MW01100206

Knobil and Neill's Physiology of Reproduction

THIRD EDITION

Knobil and Neill's Physiology of Reproduction

THIRD EDITION

Volume 2

Editor-in-Chief

Jimmy D. Neill, Ph.D.
Distinguished Professor
Department of Physiology and Biophysics
The University of Alabama at Birmingham
Birmingham, Alabama

Section Editors

Tony M. Plant, Ph.D.
University of Pittsburgh
School of Medicine
Department of Cell Biology
and Physiology
Pittsburgh, Pennsylvania

Donald W. Pfaff, Ph.D.
Department of Neurobiology
and Behavior
The Rockefeller University
New York, New York

John R. G. Challis, D.Sc., F.R.S.C.
Departments of Physiology,
Ob/Gyn, and Medicine
University of Toronto
Toronto, Ontario
Canada

David M. de Kretser, M.D., A.O.
Monash Institute of Medical Research
Monash University
Melbourne, Australia

JoAnne S. Richards, Ph.D.
Department of Molecular
and Cellular Biology
Baylor College of Medicine
Houston, Texas

Paul M. Wassarman, Ph.D.
Brookdale Department of Molecular,
Cell and Developmental Biology
Mount Sinai School of Medicine
New York, New York

AMSTERDAM • BOSTON • HEIDELBERG • LONDON
NEW YORK • OXFORD • PARIS • SAN DIEGO
SAN FRANCISCO • SINGAPORE • SYDNEY • TOKYO

Academic Press is an imprint of Elsevier

ELSEVIER
ACADEMIC
PRESS

11830 Westline Industrial Drive, St. Louis, MO 63146, USA
525 B Street, Suite 1900, San Diego, CA 92101-4495, USA
84 Theobald's Road, London WC1X 8RR, U.K.

KNOBIL AND NEILL'S PHYSIOLOGY OF REPRODUCTION
ISBN-13: 978-0-12-515402-4
ISBN-10: 0-12-515402-X

ISBN-13: 978-0-12-515402-4
ISBN-10: 0-12-515402-X

Acquisitions Editor: *Tari Paschall*
Developmental Editor: *Kirsten Funk*
Publishing Services Manager: *Julie Eddy*
Project Manager: *Kelly E. M. Steinmann*
Marketing Manager: *Patricia Howard*
Book Designer: *Bill Drone*

For all information on all Elsevier Academic Press Publications
visit our Web site at www.books.elsevier.com

Printed in the United States of America.

Last digit is the print number: 9 8 7 6 5 4 3 2 1

Contents

Contributors ix
Foreword xv
In Memoriam: Ernst Knobil (1926–2000) xxi
Preface to the First Edition xxv
Preface to the Second Edition xxvii
Preface xxix

VOLUME 1

Gametes, Fertilization, and Embryogenesis
Paul M. Wassarman, Editor

1. The Spermatozoon 3
E. M. Eddy

2. Fertilization in Mammals 55
Harvey M. Florman and Tom Ducibella

3. Gamete and Zygote Transport 113
Susan S. Suarez

4. Implantation 147
Sudhansu K. Dey and Hyunjung Lim

5. Anatomy and Genesis of the Placenta 189
Graham J. Burton, Peter Kaufmann, and Berthold Huppertz

6. Sex Determination and Differentiation 245
Amanda Swain

7. Preimplantation Embryogenesis 261
Kathleen H. Burns and Martin M. Matzuk

Female Reproductive System
JoAnne S. Richards, Editor

8. Embryology and Genetics of the Mammalian Gonads and Ducts 313
Keith L. Parker and Bernard P. Schimmer

9. Oviduct and Endometrium: Cyclic Changes in the Primate Oviduct and Endometrium 337
Alexandra P. Hess, Nihar R. Nayak, and Linda C. Giudice

10. Follicular Development: Mouse, Sheep, and Human Models 383
Aleksandar Rajkovic, Stephanie A. Pangas, and Martin M. Matzuk

11. Ovulation 425
Lawrence L. Espey and JoAnne S. Richards

12. Structure, Function, and Regulation of the Corpus Luteum 475
Richard L. Stouffer

13. Growth Hormone, Insulin-Like Growth Factors, and the Ovary 527
Carolyn A. Bondy, Jian Zhou, and Jose A. Arraztoa

14. Gonadotropin Signaling in the Ovary 547
Mary Hunzicker-Dunn and Kelly Mayo

15. Steroid Receptors in the Ovary and Uterus 593
John F. Couse, Sylvia C. Hewitt, and Kenneth S. Korach

16. Physiology and Molecular Biology of the Relaxin Peptide Family 679
Ross A. D. Bathgate, Aaron J. W. Hsueh, and O. David Sherwood

Male Reproductive System
David M. de Kretser, Editor

17. Anatomy, Vasculature, and Innervation of the Male Reproductive Tract 771
B. P. Setchell and W. G. Breed

18. Cytology of the Testis and Intrinsic Control Mechanisms 827
J. B. Kerr, K. L. Loveland, M. K. O'Bryan, and D. M. de Kretser

19. The Sertoli Cell 949
Michael D. Griswold and Derek McLean

20. Physiology of Testicular Steroidogenesis 977
Douglas M. Stocco and Michael J. McPhaul

21. Endocrine Regulation of Spermatogenesis 1017
Liza O'Donnell, Sarah J. Meachem, Peter G. Stanton, and Robert I. McLachlan

22. The Epididymis 1071
Bernard Robaire, Barry T. Hinton, and Marie-Claire Orgebin-Crist

23. Physiology of the Male Accessory Sex Structures: The Prostate Gland, Seminal Vesicles, and Bulbourethral Glands 1149
Gail P. Risbridger and Renea A. Taylor

24. Male Sexual Function 1173
Shalender Bhasin and George S. Benson

25. Immunophysiology of the Male Reproductive Tract 1195
M. P. Hedger and D. B. Hales

Pituitary and Hypothalamus
Jimmy D. Neill, Editor

26. Pituitary and Hypothalamus: Perspectives and Overview 1289
*John W. Everett**

27. Anatomy of the Hypothalamo–Hypophysial Complex 1309
Robert B. Page

28. Physiology of the Gonadotropin-Releasing Hormone Neuronal Network 1415
Allan E. Herbison

29. Gonadotropes and Lactotropes 1483
Gwen V. Childs

30. Gonadotropins: Chemistry and Biosynthesis 1581
George R. Bousfield, Li Jia, and Darrell N. Ward

**Deceased.*

31. Gonadotropin-Releasing Hormone Regulation of Gonadotropin Biosynthesis and Secretion . 1635
Kyeong-Hoon Jeong and Ursula B. Kaiser

32. Prolactin: Structure, Function, and Regulation of Secretion 1703
Karen A. Gregerson

VOLUME 2

Reproductive Behavior and Its Control
Donald W. Pfaff, Editor

33. Neurobiology of Male Sexual Behavior 1729
Elaine M. Hull, Ruth I. Wood, and Kevin E. McKenna

34. Hormonal, Neural, and Genomic Mechanisms for Female Reproductive Behaviors, Motivation, and Arousal 1825
D. W. Pfaff, Y. Sakuma, L.-M. Kow, A. W. L. Lee, and A. Easton

35. Maternal Behavior 1921
M. Numan, A. S. Fleming, and F. Levy

36. Communicative Behaviors, Hormone–Behavior Interactions, and Reproduction in Vertebrates 1995
John C. Wingfield

37. Pheromones and Mammalian Reproduction 2041
John G. Vandenbergh

Reproductive Processes and Their Control
Tony M. Plant, Editor

38. Puberty in the Rat 2061
Sergio R. Ojeda and Michael K. Skinner

39. Puberty in the Sheep 2127
Douglas L. Foster and Leslie M. Jackson

40. Puberty in Nonhuman Primates and Humans 2177
Tony M. Plant and Selma F. Witchel

41. Seasonal Regulation of Reproduction in Mammals 2231
Benoît Malpaux

42. Neuroendocrine Control of Mating-Induced Ovulation 2283
Alexander S. Kauffman and Emilie F. Rissman

43. Neuroendocrine Control of the Ovarian Cycle of the Rat 2327
Marc E. Freeman

44. Neuroendocrine Control of the Ovarian Cycle of the Sheep 2389
Robert L. Goodman and E. Keith Inskeep

45. Control of Follicular Development, Corpus Luteum Function, the Maternal Recognition of Pregnancy, and the Neuroendocrine Regulation of the Menstrual Cycle in Higher Primates 2449
Anthony J. Zeleznik and Clifford R. Pohl

46. Suckling and the Control of Gonadotropin Secretion 2511
Alan S. McNeilly

47. Physiological Mechanisms Integrating Metabolism and Reproduction 2553
Stephanie M. Krasnow and Robert A. Steiner

48. Stress and the Reproductive System 2627
Michel Ferin

49. Aging in the Hypothalamic–Pituitary–Testicular Axis 2697
David J. Handelsman

50. Aging in the Hypothalamic–Pituitary–Ovarian Axis 2729
Genevieve Neal-Perry and Nanette F. Santoro

Pregnancy
John R. G. Challis, Editor

51. Immunobiology of Human Pregnancy 2759
Joan S. Hunt, Ramsey H. McIntire, and Margaret G. Petroff

52. Placental Transfer 2787
D. E. Atkinson, R. D. H. Boyd, and C. P. Sibley

53. Placental Endocrine Function 2847
Felice Petraglia, Pasquale Florio, and Michela Torricelli

54. Maternal Adaptation to Pregnancy 2899
Kent L. Thornburg, Susan P. Bagby, and George D. Giraud

55. Parturition 2925
William Gibb, Stephen J. Lye, and John R. G. Challis

56. Developmental Origins of Health and Disease: Theoretical Considerations, Mechanisms, and Implications 2975
Peter D. Gluckman and Mark A. Hanson

Lactation
Jimmy D. Neill, Editor

57. Lactation and Its Hormonal Control 2993
Margaret C. Neville

58. Oxytocin: Synthesis, Secretion, and Reproductive Functions 3055
J. Peter H. Burbach, Larry J. Young, and John A. Russell

59. Milk Ejection and Its Control 3129
J. B. Wakerley

Index 3191

Contributors

Jose A. Arraztoa, M.D.
Developmental Endocrinology Branch, National Institute of Child Health and Human Development, Bethesda, Maryland

D. E. Atkinson, Ph.D.
School of Biological Science, University of Manchester, Manchester, U.K.

Susan P. Bagby, M.D.
Division of Nephrology & Hypertension and Heart Research Center, Department of Medicine, Oregon Health & Science University, Portland, Oregon

Ross A. D. Bathgate, Ph.D.
Howard Florey Institute, University of Melbourne, Victoria, Australia

George S. Benson, M.D.
6414 Fannin Street, Suite g150, Houston, Texas

Shalender Bhasin, M.D.
Division of Endocrinology, University of the City of Los Angeles, Los Angeles, California

Carolyn A. Bondy, M.D.
Developmental Endocrinology Branch, National Institute of Child Health and Human Development, Bethesda, Maryland

George R. Bousfield, Ph.D.
Department of Biological Sciences, Wichita State University, Wichita, Kansas

R. D. H. Boyd, M.B.
School of Biological Science, University of Manchester, Manchester, U.K.

W. G. Breed, D.Phil.
Department of Anatomical Sciences, University of Adelaide, Adelaide, Australia

J. Peter H. Burbach, Ph.D.
Department of Pharmacology and Anatomy, Rudolf Magnus Institute of Neurosciences, Utrecht, The Netherlands

Kathleen H. Burns, M.D., Ph.D.
Department of Pathology, The Johns Hopkins Hospital, Baltimore, Maryland

Graham J. Burton, M.D., D.Sc.
Department of Anatomy, University of Cambridge, Cambridge, U.K.

John R. G. Challis, D.Sc., F.R.S.C.
Research and Associate Provost Professor of Physiology, Ob/Gyn and Medicine, University of Toronto, Toronto, Ontario, Canada

Gwen V. Childs, Ph.D.
Department of Neurobiology and Developmental Sciences, College of Medicine, University of Arkansas for Medical Sciences, Little Rock, Arkansas

John F. Couse, Ph.D.
Laboratory of Reproductive and Developmental Toxicology, National Institute of Environmental Health Sciences, National Institutes of Health, Research Triangle Park, North Carolina

D. M. de Kretser, M.D., A.O.
Monash Institute of Reproduction and Development, Monash Medical Centre, Melbourne, Australia

Sudhansu K. Dey, Ph.D.
Department of Pediatrics, Cell and Developmental Biology and Pharmacology, Director, Division of Reproductive and Developmental Biology, Vanderbilt University Medical Center, Nashville, Tennessee

Tom Ducibella, Ph.D.
Department of Obstetrics and Gynecology, Tufts University, New England Medical Center, Boston, Massachusetts

A. Easton, Ph.D.
Department of Neurobiology and Behavior, The Rockefeller University, New York, New York

E. M. Eddy, Ph.D.
Gamete Biology Section, Laboratory of Reproductive and Developmental Toxicology, National Institute of Environmental Health Sciences, National Institutes of Health, Research Triangle Park, North Carolina

Lawrence L. Espey, Ph.D.
Department of Biology, Trinity University, San Antonio, Texas

John W. Everett*
Department of Neurobiology, Duke University School of Medicine, Durham, North Carolina

Michel Ferin, M.D.
Department of Obstetrics/Gynecology, Columbia University, College of Physicians and Surgeons, New York, New York

A. S. Fleming, Ph.D.
Department of Psychology, University of Toronto at Mississauga, Mississauga, Ontario, Canada

Pasquale Florio, Ph.D.
Department of Pediatrics, Obstetric and Reproductive Medicine, University Siena, Siena, Italy

Harvey M. Florman, Ph.D.
Department of Cell Biology, University of Massachusetts Medical School, Worcester, Massachusetts

Douglas L. Foster, Ph.D.
Reproductive Sciences Program, Department of Obstetrics and Gynecology, Department of Ecology and Evolutionary Biology, University of Michigan, Ann Arbor, Michigan

Marc E. Freeman, Ph.D.
Department of Biological Science, Florida State University, Biomedical Research Facility, Tallahassee, Florida

William Gibb, Ph.D.
Department of Obstetrics and Gynecology, and Department of Cellular and Molecular Medicine, University of Ottawa, Ottawa, Ontario, Canada

George D. Giraud, M.D., Ph.D.
Department of Medicine, Heart Research Center, Oregon Health & Science University, Portland, Oregon

Linda C. Giudice, M.D., Ph.D.
Center for Research on Reproduction and Women's Health and Genomic Medicine, Department of Obstetrics and Gynecology, Stanford University School of Medicine, Stanford, California

Peter D. Gluckman, F.R.S.
Liggins Institute, University of Auckland, Grafton, Auckland, New Zealand

Robert L. Goodman, Ph.D.
Department of Physiology and Pharmacology, West Virginia University, Morgantown, West Virginia

Karen A. Gregerson, Ph.D.
University of Cincinnati, College of Pharmacy, Cincinnati, Ohio

Michael D. Griswold, Ph.D.
School of Molecular Biosciences, Washington State University, Pullman, Washington

D. B. Hales, Ph.D.
Department of Physiology and Biophysics, University of Illinois at Chicago, Chicago, Illinois

David J. Handelsman, M.B., B.S., Ph.D.
ANZAC Research Institute, Sydney, Australia

Mark A. Hanson, D.Phil.
Centre for Developmental Origins of Health and Disease, University of Southampton, Princess Anne Hospital Level F (887), Southampton, U.K.

M. P. Hedger, Ph.D.
Monash Institute of Medical Research, Monash University, Melbourne, Australia

Allan E. Herbison, M.B.Ch.B., Ph.D.
Center for Neuroendocrinology and Department of Physiology, Otago School of Medical Sciences, Dunedin, New Zealand

Alexandra P. Hess, M.D.
Center for Research on Reproduction and Women's Health and Genomic Medicine, Department of Obstetrics and Gynecology, Stanford University School of Medicine, Stanford, California

---

*Deceased.

Sylvia C. Hewitt, M.S.
National Institute of Environmental Health Sciences, National Institutes of Health, Research Triangle Park, North Carolina

Barry T. Hinton, Ph.D.
Department of Cell Biology, University of Virginia Health System, Charlottesville, Virginia

Aaron J. W. Hsueh, Ph.D.
Division of Reproductive Biology, Department of Obstetrics and Gynecology, Stanford University School of Medicine, Stanford, California

Elaine M. Hull, Ph.D.
Psychology Department, Florida State University, Tallahassee, Florida

Joan S. Hunt, Ph.D.
Department of Anatomy and Cell Biology, University of Kansas Medical Center, Kansas City, Kansas

Mary Hunzicker-Dunn, Ph.D.
Department of Cell and Molecular Biology, Center for Reproductive Science, Northwestern University Medical School, Chicago, Illinois

Berthold Huppertz, Ph.D.
Department of Anatomy II, University Hospital RWTH Aachen, Aachen, Germany

E. Keith Inskeep, Ph.D.
Division of Animal and Veterinary Sciences, West Virginia University, Morgantown, West Virginia

Leslie M. Jackson, Ph.D.
Reproductive Sciences Program, Department of Obstetrics and Gynecology, Department of Ecology and Evolutionary Biology, University of Michigan, Ann Arbor, Michigan

Kyeong-Hoon Jeong, Ph.D.
Division of Endocrinology, Diabetes, and Hypertension, Brigham and Women's Hospital and Harvard Medical School, Boston, Massachusetts

Li Jia, Ph.D.
Department of Biological Sciences, Wichita State University, Wichita, Kansas

Ursula B. Kaiser, M.D.
Division of Endocrinology, Diabetes, and Hypertension, Brigham and Women's Hospital and Harvard Medical School, Boston, Massachusetts

Alexander S. Kauffman, Ph.D.
Department of Biochemistry and Molecular Genetics, University of Virginia School of Medicine, Center for Research in Reproduction, Charlottesville, Virginia

Peter Kaufmann, M.D.
Department of Anatomy II, University Hospital RWTH Aachen, Aachen, Germany

J. B. Kerr, Ph.D.
Department of Anatomy and Cell Biology, Faculty of Medicine, Nursing and Health Sciences, Monash University, Victoria, Australia

Kenneth S. Korach, Ph.D.
Program Director, Environ Disease Medicine Program Chief, National Institute of Environmental Health Sciences, Research Triangle Park, North Carolina

L.-M. Kow, Ph.D.
Department of Neurobiology and Behavior, The Rockefeller University, New York, New York

Stephanie M. Krasnow, Ph.D.
Departments of Physiology and Biophysics, Obstetrics and Gynecology, and Biology, University of Washington, Seattle, Washington

A. W. L. Lee, Ph.D.
Department of Neurobiology and Behavior, The Rockefeller University, New York, New York

F. Levy, Ph.D.
Responsable de l équipe de Comportement, Physiologie de la Reproduction et des Comportements, INRA-CNRS-Université de Tours, Centre INRA de Tours, Nouzilly, France

Hyunjung Lim, Ph.D.
Departments of Obstetrics and Gynecology and Cell Biology and Physiology, Washington State University School of Medicine, St. Louis, Missouri

K. L. Loveland, Ph.D.
Monash Institute of Reproduction and Development, Monash Medical Centre, Monash University, Victoria, Australia

Stephen J. Lye, Ph.D.
Department of Physiology, Department of Obstetrics and Gynecology, University of Toronto; Division of Child Health and Human Development, Samuel Lunenfeld Research Institute, Mount Sinai Hospital, Toronto, Ontario, Canada

Benoît Malpaux, Ph.D.
UMR INRA-CNRS-Université de Tours, Physiologie de la Reproduction et des Comportements, Centre INRA de Tours, Nouzilly, France

Martin M. Matzuk, M.D., Ph.D.
Department of Pathology, Baylor College of Medicine, Houston, Texas

Kelly Mayo, Ph.D.
Department of Biochemistry, Cellular and Molecular Biology, Center for Reproductive Science, Northwestern University, Evanston, Illinois

Ramsey H. McIntire, Ph.D.
Department of Anatomy and Cell Biology, University of Kansas Medical Center, Kansas City, Kansas

Kevin E. McKenna, Ph.D.
Northwestern University Medical School, Chicago, Illinois

Robert I. McLachlan, Ph.D.
Prince Henry's Institute of Medical Research, Monash Medical Centre, Victoria, Australia

Derek McLean, Ph.D.
Department of Animal Sciences, Washington State University, Pullman, Washington

Alan S. McNeilly, D.Sc., F.R.S.E.
Centre for Reproductive Biology, The University of Edinburgh, Edinburgh, U.K.

Michael J. McPhaul, M.D.
Department of Internal Medicine, University of Texas Southwestern Medical Center, Dallas, Texas

Sarah J. Meachem, Ph.D.
Prince Henry's Institute of Medical Research, Monash Medical Centre, Victoria, Australia

Nihar R. Nayak, Ph.D.
Center for Research on Reproduction and Women's Health and Genomic Medicine, Department of Obstetrics and Gynecology, Stanford University School of Medicine, Stanford, California

Genevieve Neal-Perry, M.D., Ph.D.
Department of Obstetrics and Gynecology, Albert Einstein College of Medicine, The Bronx, New York

Margaret C. Neville, Ph.D.
University of Colorado Health Sciences Center at Fitzsimmons, Aurora, Colorado

M. Numan, Ph.D.
Department of Psychology, Boston College, Chestnut Hill, Massachusetts

M. K. O'Bryan, Ph.D.
Monash Institute of Reproduction and Development, Monash University, Monash Medical Centre, Victoria, Australia

Liza O'Donnell, Ph.D.
Prince Henry's Institute of Medical Research, Monash Medical Centre, Victoria, Australia

Sergio R. Ojeda, D.V.M.
Division of Neuroscience, Oregon Regional Primate Research Center, Beaverton, Oregon

Marie-Claire Orgebin-Crist, Ph.D.
Center for Reproductive Biology Research, Vanderbilt School of Medicine, Nashville, Tennessee

Robert B. Page, M.D.
Emeritus Professor of Neurosurgery/Department of Neurosurgery, Milton S. Hershey Medical Center, Pennsylvania State University, Hershey, Pennsylvania

Stephanie A. Pangas, Ph.D.
Departments of Pathology and Molecular and Cellular Biology, Baylor College of Medicine, Houston, Texas

Keith L. Parker, M.D., Ph.D.
Department of Internal Medicine, Division of Endocrinology and Metabolism, University of Texas Southwestern Medical Center, Dallas, Texas

Felice Petraglia, M.D.
Department of Pediatrics, Obstetrics and Reproductive Medicine, University Siena, Siena, Italy

Margaret G. Petroff, Ph.D.
Department of Anatomy and Cell Biology, University of Kansas Medical Center, Kansas City, Kansas

D. W. Pfaff, Ph.D.
Department of Neurobiology and Behavior, The Rockefeller University, New York, New York

Tony M. Plant, Ph.D.
Departments of Cell Biology and Physiology and Obstetrics, Gynecology, and Reproductive Sciences, University of Pittsburgh School of Medicine, Pittsburgh, Pennsylvania

Clifford R. Pohl, Ph.D.
School of Health Sciences, Duquesne University, Pittsburgh, Pennsylvania

Aleksandar Rajkovic, M.D., Ph.D.
Departments of Obstetrics and Gynecology, Baylor College of Medicine, Houston, Texas

JoAnne S. Richards, Ph.D.
Department of Molecular and Cellular Biology, Baylor College of Medicine, Houston, Texas

Gail P. Risbridger, Ph.D.
Centre for Urology Research, Monash Institute of Reproduction and Development, Monash University, Victoria, Australia

Emilie F. Rissman, Ph.D.
Department of Biochemistry and Molecular Genetics, University of Virginia School of Medicine, Charlottesville, Virginia

Bernard Robaire, Ph.D.
Department of Pharmacology and Therapeutics, Department of Obstetrics and Gynecology, McGill University, Montréal, Québec, Canada

John A. Russell, M.B.Ch.B., Ph.D.
Division of Biomedical Sciences, University of Edinburgh, Edinburgh, U.K.

Y. Sakuma, M.D., Ph.D.
Department of Physiology, Nippon University School of Medicine, Tokyo, Japan

Nanette F. Santoro, M.D.
Department of Obstetrics and Gynecology and Women's Health, Albert Einstein College of Medicine, The Bronx, New York

Bernard P. Schimmer, Ph.D.
Banting and Best Department of Medical Research and Department of Pharmacology, University of Toronto, Toronto, Ontario, Canada

B. P. Setchell, Ph.D., Sc.D.
Department of Anatomical Sciences, University of Adelaide, Adelaide, Australia

O. David Sherwood, Ph.D.
Department of Molecular and Integrative Physiology and College of Medicine, University of Illinois at Urbana-Champaign, Urbana, Illinois

C. P. Sibley, Ph.D.
School of Biological Science, University of Manchester, Manchester, U.K.

Michael K. Skinner, Ph.D.
Center for Reproductive Biology, School of Molecular Biosciences, Washington State University, Pullman, Washington

Peter G. Stanton, Ph.D.
Prince Henry's Institute of Medical Research, Monash Medical Centre, Victoria, Australia

Robert A. Steiner, Ph.D.
Department of Physiology and Biophysics, School of Medicine, University of Washington, Seattle, Washington

Douglas M. Stocco, Ph.D.
Department of Cell Biology and Biochemistry, Texas Tech University Health Sciences Center, Lubbock, Texas

Richard L. Stouffer, Ph.D.
Senior Scientist and Head, Division of Reproductive Sciences, Oregon National Primate Research Center/Oregon Health & Science University, Beaverton, Ohio

Susan S. Suarez, Ph.D.
Department of Biomedical Sciences, Cornell University, College of Veterinary Medicine, Ithaca, New York

Amanda Swain, Ph.D.
Institute of Cancer Research, University of London, London, U.K.

Renea A. Taylor, Ph.D.
Centre for Urology Research, Monash Institute of Reproduction and Development, Monash University, Victoria, Australia

Kent L. Thornburg, Ph.D.
Department of Medicine, Heart Research Center, Oregon Health & Science University, Portland, Oregon

Michela Torricelli, M.D.
Department of Pediatrics, Obstetrics and Reproductive Medicine, University Siena, Siena, Italy

John G. Vandenbergh, Ph.D.
Department of Zoology, North Carolina State University, Raleigh, North Carolina

J. B. Wakerley, Ph.D.
Department of Anatomy, University of Bristol, School of Medical Sciences, Bristol, U.K.

Darrell N. Ward, Ph.D.
Department of Biological Sciences, Wichita State University, Wichita, Kansas

John C. Wingfield, Ph.D.
Department of Zoology, University of Washington, Seattle, Washington

Selma F. Witchel, M.D.
Department of Pediatrics, University of Pittsburgh School of Medicine, Pittsburgh, Pennsylvania

Ruth I. Wood, Ph.D.
Department of Cell and Neurobiology, USC School of Medicine, Los Angeles, California

Larry J. Young, Ph.D.
Yerkes Primate Center, Emory University, Atlanta, Georgia

Anthony J. Zeleznik, Ph.D.
Department of Cell Biology and Physiology, University of Pittsburgh School of Medicine, Pittsburgh, Pennsylvania

Jian Zhou, M.D., Ph.D.
Developmental Endocrinology Branch, National Institute of Child Health and Human Development, National Institutes of Health, Bethesda, Maryland

Foreword*

I am pleased and honored to have been asked to prepare the Foreword to this volume of work depicting the progress in research on the physiology of reproduction as well as the resulting gains in understanding made over the past few years. The expertise that is represented by the numerous contributors to this work is so impressive that I am humbled even to contemplate adding anything of note. It is only by virtue of having personally witnessed a very large segment of twentieth century research on reproduction that I am emboldened to reflect on the byways and the trail blazings that have brought this field to its present proud state of enlightenment with regard to the long sought-after means of controlling the procreative process in humankind. Clearly, there are many important and knotty problems yet to be resolved, but the pace of progress over the past several years has quickened to the extent that one is left in expectant wonderment as to where and when the next revolutionizing development will occur.

The experimental method of studying reproduction was initiated in 1849 with Berthold's discovery of a blood-borne activity that came from the testis and stimulated growth of distant organs such as the comb and wattles. In so doing he utilized one of the most fundamental means of demonstrating the function of an endocrine organ, namely, surgical removal to determine what deficiencies follow, coupled with implantation or transplantation to ascertain whether the deficiencies were repaired. At that time it was not possible to take the next step, namely, preparation of an active extract of the testes, because nothing was known about the nature of the bioactivity. Forty years later, Brown-Séquard claimed to have prepared an active extract of dog testes; however, as is well known, his enthusiastic claims for restoration of his own sexual activity at an advanced age were not substantiated. Actually, these simple means of studying reproductive physiology persisted well into the twentieth century, including the studies of such pioneering stalwarts as Marshall, Heape, Prenant, Bouin, Ancel, Loeb, Cushing, and Aschner. Observations otherwise were limited to cyclic and seasonal changes in sexual behavior among common laboratory and small domestic animals. This type of eyeball research remained in vogue through the early 1920s and overlapped the extension of visualization to the microscopic level. The latter revealed, for the first time, the precise timing of events in the ovarian cycle through microscopically observable cellular changes in the vaginal fluid. My point in mentioning these early studies is to emphasize that although the tools and techniques were inordinately primitive by present standards, the results established a firm base of knowledge on which to build.

The study of cyclic changes in the vaginal smear in rats and the findings of estrogenic activity in follicular fluid during the early 1920s led to an explosion of interest in the study of reproduction. The field was fortunate in attracting to its ranks a small band of exceedingly able biologists and biochemists who, in 1932, were to become authors of the classic first edition compendium, *Sex and Internal Secretions*, a volume overwhelmingly devoted to reproductive endocrinology. It was this landmark of progress that finally gave propriety to the study of reproduction and put it on a par with the study of other major bodily systems. Incredible as it may seem, it was only a decade earlier that a distinguished panel of the National Research Council had declared that sex research was not a fitting topic for scientific study.

Lest our pride in today's spectacular pace of progress unduly bedazzle the mind, it should not be overlooked that the developments recorded in the ten-year span from 1926 to 1936 may never be equaled. Among those monumental achievements, all of the native sex steroid hormones were

*This Foreword by Roy O. Greep is reprinted from the 1988 (first) and 1994 (second) editions of this work (E. Knobil and J. D. Neill, *The Physiology of Reproduction*. Raven Press, New York. Copyright Elsevier). This now-deceased author wrote a remarkably prescient account of the future of research in the field that remains as relevant today as it did in the first two editions.

brought to light, their structures were determined, their functions were defined, and they were made available in pure form for research and therapy. Similarly, all of the pituitary, placental, and urinary tropic hormones were identified, and their functions were defined. Like today's competition for priority rights, publicity, and potential financial gain, these earlier periods also were times for intense rivalries, but rarely with prospects for financial rewards. It would be difficult to overstate the boost that was given to basic and clinical research in reproduction as a result of the availability of estradiol-17β, testosterone, and progesterone in pure form and of known potency. The replacement of homemade extracts and such elastic entities as rat units, mouse units, capon units, and so forth, with micrograms of pure hormone was revolutionizing and allowed the study of reproduction on a quantitative basis.

Prior to World War II the thrust of research on reproduction dealt predominantly with the steroid hormones. This was the heyday of steroid biochemistry. After World War II the emphasis shifted to the protein and peptide hormones, where it still remains strong. This prolonged and difficult effort yielded many biochemical triumphs. Most notable among these were the isolation of the pituitary, placental, and urinary gonadotropins, as well as the determination of their primary structure as glycoproteins comprised of two dissimilar and covalently bonded subunits, the isolation and synthesis of the gonadotropin-releasing hormone (GnRH) of hypothalamic origin, and the isolation and structural characterization of relaxin.

The availability of pure protein and polypeptide hormones made possible the production of hormone-specific antibodies as well as the application of immunological techniques to the study of reproduction. An outcome of great consequence was the development of radioimmunoassay as the new means of measuring all of the hormones relating to reproduction. The sensitivity of this new technique was so great that it made possible, for the first time, the measurement of all these hormones in the body fluids. It had the further distinct advantage of requiring such a small amount of fluid that the monitoring of blood levels of the hormones of reproduction could be done throughout an estrous or menstrual cycle by close serial sampling. This revealed still another and most unexpected finding, the pulsatile pattern of secretion.

Identifying the homeostatic mechanism(s) responsible for maintaining a steady state in various physiologic systems of the body has been fraught with many challenging problems, but these pale in comparison with the difficulties encountered in trying to elucidate the mechanisms maintaining a constantly changing system, a characteristic of the reproductive system of female mammals. The earliest piece of evidence suggested the existence of a "push-pull" mechanism that later came to be known as *negative feedback*. It was based on the demonstration that an estrogenic extract administered to immature rats would maintain the ovaries in an infantile state. This was quickly followed by conclusive evidence that estrogen acted to inhibit pituitary follicle-stimulating hormone (FSH) stimulation of follicular growth and maturation; however, the effect on luteinizing hormone (LH), ovulation, and luteinization remained unsettled. Gaps continued to exist in all proposed explanations of reproductive cycles. None of these explanations took into account the influence of photoperiodicity on seasonal breeders, nor did they account for the role of the stimulus of mating in nonspontaneous ovulators. Following the discovery of the hypothalamic control of pituitary function, estrogen was shown to exert its action on both the pituitary and the hypothalamus; however, the problem of accounting for cyclicity remained. Adding to the complexity, radioimmunoassay revealed an unexpectedly high level of blood estrogen just prior to ovulation, an event not in keeping with the negative feedback concept.

Finally, after many years of searching for a way out of this frustrating situation, a glimmer of light appeared at the end of this long dark tunnel—light that soon turned to brilliance. In 1969, Goding and associates found that the administration of large doses of estrogen to ewes at the time of estrus did not block, but instead entrained, ovulation. Shortly thereafter, in more elaborate examination of the relationship of blood estrogen levels and ovulation in rhesus monkeys in Knobil's laboratory, it was revealed that elevated estrogen levels preceded and appeared to trigger ovulation. On further examination, Knobil and colleagues found that when blood estrogen reached a critical level the feedback mechanism switched from a negative to a positive, or stimulative, action. This utterly new finding greatly advanced our understanding of the endocrine mechanism governing reproductive cycles. There still remain, however, some uncertainties: Why does the switch in feedback action occur; to what extent and at what stage of the cycle does estrogen act at the level of the pituitary or the hypothalamus, or both; and lastly, what role, if any, do the ovarian peptides, especially inhibin, play in controlling reproductive cycles?

The progress of research on reproduction has been chronicled in numerous review articles by individual authors. Many have appeared in *Recent Progress in Hormone Research, Volumes 1 to 42*. Other major sources include the multiple editions of such titles as: *Marshall's Physiology of Reproduction, Fourth Edition* (1990); *Sex and Internal Secretions*, whose third and last edition was issued in 1961; two volumes on the *Female Reproductive System* (1973), and one on the *Male Reproductive System* (1975) in *Section 7 of the Handbook of Physiology*, published by the American Physiological Society; and four serial volumes on reproductive physiology in the *International Review of Physiology*, the last one being issued in 1983. The present volume will provide comprehensive coverage and meet the current needs of the field of reproductive physiology, a field that is rapidly gathering momentum from the application of new and highly sophisticated tools and techniques.

In viewing the vast literature dealing with research on the male and female reproductive systems and considering the rate at which it is accumulating, one might ask whether this staggering proliferation of books and articles is essential to progress; the answer is an emphatic "Yes!" The yardstick by which progress is measured in this or any other field is not in the number of articles published or the amount of financial support but in improved understanding. Such gains are generally marked by sharp peaks at indeterminate intervals, separated by avalanches of incremental gains, as recorded in an ever-growing list of journals. The point to remember is that without this persistent chipping away at a major problem there would be no solutions and no quantum leaps forward. In research very little comes from out of the blue. Part of the driving force in research is its adventuresome nature and ever-present possibility that one's efforts will pay off in an important manner. It may not be entirely fair, but in research (as in most human activities), the spoils go to the victor in the form of kudos, prizes, awards, public attention, and, increasingly in the present technological age, monetary gains—sometimes of great magnitude. What effect this latter may have, if any, on the long-cherished sanctity of science has not been determined, but it has become a matter of concern.

This volume bears the title *The Physiology of Reproduction*. Physiology, by traditional consensus, is that branch of science which studies the functions of a living organism or any of its parts and includes the basic underlying processes. It will be understood that most of the studies reviewed here will be based more on holistic research than on research at the submicroscopic or molecular level. It is unfortunate that the excitement generated by recent fantastic advances in molecular biology and development has tended to downgrade the value of whole-animal research, and physiology in particular is sometimes looked upon as passé. Actually, the two categories of research are complementary, and both are essential for maximum advancement of knowledge. Whole-animal research cannot become outdated because it is the quintessence of biological relevance and the means by which molecular findings must ultimately be evaluated.

In the same vein, no one immersed in reproductive endocrinology can be unaware of the current tendency to regard research at the molecular level as representative of exceptional scientific talent. This is a common consequence of the opening of a new arena of investigation. I recall an incident that happened at a scientific meeting back in the 1930s. The first three papers in a session chaired by an eminent embryologist were on endocrine topics—mine was the third. That being ended, the chairman took pains to assure the audience that the meeting could now turn to considerations of more fundamental nature. One of the other three papers was given by Herbert M. Evans, who bristled noticeably but held his fire. There was also an earlier period when one either worked on steroid biochemistry or something of lesser appeal like biology. Anyone who remembers the 1950s will recall a flash-in-the-pan ignited by cybernetics, a study of automatic control systems both neural and physical. The gurus of cybernetics captured the attention of the press and of audiences throughout the land, but eventually this obsession suffered the fate of other passing preoccupations. My own observation is that the closer one approaches the molecular level of research, the more one becomes dependent on highly sophisticated instrumentation to make the observations and to read out results that are often quite free of extraneous variables. Toward the obverse situation, one's dependence on an extensive background of experience and physiological increases as does the unavoidable complex of *in vivo* variables that must be taken into account. In either case we have today the availability of far more diverse approaches to a given problem in any field of biomedical research than has ever existed before. In Berthold's day there was only one experimental method available; today's number is untold but is probably in the hundreds, perhaps thousands. This is an exceedingly promising situation and one to which investigators of all persuasions must adjust. Open minds will experience exhilaration over substantive achievements at any point on this observational spectrum.

One of the major factors influencing research on reproduction has been the availability of funds or lack thereof. Prior to the institution of federal funding (i.e., prior to the middle of the twentieth century), reproductive research was sparsely supported by university departmental funds, industry, small grants from the Committee for Research in Problems of Sex within the National Research Council, and some aid from the Rockefeller Foundation. The National Institutes of Health were slow in providing significant support of research on reproduction because of restrictions on the support of work related in any way to birth control. This occurred despite the simultaneous postwar baby boom. What kept research afloat during this critical period was major support by the Ford Foundation plus lesser contributions by other major foundations. It was not until the establishment in 1968 of the Center for Population Research in the NICHD that major governmental funding in this area became available, but the boost was short-lived. As a result of the imposition of fiscal restraints in the early 1970s, federal support dwindled and has remained at a minimal level ever since. Support from all sources is woefully incommensurate with the distressing expansion of the human population and the need for safe, effective, economical, and readily available means of limiting human fertility.

The physiology of reproduction is predominantly under hormonal control. The first essential step in studying reproduction was identification of the hormones involved and the functions they serve. This having been accomplished, efforts turned to a detailed analysis as to how hormones act within the body. During the 1980s there has been a rising tide of interest in the binding of steroid, protein, or peptide hormones to receptors on specific target cells. Much effort is currently being directed toward the isolation and chemical characterization of these receptors. They are known to be composed of a protein or proteins, and some information has already been gained as to their partial or provisional structure. This, however, is only a preliminary step in the complex process whereby hormone action results in an end response such as growth, secretion of a target cell hormone, or altered behavior. The curtain has already been raised on the climatic and final chapter of the story on how hormones act. This involves linkage of the hormone-receptor complex with the nuclear genetic apparatus leading through a now well-defined series of processes to the manifestation of a physiological response in the living organism. Genes that bring about the expression of certain hormonal signals are being isolated, modified, transferred between species, and also inserted into bacteria where they direct the biosynthesis of specific hormones in large quantity. Thus genes are being manipulated in ways that raise the potential of altering the reproductive process. It is largely as a result of developments in endocrinology at the molecular level that bewildering possibilities loom on the horizons of reproductive research—they are within reach; they are science, not fiction; and they stagger the imagination.

It being granted that nothing succeeds like success, then the new edition of this highly successful two-volume compendium on *The Physiology of Reproduction* is destined for an illustrious fate. This second edition will maintain the same high standards of the first and again fulfill an existing need in a field that is experiencing rapid growth and exhilarating progress. Like the first edition, this one will provide a critical assessment of the state-of-the-art in every aspect of research on the physiology of reproduction by eminent authorities.

In the years intervening between this edition and the last, notable changes have taken place in the study of reproduction. These stem largely from major advances in technology. Remarkable new instruments, techniques, and methods have enabled investigators to probe ever deeper into the interaction between hormones and genes, thereby eliciting *in vivo* responses. New parameters are being added to the target tissues of the classical reproductive hormones as revealed by the presence of receptor sites in tissues, the physiologic significance of which often remains tantalizingly obscure. Similarly, newly identified substances of endocrine or paracrine nature are being added to this domain of research with persisting frequency. Some of these substances—the endothelins, interleukins, activins, inhibins and prorenin, to name a few—also exhibit a puzzling array of effects on extraneous tissues. Their study is being aided by the fact that their structure is known and, though rare, they are available.

Great strides are also being made in many other aspects of research on reproduction. Much work is being done on the structure of receptors and the loci of binding sites on segments of the folded gonadotropic molecules. A full-scale effort is underway seeking an elucidation of the neural mechanism underlying pulsatile secretion. Neuroendocrinologists are closing in on an elusive pulse generator located in the central nervous system. This looms as another landmark discovery in reproductive biology.

Research on reproduction is flourishing and the future appears bright. The taboos are gone. All aspects of the reproductive process are an open book. One area that has taken a quantum leap forward is the clinical application of an important body of relevant new knowledge gained in both basic and clinical spheres. Expanded opportunities have been opened by greatly improved diagnostic procedures, more effective treatment of disorders, and new methods of controlling fertility. Contributing greatly to this explosive development is the dissemination of information on reproductive matters to the lay public by the mass media. Concerned individuals have been made aware of the existing new means of manipulating the male and female reproductive systems for enhancement or inhibition of fertility. The joys and comforts that accrue respectively to these opposing modes of fertility control have enriched the lives of a grateful public. To that end I may note that it was by virtue of these frontier reproductive measures that my own progeny includes a new grandson and namesake.

Roy O. Greep

In Memoriam: Ernst Knobil (1926–2000)

Ernst Knobil died on April 13, 2000, in his 73rd year of life, thus ending a remarkable career of outstanding scientific accomplishments, of leadership positions in physiology and endocrinology, and of mentorship to numerous students and fellows. He was noted for his clarity of thought, his endless pursuit of excellence, and his abiding interest in integrative biology.

Dr. Knobil was a visionary leader and a pioneer in many areas of endocrinology, including growth and reproduction. Dr. Knobil's now classic contributions include the species-specific effects of GH, a model for positive and negative estrogen feedback control of the menstrual cycle, and elucidation of the hypothalamic GnRH pulse generator. His discovery that pulsatile GnRH stimulates LH, whereas continuous GnRH desensitizes pituitary LH secretion, has forever altered the field of reproductive endocrinology. This remarkable experimental observation unmasked a pivotal role for pulsatile secretion as a mechanism of hormonal control.

The elder son of an Austrian father (Jakob Knobil) and a German mother (Regina Seidmann), Ernst was born in Berlin, Germany, on September 20, 1926. When Ernst was about six years old, the Knobil family moved to Paris due to the deteriorating political conditions in Germany. Then, in 1940, when the Germans invaded Paris and Ernst was 13, the family emigrated to New York City.

Ernst entered the New York State College of Agriculture at Cornell in 1942, at the age of 15. He chose Animal Science as his major due to interests developed from time spent on farms in France during the summers, and from attending the Kinderhook Farm Camp, after moving to New York. Upon graduating from Cornell in 1948 (including a two-year interruption for service in the U.S. Army), he entered graduate school in zoology upon recommendation of the late Sidney Asdell, author of the well-known reference work, *Patterns of Mammalian Reproduction*. He worked in the laboratory of Professor Samuel L. Leonard, who, as a member of Hisaw's laboratory at the University of Wisconsin, had participated in the initial identification of LH. After completing his Ph.D., Ernst accepted a postdoctoral position with Roy O. Greep at the Harvard School of Dental Medicine from 1951 to 1953. In Greep's laboratory, he was introduced to the

Endocrine Reviews 22(6):721–723

rhesus monkey as an experimental animal, and he performed studies on adrenal cortical function and on fetal–maternal interrelationships. While a fellow, he assumed Greep's teaching duties in endocrinology and rapidly gained recognition as a gifted and scholarly teacher. Because of this, in 1953, he was appointed Instructor in the Physiology Department of the Harvard Medical School. In 1957, he was promoted to Assistant Professor after having been selected by Harvard Medical School for the prestigious Markle Scholar in Academic Medicine for the years 1956–1961.

Ernst's research first gained international recognition for his discovery of the "species specificity" of GH during his research in 1954–1959. His initial studies in hypophysectomized monkeys confirmed that bovine and porcine GH preparations were inert in the monkey, as had been reported for humans. However, he found that GH extracted from monkey pituitary glands was fully effective in stimulating various parameters of growth (1). These findings were quickly confirmed in humans and gave birth to the program of the National Pituitary Agency to collect human pituitary glands from which GH was purified for treatment of GH deficiency. This program was later discontinued when human GH became the first protein to be produced for clinical use using recombinant DNA technology.

In 1961, Ernst was appointed Chairman of the Department of Physiology at the University of Pittsburgh School of Medicine, a position he held for 20 years. There he initiated studies on regulation of reproductive cycles in rhesus monkeys because primates had been little studied in this regard, and because new methodologies for measuring hormones in blood were becoming available. First, he established the Center for Research in Primate Reproduction and built the Pittsburgh Primate Center to supply animals and to train postdoctoral fellows. Next, he developed RIAs for progesterone, LH, FSH, and E2 "after long and arduous methodological studies" (2) to permit description of the time courses of circulating hormones during the menstrual cycle of the rhesus monkey. These and subsequent studies revealed that E2, rather than progesterone, was the ovarian hormone that stimulated the mid-cycle LH surge. Moreover, he observed that low tonic levels of E2 were inhibitory to LH secretion, whereas prolonged elevations of E2 were stimulatory to LH secretion. Progesterone was shown to be primarily inhibitory, synergizing with low levels of E2 and at higher levels, antagonizing the stimulatory effect of estradiol (2). These studies were monumentally important because they established for the first time the exact roles of E2 and progesterone in regulating gonadotropin secretion and, hence, the reproductive cycle.

Another crucial discovery was the episodic nature of the interaction between the hypothalamus and pituitary. Pulses of LH secretion were observed at about hourly intervals and, hence, were denominated "circhoral oscillations." The LH pulses were correctly assumed to be the consequence of pulsatile GnRH secretion by the hypothalamus into the pituitary portal circulation which, in turn, gave rise to the notion of an oscillator or signal generator in the central nervous system ("GnRH pulse generator") (2,3). A decade or more of study revealed that pulsatile LH secretion was a crucial regulator of the menstrual cycle. For prolonged LH secretion, GnRH had to be administered in a pulsatile fashion; continuous administration resulted in desensitization of LH secretion (3), an observation that forms the basis of a currently popular treatment of prostate cancer in humans. Knobil further observed that the hourly pulsatile infusions into monkeys deprived of endogenous GnRH by lesioning of the arcuate nucleus reinitiated menstrual cycles. These cycles were characterized by normal LH and FSH levels and were accompanied by normal follicular growth, ovulation, and corpus luteum function as signified by normal patterns of E2 and progesterone secretion. From these studies, he argued that the ovary rather than the hypothalamus was the "zeitgeber" of the menstrual cycle; an extension of this notion was that, unlike the rat, the monkey did not require a midcycle surge of GnRH secretion to stimulate LH secretion, the heightened responsiveness of the pituitary to GnRH occasioned by E2 being sufficient (3,4).

Extension of these studies to normal prepubertal monkeys led to the initiation of menstrual cycles characterized by normal gonadotropin and E2 and progesterone levels. From these studies he concluded "that neither adenohypophysial nor ovarian competence is limiting in the initiation of puberty in the rhesus monkey." Rather, "puberty is normally initiated by the activation of hypothalamic mechanisms that control the circhoral pulsatile release of GnRH into the pituitary portal circulation" (3).

A final breakthrough in the understanding of the primate menstrual cycle was the multiunit recording in the medial basal hypothalamus of electrical activity associated with each pulse of LH secretion. Initially, these studies were confined to ovariectomized animals because restraint in the apparatus necessary for electrical recordings from the hypothalamus suppressed menstrual cycles in intact monkeys. This problem was circumvented by adopting telemetry that permitted electrical recordings of hypothalamic activity in unrestrained monkeys exhibiting normal menstrual cycles (4,5).

This very brief description of his studies hardly conveys the transforming nature of the findings. To understand their impact, it is only sufficient to note that we understand the regulation of reproductive cycles in rhesus monkeys better than for any other species. This monumental result can be attributed to several personal characteristics: 1) the logical force that is evident throughout his research career; 2) a level of clear and incisive thought that makes Occam's razor seem dull by comparison; 3) insistence on the development of methods appropriate to the experimental question being asked, no matter how arduous or time consuming; and 4) a high level of scholarship in all matters, at all times, and in all places.

In 1981, Ernst accepted the Deanship of the University of Texas Medical School at Houston. There he continued to explore regulation of the reproductive cycles in macaques; indeed, most of the studies described above on electrical recordings of the hypothalamus were conducted in Houston. After his tenure as Dean ended in 1984, he continued as Director of the Laboratory of Neuroendocrinology until 1997 when he closed his laboratory. More than 80 fellows and students studied in his laboratories in Boston, Pittsburgh, and Houston, and contributed to the studies described earlier.

Ernst received many awards, including the highest ones awarded by the Society for the Study of Reproduction (Carl G. Hartman Award, 1983), The Endocrine Society (Fred Conrad Koch Award, 1982), and the American Physiological Society (Walter B. Cannon Memorial Lecture, 1997). He was elected to numerous positions of leadership including the Presidencies of The Endocrine Society (1976), the American Physiological Society (1979), and the International Society of Endocrinology (1984–1988). A noteworthy achievement of his Endocrine Society presidency was the establishment of *Endocrine Reviews*. He was a member of many other U.S. and foreign scientific societies' review boards, NIH study sections, and the editorial boards of numerous scientific journals.

Dr. Knobil was a member of the U.S. National Academy of Science (1986), the American Academy of Arts and Sciences, a foreign associate of the French Academy of Science, the Italian National Academy of Science and the Belgian Royal Academy of Medicine. He received several honorary degrees, including the University of Bordeaux (1980), the Medical College of Wisconsin (1983), the University of Liege (1994), and the University of Milan (2000). In addition to being the author of 217 scientific papers, he was the editor of several reference books in endocrinology and reproduction, including *The Handbook of Physiology* (1974), *The Physiology of Reproduction* (1988, 1994), and *The Encyclopedia of Reproduction* (1998). In addition to his numerous scientific contributions, Ernst was a visionary leader in the field of endocrinology and a wonderful mentor and friend to the many people who were privileged to work closely with him.

He is survived by his wife, Julane Hotchkiss, four children, and three grandchildren.

ACKNOWLEDGMENTS

The editorial assistance of Dr. Adolph Friedman, Staff Consultant for the History Project of The Endocrine Society, is gratefully acknowledged.

Jimmy D. Neill
Distinguished Professor
Department of Physiology and Biophysics
University of Alabama School of Medicine
Birmingham, Alabama 35294

Address all correspondence and requests for reprints to: Jimmy D. Neill, Department of Physiology and Biophysics, University of Alabama School of Medicine, Birmingham, Alabama 35294.

REFERENCES

1. **Knobil E, Greep RO** 1959 The physiology of growth hormone with particular reference to its action in the rhesus monkey and the "species specificity" problem. Recent Prog Horm Res 15:1–69
2. **Knobil E** 1974 On the control of gonadotropin secretion in the rhesus monkey. Recent Prog Horm Res 30:1–46
3. **Knobil E** 1980 The neuroendocrine control of the menstrual cycle. Recent Prog Horm Res 36:53–88
4. **Hotchkiss J, Knobil E** 1994 The menstrual cycle and its neuroendocrine control. In: Knobil E, Neill JD, eds. The Physiology of Reproduction. Vol. 2, ed. 2 New York: Raven Press; 711–749
5. **Knobil E** 1997 The wisdom of the body revisited (1997 Walter B. Cannon Memorial Lecture). News Physiol Sci 14:1–11

Preface to the First Edition

This work was undertaken, after much deliberation, in an attempt to fill a need for a comprehensive, scholarly treatise on the physiology of mammalian reproduction. A major inspiration for this effort has been the volume *Sex and Internal Secretions,* a factual and conceptual beacon which guided generations of reproductive biologists from the time of its first publication in 1932 to the appearance of its last edition over a quarter of a century ago.

The book is divided into five major sections, and these, in turn, are loosely arrayed in two domains. The first covers the components of the reproductive system, and the second discusses reproductive processes and their physiological control. In the latter, we have included reproductive behavior in the conviction that this fundamental aspect of reproduction clearly belongs in the physiological realm and will remain a demanding challenge long after all the other mysteries in the field have been resolved.

In our discussion of reproductive systems, we have been aware of the profound differences among mammals in the way some fundamental processes, such as the ovarian cycle, are controlled. We have addressed this issue, in part, by providing separate coverage of major mammalian groups where this seemed appropriate. It has been left to the reader to ascertain the similarities and differences among them. In any case, we must ever be mindful in considering reproductive processes, from the control of ovulation to the initiation of parturition, not to extrapolate from one species to another without due reflection.

It is hoped that this book will be useful to all serious students of reproductive physiology be they scientists, teachers, or physicians.

THE EDITORS

Preface to the Second Edition

The second edition of this work represents the labor of some 61 groups of authors and is, therefore, marked by inevitable redundancies and lacunae. The six years that have elapsed between the first and second editions have seen dramatic and often unanticipated developments in some aspects of reproductive biology, with only little new understanding in others. But, as expected, the quantity and difficulty of the questions raised has increased manifold. We remain markedly ill-informed of the complex control systems that govern reproductive processes and surprised by the striking species differences in the accomplishment of common, fundamental reproductive tasks. The control of ovulation, the advent of puberty, and the initiation of parturition are but three cases in point.

We are deeply saddened by the untimely loss of Larry L. Ewing who contributed so much to the construction of the first edition with his wisdom and good humor, to say nothing of his hard work and devotion to the enterprise.

As these volumes were in the final stages of completion, the field also lost one of its great pioneers and most sagacious contributors, John W. Everett. We are privileged to count one of his last publications among the chapters of this work.

We hope that our expectations for the first edition have also been achieved in the second: scholarly, comprehensive examinations of the principal issues in reproductive physiology that will remain useful for a decade, at least, to all serious students of reproductive physiology.

THE EDITORS

Preface

In the decade that has passed since the last appearance of this work, the study of reproductive physiology has undergone monumental changes. Chief among these advances are organismal cloning with the attendant discovery of a diverse set of stem cells, and the cloning and sequencing of genomes with the resulting identification of proteomes. Therefore, the demanding challenge to our authors was incorporating the effect of these new findings into our understanding of the physiology of reproduction. Indeed, the 59 groups of authors of the current work have met this challenge in splendid fashion, providing a synthesis of the new information at molecular, cellular, and organismal levels of organization. The bibliographies of the chapters are extensive, and taken together with earlier editions of the work, provide coverage of all of the relevant literature of the topics being reviewed.

Given the explosion of new information, how can we claim the current work is comprehensive, when earlier editions had similar lengths (60 and 61 chapters) to the present one (59 chapters)? As noted by others (B. Alberts, et al. [1983]. *Molecular Biology of the Cell.* Garland Publishing, New York and London), this is because "there is a paradox in the growth of scientific knowledge. As information accumulates in ever more intimidating quantities, disconnected facts and impenetrable mysteries give way to rational explanations, and simplicity emerges from chaos." Whether this dictum can survive studies of proteomes with their large numbers of interacting proteins in regulatory networks remains to be seen.

The major changes in our understanding of reproduction and the passage of time have necessitated changes in the list of authors and editors. Of the 59 groups of authors, 44 are new to this edition. Only two editors remain from the first edition (Pfaff and Neill) so that five of seven section editors are new; separate sections and editors were also added to cover "Pregnancy" and "Lactation." Perhaps the most substantial change in this edition is the absence of Ernst Knobil as the editor-in-chief, due to his untimely death in the year 2000. This edition is dedicated to his memory, and is so signified by including a memorial article about him, and renaming the work *Knobil and Neill's Physiology of Reproduction*. His wise counsel has been missed, but we have endeavored to achieve his expectations expressed in the second edition that the work be a "scholarly, comprehensive examination of the principal issues in reproductive physiology" that will be useful "to all serious students of reproductive physiology."

A feature of the new edition that has not changed is the Foreword by Roy O. Greep. This pioneer in the discovery and characterization of pituitary hormones wrote a remarkably prescient account of the future of research in the field in 1988 that remains relevant even now eighteen years since it was written.

We welcome a new publisher, Academic Press, an imprint of Elsevier, to this third edition of the work. Particular thanks go to Dr. Jasna Markovac, Senior Vice-President of the Science and Technology Division at Elsevier, whose interest in this work dates to preparation of the second edition as its executive editor. The current edition owes its existence to her.

Finally, we thank the Senior Development Editor of this edition, Kirsten Funk, whose day-to-day work on these volumes has brought them to fruition. Her remarkable persistence and constant cheerfulness have sustained the prodigious effort required to produce such a voluminous work.

Jimmy D. Neill
Editor-in-Chief

Reproductive Behavior and Its Control

Knobil and Neill's Physiology of Reproduction, Third Edition
edited by Jimmy D. Neill,

CHAPTER 33

Neurobiology of Male Sexual Behavior

Elaine M. Hull,[1] Ruth I. Wood,[2] and Kevin E. McKenna[3]

Overview, 1730
Patterns of Sexual Behavior of Male Mammals, 1730
Description of Behavioral Elements, 1730
Sexual Motivation vs. Performance, 1732
Sexual Experience, 1733
Sexual Behavior Throughout the Lifespan: Puberty and Aging, 1734
Measures of Penile Function, 1734
Role of the Penis in Male Sexual Behavior, 1736
Anatomy of the Penis and the Mechanics of Erection, 1736
Neural Innervation of the Penis, 1738
Autonomic and Somatic Control of Ejaculation, 1740
Tactile Receptors on the Penis, 1741
Cellular Mediators of Erection, 1741
Role of Gonadal Steroids in the Control of Male Sexual Behavior, 1744
Time Course of Changes in Copulation Following Castration, 1744
Time Course of Changes in Copulation Following T Restoration, 1744
Role of T Metabolites in Maintaining and Restoring Copulation, 1745
Effects of Castration and Hormone Replacement on *Ex Copula* Penile Responses, 1746
Steroid Action on Steroid Hormone Receptors, 1747
Role of Progesterone in Male Sexual Behavior, 1748
Effects of Systemically or Intracerebroventricularly Administered Drugs, 1748
Dopamine, 1748
Norepinephrine, 1752
Serotonin, 1752
Acetylcholine, 1754
Gamma Amino Butyric Acid, 1755
Opioids, 1755
Oxytocin, 1755
Nitric Oxide, 1756
Gonadotropin-Releasing Hormone, 1757
α-Melanocyte–Stimulating Hormone and Related Peptides, 1757
Other Peptides, 1758
Brain Areas and Circuitry Implicated in the Control of Masculine Sexual Behavior, 1759
Sensory Inputs, 1759
Major Integrative Sites, 1768
Effects of Direct Application of Drugs Affecting Specific Transmitters, 1772, 1782
Major Motor Outputs, 1785
Other Brain Areas, 1786
Spinal Cord, 1787
Circuitry and Anatomical Interconnections, 1789
The Chemosensory Circuits, 1790
Somatosensory Input from the Genitals, 1791
An Ejaculation-Related Circuit, 1791
Efferents from the MPOA, 1793
Sexual Behavior in the Context of Mammalian Social Behavior, 1794
Unanswered Questions and Future Directions, 1794
References, 1795

[1]Department of Psychology, Florida State University, Tallahassee, Florida;
[2]Department of Cell and Neurobiology, University of Southern California School of Medicine, Los Angeles, California;
[3]Department of Physiology, Northwestern University Medical School, Chicago, Illinois.

This chapter summarizes recent contributions to our understanding of how hormones, the central nervous system, and peripheral structures interact to regulate male sexual behavior. Copulatory behaviors of the most commonly studied species are described, including the measures used to study them and the conceptual contexts that provide the rationales for those studies. Next, the innervation and functional mechanisms of penile erection and ejaculation are reviewed. Then the behavioral and physiological effects of gonadal hormones and systemically or intraventricularly administered drugs are summarized. Next, the functions of the brain areas most clearly implicated in the control of male sexual behavior, including effects of lesions and stimulation, local hormonal and pharmacological manipulations, and measures of neural activity, are reviewed. Finally, the functional interactions among the structures that mediate male sexual behavior are described, as well as similarities to neural mechanisms controlling other social behaviors. A set of questions raised by previous research that remain unanswered provide the final section.

OVERVIEW

Sexual behavior in animals, including the courtship that precedes it, is characterized by enormous diversity. This diversity assures that mating will occur with the optimal partner at the most appropriate time and place, in order to pass parental genotypes on to the next generation. Sexual behavior is an expression of reproductive physiology that developed days, months, or years earlier, in complex neuroendocrine interactions. Because this chapter focuses on male reproductive physiology, we will emphasize his behavior; however, the contributions of the female cannot be overemphasized. Mating requires the active participation of two individuals, and when mating is viewed from the female perspective, a different interpretation often emerges. However, a male-oriented perspective is warranted in the present endeavor, as long as one is mindful that neither the male- nor the female-centered perspective provides a full description of the complex reproductive interaction.

In this chapter, we summarize recent contributions to our understanding of how hormones, the central nervous system, and peripheral structures interact to regulate male sexual behavior. We begin by describing copulatory behaviors of the most commonly studied species, including the measures used to study them and the conceptual contexts that provide the rationales for those studies. We next review the innervation and functional mechanisms of penile erection and ejaculation. We then turn to the behavioral and physiological effects of gonadal hormones and systemically or intraventricularly administered drugs. Next, we summarize the functions of the brain areas most clearly implicated in the control of male sexual behavior, including effects of lesions and stimulation, local hormonal and pharmacological manipulations, and measures of neural activity. Finally, we describe the functional interactions among the structures that mediate male sexual behavior. We conclude with a set of questions raised by previous research that remain unanswered. This chapter is an updating of Hull et al. (438) and of Meisel and Sachs (630); we urge readers to consult those chapters for more in-depth coverage of some of the older research.

PATTERNS OF SEXUAL BEHAVIOR OF MALE MAMMALS

Description of Behavioral Elements

Precopulatory Behaviors

Male animals use species-specific displays to advertise their fitness and suitability as a partner. As they approach the female, they also gain information about her desirability and willingness to mate. The male often begins his approach by investigating the female's face and then her anogenital region, which exposes him to chemosensory or, in the case of some primates, visual stimuli that may arouse him and signal the female's receptivity. In rodents both partners may emit ultrasonic vocalizations, which are mutually arousing (306,322,779,1009).

Copulatory Behavior

Mounts. A receptive female typically remains immobile while the male approaches from her rear, clasps her flanks with his forepaws, and initiates a series of shallow thrusts with his pelvis (Fig. 1). The male's penis is usually at least partially erect during this thrusting. The female may assume a rigid lordosis posture, with her back flat or concave and her tail deflected to one side, exposing her vagina. The male usually has one or both hindfeet on the ground, although male macaques use a double foot-clasp mount, in which their hindfeet grasp the female's hindlegs (173). Male felids (517) and mustelids (160) grasp the female's neck with their teeth before or during the mount. A female may engage in proceptive behaviors before a mount, such as hopping and darting, followed by an abrupt stop, which may cause the pursuing male to run into her. Alternatively, a female

A

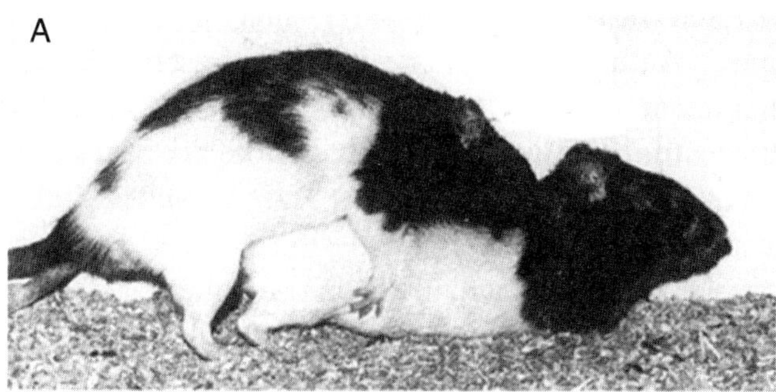

B

C

FIG. 1. Three views of copulating rats. Usually, as in **A,** one is unable to see the genitals. Therefore, without a physiological means of detecting intromission, scoring of copulatory behavior in rats and other small mammals relies upon relatively subtle differences in the patterns of movement to discriminate among mounts, intromissions, and ejaculations. Occasionally the genitals can be seen. In **B,** insertion is in progress, and the tight clasp suggests that this is an ejaculatory intromission. In **C,** the engorged glans penis can be seen extending forward just in front of the male's right hindleg. The position of his forelegs indicates that he is dismounting after ejaculation. In all three figures the female is in the characteristic receptive posture of lordosis, with body lowered and rump and head elevated. (We thank Ronald J. Barfield for contributing these photographs.)

may reject the male's advances by escaping, keeping her tail and hindquarters low, or kicking him.

Intromissions. Intromission is the defining event of copulation. If the male detects the female's vagina, he will perform a deeper, intravaginal thrust, followed by a rapid, springing dismount. This behavior pattern is reliably associated with penile insertion (778) and is used as the measure of intromission in rats and many other rodents. In mice and some other species, the male shows repeated intravaginal thrusts throughout the intromission (616), whereas in ungulates, the male may ejaculate immediately after vaginal insertion (105,566). Following an intromission, the male will typically groom his genitalia.

The number of intromissions preceding ejaculation may be an important factor in promoting fertility. Increased numbers of intromissions preceding ejaculation can increase the number of sperm in the ejaculate (3,952,953) and facilitate sperm transport in the female (3). Furthermore, females of some species require sufficient vaginocervical stimulation in order to ovulate or to trigger a progestational state [reviewed in (244)]. However, rats typically mate in a group setting, with multiple partners, so moderate decreases in the number of intromissions preceding ejaculation is unlikely to impair a male's ability to impregnate a female.

Ejaculation. Most male mammals ejaculate only after receiving stimulation from multiple intromissions. Ejaculation is characterized behaviorally by a deeper, longer thrust and a slow, relaxed dismount. Among canids ejaculation usually begins soon after penile insertion and is followed by swelling of the base of the penis, which results in a prolonged "lock" of the male to the female (89). Technically, seminal emission refers to the transfer of semen into the proximal urethra, whereas ejaculation refers to the forceful expulsion of the semen from the distal urethra. Rhythmic contractions of skeletal and striated perineal muscles, including the bulbospongiosus, ischiocavernosus, and anal sphincter, usually accompany ejaculation. In human males and females, such muscle contractions are associated with orgasm, the subjective correlate of the culmination of sexual excitement.

Postejaculatory Behavior and Satiety

Postejaculatory Interval. After ejaculation, the male typically grooms himself and enters a period of sexual quiescence. The postejaculatory interval may be as short as 30 seconds in Syrian hamsters or as long as hours to days in other species (243); in rats it lasts 5 to 10 minutes. During the postejaculatory interval male gerbils footstomp and rats emit ultrasonic (22 kHz) vocalizations. The first 50% to 75% of this interval is regarded as the absolute refractory period, when the male will not copulate again, regardless of the amount of stimulation. The ultrasonic vocalizations occur during the absolute refractory period, as does a pattern of synchronized EEG waves, which are otherwise characteristic of sleep (78). Blumberg and Moltz (124,125) suggested that the vocalizations may cool the lower forebrain, including the medial preoptic area (MPOA), which increases in temperature by 1°C during copulation. Blood may be redirected to the nose while air is being inhaled and then forcefully expelled, thereby producing the vocalization. Thus, the vocalization may be an artifact of the cooling process. During the relative refractory period, which occupies the remainder of the postejaculatory interval, the male may resume copulation if presented with a novel female or if subjected to mildly painful stimuli. In rats, the failure to copulate during the postejaculatory interval does not result from a failure of erectile function. In fact, *ex copula* touch-based erections may even be enhanced after ejaculation (716), and noncontact erections elicited by remote cues from a female can be observed during the period of ultrasonic vocalizations (844).

Sexual Satiety. After achieving seven to eight ejaculations most male rats reach sexual satiety, during which they may not copulate at all for 1 to 3 days. However, some males can resume copulation with a novel female, a phenomenon referred to as the "Coolidge effect." This refers to an apocryphal occasion when President and Mrs. Coolidge visited a chicken farm. Mrs. Coolidge was impressed that one rooster copulated repeatedly and asked the farmer to call this to Mr. Coolidge's attention when he later visited the facility. Upon hearing her observation, Mr. Coolidge observed that the rooster was mating with many different hens, and he asked the farmer to point that out to Mrs. Coolidge. Recent data suggest that a reduction in androgen receptors in the medial preoptic area and related areas may underlie the exhaustion (288).

Sexual Motivation vs. Performance

It is sometimes useful to measure the motivation to engage in sexual activity separately from the performance of the behavior. This distinction is similar to the classic distinction between appetitive and consummatory behaviors (951). Appetitive behaviors are considered to be those variable activities that bring an organism into contact with a certain stimulus, which would then elicit relatively stereotypic consummatory behaviors. Appetitive behaviors are thought to reflect an underlying, inferred motivational state. There have been arguments as to whether motivational states include a general undifferentiated arousal, analogous to an engine that can activate numerous different goal-directed behaviors, with sensory cues providing the steering wheel, or whether different energizing forces, with separate physiological bases, promote different aspects of behavior. In this review we agree with the position that sexual motivation includes conceptually separable influences on sensory and motor processes, which may be sensitized by gonadal hormones, as well as more general activational processes that may affect other behaviors as well. Furthermore, we acknowledge that it may be difficult to distinguish appetitive from consummatory behaviors. For example, some may consider anogenital sniffing of a female to be an appetitive behavior; however, it usually blends seamlessly into mounting, which in turn provides a stimulus for intromission, which is a precursor to ejaculation and postejaculatory behaviors. Thus, it may be difficult to parse the behavioral sequence into strictly appetitive and consummatory elements. Another problem is that measures of sexual motivation may be confounded with other factors. For example, latencies to mount or intromit have sometimes been used as measures of sexual motivation. However, drugs or lesions may impair sensory processing, performance of motor patterns, or the ability to achieve an erection, all of which may delay the onset of copulation without affecting sexual motivation. Therefore, it is preferable to use tests that are specifically designed to assess motivation and that do not confound sensory, motor, or learning ability with motivation per se.

Several measures of sexual motivation do not rely directly on copulatory behavior. Place preference tests allow the male to copulate with a female in one of two interconnected areas and, on other trials, to remain alone in the other. Later the male is placed into the center of the apparatus and allowed to spend time in either of the two areas. The amount of time he spends in the area previously associated with copulation, compared to time in the area that he inhabited alone, is used as an indication of sexual motivation. An advantage of this test is that little motor ability is required on the test day, and drugs administered during the conditioning phase should have been metabolized before the test day. However, the animal must be able to remember the stimuli in the area in which he copulated and associate them with the reinforcing effects of copulation. A similar procedure has been used with sexually naïve males, which are allowed to spend time near either of two screened areas containing an estrous or a nonestrous female (5,38a). No more motor activity is needed to approach the estrous female, compared to the nonestrous female; thus motor activity is not confounded with motivation. However, drugs administered on the test day may influence sensory processing. A third test uses an obstruction apparatus, in which a male must cross a barrier or electrified grid in order to reach a female. However, this test confounds sexual motivation with motor ability. A fourth measure uses a bilevel apparatus in which a male and female initially copulate (653). Later, the male is placed into the apparatus alone, and the number of level changes that he makes, presumably in search of the female, is used as the measure of his sexual motivation. This test is easily scored, but it, too, confounds motor ability with sexual motivation. A fifth measure is lever pressing for a secondary reinforcer that was previously paired with copulation [reviewed in (276)]. The difficulty of this test can be adjusted by increasing the number of bar presses required for presentation of the secondary reinforcer. However, as with tests described above, it confounds both motor ability and the ability to learn the secondary reinforcement with sexual motivation. One measure that does not confound motor and motivational factors is the cross-maze or X-maze (441,995). A female is placed into a goal box at the end of one of four arms of a maze; other stimuli are placed into the other three arms, or they remain empty. The male is placed in the center start box, and the latency to reach an arm on each of 10 trials is recorded. Sexual motivation is measured as the percentage of trials on which the male chooses the female's goal box, out of the total number of trials on which he runs. Motor ability is assessed by the speed of running to all goal boxes and the number of trials on which the male does not run at all. Again, drugs administered on the test day may influence sensory processing or the ability to remember the cues that mark the female's compartment. A somewhat similar procedure uses a goal compartment/runway apparatus, in which a male can enter the female's chamber through a one-way door, copulate at will, and leave through a second one-way door that leads to a runway that connects to the first door (96). Time spent in the female's compartment, run latency, run duration, number of copulatory behaviors, and latency to depart are recorded. Time in the female's compartment and latency to depart are relatively pure measures of sexual motivation, whereas run latency, run duration, and number of copulatory behaviors may reflect both motivation and motoric ability. Furthermore, as with the X-maze, drugs administered on the test day may influence sensory processes. Therefore, there are no "pure" tests of sexual motivation that are devoid of any potential confounds. It would be desirable to perform multiple tests that assess sensory and motor factors and the ability to remember conditioned cues, in addition to the measure of sexual motivation. At least, reports of effects on sexual motivation should include the caveat that alterations in sensory, motor, or memory abilities may influence the measure of motivation.

Sexual Experience

Previous sexual experience increases sexual "efficiency." Experienced males copulate with shorter latencies to mount, intromit, and ejaculate and with fewer mounts and intromissions preceding ejaculation (113,242,518,762). (However, as noted above, too few intromissions before ejaculation may result in lower fertility.) Precontact ultrasonic vocalizations also increase with experience (113). Furthermore, experienced males may be less susceptible to the effects of castration, various brain lesions, and exposure to a novel environment (46,110,196,238,492,549, 656,762,832,856). Sexually experienced, but not sexually naïve, male rats showed decreased latencies and increased numbers of touch-based erections following exposure to an inaccessible female (851). This effect was not blocked by section of the hypogastric nerve, which is thought to mediate "psychogenic" erections. In addition, ejaculation elicited more cells that were immunoreactive for Fos, the protein product of the immediate-early gene *c-fos*, in the MPOA (574) and in the shell region of the nucleus accumbens (558) of sexually experienced male rats, compared to sexually naïve males. In the latter study, even the odor of an estrous female elicited more Fos immunoreactivity in the nucleus accumbens shell of experienced males than in naïve animals. Therefore, one mechanism for the greater copulatory efficiency and resistance to various disruptions may be increased sensitivity to sexually relevant stimuli.

Sexually naïve males, pre-exposed for 5 minutes to stimulus animals behind a Plexiglas barrier, showed faster running times in a straight runway to an estrous female than to a nonestrous female (560). Therefore, even naïve males show unconditional motivation for female precopulatory cues. However, only males that had previously copulated to one ejaculation showed faster run times to both hormone-treated and untreated females on a second trial; one previous intromission did not affect running time on the second trial, nor did lack of exposure to a female between the trials (560). Thus, the reinforcement of ejaculation was necessary to increase the motivational impact of cues from the female. However, timing of copulation may also be important for its rewarding effects. Male rats copulating with females that controlled the pace of copulation did not show conditioned place preferences, unlike males that controlled the copulatory pace (155a).

Male rats that had previously copulated in the context of a neutral odor ejaculated preferentially with the female scented with the conditioned stimulus, although they intromitted indiscriminately with females bearing that odor or no odor (481). Both ejaculation and the postejaculatory interval, with the female present, were necessary unconditioned stimuli for the preference to develop (477) [reviewed in (764)]. Furthermore, exposure to the conditioned odor resulted in an increase in Fos immunoreactivity in areas different from those that were activated by estrous odors, but activation by both types of odor converged on the nucleus accumbens core (478).

Sexual Behavior Throughout the Lifespan: Puberty and Aging

Copulatory ability is acquired at variable rates in pubertal male rats, with the first mounts occurring between 40 and 50 days of age, the first intromissions between 44 and 75 days, and the first ejaculations between 48 and 75 days [reviewed in (630)]. Gonadal hormones are required, as prepubertal castration prevented the development of sexual behavior (520). Exogenous testosterone (T) or estrogen (E) hastened the onset of copulation (81,915). However, the natural onset of the behavior occurred several days before the pubertal rise in T (852,914). Touch-based erections also occurred before the hormonal surge (852). However, the maturation of genital structures or their innervation was not the limiting factor, as spinally transected male rats showed erections and anteroflexions approximately 10 days before spinally intact animals.

Male hamsters, unlike rats, begin to show copulatory behavior only after the pubertal increase has begun, although before it reaches its zenith (827). Lack of T during puberty resulted in decreases in the numbers of mounts, intromissions, and ejaculations when T was replaced in adulthood, compared with castrates that had T replacement during puberty (879). These animals also showed a facilitation of female-typical behavior when implanted with E and progesterone in adulthood, compared with males with T replacement during puberty. Repeated sexual experiences did not compensate for the behavioral deficits, relative to castrates that had T during puberty.

In male hamsters the odor of a receptive female activated Fos expression in the MPOA even before puberty (826); however, an increase in dopamine (DA) activity in the MPOA (measured as an increase in the DA metabolite DOPAC) in response to female odor occurred only after puberty (878). Therefore, some apparently hormonal factor is permissive for this important area to become fully responsive to stimuli from the female. Sisk and her colleagues have suggested that puberty is a second organizational period during which gonadal hormones permanently alter neural processing in areas important for sexual behavior (827,879,906).

Copulatory ability in aged males is very variable. In aged male monkeys sexual behavior was not correlated with T levels (164,165), and exogenous T did not improve behavior in either castrated or intact old monkeys, compared with young males (769,770). In 30-month-old mice, T levels, gonadal mass, and spermatogenesis were all lower compared to 6-month-old animals; however, some aged males could copulate, whereas others could not, despite a similarity in T levels (142). A thorough study of young and old gonadally intact rats and young and old T-replaced castrates also suggested that diminished T levels did not account for the inability of old males to ejaculate (166). In that study old gonadally intact males had lower levels of serum T, lower levels of nuclear androgen receptors (AR) in the preoptic area/anterior hypothalamus, and lower aromatase activity in the preoptic area. Testosterone replacement in old castrates resulted in similar T levels as in T-replaced young males and higher levels than in young intact males; however, only 25% of the old males ejaculated, whereas all of the young males did. Although there were fewer nuclear ARs in the old T-replaced castrates than in young T-replaced castrates, the numbers were similar to those of young intact males. Thus, neither serum T nor nuclear ARs could account for the behavioral deficits in the old males. However, a subsequent study reported that nuclear estrogen receptors (ER) in the amygdala were consistently lower in old males than young males, and that T replacement did not restore receptors to the levels of young males (830). They suggested that the ability of old males to bind E, derived from T, may be a limiting factor in their androgen responsiveness.

Measures of Penile Function

Observations during Copulation

The most direct method to study penile erection is observation during copulatory behavior. This is relatively straightforward in many species. However, this can be difficult in rodents, because erections are very brief and often obscured from view. Erection, intromission, and ejaculation are often assumed based on the animal's behavior. For example, rodents display characteristic behaviors at ejaculation. However, in experimental situations using brain lesions or administration of a drug, ejaculation could be affected without obvious changes in the behavior. In such cases, experimenters may inspect the female's vagina immediately after copulation for evidence of ejaculated sperm or examine the cage floor for a copulatory plug of coagulated semen. This may underestimate ejaculation, because males often eat their ejaculate while grooming their genitalia immediately after ejaculation (724). Direct observations of penile erections can be made using a mirror placed beneath a clear cage. Because genital reflexes are difficult to measure while the male is copulating, paradigms have been developed for monitoring them *ex copula*. However, evidence has been presented that *in copula* and *ex copula* erections may differ significantly in their physiological, CNS, neurochemical or hormonal control (843). Thus, extreme caution is necessary in predicting the effects of a manipulation on copulation based on *ex copula* tests. To overcome these problems, techniques have been developed to measure physiological responses during copulation. These techniques have included chronic implantation of electrodes to measure the electrical activity of the striated perineal muscles (418). Telemetric measurements of penile pressures have also been used in a variety of contexts (337). These methods are technically challenging, but provide robust physiological measurements during normal sexual behavior.

Ex Copula Measures of Penile Function

Spontaneous or Drug-induced Erections. Erections can often be observed when a male is alone in his home cage or a neutral arena, in the absence of an apparent sexual stimulus. Therefore, these erections are referred to as spontaneous. The number of spontaneous erections can be increased by administration of various drugs systemically or into specific CNS sites. These erections are referred to as drug-induced erections. These usually involve extension of the engorged glans beyond the sheath and are typically accompanied by genital grooming. With some drug administration, (e.g., the DA agonist apomorphine) a characteristic pattern of penile erection, yawning and stretching is observed (639). The obvious advantage of this model for the study of erection is its ease and simplicity. However, the probability of drug induced erections may be modified by prior copulation (853).

Noncontact Erections. Another model for studying erection in the unanesthetized rat is the noncontact erection (NCE). Male rats develop penile erections in response to the presence of an estrous female even when physical contact is prevented (854). Volatile odors from the estrous female have been shown to be the necessary and sufficient stimulus for this response (499,842). These erections are generated by the central nervous system without pelvic sensory stimulation. Their motor pattern is very similar to spontaneous and drug induced erections and may be the closest model for psychogenic erections in humans. One concern is that the dependence on olfactory stimuli of this response is unlike human sexual response, which is more dependent on visual and auditory cues. Thus, NCE in rats and psychogenic erection may have very different CNS pathways, at least at the level of forebrain sensory mechanisms. Nevertheless, this model is an important contribution, because the response is produced by supraspinal mechanisms and not spinal reflex pathways. Further, compared to copulatory behavior, it does not require complex motor behaviors and social interactions. Analysis of NCEs following discrete brain lesions has demonstrated important distinctions in the neural control of copulatory performance and erectile capability (554). It should also be noted that this response is seen only in pigmented rats, not in albino strains (841).

Touch-based Erections, Anteroflexions, and Seminal Emissions. Reflexive, or touch-based, erections have been elicited by manually stimulating the penis, primarily in the dog (90,381). However, penile erection is inhibited, not stimulated, by tactile stimulation of the penis in rats (383). A different method for evoking erection in rats was developed by Hart (383). A similar technique has been described in mice (838). With the rat restrained on his back, the penile sheath is retracted, exposing the glans penis. This gives rise to a series of erections, which are seen as tumescence of the glans penis due to engorgement of the corpus spongiosum (Fig. 2). Penile anteroflexions (previously called “flips”) also occur. These are caused by erection of the corpus cavernosum and contractions of the ischiocavernosus muscles, causing the penis to rise up from its normal posteroflexed position. There are three gradations of glans erection: in the first, the penile body elongates and rises; in the second, the glans becomes engorged and slightly flared; and in the third, the glans flares intensely into a cup. During copulation the cup is important for depositing the ejaculated semen around the cervix, where it coagulates to form a solid “copulatory plug.” Without this plug, much of the semen seeps out of the vagina, and pregnancy rarely occurs (839). Seminal emission occasionally occurs in the *ex copula* setting.

The *ex copula* and *in copula* erections are similar in physiological terms, based on parasympathetically mediated vasodilation in the penis and contractions of the striated penile muscles (383,418,846,990). However, somewhat different hormonal and neurochemical regulation of the two types of erections have been noted (840) (see below). The sensory stimulus responsible for this *ex copula* erection is unclear. It has been postulated that the continuing pressure of the sheath around the base of the penis provides the sensory basis.

Urethrogenital Reflex. A model has been described for studying sexual reflexes in anesthetized male and female rats (626). In urethane anesthetized, acutely spinalized rats, complex coordinated sexual responses can be elicited by a variety of pelvic stimuli, including stimulation of the dorsal nerve of the penis. It was shown that urethral distension is a quantitative and highly reproducible stimulus. Hence, this response has been referred to as the urethrogenital (UG) reflex. In male rats, the UG reflex consists of clonic contractions of the perineal muscles, rhythmic firing in the cavernous nerve, penile erections, and ejaculation.

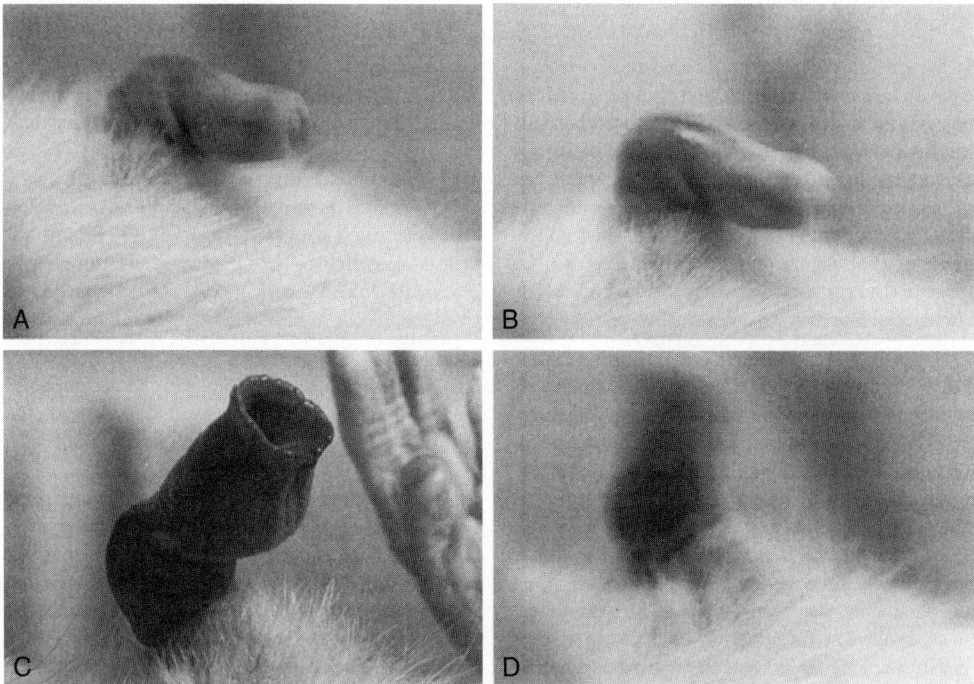

FIG. 2. Reflexive erections displayed by supine male rats after retraction of the penile sheath. Several days earlier the suspensory ligaments had been surgically removed, thereby preventing retraction of the penis during the test and permitting better visualization of the distal penile body (perpendicular to the torso in B) and of the entire glans (parallel to the torso in B). In A–C, the glans is directed toward the tail, its normal orientation. In D the glans points rostrally, the orientation necessary to achieve intromission. **A:** The quiescent (flaccid) penis. **B:** Tumescence and elevation of the penile body without glans erection. **C:** Intense erection of the glans ("cup") and of the penile body. **D:** Anteroflexion ("flip") of the penis due to a straightening of the penile body. The blurring is due to the speed of the response relative to the shutter speed (1/30 sec). [The photographs are from (537a), with permission from Elsevier.]

Thus, this reflex has been viewed as a model for both penile erection and ejaculatory reflexes. The perineal muscles were all activated simultaneously, as is seen in human climax (319,755) and in rats during copulation (418,677). The somatic muscle bursts were synchronized with bursts in the cavernous nerve, driven by both pelvic (parasympathetic) and hypogastric (sympathetic) nerve bursting, indicating a multisegmental organization. This reflex can be evoked only in animals that have high spinal transections or lesions in certain brain areas. A group of serotonergic neurons in the medullary reticular nucleus has been identified as a source of descending inhibitory effects on spinal sexual reflexes in this animal model (598) (see below).

ROLE OF THE PENIS IN MALE SEXUAL BEHAVIOR

Anatomy of the Penis and the Mechanics of Erection

Successful fertilization in mammals is dependent on an erect penis to deliver sperm into the female reproductive tract. The anatomy of mammalian penes varies between a fibroelastic organization and a highly vascular organ. In ungulates, such as sheep and goats, erection occurs primarily by extrusion of the fibroelastic penis. This is largely accomplished by the action of penile muscles. There is little engorgement of this type of penis during erection. In other animals, such as the rodent, cat, dog, monkey, and human, erection involves penile enlargement and stiffening. This is accomplished by coordinated vascular relaxation and contraction of striated penile muscles. The relative importance of these two components differs across species.

The general plan of the penis is common among all mammalian species (Fig. 3). The erectile structures of the penis consist of the paired corpora cavernosa, which occupy most of the shaft of the penis, and the corpus spongiosum, which surrounds the urethra and is expanded proximally as the penile bulb and distally as the glans penis. In most species, the two corpora cavernosa are fused, creating a single vascular space. Experimentally and clinically, this allows a drug injected anywhere into the corpus cavernosum to reach the entire structure. Likewise, pressure throughout the corpora can be monitored from a

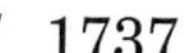

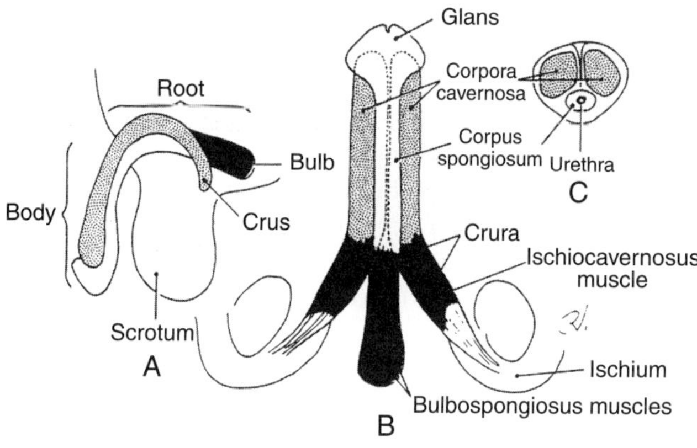

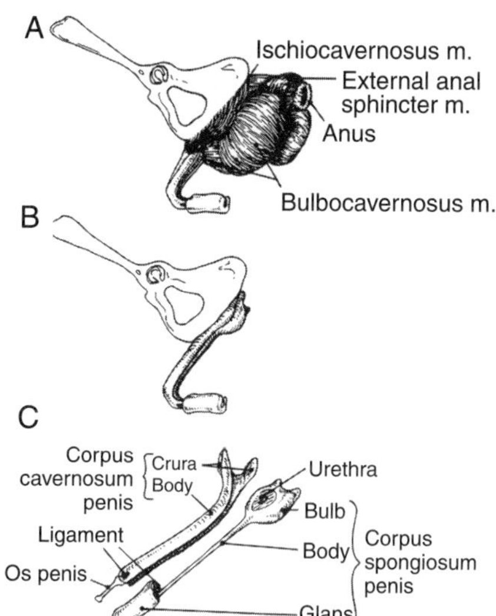

FIG. 3. Some basic structural similarities in penile anatomy illustrated from the human (top: **A**, lateral view; **B**, inferior view with penile body lifted; C, cross section) and rat (bottom: **A**, lateral view including all muscles; **B**, lateral view with muscles removed; **C**, "exploded" view). In both species, the penile crura, attached to the ischium, are covered by the ischiocavernosus muscles and are continuous with the body of the penis, composed primarily of the paired corpora cavernosa. The penile bulb, wrapped by the bulbospongiosus (=bulbocavernosus) muscles, gives rise to the corpus spongiosum, which terminates in the glans penis. The rat muscle identified as the external anal sphincter is also known as the levator ani or the dorsal bulbospongiosus. [Figures of human penis from (1014a) are reprinted by permission of Oxford University Press. Those of the rat penis from (391a) are reprinted by permission of Pergamon Press.]

single site. The corpora cavernosa are composed of vascular tissue that consists of sinuses (the trabeculae) whose arterial blood supply arises from the resistance helicine arteries. The helicine arteries are fed from the deep penile cavernosal artery. The trabeculae are drained by the subtunical venules that join to form the emissary venules and eventually form the cavernosal veins. The corpora cavernosa are surrounded by a tough capsule, the tunica albuginea.

The proximal ends of the corpora taper into tails called crura. These are anchored onto the ischiopubic ramus and surrounded by the striated ischiocavernosus muscles. The corpus spongiosum is a spongy erectile tissue with a similar histologic appearance to the corpus cavernosum. The proximal portion of the corpus spongiosum, the urethral bulb, is surrounded by the striated bulbospongiosus muscle. This was previously referred to as the bulbocavernosus muscle. Some animals' penes (but not humans) contain a bone (called the os penis), which lies centrally in the body and extends forward to the glans.

In normal behavioral and physiological settings, the erection and detumescence of the corpora cavernosa and corpus spongiosum are usually coordinated. However, in certain experimental and pathological conditions, they may act quite independently, in part because there is at least some independence of their vascular (282) and neural (168,850) effector systems. Thus, the term erection may be imprecise or inadequate, without further characterization of specific physiological processes in the penis. This distinction is essential when comparing literature discussing different species. In humans, erection almost always refers only to the penile shaft, without specifying whether the glans is erect. However, in rats and other rodents, the term erection in *ex copula* settings is used to refer to engorgement of the glans, whereas changes in the penile body are termed anteroflexions or flips. To further complicate matters, engorgement of the corpora cavernosa in experiments with anesthetized rats is also called erection.

Erection results from a combination of vascular, smooth muscle, and striated muscle actions. The relative contribution of each of these factors varies from one species to another. Erection is caused by relaxation of smooth muscles in the arterial supply to the erectile tissue, as well as in the trabeculae. This results in an increased inflow of blood that fills the corpora and causes expansion of the erectile tissue. The increased intracavernous pressure compresses and occludes the venous outflow, trapping blood in the corpora. Contraction of the striated muscles overlying the vascular reservoirs (penile bulb and crura) increases penile rigidity (319,526). In some species, such as the rat, the striated penile muscles are critical for erection. This is probably less true in the human, although there is some controversy over the role of striated penile muscles in human erection (319,527,870).

Physiological measures of intracavernous (107) and intraspongiosus (871) pressures, and of the electromyographic activity of the ischiocavernosus and bulbospongiosus muscles (393,418,871) have revealed that reflexive erections in the rat consist of plateaus of increased intracavernous and intraspongiosus pressures. The penile pressures are below systolic blood pressure for several seconds. Peaks of suprasystolic pressure occur on top of these pressure plateaus. The occurrence of pressure plateaus often occurred before any visible behavioral event. The peaks, lasting about 1 second, are generated by bursts of activity of the ischiocavernosus and bulbospongiosus muscles and are seen during glans erections, cups, and anteroflexions, with the highest pressures occurring during anteroflexions. Excision of the bulbospongiosus muscle (839) or denervation of this muscle (681) suppressed the cups. Excision of the ischiocavernosus muscle eliminated the anteroflexions (839). Section of the cavernous nerves eliminated corpus cavernosum erections (cups and anteroflexions) but not glans erections (850). The explanation for this effect is unclear. Vasodilator pathways to the corpus spongiosum other than the cavernous nerve may exist. Alternatively, the withdrawal of sympathetic tone in the glans or sympathetically mediated shift of blood flow to the glans may be sufficient to induce glans erections. In anesthetized dogs, monkeys, and rats, plateau increases of penile pressure were elicited by pelvic and cavernous nerve stimulation, and stimulation of the pudendal nerve was required to record suprasystolic penile pressure rises (339,569,570). Therefore, erections need a strong coordination of the sacral parasympathetic outflow to the penis with the pudendal outflow to the perineal striated muscles.

Neural Innervation of the Penis

Penile erection depends on the coordination of sympathetic, parasympathetic and somatic innervation of the penis and related structures (Fig. 4). The pelvic nerves (primarily parasympathetic and proerectile), hypogastric nerves (primarily sympathetic and antierectile), and pudendal nerves (somatic sensory and motor) comprise the three major pathways. The pelvic nerve fibers originate in the lumbosacral spinal cord and travel via the pelvic plexus and cavernous nerve to the penile corpora and vasculature. Although this nerve is primarily parasympathetic, it apparently also carries some sympathetic postganglionic fibers, at least in rats (226,227). This is the major proerectile pathway in all mammals studied so far (400,516,568, 794,885,941,978). The pelvic ganglion, or pelvic plexus, provides sympathetic and parasympathetic innervation to the penis via the cavernous nerve. The pelvic ganglion of rats is sexually dimorphic; the male ganglion contains two to four times as many neurons as the female's (360).

The perineal striated muscles (bulbospongiosus and ischiocavernosus muscles) are innervated by the pudendal nerve. Pudendal motoneurons are located in Onuf's (Onufrowicz) nucleus, located in the ventral sacral cord. In the rat, pudendal neurons are located in two nuclei: the dorsomedial (DM; also named the spinal nucleus of the bulbocavernosus, or SNB) and dorsolateral (DL) nuclei in lumbosacral segments (721,876). The pudendal nerve divides into distinct sensory and motor branches near the ischium (625,963). Both branches of the pudendal nerve also carry sympathetic efferent fibers (442,625). The sensory branch is primarily composed of fibers from the dorsal nerve of the penis. The motor branch of this nerve has primarily somatic fibers that originate in the lumbosacral cord and innervate the striated penile muscles (136,137,625,862,876,962,964). The contraction of these muscles on a flaccid penis does not elicit erection. However, when the muscles contract on an erect penis, penile rigidity and intrapenile pressure dramatically increase to suprasystolic levels (870).

The third pathway is the sympathetic innervation of the penis. Preganglionic sympathetic fibers reach the penis via the lumbar splanchnic nerves or via the paravertebral sympathetic chain (449). They relay on ganglionic neurons in the hypogastric plexus or in the paravertebral sympathetic chain ganglia. Sympathetic postganglionic axons reach the penis through a variety of routes. One route begins with the hypogastric nerve, which extends to the pelvic plexus, from which the cavernous nerve extends to the penis. Another route is the paravertebral sympathetic chain, which connects via the pelvic nerve to the pelvic plexus and then via the cavernous nerve to the penis. In the rat, this nerve provides the majority of adrenergic sympathetic fibers to the penis (128,225,226,332,339). Numerous experiments performed so far in rats, rabbits, cats, dogs, and monkeys have identified an antierectile sympathetic contingent in the paravertebral sympathetic chain (247,339,460,569,570,907). It is likely that a basal sympathetic tone maintains the penis in a flaccid state. Consistent with this hypothesis is the observation that the intracavernous injection of the α-adrenergic antagonists phenoxybenzamine and phentolamine elicits erection in humans (140).

A potential proerectile role of sympathetic axons in the hypogastric nerve has been debated. Electrical stimulation of the hypogastric nerve decreased penile blood flow in intact anesthetized rabbits (516). In rabbits (907), rats (228), and dogs (159,247) stimulation of the hypogastric nerve had no effect on intracavernous pressure unless the pelvic nerve had been previously sectioned. However, stimulation of the hypogastric nerve in anesthetized cats increased penile volume (43). In some studies, erections persisted following lesions of the sacral spinal cord or the parasympathetic outflow in cats (828). Noncontact erections were also unaffected by section of the hypogastric nerve (224).

A proerectile role for sympathetic pathways has been identified (332). Stimulation of the MPOA elicited

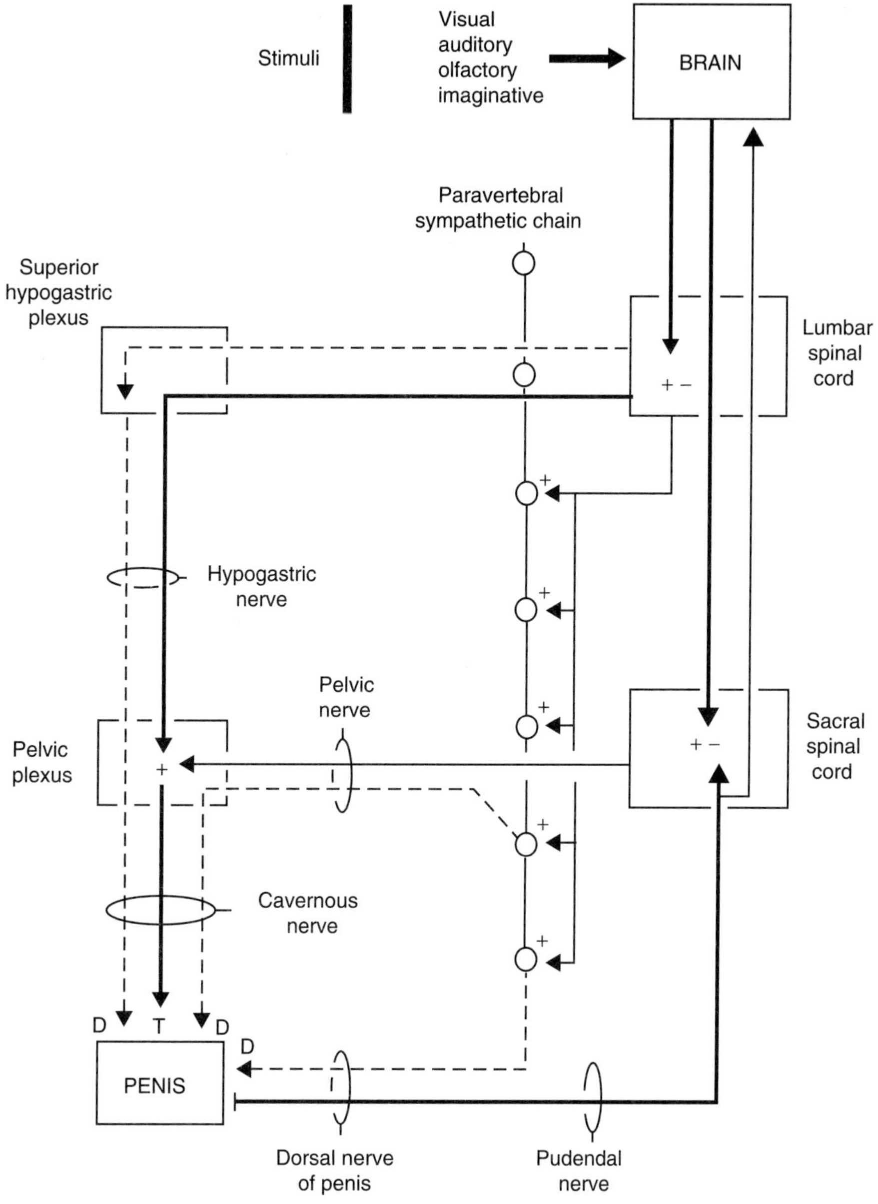

FIG. 4. Schematic diagram of the excitatory (+) and inhibitory (–) innervation of the penis for the regulation of tumescence (T) and detumescence (D). Note that the motor branch of the pudendal nerve is not depicted, nor are the striated muscles to which the pudendal nerve projects. Also, the penis is shown as a single structure, although recent evidence suggests that the innervation of the penile body and glans are not identical. The depicted innervation may better represent that of the body than that of the glans. [Diagram is from (237a) and is reprinted by permission of Harwood Academic Publishers.]

penile erection by both parasympathetic and sympathetic mechanisms (Fig. 5). The parasympathetic mechanism is the familiar vasodilation mediated by activation of the cavernous nerve. Bilateral section of the parasympathetic input to the penis abolished the erectile response. These data confirm that the primary proerectile influence is via the parasympathetic innervation. However, sympathetic fibers also contribute to erectile function. Acute or chronic section of the paravertebral sympathetic chain or destruction of sympathetic fibers by the use of the catecholamine neurotoxin 6-hydroxydopamine (6-OHDA), resulted in a significant decrease in the erectile response. The investigators suggested that proerectile CNS stimulation resulted in an increase in vascular tone in other pelvic structures, thereby diverting blood flow away from those structures and toward the penis. Therefore, stimulation of the MPOA activates a coordinated response that is mediated by both parasympathetic and sympathetic outflow.

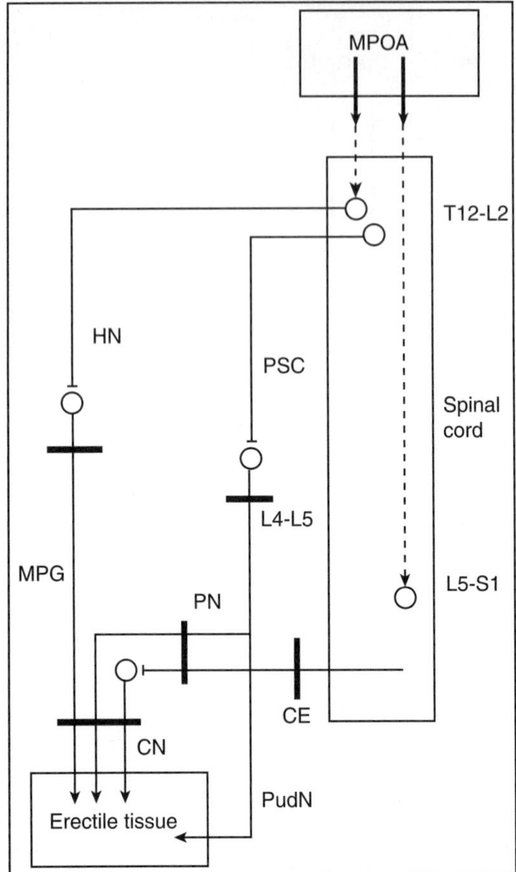

FIG. 5. Diagrammatic representation of peripheral autonomic pathways potentially involved in erectile response elicited by medial preoptic area (MPOA) stimulation. Sites of neural lesions are represented by black bars. No direct projections from MPOA to spinal autonomic nuclei have been reported; dotted lines represent hypothetical pathways. CE, cauda equina; CN, cavernous nerve; HN, hypogastric nerve; L4-L5, 4th and 5th lumbar levels of the paravertebral sympathetic chain; L6-S1, 6th lumbar and 1st sacral level of the spinal cord; MPG, major pelvic ganglion; PN, pelvic nerve; PSC, paravertebral sympathetic chain; PudN, pudendal nerve; T12-L2, 12th thoracic to 2nd lumbar level of the spinal cord. [Figure is from (335) and is reproduced with permission from the American Physiological Association.]

Autonomic and Somatic Control of Ejaculation

Ejaculation consists of two phases: emission and expulsion. Emission involves parasympathetic responses, such as secretion of seminal fluids from epithelial cells and the accessory sex glands, as well as sympathetic responses that move the seminal fluids to the proximal urethra. The coordinated emission pattern consists of closure of the bladder neck, contraction of the seminal vesicles, and contraction of prostate and ductus deferens (126,246,802). Bladder neck closure is elicited by sympathetic nerve stimulation (472,474). Stimulation of the sympathetic lesser splanchnic nerves induces contraction of the seminal vesicle in rats (427). Stimulation of sympathetic (and also parasympathetic) nerves elicits a strong contractile response in the ductus deferens, promoting transport of sperm and epididymal fluid towards the urethra (471,488,1010). In addition, stimulation of the sympathetic nerves (and also the parasympathetic nerves to a lesser extent) causes expulsion of prostatic fluid into the urethra (992). Thus, emission is primarily controlled by sympathetic mechanisms, with some contribution from parasympathetic nerves (Fig. 6).

Expulsion of the semen, which involves both sympathetic and somatic outputs, is a result of rhythmic contractions of smooth muscles of the urethra and striated perineal muscles, primarily the bulbospongiosus muscle. The ejaculatory reflex is initiated primarily by sensory receptors in the glabrous skin of the glans penis [reviewed in (402)]. However, ejaculation occurring during nocturnal penile tumescence or certain other conditions may not require direct stimulation of the glans. Sensory input travels via the sensory fibers of the dorsal penile nerve and pudendal nerve to the spinal cord and also via the hypogastric nerve

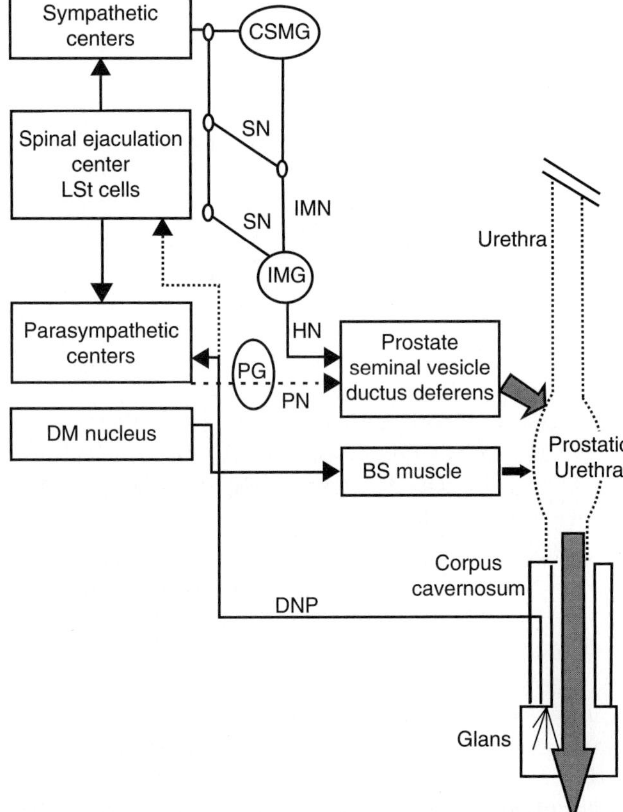

FIG. 6. Schematic presentation of the spinal ejaculation generator, demonstrating the spinal cord components and autonomic, somatomotor, and sensory efferents and afferents. CSMG, celiac superior mesenteric ganglia; IMG, inferior mesenteric ganglia; PSC, paravertebral sympathetic chain; IMN, intermesenteric nerve; HN, hypogastric nerve; PG, pelvic ganglia; PN, pelvic nerve; SN, splanchnic nerve; DNP, dorsal nerve of the penis; DM, dorsomedial; BS bulbospongiosus. [Reprinted from (206) with permission from Elsevier Press.]

to the paravertebral chain of sympathetic ganglia. Expulsion of semen is controlled by somatic fibers from the ventral horn of S2-S4 spinal segments, which travel in the motor branch of the pudendal nerve to the muscles of the pelvic floor, including the bulbospongiosus and ischiocavernosus muscles. A group of neurons in the lumbar spinal cord has been identified as responsible for generating ejaculation (958) [reviewed in (206)].

Tactile Receptors on the Penis

The mammalian penis is equipped with both slowly adapting (SA) and rapidly adapting (RA) mechanoreceptors, which are differentially distributed on the glans and may have different functions. Both types of receptors were found throughout the length of the glans of cats, but the SA type predominated at the distal end, and RA receptors were concentrated at the proximal end, just distal to the anterior margin of preputial attachment (456). The glans has a high concentration of free nerve endings, which may mediate low force, slowly adapting afferent responses (155,482) as well as high threshold mechanoreceptors (450). These latter afferents have an inhibitory effect on reflex contractions of the penile muscles. The glans of several species has a unique sensory ending, the "genital end bulb" (352) or "lamellated corpuscles" (452). These are prominent under the penile spines of the rat and may mediate the rapidly adapting response to stroking of the glans and manipulation of the spines (457). It has recently been described that these sensory structures contain melanocortin type IV receptors (969). The specific physiological significance of this finding is unknown, although melanocortin IV agonists have been reported to increase erections (see section on α-Melanocyte–Stimulating Hormone and Related Peptides for a discussion of systemically administered melanocortin drugs). Ruffini corpuscles in the glans may mediate deep pressure responses (482). The penis is remarkable for its relative lack of Merkel or Meissner corpuscles (368,452), which probably accounts for the relatively high tactile threshold of the penis.

It has been proposed (57,385) that the mechanoreceptors of the glans serve to detect the vagina, but Johnson et al. (456) suggested that SA receptors could also serve to transmit information concerning skin movement, pressure, and state of erection. Penile mechanoreceptors are more responsive when the penis is erect (451) or near core body temperature (455). The role of the penile spines in sensory processes is unclear. Male marmosets whose penile spines had been removed with a depilatory required more thrusts to achieve intromission, presumably because of difficulty in detecting the vaginal orifice (248). Neither intravaginal thrusting nor ejaculation was significantly affected. However, a sensory role for penile spines in the copulatory behavior of rats is unlikely. Castrated male rats whose penile spines have regressed, but whose copulation is maintained by estradiol, display the ejaculatory pattern after the normal number of intromissions, rather than the increase that would be expected if the spines have an important sensory role (86,717,912).

Cellular Mediators of Erection

Penile tumescence is the result of relaxation of smooth muscle within the penis. Both arterial and trabecular smooth muscle are involved. Relaxation of the arterial smooth muscle leads to increases in penile blood flow, and trabecular smooth muscle contractions lead to opening of the sinusoids in the penile erectile tissue. Trabecular smooth muscle cells are functionally coordinated by means of gap junctions, which allow for the electrotonic spread of current and second messenger signals from cell to cell. This system causes the trabecular smooth muscle to behave as a syncytial tissue. A consequence of this is that the autonomic input does not have to innervate each cell, but can exert its effects by synapsing with relatively few cells, which are, in turn, functionally connected in a tissue-wide network (176).

The state of contractile tone is determined by a dynamic balance of contractile and relaxant mechanisms (40,955). This contractile tone is largely mediated by the levels of intracellular calcium concentration. Decreases in intracellular calcium lead to relaxation; increases lead to contraction. Increases in intracellular calcium then activate a sequence of biochemical events, which require ATP and result in actin and myosin cross bridge formation, resulting in contraction. Trabecular smooth muscle can also exist in a state of contraction that requires very low levels of ATP, referred to as a latch state, whose exact mechanisms are unknown. Trabecular smooth muscle probably spends a considerable portion of time in this state, and it is from this state that erectile relaxation takes place.

Nitric oxide (NO) is the primary mediator of vascular relaxation of the corpora cavernosa [reviewed in (41,118,446)] (Fig. 7A). Nitric oxide is a short-lived gaseous messenger that is produced by the enzyme nitric oxide synthase (NOS). In the penis, there are two different isoforms responsible for the production of nitric oxide. NOS type I (NOS I, sometimes referred to as neuronal or nNOS) is present in the parasympathetic nerves innervating the erectile tissue. Another NOS isoform, NOS type III (NOS III, also referred to as endothelial or eNOS) is present in the endothelium

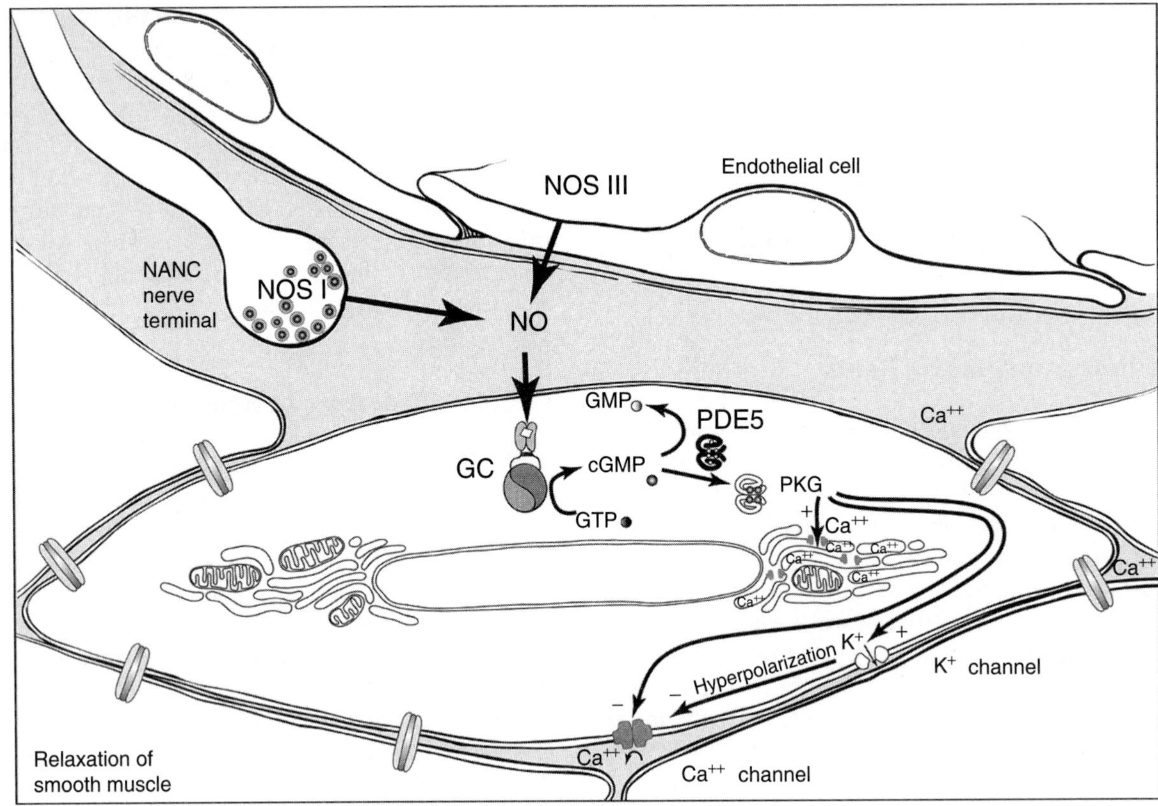

A

FIG. 7A. Relaxation of penile smooth muscle: cGMP mechanisms. Nitric oxide (NO) produced by nitric oxide synthase type I (NOS I) in NANC nerves and nitric oxide synthase type III (NOS III) in endothelial cells activates guanylyl cyclase (GC) in smooth muscle cells. This results in the production of cGMP from GTP. cGMP is metabolized by phosphodiesterase type 5 (PDE5). cGMP causes the activation of protein kinase G (PKG). PKG causes an increased uptake of calcium (Ca^{++}) into intracellular stores and a reduction of calcium entry into the cell through calcium channels. PKG also opens potassium (K^+) channels, leading to hyperpolarization, which also closes the calcium channels. The resulting decrease in intracellular calcium concentration from these mechanisms leads to relaxation of the smooth muscle. NANC, non-adrenergic, non-cholinergic. [Reprinted from (831a), with permission from the Society for the Scientific Study of Sexuality.]

of the erectile tissue. NOS I produces nitric oxide that is released from the parasympathetic nerves. This NO may affect smooth muscle directly to cause relaxation. In addition, it may stimulate the production of NO by NOS III in the endothelial cells. NO released from the endothelial cells could then relax the penile smooth muscle. This cascade of NOS I and NOS III could provide an amplification mechanism whereby neural activation leads to a greatly enhanced nitric oxide release in the erectile tissue (444).

NO diffuses into smooth muscle cells and activates guanylyl cyclase to produce cGMP. Increases in cGMP levels lead to activation of protein kinase G (PKG) and, to a lesser extent, protein kinase A (PKA). These enzymes exert regulatory effects by phosphorylating certain proteins within the cell. One such protein is phospholamban, which inhibits the calcium pump that sequesters calcium into the sarcoplasmic reticulum. Phosphorylation of phospholamban decreases its inhibitory function on the calcium pump, leading to a decrease in intracellular calcium concentration (425). Lower calcium levels lead to relaxation. The cGMP induces smooth muscle relaxation via several mechanisms: (a) activation of protein kinase G (PKG); (b) activation of ion channels that extrude or sequester calcium; (c) opening of potassium channels to hyperpolarize smooth muscle cells; (d) activation of myosin lightchain kinases; (e) cGMP may also inhibit the contractile IP_3 pathway (446,530,686,925). The activity of cGMP is terminated by phosphodiesterase type 5 (PDE_5). Sildenafil citrate (Viagra) prolongs cGMP activity by inhibiting PDE_5.

Other processes that may contribute to erection include activation of adenylyl cyclase by vasoactive intestinal peptide (VIP), calcitonin gene-related peptide (CGRP), and prostaglandin E_1 (PGE_1) (41,118,119,669) (Fig. 7B). These elevate intracellular levels of cAMP. Cyclic AMP activates PKA, and to a lesser extent, PKG.

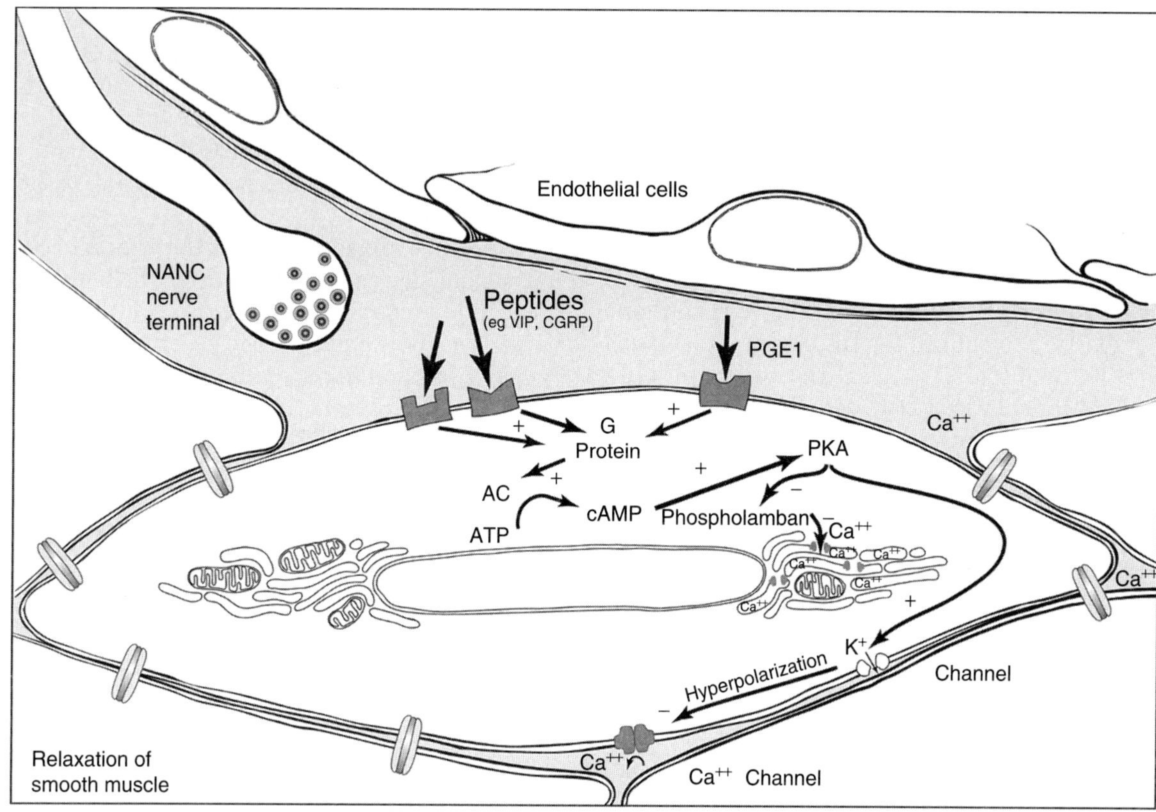

FIG. 7B. Relaxation of penile smooth muscle: cAMP mechanisms. Peptides such as vasoactive intestinal polypeptide (VIP) and calcitonin gene-related peptide (CGRP), and prostaglandins such as PGE1 bind to specific receptors on the smooth muscle cell membrane. These are linked to G protein that activates adenylyl cyclase (AC) in smooth muscle cells. This results in the production of cAMP from ATP. cAMP causes the activation of protein kinase A (PKA). PKA causes an increased uptake of calcium (Ca^{++}) into intracellular stores by inhibiting an inhibitor of calcium uptake, phospholamban. PKA also opens potassium (K^+) channels, leading to hyperpolarization, which closes the calcium channels. The resulting decrease in intracellular calcium concentration from these mechanisms leads to relaxation of the smooth muscle. [Reprinted from (831a), with permission from the Society for the Scientific Study of Sexuality.]

These decrease intracellular calcium concentrations by calcium sequestration and hyperpolarizing the cell. At least some of the acute proerectile effects of PGE_1 are mediated by an increase in the production of NO, and repeated administration of PGE_1 increases the content of both neuronal and endothelial forms of NOS (275).

Detumescence of the penis is induced and maintained by sympathetic adrenergic nerves, whose stimulation of α_1 receptors on smooth muscle cells in the corpora cavernosa increases intracellular calcium concentration via activation of the inositol triphosphate (IP_3) second messenger cascade. The IP_3 cascade causes a release of calcium from intracellular stores and allows entry of calcium through plasma membrane calcium channels. The increase in intracellular calcium then activates a sequence of events, which require ATP and result in actin and myosin cross bridge formation, resulting in contraction of smooth muscle cells, decreased blood inflow, and detumescence (40,175). Stimulation of α_1 receptors can also increase the sensitivity of smooth muscle to calcium, without affecting calcium levels, via the Rho-kinase pathway (917). A Rho-kinase antagonist [(+)-(R)-trans-4-(1-aminoethyl)-N-(4-pyridyl) cyclohexanecarboxamide dihydrochloride monohydrate, Y-27632] applied to strips of human or rabbit corpus cavernosum inhibited smooth muscle contraction that had been elicited by either the α receptor agonist phenylephrine or electrical field stimulation (803). Y-27632 also stimulated penile erection in rats (174). Castration increased RhoA and Rho-kinase protein levels in cavernous tissue of male rats; intracavernosal injections of Y-27632 dose-dependently increased cavernosal pressure in castrated, gonadally intact, and T-replaced rats (1016). Therefore, inhibition of Rho-kinase may provide a new treatment for erectile dysfunction. Other factors that contribute to detumescence include endothelin, angiotensin, and thromboxane A_2.

ROLE OF GONADAL STEROIDS IN THE CONTROL OF MALE SEXUAL BEHAVIOR

Male sexual behavior is heavily dependent on T and its metabolites. T is secreted by the Leydig cells of the testes and reaches its nontesticular targets via the blood. In all mammals studied, the display of mating behavior in adult males is promoted by circulating T and/or its metabolites, estradiol (E_2) and dihydrotestosterone (DHT). Increasing production of T at puberty is essential for the increased sexual activity of maturing males, and if the source of T is removed by surgical castration, sex drive and sexual activity typically decline. However, the importance of gonadal steroids in male sexual behavior varies among species (6). In most rodents, testicular steroids are essential, and castration eliminates mating. By contrast, T plays a more modulatory role in male sexual activity for humans (397) and other species such as the big brown bat (*Eptesicus fuscus*) (654).

The stimulatory effects of T on sexual behavior of adult males are termed activational effects, to distinguish them from T's organizational effects during sexual differentiation. In adult males, T is generally thought to have a long-term permissive effect that prepares the male to respond to cues from receptive females. T is usually present in considerably greater quantities than is necessary to stimulate sexual behavior (236). In most species, relatively small fluctuations in circulating T levels should not affect behavior (6).

Time Course of Changes in Copulation Following Castration

Although plasma hormone levels decline to unmeasurable levels within 24 hours after castration (505), male rats may continue to exhibit the complete copulatory pattern for days or weeks (233). However, intromission latency begins to lengthen soon after castration. On the other hand, the number of intromissions preceding ejaculation actually decreases in the early days after castration (233). Therefore, one of the normal functions of T may be to increase the number of intromissions preceding ejaculation, thereby increasing the number of sperm in the ejaculate, facilitating sperm transport, and triggering a progestational state in the female (952, 953).

The post-castration decline in sexual behavior follows a distinct sequence, which is consistent across species: ejaculation stops first, then intromission, and later mounting (92,233,361,523,664,832). The differential loss of copulatory elements following castration occurs, in part, because the different elements depend on different peripheral target cells or mechanisms. For example, mounting does not depend on the tactile sensitivity of the penis or its ability to become erect, as intromission does, and ejaculation requires even more penile sensory input and motor output than does intromission. The various behaviors may also depend on different neuronal circuits in the central nervous system, which have different degrees of hormonal dependence.

There are mixed reports of the effects of castration on sexual desire and copulation in men. Based on anecdotal accounts, Kinsey et al. (476) concluded that there was little basis for the assertion that castration impaired sexual function in most men. Subsequently, more detailed reports revealed considerable variability. Heim and Hursch (397) reviewed the results of prospective studies of men who had been castrated as "treatment" for sexual offenses. In these studies, half to two-thirds of the men reported a rapid loss of sexual desire and interest, whereas in the remaining men sexual activity waned gradually, with as many as 10% reporting sexual intercourse for up to 20 years after castration. Depending on the individual, castration would either eliminate intercourse immediately, produce a progressive loss of sexual activity over the period of a year, or in some cases have little effect on sexual behavior. There was a greater effect the older the individual was at the time of castration (397).

Time Course of Changes in Copulation Following T Restoration

Exogenous T restores copulatory behaviors in the reverse order in which they were lost, with mounting occurring first, followed by intromission and, finally, ejaculation (523). The effectiveness of T replacement suggests that it is the loss of T that leads to a reduction in sexual activity, rather than other effects of castration, such as the consequent rise in pituitary gonadotropins. Higher doses of hormone are required to restore copulatory behavior in long-term castrates than if hormones are administered soon after castration to maintain the behavior (233). Presumably, continuity in the exposure of neural and peripheral tissues to T maintains the responsiveness of these tissues, whereas a greater input is needed to reactivate dormant copulatory systems. Increasing doses of T accelerate the recovery of copulatory activity in long-term castrates (233,414). Although individual males vary widely in their copulatory performance, once an individual's pre-castration copulatory level is reached, additional T treatment produces little benefit (361,519).

In long-term castrates, 5 to 10 days of T exposure is typically required to reinstate copulation (91,233, 620,790). On the other hand, T increased neural firing in the MPOA within minutes (756) and increased firing in the lateral hypothalamus within seconds (1043).

Moreover, E stimulated chemo-investigation and mounting in castrated rats within 35 minutes (222), and T facilitated mounting within 60 minutes in castrated mice (448). Therefore, steroids begin to act immediately within the brain, but longer-term genomic effects are necessary for the complete restoration of copulatory behavior. This conclusion was supported by the finding that the protein synthesis inhibitor anisomycin blocked the effects of T on copulatory behavior (618). On the other hand, anisomycin did not disrupt, and in some conditions actually facilitated, the restoration of touch-based erections by T (632).

Role of T Metabolites in Maintaining and Restoring Copulation

T is the principal steroid hormone produced by the testes and present in systemic circulation. However, T is converted in target organs to potent estrogenic or androgenic metabolites. E formed by the P450 enzyme aromatase binds to ERs, of which there are at least two types: ER-α and ER-β. The major androgenic metabolite of T is DHT, which is formed by the enzyme 5α-reductase. Although both T and DHT bind to the AR, DHT has approximately fivefold greater affinity for the AR (1012). Moreover, DHT cannot be aromatized to E_2, and is therefore considered to have a "pure" androgenic action. Some target cells may produce both E_2 and DHT and may have both ERs and ARs.

Together, E_2 and DHT maintain copulation and *ex copula* reflexes. However, the balance of estrogenic and androgenic stimulation of male sexual activity is complex and may be species-specific. In rats, exogenous E_2 is sufficient to reinstate most copulatory behavior in castrated males [234,912; reviewed in (201,523)]. In addition, systemic administration of aromatase inhibitors (109,127,829,965) or ER antagonists (109) to castrates inhibited or prevented the subsequent restoration of copulation by T administration. However, the effects of hormone manipulations may depend on the specifics of drugs and testing regimens, as other investigators have found no effect of ER antagonists on copulation (966).

By contrast, neither DHT (108,615,1007) nor the nonaromatizable androgen methyltrienolone (R1881) (84) was sufficient to restore or maintain copulation in castrated male rats. Furthermore, synthetic androgens (5α-androstanediols) that can be aromatized to E_2, but not 5α-reduced to DHT, were even more effective than T in restoring sexual behavior in castrated rats (685) or mice (714). The effectiveness of E_2, and the ineffectiveness of DHT, in restoring or maintaining copulation have given rise to the "aromatization hypothesis," which states that it is the aromatization of T to E that is critical for maintaining or restoring copulation in male rats.

However, E is not sufficient to maintain all aspects of sexual behavior. Frequently, E-treated castrates were unable to ejaculate or had fewer ejaculations than males treated with T or with a combination of E_2 and DHT (281,462,617,791,792,913,965). In addition, administration of the nonsteroidal antiandrogen flutamide reduced the ability of T to restore copulation in castrated male rats (341,353,966). An antiandrogen with greater affinity for the AR ($\alpha\alpha\alpha$-trifluoro-2-methyl-4′-nitro-m-lactoluidide, SCH-16423) eliminated all mating behavior, including mounting, in 80% of male rats treated with a dose of T that restored ejaculation in all of the control males (619). Androgens also appear to preferentially stimulate appetitive aspects of sexual behavior. In this regard, DHT treatment in castrated male rats stimulated noncontact erections and ultrasonic vocalizations, but E_2 did not (593, 201). Likewise, flutamide impaired restoration of partner preference, scent marking, and 50-kHz vocalizations in gonadectomized males with T replacement, while the ER antagonist RU-58668 inhibited only scent marking and ultrasound (966). In related studies, the aromatase inhibitor fadrozole did not prevent partner preference in castrated males with T replacement (965). Using a level-searching test of sexual motivation, Roselli et al. (829) reported that fadrozole treatment reduced anticipatory level searching, but that E treatment only partially restored the behavior. Taken together, these studies suggest that stimulation of ER is not sufficient, and in some cases may not be necessary, to restore or maintain full copulatory behavior in male rats or Syrian hamsters. Rather, conversion of T to DHT, and subsequent binding to AR, may be necessary to promote central integrative processes and to maintain peripheral sensory and motor functions that contribute to copulatory performance. Thus, stimulation of both ERs and ARs may be necessary for full restoration of copulatory behavior.

Studies of genetically altered mice lacking ERα or ERβ have furthered our understanding of the roles of androgens and estrogens in male sexual activity. Both forms of ER are normally expressed in the brain (889). Gonadally intact male mice lacking the ER-α (ERα "knock out" mice, or ERαKO) mounted normally, but had fewer intromissions than wild-type mice, and almost no ejaculations (713,715). This is not due to endocrine dysfunction, because ERαKO males secrete more T than do wild-type mice, due to diminished ER-mediated negative feedback (1002). Moreover, castration of ERαKO males and replacement with normal levels of T (1002) or with higher than normal levels of DHT (715) increased mounting, but did not restore ejaculation. However, concurrent treatment with T and a DA agonist instated masculine sexual behavior in ERαKO males (1000). There appeared to be no compensation by up-regulation of other steroid receptors, because AR immunoreactivity (713) and ER-β mRNA (221) were comparable in ERαKO and wild-type mice. Recently, deficits in sexual behavior have been reported in aromatase knock out (ArKO) mice (64, 610). Although ArKO mice are capable of siring litters when housed with females, they appear to be less able to discriminate odors from conspecifics (64). In contrast to the dramatic effects of ERαKO or ArKO on male sexual behavior, there were no deficits in adult mating behavior of βERKO mice (712), although the pubertal onset of ejaculation was delayed in βERKO males (948). These data suggest that ERα and aromatase, perhaps in concert with androgens, play a substantial role in male sexual behavior in mice.

In a number of species, the aromatization hypothesis has little or no support. DHT is able to maintain or restore copulation in rabbits (11), guinea pigs (36,146), hamsters [reviewed in (890); see also (825)], deer mice (198), monkeys (199,666,768), and mice (575). Fadrozole did not suppress sexual behavior in hamsters (215) or green anole lizards (1017). On the other hand, DHT and R1881 were ineffective in stimulating copulation in male gerbils (1037), pigs (538), or sheep (745). In ferrets, either T or E facilitated partner preference in sexually experienced males (466). The anti-estrogen tamoxifen reduced libido in dogs (218), and fadrozole reduced sexual activity in castrated, T-treated monkeys (1055). In the final analysis, although the balance of androgenic and estrogenic stimuli that optimizes sexual behavior may vary among species, males are normally exposed to both androgenic and estrogenic metabolites of T, which act on ERs and ARs to induce appetitive and consummatory aspects of mating.

Effects of Castration and Hormone Replacement on *Ex Copula* Penile Responses

Ex copula reflexes are lost more rapidly after castration, and restored more rapidly with T restoration, than is copulation. In addition, restoration of these two functions is mediated by different T metabolites.

Animal Studies

In spinally transected male rats, touch-based erections were reduced within 24 hours of castration, and reached a nadir after 12 days (393). Castration in spinally intact males reduced cups within 4 days, the earliest time tested, with anteroflexions decreasing by 7 days, and glans erections, at 11 days (635). T replacement maintained touch-based erections at normal levels, even though the lowest replacement plasma T levels were much lower (0.79 ng/ml) than those in intact males (1.95 ng/ml) (236). Therefore, as with copulation, adult levels of plasma T are higher than needed to activate penile reflexes.

Touch-based erections can also be reinstated more rapidly than copulation, following T replacement. Indeed, an increase in erections was seen in spinally transected castrated rats 6 hours after implantation of a T-containing capsule; maximal stimulation was achieved by 24 hours (393). Facilitation was observed in spinally intact males 24 hours after T replacement, with maximal increases at 48 hours (354). Only 1 of 10 of these same males mounted in copulation tests 52 hours after T replacement. Thus, penile reflexes decline more rapidly after castration in spinally transected males and are also restored more rapidly by T replacement. The slower restoration in intact males may reflect the longer time required for supraspinal inhibition to be reduced sufficiently for the already-primed spinal effectors to act.

Noncontact erections also show more rapid post-castration loss and post-replacement restoration than does copulation. Noncontact erections were completely lost on the first test, 3 days after castration, and were completely restored on the first test, 3 days after hormone restoration (593). Thus, hormonal mechanisms controlling noncontact and touch-based erections differ temporally, at least, from those controlling copulation.

Whereas E_2 is the primary metabolite for activation of copulation, DHT is necessary and sufficient for maintaining and restoring *ex copula* reflexes in male rats (354,386,635). The DHT regimens that maintained or restored reflexes were completely ineffective in activating mounting (354,635). DHT is also the active androgen for maintenance of nitric oxide-mediated erections (571; see below). Co-administration of either the AR antagonist flutamide (354) or a 5α-reductase inhibitor (135) with T blocked the restorative effects of T. Noncontact erections were also maintained by DHT, but not E_2 (201,593).

Although E is ineffective in maintaining *ex copula* erections, it can maintain *in copula* erections. E-treated castrates had as high a percentage of intromission patterns on which they actually achieved insertion, verified by means of nontoxic tempera paint placed in the female's vagina, as did control males (718). EMG recordings from the bulbospongiosus muscle during intromission also confirmed that E was at least as effective as T in maintaining the duration, frequency, and average amplitude of EMG bursts (417) (Fig. 8). Sachs (840) proposed that a "behavioral cascade," organized in the brain and primed by E, can activate

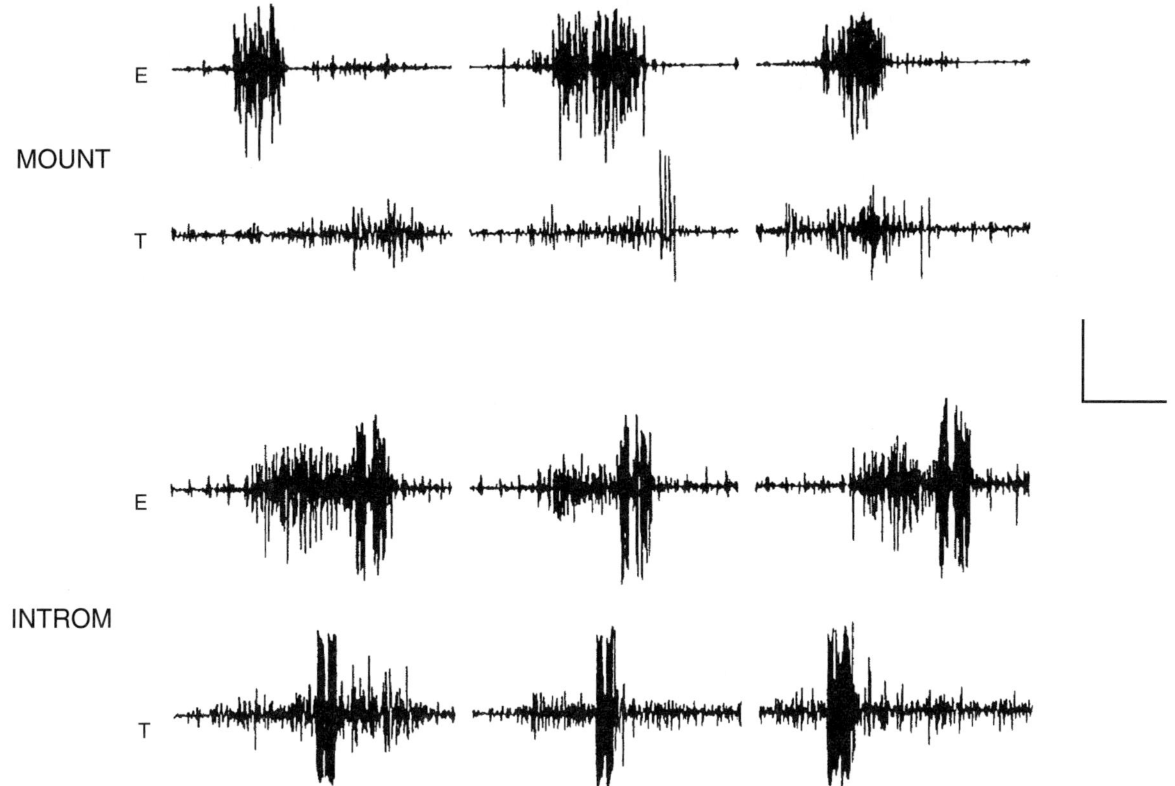

FIG. 8. Electromyographic (EMG) recordings from bulbospongiosus (=bulbocavernosus) muscle during three mounts and three intromissions (INTROM) by rats castrated 6 to 8 weeks earlier and maintained since that time with subcutaneous Silastic capsules filled with testosterone (T) or estradiol benzoate (E_2). Quantitative analyses confirmed the impression from these examples that the duration, frequency, and average amplitude of the EMG bursts were at least as great for the E_2-treated males as for the T-treated males. [Figure is reprinted from (417) with permission.]

reflexes in the context of copulation but cannot disinhibit those reflexes *ex copula*.

Studies on Human Males

T levels in men have shown little (97,796,808) or no (44,528,772) difference between men with erectile dysfunction and normally functioning men. However, there was a correlation between resistive index in cavernous arteries and free T levels in one study of men with erectile dysfunction (61). Furthermore, in aging men with moderate declines in T levels, exogenous T treatment improved erectile function and nocturnal penile tumescence (162,877). In addition, double-blind studies of hypogonadal men reported that T treatment dose-dependently increased the number of erections in diary reports (235,710,860), as well as sexual thoughts accompanying those erections (710). However, only nocturnal erections (710) or those in response to fantasy (73) were hormone-sensitive, not those in response to films (73). Exogenous T in eugonadal men (which resulted in supraphysiological levels) or in hypogonadal men (which resulted in normal levels) increased subjective sexual arousal in response to sexual audiotapes and also increased the bias toward sexual auditory stimuli in a dichotic listening task (30). T and DHT were equally effective in activating sexual activity in agonadal men, and administration of either an E_2 receptor antagonist or an aromatase inhibitor did not affect sexual function (348). However, there is not a strong correlation between endogenous T levels and sexual behavior in eugonadal men (143,503,795).

Steroid Action on Steroid Hormone Receptors

Gonadal steroids are thought to promote male sexual behavior largely by acting on ARs and ERs, which are widely but selectively distributed throughout the brain. The distribution of steroid hormone receptors in the CNS has been determined through receptor-binding, autoradiography, immunocytochemistry, and in situ hybridization. Classical steroid receptors visualized by these techniques are present in large numbers in brain areas important for male sexual behavior, including the MPOA [see (892)]. ARs and ERs are typically concentrated in the neuronal cell nucleus. Once bound to the appropriate ligand, classical steroid receptors act as transcription factors to promote synthesis of new proteins. These actions are consistent with the typical lengthy time-course through which steroids promote male sexual behavior (618,1038).

There is also increasing evidence for nongenomic effects of steroids on male sexual behavior. These include rapid (<1 h) effects of androgens and estrogens (see Time Course of Changes in Copulation Following T Restoration above), particularly in brain regions with few classical steroid receptors. Nongenomic effects of steroids may be mediated by membrane receptors that alter ion channel activity and second-messenger systems (195,314,469,661,683,886,954,997,1034). T has affected firing rates of MPOA neurons within minutes (756) and activity of hypothalamic neurons within seconds (1043) in castrated rats. Neurons in MPOA slices responded to either T or E in the perfusion medium, but not both, and the E- and T-sensitive cells were differentially distributed (891). Recordings of bulbospongiosus muscle activity during tests of touch-based erections showed that T activated some EMG activity in castrated rats as early as 5 minutes after injection, although no erections were seen at that time (849). In support of a nongenomic action of T on erection, administration of anisomycin, a potent inhibitor of protein synthesis, did not affect T's short-latency (within 24 hours) activation of touch-based erections (632). Thus, some effects of T on penile reflexes may be mediated by nongenomic membrane effects.

Role of Progesterone in Male Sexual Behavior

Progesterone (P) is secreted by the adrenals and the testes and is a precursor to T. P stimulated mounting in unisexual lizards (351) and in some males of the ancestral species (544). P implants partially overcame the effects of castration on intromission, but not ejaculation, in naïve rats (1022, 1023). Sexually naïve male mice lacked the P receptor (PRKO) showed deficits on their first copulatory test, but not on subsequent tests (766). However, P appears to be inhibitory to male primate sexual behavior. The synthetic progestin Depo-Provera (medroxyprogesterone acetate) has been used to inhibit sexual interest in sex offenders. Both P and Depo-Provera decreased sexual motivation and copulation in cynomologus monkeys (1055,1056). It is not clear whether P's effects are mediated by the classic P receptor or by nongenomic effects (589–592).

EFFECTS OF SYSTEMICALLY OR INTRACEREBROVENTRICULARLY ADMINISTERED DRUGS

Steroid hormones regulate sexual behavior primarily by slow, genomically mediated processes. However, sexual behavior requires rapid sensory and motor processing and intricate moment-to-moment interactions with a partner. Therefore, the slow hormonal effects must be translated into rapid neural responses. Hormones bias sensory inputs to favor processing of sexually relevant stimuli and also prime central integrative areas and motor outputs to promote appropriate sexual responses. Major targets of steroid priming are enzymes that synthesize (or catabolize) neurotransmitters or neuromodulators, mechanisms that govern neurotransmitter release, receptors, or second messenger systems. Drugs that influence any of these targets may be administered to study the influence of a neurotransmitter on behavior. No drug is entirely specific for its major effect; therefore, it is necessary to block the effects of an agonist with an appropriate antagonist or to administer a drug of a different class that targets the same receptor or enzyme.

There are several advantages to systemic or intracerebroventricular (icv) drug administration. First, the sites in which a given neurotransmitter influences a behavior may not be known a priori. Second, the same neurotransmitter may act synergistically in several different central and/or peripheral sites. Finally, drugs under consideration for possible clinical use must be effective with systemic administration. However, drugs may also have competing effects at different sites. Therefore, studies utilizing both systemic and site-directed routes of administration must be used to fully explicate a drug's effects on behavior [see Table 2 of (438) for a summary of systemically or intraventricularly administered drug effects].

Dopamine

The catecholamines DA, norepinephrine (NE), and epinephrine, are formed from the amino acid tyrosine; the rate-limiting enzyme, tyrosine hydroxylase, produces L-DOPA, which is then decarboxylated by aromatic amino acid decarboxylase to form DA. In noradrenergic neurons DA is converted to NE by dopamine beta-hydroxylase. In a few neurons NE is converted to epinephrine by phenylethanolamine N-methyl transferase. The major dopaminergic cell groups are the A9 group in the substantia nigra of the midbrain and the A10 group in the adjacent ventral tegmental area (VTA). Axons of the A9 group ascend to the dorsal striatum (caudate-putamen) via the nigrostriatal tract, while A10 axons ascend via the mesocorticolimbic tract to the ventral striatum (nucleus accumbens), prefrontal cortex, and other limbic structures [reviewed in (280)]. Several dopaminergic cell groups reside in the hypothalamus, including the A14 periventricular cell bodies, which lie along the third ventricle and send axons laterally into the adjacent MPOA and anterior hypothalamus [reviewed in (684)].

DA has long been known to facilitate sexual function. Administration of L-DOPA to Parkinsonian patients often resulted in increased libido and sexual potency (75,130), and the classic DA agonist apomorphine has been used to treat erectile dysfunction (513,514,725). A new sublingual form of apomorphine appears to be clinically effective, with fewer side effects (262,331) [reviewed in (728)].

In rats, too, systemically administered DA agonists have facilitated male sexual behavior [reviewed in (114,330,637)]. L-DOPA and/or apomorphine has induced sexually sluggish males to copulate, increased the numbers of ejaculations, and decreased ejaculation latency and the numbers of intromissions preceding ejaculation. Apomorphine has also elicited copulation in sexually satiated male rats (606,819) and in socially stressed mice (931) and rats (704). In male rats that were sexually sluggish because of medial frontal cortex damage, DA releasers decreased their usual long mount and intromission latencies (12); and in short-term (586) and long-term (869) castrated rats, apomorphine partially restored copulation. Even mice lacking the gene for ERα, which usually show little copulatory behavior, could copulate normally after receiving systemic injections of apomorphine (Fig. 9) (1000). Therefore, stimulation of DA receptors appears to render both organizational and activational effects of ER-α unnecessary for male copulatory behavior,

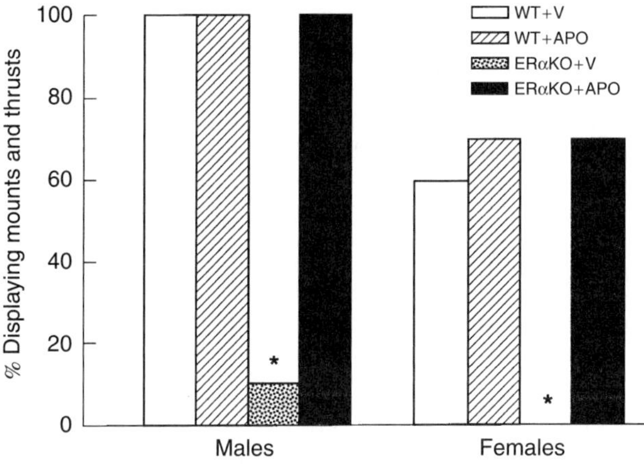

FIG. 9. The percentage of wild-type (WT) and estrogen receptor α knock-out (ERαKO) mice treated with either vehicle (V) or apomorphine (APO) showing mounting and thrusting behavior toward a receptive female. *Significantly lower than the other groups, $P < 0.05$. [Figure is reprinted from (1000), with permission.]

at least in some animals. Systemic administration of apomorphine also elicited penile erections and genital grooming in mice (797). Its effects were blocked by the centrally active antagonist haloperidol, but not by domperidone, which does not cross the blood–brain barrier. Cocaine, which inhibits DA reuptake, increased the percentage of rats that copulated and produced additive effects with paradoxical sleep deprivation, which the authors suggested may increase sensitivity of DA receptors (39). Apomorphine administered systemically to hamsters produced only slight decreases in ejaculation latency and postejaculatory interval, although it also abolished long intromissions (a sign of approaching sexual satiety) in 60% of the animals (59).

Daily injections of amphetamine, which releases endogenous DA, result in increased locomotor responsiveness to amphetamine (sensitization). Similar injections in sexually naïve male rats facilitated copulation on their first test (301). Therefore, the presumably dopaminergic mechanisms that result in drug sensitization, also "cross-sensitize" to a natural motivated behavior. Furthermore, similar mechanisms may at least partially explain the enhanced copulatory ability and resistance to surgical and environmental insults of sexually experienced animals.

As expected, antagonists that block the effects of endogenous DA interfere with sexual behavior of experienced (7,17,19,760) and inexperienced (10) male rats. The effects of DA antagonists ranged from failure to copulate to increased latencies to intromit and ejaculate. The numbers of touch-based erections were also decreased by a centrally active antagonist, haloperidol, but not by an antagonist that does not cross the blood–brain barrier, domperidone (751). Both a DA and an opioid antagonist blocked the facilitative effects of mild tail pinch in sexually sluggish castrates maintained on a suboptimal T regimen (541). Therefore, tail pinch may have activated behavior by increasing activity of midbrain DA systems via an opioid mechanism. The facilitation was also seen on subsequent days when tail pinch was no longer administered, suggesting that possible acute increases in motor activity did not explain the sensitization to sexual cues.

The DA antagonist haloperidol has inhibited measures of sexual motivation, as well as copulation. Sexually naïve male rats were given 10 daily trials on which they traversed a runway to be near an estrous or a nonestrous female on different days (556). They were then injected with vehicle or one of three doses of haloperidol and allowed to copulate to one ejaculation; no copulatory measure was significantly affected by the drug. Two days later vehicle treated animals ran faster after sexual experience than before and also ran faster for the estrous than for the nonestrous female. Males treated with the low and moderate doses of haloperidol showed no increase in running speed following sexual experience and did not prefer the estrous over the nonestrous female. Males treated with the highest dose traversed the runway more slowly after sexual experience than before and did not prefer the estrous female. In addition, they showed more retreats as they approached the goal box, as if it were aversive. The drug effects on the measure of motivation in this experiment do not appear to be confounded with motor side effects. First, the drug was given two days before the runway test and had sufficient time to be metabolized before that test. Second, no dose significantly affected copulation on the experience day; all males ejaculated. Thus, a systemically administered DA antagonist selectively impaired the experience-induced enhancement of sexual motivation.

A similar experiment from the same lab reinforced their earlier conclusion. In the second experiment two doses of haloperidol or vehicle were given to sexually naïve males before traversing a runway for an empty goal box, an estrous female, or a nonestrous female, following a 4-minute exposure to the target through a perforated barrier (557). Run times for the empty goal box and for the nonestrous female were not affected by the drug, suggesting that motoric ability was not compromised. However, run times for the estrous female were slower, compared to vehicle-treated animals, and were not significantly different from their own run times to the other two targets. Two additional experiments from that lab used conditioned olfactory stimuli to test sexual motivation (559). Males copulated on five occasions in the presence of either orange or almond odors and spent a similar amount of time alone in the presence of the other odor. Following the 10 conditioning days, all males ran faster to the goal box scented with the odor previously associated with copulation. A second group of males experienced the same conditioning protocol, but were injected with one of three doses of haloperidol or vehicle before the runway test. Vehicle-treated animals showed the expected increase in running speed to the odor associated with copulation and slower running to the control odor. The low and middle doses of haloperidol were associated with no differences in run times to the two odors and no differences relative to baseline. The highest dose resulted in slower running to both targets, suggesting that motor ability had been impaired. However, the fact that the two lower doses selectively blocked the preference for the conditioned stimulus, without slowing running, as compared to baseline, suggests that antagonism of DA receptors can inhibit motivational processes independently of motor impairment. There is one report of a lack of effect of dopaminergic drugs on a measure of sexual motivation in sexually naïve male rats (4). In that experiment the DA agonists apomorphine and amphetamine failed to affect the time spent in front of a receptive female behind a barrier, and only a high dose of the DA antagonist (cis-flupenthixol), which also impaired motor behavior, decreased this measure. However, as the author noted, it took no more motor activity to stand in front of the estrous female than in front of the male, so it is not clear that the motoric effects accounted for the apparent decrease in motivation by the high dose.

Mice that lacked the gene for tyrosine hydroxylase only in dopaminergic, and not noradrenergic, neurons, needed daily injections of L-DOPA in order to survive (939). These DA-deficient males mounted other males and nonestrous females, as well as estrous females, following their daily L-DOPA injections. After castration they continued to mount receptive females for 21 days, whereas wild-type castrates continued mounting for only 7 days. Furthermore, DA–/– castrates were more sensitive than wild-type castrates to both T and L-DOPA, in that lower doses were sufficient to restore copulation. Thus, DA receptors may have been more efficiently coupled to second messenger systems in the DA-deficient animals.

DA receptors are classified into two families. The D_1 family comprises D_1 and D_5 receptors, while the D_2 family consists of D_2, D_3, and D_4 receptors. The D_1-like receptors stimulate adenylyl cyclase, whereas the D_2 family members inhibit adenylyl cyclase and also influence certain ion channels and the phosphoinositide system. DA and apomorphine activate both families, but drugs that are relatively selective for a given subtype have been used to analyze the roles of the subtypes in various aspects of copulation.

There is some agreement that stimulation of D_2-like receptors facilitates ejaculation. A D_3-selective agonist (7-OH-DPAT) and two D_2/D_3 agonists (SND 919 and B-HT 920) decreased ejaculation latencies and the numbers of intromissions that preceded ejaculation in sexually active male rats, but not in sexually inactive animals (293,294,328). Conversely, a D_2/D_3 antagonist (eticlopride), blocked the pro-ejaculatory effects of 7-OH-DPAT (21) and of SND 919 (292), confirming their receptor specificity. However, cholinergic muscarinic receptors may mediate the pro-ejaculatory effects of D_2-like receptors, as the muscarinic antagonist atropine inhibited the effects of both dopaminergic and muscarinic agonists (1052). The effects of dopaminergic agonists are dose-dependent, with low doses facilitating and high doses inhibiting copulation, probably as a result of stereotypic behaviors elicited by the high doses (308). Similarly, a D_1 agonist (SKF 81292) facilitated copulation in the DA-deficient mice mentioned above, whereas a D_2 agonist (quinpirole) inhibited copulation as it increased stereotypic behavior (939). A different D_1 agonist (SKF 38393) also increased a measure of sexual motivation, the time spent in a goal compartment with a receptive female (96). Contributing to this effect was an increase in the number of copulatory behaviors; the latency to run to the female and run duration were not affected. Thus, a measure of sexual motivation was again affected independently of obvious motor confounds.

There are apparently contradictory effects of D_1- and D_2-like agonists on *ex copula* genital reflexes. In one series of experiments a presumed-selective D_2 agonist elicited erections in a neutral arena, whereas a D_1 agonist inhibited such erections (1053). However, eticlopride, a D_2/D_3 antagonist, administered either with cocaine or with presumed-selective D_2 agonists, actually increased drug-induced erections, suggesting that stimulation of D_2-like receptors may inhibit, rather than stimulate, erections (295). Cocaine would have increased extracellular levels of DA by decreasing reuptake, thereby resulting in stimulation of both D_1- and D_2-like receptors. It is likely that the presumed-selective D_2-like agonists also had substantial activity at D_1 receptors. D_2-like receptors may also inhibit touch-based erections in restrained male rats. A D_2/D_3 agonist (quinelorane) decreased touch-based erections following either systemic or intra-MPOA administration (117). As will be discussed below in the section on the MPOA, intense stimulation of D_2-like receptors appears to shift autonomic balance to favor sympathetically mediated seminal emission and ejaculation, and away from parasympathetically mediated erection. However, two selective D_2/D_3 agonists (7-OH-DPAT and B-HT 920) elicited erections, as well as stretching, yawning, and sedation in sexually naïve male rats in a neutral arena (296). It is not clear whether different dose ranges or different test regimens (touch-based erections vs. drug-induced erections in a neutral arena) may account for these divergent findings.

The facilitative influence of D_1-like receptors on male sexual behavior appears to be conserved across phyla. Systemic administration of a D_1 agonist (SKF 81297) to two species of lizards (little striped whiptail lizards and parthenogenetic desert grassland lizards) increased the proportion of animals that mated after gonadectomy and decreased mount latencies (1033). Conversely, systemic injections of a DA receptor antagonist in leopard geckos inhibited courtship behavior in castrated, T-implanted males [reviewed in (1032)]. Similarly, in quail systemic injections of a D_1 agonist (SKF 38393) increased two measures of appetitive behavior: the amount of time near a window through which a female could be seen and the time looking through the window; it also increased two copulatory measures: the numbers of mount attempts and of cloacal contact movements (69). Conversely, a D_1 antagonist (SCH 23390) decreased the approaches to the female's window and the number of mount attempts. Administration of D_2-selective drugs produced the opposite pattern of results: the D_2 agonist (quinpirole) decreased both appetitive and copulatory measures, and a D_2 antagonist (spiperone) increased mount attempts but decreased the number of times the male approached the area of the window. A somewhat earlier article reported that the classic DA agonist apomorphine inhibited both appetitive and copulatory measures, apparently by stimulating primarily D_2 receptors (161). The inhibition by the two higher doses was associated with intense stereotyped pecking; however, the lowest dose inhibited sexual behavior without significant increases in pecking. Two indirect DA agonists (nomifensine, which inhibits reuptake, and amfonelic acid, which primarily increases release) produced mixed facilitative and inhibitory effects. A recent article from the same lab reported that icv injections of DA itself inhibited sexual behavior in quail, and suggested that $\alpha 2$ noradrenergic receptors, rather than DA receptors, mediated the inhibitory effects (217). However, the earlier studies using selective DA agonists and antagonists probably avoided the problem of cross-reactivity with $\alpha 2$ receptors.

In summary, there is impressive agreement that systemically administered DA agonists facilitate several aspects of mammalian sexual behavior, including sexual motivation, genital reflexes, and copulation itself, and that DA antagonists inhibit those same measures. Systemically administered DA agonists have restored sexual behavior in castrates, in stressed or sexually satiated males, in males with prefrontal cortex lesions, in males lacking ERα, and in males lacking endogenous DA (reviewed above). DA agonists have also facilitated erections in several contexts, although the relative importance of D_1- and D_2-like receptors for genital reflexes is not clear. Conversely, DA antagonists have inhibited both conditioned and unconditioned sexual motivation, in several cases without confounding motor effects, have decreased genital reflexes, and have decreased the incidence, rate, and efficiency of copulation. There is also conservation of the facilitative role of D_1-like receptors in reptiles, birds, and mammals.

A recent critical review has suggested that DA has no importance for sexual motivation, because several of the studies that reported inhibition by DA antagonists either used measures of approach to a female, which can confound motoric and motivational factors, or studied sexually experienced males (732). However, several of the studies using approach to a female specifically ruled out motoric effects, as noted above. Furthermore, the increase in sexual motivation that follows copulatory experience is an important issue, and experiments have separately studied dopaminergic effects on both the initial conditioning and the subsequent expression of sexual motivation. DA antagonists have inhibited both the initial sensitization and the subsequent expression of the enhanced motivation, without confounding motor effects. While no procedure has been devised to measure motivation selectively, without any possibility of sensory, learning, or motoric contributions, the preponderance of evidence suggests that DA contributes to sexual motivation, as well as to genital reflexes and copulatory performance.

Norepinephrine

Noradrenergic cell bodies in the central nervous system are located primarily in the locus coeruleus (A6) and in nuclei (A1-5,7) throughout the medulla and pons [reviewed in (509)]. Axons from the locus coeruleus project to the cerebral cortex and hippocampus via the dorsal noradrenergic bundle; axons from the remaining cell groups project to the hypothalamus, amygdala, and septum via the ventral noradrenergic bundle. NE is also the major postganglionic neurotransmitter for the sympathetic nervous system.

Experiments using noradrenergic drugs and lesions of NE pathways have produced inconsistent results. Electrolytic lesions of the ascending dorsal noradrenergic bundle in the midbrain resulted in decreased postejaculatory intervals and increased ejaculations in a 1-hour test (193,194) However, cytotoxic (6-OHDA) lesions of the same area were without effect on copulation (194). Electrolytic lesions of the locus coeruleus or systemic inhibition of NE synthesis inhibited copulation in one experiment (624), but not in a more recent one (287). However, lesions of the locus coeruleus blocked the "Coolidge effect" (the facilitative effect of introduction of a new female to a sexually satiated male) and decreased the facilitative effects of 8-OH-DPAT, a 5-HT_{1A} agonist in satiated and nonsatiated males (287).

Presynaptic α_2 autoreceptors exert inhibitory control of NE release. Inhibition of α_2 autoreceptors with yohimbine increased mounts and intromissions by sexually sluggish males (191,909), castrated males (192), and males with anesthetized penes (191). Yohimbine also reversed sexual satiety (819–821). However, yohimbine's effects on satiety were inhibited by the DA antagonist haloperidol, suggesting that activation of the DA system mediated yohimbine's effects (819). A more selective α_2 antagonist (delaquamine) also facilitated copulation in sexually naïve males and increased mounting by castrates (942). Copulatory behavior was facilitated by two other α_2 antagonists; rauwalscine decreased the intercopulatory and postejaculatory intervals, and phentolamine decreased the number of intromissions and time preceding ejaculation (187). In hamsters, too, yohimbine appears to facilitate copulation, increasing the number of ejaculations and the numbers of long intromissions (usually a sign of approaching satiety) (59). Therefore, increasing NE release by blocking α_2 autoreceptors appears to enhance sexual performance in both rats and hamsters. Conversely, stimulating α_2 autoreceptors with clonidine inhibited copulation and touch-based *ex copula* erections in rats (188,192). However, high doses of yohimbine inhibited copulatory behavior (857), and the NE precursor dl-threodihydroxyphenylserine increased mount and intromission latencies (12). Therefore, increased NE activity may increase or decrease sexual behavior. Moderate doses of yohimbine may facilitate sexual behavior by increasing NE release and thereby stimulating α_1 receptors; blocking α_1 receptors with prazosin or methoxamine impaired copulation (190,192). Yohimbine also facilitated erections in men with moderate erectile dysfunction, with greater success in men with less severe dysfunction and higher levels of T (362).

As the major postganglionic neurotransmitter of the sympathetic nervous system, NE promotes penile detumescence and inhibits penile reflexes. In contrast to its facilitation of copulation, yohimbine inhibited *ex copula* touch-based erections and seminal emission (910). However, the α_2 agonist clonidine also inhibited touch-based erections (188). Because both stimulation and inhibition of α_2 receptors decreased penile reflexes, it appears that low to moderate levels of NE may facilitate erections, whereas high levels are inhibitory.

Systemically administered drugs affect both central and peripheral sites that control genital reflexes. However, guanethidine, which depletes peripheral noradrenergic nerve terminals without affecting central sites, has been used to separate these influences. A single injection increased the number of anteroflexions in restrained supine males; however, treatments over 4 to 8 weeks decreased touch-based erections and especially seminal emissions (922). Therefore, there appears to be a facilitative influence of the sympathetic nervous system on erection, as also demonstrated by Giuliano and colleagues (332,335), in addition to its major role in seminal emission. The mixed excitatory and inhibitory influences of NE on male sexual behavior may be relevant to human sexual function, as reviewed insightfully by Bancroft and Janssen (72).

Serotonin

Serotonin (5-hydroxytryptamine, or 5-HT) is synthesized from tryptophan by the rate-limiting enzyme, tryptophan hydroxylase, thereby producing 5-hydroxytryptophan, which is converted to 5-HT by aromatic amino acid decarboxylase. Nine groups of serotonergic neurons (B1-9) are scattered throughout the medulla, pons, and midbrain. Axons from these nuclei ascend and descend to nearly every part of the brain and spinal cord. However, some cell bodies lie outside these nuclei, and fewer than half the neurons in these nuclei are serotonergic [reviewed in (312,407)].

5-HT is generally inhibitory to sexual behavior. A common side effect of selective serotonin reuptake inhibitor (SSRI) antidepressants is difficulty achieving orgasm or ejaculation, accompanied sometimes by decreased sexual interest [reviewed in (831,884)]. These inhibitory effects typically increase over the same time course as the antidepressant effects. The physiological mechanisms that mediate these effects are unclear, but they may include increased prolactin secretion, anticholinergic effects, or decreased nitric oxide production (831) or decreased penile sensitivity (1048). SSRIs have also been used to treat premature ejaculation (367,473,627,989,1048), although unwanted side effects have included anorgasmia and decreased sexual interest (627).

Chronic administration of the SSRI fluoxetine to rats also resulted in decreased sexual motivation and ejaculation (158,311,946,981). Oxytocin injections reversed the inhibition of ejaculation but not of sexual motivation, measured as level changes in a bi-level apparatus (158). In hamsters, too, a 5-HT reuptake inhibitor, clomipramine, administered for 14 days in sugar water, decreased the number of ejaculations (129). One mechanism of the inhibition of ejaculation may be a decreased pressure response of the seminal vesicle to splanchnic nerve stimulation (427). Ejaculation was not inhibited in most studies of acute, as opposed to chronic, administration of fluoxetine (158,690,981). However, in one study acute injections slowed copulation, an effect that was reversed by amantadine, an NE releaser, suggesting an interaction between 5-HT and NE influences (1047). Acute injections of fluoxetine inhibited one measure of sexual motivation (level changing in search of a female) but not time spent near a receptive female (981). Injections of the 5-HT precursor, 5-hydroxytryptophan, also increased the latencies to intromit and ejaculate and increased the intromissions preceding ejaculation (20). Similarly, administration of 5-HT intrathecally (around the spinal cord) abolished the urethrogenital reflex, which was restored by systemic injections of methysergide, a 5-HT antagonist (599).

Copulation has been facilitated by drugs that deplete 5-HT (24,676,859) and by lesions of the medial, but not the dorsal, raphe nucleus (27,496), a major source of 5-HT in the brain. Facilitative effects were found most often in animals that were originally noncopulators (229,325,859), in sexually naïve males (818), or in castrates maintained on suboptimal T replacement (525,587,818,914,916). Similarly, 5-HT depletion resulting from lesions of the median and pontine raphe nuclei increased intense touch-based penile erections (cups) and anteroflexions (682), and neurotoxic lesions of 5-HT terminals in the brain and spinal cord disinhibited the UG reflex (600). Inhibition of 5-HT synthesis by p-chlorophenylalanine (pCPA) decreased the latency to the first touch-based erection, but also decreased the numbers of erections (611). pCPA-treated males were more likely to have noncontact erections in the presence of a female, but not when tested alone (611). Thus, 5-HT inhibits erections in response to a female, and has mixed, but mostly inhibitory effects on touch-based erections.

In contrast to 5-HT's generally inhibitory effects on sexual behavior, stimulation of one receptor subtype, the 5-HT_{1A} receptor, facilitates ejaculation. The classic 5-HT_{1A} agonist 8-hydroxy-2-(di-*n* propylamino)tetralin (8-OH-DPAT) decreased both the time and number of intromissions preceding ejaculation (18,25,26,804,874), with some rats ejaculating on their first intromission (20,25,208,366). Flesinoxan, another 5-HT_{1A} agonist, also facilitated ejaculation, although no animals ejaculated on their first intromission (366). Because the effects of 8-OH-DPAT are opposite those of 5-HT itself, and because some 5-HT_{1A} receptors are autoreceptors on serotonergic somas and dendrites, it was initially thought that 8-OH-DPAT facilitated sexual behavior by inhibiting 5-HT release. However, lesions of serotonergic nuclei did not alter the effects of 8-OH-DPAT, suggesting that postsynaptic receptors mediated its effects (286). In addition, microinjections of the drug into terminal fields (MPOA and nucleus accumbens), where all 5-HT_{1A} receptors are postsynaptic, also facilitated sexual behavior (289,613). 8-OH-DPAT's effects are dependent on normal levels of T and on sexual experience. In castrates with subnormal T replacement 8-OH-DPAT did not facilitate copulation, and in those with threshold T levels, only sexually experienced males copulated after 8-OH-DPAT administration (836). Males with high levels of T showed a strong effect of the drug.

A likely mechanism of 8-OH-DPAT's effects is an increase in noradrenergic activity in some brain area(s). 8-OH-DPAT, the α_2-adrenergic receptor antagonist yohimbine, and the opioid antagonist naloxone all increased the percentage of sexually satiated males that copulated on the following day (820,821). Furthermore, neurotoxic lesions of noradrenergic neurons blocked the facilitation by 8-OH-DPAT and naloxone in satiated males, but not that of yohimbine (821). Similar lesions also decreased the effects of 8-OH-DPAT in nonexhausted males (287).

The most common effect of 8-OH-DPAT is facilitation of ejaculation, which is mediated in part by the sympathetic nervous system. In agreement with a pro-sympathetic, anti-parasympathetic influence,

8-OH-DPAT inhibited touch-based erections, cups, and anteroflexions in male rats, in addition to decreasing the time and number of intromissions preceding ejaculation (804). A moderate dose of 8-OH-DPAT also decreased noncontact erections (780) and facilitated ejaculation in monkeys, although a high dose delayed ejaculation (781). Thus, stimulation of 5-HT_{1A} receptors appears to shift autonomic balance to favor ejaculation, in addition to promoting copulation in sexually inactive males. However, the effects of 8-OH-DPAT show some species specificity. Low doses had no effect in male ferrets, and the same doses that facilitated ejaculation in male rats inhibited copulation in ferrets (739).

Stimulation of other 5-HT receptor subtypes results in generally inhibitory effects on copulation, as expected from the inhibitory effects of 5-HT itself. In rats the 5-HT_{1B} agonist anpirtoline inhibited ejaculation; its effects were blocked by two selective 5-HT_{1B} antagonists (408). Co-administration of an SSRI (citalopram) and a 5-HT_{1A} antagonist lengthened ejaculation latency, and that effect was blocked by a 5-HT_{1B} antagonist (23). A 5-HT_{1B} antagonist also blocked the increased ejaculation latency resulting from administration of the 5-HT precursor 5-hydroxytryptophan (plus a peripheral decarboxylase inhibitor to restrict the increase in 5-HT to the CNS) (22). 5-HT_{1B} knock-out mice were less debilitated by several serotonergic drugs than were wild-type mice, again suggesting an inhibitory effect of the 5-HT_{1B} receptor (823). Consistent with a facilitative effect of 5-HT_{1A} receptors on ejaculation, a 5-HT_{1A} antagonist, co-administered with the SSRI, produced a further lengthening of ejaculation latency. Therefore, 5-HT_{1A} and 5-HT_{1B} receptors have opposite effects on ejaculation, with 5-HT_{1A} receptors facilitating ejaculation and 5-HT_{1B} receptors inhibiting it.

In contrast to the opposing effects of 5-HT_{1A} and 5-HT_{1B} receptors in rats, two studies of mice reported inhibitory effects of both the 5-HT_{1A} agonist 8-OH-DPAT and two 5-HT_{1B} agonists. 8-OH-DPAT decreased both the time spent near a receptive female (782) and copulatory performance (823). The 5-HT_{1B} agonist CGS-12066A also decreased time spent near a receptive female (782), and the 5-HT_{1B} agonist TFMPP inhibited the execution of copulation. On the other hand, 5-HT_{1B} knockout mice required more stimulation to achieve ejaculation, but were less sensitive to the inhibitory effects of 5-hydroxytryptophan, the 5-HT precursor (823). The authors suggested that compensatory mechanisms may have rendered the knockout mice less sensitive to the inhibitory effects of high levels of 5-HT.

Stimulation of 5-HT_{2C} receptors may have effects opposite those of 5-HT_{1A} agonists: facilitation of erection and inhibition of ejaculation. The 5-$HT_{2C/1D}$ agonist mCPP increased the number of male monkeys that had drug-induced and noncontact erections, but decreased the number of males that initiated copulation and achieved ejaculation (780). [mCPP was classified as a 5-$HT_{1C/1D}$ agonist when these studies were conducted. More recently the 5-HT_{1C} receptor has been renamed the 5-HT_{2C} receptor (426). We have changed the originally published nomenclature to be consistent with the current classification.] mCPP has also increased firing of the cavernous nerves and increased intracavernous pressure in anesthetized rats (920) and resulted in drug-induced erections in awake rats (102). A more selective 5-HT_{2C} agonist (RO60-0175) also increased erections in awake rats (668). Therefore, 5-HT_{1A} receptors appear to promote sympathetically mediated ejaculation, whereas 5-HT_{2C} receptors facilitate parasympathetic effects on erection, but may inhibit ejaculation.

A relatively nonselective 5-HT_2 agonist, DOI, has inhibited copulation (309,347,486,729), and a 5-HT_2 antagonist, facilitated copulation (347). To summarize, 5-HT_{1B}, 5-HT_{2C}, and other 5-HT_2 agonists inhibit ejaculation and may impair copulation, although 5-HT_{2C} agonists may also promote erection. 5-HT_{1A} agonists, on the other hand, facilitate ejaculation and, in otherwise inactive males, stimulate copulation.

Acetylcholine

Acetylcholine (ACh) is produced from choline and acetyl coenzyme A by choline acetyltransferase. Eight major groups of cholinergic cell bodies (Ch 1-8) lie in the basal forebrain and brainstem [reviewed in (662)]. ACh is also the neurotransmitter for peripheral motor nerves, preganglionic autonomic neurons, and postganglionic parasympathetic neurons. There are five types of muscarinic receptors, coupled to either $G_{q/11}$ or $G_{I/O}$ G-proteins (31). Their effects may be either excitatory or inhibitory, depending on the tissue (947). Nicotinic receptors have been divided into nine subtypes (31) and are excitatory, acting on sodium channels (947).

Increasing levels of ACh by slowing its degradation (using the acetylcholinesterase antagonist physostigmine) or administration of the muscarinic agonist pilocarpine increased the number of ejaculations; these effects were blocked by the muscarinic antagonist atropine (1052). Pilocarpine also induced erections in a neutral chamber (580). The muscarinic agonist oxotremorine also facilitated copulation in both experienced and naïve castrates with T replacement, but not in males without T (807). Therefore, systemic administration of muscarinic drugs appears to facilitate copulation and *ex copula* erections. However, because systemically administered cholinergic drugs affect striated muscles and parasympathetic targets, it is difficult to discern whether their effects are centrally or peripherally mediated.

Gamma Amino Butyric Acid

Gamma amino butyric acid (GABA) is the most common inhibitory neurotransmitter in the brain [reviewed in (746)]. It is formed from glutamate by glutamic acid dehydroxylase (GAD). Stimulation of $GABA_A$ receptors hyperpolarizes neurons by opening chloride channels. $GABA_B$ receptors act through the $G_{i/o}$ G-protein to inhibit several intracellular processes.

Systemic injections of both $GABA_A$ (15) and $GABA_B$ (731) agonists impaired copulation without obvious effects on motor activity. Similar effects of the $GABA_A$ agonist were seen in rabbits (734). However, a different $GABA_A$ agonist, as well as an inhibitor of GABA synthesis and an inhibitor of GABA catabolism all impaired social and drinking behaviors, as well as sexual behavior, suggesting that these treatments had nonspecific effects on motivated behaviors (738). Furthermore, some inhibitory effects on copulation may have been mediated by decreased EMG activity in ischiocavernosus muscles during thrusting, a peripheral, rather than central effect (737).

Opioids

There are three major classes of endogenous opioids: endorphins, enkephalins, and dynorphins [reviewed in (584)]. Endorphins are synthesized from proopiomelanocortin (POMC) in the anterior and intermediate lobes of the pituitary and in the hypothalamus. Met- and Leu-enkephalin are distributed throughout the central and peripheral nervous systems. Dynorphins comprise four major classes, including dynorphins A and B and neoendorphins α and β, which also have a wide distribution. A more recently discovered peptide, orphanin FQ (also called nociceptin), binds to an "orphan" receptor that has structural similarity with other opioid receptors. Besides the orphanin FQ receptor, three major types of opioid receptors have been identified: μ (activated primarily by β-endorphin), δ (activated by enkephalins), and κ (activated by dynorphins) (584). Opioids and other neuropeptides are usually co-localized with classical transmitters and are released under conditions of high neural activity.

Opiates and endogenous opioids have mixed, dose-dependent effects on male sexual behavior [reviewed in (438)]. The suggestion that opioids and opiates generally inhibit sexual behavior is based in part on the facilitative effects of antagonists, especially in sexually sluggish males (321), sexually satiated males (821), or sexually naïve males tested in a novel environment (762). Male quail also showed facilitation of copulatory behaviors, but not anticipatory behaviors, by centrally administered naloxone (811). However, systemic administration of the opiate antagonists naloxone and naltrexone have also had inhibitory effects, including increased postejaculatory intervals (855,973) and decreased ability of mild tail pinch to stimulate copulation in castrates with subnormal T replacement (541).

A second line of evidence for inhibitory effects of opiates is that morphine or the κ agonists U-50,488H or bremazocine inhibited sexual behaviors in active copulators (9,14,540). However, inhibitory effects of the opioid antagonist naloxone on active copulators (14) suggests that at least a minimal amount of activity at opioid receptors can facilitate copulation, but that an excess may impair sexual behavior. Some inhibitory effects of systemically administered morphine may result from inhibition of genital reflexes. All doses of morphine suppressed seminal emissions, and touch-based erections were also inhibited (346). However, naloxone also inhibited touch-based erections in those experiments, again suggesting that some opioid activity can facilitate sexual function. Furthermore, naloxone also inhibited anticipatory level-changing in a bi-level apparatus (973,974,976), suggesting that some endogenous opioid activity facilitates sexual motivation. Opioid facilitation may result from disinhibition of the mesolimbic DA system, as there are μ receptors on inhibitory GABAergic neurons in the VTA. Increasing levels of enkephalins, by intraventricular administration of an enkephalinase inhibitor, decreased both the time and number of intromissions preceding ejaculation, but also increased mount and intromission latencies (13). Thus, low levels of opioid activity may facilitate genital reflexes and sexual motivation; however, excess activity can interfere with sexual behavior, and some measures of behavior may be more sensitive than others to such interference. Enkephalins were increased in tissue from several brain areas following copulation either to one ejaculation or to exhaustion, with no significant difference between the two groups (822). These authors also reported that sexually inactive males had generally higher enkephalin content than did sexually active males, again supporting an inhibitory effect of relatively high opioid levels.

Oxytocin

Oxytocin (OT) is a nonapeptide that is synthesized in the paraventricular (PVN) and supraoptic nuclei of the hypothalamus and released both from terminals in the posterior pituitary and into the CNS. Circulating OT from the PVN and posterior pituitary stimulates smooth muscles, thereby facilitating seminal emission and, in humans, contributing to orgasm. In female mammals it stimulates milk let-down and parturition. Plasma OT was increased by copulation in naïve, but not sexually experienced male rats (410). In the naïve males plasma OT was correlated with intensity of copulation. Systemically administered OT also restored copulation in males whose copulatory behavior had been impaired by chronic fluoxetine (158). Facilitative effects of low doses of OT have been observed in rats (55) and dominant squirrel monkeys (1018), but high doses have inhibited copulation in prairie voles (582) and rats (927) [reviewed in (1020)]. A facilitative effect of endogenous OT on copulation was supported by the dose-related inhibitory effects of an OT antagonist, administered icv (50).

Intraventricular injections of OT also increased spontaneous erections, and this increase was blocked by an OT antagonist or an anticholinergic drug (atropine), but not by a DA antagonist (51,52). The OT antagonist also blocked erections elicited by the DA agonist apomorphine (52), suggesting that DA is "upstream" of OT in the control of drug-induced erections; ACh may, in turn, be "downstream" from OT. Finally, an OT antagonist, administered icv, inhibited noncontact erections, suggesting a role for OT in the control of erections elicited by normal physiological stimuli, as well as drug-induced erections (641).

Nitric Oxide

Nitric oxide (NO) is a soluble gas produced in the conversion of L-arginine to citrulline by NO synthase (NOS). NO contributes both peripherally and centrally to vasodilation, erection, and other parasympathetic functions. Both neuronal (nNOS) and endothelial (eNOS) isoforms of NOS are present in the corpus cavernosum and dorsal penile nerve and are up-regulated by T or DHT (63,571,595,805,1057). Castration decreased, and exogenous T restored, erectile responses elicited by cavernous nerve stimulation, systemically administered apomorphine, and/or injection of the nonspecific phosphodiesterase inhibitor papaverine into the penis (63,571,595,671,672,1057) [reviewed in (670)]. Phosphodiesterases inactivate cGMP, the most common second messenger of NO, as well as cAMP; therefore, their inhibition would prolong NO's effects. Indeed, sildenafil (Viagra), vardenafil (Levitra), tadalafil (Cialis), and other drugs used to treat erectile dysfunction inhibit phosphodiesterase V, which is more selective for cGMP (180,338,343,663,783). Sildenafil did not affect erections in healthy men but did decrease the interval between ejaculation and a subsequent erection (62). Therefore, sildenafil can increase penile vasodilation in men with erectile dysfunction and can facilitate postejaculatory erections in healthy men, but not pre-ejaculatory erections. In rats, too, sildenafil increased the occurrence of penile erections, as well as genital grooming and homosexual mounting (296), and vardenafil increased intracavernous pressure in anesthetized rats to a greater extent than did sildenafil (334).

Although castration decreased NOS in the penis and epididymis, it actually increased NOS in the seminal vesicles and lateral prostate; exogenous T in castrates decreased NOS to normal levels in the latter structures (167). Furthermore, systemic administration of L-arginine impaired fertility in male rats, without inhibiting sexual behavior (800). Thus, excessive NO in some parts of the reproductive tract may impair aspects of ejaculation or sperm motility.

Systemically administered NOS inhibitors have decreased the numbers of sexually experienced (112,437) and inexperienced (99) males that could ejaculate, increased their mounts and decreased their intromissions. NOS inhibitors have also decreased the number of noncontact erections (638) and touch-based erections (437), but increased the number of seminal emissions (437), consistent with a facilitative role for NO on parasympathetic function and inhibition of sympathetic activity. Sexual motivation, measured as percent choice of a female in an X-maze (437) or as increased mount latency (112) was not affected by NOS inhibitors, although another study reported a decrease in motivation, measured as decreased precoital activity and decreased numbers of animals that mounted (801). In the latter study, the numbers of spermatozoa were normal, but the number of pregnancies and the number of implanted embryos per pregnancy were lower than in controls. One explanation for the impaired fertility may be the requirement for NO to be injected by the sperm into the egg to elicit release of internal stores of calcium, which in turn triggers cell division (510). nNOS "knock out" mice had normal penile function, as a result of increased production of eNOS (144). Indeed, nNOS–/– mice required fewer mounts and intromissions to ejaculate (506), again suggestive of NO's normal role in activating parasympathetic, and inhibiting sympathetic function. Thus, NO may help to prevent "premature ejaculation."

There have been inconsistent results of systemically administered NO donors or precursor. L-arginine increased the number of naïve male rats that copulated and improved the performance of sexually experienced males (99). Production of NO in the corpus cavernosum was also increased by L-arginine (274). However, the NO donor sodium nitroprusside increased intracavernosal pressure only in untreated castrates, not in T-replaced males (805). Also, there were only marginal facilitative effects of L-arginine on touch-based erections (437), and no effects on intracavernosal pressure elicited by electrical stimulation in castrates, with or without T replacement (805). Therefore, NOS activity in many gonadally intact males may be sufficient for normal copulation and may not benefit from excess precursor or NO donors.

Gonadotropin-Releasing Hormone

Gonadotropin-releasing hormone (GnRH), a decapeptide, is produced by neurons in the mediobasal hypothalamus, the lamina terminalis, and the MPOA (80). After release in the median eminence, GnRH is carried into the hypophyseal portal system to the anterior pituitary, where it stimulates gonadotropin release. It also acts as a neuromodulator in several brain areas, including the MPOA and amygdala (80,502). It is released in response to female pheromones in mice and hamsters [reviewed in (657,658,1006)].

Administration of GnRH has resulted in inconsistent effects. Moss and his colleagues reported that GnRH, administered to castrated male rats maintained on a low dose of T, decreased their intromission and ejaculation latencies; gonadally intact males were not affected (694). Myers and Baum (700), however, reported facilitative effects only in intact males, not in T-replaced castrates; indeed, GnRH lengthened the PEI in T-treated castrates. Lengthened PEI was also observed in old rhesus monkeys (771). GnRH had no effects in male rats in another study (837), but increased the mount rate in male rats with anesthetized genitalia in another experiment (255) and decreased intromission and ejaculation latencies in gray-tailed voles (131). GnRH administered icv also promoted the mating behavior of sexually naïve male hamsters whose vomeronasal organs had been removed (VNX) (284). Previous sexual experience is usually necessary for VNX male hamsters to copulate normally; thus, administration of GnRH compensated for sexual experience. Similarly administered GnRH increased Fos-ir in the MPOA of male hamsters in response to female pheromones; it also increased the MPOA Fos response in sexually experienced VNX hamsters (1006). Sexually naïve VNX males showed enhanced Fos-ir in the Me, but not the MPOA, in response to pheromone plus icv GnRH. In hypogonadal men with decreased LH release GnRH improved sexual function (689). However, it resulted in little (100,270) or no (754) improvement in men with erectile dysfunction.

Chronic administration of a synthetic GnRH agonist [(6-D-(2-napthyl)-alanine)GnRH] disrupted copulation in gonadally intact male rats and also decreased levels of T (256). The likely explanation for this inhibitory effect is that endogenous GnRH is released in a pulsatile fashion, and continuous high doses of GnRH inhibit LH release and gonadal function (223,487). Thus, GnRH agonists have been suggested as anti-fertility drugs (141). Apparently, they also inhibit sexual behavior, perhaps as a result of the decrease in T.

α-Melanocyte–Stimulating Hormone and Related Peptides

α-Melanocyte–stimulating hormone (α-MSH) is secreted into the systemic circulation from the intermediate lobe of the pituitary; it and related melanocortin peptides, derived from the precursor proopiomelanocortin (POMC), are also expressed in numerous CNS and peripheral structures. Besides α-MSH, these peptides include adrenocorticotropic hormone (ACTH) and β-lipotropin, released from the anterior pituitary, the opioid β-endorphin, and other variants formed by post-translational processing. They bind to a family of five G-protein–coupled receptors (MC-1 to MC-5), all of which activate adenylyl cyclase. [See (329) for an excellent review.]

α-MSH and ACTH, administered icv, have been shown to increase erections in dogs, cats, rats, mice, and rabbits as early as the 1960s [reviewed in (47,329)]. Other commonly reported effects have been decreases in the numbers of mounts and intromissions preceding ejaculation and in latency to ejaculate. Both α-MSH and ACTH elicited penile erections, stretching and yawning in rats, when injected into the third ventricle; the stretching and yawning were blocked by an MC-4 receptor antagonist, but penile erection was not (54), suggesting that a different MC receptor was important for facilitation of erection (but see below). However, ACTH-induced erections were inhibited by either icv or systemic injections of the NOS antagonist L-NAME, suggesting that the effects of icv ACTH were mediated by increased central production of NO (775).

There has been a resurgence of interest in these peptides with the production of several synthetic peptides that are selective for subtypes of melanocortin (MC) receptors. Two of these have been reported to increase erections and sexual interest in men [Melanotan-II, MT-II: (254,969,1003,1004); PT-141: (679)], with nausea, stretching, yawning, and decreased appetite reported as side effects of MT-II. MT-II is a selective melanocortin-4 (MC-4) agonist that has also increased erections in rats, especially after intrathecal injections, although icv and iv injections were also effective (1005). The MC-3/MC-4 receptor antagonist

SHU 9119 blocked the effects of MT-II, confirming its receptor specificity. Intracavernosal injections had no effect, suggesting that MC receptors in the spinal cord and brain mediate the proerectile activity. In rabbits, too, iv injections of MT-II increased cavernosal pressure, and again isolated cavernosal tissue was unresponsive (982). Both SHU 9119 and the NOS inhibitor L-NAME abolished the effects of MT-II, indicating that stimulation of central MC-3/MC-4 receptors achieves the proerectile effects by releasing NO. L-NAME also blocked the proerectile effects of icv administration of ACTH (775) and of α-MSH and OT (678) in rats. Furthermore, MT-II enhanced copulatory behavior and augmented erections elicited by cavernous nerve stimulation in wild-type, but not MC-4 receptor knockout mice (969). In addition, MC-4 receptors were expressed in nerve fibers and mechanoreceptors in both rat and human penis, but not in cavernosal smooth muscle; expression was also observed in rat spinal cord, brain stem, hypothalamus, and pelvic ganglion (969). Intranasal administration of the MC-4 agonist PT-141 also elicited drug-induced erections in rats, as well as in men with erectile dysfunction (679). Finally, both systemic and icv administration of a different, highly selective, MC-4 receptor agonist, THIQ, increased intracavernous pressure and elicited reflexive erections in rats; these effects were blocked by both the endogenous nonselective antagonist, agouti-related protein, and a synthetic MC-4–preferring antagonist, MBP10 (604). These effects were also attenuated by an OT antagonist.

In summary, endogenous α-MSH and ACTH, as well as synthetic MC-3 and MC-4 receptor agonists, may promote erection by acting synergistically at nerve endings in the penis, in the lower spinal cord, and in several areas of the brain. Some of these effects are mediated by NO, and perhaps also by OT.

Other Peptides

Vasoactive Intestinal Peptide (VIP)

VIP contributes to penile erection through stimulation of adenylyl cyclase in the corpus cavernosum, thereby relaxing smooth muscle [reviewed in (45,118)]. Antagonists to VIP and ACh additively reduced erection elicited by pelvic nerve stimulation; however, even the additive facilitative effects of VIP and ACh did not elicit a full erectile response (932). Thus, while both VIP and ACh contribute to erection, they do not appear to be principal neurotransmitters for that effect. VIP may facilitate copulation, apart from its effects on erection. Systemic injection of VIP to castrated male rats, maintained on suboptimal T replacement, reduced intromission latency and interintromission interval; these effects were blocked by a VIP antagonist administered with VIP (350).

Prostaglandin E_2 (PGE_2)

Injection of PGE_2 into the fourth ventricle of male rats decreased the postejaculatory interval (123) [reviewed in (680)]. This treatment also increased basal forebrain temperature, measured in the MPOA. As noted above, temperature in this area increases during copulation and decreases during the postejaculatory interval. It is not clear whether PGE_2 influences copulation directly, or through the increase in brain temperature. It is interesting that manipulations of PGE_2 during early development influenced masculinization of rats (38).

Galanin and Galanin-Like Peptide

Better known for its role in stimulating feeding, galanin, microinjected icv, decreased mounts and intromissions in male rats (776). A galanin antagonist (galantide) increased those behaviors (98). In contrast to these apparently inhibitory effects of galanin, microinjection of the peptide into the MPOA of castrated, T-replaced rats facilitated their copulatory behavior (122), and icv injections of a galanin antagonist in ferrets decreased their time spent near a receptive female, although it did not affect copulation (744). It is not clear whether these contradictory findings reflect site, species, dosage, or other differences. Galanin-like peptide (GALP) shares partial sequence homology with galanin but is derived from a different gene. Icv administration of GALP facilitated several measures of male sexual behavior in male rats, whereas similar administration of galanin inhibited those same measures (310). GALP is known to stimulate GnRH secretion; however, GALP increased sexual behavior even in castrated males, suggesting that effects on testosterone secretion did not mediate the behavioral effects.

Neuropeptide Y (NPY)

NPY is widely distributed throughout the brain and is often co-localized with NE (181). NPY, administered icv, stimulated feeding, but inhibited copulation in male rats (189,777). Consistent with an inhibitory effect of NPY, icv administration of an NPY antagonist facilitated copulation (774). However, systemically administered NPY failed to affect reflexive erections (189).

Prolactin

A large number of studies in the 1980s revealed inhibitory effects of chronic prolactinemia, induced by implanting excess pituitary glands or prolactin-secreting tumors, by injecting drugs that stimulate PRL release, or by direct injection of PRL [reviewed in (438)]. The most commonly reported effects were increased mount and ejaculation latencies and general slowing of copulation. However, there were some contradictory effects, and the mechanism by which chronically elevated prolactin may impair sexual behavior is not clear.

Corticotropin-Releasing Hormone (CRH)

Injections of CRH into the third ventricle increased mount, intromission, and ejaculation latencies and increased the numbers of mounts and intromissions preceding ejaculation (905). These inhibitory effects of CRH were attenuated by co-administration of the opioid antagonist naloxone. Furthermore, CRH can inhibit GnRH release, possibly acting through endogenous opioids (35). The promotion of freezing and flight behaviors by CRH, in addition to its inhibition of sexual behavior, suggests that it organizes a constellation of behaviors appropriate to stressful or threatening situations.

Cholecystokinin (CCK)

There have been mixed reports of CCK effects on male sexual behavior. One study found that systemic administration of sulfated CCK decreased the numbers of intromissions and the time preceding ejaculation; these effects were reversed by the CCK antagonist proglumide (759). However, other studies found no effect of systemic (121,543) or icv (121,253) administration of CCK. On the other hand, there are at least two subtypes of CCK receptor (37), which may have opposite effects (231,673). Thus, the failure to find significant effects of CCK may be due to opposing excitatory and inhibitory effects. (See section on the Mesocorticolimbic DA System.)

Angiotensin II

Angiotensin II, microinjected into the third ventricle increased intromission latency, the number of intromissions preceding ejaculation, and postejaculatory interval (185). The most effective dose also increased drinking during the test. A more recent study confirmed the inhibitory effects, which were blocked by an angiotensin I receptor antagonist administered in the drinking water (463). However, it is not clear whether the sexual impairment was a direct effect or a result of thirst.

BRAIN AREAS AND CIRCUITRY IMPLICATED IN THE CONTROL OF MASCULINE SEXUAL BEHAVIOR

Methods used to investigate the neural organization of copulation include lesions, electrophysiological stimulation and recording, hormonal and drug manipulations, microdialysis and measures of cellular activation. As our understanding of the relevant brain architecture has expanded, the techniques have been refined to focus on discrete subregions within the MPOA and its afferent and efferent targets. The MPOA is an essential organizing center for male copulatory behavior across a wide variety of species. For this reason, our discussion is organized around afferent and efferent pathways of the MPOA, beginning rostrally. However, it should be noted that we have not included all sensory and motor circuits that contribute to the neural organization of male sexual behavior. Visual and auditory stimuli clearly contribute to the elicitation of male sexual behavior, and somatomotor pathways are necessary for the patterned motor output. However, relatively little research has been devoted to these connections. Therefore, we will focus on the primary directions of research in this area. In our discussion of the neural circuits for male sexual behavior, we have followed the anatomical parcellation and nomenclature of Paxinos and Watson (748).

Sensory Inputs

Olfactory Bulbs

The olfactory bulbs transmit chemosensory signals transduced in the olfactory mucosa and vomeronasal organ to higher centers in the forebrain. Chemosensory cues contribute to male sexual behavior across a wide variety of species (rodents to primates). However, the relative importance of chemosensory stimuli, compared with other sensory cues (vision, audition), is tailored to the biology of individual species. Chemosensory cues are particularly important for mating in rodents and other nocturnal macrosmatic species that mate in underground burrows.

The olfactory system includes two functionally and anatomically distinct components (Fig. 10). In the main olfactory system, volatile odors are transduced by receptor neurons in the olfactory mucosa lining

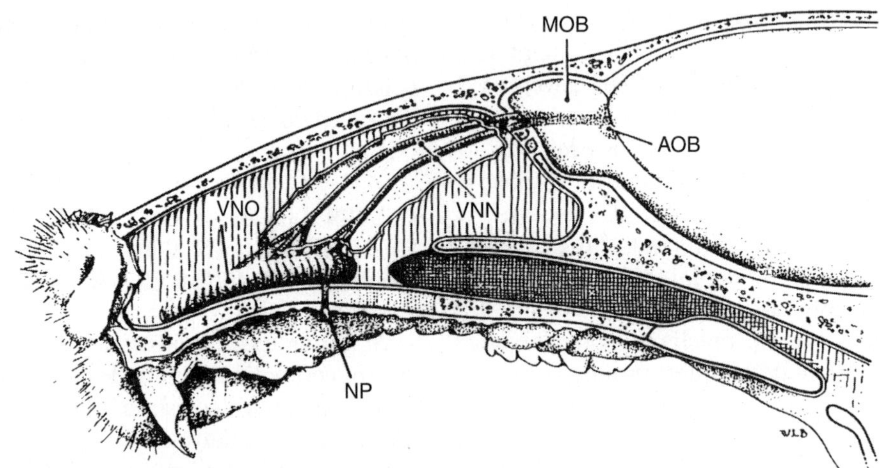

FIG. 10. Relation between the vomeronasal organ and the olfactory bulbs of the Syrian hamster. The vomeronasal organ (VNO) projects to the accessory olfactory bulb (AOB) via the vomeronasal nerve (VNN). The vomeronasal nerve travels medial to the main olfactory bulb on the way to the accessory olfactory bulb. Olfactory information is presumed to reach the vomeronasal organ through the nasopalatine duct. In contrast, the main olfactory bulb receives innervation from the olfactory nerve originating from receptors in the nasal mucosa (pathway not shown). Efferent connections of the main and accessory olfactory bulbs as they relate to copulation are illustrated in Figs. 14, 25, and 27. AOB, accessory olfactory bulb; MOB, main olfactory bulb; NP, nasopalatine duct; VNN, vomeronasal nerve; VNO, vomeronasal organ. [We appreciate the generosity of Sarah W. Newman for providing this illustration adapted from one previously published (1015).]

the nasal cavity. Olfactory receptor neurons project through the cribriform plate to the main olfactory bulb. By contrast, the accessory olfactory bulb receives projections from the vomeronasal organ, a separate sensory organ at the base of the nasal cavity, which responds preferentially to nonvolatile chemosensory cues from conspecifics. Recent studies suggest that the vomeronasal organ may also detect volatile odor cues (957,1054). Stimuli transduced in the vomeronasal organ are particularly important for social behavior, including mating, aggression, affiliation, and maternal behavior. However, in microsmatic species including humans, the vomeronasal organ and accessory olfactory bulbs are greatly diminished and may be nonfunctional.

Effects of Lesions. Bilateral destruction of the olfactory bulbs eliminates all chemosensory input; selective lesions of the main or accessory olfactory systems interrupt transmission of chemosensory cues transduced in the olfactory mucosa and vomeronasal organ, respectively. Deficits in sexual behavior following olfactory bulbectomy depend on the species and prior sexual experience of the males under study.

In male Syrian hamsters, bilateral removal of the olfactory bulbs immediately and permanently abolished mating (698). This effect was largely due to damage to the accessory olfactory system. Deafferentation of the accessory olfactory bulb through vomeronasal nerve cuts (which also damage part of the main olfactory bulb) severely impaired mating in up to 44% of sexually experienced male hamsters (711,786, 787,1015), but removal of the vomeronasal organ without damage to the main olfactory system had little effect in experienced males (656). The reduction in sexual behavior after vomeronasal organ removal was more pronounced in sexually naïve males (656). By contrast, there was minimal change in sexual activity following deafferentation of the main olfactory bulb by irrigation of the nasal cavity with zinc sulfate to destroy olfactory receptor neurons (241,550,711,785,786,1015). Nonetheless, combined deafferentation of the main and accessory olfactory systems eliminated mating, similar to the effects of bilateral olfactory bulbectomy (660,786,1015). The main and accessory olfactory bulbs have parallel yet overlapping projections to amygdaloid nuclei through the lateral olfactory tract. Bilateral transection of the lateral olfactory tract blocked mating, presumably by interrupting olfactory bulb efferents to the amygdala (240).

In rats, olfactory bulbectomy reduced the percentage of males that copulated to ejaculation (521,522,573,633, 634,993,1013) [reviewed in (636)]. The failure to ejaculate stems from both an inability to initiate copulation (268,521,522,633,993,1013) as well as a failure to sustain copulation (633). In those males that continued to mate after olfactory bulbectomy, intromission and ejaculation were delayed (104,154,265,269,399,521), and anogenital investigation of females was reduced. However, preference for a receptive female over a nonreceptive female was unaffected (269). In rats, olfactory bulbectomy also eliminated noncontact erections elicited by female rat urine (499). As in male hamsters, sexually naïve male rats may be more sensitive to anosmia (104), although severe copulatory dysfunction has also been reported following olfactory bulbectomy in sexually experienced males (521,633,993).

Supplementation with exogenous T or gonadotropins did not restore sexual behavior in bulbectomized male rats (521,522,633), suggesting that the copulatory deficit was not a secondary consequence of diminished gonadal output. However, increasing general arousal could activate copulation in male rats after olfactory bulbectomy (79,152). Mildly painful flank shock (633) or tail pinch (993) stimulated copulation in olfactory bulbectomized male rats, but these anosmic animals still exhibited longer latency to intromission and ejaculation compared with controls (633). Moreover, the effects of tail pinch and flank shock are transient, as olfactory bulbectomized males failed to copulate when retested the following week without exogenous stimulation (633,993).

In view of the modest effects of olfactory bulbectomy in male rats, few studies have investigated selective deafferentation of the main and accessory olfactory system. Zinc sulfate lesions of the olfactory mucosa reduced the percentage of isolated male rats that copulated (949), yet group-housed male rats in other studies were unaffected (154,949). Prior copulatory experience may also compensate for destruction of the olfactory mucosa, since sexually experienced male rats were less affected by zinc sulfate lesions. Likewise, removal of the vomeronasal organ in sexually experienced rats decreased the intromission rate and increased the latency to the first intromission, but did not block ejaculation (856). Interestingly, noncontact erections appear to be mediated by odors transmitted via the main olfactory system, since lesions of the vomeronasal organ had no effect, but zinc sulfate lesions significantly decreased noncontact erection (499). As with hamsters, simultaneous deafferentation of the main and accessory olfactory system produced more severe deficits in sexual activity than damage to either chemosensory system alone (265).

Chemosensory cues are also critically important for mating in male mice. Olfactory bulbectomy eliminated mounting in essentially all male mice (264,834,835). Zinc sulfate treatment did not interrupt mating (264). Although early studies reported that vomeronasal organ lesions sharply decreased the proportion of males that intromitted and ejaculated (182), recent studies have found no significant reduction in mating (730). Indeed, male mice with null mutations of the transient receptor potential 2 (TRP2) cation channel to disrupt vomeronasal neuron function displayed high levels of mounting towards male intruders (539,928). Based on these recent results, it appears that the vomeronasal system in mice has an important, but not critical, role in copulation.

The role of the olfactory system in male sexual behavior has received less attention in other species. Olfactory bulbectomy produced variable deficits in the behavior of male guinea pigs (94), while vomeronasal organ removal had no effect (95). Olfactory bulbectomy or damage to the nasal epithelium failed to block copulation in male dogs (388), cats (58), rhesus monkeys (342), gerbils (169), sheep (304), or ferrets (465), although nares occlusion blocked heterosexual partner preference in ferrets.

In early studies, the effects of olfactory bulbectomy were attributed to the resulting anosmia. However, it has been argued that the olfactory bulbs have additional integrative functions other than simple sensory processing [e.g., (153)]. In particular, the olfactory bulbs receive ascending inputs from olfactory-related structures and from nonolfactory subcortical areas. Undoubtedly, both olfactory and nonsensory functions of the olfactory bulbs are important.

Activation of c-fos or Other Immediate-early Genes. Although lesions provide information about brain areas that are critical for behavior, they may fail to identify other areas that contribute to behavioral performance. Activation of immediate-early genes provides a complementary approach to map brain circuits that are activated during behavior [reviewed in (758)] (Fig. 11). Immediate-early genes are expressed transiently in response to stimulation [reviewed in (257)]. Their protein products serve as transcription regulators for other genes, which may in turn affect neural activity or responsiveness to certain stimuli.

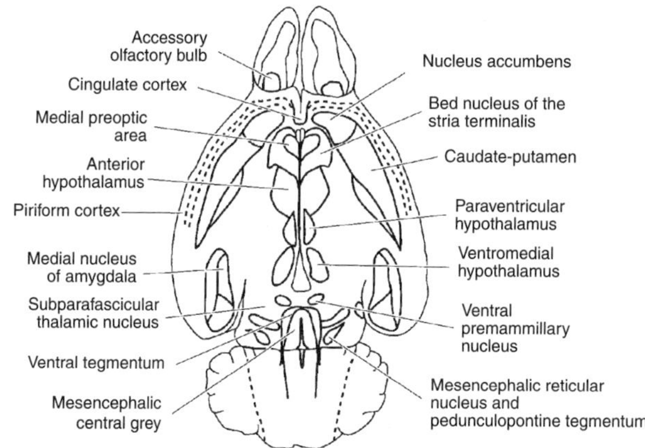

FIG. 11. Regions within a horizontal section of rat brain in which Fos induction is observed after sexual stimulation in females and males. With some notable exceptions, Fos has been observed in similar regions in hamsters and gerbils. [Figure is reprinted from (758) with permission.]

The distribution of mRNA or protein for immediate-early gene products visualized with in situ hybridization or immunocytochemistry, respectively, provides a map of neuronal activation. In studies of male sexual behavior, the *c-fos* gene and its Fos protein product are widely used.

Copulation in male hamsters increased Fos in the main and accessory olfactory bulbs. The increase in Fos in the olfactory bulbs was specific to chemosensory stimuli rather than mating, because female odors alone induced Fos (283,299,491,933). Although T is required for sexual activity, Fos expression in the olfactory bulb was unaffected by castration (933), suggesting that castration may alter downstream responses to sex-related chemosensory cues, but not the detection of those cues. Moreover, Fos was expressed in the accessory olfactory bulbs and its efferent targets even after ablation of the main olfactory system with zinc sulfate (283). On the other hand, Fos expression in main olfactory bulb efferent targets was maintained after vomeronasal lesions in sexually experienced males (298,934). In sexually naïve males with vomeronasal lesions, odors alone were unable to stimulate Fos (298). Therefore, pheromonal stimulation of the vomeronasal organ and accessory olfactory bulb is sufficient for Fos induction in the accessory olfactory bulb and its central targets, but odors transmitted through the main olfactory system may also induce Fos, particularly in males with prior sexual experience.

In male rats and mice, mating or exposure to estrous female bedding stimulates Fos in the accessory olfactory bulbs and downstream structures (82,210, 370,816); mating elicits more Fos than does chemosensory stimulation alone (468). However, noncontact erections in male rats fail to enhance Fos expression. Within the accessory olfactory bulb in male mice, Fos induction by female stimuli is concentrated in the rostral region, and is blocked by removal of the vomeronasal organ (261,612). This supports the existence of functional heterogeneity in the accessory olfactory system.

Although the accessory olfactory system appears to be more important than the main olfactory system for Fos induction in rodents, male ferrets, which are carnivores, rely on the main olfactory system (467). Both estrous female bedding and peppermint odor increased Fos in the main olfactory bulb of castrated male ferrets. However, female pheromones failed to stimulate Fos in the accessory olfactory bulb. In male ferrets, unlike male hamsters, T treatment increased the Fos response to female odors, but not to peppermint. Moreover, the ferret olfactory bulb has ARs. Thus, in this species, T has the potential to increase the responsiveness of olfactory bulb neurons to sex-related chemosensory stimuli.

Amygdala

When considering the connections and functions of "the amygdala," it is important to differentiate between the different amygdaloid nuclei. Indeed, there is controversy as to whether the amygdala should be considered an entity at all, rather than a conglomeration of dissimilar areas related to learning, motivation and fear (central nucleus and basolateral division) or to chemosensory stimuli and social behavior (corticomedial division) (Fig. 12) (937). Studies in rodents highlight the importance of the corticomedial amygdala in male sexual activity. The corticomedial amygdala is a key nodal point for integration of chemosensory, somatosensory and hormonal stimuli. In particular, the medial amygdala (Me) relays chemosensory stimuli from the olfactory bulbs to midline nuclei of the preoptic area and hypothalamus.

During mating, the amygdala receives chemosensory, somatosensory, and hormonal stimuli via the main and accessory olfactory bulbs, the thalamic subparafascicular nucleus, and steroid receptors, respectively. Chemosensory cues are transmitted by the projections of mitral and tufted cells in the olfactory bulbs through the lateral olfactory tract, terminating in the medial and cortical amygdaloid nuclei (anterior, posterolateral, and posteromedial). In particular, efferent projections of the accessory olfactory bulb target the medial and posteromedial nuclei, while the main olfactory bulbs project widely to the anterior and posterolateral cortical nuclei and to olfactory processing areas in ventral allocortex (Fig. 13). Lesions that interfere with the transmission of chemosensory cues disrupt male sexual activity according to the

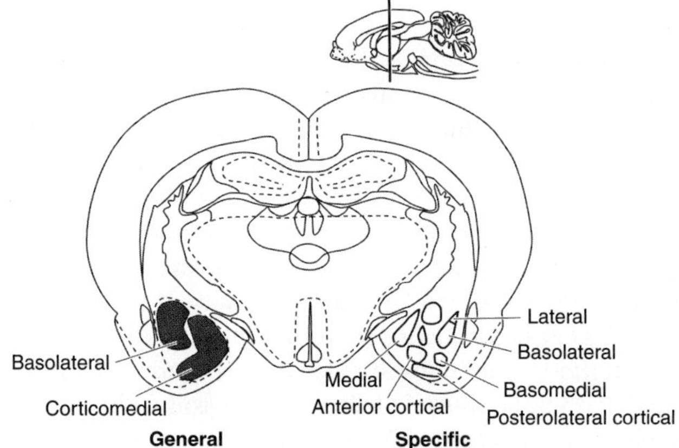

FIG. 12. The amygdala, an almond-shaped structure located in each temporal lobe, seen in a schematic drawing of a coronal section of the rat brain. Two general amygdaloid regions, the basolateral and corticomedial nuclei (left), have been studied extensively. Destruction of the corticomedial nuclei, but not the basolateral nuclei, severely affects male copulatory behavior in rodents. The specific nuclei of the basolateral and corticomedial amygdala are depicted (right). [Figure is reprinted from (702), with permission.]

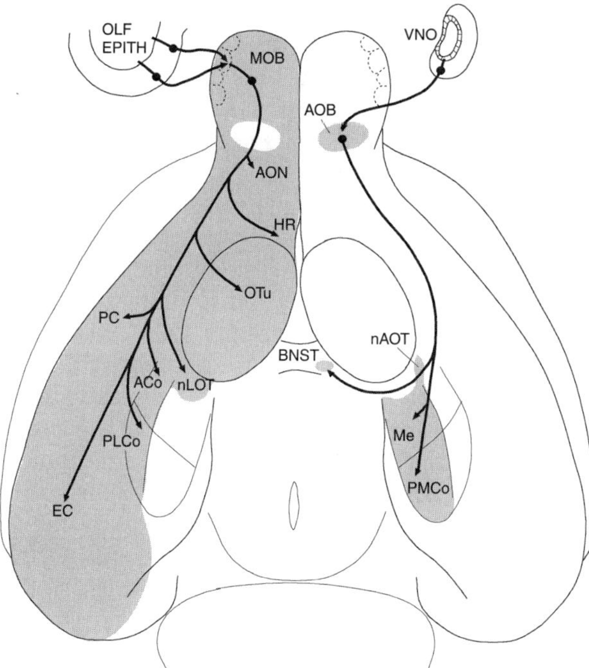

FIG. 13. Diagram of the ventral surface of the hamster brain. Shaded areas indicate brain regions that receive efferent projections of the olfactory system, via the main olfactory bulb (left), and of the vomeronasal system, via the accessory olfactory bulb (right). ACo, anterior cortical nucleus of the amygdala; AOB, accessory olfactory bulb; AON, anterior olfactory nucleus; BNST, bed nucleus of the stria terminalis; EC, entorhinal cortex; HR, hippocampal rudiment; Me, medial nucleus of the amygdala; MeA, medial nucleus of the amygdala, anterior division; MeP, medial nucleus of the amygdala, posterior division; MOB, main olfactory bulb; nAOT, nucleus of the accessory olfactory tract; nLOT, nucleus of the lateral olfactory tract; OLF EPITH, olfactory epithelium; OTu, olfactory tubercle; PC, piriform cortex; PLCo, posterolateral cortical nucleus of the amygdala; PMCo, posteromedial cortical nucleus of the amygdala; VNO, vomeronasal organ. [Figure is reprinted from (1028), with permission.]

specific chemosensory requirements for mating in that species.

In addition, the Me receives ascending somatosensory cues from the penis and perineal region relayed through the subparafascicular nucleus. However, projections of the subparafascicular nucleus are not limited to the amygdala: somatosensory cues are conveyed directly to the MPOA, as well.

The medial and posteromedial cortical nuclei also have abundant receptors for androgens and estrogens. Testosterone promotes male sexual activity, at least in part, through binding to amygdaloid ARs or to ERs after local aromatization. However, steroid receptors are not limited to the amygdala, but are also concentrated in midline nuclei of the preoptic area and hypothalamus, including MPOA. Within the Me, ARs and ERs are localized preferentially to the posterior subnucleus, whereas olfactory bulb efferents terminate preferentially in the anterior subnucleus.

Effects of Radiofrequency or Cell-body Lesions. The impact of amygdala lesions on male sexual behavior depends on the specific sites within the amygdala. In general, lesions of the central nucleus or the basolateral division (basolateral, basomedial, and lateral nuclei) do not disrupt copulation in male rats (323,380,492) or hamsters (535), although there is some indication that male rats copulate more rapidly following basolateral amygdaloid lesions (380). The central and basolateral regions of the amygdala have connections with the frontal cortex, brainstem areas for autonomic control, and ventral striatum [see (937)]. Through their connections, nuclei of the central and basolateral amygdala are important for conditioned responses during fear and reinforcement. In their landmark studies of sexual motivation in male rats, Everitt and co-workers (277–279) showed that basolateral amygdala lesions abolished bar-pressing for a secondary reinforcer previously paired with access to a receptive female. However, as Everitt (276) indicates, lesions of the basolateral amygdala produced deficits that were not specific to sexual motivation. Rather, the basolateral amygdala mediates the effects of conditioned cues on motivated behaviors, including sexual motivation.

In contrast to lesions of the basolateral or central amygdala, lesions of the corticomedial amygdala impair copulation. The behavioral deficits depend on the species and the precise location of the lesions. In particular, lesions of the anterior subdivision of the Me abolish copulation in hamsters (537). This observation is consistent with the essential role of chemosensory cues for male hamster sexual behavior (703). By contrast, lesions of posterior Me impair but do not eliminate mating (537). In male rats (239,250,323, 380,492,623,960), and gerbils (395), Me lesions reduce copulatory efficiency, producing longer mount and ejaculation latencies, more intromissions preceding ejaculation, longer interintromission intervals, and/or fewer ejaculations to sexual exhaustion. Lesions restricted to the dorsolateral part of the Me of male rats reduced the number of ejaculations and increased ejaculation latency, but did not affect mounting (323,380), suggesting that this subregion is especially important for the control of ejaculation.

Through its efferent projections to the MPOA, the Me facilitates appetitive sexual behavior and subsequent copulation. Unilateral lesions of the MPOA impaired, but did not eliminate mounting in male rats (493). However, when combined with lesion of the contralateral Me, copulatory behavior was severely disrupted. Similar effects have been reported in gerbils (395). Likewise, chemical stimulation of the Me increased DA release in the MPOA (251), and microinjection of the DA agonist apomorphine into the MPOA restored sexual behavior inhibited by

Me lesions (250). As measured by microdialysis, basal DA levels in MPOA were unaffected in males with Me lesions, but there was no DA release in response to a receptive female or during copulation (250). These observations suggest that the Me is critical for copulatory behavior organized through the MPOA. Because there are no DA-containing neurons in the Me of rats, amygdala efferents apparently synapse, either directly or indirectly, on dopaminergic cell bodies or terminals in the MPOA. Using a somewhat different model for sexual motivation, radiofrequency lesions of the posterior Me abolished noncontact erections and reduced preference for an estrous female in a partner preference test, but had no effect on copulatory behavior (494,497). Finally, a subregion of the posterodorsal Me appears to be associated with sexual satiety. Lesions of this area in male hamsters increased the number of ejaculations preceding satiety (743), and Fos expression correlated with the onset of satiety (742).

Activation of c-fos or Other Immediate-early Genes. In view of the connections of the corticomedial amygdala, it is not surprising that mating activates Fos in amygdaloid nuclei of male rats (82,209,980), hamsters (489–491), gerbils (394), prairie voles (994), and musk shrews (324). Increasing amounts of copulation, from mounting to several ejaculations, elicited increasing amounts of Fos-ir in these sites (82,209,980). The pattern of mating-induced Fos within the amygdala mirrors the importance of different amygdaloid nuclei as determined in lesion studies. Thus, the largest number of Fos-positive neurons were present in the Me, with additional expression in the anterior and posteromedial cortical nuclei (489). Mating did not induce significant Fos expression in the central and basolateral nuclei. Interestingly, the distribution of mating-induced Fos within the medial nucleus does not precisely accord with the results of tract tracing and lesion studies. Although the anterior subdivision receives the majority of direct olfactory bulb efferents, and lesions of this area produce greater deficits in copulatory behavior (537), the largest concentration of Fos-positive neurons is found in the posterior subnucleus of the Me (489). Furthermore, activation of the posterodorsal subdivision of the Me (MePD) appears to be associated specifically with ejaculation (see section on An Ejaculation-Related Circuit, below). Co-localization of Fos-ir with either GAD or glutamate revealed that half of the activated cells in the lateral MePD were GABAergic, while a quarter were glutamatergic (900). Finally, Fos in the Me, but not in the MPOA, was positively correlated with the duration of the postejaculatory interval, suggesting that this area may contribute to postejaculatory quiescence (574).

Chemosensory cues alone induce substantial Fos expression in the corticomedial amygdala. Administration of lidocaine anesthetic to the penis of male rats resulted in few intromissions, but a significant increase in Fos-ir in both the Me and MPOA (709), suggesting that chemosensory stimuli are the primary determinants of the mounting-induced increments in neuronal Fos expression. Likewise, exposure of male rats to bedding from estrous females increased Fos-ir at each level of the vomeronasal circuit, including the Me (139,210,468). In hamsters, exposing males to a cotton swab containing female hamster vaginal secretions elicited similar results (283,299,489). In male gerbils, Heeb and Yahr (394) identified the medial part of the posterodorsal Me as activated by exposure to sex-related odors, whereas the lateral portion was activated only following ejaculation (see Circuitry and Anatomical Connections, below).

Chemosensory cues for Fos induction in the Me can be further subdivided into those transduced by the main and accessory olfactory systems. Intranasal zinc sulfate treatment to damage the olfactory mucosa without affecting the vomeronasal system decreased copulation-induced Fos-ir in the anterior Me (285), suggesting that input from the main olfactory system contributes to the activation of neurons in this area. In sexually experienced male hamsters, vomeronasal lesions had minimal effect on Fos expression in response to female stimuli, but similar lesions practically eliminated this Fos in the Me in naïve males (298).

As determined by colocalization of Fos with steroid receptors, many hormone-sensitive neurons in the Me are activated during mating. In male rats (358) and hamsters (1026), AR-containing neurons in posterior Me express Fos after mating. Furthermore, neurons that co-localize Fos-ir and AR-ir are interconnected, as determined in tract tracing studies (358) (see circuitry section below). On the other hand, castration does not appear to reduce Fos in the Me and other central chemosensory targets of male rats (83) or hamsters (933). Likewise, the pattern of Fos expression in Me and MPOA was similar in wild-type mice and in those lacking the aromatase gene (60). From these observations, it appears that gonadal steroids have a long-term permissive action on mating behavior circuitry, but are not required acutely for Fos activation.

Effects of Hormone Implants. There is considerable evidence that the corticomedial amygdala is sensitive to gonadal steroid hormones. Receptors for androgens and estrogens are concentrated in the posterior subdivision of the Me (542,892,896,1026). This subnucleus and the encapsulated area of the bed nucleus of the stria terminalis (BST) are approximately twice as large in male rats as in females, and contain more substance P, CCK, and vasopressin (411). Likewise, the posteromedial cortical amygdaloid nucleus is also larger in males (986). The posterodorsal

Me also shows a different pattern of synaptic contacts, compared with females (706,707). Castration in male rats (203) or exposure to short-day photoperiod in Siberian hamsters (202) caused the posterodorsal Me to shrink. Androgen treatment of adult female rats increased the posterodorsal Me to male-typical size, through an enlargement in soma size. Therefore, hormone treatment in adulthood can completely reverse a sexual dimorphism in a brain area important for sexual behavior.

Functionally, hormone implants in the Me can facilitate the maintenance or restoration of male sexual behavior. Bilateral T implants into the Me of castrated male rats delayed the loss of both copulation and noncontact erections (111). Similarly, either T (1024) or E (205,1025) implants into the Me restored copulatory behavior in castrated male hamsters. Likewise, E stimulated sexual activity in male rats (799), including those treated systemically with an aromatase blocker (429). However, DHT implants were ineffective in facilitating sexual behavior, suggesting that stimulation of ARs in the Me is not sufficient for behavioral restoration in male hamsters (1024). On the other hand, DHT implants into the Me were sufficient for behavioral activation in castrated rats receiving subthreshold systemic injections of E (85). Implants of the AR antagonist hydroxyflutamide into the Me of T-replaced castrated male rats partially inhibited copulation (622), but the E antagonist tamoxifen likewise failed to block mating in castrated male hamsters with systemic T replacement (1030). Therefore, stimulation of ARs in the Me contributes to male copulatory behavior in both rats and hamsters. However, androgens in Me are not necessary, and may not be sufficient for copulation in male rats or hamsters.

Because the Me receives chemosensory and hormonal stimuli, both of which are essential for mating in male hamsters, the interactions between hormonal and chemosensory stimulation were probed. Removal of chemosensory input by olfactory bulbectomy, either ipsilateral or contralateral to a T implant into the Me, rendered the implant ineffective (Fig. 14) (1025). Similar implants into the MPOA were less susceptible to the loss of olfactory input (1029). Cottingham and Pfaff (219) suggested that steroid-responsive networks should exhibit redundancy, amplification, stability and selective filtering. Because both the Me and the MPOA transduce steroid cues and receive chemosensory input, Coolen and Wood (205) tested whether steroid implants into these two areas exhibit steroid amplification. Although there is some redundancy, in that steroid implants into either area were sufficient to activate copulation, there appears to be no amplification, in that males with dual implants into the Me and MPOA showed no greater copulatory ability than did males with single implants.

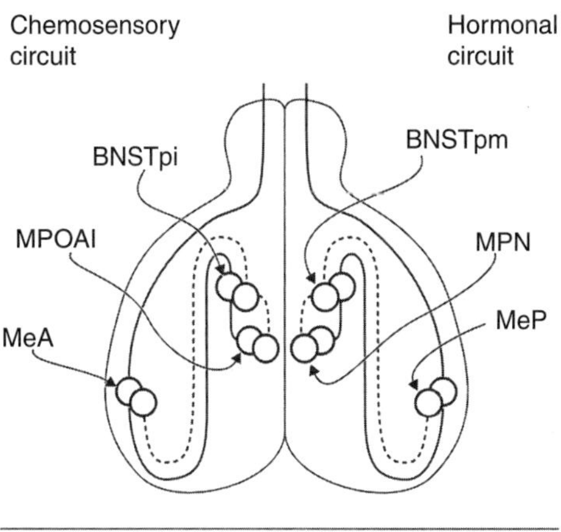

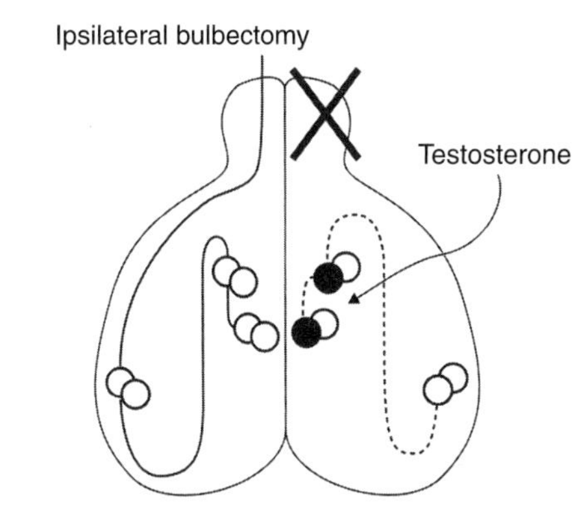

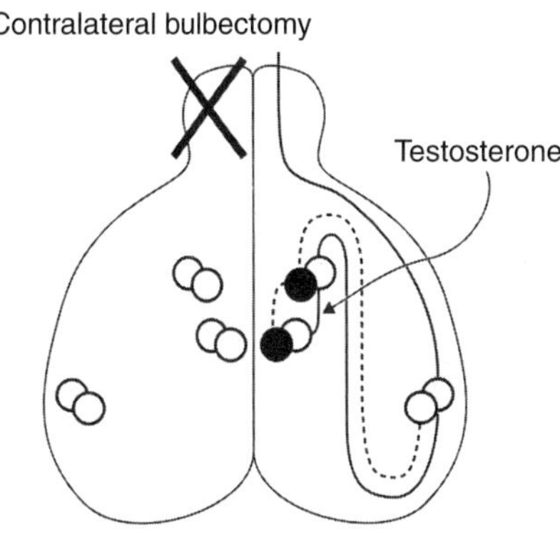

FIG. 14. Schematic diagram of mating behavior circuitry through the medial amygdaloid nucleus (Me), bed nucleus of the stria terminalis (BNST), and medial preoptic area (MPOA) in the male Syrian hamster brain, illustrating the separate pathways for receipt of chemosensory and hormonal cues. **Top:** Normal male. Elements of the chemosensory circuit (*left*) include the anterior subdivision of the Me (MeA), posterointermediate subdivision of BNST (BNSTpi), and the lateral subdivision of MPOA (MPOA1). The hormonal circuit (*right*) consists of the posterior subdivision of the Me (MeP), the posteromedial subdivision of BNST (BNSTpm), and the medial preoptic nucleus (MPN). **Middle:** Testosterone stimulation of BNSTpm and MPN (*solid circles*) combined with removal of the ipsilateral olfactory bulb prevents communication between the chemosensory and hormonal circuits. **Bottom:** Removal of the contralateral olfactory bulb permits chemosensory and hormonal integration. [Figure is reprinted from (1029) with permission.]

Microinjections of Drugs. Microinjection of angiotensin II into the Me impaired sexual behavior in male rats (138). This effect was prevented by antagonists to both angiotensin I and II receptors. However, as with icv microinjections, it is not clear whether increased thirst interfered with copulation. (See section on Systemically and Intracerebroventricularly Injected Drugs.)

Amygdaloid Efferents. Efferents from the corticomedial amygdala travel via the stria terminalis and ventral amygdalofugal pathway to innervate the BST, MPOA, VMH, and other midline hypothalamic nuclei (504). In male hamsters, cutting the stria terminalis increased mount latency, ejaculation latency, and interintromission interval for 2 weeks postoperatively (536). Knife cuts of the ventral pathway produced a less severe loss of mating behavior (534). However, combined knife cuts of both the strial and ventral projections of the corticomedial amygdala eliminated copulation (536). Similar pathways probably contribute to male sexual behavior in rats as well. In this regard, lesions of the Me produced greater mating deficits than cuts of the stria terminalis alone (495). Giantonio et al. (323) reported that lesions of the stria terminalis significantly increased ejaculation latency in male rats and that these males displayed a "distinct, though nonsignificant" (p. 42) increase in the number of mounts and intromissions prior to ejaculation.

Through the stria terminalis and ventral pathway, the anterior and posterior subdivisions of the Me maintain distinct, yet parallel connections with subnuclei of BST. While the posterior subnucleus of Me projects to the posteromedial BST, the anterior Me sends projections preferentially to the intermediate subdivisions of posterior BST (157,204,345). These connections are part of the extended amygdala, a ring of interconnected nuclei linking the central and Me with BST (32).

Bed Nucleus of the Stria Terminalis (BST)

In the control of male sexual behavior, nuclei of the BST play a modulatory role. Like the amygdala, the BST is a heterogeneous brain region with functionally and anatomically distinct subnuclei [reviewed in (33)]. The posteromedial regions of BST are particularly relevant for control of male mating behavior (703). As part of the extended amygdala, the Me and BST share overlapping functions and connections (32). In particular, the posteromedial BST and posterior Me each have abundant steroid hormone receptors (542,892,896,1026). Anatomically, the BST is positioned as an intermediate between the Me and MPOA (157,204,345,594). Nonetheless, chemosensory cues transmitted through Me can also bypass BST to reach the MPOA directly.

Effects of Radiofrequency or Cell-body Lesions. Lesions of the BST do not necessarily block mating in rats (272,967) or hamsters (788). Instead, the behavioral deficits suggest that BST contributes to preparation for mating. Thus, BST lesions reduce noncontact erections in male rats (554) and anogenital investigation in male hamsters (788). However, in gerbils BST plays a more central role in copulatory behavior (868).

In male rats, lesions of BST have produced copulatory deficits similar to those seen following corticomedial amygdala lesions. Hence, lesions of BST significantly increased the number of intromissions preceding ejaculation, the interintromission interval, and, consequently, the ejaculation latency (272,967). Touch-based erections were not affected (967). On the other hand, smaller radiofrequency lesions of the BST greatly decreased the number of noncontact erections (554). These lesions also slowed the rate of copulation, but did not affect intromission ratio, suggesting that *in copula* erections were not affected. These results are in contrast to the effects of lesions of the MPOA, which had little effect on noncontact erections, but dramatically impaired copulation (554). In male hamsters, large lesions of BST did prevent copulation, but smaller lesions centered in posteromedial BST selectively reduced anogenital investigation without preventing ejaculation (788). On the other hand, Finn and Yahr (300a) have suggested that a more ventral area of the BST (vBST) in rats may correspond to the gerbil's lateral sexually dimorphic area (LSDA) of the MPOA. Bilateral cell-body lesions of the vBST produced severe and persistent copulatory impairments, although damage to the adjacent posterodorsal preoptic nucleus (PdPN) may have contributed to these deficits. These authors suggested that axons from the vBST to the retrorubral field of the midbrain are important for copulatory behavior.

In male gerbils bilateral lesions of either the caudal medial BST or the sexually dimorphic area (SDA) of the preoptic area severely impaired copulation (868). Furthermore, unilateral lesions of the caudal medial BST and the contralateral SDA also decreased copulation. Ipsilateral lesions of the Me and SDA had much less effect. Therefore, the caudal medial BST does not simply relay input from the Me to the SDA. Rather, it appears to perform an important integrative function. It is not clear whether the different effects of BST lesions in rats (272,300a,747,967), hamsters (788), and gerbils (868) reflect differences among species, experimental procedures, or the specific sub-areas targeted.

Activation of c-fos or Other Immediate-early Genes. The pattern of mating-induced Fos expression in BST is similar to that for Me. Thus, in male rats, hamsters and gerbils, mating stimulates Fos in BST, particularly in the posterior and medial subdivisions

(283,394,468,490). By contrast, mating reduced Fos in BST of male macaques (665). In rodents, exposure to chemosensory cues from estrous females [hamsters: (283,299,490)] or their bedding [rats: (468); gerbils: (394)] also stimulated Fos, although mating to ejaculation induced further increases in Fos (209). In contrast, there was no increase in Fos-ir in the BST of male ferrets exposed to estrous female bedding (467). In hamsters, Fos induction in posteroventral BST and MPOA appears to be selective for mating and sexual stimuli, because aggression failed to induce Fos in these regions (490).

Hormonal Effects. Posteromedial BST has abundant steroid receptors which extend ventromedially from the stria terminalis to the medial preoptic nucleus (542,892,896,1026). While the role of androgens and estrogens in the BST has not been extensively tested using intracerebral implants, gonadal steroids do modify neurotransmitters within BST. In particular, castration decreased immunoreactivity for CCK in the encapsulated portion of BST (894). Testosterone treatment restored CCK to normal levels. Similar effects were observed in sexually dimorphic areas of the Me and medial preoptic nucleus.

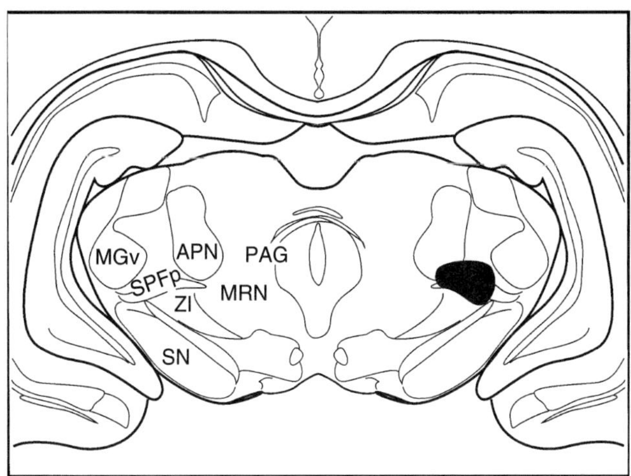

FIG. 15. Site of FluoroGold (FG) injection into the lateral central tegmental field (CTF) (black area on right side of section). The site of injection was defined as the area of brilliant FG surrounding the region of necrosis induced by the injection. APN, anterior pretectal nucleus; MGv, ventral medial geniculate nucleus of the thalamus; MRN, median raphe nucleus; PAG, periaqueductal gray; SN, substantia nigra; SPFp, parvocellular portion of the subparafascicular nucleus; ZI, zona incerta. [Figure is redrawn from (357), with permission.]

Central Tegmental Field/Subparafascicular Nucleus of the Thalamus

The midbrain tegmentum is reciprocally connected to the MPOA, Me, and anterior hypothalamus (211,214, 357,358,696,893,895). Subregions of the midbrain tegmentum have been referred to as the central tegmental field [CTF; e.g., (82,355,357–359,893)] or the dorsolateral tegmentum [DLT; e.g., (132,133,266, 327,583,741)]. This region lies immediately dorsal to the lateral half of the substantia nigra and includes the subparafascicular nucleus, part of the caudal zona incerta, the peripeduncular nucleus, the mesencephalic reticular nucleus, and the anterior pretectal nucleus (Fig. 15). In particular, the medial parvocellular portion of the subparafascicular nucleus (SPFp) is positioned to relay somatosensory cues from the penis and perineum to the MPOA and Me. The SPFp has been implicated as part of an ejaculation circuit in rats, hamsters, and gerbils (see Circuitry and Anatomical Connections, below).

Effects of Electrolytic or Cell-body Lesions. Lesion studies in male rats suggest that both sexual motivation and copulatory performance are affected by connections between the MPOA and CTF. Bilateral electrolytic lesions of CTF virtually eliminated the mating behavior of male rats (132,133,372). Furthermore, CTF lesions impaired copulatory behavior, as well as sniffing, pursuit, self-grooming, and genital exploration of the female (327). Similarly, bilateral lesions of either MPOA or CTF, or unilateral lesions of MPOA and contralateral CTF inhibited both copulation and the preference for an estrous female (266), suggesting that connections between these two areas are essential for copulation (132).

Subsequent lesion studies emphasized the importance of CTF in general, and SPFp in particular, in transmitting ascending genitosensory information to MPOA. Combined ipsilateral excitotoxic lesions of the CTF and Me in male rats abolished copulation-induced Fos-expression in the MPOA (82). Unilateral lesions of either the CTF or Me alone did not affect Fos in the MPOA. Likewise, unilateral MPOA lesions did not affect Fos in the CTF or Me. In gerbils, bilateral lesions of the SPFp did not affect copulatory behavior, suggesting that ejaculation-induced Fos expression represents a response to ejaculation-related stimuli, rather than participation in the control of ejaculation (395). However, in tract tracing studies with male rats, SPFp receives projections from lumbar spinothalamic neurons (214), which are essential for sexual behavior (958) and express ejaculation-induced Fos (959).

Expression of c-fos or Other Immediate-early Genes. Ejaculation selectively increased Fos in the SPFp of rats (82,208–210,1001), gerbils (394), hamsters (491), and musk shrews (324). Fos was not induced by chemosensory investigation, mounts or intromissions. In human male volunteers, ejaculation stimulated cerebral blood flow in SPFp as measured by positron emission tomography (423). Likewise, electrical stimulation of CTF accelerated mating in male rats (887).

To test whether SPFp Fos expression was selective for ejaculation, Coolen et al. (208) administered the 5-HT_{1A} agonist, 8-OH-DPAT, to male rats to decrease the number of intromissions preceding ejaculation. Indeed, some animals ejaculated on their first intromission. Males that ejaculated after few or no mounts or intromissions had almost as many Fos-ir neurons in the SPFp as did those that ejaculated following the normal numbers of mounts and intromissions. Additional support for the suggestion that the SPFp responds to ejaculation-related sensory input comes from the observation that male and female rats have very similar patterns of Fos-ir following ejaculation (980).

Presence of Steroid Receptors. SPFp neurons have ARs (355), and many androgen-responsive neurons in SPFp that project to MPOA express Fos following ejaculation (358).

Connections. The SPFp consists of medial and lateral subdivisions. Neurons of the lateral subdivision express calcitonin gene-related peptide and transmit visual and auditory stimuli for conditioned fear (213). The medial portion of the SPFp contains a high concentration of galanin-ir fibers (980). Mating induces Fos in the medial subdivision, and Fos-positive nuclei are surrounded by galanin-ir fibers. Medial SPFp receives projections from neurons in laminae 7 and 10 of the lumbosacral spinal cord (214,357,959), and has bidirectional connections with forebrain nuclei controlling sexual behavior (214). Many of these connections also contained AR and mating-induced Fos (358). Thus, galanin may be used to relay ejaculation-related somatosensory information through SPFp.

Major Integrative Sites

Medial Preoptic Area (MPOA)

The MPOA is a critical integrative site for male sexual behavior in all vertebrate species that have been tested. This is remarkable considering the range of sensory stimuli that elicit mating and the diversity of motor patterns that express copulation. Such behavioral flexibility is possible because the MPOA receives indirect input from every sensory modality and sends reciprocal connections back to those sources (893). These reciprocal connections allow the MPOA to modify the processing of sensory input. Furthermore, the MPOA and its principal afferents contain steroid receptors, which provide the means to bias sensory input to favor sexually relevant stimuli. Efferent connections from the MPOA are critical for the initiation and patterning of copulation. They project to hypothalamic, midbrain, and brain stem nuclei that regulate autonomic or somatomotor patterns and motivational states [reviewed in (895,1035)].

Like other elements of the circuitry for male sexual behavior (e.g., Me and BST), the MPOA is a heterogeneous region comprising separate subnuclei, each with a distinct profile of functions, connections (594,893), and neurotransmitter content (897). The MPOA includes a medial periventricular zone chiefly concerned with neuroendocrine regulation, and a medial zone that functions in male sexual behavior and maternal behavior. Subnuclei within the MPOA are heavily interconnected. In particular, there is potential for reciprocal interaction of sexual behavior and neuroendocrine function via connections between nuclei of the medial and periventricular zones. Within the medial zone of the MPOA, the medial preoptic nucleus (MPN) and posterodorsal preoptic nucleus (PdPN) are particularly important for sexual behavior. Based on a variety of experimental approaches, it appears that the MPOA is essential for copulation, and contributes to sexual motivation, as well.

Effects of Electrolytic and Cell-body Lesions. Lesions of the MPOA severely impair copulation in male rats (16,106,132,170,238,266,323,326,373,398, 401,461,498,524,628,629,837,901,961,968), hamsters (305,788), mice (93), guinea pigs (767), gerbils (200, 1039,1040), dogs (384), cats (392), ferrets (172), goats (387), marmosets (555), rhesus monkeys (908), birds (67), lizards (269a,1008), snakes (508), and fish (576). Large lesions of the MPOA, extending into the rostral anterior hypothalamus, eliminated copulation in sexually experienced male rats; indeed, none of the lesioned males mounted a receptive female rat in periodic tests up to 3 (398) or even 8 months (326) after surgery. It is important to note that similar results were obtained using either electrolytic lesions, which destroy both cell bodies and fibers of passage, or axon-sparing excitotoxic lesions. This indicates that neurons within the MPOA play a critical role in the activation of copulation (375).

A number of attempts have been made to reactivate copulation in male rats that have MPOA lesions. In the early studies of Heimer and Larsson (398), stimulation of lesioned males by handling or exposure to different females failed to elicit copulation. Chronic T administration was similarly ineffective, an indication that the effects of the MPOA lesions were not an indirect result of diminished gonadal output. Even more extreme forms of excitation, such as repeated flank shock, could not compensate for MPOA damage (629). However, systemic injection of lisuride, a nonspecific monoamine receptor agonist, transiently activated copulation in 100% of male rats with MPOA lesions, with 50% copulating to ejaculation (375). When these animals were subsequently treated with saline, they once again failed to copulate. Under similar conditions, administration of GnRH (837) or naloxone (375) was ineffective.

Smaller lesions of the MPOA typically produced less severe deficits. In marmosets, lesions confined to the MPOA decreased copulation, while lesions at the junction of the MPOA and anterior hypothalamus eliminated intromission and ejaculation (555). Similarly, in rats, lesions placed in rostral MPOA produced less severe copulatory deficits, with many of the males copulating to ejaculation (46,326,398). Males that were still able to copulate after small preoptic lesions had prolonged intromission and ejaculation latencies (46), were inconsistent from test to test (398), and, in some cases, recovered normal copulatory ability over time (326). Small lesions targeting the posterodorsal preoptic nucleus, which is selectively activated by ejaculation in male gerbils (394) decreased mounting and delayed ejaculation, suggesting that this area normally facilitates copulation (395). By contrast, damage to caudal MPOA, which included the rostral anterior hypothalamus, was associated with a severe copulatory deficit (968). In male hamsters, lesions of caudal MPOA centered in the magnocellular region of the medial preoptic nucleus eliminated copulation (788).

In rats and gerbils, the MPOA has sexually dimorphic subregions that are attractive candidates to regulate male sexual behavior. In gerbils, the sexually dimorphic area consists of three interconnected subnuclei that are part of a critical circuit for the control of male sexual behavior [reviewed in (1035)]. Unilateral excitotoxic lesions of the medial SDA, paired with similar lesions of the contralateral lateral SDA, were as effective as bilateral lesions of each (1036). Conversely, Arendash and Gorski (46) found that lesions of the sexually dimorphic nucleus of MPOA [SDN-POA; (349)] in the male rat had no effect on copulation in experienced rats, and produced only a transient mating delay in sexually naïve males (238). However, lesions dorsal to SDN-POA reduced the percentage of male rats that ejaculated (46).

It was surprising that MPOA lesions in juvenile male rats reared in isosexual or heterosexual groups produced only minimal copulatory deficits when these animals were tested as adults (485,533,961). The only discernible effects were that lesioned males had fewer ejaculations before sexual exhaustion and showed a more rapid decline in copulation following castration (533). In contrast, males lesioned as juveniles and reared in social isolation did not copulate as adults (485,961). However, similar detrimental effects of social isolation in male rats without brain damage have also been reported [e.g., (318)]. Furthermore, copulation was observed in male rats that received lesions as juveniles and were socially isolated, but were handled daily until adulthood (628), suggesting that interactions with conspecifics are not critical for the development of copulatory behavior in these animals. Two other studies in rats reported incomplete behavioral recovery following MPOA lesions in juvenile male rats (628,629). However, the MPOA lesions in these studies were larger and more caudal, and would be expected to produce the most severe deficits in adult male rats. Thus far, juvenile plasticity in copulatory behavior after MPOA lesion seems to be specific to rats, as MPOA lesions in juvenile dogs (390,391) and cats (532) eliminated copulation, despite extensive social experience.

It is significant that large MPOA lesions eliminate not only ejaculation, but also the initiation of copulation. This has been interpreted as a reduction in sexual arousal or motivation [e.g., (170,266,326)]. However, male rats with MPOA lesions pursued estrous female rats and investigated their anogenital region, climbed over them, and occasionally clasped and mounted them without thrusting (373,398,629). A similar pattern of behavior has been described for male cats (392) and dogs (384) with MPOA lesions. Indeed, male dogs with MPOA lesions mounted and sometimes showed shallow thrusting, but did not copulate effectively (384). Although monkeys or rats with preoptic lesions rarely showed female-directed sexual contact, they would press a bar for access to a receptive female (277,908) or for a secondary reinforcer that had previously been associated with a receptive female [reviewed in (276)]. Furthermore, the frequency of masturbation by the monkeys was unaffected by MPOA lesion; although they were clearly sexually aroused, they did not appear to recognize or respond to the female as a sexual partner. Similarly, MPOA lesions that dramatically impaired copulation had little or no effect on noncontact erections (554). Based on this apparent dissociation between sexual motivation and copulatory performance, Everitt (276) suggested that the MPOA controls only copulatory performance and does not influence sexual motivation.

However, lesions of the MPOA have diminished sexual motivation in other contexts, including the preference for a female partner in rats (266,269,740) and ferrets (475,733), pursuit of a female by male rats (736), or precopulatory behavior (anticipatory erections, tongue flicking, and anogenital investigations) in marmosets (555). Therefore, while sexual motivation is not eliminated by MPOA lesions, it is clearly diminished. The evolutionary conservation of MPOA influence on both sexual motivation and performance is apparent in studies of Japanese quail and starlings. In quail, small lesions of subregions of the MPOA differentially impaired copulatory performance and sexual motivation, measured as time spent in front of a window through which the male could view a female (68). In starlings, MPOA lesions selectively reduced singing and the gathering of green nesting materials (810).

The specificity of effects of MPOA lesions is highlighted by contrast with the effects of lesions in the lateral preoptic area (LPOA). Like humans, rats have erections during episodes of paradoxical sleep (873). Schmidt et al. (872) showed that rats with MPOA lesions experience no disruption of sleep or of erections during sleep. In contrast, LPOA lesions dramatically reduced the occurrence of erection during paradoxical sleep, whereas erections during the waking state were unchanged. The reduction in erections was not attributable to impaired paradoxical sleep, although males with LPOA lesions had extended insomnia and less slow-wave sleep. Although most of the affected males also had some damage to the ventral BST, case analysis suggested that it was LPOA damage that contributed most to the observed effects. This study also exemplifies the context-specificity of the regulation of erection.

Effects of Electrical or Chemical Stimulation. As expected from the inhibitory effects of lesions, stimulation of the MPOA has resulted in facilitation of copulation in male rats (588,655,824,972), guinea pigs (603), and opossums (103,815). The typical effects included decreases in mounts and intromissions preceding ejaculation, ejaculation latency, and postejaculatory intervals. Lesions of the medial forebrain bundle, which carries the major efferents from the MPOA, diminished the effects of stimulation in opossums (815). Indeed, stimulation of the hypothalamic area that encompasses the medial forebrain bundle produced facilitation of the same measures as MPOA stimulation (147,148,924,979).

Electrical stimulation of the MPOA failed to restore copulation in male rats 24 hours after they had copulated to sexual satiety, although it facilitated several measures in nonsatiated males (824). Furthermore, combining the electrical stimulation with subthreshold doses of the DA agonist apomorphine or the α_2-adrenoceptor antagonist yohimbine was also ineffective in satiated males. Higher doses of both drugs had previously restored copulation in satiated animals (819). The authors suggested that different mechanisms may underlie sexual satiety and the postejaculatory interval, and that stimulation of an excitatory site is not sufficient to reverse sexual satiety.

Repeated electrical stimulation (four per day, separated by two hours) of the MPOA of previously noncopulating rats led to "kindling" (increased afterdischarges, leading eventually to a seizure) and the subsequent ability of most males to copulate on tests without stimulation (735). Males that initially copulated without stimulation were not further facilitated. Therefore, kindling in the MPOA decreased the threshold for initiating copulation but did not affect normal copulatory performance. Kindling-like stimulation that did not result in seizure activity resulted in permanent facilitation of copulation in previously noncopulating rats, even after 8 months (784). Similarly, MPOA kindling in T-treated female rats induced male-like copulatory behavior (252).

Electrical stimulation of the MPOA or medial forebrain bundle has also elicited *ex copula* erections and/or seminal emissions in monkeys (578,579,817) and rats (403,971); however, other researchers, using different stimulation parameters, reported erections only following the end of stimulation trains (220). Either electrical or glutamatergic stimulation of the MPOA elicited increases in intracavernous pressure in anesthetized rats (340,867) and also elicited the UG reflex, even without genital stimulation (Fig. 16) (601). The effects of MPOA stimulation were enhanced by intrathecal application of an NO donor, a cGMP analog, and sildenafil, and were decreased by the NOS inhibitor L-NAME (867). MPOA neurons do not project directly to the lower spinal cord, where genital reflexes are controlled; therefore, MPOA efferents must activate downstream sites that in turn regulate erection and seminal emission. Furthermore, the MPOA appears not to be necessary for penile erection or seminal emission, since MPOA lesions have had little effect on noncontact erections (554), touch-based erections (940), or spontaneous seminal emission (16).

Effects of Direct Applications of Steroids or Steroid Antagonists. In view of the substantial numbers of steroid receptors in the MPOA, it is not surprising that intracerebral application of T facilitates mating in castrated males. Fisher (303) first tested brain localization of T effects on copulation, using a water-soluble form of T. Although detailed methods and data are not available, Fisher indicated that the preoptic area was the most effective site for the activation of copulation in castrated male rats. A more detailed study, using crystalline T implants, reported that 100% of the males receiving MPOA

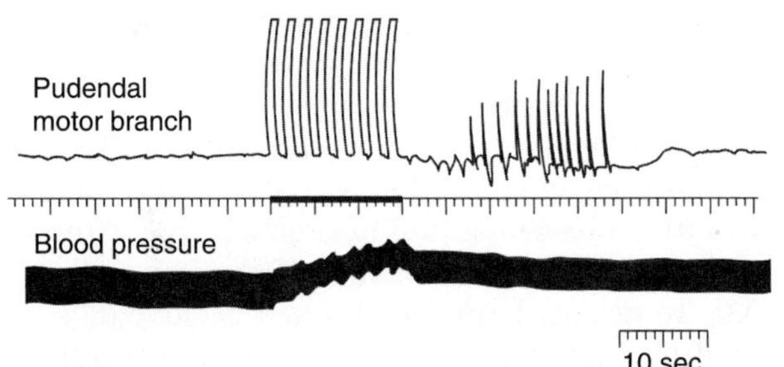

FIG. 16. Polygraph tracing illustrating that in many trials the UG reflex was initiated by hypothalamic stimulation in the absence of any genital stimulation. Bilateral stimulation of the MPOA (400 μA/0.2 ms pulses delivered at 50 Hz, 1 sec on/off) is indicated by the black bar on the timer trace. After termination of the stimulus, the UG reflex was elicited as shown by the rhythmic activity in the pudendal motor branch. Increases in blood pressure often accompanied hypothalamic stimulation. [Figure is reprinted from (601), with permission.]

implants copulated to ejaculation (232). Cholesterol control implants were ineffective, and T implants in hypothalamic, hippocampal, and "other" sites yielded 0% to 81% of animals copulating to ejaculation. The effectiveness of T implants in the MPOA has been confirmed in studies of castrated male rats (177,458, 470,546), ferrets (944), birds (76,77,445,996), and anole lizards (688). Therefore, androgenic mechanisms in the MPOA that stimulate copulation are highly conserved in evolution. Furthermore, mice lacking the aromatase gene (and therefore estrogen) showed increased loss of dopaminergic neurons in the MPOA, compared to wild-type mice (406a).

Nonetheless, stimulation of sexual activity in castrated males by intracerebral steroid implants is variable. For instance, although 100% of the males implanted with T in the MPOA copulated to ejaculation in one study (232), these males ejaculated in only 57% of the tests. Even among males that copulated to ejaculation, mating behavior was not always normal [e.g., (458)]. Similar observations have been reported for male hamsters (547,1026,1029) and mice (609,708). Intracerebral T implants in mice increased ultrasonic vocalizations, preference for a female, and urine marking in response to female urine; however, little copulation was observed (609,708). Most likely, this is due to selective stimulation of only a limited portion of the male mating behavior circuitry, and the absence of systemic androgens. In this regard, combined implants of T in the MPOA and Me did not further stimulate sexual activity above that induced by single MPOA implants alone in male mice (904) or hamsters (205).

Estrogen in the MPOA stimulated sexual activity in castrated male rats (178) and hamsters (547). Some investigators have not observed ejaculation with E implants (237). However, when Christensen and Clemens (177) compared the efficacy of T or E in the MPOA to activate copulation, only 30% of T-implanted males copulated to ejaculation, whereas 70% of males implanted with E ejaculated. In a subsequent study, infusion of T or E into the rostral anterior hypothalamus stimulated mounting (178). However, concurrent treatment with ATD (an aromatase inhibitor) blocked the increase in mounting induced by T, without affecting E-stimulated mounts. In complementary studies, the nonsteroidal aromatase inhibitor fadrozole impaired sexual behavior when delivered into the MPOA (183) or the lateral ventricle (965). Moreover, Clancy et al. (184) implanted E into the MPOA of gonadally intact males treated systemically with an aromatase inhibitor. Thus, E was available only to the MPOA, although androgens were available throughout the body. Estrogen-treated animals regained about half the numbers of mounts, intromissions, and ejaculations that they had exhibited before surgery, while copulatory behavior of controls remained low. Therefore, E in the MPOA combined with systemic androgen is sufficient to maintain at least some copulatory and ejaculatory behavior. Collectively, these results are consistent with the idea that E is able to stimulate copulation in males, and that the effects of T require aromatization to E within cells of the central nervous system.

The aromatization of T to E in the MPOA may also be important for sexual preference in rams. Rams that prefer to mate with other rams have reduced capacity to aromatize T to E in the MPOA and anterior hypothalamus, in addition to lower circulating levels of T and E (806). High sexually performing rams also had more ERs than did low sexually performing rams (29).

There have been relatively few experiments testing implants of DHT in the MPOA to activate male sexual behavior. DHT implants in the MPOA and anterior hypothalamus induced copulation in only 33% to 35% of male rats, a percentage not significantly different from cholesterol-treated controls (237,458). For the DHT-treated males that copulated to ejaculation, intromission latency and the number of mounts to intromission were significantly elevated compared with precastration levels, although other measures of copulation were normal (458). Implantation of DHT into the anterior hypothalamus of male hamsters had no detectable effect on copulation (547). Baum et al. (85) were somewhat more successful in stimulating copulation with DHT implants in the MPOA/anterior hypothalamus of male rats. In this study, castrated males ejaculated on about 35% of the tests prior to implantation, whereas the percentage increased to about 70% after the males received intracranial DHT. Likewise, implants of the anti-androgen flutamide into the MPOA blocked sexual behavior in castrated rats treated systemically with T (379). Male guinea pigs have been shown to be particularly responsive to intrapreoptic area implants of DHT, with full copulatory behavior restored in virtually all males tested (146).

Another approach to study the effects of intracranial steroid administration on copulation has been to combine a brain implant with systemic hormone treatment subthreshold for the elicitation of behavior. In one study, MPOA implants of DHT combined with subthreshold systemic DHT treatment in castrated rats promoted ejaculation in 40% of males (145). When Davis and Barfield (237) gave brain implants of E to castrated male rats treated systemically with DHT, 60% copulated to ejaculation. Baum et al. (85) performed the complementary experiment by implanting DHT into the brain of male rats receiving systemic E_2. DHT implants in the MPOA/AH were ineffective, but DHT did stimulate mating when implanted in the lateral septum or Me.

Yet another approach is to block either steroid receptors or enzymes that produce E_2 or DHT locally. Fadrozole, a nonsteroidal aromatase inhibitor, delivered into either the MPOA (183) or the lateral ventricle (965), impaired copulation in male rats. In contrast, an implant of E_2 in the MPOA partially reversed the impairment by systemic administration of fadrozole (184). These studies support the importance of E_2 in the MPOA for male sexual behavior. However, MPOA androgen also contributes to sexual behavior. Administration of the antiandrogen flutamide into the anterior MPOA impaired copulation, but not sexual motivation (partner preference), whereas flutamide in the posterior MPOA had the reverse effects: impairment of motivation, but not copulatory performance (621).

Effects of Direct Application of Drugs Affecting Specific Transmitters

The MPOA plays a central role in coordinating many facets of sexual behavior, so it is not surprising that drugs administered into the MPOA have affected numerous behavioral measures. Most effects of locally administered drugs resemble those of the same drug injected systemically. However, there are a few site-specific differences.

Dopamine. MPOA DA facilitates copulation, genital reflexes, and sexual motivation. Dopaminergic neurons of the periventricular system lie along the third ventricle and send widely branching axons laterally into the MPOA and anterior hypothalamus [reviewed in (684)]. The classic D_1/D_2 agonist apomorphine, microinjected into the MPOA, facilitated copulation in intact male rats (433) and in long-term castrates (869), and also increased the numbers of touch-based erections and seminal emissions (749). Similarly, reverse-dialysis of the DA reuptake inhibitor bupropion into the MPOA increased levels of extracellular DA and increased both touch-based and noncontact erections (2). Conversely, the D_1/D_2 antagonist *cis*-flupenthixol in the MPOA impaired copulation (995) and genital reflexes (752) and decreased the number of choices of a female in an X-maze, without affecting running speed or the number of no-choice trials (995). The decrease in sexual motivation (choice of the female) was relatively modest, probably because the drug was injected only unilaterally. However, there was spatial specificity with both apomorphine's and *cis*-flupenthixol's effects; microinjections anterior, dorsal, or lateral to the MPOA were ineffective in all experiments. There was also behavioral specificity, in that neither apomorphine nor *cis*-flupenthixol affected eating, drinking, locomotion, or rearing in the home cage, although flupenthixol slightly increased time spent inactive (995). MPOA microinjections of a different DA antagonist, haloperidol, also resulted in increased intromission latency and decreased ejaculations in a bilevel apparatus, although motoric slowing may have contributed to these deficits (761). In a more recent study, microinjection of apomorphine into the MPOA of males whose copulatory ability had been severely compromised by large excitotoxic lesions of the amygdala, fully restored their copulatory ability (Fig. 17) (250). Trends toward apomorphine-induced increases in intromissions and ejaculations in sham animals were not statistically significant. These data suggest that a major consequence of amygdala lesions is

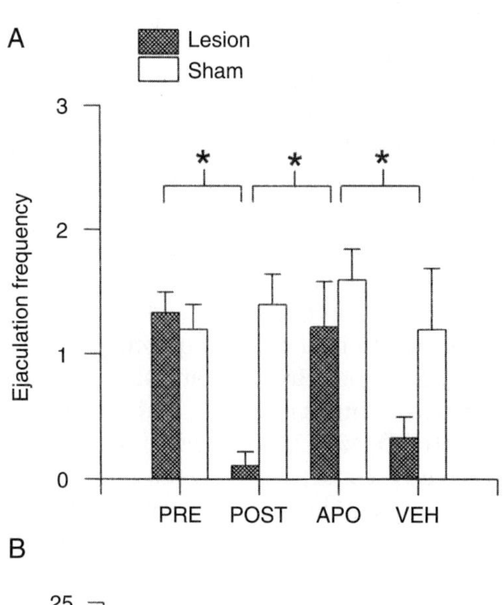

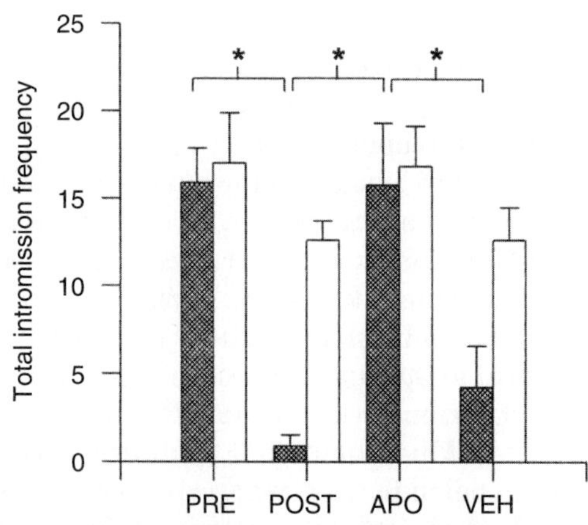

FIG. 17. Ejaculation frequency and total intromission frequency for animals with amygdala lesions or sham lesions. There were no significant differences in ejaculation frequency or total intromission frequency for animals with sham lesions. Animals with amygdala lesions displayed significantly fewer ejaculations and fewer total intromissions after surgery (POST) compared with before surgery (PRE). Microinjections of apomorphine (APO), but not vehicle (VEH), restored measures of copulation for animals with amygdala lesions. Values are expressed as mean + SEM (*$P < 0.05$). [Figure is reprinted from (250), with permission.]

impaired DA activity in the MPOA (see section on Chemical Changes Detected by Microdialysis or from Tissue Punches).

6-OHDA neurotoxic lesions of DA neurons in the MPOA impaired copulation only within the first 24 hours after the lesion (88) or if a subthreshold dose of a DA synthesis inhibitor was administered (115). There was only a 23% depletion of DA, a likely result of the sparse population of DA transporters in the MPOA, which are necessary to transport the toxin into the axon terminals. Thus, increased DA synthesis in remaining neurons occurred during the first 24 hours after the lesion and was able to restore copulation to normal. This effect was specific for DA, since the 6-OHDA was co-administered with desipramine, an NE transporter inhibitor; and NE levels were not affected.

Stimulation of D_1 receptors has consistently facilitated copulation and touch-based erections, whereas D_2 agonists have produced biphasic effects. A low dose of the D_2/D_3 agonist quinelorane, microinjected into the MPOA, decreased the latency to the first genital reflex, without affecting the numbers of erections, anteroflexions, or seminal emissions (87). Thus, small increases in dopaminergic stimulation may disinhibit reflexes via D_2-like receptors. However, a large dose of quinelorane, or of the D_1 antagonist SCH-23390, decreased erections but increased seminal emissions. Thus, intense stimulation of D_2-like receptors or antagonism of D_1-like receptors appears to facilitate sympathetically controlled seminal emission, while inhibiting parasympathetically controlled erections. Conversely, a D_1 full agonist, tetrahydrothienopyridine (THP), increased touch-based erections but inhibited seminal emission (597). The D_1 partial agonist SKF-38393 was ineffective (unpublished observations). Experiments using low and high doses of apomorphine, together with D_1 and D_2 antagonists, also support the opposing effects of D_1 and D_2 receptors on autonomic function. The low dose of apomorphine (1 μg) increased touch-based erections, but not seminal emissions; this effect was blocked by both D_1 (SCH-23390) and D_2 (raclopride) antagonists, suggesting that both disinhibition by D_2 receptors and facilitation by D_1 receptors contributed to apomorphine's increase (435) (Fig. 18). Unexpectedly, a high dose (10 μg) of apomorphine failed to increase erections significantly; however, co-administration of the high dose with a D_2 antagonist unmasked the facilitative effects apparently mediated by D_1 receptors. The high dose of apomorphine, administered alone, did increase seminal emissions, an effect that was blocked by a high dose of the D_2 antagonist, but not by the D_1 antagonist.

The association of D_1 receptor activation in the MPOA with erections, and of intense D_2 stimulation with ejaculation, was supported in copulation tests.

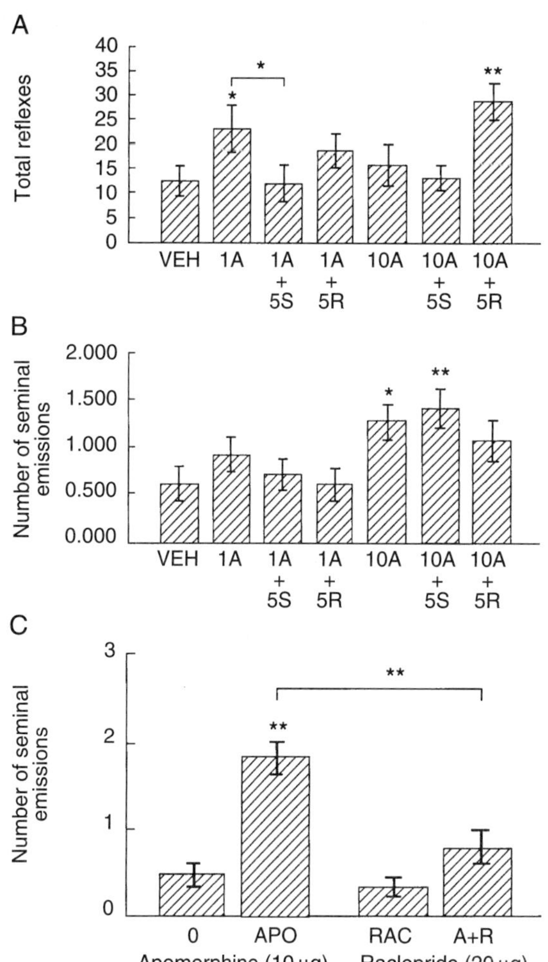

FIG. 18. Effects of low and high doses of apomorphine, alone or with a D1 or D2 antagonist, on ex copula genital reflexes. **A:** Effects on touch-based erections and anteroflexions (Total Reflexes). The low dose (1 μg) of apomorphine (1A) increased total erections plus anteroflexions. Addition of 5 μg of the D1 antagonist SCH-23390 (1A+5S) blocked this effect. Addition of 5 μg of the D2 antagonist raclopride partially blocked the effect. The high dose (10 μg) of apomorphine alone (10A) did not facilitate this measure; addition of 5 μg of the D2 antagonist raclopride to apomorphine (10A + 5R) significantly increased erections plus anteroflexions. **B:** Effects on seminal emissions. The high dose (10 μg) of apomorphine (10A) increased the number of seminal emissions. Addition of 5 μg of the D1 antagonist SCH-23390 (1A + 5S) resulted in slightly more seminal emissions than did apomorphine alone. Addition of 5 μg of the D2 antagonist raclopride to 10 μg apomorphine (10A + 5R) reduced the effect of apomorphine. **C:** Effects of the high dose (10 μg) of apomorphine and a high dose (20 μg) of the D2 antagonist raclopride, alone or together, on seminal emissions. Apomorphine again increased seminal emissions, and that increase was blocked by co-administration of the high dose of the D2 antagonist raclopride. [Figure is reprinted from (435) with permission.]

The D_1 agonist THP, microinjected into the MPOA, increased copulatory efficiency, resulting in more ejaculations per 30-minute test (597). Conversely, the D_1 antagonist SCH-23390 increased intromission latency but decreased the threshold for ejaculation (440). It also decreased the choice of the female in an X-maze, without impairing motor activity (693). A high dose of the D_2/D_3 agonist quinelorane also increased intromission latency and decreased ejaculatory threshold (440,693). However, it did not decrease the percentage of X-maze trials on which the male chose the female, but rather increased the latency to reach the female's goal box (693). This delay was not the result of slowed motor function, but of apparent reluctance to cross the line in front of the female's compartment, beyond which the male would be placed into that compartment. In summary, D_1 and D_2 receptors in the MPOA appear to exert reciprocal effects on copulation and genital reflexes. Activation of D_1 receptors facilitates parasympathetically mediated erections and copulatory performance, whereas activation of D_2 receptors promotes sympathetically mediated ejaculation and the post-ejaculatory state.

Serotonin. 5-HT is generally inhibitory to male sexual behavior, and large doses of 5-HT microinjected into the MPOA did impair copulation (289,983); furthermore, a 5-HT_{1B} agonist delayed ejaculation (289). In accord with the facilitative effects of systemically administered 5-HT_{1A} agonists, 8-OH-DPAT microinjected into the MPOA facilitated ejaculation (286). 8-OH-DPAT, administered by reverse dialysis into the MPOA, also facilitated copulation (613) and increased both DA and 5-HT levels (563). The facilitative effects of 8-OH-DPAT were partially blocked by the D_2 antagonist raclopride, but not by the 5-HT_{1A} antagonist p-MPPI, suggesting that the elevation of DA levels, resulting in stimulation of D_2 receptors, mediated much of the facilitation (613).

GABA. Gamma-amino butyric acid (GABA) is a ubiquitous inhibitory neurotransmitter in adult mammals. As expected, the GABAergic drugs muscimol and ethanolamine-O-sulfate, microinjected into the MPOA, decreased the numbers of male rats that mounted, intromitted, or ejaculated, whereas the $GABA_A$ antagonists bicuculline or 3-mercaptopropionic acid (3-MPA) dramatically shortened ejaculation latency and postejaculatory intervals and decreased the numbers of intromissions preceding ejaculation (290,291). However, MPOA administration of the same two antagonists failed to restore copulation in satiated males, again suggesting that the postejaculatory interval is governed by mechanisms separate from those that regulate sexual satiety.

Opioids. As with systemic manipulations, studies using microinjections of opiates or opioids into the MPOA suggest that low doses of agonists facilitate, and high doses inhibit copulation. The lowest doses of the μ agonist morphine and of the κ agonist dynorphin (1–13) decreased the time and numbers of intromissions before ejaculation, and the highest dose of morphine resulted in failure to resume copulation after the second ejaculation (74). The selective μ agonist morphiceptin delayed the onset of copulation but had no effect on copulatory performance, touch-based erections, sexual motivation, or locomotion (614). However, β-endorphin microinjected into the MPOA both delayed copulation and inhibited its performance (430,975). In support of the importance of low levels of endogenous opioids, naloxone in the MPOA prevented induction of sexual reinforcement (8).

NE. Administration of NE into the MPOA facilitated both sexual arousal and copulatory performance; conversely, the β-noradrenergic antagonist propranolol and the a antagonist phenoxybenzamine inhibited copulation (585). The α_2 agonist clonidine, which would have decreased NE release by stimulating autoreceptors, decreased the number of animals that copulated and increased ejaculation latency and interintromission interval; however, intromissions preceding ejaculation decreased (186). In addition, the α_2 antagonist yohimbine (which would have disinhibited NE release) in the MPOA blocked the inhibitory effects of systemically injected clonidine (186). Therefore, stimulation of α_2 receptors in the MPOA appears to facilitate copulation.

Nitric Oxide (NO). As with systemic manipulations, local increases in NO facilitate some aspects of copulation. Reverse dialysis of its precursor, L-arginine, into the MPOA increased mount rate, whereas the NOS inhibitor L-NMMA decreased that measure (865). L-NMMA microinjections increased the number of *ex copula* seminal emissions and decreased the number of inexperienced males that copulated, although it did not affect experienced males (691,692). More recently, L-NAME microinjections into the MPOA prevented intromissions and ejaculations in sexually naïve males and decreased the numbers of intromissions and ejaculations in experienced males (512). Furthermore, in naïve males, L-NAME microinjections before each of seven exposures to an inaccessible female prevented the facilitative effect of those exposures, which was observed in saline-treated males, compared to nonexposed animals (512).

Nitrergic function in the MPOA is hormonally regulated. Castration of male rats decreased production of cGMP (the most common second messenger of NO) and blocked NMDA-induced increases in plasma GnRH and in cGMP in MPOA slices (789). Castration also decreased nNOS-ir in the MPN of male hamsters (365) and rats (258,792). T replacement for 2, 5, or 10 days in castrated rats resulted in progressive improvement in copulation and increases in NOS-ir; shorter mount latencies were correlated with greater NOS-ir density (863). Either E or T replacement, but not DHT, was able to restore NOS-ir (792). Furthermore, ERa knockout mice had less nNOS-ir than wild-type mice or mice lacking AR because of the testicular feminization mutation (*tfm*) (882). Thus, the aromatization of T to E likely mediates the upregulation of NOS-ir. Gonadal steroid receptors are co-localized with nNOS-ir in mice (882), rats, (863), and hamsters (365). An NO donor, sodium nitroprusside, reverse-dialyzed into the MPOA of castrated male rats with systemic DHT implants (to maintain genital and sensory structures) was able to restore copulation fully in half of the animals (863a). At least some of NO's effects are mediated by cGMP, since a guanylyl cyclase inhibitor decreased the NO donor's facilitative effects on copulation and MPOA DA release in gonadally intact males (862a). In contrast to the castration-induced decreases in NOS-ir in the above studies, there is one report of increased NOS-ir and activity after castration and reduction by either T or DHT (902). The reason for this apparent discrepancy is not clear, but may be related to the more posterior location examined by Singh and colleagues or their use of an antibody that labeled very few cells. In summary, NO in the MPOA enhances copulation and increases GnRH release. T, probably through aromatization to E and activation of ERα, increases production of NOS in the MPN, although apparently not in more posterior areas.

Glutamate. Microinjection of the N-methyl-D-aspartate (NMDA) receptor antagonist MK-801 into the MOA of sexually naïve or experienced males impaired copulation (249a,985). Furthermore, in sexually naïve males microinjection of MK-801 before each of seven noncopulatory exposures to an estrous female, prevented the facilitation of such exposures on a drug-free test on day 8 that was observed in saline-treated males, compared to unexposed males. Therefore, endogenous glutamatergic stimulation of NMDA receptors in the MPOA facilitates copulation in both naïve and experienced males and is permissive for the sensitization to female cues by repeated exposures to a female in sexually naïve males. In tests of genital reflexes in anesthetized male rats, microinjection of glutamate into the MPOA-elicited erectile responses (340) and the UG reflex (601).

ACh. The nonspecific ACh agonist carbachol or the muscarinic agonist oxotremorine, microinjected into the MPOA, decreased the numbers of intromissions preceding the first ejaculation (432). This effect was blocked by the muscarinic antagonist scopolamine, which also decreased the numbers of animals that mounted, intromitted, or ejaculated (439). Acetylcholinesterase immunoreactivity in the SDA was more dense in male gerbils than in females; castration decreased staining intensity, which was restored by T replacement. Therefore, ACh in the MPOA may contribute to the hormonal control of sexual behavior.

PGE$_2$. Bilateral administration of PGE$_2$ into the MPOA of castrated male rats, maintained on a subthreshold T replacement regimen, activated copulation within 30 minutes (197). Half the animals copulated and 40% copulated to ejaculation. Thus, PGE$_2$ in the MPOA facilitates copulation, possibly by increasing MPOA temperature (see section on systemic manipulation of PGE$_2$).

Orexin/Hypocretin. Orexin/hypocretin is better known for its facilitation of feeding and prolonging wakefulness. However, microinjection of orexin/hypocretin into the MPOA of male rats also facilitated copulation (364).

Electrophysiological Recordings. Electrophysiological recordings have revealed that different neurons in the MPOA participate in the control of sexual motivation and copulatory performance. Some neurons were activated only before male rats began to copulate, while others were activated only during copulation (888). Similar results were obtained in monkeys, in which neural activity was highest while the animal was pressing a lever to bring a female closer; activity decreased slightly during copulation and ceased after ejaculation (722). In slice preparations from the MPOA of Japanese quail, intra- and extracellular recordings revealed that both applications of DA inhibited most cells (52%–80%), but excited a few (10%–24%). These effects were not blocked by either D_1 or D_2 antagonists, but were blocked by α_1 or α_2 antagonists, respectively (216). The effects were not blocked by dopamine-β-hydroxylase inhibitors, indicating that DA was not being converted to NE to produce the effects. Therefore, in quail DA affects neuronal activity via cross-talk with NE receptors.

Chemical Changes Detected by Microdialysis or from Tissue Punches. There is a close association between extracellular MPOA DA levels and sexual behavior in male rats. DA increased as soon as a receptive female was placed behind a perforated barrier; DA levels remained high, or increased further, when the animals were allowed to copulate (Fig. 19) (434,866). The recent presence of T is required for both copulation and the DA response to the female. All gonadally intact males, as well as two-thirds of animals castrated one week before, showed a precopulatory DA increase and copulated when the barrier was removed (434). The remaining one-third of the 1-week castrates, and all 2-week castrates did not show a DA response to the female and did not copulate. There was behavioral specificity, in that neither voluntary exercise in a running wheel nor presentation of another male behind the barrier elicited a DA increase (434), and eating a palatable food did not increase the DA metabolite DOPAC in a previous study (436). Furthermore, probes anterior or lateral to the MPOA did not detect a response to the female. The increase in DA before copulation began provides evidence that the DA increase was not caused by

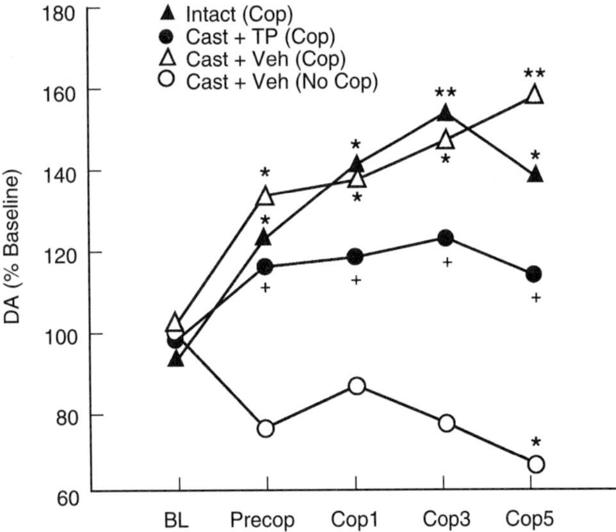

FIG. 19. Extracellular dopamine in the MPOA of male rats during baseline, a precopulatory period (estrous female behind a perforated barrier), and three 6-minute periods after the barrier was removed and the animals were free to copulate. All gonadally intact males and all castrates treated with testosterone propionate (200 μg/day) showed a significant increase in dopamine during the precopulatory period and during copulation; all of these animals did copulate. A total of nine of 14 oil-treated 1-week castrates also showed the precopulatory dopamine response and copulated after the barrier was removed. The remaining 1-week and all four 2-week oil-treated castrates failed to show the precopulatory dopamine response and failed to copulate; data from these two groups are combined. *P <0.05 compared to final baseline for intact males or for 1-week vehicle-treated castrates that copulated. **P <0.01 compared to final baseline for intact males or for 1-week vehicle-treated castrates that copulated. +P <0.05 compared to baseline for testosterone-treated castrates. #P <0.05 compared to final baseline for vehicle-treated castrates that failed to copulate. [Figure is reprinted from (434) with permission.]

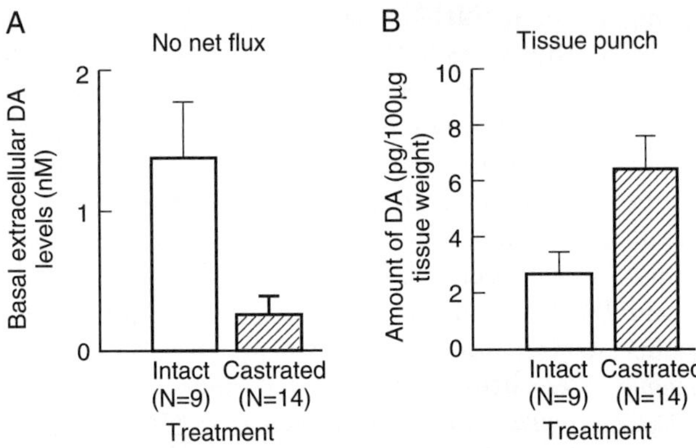

FIG. 20. A: Basal levels of extracellular dopamine in MPOA of 1-month castrates and gonadally intact males. Absolute levels were determined, using the no-net-flux method (see text). Intact males had significantly higher dopamine levels than did castrates. **B:** Dopamine levels in tissue punches from the MPOA of 1-month castrates or gonadally intact males. Castrates had significantly more dopamine in tissue than did intact males. Because almost all dopamine in tissue is stored in vesicles, this suggests that castrates synthesize and store dopamine normally, or even excessively; the deficit in extracellular dopamine of castrates may be due to decreased release. [Data are from (259); figure is from (434a), with permission.]

copulation, but was more likely associated with sexual motivation.

In the original studies described above, basal levels of extracellular DA were not measured; therefore, it was not clear whether basal levels were also affected by castration, or only the response to a female. Therefore, the no-net-flux technique was used to measure absolute basal levels in castrates, compared to gonadally intact males (259). Briefly, different amounts of DA are added to the dialysate; if there is less DA in the brain than in the dialysate, some will diffuse out of the dialysate into the brain, and the loss can be detected (720). Conversely, if there is less DA in the dialysate, or none, as in standard dialysis, DA will diffuse from the brain into the dialysate, and the gain can be detected. A regression line is drawn, and the point at which it crosses from loss to gain represents the basal extracellular DA level. In castrates the absolute extracellular DA levels were only about one-fifth those of gonadally intact males [0.3 vs. 1.3 nM; (259)] (Fig. 20). Therefore, the deficit in castrates was a general one, not limited to the increase in response to a female. However, systemically administered amphetamine elicited greater DA release in castrates than in intact males, and there was more DA in tissue punches from castrates than in those from intact males (259). Therefore, synthesis and storage of DA were at least as great in castrates as in intact males, but castrates were unable to release their abundant stores.

Further evidence for an association between DA release and the ability of castrates to copulate was provided by a study of restoration of copulation by 2-, 5-, or 10-day regimens of T treatment (790). No 2-day T-treated castrate showed a DA response to the female and none copulated. Eight of nine 5-day T-treated castrates showed a DA response and those eight copulated, with five also able to ejaculate. All 10-day T-treated castrates showed a precopulatory DA response, and all copulated. There were numerous correlations between DA release and copulatory measures. Therefore, both loss of copulation following castration and its restoration following T replacement were closely associated with female-elicited DA release in the MPOA.

A major factor regulating DA release in the MPOA is NO. Both basal (561) and copulation-induced (564) DA release were dependent, in part, on NO. Furthermore, NOS-ir in the MPN is positively regulated by T and E (163,258,365,792,882) [but see (902)]. Therefore, T may maintain both basal and female-stimulated DA release by upregulating NOS in the MPOA.

A similar mechanism was proposed for the regulation of GnRH release (789). Systemic injections of the glutamate agonist NMDA increased the production of NO in the MPOA of gonadally intact male and female rats, as inferred from increases in cGMP, a common second messenger for NO. In both sexes gonadectomy resulted in lower basal cGMP levels, a lack of response to systemically injected NMDA, and a lack of increase in LH. The increases in intact animals were blocked by both an NMDA antagonist and a NOS antagonist, verifying the specificity of the effect to NMDA receptors and NOS.

T's metabolites were differentially effective in restoring basal and female-stimulated DA release in long-term castrates (791). E_2 restored high basal levels of DA, but not the increase in response to a female. E_2-treated males all intromitted, but none showed a behavioral ejaculation pattern. Neither DHT nor oil vehicle maintained basal DA levels or a female-stimulated increase, and no DHT- or oil-treated animal copulated. Although DHT was ineffective by itself, when it was administered together with E_2, the combination effectively restored basal and female-stimulated DA responses, as well as copulation.

Although extracellular DA levels were positively correlated with copulatory measures, tissue (stored) DA levels were negatively associated with copulation (792). The noncopulating groups (DHT and oil) had higher tissue levels of DA than did the groups that copulated (E_2, E_2 +DHT, and T). As in the Du et al. (259) study above, DHT- and oil-treated animals could synthesize and store DA, but had difficulty releasing it in basal conditions and also in response to a female. Furthermore, a deficiency in NO may have led to the impaired DA release; DHT- and oil-treated males had less NOS-ir than did the three groups that were able to copulate (792). A similar finding regarding an apparent DA response to stimuli from a female was recently reported regarding juvenile vs. adult male hamsters (878). Adult hamsters showed an increase in DOPAC in punches from the MPOA in response to presentation of vaginal secretions from an estrous female, suggesting release of DA and subsequent uptake and metabolism, whereas juveniles showed no such response and were unable to copulate (see section on Sexual Behavior Throughout the Lifespan: Puberty and Aging).

A major factor regulating the MPOA DA response to a female is input from the Me. Small radiofrequency lesions of this area in highly experienced rats increased both the number of intromissions and the time preceding ejaculation and decreased the number of ejaculations (250) (see section on Effects of Direct Application of Drugs Affecting Specific Transmitters). These lesions did not affect basal DA levels, but they abolished the DA increase in response to an estrous female (Fig. 21). Thus, as with E_2 treatment of castrates (791), normal basal levels of MPOA DA were compatible with suboptimal copulatory ability.

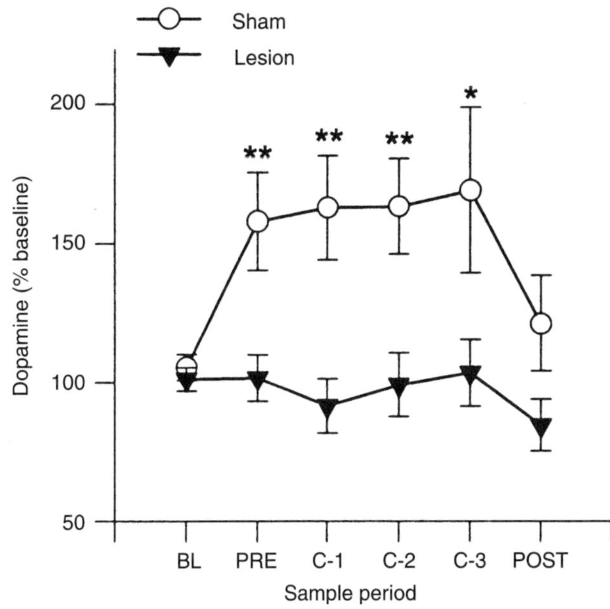

FIG. 21. Levels of dopamine in dialysate from the MPOA of animals with medial amygdala lesions. Levels represent percent changes from baseline (BL) in response to precopulatory exposure to an estrous female (PRE), during copulation (C1-C3), and after copulation (POST). Extracellular levels of dopamine significantly increased during the precopulatory and copulatory stages of testing for animals with sham lesions but not for animals with medial amygdalar lesions. The baseline value used for with sham lesions, but not for animals with medial amygdalar lesions. The baseline value used for computation was obtained by dividing the value of the last baseline by the mean of all three baselines. Values are expressed as mean + SEM (*$P<0.05$; ** $P<0.01$). [Figure is reprinted from (259), with permission.]

However, an increase in DA before and during copulation appears to contribute to the ability to ejaculate. As noted above, apomorphine microinjected into the MPOA of males with large excitotoxic lesions of the amygdala completely restored their copulatory ability, which had been almost abolished by the lesions (250). Therefore, input from the Me is important for increasing DA release in the MPOA in response to a female and during copulation, and this is a major way in which the Me facilitates copulatory ability. A recent critical review observed that "there was no delay whatsoever with regard to the initiation of sexual behavior despite the absence of preoptic DA release" in the Dominguez et al. (250) article (732). They suggested that, therefore, DA has little or nothing to do with sexual motivation. However, the data do show trends that approach statistical significance for both mount latency and intromission latency; the relatively high variability among animals with lesions may have contributed to the lack of statistical significance. Furthermore, there was not an "absence of DA release," but rather a lack of female-stimulated increase in DA; basal DA levels were normal and, as we have suggested, are sufficient for suboptimal copulation. Therefore, the data from both microinjection and microdialysis experiments suggest that MPOA DA does contribute to sexual motivation, copulatory ability, and genital reflexes, although other structures and other neurotransmitters also contribute to these processes.

The olfactory bulbs apparently provide the signal relayed by the amygdala to the MPOA. Male hamsters with sham bulbectomy, or with unilateral bulbectomy contralateral to the MPOA microdialysis site, showed increased extracellular DA in the MPOA when presented with an estrous female (956). However, bilateral or ipsilateral bulbectomy inhibited the DA response and impaired copulatory behavior, similar to the results of Dominguez et al. (250). In sexually mature male hamsters, pheromones from estrous females elicited an apparent increase in DA release, inferred from an increase in the DA metabolite DOPAC, in tissue from the MPOA (878). However, tissue from prepubertal males did not show such an increase. Therefore, the neural processing of sexually relevant chemosensory stimuli matures during puberty.

The 5-HT metabolite 5-hydroxyindole acetic acid (5-HIAA) increased in dialysates from the POA of male rats that had ejaculated (316,605), and 5-HT itself was increased in tissue punches from the POA following ejaculation (608). These authors suggested that postejaculatory 5-HT increases in the POA contributed to postejaculatory quiescence. However, a more recent study, based on analysis of 5-HT itself, concluded that MPOA and, more laterally, POA 5-HT levels were constant throughout copulation and the postejaculatory interval (562). However, 5-HT was increased in the anterior lateral hypothalamic area (LHA) at the time of ejaculation. The increase in 5-HIAA observed by Fumero et al. (316) and Mas et al. (605) may have resulted from diffusion from the adjacent LHA. Therefore, it seems unlikely that 5-HT in the MPOA contributes to postejaculatory quiescence.

Expression of c-fos and/or Other Immediate-early Genes. In keeping with the pivotal role of the MPOA in the control of male sexual behavior, numerous studies in a variety of species have documented Fos induction in the MPOA with sexual activity. Fos is expressed in the MPOA following copulation in male rats (82,816,980), gerbils (394), mice (370), and hamsters (283,489). In particular, neuronal activation in the MPOA appears to be selective for sexual activity, because mating, but not aggressive behavior, increased Fos-ir in the MPOA of male hamsters (490). Sexual experience may also prime MPOA neurons to be more responsive to sexual stimuli. The number of Fos-ir neurons in the MPN following one ejaculation was greater in sexually experienced male rats compared to naïve males, even though experienced males required fewer intromissions to reach ejaculation (574). Likewise, chemosensory cues were more effective in stimulating Fos expression in MPOA of sexually experienced male hamsters (298).

Fos is expressed in androgen receptor-containing neurons of the MPOA in rats (356), but Fos does not colocalize with aromatase in quail MPOA (307). Using retrograde tract-tracing, Gréco et al. (358) demonstrated Fos-positive neurons in Me and CTF that project to the MPOA. Heeb and Yahr (396) reported similar findings in male gerbils. Moreover, mating-induced Fos expression in the MPOA requires afferent input from both Me and CTF, because combined ipsilateral lesions of Me and CTF blocked copulation-induced Fos in the MPN, though single lesions did not affect Fos (82). In turn, Fos-positive neurons in the MPOA send their projections to the PAG (929).

Fos may be expressed in the MPOA in response to female odors alone, although there is less consensus on this point. Sexually experienced male rats exposed to estrous-female bedding showed an increase in Fos-ir in the MPOA (139,468). Likewise, chemosensory cues from females activated Fos in MPOA of male hamsters (299,491), macaques (665), gerbils (394), and aromatase-knockout mice (60). Similar results were obtained using fMRI in marmosets (297) or with brain temperature recordings in male rats (484). In addition, electrical stimulation of the vomeronasal organ activated Fos in the MPOA of male hamsters (659). However, other investigators did not report an MPOA Fos response to female odors in rats (82,210), mice (370), hamsters (283), or ferrets (467). It may be that certain subregions of the MPOA are activated selectively by ejaculation while others respond preferentially to chemosensory cues. In this regard, chemosensory cues selectively activated the medial and lateral sexually dimorphic areas of the male gerbil (394), as well as the magnocellular subdivision of the MPOA in hamsters (299,491). However, ejaculation activated Fos within the medial preoptic nucleus of both hamsters (491) and rats (210). In a similar manner, the posterodorsal preoptic nucleus (PdPN) in gerbils expressed Fos only after ejaculation (394). Whether chemosensory cues inducing Fos in MPOA are mediated by the main or accessory olfactory systems remains unclear (934,298).

The neurochemical identity of Fos-positive neurons has only recently been studied. In male gerbils, approximately half of the Fos-positive cells in the medial sexually dimorphic area and PdPN are GABAergic (900). Nearly 25% of cells in the medial sexually dimorphic area are glutamatergic, but PdPN neurons do not express glutamate. In male rats, galanin neurons in the MPOA are activated selectively by ejaculation (65). Administration of a DA D_1 antagonist before copulation decreased ejaculation-induced

Fos (574), suggesting that at least some of the copulation-induced Fos-ir may result from stimulation of D_1 receptors.

Effects of Intracerebral Grafts. Sexual arousal and copulatory ability decline with advancing age in males of many species. Injection of suspensions of fetal MPOA neurons into the MPOA of 19- to 24-month-old rats improved their copulatory behavior between 21 and 45 days after implantation (443). The improvements lasted until the end of the experiment 4.5 months later. Grafts from cerebral cortex to MPOA or from MPOA to the ventromedial hypothalamus (VMH) were ineffective. Two hormonal measures, serum T and the postcastration increase in luteinizing hormone, were also increased by the grafts. Therefore, some age-related declines in copulatory ability, sexual motivation, and neuroendocrine function may result at least partially from dysfunction in the MPOA.

Activation of Neurotransmitter Receptors. Endocytosis of μ opioid receptors in the MPOA was observed within 30 minutes of copulation and was still evident 6 hours later (207). The opioid antagonist naloxone prevented receptor internalization. Microinjections of a μ agonist produced similar endocytosis, and mating resulted in increased Fos-ir in neurons that contained μ receptors. However, naloxone pretreatment did not prevent the Fos response in μ receptor-containing neurons, suggesting that the Fos response did not result from μ activation during mating.

Summary of MPOA Functional Roles. The MPOA receives both chemosensory and genitosensory input, as well as information from other sensory modalities. Reciprocal connections with the sources of the sensory input allow hormone-concentrating neurons to bias the sensory input to favor sexually relevant information. Output from the MPOA to other hypothalamic, midbrain, and brainstem nuclei that control somatomotor patterns and genital reflexes contribute to the initiation and orchestration of complex, interactive copulatory behaviors.

Mesocorticolimbic DA Tract

Cell bodies in the VTA give rise to axons of the mesocorticolimbic DA tract that innervate the nucleus accumbens (Acb), and other regions in the ventral forebrain (249). This system is critical for appetitive behavior and reinforcement and is activated before and/or during a variety of motivated behaviors, including eating, drinking, sexual activity, drug self-administration, and intracranial self-stimulation (447,483). It is not clear whether this system is more important for reward processes (1019), behavioral activation (501,858,1014), "wanting" (as opposed to "liking," (464,817a), or incentive learning (245,918). However, it is clearly essential for appetitive behavior. The mesocorticolimbic tract is activated both before and during copulation in male rats. DA released in the Acb is important for the activation of numerous motivated behaviors, including sexual behavior, but appears to have little influence on specifically sexual motivation or on copulation per se. Because most research on this system has focused on the Acb and VTA, this review will also concentrate on these areas.

Effects of Electrolytic or Cell-body Lesions or of Electrical Stimulation. Few recent studies have tested the effects of lesions of the Acb or VTA on sexual behavior. In general, elements of the mesolimbic DA tract appear to be important for sexual arousal in response to remote cues from females, but play a relatively less critical role in copulatory performance per se. Lesions of the VTA and the adjacent interpeduncular nucleus increased the postejaculatory interval, but did not affect other copulatory measures (133). Likewise, male rats with DA-depleting lesions of the Acb had a lower incidence of noncontact erections, fewer erections, and longer latency to display erection; radiofrequency lesions of the Acb also increased the latency to the first noncontact erection (552). Nonetheless, copulation was normal in both groups of lesioned rats. Using excitotoxic lesions of the Acb, Kippin et al. (480) observed a reduction in noncontact erection, as well as intromission and ejaculation. However, lesioned male rats maintained a normal preference for receptive females.

Electrical stimulation of the VTA decreased the latencies to mount, intromit, and ejaculate (271,596) and increased the number of ejaculations in a 30-minute test (596). These results are consistent with a general activating effect of the mesocorticolimbic tract.

Effects of Castration or of Direct Applications of Hormones. Although the Acb and VTA are logical sites for gonadal steroid hormones to stimulate sexual motivation, there are relatively few data to support this hypothesis. Intracerebral implants of T into the VTA of castrated male house mice did not restore mounting, ultrasonic vocalizations, urine marking, or attraction to female urine (903). However, the combined application of T to the MPOA and VTA was more effective than implants in the MPOA alone. Testosterone may also be reinforcing, independent of its effects on sexual motivation. Male rats formed a conditioned place preference for intra-Acb injections of T (726), and T-induced conditioned place preference was blocked by intra-Acb injections of a DA antagonist (727). This effect appears to be due to androgenic metabolites of T acting on the shell of the Acb, because intracerebral implants of DHT into the Acb shell also induced conditioned place preference (315).

The mechanisms through which steroid hormones may act on the mesolimbic DA system are unclear. There are contradictory reports on the effects of castration on Acb DA levels. In male rats, castration depressed DA concentrations in the Acb, and this effect was reversed by replacement with T, E, or DHT (28,675). However, Hernandez et al. (405) found that amphetamine released more DA in dialysates from prepubertally castrated males, suggesting that there may be more DA stored in tissue. Likewise, the presence of steroid receptor-containing neurons in the Acb or VTA remains controversial. Early studies reported no binding of hormones to classic steroid receptors in the VTA, though low levels of binding were found in the Acb (757,861). A more recent study found small patches of AR neurons in the VTA that were also immunoreactive for tyrosine hydroxylase, the rate-limiting enzyme for catecholamine synthesis (507). Despite the scarcity of AR and ER in the mesolimbic DA system, there is the additional possibility that gonadal steroids may act on Acb and VTA via nongenomic receptors.

Expression of c-fos or Other Immediate-early Genes or Receptor Activation. Mating stimulated Fos in the Acb of male rats (82,816). Moreover, sexual experience enhanced Fos-ir in the Acb induced by an estrous female, particularly in the Acb shell (558). Fos was expressed in the Acb in response to female odors, but was also conditioned to a nonsexual odor paired with mating (478). Conditioned odor cues also stimulated Fos-ir in the VTA (479). Increased Fos-ir in the VTA of male rats that were exposed to sex-related cues or that copulated was found in both dopaminergic (tyrosine hydroxylase-ir) and nondopaminergic neurons (66). Both types of stimuli also activated neurons in the Acb core and shell. In addition, mu opioid receptors in the VTA were internalized following both sex-related cues and copulation. The authors suggested that mu receptors on GABAergic neurons in the VTA release dopaminergic neurons from tonic inhibition. Tail pinch, an arousing stimulus that can promote copulation (911) in olfactory bulbectomized males (993) or castrates maintained on subnormal T replacement (541) (reviewed above), also increased Fos-ir in the VTA, as well as the MPOA, BST, PVN, Me, and several sites not usually associated with copulation. However, Fos expression in the Acb in response to sexual activity be may limited to rats. In hamsters and gerbils, no mating-induced increase in Acb Fos-ir was observed (394,489).

Effects of Direct Applications of Drugs Affecting Specific Transmitters.

DA. Stimulation of impulse-regulating autoreceptors in the VTA with apomorphine delayed the onset of copulation and slowed its rate (431). Similar microinjections slowed the speed of running in all four arms of an X-maze and increased the number of trials on which the male failed to leave the start area; however, neither the percentage of trials on which the male chose the female's goal box nor *ex copula* genital reflexes were altered (441). Thus, inhibiting mesocorticolimbic DA activity decreased behavioral activation but had no effect on specifically sexual motivation or on genital reflexes.

Application of drugs to the Acb also affected primarily activity measures. The onset of copulation was speeded by amphetamine and delayed by *cis*-flupenthixol or the DA neurotoxin 6-OHDA [reviewed in (276)]. Amphetamine also increased responding for a secondary reinforcer that had been paired with copulation (276). The D_2/D_3 agonist quinelorane increased the number of trials on which the male failed to leave the start area but did not affect the percentage of trials on which he chose the female's goal box or his copulatory behavior when he reached the female (693). It is not clear whether quinelorane's effects resulted from stimulation of terminal autoreceptors or postsynaptic receptors. However, drug injections into either the VTA or the Acb have affected general activation, with little influence on specifically sexual motivation or on copulatory measures.

Other Neurotransmitters. A cholecystokinin A (CCK_A) antagonist, microinjected into the posteromedian Acb, attenuated the facilitation that resulted from electrical stimulation of the VTA, as did either a CCK_A or CCK_B antagonist into the anterolateral Acb (596). CCK is located in dopaminergic terminals in the posteromedian Acb, but is in nondopaminergic terminals in the anterolateral portion (416). These data suggest that CCK potentiates the facilitative effects of mesocorticolimbic DA activity.

5-HT, microinjected into the Acb inhibited ejaculation (increased mounts, intromissions, and time preceding ejaculation); the 5-HT_{1A} agonist 8-OH-DPAT decreased those same measures (409). Injection of these compounds into the dorsal striatum was without effect. The authors suggested that 8-OH-DPAT's effects may have resulted from antagonism of 5-HT_2 receptors.

Microinjection of the opioid antagonist naloxone into the VTA decreased anticipatory level changing in a bi-level apparatus but had no effect on sexual performance; β-endorphin had no effect (974). Stimulation of opioid receptors in the VTA may disinhibit DA neurons there and increase behavioral activation.

Changes Detected by Electrophysiological Recording. A larger proportion of single cells in the accumbens (Acb) increased their firing rates in response to vaginal estrous odors from unfamiliar females than from familiar females (1023a).

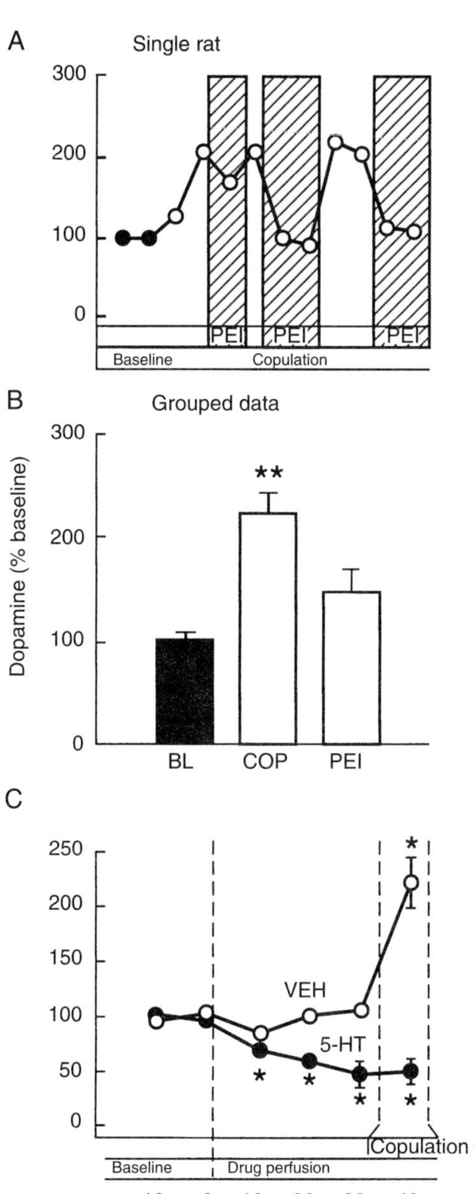

FIG. 22. Levels of dopamine in microdialysate collected from the nucleus accumbens (Acb) of male rats during sexual behavior sessions. **A:** Representative example of temporal changes in a single rat. **B:** Mean change in DA levels (±SE) for a group of six rats. Samples were collected at 3-minute intervals during three ejaculatory series and are designated precopulatory baseline (BL; filled circles), copulation (COP; open circles), or postejaculatory interval (PEI; hatched circles). Dopamine content was significantly greater in samples collected during copulation, compared with samples collected under baseline and postejaculatory conditions (**$P<0.01$). **C:** Temporal change in dopamine concentration of the dialysates collected from the Acb before and during 40 minutes of 5-HT perfusion into the LHA of six male rats. Samples were collected at 10-minute intervals. An estrous female was introduced to the male, and copulation was allowed during collection of the final sample during 5-HT perfusion. A significant decrease in dopamine occurred throughout the entire 5-HT perfusion period. This treatment blocked the increase in Acb DA seen in control animals (*$P<0.05$, perfusion times vs. time 0 [baseline]). [Figure is from (565) with permission.]

Similarly, the magnitudes of the responses to novel female odors were greater than to those of familiar females or control odors.

Chemical Changes Detected by Microdialysis or Voltammetry. Exposure of a male rat to the odor of an estrous female, but not of a nonestrous female, increased extracellular DA in the Acb (230,316,567,763,999). DA levels also increased during copulation (316,607,773,999). The DA response to the female's odor occurred on the first exposure to the odor (999) and was attenuated by systemically injected naloxone, which may have blocked the disinhibition of DA neurons by endogenous opioids (674).

Finer temporal analysis, using in vivo voltammetry (120, 607) or microdialysis with capillary chromatography (565), revealed that Acb DA rose in response to the presentation of the female, increased further during copulation, fell following each ejaculation, and increased again when the male resumed copulation (Fig. 22). Reverse dialysis of 5-HT into the lateral hypothalamic area (LHA) decreased basal DA in the Acb and prevented the DA response to the female (565). Because 5-HT is increased in the LHA with ejaculation (562), one factor contributing to the post-ejaculatory interval may be decreased Acb DA, as a result of 5-HT release in the LHA.

Sexually satiated male rats, presented with a novel estrous female behind a barrier, showed a slight increase in extracellular DA in the Acb; DA rose significantly when the barrier was removed and the animals copulated (302). Furthermore, sensitization to the motor-activating effects of amphetamine, following repeated systemic injections of amphetamine, also increased the DA response to a female (301). Therefore, there is cross-sensitization between a psychostimulant drug and a natural motivated behavior. On the other hand, prenatal stress decreased the number of rats that were able to copulate, but did not affect the DA response to an inaccessible female or DA levels during copulation (991).

Nigrostriatal DA Tract

The nigrostriatal tract arises from the substantia nigra of the midbrain (A9 cell group) and innervates the dorsal striatum (caudate putamen). Nigrostriatal activity enhances the readiness to respond to external stimuli (814). Loss of nigrostriatal neurons in Parkinson's disease leads to difficulty in initiating movements and to slowing of movements. Nigrostriatal neurons respond rapidly to any sudden stimulus, but do not contribute to specific information about those stimuli (880,881).

Effects of Electrolytic or Cell-body Lesions. Bilateral lesions of the substantia nigra slowed copulation and decreased the number of ejaculations (133). These data support the proposal that nigrostriatal activity contributes to the somatomotor patterns of copulation.

Chemical Changes Detected by Microdialysis or Voltammetry. DA was released in the dorsal striatum only during copulation, in contrast to the DA release in the Acb and MPOA (230). Therefore, nigrostriatal DA appears to be more important for motor activation than for sexual motivation.

Expression of c-fos and/or Other Immediate-early Genes. Copulation elicited no increase in Fos-ir in the dorsal striatum, in contrast to the significant increases in the Acb, MPOA, BST, and piriform cortex (816).

Paraventricular Nucleus (PVN) of the Hypothalamus

The paraventricular nucleus (PVN) consists of a magnocellular division, which releases oxytocin and vasopressin from the posterior pituitary, and a parvocellular division, which projects to several brain areas and the spinal cord [reviewed in (938)]. Axons projecting to the spinal cord contain oxytocin, vasopressin, somatostatin, DA, or other, undetermined neurotransmitters. The parvocellular division receives monoaminergic input from noradrenergic and serotonergic brainstem nuclei, as well as from periventricular DA neurons. The PVN is an important integrative site for autonomic and neuroendocrine functions. It is important for both noncontact erections and seminal emission, but provides less critical contributions to touch-based erections and copulation. Oxytocinergic neurons projecting to several areas of the brain and spinal cord mediate much of the PVN's influence, and DA, glutamate, hexarelin analog peptides, and NO provide excitatory drive to those oxytocinergic efferents. GABA and opioid peptides and opiate drugs inhibit these neurons.

Effects of Electrolytic or Cell-body Lesions. Excitotoxic lesions of the parvocellular PVN decreased noncontact erections but did not affect copulation (553). Similar parvocellular lesions also failed to affect copulation but did decrease the amount of semen ejaculated and also decreased the number of oxytocin-immunoreactive fibers in the lumbosacral spinal cord (1). Larger lesions encompassing both the parvo- and magnocellular divisions inhibited both noncontact and touch-based erections and, in copulation tests, decreased intromission ratio and increased ejaculation latency (553). On the other hand, electrolytic lesions of the PVN only decreased the latency to the first touch-based erection (i.e., facilitated erection) (682). Therefore, the PVN appears to contribute to both noncontact erections and seminal emission, but PVN lesions have inconsistent effects on touch-based erections and copulation.

Lesions of the lateral parvocellular portion of the PVN destroyed the neurophysin-containing axons to the sexually dimorphic spinal nucleus of the bulbocavernosus (SNB) (987). Neurophysin is used as a marker for oxytocin and vasopressin. Neurophysin-containing terminals appeared to be in contact with somas and dendrites of the SNB. Thus, PVN is apparently the source of oxytocinergic neurons to the SNB, which promote erection, seminal emission, and ejaculation.

Effects of Direct Applications of Drugs Affecting Specific Transmitters.

DA and Oxytocin. Microinjections of the DA agonist apomorphine (639), the D_2 agonist quinpirole, or oxytocin (53) into the PVN elicited drug-induced erections, which were blocked by an oxytocin antagonist microinjected icv (640). PVN microinjections of apomorphine (749) or the D_2/D_3 agonist quinelorane (263) also increased touch-based erections and *ex copula* seminal emissions. In addition, apomorphine in the PVN elicited increases in intracavernous pressure in anesthetized rats, a response that was enhanced by systemic administration of the monoamine oxidase B inhibitor, selegiline (34). An oxytocin antagonist administered icv, but not into

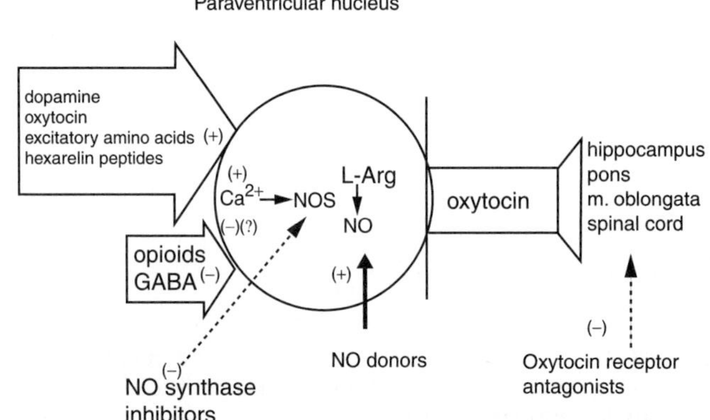

FIG. 23. A schematic representation of oxytocinergic neurons originating in the paraventricular nucleus of the hypothalamus (PVN) and projecting to extrahypothalamic brain areas and the spinal cord involved in sexual activity. Activation of these neurons by dopamine, excitatory amino acids, oxytocin itself, and hexarelin analog peptides facilitates penile erection and sexual activity. The direct or indirect activation of these neurons and the facilitative effects on erection and sexual activity induced by the above substances can be reduced or abolished by the stimulation of opioid and GABAergic receptors. These oxytocinergic neurons are activated by the activation of nitric oxide synthase (NOS), present in their cell bodies. Endogenous NO, formed by stimulation of dopamine, excitatory amino acid, or oxytocin receptors, or exogenous NO, derived from NO donors delivered to the PVN, activates oxytocinergic neurons by a yet-unidentified mechanism. This causes the release of oxytocin in brain areas distant from the PVN, as well as the spinal cord, inducing penile erection. Similar mechanisms operate during noncontact erections and copulation. [Figure reproduced from (49), with permission.]

the PVN, inhibited noncontact erections (650), suggesting that axons from the PVN, terminating perhaps in the hippocampus (644), facilitate noncontact erections. In summary, DA, acting via D_2 receptors, apparently increases oxytocin release, which in turn promotes drug-induced and noncontact erections [reviewed in (48,49)] (Fig. 23).

NO. PVN administration of the NOS inhibitor N(G)-nitro-L-arginine methyl ester)L-NAME) decreased noncontact erections and impaired copulation in male rats (649). Another NOS inhibitor, N(G)-monomethyl L-arginine (L-NMMA), reverse-dialyzed into the PVN, decreased touch-based erections, whereas the NO precursor L-arginine increased erections; however, copulation was not affected by either drug (864). These authors noted that similar administration of L-NMMA into the MPOA did decrease the rate of mounting in copulation tests (865). Therefore, manipulations of NO in the PVN have had inconsistent effects on copulation, but NO antagonists have consistently inhibited noncontact and touch-based erections.

Amino Acids. The glutamate agonist N-methyl-D-aspartic acid (NMDA), microinjected into the PVN, elicited erections; this effect was blocked by the NMDA antagonist MK-801 (dizocilpine) and by the NOS antagonist L-NAME (646). Therefore, glutamate, by opening calcium channels associated with NMDA receptors, may lead to calcium activation of calcium-calmodulin kinase (CAM kinase), which in turn activates NOS. Conversely, microinjections of the $GABA_A$ agonist muscimol into the PVN inhibited erections elicited by apomorphine, oxytocin, or NMDA (642). The $GABA_B$ agonist baclofen was ineffective. Extracellular levels of glutamate and aspartic acid in the PVN increased during exposure to a receptive female and increased further during copulation (648). Furthermore, microinjection of an NMDA receptor antagonist (MK-801) into the PVN inhibited both noncontact erections and copulation; an AMPA receptor antagonist (CNQX) also impaired copulation, but to a lesser extent (648).

Adrenocorticotropin. Adrenocorticotropin (ACTH [1-24]), microinjected into the periventricular hypothalamus, including the PVN, elicited erections (54). Similar injections into the POA, caudate nucleus, or hippocampus were ineffective.

Opiates. Morphine, administered unilaterally into the PVN before introduction of an inaccessible estrous female prevented noncontact erections, as well as the increases in production of NO (see section on microdialysis below) observed with control injections (641).

Hexarelin Analog Peptides. Hexarelin analog peptides, microinjected into the PVN elicited erections, probably by stimulating specific receptors on oxytocinergic cell bodies, which increase Ca^{2+} influx and activate NOS (643,652); reviewed in (49).

Chemical Changes Detected by Microdialysis. An increase in NO, inferred from increases in NO_2 and NO_3, was detected in microdialysate from the PVN during both noncontact erections and copulation (641,649). This increase was blocked by icv injection of hemoglobin, an NO scavenger, but not by injections into the PVN; PVN injections of hemoglobin also failed to inhibit noncontact erections (649). Therefore, NO appears to work intracellularly in the PVN, but to have additional effects in other areas that are important for erections. PVN microinjections of morphine inhibited both copulation and the copulation-induced increase in NO (641). L-arginine, reverse-dialyzed into the PVN, increased both NO production and touch-based erections, and L-NMMA had the opposite effects (864). Both NO production and noncontact erections were increased by apomorphine or the D_2 agonist LY-171555, but not by the D_1 agonist SKF-38393 (645). The increase in NO production was blocked by a D_2 antagonist (L-sulpiride) but not by a D_1 antagonist (SCH-23390). However, the increase in erections was blocked by both antagonists, suggesting that both D_1 and D_2 receptors contribute to erectile function. On the other hand, a later study failed to find an inhibitory effect of a D_1 or D_2 antagonist, or of an oxytocin antagonist, administered into the PVN, on noncontact erections or on the production of NO in the PVN, although an NMDA antagonist reduced both (651). Furthermore, the NMDA microinjections that elicited erections (see above) also increased NO in the PVN (646).

Microinjection of omega-conotoxin, a potent inhibitor of N-type voltage-activated calcium channels, prevented apomorphine- and oxytocin-induced erections and NO production (930). Exogenous provision of NO by the NO donors sodium nitroprusside or hydroxylamine overcame the effects of the omega-conotoxin, bypassing the need for activation of N-type channels that were indirectly activated by apomorphine and oxytocin. Omega-conotoxin also failed to block NMDA facilitation of NO production and erections, because NMDA stimulates ligand-activated calcium channels, not voltage-gated channels. Consistent with the facilitative effects of the DA agonist apomorphine and of D_2 agonists, increases in DA release in the PVN were observed during noncontact erections and, to an even greater extent, with copulation (647).

Presence of Steroid Receptors. The lateral parvocellular subdivision of the PVN projects to the L5-L6 region of the spinal cord; 30% of those neurons concentrated E, and almost half of the E-concentrating neurons projected to the lumbar spinal cord (988). Therefore, some of the effects of steroids on the SNB may be indirect, mediated by steroid-sensitive afferents.

Expression of c-fos and/or Other Immediate-early Genes. Intromission and ejaculation increased

Fos-ir in the parvocellular regions of the PVN, with ejaculation eliciting 2.5 times as many Fos-ir neurons (1021). Ejaculation also elicited more Fos-ir neurons in the magnocellular division than did intromission, although overall numbers were lower than in the parvocellular regions. Half the Fos-ir neurons contained oxytocin, even in noncopulating males. Indeed, copulation did not affect the percentage of double-labeled cells, except in the caudal-most PVN, where one-third of the lateral parvocellular neurons were double-labeled following ejaculation, but no cells were double-labeled in noncopulating animals. However, use of a more selective antibody for Fos, as opposed to Fos plus Fos-related antigens (Fras), revealed no double labeling of Fos and oxytocin following ejaculation in male rats (L. Coolen, personal communication). In gerbils neither exposure to an arena previously associated with copulation, nor copulation itself, increased PVN Fos expression (394). On the other hand, in a more recent report, copulation did increase the number of neurons double-labeled for OT and Fos-ir (705). Exposure to an anesthetized female also increased Fos-ir in OT-containing cells in sexually experienced males, but not in naïve males. Perhaps variability in Fos-ir in non-OT cells contributed to a lack of statistical significance in the earlier studies that found no Fos increases.

Expression of mRNA or Immunocytochemistry for Neuropeptides. Sexually impotent male rats had less expression of oxytocin mRNA in the PVN, and greater expression of mRNA for proenkephalin and prodynorphin, than did sexually competent males (56). This finding fits well with the observation that morphine in the PVN inhibited NO production and copulation (641). In addition, NOS is co-localized with oxytocin in the PVN (1042). Activation of these neurons facilitates *ex copula* erections and may also contribute to copulation. The oxytocin is released in the hippocampus, where it promotes erections (644), and in the spinal cord, where it promotes seminal emission (1).

Efferent Projections. Axons project from the PVN to the nucleus paragigantocellularis (nPGi) (70), a major source of inhibition of genital reflexes and copulation (see section on nPGi below). Axon terminals formed close appositions to serotonergic neurons that projected to the lumbosacral spinal cord and indirectly to the corpus cavernosum. Thus, the PVN is positioned to exert control over inhibitory serotonergic projections to the spinal areas that control erection.

Lateral Hypothalamus (LH)

Reciprocal connections of the MPOA with downstream structures pass through the lateral hypothalamus (LH) as part of the medial forebrain bundle (895), and LH lesions that sever these connections are as effective as MPOA lesions in abolishing copulation. However, LH cell bodies may also influence copulation.

Effects of Direct Applications of Drugs. Microinjection into the anterior LH (LHA) of a selective serotonin reuptake inhibitor (SSRI) delayed the onset of copulation and delayed ejaculation after the animals did begin to copulate (562). Thus, the impairment by SSRIs of sexual motivation and ejaculation [reviewed in (831)] may be mediated, in part, by increased 5-HT in the LH.

Changes Detected by Microdialysis. 5-HT is released in the LHA when male rats ejaculate (Fig. 24) (562). Reverse dialysis of 5-HT into the LHA decreased basal DA levels in the Acb and prevented the usual DA increase during copulation (565). Because DA release in the Acb contributes to numerous motivated behaviors, the inhibition induced by 5-HT in the LHA may at least partially explain the diminished sexual motivation associated with clinical use of SSRI antidepressants [reviewed in (831)].

Orexin/Hypocretin Neurons. Orexin/hypocretin neurons are located in the perifornical lateral hypothalamus. Exposure to a receptive female or copulation increased the number of orexin/hypocretin neurons that were Fos-immunoreactive, compared to exposure to another male or placement into an empty cage, respectively (699).

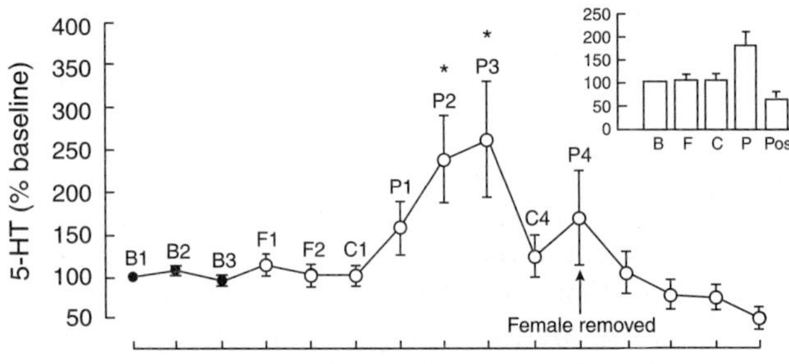

FIG. 24. Temporal changes in extracellular serotonin (5-HT) collected from the lateral hypothalamic area (LHA) of male rats before and during copulation. Each data point is the mean (+SE) for 6-minute dialysate samples collected during baseline (B), in the presence of an estrous female (F), during copulation (C), during the postejaculatory interval (P), and after the female was removed (expressed as a % of mean baseline levels). A total of four samples were analyzed after the female was removed, at 30-minute intervals. 5-HT levels increased during the second (P2) and third (P3) postejaculatory intervals, compared to final baseline (B3), female behind barrier (F2), and first copulation period (C1). 5-HT during P3 was also higher than during the fourth copulation interval (C4). Samples collected during the second and third copulation series were not analyzed, because most males ejaculated before a full 6-minute sample could be collected. The summary graph (inset) represents the mean (+SE) for data from the 15 sample periods collapsed into five groups, based on behavioral condition. Samples collected during the postejaculatory intervals showed higher 5-HT levels compared to all other conditions. Basal extracellular concentrations of 5-HT in the LHA were calculated to be 1.6 + 0.1 nM. [Figure from (562), with permission.]

Ventromedial Hypothalamus (VMH)

The VMH is recognized primarily for its role in the control of female rat sexual behavior. (See Chapter 35 on female sexual behavior.) However, the VMH may also modify male sexual behavior. Like the MPOA, the VMH has abundant receptors for androgens and estrogens (896,1031) and receives genitosensory input through midbrain projections (214), as well as chemosensory stimuli via projections from the Me (157,204,345).

Effects of Electrolytic or Cell-body Lesions. Lesions of VMH often enhance aspects of male sexual activity, suggesting that VMH may tonically inhibit masculine sexual behavior. In this regard, small electrolytic lesions of the VMH in castrated male rats treated with exogenous T reduced the latency to express copulatory behavior (179). However, in male hamsters, VMH lesions increased ultrasonic vocalizations, but did not alter copulation (305).

Effects of Hormonal Manipulations. Both ARs and ERs are present in the VMH of male rats (896) and hamsters (1031). Nonetheless, intracerebral implants of T into the VMH in castrated male rats or mice failed to stimulate copulation or ultrasonic vocalizations (378,708). However, T in the VMH restored partner preference in male rats (378), and increased urine marking in male mice (708), suggesting that the VMH contributes to hormone-sensitive sexual motivation. Conversely, intracerebral implants of the AR antagonist hydroxyflutamide into the VMH of castrated male rats with T replacement blocked both sexual motivation (379) and sexual behavior (379,622). Therefore, it appears that the VMH transduces androgen stimuli important for sexual motivation and copulation. The inhibitory effects of hydroxyflutamide in the VMH stand in contrast to the facilitative effects of VMH lesions described above. It may be that AR-containing neurons inhibit other VMH neurons, or may themselves be inhibited during mating, thereby facilitating male copulatory behavior. Androgen-responsive neurons in the VMH may also affect sexual activity through projections to other sites, including the midbrain (545).

Expression of c-fos or Other Immediate-early Genes. Mating induced Fos-ir in the VMH in male rats (209) and gerbils (394). In rats, bedding from estrous females failed to increase Fos-ir in the VMH, suggesting that mating is required to activate Fos expression (139). However, Fos expression was unaffected by pelvic nerve hemisection in gonadectomized male rats primed with E and progesterone (1001). This suggests that Fos expression in the VMH of males is more dependent on chemosensory input or genitosensory input carried through other pathways, such as the pudendal nerve. By contrast, copulatory activity was not required for the increase in Fos expression in male gerbils (394).

In contrast to the mating-induced Fos induction observed in the VMH of male rats and gerbils, mating may inhibit neural activity in the VMH of other species. Mating did not increase Fos-ir in the VMH of male musk shrews, although Fos-ir was significantly increased in the VMH of females (324). Similar results were observed in male hamsters (491), mice (370), and ferrets (467,515). In male macaques, mating actually elicited less Fos-ir in the VMH than did exposure to an inaccessible female (665).

Major Motor Outputs

Ventral Premammillary Nucleus

The ventral premammillary nucleus receives input from the MPOA, the VMH, and the posterodorsal part of the Me (156). ARs are located in this nucleus in rats (896,1049,1051) and Syrian hamsters (1027). Fos-ir was increased in this nucleus in male rats following copulation (209), and Fos-ir neurons also contained AR (358). Female-soiled bedding elicited Fos-ir in the ventral premammillary nucleus of mice, although no increase was observed in the VMH, posterodorsal medial amygdala, or posteromedial cortical amygdala of these animals (1050). Gerbils did not show increased Fos-ir following either copulation or exposure to an arena previously associated with copulation (394).

Midbrain Periaqueductal Gray (PAG)

In the context of sexual behavior, the PAG is primarily associated with vocalization (511) and with lordosis in females (via extensive connections with VMH; see Chapter 35). Nonetheless, the PAG has steroid receptors (695,697) and extensive reciprocal connections with the MPOA (422,695,812,813,893,).

Effects of Electrolytic or Cell-body Lesions. The effects of PAG lesions in male rodents are similar to the effects of VMH lesions. That is, lesions of the PAG stimulate sexual activity in male rats (133,376). Electrical stimulation of caudal PAG in male guinea pigs produced vocalizations related to sexual behavior (511). However, combined unilateral lesions of the ventrolateral PAG and the contralateral sexually dimorphic area of the MPOA in male gerbils did not affect mating (300). On the other hand, PAG lesions did block the elicitation of the UG reflex by electrical stimulation of the MPOA (597a).

Steroid Receptors and Connections. The caudal two-thirds of the PAG contains a large population of ERα and AR receptors (697). In the middle third of the PAG, steroid-sensitive neurons were distributed primarily in the dorsomedial and lateral PAG; in caudal PAG, immunoreactive cells were located primarily in the dorsal half. Afferents from the MPOA terminated in close apposition to ERα-ir and AR-ir neurons in the PAG, and between 17% and 54% of PAG neurons that projected to the nucleus paragigantocellularis (nPGi, see below) were also immunoreactive for ERα or AR (695). These data suggest the existence of a hormone-sensitive output pathway for male sexual behavior that extends from the MPOA through the PAG to the nPGi. Similarly, in rhesus monkeys, the PAG projects to the nucleus retroambiguus, a premotor area in the medulla (977).

Nucleus Paragigantocellularis of the Medulla (nPGi)

Spinal mechanisms that control erection are under tonic inhibitory control. In rats (382,847,848), mice (838), and dogs (381,389) spinal transection results in an increase in the number or intensity of erections or UG reflexes or a decrease in the amount of stimulation required to elicit such responses. There is increasing evidence that a major source of that inhibition is the nPGi, located in the ventrolateral medulla.

Effects of Electrolytic or Cell-body Lesions. Bilateral electrolytic lesions of the nPGi in sexually naïve male rats increased the number of animals that ejaculated on their first exposure to an estrous female (1046). In the males that did ejaculate, nPGi lesions decreased the time and the numbers of mounts and intromissions preceding ejaculation and increased copulatory efficiency. In sexually experienced males nPGi lesions increased the latency to sexual satiety and increased the number of ejaculations preceding satiety (1046). Similar lesions decreased the latency to the onset of touch-based penile reflexes and increased the number of anteroflexions (419,602). External anal reflexes were unaffected by nPGi lesions. Instead, anal reflexes were facilitated by nucleus raphe obscurus lesions, which had no effect on erection, indicating a differential control of sexual and defecatory function (419). Lesions of the nPGi also allowed the UG reflex to be elicited without spinal transection (598). Indeed, nPGi lesions were as effective as spinal transection in disinhibiting the UG reflex, suggesting that the nPGi is a major source of the tonic inhibition.

Effects of Electrical Stimulation. Electrical stimulation of the nPGi of male rats has been demonstrated to activate and inhibit pudendal motoneurons (577). It elicited field potentials in the lumbosacral cord in regions containing spinal cord neurons projecting to motoneurons of the pudendal nerve (943). Sympathetic fibers in the pudendal nerve were specifically activated by nPGi stimulation (454). Spinal reflex activation of pudendal neurons by stimulation of the dorsal nerve of the penis was inhibited by nPGi stimulation (453).

Immunocytochemistry. Anatomical tracing studies demonstrated that neurons in the nPGi project directly to pudendal motoneurons in the dorsolateral and dorsomedial nuclei, sympathetic and parasympathetic preganglionic neurons in the T13-L2 and L6-S1 segments, respectively, and to the interneuronal regions of the medial gray lumbosacral cord (404,602). A majority (78%) of nPGi neurons projecting to the lumbosacral spinal cord contain 5-HT (599). These serotonergic neurons receive a direct projection from the PVN (70). Animals pretreated with the serotonergic neurotoxin 5,7 dihydroxytryptamine demonstrated a decreased descending inhibition in UG reflex tests, and application of serotonin to the spinal cord suppressed the UG reflex in spinalized rats (600). Therefore, serotonergic input could mediate the tonic inhibition of genital reflexes.

Other Brain Areas

In a classic study, electrical stimulation of the cingulate gyrus, the medial dorsal nucleus of the thalamus, the periventricular region, and the VTA elicited erections in conscious monkeys (578). Stimulation of the hippocampus in anesthetized rats also elicited erections (171). The hippocampus receives excitatory oxytocinergic inputs from the PVN (48), as well as serotonergic input from the raphe nuclei (580,581).

Lesions of the lateral septum decreased the numbers of mounts and intromissions (500). However, the AR antagonist hydroxyflutamide, implanted in the lateral septum of male rats, failed to prevent the T-induced restoration of copulation in castrated male rats, although similar implants into the MPOA did prevent such restoration (622). Thus, the lateral septum may facilitate copulation, but ARs in that area appear not to contribute to its effects. Microinjection of vasopressin into the lateral septum of monogamous male prairie voles promoted pair bonding, and either a vasopressin or an oxytocin antagonist in the lateral septum blocked pair bonding elicited by either mating or microinjection of vasopressin (551).

Lesions of the caudal zona incerta, in the subthalamic region, abolished copulation in male rats, without affecting the preference for a receptive female (267,583). Thus, the zona incerta appears to be an important node in the circuit that controls copulation; however, little is known of its specific contributions.

Several cortical regions were activated during positron emission tomography (PET) scans when men were shown sexually explicit films, compared to neutral or humorous films (926). These areas included the inferior temporal cortex, a visual association area; the right insula and right inferior frontal cortex, paralimbic areas that associate highly processed sensory information with emotional states; and the left anterior cingulate cortex, which is associated with autonomic and neuroendocrine states. Plasma T levels were correlated positively with activity in some of these areas.

Spinal Cord

The spinal cord contains the autonomic and somatic nuclei that control the hemodynamic and striated muscle effectors of erection, ejaculation, and detumescence and that provide initial processing of somatic and visceral information from the genital region. Both descending axons from the brain and local reflex loops are important for these processes [see (332) for review].

Effects of Lesions or Transections

There is convincing evidence of descending inhibition from supraspinal sources, which can be eliminated by transection of the spinal cord. Spinal cord transection releases the UG reflex from descending inhibition, allowing it to he elicited by urethral stimulation (626). Early postnatal spinal cord transection also accelerates the developmental expression of touch-based erections in male rats from about 38 to about 28 days of age, without altering androgen titers or their morphological correlates (631). Spinal cord transection also reversed the inhibition of penile reflexes resulting from dorsal penile nerve transection (848), application of a topical anesthetic to the glans penis (921), or excision of the penile sheath, which removes the tonic stimulation necessary for penile reflexes in neurally intact rats (572).

Effects of Direct Application of Drugs Selective for a Neurotransmitter

The need for caution in interpreting the effects of spinal transection is emphasized by a study in which spinal anesthesia was induced in rats at the thoracic level via an intrathecal cannula (845). After induction of spinal block, supine males exhibited the abbreviated erection latencies characteristic of spinally transected males, but unlike transected males, they had fewer and weaker erections. The reduction in erection latency may truly be due to the removal of inhibitory influences from the brain, but the decreased number and intensity of erections in spinally blocked males implies interference with excitatory influences from the brain. The increase in touch-based erections after surgical transection may therefore be secondary to neural reorganization within the cord, rather than a direct effect of removing supraspinal inhibition.

5-HT. The UG reflex typically can not be elicited unless the spinal cord is transected (626), the nPGi is ablated (598,602), or the MPOA is stimulated (601). However, either intrathecal or icv administration of the 5-HT neurotoxin 5,7-DHT allowed the UG reflex to be elicited without spinal transection or nPGi ablation or MPOA stimulation (600). This suggests that descending serotonergic axons are responsible for a major part of the central inhibition of this reflex. As noted above, a majority (78% ipsilateral) of the axons from the nPGi contain 5-HT (599). Intrathecal administration of 5-HT abolished the UG reflex, and the 5-HT antagonist methysergide prevented the 5-HT induced blockade of the reflex (599). In addition, lesions of the median raphe nuclei (a major source of serotonergic neurons) significantly increased the numbers of anteroflexions (penile body erections) and cups (intense glans erections) in tests of touch-based erections (682).

Intrathecal administration of either 5-HT or thyrotropin releasing hormone (TRH) resulted in only slight increases in mount and intromission latencies. However, co-administration of 5-HT and TRH caused a marked increase in mount and intromission latencies (377). However, in a test of touch-based penile reflexes, intrathecal administration of TRH alone decreased the proportion of responders, decreased the numbers of reflexes, and increased their latencies (420). TRH and 5-HT are co-localized in some bulbospinal neurons (415). Furthermore, methiothepin, a relatively nonspecific 5-HT antagonist, partially blocked some of the inhibitory effects of TRH (421). Therefore, either 5-HT or TRH may inhibit copulation and genital reflexes, but their effects may be enhanced by co-administration or endogenous co-release.

In contrast to the demonstrated inhibitory effects of 5-HT on genital reflexes, stimulation of one receptor subtype, the 5-HT_{2C} receptor, appears to facilitate erectile function. (See section on systemically or intraventricularly administered serotonergic drugs.) Consistent with such a facilitative effect, a systemically administered 5-HT_{2C} agonist (mCPP) facilitated erections in monkeys (780) and rats (101,102,668). In anesthetized animals, the reflex response of the cavernous nerve to stimulation of the dorsal penile nerve is facilitated after the complete section of the spinal cord at the thoracic level and after systemic injection of mCPP (920).

Intrathecal administration of 8-OH-DPAT decreased the percentage of male rats displaying touch-based erections and seminal emissions (531). This is in dramatic contrast to the facilitative effects of intrathecally administered 8-OH-DPAT on copulation, and especially ejaculatory threshold. In the copulation tests,

8-OH-DPAT decreased ejaculation latency, intercopulatory interval, and the number of intromissions preceding ejaculation (531). The basis for the apparent discrepancy between *in copula* and *ex copula* measures is not clear. However, a similar discrepancy was observed with intrathecal apomorphine [(750); see below].

GABA. In the lower spinal cord, GABA inhibits sexual reflexes. Intrathecal injection of the $GABA_B$ agonist baclofen into the lumbosacral (L5-S1) spinal cord decreased the number of touch-based erections and increased the latency to the first glans erection. The highest dose completely blocked penile responses (116). However, none of the doses used prevented males from copulating to ejaculation. *In copula* ejaculation shortly before the reflex test facilitated the onset of touch-based erections in saline controls and also blocked the inhibitory effects of the lower two doses of baclofen, but not the highest dose. Therefore, as in the experiment using 8-OH-DPAT (531), different neural mechanisms appear to control copulation and *ex copula* reflexes. In contrast to the inhibitory effects of intrathecal baclofen on *ex copula* reflexes, the $GABA_A$ agonist THIP produced only slight inhibitory effects at the highest dose. Therefore, stimulation of spinal $GABA_B$, but not $GABA_A$, receptors inhibits *ex copula* reflexes.

DA. Intrathecal infusion of lisuride, a nonspecific monoamine agonist, into the region of the rostral lumbar spinal cord of male rats produced the same effects on copulation as seen following systemic injection, namely, decreased time and number of intromissions preceding ejaculation (371). Therefore, stimulation of some type(s) of monoamine receptors appears to facilitate ejaculation, perhaps by increasing sensitivity to peripheral stimulation. Furthermore, intrathecal administration of the catecholamine neurotoxin 6-OHDA increased postejaculatory intervals of male rats and rendered the animals more sensitive to androgen deprivation (374).

Intrathecal administration of the DA agonist apomorphine also decreased the number of intromissions preceding ejaculation, suggesting that stimulation of spinal cord DA receptors facilitated ejaculation (750). However, intrathecal apomorphine also increased the interintromission interval, the ejaculation latency, and the postejaculatory interval. In touch-based reflex tests, the highest dose of intrathecal apomorphine decreased the numbers of touch-based erections and anteroflexions but did not affect seminal emissions (750). These data stand in contrast to the effects of systemic injection of apomorphine, which facilitated erections at low doses, inhibited erections at high doses, and facilitated seminal emissions at both low and high doses (751). However, in anesthetized rats, intracavernous pressure was increased by apomorphine administered intrathecally (333). The explanation for the different results from these studies is unclear.

NE. Intrathecal administration of NE increased the frequency of pelvic thrusting during intromission, whereas intrathecal administration of both the α_2 agonist clonidine and the β agonist isoproterenol decreased the frequency of thrusting (406). It is likely that clonidine's inhibitory effect resulted from stimulation of α_2 autoreceptors, thereby decreasing NE release. These results are consistent with the findings that α_1 receptors probably mediate NE's facilitative effects on copulation.

Oxytocin. Administration of oxytocin at the lumbosacral, but not the thoracolumbar, level increased intracavernous pressure in dose-dependent fashion (336). The proerectile effects of intrathecal oxytocin were blocked by an oxytocin antagonist, pelvic nerve section, and a nicotinic antagonist, which blocked transmission between the pre- and postganglionic parasympathetic nerves. An oxytocin agonist also produced proerectile effects. Intrathecal administration of a nitric oxide inhibitor blocked the proerectile effects of intrathecally delivered oxytocin, indicating that the effects of oxytocin in the spinal cord are mediated through nitric oxide interneurons. Vasopressin, a nonapeptide that differs from oxytocin by only one amino acid, was ineffective, as were lumbosacral saline and systemic oxytocin. Blockade of striated muscle activation failed to inhibit oxytocin's facilitative effects, suggesting that oxytocin's effects were mediated by the parasympathetic nervous system, and not by striated penile muscles. These data provide compelling evidence for the proerectile role of oxytocin-containing fibers descending from the PVN to the lumbosacral cord.

Opiates. Intrathecally administered naloxone decreased the number of intromissions preceding ejaculation, whereas morphine administered in a similar manner increased the number of intromissions (1011). These treatments had no other effects on copulation. The authors suggested that naloxone and morphine influenced the strength of the sensory signal from each intromission.

NO. As noted in the section on MPOA stimulation, intrathecal administration of an NO donor, a cGMP analog, or the phosphodiesterase V inhibitor sildenafil enhanced the increase in intracavernous pressure elicited by electrical stimulation of the MPOA; the NOS inhibitor L-NAME reduced the response (867). None of these treatments affected blood pressure or intracavernous responses to cavernous nerve stimulation. The authors concluded that central NO/cGMP can affect erectile capacity.

Chemical Changes Detected by Analysis of Cerebrospinal Fluid

The concentration of GABA in cerebrospinal fluid of male rats increased by more than 1,000% following ejaculation; concentrations of aspartate and glutamate increased by about 200% (793). Therefore, the concentration of a major inhibitory neurotransmitter increased markedly during the absolute refractory interval.

Presence of Steroid Receptors

AR-ir was found in three areas of the male rat lumbosacral spinal cord: the spinal nucleus of the bulbocavernosus (SNB), dorsolateral nucleus (DLN), and retrodorsolateral nucleus (RDLN) (313). Virtually no AR-ir was seen in long-term castrated males or males with the Tfm mutation, which renders the AR inactive. Furthermore, unilateral capsules filled with T and placed next to the bulbospongiosus muscle, increased the dendritic arborization of SNB neurons of castrated male rats. In contrast, testosterone administered to the spinal cord of castrated rats, was ineffective (798). Thus, despite the presence of androgen receptors in the pudendal motoneurons, the trophic effect of androgens is mediated by actions on the target muscle.

Gréco et al. (359), also found dense labeling for AR in the L5-S1 segments of the spinal cord, especially in the dorsal part of lamina X. Furthermore, mating-induced Fos-ir was predominantly located in AR-ir neurons. Therefore, the lumbosacral neurons that are activated by mating are androgen sensitive.

Immunocytochemistry

The 5-HT_{2C} receptor at the spinal level appears to facilitate erectile function. Consistent with such a facilitative effect, 5-HT_{2C} receptor immunoreactivity was found on motoneurons of the sacral parasympathetic nucleus, the dorsal gray commissure, and the motoneurons of the ventral horn (71). Furthermore, 5-HT_{2C} receptor immunoreactivity was exhibited by all neurons retrogradely labeled from the corpus cavernosum to the sacral parasympathetic nucleus and the dorsal gray commissure of L5-L6, and in ventral horn motoneurons retrogradely labeled from the ischiocavernosus and bulbospongiosus muscles. Therefore, the neurons that are known to promote erection possess 5-HT_{2C} receptors.

Immunocytochemical studies revealed that dopaminergic (DA) fibers and terminals exist in virtually all laminae throughout the spinal cord (424,809). D_2 receptors identified by immunohistochemical and in situ hybridization techniques were noted in the parasympathetic area of the lumbosacral spinal cord and found to be particularly abundant in the dorsomedial and dorsolateral motoneurons innervating the bulbospongiosus and ischiocavernosus muscles (970).

Neurons and fibers immunoreactive for α_{2a} and α_{2c} adrenoceptor subtypes were identified in the intermediolateral cell column, the dorsal gray commissure and the ventral horn of the T12-L2 and L5-S1 spinal cord. The dorsal horn displayed only immunoreactive fibers. Pseudorabies virus-infected neurons in the autonomic nuclei were both immunoreactive for α_{2a} and α_{2c} adrenoceptor subtypes and closely apposed by α_{2a} and α_{2c} immunoreactive fibers (1041). These results suggest an intraspinal modulation of the noradrenergic and adrenergic control of the autonomic outflow to the penis by pre- and postsynaptic α_2 adrenoceptors.

Oxytocinergic axons from the PVN densely innervate clusters of sympathetic preganglionic neurons in the intermediolateral cell column of the thoracolumbar spinal cord (667,936). There is also immunoreactivity for oxytocin in the sacral parasympathetic nucleus of the rat spinal cord, the dorsal horn and the dorsal gray commissure. Oxytocin fibers make synaptic contacts on preganglionic neurons in the sacral parasympathetic nucleus (875,936,945). Oxytocin receptor binding was found in the superficial dorsal horn, sacral parasympathetic nucleus, and dorsal gray commissure of the male rat lumbosacral spinal cord (984).

Neurons in segments L3 and L4 of the lumbar spinal cord contain both galanin and CCK (459) and project to the SPFp (958,998). These neurons are thought to relay somatosensory information from the genitals to the brain. Galanin and CCK immunoreactive neurons showed increased Fos expression only in males that had ejaculated one or two times and not in animals that only mounted or intromitted (except for a few Fos-ir neurons in one male that only intromitted) (958). Furthermore, lesions of galanin-containing neurons in L3 and L4 severely impaired ejaculatory ability of male rats (958). Therefore, these neurons not only relay ejaculation-specific sensory information to the brain, but also contribute to the execution of the behavior.

CIRCUITRY AND ANATOMICAL INTERCONNECTIONS

Thus far, we have considered separately the functions and connections of different brain nuclei in the control of male sexual behavior. However, it is probably more accurate to view each region of the brain as belonging to one or more functional circuits that converge on the MPOA. In terms of afferent input, MPOA receives chemosensory cues from the vomeronasal organ and olfactory mucosa, somatosensory stimuli from the genitals, as well as direct and indirect hormonal input via steroid receptor-containing neurons. There are also excitatory and disinhibitory pathways from the MPOA to the midbrain and brain stem, from which neurons descend to the spinal cord nuclei to regulate genital reflexes and copulatory motor patterns.

A great deal of research is directed at understanding the signals and pathways through which individual sensory and endocrine stimuli influence mating behavior. However, sensory and humoral cues are seldom presented in isolation. Instead, an organism in a natural setting is confronted with multiple signals from its internal and external environment that together are interwoven to yield an experience richer than the sum of the individual modalities (923).

To understand complex motivated behaviors, we must determine how different cues are weighed, prioritized, and integrated. The specific signals to guide sexual activity are tailored to the biology of the species. While rodents rely principally on chemosensory stimuli to initiate behavior, other species use visual, auditory, or somatosensory stimuli, or a combination of sensory cues. Nonetheless, multimodal integration of sensory and humoral signals is ubiquitous, and further studies may uncover common mechanisms that guide the behavior of diverse species.

The Chemosensory Circuits

Lesions along the chemosensory circuit that eliminate mating have established the importance of the Me, BST, and MPOA in male sexual behavior (Fig. 25). Our understanding of this circuit is enhanced through the use of Fos to identify activated neurons, including steroid-responsive neurons. However, the activity of the chemosensory circuit is also modified by hormonal stimuli. Me, BST, and MPOA each contain large numbers of steroid receptor-containing neurons, and intracerebral implants of T in either Me or MPOA

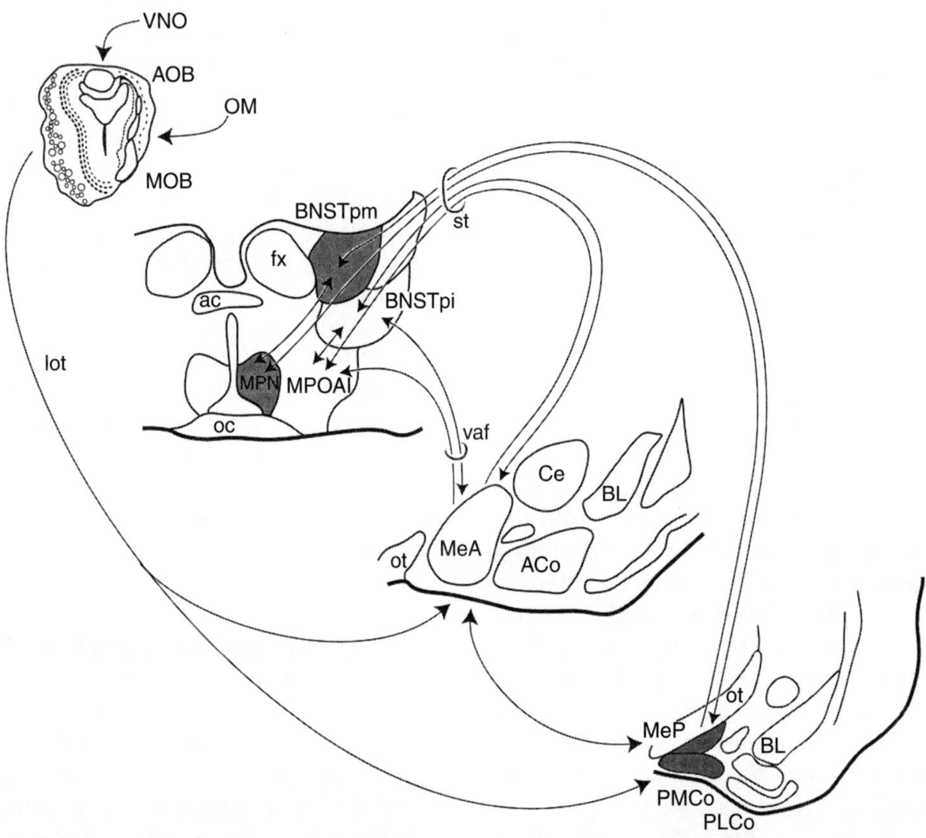

FIG. 25. Schematic diagram of principal limbic nuclei and connections in hamster brain that transmit chemosensory and hormonal cues to control male sexual behavior. Shading indicates areas with abundant steroid receptor-containing neurons. ac, anterior commissure; ACo, anterior cortical amygdaloid nucleus; AOB, accessory olfactory bulb; BL, basolateral amygdaloid nucleus; BNST, bed nucleus of the stria terminalis; BNSTpi, posterointermediate subdivision of BNST; BNSTpm, posteromedial subdivision of BNST; Ce, central amygdaloid nucleus; fx, fornix; lot, lateral olfactory tract; Me, medial amygdaloid nucleus; MeA, anterior subdivision; MeP, posterior subdivision; MOB, main olfactory bulb; MPN, medial preoptic nucleus; MPOAl, lateral subdivision of the medial preoptic area; oc, optic chiasm; OM, olfactory mucosa; ot, optic tract; PLCo, posterolateral cortical amygdaloid nucleus; PMCo, posteromedial cortical amygdaloid nucleus; st, stria terminalis; vaf, ventral amygdalofugal pathway; VNO, vomeronasal organ. [Reprinted from (1024a) with permission.]

enhance sexual activity above castrate levels (547, 799,1028). These effects are mediated by aromatization to E, for implants of DHT in either area are without effect (548,1024). Steroid implants not only facilitate male copulatory behavior, they also enhance the response to odor cues. Specifically, male mice with intracerebral implants in the MPOA will make ultrasonic vocalizations to female urine (903). Likewise, intracerebral T implants in Me or BST/MPOA of male hamsters stimulate anogenital investigation (1028). Together, these studies suggest that steroid action along the chemosensory pathway facilitates sexual behavior and responsiveness to sexually relevant chemosensory stimuli.

The challenge is to understand how steroid hormones interact with chemosensory cues to regulate male sexual behavior. Cottingham and Pfaff (219) have proposed that steroids act as a gating mechanism to permit or enhance transmission of chemosensory cues through steroid-sensitive brain nuclei. At the cellular level, castration may reduce synaptic efficiency in the transmission of odor cues by inhibiting release of neurotransmitters and/or suppressing postsynaptic receptors. Gonadal steroids also promote growth and branching of neurons. Although these effects are most dramatic during development, steroids also stimulate neuronal growth in the adult, as detailed by Garcia-Segura et al. (317). In the male Syrian hamster, castration reduced the mean somal area, highest dendritic branch, and percentage of neurons in Me with tertiary branches (344). There is also potential for steroid effects on synaptic connections. Changes in connectivity can take place rapidly, in the absence of substantial alterations in gross neuronal structure. This has been best-studied in females, where synapses in the arcuate nucleus of the hypothalamus are remodeled over the course of the 4-day estrous cycle (719). In the male, loss of synaptic connections after castration could have a major impact on transmission of chemosensory cues through Me, BST, and MPOA. It is likely that both structural and neurochemical mechanisms contribute to the decline in sexual behavior after castration.

Somatosensory Input from the Genitals

The dorsal nerve of the penis (DNP), a branch of the pudendal nerve, carries sensory input from the penile skin, prepuce, and glans penis (919). It is the major source of afferent input from the penis, although the cavernous nerve also contributes input from deeper structures (919). The afferents terminate primarily in the medial portions of the dorsal horn and in the medial central gray matter (dorsal gray commissure) of the lumbosacral spinal cord (625,687,701,950). Sensory information is relayed to supraspinal sites via both spinothalamic and spinoreticular pathways. The spinothalamic pathways primarily convey the fastest fibers related to the encapsulated nerve endings of the penis. They travel in the dorsal columns and consist primarily of fast myelinated fibers. These fibers terminate in the posterolateral nucleus of the thalamus, and input is subsequently relayed to the medial thalamus. Spinoreticular fibers tend to be slower than the spinothalamic fibers. They travel in the contralateral (and to a lesser extent ipsilateral) lateral spinal columns and terminate in the brainstem reticular formation.

Electrical stimulation of the DNP elicited activity in the nPGi, the PVN, the MPOA, and the cortex (369,428,1044). The CTF of male rats (82) and SPFp of rats (208,209,210), gerbils (394), and hamsters (491) show increased Fos-ir following copulation and also project to the MPOA. Finally, a combined anterograde and retrograde tracer study revealed that axons from the L3 and L4 segments of the lumbar spinal cord terminated in close proximity to SPFp neurons that projected to the MPOA or BNSTpm (998).

Studies in humans have examined cortical evoked potentials following electrical stimulation of the dorsal nerve of the penis/clitoris (a division of the pudendal nerve). Evoked potentials were recorded bilaterally from cortical areas, with the highest amplitude in the midline (Cz-2) over the sensory cortex (320,363). This distribution is consistent with pudendal representation deep in the midline interhemispheric fissure in humans (753) and cats (134). The amplitudes of cortical evoked responses were larger in men than in women (369) although slightly shorter in latency in women (723). The smaller size in women may be related to fewer fibers innervating the clitoris relative to the penis or to the greater accessibility of stimulation of the male dorsal nerve.

An Ejaculation-Related Circuit

Ejaculation, but not copulation without ejaculation, elicited small clusters of Fos-ir in the posterodorsal medial amygdala (MePD), the posteromedial bed nucleus of the stria terminalis (BSTpm), posterodorsal preoptic nucleus (PdPN), and the medial portion of the parvocellular subparafascicular thalamic nucleus (SPFp) of rats (208–211), hamsters (283,491,742), and gerbils (394,396,899,900) (Fig. 26). Using anterograde and retrograde tracing techniques, it has been demonstrated that neurons in each of these areas expressing ejaculation-induced Fos are reciprocally connected with the MPOA (211,356,396) (Fig. 27). In addition, combined unilateral lesions of MePD and SPFp in rats (82) and lesions of the SPFp in gerbils (395)

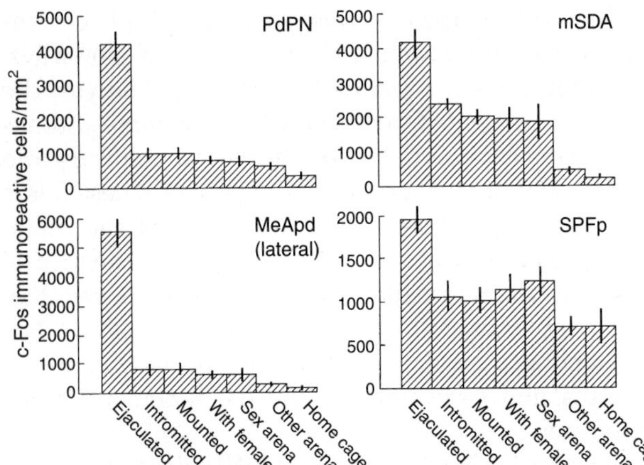

FIG. 26. Increased densities of Fos-ir cells in the PdPN, mSDA, lateral MeApd and SPFp of male gerbils that copulated to ejaculation. In the mSDA and SPFp, Fos-ir densities also increased when males were placed in the sex arena. Bar heights indicate group means. Vertical lines denote standard errors. N = 7 in all four areas for the Home Cage group and in the SPFp for the ejaculated group. In all other cases, n = 8. Fos-ir, immunoreactivity to Fos, the protein product of the immediate-early gene c-fos; PdPN, Posterodorsal preoptic nucleus; mSDA, medial sexually dimorphic area; MeApd, posterodorsal region of the medial amygdala; SPFp, subparafascicular nucleus. [Reprinted from (394), with permission.]

resulted in decreased mating-induced Fos expression in the MPOA, demonstrating that these inputs contribute to the neural activation of the MPOA. However, the majority of ejaculation-activated cells in the MePD, PdPN, or mSDA of gerbils did not project to other members of the ejaculation-related circuit; the authors suggested that the majority of the Fos-ir neurons either were interneurons or projected to other brain areas (396). Indeed, a later article reported that ejaculation-activated neurons in the PdPN and MePD projected to the anteroventral periventricular nucleus (AVPV), the dorsal portion of the principal part of the BST, and the retrorubral field (899).

It is not clear whether the activated neurons primarily received sensory input or activated motor patterns leading to ejaculation. The similar pattern of Fos-ir in female rats following ejaculation by the male (209,765,833) suggests that these neurons are activated by sensory input, rather than programming motor output. However, lesions of the PdPN or MePD decreased mounting and delayed ejaculation (395). Thus, these data suggest a contribution of these structures to the motor control of copulation, but not a major specific role in triggering ejaculation. Lesions of

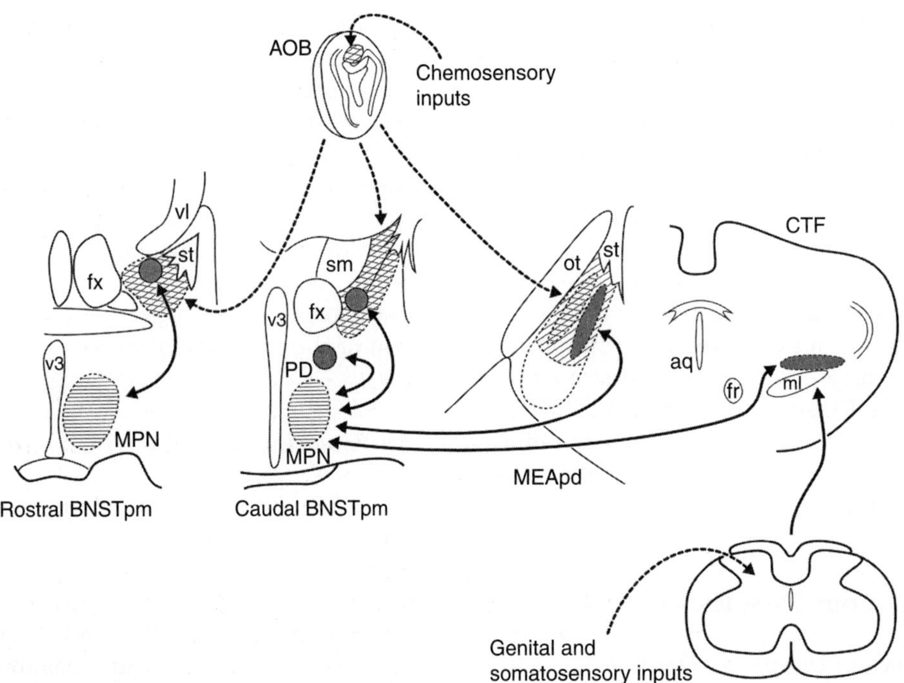

FIG. 27. Schematic overview of neural activation in circuits underlying male sexual behavior. Areas where Fos is induced following chemosensory cues or chemosensory investigation are illustrated by diagonal stripes from upper left to lower right. Areas where Fos is induced primarily following ejaculation are illustrated in dark shading. Areas where Fos is induced by all consummatory elements of behavior are illustrated by diagonal stripes from lower left to upper right. AOB, accessory olfactory bulbs; MPN, medial preoptic nucleus; PD, posterodorsal preoptic nucleus; BNSTpm, posteromedial bed nucleus of the stria terminalis; MEApd, posterodorsal medial amygdala; CTF, central tegmental field; LSSC, lumbosacral spinal cord; v3, third ventricle; fx, fornix; vl, lateral ventricle; st, stria terminalis; sm, stria medularis; ot, optic tract; aq, aqueduct; fr, fasciculus retroflexus; ml, medial lemniscus. [Figure reproduced from (206), with permission.]

the posterior thalamus that included SPFp and zona incerta completely disrupted ejaculatory behavior in male rats (583), supporting a role for the posterior thalamus in control of ejaculation. In contrast, small lesions restricted to the SPFp in gerbils had no effect on copulation (395), suggesting that the SPFp may relay copulation-related information, rather than triggering ejaculation. The SPFp receives ascending sensory inputs, either directly or indirectly, from the lumbosacral and cervical regions of the spinal cord (459,529) and, in turn, projects to the BSTpm and MePD of rats (212,1045). In gerbils the SPFp projects to the mSDA, PdPN, MePD, but does not receive reciprocal input from those structures (899).

Yahr and her colleagues have provided information about the neurotransmitters used by ejaculation-activated neurons in gerbils. One-third to one-half of Fos-ir neurons in the lateral PdPN and MePD expressed NOS (898). In addition, half of the Fos-ir neurons in the MePD, lateral PdPN, and medial sexually dimorphic nucleus of the MPOA (mSDA) were GABAergic, and a quarter of those in the MePD and mSDA were glutamatergic; the PdPN appeared to have no glutamatergic cells (900). The authors suggested that these neurons may influence the timing of ejaculation, the length of the postejaculatory interval, or release of LH or prolactin at the time of ejaculation.

Male hamsters show Fos activation in the BSTpm and MePD only after multiple ejaculations, or when they are reaching sexual satiety after only a few ejaculations, due to mating on previous consecutive days (742). Moreover, small lesions directed at the lateral MePD increased the number of ejaculations necessary for satiety (743). Hence, the MePD may contribute to ejaculatory behavior and promote sexual satiety in addition to other aspects of male sexual behavior.

Lesions of the lateral hypothalamus severely affect display of ejaculations, but do not affect the ability to mount and intromit (480). Thus, this area may play an excitatory role in the regulation of ejaculation. Interestingly, serotonin is released in the anterior lateral hypothalamus (LHA) at the time of ejaculation (562), and microinjection of a SSRI into the LHA increased the latencies to mount, intromit, and ejaculate (562). Hence, serotonin in the LHA may influence sexual motivation as well as ejaculatory behavior.

A recent study by Holstege and co-workers (423) used positron emission tomography to study increases in regional cerebral blood flow during ejaculation in men. Increased blood flow was observed following ejaculation compared to sexual stimulation in the mesodiencephalic junction, which includes the location of the SPFp. In addition, increased blood flow was observed in cerebellum, lateral putamen and claustrum, and several cortical areas. Increased blood flow was not detected in the MPOA or BNST, which contrasts with studies in rodents but is in agreement with a study in male macaques that failed to detect Fos activation in the MPOA following ejaculation (665).

Efferents from the MPOA

The primary efferents from the MPOA travel through the medial forebrain bundle to the midbrain (935). However, we should emphasize that most of the afferent and efferent connections of the MPOA with other brain regions are reciprocal (893). Therefore, input to the MPOA from "downstream" structures can modulate MPOA processing, as well as receive its output. Evidence for the importance of axons leaving the MPOA laterally to join the medial forebrain bundle include reports that knifecuts lateral to the MPOA disrupted copulation similarly to lesions of the MPOA itself, whereas knife cuts either rostral or caudal to the MPOA were ineffective (883,940). Similarly, lesions along the extent of the medial forebrain bundle caudal to the MPOA permanently eliminated copulation, whereas lesions rostral to the MPOA did not affect copulation (149–151,401,412,413). Furthermore, contralateral lesions of the MPOA and medial forebrain bundle were as effective as bilateral lesions of either structure (401).

The critical axons in the medial forebrain bundle provide reciprocal connections between the MPOA and the ventral pons and medulla (696) (Fig. 28). Dense projections from the MPOA reach the nPGi and midline raphe nuclei. Lesions of the nPGi facilitated copulation (1046), touch-based erections and anteroflexions (602), and the UG reflex (598). 5-HT from the raphe nuclei is also generally inhibitory to male sexual behavior [reviewed in (409)]. Therefore, one means by which the MPOA can facilitate sexual behavior is by inhibiting these two areas. However, these areas may also send reciprocal input to the MPOA to inhibit sexual behavior. The MPOA also provides dense innervation to the PAG, which in turn projects to the nPGi. Thus, the MPOA can influence the nPGi through both direct and indirect connections. The PAG has numerous AR- and/or ER-containing neurons (697). Furthermore, both the MPOA efferents to the PAG and the PAG efferents to the nPGi contain AR and/or ERa, and axons from the MPOA end in close juxtaposition to PAG-to-nPGi efferents (695). Therefore, gonadal hormones could affect the ability of the MPOA to disinhibit sexual behavior. However, it is unlikely that mere disinhibition can account for the elicitation of reflexes that has resulted from electrical or chemical stimulation of the MPOA (335,340,578,579,601,817). It seems likely that there is also a pathway that provides excitation to the spinal neurons that control the reflexes. Peripheral pathways

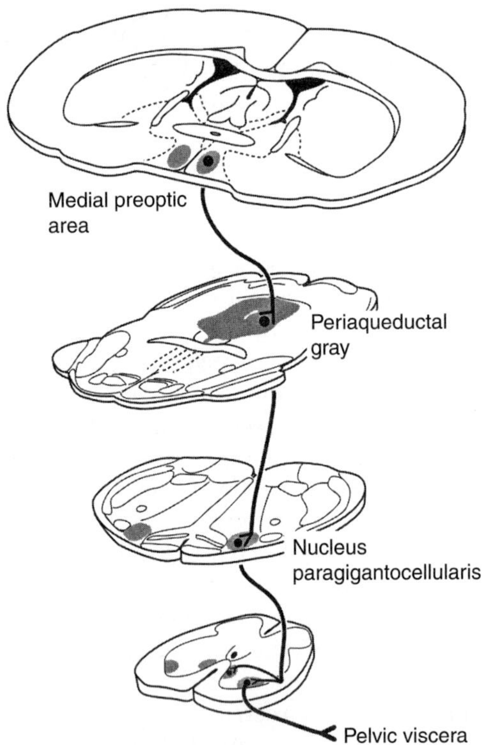

FIG. 28. Summary figure showing the MPOA-PAG-nPGi-spinal cord circuit. MPOA projections to the periaqueductal gray (PAG) terminate preferentially among PAG neurons projecting to the nucleus paragigantocellularis (nPGi). Descending projections from the nPGi terminate within the dorsomedial and dorsolateral motor pools of the ventral horn of the lumbosacral spinal cord. Motoneurons from these pools innervate the bulbocavernosus and ischiocavernosus muscles, which are essential for penile erection and ejaculation. [Figure is from (695), with permission.]

that mediate that excitatory influence appear to include activation of both parasympathetic outflow, via the pelvic and cavernous nerves, and sympathetic efferents, via the paravertebral sympathetic chain and possibly the hypogastric nerve (335). Although the parasympathetic system provides the main proerectile force, the sympathetic system may also contribute (43,335,907), perhaps by producing vasoconstriction in nongenital areas, thereby diverting blood to the penis (335). (See section on neural innervation of the penis, above.)

Sexual Behavior in the Context of Mammalian Social Behavior

The same brain areas that control male sexual behavior also regulate other social behaviors, including female sexual behavior, maternal behavior, aggression, and territorial marking. Newman (703) has suggested that these structures form a richly and reciprocally interconnected circuit subserving mammalian social behaviors. Most of the areas, except the midbrain, are copiously populated with steroid receptors, and all contribute to more than one behavior. Perinatal, adolescent, and adult hormones mold sexually dimorphic responsiveness to social stimuli. Substructures of these brain areas may be specialized for specific behavioral components. However, the circuit that is activated specifically by ejaculation in male rats (208,209), hamsters (491,742), and gerbils (394), also responds in female rats to ejaculation by a male (260, 273,765). Therefore, this circuit may either "count" stimuli or regulate timing in both males and females to promote a successful pregnancy. It is unclear whether the same neurons contribute to different behaviors or whether behavior-specific neurons within each structure mingle with neurons specific for other behaviors. Organizational and activational effects of hormones may confer specificity in response to stimuli and thereby promote diversity in the network. However, appreciation of the common themes underlying diverse social behaviors may provide insights into the mechanisms that control them.

UNANSWERED QUESTIONS AND FUTURE DIRECTIONS

Answers to previous questions inevitably raise new questions to guide future research. These questions fall into several broad categories: anatomical interconnections, immediate and long-term neurotransmitter effects and their interactions, electrophysiological responses, roles of intracellular messengers and gene expression, similarities and differences within and between species in the control of male sexual behavior, and in the control of other social behaviors.

Questions concerning anatomical interconnections include the following: By which pathways do the main and vomeronasal olfactory systems promote sexual arousal? How do the pathways that control noncontact, touch-based, and *in copula* erections differ and where do they converge? Similarly, what are the similarities and differences between pathways that elicit the UG reflex and *in copula* ejaculation? Which neural efferents from the MPOA carry the disinhibitory and the excitatory influences on genital reflexes? What pathways mediate the parasympathetic and sympathetic proerectile effects of MPOA stimulation?

Issues related to neurotransmitters include: Which neurotransmitters, and which receptors, mediate the major effects in each of the sensory, integrative, and motor control sites? Which neuropeptides are co-released with classic transmitters, and under which conditions? Do changes in neurotransmitter release reflect primarily changes in cell firing or influences on presynaptic terminals? Which (combinations of) neurotransmitters elicit long-term changes in sensitivity, as opposed to immediate responses to stimuli?

Questions related to intracellular and genetic influences include the following: Which ion channels and/or second messengers mediate the immediate and long-term effects of neurotransmitters in brain areas that respond to sexually relevant stimuli? What changes in gene transcription are elicited by steroid hormones binding to classic steroid receptors? What are the effects on sexual behavior of rapid membrane responses to steroid hormones? How does sexual experience affect gene transcription? How does activation of immediate-early genes facilitate subsequent sexual behavior? Are new neurons recruited by sexual experience, or are the facilitative effects of experience mediated by strengthening of previous connections? What do patterns of electrophysiological responses contribute to the neural control of copulation?

Questions concerning broader issues include: How do general activational systems, such as the mesocorticolimbic system, mesh with systems that promote specific behaviors? What *does* the MPOA do? What differences in neuroanatomical and physiological processes result in species differences in behavior? How are behavioral differences between species related to ecological niches? How are the neural elements of male sexual behavior similar to and different from those that control other social behaviors?

REFERENCES

1. Ackerman, A. E., Lange, G. M., and Clemens, L. G. (1997). Effects of paraventricular lesions on sex behavior and seminal emissions in male rats. *Physiol. Behav.* 63, 49–53.
2. Adachi, H., Sato, Y., Kato, R., Hisasue, S., Suzuki, K., Masumori, N., Itoh, N., and Tsukamoto, T. (2003). Direct evidence of facilitative actions of dopamine in the medial preoptic area on reflexive and noncontact erections in male rats. *J. Urol.* 169, 386–389.
3. Adler, N. T., and Toner, J. P. (1986). The effects of copulatory behavior on sperm transport and fertility in rats. *Ann. N Y Acad. Sci.* 474, 21–32.
4. Ågmo, A. (2003a). Lack of opioid or dopaminergic effects on unconditioned sexual incentive motivation in male rats. *Behav. Neurosci.* 117, 55–68.
5. Ågmo, A. (2003b). Unconditioned sexual motivation in the Norway rat (*Rattus norvegicus*). *J. Comp. Psychol.* 117, 3–14.
6. Ågmo, A., and Ellingsen, E. (2003). Relevance of non-human animal studies to the understanding of human sexuality. *Scand. J. Psychol.* 44, 293–301.
7. Ågmo, A., and Fernandez, H. (1989). Dopamine and sexual behavior in the male rat: a reevaluation. *J. Neural Transm.* 77, 21–37.
8. Ågmo, A., and Gomez, M. (1993). Sexual reinforcement is blocked by infusion of naloxone into the medial preoptic area. *Behavior. Neurosci.* 107, 812–818.
9. Ågmo, A., and Paredes, R. (1988). Opioids and sexual behavior in the male rat. *Pharmacol. Biochem. Behav.* 30, 1021–1034.
10. Ågmo, A., and Picker, Z. (1990). Catecholamines and the initiation of sexual behavior in male rats without sexual experience. *Pharmacol. Biochem. Behav.* 35, 327–334.
11. Ågmo, A., and Södersten, P. (1975). Sexual behavior in castrated rabbits treated with testosterone, oestradiol, dihydrotestosterone or oestradiol in combination with dihydrotestosterone. *J. Endocrinol.* 67, 327–332.
12. Ågmo, A., and Villalpando, A. (1995). Central nervous stimulants facilitate sexual behavior in male rats with medial prefrontal cortex lesions. *Brain Res.* 696, 187–193.
13. Ågmo, A., Gomez, M., and Irazabal, Y. (1994a). Enkephalinase inhibition facilitates sexual behavior in the male rat but does not produce conditioned place preference. *Pharmacol. Biochem. Behav.* 47, 771–778.
14. Ågmo, A., Paredes, R. G., and Contreras, J. L. (1994b). Opioids and sexual behaviors in the male rabbit: the role of central and peripheral opioid receptors. *J. Neural Transm. Gen. Sec.* 97, 211–223.
15. Ågmo, A., Paredes, R. G., Sierra, L., and Garces, I. (1997). The inhibitory effects on sexual behavior and ambulatory activity of the mixed $GABA_A/GABA_B$ agonist progabide are differentially blocked by GABA receptor antagonists. *Psychopharmacology* 129, 27–34.
16. Ågmo, A., Soulairac, M.-L., and Soulairac, A. (1977). Preoptic lesions, sexual behavior, and spontaneous ejaculation in the rat. *Scand. J. Psychol.* 18, 345–347.
17. Ahlenius, S., and Larsson, K. (1984a). Apomorphine and haloperidol-induced effects on male rat sexual behavior: no evidence for actions due to stimulation of central dopamine autoreceptors. *Pharmacol. Biochem. Behav.* 21, 463–466.
18. Ahlenius, S., and Larsson, K. (1984b). Lisuride, LY-141865, and 8-OH-DPAT facilitate male rat sexual behavior via a non-dopaminergic system. *Psychopharmacology* 83, 330–334.
19. Ahlenius, S., and Larsson, K. (1990). Effects of selective D_1 and D_2 antagonists on male rat sexual behavior. *Experentia* 46, 1026–1028.
20. Ahlenius, S., and Larsson, K. (1991). Opposite effects of 5-methoxy-N, N-di-methyl-tryptamine and 5-hydroxytryptophan on male rat sexual behavior. *Pharmacol. Biochem. Behav.* 38, 201–205.
21. Ahlenius, S., and Larsson, K. (1995). Effects of the dopamine D_3 receptor ligand 7-OH-DPAT on male rat ejaculatory behavior. *Pharmacol. Biochem. Behav.* 51, 545–547.
22. Ahlenius, S., and Larsson, K. (1998). Evidence for an involvement of $5\text{-}HT_{1B}$ receptors in the inhibition of male rat ejaculatory behavior produced by 5-HTP. *Psychopharmacology* 137, 374–382.
23. Ahlenius, S., and Larsson, K. (1999). Synergistic actions of the $5\text{-}HT_{1A}$ receptor antagonist WAY-100635 and citalopram on male rat ejaculatory behavior. *Eur J. Pharm.* 379, 1–6.
24. Ahlenius, S., Eriksson, H., Larsson, K., Modigh, K., and Södersten, P. (1971). Mating behavior in the male rat treated with p-chlorophenylalanine methyl ester alone and in combination with pargyline. *Psychopharmacologia* 20, 383–388.
25. Ahlenius, S., Larsson, K., and Arvidsson, L. E. (1989). Effects of stereoselective $5\text{-}HT_{1A}$ agonists on male rat sexual behavior. *Pharmacol. Biochem. Behav.* 33, 691–695.
26. Ahlenius, S., Larsson, K., Svensson, L., Hjorth, S., Carlsson, A., Lindberg, P., Wikstrom, H., Sanchez, D., Arvidsson, L. E., Hacksell, U., and Nilsson, J. L. (1981). Effects of a new type of 5-HT receptor agonist on male rat sexual behavior. *Pharmacol. Biochem. Behav.* 15, 785–792.
27. Albinsson, A., Andersson, G., Andersson, K., Vega-Matuszczyk, J., and Larsson, K. (1996). The effects of lesions in the mesencephalic raphe systems on male rat sexual behavior and locomotor activity. *Behav. Brain Res.* 80, 57–63.

28. Alderson, L. M., and Baum, M. J. (1981). Differential effects of gonadal steroids on dopamine metabolism in mesolimbic and nigro-striatal pathways of male rat brain. *Brain Res.* 218, 189–206.
29. Alexander, B. M., Perkins, A., Van Kirk, E. A., Moss, G. E., and Fitzgerald, J. A. (1993). Hypothalamic and hypophyseal receptors for estradiol in high and low sexually performing rams. *Horm. Behav.* 27, 296–307.
30. Alexander G. M., Swerdloff, R. S., Wang, C., Davidson, T. McDonald, V., Steiner, B., and Hines, M. (1997). Androgen-behavior correlations in hypogonadal men and eugonadal men: I. Mood and response to auditory sexual stimuli. *Horm. Behav.* 31, 110–119.
31. Alexander, S. P. H., and Peters, J. A., Eds. (2000). *TiPS Receptor and Ion Channel Nomenclature Supplement* 11, Elsevier.
32. Alheid, G. F. (2003). Extended amygdala and basal forebrain. *Ann. N Y Acad. Sci.* 985, 185–205.
33. Alheid, G. F., de Olmos, J. S., and Beltramino, C. A. (1995). Extended amygdala and basal forebrain. In *The Rat Nervous System* (G. E. Paxinos, Ed.), pp. 495–578. Academic Press, San Diego.
34. Allard, J., Bernabe, J., Derdinger, F., Alexantre, L., McKenna, K., and Giuliano, F. (2002). Selegiline enhances erectile activity induced by dopamine injection in the paraventricular nucleus of the hypothalamus in anesthetized rats. *Int. J. Impot. Res.* 14, 518–522.
35. Almeida, O. F. X., Nikolarakis, K. E., and Herz, A. (1988). Evidence for the involvement of endogenous opioids in the inhibition of luteinizing hormone by corticotropin-releasing factor. *Endocrinology* 122, 1034–1041.
36. Alsum, P., and Goy, R. (1974). Actions of esters of testosterone, dihydrotestosterone, or estradiol on sexual behavior in castrated male guinea pigs. *Horm. Behav.* 5, 207–217.
37. Altar, C. A. (1989). Cholecystokinin receptor subtypes and neuromodulation. *Prog. Neuro-Psychopharmacol Biol. Psychiatry* 13, 381–393.
38. Amateau, S. K., and McCarthy, M. M. (2004). Induction of PGE2 by estradiol mediates developmental masculinization of sex behavior. *Nature Neurosci.* 7, 643–650.
38a. Amstislavskaya, T. G., and Popova, N. K. (2004). Female-induced sexual arousal in male mice and rats: behavioral and testosterone response. *Horm. Behav.* 46, 544–550.
39. Andersen, M. L., and Tufik, S., (2002). Distinct effects of paradoxical sleep deprivation and cocaine administration on sexual behavior in male rats. *Addict. Biol.* 7, 2511–253.
40. Andersson, K. E., and Stief, C. G. (1997). Neurotransmission and the contraction and relaxation of penile erectile tissues. *World J. Urol.* 15, 14–20.
41. Andersson, K. E., and Wagner, G. (1995). Physiology of penile erection. *Physiol. Rev.* 75, 191–236.
42. Andersson, K. E., Hedlund, P., and Alm, P. (2000). Sympathetic pathways and adrenergic innervation of the penis. *Int. J. Impot. Res.* 12, S5–S12.
43. Andersson, P. O., Björnberg, J., Bloom S. R., and Mellander, S. (1987). Vasoactive intestinal polypeptide in relation to penile erection in the cat evoked by pelvic and by hypogastric nerve stimulation. *J. Urol.* 138, 419–422.
44. Ansari, J. M. (1975). A study of 65 impotent males. *J. Psychiat.* 127, 337–341.
45. Aoki, H., Matsuzaka, J., Yeh, K. H., Sato, F., Fujioka, T., Kubo, T., Ohhori, T., and Yasuda, N. (1994). Involvement of vasoactive intestinal peptide (VIP) as a humoral mediator of penile erectile function in the dog. *J. Androl.* 15, 174–182.
46. Arendash, G. W., and Gorski, R. A. (1983). Effects of discrete lesions of the sexually dimorphic nucleus of the preoptic area or other medial preoptic regions on the sexual behavior of male rats. *Brain Res. Bull.* 10, 147–154.
47. Argiolas, A. (1999). Neuropeptides and sexual behaviour. *Neurosci. Biobehav. Rev.* 23, 1127–1142.
48. Argiolas, A., and Melis, M. R. (1995). Neuromodulation of penile erection: an overview of the role of neurotransmitters and neuropeptides. *Prog. Neurobiol.* 47, 235–255.
49. Argiolas, A., and Melis, M. R. (2004). The role of oxytocin and the paraventricular nucleus in the sexual behavior of male mammals. *Physiol. Behav.* 83, 309–317.
50. Argiolas, A., Collu, M., Gessa, G. L., Melis, M. R., and Serra, G. (1988a). The oxytocin antagonist d(CH2)5Tyr(Me)-Om8-vasotocin inhibits male copulatory behaviour in rats. *Eur. J. Pharmacol.* 149, 389–392.
51. Argiolas, A., Melis, M. R., and Gessa, G. L. (1986). Oxytocin: an extremely potent inducer of penile erection and yawning in male rats. *Eur. J. Pharmacol.* 130, 265–272.
52. Argiolas, A., Melis, M. R., and Gessa, G. L. (1988b). Yawning and penile erection: central dopamine-oxytocin-adrenocorticotropin connection. *Ann. N Y Acad. Sci.* 525, 330–337.
53. Argiolas, A., Melis, M. R., Mauri, A., and Gessa, G. L. (1987). Paraventricular nucleus lesion prevents yawning and penile erection induced by apomorphine and oxytocin, but not by ACTH in rats. *Brain Res.* 421, 349–352.
54. Argiolas, A., Melis, M. R., Murgia, S., and Schioth, H. B. (2000). ACTH- and alpha-MSH-induced grooming, stretching, yawning and penile erection in male rats: site of action in the brain and role of melanocortin receptors. *Brain Res. Bull.* 51, 425–431.
55. Arletti, R., Bazzani, C., Castelli, M., and Bertolini, A. (1985). Oxytocin improves male copulatory performance in rats. *Horm. Behav.* 19, 14–20.
56. Arletti, R., Calza, L., Giardino, L., Benelli, A., Cavazzuti, E., and Bertolini, A. (1997). Sexual impotence is associated with a reduced production of oxytocin and with an increased production of opioid peptides in the paraventricular nucleus of male rats. *Neurosci. Lett.* 233, 65–68.
57. Aronson, L. R., and Cooper, M. L. (1968). Desensitization of the glans penis and sexual behavior in cats. In *Reproduction and Sexual Behavior* (M. Diamond, Ed.), pp. 51–82. Indiana University Press, Bloomington.
58. Aronson, L. R., and Cooper, M. L. (1974). Olfactory deprivation and mating behavior in sexually experienced male cats. *Behav. Biol.* 11, 459–480.
59. Arteaga, M., Motte-Lara, J., and Velazquez-Moctezuma, J. (2002). Effects of yohimbine and apomorphine on the male sexual behaviour pattern of the golden hamster (Mesocricetus auratus). *Eur. Neuropsychopharmacol.* 12, 39–45.
60. Aste, N., Honda, S., and Harada, N. (2003). Forebrain Fos responses to reproductively related chemosensory cues in aromatase knockout mice. *Brain Res. Bull.* 60, 191–200.
61. Aversa, A., Isidori, A. M., De Martino, M. U., Caprio, M., Fabbrini, E., Rocchietti-March M., Frajese, G., and Fabbri, A. (2000a). Androgens and penile erection: evidence for a direct relationship between free testosterone and cavernous vasodilation in men with erectile dysfunction. *Clin. Endocrinol.* 53, 517–522.
62. Aversa, A., Mazzilli, F., Rossi, T., Delfino, M., Isidori, A. M., and Fabbri, A. (2000b). Effects of sildenafil (Viagra) administration on seminal parameters and post-ejaculatory refractory time in normal males. *Hum. Reprod.* 15, 131–134.
63. Baba, K., Yajima, M., Carrier, S., Akkus, E., Reman, J., Nunes, L., Lue, T. F., and Iwamoto, T. (2000). Effect of testosterone on the number of NADPH diaphorase-stained nerve fibers in the rat corpus cavernosum-dorsal nerve. *Urology* 56, 533–538.
64. Bakker, J., Honda, S., Harada, N., and Balthazart, J. (2002a). Sexual partner preference requires a functional aromatase (cyp19) gene in male mice. *Horm. Behav.* 42, 158–171.

65. Bakker, J., Woodley, S. K., Kelliher, K. R., and Baum, M. J. (2002b). Sexually dimorphic activation of galanin neurones in the ferret's dorsomedial preoptic area/anterior hypothalamus after mating. *J. Neuroendocrinol.* 14, 116–125.
66. Balfour, M. E., Yu, L., and Coolen, L. (2004). Sexual behavior and sex-associated environmental cues activate the mesolimbic system in male rats. *Neuropsychopharmacology* 29, 718–730.
67. Balthazart, J., and Surlemont, C. (1990). Copulatory behavior is controlled by the sexually dimorphic nucleus of the quail POA. *Brain Res. Bull.* 25, 7–14.
68. Balthazart, J., Absil, P., Gerard, M., Appeltants, D., and Ball, G. F. (1998). Appetitive and consummatory male sexual behavior in Japanese quail are differentially regulated by subregions of the preoptic medial nucleus. *J. Neurosci.* 18, 6512–6527.
69. Balthazart, J., Castagna, C., and Ball, G. F. (1997). Differential effects of D_1 and D_2 dopamine-receptor agonists and antagonists on appetitive and consummatory aspects of male sexual behavior in Japanese quail. *Physiol. Behav.* 62, 571–580.
70. Bancila, M., Giuliano, F., Rampin, O., Mailly, P., Brisorgueil, M. J., Calas, A., and Verge, D. (2002). Evidence for a direct projection from the paraventricular nucleus of the hypothalamus to putative serotoninergic neurons of the nucleus paragigantocellularis involved in the control of erection in rats. *Eur. J. Neurosci.* 16, 1240–1248.
71. Bancila, M., Verge, D., Rampin, O., Backstrom, J. R., Sanders-Bush, E., McKenna, K. E., Marson, L., Calas, A., and Giuliano, F. (1999). 5-Hydroxytryptamine 2C receptors on spinal neurons controlling penile erection in the rat. *Neuroscience* 92, 1523–1537.
72. Bancroft, J., and Janssen, E. (2000). The dual control model of male sexual response: a theoretical approach to centrally mediated erectile dysfunction. *Neurosci. Biobehav. Rev.* 24, 571–519.
73. Bancroft, J., and Wu, F. C. (1983). Changes in erectile responsiveness during androgen replacement therapy. *Arch. Sex. Behav.* 12, 59–66.
74. Band, L. C., and Hull, E. M. (1990). Morphine and dynorphin (1-13) microinjected into the medial preoptic area and nucleus accumbens: effects on sexual behavior in male rats. *Brain Res.* 524, 77–84.
75. Barbeau, A. (1969). L-DOPA therapy in Parkinson's disease: a critical review of nine years' experience. *Can. Med Assoc. J.* 101, 791–800.
76. Barfield, R. J. (1969). Activation of copulatory behavior by androgen implanted into the preoptic area of the male fowl. *Horm. Behav.* 1, 37–52.
77. Barfield, R. J. (1971). Activation of sexual and aggressive behavior by androgen implanted into the male ring dove brain. *Endocrinology* 89, 1470–1476.
78. Barfield, R. J., and Geyer, L. A. (1975). The ultrasonic postejaculatory vocalization and the postejaculatory refractory period of the male rat. *J. Comp. Physiol. Psychol.* 88, 723–734.
79. Barfield, R. J., and Sachs, B. D. (1968). Sexual behavior: stimulation by painful electrical shock to skin in male rats. *Science* 161, 392–395.
80. Barry, J., Hoffman, G. E., and Wray, S. (1985). LHRH containing systems. In *Handbook of Chemical Neuroanatomy* (A. Bjorklund and T. Hokfelt, Eds.), pp. 116–215. Elsevier, Amsterdam.
81. Baum, M. J. (1972). Precocious mating in male rats following treatment with androgen or estrogen. *J. Comp. Physiol. Psychol.* 78, 356–367.
82. Baum, M. J., and Everitt, B. J. (1992). Increased expression of *c-fos* in the medial preoptic area after mating in male rats: role of afferent inputs from the medial amygdala and midbrain central tegmental field. *Neuroscience* 50, 627–646.
83. Baum, M. J., and Wersinger, S. R. (1993). Equivalent levels of mating-induced neural *c-fos* immunoreactivity in castrated male rats given androgen, estrogen, or no steroid replacement. *Biol. Reprod.* 48, 1341–1347.
84. Baum, M. J., Kingsbury, P. A., and Erskine, M. S. (1987). Failure of the synthetic androgen 17 beta-hydroxyl-17alpha-methyl-estra-4,9,11-triene-3-one (methyltrienolone, R1881) to duplicate the activational effect of testosterone on mating in castrated male rats. *J. Endocrinol.* 113, 15–20.
85. Baum, M. J., Tobet, S. A., Starr, M. S., and Bradshaw, W. G. (1982). Implantation of dihydrotestosterone propionate into the lateral septum or medial amygdala facilitates copulation in castrated male rats given estradiol systemically. *Horm. Behav.* 16, 208–223.
86. Baumgardner, D. J., and Dewsbury, D. A. (1980). Pseudopregnancy in female rats: effects of hormonal manipulations of the male. *Physiol. Behav.* 24, 693–697.
87. Bazzett, T. J., Eaton, R. C., Thompson, J. T., Markowski, V. P., Lumley, L. A., and Hull, E. M. (1991). Dose dependent D_2 effects on genital reflexes after MPOA injections of quinelorane and apomorphine. *Life Sci.* 48, 2309–2315.
88. Bazzett, T., Lumley, L., Bitran, D., Markowski, V., Warner, R., and Hull, E. (1992). Male rat copulation following 6-OHDA lesions of the medial preoptic area: resistance to repeated administration and rapid behavioral recovery. *Brain Res.* 580, 164–170.
89. Beach, F. A. (1969). Locks and beagles. *Am. Psychol.* 24, 971–989.
90. Beach, F. A. (1984). Hormonal modulation of genital reflexes in male and masculinized female dogs. *Behav. Neurosci.* 98, 325–332.
91. Beach, F. A., and Holz-Tucker, A. M. (1949). Effects of different concentrations of androgen upon sexual behavior in castrated male rats. *J. Comp. Physiol. Psychol.* 42, 433–453.
92. Beach, F. A., and Pauker, R. S. (1949). Effects of castration and subsequent androgen administration upon mating behavior in the male hamster (*Cricetus auratus*). *Endocrinology* 45, 211–221.
93. Bean, N. J., Núñez, A. A., and Conner, R. (1981). Effects of medial preoptic lesions on male mouse ultrasonic vocalizations and copulatory behavior. *Brain Res. Bull.* 6, 109–112.
94. Beauchamp, G. K., Magnus, J. G., Shmunes, N. T., and Durham, T. (1977). Effects of olfactory bulbectomy on social behavior of male guinea pigs (*Cavia porcellus*). *J. Comp. Physiol. Psychol.* 91, 336–346.
95. Beauchamp, G. K., Martin, I. G., Wysocki, C. J., and Wellington, J. L. (1982). Chemoinvestigatory and sexual behavior of male guinea pigs following vomeronasal organ removal. *Physiol. Behav.* 29, 329–336.
96. Beck, J., Bialy, M., and Kostowski, W. (2002). Effects of D(1) receptor agonist SKF 38393 on male rat sexual behavior and postcopulatory departure in the goal compartment-runway paradigm. *Physiol. Behav.* 76, 91–97.
97. Becker, A. J., Uckert, S., Stief, C. G., Scheller, F., Knapp, W. H., Hartmann, U., and Jonas, U. (2001). Cavernous and systemic testosterone plasma levels during different penile conditions in healthy males and patients with erectile dysfunction. *Urology* 58, 435–440.
98. Benelli, A., Arletti, R., Bertolini, A., Menozzi, B., Basaglia, R., and Poggioli, R. (1994). Galantide stimulates sexual behavior in male rats. *Eur. J. Pharmacol.* 260, 279–282.
99. Benelli, A., Bertolini, A., Poggioli, R., Cavazzuti, E., Calza, L., Giardino, L., and Arletti, R. (1995). Nitric oxide is involved in male sexual behavior of rats. *Eur. J. Pharmacol.* 294, 505–510.
100. Benkert, O. (1975). Effects of hypothalamic releasing hormones in depression and sexual impotence.

In *Neuropsychopharmacology* (J. R. Boissier, H. Hippius, and P. Pichot, Eds.), pp. 663–671. Excerpta Medica, Amsterdam.

101. Berendsen, H. H., and Broekkamp, C. L. (1987). Drug-induced penile erections in rats, indications of serotonin$_{1B}$ receptor mediation. *Eur. J. Pharmacol.* 135, 279–287.
102. Berendsen, H. H., Jenck, F., and Broekkamp, C. L. (1990). Involvement of 5-HT$_{1C}$-receptors in drug-induced penile erections in rats. *Psychopharmacology* 101, 57–61.
103. Bergquist, E. H. (1970). Output pathways of hypothalamic mechanisms for sexual, aggressive, and other motivated behaviors in opossum. *J. Comp. Physiol. Psychol.* 70, 389–398.
104. Bermant, G., and Taylor, L. (1969). Interactive effects of experience and olfactory bulb lesions in male rat copulation. *Physiol. Behav.* 4, 13–17.
105. Bermant, G., Clegg, M. T., and Beamer, W. (1969). Copulatory behaviour of the ram, *Ovis aries*: I. A normative study. *Anim. Behav.* 17, 700–705.
106. Bermond, B. (1982). Effects of medial preoptic hypothalamus anterior lesions on three kinds of behavior in the rat: intermale aggressive, male sexual, and mouse killing behavior. *Aggres. Behav.* 8, 335–354.
107. Bernabé, J., Rampin, O., Sachs, B. D., and Giuliano, F. (1999). Intracavernous pressure during erection in rats: an integrative approach based on telemetric recording. *Am. J. Physiol.* 276, R441–R449.
108. Beyer, C., Larsson, K., Perez-Palacios, G., and Morali, G. (1973). Androgen structure and male sexual behavior in the castrated rat. *Horm. Behav.* 4, 99–108.
109. Beyer, C., Morali G., Naftolin, F., Larsson, K., and Perez-Palacios, G. (1976). Effect of some antiestrogens and aromatase inhibitors on androgen induced sexual behavior in castrated male rats. *Horm. Behav.* 7, 353–363.
110. Bialy, M., and Beck, J. (1993). The influence of vibrissae removal on copulatory behavior in male rats. *Acta Neurobioligiae Experimentalis* 53, 415–419.
111. Bialy, M., and Sachs, B. D. (2002). Androgen implants in medial amygdala briefly maintain noncontact erection in castrated male rats. *Horm. Behav.* 42(3), 345–355.
112. Bialy, M., Beck, J., Abramczyk, P., Trzebski, A., and Przybylski, J. (1996). Sexual behavior in male rats after nitric oxide synthesis inhibition. *Physiol. Behav.* 60, 139–143.
113. Bialy, M., Rydz, M., and Kacamarek, L. (2000). Precontact 50-kHz vocalizations in male rats during acquisition of sexual experience. *Behav. Neurosci.* 114, 983–990.
114. Bitran, D., and Hull, E. M. (1987). Pharmacological analysis of male rat sexual behavior. *Neurosci. Biobehav. Rev.* 11, 365–389.
115. Bitran, D., Hull, E. M., Holmes, G. M., and Lookingland, K. J. (1988a). Regulation of male rat copulatory behavior by preoptic incertohypothalamic dopamine neurons. *Brain Res. Bull.* 20, 323–331.
116. Bitran, D., Miller, S. A., McQuade, D. B., Leipheimer, R. E., and Sachs, B. D. (1988b). Inhibition of sexual reflexes by lumbosacral injection of a GABA$_B$ agonist in the male rat. *Pharmacol. Biochem. Behav.* 31, 657–666.
117. Bitran, D., Thompson, J. T., Hull, E. M., and Sachs, B. D. (1989). Quinelorane (LY163502), a D_2 dopamine receptor agonist, facilitates seminal emission but inhibits penile erection in the rat. *Pharmacol. Biochem. Behav.* 34, 453–458.
118. Bivalacqua, T. J., Champion, H. C., Hellstrom, W. J. G., and Kadowitz, P. J. (2000). Pharmacotherapy for erectile dysfunction. *Trends Pharmacol. Sci.* 21, 484–489.
119. Bivalacqua, T. J., Rajasekaran, M., Champion, H. C., Wang, R., Sikka, S. C., Kadowitz, P. J., and Hellstrom, W. J. (1998). The influence of castration on pharmacologically induced penile erection in the cat. *J. Androl.* 19, 551–557.
120. Blackburn, J. R., Pfaus, J. G., and Phillips A. G. (1992). Dopamine functions in appetitive and defensive behaviours. *Prog. Neurobiol.* 39, 247–279.
121. Bloch, G. J., Babcock, A. M., Gorski, R. A., and Micevych, P. E. (1988). Effects of cholecystokinin on male copulatory behavior and lordosis behavior in male rats. *Physiol. Behav.* 43, 351–357.
122. Bloch, G. J., Butler, P. C., Kohlert, J. G., and Bloch, D. A. (1993). Microinjection of galanin into the medial preoptic nucleus facilitates copulatory behavior in the male rat. *Physiol. Behav.* 54, 615–624.
123. Blumberg, M. S. (1991). Prostaglandin E accelerates sexual behavior in male rats. *Physiol. Behav.* 50, 95–99.
124. Blumberg, M. S., and Moltz, H. (1987). Hypothalamic temperature and the 22 kHz vocalization of the male rat. *Physiol. Behav.* 40, 637–640.
125. Blumberg, M. S., and Moltz, H. (1988). How the nose cools the brain during copulation in the male rat. *Physiol. Behav.* 43, 173–176.
126. Bohlen, D., Hugonnet, C. L., Mills, R. D., Sonntag R. W., and Studer, U. E. (2000). Five meters of H(2)O: the pressure at the urinary bladder neck during human ejaculation. *Prostate* 44, 339–341.
127. Bonsall, R. W., Clancy, A. N., and Michael, R. P. (1992). Effects of the nonsteroidal aromatase inhibitor, Fadrozole, on sexual behavior in male rats. *Horm. Behav.* 26, 240–254.
128. Booth, A. M. Jr. (1976). Nervous control of penile erection and copulatory behavior in the rat. Unpublished doctoral dissertation, University of Pittsburgh.
129. Boscarino, B. T., and Parfitt, D. B. (2002). Chronic oral administration of clomipramine decreases sexual behavior in the male Syrian hamster (*Mesocricetus auratus*). *Physiol Behav.* 75, 361–366.
130. Bowers, M. B. Jr., Van Woert, M., and Davis, L. (1971). Sexual behavior during L-dopa treatment for Parkinsonism. *Am. J. Psychiatry* 127, 1691–1693.
131. Boyd, S. K., and Moore, F. L. (1985). Luteinizing hormone-releasing hormone facilitates the display of sexual behavior in male voles (*Microtus canicaudus*). *Horm. Behav.* 19, 252–264.
132. Brackett, N. L., and Edwards, D. A. (1984). Medial preoptic connections with the midbrain tegmentum are essential for male sexual behavior. *Physiol. Behav.* 32, 79–84.
133. Brackett, N. L., Iuvone, P. M., and Edwards, D. A. (1986). Midbrain lesions, dopamine and male sexual behavior. *Behav. Brain Res.* 20, 231–240.
134. Bradley, W. E. (1980). Cerebrocortical innervation of the urinary bladder. *Tohoku J. Exp. Med.* 31, 7–13.
135. Bradshaw, W. G., Baum, M. J., and Awh, C. C. (1981). Attenuation by a 5a-reductase inhibitor of the activational effect of testosterone propionate on penile erections in castrated male rats. *Endocrinology* 109, 1047–1051.
136. Breedlove, S. M. (1984). Steroid influences on the development and function of a neuromotor system. *Prog. Brain Res.* 61, 147–170.
137. Breedlove, S. M. (1985). Hormonal control of the anatomical specificity of motoneuron-muscle innervation in rats. *Science* 227, 1357–1359.
138. Breigeiron, M. K., Morris, M., Lucion, A. B., and Sanvito, G. L. (2002). Effects of angiotensin II microinjected into medial amygdala on male sexual behavior in rats. *Horm. Behav.* 41, 267–274.
139. Bressler, S. C., and Baum, M. J. (1996). Sex comparison of neuronal Fos immunoreactivity in the rat vomeronasal projection circuit after chemosensory stimulation. *Neuroscience* 71, 1063–1072.

140. Brindley, G. S. (1986). Pilot experiments on the actions of drugs injected into the human corpus cavernosum penis. *Br. J. Pharmacol.* 87, 495–500.
141. Brodie, T. D., and Crowley, W. F. (1984). Neuroendocrine control of reproduction and its manipulation with LHRH and its analogs. *Trends Neurosci.* 7, 340–342.
142. Bronson, F. H., and Desjardins, C. (1982). Endocrine responses to sexual arousal in male mice. *Endocrinology* 111, 1286–1291.
143. Brown, W. A., Monti, P. M., and Corriveau, D. P. (1978). Serum testosterone and sexual activity and interest in men. *Arch. Sex. Behav.* 7, 97–103.
144. Burnett, A. L., Nelson, R. J., Calvin, D. C., Liu, J. X., Demas, G. E., Klein, S. L., Kriegsfeld, L. J., Dawson, T. M., and Snyder, S. H. (1996). Nitric oxide-dependent penile erection in mice lacking neuronal nitric oxide synthase. *Mol. Med.* 2, 288–296.
145. Butera, P. C., and Czaja, J. A. (1989a). Activation of sexual behavior in male rats by combined subcutaneous and intracranial treatments of 5-dihydrotestosterone. *Horm. Behav.* 2, 92–105.
146. Butera, P. C., and Czaja, J. A. (1989b). Effects of intracranial implants of dihydrotestosterone on the reproductive physiology and behavior of male guinea pigs. *Horm. Behav.* 23, 424–431.
147. Caggiula, A. R., and Hoebel, B. (1966). "Copulation-reward site" in the posterior hypothalamus. *Science* 153, 1284–1285.
148. Caggiula, A. R., and Szechtman, H. (1972). Hypothalamic stimulation: a biphasic influence on copulation of the male rat. *Behav. Biol.* 7, 591–598.
149. Caggiula, A. R., Antelman, S. M., and Zigmond, M. J. (1973). Disruption of copulation in male rats after hypothalamic lesions: a behavioral, anatomical and neurochemical analysis. *Brain Res.* 59, 273–287.
150. Caggiula, A. R., Antelman, S. M., and Zigmond, M. J. (1974). Ineffectiveness of sexually arousing stimulation after hypothalamic lesions in the rat. *Physiol. Behav.* 12, 313–316.
151. Caggiula, A. R., Gay, V. L., Antelman, S. M., and Leggens, J. (1975). Disruption of copulation in male rats after hypothalamic lesions: a neuroendocrine analysis. *Neuroendocrinology* 17, 193–202.
152. Caggiula, A. R., Shaw, D. H., Antelman, S. M., and Edwards, D. J. (1976). Interactive effects of brain catecholamines and variations in sexual and non-sexual arousal on copulatory behavior of male rats. *Brain Res.* 111, 321–336.
153. Cain, D. P. (1974). The role of the olfactory bulb in limbic mechanisms. *Psychol. Bull.* 81, 654–671.
154. Cain, D. P., and Paxinos, G. (1974). Olfactory bulbectomy and mucosa damage: effects on copulation, irritability, and interspecific aggression in male rats. *J. Comp. Physiol. Psychol.* 86, 202–212.
155. Calaresu, F. R., and Mitchell, R. (1969). Cutaneous mechanoreceptors in the glans penis of the rat. *Brain Res.* 15, 295–297.
155a. Camacho, F., Sandoval, C, and Paredes, R. G. (2004). Sexual experience and conditioned place preference in male rats. *Pharmacol. Biochem. Behav.* 78, 419–425.
156. Canteras, N. S., Simerly, R. B., and Swanson, L. W. (1992). Connections of the posterior nucleus of the amygdala. *J. Comp. Neurol.* 324, 143–179.
157. Canteras, N. S., Simerly, R. B., and Swanson, L. W. (1995). Organization of projections from the medial nucleus of the amygdala: a PHAL study in the rat [published erratum appears in *J. Comp. Neurol.* (1996). 369, 328–330]. *J. Comp. Neurol.* 360, 213–245.
158. Cantor, J. M., Binik, Y. M., and Pfaus, J. G. (1999). Chronic fluoxetine inhibits sexual behavior in the male rat: reversal with oxytocin. *Psychopharmacology* 144, 355–362.
159. Carati, C. J., Creed, K. E., and Keogh, E. J. (1987). Autonomic control of penile erection in the dog. *J. Physiol.* 384, 525–538.
160. Carroll, R. S., Erskine, M. S., Doherty, P. C., Lundell, L. A., and Baum, M. J. (1985). Coital stimuli controlling luteinizing hormone secretion and ovulation in the female ferret. *Biol. Reprod.* 32, 925–933.
161. Castagna, C., Ball, G. F., and Balthazart, J. (1997). Effects of dopamine agonists on appetitive and consummatory male sexual behavior in Japanese quail. *Pharmacol. Biochem. Behav.* 58, 403–414.
162. Cavallini, G., Caracciolo, S., Vitali, G., Modenini, F., and Biagiotti, G. (2004). Carnitine versus androgen administration in the treatment of sexual dysfunction, depressed mood, and fatigue associated with male aging. *Urology* 63, 641–646.
163. Ceccatelli, S., Grandison, L., Scott, R. E., Pfaff, D. W., and Kow, L. M. (1996). Estradiol regulation of nitric oxide synthase mRNAs in rat hypothalamus. *Neuroendocrinology* 64, 357–363.
164. Chambers, K. C., and Phoenix, C. H. (1981). Diurnal patterns of testosterone, dihydrotestosterone, estradiol, and cortisol in serum of rhesus males: relationship to sexual behavior in aging males. *Horm. Behav.* 15, 416–426.
165. Chambers, K. C., Hess, D. L., and Phoenix, C. H. (1981). Relationship of free and bound testosterone to sexual behavior in old rhesus males. *Physiol. Behav.* 27, 615–620.
166. Chambers, K. C., Thornton, J. E., and Roselli, C. E. (1991). Age-related deficits in brain androgen binding and metabolism, testosterone, and sexual behavior of male rats. *Neurobiol. Aging* 12, 123–130.
167. Chamness, S. L., Ricker, D. D., Crone, J. K., Dembeck, C. L., Maguire, M. P., Burnett, A. L., and Chang, T. S. (1995). The effect of androgen on nitric oxide synthase in the male reproductive tract of the rat. *Fert. Ster.* 63, 1101–1107.
168. Chapelle, P. A., Durand, J., and Lacert, P. (1980). Penile erection following complete spinal cord injury in man. *Br. J. Urol.* 52, 216–219.
169. Cheal, M., and Domesick, V. B. (1979). Mating in male Mongolian gerbils after olfactory bulbectomy. *Physiol. Behav.* 22, 199–202.
170. Chen, J. J., and Bliss, D. K. (1974). Effects of sequential preoptic and mammillary lesions on male rat sexual behavior. *J. Comp. Physiol. Psychol.* 87, 841–847.
171. Chen, K. K., Chan, J. Y. H. Chang, L. S., Chen, M. T., and Chan, S. H. H. (1992). Elicitation of penile erection following activation of the hippocampal formation in the rat. *Neurosci. Lett.* 141, 218–222.
172. Cherry, J. A., and Baum, M. J. (1990). Effects of lesions of a sexually dimorphic nucleus in the preoptic/anterior hypothalamic area on the expression of androgen- and estrogen-dependent sexual behaviors in male ferrets. *Brain Res.* 522, 191–203.
173. Chevalier-Skolnikoff, S. (1975). Heterosexual copulatory patterns in stumptail macaques (Macaca arctoides) and in other macaque species. *Arch. Sex. Behav.* 4, 199–220.
174. Chitaley, K, Wingard, C. J., Webb, R. C., Branam, H., Stopper, V. S., Lewis, R. W., and Mills, T. M. (2001). Antagonism of Rho-kinase stimulates rat penile erection via a nitric oxide-independent pathway. *Nature Med.* 7, 119–122.
175. Christ, G. J. (1995). The penis as a vascular organ: the importance of corporal smooth muscle tone in the control of erection. *Uro. Clin. North Am.* 22, 727–745.
176. Christ, G. J., Wang, H. Z., Venkateswarlu, K., Zhao, W., and Day, N. S. (1999). Ion channels and gap junctions: their role in erectile physiology, dysfunction, and future therapy. *Mol. Urol.* 3, 61–73.
177. Christensen, L. W., and Clemens, L. G. (1974). Intrahypothalamic implants of testosterone or estradiol and

resumption of masculine sexual behavior in long-term castrated male rats. *Endocrinology* 95, 984–990.

178. Christensen, L. W., and Clemens, L. G. (1975). Blockade of testosterone-induced mounting behavior in the male rat with intracranial application of the aromatization inhibitor, androst-triene-dione. *Endocrinology* 97, 1545–1551.
179. Christensen, L. W., Nance, D. M., and Gorski, R. A. (1977). Effects of hypothalamic and preoptic lesions on reproductive behavior in male rats. *Brain Res. Bull.* 2, 137–141.
180. Christiansen, E., Guirguis, W. R., Cox, D., and Osterloh, I. H. (2000). Long-term efficacy and safety of oral Viagra (sildenafil citrate) in men with erectile dysfunction and the effect of randomized treatment withdrawal. *Int. J. Impot. Res.* 12, 177–182.
181. Chronwall, B. M., DiMaggio, D. A., Massari, V. J., Pickel, V. M., and O'Donohue, T. L. (1985). The anatomy of neuropeptide Y-containing neurons in rat brain. *Neuroscience* 15, 1159–1181.
182. Clancy, A. N., Coquelin, A., Macrides, F., Gorski, R. A., and Noble, E. P. (1984). Sexual behavior and aggression in male mice: involvement of the vomeronasal system. *J. Neurosci.* 4, 2222–2229.
183. Clancy, A. N., Zumpe, D., and Michael, R. P. (1995). Intracerebral infusion of an aromatase inhibitor, sexual behavior and brain estrogen receptor-like immunoreactivity in intact male rats. *Neuroendocrinology* 61, 98–111.
184. Clancy, A. N., Zumpe, D., and Michael, R. P. (2000). Estrogen in the medial preoptic area of male rats facilitates copulatory behavior. *Horm. Behav.* 38, 86–93.
185. Clark, J. T. (1989). A possible role for angiotensin II in the regulation of male sexual behavior in rats. *Physiol. Behav.* 45, 221–246.
186. Clark, J. T. (1991). Suppression of copulatory behavior in male rats following central administration of clonidine. *Neuropharmacology* 30, 373–382.
187. Clark, J. T. (1995). Sexual arousal and performance are modulated by adrenergic-neuropeptide-steroid interactions. In *The Pharmacology of Sexual Function and Dysfunction* (J. Bancroft, Ed.), pp. 55–68. Proceedings of the Esteve Foundation Symposium VI, Son Vida, Mallorca, October 9–12, 1994. Excerpta Medica, Amsterdam.
188. Clark, J. T., and Smith, E. R. (1990). Clonidine suppresses copulatory behavior and erectile reflexes in male rats: lack of effect of naloxone pretreatment. *Neuroendocrinology* 51, 357–364.
189. Clark, J. T., Kalra, P. S., and Kalra, S. P. (1985a). Neuropeptide Y stimulates feeding but inhibits sexual behavior in rats. *Endocrinology* 117, 2435–2442.
190. Clark, J. T., Kalra, S. P., and Kalra, P. S. (1987). Effects of a selective alpha-1-adrenoceptor agonist, methoxamine, on sexual behavior and penile reflexes. *Physiol. Behav.* 40, 747–753.
191. Clark, J. T., Smith, E. R., and Davidson, J. M. (1984). Enhancement of sexual motivation in male rats by yohimbine. *Science* 225, 847–849.
192. Clark, J. T., Smith, E. R., and Davidson, J. M. (1985b). Evidence for the modulation of sexual behavior by α-adrenoceptors in male rats. *Neuroendocrinology* 41, 36–43.
193. Clark, T. K. (1975). Sexual inhibition is reduced by rostral midbrain lesions in the male rat. *Science* 190, 169–171.
194. Clark, T. K. (1980). Male rat sexual behavior compared after 6-OHDA and electrolytic lesions in the dorsal NA bundle region of the midbrain. *Brain Res.* 202, 429–443.
195. Clarke, C. H., Norfleet, A. M., Clarke, M. S., Watson, C. S., Cunningham, K. A., and Thomas, M. L. (2000). Perimembrane localization of the estrogen receptor alpha protein in neuronal processes of cultured hippocampal neurons. *Neuroendocrinology* 71, 34–42.
196. Claro, F., Segovia, S., Guillamon, A., and Del Abril, A. (1995). Lesions in the medial posterior region of the BST impair sexual behavior in sexually experienced and inexperienced male rats. *Brain Res. Bull.* 36, 1–10.
197. Clemens, L. G., and Gladue, B. A. (1977). Effect of prostaglandin E2 on masculine sexual behaviour in the rat. *J. Endocrinol.* 75, 383–389.
198. Clemens, L. G., and Pomerantz, S. M. (1982). Testosterone acts as a prohormone to stimulate male copulatory behavior in male deer mice (*peromyscus maniculatus bairdi*). *J. Comp. Physiol. Psychol.* 96, 114–122.
199. Cochran, C. A., and Perachio, A. A. (1977). Dihydrotestosterone propionate effects on dominance and sexual behaviors in gonadectomized male and female rhesus monkeys. *Horm. Behav.* 8, 175–187.
200. Commins, D., and Yahr, P. (1984). Lesions of the sexually dimorphic area disrupt mating and marking in male gerbils. *Brain Res. Bull.* 13, 185–193.
201. Cooke, B. M., Breedlove, S. M., and Jordan, C. (2003). Both estrogen receptors and androgen receptors contribute to testosterone-induced changes in the morphology of the medial amygdala and sexual arousal in male rats. *Horm. Behav.* 43, 335–346.
202. Cooke, B. M., Hegstrom, C. D., Keen, A., and Breedlove, S. M. (2001). Photoperiod and social cues influence the medial amygdala but not the bed nucleus of the stria terminalis in the Siberian hamster. *Neurosci. Lett.* 312, 9–12.
203. Cooke, B. M., Tabibnia, G., and Breedlove, S. M. (1999). A brain sexual dimorphism controlled by adult circulating androgens. *Proc. Nat. Acad. Sci. U S A* 96, 7538–7540.
204. Coolen, L. M., and Wood, R. I. (1998). Bidirectional collections of the medial amygdaloid nucleus in the Syrian hamster brain: simultaneous anterograde and retrograde tract tracing. *J. Comp. Neurol.* 399, 189–209.
205. Coolen, L. M., and Wood, R. I. (1999). Testosterone stimulation of the medial preoptic area and medial amygdala in the control of male hamster sexual behavior: redundancy without amplification. *Behav. Brain Res.* 98, 143–153.
206. Coolen, L. M., Allard, J., Truitt, W. A., and McKenna, K. E. (2004a). Central regulation of ejaculation. *Physiol. Behav.* 83, 203–215.
207. Coolen, L. M., Fitzgerald, M. E., Yu, L., and Lehman, M. N. (2004b). Activation of mu opioid receptors in the medial preoptic area following copulation in male rats. *Neuroscience* 124, 11–21.
208. Coolen, L. M., Olivier, B., Peters, H. J., and Veening, J. G. (1997a). Demonstration of ejaculation-induced neural activity in the male rat brain using 5-HT_{1A} agonist 8-OH-DPAT. *Physiol. Behav.* 62, 881–891.
209. Coolen, L. M., Peters, H. J., and Veening, J. G. (1996). Fos immunoreactivity in the rat brain following consummatory elements of sexual behavior: a sex comparison. *Brain Res.* 738, 67–82.
210. Coolen, L. M., Peters, H. J., and Veening, J. G. (1997b). Distribution of Fos immunoreactivity following mating versus anogenital investigation in the male rat brain. *Neuroscience* 77, 1151–1161.
211. Coolen, L. M., Peters, H. J., and Veening, J. G. (1998). Anatomical interrelationships of the medial preoptic area and other brain regions activated following male sexual behavior: a combined *c-fos* and tract-tracing study. *J. Comp. Neurol.* 397, 421–435.
212. Coolen, L. M., Veening, J. G., Murphy, A. Z., and Shipley, M. T. (1997c). Projections of subparafascicular nucleus: evidence for a discrete spino-thalamo-forebrain circuit activated by mating. *Soc. Neurosci. Abst.* 27, no. 532.3.
213. Coolen, L. M., Veening, J. G., Petersen, D. W., and Shipley, M. T. (2003a). Parvocellular subparafascicular thalamic

nucleus in the rat: anatomical and functional compartmentalization. *J. Comp. Neurol.* 463(2), 117–131.
214. Coolen, L. M., Veening, J. G., Wells, A. B., and Shipley, M. T. (2003b). Afferent connections of the parvocellular subparafascicular thalamic nucleus in the rat: evidence for functional subdivisions. *J. Comp. Neurol.* 463(2), 132–156.
215. Cooper, T. T., Clancy, A. N., Karom, M., Moore, T. O., and Albers, H. E. (2000). Conversion of testosterone to estradiol may not be necessary for the expression of mating behavior in male Syrian hamsters (*Mesocricetus auratus*). *Horm. Behav.* 37, 237–245.
216. Cornil, C. A., Balthazart, J., Motte, P., Massotte, L., and Seutin, V. (2002). Dopamine activates noradrenergic receptors in the preoptic area. *J. Neurosci.* 22, 9320–9330.
217. Cornil, C. A., Seutin, V., and Balthazart, J. (2004). Interaction of dopamine with noradrenergic receptors in the preoptic area and sexual behavior in male quail. *Horm. Behav.* 46, 128–129.
218. Corrada, Y., Arias, D., Rodriguez, R., Spaini, E., Fava, F., and Gobello, C. (2004). Effect of tamoxifen citrate on reproductive parameters of male dogs. *Theriogenology* 61, 1327–1341.
219. Cottingham, S. L., and Pfaff, D. W. (1986). Interconnectedness of steroid hormone-binding neurons: existence and implications. *Curr. Top. Neuroendocrinol.* 7, 223–250.
220. Courtois, F. J., and MacDougall, J. C. (1988). Higher CNS control of penile responses in rats: the effect of hypothalamic stimulation. *Physiol. Behav.* 44, 165–171.
221. Couse, J. F., Lindzey, J., Grandien, K., Gustafsson, J. A., and Korach, K. S. (1997). Tissue distribution and quantitative analysis of estrogen receptor-alpha (ERalpha) and estrogen receptor-beta (ERbeta) messenger ribonucleic acid in the wild-type and ERalpha-knockout mouse. *Endocrinology* 138, 4613–4621.
222. Cross, E., and Roselli, C. E. (1999). 17beta-estradiol rapidly facilitates chemoinvestigation and mounting in castrated male rats. *Am. J. Physiol.* 276(5 Pt 2), R1346–R1350.
223. Crowley, W. F. Jr., Filicori, M., Spratt, D. I., and Santoro, N. F. (1985). The physiology of gonadotropin-releasing hormone (GnRH) secretion in men and women. *Recent Prog. Horm. Res.* 41, 473–531.
224. Cruz, M. R., Liu, Y. C., Manzo, J., Pacheco, P., and Sachs, B. D. (1999). Peripheral nerves mediating penile erection in the rat. *J. Auton. Nerv. Sys.* 76, 15–27.
225. Dail, W. G., and Minorsky, N. (1986). Composition of the pelvic nerve. *Exp. Neurol.* 92, 278–283.
226. Dail, W. G., Manzanares, K., Moll, M. A., and Minorsky, N. (1985). The hypogastric nerve innervates a population of penile neurons in the pelvic plexus. *Neuroscience* 16, 1041–1046.
227. Dail, W. G., Minorsky, N., Moll, M. A., and Manzanares, K. (1986). The hypogastric nerve pathway to penile erectile tissue: histochemical evidence supporting a vasodilator role. *J. Auton. Nerv. Sys.* 15, 341–349.
228. Dail, W. G., Walton, G., and Olmstead, M. P. (1989). Penile erection in the rat: stimulation of the hypogastric nerve elicits increases in penile pressure after chronic interruption of the sacral parasympathetic outflow. *J. Auton. Nerv. Sys.* 28, 251–258.
229. Dallo, J. (1977). Effect of two brain serotonin depletors on the sexual behavior of male rats. *Pol. J. Pharmacol. Pharm.* 29, 247–251.
230. Damsma, G. Pfaus, J. G., Wenkstern, D., Philips, A. G., and Fibiger, H. C. (1992). Sexual behavior increases dopamine transmission in the nucleus accumbens and striatum of male rats: a comparison with novelty and locomotion. *Behav. Neurosci.* 106, 181–191.
231. Dauge, V., Steimes, P., Derrien, M., Beau, N., Roques, B. P., and Feger, J. (1989). CCK8 effects on motivational and emotional states of rats involve CCKA receptors of the postero-median part of the nucleus accumbens. *Pharmacol. Biochem. Behav.* 34, 157–163.
232. Davidson, J. M. (1966a). Activation of the male rat's sexual behavior by intracerebral implantation of androgen. *Endocrinology* 79, 783–794.
233. Davidson, J. M. (1966b). Characteristics of sex behaviour in male rats following castration. *Anim. Behav.* 14, 266–272.
234. Davidson, J. M. (1969). Effects of estrogen on the sexual behavior of male rats. *Endocrinology* 84, 1365–1372.
235. Davidson, J. M., Camargo, C. A., and Smith, E. R. (1979). Effects of androgen on sexual behavior in hypogonadal men. *J. Clin. Endocrinol. Metab.* 48, 955–958.
236. Davidson, J. M., Stefanick, M. L., Sachs, B. D., and Smith, E. R. (1978). Role of androgen in sexual reflexes of the male rat. *Physiol. Behav.* 21, 141–146.
237. Davis, P. G., and Barfield, R. J. (1979). Activation of masculine sexual behavior by intracranial estradiol benzoate implants in male rats. *Neuroendocrinology* 28, 217–227.
237a. DeGroat, W. C., and Steers, W. D. (1988). Neuroanatomy and neurophysiology of penile erection. In *Contemporary Management of Impotence and Infertility* (E. A. Tanagho, T. F. Lue, and R. D. McClure, Eds.), pp. 3–27. Williams & Wilkins, Baltimore.
238. de Jonge, F. H., Louwerse, A. L., Ooms, M. P., Evers, P., Endert, E., and van de Poll, N. E. (1989). Lesions of the SDN-POA inhibit sexual behavior of male Wistar rats. *Brain Res. Bull.* 23, 483–492.
239. de Jonge, F. H., Oldenburger, W. P., Louwerse, A. L., and Van de Poll, N. E. (1992). Changes in male copulatory behavior after sexual exciting stimuli: effects of medial amygdala lesions. *Physiol. Behav.* 52, 327–332.
240. Devor, M. (1973). Components of mating dissociated by lateral olfactory tract transection in male hamsters. *Brain Res.* 64, 437–441.
241. Devor, M., and Murphy, M. R. (1973). The effect of peripheral olfactory blockade on the social behavior of the male golden hamster. *Behav. Biol.* 9, 31–42.
242. Dewsbury, D. A. (1969). Copulatory behavior of rats (*Rattus norvegicus*) as a function of prior copulatory experience. *Animal Behav.* 17, 217–223.
243. Dewsbury, D. A. (1972). Patterns of copulatory behavior in male mammals. *Q. Rev. Biol.* 47, 1–33.
244. Dewsbury, D. A. (1990). Modes of estrus induction as a factor in studies of the reproductive behavior of rodents. *Neurosci. Biobehav. Rev.* 14, 147–155.
245. Di Chiara, G. (1995). The role of dopamine in drug abuse viewed from the perspective of its role in motivation. *Drug Alcoh. Dep.* 38, 95.
246. Dieckmann K. P., Huland H., and Gross A. J. (1992). A test for the identification of relevant sympathetic nerve fibers during nerve sparing retroperitoneal lymphadenectomy. *J. Urol.* 148, 1450–1452.
247. Diederichs, W., Stief, C. G., Benard, F., Bosch, R., Lue, T. F., and Tanagho, E. A. (1991). The sympathetic role as an antagonist of erection. *Urol. Res.* 19, 123–126.
248. Dixson, A. F. (1991). Sexual selection, natural selection and copulatory patterns in male primates. *Folia Primatol.* 57, 96–101.
249. Domesick, V. B. (1988). Neuroanatomical organization of dopamine neurons in the ventral tegmental area. In *The Mesocorticolimbic Dopamine System* (P. W. Kalivas and C. B. Nemeroff, Eds.). *Ann. N Y Acad. Sci.* 537, 10–26.
249a. Dominguez, J. M., and Coolen, L. M. (2004). Activation of NMDA receptors in the medial preoptic nucleus is essential for the expression of Fos and male sexual behavior. *Soc. Behav. Neuroendocrinol. Abst.*, in *Horm. Behav.* 46, 129.

250. Dominguez, J., Riolo, J. V., Xu, Z., and Hull, E. M. (2001). Regulation by the medial amygdala of copulation and medial preoptic dopamine release. *J. Neurosci.* 21, 349–355.
251. Dominguez, J. M., and Hull, E. M. (2001). Stimulation of the medial amygdala enhances medial preoptic dopamine release: implications for male rat sexual behavior. *Brain Res.* 917, 225–229.
252. Dominguez-Salazar, E., Portillo, W., Velazquez-Moctezuma, J., and Paredes, R. G. (2003). Facilitation of male-like coital behavior in female rats by kindling. *Behav. Brain Res.* 140, 57–64.
253. Dornan, W. A., and Malsbury, C. W. (1989). Peptidergic control of male rat sexual behavior: the effects of intracerebral injections of substance P and cholecystokinin. *Physiol. Behav.* 46, 547–556.
254. Dorr, R. T., Lines, R., Livine, N., Brooks, C., Xiang, L., Hruby, V. J., and Hadley, M. E. (1996). Evaluation of melanotan-II, a superpotent cyclic melanotropic peptide in a pilot phase-I clinical study. *Life Sci.* 58, 1777–1784.
255. Dorsa, D. M., and Smith, E. R. (1980). Facilitation of mounting behavior in male rats by intracranial injections of luteinizing hormone-releasing hormone. *Peptides* 1, 147–155.
256. Dorsa, D. M., Smith, E. R., and Davidson, J. M. (1981). Endocrine and behavioral effects of continuous exposure of male rats to a potent luteinizing hormone-releasing hormone (LHRH) agonist: evidence for central nervous system actions of LHRH. *Endocrinology* 109, 729–735.
257. Dragunow, M., and Faull, R. (1989). The use of *c-fos* as a metabolic marker in neuronal pathway tracing. *J. Neurosci. Meth.* 29, 261–265.
258. Du, J., and Hull, E. M. (1999). Effects of testosterone on neuronal nitric oxide synthase and tyrosine hydroxylase. *Brain Res.* 836, 90–98.
259. Du, J., Lorrain, D. S., and Hull, E. M. (1998). Castration decreases extracellular, but increases intracellular, dopamine in medial preoptic area of male rats. *Brain Res.* 782, 11–17.
260. Dudley, A., Rajendren, G., and Moss R. L. (1996). Signal processing in the vomeronasal system: modulation of sexual behavior in the female rat. *Crit. Rev. Neurobiol.* 10, 265–290.
261. Dudley, C. A., and Moss, R. L. (1999). Activation of an anatomically distinct subpopulation of accessory olfactory bulb neurons by chemosensory stimulation. *Neuroscience* 91, 1549–1556.
262. Dula, E., Keating, W., Siami, P. F., Edmonds, A., O'Neil, J., and Buttler, S. (2000). Efficacy and safety of fixed-dose and dose-optimization regimens of sublingual apomorphine versus placebo in men with erectile dysfunction. The Apomorphine Study Group. *Urology* 56, 130–135.
263. Eaton, R. C., Markowski, V. P., Lumley, L. A., Thompson, J. T., Moses, J., and Hull, E. M. (1991). D_2 receptors in the paraventicular nucleus regulate genital responses and copulation in male rats. *Pharm. Biochem. Behav.* 39, 177–181.
264. Edwards, D. A., and Burge, K. G. (1973). Olfactory control of the sexual behavior of male and female mice. *Physiol. Behav.* 11, 867–872.
265. Edwards, D. A., and Davis, A. B. (1997). Deafferentation of the olfactory bulbs of male rats reduces erection to remote cues from females. *Physiol. Behav.* 62, 145–149.
266. Edwards, D. A., and Einhorn, L. C. (1986). Preoptic and midbrain control of sexual motivation. *Physiol. Behav.* 37, 329–335.
267. Edwards, D. A., and Isaacs, S. (1991). Zona incerta lesions: effects on copulation, partner-preference and other socio-sexual behaviors. *Behav. Brain Res.* 44, 145–150.
268. Edwards, D. A., Griffis, K. T., and Tardival, C. (1990). Olfactory bulb removal: effects on sexual behavior and partner-preference in male rats. *Physiol. Behav.* 48, 447–450.
269. Edwards, D. A., Walter, B., and Liang, P. (1996). Hypothalamic and olfactory control of sexual behavior and partner preference in male rats. *Physiol. Behav.* 60, 1347–1354.
269a. Edwards, N., Kriegsfeld, L., and Crews, D. (2004). Neural substrates for sexual and thermoregulatory behavior in the male leopard gecko, Eublepharis macularius. *Brain Res.* 1029, 77–83.
270. Ehrensing, R. H., Kastin, A. J., and Schally, A. V. (1981). Behavioral and hormonal effects of prolonged high doses of LHRH in male impotency. *Peptides* 1, 147–155.
271. Eibergen, R. D., and Caggiula, A. R. (1973). Ventral midbrain involvement in copulatory behavior of the male rat. *Physiol. Behav.* 10, 435–441.
272. Emery, D. E., and Sachs, B. D. (1976). Copulatory behavior in male rats with lesions in the bed nucleus of the stria terminalis. *Physiol. Behav.* 17, 803–806.
273. Erskine, M. S., and Hanrahan, S. B. (1997). Effects of paced mating on *c-fos* gene expression in the female rat brain. *J. Neuroendocrinol.* 9, 903–912.
274. Escrig, A., Gonzalez-Mora, J. L., and Mas, M. (1999a). Nitric oxide release in penile corpora cavernosa in a rat model of erection. *J. Physiol.* 516 (Pt 1), 261–269.
275. Escrig, A., Marin, R., and Mas, M. (1999b). Repeated PGE1 treatment enhances nitric oxide and erection responses to nerve stimulation in the rat penis by upregulating constitutive NOS isoforms. *J. Urol.* 162, 2205–2210.
276. Everitt, B. J. (1990). Sexual motivation: a neural and behavioral analysis of the mechanisms underlying appetitive and copulatory responses of male rats. *Neurosci. Biobehav. Rev.* 14, 217–232.
277. Everitt, B. J., and Stacey, P. (1987). Studies of instrumental behavior with sexual reinforcement in male rats (*Rattus norvegicus*): II. Effects of preoptic lesions, castration, and testosterone. *J. Comp. Psychol.* 101, 407–419.
278. Everitt, B. J., Cador, M., and Robbins, T. W. (1989). Interactions between the amygdala and ventral striatum in stimulus-reward associations: studies using a second-order schedule of sexual reinforcement. *Neuroscience* 30(1), 63–75.
279. Everitt, B. J., Fray, P., Kostarczyk, E., Taylor, S., and Stacey, P. (1987). Studies of instrumental behavior with sexual reinforcement in male rats (*Rattus norvegicus*): I. Control by brief visual stimuli paired with a receptive female. *J. Comp. Psychol.* 101(4), 395–406.
280. Fallon, J. H. (1986). Topographic organization of ascending dopaminergic projections. In *The Mesocorticolimbic Dopamine System* (P. W. Kalivas and C. B. Nemeroff, Eds.). *Ann. N Y Acad. Sci.* 537, 1–9.
281. Feder, H. H., Naftolin, F., and Ryan, K. J. (1974). Male and female sexual responses in male rats given estradiol benzoate and 5 alpha-androstan-17 beta-ol-3-one propionate. *Endocrinology* 94, 136–141.
282. Fernandez, E., Dail, W. G., Walton, G., and Martinez, G. (1991). The vasculature of the rat penis: a scanning electron microscopic and histologic study. *Am. J. Anat.* 192, 307–318.
283. Fernandez-Fewell, G. D., and Meredith, M. (1994). *C-fos* Expression in vomeronasal pathways of mated or pheromone-stimulated male golden hamsters: contribution from vomeronasal sensory input and expression related to mating performance. *J. Neurosci.* 14, 3643–3654.
284. Fernandez-Fewell, G. D., and Meredith, M. (1995). Facilitation of mating behavior in male hamsters by LHRH and AcLHRH5-10: interaction with the vomeronasal system. *Physiol. Behav.* 57, 213–221.
285. Fernandez-Fewell, G. D., and Meredith, M. (1998). Olfactory contribution to Fos expression during mating in inexperienced male hamsters. *Chem. Sens.* 23, 257–267.

286. Fernandez-Guasti, A., and Escalante, A. L. (1991). Role of presynaptic serotonergic receptors on the mechanism of action of 5-HT_{1A} and 5-HT_{1B} agonists on masculine sexual behaviour: physiological and pharmacological implications. *J. Neural Transm.* 85, 95–107.
287. Fernandez-Guasti, A., and Rodriguez-Manzo, G. (1997). 8-OH-DPAT and male rat sexual behavior: partial blockade by noradrenergic lesion and sexual exhaustion. *Pharmacol. Biochem. Behav.* 56, 111–116.
288. Fernandez-Guasti, A., and Rodriguez-Manzo, G. (2003). Pharmacological and physiological aspects of sexual exhaustion in male rats. *Scand. J. Psychol.* 44, 257–263.
289. Fernandez-Guasti, A., Escalante, A. L., Ahlenius, S., Hillegaart, V., and Larsson, K. (1992). Stimulation of 5-HT_{1A} and 5-HT_{1B} receptors in brain regions and its effects on male rat sexual behavior. *Eur. J. Pharmacol.* 210, 121–129.
290. Fernandez-Guasti, A., Larsson, K., and Beyer, C. (1985). Comparison of the effects of different isomers of bicuculline infused in the preoptic area on male rat sexual behavior. *Experientia* 41, 1414–1416.
291. Fernandez-Guasti, A., Larsson, K., and Beyer, C. (1986). GABAergic control of masculine sexual behavior. *Pharmacol. Biochem. Behav.* 24, 1065–1070.
292. Ferrari, F., and Giuliani, D. (1994). The selective D_2 dopamine receptor antagonist eticlopride counteracts the *ejaculatio praecox* induced by the selective D_2 dopamine agonist SND 919 in the rat. *Life Sci.* 55, 1155–1162.
293. Ferrari, F., and Giuliani, D. (1995). Sexual attraction and copulation in male rats: effects of the dopamine agonist SND 919. *Pharmacol. Biochem. Behav.* 50, 29–34.
294. Ferrari, F., and Giuliani, D. (1996a). Behavioral effects induced by the dopamine D_3 agonist 7-OH-DPAT in sexually-active and -inactive male rats. *Neuropharmacology* 35, 279–284.
295. Ferrari, F., and Giuliani, D. (1996b). Influence of eticlopride on cocaine and DA D_2 agonist-induced behavioral effects in rats. *Pharmacol. Biochem. Behav.* 53, 525–530.
296. Ferrari, F., Ottani, A., and Giuliani, D. (2002). Influence of sildenafil on central dopamine-mediated behaviour in male rats. *Life Sci.* 70, 1501–1508.
297. Ferris, C. F., Snowdon, C. T., King, J. A., Duong, T. Q., Ziegler, T. E., Ugurbil, K., Ludwig, R., Schultz-Darken, N. J., Wu, Z., Olson, D. P., Sullivan, Jr J. M., Tannenbaum, P. L., and Vaughan, J. T. (2001). Functional imaging of brain activity in conscious monkeys responding to sexually arousing cues. *NeuroReport* 12, 2231–2236.
298. Fewell, G. D., and Meredith, M. (2002). Experience facilitates vomeronasal and olfactory influence on Fos expression in medial preoptic area during pheromone exposure or mating in male hamsters. *Brain Res.* 941, 91–106.
299. Fiber, J. M., Adages, P., and Swain, J. M. (1993). Pheromones induce *c-fos* in limbic areas regulating male hamster mating behavior. *NeuroReport* 4, 871–874.
300. Finn, P. D., and Yahr, P. (1994). Projection of the sexually dimorphic area of the gerbil hypothalamus to the retrorubral field is essential for male sexual behavior: role of A8 and other cells. *Behav. Neurosci.* 108, 362–378.
300a. Finn, P. D., and Yahr, P. (2005). Projection from the ventral bed nucleus of the stria terminalis to the retrorubral field in rats and the effects of cells in these areas on mating in male rats versus gerbils. *Horm. Behav.* 47, 123–138.
301. Fiorino, D. F., and Phillips, A. G. (1999). Facilitation of sexual behavior and enhanced dopamine efflux in the nucleus accumbens of male rats after D-amphetamine-induced behavioral sensitization. *J. Neurosci.* 19, 456–463.
302. Fiorino, D. F., Coury, A., and Phillips, A. G. (1997). Dynamic changes in nucleus accumbens dopamine efflux during the Coolidge effect in male rats. *J. Neurosci.* 17, 4849–4855.
303. Fisher, A. E. (1956). Maternal and sexual behavior induced by intracranial chemical stimulation. *Science* 124, 228–229.
304. Fletcher, I. C., and Lindsay, D. R. (1968). Sensory involvement in the mating behaviour of domestic sheep. *Anim. Behav.* 16, 410–414.
305. Floody, O. R. (1989). Dissociation of hypothalamic effects on ultrasound production and copulation. *Physiol. Behav.* 46, 299–307.
306. Floody, O. R., and Pfaff, D. W. (1977). Communication among hamsters by high-frequency acoustic signals: III. Responses evoked by natural and synthetic ultrasounds. *J. Comp. Physiol. Psychol.* 91, 820–829.
307. Foidart, A., Meddle, S. L., and Balthazart, J. (1999). Mating-induced Fos matase are not co-localized in the preoptic area. *NeuroReport* 10, 907–912.
308. Foreman, M. M., and Hall, J. L. (1987). Effects of D_2-dopaminergic receptor stimulation on male rat sexual behavior. *J. Neural Transm.* 68, 153–170.
309. Foreman, M. M., Hall, J. L., and Love, R. L. (1989). The role of the 5-HT_2 receptor in the regulation of sexual performance of male rats. *Life Sci.* 45, 1263–1270.
310. Fraley, G. S., Thomas-Smith, S. E., Acohido, B. V., Steiner, R. A., and Clifton, D. K. (2004). Stimulation of sexual behavior in the male rat by galanin-like peptide. *Horm. Behav.* 46, 551–557.
311. Frank, J. L., Hendricks, S. E., and Olson, C. H. (2000). Multiple ejaculations and chronic fluoxetine: effects on male rat copulatory behavior. *Pharmacol. Biochem. Behav.* 66, 337–342.
312. Frazer, A., and Hensler, J. G. (1999). Serotonin. In *Basic Neurochemistry* (G. J. Siegel, B. W. Agranoff, R. W. Albers, S. K. Fisher, and M. D. Uhler, Eds.), pp. 263–292. Lippencott-Raven Press, Philadelphia.
313. Freeman, L. M., Padgett, B. A., Prins, G. S., and Breedlove, S. M. (1995). Distribution of androgen receptor immunoreactivity in the spinal cord of wild-type, androgen-insensitive and gonadectomized male rats. *J. Neurobiol.* 27, 51–59.
314. Frye, C. A. (2001). The role of neurosteroids and non-genomic effects of progestins and androgens in mediating sexual receptivity of rodents. *Brain Res. Rev.* 37, 201–222.
315. Frye, C. A., Rhodes, M. E., Rosellini, R., and Svare, B. (2002). The nucleus accumbens as a site of action for rewarding properties of testosterone and its 5alpha-reduced metabolites. *Pharmacol. Biochem. Behav.* 74, 119–127.
316. Fumero, B., Fernandez-Vera, J. R., Gonzalez-Mora, J. L., and Mas, M. (1994). Changes in monoamine turnover in forebrain areas associated with masculine sexual behavior: a microdialysis study. *Brain Res.* 662, 233–239.
317. Garcia-Segura, L. M., Chowen, J. A., Parducz, A., and Naftolin, F. (1994). Gonadal hormones as promoters of structural synaptic plasticity: cellular mechanisms. *Prog. Neurobiol.* 44, 279–307.
318. Gerall, H. D., Ward, I. L., and Gerall, A. A. (1967). Disruption of the male rat's sexual behaviour induced by social isolation. *Anim. Behav.* 15, 54–58.
319. Gerstenberg, T. C., Levin, R. J., and Wagner, G. (1990). Erection and ejaculation in Man. Assessment of the electromyographic activity of the bulbocavernosus and ischiocavernosus muscles. *Brit. J. Urol.* 65, 395–402.
320. Gerstenberg, T. C., Nordling, J., and Hald, T. (1991). Evoked potentials from the lower urinary tract. II. The spino-cortical neuraxis. A methodological study. *Scand. J. Urol. Nephrol. Suppl.* 138, 41–46.
321. Gessa, G. L., Paglietti, E., and Quarantotti, B. P. (1979). Induction of copulatory behavior in sexually inactive rats by naloxone. *Science* 204, 203–205.
322. Geyer, L. A., and Barfield, R. J. (1978). Influence of gonadal hormones and sexual behavior on ultrasonic vocalization

in rats: I. Treatment of females. *J. Comp. Physiol. Psychol.* 92, 438–446.

323. Giantonio, G. W., Lund, N. L., and Gerall, A. A. (1970). Effect of diencephalic and rhinencephalic lesions on the male rat's sexual behavior. *J. Comp. Physiol. Psychol.* 73, 38–46.
324. Gill, C. J., Wersinger, S. R., Veney, S. L., and Rissman, E. F. (1998). Induction of *c-fos*-like immunoreactivity in musk shrews after mating. *Brain Res.* 811, 21–28.
325. Ginton, A. (1976). Copulation in noncopulators: effect of PCPA in male rats. *Pharmacol. Biochem. Behav.* 4, 357–359.
326. Ginton, A., and Merari, A. (1977). Long range effects of MPOA lesion on mating behavior in the male rat. *Brain Res.* 120, 158–163.
327. Giordano, M., Guemes, M., Lopez-Arias, V., and Paredes, R. G. (1998). Socio-sexual behavior in male rats after lesions of the dorsolateral tegmentum. *Physiol. Behav.* 65, 89–94.
328. Giuliani, D., and Ferrari, F. (1996). Differential behavioral response to dopamine D_2 agonists by sexually naïve, sexually active, and sexually inactive male rats. *Behav. Neurosci.* 110, 802–808.
329. Giuliano, F. (2004). Control of penile erection by the melanocortinergic system: experimental evidences and therapeutic perspectives. *J. Androl.* 25, 683–694.
330. Giuliano, F., and Allard J. (2001). Dopamine and sexual function. *Int. J. Impotence Res.* 13, Suppl 3, S18–S28.
331. Giuliano, F., and Allard J. (2002). Apomorphine SL (Uprima): preclinical and clinical experiences learned the first central nervous system-acting ED drug. *Int. J. Impot. Res.* 14, Suppl 1, S53–S56.
332. Giuliano, F., and Rampin, O. (2000). Central neural regulation of penile erection. *Neurosci. Biobehav. Rev.* 24, 517–533.
333. Giuliano, F., Allard, J., Rampin, O., Droupy, S., Benoit, G., Alexandre, L., and Bernabe, J. (2001a). Spinal proerectile effect of apomorphine in the anesthetized rat. *Int. J. Impotence Res.* 13, 110–115.
334. Giuliano, F., Bernabe, J., Alexandre, L., Niewoehner, U., Haning, H., and Bischoff, E. (2003). Pro-erectile effect of vardenafil: in vitro experiments in rabbits and in vivo comparison with sildenafil in rats. *Eur. Urol.* 44, 731–736.
335. Giuliano, F., Bernabé, J., Brown, K., Droupy, S. Benoit, G., and Rampin, O. (1997). Erectile response to hypothalamic stimulation in rats: role of peripheral nerves. *Am. J. Physiol.* 273, R1990–R1997.
336. Giuliano, F., Bernabe, J., McKenna, K., Longueville, F., and Rampin, O. (2001b). Spinal proerectile effect of oxytocin in anesthetized rats. *Am. J. Physiol. – Reg. Integ. Comp. Physiol.* 280, R1870–R1877.
337. Giuliano, F., Bernabe, J., Rampin, O., Courtois, F., Benoit, G., and Rousseau, J. P. (1994). Telemetric monitoring of intracavernous pressure in freely moving rats during copulation. *J. Urol.* 152, 1271–1274.
338. Giuliano, F., Montorsi, F., Mirone, V., Rossi, D., and Sweeney, M. (2000). Switching from intracavernous prostaglandin E1 injections to oral sildenafil citrate in patients with erectile dysfunction: results of a multicenter European study. *J. Urol.* 164, 708–711.
339. Giuliano, F., Rampin, O., Bernabé, J., and Rousseau, J. P. (1995). Neural control of penile erection in the rat. *J. Autonom. Nerv. Sys.* 55, 36–44.
340. Giuliano, F., Rampin, O., Brown, K., Courtois, F., Benoit, G., and Jardin, A. (1996). Stimulation of the medial preoptic area of the hypothalamus in the rat elicits increases in intracavernous pressure. *Neurosci. Lett.* 209, 1–4.
341. Gladue, B. A., and Clemens, L. G. (1980). Flutamide inhibits testosterone-induced masculine sexual behavior in male and female rats. *Endocrinology* 106, 1917–1922.
342. Goldfoot, D. A., Essock-Vitale, S. M., Asa, C., Thornton, J. E., and Leshner, A. L. (1978). Anosmia in male rhesus monkeys does not alter copulatory activity with cycling females. *Science* 199, 1095–1096.
343. Goldstein, I., Lue, T. F., Padma-Nathan, H., Rosen, R. C., Steers, W. D., Wicker, P. A. (1998). Oral sildenafil in the treatment of erectile dysfunction. *N. Engl. J. Med.* 338, 1397–1404.
344. Gomez, D. M., and Newman, S. W. (1991). Medial nucleus of the amygdala in the adult Syrian hamster: a quantitative Golgi analysis of gonadal hormonal regulation of neuronal morphology. *Anat Rec.* 231, 498–509.
345. Gomez, D. M., and Newman, S. W. (1992). Differential projections of the anterior and posterior regions of the medial amygdaloid nucleus in the Syrian hamster. *J. Comp. Neurol.* 317, 195–218.
346. Gomez-Marrero, J., Feria, M., and Mas, M. (1988). Stimulation of opioid receptors suppresses penile erectile reflexes and seminal emission in rats. *Pharmacol. Biochem. Behav.* 31, 393–396.
347. Gonzalez, M. I., Farabollini, F., Albonetti, E., and Wilson, C. A. (1994). Interactions between 5-hydroxytryptamine (5-HT) and testosterone in the control of sexual and nonsexual behaviour in male and female rats. *Pharm. Biochem. Behav.* 47, 591–601.
348. Gooren, L. J. (1985). Human male sexual functions do not require aromatization of testosterone: a study using tamoxifen, testolactone, and dihydrotestosterone. *Arch. Sex. Behav.* 14, 539–548.
349. Gorski, R. A., Gordon, J. H., Shryne, J. E., and Southam, A. M. (1978). Evidence for a morphological sex difference within the medial preoptic area of the rat brain. *Brain Res.* 148, 333–346.
350. Gozes, I., Meltzer, E., Rubinrout, S., Brenneman, D. E., and Fridkin, M. (1989). Vasoactive intestinal peptide potentiates sexual behavior: inhibition by a novel antagonist. *Endocrinology* 125, 2945–2949.
351. Grassman, M., and Crews, D. (1986). Progesterone induction of pseudocopulatory behavior and stimulus-response complementarity in an all-female lizard species. *Horm. Behav.* 20, 327–335.
352. Gray, B. W., Beckett, S. D., and Henry, D. F. (1985). Microscopic characteristics of genital end bulbs in the penis of bulls. *Am. J. Vet. Res.* 46, 2393–2398.
353. Gray, G. D. (1977). Differential effects of the antiandrogen flutamide on aspects of sexual behavior in castrated, androgen-treated male rats. *Psychoneuroendocrinology* 2, 315–320.
354. Gray, G. D., Smith, E. R., and Davidson, J. M. (1980). Hormonal regulation of penile erection in castrated male rats. *Physiol. Behav.* 24, 463–468.
355. Gréco, B., Edwards, D. A., Michael, R. P., and Clancy, A. N. (1996). Androgen receptor immunoreactivity and mating-induced Fos expression in the forebrain and midbrain structures in the male rat. *Neuroscience* 75, 161–171.
356. Gréco, B., Edwards, D. A., Michael, R. P., and Clancy, A. N. (1998a). Androgen receptors and estrogen receptors are colocalized in male rat hypothalamic and limbic neurons that express Fos immunoreactivity induced by mating. *Neuroendocrinology* 67, 18–28.
357. Gréco, B., Edwards, D. A., Michael, R. P., Zumpe, D., and Clancy, A. N. (1999). Colocalization of androgen receptors and mating-induced FOS immunoreactivity in neurons that project to the central tegmental field in male rats. *J. Comp. Neurol.* 408, 220–236.
358. Gréco, B., Edwards, D. A., Zumpe, D., and Clancy, A. N. (1998b). Androgen receptor and mating-induced Fos immunoreactivity

are co-localized in limbic and midbrain neurons that project to the male rat medial preoptic area. *Brain Res.* 781, 15–24.

359. Gréco, B., Edwards, D. A., Zumpe, D., Michael, R. P., and Clancy, A. N. (1998c). Fos induced by mating or noncontact sociosexual interaction is colocalized with androgen receptors in neurons within the forebrain, midbrain, and lumbosacral spinal cord of male rats. *Horm. Behav.* 33, 125–138.
360. Greenwood, D., Coggeshall, R. E., and Hulsebosch, C. E. (1985). Sexual dimorphism in the numbers of neurons in the pelvic ganglia of adult rats. *Brain Res.* 340, 160–162.
361. Grunt, J. A., and Young, W. C. (1953). Consistency of sexual behavior patterns in individual male guinea pigs following castration and androgen therapy. *J. Comp. Physiol. Psychol.* 46, 138–144.
362. Guay, A. T., Spark, R. F., Jacoobson, J., Murray, F. T., and Geisser, M. E. (2002). Yohimbine treatment of organic erectile dysfunction in a dose-escalation trial. *Int. J. Impot. Res.* 14, 25–31.
363. Guerit, J. M., and Opsomer R. J. (1991). Bit-mapped imaging of somatosensory evoked potentials after stimulation of the posterior tibial nerves and dorsal nerve of the penis/clitoris. *Electroencephalogr. Clin. Neurophysiol.* 80, 228–237.
364. Gulia, K. K., Mallick, H. N., and Kumar, V. M. (2003). Orexin A (hypocretin-1) application at the medial preoptic area potentiates male sexual behavior in rats. *Neuroscience* 116, 921–923.
365. Hadeishi, Y., and Wood, R. I. (1996). Nitric oxide synthase in mating behavior circuitry of male Syrian hamster brain. *J. Neurobiol.* 30, 480–492.
366. Haensel, S. M., and Slob, A. K. (1997). Flesinoxan: a prosexual drug for male rats. *Eur. J. Pharmacol.* 330, 1–9.
367. Haensel, S. M., Rowland, D. L., and Slob, A. K. (1995). Serotonergic drugs and masculine sexual behavior in laboratory rats and men. In *The Pharmacology of Sexual Function and Dysfunction* (J. Bancroft, Ed.), pp. 235–247. Proceedings of the Esteve Foundation Symposium VI, Son Vida, Mallorca, October 9–12, 1994. Excerpta Medica, Amsterdam.
368. Halata, Z., and Munger, B. L. (1986). The neuroanatomical basis for the protopathic sensibility of the human glans penis. *Brain Res.* 371, 205–230.
369. Haldeman, S., Bradley, W. E., Bhatia, N. N., and Johnson, B. K. (1982). Pudendal evoked responses. *Arch. Neurol.* 39, 280–283.
370. Halem, H. A., Cherry, J. A., and Baum, M. J. (1999). Vomeronasal neuroepithelium and forebrain Fos responses to male pheromones in male and female mice. *J. Neurobiol.* 39, 249–263.
371. Hansen, S. (1982). Spinal control of sexual behavior: effects of intrathecal administration of lisuride. *Neurosci. Lett.* 33, 329–332.
372. Hansen, S., and Gummesson, B. M. (1982). Participation of the lateral midbrain tegmentum in the neuroendocrine control of sexual behavior and lactation in the rat. *Brain Res.* 251, 319–325.
373. Hansen, S., and Hagelsrum, L. J. af. (1984). Emergence of displacement activities in the male rat following thwarting of sexual behavior. *Behav. Neurosci.* 98, 868–883.
374. Hansen, S., and Ross, S. B. (1983). Role of descending monoaminergic neurons in the control of sexual behavior: effects of intrathecal infusions of 6-hydroxydopamine and 5,7-dihydroxytryptamine. *Brain Res.* 268, 285–290.
375. Hansen, S., Köhler, C., Goldstein, M., and Steinbusch, H. V. (1982a). Effects of ibotenic acid-induced neuronal degeneration in the medial preoptic area and the lateral hypothalamic area on sexual behavior in the male rat. *Brain Res.* 239, 213–232.
376. Hansen, S., Köhler, C., and Ross, S. B. (1982b). On the role of the dorsal mesencephalic tegmentum in the control of masculine sexual behavior in the rat: effects of electrolytic lesions, ibotenic acid and DSP 4. *Brain Res.* 240, 311–320.
377. Hansen, S., Svensson, L., Hokfelt, T., and Everitt, B. J. (1983). 5-Hydroxytryptamine-thyrotropin releasing hormone interactions in the spinal cord: effects on parameters of sexual behaviour in the male rat. *Neurosci. Lett.* 42, 299–304.
378. Harding, S. M., and McGinnis, M. Y. (2003). Effects of testosterone in the VMN on copulation, partner preference, and vocalizations in male rats. *Horm. Behav.* 43, 327–335.
379. Harding, S. M., and McGinnis, M. Y. (2004). Androgen receptor blockade in the MPOA or VMN: effects on male sociosexual behaviors. *Physiol. Behav.* 81, 671–680.
380. Harris, V. S., and Sachs, B. D. (1975). Copulatory behavior in male rats following amygdaloid lesions. *Brain Res.* 86, 514–518.
381. Hart, B. L. (1967). Sexual reflexes and mating behavior in the male dog. *J. Comp. Physiol. Psychol.* 64, 388–399.
382. Hart, B. L. (1968a). Alteration of quantitative aspects of sexual reflexes in spinal male dogs by testosterone. *J. Comp. Physiol. Psychol.* 66, 726–730.
383. Hart, B. L. (1968b). Sexual reflexes and mating behavior in the male rat. *J. Comp. Physiol. Psychol.* 65, 453–460.
384. Hart, B. L. (1974). The medial preoptic-anterior hypothalamic area and sociosexual behavior of male dogs: a comparative neuropsychological analysis. *J. Comp. Physiol. Psychol.* 86, 328–349.
385. Hart, B. L. (1978). Hormones, spinal reflexes, and sexual behaviour. In *Biological Determinants of Sexual Behaviour* (J. B. Hutchison, Ed.), pp. 319–347. John Wiley & Sons, Chichester, UK.
386. Hart, B. L. (1979). Activation of sexual reflexes of male rats by dihydrotestosterone but not estrogen. *Physiol. Behav.* 23, 107–109.
387. Hart, B. L. (1986). Medial preoptic-anterior hypothalamic lesions and sociosexual behavior of male goats. *Physiol. Behav.* 36, 301–305.
388. Hart, B. L., and Haugen, C. M. (1972). Scent marking and sexual behavior maintained in anosmic dogs. *Commun. Behav. Biol.* 6, 131–135.
389. Hart, B. L., and Kitchell, R. L. (1966). Penile erection and contraction of penile muscles in the spinal and intact dog. *Am. J. Physiol.* 210, 257–262.
390. Hart, B. L., and Ladewig, J. (1979). Effects of medial preoptic-anterior hypothalamic lesions on development of sociosexual behavior in dogs. *J. Comp. Physiol. Psychol.* 93, 566–573.
391. Hart, B. L., and Ladewig, J. (1980). Accelerated and enhanced testosterone secretion in juvenile male dogs following medial preoptic-anterior hypothalamic lesions. *Neuroendocrinology* 30, 20–24.
391a. Hart, B. L., and Melese-d'Hospital, P. (1983). Penile mechanisms and the role of the striated penile muscles in penile reflexes. *Physiol. Behav.* 31, 807–813.
392. Hart, B. L., Haugen, C. M., and Peterson, D. M. (1973). Effects of medial preoptic-anterior hypothalamic lesions on mating behavior of male cats. *Brain Res.* 54, 177–191.
393. Hart, B. L., Wallach, S. J. R., and Melese-D'Hospital, P. Y. (1983). Differences in responsiveness to testosterone of penile reflexes and copulatory behavior of male rats. *Horm. Behav.* 17, 274–283.
394. Heeb, M. M., and Yahr, P. (1996). *c-fos* immunoreactivity in the sexually dimorphic area of the hypothalamus and related brain regions of male gerbils after exposure to sex-related stimuli or performance of specific sexual behaviors. *Neuroscience* 72, 1049–1071.

395. Heeb, M. M., and Yahr, P. (2000). Cell-body lesions of the posterodorsal preoptic nucleus or posterodorsal medial amygdala, but not the parvicellular subparafascicular thalamus, disrupt mating in male gerbils. *Physiol. Behav.* 68, 317–331.
396. Heeb, M. M., Yahr. P. (2001). Anatomical and functional connections among cell groups in the gerbil brain that are activated with ejaculation. *J. Comp. Neurol.* 439, 248–258.
397. Heim, N., and Hursch, C. J. (1979). Castration for sex offenders: treatment or punishment? A review and critique of recent European literature. *Arch. Sex. Behav.* 8, 281–304.
398. Heimer, L., and Larsson, K. (1966/1967). Impairment of mating behavior in male rats following lesions in the preoptic-anterior hypothalamic continuum. *Brain Res.* 3, 248–263.
399. Heimer, L., and Larsson, K. (1967). Mating behavior of male rats after olfactory bulb lesions. *Physiol. Behav.* 2, 207–209.
400. Henderson, V. E., and Roepke, M. H. (1933). On the mechanism of erection. *Am. J. Physiol.* 106, 441–448.
401. Hendricks, S. E., and Scheetz, H. A. (1973). Interaction of hypothalamic structures in the mediation of male sexual behavior. *Physiol. Behav.* 10, 711–716.
402. Hendry, W. F., Althof, S. E., Benson, G. S., Haensel, S. M., Hull, E. M., Kihara, K., and Opsomer, R. J. (2000). Male orgasmic and ejaculatory disorders. In *Erectile Dysfunction* (A. Jardin, G. Wagner, S. Khoury, F. Giuliano, H. Padma-Nathan, and R. Rosen, Eds.), pp. 477–506. Geneva, Switzerland: World Health Organization.
403. Herberg, L. J. (1963). Seminal ejaculation following positively reinforcing electrical stimulation of the rat hypothalamus. *J. Comp. Physiol. Psychol.* 56, 679–685.
404. Hermann, G. E., Holmes, G. M., Rogers, R. C., Beattie, M. S., Bresnahan, J. C. (2003). Descending spinal projections from the rostral gigantocellular reticular nuclei complex. *J. Comp. Neurol.* 455, 210–221.
405. Hernandez, L., Gonzalez, L., Murzi, E., Paez, X., Gottberg, E., and Baptista, T. (1994a). Testosterone modulates mesolimbic dopaminergic activity in male rats. *Neurosci. Lett.* 171, 172–174.
406. Hernandez, L., Gonzalez, M., Oropeza, M. V., Guevara, M. A., Cervantes, M., and Morali, G. (1994b). Effects of intrathecal administration of adrenergic agonists on the frequency of copulatory pelvic thrusting of the male rat. *Arch. Med. Res.* 25, 419–425.
406a. Hill, R. A., Pompolo, S., Jones, M. E., Simpson, E. R., and Boon, W. C. (2004). Estrogen deficiency leads to apoptosis in dopaminergic neurons in the medial preoptic area and arcuate nucleus of male mice. *Molec. Cell. Neurosci.* 27, 466–476.
407. Hillegaart, V. (1991). Functional topography of brain serotonergic pathways in the rat. *Acta Physiol. Scand. Suppl.* 598, 1–54.
408. Hillegaart, V., and Ahlenius, S. (1998). Facilitation and inhibition of male rat ejaculatory behaviour by the respective 5-HT_{1A} and 5-HT_{1B} receptor agonists 8-OH-DPAT and anpirtoline, as evidenced by use of the corresponding new and selective receptor antagonists NAD-299 and NAS-181. *Br. J. Pharm.* 125, 1733–1743.
409. Hillegaart, V., Ahlenius, S., and Larsson, K. (1991). Region-selective inhibition of male rat sexual behavior and motor performance by localized forebrain 5-HT injections: a comparison with effects produced by 8-OH-DPAT. *Behav. Brain Res.* 42, 169–180.
410. Hillegaart, V., Alster, P., Uvnas-Moberg, K., and Ahlenius, S. (1998). Sexual motivation promotes oxytocin secretion in male rats. *Peptides* 19, 39–45.
411. Hines, M. Allen, L. S., and Gorski, R. A. (1992). Sex differences in subregions of the medial nucleus of the amygdala and the bed nucleus of the stria terminalis of the rat. *Brain Res.* 579, 321–326.
412. Hitt, J. C., Bryon, D. M., and Modianos, D. T. (1973). Effects of rostral medial forebrain bundle and olfactory tubercle lesions upon sexual behavior of male rats. *J. Comp. Physiol. Psychol.* 82, 30–36.
413. Hitt, J. C., Hendricks, S. E., Ginsberg, S. I., and Lewis, J. H. (1970). Disruption of male, but not female, sexual behavior in rats by medial forebrain bundle lesions. *J. Comp. Physiol. Psychol.* 73, 377–384.
414. Hlinak Z., and Madlafousek J. (1982). Initiation of copulatory behaviour in castrated male rats implanted with very small testosterone-filled silastic capsules. *Endokrinologie* 79, 35–43.
415. Hokfelt, T., Fried, G., Hansen, S., Holets, V., Lundberg, J. M., and Skirboll, L. (1986). Neurons with multiple messengers: distribution and possible functional significance. *Prog. Brain Res.* 65, 115–137.
416. Hokfelt, T., Skirboll, L., Rehfeld, J. F., Goldstein, M., Markey, K., and Dann, O. (1980). A subpopulation of mesencephalic dopamine neurons projecting to limbic areas contains a cholecystokinin-like peptide: evidence from immunohistochemistry combined with retrograde tracing. *Neuroscience* 5, 2093–2124.
417. Holmes, G. M., and Sachs, B. D. (1992). Erectile function and bulbospongiosus EMG activity in estrogen-maintained castrated rats vary with behavioral context. *Horm. Behav.* 26, 406–419.
418. Holmes, G. M., Chapple, W. D., Leipheimer, R. E., and Sachs, B. D. (1991). Electromyographic analysis of male rat perineal muscles during copulation and reflexive erections. *Physiol. Behav.* 49, 1235–1246.
419. Holmes, G. M., Hermann, G. E., Rogers, R. C., Bresnahan, J. C., and Beattie, M. S. (2002). Dissociation of the effects of nucleus raphe obscurus or rostral ventrolateral medulla lesions on eliminatory and sexual reflexes. *Physiol. Behav.* 75, 49–55.
420. Holmes, G. M., Rogers, R. C., Bresnahan, J. C., and Beattie, M. S. (1997). Differential effects of intrathecal thyrotropin-releasing hormone (TRH) on perineal reflexes in male rats. *Physiol. Behav.* 61, 57–63.
421. Holmes, G. M., Stephens, R. L., Breshnahan, J. C., and Beattie, M. S. (2001). Intrathecal methiothepin and thyrotropin-releasing hormone (TRH): effects on penile erection. *Physiol. Behav.* 73, 59–64.
422. Holstege, G. (1987). Some anatomical observations on the projections from the hypothalamus to brainstem and spinal cord: an HRP and autoradiographic tracing study in the cat. *J. Comp. Neurol.* 260, 98–126.
423. Holstege, G., Georgiadis, J. R., Paans, A. M., Meiners, L. C., van der Graaf, F. H., and Reinders, A. A. (2003). Brain activation during human male ejaculation. *J. Neurosci.* 23, 9185–9193.
424. Holstege, J. C., Van Dijken, H., Buijs, R. M., Goedknegt, H., Gosens, T., Bongers, C. M. (1996). Distribution of dopamine immunoreactivity in the rat, cat and monkey spinal cord. *J. Comp. Neurol.* 37, 631–52.
425. Horowitz, A., Menice, C. B., Laporte, R., and Morgan, K. G. (1996). Mechanisms of smooth muscle contraction. *Physiol. Rev.* 76, 967–1003.
426. Hoyer, D., Clarke, D. E., Fozard, J. R., Hartig, P. R., Martin, G. R., Mylecharane, E. J., Saxena, P. R., and Humphrey, P. P. (1994). International Union of Pharmacology classification of receptors for 5-hydroxytryptamine (serotonin). *Pharmacol. Rev.* 46, 157–203.
427. Hsieh, J. T., Chang, H. C., Law, H. S., Hsieh, C. H., and Cheng, J. T. (1998). In vivo evaluation of serotonergic agents and alpha-adrenergic blockers on premature ejaculation by inhibiting the seminal vesicle pressure response to electrical nerve stimulation. *Brit. J. Urol.* 82, 237–240.
428. Hubscher, C. H., and Johnson, R. D. (1996). Responses of medullary reticular formation neurons to input from the male genitalia. *J. Neurophysiol.* 76, 2474–2482.

429. Huddleston, G. G., Michael, R. P., Zumpe, D., and Clancy, A. N. (2003). Estradiol in the male rat amygdala facilitates mounting but not ejaculation. *Physiol. Behav.* 79, 239–246.

430. Hughes, A. M., Everitt, B. J., and Herbert, J. (1987). Selective effects of beta-endorphin infused into the hypothalamus, preoptic area and bed nucleus of the stria terminalis on the sexual and ingestive behaviour of male rats. *Neuroscience* 23, 1063–1073.

431. Hull, E. M., Bazzett, T. J., Warner, R. K., Eaton, R. C., and Thompson, J. T. (1990). Dopamine receptors in the ventral tegmental area modulate male sexual behavior in rats. *Brain Res.* 512, 1–6.

432. Hull, E. M., Bitran, D., Pehek, E. A., Holmes, G. M., Warner, R. K., Band, L. C., and Clemens, L. G. (1988a). Brain localization of cholinergic influence on male sex behavior in rats: antagonists. *Pharm. Biochem. Behav.* 31, 169–174.

433. Hull, E. M., Bitran, D., Pehek, E. A., Warner, R. K., and Band, L. C. (1986). Dopaminergic control of male sex behavior in rats: effects of an intracerebrally infused agonist. *Brain Res.* 370, 73–81.

434. Hull, E. M., Du, J., Lorrain, D. S., and Matuszewich, L. (1995). Extracellular dopamine in the medial preoptic area: implications for sexual motivation and hormonal control of copulation. *J. Neurosci.* 15, 7465–7471.

434a. Hull, E. M., Du, J., Lorrain, D. S., and Matuszewich, L. (1997). Testosterone, preoptic dopamine, and copulation in male rats. *Brain Res. Bull.* 44, 327–333.

435. Hull, E. M., Eaton, R. C., Markowski, V. P., Moses, J., Lumley, L. A., and Loucks, J. A. (1992). Opposite influence of medial preoptic D_1 and D_2 receptors on genital reflexes: implications for copulation. *Life Sci.* 51, 1705–1713.

436. Hull, E. M., Eaton, R. C., Moses, J., and Lorrain, D. (1993). Copulation increases dopamine activity in the medial preoptic area of male rats. *Life Sci.* 52, 935–940.

437. Hull, E. M., Lumley, L. A., Matuszewich, L., Dominguez, J., Moses, J., and Lorrain, D. S. (1994). The roles of nitric oxide in sexual function of male rats. *Neuropharmacology* 33, 1499–1504.

438. Hull, E. M., Meisel, R. L., & Sachs, B. D. (2002). Male sexual behavior. In *Hormones, Brain and Behavior* (D. W. Pfaff, A. P. Arnold, A. M. Etgen, S. E. Fahrbach, and R. T. Rubin, Eds.), pp. 3–137. Academic Press, San Diego.

439. Hull, E. M., Pehek, E. A., Bitran, D., Holmes, G. M., Warner, R. K., Band, L. C., Bazzett, T., and Clemens, L. G. (1988b). Brain localization of cholinergic influence on male sex behavior in rats: agonists. *Pharm. Biochem. Behav.* 31, 175–178.

440. Hull, E. M., Warner, R. K., Bazzett, T. J., Eaton, R. C., Thompson, J. T., and Scaletta, L. L. (1989). D_2/D_1 ratio in the medial preoptic area affects copulation of male rats. *J. Pharmacol. Exp. Ther.* 251, 422–427.

441. Hull, E. M., Weber, M. S., Eaton, R. C., Dua, R., Markowski, V. P., Lumley, L., and Moses, J. (1991). Dopamine receptors in the ventral tegmental area affect motor, but not motivational or reflexive, components of copulation in male rats. *Brain Res.* 554, 72–76.

442. Hulsebosch, C. E., and Coggeshall, R. E. (1982). An analysis of the axon populations in the nerves to the pelvic viscera in the rat. *J. Comp. Neurol.* 211, 1–10.

443. Hung, S. H., Pi, W. P., Tsai, Y. F., and Peng, M. T. (1997). Restoration of sexual behavior in aged male rats by intracerebral grafts of fetal preoptic area neurons. *J. Form. Med. Assoc.* 96, 812–818.

444. Hurt K. J., Musicki B., Palese M. A., Crone J. K., Becker R. E., Moriarity J. L., Snyder S. H., and Burnett A. L. (2002). Akt-dependent phosphorylation of endothelial nitric-oxide synthase mediates penile erection. *Proc. Natl. Acad. Sci. U S A* 99, 4061–4066.

445. Hutchison, J. B. (1978). Hypothalamic regulation of male sexual responsiveness to androgen. In *Biological Determinants of Sexual Behaviour* (J. B. Hutchison, Ed.), pp. 277–317. John Wiley & Sons, Chichester, UK.

446. Ignarro, L. J., Cirino, G., Casini, A., Napoli, C. (1999). Nitric oxide as a signaling molecule in the vascular system: an overview. *J. Cardiovasc. Pharmacol.* 34, 879–886.

447. Ikemoto, S., and Panksepp, J. (1999). The role of nucleus accumbens dopamine in motivated behavior: a unifying interpretation with special reference to reward-seeking. *Brain Res. Rev.* 31, 6–41.

448. James, P. J., and Nyby, J. G. (2002). Testosterone rapidly affects the expression of copulatory behavior in house mice (*Mus musculus*). *Physiol. Behav.* 75, 287–294.

449. Janig, W., and McLachlan, E. M. (1987). Organization of lumbar spinal outflow to distal colon and pelvic organs. *Physiol. Rev.* 67, 1332–1404.

450. Johnson, R. (1989). The physiology of single spinal cord neurons responding to penile stimulation in the rat. *Soc. Neurosci. Abstr.* 15, 756.

451. Johnson, R. D. (1988). Efferent modulation of penile mechanoreceptor activity. In *Progress in Brain Research*, Vol. 74, *Transduction and Cellular Mechanisms in Sensory Receptors* (A. Iggo, and W. Hamann, Eds.), pp. 319–324. Elsevier, Amsterdam.

452. Johnson, R. D., and Halata, Z. (1991). Topography and ultrastructure of sensory nerve endings in the glans penis of the rat. *J. Comp. Neurol.* 312, 299–310.

453 Johnson, R. D., and Hubscher, C. H. (1998). Brainstem microstimulation differentially inhibits pudendal motoneuron reflex inputs. *NeuroReport* 9, 341–345.

454. Johnson, R. D., and Hubscher, C. H. (2000). Brainstem microstimulation activates sympathetic fibers in pudendal nerve motor branch. *NeuroReport* 11, 379–382.

455. Johnson, R. D., and Kitchell, R. L. (1987). Mechanoreceptor response to mechanical and thermal stimuli in the glans penis of the dog. *J. Neurophysiol.* 57, 1813–1836.

456. Johnson, R. D., Kitchell, R. L., and Gilanpour, H. (1986). Rapidly and slowly adapting mechanoreceptors in the glans penis of the cat. *Physiol. Behav.* 37, 69–78.

457. Johnson, R., and Dugan, V. P. (1988). Responses of spine associated penile mechanoreceptors to tangential mechanical stimulation. *Soc. Neurosci. Abst.* 14, 726.

458. Johnston, P., and Davidson, J. M. (1972). Intracerebral androgens and sexual behavior in the male rat. *Horm. Behav.* 3, 345–357.

459. Ju, G., Ceccatelli, S., Hokfelt, T., and Frey, P. (1987). Immunohistochemical evidence for a spinothalamic pathway co-containing cholecystokinin- and galanin- like immunoreactivities in the rat. *Neurosci.* 20, 439–456.

460. Junemann, K. P., Persson-Junemann, C., Lue, T. F., Tanagho, E. A., and Alken, P. (1989). Neurophysiological aspects of penile erection: the role of the sympathetic nervous system. *Br. J. Urol.* 64, 84–92.

461. Kamel, F., and Frankel, A. I. (1978). The effect of medial preoptic area lesions on sexually stimulated hormone release in the male rat. *Horm. Behav.* 10, 10–21.

462. Kaplan, M. E., and McGinnis, M. Y. (1989). Effects of ATD on male sexual behavior and androgen receptor binding: a reexamination of the aromatization hypothesis. *Horm. Behav.* 23, 10–26.

463. Keaton, A. K., and Clark, J. T. (1998). Effects of angiotensin II on sexual function, blood pressure, and fluid intake are differentially affected by AT-1 receptor blockade. *Physiol. Behav.* 64, 339–346.

464. Kelley, A. E., and Berridge, K. C. (2002). The neuroscience of natural rewards: relevance to addictive drugs. *J. Neurosci.* 22, 3306–3311.

465. Kelliher, K. R., and Baum, M. J. (2001). Nares occlusion eliminates heterosexual partner selection without disrupting coitus in ferrets of both sexes. *J. Neurosci.* 21, 5832–5840.

466. Kelliher, K., and Baum, M. (2002). Effect of sex steroids and coital experience on ferrets' preference for the smell, sight and sound of conspecifics. *Physiol. Behav.* 76, 1–7.

467. Kelliher, K. R., Chang, Y. M., Wersinger, S. R., and Baum, M. J. (1998). Sex difference and testosterone modulation of pheromone-induced neuronal Fos in the ferret's main olfactory bulb and hypothalamus. *Biol. Reprod.* 59, 1454–1463.

468. Kelliher, K. R., Liu, Y. C., Baum, M. J., and Sachs, B. D. (1999). Neuronal Fos activation in olfactory bulb and forebrain of male rats having erections in the presence of inaccessible estrous females. *Neuroscience* 92, 1025–1033.

469. Kelly, M. J., Qiu, J., Wagner, E. J., Ronnekleiv, O. K. (2002). Rapid effects of estrogen on G protein-coupled receptor activation of potassium channels in the central nervous system (CNS). *J. Steroid Biochem. Mol. Biol.* 83, 187–193.

470. Kierniesky, N. C., and Gerall, A. A. (1973). Effects of testosterone propionate implants in the brain on the sexual behavior and peripheral tissue of the male rat. *Physiol. Behav.* 11, 633–640.

471. Kihara, K., and de Groat, W. C. (1997a). Sympathetic efferent pathways projecting bilaterally to the vas deferens in the rat. *Anat. Rec.* 248, 291- 299.

472. Kihara, K., and de Groat, W. C. (1997b). Sympathetic efferent pathways projecting to the bladder neck and proximal urethra in the rat. *J. Auton. Nerv. Syst.* 62, 134–142.

473. Kim, S. C., and Seo, K. K. (1998). Efficacy and safety of fluoxetine, sertraline and clomipramine in patients with premature ejaculation: a double-blind, placebo controlled study. *J. Urol.* 159, 425–427.

474. Kimura, Y., Adachi, K., Kisaki, N., and Ise K. (1975). Role of alpha-adrenergic receptor mechanism in closure of the internal urethral orifice during ejaculation. *Urol Int.* 30, 341–349.

475. Kindon, H. A., Baum M. J., and Paredes, R. J. (1996). Medial preoptic/anterior hypothalamic lesions induce a female-typical profile of sexual partner preference in male ferrets. *Horm. Behav.* 30, 514–527.

476. Kinsey, A. C., Pomeroy, W. B., and Martin, C. E. (1948). "Sexual Behavior in the Human Male." W. B. Saunders Co., Philadelphia.

477. Kippin, T. E., and Pfaus, J. G. (2001). The development of olfactory conditioned ejaculatory preferences in the male rat. I. Nature of the unconditioned stimulus. *Physiol. Behav.* 73, 457–469.

478. Kippin, T. E., Cain, S. W., and Pfaus, J. G. (2003). Estrous odors and sexually conditioned neutral odors activate separate neural pathways in the male rat. *Neuroscience* 117, 971–979.

479. Kippin, T., E., Manitt, C., Talianakis S., and Pfaus, J. G. (1996). Fos immunoreactivity in male rat brain following exposure to conditioned and unconditioned sex odors. *Soc. Neurosci. Abst.* 22, 154.

480. Kippin, T. E., Sotiropoulos, V., Badih, J., and Pfaus, J. G. (2004). Opposing roles of the nucleus accumbens and anterior lateral hypothalamic area in the control of sexual behaviour in the male rat. *Eur. J. Neurosci.* 19, 698–704.

481. Kippin, T. E., Talianakis, S., Schattmann, L., Bartholomew, S., and Pfaus, J. G. (1998). Olfactory conditioning of sexual behavior in the male rat. *J. Comp. Psychol.* 112, 389–399.

482. Kitchell, R. L., Gilanpour, H., and Johnson, R. D. (1982). Electrophysiologic studies of penile mechanoreceptors in the rats. *Exp. Neurol.* 75, 229–244.

483. Kiyatkin, E. A. (2002). Dopamine in the nucleus accumbens: cellular actions, drug- and behavior-associated fluctuations, and a possible role in an organism's adaptive activity. *Behav. Brain Res.* 137, 27–46.

484. Kiyatkin, E. A., and Mitchum, R. D. Jr. (2003). Fluctuations in brain temperature during sexual interaction in male rats: an approach for evaluating neural activity underlying motivated behavior. *Neuroscience* 119, 1169–1183.

485. Klaric, J. S. (1990). Prepubertal MPOA lesions and housing condition: effects on rats' male sex behaviors. *Physiol. Behav.* 47, 1287–1289.

486. Klint, T., and Larsson, K. (1995). Clozapine acts as a 5-HT_2 antagonist by attenuating DOI-induced inhibition of male rat sexual behaviour. *Psychopharmacology* 119, 291–294.

487. Knobil, E. (1980). The neuroendocrine control of the menstrual cycle. *Recent Prog. Horm. Res.* 36, 53–88.

488. Kolbeck, S. C., and Steers, W. D. (1992). Neural regulation of the vas deferens in the rat: an electrophysiological analysis. *Am. J. Physiol.* 263, R331–R338.

489. Kollack, S. S., and Newman, S. W. (1992). Mating behavior induces selective expression of Fos protein within the chemosensory pathways of the male Syrian hamster brain. *Neurosci. Lett.* 143, 223–228.

490. Kollack-Walker, S., and Newman, S. W. (1995). Mating and agonistic behavior produce different patterns of Fos immunolabeling in the male Syrian hamster brain. *Neuroscience* 66, 721–736.

491. Kollack-Walker, S., and Newman, S. W. (1997). Mating-induced expression of *c-fos* in the male Syrian hamster brain: role of experience, pheromones, and ejaculations. *J. Neurobiol.* 32, 481–501.

492. Kondo, Y. (1992). Lesions of the medial amygdala produce severe impairment of copulatory behavior in sexually inexperienced male rats. *Physiol. Behav.* 51, 939–943.

493. Kondo, Y., and Arai, Y. (1995). Functional association between the medial amygdala and the medial preoptic area in regulation of mating behavior in the male rat. *Physiol. Behav.* 57, 69–73.

494. Kondo, Y., and Sachs, B. D. (2002). Disparate effects of small medial amygdala lesions on noncontact erection, copulation, and partner preference. *Physiol. Behav.* 76, 443–447.

495. Kondo, Y., and Yamanouchi, K. (1995). The possible involvement of the nonstrial pathway of the amygdala in neural control of sexual behavior in male rats. *Brain Res. Bull.* 38, 37–40.

496. Kondo, Y., and Yamanouchi, K. (1997). Potentiation of ejaculatory activity by median raphe nucleus lesions in male rats: effect of p-chlorophenylalanine. *Endocr. J.* 44, 873–879.

497. Kondo, Y., Sachs, B. D., and Sakuma, Y. (1998). Importance of the medial amygdala in rat penile erection evoked by remote stimuli from estrous females. *Behav. Brain Res.* 91, 215–221.

497a. Marson, L. (2004). Lesions of the periaqueductal gray block the medial preoptic-induced activation of the urethrogenital reflex in male rats. *Neurosci. Lett.* 367, 278–282.

498. Kondo, Y., Shinoda, A., Yamanouchi, K., and Arai, Y. (1990). Role of septum and preoptic area in regulating masculine and feminine sexual behavior in male rats. *Horm. Behav.* 24, 421–434.

499. Kondo, Y., Tomihara, K., and Sakuma, Y. (1999). Sensory requirements for noncontact penile erection in the rat. *Behav. Neurosci.* 113, 1062–1070.

500. Kondo, Y., Yamanouchi, K., and Arai, Y. (1993). P-chlorophenylalanine facilitates copulatory behavior in septal lesioned but not in preoptic lesioned male rats. *J. Neuroendocrinol.* 5, 629–633.

501. Koob, G. F. (1992). Drugs of abuse: anatomy, pharmacology and function of reward pathways. *Trends Pharmacol. Sci.* 13, 177–184.

502. Kostarczyk, E. M. (1986). The amygdala and male reproductive functions: I. Anatomical and endocrine bases. *Neurosci. Biobehav. Rev.* 10, 67–77.

503. Kraemer, H. C., Becker, H. B., Brodie, H. K. H., Doering, C. H., Moos, R. H., and Hamburg, D. A. (1976). Orgasmic frequency and plasma testosterone levels in normal human males. *Arch. Sex. Behav.* 5, 125–132.
504. Krettek, J. E., and Price, J. L. (1978). Amygdaloid projections to subcortical structures within the basal forebrain and brainstem in the rat and cat. *J. Comp. Neurol.* 178, 225–254.
505. Krey, L. C., and McGinnis, M. Y. (1990). Time-courses of the appearance/disappearance of nuclear androgen + receptor complexes in the brain and adenohypophysis following testosterone administration/withdrawal to castrated male rats: relationships with gonadotropin secretion. *J. Steroid Biochem.* 35, 403–408.
506. Kriegsfeld, L. J., Demas, G. E., Huang, P. L., Burnett, A. L., and Nelson, R. J. (1999). Ejaculatory abnormalities in mice lacking the gene for endothelial nitric oxide synthase (eNOS–/–). *Physiol. Behav.* 67, 561–566.
507. Kritzer, M. F. (1997). Selective colocalization of immunoreactivity for intracellular gonadal hormone receptors and tyrosine hydroxylase in the ventral tegmental area, substantia nigra, and retrorubral fields in the rat. *J. Comp. Neurol.* 379, 247–260.
508. Krohmer, R. W., and Crews, D. (1987). Temperature activation of courtship behavior in the male red-sided garter snake (*Thamnophis sirtalis parietalis*): role of the anterior hypothalamus-preoptic area. *Behav. Neurosci.* 101, 228–236.
509. Kuhar, M. J., Couceyro, P. R., and Lambert, P. D. (1999). Catecholamines. In *Basic Neurochemistry* (G. J. Siegel, B. W. Agranoff, R. W. Albers, S. K. Fisher, and M. D. Uhler, Eds.), pp. 243–262. Lippincott-Raven, Philadelphia.
510. Kuo, R. C., Baxter, G. T., Thompson, S. H., Stricker, S. A., Patton, C., Bonaventura, J., and Epel, D. (2000). NO is necessary and sufficient for egg activation at fertilization. *Nature* 406, 633–636.
511. Kyuhou, S., and Gemba, H. (1998). Two vocalization-related subregions in the midbrain periaqueductal gray of the guinea pig. *NeuroReport* 9, 1607–1610.
512. Lagoda, G., Muschamp, J. M., Vigdorchik, A., and Hull, E. M. (2004). A nitric oxide synthase inhibitor in the medial preoptic area inhibits copulation and stimulus sensitization in male rats. *Behav. Neurosci.* 118, 1317–1323.
513. Lal, S., Ackman, D., Thavundayil, J. X., Kiely, M. E., and Etienne, P. (1984). Effect of apomorphine, a dopamine receptor agonist, on penile tumescence in normal subjects. *Prog. Neuropsychopharm. Biol. Psychiat.* 8, 695–699.
514. Lal, S., Laryea, E., Thavundayil, J. X., Nair, N. P., Negrete, J., Ackman, D., Blundell, P., and Gardiner, R. J. (1987). Apomorphine-induced penile tumescence in impotent patients: preliminary findings. *Prog. Neuropsychopharm. Biol. Psychiat.* 11, 235–242.
515. Lambert, G. M., Rubin, B. S., and Baum, M. J. (1992). Sex difference in the effect of mating on *c-fos* expression in luteinizing hormone-releasing hormone neurons of the ferret forebrain. *Endocrinology* 131, 1473–1480.
516. Langley, J. N., and Anderson, H. K. (1895). The innervation of the pelvic and adjoining viscera. *J. Physiol. (Lond.)* 19, 71–130.
517. Lanier, D. L., and Dewsbury, D. A. (1976). A quantitative study of copulatory behaviour of large *felidae*. *Behav. Proc.* 1, 327–333.
518. Larsson, K. (1959). Experience and maturation in the development of sexual behaviour in the male puberty rat. *Behaviour* 14, 101–107.
519. Larsson, K. (1966). Individual differences in reactivity to androgen in male rats. *Physiol. Behav.* 1, 255–258.
520. Larsson, K. (1967). Testicular hormone and developmental changes in mating behavior of the male rat. *J. Comp. Physiol. Psychol.* 63, 223–230.
521. Larsson, K. (1969). Failure of gonadal and gonadotrophic hormones to compensate for an impaired sexual function in anosmic male rats. *Physiol. Behav.* 4, 733–737.
522. Larsson, K. (1975). Sexual impairment of inexperienced male rats following pre- and postpubertal olfactory bulbectomy. *Physiol. Behav.* 14, 195–199.
523. Larsson, K. (1979). Features of the neuroendocrine regulation of masculine sexual behavior. In *Endocrine Control of Sexual Behavior* (C. Beyer, Ed.), pp. 77–163. Raven Press, New York.
524. Larsson, K., and Heimer, L. (1964). Mating behaviour of male rats after lesions in the preoptic area. *Nature* 202, 413–414.
525. Larsson, K., Füxe, K., Everitt, B. J., Holmgren, M., and Södersten, P. (1978). Sexual behavior in male rats after intracerebral injection of 5,7-dihydroxytryptamine. *Brain Res.* 141, 293–303.
526. Lavoisier, P., Courtois, F., Barres, D., and Blanchard, M. (1986). Correlation between intracavernous pressure and contraction of the ischiocavernosus muscle in man. *J. Urol.* 136, 936–939.
527. Lavoisier, P., Proulx, J., and Courtois, F. (1988). Reflex contractions of the ischiocavernosus muscles following electrical and pressure stimulations. *J. Urol.* 139, 396–399.
528. Lawrence, D. M., and Sawyer, G. I. M. (1974). Plasma testosterone and testosterone binding affinities in men with impotence, oligospermia and hypogonadism. *Brit. Med. J.* 1, 349–351.
529. LeDoux, J. E., Ruggiero, D. A., Forest, R., Stornetta, R., and Reis, D. J. (1987). Topographic organization of convergent projections to the thalamus from the inferior colliculus and spinal cord in the rat. *J. Comp. Neurol.* 264, 123–146.
530. Lee, M. R., Li, L., and Kitazawa, T. (1997). Cyclic GMP causes Ca2+ desensitization in vascular smooth muscle by activating the myosin light chain phosphatase. *J. Biol. Chem.* 272, 5063–5068.
531. Lee, R. L., Smith, E. R., Mas, M., and Davidson, J. M. (1990). Effects of intrathecal administration of 8-OH-DPAT on genital reflexes and mating behavior in male rats. *Physiol. Behav.* 47, 665–669.
532. Leedy, M. G., and Hart, B. L. (1986). Medial preoptic-anterior hypothalamic lesions in prepubertal male cats: effects on juvenile and adult sociosexual behaviors. *Physiol. Behav.* 36, 501–506.
533. Leedy, M. G., Vela, E. A., Popolow, H. B., and Gerall, A. A. (1980). Effect of prepubertal medial preoptic area lesions on male rat sexual behavior. *Physiol. Behav.* 24, 341–346.
534. Lehman, M. N. (1982). Neural pathways of the vomeronasal and olfactory systems controlling sexual behavior in the male golden hamster. Unpublished Ph.D. dissertation, University of Michigan, Ann Arbor.
535. Lehman, M. N., and Winans, S. S. (1982). Vomeronasal and olfactory pathways to the amygdala controlling male hamster sexual behavior: autoradiographic and behavioral analyses. *Brain Res.* 240, 27–41.
536. Lehman, M. N., Powers, J. B., and Winans, S. S. (1983). Stria terminalis lesions alter the temporal pattern of copulatory behavior in the male golden hamster. *Behav. Brain Res.* 8, 109–128.
537. Lehman, M. N., Winans, S. S., and Powers, J. B. (1980). Medial nucleus of the amygdala mediates chemosensory control of male hamster sexual behavior. *Science* 210, 557–560.

537a. Leipheimer, R. E., and Sachs, B. D. (1988). GABAergic regulation of penile reflexes and copulation in rats. *Physiol. Behav.* 42, 351–357.

538. Levis, D. G., and Ford, J. J. (1989). The influence of androgenic and estrogenic hormones on sexual behavior in castrated adult male pigs. *Horm. Behav.* 23, 393–411.
539. Leypold, B. G., Yu, C. R., Leinders-Zufall, T., Kim, M. M., Zufall, F., Axel, R. (2002). Altered sexual and social behaviors in trp2 mutant mice. *Proc. Natl. Acad. Sci. U S A* 99, 6376–6381.

540. Leyton, M., and Stewart, J. (1992). The stimulation of central kappa opioid receptors decreases male sexual behavior and locomotor activity. *Brain Res.* 594, 56–74.
541. Leyton, M., and Stewart, J. (1996). Acute and repeated activation of male sexual behavior by tail pinch: opioid and dopaminergic mechanisms. *Physiol Behav.* 60, 77–85.
542. Li, H. Y., Blaustein, J. D., De Vries, G. J., and Wade, G. N. (1993). Estrogen-receptor immunoreactivity in hamster brain: preoptic area, hypothalamus and amygdala. *Brain Res.* 631, 304–312.
543. Linden, A., Hansen, S., Bednar, I., Forsberg, G., Södersten, P., and Uvnas-Moberg, K. (1987). Sexual activity increases plasma concentrations of cholecystokinin octapeptide and offsets hunger in male rats. *J. Endocrinol.* 115, 91–95.
544. Lindzey, J., and Crews, D. (1988). Effects of progestins on sexual behaviour in castrated lizards (*Cnemidophorus inornatus*). *J. Endocrinol.* 119, 265–273.
545. Lisciotto, C. A., and Morrell, J. I. (1994). Sex differences in the distribution and projections of testosterone target neurons in the medial preoptic area and the bed nucleus of the stria terminalis of rats. *Horm. Behav.* 28, 492–502.
546. Lisk, R. D. (1967). Neural localization for androgen activation of copulatory behavior in the male rat. *Endocrinology* 80, 754–761.
547. Lisk, R. D., and Bezier, J. L. (1980). Intrahypothalamic hormone implantation and activation of sexual behavior in the male hamster. *Neuroendocrinology* 30, 220–227.
548. Lisk, R. D., and Greenwald, D. P. (1983). Central plus peripheral stimulation by androgen is necessary for complete restoration of copulatory behavior in the male hamster. *Neuroendocrinology* 36, 211–217.
549. Lisk, R. D., and Heimann, J. (1980). The effects of sexual experience and frequency of testing on retention of copulatory behavior following castration in the male hamster. *Behav. Neural. Biol.*, 28, 156–171.
550. Lisk, R. D., Zeiss, J., and Ciaccio, L. A. (1972). The influence of olfaction on sexual behavior in the male golden hamster (*Mesocricetus auratus*). *J. Exp. Zool.* 181, 69–78.
551. Liu, Y, Curtis, J. T., and Wang Z. (2001). Vasopressin in the lateral septum regulates pair bond formation in male prairie voles (*Microtus ochrogaster*). *Horm. Behav.* 115, 910–919.
552. Liu, Y. C., Sachs, B. D., and Salamone J. D. (1998). Sexual behavior in male rats after radiofrequency or dopamine-depleting lesions in nucleus accumbens. *Pharmacol. Biochem. Behav.* 60, 585–592.
553. Liu, Y. C., Salamone, J. D., and Sachs, B. D. (1997a). Impaired sexual response after lesions of the paraventricular nucleus of the hypothalamus in male rats. *Behav. Neurosci.* 111, 1361–1367.
554. Liu, Y. C., Salamone, J. D., and Sachs, B. D. (1997b). Lesions in medial preoptic area and bed nucleus of stria terminalis: differential effects on copulatory behavior and noncontact erection in male rats. *J. Neurosci.* 17, 5245–5253.
555. Lloyd, S. A., and Dixson, A. F. (1988). Effects of hypothalamic lesions upon the sexual and social behaviour of the male common marmoset (*Callithrix jacchus*). *Brain Res.* 463, 317–329.
556. Lopez, H. H., and Ettenberg, A. (2000). Haloperidol challenge during copulation prevents subsequent increase in male sexual motivation. *Pharmacol. Biochem. Behav.* 67, 387–393.
557. Lopez, H. H., and Ettenberg, A. (2001). Dopamine antagonism attenuates the unconditioned incentive value of estrous female cues. *Pharmacol. Biochem. Behav.* 68, 411–416.
558. Lopez, H. H., and Ettenberg, A. (2002a). Exposure to female rats produces differences in c-fos induction between sexually-naïve and experienced male rats. *Brain Res.* 947, 57–66.
559. Lopez, H. H., and Ettenberg, A. (2002b). Sexually conditioned incentives: attenuation of motivational impact during dopamine receptor antagonism. *Pharmacol. Biochem. Behav.* 72, 65–72.
560. Lopez, H. H., Olster, D. H., and Ettenberg, A. (1999). Sexual motivation in the male rat: the role of primary incentives and copulatory experience. *Horm. Behav.* 36, 176–185.
561. Lorrain, D. S., and Hull, E. M. (1993). Nitric oxide increases dopamine and serotonin release in the medial preoptic area. *NeuroReport* 5, 87–89.
562. Lorrain, D. S., Matuszewich, L., Friedman, R. D., and Hull, E. M. (1997). Extracellular serotonin in the lateral hypothalamic area is increased during postejaculatory interval and impairs copulation in male rats. *J. Neurosci.* 17, 9361–9366.
563. Lorrain, D. S., Matuszewich, L., and Hull, E. M. (1998). 8-OH-DPAT influences extracellular levels of serotonin and dopamine in the medial preoptic area of male rats. *Brain Res.* 790, 217–223.
564. Lorrain, D. S., Matuszewich, L., Howard, R. V., Du, J., and Hull, E. M. (1996). Nitric oxide promotes medial preoptic dopamine release during male rat copulation. *NeuroReport* 8, 31–34.
565. Lorrain, D. S., Riolo, J. V., Matuszewich, L., and Hull, E. M. (1999). Lateral hypothalamic serotonin inhibits nucleus accumbens dopamine: implications for sexual refractoriness. *J. Neurosci.* 19, 7648–7652.
566. Lott, D. F. (1981). Sexual behavior and intersexual strategies in American bison. *Z. Tierpsychol.* 56, 97–114.
567. Louilot, A., Gonzalez-Mora, J. L., Guadalupe, T., and Mas, M. (1991). Sex-related olfactory stimuli induce a selective increase in dopamine release in the nucleus accumbens of male rats: a voltammetric study. *Brain Res.* 553, 313–317.
568. Lue, T. F., Gleason, C. A., Brock, G. B., Carroll, P. R., and Tanagho, E. A. (1995). Intraoperative electrostimulation of the cavernous nerve: technique, results and limitations. *J. Urol.* 154, 1426–1428.
569. Lue, T. F., Takamura, T., Schmidt, R. A., Palubinskas, A. J., and Tanagho, E. A. (1983). Hemodynamics of erection in the monkey. *J. Urol.* 130:1237–1241.
570. Lue, T. F. Takamura, T., Umraiya, M., Schmidt, R. A., and Tanagho, E. A. (1984). Hemodynamics of canine corpora cavernosa during erection. *Urology* 24, 347–352.
571. Lugg, J. A., Rajfer, J., and Gonzalez-Cadavid, N. F. (1995). Dihydrotestosterone is the active androgen in the maintenance of nitric oxide-mediated penile erection in the rat. *Endocrinology* 136, 1495–1501.
572. Lumia, A. R., Sachs, B. D., and Meisel, R. L. (1980). Spinal transection restores sexual reflexes of rats following suppression by penile sheath removal. *Physiol. Behav.* 25, 89–92.
573. Lumia, A. R., Zebrowski, A. F., and McGinnis, M. Y. (1987). Olfactory bulb removal decreases androgen receptor binding in amygdala and hypothalamus and disrupts masculine sexual behavior. *Brain Res.* 404, 121–126.
574. Lumley, L. A., and Hull, E. M. (1999). Effects of a D_1 antagonist and of sexual experience on copulation-induced Fos-like immunoreactivity in the medial preoptic nucleus. *Brain Res.* 829, 55–68.
575. Luttge, W. G., and Hall, N. R. (1973). Differential effectiveness of testosterone and its metabolites in the induction of male sexual behavior in two strains of albino mice. *Horm. Behav.* 4, 31–43.
576. Macey, M. J., Pickford, G. E., and Peter, R. E. (1974). Forebrain localization of the spawning reflex response to exogenous neurohypophysial hormones in the killfish, *Fundulus heteroclitus*. *J. Exp. Zool.* 190, 269–280.
577. Mackel, R. (1979). Segmental and descending control of the external urethral and anal sphincters in the cat. *J. Physiol.* 294, 105–122.
578. MacLean, P. D., and Ploog, D. W. (1962). Cerebral representation of penile erection. *J. Neurophysiol.* 25, 29–55.

579. MacLean, P. D., Dua, S., and Denniston, R. H. (1963). Cerebral localization for scratching and seminal discharge. *Arch. Neurol.* 9, 485–497.
580. Maeda, N., Matsuoka, N., and Yamaguchi, I. (1990). Septohippocampal cholinergic pathway and penile erections induced by dopaminergic and cholinergic stimulants. *Brain Res.* 537, 163–168.
581. Maeda, N., Matsuoka, N., and Yamaguchi, I. (1994). Role of the dopaminergic, serotonergic and cholinergic link in the expression of penile erection in rats. *Jap. J. Pharmacol.* 66, 59–66.
582. Mahalati, K., Okanoya, K., Witt, D. M., and Carter, C. S. (1991). Oxytocin inhibits male sexual behavior in prairie voles. *Pharmacol. Biochem. Behav.* 39, 219–222.
583. Maillard, C. A., and Edwards, D. A. (1991). Excitotoxin lesions of the zone incerta/lateral tegementum continuum: effects on male sexual behavior in rats. *Behav. Brain Res.* 46, 143–149.
584. Mains, R. E., and Eipper, B. A. (1999). Peptides. In *Basic Neurochemistry* (G. J. Siegel, B. W. Agranoff, R. W. Albers, S. K. Fisher, and M. D. Uhler, Eds.), 6th ed., pp. 363–382. Lippencott-Raven Publishers, Philadelphia.
585. Mallick, H., Manchanda, S. K., and Kumar, V. M. (1996). Beta-adrenergic modulation of male sexual behavior elicited from the medial preoptic area in rats. *Behav. Brain Res.* 74, 181–187.
586. Malmnas, C. O. (1976). The significance of dopamine, versus other catecholamines, for L-dopa induced facilitation of sexual behavior in the castrated male rat. *Pharmacol. Biochem. Behav.* 4, 521–526.
587. Malmnas, C. O., and Meyerson, B. J. (1971). p-Chlorophenylalanine and copulatory behaviour in the male rat. *Nature* 232, 398–400.
588. Malsbury, C. W. (1971). Facilitation of male rat copulatory behavior by electrical stimulation of the medial preoptic area. *Physiol. Behav.* 7, 797–805.
589. Mani, S, K., Blaustein, J. D., and O'Malley, B. W. (1997). Progesterone receptor function from a behavioral perspective. *Horm. Behav.* 31, 244–255.
590. Mani, S. K., Allen, J. M., Clark, J. H., Blaustein, J. D., and O'Malley, B. W. (1994a). Convergent pathways for steroid hormone- and neurotransmitter-induced rat sexual behavior [published erratum appears in *Science* (1995). 268, 1833]. *Science* 265, 1246–1249.
591. Mani, S. K., Blaustein, J. D., Allen, J. M., Law, S. W., O'Malley, B. W., and Clark, J. H. (1994b). Inhibition of rat sexual behavior by antisense oligonucleotides to the progesterone receptor. *Endocrinology* 135, 1409–1414.
592. Mani, S. K., Fienberg, A. A., O'Callaghan, J. P., Snyder, G. L., Allen, P. B., Dash, P. K., Moore, A. N., Mitchell, A. J., Bibb, J., Greengard, P., and O'Malley, B. W. (2000). Requirement for DARPP-32 in progesterone-facilitated sexual receptivity in female rats and mice. *Science* 287, 1053–1056.
593. Manzo, J., Cruz, M. R., Hernandez, M. E., Pacheco, P., and Sachs, B. D. (1999). Regulation of noncontact erection in rats by gonadal steroids. *Horm. Behav.* 35, 264–270.
594. Maragos, W. F., Newman, S. W., Lehman, M. N., and Powers, J. B. (1989). Neurons of origin and fiber trajectory of amygdalofugal projections to the medial preoptic area in Syrian hamsters. *J. Comp. Neurol.* 280, 59–71.
595. Marin, R., Escrig, A., Abreu, P., and Mas, M. (1999). Androgen-dependent nitric oxide release in rat penis correlates with levels of constitutive nitric oxide synthase isoenzymes. *Biol. Reprod.* 61, 1012–1016.
596. Markowski, V. P., and Hull, E. M. (1995). Cholecystokinin modulates mesolimbic dopaminergic influences on male rat copulatory behavior. *Brain Res.* 699, 266–274.
597. Markowski, V. P., Eaton, R. C., Lumley, L. A., Moses, J., and Hull, E. M. (1994). A D_1 agonist in the MPOA area facilitates copulation of male rats. *Pharm. Biochem. Behav.* 47, 483–486.
597a. Marson, L. (2004). Lesions of the periaqueductal gray block the medial preoptic area-induced activation of the urethrogenital reflex in male rats. *Neurosci. Lett.* 367, 278–282.
598. Marson, L., and McKenna, K. E. (1990). The identification of a brainstem site controlling spinal sexual reflexes in male rats. *Brain Res.* 515, 303–308.
599. Marson, L., and McKenna, K. E. (1992). A role for 5-hydroxytryptamine in descending inhibition of spinal sexual reflexes. *Exp. Brain Res.* 88, 313–320.
600. Marson, L., and McKenna, K. E. (1994a). Serotonergic neurotoxic lesions facilitate male sexual reflexes. *Pharmacol. Biochem. Behav.* 47, 883–888.
601. Marson, L., and McKenna, K. E. (1994b). Stimulation of the hypothalamus initiates the urethrogenital reflex in male rats. *Brain Res.* 638, 103–108.
602. Marson, L., List, M. S., and McKenna, K. E. (1992). Lesions of the nucleus paragigantocellularis alter ex copula penile reflexes. *Brain Res.* 592, 187–192.
603. Martin, J. R. (1976). Motivated behaviors elicited from hypothalamus, midbrain, and pons of guinea pig (*Cavia porcellus*). *J. Comp. Physiol. Psychol.* 90, 1011–1034.
604. Martin, W. J., McGowan, E., Cashen, D. E., Gantert, L. T., Drisko, J. E., Hom, G. J., Nargund, R., Sebhat, I., Howard, A. D., Van der Ploeg, L. H., and MacIntyre, D. E. (2002). Activation of melanocortin MC(4) receptors increases erectile activity in rats ex copula. *Eur. J. Pharmacol.* 454, 71–79.
605. Mas, M., Fernandez-Vera, J. R., and Gonzalez-Mora, J. L. (1995a). Neurochemical correlates of sexual exhaustion and recovery as assessed by in vivo microdialysis. *Brain Res.* 675, 13–19.
606. Mas, M., Fumero, B., and Perez-Rodriguez, I. (1995b). Induction of mating behavior by apomorphine in sexually sated rats. *Euro. J. Pharmacol.* 280, 331–334.
607. Mas, M., Gonzalez-Mora, J. L., Louilot, A., Sole, C., and Guadalupe, T. (1990). Increased dopamine release in the nucleus accumbens of copulating male rats as evidenced by in vivo voltammetry. *Neurosci. Lett.* 110, 303–308.
608. Mas, M., Rodriguez del Castillo, A., Guerra, M., Davidson, J. M., and Battaner, E. (1987). Neurochemical correlates of male sexual behavior. *Physiol. Behav.* 41, 341–345.
609. Matochik, J. A., Sipos, M. L., Nyby, J. G., and Barfield, R. J. (1994). Intracranial androgenic activation of male-typical behaviors in house mice: motivation versus performance. *Behav. Brain Res.* 60, 141–149.
610. Matsumoto, T., Honda, S., and Harada, N. (2003). Neurological effects of aromatase deficiency in the mouse. *J. Steroid Biochem. Mol. Biol.* 86, 357–365.
611. Matsumoto, T., Kondo, Y., Sachs, B. D., and Yamanouchi, I. (1997). Effects of p-chlorophenylalanine on reflexive and noncontact penile erections in male rats. *Physiol. Behav.* 61, 165–168.
612. Matsuoka, M., Yokosuka, M., Mori, Y., and Ichikawa, M. (1999). Specific expression pattern of Fos in the accessory olfactory bulb of male mice after exposure to soiled bedding of females. *Neurosci. Res.* 35, 189–195.
613. Matuszewich, L., Lorrain, D. S., Trujillo, R., Dominguez, J., Putnam, S. K., and Hull, E. M. (1999). Partial antagonism of 8-OH-DPAT's effects on male rat sexual behavior with a D_2, but not a 5-HT_{1A}, antagonist. *Brain Res.* 820, 55–62.
614. Matuszewich, L., Ormsby, J. L., Moses, J., Lorrain, D. S., and Hull, E. M. (1995). Effects of morphiceptin in the medial preoptic area on male sexual behavior. *Psychopharmacology* 122, 330–335.
615. McDonald, P. G., Beyer, C., Newton, F., Brien, B., Baker, R., Tan, H. S., Sampson, C., Kitching, P., Greenhill, R., and Pritchard, D. (1970). Failure of 5a-dihydrotestosterone to initiate sexual behavior in the castrated male rat. *Nature* 227, 964–965.

616. McGill, T. E. (1962). Sexual behavior in three inbred strains of mice. *Behaviour* 19, 341–350.
617. McGinnis, M. Y., and Dreifuss, R. M. (1989). Evidence for a role of testosterone-androgen receptor interactions in mediating masculine sexual behavior in male rats. *Endocrinology* 124, 618–626.
618. McGinnis, M. Y., and Kahn, D. F. (1997). Inhibition of male sexual behavior by intracranial implants of the protein synthesis inhibitor anisomycin into the medial preoptic area of the rat. *Horm Behav.* 31, 15–23.
619. McGinnis, M. Y., and Mirth, M. C. (1986). Inhibition of cell nuclear androgen receptor binding and copulation in male rats by an antiandrogen, Sch 16423. *Neuroendocrinology* 43, 63–68.
620. McGinnis, M. Y., Mirth, M. C., Zebrowski, A. F., and Dreifuss, R. M. (1989). Critical exposure time for androgen activation of male sexual behavior in rats. *Physiol. Behav.* 46, 159–165.
621. McGinnis, M. Y., Montana, R. C., and Lumia, A. R. (2002). Effects of hydroxyflutamide in the medial preoptic area or lateral septum on reproductive behaviors in male rats. *Brain Res. Bull.* 59, 227–234.
622. McGinnis, M. Y., Williams, G. W., and Lumia, A. R. (1996). Inhibition of male sex behavior by androgen receptor blockade in preoptic area or hypothalamus, but not amygdala or septum. *Physiol Behav.* 60, 783–789.
623. McGregor, A., and Herbert, J. (1992). Differential effects of excitotoxic basolateral and corticomedial lesions of the amygdala on the behavioural and endocrine responses to either sexual or aggression-promoting stimuli in the male rat. *Brain Res.* 574, 9–20.
624. McIntosh, T. K., and Barfield, R. J. (1984). Brain monoaminergic control of male reproductive behavior. III. Norepinephrine and the postejaculatory refractory period. *Behav. Brain Res.* 12, 275–281.
625. McKenna, K. E., and Nadelhaft, I. (1986). The organization of the pudendal nerve in the male and female rat. *J. Comp. Neurol.* 248, 532–549.
626. McKenna, K. E., Chung, S. K., and McVary, K. T. (1991). A model for the study of sexual function in anesthetized male and female rats. *Am. J. Physiol.* 30, R1276–R1285.
627. McMahon, C. G., and Touma, K. (1999). Treatment of premature ejaculation with paroxetine hydrochloride. *Int. J. Impot. Res.* 11, 241–245.
628. Meisel, R. L. (1982). Effects of postweaning rearing condition on recovery of copulatory behavior from lesions of the medial preoptic area in rats. *Dev. Psychobiol.* 15, 331–338.
629. Meisel, R. L. (1983). Recovery of masculine copulatory behavior from lesions of the medial preoptic area: effects of age versus hormonal state. *Behav. Neurosci.* 97, 785–793.
630. Meisel, R. L., and Sachs, B. D. (1994). The physiology of male sexual behavior. In *Physiology of Reproduction* (E. Knobil, and J. D. Neill, Eds.), pp. 3–106. Raven Press, New York.
631. Meisel, R. L., and Sachs, B. D. (1980). Spinal transection accelerates the developmental expression of penile reflexes in male rats. *Physiol. Behav.* 24, 289–292.
632. Meisel, R. L., Leipheimer, R. E., and Sachs, B. D. (1986). Anisomycin does not disrupt the activation of penile reflexes by testosterone in rats. *Physiol. Behav.* 37, 951–956.
633. Meisel, R. L., Lumia, A. R., and Sachs, B. D. (1980). Effects of olfactory bulb removal and flank shock on copulation in male rats. *Physiol. Behav.* 25, 383–387.
634. Meisel, R. L., Lumia, A. R., and Sachs, B. D. (1982). Disruption of copulatory behavior of male rats by olfactory bulbectomy at two, but not ten days of age. *Exp. Neurol.* 77, 612–624.
635. Meisel, R. L., O'Hanlon, J. K., and Sachs, B. D. (1984a). Differential maintenance of penile responses and copulatory behavior by gonadal hormones in castrated male rats. *Horm. Behav.* 18, 56–64.
636. Meisel, R. L., Sachs, B. D., and Lumia, A. R. (1984b). Olfactory bulb control of sexual function. In *Early Brain Damage* (S. Finger and C. R. Almli, Eds.), Vol. 2, pp. 253–268. Academic Press, Orlando, FL.
637. Melis, M. R., and Argiolas, A. (1995). Dopamine and Sexual Behavior. *Neurosci. Biobehav. Rev.* 19, 19–38.
638. Melis, M. R., and Argiolas, A. (1997). Role of central nitric oxide in the control of penile erection and yawning. *Prog. Neuro-Psychopharmacol. Biol. Psychiat.* 21, 899–922.
639. Melis, M. R., Argiolas, A., and Gessa, G. L. (1987). Apomorphine-induced penile erection and yawning: site of action in brain. *Brain Res.* 415, 98–104.
640. Melis, M. R., Argiolas, A., and Gessa, G. L. (1989). Evidence that apomorphine induces penile erection and yawning by releasing oxytocin in the central nervous system. *Eur. J. Pharmacol.* 164, 565–572.
641. Melis, M. R., Spano, M. S., Succu, S., and Argiolas, A. (1999a). The oxytocin antagonist d(CH2)5Tyr(Me)2-Orn8-vasotocin reduces non-contact penile erections in male rats. *Neurosci. Lett.* 265, 171–174.
642. Melis, M. R., Spano, M. S., Succu, S., and Argiolas, A. (2000a). Activation of gamma-aminobutyric acid(A) receptors in the paraventricular nucleus of the hypothalamus reduces apomorphine-, N-methyl-D-aspartic acid- and oxytocin-induced penile erection and yawning in male rats. *Neurosci. Lett.* 281, 127–130.
643. Melis, M. R., Spano, M. S., Succu, S., Locatelli, V., Torsello, A., Muller, E. E., Deghenghi R. Argiolas A. (2000b). EP 60761- and EP 50885-induced penile erection: structure–activity studies and comparison with apomorphine, oxytocin and N-methyl-D-aspartic acid. *Int. J. Impot. Res.* 12, 255–262.
644. Melis, M. R., Stancampiano, R., and Argiolas, A. (1992). Hippocampal oxytocin mediates apomorphine-induced penile erection and yawning. *Pharm. Biochem. Behav.* 42, 61–66.
645. Melis, M. R., Succu, S., and Argiolas, A. (1996). Dopamine agonists increase nitric oxide production in the paraventricular nucleus of the hypothalamus: correlation with penile erection and yawning. *Eur. J. Neurosci.* 8, 2056–2063.
646. Melis, M. R., Succu, S., Iannucci, U., and Argiolas, A. (1997). N-methyl-D-aspartic acid-induced penile erection and yawning: role of hypothalamic paraventricular nitric oxide. *Eur. J. Pharmacol.* 328,115–123.
647. Melis, M. R., Succu, S., Mascia, M. S., Cortis, L., and Argiolas, A. (2003). Extra-cellular dopamine increases in the paraventricular nucleus of the hypothalamus: correlation with penile erection and yawning. *Eur. J. Neurosci.* 17, 1266–1272.
648. Melis, M. R., Succu, S., Mascia, M. S., Cortis, L., and Argiolas, A. (2004). Extracellular excitatory amino acids increase in the paraventricular nucleus of male rats during sexual activity: main role of N-methyl-d-aspartic acid receptors in erectile function. *Eur. J. Neurosci.* 19, 2569–2575.
649. Melis, M. R., Succu, S., Mauri, A., and Argiolas, A. (1998). Nitric oxide production is increased in the paraventricular nucleus of the hypothalamus of male rats during non-contact penile erections and copulation. *Eur. J. Neurosci.* 10, 1968–1974.
650. Melis, M. R., Succu, S., Spano, M. S., and Argiolas, A. (1999b). Morphine injected into the paraventricular nucleus of the hypothalamus prevents noncontact penile erections and impairs copulation: involvement of nitric oxide. *Eur. J. Neurosci.* 11, 1857–1864.
651. Melis, M. R., Succu, S., Spano, M. S., and Argiolas, A. (2000c). Effect of excitatory amino acid, dopamine, and oxytocin receptor antagonists on noncontact penile erections and paraventricular nitric oxide production in male rats. *Behav. Neurosci.* 114, 849–857.

652. Melis, M. R., Succu, S., Spano, M. S., Deghenghi, R., and Argiolas, A. (2001). EP 91073 prevents EP 80661-induced penile erection: new evidence for the existence of specific EP peptide receptors mediating penile erection. *Neuropharmacology* 41, 254–262.
653. Mendelson, S. D., and Pfaus, J. G. (1989). Level searching: a new assay of sexual motivation in the male rat. *Physiol. Behav.* 45, 337–341.
654. Mendonca, M. T., Chemetsky, S. D., Nester, K. E, and Gardner, G. L. (1996). Effects of gonadal sex steroids on sexual behavior in the big brown bat, *Eptesicus fuscus*, upon arousal from hibernation. *Horm. Behav.* 30, 153–161.
655. Merari, A., and Ginton, A. (1975). Characteristics of exaggerated sexual behavior induced by electrical stimulation of the medial preoptic area in male rats. *Brain Res.* 86, 97–108.
656. Meredith, M. (1986). Vomeronasal organ removal before sexual experience impairs male hamster mating behavior. *Physiol. Behav.* 36, 737–743.
657. Meredith, M. (1998). Vomeronasal, olfactory, hormonal convergence in the brain. Cooperation or coincidence? *Ann. N Y Acad. Sci.* 855, 349–361.
658. Meredith, M., and Fernandez-Fewell, G. (1994). Vomeronasal system, LHRH, and sex behaviour. *Psychoneuroendocrinology* 19, 657–672.
659. Meredith, M., and Fewell, G. (2001). Vomeronasal organ: electrical stimulation activates Fos in mating pathways and in GnRH neurons. *Brain Res.* 922, 87–94.
660. Meredith, M., Marques, D. M., O'Connell, R. J., and Stem, F. L. (1980). Vomeronasal pump: significance for male hamster sexual behavior. *Science* 207, 1224–1226.
661. Mermelstein, P. G., Becker, J. B., and Surmeier, D. J. (1996). Estradiol reduces calcium currents in rat neostriatal neurons via a membrane receptor. *J. Neurosci.* 16, 595–604.
662. Mesulam, M. M. (1995). Structure and function of cholinergic pathways in the cerebral cortex, limbic system, basal ganglia, and thalamus of the human brain. In *Psychopharmacology: The Fourth Generation of Progress* (F. E. Bloom, and D. J. Kupfer, Eds.), pp. 135–146. Raven Press, New York.
663. Meuleman, E., Cuzin, B., Opsomer, R. J., Hartmann, U., Bailey, M. J., Maytom, M. C., Smith, M. D., and Osterloh, I. H. (2001). A dose-escalation study to assess the efficacy and safety of sildenafil citrate in men with erectile dysfunction. *BJU Int.* 87, 75–81.
664. Michael, R. P., and Wilson, M. (1974). Effects of castration and hormone replacement in fully adult male rhesus monkeys (*Macaca mulatta*). *Endocrinology* 95, 150–159.
665. Michael, R. P., Clancy, A. N., and Zumpe, D. (1999). Effects of mating on *c-fos* expression in the brains of male macaques. *Physiol. Behav.* 66, 591–597.
666. Michael, R. P., Zumpe, D., and Bonsall, R. W. (1986). Comparison of the effects of testosterone and dihydrotestosterone on the behavior of male cynomologus monkeys (*Macaca fascicularis*). *Physiol. Behav.* 36, 349–355.
667. Millan, M. J., Millan, M. H., Czlonkowski, A., and Herz, A. (1984). Vasopressin and oxytocin in the rat spinal cord: distribution and origins in comparison to (Met)enkephalin, dynorphin and related opioids and their irresponsiveness to stimuli modulating neurohypophyseal secretion. *Neuroscience* 13, 179–187.
668. Millan, M. J., Peglion, J. L, Lavielle, G., and Perrin-Monneyron, S. (1997). 5-HT_{2C} receptors mediate penile erections in rats, actions of novel and selective agonists and antagonists. *Eur. J. Pharmacol.* 325, 9–12.
669. Miller, M. A., Morgan, R. J., Thompson, C. S., Mikhailidis, D. P., and Jeremy, J. Y. (1995). Effects of papaverine and vasointestinal polypeptide on penile and vascular cAMP and cGMP in control and diabetic animals: an in vitro study. *Int. J. Impot. Res.* 7, 91–100.
670. Mills, T. M., Reilly, C. M., Lewis, R. W. (1996). Androgens and penile erection: a review. *J. Androl.* 17, 633–638.
671. Mills, T. M., Stopper, V. S., and Wiedmeier, V. T. (1994). Effect of castration and androgen replacement on the hemodynamics of penile erection in the rat. *Biol. Reprod.* 51, 234–238.
672. Mills, T. M., Wiedmeier, V. T., and Stopper, V. S. (1992). Androgen maintenance of erectile function in the rat penis. *Biol. Reprod.* 46, 342–348.
673. Minabe, Y., Ashby, C. R., Jr., and Wang, R. Y. (1991). The CCK-A receptor antagonist devazepide but not the CCK-B antagonist L-365,260 reverses the effects of chronic clozapine and haloperidol on midbrain dopamine neurons. *Brain Res.* 549, 151–154.
674. Mitchell, J. B., and Gratton, A. (1991). Opioid modulation and sensitization of dopamine release elicited by sexually relevant stimuli: a high speed chronoamperometric study in freely behaving rats. *Brain Res.* 551, 20–27.
675. Mitchell, J. B., and Stewart, J. (1989). Effects of castration, steroid replacement, and sexual experience on mesolimbic dopamine and sexual behaviors in the male rat. *Brain Res.* 491, 116–127.
676. Mitler, M. M., Morden, B., Levine, S., and Dement, W. (1972). The effects of parachlorophenylalanine on the mating behavior of male rats. *Physiol. Behav.* 8, 1147–1150.
677. Miura, T., Kondo, Y., Akimoto, M., and Sakuma, Y. (2001). Electromyography of male rat perineal musculature during copulatory behavior. *Urol. Int.* 67, 240–245.
678. Mizusawa, H., Hedlund, P., and Andersson, K. E. (2002). Alpha-melanocyte stimulating hormone and oxytocin induced penile erections, and intracavernous pressure increases in the rat. *J. Urol.* 167, 757–760.
679. Molinoff, P. B., Shadiack, A. M., Earle, D., Diamond, L. E., and Quon, C. Y. (2003). PT-141: a melanocortin agonist for the treatment of sexual dysfunction. *Ann. N Y Acad. Sci.* 994, 96–102.
680. Moltz, H. (1990). E-series prostaglandins and arginine vasopressin in the modulation of male sexual behavior. *Neurosci. Biobehav. Rev.* 14, 109–115.
681. Monaghan, E. P., and Breedlove, S. M. (1992). The role of the bulbocavernosus in penile reflex behavior in rats. *Brain Res.* 587, 178–180.
682. Monaghan, E. P., Arjomand, J., and Breedlove, S. M. (1993). Brain lesions affect penile reflexes. *Horm. Behav.* 27, 122–131.
683. Moore, F. L., and Evans, S. J. (1999). Steroid hormones use non-genomic mechanisms to control brain functions and behaviors: a review of evidence. *Brain Behav. Evol.* 54(1), 41–50.
684. Moore, K. E., and Lookingland, K. J. (1995). Dopaminergic neuronal systems in the hypothalamus. In *Psychopharmacology: The Fourth Generation of Progress* (F. E. Bloom and D. J. Kupfer, Eds.), pp. 245–246. Raven Press, New York.
685. Morali, G., Lemus, A. E., Munguia, R., Arteaga, M., Perez-Palacios, G., Sundaram, K., Kumar, N., and Bardin, C. W. (1993). Induction of male sexual behavior in the rat by 7 alpha-methyl-19-nortestosterone, an androgen that does not undergo 5 alpha-reduction. *Biol. Reprod.* 49, 577–581.
686. Moreland, R. B., Goldstein, I., Kim, N. N., and Traish, A. (1999). Sildenafil citrate, a selective phosphodiesterase type 5 inhibitor: research and clinical implications in erectile dysfunction. *Trends Endocrinol. Metab.* 10, 97–104.
687. Morgan, C., Nadelhaft, I., and de Groat, W. C. (1981). The distribution of visceral primary afferents from the pelvic nerve to Lissauer's tract and the spinal gray matter and its relationship to the sacral parasympathetic nucleus. *J. Comp. Neurol.* 201, 415–440.
688. Morgantaler, A., and Crews, D. (1978). Role of the anterior hypothalamus-preoptic area in the regulation of reproductive behavior in the lizard, *Anolis carolinensis*: implantation studies. *Horm. Behav.* 11, 61–73.

689. Mortimer, C. H., McNeilly, A. S., Fisher, R. A., Murray, M. A. F., and Besser, G. M. (1974). Gonadotrophin-releasing hormone therapy in hypogonadal males with hypothalamic or pituitary dysfunction. *Br. Med. J.* 4, 617–621.
690. Mos, J., Mollet, I., Tolboom, J. T., Waldinger, M. D., and Olivier, B. (1999). A comparison of the effects of different serotonin reuptake blockers on sexual behaviour of the male rat. *Euro. J. Neuropsychopharmacol.* 9, 123–135.
691. Moses, J., and Hull, E. M. (1994). Inhibition of nitric oxide synthase in the MPOA increases seminal emissions in restrained supine rats and reduces the incidence of copulation in the copulation test. *Conf. Reprod. Behav. Abst.* 26, 37.
692. Moses, J., and Hull, E. M. (1999). A nitric oxide synthesis inhibitor administered into the medial preoptic area increases seminal emissions in an *ex copula* reflex test. *Pharmacol. Biochem. Behav.* 63, 345–348.
693. Moses, J., Loucks, J. A., Watson, H. L., Matuszewich, L., and Hull, E. M. (1995). Dopaminergic drugs in the medial preoptic area and nucleus accumbens: effects on motor activity, sexual motivation, and sexual performance. *Pharmacol. Biochem. Behav.* 51, 681–686.
694. Moss, R. L., McCann, S. M., and Dudley, C. A. (1975). Releasing hormones and sexual behavior. *Prog. Brain. Res.* 42, 37–46.
695. Murphy, A. Z., and Hoffman, G. E. (2001). Distribution of gonadal steroid receptor-containing neurons in the preoptic-periaqueductal gray-brainstem pathway: a potential circuit for the initiation of male sexual behavior. *J. Comp. Neurol.* 438, 191–212.
696. Murphy, A. Z., Rizvi, T. A., Ennis, M., and Shipley, M. T. (1999a). The organization of preoptic-medullary circuits in the male rat, evidence for interconnectivity of neural structures involved in reproductive behavior, antinociception and cardiovascular regulation. *Neuroscience* 91, 1103–1116.
697. Murphy, A. Z., Shupnik, M. A., and Hoffman, G. E. (1999b). Androgen and estrogen (alpha) receptor distribution in the periaqueductal gray of the male rat. *Horm. Behav.* 36, 98–108.
698. Murphy, M. R., and Schneider, G. E. (1970). Olfactory bulb removal eliminates mating behavior in the male golden hamster. *Science* 167, 302–304.
699. Muschamp, J. W., Lagoda, G., and Hull, E. M. (2004) Sexual stimuli increase Fos immunoreactivity in hypocretin/orexon neurons of male rats. *Soc. Neurosci. Abst.* 34. 888.20.
700. Myers, B. M., and Baum, M. J. (1980). Facilitation of copulatory performance in male rats by naloxone: effects of hypophysectomy, 17 alpha-estradiol, and luteinizing hormone releasing hormone. *Pharmacol. Biochem. Behav.* 12, 365–370.
701. Nadelhaft I., Roppolo J., Morgan, C., and de Groat, W. C. (1983). Parasympathetic preganglionic neurons and visceral primary afferents in monkey sacral spinal cord revealed following application of horseradish peroxidase to pelvic nerve. *J. Comp. Neurol.* 216, 36–52.
702. Nelson, R. J. (1998). *An Introduction to Behavioral Endocrinology*, p. 224. Sinauer Associates, Inc., Sunderland, MA.
703. Newman, S. W. (1999). The medial extended amygdala in male reproductive behavior: a node in the mammalian social behavior network. *Ann. N Y Acad. Sci.* 877, 242–257.
704. Niikura, S., Yokoyama, O., Komatsu, K., Yotsuyanagi, S., Mizuno, T., and Namiki, M. (2002). A causative factor of copulatory disorder in rats following social stress. *J. Urol.* 168, 843–849.
705. Nishitani, S., Moriya, T., Kondo, Y., Sakuma, Y., and Shinohara, K. (2004). Induction of Fos immunoreactivity in oxytocin neurons in the paraventricular nucleus after female odor exposure in male rats: effects of sexual experience. *Cell. Molec. Neurobiol.* 24, 283–291.
706. Nishizuka, M., and Arai, Y. (1981). Sexual dimorphism in synaptic organization in the amygdala and its dependence on neonatal hormone environment. *Brain Res.* 212, 31–38.
707. Nishizuka, M., and Arai, Y. (1983). Regional difference in sexually dimorphic synaptic organization of the medial amygdala. *Exp. Brain Res.* 49, 462–465.
708. Nyby, J., Matochik, J. A., and Barfield, R. J. (1992). Intracranial androgenic and estrogenic stimulation of male-typical behaviors in house mice (*Mus domesticus*). *Horm. Behav.* 26, 24–45.
709. Oboh, A. M., Paredes, R. G., and Baum M. J. (1995). A sex comparison of increments in Fos immunoreactivity in forebrain neurons of gonadectomized, testosterone-treated rats after mounting an estrous female. *Neurobiol. Learn. Mem.* 63, 66–73.
710. O'Carroll, R., Shapiro, C., and Bancroft, J. (1985). Androgens, behaviour and nocturnal erection in hypogonadal men: the effects of varying the replacement dose. *Clin. Endocrinol. (Oxf.)* 23, 527–538.
711. O'Connell, R. J., and Meredith, M. (1984). Effects of volatile and nonvolatile chemical signals on male sex behaviors mediated by the main and accessory olfactory systems. *Behav. Neurosci.* 98, 1083–1093.
712. Ogawa, S., Chan, J., Chester, A. E., Gustafsson, J. A., Korach, K. S., and Pfaff, D. W. (1999). Survival of reproductive behaviors in estrogen receptor beta gene-deficient (betaERKO) male and female mice. *Proc. Natl. Acad. Sci. U S A* 96, 12887–12892.
713. Ogawa, S., Lubahn, D. B., Korach, K. S., and Pfaff, D. W. (1997). Behavioral effects of estrogen receptor gene disruption in male mice. *Proc. Natl. Acad. Sci. U S A* 94, 1476–1481.
714. Ogawa, S. Robbins, A., Kumar, N., Pfaff, D. W., Sundaram, K., and Bardin, C. W. (1996). Effects of testosterone and 7 alpha-methyl-19-nortestosterone (MENT) on sexual and aggressive behaviors in two inbred strains of male mice. *Horm. Behav.* 30, 74–84.
715. Ogawa, S., Washburn, T. F., Taylor, J, Lubahn, D. B., Korach, K. S., and Pfaff, D. W. (1998). Modification of testosterone-dependent behaviors by estrogen receptor-alpha gene disruption in male mice. *Endocrinology* 139, 5058–5069.
716. O'Hanlon, J. K., and Sachs, B. D. (1980). Penile reflexes in rats after different numbers of ejaculations. *Behav. Neurol. Biol.* 29, 338–348.
717. O'Hanlon, J. K., and Sachs, B. D. (1986). Fertility of mating in rats (*Rattus norvegicus).* Contributions of androgen-dependent morphology and actions of the penis. *J. Comp. Psychol.* 100, 178–187.
718. O'Hanlon, J. K., Meisel, R. L., and Sachs, B. D. (1981). Estradiol maintains castrated male rats' sexual reflexes *in copula*, but not *ex copula*. *Behav. Neur. Biol.* 32, 269–273.
719. Olmos, G., Naftolin, F., Perez, J., Tranque, P. A., and Garcia-Segura, L. M. (1989). Synaptic remodeling in the rat arcuate nucleus during the estrous cycle. *Neuroscience* 32, 663–667.
720. Olson, R. J., and Justice, J. B., Jr. (1993). Quantitative microdialysis under transient conditions. *Anal. Chem.* 65, 1017–1022.
721. Onufrowicz, B. (1899). Notes on the arrangement and function of the cell groups in the sacral region of the spinal cord. *J. Nerv. Ment. Dis.* 26, 498–504.
722. Oomura, Y., Aou, S., Koyama, Y., and Yoshimatsu, H. (1988). Central control of sexual behavior. *Brain Res. Bull.* 20, 863–870.
723. Opsomer, R. J., Guerit, J. M., Wese, F. X., and Van Cangh, P. J. (1986). Pudendal cortical somatosensory evoked potentials. *J. Urol.* 135, 1216–1218.
724. Orbach, J. (1961). Spontaneous ejaculation in rat. *Science* 134, 1072–1073.

725. O'Sullivan, J. D., and Hughes, A. J. (1998). Apomorphine-induced penile erections in Parkinson's disease. *Mov. Disord.* 13, 536–539.
726. Packard, M. G., Cornell, A. H., and Alexander, G. M. (1997). Rewarding affective properties of intra-nucleus accumbens injections of testosterone. *Behav. Neurosci.* 111, 219–224.
727. Packard, M. G., Schroeder, J. P., and Alexander, G. M. (1998). Expression of testosterone conditioned place preference is blocked by peripheral or intra-accumbens injection of alpha-flupenthixol. *Horm. Behav.* 34, 39–47.
728. Padma-Nathan, H., and Giuliano, F. (2001). Oral drug therapy for erectile dysfunction. *Urol. Clin. N. Am.* 28, 321–334.
729. Padoin, M. J., and Lucion, A. B. (1995). The effect of testosterone and DOI (1-(2,5-dimethoxy-4-iodophenyl)-2-aminopropane) on male sexual behavior of rats. *Eur. J. Pharmacol.* 277, 1–6.
730. Pankevich, D. E., Baum, M. J., and Cherry, J. A. (2004). Olfactory sex discrimination persists, whereas the preference for urinary odorants from estrous females disappears in male mice after vomeronasal organ removal. *J. Neurosci.* 24, 9451–9457.
731. Paredes, R. G., and Ågmo, A. (1995). The $GABA_B$ antagonist CGP 35348 inhibits the effects of baclofen on sexual behavior and motor coordination. *Brain Res. Bull.* 36, 495–497.
732. Paredes, R. G., and Ågmo, A. (2004). Has dopamine a physiological role in the control of sexual behavior? A critical review of the evidence. *Prog. Neurobiol.* 73, 179–226.
733. Paredes, R. G., and Baum, M. J. (1995). Altered sexual partner preference in male ferrets given excitotoxic lesions of the preoptic area/anterior hypothalamus. *J. Neurosci.* 15, 6619–6630.
734. Paredes, R. G., Contreras, J. L., and Ågmo, A. (1998a). GABAergic drugs and sexual behaviour in the rabbit: evidence for species-specific effects. *J. Psychopharmacol.* 12, 186–191.
735. Paredes, R. G., Haller, A. E., Manero, M. C., Alvaradom, R., and Ågmo, A. (1990). Medial preoptic area kindling induces sexual behavior in sexually inactive male rats. *Brain Res.* 515, 20–26.
736. Paredes, R. G., Highland, L., and Karam, P. (1993a). Socio-sexual behavior in male rats after lesions of the medial preoptic area: evidence for reduced sexual motivation. *Brain Res.* 618, 271–276.
737. Paredes, R. G., Holmes, G. M., Sachs, B. D., and Ågmo, A. (1993b). Electromyographic activity of rat ischiocavernosus muscles during copulation after treatment with a GABA-transaminase inhibitor. *Behav. Neurol. Biol.* 60, 118–122.
738. Paredes, R. G., Karam, P., Highland, L., and Ågmo, A. (1997). GABAergic drugs and socio-sexual behavior. *Pharmacol. Biochem. Behav.* 58, 291–298.
739. Paredes, R. G., Kica, E., and Baum, M. J. (1994). Differential effects of the serotonin1A agonist, 8-OH-DPAT, on masculine and feminine sexual behavior of the ferret. *Psychopharmacology* 114, 591–596.
740. Paredes, R. G., Tzschentke, T., and Nakach, N. (1998b). Lesions of the medial preoptic area/anterior hypothalamus (MPOA/AH) modify partner preference in male rats. *Brain Res.* 813, 81–83.
741. Paredes, R. G., Pina, A. L., and Bermudez-Rattoni, F. (1993c). Hypothalamic but not cortical grafts induce recovery of sexual behavior and connectivity in medial preoptic area-lesioned rats. *Brain Res.* 620, 351–355.
742. Parfitt, D. B., and Newman, S. W. (1998). Fos-immunoreactivity within the extended amygdala is correlated with the onset of sexual satiety. *Horm. Behav.* 34, 17–29.
743. Parfitt, D. B., Coolen, L. M., Newman, S. W., and Wood, R. I. (1996). Lesions of the posterior medial nucleus of the amygdala delay sexual satiety. *Soc. Neurosci. Abstr.* 22, 155.
744. Park, J. J., and Baum, M. J. (1999). Intracerebroventricular infusion of the galanin antagonist M40 attenuates heterosexual partner preferences in ferrets. *Behav. Neurosci.* 113, 391–400.
745. Parrott, R. F. (1986). Minimal effects of 17 beta-hydroxy-17 alpha-methyl-estra-4,9,11-triene-3-one (R1881) on sexual behaviour in prepubertally castrated rams. *J. Endocrinol.* 110, 481–487.
746. Paul, S. M. (1995). GABA and glycine. In *Psychopharmacology: The Fourth Generation of Progress* (F. E. Bloom, and D. J. Kupfer, Eds.), pp. 87–94. Raven Press, New York.
747. Paxinos, G. (1976). Interruption of septal connections: effects on drinking, irritability, and copulation. *Physiol. Behav.* 17, 81–88.
748. Paxinos, G., and Watson, C. (1998). *The Rat Brain in Stereotaxic Coordinates*, 4th ed. Academic Press, New York.
749. Pehek, E. A., Thompson, J. T., and Hull, E. M. (1989a). The effects of intracranial administration of the dopamine agonist apomorphine on penile reflexes and seminal emission in the rat. *Brain Res.* 500, 325–332.
750. Pehek, E. A., Thompson, J. T., and Hull, E. M. (1989b). The effects of intrathecal administration of the dopamine agonist apomorphine on penile reflexes and copulation in the male rat. *Psychopharmacology* 99, 304–308.
751. Pehek, E. A., Thompson, J. T., Eaton, R. C., Bazzett, T. J., and Hull, E. M. (1988a). Apomorphine and haloperidol, but not domperidone, affect penile reflexes in rats. *Pharmacol. Biochem. Behav.* 31, 201–208.
752. Pehek, E. A., Warner, R. K., Bazzett, T., Bitran, D., Band, L. C., Eaton, R. C., and Hull, E. M. (1988b). Microinjection of cis-flupenthixol, a dopamine antagonist, into the medial preoptic area impairs sexual behavior of male rats. *Brain Res.* 443, 70–76.
753. Penfield, W., and Rasmussen A. T. (1950). *The Cerebral Cortex of Man: A Clinical Study of Localization of Function.* Macmillan Publishing Co, New York.
754. Perras, B., Smolnik, R., Fehm, H. L., and Born, J. (2001). Signs of sexual behaviour are not increased after subchronic treatment with LHRH in young men. *Psychoneuroendocrinology* 26, 1–15.
755. Petersén, I., and Stener, I. (1970). An electromyographical study of the striated urethral sphincter, the striated anal sphincter, and the levator ani muscle during ejaculation. *Electromyography* 1, 23–68.
756. Pfaff, D. W., and Pfaffmann, C. (1969). Olfactory and hormonal influences on the basal forebrain of the male rat. *Brain Res.* 15, 137–156.
757. Pfaff, D., and Keiner, M. (1973). Atlas of estradiol-concentrating cells in the central nervous system of the female rat. *J. Comp. Neurol.* 151, 121–158.
758. Pfaus, J. G., and Heeb, M. M. (1997). Implications of immediate-early gene induction in the brain following sexual stimulation of female and male rodents. *Brain Res. Bull.* 44, 397–407.
759. Pfaus, J. G., and Phillips, A. G. (1987). Cholecystokinin facilitates ejaculation in male rats: blockade with proglumide and apomorphine. *Eur. J. Pharmacol.* 141, 331–338.
760. Pfaus, J. G., and Phillips, A. G. (1989). Differential effects of dopamine receptor antagonists on the sexual behavior of male rats. *Psychopharmacology* 98, 363–368.
761. Pfaus, J. G., and Phillips, A. G. (1991). Role of dopamine in anticipatory and consummatory aspects of sexual behavior in the male rat. *Behav. Neurosci.* 105, 727–743.
762. Pfaus, J. G., and Wilkins, M. F. (1995). A novel environment disrupts copulation in sexually naïve but not experienced male rats: reversal with naloxone. *Physiol. Behav.* 57, 1045–1049.

763. Pfaus, J. G., Damsma, G., Nomikos, G. G., Wenkstern, D. G., Blaha, C. D., Philips, A. G., and Fibiger, H. C. (1990). Sexual behavior enhances central dopamine transmission in the rat. *Brain Res.* 530, 345–348.
764. Pfaus, J. G., Kippin, T. E., and Centeno, S. (2001). Conditioning and sexual behavior: a review. *Horm. Behav.* 40, 291–321.
765. Pfaus, J. G., Kleopoulus, S. P., Mobbs, C. V., Gibbs, R. B., and Pfaff, D. W. (1993). Sexual stimulation activates *c-fos* within estrogen-concentrating regions of the female rat forebrain. *Brain Res.* 624, 253–267.
766. Phelps, S. M., Lydon, J. P., O'malley, B. W., and Crews, D. (1998). Regulation of male sexual behavior by progesterone receptor, sexual experience, and androgen. *Horm. Behav.* 34, 294–302.
767. Phoenix, C. H. (1961). Hypothalamic regulation of sexual behavior in male guinea pigs. *J. Comp. Physiol. Psychol.* 54, 72–77.
768. Phoenix, C. H. (1974). Effects of dihydrotestosterone on sexual behavior of castrated male rhesus monkeys. *Physiol. Behav.* 12, 1045–1055.
769. Phoenix, C. H., and Chambers, K. C. (1986). Threshold for behavioral response to testosterone in old castrated male rhesus monkeys. *Biol. Reprod.* 35, 918–926.
770. Phoenix, C. H., and Chambers, K. C. (1988). Testosterone therapy in young and old rhesus males that display low levels of sexual activity. *Physiol. Behav.* 43, 479–484.
771. Phoenix, C. H., and Chambers, K. C. (1990). Sexual performance of old and young male rhesus macaques following treatment with GnRH. *Physiol. Behav.* 47, 513–517.
772. Pirke, K. M., Kockott, G., Aldenhoff, J., Besinger, U., and Feil, W. (1979). Pituitary gonadal system function in patients with erectile impotence and premature ejaculation. *Arch. Sex. Behav.* 8, 41–48.
773. Pleim, E. T., Matochik, J. A., Barfield, R. J., and Auerbach, S. B. (1990). Correlation of dopamine release in the nucleus accumbens with masculine sexual behavior in rats. *Brain Res.* 524, 160–163.
774. Poggioli, R., Benelli, A, Arletti, R., Menozzi, B., and Bertolini, A. (1994). Benextramine, an NPY antagonist, improves sexual behavior in male rats. *Neuropeptides* 26, 355–358.
775. Poggioli, R., Benelli, A., Arletti, R., Cavazzuti, E., and Bertolini, A. (1995). Nitric oxide is involved in the ACTH-induced behavioral syndrome. *Peptides* 16, 1263–1268.
776. Poggioli, R., Rasori, E., and Bertolini, A. (1992). Galanin inhibits sexual behavior in male rats. *Eur. J. Pharmacol.* 213, 87–90.
777. Poggioli, R., Vergoni, A. V., Marrama, D., Giuliani, D., Bertolini, A. (1990). NPY-induced inhibition of male copulatory activity is a direct behavioural effect. *Neuropeptides*, 16, 169–172.
778. Pollak, E. I., and Sachs, B. D. (1975). Masculine sexual behavior and morphology: paradoxical effects of perinatal androgen treatment in male and female rats. *Behav. Biol.* 13, 401–411.
779. Pomerantz, S. M., and Clemens, L. G. (1981). Ultrasonic vocalizations in male deer mice (*Peromyscus maniculatus bairdi*), their role in male sexual behavior. *Physiol. Behav.* 27, 869–872.
780. Pomerantz, S. M., Hepner, B. C., and Wertz, J. M. (1993a). 5-HT$_{1A}$ and 5-HT$_{1C}$/$_{1D}$ receptor agonists produce reciprocal effects on male sexual behavior of rhesus monkeys. *Eur. J. Pharmacol.* 243, 227–234.
781. Pomerantz, S. M., Hepner, B. C., and Wertz, J. M. (1993b). Serotonergic influences on male sexual behavior of rhesus monkeys: effects of serotonin agonists. *Psychopharmacology* 111, 47–54.
782. Popova, N. K., and Amstislavskaya, T. G., (2002). Involvement of the 5-HT(1A) and 5-HT(1B) serotonergic receptor subtypes in sexual arousal in male mice. *Psychoneuroendocrinology* 27, 609–618.
783. Porst, H., Rosen, R., Padma-Nathan, H., Goldstein, I., Giuliano, F., Ulbrich, E., and Bandel, T. (2001). The efficacy and tolerability of vardenafil, a new, oral, selective phosphodiesterase type 5 inhibitor, in patients with erectile dysfunction: the first at-home clinical trial. *Int. J. Impotence Res.* 13, 192–199.
784. Portillo, W., Basanez, E., and Paredes, R. G. (2003). Permanent changes in sexual behavior induced by medial preoptic area kindling-like stimulation. *Brain Res.* 96, 10–14.
785. Powers, J. B., and Winans, S. S. (1973). Sexual behavior in peripherally anosmic male hamsters. *Physiol. Behav.* 10, 361–368.
786. Powers, J. B., and Winans, S. S. (1975). Vomeronasal organ: critical role in mediating sexual behavior of the male hamster. *Science* 187, 961–963.
787. Powers, J. B., Fields, R. B., and Winans, S. S. (1979). Olfactory and vomeronasal system participation in male hamsters attraction to female vaginal secretions. *Physiol. Behav.* 22, 77–84.
788. Powers, J. B., Newman, S. W., and Bergondy, M. L. (1987). MPOA and BNST lesions in male Syrian hamsters: differential effects on copulatory and chemoinvestigatory behaviors. *Behav. Brain Res.* 23, 181–195.
789. Pu, S., Xu, B., Kalra, S. P., and Kalra, P. S. (1996). Evidence that gonadal steroids modulate nitric oxide efflux in the medial preoptic area: effects of N-methyl-D-aspartate and correlation with luteinizing hormone secretion. *Endocrinology* 137, 1949–1955.
790. Putnam, S. K., Du, J., and Hull, E. M. (2001). Testosterone restoration of copulation and medial preoptic dopamine release in castrated male rats: 2-, 5-, and 10-day treatments. *Horm. Behav.* 39, 216–224.
791. Putnam, S. K., Sato, S., and Hull, E. M. (2003). Hormonal maintenance of copulation in castrates: association with extracellular dopamine in MPOA. *Horm. Behav.* 44, 419–426.
792. Putnam, S. K., Sato, S., and Hull, E. M. (2005). Effects of testosterone metabolites on copulation, medial preoptic dopamine content, and nitric oxide synthase. *Horm. Behav.* 47, 513–522.
793. Quershi, G. A., and Södersten, P. (1986). Sexual activity alters the concentration of amino acids in cerebrospinal fluid of male rat. *Neurosci. Lett.* 70, 374–378.
794. Quinlan, D. M., Nelson, R. J., Partin, A. W., Mostwin, J. L., and Walsh, P. C. (1989). The rat as a model for the study of penile erection. *J. Urol.* 141, 656–661.
795. Raboch, J., and Starka, L. (1973). Reported coital activity of men and levels of plasma testosterone. *Arch. Sex. Behav.* 2, 309–315.
796. Raboch, J., Mellan, J., and Starka, L. (1975). Plasma testosterone in male patients with sexual dysfunction. *Arch. Sex. Behav.* 4, 541–545.
797. Rampin, O., Jerome, N., and Suaudeau, C. (2003). Proerectile effects of apomorphine in mice. *Life Sci.* 72, 2329–2336.
798. Rand, M. N., and Breedlove, S. M. (1995). Androgen alters the dendritic arbors of SNB motoneurons by acting upon their target muscles. *J. Neurosci.* 15, 4408–4416.
799. Rasia-Filho, A. A., Peres, T. M., Cubilla-Gutierrez, F. H., and Lucion, A. B. (1991). Effect of estradiol implanted in the corticomedial amygdala on the sexual behavior of castrated male rats. *Braz. J. Med. Biol. Res.* 24, 1041–1049.
800. Ratnasooriya, W. D., and Dharmasiri, M. G. (2001). L-arginine, the substrate of nitric oxide synthase, inhibits fertility of male rats. *Asian J. Androl.* 3, 97–103.

801. Ratnasooriya, W. D., Dharmasiri, M. G., and Wadsworth, R. M. (2000). Reduction in libido and fertility of male rats by administration of the nitric oxide (NO) synthase inhibitor N-nitro-L-arginine methylester. *Int. J. Androl.* 23, 187–191.
802. Recker, F., and Tscholl, R. (1993). Monitoring emission as direct intraoperative control for nerve sparing retroperitoneal lymphadenectomy. *J. Urol.* 150, 1360–1364.
803. Rees, R. W., Ralph, D. J., Royle, M., Moncada, S., and Cellek, S. (2001). Y-27632, an inhibitor of Rho-kinase, antagonizes noradrenergic contractions in the rabbit and human corpus cavernosum. *Brit. J. Pharmacol.* 133, 455–458.
804. Rehman, J., Kaynan, A., Christ, G., Valcic, M., Maayani, S., and Melman, A. (1999). Modification of sexual behavior of Long-Evans male rats by drugs acting on the 5-HT_{1A} receptor. *Brain Res.* 821, 414–425.
805. Reilly, C. M., Zamorano, P., Stopper, V. S., and Mills, T. M. (1997). Androgenic regulation of NO availability in rat penile erection. *J. Androl.* 18, 110–115.
806. Resko, J. A., Perkins, A., Roselli, C. E., Stellflug, J. N., and Stormshak, F. K. (1999). Sexual behaviour of rams: male orientation and its endocrine correlates. *J. Reprod. Fert. Suppl.* 54, 259–269.
807. Retana-Marquez, S., and Velazquez-Moctezuma, J. (1997). Cholinergic-androgenic interaction in the regulation of male sexual behavior in rats. *Pharmacol. Biochem. Behav.* 56, 373–378.
808. Rhoden, E. L., Teloken, C., Mafessoni, R., and Souto, C. A. (2002). Is there any relation between serum levels of total testosterone and the severity of erectile dysfunction? *Int. J. Impot. Res.* 14, 167–171.
809. Ridet J. L., Sandillon, F., Rajaofetra, N., Geffard, M., and Privat, A. (1992). Spinal dopaminergic system of the rat: light and electron microscopic study using an antiserum against dopamine, with particular emphasis on synaptic incidence. *Brain Res.* 598, 233–241.
810. Riters, L. V., and Ball, G. F. (1999). Lesions to the medial preoptic area affect singing in the male European starling (*Sturmus vulgaris*) *Horm. Behav.* 36, 276–286.
811. Riters, L. V., Absil, P., and Balthazart, J. (1999). Effects of naloxone on the acquisition and expression of appetitive and consummatory sexual behavior in male Japanese quail. *Physiol. Behav.* 66, 763–773.
812. Rizvi, T. A., Ennis, M., and Shipley, M. T. (1992). Reciprocal connections between the medial preoptic area and the midbrain periaqueductal gray in rat: a WGA-HRP and PHA-L study. *J. Comp. Neurol.* 315, 1–15.
813. Rizvi, T. A., Murphy, A. Z., Ennis, M., Behbehani, M. M., and Shipley, M. T. (1996). Medial preoptic area afferents to periaqueductal gray medullo-output neurons: a combined Fos and tract tracing study. *J. Neurosci.* 16, 333–344.
814. Robbins, T. W., and Everitt, B. J. (1992). Functions of dopamine in the dorsal and ventral striatum. *Sem. Neurosci.* 4, 119–128.
815. Roberts, W. W., Steinberg, M. L., and Means, L. W. (1967). Hypothalamic mechanisms for sexual, aggressive, and other motivational behaviors in the opossum, *Didelphis virginiana. J. Comp. Physiol. Psychol.* 64, 1–15.
816. Robertson, G. S., Pfaus, J. G., Atkinson, L. J., Matsumura, H., Phillips, A. G., and Fibiger, H. C. (1991). Sexual behavior increases *c-fos* expression in the forebrain of the male rat. *Brain Res.* 564, 352–357.
817. Robinson, B. W., and Mishkin, M. (1966). Ejaculation evoked by stimulation of the preoptic area in monkeys. *Physiol. Behav.* 1, 269–272.
817a. Robinson, S., Sandstrom, S. M., Denenberg, V. H., and Palmiter, R. D. (2005). Distinguishing whether dopamine regulates liking, wanting, and/or learning about rewards. *Behav. Neurosci.* 119, 5–15.
818. Rodriguez, M., Castro, R., Hernandez, G., and Mas, M. (1984). Different roles of catecholaminergic and serotoninergic neurons of the medial forebrain bundle on male rat sexual behavior. *Physiol. Behav.* 33, 5–11.
819. Rodriguez-Manzo, G. (1999). Yohimbine interacts with the dopaminergic system to reverse sexual satiation: further evidence for a role of sexual motivation in sexual exhaustion. *Eur. J. Pharmacol.* 372, 1–8.
820. Rodriguez-Manzo, G., and Fernandez-Guasti, A. (1994). Reversal of sexual exhaustion by serotonergic and noradrenergic agents. *Behav. Brain Res.* 62, 127–134.
821. Rodriguez-Manzo, G., and Fernandez-Guasti, A. (1995). Participation of the central noradrenergic system in the reestablishment of copulatory behavior of sexually exhausted rats by yohimbine, naloxone, and 8-OH-DPAT. *Brain Res. Bull.* 38, 399–404.
822. Rodriguez-Manzo, G., Asai, M., and Fernandez-Guasti, A. (2002). Evidence for changes in brain enkephalin contents associated to male rat sexual activity. *Behav. Brain Res.* 131, 47–55.
823. Rodriguez-Manzo, G., Lopez-Rubalcava, C., Hen, R., and Fernandez-Guasti, A. (2002). Participation of 5-HT(1B) receptors in the inhibitory actions of serotonin on masculine sexual behaviour of mice: pharmacological analysis in 5-HT(1B) receptor knock-out mice. *Brit. J. Pharmacol.* 136, 1127–1134.
824. Rodríguez-Manzo, G., Pellicer, F., Larsson, K., and Fernandez-Guasti, A. (2000). Stimulation of the medial preoptic area facilitates sexual behavior but does not reverse sexual satiation. *Behav. Neurosci.* 114, 553–560.
825. Romeo, R. D., Cook-Wiens, E., Richardson, H. N., and Sisk, C. L. (2001). Dihydrotestosterone activates sexual behavior in adult male hamsters but not in juveniles. *Physiol. Behav.* 73, 579–584.
826. Romeo, R. D., Parfitt, D. B., Richardson, H. N., and Sisk, C. L. (1998). Pheromones elicit equivalent levels of Fos-immunoreactivity in prepubertal and adult male Syrian hamsters. *Horm. Behav.* 34, 48–55.
827. Romeo, R. D., Richardson, H. N., and Sisk, C. L. (2002). Puberty and the maturation of the male brain and sexual behavior: recasting a behavioral potential *Neurosci. Biobehav. Rev.* 26, 381–391.
828. Root, W. S., and Bard, P. (1947). The mediation of feline erection through sympathetic pathways with some remarks on sexual behavior after deafferentation of the genitalia. *Am. J. Physiol.* 151, 80–90.
829. Roselli, C. E., Cross, E., Poonyagariyagorn, H. K., and Stadelman, H. L. (2003). Role of aromatization in anticipatory and consummatory aspects of sexual behavior in male rats. *Horm. Behav.* 44, 146–151.
830. Roselli, C. E., Thornton, J. E., Chambers, K. C. (1993). Age-related deficits in brain estrogen receptors and sexual behavior of male rats. *Behav. Neurosci.* 107, 202–209.
831. Rosen, R. C., Lane, R. M., and Menza, M. (1999). Effects of SSRIs on sexual function: a critical review. *J. Clin. Psychopharmacol.* 19, 67–85.
831a. Rosen, R. R., and McKenna, K. E. (2002). PDE-5 inhibition and sexual response: pharmacological mechanisms and clinical outcomes. *Ann. Rev. Sex. Res.* 13, 36–88.
832. Rosenblatt, J. S., and Aronson, L. R. (1958). The decline of sexual behavior in male cats after castration with special reference to the role of prior sexual experience. *Behaviour,* 12, 285–338.
833. Rowe, D. W., and Erskine, M. S. (1993). *c-fos* proto-oncogene activity induced by mating in the preoptic area, hypothalamus and amygdala in the female rat: role of afferent input via the pelvic nerve. *Brain Res.* 621, 25–34.

834. Rowe, F. A., and Edwards, D. A. (1972). Olfactory bulb removal: influences on the mating behavior of male mice. *Physiol. Behav.* 8, 37–41.
835. Rowe, F. A., and Smith, W. E. (1973). Simultaneous and successive olfactory bulb removal: influences on the mating behavior of male mice. *Physiol. Behav.* 10, 443–449.
836. Rowland, D. L., and Houtsmuller, E. J. (1998). 8-OH-DPAT interacts with sexual experience and testosterone to affect ejaculatory response in rats. *Pharmacol. Biochem. Behav.* 60, 143–149.
837. Ryan, E. L., and Frankel, A. I. (1978). Studies on the role of the medial preoptic area in sexual behavior and hormonal response to sexual behavior in the mature male laboratory rat. *Biol. Reprod.* 19, 971–983.
838. Sachs, B. D. (1980). Sexual reflexes of spinal male house mice. *Physiol. Behav.* 24, 489–492.
839. Sachs, B. D. (1982). Role of striated penile muscles in penile reflexes, copulation, and induction of pregnancy in the rat. *J. Reprod. Fertil.* 66, 433–443.
840. Sachs, B. D. (1983). Potency and fertility: hormonal and mechanical causes and effects of penile actions in rats. In *Hormones and Behaviour in Higher Vertebrates* (J. Balthazart, E. Pröve, and R. Gilles, Eds.), pp. 86–110. Springer-Verlag, Berlin.
841. Sachs, B. D. (1996). Penile erection in response to remote cues from females: albino rats severely impaired relative to pigmented strains. *Physiol. Behav.* 60, 803–808.
842. Sachs, B. D. (1997). Erection evoked in male rats by airborne scent from estrous females. *Physiol. Behav.* 62, 921–924.
843. Sachs, B. D. (2000). Contextual approaches to the physiology and classification of erectile function, erectile dysfunction, and sexual arousal. *Neurosci. Biobehav. Rev.* 24, 541–560.
844. Sachs, B. D., and Bialy, M. (2000). Female presence during the postejaculatory interval facilitates penile erection and 22 kHz vocalization in male rats. *Behav. Neurosci.* 114, 1203–1208.
845. Sachs, B. D., and Bitran, D. (1990). Spinal block reveals roles for brain and spinal cord in the mediation of reflexive erection in rats. *Brain Res.* 528, 99–108.
846. Sachs, B. D., and Garinello, L. D. (1978). Interaction between penile reflexes and copulation in male rats. *J. Comp. Physiol. Psychol.* 92, 759–767.
847. Sachs, B. D., and Garinello, L. D. (1979). Spinal pacemaker controlling sexual reflexes in male rats. *Brain Res.* 171, 152–156.
848. Sachs, B. D., and Garinello, L. D. (1980). Hypothetical spinal pacemaker regulating penile reflexes in rats: evidence from transection of spinal cord and dorsal penile nerves. *J. Comp. Physiol. Psychol.* 94, 530–535.
849. Sachs, B. D., and Leipheimer, R. E. (1988). Rapid effect of testosterone on striated muscle activity in rats. *Neuroendocrinology* 48, 453–458.
850. Sachs, B. D., and Liu, Y. C. (1991). Maintenance of erection of penile glans, but not penile body, after transection of rat cavernous nerves. *J. Urol.* 146, 900–905.
851. Sachs, B. D., and Liu, Y. C. (1998). Mounting and brief noncontact exposure of males to receptive females facilitate reflexive erection in rats, even after hypogastric nerve section. *Physiol. Behav.* 65, 413–421.
852. Sachs, B. D., and Meisel, R. L. (1979). Pubertal development of penile reflexes and copulation in male rats. *Psychoneuroendocrinology* 4, 287–296.
853. Sachs, B. D., Akasofu, K., and McEldowney, S. (1994a). Effects of copulation on apomorphine-induced erection in rats. *Pharmacol. Biochem. Behav.* 48, 423–428.
854. Sachs, B. D., Akasofu, K., Citron, J. H., Daniels, S. B., and Natoli, J. H. (1994b). Noncontact stimulation from estrous females evokes penile erection in rats. *Physiol. Behav.* 55, 1073–1079.
855. Sachs, B. D., Valcourt, R. J., and Flagg, H. C. (1981). Copulatory behavior and sexual reflexes of male rats treated with naloxone. *Pharmacol. Biochem. Behav.* 14, 251–253.
856. Saito, T. R., and Moltz, H. (1986). Copulatory behavior of sexually naïve and sexually experienced male rats following removal of the vomeronasal organ. *Physiol. Behav.* 37, 507–510.
857. Sala, M., Braida, D., Leone, M. P., Calcaterra, P., Monti, S., and Gori, E. (1990). Central effect of yohimbine on sexual behavior in the rat. *Physiol. Behav.* 47, 165–173.
858. Salamone, J. D., and Correa, M. (2002). Motivational views of reinforcement: implications for understanding the behavioral functions of nucleus accumbens dopamine. *Behav. Brain Res.* 137, 3–25.
859. Salis, P. S., and Dewsbury, D. A. (1971). p-Chlorophenylalanine facilitates copulatory behaviour in male rats. *Nature* 232, 400–401.
860. Salmimies, P., Kockott, G., Pirke, K. M., Vogt, H. J., and Schill, W. B. (1982). Effects of testosterone replacement on sexual behavior in hypogonadal men. *Arch. Sex. Behav.* 11, 345–353.
861. Sar, M., and Stumpf, W. E. (1975). Distribution of androgen-concentrating neurons in rat brain. In *Anatomical Neuroendocrinology* (W. E. Stumpf and L. D. Grant, Eds.), pp. 120–133. Karger, Basel.
862. Sato, M., Mizuno, N., and Konishi, A. (1978). Localization of motoneurons innervating perineal muscles: a HRP study in cat. *Brain Res.* 140, 149–154.
862a. Sato, S., and Hull, E. M. (2004). The effects of cGMP analog on the MPOA dopamine levels and copulation in male rats. *Soc. Neurosci. Abst.* 34, 423.12.
863. Sato, S., Braham, C. S., Putnam, S. K., and Hull, E. M. (2005). Neuronal nitric oxide synthase and gonadal steroid interaction in the MPOA of male rats: co-localization and testosterone-induced restoration of copulation and nNOS immunoreactivity. *Brain Res.* 1043, 205–213.
863a. Sato, S., Wersinger, S. W., and Hull, E. M. (2005). Effects of NO donor and cGMP analog in the MPOA of DHT-treated castrates. *Soc. Behav. Neuroendocrinol. Abst.* 9, #211, in *Horm. Behav.* 48, 126.
864. Sato, Y., Christ, G. J., Horita, H., Adachi, H., Suzuki, N., and Tsukamoto, T. (1999). The effects of alterations in nitric oxide levels in the paraventricular nucleus on copulatory behavior and reflexive erections in male rats. *J. Urol.* 162, 2182–2185.
865. Sato, Y., Horita, H., Kurohata, T., Adachi, H., and Tsukamoto, T. (1998). Effect of the nitric oxide level in the medial preoptic area on male copulatory behavior in rats. *Am. J. Physiol.* 274, R243–R247.
866. Sato, Y., Wada, H., Horita, H., Suzuki, N., Shibuya, A., Adachi, H., Kato, R., Tsukamoto, T., and Kumamoto, Y. (1995). Dopamine release in the medial preoptic area during male copulatory behavior in rats. *Brain Res.* 692, 66–70.
867. Sato, Y., Zhao, W., and Christ, G. J. (2001). Central modulation of the NO-cGMP pathway affects the MPOA-induced intracavernous pressure response. *Am. J. Physiol. Reg. Integ. Comp. Physiol.* 281, R269–R278.
868. Sayag, N., Hoffman, N. W., and Yahr, P. (1994). Telencephalic connections of the sexually dimorphic area of the gerbil hypothalamus that influence male sexual behavior. *Behav. Neurosci.* 108, 743–757.
869. Scaletta L. L., and Hull, E. M. (1990). Systemic or intracranial apomorphine increases copulation in long-term castrated male rats. *Pharmacol. Biochem. Behav.* 37, 471–475.
870. Schmidt, M. H., and Schmidt, H. S. (1993). The ischiocavernosus and bulbospongiosus muscles in mammalian penile rigidity. *Sleep* 16, 171–183.
871. Schmidt, M. H., Valatx, J. L., Sakai, K., Debilly, G., and Jouvet, M. (1995). Corpus spongiosum penis pressure and

perineal muscle activity during reflexive erections in the rat. *Am. J. Physiol.* 269, R904–R913.

872. Schmidt, M. H, Valatx, J. L., Sakai, K., Fort, P., and Jouvet, M. (2000). Role of the lateral preoptic area in sleep-related erectile mechanisms and sleep generation in the rat. *J. Neurosci.* 20, 6640–6647.
873. Schmidt, M. H, Valatx, J. L, Schmidt, H. S, Wauquier, A, Jouvet, M. (1994). Experimental evidence of penile erections during paradoxical sleep in the rat. *NeuroReport* 5, 561–564.
874. Schnur, S. L., Smith, E. R., Lee, R. L., Mas, M., and Davidson, J. M. (1989). A component analysis of the effects of DPAT on male rat sexual behavior. *Physiol. Behav.* 45, 897–901.
875. Schoenen, J., Lotstra, F., Vierendeels, G., Reznik, M., and Vanderhaeghen, J. J. (1985). Substance P, enkephalins, somatostatin, cholecystokinin, oxytocin, and vasopressin in human spinal cord. *Neurology* 35, 881–890.
876. Schrøder, H. D. (1980). Organization of the motoneurons innervating the pelvic muscles of the male rat. *J. Comp. Neurol.* 192, 567–587.
877. Schultheiss, D., Hiltl, D. M., Meschi, M. R., Machtens, S. A., Truss, M. C., Stief, C. G., and Jonas, U. (2000). Pilot study of the transdermal application of testosterone gel to the penile skin for the treatment of hypogonadotropic men with erectile dysfunction. *World J. Urol.* 18, 431–435.
878. Schultz, K. M., Richardson, H. N., Romeo, R. D., Morris, J. A., Lookingland, K. J., and Sisk, C. L. (2003). Medial preoptic area dopaminergic responses to female pheromones develop during puberty in the male Syrian hamster. *Brain Res.* 988, 139–145.
879. Schultz, K. M., Richardson, H. N., Zehr, J. L., Osetek, A. J., Menard, T. A., and Sisk, C. L. (2004). Gonadal hormones masculinize and defeminize reproductive behaviors during puberty in the male Syrian hamster. *Horm. Behav.* 45, 242–249.
880. Schultz, W. (1992). Activity of dopamine neurons in the behaving primate. *Sem. Neurosci.* 4, 129–138.
881. Schultz, W. (2002). Getting formal with dopamine and reward. *Neuron.* 36, 241–263.
882. Scordalakes, E. M., Shetty, S. J., and Rissman, E. F. (2002). Roles of estrogen receptor alpha and androgen receptor in the regulation of neuronal nitric oxide synthase. *J. Comp. Neurol.* 453, 336–344.
883. Scouten, C. W., Burrell, L., Palmer, T., and Cegavske, C. F. (1980). Lateral projections of the medial preoptic area are necessary for androgenic influence on urine marking and copulation in rats. *Physiol. Behav.* 25, 237–243.
884. Seagraves, R. T. (1990). Effects of psychotropic drugs on human erection and ejaculation. *Arch. Gen. Psychiatry* 46, 275–284.
885. Semans, J. H., and Langworthy, O. R. (1938). Observations on the neurophysiology of sexual function in the male cat. *J. Urol.* 40, 836–846.
886. Shakil, T., Hoque, A. N., Husain, M., and Belsham, D. D. (2002). Differential regulation of gonadotropin-releasing hormone secretion and gene expression by androgen: membrane versus nuclear receptor activation. *Molec. Endocrinol.* 16, 2592–1602.
887. Shimura, T., and Shimokochi, M. (1991). Modification of male rat copulatory behavior by lateral midbrain stimulation. *Physiol. Behav.* 50, 989–994.
888. Shimura, T. Yamamoto, T., and Shimokochi, M. (1994). The medial preoptic area is involved in both sexual arousal and performance in male rats: re-evaluation of neuron activity in freely moving animals. *Brain Res.* 640, 215–222.
889. Shugrue, P. J., Lane, M. V., and Merchenthaler, I. (1997). Comparative distribution of estrogen receptor-alpha and -beta mRNA in the rat central nervous system. *J. Comp. Neurol.* 388, 507–525.
890. Siegel, H. I. (1985). Male sexual behavior. In *The Hamster: Reproduction & Behavior* (H. I. Siegel, Ed.), pp. 191–206. Plenum Press, New York.
891. Silva, N. L., and Boulant, J. A. (1986). Effects of testosterone, estradiol, and temperature on neurons in preoptic tissue slices. *Am. J. Physiol.* 250, R625–R632.
892. Simerly, R. B. (1995). Hormonal regulation of limbic and hypothalamic pathways. In *Neurobiological Effects of Sex Steroid Hormones* (P. E. Micevych and R. P. Hammer Jr., Eds.), pp. 85–116. Cambridge University Press, Cambridge.
893. Simerly, R. B., and Swanson, L. W. (1986). The organization of neural inputs to the medial preoptic nucleus of the rat. *J. Comp. Neurol.* 246, 312–342.
894. Simerly, R. B., and Swanson, L. W. (1987). Castration reversibly alters levels of cholecystokinin immunoreactivity in cells of three interconnected sexually dimorphic fore-brain nuclei in the rat. *Proc. Natl. Acad. Sci. U S A* 84, 2087–2091.
895. Simerly, R. B., and Swanson, L. W. (1988). Projections of the medial preoptic nucleus: a *Phaseolis vulgaris* leucoagglutinin anterograde tract-tracing study in the rat. *J. Comp. Neurol.* 270, 209–242.
896. Simerly, R. B., Chang, C., Muramatsu, M., and Swanson, L. W. (1990). Distribution of androgen and estrogen mRNA-containing cells in the rat brain: an in situ hybridization study. *J. Comp. Neurol.* 294, 76–95.
897. Simerly, R. B., Gorski, R. A., and Swanson, L. W. (1986). Neurotransmitter specificity of cells and fibers in the medial preoptic nucleus: an immunohistochemical study in the rat. *J. Comp. Neurol.* 246, 343–363.
898. Simmons, D., and Yahr, P. (2001). Posterodorsal preoptic nucleus (PdPN) and posterodorsal medial amygdala (MeApd) cells that are activated with ejaculation: additional projections and presence of nitric oxide synthase (NOS). *Soc. Behav. Neuroendocrinol. Abst.* 5, 349.
899. Simmons, D., and Yahr, P. (2002). Projections of the posterodorsal preoptic nucleus and the lateral part of the posterodorsal medial amygdala in male gerbils, with emphasis on cells activated with ejaculation. *J. Comp. Neurol.* 444, 75–94.
900. Simmons, D. A., and Yahr, P. (2003). GABA and glutamate in mating-activated cells in the preoptic area and medial amygdala of male gerbils. *J. Comp. Neurol.* 459, 290–300.
901 Singer, J. J. (1968). Hypothalamic control of male and female sexual behavior in female rats. *J. Comp. Physiol. Psychol.* 66, 738–742.
902. Singh, R. Pervin, S. Shryne, J. Gorski, R., and Chaudhuri, G. (2000). Castration increases and androgens decrease nitric oxide synthase activity in the brain: physiologic implications. *Proc. Natl. Acad. Sci. U S A* 97, 3672–3677.
903. Sipos, M. L., and Nyby, J. G. (1996). Concurrent androgenic stimulation of the ventral tegmental area and medial preoptic area, synergistic effects on male-typical reproductive behaviors in house mice. *Brain Res.* 729, 29–44.
904. Sipos, M. L., and Nyby, J. G. (1998). Intracranial androgenic activation of male-typical behaviours in house mice: concurrent stimulation of the medial preoptic area and medial nucleus of the amygdala. *J. Neuroendocrinol.* 10, 577–586.
905. Sirinathsinghji, D. J. (1987). Inhibitory influence of corticotropin releasing factor on components of sexual behaviour in the male rat. *Brain Res.* 407, 185–190.
906. Sisk, C. L., Schultz, K. M., and Zehr, J. L. (2003). Puberty: a finishing school for male social behavior. *Ann. N Y Acad. Sci.* 1007, 189–198.
907. Sjöstrand, N., and Klinge, E. (1979). Principal mechanisms controlling penile retraction and protrusion in rabbits. *Acta Physiol. Scand.* 106, 199–214.
908. Slimp, J. C., Hart, B. L., and Goy, R. W. (1978). Heterosexual, autosexual and social behavior of adult male

rhesus monkeys with medial preoptic-anterior hypothalamic lesions. *Brain Res.* 142, 105–122.

909. Smith, E. R., Lee, R. L., Schnur, S. L., and Davidson, J. M. (1987a). Alpha 2-adrenoceptor antagonists and male sexual behavior: I. Mating behavior. *Physiol. Behav.* 41, 7–14.
910. Smith, E. R., Lee, R. L., Schnur, S. L., and Davidson, J. M. (1987b). Alpha 2-adrenoceptor antagonists and male sexual behavior: II. Erectile and ejaculatory reflexes. *Physiol. Behav.* 41, 15–9.
911. Smith, W. J., Stewart, J., and Pfaus, J. G. (1997). Tail pinch induces Fos immunoreactivity within several regions of the male rat brain: effects of age. *Physiol. Behav.* 61, 717–723.
912. Södersten, P. (1973). Estrogen-activated sexual behavior in male rats. *Horm. Behav.* 4, 247–256.
913. Södersten, P., and Gustafsson, J. A. (1980). Activation of sexual behaviour in castrated rats with the synthetic androgen 17 beta-hydroxy-17 alpha-methyl-estra-4,9,11-triene-3-one 7 1881. *J. Endocrinol.* 87, 279–283.
914. Södersten, P., Damassa, D. A., and Smith, E. R. (1977). Sexual behavior in developing male rats. *Horm. Behav.* 8, 320–341.
915. Södersten, P., Eneroth, P., Mode, A., and Gustafsson, J. A. (1985). Mechanisms of androgen-activated sexual behaviour in rats. In *Neurobiology* (R. Gilles, and J. Balthazart, Eds.), pp. 48–59. Springer-Verlag, Berlin.
916. Södersten, P., Larsson, K., Ahlenius, S., and Engel, J. (1976). Stimulation of mounting behavior but not lordosis behavior in ovariectomized female rats by p-chlorophenylalanine. *Pharmacol. Biochem. Behav.* 5, 329–333.
917. Somlyo, A. P., and Somlyo, A. V. (2000). Signal transduction by G-proteins, rho-kinase and protein phosphatase to smooth muscle and non-muscle myosin II. *J. Physiol. (Lond.)* 522, 177–185.
918. Spanagel, R., and Weiss, F. (1999). The dopamine hypothesis of reward: past and current status. *Trends Neurosci.* 22, 521–527.
919. Steers, W. D. (2000). Neural pathways and central sites involved in penile erection: neuroanatomy and clinical implications. *Neurosci. Biobehav. Rev.* 24, 507–516.
920. Steers, W. D, and. de Groat, W. C. (1989). Effects of m-chlorophenylpiperazine on penile and bladder function in rats. *Am. J. Physiol.* 257, R1441–R1449.
921. Stefanick, M. L., Smith, E. R., and Davidson, J. M. (1983). Penile reflexes in intact rats following anesthetization of the penis and ejaculation. *Physiol. Behav.* 31, 63–65.
922. Stefanick, M. L., Smith, E. R., Szumowski, D. A., and Davidson, J. M. (1985). Reproductive physiological behavior in the male rat following acute and chronic peripheral adrenergic depletion by guanethidine. *Pharmacol. Biochem. Behav.* 23, 55–63.
923. Stein, B. E., and Meredith, M. A. (1994). *The Merging of the Senses.* MIT Press, Cambridge, MA.
924. Stephan, F. K., Valenstein, E. S., and Zucker, I. (1971). Copulation and eating during electrical stimulation of the rat hypothalamus. *Physiol. Behav.* 7, 587–593.
925. Stief, C. G., Noack, T., and Andersson, K. E. (1997). Signal transduction in cavernous smooth muscle. *World J. Urol.* 15, 27–31.
926. Stoleru, S., Gregoire, M. C., Gerard, D., Decety, J., Lafarge, E., Cinotti, L., Lavenne, F., Le Bars, D., Vernet-Maury, E., Rada, H., Collet, C., Mazoyer, B., Forest, M. G., Magnin, F., Spira, A., and Comar, D. (1999). Neuroanatomical correlates of visually evoked sexual arousal in human males. *Arch. Sex. Behav.* 28, 1–21.
927. Stoneham, M. D., Everitt, B. J., Hansen, S., Lightman, S. L., and Todd, K. (1985). Oxytocin and sexual behaviour in the male rat and rabbit. *J. Endocrinol.* 107, 97–106.
928. Stowers, L., Holy, T. E, Meister, M., Dulac, C., and Koentges, G. (2002). Loss of sex discrimination and male-male aggression in mice deficient for TRP2. *Science* 295, 1493–1500.
929. Struthers, W. M. (2001). Sex-induced fos in the medial preoptic area: projections to the midbrain. *NeuroReport* 12, 3065–3068.
930. Succu, S., Spano, M. S., Melis, M. R., and Argiolas, A. (1998). Different effects of omega-conotoxin on penile erection, yawning and paraventricular nitric oxide in male rats. *Eur. J. Pharmacol.* 359, 19–26.
931. Sugiura, K., Yoshimura, H., and Yokoyama, M. (1997). An animal model of copulatory disorder induced by social stress in male mice: effects of apomorphine and L-dopa. *Psychopharmacology* 133, 249–255.
932. Suh, J. K., Mun, K. H., Cho, C. K., Shin, H. C., Kim, Y. S., and Park, T. C. (1995). Effect of vasoactive intestinal peptide and acetylcholine on penile erection in the rat in vivo. *Int. J. Imp. Res.* 7, 111–118.
933. Swann, J. M. (1997). Gonadal steroids regulate behavioral responses to pheromones by actions on a subdivision of the medial preoptic nucleus. *Brain Res.* 750, 189–194.
934. Swann, J., Rahaman, F., Bijak, T., and Fiber, J. (2001). The main olfactory system mediates pheromone-induced fos expression in the extended amygdala and preoptic area of the male Syrian hamster. *Neuroscience* 105, 695–706.
935. Swanson, L. W. (1976). An autoradiographic study of the efferent connections of the preoptic region in the rat. *J. Comp. Neurol.* 167, 227–256.
936. Swanson, L. W., and McKellar, S. (1979). The distribution of oxytocin- and neurophysin-stained fibers in the spinal cord of the rat and monkey. *J. Comp. Neurol.* 188, 87–106.
937. Swanson, L. W., and Petrovich, G. D. (1998). What is the amygdala? *Trends Neurosci.* 21, 323–331.
938. Swanson, L. W., and Sawchenko, P. E. (1980). Paraventricular nucleus: a site for the integration of neuroendocrine and autonomic mechanisms. *Neuroendocrinology* 31, 410–417.
939. Szczypka, M. S., Zhou, Q. Y., and Palmiter, R. D. (1998). Dopamine-stimulated sexual behavior is testosterone dependent in mice. *Behav. Neurosci.* 112, 1229–1235.
940. Szechtman, H., Caggiula, A. R., and Wulkan, D. (1978). Preoptic knife cuts and sexual behavior in male rats. *Brain Res.* 150, 569–591.
941. Tai, C., Booth, A. M., deGroat, W. C., and Roppolo, J. R. (1998). Penile erection produced by microstimulation of the sacral spinal cord of the cat. *IEEE Trans. Rehabil. Eng.* 6, 374–381.
942. Tallentire, D., McRae, G., Spedding, R., Clark, R., and Vickery, B. (1996). Modulation of sexual behaviour in the rat by a potent and selective 2-adrenoceptor antagonist, delequamine (RS-15385-197). *Brit. J. Pharm.* 118, 63–72.
943. Tanaka, J., and Arnold, A. P. (1993). An electrophysiological study of descending projections to the lumbar spinal cord in adult male rats. *Exp. Brain Res.* 96, 117–124.
944. Tang, Y. P., and Sisk, C. L. (1991). Testosterone in MPOA elicits behavioral but not neuroendocrine responses in ferrets. *Brain Res. Bull.* 26, 373–378.
945. Tang, Y., Rampin, O., Calas, A., Facchinetti, P., and Giuliano, F. (1998). Oxytocinergic and serotoninergic innervation of identified lumbosacral nuclei controlling penile erection in the male rat. *Neuroscience* 82, 241–254.
946. Taylor, G., Bardgett, M., Csernansky, J., Early, T., Haller, J., Scherrer, J., and Womack, S. (1996). Male reproductive systems under chronic fluoxetine or trimipramine treatment. *Physiol. Behav.* 59, 479–485.
947. Taylor, P., and Brown, J. H. (1999). Acetycholine. In *Basic Neurochemistry* (G. J. Siegel, B. W. Agranoff, R. W. Albers, S. K. Fisher, and M. D. Uhler, Eds.), pp. 213–242. Lippincott-Raven, Philadelphia.

948. Temple, J. L., Scordalakes, E. M., Bodo, C., Gustafsson, J. A., and Rissman, E. F. (2003). Lack of functional estrogen receptor beta gene disrupts pubertal male sexual behavior. *Horm. Behav.* 44, 427–434.
949. Thor, D. H., and Flannelly, K. J. (1977). Social-olfactory experience and initiation of copulation in the virgin male rat. *Physiol. Behav.* 19, 411–417.
950. Thor, K. B., Morgan, C., Nadelhaft, I., Houston, M., and de Groat, W. C. (1989). Organization of afferent and efferent pathways in the pudendal nerve of the female cat. *J. Comp. Neurol.* 288, 263–279.
951. Tinbergen, N. (1951). *The Study of Instinct.* Oxford University Press, Oxford, UK.
952. Toner, J. P, and Adler, N. T. (1986). The pre-ejaculatory behavior of male and female rats affects the number of sperm in the vagina and uterus. *Physiol. Behav.* 36, 363–367.
953. Toner, J. P., Attas, A. I., and Adler, N. T. (1987). Transcervical sperm transport in the rat: the roles of pre-ejaculatory behavior and copulatory plug fit. *Physiol. Behav.* 39, 371–375.
954. Toran-Allerand, C. D. (2004). Minireview: a plethora of estrogen receptors in the brain: where will it end? *Endocrinology* 145, 1069–1074.
955. Traish, A., Kim, N. N., Moreland, R. B., and Goldstein, I. (2000). Role of alpha adrenergic receptors in erectile function. *Int. J. Impot. Res.* 12 (Suppl 1), S48–S63.
956. Triemstra, J. L., Nagatani, S., and Wood, R. I. (2005). Chemosensory cues are essential for mating-induced dopamine release in MPOA of male Syrian hamsters *Neuropsychopharmacology* 30, 1436–1442.
957. Trinh, K., and Storm, D. R. (2003). Vomeronasal organ detects odorants in absence of signaling through main olfactory epithelium. *Nat. Neurosci.* 6, 519–525.
958. Truitt, W. A., and Coolen, L. M. (2002). Identification of a potential ejaculation generator in the spinal cord. *Science* 297, 1566–1569.
959. Truitt, W. A., Shipley, M. T., Veening, J. G., and Coolen, L. M. (2003). Activation of a subset of lumbar spinothalamic neurons after copulatory behavior in male but not female rats. *J. Neurosci.* 23, 325–331.
960. Tsutsui, Y., Shinoda, A., and Kondo, Y. (1994). Facilitation of copulatory behavior by pCPA treatments following stria terminalis transection but not medial amygdala lesion in the male rat. *Physiol. Behav.* 56, 603–608.
961. Twiggs, D. G., Popolow, H. B., and Gerall, A. A. (1978). Medial preoptic lesions and male sexual behavior: age and environmental interactions. *Science* 200, 1414–1415.
962. Ueyama, T., Arakawa, H., and Mizuno, N. (1985). Contralateral termination of pudendal nerve fibers in the gracile nucleus of the rat. *Neurosci. Lett.* 62, 113–117.
963. Ueyama, T., Arakawa, H., and Mizuno, N. (1987). Central distribution of efferent and afferent components of the pudendal nerve in rat. *Anat. Embryol.* 177, 37–49.
964. Ueyama, T., Mizuno, N., Nomura, S., Konishi, A., Itoh, K., and Arakawa, H. (1984). Central distribution of afferent and efferent components of the pudendal nerve in cat. *J. Comp. Neurol.* 222, 38–46.
965. Vagell, M. E., and McGinnis, M. Y. (1997). The role of aromatization in the restoration of male rat reproductive behavior. *J. Neuroendocrinol.* 9, 415–421.
966. Vagell, M. E., and McGinnis, M. Y. (1998). The role of gonadal steroid receptor activation in the restoration of sociosexual behavior in adult male rats. *Horm. Behav.* 33, 163–179.
967. Valcourt, R. J., and Sachs, B. D. (1979). Penile reflexes and copulatory behavior in male rats following lesions in the bed nucleus of the stria terminalis. *Brain Res. Bull.* 4, 131–133.
968. Van de Poll, N. E., and Van Dis, H. (1979). The effect of medial preoptic-anterior hypothalamic lesions on bisexual behavior of the male rat. *Brain Res. Bull.* 4, 505–511.
969. Van der Ploeg, L. H. T., Martin, W. J., Howard, A. D., Nargund, R. P., Austin, C. P., Guan, X., Drisko, J., Cashen, D., Sebhat, I., Patchett, A. A., Figueroa, D. J., DiLella, G., Connolly, B. M., Weinberg, D. H., Tan, C. P., Pallyha, O. C., Pong, S. S., MacNeil, T., Rosenblum, C., Vongs, A., Tang, R., Yu, H., Sailer, A. W., Fong, T. M., Huang, C., Tota, M. R., Chang, R. S., Stearns, R., Tamvakopoulos, C., Christ, G., Drazen, D. L., Spar, B. D., Nelson, R. J., and NacIntyre, D. E. (2002). A role for the melanocortin 4 receptor in sexual function. *Proc. Nat. Acad. Sci. U S A* 99, 11381–11386.
970. van Dijken, H., Dijk, J., Voom, P., and Holstege, J. C. (1996). Localization of dopamine D2 receptor in rat spinal cord identified with immunocytochemistry and in situ hybridization. *Eur. J. Neurosci.* 8, 621–628.
971. van Dis, H., and Larsson, K. (1970). Seminal discharge following intracranial electrical stimulation. *Brain Res.* 23, 381–386.
972. van Dis, H., and Larsson, K. (1971). Induction of sexual arousal in the castrated male rat by intracranial stimulation. *Physiol. Behav.* 6, 85–86.
973. van Furth, W. R., and van Ree, J. M. (1994). Endogenous opioids and sexual motivation and performance during the light phase of the diurnal cycle. *Brain Res.* 636, 175–179.
974. van Furth, W. R., and van Ree, J. M. (1996). Sexual motivation: involvement of endogenous opioids in the ventral tegmental area. *Brain Res.* 729, 20–28.
975. van Furth, W. R., van Ernst, M. G., and van Ree, J. M. (1995). Opioids and sexual behavior of male rats: involvement of the medial preoptic area. *Behav. Neurosci.* 109, 123–134.
976. van Furth, W. R., Wolterink-Donselaar, I. G., and van Ree, J. M. (1994). Endogenous opioids are differentially involved in appetitive and consummatory aspects of sexual behavior of male rats. *Amer. J. Physiol.* 266, R606–R613.
977. Vanderhorst, V. G., Terasawa, E., Ralston, H. J. III, and Holstege, G. (2000). Monosynaptic projections from the lateral periaqueductal gray to the nucleus retroambiguus in the rhesus monkey: implications for vocalization and reproductive behavior. *J. Comp. Neurol.* 424, 251–268.
978. Vardi, Y., Youdim, M. B. H., Levin, R. D., and Finberg, J. P. M. (1989). A model for the study of penile erection in the rat. *Int. J. Impot. Res.* 1, 11–17.
979. Vaughan, E., and Fisher, A. E. (1962). Male sexual behavior induced by intracranial electrical stimulation. *Science* 137, 758–760.
980. Veening, J. G., and Coolen, L. M. (1998). Neural activation following sexual behavior in the male and female rat brain. *Behav. Brain. Res.* 92, 181–193.
981. Vega Matuszcyk, J., Larsson, K., and Eriksson, E. (1998). The selective serotonin reuptake inhibitor fluoxetine reduces sexual motivation in male rats. *Pharmacol. Biochem. Behav.* 60, 527–532.
982. Vemulapalli, R., Kurowski, S., Salisbury, B., Parker, E., and Davis, H. (2001). Activation of central melanocortin receptors by MT-II increases cavernosal pressure in rabbits by the neuronal release of NO. *Brit. J. Pharmacol.* 134, 1705–1710.
983. Verma, S., Chhina, G. S., Kumar, M. V., Singh, B. (1989). Inhibition of male sexual behavior by serotonin application in the medial preoptic area. *Physiol. Behav.* 46, 327–330.
984. Véronneau-Longueville. F., Rampin, O., Freund-Mercier, M. J., Tang, Y., Calas, A., Marson, L., McKenna, K. E., Stoeckel, M. E., Benoit, G., and Giuliano, F. (1999). Oxytocinergic innervation of automatic nuclei controlling penile erection in the rat. *Neuroscience* 93, 1437–1447.

985. Vigdorchik, A., Lagoda, G., Boje, K. M. K., and Hull, E. M. (2003). A glutamate NMDA receptor antagonist, microinjected into the MPOA, impairs male sexual behavior. *Soc. Neurosci. Abst.* 33, 404.11.
986. VinanderCaerols, C., Collado, P., Segovia, S., and Buillamon, A. (1998). Sex differences in the posteromedial cortical nucleus of the amygdala in the rat. *NeuroReport* 9, 2653–2656.
987. Wagner, C. K., and Clemens, L. G. (1993). Neurophysin-containing pathway from the paraventricular nucleus of the hypothalamus to a sexually dimorphic motor nucleus in lumbar spinal cord. *J. Comp. Neurol.* 336, 106–116.
988. Wagner, C. K., Sisk, C. L., and Clemens, L. G. (1993). Neurons in the paraventricular nucleus of the hypothalamus that project to the sexually dimorphic lower lumbar spinal cord concentrate 3H-estradiol in the male rat. *J. Neuroendocrinol.* 5, 545–551.
989. Waldinger, M. D., Hengeveld, M. W., Zwinderman, A. H., and Olivier, B. (1998). Effects of SSRI antidepressants on ejaculation: a double-blind, randomized, placebo-controlled study with fluoxetine, fluvoxamine, paroxetine, and sertraline. *J. Clin. Psychopharmacol.* 18, 274–281.
990. Wallach, S. J. R., and Hart, B. L. (1983). The role of the striated penile muscles of the male rat in seminal plug dislodgement and deposition. *Physiol. Behav.* 31, 815–821.
991. Wang, C. T., Huang, R. L., Tai, M. Y., Tsai, Y. F., and Peng, M. T. (1995). Dopamine release in the nucleus accumbens during sexual behavior in prenatally stressed adult male rats. *Neurosci. Lett.* 200, 29–32.
992. Wang, J. M., McKenna, K. E., and Lee, C. (1991). Determination of prostatic secretion in rats: effect of neurotransmitters and testosterone. *Prostate* 18, 289–301.
993. Wang, L., and Hull, E. M. (1980). Tail pinch induces sexual behavior in olfactory bulbectomized male rats. *Physiol. Behav.* 24, 211–215.
994. Wang, Z., Hulihan, T. J., and Insel, T. R. (1997). Sexual and social experience is associated with different patterns of behavior and neural activation in male prairie voles. *Brain Res.* 767, 321–332.
995. Warner, R. K., Thompson, J. T., Markowski, V. P., Loucks, J. A., Bazzett, T. J., Eaton, R. C., and Hull, E. M. (1991). Microinjection of the dopamine antagonist cis-flupenthixol into the MPOA impairs copulation, penile reflexes and sexual motivation in male rats. *Brain Res.* 540, 177–182.
996. Watson, J. T., and Adkins-Regan, E. (1989). Activation of sexual behavior by implantation of testosterone propionate and estradiol benzoate into the preoptic area of the male Japanese Quail (*Cotumix japonica*). *Horm. Behav.* 23, 251–268.
997. Watters, J. J., and Dorsa, D. M. (1998). Transcriptional effects of estrogen on neuronal neurotensin gene expression involve cAMP/protein kinase A-dependent signaling mechanisms. *J. Neurosci.* 18, 6672–6680.
998. Wells, A. B., and Coolen, L. M. (2001). Lumbar spinal cord efferents directly contact thalamic neurons that project to MPOA or BNST. *Soc. Behav. Neuroendocrinol. Abst.* 5, 354.
999. Wenkstern, D., Pfaus, J. G., and Fibiger, H. C. (1993). Dopamine transmission increases in the nucleus accumbens of male rats during their first exposure to sexually receptive female rats. *Brain Res.* 618, 41–46.
1000. Wersinger, S. R., and Rissman, E. F. (2000). Dopamine activates masculine sexual behavior independent of the estrogen receptor alpha. *J. Neurosci.* 20, 4248–4254.
1001. Wersinger, S. R., Baum, M. J., and Erskine, M. S. (1993). Mating-induced FOS-like immunoreactivity in the rat forebrain, a sex comparison and a dimorphic effect of pelvic nerve transection. *J. Neuroendocrinol.* 5, 557–568.
1002. Wersinger, S. R., Sannen, K., Villalba, C., Lubahn, D. B., Rissman, E. F., and De Vries, G. J. (1997). Masculine sexual behavior is disrupted in male and female mice lacking a functional estrogen receptor alpha gene. *Horm. Behav.* 32, 176–183.
1003. Wessels, H., Fuciarelli, K., Hansen, J., Hadley, M. E., Hruby, V. J., Dorr, R., and Levine, N. (1998). Synthetic melanotropic peptide initiates erections in men with psychogenic erectile dysfunction: double-blind, placebo controlled crossover study. *J. Urol.* 160, 389–393.
1004. Wessels, H., Gralnek, D., Dorr, R., Hruby, V. J., Hadley, M. E., and Levine, N. (2000). Effect of an alpha-melanocyte stimulating hormone analog on penile erection and sexual desire in men with organic erectile dysfunction. *Urology* 56, 641–646.
1005. Wessels, H., Hruby, V. J., Hackett, J., Han, G., Balse-Sriinivasan, P., and Vanderah, T. W. (2003). Ac-Nle-c(Asp-His-DPhe-Arg-Trp-Lys)-NH2 induces penile erection via brain and spinal melanocortin receptors. *Neuroscience* 118, 755–762.
1006. Westberry, J., and Meredith, M. (2003). The influence of chemosensory input and gonadotropin releasing hormone on mating behavior circuits in male hamsters. *Brain Res.* 974, 1–16.
1007. Whalen, R. E., and Luttge, W. G. (1971). Testosterone, androstenedione and dihydrotestosterone: effects on mating behavior of male rats. *Horm. Behav.* 2, 117–125.
1008. Wheeler, J. M., and Crews, D. (1978). The role of the anterior hypothalamus-preoptic area in the regulation of male reproductive behavior in the lizard, *Anolis carolinensis*: lesion studies. *Horm. Behav.* 11, 42–60.
1009. White, N. R., Cagiano, R., Moises, A. U., and Barfield, R. J. (1990). Changes in mating vocalizations over the ejaculatory series in rats (*Rattus norvegicus*). *J. Comp. Psychol.* 104, 255–262.
1010. Whitelaw, G. P., and Smithwick, R. H. (1951). Some secondary effects of sympathectomy with particular reference to disturbance of sexual function. *N. Engl. J. Med.* 245, 121–130.
1011. Wiesenfeld-Hallin, Z., and Sodersten, P. (1984). Spinal opiates affect sexual behaviour in rats. *Nature* 309, 257–258.
1012. Wilbert, D. M., Griffin, J. E., and Wilson, J. D. (1983). Characterization of the cytosol androgen receptor of the human prostate. *J. Clin. Endocrinol. Metabol.* 56, 113–120.
1013. Wilhelmsson, M., and Larsson, K. (1973). The development of sexual behavior in anosmic male rats reared under various social conditions. *Physiol. Behav.* 11, 227–232.
1014. Wilson, C., Nomikos, G. G., Collu, M., and Fibiger, H. C. (1995). Dopaminergic correlates of motivated behavior: importance of drive. *J. Neurosci.* 15, 5169–5178.
1014a. Wilson, D. B., and Wilson, W. J. (1978). *Human Anatomy*. Oxford University Press, New York.
1015. Winans, S. S., and Powers, J. B. (1977). Olfactory and vomeronasal deafferentation of male hamsters: histological and behavioral analyses. *Brain Res.* 126, 325–344.
1016. Wingard, C. J., Husain, S., Williams, J., and James, S. (2003). RhoA-Rho kinase mediates synergistic ET-1 and phenylephrine contraction of rat corpus cavernosum. *Am. J. Physiol. Reg. Integ. Comp. Physiol.* 285, R1145–R1152.
1017. Winkler, S. M., and Wade, J. (1998). Aromatase activity and regulation of sexual behaviors in the green anole lizard. *Physiol. Behav.* 64, 723–731.
1018. Winslow, J. T., and Insel, T. R. (1991). Social status in pairs of male squirrel monkeys determines the behavioral response to central oxytocin administration. *J. Neurosci.* 24, 761–765.

1019. Wise, R. A., and Bozarth, M. A. (1984). Brain reward circuitry: four circuits "wired" in apparent series. *Brain Res. Bull.* 12, 203–208.

1020. Witt, D. M. (1995). Oxytocin and rodent sociosexual responses: from behavior to gene expression. *Neurosci. Biobehav. Rev.* 19, 315–324.

1021. Witt, D. M., and Insel, T. R. (1994). Increased Fos expression in oxytocin neurons following masculine sexual behavior. J. *Neuroendocrinol.* 6, 13–18.

1022. Witt, D. M., Young, L. J., and Crews, D. (1994). Progesterone and sexual behavior in males. *Psychoneuroendocrinology* 19, 553–562.

1023. Witt, D. M., Young, L. J., and Crews, D. (1995). Progesterone modulation of androgen-dependent sexual behavior in male rats. *Physiol. Behav.* 57, 307–313.

1023a. Wood, D. A., Kosobud, A. E., and Rebec, G. V. (2004). Nucleus accumbens single-unit activity in freely behaving male rats during approach to novel and non-novel estrus. *Neurosci. Lett.* 368, 29–32.

1024. Wood, R. I. (1996). Estradiol, but not dihydrotestosterone, in the medial amygdala facilitates male hamster sex behavior. *Physiol. Behav.* 59, 833–841.

1024a. Wood, R. I. (1997). Thinking about networks in the control of male hamster sexual behavior. *Horm. Behav.* 32, 40–45.

1025. Wood, R. I., and Coolen, L. M. (1997). Integration of chemosensory and hormonal cues is essential for sexual behaviour in the male Syrian hamster: role of the medial amygdaloid nucleus. *Neuroscience* 78, 1027–1035.

1026. Wood, R. I., and Newman, S. W. (1993). Mating activates androgen receptor-containing neurons in the chemosensory pathways of the male Syrian hamster brain. *Brain Res.* 614, 65–77.

1027. Wood, R. I., and Newman, S. W. (1995a). Androgen and estrogen receptors coexist within individual neurons in the brain of the Syrian hamster. *Neuroendocrinology* 62, 487–497.

1028. Wood, R. I., and Newman, S. W. (1995b). Hormonal influence on neurons of the mating behavior pathway in male hamsters. In *Neurobiological Effects of Sex Steroid Hormones* (P. E. Micevych, and R. P. Hammer, Jr., Eds.), pp. 3–39. Cambridge University Press, New York.

1029. Wood, R. I., and Newman, S. W. (1995c). Integration of chemosensory and hormonal cues is essential for mating in the male Syrian hamster. *J. Neurosci.* 15, 7261–7269.

1030. Wood, R. I., and Williams, S. J. (2001). Steroidal control of male hamster sexual behavior: redundancy without hierarchy. *Physiol. Behav.* 72, 727–733.

1031. Wood, R. I., Brabec, R. K., Swain, J. M., and Newman, S. W. (1992). Androgen and estrogen receptor concentrating neurons in chemosensory pathways of the male Syrian hamster brain. *Brain Res.* 596, 89–98.

1032. Woolley, S. C., Sakata, J. T., and Crews, D. (2004). Evolutionary insights into the regulation of courtship behavior in male amphibians and reptiles. *Physiol. Behav.* 83, 347–360.

1033. Woolley, S. C., Sakata, J. T., Gupta, A., and Crews, D. (2001). Evolutionary changes in dopaminergic modulation of courtship behaviors in *Cnemidophorus* whiptail lizards. *Horm. Behav.* 40, 483–489.

1034. Xiao, L., and Becker, J. B. (1998). Effects of estrogen agonists on amphetamine-stimulated striatal dopamine release. *Synapse* 29, 379–391.

1035. Yahr, P. (1995). Neural circuitry for the hormonal control of male sexual behavior. In *Neurobiological Effects of Sex Steroid Hormones* (P. E. Micevych, and R. P. Hammer, Jr., Eds.), pp. 40–56. Cambridge University Press, New York.

1036. Yahr, P., and Gregory, J. E. (1993). The medial and lateral cell groups of the sexually dimorphic area of the gerbil hypothalamus are essential for male sex behavior and act via separate pathways. *Brain Res.* 631, 287–296.

1037. Yahr, P., and Stephens, D. R. (1987). Hormonal control of sexual and scent marking behaviors of male gerbils in relation to the sexually dimorphic area of the hypothalamus. *Horm. Behav.* 21, 331–346.

1038. Yahr, P., and Ulibarri, C. (1987). Polyadenylated and non-adenylated messenger RNA and androgen control of sexual behavior and scent marking in male gerbils. *Horm. Behav.* 21, 53–64.

1039. Yahr, P., Commins, D., Jackson, J. C., and Newman, A. (1982). Independent control of sexual and scent marking behaviors of male gerbils by cells in or near the medial preoptic area. *Horm. Behav.* 16, 304–322.

1040. Yahr, P., Finn, P. D., Hoffman, N. W., and Sayag N. (1994). Sexually dimorphic cell groups in the medial preoptic area that are essential for male sex behavior and the neural pathways needed for their effects. *Psychoneuroendocrinology* 19, 463–470.

1041. Yaîci, E. D., Rampin, O., Calas, A., Jestin, A., McKenna, K. E., Leclerc, P., Benoit, G., and Giuliano, F. (2002). Alpha(2a) and alpha(2c) adrenoceptors on spinal neurons controlling penile erection. *Neuroscience* 114, 945–960.

1042. Yamada, K., Emson, P., and Hokfelt, T. (1996). Immunohistochemical mapping of nitric oxide synthase in the rat hypothalamus and colocalization with neuropeptides. *J. Chem. Neuroanat.* 10, 295–316.

1043. Yamada, Y. (1979). The effects of testosterone on unit activity in rat hypothalamus and septum. *Brain Res.* 172, 165–169.

1044. Yanagimoto, M. Honda, K. Goto, Y., and Negoro, H. (1996). Afferents originating from the dorsal penile nerve excite oxytocin cells in the hypothalamic paraventricular nucleus of the rat. *Brain Res.* 733, 292–296.

1045. Yasui, Y., Saper, C. B., and Cechetto, D. F. (1991). Calcitonin gene-related peptide (CGRP) immunoreactive projections from the thalamus to the striatum and amygdala in the rat. *J. Comp. Neurol.* 308, 293–310.

1046. Yells, D. P., Hendricks, S. E., and Prendergast, M. A. (1992). Lesions of the nucleus paragigantocellularis effects on mating behavior in male rats. *Brain Res.* 596, 73–79.

1047. Yells, D. P., Prendergast, M. A., Hendricks, S. E., and Miller, M. E. (1995). Monoaminergic influences on temporal patterning of sexual behavior in male rats. *Physiol. Behav.* 58, 847–852.

1048. Yilmaz, U., Tatlisen, A., Turan, H., Arman, F., and Ekmekcioglu, O. (1999). The effects of fluoxetine on several neurophysiological variables in patients with premature ejaculation. *J. Urol.* 161, 107–111.

1049. Yokosuka, M., and Hayashi, S., (1996). Co-localization of neuronal nitric oxide synthase and androgen receptor immunoreactivity in the premammillary nucleus in rats. *Neurosci. Res.* 26, 309–314.

1050. Yokosuka, M., Matsuoka, M., Ohtani-Kaneko, R., Iigo, M., Hara, M., Hirata, K., and Ichikawa, M. (1999). Female-soiled bedding induced fos immunoreactivity in the ventral part of the premammillary nucleus (PMv) of the male mouse. *Physiol. Behav.* 68, 257–261.

1051. Yokosuka, M., Prins, G. S., and Hayashi, S. (1997). Co-localization of androgen receptor and nitric oxide synthase in the ventral premammillary nucleus of the newborn rat: an immunohistochemical study. *Brain Res. Dev. Brain Res.* 99, 226–233.

1052. Zarrindast, M. R., Mamanpush, S. M., and Rashidy-Pour, A. (1994). Morphine inhibits dopaminergic and cholinergic induced ejaculation in rats. *Gen. Pharmacol.* 25, 803–808.

1053. Zarrindast, M. R., Shokravi, S., and Samini, M. (1992). Opposite influences of dopaminergic receptor subtypes on penile erection. *Gen. Pharmacol.* 23, 671–675.
1054. Zufall, F., Kelliher, K. R., and Leinders-Zufall, T. (2002). Pheromone detection by mammalian vomeronasal neurons. *Microsc. Res. Tech.* 58, 251–260.
1055. Zumpe, D., Clancy, A. N., Bonsall, R. W., and Michael, R. P. (1996). Behavioral responses to Depo-Provera, Fadrozole, and estradiol in castrated, testosterone-treated cynomolgus monkeys (*Macaca fascicularis*): the involvement of progestin receptors. *Physiol. Behav.* 60, 531–540.
1056. Zumpe, D., Clancy, A. N., and Michael, R. P. (1997). Effects of progesterone on the sexual behavior of castrated, testosterone-treated male cynomolgus monkeys (*Macaca fascicularis*). *Physiol. Behav.* 62, 61–67.
1057. Zvara, P., Sioufi, R., Schipper, H. M., Begin, L. R., and Brock, G. B. (1995). Nitric oxide mediated erectile activity is a testosterone dependent event: a rat erection model. *Int. J. Impot. Res.* 7, 209–219.

Knobil and Neill's Physiology of Reproduction, Third Edition
edited by Jimmy D. Neill,
Elsevier © 2006

CHAPTER 34

Hormonal, Neural, and Genomic Mechanisms for Female Reproductive Behaviors, Motivation, and Arousal

D. W. Pfaff, Y. Sakuma, L.-M. Kow, A. W. L. Lee, and A. Easton

Background, 1825
Theoretical Questions, 1826
Hormone Actions, 1828
Mechanisms Underlying the Primary Mating Behavior, Lordosis, 1863
Molecular Mechanisms, from Hormone to Gene to Behavior, 1865
Sexual Motivation and Courtship Behaviors, 1871
Preoptic Neurons as Nodes for Reproductive Neuroendocrinology and Sex and Courtship Behaviors, 1872
Courtship and Sexual Behaviors of Female Rats, 1877
Preoptic Area Neurons Foster Courtship Behaviors, 1880
Social Recognition and Affiliation, 1881
Arousal, 1882
BBURP Theory, 1882
Background and Operational Definition of Arousal, 1883
Hormone-Dependent Genes Influencing Arousal, 1886
Biophysics of Arousal Transmitters Affecting Hypothalamic Neurons that Control Sexual Behavior, 1887
Overall Summary, 1891
References, 1892

After a general introduction (Background), from the second edition of this text, which covers the historical highlights of hormone–behavior relations in females, we review mechanisms for lordosis behavior itself (Mechanisms Underlying the Primary Mating Behavior, Lordosis). By rigorous logic, lordosis fulfills the criteria for a motivated behavior. Prominent signs of female sexual motivation are found in their courtship behaviors, which are covered in Sexual Motivation and Courtship Behaviors. In turn, all theories of motivation say that the expression of vigorous motivated behavior depends on the arousal (activation) of brain and behavior. We discuss genomic contributions to the expression of hormone-dependent arousal in Arousal. All of the basic characteristics of hormone-facilitated mechanisms covered here conform to more general principles of hormone–behavior relations (covered elsewhere in textbook form). Finally, the entire chapter is summarized (Overall Summary). To keep this chapter to a reasonable length, we have left out tremendous amounts of electrophysiological detail about female sex behavior circuitry because they were well covered in the first and second editions of this text.

BACKGROUND

The first edition of *The Physiology of Reproduction* included neurophysiological details of lordosis behavior and related circuitry. The second edition, with Margaret McCarthy and Susan Schwartz-Giblin, added much neurochemistry. A tribute to the reliability and durability of knowledge in this field is

Department of Neurobiology and Behavior, The Rockefeller University, New York, New York.

that the older literature can be included in capsule form in this introductory section to our chapter in the third edition. Only those segments of the second edition relevant to our present arguments have been reproduced here.

The modern era of research on mechanisms of reproductive behavior began when Frank Beach (1) and William C. Young (2) collected and summarized the data showing the dependence of mating behaviors on gonadal steroid hormones. A variety of female reproductive behaviors clearly required circulating estrogens, and the behavioral effects of estrogens could be amplified by progestins. Clear explanations of the cellular mechanisms for female reproductive behavior waited (a) until we could achieve a more comprehensive description of steroid hormone actions at the cellular level, and (b) until a complete neural circuit for a single reproductive behavior could be described. Steroid hormone actions on cells have been analyzed most easily in those tissues that have specific hormone receptors. The exact locations in the vertebrate brain of nerve cells with such receptors for sex steroid hormones eventually were described with steroid hormone autoradiography (3–5). From there, receptor-mediated effects of estrogens on electrical and synthetic capacities of hypothalamic and preoptic neurons could more efficiently be studied. Estrogen- or progesterone-induced changes in these neurons might control primary reproductive behaviors but might also be relevant for other hormone-dependent functions such as ovulation, communicative behaviors, locomotion, feeding, and aggression. With an emphasis on mechanisms that are related to the control of the hormone-dependent female reproductive behavior, lordosis, the first section of this chapter treats hormone actions on nerve cells: receptors, electrophysiological effects, and changes in RNA and protein synthesis. Results now available open the way for the application of certain molecular biological and protein chemical techniques to hypothalamic neurons to probe their function with chemically specific driving forces (estrogens and progestins) and clear functional end points (female reproductive behaviors).

The other requirement for launching detailed cellular investigations from the platform provided by the early work on hormone–behavior relations was knowing enough about a behavioral circuit that the functional meaning of particular cellular changes could be deduced. Knowing the characteristics of an entire behavioral circuit could be necessary, for example, to see exactly how hormone-stimulated changes in hypothalamic neurons could play on other, extra-hypothalamic neural mechanisms to effect a behavioral change. In this regard, the neural circuit for lordosis behavior (6) provides a paradigm for studying how genomic and electrical changes in hypothalamic and forebrain neurons can be read out through changes in behavior-controlling circuitry. Besides providing the first complete circuit for a mammalian behavior, the increasingly detailed description of the mechanisms for lordosis behavior offers two opportunities. The properties of models of neurons at different levels of the neuraxis can be summarized and conceptualized in such a way as to account completely for the properties of a reproductive behavior. Further, within each neural module, our behavioral and endocrine knowledge, as well as the constraints placed by the other modules, can allow us to predict the properties of individual cell types. The second section of this chapter summarizes the neural mechanisms of lordosis as the most intensively studied female reproductive behavior.

Theoretical Questions

Mammalian Brain

Hormone-driven reproductive behaviors have given us the opportunity to answer global theoretical questions about how the mammalian brain manages the mechanisms for behaviors of adult organisms. At least four fundamental questions can be asked. First, what causes a particular behavioral response to occur? Second, given that this behavioral response can occur, what regulates its amplitude? Third, as that response is occurring, what prevents other responses? Fourth, as Sherrington asked in 1906, what executes the transition from the behavioral response in question to the next?

The literature reviewed here answers, for female reproductive behavior, the first and second questions in considerable detail. Completion of the neural circuit for lordosis behavior (6) proved that it is in fact possible to determine mechanisms for a mammalian behavior. Success in providing this paradigm depended on the relative simplicity of the stimuli driving the behavior and the motor topography of the response itself. Most exquisitely, the strong steroid hormonal controls have allowed us, and continue to allow us, to exploit the experimental tools of molecular endocrinology.

The clarity of the neural mechanisms at the systems and cellular levels that regulate reproductive behavior has allowed the formation of certain other global conclusions. It is clear that even with a well-defined steroid hormonal effect and a simple vertebrate behavior like lordosis, we are not dealing with a one-hormone-one-gene-one-behavior system. In fact, early responses to estrogen by cells in the ventromedial nucleus of the hypothalamus (VMN) appear to prepare the way for later responses. Some of the early responses are closely related to growthlike neuronal mechanisms, whereas some of the later responses can be tied specifically to female reproductive behavior. Where estrogen leads to increases in specific hypothalamic neuronal products, and, in turn, those products can be shown to facilitate female reproductive behavior, molecular manufacture of that product can be assumed to be one way in which the hormone facilitates the behavior.

Axiomatic Requirements for Reproduction

Theoretical approaches to physics and chemistry have been much different from those for neurobiology. In the former cases, entire fields can be governed by a small number of equations: mechanics by Newton's laws, electricity and magnetism by Maxwell's equations, and thermodynamics by Gibbs's phase rule are examples. No part of neurobiological theory partakes of this simplicity. Instead, the safest approach to neurobiological theory may be one that has fared well in mathematics: the axiomatic approach.

It is part of the very definition of life that animals must reproduce. Following from this axiom, what are the obvious requirements for reproduction? We emphasize here only those clear requirements that involve the central nervous system (CNS) and assume that gametogenesis and implantation will be managed well by peripheral mechanisms. Obviously, the sperm and egg need to achieve contact, thus requiring reproductive behavior. Assuming that circulating hormones are driving this behavior to coordinate central neural mechanisms with peripheral preparations for reproduction, nerve cells executing reproductive behavior must in addition receive inputs showing the existence of the mating partner. Other permissive influences that help to guarantee the success of reproduction include adequate nutrition, adequate water and salt supplies, appropriate temperature and day length, appropriate time of day, the existence of nesting material, and the absence of stress.

Cellular and molecular mechanisms by which the steroid hormones themselves induce lordosis behavior have been unraveled in considerable detail and are presented in the following sections. Analyses of the promoters of hormone-responsive genes are now approachable for hormone-dependent neurons and can add to behavioral explanation. Our own work has shown that the gene coding for estrogen receptor-α (ERα) is required for the entire chain (Fig. 1) of behaviors associated with reproduction, from the locomotion of courtship behaviors, through mating behavior itself, to maternal behaviors.

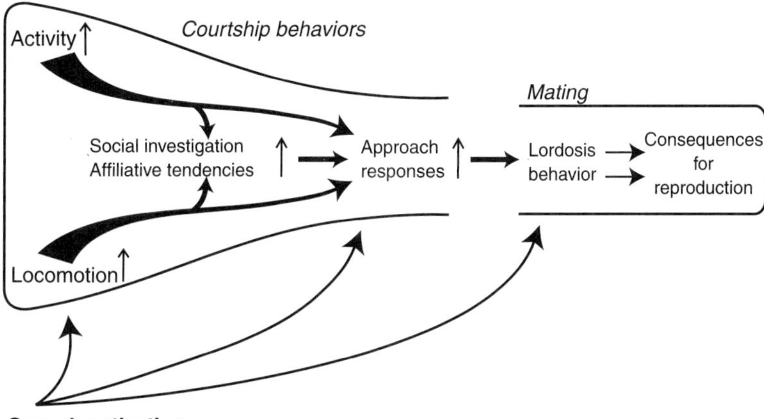

FIG. 1. Even in lower animals there are substantial sets and sequences of social behaviors preceding mating responses. Uncovering their mechanisms gives us the chance to address the problem stated by the great British neurophysiologist, Sir Charles Sherrington, who said that the neurobiologist must account for "the flow of behaviour through time." Motivational influences, fueled by sex hormones, facilitate not only copulatory responses but components of the courtship responses that precede them. Conversely, the depth of knowledge we have about the concrete mating and courtship responses helps us explain motivation, the underlying abstract concept. The estrogenic effect on locomotor behavior—a mechanism for motivational actions on early courtship behaviors—depends on the gene for estrogen receptor-α (821). Likewise, the estrogenic effects on lordosis behavior itself and on maternal behavior.

Hormone Actions

Hormones and Behavior

The dependence of female reproductive behavior on estrogen and progestins was elucidated thoroughly by Beach (1), Young (2), and their collaborators. We can use these steroid hormones to probe molecular mechanisms for a mammalian behavior (7).

The interactions of estrogens and progestins administered experimentally with the effects of naturally circulating hormones and the comparisons of effective doses to physiological levels have been reviewed (8). Serum levels of estradiol in the physiological range promote female sexual behavior in ovariectomized rats, and the estrogen effect on behavior can be facilitated by progesterone levels clearly below those that are physiologically available (9). Estradiol by itself can support high levels of lordosis behavior in ovariectomized female rats (10,11), but optimal combinations of estrogens followed by progesterone produce much more complete displays of reproductive behavior. In estrogen-primed females, progesterone, as well as luteinizing hormone–releasing hormone (LHRH) (12,13), offers the opportunity of synchronizing behavioral with gametogenic events. The critical role of gene expression for the progesterone receptor (PR) in this process has been demonstrated by the use of direct hypothalamic administration of antisense DNA for PR messenger RNA (mRNA) (14). When applied 12 but not 24 hours after estradiol, hypothalamic PR antisense treatment significantly reduced lordosis behavior measured either as a reflex or in a mating behavior test (Fig. 2). Notably, preceptive behaviors, strongly progesterone dependent, were greatly reduced in their occurrence.

The behavioral actions of estrogen are clearly on neurons directly rather than through an obligatory pituitary mechanism because estrogen and progesterone can support female reproductive behavior in hypophysectomized ovariectomized rats (15). Some of the behaviorally active progesterone may come from the adrenals because suppression of adrenal activity has been shown to result in a blockade of ovulation, but the major rise in progesterone levels after the critical period on the day of proestrus is from the ovary (16). In ovariectomized rats and guinea pigs, the adrenal progesterone is behaviorally effective enough that administration of corticotropin (adrenocorticotropic hormone [ACTH]) can stimulate lordosis (17). In some circumstances, higher levels of estrogen, achieved just before behavioral tests in ovariectomized estrogen-primed female rats, can themselves produce high levels of lordosis behavior, as though the high levels of estrogen were mimicking the action of progesterone (18). In these circumstances estradiol probably is not restricted to actions

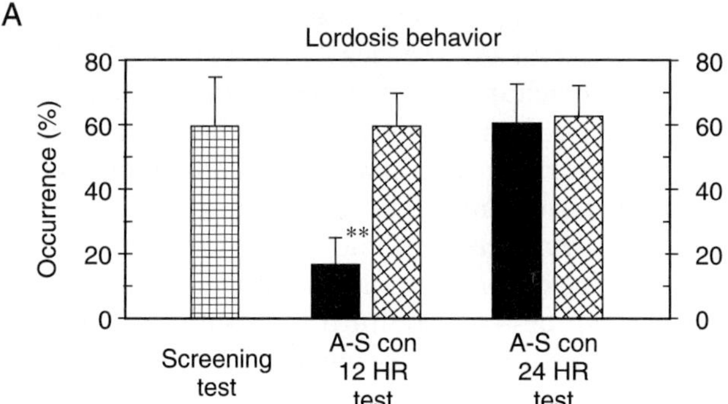

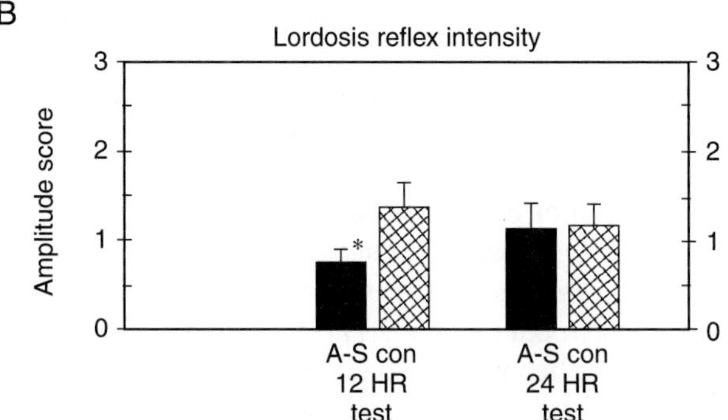

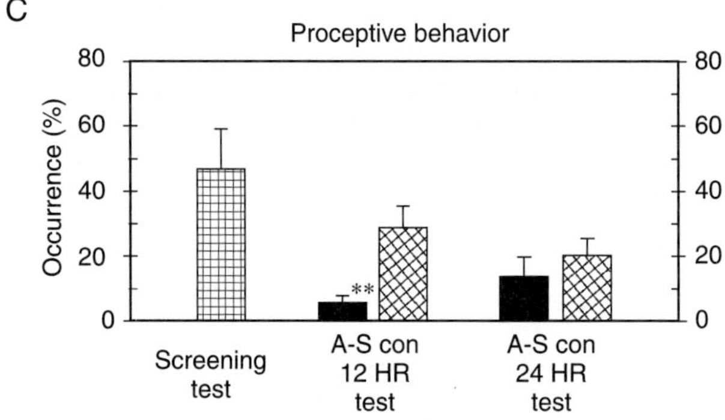

FIG. 2. Antisense DNA directed against mRNA for progesterone receptor (PR) reduces female rat reproductive behavior. Ovariectomized estrogen-treated female rats were microinjected into the ventromedial hypothalamus with antisense DNA constructs (A-S, *dark bars*) or with control DNA constructs composed of the same nucleotide bases in scrambled order (Con, *crosshatch bars*). Treated at 12 but not 24 hours after estrogen, rats with antisense against PR displayed significantly less lordosis behavior (**A**), with smaller reflex amplitude (**B**). The largest reduction in behavioral frequency from antisense was in proceptive behavior (**C**), which is known to be strongly progesterone sensitive. (From Ogawa, S., Olazabal, U. E., and Pfaff, D. W. [1994]. Effects of intrahypothalamic administration of antisense DNA for progesterone receptor mRNA on reproductive behavior and progesterone receptor immunoreactivity in female rat. *J. Neurosci.* 14, 1766–1774, with permission.)

through ERs, and the behavioral effects may be attributed partly to actions of estradiol through PRs (19).

Comparisons across Species

The generality of the conclusion that steroid hormones from the ovary facilitate female reproductive behavior among vertebrates has been documented in the older literature and well reviewed (1,2). Of special interest are the comparisons between widely used laboratory animals, such as rats and primates. In female monkeys, attractivity and proceptivity clearly change according to the stage of the menstrual cycle or the steroid injected, but there is less agreement among results from different laboratories on the effects of steroid hormones on receptivity [reviewed in Clemens and Weaver (20)]). Among subhuman primates, the degree to which copulation is restricted to the periovulatory period depends on the species studied, but in adult human females, attempts to show that ovarian steroids are required for normal sexual activity are either negative or fraught with methodological problems [reviewed in Feder (21)]. In general, although the mechanisms for steroid hormone–behavioral relations can be assumed to be in place, the increasing power of social and other environmental factors in affecting reproductive behavior would give steroid hormones less of an explanatory role in humans. A neuroendocrine component related to sexual orientation among humans has been claimed, based on the demonstration that the secretory pattern of luteinizing hormone (LH) in male homosexuals was intermediate between that typical of heterosexual men and that of women (22). In women, premenstrual changes in susceptibility to depression, anxiety, irritability, or other mood changes are prominent and may depend in part on changes in progesterone levels (23).

Gene Memory

Time Courses of Hormone Action. Developmental effects of steroid sex hormones on sexual differentiation have been widely reviewed. Behaviorally, there are critical androgen-sensitive periods in which the eventual capacity for feminine-typical behavior can be disrupted by administration of testosterone (24–26). Androgenic hormones given prenatally or postnatally to developing female rats decrease the rate of lordosis behavior (27). In genetic males, prenatal stress can increase the eventual rate of female-typical lordotic responding (28), probably because of an alteration in fetal testicular enzyme activity (29). The behavioral aspects of sexual differentiation have been reviewed (30). There are presumed to be two aspects of behavioral differentiation: defeminization, dependent primarily on aromatization of testosterone to produce estradiol (31–34), and masculinization, which may be achieved partly through aromatization but also through androgen receptors (34). Prevention of aromatization not only allows greater levels of female-typical behavior in genetic males, it allows higher levels of hypothalamic PRs (35).

Based on the foregoing notion that defeminization depends on aromatization of testosterone to produce estradiol, which in turn works through hypothalamic and preoptic ERs, we hypothesized that antisense DNA sequences against the mRNA for the ER would interfere with behavioral effects of neonatal androgenization in female rats. On postnatal day 3, female rats were given bilateral hypothalamic infusion of either antisense DNA spanning the translation start site for the ER or a control (scrambled) DNA sequence, and were also lightly androgenized with 50 μg testosterone propionate (37). In terms of lordosis behavior in adulthood, the neonatal antisense treatment significantly protected against the effect of this neonatal androgenization treatment, thus confirming the importance of the ER for this sexual differentiation process and demonstrating for the first time a long-term behavioral impact from the interruption of expression of a specific gene (Fig. 2).

At the other end of the life span, chronic exposure to high levels of estradiol may mimic some aspects of normal reproductive aging in the female (see Chapter 59).

This chapter concentrates on temporal aspects of estrogen and progesterone actions that bear on cellular mechanisms for behavior in the normal adult.

Gene Memory: Precedents. Some aspects of hormone actions on behavior raise the mechanistic possibility that steroid influences on nerve cells cause changes in genomic structure such that subsequent transcriptional responses to normal controlling signals are altered for some time. These alterations might persist long after the initial cause disappeared and may result from changes in DNA itself or from protein–DNA interactions. Previous studies on other cell types lend precedents for this notion and lead to the concept of "estrogen memory" or, more generally, "gene memory" (38). Work with a human hepatoma cell line (39) demonstrated that a brief estrogen treatment could alter responses to subsequent hormonal treatment: enhanced apolipoprotein synthesis as well as induced type II estrogen binding sites could be propagated in these cells for several generations in the absence of further hormone treatment. This type of effect was not restricted to an individual cell line. A classical estrogen-dependent system, that for the expression of vitellogenin, also demonstrated more rapid responses and higher levels of expression after an initial exposure to estrogen (40–45).

Time Course of Estrogen Action in Adult Brain

Long-term Priming Actions. Several studies have shown that ovariectomized female rats that receive periodic injections of estradiol benzoate are behaviorally more sensitive to a subsequent treatment with estradiol benzoate several days later (36,46,48). Although the interpretation favored was that there existed a long-term priming action of estrogen that would make neurons more sensitive to subsequent estrogen effects, it was also possible that estradiol benzoate in an oil vehicle simply persisted in the subcutaneous depot and circulation for several days. Thus, for studying long-term estrogen actions, we used estradiol-17β in Silastic capsules so that, on capsule removal, blood levels of estradiol would quickly return to control levels. Under these circumstances as well, animals pretreated with estradiol showed significantly better lordosis scores when subsequently challenged with estradiol or estradiol plus progesterone than did animals pretreated with cholesterol (49). We found no evidence that the long-term potentiation of lordosis behavior by estradiol was caused by a change in ER or PR dynamics. In ovariectomized guinea pigs, too, subthreshold doses of estradiol benzoate facilitate the behavioral actions of subsequent estrogen-plus-progesterone treatments (50,51). Wilcox and Feder (52) inferred that the behavioral action was correlated with an increased induction of cytoplasmic PRs.

Czaja and Butera (53) looked at long-term estrogen effects by contrasting short-term ovariectomized with long-term ovariectomized females and found that whereas long-term estrogen-deprived females had lower lordosis quotients in response to subsequent estrogen treatment, the estrogenic suppression of food intake was not affected by long-term estrogen deprivation. In guinea pigs, too, long-term ovariectomy caused a decrease in the proportion of females from which lordosis could be elicited in response to subsequent estrogen–progesterone treatment, but the effects of estradiol on food intake, body weight, and vaginal membrane condition had not diminished (54). Thus, there exists an action of estrogen, probably directly on nerve cells, whose time course can be measured in weeks, and whose effects are particularly relevant for lordosis behavior.

Estrogen Required before Progesterone Facilitation. In terms of the minimal latency required between the beginning of estrogen treatment and reproductive behavior testing, the intravenous administration of 17β-estradiol has been shown to facilitate lordosis response 23 or 24 hours later, and lordosis will not appear earlier than 20 hours (55,56). To study the minimal duration from first estrogen exposure in ovariectomized animals to a positive behavioral result, while at the same time studying nuclear receptor occupation, we (57) gave ovariectomized female rats an intraperitoneal injection of 10 μg of 17β-estradiol (free alcohol) and tested at various intervals for the lordosis reflex. Mean lordosis responses did not rise significantly above pre-estradiol levels 2, 12, or 23 hours after estradiol. However, after a progesterone injection 23.5 hours after estradiol, lordosis behavior rose significantly at the 25- and 28-hour tests (57). When the same amount of 17β-estradiol was given intraperitoneally, radioactively labeled, peak cell nuclear binding in the hypothalamus was seen 1 and 2 hours after injection and had declined almost to background levels by 12 hours. In the hypothalamic samples obtained 12, 18, and 24 hours after isotope injection, total cell nuclear radioactivity was barely twice the background counts. Thus, little if any labeled estradiol was retained in neuronal cell nuclear receptor sites beyond 12 hours, so behavioral effects of estrogen apparently can be manifest in the virtual absence of occupied cell nuclear receptor sites.

Furthermore, continuous estrogen application during the interval from first estrogen exposure to behavioral test is not necessary for a positive behavioral result. With estradiol-filled implants that could be taken in and out of ovariectomized female rats, Sodersten et al. (58) were able to induce sexual receptivity with two discontinuous estrogen exposures. With the obligatory 24-hour estrogen exposure interval finely dissected, we (59,60) demonstrated that surprisingly short and discontinuous periods of estradiol treatment were sufficient to induce progesterone-facilitated female sexual behavior. Using various temporal schedules with estradiol-filled subcutaneous Silastic capsules, we found that two periods of estradiol treatment with a total exposure time of 2 hours were sufficient for both activating the lordosis reflex and elevating cytosol PR levels. The second, discontinuous estradiol treatment hour had to begin not less than 4 or more than 13 hours after the end of the first estradiol treatment period for a positive behavioral result (60). Cell nuclear estradiol receptor measurements also showed that, besides the discontinuous estradiol treatment, an essentially discontinuous pattern of nuclear ER levels (3.6% receptor occupation between two peaks) was associated with positive lordosis behavior tests and elevated PR levels (60). These results are analogous to the requirement for discontinuous estrogen to evoke increases in DNA levels in the uterus (61).

Nevertheless, it is also true that for reproductive behavior to occur, ER occupation has to be long enough to be measured in hours rather than just a few minutes. In the same series of experiments where relatively brief pulsed administrations of 17β-estradiol led to sexual receptivity, estriol-filled implants were behaviorally ineffective (58). Analogously, estriol, usually regarded as a weak estrogen (62), is probably not as potent because its nuclear receptor occupation is not long enough, for example, to promote uterine growth, and if it is injected repeatedly, its estrogenic potency can be significantly increased (63,64). All of these results can be explained by the shorter time of retention in cell nuclei of the estriol–receptor complex than the estradiol–receptor complex.

Implications for Gene Memory. The very existence of a long-term decline of hormone responsivity after ovariectomy and the ability of intervening injections of estradiol to reverse that decline over a similarly long period (36,46,48,49) prove that there are actions of estrogens whose time courses are long and that outlast the elevated estrogen levels in blood during the treatment. The "estrogen memory" presumably is carried by a hormone-dependent hypothalamic mechanism, perhaps an alteration in genomic structure. Similarly, during pulsatile estrogen treatments, the ability of the first pulse to set up a condition whereby the second pulse could participate effectively in behavioral facilitation (58–60) indicates that the first pulse had effects lasting beyond the time of elevated estrogen levels. This is another type of proof of an "estrogen memory" phenomenon, perhaps residing in molecular alterations in hypothalamic neurons.

Trigger versus Maintenance Hypothesis. The conflict has been resolved between the "trigger" and "maintenance" hypotheses on the requirements for estrogen administration before a positive reproductive behavior result. The trigger hypothesis stated that a single brief exposure to estradiol would set off a chain of events that continued ballistically, with no further requirement for estradiol presence. This hypothesis was inspired by the reports (65,66) that a single, brief (approximately 30-minute) exposure to estradiol was sufficient priming for lordosis behavior. The bulk of the data with respect to the trigger hypothesis [reviewed in Clemens and Weaver (20)] shows clearly that it is wrong. For example, in ovariectomized guinea pigs, repeated pulse administration of estradiol clearly was more effective than a single large dose for facilitating lordosis behavior (67). The maintenance hypothesis stated that estrogen must be present continuously through the interval between the beginning of estrogen action and the behavioral tests for lordosis to occur. Evidence against this hypothesis includes discontinuous treatments with estradiol in Silastic capsules (59,60), which allowed lordosis behavior to occur. Further, work with guinea pigs using pulsatile administration of unesterified estradiol (unlikely to remain long in the blood) showed that positive behavioral tests could follow such treatments (67), and ovariectomized rats given low estradiol doses in a pulsatile fashion did display lordosis responses (68). Thus, for lordosis behavior to occur, it is extremely unlikely that estrogen must be present continuously in the brain from the beginning of hormone treatment to the time of the behavioral test.

All of these data can be encompassed by a "cascade" hypothesis, stating that for reproductive behavior to occur, discrete events must occur in medial hypothalamic neuronal cell nuclei as a result of ER occupation, and that earlier estrogen-stimulated events are required for later estrogen-stimulated events to occur. This hypothesis would not require that ERs be occupied continuously up to the time of the behavioral test (and thus not fall prey to the shortcomings of the maintenance hypothesis), but it would require that estrogen be present and estrogen nuclear receptors be occupied at certain critical periods during the interval from the beginning of estrogen treatment to the behavioral test (and thus be much more demanding than the trigger hypothesis). The data on pulsed administrations of estradiol for stimulating female reproductive behavior (58–60) are consistent with this hypothesis and suggest that at least two phases of estrogen action are required. Ultrastructural results from the study of neuronal cell nuclei in the ventromedial hypothalamus, showing a difference after the second phase of two pulsed estradiol treatments, are also consistent with this notion (69). Protein synthesis inhibition just before either pulse of estradiol administration (in a two-pulsed treatment) or between the two pulses disrupted the effects of the hormone on lordosis behavior, whereas protein synthesis inhibition several hours after the second pulse had no effect (70). Moreover, anesthesia during estrogen administration can inhibit the effect of the hormone on lordosis behavior (71). Thus, electrical activity, synaptic input, and de novo protein synthesis all are likely to be essential components of the cascade of events required after the onset of estradiol treatment.

Progesterone. Whereas progesterone multiplies the effectiveness of estrogen treatment for promoting lordosis behavior—under some circumstances estrogen alone is sufficient (10,11)—progesterone is required for the normal occurrence of proceptive behavior (72,73). When progesterone is given intravenously, significant amounts of lordosis behavior can be seen as early as 0.5 hour, with high levels of proceptive behavior seen by 2 hours after progesterone administration (74). McGinnis et al. (75) found a very high percentage of female rats showing lordosis and proceptivity by 1 hour after intravenous progesterone treatment and correlated their results with PR occupation. Therefore, if progesterone's action on reproductive behavior includes a requirement for new protein synthesis, these proteins must turn over rapidly.

A second kind of progesterone action also remains to be explained. In ovariectomized guinea pigs, after estrogen- and progestin-induced sexual receptivity, the animals were refractory to further hormonal stimulation (76). During the luteal phase of the normal guinea pig estrous cycle, the inhibition of lordosis disappeared after ovariectomy, yet during this luteal period estradiol could still prepare the neural substrate of lordosis behavior (77). Refractoriness to hormonal stimulation does not occur when receptive behavior is induced with estrogen alone. All of these facts are consistent with the interpretation that progesterone can exert a biphasic action on female reproductive behavior, first facilitating and then inhibiting receptivity (76).

Conditions at Onset of Hormone Treatment. Several studies indicate that physiological or pharmacological conditions at the onset of estradiol treatment can influence subsequent hormone binding or behavioral effectiveness. Striking was the demonstration by Roy and colleagues (78) that anesthetizing rats during exposure to estradiol could inhibit the induction of behavioral receptivity. For an effect in the opposite direction, Axelson (79,80) reported that access to opportunities for wheel running would actually enhance the behavioral effectiveness of estradiol treatment. Such work may be related to the neuropharmacological tradition of study in which transmitter agonists or antagonists can influence hormone responsivity. Thus, the muscarinic agonist bethanechol can increase estrogen binding by hypothalamic cytosols from female but not male rats (81), and, counterintuitively, atropine can also increase estrogen binding, but by female and not male hypothalamic cytosols (82). Likewise, a large number of experiments have demonstrated the effectiveness of manipulations of catecholaminergic transmitters for altering hormone binding or reproductive behavior (83–90). In one study, Montemayer et al. (91) showed that the α_1-noradrenergic antagonist prazosin, administered at the time of estradiol treatment, reduced not only sexual receptivity but levels of estrogen binding in hypothalamic neuronal nuclei.

An obvious aspect of the animal's physiological condition at the time of hormone administration is genetic sex. References cited later report that estrogen can increase preproenkephalin gene expression in a tissue-specific manner in genetic females but not males. In a similar manner, estradiol treatment can increase PR mRNA in a brain region–specific manner in females but not males. It is not the case that males simply lack estradiol receptors in the hypothalamus or preoptic area (POA). It is also not the case that males are incapable of expressing the gene for the PR, because by in situ hybridization we have demonstrated that intact male rats have significant levels of PR mRNA in hypothalamic neurons (92). Therefore, it is likely that males lack adequate concentrations of other important nuclear proteins that help the DNA-bound ER on its estrogen response element to effect changes in transcription. A complementary hypothesis is that males have masking or squelching proteins that prevent the ER from binding to estrogen response elements in hormone-responsive genes. The other nuclear proteins referred to could result from the signal transduction pathways activated under the physiological or pharmacological conditions referred to in preceding paragraphs. Changes in kinases or phosphatases active on hormone receptor proteins could be important in this regard.

Overall, these results emphasizing the importance of genetic, physiological, and pharmacological conditions at the time of hormone administration underline the fact, elaborated in the section called Outlook, that for activating hormone-responsive genes and behavior, sex steroids do not work alone. In line with the multiple conditions required for successful reproduction enumerated previously, *combinations* of events are necessary for optimal activation of the circuitry for reproductive behavior. Although the various requirements for reproduction can affect behavior circuitry at levels other than the hypothalamus, and although they can use transmitter and peptide encoding mechanisms separate from any individual gene under study, it is possible that the combinatorial requirements of transcription factors—even for genes dominated by hormone influences—might also underlie some of the integration among influences on hypothalamic neurons, signaling that the basal requirements for reproduction are present.

Implications for Study. The primary importance of these hormone–behavioral descriptions is to set up a well-studied system for the systematic investigation of cellular mechanisms of a hormone-driven behavioral system, including electrical and molecular-synthetic mechanisms (7) (and see II.A, Molecular Mechanisms, from Hormone to Gene to Behavior). In molecular terms, this is a kind of "expression system." With this in mind, it is important to determine whether the lordosis reflex occurs. In a more general context, however, it may also be that these effects of estrogens and progesterone explain a mechanism of sexual motivation (93). As part of that motivational change, a major theme may be a changed response to somatosensory input—that the irritability of the ovariectomized animal and the tendency to treat somatosensory inputs as noxious may be reduced by estrogen–progesterone treatment, and this may have implications as well for reproductive behavior (94,95). If so, it would be an example of how a specific, molecularly explicable behavioral reflex can be affected by a general disposition to somatosensory input.

Hormone Receptors in Brain

In the vertebrate brain, as in other hormone-responsive tissues, nuclear hormone receptors are expected to act as transcription factors involved not only in neuronal growth and development but in the facilitation of specific behaviors. Our development and use of the histochemical approach to hormone receptors—steroid hormone autoradiography—have been designed to address the crux of the issue of exactly where hormones act in the brain to influence behaviorally relevant neuronal events.

Lawfulness of Hormone Receptor Anatomy in Vertebrate Brains. Steroid hormone autoradiography has allowed precise locations of neurons with steroid sex hormone receptors to be mapped. In particular, estrogen-concentrating neurons, in many cases androgen-concentrating neurons, and in a few cases progesterone-concentrating neurons have been studied across a wide range of vertebrate species. In our laboratory as well as others, all the major vertebrate classes have been covered.

Among fish, paradise fish (96), goldfish (97), and sunfish (98) have been studied with steroid autoradiography. Among amphibians, *Xenopus* (99–101) and *Rana* (102) have been studied for hormone-concentrating neurons. Among reptiles, the lizard *Anolis* has been studied (103,104), as well as garter snakes (105). Among birds, both passerine and non-passerine forms have been covered. The brains of chickens (106–108), doves (109), chaffinches (110,111), and zebra finches (112) all have been mapped for estrogen- or androgen-concentrating neurons.

The greatest amount of autoradiographic work has been done with mammals. The initial work was done with rodents, particularly with estrogen- and androgen-concentrating cells in the rat brain (3,4, 113,114). Subsequent work with rats amplified these findings (115,116). Similar distributions of hormone-concentrating cells have been seen in other rodents: hamsters (117), guinea pig (118), and gerbils (119). A lagomorph, the rabbit, has also been studied (120). In rodents, progesterone-concentrating cells were covered by Warembourg (121,122). Among carnivores, mink (123) and cats (124) have been studied. Among primates, hormone-concentrating neurons have been mapped autoradiographically in the brains of rhesus monkeys (125–128) as well as squirrel monkeys (129).

In all of these cases, certain findings about the neuroanatomical distribution of sex steroid–concentrating neurons (estrogen-concentrating, androgen-concentrating, and progesterone-concentrating cells) remain true, suggesting an orderly, lawful development of such hormone-concentrating cells in the vertebrate train (Table 1).

First, in representative species from all major vertebrate classes, neurons specifically concentrating steroid sex hormones have been detected autoradiographically. Second, in all species, these hormone-concentrating cells can be found in the medial POA, in cell groups of the basomedial (tuberal) hypothalamus, in specific limbic forebrain structures, and in a specific area of the mesencephalon deep to the tectum [reviewed in Morrell et al. (130), Pfaff and Conrad (131), and Morrell and Pfaff (132)]. Third, papers on the autoradiographic findings in each species include evidence from the endocrine and behavioral physiology of that species that nerve cell groups that bind estrogenic or androgenic hormones participate in the control of hormone-modulated functions. The physiological findings usually emphasize the participation of those cell groups in the regulation of gonadotropin release or mating behavior. Finally, in the vast majority of cases, genetic sex of the animal studied by autoradiography does not appear to be an important factor in determining the neuroanatomical pattern of sex steroid–binding cells in the brain.

TABLE 1. *General conclusions from autoradiographic studies of steroid sex hormone–binding cells in the brains of vertebrates*

1. All species have hormone-concentrating nerve cells in specific locations.
2. Hormone-concentrating cells are in the medial preoptic area, medial (tuberal) hypothalamus, and limbic forebrain structures.
3. Nerve cells groups that bind hormones participate in the control of hormone-modulated functions.

From Pfaff, D. W., and Conrad, L. C. A. (1978). Hypothalamic neuroanatomy: steroid hormone binding and patterns of axonal projections. *Int. Rev. Cytol.* 54, 245–265.

Properties of Individual Sex Steroid Hormone–Binding Neurons. Quantitative analysis of estrogen autoradiograms from rat brain revealed that, even under conditions designed to maximize hormone uptake and analytic sensitivity, unlabeled cells can be found in the same cell groups as hormone-binding neurons, even in the neuronal cell groups with the highest density of hormone receptors (133,134). Thus, the heterogeneity long suspected within nerve cell groups with respect to this hormone-binding property is proven. Moreover, the quantitative distribution of the numbers of grains per cell differed significantly from the Poisson distribution, indicating that the kinetics of binding are not dominated by a random process such as diffusion through lipid. There were no indications of discrete modes of uptake. Finally, the distributions of numbers of grains per cell did not differ across four hormone-binding brain regions, suggesting that functional differences in the consequences of hormone uptake are not caused by fine quantitative features of the uptake itself but rather by differences among the synthetic products and axonal connections of different hormone-binding neurons.

Because of the heterogeneity within nerve cell groups with respect to hormone binding, it is wise for the analysis of any particular neuroendocrine problem to include experimental techniques that can examine the cells involved on an individual cell-by-cell basis. For studying the properties of steroid-binding neurons, therefore, combinations of steroid hormone autoradiography with other histochemical and neuroanatomical techniques in the same tissue are useful for revealing the properties of individual nerve cells. Combination of steroid autoradiography in the same tissue with immunocytochemistry can reveal the peptides produced by hormone-concentrating neurons (see later), whereas combinations of steroid autoradiography with retrograde neuroanatomical measures can reveal the destinations of the axons of individual hormone-binding neurons (see section on Axonal Projections of Hormone-Concentrating Neurons).

Peptides Produced in Hormone-concentrating Neurons. A method for combining steroid hormone autoradiography and peptide immunocytochemistry in the same sections of neural tissue has been described (135). The magnocellular groups in the anterior hypothalamus present an interesting problem for this technique in that they contain estrogen-binding neurons and produce neurophysins, parent proteins for oxytocin and vasopressin. Because estrogen treatment can stimulate the release of oxytocin (136) as well as vasopressin (137), we might predict that estrogen directly addresses the cell nuclei of oxytocin- and vasopressin-producing neurons. In fact, the decline of pituitary oxytocin content during proestrus and estrus in normally cycling female rats suggests that physiological changes in ovarian steroid levels can affect the release of posterior pituitary peptides (138). With combined autoradiography–immunocytochemistry in the normal Long-Evans rat, we found estrogen-concentrating, neurophysin-containing neurons, as predicted, in the hypothalamus (139). These were almost exclusively in the paraventricular nucleus (PVN), especially in the posterior subnucleus. Especially interesting is the comparison of these results with a similar experiment in the Brattleboro rat, a strain that is congenitally unable to produce vasopressin or its associated neurophysin but that produces oxytocin normally (140). Here also, a majority of magnocellular neurons that concentrate estradiol in the cell nucleus were found in the PVN, ventral and medial to this lateral subnucleus, and especially in the posterior subnucleus. Quantitative analysis of these results compared with the genetically normal animal suggested that, in the normal rat, both oxytocin- and vasopressin-producing neurons concentrate radioactive estradiol. Further, the anatomical distribution of estrogen-concentrating neurons in the PVN suggests that estrogen addresses those cells that project to the medulla or spinal cord, indicating that estrogen-influenced oxytocin or vasopressin production could serve in the hypothalamic regulation of autonomic or behavioral functions.

Estrogen-concentrating neurons in the arcuate nucleus of the hypothalamus also are beginning to be characterized. Clearly, some of the neurons containing the enzyme tyrosine hydroxylase concentrate radioactive estrogen in the cell nucleus (141,142). Dramatically, a high proportion of tyrosine hydroxylase–producing arcuate neurons concentrate the radioactive progestin [^{3}H]R-5020 (143). Small numbers of arcuate neurons that produce the pro-opiomelanocortin (POMC) peptide β-endorphin, or dynorphin, concentrate radioactive estrogens (144), and these could be relevant both for opiate-influenced neuroendocrine functions and for behavioral control. Leranth et al. (145) have presented evidence that estrogen-responsive arcuate neurons can also contain glutamic acid decarboxylase. In both the arcuate nucleus and in the cell-poor zone of the basomedial hypothalamus, neurons with immunoreactive prolactin concentrate radioactive estrogen (146).

Combinations of steroid autoradiography with peptide immunocytochemistry have also revealed a surprising disjunction between estrogen binding and peptide content. Contrary to prediction, we could not find a significant number of LHRH-containing immunoreactive neurons in the POA or diagonal bands of Broca, which also gave evidence of hormone concentration in the cell nucleus (147). Indeed, progesterone-binding neurons in the POA also were found *not* to contain immunoreactive LHRH (148). Other laboratories have extended this finding that ERs or PRs are not demonstrable in LHRH-expressing neurons, leading to the general theory that many actions of sex steroids on LHRH dynamics, insofar as they involve nuclear receptors, must be routed indirectly through other cells.

The use of double immunocytochemistry to discern properties of hormone-binding cells has been especially successful in the guinea pig brain. For example, estrogen-induced PRs were found to be present exclusively in cells that contained ERs (149). Importantly, in guinea pigs, enkephalin immunoreactivity was colocalized impressively with PR immunoproduct only in cells of the hypothalamic arcuate nucleus and the ventromedial/ventrolateral area of the hypothalamus (150). In rat brain, combining estrogen autoradiography with opioid peptide immunocytochemistry revealed double-labeled cells in the ventrolateral subdivision of the VMN (151). Of similar behavioral importance is the fact that substance P and PR immunoreactivity colocalized in neurons in the ventrolateral area of the hypothalamus in guinea pigs (152). In turn, in rats these substance P neurons can project to the midbrain central gray and can also contain estrogen-stimulated enkephalin (153). The potential implications for female reproductive behavior are clear. In fact, behaviorally effective estradiol pulses can induce PR immunoreactivity selectively in substance P–containing hypothalamic neurons (154). Finally, in the POA, estrogen-receptive neurons receive enkephalin-immunoreactive synapses in a manner that is sexually dimorphic (155), indicating that behaviorally important hormone-sensitive peptide production by hypothalamic neurons may affect not only the midbrain but the POA and basal forebrain.

It is expected that continued work with tritiated estrogens, androgens, and progestins, using autoradiography in combination with immunocytochemistry and using the increased number of antisera available, will produce a catalog of the peptide-producing properties of hormone-concentrating neurons that is both complete and functionally suggestive.

Axonal Projections of Hormone-concentrating Neurons. Methods for combining retrograde neuroanatomical techniques with steroid hormone autoradiography to chart axonal projections of steroid-binding neurons have been described (156–158).

Of greatest importance for the control of female reproductive behavior is whether estrogen-binding cells in the VMN, the primary site for estrogenic action on lordosis (see later), project to the midbrain central gray (a primary midbrain site for transmitting hypothalamic effects to the lower brain stem). To answer this question, we delivered microinjections of the fluorescent dye primuline to the dorsal midbrain in rats that would later be given a nuclear saturating dose of tritiated estradiol (157). Indeed, certain estrogen-concentrating cells in the ventrolateral subdivision of the VMN sent axons to the dorsal midbrain. Whereas 25% to 35% of estrogen-concentrating ventromedial hypothalamic neurons projected to the region of the midbrain central gray, other ventromedial hypothalamic cells concentrated estrogen only or only projected to the midbrain. Again, heterogeneity among neurons with respect to neuroanatomical as well as neuroendocrine properties showed the usefulness of a cell-by-cell histochemical approach.

Of importance for maternal behavior are projections from estrogen-concentrating neurons in the POA through the medial forebrain bundle toward or into the ventral midbrain. Estrogen affects maternal behavior through POA neurons, lesions of which can abolish maternal behavior (159). POA efferents through the medial forebrain bundle are crucial (see Chapter 36), and lesions in the ventral midbrain can reduce or abolish maternal behavior (160,161). With fluorescent dyes applied locally in the ventral midbrain, followed by treatment of the same rats with tritiated estradiol and tissue processing for combined retrograde neuroanatomical and autoradiographic techniques, we found that the greatest number of estrogen-concentrating neurons that projected to the ventral midbrain were in the caudal medial POA, approximately at the level of the anterior pole of the suprachiasmatic nucleus (162,163). The medial POA also contains estrogen-concentrating neurons that project to the medial amygdala (158,164).

When the neuroanatomical literature is compared with the literature on hormone-binding neurons, there is an obvious, distinct tendency for steroid hormone–binding neurons to project to cell groups that also contain hormone-binding cells—there appear to be "webs" of steroid hormone–binding systems (4,165,166). One implication of this type of hormone-binding neuronal system would be the opportunity for cascades of steroid hormone effects in which a given chemical or electrical effect of a given steroid early in the system can be multiplied by an effect later in the system. A second obvious possibility is the opportunity for feedback within hormone-concentrating systems. The implications of the interconnectedness of steroid hormone–binding cell groups will be explored with physiological and neurochemical studies, but already its theoretical relationship to the possible requirement for certain forms of synaptic input to obtain some estrogen effects is clear (see later). It is expected that continued work with combinations of retrograde neuroanatomical methodology with steroid hormone autoradiography will extend our knowledge of the projections of estrogen-, androgen-, and progestin-concentrating neurons. In turn, this information is crucial for explaining how hormone-binding neurons control neuroendocrine or behavioral functions.

Cell Fractionation. The modern era of work on mechanisms of hormone action began with the demonstration by Jensen and Jacobson (167) that the isotopically labeled estradiol was retained specifically in target tissues. The implications of this and subsequent studies of ERs and PRs for the control of gene expression have been explored most thoroughly in chick oviduct (168). In brain, scintillation counting of dissected regions of the nervous system showed similar estrogen retention in neural regions construed to be related to the control of mating behavior and gonadotropin release (169–171). Estrogen concentration by cell nuclei in the hypothalamus was clearly demonstrated (172). Many aspects of the studies regarding the chemical structure of ERs and PRs have remained controversial, including the nature of the transformation on steroid binding, but there are also reported to be important similarities of receptor structure across tissues and steroid systems (173). Estrogen receptors exist in at least two states. Before steroid binding, the receptor probably is a monomer, with a molecular weight of approximately 70,000 Da, whereas the activated receptor is probably a dimer, at least based on the results from calf uterine cytosol receptors (174). The Scatchard plot of estrogen binding in calf uterine cytosol at equilibrium shows positive cooperativity (174). Estriol is a weaker estrogen than estradiol in a variety of biological systems (175), including behavioral studies (see earlier). Not only does estriol interfere with the biologically effective nuclear occupancy of the estradiol–receptor complex, it inhibits the cooperative aspect of the uterine estradiol receptor binding (175).

Estrogen Receptors. Properties of ERs in brain have been reviewed (176), and hypothalamic receptors compared with ERs in other target tissues (177–179). Uniformly, high-affinity, limited-capacity sites are found, as expected from studies of ERs in other target tissues.

Studies on the ontogeny of ERs in the brain have demonstrated a rise in receptor content before puberty (180) and indicate that ERs in the male hypothalamus are occupied by estrogen derived from testosterone by aromatization (181,182). In mouse brain, a sharp perinatal increase in ER levels has been documented (183,184). It should be noted that, at least from studies in guinea pigs, changes in lordosis behavior during development are not necessarily tightly correlated with changes in hypothalamic ER concentration or affinity (185).

To study ERs in brain and pituitary during aging, Wise and Camp (186) measured nuclear ER concentrations in young adult, middle-aged, or old (16 to 18 months) cycling female rats. Saturated nuclear receptor concentrations were assessed in POA, medial basal hypothalamus, amygdala, and pituitary, and in the old rats there were decreased receptor concentrations in the POA, medial basal hypothalamus, and pituitary. There was no change in receptor affinity with age. Data were consistent with the notion that changes in nuclear ER concentrations with age might be related to the decline in reproductive function. In related experiments, microdissection of cell groups demonstrated a decrease in the number of cytoplasmic estradiol receptors in the preoptic region and in the VMN (187), and this decline was associated with deficits in female mating behavior.

Correlations of Estrogen Receptors with Lordosis Behavior. Across a variety of experimental approaches there are strong correlations between ER function and lordosis behavior levels along three dimensions: spatial correlations (high ER levels in brain locations where estrogen implants facilitate reproductive behavior), temporal correlations, and functional correlations (interference with receptor function leading to interference with reproductive behavior) (176,188). Occupancy of approximately 50% of nuclear ERs is consistent with the occurrence of lordosis behavior (189). It appears that prolonged occupation of the nuclear receptor is correlated with estrogen effects on other biological functions (190). Correlations between ERs and reproductive behavior during aging were summarized in the previous paragraph. With respect to sex differences and to behavioral changes associated with perinatal hormone treatments, there are changes in ER levels in forebrain structures related to mating behavior (191,192), although the magnitude of the changes in receptor levels certainly does not suggest that these are the only reasons for behavioral alterations. Certain comparisons across species in behavior and receptor levels are striking. Feder et al. (193) found that rats and guinea pigs, behaviorally more responsive to estradiol than hamsters, also had much bigger levels of estrogen uptake in the hypothalamus than did hamsters. Further data indicating correlations between ER levels in the brain and mating behavior, deriving from studies with intracerebral hormone implants and with the use of antiestrogens, are reviewed in the section on Hypothalamic Hormone Implants Facilitate Behavior and the section on Effects of Receptor Blockers: Behavior Requires Hormone Binding, respectively.

Relatively new lines of investigations treat the effects of various neural, pharmacological, or environmental changes on ER levels. For example, prazosin, an α_1-noradrenergic antagonist, reduced nuclear ER binding in the medial basal hypothalamus and in the POA of estrogen-primed female guinea pigs, but no effect on cytosol ER levels was detected (194). A dopamine β-hydroxylase (DBH) inhibitor was reported to decrease the concentration of cytosol ERs in the mediobasal hypothalamus of ovariectomized rats, coupled with a surprising increase in nuclear ER levels (195). These results were interpreted as suggesting that inhibition of DBH could modify unoccupied ERs in such a way that they had a higher affinity for cell nuclear components. The subsequent discovery that DBH-immunoreactive varicosities are closely associated with a high percentage of ER neurons in the hypothalamus further supports the involvement of noradrenergic transmitters with estrogen-binding cells (196). Variations through the light cycle in levels of cytoplasmic ERs in the brain, in the absence of circulating estrogens in ovariectomized female rats, indicate changes in the number of estrogen-binding sites but not in the affinity of the receptors for the hormone (197). These diurnal variations in ER levels have not been seen in the brains of castrated male rats and, in the female, do not depend on the pineal (198).

Whereas a large amount of work on ERs in the brain and pituitary depended on the use of isotopically labeled estradiol for both cell fractionation and autoradiographic approaches, in the mid-1980s it became possible to apply immunochemical techniques using antibodies to ERs (199). When immunocytochemical and steroid-binding assays were compared for ERs in human breast tumors, the nuclear immunocytochemical staining was significantly associated with the concentration of cytosol ERs (200). Importantly, monoclonal antibodies to ERs in various estrogen-sensitive tissues indicate that a very high portion of the ERs are contained in the cell nucleus (201), a conclusion consistent with some approaches to cell fractionation (202). In the brains of ovariectomized, estrogen-primed female rats, using an anti-ER antibody, we have detected cell nuclear staining in the medial basal hypothalamus (203), but have not optimized a technical approach.

Progesterone Receptors. Properties of PRs in brain have been reviewed, and limited-capacity receptors with high affinity similar to those found in the pituitary are found in medial hypothalamus and POA (204). PRs not dependent on estrogen are found through much of the rat brain, including midbrain, cortex, and cerebellum (205). A feature common across the species studied so far, including monkeys, is a high level of PRs in the mediobasal hypothalamus and POA (204).

For the study of PRs during development, the intact female rat brain was investigated for cytosol binding and for a nuclear exchange assay during fetal and immediate postnatal periods (206). Receptors in the hypothalamus and POA appeared at day 1, increased during the first postnatal week, and then remained constant through days 10 to 28. Cytosol receptors in the cortex, detected first on the day of birth, rapidly increased through days 1 to 7 and declined after day 10 (206). In guinea pigs, cytosol PRs were studied, and the neonatal brain was found to contain both high- and low-affinity binders similar to those found in the adult (207). In the neonatal guinea pig brain, sex differences in cytoplasmic PRs were not found. Nuclear PRs were also measured, with an exchange assay, and the neonatal brain contained a high-affinity (0.1 nM) binder (208). Here, there was a sex difference in the hypothalamus, with females having higher levels of nuclear receptors than males. To see whether PR concentrations were changing during aging, Wise and Parsons (209) measured cytosol receptor levels in various brain areas and in the pituitary of young and middle-aged ovariectomized rats, 2 and 4 days after implantation of capsules containing estradiol. Cytoplasmic PR concentrations in the middle-aged animals were significantly lower both in the POA and in the medial basal hypothalamus than in the young animals on day 2 of estradiol treatment, but not on day 4. This was perfectly correlated with deficits in the middle-aged rats with respect to all aspects of reproductive behavior tested (209).

After the discovery of specific PRs in brain (210), a question of immediate importance was whether these receptors could be induced by estrogen. MacLusky and McEwen (211) found that there were two anatomically distinct PR systems in the rat brain, one of which is responsive to estrogen and the other of which is not. PRs in the cytosol fractions from uterus, pituitary, hypothalamus, and POA were markedly increased after estrogen treatment, whereas in the midbrain and cerebral cortex, concentrations of receptor in the cytosol were approximately equal in untreated and estrogen-primed ovariectomized female rats (205,211). In fact, the estrogen-sensitive type of PR is a good example of an estrogen-induced protein in brain tissue. The time course after estradiol of the activation of the cytoplasmic PR and its subsequent decay remarkably parallels the increase and decrease of female reproductive behavior (212). A significant increase in progesterone-facilitated sexual receptivity was first seen 18 hours after estradiol capsule insertion, at which point there was a significant increase in cytosol PRs. After 48 hours of estradiol implantation, both inducible PRs in basal medial hypothalamus and POA and progesterone-facilitated behavior were approaching maximal levels. A similar parallelism was seen after estradiol capsule removal in the decay of PR and behavioral levels. There appeared to be a rough "threshold level" of estrogen-inducible PRs (25% to 35% of maximal induction) that would be associated with the appearance of progesterone-facilitated lordosis (212).

Correlation with Behavior. There are strong correlations between the induction of hypothalamic PRs and reproductive behavior in female rodents (188), but there is also evidence that the induction of PRs is not sufficient for the behavior to occur (213). Some of the correlations are spatial, based on the fact that high levels of PRs have been discovered where implants of progestin can lead to female reproductive behavior (implant data reviewed in following section) (214). Further, within these hypothalamic–preoptic structures, the presence of sex differences in PR levels correlates with sex differences in female reproductive behavior (215). Other correlations are temporal. After the onset of a pulse of estradiol administration, the rises in PR levels and reproductive behavior are parallel, as are the declines after termination of estrogen treatment (212). Striking was the rapid onset of lordosis and preceptive behaviors tightly correlated with the rapid increase in nuclear PR occupation in the hypothalamus and POA (216). Further, in their requirements for particular temporal schedules of discontinuous estrogen treatment, hypothalamic progesterone levels and lordosis behavior are also very similar (59).

In ovariectomized guinea pigs, cytoplasmic PRs in the brain are specific: biologically active progestins compete well for the binding site, whereas androgens, glucocorticoids, and estrogens do not. After a single injection of estradiol used to facilitate behavioral receptivity, the latency for increases in cytoplasmic progestin receptor levels in the hypothalamus correlates well with behavioral latencies (217). Moreover, with respect to subsequent refractoriness in female reproductive behavior, a high but not a low dose of progesterone suppressed behavioral responsiveness to a second progesterone injection 40 hours later, and a similar dose dependency was observed when looking at the concentration of available cytoplasmic PRs 40 hours later (218). Other correlations between hypothalamic PRs and female reproductive behavior are based on the assumption that part of the mechanism by which progesterone stimulates behavior involves new protein synthesis because applications of protein synthesis inhibitors can block the effects of progesterone on lordosis [reviewed in Parsons and Pfaff (188)].

Antisense DNA to Test Essential Role of Progesterone Receptor Biosynthesis. The PR is a transcription factor that has been cloned and analyzed extensively. Cloning and sequencing genes expressed in the nervous system, in general, followed by expression and regulation studies, provide molecular measurements whose connections to nervous system function are only correlational. We require active manipulation of gene expression pathways to prove the importance of a given molecular event for a specific behavior. We hypothesized that if PR mRNA increases correlated with facilitation of reproductive behaviors (see section on Molecular Mechanisms, from Hormone to Gene to Behavior) or actually help drive them, then effective PR antisense DNA sequences might reduce those behaviors (14). Antisense oligonucleotides spanning the translation start site in the mRNA for PR were microinjected directly among ventromedial hypothalamic neurons, and their behavioral effects were compared with control oligomers composed of the same nucleotide bases in scrambled order. When applied 12 but not 24 hours after estradiol, the PR antisense treatment significantly reduced lordosis behavior measured either as a reflex or in a mating behavior test (Fig. 2). Notably, proceptive behaviors, strongly progesterone dependent, were greatly reduced in their occurrence. To see if PR protein was also reduced, antisense oligomers were administered near the ventromedial hypothalamus on one side of the brain while the other side received the scrambled control sequence or vehicle. PR-immunoreactive cells on the antisense side were significantly reduced. Thus, for the first time it can be claimed that interrupting gene expression for a transcription factor, in this case PR, in a particular group of neurons can have behavioral effects, in this case the reduction of lordosis behavior and preceptive behavior (14).

Hypothalamic Hormone Implants Facilitate Behavior. Part of the tight set of correlations between ERs or PRs and reproductive behavior control is the body of data showing that localized hormone implants in sites that have high levels of receptors facilitate female reproductive behavior.

Estrogen. Early experiments showed that increased lordosis behavior can result from intrahypothalamic implants of estrogen alone (219). Studies using estradiol benzoate implicated the VMN as an especially important target site (220). Definitive results were collected by Rubin and Barfield (221), who used bilateral implants of 17β-estradiol diluted 1:250 with cholesterol and discovered that lordosis behavior could occur as long as both cannulae were in or very close to the VMN. Diluted implants prepared with tritiated estradiol proved that the behavioral results did not depend on spread away from the ventromedial hypothalamus (222,223). In the female hamster as well, estrogen implants in and around the VMN can induce sexual receptivity (224). In sheep, bilateral hypothalamic implants of crystalline estradiol were able to induce proceptivity in a high percentage of ewes, approximately equal in effectiveness to a peripheral injection of estrogen (225).

In castrated male rats, estradiol benzoate implants in the region of the ventromedial hypothalamus were more effective for stimulating lordosis than implants in the POA (226), whereas for the activation of masculine copulatory behavior, estradiol benzoate implants in the anterior hypothalamic-preoptic zone were more effective (227). Other behavioral systems also can be affected by hypothalamic estrogen implants. Estradiol benzoate in the vicinity of the ventromedial hypothalamus significantly depresses food intake of ovariectomized female rats within 12 hours (228). In contrast, estrogen in the POA is more effective for increasing locomotor activity of female rats (229), a finding that was replicated (230).

Progesterone. Sites of progesterone action on female reproductive behavior proved more difficult to pin down than was the case for estradiol. Whether that was because of the lipid solubility of progesterone, questions of metabolism, or the actual modes of progesterone's action is not yet clear. Progesterone in the mesencephalic reticular formation can facilitate lordosis behavior with short latency (231), and this may be related to high uptake of radioactive progesterone in the midbrain (232). Limbic sites of progesterone administration, such as the hippocampus or amygdala, also can facilitate lordosis behavior in estrogen-primed female rats (233). The early findings on progesterone implantation in the mesencephalic reticular formation have been repeated and extended to the conclusion that both the habenula and the midbrain reticular formation are progesterone-sensitive regions—progestin action in these areas facilitates both perceptive and receptive components of female reproductive behavior (234).

The first progesterone implant studies with female guinea pigs suggested that progesterone might be acting in the medial basal hypothalamus to facilitate lordosis (235). Definitive work by Rubin and Barfield (236) used bilateral implantation of double-barreled cannula assemblies in estrogen-primed, ovariectomized female rats. They found that a high percentage of females with progesterone implanted in the ventromedial hypothalamus showed high levels of lordosis behavior. It was even possible to activate lordosis with sequential implantation of diluted estradiol followed by progesterone in the ventromedial hypothalamus (237), indicating that estrogen and progesterone stimulation of the ventromedial hypothalamus is sufficient for facilitating reproductive behavior in ovariectomized female rats. Because progesterone has a biphasic effect on reproductive behavior in many female rodents, it was interesting that progesterone implants in the ventromedial hypothalamus could inhibit the subsequent hormonal facilitation of lordosis behavior (238), suggesting that the ventromedial hypothalamus mediates both facilitative and inhibitory effects of this hormone. The behavior-facilitating action is not likely to result from progesterone metabolites because implants of two metabolites, 5α-dihydroprogesterone and 20α-hydroxyprogesterone, were considerably less effective than the parent hormone (239). This same set of results suggested a poor correlation between behavioral efficacy and PR-binding properties of the steroids used. In summary, although large bodies of data show a close relationship between the induction of cytosol PRs by estrogen and the activation of female reproductive behavior [reviewed in Parsons and Pfaff (188)], there are also indications that the PR by itself is not sufficient for behavioral activation, nor is it absolutely necessary [reviewed in Barfield et al. (213)].

Effects of Receptor Blockers: Behavior Requires Hormone Binding. In the preceding sections, correlations between ERs or PRs in brain and the facilitation of female reproductive behavior were established. Autoradiography shows that estrogen- and progestin-binding cells can be found in neuronal groups involved in the control of the behavior. Some of the axonal projections of individual estrogen-binding neurons, for example from the ventromedial hypothalamus to the midbrain central gray, clearly can help to control the behavior. The temporal characteristics of hormone receptor induction or occupation, studied with cell fractionation techniques, fit with lordosis behavior control. And, crucially, local implants of estrogen or progestins in specific groups of estrogen- or progestin-binding cells can facilitate female reproductive behavior, even when the steroid can be shown not to have leaked to other parts of the brain or to other organs. If steroid hormone receptors, studied autoradiographically and with cell fractionation, are *required* for behavior, receptor blockers should inhibit the effects of estrogens or progestins on female reproductive behavior.

Antiestrogens. A group of compounds classified as antiestrogens block nuclear binding of estradiol in the uterus (240,241). Antiestrogens (e.g., CI-628) can also block accumulation of radioactive estradiol in cell nuclei in the brain (242–245). Functionally, antiestrogens prevent the long-term uterine responses to estradiol that provide estrogen-stimulated uterine growth (246), presumably by interfering with the nuclear action of an estrogen–receptor complex. Similarly, with respect to brain function, systemic injections of antiestrogens diminish estrogen-stimulated lordosis behavior. Decreased lordosis follows the blocking action of CI-628 (245,247,248) or clomiphene (249). Decreased lordosis also followed treatment with the antiestrogen MER-25 (250). In general, experimental manipulations that interfere with nuclear binding of estradiol can also decrease the occurrence of lordosis behavior (251,252).

What about blocking ERs in the VMN itself? Because implant studies (see the section on Hypothalamic Hormone Implants Facilitate Behavior) have demonstrated that estradiol stimulation of these neurons is sufficient estrogen treatment for the display of lordosis behavior, similar implants of an antiestrogen could test the hypothesis that these receptors are necessary for lordosis in female rats. Indeed, implants of the antiestrogen tamoxifen in the ventromedial hypothalamus can inhibit lordosis behavior (253,254). In our laboratory, application of tamoxifen or a second antiestrogen, LY156758, to the VMN before estradiol treatment significantly reduced lordosis responding, whereas tamoxifen in the ventromedial hypothalamus 12 hours after estradiol was without effect (254,255). For significant effects on lordosis behavior, both bilateral cannula tips had to be in or adjacent to the VMN, and under these conditions higher concentrations of tamoxifen essentially eliminated female reproductive behavior (254,255). It is reasonable to conclude that estrogen binding in the VMN is necessary for the normal estrogenic stimulation of lordosis behavior.

Antiprogestins. When progesterone, R-5020, progesterone metabolites, and certain other compounds were compared, there was a good correlation between their potency for facilitating lordosis and their relative binding affinities for PRs in the hypothalamus (256).

This led the way to correlations between receptor binding and behavior along other dimensions (188). For example, in ovariectomized guinea pigs, the latencies to the increase and subsequent decrease of cytoplasmic PRs correlated well with the time course of sexual receptivity (257). Results of an assay for nuclear PRs also were consistent with a parallelism to lordosis behavior on the basis of estrogen priming dose, time course, and responsivity to a second progesterone injection (258). In rats, nuclear translocation of PRs just precedes the earliest appearance of lordosis behavior (259). Nevertheless, although male guinea pigs respond to estrogen and progesterone with much less frequency than females, and nuclear PR levels were lower in males then females after estrogen priming, Blaustein et al. (260) concluded that sex differences in the PR system alone would not account for behavioral sex differences. Crucial evidence has come from the use of the PR antagonist RU38486 (261–263). A high dose but not a low dose of RU38486 inhibited lordosis in estrogen- and progesterone-treated female guinea pigs, and, in turn, this inhibition could be overcome by a large dose of progesterone, suggesting that the antagonist effect really resulted from the blocking of PRs. Therefore, under these conditions in female guinea pigs, progesterone binding with hypothalamic PRs was required for female sexual behavior. Indeed, conclusions from the early work with RU486 fit perfectly with the antisense DNA findings cited earlier.

Summary. Receptors for estradiol and progesterone, chemically similar to those in other target tissues, are found in nerve cells that have a lawful neuroanatomical distribution in the vertebrate brain. Hormone-receiving nerve cell bodies are found among cell groups critical for female reproductive behavior, and some estradiol-concentrating cells send axons along trajectories known to be important for lordosis, such as to the midbrain central gray. Temporal as well as spatial properties allow tight correlations between estrogen and PRs and female reproductive behavior. Importantly, the behavioral effects of local implants of estradiol and progesterone in the VMN prove that occupation of these receptors is sufficient for lordosis. Because antiestrogens and an antiprogestin can block female reproductive behavior, receptor occupation must be necessary for lordosis. Mechanisms through which estrogens and progestins act in hypothalamic neurons with steroid hormone receptors include electrophysiological changes (see Electrophysiological Actions subsection) and increases in rate of synthesis and, in turn, hypothalamic changes are read out through known circuits in the brain stem and spinal cord (see Neural Circuit and Mechanisms, below).

Neurotransmitters Influenced

Estrogen effects on hypothalamic neurons probably require those neurons to have a normal synaptic input for the full elaboration of the hormone effect on hypothalamic output. In vivo, a major electrophysiological action of estradiol on ventromedial hypothalamic neurons allows resting discharge in formerly silent cells (264). In vitro, where synaptic input has largely been cut off by the preparation of a hypothalamic tissue slice, estrogen does not cause spontaneous activity but specifically increases the responses of formerly silent cells to orthodromic excitations (265). Thus, the way in which estrogen influences the background activity of ventromedial hypothalamic neurons is probably to increase their responses to neurotransmitters and neuropeptides. Hormone effects on neurotransmitters and the participation of those transmitters in controlling reproductive behavior are reviewed in this section, and neuropeptides in the next section.

Rather than cover every possible transmitter candidate, we have focused on those transmitter systems in which the greatest amount of evidence has accumulated for steroid hormone effects and for participation in reproductive behavior: the cholinergic, adrenergic, and serotonergic systems. In particular, we look for situations in which the existence and direction of hormone effects on transmitter synthesis, turnover, and receptors, and the electrophysiological effects of the transmitter, are consistent with the existence and direction of that transmitter's effect on lordosis behavior.

Acetylcholine

Hormone Effects. Treatment of ovariectomized rats with estradiol can significantly increase muscarinic cholinergic receptors in medial basal hypothalamus (266–268). For example, Dohanich et al. (266) used homogenized dissected mediobasal hypothalamic, preoptic, and other samples and, with the high-affinity muscarinic antagonist [^{3}H]QNB, found that estradiol benzoate increased muscarinic binding in the medial basal hypothalamus and decreased it in the medial POA in a dose-dependent fashion. With quantitative receptor autoradiography, Rainbow et al. (269) determined that the increased [^{3}H]QNB binding results from an increase in the number of receptors rather than from an increased affinity. The increase in ventromedial hypothalamic muscarinic receptors in the region including the cells through which estrogen can activate lordosis behavior occurred by 18 to 24 hours after exposure to estradiol, roughly matching the latency for sexual behavior increases. The estrogen effect could not be shown in the ventromedial hypothalamus of male rats, and in females it depended on protein synthesis. In terms of cholinergic cell bodies giving rise to terminals in the medial basal hypothalamus, it may be relevant that estrogen can increase choline acetylase in the nucleus of the horizontal limb of the diagonal band in ovariectomized females but not in castrated males (270). Thus, estrogen effects on muscarinic cholinergic receptors may be one of the ways in which the hormone influences the electrical activity of ventromedial hypothalamic neurons, in a behaviorally relevant fashion.

Cholinergic influences on the electrical activity of neurons in the VMN were studied by recording in vitro from hypothalamic tissue slices (265). Application of acetylcholine (ACh) excited ventromedial neurons, and estrogen treatment of ovariectomized animals increased the number of cells responding to ACh. These electrophysiological findings are consistent with muscarinic receptor induction by estrogen and with an excitatory effect of cholinergic influences operating through these neurons on lordosis behavior.

Reproductive Behavior. Clemens and colleagues explored the role of cholinergic influences on female mating behavior by infusing cholinergic agonists and antagonists into preoptic, medial hypothalamic, and other brain regions. After carbachol infusion into the medial POA, lordosis performance was elevated within 15 minutes (271). Maximum cholinergic facilitation required estrogen because ovariectomized females without hormone treatment before the intracerebral infusion showed very poor carbachol effects (272). Because Dohanich and Clemens (273) found good carbachol effects on lordosis with bilateral infusions in the medial POA or the ventromedial hypothalamus but not in the mesencephalic reticular formation or frontal cortex, they suggested that the lordosis effects of cholinergic stimulation may be related to the ability of that area to concentrate estrogen. Indeed, we found correlations between responses to estrogen and responses to cholinergically active drugs when measuring lordosis reflexes after hormonal or cholinergic manipulation of the VMN (274,275).

The ACh synthesis inhibitor hemicholinium and the receptor blocker atropine sulfate both reduced lordotic behavior (276). Increases in lordosis behavior after carbachol or oxotremorine (primarily a muscarinic agonist) could be prevented by pretreatment with atropine or scopolamine, muscarinic receptor antagonists (277). Further behavioral pharmacological analyses showed that the lordosis-facilitating effect of ACh was not blocked by the selective M_1 antagonist, pirenzepine, infused into the ventromedial hypothalamus (278) or the cerebral ventricle (279). Moreover, by using a battery of selective M_1 and M_2 agonists and antagonists and comparing the potencies of their effects on lordosis to their capacity to bind M_1 and M_2 receptors, Dohanich et al. (279) concluded that the behavioral effect was likely to be mediated by M_2 receptors. However, at the time of this study, only M_1 and M_2 receptors were examined, and the possible involvement of other muscarinic receptors was not considered. Such a possibility is substantiated by the presence of non-M_1, non-M_2 receptors in the hypothalamus (280–282) and further substantiated by the electrophysiological findings discussed in the next section.

Where do the cells stimulated by hypothalamic cholinergic influences send their information for the control of female reproductive behavior? Richmond and Clemens (283,284) gave infusions of carbachol into the midbrain central gray in estrogen-treated female rats and found a dose-related lordosis behavior facilitation. In similar experiments, a dose-related facilitation by oxotremorine could be blocked by pretreatment with scopolamine. Finally, rats with midbrain central gray lesions showed not only lower lordosis behavior but lower cholinergic facilitation. Richmond and Clemens (284) suggested that the midbrain central gray plays a critical role in the cholinergic mediation of sexual receptivity.

All of these results indicate that, in the VMN, estrogen-induced facilitations of cholinergic functions (reviewed earlier) participate in the mechanisms by which estrogen facilitates lordosis behavior, and that some aspects of the cholinergically facilitated outputs are routed through the midbrain central gray.

Electrophysiological Actions. In vitro experiments using hypothalamic slices containing the VMN were conducted to elucidate the neuronal mechanism(s) of the facilitatory effect of ACh. These experiments showed, first, that most (up to 80%) of the ACh-responsive VMN neurons responded with excitation, only a few with inhibition, and even fewer with biphasic responses (265,285,286). This implies that the net response to ACh from the entire VMN would be an excitation. Also, in vivo estrogen treatment increased the responsiveness of VMN neurons to ACh (265). Therefore, in female rats, estrogen priming is likely to potentiate the excitation of the VMN evoked by ACh. These electrophysiological findings, together with the behavioral observation that ACh can act through the VMN to facilitate the estrogen-dependent lordosis (273,277,278), indicate that the lordosis-facilitating effect of ACh is associated with the excitatory neuronal action of ACh.

In the VMN, neuronal actions of ACh appeared to be mediated by muscarinic receptors because both the excitatory and the inhibitory actions were blocked by scopolamine (287). In our ongoing electropharmacological analyses (286), the inhibitory action of ACh was blocked by the M_2-selective antagonist AF-DX 116 but was not affected by pirenzepine or mimicked by the selective M_1 agonist McN-A343. Obviously, the inhibition is mediated through M_2 and not M_1 receptors. This finding, together with the conclusion from behavioral observation (279), implies that the lordosis-facilitating effect is associated with the inhibitory action. This is in conflict with the results from the electrophysiological studies described previously.

This apparent conflict, however, is resolved by additional electrophysiological findings. First, in exciting VMN neurons, oxotremorine-M (O-M) and carbachol, the two muscarinic agonists that were found to be more potent than McN-A343 in facilitating lordosis (279), were also greater than 1,000-fold and greater than fivefold, respectively, more potent than ACh itself, which in turn was more potent than McN-A343. This positive correlation between the potency of muscarinic agonists in facilitating lordosis and in exciting VMN neurons indicates that the facilitatory effect is associated with the excitatory action. It is also interesting to note that, although the order of the potency of the excitatory action is O-M > carbachol > McN-A343, that of the binding capacity for M_1 receptors is O-M > McN-A343 > carbachol (279). The inconsistency in the orders suggests that the excitatory action of muscarinic agonists was not mediated entirely by M_1 receptors. Indeed, our preliminary results show that pirenzepine (1 to 50 μM) blocked excitation only in some neurons and often incompletely. In the units tested, pirenzepine abolished or attenuated the excitation by McN-A343, ACh, carbachol, or O-M in 1/1, 6/8, 1/3, or 1/4 units, respectively. In contrast, 4-DAMP, which has been shown to bind to M_1 as well as M_3 receptors (288,289), abolished the excitation by ACh or O-M in 3/3 units. In two of these three units, the excitation was either not affected or only attenuated by pirenzepine.

These findings, although sketchy, suggest that the excitation of VMN neurons by muscarinic agonists, especially O-M and carbachol, is mediated partially or possibly entirely by M_3 receptors. However, the involvement of other non-M_1, non-M_2 receptors cannot be ruled out because the existence of such receptors (280), as well as genetically defined M_3 (282) and M_5 receptor gene expression (281) in the hypothalamus, has been demonstrated. Thus, electrophysiological analyses showed that the lordosis-facilitating effect of ACh in the VMN does not appear to be mediated by M_2 receptors, which mediate the inhibitory neuronal action, but rather by M_3 or other non-M_1, non-M_2 receptors that also mediate the excitatory action of this transmitter.

Adrenergic Transmitters

Hormone Effects. Adrenergic transmitters traditionally were difficult to study selectively—their turnovers, receptors, effects—primarily because multiple actions of some pharmacological agents frustrated the analysis of individual catecholamines. A clear picture then emerged slowly, as reviewed by Meyerson et al. (290).

The major supply of norepinephrine (NE) to the hypothalamus and POA comes from the NE neurons in the medulla, including those in the A1 and A2 groups (291). In ovariectomized rats, treatment with estrogen has been shown to increase the resting activity and the excitability of the NE neurons in the A1 region (292). Because increases in neuronal activity are generally associated with increases in transmitter release, estrogen is likely to act on A1 noradrenergic neurons to boost the supply of NE to the hypothalamus. In noradrenergic A2 neurons, Jennes et al. (293) found that the proportion of neurons expressing the c-*fos* gene was highest at proestrus and was also increased by exogenous estrogen. Because the evidence they discussed indicated that activation of c-*fos* gene would lead to increases in the synthesis of tyrosine hydroxylase and DBH, their findings suggest that estrogen can stimulate these two enzymes to increase the supply of NE to the hypothalamus. Indeed, the activity of DBH in rat hypothalamus has been found to be increased at estrus compared with the diestrus of the estrous cycle (294).

Treatment of ovariectomized female rats with estradiol benzoate decreases monoamine oxidase (MAO) activity in the basal medial hypothalamus (295). Because more detailed dissections allowed Luine and Rhodes (296) to show that MAO activity is cut by half in the lateral portion of the VMN, we can conclude that the rate of degradation of some monoamines is lower after estrogen treatment in hypothalamic regions relevant to lordosis behavior. Further, the decreased MAO activity holds only for type A MAO (type B, with phenylethylamine as a substrate, is not affected), and the decreased enzyme activity cannot be caused by a single large dose of estrogen given intravenously (297).

NE turnover in the hypothalamus and POA has been correlated with the preovulatory LH surge; estrogen treatment consistent with LH release is accompanied by increased NE turnover in the median eminence, the arcuate nucleus of the hypothalamus, and the medial POA (298–300). Further, the increase of turnover rates of NE in the POA after 3 days of estrogen priming, apparently a prerequisite for the beginning of the LH surge, can be observed in female rats but not in males (301). Changes in catecholamine content in the ventromedial hypothalamus (302) and effects of estrogen and progesterone on NE turnover in the ventromedial hypothalamus (303) are hard to relate directly to lordosis behavior.

Insight regarding the effects of ovarian hormones on adrenergic receptors can be gained from the control of the sympathetic contractile response of uterine smooth muscle. There are two types of contractile responses, stimulatory and inhibitory, which depend, respectively, on α- and β-adrenergic agonists (304). Treatment with estrogen causes the α-adrenergic response in uterine smooth muscle, whereas progesterone after estrogen treatment results in the β-adrenergic response (305). Direct receptor measurement showed that estrogen increases the uterine concentration of α-adrenergic receptors (306). In agreement with this, estrogen was found to increase the sensitivity of the electrophysiological response of urethral strips to the α agonist phenylephrine (307). An estrogen effect on α-adrenoceptors in the brain has also been examined. Estradiol treatment caused an increase in the apparent number of α_1 but not α_2 receptors in the nucleus tractus solitarius (308). More relevant to reproductive behaviors, estrogen treatment has been reported to increase the α_{1B} subtype of α_1 receptors by five- to sixfold in the hypothalamus and POA. The other subtype, α_{1A}, did not seem to be affected (309). This observation, that estrogen acts specifically on α_{1B} receptors, fits very well with findings that these receptors are required for mediating the excitatory neuronal action and the lordosis-facilitating effect of NE (see later).

In the brain, the clearest set of results has related noradrenergic stimulation of female guinea pig lordosis behavior to noradrenergic effects on estrogen-inducible PRs in hypothalamic tissue (310–313). Females injected with a DBH inhibitor or an α-receptor antagonist had fewer hypothalamic PRs and did not show lordosis behavior after estrogen and progesterone treatment. The effective receptor blockers decreased binding of α_1 ligands, but not α_2 ligands (314). Other experiments showing estrogen-induced increases in β-adrenergic receptors (315,316) are harder to relate to other data. The clearest implication of work on adrenergic receptors is that increases in α_1 receptors in the hypothalamus are correlated with the occurrence of lordosis behavior.

Adrenergic effects on neurons of the VMN have been studied electrophysiologically in hypothalamic tissue slices (265). The biphasic response of individual hypothalamic neurons after application of NE is caused by a short-lived α_2 inhibition superimposed on a longer-lasting α_1 excitation (317). Estrogen increases the number of neurons responding to the NE (265). An increased α_1 excitatory drive on these neurons after estrogen is consistent with α_1-adrenergic facilitation of female of reproductive behavior.

Reproductive Behavior. Amphetamine given to female rats treated with the dopamine receptor blocker pimozide can increase lordosis behavior (318), suggesting a behavioral role for NE. Infusing NE to the medial POA or the arcuate-ventromedial region of the hypothalamus also can activate lordosis in female rats given low doses of estrogen (319). Consistent with these findings, the administration of a DBH inhibitor or pretreatment with the α-adrenergic antagonist phenoxybenzamine can reduce the performance of lordosis (320). Because phenoxybenzamine decreases binding of α_1 ligands but not α_2 ligands, a role for α_1-adrenergic receptors in the control of lordosis behavior through hypothalamic neurons can be suggested (321). If α_1 receptors facilitate lordosis, as discussed by Michanek (322), then behavioral and electrophysiological results fit perfectly. Indeed, it was demonstrated subsequently that infusion of NE into the ventromedial hypothalamic area of estrogen-primed female rats could induce lordosis, and this NE effect was blocked by systemic administration of the α_1 antagonist prazosin (323). Conversely, infusion of prazosin but not antagonists specific for α_2 or β receptors into the VMN attenuated the lordosis-facilitating effect of subsequently administered progesterone (324). To complete the story, we infused selective adrenergic agonists instead of antagonists into the VMN of female rats (325). Infusions of two α_1 agonists, phenylephrine and methoxamine, but not the α_2 agonist clonidine or the β agonist isoproterenol, facilitated lordosis in estrogen-primed rats. The specificity of the facilitatory effect of α_1 agonists was also indicated by the fact that they were not effective if infused into brain regions outside the VMN or into the VMN of rats unprimed with estrogen. Furthermore, in a double-infusion experiment, the lordosis-facilitating effect of methoxamine was found to be blocked by chloroethylclonidine, a specific antagonist for the α_{1B} receptor subtype. Thus, from results obtained with agonists as well as antagonists, it is clear that, in the VMN at least, the lordosis-facilitating effect of NE is mediated by α_1 receptors involving the α_{1B} subtype.

The anatomical basis for the implication of neuronal α-adrenergic receptors in lordosis behavior probably resides in the responses of medial hypothalamic neurons to ascending noradrenergic fibers distributed through the ventral noradrenergic bundle (326). Damage to the ascending noradrenergic bundles is associated with loss of lordosis (327). Surprisingly, adrenergic mechanisms in the spinal cord are not likely to be closely related to lordosis behavior because monoamine agonists were not able to reinstate lordosis behavior in estrogen-primed spinal female rats (328), nor are detectable monoamine levels in spinal cord required for normal lordosis behavior in the female rat (328).

The situation with β-adrenergic receptors is less clear. The β blocker LB-46, implanted in the hypothalamus, facilitated lordosis behavior in female rats in one experiment (329), but the β-adrenergic receptor blocker propranolol implanted in the medial POA or in the arcuate-ventromedial hypothalamic region decreased lordosis quotients in estrone-primed females (319).

Electrophysiological Actions. Adrenergic effects on neurons of the VMN have been studied electrophysiologically in hypothalamic tissue slices (265,317,325). Estrogen increases the number of neurons responding to NE, which can evoke excitation, inhibition, or biphasic inhibition–excitation in individual neurons (265). Receptor subtypes responsible for mediating these responses were determined by in vitro neuropharmacological analyses using both selective agonists and antagonists (317). The main finding from these analyses on VMN neurons is that α_1 receptors, which mediate the lordosis-facilitating effect of NE (see earlier), are responsible for mediating the excitatory action of NE. Activation of α_2 receptors, not facilitating lordosis, resulted in the inhibition of neuronal activity; and β receptors, which also are not involved in the facilitation of lordosis, affected neuronal activity only occasionally. Obviously, the lordosis-facilitating effect of NE is associated with its excitatory neuronal action. This association is illustrated by the fact that the α_1 agonists phenylephrine and methoxamine both excited VMN neurons and facilitated lordosis (325). Furthermore, both the neuronal action and the behavioral effect of these agonists were blocked by the α_{1B} antagonist chloroethylclonidine, indicating that α_{1B} receptors are required in both cases (325). In view of these closely parallel findings, it is probable that, in the VMN, NE facilitates lordosis by exciting VMN neurons through α_1 receptors. Further electrophysiological investigations showed that the excitatory neuronal action of NE requires both α_{1B} *and* α_{1A} receptors (330). It would be interesting to see if this is also true for lordosis facilitation.

In summary, the only clear configuration of results with adrenergic transmitters is related to α_1 receptors. Under the influence of estrogen, there is a slower rate of degradation of NE in the ventromedial hypothalamic region. Through α_1 receptors, estrogen-inducible PRs are increased, fostering lordosis behavior, and there is an electrophysiological excitation of ventromedial hypothalamic neurons, similarly excitatory to lordosis behavior (Table 2). Thus, it is consistent that an α_1 receptor blocker can decrease lordosis. Anatomically, the train of events probably is that NE released from terminals of the axons in the ventral noradrenergic bundle excites ventromedial hypothalamic neurons through α_1 receptors, and the output of the VMN is responsible for the reproductive behavior effects.

Serotonin. Various aspects of the relationship between serotonin and lordosis have been discussed by Mendelson in a thorough, scholarly review (331).

Hormone Effects. From the original work of Fuxe (332), it was clear that serotonin-containing fibers were distributed widely in the hypothalamus. Steinbusch (333) and Kawata et al. (334), using immunocytochemistry, described serotonin fibers widespread in the hypothalamus. For the control of lordosis behavior, the dense serotonin innervation of the VMN is an important feature (335,336).

TABLE 2. *Comparisons of effects of various neurotransmitters and neuropeptides on lordosis behavior, neuronal electrical activity, and second messengers*

Agents (type of receptor)	Effects on: Lordosis behavior	Neuronal activity	Membrane enzymes
Lordosis-facilitating agents			
Acetylcholine (M_3?)	↑	↑	↑ on PLC
Norepinephrine (α_1)	↑	↑	↑ on PLC, PLA_2, PLD
Serotonin (5-HT_2)	↑	↑	↑ on PLC,
LHRH	↑	~	↑ on PLC, PLA_2
PRL	↑	~	↑ on PLC
Oxytocin	↑	↑	↑ on PLC
Substance P (NK1)	↑	↑	↑ on PLC
$GABA_A$	↑	↓	
Lordosis-inhibiting agents			
Serotonin (5-HT_{1A})	↓	↓	↓ on AC
Dopamine (D_1? D_2?)	↓	↑or↓	↑(D_1) or ↓ (D_2) on AC
CRF	↓	?	↑ on AC
β-Endorphin	↓	↓	↓ on AC
Neuropeptide Y	↓	↓?	↓ on AC
Cholecystokinin	↓	↑	↑ on AC, PLC
Glutamate	↓	↑	

From Kow, L.-M., Paden, C. M., and Pfaff, D. W. (1981). Dispensability of spinal monoaminergic systems in mediating the lordosis reflex of the female rat. *Pharmacol. Biochem. Behav.* 14, 707–711, with permission; and unpublished data.

↑, stimulatory; ↓, inhibitory; ~, no direct action; AC, adenylyl cyclase; PLA, PLC, PLD, phospholipase A, C, D; GABA, γ-aminobutyric acid; 5-HT, 5-hydroxytryptamine.

Estrogen treatment increases the number of serotonin receptors in the hypothalamus (337,338). There are, however, no sex differences reported in serotonin receptor binding in the ventromedial hypothalamus, even though they have been reported in other regions of the brain (339). In vitro electrophysiological recording from hypothalamic slices shows an increased responsivity of ventromedial hypothalamic neurons to serotonin after estrogen treatment (265). This fits with increased serotonin receptor binding after estrogen (see earlier). It is not yet possible, however, to reason in a syllogistic fashion from the estrogen treatment through the various effects on hypothalamic neurons to deduce the observed behavioral effects.

Reproductive Behavior. Treatment with serotonin precursors inhibits lordosis behavior (340,341). Similarly, raising serotonin levels by blocking reuptake decreases reproductive behavior in female rats (342,343). Manipulating the serotonin system in the opposite direction, inhibiting serotonin synthesis with parachlorophenylalanine (PCPA), increases lordosis behavior (344–348). Similarly, serotonin antagonists increase lordosis in estrogen-primed rats when given systemically (349) or when implanted in the brain (350). Female reproductive behavior facilitation evoked by PCPA is accompanied by a specific depletion of serotonin and is not associated with the release of adrenal progesterone (351).

Destruction of serotonin fibers in the hypothalamus by injection of 5,7-dihydroxytryptamine (5,7-DHT) facilitated lordosis behavior in estrogen-treated female rats (352), and this effect could be reversed by transplanting serotonin-containing fetal raphe cells into the hypothalamus (353). The release of lordosis behavior from serotonergic inhibition by hypothalamic application of 5,7-DHT has been confirmed, and subsequent experiments suggest that one aspect of behavioral sex differences depends on this serotonergic inhibition of ventromedial hypothalamic cells (354,355).

With the advent of a more comprehensive serotonin pharmacology, the inhibitory effect of serotonin on lordosis has been subjected to intense scrutiny to determine the responsible receptor type. There is now an overwhelming consensus that the lordosis-inhibiting effect of serotonin is mediated by post-synaptic 5-hydroxytryptamine 1A (5-HT_{1A}) receptors (356,357). Such receptors obviously exist in the ventromedial hypothalamus because in intact, pro-estrous rats, lordosis was inhibited by the 5-HT_{1A} agonist 8-hydroxy-2-(di-*n*-propylamino)tetralin (8-OH-DPAT) not only when it was given systemically but when it was infused directly into the ventromedial hypothalamus (358). This effect is site specific because infusions into the posterior hypothalamus were ineffective.

In addition to inhibiting, serotonin can also facilitate lordosis (359). This facilitatory effect is obviously mediated mainly by 5-HT_2 receptors (360–363), although presynaptic 5-HT_{1A} and 5-HT_{1B} receptors may also be involved (356).

Serotonergic inhibition of female reproductive behavior is not necessarily a specific action on lordosis. The "serotonin behavioral syndrome" (364) consists of a variety of stereotyped postural responses, including hypersensitivity to touch, and shows a sex difference in that females respond at lower doses of L-tryptophan. The manner in which specific reproductive effects of serotonin might be integrated with other serotonergic effects in the forebrain remains to be explored.

Electrophysiological Actions. Recorded from hypothalamic slices maintained in vitro, responses of VMN neurons to serotonin include inhibition, excitation, and biphasic inhibition–excitation, with the majority of neurons showing inhibition (265). Correlations of these neuronal responses to the behavioral effects of serotonin were investigated with electropharmacological analyses using selective agonists and antagonists in behavioral pharmacological studies (365). These investigations revealed that the inhibition of neuronal activity is mediated by the same receptor subtype, 5-HT_{1A}, that also mediates the lordosis-inhibiting effect, with the excitatory neuronal action by 5-HT_2, the subtype responsible for mediating lordosis facilitation. The inhibitory and the excitatory phases of the biphasic response are mediated independently by 5-HT_{1A} and 5-HT_2 receptors, respectively. Thus, serotonin has parallel dual effects on neuronal activity and lordosis behavior. From these relationships, it is again tempting to propose that neurotransmitters affect lordosis by altering the activity of VMN neurons.

Coordination of Lordosis and Feeding Behaviors. Electrophysiological analyses of the serotonergic (365) and noradrenergic (317) systems in the VMN also revealed that these two transmitter systems have several similar characteristics. First, in each of the two systems there are receptor subtypes, 5-HT_{1A} and 5-HT_2 or α_1 and α_2, respectively, that mediate opposite neuronal actions, and these "opposite" types of receptors colocalize in the majority of individual VMN neurons. Second, in such neurons, the opposite actions mediated by the two opposite types of serotonergic or adrenergic receptors can neutralize each other. Third, the two opposite types of receptors in each case are differentially sensitive to their respective transmitter, with the receptor subtype mediating inhibitory neuronal actions (i.e., 5-HT_{1A} and α_2 receptors) being more sensitive. Therefore, the net response of individual VMN neurons or the entire VMN to these transmitters depends on two factors: the makeup of receptor subtypes and the amount of the transmitter released. Thus, by altering either or both of these factors, one can modulate VMN output and, subsequently, lordosis.

Both serotonin and NE regulate not just sexual behaviors but feeding. Serotonin is known to stimulate feeding through 5-HT_{1A} receptors (357,366,367) and to inhibit feeding through 5-HT_2 receptors (357). Similarly, NE can elicit feeding through α_2 receptors (368) and inhibit feeding by activating β receptors (369). These observations, combined with results from electropharmacological analyses (317,365), indicate that stimulation and inhibition of food intake are associated, respectively, with reduction and increase of neuronal activity in the VMN, just the opposite of the associations between lordosis and VMN activity. This in turn, further implies that when a change occurs to either or both of these transmitter systems in the VMN that promotes one behavior, it will also simultaneously depress the other behavior. Indeed, estrogen, which is capable of inducing lordosis and probably does so through its effects on serotonergic and NE systems in the VMN (see previous discussions), can also inhibit food intake (370–373).

Similarly, but in a converse way, food deprivation, which can be expected to foster the motivation for feeding, decreases NE activity in the hypothalamus (374), alters serotonin synthesis in the brain (375), causes rats to stop cycling (376), and blocks both ovulation and lordosis behavior in hamsters (377). Therefore, it is likely that, in addition to regulating reproductive behaviors, the serotonergic and adrenergic systems in the VMN also serve as mediators in coordinating reproduction and feeding.

The earlier literature suggested that dopaminergic influences could inhibit lordosis behavior (290). A more recent hypothesis is that dopaminergic effects on reproductive behavior can be understood in terms of differential actions on courtship behavior, which involves activity and locomotion, versus lordosis behavior itself, which requires immobility (378). Caggiula et al. (378) found that intraventricular treatment with 6-hydroxydopamine, which depletes dopamine in the forebrain, actually increased the lordosis reflex but severely reduced courtship responses.

Amino Acid Transmitters. The pioneering work of Eugene Roberts in discovering γ-aminobutyric acid (GABA) in the brain and verifying its role as a classic neurotransmitter was pivotal in leading the way to the eventual discovery of all the amino acid transmitters. This class of transmitter is easily segregated into two groups by their actions on neuronal activity: the inhibitory transmitters (GABA and glycine) and the excitatory amino acids (glutamate and aspartate).

The role of the amino acid transmitters in the control of reproductive behavior has been left relatively uninvestigated, particularly in light of their preponderance in the hypothalamus. Quantitative analysis has indicated that half (49%) of all synapsing boutons examined in the hypothalamus were immunoreactive for GABA, suggesting that GABA is the numerically dominant neurotransmitter there and emphasizing the importance of inhibitory circuits in this region (379). Indeed, by electron microscopy a large number of intrinsic synapses are detected in VMN (380) and POA (381). There is also a substantial and separate population of presynaptic boutons immunoreactive for glutamate, and to a lesser extent for aspartate, in nuclei of the hypothalamus, including the VMN and arcuate (382). Evidence based on dense-core vesicles suggests that colocalization with peptides may be the rule rather than the exception for both inhibitory and excitatory amino acid axon terminals in the hypothalamus (382), further underlining the importance of the amino acid transmitters in controlling neural activity in this brain structure.

GABA Receptors. The receptors for GABA are of two general categories, the $GABA_B$ receptors, which are members of the G-protein–linked family and are coupled to slow inhibitory currents (383), and $GABA_A$ receptors, which are members of the ligand-gated ion channel family and mediate rapid inhibitory chloride currents (384–386). The $GABA_A$ receptor is believed to be a heteroligomer consisting of up to five subunits and showing strong homology in both structure and amino acid sequence to the ACh receptor (386). Several distinct clones have been isolated for each of four major subunit types (384), and the pharmacological specificity and channel properties of native receptors depend on subunit composition (387–389). Tremendous variety exists in the pharmacological profile of the $GABA_A$ receptor because of the presence of additional binding sites for benzodiazepines, barbiturates, convulsants, and steroids, which may or may not be present on a particular receptor.

Hormonal modulation of the GABA system has been assessed in terms of both synthesis of the neurotransmitter and affinity and number of receptors. Some steroids have been found to have strikingly fast, presumably nongenomic effects on the affinity of the receptor. Naturally occurring steroid metabolites of progesterone and deoxycorticosterone are potent ligands of the $GABA_A$ receptor complex and can stimulate chloride ion flux at physiological concentrations (390,391). Progesterone itself has also been found to enhance GABAergic responses under some assay conditions (392,393). Some progesterone metabolites that potentiate GABA action are actually synthesized in the brain and hence are referred to as *neurosteroids* (394,395). Other neurosteroids, such as pregnenolone sulfate, can act as antagonists at the $GABA_A$ receptor (396), thus allowing the possibility of a high degree of local control of the GABA system. Although the mechanism of direct steroid modulation of the $GABA_A$ receptor is not entirely clear (i.e., is there a protein to which the steroid binds, or does the steroid modulate the phospholipid environment around the receptor?) (397–400), there is evidence that a particular subunit of the receptor confers sensitivity to steroids (401) and neurosteroids (386).

The ovarian steroid estrogen has also been found to have effects on the $GABA_A$ receptor, although the direction of the effect appears to depend on the precise conditions of the assay, the level of estrogen treatment, and the particular brain region examined (402–405). However, when hormone treatment is given in a regimen that reliably induces female sexual receptivity, the pattern of effects is informative. Two days of exposure to estrogen alone reduces high-affinity muscimol binding in many estrogen-concentration brain areas, including the VMN, arcuate nucleus, amygdala, and midbrain central gray (405,406). Two days of estrogen followed by 4 hours of exposure to progesterone results in a reversal of the effects of estrogen selectively in the VMN and midbrain central gray (406), and this hormonal modulation does not occur in similarly treated males (407). It is notable that estrogen treatment has been reported not to effect the level of muscimol binding in the POA (405), despite the fact that a high percentage of the estrogen-concentrating cells in this region are reportedly GABAergic (408). Thus, steroid hormone treatment has consistent effects on muscimol binding in two brain areas critical to the control of female sexual behavior, the VMN and midbrain central gray, and these effects are not seen in males or in other estrogen-concentrating brain areas that are not critical to the behavior.

Glutamic Acid Decarboxylase. The effects of estrogen treatment on the GABA-synthetic enzyme glutamic acid decarboxylase (GAD) have also been somewhat contradictory. Estrogen and progesterone have both been reported to be potent inhibitors of GAD activity by some authors (409), although others have reported increases (410,411) or no change in GAD activity (405). It has become apparent that there are two separate forms of GAD, coded by two distinct genes (412–414), that have distinct intraneuronal distributions and different dependence on the pyroxidal phosphate cofactor (415). The enzyme referred to as GAD_{67} is distributed throughout the neurons, including the cell body, and is not dependent on the pyroxidal phosphate cofactor for its activity. In contrast, GAD_{65} is found predominantly in the nerve terminal and frequently exists in the inactive or apoenzyme form (415,416). Lack of knowledge concerning these two different enzymes may have generated unexpected variance in experimental design. An analysis of the effects of hormonal treatments that induce LH and prolactin release and female sexual behavior on mRNA levels of GAD indicate that there is a decrease in synthesis of GAD_{67} in the POA, zona incerta, arcuate nucleus, and median eminence, but not in the vicinity of the VMN (417).

An intriguing aspect of the GABA system in the hypothalamus is the degree of species differences. It has been reported that virtually all progesterone-concentrating cells in the hypothalamus of primates (418) and guinea pigs (419) use GABA as a transmitter. The fact that the great majority of PRs induced by estrogen are found in estrogen-concentrating neurons (420) further suggests that a large number of estrogen-concentrating neurons in the hypothalamus are GABAergic. A highly notable exception to this rule, however, appears to be rats, in which GAD is *not* found in the VMN of the hypothalamus (421), although there are estrogen-responsive cells in the arcuate nucleus of the rat that use GABA as a transmitter and receive GABAergic input (422). The potential physiological significance of the lack of GABAergic neurons in the VMN of rats remains unclear. However, there is a high concentration of $GABA_A$ receptors in the VMN, and binding levels are significantly higher in the posterior VMN of males than females (407). Therefore, it is possible in rats that GABAergic neurons surrounding the VMN project into this nucleus to regulate neuronal output and control lordosis behavior.

Behavior. GABA was first implicated in the control of lordosis in septal-lesioned animals. Picrotoxin, a $GABA_A$ antagonist, inhibited the high level of lordosis seen in these females when infused into the substantia nigra (423). The behavioral effect of picrotoxin was correlated with an increase in dopamine turnover in the striatum, suggesting that GABAergic inhibition of dopaminergic transmission facilitates lordosis.

GABAergic transmission in the hypothalamus has alternately been reported to increase (424–426), decrease (427–429), or have no effect on lordosis (427,429,430). However, this discrepancy is more apparent than real and is clarified by an examination of the behavioral effects of GABA in specific brain regions. Infusion of the $GABA_A$ agonist muscimol directly into the hypothalamus or the midbrain central gray of subthreshold hormonally primed females induces a significant facilitation of lordosis (425,426).

More important, infusion of the $GABA_A$ antagonist bicuculline into the hypothalamus or midbrain central gray significantly reduces lordosis in previously receptive females (425,430a), indicating that tonically active GABA transmission regulates the behavior. Local benzodiazepine effects in midbrain central gray on lordosis [unpublished data (430b)] are also consistent with the importance of the GABA–benzodiazepine receptor complex. In dramatic contrast, when $GABA_A$ drugs are infused into the POA, the opposite effects on behavior are seen (425). The effect of a neuroactive substance that modulates lordosis having opposite actions in the POA from the hypothalamus and midbrain central gray is internally consistent with the behavior of many other neuroactive substances (see previous sections of this chapter). This consistency of GABA actions is also exceptional in that it is contradictory to what might be predicted for an inhibitory transmitter because, in general, agents that increase neuronal excitability in the VMN and midbrain central gray facilitate lordosis, whereas the opposite is true for the POA.

Further evidence for the potential role of endogenous GABA in controlling lordosis is based on the finding that during mating, ejaculation by the male rat causes an abrupt and marked increase in the concentration of GABA in the cerebrospinal fluid of the female rat and an equally abrupt and marked decrease in receptivity (428). When GABA levels are increased nonspecifically in the cerebrospinal fluid or specifically in the POA, there is a decrease in receptivity.

This endogenous increase in GABA levels being correlated with behavior is corroborated by changes in tissue GABA levels over the estrous cycle and suggests that GABA may be an important modulator of the natural onset and termination of receptivity. When naturally cycling females are assayed for GABA levels in specific brain areas, there is an increase in GABA in the hypothalamus and a decrease in the POA during proestrus, when females are sexually receptive. The opposite relationship exists when females are in late estrus—that is, GABA levels are high in the POA and low in the hypothalamus (426,431).

Whether GABA exerts its behavioral effects by a widespread inhibitory action or through specific subsets of ventromedial hypothalamic, POA, or midbrain central gray neurons is not yet clear. If the latter, does GABA locally disinhibit particular subsets, or does it dampen competing behavioral responses? Finally, with our pharmacological local applications of GABA-active agents, the possible role of desensitization cannot be ignored. From these notions, one important parameter that must be controlled for in studies of behavioral effects of GABA is the alteration of competing behaviors that would then secondarily alter sexual receptivity. For example, antagonizing GABAergic transmission in either the hypothalamus or midbrain central gray can induce escape behavior (432), defensive behavior (433,434), and changes in nociception (435). Clearly, induction of defensive or escape behaviors and changes in the sensitivity to somatosensory stimuli could indirectly affect sexual receptivity. Experiments designed to control for this possibility by limiting the access of the male to the female and by simultaneously measuring behavioral indices of locomotion, aversion, rejection, and nociception indicate that when GABAergic transmission in the hypothalamus and midbrain central gray is blocked by very low doses of bicuculline, only lordosis responding is altered (430a).

A novel approach using antisense technology further demonstrated the significance of GABA in the control of female reproductive behavior. Synthetic oligonucleotides constructed to be antisense to the vicinity of the translation-start codon for GAD exhibited a significant inhibitory effect on lordosis behavior when administered into the hypothalamus and midbrain central gray, but not the POA (435a,b). When antisense oligonucleotides were constructed specifically to block the translation of mRNA for GAD_{65} and GAD_{67}, it was apparent that both enzymes were providing GABA that was important to the control of lordosis, but with a different time course.

The role of $GABA_B$-mediated synaptic transmission in the control of lordosis has also been investigated. Peripheral administration of the $GABA_B$ agonist baclofen significantly reduced lordosis responding in hormonally primed female rats (436), and this dose-dependent effect is attenuated after hypothalamic serotonin (5-HT) lesions (437). Baclofen administration is also associated with decreased 5-HT activity in the medial POA and VMN and enhanced activity in the midbrain central gray, suggesting that interactions between GABA acting at the $GABA_B$ receptor and 5-HT may be important for the activation of lordosis (437).

Glycine. In contrast to the substantial body of evidence for steroid modulation of GABAergic transmission, relatively little is known of the effect of steroids on glycine responses. Voltage clamping of spinal cord neurons in culture indicated that whereas progesterone enhances GABA-induced chloride currents, it also antagonized those induced by glycine (392). Progesterone and its active metabolites appear to act on the strychnine-sensitive form of the glycine receptor.

Despite the presence of some immunoreactivity for strychnine-sensitive glycine receptors in the magnocellular portions of the PVN and supraoptic nucleus (438) and strychnine-insensitive sites in the amygdala (439), there seems to be very little glycine present in the hypothalamus. Nonetheless, the possible role of inhibitory glycinergic neurons in the ventromedial hypothalamus on lordosis responding was investigated. Infusion of the glycinergic antagonist strychnine into the ventromedial hypothalamus significantly induced lordosis in subthreshold-primed, ovariectomized, adrenalectomized rats (440). However, infusion of glycine into this same brain area failed to inhibit lordosis responding in fully receptive females. The inability of glycine to inhibit the behavior may result from its rapid removal from the synaptic region or insufficient inhibition of neurons after a restricted injection (440). A third explanation is that glycine was ineffective because of interference with activation of the N-methyl-D-aspartate (NMDA) receptor, which has the opposite effect on lordosis in this brain area (457,457a). The latter possibility exists because, in addition to its role as an inhibitory transmitter, glycine may facilitate excitatory transmission in the brain through an allosteric activation of the NMDA receptor at concentrations as low as 10 nM (441).

Excitatory Amino Acids. *Receptors.* The predominant excitatory amino acid transmitter is glutamate, and it is capable of activating at least three distinct classes of receptors: the NMDA ionotropic receptors, the non-NMDA ionotropic receptors, and the glutamate metabotropic receptor. Although NMDA receptors are important in synaptic function, they do not mediate postsynaptic potentials resulting directly from chemical synaptic activation; rather, their action is voltage dependent, and the non-NMDA receptors are responsible for the voltage-independent portion of the synaptic response to chemical transmission in many neuronal pathways (442).

An NMDA receptor subunit has been cloned (443) and exhibits no substantial homology with the non-NMDA receptor. Three separate subunits have been cloned for the non-NMDA receptor (444,445), and all three are expressed to varying degrees in the VMN (445). There are large amounts of glutamate in boutons making synaptic contact with neuroendocrine neurons in the arcuate nucleus, PVN, and supraoptic nucleus (446), suggesting that this transmitter is important in neuroendocrine regulation. Endogenous excitatory amino acid neurotransmission has consistently been reported to elevate serum LH levels in male and female rats (447–449), an effect attributed to a direct action on LHRH cells in the hypothalamus (450). In the POA, [^{3}H]glutamate binding to the non-NMDA binding site is increased by 2 days of estradiol treatment and further increased by estradiol followed by progesterone. Estrogen treatment did not change the density of NMDA binding in this area (451). Interestingly, in the same animals, estrogen treatment increased the density of NMDA agonist binding sites exclusively in the CA1 region of the hippocampus. Whether the influences of ovarian steroids on [^{3}H]glutamate binding are mediated genomically or through membrane effects has not been determined.

Smith and colleagues have reported that estrogen and progesterone have antagonistic affects on the excitatory amino acids. For example, progesterone administered either locally or systemically suppressed glutamate excitation of cerebellar Purkinje neurons within 3 to 10 minutes, whereas estrogen treatment significantly increased the Purkinje cell excitatory responses to glutamate (452,453). The effects of progesterone do not depend on prior treatment with estrogen, and the mechanisms of action for both steroids appear to be nongenomic (452–455). Both the generalizability of these phenomena to other brain areas and the significance of such effects to excitatory amino acid modulation of reproductive processes await further investigation.

In contrast to its inhibitory effect on GABA and glycine, the neurosteroid pregnenolone sulfate is a positive allosteric modulator at the NMDA receptor (456). Because the antagonist and positive effects of this neuromodulator occur at approximately the same concentration, this neurosteroid may play an important role in increasing excitatory tone.

Behavior. The excitatory amino acid transmitter glutamate markedly and transiently inhibits lordosis when infused into the hypothalamus of receptive female rats (457), suggesting that excitation of certain medial hypothalamic neurons inhibits the lordosis reflex. Furthermore, infusion of NMDA into the hypothalamus also inhibits lordosis, suggesting a tonically active non-NMDA system. However, not only is lordosis reduced, there is a significant increase in the number of females that vocalize in response to mountings by the male, suggesting an increased sensitivity to somatosensory input (457a). It has been suggested that for a female rat to be highly receptive she must also be partially analgesic so that high-intensity stimulation from the male is not perceived as noxious. Therefore, the decrease in receptivity after NMDA injection into the hypothalamus may result from a partial loss of this analgesia, so that the female now attempts to avoid being mounted by the male. In this view, the effect of NMDA may not directly involve the neural circuitry controlling lordosis.

When NMDA is infused into the POA of fully receptive rats, there is no effect on lordosis responding, but administration of the NMDA antagonist 2-amino-5-phosphonopentoic acid (AP-5) reduces lordosis responding (457a). Conversely, administration of AP-5 into the hypothalamus has no effect on fully receptive females.

When tissue glutamate levels were measured in intact cycling rats, there were significantly higher levels of glutamate in the posterior medial basal hypothalamus during diestrus and proestrus and 24 hours after estrogen treatment compared with ovariectomized females. Notably, there was no change in glutamate concentration in the POA (458).

There is evidence for an excitatory amino acid projection from the ventromedial hypothalamus to the medial zona incerta (459), and the ventromedial hypothalamus and zona incerta both project to the midbrain central gray, where there is a high concentration of binding sites for all the excitatory amino acid receptor types (460). Taken together, the evidence suggests that this excitatory amino acid neurocircuitry may be important in mediating the effects of zona incerta projections to the midbrain central gray on reproductive behavior (461,462).

Outlook. It has been suggested that amino acid transmitters participate in large numbers of synapses involved in neuroendocrine regulation, outnumbering individual amines, peptides, or other neuroactive substances. The results reviewed here concerning the effect of amino acid transmitters on female sexual behavior indicate that in brain areas where GABA decreases receptivity (POA), excitatory amino acids facilitate lordosis. In sites at which GABA facilitates receptivity, such as the medial hypothalamus, excitatory amino acids are inhibitory. This raises the question of whether the actions of GABA and excitatory amino acids are functionally related. For example, does GABA in the hypothalamus increase receptivity by inhibiting an excitatory amino acid projection to this area that is itself inhibitory to lordosis? Alternatively, GABA and excitatory amino acids may be acting through independent neuronal pathways that do not interact directly. More behavioral pharmacological work will answer this question.

Neuropeptides Influenced

Luteinizing Hormone–releasing Hormone

Mechanisms of Hormonal Control. Locations of LHRH-immunoreactive cells have been reviewed [(463); and see Chapter 28 of the second edition of this text)]. With a variety of antisera and a number of species used, there are still some features of the anatomy of LHRH-containing cells that are constant. LHRH cell bodies can always be found distributed sparsely in a continuum from the ventral septum through the diagonal bands and, importantly, the medial POA to the anterior hypothalamus (especially just dorsal to the supraoptic nucleus). In both males and females, quantitative estimations show that the number of LHRH cell bodies is greatest in the POA (464). In the rat, although occasional reports have placed LHRH cell bodies in the medial basal hypothalamus, the validity of this finding has been questioned (465). Differential immunoreactivity in LHRH neurons to antisera with different binding requirements has been used to infer some features of LHRH processing; the decapeptide is synthesized as part of a larger protein extended at both the N and carboxyl terminals, and processing occurs as the material is transported (466). Comparisons across species have been reviewed (463). In subhuman primates such as the baboon (467), LHRH neurons were found in the medial basal hypothalamus as well as in the medial POA and in the region dorsal to the supraoptic nucleus. In humans (468,469) immunoreaction product can be seen associated with the rough endoplasmic reticulum and the Golgi apparatus in cells from fetal tissue fixed with acrolein; overall, immunopositive neurons were abundant in the infundibular hypothalamus as well as in the POA.

An exciting finding with respect to POA LHRH neurons is their propensity to form anatomical connections with each other [(470,471); and see Chapter 28 of the second edition of this text)] because such connections could be involved in the formation of LHRH pulses (472). Indeed, in cultures of LHRH neurons derived from the olfactory placode we observed that the cells would survive much better if they were touching each other (473). We sought to determine if LHRH receptors on preoptic neurons might be of physiological importance for normal estrous cyclicity in the female rat. Local preoptic administration of a long-acting LHRH antagonist, antide, was able to interrupt the estrous cycle for periods ranging from 11 days to more than 100 days (474), but control infusions into other sites were not effective. Therefore, functional LHRH receptors on preoptic neurons, as might be indicated by anatomical connections observed between LHRH neurons, are important for mediating the LHRH release required to drive the estrous cycle.

The classic distribution of LHRH-containing axons is to the median eminence. Analyzed quantitatively (475), their appearance in the lateral palisade zone can be well documented, as can their partial overlap with other peptides and their relationship to dopamine terminals. Perhaps most important for reproductive behavior are those LHRH axonal terminals found in the midbrain central gray (464).

In conjunction with studies on the development of the LHRH neuronal system, the peptide was discovered in the nervus terminalis (476). Fibers of the nervus terminalis connect the nasal septum and the septal mucosa with the basal forebrain, including the POA, through a ganglionated nerve that is distinct from the main olfactory and vomeronasal nerves. Immunoreactive LHRH is present in the cell bodies and fibers of nervus terminalis in both fetal and adult guinea pig. Because, in the fetal rat, there are periods when the immunoreactive LHRH of the nervus terminalis is the only LHRH detected in the CNS (477), there is a possibility that it has some unique function during development. Transection studies and work with fetal applications of LHRH antisera ruled out the hypothesis that nervus terminalis LHRH is required for olfactory preferences in the neonate (478).

What is the cellular nature of LHRH neurons potentially sensitive to steroid hormones and important for reproductive behavior? Clearly, they are dissociated from estrogen-concentrating neurons as shown by combinations of steroid hormone autoradiography with LHRH immunocytochemistry (see later). Our ideas about the cellular history of LHRH neurons were revolutionized by the finding that they are born on the medial side of the olfactory placode and migrate up the nasal septum into the basal forebrain (479). Following up this discovery with electron microscopic investigations, we saw an unusual distribution of immunoreactive material in LHRH neurons on the olfactory placode, a marked change of shape during migration, and an apparent average migration rate of about 40 μm/h (480). Positive in situ hybridization results for LHRH mRNA among cells in the olfactory placode of the fetal mouse proved that the LHRH gene is actually being expressed in these cells (481). So far, no other neuroendocrine peptide we have tested exhibits such a developmental history and migration route (481). Another confirmation of the olfactory placode origin of LHRH neurons came from the demonstration that LHRH-immunoreactive cells can be cultured from the olfactory placode (473). The LHRH neuronal migration route had implications for the pathogenesis of Kallmann's syndrome: hypogonadotropic hypogonadism coupled with anosmia. By studying LHRH neuronal immunoreactivity in the nose and brain of a fetus with Kallmann's syndrome compared with normal fetuses of the same age and sex, we were able to demonstrate that Kallmann's syndrome can result from a failure of LHRH neuronal migration: at a developmental stage where normal fetuses had LHRH cells in the POA and hypothalamus, all LHRH cells in the Kallmann's fetus were "stuck" out in the nose (482). This finding also confirmed that the migration normally occurs during human brain development.

With respect to mechanisms of LHRH neuronal migration into the brain from the nose, a nerve cell adhesion molecule (NCAM) evidently plays an important role (483). It forms a scaffolding along and within which LHRH neurons migrate toward the brain. The important role for NCAM in the migration is demonstrated by the fact that local microinjections of antiserum to NCAM penetrating the olfactory placode at 10 days of fetal age in the mouse can effectively reduce the number of gonadotropin-releasing hormone (GnRH) neurons on the migration pathway (484). Because the "pioneer cells" that move out early from the olfactory placode and form a cellular aggregate linking the nasal cavity to the developing forebrain are themselves glial cells (485), we speculate that initially these cells lay down the NCAM pathway, followed in turn by the axons of nervus terminalis and the vomeronasal nerve, and then by LHRH neurons themselves, which will eventually migrate to the basal forebrain.

In the development of the entire LHRH system in the guinea pig brain (486), at one stage (50 days postfertilization) immunoreactive fibers could be seen as far posterior as the pontine central gray; because Yamanouchi and Arai (487) have reported lesion effects there on reproductive behavior, these LHRH fibers conceivably could be involved in that function.

Ultrastructurally, in the median eminence, LHRH immunoreactivity is found in membrane-bound granules in terminals, consistent with a classic neurosecretory function (488,489). In the POA, electron microscopic–immunocytochemical studies show LHRH neurons often to be bipolar [reviewed in Krisch (490) and Kozlowski and Hostetter (491)]. Preoptic LHRH neurons, often seen near blood vessels, have neurosecretory granules distributed throughout the cell, with many lying directly under the plasmalemma (492). In these neurons the greatest amount of rough endoplasmic reticulum was often distinctly separated from the greatest concentrations of Golgi, and no axosomatic synapses could be detected. Electron microscopic examination of LHRH terminals in the midbrain central gray (493) may reveal features of LHRH secretion peculiarly related to the control of reproductive behavior.

Electrical stimulation in the neighborhood of these preoptic LHRH neurons, in either male or female rats, causes LHRH secretion leading to LH release, dependent on the electrical stimulation parameters. Among square-wave stimuli, 60 pulses per second was an optimal frequency, but direct-current stimulation was even more effective (494). Optimal square-wave electrical stimuli produced significant increases in LH release after 5 minutes. After electrical stimulation in the medial POA, the increase in the concentration of LHRH in portal blood was linearly related to the strength of the stimulating current, and the amount of LHRH secreted decreased when the stimulation was turned off (495). The importance of the LHRH cells stimulated can be deduced from the fact that transplantation of normal preoptic tissue to hypogonadal mice with a genetic deficiency in LHRH can allow mating, pregnancy, and delivery in animals that otherwise would have been infertile (496). In rhesus monkeys, the role of LHRH in causing LH release appears to be permissive (497)—the precise temporal pattern of gonadotropin secretion is not directed by a corresponding pattern of hypothalamic LHRH secretion. The optimal LHRH release pattern for fulfilling this role is not a continuous LHRH release but instead is a pulsatile pattern of secretion (498). Corresponding tests of pulsatile applications of LHRH to brain tissue for stimulating reproductive behavior have not yet been carried out.

In studies of the role of LHRH in humans (499), it is clear that gonadotropins are discharged episodically (500–504), and pulsatile administration of LHRH can reverse hypogonadotropic states in humans (505,506) as in monkeys (507,508). Crowley and McArthur (509) found that giving LHRH at 2-hour intervals over 27 days to an amenorrheic patient with Kallmann's syndrome resulted in ovulation. Desensitization of the pituitary with long-acting LHRH agonists may be a useful approach to the treatment of precocious puberty in the human (510). For the opposite problem, failure of normal puberty in hypogonadotropic hypogonadism, pulsatile administration of LHRH subcutaneously led to endocrine and behavioral signs of gonadotropin release (511).

Control. After castration, hypothalamic LHRH content declines; this can be detected by radioimmunoassay (512–515) or by immunocytochemistry (512,516,517). Because, on castration, LH secretion increases—released from the negative feedback of gonadal steroids—the decline in hypothalamic LHRH content may reflect an increased rate of LHRH secretion with which changes in LHRH synthesis cannot keep pace. Indeed, in the male rat, quantitative immunocytochemistry shows that castration is followed not only by a decrease in the number and mean optical density of detectable LHRH cell bodies but by a decreased density of immunocytochemically stained fibers in the median eminence, which may be correlated to LHRH secretion leading to LH release (464). Increased LHRH release apparently exceeded any change in LHRH synthesis. In the female rat the situation may be more complicated. In the ovariectomized rat, in contrast to the female given estrogen treatment, the decline in immunocytochemically stained fibers in the median eminence was slight, and in the midbrain central gray the difference in the immunocytochemical result was in the opposite direction—the less intensely stained fibers were actually found in the estrogen-treated animals—consistent with a role for midbrain LHRH secretion in lordosis behavior (464).

Thus, in the case of the female, the increased number of LHRH cell bodies in the POA after estrogen treatment may reflect an effect of estradiol on LHRH synthesis. In fact, estrogen effects on LHRH release can be blocked by inhibition of RNA or protein synthesis (518,519). If possible estrogen effects on LHRH synthesis were mediated by ERs in preoptic neurons, through the classic mechanism of cell nuclear concentration, we would expect to find autoradiographic evidence of cell nuclear concentration of tritiated estradiol in LHRH-producing cells. We tested this hypothesis (520) by combining steroid hormone autoradiography with immunocytochemistry in the same tissue section, looking for estrogen-concentrating cells that had immunoreactive LHRH in the cytoplasm. Surprisingly, virtually none of the estrogen-concentrating neurons had immunoreactive LHRH in the cytoplasm. Thus, in the POA, any genomic effects of estradiol on LHRH synthesis would not be exerted on the LHRH neurons themselves but must be mediated by another class of neurons.

Transcription. Another way of evaluating possible effects of estrogen on LHRH synthesis is to measure levels of messenger RNA for LHRH in preoptic neurons. Using in situ DNA/RNA hybridization (521,522), we localized neurons in the rat brain containing messenger RNA that hybridizes with labeled synthetic oligomers complementary to the human LHRH sequence. Most of the cells with detectable LHRH messenger RNA were in the diagonal bands of Broca and the medial POA. They were in cell groups previously shown by immunocytochemistry to contain LHRH cell bodies, thus arguing against the theory that many hitherto undetected LHRH neurons were depleted of immunocytochemically recognizable LHRH because of quick release of the decapeptide.

For in situ hybridization studies to test possible stimulatory actions of steroid hormones on LHRH gene expression, we chose a long, 7-day treatment with estrogen of the sort that previous workers had shown would lead to repeated LH surges and thus exercise the LHRH and LH transcriptional systems maximally. Under these circumstances, a long and strong estrogen treatment of ovariectomized female rats led to significant increases in LHRH mRNA (523). Roberts and coworkers, using solution hybridization assays and a different protocol for estrogen treatment, similarly showed a stimulatory action of estrogen on LHRH mRNA (524). Rosie et al. (525) ovariectomized female rats on diestrus and gave estrogen in a previously established "positive-feedback" mode. Sacrificing the animals on the day of presumptive proestrus, they showed a stimulatory action of estrogen on LHRH mRNA levels in the POA. Similarly, Park et al. (526) and Zoeller and Young (527), following animals through the normal estrous cycle, reported highest levels of LHRH mRNA about the time of the LH surge. All of these findings can be subsumed by the idea that maximum LHRH gene expression will be found on occasions of accelerated LHRH release and that indirect sex steroid effects on LHRH mRNA might be mediated in this way. Overall, so far, controls in the genetic female over LHRH gene expression fit well with what might be expected for a transcriptional system centrally important for management of the LH ovulatory surge and female reproductive behavior.

It is also conceivable that neurotransmitters and neuropeptides, besides working through conventional electrophysiological mechanisms, might exert influences on LHRH neurons through controls over levels of LHRH mRNA. We have tested such hypotheses using in situ hybridization. An obvious example for initial experiments is the neurotransmitter GABA. This transmitter, working primarily through $GABA_B$ receptors, can inhibit gonadotropin release (528–530). The rapid disruption of an ongoing LH surge after intraventricular injection of the selective $GABA_B$ receptor agonist baclofen suggests a direct involvement of $GABA_B$ receptors in an inhibitory component in the neuroendocrine control over the LH surge (531). With in situ hybridization we found that local preoptic administration of $GABA_B$ but not $GABA_A$ receptor agonists could quantitatively reduce LHRH expression (532), indicating that controls over LHRH gene expression may supplement ordinary electrophysiological mechanisms. Likewise, we predicted that α_1-adrenergic inputs might support normal LHRH gene expression, reasoning from the role of α_1-adrenergic inputs in the facilitation of the ovulatory LH surge. Weesner et al. found that local preoptic administration of an α_1 blocker could reduce LHRH mRNA (533). In fact, during an experiment with estrogen-stimulated increases in LHRH mRNA in which we replicated and extended the work of Rosie et al. (525), Weesner et al. found that an α_1-adrenergic blocker could interfere both with the production of the LH surge and with the estrogen-stimulated increase in LHRH mRNA (534).

Reproductive Behavior. LHRH can facilitate lordosis behavior in female rats (535,536). Because this effect can be observed in hypophysectomized animals, the decapeptide is not working through LH release but must be acting on other nerve cells (536). Some estrogen priming is required for the LHRH effect on behavior. In the sequence of hormone treatments before the behavioral test, therefore, one could compare estradiol plus LHRH with estradiol plus progesterone. There is no evidence, however, that LHRH shares mechanisms of behavioral action with progesterone or that it is part of the progesterone effect; they may be independent actions because under some circumstances their behavioral effects are roughly additive (537). During aging, reduction of LHRH content in the septum and the midbrain is correlated in male rats with reproductive behavior loss (538). In humans, one study showed that men receiving daily LHRH had greater sexual arousal (539), but the size of the effect may depend greatly on the exact conditions of LHRH administration because a thorough follow-up study showed only small, nonsignificant trends toward increases in various sexual behavior parameters (540).

Studies from various laboratories show a consensus that one important site of action for LHRH on female reproductive behavior is the midbrain central gray. Not only can LHRH administered there potentiate lordosis behavior, but an LHRH antiserum given there can block lordosis behavior, indicating that normal LHRH action may be an important part of behavioral mechanisms in the central gray (541). Some results indicate that the ability of LHRH in the midbrain central gray to facilitate female reproductive behavior can be antagonized by opiate peptidergic mechanisms there (542). Behavioral effects of LHRH were reported to be similar to those obtained from hypothalamus (543), but that is not because LHRH infused into the midbrain leaks to the hypothalamus; animals with serotonin infusions in hypothalamus such that LHRH in the arcuate-ventromedial area would have no behavioral effect still showed an enhancement of lordosis with LHRH infusions in the midbrain central gray (544).

It was surprising that administration of an LHRH antiserum to the midbrain central gray could actually block lordosis behavior (541). This indication that LHRH action might be required for normal behavioral performance enlarged on the results of Kozlowski and Hostetter (491) in which anti-LHRH injected into the lateral ventricles caused a reduction of sexual behavior. If, indeed, LHRH in the midbrain is required for lordosis behavior, the behavioral importance of LHRH-containing terminals, whose LHRH content can be influenced by estrogen treatment, is confirmed (464).

Modifications of the LHRH decapeptide indicate that the pituitary effects and the behavioral potentiation by LHRH can be manipulated independently (545). For example, an LHRH analog that inhibits ovulation can facilitate female reproductive behavior (546). With local infusion of LHRH or its analogs into the midbrain central gray, also, there was no correlation between the sign or magnitude of behavioral effects and the previously reported effects on LH release (547). Dudley et al. (548) used fragments of LHRH administered intraventricularly and found that the $LHRH_{5\text{-}10}$ fragment, although inactive for releasing LHRH, would still potentiate lordosis behavior. Whether brain tissue and pituitary differ in their receptor requirements or in the means necessary to protect the decapeptide from breakdown, or both, remains to be discovered.

Presumably, different peptides acting on female reproductive behavior fill different physiologically integrative requirements. Obviously, LHRH could serve to synchronize female reproductive behavior with ovulation so that mating behavior would result in fertilization. In cellular terms, how LHRH affects a target nerve cell is still obscure. Intracellular recording in vitro indicates that LHRH has an effect similar to the late, slow muscarinic excitatory postsynaptic potential, which reduces a potassium current, and the net effect of LHRH action is to encourage bursts of action potentials (549). In the hypothalamus, iontophoresis of LHRH does result in changed electrical activity (550). In vitro electrophysiological experiments are especially convenient for peptides because they allow better control over concentration and better protection from degradation—single-unit activity in the hypothalamus and POA reveals both direct responses to LHRH and neuromodulatory actions of the decapeptide (551). The neuromodulatory mode dominates. Indeed, in midbrain central gray, where LHRH is most effective in facilitating lordosis, LHRH was found to be without effect on neuronal activity when the peptide was administered alone, even at a dose 10 to 100 times that of other effective peptides (552). But, when administered together with neurotransmitters, LHRH can modulate responses of neurons in the POA to NE or serotonin (551,553) or potentiate responses of midbrain central gray neurons to GABA (552), which itself can facilitate lordosis when infused into the midbrain central gray (435b). Results with analogs so far support an amphiphilic theory of protein hormone action in which the β-pleated sheet of the decapeptide allows lipophilic side chains in one direction and hydrophilic side chains in the other direction to foster action on the cell membrane (553). A larger series of structure–function studies is required.

Prolactin

Synthesis. For determining locations of possible prolactin-synthesizing nerve cells, we were stimulated by reports (554,555) of prolactin immunoreactivity in brain. Using immunocytochemistry with the antiserum to prolactin from the National Hormone and Pituitary Program (and two other antisera) as an end point, we found immunoreactive prolactin neurons in the mediobasal hypothalamus (556). Immunopositive cell bodies were on the lateral side of the arcuate nucleus, extending near the bottom of the brain, ventral and lateral to the ventromedial nuclei. Fibers from these cells were distributed widely throughout the brain, and the midbrain central gray had a particularly dense distribution of immunopositive fibers. This may be important for the behavioral effect summarized later. Electron microscopic immunocytochemistry of the mediobasal hypothalamic neurons in question revealed immunoreactive prolactin in large granules in the cytoplasm of the soma (557).

Is the immunoreactive prolactin actually synthesized in hypothalamic cell bodies? Because the immunocytochemical results can be obtained in hypophysectomized rats, it is not the case that the immunoreactive protein simply comes up from the pituitary through the portal vessels (556). Moreover, hypothalamic explants can synthesize a protein that, after immunoprecipitation, travels on a sodium dodecyl sulfate–polyacrylamide gel similar to pituitary prolactin (558). Following up this evidence with extensive immunocytochemical and Western blot studies, we found that even though the great preponderance of the results indicated that an immunoreactive product very similar to rat pituitary prolactin was being expressed by hypothalamic neurons, the National Institutes of Health prolactin antiserum is not monospecific, and not every control condition could be explained by the assumption that only prolactin was being detected (559). Evidence related to transcription of the prolactin gene is reviewed in the following section.

If estrogen participates in the control of prolactin synthesis in hypothalamic neurons, one would expect that estrogen could be localized in the cell nuclei of at least a portion of the prolactin-producing cells. Combining steroid hormone autoradiography with prolactin immunocytochemistry in the same hypothalamic tissue sections, Shivers et al. (560) found that a substantial portion of prolactin-immunopositive cells in the medial basal hypothalamus concentrate radioactive estradiol in the cell nucleus. Thus, for at least a portion of the neurons, it is possible that estrogen effects on prolactin transcription can be studied in a manner analogous to estrogen effects on pituitary cell lines (Haisenleder, Chapter 31, second edition).

Transcription. Using in situ complementary DNA/RNA hybridization, we obtained preliminary evidence of cells containing the prolactin mRNA in the mediobasal hypothalamus (561) in anatomical locations similar to the distribution of neurons with immunoreactive prolactin. Further, Northern blots revealed hybridizable hypothalamic mRNA indistinguishable, according to sizing on agarose, from mature pituitary prolactin mRNA (562), indicating that the prolactin gene may be expressed in hypothalamic neurons. Using reverse transcriptase followed by the polymerase chain reaction (RT-PCR), Emanuele and colleagues (563) obtained findings that confirmed and extended their earlier results from radioimmunoassays: prolactin mRNA can be detected in brain tissue. Likewise, during experiments in our laboratory that revealed effects of pulsatile estrogen administration on prolactin mRNA in the pituitary, Dutt (564) also used in situ hybridization to search for prolactin gene-expressing cells in the brain. Small numbers of cells containing prolactin mRNA were found in the mediobasal hypothalamus and also in the amygdala. Further, the use of RT-PCR gave results supporting the in situ hybridization evidence. In summary, results from various laboratories with radioimmunoassay, immunocytochemistry, bioassay, Northern blot, in situ hybridization, and RT-PCR indicate that a transcriptional system very similar to rat prolactin is operating in hypothalamic neurons and that hybridization with rat prolactin reagents and immunoreactivity with prolactin antisera can detect hypothalamic neuronal products.

Reproductive Behavior. Microinfusion of prolactin through cannulae aimed at the midbrain central gray can facilitate female reproductive behavior (556). Estrogen-pretreated females with low levels of lordosis showed significantly more reproductive behavior approximately 40 minutes after prolactin microinfusion, and lordosis performance returned to baseline levels by approximately 6 hours after prolactin treatment. Neither vehicle infusion nor microinfusion of growth hormone, oxytocin, vasopressin, or ACTH had this effect on lordosis, nor did infusion of prolactin into the cerebral cortex. In females given enough estrogen to show high levels of lordosis behavior, an antiprolactin antibody decreased lordosis performance.

Substance P. Substance P–immunoreactive perikarya (substance P neurons) have been found in the VMN (565,566) and are almost exclusively in the ventrolateral portion of the nucleus (567), where estrogen-concentrating neurons are located (4). Based on their location, it is clear that substance P neurons also contain ERs (568). In rats (569–571) and hamsters as well (571), substance P content in the hypothalamus varies with estrous cycle.

These observations are consistent with the notion that substance P is involved in the regulation of reproductive behavior. This is further supported by the findings that substance P neurons in the VMN project to the midbrain central gray (572) and that this brain region contains substance P receptors (573–576).

The preceding discussion suggests that substance P transmits a lordosis-promoting effect originated by estrogen in the VMN to midbrain central gray. Indeed, Dornan and Malsbury (577,578) studied its role in female sexual receptivity using ovariectomized female rats with just enough estradiol to display moderate levels of lordosis. Then, acidified substance P or LHRH was infused through cannulae directed at the dorsal midbrain central gray. Both acidified substance P and LHRH promptly facilitated lordosis behavior, and use of different doses of substance P yielded an orderly dose-related behavioral facilitation. Infusion of an antiserum to substance P, on the other hand, inhibited lordosis (578), indicating that the endogenous substance P is important physiologically.

In vitro electrophysiological studies show that substance P has two kinds of actions on neurons in midbrain central gray (552). It can act as a neurotransmitter to exert an exclusively excitatory action on responsive neurons. Also, at a concentration too low to have the direct action, substance P can potentiate the inhibitory action of GABA and the excitatory action of NE, both actions being associated with lordosis facilitation. It is not known whether substance P facilitates lordosis by its direct or indirect neuromodulatory action, or both. However, in the midbrain central gray, the direct excitatory action of substance P is very similar to those of thyrotropin-releasing hormone (TRH) and oxytocin (552), which did not facilitate lordosis when infused into this brain region (556,579). On the other hand, its modulatory action is very similar to that of LHRH, which is lordosis facilitating at the midbrain central gray (see earlier). For LHRH, prolactin, and substance P, the neuromodulatory action appears to be an important neural mechanism underlying the facilitation of lordosis and perhaps other behavioral effects (580) of neuropeptides.

Corticotropin-releasing Factor. Because there are close neuroanatomical relationships between neuronal systems containing opioid peptides and corticotropin-releasing factor (CRF) (581), and since β-endorphin can inhibit lordosis behavior (see later), Sirinathsinghji et al. (582) tested the hypothesis that CRF also could affect female reproductive behavior in rats. Infusions of CRF in the arcuate-ventromedial region of estrogen-primed female rats markedly reduced lordosis behavior, and this CRF effect could be blocked by an antiserum to β-endorphin. Combined infusion of CRF and an antiserum to ACTH left the CRF behavioral effect intact. In the midbrain central gray, CRF also was able promptly to inhibit lordosis behavior, and the anti–β-endorphin antiserum actually reversed this effect: reproductive behavior performance was higher than for saline-infused controls. LHRH in the midbrain central gray massively facilitated lordosis, easily overcoming the CRF-caused inhibition. CRF might act directly on neurons involved in the circuitry for lordosis, but most of these results fit with the notion of CRF operating through an interaction with POMC-derived peptides. Sirinathsinghji et al. (582) also considered the possible involvement of α-melanocyte–stimulating hormone (α-MSH). α-MSH is released from the pituitary in response to CRF, immunoreactive α-MSH neurons can be found in the arcuate nucleus (583,584), and, indeed, α-MSH can facilitate lordosis behavior (585,586). Electrophysiological work and studies of peptide release and membrane receptors will be required to unravel the situation in which several peptides apparently act and interact in one brain region, such as the midbrain central gray, to control a specific behavioral response.

Opioid Peptides. Perhaps the most significant effect of gonadal hormones on opioid peptide gene expression is the facilitation by estradiol of preproenkephalin mRNA synthesis. This was covered previously (second edition) and is tied into behavioral explanations. Notably, acting through δ receptors, enkephalin can stimulate lordosis behavior (587,588). Therefore, the behavioral meaning of changes in gene expression in the case of preproenkephalin depends on the localization of a specific opioid receptor type.

β-Endorphin. Long-standing evidence shows that opioid peptides can inhibit LH secretion and that this effect can be reversed by naloxone (589–591). It was natural, therefore, to investigate the possible roles for opioid peptides in the control of female reproductive behavior.

Intraventricular infusion of β-endorphin can suppress lordosis behavior (592). Because an opioid peptide might be expected to have many nonspecific behavior effects, Wiesner and Moss (593) used a variety of measures to see if the reproductive behavioral effects were secondary to more generalized opioid actions. Intraventricular β-endorphin in doses of 0.5 to 4 μg suppressed lordosis behavior. Courtship behaviors were also abolished. There was no indication of massive behavioral effects such as catalepsy, nor were there large effects on general responsiveness to somatosensory stimuli. Blood pressure was not altered by β-endorphin. There was a significant reduction on some measures of open-field behavior, but this was not correlated with changes in sex behavior. Wiesner and Moss (593) concluded that there was little indication that lordosis effects were secondary, in a trivial sense, to generalized behavioral suppressions. Because the opioid peptide effect has not been obtained routinely in some laboratories, Alien et al. (594) used the longer-acting antagonist naltrexone. Subcutaneous injection of naltrexone hydrochloride facilitated lordosis behavior in estrogen-primed ovariectomized female rats. Only an intermediate dose of 3 mg/kg was effective in this regard.

The first report of a site of action of opioid peptides for reproductive behavior control was from Sirinathsinghji et al. (595), who infused naloxone in the midbrain central gray and found enhanced lordosis behavior. Lordosis increases were noted 5 minutes after the infusion, and they lasted for approximately 2 hours. Working with the theory that opioid peptides in the midbrain have behavioral effects partly through their relationships to LHRH terminals (596), Sirinathsinghji (542) used estrogen-treated ovariectomized female rats to look for β-endorphin effects in midbrain central gray. Infusion of β-endorphin inhibited lordosis behavior within 30 minutes, and this effect could be overcome by LHRH infusion. Naloxone infused in the midbrain markedly facilitated lordosis as quickly as the animals could be tested. This type of midbrain infusion of β-endorphin does not affect LH release in the manner that, for example, a medial hypothalamic infusion of β-endorphin would (597). One report (598) investigated possible sites of action in the spinal cord. Based on the observations of enkephalin-containing cells in the substantia gelatinosa and the possibility that opioid peptides could be involved in the control of somatosensory stimuli related to reproduction, they used intrathecal injections of morphine in ovariectomized rats pretreated with estradiol and progesterone. Under these conditions, morphine markedly inhibited lordosis in an orderly, dose-dependent fashion, whereas the receptor antagonist naloxone enhanced reproductive behavior.

Across a number of neuroendocrine measures, the actions of opiate antagonists are like those of progesterone (599). Sensibly, the common actions of reductions in opiate peptide action would serve to synchronize LH release and lordosis behavior. This is a complex situation, however, with room for the possibilities of antagonisms among different peptide systems controlling somatosensory responsiveness and biphasic actions (e.g., of progesterone). It appears that for maximum receptivity, female rats must be both highly aroused motorically and particularly analgetic in the sense that they do not treat high-intensity somatosensory input as noxious (600). Also, a preexisting progestin action is required for naloxone to be effective on some neuroendocrine measures. One possibility is that progesterone first increases opioid tone, leading to a partial analgesia, which allows the lordosis reflex. The opposite possibility is that the immediate progestin action would be like naloxone (increasing lordosis behavior) but that subsequently it would increase the synthesis of opioid peptides (the second limb of a biphasic action on behavior).

Control of Synthesis. Estrogen decreases radioimmunoassayed endorphin levels in the mediobasal hypothalamus in rats, and it is not yet clear how effects on synthesis versus release interact in causing this result (601). The possibility of genomic actions of estradiol in this respect is raised by the demonstration—using the combination of estrogen autoradiography and β-endorphin immunocytochemistry in the same tissue—that estradiol is concentrated in the nuclei of some β-endorphin–producing cells (144). Possible effects of estrogen on transcription of the POMC gene are tested using hybridization with mediobasal hypothalamic tissue (602).

Oxytocin. The relationship of gonadal steroid hormones to oxytocin synthesis and action appears so far to be unique. Estrogen can increase both the synthesis and release of the peptide and also induce the production of its receptors. In turn, oxytocin can promote lordosis behavior. It is conceivable that some of the mechanisms by which gonadal hormones prime the oxytocin components of reproductive behavior control have themselves a self-multiplying character (603).

Molecular Synthesis. Beginning with classic observations of estrogen-induced oxytocin release (604,605), it became of interest to determine how hormone-induced synthesis could affect behavioral mechanisms. After cloning of the oxytocin gene and sequencing of its promoter, it became evident that potential estrogen response elements existed in the promoter (606,607). That, coupled with the observation that small numbers of neurophysin-immunoreactive neurons in the posterior PVN and the anterior commissural nucleus accumulate estrogen in the cell nucleus (608–610), suggested that, at least in some hypothalamic neurons, estrogen might directly turn on oxytocin gene expression. We used in situ hybridization to determine that in the anterior commissural nucleus and the supraoptic nucleus, estrogen caused the appearance of some neurons with higher amounts of oxytocin mRNA per cell than would normally be found in control ovariectomized rats (611). These results were consistent with Burbach's finding that estrogen, acting through its receptor on the rat oxytocin promoter, could indeed promote transcription of a reporter gene (612), and with other approaches to oxytocin mRNA measurements. Using Northern blot and dot-blot analyses, Van Tol et al. (613) demonstrated a significant increase in oxytocin mRNA levels during the estrus phase of the female rat's cycle: almost twice the levels of oxytocin mRNA were achieved compared with diestrus.

Similarly, oxytocin mRNA in the brain of the female rat is increased concomitant with puberty, and ovariectomy in the adult female decreases mRNA values (614).

In a subsequent set of studies, we targeted those neurons that are most likely to be involved directly with reproductive behavior, the cells with axons traveling to the spinal cord (615). In this work we combined in situ hybridization with a retrograde tracer technique to study effects of estrogen on oxytocin mRNA–expressing cells that had been back-filled by application of fluorogold pellets to the spinal cord. Oxytocin mRNA–containing neurons that project to the cord were found predominantly in middle and posterior levels of the paraventricular hypothalamic nucleus of female rats, comprising 12% to 14% of the retrograde labeled cells there. High but not low doses of estradiol were followed by significantly greater amounts of oxytocin mRNA in a subpopulation of these hypothalamic neurons that project to the cord. Thus, it is possible that estrogens affect a small population of oxytocin gene–expressing cells of particular relevance for female reproductive behavior. Progesterone may further increase oxytocin mRNA levels in the PVN (616), a possibility that was tested further in our laboratory with in situ hybridization for intronic probes generated by single-sided PCR (617). It is also likely that there are indirect means of hormone action, perhaps routed through other cells and perhaps consequent to increased oxytocin release.

Receptors. Just as it induces PRs, estrogen can also induce oxytocin receptors in the VMN (618–620). Initially, oxytocin receptors are induced in perikarya located in the VMN, and with time, and especially with subsequent progesterone treatment, they spread laterally by transportation along dendrites to an area lateral to the VMN (621–623). This lateral spread brings the induced oxytocin receptors into contact with oxytocin fibers and terminals, which are also located in the area lateral to the VMN and whose oxytocin content is increased by estrogen and progesterone (624). These effects by estrogen and progesterone, as discussed in the following section, are critical for the facilitatory effect of oxytocin in lordosis.

Reproductive Behavior. Like LHRH, oxytocin can facilitate lordosis at several brain sites. It does so when infused into the cerebral ventricles (625–629) or ventromedial hypothalamus (579,621,630), although not the midbrain central gray (579). This lordosis-facilitating effect requires mediation by oxytocin receptors, as indicated by the use of selective oxytocin receptor antagonists (629,631,632). It also requires priming with estrogen (633), indicating that the estrogen-induced synthesis of oxytocin or oxytocin receptors is necessary. The facilitatory effect of oxytocin can also be progesterone dependent. This is true for exogenous (624,628) as well as endogenous oxytocin (632). Indeed, in the ventromedial hypothalamus, oxytocin facilitated lordosis only when it was infused into the posterior portion, where the progesterone effect in inducing the spread of estrogen-induced oxytocin receptors was observed (621).

It is not necessarily the case that the sole job of oxytocin released in the hypothalamus is to facilitate reproductive behavior. Indeed, a coherent theory of how oxytocin actions on a variety of instinctive behaviors should be conceptualized (634,635) points out that the main mode of the oxytocin effect may be to protect biologically adaptive responses from disruptions caused by mild stress. Thus, hormone-stimulated oxytocin gene expression, release, and binding all may satisfy just one means by which hypothalamic provide biologically important interactions between external and internal environments.

Electrophysiological Action. In hypothalamic slices maintained in vitro, oxytocin activated VMN neurons predominantly [up to 94% of the neurons tested by excitation (636,637)]. Further electrophysiological and pharmacological analyses of this excitatory action uncovered several interesting findings (637). First, like the behavioral effect and the effect on oxytocin receptors in the VMN, the number of VMN neurons excited by oxytocin is also increased by estrogen. Furthermore, estrogen treatment resulted in the significant correlations of neuronal responses to oxytocin with those to NE and ACh, which also can facilitate lordosis. Therefore, estrogen not only makes oxytocin excite more VMN neurons but makes it stimulate the right kind of neurons. Second, corresponding nicely to the facilitatory behavioral effect, the excitatory electrophysiological action is also mediated by oxytocin receptors. It is mimicked by a selective oxytocin agonist but not by a metabolite of the peptide; and conversely, it is blocked by selective oxytocin antagonists, including the one that inhibited the facilitatory behavioral effect of the endogenous oxytocin. Thus, the excitatory neuronal action of oxytocin in the VMN is obviously associated with and may be responsible for its lordosis-facilitating effect.

The importance of oxytocin receptor gene expression has been made apparent through both the behavioral and the electrophysiological experiments reviewed earlier. Now that this gene has been cloned (638), we are using antisense DNA oligomers spanning the translation start site to see if interrupting oxytocin receptor synthesis has implications for binding or behavior (639). Likewise, in situ hybridization with PCR-generated probes will determine whether gonadal steroid treatments are followed by increased oxytocin receptor gene expression (640).

Second Messenger Systems

Correlation with Effects on Lordosis. As a prelude to investigating the intracellular mechanisms underlying the inhibitory and facilitatory effects of peptides on lordosis, a survey was made to identify the second messenger system coupled to each transmitter and neuropeptide that can affect lordosis. The results of the survey and the corresponding electrophysiological actions are listed in Table 2. Because most knowledge about second messengers is gathered from non-nervous systems, the relevance of the second messenger system summary in Table 2 to the regulation of lordosis remains to be proven. Nevertheless, the list reveals two patterns: a striking correlation between lordosis-facilitating agents and the phosphoinositide system and a less monolithic connection between lordosis-inhibiting agents and the adenylate cyclase system. These correlations suggest that two major second messenger systems mediate opposite effects on lordosis. They also indicate that the phosphoinositide system serves as a common pathway for the facilitation of lordosis. Indeed, one can facilitate lordosis by activating the phosphoinositide system directly with phorbol esters without involving membrane receptors (641). The role of the adenylate cyclase system is less clear because it can be affected in opposite ways by lordosis-inhibiting agents.

Hormone Effects. One important characteristic of the lordosis-facilitating effect is the dependence on estrogen. This is concluded from the observations on all the agents examined to date, including α-adrenergic agonists (325), muscarinic agonists (642,643), LHRH (535,644), oxytocin (633), and $GABA_A$ agonists (645). Without estrogen priming, these lordosis-facilitating agents did not facilitate lordosis, even at supramaximal doses. It is exciting, therefore, that a strong action of estrogen on the phosphoinositide system has been observed. Estrogen priming at a level that would induce lordosis also induces the production of several proteins in the VMN (646). The most prominent protein induced fulfills all the criteria to be an isoform of a key enzyme in the phosphoinositide system, phospholipase C-α (PLC-α) (647). Estrogen can also modulate other aspects of the phosphoinositide turnover (648–650). In MCF-7 breast cancer cells, estradiol has been found to potentiate phosphoinositide turnover (651) and to cause an enhancement (652) or a biphasic change in PLC activity (653). Regarding another key enzyme of the phosphoinositide system, protein kinase C (PKC), estrogen has been shown to promote its translocation from the cytosol to the membrane (654,655), a process that would activate the enzyme, to increase the activity of PKC in the rat pituitary in vivo and in vitro (656), and to increase remarkably the PKC-dependent phosphorylation of PKC substrates in luteal tissue (657).

Given these findings, it is no wonder that estrogen also potentiates the stimulation of the phosphoinositide second messenger system by lordosis-stimulating agents. This was in fact reported for the muscarinic agonist carbachol in the uterus (658), for adrenergic agonists in the uterus (659) and brain (660), for LHRH on ovarian cells (661), and for oxytocin (648). It is clear that alteration of the phosphoinositide system provides one mechanism for the estrogen induction of lordosis.

The findings on agents that can affect lordosis are summarized in Table 2. For lordosis-facilitating agents, an agent's effect invariably is associated with its activation of the phosphoinositide second messenger system and its excitatory action on neuronal activity. These associations elucidate a mechanism for the facilitation of lordosis: because the activation of the phosphoinositide system can lead to increases in neuronal activity (662), and because an increase in neuronal activity is required for the facilitation of lordosis (discussed later), the associations indicate that there are causal relationships between the cellular actions and the behavioral effect. Such a relationship also indicates that the phosphoinositide system is a common pathway for the facilitatory effects of various agents. This common pathway can be regulated by estrogen, and estrogen is required for the lordosis-facilitating agents to work.

Lordosis-inhibiting agents may depress lordosis in part by inhibiting the adenylate cyclase system and neuronal activity. Our electrophysiological studies show that, in the VMN at least, a neuron can receive both inhibitory and the excitatory actions from different transmitters (265) or from a given transmitter through different subtypes of receptors (317,365), and that the opposite actions can neutralize each other (317,365). Therefore, lordosis-inhibiting agents may do so by bringing down VMN neuronal activity.

Electrophysiological Actions

Electrical Recording In Vivo. Electrical activity of nerve cells in the hypothalamus and POA, as well as in limbic forebrain structures, varies during the estrous cycle of the rat. Elevated firing rates can be correlated with ovulation (663,664). In the ventral portion of the medial anterior hypothalamus, single-unit activity was highest on the day of proestrus (665). Similar conclusions had been reported by Moss and Law (666). In these experiments, circulating hormones were probably acting directly on hypothalamic tissue because, in studies in which an island of hypothalamic tissue was prepared surgically, similar variations in electrical activity were seen throughout the estrous cycle (667). Antidromic identification of ventral anterior hypothalamic neurons by electrical stimulation in the ventromedial-arcuate region suggested that those neurons project to the arcuate nucleus region (668). Estrogen certainly plays a role in electrophysiological variations through the estrous cycle. Recording in ovariectomized female rats, Kawakami et al. (669) showed increases caused by estrogen in the firing rates of arcuate neurons. Similarly, in the medial anterior hypothalamus, Cross and Dyer (670) showed increases in resting discharge rates after estradiol injection. Multiunit activity in the VMN and the medial anterior hypothalamus tended to be more responsive to peripheral stimuli after estrogen treatment, with a shift in response direction toward excitation (671). The only caution here is that for this type of study single-unit recording should be used because adjacent hypothalamic neurons can have different functional properties (672).

The predominant effect of estradiol in the POA and in the most anterior neurons of the anterior hypothalamus is different from that in the basal medial hypothalamus and in the anterior hypothalamus near the VMN. Lincoln (673) showed that estradiol treatment of ovariectomized rats is followed by lower spontaneous activity of units in the preoptic region and in the lateral septum. Probing the vaginal cervix produced inhibition of preoptic neurons in estrogen-treated animals (674). In the same experiments where Kawakami et al. (669) found estrogen-dependent increases in arcuate electrical activity, they saw no changes or mixed changes in preoptic neuron discharge following estrogen treatment. Yagi (675,676) looked for changes in single-unit activity in the POA at short latencies after systemic estradiol injections. Some of the units showed complicated biphasic responses, but many preoptic units decreased their activity to very low levels after estradiol injection, and in some cases activity remained low for the duration of electrical recording.

In a set of single-unit recording experiments (264), conditions were intended to be relevant for the control of mating behavior. To simplify the endocrine analysis, ovariectomized animals were used, one half of them given long-term estrogen treatment and the other half used as estrogen-free controls. All of the estrogen-treated animals performed strong lordosis reflexes, and none of the controls did. The number of single units with recordable spontaneous electrical activity in the basomedial hypothalamus was significantly greater in the estrogen-treated animals. In the medial POA, the opposite was the case: the number of single units with recordable spontaneous activity was significantly less in the estrogen-treated animals. These findings and the entire distributions of firing rates in the two anatomical regions supported the interpretation that in the basal medial hypothalamus estradiol stimulates silent neurons to fire at a low rate and that it has the opposite effect in the medial POA (264). Many ventromedial hypothalamic neurons send axons to the midbrain central gray, as shown by responses to antidromic stimulation (677). Similar to the results of Bueno and Pfaff (264), a high percentage of these cells have no resting discharge. The stimulation threshold for antidromic activation was significantly lower in the estrogen-treated animals (677).

Some analogous recording experiments have been done in female monkeys, not with respect to behavior but based on evidence that electrical activity is involved in the progesterone-induced LH surge. Because the surge is blocked by anesthesia (678) but not by hypothalamic deafferentation (679), there is probably a medial basal hypothalamic action of progesterone where there are PRs (680) and, in the monkey, LHRH neurons (681,682). Similar to the results of Bueno and Pfaff (264), Yeoman and Terasawa (683) found that the firing rate of single units in the ventral hypothalamus was slow. Progesterone injection increased unit activity in the ventral hypothalamus but not the dorsal hypothalamus, and the increases in circulating LH and ventral hypothalamic unit activity were significantly correlated over time. Overall, electrophysiological results from several laboratories indicate that estrogen, or estrogen plus progesterone, can facilitate various indices of electrical activity in and around the VMN in a manner that could be involved in the stimulation of female mating behavior. In the medial POA, which can inhibit female mating behavior, estrogen appears to decrease single-unit electrical activity.

Responsiveness of neurons to somatosensory stimulation was generally low (264). In the medial POA, if present, it was depressed by estrogen treatment, whereas it tended to be elevated in the medial anterior hypothalamus and basomedial hypothalamus by estrogen treatment (264). Similarly, Chan et al. (684) found a small percentage of ventromedial hypothalamic neurons specifically responsive to vaginocervical stimulation. Estrogen and progesterone significantly increased the percentage of neurons responding specifically to this stimulation but had no effect on the nonspecifically responding neurons. In these experiments, iontophoretic applications of prolactin excited the majority of neurons, whereas the type of responses to LHRH tended to be related to the type of response evoked by peripheral stimulation. Overall, experiments from different laboratories suggest that estrogen or estrogen plus progesterone priming can increase the probability that neurons in and around the VMN will respond to somatosensory stimulation, but the percentage of neurons responding, the amplitude of the response, and their time course make it unlikely that such somatosensory responsiveness would control lordosis behavior in a reflex-like manner.

It should be emphasized that neurons in the basomedial hypothalamus have low resting discharge rates compared with other brain regions (264,665, 666,685). Neurons antidromically identified from the midbrain tended, in most cases, to have no spontaneous activity (677), and estrogen effects on basomedial hypothalamic neurons were interpreted as stimulating silent neurons to show low resting discharge rates (264). This is consistent with the fact that a substantial percentage of estrogen-binding neurons send their axons to the midbrain central gray (157). These facts, together with the sluggish responsiveness to peripheral stimuli by medial hypothalamic neurons, further support the interpretation that ventromedial hypothalamic neurons control female reproductive behavior by a descending tonic output and that reflex loops primarily are completed at lower levels of the neuraxis.

Estrogen effects on nerve cells at lower levels of the neuraxis may well be indirect, depending on a hypothalamic output, even though there are a small number of estrogen-binding cells in the dorsal horn of the spinal cord (686). In any case, the nature of these effects fits well with a participation in the control of lordosis behavior. Estrogen treatment causes an expansion of the pudendal nerve receptive field (687,688) as well as an expansion of the receptive field of dorsal root L6 (689). These hormone effects depend on the inherent averaging of recording techniques that take into account the electrical activity of the entire pudendal nerve or the entire dorsal root. Recording from neurons deep in the dorsal horn of the lumbosacral spinal cord, Kow et al. (690) found that the ratio of excitatory to inhibitory responses, when lordosis-eliciting stimuli were applied, was increased by estrogen treatment, whereas the excitatory/inhibitory ratio to behaviorally irrelevant stimulation was not affected. Indeed, estrogen lowers the threshold for cutaneous pressure in eliciting lordosis (691).

In the medial medullary reticular formation there are neurons whose inputs reveal an extensive and specific convergence between inputs from lordosis-eliciting stimuli and the midbrain central gray (which relays lordosis-relevant estrogen influences from the hypothalamus) (692). In such neurons, estrogen can increase the ratio of excitatory responses to inhibitory responses after lordosis-eliciting stimulation, thus facilitating behaviorally relevant neuronal excitability. It was striking that in the medullary reticular neurons most easily related to lordosis behavior—reticulospinal units, which gave prompt responses to flank-perineal stimuli—estrogen increased the percentage of excitatory responses (692). Further, estrogen increased the rate of resting discharge in reticulospinal neurons that responded to flank-perineal stimuli but not in neurons that did not respond to behaviorally relevant stimuli and not in neurons that failed to project back to the spinal cord. It is particularly important for a behavioral mechanism that these medullary reticular neurons, whose resting discharge and responsiveness are facilitated by estrogen and that demonstrate a convergence between lordosis-eliciting peripheral stimuli in midbrain central gray, have been implicated further in a mechanism by which initial perineal stimuli (during mating behavior) could facilitate lumbar axial muscular responses to subsequent stimuli (693,694). All of these facts are consistent with the following interpretation (287): estrogen effects on medial hypothalamic neurons, relayed through the midbrain central gray, increase the excitability of certain medial medullary reticular neurons. These neurons receive an input from skin areas involved in triggering lordosis behavior. In the manner of a spinobulbospinal reflex, their responses to the first inputs from flank or perineal skin regions excite such medial medullary reticular neurons projecting back to the lumbar spinal cord, facilitating segmental responses to subsequent perineal inputs. These segmental responses include the lumbar axial motor neuronal output that is responsible for lordosis behavior.

In the midbrain, during single-unit recording from a population of neurons probably including many short-axoned cells, extensive convergence was observed between somatosensory and hypothalamic effects on individual neurons (695). Central gray neurons were found whose response profiles matched exactly what would be required for facilitating lordosis behavior: excitation by the ventromedial hypothalamus, by nucleus gigantocellularis of the medulla, and by touch or pressure on the skin, with inhibition by painful stimuli and by electrical stimulation of the POA (695). Other midbrain central gray neurons were antidromically identified by electrical stimulation of the medullary reticular formation as having descending axons (696). A particularly large group of cells was noted in the ventrolateral central gray, and these antidromically identified neurons had very low rates of spontaneous activity or none at all (696). Estrogen treatment increased resting discharge rates in a subpopulation of these neurons by creating a group of cells with firing rates greater than three per second; no such cells existed in ovariectomized control preparations (697). Particularly important is that this sort of estrogen effect on spontaneous activity was not seen among the cells that did not have descending axons, suggesting that the estrogen effect is specific to those nerve cells with central roles in reproductive behavior control.

During the estrous cycle, central gray unit activity is higher during late proestrus and estrus than during metestrus (698). Electrical excitability of the cell membrane of these antidromically identified central gray neurons was measured by the ability of the antidromic spike to invade the cell body. Electrical stimulation of the VMN increased excitability according to this measure, whereas lesions decreased it; POA manipulations had the opposite effects—electrical stimulation of the POA reduced central gray neuronal excitability, whereas lesions facilitated it (697). Thus, the electrical activity of these midbrain central gray neurons fits perfectly what would be expected for the excitation of lordosis: they are estrogen sensitive, and neural manipulations that increase lordosis also increase excitability, whereas those that decrease lordosis behavior also reduce cell membrane excitability. Similarly, in hamsters, midbrain neurons responsive to lordosis-triggering stimuli can be recorded, even in urethane-anesthetized animals (699). Such neurons tended to cluster in the deep layers of the superior colliculus, the adjacent intercollicular zone, and in the central gray. Although there is a trend toward increasing responsiveness of these midbrain neurons toward any somatosensory stimulus with estrogen–progesterone treatment, by far the most quantitatively significant result is an increased responsiveness to stimuli that will trigger lordosis after estrogen plus progesterone treatment (700). Rose and Michael (701) recorded from midbrain and pontine units in squirrel monkeys. In estrogen-treated monkeys, units responding to tactile stimuli tended to have larger receptive fields, and responsive units comprised a larger portion of the local recording sample than in the control untreated monkeys. A greater portion of the units responded to vaginal stimulation in the estrogen-treated animals, but responsiveness to painful stimuli was not affected.

Electrical Recording In Vitro. For increasing precision of control over the effective concentration and time of action of agents applied, and for decreasing the possibilities of metabolism of the agents applied or for their indirect actions, in vitro electrophysiological recording, for example in hypothalamic tissue slices, is a good approach. Recording from hypothalamic cells in female rats that had been treated chronically with estrogen or had been in the control, ovariectomized state allows the discernment of estrogen effects after the normal process of nuclear concentration of estradiol in the living animal. In the suprachiasmatic nucleus of the hypothalamus, estrogen specifically facilitates neuronal responses to ACh and serotonin in nerve cells firing with an irregular pattern, but not in neurons firing at a constant frequency (702).

Recording in vitro from the VMN is likely to give results relevant for the physiological control of lordosis behavior (see later). Here, estradiol facilitates not only excitatory responses to electrically stimulated orthodromic input but responses to ACh, serotonin, and NE (265). Responses to glutamate, dopamine, and GABA were not affected by hormone treatment. Estradiol also potentiates the neuromodulatory action of TRH on the responses of ventromedial hypothalamic neurons to glutamate and ACh (285). This is especially interesting in view of the effects of estradiol on TRH receptors (703–705). These estrogen effects on ventromedial hypothalamic neurons are specific in that many types of responses are not affected—responses to glucose (706), to vasopressin (636), and also by arcuate neurons to anterior pituitary hormone (708).

Fast electrophysiological effects of estrogen, as might be achieved with in vitro applications of the steroid, might be relevant for the negative feedback effects of this hormone. The spontaneous activity of tuberoinfundibular neurons can be inhibited by iontophoresis of estradiol (709), which fits with the finding that actinomycin D cannot prevent the inhibitory actions of estradiol on LH secretion (710,711). In hypothalamic slices from female guinea pigs, neurons with an average resting membrane potential of −35 mV were recorded, and several of them were hyperpolarized by estradiol application in the bathing medium (712). In later experiments recordings from neurons in this type of hypothalamic slice showed not only that estradiol could hyperpolarize or inhibit the spontaneous activity of cells in the arcuate region of the hypothalamus, but that a few of the estrogen-responsive neurons contained immunoreactive LHRH (713). Pan et al. (708) did not find estrogen effects on responses of arcuate neurons to anterior pituitary hormones (possibly relevant for short-loop feedback), but did find a slight estrogen facilitation of arcuate neuronal response to cholecystokinin-8.

Comparisons of In Vivo with In Vitro Hormone Effects. In vivo and in vitro recording situations offer different advantages for the electrophysiologist in that many of the hormone effects testable in one would not be directly relevant for the other, but certain differences among results might have important implications. In vivo, estrogen increased the resting discharge rate of slowly firing or silent neurons (264). In vitro, estrogen did not increase simple resting discharge rate, suggesting that synaptic inputs not present in the hypothalamic slice are required for the estrogen effect on firing rate (265). Consistent with this are the findings that, in vitro, estrogen does facilitate the response to orthodromic electrically stimulated input and to neurotransmitters (265). In a set of neurochemical experiments, a similar comparison may obtain. With labeled amino acid precursors applied in vivo to ventromedial hypothalamic neurons, estrogen could be shown to induce a protein with molecular weight about 70,000 Da (714), but with the entire experiment carried out in vitro we have not seen this protein. Conceivably, the synthetic response of a hypothalamic neuron to estrogen could depend on the barrage of synaptic input. In turn, synaptic input to hypothalamic neurons may be important for the estrogen effect on the overall behavior, lordosis, because anesthesia during estrogen priming inhibited the induction of behavioral receptivity (715).

Thus, synaptic input may be required for estrogen effects on spontaneous activity, on an estrogen-induced protein, and perhaps even on lordosis behavior itself. This suggests that in the cascade of hormone-induced changes after ER occupation (69,70), electrical activity dependent on synaptic input is an essential element. In turn, estrogen effects on the neurotransmitter and neuropeptide systems, reviewed previously, probably have, as one of their functions, to maintain this synaptically driven spontaneous activity in ventromedial hypothalamic neurons.

Role in Reproductive Behavior Control

Hypothalamic Electrical Activity Required. With such a large body of data on electrical actions of estrogens, one normally would presume that they had an important role in the control of female reproductive behavior, even if primarily in the stimulus–secretion coupling mode. Importantly, at several levels of the neuraxis, the nature of estrogen's electrical effects matches the requirements for behavior control—in the spinal cord, in the medullary reticular formation (692,693), in the midbrain (697), and in the hypothalamus (264).

Interfering with electrical activity in a manner that does not produce overall cellular destruction provides evidence that hypothalamic electrical activity is actually required for estrogen-driven lordosis behavior. We used intrahypothalamic infusions of the sodium channel blocker tetrodotoxin to see if electrical activity mediated by voltage-dependent sodium channels is needed for the occurrence of the behavior (716,717). Bilateral application of tetrodotoxin near the ventromedial hypothalamus caused a dose-dependent decrease in the magnitude of lordosis behavior. Interestingly, the first significant decrease in lordosis was not reached until 40 minutes after tetrodotoxin, and lowest lordosis scores occurred 2 to 4 hours after infusion. In electrophysiological preparations, the same type of intrahypothalamic tetrodotoxin infusion suppressed action potentials completely within 5 or 6 minutes and lasted for hours. In contrast, intrahypothalamic infusion of local anesthetics usually depressed electrical activity only temporarily or incompletely and had no effect on lordosis behavior. Thus, large and prolonged decreases in electrical activity in the basomedial hypothalamus can cause a decline in lordosis behavior (7).

The time course of the behavioral, as opposed to the electrophysiological, changes has several implications. Because electrical activity was suppressed within 5 or 6 minutes but the behavior could persist as long as 40 minutes or more after tetrodotoxin (716,717), it could not be the case that hypothalamic electrical responses to somatosensory input are required for the behavior on an event-by-event basis. Instead, a tonic hypothalamic output must be facilitating the behavior over a much longer period than an individual mount by the male rat. Insofar as one meaning for the electrical activity is a stimulus-secretion–coupled release of important hypothalamic peptides in the midbrain or at another level of the neuraxis, the long time course of behavioral survival after tetrodotoxin infusion is a measure of time course of action of peptides that had been released just before tetrodotoxin. Finally, it is not always the case that tetrodotoxin action on lordosis behavior is temporally sluggish: when tetrodotoxin is applied in the dorsal midbrain, lordosis behavior loss is almost immediate (718,719). This in turn is consistent with the rapid responses by neurons in the midbrain central gray to electrical stimulation and their prompt responses to somatosensory input.

Because tetrodotoxin can block a sodium channel whether it is open or closed (720,721), whereas local anesthetics, operating by a different mechanism, act preferentially on open channels (722), it might be inferred that the most important cells in the ventromedial hypothalamus for reproductive behavior control are the silent or slowly firing cells (7). This implication fits well with the electrophysiological finding (264) that estrogen in the basal medial hypothalamus facilitates the resting discharge of silent or slowly firing neurons.

If an experimental manipulation of a hormone-dependent cell decreases the cell's output, was that because the experimental manipulation interfered with a cellular function that is part of the hormone's effect, or did it simply disrupt some "housekeeping activities" of that cell? Because sodium channel blocking can decrease lordosis (717) in the ventromedial hypothalamus, because electrical stimulation (723) there can increase lordosis behavior, and because estrogen leads to increased electrical activity in the ventromedial hypothalamic region (see earlier reference), by far the most parsimonious interpretation is that one means by which estrogens foster lordosis behavior is through the increase in ventromedial hypothalamic electrical activity, particularly in slowly firing neurons (7).

Electrophysiological Changes Are Not the Only Behaviorally Important Mechanisms. In addition to the good match between estrogen effects on electrical activity at various levels of the CNS and the requirements for lordosis behavior, there are parallels between detailed electrophysiological results in the midbrain and behavioral requirements. Recording from midbrain central gray neurons that were antidromically identified from the medullary reticular formation (good candidates for carrying hormone-dependent behaviorally relevant information to the lower brainstem), we measured the effects of hypothalamic or preoptic manipulations on the electrical excitability of these midbrain cells (696,697). Throughout, there were striking parallels; for example, electrical stimulation of the VMN increased central gray neuronal electrical excitability even as it increased lordosis behavior. Lesions of the ventromedial hypothalamus had the opposite effect (7). This pattern of results is consistent with the notion that changes in hypothalamic electrical activity are an important part of the means by which estrogen effects on hypothalamic cells alter lordosis behavior.

Nevertheless, there were important differences between certain aspects of estrogen effects on midbrain central gray electrophysiology and the effect of that hormone on lordosis behavior as a whole (7). Whereas the effect of ventromedial hypothalamic electrical stimulation on central gray neuronal electrical excitability did not require a background of estrogen action and could be registered immediately, the effect of ventromedial stimulation with the very same parameters on lordosis behavior as a whole did require subthreshold estrogen priming and required at least 25 to 50 minutes for a full behavioral effect (7,723). Therefore, to go from a well-documented electrical effect in the midbrain to a full behavioral mechanism requires a priming action of estrogen that takes at least 25 to 50 minutes to be registered.

Because the estrogen in these experiments was given over a period of days, there was plenty of time for new protein synthesis before the electrophysiological or behavioral measures were taken. In view of the fact that new RNA and protein synthesis is required for the action of steroid hormones on female reproductive behavior (see later), the time-consuming process required for the effects of ventromedial hypothalamic electrical stimulation on lordosis behavior must involve release, diffusion, and action of peptides resulting from this synthesis. Thus, in addition to the fast electrical effects of hypothalamic stimulation on midbrain neurons (themselves hormone dependent), the temporal delays between the electrical responses and the behavioral results can be taken as a measure of the time course of release, diffusion, and action of the transmitters and peptides involved. With respect to the role of the electrical changes caused by hormone action in ventromedial hypothalamus, if they are mainly in a tonic stimulus–secretion coupling mode (as opposed to a reflex mode), it is particularly clear that the fast electrical responses in midbrain central gray neurons would not adequately account for the entire behaviorally relevant transmission through the midbrain. In turn, we infer that midbrain central gray neurons, in addition to their fast, hormone-sensitive electrical responses, have a longer-acting type of electrical response that we did not record, or they have a behaviorally important nonelectrical response mode. The same implication about the possibilities of longer-acting hypothalamic effects important for the entire behavioral mechanism would hold for neurons in the lordosis circuit at lower levels of the neuraxis.

In summary, at several levels of the CNS estrogen effects on neuronal resting discharge or responsiveness match the requirements for participation in the behavioral mechanism. Moreover, effects of estradiol on ventromedial hypothalamic electrical activity can be proven to be part of the essential mechanism (7). Especially important is the participation of slowly firing ventromedial hypothalamic neurons with little or no resting discharge. The slow registration of the effects of various electrical manipulations on lordosis behavior requires that the mode of action of hypothalamic output be a facilitating background "tonic" action rather than a rapid gate-closing reflex control that operates on a mount-by-mount basis.

Not all of the characteristics of hypothalamic participation in reproductive behavior control can be accounted for by these electrophysiological results. Therefore, modes of hormone action in the medial hypothalamus that include molecular synthetic and secretory steps are included in the next section.

MECHANISMS UNDERLYING THE PRIMARY MATING BEHAVIOR, LORDOSIS

Neuronal circuitry sufficient for producing lordosis behavior has been reviewed extensively (724,725) and is depicted in Fig. 3. Three main features stand out. First, despite the fact that there are many opportunities for sensory information relevant for triggering the behavior to bridge over to motor control pathways, the circuit does not operate unless an estrogen-dependent signal is carried from the VMN to the midbrain. This estrogen-dependent signal in turn depends on estrogen-dependent gene expression in the hypothalamus, reviewed later.

Second, the circuitry is arranged in modules (Fig. 3), which, by good fortune, match the embryological divisions of the developing neuraxis. Third, for reasons we have guessed at (726) but not proven, there is an extensive overlap between lordosis behavior–controlling neurons and the pathways signaling pain and analgesia. We have hypothesized that control over pain is centrally involved in the female's ability to respond sexually to tactile and interoceptive stimuli that otherwise might prove noxious.

Because of the dominating role of estrogens (and after estrogens, progestins) in the neural circuitry pictured in Fig. 3 and in the gene expression studies reviewed in the following sections, one might have been tempted to talks about hormones "causing behavior." As detailed in textbook form [Chapter 6 in Pfaff et al. (727)], hormones do not "cause" behavior. They alter probabilities of responses to given stimuli. For example, in our experiments, when adequate schedules of estradiol and progesterone are given

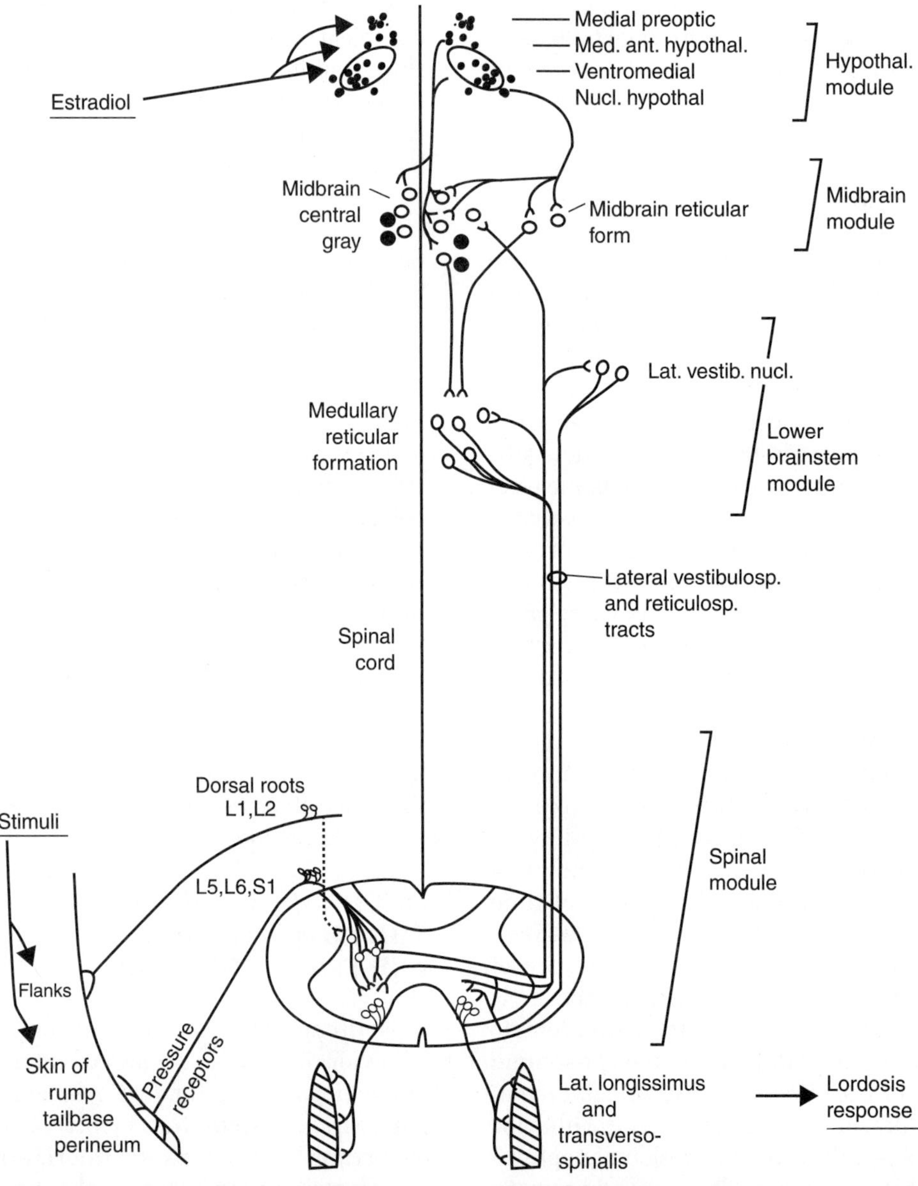

FIG. 3. The neural circuit for producing lordosis behavior. It is bilaterally symmetrical, but is drawn here on one side to maintain visual clarity. At lower levels of the lordosis circuit there are multiple opportunities for sensory signals to affect motor control pathways. However, the circuit does not operate unless there is an estrogen-dependent enabling signal from the ventromedial nucleus of the hypothalamus down to the midbrain. The modules, deduced from neuroanatomical and neurophysiological data, turn out to match the embryological divisions of the developing neuraxis. (From Pfaff, D. W. [1999]. *Drive: Neural and Molecular Mechanisms for Sexual Motivation.* MIT Press, Cambridge, MA.) As a side point, overlaps between sex behavior circuitry and neural pathways involved with pain signaling and pain suppression are remarkable (726).

to female rats or mice, to mimic their normal sex hormone conditions that lead to ovulation, they do not simply *spring* into the sway-backed posture called lordosis that allows the male to fertilize the newly released eggs. Nor does the naturally estrous female, "in heat," do lordosis in a vacuum. Instead, the behavior is triggered by tactile stimuli on the flanks and rump consequent to mounting by the male. In the absence of estrogens plus progestins, lordosis will not occur. Systematically, estrogens followed by progesterone raise the probability that such tactile stimuli will cause lordosis to occur (Fig. 4). We can see the rising amplitude of response to an adequate stimulus and the rising probability of lordosis as a function of hormone dose.

Molecular Mechanisms, from Hormone to Gene to Behavior

Several genes have the following two properties: that estrogen administration to female rats or mice significantly elevates their transcript levels, and that their gene products promote female reproductive behavior (Fig. 5). They would seem to provide a quite adequate basis for saying that we know the molecular basis of this particular type of hormone action on this particular behavior. However, the list of genes in Fig. 5 leads immediately to two new questions: (a) can we formulate the genes we already know in a manner that makes biological sense, and (b) is the list curtailed by virtue of a lack of reagents up until the recent past? The answer to both questions is "yes."

The first question has been answered using a modular perspective, which was stated by Mong and Pfaff (728), upon which this section depends. In fact, the functional genomics of hormone/gene/behavior relations in the CNS provide a theoretical paradigm for linking genes to neural circuits to behavior. We propose that estrogen-induced genes are organized in modules: *G*rowth of hypothalamic neurons; *A*mplification of the estrogen effect by progesterone; *P*reparative behaviors; *P*ermissive actions on sex behavior circuitry; and *S*ynchronization of mating behavior with ovulation (GAPPS; Fig. 6).

Here we theoretically propose a gene network—really, a micronet—downstream of estrogen action in the forebrain responsible for courtship and lordosis behavior in females. Not a massive review of the literature or an original data report, this section comprises an attempt to draw different transcriptional systems into a new theoretical formulation. These efforts to explain hormone-driven behaviors have benefited from Rosenfeld's example of genetic network control over pituitary gland development (729). Both direct and indirect causal routes are depicted in the following.

Causal Routes, Downstream from Genomic Action: A Modular System Emergent

The primary sex behavior of female quadrupeds, lordosis, depends on defined physical signals: cutaneous stimuli and estrogens plus progestins (729). The neural circuit has been worked out; estrogen-dependent transcription in ventromedial hypothalamic cells allows permissive signals to the midbrain central grey, thus enabling the rest of the circuit. In the absence of fear or anxiety-provoking conditions, females under the influence of estrogens plus progestins will demonstrate courtship and then mating behaviors. During the normal female cycle, these behavioral components of reproduction are synchronized with ovulation. Thus, with the mediation of estrogens and progestins, the neural, behavioral, and endocrine preparations for reproduction are harmonized.

Because (a) the hormone receptors discovered in neurons turned out to be transcription factors, (b) mating behaviors follow estrogen administration by more than 18 hours, and (c) inhibitors of RNA and protein synthesis disrupt the estrogen effect, it was natural to look for genes whose induction comprised important mechanisms for the behavioral facilitation. In the following sections are listed gene/neuronal and gene/glial modules found so far.

Throughout these studies, we compare two extremely similar transcription factors, ERs α versus β. Likely gene duplication products, both have high affinity for estradiol and for genomic estrogen response elements in vertebrate gene promoters (730–733). Yet, they have distinctly different neuroanatomical patterns of expression (734) and different functional consequences.

All of the genetic systems discussed here have the character that estrogen treatment turns them on in vivo, and that they participate in facilitating estrogen-dependent female reproductive behaviors.

Direct Effects, from Gene Induction to Neural Circuit to Behavioral Change. Hormone effects on neurotransmitter receptors in ventromedial hypothalamic neurons directly trigger the rest of the lordosis circuit to operate.

Noradrenergic α_{1B} *receptors* are induced (735) in vivo in female rats by estrogen treatment in ventromedial hypothalamic cells that govern the rest of the lordosis behavior circuit (736,737). Noradrenergic ascending afferents synapse on ventromedial hypothalamic neurons, coming in from the ventral noradrenergic bundle, which originates in arousal-related neuronal groups A1 and A2, and signals heightened arousal upon stimulation from the male. In biophysical studies, directly applied noradrenergic stimulation increases the electrical activity of ventromedial hypothalamic neurons (736). Beginning with G_q or G_{11} proteins activating PLC, noradrenergic action produces

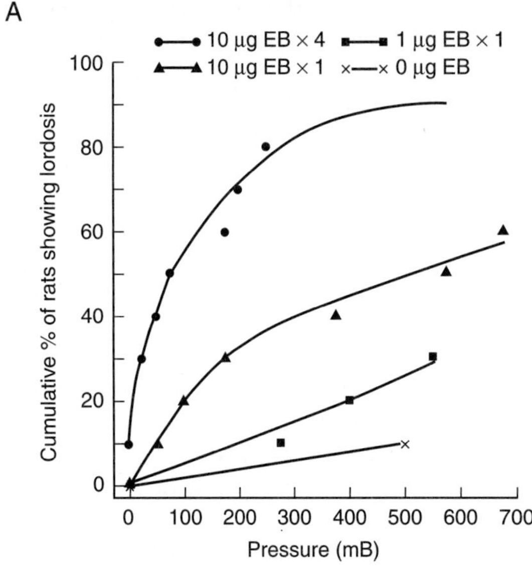

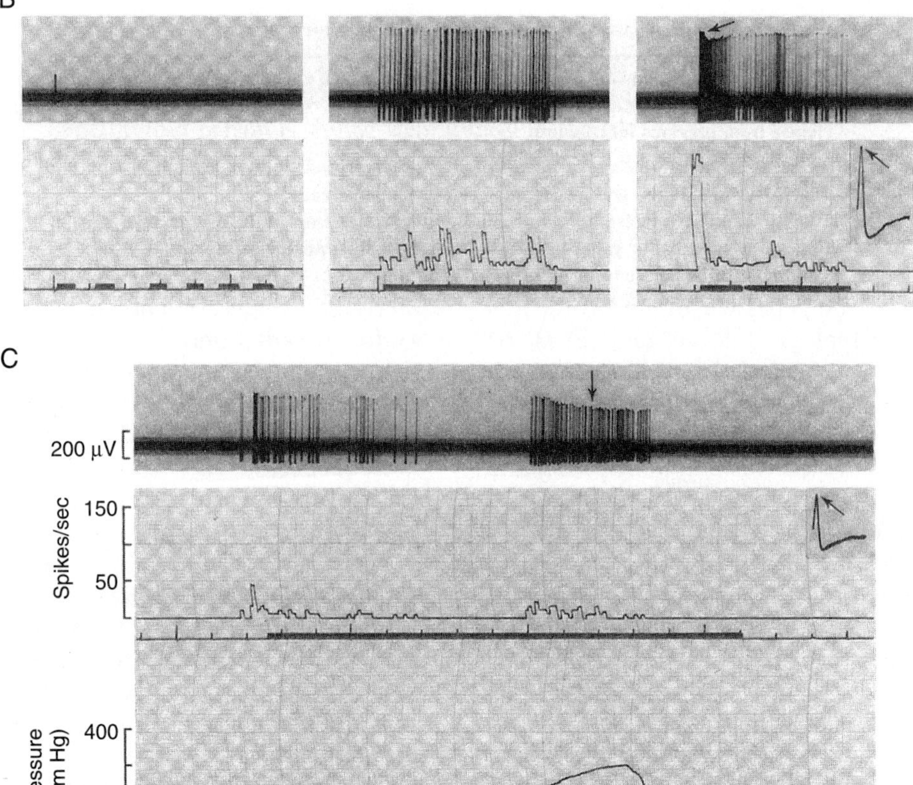

FIG. 4. A: Estradiol benzoate (EB) administered to an ovariectomized rat in increasing doses raises the probability of the lordosis response, as a function of the amount of pressure on the relevant skin area of the female rat (From Kow, L.-M., Montgomery, M. O., and Pfaff, D. W. [1979]. Triggering of lordosis reflex in female rats with somatosensory stimulation: quantitative determination of stimulus parameters. *J. Neurophysiol.* 42, 195–202.) **B, C:** Electrophysiological responses of individual primary sensory neurons in the relevant dorsal root ganglion of an anesthetized female rat to pressure stimuli on the relevant portion of the skin. Pressure intensity is shown quantitatively on the bottom trace of each figure. Sustained pressure led to action potentials in a neuron that had no spontaneous discharge. A sudden peak of pressure, as would be caused by stimulation from a male rat during mounting (**B,** *upper right*), caused a high rate of firing. (From Kow, L.-M., and Pfaff, D. W. [1979]. Responses of single units in sixth lumbar dorsal root ganglion of female rats to mechanostimulation of relevant for lordosis reflex. *J. Neurophysiol.* 42, 203–213.)

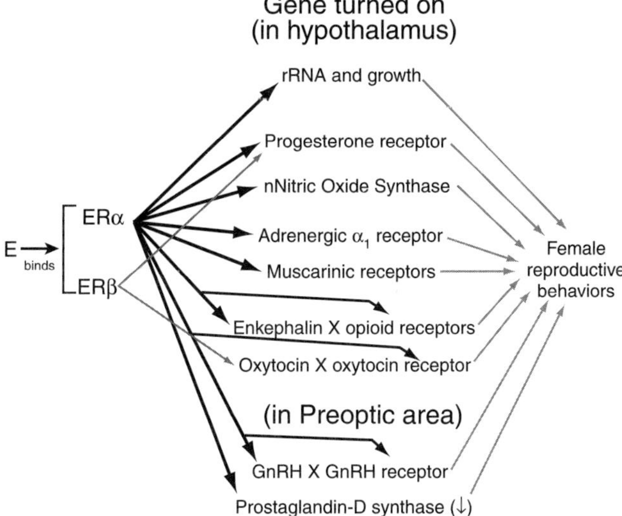

FIG. 5. Estrogens (E) working through estrogen receptor-α (ERα) or ERβ increase certain transcript levels in hypothalamic neurons (or, in the case of prostaglandin D synthase [PGDS] in the preoptic area [POA], decrease), thus facilitating female reproductive behaviors. (Updated from Pfaff, D. W. [1999]. *Drive: Neural and Molecular Mechanisms for Sexual Motivation.* MIT Press, Cambridge, MA.) Details of the inductions reviewed in *Hormones, Brain and Behavior*, Chapter 47 show how the *specificity* of genomic change according to neuroanatomical location and gender makes sense in terms of estrogenic controls over behavior. Considering all of the effects of estrogens in hypothalamus and POA, at this point they can be conceived in the modules proposed in Fig. 6.

both diacylglycerol, a PKC activator, and inositol-3-phosphate, which mobilizes intracellular calcium. This signal transduction route is predicted to lead to L-type calcium channel opening as in the heart, but this needs to be established. The induction of α_{1B} receptors by estrogen treatment is consistent with the greater electrophysiological effectiveness of noradrenergic stimulation after estrogen, but the detailed step-by-step transduction route to the channel now provides a timely subject for analysis. Because these

Modular systems downstream from hormone-facilitated transcription responsible for a mammalian social behavior: "GAPPS".

- *Growth* (rRNA, cell body, synapses).
- *Amplify* (pgst/PR → → downstream genes).
- *Prepare* (indirect behavioral means; analgesia)

  (ENK gene) and anxiolysis (OT gene).
- *Permit* (NE alpha-1b; muscarinic receptors).
- *Synchronize* (GnRH gene, GnRH rcptr gene)

FIG. 6. Functional modules through which molecular alterations caused by estrogens could facilitate female mouse courtship and mating behaviors: GAPPS. Details are given in the text and in Mong and Pfaff (728).

ventromedial hypothalamic neurons are at the top of the lordosis behavior circuit, the noradrenergic effect fosters reproductive behavior.

Muscarinic receptors responding to the neurotransmitter ACh are also expressed in ventromedial hypothalamic neurons (738). Estrogen treatment increases their activities as determined electrophysiologically. Inputs to ventromedial hypothalamic neurons come from, among other places, the lateral dorsal nucleus of the tegmentum. Neurons there are part of the ascending arousal pathways, and would signal stimulation from the male upon mounting the female. We note that apparent redundancy between ascending noradrenergic and muscarinic cholinergic afferents to the hypothalamus helps to guarantee that the system will not fail, and so exemplifies a design characteristic prominent in brain stem arousal system neurobiology. In any case, inducing muscarinic receptors increases the ventromedial hypothalamic cellular electrophysiological response to ACh. Whether the estrogen effect uses a membrane receptor, a signal transduction mechanism, or a classic genomic facilitation is not yet known. However, it is known that the enhanced ventromedial hypothalamic neuronal output primes lower pathways in the circuit for lordosis behavior (739).

Indirect Effects, from Gene Induction to Downstream Genes to Behavioral Change. Some hormone effects occur early, long before the onset of reproductive behaviors, and set the stage for later developments.

Neuronal Growth. Growth promotion by estrogens in ventromedial hypothalamic neurons, in female rats, follows from the stimulation of synthesis of ribosomal RNA, which precedes the elaboration of dendrites and synapses on ventromedial hypothalamic neurons observed after hormonal treatment. The earliest estrogen effect is the increase of transcription of ribosomal RNA (740), followed rapidly by morphological effects, including those in the nucleolus itself (741), and a striking elaboration of rough endoplasmic reticulum in the cytoplasm (742). Catherine Woolley (743) and her colleagues have shown, probably consequent to the aforementioned phenomena, a stimulatory effect of estrogen treatment on dendritic growth. In the female rat hypothalamus, Maya Frankfurt (744) and Lori Flanagan (745,746) have reported that estrogens foster dendritic growth and an increased number of synapses (747).

Therefore, in ventromedial hypothalamic cells that control lordosis behavior circuitry, estrogens apparently provide the structural basis for increased synaptic activity and, therefore, greater sex behavior–facilitating output. Although the greater signaling capacity of these ventromedial hypothalamic cells thus proposed is consistent with the actual electrophysiological activity of such cells after estrogen treatment, the causal relation of structure to function still needs direct proof.

Amplification by Progesterone. Administration of progesterone 24 or 48 hours after estrogen priming greatly amplifies the effect of estrogen on mating behavior. This effect requires the nuclear PR because it disappears after antisense DNA against PR mRNA has been administered onto ventromedial hypothalamic neurons (748–750). It also disappears in PR knockout mice (751). Because PR, itself, is a transcription factor, its induction by estrogen might imply that certain downstream genes would consequently be upregulated. With molecular probes directed to specific genes, studied primarily in female rats, several have been revealed as upregulated by progesterone: these include neuropeptide Y receptor (752), galanin (753,754), oxytocin (755), GnRH (756,757), μ opioid receptors (758), POMC (759), GAD (760), a glutamate receptor (761) and tyrosine hydroxylase (762,763). The manners in which these particular downstream genes contribute to reproductive behaviors will be exciting to explore.

Gonadotropin-Releasing Hormone. The physiological importance of estrogenic elevation of GnRH (LHRH) mRNA levels under positive feedback conditions—as well as elevation of the receptor mRNA for GnRH—would be to *synchronize* reproductive behavior with the ovulatory surge of LH. The same GnRH decapeptide that stimulates the ovulatory release of gonadotropins also facilitates mating behavior (764,765). In many small animals, synchrony of sex behavior with ovulation would be biologically adaptive because it eliminates unnecessary exposure to predation. In this respect, the behavioral effect of this neuropeptide is consonant with its peripheral physiological action.

GnRH also brings up the unusual case of an individual gene causally related to a human social behavior. During development in vertebrates ranging from fish to humans, GnRH neurons migrate from their birthplace in the olfactory placode into the brain (766). A human with damage at the Kallmann's syndrome (767) locus on the X chromosome did not fail to express the GnRH gene in the appropriate neurons. Instead, the neurons failed to migrate out of the olfactory placode (768). A single gene for the Kall protein (769,770) accounts for the deficit. It is for an extracellular matrix protein that is necessary for GnRH neuronal migration and which, in fact, decorates the migration route (771). A striking feature of the phenotype in men is important to note: they have no libido. Here is the causal route. The men have no sexual drive *because* they have little testosterone *because* they have little LH and follicle-stimulating hormone circulating from the pituitary gland, *because* no GnRH is coming down the portal circulation to the pituitary from the hypothalamus *because* there is no GnRH in the hypothalamus, *because* the GnRH neurons did not migrate during development into the brain, *because* of a mutation in the gene for the Kall protein. Therefore, we can causally connect, step by step, an individual gene to an important human social behavior, but at least six causal links are required. This causal route illustrates the complexity of gene–behavior relationships in humans.

Indirect Effects, from Gene Induction to Intermediate Behaviors. Some of the genes affected by estrogens work by altering other behaviors that then prepare the animal for the behavior in question, in this case mating.

Analgesia. The enkephalin gene is turned on rapidly in female mouse and rat hypothalamic neurons by estrogens (772–774), within approximately 30 minutes, and this is proven by in vitro transcription assays to represent a hormone-facilitated transcriptional facilitation (775). The route of action upon lordosis of the enkephalin gene product would theoretically be indirect, through other behaviors. That is, we propose that through the reduction of pain, enkephalins help allow the female to engage in mating behavior despite the mauling she receives from the male. The strong somatosensory and interoceptive stimuli that ordinarily would be treated by the female as noxious are now tolerable and allow successful mating to proceed (776). Hypothetically, the ability of estrogens to turn on genes for opioid receptors as well (777) has the potential of multiplying the hormone's effect on mating behavior sequences. Specificity among opioid receptors, as well as neuroanatomical site specificity, is observed in this course of action (778–781). The indirect route of action of this multiplicative set of gene inductions is likely to allow the female to participate in sex behavior sequences.

Anxiety Reduction. The oxytocin gene and the gene for its receptor are both expressed by hypothalamic neurons at higher levels in the presence of estrogens. The indirect route of action of this multiplicative set of gene inductions on mating behavior is likely through a behavioral link: anxiety reduction allows courtship and mating. This proposal is consistent with previous formulations; oxytocin has been conceived as protecting instinctive behaviors connected with reproduction, maternity, and other social behaviors from the disruptive effects of stress (782). Indeed, oxytocin has an anxiolytic action in the presence of estrogens (which presumably elevate the oxytocin receptor gene product) (783).

Social Recognition and Aggression. The induction of the oxytocin gene by estrogens is an ERβ-dependent (784), behaviorally significant (785) phenomenon. This makes sense because only ERβ gene expression is found in oxytocinergic cells (734). In turn, oxytocinergic projections to the amygdala are thought to be important for social recognition in mice, which helps to prevent aggression (786–788). Thus, the lack of social recognition by ERβ knockout mice (789) could explain the hyperaggressiveness displayed by ERβ knockout male mice (790). Altogether, these data invoke the idea of a four-gene micronet important for social behaviors.

Orchestrated Genomic Responses to Hormones (GAPPS). From this series of individual gene inductions by estrogens acting in the basal forebrain, and the preceding recounting of downstream genes and their physiological routes of action, there emerges a theoretical molecular "formula" that appears to account for some of the causal relations between sex hormones and female sex behaviors. First, there is a hormone-dependent *growth* response, which permits hormone-facilitated, behavior-directing hypothalamic neurons a greater range of input/output connections and, thus, physiological power. Secondly, progesterone can *Amplify* the estrogen effect, in part through the downstream genes listed previously. Then, through indirect behavioral means—the reduction of anxiety and a partial analgesia—the female as an organism is *Prepared* for engaging in reproductive behavior sequences. Here, the genes for oxytocin (and its receptor) as well as for the opioid peptide enkephalin (and its receptors) are important. Next, neurotransmitter receptor induction by estradiol *Permits* the neural circuit for lordosis behavior to be activated. The noradrenergic α_1 receptor and the muscarinic ACh receptors are crucial here, in the VMN. Finally, induction of the decapeptide that triggers ovulation, GnRH, as well as its cognate receptor acts to *Synchronize* mating behavior with ovulation in a biologically adaptive fashion.

This theoretical formulation is intended to tie together disparate results from several transcriptional systems into one set of modules. Even so, the genomic mechanisms uncovered so far probably represent only a subset of the full range of neurochemical steps underlying sex behaviors.

A Note on Technology. New technological approaches have been used to extend the demonstration of hormone effects on the enkephalin gene. Viral vectors have been constructed that are sufficient for expression of foreign genes in neuronal tissue (791,792). In this application, the rat enkephalin gene promoter was used to drive the reporter gene β-galactosidase in an effort to see if estrogens operating on the promoter could increase expression of the reporter gene when the viral vector had been transfected to hypothalamic neurons (793). Indeed, a significant increase in the number of cells expressing the reporter gene was seen after estrogen administration. This fact, plus the ability of this viral vector technique to reveal anatomically correct expression of preproenkephalin (794), provides a valid confirmation that an estrogen-induced increase is a transcriptional facilitation.

In another set of technically interesting investigations we collaborated with David Borsook and Stephen Hyman to use transgenic mice bearing a reporter gene driven by the enkephalin promoter (i.e., β-galactosidase) (795,796). Here, hormone effects on the enkephalin promoter were revealed in females but not males and depended on intermediate levels of stress. Thus, the hormone effect was extended to mice, but significant interactions with stress levels were also indicated (795,796).

The *second question* posed earlier (at the beginning of Molecular Mechanisms, from Hormone to Gene to Behavior) has to do with adding efficiently to the list of genes turned on by estrogens in the brain. Our approach has been to do four microarray studies under different conditions, with transcripts from thousands of genes semiquantitatively examined, to break open new questions about steroid hormone actions in the brain. One of the first transcripts discovered with microarrays, that coding for prostaglandin D synthase, is highlighted later in Hormone-Dependent Genes Influencing Arousal.

From Genes to Behavior

Here we update molecular genetic and behavior information introduced in Chapter 47 of *Hormones, Brain and Behavior* (2002). It is clear that connecting individual genes to specific mammalian behaviors will be a difficult task (797). Three sets of reasons can be illustrated. First, learning from basic genetics, the pleiotropy of genetic action indicates that many genes can have several effects, especially when different portions of the life cycle are considered. Second, consider the task of the 21st century neurobiologist. Many wish to study neuronal and genetic mechanisms underlying integration in the adult brain. However, gene actions important for behavior include a variety of especially obvious indirect routes of action, with several types of developmental effects. In addition, during adulthood, genes expressed and acting outside the brain may be important for certain behaviors. As well, discerning genetic influences on elementary sensory and motor capacities must be distinguished from genetic influences on truly integrative mechanisms.

A third way of expressing the potential difficulty of interpreting neural and behavioral data in transgenic animals and gene knockout mice is presented now. Most of us working with such animals try to imagine genetic manipulations as punctate in space, singular in time, and specific on the genome. The actual situation is more complicated than that. Genetic alterations are manifested in parts of the brain that were not intended by the experimenter, and because they are permanent throughout the life cycle, allow time for complex feedbacks and compensations. Finally, damage to one gene can have implications for the expression of a different gene. For all of these reasons, genes to be studied and behaviors to be analyzed must be selected with care.

Nevertheless, in some circumstances an investigator "gets lucky" and a gene–behavior relationship is revealed. For example, after the discovery that GnRH neurons are born in the olfactory placode and migrate during development into the basal forebrain, a finding reliable across species from fish to human beings (798–800), it occurred to us that the migration phenomenon might be related to Kallmann's syndrome (hypogonadotropic hypogonadism frequently coupled with anosmia). This human syndrome provides a superb example of how subtle the causal routes can be, between an individual gene and a mammalian behavior—in this case, a human social behavior. Men with Kallmann's syndrome demonstrate an absence of libido and, generally, a disinterest in the opposite sex. Genetic influence on X-linked Kallmann's syndrome can be understood now because genetic damage at Xp22.3 has been correlated with the failure of GnRH neuronal migration (799). In an individual with such genetic damage, the GnRH neurons were dammed up in the olfactory apparatus and never reached the brain. In turn, the disorder of neuronal migration is understood by the fact that damage on chromosome Xp22.3 disrupts a specific gene, which codes for a cell surface protein. This cell surface protein is present (799a) during migration of GnRH neurons into the brain.

Putting these and related facts in a logical order leads to a clear example of complex participation by an individual gene in a human social behavior. The formulation is as follows. In men with X-linked Kallmann's syndrome, behavioral libido is reduced because of low testosterone levels. These are low because of reduced gonadotropins output from the pituitary. In turn, gonadotrophins are low because no GnRH is coming from the brain through the portal circulation to the pituitary. This, in turn, is because there are no GnRH cells in the basal forebrain, because of the failure of GnRH neuronal migration. The migration failed because of the absence of the Kallmann protein that is produced from a gene at Xp22.3. In summary, these data prove a genetic influence on an important human social behavior, but at the same time illustrate the complicated, multistep nature of that influence.

The difficulties in discovering mechanisms of genetic contributions to behavior are ameliorated in the case of many hormone-dependent behaviors because of several strategic advantages. Systematic experimental approaches to well-chosen hormone-dependent behaviors have proceeded quickly for the following reasons. They are triggered by simple stimuli and their responses have a simple topography. With steroid-dependent behaviors, we have all the tools of experimental endocrinology at our command, including a wealth of information about steroid chemistry and steroid pharmacology. Most important of all for studying the functional genomics of behavior is the fact that the types of nuclear receptors studied in this laboratory for a long time have turned out to be ligand-activated transcription factors.

Additional advantages include the fact that we can evoke many hormone-dependent responses in the laboratory in their natural form. Moreover, many of these responses are conserved across species. Furthermore, several hormone-dependent responses are associated with interesting behavioral *chains* and *choices*. For example, leading to female reproductive behaviors are series of hormone-dependent communications and locomotion responses, which act as "hormone dependent behavioral funnels" (725). These serve to bring reproductively competent conspecifics, and only they, into proximity with each other. These are important to study because, for example, failure to exhibit approach responses involving locomotion toward the potential mate can account for the absence of reproductive behavior during severe environmental cold (801). When studied in a seminatural environment, proceptive behaviors of female mice can be quite dramatic, sometimes comprising rapid, systematic locomotor responses between nests across a large open region (802,802a). Among the behavioral *choices* available to a female mammal would be those between reproductive responses and aggression. This is especially impressive in female hamsters (803–806). In all of these respects, hormone-dependent behaviors can lead the way in connecting genetic influences to a thoroughgoing understanding of animal and human social behaviors.

In summary, these patterns of gene expression with gene products serving individual biological roles, including the "formula" for arousal plus analgesia (807), have the combined effects of integrating reproductive behaviors with the endocrine system and with the environment.

Specificity in the Actions of Individual Genes on Aggressive and Sociosexual Behaviors. The functional genomics of female mouse sociosexual behaviors have been easy enough to study that a large body of data has led to several "lessons," numbered and italicized in the following:

1. If the gene that codes for the classic ER, ERα, is knocked out, the female mouse's behavior is masculinized. She behaves more like a male and,

in fact, is treated like a male by other mice (808). In such an experiment, the gene's product is absent throughout the body and throughout the life of the mouse. In dramatic contrast, if the ERα function is interrupted during the neonatal critical period by the microinjection of an antisense DNA oligomer directed against the translation start site of ERα mRNA, only in the hypothalamus, then the masculinizing influence of testosterone during the neonatal critical period is opposed (809). Reasoning from the comparisons between these two sets of results, we infer that *the effect of a given gene on a given type of behavior depends on where in the brain and exactly when that gene is expressed.*

2. Deleting the ERα gene permanently in a male mouse abolishes aggression (810). However, in a female mouse, the same type of mutation increases aggression (811). The results for the two sexes are opposite. Now consider deletion of the gene coding for ERβ. In males, such a gene knockout can increase aggression (812). However, in female mice, knocking out ERβ reduces testosterone-facilitated aggression (813). From these two sets of contrasts between male and female mice, we infer that *the effect of a given gene on aggression depends on the gender in which that gene is expressed.*

3. With respect to aggressive behaviors, the magnitude of the phenotype in the ERβ knockout male mouse declines with age (812). We do not know why. Further, when measuring locomotor behaviors in ERα knockout female mice compared with their littermate wild-type controls, we see that the phenotype (failure to respond to estrogens with increased locomotion) gets stronger with age (814). We also do not know the mechanism of that age-related change. Nevertheless, it is clear that *the effect of a specific gene on aggressive (or locomotor) behaviors can depend on the age of the animals at which the behavioral assay is conducted.*

4. During the resident–intruder paradigm for testing aggression, ERα knockout female mice display high levels of increased aggression toward female intruders. Their aggression persists well beyond that shown by their wild-type littermate controls, regardless of whether the intruder is a female mouse treated with estrogens and progestins. In contrast, if the intruder is an olfactory bulbectomized male mouse, the ERα knockout female's aggression is at a very low level, not any different from that of the wild-type female mouse. This comparison shows that *the effect of a specific gene on aggressive behavior can depend on the nature of the opponent.*

5. Female mice tested for their vigor in nest defense, during preliminary experiments (815) showed a marked effect of the null deletion of ERβ. At the beginning of each test, ERβ knockouts showed a much greater number of attacks than wild-type controls. On the other hand, when tested for testosterone-facilitated aggression (813) ERβ knockouts responded with significantly less frequent aggression for a given dose of testosterone. This comparison suggests that *the effect of a specific gene on aggressive behavior can depend on the type of aggression tested.*

In all of these five aspects, and probably in other respects waiting to be discovered, we see specificity in hormone–gene–behavior relations. An even more important property of the data was brought out in Chapter 47 of *Hormones, Brain and Behavior* (2002), which can be expressed here in capsule form as follows. The history of functional genomics was dominated by George Beadle's and Edward Tatum's discoveries in *Neurospora*, which resulted in the "one gene–one enzyme" concept (816). In dramatic contrast, in our field the data (Fig. 7) show that the proper concepts are more complicated than that. We see, in both male and female mice, that *patterns of gene expression* are required for *patterns of sociosexual behaviors.*

Rapid Actions of Steroid Sex Hormones. All of the emphasis in discussing steroid sex hormone actions on neurons and reproductive behaviors up to this point has been on the transcriptional mechanisms. However, rapid membrane-limited actions have also been proven to occur [reviewed in Kelly and Ronnekleiv (817)]. In molecular biological experiments with a neuroblastoma cell line, they have proven effective in facilitating later, transcriptional actions (Fig. 8A), and involve more than one signal transduction pathway (Fig. 8B). Applied to hypothalamic neurons, they are effective in facilitating lordosis behavior (Fig. 8C), using mechanisms which include both protein kinase A (PKA) and PKC signaling (Fig. 8D).

SEXUAL MOTIVATION AND COURTSHIP BEHAVIORS

The status of sex behavior as a biologically regulated motivated behavior has been argued rigorously (818) [summarized in Bodnar et al. (819)]. If a female laboratory animal is tested for mating behaviors in a strictly controlled environment, in the absence of estrogens and progesterone, and she demonstrates no mating behaviors at all—and then, an adequate schedule of estrogens and progesterone is administered (with all other conditions held absolutely constant), and she shows mating behaviors—a variable is required in the "equation" for this behavior to account for the appearance of the behavior. This variable is named *motivation.* In our case, it is a hormone-sensitive motivational change to account for the appearance of this biologically regulated type of behavior.

Some of the most striking natural examples of female laboratory animals motivated to seek males are illustrated in their courtship behaviors. Well known

Comparisons of ERα and ERβ functions in CNS				
Necessary?	Sufficient?		Assays in the ♀	Assays in the ♂
ER α and β	Neither α nor β	"ER α and β must synergize"	Social recognition	none
Neither α nor β	Either α or β	"ER α and β can subst. for each other"	E induction of PR (ICC) E reduction of ER (ICC)	E induction of PR (ICC) E reduction of ER (ICC) simple mounting
α, β each for its own	α, β each for its own	"ER α vs β contribs. not related"	Maternal behavior (α) Suppression of aggression (α) Reduction of food intake (α) Reduction of anxiety (α)	Ejaculation (α) Anxiety response (α)
ER α absent β	ER α absent β	"ER β can reduce α effect"	Lordosis behavior	Aggression
Optimal α/β balance	Optimal α/β balance	"ER α vs β always opposed"	none	none

FIG. 7. Contributions of two different estrogen receptor (ER) genes to sociosexual behaviors and other functions. A large number of behavioral measurements have been carried out with ERα knockouts, ERβ knockouts, and their respective wild types, both in genetic females and genetic males (right two columns). Theoretical scenarios as to the functional relations between ERα and ERβ gene products are portrayed in the left three columns. Different combinations of ERα and ERβ genes participate in different behaviors. In females, the two gene products must synergize for social recognition; oppose each other for lordosis control; and are unrelated to each other for maternal behavior. In males, the two gene products can substitute for each other with simple mounting; are opposed to each other for regulation of aggression; and are unrelated with respect to ejaculation. As opposed to the classic Beadle and Tatum formulation (one gene–one enzyme) from their work with the fungus *Neurospora*, for the mammalian central nervous system, different *patterns* of gene expression are required for different *patterns* of behavior. (From Pfaff, D. W., et al. [2002]. Chapter 47. In *Hormones, Brain and Behavior* [D. W. Pfaff, A. Arnold, A. Etgen, S. Fahrbach, and R. Rubin, Eds.]. Academic Press, San Diego.)

is the "hopping and darting" seen during experiments with sexually motivated female rats. However, for the genetics, experiments with mice are becoming increasingly popular. Informal reports of laboratory work with mice held that female mice do not show courtship behaviors. An alternative point of view was that sufficiently natural environments for eliciting these behaviors had not yet been achieved. Therefore, we constructed "seminatural environments" and videotaped female mouse behavior during mating encounters. Under these conditions, female mice showed at least two kinds of striking courtship behaviors: (a) direct approaches to the stud male, and (b) dramatic withdrawal and approach series of responses reminiscent of the pacing behaviors demonstrated by Erskine and others [reviewed in Blaustein and Erskine (820)] in female rats. In these seminatural environments (802), a female mouse, having mated in a nest box, would bolt from that box and run at unprecedented speeds across the environment. Later she would return, sometimes simply retracing her steps and other times by a completely different route (Fig. 9A). These behaviors appear only if the female has estrogen and progesterone treatment. Two conclusions emerge. First, female mice do have courtship behaviors. Second, under some circumstances a biologically regulated behavior may occur only if it is examined under conditions reasonably close to those for which the mouse genome evolved.

All of the female-typical courtship behaviors mentioned in the following sections involve locomotion. Knowing that estrogens foster locomotion in these animals, at least under safe, unthreatening conditions, we examined the potential participation of the two ER genes (821) in locomotion (Fig. 9B).

Preoptic Neurons as Nodes for Reproductive Neuroendocrinology and Sex and Courtship Behaviors

Preoptic Neurons as Targets of Estrogen Action

Interconnected webs of estrogen-sensitive neuron groups have been implicated in the regulation of

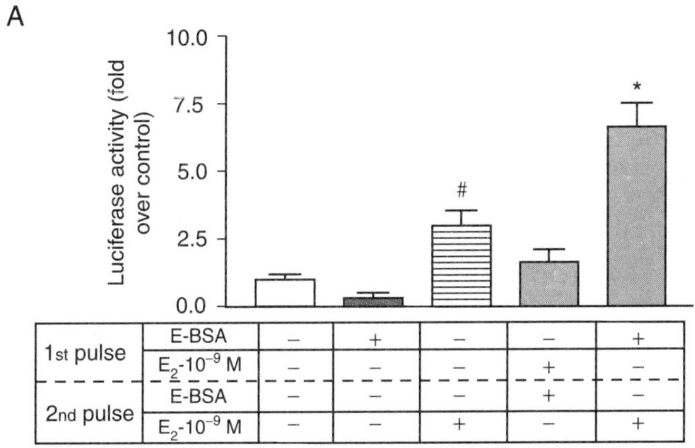

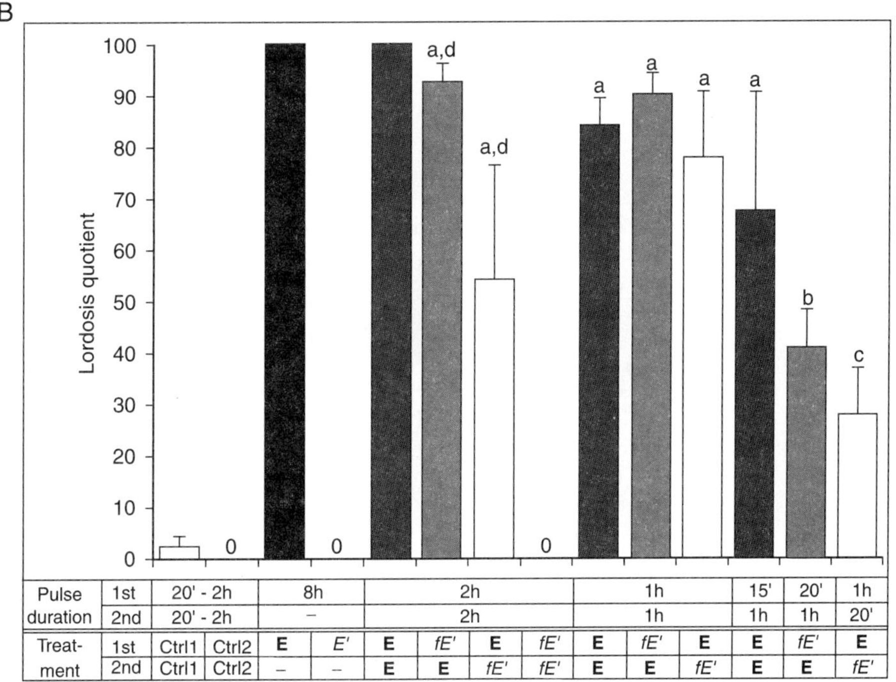

FIG. 8. A: Molecular data. In transient transfection experiments with neuroblastoma cells, using a two-pulse hormone administration protocol, the first pulse comprising membrane-limited estradiol can potentiate the transcriptional facilitation of the second pulse of pure estradiol. (Data from Vasudevan, N., Kow, L.-M., and Pfaff, D. W. [2001]. Rapid membrane estrogenic effects required for full expression of slower genomic actions in a nerve cell line. *Proc. Natl. Acad. Sci. U S A* 98, 12267–12271.) **B:** Behavioral data: estrogens applied directly to the ventromedial hypothalamus of ovariectomized female rats. Using a two-pulse hormone administration protocol, we showed that a first pulse of membrane-limited estradiol could potentiate lordosis behavior after a second pulse of pure estradiol. (Data from Kow, L.-M., et al. [2004]. *Proc. Natl. Acad. Sci. U S A.*)

Continued

multiple facets of reproductive behavior in the rat, based on results from morphological and tract-tracing studies as well as immunohistochemistry and in situ hybridization for ER molecules (822), and the POA is the nodal point in this network. In this network, estrogen allocates neural signals to particular routes relevant for estrogen-dependent functions; in the absence of estrogen, neural signals flow otherwise. The allocation of neural signals can be accomplished by either facilitation or inhibition of neural transmission in individual neural connections by estrogen. Electrophysiological results support such a theory by showing site- and sex-specific regulation of neuronal excitability by estrogen (823). As in other structures in the basal forebrain and hypothalamus, the POA is heterogeneous in neuronal composition, with neurons with different projection targets and sensitivities to estrogen (823). Projection targets of POA neurons include structures relevant to different components of sexual behavior (823,824).

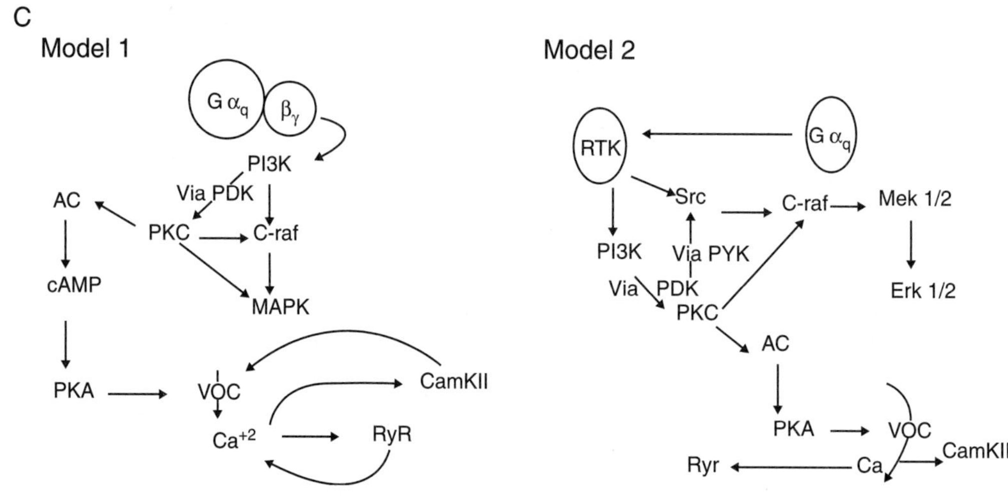

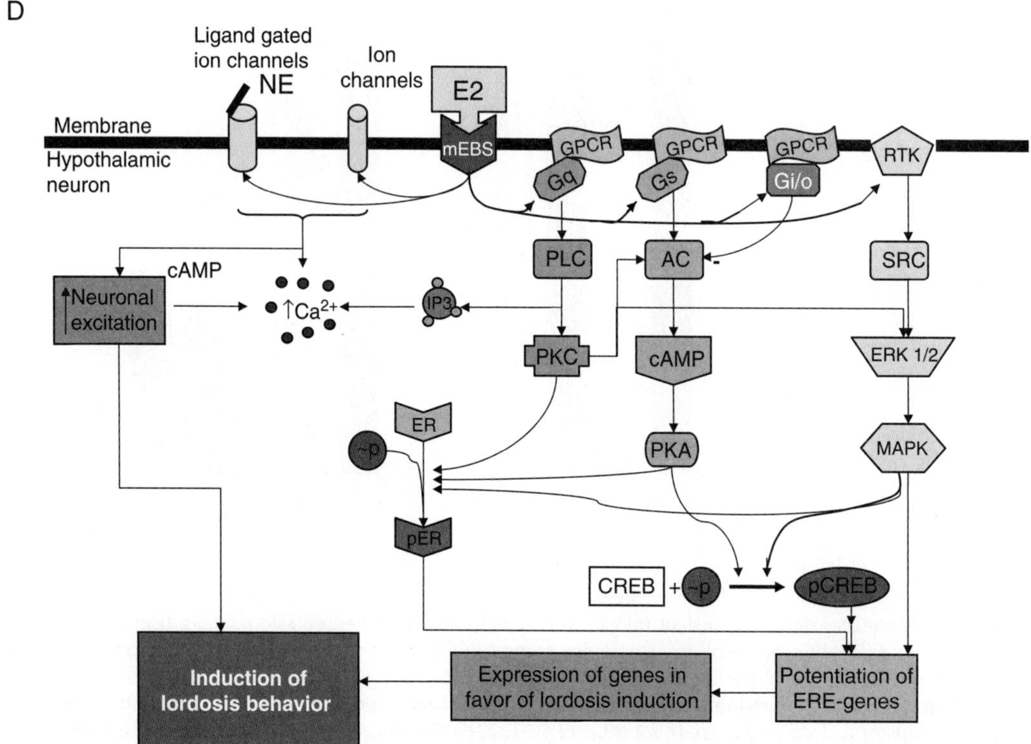

FIG. 8—Cont'd. C: Theory applied to molecular data. From all of the transient transfection experiments we have done so far, and from the literature on signal transduction alterations after rapid estrogen actions, we derive two models that both seem to fit the facts to date. (From Vasudevan, N., Kow, L.-M., and Pfaff, D. [2005]. Integration of steroid hormone initiated membrane action to genomic function in the brain. *Steroids* 70, 388–396.) **D:** Theory applied to behavioral data: model of potential signaling pathway for estrogen's membrane actions that facilitate its lordosis-inducing genomic actions. The model begins with the fact that brief applications of a membrane-limited estrogen in the hypothalamus in vivo can facilitate estrogen induction of lordosis behavior (**B**). Also, during electrophysiological studies of ventromedial hypothalamic neurons, which are critical for the induction of lordosis behavior, estrogenic membrane actions were shown to potentiate (a) the excitatory effects of histamine (activator for G-protein–coupled receptor [GPCR]-G_q and GPCR-G_s) and (b) N-methyl-D-aspartate (NMDA; ligand-gated ion channels), as well as (c) enhancing an inward Na^+ current; estrogen also (d) rapidly attenuated inhibition by histamine (GPCR-$G_{i/o}$ agonist) and (e) reduced outward K^+ currents. These effects would lead both to increases in intracellular [Ca^{2+}] and to the excitation of hypothalamic neurons. All other effects of estrogen's acute membrane actions shown here, such as increase in intracellular Ca^{+2}, activations of signaling pathways—including but not limited to cyclic adenosine monophosphate (cAMP)/protein kinase A (PKA)/cAMP response element–binding protein (CREB), phospholipase C (PLC)/protein kinase C (PKC), and steroid receptor coactivator (SRC)/mitogen-activated protein kinase (MAPK)—have been well documented in the hormone action literature (Kow, L.-M., et al., unpublished data). E2, estradiol; mEBS, membrane estrogen-binding site; RTK, receptor tyrosine kinase.

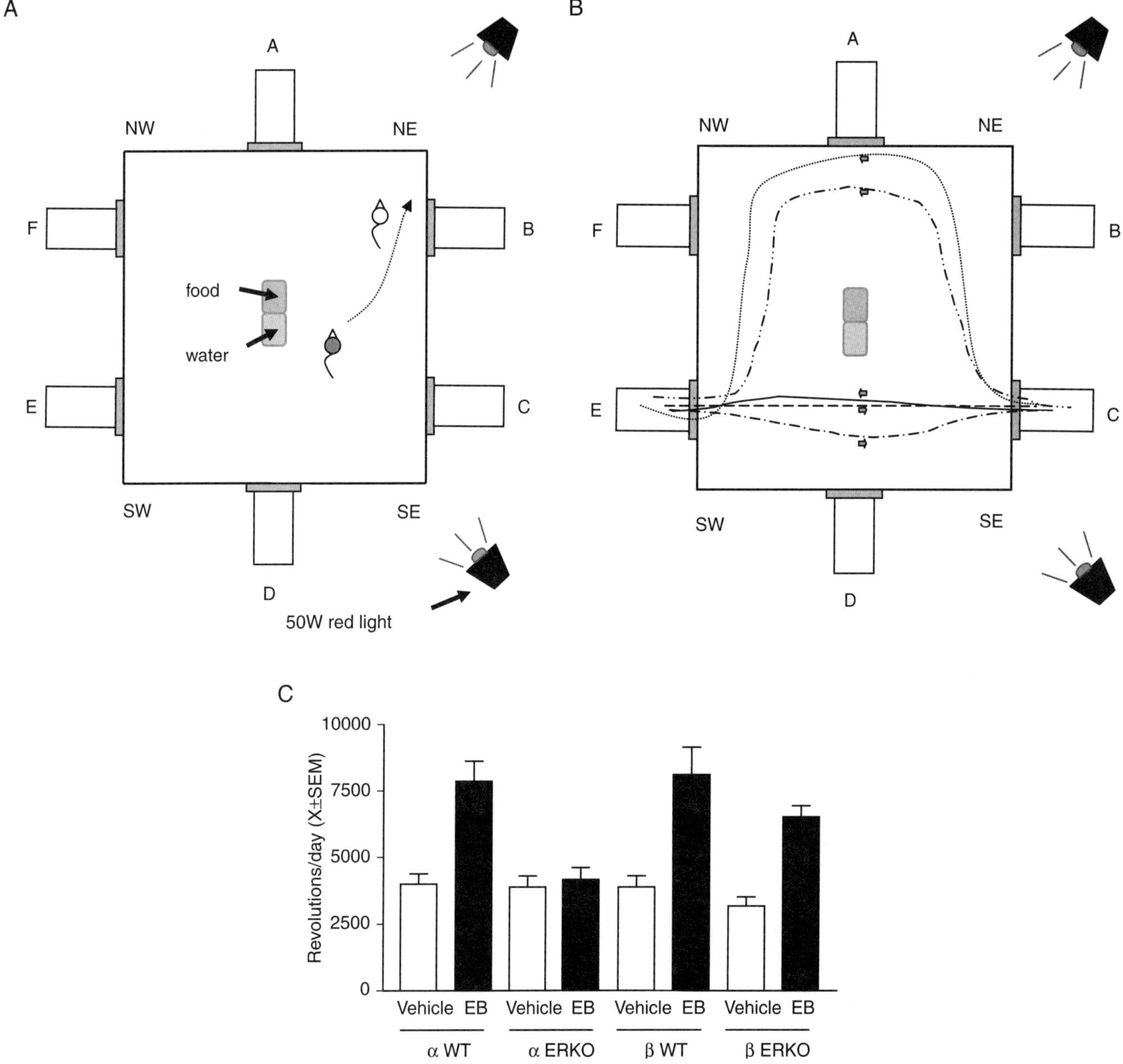

FIG. 9. Tracings of courtship and approach responses by female mice in a seminatural environment (From Garey, J., Kow, L.-M., Huynh, W., Ogawa, S., and Pfaff, D. W. [2002]. Temporal and spatial quantitation of nesting and mating behaviors among mice housed in a semi-natural environment. *Horm. Behav.* 42, 294–306.) **A:** A direct approach by the female (*gray*) to the male (*open, white*), in which she locomotes immediately to sit by his side and just in front of him. **B:** A more complex pattern of courtship and pacing behaviors in which the female mouse approaches the male to mate in a nest box, and then, after each mount by the male, does complex "round trips" to and from that nest box, thus to be mounted again there. Her exits from the nest box after mounts are characterized by extremely high velocities. **C:** Estrogenic effects on running-wheel activity in female mice depend on the gene for estrogen receptor-α, not the gene for estrogen receptor-β. (From Ogawa, S., Chan, J., Gustafsson, J. A., Korach, K. S., and Pfaff, D. W. [2003]. Estrogen increases locomotor activity in mice through estrogen receptor alpha: specificity for the type of activity. *Endocrinology* 144, 230–239.)

Neurons with ERα, which is irreplaceably involved in the accomplishment of estrogen-dependent sexual differentiation of the developing brain (826) and in the regulation of reproductive endocrinology and behavior in sexually mature mammals (827), or those with recently discovered ERβ, which provides alternate route for certain important actions of estrogen (828–830), are intermingled in the POA. Studies in ER knockout mice suggest that several aspects of rodent sexual behavior depend on estrogen not only in females but males (827). Although the mechanism of estrogen action awaits further scrutiny, the VMN has been considered as the major site of estrogen action to facilitate the lordosis reflex, the major receptive component in female rat sexual behavior (831). Although the VMN is the most effective site for the estrogen priming of sexual receptivity, the POA is another structures from which local implants of estrogen induce receptivity in ovariectomized female hamster (832). Electrical stimulation of the POA (833,834), as well as the ventral tegmental area (VTA) (835), disrupts the lordosis reflex without diminishing proceptivity. The neural substrate for this effect is a direct excitatory innervation of the VTA by a subset of POA neurons, which decrease their excitability by the actions of estrogen (836,837). On the other hand, an excitotoxic lesion of the POA enhances lordosis reflex and diminishes proceptivity (833). Thus, the POA appears to contain neural substrates that exert diametric effects on the receptive and proceptive components of female rat sexual behavior. Recordings of neuronal activity in the POA implicated different subsets of POA neurons in the regulation of individual behavioral components (838). Similarly, facets of male rat sexual behavior, such as mounting (839), sexual preference, and motivation (840,841), are regulated separately by different subsets of POA neurons under the influence of estrogen (842) (Fig. 10).

Sex Differences in the Preoptic Area

The POA is among the brain structures most closely studied for sexual difference in the rat and other

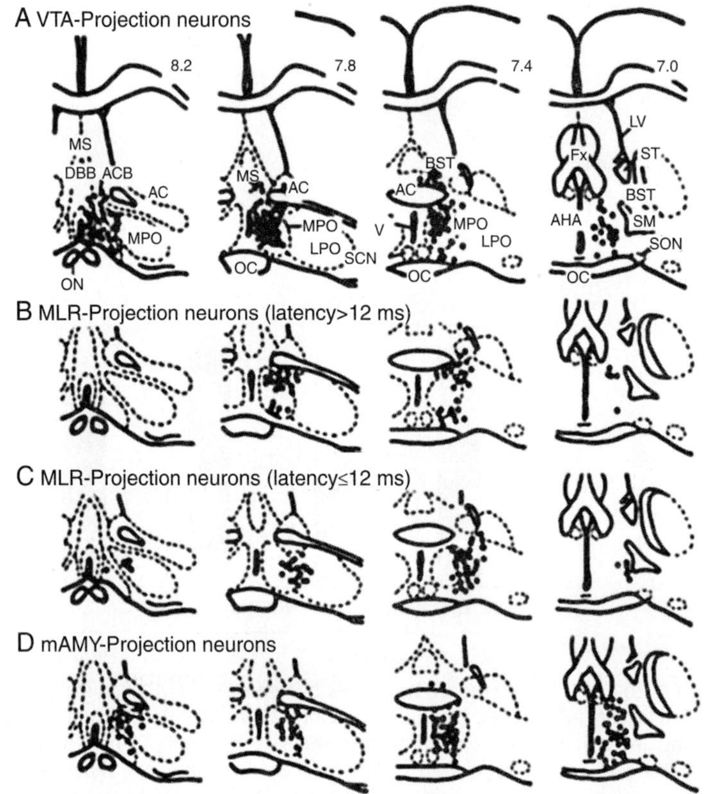

FIG. 10. Results from the laboratory of Yasuo Sakuma (Tokyo), as well as other laboratories, have shown that neurons in different parts of the female preoptic area likely make different types of contributions to female reproductive physiology. Shown here are the neuroanatomical distributions of neurons in the preoptic region of the ovariectomized female rats that were antidromically activated from the ventral tegmental area (VTA) (**A**), midbrain locomotor region (MLR) (**B, C**), or the medial amygdala (mAMY) (**D**). Frontal sections of the brain are 0.4 mm apart. The numbers in the *top panels* indicate distances from the interaural line in millimeters. Neurons projecting to the MLR are of likely importance for the locomotor aspects of courtship behaviors. Estrogen effects on the excitability of cell bodies of neurons shown in **B** are opposite to those on neurons shown in **C**—that is, estrogens inhibited neurons in **A, B,** and **D,** but excited those in **C**. AC, anterior commissure; ACB, nucleus accumbens; AHA, anterior hypothalamic area; BST, bed nucleus of the stria terminalis; DBB, diagonal band of Broca; Fx, fornix; LPO, lateral preoptic area; LV, lateral ventricle; MPO, medial preoptic area; MS, medial septum; OC, optic chiasm; ON, optic nerve; SCN, suprachiasmatic nucleus; SM, stria medullaris; SO, supraoptic nucleus; ST, stria terminalis; V, third ventricle. (From Takeo, T., and Sakuma, Y. [1995]. Diametrically opposite effects of estrogen on the excitability of female rat medial and lateral preoptic neurons with axons to the midbrain locomotor region. *Neurosci. Res.* 22, 73–80.)

mammals, including humans (843,844). Female and male POAs differ in synaptic organization (845,846), dendritic branching patterns (847), and the size or the number of neurons they contain (848). A portion of the POA that contains a larger number of neurons and synaptic contacts in the male than in the female has therefore been designated as the sexually dimorphic nucleus of the POA (SDN-POA) (849). On the other hand, a larger number of ERβ-expressing neurons are found in females than in males in the anteroventral periventricular nucleus, which occupies a more rostral portion of the POA (829).

The morphometric sex difference is the result of the developmental effect of testosterone, which organizes the brain to develop toward the male phenotype (848,849). Therefore, sexual differentiation of POA morphology as well as various functions can be manipulated in the rat by castration or testosterone injection during the neonatal period. The female-type brain develops in the absence of estradiol because estradiol-binding protein in the fetal and neonatal blood does not allow estradiol to penetrate the blood–brain barrier. In males, testosterone enters the brain and is aromatized locally to estrogen by enzymatic action (846). The sex difference can be reversed experimentally by the castration of male rats or the administration of testosterone to female rats during the neonatal period. Administration of estrogen in large amounts or nonsteroidal estrogenic compounds also masculinizes female rat brain, presumably by saturating or evading estrogen-binding protein (850).

The SDN-POA in the male rat has a volume five times larger compared with that of the adult female (849). Earlier lesion experiments showed that number of mounts, intromissions, and ejaculations endures lesions confined to the SDN-POA, whereas these behaviors diminished substantially after small electrolytic lesions in the dorsal POA (851,852). In the ferret, the SDN-POA has been associated not with the implementation but with motivational or preferential aspects of male sexual behavior (851,853). Indeed, in the ram with innately different aromatase mRNA levels, which is characteristic of this species, the mRNA levels correlated significantly and positively with the volume of the SDN-POA and preference for a female over a male (854). In the female rat given testosterone during neonatal period, the positive correlation with partner preference toward a female and the volume of the SDN-POA is maintained when the females are both sexually experienced and in the presence of circulating testosterone (854,855). Interestingly, not only aromatase inhibitors, but post-weaning social isolation of gonadally intact male rats reduces the volume or neuronal size in the SDN-POA and the medial amygdala and leads to deficits in adult sexual behavior, presumably through lower testosterone levels caused by isolation (856). These results suggest that sexual experience interacts with steroid exposure to shape partner preference (857). The role of social experience cannot be overlooked as a factor in the sexual differentiation of the brain and the display of sexual behaviors.

Sexual Partner Preferences

Odor-dependent sexual partner preference can be determined by the time spent sniffing odors of conspecific potential partners. When estrous female rats are allowed to make alternative choices between odors of sexually active and inactive males in an apparatus that restricts physical contacts or copulation, females prefer sexually active males to inactive ones (858,859). This bias, together with the total time spent for sniffing, diminishes after lesions of the POA, showing loss of preference and sexual motivation as well (Xiao, Kondo and Sakuma, in press). Different from the males, preference in estrous females does not depend on the SDN-POA: small lesions confined to the SDN-POA region cause no change in female-typical partner preference or proceptive behavior (860). In a striking contrast, a bilateral lesion restricted to the dorsal portion of the medial amygdala, which presumably disrupts the ascending neural pathway for the perception of odor, eliminates preference of estrous female rats for sexually active males without affecting the time spent sniffing: they apparently cannot discriminate odor of sexually active males from those of sexually inactive males, but their sexual motivation persists (861).

Sexual partner preference in the rat is sex specific and depends on sex hormones, but in a manner different from receptivity in females or territorial aggression in males. Male rats castrated as adults, when treated with a combination of estrogen and progesterone, prefer conspecific males to estrous females, whereas ovariectomized, testosterone-treated females maintained female-typical preference toward sexually active males (862). Testosterone might have been aromatized to estrogen to induce female preference toward males. However, conventional understanding of brain sex differentiation does not explain induction of female-typical preference in castrated males by sex hormones.

Courtship and Sexual Behaviors of Female Rats

Increased locomotor activity in estrous female rats (863–865) increases their chances of running into males (866). When encountering a male, the female adopts a characteristic gait, a hopping-darting sequence immediately in front of the male, then abruptly halts and usually assumes a crouching posture. This presents her hindquarters directly toward the male, soliciting him to pursue and subsequently to mount her from the back. At this time or between each hopping-darting sequence, the highly receptive female often exhibits ear-wiggling, a rapid oscillatory movement of her head. These movements comprise solicitations because they usually evoke mounting by her male partner (867).

The lordosis reflex, a postural reflex with a dorsiflexion of the vertebral column, represents the consummatory phase of female reproductive behavior in the rat. Palpation of the flank of female rats, followed immediately by pressure stimuli on the perineum applied by the male partner, elicits this reflex. As a result of the spinal dorsiflexion and extension of the legs, the female elevates her head and rump, and reveals her perineal region. This posture enables penile insertion by the male and insemination to occur. In the laboratory, similar cutaneous stimuli applied manually by the experimenter elicit the lordosis reflex in female rats in estrus. There are no other sensory requirements for the reflex, such as visual, olfactory, or acoustic inputs, although some of them play important roles in more natural settings to initiate the proceptive components of sexual interaction (868).

Hormonal Regulation of Behavior, and Preoptic Neurons

Building on the introduction to this chapter (see Background), and thinking especially about temporal parameters, we note that the lordosis reflex can be reinstated within 12 to 48 hours in ovariectomized female rats after estrogen treatment (869). Systemic progesterone supplementation at this time induces solicitatory behavior and potentates the reflex approximately 6 hours after the injection (870). Estrogen in a large amount, or other compounds such as blockers of neural transmission or certain hypothalamic peptides, can replace the action of progesterone (870), but the latency for progesterone action can be shortened to 10 to 60 minutes by applying it intravenously or implanting it directly into the brain (871); the time needed for estrogen action does not depend on the route of administration (872).

Both receptor-mediated genomic activation and direct membrane effects are implicated in the actions of estrogen and progesterone in inducing sexual behavior. The long latency in promoting the lordosis reflex implicates genomic activation by estrogen. Indeed, the effect can be blocked by inhibitors of protein synthesis (872). Estrogen has been shown to induce molecules that modify neuronal excitability, including Na^+-K^+ adenosine triphosphatase in the POA (873). Other estrogen-inducible molecules include ion channel proteins (874) and enzymes that modify the kinetic properties of the existing channel (875–877). Estrogen can also cause rapid changes in neuronal excitability by activating kinase pathways to effect multiple downstream actions in hypothalamic neurons (878). An emerging hypothesis implicates the binding of estrogen to presumed membrane receptors, or interactions with the binding of ligands to their receptors (878,879).

The rapid effect of progesterone on the lordosis reflex suggests that this steroid affects the neuronal membrane, although receptor-mediated actions cannot be ruled out. PR molecules are induced in the VMN by estrogen in a time course that coincides with the emergence of the lordosis reflex (872). On the other hand, membrane-active metabolites of this steroid enhance the lordosis reflex when provided to the medial POA (880). Albumin-bound progesterone, which does not penetrate the cell membrane, is capable of releasing GnRH from hypothalamic slices (881). These substances either directly modulate ion flux across the neuronal membrane or activate the intracellular messenger system. Tetrahydroprogesterone, an A-ring reduced metabolite of progesterone, is a $GABA_A$ receptor agonist (882). Progesterone also increases the levels of adenylate cyclase and cyclic adenosine monophosphate (AMP) in the ventromedial hypothalamus and POA (883). Cyclic AMP (884) and forskolin (885), an adenylate cyclase activator, are among substances that enhance the lordosis reflex in ovariectomized, estrogen-treated female rats by replacing progesterone.

Inhibitory Effects of the Preoptic Area on Lordosis

The role of the POA in the control of the lordosis reflex is evidently more complex than would appear from the results of earlier studies that showed enhanced behavioral sensitivity to estrogen after POA lesions (886) and disruption of the reflex by stimulation of this structure (887). For example, POA lesions in nonovariectomized rats disrupted the lordosis reflex, estrous cyclicity, and ovulation altogether (888,889). Obliteration of GnRH neurons in the POA, which would otherwise promote the lordosis reflex (890), may be responsible for these effects. Indeed, transplants of preoptic grafts containing GnRH neurons resuscitated the reflex in hypogonadal mice (891), although it is not known whether the graft established physiological connections to accomplish this effect. These results suggest that GnRH neurons in the POA promote the lordosis reflex, but because GnRH neurons themselves lack ERs (892), they cannot account for the changes in behavioral sensitivity to estrogen that follow lesions of the POA, unless the effect is mediated by a nongenomic mechanism (893).

The notion that neurons in the POA, with the obvious exception of GnRH cells scattered in this region, inhibit the lordosis reflex has left room for further scrutiny. There are fibers of passage in the POA with septal origins (894). These fibers presumably originate in septal neurons that exert powerful inhibitory effects on the lordosis reflex (895). This led to speculation that disruption of fibers of passage by POA lesions, and not the destruction of local cell bodies, decreases the amount of estrogen needed to induce the lordosis reflex (896,897). To dissociate the effects of neurons and fibers of passage in the POA on the lordosis reflex (898), we removed descending axons of passage in the female rat POA by a knife cut between the septum and POA and stimulated the deafferented POA (834). The roof-cut animals showed an enhanced lordosis reflex, as in the earlier literature (897), presumably because of the removal of septal inhibition of the reflex. Notwithstanding the enhanced sensitivity to estrogen, electrical stimulation of the POA in these rats caused a prompt and strong suppression of the lordosis reflex. The POA therefore contains neurons that inhibit the reflex.

The roof-cut animals and intact controls showed a marked difference in the time course of the POA-induced inhibition of the lordosis reflex. Because adjunct bilateral cuts of the stria terminalis caused the animals to respond to the POA stimulation as rapidly as the roof-cut animals, the slow time course of inhibition of the lordosis reflex in the controls was due to a simultaneous activation of both inhibitory and facilitatory POA components for the reflex. The stria terminalis contains an amygdalofugal, facilitatory neural component for the lordosis reflex and descends through the POA (899). The roof cut as well as the transection of the stria terminalis were likely to remove this component, which would otherwise antagonize the inhibitory effects of the local POA neurons on the lordosis reflex. As a corollary, ascending preoptic projections into the septum, which originate in cells in the ventral part of the POA (900), are not involved in the suppression of the reflex.

The estrogen-sensitive projections of the POA to the VTA (836,901) are the neural substrate for the preoptic suppression of the lordosis reflex. Electrical stimulation of the VTA causes powerful and prompt disruption of the lordosis reflex in estrogen-treated ovariectomized female rats (835), in the same way as does POA stimulation in the roof-cut animals. The integrity of the dopaminergic system is not necessary for interruption of the lordosis reflex by VTA stimulation because pretreatment with pimozide scarcely affected the stimulation effect. Nondopaminergic VTA efferents to the midbrain or pontine structures are likely to mediate the behavioral effect (902). The axons from the VTA descend in the ventral part of the central gray and extend as far caudally as the dorsal pontine tegmentum to the level of the locus coeruleus.

Estrogen alters the excitability of POA neurons with axons to the VTA in a sexually dimorphic manner, in close association with the behavioral sensitivity of each animal to this hormone (836,901). In ovariectomized females and neonatally orchidectomized males, estrogen increased the threshold and prolonged the refractory period for the antidromic activation of POA neurons by VTA stimulation. Estrogen had no effect on neuronal excitability in androgenized females. Another series of antidromic activation experiments also showed that estrogen increased the threshold for electrical stimulation of the ventral central gray to activate these VTA neurons (837). At the same time, estrogen significantly decreased the probability of propagation of antidromic action potentials into the neuronal soma and suppressed the generation of somatodendritic spikes in VTA neurons (837). Electrical stimulation of the POA with a train of pulses promoted the generation of somatodendritic spikes; conversely, an electrolytic preoptic lesion blocked antidromic propagation, similar to estrogen treatment. In this way, preoptic stimulation has a facilitatory influence on the soma of VTA neurons. These effects of estrogen would culminate in decreasing the efficacy of transmission from the POA to the ventral central gray. The pattern of estrogen-induced changes in the excitability of VTA neurons is that required for the behavioral disinhibition of the lordosis reflex. Electrical stimulation of the ventral part of the central gray causes a disruption of the lordosis reflex (851,852) contrary to that of the dorsal subdivision (697). Functional dichotomy between the ventral and dorsal subdivisions of the central gray has also been noted in autonomic regulation.

Electrical stimulation or local infusion of carbachol to the dorsal pontine tegmentum produces postural atonia because of a generalized inhibition of spinal motoneurons, an effect presumably implemented by activation of reticulospinal neurons in the medullary inhibitory center. This postural atonia would suffice to dissolve the lordosis posture in the female rat. It is of interest that the POA and the dorsal pontine tegmentum are implicated in the induction of the paradoxical sleep that accompanies postural atonia. This phase of sleep is now considered to be equivalent to the "electroencephalographic after-reaction" in the postcopulatory female rabbit.

Preoptic Area Neurons Foster Courtship Behaviors

Whereas activation of the estrogen-sensitive preoptic efferents to the VTA suppresses the reflexive display of lordosis, initiation of sexual interactions depends on the integrity of the POA. Preoptic lesions compel female rats to avoid contact with males, despite their enhanced ability to show the lordosis reflex (921). In nonrestrained, free-moving female rats in estrus, a transitory activation of some preoptic neurons precedes the commencement of reproductive interactions with the males (903).

Sexual odors are important in initiating copulatory interactions in several species. In the rat, the effect depends on the accessory olfactory system (904), and the POA is a major target of accessory olfactory inputs (905). Damage to the accessory olfactory system might diminish olfactory cues that are necessary for initiation of copulatory interaction. It seems worthwhile to explore this hypothesis, despite the possibility that the amygdala, which relays accessory olfactory inputs to the POA, also may affect the lordosis reflex through its projection to the VMN (906).

Increased locomotor activity in female rats in estrus embodies enhanced sexual motivation (907). The spinal stepping mechanism that induces locomotion is regulated by the midbrain locomotor region (MLR) in the rat (908). The rat MLR has been identified in the rostral part of the pedunculopontine and cuneiform nuclei of the midbrain (909). Estrogen is clearly responsible for the increased locomotion in female rats in estrus because systemic administration of estrogen to ovariectomized rats increases both open-field (910,911) and wheel-running (912) activities. The brain structure responsible for the action of estrogen on the open-field activity is not known, but the medial POA has been positively identified as a site for the estrogen-induced activation of wheel running (913). The POA contributes to the rostrocaudal neural axis for the locomotor synergy (914) with its heavy projections to the MLR (909,915). The preoptic locomotor region (916), from which stepping can be initiated by electrical (917) or chemical (918) stimulation, is situated in the medial portion of the lateral POA. In contrast, locomotor activity can be consistently reduced by cholinergic activation of the periventricular medial POA (919,920).

Antidromic activation experiments have identified two groups of preoptic neurons that project directly to the MLR. The antidromic activation latency and the histological localization differed between the two populations (825). Even though both overlapped in part, many responses with latencies shorter than 12.0 msec were recorded from the medial portion of the lateral POA, lateral to the recording sites of the longer-latency potentials in the medial POA. Estrogen lowered the antidromic activation thresholds in neurons with shorter latencies, but had the opposite effect on neurons with latencies longer than 12.0 msec.

Electrical stimulation of the POA disrupts receptivity but not proceptivity (834). Electrolytic (921) or excitotoxin-induced (833) lesions of the POA have the opposite effect. Preoptic innervation of the MLR has been implicated in the regulation of motivational behavior (922). Disruption of neural transmission in the peripeduncular nucleus, through which fibers from the preoptic and other subpallidal structures descend to the MLR (915), diminishes the lordosis reflex but spares sexual motivation in female rats (923–925). Because a no-exit paradigm (921) was used in these studies, a lesion-induced rejection of males might have contributed to the effect. Females with the peripeduncular lesions spent more time with males, but characteristically failed to engage in darting and other solicitatory behavior (925). It is possible that decreased locomotion, as indicated by the lack of darting, might have made the lesioned females spend more time with males than the controls. Although the nucleus accumbens activates locomotion through innervation of the POA (926), lesions of the accumbens do not modify soliciting activity (865). The POA may be an independent entity for this behavior. Because estrogen inhibits neural impulse flow from the medial POA and facilitates that from the lateral POA, when combined, the effects of estrogen would culminate in an increased locomotor activity. It has been shown that the medial POA projection suppresses locomotion (919,927), whereas that from the lateral POA activates it (918,928). Such selective filtration of the neural impulse flow would result in the increased locomotor activity that is typical of female rats in estrus.

With respect to estrogenic actions in POA , it should be noted that facilitatory effects of systemically administered estradiol have been found on running-wheel behavior; for example, Blizard (929), Gentry and Wade (930), and Fahrbach et al. (931) found preoptic implants of estradiol to increase running-wheel but not open-field activity of female rats. This discovery was further illuminated by Sakuma's (932) finding that estrogen inhibited neurons in the dorsal part of the medial POA. In contrast, Slater and Blizard (933) found estradiol significantly increased open-field activity in the Maudsley Reactive strain of rats. Estradiol-treated mice have also been shown to exhibit increased fearfulness on a dark–light transition task and on tone-shock–conditioned fear learning (934). The facilitation of running-wheel activity by estrogens depends on the gene for ERα (821).

The neurochemistry of estrogenic effects in the preoptic neurons will become clearer in the future. Already, microarray results comparing gene inductions and repressions in POA compared with hypothalamus have been reported (935), and the work on prostaglandin D synthase, reviewed later, exemplifies the high yield of this approach. On the behavioral level, a selective facilitation of melanocortin receptors has been reported selectively to facilitate sexual courtship behaviors in female rats (936), but the site of action for this effect has not yet been determined.

In summary, different neuronal groups in the POA have different modes of action on reproductive behaviors, some of them primarily to inhibit lordosis (and so their neuronal inhibition would facilitate lordosis), and others to foster the locomotor aspects of courtship behaviors.

Social Recognition and Affiliation

As depicted in Fig. 11, another aspect of premating behavior follows from locomotor/approach behaviors and has to do with social recognition and affiliation. At least four genes, those coding for ERα, ERβ, oxytocin, and the oxytocin receptor, appear to be crucial for social recognition in female mice (789). The coordinated action on social behavior of these four gene products brings together a tremendous amount of data from molecular endocrinology, neuroanatomy, neurophysiology, and ethology in the following way (Fig. 11): estrogens acting through ERβ in the PVN facilitate transcription of the oxytocin gene, and the oxytocin is then transported to the amygdala, where estrogens acting through ERα turn on the gene for the oxytocin receptor. It is in the amygdala (937) that oxytocin has been shown to foster

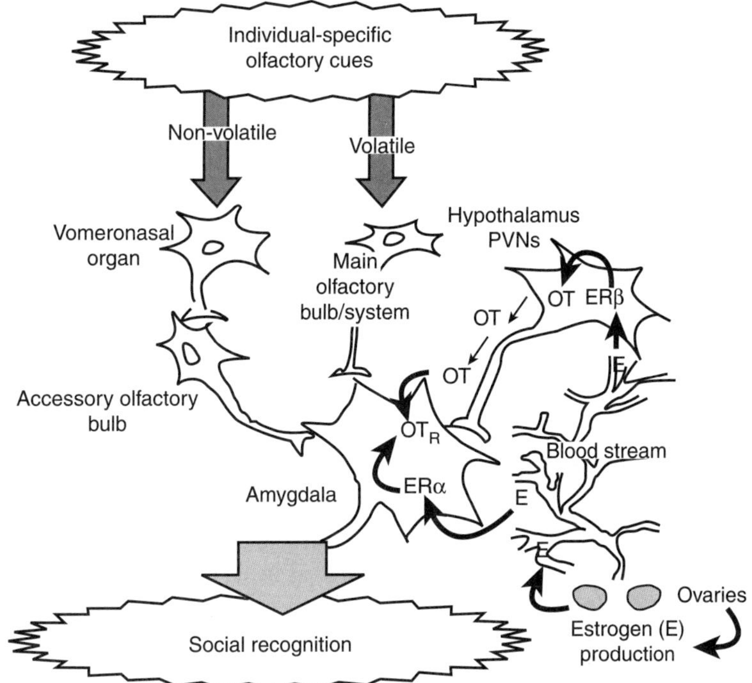

FIG. 11. Four genes involved in a micronet supporting social recognition. These are the genes for estrogen receptor-α (ERα), ERβ, oxytocin (OT), and oxytocin receptor (OT-R). Circulating estrogens (E) facilitate transcription from the OT gene in the paraventricular nucleus (PVN) as a result of binding to ERβ. This OT is transported to the amygdala. Meanwhile, in the amygdala, circulating estrogens facilitate transcription from the OT-R gene as a result of binding to ERα. Thus, OT-R is there to receive OT. This is important because Ferguson and Insel (787,788) showed that in the amygdala, OT can facilitate social recognition. This theory makes sense because of the importance of pheromonal and olfactory information for mouse social recognition—and these signals pour into the amygdala. (From Choleris, E., Gustafsson, J. A., Korach, K. S., Muglia, L. J., Pfaff, D. W., and Ogawa, S. [2003]. An estrogen-dependent four-gene micronet regulating social recognition: a study with oxytocin- and estrogen receptor alpha and beta-knockout mice. *Proc. Natl. Acad. Sci. U S A* 100, 6192–6197.)

social recognition. Thus, we have four-gene micronet acting in at least two brain regions to support this sociosexual behavior. In addition, Young and colleagues have shown the efficacy of vasopressin-1a expression in neuroanatomically discrete locations of the basal forebrain for enhancing social behaviors (938).

AROUSAL

Elementary brain arousal underlies sexual arousal and motivation, which in turn underlie sexual behavior. During a review of the literature on neural and genetic mechanisms for CNS arousal [see Chapter 7 in Pfaff (939), on which this section depends], we were able to deduce the universal operating features of arousal systems in the mammalian brain. What is the major, essential feature of arousal systems in the mammalian CNS? The answer is simply that they are not allowed to fail. Therefore, overlapping functions among genes, neuropeptides, neurotransmitters, individual neurons, and nerve cell groups are required. The system's function must be protected against consequences of the loss of individual components. Redundancy is expected in neuroanatomical circuitry, neurophysiological mechanisms, and especially among genes for receptors and so forth.

BBURP Theory

Beyond the simple requirement for reliability, arousal systems in mammalian brains have the following universal properties, briefly summarized here.

Bilaterality

Are they symmetrical, left for right? Yes, in every major respect, ascending arousal systems in the brain stem and descending controls (e.g., from the PVN) are bilaterally symmetrical. Even in circumstances where connections look unbalanced left for right, eventual functional effects look closely coupled.

Are both left and right sides of the neural circuits required? No. Unilateral damage can cause asymmetry in the electroencephalogram (EEG), specifically during rapid eye movement sleep (940), but not coma. Comatose states in patients only occur after serious bilateral damage to the systems specified in Pfaff (939).

Bipolarity (Bidirectionality)

Much of the neuroanatomy and neurophysiology reviewed in Pfaff (939) emphasized systems ascending from the brain stem and following either a low road through the basal forebrain or a high road through the thalamus to affect the cerebral cortex. But are arousal systems essentially bipolar? Do they work in both directions essentially at the same time? Yes. An important hypothalamic cell group, the PVN, provides a wonderful example because it is involved in all four forms of arousal, cortical, autonomic, endocrine, and behavioral. Exactly how PVN neurons have their wide range of effects remains a bit of a mystery. Projections to the locus ceruleus could have something to do with the PVN's alerting effects on the cortical EEG, and projections to autonomic control centers in the medulla and spinal cord surely explain some of the PVN's effects on autonomic arousal, through vasopressinergic, oxytocinergic, and corticotropin-releasing hormone synapses. But we need much more neurochemical and biophysical detail than this to understand how PVN neurons can change the state of a mammalian brain and body.

Tucked within these long-distance connections from brain stem to basal forebrain or hypothalamus, and the reverse, is a much smaller set of reciprocal connections that comprise what Clifford Saper has dubbed the "sleep switch." It appears that sleep-promoting neurons in the ventrolateral POA not only receive ascending connections from wake-promoting histaminergic neurons in the tuberomammillary neurons of the hypothalamus, but send a descending projection back to those neurons. Saper hypothesizes that these connections provide a reciprocally opposed "flip-flop" control over the sleep–wake cycle. At issue is whether temperature controls in the POA play an important causal role in governing waking and sleeping or whether they work simply as a parallel, correlated system.

Another example of descending pathways related to arousal would be those emanating from the brain's "circadian clock," the suprachiasmatic nucleus of the hypothalamus (941). Taking in signals about daily changes in ambient light from the retina and feeding them to non–image-forming visual systems and elsewhere, the suprachiasmatic nucleus both sends axons down the neuraxis and receives ascending signals. In a following section, we face the question of what happens when rhythmic circadian signals encounter homeostatic sleep drive forces.

POA neurons, besides their likely participation in Saper's "sleep switch," have longer-distance descending projections related to autonomic nervous system controls, as well as temperature controls and sexual arousal (942). Extensive descending connections controlling autonomic nervous system reactions, as determined from the neuroanatomical studies of Arthur Loewy and colleagues (943) [reviewed in Loewy (944); see also Luppi et al. (945) and

Condes-Lara et al. (946)]. In fact, I suspect that the reciprocal connections between brain stem arousal-related sites like the locus ceruleus and forebrain sites like the amygdala could provide the regulatory negative feedback required to bring rapid arousal changes under control (see later).

Finally, the "master cells" themselves, large reticular neurons in the arousal crescent pictured in Pfaff (939), have descending as well as ascending axons. In all of the preceding examples, the bipolar, bidirectional nature of primitive arousal systems comes through clearly. Neural systems supporting arousal traverse the entire neural axis.

Response Potentiation

In all cases, the major effect of a mammalian CNS arousal system is to enable motoric and autonomic responses to sensory stimuli, especially if those stimuli have emotional import.

Thus, we have proposed a *B*ilaterally symmetric, *B*ipolar (bidirectional), *U*niversal *R*esponse-*P*otentiating (BBURP) system (Fig. 12). Among all vertebrates, this system readies the organism to respond to stimuli of all modalities, initiate voluntary locomotion, and react with feeling. Especially important for our purposes in this chapter is that it contributes to the intensity and expression of sexual arousal.

Experimental progress on the relation of generalized arousal systems to sexual arousal is being made on at least three fronts: crisp logical definition of generalized arousal (see the section on Background and Operational Definition of Arousal); discovery of hormone-response genes contributing to arousal (see the section on Hormone-Dependent Genes Influencing Arousal); and unraveling of signal transduction chains that allow generalized arousal forces (transmitter effects) to influence neurons involved in sexual arousal and behavior (see the section on Biophysics of Arousal Transmitters Affecting Hypothalamic Neurons that Control Sexual Behavior).

Background and Operational Definition of Arousal

If mechanisms of brain arousal have certain universal characteristics in the mammalian brain, and if they also underlie certain facets of sexual behaviors, then it is interesting to delve into their conceptual foundations. Here, relying on the historical background and logical approach to arousal theory contributed by Jonathan Frohlich to Chapter 47 in *Hormones, Brain and Behavior*, we introduce ideas of how generalized CNS arousal could be related to reproductive physiology.

In their investigations of hormone effects on animal behavior, researchers typically have studied the effects of given hormonal manipulations on specific behavioral outcome measures. In addition to standard reproductive, parental, and progressive behavior assays, common behavioral indices include running-wheel and open-field behavior, light–dark exploration, radial arm maze, elevated plus maze, acoustic and tactile startle, Morris water maze, fear conditioning, active and passive avoidance, and exploratory behaviors. In these studies, the independent variable is hormone type/level, and the dependent variable is the amount, intensity, or duration of performance on the specific behavioral task.

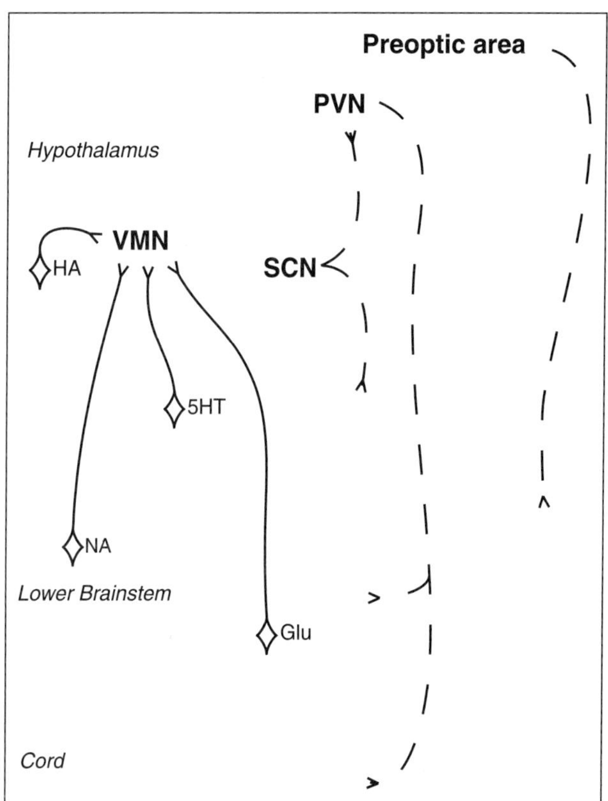

FIG. 12. Diagram of some of the components of BBURP (bilateral, bidirectional universal response potentiation) theory (Pfaff, 2005) underlying generalized and certain specific forms of central nervous system arousal. All of these mechanisms are bilaterally symmetrical. Catecholaminergic, serotonergic, and glutaminergic influences *ascending* from the brain stem are among those helping to "wake up" the forebrain. *Descending* influences related to arousal include those from the paraventricular nucleus of the hypothalamus (PVN), suprachiasmatic nucleus of the hypothalamus (SCN), and the preoptic area. Importantly, histaminergic, noradrenergic, serotonergic, and glutaminergic inputs to the ventromedial nucleus of the hypothalamus show how ascending arousal influences can potentiate lordosis behavior and other female-typical responses related to reproduction. This is generalized arousal potentiating a specific, sexual arousal. Glu, glutamate; HA, histamine; 5HT, serotonin; NA, noradrenaline; VMN, ventromedial nucleus of the hypothalamus.

In our search for novel measures and indices of hormonal and genetic effects on behavior, however, we are trying an alternative approach to quantifying the effects of hormones on behavior (947). That is, rather than investigating the effect of a given hormonal manipulation on a specific behavioral task, we subjected experimental animals to an array of behavioral paradigms from a variety of domains to tap into a conceptually larger, more fundamental dependent variable. Toward this end, we are investigating the hormonal and genetic foundations of *arousal* because it lies at the base of specific states of sexual motivation (948). This section deals with two aspects of our study. First, what is the empirical structure of arousal in mice, including its operational definition and mathematical formulation? Second, to what extent are aspects of this structure hormone sensitive?

Development of the Arousal Concept

> *Arouse: to move toward (readiness for) action from a state of inactivity.*
>
> *—Condensed from Webster's Third New International Dictionary*

The history of the arousal construct is rooted in Immanuel Kant's *Lebenskraft*, or "life energy," Gottfried Ewald's *biotonus*, a metabolically regulated temperament, and Pavlov's "excitatory substance" of the CNS [see Strelau (949) for a review]. Hebb (950), in his theory of motivation, postulated the arousal system to be the essence of drive, which he further defined as "the energy of movement." As such, sensory stimulation was to play a dual role in behavior, exerting both a "cue function" that determined the precise form of behavioral response, and an "arousal function" that provided the requisite neurological support for all activity. Moreover, in his notion of the "conceptual nervous system," Hebb (951) furthered this idea by proposing two frameworks of sensory projections to the cortex. Whereas the first system encompassed quick and efficient projections through the neothalamus, the "arousal system" is characterized by widespread nonspecific projections.

In contrast to Hebb's (950,951) focus on the stimulatory aspects of arousal, Duffy (952–954) highlighted the behavioral response elements. Duffy noted that because living beings are, essentially, energy systems, they can be characterized by their degree of activation, which was defined as "the extent of release of potential energy stored in the tissues of the organism as this is shown in activity or response." Both overt (i.e., locomotion) and covert (i.e., thinking) activities were proposed to depend on activation. Furthermore, activation was proposed to vary along dimensions of direction (approach or withdrawal from stimuli) and degree (intensity or force of response), and to range potentially from deep coma to extreme excitement. The various behavioral measures proposed to quantify this dimension included metabolic rates and patterns, skeletal muscle tensions, electrical resistance of the skin, EEG, blood pressure, and respiration. Other theories of arousal have been reviewed in Frohlich et al. (955).

Arousal has been also understood as a neurophysiological construct. The origins of this notion lie in Moruzzi and Magoun's (956) groundbreaking experiments that found stimulation of the cat brain stem to elicit EEG changes strikingly similar to those observed in normal arousal reactions. Moreover, lesioning or anesthetizing the reticular activating system inhibited behavior and engendered EEG characteristics of sleep (957–959). Thus, "arousal" was proposed to be an emergent property of these ascending brain stem pathways.

More recent neuroanatomical research, however, has pointed to the physiological and neurochemical fractionation of the arousal construct, with discrete arousal pathways isolated within distinct noradrenergic, dopaminergic, serotonergic, and cholinergic systems [see Robbins and Everitt (960) for a review]. Robbins and Everitt (960) also point out, however, that "in many situations these reticular activating systems are simultaneously active in varying degrees, to optimize processing capacity and facilitate response output." Therefore, despite such evidence of physiological multiplicity, the arousal system may yet function, on a behavioral level, in a unitary manner.

The aforementioned literature seemed to indicate two extreme, falsely dichotomous possibilities for the structure of arousal: that arousal functions purely as a single, unitary construct, or it must be fractionated into a large number of separately delineated categories. Our own research resolved this problem by using a mathematical/statistical approach (961), principal components analysis (PCA), applied to arousal-related behavioral results with mice. Although behaviors did not cluster neatly into the predicted sensory, motor, and affective dimensions, an interesting formulation was seen, and the similarities of arousal components between cluster analysis and PCA gave us further confidence in the calculations and their implications.

We (814) have also identified a generalized arousal component in the behavior of mice. Analyzed by mathematical/statistical approaches across experiments, investigators, and mouse populations, it accounts for approximately one third of the variance in arousal-related measures. A combination of mathematical and genetic approaches to arousal in an experimentally tractable mammal will open up analysis of a CNS function of considerable theoretical significance with obvious contributions to the biological regulation of sexual arousal and behavior.

Hormonal Influences on Arousal

Having begun to work out the structure of arousal in mice, we then sought to determine the extent to which this set of functions was hormone sensitive. Research conducted in the mid-1990s has illuminated our understanding of hormone–behavior relationships (962–964). We were primarily interested in the roles of estrogen and thyroid hormones in arousal-related behaviors. As such, we first offer a brief review of the effects of estrogen and thyroid hormones on behavior.

Estrogens. Numerous studies have investigated the role of estrogen in mental states. For example, symptoms of depression, sleep disturbance, irritability, anxiety, panic, memory difficulties, and cognitive dysfunction in menopausal women have been proposed to result from decreased brain estrogen levels (see Arpels (965) for a review]. Clearly, estrogen can heighten mood (966,967). In female rats (968–970) and mice (Blackman and Pfaff, unpublished observations), it can act as an *antidepressant*. Similarly, in forced swimming tests, Okada et al. (971) found ovariectomy-induced immobility to be significantly decreased in estradiol-treated rats. A review by Fink et al. (972) outlined the modulatory effects of estrogens on brain neuropeptide and monoamine neurotransmitter systems, as an approach to understand the influence of estrogen on mood and cognition. In their study, estradiol treatment significantly increased the density of 5-HT_{2A} binding sites in the anterior frontal, anterior cingulate, and piriform cortex, olfactory tubercle, nucleus accumbens, and lateral dorsal raphe nucleus. This finding was taken to explain why fluoxetine (Prozac), which inhibits the reuptake of serotonin, alleviates depressive symptoms of premenstrual syndrome and postnatal depression, because these syndromes coincide with decreased estrogen concentrations. Estrogenic regulation of the PR gene in serotonergic midbrain raphe neurons in ER knockout as well as wild-type mice (973) suggests a role for ERβ.

In rodents, estrogens have been demonstrated to amplify the functioning of the hypothalamic–pituitary–adrenal axis, possibly through glucocorticoid receptor–mediated negative feedback mechanisms (974). Research has also demonstrated estrogen administration to facilitate stress-induced ACTH and corticosterone secretion from the anterior pituitary and adrenal cortex, respectively (974,975). Becker and Breedlove (976) further noted the significance of the hypothalamic–pituitary system in that the hypothalamus is strongly involved in the mechanism of secretion of almost every hormone. They remind us that hypothalamic–pituitary-controlled hormones influence a plethora of target organs, including gonads, adrenal cortex, thyroid gland, mammary gland, and body cells. Moreover, the hypothalamic–pituitary–adrenal axis plays an important role in maintaining homeostasis (977). Thus, estrogenic facilitation of this system could exert significant systemic effects with broad behavioral consequences.

Thyroid Hormones. In human clinical syndromes, hyperthyroidism leads to the etiology of Graves' disease, an illness characterized by tremors, restlessness, and excitability (978). Conversely, hypothyroidism often engenders symptoms of weakness, fatigue, weight gain, and bradycardia (979). Extensive research has investigated the relationship between thyroid hormone and psychopathology (980, 981), sensation seeking, and anxiety (982,983), highlighting the relationship between the hypothalamic–pituitary–thyroid axis and affect and cognition, noting that hypothyroid patients reported depression more than any other psychiatric illness. Thyroid hormones have been used effectively to treat mood disorders, as well as to augment the effects of antidepressants (983–986). In an interesting study, Joffe et al. (987) found that cognitive–behavioral psychotherapy manipulated thyroid hormone levels in a manner similar to antidepressant medication.

Thyroid hormones have been shown to be significant in early rodent growth (988). In a study by McCarthy et al. (989), neonatal administration of thyroxine to rats caused hyperresponsiveness of the sympathoadrenal medullary system to acute stress. This hyperresponsiveness remained through adulthood. Sjoden and Soderberg (990) administered high doses of triiodothyronine and L-thyroxine sodium administered to 4-day-old rats and found increased open-field behavior. Rastogi and Singhal (991) demonstrated that administration of triiodothyronine for 30 days to neonatal rats engendered hyperthyroidism, which hastened the genesis of assorted behavioral characteristics at maturation. Spontaneous motor activity, for example, increased 69%. Such hyperthyroidism was also associated with increased synthesis and utilization of catecholamines and 5-HT in the brain.

Thyroid hormones have been shown to influence the acquisition and retention of behavioral responses (992). In a study by Deshimaru et al. (993), hyperthyroidism was induced in 10 adult male rats by oral doses of thyroideum siccatum. These rats lost weight and became hyperactive, irritable, and aggressive. Electron microscopy revealed various neuronal effects of induced hyperthyroidism, including synaptic changes in frontal and temporal cortex, hippocampus, and hypothalamus, abnormalities in the presynaptic membrane, and increased mitochondria in nerve cells of the frontal cortex. Other studies have found thyroid hormones to modify the sensitivity of dopamine receptors (994).

Additional effects of thyroid hormone on animal performance in experimental behavioral paradigms have been noted. For example, hypothyroid mice have been found to evince decreased locomotion in both center and margin regions of the open-field test, with hyperthyroid mice showing the opposite pattern (995). McEachron et al. (996) noted significantly decreased free-running activity cycles (in running wheel) in thyroidectomized rats compared with controls. Upon thyroxine treatment, however, circadian activity of the former significantly increased while controls remained unchanged. TRH administration has also been demonstrated to elicit increased spontaneous locomotor activity in rats placed in a plastic cage resting on an activity meter. Finally, administration of uncontrolled foot shocks to rats has been found to increase brain levels of triiodothyronine in males and females compared with controls (997).

Hormone-Dependent Genes Influencing Arousal

How might one investigate genetic determinants of arousal in mice? Advances in molecular biology have enabled targeted disruptions, or knockouts, of specific genes in mice (998–1000). Behavioral studies of such knockout mice promise to discern functional roles of the identified genes. A review by Nelson (1001) highlighted numerous advantages and limitations of this technique, and detailed the production of knockout mice. Specific methodological and behavioral

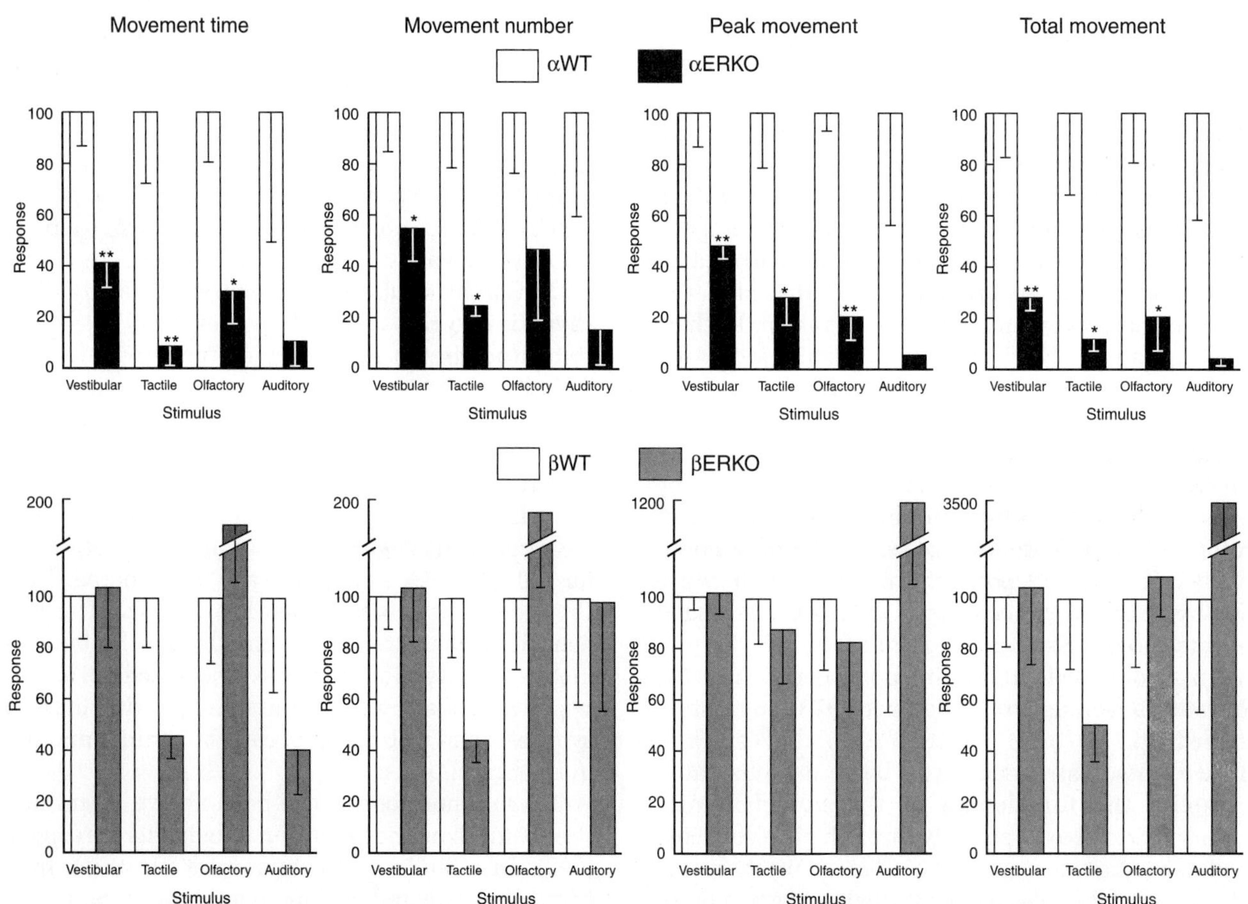

FIG. 13. Genetic manipulation of generalized arousal. Sensory alertness: knockout of the gene for estrogen receptor-α (ERα) significantly reduced all four measures of responses to all four stimulus modalities tested: vestibular, tactile, olfactory, and auditory. The perfect genetic control, knockout of the likely gene duplication product ERβ, had no significant effect. (From Garey, J., Goodwillie, A., Frohlich, J., Morgan, M., Gustafsson, J. A., Smithies, O., Korach, K. S., Ogawa, S., and Pfaff, D. W. [2003]. Genetic contributions to generalized arousal of brain and behavior. *Proc. Natl. Acad. Sci. U S A* 100, 11019–11022.)

paradigms have been proposed to evaluate cognitive, emotional, and behavioral dimensions in knockout mice (1002,1003).

We are currently extending our investigation of the roles of estrogen and thyroid hormone in arousal by studying mice with targeted deletions of ERα, ERβ, thyroid hormone receptor-α (TRα), and thyroid hormone receptor-β (TRβ) genes.

For example, comparing the roles for the genes coding for ERα versus ERβ (likely gene duplication products), we studied female mice, individually housed, sleeping in their home cages as they do during the *light phase* of the daily light cycle. Thus, imprecision due to ongoing activities and biases due to fear and anxiety were removed from the experimental protocol. For all sensory modalities tested, ERα knockout female mice were less responsive to sensory stimuli than their wild-type female littermate controls (Fig. 13). Surprisingly, disruption of the gene for the closely related ER-β, a likely gene duplication product and therefore a perfect genetic control group, did not have the same effect. Likewise, in terms of locomotor activity in running wheels, highest during the *dark phase* of the daily light cycle, ERα knockout females were less active. Interestingly, this phenotype was dependent on age; older ERα knockout females were subject to the genetic effect, whereas the younger ERα knockout females were not (Fig. 14). Again, the differences between ERβ knockout females and their wild-type littermate controls were not significant (data not shown). The graphs for running-wheel activity of ERβ knockout mice and their controls were virtually identical. In sum, ERα knockout females were less responsive to external stimuli and less motorically active.

Up to this point, much of the discussion has centered on the effects of gene knockouts on behavior. What about the effects of sex hormones on gene expression yielding transcripts related to arousal physiology? Mong et al. (1004) used microarrays to study more than 11,000 transcript levels in female mice, looking for estrogen-influenced genes (Fig. 15A). They were struck by the obvious hormone effect on mRNA levels for lipocalin-type prostaglandin D synthase (L-PGDS). Surprisingly, estrogen treatment regulated its gene expression in opposite directions when two neighboring brain regions were compared: mRNA levels were increased in the basomedial hypothalamus, but decreased in the POA (Fig. 15B). Then, Mong et al. (1005) used a novel antisense DNA oligomer approach to test the effect of knocking down L-PGDS mRNA function specifically in the POA. They not only caused elevated activity levels by this maneuver, but also were able to discover the performance of lordosis behavior in ovariectomized female mice that had not received estrogen treatment. Thus, they revealed a powerful effect of a generalized arousal manipulation on a specific (sexual) arousal mechanism.

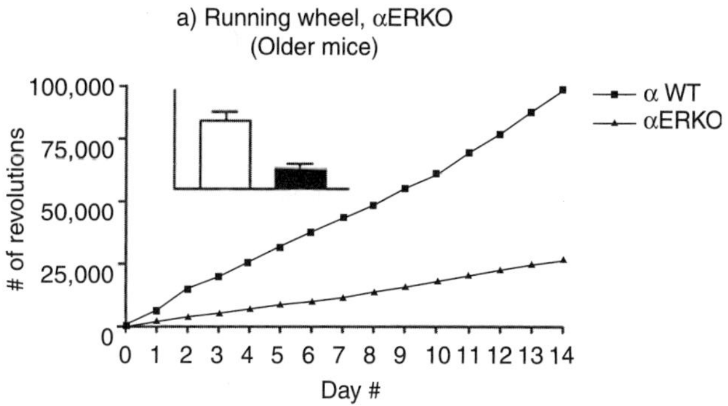

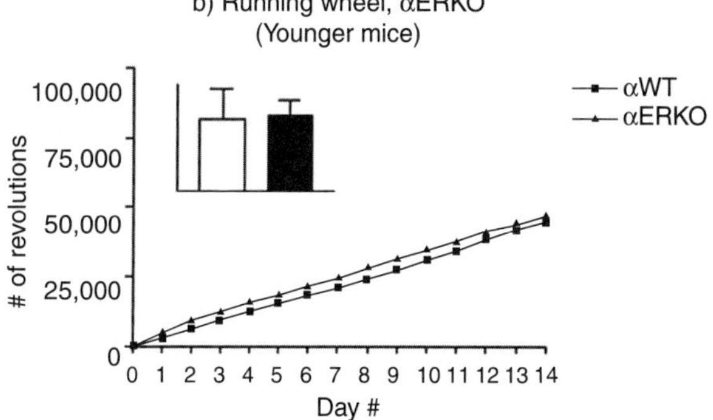

FIG. 14. Genetic manipulation of generalized arousal: voluntary motor activity. In older (35-week) female mice, knocking out the gene for estrogen receptor-α significantly reduced running-wheel activity. For reasons we do not understand, in younger (15-week) animals there was no phenotype. There was no effect of knocking out estrogen receptor-β on running wheel activity (data not shown). (From Garey, J., Goodwillie, A., Frohlich, J., Morgan, M., Gustafsson, J. A., Smithies, O., Korach, K. S., Ogawa, S., and Pfaff, D. W. [2003]. Genetic contributions to generalized arousal of brain and behavior. *Proc. Natl. Acad. Sci. U S A* 100, 11019–11022.)

Biophysics of Arousal Transmitters Affecting Hypothalamic Neurons that Control Sexual Behavior

In addition to the effects of sex hormones on genes related to generalized arousal, just mentioned, what are some of the mechanisms by which generalized arousal could affect a specific sex behavior such as lordosis? The electrophysiological activation of neurons in the VMN by arousal-related transmitters represents one possibility. Not only NE (1006) but histamine (Fig. 16) can elevate the electrical discharge rate in VMN neurons. These electrophysiological data are consistent with the notion that elevated arousal could increase sex behavior levels. However, variations in the nature of the environment make things somewhat more complicated. Morgan et al. (1007) have made the following point: hormone-facilitated arousal in a

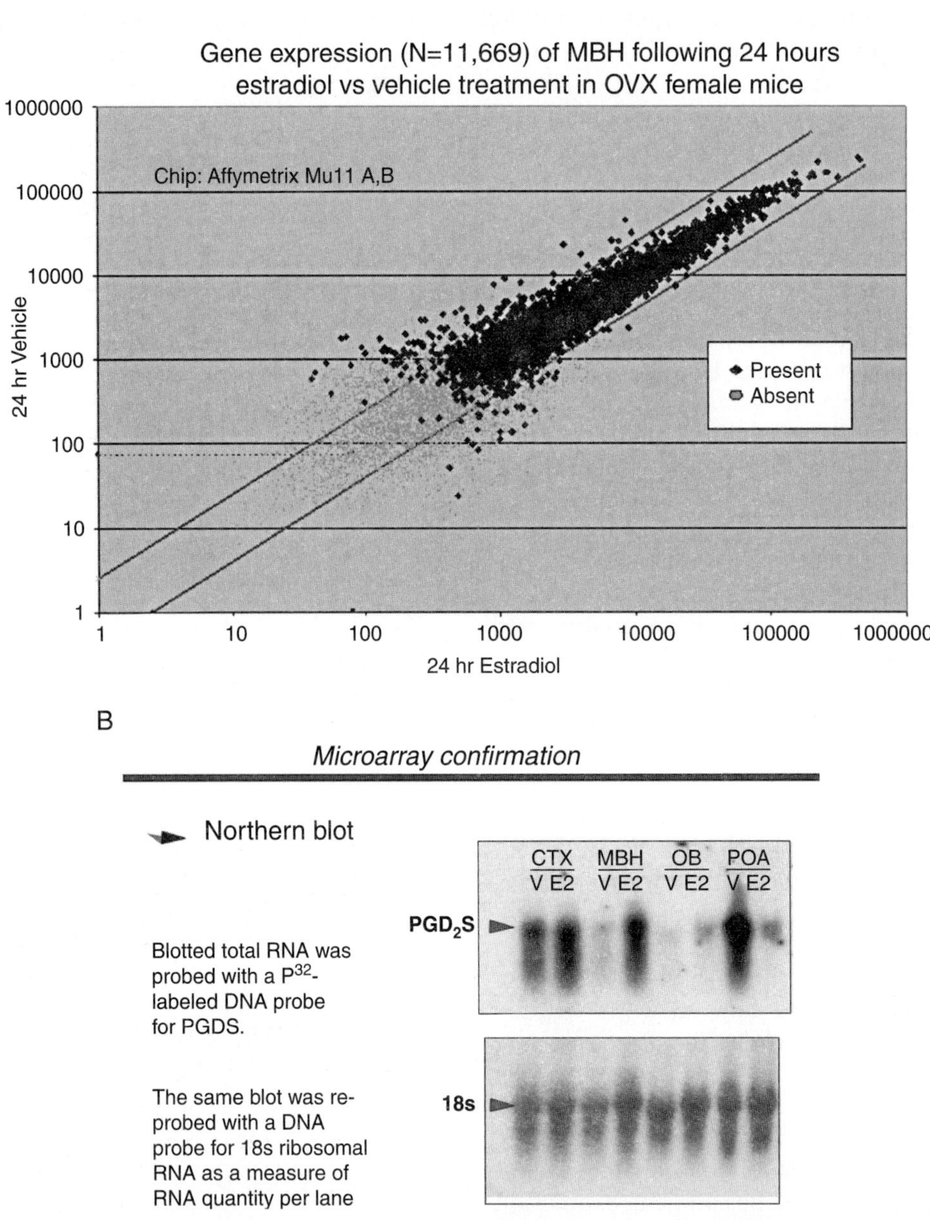

FIG. 15. Estrogen effects in a microarray experiment lead to a gene whose expression influences arousal and sex behavior. **A:** The microarray data display, in which more than 11,000 transcripts were measured. The surprising finding that in the medial basal hypothalamus (MBH) the transcript encoding lipocalin-type prostaglandin D synthase (PGDS) was revealed by one of the dots below and to the right of the straight lines (which represent doubling and halving of transcript levels due to 24 hours of estradiol treatment). **B:** Microarray findings were followed up by Northern blot analysis. We confirmed that in the MBH, estradiol (E2) markedly raises PGDS transcript levels. Surprisingly, in the preoptic area (POA) the opposite regulation was observed. Based on current neuroanatomy and neurophysiology, the POA finding was functionally provocative, and we followed it up (**C, D**).

Continued

safe environment may indeed encourage locomotion, courtship behavior, and sex. However, in a frightening environment, the same high arousal levels might increase fear, and militate against sex. In humans, hormones influence mood, and the exact nature and strength of the hormone effects depend on context (967,1008). Mechanisms by which the simplest mood alterations occur may be explored by the forced swim test in animals, in which estrogen treatment increases swimming time and decreases immobility ("giving up") (968–970). In turn, in human patients, there is evidence of generalized arousal effects on sex behavior: depressed women and men suffer from decreased libido and have a harder time reaching orgasms.

C

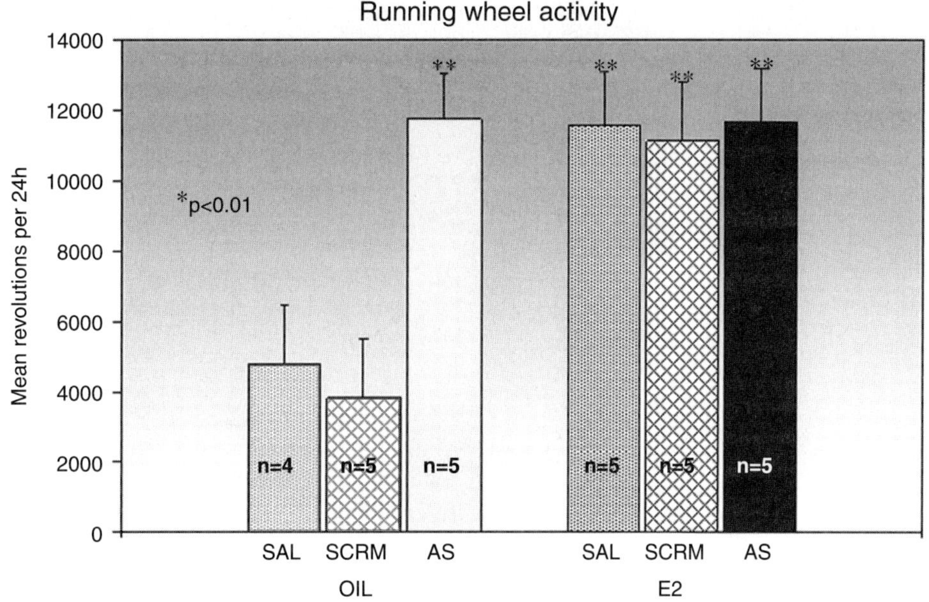

D

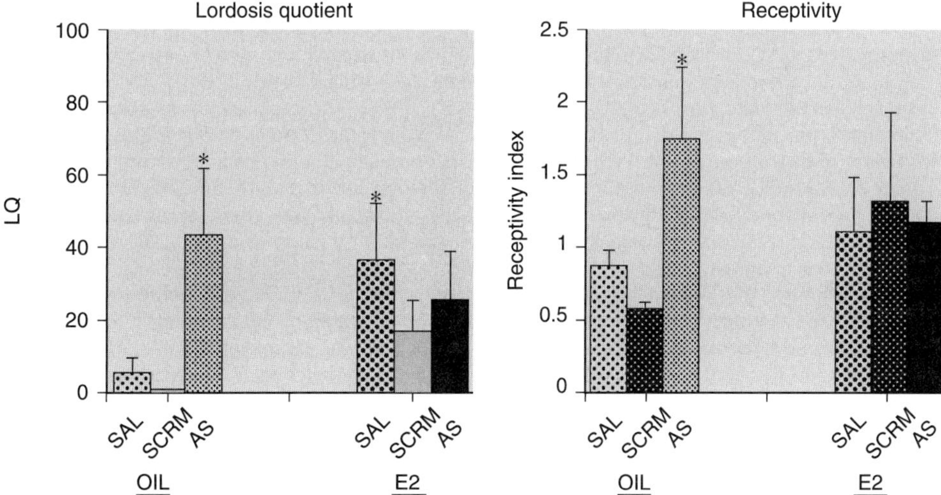

FIG. 15—Cont'd. C: Microinjection of locked nucleic acid antisense oligomers directed against the mRNA coding for PGDS, directly into the POA. As predicted from the POA Northern blot results in **B,** interfering with PGDS mRNA function in the POA elevated the performance of voluntary motor activity, similar to the effect of E2 itself. The scrambled sequence control (SCRM) had no effect compared with the saline vehicle control (SAL). **D:** Microinjection of locked nucleic acid antisense oligomers directed against the mRNA coding for PGDS, directly into the POA. Surprisingly, this molecular manipulation permitted lordosis behavior in ovariectomized female mice without estrogen treatment, as well as elevating receptivity levels. This experiment represents an extreme example of a generalized arousal force able to facilitate a specific form of arousal, in this case sexual arousal. (From Mong, J. A., Devidze, N., Frail, D. E., O'Connor, L. T., Samuel, M., Choleris, E., Ogawa, S., and Pfaff, D. W. [2003]. Estradiol differentially regulates lipocalin-type prostaglandin D synthase transcript levels in the rodent brain: evidence from high-density oligonucleotide arrays and in situ hybridization. *Proc. Natl. Acad. Sci. U S A* 100, 318–323, and Mong, J. A., Devidze, N., Goodwillie, A., and Pfaff, D. W. [2003]. Reduction of lipocalin-type prostaglandin D synthase in the preoptic area of female mice mimics estradiol effects on arousal and sex behavior. *Proc. Natl. Acad. Sci. U S A* 100, 15206–15211.)

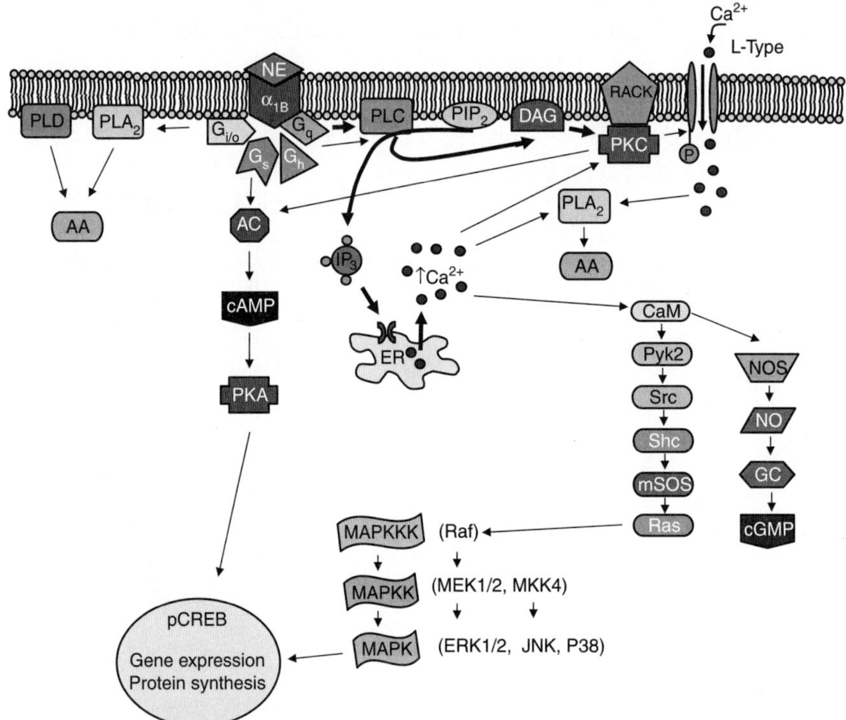

FIG. 16. Schematic derived from the literature on α-adrenergic signaling. This is an important signaling system because it specifies how a generalized arousal system could influence specific, sexual arousal, in this case through norepinephrine increasing the electrical activity of the ventromedial hypothalamic neurons that control lordosis behavior. α_{1B}-Adrenergic receptors couple primarily to the G-protein–coupled receptor G_q, stimulating phospholipase C (PLC). PLC cleaves phosphatidylinositol 4,5-bisphosphate (PIP_2), generating inositol 1,4,5-trisphosphate (IP_3) and diacylglycerol (DAG). IP_3 mobilizes intracellular Ca^{2+} and DAG activates protein kinase C (PKC). The mobilization of cytoplasmic Ca^{2+} stimulates phospholipase A_2 (PLA_2), releasing arachidonic acid (AA), and activates nitric oxide synthase (NOS), producing nitric oxide (NO), guanylyl cyclase (GC), and cyclic guanosine monophosphate (cGMP). PKC has been proposed to bind to receptor for activated C kinases (RACK) near the L-type Ca^{2+} channel, which it phosphorylates, causing an influx of Ca^{2+}. There is some evidence that α_{1B}-adrenergic receptors may also couple to G_h, leading to production of IP_3 and DAG, to G_s, leading to stimulation of protein kinase A (PKA), and to G_i/G_o, increasing AA through PLA_2 or phospholipase D (PLD). Although the precise signaling mechanism is not known, α_{1B}-adrenergic receptors can stimulate the mitogen-activated protein kinase (MAPK) pathway. α_{1B}-Adrenergic receptors can activate extracellular signal-regulated kinases (ERKs), P38, or c-Jun N-terminal kinases (JNKs), depending on the particular cell line or tissue in which it is expressed. AC, adenylyl cyclase; CaM, calmodulin; cAMP, cyclic 3′,5′-adenosine monophosphate; ER, endoplasmic reticulum; NE, norepinephrine; pCREB, cAMP response element–binding protein. Sources include Etgen (735), Minneman (1011), Michelotti et al. (1012), Koshimizu et al. (1013), and Kreda et al. (1014).

The mechanisms brought forward here in the section on Hormone-Dependent Genes Influencing Arousal and the section on Biophysics of Arousal Transmitters Affecting Hypothalamic Neurons that Control Sexual Behavior may be related to this effect.

As of August 2004, we are beginning to be able to picture the chain of events from the occupation of the α_{1B}-adrenergic receptor, the subtype most relevant for female reproductive behavior (735), to the stimulation of hypothalamic electrical activity. α_{1B}-Adrenergic receptors couple primarily to the G-protein–coupled receptor Ga_q, stimulating PLC. PLC cleaves phosphatidylinositol bisphosphate (PIP_2), generating inositol trisphosphate (IP_3) and diacylglycerol (DAG). IP_3 mobilizes intracellular Ca^{2+} and DAG activates PKC. PKC has been proposed to bind to receptor for activated C kinases (RACK) near the L-type Ca^{2+} channel, which it phosphorylates, causing an influx of Ca^{2+}. There is some evidence that α_{1B}-adrenergic receptors may also couple to G_h, leading to production of IP_3 and DAG, and to G_s, leading to stimulation of PKA. Although the precise signaling mechanism is not known, α_{1B}-adrenergic receptors can stimulate the mitogen-activated protein kinase (MAPK) pathway.

α_{1B}-adrenergic receptors can activate extracellular signal–regulated kinases (ERKs), P38, or c-Jun N-terminal kinases (JNKs), depending on the particular cell line or tissue it is expressed in (Fig. 16).

Likewise, the physical series of mechanisms joining the occupation of the histamine-1 (H_1) receptor to the facilitation of electrical activity in hypothalamic nerve cells is beginning to be elucidated (1009) (Fig. 17). Both of these nascent lines of research offer concrete ideas as to how generalized arousal forces increased sexual arousal in experimental animals.

OVERALL SUMMARY

The general introduction (see Background) showed how current mechanistic findings fit into a long and successful history of results on hormone–behavior relations in females. In turn, sex behavior findings form one part of a quickly growing field of hormone/brain/behavior studies (1010). Although all the data reviewed here are derived from experiments with animals, a large number of mechanistic steps have been preserved in human reproductive physiology and the

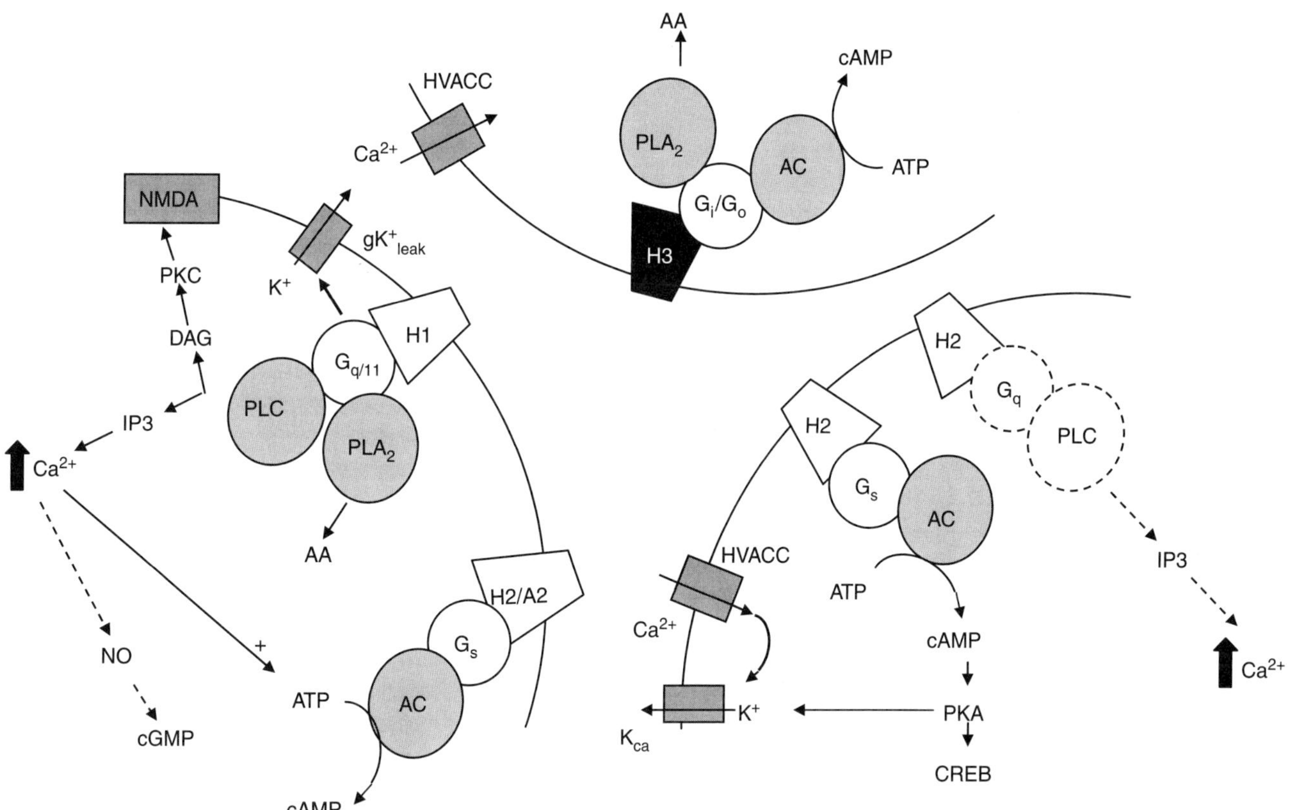

FIG. 17. Schematic representation, based on the literature regarding histamine neurobiology, of membrane and cellular responses to histamine receptor activation. Like Fig. 16, this step-by-step formulation is important because it specifies how a generalized arousal system could influence specific, sexual arousal by increasing the electrical activity of the ventromedial hypothalamic neurons that control lordosis behavior. Figures drawn with *dashed lines* indicate possible, but as yet unverified, signal transduction pathways. Here we concentrate on histamine-1 (H_1) receptors, likely the most important for causing ventromedial nucleus (VMN) nerve cells to fire. H_1 receptors, when activated, bind to the $G_{q/11}$ protein, stimulating PIP_2 (phosphatidylinositol 4,5-bisphosphate) hydrolysis, which results in the formation of inositol 1,4,5-trisphosphate (IP_3) and diacylglycerol (DAG). IP_3 induces the release of calcium from intracellular stores, which can initiate the augmentation of cyclic adenosine monophosphate (cAMP) production by neighboring H2 or adenosine receptors. DAG potentiates the activity of protein kinase C (PKC), which in turn phosphorylates the N-methyl-D-aspartate (NMDA) receptor, weakening the Mg^{+} block. H_1 receptor activation also leads to an increase in arachidonic acid (AA) and cGMP, and the blockade of a leak K^{+} conductance. Sources include Brown et al. (1009), Hill (1015), and Greene and Haas (1016).

human CNS (see Mechanisms Underlying the Primary Mating Behavior, Lordosis). Tracing the neural circuit for lordosis behavior [reviewed in Pfaff (6)] permitted serious work on biophysical and genomic mechanisms for this behavior itself [reviewed in Pfaff (725)]. Several genes have the properties (a) that estrogens turn them on; and (b) that their products foster female reproductive behaviors. Functional modules for *G*rowth of hypothalamic neurons, *A*mplification of the estrogen effect, *P*reparatory behaviors, *P*ermissive actions of hypothalamic neurons, and *S*ynchronization of mating with ovulation (GAPPS) reveal an organized "gene network" emerging for the regulation of mating behavior.

Courtship behaviors are indices of sexual motivation and depend on specific sets of neurons in the POA, as opposed to lordosis, which uses basomedial hypothalamic mechanisms. These preoptic mechanisms were spelled out in the section on Sexual Motivation and Courtship Behaviors. Going conceptually deeper, all motivated behaviors depend on arousal, the elementary activation of brain and behavior (939). A BBURP theory (*B*ilateral, *B*idirectional *U*niversal *R*esponse *P*otentiation) of generalized arousal has been proposed. Both the mathematical structure of arousal and some of the genomic contributions to the expression of hormone-dependent arousal were introduced in the section on Arousal.

Many of the "growth points" in this field show up at the extremes of the material reviewed here. For example, some of the most exciting questions try for greater molecular and biochemical detail: given that we know a lot about nuclear receptor genes required for female reproductive behavior, what about the coactivators that must cooperate with ERs to facilitate crucial transcriptional steps? Another question: given that rapid actions of estrogens at the membrane can facilitate later transcriptional events as well as sex behavior, what is the exact chain of physical steps from the arrival of the steroid at the membrane to the eventual completion of signal transduction? In addition, what channel properties are altered by sex hormones when we record estrogen-altered electrical activity in mouse brain neurons? Other interesting questions cluster at the opposite, behavioral extreme: the abstract and conceptual basis of motivational and arousal influences that are important for mating behaviors in both sexes. Here we can ask, what is the structure of such arousal influences—when is a generalized arousal state important, and when would it be specific to sexuality itself? Also, when is there just too much arousal, such as during high levels of environmental stress? Such behavioral and ecological questions lie at the heart of gene–environment interactions. Finally, all of the basic characteristics of hormone-facilitated mechanisms covered here conform to more general principles of hormone–behavior relations (727).

REFERENCES

1. Beach, F. (1948). *Hormones and Behavior*. Hoeber, New York.
2. Young, W. C. (1961). The hormones and mating behavior. In *Sex and Internal Secretions* (W. C. Young, Ed.), Vol. 2. Williams & Wilkins, Baltimore.
3. Pfaff, D. W. (1968). Uptake of estradiol-17β-H^3 in the female rat brain: an autoradiographic study. *Endocrinology* 82, 1149–1155.
4. Pfaff, D. W., and Keiner, M. (1973). Atlas of estradiol-concentrating cells in the central nervous system of the female rat. *J. Comp. Neurol.* 151, 121–158.
5. Morrell, D. W., and Pfaff, D. W. (1978). A neuroendocrine approach to brain functions: localization of sex steroid concentrating cells in vertebrate brains. *Am. Zool.* 18, 447–460.
6. Pfaff, D. W. (1980). *Estrogens and Brain Function: Neural Analysis of a Hormone-Controlled Mammalian Reproductive Behavior*. Springer-Verlag, New York.
7. Pfaff, D. W. (1983). Impact of estrogens on hypothalamic nerve cells: ultrastructural, chemical, and electrical effects. *Recent Prog. Horm. Res.* 39, 127–179.
8. Sodersten, P. (1985). Estradiol-progesterone interactions in the reproductive behavior of female rats. In *Current Topics in Neuroendocrinology* (D. Ganten and D. Pfaff, Eds.), Vol. 5. Springer-Verlag, Berlin.
9. Sodersten, P., and Eneroth, P. (1981). Serum levels of oestradiol-17β and progesterone in relation to sexual receptivity in intact and ovariectomized rats. *J. Endocrinol.* 89, 45–54.
10. Davidson, J. M., Rodgers, C. H., Smith, E. R., and Bloch, G. J. (19868). Stimulation of female sex behavior in adrenalectomized rats with estrogen alone. *Endocrinology* 82, 193–195.
11. Pfaff, D. W. (1970). Nature of sex hormone effects on rat sex behavior: specificity of effects and individual patterns of response. *J. Comp. Physiol. Psychol.* 73, 349–358.
12. Moss, R. L., and McCann, S. M. (1973). Induction of mating behavior in rats by luteinizing hormone-releasing factor. *Science* 181, 177–179.
13. Pfaff, D. W. (1973). Luteinizing hormone releasing factor (LRF) potentiates lordosis behavior in hypophysectomized ovariectomized female rats. *Science* 182, 1148–1149.
14. Ogawa, S., Olazabal, U. E., and Pfaff, D. W. (1994). Effects of intrahypothalamic administration of antisense DNA for progesterone receptor mRNA on reproductive behavior and progesterone receptor immunoreactivity in female rat. *J. Neurosci.* 14, 1766–1774.
15. Pfaff, D. W. (1970). Mating behavior of hypophysectomized rats. *J. Comp. Physiol. Psychol.* 72, 45–50.
16. Feder, H. H., Brown-Grant, K., and Corker, C. S. (1971). Preovulatory progesterone, the adrenal cortex and the "critical period" for luteinizing hormone release in rats. *J. Endocrinol.* 50, 29–39.
17. Feder, H. H., and Ruf, K. B. (1969). Stimulation of progesterone release and estrous behavior by ACTH in ovariectomized rodents. *Endocrinology* 84, 171–174.
18. Kow, L.-M., and Pfaff, D. W. (1975). Induction of lordosis in female rats: two modes of estrogen action and the effect of adrenalectomy. *Horm. Behav.* 6, 259–276.
19. Parson, B., Rainbow, T. C., Snyder, L., and McEwen, B. S. (1984). Progesterone-like effects of estradiol on reproductive behavior and hypothalamic progestin receptors in the female rat. *Neuroendocrinology* 39, 25–30.

20. Clemens, L. G., and Weaver, D. R. (1985). The role of gonadal hormones in the activation of feminine sexual behavior. In *Handbook of Behavioral Neurobiology* (N. Adler, D. W. Pfaff, and R. Goy, Eds.), Vol. 7, pp. 183–227. Plenum Press, New York.
21. Feder, H. H. (1984). Hormones and sexual behavior. *Annu. Rev. Psychol.* 35, 165–200.
22. Gladue, B. A., Green, R., and Hellman, R. E. (1984). Neuroendocrine response to estrogen and sexual orientation. *Science* 225, 1496–1499.
23. Sanders, S. A., and Reinisch, J. M. (1985). Behavioral effects on humans of progesterone-related compounds during development and in the adult. In *Current Topics in Neuroendocrinology* (D. Ganten and D. Pfaff, Eds.), Vol. 5, pp. 175–205. Springer-Verlag, Berlin.
24. Gerall, A. A. (1966). Hormonal factors influencing masculine behavior of female guinea pigs. *J. Comp. Physiol. Psychol.* 62, 356–369.
25. Goy, R. W., Bridson, W. E., and Young, W. C. (1964). Period of maximal susceptibility of the prenatal female guinea pig to masculinizing actions of testosterone propionate. *J. Comp. Physiol. Psychol.* 57, 166–174.
26. Phoenix, C. H., Goy, R., Gerall, A. A., and Young, W. C. (1959). Organizing action of prenatally administered testosterone propionate on the tissues mediating mating behavior in the female guinea pig. *Endocrinology* 65, 369–382.
27. Ward, I. L., and Renz, F. J. (1972). Consequences of perinatal hormone manipulation of the adult sexual behavior of female rats. *J. Comp. Physiol. Psychol.* 78, 349–355.
28. Ward, I. L. (1972). Prenatal stress feminizes and demasculinizes the behavior of males. *Science* 175, 82–84.
29. Ward, I. L. (1984). The prenatal stress syndrome: current status. *Psychoneuroendocrinology* 9, 3–11.
30. Ward, I. L., and Ward, B. (1985). Sexual behavior differentiation: effects of prenatal manipulations in rats. In *Handbook of Behavioral Neurobiology* (N. Adler, D. W. Pfaff, and R. W. Goy, Eds.), pp. 77–98. Plenum Press, New York.
31. Beatty, W. W. (1975). Gonadal hormones and sex differences in nonreproductive behavior in rodents: organizational and activational influences. *Horm. Behav.* 12, 112.
32. Baum, M. J. (1979). Differentiation of coital behavior in mammals: a comparative analysis. *Neurosci. Biobehav. Rev.* 3, 265.
33. Plapinger, L., and McEwen, B. S. (1978). Gonadal steroid-brain interactions in sexual differentiation. In *Biological Determinations of Sexual Behavior* (J. Hutchinson, Ed.), p. 153. John Wiley & Sons, New York.
34. McEwen, B. S. (1983). Gonadal steroid influences on brain development and sexual differentiation. In *Reproductive Physiology IV* (R. O. Greep, Ed.), p. 99. University Park Press, Baltimore.
35. Parsons, B., Rainbow, T. C., and McEwen, B. S. (1984). Organizational effects of testosterone via aromatization on feminine reproductive behavior and neural progestin receptors in rat brain. *Endocrinology* 115, 1412–1417.
36. Whalen, R. E., and Nakayama, K. (1965). Induction of estrous behavior: facilitation by repeated hormone treatment. *J. Endocrinol.* 33, 525–526.
37. McCarthy, M. M., Schlenker, E., and Pfaff, D. W. (1993). Neonatal intracerebral infusion of antisense DNA to estrogen receptor mRNA alters estrogen-dependent parameters in adult rats. *Endocrinology* 133, 433–439.
38. Pfaff, D. W., Brooks, P. J., Funabashi, T., Pfaus, J. G., and Mobbs, C. V. (1992). Gene memory in neuroendocrine and behavioral systems. *Ciba Found. Symp.* 168, 165–186.
39. Tarn, S.-O., Hache, R. J. G., and Deeley, R. G. (1986). Estrogen memory effect in human hepatocytes during repeated cell division without hormone. *Science* 234, 1234–1237.
40. Deeley, R. G., Udell, D. S., Burnes, A. T. H., Gordon, J. I., and Goldberger, R. F. (1977). Kinetics of avian vitellogenin messenger RNA induction. *J. Biol. Chem.* 252, 7913–7915.
41. Jost, J.-O., Ohno, T., Panyim, S., and Schuerch, A. R. (1978). Appearance of vitellogenin mRNA sequences and rate of vitellogenin synthesis in chicken liver following primary and secondary stimulation by 17β-estradiol. *Eur. J. Biochem.* 1978; 84:355–361.
42. Searle, P. F., and Tata, J. R. (1981). Vitellogenin gene expression in male *Xenopus* hepatocytes during primary and secondary stimulation with estrogen in cell cultures. *Cell* 23, 741–746.
43. Baker, H. J., and Shapiro, D. J. (1977). Kinetics of estrogen induction of *Xenopus laevis* vitellogenin messenger RNA as measured by hybridization to complementary DNA. *J. Biol. Chem.* 252, 8428–8434.
44. Burch, J. B. E., and Evans, M. I. (1986). Chromatin structural transitions and the phenomenon of vitellogenin gene memory in chickens. *Mol. Cell. Biol.* 6, 1886–1893.
45. Hache, R. J. G., Shui-Pang, T., Cochrane, A., Nesheim, M., and Deeley, R. G. (1987). Long-term effects of estrogen on avian liver: estrogen-inducible switch in expression of nuclear, hormone-binding proteins. *Mol. Cell. Biol.* 7, 3538–3547.
46. Gerall, A. A., and Dunlap, J. L. (1973). The effect of experience and hormones on the initial receptivity in female and male rats. *Physiol. Behav.* 10, 851–854.
48. Beach, F. A., and Orndoff, R. K. (1974). Variation in the responsiveness of female rats to ovarian hormones as a function of preceding hormonal deprivation. *Horm. Behav.* 5, 202–205.
49. Parsons, B., MacLusky, N. J., Krieger, M. S., McEwen, B. S., and Pfaff, D. W. (1979). The effects of long-term estrogen exposure on the induction of sexual behavior and measurements of brain estrogen and progestin receptors in the female rat. *Horm. Behav.* 13, 301–313.
50. Walker, W. A., and Feder, H. H. (1977). Inhibitory and facilitatory effects of various anti-estrogens on the induction of female sexual behavior by estradiol benzoate in guinea pigs. *Brain Res.* 134, 455–465.
51. Walker, W. A., and Feder, H. H. (1979). The comparative potency of various steroids to complete the priming process for lordosis in guinea pigs. *Horm. Behav.* 12, 299–308.
52. Wilcox, J. N., and Feder, H. H. (1983). Long-term priming with a low dosage of estradiol benzoate or an antiestrogen (enclomiphene) increases nuclear progestin receptor levels in brain. *Brain Res.* 266, 243–251.
53. Czaja, J. A., and Butera, P. C. (1985). Behavioral consequences of hormonal deprivation on the responsiveness of female rats to estradiol. *Physiol. Behav.* 35, 873–877.
54. Czaja, J. A., Butera, P. C., and McCaffrey, T. A. (1985). Duration of hormonal deprivation: influences on physiological and behavioral responsiveness to estradiol. *Horm. Behav.* 19, 52–63.
55. Green, R., Luttge, W. G., and Whalen, R. E. (1970). Induction of receptivity in ovariectomized rats by a single intravenous injection of estradiol-17-beta. *Physiol. Behav.* 5, 137–141.
56. Whalen, R. E., and Gorzalka, B. B. (1973). Effects of an estrogen antagonist on behavior and on estrogen retention in neural and peripheral target tissues. *Physiol. Behav.* 10, 35–40.
57. McEwen, B. S., Pfaff, D. W., Chaptal, C., and Luine, V. N. (1975). Brain cell nuclear retention of [^{3}H]estradiol doses able to promote lordosis: temporal and regional aspects. *Brain Res.* 86, 155–161.
58. Sodersten, O., Eneroth, O., and Hansen, S. (1981). Induction of sexual receptivity in ovariectomized rats by pulse administration of oestradiol-17β. *J. Endocrinol.* 89, 55–62.
59. Parsons, B., Rainbow, T. C., Pfaff, D. W., and McEwen, B. S. Oestradiol, sexual receptivity and cytosol progestin receptors in rat hypothalamus. *Nature* 292, 58–59.

60. Parsons, B., McEwen, B. S., and Pfaff, D. W. (1982). A discontinuous schedule of estradiol treatment is sufficient to activate progesterone-facilitated feminine sexual behavior and to increase cytosol receptors for progestins in the hypothalamus of the rat. *Endocrinology* 110, 613–619.
61. Harris, J., and Gorski, J. (1978). Evidence for a discontinuous requirement for estrogen in stimulation of deoxyribonucleic acid synthesis in the immature rat uterus. *Endocrinology* 103, 240–245.
62. Huggins, C., and Jensen, E. V. (1955). The depression of estrone-induced uterine growth by phenolic estrogens with oxygenated functions at position 6 or 16: the impeded estrogens. *Exp. Med.* 102, 335–342.
63. Anderson, J., Peck, E. J. Jr., and Clark, J. H. (1975). Estrogen-induced uterine responses and growth: relationship to estrogen binding by uterine nuclei. *Endocrinology* 96, 160–167.
64. Clark, J. H., Paszko, Z., and Peck, E. J. Jr. (1977). Nuclear binding and retention of the receptor estrogen complex: relation to the agonistic and antagonistic properties of estriol. *Endocrinology* 100, 91–96.
65. Bullock, D. W. (1970). Induction of heat in ovariectomized guinea pigs by brief exposure to estrogen and progesterone. *Horm. Behav.* 1, 137–143.
66. Johnston, P. G., and Davidson, J. M. (1979). Priming action of estrogen: minimum duration of exposure for feedback and behavioral effects. *Neuroendocrinology* 28, 155–159.
67. Wilcox, J. N., Barclay, S. R., and Feder, H. H. (1984). Administration of estradiol-17β in pulses to female guinea pigs: self-priming effects of estrogen on brain tissues mediating lordosis. *Physiol. Behav.* 32, 483–488.
68. Clark, A. S., and Roy, E. J. (1983). Behavioral and cellular responses to pulses of low doses of estradiol-17β. *Physiol. Behav.* 30, 561–565.
69. Jones, K. J., Pfaff, D. W., and McEwen, B. S. (1985). Early estrogen-induced nuclear changes in rat hypothalamic ventromedial neurons: an ultrastructural and morphometric analysis. *J. Comp. Neurol.* 239, 255–266.
70. Parsons, B., Rainbow, T. C., Pfaff, D. W., and McEwen, B. S. (1982). Hypothalamic protein synthesis essential for the activation of the lordosis reflex in the female rat. *Endocrinology* 110, 620–624.
71. Roy, E. J., Lynn, D. M., and Clark, A. S. (1985). Inhibition of sexual receptivity by anesthesia during estrogen priming. *Brain Res.* 337, 163–166.
72. Fadem, B. H., Barfield, R. J., and Whalen, R. E. (1979). Dose-response and time-response relationships between progesterone and the display of patterns of receptive and preceptive behavior in the female rat. *Horm. Behav.* 13, 40–48.
73. Tennent, B. J., Smith, E. R., and Davidson, J. M. (1980). The effects of estrogen and progesterone on female rat preceptive behavior. *Horm. Behav.* 14, 65–75.
74. Glaser, J. H., Rubin, B. S., and Barfield, R. J. (1983). Onset of the receptive and proceptive components of feminine sexual behavior in rats following the intravenous administration of progesterone. *Horm. Behav.* 17, 18–27.
75. McGinnis, M. Y., Parsons, B., Rainbow, T. C., Krey, L. C., and McEwen, B. S. (1981). Temporal relationship between cell nuclear progestin receptor levels and sexual receptivity following intravenous progesterone administration. *Brain Res.* 218, 365–371.
76. Zucker, I., and Goy, R. W. (1967). Sexual receptivity in the guinea pig: inhibitory and facilitatory actions of progesterone and related compounds. *J. Comp. Physiol. Psychol.* 64, 378–383.
77. Zucker, I. (1968). Biphasic effects of progesterone on sexual receptivity in the female guinea pig. *J. Comp. Physiol. Psychol.* 65, 472–478.
78. Roy, E. J., Lynn, D. M., and Clark, A. S. (1985). Inhibition of sexual receptivity by anesthesia during estrogen priming. *Brain Res.* 337, 163–166.
79. Axelson, J. (1983). Effect of voluntary exercise on sexual behavior in female rats. *Physiol. Behav.* 30, 963–966.
80. Axelson, J. (1987). Effects of voluntary and forced exercise on lordosis behavior in female rats. *Horm. Behav.* 21, 384–392.
81. Lauber, A. H., and Whalen, R. E. (1988). Muscarinic cholinergic modulation of hypothalamic estrogen binding site. *Brain Res.* 443, 21–26.
82. Lauber, A. H. (1988). Atropine sulfate modulates estrogen receptor binding by female, but not male, rat hypothalamus. *Brain Res. Bull.* 20, 273–276.
83. Blaustein, J. D., Brown, T. J., and McElroy, J. F. (1986). Some catecholamine inhibitors do not cause accumulation of nuclear estrogen receptors in rat hypothalamus and anterior pituitary gland. *Neuroendocrinology* 43, 143–149.
84. Blaustein, J. D., Brown, T. J., and Sweargengen, E. (1986). Dopamine-beta-hydroxylase inhibitors modulate the concentration of functional estrogen receptors in female rat hypothalamus and pituitary. *Neuroendocrinology* 43, 150–158.
85. Blaustein, J. D. (1987). The alpha-1-noradrenergic antagonist prazosin decreases the concentration of estrogen receptors in the female rat hypothalamus. *Brain Res.* 404, 39–50.
86. Clark, A. S., Nock, B., Feder, H. H., and Roy, E. J. (1985). Alpha-1-noradrenergic receptor blockade decreases nuclear estrogen receptor binding in guinea pig hypothalamus and preoptic area. *Brain Res.* 330, 197–199.
87. Nock, B., and Feder, H. H. (1979). Noradrenergic transmission and female sexual behavior of guinea pigs. *Brain Res.* 166, 369–380.
88. Nock, B., and Feder, H. H. (1981). Neurotransmitter modulation of steroid action in target cells that mediate reproduction and reproductive behavior. *Neurosci. Biobehav. Rev.* 5, 437–447.
89. Nock, B., and Feder, H. H. (1984). Alpha-1 noradrenergic regulation of hypothalamic progestin receptors and guinea pig sexual behavior. *Brain Res.* 310, 77–85.
90. Nock, B., Blaustein, J. D., and Feder, H. H. (1981). Changes in noradrenergic transmission alter the concentration of cytoplasmic progestin receptors in hypothalamus. *Brain Res.* 207, 371–396.
91. Montemayer, M. M., Clark, A. S., Lynn, D. M., and Roy, E. J. (1990). Modulation by norepinephrine of neural responses to estradiol. *Neuroendocrinology* 52, 473–480.
92. Lauber, A. H., Romano, G. J., and Pfaff, D. W. (1991). Sex differences in estradiol regulation of progestin receptor mRNA in rat mediobasal hypothalamus as demonstrated by in situ hybridization. *Neuroendocrinology* 53, 608–613.
93. Pfaff, D. W. (1982). Motivational concepts: definitions and distinctions. In *The Physiological Mechanisms of Motivation* (D. W. Pfaff, Ed.), pp. 3–24. Springer-Verlag, New York.
94. Emery, D. E., and Moss, R. L. (1984). Lesions confined to the ventromedial hypothalamus decrease the frequency of coital contacts in female rats. *Horm. Behav.* 18, 313–329.
95. Schwartz-Giblin, S., and Pfaff, D. W. (1985). Intrathecal bicuculline in rats evokes hyperalgesia and exaggerates avoidance of contact by other rats. *Soc. Neurosci. Abstr.* 83, 5.
96. Davis, R. E., Morrell, J. I., and Pfaff, D. W. (1977). Autoradiographic localization of sex steroid-concentrating cells in the brain of the teleost *Marcropodus opercularis* (Osteichthyes: Belontiidae). *Gen. Comp. Endocrinol.* 33, 496–505.
97. Kirn, Y. S., Stumpf, W. E., and Star, M. (1978). Topography of estrogen target cells in the forebrain of goldfish, *Carrassius auratus*. *J. Comp. Neurol.* 182, 611–620.
98. Morrell, J. I., and Pfaff, D. W. (1981). Autoradiographic technique for steroid hormone localization: application to the

vertebrate brain. In *Neuroendocrinology of Reproduction: Physiology and Behavior* (N. T. Adler, Ed.), pp. 519–531. Plenum Press, New York.

99. Kelley, D. B., Morrell, J. I., and Pfaff, D. W. (1975). Autoradiographic localization of hormone-concentrating cells in the brain of the amphibian, *Xenopus laevis*: I. Testosterone. *J. Comp. Neurol.* 164, 47–62.
100. Morrell, J. I., Kelley, D. B., and Pfaff, D. W. (1975). Autoradiographic localization of hormone-concentrating cells in the brain of an amphibian, *Xenopus laevis*: II. Estradiol. *J. Comp. Neurol.* 164, 63–78.
101. Roy, E. J., Wilson, M. A., and Kelley, D. B. (1986). Estrogen-induced progestin receptors in the brain and pituitary of the South African clawed frog, *Xenopus laevis*. *Neuroendocrinology* 42, 51–56.
102. Kelley, D. B., Lieberburg, I., McEwen, B. S., and Pfaff, D. W. (1978). Autoradiographic and biochemical studies of steroid hormone-concentrating cells in the brain of *Rana pipiens*. *Brain Res.* 140, 287–305.
103. Morrell, J. L., Crews, D., Ballin, A., Morgentaler, A., and Pfaff, D. W. (1979). ^{3}H-estradiol, ^{3}H-testosterone and ^{3}H-dihydrotestosterone localization in the brain of the lizard *Anolis carolinensis*: an autoradiographic study. *J. Comp. Neurol.* 188, 201–224.
104. Martinez-Vargas, M. C., Keefer, D. A., and Stumpf, W. E. (1978). Estrogen localization in the brain of the lizard, *Anolis carolinensis*. *J. Exp. Zool.* 205, 141–147.
105. Halpern, M., Morrell, J. I., and Pfaff, D. W. (1982). Cellular ^{3}H-estradiol and ^{3}H-testosterone localization in the brains of garter snakes: an auto-radiographic study. *Gen. Comp. Endocrinol.* 46, 211–224.
106. Meyer, C. (1973). Testosterone concentration in the male chick brain: an autoradiographic survey. *Science* 180, 1381–1383.
107. Wood-Gush, D. G. M., Langley, G. A. S., Leitch, A. F., Gentle, M. J., and Gilbert, A. B. (1977). An autoradiographic study of sex steroids in the chicken telencephalon. *Gen. Comp. Endocrinol.* 31, 161–168.
108. Barfield, R. J., Ronay, G., and Pfaff, D. W. (1978). Autoradiographic localization of androgen-concentrating cells in the brain of the male domestic fowl. *Neuroendocrinology* 26, 297–311.
109. Martinez-Vargas, M. C., Stumpf, W. E., and Star, M. (1976). Anatomical distribution of estrogen target cells in avian CNS: a comparison with the mammalian CNS. *J. Comp. Neurol.* 167, 83–104.
110. Zigmond, R. E., Nottebohm, F., and Pfaff, D. W. (1973). Androgen-concentrating cells in the midbrain of a songbird. *Science* 179, 1005–1007.
111. Zigmond, R. E., Detrick, R. A., and Pfaff, D. W. (1980). An autoradiographic study of the localization of androgen concentrating cells in the chaffinch. *Brain Res.* 182, 369–381.
112. Arnold, A., Nottebohm, F., and Pfaff, D. W. (1976). Hormone-concentrating cells in vocal control and other areas of the brain of the zebra finch *(Poephila guttala)*. *J. Comp. Neurol.* 165, 487–512.
113. Pfaff, D. W. (1968). Autoradiographic localization of radioactivity in rat brain after injection of tritiated sex hormones. *Science* 161, 1355–1356.
114. Stumpf, W. E. (1968). Estradiol concentrating neurons: topography in the hypothalamus by dry-mount autoradiography. *Science* 162, 1001–1003.
115. Star, M., and Stumpf, W. E. (1972). Cellular localization of androgen in the brain and pituitary after the injection of tritiated testosterone. *Experientia* 28, 1364–1366.
116. Sheridan, P. J. (1979). The nucleus interstitialis striae terminalis and the nucleus amygdaloideus medialis: prime targets for androgen in the rat forebrain. *Endocrinology* 104, 130–136.
117. Krieger, M. S., Morrell, J. I., and Pfaff, D. W. (1976). Autoradiographic localization of estradiol-concentrating cells in the female hamster brain. *Neuroendocrinology* 22, 193–205.
118. Warembourg, M. (1977). Topographical distribution of estrogen-concentrating cells in the brain and pituitary of the guinea pig. *Neurosci. Lett.* 5, 315–319.
119. Commins, D., and Yahr, P. (1985). Autoradiographic localization of estrogen and androgen receptors in the sexually dimorphic area and other regions of the gerbil brain. *J. Comp. Neurol.* 231, 473–489.
120. Anderson, J., Peck, E. J. Jr., and Clark, J. H. (1975). Estrogen-induced uterine responses and growth: relationship to estrogen binding by uterine nuclei. *Endocrinology* 96, 160–167.
121. Warembourg, M. (1978). Radioautographic study of the brain and pituitary after (^{3}H)progesterone injection into estrogen-primed ovariectomized guinea pigs. *Neurosci. Lett.* 7, 1–5.
122. Warembourg, M. (1978). Radioautographic study of the rat brain, uterus and vagina after (^{3}H)R5020 injection. *Mol. Cell. Endocrinol.* 12, 67–79.
123. Morrell, J. I., Ballin, A., and Pfaff, D. W. (1977). Autoradiographic demonstration of the pattern of ^{3}H-estradiol concentrating cells in the brain of a carnivore, the mink, *Mustela vison*. *Anat. Rec.* 189, 609–624.
124. Rees, H. D., Switz, G. M., and Michael, R. P. (1980). The estrogen-sensitive neural system in the brain of female cats. *J. Comp. Neurol.* 193, 789–804.
125. Pfaff, D. W., Gerlach, J., McEwen, B. S., Ferin, M., Carmel, O., and Zimmerman, E. (1976). Autoradiographic localization of hormone-concentrating cells in the brain of the female rhesus monkey. *J. Comp. Neurol.* 170, 279–294.
126. Gerlach, J., McEwen, B. S., Pfaff, D., Moskovitz, S., Ferin, M., Carmel, P. W., and Zimmerman, E. A. (1976). Cells in regions of rhesus monkey brain and pituitary retain radioactive estradiol, corticosterone and cortisol differentially. *Brain Res.* 103, 603–612.
127. Rees, H. D., and Michael, R. P. (1982). Brain cells of the male rhesus monkey accumulate ^{3}H-testosterone or its metabolites. *J. Comp. Neurol.* 206, 273–277.
128. Michael, R. O., and Rees, H. D. (1982). Autoradiographic localization of ^{3}H-dihydrotestosterone in the preoptic area, hypothalamus, and amygdala of a male rhesus monkey. *Life Sci.* 30, 2087.
129. Warembourg, M. (1977). Topographical distribution of estrogen-concentrating cells in the brain and pituitary of the squirrel monkey. *Neurosci. Lett.* 5, 315–319.
130. Morrell, J. I., Kelley, D., and Pfaff, D. W. (1975). Sex steroid binding in the brains of vertebrates: studies with light microscopic autoradiography. In *Brain–Endocrine Interaction II* (K. Kingge, D. Scott, H. Kobayashi, and S. Ishii, Eds.), pp. 230–256. Karger, Basel.
131. Pfaff, D. W., and Conrad, L. C. A. (1978). Hypothalamic neuroanatomy: steroid hormone binding and patterns of axonal projections. *Int. Rev. Cytol.* 54, 245–265.
132. Morrell, J. I., and Pfaff, D. W. (1981). Autoradiographic technique for steroid hormone localization: application to the vertebrate brain. In *Neuroendocrinology of Reproduction: Physiology and Behavior* (N. T. Adler, Ed.), pp. 519–531. Plenum Press, New York.
133. Krieger, M. S., Morrell, J. I., and Pfaff, D. W. (1978). Quantitative autoradiographic analysis of estrogen concentrating cells in hypothalamus, preoptic area and amygdala. *Soc. Neurosci. Abstr.* 4, 347.
134. Morrell, J. I., Krieger, M. S., and Pfaff, D. W. (1986). Quantitative autoradiographic analysis of estradiol retention by cells in the preoptic area, hypothalamus and amygdala. *Exp. Brain Res.* 62, 343–354.

135. Morrell, J. T., and Pfaff, D. W. (1983). Immunocytochemistry of steroid hormone receiving cells in the central nervous system. *Methods Enzymol.* 103, 639–662.
136. Yamaguchi, K., Akaishi, T., and Negoro, H. (1979). Effect of estrogen treatment on plasma oxytocin and vasopressin in ovariectomized rats. *Endocrinol. Jpn.* 26, 197–205.
137. Skowsky, W. R., Swan, L., and Smith, P. (1979). Effects of sex steroid hormones on arginine vasopressin in intact and castrated male and female rats. *Endocrinology* 104, 105–108.
138. Crowley, W. R., O'Donohue, T. L., George, J. M., and Jacobowitz, D. M. (1978). Changes in pituitary oxytocin and vasopressin during the estrous cycle and after ovarian hormones: evidence for mediation by norepinephrine. *Life Sci.* 23, 2579–2586.
139. Rhodes, C. H., Morrell, J. I., and Pfaff, D. W. (1981). Distribution of estrogen-concentrating, neurophysin-containing magnocellular neurons in the rat hypothalamus as demonstrated by a technique combining steroid autoradiography and immunohistology in the same tissue. *Neuroendocrinology* 33, 18–23.
140. Rhodes, C. H., Morrell, J. I., and Pfaff, D. W. (1982). Estrogen-concentrating neurophysin-containing hypothalamic magnocellular neurons in the vasopressin-deficient (Brattleboro) rat: a study combining steroid autoradiography and immunocytochemistry. *J. Neurosci.* 2, 1718–1724.
141. Sar, M. (1984). Estradiol is concentrated in tyrosine hydroxylase-containing neurons of the hypothalamus. *Science* 223, 938–940.
142. McCabe, J., Morrell, J., and Pfaff, D. W. Unpublished observations.
143. Fox, S., Harlan, R., Shivers, B., and Pfaff, D. W. (1990). Chemical characterization of neuroendocrine targets for progesterone in the female rat brain and pituitary. *Neuroendocrinology* 51, 276–283.
144. Morrell, J. I., McGinty, J. F., and Pfaff, D. W. (1985). A subset of β-endorphin- or dynorphin-containing neurons in the medial basal hypothalamus accumulates estradiol. *Neuroendocrinology* 41, 417–426.
145. Leranth, C., Sakamoto, H., Maclusky, N. J., Shanabrough, M., and Naffolin, F. (1985). Estrogen responsive cells in the arcuate nucleus of the rat contain glutamic acid decarboxylase (GAD): an electron microscopic immunocytochemical study. *Brain Res.* 331, 376–381.
146. Shivers, B. D., Harlan, R. E., and Pfaff, D. W. (1989). A subset of neurons containing immunoreactive prolactin is a target for estrogen regulation of gene expression in rat hypothalamus. *Neuroendocrinology* 49, 23–27.
147. Shivers, B., Harlan, R., Morrell, J., and Pfaff, D. (1983). Absence of oestradiol concentration in cell nuclei of LHRH-immunoreactive neurons. *Nature* 304, 345–347.
148. Fox, S., Harlan, R., Shivers, B., and Pfaff, D. W. (1990). Chemical characterization of neuroendocrine targets for progesterone in the female rat brain and pituitary. *Neuroendocrinology* 51, 276–283.
149. Blaustein, J. D., and Torcotte, J. C. (1989). Estradiol-induced progestin receptor immunoreactivity is found only in estrogen receptor-immunoreactive cells in guinea pig brain. *Neuroendocrinology* 49, 454–461.
150. Olster, D. H., and Blaustein, J. D. (1990). Immunocytochemical colocalization of progestin receptors and β-endorphin or enkephalin in the hypothalamus of female guinea pigs. *J. Neurobiol.* 21, 768–780.
151. Akesson, T. R., and Micevych, P. E. (1991). Endogenous opioid-immunoreactive neurons of the ventromedial hypothalamic nucleus concentrate estrogen in male and female rats. *J. Neurosci. Res.* 28, 359–366.
152. Nielsen, K. H., and Blaustein, J. D. (1990). Many progestin receptor-containing neurons in the guinea pig ventrolateral hypothalamus contain substance P: immunocytochemical evidence. *Brain Res.* 517, 175–181.
153. Priest, C., and Micevych, P. E. (1992). Unpublished data.
154. Olster, D. H., and Blaustein, J. D. (1992). Estradiol pulses induce progestin receptors selectively in substance P-immunoreactive neurons in the ventrolateral hypothalamus of female guinea pigs. *J. Neurobiol.* 23, 293–301.
155. Langub, M. C. Jr., and Watson, R. E. Jr. (1992). Estrogen receptive neurons in the preoptic area of the rat are post-synaptic targets of a sexually dimorphic enkephalinergic fiber plexus. *Brain Res.* 573, 61–69.
156. Arnold, A. P. (1980). A technique for simultaneous steroid autoradiography and retrograde labeling of neurons. *Brain Res.* 194, 210–212.
157. Morrell, J. I., and Pfaff, D. W. (1982). Characterization of estrogen-concentrating hypothalamic neurons by their axonal projections. *Science* 217, 1273–1276.
158. Morrell, J. I., Schwanzel-Fukuda, M., Fahrbach, S. E., and Pfaff, D. W. (1984). Axonal projections and peptide content of steroid hormone concentrating neurons. *Peptides* 5, 227–239.
159. Numan, M., and Callahan, E. C. (1980). The connections of the medial preoptic region and maternal behavior in the rat. *Physiol. Behav.* 25, 653–665.
160. Gaffori, O., and Le Moal, M. (1979). Disruption of maternal behavior and appearances of cannibalism after ventral mesencephalic tegmentum lesions. *Physiol. Behav.* 23, 317–323.
161. Numan, M., and Nagle, D. S. (1983). The preoptic area and the substantia nigra interact in the control of maternal behavior in the rat. *Behav. Neurosci.* 97, 120–139.
162. Fahrbach, S., Morrell, J. I., and Pfaff, D. W. (1983). Estrogen-concentrating neurons in the medial preoptic area send axons to the ventral tegmental area and amygdala. *Soc. Neurosci. Abstr.* 9, 516.
163. Fahrbach, S., Morrell, J. I., and Pfaff, D. W. (1986). Identification of medial preoptic neurons that concentrate estradiol and project to the midbrain in the rat. *J. Comp. Neurol.* 247, 364–382.
164. Morrell, J. I., and Pfaff, D. W. (1984). Hypothalamic and limbic system estradiol (E_2) concentrating neurons that project to the amygdala. *Soc. Neurosci. Abstr.* 10.
165. Rees, H. D., and Michael, R. P. (1980). The estrogen-sensitive neural system in the brain of female cats. *J. Comp. Neurol.* 193, 789–804.
166. Cottingham, S., and Pfaff, D. W. (1986). Interconnectedness of steroid hormone binding neurons: existence and implications. In *Current Topics in Neuroendocrinology* (D. Ganten and D. Pfaff, Eds.), Vol. 7, pp. 223–250. Springer-Verlag, Berlin.
167. Jensen, E. V., and Jacobson, H. I. (1962). Basic guides to the mechanism of estrogen action. *Recent Prog. Horm. Res.* 18, 387–414.
168. Buller, W., and O'Malley, B. (1976). The biology and mechanism of steroid hormone receptor interaction with the eukaryotic nucleus. *Biochem. Pharmacol.* 25, 1–12.
169. Eisenfeld, A. J., and Axelrod, J. (1966). Effect of steroid hormones, ovariectomy, estrogen pretreatment, sex and immaturity on the distribution of ^{3}H-estradiol. *Endocrinology* 79, 38–42.
170. Kato, J., and Villee, C. A. (1967). Factors affecting uptake of estradiol-6, 7-^{3}H by the hypophysis and hypothalamus. *Endocrinology* 80, 1113–1138.
171. Kato, J., and Villee, C. A. (1967). Preferential uptake of estradiol by the anterior hypothalamus of the rat. *Endocrinology* 80, 567–575.
172. Zigmond, R. E., and McEwen, B. S. (1970). Selective retention of estradiol by cell nuclei in specific brain regions of the ovariectomized rat. *J. Neurochem.* 17, 889–899.

173. Sherman, M., and Stevens, T. (1984). Structure of mammalian steroid receptors: Evolving concepts and methodological development. *Annu. Rev. Physiol.* 46, 83–105.
174. Notides, A. C., Lerner, N., and Hamilton, D. E. (1981). Positive cooperativity of the estrogen receptor. *Proc. Natl. Acad. Sci. U S A* 78, 4926–4930.
175. Sasson, S., and Notides, A. C. (1984). The estriol-induced inhibition of the estrogen receptor's positive cooperativity. *J. Steroid Biochem.* 20, 1021–1026.
176. McEwen, B. S., Davis, P. G., Parsons, B., and Pfaff, D. W. (1979). The brain as a target for steroid hormone action. *Annu. Rev. Neurosci.* 2, 65–112.
177. Peck, E. J., and Kelner, K. L. (1983). A comparison of central and peripheral estrogen targets. In *Gene Regulation by Steroid Hormones II* (A. K. Roy and J. H. Clark, Eds.), pp. 201–217. Springer-Verlag, New York.
178. Kelner, K. L., Kirchick, H. J., and Peck, E. J. Jr. (1982). Differential sensitivity of estrogen target tissues: the role of the receptor. *Endocrinology* 111, 1986–1995.
179. Kelner, K. L., and Peck, E. J. Jr. (1981). Resolution of estrogen binding species in the hypothalamus and pituitary. *J. Receptor Res.* 2, 47–62.
180. Plapinger, L., and McEwen, B. S. (1973). Ontogeny of estradiol- binding sites in rat brain: I. appearance of presumptive adult receptors in cytosol and nuclei. *Endocrinology* 93, 1119–1128.
181. Lieberburg, I., Krey, L. C., and McEwen, B. S. (1979). Sex differences in serum testosterone and in exchangeable brain cell nuclear estradiol during the neonatal period in rats. *Brain Res.* 178, 207–212.
182. Westley, B. R., and Salaman, D. F. (1976). Role of oestrogen receptor in androgen-induced sexual differentiation of the brain. *Nature* 262, 407–408.
183. Friedman, W. J., McEwen, B. S., Toran-Allerand, C. D., and Gerlach, J. L. (1983). Perinatal development of hypothalamic and cortical estrogen receptors in mouse brain: methodological aspects. *Dev. Brain Res.* 11, 19–27.
184. Gerlach, J. L., McEwen, B. S., Toran-Allerand, C. D., and Friedman, W. J. (1983). Perinatal development of estrogen receptors in mouse brain assessed by radioautography, nuclear isolation and receptor assay. *Dev. Brain Res.* 11, 7–18.
185. Ryer, H. I., and Feder, H. H. (1984). Development of steroid-receptor systems in guinea pig brain: II. Cytoplasmic progestin receptors. *Dev. Brain Res.* 13, 15–21.
186. Wise, P. M., and Camp, P. (1984). Changes in concentrations of estradiol nuclear receptors in the preoptic area, medial basal hypothalamus, amygdala, and pituitary gland of middle-aged and old cycling rats. *Endocrinology* 114, 92–98.
187. Wise, P. M., McEwen, B. S., Parsons, B., and Rainbow, T. C. (1984). Age-related changes in cytoplasmic estradiol receptor concentrations in microdissected brain nuclei: correlations with changes in steroid-induced sexual behavior. *Brain Res.* 321, 119–126.
188. Parsons, B., and Pfaff, D. W. (1985). Hypothalamic progestin receptors and female reproductive behavior. In *Current Topics in Neuroendocrinology* (D. Ganten and D. Pfaff, Eds.), Vol. 5, pp. 103–140. Springer-Verlag, Berlin.
189. McGinnis, M. Y., Krey, L. C., MacLusky, N. J., and McEwen, B. S. (1981). Steroid receptor levels in intact and ovariectomized estrogen-treated rats: an examination of quantitative temporal and endocrine factors influencing the efficacy of an estradiol stimulus. *Neuroendocrinology* 33, 158–165.
190. Anderson, J. N., Clark, J. H., and Peck, E. J. Jr. (1972). The relationship between nuclear receptor estrogen binding and uterotrophic responses. *Biochem. Biophys. Res. Commun.* 48, 1460–1468.
191. Rainbow, T. C., Parsons, B., MacLusky, N., and McEwen, B. (1982). Estradiol receptors in rat hypothalamic and limbic nuclei. *J. Neurosci.* 2, 1439–1445.
192. Nordeen, E. J., and Yahr, P. (1983). A regional analysis of estrogen binding to hypothalamic cell nuclei in relation to masculinization and de-feminization. *J. Neurosci.* 3, 933–941.
193. Feder, H. H., Siegel, H., and Wade, G. N. (1974). Uptake of [6,7-^{3}H]estradiol-17β in ovariectomized rats, guinea pigs, and hamsters: correlation with species differences in behavioral responsiveness to estradiol. *Brain Res.* 71, 93–103.
194. Clark, A. S., Nock, B., Feder, H. H., Roy, E. J. (1985). α_1-Noradrenergic receptor blockade decreases nuclear estrogen receptor binding in guinea pig hypothalamus and preoptic area. *Brain Res.* 330, 197–199.
195. Blaustein, J. D. (1986). Cell nuclear accumulation of estrogen receptors in rat brain and pituitary gland after treatment with a dopamine-β-hydroxylase inhibitor. *Neuroendocrinology* 42, 44–50.
196. Tetel, M. J., and Blaustein, J. D. (1991). Immunocytochemical evidence for noradrenergic regulation of estrogen receptor concentrations in the guinea pig hypothalamus. *Brain Res.* 565, 321–329.
197. Roy, E. J., and Wilson, M. A. (1981). Diurnal rhythm of cytoplasmic estrogen receptors in the rat brain in the absence of circulating estrogen. *Science* 213, 1525–1527.
198. Wilson, M. A., Clark, A. S., Clyde, V., and Roy, E. J. (1983). Characterization of a pineal-independent diurnal rhythm in neural estrogen receptors and its possible behavioral consequences. *Neuroendocrinology* 37, 14–22.
199. Greene, G. L. (1984). Application of immunochemical techniques to the analysis of estrogen receptor structure and function. In *Biochemical Actions of Hormones* (G. Litwack, Ed.), Vol. XL, pp. 207–239. Academic Press, New York.
200. King, W. J., DeSombre, E. R., Jensen, E. V., and Greene, G. L. (1985). Comparison of immunocytochemical and steroid-binding assays for estrogen receptor in human breast tumors. *Cancer Res.* 45, 293–304.
201. King, W. J., and Greene, G. L. (1984). Monoclonal antibodies localize oestrogen receptor in the nuclei of target cells. *Nature* 307, 745–747.
202. Welshons, W. V., Lieberman, M. E., and Gorski, J. (1984). Nuclear localization of unoccupied oestrogen receptors: cytochalasin enucleation of GH3 cells. *Nature* 307, 747–749.
203. Meisel, R., Morrell, J. I., and Pfaff, D. W. Unpublished observations.
204. Kato, J. (1985). Progesterone receptors in brain and hypophysis. In *Current Topics in Neuroendocrinology* (D. Ganten and D. Pfaff, Eds.), Vol. 5, p. 221. Springer-Verlag, Berlin.
205. MacLusky, N. J., and McEwen, B. S. (1980). Progestin receptors in rat brain: distribution and properties of cytoplasmic progestin binding sites. *Endocrinology* 106, 192–202.
206. Kato, J., and Onouchi, T. (1983). Progestin receptors in female rat brain and hypophysis in the development from fetal to postnatal stages. *Endocrinology* 113, 29–36.
207. Ryer, H. I., and Feder, H. H. (1984). Development of steroid-receptor systems in guinea pig brain: I. Cytoplasmic estrogen receptor. *Dev. Brain Res.* 13, 9–14.
208. Ryer, H. I., and Feder, H. H. (1984). Development of steroid-receptor systems in guinea pig brain: III. Nuclear progestin receptors. *Dev. Brain Res.* 13, 23–27.
209. Wise, P. M., and Parsons, B. (1984). Nuclear estradiol and cytosol progestin receptor concentrations in the brain and the pituitary gland and sexual behavior in ovariectomized estradiol-treated middle-aged rats. *Endocrinology* 115, 810–816.
210. Kato, J., and Onouchi, T. (1977). Specific progesterone receptors in the hypothalamus and anterior hypophysis of the rat. *Endocrinology* 101, 920–928.

211. MacLusky, N. J., and McEwen, B. S. (1978). Oestrogen modulates progestin receptor concentrations in some rat brain regions but not in others. *Nature* 274, 276–278.
212. Parsons, B., MacLusky, N. J., Krey, K., Pfaff, D. W., and McEwen, B. S. (1980). The temporal relationship between estrogen-inducible progestin receptors in the female rat brain and the time course of estrogen activation of mating behavior. *Endocrinology* 107, 774–779.
213. Barfield, R. J., Glaser, J. H., Rubin, B. S., and Etgen, A. M. (1984). Behavioral effects of progestin in the brain: a review. *Psychoneuroendocrinology* 9, 217–231.
214. Moguilevsky, M., and Raynaud, J.-P. (1979). The relevance of hypothalamic and hypophyseal progestin receptor regulation in the induction and inhibition of sexual behavior in the female rat. *Endocrinology* 105, 516–522.
215. Parsons, B., Rainbow, T., MacLusky, N., and McEwen, B. (1982). Progestin receptor levels in rat hypothalamic and limbic nuclei. *J. Neurosci.* 2, 1446–1452.
216. Rainbow, T. C., McGinnis, M. Y., Krey, L. C., and McEwen, B. S. (1982). Nuclear progestin receptors in rat brain and pituitary. *Neuroendocrinology* 34, 426–432.
217. Blaustein, J. D., and Feder, H. H. (1979). Cytoplasmic progestin-receptors in guinea pig brain: characteristics and relationship to the induction of sexual behavior. *Brain Res.* 169, 681–697.
218. Blaustein, J. D., and Feder, H. H. (1979). Cytoplasmic progestin receptors in female guinea pig brain and their relationship to refractoriness in expression of female sexual behavior. *Brain Res.* 177, 489–498.
219. Lisk, R. D. (1962). Diencephalic placement of estradiol and sexual receptivity in the female rat. *Am. J. Physiol.* 203, 493.
220. Barfield, R. J., and Chen, J. J. (1977). Activation of estrous behavior in ovariectomized rats by intracerebral implants of estradiol benzoate. *Endocrinology* 101, 1716.
221. Rubin, B. S., and Barfield, R. J. (1980). Priming of estrous responsiveness by implants of 17 β-estradiol in the ventromedial hypothalamic nucleus of female rats. *Endocrinology* 106, 504–509.
222. Davis, P. G., McEwen, B., and Pfaff, D. W. (1979). Localized behavioral effects of tritiated estradiol implants in the ventromedial hypothalamus of female rats. *Endocrinology* 104, 898–903.
223. Davis, P. G., Krieger, M. S., Barfield, R. J., McEwen, B. S., and Pfaff, D. W. (1982). The site of action of intrahypothalamic estrogen implants in feminine sexual behavior: an autoradiographic analysis. *Endocrinology* 111, 1581–1586.
224. Malsbury, C. W., Miceli, M. O., and Scouten, C. W. Neural basis of reproductive behavior. In *The Hamster* (H. I. Siegel, Ed.), pp. 229–259. Plenum Press, New York.
225. Blache, D., Fabre-Nys, C. J., and Venier, G. (1991). Ventromedial hypothalamus as a target for oestradiol action on proceptivity and luteinizing hormone surge of the ewe. *Brain Res.* 546, 241–249.
226. Davis, P. G., and Barfield, R. J. (1979). Activation of feminine sexual behavior in castrated male rats by intrahypothalamic implants of estradiol benzoate. *Neuroendocrinology* 28, 228–233.
227. Davis, P. G., and Barfield, R. J. (1979). Activation of masculine sexual behavior by intracranial estradiol benzoate implants in male rats. *Neuroendocrinology* 28, 217–227.
228. Wade, G. N., and Zucker, I. (1970). Modulation of food intake and locomotor activity in female rats by diencephalic hormone implants. *J. Comp. Physiol. Psychol.* 72, 328–336.
229. Wade, G. N., and Zucker, I. (1970). Modulation of food intake and locomotor activity in female rats by diencephalic hormone implants. *J. Comp. Physiol. Psychol.* 72, 328–336.
230. Fahrbach, S., Meisel, R. L., and Pfaff, D. W. (1985). Preoptic implants of estradiol increase wheel running but not the open field activity of female rats. *Physiol. Behav.* 35, 985–992.
231. Ross, J., Claybaugh, C., Clemens, L. G., and Gorski, R. A. (1971). Short latency induction of estrous behavior with intracerebral gonadal hormones in ovariectomized rats. *Endocrinology* 89, 32–38.
232. Wade, G. N., and Feder, H. H. (1972). [1,2-^{3}H]Progesterone uptake by guinea pig brain and uterus: differential localization, time-course of uptake and metabolism, and effects of age, sex, estrogen-priming and competing steroids. *Brain Res.* 45, 525–543.
233. Franck, J. E., and Ward, I. L. (1981). Intralimbic progesterone and methysergide facilitate lordotic behavior in estrogen-primed female rats. *Neuroendocrinology* 32, 50–56.
234. Tennent, B. J., Smith, E. R., and Davidson, J. M. (1982). Effects of progesterone implants in the habenula and midbrain on preceptive and receptive behavior in the female rat. *Horm. Behav.* 16, 352–363.
235. Morin, L. O., and Feder, H. H. (1974). Hypothalamic progesterone implants and facilitation of lordosis behavior in estrogen-primed ovariectomized guinea pigs. *Brain Res.* 70, 81–93.
236. Rubin, B. S., and Barfield, R. J. (1983). Progesterone in the ventromedial hypothalamus facilitates estrous behavior in ovariectomized estrogen-primed rats. *Endocrinology* 113, 797–804.
237. Rubin, B. S., and Barfield, R. J. (1983). Induction of estrous behavior in ovariectomized rats by sequential replacement of estrogen and progesterone to the ventromedial hypothalamus. *Neuroendocrinology* 37, 218–224.
238. Rubin, B. S., and Barfield, R. J. (1984). Progesterone in the ventromedial hypothalamus of ovariectomized, estrogen-primed rats inhibits subsequent facilitation of estrous behavior by systemic progesterone. *Brain Res.* 294, 1–8.
239. Glaser, J. H., Etgen, A. M., and Barfield, R. J. (1985). Intrahypothalamic effects of progestin agonists on estrous behavior and progestin receptor binding. *Physiol. Behav.* 34, 871–877.
240. Capony, E., and Rochefort, H. (1975). *In vivo* effect of anti-estrogens on the localization of replenishment of estrogen receptor. *Mol. Cell. Endocrinol.* 3, 213–251.
241. Ruh, T. S., and Ruh, M. E. (1974). The effect of anti-estrogens on the nuclear binding of the estrogen receptor. *Steroids* 24, 209–224.
242. Chazal, G., Faudon, M., Gogan, E., and Rotsztejn, W. (1975). Effects of two estradiol antagonists upon the estradiol uptake in the rat brain and peripheral tissues. *Brain Res.* 89, 245–254.
243. Luine, V. N., and McEwen, B. S. (1977). Effects of an estrogen antagonist on enzyme activities and [^{3}H]-estradiol nuclear binding in uterus, pituitary and brain. *Endocrinology* 100, 903–910.
244. Luttge, W. G., Gray, H. E., and Hughes, J. R. (1976). Regional and subcellular [^{3}H]estradiol localization in selected brain regions and pituitary of female mice: effects of unlabelled estradiol and various anti-hormones. *Brain Res.* 104, 273–281.
245. Roy, E. J., and Wade, G. N. (1977). Binding of ^{3}H-estradiol by brain cell nuclei and female rat sexual behavior: inhibition by antiestrogens. *Brain Res.* 126, 73–87.
246. Katzenellenbogen, B. S., and Ferguson, E. R. (1975). Antiestrogen action in the uterus: biological ineffectiveness of nuclear bound estradiol after antiestrogen. *Endocrinology* 97, 1–12.
247. Arai, Y., and Gorski, R. A. (1986). Effect of anti-estrogen on steroid induced sexual receptivity in ovariectomized rats. *Physiol. Behav.* 3, 351–353.

248. Powers, J. B. (1975). Anti-estrogenic suppression of the lordosis response in female rats. *Horm. Behav.* 6, 379–392.
249. Ross, J. W., Paup, D. C., Brant-Zawadski, M., Marshall, J. R., and Gorski, R. A. (1973). Effects of *cis-* and *trans*-clomiphene in the induction of sexual behavior. *Endocrinology* 93, 681–685.
250. Meyerson, B. J., and Lindstrom, L. (1986). Effect of an oestrogen antagonist ethamoxytriphetol [MER-25] on oestrus behavior in rats. *Acta Endocrinol (Kbh)* 59, 41–48.
251. Blaustein, J. D., and Wade, G. N. (1977). Sequential inhibition of sexual behavior by progesterone in female rats: comparison with a synthetic antiestrogen. *J. Comp. Physiol. Psychol.* 91, 752–760.
252. Walker, W. A., and Feder, H. H. (1977). Anti-estrogen effects on estrogen accumulation in brain cell nuclei: neurochemical correlates of estrogen action on female sexual behavior in guinea pigs. *Brain Res.* 134, 467–480.
253. Howard, S. B., Etgen, A. M., and Barfield, R. J. (1984). Antagonism of central estrogen action by intracerebral implants of tamoxifen. *Horm. Behav.* 18, 256–266.
254. Meisel, R. L., Dohanich, G. O., McEwen, B. S., and Pfaff, D. W. (1985). Brain region specificity in antiestrogen inhibition of lordosis in female rats. *Soc. Neurosci. Abstr.* 11, 161.
255. Meisel, R. L., Dohanich, G. O., McEwen, B. S., and Pfaff, D. W. (1987). Antagonism of sexual behavior in female rats by ventromedial hypothalamic implants of antiestrogen. *Neuroendocrinology* 45, 201–207.
256. Feder, H. H., Blaustein, J. D., and Nock, B. L. (1979). Oestrogen-progestin regulation of female sexual behavior in guinea pigs. *J. Steroid Biochem.* 11, 873–877.
257. Blaustein, J. D., and Feder, H. H. (1979). Cytoplasmic progestin-receptors in guinea pig brain: characteristics and relationship to the induction of sexual behavior. *Brain Res.* 169, 481–497.
258. Blaustein, J. D., and Feder, H. H. (1980). Nuclear progestin receptors in guinea pig brain measured by an *in vitro* exchange assay after hormonal treatments that affect lordosis. *Endocrinology* 106, 1061–1069.
259. McGinnis, M. Y., Parsons, B., Rainbow, T. C., Krey, L. C., and McEwen, B. S. (1981). Temporal relationship between cell nuclear progestin receptor levels and sexual receptivity following intravenous progesterone administration. *Brain Res.* 218, 365–371.
260. Blaustein, J. D., Ryer, H. I., and Feder, H. H. (1980). A sex difference in the progestin receptor system of guinea pig brain. *Neuroendocrinology* 31, 403–409.
261. Brown, T. J., and Blaustein, J. D. (1984). Inhibition of sexual behavior in female guinea pigs by a progestin receptor antagonist. *Brain Res.* 301, 343–349.
262. Etgen, A. M., and Barfield, R. J. (1986). Antagonism of female sexual behavior with intracerebral implants of antiprogestin RU 38486: correlation with binding to neural progestin receptors. *Endocrinology* 119, 1610–1617.
263. Vathy, I. U., Etgen, A. M., and Barfield, R. J. (1989). Actions of RU 38486 on progesterone facilitation and sequential inhibition of rat estrous behavior: correlation with neural progestin receptor levels. *Horm. Behav.* 23, 43–56.
264. Bueno, J., and Pfaff, D. W. (1976). Single unit recording in hypothalamus and preoptic area of estrogen-treated and untreated ovariectomized female rats. *Brain Res.* 101, 67–78.
265. Kow, L.-M., and Pfaff, D. W. (1985). Estrogen effects on neuronal responsiveness to electrical and neurotransmitter stimulation: an *in vitro* study on the ventromedial nucleus of the hypothalamus. *Brain Res.* 347, 1–10.
266. Dohanich, G. O., Witcher, J. A., Weaver, D. R., and Clemens, L. G. (1982). Alteration of muscarinic binding in specific brain areas following estrogen treatment. *Brain Res.* 241, 347–350.
267. Olsen, K. L., Edwards, E., McNally, W., Schechter, N., and Whalen, R. E. (1982). Muscarinic receptors in hypothalamus: effects of cyclicity, sex and estrogen treatment. *Soc. Neurosci. Abstr.* 8, 423.
268. Rainbow, T. C., DeGroff, V., Luine, V. N., and McEwen, B. S. (1980). Estradiol-17β increases the number of muscarinic receptors in hypothalamic nuclei. *Brain Res.* 198, 239–243.
269. Rainbow, T. C., Synder, L., Berck, D. J., and McEwen, B. S. (1984). Correlation of muscarinic receptor induction in the ventromedial hypothalamic nucleus with the activation of feminine sexual behavior by estradiol. *Neuroendocrinology* 39, 476–480.
270. Luine, V. N., and McEwen, B. S. (1983). Sex differences in cholinergic enzymes of diagonal band nuclei in the rat preoptic area. *Neuroendocrinology* 36, 475–482.
271. Clemens, L. G., Dohanich, G. O., and Witcher, J. A. (1981). Cholinergic influences on estrogen-dependent sexual behavior in female rats. *J. Comp. Physiol. Psychol.* 95, 763–770.
272. Clemens, L. G., Dohanich, G., and Barr, P. (1983). Cholinergic regulation of feminine sexual behavior in laboratory rats. In *Hormones and Behaviour in Higher Vertebrates* (J. Balthazart, E. Prove, and R. Gilles, Eds.), pp. 56–68. Springer-Verlag, Berlin.
273. Dohanich, G. O., and Clemens, L. G. (1981). Brain areas implicated in cholinergic regulation of sexual behavior. *Horm. Behav.* 15, 157–167.
274. Kaufman, L., McEwen, B. S., and Pfaff, D. W. (1988). Cholinergic mechanisms of lordotic behavior in rats. *Physiol. Behav.* 43, 507–514.
275. Kaufman, L. S., Pfaff, D. W., and McEwen, B. S. (1985). Effects of intracranial application of various cholinergic drugs and estrogen on lordotic behavior in Long Evans rats. *Soc. Neurosci. Abstr.* 11, 1294.
276. Clemens, L. G., and Dohanich, G. P. (1980). Inhibition of lordotic behavior in female rats following intracerebral infusion of anticholinergic agents. *Pharmacol. Biochem. Behav.* 13, 89–95.
277. Dohanich, G. O., Barr, P. J., Witcher, J. A., and Clemens, L. G. (1984). Pharmacological and anatomical aspects of cholinergic activation of female sexual behavior. *Physiol. Behav.* 32, 1021–1026.
278. Kaufman, L. S., Pfaff, D. W., and McEwen, B. S. (1988). Cholinergic mechanisms of lordotic behavior in rats. *Physiol. Behav.* 43, 507–514.
279. Dohanich, G. O., McMullan, D. M., Cada, D. A., and Mangum, K. A. (1991). Muscarinic receptor subtypes and sexual behavior in female rats. *Pharmacol. Biochem. Behav.* 38, 115–124.
280. Ehlert, F. J., and Tran, L. L. P. (1990). Regional distribution of M_1, M_2 and non-M_1, non-M_2 subtypes of muscarinic binding sites in rat brain. *Pharmacol. Exp. Ther.* 255, 1148–1157.
281. Vilaro, M. T., Palacios, J. M., and Mengod, G. (1990). Localization of M_5 muscarinic receptor mRNA in rat brain examined by *in situ* hybridization histochemistry. *Neurosci. Lett.* 114, 154–159.
282. Wall, S. J., Yasuda, R. O., Li, M., and Wolfe, B. B. (1991). Development of an anti-serum against M_3 muscarinic receptors: distribution of m_3 receptors in rat tissues and clonal cell lines. *Mol. Pharmacol.* 40, 783–789.
283. Richmond, G., Clemens, L. (1984). Midbrain involvement in cholinergic facilitation of feminine sexual behavior. *Soc. Neurosci. Abstr.*
284. Richmond, G., and Clemens, L. (1986). Evidence for involvement of midbrain central gray in the cholinergic mediation of feminine sexual receptivity in rats. *Behav. Neurosci.* 100, 376–380.

285. Kow, L.-M., and Pfaff, D. W. (1987). Neuropeptides TRH and cyclo(His-Pro) share neuromodulatory, but not stimulatory, action on hypothalamic neurons *in vitro*: implication for the regulation of feeding. *Exp. Brain Res.* 67, 93–99.
286. Tsai, J.-T., Kow, L.-M., and Pfaff, D. W. Unpublished observations.
287. Kow, L.-M., and Pfaff, D. W. (1984). Neuronal responses to somatosensory and neuropharmacological stimuli *in vivo* and *in vitro*: effects of hormonal conditions. In *Modulation of Sensorimotor Activity During Alterations in Behavioral States* (R. Bandler, Ed.), pp. 217–230. Alan R. Liss, New York.
288. Doods, H. N., Mathy, M.-J., Davidesko, D., van Charldorp, K. J., de Jonge, A., and van Zwieten, P. A. (1987). Selectivity of muscarinic antagonists in radioligand and *in vivo* experiments for the putative m_1, m_2 and m_3 receptors. *J. Pharmacol. Exp. Ther.* 242, 257–262.
289. Lucot, J. B., van Charldorp, K. J., and Tulp, M. T. M. (1991). Idaverine, an m_2- vs. m_3-selective muscarinic antagonist, does not prevent motion sickness in cats. *Pharmacol. Biochem. Behav.* 40, 345–349.
290. Meyerson, B. J., Malmnas, G. O., and Everitt, B. J. (1985). Neuropharmacology, neurotransmitters, and sexual behavior in mammals. In *Handbook of Behavioral Neurobiology* (N. Adler, D. W. Pfaff, and R. Goy, Eds.), Vol. 7, pp. 495–536. Plenum Press, New York.
291. Moore, R. Y., and Bloom, F. E. (1979). Central catecholamine neuron systems: anatomy and physiology of the norepinephrine and epinephrine systems. *Annu. Rev. Neurosci.* 2, 113–168.
292. Kaba, H., Saito, H., Otsuka, K., Seto, K., and Kawakami, M. (1983). Effects of estrogen on the excitability of neurons projecting from the noradrenergic a_1 region to the preoptic and anterior hypothalamic area. *Brain Res.* 274, 156–159.
293. Jennes, L., Jennes, M. E., Purvis, C., and Nees, M. (1992). *c-fos* expression in noradrenergic a_2 neurons of the rat during the estrous cycle and after steroid hormone treatments. *Brain Res.* 586, 171–175.
294. Simpson, C. W., Cummins, C., and Dicara, L. V. (1974). Dopamine-β-hydroxylase activity in rat hypothalamus during the estrus cycle. *Pharmacol. Biochem. Behav.* 3, 693–696.
295. Luine, V. N., Khylchevskaya, R. I., and McEwen, B. S. (1975). Effects of gonadal steroids on activities of monoamine oxidase and choline acetylase in rat brain. *Brain Res.* 86, 293–306.
296. Luine, V. N., and Rhodes, J. C. (1983). Gonadal hormone regulation of MAO and other enzymes in hypothalamic area. *Neuroendocrinology* 36, 235–241.
297. Luine, V. N., and McEwen, B. S. (1977). Effect of oestradiol on turnover of type A monoamine oxidase in brain. *J. Neurochem.* 28, 1221–1227.
298. Wise, P. M., Rance, N., Selmanoff, M., and Barraclough, C. A. (1981). Changes in radioimmunoassayable luteinizing hormone-releasing hormone in discrete brain areas of the rat at various times on proestrus, diestrous day 1, and after phenobarbital administration. *Endocrinology* 108, 2179–2185.
299. Wise, P. M., Rance, N., and Barraclough, C. A. (1981). Effects of estradiol and progesterone on catecholamine turnover rates in discrete hypothalamic regions in ovariectomized rats. *Endocrinology* 108, 2186–2193.
300. Rance, N., Wise, P. M., and Barraclough, C. A. (1981). Negative feedback effects of progesterone correlated with changes in hypothalamic norepinephrine and dopamine turnover rates, median eminence luteinizing hormone-releasing hormone, and peripheral plasma gonadotropins. *Endocrinology* 108, 2194–2199.
301. Hiemke, J. (1985). Sex-specific effects of estradiol on hypothalamic noradrenaline turnover in gonadectomized rats. *Exp. Brain Res.* 59, 68–72.
302. Crowley, W. R., O'Donohue, T. L., and Jacobowitz, D. M. (1978). Changes in catecholamine content in discrete brain nuclei during the estrous cycle of the rat. *Brain Res.* 147, 315–326.
303. Crowley, W. R., O'Donohue, T. L., Wachslicht, H., and Jacobowitz, D. M. (1978). Effects of estrogen and progesterone on plasma gonadotropins and on catecholamine levels and turnover in discrete brain regions of ovariectomized rats. *Brain Res.* 154, 345–357.
304. Diggs, K. G. (1982). Adrenoceptors in the uterus. *J. Auton. Pharmacol.* 2, 53–67.
305. Miller, M. D., and Marshall, J. M. (1965). Uterine response to nerve stimulation: relation to hormonal status and catecholamines. *Am. J. Physiol.* 209, 859–865.
306. Hoffman, B. B., Lavin, T. N., Lefkowitz, R. J., and Ruffolo, R. R. (1981). Alpha adrenergic receptor subtypes in rabbit uterus: mediation of myometrial contraction and regulation by steroids. *J. Pharmacol. Exp. Ther.* 219, 290–295.
307. Callahan, S. M., and Creed, K. E. (1985). The effects of oestrogens on spontaneous activity and responses to phenylephrine of the mammalian urethra. *J. Physiol. (Lond.)* 358, 35–46.
308. Shackelford, D. P. Jr., McConnaughey, M. M., and Iams, S. G. (1988). The effects of estradiol and mestranol on alpha-adrenoceptors in select regions of the rat brain. *Brain Res. Bull.* 21, 329–333.
309. Petitti, N., Karkanias, G. B., and Etgen, A. M. (1992). Estradiol selectively regulates α_{1B}-noradrenergic receptors in the hypothalamus and preoptic area. *J. Neurosci.* 12, 3869–3876.
310. Crowley, W. R., Feder, H. H., and Morin, L. P. (1976). Role of monoamines in sexual behavior of the female guinea pig. *Pharmacol. Biochem. Behav.* 4, 67–71.
311. Crowley, W. R., Nock, B., and Feder, H. H. (1978). Facilitation of lordosis behavior by clonidine in female guinea pigs. *Pharmacol. Biochem. Behav.* 8, 207–209.
312. Nock, B., and Feder, H. H. (1979). Noradrenergic transmission and female sexual behavior of guinea pigs. *Brain Res.* 166, 369–380.
313. Nock, B., Blaustein, J. D., and Feder, H. H. (1981). Changes in noradrenergic transmission alter the concentration of cytoplasmic progestin receptors in hypothalamus. *Brain Res.* 207, 371–396.
314. Nock, B., and Feder, H. H. (1984). α_1-Noradrenergic regulation of hypothalamic progestin receptors and guinea pig lordosis behavior. *Brain Res.* 310, 77–85.
315. Vacas, M. I., and Cardinali, D. P. (1980). Effect of estradiol on α- and β-adrenoceptor density in medial basal hypothalamus, cerebral cortex and pineal gland of ovariectomized rats. *Neurosci. Lett.* 17, 73–77.
316. Wilkinson, M., Herdon, H., Pearce, M., and Wilson, C. (1979). Radioligand binding studies on hypothalamic noradrenergic receptors during the estrous cycle or after steroid injection in ovariectomized rats. *Brain Res.* 168, 652–655.
317. Kow, L.-M., and Pfaff, D. W. (1987). Responses of ventromedial hypothalamic neurons *in vitro* to norepinephrine: dependence on dose and receptor type. *Brain Res.* 413, 220–228.
318. Everitt, B. J., Fuxe, K., and Hokfelt, T. (1974). Inhibitory role of dopamine and 5-hydroxytryptamine in the sexual behavior of female rats. *Eur. J. Pharmacol.* 29, 187–191.
319. Foreman, M. M., and Moss, R. L. (1978). Role of hypothalamic alpha and beta adrenergic receptors in the control of lordotic behavior in the ovariectomized estrogen-primed rat. *Pharmacol. Biochem. Behav.* 9, 235–241.
320. Nock, B., and Feder, H. H. (1979). Noradrenergic transmission and female sexual behavior of guinea pig. *Brain Res.* 166, 369–380.

321. Nock, B., and Feder, H. H. (1984). α_1-Noradrenergic regulation of hypothalamic progestin receptors and guinea pig lordosis behavior. *Brain Res.* 310, 77–85.
322. Michanek, A. (1979). Potentiation of D- and L-amphetamine effects on copulatory behavior in female rats by treatment with α-adrenoreceptor blocking drugs. *Arch. Int. Pharmacodyn. Ther.* 239, 241–256.
323. Fernandez-Guasti, A., Larsson, K., and Beyer, C. (1985). Potentiative action of α- and β-adrenergic receptor stimulation in inducing lordosis behavior. *Pharmacol. Biochem. Behav.* 22, 613–617.
324. Etgen, A. M., Vathy, I., Petitti, N., Ungar, S., and Karkanias, G. B. (1990). Ovarian steroids, female reproductive behavior, and norepinephrine neurotransmission in the hypothalamus. *Comp. Physiol.* 8, 116–128.
325. Kow, L.-M., Weesner, G. D., and Pfaff, D. W. (1992). α_1-Adrenergic agonists act on the ventromedial hypothalamus to cause neuronal excitation and lordosis facilitation: electrophysiological and behavioral evidence. *Brain Res.* 588, 237–245.
326. Hansen, S., Stanfiled, E. J., and Everitt, B. J. (1980). The role of ventral bundle noradrenergic neurones in sensory components of sexual behaviour and coitus-induced pseudopregnancy. *Nature* 286, 152–154.
327. Herndon, J. G. Jr. (1976). Effects of midbrain lesions on female sexual behavior in the rat. *Physiol. Behav.* 17, 143.
328. Kow, L.-M., Paden, C. M., and Pfaff, D. W. (1981). Dispensability of spinal monoaminergic systems in mediating the lordosis reflex of the female rat. *Pharmacol. Biochem. Behav.* 14, 707–711.
329. Ward, I. L., Crowley, W. R., Zemlan, F. O., and Margules, D. L. (1975). Monoaminergic mediation of female sexual behavior. *J. Comp. Physiol. Psychol.* 88, 53–61.
330. Kow, L.-M., and Pfaff, D. W. (1992). Relations between α_{1a}- and α_{1b}-adrenoceptor subtypes: an electrophysiological assessment in hypothalamic neurons. *Soc. Neurosci. Abstr.* 18, 589.
331. Mendelson, S. D. (1992). A review and reevaluation of the effects of serotonin on lordosis behavior in the female rat. *Neurosci. Biobehav. Rev.* 16, 309–350.
332. Fuxe, K. (1965). Evidence for the existence of monoamine neurons in the central nervous system: IV. Distribution of monoamine nerve terminals in the central nervous system. *Acta Physiol. Scand.* 64(Suppl. 247), 38–85.
333. Steinbusch, H. W. M. (1981). Distribution of serotonin-immunoreactivity in the central nervous system of the rat-cell bodies and terminals. *Neuroscience* 6, 557–618.
334. Kawata, M., Takeuchi, Y., Ueda, S., Matsuura, T., and Sano, Y. (1984). Immunohistochemical demonstration of serotonin nerve fibers in the hypothalamus including posterior lobe of the rat and monkey. *Cell. Tissue Res.* 236, 495–503.
335. Ueda, S., Kawata, M., Takeuchi, Y., and Sano, Y. (1983). Immunohistochemical demonstration of serotonin nerve fibers in the hypothalamus of the rat. *Anat. Embryol.* 168, 315–330.
336. Kawata, M., Takeuchi, Y., Ueda, S., Matsuura, T., and Sano, Y. (1984). Immunohistochemical demonstration of serotonin-containing nerve fibers in the hypothalamus of the monkey, *Macaca fuscata*. *Cell. Tissue Res.* 236, 495–503.
337. Biegon, A., Fischette, C. T., Rainbow, T. C., and McEwen, B. S. (1982). Serotonin receptor modulation by estrogen in discrete brain nuclei. *Neuroendocrinology* 35, 287–291.
338. Biegon, A., and McEwen, B. S. (1982). Modulation by estradiol of serotonin receptors in brain. *J. Neurosci. 2,* 199–205.
339. Fischette, C. T., Biegon, A., and McEwen, B. S. (1983). Sex differences in serotonin I receptor binding in rat brain. *Science* 222, 333–335.
340. Meyerson, B. J. (1964). Central nervous monoamines and hormone induced estrus behaviour in the spayed rat. *Acta Physiol. Scand.* 63(Suppl. 241), 3–32.
341. Meyerson, B. J. (1975). Drugs and sexual motivation in the female rat. In *Sexual Behavior, Pharmacology and Biochemistry* (M. Sandier and G. L. Gessa, Eds.). Raven Press, New York.
342. Meyerson, B. J. (1966). The effect of imipramine and related antidepressive drugs on estrus behaviour in ovariectomized rats activated by progesterone, reserpine or tetrabenzine in combination with estrogen. *Acta Physiol. Scand.* 67, 411–422.
343. Everitt, B. J., and Fuxe, K. (1977). Serotonin and the sexual behavior of female rats: effects of hallucinogenic indolealkylamines and phenylethylamines. *Neurosci. Lett.* 4, 213–220.
344. Meyerson, B. J., and Lawander, T. (1970). Serotonin synthesis inhibition and estrous behavior in female rats. *Life Sci.* 9, 661–671.
345. Ahlenius, S., Engel, J., Eriksson, J., Modigh, K., and Sodersten, P. (1972). Importance of central catecholamines in the mediation of lordosis behavior in ovariectomized rats treated with estrogen and inhibitors of monoamine synthesis. *J. Neural Transm.* 33, 247–256.
346. Zemlan, F. O., Ward, I. L., Crowley, W. R., and Margules, D. L. (1973). Activation of lordotic responding in female rats by suppression of serotonergic activity. *Science* 179, 1010–1011.
347. Everitt, B. J., Fuxe, K., Hokfelt, T., and Jonsson, G. (1975). Role of monoamines in the control of hormones of sexual receptivity in the female rat. *J. Comp. Physiol. Psychol.* 89, 556–572.
348. Gradwell, P. B., Everitt, B. J., and Herbert, J. (1975). 5-Hydroxytryptamine in the central nervous system and sexual receptivity of female rhesus monkeys. *Brain Res.* 88, 281–293.
349. Fuxe, K., Everitt, B. J., Agnati, L., Fredholm, B., and Jansson, B. (1976). On the biochemistry and pharmacology of hallucinogens. In *Schizophrenia Today* (D. Kermal, G. Bartholini, and P. Richter, Eds.). Pergamon Press, New York.
350. Ward, I. L., Crowley, W. R., Zemlan, F. O., and Margules, D. L. (1975). Monoaminergic mediation of female sexual behavior. *J. Comp. Physiol. Psychol.* 88, 51–61.
351. Zemlan, F. O., Trulson, M. E., Howell, R., and Hoebel, B. G. (1977). Influence of *p*-chloroamphetamine on female sexual reflexes and brain monoamine levels. *Brain Res.* 123, 347–356.
352. Luine, V. N., Frankfurt, M., Rainbow, T. C., Biegon, A., and Azmitia, E. (1983). Intrahypothalamic 5,7-dihydroxytryptamine facilitates feminine sexual behavior and decreases [^{3}H]-imipramine binding and 5-HT uptake. *Brain Res.* 264, 344–348.
353. Luine, V. N., Renner, K. J., Frankfurt, M., and Azmitia, E. C. (1984). Facilitated sexual behavior reversed and serotonin restored by raphe nuclei transplanted into denervated hypothalamus. *Science* 226, 1436–1438.
354. Moreines, J., Pfaff, D., and McEwen, B. S. (1984). Serotonergic inhibition of female rat sexual behavior in the basomedial hypothalamus. *Soc. Neurosci. Abstr.* 10, 823.
355. Moreines, J., Kelton, M., Luine, V. N., Pfaff, D. W., and McEwen, B. S. (1988). Hypothalamic serotonin lesions unmask hormone responsiveness of lordosis behavior in adult male rats. *Neuroendocrinology* 47, 453–458.
356. Mendelson, S. D., and Gorzalka, B. B. (1989). Differential roles of 5-HT receptor subtypes in female sexual behavior. In *Behavioural Pharmacology of 5-HT* (O. Bevan, A. R. Cools, and T. Archer, Eds.), pp. 55–71. Lawrence Erlbaum Associates, London.
357. Wilkinson, L. O., and Dourish, C. T. (1991). Serotonin and animal behavior. In *Serotonin Receptor Subtypes: Basic and*

Clinical Aspects (S. J. Peroutka, Ed.), pp. 147–210. Wiley-Liss, New York.

358. Uphouse, L., Montanez, S., Richards-Hill, R., Caldarola-Pastuszka, M., and Droge, M. (1991). Effects of the 5-HT_{1A} agonist, 8-OH-DPAT, on sexual behaviors of the proestrous rat. *Pharmacol. Biochem. Behav.* 39, 635–640.
359. Wilson, C. A., and Hunter, A. J. (1985). Progesterone stimulates sexual behavior in female rats by increasing 5-HT activity on 5-HT_2 receptors. *Brain Res.* 333, 223–229.
360. Mendelson, S. D., and Gorzalka, B. B. (1985). A facilitatory role for serotonin in the sexual behavior of the female rat. *Pharmacol. Biochem. Behav.* 22, 1025–1033.
361. Mendelson, S. D., and Gorzalka, B. B. (1986). Serotonin type 2 antagonists inhibit lordosis behavior in the female rat: reversal with quipazine. *Life Sci.* 38, 33–39.
362. Hunter, A. J., Hole, D. R., and Wilson, C. A. (1985). Studies into the dual effects of serotonergic pharmacological agents on female sexual behaviour in the rat: preliminary evidence that endogenous 5HT is stimulatory. *Pharmacol. Biochem. Behav.* 22, 5–13.
363. James, M. D., Lane, S. M., Hole, D. R., and Wilson, C. A. (1989). Hypothalamic sites of action of the dual effect of 5-HT on female sexual behaviour in the rat. In *Behavioural Pharmacology of 5-HT* (O. Bevan, A. R. Cools, and T. Archer, Eds.), pp. 73–77. Lawrence Erlbaum Associates, London.
364. Fischette, C. T., Biegon, A., and McEwen, B. S. (1984). Sex steroid modulation of the serotonin behavioral syndrome. *Life Sci.* 35, 1197–1206.
365. Kow, L.-M., Tsai Y-F., Wang, L., and Pfaff, D. W. (1992). Electrophysiological analyses of serotonergic actions on neurons in hypothalamic ventromedial nucleus *in vitro*: receptor subtypes involved and implications for regulation of feeding and lordosis behaviors. *Clin. J. Physiol.* 35, 105–121.
366. Bendotti, C., and Samanin, R. (1986). 8-Hydroxy-2-(di-*n*-propylamino)tetralin (8-OH-DPAT) elicits eating in free-feeding rats by acting on central serotonin neurons. *Eur. J. Pharmacol.* 121, 147–150.
367. Huston, P. H., Dourish, C. T., and Curzon, G. (1988). Evidence that the hyperphagic response to 8-OH-DPAT is mediated by 5-HT_{1A} receptors. *Eur. J. Pharmacol.* 150, 361–366.
368. Goldman, C. K., Marino, L., and Leibowitz, S. F. (1985). Postsynaptic α_2-noradrenergic receptors mediate feeding induced by paraventricular nuclear injection of norepinephrine and clonidine. *Eur. J. Pharmacol.* 115, 11–19.
369. Wellman, P. J., and Davies, B. T. (1991). Suppression of feeding induced by phenylephrine microinjections within the paraventricular hypothalamus in rats. *Appetite* 17, 121–128.
370. Beatty, W. W., O'Briant, D. A., and Vilberg, T. R. (1974). Suppression of feeding by intrahypothalamic implants of estradiol in male and female rats. *Bull. Psychonom. Soc.* 3, 273–274.
371. Butera, P. C., and Czaja, J. A. (1984). Intracranial estradiol in ovariectomized guinea pigs: effects on ingestive behaviors and body weight. *Brain Res.* 322, 41–48.
372. Donohoe, T. O., and Stevens, R. (1982). Modulation of food intake by hypothalamic implants of estradiol benzoate, estrone, estriol, and CI-628 in female rats. *Pharmacol. Biochem. Behav.* 16, 93–99.
373. Wade, G. N., and Zucker, I. (1970). Modulation of food intake and locomotor activity in female rats by diencephalic hormone implants. *J. Comp. Physiol. Psychol.* 72, 328–336.
374. Wurtman, R. J., and Wurtman, J. J. (1983). Underfeeding decreases brain and hypothalamic norepinephrine activity. In *Nutrition and the Brain* (R. J. Wurtman and J. J. Wurtman, Eds.), Vol. 2, pp. 177–181. Raven Press, New York.
375. Blundell, J. E. (1984). Serotonin and appetite. *Neuropharmacology* 23, 1537–1551.
376. Quigley, K., Goya, R., and Meites, J. (1987). Rejuvenating effects of 10-week underfeeding period on estrous cycles in young and old rats. *Biol. Aging* 8, 225–232.
377. Dickerman, R. W., Li, H. Y., and Wade, G. N. (1991). Food deprivation suppresses steroid-induced estrous behavior in ovariectomized Syrian hamsters. *Soc. Neurosci. Abstr.* 17, 500.
378. Caggiula, A. R., Herndon, J. G. Jr., Scanlon, R., Greenstone, D., Bradshaw, W., and Sharp, D. (1979). Dissociation of active from immobility components of sexual behavior in female rats by central 6-hydroxydopamine: implications for CA involvement in sexual behavior and sensorimotor responsiveness. *Brain Res.* 172, 505–520.
379. Decavel, C., and van den Pol, A. N. (1990). GABA: a dominant neurotransmitter in the hypothalamus. *J. Comp. Neurol.* 302, 1019–1037.
380. Nishizuka, M., and Pfaff, D. W. (1989). Intrinsic synapses in the ventromedial nucleus of the hypothalamus: an ultrastructural study. *J. Comp. Neurol.* 286, 260–268.
381. Nishizuka, M., and Pfaff, D. W. (1989). Medial preoptic islands in the rat brain: electron microscopic evidence for intrinsic synapses. *Exp. Brain Res.* 77, 295–301.
382. van den Pol, A. N. (1991). Glutamate and aspartate immunoreactivity in hypothalamic presynaptic axons. *J. Neurosci.* 11, 2087–2101.
383. Moratalla, R., and Bowery, N. G. (1991). Chronic lesion of corticostriatal fibers reduces $GABA_B$ but not $GABA_A$ binding in rat caudate putamen: an autoradiographic study. *Neurochem. Res.* 16, 309–315.
384. Olsen, R. W., Bureau, M. H., Endo, S., and Smith, G. (1991). The $GABA_A$ receptor family in the mammalian brain. *Neurochem. Res.* 16, 317–325.
385. Olsen, R. W., and Tobin, A. J. (1990). Molecular biology of $GABA_A$ receptors. *FASEB J.* 4, 1469–1480.
386. Schofield, P. R., Darlison, M. G., Fujita, N., Burt, D. R., Stephenson, F. A., Rodriguez, H., Rhee, L. M., Ramachandran, J., Reale, V., Glencorse, T. A., Seeburg, P. H., and Barnard, E. A. (1987). Sequence and functional expression of the GABA-A receptor shows a ligand-gated receptor superfamily. *Nature* 328, 221–227.
387. Shingai, R., Sutherland, M. L., and Barnard, E. A. (1991). Effects of subunit types of the cloned $GABA_A$ receptor on the response to a neurosteroid. *Eur. J. Pharmacol.* 206, 77–80.
388. Verdoorn, T. A., Draguhn, A., Ymer, S., Seeburg, P. H., and Sakmann, B. (1990). Functional properties of recombinant rat $GABA_A$ receptors depend upon subunit composition. *Neuron* 4, 919–928.
389. Pritchett, D., Sontheimer, J., Shivers, B. D., Ymer, S., Kettenmann, H., Schofield, P. R., and Seeburg, P. H. (1989). Importance of a novel $GABA_A$ receptor subunit for benzodiazepine pharmacology. *Nature* 338, 582–585.
390. Majewska, M. D., Harrison, N. L., Schwartz, R. D., Barker, J. L., and Paul, S. M. (1986). Steroid hormone metabolites are barbiturate-like modulators of the GABA receptor. *Science* 232, 1004–1007.
391. Morrow, A. L., Pace, J. R., Purdy, R. H., and Paul, S. M. (1990). Characterization of steroid interactions with gamma-aminobutyric acid receptor-gated chloride ion channels: evidence for multiple steroid recognition sites. *Mol. Pharmacol.* 37, 263–270.
392. Wu, F.-S., Gibbs, T. T., and Farb, D. H. (1990). Inverse modulation of gamma-aminobutyric acid- and glycine-induced currents by progesterone. *Mol. Pharmacol.* 37, 597–602.
393. Lopez-Colome, A.-M., McCarthy, M. M., and Beyer, C. (1990). Enhancement of [^{3}H]muscimol binding to brain synaptic membranes by progesterone and related pregnanes. *Eur. J. Pharmacol.* 176, 297–303.

394. Hu, Z. Y., Bourreau, E., Testa, I. J., Robel, O., and Baulieu, E. E. (1987). Neurosteroids: oligodendrocyte mitochondria convert cholesterol to pregnenolone. *Proc. Natl. Acad. Sci. U S A* 84, 8215–8219.

395. Le Goascogne, C., Robel, O., Gouezou, M., Sananes, N., Baulieu, E. E., and Waterman, M. (1988). Neurosteroids: cytochrome P-450scc in rat brain. *Science* 237, 1212–1215.

396. Majewska, M. D., and Schwartz, R. D. (1987). Pregnenolone-sulfate: an endogenous antagonist of the gamma-aminobutyric acid receptor complex in brain? *Brain Res.* 404, 355–360.

397. Gee, K. W., Bolger, M. B., Brinton, R. E., Coirini, H., and McEwen, B. S. (1988). Steroid modulation of the chloride ionophore in rat brain: structure-activity requirements, regional dependence and mechanism of action. *J. Pharmacol. Exp. Ther.* 246, 803–812.

398. Turner, J. O., and Simmonds, M. A. (1989). Modulation of the $GABA_A$ receptor complex by steroids in slices of rat cuneate nucleus. *Br. J. Pharmacol.* 96, 409–417.

399. Kirkness, E. F., and Turner, A. J. (1988). The stimulatory effects of secobarbital and pregnenolone on the $GABA_A$ receptor can be blocked selectively. *Eur. J. Pharmacol.* 150, 385–388.

400. Peters, J. A., Kirkness, E. F., Callachan, H., Lambert, J. J., and Turner, A. J. (1988). Modulation of the $GABA_A$ receptor by depressant barbiturates and pregnane steroids. *Br. J. Pharmacol.* 94, 1257–1269.

401. Lan, N. C., Gee, K. W., Bolger, M. B., and Chen, J. S. (1991). Differential responses of expressed recombinant human gamma-aminobutyric acid receptors to neurosteroids. *J. Neurochem.* 57, 1818–1821.

402. Hamon, M., Goetz, C., Euvrard, C., Pasqualini, C., Le Dafniet, M., Kerdelhue, B., Cesselin, F., and Peillon, F. (1983). Biochemical and functional alterations of central GABA receptors during chronic estradiol treatment. *Brain Res.* 279, 141–152.

403. Maggi, A., and Perez, J. (1986). Estrogen-induced up-regulation of gamma-aminobutyric acid receptors in the CNS of rodents. *J. Neurochem.* 47, 1793–1797.

404. Perez, J., Zucchi, I., and Maggi, A. (1986). Sexual dimorphism in the response of the GABAergic system to estrogen administration. *J. Neurochem.* 47, 1798–1803.

405. O'Conner, L. H., Nock, B., and McEwen, B. S. (1988). Regional specificity of gamma-aminobutyric acid receptor regulation by estradiol. *Neuroendocrinology* 47, 473–481.

406. Schumacher, M., Coirini, H., and McEwen, B. S. (1989). Regulation of high affinity $GABA_A$ receptors in specific brain regions by ovarian hormones. *Neuroendocrinology* 50, 315–320.

407. McCarthy, M. M., Coirini, H., Schumacher, M., Johnson, A. E., and Pfaff, D. W. (1992). Steroid regulation and sex differences in [^{3}H]muscimol binding in hippocampus, hypothalamus and midbrain in rats. *J. Neuroendocrinol.* 4, 1-7.

408. Flugge, G., Oertel, W. H., Wuttke, W. (1986). Evidence for estrogen-receptive GABAergic neurons in the preoptic/anterior hypothalamic area of the rat brain. *Neuroendocrinology* 43, 1–5.

409. Wallis, C. J., and Luttge, W. G. (1980). Influence of estrogen and progesterone on glutamic acid decarboxylase activity in discrete regions of rat brain. *J. Neurochem.* 34, 609–613.

410. Nicolletti, F., Grandison, L., and Meek, J. L. (1985). Effects of repeated administration of estradiol benzoate on tubero-infundibular GABAergic activity in male rats. *J. Neurochem.* 44, 1217–1220.

411. Duvilanski, B., Maines, V. M., and Debeljuk, L. (1983). GABA-related enzymes in the hypothalamus of rats treated with estradiol. *Eur. J. Pharmacol.* 89, 259–264.

412. Julien, J.-F., Legay, F., Dumas, S., Tappaz, M., and Mallet, J. (1987). Molecular cloning, expression and in situ hybridization of rat brain glutamic acid decarboxylase messenger RNA. *Neurosci. Lett.* 73, 173–180.

413. Kobayashi, Y., Kaufman, D. L., and Tobin, A. J. (1987). Glutamic acid decarboxylase cDNA: nucleotide sequence encoding an enzymatically active fusion protein. *J. Neurosci.* 7, 2768–2772.

414. Wyborski, R. J., Bond, R. W., and Gottlieb, D. I. (1990). Characterization of a cDNA coding for rat glutamic acid decarboxylase. *Brain Res. Mol. Brain Res.* 8, 193–198.

415. Kaufman, D. L., Houser, C. R., and Tobin, A. J. (1991). Two forms of the (gamma)-aminobutyric acid synthetic enzyme glutamate decarboxylase have distinct intraneuronal distributions and cofactor interactions. *J. Neurochem.* 56, 720–723.

416. Martin, D. L., Martin, S. B., Wu, S. J., and Espina, N. (1991). Cofactor interactions and the regulation of glutamate decarboxylase activity. *Neurochem. Res.* 16, 243–249.

417. Leigh, A. J., Carter, N. D., Horton, R., Silverlight, J. J., and Wilson, C. A. (1990). Ovarian steroid regulation of glutamic acid decarboxylase gene expression in individual hypothalamic nuclei. *J. Neuroendocrinol.* 2, 433–438.

418. Leranth, C., Shanabrough, M., and Naftolin, F. (1991). Estrogen induces ultrastructural changes in progesterone receptor-containing GABA neurons of the primate hypothalamus. *Neuroendocrinology* 54, 571–579.

419. Brown, T. J., MacLusky, N. J., Leranth, C., Shanabrough, M., and Naftolin, F. (1990). Progestin receptor-containing cells in guinea pig hypothalamus: afferent connections, morphological characteristics, and neurotransmitter content. *Mol. Cell. Neurosci.* 1, 58–77.

420. Blaustein, J. D., and Turcotte, J. C. (1989). Estradiol-induced progestin receptor immunoreactivity is found only in estrogen receptor immunoreactive cells in guinea pig brain. *Neuroendocrinology* 49, 454–461.

421. Okamura, J., Abitbol, M., Julien, J.-F., Dumas, S., Berod, A., Geffard, M., Kitahama, K., Bobillier, P., Mallet, J., and Wiklund, L. (1990). Neurons containing messenger RNA encoding glutamate decarboxylase in rat hypothalamus demonstrated by in situ hybridization, with special emphasis on cell groups in medial preoptic area, anterior hypothalamic area and dorsomedial hypothalamic nucleus. *Neuroscience* 39, 675–699.

422. Leranth, C., Sakamoto, H., MacLusky, N. J., Shanabrough, M., and Naftolin, F. (1985). Estrogen responsive cells in the arcuate nucleus of the rat contain glutamate acid decarboxylase (GAD): an electron microscopic immunocytochemical study. *Brain Res.* 331, 376–381.

423. McGinnis, M. Y., Gordon, J. H., and Gorski, R. A. (1980). Influence of gamma-aminobutyric acid on lordosis behavior and dopamine activity in estrogen primed spayed female rats. *Brain Res.* 184, 179–191.

424. Donoso, A. O., and Zarate, M. B. (1981). Enhancement of lordotic behavior by injection of GABA into the ventromedial nucleus of hypothalamus. *IRCS J. Med. Sci.* 9, 1039.

425. McCarthy, M. M., Malik, K., and Feder, J. J. (1990). Increased GABAergic neurotransmission in medial hypothalamus facilitates lordosis but has opposite effects in the preoptic area/anterior hypothalamus. *Brain Res.* 501, 40–43.

426. McCarthy, M. M., Masters, D. B., Fiber, J. M., Lopez-Colome, A. M., Beyer, C., Komisaruk, B. R., and Feder, H. H. (1991). GABAergic control of receptivity in the female rat. *Neuroendocrinology* 53, 473–479.

427. Masco, D., Weigel, R., and Carrer, J. F. (1986). Gamma-aminobutyric acid mediates ventromedial hypothalamic mechanisms controlling the execution of lordotic responses in the female rat. *Behav. Brain. Res.* 19, 153–162.

428. Quershi, G. A., Bednar, I., Forsberg, G., and Sodersten, P. (1988). GABA inhibits sexual behavior in female rats. *Neuroscience* 27, 169–174.

429. Agmo, A., Soria, O., and Paredes, P. (1989). GABAergic drugs and lordosis behavior in the female rat. *Horm. Behav.* 23, 368–380.
430. Fernandez-Guasti, A., Larsson, K., and Beyer, C. (1986). Lordosis behavior and GABAergic neurotransmission. *Pharmacol. Biochem. Behav.* 24, 673–676.
430a. McCarthy, M. M., Pfaff, D. W., and Schwartz-Giblin, S. (1993). The role of steroid modulation of amino acid transmitters in the regulation of female reproduction. *Am. Zool.* 33, 275–284.
430b. McCarthy, M, M. Unpublished data.
431. Frankfurt, M., Fuchs, E., and Wuttke, W. (1984). Sex differences in gamma-aminobutyric acid and glutamate concentration in discrete brain nuclei. *Neurosci. Lett.* 50, 245–250.
432. Brandao, M. L., Di Scala, G., Bouchet, M. J., and Schmitt, P. (1986). Escape behavior produced by the blockade of glutamic acid decarboxylase (GAD) in mesencephalic central gray or medial hypothalamus. *Pharmacol. Biochem. Behav.* 24, 497–501.
433. Bandler, R., Depaulis, A., and Vergnes, M. (1986). Identification of midbrain neurons mediating defensive behavior in the rat by microinjections of excitatory amino acids. *Behav. Brain Res.* 15, 107–109.
434. Depaulis, A., and Vergnes, M. (1986). Elicitation of intraspecific defensive behaviors in the rat by microinjection of picrotoxin, a gamma-aminobutyric acid antagonist, into the midbrain periaqueductal gray matter. *Brain Res.* 367, 87–95.
435. Moreau, J.-L., and Fields, H. L. (1986). Evidence for GABA involvement in midbrain control of medullary neurons that modulate nociceptive transmission. *Brain Res.* 397, 37–46.
435a. McCarthy, M. M., Schwartz-Giblin, S., and Pfaff, D. W. (1991). Intracerebral administration of glutamic acid decarboxylase (GAD) antisense oligodeoxynucleotide reduces lordosis behavior in the rat. *Soc. Neurosci. Abstr.* 17, 197.5.
435b. McCarthy, M. M., Pfaff, D. W., and Schwartz-Giblin, S. (1991). Midbrain central gray $GABA_A$ receptor activation enhances, and blockade reduces, sexual behavior in the female rat. *Exp. Brain Res.* 86, 108–116.
436. Agmo, A., Soria, O., and Paredes, P. (1989). GABAergic drugs and lordosis behavior in the female rat. *Horm. Behav.* 23, 368–380.
437. Luine, V., Cowell, J., and Frankfurt, M. (1991). GABAergic-serotonergic interactions in regulating lordosis. *Brain Res.* 556, 171–174.
438. van den Pol, A. N., and Gores, T. (1988). Glycine and glycine receptor immunoreactivity in brain and spinal cord. *J. Neurosci.* 8, 472–492.
439. Bristow, D. R., Bowery, N. G., and Woodruff, G. N. (1986). Light microscopic autoradiographic localization of [^{3}H]glycine and [^{3}H]strychnine binding sites in rat brain. *Eur. J. Pharmacol.* 126, 303–308.
440. Sandoval, Y., Komisaruk, B., and Beyer, C. (1988). Possible role of inhibitory glycinergic neurons in the regulation of lordosis behavior in the rat. *Pharmacol. Biochem. Behav.* 29, 303–307.
441. Johnson, J. W., and Ascher, P. (1987). Glycine potentiates the NMDA response in cultured mouse brain neurons. *Nature* 325, 529–531.
442. Monaghan, D. T., Bridges, R. J., and Cotman, C. W. (1989). The excitatory amino acid receptors: their classes, pharmacology and distinct properties in the function of the central nervous system. *Annu. Rev. Pharmacol. Toxicol.* 29, 365–402.
443. Kumar, K. N., Tilakaratne, M., Johnson, P. S., Alien, A. E., and Michaelis, E. K. (1991). Cloning of cDNA for the glutamate-binding subunit of an NMDA receptor complex. *Nature* 354, 31–37.
444. Hollman, M., O'Shea-Greenfield, A., Rogers, S. W., and Heinemann, S. (1989). Cloning by functional expression of a member of the glutamate receptor family. *Nature* 342, 643–648.
445. Boulter, J., Hollman, M., O'Shea-Greenfield, A., Hartley, M., Deneris, E., Maron, C., and Heinemann, S. (1990). Molecular cloning and functional expression of glutamate receptor subunit genes. *Science* 249, 1033–1037.
446. van den Pol, A. N., Wuarin, J.-O., and Dudek, F. E. (1990). Glutamate, the dominant excitatory transmitter in neuroendocrine regulation. *Science* 250, 1276–1278.
447. Arslan, M., Pohl, C. R., and Plant, T. M. (1988). DL-2-Amino-5-phosphonopen-tanoic acid, a specific *N*-methyl-*d*-aspartic acid receptor antagonist, suppresses pulsatile LH release in the rat. *Neuroendocrinology* 47, 465–468.
448. Donoso, A. O., Francisco, F. J., and Negro-Vilar, A. (1990). Glutamate receptors of the non-*N*-methyl-*d*-aspartic acid type mediate the increase in luteinizing hormone-releasing hormone release by excitatory amino acids *in vitro*. *Endocrinology* 126, 414–420.
449. Lopez, F. J., Donoso, A. O., and Negro-Vilar, A. (1990). Endogenous excitatory amino acid neurotransmission regulates the estradiol-induced LH surge in ovariectomized rats. *Endocrinology* 126, 1771–1773.
450. Bourguignon, J.-O., Gerard, A., and Franchimont, P. (1989). Direct activation of gonadotropin-releasing hormone secretion through different receptors to neuroexcitatory amino acids. *Neuroendocrinology* 49, 402–408.
451. Weiland, N. G. (1993). Estradiol selectively regulates agonist binding sites on *N*-methyl-*d*-aspartate receptor complex in the CA1 region of the hippocampus. *Endocrinology* 131, 662–668.
452. Smith, S. S., Waterhouse, B. D., Chapin, J. K., and Woodward, D. J. (1987). Progesterone alters GABA and flutamate responsiveness: a possible mechanism for its anxiolytic action. *Brain Res.* 400, 353–359.
453. Smith, S. S., Waterhouse, B. D., and Woodward, D. J. (1987). Sex steroid effects on extrahypothalamic CNS: I. Estrogen augments neuronal responsiveness to iontophoretically applied glutamate in the cerebellum. *Brain Res.* 422, 40–51.
454. Smith, S. S., Waterhouse, B. D., and Woodward, D. J. (1987). Sex steroid effects on extrahypothalamic CNS: II. Progesterone, alone and in combination with estrogen, modulates cerebellar responses to amino acid neurotransmitters. *Brain Res.* 422, 52–62.
455. Smith, S. S., Waterhouse, B. D., and Woodward, D. J. (1988). Locally applied estrogens potentiate glutamate-evoked excitation of cerebellar Purkinje cells. *Brain Res.* 475, 272–282.
456. Wu, F.-S., Gibbs, T. T., and Farb, D. H. (1991). Pregnenolone sulfate: a positive allosteric modulator at the N-methyl-D-aspartate receptor. *Mol. Pharmacol.* 40, 333–336.
457. Kow, L.-M., Harlan, R. E., Shivers, B. D., and Pfaff, D. W. (1985). Inhibition of the lordosis reflex in rats by intrahypothalamic infusion of neural excitatory agents: evidence that the hypothalamus contains separate inhibitory and facilitatory elements. *Brain Res.* 341, 26–34.
457a. McCarthy, M. M., Curan, G. H., and Feder, H. H. (1991). Excitatory amino acid modulation of lordosis in the rat. *Neurosci. Lett.* 126, 94–97.
458. Mansky, T., and Wuttke, W. (1983). Glutamate in hypothalamic and limbic structures of diestrous, proestrous, ovariectomized and ovariectomized estrogen-treated rats. *Neurosci. Lett.* 38, 51–56.
459. Eaton, M. J., and Moss, R. L. (1990). Electrophysiological evidence for an excitatory amino acid projection from the ventromedial hypothalamus to the medial zona incerta. *Synapse* 5, 77–79.
460. Albin, R. L., Makowiec, R. L., Hollingsworth, Z., Dure, L. S. I. V., Penney, J. B., and Young, A. B. (1990). Excitatory amino

acid binding sites in the peri-aqueductal gray of the rat. *Neurosci. Lett.* 118, 112–115.

461. Dornan, W. A., Peterson, M., Matuszewich, L., and Malen, P. (1991). Ibotenic acid-induced lesions of the medial zona incerta decrease lordosis behavior in the female rat. *Behav. Neurosci.* 105, 210–214.
462. Wilson, C. A., James, M. D., Grierson, J. O., and Hole, D. R. (1991). Involvement of catecholaminergic systems in the zona incerta in the steroidal control of gonadotrophin release and female sexual behavior. *Neuroendocrinology* 53, 113–123.
463. Krey, L. C., and Silverman, A. J. (1983). Luteinizing hormone releasing hormone (LHRH). In *Brain Peptides* (D. T. Krieger, M. J. Brownstein, and J. B. Martin, Eds.), pp. 687–709. Wiley-Interscience, New York.
464. Shivers, B. D., Harlan, R., Morrell, J. I., and Pfaff, D. W. (1983). Immunocytochemical localization of luteinizing hormone-releasing hormone in male and female rat brains. *Neuroendocrinology* 36, 1–12.
465. Kozlowski, G. O., and Dees, W. L. (1984). Immunocytochemistry for LHRH neurons in the arcuate nucleus area of the rat: fact or artifact? *J. Histochem. Cytochem.* 32, 83–91.
466. King, J. C., and Anthony, E. L. P. (1983). LHRH neurons and their projections in humans and other mammals: species comparisons. *Peptides* 5, 195–207.
467. Marshall, P. E., and Goldsmith, P. C. (1980). Neuroregulatory and neuroendocrine GnRH pathways in the hypothalamus and forebrain of the baboon. *Brain Res.* 193, 353–372.
468. King, J. C., Anthony, E. L. P., Fitzgerald, D. M., and Stopa, E. G. (1985). Luteinizing hormone-releasing hormone neurons in human preoptic/hypothalamus: differential intraneuronal localization of immunoreactive forms. *J. Clin. Endocrinol. Metab.* 60, 88–97.
469. King, J. C., and Anthony, E. L. P. (1984). Biosynthesis of LHRH: inferences from immunocytochemical studies. *Peptides* 4, 963–970.
470. Leranth, C. S., Segura, L. M. G., Palkovits, M., MacLusky, N. J., Shanabrough, M., and Naftolin, F. (1985). The LH-RH-containing neuronal network in the preoptic area of the rat: demonstration of LH-RH-containing nerve terminals in synaptic contact with LH-RH neurons. *Brain Res.* 345, 332–336.
471. Witkin, J. W., Ferlin, M., Popilskis, S. J., and Silverman, A.-J. (1991). Effects of gonadal steroids on the ultrastructure of GnRH neurons in the rhesus monkey: synaptic input and glial apposition. *Endocrinology* 129:1083–1088.
472. Knobil, E. (1991). Electrophysiologic approaches to the hypothalamic GnRH pulse generator. In *Neuroendocrine Regulation of Reproduction* (S. S. C. Yen and W. W. Vale, Eds.) Serono Symposia, Norwell, MA.
473. Schwanzel-Fukuda, M., Jorgenson, K., Bergen, H., Weesner, G., and Pfaff, D. W. (1992). Biology of normal LHRH neurons during and after their migration from olfactory placode. *Endocr. Rev.* 13, 623–633.
474. Weesner, G. D., and Pfaff, D. W. (1992). Disruption of estrous cyclicity following administration of an LHRH antagonist to the preoptic area of the rat. *Soc. Neurosci. Abstr.* 18, 118.
475. Calza, L., Agnati, L. F., Fuxe, K., Giardino, L., and Goldstein, M. (1983). Morphometrical analysis of the distribution of luteinizing hormone-releasing hormone and tyrosine hydroxylase-immunoreactive nerve terminals within the lateral palisade zone of the median eminence of the male rat. *Neurosci. Lett.* 43, 179–183.
476. Schwanzel-Fukuda, M., and Silverman, A. J. (1980). The nervus terminalis of the guinea pig: a new luteinizing hormone-releasing hormone (LHRH) neuronal system. *J. Comp. Neurol.* 191, 213–225.
477. Schwanzel-Fukuda, M., Morrell, J. I., and Pfaff, D. W. (1985). Ontogenesis of neurons producing luteinizing hormone-releasing hormone (LHRH) in the nervus terminalis of the rat. *J. Comp. Neurol.* 238, 348–364.
478. Schwanzel-Fukuda, M., and Pfaff, D. W. (1987). Passive immunization of fetal rats with antiserum to luteinizing hormone-releasing hormone (LHRH) or transection of the central roots of the nervus terminalis does not affect rat pups' preference for home nest. *Physiol. Behav.* 41, 613–619.
479. Schwanzel-Fukuda, M., and Pfaff, D. W. (1989). Origin of luteinizing hormone-releasing hormone neurons. *Nature* 338, 161–164.
480. Zheng, L.-M., Pfaff, D. W., and Schwanzel-Fukuda, M. (1992). Electron microscopic identification of luteinizing hormone-releasing hormone-immunoreactive neurons in the medial olfactory placode and basal forebrain of embryonic mice. *Neuroscience* 46, 407–418.
481. Schwanzel-Fukuda, M., Zheng, L.-M., Bergen, H., Weesner, G., and Pfaff, D. W. (1992). LHRH neurons: functions and development. *Prog. Brain Res.* 93, 189–203.
482. Schwanzel-Fukuda, M., Bick, D., and Pfaff, D. W. (1989). Luteinizing hormone-releasing hormone (LHRH)-expressing cells do not migrate normally in an inherited hypogonadal (Kallmann) syndrome. *Brain Res. Mol. Brain Res.* 6, 311–326.
483. Schwanzel-Fukuda, M., Abraham, S., Crossin, K. L., Edelman, G. M., and Pfaff, D. W. (1992). Immunocytochemical demonstration of neural cell adhesion molecule (NCAM) along the migration route of luteinizing hormone-releasing hormone (LHRH) neurons in mice. *J. Comp. Neurol.* 321, 1–18.
484. Schwanzel-Fukuda, M., Abraham, S., Reinhard, G. R., Corssin, K. L., Edelman, G. M., and Pfaff, D. W. (1991). Antibody to neural cell adhesion molecule (NCAM) can disrupt the migration of luteinizing hormone-releasing hormone (LHRH) neurons into the mouse brain. *Soc. Neurosci. Abstr.* 17, 547.
485. Schwanzel-Fukuda, M., and Pfaff, D. W. (1992). Neuron specific enolase (NSE)-immunoreactivity is present in cells on the migration route of luteinizing hormone-releasing hormone (LHRH) neurons originating from the olfactory placode in mice. *Soc. Neurosci. Abstr.* 18, 233.
486. Schwanzel-Fukuda, M., Robinson, J. S., and Silverman, A. J. (1981). The fetal development of the luteinizing hormone-releasing hormone (LHRH) neuronal systems of the guinea pig brain. *Brain Res. Bull.* 7, 293–315.
487. Yamanouchi, K., and Arai, Y. (1983). Forebrain and lower brainstem participation in facilitatory and inhibitory regulation of the display of lordosis in female rats. *Physiol. Behav.* 30, 155–159.
488. Goldsmith, P. C., and Ganong, W. F. (1975). Ultrastructural localization of luteinizing hormone-releasing hormone in the median eminence of the rat. *Brain Res.* 97, 181–193.
489. Silverman, A. J., and Desnoyers, P. (1976). Ultrastructural immunocytochemical localization of luteinizing hormone-releasing hormone (LH-RH) in the median eminence of the guinea pig. *Cell. Tissue Res.* 169, 157–166.
490. Krisch, B. (1986). In *Current Topics in Neuroendocrinology* (D. Ganten and D. Pfaff, Eds.), Vol. 7. Springer-Verlag, Berlin.
491. Kozlowski, G. O., and Hostetter, G. (1978). Cellular and subcellular localization and behavioral effects of gonadotropin-releasing hormone (Gn-RG) in the rat. In *Brain-Endocrine Interaction III: Neural Hormones and Reproduction* (D. E. Scott, G. O. Kozlowski, and A. Weindl, Eds.), pp. 138–153. Karger, Basel.
492. Kozlowski, G. O., and Chu, L. (1980). Cellular characteristics of immunolabeled luteinizing hormone releasing hormone (LHRH) neurons. *Peptides* 1, 37–46.
493. Nishizuka, M. (1986). Unpublished observations.

494. Jamieson, M. G., and Fink, G. (1976). Parameters of electrical stimulation of the medial preoptic area for release of gonadotrophins in male rats. *J. Endocrinol.* 68, 57–70.
495. Fink, G., and Jamieson, M. G. (1976). Immunoreactive luteinizing hormone releasing factor in rat pituitary stalk blood: effects of electrical stimulation of the medial preoptic area. *J. Endocrinol.* 68, 71–87.
496. Gibson, M. J., Krieger, D. T., Charlton, H. M., Zimmerman, E. A., Silverman, A.-J., and Perlow, M. J. (1984). Mating and pregnancy can occur in genetically hypogonadal mice with preoptic area brain grafts. *Science* 225, 949–951.
497. Knobil, E., Plant, T. M., Wildt, L., Belchetz, P. E., and Marshall, G. (1980). Control of the rhesus monkey menstrual cycle: permissive role of hypothalamic gonadotropin-releasing hormone. *Science* 207, 1371–1373.
498. Knobil, E. (1980). The neuroendocrine control of the menstrual cycle. *Recent Prog. Horm. Res.* 36, 53–88.
499. Crowley, W. F., Filicori, M., Spratt, D. I., and Santoro, N. F. (1985). The physiology of gonadotropin-releasing hormone (GnRH) secretion in men and women. *Recent Prog. Horm. Res.* 41, 473–531.
500. Santen, R. J., and Bardin, C. W. (1973). Episodic luteinizing hormone secretion in man: pulse analysis, clinical interpretation, physiologic mechanisms. *J. Clin. Invest.* 52, 2617.
501. Wentz, A. C., Seegar-Jones, G., and Sapp, K. (1976). Pulsatile gonadotropin output in menstrual dysfunction. *Obstet. Gynecol.* 47, 309.
502. Strickler, R. C., Borth, R., and Woolever, A. (1977). The pulsatile pattern of gonadotropin release in normal men, normal women and amenorrheic women. *Obstet. Gynecol.* 50, 340.
503. Naftolin, F., Yen, S. S. C., and Perlman, D. (1973). Nocturnal patterns of serum gonadotropins during the menstrual cycle. *J. Clin. Endocrinol. Metab.* 37, 6.
504. Root, A., DeCherney, A., Russ, D., Duckett, G., Garcia, C.-R., and Wallach, E. (1972). Episodic secretion of luteinizing and follicle stimulating hormone in agonadal and hypogonadal adolescents and adults. *J. Clin. Endocrinol. Metab.* 35, 700.
505. Leyendecker, G. (1979). The pathophysiology of hypothalamic ovarian failure. *Eur. J. Obstet. Gynecol. Reprod. Biol.* 9, 175.
506. Leyendecker, G., Struve, T., and Plotz, E. J. Induction of ovulation with chronic intermittent administration of LH-RH in women with hypothalamic and hyperprolactinemic amenorrhea. In *61st Annual Endocrine Society Meeting,* Anaheim, CA, Abstr. 926.
507. Belchetz, P. E., Plant, T. M., Nakai, Y., Keogh, E. J., and Knobil, E. (1979). Hypophysial response to continuous and intermittent delivery of hypothalamic gonadotropin-releasing hormone. *Science* 202, 631.
508. Wildt, L., Marshall, G., and Knobil, E. (1980). Experimental induction of puberty in the infantile female rhesus monkey. *Science* 207, 1373.
509. Crowley, W. F., and McArthur, J. W. (1980). Stimulation of the normal menstrual cycle in Kallmann's syndrome by pulsatile administration of luteinizing hormone-releasing hormone (LHRH). *J. Clin. Endocrinol. Metab.* 51, 173–175.
510. Crowley, W. F., Comite, F., Vale, W., Rivier, J., Loriaux, D. L., and Cutler, G. B. (1981). Therapeutic use of pituitary desensitization with a long-acting LHRH agonist: a potential new treatment for idiopathic precocious puberty. *J. Clin. Endocrinol. Metab.* 52, 370–372.
511. Hoffman, A. R., and Crowley, W. F. (1982). Induction of puberty in men by long-term pulsatile administration of low-dose gonadotropin-releasing hormone. *N. Engl. J. Med.* 307, 1237–1241.
512. Culler, M. D., McArthur, N. H., Dees, W. L., Owens, R. E., and Harms, P. G. (1982). Inhibition of the postovariectomy depletion of hypothalamic luteinizing hormone-releasing hormone (LHRH) by suckling. *Biol. Reprod.* 26, 633–639.
513. Wheaton, J. E., and McCann, S. M. (1976). Luteinizing hormone-releasing hormone in peripheral plasma and hypothalamus of normal and ovariectomized rats. *Neuroendocrinology* 20, 296–310.
514. Root, A. W., Reiter, E. O., Duckett, G. E., and Sweetland, M. L. (1975). Effects of short-term castration and starvation upon hypothalamic content of luteinizing hormone-releasing hormone in adult male rats. *Proc. Soc. Exp. Biol. Med.* 150, 602–609.
515. Moguilevsky, J. A., Scachi, O., Debeljuk, L., and Faigon, M. R. (1975). Effect of castration upon hypothalamic luteinizing hormone releasing factor (LH-RF). *Neuroendocrinology* 17, 189–192.
516. Gross, D. S. (1980). Effect of castration and steroid replacement on immunoreactive gonadotropin-releasing hormone in the hypothalamus and preoptic area. *Endocrinology* 106, 1442–1450.
517. Campbell, G. T., and Ramaley, J. A. (1978). Immunocytochemistry of hypothalamic GnRH following manipulation of sex steroid levels: evidence that androgens have divergent effects on GnRH stores and serum follicle stimulating hormone. *Biol. Reprod.* 19, 620–627.
518. Jackson, G. L. (1972). Effect of actinomycin D on estrogen-induced release of luteinizing hormone in ovariectomized rats. *Endocrinology* 91, 1284–1287.
519. Jackson, G. L. (1973. Time interval between injection of estradiol benzoate and LH release in the rat and effect of actinomycin D or cycloheximide. *Endocrinology* 93, 887–892.
520. Shivers, B., Harlan, R., Morrell, J., and Pfaff, D. W. (1983). Absence of oestradiol concentration in cell nuclei of LHRH-immunoreactive neurons. *Nature* 304, 345–347.
521. Shivers, B. D., Harlan, R. E., Hejtmancik, J. F., Conn, P. M., and Pfaff, D. W. (1985). Localization of cells containing LHRH-like mRNA in rat brain. *Soc. Neurosci. Abstr.* 11, 143.
522. Shivers, B. D., Harlan, R. E., Hejtmancik, J. F., Conn, P. M., and Pfaff, D. W. (1986). Localization of cells containing LHRH-like mRNA in rat fore-brain using *in situ* hybridization. *Endocrinology* 118, 883–885.
523. Rothfeld, J., Hejtmancik, J. F., Conn, P. M., and Pfaff, D. W. (1989). *In situ* hybridization for LHRH mRNA following estrogen treatment. *Brain Res. Mol. Brain Res.* 6, 121–126.
524. Roberts, J. L., Dutlow, C. M., Jakubowski, M., Blum, M., and Millar, R. P. (1989). Estradiol stimulates preoptic area-anterior hypothalamic proGnRH-GAP gene expression in ovariectomized rats. *Brain Res. Mol. Brain Res.* 6, 127–132.
525. Rosie, R., Phomson, E., and Fink, G. (1990). Oestrogen positive feedback stimulates the synthesis of LHRH mRNA in neurons of the rostral diencephalon of the rat. *J. Endocrinol.* 124, 285–293.
526. Park, O.-K., Gugneja, S., and Mayo, K. E. (1990). Gonadotropin-releasing hormone gene expression during the rat estrous cycle: effects of pentobarbital and ovarian steroids. *Endocrinology* 127, 365–375.
527. Zoeller, R. T., and Young, S. W., III. (1988). Changes in cellular levels of messenger ribonucleic acid encoding gonadotropin-releasing hormone in the anterior hypothalamus of female rats during the estrous cycle. *Endocrinology* 123, 1688–1689.
528. Adler, B. A., and Crowley, W. R. (1986). Evidence for gamma-aminobutyric acid modulation of ovarian hormonal effects on luteinizing hormone secretion and hypothalamic catecholamine activity in the female rat. *Endocrinology* 118, 91–97.
529. Masoto, C., and Negro-Vilar, A. (1987). Activation of gamma-aminobutyric acid B-receptors abolishes naloxone-stimulated luteinizing hormone release. *Endocrinology* 121, 2251–2255.

530. Masoto, C., Wisniewski, G., and Negro-Vilar, A. (1989). Different gamma-aminobutyric acid receptor subtypes are involved in the regulation of opiate-dependent and independent luteinizing hormone-releasing hormone secretion. *Endocrinology* 125, 548–553.
531. Akema, T., and Kimura, F. (1991). The mode of $GABA_B$ receptor-mediated inhibition of the preovulatory luteinizing hormone surge in female rats. *Brain Res.* 562, 169–172.
532. Bergen, H. T., Hejtmancik, J. F., and Pfaff, D. W. (1991). Effects of gamma-aminobutyric receptor agonists and antagonists on LHRH-synthesizing neurons as detected by immunocytochemistry and *in situ* hybridization. *Exp. Brain Res.* 87, 46–56.
533. Weesner, G. D., Bergen, H. T., and Pfaff, D. W. (1992). Differential regulation of luteinizing hormone-releasing hormone and galanin mRNA levels by alpha-1 adrenergic agents in the rat. *J. Neuroendocrinol.* 4, 331–336.
534. Weesner, G. D., Krey, L. C., and Pfaff, D. W. (1993). Alpha-1 adrenergic regulation of estrogen-induced increases in luteinizing hormone-releasing hormone (LHRH) mRNA and release. *Brain Res. Mol. Brain Res.* 17, 77–82.
535. Moss, R. L., and McCann, S. M. (1973). Induction of mating behavior in rats by luteinizing hormone-releasing factor. *Science* 181, 177–179.
536. Pfaff, D. W. (1973). Luteinizing hormone releasing factor (LRF) potentiates lordosis behavior in hypophysectomized ovariectomized female rats. *Science* 182, 1148–1149.
537. Pfaff, D. W., and Modianos, D. (1976). Steroid and peptide hormones, and the neural mechanisms for reproductive behavior. In *Proceedings of the V International Congress of Endocrinology, Hamburg, Vol. 1.* (V. H. T. James, Ed.), pp. 67–71. Excerpta Medica, Amsterdam.
538. Dorsa, D. M., Smith, E. R., and Davidson, J. M. (1984). Immunoreactive-β-endorphin and LHRH levels in the brains of aged male rats with impaired sex behavior. *Neurobiol. Aging* 5, 115–120.
539. McAdoo, B. C., Doering, C. H., Kraemer, H. C., Dessert, N., Brodie, H. K. H., and Hamburg, D. A. (1978). A study of the effects of gonadotropin-releasing hormone on human mood and behavior. *Psychosom. Med.* 40, 199–209.
540. Ehrensing, R. H., Kastin, A. J., and Schally, A. V. (1981). Behavioral and hormonal effects of prolonged high doses of LHRH in male impotency. *Peptides* 2, 115–121.
541. Sakuma, Y., and Pfaff, D. W. (1980). LH-RH in the mesencephalic central grey can potentiate lordosis reflex of female rats. *Nature* 283, 566–567.
542. Sirinathsinghji, D. J. S. (1984). Modulation of lordosis behavior of female rats by naloxone, β-endorphin and its antiserum in the mesencephalic central gray: possible mediation via GnRH. *Neuroendocrinology* 39, 222–230.
543. Riskind, O., and Moss, R. L. (1979). Midbrain central gray: an extrahypothalamic site for LHRH potentiation of lordotic behavior in ovariectomized estrone-primed rats. *Brain Res. Bull.* 4, 203–205.
544. Riskind, O., and Moss, R. L. (1983). Midbrain LHRH infusions enhance lordotic behavior in ovariectomized estrogen-primed rats independently of a hypothalamic responsiveness to LHRH. *Brain Res. Bull.* 11, 481–485.
545. Kastin, A. J., Coy, D. H., Schally, A. V., and Zadina, J. E. (1980). Dissociation of effects of LH-RH analogs on pituitary regulation and reproductive behavior. *Pharmacol. Biochem. Behav.* 13, 913–914.
546. Zadina, J. E., Kastin, A. J., Fabre, L. A., and Coy, D. H. (1981). Facilitation of sexual receptivity in the rat by an ovulation-inhibiting analog of LHRH. *Pharmacol. Biochem. Behav.* 15, 961–964.
547. Sakuma, Y., and Pfaff, D. W. (1983). Modulation of the lordosis reflex of female rats by LHRH, its antiserum and analogs in the mesencephalic central gray. *Neuroendocrinology* 36, 218–224.
548. Dudley, C. A., Vale, W., Rivier, J., and Moss, R. L. (1983). Facilitation of sexual receptivity in the female rat by a fragment of the LHRH decapeptide, Ac-$LHRH^{5\text{-}10}$. *Neuroendocrinology* 36, 486–488.
549. Adams, P. R., Brown, D. A., and Constanti, A. (1982). Pharmacological inhibition of the M-current. *J. Physiol. (Lond.)* 332, 223–262.
550. Chan, A., Dudley, C. A., and Moss, R. L. (1983). Action of prolactin, dopamine and LHRH on ventromedial hypothalamic neurons as a function of ovarian hormones. *Neuroendocrinology* 36, 397–403.
551. Pan, J.-T., Kow, L.-M., and Pfaff, D. W. (1988). Modulatory actions of luteinizing hormone-releasing hormone on electrical activity of preoptic neurons in brain slices. *Neuroscience* 27, 623–628.
552. Ogawa, S., Kow, L.-M., and Pfaff, D. W. (1992). Effects of lordosis-relevant neuropeptides on midbrain periaqueductal gray neuronal activity *in vitro. Peptides* 13, 965–975.
553. Pan, J.-T., Kow, L.-M., Kendall, D. A., Kaiser, E. T., and Pfaff, D. W. (1986). Electrophysiological test of an amphiphilic β-structure in LHRH action. *Mol. Cell. Endocrinol.* 48, 161–166.
554. Toubeau, G., Desclin, J., Parmentier, M., and Pasteels, J. L. (1979). Cellular localization of a prolactin-like antigen in the rat brain. *J. Endocrinol.* 83, 261–266.
555. Fuxe, K., Kokfelt, T., Eneroth, O., Gustafsson, J.-A., and Skett, P. (1977). Prolactin-like immunoreactivity: localization in nerve terminals of rat hypothalamus. *Science* 196, 899–900.
556. Harlan, R., Shivers, B., and Pfaff, D. W. (1983). Midbrain microinfusions of prolactin increase the estrogen-dependent behavior, lordosis. *Science* 219, 1451–1453.
557. Nishizuka, M., Shivers, B. D., Leranth, C., and Pfaff, D. W. (1990). Ultrastructural characterization of prolactin-like immunoreactivity in rat medial basal hypothalamus. *Neuroendocrinology* 51, 249–254.
558. Harlan, R. E., Shivers, B. D., Kalamaridis, M., and Pfaff, D. W. (1984). Synthesis of immunoreactive prolactin in the rat hypothalamus. *Soc. Neurosci. Abstr.* 10, 286.
559. Harlan, R. E., Shivers, B. D., Fox, S. R., Kaplove, K. A., Schacter, B. S., and Pfaff, D. W. (1989). Distribution and partial characterization of immunoreactive prolactin in the rat brain. *Neuroendocrinology* 49, 7–22.
560. Shivers, B. D., Harlan, R. E., and Pfaff, D. W. (1989). A subset of neurons containing immunoreactive prolactin is a target for estrogen regulation of gene expression in rat hypothalamus. *Neuroendocrinology* 49, 23–27.
561. Schachter, B., Shivers, B., Harland, R., and Pfaff, D. W. (1983). Evidence for prolactin messenger RNA in the rat brain. *Endocrinol. Soc. Abstr.* 346, 167.
562. Schachter, B. S., Durgerian, S., Harlan, R. E., Pfaff, D. W., and Shivers, B. D. (1984). Prolactin mRNA exists in rat hypothalamus. *Endocrinology* 114, 1947–1949.
563. Emanuele, N. V., Jurgens, J. K., Halloran, M. M., Tentler, J. J., Lawrence, A. M., and Kelley, M. R. (1992). The rat prolactin gene is expressed in brain tissue: detection of normal and alternatively spliced prolactin messenger RNA. *Mol. Endocrinol. 6,* 35–42.
564. Dutt, A. (1992). Unpublished data.
565. Ljungdahl, A., Hokfelt, T., and Nilsson, G. (1978). Distribution of substance P immunoreactivity in the central nervous system of the rat: I. Cell bodies and nerve terminals. *Neuroscience* 3, 861–943.
566. Panula, O., Yang, H.-Y. T., and Cost, E. (1984). Comparative distribution of bombesin/GRP- and substance P-like immunoreactivities in rat hypothalamus. *J. Comp. Neurol.* 224, 606–617.

567. Larsen, P. J. (1992). Distribution of substance P-immunoreactive elements in the preoptic area and the hypothalamus of the rat. *J. Comp. Neurol.* 316, 287–313.
568. Akesson, T. R., and Micevych, P. E. (1988). Estrogen concentration by substance P-immunoreactive neurons in the medial basal hypothalamus of the female rat. *J. Neurosci. Res.* 19, 412–419.
569. Antonowicz, U., Jakubowska-Naziemblo, B., Cannon, D., and Powell, D. (1981). Immunoreactive substance P content in median eminence and pituitary gland during oestrus, dioestrus and after anterior hypothalamic deafferentiation. *Endokrinologie* 79, 25–34.
570. Kerdelhue, B., Tartar, A., Lenoir, V., El Abed, A., Hublau, O., and Millar, R. P. (1985). Binding studies of substance P anterior pituitary binding sites: changes in substance P binding sites during the rat estrous cycle. *Regul. Pept.* 10, 133–143.
571. Debeljuk, L., Villanua, M. A., and Bartke, A. (1992). Substance P variations in the hypothalamus of golden hamsters at different stages of the estrous cycle. *Neurosci. Lett.* 137, 178–180.
572. Dornan, W. A., Akesson, T. R., and Micevych, P. E. (1990). A substance P projection from the VMH to the dorsal midbrain central gray: implication for lordosis. *Brain Res. Bull.* 25, 791–796.
573. Dam, T.-V., Martinelli, B., and Quirion, R. (1990). Autoradiographic distribution of brain neurokinin-1/substance P receptors using a highly selective ligand [^{3}H]-[Sar^9,Met(O$_2$)^{11}]-substance P. *Brain Res.* 531, 333–337.
574. Mantyh, P. W., Hunt, S. O., and Maggio, J. E. (1984). Substance P receptors: localization by light microscope autoradiography in rat brain using [^{3}H]SP as the radioligand. *Brain Res.* 307, 147–165.
575. Wolf, S. S., Moody, T. W., Quirion, R., and O'Donohue, T. L. (1985). Biochemical characterization and autoradiographic localization of central substance P receptors using [^{125}]physalaemin. *Brain Res.* 332, 299–307.
576. Rothman, R. B., Herkenham, M., Pert, C. B., Liang, T., and Cascier, M. A. (1984). Visualization of rat brain receptors for the neuropeptide, substance P. *Brain Res.* 309, 47–54.
577. Dornan, W. A., and Malsbury, C. W. (1984). Facilitation of lordosis by infusion of substance P in the midbrain central gray. *Soc. Neurosci. Abstr.* 10, 172.
578. Dornan, W. A., Malsbury, C. W., and Penney, R. B. (1987). Facilitation of lordosis by injection of substance P into the midbrain central gray. *Neuroendocrinology* 45, 498–506.
579. Caldwell, J. D., Jirikowski, G. F., Greer, E. R., and Pedersen, C. A. (1989). Medial preoptic area oxytocin and female sexual receptivity. *Behav. Neurosci.* 103, 655–662.
580. Kow, L.-M., and Pfaff, D. W. (1988). Neuromodulatory actions of peptides. *Annu. Rev. Pharmacol. Toxicol.* 28, 163–188.
581. Roth, K. A., Weber, E., Barchas, J. D., Chang, D., and Chang, J. K. (1983). Immunoreactive dynorphin-(1-8) and corticotropin-releasing factor in subpopulation of hypothalamic neurons. *Science* 219, 189–191.
582. Sirinathsinghji, D. J. S., Whittington, P. E., Audsley, A. R., and Fraser, H. M. (1983). β-Endorphin regulates lordosis in female rats by modulating LH-RH release. *Nature* 301, 62–64.
583. O'Donohue, T. L., Miller, R. L., and Jacobowitz, D. M. (1979). Identification, characterization and stereotaxic mapping of intraneuronal α-melanocyte stimulating hormone-like immunoreactive peptides in discrete regions of the rat brain. *Brain Res.* 176, 101–123.
584. Watson, S. J., and Akil, H. (1980). α-MSH in rat brain: occurrence within and outside of β-endorphin neurons. *Brain Res.* 182, 217–223.
585. Thody, A. J., Wilson, C. A., and Everard, D. (1981). α-Melanocyte stimulating hormone stimulates sexual behaviour in the female rat. *Psychopharmacology (Berl.)* 74, 153–156.
586. Thody, A. J., Wilson, C. A., and Everard, D. (1979). Facilitation and inhibition of sexual receptivity in the female rat by α-MSH. *Physiol. Behav.* 22, 447–450.
587. Pfaus, J. G., and Gorzalka, B. B. (1987). Selective activation of opioid receptors differentially affects lordosis behavior in female rats. *Peptides* 8, 309–317.
588. Pfaus, J. G., and Pfaff, D. W. (1992). Mu, delta, and kappa opioid receptor agonists selectively modulate sexual behaviors in the female rat: differential dependence on progesterone. *Horm. Behav.* 26, 457–473.
589. Barraclough, C. A., and Sawyer, C. H. (1955). Inhibition of the release of pituitary ovulatory hormone in the rat by morphine. *Endocrinology* 57, 329–337.
590. Ieri, T., Chen, H. T., Campbell, G. A., and Meites, J. (1980). Effects of naloxone and morphine on the pro-oestrous surge of prolactin and gonadotrophins in the rat. *Endocrinology* 106, 1568–1570.
591. Pang, C. N., Zimmerman, E., and Sawyer, C. H. (1977). Morphine inhibition of the preovulatory surges of plasma luteinizing hormone and follicle stimulating hormone in the rat. *Endocrinology* 101, 1726–1732.
592. Wiesner, J. B., and Moss, R. L. (1984). Beta-endorphin suppression of lordosis behavior in female rats: lack of effect of peripherally-administered naloxone. *Life Sci.* 34, 1455–1462.
593. Wiesner, J. B., and Moss, R. L. (1986). Suppression of receptive and preceptive behavior in ovariectomized, estrogen-progesterone-primed rats by intraventricular beta-endorphin: studies of behavioral specificity. *Neuroendocrinology* 43, 57–62.
594. Alien, D. L., Renner, K. J., and Luine, V. N. (1985). Naltrexone facilitation of sexual receptivity in the rat. *Horm. Behav.* 19, 98–103.
595. Sirinathsinghji, D. J. S., Audsley, A., and Whittington, P. E. (1981). Naloxone administration in the mesencephalic central gray can potentiate lordosis reflex of female rats. *Br. Soc. Endocrinol.*
596. Sirinathsinghji, D. J. S., Rees, L. H., Rivier, J., and Vale, W. (1983). Corticotropin releasing factor is a potent inhibitor of sexual receptivity in the female rat. *Nature* 305, 232–235.
597. Wiesner, J. B., Koenig, J. I., Krulich, L., and Moss, R. L. (1984). Site of action for β-endorphin-induced changes in plasma luteinizing hormone and prolactin in the ovariectomized rat. *Life Sci.* 34, 1463–1473.
598. Wiesenfeld-Hallin, Z., and Sodersten, P. (1984). Spinal opiates affect sexual behavior in rats. *Nature* 309, 257–258.
599. Crowley, W. R. (1986). Reproductive neuroendocrine regulation in the female rat by central catecholamine-neuropeptide interactions: a local control hypothesis. *Ann. N Y Acad. Sci.* 474, 423–436.
600. Pfaff, D. W. (1982). Motivational concepts: definitions and distinctions. In *The Physiological Mechanisms of Motivation* (D. W. Pfaff, Ed.), pp. 3–24. Springer-Verlag, New York.
601. Wardlaw, S. L., Thoron, L., and Frantz, G. A. (1982). Effects of sex steroids on brain beta endorphin. *Brain Res.* 245, 327.
602. Shivers, B. D., Harlan, R. E., and Pfaff, D. W. (1989). A subset of neurons containing immunoreactive prolactin is a target for estrogen regulation of gene expression in rat hypothalamus. *Neuroendocrinology* 49, 23–27.
603. Pfaff, D. W. (1988). Multiplicative responses to hormones by hypothalamic neurons. In *Recent Progress in Posterior Pituitary Hormones* (S. Yoshida and L. Share, Eds.), pp. 257–267. Elsevier, Amsterdam.
604. Robinson, A. G. (1974). Elevation of plasma neurophysin in women on oral contraceptives. *J. Clin. Invest.* 54, 209–212.

605. Robinson, A. G., Halusczak, C., Wilkins, J. A., Huellmantel, A. B., and Watson, C. G. (1977). Physiological control of two neurophysins in humans. *J. Clin. Endocrinol. Metab.* 44, 330–339.
606. Mohr, E., Bahnsen, U., Kiessling, C., and Richter, D. (1988). Expression of the vasopressin and oxytocin genes in rats occurs in mutually exclusive sets of hypothalamic neurons. *FEBS Lett.* 242, 144–148.
607. Richard, S., and Zingg, H. H. (1990). The human oxytocin gene promoter is regulated by estrogens. *J. Biol. Chem.* 265, 6098–6103.
608. Sar, M., and Stumpf, W. E. (1980). Simultaneous localization of ^{3}H estradiol and neurophysin 1 or arginine vasopressin in hypothalamic neurons demonstrated by a combined technique of dry-mount autoradiography and immunocytochemistry. *Neurosci. Lett.* 17, 179–184.
609. Rhodes, C. H., Morrell, J. I., and Pfaff, D. W. (1981). Distribution of estrogen-concentrating, neurophysin-containing magnocellular neurons in the rat hypothalamus as demonstrated by a technique combining steroid autoradiography and immunohistology in the same tissue. *Neuroendocrinology* 33, 18–23.
610. Rhodes, C. H., Morrell, J. I., and Pfaff, D. W. (1982). Estrogen concentrating neurophysin containing hypothalamic magnocellular neurons in the vasopressin deficient (Brattleboro) rat: a study combining steroid autoradiography and immunohistochemistry. *J. Neurosci.* 2, 1718–1724.
611. Chung, S. K., McCabe, J. T., and Pfaff, D. W. (1991). Estrogen influences on oxytocin mRNA expression in preoptic and anterior hypothalamic regions studied by in situ hybridization. *J. Comp. Neurol.* 307, 281–295.
612. Burbach, J. P. H., Adan, R. A. H., van Tol, H. H. M., Verbeeck, M. A. E., Axelson, J. F., van Leeuwen, F. W., Beekman, J. M., and Ab, G. (1990). Regulation of the rat oxytocin gene by estradiol. *J. Neuroendocrinol.* 2, 633–639.
613. Van Tol, H. H. M., Bolwerk, E. L. M., Liu, B., and Burbach, J. P. H. (1988). Oxytocin and vasopressin gene expression in the hypothalamo-neurohypophyseal system of the rat during the estrous cycle, pregnancy and lactation. *Endocrinology* 122, 945–951.
614. Miller, F. R., Ozimek, G., Milner, R. J., and Bloom, F. E. (1989). Regulation of neuronal oxytocin mRNA by ovarian steroids in the mature and developing hypothalamus. *Proc. Natl. Acad. Sci. U S A* 86, 2468–2472.
615. Chung, S. K., Haldar, J., and Pfaff, D. W. (1991). Combination of in situ hybridization with retrograde marker to study hormone effect on oxytocin mRNA-expressing neurons which project to spinal cord. *Soc. Neurosci. Abstr.* 17, 715.
616. Kawata, M., McCabe, J. T., Chung, S. K., Dutt, A., Yuri, K., Hirakawa, M., Kumamoto, K., Hirayama, Y., and Pfaff, D. W. (1991). The effect of progesterone on oxytocin messenger RNA in hypothalamic neurons of estrogen-treated female rats studied with quantitative in situ hybridization histochemistry. *Biomed. Res.* 12, 405–415.
617. Brooks, P. J., Kaplitt, M. G., Kleopoulos, S. P., Funabashi, T., Mobbs, C. V., and Pfaff, D. W. (1993). Detection of messenger RNA and low-abundance heteronuclear RNA with single-stranded DNA probes produced by amplified primer extension labeling *J. Histochem. Cytochem.* 41, 1761–1766.
618. De Kloet, E. R., Rotteveel, F., Voorhuis, T. A. M., and Terlou, M. (1985). Topography of binding sites for neurohypophyseal hormones in rat brain. *Eur. J. Pharmacol.* 110, 113–119.
619. Coirini, H., Johnson, A. R. E., and McEwen, B. S. (1989). Estradiol modulation of oxytocin binding in the ventromedial hypothalamus nucleus of male and female rats. *Neuroendocrinology* 50, 193–198.
620. Johnson, A. E., Coirini, H., Ball, G. F., and McEwen, B. S. (1989). Anatomical localization of the effects of 17β-estradiol on oxytocin receptor binding in the ventromedial hypothalamic nucleus. *Endocrinology* 124, 207–211.
621. Schumacher, M., Coirini, H., Pfaff, D. W., and McEwen, B. S. (1990). Behavioral effects of progesterone associated with rapid modulation of oxytocin receptors. *Science* 250, 691–694.
622. Coirini, H., Schumacher, M., Flanagan, L. M., and McEwen, B. S. (1991). Transport of estrogen-induced oxytocin receptors in the ventromedial hypothalamus. *J. Neurosci.* 11, 3317–3324.
623. Johnson, A. R. E., Harbaugh, C. R., and Gelhard, R. E. (1991). Projections from the ventromedial nucleus of the hypothalamus contain oxytocin binding sites. *Brain Res.* 567, 332–336.
624. Schumacher, M., Coirini, H., Frankfurt, M., and McEwen, B. S. (1989). Localized actions of progesterone in hypothalamus involve oxytocin. *Proc. Natl. Acad. Sci. U S A* 86, 6798–6801.
625. Arletti, R., and Bertolini, A. (1985). Oxytocin stimulates lordosis behavior in female rats. *Neuropeptides* 6, 247–253.
626. Caldwell, J. D., Prange, A. J. Jr., and Pedersen, C. A. (1984). Oxytocin facilitates sexual behavior in estrogen-treated ovariectomized rats. *J. Steroid Biochem.* 20, 1510.
627. Caldwell, J. D., Prange, A. J. Jr., and Pedersen, C. A. (1986). Oxytocin facilitates the sexual receptivity of estrogen-treated female rats. *Neuropeptides* 7, 175.
628. Gorzalka, B. B., and Lester, G. L. L. (1987). Oxytocin-induced facilitation of lordosis behavior in rats is progesterone-dependent. *Neuropeptides* 10, 55–65.
629. Elands, J., von Doremalen, E., Spruyt, B., and de Kloet, E. R. (1991). Oxytocin receptors in the rat hypothalamic ventromedial nucleus: a study of possible mediators of female sexual behavior. *Coll. INSERM* 208, 311–319.
630. Schulze, H. G., and Gorzalka, B. B. (1991). Oxytocin effects on lordosis frequency and lordosis duration following infusion into the medial preoptic area and ventromedial hypothalamus of female rats. *Neuropeptides* 18, 99–106.
631. Caldwell, J. D., Barakat, A. S., Smith, D. D., Hruby, V. J., and Pedersen, C. A. (1990). A uterotonic antagonist blocks the oxytocin-induced facilitation of female sexual receptivity. *Brain Res.* 512, 291–296.
632. Witt, D. M., and Insel, T. R. (1991). A selective oxytocin antagonist attenuates progesterone facilitation of female sexual behavior. *Endocrinology* 128, 3269–3276.
633. Caldwell, J. D. (1992). Central oxytocin and female sexual behavior. *Ann. N Y Acad. Sci.* 652, 166–179.
634. McCarthy, M. M., Chung, S. K., Ogawa, S., Kow, L.-M., and Pfaff, D. W. (1991). Behavioral effects of oxytocin: is there a unifying principle? In *Vasopressin* (S. Jard and R. Jamison, Eds.), pp. 195–212. John Libbey Eurotext, Montpellier, France.
635. McCarthy, M. M., Kow, L.-M., and Pfaff, D. W. (1992). Speculations concerning the physiological significance of central oxytocin in maternal behavior. *Ann. N Y Acad. Sci.* 652, 70–82.
636. Kow, L.-M., and Pfaff, D. W. (1986). Vasopressin excites ventromedial hypothalamic glucose-responsive neurons *in vitro*. *Physiol. Behav.* 37, 153–158.
637. Kow, L.-M., Johnson, A. E., Ogawa, S., and Pfaff, D. W. (1991). Electrophysiological actions of oxytocin on hypothalamic neurons *in vitro*: neuropharmacological characterizations and effects of ovarian steroids. *Neuroendocrinology* 54, 526–535.
638. Kimura, T., Tanizawa, O., Mori, K., Brownstein, M. J., and Okayama, H. (1992). Structure and expression of a human oxytocin receptor. *Nature* 356, 526–529.
639. McCarthy, M. M., et al. (1992). Unpublished data.
640. Brown, H., et al. (1992). Unpublished data.
641. Mobbs, C. V., Rothfeld, J. M., Saluja, R., and Pfaff, D. W. (1989). Phorbol esters and forskolin infused into midbrain

central gray facilitate lordosis. *Pharmacol. Biochem. Behav.* 34, 665–667.

642. Clemens, L. G. (1983). Neural plasticity and feminine sexual behaviour in the rat. In *Sex and Behavior* (T. E. McGill, D. A. Dewsbury, and B. D. Sachs, Eds.), pp. 243–266. Plenum Press, New York.
643. Menard, C. S., and Dohanich, G. P. (1990). Physostigmine facilitation of lordosis in naturally cycling female rats. *Pharmacol. Biochem. Behav.* 36, 853–858.
644. Rodriguez-Sierra, J. F., and Komisaruk, B. R. (1982). Common hypothalamic sites for activation of sexual receptivity in female rats by LHRH, PGE_2 and progesterone. *Neuroendocrinology* 35, 363–369.
645. Pfaus, J., et al. (1992). Unpublished data.
646. Mobbs, C. V., Harlan, R. E., Burrows, M., and Pfaff, D. W. (1988). An estradiol-induced protein synthesized in the ventromedial hypothalamus and transported to the midbrain central gray. *J. Neurosci.* 8, 113–118.
647. Mobbs, C. V., Fink, G., and Pfaff, D. W. (1990). HIP-70: an isoform of phosphoinositol-specific phospholipase C-α. *Science* 249, 566.
648. Hixon, J. E., and Flint, A. P. F. (1987). Effects of luteolytic does of oestradiol benzoate on uterine oxytocin receptor concentrations, phosphoinositide turnover and prostaglandin $F_{2\alpha}$ secretion in sheep. *J. Reprod. Fertil.* 79, 457–467.
649. Grove, R. E., and Korach, K. S. (1987). Estrogen stimulation of phosphatidylinositol metabolism in mouse uterine tissue. *Endocrinology* 121, 1083–1088.
650. Ignar-Trowbridge, D. M., Hughes, A. R., Putnery, J. W. Jr., McLachlan, J. A., and Korach, K. S. (1991). Diethylstilbestrol stimulates persistent phosphatidylinositol lipid turnover by an estrogen receptor-mediated mechanism in immature mouse uterus. *Endocrinology* 129, 2423–2430.
651. Freter, C. E., Lippman, M. E., Cheville, A., Zinn, S., and Gelmann, E. P. (1988). Alterations in phosphoinositide metabolism associated with 17β-estradiol and growth factor treatment of MCF-7 breast cancer cells. *Mol. Endocrinol.* 3, 159–166.
652. Taylor, J. A., and Welshons, W. V. (1992). MCF-7 breast cancer cell proliferation and membrane-associated phospholipase C activity: implications for mitogenic signalling mechanisms. In *Endocrine Society Annual Meeting Abstracts*.
653. Graber, R., Sumida, C., Vallette, G., and Nunez, E. A. (1992). Estradiol-induced nongenomic and long-term effects on the PIP_2-phospholipase C activity of MCF-7 cells. *Endocrine Society Annual Meeting Abstracts*.
654. Sidorkina, O. M., Morozova, T. M., and Rau, V. A. (1988). Translocation of protein kinase C under the action of estradiol from the cytosol into the cell membranes and activation of the enzyme in the target cells. *Biochemistry (Mosc.)* 53, 347–352.
655. Morozova, T. M., Mitina, R. L., Rau, V. A., and Sidorkina, O. M. (1989). Mechanisms of the stimulating action of estradiol on protein kinase C in plasma membranes of target cells. *Biochemistry (Mosc.)* 54, 475–482.
656. Drouva, S. V., Gorenne, I., Laplante, E., Rerat, E., Enhalbert, A., and Kordon, C. (1990). Estradiol modulates protein kinase C activity in the rat pituitary *in vivo* and *in vitro*. *Endocrinology* 126, 536–544.
657. Steinschneider, A., Rao, M. C., Khan, I., McLean, M. O., and Gibori, G. (1991). Calcium-calmodulin and calcium-phospholipid dependent phosphorylation of membranous fraction proteins related to the tropic regulation by estradiol in the corpus luteum. *Endocrinology* 128, 263–272.
658. Ruzycky, A. L., and Crankshaw, D. J. (1988). Role of inositol phospholipid hydrolysis in the initiation of agonist-induced contractions of rat uterus: effects of domination by 17β-estradiol and progesterone. *Can. J. Physiol. Pharmacol.* 66, 10–17.
659. Riemer, R. K., Goldfien, A., and Roberts, J. M. (1987). Estrogen increases adrenergic- but not cholinergic-mediated production of inositol phosphates in rabbit uterus. *Mol. Pharmacol.* 32, 663–668.
660. Favit, A., Fiore, L., Nicoletti, F., and Canonico, P. L. (1991). Estrogen modulates stimulation of inositol phospholipid hydrolysis by norepinephrine in rat brain slices. *Brain Res.* 555, 65–69.
661. Lida, K., Imai, A., and Tamaya, T. (1991). Stimulatory effects of estrogen on gonadotropin-releasing hormone-induced phosphoinositide turnover in granulosa cells. *J. Steroid Biochem. Mol. Biol.* 38, 583–586.
662. Kaczmarek, L. K. (1987). The role of protein kinase C in the regulation of ion channels and neurotransmitter release. *Trends Neurosci.* 10, 30–34.
663. Kawakami, M., Terasawa, E., and Ibuki, T. (1970). Changes in multiple unit activity of the brain during the estrous cycle. *Neuroendocrinology* 6, 30.
664. Terasawa, E., and Sawyer, C. H. (1969). Changes in electrical activity in the rat hypothalamus related to electrochemical stimulation of adenohypophyseal function. *Endocrinology* 85, 143.
665. Dyer, R. G., Pritchett, C. J., and Cross, B. A. (1972). Unit activity in the diencephalon of female rats during the oestrous cycle. *J. Endocrinol.* 53, 151.
666. Moss, R. L., and Law, O. T. (1971). The estrous cycle: its influence on single unit activity in the forebrain. *Brain Res.* 30, 435.
667. Cross, B. A., and Dyer, R. G. (1971). Cyclic changes in neurons of the anterior hypothalamus during the rat estrous cycle and the effect of anesthesia. In *Steroid Hormones and Brain Function: UCLA Forum in Medical Sciences, Number 15* (C. H. Sawyer and R. A. Gorski, Eds.), pp. 95–102. University of California Press, Los Angeles.
668. Dyer, R. G. (1973). An electrophysiological dissection of the hypothalamic regions which regulate the pre-ovulatory secretion of luteinizing hormone in the rat. *J. Physiol. (Lond.)* 234, 421.
669. Kawakami, M., Terasawa, E., Ibuki, T., and Manaka, M. (1971). Effects of sex hormones and ovulation-blocking steroids and drugs on electrical activity of the rat brain. In *Steroid Hormones and Brain Function: UCLA Forum in Medical Sciences, Number 15* (C. H. Sawyer and R. A. Gorski, Eds.), pp. 79–93. University of California Press, Los Angeles.
670. Cross, B. A., and Dyer, R. G. (1972). Ovarian modulation of unit activity in the anterior hypothalamus of the cyclic rat. *J. Physiol. (Lond.)* 222, 25.
671. Alcaraz, M., Guzman-Rores, C., Salas, M., and Beyer, C. (1969). Effect of estrogen on the responsivity of hypothalamic and mesencephalic neurons in the female cat. *Brain Res.* 15, 439.
672. Pfaff, D. W., and Gregory, E. (1971). Correlation between preoptic area unit activity and the cortical EEG: difference between normal and castrated male rats. *Electroencephalogr. Clin. Neurophysiol.* 31, 223.
673. Lincoln, D. W. (1967). Unit activity in the hypothalamus, septum and preoptic area of the rat: characteristics of spontaneous activity and the effect of oestrogen. *J. Endocrinol.* 37, 177.
674. Lincoln, D. W., and Cross, B. A. (1967). Effect of oestrogen on the responsiveness of neurons in the hypothalamus, septum and preoptic area of rats with light-induced persistent oestrus. *J. Endocrinol.* 37, 191.
675. Yagi, K. (1970). Effects of estrogen on the unit activity of the rat hypothalamus. *J. Physiol. Soc. Jpn.* 32, 692.
676. Yagi, K. (1973). Changes in firing rates of single preoptic and hypothalamic units following an intravenous administration of estrogen in the castrated female rat. *Brain Res.* 53, 343.

677. Sakuma, Y., and Pfaff, D. W. (1982). Properties of ventromedial hypothalamic neurons with axons to midbrain central gray. *Exp. Brain Res.* 46, 292–300.
678. Terasawa, E., Noonan, J., and Bridson, W. E. (1982). Anesthesia with pentobarbitone blocks the progesterone-induced luteinizing hormone surge in the ovariectomized rhesus monkey. *J. Endocrinol.* 92, 327.
679. Terasawa, E., Yeoman, R. R., and Bridson, W. E. (1983). Effects of complete neural deafferentation of the medial basal hypothalamus (MBH) during the progesterone (P)-induced LH surge in the female rhesus monkey. In *65th Annual Meeting of the Endocrine Society*, p. 157.
680. Garris, D. R., Billiar, R. B., Takaoka, Y., White, R., and Little, B. (1982). Autoradiographic analysis of progesterone-concentrating cells in the isolated rhesus monkey hypothalamus. *Neuroendocrinology* 35, 388.
681. Goldsmith, P. C., Lamberts, R., and Bregina, L. R. (1983). Gonadotropin-releasing hormone neurons and pathways in the primate hypothalamus and forebrain. In *Neuroendocrine Aspects of Reproduction* (R. L. Norman, Ed.), pp. 7–45. Academic Press, New York.
682. Silverman, A. J., Antunes, J. L., Abrams, G. M., Nilaver, G., Thau, R., Robinson, J. A., Ferin, M., and Krey, L. C. (1982). The luteinizing hormone-releasing hormone pathways in rhesus *(Macaca mulatta)* and pig tailed *(Macaca nemestrina)* monkeys: new observations on thick unembedded sections. *J. Comp. Neurol.* 211, 309.
683. Yeoman, R. R., and Terasawa, E. (1984). An increase in single unit activity of the medial basal hypothalamus occurs during the progesterone-induced luteinizing hormone surge in the female rhesus monkey. *Endocrinology* 115, 2445–2452.
684. Chan, A., Dudley, C. A., and Moss, R. L. (1984). Hormonal and chemical modulation of ventromedial hypothalamic neurons responsive to vaginocervical stimulation. *Neuroendocrinology* 38, 328–336.
685. Yagi, K., and Sawaki, Y. (1973). Feedback of estrogen in the hypothalamic control of gonadotrophin secretion. In *Neuroendocrine Control* (K. Yagi and S. Yoshida, Eds.), pp. 297–325. University of Tokyo Press, Tokyo.
686. Morrell, J. I., Wolinsky, T. D., Krieger, M. S., and Pfaff, D. W. (1982). Autoradiographic identification of estradiol-concentrating cells in the spinal cord of the female rat. *Exp. Brain Res.* 45, 144–150.
687. Komisaruk, B. R., Adler, N. T., and Hutchison, J. (1972). Genital sensory field: enlargement by estrogen treatment in female rats. *Science* 178, 1295–1298.
688. Kow, L.-M., and Pfaff, D. W. (1973). Effects of estrogen treatment on the size of receptive field and response threshold of pudendal nerve in the female rat. *Neuroendocrinology* 13, 299–313.
689. Kow, L.-M., and Pfaff, D. W. (1979). Responses of single units in sixth lumbar dorsal root ganglion of female rats to mechanostimulation of relevant for lordosis reflex. *J. Neurophysiol.* 42, 203–213.
690. Kow, L.-M., Zemlan, F. O., and Pfaff, D. W. (1980). Responses of lumbosacral spinal units to mechanical stimuli related to analysis of lordosis reflex in female rates. *J. Neurophysiol.* 43, 27–45.
691. Kow, L.-M., Montgomery, M. O., and Pfaff, D. W. (1979). Triggering of lordosis reflex in female rats with somatosensory stimulation: quantitative determination of stimulus parameters. *J. Neurophysiol.* 42, 195–202.
692. Kow, L.-M., and Pfaff, D. W. (1982). Responses of medullary reticulospinal and other reticular neurons to somatosensory and brainstem stimulation in anesthetized or freely moving ovariectomized rats with or without estrogen treatment. *Exp. Brain Res.* 47, 191–202.
693. Sullivan, J., Schwartz-Giblin, S., and Pfaff, D. (1985). Lumbar axial muscle activity is correlated with electroencephalogram arousal. *Soc. Neurosci. Abstr.* 11, 882.
694. Sullivan, J. M., Schwartz-Giblin, S., and Pfaff, D. W. (1986). Correlations between EEG state and spontaneous and evoked axial muscle. *Brain Res.* 368, 197–200.
695. Sakuma, Y., and Pfaff, D. W. (1980). Convergent effects of lordosis-relevant somatosensory and hypothalamic influences on central gray cells in the rat mesencephalon. *Exp. Neurol.* 70, 269–281.
696. Sakuma, Y., and Pfaff, D. W. (1980). Cells of origin of medullary projections in central gray of rat mesencephalon. *J. Neurophysiol.* 44, 1002–1011.
697. Sakuma, Y., and Pfaff, D. W. (1980). Excitability of female rat central gray cells with medullary projections: changes produced by hypothalamic stimulation and estrogen treatment. *J. Neurophysiol.* 44, 1012–1023.
698. Haskins, J. T., and Moss, R. L. (1983). Action of estrogen and mechanical vaginocervical stimulation on the membrane excitability of hypothalamic and midbrain neurons. *Brain Res. Bull.* 10, 489–496.
699. Rose, J. D. (1982). Midbrain distribution of neurons with strong, sustained responses to lordosis trigger stimuli in the female golden hamster. *Brain Res.* 240, 364–367.
700. Rose, J. D., and Bieber, S. L. (1984). Joint and separate effects of estrogen and progesterone on responses of midbrain neurons to lordosis-controlling somatic stimuli in the female golden Syrian hamster. *J. Neurophysiol.* 54, 1040–1054.
701. Rose, J. D., and Michael, R. P. (1978). Facilitation by estradiol of midbrain and pontine unit responses to vaginal and somatosensory stimulation in the squirrel monkey *(Saimiri sciureus)*. *Exp. Neurol.* 58, 46–58.
702. Kow, L.-M., and Pfaff, D. W. (1984). Suprachiasmatic neurons in tissue slices from ovariectomized rats: electrophysiological and neuropharmacological characterization and the effects of estrogen treatment. *Brain Res.* 297, 275–286.
703. De Lean, A., Ferland, L., Drouin, J., Kelly, P. A., and Labrie, F. (1977). Modulation of pituitary thyrotropin releasing hormone receptor levels by estrogens and thyroid hormones. *Endocrinology* 100, 1496–1504.
704. De Lean, A., Garon, M., Kelly, P. A., and Labrie, F. (1977). Changes of pituitary thyrotropin releasing hormone (TRH) receptor level and prolactin response to TRH during the rat estrous cycle. *Endocrinology* 100, 1505–1510.
705. Gershengorn, M. C., Marcus-Samuels, B. E., and Geras, E. (1979). Estrogens increase the number of thyrotropin-releasing hormone receptors on mammotropic cells in culture. *Endocrinology* 105, 171–176.
706. Kow, L.-M., and Pfaff, D. W. (1985). Actions of feeding-relevant agents on hypothalamic glucose-responsive neurons *in vitro*. *Brain Res. Bull.* 15, 509–513.
708. Pan, J.-T., Kow, L.-M., and Pfaff, D. W. (1986). Single-unit activity of hypothalamic arcuate neurons in brain tissue slices. *Neuroendocrinology* 43, 189–196.
709. Kelly, M. J., Moss, R. L., Dudley, C. A., and Fawcett, C. P. (1977). The specificity of the response of preoptic septal area neurons to estrogen: 17α-estradiol versus 17β-estradiol and the response of extrahypothalamic neurons. *Exp. Brain Res.* 30, 43–52.
710. Dufy, B., Partouche, C., Poulain, D., Dufy-Barbe, L., and Vincent, J. D. (1976). Effects of oestrogen on the electrical activity of identified and unidentified hypothalamic units. *Neuroendocrinology* 22, 38–47.
711. Dufy, B., Dufy-Barbe, L., and Vincent, J. D. (1978). Effects of protein synthesis inhibitors on the negative feedback effect of estrogen on LH release. *Horm. Res.* 9, 279–291.

712. Kelly, M. J., Kuhnt, U., and Wuttke, W. (1980). Hyperpolarization of hypothalamic parvocellular neurons by 17β-estradiol and their identification through intracellular staining with procion yellow. *Exp. Brain Res.* 40, 440–447.
713. Kelly, M. J., Ronnekleiv, O. K., and Eskay, R. L. (1984). Identification of estrogen-responsive LHRH neurons in the guinea pig hypothalamus. *Brain Res. Bull.* 12, 399–407.
714. Mobbs, C. V., Harlan, R. E., and Pfaff, D. W. (1985). An estradiol-induced protein in the hypothalamus. *Endocrinology Society Abstracts*.
715. Roy, E. J., Lynn, D. M., and Clark, A. S. (1985). Inhibition of sexual receptivity by anesthesia during estrogen priming. *Brain Res.* 337, 163–166.
716. Harlan, R. E., Shivers, B. D., and Pfaff, D. W. (1981). Requirement of hypothalamic sodium currents for the lordosis reflex. *Soc. Neurosci. Abstr.* 7, 615.
717. Harlan, R., Shivers, B. D., Kow, L.-M., and Pfaff, D. W. (1983). Estrogenic maintenance of lordotic responsiveness: requirement for hypothalamic action potentials. *Brain Res.* 268, 67–78.
718. Harlan, R. E., Shivers, B. D., and Pfaff, D. W. (1982). Dorsal midbrain sodium currents are required for the lordosis reflex in female rats. *Soc. Neurosci. Abstr.* 8, 930.
719. Rothfeld, J. M., Harlan, R. E., Shivers, B. D., and Pfaff, D. W. (1986). Reversible disruption of lordosis via midbrain infusions of procaine and tetrodotoxin. *Pharmacol. Biochem. Behav.* 25, 857–863.
720. Narahashi, R., Anderson, N., and Moore, J. (1966). Tetrodotoxin does not block excitation from inside the nerve membrane. *Science* 153, 765–767.
721. Ritchie, J., and Rogart, R. (1977). The binding of saxitoxin and tetrodotoxin to excitable tissue. *Rev. Physiol. Biochem. Pharmacol.* 79, 1–49.
722. Courtney, K. R., Kendig, J., and Cohen, E. (1978). The rates of interaction of local anesthetics with sodium channels in nerve. *J. Pharmacol. Exp. Ther.* 207, 594–604.
723. Pfaff, D. W., and Sakuma, Y. (1979). Facilitation of the lordosis reflex of female rats from the ventromedial nucleus of the hypothalamus. *J. Physiol. (Lond.)* 288, 189–202.
724. Pfaff, D. W. (1980). *Estrogens and Brain Function: Neural Analysis of a Hormone Controlled Mammalian Reproductive Behavior*. Springer-Verlag, New York.
725. Pfaff, D. W. (1999). *Drive: Neural and Molecular Mechanisms for Sexual Motivation*. MIT Press, Cambridge, MA.
726. Bodnar, R. J., Commons, K., and Pfaff, D. W. (2002). *Central Neural States Relating Sex and Pain*. Johns Hopkins University Press, Baltimore.
727. Pfaff, D. W., Phillips, M. I., and Rubin, R. T. (2004). *Principles of Hormone/Behavior Relations*. Academic Press/ Elsevier, San Diego.
728. Mong,, J. A., and Pfaff, D. W. (2004). Hormonal symphony: steroid orchestration of gene modules for sociosexual behaviors. *Mol. Psychiatry* 9, 550–556.
729. Pfaff, D. W. (1999). *Drive: Neural and Molecular Mechanisms for Sexual Motivation*. MIT Press, Cambridge, MA.
730. Jensen, E. V., and Jacobson, H. I. (1962). Basic guides to the mechanism of estrogen action. *Recent Prog. Horm. Res.* 18, 387–414.
731. Walter, P., Green, S., Greene, G., Krust, A., Bornert, J.-M., Jeltsch, J.-M., Staub, A., Jensen, E., Scrace, G., Waterfield, M., and Chambon, P. (1985). Cloning of the human estrogen receptor cDNA. *Proc. Natl. Acad. Sci. U S A* 82, 7889–7893.
732. Kuiper, C. G., Enmark, E., Pelto-Huikko, M., Nilsson, S., and Gustafsson, J. A. (1996). Cloning of a novel estrogen receptor expressed in rat prostate and ovary. *Proc. Natl. Acad. Sci. U S A* 93, 5925–5230.
733. Nilsson, S., and Gustafsson, J. A. (2002). Biological role of estrogen and estrogen receptors. *Crit. Rev. Biochem. Mol. Biol.* 37, 1–28.
734. Shughrue, P. J., Lane, M. V., and Merchanthaler, I. (1997). Comparative distribution of estrogen receptor alpha and beta RNA in the rat central nervous system. *J. Comp. Neurol.* 388, 507–525.
735. Etgen, A. M. (2002). Estrogen regulation of neurotransmitter and growth factor signaling in the brain. In *Hormones, Brain and Behavior* (D. W. Pfaff, A. Arnold, A. Etgen, S. Fahrbach, and R. Rubin, Eds.), pp. 381–440. Academic Press, San Diego.
736. Kow, L., Weesner, G., and Pfaff, D. (1992). Alpha 1-adrenergic agonists act on the ventromedial hypothalamus to cause neuronal excitation and lordosis facilitation: electrophysiological and behavioral evidence. *Brain Res.* 588, 237–245.
737. Petitti, N., Karkanias, G. B., and Etgen, A. M. (1992). Estradiol selectively regulates alpha 1B-noradrenergic receptors in the hypothalamus and preoptic area. *J. Neurosci.* 12, 3869–3876.
738. Kow, L. M., Tsai, Y. F., Weiland, N. G., McEwen, B. S., and Pfaff, D. W. (1995). In vitro electro-pharmacological and autoradiographic analyses of muscarinic receptor subtypes in rat hypothalamic ventromedial nucleus: implications for cholinergic regulation of lordosis. *Brain Res.* 694, 29–39.
739. Menard, C. S., and Dohanich, G. P. Physostigmine facilitation of lordosis in naturally cycling female rats. *Pharmacol. Biochem. Behav.* 36, 853–858.
740. Jones, K., Harrington, C., and Pfaff, D. W. (1990). Steroid hormone regulation of ribosomal RNA in rat hypothalamus: early detection using in situ hybridization and precursor-product ribosomal DNA probes. *J. Neurosci.* 10, 1513–1521.
741. Cohen, R., Chung, S., and Pfaff, D. (1984). Alteration by estrogen of the nucleoli in nerve cells of the rat hypothalamus. *Cell. Tissue Res.* 235, 485–489.
742. Cohen, R., and Pfaff, D. (1981). Ultrastructure of neurons in the ventromedial nucleus of the hypothalamus in ovariectomized rats with or without estrogen treatment. *Cell. Tissue Res.* 217, 451–470.
743. Woolley, C. S., and Cohen, R. S. (2002). Sex steroids and neuronal growth in adulthood. In *Hormones, Brain and Behavior* (D. W. Pfaff, A. Arnold, A. Etgen, S. Fahrbach, and R. Rubin, Eds.), pp. 717–777. Academic Press, San Diego.
744. Frankfurt, M. (1994). Gonadal steroids and neuronal plasticity: Studies in the adult rat hypothalamus. *Annu. N Y Acad. Sci.* 743, 45–60.
745. Flanagan-Cato, L. (2004). Estrogen-induced remodeling of hypothalamic neural circuitry [review]. *Front. Neuroendocrinol.* 4, 309–329.
746. Flanagan-Cato, L. M., Calizo, L. H., and Daniels, D. (2001). The synaptic organization of VMH neurons that mediate the effects of estrogen on sexual behavior. *Horm. Behav.* 40, 178–182.
747. Carrer, H., and Aoki, A. (1982). Ultrastructural changes in the hypothalamic ventromedial nucleus of ovariectomized rats after estrogen treatment. *Brain Res.* 240, 221–233.
748. Pollio, G., Xue, P., Zanisi, M., Nicolin, A., and Maggi, A. (1993). Antisense oligonucleotide blocks progesterone-induced lordosis behavior in ovariectomized rats. *Brain Res. Mol. Brain Res.* 19, 135–139.
749. Ogawa, S., Olazabal, U. E., Parhar, I. S., and Pfaff, D. W. (1994). Effects of intrahypothalamic administration of a ntisense DNA for progesterone receptor mRNA on reproductive behavior and progesterone receptor immunoreactivity in female rat. *J. Neurosci.* 14, 1766–1774.
750. Mani, S. K., Blaustein, J. D., Allen, J. M., Law, S. W., O'Malley, B. W., and Clark, J. H. (1994). Inhibition of rat sexual behavior by antisense oligonucleotides to the progesterone receptor. *Endocrinology* 135, 1409–1414.
751. Lydon, J. P., DeMayo, F. J., Funk, C. R., Mani, S. K., Hughes, A. R., Montgomery, C. A. Jr., Shyamala, G.,

Conneely, O. M., and O'Malley, B. W. (1995). Mice lacking progesterone receptor exhibit pleiotropic reproductive abnormalities. *Genes Dev.* 9, 2266–2278.

752. Xu, M., Urban, J. H., Hill, J. W., and Levine, J. E. (2000). Regulation of hypothalamic neuropeptide Y1 receptor gene expression during the estrous cycle: role of progesterone receptors. *Endocrinology* 141, 3319–3327.
753. Brann, D. W., Chorich, L. P., and Mahesh, V. B. (1993). Effects of progesterone on galanin mRNA levels in the hypothalamus and the pituitary: correlation with the gonadotrophin surge. *Neuroendocrinology* 58, 531–538.
754. Rossmanith, W. G., Marks, D. L., Clifton, D. K., and Steiner, R. A. (1996). Induction of galanin mRNA in GNRH neurons by estradiol and its facilitation by progesterone. *J. Neuroendocrinol.* 8, 185–191.
755. Thomas, A., Shughrue, P. J., Merchenthaler, I., and Amico, J. A. (1999). The effects of progesterone on oxytocin mRNA levels in paraventricular nucleus of the female rat can be altered by the administration of diazepam or RU486. *J. Neuroendocrinol.* 11, 137–144.
756. Cho, B. N., Seong, J. Y., Cho, H., and Kim, K. (1994). Progesterone stimulates GnRH gene expression in the hypothalamus of ovariectomized, estrogen treated adult rats. *Brain Res.* 652, 177–180.
757. Petersen, S. L., McCrone, S., Keller, M., and Shores, S. (1995). Effects of estrogen and progesterone on luteinizing hormone-releasing hormone messenger ribonucleic acid levels: consideration of temporal and neuroanatomical variables. *Endocrinology* 136, 3604–3610.
758. Petersen, S. L., and LaFlamme, K. D. (1997). Progesterone increases levels of mu-opioid receptor mRNA in the preoptic area and arcuate nucleus of ovariectomized, estradiol-treated female rats. *Brain Res. Mol. Brain Res.* 52, 32–37.
759. Petersen, S. L., Keller, M. L., Carder, S. A., and McCrone, S. (1993). Differential effects of estrogen and progesterone on levels of POMC mRNA levels in the arcuate nucleus: relationship to the timing of LH surge release. *J. Neuroendocrinol.* 5, 643–648.
760. Unda, R., Brann, D. W., and Mahesh, V. B. (1995). Progesterone suppression of glutamic acid decarboxylase (GAD_{67}) mRNA levels in the preoptic area: correlation to luteinizing hormone surge. *Neuroendocrinology* 62, 562–570.
761. Gu, G., Varoqueaux, F., and Simerly, R. B. (1999). Hormonal regulation of glutamate receptor gene expression in the anteroventral periventricular nucleus of the hypothalamus. *J. Neurosci.* 19, 3213–3222.
762. Arbogast, L. A., and Voogt, J. L. (1993). Progesterone reverses the estradiol-induced decrease in tyrosine hydroxylase mRNA levels in the arcuate nucleus. *Neuroendocrinology* 58, 501–510.
763. Arbogast, L. A., and Voogt, J. L. (1994). Progesterone suppresses tyrosine hydroxylase messenger ribonucleic acid levels in the arcuate nucleus on proestrus. *Endocrinology* 135, 343–350.
764. Moss, R., and McCann, S. (1973). Induction of mating behavior in rats by luteinizing hormone-releasing factor. *Science* 181, 177–179.
765. Pfaff, D. (1973). Luteinizing hormone releasing factor (LRF) potentiates lordosis behavior in hypophysectomized ovariectomized female rats. *Science* 182, 1148–1149.
766. Schwanzel-Fukuda, M., and Pfaff, D. W. (1989). Origin of luteinizing hormone-releasing hormone neurons. *Nature* 338, 161–164.
767. MacColl, G., Quinton, R., and Bouloux, P. M. G. (2002). GnRH neuronal development: insights into the mechanisms of hypogonadotropic hypogonadism. *Trends Endocrinol. Metab.* 13, 112–118.
768. Schwanzel-Fukuda, M., Bick, D., and Pfaff, D. W. (1989). Luteinizing hormone releasing hormone (LHRH) expressing cells do not migrate normally in an inherited hypogonadal (Kallmann) syndrome. *Brain Res. Mol. Brain Res.* 6, 311–326.
769. Legouis, R., Hardelin, J. P., Levilliers, J., Claveric, J. M., Compain, S., Wunderle, V., Millasseau, P., Le Paslier, D., Cohen, D., Caterina, D., Bougueleret, L., Delmarre-van der Waal, H., Lutfalla, G., Weissenbach, J., and Petit, C. (1991). The candidate gene for the X-linked Kallmann syndrome encodes a protein related to adhesion molecules. *Cell* 67, 423–435.
770. Ballabio, A., and Camerino, G. (1992). The gene for X-linked Kallmann syndrome: a human neuronal migration defect. *Curr. Opin. Genet. Dev.* 2, 417–421.
771. Dellovade, T., Hardelin, J. P., Pfaff, D. W., Schwanzel-Fukuda, M. (2003). Anosmin-I immunoreactivity during embryogenesis in a primitive eutherian mammal. *Dev. Brain Res.* 140, 157–167.
772. Romano, G. J., Harlan, R. E., Shivers, B. D., Howells, R. D., and Pfaff, D. W. (1988). Estrogen increases proenkephalin messenger ribonucleic acid levels in the ventromedial hypothalamus of the rat. *Mol. Endocrinol.* 2, 1320–1328.
773. Zhu, Y.-S., Cai, L. Q., You, X., Duan, Y., Imperato-McGinley, J., Chin, W. W., and Pfaff, D. W. (2001). Molecular analysis of estrogen induction of preproenkephalin gene expression and its modulation by thyroid hormones. *Brain Res. Mol. Brain Res.* 91, 23–33.
774. Priest, C. A., Eckersell, C. B., and Micevych, P. E. (1995). Estrogen regulates preproenkephalin-A mRNA levels in the rat ventromedial nucleus: temporal and cellular aspects. *Brain Res. Mol. Brain Res.* 28, 251–262.
775. Vasudevan, N., Zhu, Y.-S., Daniel, S., Koibuchi, N., Chin, W. W., and Pfaff, D. (2001). Crosstalk between oestrogen receptors and thyroid hormone receptor isoforms results in differential regulation of the preproenkephalin gene. *J. Neuroendocrinol.* 13, 779–790.
776. Bodnar, R., Commons, K., and Pfaff, D. (2002). *Central Neural States Relating Sex and Pain.* Johns Hopkins University Press, Baltimore.
777. Quinones-Jenab, V., Jenab, S., Ogawa, S., Inturrisi, C., and Pfaff, D. W. (1997). Estrogen regulation of mu-opioid receptor mRNA in the forebrain of female rats. *Brain Res. Mol. Brain Res.* 47, 134–138.
778. Acosta-Martinez, M., and Etgen, A. M. (2002). Activation of mu-opioid receptors inhibits lordosis behavior in estrogen and progesterone-primed female rats. *Horm. Behav.* 41, 88–100.
779. Pfaus, J., and Pfaff, D. (1992). μ-, δ- and κ-Opioid receptor agonists selectively modulate sexual behaviors in the female rat: differential dependence on progesterone. *Horm. Behav.* 26, 457–473.
780. Pfaus, J., and Gorzalka, B. (1987). Opioids and sexual behavior [review]. *Neurosci. Biobehav. Rev.* 11, 1–34.
781. Pfaus, J. G., and Gorzalka, B. B. (1987). Selective activation of opioid receptors differentially affects lordosis behavior in female rats. *Peptides* 8, 309–317.
782. McCarthy, M., Chung, S., Ogawa, S., and Pfaff, D. W. (1991). Behavioral effects of oxytocin: is there a unifying principle? In *Vasopressin* (S. Jard and J. Ramison, Eds.), Colloque INSERM 208, pp. 195–212. INSERM/John Libbey Eurotext, Montrouge, France.
783. McCarthy, M. M., McDonald, C. H., Brooks, P. J., and Goldman, D. (1996). An anxiolytic action of oxytocin is enhanced by estrogen in the mouse. *Physiol. Behav.* 60, 1209–1215.
784. Nomura, M., McKenna, E., Korach, K. S., Pfaff, D. W., and Ogawa, S. (2002). Estrogen receptor-beta regulates transcript levels for oxytocin and arginine vasopressin in the

hypothalamic paraventricular nucleus of male mice. *Brain Res. Mol. Brain Res.* 109, 84–94.

785. Krezel, W., Dupont, S., Krust, A., Chambon, P., and Chapman, P. F. (2001). Increased anxiety and synaptic plasticity in estrogen receptor β-deficient mice. *Proc. Natl. Acad. Sci. U S A* 98, 12278–12282.
786. Insel, T., and Young, L. (2000). Neuropeptides and the evolution of social behavior [review]. *Curr. Opin. Neurobiol.* 10, 784–789.
787. Ferguson, J., Young, L, and Insel, T. (2002). The neuroendocrine basis of social recognition [review]. *Front. Neuroendocrinol.* 2, 200–224.
788. Ferguson, J., Aldag, J. M., Insel, T. R., Young, L. J. (2001). Oxytocin in the medial amygdala is essential for social recognition in the mouse. *J. Neurosci.* 21, 8278–8285.
789. Choleris, E., Gustafsson, J. A., Korach, K. S., Muglia, L. J., Pfaff, D. W., and Ogawa, S. (2003). An estrogen-dependent four-gene micronet regulating social recognition: a study with oxytocin- and estrogen receptor alpha and beta-knockout mice. *Proc. Natl. Acad. Sci. U S A* 100, 6192–6197.
790. Nomura, M., Durback, L., Chan, J., Gustafsson, J.-A., Smithies, O., Korach, K. S., Pfaff, D. W., and Ogawa, S. (2002). Genotype/age interactions on aggressive behavior in gonadally intact estrogen receptor β knockout (βERKO) male mice. *Horm. Behav.* 41, 288–296.
791. Kaplitt, M. G., Pfaus, J., Kleopoulos, S. P., Hanlon, B. A., Rabkin, S. D., and Pfaff, D. W. (1991). Expression of a functional foreign gene in adult mammalian brain following in vivo transfer via a herpes simplex virus type 1 defective viral vector. *Mol. Cell. Neurosci.* 2, 320–330.
792. Kaplitt, M. G., Rabkin, S. D., and Pfaff, D. W. (1993). Molecular alterations in nerve cells: direct manipulation and physiological mediation. *Curr. Top. Neuroendocrinol.* 11, 169–191.
793. Yin, J., Kaplitt, M. G., and Pfaff, D. W. (1995). In vivo promoter analysis in the adult central nervous system using viral vectors. In *Viral Vectors: Gene Therapy and Neuroscience Applications* (M. Kaplitt and A. Loewy, Eds.), pp. 157–171. Academic Press, San Diego.
794. Kaplitt, M. G., Kwong, A. D., Kleopoulos, S. P., Mobbs, C. V., Rabkin, S. D., and Pfaff, D. W. (1994). Preproenkephalin promoter yields region-specific and longterm expression in adult brain following direct in vivo gene transfer via a defective herpes simplex viral vector. *Proc. Natl. Acad. Sci. U S A* 91, 8979–8983.
795. Priest, C. A., Eckersell, C. B., and Micevych, P. E. (1995). Estrogen regulates preproenkephalin-A mRNA levels in the rat ventromedial nucleus: temporal and cellular aspects. *Brain Res. Mol. Brain Res.* 28, 251–262.
796. Priest, C. A., Borsook, D., and Pfaff, D. W. (1996). Estrogen and stress interact to regulate the hypothalamic expression of a human proenkephalin promoter-β-galactosidase fusion gene in a site-specific and sex-specific manner. *J. Neuroendocrinol.* 9, 317–326.
797. Pfaff, D., Berrettini, W., Joh, T. H., and Maxson, S. C., Eds. (1999). *Genetic Influences on Neural and Behavioral Functions*. CRC Press, Boca Raton, FL.
798. Schwanzel-Fukuda, M., and Pfaff, D. W. (1989). Origin of luteinizing hormone-releasing hormone neurons. *Nature* 338, 161–164.
799. Schwanzel-Fukuda, M., Bick, D., and Pfaff, D. W. (1989). Luteinizing hormone releasing hormone (LHRH) expressing cells do not migrate normally in an inherited hypogonadal (Kallmann) syndrome. *Brain Res. Mol. Brain Res.* 6: 311–326.

799a. Dellovade, T., et al. (2005). Manuscript submitted.

800. Parhar, I. S., Soga, T., Ogawa, S., Ogawa, S., Pfaff, D., and Sakuma, Y. (2005). Nonmammalian gonadotropin-releasing hormone molecules in the brain of promoter transgenic rats. *Proc. Natl. Acad. Sci. U S A* 102, 5880–5885.
801. Chan, J., Ogawa, S., and Pfaff, D. W. (2001). Reproductive-related behaviors of Swiss-Webster female mice living in a cold environment. *Proc. Natl. Acad. Sci. U S A* 98, 700–704.
802. Garey, J., Kow, L.-M., Huynh, W., Ogawa, S., and Pfaff, D. W. (2002). Temporal and spatial quantitation of nesting and mating behaviors among mice housed in a semi-natural environment. *Horm. Behav.* 42, 294–306.

802a. Garey, J., et al. (2005). Manuscript in preparation.

803. Floody, O. R, and Pfaff, D. W. (1977). Aggressive behavior in female hamsters: the hormonal basis for fluctuations in female aggressiveness correlated with estrous state. *J. Comp. Physiol. Psychol.* 91, 443–464.
804. Floody, O. R, and Pfaff, D. W. (1977). Communication among hamsters by high-frequency acoustic signals: I. Physical characteristics of hamster calls. *J. Comp. Physiol. Psychol.* 91, 794–806.
805. Floody, O. R, Pfaff, D. W, and Lewis, C. D. (1977). Communication among hamsters by high-frequency acoustic signals: II. Determinants of calling by females and males. *J. Comp. Physiol. Psychol.* 91, 807–819.
806. Floody, O. R., and Pfaff, D. W. (1977). Communication among hamsters by high-frequency acoustic signals: III. Responses evoked by natural and synthetic ultrasound. *J. Comp. Physiol. Psychol.* 91, 820–829.
807. Bodnar, R., Commons, K., and Pfaff, D. (2001). *Central Neural States Relating Sex and Pain*. Johns Hopkins University Press, Baltimore.
808. Ogawa, S., Taylor, J., Lubahn, D. B., Korach, K. S., and Pfaff, D. W. (1996). Reversal of sex roles in genetic female mice by disruption of estrogen receptor gene. *Neuroendocrinology* 64, 467–470.
809. McCarthy, M. M., Schlenker, E. H., and Pfaff, D. W. (1993). Enduring consequences of neonatal treatment with antisense oligodeoxynucleotides to estrogen receptor messenger ribonucleic acid on sexual differentiation of rat brain. *Endocrinology* 133, 433–439.
810. Ogawa, S., Washburn, T. F., Taylor, J., Lubahn, D. B., Korach, K. S., and Pfaff, D. W. (1998). Modifications of testosterone-dependent behaviors by estrogen receptor-alpha gene disruption in male mice. *Endocrinology* 139, 5058–5069.
811. Ogawa, S., Eng, V., Taylor, J., Lubahn, D. B., Korach, K. S., and Pfaff, D. W. (1998). Roles of estrogen receptor-alpha gene expression in reproduction-related behaviors in female mice. *Endocrinology* 139, 5070–5081.
812. Nomura, M., Durbak, L., Chan, J., Smithies, O., Gustafsson, J. A., Korach, K. S., Pfaff, D. W., and Ogawa, S. (2002). Genotype/age interactions on aggressive behavior in gonadally intact estrogen receptor β-knockout (βERKO) male mice. *Horm. Behav.* 41, 288–296.
813. Durbak, L., Pfaff, D., and Ogawa, S. (2002). Role of the estrogen receptor beta gene in testosterone inducible aggression in female mice. *Soc. Neurosci. Abstr.* 288.
814. Garey, J., Goodwillie, A., Frohlich, J., Morgan, M., Gustafsson, J. A., Smithies, O., Korach, K. S., Ogawa, S., and Pfaff, D. W. (2003). Genetic contributions to generalized arousal of brain and behavior. *Proc. Natl. Acad. Sci. U S A* 100, 11019–11022.
815. Le, T., Mirasol, E., Pfaff, D., and Ogawa, S. (2002). Role of the estrogen receptor beta gene in postpartum aggression in mice. *Soc. Neurosci. Abstr.* 288.
816. Beadle, G. W., and Tatum, E. L. (1941). Genetic control of biochemical reactions in *Neurospora*. *Proc. Natl. Acad. Sci. U S A* 27, 499–506.
817. Kelly, M. J., and Ronnekleiv, O. K. (2002). Rapid membrane effects of estrogen in the central nervous system. In *Hormones, Brain and Behavior* (D. W. Pfaff, A. Arnold, A. Etgen, S. Fahrbach, and R. Rubin, Eds.), pp. 361–381. Academic Press, San Diego.

818. Pfaff, D. (1982). Neurobiological mechanisms of sexual motivation. In *The Physiological Mechanisms of Motivation* (D. W. Pfaff, Ed.), pp. 287–317. Springer-Verlag, New York.
819. Bodnar, R., Commons, K., and Pfaff, D. (2002). *Central Neural States Relating Sex and Pain*. Johns Hopkins University Press, Baltimore.
820. Blaustein, J. D., and Erskine, M. S. (2002). Feminine sexual behavior: cellular integration of hormonal and afferent information in the rodent brain. in *Hormones, Brain and Behavior* (D. W. Pfaff, A. Arnold, A. Etgen, S. Fahrbach, and R. Rubin, Eds.), pp. 361–381. Academic Press, San Diego.
821. Ogawa, S., Chan, J., Gustafsson, J. A., Korach, K. S., and Pfaff, D. W. (2003). Estrogen increases locomotor activity in mice through estrogen receptor alpha: specificity for the type of activity. *Endocrinology* 144, 230–239.
822. Cottingham, S. L., and Pfaff, D. W. (1986). Interconnectedness of steroid hormone-binding neurons: existence and implications. *Curr. Top. Neuroendocrinol.* 7, 223–249.
823. Sakuma, Y. (1984). Influences of neonatal gonadectomy or androgen exposure on the sexual differentiation of the rat ventromedial hypothalamus. *J. Physiol. (Lond.).* 349, 273–286.
824. Nakano, K., Suga, S., Kondo, Y., Sato, T., and Sakuma, Y. (1997). Estrogen-excitable forebrain projections to the ventral premammillary nucleus of the female rat. *Neurosci. Lett.* 225, 17–20.
825. Takeo, T., and Sakuma, Y. (1995). Diametrically opposite effects of estrogen on the excitability of female rat medial and lateral preoptic neurons with axons to the midbrain locomotor region. *Neurosci. Res.* 22, 73–80.
826. Wagner, C. K., Pfau, J. L., de Vries, G. J., and Merchenthaler, I. J. (2001). Sex differences in progesterone receptor immunoreactivity in neonatal mouse brain depend on estrogen receptor alpha expression. *J. Neurobiol.* 47, 176–182.
827. Ogawa, S., Eng, V., Taylor, J., Lubahn, D. B., Korach, K. S., and Pfaff, D. W. (1998). Roles of estrogen receptor alpha gene expression in reproduction-related behaviors in female mice. *Endocrinology* 139, 5070–5081.
828. Kuiper, G. G., Shughrue, P. J., Merchenthaler, I., and Gustafsson, J. A. (1998). The estrogen receptor beta subtype: a novel mediator of estrogen action in neuroendocrine systems [review]. *Front. Neuroendocrinol.* 19, 253–286.
829. Orikasa, C., Kondo, Y., Hayashi, S., McEwen, B. S., and Sakuma, Y. (2002). Sexually dimorphic expression of estrogen receptor beta in the anteroventral periventricular nucleus of the rat preoptic area: implication in luteinizing hormone surge. *Proc. Natl. Acad. Sci. U S A* 99, 3306–3311.
830. Choleris, E., Gustafsson, J. A., Korach, K. S., Muglia, L. J., Pfaff, D. W., and Ogawa, S. (2003). An estrogen-dependent four-gene micronet regulating social recognition: a study with oxytocin- and estrogen receptor alpha and beta-knockout mice. *Proc. Natl. Acad. Sci. U S A* 100, 6192–6197.
831. Pfaff, D. W., and Sakuma, Y. (1979). Facilitation of the lordosis reflex of female rats from the ventromedial nucleus of the hypothalamus. *J. Physiol. (Lond.)* 288, 189–202.
832. Floody, O. R., Blinn, N. E., Lisk, R. D., and Vomachka, A. J. (1987). Localization of hypothalamic sites for the estrogen-priming of sexual receptivity in female hamsters. *Behav. Neurosci.* 101, 309–314.
833. Hoshina, Y., Takeo, T., Nakano, K., Sato, T., and Sakuma, Y. (1994). Axon-sparing lesion of the preoptic area enhances receptivity and diminishes proceptivity among components of female rat sexual behavior. *Behav. Brain Res.* 61, 197–204.
834. Takeo, T., Chiba, Y., and Sakuma, Y. (1993). Suppression of the lordosis reflex of female rats by efferents of the medial preoptic area. *Physiol. Behav.* 53, 831–838.
835. Hasegawa, T., Takeo, T., Akitsu, H., Hoshina, Y., and Sakuma, Y. (1991). Interruption of the lordosis reflex of female rats by ventral midbrain stimulation. *Physiol. Behav.* 50, 1033–1038.
836. Hasegawa, T., and Sakuma, Y. (1993). Developmental effect of testosterone on estrogen sensitivity of the rat preoptic neurons with axons to the ventral tegmental area. *Brain Res.* 611, 1–6.
837. Sakamoto, Y., Suga, S., and Sakuma, Y. (1993). Estrogen-sensitive neurons in the female rat ventral tegmental area: a dual route for the hormone action. *J. Neurophysiol.* 70, 1469–1475.
838. Kato, A., and Sakuma, Y. (2000). Neuronal activity in female rat preoptic area associated with sexually motivated behavior. *Brain Res.* 862, 90–102.
839. Hansen, S., Kohler, C., Goldstein, M., and Steinbusch, H. V. (1982). Effects of ibotenic acid-induced neuronal degeneration in the medial preoptic area and the lateral hypothalamic area on sexual behavior in the male rat. *Brain Res.* 239, 213–232.
840. Paredes, R. G., Tzschentke, T., and Nakach, N. (1998). Lesions of the medial preoptic area/anterior hypothalamus (MPOA/AH) modify partner preference in male rats. *Brain Res.* 813, 1–8.
841. Paredes, R. G., Highland, L., and Karam, P. (1993). Socio-sexual behavior in male rats after lesions of the medial preoptic area: evidence for reduced sexual motivation. *Brain Res.* 618, 271–276.
842. Clancy, A. N., Zumpe, D., and Michael, R. P. (2000). Estrogen in the medial preoptic area of male rats facilitates copulatory behavior. *Horm. Behav.* 38, 86–93.
843. Gorski, R. A. (2002). Hypothalamic imprinting by gonadal steroid hormones. *Adv. Exp. Med. Biol.* 511, 57–70.
844. Tobet, S. A., and Hanna, I. K. (1997). Ontogeny of sex differences in the mammalian hypothalamus and preoptic area. *Cell. Mol. Neurobiol.* 17, 565–601.
845. Raisman, G., and Field, P. M. (1971). Sexual dimorphism in the preoptic area of the rat. *Science* 173, 731–733.
846. Maclusky, N. J., and Naftolin, F. (1981). Sexual differentiation of the central nervous system. *Science* 211, 1294–1302.
847. Greenough, W. T., Carter, C. S., Steerman, C., and DeVoogd, T. J. (1977). Sex differences in dentritic patterns in hamster preoptic area. *Brain Res.* 126, 63–72.
848. Gorski, R. A., Gordon, J. H., Shryne, J. E., and Southam, A. M. (1978). Evidence for a morphological sex difference within the medial preoptic area of the rat brain. *Brain Res.* 148, 333–346.
849. Gorski, R. A., Harlan, R. E., Jacobson, C. D., Shryne, J. E., and Southam, A. M. (1980). Evidence for the existence of a sexually dimorphic nucleus in the preoptic area of the rat. *J. Comp. Neurol.* 193, 529–539.
850. Gorski, R. A. (1986). Sexual differentiation of the brain: a model for drug-induced alterations of the reproductive system. *Environ. Health Perspect.* 70, 163–175.
851. Arendash, G. W., and Gorski, R. A. (1983). Effects of discrete lesions of the sexually dimorphic nucleus of the preoptic area or other medial preoptic regions on the sexual behavior of male rats. *Brain Res. Bull.* 10, 147–154.
852. Arendash, G. W., and Gorski, R. A. (1983). Effects of discrete lesions of the sexually dimorphic nucleus of the preoptic area or other medial preoptic regions on the sexual behavior of male rats. *Brain Res. Bull.* 10, 147–154.
853. Kindon, H. A., Baum, M. J., and Paredes, R. J. (1996). Medial preoptic/anterior hypothalamic lesions induce a female-typical profile of sexual partner preference in male ferrets. *Horm. Behav.* 30, 514–527.
854. Roselli, C. E., Larkin, K., Resko, J. A., Stellflug, J. N., and Stormshak, F. (2004). The volume of a sexually dimorphic nucleus in the ovine medial preoptic area/anterior hypothalamus varies with sexual partner preference. *Endocrinology* 145, 478–483.
855. Woodson, J. C., Balleine, B. W., and Gorski, R. A. (2002). Sexual experience interacts with steroid exposure to

shape the partner preferences of rats. *Horm. Behav.* 42, 148–157.
856. Cooke, B. M., Chowanadisai, W., and Breedlove, S. M. (2000). Post-weaning social isolation of male rats reduces the volume of the medial amygdala and leads to deficits in adult sexual behavior. *Behav. Brain Res.* 117, 107–113.
857. Houtsmuller, E. J., Brand, T., de Jonge, F. H., Joosten, R. N., van de Poll, N. E., and Slob, A. K. (1994). SDN-POA volume, sexual behavior, and partner preference of male rats affected by perinatal treatment with ATD. *Physiol. Behav.* 56, 535–541.
858. Slob, A. K., and Van der Schoot, P. (1982). Testosterone induced mounting behavior in adult female rats born in litters of different female to male ratios. *Physiol. Behav.* 28, 1007–1010.
859. Broekman, M., de Bruin, M., Smeenk, J., Slob, A. K., and Van der Schoot, P. (1988). Partner preference behavior of estrous female rats affected by castration of tethered male incentives. *Horm. Behav.* 22, 324–337.
860. Turkenburg, J. L., Swaab, D. F., Endert, E., Louwerse, A., and Van de Poll, N. E. (1988). Effects of lesions of the sexually dimorphic nucleus on sexual behavior of testosterone-treated female Wistar rats. *Brain Res. Bull.* 21, 215–224.
861. Kondo, Y., and Sakuma, Y. (2004). The medial amygdala controls coital access of female rats by altering emotional response toward conspecific males. *Eur. J. Neurosci.* (in press).
862. Xiao, K., Kondo, Y., and Sakuma, Y. (2004). Sex-specific effects of gonadal steroids on odor-prompted sexual partner preference in the rat. *Horm. Behav.* (in press).
863. Edwards, D. A., and Einhorn, L. C. (1986). Preoptic and midbrain control of sexual motivation. *Physiol. Behav.* 37, 329–335.
864. Quadagno, D. M., Shryne, J., Anderson, C., and Gorski, R. A. (1972). Influence of gonadal hormones on social, sexual, emergence, and open field behaviour in the rat (*Rattus norvegicus*). *Anim. Behav.* 20, 732–740.
865. Rivas, F. J., and Mir, D. (1990). Effects of nucleus accumbens lesion on female rat sexual receptivity and proceptivity in a partner preference paradigm. *Behav. Brain Res.* 41, 239–249.
866. Fahrbach, S. E., Meisel, R. L., and Pfaff, D. W. (1985). Preoptic implants of estradiol increase wheel running but not the open field activity of female rats. *Physiol. Behav.* 35, 985–992.
867. Sakuma, Y. (1992). Brain control of female sexual behavior. In *Brain Control of Reproductive System* (A. Yokoyama, Ed.), pp. 141–163. Japan Scientific Societies Press/CRC Press, Tokyo/Boca Raton.
868. Kow, L. M., and Pfaff, D. W. (1981). Physiology of somatosensory and estrogenic control over the lordosis reflex. *Exp. Brain Res.* Suppl. 3, 262–273.
869. Parsons, B., Maclusky, N. J., Krey, L., Pfaff, D. W., and McEwen, B. S. (1980). The temporal relationship between estrogen-inducible progestin receptors in the female rat brain and the time course of estrogen activation of mating behavior. *Endocrinology* 107, 774–779.
870. Parsons, B., Rainbow, T. C., Snyder, L., and McEwen, B. S. (1984). Progesterone-like effects of estradiol on reproductive behavior and hypothalamic progestin receptors in the female rat. *Neuroendocrinology* 39, 25–30.
871. McEwen, B. S. (1988). Genomic regulation of sexual behavior. *J. Steroid Biochem.* 30, 179–183.
872. Parsons, B., McEwen, B. S., and Pfaff, D. W. (1982). A discontinuous schedule of estradiol treatment is sufficient to activate progesterone-facilitated feminine sexual behavior and to increase cytosol receptors for progestins in the hypothalamus of the rat. *Endocrinology* 110, 613–619.
873. del Castillo, A. R., Battaner, E., Guerra, M., Alonso, T., and Mas, M. (1987). Regional changes of brain Na^+, K^+-transporting adenosine triphosphatase related to ovarian function. *Brain Res.* 416, 113–118.
874. Pragnell, M., Snay, K. J., Trimmer J. S., MacLusky, N. J., Naftolin, F., Kaczmarek, L. K., and Boyle, M. B. (1990). Estrogen induction of a small, putative K^+ channel messenger RNA in rat uterus. *Neuron* 4, 807–812.
875. Minami, T., Oomura, Y., Nabekura, J., and Fukuda, A. (1990). 17 Beta-estradiol depolarization of hypothalamic neurons is mediated by cyclic AMP. *Brain Res.* 519, 301–307.
876. Mobbs, C. V., Fink, G., and Pfaff, D. W. (1990). HIP-70: a protein induced by estrogen in the brain and LH-RH in the pituitary. *Science* 247, 1477–1479.
877. Olazabal, U. E., Pfaff, D. W., and Mobbs, C. V. (1992). Estrogenic regulation of heat shock protein 90 kDa in the rat ventromedial hypothalamus and uterus. *Mol. Cell. Endocrinol.* 84, 175–183.
878. Qiu, J., Bosch, M. A., Tobias, S. C., Grandy, D. K., Scanlan, T. S., Ronnekleiv, O. K., and Kelly, M. J. (2003). Rapid signaling of estrogen in hypothalamic neurons involves a novel G-protein-coupled estrogen receptor that activates protein kinase C. *J. Neurosci.* 23, 9529–9540.
879. Kelly, M. J., and Levin, E. R. (2001). Rapid actions of plasma membrane estrogen receptors. *Trends Endocrinol. Metab.* 12, 152–156.
880. Rodriguez-Manzo, G., Cruz, M. L., and Beyer, C. (1986). Facilitation of lordosis behavior in ovariectomized estrogen-primed rats by medial preoptic implantation of 5 beta, 3 beta, pregnenolone: a ring A reduced progesterone metabolite. *Physiol. Behav.* 36, 277–281.
881. Ke, F. C., and Ramirez, V. D. (1987). Membrane mechanism mediates progesterone stimulatory effect on LHRH release from superfused rat hypothalamic in vitro. *Neuroendocrinology* 45, 514–517.
882. Majewska, M. D., Harrison, N. L., Schwartz, R. D., Barker, J. L., and Paul, S. M. (1986). Steroid hormone metabolites are barbiturate-like modulators of the GABA receptor. *Science* 232, 1004–1007.
883. Collado, M. L., Rodriguez-Manzo, G., and Cruz, M. L. (1985). Effect of progesterone upon adenylate cyclase activity and cAMP levels on brain areas. *Pharmacol. Biochem. Behav.* 23, 501–504.
884. Beyer, C., Canchola, E., and Larsson, K. (1981). Facilitation of lordosis behavior in the ovariectomized estrogen primed rat by dibutyryl cAMP. *Physiol. Behav.* 26, 249–251.
885. Mobbs, C. V., Rothfeld, J. M., Saluja, R., and Pfaff, D. W. (1989). Phorbol esters and forskolin infused into midbrain central gray facilitate lordosis. *Pharmacol. Biochem. Behav.* 34, 665–667.
886. Powers, B., and Valenstein, E. S. (1972). Sexual receptivity: Facilitation by medial preoptic lesions in female rats. *Science* 175, 1003–1005.
887. Moss, R. L., Paloutzian, R. F., and Law, O. T. (1974). Electrical stimulation of forebrain structures and its effect on copulatory as well as stimulus-bound behavior in ovariectomized hormone-primed rats. *Physiol. Behav.* 12, 997–1004.
888. Leedy, M. G. (1984). Effects of small medial preoptic lesions on estrous cycles and receptivity in female rats. *Psychoneuroendocrinology* 9, 189–196.
889. Popolow, H. B., King, J. C., and Gerall, A. A. (1981). Rostral medial preoptic area lesions' influence on female estrous processes and LHRH distribution. *Physiol. Behav.* 27, 855–861.
890. Dudley, C. A., and Moss, R. L. (1988). Facilitation of lordosis in female rats by CNS-site specific infusions of an LH-RH fragment-Ac-LH-RH-(5-10). *Brain Res.* 441, 161–167.
891. Gibson, M. J., Moscovitz, H. C., Kokoris, G. J., and Silverman, A.-J. (1987). Female sexual behavior in hypogonadal mice with GnRH-containing brain grafts. *Horm. Behav.* 21, 211–222.

892. Shivers, B. D., Harlan, R. E., Morrell, J. I., and Pfaff, D. W. (1983). Absence of oestradiol concentration in cell nuclei of LHRH-immunoreactive neurones. *Nature* 304, 345–347.

893. Rosie, R., Sumner, B. E., and Fink, G. (1994). An alpha 1 adrenergic mechanism mediates estradiol stimulation of LHRH mRNA synthesis and estradiol inhibition of POMC mRNA synthesis in the hypothalamus of the prepubertal female rat. *J. Steroid Biochem. Mol. Biol.* 49, 399–406.

894. Risold, P. Y., and Swanson, L. W. (1997). Connections of the rat lateral septal complex. *Brain Res. Rev.* 24, 115–195.

895. King, T. R., and Nance, D. M. (1985). The effects of unilateral frontolateral hypothalamic knife cuts and asymmetrical unilateral septal lesions on lordosis behavior of rats. *Physiol. Behav.* 35, 955–959.

896. Clough, R. W., Rodriguez Sierra, J. F., and Hoppe, L. B. (1986). Kainic acid lesioning of the preoptic area alters positive feedback of estrogen in prepubertal female rats. *Biol. Reprod.* 35, 1269–1276.

897. Yamanouchi, K., and Arai, Y. (1983). Forebrain and lower brainstem participation in facilitatory and inhibitory regulation of the display of lordosis in female rats. *Physiol. Behav.* 30, 155–159.

898. Tsukahara, S., and Yamanouchi, K. (2001). Neurohistological and behavioral evidence for lordosis-inhibiting tract from lateral septum to periaqueductal gray in male rats. *J. Comp. Neurol.* 431, 293–310.

899. Takeo, T., Kudo, M., and Sakuma, Y. (1995). Stria terminalis conveys a facilitatory estrogen effect on female rat lordosis reflex. *Neurosci. Lett.* 184, 79–81.

900. Luiten, P. G., Kuipers, F., and Schuitmaker, H. (1982). Organization of diencephalic and brainstem afferent projections to the lateral septum in the rat. *Neurosci. Lett.* 30, 211–216.

901. Hasegawa, T., and Sakuma, Y. (1990). Estrogen-induced suppression of female rat forebrain neurons with axons to ventral midbrain. *Neurosci. Lett.* 119, 171–174.

902. Fallon, J. H., Schmued, L. C., Wang, C., Miller, R., and Banales, G. (1984). Neurons in the ventral tegmentum have separate populations projecting to telencephalon and inferior olive, are histochemically different, and may receive direct visual input. *Brain Res.* 321, 332–336.

903. Sakuma, Y. (1994). Estrogen-induced changes in the neural impulse flow from the female rat preoptic region. *Horm. Behav.* 28, 438–444.

904. Rajendren, G., Dudley, C. A., and Moss, R. L. (1993). Influence of male rats on the luteinizing hormone-releasing hormone neuronal system in female rats: role of the vomeronasal organ. *Neuroendocrinology* 57, 898–906.

905. Masco, D. H., and Carrer, H. F. (1984). Pathways conducting amygdaloid influence on feminine sexual behavior in the rat. *Behav. Brain Res.* 11, 205–212.

906. Renaud, L. P. (1976). An electrophysiological study of amygdalohypothalamic projections to the ventromedial nucleus of the rat. *Brain Res.* 105, 45–58.

907. Edwards, D. A., and Maillard, C. (1988). Subthalamic and mesencephalic locomotor regions: brain damage augments the importance of female movement for the display of sexual behavior in male rats. *Physiol. Behav.* 44, 803–809.

908. Garcia-Rill, E., Kinjo, N., Atsuta, Y., Ishikawa, Y., Webber, M., and Skinner, R. D. (1990). Posterior midbrain-induced locomotion. *Brain Res. Bull.* 24, 499–508.

909. Swanson, L. W., Mogenson, G. J., Gerfen, C. R., and Robinson, P. (1984). Evidence for a projection from the lateral preoptic area and substantia innominata to the "mesencephalic locomotor region" in the rat. *Brain Res.* 295, 161–178.

910. Slater, J., and Blizard, D. A. (1976). A reevaluation of the relation between estrogen and emotionality in female rats. *J. Comp. Physiol. Psychol.* 90, 755–764.

911. Blizard, D. A., Lippman, H. R., and Chen, J. J. (1975). Sex differences in open-field behavior in the rat: the inductive and activational role of gonadal hormones. *Physiol. Behav.* 14, 601–608.

912. Fahrbach, S. E., Meisel, R. L., and Pfaff, D. W. (1985). Preoptic implants of estradiol increase wheel running but not the open field activity of female rats. *Physiol. Behav.* 35, 985–992.

913. Fahrbach, S. E., Morrell, J. I., and Pfaff, D. W. (1986). Identification of medial preoptic neurons that concentrate estradiol and project to the midbrain in the rat. *J. Comp. Neurol.* 247, 364–382.

914. Mori, S., Matsuyama, K., Kohyama, J., Kobayashi, Y., and Takakusaki, K. (1992). Neuronal constituents of postural and locomotor control systems and their interactions in cats [review]. *Brain Dev.* 14 (Suppl.), S109–S120.

915. Swanson, L. W., Mogenson, G. J., Simerly, R. B., and Wu, M. (1987). Anatomical and electrophysiological evidence for a projection from the medial preoptic area to the "mesencephalic and subthalamic locomotor regions" in the rat. *Brain Res.* 405, 108–122.

916. Brudzynski, S. M., Mclachlan, R. S., and Girvin, J. P. (1991). Involvement of M1 and M2 muscarinic receptors of the basal forebrain in cholinergically mediated changes in the rat locomotion. *Prog. Neuropsychopharmacol. Biol. Psychiatry* 15, 279–284.

917. Sinnamon, H. M. (1993). Preoptic and hypothalamic neurons and the initiation of locomotion in the anesthetized rat [review]. *Prog. Neurobiol.* 41, 323–344.

918. Marciello, M., and Sinnamon, H. M. (1990). Locomotor stepping initiated by glutamate injections into the hypothalamus of the anesthetized rat. *Behav. Neurosci.* 104, 980–990.

919. Brudzynski, S. M., and Eckersdorf, B. (1984). Inhibition of locomotor activity during cholinergically-induced emotional-aversive response in the cat. *Behav. Brain Res.* 14, 247–253.

920. Brudzynski, S. M., and Mogenson, G. J. (1984). The role of the nucleus reticularis tegmenti pontis in locomotion: a lesion study in the rat. *Brain Res. Bull.* 12, 513–520.

921. Whitney, J. F. (1986). Effect of medial preoptic lesions on sexual behavior of female rats is determined by test situation. *Behav. Neurosci.* 100, 230–235.

922. Mink, J. W., Sinnamon, H. M., and Adams, D. B. (1983). Activity of basal forebrain neurons in the rat during motivated behaviors. *Behav. Brain Res.* 8, 85–108.

923. Hansen, S., and Kohler, C. (1984). The importance of the peripeduncular nucleus in the neuroendocrine control of sexual behavior and milk ejection in the rat. *Neuroendocrinology* 39, 563–572.

924. Lorenzo, A., and Carrer, H. F. (1988). Verapamil injections in the peripeduncular nucleus suppress multiunit evoked activity and sexual receptivity in female rats. *Physiol. Behav.* 42, 417–422.

925. Pfeifle, J. K., and Edwards, D. A. (1983). Midbrain lesions eliminate sexual receptivity but spare sexual motivation in female rats. *Physiol. Behav.* 31, 385–389.

926. Swerdlow, N. R., Swanson, L. W., and Koob, G. F. (1984). Electrolytic lesions of the substantia innominata and lateral preoptic area attenuate the "supersensitive" locomotor response to apomorphine resulting from denervation of the nucleus accumbens. *Brain Res.* 306, 141–148.

927. Brudzynski, S. M., and Mogenson, G. J. (1986). Decrease of locomotor activity by injections of carbachol into the anterior hypothalamic/preoptic area of the rat. *Brain Res.* 376, 38–46.

928. Sinnamon, H. M. (1992). Microstimulation mapping of the basal forebrain in the anesthetized rat: the "preoptic locomotor region." *Neuroscience* 50, 197–207.

929. Blizard, D. A. (1983). Sex differences in running wheel behavior in the rat: the inductive and activational effects of gonadal hormones. *Anim. Behav.* 31, 378–384.
930. Gentry, R. T., and Wade, G. N. (1976). Sex differences in sensitivity of food intake, body weight, and running wheel activity to ovarian steroids in rats. *J. Comp. Physiol. Psychol.* 90, 747–754.
931. Fahrbach, S., Meisel, R., and Pfaff, D. W. (1985). Preoptic implants of estradiol increase wheel running but not the open field activity of female rats. *Physiol. Behav.* 35, 985–992.
932. Sakuma, Y. (1994). Estrogen-induced changes in the neural impulse flow from the female rat preoptic region. *Horm. Behav.* 28, 438–444.
933. Slater, J., and Blizard, D. A. (1976). A reevaluation of the relation between estrogen and emotionality in female rats. *J. Comp. Physiol. Psychol.* 90, 755–764.
934. Morgan, M., and Pfaff, D. (1999). Effects of estrogen on fear-related assays in female mice. *Soc. Neurosci. Abstr.* 350, 873.
935. Mong, J. A., Devidze, N., Frail, D. E., O'Connor, L. T., Samuel, M., Choleris, E., Ogawa, S., and Pfaff, D. W. (2003). Estradiol differentially regulates lipocalin-type prostaglandin D synthase transcript levels in the rodent brain: evidence from high-density oligonucleotide arrays and in situ hybridization. *Proc. Natl. Acad. Sci. U S A* 100, 318–323.
936. Pfaus, J. G., Shadiack, A., Van Soest, T., Tse, M., and Molinoff, P. (2004). Selective facilitation of sexual solicitation in the female rat by a melanocortin receptor agonist. *Proc. Natl. Acad. Sci. U S A* 101, 10201–10204.
937. Ferguson, J., Aldag, J. M., Insel, T. R., and Young, L. J. (2001). Oxytocin in the medial amygdala is essential for social recognition in the mouse. *J. Neurosci.* 21, 8278–8285.
938. Pitkow, L. J., Sharer, C. A., Ren, X., Insel, T. R., Terwilliger, E. F., and Young, L. J. (2001). Facilitation of affiliation and pair-bond formation by vasopressin receptor gene transfer into the ventral forebrain of a monogamous vole. *J. Neurosci.* 21, 7392–7396.
939. Pfaff, D. W. (2005). *Brain Arousal and Information Theory: Neural, Hormonal and Genetic Analyses*. Harvard University Press, Cambridge, MA.
940. Kushida, C. A., Rye, D. B., Nummy, D., Milton, J. G., Spire, J. P., and Rechtschaffen, A. (1991). Cortical asymmetry of REM sleep EEG following unilateral pontine hemorrhage. *Neurology* 41, 598–601.
941. Kriegsfeld, L. J., LaSauter, J., Hamada, T., and Silver, R. (2002). Circadian rhythms in the endocrine system. In *Hormones, Brain and Behavior* (D. W. Pfaff, A. Arnold, A. Etgen, S. Fahrbach, and R. Rubin, Eds.), pp. 33–93. Academic Press, San Diego.
942. Hull, E., Meisel, R., and Sachs, B. D. (2002). Male sexual behavior. In *Hormones, Brain and Behavior* (D. W. Pfaff, A. Arnold, A. Etgen, S. Fahrbach, and R. Rubin, Eds.), pp. 1–139. Academic Press, San Diego.
943. Westerhaus, M. J., and Loewy, A. D. (1999). Sympathetic-related neurons in the preoptic region of the rat identified by viral transneuronal labeling. *J. Comp. Neurol.* 414, 361–378.
944. Loewy, A. D. (1991). Forebrain nuclei involved in autonomic control. *Prog. Brain Res.* 87, 253–268.
945. Luppi, P. H., Aston-Jones, G., Akaoka, H., Chouvet, G., and Jouvet, M. (1995). Afferent projections to the rat locus coeruleus demonstrated by retrograde and anterograde tracing with cholera-toxin B subunit and *Phaseolus vulgaris* leucoagglutinin. *Neuroscience* 65, 119–160.
946. Condes-Lara, M., Gonzalez, N. M., Martinez-Lorenzana, G., Delgado, O. L., and Freund-Mercier, M. J. (2003). Actions of oxytocin and interactions with glutamate on spontaneous and evoked dorsal spinal cord neuronal activities. *Brain Res.* 976, 75–81.
947. Frohlich, J., Ogawa, S., Morgan, M., Burton, L., and Pfaff, D. (1999). Hormones, genes and the structure of sexual arousal. *Behav. Brain Res.* 105, 5–27.
948. Bodnar, R., Commons, K., and Pfaff, D. W. (2001). *Central Neural States Relating Sex and Pain*. Johns Hopkins University Press, Baltimore.
949. Strelau, J. (1994). *The Concepts of Arousal and Arousability as Used in Temperament Studies*. American Psychological Association, Washington, DC.
950. Hebb, D. O. (1958). *A Textbook of Psychology*. WB Saunders, Philadelphia.
951. Hebb, D. O. (1955). Drives and the C. N. S. *Psychol. Rev.* 62, 243–254.
952. Duffy, E. (1951). The concept of energy mobilization. *Psychol. Rev.* 58, 30–40.
953. Duffy, E. (1957). The psychological significance of the concept of arousal or activation. *Psychol. Rev.* 64, 265–275.
954. Duffy, E. (1962). *Activation and Behavior*. Wiley, New York.
955. Frohlich, J., Ogawa, S., Morgan, M., Burton, L., and Pfaff, D. (1999). Hormones, genes and the structure of sexual arousal. *Behav. Brain Res.* 105, 5–27.
956. Moruzzi, G., and Magoun, H. W. (1949). Brain stem reticular formation and activation of the EEG. *Electroencephalogr. Clin. Neurophysiol.* 1, 455–473.
957. Lindsley, D. B. (1956). Physiological psychology. *Annu. Rev. Psychol.* 7, 323–348.
958. Lindsley, D. B., Bowden, J. W., and Magoun, H. W. (1949). Effect upon the EEG of acute injury to the brain stem activating system. *Electroencephalogr. Clin. Neurophysiol.* 1, 475–486.
959. Lindsley, D. B., Schreiner, L. H., Knowles, W. B., and Magoun, H. W. (1950). Behavioral and EEG changes following chronic brain stem lesions in the cat. *Electroencephalogr. Clin. Neurophysiol.* 1, 483–498.
960. Robbins, T., and Everitt, B. (1996). Arousal systems and attention. In *Handbook of Cognitive Neuroscience* (M. Gazzaniga, Ed.), pp. 703–720. MIT Press, Cambridge, MA.
961. Frohlich, J., Morgan, M., Ogawa, S., Burton, L., and Pfaff, D. (2002). Statistical analysis of hormonal influences on arousal measures in ovariectomized female mice. *Horm. Behav.* 42, 414–423.
962. Becker, J. B., and Breedlove, S. M. (1992). Introduction to behavioral endocrinology. In *Behavioral Endocrinology* (J. B. Becker, S. M. Breedlove, and D. Crews, Eds.), pp. 3–37. MIT Press, Cambridge, MA.
963. Knobil, E., and Neill, J., Eds. (1994). *The Physiology of Reproduction*, 2nd ed. Raven Press, New York.
964. Nelson, R. J. (1995). *An Introduction to Behavioral Endocrinology*. Sinauer, Sunderland, MA.
965. Arpels, J. (1996). The female brain hypoestrogenic continuum from the premenstrual syndrome to menopause. *Reprod. Med.* 41, 633–639.
966. Schmidt, P., Nieman, L., et al. (1998). Differential behavioral effects of gonadal steroids in women with and in those without premenstrual syndrome. *N. Engl. J. Med.* 338, 209–216.
967. Soares, C. N., Poitras, J. R. and Prouty, J. (2003). Effect of reproductive hormones and selective estrogen receptor modulators on mood during menopause. *Drugs Aging* 20, 85–100.
968. Rachman, I., Unnerstall, J., Pfaff, D. W., and Cohen, R. S. (1998). Estrogen alters behavior and forebrain c-fos expression in ovariectomized rats subjected to the forced swim test. *Proc. Natl. Acad. Sci. U S A* 95, 13941–13946.
969. Marvan, M. L., Santana, S., Chavez-Chavez, L., and Bertran, M. (1997). Inescapable shocks accentuate fluctuations of forced swimming immobility in different phases of the rat estrous cycle. *Arch. Med. Res.* 28, 369–372.

970. Marvan, M. L., Chavez-Chavez, L., and Santana, S. (1996). Clomipramine modifies fluctuations of forced swimming immobility in different phases of the rat estrous cycle. *Arch. Med. Res.* 27, 83–86.
971. Okada, M., Hayashi, N., Kometani, M., Nakao, K., and Inukai, T. (1997). Influences of ovariectomy and continuous replacement of 17 beta-estradiol on the tail skin temperature and behavior in the forced swimming test in rats. *Jpn. J. Pharmacol.* 73, 93–96.
972. Fink, G., Sumner, B. E., Rosie, R., Grace, O., and Quinn, J. P. (1996). Estrogen control of central neurotransmission: effect on mood, mental state, and memory. *Cell. Mol. Neurobiol.* 16, 325–344.
973. Alves, S., McEwen, B., Hayashi, S., Korach, K. S., Pfaff, D. W., and Ogawa, S. (2000). Estrogen regulated progestin receptors are found in the midbrain raphe but not hippocampus of estrogen receptor alpha (ERα) gene disrupted mice. *J. Comp. Neurol.* 427, 185–195.
974. Handa, R. J., Burgess, L. H., Kerr, J. E., and O'Keefe, J. A. (1994). Gonadal steroid hormone receptors and sex differences in the hypothalamo-pituitary-adrenal axis. *Horm. Behav.* 28, 464–476.
975. Burgess, L. H., and Handa, R. J. (1992). Estrogen alters adrenocorticotropic hormone and corticosterone secretion and glucocorticoid receptor mediated function. *Endocrinology* 131, 1261–1269.
976. Becker, J. B., and Breedlove, S. M. (1992). Introduction to behavioral endocrinology. In *Behavioral Endocrinology* (J. B. Becker, S. M. Breedlove, and D. Crews, Eds.), pp. 3–37. MIT Press, Cambridge, MA.
977. Handa, R. J., Burgess, L. H., Kerr, J. E., and O'Keefe, J. A. (1994). Gonadal steroid hormone receptors and sex differences in the hypothalamo-pituitary-adrenal axis. *Horm. Behav.* 28, 464–476.
978. McKenzie, J. M., and Zakarija, M. (1989). Hyperthyroidism. In *Endocrinology* (L. J. DeGroot, Ed.), Vol. 1, pp. 646–682. WB Saunders, Philadelphia.
979. Utiger, R. D. (1989). Hypothyroidism. In *Endocrinology* (L. J. DeGroot, Ed.), Vol. 1, pp. 702–721. WB Saunders, Philadelphia.
980. Dorn, L. D., Burgess, E. S., Dichek, H. L., Putnam, F. W., Chrousos, G. P., and Gold, P. W. (1996). Thyroid hormone concentrations in depressed and nondepressed adolescents: group differences and behavioral reactions. *J. Am. Acad. Child Adolesc. Psychiatry* 35, 299–306.
981. Gorman, J. M., and Hatterer, J. A. (1994). The role of thyroid hormone in refractory depression. In *Refractory Depression: Current Strategies and Future Directions* (W. Anton, J. Zohar, S. P. Roose, and J. D. Amsterdam, Eds.), pp. 121–128. John Wiley, Chichester, United Kingdom.
982. Balada, F., Torrubia, R., and Arque, J. M. (1992). Thyroid hormone correlates of sensation seeking and anxiety in healthy human females. *Neuropsychobiology* 25, 208–213.
983. Loosen, P. T. (1986). Hormones of the hypothalamic-pituitary-thyroid axis: a psychoneuroendocrine perspective. *Pharmacopsychiatry* 19, 401–415.
984. Gupta, S., Masand, P. S., and Tanquary, J. F. (1991). Thyroid hormone supplementation of fluoxetine in the treatment of major depression. *Br. J. Psychiatry* 159, 866–867.
985. Sullivan, G. M., and Gorman, J. M. (2000). The use of thyroid hormones in mood disorders. *Psychiatr. Ann.* 30, 129–136.
986. Shelton, R. C. (1999). Treatment options for refractory depression. *J. Clin. Psychiatry* 60, 57–63.
987. Joffe, R., Segal, Z., and Singer, W. (1996). Change in thyroid hormone levels following response to cognitive therapy for major depression. *Am. J. Psychiatry* 153, 411–413.
988. Wallace, H., Pate, A., and Bishop, J. O. (1995). Effects of perinatal thyroid hormone deprivation on the growth and behavior of newborn mice. *J. Endocrinol.* 145, 251–262.
989. McCarty, R., Kirby, R. F., and Brunjes, P. C. (1983). Neonatal hyperthyroidism: effects on sympathetic responses to stress in adult rats. *Am. J. Physiol.* 245, R95–R99.
990. Sjoden, P. O., and Soderberg, U. (1976). Effects of neonatal thyroid hormone stimulation and differential preweaning rearing on open-field behavior in adult rats. *Dev. Psychobiol.* 9, 413–424.
991. Rastogi, R. B., and Singhal, R. L. (1976). Influence of neonatal and adult hyperthyroidism on behavior and biosynthetic capacity for norepinephrine, dopamine and 5-hydroxytryptamine in rat brain. *J. Pharmacol. Exp. Ther.* 198, 609–618.
992. Jefferys, D., and Funder, J. W. (1989). Thyroid hormones and the acquisition and retention of behavioral responses. *Endocrinology* 125, 1103–1105.
993. Deshimaru, M., Miyakawa, T., Hattori, E., Shikai, I., Shimoji, A., and Nagatoshi, K. (1979). The effects of hyperthyroidism induced by thyroideum siccatum on the central nervous system: an electron microscope study. *Folia Psychiatr. Neurol. Jpn.* 33, 517–523.
994. Crocker, A. D., and Overstreet, D. H. (1984). Modifications of the behavioral effects of haloperidol and of dopamine receptor regulation by altered thyroid status. *Psychopharmacology* 82, 102–106.
995. Rial, R. V., Tur, J. A., Palmer, A. M., and Tur, J. (1987). Altered responsiveness to ambiental stimuli in altered thyroidal states. *Physiol. Behav.* 41, 119–123.
996. McEachron, D. L., Lauchlan, C. L., and Midgley, D. E. (1993). Effects of thyroxine and thyroparathyroidectomy on circadian wheel running in rats. *Pharmacol. Biochem. Behav.* 46, 243–249.
997. Friedman, Y., Bacchus, R., Joffe, R. T., and Nobrega, J. N. (1999). Acute stress increases thyroid hormones levels in rat brain. *Biol. Psychiatry* 45, 234–237.
998. Bernstein, A., and Breitman, M. (1989). Genetic ablation in transgenic mice. *Mol. Biol. Med.* 6, 523–530.
999. Bolivar, V., Cook, M., and Flaherty, L. (2000). List of transgenic and knockout mice: behavioral profiles. *Mamm. Genome* 11, 260–274.
1000. Nelson, R. J., and Young, K. A. (1998). Behavior in mice with targeted disruption of single genes. *Neurosci. Biobehav. Rev.* 22, 453–462.
1001. Nelson, R. J. (1997). The use of genetic "knockout" mice in behavioral endocrinology research. *Horm. Behav.* 31, 188–196.
1002. Crawley, J. N. (1999). Behavioral phenotyping of transgenic and knockout mice: experimental design and evaluation of general health, sensory functions, motor abilities and specific behavioral tests. *Brain Res.* 835, 18–26.
1003. Crawley, J. N., and Paylor, R. (1997). A proposed test battery and constellations of specific behavioral paradigms to investigate the behavior phenotypes of transgenic and knockout mice. *Horm. Behav.* 31, 197–211.
1004. Mong, J. A., Devidze, N., Frail, D. E., O'Connor, L. T., Samuel, M., Choleris, E., Ogawa, S., and Pfaff, D. W. (2003). Estradiol differentially regulates lipocalin-type prostaglandin D synthase transcript levels in the rodent brain: evidence from high-density oligonucleotide arrays and in situ hybridization. *Proc. Natl. Acad. Sci. U S A* 100, 318–323.
1005. Mong, J. A., Devidze, N., Goodwillie, A., and Pfaff, D. W. (2003). Reduction of lipocalin-type prostaglandin D synthase in the preoptic area of female mice mimics estradiol effects on arousal and sex behavior. *Proc. Natl. Acad. Sci. U S A* 100, 15206–15211.

1006. Kow, L., Weesner, G., and Pfaff, D. W. (1992). Alpha 1-adrenergic agonists act on the ventromedial hypothalamus to cause neuronal excitation and lordosis facilitation: electrophysiological and behavioral evidence. *Brain Res.* 588, 237–245.

1007. Morgan, M. A., Schulkin, J., and Pfaff, D. W. (2004). Estrogens and non-reproductive behaviors related to activity and fear. *Neurosci. Biobehav. Rev.* 28, 55–63.

1008. Rubinow, D. R., Schmidt, P. J., Roca, C. A., and Daly, R. C. (2002). Gonadal hormones and behavior in women: concentrations versus context. In *Hormones, Brain and Behavior* (D. W. Pfaff, A. Arnold, A. Etgen, S. Fahrbach, and R. Rubin, Eds.), pp. 37–75. Academic Press, San Diego.

1009. Brown, R. E., Stevens, D. R., and Haas, H. L. (2001). The physiology of brain histamine. *Prog. Neurobiol.* 63, 637–672.

1010. Pfaff, D. W., Arnold, A., Etgen, A., Fahrbach, S., and Rubin, R., Eds. (2002). *Hormones, Brain and Behavior*. Academic Press, San Diego.

1011. Minneman, K. (1988). Alpha 1-adrenergic receptor subtypes, inositol phosphates, and sources of cell Ca2+. *Pharmacol. Rev.* 40, 87–119.

1012. Michelotti, G. A., Price, D. T., and Schwinn, D. A. (2000). Alpha 1-adrenergic receptor regulation: basic science and clinical implications. *Pharmacol. Ther.* 88, 281–309.

1013. Koshimizu, T. A., Yamauchi, J., Hirasawa, A., Tanoue, A., and Tsujimoto, G. (2002). Recent progress in alpha 1-adrenoceptor pharmacology. *Biol. Pharm. Bull.* 25, 401–409.

1014. Kreda, S. M., Sumner, M., Fillo, S., Ribeiro, C. M., Luo, G. X., Xie, W., Daniel, K. W., Shears, S., Collins, S., and Wetsel, W. C. (2001). α_1-Adrenergic receptors mediate LH-releasing hormone secretion through phospholipases C and A_2 in immortalized hypothalamic neurons. *Endocrinology* 132, 4839–4851.

1015. Hill, S. J. (1991). Histamine receptors and interactions between second messenger transduction systems. *Agents Actions Suppl.* 33, 145–159.

1016. Greene, R. W., and Haas, H. L. (1990). Effects of histamine on dentate granule cells in vitro. *Neuroscience* 34, 299–303.

Knobil and Neill's Physiology of Reproduction, Third Edition
edited by Jimmy D. Neill,
Elsevier © 2006

CHAPTER 35

Maternal Behavior

M. Numan,[1] A. S. Fleming,[2] and F. Levy[3]

Introduction, 1922
Hormonal Bases of Maternal Behavior, 1924
Endocrine Changes Associated with Parturition, 1924
Hormonal Determinants of the Onset of Maternal Behavior, 1925
Hormonal Bases of Maternal Aggression, 1931
Motivational Analysis of Maternal Behavior, 1932
Evidence in Support of an Approach-Avoidance Motivational Analysis of the Occurrence of Maternal Behavior, 1933
Maternal Aggression, 1936
Conclusions, 1937
Neuroanatomy and Neurochemistry of Maternal Behavior, 1937
Sensory Mechanisms Mediating Maternal Behavior, 1938
Neural Circuits Regulating Maternal Behavior, 1940
Neurochemistry and Molecular Neurobiology of Maternal Behavior, 1949
Conclusions, 1955
Experiential Influences on Maternal Behavior, 1956
Maternal Experience and Parity Effects in Adult Females, 1956
Pup Experience and Female Allomothering, 1957
Postpartum Maternal Experience, 1958
Hormonal Effects on Robustness of Maternal Experience, 1960
Sensory and Psychological Features of Maternal Experience, 1961
Individual Recognition, 1961
Mechanisms of Maternal Experience, 1962
Neurochemical Bases of Maternal Experience, 1963
Neuroanatomical Bases of Maternal Experience and Olfactory Learning, 1963
Conclusions: Maternal Experience and Individual Recognition, 1966
Effects of Early Preweaning Influences on Adult Maternal Behavior, 1966
Sensory and Psychological Mediation of Early Experience, 1966
Maternal Deprivation and Isolation Studies, 1967
Hormonal Effects of Early Experience, 1971
Neural and Neurochemical Effects of Early Experience, 1973
General Conclusions, 1974
References, 1975

[1]Department of Psychology, Boston College, Chestnut Hill, Massachusetts;
[2]Department of Psychology, University of Toronto at Mississauga, Mississauga, Ontario, Canada;
[3]Responsable de l équipe de Comportement, Physiologie de la Reproduction et des Comportements, INRA-CNRS-Université de Tours-Haras Nationaux, Nouzilly, France.

INTRODUCTION

Survival of newborn young relies on the mother and her ability to provide food, warmth, shelter, and protection from predators and conspecifics. In placental mammals, the synchronization of maternal behavior with parturition and lactation ensures that the mother responds to the needs of the young at the appropriate time. This temporal synchrony is accomplished by finely tuned physiological changes that underlie both the onset of maternal behavior and the preparation for and induction of parturition and lactation. The study of the biological mechanisms that influence and regulate pregnancy, parturition, and lactation is therefore important for a better understanding of maternal responsiveness.

The aim of this chapter is to describe the hormonal, sensory, neural, and experiential mechanisms that regulate the onset of maternal behavior across a variety of placental mammals that represent different behavioral strategies. In this context, we use a definition of maternal behavior, which refers to behaviors displayed during the very first days immediately before and after parturition that are preparatory to the arrival of the young (nest building) or are in response to the young (licking, nursing, carrying) or to threatening conspecifics (maternal aggression). We also consider how these biological mechanisms affect other behavioral systems that impact on mothers' responses to their young, including their affect, attention, perceptions, behavioral flexibility and learning. Maternal behavior usually emerges at or close to parturition. Just after birth, the female shows a very rapid interest in the newborn. Cleaning of the neonate and the consumption of amniotic fluids and placenta are widespread behaviors among mammalian orders, except in fully aquatic mammals (*Cetacea*) or semi-aquatic mammals (*Pinnipedia*). Mothers of many mammals also emit characteristic vocalizations in response to their young and show retrieval, gathering, herding, or carrying behaviors that protect the young from predation and tend to keep the young in close proximity to the mother. As well, most new mothers protect their young from predators and conspecifics. However, the most important and common pattern of maternal behavior in mammals is nursing, which occurs shortly after the young are born.

In so-called "altricial" species (most rodents, canids, felids), the mother builds a nest in which she gives birth to a large litter of young that are not fully developed and have limited sensory and locomotor abilities. Nests can be composed of natural materials found in the environment (e.g., leaves, straw, twigs, string) or, as in the case of rabbits, the mothers own hair which she begins to pull out before parturition. In rodents, once the mother has retrieved and gathered the pups together into the nest, she spends most of her time in a nursing posture (1–3) and she licks the young, a behavior that functions to promote urination and elimination by the offspring and to maintain the mother's fluid balance (4). In rabbits, a very different pattern is seen where the mother (doe) remains with and nurses her bunnies only once or twice a day for less than 10 minutes at a time (5). In most altricial species, the young stay huddled together within the nest for the first 10 days, hence, there is little need for the mother to recognize individual members of her litter, as long as they remain in the nest. However, among rodents that live in colonies there may be a need for the mother to recognize the specific characteristics of her litter. In contrast to the mother, virgin female rats do not exhibit maternal behavior towards newborn young (6,7). However, with continuous contact with foster young, virgins will also eventually become maternal (after 2–10 days, depending on the strain and other conditions) and when they do their pattern of behavior is very similar to that of the new mother, although they do not lactate (6,7). This process of sensitization to pup stimulation is often referred to as the induction or sensitization of maternal behavior. In contrast to laboratory rats, some strains of laboratory mice show extremely short sensitization latencies, and therefore appear to depend less on parturitional hormones to exhibit maternal behavior; in these cases, virgin mice are immediately responsive to young on first exposure (8,9). In the so-called "precocial" species (most ungulates), the late pregnant ewe shows a shift from gregarious behavior to a tendency to seek isolation from the flock. This allows the mother to give birth and become familiarized with her neonate without being disturbed by other conspecifics. Immediately after parturition, the mother licks her newborn until the young are cleaned of the amniotic fluid. She also emits maternal bleats and, as the lamb can stand, it searches for the udder; in turn, the mother responds by arching her body, thereby facilitating access to the udder. From these behavioral observations, licking behavior, emission of maternal bleats, udder acceptance and suckling are defined as "maternal acceptance" while aggressive behavior, emission of protest bleats and udder refusal are indicative of "rejection" of the lamb. These precocial species tend to have a small litter of fully developed young who are capable of following the mother shortly after birth. Indeed, they are constantly on the move in search of food, hence, it is vital that young are able to follow their mothers. Because own and alien young co-occur in the same flock, nursing ungulate females potentially risk having their limited milk supply usurped by young that are not their offspring. Mothers in these species develop discriminative maternal care favoring their own young, allowing them to suck while rejecting any alien young that may approach the udder. In this respect, the establishment of a selective bond within the first few hours after parturition represents one of the essential characteristics of maternal behavior in precocial species. Individual recognition of young has been demonstrated in sheep, goat, cattle, horse (10–14). This characteristic is different from maternal responsiveness, which is a nonspecific interest toward any newborn and occurs immediately at birth in both altricial and precocial species. In these ungulate species, virgin animals tend not to exhibit maternal behavior towards young.

A continuum of mothering styles can be found between the two extremes of altricial and precocial species. In some types of mammals, neonates have functional sensory systems but limited thermoregulation (e.g., pigs) or locomotion (e.g., primates). Sows build a nest and do not develop selectivity toward their piglets immediately after parturition, although they are able to recognize their young within one day of parturition (14). Most primates do not build nests but instead the mother carries the offspring on her body to which the infants attach, using their strong prehensile hands. In some primates, the young are carried on the mother's back (15); and in some, fathers, siblings, and other group members participate in the carrying and caregiving (15). Mothers act maternally only towards their own infants but this selectivity may develop later, e.g., at about 2 months in crab-eating monkeys (16).

A cross-cultural survey of mothering practices in humans indicates that, although the biological mother is typically the primary caregiver, and in that role usually breastfeeds her infant (17), there is enormous variability in how mothers feel about, transport, and communicate with their infants and the extent of touching, vocalizing, and visual attention that is involved in those interactions (17–22). As well, there are large variations in the extent to which siblings, fathers, other relatives, and friends participate in caregiving (22). Despite this range of cultural variation in early maternal behavior and surrounding social systems, universal patterns and common functions of human maternal behavior are also suggested by a number of studies. For instance, particular maternal responses tend to follow particular infant behaviors: Nurturant responses follow vocal distress, imitation follows nondistress vocalizations, and encouragement to the environment follows explorations. These imitative responses to infant vocalizations reflect and foreshadow the special human capacity for linguistic communication (23). In fact, sensitive maternal responsiveness can be operationally defined as the "mothers" prompt contingent and appropriate behaviors that have identifiable, immediate, and direct antecedents in the behaviors of their infants (23). Moreover, like other primates, humans are able to discriminate their own infants and may show a preference for them, but, unlike the ungulates, they do not necessarily reject infants belonging to other mothers.

Whatever the commonalities and variations in patterns of maternal behavior, adaptive mothering necessarily involves behaving so as to ensure the infant's survival and healthy development; thus, instrumental attitudes and behaviors (like nursing) that facilitate the infant's growth, as well as emotional/nurturant attitudes and behaviors that promote the infant's emotional development, comprise some of the more frequently used measures of maternal behavior in human beings. Given the length of dependence of the human infant and of prepubertal development, it is not surprising that in primates, especially humans, affect and experience play a considerable role in socialization and the development of a mothering style.

Considering these species similarities and differences, one would expect that some immediate factors (hormonal, neural, and sensory) that control the onset of maternal behavior would be shared across mammalian species and other factors may differ between species. Those neuroendocrine events that are triggered by the onset of parturition and involved in the expression of maternal behavior are likely to have much in common across species since neuroendocrine control of pregnancy and parturition is relatively similar between species. Differences likely emerge after birth when the socialization of the young commence and, not surprisingly, are most influential among the primates and humans.

The first section of this chapter will compare, across altricial and precocial species, the hormonal factors regulating the onset of maternal behavior. The second section will focus on physiological factors that control the initial motivation to interact with young and the heightened aggressiveness displayed by lactating females. The third section will emphasize sensory factors, the neural circuits, and the various neurochemical systems involved in the onset of maternal behavior. In the fourth and final section, we will evaluate how experience throughout the life of the animal (infancy, juvenile, adult) can affect maternal behavior. A particular emphasis will be given to sensory, endocrine, and neural mechanisms that underlie memory for the litter (altricial species) and for individual offspring as found in a precocial species. Where there is adequate information for a class of mammal, in each section the different classes are indicated as a heading. For some topics, especially those related to genetics or neural mechanisms, primary emphasis is on the very extensive work on rodents and, sometimes, sheep.

HORMONAL BASES OF MATERNAL BEHAVIOR

Endocrine Changes Associated with Parturition

There is now substantial evidence that the hormones associated with late pregnancy and the parturitional period acting on brain receptors account for the rapid activation of maternal responsiveness seen at parturition; these include the steroid hormones, estradiol and progesterone which are synthesized by the ovaries and released into the circulatory system as well as the protein hormones, prolactin and oxytocin that are released within the brain and from cells or nerve terminals within the pituitary gland (Fig. 1) (24–31).

In addition, activity of the hypothalamic–pituitary–adrenal axis (HPA) and associated corticosterone or cortisol release is elevated during the peripartum period (32–34). Associated with gestation and/or parturition, there also occur important changes in the neurohormones, oxytocin and β-endorphins as well as in the neurotransmitters, dopamine, norepinephrine, glutamate and GABA (29,35) in rat, rabbit, and sows (30,31,36–39).

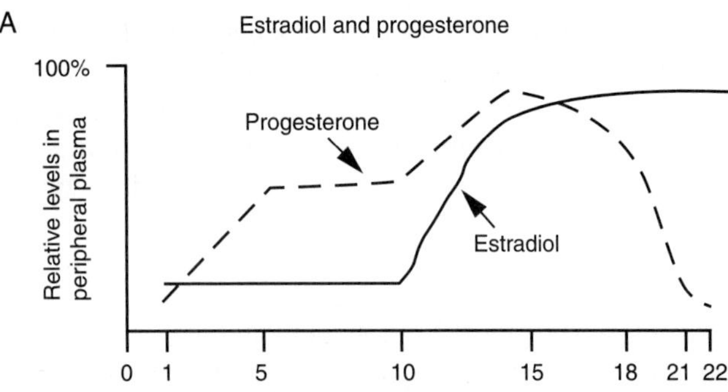

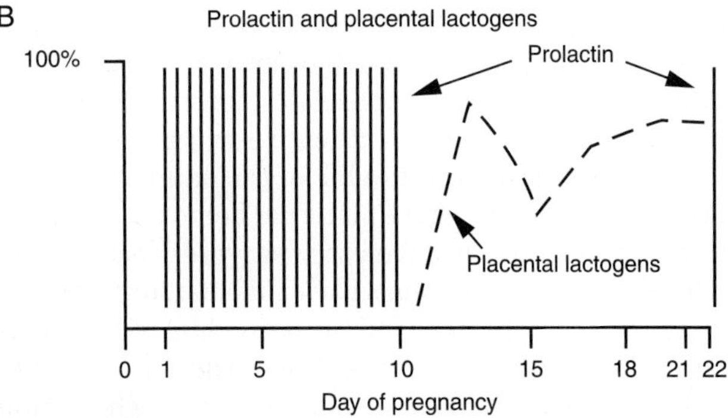

FIG. 1. Relative blood plasma levels of estradiol, progesterone (shown in panel **A**), prolactin, and placental lactogens (shown in panel **B**) over the course of the rat's 22-day pregnancy. For each hormone, data are depicted as the percentage of the maximum value detected during pregnancy. Prolactin is secreted in daily surges over the first half of pregnancy and on day 22. We show surges as vertical lines. [Modified from Bridges (105) and Grattan (55).]

Hormonal changes associated with parturition show similarities but also some differences across mammalian species. In most mammals (rat, mice, ungulates, carnivorous, some primates), parturition is characterized by a rapid increase in the ratio of estradiol (E) to progesterone (P) concentrations. In rats and rabbits, plasma levels of progesterone are high during pregnancy and drop abruptly after day 20 while estradiol levels are low throughout pregnancy and rise from day 15 up to the day of parturition (40,41). In sheep, 2 to 4 days before parturition plasma levels of progesterone drop whereas estradiol levels rise one day before parturition, peak at parturition and then decrease to basal levels within the first four hours postpartum (42–44). The common marmoset is the only primate for which progesterone decline occurs 1 week prior to parturition and is associated with a rise in estradiol towards the end of gestation; this pattern is similar to the pattern found in rats and sheep (15).

In contrast to the pattern of changing E to P ratios, hamsters, most female primates, and humans, show a different pattern of endocrine changes during late gestation. In the hamster, the rhesus monkey, the chimpanzee and the gorilla, both progesterone and estrogen rise together during gestation, peak at parturition and then decline (45–49). In humans, there is much controversy about the change of the estradiol/progesterone ratio. A fall in progesterone and a rise in estradiol plasma levels were reported during the last weeks of pregnancy (50), whereas other studies showed no significant change (51). However, in the myometrium, a 40% decrease of this ratio has been reported at parturition (51); and in the amniotic fluid labor and delivery are associated with a significant rise in 17β-estradiol levels (52), and in other estrogens such as estrone and estriol (53).

Lactogenic hormones are secreted from either the anterior pituitary (pituitary prolactin) or the placenta (placental lactogens) during most of pregnancy (54,55). In the rat, prolactin levels are low during the second half of pregnancy while plasma levels of placental lactogens increase. The change of steroid balance induces a rise in prolactin levels 24 hours to 48 hours before parturition and this high level is maintained after parturition due to the mammary stimulation (56,57) (Fig. 1B). Similar patterns of prolactin secretion have been reported in mice, rabbit (41,58), sheep (59), and primates (60). An exception is the hamster, which shows the highest prolactin levels in early pregnancy and maintains medium levels from late gestation into lactation (61).

Elevated plasma levels of glucocorticoids, especially corticosterone or cortisol, measured around parturition, have been reported in rodents (62), also in cattle (63,64), goats (65), and sheep (66). In cattle and goats, cortisol concentration increases in association with labor and peaks at birth (67). Both nonhuman and human primates show a state of hypercortisolism during pregnancy followed by a reduction after parturition (32,68,69). However, as in other species, the highest plasma concentrations of cortisol have been demonstrated at the beginning of labor (70).

In contrast to the steroids, the rise of plasma oxytocin is strictly contemporaneous with the actual phase of expulsion, for all species that have been investigated (71). This rise is induced by the decrease of plasma progesterone and/or an increase in estrogen and the vaginocervical stimulation occurring at this time. Oxytocin plays a role in the initiation and expulsion phase of labor (71).

All of these hormonal changes reflect changes in the peripheral compartment. Because of the existence of a blood–brain barrier, one cannot infer the existence of similar changes in the cerebral compartment, the actual site of hormonal effects on behavior. Parturition induces physiological changes within several brain areas. In sheep, as well as in cattle, oxytocin concentrations are greatly increased in the cerebrospinal fluid (CSF) within the first few minutes after expulsion of the neonate (72), but not in animals that have been given an epidural anesthesia, which blocks the feedback effects of vaginal stimulation (73,74). This increase of oxytocin in the ventricular system appears to be a common characteristic of mammals, although in humans, CSF oxytocin concentrations do not change at parturition (75). Oxytocin release within the brain has also been found in rats and sheep using in vivo microdialysis sampling techniques. Oxytocin levels increase at parturition in the paraventricular nucleus (PVN) of the hypothalamus (76), the main source of oxytocinergic projections in the brain, in the supraoptic nucleus of the hypothalamus (SON) (77), and in the oxytocinergic terminals within the olfactory bulbs (78,79), the medial preoptic area (MPOA), the mediobasal hypothalamus (80), the bed nucleus of the stria terminalis (BST) (80), and the septum (81). (For a description of relevant neuroanatomy, see section on Neural Circuits Regulating Maternal Behavior.) These increases are reflected in an increase of levels of oxytocin immunoreactivity, mRNA expression and/or receptor mRNA expression in the different brain sites (82–84). Thus, parturition is characterized by a synchronized synthesis and release of oxytocin in various brain regions; an activation that is amplified by an increase of the number of oxytocin receptors (85).

Other peptides show important variations within the brain at the time of parturition. In the rat, β-endorphin concentration is elevated during gestation and drops at parturition, especially in the MPOA (86). In this structure, the concentration of μ-opiate receptor increases during gestation and peaks at parturition (87). In sheep, pre-proenkephalin mRNA expression, similar to oxytocin mRNA expression, increases at parturition in the PVN (82,88). In contrast, the concentration of β-endorphin is elevated at parturition only in the plasma but not in the cerebrospinal fluid in humans (89,90). Classical neurotransmitters also show increased release associated with parturition. A general activation of the noradrenergic system is reported mainly in the hypothalamus, the hippocampus and the cortex (91,92). In sheep, noradrenaline release increases in the same brain areas as found for oxytocin (76). This similarity in pattern of release suggests that these two systems can cooperate to induce maternal responsiveness. In humans, a rise of catecholamines, especially noradrenaline, has been reported both in the plasma and in cerebral compartments (93,94). Therefore, in all species studied, the peripartum period is characterized by a gradual increase in plasma estradiol, prolactin, cortisol, and β-endorphins and by an activation of mainly the oxytocinergic and noradrenergic systems at the time of parturition. These changes are also found in the cerebral compartment although in humans this has not been established. The fact that these simultaneous activations occur in different brain areas that are not directly connected suggests that the regulation of maternal behavior is under multifactorial control. For a discussion of the effects of brain hormones and peptides on maternal behavior, see section on Neurochemistry and Molecular Neurobiology of Maternal Behavior.

Hormonal Determinants of the Onset of Maternal Behavior

Hormones and neurochemicals associated with parturition serve multiple functions. They prepare the prospective mother physiologically, by acting on mammary tissue prior to the initiation of lactation (95), by acting on the uteri, and to produce analgesia during the birth process (96–98). As summarized by Rosenblatt (27), these hormones also contribute to elevated maternal responsiveness shown by the newly parturient mother (Table 1) (28,29,31, 99,100).

Steroids

Estrogen and Progesterone. The steroid balance around parturition plays a critical role in the hormonal induction of maternal behavior in almost all mammals that have been studied (99,100).

TABLE 1. *Effects of different hormones and vagino-cervical stimulation on maternal behavior in mammals*

	Progesterone and estradiol	Cortisol	Prolactin (under steroid priming)	Vagino-cervical stimulation	Oxytocin
Rat	Full maternal behavior Maternal aggression	Licking, time spent in the nest	Full maternal behavior	Full maternal behavior	Full maternal behavior
Mice	Nest building Maternal aggression	Undetermined	Retrieving, crouching	Undetermined	Inhibition of infanticide
Rabbit	Digging, straw carrying, hair-plucking	Undetermined	Straw carrying, hair-plucking; emission of "nipple-search pheromone"; inhibition on digging	Undetermined	Undetermined
Sheep	Partial maternal	Full maternal behavior (CRH)	No effect on maternal behavior	Full maternal behavior	Full maternal behavior
Non-human primate	Instrumental bar-press to infants (marmosets)	Negatively correlated with infant carrying (gorillas)	Undetermined	No effect on maternal behavior	Undetermined
Human	Positive correlation with attachment feelings	Positive correlation with attraction to infant stimuli	Undetermined	Undetermined	Undetermined

Rats. This had been clearly shown in rats using two models of females: virgin rats given injections or silastic capsules containing progesterone and estrogen and rats whose pregnancy were terminated by removal of the pups and uteri (hysterectomy). Injection of high pharmacologic doses of estradiol benzoate in virgin female rats induced maternal behavior after 1 or 2 days of pup exposure, whereas females without hormonal treatment responded maternally after about 5 days of pup exposure (101). Pretreatment with progesterone reduced the effective dose estrogen (estradiol benzoate) (102). Moreover, when pregnant rats were hysterectomized on day 16 of gestation a high percentage of females displayed maternal behavior after one day of pup exposure. This high level of responsiveness appears to be the result of the endocrine effects of hysterectomy, including a decline in ovarian progesterone followed by a rise in estrogen. Siegel and Rosenblatt (101) showed that this progesterone to estrogen profile stimulated maternal behavior. If animals that undergo pregnancy termination through hysterectomy also have their ovaries removed (ovariectomy) at the same time, they show a decline in both progesterone and estrogen; when tested with foster pups, these animals did not show rapid maternal behavior. However, injection with estrogen reverses theses effects and animals showed an immediate maternal response on their first exposure to pups. On the other hand, high levels of progesterone artificially maintained during estradiol treatment inhibited the facilitatory effect of the estrogen (103). In virgin rats, using physiological doses of steroids Bridges (104) showed that with a progesterone and estradiol treatment for 12 days followed by progesterone withdrawal, females responded maternally after 1 or 2 days of pup exposure. Progesterone withdrawal, in the absence of E treatment had no effect and concomitant progesterone and estradiol treatment was not effective if progesterone was not withdrawn prior to behavioral testing (29,104,105). Overall, these results show that estradiol is essential to stimulate maternal care and progesterone has two main functions. A prolonged exposure to progesterone primes the female to respond to the rise of estradiol occurring during late pregnancy and at parturition. Second, the decline in progesterone secretion at the end of pregnancy synchronizes the onset of maternal behavior with parturition.

Rabbits. The participation of steroids in the expression of specific components of maternal behavior has been reported. In rabbits, ovariectomized females given estradiol benzoate plus progesterone display digging, while those given only estradiol benzoate do not (106). This steroid treatment also induces straw-carrying but only after progesterone withdrawal. On the other hand, progesterone withdrawal in such estradiol benzoate/progesterone treated females induces hair-plucking in no more than 40% of the animals. Additional factors are required to promote the loosening of body hair and the participation of androgens that are in high concentration during pregnancy has been reported (107).

Other Rodents. In mice, nest building is under the influence of both high progesterone levels and low estradiol levels (108). In contrast to rats, rabbits, and mice, in hamsters nest building begins during midpregnancy when both steroids are elevated. Concurrent treatment with progesterone and estradiol induced nest building while treatment with one hormone alone was without effect (109). Steroids do not induce the onset of pup-directed maternal care in the hamster and a decrease in the progesterone/estradiol ratio is presumably not involved since both steroids decline over the last 2 days of pregnancy (110). Also, treatment with only estradiol benzoate given to ovariectomized females induced pup-retrieving, but only in 40% of the animals (111). It seems therefore that steroids are not sufficient to induce maternal response to pups in hamsters and other physiological factors are likely involved.

Sheep. In sheep, acceptance of a lamb occurs only when estrogen levels are high, at estrus and the very end of pregnancy (112). In addition, correlations have been found between plasma estradiol levels and latency to groom (44,113). Finally, a single injection of estradiol can extend by one day the period during which parturient ewes are responsive to lambs (114) and, at pharmacologic doses, can induce maternal responsiveness in a high proportion of nonpregnant ewes (112,114,115). However, this hormonal effect in nonparturient ewes is unusual; excessively high doses of estradiol in these animals also produce a variety of abnormal behaviors, prolonged estrus and, even, male sexual behavior. Injections at more physiological levels are essentially without effect (116,117). It seems that in contrast to rodents and rabbits, in sheep steroids alone are not sufficient to induce maternal responsiveness; as described below (Neurochemistry and Molecular Neurobiology of Maternal Behavior), where they are effective, it is in combination with other physiological factors to trigger the behavior.

Primates. In nonhuman primates many contradictory results exist regarding the role of steroids in the display of maternal behavior. Evidence against a role for estradiol or progesterone in maternal behavior has been reported in a number of species. For instance, in gorillas, no significant association was found between prepartum urinary estrone conjugates or pregnanediol-3α-glucoronide and maternal behavior (49). As well, in rhesus macaque, whether mothers were accepting or abusive was found to be unrelated to plasma levels of estradiol and progesterone (48); and, in marmosets, prepartum urinary levels of estradiol were negatively correlated with maternal carrying effort and poor infant survivorship (118).

On the positive side, a host of recent studies have reported positive correlations between sex steroid profiles and maternal behavior at parturition. In red-bellied tamarins maternal behavior was related to prepartum urinary estradiol profile: females with high prepartum urinary estradiol titers showed positive behaviors toward their neonates and those with low prepartum estradiol titers reject them (119). In a study on pigtail macaques, females were found to increase their rate of interactions with other females' infants as pregnancy advanced and this behavioral change was correlated with increases in plasma estradiol and progesterone levels (120). In common marmosets, operant studies have provided evidence indicating a close relation between circulating levels of estradiol and progesterone and maternal motivation (15,121). Nonpregnant nulliparous female marmosets learned a bar-press task to have access to infant stimuli. Bar pressing varied with hormonal state over natural pregnancy. Bar pressing increased and peaked in the final 10 days of pregnancy when plasma estradiol levels were at a maximum and progesterone levels had declined.

More recently, a significant inverse correlation was reported between fecal estrone conjugates and rejection behavior observed during the first 12 weeks of lactation in Japanese macaques (32,122). Direct evidence for steroid regulation of maternal behavior was reported in studies using exogenous administration of estradiol and progesterone. Rhesus macaque females whose ovaries had been removed increased their rate of interactions with other females' infants after receiving estradiol in doses similar to those of middle-late pregnancy (120). Further, nonpregnant female marmosets treated with an estradiol and progesterone regimen that mimics the hormonal profile observed late in pregnancy (high estradiol and progesterone followed by a decline in progesterone) displayed the greatest level of operant responses to infant stimuli (15,121).

Humans. In humans, there is some evidence that as in other mammals, hormones may also exert important influences on mothers' early interactions with their infants. There is considerable variability among mothers in the pattern of their maternal feelings across pregnancy and the postpartum period. However, for many mothers there is a growth of attachment-related feelings during the second trimester and once again after the birth of the baby (123). The role of steroids in this elevated responsiveness shown by new mothers has received some attention. In one study, Fleming and colleagues (124) assessed prospective mothers longitudinally across pregnancy and the early postpartum period for circulating hormones and self-report feelings or attachment. Although few hormone-attitude correlations were found during pregnancy, mothers who exhibited a smaller pre-to-postpartum decline in the estradiol/progesterone ratio expressed higher feelings of attachment to their infants during the puerperium. However, when behavioral, as opposed to attitudinal measures of responsiveness are considered, the adrenal hormones rather than the ovarian or placental steroids seem to be important for maternal responsiveness (23).

Cortisol and Corticosterone

Rats. Although corticosterone levels are high at parturition, it has been unclear as to how, or even if, corticosterone affects the expression of maternal behavior during the early postpartum period. In the rat, disparity in effects of removal of the adrenal glands, the source of glucocorticoids, on maternal behavior had been reported. In some cases, adrenalectomy appears to be facilitatory with adrenalectomized mother rats spending more time in the nest and in a lactating position (125), and in other cases inhibitory or disruptive (126). More recently, the involvement of the adrenal glands and corticosterone replacement on maternal behavior in primiparous rats has been reconsidered (127). Adrenalectomy had no effect on the onset of maternal behavior at parturition. All animals exhibited maternal behavior. However, adrenalectomy did decrease some ongoing maternal behaviors, including licking and time in the nest; and high doses of corticosterone replacement reversed these effects. Interestingly, in virgin animals induced to become maternal through sensitization techniques, adrenalectomy and corticosterone replacement produced opposite effects; while having no effect on the latency to become maternal, adrenalectomy increased licking in maternal virgins and corticosterone reversed this effect.

Corticosterone is, therefore, not necessary for the onset of the motivation to respond to pups in either the postpartum animal or the virgin animals; however, corticosterone does play a modulating role on the intensity of licking behavior, which differs as a function of the endocrine background of the animal. These virgin-postpartum differences in maternal responses to adrenalectomy and corticosterone replacement are consistent with differences between the two types of animal in the HPA responses to stress, in which postpartum animals show a clear hyporesponsiveness (128). Based on these data it is not clear whether the effects of adrenalectomy and corticosterone replacement in the above studies reflect changes in the glucocorticoid itself, or in its feedback effects on corticotrophin-releasing hormone (CRH) or adrenocorticotropic hormone (ACTH). Of possible relevance here are the observations that CRH produce different neural effects in virgins and in postpartum animals (129), and endogenous levels within limbic system are also different in virgins and lactating animals (130).

Sheep. In sheep, there is some evidence showing that the HPA axis may be involved in the display of maternal behavior (131). Intracerebroventricular infusions of CRH facilitated the expression of acceptance behaviors in nonpregnant females receiving concurrent vaginocervical stimulation. The increase of CRH mRNA expression at parturition in brain sites that have been implicated in the onset of maternal behavior (132) strengthens the possible role of corticosterone, although these data are interpreted in terms of CRH potentiation on the functioning of other neuropeptide systems (e.g., oxytocin).

Humans. Consistent with a role of the glucocorticoids in the modulation of maternal behavior, Fleming and colleagues found in human mothers a significant positive correlation between postpartum plasma levels of cortisol and the amount of contact and approach responses shown by mothers toward their nursing infants (133). In addition, postpartum concentrations of cortisol were associated with greater attraction to infant odors, with greater discriminative responding to their own infant odors (134), and with more sympathy in response to infant cries (Fig. 2) (135).

These studies, although only showing correlations (*not* causal relations) between hormonal levels and behavior, suggest that in humans the adrenal system may play a role in mothers' initial responses to their infants. There is evidence that the adrenal corticoids are activated during high arousal states and that they enhance attention and acuity in a variety of sensory systems (136–139). The enhancement induced by the adrenal system may influence maternal responsiveness by enhancing the salience of infant stimulation.

Vaginocervical Stimulation. For some species (sheep), ovarian steroids are not able by themselves to stimulate maternal care and for other species (rat), steroid treatment facilitates maternal behavior, but the immediate maternal behavior normally shown at parturition has not been observed. Therefore, steroids have a priming effect allowing other factors to trigger the behavior. Such physiological factors include additional hormones such as prolactin and oxytocin (see Neuroanatomy and Neurochemistry of Maternal Behavior) and somatosensory stimulation that is associated with parturition. Indeed, the synchrony between parturition and the various physiological changes cited above suggest that the actual process of expulsion of the fetus is of primary importance.

Rats. In rats, there is indication that vaginocervical stimulation is important for the activation of maternal care. Nonpregnant intact multiparous females receiving vaginocervical stimulation show immediate maternal behavior (140). However, this effect is synergistic with that of steroids because vaginocervical stimulation was ineffective in ovariectomized females.

Sheep. It is in sheep that vaginocervical stimulation plays a preeminent role for triggering the onset of maternal responsiveness. In nonpregnant ewes primed with progesterone and estradiol, the induction of maternal responsiveness in response to neonates was obtained in 80% of the cases when ewes received 5 minutes of vaginocervical stimulation. By contrast, only 20% of females responded maternally without this stimulation (141). These results have been further confirmed in other breeds of sheep, with low doses of estradiol and even in ewes at estrus (115,117,142,143). This stimulation not only increased the number of maternal ewes but induced the full complement of

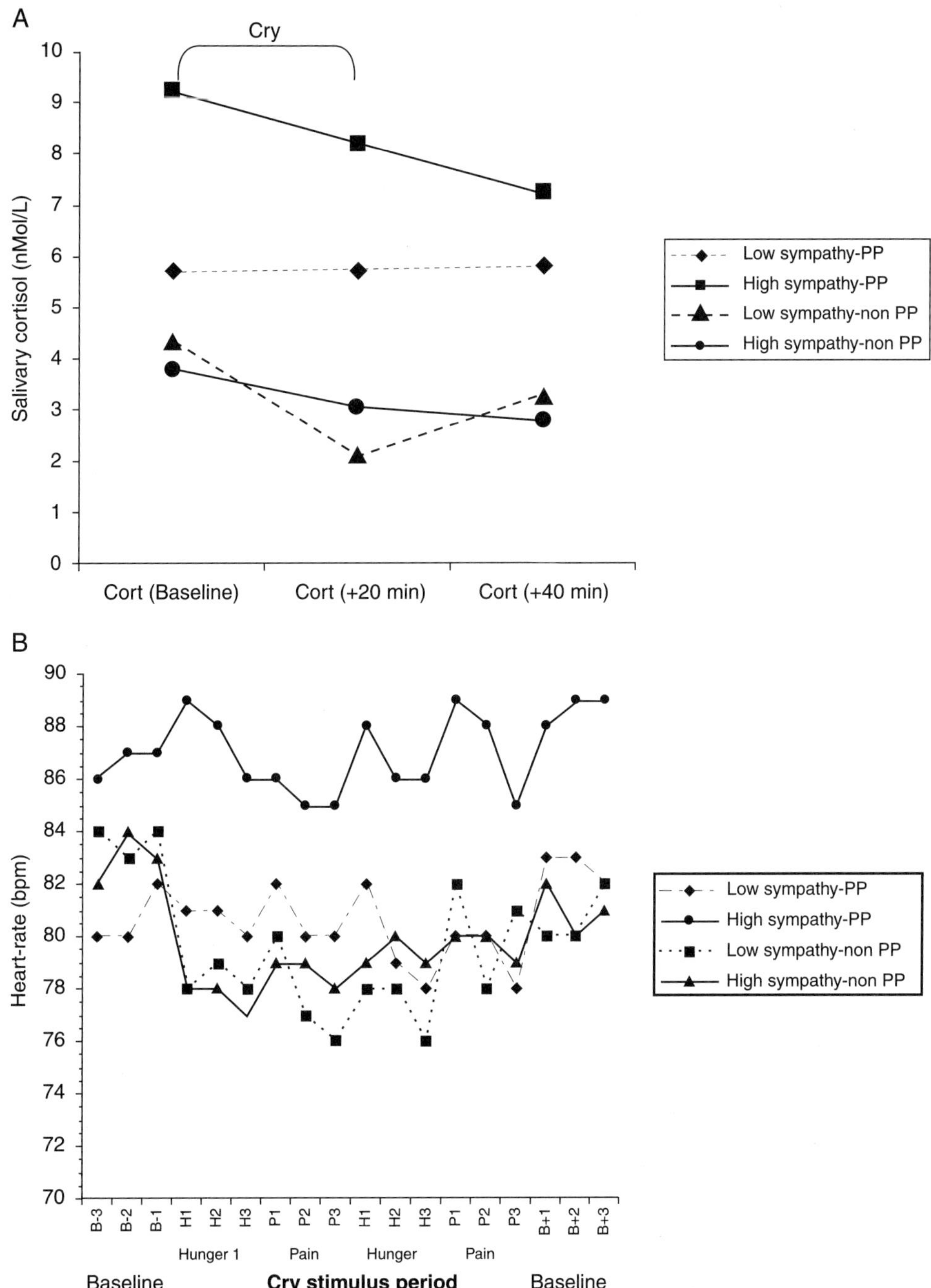

FIG. 2. A: Salivary Cortisol (nMol/L) at Baseline (0 min), During (20 min), and After Cry (40 min) Stimuli in Mothers (PP) and Nonpostpartum (Non PP) Women Experiencing High and Low Sympathy in Response to Cries (High and Low Sympathy established by median split on Sympathy Emotion Response Scale [ERS]). Numbers above symbols reflect standard errors. Cries were 30–40 second recorded hunger cries played to subjects through earphones. Cries were recorded from newborn infants in a maternity ward nursery. [(Modified from Stallings et al. (135).] **B:** Heart Rate Responses to Infant Hunger Cries in New Mothers (PP) and Nonpostpartum (Non PP) Women Experiencing High and Low Sympathy in Response to Cries. (High and Low Sympathy established by median split on a visual analogue scale for sympathy) B-3, B-2, B-1 (baseline, last three 5-sec intervals); H1-H3 (first and second Hunger cry, first three 5-sec intervals); P1-P3 (first and second Pain cry, first three 5-sec intervals). Pain and hunger cries were recorded from newborn infants in a maternity ward nursery. [Modified from Stallings et al. (135).]

maternal responsiveness, licking, maternal bleats, acceptance at the udder, and reduction in aggressive behavior. Vaginocervical stimulation stimulated even the attraction toward amniotic fluid that is a characteristic trait of maternal ewes at parturition (144) (see section on Sensory Mechanisms Mediating Maternal Behavior). On the other hand, vaginocervical stimulation without a pretreatment of steroids was ineffective, which confirms the importance of priming by steroid hormones (115,142). This facilitatory action of vaginocervical stimulation was also established in parturient ewes. Five minutes of vaginocervical stimulation 1 hour after parturition re-induced maternal responses, especially licking a new-born lamb (141). Vaginocervical stimulation was even effective at 24-hour postpartum (145). If stimulation of the genital tract was prevented at parturition by peridural anaesthesia, maternal behavior was greatly disturbed and a loss of preference for amniotic fluid was observed (146,147).

Primates. In primates, cesarean delivery prevents acceptance of the neonate in rhesus monkeys (148,149). However, presence of vaginal stimulation is not prerequisite to maternal care, since in the c-sectioned mother swabbing the newborn with amniotic fluids and vaginal secretions allowed its acceptance (149). In humans, the impact of cesarean delivery or peridural anesthesia is poorly documented and whether these medical procedures would disturb the first mother–infant interactions or, on the contrary, would be advantageous as a result of the reduced pain remains an interesting question to investigate.

The neural mechanisms by which vaginocervical stimulation initiates of the onset of maternal responsiveness involved mainly oxytocin release in the brain, and we review the importance of this neuropeptide for the onset of maternal behavior both in rats and sheep (see Oxytocin).

Prolactin

Rats and Other Rodents. In rats, robust demonstrations of prolactin effects on maternal care have been reported in a series of thoughtful studies by Bridges and his group. In a first series of studies, they found that hypophysectomized virgin rats (with removal of the pituitary gland) failed to respond maternally to the stimulatory effects of a steroid treatment; replacement with prolactin reversed this effect (150). In a second series of studies, ovariectomized steroid-treated nulliparous rats were used in which endogenous prolactin was suppressed by bromocriptine (Fig. 3B). Suppression of estradiol-induced rise in prolactin delayed for 4 to 5 days the expression of maternal behavior, mainly retrieving and crouching (151). The negative effects of bromocriptine can be restored by a concurrent administration of prolactin (151).

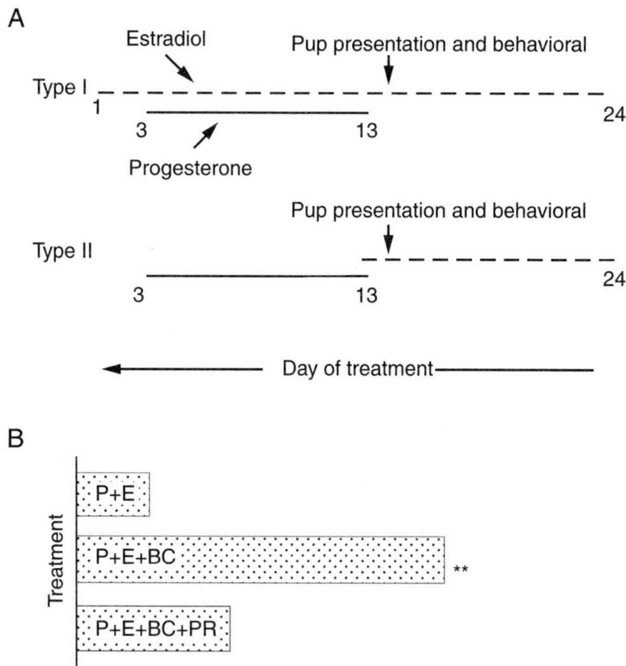

FIG. 3. A: Two steroid hormone treatment schedules used by Bridges's group to stimulate maternal behavior in ovariectomized virgin (nulliparous) female rats. In the type I concurrent schedule, females receive subcutaneous implants of Silastic capsules containing estradiol for 24 days. Over days 3–13 they are also implanted subcutaneously with progesterone. Twenty-four hours following the removal of the progesterone capsules, on day 14 of treatment, females are presented with pups and daily observations of maternal behavior commence, with a freshly nourished litter being presented on each day. In the type II sequential schedule, females are first exposed to progesterone for 10 days, the progesterone implant is removed and replaced with estradiol containing capsules, and 24 hours after the onset of estradiol treatment, the females are exposed to pups and the onset of maternal responsiveness is measured. **B:** Median latencies to the onset of maternal behavior in ovariectomized virgin female rats treated with the type II hormone regimen. One group was treated with progesterone (P) and estradiol (E), and another was treated similarly, except that bromocriptine (BC, a dopamine agonist drug that inhibits the release of prolactin from the pituitary) was administered subcutaneously throughout the period of estradiol treatment. In addition to treatment with E, P, and BC, the third group also received subcutaneous prolactin injections during the period of estradiol and bromocriptine treatment. In relevant groups, appropriate vehicle control injections were also administered. Rats treated with P+E+BC have significantly longer latencies to the onset of maternal behavior than each of the remaining two groups. [Panels A and B are modified from Bridges and Ronsheim (151).]

Since prolactin can gain access to the brain through active transport across the blood–brain barrier and considerable prolactin is found within the cerebrospinal fluid, the primary site of its behavioral action has been investigated and appears to involve the medial preoptic area (see section on Neurochemistry and Molecular Neurobiology of

Maternal Behavior and section on MPOA/vBST and Maternal Behavior).

In mice, prolactin further enhances the already high levels of maternal behavior shown by virgin females. Prolactin-treated mice showed more maternal motivation, measured by the frequency of bar-pressing for pups (152). They also retrieve and crouch over pups more rapidly (153). Related to this behavioral function are studies with prolactin-receptor knockout mice (154) (see section Neurochemistry and Molecular Neurobiology of Maternal Behavior).

Rabbits. In rabbits, prolactin plays an important role in relation to nest building, reflected in the finding that injections of bromocriptine at the end of gestation reduced the incidence of nest building (155–157). Bromocriptine treatment also prevented the stimulatory effect of an estradiol and progesterone regime on straw-carrying and hair-plucking in ovariectomized does (106). However, prolactin seems to have an inhibitory effect on another item of nest building: injection of prolactin antagonizes the stimulatory effect of steroids on digging (106). Prolactin, together with estradiol and progesterone, play a role in stimulating the emission of an olfactory signal by the nipple, "the nipple search pheromone" that allows the pup to locate maternal nipples (158,159). Estradiol and progesterone treatment reinstates the emission of this pheromone in ovariectomized does (160). Progesterone withdrawal induces a decrease of pheromonal emission while prolactin administration maintains a high level (40). Prolactin also appears to influence behavior toward pups. When bromocriptine was injected at the end of pregnancy, the percentage of primiparous rabbits that entered the nest box and crouched over the litter decreased dramatically (99,157). Where prolactin acts on the brain to promote maternal behaviors is not determined yet.

Sheep. In sheep, attempts to show a possible influence of prolactin on maternal responsiveness have failed so far. In parturient ewes, injection of dibromoergocryptine to block prolactin release did not prevent the onset of maternal responsiveness (112). In nongestant ewes primed with steroids, icv infusions of 50 or 100 μg of prolactin did not induce maternal responsiveness (79). This discrepancy could reflect an important difference in the physiological control of maternal responsiveness between rodents and ungulates.

Although the parturitional hormones might appear to activate maternal behavior in a unitary fashion, the different hormones exert somewhat different behavioral effects, and any one hormone exerts multiple effects. Moreover, their varied effects probably result from their action on a variety of different neural pathways (see section on Neuroanatomy and Neurochemistry of Maternal Behavior). Hormones augment maternal responsiveness during the periparturitional period by reducing the competing effects of alternative nonmaternal behaviors, and by ensuring that dams will continue to respond to pups when the period of hormonal priming ends. How hormonal factors act on behavioral processes and neural substrates is the focus in the following sections (Motivational Analysis of Maternal Behavior; Neuroanatomy and Neurochemistry of Maternal Behavior).

Hormonal Bases of Maternal Aggression

Another important aspect of maternal behavior is the heightened aggressiveness displayed by the mother towards intruders (161,162). Such behavior ensures the protection of the offspring from infanticide by conspecific males and females. Although maternal aggression has been reported in a variety of species (rodents, primates, ungulates), most of the data has been obtained from laboratory rodent studies. To examine maternal aggression, an intruder (male or female) is introduced into the home cage of a lactating female; sometimes pups are present, other times they are not. The latency to attack and the number and duration of attacks, threats, genital sniffing, and intruder defensive behaviors are the primary behaviors that are quantified.

Rats and Other Rodents

Maternal aggression in rats increases during the last week of gestation, rises around parturition and remains high during the first two postpartum weeks (163–166). Aggressiveness then gradually decreases, reaching nonpregnant levels by weaning. This, together with the fact that virgin rats display a low level of aggressiveness towards intruders suggests that the hormonal events associated with pregnancy and parturition are likely involved in the control of the behavior. Hysterectomy and ovariectomy on day 16 of pregnancy, which mimics the changes of steroids that occurs at parturition, and injection of estradiol benzoate induces high aggressive behavior toward male intruders, even when females had not had any exposure to pups (167). Estradiol may be more critical than progesterone for the onset of maternal aggression because estradiol treatment alone, in the absence of progesterone induces aggression in virgin rats (169,170). However, progesterone can stimulate aggression if administered levels are high enough, or if rats are exposed to progesterone for a long period (16 days), followed by estradiol (168,169).

Although aggression can be stimulated by hormones alone, both hormonal priming and exposure to pups is required for the full expression of aggressive behavior.

The full spectrum of maternal aggression can be induced in virgin rats if they are treated with the steroid regimen that mimics late pregnancy in virgin rats who have also been exposed to pups (165). Steroid facilitation of maternal aggression declines quickly in the absence of pups but is maintained in their presence (165,170). Finally, hormonal treatments of nonpregnant rats that activate maternal aggression are not sufficient to induce maternal behavior. Steroid treatment in hypophysectomized females elicits maternal aggression without producing short-latency maternal responses, while short latency maternal behavior can only be induced if the steroid treatment is administered to animals retaining their pituitary glands and, hence, prolactin function (170). These data indicate that pituitary products (especially prolactin) are necessary for maternal behavior, but not for maternal aggression.

In mice, the development of aggressive behavior across pregnancy, parturition and lactation appears to be different from rats. Levels of aggression are low during pregnancy and during the postpartum estrus, but rise to high levels by 24 to 48 hours after parturition (166,171). Contrary to rats, estrogens inhibit the display of maternal aggression. When females are ovariectomized just prior to parturition, thereby eliminating the surge of estradiol, they exhibit maternal aggression more quickly than do sham animals. Estrogen treatment reverses this effect (171). Although the administration of progesterone during lactation does not affect the full repertoire of aggressive behaviors (172), progesterone plays a role in stimulating threats, a less-intense aggressive behavior, in pregnant females (166). Peak levels of threats are high at the end of gestation when circulating progesterone levels are high (166). Progesterone treatment induces threats and its removal attenuates this behavior (173). Further experiments conducted in order to assess the role of progesterone in the pregnant mice use the hysterectomy paradigm that reduces progesterone levels. Hysterectomy reduced the aggressive behavior displayed by pregnant mice toward intruders, while implants of progesterone stimulated aggression in hysterectomized animals (174). However, this treatment does not fully restore the behavior to the level of fighting normally observed in pregnant animals, suggesting that other neuroendocrine factors may be involved.

MOTIVATIONAL ANALYSIS OF MATERNAL BEHAVIOR

Species-typical behaviors are often divided into an appetitive and a consummatory phase (175). The appetitive stage is made up of behaviors that bring an organism in contact with a desired stimulus or goal; it reflects an underlying motivation, appetite, or drive to interact with a particular stimulus. The consummatory phase consists of behaviors that are performed once the goal is achieved. For rodent maternal behaviors (176,177), nursing behavior, because it is the primary goal, is the major consummatory component. The appetitive component of maternal behaviors in rodents would then consist of approaching a displaced pup and retrieving it back to the nest site. The appetitive stage is usually composed of active, voluntary motor responses, while the consummatory phase is more reflexive in nature. The reflexive nature of rodent nursing behavior has clearly been shown (177).

Hormonal and other physiological factors that occur near the time of parturition may promote maternal behaviors by influencing both appetitive and consummatory processes. Also, in order for maximal maternal responsiveness to occur, it may be necessary for hormonal factors to depress competing behaviors. Motivational models have been proposed which argue that maternal behavior occurs when the tendency to approach infant stimuli and engage in maternal behavior is greater than the tendency to avoid or withdraw from infants (178–180). Such a view may be particularly relevant to first-time mothers who may be wary of novel infant stimuli. Therefore, in terms of motivational processes, the hormonal and other physiological events of late pregnancy, parturition, and lactation may promote maternal responsiveness by acting on the brain to decrease fear/aversion of infant stimuli (a competing motivational state), increase attraction/approach toward infant stimuli (appetitive maternal motivation), or both. These possibilities are shown in Fig. 4. By focussing on underlying motivational processes rather than on particular behaviors, one may ultimately be able to identify neural mechanisms that have wide species generality.

In the section on the hormonal control of maternal behaviors we noted that some species show high levels of maternal responsiveness in the absence of exposure to pregnancy hormones [also see Numan and Insel (100)]: several species show high levels of alloparental behavior (certain vole species, primates, and humans) and many strains of laboratory mice (but not wild mice) also show high levels of maternal behavior while they are nulliparous and estrous cycling. Alloparental behavior is said to occur when members of the social group who are not the biological mother give maternal-like care to young. This can include fathers, siblings, other relatives and even non-relatives. What this data suggests is that natural selection, selective breeding, or inbreeding can modify central neural mechanisms so that the level of attraction toward, and avoidance of, infant stimuli

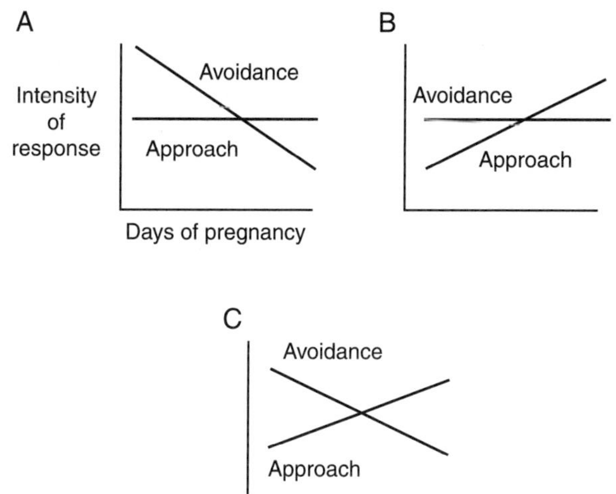

FIG. 4. Three alternative approach-avoidance models of the onset of maternal behavior at parturition. Maternal behavior occurs when central neural approach systems are more active than central neural avoidance systems with respect to infant-related stimuli. In **A** the physiological events of pregnancy primarily decrease avoidance; in **B** these physiological factors promote approach responses; and in **C** avoidance systems are depressed and approach systems are activated toward the end of pregnancy. Research on the rat is most consistent with model **C**. (Reproduced with permission from Numan M, Sheehan TP. [1997]. Neuroanatomical circuitry for mammalian maternal behavior. *Ann. N Y Acad. Sci.* 807, 101–125. Copyright 1997 by New York Academy of Sciences.)

is modified. Nulliparous human females are capable of adopting and caring for young and this may be the result of a neural functional organization which results in a high level of attraction and a low level of avoidance with respect to infant-related stimuli in such females. However, as reviewed in section on Hormonal Bases of Maternal Behavior, the hormonal events of late pregnancy and parturition are still capable of acting on the relevant neural circuitry in humans to further increase maternal responsiveness.

Evidence in Support of an Approach-Avoidance Motivational Analysis of the Occurrence of Maternal Behavior

Sensitization of Maternal Behavior in Nulliparous Females

Rats. Rat mothers often become pregnant soon after birth of a litter. They nurse the senior litter while gestating a junior litter. Under these circumstances, the older siblings are in a position to be exposed to and perhaps care for junior siblings. They show both nonmaternal, play-like or affiliative responses, and maternal responses to pups, with no evidence of undergoing an endocrine change (181–183). The initial responses, seen within a few days of contact with foster pups, involve licking and even crouching over pups, followed after 4 to 6 days by pup-retrieval (A. S. Fleming, personal observation). They also engage in "play" behaviors with pups (182,183). Hence, juveniles of both sexes are very responsive to young, in fact, more so than are adult nulliparous animals (184,185). They are also considerably less fearful and neophobic (184) and withdraw less from pups, which may permit expression of affiliative responses (181).

In contrast to the juvenile animal, adult virgin female rats do not show maternal behavior upon initial exposure to young, but after several days of constant exposure to pups a sensitization process occurs which results in the initiation of maternal behavior. Rosenblatt and Mayer (180) have provided evidence for an approach/withdrawal view of sensitization. According to this view, when a nulliparous female is first exposed to pups, both avoidance and approach tendencies are activated, but avoidance is much higher than attraction. Maternal behavior eventually occurs (after 5–7 days of constant pup exposure) because constant association with pups results in a habituation of neophobic withdrawal responses (familiarization process), allowing approach responses to become dominant. Once approach responses prevail, proximal pup stimulation further increases appetitive responses (maternal motivation) until a threshold is reached which allows for the occurrence of full maternal behaviors. Support comes from Fleming and Luebke (186). Female rats usually have a preferred sleeping/resting corner in their cages. When Fleming and Luebke (186) placed pups in the preferred area of virgin females' cages, such females switched their sleeping/resting area to another part of the cage. This switching behavior occurred for the first 3 to 4 days of pup exposure. After this time, a period of tolerance occurred where the females stopped the switching response, remaining in the same area with the pups, but still not showing maternal behavior. After a day or two of tolerance, and proximal pup stimulation, maternal behavior began to occur.

Evidence that during the sensitization process the virgin female must first overcome an inhibitory effect of pup stimulation before she can develop maternal behavior is provided by research which shows that disruption of the ability of the virgin to detect olfactory-related chemosensory stimuli actually facilitates maternal behavior by shortening the sensitization latencies. Both the primary olfactory system (olfactory nerve projections to the main olfactory bulb) and the vomeronasal system (vomeronasal organ projections to the accessory olfactory bulb) are involved in this inhibitory process (187). Naive virgin females who are anosmic or hyposmic as a result of damage to both of these chemosensory systems become maternal after about 24 hours of pup exposure, which contrasts with the standard sensitization latency of about

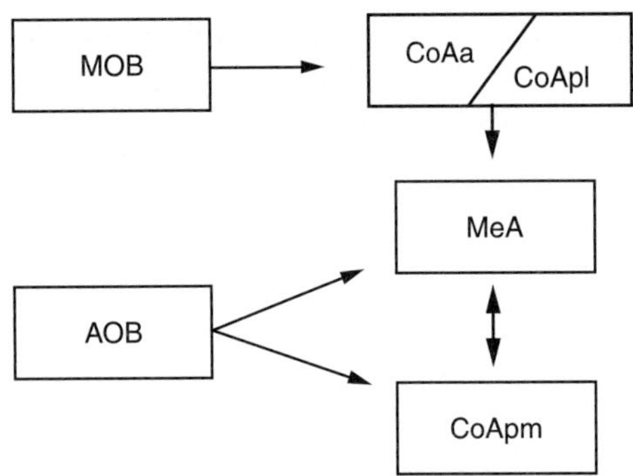

FIG. 5. Schematic representation of the neural connections between the main olfactory bulb (MOB), accessory olfactory bulb (AOB), and corticomedial amygdala to show how vomeronasal and primary olfactory input can converge onto the medial amygdaloid nucleus (MeA). CoAa, anterior part of the cortical amygdaloid nucleus; CoApl, posterolateral part of the cortical amygdaloid nucleus; CoApm, posteromedial part of the cortical amygdaloid nucleus. [Reproduced with permission from Numan and Insel (100) (Fig. 4.4, p. 74). Copyright 2003 by Springer-Verlag.]

1 week. It has been proposed that by disrupting chemosensory inputs, novel odors/pheromones from pups, which are considered to be aversive, are not detected, and this allows the female to respond more quickly to the positive aspects of pups which foster maternal responsiveness.

Both the main and accessory olfactory bulbs send efferents to the corticomedial amygdala (Fig. 5), with main olfactory and vomeronasal input converging on the medial nucleus of the amygdala (MeA) (188–191). Because of the involvement of the amygdala in fear and anxiety-related processes (192,193), it can be proposed that novel olfactory stimuli from pups suppress maternal responsiveness in virgins because such stimuli activate fear and withdrawal responses, which compete with the development of maternal responses. Importantly, electrical or excitotoxic amino acid lesions of the corticomedial amygdala shorten sensitization latencies in virgins to about 3 days of pup exposure (194–197).

Corticomedial amygdala-lesioned females do not show withdrawal responses to pups (they do not switch sleeping areas), and this would allow the positive effects of proximal pup stimulation to occur sooner. Given that corticomedial amygdala lesions facilitate maternal behavior in virgins, one would predict that stimulation of this nucleus would lengthen sensitization latencies, presumably by enhancing fear-related processes, and this is exactly what was found by Morgan and colleagues (198).

Approach-Avoidance in Postpartum Mothers

Rats. How do the physiological events of late pregnancy, parturition, and lactation promote maternal behaviors? Although experimental procedures which decrease avoidance behavior (e.g. amygdala lesions) facilitate sensitization, such data do not prove that the physiological events which naturally promote maternal behavior act in that way. Maternal behaviors could be triggered near the time of parturition simply because appetitive motivation is increased (Fig. 4). Actually, maternal behavior in postpartum rats is associated with both increases in appetitive maternal motivation and decreases in fearfulness.

One might predict, based on the effects of anosmia on the sensitization process, that parturient rats are anosmic or hyposmic. This is not the case: Bauer (199) and Kinsley and Bridges (200) have found that late pregnant female rats (day 22 of pregnancy), when given a 2-choice test, prefer (spend more time in) a cage compartment that contains bedding soiled by pups over clean bedding. Such a preference is not shown by virgin rats or rats in earlier stages of pregnancy. Fleming and colleague (201) reported that when ovariectomized virgin female rats are treated with a hormone regimen that would facilitate maternal behavior, such females are more attracted to pup-related odors than are untreated females. These results support the view that the hormonal events of late pregnancy promote maternal behavior by making pup-stimuli more attractive. The increased attractiveness of pup-related odors is undoubtedly involved in the natural onset of maternal behavior at parturition. However, since anosmic females show maternal behavior, the functional importance of the increased attractiveness of pup odors is that the inhibitory effects of such odors are removed.

Other aspects of pup stimulation, in addition to olfactory input, must be attractive to postpartum females, since anosmic females show maternal behavior. Research using the conditioned place preference paradigm has indicated that pups can serve as strong reinforcers to postpartum females, but not to virgin (nonmaternal) females (202–204). Postpartum females, but not naive virgins, develop of preference for a cage compartment that had been paired with pups; and tactile input from the pups to the mother's perioral and ventral trunk region plays a role in this reinforcing effect (205). Significantly, when one compartment of a conditioned place preference apparatus is associated with pups while the other is associated with cocaine administration, day 8 postpartum females develop a preference for the pup-associated compartment (203,204), suggesting that pups are highly reinforcing during the early postpartum period. In a related study, Lee and colleagues (206) found that postpartum rats will bar-press in an operant chamber at a high rate when rat pups are presented as the reward, but only if the mother is also allowed to interact with the pups after their presentation. Importantly, pregnant females, who were not maternally responsive, did not bar-press at a high rate for pups, although they did so when a palatable food was used as a reward.

With respect to aversion processes, several pieces of evidence show that postpartum females exhibit a general reduction in fear-related behaviors when compared to pregnant females and naive virgins (168,186,201,207–211). Postpartum females emerge more quickly into a novel open-field arena and spend more time in the open (less protected) arms of an elevated plus maze than do virgin and pregnant females. Importantly, when virgin females are treated with hormones that are capable of facilitating maternal behavior, they also show a decrease in general fearfulness (168,201). These results indicate that hormonal and other physiological events associated with parturition do not simply cause a specific reduction in fearfulness to novel pup-related stimuli, but instead cause a more general reduction in fearfulness to a large variety of stimuli. Perhaps such effects not only promote maternal responsiveness to pup stimuli, but also allow the female to effectively care for her young under stressful or demanding conditions. Some support for this idea comes from a study by Fride and colleagues (212), who reported on the adult maternal behavior of females that were stressed prenatally, and therefore presumably more fearful than control females (213). Although prenatally stressed females did not differ from controls when standard retrieving tests were administered, such females were much less likely than controls to retrieve pups when they had to pass through an airstream in an alley in order to reach the displaced pups (214). Interestingly, there also appears to be a relationship between the level of fearfulness in postpartum rats and the nature of nursing behavior shown by such females (56,215). One possibility is that as fearfulness increases, the female is more easily distracted by insignificant environmental stimuli and that this, in turn, decreases the incidence of active nursing behavior. Finally, as we will review below, a general reduction in fearfulness may permit the effective display of maternal aggression.

A related body of literature has shown that the HPA axis is hyporesponsive to stressful stimuli during lactation. HPA activation has often been used as a neuroendocrine index of fear or anxiety. When one compares lactating rats with nonlactating rats, stress-induced HPA activation is reduced in lactators (210, 216–218). Is this reduction in HPA neuroendocrine responsiveness to stressors an essential component of the reduction in general fearfulness shown by lactating females? The answer appears to be no. HPA hyporesponsiveness to stress can first be detected during the last week of pregnancy, before a general reduction in behavioral fearfulness can be detected, leading Neumann and colleagues (210) to conclude that there is an independent regulation of HPA axis activity and fearfulness.

Other Rodents. Not much work on a motivational analysis of maternal behavior has been done on other species. For hamsters (219), rabbits (220), gerbils (221), and sheep (79,222,223), there is evidence, as in the rat, that the physiological events occurring near the time of parturition modify the effects of olfactory input from young so that such input is re-routed away from brain centers mediating aversion and toward those centers mediating approach behavior.

In laboratory mice, although high levels of maternal behavior are observed in virgin females, Hauser and Gandelman (152) found that in an operant conditioning paradigm, postpartum females press a lever at a much higher rate than do maternal virgins when pups are used as a reinforcing stimulus. Therefore, although the baseline level of maternal responsiveness in laboratory mice is relatively high, the physiological events associated with the natural maternal condition appear to further increase the appetitive aspects of maternal behavior, and may also cause a general reduction in fearfulness (see section on Maternal Aggression).

Sheep. Among sheep, there is also some evidence that the physiological events occurring at parturition induce a reduction of fearfulness. Pregnant ewes display lower fear reactions than do nonpregnant ewes. In ewes that are isolated and presented with a novel object, there occurs a negative correlation between plasma progesterone and level of fear (224).

Primates. Similar findings are available for marmoset monkeys, a species that shows a high level of alloparental behavior. Pryce and colleagues (121) found that adult nulliparous females who had pre-adult alloparental experience and were also treated with progesterone and estradiol to simulate late-pregnancy endocrine patterns showed higher operant bar-pressing rates for infant stimuli than did control females who were not treated with steroids, but who also had preadult alloparental experience. There is also some work on rhesus monkeys which suggests that the hormonal changes associated with parturition decrease neophobic responses to infant stimuli while also increasing the appetitive aspects of maternal behavior (225,226).

Humans. The relation between maternal behavior and affect in humans has primarily been studied in the context of postpartum depression. Epidemiologic studies have found that between 10% and 15% of women exhibit depressive symptoms in the first weeks following delivery (227), that the rise in incidence is within 30 days postpartum but persists for up to 2 years, and that the great majority of these depressive episodes resolve spontaneously within 3 to 6 months (228,229). The symptom profile of postpartum depression (PPD) resembles that of a major depressive episode experienced at other times in life, but it is unique in its timing and it always involves at least the mother–baby dyad and in most cases an entire family unit.

Comparisons between depressed and nondepressed mothers indicate that depressed mothers are more intrusive and irritated and respond less sensitively, less contingently, and more negatively to their infants than do nondepressed mothers (230–233). When observed at 3 months postpartum, depressed dyads exhibited less mutual attentiveness, less vocal and visual communications, less touch interactions and less smiling compared to nondepressed dyads (232–235). These relations are now well established.

As with other species, mothers' maternal responsiveness is related to other behavioral systems, in this case, her affective state as well as to how focused the mother is on the infant, how easily distracted she is, and how well she attends to the infant. This relation has been well-established in the animal literature in which rat mothers who show reduced prepulse inhibition and attention set-shifting performance, are more easily distracted in the maternal context, and spend less time in contact with and actively licking their young (236). What might be the endocrine underpinnings of this affect profile in new mothers?

Few consistent relations have been found between levels of hormones during either the prepartum or the postpartum periods and postpartum affect—and it is not for want of trying to find the relation (237). In terms of a role for the HPA axis in mothers with affective dysphoria, results are mixed. Increased cortisol plasma levels correlated with PPD in some studies (238) but not in others (239,240) and has even been associated with positive affect (241). Fleming and colleagues (242,243) find that the hormone most consistently associated with mothers positive *and* negative affect is the glucocorticoid, cortisol; however, whether the relation between cortisol levels and affect is positive or negative depends on (a) The mothers' age; for teen mothers, high cortisol is associated with more energy, well-being, whereas for older mothers it is associated with negative affect, fatigue, etc. (b) The postpartum period; in older mothers cortisol has positive relations with mood at 2 to 4 days postpartum and negative relations at 3 to 4 months postpartum. (c) The context for cortisol elevation; if associated with prior stressors, cortisol is associated with negative mood; in the absence of stressors, the relation may be positive. These different sets of correlations indicate that in different contexts, ages, possibly other "stress" hormones (parasympathetic and sympathetic hormones/neurochemicals) are likely also released along with cortisol, and the profile of these may contribute to differences in mood state. With respect to PPD in humans, which is usually associated with anxiety, this emotional profile is antagonistic to adaptive maternal responding and is opposite to the decreased fearfulness that is typically shown by postpartum female mammals. Future research should be aimed at clarifying the abnormal physiological states that give rise to this dysfunctional emotional response.

In addition, among primiparous mothers cortisol has been positively associated both with the degree of expressed attachment to their infants, and to their ratings of infant odors (123,244). Finally, Altemus (245) and Uvnas-Moberg (246) provide evidence that lactation in nondepressed women is associated with decreases in general fearfulness and stress reactivity.

Maternal Aggression

Maternal aggression, or the heightened aggressiveness displayed by lactating females in a variety of species (161,166), may be promoted by the decrease in general fearfulness that occurs postpartum. Although maternal aggression has been documented in a variety of species, most of the experimental work has been done on rodents (162).

Rodents

The evidence from rodents indicates that the hormonal events of late pregnancy increase aggression in both rats and mice and also allow pup stimulation to further increase and maintain the aggression during the postpartum period in the absence of further direct endocrine involvement. In other words, hormonal factors induce, but do not maintain, maternal aggression. The most important pup stimuli are ventral somatic sensory stimuli from nuzzling pups, which include suckling stimulation (172,247). That such stimulation is not inducing hormone secretions, for example, prolactin release, which then maintains aggression, is shown by the fact that postpartum hypophysectomy does not decrease the aggressiveness of such females, while pup removal (for 5 to 24 hr) does (248–250). Finally, note that the maternal behavior shown by virgin rats and mice (sensitized maternal behavior) is not associated with high levels of maternal aggression (172,251–253). Therefore, the endocrine events of late pregnancy and parturition appear to set up or prime the relevant neural circuits so that postpartum pup stimulation can then promote the appropriate occurrence of maternal aggression toward intruders at the nest site.

What is the evidence that the postpartum decrease in general fearfulness is a factor that promotes maternal aggression? Hansen and colleagues (254) were the first to draw attention to this relationship when they found that virgin rats that are treated with antianxiety agents such as diazepam show both decreased fearfulness and increased aggression. Other work has confirmed these results (255–257). Since diazepam and other benzodiazepines act on $GABA_A$ receptors to potentiate GABA transmission, in this way presumably increasing the inhibitory tone on neural areas controlling anxiety, they suggested that motherhood may be associated with increased GABAergic transmission at certain central neural sites. In support, it has been found that when postpartum rats are treated systemically with benzodiazepine antagonists, maternal aggression is decreased (254). Additional evidence can be outlined as follows: (a) In laboratory mice, there is a strong negative correlation between the level of anxiety or fearfulness and the degree of postpartum maternal aggression (258–260). (b) Brain lesions which further decrease fearfulness in postpartum rats cause an increase in the already high levels of maternal aggression shown by these females (261). (c) The higher levels of maternal aggression shown by lactating rats in comparison to sensitized maternal virgins is inversely related to the level of fearfulness shown by these two groups (252).

Conclusions

Taken together, evidence, primarily from rodents, has been presented which shows that the physiological events associated with late pregnancy and parturition act on the brain to increase the attractiveness and decrease the aversiveness of infant-related stimuli. These events also set up a neural condition that results in a general decrease in postpartum fearfulness. Importantly, although hormonal and parturitional factors appear to set up the neural condition that allows these behavioral changes to take place, the subsequent maintenance of these motivational changes seems to be dependent on continued mother–young interaction, and appears to be relatively independent of hormonal mediation (hypophysectomized females continue to show both maternal behavior toward young and maternal aggression toward intruders).

Just where and how hormonal factors act to produce the shift from avoidance to approach with respect to infant stimuli is the subject of subsequent sections. It is relevant to point out, however, that recent work has shown that prolactin increases neurogenesis in the olfactory bulb (262,263). In addition, with respect to the possible underlying physiological mechanisms related to the decrease in general fearfulness that occurs postpartum, the following can be noted. First, although postpartum neuroendocrine stress hyporesponsiveness appears to be independent of the decrease in fearfulness, there may be some points of commonality. CRH is not only a neurohormone that stimulates ACTH release, but it is also a neurotransmitter/neuromodulator within central neural circuits that mediate anxiety-related behaviors (264). Importantly, there is evidence that decreased activity within central CRH systems may mediate the decrease in fearfulness that occurs in lactating females: (a) Intracerebroventricular (icv) injection of CRH causes greater neural activation in virgin female rats than in lactating females, and some of the activated neural areas are involved in regulating fearfulness, such as the corticomedial amygdala (which is also involved in inhibiting maternal behavior) (129). (b) CRH mRNA levels in the amygdala are lower in lactating rats than in virgins (130).

Brain prolactin and oxytocin systems may also play a role in reduced fearfulness. Torner et al. (265) found that icv administration of prolactin caused an anxiolytic effect, as measured on the elevated plus maze, in virgin female rats. In postpartum females, it is unlikely that suckling-induced prolactin release from the pituitary causes anxiolysis since hypophysectomized postpartum females show maternal aggression. However, as will be shown, there are neurons within the brain which release prolactin as a neurotransmitter/neuromodulator. Perhaps the physiological events of late pregnancy and parturition stimulate the development of brain prolactin circuits, whose subsequent activity is then maintained by infant stimulation, and perhaps this activated system decreases fear-related behaviors. A similar relationship might apply to brain oxytocin systems, since research has shown that the action of oxytocin at central neural sites produces an anxiolytic effect (266,267). Although oxytocin released from the neurohypophysis is required for the milk-ejection reflex, it is unlikely that significant amounts of systemic oxytocin can cross the blood–brain barrier (268). However, activation of brain oxytocin circuits postpartum may contribute to fear reduction (see below).

NEUROANATOMY AND NEUROCHEMISTRY OF MATERNAL BEHAVIOR

Most of the research on the neural basis of maternal behavior has been done on rodents and sheep. However, since maternal behavior is a defining feature of mammals, and the critical brain regions include the limbic system and hypothalamus, which are phylogenetically older parts, the data is likely to define a core neural circuitry. This section will review sensory factors involved in maternal behavior, followed by a review of central neural mechanisms. Two major neural circuits will be outlined: one which activates maternal behavior, presumably by promoting approach responses and maternal motivation, and one which inhibits maternal behavior, presumably by activating withdrawal responses and fearfulness.

Sensory Mechanisms Mediating Maternal Behavior

Laboratory strains of maternal rats and mice, which give birth to altricial young, do not form specific attachments to their young; they will care for their own young and young from another mother. In contrast, maternal sheep, which give birth to precocial young, do learn to recognize their young, and will not care for alien young. Therefore, the discussion of rodents will focus on those sensory cues used by mothers to generally recognize pups so that they can perform maternal responses, while the discussion of sheep will also focus on those sensory cues that mothers utilize to specifically identify their own lamb.

Rodents

An influential study by Beach and Jaynes (269) developed the concept of multisensory control of maternal behavior in rats. This view proposed that although many infant-related stimuli may influence maternal responses, no single sensory modality is essential for maternal behavior. A crucial experiment investigated the sensory cues from pups that are necessary for the retrieval response in lactating rats. It was found that surgical elimination of vision, olfaction, or tactile sensitivity of the snout and perioral region did not prevent the response. Furthermore, elimination of any two or all three of these sensory modalities, although causing deficits, did not prevent retrieval of pups. A study by Herrenkohl and Rosenberg (270) provided additional support by showing that deafened postpartum rats were capable of retrieving their young.

Subsequent research has basically confirmed the nonessential role of vision and hearing for maternal behavior in rats (271,272). But the roles of olfaction and tactile sensitivity in maternal behavior control need to be reevaluated in the light of more recent research. Concerning the two sensory structures in the nasal cavity, the vomeronasal sensory neurons that project to the accessory olfactory bulb and the primary olfactory sensory neurons that project to the main olfactory bulb, subsequent work was in accord with Beach and Jaynes (269) in that damage to these sensory inputs does not disrupt maternal behavior in rats (273–275). However, damage to the primary olfactory system, but not the vomeronasal system, does severely disrupt maternal behavior in laboratory mice (276–278), indicating an important species difference.

Although olfactory input does not play an essential positive function in the maternal behavior of rats, recall that such input plays an inhibitory role in nonhormonally primed (virgin) female rats. Recall that an essential shift in the valence of pup odors from negative (within the context of other pup stimuli) to positive occurs during the periparturient period in rats. Therefore, although olfactory input per se is not necessary for maternal behavior, because anosmic females show such behavior, a modification in the central neural effects of olfactory input does appear essential for the onset of maternal behavior in primiparous females. The suppressive effects of olfactory input need to be removed in order for maternal behavior to occur, and the way this mechanism appears to operate under natural conditions is through a change in the motivational impact of pup olfactory stimuli from fear arousing to attracting.

Beach and Jaynes (269) argued that perioral input was not important for retrieving in rats. However, they did not indicate the time interval between their trigeminal nerve transections and their retrieval tests, nor did they verify that their deafferentation procedure actually caused a lack of perioral somatosensation. Subsequent research has shown that tactile sensitivity of the perioral region is important for retrieving. The injection of lidocaine into the mystacial pads, which anesthetizes the snout region, or section of the infraorbital branch of the trigeminal nerve, which disrupts tactile input from the snout and upper lip, blocks the retrieval response in lactating rats (177,279–282). The mothers approach displaced pups and nose them but do not retrieve them. Interestingly, when the desensitized rats are tested over a series of days, retrieval behavior recovers even though tactile sensitivity of the snout region does not. Stern and Kolunie (282) have presented evidence that such recovery involves the use of somatosensory inputs from the lower lip and chin, mediated by the mandibular branch of the trigeminal nerve. These results indicate that perioral somatic sensory inputs play a major role in regulating the retrieval response, and it is likely that such inputs operate at the level of the brainstem, where they trigger mouth opening in response to perioral tactile input from pups.

Beach and Jaynes' (269) multisensory view of maternal behavior in rats examined retrieving behavior. What about the role of sensory factors in other aspects of maternal behavior? In a series of studies, Stern's group has shown that ventral trunk somatic sensory inputs play a crucial role in regulating aspects of nursing behavior in rats. After a female retrieves her young to a nest, she begins to nurse them.

As described by Stern and Johnson (283), first she hovers over them while engaging in other activities, such as pup licking. A transition is eventually made by the mother to a low crouch and then to a high crouch position, both of which are relatively quiescent or immobile responses, which Stern (177) refers to as kyphosis. Kyphosis appears to be important in allowing all members of a young litter access to the nipple region (284). Suckling stimulation from pups is necessary for the maternal kyphosis response to be maintained over time; in the absence of such stimulation, kyphosis either does not occur or is of very short duration (3,283,285). It is likely that kyphosis is a brainstem reflex that is triggered by appropriate ventral somatic sensory inputs (261,286,287).

In rats, although tactile inputs operating at the level of the brainstem appear to trigger retrieving and kyphosis, an important question is whether such inputs also reach higher levels of the nervous system in order to more generally influence maternal responsiveness. It has been reported that postpartum females who are not retrieving because of either infraorbital nerve transection or lidocaine injections into the mystacial pads are also not likely to nurse their young, even when the pups are placed back into the mother's nest (281,282). Such females briefly investigate the pups and then move away from them. An interpretation of these results is that trigeminal sensation reaches higher levels of the brain (above the sensory and motor trigeminal nuclei in the brainstem) to influence appetitive aspects of maternal behavior, fostering interest in pups and proximity-seeking behavior. Unfortunately, other researchers have not replicated this finding (205,284,288). However, Morgan and colleagues (288) report that although postpartum rats with lidocaine injections into their mystacial pads nursed young that were placed in their nest, females that received similar lidocaine injections and also wore nylon jackets that prevented the female from receiving ventral somatic sensory inputs from pups showed very little maternal behavior. Since females that just wore nylon jackets did remain with their pups and hovered over them, it may be that tactile inputs from both the perioral and ventral regions play positive roles in maintaining overall maternal responsiveness, and only when both are eliminated is maternal motivation disrupted. If this were the case, and more research is necessary, it would mean, at least for rats, that overall tactile inputs are essential to maintain maternal motivation.

In summary [also see (289)], for rats, in order for postpartum maternal behavior to occur, olfactory inhibition over maternal responsiveness must be removed. Distal stimuli (visual, olfactory, auditory) attract the mother to her young, but none of these stimuli, acting alone, is essential. Once contact with pups is made, tactile stimuli appear to be primary for the reflexive aspects of retrieving and nursing, and for the motivational aspects of maternal behavior. Note, however, that within rodents, species differences exist. Postpartum mice rely heavily on olfactory input for their maternal behavior. An understanding of the sensory factors involved in maternal behavior may help to guide us to crucial central regulatory sites.

Sheep

Sheep form a selective bond with their lamb at birth and will then only care for that lamb, while rejecting the advances of others (99,112,290). Specifically, at parturition, a ewe will accept any lamb and care for it, but after several hours of contact with a particular lamb, the female learns its characteristics and then will only accept that lamb and will reject others. Importantly, if a mother is separated from her lamb at birth before an exclusive bond forms, her maternal interest declines so that by 12 to 24 hours she will revert to the typical nonmaternal responsiveness of nonpregnant estrous-cycling ewes, and she will reject the advances of all lambs. In other words, there is a sensitive period in the immediate postpartum period when physiological factors have changed the brain of the ewe so that maternal behavior occurs, and the subsequent mother–lamb interaction during this period is associated with an important learning mechanism which allows for the formation of a selective bond.

There are some parallels with the rat literature which indicate that in sheep some sort of olfactory inhibition over maternal behavior needs to be removed. Prior to parturition, ewes avoid amniotic fluid, but near parturition they prefer it, and these choices are dependent on the primary (and not the vomeronasal) olfactory system (223). Therefore, prior to parturition, lamb stimuli, in this case amniotic fluid, are repulsive to ewes, but at parturition they are attractive. Attraction to amniotic fluid is necessary for the development of maternal responsiveness in females without maternal experience. Washing the neonate to remove amniotic fluids disrupted the onset of maternal care in inexperienced ewes (222). Amniotic fluid is also sufficient by itself to induce maternal responsiveness in a context wherein females reject young. Parturient experienced ewes accepted 1-day-old lambs whose coats were treated with amniotic fluid. (291,292). However, in a manner analogous to rats, such an attraction to lamb olfactory stimuli is not essential for maternal behavior in maternal experienced females since multiparous parturient ewes that are made anosmic prior to parturition will still show maternal behavior (79,293,294). Therefore, the valence shift from repulsion to attraction with respect to amniotic fluid seems to primarily be a mechanism for removing inhibition over maternal responsiveness.

Importantly, olfaction is also important for the formation of a selective mother–infant bond. If a ewe receives an olfactory bulbectomy or intranasal application of zinc sulfate (which destroys the primary olfactory sensory receptors) prior to parturition, then at parturition and subsequently she will show nonselective maternal responsiveness, acting in a positive manner to any lamb she is exposed to (79,293,294). Note that vomeronasal nerve cuts performed prior to parturition do not interfere with the formation of a selective bond (79), which indicates the essential role of the primary olfactory system. In support of this idea, manipulation of the lambs' sensory cues, rather than of the mothers' sensory modality, produced similar effects (112). Ewes exposed for the first 12 hours to their lambs who were confined in a double-walled mesh cage (providing olfactory, but not physical, contact) nonetheless developed a selective bond with that lamb. In contrast, ewes that were exposed to lambs in an airtight transparent box (thereby eliminating olfactory cues) did not develop a selective bond. Also, research has shown that if a ewe can not detect the olfactory cues from her lamb after a selective bond has already been formed, then she will reject that lamb as if it were alien (295–297). To summarize, as a result of the physiological events associated with parturition, the ewe is primed to accept lambs; although the sensory cues from the lamb remain undefined, they are general in nature, since the ewe will accept any lamb. As a result of mother– infant interaction, the ewe learns the specific olfactory characteristics of the lamb. This odor(s) now has selective access to the central mechanisms regulating maternal behavior, and if the odor is not detected maternal behavior will not occur. These findings allow us to conclude that after a female forms an exclusive bond, primary olfactory input becomes essential for maternal behavior.

In addition to being important for the onset of maternal behavior, vaginocervical stimulation also modifies central neural mechanisms so as to allow for the formation of a selective maternal bond. For example, although a ewe that has already formed a selective bond with one lamb will not normally accept alien lambs, experimenter-applied vaginocervical stimulation can induce the formation of a new selective bond even when it is administered as late as 27.5 hours postpartum (141,145) As will be shown in a later section, a modification of olfactory bulb processing by vaginocervical stimulation is a likely participant in this mechanism (298).

Humans

In contrast to many other mammalian mothers, for human mothers no single sensory modality is essential for adequate maternal responsiveness although each may make a contribution (23). With respect to maternal responses to infant odors, there is substantial evidence that mothers are more attracted to newborn infant body odors than are nonmothers (244) and they are able to recognize their infants based on their odors following very little experience with them (299–301). In terms of the importance of other sensory modalities, there is evidence that although blind and deaf mothers may take longer, they can achieve normal levels of responsiveness. Human mothers are highly responsive to both auditory and visual stimuli from their infants, whereas the influence of olfactory cues is more subtle. In fact, there is surprisingly little research on the role of olfactory, tactile, and thermal modalities in human maternal responsiveness. Heart rate monitoring has provided substantial insight into the arousal properties of infant signaling, especially crying, showing that heart rates vary according to the populations studied as well as characteristics of the cry itself (23). Although the findings are equivocal (302,303) heart rate accelerations are generally interpreted as preparatory responses to stimuli and indicative of states of arousal, interest, and motivation (304,305). Heart rate decelerations have been associated with attentive or orienting behavior (302). Using heart rate measures, Bleichfeld and Moely (306) found, for instance, that hearing unfamiliar infant crying elicited heart rate accelerations among newly parturient women and decelerations among nonpregnant women. In contrast, pregnant women demonstrated both accelerations and decelerations.

How effective infant stimuli are depends, however, on a multitude of factors including mothers parity, intensity of the stimulus, whether infant cues is from own infant or another infant as well as to complex attributes of the infants such as temperament, sex, illness, and prematurity, all of which have also been shown to influence maternal responsiveness (23) and may well affect autonomic responsivity.

Neural Circuits Regulating Maternal Behavior

Figure 6, an elaboration of Fig. 5, shows that olfactory input reaches MeA, and then outlines two of the major efferents from MeA (188). One projection is to the anterior hypothalamic area (AHA) and ventromedial hypothalamic nucleus (VMN), and the other MeA projection is to the MPOA and BST. Although it is possible that pup-related odors suppress maternal behavior in virgin rats via the activation of MeA efferents that directly inhibit MPOA/BST, brain regions which play an essential positive role in maternal behavior, current evidence supports an alternative model.

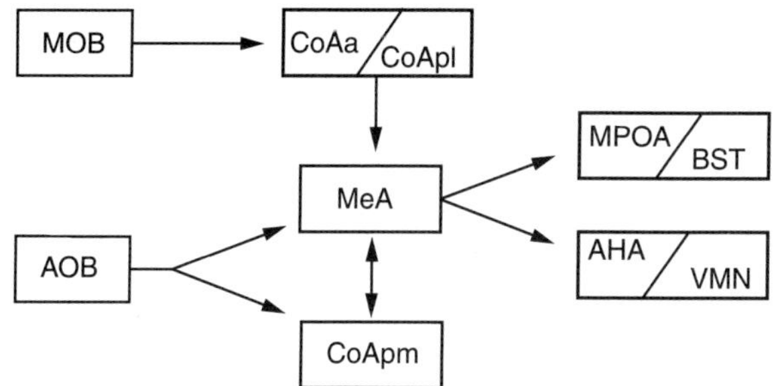

FIG. 6. An elaboration of Fig. 5 to show neural pathways through which olfactory input from the main and accessory olfactory bulbs (MOB and AOB, respectively) can reach the hypothalamus. The medial amygdaloid nucleus (MeA), which receives AOB and MOB input, is shown as projecting to the medial preoptic area (MPOA) and the bed nucleus of the stria terminalis (BST, which adjoins the MPOA, but is part of the telencephalon), and to the anterior hypothalamic area (AHA) and ventromedial nucleus (VMN) of the hypothalamus. We propose that MeA projections to MPOA may play a positive role in maternal behavior, while MeA projections to AHA/VMN may be inhibitory. CoAa, anterior part of the cortical amygdaloid nucleus; CoApl, posterolateral part of the cortical amygdaloid nucleus; CoApm, posteromedial part of the cortical amygdaloid nucleus. [Reproduced (with some modifications) with permission from Numan and Insel (100) (Fig. 5.5, p. 125). Copyright 2003 by Springer-Verlag.]

Certain MeA efferents appear to suppress maternal behavior by activating AHA/VMN. In contrast, MeA stimulatory efferents to MPOA/BST may underlie the attraction to pup-related odors that occurs in parturient females. In other words, MeA-to-AHA/VMN projections may be functionally dominant in virgin females, while MeA-to-MPOA/ BST projections may be dominant in parturient females, and this switch may be one of the mechanisms through which the onset of maternal behavior at parturition is facilitated.

Knowledge of the names and locations of certain neural structures will be important. Figure 7 shows a frontal section through the rat brain at the level of MeA and anterior hypothalamic nucleus (AHN: the main nucleus in AHA). Note the locations of the central nucleus of the amygdala (CNA) and the PVN. Figure 7B is a frontal section through MPOA and ventral BST (vBST), and also shows the ventral pallidum (VP) and the lateral septum (ventral part, LSv; intermediate part, LSi). Figure 7C shows a sagittal section through the medial part of rat brain. In addition to the nuclei already mentioned, the anterior-posterior locations of the nucleus accumbens (NA), VMN, ventral tegmental area (VTA), and periaqueductal gray (PAG) are shown.

MPOA/vBST and Maternal Behavior

MPOA/vBST neurons and their lateral connections with other brain regions need to be intact for normal maternal behavior to occur in rats and other rodents. Electrical lesions of the MPOA and adjoining vBST

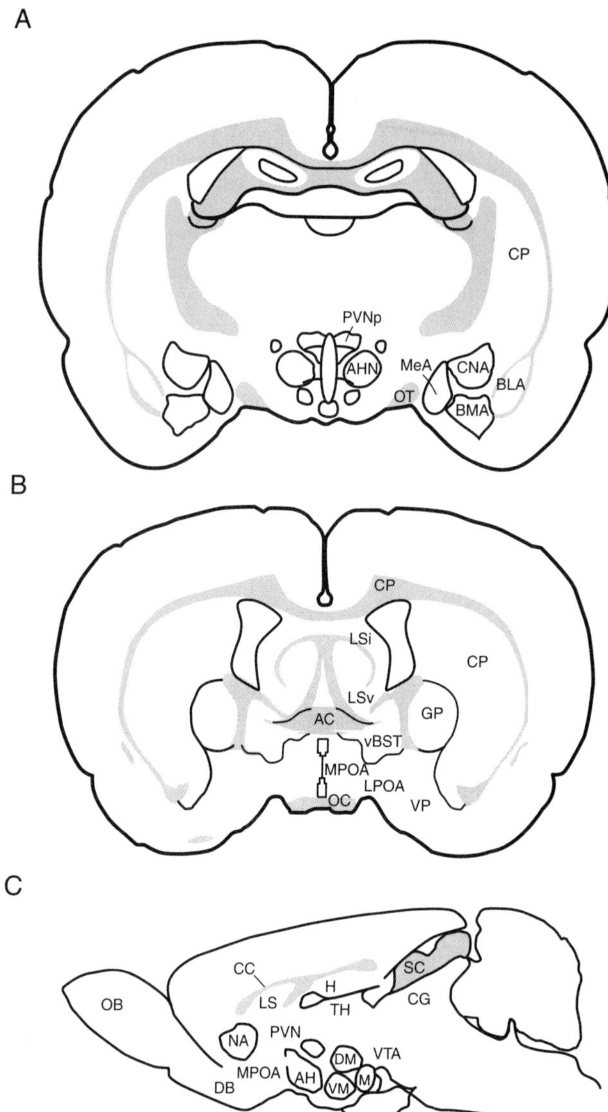

FIG. 7. Frontal section (**A**) through the rat brain showing the location of some of the amygdaloid nuclei. Frontal (**B**) and sagittal (**C**) sections of the rat brain at the level of the medial preoptic area. AC, anterior commissure; AH/AHN, anterior hypothalamic nucleus; BLA, basolateral amygdaloid nucleus; BMA, basomedial amygdaloid nucleus; CNA, central nucleus of the amygdala; CC, corpus callosum; CG, central gray (periaqueductal gray); CP, caudate-putamen; DB, nucleus of the diagonal band of Broca; DM, dorsomedial hypothalamic nucleus; GP, globus pallidus; H, hippocampus; IC, internal capsule; LPOA, lateral preoptic area; LS, lateral septum; LSi, intermediate nucleus of lateral septum; LSv, ventral nucleus of lateral septum; M, mammillary bodies; MeA, medial amygdaloid nucleus; MPOA, medial preoptic area; NA, nucleus accumbens; OB, olfactory bulb; OC, optic chiasm; OT, optic tract; PVN, paraventricular hypothalamic nucleus; PVNp, parvocellular division of the paraventricular hypothalamic nucleus; SC, superior colliculus; TH, thalamus; vBST, ventral bed nucleus of stria terminalis; VM, ventromedial hypothalamic nucleus; VP, ventral pallidum; VTA, ventral tegmental area. [Adapted from Swanson, L. W. (1992). Brain maps: structure of the rat brain. Elsevier, Amsterdam, and reproduced with permission from Numan and Insel (100) (Figs. 5.6, 5.7, pp. 127, 130). Copyright 2003 by Springer-Verlag.]

disrupt maternal behavior, as do excitotoxic amino acid lesions that spare fibers of passage and only destroy MPOA/vBST neurons (206,308–319). Finally, knife cuts that sever the lateral connections of MPOA/vBST, particularly the dorsolateral connections, are also disruptive (314,320,321).

Evidence clearly indicates that the disruption of maternal behavior by MPOA/vBST damage is the result of neurobehavioral, rather than neuroendocrine, dysfunction: such damage disrupts maternal behavior in hormonally primed females (315), in postpartum females who have already initiated maternal behavior, and in sensitized virgins; maternal behavior in the latter two cases is considered to be free from hormonal control (311,314,316,318,319) Furthermore, the maternal behavior deficit resulting from preoptic damage appears relatively specific to maternal behavior. Such females show normal body weight and temperature regulation, activity levels, sexual behavior, and hoarding behavior (311,314,315,322). The hoarding data indicate that the requisite oral motor responses that would be involved in pup retrieval are intact.

With respect to the nature of maternal deficits that occur after MPOA/vBST damage, all studies agree that retrieval behavior and nest building are abolished or severely depressed. Although there are some differences between studies with respect to nursing behavior, the best conclusion is that nursing behavior is disrupted, but not eliminated. It has been suggested that the MPOA may be particularly important for mother-initiated active/voluntary maternal responses, such as retrieval behavior, and may be less important for the more passive/reflexive aspects, such as nursing, which can be initiated by the pups and triggered by ventral somatic sensory and suckling stimulation (100,177,310,321). Indeed, the MPOA may be particularly concerned with appetitive maternal motivation, or the desire to voluntarily approach and interact with pups. In support, Lee and colleagues (206) found that MPOA lesions not only disrupted retrieving, nest building, and nursing, but also disrupted an operant bar-press response when pups were used as the reinforcing stimulus, but not when a palatable food item was used as a reward.

The importance of MPOA/vBST in maternal motivation gathers further support from findings that stimulation of this region can activate full maternal behavior. First, kindling electrical stimulation of MPOA facilitates sensitization in virgin rats (198). Second, MPOA/vBST neurons are one of the sites where hormones act to facilitate the onset of maternal behavior in rats. MPOA/vBST neurons contain estrogen receptors (both alpha and beta types) (323), progesterone receptors (324), and prolactin receptors (325,326), and research has shown that estradiol (318) and lactogenic hormones (prolactin and placental lactogens) (327,328) act locally on MPOA/vBST to stimulate the onset of maternal behavior. Importantly, estrogen receptor mRNA (329), estrogen receptors (329–331), and mRNA for the long form of the prolactin receptor (55,325) have been found to increase in MPOA as pregnancy advances in the rat. Finally, the injection of either an estrogen (332) or prolactin receptor antagonist (333) into MPOA has been found to disrupt the hormonally induced onset of maternal behavior in rats. Interestingly, the neural site(s) where progesterone and progesterone withdrawal act to influence maternal behavior have yet to be elucidated (324,334–336).

Finally, during maternal behavior in rats (and other species), there is an increase in the number of MPOA/vBST neurons that contain Fos proteins. Fos genes, which are referred to as immediate early genes, are activated by intracellular second messenger cascades that are triggered by a variety of extracellular stimuli (e.g., hormones, neurotransmitters, neuromodulators). The activated fos genes produce their protein products (cFos, Fos B, Fra-1, Fra-2), which serve as transcription factors that, in turn, activate late-responding structural genes, the protein products of which are presumed to alter neuron function (e.g., by possibly modifying neurotransmitter or neurotransmitter receptor levels) (337,338). Since the promoter region of Fos genes contains regulatory elements which allow Fos protein expression to be influenced by a variety of extracellular signals (339–341), it is not surprising that Fos expression has been used as a neuronal activation mapping procedure.

cFos and Fos B expression increase in the MPOA/vBST during maternal behavior in rats (25,342–347). Importantly, such expression not only occurs in postpartum females during maternal behavior, but also occurs in sensitized maternal virgin females (25,348). Furthermore, control stimuli, such as candy, other social stimuli, or pup presentation to nonmaternal females, are much less effective inducers of Fos protein expression in MPOA/vBST neurons than is pup presentation to maternal females who show appropriate maternal responsiveness (25,202,349,350).

One might predict that during maternal behavior Fos proteins, in their role as transciption factors, maintain the functional integrity of MPOA/vBST neurons. During maternal behavior, important structural proteins within MPOA/vBST neurons may be turned over or used up, and Fos proteins may signal the genetic machinery to renew the synthesis of these proteins, which may be necessary for the continuance of maternal behavior. Therefore, one might expect that Fos proteins would be expressed in MPOA/vBST for as long as females are exhibiting maternal behavior, and this appears to be the case (Fig. 8) (345).

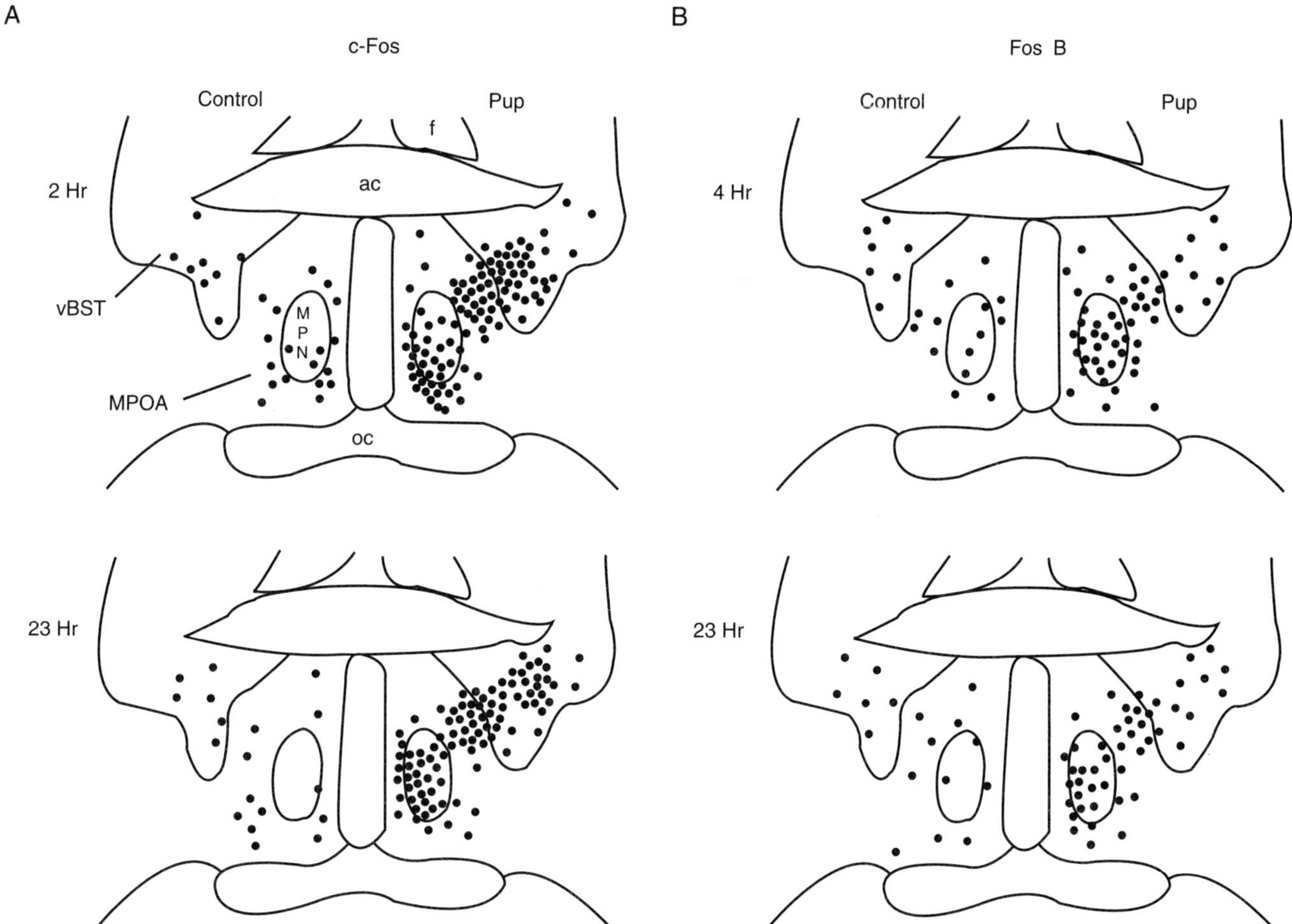

FIG. 8. Representative frontal sections through the level of the medial preoptic area (MPOA) on which is plotted the location of cFos (panel **A**) and Fos B immunoreactive cells (panel **B**) in postpartum primiparous female rats who were exposed to pups and showed maternal behavior for either 2, 4, or 23 hr, and for control females who were not exposed to pups. ac, anterior commissure; f, fornix; MPN, medial preoptic nucleus; oc, optic chiasm; vBST, ventral part of the bed nucleus of the stria terminalis. [Reproduced with permission from Stack and Numan (345). Copyright 2000 by the American Psychological Association, Inc.]

Stack and colleagues (346) took advantage of the fact that unilateral damage to MPOA/vBST, unlike bilateral, does not disrupt maternal behavior. They found that the occurrence of maternal behavior in postpartum females who received unilateral knife cuts severing the lateral connections of MPOA/vBST was associated with a significant decrease in cFos and Fos B expression in the ipsilateral, but not the contralateral, MPOA/vBST (Fig. 9). Since bilateral knife cuts of the lateral MPOA/vBST connections severely disrupt maternal behavior, these results support the contention that Fos proteins are marking activated neurons within the MPOA/vBST that contribute to neural circuits regulating maternal behavior.

If Fos proteins are involved in activating the synthesis of other proteins that are necessary for the proper functioning of MPOA/vBST neurons in maternally relevant circuits, then Fos may also play a role in the hormonal onset of maternal behavior in rats. Hormonal events during late pregnancy may activate Fos synthesis in MPOA/vBST neurons so that these neurons are already prepared for maternal functions upon the birth of the young. With respect to the continuance of maternal behavior during the postpartum period after the hormonal stimulus has waned, pup stimulated maternal behavior would be expected to maintain Fos expression and MPOA/vBST function (345). There is some evidence for the hypothesized involvement of Fos proteins in the hormonal onset of maternal behavior. First, the anatomical location of Fos expressing neurons in MPOA/vBST (Fig. 8) matches the distribution of estradiol receptor and prolactin receptor-containing neurons (325,351). Second, Lonstein and colleagues (352) found that cFos is co-localized with estrogen receptor alpha within MPOA/vBST neurons during maternal behavior.

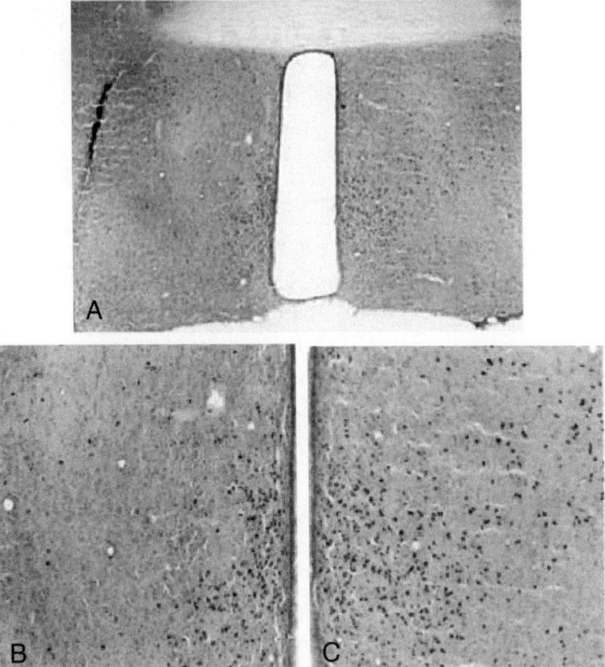

FIG. 9. Photomicrographs showing the effect of a unilateral knife cut severing the lateral connections of the medial preoptic area (MPOA) and adjoining ventral bed nucleus of the stria terminalis (vBST) on the expression of cFos within cells in the ipsilateral and contralateral MPOA and vBST of maternal rats. **A:** A low power magnification showing both the ipsilateral and contralateral regions. The knife cut is visible on the left side, the anterior commissure is located dorsally, and the third ventricle is in the middle of the photomicrograph. **B** and **C** show high power magnifications of the ipsilateral and contralateral regions, respectively. The knife cut clearly decreases cFos expression on the ipsilateral side. [Modified from Stack et al. (346) and reproduced with permission from Numan and Insel (100) (Fig. 5.15, p. 158). Copyright 2003 by Springer-Verlag.]

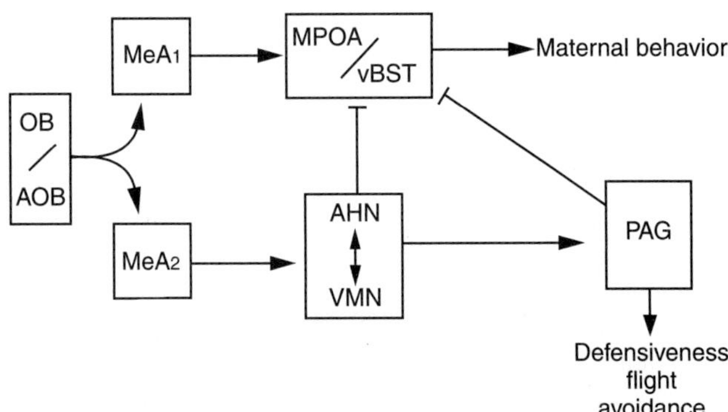

FIG. 10. Hypothetical model to explain the neural inhibition of maternal behavior. Olfactory input from the main olfactory bulb (OB) and accessory olfacotory bulb (AOB) is shown as reaching two populations of neurons in the medial amygdaloid nucleus (MeA1 and MeA2). MeA1 is shown as sending chemosensory input to the medial preoptic area and the ventral part of the bed nucleus of the stria terminalis (MPOA/vBST), and such input is hypothesized to play a positive role in maternal behavior. In suboptimally hormone-primed nonmaternal females, however, we propose that MeA2 output is dominant over MeA1 output, and that MeA2 relays novel chemosensory inputs to a central aversion system which depresses maternal behavior by directly inhibiting MPOA/vBST and by also activating defensive behavior (escape and avoidance). The central aversion network is shown as consisting of the anterior and ventromedial hypothalamic nuclei (AHN and VMN, respectively), and the midbrain periaqueductal gray (PAG). Lines ending in arrows signify excitatory projections, and those ending in a vertical bar are inhibitory. [Reproduced (with some modifications) with permission from Numan and Insel (100) (Fig. 5.14, p. 153). Copyright 2003 by Springer-Verlag.]

Third, the cfos gene contains an estrogen response element in its promoter region (353), and systemic estradiol treatment increases cFos expression in MPOA (31,354). Fourth, and most importantly, there is evidence that the endocrine events which stimulate immediate maternal responsiveness also participate in the activation of cFos synthesis in MPOA neurons of rats and sheep prior to any exposure to infant-related stimuli (355,356). Sheehan and Numan (356) found that systemic estradiol was more effective in activating cFos expression in MPOA of pregnancy-terminated female rats (who experienced pregnancy levels of lactogens and progesterone, followed by progesterone withdrawal) than it was in hysterectomized and ovariectomized virgin rats who did not have the same endogenous progesterone and lactogen exposure. Significantly, the dose of estradiol used in this study is capable of inducing maternal behavior in pregnancy-terminated females but not in virgins. Therefore, it is possible that exposure to the hormone profile of late pregnancy, which includes progesterone dominance followed by progesterone withdrawal, sensitizes the MPOA to the effects of estradiol so that more neurons express cFos, which, in turn, acts to alter the phenotype of MPOA neurons in ways that might facilitate maternal responsiveness to pup-related stimuli.

Inhibitory and Excitatory Neural Circuits Regulating Maternal Behavior

In an ingenious series of studies, Sheehan and colleagues (350) presented pups to female rats that either showed maternal behavior or did not (because of hormonal manipulations). Control females were similarly treated, but were not exposed to pups. A subsequent cFos mapping analysis of selected neural regions was used to determine brain areas that were most active in females that were either not showing maternal behavior, but instead were avoiding pups, or were showing maternal behavior. The former sites would be candidates in a neural circuit that might inhibit maternal behavior, while the latter areas might be components of an excitatory neural circuit. More attention was paid to putative inhibitory areas. As expected, the MPOA/vBST region was most active in maternal females, and this confirms the work already reviewed. LSi was also most active in maternal females.

The brain regions that were most active in nonmaternal females included the caudal part of AHN, just rostral to the VMN, LSv, and the dorsal premammillary nucleus (PMd: this nucleus is located just rostral to the mammillary bodies shown in Fig. 7C). Interestingly, MeA was active when pups were presented to either maternal or nonmaternal females (25,344). This finding fits with the model shown in Fig. 10, which suggests that there are two populations of MeA neurons, one projecting to MPOA/vBST and playing a positive role, and the other projecting to AHN/VMN and playing an inhibitory role.

Inhibitory Circuits Arising from the Corticomedial Amygdala. Although a population of MeA neurons may be neurally linked with AHN, LSv, and PMd to suppress maternal behavior (188,307,350), research has focussed on the suppressive effects of MeA projections to AHN. First, unilateral NMDA lesions of the corticomedial amygdala, which include MeA, depress cFos activation in the ipsilateral, but not contralateral, AHN of nonmaternal female rats who are exposed to pups and show avoidant responses (197). Second, excitotoxic amino acid lesions centered in the region of caudal AHN and rostral VMN have been found to facilitate maternal behavior in females rats treated with a suboptimal hormone preparation which does not facilitate maternal behavior in control females (197,357). In other words, such lesions mimicked the effects of MeA lesions, supporting anatomical data, suggesting that these areas are linked in an inhibitory neural circuit.

There is good evidence that the medial hypothalamic region which includes the AHN/VMN region is part of a central aversion system. Electrical or chemical stimulation of this region has been found to promote defensive aggression and escape/flight responses in mammals (358–360). Also, AHN and VMN (and PMd) project strongly to PAG in the midbrain, and PAG is involved in regulating a variety of responses related to anxiety, fear, escape behavior, and defensive aggression (358,361,362). Therefore, in a female who is not hormonally primed to show maternal behavior, we propose that novel olfactory input from pups activates a MeA-to-AHN/VMN-to-PAG circuit, and such activation leads to avoidance/escape responses and other processes which prevent immediate maternal responsiveness. Since AHN, VMN, and PAG also have neural projections to MPOA/vBST (307,363). when these areas are highly active they may not only promote avoidance behavior, but may also directly inhibit MPOA/vBST and maternal behavior (Fig. 10). It should be noted that a high proportion of MeA neurons are GABAergic (191). If such neurons project to AHN/VMN, the mechanism through which they could activate the aversion system needs to be elucidated.

Although a neurobehavioral model of the inhibition of maternal behavior has been presented, whenever studying the hormonal onset phase of maternal behavior, neuroendocrine explanations should not be excluded. Perhaps excitotoxic amino acid lesions of MeA and/or AHN/VMN facilitate maternal behavior not because of depression of a central aversion circuit, but instead because the lesions caused hormone changes that facilitated maternal behavior. Although neurons in MeA and AHN/VMN are involved in the control of prolactin release (196,364), the overall evidence supports a neurobehavioral over a neuroendocrine explanation of the involvement of these structures in maternal behavior (197).

MPOA/vBST Facilitatory Circuits. Anatomical studies have indicated regions to which MPOA/vBST neurons might project to promote maternal behavior. Numan and Numan (369) iontophoretically injected PHAL, an anterograde tracer, into the MPOA and vBST of postpartum rats, and the following regions were among those that received terminal input: lateral septum (LS), shell region of the nucleus accumbens (NAs: the NA is divided into a medial shell region and a lateral core region), AHN, VMN, ventral tegmental area (VTA), PAG, and retrorubral field (RRF). This study was followed up by a double-labeled anatomical study, which determined some of the sites to which cFos-expressing MPOA and vBST neurons project during maternal behavior (317). Such MPOA neurons projected to VTA, LS, the medial hypothalamic region (MH) that contains AHN and VMN, and PAG. vBST cFos-expressing neurons projected to MH, PAG, VTA, and RRF. Projections to NAs were not explored in this study (Fig. 11).

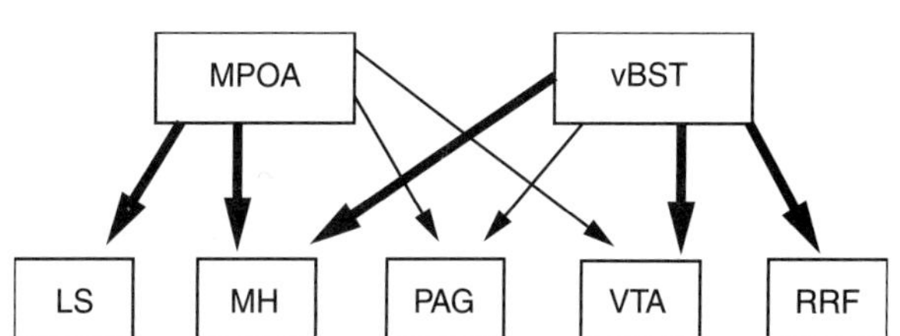

FIG. 11. A double-labeled immunocytochemical procedure was used to explore the neural projection sites of medial preoptic area (MPOA) and ventral bed nucleus of the stria terminalis (vBST) neurons that are active during the occurrence of maternal behavior in primiparous postpartum rats. MPOA neurons which were active during maternal behavior projected to lateral septum (LS), medial hypothalamus centered in the region of the ventromedial nucleus (MH), periaqueductal gray (PAG), and ventral tegmental area (VTA). vBST neurons that were active during maternal behavior projected to MH, PAG, VTA, and retrorubral field (RRF). The strength of each projection (number of cells double-labeled with Fos and a retrograde tracer) is indicated by the thickness of the line and arrow representing a particular projection route. [Reproduced with permission from Numan and Insel (100) (Fig. 5.17B, p. 162). Copyright 2003 by Springer-Verlag.]

MPOA/vBST projections to AHN, VMN, and PAG are intriguing with respect to the suggested role of these latter three regions in the inhibition of maternal behavior. Perhaps, as a result of the modification of MPOA/vBST function by pregnancy hormones (steroids and lactogens), MPOA/vBST acts to inhibit AHN, VMN, and PAG, and this depression of the proposed central aversion system decreases avoidance responses to pup stimuli. In reference to Fig. 4, such an effect would be one element (decreased avoidance of infant-related stimuli) of the dual mechanism that has been proposed as being involved in the onset of maternal behavior at parturition. Such projections may also influence the general reduction of fearfulness that is observed in postpartum females. Research has not yet focussed on testing these ideas, except for the important finding that a substantial proportion of MPOA and vBST neurons that express cFos during maternal behavior also contain glutamate decarboxylase (GAD), the synthetic enzyme for GABA, the brain's major inhibitory neurotransmitter (365).

More research has focussed on the role of MPOA/vBST projections to NAs, VTA, and RRF in the positive control of maternal behavior, and these circuits may underlie the other element in the dual mechanism: increased attraction to infant-related stimuli in the parturient and postpartum female. The VTA and RRF give rise to the mesolimbic dopamine (DA) system, which projects to NA (366–368). MPOA/vBST can influence NA function through a direct projection to NAs, or indirectly through projections to VTA/RRF (317,369). The mesolimbic DA system and the NA have a central role in the control of general motivational processes (370–373). For example, bilateral interference with the normal operation of this system has been found to disrupt brain stimulation reward (374), the rewarding effects of addictive drugs (375), sexual behavior (370,376,377), approach behavior and general locomotor activity (373), food intake (378), and hoarding behavior (373). It is not surprising, therefore, that the mesolimbic DA system and NA are also involved in maternal motivation, and this research will be described below. We propose that the mesolimbic DA system and NA are a nonspecific motivational system that can be accessed by a variety of specific motivational systems. When a particular specific motivational system activates the mesolimbic DA system, the organism shows increased and appropriate behavioral responses to the set of stimuli related to the specific system (100). Therefore, to the extent that a population of MPOA/vBST neurons is specifically related to maternal behavior, the neural connections between MPOA/vBST and NAs, VTA, and RRF may represent an interaction between a specialized and a general motivational system which allows the organism to show increased responsiveness to infant-related stimuli (100).

The evidence in favor of the involvement of the mesolimbic DA system in the maternal behavior of rats can be outlined as follows. Both electrical and 6-hydroxydopamine (6-HD: a neurotoxin that destroys DA-containing neurons) lesions of the VTA disrupt maternal behavior (176,379), as does microinjection of dopamine receptor antagonists or 6-HD into NA, particularly the shell region (380–382). Not surprisingly, maternal interaction with pups is associated with increases in the release of DA into the mother's NA (383,384). Importantly, DA receptor levels in the NA are positively correlated with the intensity of maternal behavior (licking/grooming of pups), while DA transporter levels are negatively correlated with this behavior (383). A consensus of these studies is that decreases in mesolimbic DA function interfere most with the motorically active components of maternal behavior (retrieving), while leaving nursing behavior (motorically inactive) relatively intact. Note that this deficit mirrors the effects of MPOA/vBST damage. Finally, abnormal increases or overactivity in brain DA levels also appear to disrupt maternal behavior (385–387).

There is also evidence that MPOA/vBST exerts an influence on maternal behavior by interacting with the VTA and NA. Using an asymmetrical lesion design (388,389), it was found that a unilateral MPOA lesion paired with a contralateral lesion of the VTA severely disrupted maternal retrieval of pups, while various control lesions were relatively ineffective. Furthermore, both MPOA/vBST and NAs neurons show increased cFos expression during maternal behavior in rats (343,346) and a unilateral excitotoxic amino acid lesion of MPOA/vBST (which does not disrupt maternal behavior) disrupts the expression of Fos in the ipsilateral, but not the contralateral, NAs during mother–infant interactions (346). Therefore, during maternal behavior, the output of MPOA/vBST influences activity in NAs (Fig. 12).

A straightforward hypothesis to explain these results is that the output of MPOA/vBST to VTA/RRF and/or NA drives neural activity in NAs, and that such activity increases the attractiveness of pup-related stimuli, in this way fostering retrieval behavior (100). There are some difficulties with this hypothesis, however. First, the hypothesis would predict that lesions of NAs should disrupt maternal behavior in rats, but the existing evidence is equivocal (206,390–394). Indeed, Numan and colleagues (394) found that bilateral radiofrequency lesions of NAs, performed on day 12 of pregnancy, affected neither the onset of maternal behavior at parturition nor its continuance through day 8 postpartum. Second, one of the major efferent projections of NA is to the ventral pallidum (VP), and this projection is GABAergic (395,396). One might predict that if NA output is essential for maternal behavior, depression of VP activity would not disrupt

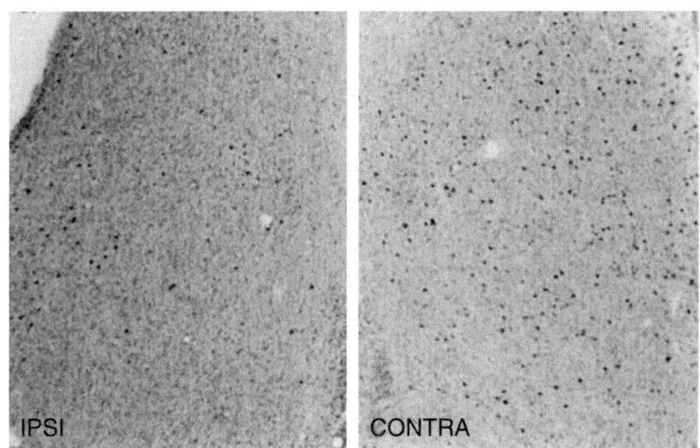

FIG. 12. Photomicrographs showing the expression of cFos within cells of the nucleus accumbens shell region (NAs). Postpartum female rats had received a unilateral lesion of the medial preoptic area and adjoining ventral bed nucleus of the stria terminalis (MPOA/vBST) and continued to show maternal behavior (bilateral lesions are needed to disrupt the behavior). The lesion decreased cFos expression in the ipsilateral (IPSI) NAs, but not in the contralateral (CONTRA) NAs. [Data from Stack et al. (346).]

maternal behavior, and might even enhance it. But the opposite appears to be the case. While depression of NAs activity does not disrupt maternal behavior (394), depression of VP activity severely disrupts the behavior (316,394). Figure 13A shows that the bilateral injection of muscimol (a GABA-A receptor agonist which depresses neural activity) into the medial NA has no effect on retrieval behavior in postpartum rats, while similar injections into VP suppressed the behavior.

Based on these findings, an alternative neural model [influenced by (372,373)] describing the interactions between MPOA, VTA, NA, and VP has been proposed (394; see Fig. 13B): MPOA efferents to NAs and/or VTA may initiate processes that inhibit the activity of the GABAergic output neurons arising from NAs, in this way disinhibiting VP, and the increased output from VP is proposed to increase the attractive properties of pup-related stimuli. Perhaps the increased cFos expression that is observed in NAs during maternal behavior represents increased activity in GABAergic interneurons (395) which function to depress the output of NAs efferents to other brain regions. The importance of VP activity for maternal behavior is significant in light of other evidence suggesting that the VP may play an important role in social motivation (397).

The model shown in Fig. 13B predicts that the linked neural activation of MPOA and VP is important for the normal expression of maternal behavior, and evidence supports this. Using an asymmetrical lesion design, Numan and colleagues (394) found that a unilateral NMDA lesion of MPOA/vBST paired with a contralateral NMDA lesion of VP disrupted maternal behavior in postpartum rats, while control lesions were relatively ineffective (Fig. 14). Importantly, the

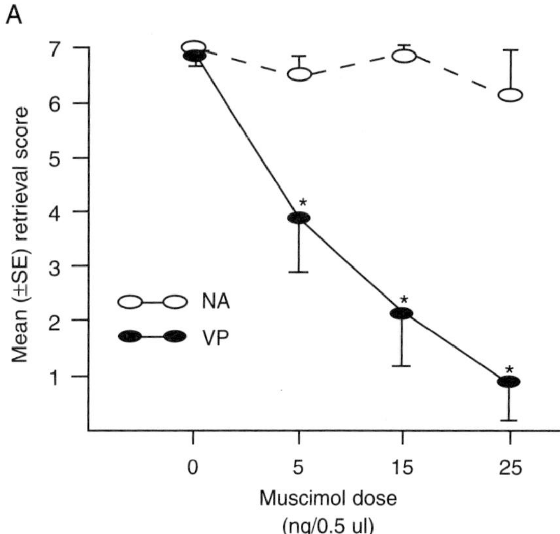

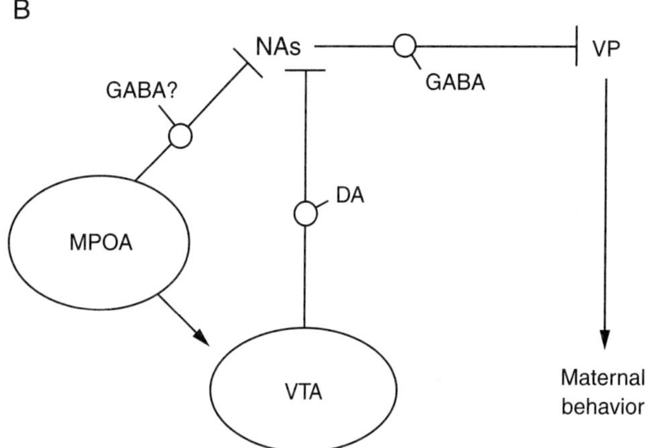

FIG. 13. A: Mean (+ SE) retrieval scores for postpartum rats that received various doses of muscimol into either the medial nucleus accumbens (M-NA) or ventral pallidum (M-VP). Retrieval rating scale: all pups retrieved to nest in 5 min = 7; 10 min = 6; 15 min = 5; 30 min = 4; 45 min = 3; 1 hr = 2; 2 hr = 1; not all pups retrieved and grouped in 2 hr = 0.* Significantly different from corresponding M-NA group. [Data from Numan et al. (394).] **B:** A hypothetical neural model to explain how medial preoptic area (MPOA), nucleus accumbens (NA), and ventral pallidum (VP) may interact in the regulation of maternal behavior. For maternal behavior to occur, the MPOA is shown as inhibiting the shell region of NA, in this way releasing the VP from GABAergic inhibition. MPOA inhibition of NA may occur via two routes: a direct inhibitory connection to NA; an excitatory connection to ventral tegmental area (VTA) which activates inhibitory dopamine (DA) input to NA. In this hypothetical circuit, connections ending in a bar are presumed to be inhibitory, and those ending in an arrow are presumed to be excitatory. [From Numan et al. (394).]

lesions that disrupted maternal behavior left general locomotor activity and hoarding behavior intact.

A Developing Neural Model

Figure 15 presents a hypothetical neural model that incorporates the reviewed data. The circuits nicely map on to the approach-avoidance model of the onset of maternal behavior shown in Fig. 4C. It is proposed

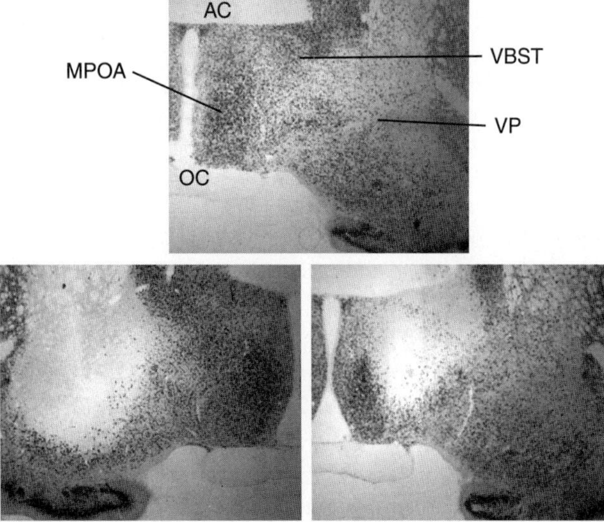

FIG. 14. Frontal sections through the preoptic region immunohistochemically processed with Neu N primary antibody. **Top:** Normal morphology of the medial preoptic area (MPOA), ventral bed nucleus of the stria terminalis (VBST) and ventral pallidum (VP). **Bottom left:** A unilateral NMDA lesion aimed at the VP. **Bottom right:** A unilateral NMDA lesion aimed at the MPOA/VBST region. The Neu N antibody recognizes a neuronal specific protein. If neurons are destroyed, the tissue appears white. AC, anterior commissure; OC, optic chiasm. [From Numan et al. (394).]

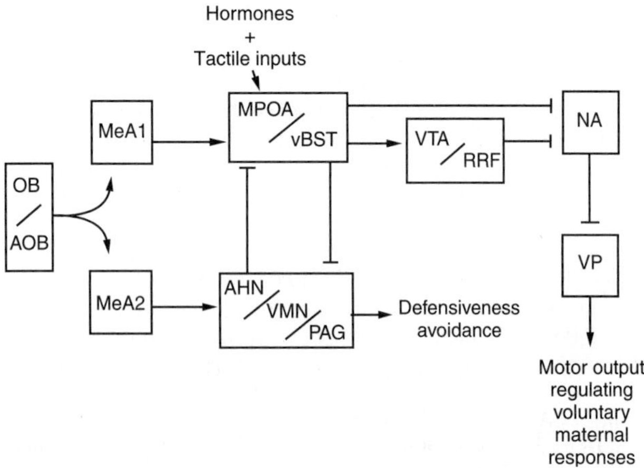

FIG. 15. A composite neural model of the regulation of maternal behavior in rats which combines the models shown in Figs. 10 and 13B. Projection lines ending in arrows signify excitatory connections and those ending in a vertical bar indicate inhibitory connections. Voluntary proactive maternal responses occur when the medial preoptic area (MPOA)/ventral bed nucleus of the stria terminalis (vBST) system is dominant over the anterior hypothalamic nucleus (AHN)/ventromedial nucleus (VMN)/periaqueductal gray (PAG) system. The latter system promotes avoidance of novel pup-related stimuli. See Figs. 7, 10, and 13 for other abbreviations. [Reproduced (with some modifications) with permission from Numan and Insel (100) (Fig. 5.21, p. 184). Copyright 2003 by Springer-Verlag.]

that in a nonhormone primed female, olfactory and MeA projections to AHN/VMN and PAG are dominant, and this promotes avoidance and also directly inhibits the positive neural circuit. As a result of the action of hormones, particularly estradiol and lactogens, at the level of MPOA/vBST, neural circuits become functional which activate maternal interest in pups and suppress the avoidance system. Olfactory (344) and tactile (343,344,398) inputs are shown as driving MPOA/vBST output. As already reviewed, although olfactory input is not essential, tactile inputs may be primary. Much more research needs to be done to test various aspects of this model, and in subsequent sections additional evidence will develop the model further. At this point, note that estradiol and lactogen receptors are not only located in MPOA/vBST, but are also located in many other nuclei as shown in Fig. 15 (323,325,399). It is possible that steroid hormones and lactogens have multiple sites of action to promote maternal responsiveness in parturient females.

Additional Neural Regions Involved in Maternal Behavior

Four other neural regions have been found to be involved in rodent maternal behavior: the PVN, the LS, lateral habenula (LHb), and the PAG.

Electrical lesions of the rat PVN, when performed during pregnancy, disrupt the onset of maternal behavior at parturition (400), but when similar lesions are performed during lactation, after maternal behavior has become established, they have no disruptive effect (322,400). Therefore, PVN neural circuits are important for the onset, but not the maintenance of maternal behavior. Importantly, the PVN is the major source of oxytocinergic neural projections to other brain regions (401,402) and oxytocin receptors are present in many of the brain nuclei that have been shown to play either a positive or negative role in maternal behavior. An interesting hypothesis, which will be developed later, is that PVN oxytocinergic projections to diverse brain regions near the time of parturition coordinate the decrease in avoidance and the increase in attraction with respect to infant stimuli, in this way facilitating the onset of maternal behavior (76,100).

The intermediate part of LS shows increased cFos expression during maternal behavior in rats, and this increased expression is driven by activity in MPOA/vBST neurons, which are known to project to LS (345,346,369). Importantly, LS lesions disrupt maternal behavior in rats, mice, and rabbits (100,403). Septal-lesioned rodent mothers are interested in their pups and they retrieve them. However, the retrieval response is confused, and the mothers persistently carry their pups around the cage. They have difficulty grouping their pups into a single nest site, but instead keep moving them from one location to another. Accordingly, nursing is disrupted because the mothers rarely settle over their pups. Since septal lesions have been associated with increased anxiety and fearfulness, it has been suggested that the constant moving of pups from one location to another is an exaggerated protective response (100,350). Other explanations for the effects of septal lesions on maternal behavior have been offered (100), but if the anxiety hypothesis is correct then it is possible that MPOA/vBST stimulatory projections to LS play a role in the fear reduction that is observed in the postpartum period. Such projections would reinforce the hypothesized inhibitory projections from MPOA/vBST to AHN/VMN and PAG, which were previously suggested as possibly promoting decreased avoidance and fearfulness.

In the section on Inhibitory and Excitatory Neural Circuits Regulating Maternal Behavior, the PAG was included as a site to which AHN/VMN projects to promote withdrawal and avoidance responses (Fig. 10). At present, there is no evidence that PAG damage can facilitate maternal behavior. However, Lonstein and colleagues (343) reported that postpartum rats with lesions of PAG show even lower levels of fearfulness than the typical low levels shown by control postpartum females and, correspondingly, higher levels of maternal aggression. These results support the literature on the PAG and fearfulness and the inverse relationship between fearfulness and maternal aggression. Since the MPOA and PAG are reciprocally interconnected (317,363), perhaps MPOA projections to PAG not only play a role in decreased fearfulness shown by maternal females, but also foster maternal aggression.

Interestingly, the caudal and ventral portion of the PAG is also necessary for the kyphotic nursing posture. Although general care of young (retrieving, nest building, hover-type nursing) in not disrupted by lesions or pharmacological depression of this region, the crouching nursing posture is suppressed, and stimulation of this region promotes kyphosis (261,284, 343,404). The output of certain PAG neurons promotes quiescence and immobility through their descending projections (358,405), and suckling stimulation appears to activate such mechanisms (406).

LHb, which has neural connections with the preoptic area and mesolimbic DA system, has also been shown to be important for rodent maternal behavior, in that lesions to this structure disrupt the behavior (407,408). More work is needed to elucidate LHb's particular role.

Primates. With respect to research on primates, lesions of the prefrontal cortex or the anterior temporal cortex in postpartum rhesus monkeys result in severe maternal behavior deficits (409,410). These findings contrast with the earlier work on rats which found that circumscribed regions of the neocortex were not essential for maternal behavior (411–413). These latter studies found that neocortical damage could disrupt maternal behavior in rats, but the critical factor was not the location of the lesion but its size: large lesions at different neocortical sites were equally disruptive, while small lesions were ineffective. In light of the primate findings, it might be worthwhile to renew an investigation of the prefrontal cortex in maternal behavior of nonprimates [see (414) for data on sheep]. The prefrontal cortex may be significant because it is a recipient of olfactory input and it also receives DA input from the VTA (mesocortical DA system) and projects to the nucleus accumbens (415,416). Importantly, the ventral pallidum, via projections to the mediodorsal thalamus, can influence the function of the prefrontal cortex (417).

Neurochemistry and Molecular Neurobiology of Maternal Behavior

This section will describe the involvement of various neurochemical systems in maternal behavior, and will review studies on transgenic mouse models and the effects of specific null mutations or gene knockouts on maternal behavior. Evidence for a positive role of the mesolimbic DA system in maternal behavior has already been reviewed. We also suggested that MPOA and NAs GABAergic efferents may be involved in maternal behavior. The focus of this section will be on neuropeptides and maternal behavior, with a major emphasis on oxytocin, opioids, and prolactin. See section on Hormonal Determinants of the Onset of Maternal Behavior for a description of changes in these neuropeptides at parturition.

Oxytocin

The PVN is a source of oxytocinergic projections to other nuclei within the brain, where it serves as a neurotransmitter or neuromodulator and regulates maternal behavior.

Rats. Similar to the conclusions derived from the effects of PVN lesions, central oxytocin neural systems are involved in the onset of maternal behavior at parturition in rats, but probably do not play an essential role in the maintenance of the behavior once it has become established [see (56,418,419) for a critical analysis of these issues]. Intracerebroventricular (icv) administration of oxytocin facilitates the onset of maternal behavior in virgin female rats, but this manipulation only works if the females have been pretreated with estradiol (420–424). Also, icv administration of an oxytocin antagonist disrupts the onset of maternal behavior in hormone-primed female rats or naturally parturient females (425–427).

Steroid hormones are essential for central oxytocin to exert positive effects on maternal behavior, which partly explains their role in maternal behavior. Steroid hormones bind to intracellular receptors and serve as transcription factors, in this way either turning on or turning off particular genes which are responsive to the steroid hormone-receptor complex. PVN neurons contain high levels of estrogen receptor beta (ERβ) (428–430) and the oxytocin gene contains estrogen response elements (431), but estradiol alone does not stimulate oxytocin synthesis (432). However, Amico and colleagues (433,434) found that systemic estradiol treatment superimposed on a background of progesterone withdrawal did increase oxytocin mRNA in the PVN. Importantly, estradiol alone, or estradiol and progesterone treatment without progesterone withdrawal was ineffective. Therefore, just those steroidal events that promote maternal behavior in rats are necessary to stimulate oxytocin synthesis in PVN.

In addition to promoting the synthesis of oxytocin, steroid hormones, particularly estradiol, also stimulate the synthesis of oxytocin receptors (the oxytocin receptor is a membrane-bound G-protein–coupled receptor) in select brain regions. Oxytocin receptors in the rat brain are present in many hypothalamic and limbic regions known to be involved in maternal behavior, and in many of these sites the expression of these receptors can be increased by estradiol [reviewed in (100); also see (435,436)]. Research on rats has shown that the hormonal and other physiological events of parturition are associated with increased expression of oxytocin receptors in the following brain regions: MPOA, VTA, the oval nucleus of the bed nucleus of the stria terminalis (BSTov, this nucleus should be distinguished from vBST; BSTov is located in the dorsal BST), and VMN (31,437,438). Note that some brain regions, such as the central nucleus of the amygdala (CNA), contain oxytocin receptors, but estradiol does not affect the expression of these receptors [see (266)].

Pedersen and colleagues (438) have provided evidence that MPOA and VTA are oxytocin-sensitive sites involved in maternal behavior. They found that direct microinjection of an oxytocin receptor antagonist into these regions disrupted the onset of maternal behavior at parturition in rats. The hormonal events of late pregnancy promote oxytocin synthesis in PVN and oxytocin receptor synthesis in MPOA and VTA. Oxytocin action at these sites may then promote the onset of maternal behavior by increasing attraction and decreasing avoidance with respect to infant-related stimuli through neural pathways already described.

In a series of studies, Yu and colleagues (439,440) presented evidence that the main olfactory bulb is another site of oxytocin action. Electrical stimulation of PVN inhibited the neural activity of mitral cells (the output cells of the olfactory bulb), presumably as a result of excitation of granule cells, which are inhibitory interneurons within the bulb. These neurophysiological effects of PVN stimulation were blocked by microinjection of an oxytocin receptor antagonist into the bulb. Also, injection of an oxytocin antagonist into the olfactory bulb disrupted the onset of maternal behavior in parturient rats, while microinjection of oxytocin into the bulb of estradiol-primed virgin female rats stimulated the onset of maternal behavior. These results suggest that oxytocin action directly on the olfactory bulb may affect maternal behavior in rats by altering olfactory function, perhaps decreasing the aversive nature of pup stimuli by affecting a select subset of cells.

It is also possible that oxytocin action at the level of CNA and BSTov is involved in the general reduction of fearfulness that occurs in postpartum female rats and which may be important for the onset of maternal behavior. CNA and BSTov are involved in promoting anxiety-related behaviors (441,442), and both of these nuclei contain CRH neurons (443). Importantly, icv administration of CRH increases fear-related behaviors and decreases maternal behavior (264,444,445). Icv administration of oxytocin, or direct application of oxytocin to CNA, has an anxiolytic effect (218,266,446,447). Finally, CRH mRNA levels in CNA are lower in lactating female rats than in virgin females (130). Perhaps oxytocin action on CNA and BSTov in postpartum female rats decreases fearfulness by inhibiting the synthesis of CRH.

It was argued that fear reduction may be essential for the occurrence of maternal aggression, and this in turn would suggest that oxytocin might potentiate maternal aggression. At present, the evidence favoring a positive role for oxytocin neural systems in maternal aggression is equivocal and inconclusive (128,448–453). Indeed, some studies have suggested that oxytocin plays a negative role in such aggression (453). Note that if it can be established that oxytocin neural systems are involved in postpartum fear reduction and maternal aggression, then there will be a continuing role for oxytocin in maternal-related processes beyond those involved in the onset of maternal behavior at parturition.

Sheep. Vaginal and cervical stimulation (VCS) that occurs at parturition in the ewe is important for the rapid onset of maternal responsiveness. Evidence has shown that VCS operates by activating central oxytocinergic neural systems. Cerebrospinal fluid (CSF) levels of oxytocin increase during parturition and as a result of VCS (72,116), and peridural anesthesia (which blocks afferent feedback from the vagina and cervix) blocks the rise in CSF oxytocin levels in parturient ewes and also disrupts the onset of maternal behavior (73). This disruptive effect of peridural anesthesia on maternal behavior is reversed by icv injections of oxytocin. Importantly, in nonpregnant multiparous steroid-primed ewes, icv injections of oxytocin activate full maternal behavior (positive maternal responses coupled with decreased aggressive responses toward the lamb) (454).

Microdialysis findings show that oxytocin is released within the olfactory bulb, BST, and MPOA during parturition in sheep (78,455). Also, increased expression of oxytocin receptor mRNA occurs at this time in several limbic and hypothalamic sites, which include MPOA, BST, LS, and MeA (83). Interestingly, oxytocin infusions directly into the MPOA of nonpregnant steroid-primed ewes inhibit aggressive responses toward test lambs but do not promote positive maternal responses (117). In contrast, infusions of oxytocin directly into the PVN have been found to promote full maternal behavior: decreased rejection and avoidance and increased acceptance of a test lamb (76). On this basis, it has been proposed that oxytocin acts at several sites to stimulate the onset of full maternal behavior at parturition in sheep and that the fully effective action of oxytocin at the level of the PVN is probably the result of a positive feedback effect of the peptide on PVN oxytocin-producing neurons which stimulates the coordinated release of oxytocin into multiple PVN target sites (76).

Opioids

Rats. There are three major endogenous opioid systems in the brain, β-endorphin, enkephalin, and dynorphin systems, and three classes of G-protein–linked opioid receptors, μ, δ, and κ. It appears that β-endorphin acting on μ receptors in the MPOA inhibits active maternal responses in rats, while unspecified endogenous opioids acting on unspecified opioid receptors in the VTA stimulate maternal behavior. Rubin and Bridges (456) reported that the application of morphine, a nonpeptide opiate agonist, directly to the MPOA of postpartum rats inhibited maternal retrieving of pups. Subsequently, it was shown that both icv injections of β-endorphin or direct application of β-endorphin to the MPOA replicated the inhibitory effect of morphine (457,458). Finally, icv administration of a highly selective μ-opioid agonist also inhibited maternal behavior (458). In support of these findings it has been found that the MPOA contains a high density of opioid axon terminals and receptors (459), that pregnancy hormones are associated with increased endorphin levels and opioid binding density in MPOA (86,460,461) and with increased synthesis of endorphin within the arcuate nucleus (462), the likely source of the endorphin input to the MPOA. Importantly, the enhanced endorphin activity observed in MPOA during pregnancy declines during the postpartum period, presumably releasing an inhibitory brake on maternal responsiveness (460). The fact that the β-endorphin synthesizing cell bodies in the arcuate nucleus of the medial basal hypothalamus are the likely source of β-endorphin input to MPOA is interesting in light of the previous discussion of a maternal inhibitory neural network in the AHN/VMN region, which lies close to the arcuate nucleus.

Interestingly, the rostral PAG may be another site for an inhibitory action of opioids on maternal behavior, since naloxone (an opiate receptor antagonist) injections into this region block the inhibitory effects of systemic morphine injections on maternal behavior (463). Since opiates activate rostral PAG neurons through a process of disinhibition, one possibility is that some aspects of opiate inhibition of maternal behavior are related to increases in avoidance behavior (463).

In contrast to the inhibitory action of opioids at the level of MPOA (and PAG), Thompson and Kristal (464) report that microinjection of morphine into VTA stimulates the onset of maternal behavior in virgin female rats exposed to pups (reduces sensitization latencies), while injection of an opiate antagonist into VTA disrupts the natural onset of maternal behavior in parturient females. The primary neurophysiological effect of opioid peptides is to produce neural inhibition. There is evidence that opiates stimulate DA release into nucleus accumbens as a result of inhibiting VTA interneurons that form inhibitory synapses on VTA DA neurons that project to NA (465,466). It is tempting to speculate that the hormonal and other physiological events associated with pregnancy termination and lactation, through differential actions on particular endogenous opioid systems and their receptors, act to potentiate VTA–opioid interactions while depressing the MPOA–opioid interactions.

Finally, it is important to note that hormonal and other physiological events of pregnancy increase dynorphin levels in the lumbar spinal cord during late pregnancy, and dynorphin action on κ and δ receptors in the spinal cord produces an analgesic effect which might reduce the pain associated with parturition (467). Importantly, Kristal's group has shown that the ingestion of amniotic fluid by parturient rats enhances this late pregnancy-induced analgesia without disrupting maternal behavior (468–470). This research is an excellent example of how opioid systems can be stimulated without inhibiting maternal behavior.

The selective effect on analgesia is dependent upon a potentiation of a particular opioid system in terms of its type, anatomical location, and receptors (μ receptors play no role in pregnancy-induced analgesia, even though they are present in the spinal cord).

Sheep and Primates. With respect to other species, in sheep there is only evidence for a positive role of opioids in maternal behavior, while research on primates has provided evidence for both stimulatory and inhibitory effects (471–476). It is interesting to note, in the context of species comparisons, that in rats opioid systems acting on PVN and SON neurons exert an inhibitory effect on oxytocin release during pregnancy and this inhibition is reduced at the end of pregnancy, facilitating the initiation of parturition (477). In contrast, in sheep, opioid systems appear to promote central oxytocin release in response to vaginocervical stimulation, and this may partly explain their facilitating effects on maternal behavior in this species (131,473).

Additional Neuropeptides

There is evidence that cholecystokinin (CCK) central neural systems are involved in maternal behavior (478–480); CCK's most important effect appears to be its ability to antagonize the inhibitory effects of β-endorphin: the coinfusion of CCK with β-endorphin into the MPOA blocks the disruptive effect of the opioid on maternal behavior in postpartum rats. Further studies on this opioid–CCK interaction, along the lines used by Micevych and Sinchak (481) in their studies of sexual behavior, are likely to yield important results.

Tachykinins, which include substance P, neuropeptide K, and neurokinin A, appear to be involved in neural systems which inhibit maternal behavior. Sheehan and Numan (482) found that microinjection of neuropeptide K into the medial hypothalamus at the level of the VMN disrupted the onset of maternal behavior in hormone-primed female rats, while similar injections into the dorsally located medial thalamus were ineffective. Since a tachykinin neural pathway exists between the MeA and the VMN, it was suggested that neuropeptide K injections into the VMN region activated the "central aversion system" which would normally be down-regulated in hormone-primed females (483).

Molecular Neurobiology and Knock-Out (KO) Mouse Models

Recent developments in molecular neurobiology have provided additional insights into the neurochemistry of maternal behavior. The development of transgenic animal lines with null mutations of certain genes (gene knockouts [KOs]) has provided evidence for the involvement of particular genes and their protein products in maternal behavior. This literature will be selectively reviewed, and the reader is referred to Leckman and Herman (484) and Numan and Insel (100) for more detail.

Some limitations of the gene KO procedure will be discussed before presenting specific findings. First, most transgenic lines are produced in mice, yet most of the work on the neural and neurochemical basis of maternal behavior has been done on rats and sheep. Therefore, species differences need to be considered in any analysis of the results from gene KO studies in mice. Indeed, evidence that does exist suggests that the control of maternal behavior in laboratory mice exhibits some striking differences from that in rats. While anosmia facilitates maternal behavior in rats, it disrupts maternal behavior in mice. Therefore, whenever a gene KO is associated with maternal deficits in mice, careful consideration must be given to the possibility that the KO disrupted the function of the primary olfactory system. Furthermore, laboratory mice are anomalous in comparison to most nonprimate mammals in that most strains of laboratory virgin female mice (which serve as the base from which transgenic strains are derived) are spontaneously maternal. In other words, they have short sensitization latencies in the absence of exposure to the hormonal events associated with pregnancy and pregnancy termination. Because wild (nonlaboratory strains) virgin female mice are not spontaneously maternal, selective breeding and inbreeding used to create the various laboratory strains probably affected the brain organization of these strains so that pup stimuli are not aversive but instead are attractive to virgins. Therefore, if one were interested in gene products that regulate the hormonally induced transition from a nonmaternal to a maternal state, transgenic mouse lines might not be the best models to use.

Other limitations of standard gene KO procedures relate to the fact that the null mutations are typically present in all cells of the body throughout the entire development of the organism. Therefore, if an alteration in maternal behavior were to be observed as a result of a particular gene KO, one would not be able to determine the site where the natural gene product acts to affect maternal behavior, nor be able to determine the time during development when expression of this gene would normally influence mechanisms or structures involved in the behavior. The future use of inducible (temporally and spatially regulated) KOs and intracerebral injections of antisense compounds aimed at specific gene products will ultimately clarify the results obtained from standard KO procedures.

Oxytocin KO Mice. In rats and sheep, oxytocin neural pathways are critically involved in the onset of maternal behavior, and this system, by acting at multiple sites, tilts the balance away from aversion and toward attraction with respect to infant stimuli. In contrast, in transgenic mouse lines with a KO mutation of the oxytocin gene, the onset of maternal behavior at parturition is perfectly normal, although the pups cannot be nourished in spite of normal nursing behavior because the milk-ejection reflex is blocked (485,486). In addition, icv injections of an oxytocin receptor antagonist do not disrupt maternal behavior in these oxytocin KO mice, indicating that some other ligand, such as vasopressin, is not binding to oxytocin receptors and compensating for the loss of endogenous oxytocin (487). The best interpretation of these results is that oxytocin neural systems are not needed for maternal behavior in laboratory mice because (as a result of selective breeding or inbreeding) motivational brain systems are already biased in favor of maternal responsiveness. Importantly, there is some evidence that oxytocin is involved in the onset of maternal behavior in wild (nonlaboratory) mice (488).

It might have been hypothesized that maternal behavior in laboratory mice is dependent upon oxytocin and virgin female mice typically show spontaneous maternal behavior because they have a chronically active brain oxytocin system that does not require stimulation by the physiological events of pregnancy and pregnancy termination. The results reviewed above allow us to discard this hypothesis.

Interestingly, although oxytocin KO mice show normal levels of maternal behavior in standard home cage tests, they show abnormally high levels of anxiety when tested in the elevated plus maze (447). This latter finding conforms to the anxiolytic role of oxytocin. Therefore, perhaps maternal behavior would be disrupted in such KO females if they were tested under more threatening situations. Koch and Ehret (111) have shown that when the maternal behavior of normal virgin female laboratory mice is tested under demanding situations, there is some evidence for the involvement of ovarian steroids. When pups were placed in a novel environment, intact virgin females retrieved pups back to a nest area while ovariectomized females did not. Systemic estradiol treatment restored normal maternal behavior in the ovariectomized females. Since estradiol can influence the synthesis of oxytocin receptors, perhaps oxytocin is involved in facilitating maternal behavior in these mice during stressful situations. Perhaps aspects of these results are related to the finding that estrogen receptor-alpha KO mice show deficits in maternal behavior (decreased retrieval behavior and increased infanticide) (489–491). Finally, in light of these findings and conjectures, it would be interesting to examine maternal aggression in oxytocin deficient mouse strains.

Brain Prolactin Systems and Prolactin KO Mice. Prolactin/lactogen receptors (in rats and mice there are two types of prolactin receptor, a long and a short form), which are membrane bound receptors belonging to the cytokine receptor family, are located in diverse hypothalamic and limbic sites that are known to be involved in maternal behavior regulation, including MPOA, AHN/VMN, and MeA, and the expression of such receptors increases during pregnancy and lactation (55,325,326). Hormonal prolactin from the anterior pituitary and placental lactogens are critically involved in the onset of maternal behavior in rats, and one of their sites of action is the MPOA. Another may be the olfactory bulbs since recent work has shown that prolactin increases neurogenesis in this region (263).

A prolactin or lactogen-like peptide is also produced by neurons in the brain, where it serves as a neurotransmitter/neuromodulator (492–496), and it is the purpose of this section to explore the possible involvement of neural, as opposed to hormonal, prolactin in maternal behavior.

Virgin female mice of a transgenic strain with a KO mutation of the prolactin gene are infertile, but when they are presented with test pups they show spontaneous maternal behavior (497). This finding should not be surprising, since pregnancy hormones are not critical for maternal behavior in laboratory mice and because hypophysectomy has no effect on maternal behavior in these strains (498). Surprisingly, however, Lucas and colleagues (154) have reported deficits in virgin females with a KO mutation of the prolactin receptor gene. Virgin female laboratory mice were exposed to test pups for 30 minutes per day and the onset of full maternal behavior was recorded. Normal females (+/+) showed maternal behavior after 2 days of pup exposure, while mice with a KO mutation of the prolactin receptor gene (–/–) were not fully maternal even after 6 days of pup exposure, at which time testing was terminated. Heterozygous females (+/–) were intermediate in their responsiveness, showing maternal behavior in 4 days (499). Since the mice were only exposed to pups for 30 minutes per day, it is not clear whether full maternal behavior would have occurred in the KO females if pup exposure had been more prolonged. Importantly, the homozygous KO females had normal spatial and olfactory abilities, attesting to the relative specificity of the behavioral effects.

How can we reconcile the facts that a prolactin KO does not disrupt maternal behavior in mice, while a prolactin receptor KO does? The most likely possibility is that some other ligand of central nervous system origin, in addition to prolactin, is capable of binding to the prolactin receptor in virgin females, in this way fostering maternal responsiveness. The search for this ligand and the neural pathways within which it is contained may shed important light on the neural processes that promote the relatively spontaneous maternal responsiveness that is observed in normal virgin laboratory female mice. To validate the importance of the prolactin receptor for maternal behavior in mice, it would be important to examine the effects of central administration of a recently developed prolactin receptor antagonist (S179D-PRL), which has been shown to disrupt the hormonal onset of maternal behavior in rats (333).

Although hormonal prolactin is necessary for the onset of maternal behavior at parturition in rats, it is not necessary for the maintenance of maternal behavior once it has been initiated: hypophysectomy or other treatments which prevent anterior pituitary prolactin release disrupt lactation but leave maternal behavior intact during the postpartum period. However, there is some evidence that a brain prolactin system may participate in the postpartum regulation of maternal behavior in rats under certain conditions. Torner and colleagues (265) report that when prolactin receptor antisense oligonucleotides are administered icv for 4 days to postpartum rats, maternal retrieval latencies are increased and this is associated with increased fearfulness as measured in the elevated plus maze [see also (500,501)]. Note that the maternal behavior of these females was tested in a novel cage, rather than the female's home cage.

Catecholamines and Dopamine β-hydroxylase (DBH) KO Mice. In the 1970s, preliminary work from Moltz's laboratory suggested that a norepinephrine system was involved in the hormonal onset of maternal behavior at parturition in rats, but was not necessary for the maintenance of the behavior after its initiation (502,503). Recent findings suggest that these ideas deserve reconsideration. Thomas and Palmiter (504) produced a transgenic mouse line with a KO mutation of the DBH gene. DBH is an enzyme that is necessary for norepinephrine (and epinephrine) synthesis, and although such mice were normal on a variety of cognitive and sensory tests (including olfactory sensitivity tests), they showed a severe disruption of maternal behavior at parturition. Additionally, virgin DBH KO females did not show spontaneous maternal behavior toward test pups while virgin females of the normal strains from which the DBH KO strain was derived did show short latency sensitized maternal behavior. Interestingly, although maternal responsiveness toward pups did not occur at parturition in the DBH KO females, these mice were capable of showing maternal behavior after a period of prolonged cohabitation with pups. It is as if the DBH KO made these mice act like rats (prolonged sensitization latencies), while at the same time preventing the stimulatory effects of pregnancy hormones that are typically observed in rats. An interesting hypothesis to explain these data is that a norepinephrine system mediates the hormonally induced transition from a nonmaternal to a maternal state in rats, and virgin laboratory mice show spontaneous maternal behavior because the hypothesized norepinephrine system is active without a hormonal trigger. In a later section we will discuss the involvement of norepinephrine systems in what has come to be called maternal memory: the effects of interactions with young on subsequent maternal behavior. A relationship between the possible role of norepinephrine in the short-latency onset of maternal behavior at parturition and its involvement in maternal memory deserves consideration.

More work needs to be done on norepinephrine's possible role in maternal behavior. A DBH KO would not only affect central norepinephrine (and epinephrine) systems, but would also affect the functioning of the autonomic nervous system. Comprehensive neuroanatomical, pharmacological, and selective lesioning and stimulation procedures are needed to study particular norepinephrine systems before we can reach any conclusions about the underlying mechanisms and the degree of specificity of norepinephrine's involvement in maternal behavior. Since brainstem norepinephrine input to the olfactory bulb modulates olfactory processing (298), and since olfaction is involved in the maternal behavior of both rats and mice, norepinephrine input to the bulb (from the locus coeruleus) may be a possible circuit to consider. Norepinephrine input to the MPOA/BST should also be considered (320).

Fos B KO Mice. Brown and colleagues (505) have reported severe deficits in the maternal behavior of both virgin female mice and postpartum female mice in a strain with a KO mutation of the Fos B gene. Pregnancy proceeded normally in these females and hormone levels (gonadal steroids, prolactin, growth hormone) and mammary gland development were normal. However, the postpartum females failed to retrieve and nurse their own young or foster replacement young. Additional tests showed that these KO females had normal spatial abilities, spatial memory, and olfactory function.

Brown and colleagues (505) also noted that Fos B increases in the MPOA of normal female mice during maternal behavior, and this conforms to the work done on rats (506). Brown and colleagues (505) suggest that Fos B in MPOA may be crucial for maternal care. In rats, it has also been argued that Fos proteins may be maintaining the functional integrity of MPOA neurons during episodes of maternal behavior since Fos expression (cFos and Fos B) remains elevated for as long as a female is nurturing her young (345). Of course, in order to provide more definitive evidence for this hypothesis, the effects on maternal behavior of a selective down-regulation of cFos and/or Fos B synthesis within MPOA, through the use of antisense technology, for example, would have to be examined.

If it can be confirmed that Fos proteins, as transcription factors, are maintaining the function of MPOA neurons during maternal behavior, then we will have an important tool to analyze the molecular neurobiology of MPOA systems relevant to maternal behavior. If we can determine the extracellular factors that activate fos genes within MPOA, and if we can also uncover the late-responding genes that are affected by Fos proteins within MPOA, we will be taking important steps toward unraveling some of the chemical neuroanatomy of preoptic neural circuits regulating maternal behavior. At this point, next to nothing is known about the actual neurotransmitters used in MPOA efferent circuits that regulate maternal behavior.

Additional Transgenic Mouse Lines. Two genomically imprinted genes that are paternally expressed (Pegl, also referred to as Mest, and Peg 3) are involved in maternal behavior (507–509). In these genes, it is only the allele inherited from the father that is expressed in offspring (male and female) during development and in adulthood, whereas the maternal allele is silent. A KO mutation of the paternally inherited Peg 1 or Peg 3 allele is associated with deficits in the maternal behavior (retrieving, nursing, and nest building) of female mice that carry these mutations, although pregnancy and parturition are normal. Virgin female mice with these mutations also exhibit a major disruption in maternal behavior. A disturbance in olfactory sensitivity is not a likely cause of these deficits. The specific functions of these genes are not known, and it is also not known whether they are producing their effects on maternal behavior as a result of actions early in development (since these genes have developmental effects) or in adulthood. These genes are expressed in the hypothalamus and limbic system of normal adult mice, which includes expression in MPOA and MeA.

The work reviewed up to this point has described the effects of selective gene deletions on pup-directed maternal behavior (retrieving and nursing) in mice. A recent study has examined both maternal behavior and maternal aggression. Gammie and Nelson (510) report that a null mutation of the neuronal nitric oxide synthase (nNOS) gene disrupts maternal aggression without disrupting pup-directed maternal behavior in this transgenic mouse line. nNOS is the enzyme necessary for the synthesis of the gaseous intracellular messenger and intercellular neurotransmitter/ neuromodulator nitric oxide (NO) (511). NOS converts L-arginine and O_2 into NO and L-citrulline, and the immunohistochemical detection of citrulline within neurons has been taken as evidence for NO production. Gammie and Nelson (510) also reported that in normal laboratory mice, L-citrulline positive neurons increased in the MPOA and in the hypothalamic region below the PVN during bouts of maternal aggression. These results suggest that NO may facilitate maternal aggression by acting within these neural systems. In prairie voles, Gammie and Nelson (512) report that elevated citrulline immunoreactivity occurs selectively within PVN neurons during maternal aggression in postpartum females. Since NO is involved in the control of oxytocin release (513), it is possible that NO influences on maternal aggression are mediated via influences on oxytocin pathways.

Interestingly, a recent study has reported that microinjection of a NOS inhibitor directly in the MPOA of postpartum rats disrupts maternal retrieving of pups (514). This result contrasts with the finding that a deletion of the nNOS gene disrupts maternal aggression but leaves retrieval behavior intact in postpartum mice. Although species differences might exist, it is also possible that compensatory mechanisms occur in MPOA neurons regulating pup care behavior when nNOS is deleted throughout development in the transgenic mouse strain, which contrasts with the acute disruption of NOS activity in rats that was produced in the Popeski and colleagues (514) study. Importantly, Dominguez and colleagues (515) have reported that NOS and NO within the MPOA play positive roles in the regulation of male sexual behavior in rats. They present evidence that this effect is mediated by the facilitation of DA release within the MPOA by NO. Since the hypothalamic regulation of maternal behavior and male sexual behavior in rats share many similarities (314), it might be that similar NO–DA interactions within MPOA are involved in facilitating retrieval behavior. There is some evidence that DA release within MPOA may be involved in the maternal behavior of rats (387,516,517). The source of such DA input would be from the periventricular and incerto-hypothalamic DA systems.

Conclusions

Our understanding of the neural basis of maternal behavior is relatively undeveloped and limited to a few species. Although more work needs to be done, we have outlined our current understanding of the neural circuits, neural sites of hormone action, neurotransmitters, and genes which are involved in maternal behavior. In the next section we will explore how experiential factors operate on these defined neural systems so that maternal behavior and its underlying neural circuits can be modified. Although most of the findings have been obtained from research on rodents and sheep, we believe that we are beginning to define core neural systems underlying maternal behavior in mammals (Fig. 16). Hopefully, this research will develop further and become relevant to

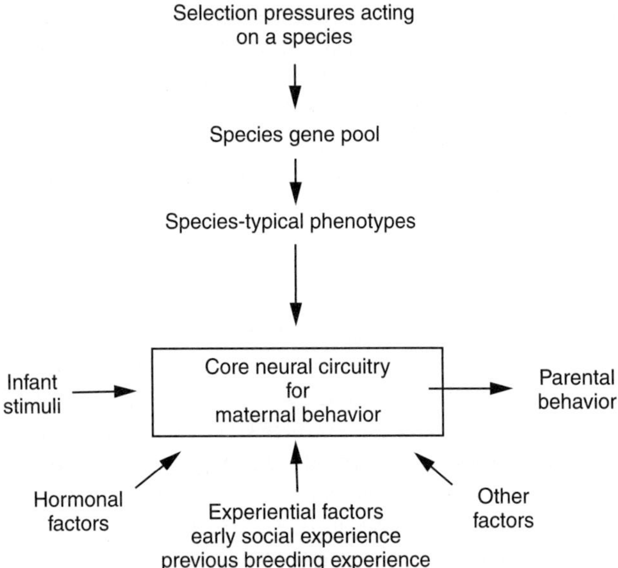

FIG. 16. A diagrammatic representation of how species differences in mammalian maternal behavior may have evolved. We show a core neural circuitry, made up of the phylogenetically older parts of the brain (limbic system, hypothalamus), which regulates parental responsiveness to infant stimuli. The operation of this core substrate, in turn, can be influenced by genetic, hormonal, experiential, and other factors. We argue that the core neural circuitry which regulates maternal behavior may be highly conserved across mammalian species. However, as a result of different species being affected by differential natural selection pressures, what might vary among species is the degree to which this circuitry is dependent on hormonal facilitation, and whether, and to what degree, this neural network can be influenced by experiential factors during ontogeny. [Reproduced with permission from Numan and Insel (100) (Fig. 2.1, p. 10). Copyright 2003 by Springer-Verlag.]

an understanding of the neural dysfunctions that occur during cases of child abuse or neglect and postpartum depression in humans. Importantly, several recent neuroimaging studies in humans have indicated that many of the brain regions that contribute to maternal behavior in rats are activated during the occurrence of normal maternal behavior in human mothers (518,519).

EXPERIENTIAL INFLUENCES ON MATERNAL BEHAVIOR

In this section the effects of caregiving experience on the expression of postpartum maternal behavior are considered. Experiences that affect how a mother interacts with her young occur throughout the life of the animal, during early life when the neonate and infant are being cared for themselves, during the juvenile and adolescent periods when young animals have the opportunity to interact with their younger siblings, and as adult, after the birth of their offspring. For each type of experience, we will discuss the phenomenology of experience and its sensory, neuroanatomical, and neurochemical foundations. We explore how experiences earlier in life affect the development of the neurobiology that underlies the final expression of maternal behavior during the postpartum period.

Maternal Experience and Parity Effects in Adult Females

Rats

First time mother rats, hamsters, and voles (raised in the laboratory with optimal food, housing, and environmental conditions) are usually maternal at parturition. Differences in behavior that occur between first-time (primiparous) and multiparous mothers are often subtle or difficult to detect, and they may reflect not only differences in experience, but also differences in the mother's maturity, age, or lactational competence (100,109,411,520,521). Mothers may show differences in the intensity of some behaviors, coordination of the different components, or in the speed and efficiency with which behaviors, like retrieval, are expressed (109,521,522).

In the postpartum mother (when hormonal influences are present) the effects of maternal experience and parity on maternal behavior are most evident when the mother is challenged in some way. Multiparous rats are less disturbed than primiparous mothers by a variety of experimental manipulations including C-section (523), endocrine manipulations (524,525), morphine administration (29,526), and brain lesions (50,187,527,528). In addition, maternal experience can affect the threshold of activation of behavior in response to pups and/or their cues. Pup cues which are initially ineffective in eliciting maternal behavior in first-time mothers, may become effective in multiparous animals (9).

The effects of hormone administration also vary as a function of parity (nulliparous versus primiparous versus multiparous). In rat strains where nulliparous females take a long time to become maternally sensitized (Sprague-Dawley rats require 8–10 days), the Bridges (104) hormone regimen of progesterone (P) and estrogen (E) produces more rapid sensitization of maternal behavior in multiparous, than in nulliparous, rats (A. S. Fleming, personal communication). Moreover, P administration toward the end of pregnancy, when P would normally decline, blocks the onset of maternal behavior in primiparous females but not in multiparous females (524,525); it seems that progesterone decline is necessary for the onset of the behavior in the first time mother, but not in the experienced mother. Hence, the multiparous rodent central nervous system (CNS) is both more sensitive to effects of parturitional hormones and less dependent on those hormones to express maternal responses to pup cues.

Sheep

In contrast to multiparous mothers, first-time mother sheep are more disturbed by parturition and take longer to accept their newborn lambs. As a result, access to the udder is delayed, which reduces the chance of lamb survival (529). As in rodents, parity influences the ewe's responsiveness to some of the physiological determinants of maternal behavior. In contrast to their effects in multiparous ewes, in inexperienced nongestant females, neither steroid priming nor vaginocervical stimulation is effective in inducing maternal responsiveness (131,115,530,531). In addition, neither oxytocin, opiates nor CRH treatment induces maternal behavior in steroid-primed nulliparous ewes as it does in multiparous females (131). Finally, deprivation of vaginocervical stimulation by means of epidural anesthesia induces greater deficits in maternal responding in primiparous, as opposed to multiparous, females (146).

Primates

In monkeys, the prior experience of giving birth, nursing and caring for own infant, clearly affects subsequent maternal adequacy and quality of maternal behavior. There is a higher incidence of infant rejection, neglect, abandonment, and resulting infant mortality in primiparous versus multiparous female marmosets (Callithrix species), tamarins (Saguinus species) and pigtail, Japanese and rhesus macaques (*Macaca nemestrina*, *fuscata*, and *mulatta*) (32,532–536). On the other hand, first-time vervet monkey (*Cercopithecus aethiops*) mothers are more likely than multiparous mothers to restrain their infants and remain in close contact with them (537,538). As found for sheep, in some monkeys parity effects are accompanied by differential effectiveness of hormones in the facilitation of maternal behavior. In red-bellied tamarins the effects of exogenous hormone replacement are only apparent in previously experienced, multiparous animals (15).

Humans

In humans, prior maternal experience is the single most reliable predictor of positive responsiveness to offspring (539,540). In comparison to primiparous mothers, during the initial postpartum period multiparous mothers are more attracted to the body odors of newborn infants even though they are not better at discriminating their own infants' odors from another infant (541). They respond more rapidly and sympathetically to the cry of their own infant (542) and more sympathetically to pain than to hunger cries (135). They also show greater heart-rate accelerations to the cry of their own infant (543) but also show lower arousal responses, suggesting a lower overall arousal state and reduced anxiety (544).

By 3 months postpartum, when primiparous mothers have become familiar with their infants, parity differences seem to be reversed. Primiparous mothers expressed more affectionate affect, showed a higher level of reciprocal interactions and provided more infant-directed vocalization and stimulation than multiparous mothers (545). These relationships between maternal experience and maternal attitudes and behavior may be mediated by mother's mood, since prior maternal experience is also a strong predictor of postpartum affect (23,233).

In conclusion, parity appears to alter the nervous system so that hormones and/or stimuli have an effect that they would not have had otherwise. It is not clear whether parity differences reflect experiences acquired while caring for young or their cues, or other factors like the mother's age, level of physiological maturation, endocrine/lactational status, or other social and psychological developmental factors (546).

Pup Experience and Female Allomothering

In some species, juveniles or sub-adults care for younger siblings. They provide stimulation and warmth to their siblings but may also gain experiences that "prepare" them for later caregiving as adults. Sibling allomothering may also enhance an animal's fitness (546a). Across species, the extent to which juveniles care for younger siblings depends on the structure of social groups.

Rats

As indicated earlier, older siblings sometimes have the opportunity to interact with their younger siblings in the nest and have the opportunity to gain pup-related experiences in the context of "play" or the exhibition of maternal retrievals, licking, and crouching over pups. Juvenile caregiving experience can have long lasting effects on sensitization latencies in adult nullipara, in both rats (185,547) and voles (365). Adult nullipara become maternal more quickly in response to exposure to foster pups if they have had prior exposure to pups during the juvenile period. However, juvenile caregiving experience in rats seems to have little effect on the quality of mothering exhibited during the postpartum period, under hormonal priming (183). Preweaning experiences, however, set the stage for effects of later juvenile experiences. If juvenile rats are reared with exposure to pups, or with a same age conspecific, they show higher levels of licking of offspring when they give birth than if they were isolated during this juvenile period, but only if they were initially reared by their own mothers, rather than experiencing repeated separations (see section on Maternal Deprivation and Isolation Studies). This interactive effect of preweaning and juvenile experiences on later maternal behavior also surfaces when pups are repeatedly separated from their mothers; juvenile "play" experience with pups can actually have negative effects on adult postpartum maternal behavior (182).

Sheep

There is no evidence among sheep that allomothering occurs nor that exposure to younger siblings in the matriarchical group in any way enhances maternal responsiveness later on. However, this has not been studied.

Primates

In new world monkeys, carrying by older siblings of younger siblings enhances survival of subsequent offspring (15). In marmoset social groups, composed of parents and older siblings, the infants spend the majority of time during the first 2 weeks being carried by the father (40%) and mother (45%), with the remaining time (15%) accounted for by carrying by older siblings and juveniles (548,549). Over the next month carrying time for the infants declines to about 10% with an increase in play and affiliative social interactions of the infants with each other and their older siblings (548,549). Hence, juveniles and young adults usually have both caregiving and play experiences with their younger siblings.

Practically speaking, successful establishment of captive bred marmosets requires that offspring have experience of rearing their younger siblings before being removed from the family for breeding purposes (535,550,551). Parents without this early experience will often kill or neglect their infants (121,552,553). In contrast to the marmosets, among rhesus monkeys most laboratory rearing environments do not provide developing females with experience with younger siblings or neonates per se, a practice which apparently has little effect on rhesus monkey maternal responsiveness to first born offspring (225).

Humans

There are no prospective longitudinal studies that explore the effects of experience "babysitting" or caring for younger siblings on the quality of maternal behavior shown by first-time mothers. Retrospective studies indicate that primiparous mothers with more babysitting experience report stronger attachments to their infants, display more maternal self confidence, and show a higher level of affectionate contact behaviors with their infants (178,554). This is particularly true of teen mothers, for whom prior experience seems to have marked effects (243).

Postpartum Maternal Experience

Rats

The mechanisms that underlie the enhancement of maternal behavior by earlier experience with pups can be identified to some extent (555–558). Once the parturitional hormones have exerted their effects and maternal behavior has been displayed during the postpartum period, hormonal effects of parturition subside within 3 to 5 days. Then a different set of neuroendocrine and brain mechanisms are activated. These mechanisms maintain responsiveness throughout the lactational period. The latter mechanisms are the ones that most likely overlap with parity effects (124).

As shown in Fig. 17 to identify the psychobiological mechanisms underlying the parity/maternal experience, a paradigm was developed in which primiparous mother rats are:

- Restricted from contact with their offspring at parturition. (The dam is caesarian sectioned or

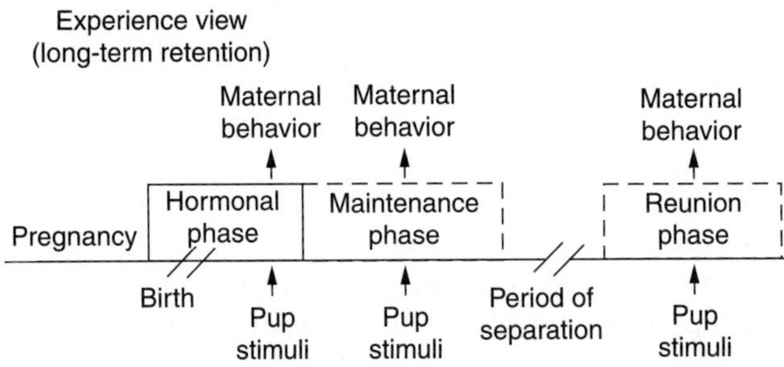

FIG. 17. One view of the mechanisms underlying the maintenance of maternal behavior in rats after its hormonally induced initiation. In the experience view (long-term retention), experience with pups and the display of the maternal behavior during the hormonal phase permanently modifies the brain so that even after a prolonged period of mother–infant separation, the brain does not revert back to a naive state, but instead pup stimuli retain the ability to have relatively immediate access to the neural circuitry regulating maternal behavior. Research on rats supports the long-term retention view of the maintenance of maternal behavior. Adapted from Numan and Insel (100).

pups are removed as they emerge from the birth canal.)

- Provided with varying amounts and kinds of contact with neonatal pups immediately after birth.
- Separated from pups for varying periods.
- Tested with foster pups to assess the retention of their maternal responsiveness.

Using this paradigm, Bridges and colleagues (555–557) and Fleming and colleagues (558) demonstrated that if pups are removed from mothers at the time of parturition, and mothers are then tested for behavior with foster pups at various postpartum times, the responsiveness to test pups wanes over the first 5 postpartum days. After a 10-day separation from young, mothers respond to pups with the same latency normally seen in virgin nulliparous animals. Experiences of pregnancy and parturition in the absence of pups do not enhance later responsiveness to pups (Fig. 18A).

In contrast, if mothers interact with their young for a brief period during the period of postpartum hormonal priming (for 1/2 to 24 hr during the first 24 hr), they exhibit a reduced latency to retrieve and crouch over foster pups for up to 3 weeks or more (555–557). The temporal parameters of the effectiveness of the experience in modifying later behavior are similar to other forms of learning, that is, (a) the longer the experience duration, the longer the memory effects are sustained; and, (b) the longer the interval between experience and test, the poorer the retention (559) (Fig. 18B). This long-term change in behavior as a result of experience interacting with pups is referred to as the maternal experience effect (MEE). The MEE also occurs in rabbits (99,560).

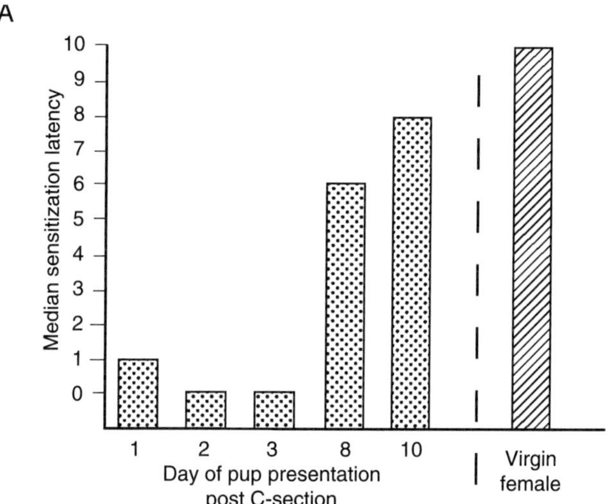

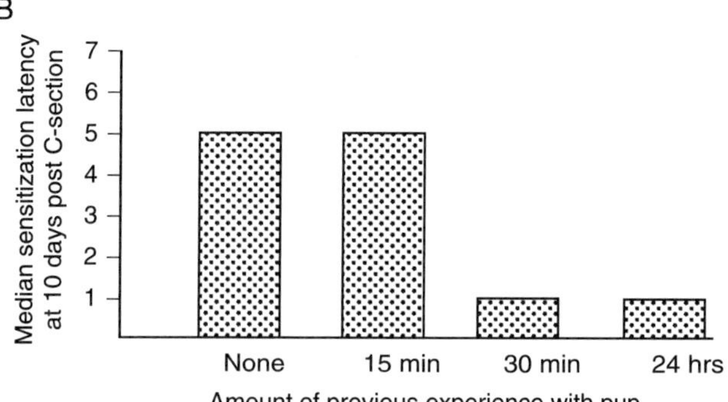

FIG. 18. A: Median sensitization latencies to onset of maternal behavior in primigravid female rats who were caesarian-sectioned on day 22 of pregnancy and then presented with pups for the first time at either 1, 2, 3, 8, or 10 days postcaesarian section. For comparison, the latency to onset of maternal behavior in virgin females is also included. [Modified from Orpen and Fleming (558); from Numan and Insel (100).] **B:** Median latencies to onset of maternal behavior at 10 days postsurgery for primigravid female rats that were caesarian sectioned on day 22 of pregnancy and allowed either no exposure to pups or 15 minutes, 30 minutes, or 24 hours' exposure to pups at 24-hours postsurgery. Maternal behavior was shown toward pups at 24 hours post-caesarian section. [Modified from Orpen and Fleming (558); from Numan and Insel (100).]

Sheep

In sheep, an increase of exposure period to young does not result in a longer retention of maternal responsiveness. Thus, 4 hours of immediate postpartum contact is sufficient to maintain maternal responsiveness following separations for up to 24 hours, but not 36 hours. Increasing the duration of mother–young contact to 7 days does not significantly restore maternal responsiveness under the 36-hour separation conditions (Fig. 19A) (147,561). Therefore, maternal responsiveness declines gradually with time of separation and independently of previous exposure duration. In contrast, with increased exposure to own lambs, the ability for individual lamb recognition can be retained for a longer period (147,561). A short exposure to the lamb for 4 hours just after birth produces good selectivity; however, after a 36-hour mother–lamb separation, this selectivity is not sustained (147,561). If ewes interact with their lambs for 7 days, then selectivity is sustained following a 36-hour separation in maternal ewes but not following a 72-hour separation (Fig. 19B). Unlike the experience based long-term retention of maternal responsiveness in rats, in sheep, the recognition memory for own lamb does not persist for very long, after separation from the lamb (561).

Nonhuman Primates

The effects of MEE has not really been studied adequately in primates. However, Harlow and colleagues (562) cite an example in which a isolation-reared

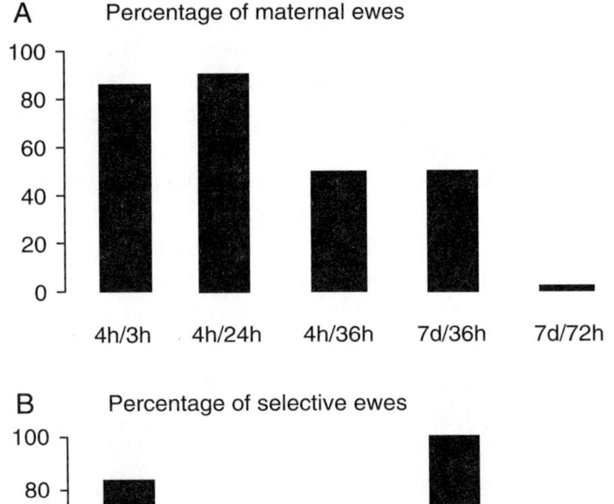

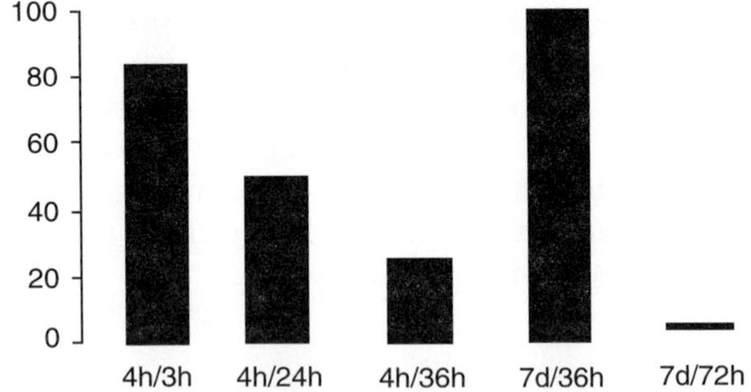

FIG. 19. A: Percentage of ewes showing maternal responsiveness after various mother–young contact (4 hr or 7 days) and separation (3, 24, 36, and 72 hr) durations. 4 hr/3 hr group: 4-hr of contact followed by 3-hr separation/4 hr/24-hr group: 4 hr of contact followed by a 24hr separation period and so on. **B:** Percentage of ewes showing individual lamb recognition after various mother–young contact (4 hr or 7 days) and separation (3, 24, 36, and 72 hr) durations. 4 hr/3 hr group: 4 hr of contact followed by 3 hr-separation/4 hr/24-hr group: 4 hr of contact followed by a 24-hr separation period and so on.

"motherless" mother monkey (see section on Maternal Deprivation and Isolation Studies), having crushed and mangled her first born, "astoundingly" (with no further experience with offspring or other infants) became an adequate mother to her second and third born (562). In this case—which is not unique—it appears that the positive/adequate response of the rhesus monkey mother to the second born was not directly dependent on the outcome of a maternal interactive experience with the first born.

That said, the best predictor of whether a rejecting primiparous "motherless" rhesus monkey mother will adequately care for subsequent offspring is how much time she spent with the fetus/neonate postpartum. Thus, if a mother was exposed to her (live) infant for at least its first 2 days (48 hours) of life, regardless of outcome (whether the infant survived or died) then the prognosis for care for subsequent offspring by the mother is excellent (563). On the other hand, if the duration of maternal exposure to the neonate was less than 24 hours then the probability that subsequent offspring will be rejected is high (563).

These effects suggest that the stimuli provided by the birth process or the neonate if these can be sustained for a period exceeding 48 hours affect the responsiveness of the mother to subsequent deliveries. Exposure to these stimuli may have effects on sensory processing, and neuroendocrine, and neurotransmitter systems which may reduce fear and aversion, and increase recognition and attraction to comparable cues associated with second born offspring. Alternately, it may be that mothers that endure a longer period of infant contact with the first birth are also the more responsive animals, reflected in enhanced performance after the second birth. Evidence from vervets (537) certainly suggests a personality dimension at work, where animals that are maternally protective as a first-time mother, will also be more protective as a multiparous mother.

Humans

There is no substantial evidence that there is a sensitive period for mother–infant attachment in humans (564). Mothers who have earlier postpartum interactions with their infants are more affectionate to them during interactions on day 2 postpartum (565). They also feel more confident and practiced. Clearly, as mothers gain experience with their infants over the first few postpartum months, they feel increasingly attached to them; they express more positive attitudes, become more efficient at feeding, and become more attuned to a variety of infant cues (545).

Hormonal Effects on Robustness of Maternal Experience

Rats

Virgin rats that become maternal through sensitization procedures show a maternal experience effect (R. S. Bridges, personal communication). Hence, the parturitional hormones are not necessary for maternal experiences to have long-term effects. However, hormones do influence the robustness of the effects of a maternal experience. Rats which acquire maternal experience under the influence of the parturitional hormones exhibit better retention of maternal behavior 30 days later than do animals that are not being stimulated with hormones at the time of the maternal experience (559). The optimal condition for the expression of maternal behavior occurs, however, when both the initial experience and the test occur during a period of hormonal priming (559,566).

Adrenal hormones are also implicated in the formation of maternal experience and in regulating the amount of licking mother rats give to their offspring. However, effects of HPA axis hormones (corticosterone in the rat) are bimodal, depending on the "reason" for their elevation and the context in which HPA system is being activated. Removal of the adrenal glands during late pregnancy reduces the MEE. Corticosterone replacement blocks the effect of adrenalectomy (567). Whether the effects are due to reduced glucocorticoid receptor activation, effects on CRH and ACTH release, or on acquisition, consolidation, or the retrieval processes and/or attention, is not known. A most striking fact is that the hormonally based postpartum learning effect is not restricted to pups or to the maternal context. In comparison to nulliparous rats or nonmothers, new mother rats also show enhanced olfactory-based social learning (568) as well as spatial learning (569).

Taken together, these studies indicate that the effect of experiences is enhanced and later behavior more easily activated by a combination of hormones and exposure to relevant pup stimuli. Hormones exert their effects on experience by altering multiple behavioral systems, including percepts, affect, attention, and memory processes that in turn influence the subsequent expression of maternal behavior (see below).

Sensory and Psychological Features of Maternal Experience

Rats

Interactive experience with pups in the context of the peripartum hormonal environment enhances the long-term retention of maternal behavior. Both somatosensory and chemosensory inputs are important. If the mother is prevented from crouching over her young her responsiveness to them declines more rapidly than is usual (570,571). If mothers were separated from their litters by wire mesh during the 1-hour postpartum exposure phase, so that they could see, hear, and smell pups, but receive no tactile or proximal chemosensory input, they showed no MEE (558). Instead, they responded to pups as nullipara would. Engaging in maternal crouching and stimulation by moving and suckling pups are probably essential features of the maternal experience in rats (288). Dams also need to receive input from the mouth region to exhibit normal maternal licking and crouching (1,572). Licking during exposure, and pup stimulation of the maternal ventrum (during suckling) are factors in subsequent maternal responsiveness (288). Either perioral or ventral trunk stimulation is sufficient for an MEE to occur; however, when both inputs were eliminated, by desensitization procedures, the MEE was eliminated (288).

In the rat olfaction is not necessary for the long-term effects of a maternal experience (274). Removal of the olfactory bulbs or of the accessory olfactory system prior to a maternal experience does not prevent the development of an experience effect, although it does reduce it somewhat (274). Nevertheless, during interactions with pups, dams learn about specific olfactory and chemosensory features of the pups. If pups are scented with an artificial odorant, mother rats respond more rapidly to pups labeled with that scent than to those labeled with a discrepant scent. Furthermore, exposure to the odor on its own (in the absence of pups) does not increase responsiveness to scented pups (573). Hence the association is related to a combination of changes in sensory processing that occurs at parturition (see section on Motivational Analysis of Maternal Behavior).

Individual Recognition

Sheep

In sheep, the cues that are learned from the lamb for the mother to remain maternal after being exposed to a lamb are clearly olfactory (112,115). In two different experiments, lambs were partially separated from their mothers at birth so that only some cues were available. Mothers were tested for retention of their maternal responses after either 8 hours or 12 hours of partial separation. Deprivation of tactile stimulation from the lamb led to a minor disturbance of maternal responsiveness. On the other hand, placing the lamb in an odor-proof box with a transparent side led to performances no better than those of mothers deprived of their neonates. In addition, when lambs were placed inside a box with a ventilation system allowing only the smell of the lamb to return to their mothers, the rate of maternal acceptance was greater than when only the visual and auditory cues were provided. Thus deprivation of olfactory cues, but not other cues, leads ewes to perform as if they had been totally deprived of their young.

Olfactory cues from the lamb are not only crucial for the consolidation of maternal behavior; they are also important for the maintenance of selectivity in the mother. Ewes exposed for 12 hours to their lambs that were confined in a double-walled mesh cage (receiving olfactory, but not physical, contact) nonetheless developed a selective bond with that lamb. In contrast, ewes that were exposed to lambs in an airtight transparent box (thereby eliminating olfactory cues) did not develop a selective bond.

Although ewes depend primarily on olfactory cues to recognize their lambs at suckling, they are also able to recognize their neonate from a short distance using auditory and visual cues (574–576). The nonolfactory recognition is already effective as early as

8 hours postpartum. The processes involved in proximal recognition appear independent of those involved in distal recognition since anosmic ewes that allow suckling any lamb can, nevertheless, discriminate their lambs at a distance (574). Interestingly, anosmic ewes can compensate for their loss of olfaction by using the auditory/visual cues from their lambs (574).

Humans

Some human mothers are able to recognize their own infants using olfactory, touch, and auditory cues (23,300,541). This identification is probably not critical during the early postpartum period when mother and altricial young remain in close contact with one another, but they may increase the mothers' feelings of attachment in general.

Nonetheless, mothers who were better at recognizing their own infants' odors had earlier and longer contact with their infants after birth and had spent more time in close proximal contact with their infants during interactions (244). The effective odors are probably derived from the sweat and sebaceous glands on the baby's body and back. Also, mothers may recognize their own infants based on similarities between their own odors and the odors of their infants. In most human societies mothers are not likely to "lose" their infants in the herd, as are ewes. Individual recognition is probably a by-product of mechanisms that enhance the mother's responsiveness to the infant and may itself enhance the mother's positive affect (23).

Recognition of a mother's own infant's cries versus other infants' cries has also been studied. Mothers can identify crying samples belonging to their own infants versus others (577,578). Patterns of maternal heart rate change differed in response to audiotapes of their own versus another's baby (543,578) also reported findings consistent with preparatory arousal to the mother's own infant's cries. The ecologic relevance of these studies is suggested by the finding that among mothers who roomed together in the hospital with their infants, and with several other infant–mother pairs, 58% reported awakening to their own infant's cries on the first few nights, whereas thereafter the percentage rose to 96%.

Mechanisms of Maternal Experience

Studies in rodents suggest that parity and experience alter underlying neuroendocrine and neural mechanisms to render them more responsive to physiological factors associated with subsequent exposure to offspring.

Rats

Circulating PRL is higher in multiparous as opposed to primiparous lactating dams through a conditioning mechanism (579). With experience with suckling young, olfactory and visual cues from the young become associative cues and can elicit the release of PRL and the maintenance of milk production (580). Hence, offspring do not need to suckle continuously or remain close to the female to obtain adequate nutrition. Instead, milk is available to pups when they reattach and they can nurse for brief bouts appropriate to their developmental age. In multiparous animals the conditioned release of hormones occurs a week earlier than first seen in primiparous mothers (580).

Parity effects on brain and circulating hormones are also seen for hormone receptors. For instance, the density of E receptors in olfactory cortices is higher in multiparous than primiparous nonpartum mice (581,582). This may explain why, in comparison to primiparous nonpartum animals, multiparous rats are more maternally responsive to exogenous steroids (99,262).

Sheep

Neuroendocrine systems are clearly changed by a previous parity and experience in ways that enhance subsequent maternal responsiveness to infant cues and/or to hormones. Parity enhances OT receptor mRNA expression in the PVN (83) and estrogen receptor immunoreactivity in the PVN and the MPOA (583) presumably promoting the rapid onset of maternal behavior (117,355). Finally, in partum sheep, the brain release of neurotransmitters and neuropeptides involved in maternal recognition learning is higher in multiparous than in primiparous animals (79,584,585).

Primates

Little is known about parity differences and neuroendocrine receptors in the nonhuman primate brain. Bardi et al. (32) reported that the sole primiparous mother in a sample of eight mothers, was the only one of the group to reject her offspring at birth. She also had the highest postpartum stress index and the highest levels of prepartum and postpartum cortisol (CORT) levels.

Humans

Primiparous human mothers also express higher levels of salivary CORT than do multiparous mothers on day 2 postpartum. This effect is likely related to their enhanced anxiety and emotional lability (135); also, correlations between CORT and affectionate behavior or attraction to infant body odors are stronger in primiparous than in multiparous mothers (244). Primiparous mothers with high CORT show the same elevated attraction to infant odors as do multiparous mothers, with both high and low levels. It seems that experience and hormones are parallel routes to the same end.

Neurochemical Bases of Maternal Experience

Rats

In addition to hormonal changes associated with experience, brain neurochemicals and peptides may also change with maternal learning. As indicated above, with parturition and during the postpartum period before access to pups, the mother undergoes changes in endocrine systems, in autonomic function, and in peptidergic systems (i.e., OT, CRH).

For example, inhibition of brain of protein synthesis reduces retention of maternal memory (124). Brain proteins must be synthesized at the time of the maternal experience for the experience to be consolidated. Inhibition of brain proteins after the consolidation phase is over is without effect on the MEE. Manipulations of the NE, DA, and opioid systems all affect maternal memory, as well as being involved in the initial motivation to be maternal (384,391,586). For instance, NE beta-receptor blocker (propranolol) immediately after exposure to pups in the postpartum animal blocked maternal memory. In contrast NE receptor agonist (isoproteronol) enhanced maternal memory (587). There is also a disruptive effect on the retention of MEE by DA receptor blockade during the exposure or experience phase (588).

Neuropeptides such as oxytocin and prolactin are also likely to be involved in the acquisition and consolidation of MEE or maternal memory. They are present at parturition, enhance maternal behavior and are important for the formation of social memories within other functional contexts (56,426,438,589–593). Together, these systems are likely involved in all phases of MEE and probably interact differently at different stages, on different aspects of the learning process, and in different sites within the maternal neural circuit.

Adequate function in hormonal and neurotransmitter systems support learning and experience effects through a number of psychobiological mechanisms. Parturition increases the salience (through-put) of associative cues, most likely unconditioned pup cues (e.g., proximal tactile or olfactory) during the learning phase. It increases the hedonic value of those cues and hence their reinforcing properties. Hormones act to facilitate or strengthen the association between the conditioned and unconditioned pup-associated cues. Finally, they produce internal cues that themselves recursively act as conditioned stimuli. This recursive mechanism could explain the fact that experiences acquired under hormones and their interoceptive cues are better retained under those same hormones, as described before. Also, neurochemicals released at the time of parturition could increase the individuals' arousal, attentional, and/or perceptual function and memory formation postpartum in much the same way as they are known to do within other behavioral contexts.

Neuroanatomical Bases of Maternal Experience and Olfactory Learning

Rats

There appears to be a distinctive brain mechanism that underlies maternal memory, i.e., the acquisition, consolidation, and retention of a maternal experience. This mechanism intersects with, or maps onto, the "maternal circuit" that mediates the expression of maternal behavior, as indicated above (see section Neural Circuits Regulating Maternal Behavior). Areas of the brain that exhibit a long-lasting cytoarchitectural and/or functional change following maternal experience include the MPOA, the basolateral amygdala, and the parietal cortex (594,595). These sites show greater neuronal Fos immunoreactivity in response to presentation of pups and the exhibition of maternal behavior in maternally experienced over inexperienced mothers (594). In comparison to inexperienced mothers, pup-exposed experienced mothers also show enhanced astroglial GFAP staining in MPOA, suggesting enhanced support for functional neurons in that area (595). Finally, when primiparous animals first interact with the pups during the postpartum period, they undergo greater elevations in the proliferation of new neurons in the subventricular zone, than do nulliparous or parturient animals whose pups had been removed (596). Moreover, comparisons between maternally experienced and inexperienced animals 30 days after parturition and separation from pups showed experienced animals to have higher levels of surviving new neurons in maternally relevant regions, especially in the vBST and shell NA, but lower levels in other regions, especially the amygdala.

With experience and a previous parity, hormone receptors in the MPOA are more numerous, likely rendering this region more sensitive to the hormonal induction of maternal behavior (Bridges–opioid receptor work). Although the MPOA/vBST is changed by experience, this region seems not to be involved in the acquisition of maternal experience. Lee and colleagues (597) and Li and Fleming (391) found that lesions of the nucleus accumbens, whole or shell region, a limbic brain site involved in reinforcement, undertaken immediately after a maternal experience, consistently blocked the formation of a maternal experience (Fig. 20). While lesions preceding the experience had similar effects on memory formation, they did not affect the quality of maternal behavior expressed during the postpartum experience. Protein synthesis inhibitors infused into the shell region of the nucleus accumbens immediately after a maternal experience produced similar effects as did post-experience lesions (391). Future research will have to reconcile data which indicates that NA shell lesions do not have a major disruptive effect on ongoing maternal behavior (see section on MPOA/vBST Facilitatory Circuits), yet have a substantial depressing effect on the development of the MEE.

Part of the maternal circuit is indeed changed by experience and also interacts with regions and systems that are activated when experiences are acquired and consolidated. The MPOA/vBST is the region of the brain that is essential for the expression of maternal behavior but is changed by experience (see section on Neural Circuits Regulating Maternal Behavior); the basolateral amygdala is the region of the brain that receives input from multiple sensory systems and, in the maternal context, is activated by olfactory and somatosensory input from the pups; the parietal cortex is involved in processing of tactile and somatosensory input; and the nucleus accumbens is involved in reward processes and associated consolidation processes associated with the formation of a maternal memory.

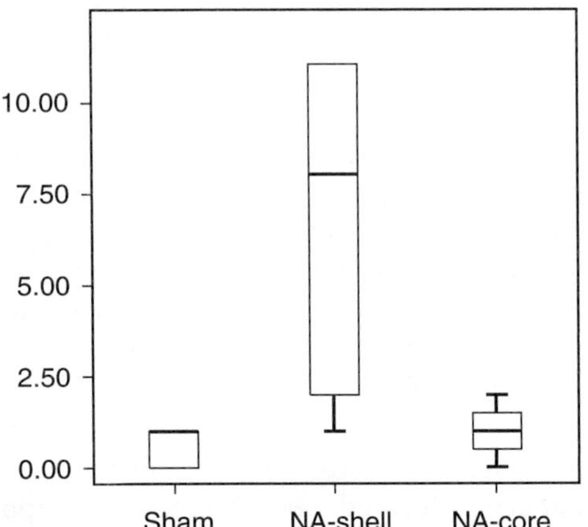

FIG. 20. Median latency (in days) to exhibit full maternal behavior in rats that were administered nucleus accumbens (NA) shell, NA core, or sham lesions during pregnancy (before maternal experience, MEE) and tested for maternal behavior 9 days after initial pup experience. The NA shell lesions significantly increased maternal latency, whereas NA core lesions did not. The box represents the interquartile range. The whiskers are lines that extend from the box to the highest and lowest values, excluding outliers. The line across the box indicates the median, *$P < 0.05$, compared with the sham control group. [From Li and Fleming (391 or 392).]

Sheep

In sheep, how the whole maternal brain circuit is changed by parturition and interactions with the lamb has not been extensively studied, as it has been in rats. However, because in sheep olfactory cues are crucial for both consolidation of maternal behavior and the learning of individual characteristics of the offspring, focus has been on the mechanism by which parturition produces changes in the olfactory system and how these changes are influenced by experience.

Electrophysiological recordings undertaken in ewes from olfactory bulb mitral cells (117) showed during pregnancy that none of these cells responded preferentially to lamb or amniotic fluid odors. Instead, the majority of cells responded preferentially to food odors. After birth, there was a dramatic increase in the number of cells that responded preferentially to lamb odors, supporting the idea that the change in salience of the lamb odor that occurs at the time of parturition is mediated by a shift in olfactory cell responsivity. However, the majority of cells that responded to lamb odors did not differentiate between the odor of the ewe's own lamb and that of an alien lamb. Nevertheless, a proportion of the cells did respond preferentially to the odor of the ewe's own lamb. These results indicate that lamb odors are a powerful olfactory stimulus, producing cell activity, only in the period after birth when attraction to lambs occurs. As well, a small proportion of cells actually show preferential responsiveness to the odor of own lamb at the time when the recognition of lamb odors is a priority.

These shifts in electrical properties of mitral cells that are associated with maternal responsiveness and with olfactory selectivity are reflected in concurrent parturitional changes in the release of amino acids (GABA and glutamate) within the olfactory bulb (117), the action of the noradrenergic system projecting into the olfactory bulb, and the release of oxytocin from the olfactory bulb (598,599). In comparison to before birth, after parturition, when ewes have established a selective bond with their lambs, the odors of these lambs, but not those of alien ones, increase the release of the excitatory amino acid (glutamate) and the inhibitory amino acid, GABA. The increase of

GABA after birth was significantly greater than that of glutamate (117) which may reflect an experience-based increase in the inhibition of activity of certain mitral cells. Experimental evidence supports these correlations. Infusion of a GABAa receptor antagonist, bicuculline, in the olfactory bulb prevents lamb recognition after the selective bonding had been formed (600). The general increase of GABA likely refines the olfactory signal by inhibiting all mitral cells, except those activated by the odor of the familiar lamb.

The selective increase in reactivity of the olfactory bulb to familiar lamb odors also results from noradrenergic input from the pons locus coeruleus (601). A dramatic increase of noradrenaline release occurs during the first 2 hours after parturition, i.e., the learning phase of lamb odor (602). Lesions of noradrenergic projections to the OB or direct infusions of a β-adrenergic antagonist significantly inhibit the number of ewes developing the olfactory memory, without affecting odor perception (603,604).

Oxytocin release within the OB (see section on Oxytocin) may also act to facilitate lamb odor memory because it modulates noradrenaline release. Infusion of oxytocin within the OB causes a significant increase in noradrenaline concentrations in the olfactory bulb (598). Although we have no direct evidence, the fact that maternal behavior induced by intracerebroventricular oxytocin infusions results in selective bonding, supports this hypothesis (474).

These neural changes result from the first experience of parturition and interactions with the young. The neurotransmitters released at parturition from intrinsic neurons within the olfactory bulb are higher in ewes that had previously experienced mothering than in those that had not (584). In ewes giving birth for the first time, there is no measurable increase in glutamate or GABA release in the olfactory bulb at parturition.

Although interactions between different cell types in the olfactory bulbs are essential to the selective recognition process, they cannot achieve this without the assistance of neural inputs into the bulbs from other sources. Experience with lambs is, however also important for the release of noradrenaline at birth itself. In fact, after the first birth, experience produces a fivefold increase in noradrenaline levels at parturition (602). This experience-dependent increase may also be facilitated by a simultaneous increase in the peptide oxytocin which is significantly higher in multiparous, than in primiparous, ewes at birth (599). However, it is not just the higher concentrations of this peptide that facilitates the release of noradrenaline, since experience also influences the magnitude of noradrenaline release to controlled, dose-dependent infusions of oxytocin directly into the olfactory bulb by retrodialysis (79).

Of course, changes that occur in the OB induced by parturition and the first interactions with the lamb only represent the first stage in the processing of lamb odors. How the rest of the brain handles this information and whether the more central olfactory and limbic structures are involved, is currently under investigation.

Using Fos expression the neuranatomical substrates activated at parturition and during the first interactions with the lamb were mapped. Postparturient ewes exposed to their lambs for 30 minutes showed increased cFos mRNA and/or zif/268 expression compared to animals not exposed to lambs in primarily in the main olfactory processing regions (OB, piriform cortex, entorhinal cortex, orbitofrontal cortex) (355). Recently, the number of Fos immunoreactive cells have been compared in intact ewes and in anosmic ewes that do not form an exclusive olfactory memory for their lambs (576). Consistent with the Da Costa and colleagues study (355), in intact ewes, as compared to anosmic ewes, significant increases of Fos labeling was found in the olfactory structures and projection sites. In addition, the cortical amygdala also shows enhanced expression in parturient intact animals but not in anosmic animals. Which of these structures is involved in the development of selectivity and the formation of the memory was addressed in a series of implant studies. Reversible inactivation of the medial frontal cortex by infusing tetracaine, a local anesthetic, did not prevent the formation of an olfactory memory for the familiar lamb but it did inhibit the aggressive rejection directed to strange lambs (414). In contrast, a role for both the medial and the cortical nuclei of the amygdala, which receives olfactory input from the olfactory bulbs, has been recently demonstrated (605,606). Infusion of the anesthetic, lidocaine, for the first 8 hours postpartum, in either of these nuclei prevents animals from learning to discriminate their own lamb from an alien lamb and, hence, both are permitted to suckle. This effect is due to the blockade of the acquisition of olfactory information and not to lidocaine effects on the retrieval of this memory or on the display of the rejection response towards the alien lamb, since the rejection response also is not expressed 2 hours later, after the lidocaine has worn off.

Primates and Humans

There are a number of fMRI studies indicating activation of the cingulate cortex and other limbic and hypothalamic regions with exposure to infant cues (519). There is also some evidence showing that in mothers brain size may increase 6 to 8 weeks postpartum when compared to the size during the third trimester of pregnancy (607). Whether the changes are in production of new neurons or glial cells, or to enlargement of cells we do not know. Also not known is whether these changes are due to interactive social experiences, the experience of hormones of pregnancy, endocrine and somatosensory experiences associated with parturition or lactation, or the suckling stimulation itself.

Conclusions: Maternal Experience and Individual Recognition

In the laboratory rat, experience has very little effect on the maternal behavior repertoire. Movements are changed only in the speed with which pups are retrieved and in where on the body the mother grasps the pup to retrieve it. More likely the MEE involves perceptual learning. The mother develops a familiarity with pup associated stimuli. This reduces her natural "neophobic" tendency to withdraw from novel, new stimuli. When she encounters pups again they no longer induce a fear-withdrawal response. With maternal experience pups acquire heightened reinforcing properties which further motivate the mother to approach and respond to them (566).

In humans, maternal experience has multiple psychological effects. There is an increased ability to both perceive and then discriminate infant cues. Successful maternal behavior is associated with heightened self-esteem, positive affect, and greater feelings of maternal competence. In fact, mothers who are better able to recognize and then discriminate their infants' cues, from others, also report more positive maternal feelings and attitudes. Finally, experience with infants also provides the mother with realized, as opposed to imagined, practice at breastfeeding, changing the diaper, and so on (23). It is not hard to imagine that the positive benefits of experience would asymptote. With too many offspring, more negative reactions could also occur.

EFFECTS OF EARLY PREWEANING INFLUENCES ON ADULT MATERNAL BEHAVIOR

Although experiences during adulthood clearly have an impact on behavior, the efficacy of the adult experience depends on the sensitivity of substrate systems that have been molded by earlier experiences acquired in interaction with mother and siblings (608).

Sensory and Psychological Mediation of Early Experience

Rats

The mother rat and maternal nest provide a host of stimuli that the young can use to guide their later behavior. Newborn pups come to recognize and preferentially approach their own mother and nest site based on odors (609). First attachments to the teats are guided and activated by the odor and taste of amniotic fluid first experienced in utero (610,611) licking provided by the mother (612). Attraction to mother's odors helps the infant orient to the mother, locate a nipple position, as well as providing a basis for later food preferences. These associations occur as a result of the pairing of mothers' odors with licking stimulation provided by the mother prior to a nursing bout (608,613). The neurobiology of this early conditioning involves neural and neurochemical systems that are also involved in later learning (608,613).

Until recently, the long-term effects of learned attraction to mother's odors on female offspring were not understood. However, mother rats respond preferentially to pups scented with the same odors that they experienced earlier on their own mothers than to pups that carried a different scent. Simple exposure to odorants early in life, not in association with mother, produced no such preference in adulthood (614,615). Whether these effects are mediated by changes in OB mechanisms and persist into adulthood and mediate adult behavior, we do not know.

When offspring receive somatosensory stimulation (through mothers' licking), there occur long-term effects on the quality of mothering they come to show towards their own offspring when they grow up. The experience of receiving more somatosensory stimulation and a more varied nest experience, while in the presence of the natural mother, siblings, and nest odor, directly affect physiology (608,613). Young that receive more licking and grooming from their mothers grow up to show higher levels of licking and grooming toward their own offspring by comparison to those who received less licking (616,617). Furthermore, maternal traits of high or low licking of offspring are not inherited. Instead, licking of offspring as a mother depends on experiences of being licked as a pup. For example, long-term licking effects can be produced by altering the characteristics of the mother and nest to enhance or reduce the amount of licking the pups receive. The amount of licking experienced by an individual pup is reduced if the mother is rendered anosmic through olfactory bulbectomy or, if pups are raised in a large litter (615,618). Total licking by mothers on day 6 postpartum and total licking by their adult offspring at the same time period postpartum were positively and significantly correlated. Reduced experience of being licked (by virtue of having a bulbectomized mother or being raised in a large litter) had effects on licking in female offspring when they became mothers, even though the offspring as adults could smell their litters and/or had litters that were of intermediate size (619). Hence, the "reasons" or causes for more or less licking in mothers and daughters changed across the generations.

The most effective period for transmission of licking effects seems to be during the first postpartum week (617,619). This cross-generational effect of licking provides a nongenomic mechanism for the transmission of behavioral phenotypes. Enhanced licking is associated with multiple receptors and hippocampal effects (in the "licking" animal and their offspring). Whether these transgenerational behavioral effects are reflected in effects on "maternal" brain systems specifically involved in licking itself or in other maternal behaviors is not known.

Maternal Deprivation and Isolation Studies

Rats

There is clearly a relation between level of licking received as a neonate and the development of physiology and behavior in offspring. One way that the relationship between mothers' licking and offspring development can be established is by depriving animals of licking stimulation and then providing "replacement" stimulation. The consistency and reliability of the mother's behavior toward her pups can be altered by separating mothers and offspring during the preweaning nesting period (615,620).

In a series of maternal deprivation studies, pups were either left with their own mothers for the entire preweaning period (controls) or were separated from their mothers for 15 minutes, or 3 hours daily (from days 1–9) or for 5 or 24 hours daily during the first 18 days of life and placed in a warm chamber. In some cases, entire litters were removed and separated together. In other cases only some pups were removed and separated either alone or together. In all cases, pup temperatures were maintained during the separation period. The most severe deprivation paradigm involved raising pups artificially in a variant on Hall's "pup-in-a-cup" environment (AR) and comparing these to mother-reared sibling controls (MR) (621,622). Unlike the limited deprivation procedures which really involve separations and reunions, the "pup-in-the-cup" procedure involves a continuous isolation, where pups never know their mothers. In all these studies, animals were pair-weaned until adulthood when they were mated, gave birth, and observed for maternal behavior.

The offspring of the artificially reared females (F1 generation) were themselves reared to adulthood, mated, and tested with their own offspring (F2 generation). Using these deprivation procedures, Fleming and colleagues (182,183,236,619) found that maternal behavior deficits in the offspring were related to the degree of earlier maternal deprivation (Fig. 21). Offspring that had experienced the least deprivation (3 hours daily for 9 days) showed no deficits in adult maternal behavior in comparison to controls (182); those that experienced somewhat longer periods of daily deprivation (5 hours daily for 18 days) showed initial deficits in licking behavior compared to controls, and once they were fully maternal these deficits

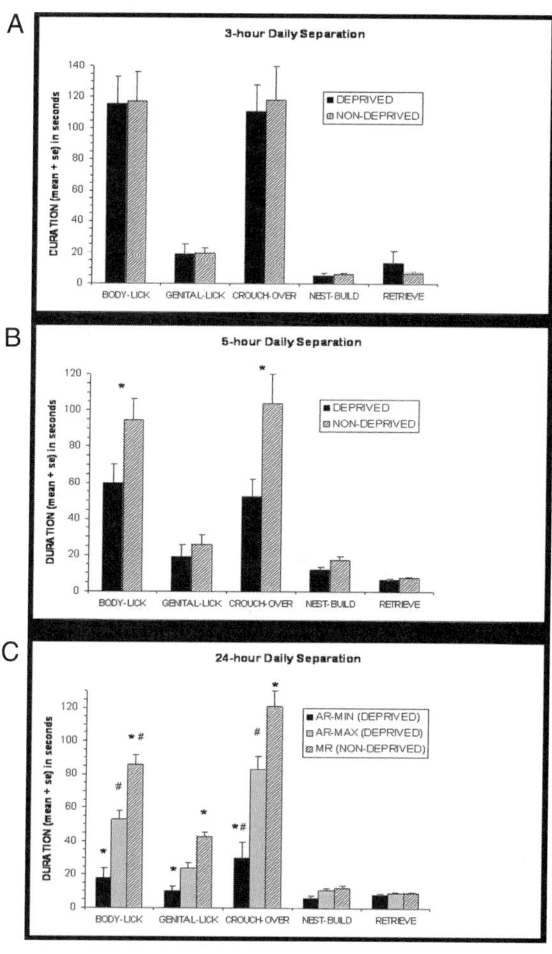

FIG. 21. Durations of different maternal behaviors averaged across 5 test days in mothers tested postpartum that had been separated from their own mothers (DEPRIVED) or not separated (NOT DEPRIVED) for 3 (**A**) or 5 (**B**) or 24 (**C**) hr daily during the preweaning period. In **C** pups were reared without mother (artificial, AR-DEPRIVED) or by mother (MR-NONDEPRIVED). AR animals either had received 6 additional "strokings" with a paintbrush daily (AR-MAX) or two strokings daily (AR-MIN). Shared symbols within a behavioral category indicate groups differ significantly, $P < 0.05$. No differences were seen between groups that had been separated for 3 hr daily, whereas groups separated for 5 or 24 hr daily showed group differences in licking and crouching. [Modified from Fleming et al. (615).]

persisted in both licking and crouching by comparison to controls (183). Artificially reared mothers (AR, "cup pups") exhibited the greatest reductions in pup body licking, genital licking, and crouching as adults (615,619).

The intergenerational effect of licking pattern was also demonstrated when the female offspring of AR and MR mothers were tested with their own offspring. In the F2 generation, offspring of MR mothers showed more licking toward their own offspring than did offspring of AR mothers (615,619).

To determine whether the deleterious effects on adult maternal behavior of more extended preweaning maternal deprivation (24 hr separation) could be remedied or reversed by providing additional licking-like somatosensory stimulation known to influence a variety of metabolic, endocrine, and homeostatic systems (608), pups were reared artificially and then the effects of more or less stimulation during the preweaning was determined. Complete privation of mothering resulted in deficits in many maternal behaviors, but not in the motivation to respond maternally. All groups retrieved young, so the usual reduction in "neophobia" seen in postpartum rats occurred in AR groups. Deficits were associated with changes in the salience to the mother of pup cues (olfactory, taste?). Deficient body licking, genital licking, and crouching produced by AR were partially restored, however, by rearing AR animals with increased licking-like stimulation produced by multiple daily swabbings with a paintbrush. Subsequent studies indicate that the addition of a conspecific into the cup environment ("Peer-Reared") further restored adult maternal licking behavior. This indicates that licking-like stimulation reduces deficits in a number of maternal behaviors. In the adult, brushing also had an effect on deficits in attentional and emotional behaviors, but not in social learning and hyperactivity induced by deprivation (236,623).

There is every likelihood that a "combination" replacement regime (simulating mothers' odors, temperature, movement, and other somatosensory characteristics), would result not only in more complete adult maternal responding and reductions in deficits in many of the other social deficits that follow early maternal deprivation. The latter may well contribute to the deficits in maternal behavior (608,613,624,625). Initial results indicate that the pattern of stimulation provided by conspecifics, by their movement, warmth, and tactile characteristics also contribute to the normal development of behavior. The parallels between this rodent model and earlier work in nonhuman primates are remarkable (625).

Reversals of deprivation effects can occur if the ecologically appropriate stimulation is provided during the preweaning period. The question also arises as to whether other "therapies" or "enrichments" following weaning can reverse the deficits in behavior produced by early deprivation. This has not been attempted in the study of maternal behavior. Post-weaning environmental enrichment does reverse the negative effects of maternal separation on HPA axis function, but without reversing many of the molecular changes that are associated with the altered HPA function (616) (see below). This suggests that postweaning "therapies" may hone or tune up the expression or development of other mechanisms that compensate for the disrupted mechanisms and that could normalize the stress system and HPA function. Whether these effects are fragile and would dissociate with subsequent stressors, we do not know.

It is clear that in the absence of added stimulation, either pre- or postweaning, maternally deprived AR mothers show deficits in their ability to gain lasting benefits from their postpartum interactions with pups, that is to show a MEE (623). Hence, their ability to learn from their pup-related experiences was compromised. It is not surprising that "learning" deficits of artificial-rearing are not restricted to the maternal context. They extend and propagate to other social experiences acquired throughout later development (623,625).

The intergenerational transmission of quantitative aspects of maternal behavior in animals seems quite clear at this point. What is less clear is how these effects are transmitted. For instance, maternal deprivation affects the development of fear and emotionality (616), attentional systems (236), conspecific social learning and learning about new odors—but does not, oddly enough—disrupt performance on numerous other cognitive, nonsocial tasks, where in some cases the deprived animals excel (623,626).

A relation between maternal-deprivation induced attentional deficits and maternal licking deficits is highlighted by the finding that that in comparison to MR animals, AR animals show less inhibition on a prepulse inhibition task (pre-attentive gating task). These deficits can be reversed by extra somatosensory stimulation and that there is a positive relation between performance on measures of attention and extent of body licking (236,615).

Primates

There is a major, but probabilistic, effect of early rearing on maternal behavior in laboratory housed rhesus monkeys. In the laboratory, rhesus offspring are usually weaned between 6 and 9 months of age (juvenile), reach endocrine sexual maturity at about 3 years of age (post-adolescence/young adulthood), and psychosocial sexual maturity between 5 and 7 years of age (age of majority/adult) (625,627). "Mother-Reared" (MR) are born to their biological mother and remain with her for 6 to 9 months in the laboratory. Among MR laboratory rhesus monkeys over 90% of primiparous mothers care for their offspring (563). If female rhesus monkeys are separated from their mother shortly after birth and reared in isolation from other monkeys for 6 to 9 months, they will not mate voluntarily as young adults or adults. If they are artificially impregnated and give birth, they usually kill the infant as part of eating the after-birth, or reject the infant vigorously if it survives beyond the first 24 hours after parturition (625). The effects of early isolation are "dose dependent" (563,628,629). Earlier and longer postnatal isolation has a greater effect on later psychobiological/social measures than later isolation.

More relevant to understanding of the human condition, is the nonhuman primate model of PR. PR monkeys are separated from their mother, usually 3 to 4 days after parturition, and thereafter reared in a nonhuman primate neonatal intensive care unit (NICU) for approximately 30 days and nights, and then put into a group (N = 4–6) of comparably reared infants. PR monkeys are not socially "isolated" in the sense of having no conspecific interaction. Instead, what they are exposed to is a relatively "unregulated" social environment because there are no adult rhesus monkey mothers or fathers in place to regulate the behaviors or socialization of the peer group.

The probability that PR monkeys will care for their first born is only 65%, but if they do care for the infant they appear to do just about as well as primiparous MR monkeys. Amalgamation of studies over time indicates that the probability that the offspring of PR monkeys will themselves care for offspring turns out to be a striking 90%, which is in the same the range as offspring that were cared for by a MR mother (562,625,627,630–634). Hence, if the PR monkey actually picks up her postpartum neonate, there is a very high probability that she will be a "good enough" mother (625,635). Overall, the experience of being mothered greatly increases the probability that nulliparous rhesus monkeys will care for their first born offspring. Nevertheless, a substantial proportion of rhesus monkeys will adequately care for offspring if they have been socialized, but have no prior experience with their own mother or younger rhesus monkeys.

In terms of style of mothering or variations in naturally occurring behavior, Fairbanks has found for vervets (537) very strong correlations between maternal behavior style of mothers and the style exhibited by their female offspring towards their grandchildren, as it were. Mothers who were more protective with their daughters have daughters who were more protective when they had offspring (Fig. 22).

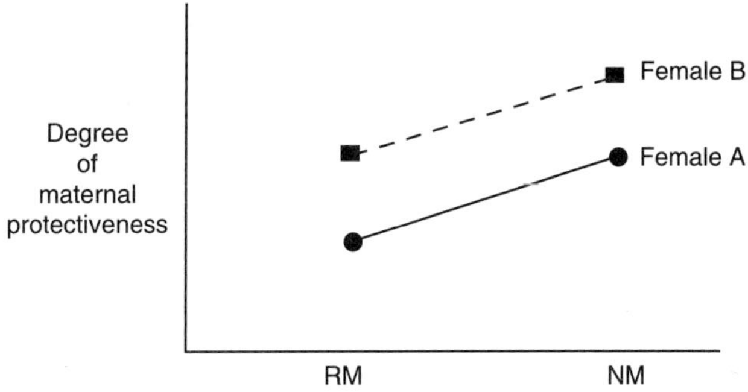

FIG. 22. The degree of maternal protectiveness shown by two postpartum vervet monkeys toward their infants under two conditions: when a long-term familiar resident male (RM) was in the group, and when a new male (NM) entered the group. Although protectiveness increases under the NM condition, individual differences are still evident with females showing protectiveness under both conditions. [Modified from Fairbanks (537); from Numan and Insel (100).]

This cross-generational correlation in maternal style likely reflects a cross-generational transmission in anxiety and timidity; hence, it may be that the more timid mothers are also more protective and restrict their offspring from exploring, leading them in turn to become more timid (636).

Humans

A widely held view is that extreme maternal deprivation or the experience of being abused by parents increases the incidence of inadequate or abusive mothering by these same infants when they become mothers (637). Much of the evidence that there are substantial effects of early maternal rearing experience on later maternal competence is based on retrospective studies, however. For example, Simons et al. (637) report that mothers who engaged in "harsh" parenting tended to have had mothers who treated them harshly and/or to have a hostile personality which itself was correlated with harsh parenting by their fathers. In low SES, high-risk groups, these intergenerational effects are even more pronounced when more severely dysfunctional parenting is evaluated.

The intergenerational effects are seen among women whose early lives were disrupted by abuse, maternal depression and poor family functioning early in adolescence (638). For instance, 30% of mothers who were abused as children abuse their own children later in life, whereas only 5% of mothers who were not abused abuse their infants (639). Consistent with the epidemiological data, Main and Goldwym (640) reported that the experience of maternal rejection as a child is related to the rejection of offspring later in life. This intergenerational effect may account for the cyclic occurrence of premature motherhood among teenagers who have often received inadequate or inappropriate parenting (641,642). These children thereafter go on to raise poorly adjusted children who tend to become teen parents themselves (643–645). In a recent series of studies on early experience correlates of maternal behavior in teen mothers (243) found that mothers who as children and adolescents received consistent care engaged in significantly more affectionate and less instrumental caretaking behaviors toward their infants as compared to mothers with multiple and changing caregivers (Fig. 23).

The two younger groups of mothers showed the early experience effects most strongly. When all age groups were combined, more frequent changes in caregivers during childhood were also related to higher levels of CORT, which in turn was related to negative affect, and then to reduced affectionate behavior. In a study of clinically depressed older mothers Gonzalez and colleagues (235) found the same basic direct relation between inconsistent early parenting received negative affect and CORT levels, as well as indirect relations between negative affect and behavioral interactions with the infants. Hence, mothers who received less early consistent care prior to age 12 years, were more likely to have high CORT levels and to become clinically depressed (235).

The implications of these studies are that early family instability has long-lasting effects on maternal behavior by altering the offspring's adult affective state. However, whether these relations are mediated by the mother in the family of origin remains an open question. Prospective evidence on the intergenerational effects of these extreme conditions not to mention of normal parenting variations simply does not exist (639,646). What has been investigated is the relationship between maternal and infant attachment style, later personality of the infant as young-adult, and attachment style of both earlier and later offspring (647). For example, infants that have been characterized in Ainsworth's Strange Situation as securely attached at 1 year of age tend to have mothers who are more responsive to their (the baby's) emotions than do the mothers of infants who are categorized as

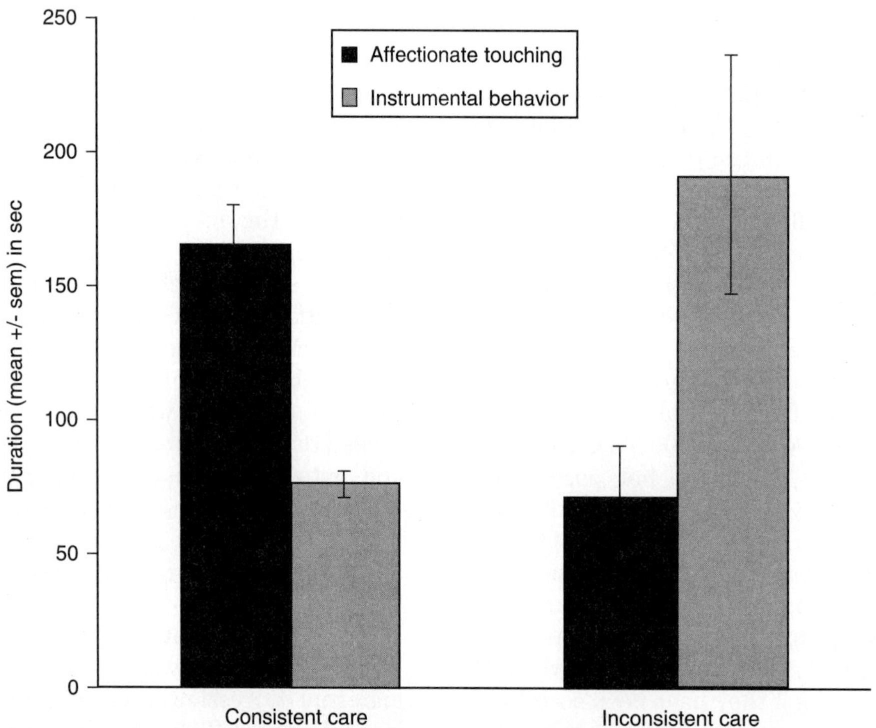

FIG. 23. Interaction between early life experiences (consistent vs. inconsistent) and type of maternal behavior (affectionate vs. instrumental) shown by 3- to 4-month-old postpartum mothers. A mother was said to have received consistent care if there was one primary caregiver present continuously during the first 12 years of life. Some examples of consistent care are having a mother figure present continuously despite multiple and changing father figures; having two parents consistently present; and having a grandparent consistently present. Some examples of inconsistent care are moving back and forth with long intervening periods between parents living in different homes, living in foster care for a period, living with other family relatives or friends for a period. Affectionate behavior consisted of stroking, patting, palming the infants during a 15 minute observation period when mother was holding, but not feeding, the infant; instrumental behavior consisted of cleaning the infant, fixing the clothes, and changing diapers. [Modified from Krpan et al. (243).]

ambivalently attached (648). Mothers of ambivalent babies were most responsive to negative affect and least responsive to positive affect (648).

Thus, the emotional quality of mothers' responsiveness to the infant may affect the infants' attachment style. Infant attachment style in turn appears to be related to adult personality. Waters and colleagues (649) report that 70% of securely attached infants (at 1 year of age) were later categorized as secure and autonomous in young adulthood (20 years of age). In contrast, only 17% of individuals who were categorized as avoidant in infancy were secure as adults, and approximately 70% of these individuals were classified as "dismissing" adults. Adult personality, in turn, appears to affect infant attachment style (650) and, likely, their sensitivity to their infants. Overall, the long-term longitudinal story has to be pieced together based on findings of multiple short-term studies, however, and there are missing pieces to the puzzle.

Intergenerational continuities are by no means fixed, immutable, or certain. Clearly parental learning, societal, and cultural influences acting between infancy and later child bearing have to be considered. In like measure, however, psychobiological factors may be significant intervening variables that contribute both to the more usual intergenerational transmission of rearing competence, and to persisting exhibition of inadequate maternal behavior when this arises.

Although there is also strong evidence to suggest transmission of attachment styles in both mothers and offspring, the psychological mechanism underlying the early experience effects have not been extensively explored. However, Belsky (545) provides a thoughtful analysis of the possible mediators of these relations, including in the equation a consideration of how mothers talk about their infants (their "mind-mindedness"), the quality of the marriage and home environment, and the acquired emotional bias used by infants to process their world. This latter "affect-memory" mechanism finds its analogy in the animal work, in early experience effects on attention, sociability, emotion, and memory of the mother-to-be.

Hormonal Effects of Early Experience

Rats

The early preweaning period has many effects on the pups' cardiac responses, HPA stress responses, and other aspects of physiological homeostasis. Some of these effects are also expressed in physiological vulnerabilities in adulthood. Whether these are direct effects of early preweaning experiences on adult maternal behavior, or the inability to gain from developmental experiences as a result of disrupted endocrine or neural responses to social and olfactory cues in general, is still an open question.

There are endocrine and endocrine receptor changes normally associated with the onset of maternal behavior in adulthood, but we are only now getting a handle on how early experiences affect development of receptor systems that underlie maternal behavior. Meaney and his colleagues (56,589) identified changes in brain receptors associated with variations in licking intensity in mother rats that show high vs low licking, grooming and arched-back nursing, (LG-ABN). High LG-ABN mothers have higher levels of OXY receptors in a number of brain regions, including the MPOA. In comparison to low LG-ABN, nulliparous females of high LG-ABN mothers became maternal more rapidly, licked pups themselves more after becoming maternal, and showed significantly higher levels of OT binding in MPOA (56). As shown in Fig. 24, offspring of high LG-ABN mothers also show a greater OT receptor induction to exogenous E treatment that is dose dependent (56). There was a dose-related increase in MPOA OT binding in nulliparous females of high, but not low, LG-ABN mothers. E did not affect OT receptor binding in these females.

Also, offspring of high LG-ABN also have higher E receptor densities in MPOA (651). Although there is evidence of variations in receptor density associated with variations in early licking received by mothers, we know very little about the hormone receptor profile associated with maternal deprivation and "replacement."

Our preliminary results indicate that the differences in AR and MR mothers in licking are probably not due to differences in either OXY or E receptors in MPOA (652). Consistent with these findings, Novakov and Fleming (653) found that if AR and MR ovariectomized nullipara are administered the Bridges Progesterone-Estrogen silastic hormone regime, the two groups are equally affected by the hormone regimen. Once maternal, the group showing the highest licking behavior was the MR females who received exogenous hormone treatment. Similar to their postpartum counterparts, AR and MR animals are equally susceptible to the activational effects of the parturitional hormones and all animals show maternal behavior; however, once maternal, the AR animals come to show a deficit in their licking and crouching, much like the AR postpartum animals. Based on these results, it is possible that the hormones are above threshold for activation of the behavior but not for the modulation of the quality of the behavior. Lower concentrations or durations of the hormones, using the Bridges regimen, might reveal differences in the activational effects of these hormones on the motivation to be maternal.

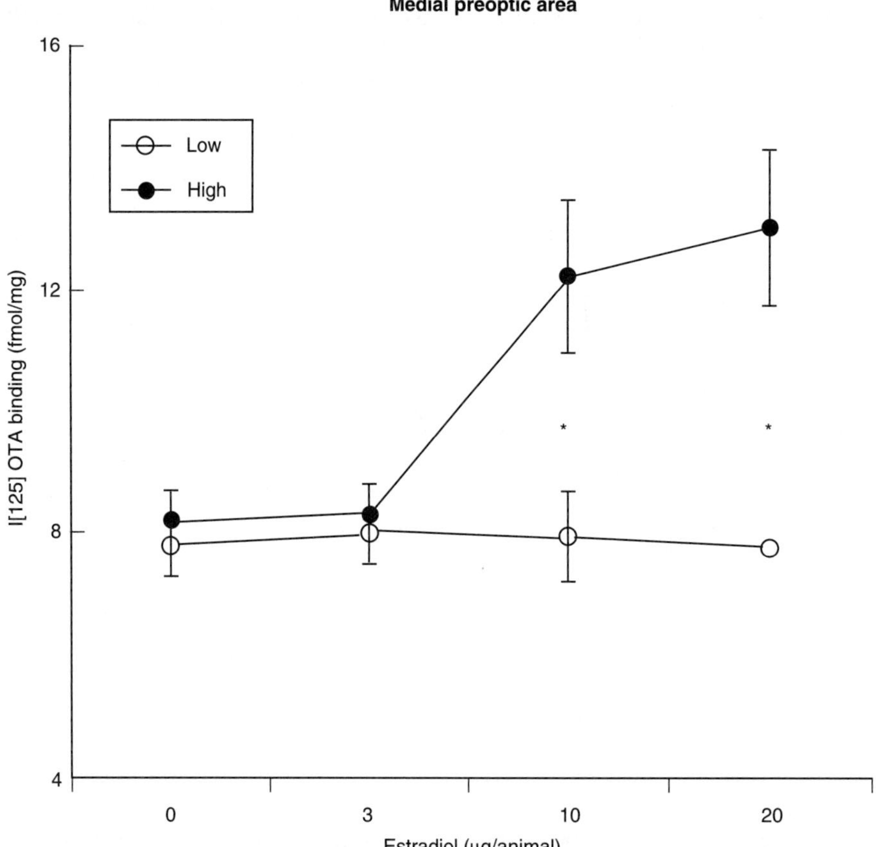

FIG. 24. Representative autoradiographic images of 125OTA (oxytocin receptor) binding (fmol/mg) in MPOA in ovariectomized virgin female offspring of high or low LG-ABN mothers as a function of estradiol replacement to the virgins. High LG-ABN and low LG-ABN animals represent animals whose mothers showed levels of licking (L), grooming (G), and arched-back nursing (ABN) that were two standard deviations above (High) and below (Low) the group mean for these behaviors.

Another possible endocrine effect of deprivation that may influence adult maternal behavior is in the demonstrated effects on the stress system and HPA axis both during the early preweaning period and later on in adulthood (618,654,655). Early deprivation may well produce dysregulations of the HPA axis, such that the basal and/or stress levels of CRH or CORT during the early postpartum period are not being maintained within the optimal regulated range to enhance maternal memory or licking. Due to early deprivation, CRH or CORT levels may be quite erratic and elevated, producing instead disrupted behavior. In a first study exploring this possibility, Rees and Fleming (in preparation) found that in comparison to MR animals, AR rearing resulted in quite elevated levels of basal CORT in adulthood during the postpartum period and that these were inversely related to pup licking shown by the mothers.

Primates

ACTH and CORT values have been measured over the first 6 months of life in 48 infant rhesus monkeys that were either mother- or peer-reared, and in response to two stressful housing transitions (656,657).

Peer-reared monkeys exhibited lower levels of ACTH over the first 6 months of life than mother-reared monkeys, although basal CORT values did not differ between groups. Mother-reared monkeys showed increases in ACTH and CORT in response to the stress of being moved to new cages, whereas the peer-reared group showed virtually no HPA axis response to this transition. The mother-reared group also exhibited higher values of ACTH and CORT in response to the more severe stress of separation from their rearing partners and housing with unfamiliar agemates. Hence, HPA axis responses to stressors are blunted in juvenile peer-reared monkeys. Despite blunting of neuroendocrine response, juvenile peer-reared monkeys display exaggerated behavioral responses to social stressors (656). This relatively early developmental effect of maternal privation on the HPA axis appears to reverse in direction as individuals move through adolescence and into adulthood (658). Older peer-reared monkeys are hyper-responsive behaviorally to social stressors such as brief social separation, and they also exhibit enhanced HPA axis responsiveness as measured in plasma CORT and ACTH increases by comparison to mother-reared monkeys (658). Overall, stressors (e.g., social separation) tend to have more severe and long lasting social consequences in peer-reared monkeys in general. Not known, of course, is whether these changes in HPA axis associated with disruptions in early rearing affect the subsequent maternal behavior shown by these monkeys.

Humans

Results with new mothers map onto these HPA findings quite directly. Mothers who are at risk, either teen or clinically depressed, are more likely to show disrupted interactions with their infants and to have elevated basal CORT levels. Although there is no direct relation between CORT levels and negative maternal behavior, there is clearly an indirect relation. Higher CORT is associated with more depression/negative affect and higher dysphoria is associated with less adapted maternal behavior. Not known is whether the elevated CORT levels that we find in our high-risk teen mothers are restricted to the postpartum period, or whether as youngsters these mothers also had elevated baseline and/or stress CORT levels (Fig. 25).

In certain contexts CORT levels are elevated in infants who are insecurely attached and/or who have intrusive caregivers (659,660). It would not be surprising to learn that these high-risk mothers have a history of an over-responsive HPA axis—much has been described for "maternally deprived" rat pups and motherless or peer-reared monkeys.

Neural and Neurochemical Effects of Early Experience

Rats

The experience of exposure to mothers' odors in association with her licking produces changes in brain lasting into the juvenile period. The association

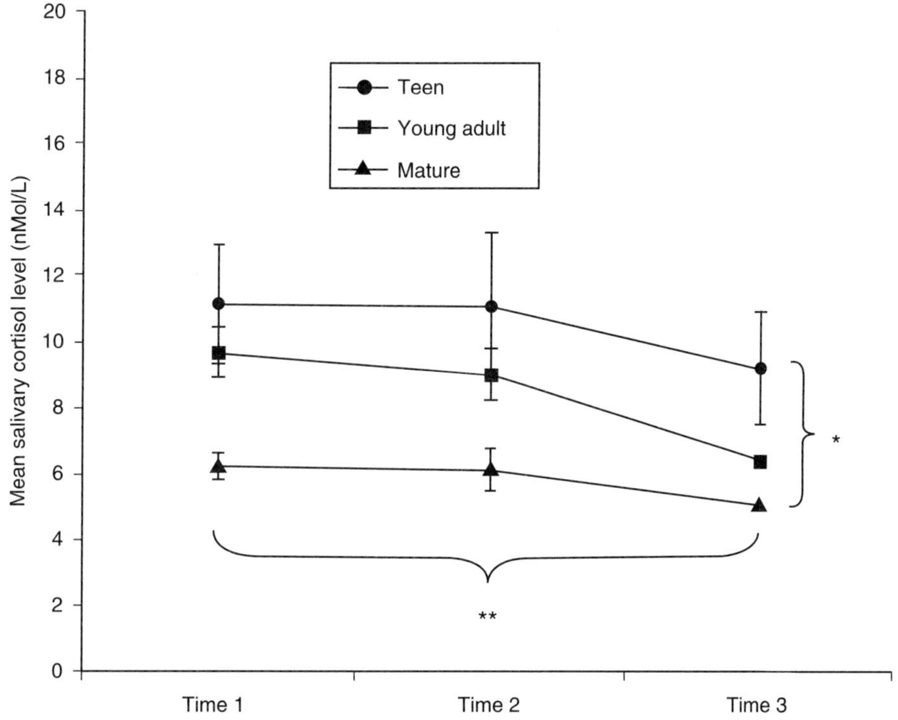

FIG. 25. Interaction between age and type of maternal behavior (affectionate vs. instrumental), $P < 0.012$. Teen, n = 41; young, n = 34; mature, n = 43. Affectionate behavior consisted of stroking, patting, palming the infants during a 15-minute observation period when mother was holding, but not feeding, the infant; instrumental behavior consisted of cleaning the infant, fixing the clothes, and changing diapers. [Modified from Krpan et al. (243).]

between licking and maternal odor produces changes in the OBs, altering their neurochemical and structural properties. This alteration depends on the activation of the noradrenergic system that originates in the pons and terminates in the OBs (613). As well, environmental manipulations during the preweaning period (including somatosensory stimulation) produce changes in cortex, hippocampus, other limbic areas, and in the MPOA (661–665) [(see also (596)]. Finally, early experience and separation have effects on NE and DA biogenic amine systems relevant to maternal behavior and MEE in adulthood (621,666,667).

Primates

In rhesus monkeys maternal privation and subsequent challenges with social stressors such as separation, produces enduring changes in behavioral and biogenic amine system development in rhesus monkeys (633,668–673). Baseline levels of CSF NE and the magnitude of response of the NE system to subsequent stressors, and baseline levels of CSF 5-hydroxyindoleacetic acid (5HIAA), the major metabolite of serotonin are affected by maternal deprivation (625,633,671–673).

Enduring or "trait-like" variation in behavioral/biological responses to social stressors appears to be related to variation in underlying neurochemical and neuroendocrine systems, and the status of these systems in the adult appears to be determined in significant respects by early rearing experience. Whether they relate directly to mechanisms that underlie adult maternal behavior, we do not know.

GENERAL CONCLUSIONS

In this chapter we have reviewed what is known about the neurobiology of maternal behavior in a variety of species representative of different levels of precocity, with varying degrees of infant dependence and with differences in developmental complexity. We discuss mechanisms at multiple levels of analysis including the endocrine, neural, neurochemical, and genetic factors on the one hand; and experiential, developmental, affective, cognitive, and other environmental factors, on the other. This exercise has led us to conclude that at all levels of mammalian complexity, from mouse to human, maternal behavior can be discussed from both a proximal and reductionistic perspective and from a psychological perspective.

We discuss similarities across species as well as their differences. Hence, we discuss the similarity in the role of the hormones, in species ranging from mice, rats, rabbit, sheep, monkeys, and humans. In all, these hormonal mechanisms involving estrogen and progesterone and oxytocin very likely play a role in mothers' responses to their offspring, although the relative importance of each, the direction of their effects, and the particular hormonal profile that is effective, may vary. For instance, estrogens induce maternal aggression in rats, but inhibit it in mice. Prolactin, on the other hand, is important for rats but not for sheep.

As well, in some species maternal behavior can be expressed in the absence of hormones; in others, this never happens. Virgin rats can be induced to become maternal through contact with pup stimulation, whereas virgin mice are immediately responsive to pup cues. Nulliparous ewes never accept a lamb at the udder.

In terms of neural circuitry of the behavior the neural mechanisms regulating the onset of maternal behavior have been most intensively studied in rats and other rodents, where the MPOA/vBST is essential for the expression of the behavior and is part of an excitatory system. Whether the role of the MPOA is conserved across species is at present unknown. For instance, in sheep, the PVN, may be more important for maternal responsiveness. Unfortunately, we know nothing about the brain or neurochemical mechanisms that mediate the human mothers' responsiveness and recognition learning of her offspring; however, based on the animal work, we can make educated guesses as to what the fMRI studies will show in terms of the pattern of cortical, limbic, and hypothalamic sites that are activated when the new mother interacts with her offspring. This constitutes an important direction for future research.

Mammals living in a changing and, sometimes, unpredictable environment possess mechanisms that permit them to alter their behavior to accommodate that variation. These are mechanisms that underlie experience, learning, and memory.

Hence, experience and behavioral plasticity seems to be a feature that characterizes the maternal behavior of most of the mammals studied to date and for many of these the olfactory system plays a prominent role. This sensory system plays a central role allowing the parturitional change in affect from avoidance to attraction to the young and is itself characterized by an intrinsic plasticity, with peripheral receptors and central projection sites undergoing extensive change throughout development and after parturition. In addition to olfaction, somatosensory and auditory cues are also important for maternal behavior in virtually all species that have been studied and these too become attached through learning and conditioning to other cues in the infant complex. The extent to which the neural mechanisms mediating these modalities also undergo enduring change has received some attention in the rat but has not been studied extensively in other species. Similarly, the visual system, so important in primates and humans has received scant attention in its role in the regulation of mothering in nonprimate species or, for that matter, in primates.

We know virtually nothing about how this combination of sensory factors is integrated by the brain to form a unique representation of a young. Is this representation distributed in different brain structures or "localized" in cortical areas and is the localization of this representation different in mammals at different phyletic levels? In this review, we also discuss for each of these different species, the role of fear, attraction, attention, reinforcement, learning, and experience in the regulation of the mothers' maternal behavior. We explore the extent to which experiences both in adulthood and in early life influence the initial expression of maternal behavior which, in most animals, is a highly stereotyped species-characteristic behavior. We find that in all species studied a host of behavioral changes occurs when the mother becomes maternal that enhances her maternal responsiveness; in their general properties, these behavioral changes are similar across species, although for some species, the data are not known. We know that like the rat and human, the mother ewe develops an attraction to the odor of the newborn; however, this has not been adequately studied in monkeys. Whether primates are also less fearful at birth as found for rat mothers, we do not know.

We make a distinction between factors influencing the "motivation" to mother or maternal responsiveness and the processes involved in the formation of a bond toward an individual offspring (in the case of selectivity, as in sheep and primates). The interactions between hormonal, sensory, and neural mechanisms underlying motivation and recognition memory are not yet understood and this issue requires further study.

We illustrate that experience effects are by no means restricted to the "higher" mammals or primates but that they occur at all phyletic levels. We also describe what is known at present about the underlying sensory, endocrine and neural mechanisms of that experience, revealing both their specificity of action and their general properties. Further, we describe the behavioral and physiological effects of experience that mediate the change in subsequent maternal responsiveness. We contrast mechanisms of attachment in the infant and in the mother; in some instances those mechanisms are the same, in others clearly different.

Finally, the role of epigenetic in the regulation of parental behavior and of the effects of experience on individual differences in development is totally unchartered territory in nonhuman mammals or humans. The question arises as to why the same early, and even later, experiences can have such different effects in different individuals. We know nothing about how subsequent postpartum experiences modulate the effects of earlier experiences in the nest and with siblings and how both interact with the genetic and neurochemical profile of the offspring to influence subsequent development and the adult maternal phenotype. Clearly we are at the early stages of our understanding of gene-environment interactions in the development of attachment and bonding processes in nonhuman and human mammals. However, these studies have now been made possible by the development of new genetic assessment technologies and we expect to see huge advances in the near future.

Irrespective of the degree of cognitive complexity, we are all biological creatures who exhibit species-characteristic behaviors that promote the production and maintenance of progeny and, hence, the species. This is accomplished by the interaction of a host of genetic and environmental influences that are expressed in the development of species-characteristic sensory, neuroendocrine, endocrine and neural structures that underlie the expression of maternal behavior at the biologically appropriate time. As this review indicates, we understand a lot about individual factors or mechanisms in the regulation of maternal behavior. Our next task is to unravel and come to understand the interaction of these mechanisms and how these interactions vary across species.

REFERENCES

1. Stern, J. M. (1989). Maternal behavior: sensory, hormonal, and neural determinants. In *Psychoendocrinology* (F. R. Brush and S. Levine S, Eds.), pp. 105–226. Academic Press, New York.
2. Stern, J. M. (1990). Multisensory regulation of maternal behavior and masculine sexual behavior: a revised view. *Neurosci. Biobehav. Rev.* 14, 183–200.
3. Stern, J. M., Dix, L., Bellomo, C., and Thramann, C. (1992). Ventral trunk somatosensory determinants of nursing behavior in Norway rats: role of nipple and surrounding sensations. *Psychobiology* 20, 71–80.
4. Friedman, M. I., Bruno, J. P., and Alberts, J. R. (1981). Physiological and behavioral consequences in rats of water recycling during lactation. *J. Comp. Physiol. Psychol.* 95, 26–35.
5. Jilge, B., and Hudson, R. (2001). Diversity and development of circadian rhythms in the European rabbit. *Chronobiol. Int.* 18, 1–26.
6. Fleming, A. S., and Rosenblatt, J. S. (1974). Maternal behavior in the virgin and lactating rat. *J. Comp. Physiol. Psychol.* 86, 957–972.
7. Rosenblatt, J. S. (1967). Nonhormonal basis of maternal behavior in the rat. *Science* 156, 1512–1514.
8. Gandelman, R. (1973). Induction of maternal nest building in virgin female mice by the presentation of young. *Horm. Behav.* 4, 191–197.
9. Noirot, E. (1972). The onset of maternal behavior in rats, hamsters, and mice: a selective review. In *Advances in the Study of Behavior* (D. S. Lehrman, R. A. Hinde, and E. Shaw, Eds.), Vol. 4, pp. 107–145. Academic Press, New York.

10. Herscher, L., Richmond, J. B., and Moore, A. U. (1963). Modifiability of the critical period for the development of maternal behavior in sheep and goats. *Behaviour* 20, 311–319.
11. Hudson, S. J., and Mullord, M. M. (1977). Investigations on maternal bonding in dairy cattle. *Appl. Anim. Ethol.* 3, 271–276.
12. Klopfer, P. H., Adams, D. K., and Klopfer, M. S. (1964). Maternal imprinting in goats. *Proc. Natl. Acad. Sci. U S A* 52, 588–592.
13. Smith, F. V., Van Toller, C., and Boyes, T. (1966). The "critical period" in the attachment of lambs and ewes. *Anim. Behav.* 14, 120–125.
14. Maletinska, J., Spinka, M., Vichova, J., and Stehulova, I. (2002). Individual recognition of piglets by sows in the early post-partum period. *Behaviour* 139, 975–991.
15. Pryce, C. R. (1996). Socializtion, hormones, and the regulation of maternal behavior in nonhuman simian primates. In *Advances in Study of Behavior* (J. Rosenblatt and C. Snowdon, Eds.), Vol. 25, pp. 432–473. Academic Press, San Diego.
16. Negayama, K., and Honjo, S. (1986). An experimental study on developmental changes of maternal discrimination of infants in crab-eating monkeys (Macaca fascicularis). *Dev. Psychobiol.* 19, 49–56.
17. Leiderman, P., and Leiderman, M. (1977). Economic change and infant care in an East African agricultural community. In *Culture and Infancy: Variations in the Human Experience* (P. H. Leiderman, S. R. Tulkin, and A. Rosenfeld, Eds.). Academic, New York.
18. Brazelton, T. B. (1977). Implications of infant development among Mayan Indians of Mexico. In *Culture and Infancy: Variations in Human Experience* (P. H. Leiderman, S. R. Tulkin, and A. Rosenfeld, Eds.), pp. 151–188. Academic, New York.
19. Konner, M. (1977). Infancy among Kalahari Desert San. In *Culture and Infancy: Variations in the Human Experience* (P. H. Leiderman, S. R. Tulkin, and A. Rosenfeld, Eds.), pp. 287–328. Academic, New York.
20. Caudill, W., and Weinstein, H. (1969). Maternal care and infant behavior in Japan and America. *Psychiatry* 32, 12–43.
21. Moss, H. (1967). Sex, age and state as determinants of mother-infant interaction. *Merrill-Palmer Q.* 128, 19–36.
22. Tronick, E. Z., Morelli, G. A., and Ivey, P. K. (1992). The Efe forager infant and toddler's pattern of social relationships: multiple and simultaneous. *Dev. Psychol.* 28, 568–577.
23. Corter, C., and Fleming, A. S. (2002). Psychobiology of maternal behavior in human beings. In *Handbook of Parenting: Biology and Ecology of Parenting* (M. H. Bornstein, Ed.), Vol. 2, pp. 141–182. Lawrence Erlbaum Associates, Mahwah, NJ.
24. Numan, M. (1994). Maternal behavior. In *The Physiology of Reproduction* (E. Knobil and J. D. Neill, Eds.), Vol. 2, pp. 221–302. Raven Press, New York.
25. Numan, M., and Numan, M. J. (1994). Expression of Fos-like immunoreactivity in the preoptic area of maternally behaving virgin and postpartum rats. *Behav. Neurosci.* 108, 379–394.
26. Rosenblatt, J. S., and Snowdon, C. T., eds. (1996). *Advances in the Study of Behavior*. Academic Press, San Diego.
27. Rosenblatt, J. S. (2002). Hormonal bases of parenting in mammals. In *Handbook of Parenting* (M. H. Bornstein, Ed.), Vol. 2, pp. 31–60. Lawrence Erlbaum Associates, Mahwah, NJ.
28. Rosenblatt, J. S. (1990). Landmarks in the physiological study of maternal behavior with special reference to the rat. In *Mammalian Parenting: Biochemical, Neurobiological, and Behavioral Determinants* (N. A. Krasnegor and R. S. Bridges, Eds.), pp. 40–60. Oxford University Press, New York.
29. Bridges, R. S. (1990). Endocrine regulation of parental behavior in rodents. In *Mammalian Parenting: Biochemical, Neurobiological, and Behavioral Determinants* (N. A. Krasnegor and R. S. Bridges, Eds.), pp. 93–117. Oxford University Press, New York.
30. Insel, T. R. (1990). Regional induction of c-fos-like protein in rat brain after estradiol administration. *Endocrinology* 126, 1849–1853.
31. Insel, T. (1990). Oxytocin and maternal behavior. In *Mammalian Parenting: Biochemical, Neurobiological, and Behavioral Determinants* (N. A. Krasnegor and R. S. Bridges, Eds.), pp. 260–280. Oxford University Press, New York.
32. Bardi, M., Shimizu, K., Barrett, G. M., Borgognini-Tarli, S. M., and Huffman, M. A. (2003). Peripartum sex steroid changes and maternal style in rhesus and Japanese macaques. *Gen. Comp. Endocrinol.* 133, 323–331.
33. Challis, J. R. G., Matthews, S. G., Gibb, W., and Lye, S. J. (2000). Endocrine and paracrine regulation of birth at term and preterm. *Endocr. Rev.* 21, 514–550.
34. Mastorakos, G., and Ilias, I. (2003). Maternal and fetal hypothalamic–pituitary–adrenal axes during pregnancy and postpartum. *Ann. N Y Acad. Sci.* 997, 136–149.
35. Caldwell, J. D., Greer, E. R., Johnson, M. F., Prange, A. J. J., and Pedersen, C. A. (1987). Oxytocin and vasopressin immunoreactivity in hypothalamic and extrahypothalamic sites in late pregnant and postpartum rats. *Neuroendocrinology* 46, 39–47.
36. Young, L. J., Muns, S., Wang, Z., and Insel, T. R. (1997). Changes in oxytocin receptor mRNA in rat brain during pregnancy and the effects of estrogen and interleukin-6. *J. Neuroendocrinol.* 9, 859–865.
37. Keverne, E. B., and Kendrick, K. M. (1990). Neurochemical changes accompanying parturition and their significance for maternal behavior. In *Mammalian Parenting: Biochemical, Neurobiological, and Behavioral Determinants* (N. A. Krasnegor and R. S. Bridges, Eds.), pp. 281–304. Oxford University Press, New York.
38. Rosenberg, P., Leidahl, L., Halaris, A., and Moltz, H. (1976). Changes in the metabolism of hypothalamic norepinephrine associated with the onset of maternal behavior in the nulliparous rat. *Pharmacol. Biochem. Behav.* 4, 647–649.
39. Stafisso-Sandoz, G., Polley, D., Holt, E., Lambert, K. G., and Kinsley, C. H. (1998). Opiate disruption of maternal behavior: morphine reduces, and naloxone restores, c-fos activity in the medial preoptic area of lactating rats. *Brain Res. Bull.* 45, 307–313.
40. Gonzalez-Mariscal, G., Chirino, R., and Hudson, R. (1994). Prolactin stimulates emission of nipple pheromone in ovariectomized New Zealand white rabbits. *Biol. Reprod.* 50, 373–376.
41. Negatu, Z., and McNitt, J. I. (2002). Hormone profiles and nest-building behavior during the periparturient period in rabbit does. *Anim. Reprod. Sci.* 72, 125–135.
42. Challis, J. R. G., and Linzell, J. L. (1971). The concentration of total unconjugated oestrogens in the plasma of pregnant goats. *Reprod. Fertil. Dev.* 26, 401–404.
43. Chamley, W. A., Buckmaster, J., Cereni, M. E., et al. (1973). Changes in the levels of progesterone, corticosteroids, estrone, oestradiol 17–beta, luteinizing hormone and prolactin in the peripheral plasma of the ewe during late pregnancy and at parturition. *Biol. Reprod.* 9, 30–35.
44. Shipka, M. P., and Ford, S. P. (1991). Relationship of circulating estrogen and progesterone concentrations during late pregnancy and the onset phase of maternal behavior in the ewe. *Appl. Anim. Behav. Sci.* 31, 91–99.
45. Baranczuk, R., and Greenwald, G. S. (1974). Plasma levels of oestrogen and progesterone in pregnant and lactating hamsters. *J. Endocrinol.* 63, 125–135.
46. Reyes, F. I., Winter, J. S. D., Faiman, C., and Hobson, W. C. (1973). Serial serum levels of gonadotropins, prolactin and sex steroids in the non pregnant and pregnant chimpanzee. *Endocrinology* 96, 1447–1455.

47. Soares, M. J., and Talamantes, F. (1982). Gestational effects on placental and serum androgen, progesterone and prolactin-like activity in the mouse. *J. Endocrinol.* 95, 29–36.
48. Maestripieri, D., and Megna, N. L. (2000). Hormones and behavior in rhesus macaque abusive and nonabusive mothers 2. Mother–infant interactions. *Physiol. Behav.* 71, 43–49.
49. Bahr, N. I., Martin, R. D., and Pryce, C. R. (2001). Peripartum sex steroid profiles and endocrine correlates of postpartum maternal behavior in captive gorillas (Gorilla gorilla gorilla). *Horm. Behav.* 40, 533–541.
50. Numan, M. (1994). A neural circuitry analysis of maternal behavior in the rat. *Acta Paediatr. Suppl.* 397, 19–28.
51. Benassayag, C., Leroy, M. J., and Rigourd, V. (1999). Estrogens receptors (ER/ERβ) in normal and pathological growth of the human myometrium: pregnancy and leiomyoma. *Am. J. Physiol.* 276, E1112–E1118.
52. Mazor, M., Hershkovitz, R., Chaim, W., et al. (1994). Human preterm birth is associated with systemic and local changes in progesterone/17 beta-estradiol ratios. *Am. J. Obstet. Gynecol.* 171, 231–236.
53. Romero, R., Emamian, M., Wan, M., Grzyboski, C., Hobbins, J. C., and Mitchell, M. D. (1987). Increased concentrations of arachidonic acid lipoxygenase metabolites in amniotic fluid during parturition. *Obstet. Gynecol.* 70, 849–851.
54. Robertson, M. C., and Friesen, H. G. (1981). Two forms of rat placental lactogen revealed by radioimmunoassay. *Endocrinology* 108, 2388–2390.
55. Grattan, D. R. (2001). The actions of prolactin in the brain during pregnancy and lactation. *Prog. Brain Res.* 133, 153–171.
56. Champagne, F., Diorio, J., Sharma, S., and Meaney, M. J. (2001). Naturally occurring variations in maternal behavior in the rat are associated with differences in estrogen-inducible central oxytocin receptors. *Proc. Natl. Acad. Sci. U S A* 98, 12736–12741.
57. Bridges, R. S., and Goldman, B. D. (1975). Diurnal rhythms in gonadotropins and progesterone in lactating and photoperiod induced acyclic hamsters. *Biol. Reprod.* 13, 617–622.
58. Gonzalez-Mariscal, G., Diaz-Sanchez, V., Melo, A. I., Beyer, C., and Rosenblatt, J. S. (1994). Maternal behavior in New Zealand white rabbits: quantification of somatic events, motor patterns, and steroid plasma levels. *Physiol. Behav.* 55, 1081–1089.
59. Davis, S. L., Reichert, L. E. Jr., and Niswender, G. D. (1971). Serum levels of prolactin in sheep as measured by radioimmunoassay. *Biol. Reprod.* 4, 145–153.
60. Ziegler, T. E., Wegner, F. H., Carlson, A. A., Lazaro-Perea, C., and Snowdon, C. T. (2000). Prolactin levels during the periparturitional period in the biparental cotton-top tamarin (*Saguinus oedipus*): Interactions with gender, androgen levels, and parenting. *Horm. Behav.* 38, 111–122.
61. Bast, J. D., and Greenwald, G. S. (1974). The daily concentrations of gonadotrophins and prolactin in the serum of pregnant or lactating hamsters. *J. Endocrinol.* 63, 527–532.
62. Neumann, I. D. (2001). Alterations in behavioral and neuroendocrine stress coping strategies in pregnant, parturient and lactating rats. *Prog. Brain Res.* 133, 143–152.
63. Hudson, S., Mullord, M., Whittlestone, W. G., and Payne, E. (1975). Diurnal variations in blood cortisol in the dairy cow. *J. Dairy Sci.* 58, 30–33.
64. Taverne, M., Bevers, M., Van der Weyden, G., Dieleman, S., and Fontijne, P. (1988). Concentration of growth hormone, prolactin, and cortisol in fetal and maternal blood and amniotic fluid during late pregnancy and parturition in cows with cannulated fetuses. *Anim. Reprod. Sci.* 17, 51–59.
65. Olsson, K., Josater-Hermelin, M., Hossaini-Hilali, J., Hydbring, E., and Dahlborn, K. (1995). Heat stress causes excessive drinking in fed and food deprived pregnant goats. *Comp. Biochem. Physiol. A Physiol.* 110, 309–317.
66. Strott, C. A., Sundel, H., and Stahlman, M. T. (1974). Maternal and fetal plasma progesterone, cortisol, testosterone and 17beta-estradiol in preparturient sheep: response to fetal ACTH infusion. *Endocrinology* 95, 1327–1339.
67. Hydbring, E., Madej, A., MacDonald, E., Drugge, B., Berglund, B., and Olsson, K. (1999). Hormonal changes during parturition in heifers and goats are related to the phases and severity of labour. *J. Endocrinol.* 160, 75–85.
68. Bahr, N. I., Christopher, C. R., Döbeli, M., and Martin, R. D. (1998). Evidence from urinary cortisol that maternal behavior is related to stress in gorillas. *Physiol. Behav.* 64, 429–437.
69. Nolten, W. E., and Rueckert, P. A. (1981). Elevated free cortisol index in pregnancy: possible regulatory mechanisms. *Am. J. Obstet. Gynecol.* 139, 492–498.
70. Chaim, W., and Mazor, M. (1998). The relationship between hormones and human parturition. *Arch. Gynecol. Obstet.* 262, 43–51.
71. Blanks, A. M., and Thornton, S. (2003). The role of oxytocin in parturition. *BJOG* 110 (Suppl 20), 46–51.
72. Kendrick, K. M., Keverne, E. B., Baldwin, B. A., and Sharman, D. F. (1986). Cerebrospinal fluid levels of acetylcholinesterase, monoamines and oxytocin during labour, parturition, vaginocervical stimulation, lamb separation and suckling in sheep. *Neuroendocrinology* 44, 149–156.
73. Levy, F., Kendrick, K. M., Keverne, E. B., Piketty, V., and Poindron, P. (1992). Intracerebral oxytocin is important for the onset of maternal behavior in inexperienced ewes delivered under peridural anesthesia. *Behav. Neurosci.* 106, 427–432.
74. Williams, G. L., Gazal, O. S., Leshin, L. S., Stanko, R. L., and Anderson, L. L. (2001). Physiological regulation of maternal behavior in heifers: Roles of genital stimulation, intracerebral oxytocin release, and ovarian steroids. *Biol. Reprod.* 65, 295–300.
75. Takagi, T., Tanizawa, O., Otsuki, Y., Sugita, N., Haruta, M., and Yamaji, K. (1985). Oxytocin in the cerebrospinal fluid and plasma of pregnant and nonpregnant subjects. *Horm. Metab. Res.* 17, 308–310.
76. Da Costa, A. P., Guevara-Guzman, R. G., Ohkura, S., Goode, J. A., and Kendrick, K. M. (1996). The role of oxytocin release in the paraventricular nucleus in the control of maternal behaviour in the sheep. *J. Neuroendocrinol.* 8, 163–177.
77. Neumann, I., Ludwig, M., Engelmann, M., Pittman, Q. J., and Landgraf, R. (1993). Simultaneous microdialysis in blood and brain: oxytocin and vasopressin release in response to central and peripheral osmotic stimulation and suckling in the rat. *Neuroendocrinology* 58, 637–645.
78. Kendrick, K. M., Keverne, E. B., Chapman, C., and Baldwin, B. A. (1988). Microdialysis measurement of oxytocin, aspartate, gamma-aminobutyric acid and glutamate release from the olfactory bulb of the sheep during vaginocervical stimulation. *Brain Res.* 442, 171–174.
79. Levy, F., Locatelli, A., Piketty, V., Tillet, Y., and Poindron, P. (1995). Involvement of the main but not the accessory olfactory system in maternal behavior of primiparous and multiparous ewes. *Physiol. Behav.* 57, 97–104.
80. Kendrick, K. M., Fabre-Nys, C., Blache, D., Goode, J. A., and Broad, K. D. (1993). The role of oxytocin release in the mediobasal hypothalamus of the sheep in relation to sexual female receptivity. *J. Neuroendocrinol.* 5, 13–21.
81. Landgraf, R., Neumann, I., and Pittman, Q. J. (1991). Septal and hippocampal release of vasopressin and oxytocin during late pregnancy and parturition in the rat. *Neuroendocrinology* 54, 378–383.
82. Broad, K. D., Kendrick, K. M., Sirinathsinghji, D. J., and Keverne, E. B. (1993). Changes in pro-opiomelanocortin and pre-proenkephalin mRNA levels in the ovine brain during pregnancy, parturition and lactation and in response to oestrogen and progesterone. *J. Neuroendocrinol.* 5, 711–719.

83. Broad, K. D., Levy, F., Evans, G., Kimura, T., Keverne, E. B., and Kendrick, K. M. (1999). Previous maternal experience potentiates the effect of parturition on oxytocin receptor mRNA expression in the paraventricular nucleus. *Eur. J. Neurosci.* 11, 3725–3737.
84. Russell, J. A., Leng, G., and Douglas, A. J. (2003). The magnocellular oxytocin system, the fount of maternity: adaptations in pregnancy. *Front. Neuroendocrinol.* 24, 27–61.
85. Neumann, I., Koehler, E., Landgraf, R., and Summy-Long, J. (1994). An oxytocin receptor antagonist infused into the supraoptic nucleus attenuates intranuclear and peripheral release of oxytocin during suckling in conscious rats. *Endocrinology* 134, 141–148.
86. Bridges, R. S., and Ronsheim, P. M. (1987). Immunoreactive beta-endorphin concentrations in brain and plasma during pregnancy in rats: possible modulation by progesterone and estradiol. *Neuroendocrinology* 45, 381–388.
87. Hammer, R. P. J., Mateo, A. R., and Bridges, R. S. (1992). Hormonal regulation of medial preoptic mu-opiate receptor density before and after parturition. *Neuroendocrinology* 56, 38–45.
88. Broad, K. D., Kendrick, K. M., Sirinathsinghji, D. J., and Keverne, E. B. (1993). Changes in oxytocin immunoreactivity and mRNA expression in the sheep brain during pregnancy, parturition and lactation and in response to oestrogen and progesterone. *J. Neuroendocrinol.* 5, 711–719.
89. Fajardo, M. C., Florido, J., Villaverde, C., Oltras, C. M., Gonzalez-Ramirez, A. R., and Gonzalez-Gomez, F. (1994). Plasma levels of beta-endorphin and ACTH during labor and immediate puerperium. *Eur. J. Obstet. Gynecol. Reprod. Biol.* 55, 105–108.
90. Olofsson, C., Ekblom, A., and Ekman-Ordeberg, G., et al. (1997). Increased cerebrospinal fluid concentration of aspartate but decreased concentration of nitric oxide breakdown products in women experiencing visceral pain during active labour. *NeuroReport.* 8, 995–998.
91. Glaser, J., Russell, V. A., and Taljaard, J. J. (1992). Rat brain hypothalamic and hippocampal monoamine and hippocampal beta-adrenergic receptor changes during pregnancy. *Brain Res.* 577, 293–299.
92. Herbison, A. E., Voisin, D. L., Douglas, A. J., and Chapman, C. (1997). Profile of monoamine and excitatory amino acid release in rat supraoptic nucleus over parturition. *Endocrinology* 138, 33–40.
93. Ohno, H., Yamashita, K., and Yahata, T., et al. (1986). Maternal plasma concentrations of catecholamines and cyclic nucleotides during labor and following delivery. *Res. Commun. Chem. Pathol. Pharmacol.* 51, 183–194.
94. Tinkanen, H., Rorarius, M., and Metsa-Ketela, T. (1993). Catecholamine concentrations in venous plasma and cerebrospinal fluid in normal and complicated pregnancy. *Gynecol. Obstet. Invest.* 35, 7–11.
95. Tucker, H. A. (1988). Lactation and its hormonal control. In The *Physiology of Reproduction* (E. Knobil, J. D. Neill, L. L. Ewing, C. L. Market, G. S. Greenwald, and D. W. Pfaff, Eds.), pp. 2235–2264. Raven Press, New York.
96. Challis, J. R. G., and Olsen, D. M. (1988). Parturition. In *The Physiology of Reproduction* (E. Knobil, J. D. Neill, L. L. Ewing, C. L. Market, G. S. Greenwald, and D. W. Pfaff, Eds.), pp. 2177–2216. Raven Press, New York.
97. Hodgen, G. D., and Itskovitz, J. (1988). Recognition and maintenance of pregnancy. In *The Physiology of Reproduction* (E. Knobil, J. D. Neill, L. L. Ewing, C. L. Market, G. S. Greenwald, and D. W. Pfaff, Eds.), pp. 1995–2022. Raven Press, New York.
98. Kristal, M. B., Thompson, A. C., Heller, S. B., and Komisaruk, B. R. (1986). Placenta ingestion enhances analgesia produced by vaginal/cervical stimulation in rats. *Physiol. Behav.* 36, 1017–1020.
99. Gonzalez-Mariscal, G., and Poindron, P. (2002). Parental care in mammals: immediate internal and sensory factors of control. In *Hormones, Brain, and Behavior* (Pfaff, D. W., Ed.), Vol. 1, pp. 215–297. Academic Press, San Diego.
100. Numan, M., and Insel, T. (2003). *The Neurobiology of Parental Behavior.* Springer-Verlag, New York.
101. Siegel, H. I., and Rosenblatt, J. S. (1975). Estrogen-induced maternal behavior in hysterectomized-overiectomized virgin rats. *Physiol. Behav.* 14, 465–471.
102. Doerr, H. K., Siegel, H. I., and Rosenblatt, J. S. (1981). Effects of progesterone withdrawal and estrogen on maternal behavior in nulliparous rats. *Behav. Neural Biol.* 32, 35–44.
103. Siegel, H. I., and Rosenblatt, J. S. (1978). Duration of estrogen stimulation and progesterone inhibition of maternal behavior in pregnancy-terminated rats. *Horm. Behav.* 11, 12–19.
104. Bridges, R. S. (1984). A quantitative analysis of the roles of dosage, sequence, and duration of estradiol and progesterone exposure in the regulation of maternal behavior in the rat. *Endocrinology* 114, 930–940.
105. Bridges, R. S. (1996). Biochemical basis of parental behavior in the rat. In *Advances in the Study of Behavior* (J. S. Rosenblatt and C. T. Snowdon, Eds.), Vol. 25, pp. 215–224. Academic Press, San Diego.
106. Gonzalez-Mariscal, G., and Rosenblatt, J. S. (1996). Maternal behavior in rabbits: a historical and multidisciplinary perspective. In *Parental Care: Evolution, Mechanisms and Adaptive Significance* (J. S. Rosenblatt and C. T. Snowdon, Eds.), Vol. 25, pp. 333–360. Academic Press, San Diego.
107. Gonzalez-Mariscal, G., Jimenez, P., Beyer, C., and Rosenblatt, J. S. (2003). Androgens stimulate specific aspects of maternal nest-building and reduce food intake in rabbits. *Horm. Behav.* 43, 312–317.
108. Lisk, R. D. (1971). Oestrogen and progesterone synergism and elicitation of maternal nest-building in the mouse. *Anim. Behav.* 17, 730–738.
109. Swanson, L. J., and Campbell, C. S. (1979). Induction of maternal behavior in nulliparous golden hamsters (Mesocricetus auratus). *Behav. Neural Biol.* 26, 364–371.
110. Siegel, H. I., and Rosenblatt, J. S. (1980). Hormonal and behavioral aspects of maternal care in the hamster: a review. *Neurosci. Biobehav. Rev.* 4, 17–26.
111. Koch, M., and Ehret, G. (1989). Estradiol and parental experience, but not prolactin are necessary for ultrasound recognition and pup-retrieving in the mouse. *Physiol. Behav.* 45, 771–776.
112. Poindron, P., and Le Neindre, P. (1980). Endocrine and sensory regulation of maternal behavior in the ewe. In *Advances in the Study of Behavior* (J. S. Rosenblatt, R. A. Hinde, and C. Beer, Eds.), Vol. 11, pp. 75–119. Academic Press, New York.
113. Dwyer, C. M., Dingwall, W. S., and Lawrence, A. B. (1999). Physiological correlates of maternal and offspring behaviour in sheep. *Physiol. Behav.* 67, 443–454.
114. Poindron, P., Martin, G. B., and Hooley, R. D. (1979). Effects of lambing induction on the sensitive period for the establishment of maternal behavior in sheep. *Physiol. Behav.* 23, 1081–1087.
115. Poindron, P., Levy, F., and Krehbiel, D. (1988). Genital, olfactory and endocrine interactions in the development of maternal behavior in the parturient ewe. *Psychoneuroendocrinology* 13, 99–125.
116. Kendrick, K. M., Keverne, E. B., Hinton, M. R., and Goode, J. A. (1991). Cerebrospinal fluid and plasma concentrations of oxytocin and vasopressin during parturition and vaginocervical stimulation in the sheep. *Brain Res. Bull.* 26, 803–807.
117. Kendrick, K. M., Levy, F., and Keverne, E. B. (1992). Changes in the sensory processing of olfactory signals induced by birth in sleep. *Science* 256, 833–836.

118. Fite, J., and French, J. (2000). Pre- and postpartum sex steroids in female marmosets (Callithrix kuhlii): is there a link with infant survivorship and maternal behavior? *Horm. Behav.* 38, 1–12.
119. Pryce, C. R., Abbott, D. H., Hodges, J. K., and Martin, R. D. (1988). Maternal behavior is related to prepartum urinary estradiol levels in red-bellied tamarin monkeys. *Physiol. Behav.* 44, 717–726.
120. Maestripieri, D., and Zehr, J. L. (1998). Maternal responsiveness increases during pregnancy and after estrogen treatment in macaques. *Horm. Behav.* 34, 223–230.
121. Pryce, C. R., Dobeli, M., and Martin, R. D. (1993). Effects of sex steroids on maternal motivation in the common marmoset (Callithrix jacchus): development and application of an operant system with maternal reinforcement. *J. Comp. Psychol.* 107, 99–115.
122. Bardi, M., Shimizu, K., Fujita, S., Borgognini-Tarli, S., and Huffman, M. A. (2001). Hormonal correlates of maternal style in captive macaques (Macaca fuscata and M-mulatta). *Int. J. Primatol.* 22, 647–662.
123. Fleming, A. S., Ruble, D., Krieger, H., and Wong, P. Y. (1997). Hormonal and experiential correlates of maternal responsiveness during pregnancy and the puerperium in human mothers. *Horm. Behav.* 31, 145–158.
124. Fleming, A. S., Morgan, H. D., and Walsh, C. (1996). Experiential factors in postpartum regulation of maternal care. In *Advances in the Study of Behavior* (J. S. Rosenblatt and C. T. Snowdon, Eds.), Vol. 25, pp. 295–332. Academic Press, San Diego.
125. Thoman, E. B., and Levine, S. (1970). Hormonal and behavioral changes in the rat mother as a function of early experience treatments of the offspring. *Physiol. Behav.* 5, 1417–1421.
126. Hennessy, M. B., Harney, K. S., Smotherman, W. P., Coyle, S., and Levine, S. (1977). Adrenalectomy-induced deficits in maternal retrieval in the rat. *Horm. Behav.* 9, 222–227.
127. Rees, S. L., Panesar, S., Fleming, A. S., and Steiner, M. (2004). The effects of adrenalectomy and corticosterone replacement on maternal behavior in the postpartum female rat. *Horm. Behav.* 46, 411–419.
128. Neumann, I. D., Toschi, N., Ohl, F., Torner, L., and Kromer, S. A. (2001). Maternal defence as an emotional stressor in female rats: correlation of neuroendocrine and behavioural parameters and involvement of brain oxytocin. *Eur. J. Neurosci.* 13, 1016–1024.
129. Da Costa, A. P., Kampa, R. J., Windle, R. J., Ingram, C. D., and Lightman, S. L. (1997). Region-specific immediate-early gene expression following the administration of corticotropin-releasing hormone in virgin and lactating rats. *Brain Res.* 770, 151–162.
130. Walker, C. D., Toufexis, D. J., and Burlet, A. (2001). Hypothalamic and limbic expression of CRF and vasopressin during lactation: implications for the control of ACTH secretion and stress hyporesponsiveness. *Prog. Brain Res.* 133, 99–110.
131. Keverne, E. B., and Kendrick, K. M. (1991). Morphine and corticotrophin-releasing factor potentiate maternal acceptance in multiparous ewes after vaginocervical stimulation. *Brain Res.* 540, 55–62.
132. Broad, K. D., Keverne, E. B., and Kendrick, K. M. (1995). Corticotrophin releasing factor mRNA expression in the sheep brain during pregnancy, parturition, and lactation and following exogenous progesterone and oestrogen treatment. *Mol. Brain Res.* 29, 310–316.
133. Fleming, A. S., Steiner, M., and Anderson, V. (1987). Hormonal and attitudinal correlates of maternal behaviour during the early postpartum period. *J. Reprod. Infant Psychol.* 5, 193–205.
134. Fleming, A. S., Steiner, M., and Corter, C. (1997). Cortisol, hedonics, and maternal responsiveness in human mothers. *Horm. Behav.* 32, 85–98.
135. Stallings, J., Fleming, A. S., Corter, C., Worthman, C., and Steiner. (2001). The effects of infant cries and odors on sympathy, cortisol, and autonomic responses in new mothers and nonpostpartum women. *Parenting: Sci. Pract.* 1, 71–100.
136. Beckwith, B. E., Lerud, K., Antes, J. R., and Reynolds, B. W. (1983). Hydrocortisone reduces auditory sensitivity at high tonal frequencies in adult males. *Pharmacol. Biochem. Behav.* 19, 431–433.
137. Melamed, S., Ugarten, U., Shirom, A., Kahana, L., Lerman, Y., and Froom, P. (1999). Chronic burnout, somatic arousal and elevated salivary cortisol levels. *J. Psychosomat. Res.* 46, 591–598.
138. Wang, S. (1997). Traumatic stress and attachment. *Acta Physiologia Scand.* 161, 164–169.
139. Henry, J. P., and Wang, S. (1998). Effects of early stress on adult affiliative behavior. *Psychoneuroendocrinology* 23, 863–875.
140. Yeo, J. A., and Keverne, E. B. (1986). The importance of vaginal-cervical stimulation for maternal behaviour in the rat. *Physiol. Behav.* 37, 23–26.
141. Keverne, E. B., Levy, F., Poindron, P., and Lindsay, D. R. (1983). Vaginal stimulation: an important determinant of maternal bonding in sheep. *Science* 219, 81–83.
142. Kendrick, K. M., and Keverne, E. B. (1991). Importance of progesterone and estrogen priming for the induction of maternal behavior by vaginocervical stimulation in sheep: effects of maternal experience. *Physiol. Behav.* 49, 745–750.
143. Rubianes, E. (1992). Genital stimulation modifies behavior towards amniotic fluid in oestrus ewes. *Appl. Anim. Behav. Sci.* 35, 35–40.
144. Lévy, F., Keverne, E. B., Piketty, V., and Poindron, P. (1990). Physiological determinism of olfactory attraction for amniotic fluids in sheep. In *Chemical Signals in Vertebrates* (D. W. MacDonald, D. Müller-Schwarze, and S. E. Natynczuck, Eds.), Vol. 5, pp. 162–165. Oxford University Press, Oxford.
145. Kendrick, K. M., Levy, F., and Keverne, E. B. (1991). Importance of vaginocervical stimulation for the formation of maternal bonding in primiparous and multiparous parturient ewes. *Physiol. Behav.* 50, 595–600.
146. Krehbiel, D., Poindron, P., Levy, F., and Prud'Homme, M. J. (1987). Peridural anesthesia disturbs maternal behavior in primiparous and multiparous parturient ewes. *Physiol. Behav.* 40, 463–472.
147. Lévy, F., Gervais, R., Kindermann, U., Litterio, M., Poindron, P., and Porter, R. (1991). Effects of early postpartum separation on maintenance of maternal responsiveness and selectivity in parturient ewes. *Appl. Anim. Behav. Sci.* 31, 101–110.
148. Meier, G. W. (1965). Maternal behaviour of feral-and laboratory-reared monkeys following the surgical delivery of their infants. *Nature* 206, 492–493.
149. Lundbald, E. G., and Hodgen, G. D. (1980). Induction of maternal-infant bonding in Rhesus and Cynomolgus monkeys after cesarean delivery. *Lab. Anim. Sci.* 30, 913.
150. Bridges, R. S., DiBiase, R., Loundes, D. D., and Doherty, P. C. (1985). Prolactin stimulation of maternal behavior in female rats. *Science* 227, 782–784.
151. Bridges, R. S., and Ronsheim, P. M. (1990). Prolactin (PRL) regulation of maternal behavior in rats: bromocriptine treatment delays and PRL promotes the rapid onset of behavior. *Endocrinology* 126, 837–848.
152. Hauser, H., and Gandelman, R. (1985). Lever pressing for pups: evidence for hormonal influence upon maternal behavior of mice. *Horm. Behav.* 19, 454–468.

153. Voci, V. E., and Carlson, N. R. (1973). Enhancement of maternal behavior and nest building following systemic and diencephalic administration of prolactin lactogen and progesterone in the mouse. *J. Comp. Physiol. Psychol.* 83, 388–393.
154. Lucas, B. K., Ormandy, C. J., Binart, N., Bridges, R. S., and Kelly, P. A. (1998). Null mutation of the prolactin receptor gene produces a defect in maternal behavior. *Endocrinology* 139, 4102–4107.
155. Anderson, C. O., Zarrow, M. X., Fuller, G. B., and Denenberg, V. H. (1971). Pituitary involvement in maternal-nest building in the rabbit. *Horm. Behav.* 2, 183–189.
156. Zarrow, M. X., Gandelman, R., and Denenberg, V. H. (1971). Prolactin: is it an essential hormone for maternal behavior in mammals? *Horm. Behav.* 2, 343–354.
157. Gonzalez-Mariscal, G., Melo, A. I., Chirino, R., Jimenez, P., Beyer, C., and Rosenblatt, J. S. (2000). Pharmacological evidence that prolactin acts from late gestation to promote maternal behavior in rabbits. *J. Neuroendocrinol.* 12, 983–992.
158. Hudson, R., and Distel, H. (1986). Pheromonal release of suckling in rabbits does not depend on the vomeronasal organ. *Physiol. Behav.* 37, 123–128.
159. Schaal, B., Coureaud, G., Langlois, D., Giniès, C., Sémon, E., and Perrier, G. (2003). Chemical and behavioural characterization of the rabbit mammary pheromone. *Nature* 424, 68–72.
160. Hudson, R., Gonzáles-Mariscal, G., and Beyer, C. (1990). Chin-marking behavior, sexual receptivity, and pheromone emission in steroid-treated, ovariectomized rabbits. *Horm. Behav.* 24, 1–13.
161. Rosenblatt, J. S., Factor, E. M., and Mayer, A. D. (1994). Relationship between maternal aggression and maternal care in the rat. *Aggress. Behav.* 20, 243–255.
162. Lonstein, J. S., and Gammie, S, C. (2002). Sensory, hormonal, and neural control of maternal aggression in laboratory rodents. *Neurosci. Biobehav. Rev.* 26, 869–888.
163. Flanelly, K. J., and Flanelly, L. (1987). Time course of postpartum aggression in rats (*Rattus norvegicus*). *J. Comp. Psychol.* 101, 101–103.
164. Mayer, A. D., and Rosenblatt, J. S. (1984). Prepartum changes in maternal responsiveness and nest defense in Rattus norvegicus. *J. Comp. Psychol.* 98, 177–188.
165. Mayer, A. D., and Rosenblatt, J. S. (1993). Peripheral olfactory deafferentation of the primary olfactory system in rats using $ZnSO_4$ nasal spray with special reference to maternal behavior. *Physiol. Behav.* 53, 587–592.
166. Svare, B. (1990). Maternal aggression: hormonal, genetic, and developmental determinants. In *Mammalian Parenting: Biochemical, Neurobiological, and Behavioral Determinants* (N. A. Krasnegor and R. S. Bridges, Eds.), pp. 118–132. Oxford University Press, New York.
167. Mayer, A. D., and Rosenblatt, J. S. (1987). Hormonal factors influence the onset of maternal aggression in laboratory rats. *Horm. Behav.* 21, 253–267.
168. Hansen, S., and Ferreira, A. (1986). Food intake, aggression, and fear behavior in the mother rat: control by neural systems concerned with milk ejection and maternal behavior. *Behav. Neurosci.* 100, 64–70.
169. Mayer, A. D., Monroy, M. A., and Rosenblatt, J. S. (1990). Prolonged estrogen-progesterone treatment of nonpregnant ovariectomized rats: factors stimulating home-cage and maternal aggression and short-latency maternal behavior. *Horm. Behav.* 24, 342–364.
170. Mayer, A. D., Ahdieh, H. B., and Rosenblatt, J. S. (1990). Effects of prolonged estrogen-progesterone treatment and hypophysectomy on the stimulation of short-latency maternal behavior and aggression in female rats. *Horm. Behav.* 24, 152–173.
171. Ghiraldi, L. L., Plonsky, M., and Svare, B. B. (1993). Postpartum aggression in mice: the role of ovarian hormones. *Horm. Behav.* 27, 251–268.
172. Svare, B., and Gandelman, R. (1976). A longitudinal analysis of maternal aggression in Rockland-Swiss albino mice. *Dev. Psychobiol.* 9, 437–446.
173. Mann, M. A., Konen, C., and Svare, B. (1984). The role of progesterone in pregnancy-induced aggression in mice. *Horm. Behav.* 18, 140–160.
174. Svare, B., Miele, J., and Kinsley, C. (1986). Mice: progesterone stimulates aggression in pregnancy-terminated females. *Horm. Behav.* 20, 194–200.
175. Wallen, K. (1990). Desire and ability: hormones and the regulation of female sexual behavior. *Neurosci. Biobehav. Rev.* 14, 233–241.
176. Hansen, S., Harthon, C., Wallin, E., Lofberg, L., and Svensson, K. (1991). Mesotelencephalic dopamine system and reproductive behavior in the female rat: effects of ventral tegmental 6-hydroxydopamine lesions on maternal and sexual responsiveness. *Behav. Neurosci.* 105, 588–598.
177. Stern, J. M. (1996). Trigeminal lesions and maternal behavior in Norway rats: II. Disruption of parturition. *Physiol. Behav.* 60, 187–190.
178. Fleming, A. S., and Orpen, G. (1986). Psychobiology of maternal behavior in rats, selected other species, and humans. In *Origins of Nurturance* (G. Melson and A. Fogel, Eds.), pp. 141–208. L. Erlbaum Associates, New Jersey.
179. Pryce, C. R. (1992). A comparative systems model of the regulation of maternal motivation in mammals. *Anim. Behav.* 43, 417–441.
180. Rosenblatt, J. S., and Mayer, A. D. (1995). An analysis of approach/withdrawal processes in the initiation of maternal behavior in the laboratory rat. In *Behavioral Development, Concepts of Approach/Withdrawal and Integrative Levels* (K. E. Hood, E. Greenberg, and E. Tobach, Eds.), pp. 177–230. Garland Press, New York.
181. Brunelli, S. A., Shindledecker, R. D., and Hofer, M. A. (1985). Development of maternal behaviors in prepubertal rats at three ages: age-characteristic patterns of responses. *Dev. Psychobiol.* 18, 309–326.
182. Rees, S., and Fleming, A. (2001). How early maternal separation and juvenile experience with pups affect maternal behavior and emotionality in adult postpartum rats. *Anim. Learn. Behav.* 29, 221–233.
183. Lovic, V., Gonzalez, A., and Fleming, A. S. (2001). Maternally separated rats show deficits in maternal care in adulthood. *Dev. Psychobiol.* 39, 19–33.
184. Mayer, A. D., Freeman, N. C., and Rosenblatt, J. S. (1979). Ontogeny of maternal behavior in the laboratory rat: factors underlying changes in responsiveness from 30 to 90 days. *Dev. Psychobiol.* 12, 425–439.
185. Gray, O., and Chesley, S. (1984). Development of maternal behavior in nulliparous rats (Rattus norvegicus): Effects of sex and early maternal experience. *J. Comp. Psychol.* 98, 91–99.
186. Fleming, A. S., and Luebke, C. (1981). Timidity prevents the virgin female rat from being a good mother: emotionality differences between nulliparous and parturient females. *Physiol. Behav.* 27, 863–868.
187. Fleming, A. S., and Rosenblatt, J. S. (1974). Olfactory regulation of maternal behavior in rats. I. Effects of olfactory bulb removal in experienced and inexperienced lactating and cycling females. *J. Comp. Physiol. Psychol.* 86, 221–232.
188. Canteras, N. S., Simerly, R. B., and Swanson, L. W. (1995). Organization of projections from the medial nucleus of the amygdala: a PHAL study in the rat. *J. Comp. Neurol.* 360, 213–245.
189. Licht, G., and Meredith, M. (1987). Convergence of main and accessory olfactory pathways onto single neurons in the hamster amygdala. *Exp. Brain Res.* 69, 7–18.

190. Scalia, F., and Winans, S. S. (1975). The differential projections of the olfactory bulb and accessory olfactory bulb in mammals. *J. Comp. Psychol.* 161, 31–55.
191. Swanson, L. W., and Petrovich, G. D. (1998). What is the amygdala? *Trends Neurosci.* 21, 323–331.
192. Davis, M. (1992). The role of the amygdala in fear-potentiated startle: implications for animal models of anxiety. *Trends Pharmacol. Sci.* 13, 35–41.
193. LeDoux, J. E. (2000). Emotion circuits in the brain. *Annu. Rev. Neurosci.* 23, 155–184.
194. Del Cerro, M. C. (1998). Role of the vomeronasal input in maternal behavior. *Psychoneuroendocrinology* 23, 905–926.
195. Fleming, A. S., Vaccarino, F., and Luebke, C. (1980). Amygdaloid inhibition of maternal behavior in the nulliparous female rat. *Physiol. Behav.* 25, 731–743.
196. Numan, M., Numan, M. J., and English, J. B. (1993). Excitotoxic amino acid injections into the medial amygdala facilitate maternal behavior in virgin female rats. *Horm. Behav.* 27, 56–81.
197. Sheehan, T., Paul, M., Amaral, E., Numan, M. J., and Numan, M. (2001). Evidence that the medial amygdala projects to the anterior/ventromedial hypothalamic nuclei to inhibit maternal behavior in rats. *Neuroscience* 106, 341–356.
198. Morgan, H. D., Watchus, J. A., Milgram, N. W., and Fleming, A. S. (1999). The long lasting effects of electrical simulation of the medial preoptic area and medial amygdala on maternal behavior in female rats. *Behav. Brain Res.* 99, 61–73.
199. Bauer, J. H. (1983). Effects of maternal state on the responsiveness to nest odors of hooded rats. *Physiol. Behav.* 30, 229–232.
200. Kinsley, C. H., and Bridges, R. S. (1990). Morphine treatment and reproductive condition alter olfactory preferences for pup and adult male odors in female rats. *Dev. Psychobiol.* 23, 331–347.
201. Fleming, A. S., Cheung, U., Myhal, N., and Kessler, Z. (1989). Effects of maternal hormones on 'timidity' and attraction to pup-related odors in female rats. *Physiol. Behav.* 46, 449–453.
202. Fleming, A. S., Korsmit, M., and Deller, M. (1994). Rat pups are potent reinforcers to the maternal animal: effects of experience, parity, hormones, and dopamine function. *Psychobiology* 22, 44–53.
203. Mattson, B. J., Williams, S., Rosenblatt, J. S., and Morrell, J. I. (2001). Comparison of two positive reinforcing stimuli: pups and cocaine throughout the postpartum period. *Behav. Neurosci.* 115, 683–694.
204. Mattson, B. J., Williams, S. E., Rosenblatt, J. S., and Morrell, J. I. (2003). Preferences for cocaine- or pup-associated chambers differentiates otherwise behaviorally identical postpartum maternal rats. *Psychopharmacol. (Berl).* 167, 1–8.
205. Magnusson, J. E., and Fleming, A. S. (1995). Rat pups are reinforcing to the maternal rat: role of sensory cues. *Psychobiology* 23, 69–75.
206. Lee, A., Clancy, S., and Fleming, A. S. (2000). Mother rats bar-press for pups: effects of lesions of the mpoa and limbic sites on maternal behavior and operant responding for pup-reinforcement. *Behav. Brain Res.* 108, 215–231.
207. Bitran, D., Hilvers, R. J., and Kellogg, C. K. (1991). Ovarian endocrine status modulates the anxiolytic potency of diazepam and the efficacy of gamma-aminobutyric acid-benzodiazepine receptor-mediated chloride ion transport. *Behav. Neurosci.* 105, 653–662.
208. Ferreira, A., Hansen, S., Nielsen, M., Archer, T., and Minor, B. G. (1989). Behavior of mother rats in conflict tests sensitive to antianxiety agents. *Behav. Neurosci.* 103, 193–201.
209. Hard, E., and Hansen, S. (1985). Reduced fearfulness in the lactating rat. *Physiol. Behav.* 35, 641–643.
210. Neumann, I. D., Johnstone, H. A., and Hatzinger, M., et al. (1998). Attenuated neuroendocrine responses to emotional and physical stressors in pregnant rats involve adenohypophysial changes. *J. Physiol.* 508 (Pt 1), 289–300.
211. Toufexis, D. J., Rochford, J., and Walker, C. D. (1999). Lactation-induced reduction in rats' acoustic startle is associated with changes in noradrenergic neurotransmission. *Behav. Neurosci.* 113, 176–184.
212. Fride, E., Dan, Y., Gavish, M., and Weinstock, M. (1985). Prenatal stress impairs maternal behavior in a conflict situation and reduces hippocampal benzodiazepine receptors. *Life Sci.* 36, 2103–2109.
213. Lehmann, J., Stohr, T., and Feldon, J. (2000). Long-term effects of prenatal stress experiences and postnatal maternal separation on emotionality and attentional processes. *Behav. Brain Res.* 107, 133–144.
214. Ohta, R., Shirota, M., Tohei, A., and Taya, K. (2002). Maternal behavior, milk ejection, and plasma hormones in Hatano high- and low-avoidance rats. *Horm. Behav.* 42, 116–125.
215. Smith, J. W., Seckl, J. R., Evans, A. T., Costall, B., and Smythe, J. W. (2004). Gestational stress induces post-partum depression-like behaviour and alters maternal care in rats. *Psychoneuroendocrinology* 29, 227–244.
216. Stern, J. M., Goldman, L., and Levine, S. (1973). Pituitary–adrenal responsiveness during lactation in rats. *Neuroendocrinology* 12, 179–191.
217. Walker, C. D., Lightman, S. L., Steele, M. K., and Dallman, M. F. (1992). Suckling is a persistent stimulus to the adrenocortical system of the rat. *Endocrinology* 130, 115–125.
218. Windle, R. J., Wood, S., and Shanks, N., et al. (1997). Endocrine and behavioural responses to noise stress: comparison of virgin and lactating female rats during non-disrupted maternal activity. *J. Neuroendocrinol.* 9, 407–414.
219. Marques, D. M. (1979). Roles of the main olfactory and vomeronasal systems in the response of the female hamster to young. *Behav. Neural Biol.* 26, 311–329.
220. Gonzalez-Mariscal, G., Chirino, R., Beyer, C., and Rosenblatt, J. S. (2004). Removal of the accessory olfactory bulbs promotes maternal behavior in virgin rabbits. *Behav. Brain Res.* 152, 89–95.
221. Clark, M. M., Spencer, C. A., and Galef, B. G. Jr. (1986). Responses to novel odors mediate maternal behavior and concaveation in gerbils. *Physiol. Behav.* 36, 845–851.
222. Levy, F., and Poindron, P. (1987). The importance of amniotic fluids for the establishment of maternal behavior in experienced and non-experienced ewes. *Anim. Behav.* 35, 1188–1192.
223. Levy, F., Poindron, P., and Le Neindre, P. (1983). Attraction and repulsion by amniotic fluids and their olfactory control in the ewe around parturition. *Physiol. Behav.* 31, 687–692.
224. Vierin, M., and Bouissou, M. (2001). Pregnancy is associated with low fear reactions in ewes. *Physiol. Behav.* 72, 579–587.
225. Gibber, J. R. (1986). Infant-directed behavior of rhesus monkeys during their first pregnancy and parturition. *Folia Primatol. (Basel)* 46, 118–124.
226. Holman, S. D., and Goy, R. W. (1995). Experiential and hormonal correlates of care-giving in rhesus macaques. In *Motherhood in Human and Nonhuman Primates* (C. R. Pryce, R. D. Martin, and D. Skuse, Eds.), pp. 87–92. Karger, Basel.
227. Carothers, A. D. (1990). Risk calculations under heterogeneity: comment on a letter by D. E. Weeks and J. Ott. *Am. J. Hum. Genet.* 47, 165–166.
228. Cox, J. L., Murray, D., and Chapman, G. (1993). A controlled study of the onset, duration and prevalence of postnatal depression. *Br. J. Psychiatr.* 163, 27–31.
229. Cooper, P. J., and Murray, L. (1995). Course and recurrence of postnatal depression: evidence for the specificity of the diagnostic concept. *Br. J. Psychiatry.* 166, 191–195.
230. Cohn, J. F., Campbell, S. B., Matias, R., and Hopkins, J. (1990). Face-to-face interactions of postpartum depressed

and nondepressed mother-infant pairs at 2 months. *Dev. Psychol.* 35, 119–123.
231. Field, T., Sandberg, D., Garcia, R., Bega-Lahr, N., Goldstein, S., and Guy, L. (1985). Pregnancy problems, postpartum depression, and early mother-infant interactions. *Dev. Psychobiol.* 21, 1152–1156.
232. Field, T., Healy, B., Goldstein, S., and Guthertz, M. (1990). Behavior-state matching and synchrony in mother-infant interactions of non-depressed versus depressed dyads. *Dev. Psychol.* 26, 7–14.
233. Fleming, A. S., Ruble, D. N., Flett, G. L., and Shaul, D. L. (1988). Postpartum adjustment in first-time mothers: relations between mood, maternal attitudes and mother-infant interactions. *Dev. Psychol.* 24, 71–81.
234. Righetti-Veltema, M., Conne-Perreard, E., Bousquet, A., and Manzano, J. (2002). Postpartum depression and mother-infant relationship at 3 months old. *J. Affect. Disord.* 70, 291–306.
235. Gonzalez, A., Steiner, M., and Fleming, A. Responsiveness to infant cries in depressed new mothers. *Child Dev.* (submitted).
236. Lovic, V., and Fleming, A. S. (2004). Artificially-reared female rats show reduced prepulse inhibition and deficits in the attentional set shifting task-reversal of effects with maternal-like licking stimulation. *Behav. Brain Res.* 148, 209–219.
237. Harris, B., Lovett, L., Newcombe, R. G., Read, G. F., Walker, R., and Riad-Fahmy, D. (1994). Cardiff puerperal mood and hormone study III. Postnatal depression at 5 to 6 weeks postpartum, and its hormonal correlates across the peripartum period. *Br. J. Psychiatry.* 168, 739–744.
238. Handley, S. L., Dunn, T. L., Waldron, G., and Baker, J. M. (1980). Tryptophan, cortisol and puerperal mood. *Br. J. Psychiatry.* 136, 498–508.
239. Smith, R., Cubis, J., and Brinsmead, M., et al. (1990). Mood changes, obstetric experience and alterations in plasma cortisol, beta-endorphin and corticotrophin releasing hormone during pregnancy and the puerperium. *J. Psychosom. Res.* 34, 53–69.
240. O'Hara, M. W., Schlechte, J. A., Lewis, D. A., and Wright, E. J. (1991). Prospective study of postpartum blues. Biologic and psychosocial factors. *Arch. Gen. Psychiatry.* 48, 801–806.
241. Harris, B., Lovett, L., Newcombe, R. G., Read, G. F., Walker, R., and Riad-Fahmy, D. (1994). Maternity blues and major endocrine changes: Cardiff puerperal mood and hormone study II. *BMJ* 308, 949–953.
242. Stallings, J., Fleming, A. S., Corter, C., Worthman, C., and Steiner, M. (2001). The effects of infant cries and odors on sympathy, cortisol, and autonomic responses in new mothers and nonpostpartum women. *Parenting, Science and Practice* 1, 71–100.
243. Krpan, K., Coombs, R., Zinga, D., Steiner, M., and Fleming, A. S. (2004). Experiential and hormonal correlates of maternal behaviour in teen and adult mothers. *Horm. Behav.*
244. Fleming, A. S., Corter, C., Franks, P., Surbey, M., Schneider, B., and Steiner, M. (1993). Postpartum factors related to mother's attraction to newborn infant odors. *Dev. Psychobiol.* 26, 115–132.
245. Altemus, M., Deuster, P. A., Galliven, E., Carter, C. S., and Gold, P. W. (1995). Suppression of hypothalamic-pituitary-adrenal axis responses to stress in lactating women. *J. Clin. Endocrinol. Metab.* 80, 2954–2959.
246. Uvnas-Moberg, K. (1997). Physiological and endocrine effects of social contact. *Ann. N Y Acad. Sci.* 807, 146–163.
247. Stern, J. M., and Kolunie, J. M. (1993). Maternal aggression of rats is impaired by cutaneous anesthesia of the ventral trunk, but not by nipple removal. *Physiol. Behav.* 54, 861–868.
248. Erskine, M. S., Barfield, R. J., and Goldman, B. D. (1978). Intraspecific fighting during late pregnancy and lactation in rats and effects of litter removal. *Behav. Biol.* 23, 206–218.
249. Erskine, M. S., Barfield, R. J., and Goldman, B. D. (1980). Postpartum aggression in rats: I. Effects of hypophysectomy. *J. Comp. Physiol. Psychol.* 94, 484–494.
250. Svare, B., Mann, M. A., Broida, J., and Michael, S. D. (1982). Maternal aggression exhibited by hypophysectomized parturient mice. *Horm. Behav.* 16, 455–461.
251. Erskine, M. S., Barfield, R. J., and Goldman, B. D. (1980). Postpartum aggression in rats: II. Dependence on maternal sensitivity to young and effects of experience with pregnancy and parturition. *J. Comp. Physiol. Psychol.* 94, 495–505.
252. Ferreira, A., Pereira, M., Agrati, D., Uriarte, N., and Fernandez-Guasti, A. (2002). Role of maternal behavior on aggression, fear and anxiety. *Physiol. Behav.* 77, 197–204.
253. Svare, B., and Gandelman, R. (1976). Postpartum aggression in mice: the influence of suckling stimulation. *Horm. Behav.* 7, 407–416.
254. Hansen, S., Ferreira, A., and Selart, M. E. (1985). Behavioural similarities between mother rats and benzodiazepine-treated non-maternal animals. *Psychopharmacol. (Berl).* 86, 344–347.
255. Olivier, B., Mos, J., and van Oorschot, R. (1985). Maternal aggression in rats: effects of chlordiazepoxide and fluprazine. *Psychopharmacol (Berl).* 86, 68–76.
256. Parmigiani, S., Ferrari, P. F., and Palanza, P. (1998). An evolutionary approach to behavioral pharmacology: using drugs to understand proximate and ultimate mechanisms of different forms of aggression in mice. *Neurosci. Biobehav. Rev.* 23, 143–153.
257. Yoshimura, H., and Ogawa, N. (1991). Ethopharmacology of maternal aggression in mice: effects of diazepam and SM-3997. *Eur. J. Pharmacol.* 200, 147–153.
258. Maestripieri, D., Badiani, A., and Puglisi-Allegra, S. (1991). Prepartal chronic stress increases anxiety and decreases aggression in lactating female mice. *Behav. Neurosci.* 105, 663–668.
259. Maestripieri, D., and D'Amato, F. R. (1991). Anxiety and maternal aggression in house mice (Mus musculus): a look at interindividual variability. *J. Comp. Psychol.* 105, 295–301.
260. Parmigiani, S., Palanza, P., Rogers, J., and Ferrari, P. F. (1999). Selection, evolution of behavior and animal models in behavioral neuroscience. *Neurosci. Biobehav. Rev.* 23, 957–969.
261. Lonstein, J. S., and Stern, J. M. (1998). Site and behavioral specificity of periaqueductal gray lesions on postpartum sexual, maternal, and aggressive behaviors in rats. *Brain Res.* 804, 21–35.
262. Bridges, R. S., and Grattan, D. R. (2003). Prolactin-induced neurogenesis in the maternal brain. *Trends Endocrinol. Metab.* 14, 199–201.
263. Shingo, T., Gregg, C., Enwere, E., et al. (2003). Pregnancy-stimulated neurogenesis in the adult female forebrain mediated by prolactin. *Science* 299, 117–120.
264. Dunn, A. J., and Berridge, C. W. (1990). Physiological and behavioral responses to corticotropin-releasing factor administration: is CRF a mediator of anxiety or stress responses? *Brain Res. Rev.* 15, 71–100.
265. Torner, L., Toschi, N., Nava, G., Clapp, C., and Neumann, I. D. (2002). Increased hypothalamic expression of prolactin in lactation: involvement in behavioural and neuroendocrine stress responses. *Eur. J. Neurosci.* 15, 1381–1389.
266. Bale, T. L., Davis, A. M., Auger, A. P., Dorsa, D. M., and McCarthy, M. M. (2001). CNS region-specific oxytocin receptor expression: importance in regulation of anxiety and sex behavior. *J. Neurosci.* 21, 2546–2552.
267. Neumann, I. D., Wigger, A., Torner, L., Holsboer, F., and Landgraf, R. (2000). Brain oxytocin inhibits basal and stress-induced activity of the hypothalamo–pituitary–adrenal axis in male and female rats: partial action within the paraventricular nucleus. *J. Neuroendocrinol.* 12, 235–243.

268. Mens, W. B., Witter, A., and van Wimersma Greidanus, T. B. (1983). Penetration of neurohypophyseal hormones from plasma into cerebrospinal fluid (CSF): half-times of disappearance of these neuropeptides from CSF. *Brain Res.* 262, 143–149.
269. Beach, F. A., and Jaynes, J. (1956). Studies of maternal retrieving in rats: III. Sensory cues involved in the lactating female's response to her young. *Behaviour* 10, 104–125.
270. Herrenkohl, L. R., and Rosenberg, P. A. (1972). Exteroceptive stimulation of maternal behavior in the naive rat. *Physiol. Behav.* 8, 595–598.
271. Ihnat, R., White, N. R., and Barfield, R. J. (1995). Pup's broadband vocalizations and maternal behavior in the rat. *Behav. Proc.* 33, 257–272.
272. Kolunie, J. M., Stern, J. M., and Barfield, R. J. (1994). Maternal aggression in rats: effects of visual or auditory deprivation of the mother and dyadic pattern of ultrasonic vocalizations. *Behav. Neural Biol.* 62, 41–49.
273. Benuck, I., and Rowe, F. A. (1975). Centrally and peripherally induced anosmia: influences on maternal behavior in lactating female rats. *Physiol. Behav.* 14, 439–447.
274. Fleming, A. S., Gavarth, K., and Sarker, J. (1992). Effects of transections to the vomeronasal nerves or to the main olfactory bulbs on the initiation and long-term retention of maternal behavior in primiparous rats. *Behav. Neural Biol.* 57, 177–188.
275. Kolunie, J. M., and Stern, J. M. (1995). Maternal aggression in rats: effects of olfactory bulbectomy, $ZnSO_4$-induced anosmia, and vomeronasal organ removal. *Horm. Behav.* 29, 492–518.
276. Gandelman, R., Zarrow, M. X., Denenberg, V. H., and Myers, M. (1971). Olfactory bulb removal eliminates maternal behavior in the mouse. *Science* 171, 210–211.
277. Lepri, J. J., Wysocki, C. J., and Vandenbergh, J. G. (1985). Mouse vomeronasal organ: effects on chemosignal production and maternal behavior. *Physiol. Behav.* 35, 809–814.
278. Vandenbergh, J. G. (1973). Effects of central and peripheral anosmia on reproduction of female mice. *Physiol. Behav.* 10, 257–261.
279. Kenyon, P., Cronin, P., and Keeble, S. (1981). Disruption of maternal retrieving by perioral anesthesia. *Physiol. Behav.* 27, 313–321.
280. Kenyon, P., Cronin, P., and Keeble, S. (1983). Role of the infraorbital nerve in retrieving behavior in lactating rats. *Behav. Neurosci.* 97, 255–269.
281. Stern, J. M., and Kolunie, J. M. (1989). Perioral anesthesia disrupts maternal behavior during early lactation in Long-Evans rats. *Behav. Neural Biol.* 52, 20–38.
282. Stern, J. M., and Kolunie, J. M. (1991). Trigeminal lesions and maternal behavior in Norway rats: I. Effects of cutaneous rostral snout denervation on maintenance of nurturance and maternal aggression. *Behav. Neurosci.* 105, 984–997.
283. Stern, J. M., and Johnson, S. K. (1990). Ventral somatosensory determinants of nursing behavior in Norway rats. I. Effects of variations in the quality and quantity of pup stimuli. *Physiol. Behav.* 47, 993–1011.
284. Lonstein, J. S., and Stern, J. M. (1997). Role of the midbrain periaqueductal gray in maternal nurturance and aggression: c-fos and electrolytic lesion studies in lactating rats. *J. Neurosci.* 17, 3364–3378.
285. Lonstein, J. S., Wagner, C. K., and De Vries, G. J. (1999). Comparison of the "nursing" and other parental behaviors of nulliparous and lactating female rats. *Horm. Behav.* 36, 242–251.
286. Lonstein, J. S., and Stern, J. M. (1997). Somatosensory contributions to c-fos activation within the caudal periaqueductal gray of lactating rats: effects of perioral, rooting, and suckling stimuli from pups. *Horm. Behav.* 32, 155–166.
287. Stern, J. M. (1991). Nursing posture is elicited rapidly in maternally naive, haloperidol-treated female and male rats in response to ventral trunk stimulation from active pups. *Horm. Behav.* 25, 504–517.
288. Morgan, H. D., Fleming, A. S., and Stern, J. M. (1992). Somatosensory control of the onset and retention of maternal responsiveness in primiparous Sprague-Dawley rats. *Physiol. Behav.* 51, 549–555.
289. Stern, J. M. (1997). Offspring-induced nurturance: animal-human parallels. *Dev. Psychobiol.* 31, 19–37.
290. Carter, C., and Keverne, B. (2002). Affiliation and pair bonding. In *Hormones, Brain and Behavior* (D. W. Pfaff, A. M. Etgen, S. E. Fahrbach, and R. T. Rubin, Eds.), Vol. 1, pp. 299–338. Academic Press, San Diego.
291. Levy, F., and Poindron, P. (1984). Influence du liquide amniotique sur la manifestation du comportement maternel chez la brebis parturiente (The influence of amniotic fluid on maternal behavior in parturient sheep). *Biol. Behav.* 9, 65–88.
292. Basiouni, G. F., and Gonyou, H. W. (1988). Use of birth fluids and cervical stimulation in lamb fostering. *J. Anim. Sci.* 66, 872–879.
293. Baldwin, B. A., and Shillito, E. E. (1974). The effects of ablation of the olfactory bulbs on parturition and maternal behaviour in Soay sheep. *Anim. Behav.* 22, 220–223.
294. Poindron, P. (1976). Mother-young relationships in intact or anosmic ewes at the time of suckling. *Biol. Behav.* 2, 161–177.
295. Alexander, G., and Stevens, D. (1981). Recognition of washed lambs by merino ewes. *Appl. Anim. Ethol.* 7, 77–86.
296. Klopfer, P. H., and Gamble, J. (1966). Maternal "imprinting" in goats: the role of chemical senses. *Z. Tierpsychol.* 23, 588–592.
297. Romeyer, A., Poindron, P., and Orgeur, P. (1994). Olfaction mediates the establishment of selective bonding in goats. *Physiol. Behav.* 56, 693–700.
298. Brennan, P. A., and Keverne, E. B. (1997). Neural mechanisms of mammalian olfactory learning. *Prog. Neurobiol.* 51, 457–481.
299. Fleming, A. S., Corter, C., Surbey, M., Franks, P., and Steiner, M. (1995). Postpartum factors related to mother's recognition of newborn infant odors. *J. Reprod. Infant Psychol.* 13, 197–210.
300. Porter, R. H., Cernoch, J. M., and McLaughlin, F. J. (1983). Maternal recognition of neonates through olfactory cues. *Physiol. Behav.* 30, 151–154.
301. Schaal, B., and Porter, R. H. (1991). "Microsomatic humans" revisited: the generation and perception of chemical signals. *Adv. Study Behav.* 20, 135–199.
302. Wisenfeld, A., and Klorman, R. (1978). The mother's psychophysiological reactions to contrasting affective expressions by her own and an unfamiliar infant. *Dev. Psychol.* 14, 294–304.
303. Leavitt, L., and Donovan, W. (1979). Perceived infant temperament, focus of control, and maternal physiological response to infant gaze. *J. Res. Personality* 13, 267–278.
304. Furedy, J., Fleming, A. S., and Ruble, D., et al. (1989). Sex differences in small-magnitude heart-rate response to sexual and infant-related stimuli: a psychophysiological approach. *Physiol. Behav.* 46, 903–905.
305. Nott, P. N., Franklin, M., Armitage, C., and Gelder, M. G. (1976). Hormonal changes and mood in the puerperium. *Br. J. Psychiatr.* 128, 379–383.
306. Bleichfeld, B., and Moely, B. E. (1984). Psychophysiological responses to an infant cry: comparison of groups of women in different phases of the maternal cycle. *Dev. Psychol.* 20, 1082–1091.
307. Canteras, N. S., Simerly, R. B., and Swanson, L. W. (1994). Organization of projections from the ventromedial nucleus of the hypothalamus: a Phaseolus vulgaris-leucoagglutinin study in the rat. *J. Comp. Neurol.* 348, 41–79.

308. Fleming, A. S., Miceli, M., and Moretto, D. (1983). Lesions of the medial preoptic area prevent the facilitation of maternal behavior produced by amygdala lesions. *Physiol. Behav.* 31, 503–510.
309. Gray, P., and Brooks, P. J. (1984). Effect of lesion location within the medial preoptic-anterior hypothalamic continuum on maternal and male sexual behaviors in female rats. *Behav. Neurosci.* 98, 703–711.
310. Jacobson, C. D., Terkel, J., Gorski, R. A., and Sawyer, C. H. (1980). Effects of small medial preoptic area lesions on maternal behavior: retrieving and nest building in the rat. *Brain Res.* 194, 471–478.
311. Kalinichev, M., Rosenblatt, J. S., and Morrell, J. I. (2000). The medial preoptic area, necessary for adult maternal behavior in rats, is only partially established as a component of the neural circuit that supports maternal behavior in juvenile rats. *Behav. Neurosci.* 114, 196–210.
312. Lee, A. W., and Brown, R. E. (2002). Medial preoptic lesions disrupt parental behavior in both male and female California mice (Peromyscus californicus). *Behav. Neurosci.* 116, 968–975.
313. Miceli, M. O., Fleming, A. S., and Malsbury, C. W. (1983). Disruption of maternal behaviour in virgin and postparturient rats following sagittal plane knife cuts in the preoptic area-hypothalamus. *Behav. Brain Res.* 9, 337–360.
314. Numan, M. (1974). Medial preoptic area and maternal behavior in the female rat. *J. Comp. Physiol. Psychol.* 87, 746–759.
315. Numan, M., and Callahan, E. C. (1980). The connections of the medial preoptic region and maternal behavior in the rat. *Physiol. Behav.* 25, 653–665.
316. Numan, M., Corodimas, K. P., Numan, M. J., Factor, E. M., and Piers, W. D. (1988). Axon-sparing lesions of the preoptic region and substantia innominata disrupt maternal behavior in rats. *Behav. Neurosci.* 102, 381–396.
317. Numan, M., and Numan, M. J. (1997). Projection sites of medial preoptic area and ventral bed nucleus of the stria terminalis neurons that express Fos during maternal behavior in female rats. *J. Neuroendocrinol.* 9, 369–384.
318. Numan, M., Rosenblatt, J. S., and Komisaruk, B. R. (1977). Medial preoptic area and onset of maternal behavior in the rat. *J. Comp. Physiol. Psychol.* 91, 146–164.
319. Oxley, G., and Fleming, A. S. (2000). The effects of medial preoptic area and amygdala lesions on maternal behavior in the juvenile rat. *Dev. Psychobiol.* 37, 253–265.
320. Numan, M., McSparren, J., and Numan, M. J. (1990). Dorsolateral connections of the medial preoptic area and maternal behavior in rats. *Behav. Neurosci.* 104, 964–979.
321. Terkel, J., Bridges, R. S., and Sawyer, C. H. (1979). Effects of transecting lateral neural connections of the medial preoptic area on maternal behavior in the rat: nest building, pup retrieval and prolactin secretion. *Brain Res.* 169, 369–380.
322. Numan, M., and Corodimas, K. P. (1985). The effects of paraventricular hypothalamic lesions on maternal behavior in rats. *Physiol. Behav.* 35, 417–425.
323. Shughrue, P. J., Lane, M. V., and Merchenthaler, I. (1997). Comparative distribution of estrogen receptor-alpha and -beta mRNA in the rat central nervous system. *J. Comp. Neurol.* 388, 507–525.
324. Numan, M., Roach, J. K., del Cerro, M. C., et al. (1999). Expression of intracellular progesterone receptors in rat brain during different reproductive states, and involvement in maternal behavior. *Brain Res.* 830, 358–371.
325. Bakowska, J. C., and Morrell, J. I. (1997). Atlas of the neurons that express mRNA for the long form of the prolactin receptor in the forebrain of the female rat. *J. Comp. Neurol.* 386, 161–177.
326. Bakowska, J. C., and Morrell, J. I. (2003). The distribution of mRNA for the short form of the prolactin receptor in the forebrain of the female rat. *Brain Res. Mol. Brain Res.* 116, 50–58.
327. Bridges, R. S., Numan, M., Ronsheim, P. M., Mann, P. E., and Lupini, C. E. (1990). Central prolactin infusions stimulate maternal behavior in steroid-treated, nulliparous female rats. *Proc. Natl. Acad. Sci. U S A* 87, 8003–8007.
328. Bridges, R. S., Robertson, M. C., Shiu, R. P., Sturgis, J. D., Henriquez, B. M., and Mann, P. E. (1997). Central lactogenic regulation of maternal behavior in rats: steroid dependence, hormone specificity, and behavioral potencies of rat prolactin and rat placental lactogen I. *Endocrinology* 138, 756–763.
329. Wagner, C. K., and Morrell, J. I. (1996). Levels of estrogen receptor immunoreactivity are altered in behaviorally-relevant brain regions in female rats during pregnancy. *Brain Res. Mol. Brain Res.* 42, 328–336.
330. Giordano, A. L., Siegel, H. I., and Rosenblatt, J. S. (1989). Nuclear estrogen receptor binding in the preoptic area and hypothalamus of pregnancy-terminated rats: correlation with the onset of maternal behavior. *Neuroendocrinology* 50, 248–258.
331. Giordano, A. L., Siegel, H. I., and Rosenblatt, J. S. (1991). Nuclear estrogen receptor binding in microdissected brain regions of female rats during pregnancy: implications for maternal and sexual behavior. *Physiol. Behav.* 50, 1263–1267.
332. Ahdieh, H. B., Mayer, A. D., and Rosenblatt, J. S. (1987). Effects of brain antiestrogen implants on maternal behavior and on postpartum estrus in pregnant rats. *Neuroendocrinology* 46, 522–531.
333. Bridges, R., Rigero, B., Byrnes, E., Yang, L., and Walker, A. (2001). Central infusions of the recombinant human prolactin receptor antagonist, S179D-PRL, delay the onset of maternal behavior in steroid-primed, nulliparous female rats. *Endocrinology* 142, 730–739.
334. Francis, K., Meddle, S. L., Bishop, V. R., and Russell, J. A. (2002). Progesterone receptor expression in the pregnant and parturient rat hypothalamus and brainstem. *Brain Res.* 927, 18–26.
335. Lonstein, J. S., and Blaustein, J. D. (2004). Immunocytochemical investigation of nuclear progestin receptor expression within dopaminergic neurones of the female rat brain. *J. Neuroendocrinol.* 16, 534–543.
336. Numan, M. (1978). Progesterone inhibition of maternal behavior in the rat. *Horm. Behav.* 11, 209–231.
337. Morgan, J. I., and Curran, T. (1991). Stimulus-transcription coupling in the nervous system: involvement of the inducible proto-oncogenes fos and jun. *Annu. Rev. Neurosci.* 14, 421–451.
338. Sheng, M., and Greenberg, M. E. (1990). The regulation and function of c-fos and other immediate early genes in the nervous system. *Neuron* 4, 477–485.
339. Bozas, E., Tritos, N., Phillipidis, H., and Stylianopoulou, F. (1997). At least three neurotransmitter systems mediate a stress-induced increase in c-fos mRNA in different rat brain areas. *Cell Mol. Neurobiol.* 17, 157–169.
340. Robertson, L. M., Kerppola, T. K., and Vendrell, M., et al. (1995). Regulation of c-fos expression in transgenic mice requires multiple interdependent transcription control elements. *Neuron* 14, 241–252.
341. Xia, Z., Dudek, H., Miranti, C. K., and Greenberg, M. E. (1996). Calcium influx via the NMDA receptor induces immediate early gene transcription by a MAP kinase/ERK-dependent mechanism. *J. Neurosci.* 16, 5425–5436.
342. Fleming, A. S., Suh, E. J., Korsmit, M., and Rusak, B. (1994). Activation of Fos-like immunoreactivity in the medial preoptic area and limbic structures by maternal and social interactions in rats. *Behav. Neurosci.* 108, 724–734.
343. Lonstein, J. S., Simmons, D. A., and Stern, J. M. (1998). Functions of the caudal periaqueductal gray in lactating

rats: kyphosis, lordosis, maternal aggression, and fearfulness. *Behav. Neurosci.* 112, 1502–1518.

344. Numan, M., and Numan, M. J. (1995). Importance of pup-related sensory inputs and maternal performance for the expression of Fos-like immunoreactivity in the preoptic area and ventral bed nucleus of the stria terminalis of postpartum rats. *Behav. Neurosci.* 109, 135–149.
345. Stack, E. C., and Numan, M. (2000). The temporal course of expression of c-Fos and Fos B within the medial preoptic area and other brain regions of postpartum female rats during prolonged mother-young interactions. *Behav. Neurosci.* 114, 609–622.
346. Stack, E. C., Balakrishnan, R., Numan, M. J., and Numan, M. (2002). A functional neuroanatomical investigation of the role of the medial preoptic area in neural circuits regulating maternal behavior. *Behav. Brain Res.* 131, 17–36.
347. Walsh, C. J., Fleming, A. S., Lee, A., and Magnusson, J. E. (1996). The effects of olfactory and somatosensory desensitization on Fos-like immunoreactivity in the brains of pup-exposed postpartum rats. *Behav. Neurosci.* 110, 134–153.
348. Kalinichev, M., Rosenblatt, J. S., Nakabeppu, Y., and Morrell, J. I. (2000). Induction of c-fos-like and fosB-like immunoreactivity reveals forebrain neuronal populations involved differentially in pup-mediated maternal behavior in juvenile and adult rats. *J. Comp. Neurol.* 416, 45–78.
349. Lonstein, J. S., Simmons, D. A., Swann, J. M., and Stern, J. M. (1998). Forebrain expression of c-fos due to active maternal behaviour in lactating rats. *Neuroscience* 82, 267–281.
350. Sheehan, T. P., Cirrito, J., Numan, M. J., and Numan, M. (2000). Using c-Fos immunocytochemistry to identify forebrain regions that may inhibit maternal behavior in rats. *Behav. Neurosci.* 114, 337–352.
351. Fahrbach, S. E., Morrell, J. I., and Pfaff, D. W. (1986). Identification of medial preoptic neurons that concentrate estradiol and project to the midbrain in the rat. *J. Comp. Neurol.* 247, 364–382.
352. Lonstein, J. S., Greco, B., De Vries, G. J., Stern, J. M., and Blaustein, J. D. (2000). Maternal behavior stimulates c-fos activity within estrogen receptor alpha-containing neurons in lactating rats. *Neuroendocrinology* 72, 91–101.
353. Schuchard, M., Landers, J. P., Sandhu, N. P., and Spelsberg, T. C. (1993). Steroid hormone regulation of nuclear proto-oncogenes. *Endocr. Rev.* 14, 659–669.
354. Auger, A. P., and Blaustein, J. D. (1995). Progesterone enhances an estradiol-induced increase in Fos immunoreactivity in localized regions of female rat forebrain. *J. Neurosci.* 15, 2272–2279.
355. Da Costa, A. P., Broad, K. D., and Kendrick, K. M. (1997). Olfactory memory and maternal behaviour-induced changes in c-fos and zif/268 mRNA expression in the sheep brain. *Brain Res. Mol.* 46, 63–76.
356. Sheehan, T., and Numan, M. (2002). Estrogen, progesterone, and pregnancy termination alter neural activity in brain regions that control maternal behavior in rats. *Neuroendocrinology* 75, 12–23.
357. Bridges, R. S., Mann, P. E., and Coppeta, J. S. (1999). Hypothalamic involvement in the regulation of maternal behaviour in the rat: inhibitory roles for the ventromedial hypothalamus and the dorsal/anterior hypothalamic areas. *J. Neuroendocrinol.* 11, 259–266.
358. Bandler, R., and Shipley, M. T. (1994). Columnar organization in the midbrain periaqueductal gray: modules for emotional expression? *Trends Neurosci.* 17, 379–389.
359. Fuchs, S. A., Edinger, H. M., and Siegel, A. (1985). The organization of the hypothalamic pathways mediating affective defense behavior in the cat. *Brain Res.* 330, 77–92.
360. Silveira, M. C., and Graeff, F. G. (1992). Defense reaction elicited by microinjection of kainic acid into the medial hypothalamus of the rat: antagonism by a GABAA receptor agonist. *Behav. Neural Biol.* 57, 226–232.
361. Bandler, R. (1988). Brain mechanisms of aggression as revealed by electrical and chemical stimulation: suggestion of a central role of the midbrain periaqueductal grey region. *Prog. Psychobiol. Physiol. Psychol.* 13, 67–153.
362. Fanselow, M. S. (1991). The midbrain periaqueductal gray as a coordinator of action in response to fear and anxiety. In *The Midbrain Periaqueductal Gray* (A. Depaulis and R. Bandler, Eds.), pp. 151–173. Plenum Press, New York.
363. Rizvi, T. A., Ennis, M., and Shipley, M. T. (1992). Reciprocal connections between the medial preoptic area and the midbrain periaqueductal gray in rat: a WGA-HRP and PHA-L study. *J. Comp. Neurol.* 315, 1–15.
364. Polston, E. K., and Erskine, M. S. (2001). Excitotoxic lesions of the medial amygdala differentially disrupt prolactin secretory responses in cycling and mated female rats. *J. Neuroendocrinol.* 13, 13–21.
365. Lonstein, J. S., and De Vries, G. J. (2001). Social influences on parental and nonparental responses toward pups in virgin female prairie voles (Microtus ochrogaster). *J. Comp. Psychol.* 115, 53–61.
366. Deutch, A. Y., Goldstein, M., Baldino, F. Jr., and Roth, R. H. (1988). Telencephalic projections of the A8 dopamine cell group. *Ann. N Y Acad. Sci.* 537, 27–50.
367. Dunnett, S. B., and Robbins, T. W. (1992). The functional role of mesotelencephalic dopamine systems. *Biol. Rev. Camb. Philos. Soc.* 67, 491–518.
368. Swanson, L. W. (1982). The projections of the ventral tegmental area and adjacent regions: a combined fluorescent retrograde tracer and immunofluorescence study in the rat. *Brain Res. Bull.* 9, 321–353.
369. Numan, M., and Numan, M. J. (1996). A lesion and neuroanatomical tract-tracing analysis of the role of the bed nucleus of the stria terminalis in retrieval behavior and other aspects of maternal responsiveness in rats. *Dev. Psychobiol.* 29, 23–51.
370. Everitt, B. J. (1990). Sexual motivation: a neural and behavioural analysis of the mechanisms underlying appetitive and copulatory responses of male rats. *Neurosci. Biobehav. Rev.* 14, 217–232.
371. Horvitz, J. C. (2002). Dopamine gating of glutamatergic sensorimotor and incentive motivational input signals to the striatum. *Behav. Brain Res.* 137, 65–74.
372. Kelley, A. E., and Berridge, K. C. (2002). The neuroscience of natural rewards: relevance to addictive drugs. *J. Neurosci.* 22, 3306–3311.
373. Mogenson, G. J., and Wu, M. (1988). Disruption of food hoarding by injections of procaine into mediodorsal thalamus, GABA into subpallidal region and haloperidol into accumbens. *Brain Res. Bull.* 20, 247–251.
374. Hunt, G. E., and McGregor, I. S. (2002). Contrasting effects of dopamine antagonists and frequency reduction on Fos expression induced by lateral hypothalamic stimulation. *Behav. Brain Res.* 132, 187–201.
375. Everitt, B. J., Dickinson, A., and Robbins, T. W. (2001). The neuropsychological basis of addictive behaviour. *Brain Res. Brain Res. Rev.* 36, 129–138.
376. Jenkins, W. J., and Becker, J. B. (2001). Role of the striatum and nucleus accumbens in paced copulatory behavior in the female rat. *Behav. Brain Res.* 121, 119–128.
377. Pfaus, J. G., and Phillips, A. G. (1991). Role of dopamine in anticipatory and consummatory aspects of sexual behavior in the male rat. *Behav. Neurosci.* 105, 727–743.
378. Stratford, T. R., Swanson, C. J., and Kelley, A. (1998). Specific changes in food intake elicited by blockade or activation

of glutamate receptors in the nucleus accumbens shell. *Behav. Brain Res.* 93, 43–50.

379. Gaffori, O., and Le Moal, M. (1979). Disruption of maternal behavior and appearance of cannibalism after ventral mesencephalic tegmentum lesions. *Physiol. Behav.* 23, 317–323.
380. Hansen, S., Harthon, C., Wallin, E., Lofberg, L., and Svensson, K. (1991). The effects of 6-OHDA-induced dopamine depletions in the ventral or dorsal striatum on maternal and sexual behavior in the female rat. *Pharmacol. Biochem. Behav.* 39, 71–77.
381. Keer, S. E., and Stern, J. M. (1999). Dopamine receptor blockade in the nucleus accumbens inhibits maternal retrieval and licking, but enhances nursing behavior in lactating rats. *Physiol. Behav.* 67, 659–669.
382. Silva, M. R., Bernardi, M. M., Cruz-Casallas, P. E., and Felicio, L. F. (2003). Pimozide injections into the nucleus accumbens disrupt maternal behaviour in lactating rats. *Pharmacol. Toxicol.* 93, 42–47.
383. Champagne, F. A., Chretien, P., Stevenson, C. W., Zhang, T. Y., Gratton, A., and Meaney, M. J. (2004). Variations in nucleus accumbens dopamine associated with individual differences in maternal behavior in the rat. *J. Neurosci.* 24, 4113–4123.
384. Hansen, S., Bergvall, A. H., and Nyiredi, S. (1993). Interaction with pups enhances dopamine release in the ventral striatum of maternal rats: a microdialysis study. *Pharmacol. Biochem. Behav.* 45, 673–676.
385. Spielewoy, C., Roubert, C., Hamon, M., Nosten-Bertrand, M., Betancur, C., and Giros, B. (2000). Behavioural disturbances associated with hyperdopaminergia in dopamine-transporter knockout mice. *Behav. Pharmacol.* 11, 279–290.
386. Stern, J. M., and Protomastro, M. (2000). Effects of low dosages of apomorphine on maternal responsiveness in lactating rats. *Pharmacol. Biochem. Behav.* 66, 353–359.
387. Vernotica, E. M., Rosenblatt, J. S., and Morrell, J. I. (1999). Microinfusion of cocaine into the medial preoptic area or nucleus accumbens transiently impairs maternal behavior in the rat. *Behav. Neurosci.* 113, 377–390.
388. Numan, M., and Smith, H. G. (1984). Maternal behavior in rats: evidence for the involvement of preoptic projections to the ventral tegmental area. *Behav. Neurosci.* 98, 712–727.
389. Numan, M., and Numan, M. J. (1991). Preoptic-brainstem connections and maternal behavior in rats. *Behav. Neurosci.* 105, 1013–1029.
390. Lee, A., Li, M., Watchus, J., and Fleming, A. (1999). Neuroanatomical basis of maternal memory in postpartum rats: selective role for the nucleus accumbens. *Behav. Neurosci.* 113, 523–538.
391. Li, M., and Fleming, A. S. (2003). Differential involvement of nucleus accumbens shell and core subregions in maternal memory in postpartum female rats. *Behav. Neurosci.* 117, 426–445.
392. Li, M., and Fleming, A. S. (2003). The nucleus accumbens shell is critical for normal expression of pup-retrieval in postpartum female rats. *Behav. Brain Res.* 145, 99–111.
393. Smith, M. O., and Holland, R. C. (1975). Effects of lesions of the nucleus accumbens on lactation and postpartum behavior. *Physiol. Psychol.* 3, 331–336.
394. Numan, M., Numan, M. J., Schwarz, J. M., Neuner, C. M., Flood, T. F., and Smith, C. D. (2005). Medial preoptic area interactions with the nucleus accumbens-ventral pallidum circuit and maternal behavior in rats. *Behav. Brain Res.* 158, 53–68.
395. Meredith, G. E., Pennartz, C. M., and Groenewegen, H. J. (1993). The cellular framework for chemical signalling in the nucleus accumbens. *Prog. Brain Res.* 99, 3–24.
396. Pennartz, C. M., Groenewegen, H. J., and Lopes da Silva, F. H. The nucleus accumbens as a complex of functionally distinct neuronal ensembles: an integration of behavioural, electrophysiological and anatomical data. *Prog. Neurobiol.* 42, 719–761.
397. Lim, M. M., and Young, L. J. (2004). Vasopressin-dependent neural circuits underlying pair bond formation in the monogamous prairie vole. *Neuroscience* 125, 35–45.
398. Malick, A., and Burstein, R. (1998). Cells of origin of the trigeminohypothalamic tract in the rat. *J. Comp. Neurol.* 400, 125–144.
399. Pfaff, D., and Keiner, M. (1973). Atlas of estradiol-concentrating cells in the central nervous system of the female rat. *J. Comp. Neurol.* 151, 121–158.
400. Insel, T. R., and Harbaugh, C. R. (1989). Lesions of the hypothalamic paraventricular nucleus disrupt the initiation of maternal behavior. *Physiol. Behav.* 45, 1033–1041.
401. De Vries, G. J., and Buijs, R. M. (1983). The origin of the vasopressinergic and oxytocinergic innervation of the rat brain with special reference to the lateral septum. *Brain Res.* 273, 307–317.
402. Lang, R. E., Heil, J., Ganten, D., Hermann, K., Rascher, W., and Unger, T. (1983). Effects of lesions in the paraventricular nucleus of the hypothalamus on vasopressin and oxytocin contents in brainstem and spinal cord of rat. *Brain Res.* 260, 326–329.
403. Fleischer, S., and Slotnick, B. M. (1978). Disruption of maternal behavior in rats with lesions of the septal area. *Physiol. Behav.* 21, 189–200.
404. Salzberg, H. C., Lonstein, J. S., and Stern, J. M. (2002). GABA(A) receptor regulation of kyphotic nursing and female sexual behavior in the caudal ventrolateral periaqueductal gray of postpartum rats. *Neuroscience* 114, 675–687.
405. Monassi, C. R., Leite-Panissi, C. R., and Menescal-de-Oliveira, L. (1999). Ventrolateral periaqueductal gray matter and the control of tonic immobility. *Brain Res. Bull.* 50, 201–208.
406. Stern, J. M., Yu, Y. L., and Crockett, D. P. (2002). Dorsolateral columns of the spinal cord are necessary for both suckling-induced neuroendocrine reflexes and the kyphotic nursing posture in lactating rats. *Brain Res.* 947, 110–121.
407. Corodimas, K. P., Rosenblatt, J. S., Canfield, M. E., and Morrell, J. I. (1993). Neurons in the lateral subdivision of the habenular complex mediate the hormonal onset of maternal behavior in rats. *Behav. Neurosci.* 107, 827–843.
408. Matthews Felton, T., Linton, L. N., Rosenblatt, J. S., and Morrell, J. I. (1998). First order and second order maternal behavior related afferents of the lateral habenula. *NeuroReport* 10, 883–887.
409. Franzen, E. A., and Myers, R. E. (1973). Neural control of social behavior: prefrontal and anterior temporal cortex. *Neuropsychologia* 11, 141–157.
410. Myers, R. E., Swett, C., and Miller, M. (1973). Loss of social group affinity following prefrontal lesions in free-ranging macaques. *Brain Res.* 64, 257–269.
411. Beach, F. A. (1937). The neural basis of innate behavior: I. The effects of cortical lesions upon the maternal behavior pattern in the rat. *J. Comp. Psychol.* 24, 393–436.
412. Davis, C. D. (1939). The effect of ablations of neocortex on mating, maternal behavior and the production of pseudopregnancy in the female rat and on copulatory activity in the male. *Am. J. Physiol.* 127, 374–380.
413. Stone, C. P. (1938). Effects of cortical destruction on reproductive behavior and maze learning in albino rats. *J. Comp. Psychol.* 26, 217–236.
414. Broad, K. D., Hinton, M. R., Keverne, E. B., and Kendrick, K. M. (2002). Involvement of the medial prefrontal cortex in mediating behavioural responses to odour cues rather than olfactory recognition memory. *Neuroscience* 114, 715–729.

415. Kalivas, P. W., Churchill, L., and Romanides, A. (1999). Involvement of the pallidal-thalamocortical circuit in adaptive behavior. *Ann. N Y Acad. Sci.* 877, 64–70.
416. Zahm, D. S. (1999). Functional-anatomical implications of the nucleus accumbens core and shell subterritories. *Ann. N Y Acad. Sci.* 877, 113–128.
417. Ray, J. P., and Price, J. L. (1992). The organization of the thalamocortical connections of the mediodorsal thalamic nucleus in the rat, related to the ventral forebrain-prefrontal cortex topography. *J. Comp. Neurol.* 323, 167–197.
418. Pedersen, C. A. (1997). Oxytocin control of maternal behavior. Regulation by sex steroids and offspring stimuli. *Ann. N Y Acad. Sci.* 807, 126–145.
419. Pedersen, C. A., and Boccia, M. L. (2003). Oxytocin antagonism alters rat dams' oral grooming and upright posturing over pups. *Physiol. Behav.* 80, 233–241.
420. Fahrbach, S. E., Morrell, J. I., and Pfaff, D. W. (1984). Oxytocin induction of short-latency maternal behavior in nulliparous, estrogen-primed female rats. *Horm. Behav.* 18, 267–286.
421. Fahrbach, S. E., Morrell, J. I., and Pfaff, D. W. (1986). Effect of varying the duration of pre-test cage habituation on oxytocin induction of short-latency maternal behavior. *Physiol. Behav.* 37, 135–139.
422. Pedersen, C. A., Ascher, J. A., Monroe, Y. L., and Prange, A. J. Jr. (1982). Oxytocin induces maternal behavior in virgin female rats. *Science* 216, 648–650.
423. Pedersen, C. A., and Prange, A. J. Jr. (1979). Induction of maternal behavior in virgin rats after intracerebroventricular administration of oxytocin. *Proc. Natl. Acad. Sci. U S A* 76, 6661–6665.
424. Wamboldt, M. Z., and Insel, T. R. (1987). The ability of oxytocin to induce short latency maternal behavior is dependent on peripheral anosmia. *Behav. Neurosci.* 101, 439–441.
425. Fahrbach, S. E., Morrell, J. I., and Pfaff, D. W. (1985). Possible role for endogenous oxytocin in estrogen-facilitated maternal behavior in rats. *Neuroendocrinology* 40, 526–532.
426. Pedersen, C. A., Caldwell, J. D., Johnson, M. F., Fort, S. A., and Prange, A. J. Jr. (1985). Oxytocin antiserum delays onset of ovarian steroid-induced maternal behavior. *Neuropeptides* 6, 175–182.
427. van Leengoed, E., Kerker, E., and Swanson, H. H. (1987). Inhibition of post-partum maternal behaviour in the rat by injecting an oxytocin antagonist into the cerebral ventricles. *J. Endocrinol.* 112, 275–282.
428. Alves, S. E., Lopez, V., McEwen, B. S., and Weiland, N. G. (1998). Differential colocalization of estrogen receptor beta (ERbeta) with oxytocin and vasopressin in the paraventricular and supraoptic nuclei of the female rat brain: an immunocytochemical study. *Proc. Natl. Acad. Sci. U S A* 95, 3281–3286.
429. Patisaul, H. B., Scordalakes, E. M., Young, L. J., and Rissman, E. F. (2003). Oxytocin, but not oxytocin receptor, is regulated by oestrogen receptor beta in the female mouse hypothalamus. *J. Neuroendocrinol.* 15, 787–793.
430. Zhang, J. Q., Su, B. Y., and Cai, W. Q. (2004). Immunolocalization of estrogen receptor beta in the hypothalamic paraventricular nucleus of female mice during pregnancy, lactation and postnatal development. *Brain Res.* 997, 89–96.
431. Burbach, J. P., van Schaick, H. S., de Bree, F. M., Lopes da Silva, S., and Adan, R. A. (1995). Functional domains in the oxytocin gene for regulation of expression and biosynthesis of gene products. *Adv. Exp. Med. Biol.* 395, 9–21.
432. Burbach, J. P. H., Adan, R. A. H., van Tol, H. H. M., et al. (1990). Regulation of rat oxytocin gene by estradiol. *J. Neuroscience.* 2, 633–639.
433. Amico, J. A., Crowley, R. S., Insel, T. R., Thomas, A., and O'Keefe, J. A. (1995). Effect of gonadal steroids upon hypothalamic oxytocin expression. *Adv. Exp. Med. Biol.* 395, 23–35.
434. Amico, J. A., Thomas, A., and Hollingshead, D. J. (1997). The duration of estradiol and progesterone exposure prior to progesterone withdrawal regulates oxytocin mRNA levels in the paraventricular nucleus of the rat. *Endocr. Res.* 23, 141–156.
435. Kremarik, P., Freund-Mercier, M. J., and Stoeckel, M. E. (1995). Oxytocin and vasopressin binding sites in the hypothalamus of the rat: histoautoradiographic detection. *Brain Res. Bull.* 36, 195–203.
436. Veinante, P., and Freund-Mercier, M. J. (1997). Distribution of oxytocin- and vasopressin-binding sites in the rat extended amygdala: a histoautoradiographic study. *J. Comp. Neurol.* 383, 305–325.
437. Insel, T. R. (1986). Postpartum increases in brain oxytocin binding. *Neuroendocrinology* 44, 515–518.
438. Pedersen, C. A., Caldwell, J. D., Walker, C., Ayers, G., and Mason, G. A. (1994). Oxytocin activates the postpartum onset of rat maternal behavior in the ventral tegmental and medial preoptic areas. *Behav. Neurosci.* 108, 1163–1171.
439. Yu, G. Z., Kaba, H., Okutani, F., Takahashi, S., and Higuchi, T. (1996). The olfactory bulb: a critical site of action for oxytocin in the induction of maternal behaviour in the rat. *Neuroscience* 72, 1083–1088.
440. Yu, G. Z., Kaba, H., Okutani, F., Takahashi, S., Higuchi, T., and Seto, K. (1996). The action of oxytocin originating in the hypothalamic paraventricular nucleus on mitral and granule cells in the rat main olfactory bulb. *Neuroscience* 72, 1073–1082.
441. Davis, M., Rainnie, D., and Cassell, M. (1994). Neurotransmission in the rat amygdala related to fear and anxiety. *Trends Neurosci.* 17, 208–214.
442. LeDoux, J. E. (1993). Emotional memory systems in the brain. *Behav. Brain Res.* 58, 69–79.
443. Sakanaka, M., Shibasaki, T., and Lederis, K. (1987). Corticotropin releasing factor-like immunoreactivity in the rat brain as revealed by a modified cobalt-glucose oxidase-diaminobenzidine method. *J. Comp. Neurol.* 260, 256–298.
444. Almeida, O. F., Yassouridis, A., and Forgas-Moya, I. (1994). Reduced availability of milk after central injections of corticotropin-releasing hormone in lactating rats. *Neuroendocrinology* 59, 72–77.
445. Pedersen, C. A., Caldwell, J. D., McGuire, M., and Evans, D. L. (1991). Corticotropin-releasing hormone inhibits maternal behavior and induces pup-killing. *Life Sci.* 48, 1537–1546.
446. Neumann, I. D., Kromer, S. A., Toschi, N., and Ebner, K. (2000). Brain oxytocin inhibits the (re)activity of the hypothalamo–pituitary–adrenal axis in male rats: involvement of hypothalamic and limbic brain regions. *Regul. Pept.* 96, 31–38.
447. Amico, J. A., Mantella, R. C., Vollmer, R. R., and Li, X. (2004). Anxiety and stress responses in female oxytocin deficient mice. *J. Neuroendocrinol.* 16, 319–324.
448. Boccia, M. L., and Pedersen, C. A. (1999). Early maternal separation alters postpartum pup-licking and grooming, lactation-associated changes in aggression and anxiety, and central oxytocin binding in female offspring. *Society for Neuroscience, 29th Annual Meeting*. Miami Beach.
449. Boccia, M. L., and Pedersen, C. A. (2001). Brief vs. long maternal separations in infancy: contrasting relationships with adult maternal behavior and lactation levels of aggression and anxiety. *Psychoneuroendocrinology* 26, 657–672.
450. Bosch, O. J., Kromer, S. A., Brunton, P. J., and Neumann, I. D. (2004). Release of oxytocin in the hypothalamic paraventricular nucleus, but not central amygdala or lateral septum in lactating residents and virgin intruders during maternal defence. *Neuroscience* 124, 439–448.

451. Consiglio, A. R., and Lucion, A. B. (1996). Lesion of hypothalamic paraventricular nucleus and maternal aggressive behavior in female rats. *Physiol. Behav.* 59, 591–596.
452. Giovenardi, M., Padoin, M. J., Cadore, L. P., and Lucion, A. B. (1998). Hypothalamic paraventricular nucleus modulates maternal aggression in rats: effects of ibotenic acid lesion and oxytocin antisense. *Physiol. Behav.* 63, 351–359.
453. Lubin, D. A., Elliott, J. C., Black, M. C., and Johns, J. M. (2003). An oxytocin antagonist infused into the central nucleus of the amygdala increases maternal aggressive behavior. *Behav. Neurosci.* 117, 195–201.
454. Kendrick, K. M., Keverne, E. B., and Baldwin, B. A. (1987). Intracerebroventricular oxytocin stimulates maternal behaviour in the sheep. *Neuroendocrinology* 46, 56–61.
455. Kendrick, K. M., Keverne, E. B., Hinton, M. R., and Goode, J. A. (1992). Oxytocin, amino acid and monoamine release in the region of the medial preoptic area and bed nucleus of the stria terminalis of the sheep during parturition and suckling. *Brain Res.* 569, 199–209.
456. Rubin, B. S., and Bridges, R. S. (1984). Disruption of ongoing maternal responsiveness in rats by central administration of morphine sulfate. *Brain Res.* 307, 91–97.
457. Mann, P. E., Kinsley, C. H., and Bridges, R. S. (1991). Opioid receptor subtype involvement in maternal behavior in lactating rats. *Neuroendocrinology* 53, 487–492.
458. Mann, P. E., Pasternak, G. W., and Bridges, R. S. (1990). Mu 1 opioid receptor involvement in maternal behavior. *Physiol. Behav.* 47, 133–138.
459. Simerly, R. B., Gorski, R. A., and Swanson, L. W. (1986). Neurotransmitter specificity of cells and fibers in the medial preoptic nucleus: an immunohistochemical study in the rat. *J. Comp. Neurol.* 246, 343–363.
460. Hammer, R. P. Jr., and Bridges, R. S. (1987). Preoptic area opioids and opiate receptors increase during pregnancy and decrease during lactation. *Brain Res.* 420, 48–56.
461. Wardlaw, S. L., and Frantz, A. G. (1983). Brain beta-endorphin during pregnancy, parturition, and the postpartum period. *Endocrinology* 113, 1664–1668.
462. Mann, P. E., Rubin, B. S., and Bridges, R. S. (1997). Differential proopiomelanocortin gene expression in the medial basal hypothalamus of rats during pregnancy and lactation. *Brain Res. Mol. Brain Res.* 46, 9–16.
463. Miranda-Paiva, C. M., Ribeiro-Barbosa, E. R., Canteras, N. S., and Felicio, L. F. (2003). A role for the periaqueductal grey in opioidergic inhibition of maternal behaviour. *Eur. J. Neurosci.* 18, 667–674.
464. Thompson, A. C., and Kristal, M. B. (1996). Opioid stimulation in the ventral tegmental area facilitates the onset of maternal behavior in rats. *Brain Res.* 743, 184–201.
465. Kalivas, P. W. (1993). Neurotransmitter regulation of dopamine neurons in the ventral tegmental area. *Brain Res. Rev.* 18, 75–113.
466. Xi, Z. X., Wu, G., Stein, E. A., and Li, S. J. (2002). GABAergic mechanisms of heroin-induced brain activation assessed with functional MRI. *Magn. Reson. Med.* 48, 838–843.
467. Gintzler, A. R., and Liu, N. J. (2001). The maternal spinal cord: biochemical and physiological correlates of steroid-activated antinociceptive processes. *Prog. Brain Res.* 133, 83–97.
468. Kristal, M. B., Thompson, A. C., and Grishkat, H. L. (1985). Placenta ingestion enhances opiate analgesia in rats. *Physiol. Behav.* 35, 481–486.
469. Kristal, M. B., Tarapacki, J. A., and Barton, D. (1990). Amniotic fluid ingestion enhances opioid-mediated but not nonopioid-mediated analgesia. *Physiol. Behav.* 47, 79–81.
470. Tarapacki, J. A., Piech, M., and Kristal, M. B. (1995). Ingestion of amniotic fluid by postpartum rats enhances morphine antinociception without liability to maternal behavior. *Physiol. Behav.* 57, 209–212.
471. Caba, M., Poindron, P., Krehbiel, D., Levy, F., Romeyer, A., and Venier, G. (1995). Naltrexone delays the onset of maternal behavior in primiparous parturient ewes. *Pharmacol. Biochem. Behav.* 52, 743–748.
472. Kalin, N. H., Shelton, S. E., and Lynn, D. E. (1995). Opiate systems in mother and infant primates coordinate intimate contact during reunion. *Psychoneuroendocrinology* 20, 735–742.
473. Kendrick, K. M., and Keverne, E. B. (1989). Effects of intracerebroventricular infusions of naltrexone and phentolamine on central and peripheral oxytocin release and on maternal behaviour induced by vaginocervical stimulation in the ewe. *Brain Res.* 505, 329–332.
474. Kendrick, K. M., Da Costa, A. P., and Broad, K. D., et al. (1997). Neural control of maternal behaviour and olfactory recognition of offspring. *Brain Res. Bull.* 44, 383–395.
475. Martel, F. L., Nevison, C. M., Rayment, F. D., Simpson, M. J., and Keverne, E. B. (1993). Opioid receptor blockade reduces maternal affect and social grooming in rhesus monkeys. *Psychoneuroendocrinology* 18, 307–321.
476. Misiti, A., Turillazzi, P. G., Zapponi, G. A., and Loizzo, A. (1991). Heroin induces changes in mother-infant monkey communication and subsequent disruption of their dyadic interaction. *Pharmacol. Res.* 24, 93–104.
477. Douglas, A. J., and Russell, J. A. (2001). Endogenous opioid regulation of oxytocin and ACTH secretion during pregnancy and parturition. *Prog. Brain Res.* 133, 67–82.
478. Lavi-Avnon, Y., Malkesman, O., Hurwitz, I., and Weller, A. (2004). Mother-infant interactions in rats lacking CCK(A) receptors. *Behav. Neurosci.* 118, 282–289.
479. Mann, P. E., Felicio, L. F., and Bridges, R. S. (1995). Investigation into the role of cholecystokinin (CCK) in the induction and maintenance of maternal behavior in rats. *Horm. Behav.* 29, 392–406.
480. Miranda-Paiva, C. M., Nasello, A. G., Yim, A. J., and Felicio, L. F. (2002). Puerperal blockade of cholecystokinin (CCK1) receptors disrupts maternal behavior in lactating rats. *J. Mol. Neurosci.* 18, 97–104.
481. Micevych, P., and Sinchak, K. (2001). Estrogen and endogenous opioids regulate CCK in reproductive circuits. *Peptides* 22, 1235–1244.
482. Sheehan, T. P., and Numan, M. (1997). Microinjection of the tachykinin neuropeptide K into the ventromedial hypothalamus disrupts the hormonal onset of maternal behavior in female rats. *J. Neuroendocrinol.* 9, 677–687.
483. Numan, M., and Sheehan, T. P. (1997). Neuroanatomical circuitry for mammalian maternal behavior. *Ann. N Y Acad. Sci.* 807, 101–125.
484. Leckman, J. F., and Herman, A. E. (2002). Maternal behavior and developmental psychopathology. *Biol. Psychiatr.* 51, 27–43.
485. Nishimori, K., Young, L. J., Guo, Q., Wang, Z., Insel, T. R., and Matzuk, M. M. (1996). Oxytocin is required for nursing but is not essential for parturition or reproductive behavior. *Proc. Natl. Acad. Sci. U S A* 93, 11699–11704.
486. Young, W. S. III, Shepard, E., Amico, J., et al. (1996). Deficiency in mouse oxytocin prevents milk ejection, but not fertility or parturition. *J. Neuroendocrinol.* 8, 847–853.
487. Insel, T. R., Gingrich, B. S., and Young, L. J. (2001). Oxytocin: who needs it? *Prog. Brain Res.* 133, 59–66.
488. McCarthy, M. M. (1990). Oxytocin inhibits infanticide in female house mice (Mus domesticus). *Horm. Behav.* 24, 365–375.
489. Ogawa, S., Taylor, J. A., Lubahn, D. B., Korach, K. S., and Pfaff, D. W. (1996). Reversal of sex roles in genetic female mice by disruption of estrogen receptor gene. *Neuroendocrinology* 64, 467–470.

490. Ogawa, S., Eng, V., Taylor, J., Lubahn, D. B., Korach, K. S., and Pfaff, D. W. (1998). Roles of estrogen receptor-alpha gene expression in reproduction-related behaviors in female mice. *Endocrinology* 139, 5070–5081.
491. Ogawa, S., Chan, J., Chester, A. E., Gustafsson, J. A., Korach, K. S., and Pfaff, D. W. (1999). Survival of reproductive behaviors in estrogen receptor beta gene-deficient (betaERKO) male and female mice. *Proc. Natl. Acad. Sci. U S A* 96, 12887–12892.
492. DeVito, W. J. (1988). Distribution of immunoreactive prolactin in the male and female rat brain: effects of hypophysectomy and intraventricular administration of colchicine. *Neuroendocrinology* 47, 284–289.
493. Dutt, A., Kaplitt, M. G., Kow, L. M., and Pfaff, D. W. (1994). Prolactin, central nervous system and behavior: a critical review. *Neuroendocrinology* 59, 413–419.
494. Paut-Pagano, L., Roky, R., Valatx, J. L., Kitahama, K., and Jouvet, M. (1993). Anatomical distribution of prolactin-like immunoreactivity in the rat brain. *Neuroendocrinology* 58, 682–695.
495. Shivers, B. D., Harlan, R. E., and Pfaff, D. W. (1989). A subset of neurons containing immunoreactive prolactin is a target for estrogen regulation of gene expression in rat hypothalamus. *Neuroendocrinology* 49, 23–27.
496. Emanuele, N. V., Jurgens, J. K., Halloran, M. M., Tentler, J. J., Lawrence, A. M., and Kelley, M. R. (1992). The rat prolactin gene is expressed in brain tissue: detection of normal and alternatively spliced prolactin messenger RNA. *Mol. Endocrinol.* 6, 35–42.
497. Horseman, N. D., Zhao, W., and Montecino-Rodriguez, E., et al. (1997). Defective mammopoiesis, but normal hematopoiesis, in mice with a targeted disruption of the prolactin gene. *EMBO J.* 16, 6926–6935.
498. Leblond, C. P., and Nelson, W. O. (1937). Maternal behavior in hypophysectomized male and female mice. *Am. J. Physiol.* 120, 167–172.
499. Ormandy, C. J., Camus, A., and Barra, J., et al. (1997). Null mutation of the prolactin receptor gene produces multiple reproductive defects in the mouse. *Genes Dev.* 11, 167–178.
500. Torner, L., Toschi, N., Pohlinger, A., Landgraf, R., and Neumann, I. D. (2001). Anxiolytic and anti-stress effects of brain prolactin: improved efficacy of antisense targeting of the prolactin receptor by molecular modeling. *J. Neurosci.* 21, 3207–3214.
501. Torner, L., Maloumby, R., Nava, G., Aranda, J., Clapp, C., and Neumann, I. D. (2004). In vivo release and gene upregulation of brain prolactin in response to physiological stimuli. *Eur. J. Neurosci.* 19, 1601–1608.
502. Moltz, H., Rowland, D., Steele, M., and Halaris, A. (1975). Hypothalamic norepinephrine: concentration and metabolism during pregnancy and lactation in the rat. *Neuroendocrinology* 19, 252–258.
503. Rosenberg, P., Halaris, A., and Moltz, H. (1977). Effects of central norepinephrine depletion on the initiation and maintenance of maternal behavior in the rat. *Pharmacol. Biochem. Behav.* 6, 21–24.
504. Thomas, S. A., and Palmiter, R. D. (1997). Impaired maternal behavior in mice lacking norepinephrine and epinephrine. *Cell* 91, 583–592.
505. Brown, J. R., Ye, H., Bronson, R. T., Dikkes, P., and Greenberg, M. E. (1996). A defect in nurturing in mice lacking the immediate early gene fosB. *Cell* 86, 297–309.
506. Mathieson, W. B., Wilkinson, M., Brown, R. E., Bond, T. L., Taylor, S. W., and Neumann, P. E. (2002). FOS and FOSB expression in the medial preoptic nucleus pars compacta of maternally active C57BL/6J and DBA/2J mice. *Brain Res.* 952, 170–175.
507. Keverne, E. B. (2001). Genomic imprinting, maternal care, and brain evolution. *Horm. Behav.* 40, 146–155.
508. Lefebvre, L., Viville, S., Barton, S. C., Ishino, F., Keverne, E. B., and Surani, M. A. (1998). Abnormal maternal behaviour and growth retardation associated with loss of the imprinted gene Mest. *Nat. Genet.* 20, 163–169.
509. Li, L., Keverne, E. B., Aparicio, S. A., Ishino, F., Barton, S. C., and Surani, M. A. (1999). Regulation of maternal behavior and offspring growth by paternally expressed Peg3. *Science* 284, 330–333.
510. Gammie, S. C., and Nelson, R. J. (1999). Maternal aggression is reduced in neuronal nitric oxide synthase-deficient mice. *J. Neurosci.* 19, 8027–8035.
511. Bredt, D. S., and Snyder, S. H. (1992). Nitric oxide, a novel neuronal messenger. *Neuron* 8, 3–11.
512. Gammie, S. C., and Nelson, R. J. (2000). Maternal and mating-induced aggression is associated with elevated citrulline immunoreactivity in the paraventricular nucleus in prairie voles. *J. Comp. Neurol.* 418, 182–192.
513. Popeski, N., Amir, S., and Woodside, B. (1999). Changes in NADPH-d staining in the paraventricular and supraoptic nuclei during pregnancy and lactation in rats: role of ovarian steroids and oxytocin. *J. Neuroendocrinol.* 11, 53–61.
514. Popeski, N., Service, G., and Woodside, B. (2003). Effects of administering a nitric oxide synthase inhibitor into the medial preoptic area on the behavior of lactating rats during exposure to pups. Abstract Viewer/Itinerary Planner, Program no. 728.3. Society for Neuroscience, Washington, DC, online.
515. Dominguez, J. M., Muschamp, J. W., Schmich, J. M., and Hull, E. M. (2004). Nitric oxide mediates glutamate-evoked dopamine release in the medial preoptic area. *Neuroscience* 125, 203–210.
516. Lonstein, J. S., Dominguez, J. M., Putnam, S. K., De Vries, G. J., and Hull, E. M. (2003). Intracellular preoptic and striatal monoamines in pregnant and lactating rats: possible role in maternal behavior. *Brain Res.* 970, 149–158.
517. Olazabal, D. E., Abercrombie, E., Rosenblatt, J. S., and Morrell, J. I. (2004). The content of dopamine, serotonin, and their metabolites in the neural circuit that mediates maternal behavior in juvenile and adult rats. *Brain Res. Bull.* 63, 259–268.
518. Seifritz, E., Esposito, F., Neuhoff, J. G., et al. (2003). Differential sex-independent amygdala response to infant crying and laughing in parents versus nonparents. *Biol. Psychiatry.* 54, 1367–1375.
519. Loberman, J., Newman, J., Horwitz, A., et al. (2002). A potential role for thalamocingulate circuitry in human maternal behavior. *Biol. Psychiatry.* 51, 431–445.
520. Moltz, H., and Robbins, D. (1965). Maternal behavior of primiparous and multiparous rats. *J. Comp. Physiol. Psychol.* 60, 417–421.
521. Carlier, C., and Noirot, E. (1965). Effects of previous experience on maternal retrieving by rats. *Anim. Behav.* 13, 423–426.
522. Wang, Z., and Novak, M. A. (1994). Parental care and litter development in primiparous and multiparous prairie voles (Microtus ochrogaster). *J. Mammal.* 75, 18–23.
523. Moltz, H., Robbins, D., and Parks, M. (1966). Caesarean delivery and maternal behavior of primiparous and multiparous rats. *J. Comp. Physiol. Psychol.* 61, 455–460.
524. Moltz, H., and Wiener, E. (1966). Effects of ovariectomy on maternal behavior of primiparous and multiparous rats. *J. Comp. Physiol. Psychol.* 61, 455–460.
525. Moltz, H., Levin, R., and Leon, M. (1969). Differential effects of progesterone on the maternal behavior of primiparous and multiparous rats. *J. Comp. Physiol. Psychol.* 67, 36–40.
526. Kinsley, C. H., and Bridges, R. S. (1988). Parity-associated reductions in behavioral sensitivity to opiates. *Biol. Reprod.* 39, 270–278.

527. Franz, J. R., Leo, R. J., Steuer, M. A., and Kristal, M. B. (1986). Effects of hypothalamic knife cuts and experience on maternal behavior in the rat. *Physiol. Behav.* 38, 629–640.
528. Schlein, P. A., Zarrow, M. X., Cohen, H. A., Denenberg, V. H., and Johnson, N. P. (1972). The differential effect of anosmia on maternal behaviour in the virgin and primiparous rat. *J. Reprod. Fertil.* 30, 139–142.
529. Poindron, P., Nowak, R., Levy, F., Porter, R. H., and Schaal, B. (1993). Development of exclusive mother-young bonding in sheep and goats. *Oxf. Rev. Reprod. Biol.* 15, 311–364.
530. Poindron, P., Le Neindre, P., and Lévy, F. (1984). Maternal behaviour in sheep and its physiological control. In *Reproduction in Sheep* (D. R. Lindsay and D. T. Pearce, Eds.), pp. 191–198. Australian Wool Corporation Technical Publication, Canberra.
531. Le Neindre, P., Poindron, P., and Delouis, C. (1979). Hormonal induction of maternal behavior in non-pregnant ewes. *Physiol. Behav.* 22, 731–734.
532. Epple, G. (1978). Reproductive and social behavior of marmosets with special reference to captive breeding. *Primates Med.* 10, 50–62.
533. Johnson, L. D., Petto, A. J., and Sehgal, P. K. (1991). Factors in the rejection and survival of captive cotton top tamarins. *Am. J. Primatol.* 25, 91–102.
534. Maestripieri, D., Wallen, K., and Carroll, K. A. (1997). Infant abuse runs in families of group-living pigtail macaques. *Child Abuse Negl.* 21, 465–471.
535. Tardif, S. D., Richter, C., and Carson, R. L. (1984). Effects if sibling-rearing experience on future reproductive success in two species of callitrichidae. *Am. J. Primatol.* 6, 377–380.
536. Snowdon, C. T. (1996). Infant care in cooperatively breeding species. In *Parental Care: Evolution, Mechanisms, and Adaptive Significance* (J. S. Rosenblatt and C. T. Snowdon, Eds.), Vol. 25, pp. 643–689. Academic Press, San Diego.
537. Fairbanks, L. A. (1996). Individual differences in maternal style. Causes and consequences for mothers and offspring. In *Advances in the Study of Behavior* (J. S. Rosenblatt and C. T. Snowdon, Eds.), Vol. 25. Academic Press, San Diego.
538. Kemps, A., Timmermans, P., and Vossen, J. (1989). Effects of mother's rearing condition and multiple motherhood on early development of mother-infant interactions in java macaques (Macaca fascicularis). *Behaviour* 111, 61–76.
539. Robson, K. M., and Kumar, R. (1980). Delayed onset of maternal affection after childbirth. *Br. J. Psychiatry.* 136, 347–353.
540. Kaitz, M., Chriki, M., Bear-Scharf, L., Nir, T., and Eidelman, A. I. (2000). Effectiveness of primiparae and multiparae at soothing their newborn infants. *J. Genet. Psychol.* 161, 203–215.
541. Schaal, B., Montagner, H., Hertling, E., Bolzoni, D., Moyse, A., and Quichon, R. (1980). (Olfactory stimulation in the relationship between child and mother). *Reprod. Nutr. Dev.* 20, 843–858.
542. Bernal, J. (1972). Crying during the first 10 days of life, and maternal responses. *Dev. Med. Child Neurol.* 14, 362–372.
543. Wisenfeld, A., and Malatesta, C. Z. (1982). Infant distress: variable affecting responses of caregivers and others. In *Parenting: Its Causes and Consequences* (L. W. Hoffman, R. Gandelman, and H. R. Schiffman, Eds.), pp. 123–139. Erlbaum Associates, Hillsdale, NJ.
544. Boukydis, C. F., and Burgess, R. L. (1982). Adult physiological response to infant cries: effects of temperament of infant, parental status, and gender. *Child Dev.* 53, 1291–1298.
545. Belsky, J., Gilstrap, B., and Rovine, M. (1984). The Pennsylvania Infant and Family Development Project, I: Stability and change in mother-infant and father-infant interaction in a family setting at one, three, and nine months. *Child Dev.* 55, 692–705.
546. Schino, G., D'Amato, F. R., and Troisi, A. (1995). Mother-infant relationships in Japanese macaques: sources of inter-individual variation. *Anim. Behav.* 49, 151–158.
546a. Hrdy, S. (2005). Evolutionary context of human development: the cooperative breeding model. In *Attachment and Bonding: A New Synthesis* (C. S. Carter, L. Ahnert, K. E. Grossman, S. B. Hrdy, M. E. Lamb, S. W. Porges, and N. Sachser, Eds.). MIT Press, Cambridge, MA.
547. Stern, J. M., and Rogers, L. (1988). Experience with younger siblings facilitates maternal responsiveness in pubertal Norway rats. *Dev. Psychobiol.* 21, 575–589.
548. Tardif, S. D., Harrison, M. L., and Simek, M. A. (1993). Communal infant care in marmosets and tamarins: relation to energetics, ecology, and social organization. In *Marmosets and Tamarins* (A. Rylands, Ed.), pp. 220–234. Oxford University Press, Oxford.
549. Locke-Haydon, J., and Chalmers, N. R. (1983). The development of infant-caregiver relationships in captive common marmosets (Callithrix jacchus). *Int. J. Primatol.* 4, 63–81.
550. Stevenson, M. F., and Rylands, A. B. (1988). The marmosets, genus Callithrix. In *Ecology and Behavior of Neotropical Primates* (R. A. Mitterminder, A. B. Rylands, A. F. Coimbra-Filhi, and G. A. B. da Fonseca, Eds.), Vol. 2. World Wildlife Fund, Washington, DC.
551. Evans, S. (1983). The pair bond of the common marmoset, Callithrix jacchus. *Anim. Behav.* 31, 651–658.
552. Pryce, C. R. (1993). The regulation of maternal behavior in marmosets and tamarins. *Behav. Process* 30, 201–224.
553. Pryce, C. R., Mutschler, T., Dobeli, M., Nievergelt, C., and Martin, R. D. (1995). Prepartum sex steroid hormones and infant-directed behavior in primiparous marmoset mothers. (callithrix jacchus). In *Motherhood in Human and Nonhuman Primates*. (C. Pryce, R. Martin, and D. Skuse, Eds.), pp. 78–86. Karger, Basel.
554. Deutsch, F., Ruble, D. N., Fleming, A. S., Brooks-Gunn, J., and Stangor, C. (1986). Becoming a mother: information-seeking and self-definitional processes. Paper presented at the annual meeting of the Eastern Psychological Association, New York.
555. Bridges, R. S. (1975). Long-term effects of pregnancy and parturition upon maternal responsiveness in the rat. *Physiol. Behav.* 14, 245–249.
556. Bridges, R. S. (1977). Parturition: its role in the long term retention of maternal behavior in the rat. *Physiol. Behav.* 18, 487–490.
557. Cohen, J., and Bridges, R. S. (1981). Retention of maternal behavior in nulliparous and primiparous rats: effects of duration of previous maternal experience. *J. Comp. Physiol. Psychol.* 95, 450–459.
558. Orpen, B. G., and Fleming, A. S. (1987). Experience with pups sustains maternal responding in postpartum rats. *Physiol. Behav.* 40, 47–54.
559. Fleming, A. S., and Sarker, J. (1990). Experience-hormone interactions and maternal behavior in rats. *Physiol. Behav.* 47, 1165–1173.
560. Gonzalez-Mariscal, G., Melo, A. I., Chirino, R., Jimenez, P., Beyer, C., and Rosenblatt, J. S. (1998). Importance of mother/young contact at parturition and across lactation for the expression of maternal behavior in rabbits. *Dev. Psychobiol.* 32, 101–111.
561. Keller, M., Meurisse, M., Jouaneau, N., Vénier, G., Archer, E., and Lévy, F. (2003). Neural networks involved in acquisition and consolidation of the lamb odour by parturient ewes. 35th Annual General Meeting of the European Brain and Behaviour Society. European Brain and Behavior Society, Barcelona.
562. Harlow, H. F., Harlow, M. K., Dodsworth, R. O., and Arling, G. L. (1966). Maternal behavior in rhesus monkeys deprived

of mothering and peer associations in infancy. *Proc. Am. Philos. Soc.* 110, 58–66.

563. Ruppenthal, G. C., Arling, G. L., Harlow, H. F., Sackett, G. P., and Suomi, S. J. (1976). A 10-year perspective of motherless-mother monkey behavior. *J. Abnorm. Psychol.* 85, 341–349.
564. Goldberg, S. (1983). Parent-infant bonding: another look. *Child Dev.* 54, 1355–1382.
565. Fleming, A. S., Ruble, D. N., Anderson, V., and Flett, G. L. (1988). Place of childbirth influences feelings of satisfaction and control in first-time mothers. *J. Psychosomat. Obstet. Gynaecol.* 8, 1–17.
566. Fleming, A. S., and Li, M. (2002). Psychobiology of maternal behavior and its early determinants in nonhuman mammals. In *Handbook of Parenting: Biology and Ecology* (M. H. Bornstein, Ed.), Vol. 2, pp. 61–97. Lawrence Ertlbaum Associates, Mahwah, NJ.
567. Graham, D., Rees, S., Steiner, M., and Fleming, A. S. (under re-review). Effects of adrenalectomy and corticosterone replacement on maternal memory in rats. *Horm. Behav.*
568. Fleming, A. S., Kuchera, C., Lee, A., and Winocur, G. (1994). Olfactory-based social learning varies as a function of parity in female rats. *Psychobiology* 22, 37–43.
569. Kinsley, C. H., Madonia, L., Gifford, G. W., et al. (1999). Motherhood improves learning and memory. *Nature* 402, 137–138.
570. Jakubowski, M., and Terkel, J. (1986). Establishment and maintenance of maternal responsiveness in postpartum Wistar rats. *Anim. Behav.* 34, 256–262.
571. Stern, J. M. (1983). Maternal behavior priming in virgin and caesarean-delivered Long-Evans rats: effects of brief contact or continuous exteroceptive pup stimulation. *Physiol. Behav.* 31, 757–763.
572. Stern, J. M., and Johnson, S. K. (1989). Perioral somatosensory determinants of nursing behavior in Norway rats (Rattus norvegicus). *J. Comp. Psychol.* 103, 269–280.
573. Malenfant, S. A., Barry, M., and Fleming, A. S. (1991). Effects of cycloheximide on the retention of olfactory learning and maternal experience effects in postpartum rats. *Physiol. Behav.* 49, 289–294.
574. Ferreira, G., Terrazas, A., Poindron, P., Nowak, R., Orgeur, P., and Levy, F. (2000). Learning of olfactory cues is not necessary for early lamb recognition by the mother. *Physiol. Behav.* 69, 405–412.
575. Terrazas, A., Ferreira, G., Levy, F., et al. (1999). Do ewes recognize their lambs within the first day postpartum without the help of olfactory cues? *Behav. Processes* 47, 19–29.
576. Keller, M., Meurisse, M., Poindron, P., et al. (2003). Maternal experience influences the establishment of visual/auditory, but not olfactory recognition of the newborn lamb by ewes at parturition. *Dev. Psychobiol.* 43, 167–176.
577. Valanne, E. H., Vuorenkoski, V., Partanen, T. J., Lind, J., and Wasz-Hockert, O. (1967). The ability of human mothers to identify the hunger cry signals of their own new-born infants during the lying-in period. *Experientia* 23, 768–769.
578. Formby, D. (1967). Maternal recognition of infant's cry. *Dev. Med. Child Neurol.* 9, 293–298.
579. Bridges, R. S., and Hammer, R. P. Jr. (1992). Parity-associated alterations of medial preoptic opiate receptors in female rats. *Brain Res.* 578, 269–274.
580. Grosvenor, C. E., Shah, G. V., and Crowley, W. R. (1990). Role of neurogenic stimuli and milk production in the regulation of prolactin secretion during lactation. In *Mammalian Parenting: Biochemical, Neurobiological, and Behavioral Determinants* (N. A. Krasnegor and R. S. Bridges, Eds.), pp. 324–342. Oxford University Press, New York.
581. Ehret, G., and Buckenmaier, J. (1994). Estrogen-receptor occurrence in the female mouse brain: effects of maternal experience, ovariectomy, estrogen and anosmia. *J. Physiol. Paris.* 88, 315–329.
582. Koch, M. (1990). Effects of treatment with estradiol and parental experience on the number and distribution of estrogen-binding neurons in the ovariectomized mouse brain. *Neuroendocrinology* 51, 505–514.
583. Meurisse, M., Gonzalez, A., Delsol, G., Caba, M., Levy, F., and Poindron, P. (2005). Estradiol receptor-alpha expression in hypothalamic and limbic regions of ewes is influenced by physiological state and maternal experience. *Horm. Behav.* 48, 34–43.
584. Keverne, E. B., Levy, F., Guevara-Guzman, R., and Kendrick, K. M. (1993). Influence of birth and maternal experience on olfactory bulb neurotransmitter release. *Neuroscience* 56, 557–565.
585. Levy, F., Guevara-Guzman, R., Hinton, M. R., Kendrick, K. M., and Keverne, E. B. (1993). Effects of parturition and maternal experience on noradrenaline and acetylcholine release in the olfactory bulb of sheep. *Behav. Neurosci.* 107, 662–668.
586. Byrnes, E. M., and Bridges, R. S. (2000). Endogenous opioid facilitation of maternal memory in rats. *Behav. Neurosci.* 114, 797–804.
587. Moffat, S. D., Suh, E. J., and Fleming, A. S. (1993). Noradrenergic involvement in the consolidation of maternal experience in postpartum rats. *Physiol. Behav.* 53, 805–811.
588. Byrnes, E. M., Rigero, B. A., and Bridges, R. S. (2002). Dopamine antagonists during parturition disrupt maternal care and the retention of maternal behavior in rats. *Pharmacol. Biochem. Behav.* 73, 869–875.
589. Francis, D. D., Young, L. J., Meaney, M. J., and Insel, T. R. (2002). Naturally occurring differences in maternal care are associated with the expression of oxytocin and vasopressin (V1a) receptors: gender differences. *J. Neuroendocrinol.* 14, 349–353.
590. Dluzen, D. E., Muraoka, S., Engelmann, M., and Landgraf, R. (1998). The effects of infusion of arginine vasopressin, oxytocin, or their antagonists into the olfactory bulb upon social recognition responses in male rats. *Peptides* 19, 999–1005.
591. Engelmann, M., Ebner, K., Wotjak, C. T., and Landgraf, R. (1998). Endogenous oxytocin is involved in short-term olfactory memory in female rats. *Behav. Brain Res.* 90, 89–94.
592. Popik, P., and Vetulani, J. (1991). Opposite action of oxytocin and its peptide antagonists on social memory in rats. *Neuropeptides* 18, 23–27.
593. Popik, P., and van Ree, J. M. (1998). Neurohypophyseal peptides and social recognition in rats. *Prog. Brain Res.* 119, 415–436.
594. Fleming, A. S., and Korsmit, M. (1996). Plasticity in the maternal circuit: effects of maternal experience on Fos-Lir in hypothalamic, limbic, and cortical structures in the post-partum rat. *Behav. Neurosci.* 110, 567–582.
595. Featherstone, R. E., Fleming, A. S., and Ivy, G. O. (2000). Plasticity in the maternal circuit: effects of experience and partum condition on brain astrocyte number in female rats. *Behav. Neurosci.* 114, 158–172.
596. Abkari, E., Sinopoli, K., Chatterjee, D., Levy, F., and Fleming, A. S. (Submitted). Effects of early and postpartum experience on neurogenesis in the mother rat. *J. Neurosci.*
597. Lee, A., Clancy, S., and Fleming, A. S. (1999). Mother rats bar-press for pups: effects of lesions of the mpoa and limbic sites on maternal behavior and operant responding for pup-reinforcement. *Behav. Brain Res.* 100, 15–31.
598. Lévy, F., Guevara-Guzman, R., Hinton, M. R., Kendrick, K. M., and Keverne, E. B. (1993). Effects of parturition and maternal experience on noradrenaline and acetylcholine release in the olfactory bulb of sheep. *Behav. Neurosci.* 107, 662–668.

599. Lévy, F., Kendrick, K. M., Goode, J. A., Guevara-Guzman, R., and Keverne, E. B. (1995). Oxytocin and vasopressin release in the olfactory bulb of parturient ewes: changes with maternal experience and effects on acetylcholine, gamma-aminobutyric acid, glutamate and noradrenaline release. *Brain Res.* 669, 197–206.
600. Kendrick, K. M. (1994). Neurobiological correlates of visual and olfactory recognition in sheep. *Behav. Process.* 33, 89–112.
601. Shipley, M. T., Halloran, F. J., and de la Torre, J. (1985). Surprisingly rich projection from the locus coeruleus to the olfactory bulb in the rat. *Brain Res.* 329, 294–299.
602. Keverne, E. B., Lévy, F., Guevara-Guzman, R., and Kendrick, K. M. (1993). Influence of birth and maternal experience on olfactory bulb neurotransmitter release. *Neuroscience* 56, 557–565.
603. Pissonnier, D., Theiry, J. C., Fabre-Nys, P., Poindron, P., and Keverne, E. B. (1985). The importance of olfactory bulb noradrenalin for maternal recognition in sheep. *Physiol. Behav.* 35, 361–363.
604. Lévy, F., Gervais, R., Kinderman, U., and Orgeur, P. (1990). Importance of b-noradrenergic receptors in the olfactory bulb of sheep for recognition of lambs. *Behav. Neurosci.* 104, 464–469.
605. Lévy, F., Meurisse, M., Perrin, G., et al. (2003). Cortical and medial nuclei of the amygdala are both involved in lamb odour memory formation by parturient mothers in sheep. In *Society EBaB ed. 35th Annual General Meeting of the European Brain and Behaviour Society*, Barcelona.
606. Keller, M., Meurisse, M., and Levy, F. (2005). Mapping of brain networks involved in consolidation of lamb recognition memory. *Neuroscience* 133, 359–369.
607. Holdcroft, A., Hall, L., Hamilton, G., Counsell, S., and Bydder, G. M. (2005). Phosphorus-31 brain MR spectroscopy in women during and after pregnancy compared with non-pregnant control subjects. *Am. J. Neuroadiol.* 26, 352–356.
608. Hofer, M. A., and Sullivan, R. (2001). Toward a neurobiology of attachment. In *Handbook of Developmental Cognitive Neuroscience* (C. A. Nelson and M. Luciana, Eds.). MIT Press, London.
609. Polan, H. J., and Hofer, M. A. (1998). Olfactory preference for mother over home nest shavings by newborn rats. *Dev. Psychobiol.* 33, 5–20.
610. Pedersen, C. A., and Blass, E. (1981). Olfactory control over suckling in albino rats. In *The Development of Perception: Psychobiological Perspectives* (R. N. Aslin, J. R. Alberts, and M. R. Petersen, Eds.), pp. 359–381. Academic Press, New York.
611. Robinson, S. R., Hoeltzel, T. C., and Smotherman, W. P. (1995). Development of responses to an artificial nipple in the rat fetus: involvement of mu and kappa opioid systems. *Physiol. Behav.* 57, 953–957.
612. Abel, R. A., Ronca, A. E., and Alberts, J. R. (1998). Perinatal stimulation facilitates suckling onset in newborn rats. *Dev. Psychobiol.* 32, 91–99.
613. Wilson, D. A., and Sullivan, R. M. (1994). Neurobiology of associative learning in the neonate: early olfactory learning. *Behav. Neural Biol.* 61, 1–18.
614. Shah, A., Oxley, G., Lovic, V., and Fleming, A. S. (2002). Effects of preweaning exposure to novel maternal odors on maternal responsiveness and selectivity in adulthood. *Dev. Psychobiol.* 41, 187–196.
615. Fleming, A. S., Kraemer, G. W., Gonzalez, A., Lovic, V., Rees, S., and Melo, A. (2002). Mothering begets mothering: the transmission of behavior and its neurobiology across generations. *Pharmacol. Biochem. Behav.* 73, 61–75.
616. Francis, D., Diorio, J., Liu, D., and Meaney, M. J. (1999). Nongenomic transmission across generations of maternal behavior and stress responses in the rat. *Science* 286, 1155–1158.
617. Champagne, F. A., Francis, D. D., Mar, A., and Meaney, M. J. (2003). Variations in maternal care in the rat as a mediating influence for the effects of environment on development. *Physiol. Behav.* 79, 359–371.
618. Liu, D., Diorio, J., Tannenbaum, B., et al. (1997). Maternal care, hippocampal glucocorticoid receptors, and hypothalamic-pituitary-adrenal responses to stress. *Science* 277, 1659–1662.
619. Gonzalez, A., Lovic, V., Ward, G. R., Wainwright, P. E., and Fleming, A. S. (2001). Intergenerational effects of complete maternal deprivation and replacement stimulation on maternal behavior and emotionality in female rats. *Dev. Psychobiol.* 38, 11–32.
620. Hofer, M. A. (2002). The riddle of development. In *Conceptions of Development: Lessons from the Laboratory* (D. J. Lewkowicz and R. Lickliter, Eds.). Psychology Press, Philadelphia.
621. Hall, F. S., Wilkinson, L. S., and Humby, T., et al. (1998). Isolation rearing in rats: pre- and postsynaptic changes in striatal dopaminergic systems. *Pharmacol. Biochem. Behav.* 59, 859–872.
622. Thoman, E. B., and Arnold, W. J. (1968). Effects of incubator rearing with social deprivation on maternal behavior in rats. *J. Comp. Physiol. Psychol.* 65, 441–446.
623. Levy, F., Melo, A. I., Galef, B. G. Jr., Madden, M., and Fleming, A. S. (2003). Complete maternal deprivation affects social, but not spatial, learning in adult rats. *Dev. Psychobiol.* 43, 177–191.
624. Moore, C. (1995). Maternal contributions to mammalian reproductive development and the divergence of males and females. In *Advances in the Study of Behavior* (P. J. B. Slater, J. S. Rosenblatt, C. T. Snowdon, and M. Milinski, Eds.), Vol. 24, pp. 47–118. Academic Press, San Diego.
625. Kraemer, G. W. (1992). A psychobiological theory of attachment. *Behav. Brain Sci.* 15, 493–511.
626. Kraemer, G. W., and Bachevalier, J. (1998). Cognitive changes associated with persisting behavioral effects of early psychosocial stress in rhesus monkeys: the view from psycobiology. In *Adversity, Stress and Psychopathology* (B. Dohrenwend, Ed.), pp. 438–462. Oxford University Press, New York.
627. Harlow, H. F. (1962). The effects of early experiences upon the personal-social, heterosexual, and maternal behavior of rhesus monkeys. *Trans. Am. Neurol. Assoc.* 87, 9–10.
628. Seay, B., Alexander, B. K., and Harlow, H. F. (1964). Maternal behavior of socially deprived rhesus monkeys. *J. Abnorm. Psychol.* 69, 345–354.
629. Arling, G. L., and Harlow, H. F. (1967). Effects of social deprivation on maternal behavior of rhesus monkeys. *J. Comp. Physiol. Psychol.* 64, 371–377.
630. Harlow, H. F. (1958). The nature of love. *Am. Psychologist* 13, 673–685.
631. Harlow, H. F., Harlow, M. K., and Hansen, E. W. (1963). The maternal affectional system of rhesus monkeys. In *Maternal Behavior in Mammals* (H. L. Rheingold, Ed.), pp. 254–281. Wiley, New York.
632. Harlow, H. F., Harlow, M. K., and Suomi, S. J. (1971). From thought to therapy: lessons from a primate laboratory. *Am. Sci.* 59, 538–549.
633. Kraemer, G. W., Ebert, M. H., Schmidt, D. E., and McKinney, W. T. (1991). Strangers in a strange land: a psychobiological study of infant monkeys before and after separation from real or inanimate mothers. *Child Dev.* 62, 548–566.
634. Suomi, S. J., and Ripp, C. (1983). A history of motherless monkey mothering at the University of Wisconsin Primate Laboratory. In *Child Abuse: The Non-human Primate Data* (M. Reite and N. Caine, Eds.), pp. 49–78. Alan R. Liss, New York.
635. Winnicott, D. W. (1965). *Ego intergation and child development.* Intl. Univ. Press, New York.

636. Fairbanks, L. A. (1989). Early experience and cross-generational continuity of mother-infant contact in vervet monkeys. *Dev. Psychobiol.* 22, 669–681.
637. Simons, R. L., Whitbeck, L. B., Conger, R. D., and Chyi-In, W. (1991). Intergenerational transmission of harsh parenting. *Dev. Psychol.* 27, 159–171.
638. Weinfield, N. S., Sroufe, L. A., and Egeland, B. (2000). Attachment from infancy to early adulthood in a high-risk sample: continuity, discontinuity, and their correlates. *Child Dev.* 71, 695–702.
639. Knutson, J. F. (1995). Psychological characteristics of maltreated children: putative risk factors and consequences. *Annu. Rev. Psychol.* 46, 401–431.
640. Main, M., and Goldwyn, R. (1984). Predicting rejection of her infant from mother's representation of her own experience: implications for the abused-abusing intergenerational cycle. *Child Abuse Negl.* 8, 203–217.
641. Rebollo, A. G., and Montero, C. M. (2000). (Perinatal variables and health inequalities in a health-care district in Caceres, Spain). *Gac. Sanit.* 14, 31–38.
642. de Paul, J., and Domenech, L. (2000). Childhood history of abuse and child abuse potential in adolescent mothers: a longitudinal study. *Child Abuse Negl.* 24, 701–713.
643. Geronimous, A. T. (1987). On teenage childbearing and neonatal mortality in the United States. *Popul. Dev. Rev.* 13, 245–279.
644. Furstenberg, F. F. Jr., Brooks-Gunn, J., and Morgan, S. P. (1987). Adolescent mothers and their children in later life. *Fam. Plann. Perspect.* 19, 142–151.
645. Maynard, R. A. (1996). *Kids having kids: a Robin Hood Foundation special report on the costs of adolescent childbearing.* Robin Hood Foundation, New York.
646. Clark, A. M., and Clark, A. D. (1977). *Early experience: myth and evidence.* Free Press, New York.
647. van Ijzendoorn, M. H. (1995). Adult attachment representations, parental responsiveness, and infant attachment: a meta-analysis on the predictive validity of adult attachment. *Psychonom. Bull.* 117, 387–403.
648. Goldberg, S., MacKay, S., and Rochester, M. (1994). Affect, attachment and maternal responsiveness. *Infant Behav. Dev.* 17, 335–339.
649. Waters, E., Merrick, S., Treboux, D., Crowell, J., and Albersheim, L. (2000). Attachment security in infancy and early adulthood: a twenty-year longitudinal study. *Child Dev.* 71(3), 684–689.
650. Fonagy, P., Steele, H., and Steele, M. (1991). Maternal representations of attachment during pregnancy predict the organization of infant-mother attachment at one year of age. *Child Dev.* 62, 891–905.
651. Champagne, F., Weaver, I. C., Diorio, J., Sharma, S., and Meaney, M. J. (2003). Natural variations in maternal care are associated with estrogen receptor alpha expression and estrogen sensitivity in the medial preoptic area. *Endocrinology* 144, 4718–4719.
652. Champagne, F., Rees, S., Meaney, M. J., and Fleming, A. S. (In preparation). The effects of maternal deprivation on estrogen and oxytocin receptors in brain of postpartum rats.
653. Novakov, M., and Fleming, A. S. (in press). Effects of early maternal deprivation on hormonal induction of maternal behavior in the virgin rat. *Horm. Behav.*
654. van Oers, H. J., de Kloet, E. R., Whelan, T., and Levine, S. (1998). Maternal deprivation effect on the infant's neural stress markers is reversed by tactile stimulation and feeding but not by suppressing corticosterone. *J. Neurosci.* 18, 10171–10179.
655. Vazquez, D. M., Van Oers, H., Levine, S., and Akil, H. (1996). Regulation of glucocorticoid and mineralocorticoid receptor mRNAs in the hippocampus of the maternally deprived infant rat. *Brain Res.* 731, 79–90.
656. Clarke, A. S. (1993). Social rearing effects on HPA axis activity over early development and in response to stress in rhesus monkeys. *Dev. Psychobiol.* 26, 433–446.
657. Clarke, A. S., Wittwer, D. J., Abbott, D. H., and Schneider, M. L. (1994). Long-term effects of prenatal stress on HPA axis activity in juvenile rhesus monkeys. *Dev. Psychobiol.* 27, 257–269.
658. Kraemer, G. W. (1997). Psychobiology of early social attachment in rhesus monkeys: clinical implications. *Ann. N Y Acad. Sci.* 836, 12–38.
659. Spangler, G., Schieche, M., Ilg, U., Maier, U., and Ackermann, C. (1994). Maternal sensitivity as an external organizer for biobehavioral regulation in infancy. *Dev. Psychobiol.* 27, 425–437.
660. Spangler, G., and Schieche, M. (1998). Emotional and adrenocortical responses of infants to the strange situation: the differential function of emotional expression. *Int. J. Behav. Dev.* 22, 681–706.
661. Post, R. M., Weiss, S. R., and Li, H., et al. (1998). Neural plasticity and emotional memory. *Dev. Psychopathol.* 10, 829–855.
662. Rosenzweig, M. R., and Bennett, E. L. (1996). Psychobiology of plasticity: effects of training and experience on brain and behavior. *Behav. Brain Res.* 78, 57–65.
663. Pascual, R., and Figueroa, H. (1996). Effects of preweaning sensorimotor stimulation on behavioral and neuronal development in motor and visual cortex of the rat. *Biol. Neonate* 69, 399–404.
664. Cramer, C. P. (1988). Experience during suckling increases weight and volume of rat hippocampus. *Brain Res.* 470, 151–155.
665. Liu, D., Diorio, J., Day, J. C., Francis, D. D., and Meaney, M. J. (2000). Maternal care, hippocampal synaptogenesis and cognitive development in rats. *Nat. Neurosci.* 3, 799–806.
666. Hansen, S. (1994). Maternal behavior of female rats with 6-OHDA lesions in the ventral striatum: characterization of the pup retrieval deficit. *Physiol. Behav.* 55, 615–620.
667. Stern, J. M., and Keer, S. E. (1999). Maternal motivation of lactating rats is disrupted by low dosages of haloperidol. *Behav. Brain Res.* 99, 231–239.
668. Suomi, S. J., Harlow, H. F., and Domek, C. J. (1970). Effect of repetitive infant-infant separation of young monkeys. *J. Abnorm. Psychol.* 76, 161–172.
669. Higley, J. D., Suomi, S. J., and Linnoila, M. (1991). CSF monoamine metabolite concentrations vary according to age, rearing, and sex, and are influenced by the stressor of social separation in rhesus monkeys. *Psychopharmacol. (Berl).* 103, 551–556.
670. Higley, J. D., Thompson, W. W., and Champoux, M., et al. (1993). Paternal and maternal genetic and environmental contributions to cerebrospinal fluid monoamine metabolites in rhesus monkeys (Macaca mulatta). *Arch. Gen. Psychiatry.* 50, 615–623.
671. Kraemer, G. W., Ebert, M. H., Lake, C. R., and McKinney, W. T. (1984). Hypersensitivity to d-amphetamine several years after early social deprivation in rhesus monkeys. *Psychopharmacol. (Berl).* 82, 266–271.
672. Kraemer, G. W., Ebert, M. H., Lake, C. R., and McKinney, W. T. (1984). Cerebrospinal fluid measures of neurotransmitter changes associated with pharmacological alteration of the despair response to social separation in rhesus monkeys. *Psychiatry Res.* 11, 303–315.
673. Kraemer, G. W., Ebert, M. H., Schmidt, D. E., and McKinney, W. T. (1989). A longitudinal study of the effect of different social rearing conditions on cerebrospinal fluid norepinephrine and biogenic amine metabolites in rhesus monkeys. *Neuropsychopharmacology* 2, 175–189.

Knobil and Neill's Physiology of Reproduction, Third Edition
edited by Jimmy D. Neill,
Elsevier © 2006

CHAPTER 36

Communicative Behaviors, Hormone–Behavior Interactions, and Reproduction in Vertebrates

John C. Wingfield

Introduction, 1996
The Diversity and Contexts of Hormone–Behavior Interactions in Reproduction, 1996
Ontogeny versus Development of Adult Life History Stages, 1996
Finite-State Machine Theory and the Reproductive Life History Stage, 1998
Classification of Environmental Factors Regulating Reproduction, 2001
Modalities of Communication, 2002
Hormones, Behavior, and Social Interactions: The Endocrine Basis of Signal Production and Reception, 2003
Synchronizing and Integrating Factors in the Reproductive Process: Dealing with a Dynamic Environment, 2006
Hormone–Behavior Interactions during the Development Phase of the Reproduction Life History Stage, 2007
Interactions with Initial Predictive Information (Photoperiod), 2008
Mate Choice and Pair Bonding, 2009
Hormone–Behavior Interactions Influencing Mature Capability (Courtship and Shedding of Gametes) of the Reproduction Life History Stage, 2010
Onset of Cycling, 2010
Parental Care, 2012
Transitions between the Sexual and Parental Phases of Reproduction, 2013
Recycling (Renesting), 2013
Hormone–Behavior Interactions and Support Functions for the Reproductive Life History Stage, 2014
Territorial Behavior, 2014
Social Status, 2015
Hormone–Behavior Interactions and Termination Phase of the Reproduction Life History Stage, 2015
Effects of Offspring and Mate, 2016
Interactions with Molt, 2016
Mating Systems, Breeding Strategies, and Alternative Reproductive Tactics, 2016
Variation in the Temporal Pattern of Breeding, 2016
Variation in Mating Systems, 2019
Variation in Breeding Strategy, 2023
An Example of the Potential for New Concepts of Hormone–Behavior Interrelationships, 2027
Finite-State Machine Theory and Testing Predictions, 2027
Avoiding the "Costs" of Testosterone, 2027
Role of Communication in Dealing with a Capricious Environment, 2027
Labile Perturbation Factors and Communication, 2028
Conclusion, 2029
Acknowledgments, 2030
References, 2030

Department of Biology, University of Washington, Seattle, Washington.

INTRODUCTION

Communication among individuals is fundamental for almost everything an organism does during its life cycle. Individuals communicate over spatial distribution, access to food, and other resources such as shelter, selection of mates, synchronization of breeding effort, and development of offspring. Although experimental evidence reveals that gametogenesis can proceed in individuals held in isolation, fertilization and production of viable young clearly require social exchange, no matter how intermittent. In other words, an organism living in total isolation has zero fitness. Some form of communication is crucial to synchronize the shedding of gametes, to establish social relationships, and for territoriality. Indeed, without efficient communication, reproduction would be chaotic and mostly unsuccessful.

Behavioral interactions important for reproduction are complex in some species and relatively simple in others. Nonetheless, it is possible to propose a framework by which the behavioral interactions important for the reproductive process can then be organized so as to investigate possible common hormone–behavior interactions and their control mechanisms. Finite-state machine theory (1) is such a heuristic theoretical framework and can be applied to any vertebrate (or indeed organism).

The Diversity and Contexts of Hormone–Behavior Interactions in Reproduction

The main purpose of this theoretical approach is to attempt comparisons across diverse vertebrate taxa so that the neuroendocrine and endocrine control systems can be compared directly. Considerable phylogenetic problems still remain when comparing mechanisms in a fish with those of a reptile or a mammal. However, it is also important to bear in mind that the hormones themselves are well conserved, at least by type, across vertebrates (Table 1). Comparing the major types of hormones such as peptides, proteins, modified amino acids, steroids, and modified fatty acids, there may be as much as 80% homology of type of hormone from fish to mammals. Obviously, the molecules testosterone and cortisol of teleost fish are identical to those of humans. On the other hand, protein hormones such as prolactin or insulin show considerable variation in amino acid sequence, but the gene families are the same [see Gorbman et al. (2) and Norris (3)]. Given this rather well-conserved "repertoire" of hormones, interesting questions emerge as to how they are used, in what contexts and which combinations, and how these patterns change with stages in the life cycle. Clearly, some form of framework for the life cycle is needed so comparisons can be made across the vertebrate classes. The framework proposed in this chapter is only a beginning and will likely change as more attention is paid to mechanisms underlying pattern and context of the reproductive processes. Nevertheless, it is important, perhaps critical at this stage, to begin these kinds of analyses. As genomics and proteomics techniques expand after the sequencing of more vertebrate genomes, the possibilities for comparisons with pattern and context of breeding loom large.

The purpose of this chapter is to present an introduction to how we might think about pattern and context of reproductive processes. This approach could be applied to all aspects of reproduction and other aspects of the life cycle, including migration, hibernation, molt, and so forth, but the emphasis here is on communication and the endocrine bases of the reproductive process. The varying degrees to which vertebrate species communicate and how this is influenced by phases in the breeding period is approached using finite-state machine theory as a possible guide. How does communication influence reproductive function and alternative reproductive strategies and tactics, and what are the neuroendocrine and endocrine bases of these interrelationships? Because the literature is now truly voluminous, even when focusing on communication, a complete review would require a document of book length. Therefore, I have tried to refer to reviews and specific examples as illustrative of the point being made.

Ontogeny versus Development of Adult Life History Stages

Before discussing finite-state machine theory, it is important to point out here that the life cycle of all vertebrates has two major components—ontogeny and the cycles of life history stages that make up annual routines in adults (Fig. 1). During ontogeny, growth and differentiation of cells gives rise to a phenotype with particular morphological, physiological, and behavioral characteristics. Usually we regard the phenotype that develops during ontogeny as irreversible, although a genome can result in different phenotypes during this developmental process [phenotypic plasticity (4)]. Sex determination, genetic and environmental, usually occurs at this time. In adult vertebrates as individuals progress from one life history stage to the next (e.g., breeding and non-breeding), there are also developmental changes that occur, resulting in morphological, physiological, and behavioral changes characteristic of each specific life history stage. However, these developmental processes

TABLE 1. *Distribution of some hormone types across vertebrate classes*

	Agnatha	Chrondrichthyeans	Osteichthyeans	Amphibians	Reptiles	Aves	Mammals
Nonpeptides							
Androgens	Yes	Yes	Yes	Yes	Yes	Yes	Yes
Estrogens	Yes	Yes	Yes	Yes	Yes	Yes	Yes
Progestins	Yes	Yes	Yes	Yes	Yes	Yes	Yes
Glucocorticoids	Yes	Yes	Yes	Yes	Yes	Yes	Yes
Mineralocorticoids	?	No	Yes/no	Yes	Yes	Yes	Yes
Vitamin D	No	?	Yes	Yes	Yes	Yes	Yes
Thyroid	Yes	Yes	Yes	Yes	Yes	Yes	Yes
Catecholamines	Yes	Yes	Yes	Yes	Yes	Yes	Yes
Indolamines	Yes	Yes	Yes	Yes	Yes	Yes	Yes
Prostaglandins	Yes	Yes	Yes	Yes	Yes	Yes	Yes
Peptides/Proteins							
Vasopressins	Yes	Yes	Yes	Yes	Yes	Yes	Yes
Oxytocins	No	Yes	Yes	Yes	Yes	Yes	Yes
GnRH	Yes/no	Yes	Yes	Yes	Yes	Yes	Yes
CRH	Yes	Yes	Yes	Yes	Yes	Yes	Yes
TRH	Yes	Yes	Yes	Yes	Yes	Yes	Yes
Prolactins	Yes	Yes	Yes	Yes	Yes	Yes	Yes
Growth hormone	?	Yes	Yes	Yes	Yes	Yes	Yes
POMC-like peptides	Yes	Yes	Yes	Yes	Yes	Yes	Yes
LH/FSH/TSH	Yes	Yes	Yes	Yes	Yes	Yes	Yes
Parathormone	No	No	Maybe	Yes	Yes	Yes	Yes
Calcitonin-like peptide	No	Maybe	Yes	Yes	Yes	Yes	Yes
Renin–angiotensin system	No	Maybe	Yes	Yes	Yes	Yes	Yes
Atrial natriuretic factor	Maybe	Yes	Yes	Yes	Yes	Yes	Yes
Secretin	Yes	Yes	Yes	Yes	Yes	Yes	Yes
Cholecystokinin	Yes	Yes	Yes	Yes	Yes	Yes	Yes
Gastrin	Yes	Yes	Yes	Yes	Yes	Yes	Yes
Ghrelin	?	?	Yes	Yes	?	Yes	Yes
Insulin	Yes	Yes	Yes	Yes	Yes	Yes	Yes
Glucagon	Yes	Yes	Yes	Yes	Yes	Yes	Yes
Somatostatin	Yes	Yes	Yes	Yes	Yes	Yes	Yes
Pancreatic polypeptide	?	?	?	?	?	Yes	Yes
Leptin	?	?	?	?	Maybe	Yes	Yes
Cytokines	Yes	Yes	Yes	Yes	Yes	Yes	Yes

CRH, corticotropin-releasing hormone; FSH, follicle-stimulating hormone; GnRH, gonadotropin-releasing hormone; LH, luteinizing hormone; POMC, pro-opiomelanocortin; TRH, thyrotropin-releasing hormone; TSH, thyroid-stimulating hormone.
See also Norris (3).

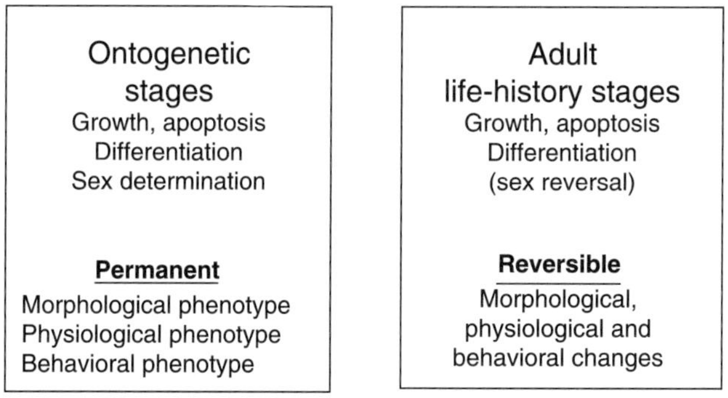

Ontogenetic stages
Growth, apoptosis
Differentiation
Sex determination

Permanent
Morphological phenotype
Physiological phenotype
Behavioral phenotype

Adult life-history stages
Growth, apoptosis
Differentiation
(sex reversal)

Reversible
Morphological, physiological and behavioral changes

FIG. 1. A comparison of ontogenetic stages and life history stages in adult vertebrates. Growth and differentiation of cells appear very similar, with one important difference: ontogenetic stages are not reversible, whereas development of a life history stage is reversible. The latter phenomenon has been referred to as *phenotypic flexibility* (4), and may have fundamental implications for hormone mechanisms operating in life history stages compared with those during ontogeny.

are reversible and occur usually on an annual basis. Furthermore, in some teleost fish, sex reversal can occur through the life cycle. These kinds of changes in morphology, physiology, and behavior *within* a phenotype have been termed *phenotypic flexibility* (4). Developmental processes during ontogeny and progression of life history stages will thus likely have very different hormonal control mechanisms and may influence communication in different ways.

Maternal (and paternal) effects influence ontogenetic stages as well as environmental cues (Fig. 2). The cycle of life history stages in adults is influenced greatly by environmental effects, including those behavioral interactions that involve varying degrees of communication. Semelparous species progress directly to reproductive maturity and then die. Iteroparous species (the vast majority of vertebrates) may undergo

FIG. 2. There are potentially many ontogenetic stages—embryo, fetus, larva, juvenile, and so forth. They can be influenced by maternal (and paternal) effects as well as environmental cues. Life history stage cycles in adults are also affected by environmental cues, but the mechanisms may be very different. Semelparous species that breed once and die may have a life cycle that is entirely ontogenetic (i.e., there is no need to repeat any stage). Iteroparous species, in contrast, may breed several times in their life span and express the full cycle of life history stages several times.

several or many cycles of life history stages, depending on longevity (Fig. 2). These comparisons may have considerable import for communication in reproduction and control mechanisms. For example, is communication in spawning Pacific salmon (*Oncorhynchus* sp.) that die after breeding under the same selection as in a trout (*Salmo trutta*), which may breed several times in its life span? Bear in mind also that endocrine disrupters in the environment may have very different effects in ontogenetic stages versus cycles of life history stages in adults.

Finite-State Machine Theory and the Reproductive Life History Stage

The life cycle of vertebrates can be organized according to a series of distinct life history stages, such as reproduction, migration (vernal and autumnal), molt, and nonbreeding (including hibernation). Vertebrates may express one or more life history stages usually on a repeated cycle of approximately 1 year (Fig. 3). Each stage is made up of a characteristic set of substages (e.g., for reproduction, see Fig. 4) that are expressed in various combinations. State at any point in the life cycle is the combination of life history stages and substages expressed. Because for each species there is a set number of stages and substages, an individual can be imagined as a machine with a finite number of states that can be expressed (1,5). Because for each individual there is a fixed number of life history stages and their substages, there is a finite number of states that can be expressed. That is, the individual can be classified as a finite-state machine. Organisms have different numbers of life history stages and thus the complexity of the finite-state machine varies markedly among taxa. It is possible that the more complex the machine, the more coordination of activities is needed among individuals. This could have considerable importance for the roles and mechanisms underlying communication involved in reproduction and other life history stages.

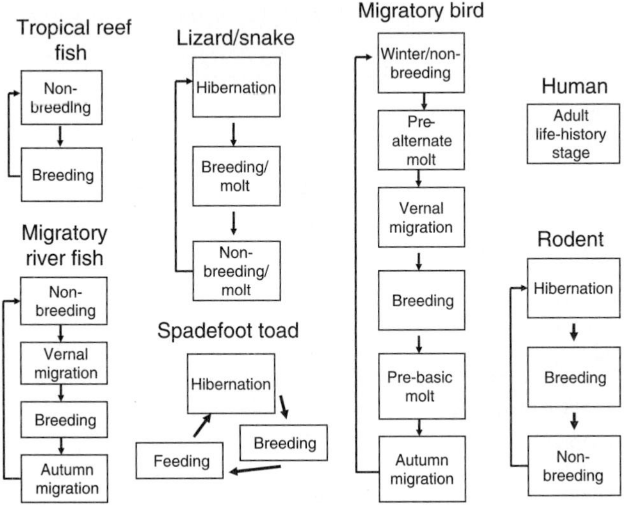

FIG. 3. Some examples of finite-state machines showing how the number of life history stages varies across taxa. Humans are unusual in showing only one adult life history stage, whereas some migratory birds may have as many as six or seven. Other species are intermediate, with either three, such as many amphibians and reptiles, or just two, as is typical of tropical reef fish. Each life history stage is a discrete life cycle event, usually, but not always, mutually exclusive of other stages. For example, in many species it is not possible to breed and migrate at the same time. There can be some degree of overlap of some stages. [Modified from Wingfield and Jacobs (5), Jacobs and Wingfield (1), and Wingfield (2006).]

Stages and Their Organization

Each life history stage is usually expressed on an annual basis, and if there are more than two, progression from one stage to the next is one-way, that is, the cycle cannot be reversed (1,5) (Fig. 3). Each life history stage is a discrete set of substages (see later), and there may be as many as six or seven, as in migratory birds, or as few as two, as in some tropical reef fish. Humans appear to be unique in having only one adult life history stage (Fig. 3); we do not migrate on a regular basis (to the exclusion of other activities), nor do we molt or show puberty more than once, as many seasonally breeding animals do (6). Life history stages can overlap, especially when in transition from one stage to the next.

Phases of Reproduction

Life history stages have three distinct phases that can be used to assess when communication is important and how it may be enhancing reproductive effort. These are the development phase; the mature capability phase, when many substages of the life history stage can be expressed; and the termination phase. Further subdivisions could be proposed for some species (Fig. 4). The development phase of one life history stage overlaps with the termination of the last and may result in very complex states in some populations (1,5). Again, it is important to note that ontogeny of morphological, physiological, and behavioral phenotypes is separate from the developmental phase of adult life history stages (Figs. 1 and 4). In birds, the first phase involves growth of the gonads from an immature state to near-functional capacity. The second phase involves rapid maturation of ovarian follicles (yolk deposition), culminating in ovulation and oviposition. In contrast, the testes progress to functional maturity without a pause (i.e., these phases are more distinct in females). Toward the end of the breeding season, the reproductive system becomes refractory to the environmental signals that stimulated the first two phases (7–9). In mammals, gonad maturation may also occur on a seasonal basis or just once (as in humans and other primates). Onset of cycling, such as estrus, may mark the mature capability phase, with anestrus marking the termination of breeding (Fig. 4). There is extensive evidence that many environmental cues, particularly social interactions (communication), can influence all three phases of breeding (see later). Further, it is clear that the number of substages of reproduction life history stages varies markedly among species, especially in the mature capability phase (Figs. 5 and 6). For example, parental care fits into this phase, but not all species show parental behavior, and in many that do it is expressed in only one sex.

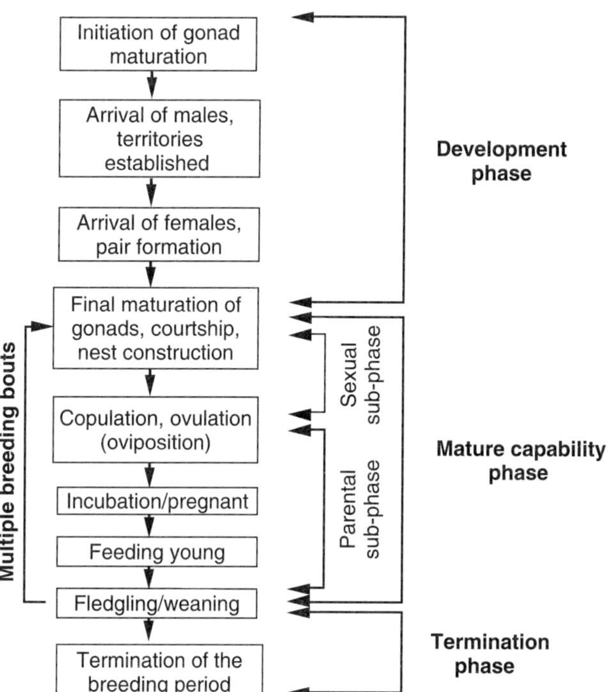

FIG. 4. The reproductive life history stage is made up of three distinct phases. First the development phase in which the gonads reach maturity and the animal is capable of reproduction. Second is the phase when actual onset of reproduction can take place. This may involve onset of estrus in mammals or yolk deposition leading to ovulation in egg-laying vertebrates. Parental care can be included in this phase. The third phase is termination of reproduction when cycling in females ceases and in many seasonally breeding animals the gonads regress.

Substages and Their Organization

Each life history stage has a unique set of substages that can be expressed in several different ways and combinations, and generally are not subject to the same "one-way" cyclic patterns. Substages of the reproductive life history stage are given in Fig. 4, but not all species express all of these substages, and thus some may be more simple than others. Substages in the development and termination phases of reproduction are in general much less complex than mature capability, when breeding actually is initiated. In general, the substages shown in Fig. 5 are not always linked (i.e., the individual cannot progress from one substage to the next in all possible connections). In other words, there are set sequences of substages, such as gonad maturation, nest building, courtship and copulation, ovulation, and oviposition/pregnancy, that are expressed in sequences characteristic of the species and gender (5). Obviously, expression of parental behavior would be inappropriate before young are born or hatched. Note that other substages such as territorial behavior may be expressed throughout the breeding life history stage and can be expressed in any combination with other substages (Figs. 5 and 6).

Not all species show parental care; in these cases, there are fewer substages in the breeding life history stage (Fig. 5). The simplest examples may be found in schooling fish, in which gametes are broadcast into the water and no parental care follows. In these cases, females may have only three substages and males just two (Fig. 5). They must communicate to coordinate timing of spawning, but other than that

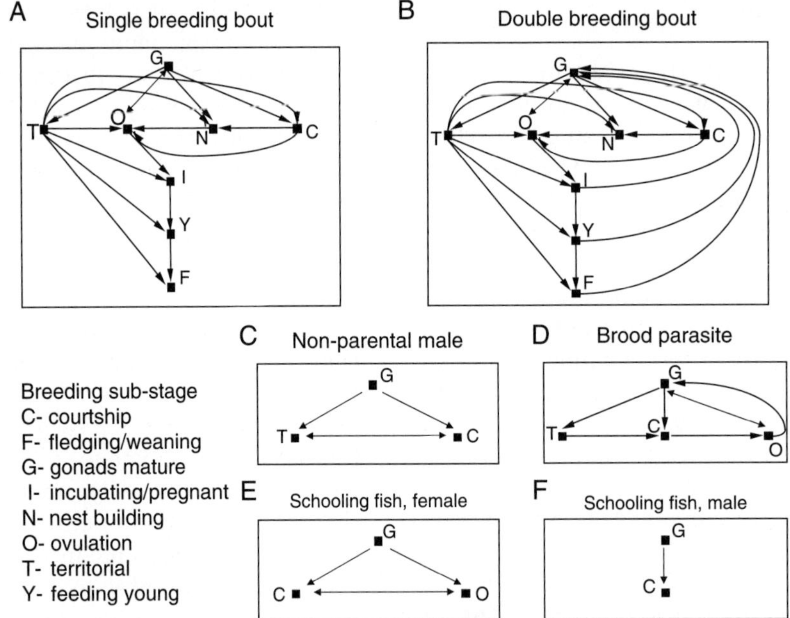

FIG. 5. Comparisons of different numbers and patterns of connectivity of substages in the reproductive life history stages of various vertebrates. **A, B:** The number of actual connections of substages (characterizing sequences of events—see text) can increase in complexity if more than one breeding bout is expressed within a breeding season. **D–F:** In contrast, those populations showing no parental care have a greatly reduced complexity of substages, as shown for brood parasites and schooling fish. **C:** Note also that nonparental males may have a much simpler set of substages than their mates that are parental (**A,B**). (Modified after Wingfield, J. C., and Jacobs, J. D. [1999]. The interplay of innate and experiential factors regulating the life history cycle of birds. In *Proceedings of the 22nd International Ornithological Congress* [N. Adams and R. Slotow, Eds.], pp. 2417–2443. BirdLife South Africa, Johannesburg, South Africa.)

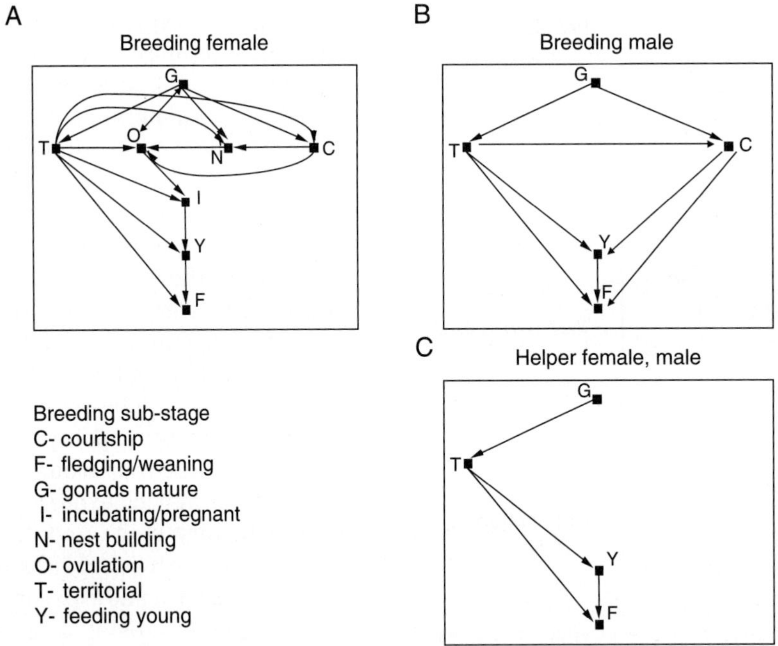

FIG. 6. Expression of substages of the reproduction life history stage can vary markedly among individuals within a population. For example, a breeding female (**A**) has a much more complex sequence of substages than males (**B**) because she is parental. In some cases of cooperative breeding, "helpers" may express parental care without showing any sexual behavior (**C**). These comparisons clearly show that communicative behaviors integrating transitions of substages and synchronizing mates and members of a group can vary markedly among individuals as well as species. (Modified after Wingfield, J. C., and Jacobs, J. D. [1999]. The interplay of innate and experiential factors regulating the life history cycle of birds. In *Proceedings of the 22nd International Ornithological Congress* [N. Adams and R. Slotow, Eds.], pp. 2417–2443. BirdLife South Africa, Johannesburg, South Africa.)

the reproductive process involves no more. Similar patterns of substages can be seen in brood parasites and nonparental males (and sometimes females; Fig. 5).

Numbers and patterns of substages may vary between males and females in a pair (Fig. 6) and, in some cases of cooperative breeding, individuals may express parental care without ovulating or showing courtship and copulation (Fig. 5). These differences in patterns among males, females, and possible "helpers" in cooperatively breeding animals are critical in determining how communication results in integration of reproductive function from one substage to the next, *and* the synchronization of individuals such as members of a pair or breeding group.

Classification of Environmental Factors Regulating Reproduction

How and why organisms breed at certain times and not others has fascinated biologists for centuries. Why do some populations have very rigid breeding periods and others show great plasticity? What environmental signals do they use to time breeding, and how are those signals transduced into neuroendocrine secretions that regulate reproductive development and onset of breeding? Although ecologists and evolutionary biologists have made significant headway at the ultimate ("why") level (10,11), physiologists still have much to learn about proximate ("how") mechanisms (8,12,13). How one environmental signal, the annual change in day length (photoperiod), regulates seasonal breeding has been studied in detail (7–9,14). However, other environmental signals that are used to time reproductive development and onset of breeding, and how they affect neuroendocrine and endocrine mechanisms that orchestrate reproductive function, are much less well understood. This is especially true of species that show great plasticity in the timing of breeding seasons [e.g., in wild birds (1,5,12,15)].

In the early 1980s, a classification of the complex array of environmental cues emanating from both the physical and social environment in relation to their major effects was proposed (12,16,17). *Initial predictive information* provides accurate, long-term predictive cues so that an individual can begin preparing for onset of breeding several weeks or even months in advance. An example is the seasonal change in day length that promotes gonadal development in anticipation of breeding. Initial predictive cues are integrated with *local predictive information* that enables short-term adjustment of reproductive function in relation to local phenology. Spring may be early and warm in one year, but late and cool in another. Examples are temperature, availability of food, nest sites, and rainfall. These two types of signals combine to coordinate reproductive development and optimize onset of breeding to maximize reproductive success.

A third group of environmental signals is called *synchronizing and integrating information* and includes all of the behavioral interactions, intersexual and intrasexual, among groups, and between adults and young (see earlier, and Figs. 5 and 6). They coordinate changes in behavior with reproductive state (18,19), and precisely synchronize the behavior of a breeding pair or group during breeding (12,20–23), including social plasticity (24), social recognition, and the ability to remember conspecifics (25). Extensive research over the past 50 years indicates that this type of information impinges on all aspects of an animal's life, and has profound implications for physiology and endocrine control mechanisms. A fourth group, known as *modifying information*, includes all the unpredictable events in the environment that have the potential to be stressful (12,26). In this chapter, the focus is kept on integrating and synchronizing information because other chapters address signals from the physical environment. Nonetheless, it is important to appreciate that none of these types of environmental signals, physical and social, act alone; they all combine to influence various phase of breeding and thus the overall reproductive process (Fig. 7). Initial predictive information triggers development of the reproductive system (in seasonally breeding animals), and these factors are necessary to maintain reproductive function and then terminate breeding at the end of the season (Fig. 7). Local predictive information acts to speed up development or termination of the reproductive system, or slow it down, and influences whether onset of cycling

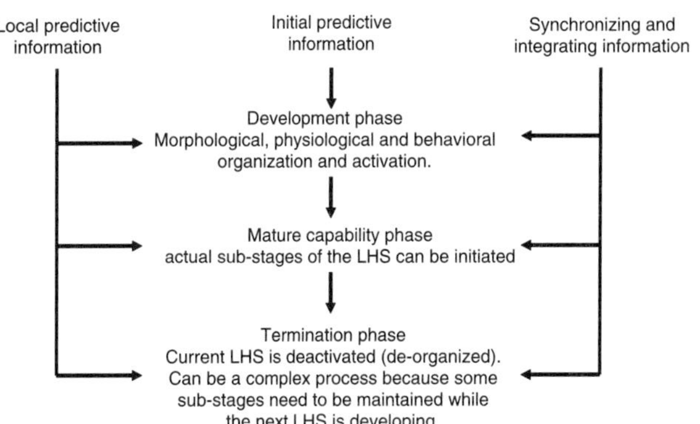

FIG. 7. Different types of environmental signals from the physical environment and from social interactions can influence all three phases of the reproduction life history stage. Initial predictive information initiates gonadal development, maintains reproductive capacity, and then terminates the breeding season. Local predictive and synchronizing and integrating information can influence all three phases as well. Thus, the phase of the reproduction life history stage is an important determinant of how, for example, a given behavioral interaction will affect neuroendocrine and endocrine mechanisms.

takes place. Synchronizing and integrating information plays similar roles (Fig. 7), but also coordinates breeding efforts of a pair or group and integrates transitions between sexual and parental behavior. Because some environmental cues act at different phases, and some behavioral interactions (e.g., between males and females) also act at different phases, it is highly likely that the mechanisms involved will also be different.

The classification described here provides a simple way of organizing all environmental signals within a framework that can be approached experimentally to identify mechanisms. Thus far, it has been possible to place all known environmental signals into one of the described categories.

Modalities of Communication

Communication systems involved in the synchronization and integration of reproductive function may be highly complex, but there are only limited pathways by which signals from an individual are transmitted and then received by another (27). Such communication systems are also involved in social modulation of hormone secretions and thus play a critical role in the integration and synchronization of all aspects of reproductive function. Depending on habitat, context, and species, communication among individuals generally involves one or several sensory modalities (23,28,29). The known sensory pathways are described in the following sections.

Visual

Visual communication can include facial expression, postures of the body, specialized adaptations of the integument (e.g., manes, plumes, antlers, fins), and color patterns (30). These visual cues can be both static and dynamic and in many cases involve a combination of colors, shapes, and postures. Some morphological signals, such as bright plumage or antlers, act as static signals. Others are used intermittently, such as making a brightly colored patch of skin or a plume more conspicuous. Usually these dynamic displays are limited to social displays and the signal can be withheld or displayed at will. The distinction between static and dynamic displays is important because of the obviously different mechanisms underlying development and production (23).

Tactile

Communication solely by tactile means is rare and may not exist in the absence of other sensory input (23). An example is induced ovulation—a reflexive response to mechanical stimulation during copulation and amplexus in amphibians when the male grasps the female during courtship, providing tactile stimuli (for males as well as females) leading to spawning (31,32). Tactile contact during grooming has important soothing and pacifying effects in birds and mammals, particularly in primates (33,34).

Auditory

Auditory communication occurs in every tetrapod vertebrate group as well as in some fish, and has reached a particularly high level of development in anurans, birds, and many mammals (35–37). Auditory communication is usually achieved by vocal production of sound, but nonvocal sounds are also important (e.g., the rattle of *Crotalus*, sounds produced by the swim bladder in some teleosts). However, the roles of sounds in reproduction are complex, such as in teleosts (38,39) and the roar of the alligator (*Alligator mississippiensis*), thought to be territorial in context (40,41).

Vibrations (Seismic Stimuli)

There is mounting evidence that some vertebrates can monitor vibrations through the substratum (41,42). Examples are some urodeles and anurans that can pick up vibrations ("thumps") through the ground and water. In the white-lipped frog (*Leptodactylus albilabris*), substrate-borne vibrations (seismic stimuli) result in responses of nerve fibers from the sacculus. Males are also able to generate substrate-borne signals by the vocal air sac striking the substrate. In other species, females may produce vibration signals by tapping their toes on substrate, allowing courting males to locate them (42–44).

The use of seismic signals has also been proposed for kangaroo rats [*Dipodomys spectabilis* (45,46)], kangaroos [*Macropodia* sp. (47)], and two species of mole rat [*Georychus* (48,49)]. The Cape mole rat (*Georychus capensis*) lives underground and apparently communicates by foot drumming on the burrow floor. Seismic signals propagate further through soil than do auditory signals (48) and may be an important sensory modality for receiving communication from others in fossorial species. Interestingly, seismic signal production is sexually dimorphic in this species, suggesting a possible endocrine basis (50). Furthermore, during an interaction involving such seismic signals individuals may synchronize footbeats (48), which may function to maintain spacing of burrows and attract mates.

Electrical

Many fish can communicate through electrical discharges [e.g., mainly teleosts, some holosteans and elasmobranchs (51–53)]. For example, in the teleost family Morymridae, each species produces a characteristic electrical wave form used in species recognition, territorial behavior, mate location, and courtship (54–58). There are many parallels in the evolution of electrical communication systems in the African Morymridae, the South American Gymnotidae, and the elasmobranch Rayidae (51,52,57). The electrical system is located in the hind part of the body musculature, and the structure of the modified muscle "plates" is correlated with the wave form generated (59,60).

Chemical

With the possible exception of the class Aves, species that communicate by chemical (pheromone) means are widespread, including most, if not all groups of fish (61–64). In pheromonal communication, one individual releases a chemical that is monitored by others at distances that can vary from a few centimeters to several kilometers. In some cases the chemical is released without any obvious behavioral display [as in passive release of pheromones by garter snakes (*Thamnophis sirtalis*) (65)]. In others there are deliberate accompanying displays such as scent marking and spraying (66–68). Some of the chemical used as pheromones may also be hormones [e.g., ovarian steroids in some teleosts (62)].

Hormones, Behavior, and Social Interactions: The Endocrine Basis of Signal Production and Reception

Throughout the life cycle, and particularly during reproduction, individuals must locate a potential mate and assess species, gender, reproductive state, age, and condition before actual mating occurs. After mate selection, the mechanics of joining egg and sperm must be coordinated. In addition, repelling competitors while avoiding the risks of combat, locating predators, and warning the mate or young are critical abilities (23). The endocrine mediation of signals involved in such sexual, social, and aggressive interactions is complex and the relationship between hormones and behavior is bidirectional. Given that communication requires some behavioral response in either the signaler or the receiver, and given that behavioral interactions are fundamental for synchronization and integration of reproductive function, it is important here to reflect on hormonal control of behavioral expression. Many books and seminal articles have been written (69–71), and other chapters in this text address these topics in detail. In relation to the action of steroid hormones on behavior, there appears to be a continuum, with organizational effects during development at one extreme and activational effects during the life cycle at the other (69,70). However, finite-state machine theory has provided some additional contexts for mechanisms and processes that may not be as well known. For example, organizational effects of hormones could occur during ontogeny that are generally irreversible, whereas similar organizational effects could occur during development of a life history stage (e.g., reproduction) that are reversible (Fig. 2). These differences, coupled with potential suppressive and interference actions of hormones on behavior, present a very plastic series of control mechanisms suited to complex behavioral interactions.

Hormones can act as true endocrine mediators of behavior, although it is important to remember that neural circuits and their neurotransmitters actually cause behavior. Peripheral endocrine secretions influence the central nervous system and expression of behavior in ways other than through a classic neurotransmitter/neuromodulator role. It is generally accepted as false to think of a stimulus resulting in secretion of a hormone that, after being taken up by the brain, triggers behavior (although this process may occur in a few specialized instances, e.g., the fight-or-flight response). Rather, hormones influence behavior by increasing the likelihood that a particular behavioral trait will be expressed over unit time, or in response to a specific external stimulus or stimuli (releasers). For example, in gonadectomized green tree frogs (*Hyla cinerea*), androgens influence the motivation to call, but under conditions of social stimulation (such as a chorus of calling male frogs), additional factors are involved (72). In other words, hormones increase the intensity and frequency of behaviors—a point that may appear obvious to many in the field, but is nonetheless crucial to bear in mind when investigating diverse species under natural conditions. Even in the absence of a particular hormone, the behavior will be expressed eventually. Some well-known and less well-known examples are given in the following sections to illustrate this point.

Organizational/Activational Effects in Relation to Finite-State Machine Theory

How, then, do circulating hormones influence behavior, as opposed to hormones produced in the brain that act as neurotransmitters or neuromodulators? Three major ways have been proposed:

1. Organizational effects: secretion of hormones at appropriate times during development can influence the development of neural circuits. Once mature, these circuits may operate independently of the hormone that influenced their development. Examples are numerous, especially the sexual dimorphism of the brain [e.g. in reptiles, rodents, and songbirds (73–76)].

2. Activational effects: seasonal or periodic increases in secretion of appropriate hormones activate specific existing neural circuits so that they fire more readily in certain contexts. Activity of these neurons depends on the immediate and continuing presence of the hormone, although only low levels may be required to maintain full function. Examples are effects of testosterone on aggressive territorial behavior and prolactin on parental behavior (70).

3. A combination of activational and organizational effects: during development (and in seasonal or periodic expression of behavior), hormones can influence both the organization of neural circuits and their immediate function. An example is the song control system of passerine (song) birds. In some species such as the canary (*Serinus canarius*) and zebra finch (*Poephila guttata*), the male sings more frequently than the female. In both species the male has larger nuclei in the brain that control production and frequency of songs (77–82). Similar mechanisms are involved in seasonal auditory plasticity in fish (39,83).

In the zebra finch, sex steroid hormones exert organizational effects during development. It appears that estradiol production just after hatching masculinizes the brain in young males so that as adults they can respond to the activational effect of testosterone and sing (84). Female zebra finches do not show an increase in estradiol production just after hatching and are unresponsive to testosterone injections when adult. As a result they rarely, if ever, sing. However, if females are given estradiol 2 to 10 days after hatching they respond to testosterone injections when adult and thus sing. In contrast, the female canary does sing spontaneously on occasion, and is always responsive to testosterone injections (often resulting in song similar to that of the male). These two species illustrate clearly the differences in how organizational and activational effects of hormones can influence behavioral patterns, although nonhormonal factors are also involved (84–90).

Of additional interest here are studies on the song control nuclei of species in which females sing normally. In some tropical species of wrens (Troglodytidae), the male and female of an established pair sing duets (91,92). The song sounds like that of a single bird but actually both male and female sing discrete parts of the song with exquisite precision. In these species the song system nuclei of both sexes are of equal size and stage of development, and both may be sensitive to testosterone. In other species there are differences in song control systems that correlate with the degree to which males and females sing (93), but most of these comparisons have phylogenetic confounds that have yet to be resolved (77).

Social traits such as those associated with reproductive aggression may differentiate along pathways separate from those of sexual traits such as singing. Individual differences can also be marked and established during ontogeny (94). In addition, organizational and activational effects may also operate for other reproductive behaviors, such as alloparental and affiliative behaviors in rodents (95). In European starlings (*Sturnus vulgaris*), expression of aggression during and outside of the breeding season varies possibly in relation to age at which testosterone exposure occurs (96).

Steroid hormones may also have major effects in sex determination in species with environmentally determined sex. In the red-eared slider (*Trachemys scripta*), sex is determined by the temperature at which the eggs incubate. Administration of estradiol to eggs at a temperature that normally produces only males can overcome the effects of temperature and produce females (97).

Suppressive Actions of Hormones

Hormone secretions may actually suppress expression of certain behaviors. For example, it has been shown in birds that elevated levels of corticosterone may suppress the expression of territorial aggression (98,99). Suppressive effects of corticosterone on sexual behavior have been described in reptiles and amphibians (100,101). Elevated levels of glucocorticoids have the potential to inhibit the reproductive system, resulting in a decrease of circulating testosterone levels in many vertebrates [see Greenberg and Wingfield (102), Moore and Jessop (103), Sapolsky (104), and Sapolsky et al. (105) for reviews].

There appears to be a more direct suppressive effect of glucocorticoids that is independent of inhibition of the reproductive system (99). For example, implants of corticosterone decreased expression of intermale aggression in the side-blotched lizard (*Uta stansburiana*), but did not affect courtship and copulation behavior when these males were exposed to females made sexually receptive by implants of estradiol (106). Furthermore, it appeared that simultaneous implants of corticosterone and testosterone also suppressed expression of aggression, suggesting that corticosterone was not acting solely by the inhibition of testosterone secretion (106). Whether this action is directly at the level of testosterone-sensitive neurons or through a different mechanism deserves further investigation. In the high-latitude–breeding white-crowned sparrow (*Zonotrichia leucophrys gambelii*), central injection of corticotropin-releasing factor rapidly decreased territorial aggression (107).

"Interference" Actions of Hormones

A fourth possible mechanism by which hormones may influence behavior involves activation of alternate behavioral patterns that override expression of other behaviors. Leshner (108) and Leshner and Politch (109) have shown that corticosterone can activate submissive behavior in agonistic encounters among mice. A decline of testosterone was irrelevant for this effect of corticosterone on submission. Androgens and glucocorticosteroids may polarize the direction of agonistic behaviors toward dominance or submission (110). If corticosterone levels rise as a result of defeat, then expression of submissive behaviors could possibly take precedence. Leshner (110) found that corticosterone treatment had little effect on the submissiveness of male mice in an aggressive encounter, but when that treatment was combined with the experience of defeat in such a paradigm, it increased submissive behavior significantly. Another example is the influence of testosterone on expression of male parental behavior in birds. Males of many passerine species provide considerable parental care, but high levels of testosterone (experimentally induced by implants) resulted in a significant decline in the rate at which they fed their young (111). High circulating concentrations of testosterone do not suppress or deactivate parental care, but the activation of aggression appears to take precedence over expression of parental care, with a decline in expression of the latter. In the Texas bobwhite quail (*Colinus virginianus texanus*), males with low circulating levels of testosterone showed alloparental care when exposed to chicks. Those males with higher levels of testosterone did not show any alloparental care (112). However, males given implants of antiandrogens such as flutamide showed reduced aggression directed at intruding males in their pens, but there was no direct effect on sexual behavior or expression of alloparental care (112).

These data suggest that high levels of testosterone do not simply suppress parental behavior in males but may activate aggressive behavior so that it is expressed in precedence over parental behavior. The etiology of this phenomenon still requires much work, but it is important to bear in mind that hormonal activation of one behavioral pattern could take precedence over expression of another at some stages of the life cycle. This could be an important additional mode of control, or just a version of suppressive actions.

Interference effects of hormones on behavior may also be related to breeding strategy, at least in birds. Lynn and Wingfield (113) showed that male parental care is essential for reproductive success in chestnut-collared longspurs (*Calcarius ornatus*) in north central Montana. As a result, it is possible that this dependence on paternal care led to the evolution of behavioral insensitivity to testosterone. Male longspurs did not show increased aggression or reduced parental care with supplemental testosterone, suggesting an insensitivity to that steroid as it relates to those behaviors (114,115).

"Persistence" Actions of Hormones

In some cases a hormone may not affect activation and expression of a behavior in a given context, but may influence persistence of the behavior after the stimulus has been removed. An example is the effect of testosterone on aggression in songbirds. Testosterone does not regulate aggression per se, but increases persistence of spontaneous aggression after a territorial intrusion. Male song sparrows (*Melospiza melodia*) with elevated testosterone after a staged territorial encounter continue spontaneously to sing, patrol the territory, and attack other birds for up to 24 hours *after* the stimulus had been removed (116–119). This is interpreted as the experimental male (staged winner) actually reinforcing his territorial status and broadcasting to other males that he is still in control. When the same experiment was performed in autumn, male song sparrows showed strong territorial behavior but testosterone levels were not detectable. Males respond to territorial intrusion in a manner similar to spring birds (with high testosterone), and even castration did not abolish this response. However, testosterone levels did not increase in response to intrusion in fall, and when the stimulus was removed, males stopped singing and patrolling within minutes. Autumn males given testosterone implants showed the persistence of aggression after an experimental simulated territorial intrusion, and the accompanying reinforcing behavior returned (117–119). There is precedent in the earlier literature for such persistence effects. Andrew (120) and Andrew and Rogers (121) showed an effect of testosterone on persistence in chicks in open-field situations. More recently, testosterone administration to male side-blotched lizards (*U. stansburiana*) increased endurance (running on a treadmill until exhaustion), activity, home range size, and control over female territories, consistent with a "persistence" effect. However, there was a cost to these effects because males with higher levels of testosterone have reduced survival (122).

Maternal Effects

It is well known that hormones of maternal origin may influence development of a phenotype during the fetal stage. Similarly, in egg-laying vertebrates, females may influence the expression of aggression in their offspring by increasing deposition of testosterone into egg yolk. In mammals, fetuses may influence each other (123,124. Maternally derived testosterone can influence development of aggressive traits in the chick [e.g. canaries (125,126)]. However, in green anoles (*Anolis carolinensis*), it appears that incubating embryos are capable of producing sex steroids, and there is no correlation between maternal testosterone and estradiol levels and those measured in eggs (127). A study in Japanese quail (*Coturnix japonica*) suggests that only a small proportion of steroids from the maternal circulation (approximately 0.1%) enter eggs, and that the majority of steroid content in the yolk reflects the production of follicular cells at the time of yolk production (128). Thus, the mechanisms at play to determine the exact nature and magnitude of maternal effects may be complicated. Nonetheless, behavioral interactions in females at the time of pregnancy and egg forming may have additional organizational effects on developing organisms that are corollaries of activational effects in the classic sense, but influenced by another individual in relation to its eventual phenotype. These perspectives may shed new light on communication and synchronization and integration of reproductive function in ecological contexts.

Synchronizing and Integrating Factors in the Reproductive Process: Dealing with a Dynamic Environment

The reproductive process involves three major types of behavior: aggression in a reproductive context (e.g. territorial aggression), sexual behavior, and parental behavior. In species in which the breeding period is very short, these behaviors are expressed briefly and may show less overlap (Fig. 8). In this case, transitions between behavioral types may be regulated by physical environmental cues and less by social interactions. If the breeding season is longer, there may be considerable overlap of potential expression of these behaviors, with complex and

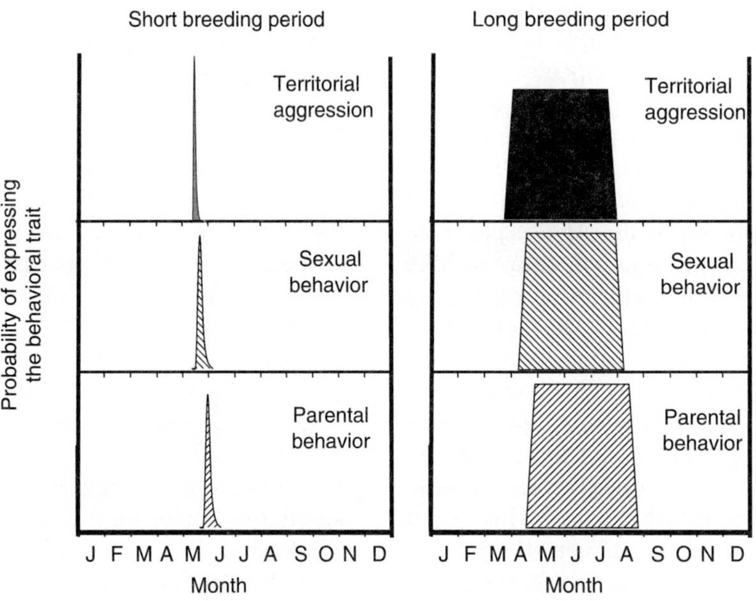

FIG. 8. The reproductive process involves three major types of behavior: aggression in reproductive context (e.g., territorial aggression), and sexual and parental behavior. In species in which the breeding period is very short, these behaviors are expressed briefly and may show less overlap. In this case, transitions between these behaviors may be regulated by physical environmental cues and less by social interactions. If the breeding season is longer, there may be considerable overlap of potential expression of these behaviors, with complex and rapid transitions among them. In this case, behavioral interactions–communication–would be much more important for synchronization and integration of reproductive function.

rapid transitions among them. In this case, behavioral interactions (i.e., communication) would be much more important for synchronization and integration of reproductive function (Fig. 8). Clearly, it is important to be able to assess the reproductive life history stage of diverse vertebrate taxa and determine how dependent the onset of breeding may be on cues from the physical environment in, for example, species with a very short season as opposed to those that have a longer breeding season and much more flexibility.

Information theory and log-linear analysis of breeding dates in the field have been used to model when individuals should integrate initial and local predictive environmental signals or rely more heavily on one or the other (129,130). For example, if the breeding season is highly predictable and restricted in time, then only one reliable environmental cue is needed to time gonadal development. On the other hand, if the breeding season is much less predictable, the animal should monitor and respond to more environmental cues to coordinate precise adjustment of reproductive development to local conditions. Predictability of seasons in a given habitat can be calculated as an "environmental information factor" (Ie). This factor reflects the degree to which a species within that habitat should integrate different environmental cues to regulate gonadal development and onset of nesting. If Ie is low, individuals should focus on one or very few reliable cues (e.g., photoperiod). When Ie is high, they should be sensitive to many environmental signals to ensure accurate timing (129,130). Demographic data, field endocrine studies, and laboratory experiments allow testing of these models (131–133) and also indicate which environmental signals are important. This theoretical framework provides a powerful model to determine mechanisms by which the brain transduces such information, and neuroendocrine/endocrine pathways can be compared in populations that have different Ie factors (degrees of responsiveness). These approaches may allow us to predict when communication is more or less important in conjunction with cues from the physical environment. Successful reproduction is always entirely dependent on some level of communication, but the complexity of such communication varies and will have an important influence on control mechanisms.

Besides the modalities by which animals communicate, the types of environmental cues regulating reproductive function, and the phases of life history stages, it is also possible to classify types of communication important for regulating reproduction. The behavioral interactions that comprise synchronizing and integrating information can be divided into four major subgroups:

1. Those systems having a direct role in regulating gonadal maturation (developmental phase) but that may play little role in the release of gametes themselves. Examples are courtship behavior, pheromones, and social facilitation.

2. Interactions regulating onset of breeding (i.e., precise timing of ovulation and release of gametes). For example, many fish spawn synchronously in large schools. Because fertilization is external, the temporal coordination of gamete shedding is critical and requires communication. Some species of mammals are induced ovulators, and females will not ovulate unless they have copulated. Once fertilization has taken place, many species show some form of parental care. This can involve communication between parents and other relatives to synchronize parental duties, and interactions between parents and offspring that coordinate (integrate) the parental phase.

3. Behavioral interactions can play support roles only indirectly related to gamete production, fusion, and rearing of young. Examples are establishment and control of a breeding territory or other resources.

4. The termination phase of the reproduction life history stage involves gonadal regression in many seasonally breeding species, often accompanied by breakdown of territories and aggression associated with defense of the territory, loss of pair bond, and dissolution of family groups. The termination phase is not always synchronous within a population and its timing can be influenced by food supply, weather, reproductive status of the mate, and age of offspring.

Now that the theoretical patterns of breeding and the organization of stages and substages of the life history cycle have been put into context with the sensory modalities by which animals communicate, the ways by which behavioral interactions influence reproductive processes, and the basic mechanisms underlying hormonal control of behavior, it is possible to go on to explore the interaction of endocrinology and behavior in synchronizing and integrating the breeding season.

HORMONE–BEHAVIOR INTERACTIONS DURING THE DEVELOPMENT PHASE OF THE REPRODUCTION LIFE HISTORY STAGE

The first critical phase of the reproductive life history stage is development of the gonad to a mature state that allows onset of actual breeding. The extent to which the gonads develop depends on the degree of seasonal changes in gonadal function, which can vary from complete regression of the testes and ovaries to an infantile state (e.g., many seasonally breeding fish, amphibians, reptiles, and birds) to very little change in gonad size, with onset of cycling (anestrus to estrus) marking the development phase (23,134,135).

Communicative behaviors underlying social interactions have dramatic effects on the development phase of reproduction and are particularly critical in species in which males and females must coordinate breeding events precisely. In addition to the gonads, accessory organs (e.g., vas deferens, oviduct) and secondary sex characters (e.g., bright plumage, antlers, pigment patterns in skin, scales) must also develop in parallel with gonadal development. Organizational effects of steroid hormones may also affect these traits during ontogeny [e.g., in fish (136)]. Hormonal control of these characters is complex (137,138), but such traits are used extensively in communicative behaviors regulating reproductive function. Changes in the brain are also essential for expression of sexual, parental, and aggressive behavioral patterns as onset of breeding takes place. There is extensive evidence here too that hormones play a major role, although genetic factors are important also (18,24,85,87,88, 139,140). Plasticity of neural mechanisms is a major focus of current research—effects of behavior on the brain in fish, includes size social status and sex thus regulation reproductive maturation. Brain affects behavior, but behavior can also affect the brain (140).

Much of this literature is beyond the scope of this chapter and is summarized elsewhere in this volume. Here I give an overview of types of communicative behaviors underlying social interactions that influence different phases of the reproduction life history stage. Specific examples are given to illustrate points. Much remains to be done concerning the regulation of morphological, physiological, and behavioral traits leading to reproductive maturity, and as the fields of genomics and proteomics expand to additional species and vertebrate groups, many exciting models will be available for future comparisons.

Interactions with Initial Predictive Information (Photoperiod)

Most experiments on the interaction of initial predictive and supplementary factors were performed in environmental chambers in which males and females were housed together. Because it is well known that the presence of the opposite sex can enhance photoperiodically induced gonadal maturation in both phases [see Wingfield et al. (12,23) for reviews], experiments were conducted on white-crowned sparrows (*Z. l. pugetensis*) in which sexes were isolated to remove intersexual social cues. Effects of temperature on testis growth persisted (15). However, isolation from males resulted in retardation of the first phase of ovarian development at temperatures of 5°C and 20°C (but not at 30°C). All temperature effects on the second phase of ovarian maturation were abolished in females isolated from males. These data clearly indicate that social cues and temperature interact to regulate both phases of gonadal maturation.

Effects of Males on Females

Much of what is known about effects of males on the reproductive state of females involves pheromones (i.e., chemical communication). For example, pheromones presumably in the urine of adult males can accelerate onset of puberty in female mice (*Mus musculus*) and other rodents [for reviews, see Vandenbergh (68,141)]. This has been demonstrated in other species as well. It is also known that pheromones produced by adult female mice can delay onset of puberty in younger females. Thus, pheromonal control can involve complex blends of potentially antagonistic signals.

It is well established for several avian species that the presence and behavior of males can accelerate reproductive development in females. Courtship behavior of males has been shown experimentally to enhance ovarian development (142). In ring doves (*Streptopelia risoria*), castration of the male resulted in less ovarian development in their mates (143). This difference was independent of contact because male and female were separated by a glass plate. If castrates were then treated with testosterone, there was a dose-dependent increase in male courtship and also the female's response (144–146). Female doves that can see a male courting them show greater ovarian development than females that could see only a male courting another female. These data suggest that male courtship must be directed at the subject female for greatest effect. Note, however, that these effects of behavioral interactions on gonadal development are effective only within a breeding season regulated by initial predictive and local predictive types of information. Subsequent evidence suggested that the mechanism for the female's response to a courting male may involve "self-stimulation" (147). Female ring doves did not respond to a courting male with increased ovarian development if they were deafened or prevented from returning appropriate courtship behavior to the male. Playback of tape-recorded female vocalizations could also mimic the effects of a courting male, indicating that the female responds to a male by undergoing active courtship displays and that these displays feed back to stimulate ovarian development (148). Thus, the male provides context, but the female's response self-stimulates.

In the green anole (*A. carolinensis*), the presence of an intact male induces a significantly higher rate of ovarian development with simultaneous exposure to springlike conditions (149,150). Isolated females or females paired with a castrated male or another female showed less ovarian growth when exposed to springlike conditions, and laid eggs without shells. Further investigations revealed that males appear to enhance gonadotropin secretion in females through displays of a pink flap of skin (dewlap) that can be extended from beneath the throat. Section of the ventrobasal hyoid abolished this display and also reduced the effect of male courtship on ovarian development. In addition, if the color of the dewlap was changed from pink to blue by injecting India ink, females responded with less ovarian growth initially, but eventually this effect was overcome (151). Similar results have been obtained in the garter snake (*T. sirtalis*). Females that mate with males on emergence from hibernation show greater rates of ovarian recrudescence than females that do not mate (152).

Effects of Females on Males

It is well known that mature, especially receptive females can influence gonadal development in males [see Wingfield et al. (23) for review]. In rhesus monkeys (*Macaca mulatta*), males exposed to estrogen-primed females in the nonbreeding season showed an increase in testis size (153). In deer mice (*Peromyscus maniculatus*), males maintained on a short day length, but housed with females, underwent off-season development of the testes. When exposed to long days, males showed more rapid testicular development when simultaneously exposed to females (154). Similar effects were seen in male musk shrews (*Suncus murinus*) (155). On the negative side, exposure of young male voles (*Microtus californicus*) to odors of their mother's bedding inhibited gonadal maturation. Whether this observation represents a mechanism to avoid inbreeding remains to be seen [see Vandenbergh (141) for review].

In the European starling (*S. vulgaris*), males exposed to females show a greater rate of photoperiodically induced testicular growth than isolated males (156,157). Essentially identical results have been obtained for male brown-headed cowbirds (*Molothrus ater*) (158–160). Presence of females also results in larger testes and greater spermatogenic activity in domestic pigeons (*Columba livia*) (161), and may influence development of song control nuclei in songbirds (90). Further effects of females on male reproductive function throughout the vertebrates have been reviewed previously (23).

Mate Choice and Pair Bonding

Mate choice has been the subject of debate for decades. Darwin (162) generally assumed females were the more discriminating sex in terms of mate choice. Moreover, the sex providing the greater share of parental care should be more discriminating in terms of mate choice (163), and this usually is the female (164). Mate choice involves species and sex identification, as well as assessment of individual characteristics, including quality of the mate (164). The endocrine bases of mate choice and the pair bonding that follows have been the focus of attention in recent years.

In red-legged partridges (*Alectoris rufa*), females in self-selected pairs (as opposed to artificial pairs) had greater egg production, and males mated to self-selected females had higher testosterone levels (165). This responsiveness of females to courting males is thus modified by "mate choice." As another example, in ducks such as the canvasback (*Aythya valisinaria*), males court females with a variety of displays that may or may not result in female receptivity. Increased male courtship induced by implantation of testosterone did not increase egg laying, serum luteinizing hormone (LH), or receptivity in unpaired females (166,167). In addition, implants of estradiol into females did not elevate receptivity to courting males or influence pair formation, nesting activity, and egg laying. Clearly, in this species, female receptivity is a key determinant of pair formation and responsiveness to subsequent synchronization factors. Only females mated to their self-chosen mates actually laid eggs. Females in experimentally forced pairs laid no eggs and showed little nesting activity, even though males courted intensely. It appears that females do not respond to males assigned to them by the investigator without a preliminary "mate-choice" phase (168). There is evidence that vocalizations from the flock may increase circulating levels of LH and egg laying in established pairs. However, unpaired females, or those in forced groups and pairs, have lower LH and prolactin levels than laying females, and their ovaries remain undeveloped. Interestingly, pituitary responsiveness to LH-releasing hormone injection in nonlaying females was similar to that of layers, suggesting that the control mechanism for responsiveness to male courtship lies at the hypothalamic level or higher (168).

Treatment of female zebra finches with estrogen may influence whether they prefer males or females as mates (169). Indeed, sex steroids may have wide-ranging effects on mate choice (170). Furthermore, estrogen treatment for the first 2 weeks post-hatch results in "masculinized" mate choice [for females, see Mansukhani et al. (171)].

Oxytocin has been implicated in pair bonding in rodents, especially in females (172,173). For example, arginine vasopressin (AVP), when administered centrally into voles (*Microtus* sp.), enhances partner preferences and pair bonding (174). Oxytocin administrated centrally can also facilitate mate choice in prairie voles (*Microtus ochrogaster*) (174), but peripheral administration of oxytocin favors mate preference in female rather than in male prairie voles (175). Because AVP has a potent water-retaining effect through V2 receptors in kidney tubules, it was predicted that environmental conditions may also effect pair bonding, particularly partner preference (176). Prairie voles from a more xeric habitat in Kansas responded more to peripheral AVP administration with increased preference for an established partner, but the effect was limited to males. Prairie voles from Illinois showed no such response to peripheral AVP. Curiously, there was no difference in distribution of V1a receptors in either population of voles (176).

HORMONE–BEHAVIOR INTERACTIONS INFLUENCING MATURE CAPABILITY (COURTSHIP AND SHEDDING OF GAMETES) OF THE REPRODUCTION LIFE HISTORY STAGE

The effects of behavioral interactions on onset of actual breeding result in synchronization of courtship, final maturation of the gonads (especially ovulation), and shedding of gametes. This is crucial if fertilization is to be achieved with maximal efficiency, and these kinds of behavioral interactions have received considerable attention.

Onset of Cycling

Once the gonads have developed to a state where actual breeding can occur (mature capability), behavioral interactions have a different role from those during gonadal development itself. By inducing onset of ovarian cycling, these effects are particularly evident in females, but may also be present in males.

Effects of Males on Females

Olfaction is an important mode of communication in mammals, and most of what is known of the effects of males on reproductive activity in females involves pheromones. These chemical signals serve as sex attractants, and can synchronize estrus in groups of females. In the mouse, presence of a strange male prevents implantation [see Vandenbergh (141) for review]. Social behavior can have marked effects on female reproductive state. In the Hanuman langur (*Presbytis entellus*), a new male that takes over the dominant position in a group kills the infants of lactating females (177). Infanticide has been identified in other species [e.g., lions; see Hausfater and Hrdy (178) for review], and it is thought that this accelerates onset of estrus in females in the group and thus enhances the potential reproductive success of the new dominant male. There is evidence for involvement of testosterone in male rodents that practice infanticide, although the underlying mechanisms remain obscure (179).

Mechanical stimulation from copulation is also important in some species for inducement of ovulation and successful pregnancy. In rabbits, rodents, mustelids, and others, vaginal stimulation with, for example, a glass rod to simulate copulation results in ovulation and in some cases an elevation of gonadal steroids (180), accompanied by increases in circulating levels of LH and prolactin, but not follicle-stimulating hormone (FSH). The enhanced estrogen secretion induced by a reflex stimulation of LH eventually results in the positive feedback loop leading to a preovulatory surge of LH. Ovulation follows within a few hours of mating, thus ensuring that sperm will be available to fertilize the ovum (or ova). In rats, section of the pudendal nerve results in loss of responsiveness to intromission by the male. Further, the size of the sensory field of this nerve, which includes the clitoris and base of the tail, is estrogen dependent. It appears that enlargement of this sensory field enhances the progestational state of the female, especially in preparation for intromission [see Vandenbergh (68) for review].

In addition to pheromones and mechanical stimulation, vocalizations also play an important role in the effects of males on reproductive state in females. Ultrasonic vocalizations occur in a wide variety of rodents and are produced mostly, but not exclusively, by males. There is much evidence that ultrasonic signals are attractive and stimulatory to females, although these sounds can also be produced as postcopulatory displays as well as during submissive behavior [see Floody (181) for review]. In rodents, two classes of signal may be produced: those associated with ejaculation and submission (20 to 30 kHz), and those used by males to attract females and stimulate sexual solicitation by them (around 50 kHz). Production of ultrasonic sounds is influenced strongly by hormonal condition of the mate and by olfactory cues (e.g., in rats). In mice, pituitary hormones play important roles in the ultrasound-eliciting potency of females (182). Because mating often occurs during postpartum estrus, male ultrasonic vocalizations may serve to facilitate mating by reducing aggression in lactating females. Typically, production of ultrasounds by males is reduced by castration and reinstated by androgen therapy (181,183,184).

Stimulatory effects of males on females are by no means universal. In rhesus monkeys, exposure of females to males made sexually active by implants of testosterone failed to induce any change in ovarian function outside the normal breeding season, despite intense signaling by these males (153).

It is well known that courtship behavior of male birds can have stimulatory effects on egg laying in females, as well as promote behaviors preparatory to egg laying. Female canaries exposed to male song build nests more actively and, furthermore, females exposed to a large repertoire of male songs show more active nest-building behavior than females exposed to small repertoires (185,186). Female Japanese quail (*C. japonica*), lay more eggs when exposed to male crowing. These females appear to be most responsive to vocalizations between dawn and dusk, when males crow most frequently in the wild (187).

In white-crowned sparrows (15), presence of males can influence responses of females to initial predictive and local predictive types of information. Although presence of males tended to increase ovarian development at warm temperatures in photostimulated birds, a much stronger effect was seen on final ovarian follicle maturation, especially at 30°C.

Male anurans (frogs) clasp females and thereby probably provide tactile stimuli for expulsion of eggs. For example, female *Rana pipiens* with a small weight tied to their backs laid eggs normally. Clasping seems to stimulate oviposition movements of the female, which in turn increases the ejaculatory response of the male (188). There is also evidence that exposure of gravid female urodeles to courting males is required for ovulation and oviposition (189). In the rough-skinned newt (*Taricha granulosa*), gravid females ovulated when exposed to courtship and insemination by males. Plasma levels of progesterone were also highest in females exposed to courting males (190).

Female cichlids (teleost fish, e.g., *Tilapia melanocephala*) rarely oviposit unless they have contact with a male (or female). Physical and chemical contact is not crucial because pure visual cues (i.e., seeing a male or female through a glass partition) are sufficient to stimulate egg laying. Nest-building behavior and courtship signals also increase ovulation rate, and once a female receives visual stimulation from a male, she can continue to ovulate and oviposit normally even if the male is removed (31). In the Japanese medaka (*Oryzias latipes*), blinded females lay eggs normally when exposed to males. However, intact females exposed to blinded males do not spawn. This may be because ovulation is stimulated by contact provided by the dorsal and anal fins of intact males, but not by blinded males (191).

Effects of Females on Males

A variety of mechanisms are involved in the regulation of behaviors by which female mammals signal receptivity, proceptivity, and attractivity. A suite of hormonal changes may occur in males in response to these signals (23). Pheromonal cues play an important part in female–male communication. In mice, perception of a receptive female by a male is accompanied by a pulse of LH (192). In the garter snake (*T. sirtalis*), females produce a pheromone on the dorsal surface of the skin that attracts courting males, even though testosterone levels in these males may be at a seasonal low (65,193).

In primates, female signals parallel reproductive state closely and clearly serve to attract the attention of males. In rhesus monkeys, sexual behavior in females is closely correlated with circulating levels of estradiol and inversely with plasma levels of progesterone (194). Male sexual behavior was associated more with fluctuating estradiol and progesterone in females than with their own testosterone levels. It is thought that female sexual behavior and other signals attract attention of males several days before ovulation and may even lead to the formation of consort pairs, thus coordinating copulation with maximum fertility (194). Sensory channels for female–male communication in primates include pheromonal mechanisms, although their precise role is still unclear (195–197). Auditory cues probably play a minimal role in most cases, whereas auditory signals are of great importance in many species. Some visual cues are generated by both postures and displays, and others by swelling of the sex skin that occurs in some primates at the peak of estrus and appears to be a strong stimulus for sexual arousal in males (198).

If nonreproductive females are brought into estrus with estradiol treatment, males respond by mounting and copulating out of season. Testicular growth is accelerated, and changes in androgen-dependent sex skin are stimulated. In addition, plasma levels of testosterone rise in males exposed to experimentally receptive females (153). Presence of estrogen-treated females also induces sexual behavior in other females, implicating female–female interactions in reproductive function. Such responsiveness of males to sexual behavior of females recurs in mammals and throughout the vertebrate classes (23).

It is also important to consider social context in responsiveness to sexual signals. In male talapoin monkeys (*Miopithecus talapoin*), the dominant male may be the only male that is sexually active within a group. As one would expect, only the dominant male shows a dramatic increase in plasma levels of testosterone in response to a sexually receptive female. Subordinate males show no increase. Testosterone therapy in castrated males increased sexual behavior of dominant males, but not subordinates. If subordinate males were given access to females in the absence of dominant males, their responsiveness was enhanced (199). Clearly, the responsiveness of males to signals from females is difficult to interpret without knowledge of dominance status.

When male ring doves are exposed to females there is an increase in plasma levels of testosterone. Exposure of males to other conspecific males had no effect on testosterone levels (200). Laboratory and field experiments on white-crowned sparrows have also indicated that implantation of females with estradiol, which induces copulation-solicitation displays, results in an increase in plasma levels of LH and testosterone in their photostimulated cagemates (201). Furthermore, if sexual activity of free-living female white-crowned sparrows and song sparrows is prolonged by estradiol implants, the decline of circulating testosterone that occurs in males as the breeding season progresses is prevented, and levels remain high (202,203).

The effects of such interactions on endocrine state are presumably mediated by visual and auditory signals. Deafened male ring doves exposed to sexually mature females have lower plasma levels of androgen than control males paired with females. In addition, plasma levels of androgens in males separated from females by a glass partition were higher than those of males held in isolation, suggesting that visual cues are also important and physical contact less so (204).

In *R. pipiens*, females emit a release call if not gravid (i.e., receptive). When clasped in the pectoral region by a sexually active male, the release call is accompanied by muscular contractions that result in the male releasing the clasp. In contrast, a receptive, gravid female remains silent when clasped. Investigations indicate that arginine-8-vasotocin (AVT) inhibits the release call, possibly by increasing water uptake, which may exert an effect by mechanical distention of the body wall or by internal pressure. Ovarian secretions are apparently not necessary for inhibition of the release call [see Diakow and Raimondi (205) for review].

In Japanese toads (*Bufo japonicus*), males and females show increases in plasma levels of LH and FSH as they emerge from hibernation and migrate to breeding ponds (206). Furthermore, males with females in amplexus had higher plasma levels of LH and FSH than solitary males (32). Even a dummy female was sufficient to trigger this response, suggesting a physical stimulation of amplexus rather than visual or other cues associated specifically with a female [i.e., tactile cues are important here (32)]. It is also important to bear in mind that this stimulation may be "self-stimulation" through some form of feedback, presumably tactile.

There is considerable evidence that female fish influence the reproductive behavior of males. In the guppy (*Lebistes reticulatus*), female behavior affects courtship in the male. If she approaches the male slowly, he will court actively; if she approaches rapidly, the male flees or attacks her (207). Females held in a glass cuvette in a male's tank provoke active courting by the male, especially if the female is actively moving. Thus, visual signals appear to be important in this species and chemical cues less so. It was also noted that males would court an anesthetized female, but this courtship was strongest in males that had been isolated previously. In the stickleback, females appear to stimulate development of nuptial coloration in the male (207).

Chemical (pheromonal) stimulation of reproductive behavior in males appears to be widespread in fish (61,62). A secretion from the ovaries of the teleost *Bathygobius soporator* provides an olfactory stimulus for full courtship by the male (208). Visual signals are also important, however. In the goldfish (*Carassius auratus*), mature males placed with either a receptive female or with stimulus pairs of spawning goldfish have gonadotropin concentrations in blood that are greater than those of males held in all-male groups. If males were separated from a spawning pair either by a solid, transparent, or perforated clear partition, no elevations of gonadotropin occurred. Thus, contact appears to be the crucial element here, as opposed to chemical cues. Even in the absence of a female, a male induced to exhibit sexual behavior by treatment with prostaglandin will elicit elevation of gonadotropin in accompanying males, whereas isolates show no increase in plasma gonadotropin. This effect occurs within 20 to 60 minutes, and in general the elevated gonadotropin level in response to a female is greater than that to another male (209).

Male goldfish can distinguish between ovulating and gravid females on the basis of chemical cues. Occlusion of the nares, or section of the olfactory nerve, results in decreased sexual behavior (210). The pheromone appears to be present in the ovarian fluid and is released after ovulation. This attracts the male and elicits persistent courtship that accompanies spawning. It is now known that this pheromone is 17α,20β-dihydroxyprogesterone, a steroid produced by the ovary that also regulates germinal vesicle breakdown (62,211). In *Poecilia reticulata*, the female also produces a pheromone to attract conspecific males. Ovariectomized females are not attractive, and hypophysectomy also decreases pheromone production. This is restored by treatment with salmon gonadotropin or estrogen (212).

Parental Care

In many, if not most, vertebrate species, some form of parental care is provided for offspring. Parental care varies greatly across taxa (213,214), and includes classic lactation by female mammals, as well as nest construction or nurturing of young by other means. In many species parental care also includes grooming and cleaning of offspring, protection (e.g., from predators or other conspecifics), play, and transmission of knowledge and skills (e.g., hunting skills and the best places for success). Some species are monogamous, with males providing parental care in the form of protection of young and nest, and keeping the young warm (215,216). In nonmammalian vertebrates, parental care is also widespread and involves a complex spectrum of behaviors, all of which must be coordinated, synchronized, and integrated. For example, when both sexes provide parental care, considerable synchronization is required to maximize reproductive effort (e.g., seabirds that alternate egg and chick care over several days or even weeks). In those species that incubate eggs, it is important to build a nest before the eggs are laid. After hatching, the shift in sedentary incubation behavior to active foraging for food must be regulated precisely.

Although we know much about the basic hormonal controls of parental care in mammals (217), much remains unknown, especially in ecological contexts, across the vertebrates. However, certain trends are apparent, and hormonal regulation of parental care can be considered in relation to certain traits as follows: construction of a nest or similar shelter; care of eggs (incubation); provision of food for offspring (escorting, active foraging, and production of specialized nutritive secretions); protective behavior (thermoregulation, antipredator, avoidance of physical disturbances); and transmission of knowledge and skills (including play behavior). In addition, these traits may have evolved independently in many taxa, and it cannot be assumed that common hormone mechanisms always exist.

The most thoroughly studied hormone in relation to parental care is prolactin, which has multiple effects on parental care throughout the vertebrates (218), including:

1. Production of milk and other nutritive secretions
2. Parental behavior, including parental aggression
3. Actions on the integument (e.g., brood patch)

A detailed review of the hormonal control of parental behavior is beyond the scope of this chapter and is treated in detail elsewhere (217,219,220). However, evidence suggests that there is great plasticity in the interrelationships of hormones and parental care, especially with regard to potential conflicts such as aggression and paternal behavior (215).

Transitions between the Sexual and Parental Phases of Reproduction

During the sexual phase of the reproductive process, expression of sexual behavior and courtship, as well as agonistic behavior associated with mate-guarding and territorial defense, are usually incompatible with expression of parental care, particularly in male birds (221). However, in other species, such as rodents, aggression and paternal care are not necessarily incompatible and can be expressed simultaneously (215). This may be particularly true in species in which males defend the territory and young from predators or conspecific intruders. Marler et al. (215) emphasize that interrelationships of aggression, parental care, and hormone control mechanisms are complex, highly plastic, and shaped by social environment. Nonetheless, those species that have a long breeding season may mate several times, resulting in a series of reproductive attempts. In these cases there are complex transitions between sexual and parental behavior that require precise synchronization and integration. In general, the sexual phase is characterized by high circulating LH and thus sex steroids, followed by a marked decline as the parental phase ensues. At least in birds, high prolactin levels are associated with the parental phase and then decrease as the next sexual phase begins (222,223). This apparent antagonism between prolactin and sex steroids (including LH) is thought to be the basis of behavioral integration in "multiple brooding." Stimuli from the nest, eggs, or young enhance prolactin secretion and thus suppress secretion of LH and sex steroids. As the young grow and become independent of the adults, stimuli for prolactin secretion are lost, allowing an increase in LH and estradiol in females (223). This precipitates development of the ovary, leading to further ovulation. The sexual receptivity of the female results in an increase in LH and testosterone in males, thus synchronizing the reproductive effort as well as integrating the transition from parental to sexual phases.

Recycling (Renesting)

Breeding animals that have lost the nest or young to a storm or predator frequently initiate another breeding attempt. Loss of the young is a considerable disruption to the normal temporal progression of the breeding season, and major reorganization of endocrine function and expression of substages is necessary to coordinate the reproductive attempt. This phenomenon has been particularly well investigated in birds such as the white-crowned sparrows, in which LH, testosterone, and testis mass in males increased dramatically after loss of the nest. This was coincident with elevated LH and estrogens in females leading to production of a replacement clutch of eggs (224). Similarly, in female mallards (*Anas platyrhynchos*), after experimental removal of the eggs (225), and in song sparrows that lost nests to extensive flooding (226), there were rapid increases in reproductive hormones just before renesting. It is important to note that in white-crowned sparrows there were also increased concentrations of testosterone when producing a replacement clutch. This is unlike multiple brooding after *successfully* raising young, and suggests different mechanisms may initiate a second clutch in these two ecological contexts (22,226).

HORMONE–BEHAVIOR INTERACTIONS AND SUPPORT FUNCTIONS FOR THE REPRODUCTIVE LIFE HISTORY STAGE

Individuals, especially males, of all vertebrate species show some form of competition for territory, trophic resources, social status within a group, and access to mates. In reproductive contexts, competition for one or all of the aforementioned may be intense and involve complex hormonal mechanisms. Experimental investigations to elucidate the mechanisms of hormone–behavior interactions in aggressive contexts have focused almost exclusively on males, although the scant literature on females also is discussed.

Territorial Behavior

The interrelationship of testosterone and aggression in mammals (indeed, all vertebrates) is complex and controversial. Although there are exceptions, the general rule is a close positive correlation of circulating testosterone and aggression in reproductive contexts. In general, an individual that gains a territory or high social rank is more likely to mate with females. Thus, one might expect an increase in circulating levels of testosterone during male–male interactions over territory and social status. In many species this is the case.

Male–Male Interactions

Trends for positive correlations of circulating testosterone and male–male aggression (especially in relation to territory establishment and defense, and mate guarding) have been well documented in birds. Plasma levels of testosterone tend to be maximal when males are establishing territories and when sexually receptive females are present (12,221).

In castrated male sparrows and quail, implantation of testosterone had no significant effect on aggression when males were held in a socially stable environment. However, on introduction of a novel male, levels of aggression directed at the intruder increased dramatically in testosterone-treated castrates (226, 250). These findings illustrate further the complex interaction of testosterone and aggression in stable versus unstable social conditions, or when an individual is challenged by a conspecific.

Free-living male song sparrows (and other species) can be "challenged" by a simulated territorial intrusion (STI). A decoy male in a cage is placed in a male's territory and tape-recorded songs played through a speaker placed alongside (227). The resident male responds by attacking the apparent intruder. This can also be done in the laboratory with males showing a dramatic and easily quantified response. If a male song sparrow is removed from his territory, a replacement male takes over within 1 to 2 days. The replacements had elevated levels of testosterone compared with control males captured in an area where territories were stable (227). Furthermore, the neighbors of the replacements (who were establishing new boundaries with the replacement males) also had higher levels of testosterone. In a further experiment, free-living male song sparrows were challenged by STI. The territorial male attacked the apparent intruder and attempted to drive him away. The attacking males were then sampled after having interacted with the intruder for at least 10 minutes. Attacking males had elevated plasma levels of testosterone compared with controls that were caught while they foraged and were not involved in an aggressive encounter (227). Further, plasma levels of LH and testosterone were elevated within approximately 10 minutes of the challenge. These data indicate strongly that aggressive encounters (challenges) can result in an increase in circulating testosterone that in turn facilitates aggressive responsiveness. The "challenge hypothesis" has been postulated as a result of these studies to explain the dynamic relationship with circulating testosterone levels. First, circulating levels of testosterone correlate with aggression only during periods of social instability. Second, male–male interactions over status and access to sexually receptive females tend to increase testosterone secretion, which in turn maintains a high level of aggression during social instability. However, this interrelationship of testosterone and aggression is by no means ubiquitous (12,221).

It is now established that males of several reptilian species defend territories aggressively and have elevated levels of testosterone. Furthermore, castration tends to reduce territorial aggression, although a strong testosterone-independent component persists, especially outside the breeding season. In a lizard (*Sceloporus jarrovi*), staged intrusions (by introducing a novel male into a free-living male's territory) result in intense fights and displays of aggression. There were no changes in plasma levels of testosterone after one of these staged intrusions (228,229). On the other hand, in the green anole, challenges from other males do increase testosterone secretion, suggesting diverse mechanisms underlying the hormonal basis of territorial aggression (230).

Many investigations have been conducted on the interrelationships of testosterone and aggression in males, and the literature has been reviewed extensively (23,231,232). Even observing fights has the potential to raise testosterone secretion in vertebrates such as fish (233) and humans (231). Interactions of testosterone, aggression, and brain peptides, especially arginine vasopressin (vasotocin) and vasoactive intestinal peptide, appear to be especially promising areas of research in relation to the ecological bases of these complex patterns (234–236).

Female–Female Interactions

Although it is well known that females of all vertebrate classes interact in reproductive contexts, the endocrine bases of female aggression and territoriality are poorly known. There is considerable evidence that adult female rodents can retard development of juvenile females by cues of pheromonal origin [e.g., mice (141)]. In golden hamsters (*Mesocricetus auratus*), females defend home territories from other females at all times and accept males only when in estrus (237). Intact, cycling females exhibit more aggressive acts than ovariectomized opponents 1 day before estrus, but no differences were found in aggression on other days of the cycle (238,239). Although the decline in aggression directed at males during estrus involves estrogen and progesterone, the regulation of female–female aggression remains equivocal.

Polyandry in birds is one of the best known examples of female–female interaction. In this mating system, females are often larger than males, have a brighter nuptial plumage, establish and defend breeding territories, compete with other females for males, and show reduced, or no parental care. Field studies on spotted sandpipers (*Actitis macularia*) and Wilson's phalaropes (*Phalaropus tricolor*) indicate that females do not have higher levels of testosterone than males, despite the apparent reversal of traditional sex roles (240,241). In the western gull (*Larus occidentalis wymani*), there is an excess of females in the population and females not mated to males can form female–female pairs that establish and defend a territory, lay eggs, and incubate. Plasma profiles of steroid hormone revealed that homosexual females had the same testosterone and estradiol levels as heterosexually paired females and males (242). Thus, it appears unlikely that homosexual females have become masculinized, but the similar levels of testosterone in males and females are consistent with territory establishment in females.

From the foregoing examples, it is clear that although testosterone is known to regulate territorial aggression in females of at least some species, measurement of circulating testosterone levels in naturally breeding populations does not always support a major role for this hormone in the regulation of female aggression.

Social Status

There is abundant evidence that in addition to territorial behavior, social status has dramatic import for reproductive success. In general, only the most dominant individuals, at least among males, breed.

Male rhesus monkeys show an increase in plasma testosterone levels on assuming alpha status in a troop. However, such high levels of testosterone are not maintained indefinitely and usually decline once the social hierarchy becomes stable (243). Removal of alpha and beta males from a group results in a dramatic decline in testosterone levels, and after return, an equally dramatic elevation. Meanwhile, males that assumed alpha and beta status during the period of removal showed high plasma testosterone levels that then declined when the original males returned and took over the alpha and beta spots (244,245).

Further evidence comes from some elegant field studies on olive baboons (*Papio anubis*). The stress associated with anesthesia and capture depressed plasma levels of testosterone, perhaps because of stress-induced elevation of cortisol secretion (246). However, suppression of testosterone titer was greatest in low-ranking males even though high-ranking males showed a similar cortisol elevation. Testosterone profiles over time after capture were generally higher in high-ranking males, again indicating individual differences in hormonal profiles and responsiveness in relation to social status (247).

There are several exceptions to this rule that may be the result of diverse mating systems and breeding strategies, or due to differences in experimental paradigms. For example, in Japanese quail, no relationship of testosterone and social status was found in established dominance hierarchies by some investigators (248,249). If naive subjects were used, however, plasma levels of testosterone were found to be higher in dominant males, especially when social relationships were being established. This correlation was lost when social hierarchies became stable (250). Similar results have also been obtained for house sparrows (*Passer domesticus*) (111).

HORMONE–BEHAVIOR INTERACTIONS AND TERMINATION PHASE OF THE REPRODUCTION LIFE HISTORY STAGE

A major phase of the breeding life history stage is its termination. Although some vertebrates spend their entire reproductive life sexually mature, most species show periodic breeding with termination of reproductive function at some stage in the annual cycle. Here too, communicative behaviors play an important role, although this is less well studied than at other times in the breeding season.

Effects of Offspring and Mate

There is evidence that the onset of photorefractoriness can be delayed by supplementary factors in relation to local conditions and to substages expressed in the breeding life history stage. Investigations of captive and naturally breeding populations of white-crowned sparrows (*Z. l. gambelii*) indicate that the timing of spontaneous gonadal regression is variable (224,226). The duration of maximum testicular size is only 10 to 12 days in captive conditions, whereas in the field, unmated males undergo gonadal regression in the fourth week of June, breeding males and females in the first week of July, and renesting birds in the third week of July (224). It obviously would not be advantageous for the parental phase to be terminated during the nestling substage because the young would not survive. However, onset of photorefractoriness prevents initiation of clutches beyond a certain date, but allows completion of a nesting phase that was initiated before this time. In the case of *Z. l. gambelii*, this period appears to be the third and fourth weeks of June (224). Exposure to females results in prolonged testicular function in starlings, but does not affect circulating LH and androgen (157).

Social interactions may also play a role. Runfeldt and Wingfield (251) showed that free-living female song sparrows given subcutaneous implants of estradiol remained sexually receptive until at least the beginning of October. Control females given empty implants became photorefractory and terminated breeding by mid-August. Males mated to control females also were refractory by mid-August, but those mated to estradiol-implanted females delayed onset of refractoriness almost 1 month later than controls. In contrast, untreated females mated to testosterone-implanted males became photorefractory at the same time as females mated to controls. The testosterone-implanted males remained on territory until the beginning of October (251). These data indicate that males adjust termination of the breeding life history stage to the reproductive state of their mates. Under natural conditions we would expect that unmated males terminate breeding before mated males (224). Females, however, do not adjust termination of breeding to the reproductive state of the males.

Interactions with Molt

In the song sparrow, implantation of females with estradiol late in the season delayed onset of molt (usually indicating that reproductive activity has ceased) up to 2 months later than control females. Males mated to females implanted with estradiol also remained on territory and delayed onset of molt compared with males mated to control females that terminated reproductive at the normal time (251). Curiously, if male reproductive behavior was prolonged by implants of testosterone, females terminated reproduction at the same time as females mated to control males and initiated molt as normal. This indicates that females time termination of breeding in response to cues other than those from males. In addition, renesting white-crowned sparrows initiated molt much later than birds that did not renest (224), suggesting that presence of offspring somehow maintains parental care while delaying onset of molt. The mechanisms involved in these adjustments of termination of breeding and onset of molt life history stages remain largely unknown.

MATING SYSTEMS, BREEDING STRATEGIES, AND ALTERNATIVE REPRODUCTIVE TACTICS

Vertebrates have evolved a bewildering array of mating systems, breeding strategies, and reproductive cycles. It is not possible here to cover all of them in detail, so I have taken an approach that considers three major categories:

1. Variation in the temporal pattern of gametogenesis and behavior. For example, cycles of gamete production and actual mating may be associated (occur simultaneously) or dissociated (occur separately). Some species may breed only once in a lifetime (semelparous), whereas most breed several times (iteroparous).
2. Variation in mating systems: monogamy, polygyny, polyandry, hermaphroditism, acquired hermaphroditism, sex reversal, and parthenogenesis.
3. Variation in breeding strategy: mimicry, "extra pair copulations," female–female pairing, cooperative breeding, brood parasitism, and eusocial systems.

Reproductive strategies are perhaps even more diverse, although the aforementioned examples are the best known.

Variation in the Temporal Pattern of Breeding

The breeding life history stage of vertebrates comprises discrete periods of gamete maturation (development); mating, when gametes are shed and fertilization occurs (mature capability); and termination of breeding. There are three major ways by which the temporal pattern of these events varies (252,253):

1. Gametes attain maturity coincident with breeding (i.e., an associated cycle). This is probably the most common tactic.

2. Gonadal recrudescence occurs after breeding activity has ceased (i.e., a dissociated cycle). Gametes are stored until the mating period.

3. Gonads develop early and remain in a functional state with viable gametes for prolonged periods (i.e., continuous breeders).

1. Associated/Dissociated Patterns

Crews (252,254) and Crews and Moore (253) point out that there are examples of species in which either or both sexes may show associated or dissociated cycles. Although there are many types of associated breeding cycles, an example of a dissociated cycle is the garter snake (*T. sirtalis*). Northern populations spend the winter in deep caves and emerge in spring. Males congregate in spectacular numbers and lie in wait for emerging females. As the females move up to the surface they emit a pheromone (vitellogenin) under the control of low levels of estrogen. Vitellogenin is emitted through the dermis of the back skin and serves to attract males for mating (65). There is intense competition among males to mate with females, and frequently a mating "ball" may be formed. Copulation acts to increase secretion of gonadotropin, which initiates ovarian development. Increased estrogen levels stimulate increased production of vitellogenin by the liver, which is then sequestered in the ovarian follicles as yolk. Sperm obtained during copulation are stored in the oviduct until ovulation later in summer (193). After ovulation and fertilization, the eggs develop in the oviduct of the female (ovoviviparity) and young are born in late summer. In males, the testes are inactive on emergence from hibernation, and they use stored sperm for mating. Later in spring they undergo spermatogenesis and store sperm until the next spring. Thus, both male and female garter snakes show a dissociated cycle of gametogenesis and mating (254).

Continuous breeders are rare because habitats in which the ultimate factor (regulating time of breeding) is always present are also rare. Examples are humans, common pigeons, and many species of tropical reef fish. Gonads are maintained at a functional level throughout much of the reproductive life of the individual and mating can occur at any time. It is possible that some opportunistic breeders may also fit this category, although good experimental data are lacking.

Semelparity

A few vertebrates are semelparous; they have a single breeding effort, and then die.

This phenomenon has been studied in fishes such as lampreys (*Petromyzon* and *Lampetra* sp.), eels (*Anguilla* sp.), and Pacific salmon [*Oncorhynchus* sp. (255)], and a group of dasyurid marsupials (256). One of the characteristics of semelparity is progression of reproductive development despite deteriorating body condition and symptoms of severe stress. Some fish species stop feeding weeks or months before spawning and development of the reproductive system occurs at the expense of body tissues (257).

It is thought that this breeding strategy evolved in Pacific salmon because of the long upstream migration from the ocean, where they spend most of their lives, during which the gonads are maturing and may comprise more than 30% of body weight. Programmed death may be selected for because body reserves are so depleted there is no energy left for a migration back to the ocean. In addition, decaying corpses of postspawned adults fertilize the waters of the spawning grounds, potentially increasing food for the hatchling salmon (255), although this is a group selection argument.

After the upstream migration to spawning grounds, male salmon fight to establish territories and attract mates, and females dig nests (redds) in the gravel of streams, both metabolically expensive behaviors (255,257). As expected, this period is accompanied by high circulating levels of cortisol (258) as well as hypertrophy of inter-renal tissue, homologous to the adrenocortical tissue of mammals (259). Such high concentrations may reduce immune responses and exhaust metabolizable energy stores, leading to massive infection by fungus and bacteria, and eventually death (255). This appears to be a programmed phenomenon because experimental force feeding of maturing and spawned sockeye salmon (*Oncorhynchus nerka*) only reversed the deterioration of body condition and prolonged life for up to 10 weeks (255). Moreover, healthy postspawned fish also die eventually without a second attempt at breeding, indicating that an endogenous program (ontogeny), or other factors, underlies the semelparous strategy.

The profile of circulating cortisol during the migration and spawning of Pacific salmon would result in suppressed reproductive development in most vertebrates. Nonetheless, both sexes shed gametes, and circulating levels of gonadotropins and sex steroids such as estradiol and 11-ketotestosterone increased in maturing salmon, peaking several weeks before spawning (255,260) despite elevated cortisol levels. Gonadal hormones may even be the stimulus for adrenocortical hypersecretion as maturation progresses because gonadectomy reversed cortisol hypersecretion (261), whereas androgen injections in gonadectomized males increased secretion (262).

The actual mechanisms by which semelparous fish are able to breed despite hyperadrenocortical function remains unknown, but one possibility is that cortisol receptors in the brain and hypothalamic–pituitary–gonadal axis are downregulated so that behavioral and reproductive traits are not directly inhibited (262a). This would allow the metabolic effects of cortisol without inhibiting breeding. On the other hand, characterization of cortisol receptors in the brains of Chinook salmon (*Oncorhynchus tshawytscha*) show that there appear to be two high-affinity, low-capacity receptors in cytosolic extracts with a differing pharmacology of binding (263). An immunohistochemical study of glucocorticoid receptor-immunoreactive staining neurons in the brains of sexually immature and mature spawning kokanee salmon indicated wide distribution of glucocorticoid receptors in the brain, similar to the mammalian pattern, but brains from spawning fish had the same pattern of glucocorticoid receptors as immature fish, and olfactory regions had greater numbers of glucocorticoid receptor- immunoreactive neurons with predominantly nuclear locations in spawning fish (264). In immature fish, glucocorticoid receptor-immunoreactive staining appeared to be largely cytoplasmic (264).

Shrewlike marsupials (*Antechinus* sp. and *Phasogale calura*) from Australia are also semelparous, reproducing once during a short mating period, after which all males die (265). Many females also die after lactation ceases and the young wean, although a few may breed through two seasons (266). The mass mortality of males may reduce competition for food when females are lactating and may also allow larger litter sizes compared with those seen in closely related species that are iteroparous, going through at least two cycles of reproductive development (256). This also is a group selection argument that needs to be reassessed in the context of more conventional evolutionary theory.

This phenomenon parallels that of the salmonids and, similarly, during the breeding period, circulating cortisol was elevated and weight of the adrenal glands increased in male *Antechinus stuarti* when mating (267). Corticosteroid-binding globulin (CBG) levels declined below the capacity to bind increasing plasma levels of cortisol (268–270), and when combined with increased cortisol levels produced an enormous increase in unbound, biologically active cortisol. Female *Antechinus*, which die later, had circulating levels of cortisol as high as those in males, but CBG capacity remained high to match circulating cortisol (268,269).

Although elevated cortisol levels were at least partially related to the aggressiveness of males during this breeding period, there were alterations in adrenocortical function at many levels. Dexamethasone treatment inhibited plasma cortisol 2 months before mating occurred, but not during the actual mating period (271). Intramuscular injection of adrenocorticotropic hormone (ACTH) significantly increased cortisol secretion at both stages, suggesting that sensitivity of the adrenal cortex did not change. Curiously, higher ACTH concentrations decreased CBG capacity (269).

Similar to Pacific salmon, cortisol hypersecretion may have a role in programmed death in these semelparous marsupials. When mating commenced, males showed obvious symptoms of chronic stress such as moderate anemia, elevated parasite load, hepatic necrosis and abscesses, ulcers, gastrointestinal tract hemorrhage, and involuted splenic follicles (269,272). Although the responsiveness of the immune system in the male *Antechinus* is greater than in females before breeding, agglutination titers used as a measure of immune function increased in females but declined markedly in males. In addition, serum immunoglobulin levels were lower in males than females, and this was accompanied by an increase in endoparasite infection in males and by the pathological processes already mentioned (269,272). In later stages, just before death, males lost fur and body mass, and were lethargic and moribund (267,273). Intramuscular injections of cortisol acetate into male *Antechinus* before the mating period increased mortality compared with controls (268).

As in Pacific salmonids, development of the breeding life history stage and an active gonad may precipitate hyperadrenocorticism and resultant pathological processes. Nonetheless, plasma levels of androgens rose in males during July to a peak in August, coincident with mating, followed by a rapid decline after mating leading to senescence (269,270,274). When males were experimentally isolated from mating, they survived longer (275), and castrated males survived even longer than intact males (269). However, gonadectomy did not alter the maximum capacity of serum CBG, which did not decline compared with intact males. However, injection of testosterone into castrated *Antechinus* males decreased CBG (270). Although in the several days to weeks surrounding the mating period, continually high levels of cortisol led to pathological states of chronic stress and finally death (275), the reproductive system remained unaffected and mating was achieved. It is of particular interest that dexamethasone treatment decreases circulating testosterone levels 2.5 months before mating, but not when testosterone is near its peak (271). This adrenocortical profile appears to be specific to semelparity because in another small marsupial that is iteroparous, the fat-tailed dunnart (*Smithsopsis crassicaudata*), males survived beyond a single season and had no marked peak in circulating cortisol, as in the semelparous species. Furthermore, circulating cortisol concentrations were always below the maximum binding capacity of CBG (270).

More recent evidence reveals that profiles of cortisol release may not be universal in semelparous marsupials because in another larger dasyurid, the quoll (*Dasyurus hallucatus*), in which all males die after mating and show the same symptoms of senescence, there was apparently no increase in plasma cortisol in males. They had profiles similar to females despite high circulating levels of testosterone, and CBG levels remained unchanged (276,277). It should be noted that male quolls can store relatively more fat than the much smaller *Antechinus* because they have a specialized fat storage organ in the tail, allowing them to prepare for the mating season by using these stores and avoiding protein breakdown. This suggests that hypersecretion of glucocorticosteroids may be a result of programmed senescence and not its cause in semelparous marsupials (276).

Variation in Mating Systems

Different mating systems can have major influences on the type and context of communicative behaviors regulating reproductive function. Compare, for example, a socially monogamous pair in which both sexes provide parental care and must coordinate reproductive effort precisely with a polygynous system in which males compete viciously for females and may have little communication with those females except for mating. Hermaphroditic and parthenogenetic species raise even more intriguing communication patterns.

Challenge Hypothesis

It is well known that testosterone activates aggression associated with male–male competition over territories and mates (20,21), although the correlation of plasma levels of testosterone with expression of territorial aggression when breeding is frequently unclear (12,221,277a). Breeding baseline levels of testosterone result in development and maintenance of morphological, physiological, and behavioral components of the male reproductive system. However, these baseline levels do not necessarily correlate with actual expression of territorial aggression. Superimposed on this breeding baseline circulating testosterone level are transient surges to much higher concentrations often tightly correlated with periods of heightened male–male competition, especially when establishing a territory, being challenged by another male, or when mate guarding. This is the "challenge hypothesis," stating that high plasma levels of testosterone, above the breeding baseline, occur during periods of social instability (12,221).

Social signals that regulate the increase in testosterone include both auditory and visual stimuli from the intruding male in song sparrows (278). Similarly, in green tree frogs (*Hyla cinerea*), chorusing results in an increase in testosterone levels in males, but this effect is mediated by social signals other than calling (279).

Territorial aggression associated with reproductive function appears to be mediated by androgens resulting from gonadal maturation in all vertebrates, and there is widespread evidence for interaction of circulating androgen levels and outcomes of social interactions (280). Urinary androgens increased as dominance relations were formed in several species of cichlid and the demoiselle fish (*Chromis dispilus*) (281). In the stoplight parrotfish (*Sparisoma viride*), androgens regulate development of male coloration, especially in terminal-phase males (282). Territorial males showed increased plasma levels of androgens when taking over a territory or when experimentally challenged by another male (283). Indeed, in a well-controlled study, it was shown that just observing fights can raise androgen levels in *Oreochromis mossambicus* (233). At the other end of the vertebrate spectrum, investigations of humans found that during the soccer World Cup in 1994, testosterone levels of Italian and Brazilian fans varied markedly before and after the game (Brazil won). The testosterone levels of the Italians decreased, whereas those of the Brazilians increased (284).

In weakly electric fish from the family Mormyridae (e.g., *Brienomyrus brachyistius*), top-ranking males undergo a large increase in electric organ discharges. Second-ranking males show a more modest increase, and lower-ranking males may actually decrease electric organ discharges (EODs). These changes were correlated with circulating levels of 11-ketotestosterone but not testosterone (285). Weakly electric fish of the family Apteronotidae show EODs in aggressive contexts. Nonaromatizable androgens raise EODs in one species (*Apteronotus leptorhynchus*) in which males produce high-frequency EODs whereas females do not, and lower EODs in another species (*Apteronotus albifrons*) in which there is no sexual dimorphism in EODs. Thus, dimorphism of EODs may be related to sensitivity to androgens (58). Together, these investigations shed light on the evolution of communication behavior, especially aggressive interactions. They also must be considered in contexts of communication networks rather than dyads. Oliveira et al. (286) show that male Siamese fighting fish (*Betta splendens*) appear to monitor aggressive interactions among neighbors.

Information on relative fighting ability may then be used in future aggressive interactions with those individuals.

Monogamy, Polygyny, and Polyandry

One of the more remarkable and consistent results of field investigations is the relationship of patterns of testosterone levels with mating systems and breeding strategies. This is potentially important and may provide insight into the evolution of hormone–behavior interrelationships. There is an increasing body of evidence suggesting a hormonal basis for mating systems, although the data thus far have been obtained almost exclusively in birds. Males of polygynous species tend to have higher plasma levels of testosterone for longer periods than monogamous males, suggesting that the temporal pattern of testosterone secretion may be related to the mating system. Indeed, if monogamous male sparrows were given implants of testosterone to mimic a polygynous-type pattern of secretion, these normally monogamous males became polygynous (12,221,287). Further, testosterone-implanted males had larger territories than controls, thus allowing settlement of more than one female (12,221). Because polygynous males potentially have much greater reproductive success, it is curious that polygyny is not as prevalent in birds as it is in other vertebrate groups. It is possible that there is a disadvantage to polygyny in some species. In the sparrows in which polygyny was induced by testosterone, there were no signs of stress as indicated by body mass, fat score, and corticosterone levels in blood. In addition, there does not appear to be reduced over-winter survival in testosterone-implanted male song sparrows, although this may be true in other species [see Ketterson et al. (288,289) and Wingfield and Silverin (13) for reviews]. However, high circulating levels of testosterone are known to inhibit parental behavior in many birds [but see Marler et al. (215) for mammals]; hence, if a male normally provides parental care, high testosterone levels could result in fewer young raised despite several mates (13,289). Testosterone implants in house sparrows reduced the number of young fledged (leaving the nest) and the number surviving, and increased the number of young that starved compared with controls (111). This reduced reproductive success appears to stem from a decrease of parental behavior in favor of enhanced aggression, such as territory defense and mate guarding. These data suggest that the mating system may be a function of male–male aggression and the degree of parental care shown by a male in a particular population. The more parental behavior shown by a male, the lower the testosterone levels must be; the more male–male aggression there is, the higher the testosterone levels tend to be. Thus, the relationship of male–male aggression and parental investment may be an important determinant of the mating system in males as well as a potent "ultimate" regulator of the temporal pattern of sex steroid hormone secretion. There are many similar studies throughout the vertebrates that support, to varying degrees, the challenge hypothesis (232), and in birds, a phylogeny-controlled analysis of many species studied in the field or in near-natural conditions showed that social modulation of testosterone was related to the mating system (290).

Two male phenotypes of the tree lizard (*Urosaurus ornatus*) are associated with alternative male reproductive tactics, one being territorial and the other nonterritorial (291,292). In response to winning territorial encounters on the previous day, the nonterritorial males display elevated plasma corticosterone and depressed plasma testosterone levels. In contrast, the territorial males show no change in plasma hormone levels the day after winning an encounter (293). It appears that one of the physiological differences between the morphs is the plasma steroid-binding globulin capacity. These lizards have a plasma steroid-binding globulin that binds both androgens and corticosterone, and the binding capacity in territorial males in significantly greater than in nonterritorial males (294). This difference in binding capacity between the morphs likely results in higher plasma levels of unbound corticosterone in the nonterritorial males. This would be especially evident when plasma corticosterone levels are elevated and free corticosterone could then trigger a decrease in plasma testosterone. The difference in free versus bound corticosterone potentially explains the morph difference in testosterone response to similar increases in corticosterone. In northern fence lizards (*Sceloporus undulatus hyacinthinus*), males also showed no change in testosterone levels after a simulated territorial intrusion, although corticosterone levels did rise after long intrusions (295).

Calling is very important for males of most anuran species and serves both as an advertisement to females and a warning in territory defense against other males. For species that call, the time spent calling is positively associated with male mating success (296). In addition, calling behavior is often considered to be one of the most energetically expensive activities for anurans (297). Data from a variety of species suggest that plasma levels of both testosterone and corticosterone are elevated during bouts of calling. A cross-species analysis suggests that corticosterone is positively correlated to both calling rate and the rank of relative energy in the call (298). Thus, sexual selection arguments would describe calling behavior as an "honest signal" (299). However, in the Tungara frog (*Physalaemus pustulosus*), it appears that endogenous plasma corticosterone levels are maintained below a threshold level. Exogenous corticosterone raises plasma corticosterone above the threshold, and calling behavior and plasma testosterone levels are reduced (300).

Emerson (299) has proposed an extension of the challenge hypothesis (221), termed the energetics–hormone vocalization (EHV) model, to explain the reported relationships between calling behavior and testosterone and corticosterone levels in male anuran amphibians. This model proposes that calling behavior drives an increase in plasma testosterone that is accompanied by an increase in plasma corticosterone levels because of the energetic demands of the behavior. Thus, over time, levels of both hormones increase until plasma corticosterone triggers a short-term stress response. Plasma testosterone levels then decline, resulting in a negative association between the two hormones (299). This appears to follow the transition from normal, seasonally variable glucocorticoid levels (baseline) to an emergency life history stage (facultative high levels) modeled by Wingfield and Ramenofsky (301). The EHV model can probably be extended to explain the apparently paradoxical differences in the relationship between testosterone, corticosterone, and aggression in a variety of taxa, especially reptiles and amphibians (103,302). In many species, it appears that when mating or courtship is energetically costly there can be a positive relationship between testosterone and corticosterone levels.

Hermaphroditism

Most vertebrates are gonochoristic (i.e., testis bearing [male] or ovary bearing [female]). However, some fishes show varying degrees of hermaphroditism (303), as described in the following.

Simultaneous Hermaphrodites that Do Not Self-fertilize. Simultaneous hermaphrodites that do not self-fertilize are individuals that have a functional ovary and testis and trade eggs and sperm during mating. The best-known examples are the sea basses (Serranidae) found in tropical and subtropical oceans worldwide. Individuals have an ovotestis (303) and during daily spawning periods form pairs or larger mating groups. Each individual performs as both male and female, reversing its sex role with its partner several times until the eggs of both have been shed. Note that the testis, ovary, and associated gonoducts are controlled separately to avoid accidental self-fertilization. In some species there are complex color pattern changes that indicate willingness to spawn as well as male phase or female phase (303). In *Serranus subligarius*, the "head-up" posture with banding pattern is given by females just before releasing eggs (to signal the male-phase hermaphrodite to follow). Female-phase fish first court another individual by advertising potential to release eggs, indicated by a swollen abdomen. The other individual then releases sperm. A male-phase animal advertises intention to release sperm with a different body posture and banding pattern. These changes occur very quickly (seconds to minutes) and are primarily under central nervous control (303).

Simultaneous Hermaphrodites that Self-fertilize. Self-fertilization in vertebrates is rare, but one well-known example is *Rivulus marmoratus*, a small killifish of neotropical mangrove swamps (303,304). Some individuals live under logs in these swamps and have variable color. Eggs are fertilized internally with sperm from the fish's own testis, and the eggs are guarded after spawning. In these gonads, mature eggs and sperm are "ovulated" into a common lumen, where fertilization occurs. Note, however, that if the eggs are exposed to temperatures below 19°C, two thirds of the young develop as primary males that will never develop ovarian tissue. These males will mate with hermaphrodites, thus reducing the tendency to homozygous clones. In addition, if young hermaphrodites are exposed to high temperatures, they develop as secondary males that can also mate with hermaphrodites.

Sequential Hermaphrodites. Sequential hermaphrodites are species in which individuals change from one sex to another. There are numerous examples of this mating strategy, especially in tropical reef fishes (305,306). There are two types, as described in the following sections.

Protogyny. Protogyny (first female, then male) may be favored where dominance of males is a major factor (307). Typically, a large male guards a number of females (harem) and protects a territory into which females enter to mate. In these instances, smaller males mate rarely, if at all, and it is an advantage to be female at this stage. Well-known examples are the tropical reef-dwelling wrasses, *Anthias* and *Thalassoma*, and the California sheephead (*Semicossyphus pulcher*). *Anthias* males dominate a group of females (306) and prevent any further sex change. However, if the male is removed, the largest (dominant) female begins to change sex (308). It is clear that behavioral domination by the male prevents sex change in others and this domination is by physical means (chasing), and does not involve purely visual signals or pheromonal cues. In *Thalassoma*, smaller fish are either male or female, with little sexual dimorphism. Females spawn daily and preferentially with the larger, terminal-phase males. However, the primary (initial-phase) males do attempt to spawn, although their rate of fertilization of eggs is probably very low. Either initial-phase males or females can change into terminal-phase males. Often the largest female changes (306,307,309,310).

Several studies have followed the endocrine events that accompany sex reversal in protogynous species. In general, estradiol levels are highest in females and 11-oxotestosterone levels highest in males, with testosterone at similar levels in both sexes. As sex reversal progresses, estradiol levels decline and 11-oxotestosterone increases. Increasing titers of androgens promote the morphological, color, and behavioral changes accompanying sex reversal (282, 311,312). More recent evidence implicates a role for AVT in sex-reversing fish (307,309,313–315). In *Thalassoma bifasciatum*, for example, AVT is involved in expression of dominance in females, whereas 11-ketotestosterone may regulate male coloration. Interactions of these two hormones, and possibly others, coordinate sex change in various contexts of social interactions. Godwin et al. (316) showed that gonadectomy did not affect the behavioral sex change, suggesting that central mechanisms such as AVT action are important. Social interactions also can increase glucocorticoid levels, which in turn can influence AVT and enzymes involved in sex steroid synthesis. Glucocorticoids may thus have a role in regulating at least the potential for sex change (309).

Protandry. Protandry (first male, then female) occurs in small, stable groups with essentially a monogamous relationship between one pair that reproduces (303,306). The largest males change sex, becoming females, and large females produce the greatest number of eggs. Thus we would expect protandry to be common in instances where small males can fertilize the eggs of larger females and, for whatever reason, large males cannot sexually monopolize groups of females. The hormonal events accompanying sex change in protandry are less well known but appear to be the reverse of the trends given in Yeung and Chan (317). In the black porgy (*Acanthopagrus schlegeli*), individuals are male for the first 2 years and then become female. In males, plasma levels of 11-ketotestosterone were high and those of estradiol low. As the transformation to females took place, estradiol levels increased while circulating levels of 11-ketotestosterone declined. Plasma concentrations of testosterone did not change in males and females (318).

Acquired Hermaphroditism. Certain deep-sea fishes are simultaneous hermaphrodites possibly because in the expanse and darkness of the deep ocean, location of mates is difficult. In one species of *Edriolychnus* (angler fish), gonochoristic females exist and secrete a pheromone that attracts the much smaller males (303). These males attach to females and anchor themselves permanently to the tissues of the female. Here the males lead a parasitic life, drawing all sustenance from the female and maintaining a large testis. These males then fertilize the female's eggs as she spawns (303). It is possible that because the female passes on nutrients through body fluids, she can also control the testes of her parasitic males by hormone secretion. If this proves to be true, then this species "acquires" a hermaphroditic state!

Parthenogenesis

True parthenogenesis, in which eggs develop without being fertilized by sperm (and with populations consisting entirely of females), is rare in vertebrates and unknown in wild populations of birds and mammals. There are several examples, although only a few have been studied in detail. We discuss two here.

Several species of whiptail lizard (*Cnemidophorus* sp.) (319) reproduce by parthenogenesis. These populations consist entirely of females. *Cnemidophorus uniparens* probably arose from a crossing of two species with males and females, *Cnemidophorus inornatus* and *Cnemidophorus gularis*, to give an F-1 diploid hybrid. Back-crossing of *C. inornatus* with the F-1 hybrid gives a triploid, all-female *C. uniparens* (254,320). A premeiotic doubling of chromosomes results in ova with a full complement of chromosomes after meiosis. Although males are absent, several species exhibit male-like copulations (termed *pseudocopulations*, which mimic the courtship behavior of sexually reproducing species) in which one female takes a male-like role and another female assumes a female-like role (319,321). Females are able to produce viable eggs in isolation, but participation in pseudocopulatory behavior enhances fecundity. Furthermore, in *C. uniparens*, female-like and male-like behavior patterns are expressed in different phases of the ovarian cycle (321). Female-like behavior is mostly restricted to the yolk deposition phase, whereas male-like behavior is most frequent in postovulatory, gravid females (319,321,322). Copulations never occurred in reproductively inactive animals. Estradiol levels were high when female-like behavior was expressed, and low with high progesterone when male- like behavior was expressed. Plasma levels of testosterone were not detectable in these females (323), although injection of testosterone activated male-like behavior (319). Progesterone injections mimic the effects of testosterone, indicating that all aspects of male morphology and physiology, but not behavior, have been eliminated. Progesterone also decreases hypothalamic estrogen and progesterone receptor expression in the ventromedial nucleus of the hypothalamus (324,325).

A more bizarre situation exists in some fish. The Amazon molly (*Poecilia formosa*) of northwestern

Mexico is parthenogenetic and the progeny are identical to the mother. They develop from eggs that are stimulated into development but not fertilized by "borrowed sperm" (326). The male genes make no contribution to the triploid "zygote." Hence, these parthenogens depend on males of a bisexual form and must spawn with these males, but are not fertilized by them. Male *Poeciliopsis lucida* provide sperm for the all-female forms. Originally, *Poeciliopsis monacha* hybridized with *P. lucida*, giving rise to a diploid unisexual form with characteristics of both parents, but which was all female. The population sustains itself by mating with male *P. lucida*. With each mating, the male chromosomes are not transmitted and the all-female hybrid persists (327).

Variation in Breeding Strategy

This is a very broad and diverse category of reproductive tactics. The great diversity of these strategies is, at first glance, difficult to comprehend and classify. However, it is extremely helpful in understanding the evolution of these diverse tactics to keep in mind the ecological or environmental conditions in which these strategies have evolved. The following is only a limited number of examples illustrating the diversity of breeding strategies among vertebrates.

Mimicry

There are examples of mimicry, morphological, physiological, and behavioral, throughout the vertebrate classes. Some mimic other individuals, or even species, and these forms serve social functions, whereas in other cases, mimicry may be used to deceive other individuals. The latter case is common among males as they compete for mating with females.

In some primates, erection of the penis and clitoris is a common social or territorial signal. In squirrel monkeys (*Saimiri scireus*), the penis can be erected in signal even when the individual is only a few days old, and both males and females of all ages use the genital display, with females exhibiting a large clitoris. This signaling is not related to reproduction directly, but rather to dominance interactions that have indirect consequences for reproductive success (328). In other species, males have evolved structures that actually mimic the sexual skin of females. In some baboons, as in some other primates, one male may dominate the entire troop, frequently attacking other males who pose a threat to his status and who could possibly mate with females. Females typically show a steroid-induced bright red swelling of the skin around the genitalia that signals to males that they are receptive (329). The bright and swollen sexual skin in females may also be presented in a wide variety of social contexts not related directly to reproduction. In the hamadryas baboon, males have bright red hindquarters as conspicuous the female's (328). These males present this signal to high-ranking males and males may mount males and females may mount females, suggesting that presentation of the skin is a general submissive posture. Males and young of several species (*Papio*, *Macaca*, and others) may develop female-like postures. Thus, males mimic female behavior and morphology in nonsexual contexts (328).

There are also many examples in fish. A well-known one is the cichlid mouth-brooders, in which females pick eggs immediately after spawning to protect them from predators within the mouth cavity. This presents a problem for fertilization. Males of these species have spots on the anal fins that mimic eggs (328). As females attempt to pick up the "eggs" into the mouth, the male releases sperm and fertilization occurs in the buccal cavity of the female.

Another striking example of mimicry is found in the spotted hyena (*Crocuta crocuta*). The animals are highly social, with a strict social hierarchy dominated by females. All members of the group, adults and cubs, greet each other with genital inspection. The female has a peniform and highly erectile clitoris complete with a false scrotum. The resemblance to male genitalia is so close that sex can be determined with confidence only by palpitation of the scrotum [the "testicles" in females consist of soft adipose tissue (330)]. This bizarre development in females has resulted in a very different development of the urinogenital system, in which the urethra and vagina exit through the peniform clitoris. Females thus copulate and give birth through this organ. Plasma levels of testosterone in females are as high as in males, but estrogen and progesterone levels are highest in females during pregnancy. There is very active production of steroids by tissue fragments of corpus luteum and stroma from the ovary of late-term pregnant females, and LH can increase production of the androgen androstenedione (331,332). This steroid is secreted and can be easily converted to testosterone or aromatized to estradiol in target tissues. In other species of hyena in which females are not masculinized (330,333), males have higher levels of testosterone than females. In spotted hyenas, the mimicry of male genitalia by females may be a result of increases in androgen during development because blood samples from two female fetuses had the same high blood levels of testosterone as adult females. However, a more detailed study showed that during the first month of life circulating levels of testosterone in males and females were similar (330) [see also Goymann et al. (334) for a useful discussion]. However, plasma levels of androstenedione were much higher in females, especially in the second part of the first month (334).

The external genitalia of male and female cubs are as well developed as those of adults. Unsexed cubs have been observed to erect the penis or clitoris during social encounters. Adults erect the penis/clitoris on rejoining the group during a greeting ceremony, although the female never erects the clitoris during sexual encounters and copulation (333). In contrast, other species of hyena use anal scent glands and not the genitalia in similar social encounters. Unlike other species, female spotted hyenas are larger than males and tend to be dominant; usually one female leads a clan. This is typically a male role in other vertebrates, and testosterone may be involved in at least the attainment of dominance. Thus, bearing in mind that the masculinized genitalia of females are not used in sexual behavior, it is possible that female spotted hyenas exhibit male sexual facies for purely aggressive and social roles [see also Goymann et al. (334)].

Mimicry is widespread in other vertebrate classes. Female mimicry has been reported in amphibians. In the red-spotted newt (*Notopthalmus viridescens*), unpaired males may mimic females. During deposition of spermatophores and the transfer stage, the male is vulnerable to cuckoldry attempts by other males. By mimicking females, unpaired males are able to transfer a spermatophore to a mating female and thus gain some fertilizations (335).

In garter snakes, females are courted by groups of 10 to 100 males as they emerge from hibernation (336). However, only one male actually mates with the female, regardless of the number of suitors. A pheromone exuded from the back of females attracts the males (65). This pheromone appears to be vitellogenin, which is under the control of estradiol. Some vitellogenin enters the dermis of the back and is released as a pheromone. Curiously, some males can mimic females by emitting this same pheromone, and are courted by other males. Female mimics (she-males) have normal hemipenes, testes, and accessory organs and have estrogen levels lower than those of females, whereas blood testosterone levels (when high after mating) are higher than in normal males. The pheromones released by she-males tend to distract normal males from real females, and thus provide an advantage that results in relatively more copulations for the female-mimics (337).

Extra-Pair Copulations

A bewildering array of morphological, physiological, and behavioral strategies have evolved to secure copulations with other individuals despite apparent pair bonds. A few examples are discussed.

In some species, many males may be much smaller than others, although all have mature testes. An example is found in salmon. Some males return to their natal streams early as "jack" salmon. They are smaller than other males but gain some fertilizations by swimming close to spawning pairs and releasing sperm quickly. Such strategies are common in other species such as the bluegill sunfish (*Lepomis macrochirus*), and there appear to be two main male strategies. Some males mature after 6 or 7 years and defend a territory on the lake bed to which they attract females to spawn (338) and compete to construct nests in densely packed colonies. The males then guard the eggs from predation until they hatch. Nesting males court and spawn with females sequentially and provide sole parental care for the developing eggs and fry in their nests. By contrast, cuckolder males do not build nests of their own or care for their offspring. Cuckolders mature precociously and steal fertilizations in the nests of parental males through two tactics: "sneakers" (age 2 to 3 years) hide behind plants and debris near the nest edge, but are visible after darting into the nest during female egg releases. "Satellites" (age 4 to 5 years) are about the size of mature females (age 4 to 8 years), and by expressing female color and behavior are able to mislead parental males into identifying them as a second female in the nest (338). These mature males have fully mature testes despite their diminutive size and will "sneak" copulations by swimming close to a spawning pair and releasing sperm. The parental male usually chases the precocial male away, but not before some eggs have been fertilized by the sneaker. However, the precocial males continue to grow and eventually become so large that they are easily spotted by the parental males and chased away. At this time, the precocial males develop a new strategy and mimic female coloration (339). In this way they can approach a spawning pair without eliciting an attack from the larger parental male. The mimic can then insert himself between the female and parental male, release sperm, and fertilize eggs (339).

Midshipman fish have two classes of reproductively active males known as parental males and sneaker males (24,83). Parental males build and guard nests from intruders and mate call ("hum") to attract females to their nests. Sneaker males neither build nests nor mate call, but instead sneak into a nest and try to steal fertilizations from the parental male when a female is present (24,83). Type 1 (parental) males have large sonic muscles and well developed testes, whereas females have a large ovary but poorly developed sonic muscles. Juvenile males have small testes and sonic muscles, but sneaker males have large testes and small sonic muscles (24). The anatomy fits very well with observed behavior: sneaker males are able to release sperm but do not hum.

Males of the Azorean rock pool blenny (*Parablennius sanguinolentus parvicornis*) show two alternate reproductive tactics. Some males hold territories and have fully expressed male secondary sex characters, whereas "floater" males do not have territories and do not express secondary sex characters. Males with territories have higher levels of 11-ketotestosterone but not testosterone (340). However, there were no differences in levels of 11-ketotestosterone and testosterone in males showing different behavioral tactics within morphs. These data suggest that the hormone differences are related to morph types and not the behavioral strategies expressed (340).

In the Seychelles warbler (*Acrocephalus sechellensis*), females in a reproductive group also copulate with males outside the group. Primary males within the group have the highest levels of testosterone, with peaks when females within the group are laying and fertilizable. However, males involved in extragroup copulations did not have higher levels of testosterone, suggesting that high testosterone may be involved more with status in the group than with courtship and sexual behavior (341).

In other species, dominance and territoriality, and thus breeding strategy, vary among morphs. The tree lizard (*U. ornatus*) shows color morphs with variations of yellow, orange, and blue pigments in the dewlap (291,293). Territorial males have orange-blue dewlaps and develop under the influence of high progesterone and testosterone levels (291,293). These males attract females for mating to their territories. Others with just orange dewlaps are satellite males that are not territorial; these develop under the influence of low progesterone and testosterone levels (291,293). These morphs may mate only by sneaking copulations or by forced copulations after intercepting a female. Plasma levels of testosterone tend to be the same among morphs, except for after a staged encounter when the less aggressive morph showed lower plasma levels of testosterone and higher levels of corticosterone (293). It is possible that social and other stresses increase corticosterone, which inhibits testosterone secretion and action (342,343).

In some instances, females may form a pair bond with another female to raise young. Several examples are known, but the best studied is in a population of western gulls (*L. occidentalis*) in southern California. In this population, there is an excess of females (adult females outnumber males by three to one), but all pairs are monogamous and no instances of polygyny have been recorded (344). As an alternative strategy, females without male mates court one another and defend a territory (there is no shortage of space on the breeding island) as well as a male–female pair, and both females actually lay eggs in the same nest. The eggs are incubated by both females and, curiously, 33% to 100% of the eggs in a clutch hatch and the young are raised as effectively by two females as they are by a male–female pair. Females of the homosexual pair actually "sneak" copulations with neighboring males, or visit "loafing areas" and actively solicit copulations from males. Thus, although females without mates have no chance of raising young alone, they can raise young if they pair with another female and then solicit extra-pair copulations from neighboring males.

Cooperative Breeding

More and more cases are being discovered of species that breed in groups, and this may be a major strategy in tropical and subtropical species of birds and mammals as well as in fish. Usually a group consists of a dominant male and female who actually breed and are assisted by one or more individuals in caring for the offspring. These helpers usually provide only parental care and do not supply gametes for reproduction. Helpers that are related to the breeding individuals may gain some reproductive fitness by helping to raise kin, whereas unrelated helpers are thought to benefit by gaining reproductive experience, and await their chance for taking over a group and eventually breeding (345).

Cooperative breeding has been observed in fish (346), especially in cichlids in African lakes. Offspring help a breeding male and female defend territory and raise young. A classic example in birds is the hornbill of tropical forests. Some species have "helpers-at-the-nest," usually related in some way to the breeding pair, that help to feed the young. In hornbills (Buderotidae), the female is "imprisoned" in a nest hole by mud daubed around the entrance hole. This is a strategy to reduce predation, but it does mean that the female and young are entirely dependent on the male and helpers for food (347). Examples in mammals are some canids (e.g., hunting dogs [*Lycaon pictus*]), the dwarf mongoose (*Helogale parvula*), and some primates (348). This strategy is widespread in birds and has been the subject of hormonal studies. It was thought that lack of reproductive activity in helpers was due to "psychological castration" (i.e., the dominant breeding pair suppress reproduction in helpers by social stress). Studies in white-browed sparrow weavers (*Plocepasser mahali*) show that although the dominant breeding female and male within a group have larger gonads and higher levels of sex steroids than helpers, there were no differences in circulating corticosterone levels, indicating that helpers were not "stressed" (349). Similar results have been obtained for Florida scrub jays (*Aphelocoma coerulescens*) and Mexican jays (*Aphelocoma ultramarina*) (350–353). The mechanisms underlying suppression of reproduction in helpers remains unknown, but hormonal correlates are complex (354–357).

Eusocial Systems

A very complex breeding strategy is illustrated by the naked mole rat (*Heterocephalus glaber*), a fossorial species that lives in large colonies throughout arid regions in East Africa. Only one female in a colony breeds, and other females have closed vaginas and immature ovaries. All males have some sperm, but some are larger and more developed than others. It is unlikely, however, that smaller males breed because they have physical problems copulating with the much larger reproductive female (358,359). There is a caste system that includes frequent workers who nest-build and forage (dig for roots and transport food), and infrequent workers who also nest-build and forage but do so less intensively than frequent workers. The latter are also larger and heavier than frequent workers. A third caste, the nonworkers, comprises the largest rats in the colony, and males of this group are most likely to mate with the dominant female and also to assist in caring for the young. Females in all castes are nonbreeding and most will never get a chance to breed. They are not sterile, but their reproductive system is suppressed. The sole breeding female may produce 1 to 4 litters of up to 12 young per year. The young are suckled by the breeding female and also fed trophic feces by other castes. All young join the frequent worker caste, but some grow faster than others and eventually join the nonworkers, whereas others remain in the frequent worker class. In an experiment in which a breeding female and five others were removed from a colony and placed in an adjacent subterranean system, two females from the infrequent worker caste had perforate vaginas and developed teats even though they were still in olfactory contact with the removed dominant female. After 12 weeks, one of the new reproductive females was pregnant, whereas the other had reverted to the imperforate vagina condition. Thus, physical contact, and not just olfactory cues, is important for maintaining dominance. In another experiment in which two colonies were combined, the dominant females fought until one died. This bizarre system has many parallels with those of social insects such as bees and ants (358,360).

Urinary progesterone levels were elevated in pregnant and reproductively active females and undetectable in nonbreeding females. Plasma levels of LH were also higher in breeding females. Furthermore, nonbreeding females had a lower sensitivity to injected gonadotropin-releasing hormone (GnRH; as assessed by an increase in LH blood levels) than breeding females, although this was reversed if nonbreeding females were given 4- to 8-hourly injections of GnRH. These data suggest that reproductive suppression of female naked mole rats involves blockade of GnRH release (361). After experimental removal of a "queen" naked mole rat, three previously nonbreeding females began showing increased aggression, evidence of ovulation, and increased urinary progesterone levels (362).

Brood Parasitism

A well-known, but uncommon, strategy is to dupe other individuals into raising young other than their own. This system is found mostly in birds and fish but may occur in other oviparous vertebrates. In birds such as the cuckoo (*Cuculus canorus*) and cowbird, females lay eggs in the nest of other species, who then incubate the parasite's eggs and raise the young. Brown-headed cowbirds (*Molothrus ater*) parasitize over 200 known species of birds, but they do not match egg color or pattern (363). The young cowbirds grow very rapidly and out-compete the chicks of their hosts. Relative fledging success of host young in nests parasitized by brown-headed cowbirds varies (363). Relative success is calculated as the number of host young fledged from all parasitized nests divided by the number of host young fledged from all unparasitized nests.

Some species, such as cuckoos of the genus *Cuculus* (363), have developed intricate mechanisms by which eggs match the color of the host's eggs. Other species in this family raise their own young, such as the New World yellow-billed cuckoo (*Coccyzus americanus*). Female European cuckoos specialize on certain passerine species and mimic their host's eggs almost perfectly (363). These "groups" of females that specialize on one host are called *gentes*, and the ability to mimic the host's egg is genetic. Once the young cuckoo hatches, it eliminates its nestmates by maneuvering eggs or young over the edge of the nest cup (363). Another group of birds are called honey guides (Indicatoridae) because they guide mammals to bee's nests to break into the nest for honey. The honey guides then feed on grubs and wax. This family is found in Africa and Asia, and most are brood parasites specializing on hole nesting barbets (Lybiidae, related to woodpeckers). In these species, the honey guide chick hatches with a hook on the upper mandible that it uses to kill its nestmates (363).

Other parasitic species are more benign. The black-headed duck (*Heteronetta atricapilla*) of South America is an obligate brood parasite mostly on gulls and coots. Because the young are precocial, they do not have to compete for feedings from adults. In yet other species, a female may parasitize females of the same species. This strategy is probably very common although difficult to study because parasitic females are so inconspicuous and secretive. In one final and remarkable example, the young have mouth markings that match those of the host's young. This is found in chicks of the brood parasite, the paradise whydah (*Vidua paradisea*) of Africa. Its host, the melba finch (*Pytila melba*), is unable to distinguish the identical mouth markings of the parasite from its own young (363). This case is so specialized that one species specializes on only one host species (i.e., there are no gentes as in cuckoos).

There are endocrinological studies of brown-headed cowbirds, particularly in relation to lack of parental care (158–160,364), but much remains to be explored in relation to communicative behaviors with their hosts.

AN EXAMPLE OF THE POTENTIAL FOR NEW CONCEPTS OF HORMONE–BEHAVIOR INTERRELATIONSHIPS

As our knowledge of the ecological evolutionary aspects of endocrinology continues to grow, and investigations of the rich resource of free-living vertebrate species expand, many new concepts of hormone–behavior interrelationships will emerge. Examples include new insights into the control of sexual behavior and indeed sex determination and gender (18,253).

Finite-State Machine Theory and Testing Predictions

One of the predictions of the finite-state machine theory is that if a life history stage such as reproduction is prolonged, then sustained high levels of hormones through that period may have deleterious effects. We have already discussed an example where testosterone interferes with paternal behavior in some species. It is also well known that high levels of glucocorticoids for prolonged periods are highly detrimental. However, many other examples may exist and many novel mechanisms may have evolved to avoid deleterious effects. One emerging example is considered in the next section.

Avoiding the "Costs" of Testosterone

In socially monogamous males, there is now extensive evidence that male–male interactions over territory and mates can result in a marked increase in testosterone secretion—the challenge hypothesis (221,365). However, parental care in males [at least in birds; see Marler et al. (215) for mammals and a different view] requires that testosterone levels be low. It has been suggested that the temporal patterns of testosterone in birds, and perhaps vertebrates in general, represent a tradeoff between social interactions that tend to increase testosterone, and paternal care that requires it decrease (221,365). Field endocrine techniques allowed experimental tests of this idea using implants of sex steroids and their blockers. The result was a series of papers that clearly indicate there may be many "costs" to high circulating levels of testosterone (289).

During the breeding life history stage, testosterone has a number of morphological, physiological, and behavioral effects that are essential for normal reproductive function in males. Experimental field studies show that elevated testosterone levels in parental males reduce parental care in pied flycatchers (366), house sparrows (111), dark-eyed juncos (289), yellow-headed blackbirds (*Xanthocephalus xanthocephalus*) (287), and the barn swallow (*Hirundo rustica*) (367). Additional costs include wounding and increased mortality (368), suppression of the immune system (369), potential increased sensitivity to stress (370), reduced over-winter survival, higher predation rates, and increased energetic costs [Fig. 6; see Ketterson et al. (289) and Beletsky et al. (287) for reviews]. On the other hand, testosterone implants may have some benefits, such as increased activity and territory size (17,371–373). Raouf et al. (374) showed that male dark-eyed juncos with testosterone implants actually gained more extra-pair fertilizations [see also Enstrom et al. (375)]. In addition, they may have higher sperm counts for longer periods (376), and energetic costs may not always be higher (373).

There is a growing literature on the potential immunosuppressive actions of testosterone in male birds [see Hillgarth and Wingfield (369)]. This is a rapidly growing field, but much controversy remains (377,378). The next few years will be critical for this topic.

ROLE OF COMMUNICATION IN DEALING WITH A CAPRICIOUS ENVIRONMENT

No individual lives in a perfectly stable environment. The life cycle is punctuated by events of regular variability—day versus night, low tide and high tide, spring, summer, autumn, and winter, rainy season and dry season, and so forth—and animals use the environmental signals discussed previously to time events in their life cycle such as reproduction. However, superimposed on this predictability are unpredictable events (perturbations) that individuals must cope with on a facultative basis. Not surprisingly, communication among individuals here also is critical, although the endocrine bases of these types of communication are largely unknown.

Labile Perturbation Factors and Communication

Direct and Indirect Perturbation Factors

A large number of unpredictable events in the environment have been described, many of which not only have the potential to be stressful but may result in high mortality rates. These numerous factors, termed labile, or transient, perturbation factors (LPFs) (1,26), can be divided into roughly two major groups. Indirect LPFs are usually very short-lived and have little long-term effect on an individual. Examples include loss of nest to a predator and short, severe storms (e.g., hail) that destroy a nest. A "flight-or-fight" response may be triggered, but the event is short-lived and results in no long-lasting effects on the adults. Resumption of normal activities, such as renesting, occur within minutes to hours (224). In contrast, direct LPFs may be manifest more slowly, their effects may be cumulative over hours or even days, and they are disruptive of normal activities (particularly breeding). Examples include prolonged severe weather, or competition with other species or individuals that may reduce access to trophic resources, thus enhancing the potential for stress. In these cases an individual may go into negative energy balance and abandon the normal life history stage. In recent years, a conservation aspect to direct perturbation factors has become clear: pollution and human disturbance have the potential to act in ways identical to natural perturbation factors (132). The direct LPFs are the focus here because they trigger marked changes in physiology and behavior that maximize survival and thus lifetime fitness.

Responses to Predators

We tend to think of communicative behaviors having direct effects on reproductive function as a result of sexual and aggressive encounters within the predictable life cycle (seasons and so forth). However, there is also a strong unpredictable component in the exposure to predators. Several rodent species (*Microtus*, *Eliomys*, *Jaculus*, and *Acomys*) respond to tape-recorded calls of owls (*Strix aluco*) by crouching and retreating to burrows (379), and in some cases (e.g., *Microtus*) fleeing behavior resulted (380). Owl calls also increased corticosterone levels in *Microtus*. Curiously, the spiny mouse (*Acomys cahirinus*) showed no behavioral response to owl vocalizations, but did have elevated corticosterone levels compared with controls (380). Female Columbian ground squirrels (*Spermophilus columbianus*) exposed to a domestic dog for 5 minutes per week showed an overall increase in plasma levels of cortisol during the active season, but this response was suppressed during lactation (381). Similarly in snowshoe hares, constant predator exposure increases cortisol levels chronically, resulting in decreased reproductive performance. Predator stress may be a major driver of the population cycle of this mammal (382,383). A striking example from animals in the field is the dramatic 10-year cycle of abundance in showshoe hares (*Lepus americanus*) of the holarctic boreal forest. In this cycle, hares increase numbers followed by a similar increase in numbers of predators, particularly the lynx (*Lynx canadensis*), after which a population decline occurs. The cycle is not a function of the increasing population outstripping food supplies, nor is it caused by increased social interactions as the population peaks (384). It appears that the cycle is driven by predation risk because as the numbers of predators increase after the rise in hare numbers, risk of predation forces hares into habitat that provides shelter but is poor in food. They are then forced into the open areas, where predation increases (384). The effects of an interaction of food and predation risk on hare behavior have major consequences for population dynamics. When showshoe hares declined in the Yukon, Canada, in the 1990s, they appeared to be chronically stressed, as indicated by higher levels of free cortisol and lower corticosteroid-binding globulin (CBG) capacity. Baseline plasma testosterone levels did not change and there was a reduced responsiveness to dexamethasone and ACTH treatment (383). Hares also showed reduced blood leukocyte counts, increased glucose mobilization, and higher over-winter loss of body weight compared with hares in years when the population was low. High predation risk, not population density or poor nutrition, resulted in chronic stress and impaired reproductive function. Reproductive success did not improve until predation risk declined (383).

In pied flycatchers, exposure of free-living birds to various predators at the nest box resulted in strong alarm calling and mobbing behavior, but only when one predator actually attacked the box was there an increase in circulating corticosterone (385). Similarly in African stonechats (*Saxicola torquata*), birds with territories next to predators had higher corticosterone levels than those not next to predators (386).

Presence of predators can be stressful to many organisms and has the potential to reduce reproductive success. Furthermore, communicative behaviors through alarm calls and the like may mediate some of these effects.

Warnings and Alarms within Reproductive Groups

Most vertebrates respond to alarm calls and warnings from other individuals, conspecific as well as heterospecific, and these responses can have dramatic effects on behavior (387–390). Sensory modalities by which these alarms are communicated can also be varied, and include auditory, visual, chemical, and electrical.

In a captive population of brown capuchin monkeys (*Cebus apella*), experimental exposure to a potential predator (human investigator) resulted in increased frequency of vocalizations that signify terrestrial predators. Although not significant, production of the vocalization was positively related to fecal and plasma levels of cortisol (391).

Parental Aggression

Many animals defend their nests and young with great intensity, especially species with altricial and defenseless young (392). Parental aggression is widespread among all vertebrates that show some form of parental care (393). Communicative behaviors are involved here in coordinating parental aggression oriented against a potential predator or conspecific intruder, as well as communicative behaviors between the parent and the aggressor. Androgens that promote territorial behavior tend to suppress paternal behavior in many species, unless the parental behavior involved includes maintaining and defending a territory (394). In laboratory rodents, many hormones have been implicated in parental aggression, including prolactin and progesterone [see Lonstein and Gammie (392) for review].

Parent–Offspring Conflict and Behavioral Cascades

Investigations of black-legged kittiwakes, Rissa tridactyla, provide an example of a unique combination of ecological and physiological factors when food shortages are prolonged, breeding black-legged kittiwakes suppress their adrenocortical response to acute stress (395). Early in reproduction, a suppressed stress response is associated with higher fat deposits and normal circulating corticosterone. At the chick substage, the suppressed stress response coincides with depleted energy reserves and an increased baseline corticosterone. In response to low food availability, adult kittiwakes allocate more time to foraging and less to guarding/brooding their chicks (396). Experimental manipulation of blood corticosterone have shown that a moderate elevation changes resource allocation by parent kittiwakes (397). This in turn results in an increase in the amount of time kittiwake chicks are unattended by parents, often resulting in predation. However, loss of offspring during food shortages increases postbreeding survival of parent kittiwakes (398). Thus, it can be argued that changing the stress response in parent black-legged kittiwakes increases lifetime fitness.

Adult black-legged kittiwakes with experimentally elevated levels of corticosterone do not alter chick-provisioning rate, and as a result may suffer long-term effects of chronically elevated levels of corticosterone (397). Pelagic-foraging kittiwakes rely on pelagic, variable food resources compared with inshore foraging species. However, chick-provisioning behavior of kittiwakes is similar to that of other species (399,400) [reviewed in Ydenberg and Bertram (401)]. Kittiwake chicks appear to be physiologically responsive to variations in daily food intake and moderate food shortages by increasing corticosterone secretion (402). Levels of corticosterone in general are thought to reflect the point when energy demands exceeds energy intake (397,402). Chronic elevation of corticosterone during early development can have long-term deleterious effects [reviewed in Welberg and Seckl (403)]. Indeed, it has been shown that high corticosterone levels increase chick begging rates and increase rates of food provisioning by parents (397). Thus, the elevated corticosterone in food-deprived kittiwake chicks represents an emergency life history strategy. This restores depleted energy reserves by modifying the behavior of their parents and thereby avoids stress.

CONCLUSION

With the possible exception of migration, the reproduction life history stage of vertebrates is the most complex component of an organism's life cycle. It is also the one life history stage that is absolutely dependent on communication of one individual with another. Even in parthenogens and one example of self-fertilization, communication among individuals is essential at some stage in the life histories of populations. It is not surprising, therefore, that patterns of reproduction, including morphological, physiological, and behavioral components, are overwhelming in their diversity and modes of communication. Nonetheless, a framework is proposed that allows an analysis of life history stages of a given organism, so that it is possible to compare reproductive function, particularly the role of communicative behaviors in social interactions, across vertebrate taxa from agnathans to mammals and in diverse habitats from totally aqueous to entirely terrestrial. Finite-state machine theory proposes that all vertebrates have a fixed number of life history stages (e.g., reproductive, nonreproductive, molt, migration), each with a characteristic set of substages (e.g., in reproduction, courtship, copulation, parental care, territoriality). The state of the individual at any point in its life cycle is determined by the combination of life history stages and substages expressed. This state includes characteristic morphology, physiology, and behavior all regulated to varying degrees by neuroendocrine and endocrine secretions. Because the number of life history stages and substages is fixed during ontogeny for each individual, there is a finite number of states. In addition, because the number of life history stages and substages varies from species to species, and possibly among individuals within a population, there may be considerable variation in possible states. In general, the more states possible (i.e., a more complex finite-state machine) means more complex patterns of social interactions, although a potential tradeoff is less flexibility in timing. However, the framework allows close comparisons, within the limits of phylogenetic confounds, that point out possible divergent and convergent evolutionary paths of mechanisms underlying hormone– behavior interactions.

Superimposed on this framework is a classification of environmental signals that regulate changes in state, as well as the ways in which the reproduction life history stage develops (in seasonal breeding species, this is separate from ontogeny)—how mature capability determines actual onset of breeding and then its termination. Support functions such as territorial behavior are interwoven into this complex mix, providing other levels of regulation, integration, and synchronization.

Thus, despite the apparently bewildering array of reproductive patterns in vertebrates, common themes emerge clearly. Phylogenetic issues are also important, but given that the hormones of the neuroendocrine and endocrine systems are remarkably well conserved from fish to mammals, the diversity of mechanisms by which they act looms large. F. Hisaw put this very aptly some years ago: "it is not the hormones that have evolved, but the uses to which they are put." As molecular techniques become more broadly available for the rich base of animal models available throughout the vertebrates, we will be able to explore those "uses" with unprecedented depth and breadth.

ACKNOWLEDGMENTS

The author has been supported by a series of grants from the National Science Foundation (currently IBN-0317141), and by grant number RO1 MH65974-01 from the National Institute of Mental Health. I also acknowledge valuable support from a Russell F. Stark University Professorship (University of Washington), and a Simon F. Guggenheim Fellowship while many of the ideas presented here were developed.

REFERENCES

1. Jacobs, J. D., and Wingfield, J. C. (2000). Endocrine control of life-cycle stages: a constraint on response to the environment? *Condor* 102, 35–51.
2. Gorbman, A., Dickhoff, W. W., Vigna, S. R., Clark, N. B., and Ralph, C. L. (1983). *Comparative Endocrinology*. Wiley, New York.
3. Norris, D. O. (1997). *Vertebrate Endocrinology*. Academic Press, New York.
4. Piersma, T., and Drent, J. (2003). Phenotypic flexibility and the evolution of organismal design. *Trends Ecol. Evol.* 18, 228–233.
5. Wingfield, J. C., and Jacobs, J. D. (1999). The interplay of innate and experiential factors regulating the life history cycle of birds. In *Proceedings of the 22nd International Ornithological Congress* (N. Adams and R. Slotow, Eds.), pp. 2417–2443. BirdLife South Africa, Johannesburg.
6. Wingfield, J. C. (2004). Allostatic load and life cycles: implications for neuroendocrine mechanisms. In *Allostasis, Homeostasis and the Costs of Physiological Adaptation* (J. Schulkin Ed.), pp. 302–342. Cambridge University Press, Cambridge.
7. Dawson, A., King, V. M., Bentley, G. E., and Ball, G. F. (2001). Photoperiodic control of seasonality in birds. *J. Biol. Rhythms* 16, 365–380.
8. Follett, B. K. (1984). Birds. In *Marshall's Physiology of Reproduction: 1. Reproductive Cycles of Vertebrates* (G. E. Lamming, Ed.), pp. 283–350. Churchill-Livingstone, Edinburgh.
9. Nicholls, T. J., Goldsmith, A. R., and Dawson, A. (1988). Photorefractoriness in birds and comparison with mammals. *Physiol. Rev.* 68, 133–176.
10. Lack, D. (1968). *Ecological Adaptations for Breeding in Birds*. Chapman and Hall, London.
11. Perrins, C. M. (1970). The timing of bird's breeding seasons. *Ibis* 112, 242–255.
12. Wingfield, J. C., Jacobs, J. D., Tramontin, A. D., Perfito, N., Meddle, S., Maney, D. L., and Soma, K. (1999). Toward and ecological basis of hormone-behavior interactions in reproduction of birds. In *Reproduction in Context* (K. Wallen and J. Schneider, Eds.), pp. 85–128. MIT Press, Cambridge, MA.
13. Wingfield, J. C., and Silverin, B. (2002). Ecophysiological studies of hormone-behavior relations birds. In *Hormones, Brain and Behavior* (D. Pfaff, A. Arnold, A. Etgen, S. Fahrbach, and R. Rubin, Eds.), Vol. 2, pp. 587–647. Academic Press, New York.
14. Gwinner, E. (1986). *Circannual Rhythms*. Springer-Verlag, Berlin.
15. Wingfield, J. C., Hahn, T. P., Wada, M., and Schoech, S. (1997). Effects of day length and temperature on gonadal development, body mass and fat depots in white-crowned sparrows, *Zonotrichia leucophrys pugetensis*. *Gen. Comp. Endocrinol.* 107, 44–62.
16. Wingfield, J. C. (1980). Fine temporal adjustment of reproductive functions. In *Avian Endocrinology* (A. Epple and M. H. Stetson, Eds.), pp. 367–389. Academic Press, New York.
17. Wingfield, J. C. (1983). Environmental and endocrine control of reproduction: an ecological approach. In *Avian Endocrinology: Environmental and Ecological Aspects* (S.-I. Mikami and M. Wada, Eds.), pp. 205–288. Japanese Scientific Societies Press, Tokyo, and Springer-Verlag, Berlin.
18. Rhen, T., and Crews, D. (2002). Variation in reproductive behavior within a sex: neural systems and endocrine activation. *J. Neuroendocrinol.* 14, 517–531.
19. Wilczynski, W., Allison, J. D., and Marler, C. A. (1993). Sensory pathways linking social and environmental cues to endocrine regions of amphibian forebrains. *Brain Behav. Evol.* 42, 252–264.

20. Balthazart, J. (1983). Hormonal correlates of behavior. In *Avian Biology* (D. S. Farner, J. R. King, and K. S. Parkes, Eds.), Vol. 7, pp. 221–365. Academic Press, New York.
21. Harding, C. F. (1983). Hormonal influences on avian aggressive behavior. In *Hormones and Aggressive Behavior* (B. Svarc, Ed.), pp. 435–467. Plenum Press, New York.
22. Wingfield, J. C., and Moore, M. C. (1987). Hormonal, social, and environmental factors in the reproductive biology of free-living male birds. In *Psychobiology of Reproductive Behavior: An Evolutionary Perspective* (D. Crews, Ed.), pp. 149–175. Prentice-Hall, Englewood Cliffs, NJ.
23. Wingfield, J. C., Whaling, C. S., and Marler, P. R. (1994). Communication in vertebrate aggression and reproduction: the role of hormones. In *The Physiology of Reproduction* (E. Knobil and J. D. Neill, Eds.), pp. 303–342. Raven Press, New York.
24. Bass, A. H., and Grober, M. S. (2001). Social and neural modulation of sexual plasticity in teleost fish. *Brain Behav. Evol.* 57, 293–300.
25. Ferguson, J. N., Young, L. J., and Insel, T. R. (2002). The neuroendocrine basis of social recognition. *Front. Neuroendocrinol.* 23, 200–224.
26. Wingfield, J. C., Breuner, C., Jacobs, J., Lynn, S., Maney, D., Ramenofsky, M., and Richardson, R. (1998). Ecological bases of hormone-behavior interactions: the "emergency life history stage." *Am. Zool.* 38, 191–206.
27. Marler, P. R., and Hamilton, W. J. (1966). *Mechanisms of Animal Behavior.* Wiley, New York.
28. Green, S., and Marler, P. R. (1979). The analysis of animal communication. In *The Handbook of Behavioral Neurobiology* (P. R. Marler and J. Vandenbergh, Eds.), Vol. 3, pp. 73–158. Plenum Press, New York.
29. Seebeok, T. A., Ed. (1977). *How Animals Communicate.* Indiana University Press, Bloomington, IN.
30. Hailman, J. J. P. (1977). *Optical Signals.* Indiana University Press, Bloomington, IN.
31. Aronson, L. R. (1965). Environmental stimuli altering the physiologic condition of the individual among lower vertebrates. In *Sex and Behavior* (F. Beach, Ed.), pp. 290–318. Wiley, New York.
32. Ishii, S., and Itoh, M. (1992). Amplexus induces surge of luteinizing hormone in male toads, *Bufo japonicus*. *Gen. Comp. Endocrinol.* 86, 34–41.
33. Marler, P. R. (1965). Communication in monkeys and apes. In: *Monkeys and Apes: Field Studies of Ecology and Behavior* (I. DeVore, Ed.), pp. 544–584. Holt, Rinehart and Winston, New York.
34. Morris, D. (1971). *Intimate Behavior.* Random House, New York.
35. Fritzsch, B., Ryan, M. J., Wilczynski, W., and Hetherington, T. E. (1988). *The Evolution of the Amphibian Auditory System.* Wiley, New York.
36. Kroodsma, D. E., and Miller, E. H. (1982). *Acoustic Communication in Birds.* Academic Press, New York.
37. Snowdon, C. T., Brown, C. H., and Peterson, M. R. (1982). *Primate Communication.* Cambridge University Press, Cambridge.
38. Bass, A. H. (1990). Sounds from the intertidal zone: vocalizing fish. *Bioscience* 40, 249–258.
39. Sisneros, J. A., and Bass, A. H. (2003). Seasonal plasticity of peripheral auditory frequency sensitivity. *J. Neurosci.* 23, 1049–1058.
40. Beach, F. A. (1944). Responses of captive alligators to auditory stimulation. *Am. Naturalist* 78, 481–505.
41. Bogert, C. M. (1960). The influence of sound on the behavior of reptiles and amphibians. In *Animal Sounds and Communication* (W. E. Lanyon and W. N. Tavolga, Eds.), pp. 137–320. American Institute of Biological Science, Washington, DC.
42. Narins, P. M. (1994). Seismic communication in amphibians. *J. Acoust. Soc. Am.* 96, 3296.
43. Lewis, E. R., and Narins, P. M. (1985). Do frogs communicate with seismic signals? *Science* 227, 187–189.
44. Narins, P. M. (1990). Seismic communication in anuran amphibians. *Bioscience* 40, 268–274.
45. Randall, J. A. (1989). Individual foot drumming signatures in banner-tailed kangaroo rats, *Dipodomys spectabilis*. *Anim. Behav.* 38, 620–630.
46. Randall, J. A., and Lewis, E. R. (1997). Seismic communication between the burrows of kangaroo rats, *Dipodomys spectabilis*. *J. Comp. Physiol. [A]* 181, 525–531.
47. Gregory, J. E., McIntyre, A. K., and Proske, U. (1986). Vibration-evoked responses from lamellated corpuscles in the legs of kangaroos. *Exp. Brain Res.* 62, 648–653.
48. Narins, P. M. (1992). Seismic signal transmission between burrows of the Cape mole rat, *Georychus capensis*. *J. Comp. Physiol. [A]* 170, 13–21.
49. Rado, R., Levi, N., Hauser, H., Witcher, J., Adler, N., Intrator, N., Wolberg, Z., and Terkel, J. (1987). Seismic signaling as a means of communication in a subterranean mammal. *Anim. Behav.* 35, 1249–1266.
50. Bennet, N. C., and Jarvis, J. U. M. (1988). The reproductive biology of the Cape mole rat, *Georychus capensis* (Rodentia, Bathyergidae). *J. Zool. (Lond).* 214, 95–106.
51. Sisneros, J. A., Tricas, T. C., and Luer, C. A. (1998). Response properties and biological function of the skate electrosensory system during ontogeny. *J. Comp. Physiol. [A]* 183, 87–99.
52. Sisneros, J. A., and Tricas, T. C. (2000). Androgen-induced changes in the response dynamics of ampullary electrosensory primary afferent neurons. *J. Neurosci.* 20, 8586–8595.
53. Zakon, H. H., and Smith, G. T. (2002). Weakly electric fish: behavior, neurobiology and neuroendocrinology. In *Hormones, Brain and Behavior* (D. Pfaff, A. Arnold, A. Etgen, S. Fahrbach, and R. Rubin, Eds.), Vol. 2, pp. 349–374. Academic Press, New York.
54. Bass, A. H., and Hopkins, C. D. (1984). Shifts in frequency tuning of electroreceptors in androgen treated mormyrid fish. *J. Comp. Physiol. [A]* 155, 713–724.
55. Hopkins, C. D. (1981). On the diversity of electric signals in a community of mormyrid electric fish in West Africa. *Am. Zool.* 21, 103–115.
56. Hopkins, C. D., and Bass, A. H. (1981). Temporal coding of species recognition in an electric fish. *Science* 21, 85–87.
57. Zakon, H. H. (2003). Insight into the mechanisms of neuronal processing from electric fish. *Curr. Opin. Neurobiol.* 13, 744–750.
58. Zakon, H. H., and Dunlap, K. D. (1999). Sex hormones and communication signals: a tale of two species. *Brain Behav. Evol.* 54, 61–69.
59. Bass, A. H. (1986). Species differences in electrical organs of mormyrids: substrates for species-typical electric organ discharge waveforms. *J. Comp. Neurol.* 244, 313–330.
60. Carr, C. E. (1990). Neuroethology of electric fish. *Bioscience* 40, 259–267.
61. Stacey, N. (1983). Hormones and pheromones in fish sexual behavior. *Biosci. Rep.* 33, 552–556.
62. Stacey, N. (2003). Hormones, pheromones and reproductive behavior. *Fish Physiol. Biochem.* 28, 229–235.
63. Stacey, N., and Sorenson, P. (2002). Hormonal pheromones in fish. In *Hormones, Brain and Behavior* (D. Pfaff, A. Arnold, A. Etgen, S. Fahrbach, and R. Rubin, Eds.), Vol. 2, pp. 375–434. Academic Press, New York.
64. Waldman, B., and Bishop, P. J. (2004). Chemical communication in an archaic anuran amphibian. *Behav. Ecol.* 15, 88–93.
65. Garstka, W. R., and Crews, D. (1981). Female sex pheromone in the skin and circulation of a garter snake. *Science* 214, 681–683.

66. Mason, R. T. (1992). Reptilian pheromones. In *Biology of the Reptilia* (C. Gans and D. Crews, Eds.), Vol. 18, pp. 114–128. University of Chicago Press, Chicago.
67. Pfeiffer, W. (1982). Chemical signals in communication. In *Chemoreception in Fishes* (T. J. Hara, Ed.), pp. 307–326. Elsevier, Amsterdam.
68. Vandenbergh, J. G., Ed. (1983). *Pheromones and Reproduction in Mammals*. Academic Press, New York.
69. Arnold, A. P., and Breedlove, S. M. (1985). Organizational and activational effects of sex steroids on brain and behavior: a reanalysis. *Horm. Behav.* 19, 469–498.
69a. Beach, F. A. (1948). *Hormones and Behavior*. P. Hoeber, New York.
70. Nelson, R. J. (1999). *An Introduction to Behavioral Endocrinology*. Sinauer Associates, Sunderland, MA.
71. Phoenix, C., Gor, R., Gerall, A., and Young, W. (1959). Organizing action of prenatally administered testosterone propionate on the tissues mediating mating behavior in the female guinea pig. *Endocrinology* 65, 369–382.
72. Burmeister, S., and Wilczynski, W. (2001). Social context influences androgenic effects on calling in the green tree frog (*Hyla cinerea*). *Horm. Behav.* 40, 550–558.
73. Balthazart, J., and Adkins-Regan, E. (2002). Sexual differentiation of brain and behavior in birds. In *Hormones, Brain and Behavior* (D. Pfaff, A. Arnold, A. Etgen, S. Fahrbach, and R. Rubin, Eds.), Vol. 3, pp. 223–302, Academic Press, New York.
74. Crews, D., Coomber, P., Baldwin, R., Azad, N., and Gonzalez-Lima, F. (1996). Brain organization in a reptile lacking sex chromosomes: effects of gonadectomy and exogenous testosterone. *Horm. Behav.* 30, 474–486.
75. De Vries, G. J., and Simerly, R. B. (2002). Anatomy, development, and function of sexually dimorphic neural circuits in the mammalian brain. In *Hormones, Brain and Behavior* (D. Pfaff, A. Arnold, A. Etgen, S. Fahrbach, and R. Rubin, Eds.), Vol. 3, pp. 137–192. Academic Press, New York.
76. Schreibman, M. P., and Magliulo-Cepriano, L. (2002). Differentiation/maturation of centers in the brain regulating reproductive function in fishes. In *Hormones, Brain and Behavior* (D. Pfaff, A. Arnold, A. Etgen, S. Fahrbach, and R. Rubin, Eds.), Vol. 2, pp. 303–324. Academic Press, New York.
77. Ball, G. F., and MacDougall-Shackleton, S. (2001). Sex differences in songbirds 25 years later: what have we learned and where do we go? *Microsc. Res. Tech.* 54, 327–334.
78. Brenowitz, E. (1997). Comparative approaches to study of the avian song control system. *J. Neurobiol.* 33, 517–531.
79. Brenowitz, E. A. (2004). Plasticity of the adult avian song control system. *Ann. N Y Acad. Sci.* 1016, 560–585.
80. Brenowitz, E. A., Margoliash, D., and Nordeen, K. W. (1997). An introduction to birdsong and the avian song control system. *J. Neurobiol.* 33, 495–500.
81. Konishi, M., and Gurney, M. E. (1982). Sexual differentiation of brain and behavior. *Trends Neurosci.* 5, 20–23.
82. Tramontin, A. D., and Brenowitz, E. A. (2000). Seasonal plasticity in the adult brain. *Trends Neurosci.* 23, 251–258.
83. Sisneros, J. A., Forlano, P. M., Deitcher, D. L., and Bass, A. H. (2004). Steroid dependent auditory plasticity leads to adaptive coupling of sender and receiver. *Science* 305, 404–407.
84. Nottebohm, F., and Arnold, A. P. (1976). Sexual dimorphism in vocal control areas of the songbird brain. *Science* 194, 211–213.
85. Agate, R. J., Grisham, W., Wade, J., Mann, S., Wingfield, J. C., Schanen, C., Palotie, A., and Arnold, A. P. (2003). Neural not gonadal origin of brain sex differences in a gynandromorphic finch. *Proc. Natl. Acad. Sci. U S A* 100, 4873–4878.
86. Arnold, A. P. (1992). Developmental plasticity in neural circuits controlling birdsong: sexual differentiation and the neural basis of learning. *J. Neurobiol.* 23, 1506–1528.
87. Arnold, A. P. (1996). Genetically triggered sexual differentiation of brain and behavior. *Horm. Behav.* 30, 495–505.
88. Arnold, A. P. (2004). Sex chromosomes and brain gender. *Nat. Rev. Neurosci.* 5, 701–708.
89. Schlinger, B. A., and Brenowitz, E. A. (2002). Neural and hormonal control of birdsong. In *Hormones, Brain and Behavior* (D. Pfaff, A. Arnold, A. Etgen, S. Fahrbach, and R. Rubin, Eds.), Vol. 2, pp. 799–839. Academic Press, New York.
90. Tramontin, A. D., Hartman, V. N., and Brenowitz, E. A. (2000). Breeding conditions induce rapid and sequential growth in adult avian song control circuits: a model of seasonal plasticity in the brain. *J. Neurosci.* 20, 854–861.
91. Brenowitz, E. A., Arnold, A. P., and Levin, R. (1985). Neural correlates of female song in tropical duetting birds. *Brain Res.* 343, 104–112.
92. Brenowitz, E. A. and Arnold, A. P. (1986). Interspecific comparisons of the size of neural song control regions and song complexity in duetting birds: evolutionary implications. *J. Neurosci.* 6, 2875–2879.
93. Gahr, M., Sonnenshcein, E., and Wickler, W. (1998). Sex differences in the size of the neural song control regions in a duetting songbird with similar song repertoire size of males and females. *J. Neurosci.* 18, 1124–1131.
94. Kendrick, A. M., and Schlinger, B. A. (1996). Independent differentiation of sexual and social traits. *Horm. Behav.* 30, 600–610.
95. Roberts, R. L., Zulio, A., Gustafson, E. A., and Carter, C. S. (1996). Perinatal steroid treatments alter alloparental and affiliative behavior in prairie voles. *Horm. Behav.* 30, 600–611.
96. Pinxten, R., De Ridder, E., Balthazart, J., Berghman, L., and Eens, M. (2000). The effect of castration on aggression in the non-breeding season is age-dependent in male European starlings. *Behaviour* 137, 647–661.
97. Crews, D., Cantu, A. R., Rhen, T., and Vohra, R. (1996). The relative effectiveness of estrone, estradiol-17β, and estriol in sex reversal in the red-eared slider (*Trachemys scripta*), a turtle with temperature dependent sex determination. *Gen. Comp. Endocrinol.* 102, 317–326.
98. Silverin, B. (1986). Corticosterone binding proteins and behavioral effects of high plasma levels of corticosterone during the breeding period in the pied flycatcher. *Gen. Comp. Endocrinol.* 64, 67–74.
99. Wingfield, J. C., and Silverin, B. (1986). Effects of corticosterone on territorial behavior of free-living song sparrows, *Melospiza melodia*. *Horm. Behav.* 20, 405–417.
100. Moore, I. T., and Mason, R. T. (2001). Behavioral and hormonal responses to corticosterone in the male red-sided garter snake, *Thamnophis sirtalis parietalis*. *Physiol. Behav.* 72, 669–674.
101. Moore, F. L., and Miller, L. J. (1984). Stress-induced inhibition of sexual behavior: corticosterone inhibits courtship behaviors of a male amphibian (*Taricha granulosa*). *Horm. Behav.* 18, 400–410.
102. Greenberg, N., and Wingfield, J. C. (1987). Stress and reproduction: reciprocal relationships. In *Reproductive Endocrinology of Fishes, Amphibians and Reptiles* (D. O. Norris and R. E. Jones, Eds.), pp. 389–426. Wiley, New York.
103. Moore, I. T., and Jessop, T. S. (2003). Stress, reproduction, and adrenocortical modulation in amphibians and reptiles. *Horm. Behav.* 43, 39–47.
104. Sapolsky, R. M. (2002). Endocrinology of the stress response. In *Behavioral Endocrinology* (2nd ed. J. B. Becker, S. M. Breedlove, D. Crews, and M. M. McCarthy, Eds.), pp. 409–450. MIT Press, Cambridge, MA.
105. Sapolsky, R., Romero, L. M. and Munck, A. U. (2001). How do glucocorticosteroids influence stress responses? Integrating permissive, suppressive, stimulatory and preparative actions. *Endocr. Rev.* 21, 55–89.
106. DeNardo, D. F., and Licht, P. (1993). Effects of corticosterone on social behavior of lizards. *Horm. Behav.* 27, 184–199.

107. Romero, L. M., Dean, S. C., and Wingfield, J. C. (1998). Neurally active peptide inhibits territorial defense in wild birds. *Horm. Behav.* 34, 239–247.
108. Leshner, A. I. (1978). *An Introduction to Behavioral Endocrinology*. Oxford University Press, New York.
109. Leshner, A. I., and Politch, J. A. (1979). Hormonal control of submissiveness in mice: irrelevance of the androgens and relevance of pituitary-adrenal hormones. *Physiol. Behav.* 22, 531–534.
110. Leshner, A. I. (1981). The role of hormones in the control of submissiveness. In *Multidisciplinary Approaches to Aggression Research* (P. F. Brain and D. Denton, Eds.), pp. 302–322. Elsevier/North Holland, Amsterdam.
111. Hegner, R. E., and Wingfield, J. C. (1987). Effects of experimental manipulation of testosterone levels on parental investment and breeding success in male house sparrows. *Auk* 104, 462–469.
112. Vleck, C. M., and Dobrott, S. J. (1993). Testosterone, antiandrogen and alloparental behavior in bobwhite quail foster fathers. *Horm. Behav.* 27, 92–107.
113. Lynn, S. E., and Wingfield, J. C. (2003). Male chestnut-collared longspurs are essential for nestling survival: a removal study. *Condor* 105, 154–158.
114. Lynn, S. E., Hayward, L. S., Benowitz-Fredericks, Z. M., and Wingfield, J. C. (2002). Behavioral insensitivity to supplementary testosterone during the parental phase in the chestnut-collared longspur (*Calcarius ornatus*). *Anim. Behav.* 63, 795–803.
115. Hunt, K. E., Hahn, T. P., and Wingfield, J. C. (1999). Endocrine influences on parental care during a short breeding season: testosterone and male parental care in Lapland longspurs (*Calcarius lapponicus*). *Behav. Ecol. Sociobiol.* 45, 360–369.
116. Wingfield, J. C., and Hahn, T. P. (1994). Testosterone and territorial behaviour in sedentary and migratory sparrows. *Anim. Behav.* 47, 77–89.
117. Wingfield, J. C. (1994). Regulation of territorial behavior in the sedentary song sparrow, *Melospiza melodia morphna*. *Horm. Behav.* 28, 1–15.
118. Wingfield, J. C. (1994). Control of territorial aggression in a changing environment. *Psychoneuroendocrinology* 19, 709–721.
119. Wingfield, J. C. (1994). Hormone-behavior interactions and mating systems in male and female birds. In *The Difference Between the Sexes* (R. V. Short and E. Balaban, Eds.), pp. 303–330. Cambridge University Press, Cambridge.
120. Andrew, R. J. (1972). Recognition processes and behavior with special reference to effects of testosterone on persistence. In *Advances in the Study of Behaviour* (D. S. Lehrman, R. A. Hinde, and E. Shaw, Eds.), Vol. 4, pp. 175–208. Academic Press, New York.
121. Andrew, R. J., and Rogers, L. (1972). Testosterone, search behavior and persistence. *Nature* 237, 343–346.
122. Sinervo, B., Miles, D. B., Frankino, W. A., Klukowski, M., and DeNardo, D. F. (2000). Testosterone, endurance and Darwinian fitness: natural and sexual selection on the physiological bases of alternative male behaviors in side-blotched lizards. *Horm. Behav.* 38, 222–233.
123. vom Saal, F. S. (1979). Prenatal exposure to androgen influence morphology and aggressive behavior of male and female mice. *Horm. Behav.* 12, 1–11.
124. vom Saal, F. S. (1983). Models of early hormone effects on intrasex aggression in mice. In *Hormones and Aggressive Behavior* (B. Svare, Ed.), pp. 197–222. Plenum Press, New York.
125. Schwabl, H. (1993). Yolk is a source of maternal testosterone for developing birds. *Proc. Natl. Acad. Sci. U S A* 90, 11446–11450.
126. Schwabl, H. (1996). Environment modifies the testosterone levels of a female bird and its eggs. *J. Exp. Zool.* 276, 157–163.
127. Lovern, M. B., and Wade, J. (2003). Sex steroids in green anoles (*Anolis carolinensis*): uncoupled maternal plasma and yolking follicle concentrations, potential embryonic steroidogenesis, and evolutionary implications. *Gen. Comp. Endocrinol.* 134, 109–115.
128. Hackl, R., Bromundt, V., Daisley, J., Kotrschal, K., and Möstl, E. (2003). Distribution and origin of steroid hormones in the yolk of Japanese quail eggs (*Coturnix coturnix japonica*). *J. Comp. Physiol. [B]* 173, 327–331.
129. Wingfield, J. C., Hahn, T. P., Levin, R., and Honey, P. (1992). Environmental predictability and control of gonadal cycles in birds. *J. Exp. Zool.* 261, 214–231.
130. Wingfield, J. C., Doak, D., and Hahn, T. P. (1993). Integration of environmental cues regulating transitions of physiological state, morphology and behavior. In *Avian Endocrinology* (P. J. Sharp, Ed.), pp. 111–122. Journal of Endocrinology, Ltd.., Bristol, United Kingdom.
131. Wingfield, J. C., Hahn, T. P., Wada, M., Astheimer, L. B., and Schoech, S. (1996). Interrelationship of day length and temperature on the control of gonadal development, body mass and fat depots in white-crowned sparrows, *Zonotrichia leucophrys gambelii*. *Gen. Comp. Endocrinol.* 101, 242–255.
132. Wingfield, J. C., Hunt, K., Breuner, C., Dunlap, K., Fowler, G. S., Freed, L., and Lepson, J. (1997). Environmental stress, field endocrinology, and conservation biology. In *Behavioral Approaches to Conservation in the Wild* (J. R. Clemmons and R. Buchholz, Eds.), pp. 95–131. Cambridge University Press, Cambridge.
133. Wingfield, J. C., Hahn, T. P., Maney, D. L., Schoech, S. J., Wada, M., and Morton, M. L. (2003). Effects of temperature on photoperiodically-induced reproductive development, circulating plasma luteinizing hormone and thyroid hormones, body mass, fat deposition and molt in mountain white-crowned sparrows, *Zonotrichia leucophrys oriantha*. *Gen. Comp. Endocrinol.* 131, 143–158.
134. Nalbandov, A. V. (1976). *Reproductive Physiology of Mammals and Birds*. Freeman Press, San Francisco.
135. Van Tienhoven, A. (1983). *Reproductive Physiology of the Vertebrates*. Cornell University Press, London.
136. Rosa-Molinar, E., Fritzsch, B., and Hendricks, S. E. (1996). Organizational-activational concept revisited: sexual differentiation in an atherinomorph teleost. *Horm. Behav.* 30, 563–575.
137. Owens, I. P. F., and Short, R. V. (1995). Hormonal basis of sexual dimorphism in birds: implications for new theories of sexual selection. *Trends Ecol. Evol.* 10, 44–47.
138. Witschi, E. (1961). Sex and secondary sexual characters. In *Biology and Comparative Physiology of Birds* (A. J. Marshall, Ed.), Vol. 2, pp. 115–168. Academic Press, New York.
139. Ball, G. F., and Balthazart, J. (2002). Neuroendocrine mechanisms regulating reproductive cycles and reproductive behavior in birds. In *Hormones, Brain and Behavior* (D. Pfaff, A. Arnold, A. Etgen, S. Fahrbach, and R. Rubin, Eds.), Vol. 2, pp. 649–798. Academic Press, New York.
140. Fernald, R. D. (2002). Social regulation of the brain: status, sex and size. In *Hormones, Brain and Behavior* (D. Pfaff, A. Arnold, A. Etgen, S. Fahrbach, and R. Rubin, Eds.), Vol. 2, pp. 435–444. Academic Press, New York.
141. Vandenbergh, J. G. (1994). Pheromones and mammalian reproduction. In *The Physiology of Reproduction* (E. Knobil and J. D. Neill, Eds.), p. 343. Raven Press, New York.
142. Lehrman, D. S., Brody, P., and Wortis, R. P. (1961). The presence of the mate and of nesting material as stimuli for the development of incubation behavior and for gonadotropin secretion in the ring dove (*Streptopelia risoria*). *Endocrinology* 68, 507–516.
143. Erickson, C. J., and Lehrman, D. S. (1964). Effects of castration of male ring doves upon ovarian activity of females. *J. Comp. Physiol. Psychol.* 58, 164–166.

144. Barfield, R. J. (1971). Gonadotropic hormone secretion in the female ring dove in response to visual and auditory stimulation. *J. Endocrinol.* 49, 305–310.
145. Erickson, C. J. (1970). Induction of ovarian activity in female ring doves by androgen treatment of castrated males. *J. Comp. Physiol. Psychol.* 71, 210–215.
146. Lott, D. S., Scholz, S. O., and Lehrman, D. S. (1967). Exteroreceptive stimulation of the reproductive system of the female ring dove (*Streptopelia risoria*) by the mate and by colony milieu. *Anim. Behav.* 15, 433–437.
147. Cheng, M. F. (1983). Behavioral "self feedback" control of endocrine states. In *Hormones and Behavior* (J. Balthazart, E. Pröve, and R. Gilles, Eds.), pp. 408–421. Springer-Verlag, Berlin.
148. Cohen, J., and Cheng, M. F. (1981). The role of the mid-brain in courtship behavior of female ring doves (*Streptopelia risoria*): evidence from radiofrequency lesion and hormone implant studies. *Brain Res.* 207, 297–301.
149. Crews, D. (1974). Effects of group stability, male-male aggression and male courtship behavior on environmentally induced ovarian recrudescence in the lizard, *Anolis carolinensis*. *J. Zool. (Lond).* 172, 419–441.
150. Crews, D., Rosenblatt, J. W., and Lehrman, D. S. (1974). Effects of unseasonal environmental regime, group presence, group composition, and male's physiologic state on ovarian recrudescence in the lizard, *Anolis carolinenesis*. *Endocrinology* 94, 541–547.
151. Crews, D. (1975). Effects of different components of male courtship behavior on environmentally induced ovarian recrudescence and mating preferences in the lizard, *Anolis carolinensis*. *Anim. Behav.* 23, 349–356.
152. Mendonça, M. T., and Crews, D. (1990). Mating-induced ovarian recrudescence in the red-sided garter snake. *J. Comp. Physiol. [A]* 166, 629–632.
153. Vandenbergh, J. G., and Drickamer, L. C. (1974). Reproductive coordination among free-ranging rhesus monkeys. *Physiol. Behav.* 13, 373–376.
154. Bronson, F. H. (1988). Seasonal regulation of reproduction in mammals. In *Physiology of Reproduction* (E. Knobil and J. D. Neill, Eds.), pp. 831–871. Raven Press, New York.
155. Wayne, N. L., Wade, G. N., and Rissman, E. F. (1991). Effects of food restriction and social cues on sexual maturation and growth in male musk shrews (*Suncus murinus*). *J. Reprod. Fertil.* 91, 385–392.
156. Burger, J. W. (1953). The effect of photic and psychic stimuli on the reproductive cycle of the male starling, *Sturnus vulgaris*. *J. Exp. Zool.* 124, 227–239.
157. Gwinner, E. (1975). Die circannuele Periodik der Fortspflanzungsactivität beim Star (*Sturnus vulgaris*) unter dem Einfluss gleichund andersgeschlectiger Artgenossen. *Z. Tierpsychol.* 38, 34–43.
158. Dufty, A. M. Jr., and Wingfield, J. C. (1986). Temporal patterns of circulating LH and steroid hormones in a brood parasite, the brown-headed cowbird, *Molothrus ater*: I. Males. *J. Zool. (Lond).* 208, 191–203.
159. Dufty, A. M. Jr., and Wingfield, J. C. (1986). Temporal patterns of circulating LH and steroid hormones in a brood parasite, the brown-headed cowbird, *Molothrus ater*: II. Females. *J. Zool. (Lond).* 208, 205–214.
160. Dufty, A. M. Jr., and Wingfield, J. C. (1986). The influences of social cues on reproductive endocrinology of male brown-headed cowbirds: field and laboratory studies. *Horm. Behav.* 20, 222–234.
161. Haase, E., Paulke, E., and Sharp. P. J. (1976). Effects of seasonal and social factors on testicular activity and hormone levels in domestic pigeons. *J. Exp. Zool.* 197, 81–96.
162. Darwin, C. (1871). *The Descent of Man, and Selection in Relation to Sex.* Murray, London.
163. Trivers, R. L. (1972). Parental investment and sexual selection. In *Sexual Selection and the Descent of Man* (B. G. Campbell, Ed.), pp. 136–179. Aldine, Chicago.
164. Searcy, W. A. (1982). The evolutionary effects of mate selection. *Annu. Rev. Ecol. Syst.* 13, 57–85.
165. Bottoni, L., Massa, R., Lea, R. W., and Sharp, P. J. (1993). Mate choice and reproductive success in the red-legged partridge (*Alectoris rufa*). *Horm. Behav.* 27, 308–317.
166. Bluhm, C. K., Phillips, R. E., and Burke, W. H. (1983). Serum levels of luteinizing hormone (LH), prolactin, estradiol and progesterone in laying and non-laying canvasback ducks. (*Aythya valisinaria*). *Gen. Comp. Endocrinol.* 52, 1–16.
167. Bluhm, C. K., Phillips, R. E., Burke, W. H., and Gupta, G. N. (1984). Effects of male courtship and gonadal steroids on pair formation, egg-laying and serum LH in canvasback ducks (*Aythya valisinaria*). *J. Zool. (Lond.)* 204, 185–200.
168. Bluhm, C. K. (1985). Social factors regulating avian endocrinology and reproduction. In *The Endocrine System and the Environment* (B. K. Follett, S. Ishii, and A. Chandola, Eds.), pp. 247–264. Japanese Scientific Societies Press, Tokyo, and Springer-Verlag, Berlin.
169. Adkins-Regan, E., and Ascenzi, M. (1987). Social and sexual behavior of male and female zebra finches treated with estradiol during the nestling period. *Anim. Behav.* 35, 1100–1112.
170. Adkins-Regan, E. (1988). Sex hormones and sexual orientation in animals. *Psychobiology* 16, 335–347.
171. Mansukhani, V., Adkins-Regan, E., and Yang, S. (1996). Sexual partner preference in female zebra finches: the role of early hormones and social environment. *Horm. Behav.* 30, 506–513.
172. Insel, T. R., and Hulihan, T. J. (1995). A gender specific mechanism for pair bonding: oxytocin and partner preference formation in monogamous voles. *Behav. Neurosci.* 109, 782–789.
173. Williams, J. R., Insel, T. J., Harbaugh, C. R., and Carter, C. S. (1992). Oxytocin administered centrally facilitates formation of a partner preference in female prairie voles (*Microtus ochrogaster*). *J. Neuroendocrinol.* 6, 247–250.
174. Cho, M. M., DeVries, A. C., Williams, J. R., and Carter, C. S. (1999). The effects of oxytocin and vasopressin on partner preferences in male and female prairie voles (*Microtus ochrogaster*). *Behav. Neurosci.* 113, 1071–1079.
175. Cushing, B. S., and Carter, C. S. (2000). Peripheral pulses of oxytocin increase partner preferences in females, but not male, prairie voles. *Horm. Behav.* 37, 49–56.
176. Cushing, B. S., Martin, J. O., Young, L. J., and Carter, C. S. (2001). The effects of peptides on partner preference formation are predicted by habitat in prairie voles. *Horm. Behav.* 39, 48–58.
177. Sugiyama, Y. (1965). On the social change of Hanuman langurs (*Presbytis entellus*) in their natural conditions. *Primates* 6, 381–418.
178. Hausfater, G., and Hrdy, S., Eds. (1984). *Infanticide: Comparative and Evolutionary Perspectives*. Aldine, New York.
179. Gandelman, R. (1983). Hormones and infanticide. In *Hormones and Aggressive Behavior* (B. B. Svare, Ed.), pp. 105–118. Plenum Press, New York.
180. Adler, N. T., and Allen, T. O. (1983). The origin of sexual behavior. In *Handbook of Behavioral Neurobiology*, Vol. 6, *Motivation* (E. Satinoff and P. Teitelbaum, Eds.). Plenum Press, New York.
181. Floody, O. R. (1981). The hormonal control of ultrasonic communication in rodents. *Am. Zool.* 21, 129–142.
182. Gould, E. (1983). Mechanisms of mammalian auditory communication. In *Advances in the Study of Mammalian Behavior* (J. F. Eisenberg and D. G. Kleiman, Eds.), Special Publication 6, pp. 265–342. American Society of Mammalogists, Shippensburg, PA.

183. Barfield, R. J., Auerbach, P., Geyer, L. A., and MacIntosh, T. K. (1979). Ultrasonic vocalizations in rat sexual behavior. *Am. Zool.* 19, 469–480.
184. Barfield, R. J., and Thomas, D. A. (1986). The role of ultrasonic vocalizations in the regulation of reproduction in rats. *Ann. N Y Acad. Sci.* 474, 33–43.
185. Hinde, R. A., and Steel, E. (1978). The influence of day length and male vocalizations on the estrogen-dependent behavior of female canaries and budgerigars with discussion of data from other species. In *Advances in the Study of Behavior* (J. Rosenblatt, R. A. Hinde, C. Beer, and M.-C. Busnel, Eds.), pp. 39–73. Academic Press, New York.
186. Kroodsma, D. E. (1976). Reproductive development in a female songbird: differential stimulation by quality of male song. *Science* 192, 574–575.
187. Guyomarc'h, C., and Guyomarc'h, J.-C. (1984). The influence of social factors on the onset of egg production in Japanese quail (*Coturnix coturnix japonica*). *Biol. Behav.*9, 333–342.
188. Noble, G. K., and Aronson, L. R. (1942). The sexual behavior of Anura: I. The normal mating pattern of *Rana pipiens. Bull. Am. Mus. Nat. Hist.* 80, 127–142.
189. Organ, J. A. (1960). The courtship and spermatophore of the salamander *Plethodon glutinosus. Copeia* 1, 34–40.
190. Moore, F. L., McCormack, C., and Swanson, L. (1979). Induced ovulation: effects of sexual behavior and insemination on ovulation and progesterone levels in *Taricha granulosa. Gen. Comp. Endocrinol.* 39, 262–269.
191. Egami, N., and Nambu, M. (1961). Factors initiating mating behavior and oviposition in the fish, *Oryzias latipes. J. Fac. Sci. Univ. Tokyo* Section IV(9 Part 2), 263–278.
192. Bronson, F. H., and Desjardins, C. (1982). Endocrine responses to sexual arousal in male mice. *Endocrinology* 111, 1286–1291.
193. Garstka, W. R., Camazine, B., and Crews, D. (1982). Interactions of behavior and physiology during the annual reproductive cycle of the red-sided garter snake (*Thamnophis sirtalis parietalis*). *Herpetology* 38, 104–123.
194. Wallen, K., Winston, L. A., Gaventa, S., Davis-Da-Silva, M., and Collins, D. C. (1984). Periovulatory changes in female sexual behavior and patterns of ovarian steroid secretion in group-living rhesus monkeys. *Horm. Behav.* 18, 431–450.
195. Goldfoot, D. A. (1981). Olfaction, sexual behavior and the pheromone hypothesis in rhesus monkeys: a critique. *Am. Zool.* 21, 153–164.
196. Keverne, E. B. (1985). Hormones and the sexual behavior of monkeys. In *Neurobiology* (R. Gilles and J. Balthazart, Eds.), pp. 37–47. Springer-Verlag, Berlin.
197. Michael, R. P., and Keverne, E. B. (1968). Pheromones and the communication of sexual status in primates. *Nature* 218, 746–749.
198. Bielert, C. (1982). Experimental examinations of baboon (*Papio ursinus*) sex stimuli. In *Primate Communication* (C. T. Snowdon, C. H. Brown, and M. R. Peterson, Eds.), pp. 373–395. Cambridge University Press, Cambridge.
199. Eberhart, J. A., Keverne, E. B., and Meller, R. E. (1980). Social influences on plasma testosterone levels in male talapoin monkeys. *Horm. Behav.* 14, 247–266.
200. Feder, H. H., Storey, A., Goodwin, D., Reboulleau, C., and Silver, R. (1977). Testosterone and 5-α-testosterone levels in peripheral plasma of male and female ring doves (*Streptopelia risoria*) during the reproductive cycle. *Biol. Reprod.* 16, 666–677.
201. Moore, M. C. (1983). Effect of female sexual displays on the endocrine physiology and behavior of male white-crowned sparrows, *Zonotrichia leucophrys. J. Zool. (Lond)* 199, 137–148.
202. Moore, M. C. (1982). Hormonal responses of free-living male white-crowned sparrows to experimental manipulation of female sexual behavior. *Horm. Behav.* 16, 323–329.
203. Wingfield, J. C., Ronchi, E., Marler, C., and Goldsmith, A. R. (1989). Interactions of steroids and prolactin during the reproductive cycle of the song sparrow, *Melospiza melodia. Physiol. Zool.* 62, 11–24.
204. O'Connell, M. E., Reboulleau, C., Feder, H. H., and Silver, R. (1981). Social interactions and androgen levels in birds: I. Female characteristics associated with increased plasma androgen levels in the male ring dove (*Streptopelia risoria*). *Gen. Comp. Endocrinol.* 44, 454–463.
205. Diakow, C., and Raimondi, D. (1981). Physiology of *Rana pipiens* reproductive behavior: a proposed mechanisms for inhibition of the release call. *Am. Zool.* 21, 295–304.
206. Itoh, M., and Ishii, S. (1990). Changes in plasma levels of gonadotropins and sex steroids in the toad, *Bufo japonicus*, in association with behavior during the breeding season. *Gen. Comp. Endocrinol.* 80, 451–464.
207. Baerends, G. P., Brower, R., and Waterbolk, H. T. (1955). Ethological studies on *Lebistes reticularis* (Peters): 1. An analysis of the male courtship pattern. *Behaviour* 8, 249–334.
208. Tavolga, W. N. (1956). Visual, chemical and sound stimuli in the sex discriminatory behavior of the gobiid fish, *Bathygobius soporator. Zoologica* 41, 49–64.
209. Kyle, A. L., Stacey, N. E., Peter, R. E., and Billard, R. (1985). Elevations in gonadotropin concentrations and milt volumes as a result of spawning behavior in the goldfish. *Gen. Comp. Endocrinol.* 57, 10–22.
210. Partridge, B. L., Liley, N. R., and Stacey, N. E. (1976). The role of pheromones in the sexual behavior of the goldfish. *Anim. Behav.* 24, 291–299.
211. Stacey, N. E., Sorenson, P., van der Kraak, G. J., and Dulka, J. G. (1989). Direct evidence that 17-α, 20-β-dihydroxy-4-pregnen-3-one functions as a goldfish primer pheromone: preovulatory release is closely associated with male endocrine responses. *Gen. Comp. Endocrinol.* 75, 62–70.
212. Meyer, J. H., and Liley, N. R. (1982). The control of production of a sexual pheromone in the female guppy, *Poecilia reticulata. Can. J. Zool.* 60, 1505–1510.
213. Clutton-Brock, T. H. (1991). *The Evolution of Parental Care.* Monographs in Behavior and Ecology. Princeton University Press, Princeton, NJ.
214. Skutch, A. F. (1976). *Parent Birds and Their Young.* University of Texas Press, Austin, TX.
215. Marler, C. A., Bester-Meredith, J. K., and Trainor, B. C. (2003). Paternal behavior and aggression: endocrine mechanisms and nongenomic transmission of behavior. *Adv. Stud. Behav.* 32, 263–323.
216. Ziegler, T. E. (2000). Hormones associated with non-maternal infant care: a review of mammalian and avian studies. *Folia Primatol.* 71, 6–21.
217. Gonzalez-Mariscal, G., and Poindron, P. (2002). Parental care in mammals: immediate internal and sensory factors of control. In *Hormones, Brain and Behavior* (D. Pfaff, A. Arnold, A. Etgen, S. Fahrbach, and R. Rubin, Eds.), Vol. 1, pp. 215–298. Academic Press, New York.
218. De Vlaming, V. L. (1979). Actions of prolactin among the vertebrates. In *Hormones and Evolution* (E. J. W. Barrington, Ed.), pp. 561–642. Academic Press, New York.
219. Buntin, J. D. (1996). Neural and hormonal control of parental behavior in birds. *Adv. Stud. Behav.* 25, 161–213.
220. Numan, M. (1994). Maternal behavior. In *The Physiology of Reproduction* (E. Knobil and J. D. Neill, Eds.), Vol. 2, pp. 221–302. Raven Press, New York.
221. Wingfield, J. C., Hegner, R. E., Dufty, A. M. Jr., and Ball, G. F. (1990). The "challenge hypothesis": theoretical implications for patterns of testosterone secretion, mating systems, and breeding strategies. *Am. Naturalist* 136, 829–846.

222. El Halawani. M. E., and Rozenboim, I. (1993). Incubation behavior in the turkey: molecular and endocrinological implications. In *Avian Endocrinology* (P. J. Sharp, Ed.), pp. 99–110. Journal of Endocrinology, Ltd., Bristol, United Kingdom.
223. Goldsmith, A. R. (1991). Prolactin and avian reproductive strategies. In *Acta 20th Congr. Int. Ornithol.*, pp. 2063–2071.
224. Wingfield, J. C., and Farner, D. S. (1979). Some endocrine correlates of renesting after loss of clutch or brood in the white-crowned sparrow, *Zonotrichia leucophrys gambelii. Gen. Comp. Endocrinol.* 38, 322–331.
225. Donham, R. S., Dane, C. W., and Farner, D. S. (1976). Plasma luteinizing hormone and the development of ovarian follicles after loss of clutch in female mallards (*Anas platyrhynchos*). *Gen. Comp. Endocrinol.* 29: 152–155.
226. Wingfield, J. C., and Farner, D. S. (1993). The endocrinology of wild species. In *Avian Biology* (D. S. Farner, J. R. King, and K. C. Parkes, Eds.), Vol. 9, pp. 163–327. Academic Press, New York.
227. Wingfield, J. C. (1985). Short-term changes in plasma levels of hormones during establishment and defense of a breeding territory in male song sparrows, *Melospiza melodia. Horm. Behav.* 19, 174–187.
228. Moore, M. C. (1986). Elevated testosterone levels during non-breeding season territoriality in a fall breeding lizard, *Sceloperous jarrovi. J. Comp. Physiol. [A]* 158, 159–163.
229. Moore, M. C. (1987). Circulating steroid hormones during rapid aggressive responses of territorial male mountain spiny lizards, *Sceloporus jarrovi. Horm. Behav.* 21, 511–521.
230. Greenberg, N., and Crews, D. (1990). Endocrine and behavioral responses to aggression and social dominance in the green anole lizard, *Anolis carolinensis. Gen. Comp. Endocrinol.* 77, 246–255.
231. Archer, J. (2005). Testosterone and human aggression: an evaluation of the challenge hypothesis. *Neurosci. Biobehav. Rev.* (*in press*).
232. Oliveira, R. F. (1998). Of fish and men: a comparative approach to androgens and social-dominance. *Behav. Brain Sci.* 21, 383.
233. Oliveira, R. F., Lopes, M., Carneiro, L. A., and Canario, A. V. M. (2001). Watching fights raises fish hormone levels. *Nature* 409, 475.
234. Goodson, J. L. (1998). Territorial aggression and dawn song are modulated by septal vasotocin and vasoactive intestinal peptide in male field sparrows (*Spizella pusilla*). *Horm. Behav.* 34, 67–77.
235. Goodson, J. L. (1998). Vasotocin and vasoactive intestinal polypeptide modulate aggression in a territorial songbird, the violet-eared waxbill (Estrildidae: *Uraeginthus granatina*). *Gen. Comp. Endocrinol.* 111, 233–244.
236. Goodson, J. L., and Bass, A. H. (2001). Social behavior functions and related anatomical characteristics of vasotocin/vasopressin systems in vertebrates. *Brain Res. Rev.* 35, 246–265.
237. Lisk, R. D., Ciaccio, L. A., and Catanzaro, D., (1983). Mating behavior of the golden hamster under semi-natural conditions. *Anim. Behav.* 31, 659–666.
238. Takahashi, L. K., and Lisk, R. D. (1983). Organization and expression of agonistic and socio-sexual behavior in golden hamsters over the estrous cycle and after ovariectomy. *Physiol. Behav.* 31, 477–482.
239. Takahashi, L. K., and Lisk, R. D. (1984). Intrasexual interactions among female golden hamsters (*Mesocricetus auratus*) over the estrous cycle. *J. Comp. Physiol.* 98, 267–275.
240. Fivizzani, A. J., and Oring, L. W. (1986). Plasma steroid hormones in the polyandrous spotted sandpiper, *Actitis macularia. Biol. Reprod.* 35, 1195–1201.
241. Fivizzani, A. J., Colwell, M. A., and Oring, L. W. (1986). Plasma steroid hormone levels in free-living Wilson's phalaropes, *Phalaropus tricolor. Gen. Comp. Endocrinol.* 62, 137–144.
242. Wingfield, J. C., Newman, A., Hunt, G. L., Jr., and Farner, D. S. (1982). Endocrine aspects of female-female pairing in the western gull (*Larus occidentalis wymani*). *Anim. Behav.* 30, 9–22.
243. Bernstein, I. S., Rose, R. M., Gordon, T. P., and Grady, C. L. (1979). Agonistic rank, aggression, social context, and testosterone in male pigtail monkeys. *Aggress. Behav.* 5, 329–399.
244. Bernstein, I. S., Gordon, T. P., and Rose, R. M. (1983). The interaction of hormones, behavior and social context in non-human primates. In *Hormones and Aggressive Behavior* (B. B. Svare, Ed.), pp. 535–562. Plenum Press, New York.
245. Rose, R. M., Bernstein, I. S., and Gordon, T. P. (1975). Consequences of social conflict on plasma testosterone levels in rhesus monkeys. *Psychosom. Med.* 37, 50–61.
246. Sapolsky, R. M. (1985). Stress-induced suppression of testicular function in the wild baboon: role of glucocorticoids. *Endocrinology* 116, 16.
247. Sapolsky, R. M. (1982). The endocrine stress-response and social status in the wild baboon. *Horm. Behav.* 16, 279–292.
248. Balthazart, J., Marra, R., and Negri-Cesi, P. (1979). Photoperiodic control of testosterone metabolism, plasma gonadotropins, cloacal gland growth, and reproductive behavior in the Japanese quail. *Gen. Comp. Endocrinol.* 39, 222–235.
249. Tsutsui, K., and Ishii, S. (1981). Effects of sex steroids on aggressive behavior of adult male Japanese quail. *Gen. Comp. Endocrinol.* 44, 480–486.
250. Ramenofsky, M. (1984). Agonistic behavior and endogenous plasma hormones in the male Japanese quail. *Anim. Behav.* 32, 698–708.
251. Runfeldt, S., and Wingfield, J. C. (1985). Experimentally prolonged sexual activity in female sparrows delays termination of reproductive activity in their untreated mates. *Anim. Behav.* 33, 403–410.
252. Crews, D. (1984). Gamete production, sex hormone secretion, and mating behavior uncoupled. *Horm. Behav.* 18, 22–28.
253. Crews, D., and Moore, M. C. (1986). Evolution of mechanisms controlling mating behavior. *Science* 231, 121–125.
254. Crews, D. (1987). Diversity and evolution of behavioral controlling mechanisms. In *Psychobiology of Reproductive Behavior* (D. Crews, Ed.), pp. 88–119. Prentice-Hall, Englewood Cliffs, NJ.
255. Dickhoff, W. W. (1989). Salmonids and annual fishes: death after sex. In *Development, Maturation, and Senescence of Neuroendocrine Systems: A Comparative Approach* (M. P. Schreibman and C. G. Scanes, Eds.), pp. 253–266. Academic Press, New York.
256. Braithwaite, R. W., and Lee, A. K. (1979). A mammalian example of semelparity. *Am. Naturalist* 113, 151–155.
257. Larsen, L. O. (1985). The role of hormones in reproduction and death in lampreys and other species which reproduce once and die. In *Current Trends in Comparative Endocrinology* (B. Lofts and W. N. Holmes, Eds.), pp. 613–616. Academic Press, New York.
258. Donaldson, E. M., and Fagerlund, U. H. M. (1968). Changes in the cortisol dynamics of sockeye salmon (*Oncorhynchus nerka*) resulting from sexual maturation. *Gen. Comp. Endocrinol.* 11, 552–561.
259. Fagerlund, U. H. M., and McBride, J. R. (1969). Suppression by dexamethasone of interrenal activity in adult sockeye salmon (*Oncorhynchus nerka*). *Gen. Comp. Endocrinol.* 12, 651–657.
260. Sower, S. A., and Schreck, C. B. (1982). Steroid and thyroid hormones during sexual maturation of Coho salmon (*Oncorhynchus kisutch*) in seawater or freshwater. *Gen. Comp. Endocrinol.* 47, 42–53.

261. Donaldson, E. M., and Fagerlund, U. H. M. (1970). Effects of sexual maturation and gonadectomy at sexual maturity on cortisol secretion rate in sockeye salmon (*Oncorhynchus nerka*). *J. Fish Res. Bd. Can.* 27, 2287–2296.
262. Fagerlund, U. H. M., and Donaldson, E. M. (1969). The effect of androgens on the distribution and secretion of cortisol in gonadectomized male sockeye salmon (*Oncorhynchus nerka*). *Gen. Comp. Endocrinol.* 12, 438–448.
262a. Wingfield, J. C., and Sapolsky, R. M. (2003). Reproduction and resistance to stress: when and how. *J. Neuroendocrinol.* 15, 711–724.
263. Knoebl, I., Fitzpatrick, M. S., and Schreck, C. B. (1996). Characterization of a glucocorticosteroid receptor in the brains of Chinook salmon, *Oncorhynchus tshawytscha*. *Gen. Comp. Endocrinol.* 101, 195–204.
264. Carruth, L. L., Jones, R. E., and Norris, D. O. (2000). Cell density and intracellular translocation of glucocorticoid receptor-immunoreactive neurons in the kokanee salmon (*Oncorhynchus nerka kennerlyi*) brain, with an emphasis on the olfactory system. *Gen. Comp. Endocrinol.* 117, 66–76.
265. Wood, D. H. (1970). An ecological study of *Antechinus stuartii* (Marsupalia) in a southeast Queensland rain forest. *Aust. J. Zool.* 18, 185–207.
266. Woolley, P. (1966). Reproduction in *Antechinus* spp. and other dasyurid marsupials. *Symp. Zool. Soc. Lond.* 15, 281–294.
267. Barnett, J. L. (1973). A stress response in *Antechinus stuartii* (MacLeay). *Aust. J. Zool.* 21, 501–513.
268. Bradley, A. J., McDonald, I. R., and Lee, A. K. (1976). Corticosteroid-binding globulin and mortality in a dasyurid marsupial. *J. Endocrinol.* 70, 323–324.
269. Bradley, A. J., McDonald, I. R., and Lee, A. K. (1980). Stress and mortality in a small marsupial (*Antechinus stuartii*, MacLeay). *Gen. Comp. Endocrinol.* 40, 188–200.
270. McDonald, I. R., Lee, A. K., Bradley, A. J., and Than, K. A. (1981). Endocrine changes in dasyurid marsupials with differing mortality patterns. *Gen. Comp. Endocrinol.* 44, 292–301.
271. McDonald, I. R., Lee, A. K., Than, K. A., and Martin, R. W. (1986). Failure of glucocorticoid feedback in males of a population of small marsupials (*Antechinus swainsonii*) during the period of mating. *J. Endocrinol.* 108, 63–68.
272. Barker, I. K., Beveridge, I., Bradley, A. J., and Lee, A. K. (1978). Observations on spontaneous stress-related mortality among males of the dasyurid marsupial *Antechinus stuartii* MacLeay. *Aust. J. Zool.* 26, 435–447.
273. Woollard, P. (1971). Differential mortality of *Antechinus stuartii* (MacLeay): nitrogen balance and somatic changes. *Aust. J. Zool.* 19, 347–353.
274. Bradley, A. J. (1987). Stress and mortality in the red-tailed phasogale, *Phasogale calura* (Marsupalia: Dasyuridae). *Gen. Comp. Endocrinol.* 67, 85–100.
275. Bradley, A. J. (1985). Steroid binding proteins and life history in marsupials. In *Current Trends in Comparative Endocrinology* (B. Lofts and W. N. Holmes, Eds.), pp. 269–270, Hong Kong University Press, Hong Kong.
276. Oakwood, M., Bradley, A. J., and Cockburn, A. (2001). Semelparity in a large marsupial. *Proc. R. Soc. Lond. B Biol. Sci.* 268, 407–411.
277. Schmidt, L. H., Bradley, A. J., Kemper, M., Kitchener, D. J., Humphreys, D. F., and How, A. (1989). Ecology and physiology of the Northern quoll, *Dasyurus hallucatus* (Marsupalia, Dasyuridae), at Mitchell Plateau, Kimberly, Western Australia. *J. Zool. (Lond.)* 217, 539–558.
277a. Wingfield, J. C., and Ramenofsky, M. (1985). Hormonal and environmental control of aggression in birds. In *Neurobiology* (R. Gilles and J. Balthazart, Eds.), pp. 92–104. Springer-Verlag, Berlin.
278. Wingfield, J. C., and Wada, M. (1989). Male-male interactions increase both luteinizing hormone and testosterone in the song sparrow, *Melospiza melodia*: specificity, time course and possible neural pathways. *J. Comp. Physiol. [A]* 166, 189–194.
279. Burmeister, S., and Wilczynski, W. (2000). Social signals influence hormones independently of calling behavior in the tree frog (*Hyla cinerea*). *Horm. Behav.* 38, 201–209.
280. Oliveira, R. F., and Almada, V. C. (1998). Androgenization of dominant males in a cichlid fish: androgens mediate the social modulation of sexually dimorphic traits. *Ethology* 104, 841–858.
281. Pankhurst, N. W., and Barnett, C. W. (1993). Relationship of population density, territorial interaction and plasma levels of gonadal steroids in spawning male demoiselles *Chromis dispilus* (Pisces, Pomacentridae). *Gen. Comp. Endocrinol.* 90, 168–176.
282. Cardwell, J. R., and Liley, N. R. (1991). Hormonal control of sex and color change in the stoplight parrotfish, *Sparisoma viride*. *Gen. Comp. Endocrinol.* 81, 7–20.
283. Cardwell, J. R., and Liley, N. R. (1991). Androgen control of social status in males of a wild population of stoplight parrotfish, *Sparisoma viride* (Scaridae). *Horm. Behav.* 25, 1–18.
284. Bernhardt, P. C., Dabbs, J. M., Jr., Fielden, J. A., and Lutter, C. D. (1998). Testosterone changes during vicarious experiences of winning and losing among fans at sporting events. *Physiol. Behav.* 65, 59–62.
285. Carlson, B. A., Hopkins, C. D., and Thomas, P. (2000). Androgen correlates of socially induced changes in the electric organ discharge waveform of a mormyrid fish. *Horm. Behav.* 38, 177–186.
286. Oliveira, R. F., McGregor, P. K., and Latruffe, C. (1998). Know thine enemy: fighting fish gather information from observing conspecific interactions. *Proc. R. Soc. Lond. B Biol. Sci.* 265, 1045–1049.
287. Beletsky, L. D., Gori, D. F., Freeman, S., and Wingfield, J. C. (1995). Testosterone and polygyny in birds. *Curr. Ornithol.* 12, 1–41.
288. Ketterson, E. D., Nolan, V., Jr., Wolf, L., and Ziegenfus, C. (1992). Testosterone and avian life histories: effects of experimentally elevated testosterone on behaviour and correlates of fitness in the dark-eyed junco (*Junco hyemalis*). *Am. Naturalist* 140, 980–999.
289. Ketterson, E. D., Nolan, V., Jr., Cawthorn, M. J., Parker, P. G., and Ziegenfus, C. (1996). Phenotypic engineering: using hormones to explore the mechanistic and functional bases of phenotypic variation in nature. *Ibis* 138, 70–86.
290. Hirschenhauser, K., Winkler, H., and Oliveira, R. F. (2003). Comparative analysis of male androgen responsiveness to social environment in birds: the effects of mating system and paternal incubation. *Horm. Behav.* 43, 508–519.
291. Hews, D. K., Thompson, C. W., Moore, I. T., and Moore, M. C. (1997). Population frequencies of alternative male phenotypes in tree lizards: geographic variation and common-garden rearing studies. *Behav. Ecol. Sociobiol.* 41, 371–380.
292. Thompson, C. W., and Moore, M. C. (1991). Throat color reliably signals status in male tree lizards, *Urosaurus ornatus*. *Anim. Behav.* 42, 745–753.
293. Knapp, R. and Moore, M. C. (1996). Male morphs in tree lizards, *Urosaurus ornatus*, have different delayed responses to aggressive encounters. *Anim. Behav.* 52, 1045–1055.
294. Jennings, D. H., Moore, M. C., Knapp, R., Matthews, L., and Orchinik, M. (2000). Plasma steroid-binding globulin mediation of differences in stress reactivity in alternative male phenotypes in tree lizards, *Urosaurus ornatus*. *Gen. Comp. Endocrinol.* 120, 289–299.

295. Klukowski, M., and Nelson, C. E. (1998). The challenge hypothesis and seasonal changes in aggression and steroids in male northern fence lizards (*Sceloporus undulatus hyacinthinus*). *Horm. Behav.* 33, 197–204.
296. Ryan, M. (1985). *The Tungara Frog.* University of Chicago Press, Chicago.
297. Bucher, T., Ryan, M., and Bartholomew, G. (1982). Oxygen consumption during resting, calling, and nest building in the frog, *Physalaemus pustulosus*. *Physiol. Zool.* 55, 10–22.
298. Emerson, S. B., and Hess, D. L. (2001). Glucocorticoids, androgens, testis mass, and the energetics of vocalization in breeding male frogs. *Horm. Behav.* 39, 59–69.
299. Emerson, S. B. (2001). Male advertisement calls: behavioral variation and physiological processes. In *Anuran Communication* (M. J. Ryan, Ed.), pp. 36–44. Smithsonian Institution Press, Washington, DC.
300. Marler, C., and Ryan, M. (1996). Energetic constraints and steroid hormone correlates of male calling behavior in the Tungara frog. *J. Zool. (Lond.)* 240, 397–409.
301. Wingfield, J. C., and Ramenofsky, M. (1999). Hormones and the behavioral ecology of stress. In *Stress Physiology in Animals* (P. H. M. Balm, Ed.), pp. 1–51. Sheffield Academic Press, Sheffield, UK.
302. Romero, L. M. (2002). Seasonal changes in plasma glucocorticoid concentrations in free-living vertebrates. *Gen. Comp. Endocrinol.* 128, 1–24.
303. Demski, L. S. (1987). Diversity in reproductive patterns and behavior in teleost fishes. In *Psychobiology of Reproduction* (D. Crews, Ed.), pp. 1–27. Prentice-Hall, Englewood Cliffs, NJ.
304. Harrington, R. W. Jr. (1961). Oviparous hermaphroditic fish with internal self fertilization. *Science* 134, 1749–1750.
305. Reinboth, R. (1980). Can sex reversion be environmentally induced? *Biol. Reprod.* 22: 49–59.
306. Shapiro, D. Y. (1984). Sex reversal and sociodemographic processes in coral reef fishes. In *Fish Reproduction* (G. W. Potts and R. J. Mouton, Eds.), pp. 103–118. Academic Press, London.
307. Reavis, R. H., and Grober, M. S. (1999). An integrative approach to sex change: social, behavioral and neurochemical changes in *Lythrypnus dalli* (Pisces). *Acta Ethol.* 2, 51–60.
308. Shapiro, D. Y., and Boulon, R. H., Jr. (1982). The influence of females on the initiation of female to male sex change in a coral reef fish. *Horm. Behav.* 16, 66–75.
309. Perry, A. N., and Grober, M. S. (2003). A model for social control of sex change: interactions of behavior, neuropeptides, glucocorticoids, and sex steroids. *Horm. Behav.* 43, 31–38.
310. Robertson, D. R. (1972). Social control of sex reversal in a coral fish . *Science* 177, 1007–1009.
311. Cochran, R. C., and Grier, H. J. (1991). Regulation of sexual succession in the protogynous black sea bass, *Centropristis striatus* (Osteichthys: Serrandiae). *Gen. Comp. Endocrinol.* 82, 69–77.
312. Hourigan, T. F., Nakamura, M., Nagahama, Y., Yamauchi, K., and Grau, G. E. (1991). Histology, ultrastructure and in vitro steroidogenesis of the testes of two male phenotypes of the protogynous fish, *Thalassoma duperrey* (Labridae). *Gen. Comp. Endocrinol.* 83, 193–217.
313. Godwin, J., Sawby, R., Warner, R. R., Crews, D., and Grober, M. S. (2000). Hypothalamic arginine vasotocin mRNA abundance variation across sexes and with sex change in a coral reef fish. *Brain Behav. Evol.* 55, 77–84.
314. Semsar, K., and Godwin, J. (2003). Social influences on the arginine vasotocin system are independent of gonads in a sex-changing fish. *J. Neurosci.* 23, 4386–4393.
315. Semsar, K., and Godwin, J. (2004). Multiple mechanisms of phenotype development in the bluehead wrasse. *Horm. Behav.* 45, 345–353.
316. Godwin, J., Crews, D., and Warner, R. R. (1996). Behavioral sex change in the absence of gonads in a coral reef fish. *Proc. R. Soc. Lond. B Biol. Sci.* 263, 1683–1688.
317. Yeung, W. S. B., and Chan, S. T. H. (1985). The in vitro metabolism of radioactive progesterone and testosterone by the gonads of the protandrous *Rhabdosargus sarba* at various sexual phases. *Gen. Comp. Endocrinol.* 59, 171–183.
318. Chang, C. F., Lin, B. Y., Lau, E. L., Lee, M. F., Yueh, W. S., Lee, Y. H., Chang, C. N., Huang, J. D., Tacon, P., Lee, F. Y., Du, J. L., and Sun, L. T. (1997). The endocrine mechanism of sex reversal in the proandrous black porgy, *Acanthopargus schlegeli*: a review. *Chin. J. Physiol.* 40, 197–205.
319. Crews, D., Gustafson, J. E., and Tokarz, R. (1983). Psychobiology of parthenogenesis in reptiles. In *Lizard Ecology* (R. B. Huey, E. R. Pianka, and T. W. Schoener, Eds.), pp. 205–231. Harvard University Press, Cambridge, MA.
320. Cuellar, O. (1971). Reproduction and mechanism of restitution in the parthenogenetic lizard *Cnemidophorus uniparens*. *J. Morphol.* 133, 139–166.
321. Moore, M. C., Whittier, J. M., and Crews, D. (1985). Sex steroid hormones during the ovarian cycle of an all-female parthenogenetic lizard and their correlation with pseudosexual behavior. *Gen. Comp. Endocrinol.* 60, 144–153.
322. Crews, D., and Fitzgerald, K. T. (1980). Sexual behavior in parthenogenetic lizards (*Cnemidophorus uniparens*). *Proc. Natl. Acad. Sci. U. S. A.* 77, 499–502.
323. Moore, M. C., Whittier, J. M., Billy, A. J., and Crews, D. (1985). Male-like behavior in an all-female lizard: relationship to ovarian cycle. *Anim. Behav.* 33, 284–289.
324. Godwin, J., Hartman, V., Grammer, M., and Crews, D. (1996). Progesterone inhibits female-typical receptive behavior and decreases hypothalamic estrogen and progesterone receptor messenger ribonucleic acid levels in whiptail lizards (genus *Cnemidophorus*). *Horm. Behav.* 30, 138–144.
325. Young, L. J., Lopreato, G. F., Horan, K., and Crews, D. (1994). Cloning and in situ hybridization analysis of estrogen receptor, progesterone receptor, and androgen receptor expression in the brain of whiptail lizards (*Cnemidophorus uniparens* and *C. inornatus*). *J. Comp. Neurol.* 347, 288–300.
326. Schultz, R. J. (1969). Hybridization, unisexuality and polyploidy in the teleost, *Poeciliopsis* (Poecililidae) and other vertebrates. *Am. Naturalist* 103, 605–619.
327. Schultz, R. J. (1973). Unisexual fish: laboratory synthesis of a "species." *Science* 179, 180–181.
328. Wickler, W. (1972). *The Sexual Code.* Doubleday, New York.
329. Ploog, D. W., Blitz, J., and Ploog, F. (1963). Studies on the social and sexual behavior of the squirrel monkey (*Saimiri sciurensis*). *Folia Primatol.* 1, 29–66.
330. Racey, P. A. and Skinner, J. D. (1979). Endocrine aspects of social mimicry in spotted hyenas, *Crocuta crocuta*. *J. Zool. (Lond).* 187, 315–316.
331. Licht, P., Frank, L. G., Yalcinkaya, T. M., Siiteri, P. K., and Glickman, S. E. (1992). Hormonal correlates of "masculinization" in female spotted hyenas (*Crocuta crocuta*): II. Maternal and fetal steroids. *J. Reprod. Fertil.* 95, 463–474.
332. Licht, P., Hayes, T., Tsai, P., Cunha, G., Kim, H., Golbus, M., Hayward, S., Martin, M. C., Jaffe, R. B., and Glickman, S. E. (1998). Androgens and masculinization of genitalia in the spotted hyena (*Crocuta crocuta*): I. Urogenital morphology and placental androgen production during fetal life. *J. Reprod. Fertil.* 113, 105–116.
333. Frank, L. G., Glickman, S. G., and Licht, P. (1991). Fatal sibling aggression, precocial development, and androgens in neonatal spotted hyenas. *Science* 252, 702–704.
334. Goymann, W., East, M. L., and Hofer, H. (2001). Androgens and the role of female "hyperaggressiveness" in spotted hyenas (*Crocuta crocuta*). *Horm. Behav.* 39, 83–92.

335. Verrell, P. A. (1983). The influence of ambient sex ratio and intermale competition on the sexual behavior of the red-spotted newt, *Notopthalmus viridescens* (Amphibia, Urodela, Salamandridae). *Behav. Ecol. Sociobiol.* 13, 307–313.
336. Crews, D., and Garstka, W. R. (1982). The ecological physiology of a garter snake. *Sci. Am.* 247, 158–168.
337. Mason, R. T., and Crews, D. (1985). Female mimicry in garter snakes. *Nature* 316, 59–60.
338. Gross, M. R. (1984). Sunfish, salmon and the evolution alternate reproductive strategies and tactics in fish. In *Fish Reproduction: Strategies and Tactics* (R. J. Wooton and G. Potts, Eds.), pp. 55–75. Academic Press, New York.
339. Dominey, W. J. (1980). Female mimicry in male bluegill sunfish: a genetic polymorphism? *Nature* 284, 546–548.
340. Oliveira, R. F., Canario, A. V. M., Grober, M. S., and Santos, R. S. (2001). Endocrine correlates of male polymorphism and alternative reproductive tactics in the Azorean rock pool blenny, *Parablennius saguinolentus parvicornis*. *Gen. Comp. Endocrinol.* 121, 278–288.
341. Van der Crommenacker, J., Richardson, D. S., Groothuis, T. G. G., Eising, C. M., Dekker, A. L. and Komdeur, J. (2004). Testosterone, cuckoldry risk and extra-pair opportunities in the Seychelles warbler. *Proc. R. Soc. Lond. B Biol. Sci.* 271, 1023–1031.
342. Knapp, R., Hews, D. K., Thompson, C. W., Ray, L. E., and Moore, M. C. (2003). Environmental and endocrine correlates of tactic switching by nonterritorial male tree lizards (*Urosaurus ornatus*). *Horm. Behav.* 43, 83–92.
343. Moore, M. C., Thompson, C. W., and Marler, C. A. (1991). Reciprocal changes in corticosterone and testosterone levels following acute and chronic handling stress in the tree lizard, *Urosaurus ornatus*. *Gen. Comp. Endocrinol.* 81, 217–226.
344. Hunt, G. L., Jr., Wingfield, J. C., Newman, A., and Farner, D. S. (1980). Sex ratio of western gulls on Santa Barbara Island. *Auk* 97, 473–479.
345. Emlen, S. T. (1991). The evolution of cooperative breeding in birds and mammals. In *Behavioral Ecology: An Evolutionary Approach* (J. R. Krebs and N. B. Davies, Eds.), pp. 301–337. Oxford University Press, Oxford.
346. Oliveira, R. F., Hirchenhauser, K., Canario, A. V. M., and Taborsky, M. (2003). Androgen levels of reproductive competitors in a co-operatively breeding cichlid. *J. Fish Biol.* 63, 1615–1620.
347. Klop, E., Curio, E., and Lastimoza, L. L. (2000). Breeding biology, nest site characteristics and nest spacing of the Visayan Tarictic hornbill *Penelopides panini panini* on Panay, Philippines. *Bird Cons. Int.* 10, 17–27.
348. MacDonald, D. W., and Moehlman, P. (1982). Cooperation, altruism and restraint in the reproduction of carnivores. In *Perspectives in Ecology* (C. P. Klopfer and P. Bateson, Eds.), pp. 433–466. Plenum Press, New York.
349. Wingfield, J. C., Hegner, R. E., and Lewis, D. (1991). Circulating levels of luteinizing hormone and steroid hormones in relation to social status in the cooperatively breeding white-browed sparrow weaver, *Plocepasser mahali*. *J. Zool. (Lond).* 225, 43–58.
350. Schoech, S. J., Mumme, R. L., and Wingfield, J. C. (1996). Prolactin and helping behavior in the cooperatively breeding Florida scrub-jay (*Aphelocoma coerulescens*). *Anim. Behav.* 52, 445–456.
351. Schoech, S. J., Mumme, R. L., and Wingfield, J. C. (1996). Delayed breeding in the cooperatively breeding Florida scrub-jay (*Aphelocoma coerulescens*): inhibition or the absence of stimulation. *Behav. Ecol. Sociobiol.* 39, 77–90.
352. Schoech, S. J., Mumme, R. L., and Wingfield, J. C. (1997). Breeding status, corticosterone, and body mass in the cooperatively breeding Florida scrub-jay (*Aphelocoma coerulescens*). *Physiol. Zool.* 70, 68–73.
353. Vleck, C. M., and Brown, J. L. (1999). Testosterone and social and reproductive behavior in *Aphelocoma* jays. *Anim. Behav.* 58, 943–951.
354. De la Cruz, C., Solis, E., Valencia, J., Chastel, O., and Sorci, G. (2003). Testosterone and helping behavior in the azure-winged magpie (*Cyanopica cyanus*): natural covariation and an experimental test. *Behav. Ecol. Sociobiol.* 55, 103–111.
355. Khan, M. Z., McNabb, F. M. A., Walters, J. R., and Sharp, P. J. (2001). Patterns of testosterone and prolactin concentrations and reproductive behavior of helpers and breeders in the cooperatively breeding red-cockaded woodpecker (*Picoides borealis*). *Horm. Behav.* 40, 1–13.
356. O'Riain, M. J., Bennett, N. C., Brotherton, P. N. M., McIlrath, G., and Clutton-Brock, T. H. (2000). Reproductive suppression and inbreeding avoidance in wild populations of co-operatively breeding meerkats (*Suricata suricata*). *Behav. Ecol. Sociobiol.* 48, 471–477.
357. Peters, A., Astheimer, L. B., and Cockburn, A. (2001). The annual testosterone profile in cooperatively breeding superb fairy-wrens, *Malurus cyaneus*, reflects their extreme infidelity. *Behav. Ecol. Sociobiol.* 50, 519–527.
358. Jarvis, J. U. M. (1981). Eusociality in a mammal: cooperative breeding in naked mole rat colonies. *Science* 212, 571–573.
359. Jarvis, J. U. M. (1991). Reproduction in naked mole rats. In *The Biology of the Naked Mole Rat* (P. Sherman, J. U. M. Jarvis, and R. D. Alexander, Eds.), pp. 384–425. Princeton University Press, Princeton, NJ.
360. Sherman, P., Jarvis, J. U. M., and Alexander, R. D., Eds. (1991). *The Biology of the Naked Mole Rat.* Princeton University Press, Princeton, NJ.
361. Faulkes, C. G., Abbott, D. H., Liddell, C. E., George, L. M., and Jarvis, J. U. M. (1991). Hormonal and behavioral aspects of reproductive suppression in female naked mole rats. In *The Biology of the Naked Mole Rat* (P. Sherman, J. U. M. Jarvis, and R. D. Alexander, Eds.), pp. 426–445. Princeton University Press, Princeton, NJ.
362. Margulis, S. W., Saltzman, W., and Abbot, D. H. (1995). Behavioral and hormonal changes in female naked mole rats (*Heterocephalus glaber*) following removal of the breeding female from a colony. *Horm. Behav.* 29, 227–247.
363. Davies, N. B. (2000). *Cuckoos, Cowbirds and Other Cheats.* Blackwell Synergy, London.
364. Dufty, A. M., Jr., Goldsmith, A. R., and Wingfield, J. C. (1987). Prolactin secretion in a brood parasite, the brown-headed cowbird, *Molothrus ater*. *J. Zool. (Lond).* 212, 669–675.
365. Wingfield, J. C., Jacobs, J. D., Soma, K. K., Maney, D. L., Hunt, K., Wisti-Peterson, D., Meddle, R. M., Ramenofsky, M., and Sullivan, K. (1999). Testosterone, aggression, and communication: ecological bases of endocrine phenomena. In *The Design of Animal Communication* (M. Hauser and M. Konishi, Eds.), pp. 255–284. MIT Press, Cambridge, MA.
366. Silverin, B. (1980). Effects of long-acting testosterone treatment on free-living pied flycatchers, *Ficedula hypoleuca*, during the breeding period. *Anim. Behav.* 28, 906–912.
367. Saino, N., and Møller, A. P. (1995). Testosterone-induced depression of male parental behavior in the barn swallow: female compensation and effects of seasonal fitness. *Behav. Ecol. Sociobiol.* 36, 151–157.
368. Dufty, A. M., Jr. (1989). Testosterone and survival. *Horm. Behav.* 23, 185–193.
369. Hillgarth, N., and Wingfield, J. C. (1997). Parasite-mediated sexual selection: endocrine aspects: In *Host–Parasite Evolution* (D. H. Clayton and J. Moore, Eds.), pp. 78–104. Oxford University Press, Oxford.

370. Schoech, S., Ketterson, E. D., and Nolan, V., Jr. (1999). Exogenous testosterone and the adrenocortical response in the dark-eyed junco, *Junco hyemalis*. *Auk* 116, 64–72.
371. Chandler, C. R., Ketterson, E. D., Nolan, V., Jr., and Ziegenfus, C. (1994). Effects of testosterone on spatial activity in free-ranging male dark-eyed juncos, *Junco hyemalis*. *Anim. Behav.* 47, 1445–1455.
372. Chandler, C. R., Ketterson, E. D., and Nolan, V., Jr. (1997). Effects of testosterone on use of space by male dark-eyed juncos when their mates are fertile. *Anim. Behav.* 54, 543–549.
373. Lynn, S. E., Houtman, A. M., Weathers, W. W., Ketterson, E. D., and Nolan, V., Jr. (2000). Testosterone increases activity but not daily energy expenditure in captive male dark-eyed juncos. *Anim. Behav.* 60, 581–587.
374. Raouf, S. A., Parker, P. G., Ketterson, E. D., Nolan, V., Jr., and Ziegenfus, C. (1997). Testosterone influences reproductive success by increasing extra-pair fertilizations in male dark-eyed juncos, *Junco hyemalis*. *Proc. R. Soc. B Biol. Sci.* 264, 1599–1603.
375. Enstrom, D. A., Ketterson, E. D., and Nolan, V. Jr. (1997). Testosterone and mate choice in the dark-eyed junco. *Anim. Behav.* 54, 1135–1146.
376. Kast, T. L., Ketterson, E. D., and Nolan, V. Jr. (1998). Variation in sperm reserves according to season, stage of reproduction, and testosterone treatment in the darkeyed junco. *Auk* 115, 684–693.
377. Hasselquist, D., Marsh, J. A., Sherman, P. W., and Wingfield, J. C. (1999). Is avian humoral immunocompetence suppressed by testosterone? *Behav. Ecol. Sociobiol.* 45, 167–175.
378. Peters, A. (2000). Testosterone treatment is immunosuppressive in superb fairy-wrens, yet free-living males with high testosterone and more immunocompetant. *Proc. Roy. Soc. Ser.* B 267, 883–889.
379. Hendrie, C. A., Weiss, S. M., and Eilam, D. (1998). Behavioral response of wild rodents to the calls of an owl: a comparative study. *J. Zool. (Lond).* 245, 439–446.
380. Eilam, D., Dayan, T., Ben-Eliyahu, S., Schulman, I., Shefer, G., and Hendrie, C. A. (1999). Differential behavioral and hormonal responses of voles and spiny mice to owl calls. *Anim. Behav.* 58, 1085–1093.
381. Hubbs, A. H., Millar, J. S., and Wiebe, J. P. (2000). Effect of brief exposure to a potential predator on cortisol concentrations in female Columbian ground squirrels (*Spermophilus columbianus*). *Can. J. Zool.* 78, 578–587.
382. Boonstra, R., and Singleton, G. R. (1993). Population declines in the snowshoe hare and the role of stress. *Gen. Comp. Endocrinol.* 91, 126–143.
383. Boonstra, R., Hik, D., Singleton, G. R., and Tinnikov, A. (1998). The impact of predator-induced stress on the snowshoe hare cycle. *Ecol. Monogr.* 79, 371–394.
384. Krebs, C. J., Boonstra, R., Boutin, S., and Sinclair, A. R. E. (2001). What drives the 10-year cycle of snowshoe hares? *BioScience* 51, 25–35.
385. Silverin, B. (1998). Behavioral and hormonal responses of pied flycatchers to environmental stressors. *Anim. Behav.* 55, 1411–1420.
386. Scheuerlein, A., Van't Hof, T. J., and Gwinner, E. (2001). Predators as stressors? Physiological and reproductive consequences of predation risk in tropical stonechats (*Saxicola torquata axillaris*). *Proc. R. Soc. Lond. B Biol. Sci.* 268, 1575–1582.
387. Fichtel, C., and Hammerschmidt, K. (2003). Responses of squirrel monkeys to their experimentally modified mobbing calls. *J. Acoust. Soc. Am.* 113, 2927–2932.
388. Marler, P. R. (1992). Functions of arousal and emotion in primate communication: a semiotic approach. In *Topics in Primatology,* Vol. 1, *Human Origins* (T. Nishida, W. S. McGrew, P. Marler, M. Pichford, and J. M. Macedonia, Eds.), pp. 225–233. University of Tokyo Press, Tokyo.
389. Marler, P. R. (2004). Bird calls: their potential for behavioral neurobiology. *Ann. N Y Acad. Sci.* 1016, 31–44.
390. Seyfarth, R., Cheney, D. L., and Marler, P. R. (1980). Vervet monkey alarm calls: semantic communication in a free-ranging primate. *Anim. Behav.* 28, 1070–1094.
391. Boinski, S., Gross, T. S., and Davis, J. K. (1999). Terrestrial predator alarm vocalizations are a valid monitor of stress in captive brown capuchins (*Cebus apella*). *Zoo Biol.* 18, 295–312.
392. Lonstein, J. S., and Gammie, S. C. (2002). Sensory, hormonal and neural control of maternal aggression in laboratory rodents. *Neurosci. Biobehav. Rev.* 26, 869–888.
393. Oliveira, R. F., and Almada, V. C. (1998). Maternal aggression during the mouthbrooding cycle in the cichlid fish, *Oreochromis mossambicus*. *Aggress. Behav.* 24, 187–196.
394. Ros, A. F., Bruintjes, R., Santos, R. S., Canario, A. V., and Oliveira, R. F. (2004). The role of androgens in the trade-off between territorial and parental behavior in the Azorean rock-pool blenny, *Parablennius parvicornis*. *Horm. Behav.* 46, 491–497.
395. Kitaysky, A. S., Wingfield, J. C., and Piatt, J. F. (1999). Dynamics of food availability, body condition and physiological stress response in breeding black-legged kittiwakes. *Funct. Ecol.* 13, 577–584.
396. Kitaysky, A. S., Hunt G. L., Flint, E. N., Rubega, M. A., and Decker, M. B. (2000). Resource allocation in breeding seabirds: responses to fluctuations in their food supply. *Mar. Ecol. Prog. Ser.* 206, 283–296.
397. Kitaysky, A. S., Kitaiskaia, E. V., Piatt, J. F., and Wingfield, J. C. (2003). Benefits and costs of increased levels of corticosterone in seabird chicks. *Horm. Behav.* 43, 140–149.
398. Golet, G. H., Irons, W. B., and Estes, J. A. (1998). Survival costs of chick rearing in black-legged kittiwakes. *J. Anim. Ecol.* 67, 827–841.
399. Harris, M. P. (1970). Breeding ecology of the Swallow-tailed Gull (*Creargus furcatus*). *Auk* 87, 215–243.
400. Henderson, B. A. (1975). Role of the chick's begging behavior in the regualtion of parental feeding behavior of *Larus glaucescens*. *Condor* 77, 488–492.
401. Ydenberg, R. C., and Bertram, D. F. (1989). Lack's clutch size hypothesis and brood enlargement studies on colonial seabirds. *Colon. Waterbirds* 12, 134–137.
402. Kitaysky, A. S., Wingfield, J. C. and Piatt, J. F. (1999). Dynamics of food availability, body condition and physiological stress response in breeding black-legged kittiwakes. *Funct. Ecol.* 13, 577–584.
403. Welberg, L. A. M., and Seckl, J. R. (2001). Prenatal stress, glucocorticoids and the programming of the brain. *J. Neuroendocrinol.* 13:113–128.

Knobil and Neill's Physiology of Reproduction, Third Edition
edited by Jimmy D. Neill,
Elsevier © 2006

CHAPTER 37

Pheromones and Mammalian Reproduction

John G. Vandenbergh

Introduction, 2041
Reception of Chemosignals, 2042
Main Olfactory System, 2043
Accessory Olfactory System, 2043
Secondary Chemical Receptors, 2043
Signaling Pheromones, 2044
Sex Attractants, 2044
Individual Identification and Mate Selection, 2046
Priming Pheromones, 2047
Acceleration of Puberty, 2047
Inhibition of Puberty, 2048
Ovarian Cyclicity, 2050
Pregnancy Block, 2051
Conclusion, 2052
Acknowledgments, 2052
References, 2052

INTRODUCTION

Working on insect chemical communication, Karlson and Butenandt (1) coined the term *pheromone* during an after-work discussion session in one of Tübingen's beer halls, according to local lore (R. Apfelbach, personal communication). The term was initially applied to insects and was associated with the notion of an innate releaser mechanism in response to a chemical signal. Wilson and Bossert (2) specifically classified pheromones as either "releasers" or "primers" in extending the pheromone concept from insects to mammals, suggesting that releasing pheromones induced an immediate behavioral response and that priming pheromones resulted in a more long-term physiological or developmental change. Bronson (3), working with rodent pheromones, proposed the term "signaling" instead of "releasing" because "releasing" denotes a rigidity of response that is probably not applicable to mammals. In this review, I follow Bronson's format. Here we deal only with signaling and priming pheromones related to reproduction. Discussion of the rich field of chemical signals relating to aggression, social status, territoriality, parental care, and other life functions of animals can be found elsewhere (4–9).

A review of chemical signals in reptiles and amphibians is provided by Madison (10) and Halpern and Martinez-Marcos (11). The important role for chemical stimuli in the reproductive behavior of fish has been reviewed by Sorensen and Stacey (11a). Birds apparently rely heavily on visual and auditory cues for mate attraction and selection, and it is unlikely that pheromones play a significant role in avian reproduction.

The term *pheromone* has been somewhat controversial, but it is well entrenched in the scientific and popular press (4). Knowledge about mammalian pheromones and their importance in reproductive processes has grown sharply in recent years. Under natural conditions, pheromones rarely operate in the absence of other sensory signals, and never out of an environmental context. An array of physical, social, and environmental factors are used by mammals, especially primates, to coordinate reproduction. Such redundancy of sensory cues from the environment helps to ensure reproductive success.

For some species of small, nocturnal, or burrowing mammals (e.g., rodents), pheromones may be particularly important as signals in a dark environment. It is not surprising, then, that much of the literature on mammalian chemical signaling concerns rodents.

Department of Zoology, North Carolina State University, Raleigh, North Carolina.

Although rodents play an important role in the discussion to follow, an attempt has been made to include a range of mammalian species.

RECEPTION OF CHEMOSIGNALS

Chemical communication involves both the production and the reception of chemosignals. Several excellent reviews of chemoreceptor systems are available (12,13). Here only selected elements of this active area of research, namely, those that relate to reproduction, are examined. The main olfactory system is well developed in most mammals and is a primary receptor for airborne chemosignals. Four other chemoreceptor systems exist, and one or more of these may be very important receptors of chemical information.

Main Olfactory System

In mammals, the receptors in the main olfactory system are located in a posterior-dorsal pocket off the nasal respiratory pathway. The flow of air containing odors can be somewhat controlled by sniffing patterns and changes in the mucosal lining. The receptors are ciliated neurons with axons coursing back to the olfactory bulb in nerve bundles that pass through the cribriform plate. The main olfactory system responds to all but the smallest molecules and without obvious specificity to biologically relevant compounds. The neural message characteristic of specific odors is carried mainly through differential responses of specific cells and in the spatial pattern of activity across the surface of the olfactory mucosa (13). The mechanism of olfactory reception is an active research area and has been reviewed (14). The basic finding is that axons from the olfactory neurons that express a specific odorant receptor project to specific sets of glomeruli in the olfactory bulb.

Accessory Olfactory System

One chemosensory organ in mammals on which remarkable progress has been made is the vomeronasal organ, the receptor organ for the accessory olfactory system. The vomeronasal organ was originally described by Frederick Ruysch in 1703 (15). It has been shown to play an important role as a receptor for priming pheromones in mammals (16). The vomeronasal organ, also known as Jacobson's organ, consists of a tubular structure above the roof of the mouth that connects to the nasal cavity or, in some species, to the oral cavity through the nasopalatine canal (16). The vomeronasal receptor neurons resemble olfactory receptors but differ in ciliary structure (16,17). The vomeronasal organ is not homogeneous but rather seems to be divided into anterior and posterior divisions, each projecting to different portions of the accessory olfactory bulb (18), and which may activate different signaling proteins (19). The receptors send axons in two or three bundles through the cribriform plate and past the main olfactory bulbs to the accessory olfactory bulb. Access of chemosignals to the vomeronasal receptors is controlled by an autonomically driven vascular pump (20). The vomeronasal system shares many characteristics with the main olfactory system, but also differs in significant ways (12). It has a more direct pathway to the areas of the brain concerned with reproduction and is specialized to respond to nonvolatile chemical signals.

Research has uncovered many of the cellular mechanisms in pheromonal reception by the vomeronasal organ. A schematic representation of the anatomy and mechanisms of the main and accessory olfactory system is given in Fig. 1 (13). Both the olfactory and vomeronasal organ receptor cells have heptahelical transmembrane receptors that belong to a group of G-protein–coupled receptors (14). The vomeronasal organ genes expressing the receptors are distinct from those in the main olfactory epithelium. The vomeronasal organ receptors contain two types of α subunits (21). Thorough reviews of the mechanisms involved in pheromonal reception and transmission at the cellular level have are available (11,22), and a diagrammatic summary is provided in Fig. 1.

The vomeronasal organ was long considered to be vestigial or absent in humans (23). Several more recent studies have presented evidence that a detectable vomeronasal organ, or at least vestigal vomeronasal pits, exist in most, if not all, humans (24,25).

Ultrastructural examination of the vomeronasal organ from patients undergoing plastic surgery, as well as cadavers and fetal specimens, revealed the presence of two potential receptor elements that may constitute a pheromone detection system (26). The status of the putative human vomeronasal organ has been extensively reviewed (16,27–29). The consensus is that the structure of a vomeronasal organ, or its evolutionary remnants, exists in humans. However, there is no strong evidence for functional sensory neurons in the epithelium, nor have any connections with the central nervous system been found.

The presence of a well-developed vomeronasal organ in mammals is often associated with a striking form of sniffing behavior called *flehmen*, in which the lips are curled and the head is often thrown up. Flehmen is particularly well expressed by ungulates and carnivores. Rodents, on the other hand, exhibit a well-developed vomeronasal organ yet are not

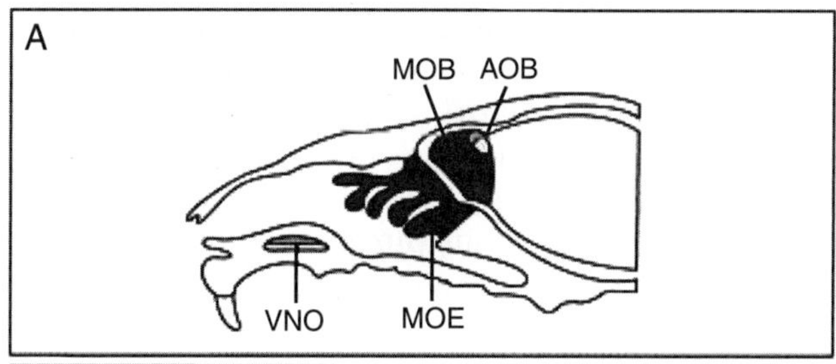

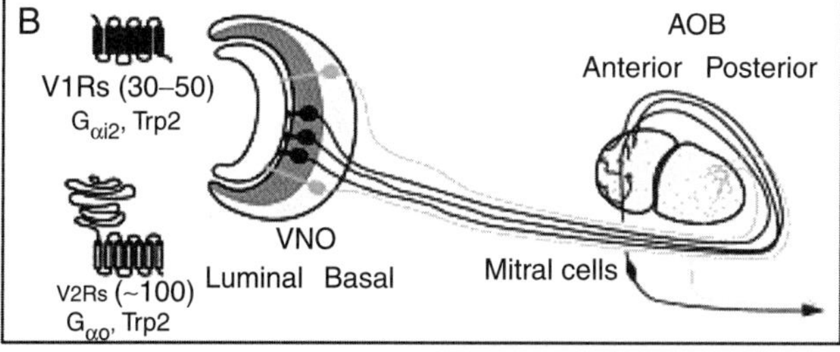

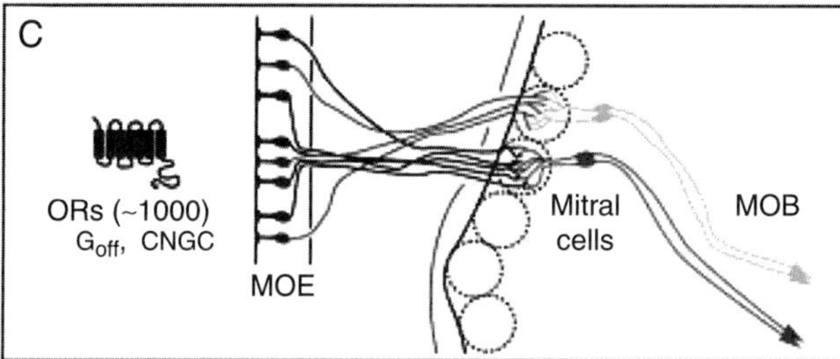

FIG. 1. A diagram summarizing the anatomical and cellular organization of the two mammalian olfactory epithelia. **A:** Location of the vomeronasal organ (VNO), the accessory olfactory bulb (AOB), the main olfactory epithelium (MOE), and the main olfactory bulb (MOB). **B:** The two categories of VNO sensory neurons. The neurons located in the apical zone of the neuroepithelium express different G proteins and receptors from those in the more basal part of the VNO neuroepithelium. Both however, express the testosterone propionate receptor-2 (TPR2) channel, a candidate component of the VNO signaling cascade. **C:** A sample of two MOE neuronal axons coursing to specific glomeruli, where they synapse with specific subsets of mitral cells. For more details, refer to Dulac (13). (From Dulac, C. [2000]. Sensory coding of pheromone signals in mammals. *Curr. Opin. Neurobiol.* 10, 511–518, with permission.)

reported to show flehmen. Wysocki (30) thoroughly reviews the comparative anatomy of this organ, and Estes (31) relates this structure to the incidence of flehmen in several mammalian taxa. Flehmen is more commonly shown by males than females, is usually shown immediately after contact with female urine, and is apparently associated with detection of estrus.

Ladewig and Hart (32) showed that when a male goat displayed flehmen after investigating female urine containing a tracer material, the urine was found throughout the vomeronasal organ. In males not showing flehmen after investigating female urine, the urine was found only in the anterior one third of the vomeronasal organ. Flehmen accompanies the pumping mechanism of the vomeronasal organ to activate into the lumen of the organ (20). The vascular operation of the vomeronasal organ is similar to that of other erectile tissues (16,33).

The clearest function of the vomeronasal organ is to serve as a receptor for signaling pheromones. It has become clear that the vomeronasal organ plays an important role in the reception of priming pheromones (12,16,34). As shown in Table 1, several reproductive events known to be primed by pheromones depend on an intact vomeronasal organ. For example, rats maintained on constant light show persistent estrus. Such rats can be induced to ovulate by exposure to male chemosignals (35,36). If, however, the rats are vomeronasalectomized, the male priming effect is significantly reduced (36). The male effect that induces earlier puberty can be eliminated by vomeronasalectomy (37), as can the pregnancy-blocking effect of a strange male (38).

Secondary Chemical Receptors

Three chemoreceptor systems that may be involved in the reception of pheromones related to reproduction are the septal organ, the nervus terminalis, and the trigeminal system. The *septal organ* is a patch of olfactory neuroepithelium on the nasal septum anterior to the main olfactory epithelium. Research indicates that the septal organ is similar to the main olfactory bulb in its signal transduction pathways, odorant response properties, and projections to the main olfactory bulb (46,47). The septal organ may play a role as a chemoreceptor related to social and sexual behavior (48), but more work is needed to clarify its role as a chemosensory organ.

The *trigeminal system* is known to respond to strong, irritating odors and may serve to modulate olfactory bulb activity (49) and, perhaps indirectly through this mechanism, influence the reception of reproductively important odors.

TABLE 1. *A selection of reproductive events influenced by the accessory olfactory system in rodents*

Reproductive event	Species	References
Male induction of estrus in		
Light-induced persistent estrus adults	Rat	(35)
Juveniles	Rat	(39,40)
Juveniles	Mouse	(40)
Female-induced testosterone surge	Mouse	(41)
Reduction of estrous cycle length	Rat	(42)
Blockage of pregnancy	Mouse	(43)
Suppression of estrus in groups	Mouse	(44)
Induction of puberty inhibition pheromone	Mouse	(10)
Activation of female reproduction	Vole	(45)

The *nervus terminalis* is another intriguing system that may be chemosensitive. All vertebrates have such a system, but its role as a chemoreceptor remains unconfirmed (12). The discovery that the neurons of the nervus terminalis contain high levels of immunoreactive luteinizing hormone–releasing hormone (LHRH) suggests that this structure may have relevance to reproduction, and thus deserves attention here. The nervus terminalis develops from the epithelium of the olfactory pit in mammals, which is also the site of origin of LHRH neurons. Most of these neurons migrate during development to their adult location in the hypothalamus. The processes from the nervus terminalis accompany those of the vomeronasal organ but terminate in the medial septal and preoptic areas of the brain (12,13). A population of LHRH-immunoreactive neurons has been localized in the ganglia of the nervus terminalis in several species (50,51).

The olfactory placode origin of LHRH neurons suggests some fascinating possibilities about the connections between this structure and the reproductive physiology mechanisms in the brain (16). The nervus terminalis might signal physical or chemical changes in the nasal septal mucosa and relay this information to the vomeronasal organ (52). This interesting finding demands additional study.

SIGNALING PHEROMONES

Mammals release to their environment an especially wide variety of odors through their excretory products (e.g., saliva and skin gland excretions, as well as interactions of these compounds with microorganisms). Some of these chemical products have evolved into chemosignals important to individual recognition, mate selection, territory marking, and reproductive coordination. Many body odors convey information about diet or are waste products that serve signaling functions. Stoddart (53) provides some provocative suggestions for the origins and functions of human odors.

Excretory products such as urine and feces serve as a rich source of glandular secretions and metabolic breakdown products that can have signaling value (4). Saliva, long known to be important in the pig (54), is receiving increased attention in other mammals as well (55). The skin contains an array of glands that are used for chemical communication (56). Sex-related steroids also serve to regulate skin gland size and secretory activity (57).

Many mammals have evolved specific behavior patterns for depositing pheromones. One of these involves "overmarking," placing a scent so that it fully or partially overlaps that of a previous mark (4,58). Overmarking can be a form of competition between rivals for a territory or a mate. For example, in voles, overmarking has been shown to result in choice of a mate based on an overmark (59). Thomas and Wolff (60), however, did not find frequent overmarking in prairie voles and provide alternative interpretations of this behavior.

Sex Attractants

Male rats are able to discriminate between male and female rat odors and display a preference for the odor of a receptive female over that of a nonreceptive diestrous female (61). Castrated or immature rats do not display this preference (62), but may discriminate between the odors when given a nonsexual reinforcement (63). These behavioral observations are supported by experiments in which electrophysiological responses of males to female rat urine odors were monitored in the olfactory bulb and hypothalamic area (64,65). These studies reveal that discrimination between estrous and diestrous female urine is independent of the male rat's androgen level. However, testosterone treatment may enhance male rat odor detection performance (66,67), suggesting that androgen treatment affects neural regions associated with preference for and attraction to female odors.

Sexual experience or learning is also important for male rats to select estrous female urine over that from diestrous females (68,69). Hayashi and Kimura (70) reported that sexual experience is important for male mice to display preferences for estrous female odors over those from diestrous females, and that a main source of the attractant involved is present in the vaginal discharge. They suggested that olfactory cues may be learned through sexual experience or that the male's hormonal state is modified with copulatory experience. Urine from estrous females and not from diestrous or ovariectomized females also serves to attract males (71). Males without previous sexual experience are attracted to the urine of intact females, although not to that of intact or castrated males or of ovariectomized or hypophysectomized females (72). Male microtine rodents exhibit attraction to conspecific female urine, though they show only a slight preference for estrous over nonestrous female odors (73). These authors suggest that behavioral responses to olfactory stimuli in microtines appear to be largely the result of learning. In contrast, sexual experience is not necessary for male lemmings to discriminate and select between estrous and nonestrous females (74,75).

Not all pheromonal effects can be readily classified as signaling or priming. An example of one that contains elements of both is the discovery of an airborne pheromone from a female rat in estrus that provokes penile erection in naive male rats (76). Here we have an airborne scent that invokes a stereotypic motor response up to several minutes after its reception (77).

The behavioral responses of golden hamsters (*Mesocricetus auratus*) to odor cues have been studied extensively. Lisk et al. (78) reported that anosmia totally suppresses courtship and copulatory behavior in male golden hamsters. Sexually experienced male hamsters prefer the odors of estrous females over those from pregnant or lactating females (79). This is highly adaptive for the male because pregnant and lactating female hamsters show high levels of aggression toward males. Murphy (80) identified hamster vaginal discharge (HVD) as containing important chemosignals that attract males and stimulate sexual behavior. Even male hamsters, anesthetized or alert, can be made sexually appealing to other males by application of HVD to their hindquarters (80). Although application to inanimate objects fails to elicit these copulatory behaviors, it tends to attract males (81). Preference tests are thought to be more sensitive and subject to less variability than attraction tests, as an evaluation of these experimental methods suggests (79).

A volatile attractant chemosignal in HVD, identified as dimethyl disulfide (DMDS) (82), actively promotes male investigative behavior, and males are attracted to even femtogram quantities of it (83). It appears that differing chemical profiles of female odors, including HVD, may be able to alert males to a female's estrous state. Males with sexual experience prefer bedding from cages of estrous females over that from diestrous females, and the males' differential responses to cage odors were eliminated by removing HVD odors (79). However, more recent studies indicate that DMDS alone is not attractive to male hamsters (84). Hamsters also rely on other olfactory signals that are not restricted to vaginal discharge. Sexually experienced males can use odors from ear glands to discriminate sex of conspecifics, and the absence of an androgen-dependent factor renders the ears of females and castrated males sexually unappealing (85).

Although male hamsters do not show preference for estrous odors of females, residence in the soiled cage of a female decreases their frequency of scent marking when the female is in estrus compared with when she is not in estrus (86). Attention has been given to the differing quantities and consistency of HVD, which vary with the female's reproductive state, peaking at estrus (87,88).

Preferences for HVD by adult male hamsters are developed by means of early androgen exposure because females receiving testosterone propionate neonatally later display significant HVD preferences after gonadectomy (89). An important role for learning has not been excluded because the responses of males to HVD can be altered by pairing exposure to HVD with induced gastrointestinal illness in an aversion-training paradigm (90).

Although other sensory cues from a female may elicit male sexual behavior (91), at least five responses of males to HVD have been identified (92):

1. Approach to and sniffing at odor
2. Licking and consumption of secretion
3. Increased sexual arousal and tendency to engage in copulatory behavior
4. Short-latency rise in plasma testosterone
5. Lowering of aggressive tendencies

Males made anosmic by a zinc sulfate nasal infusion show no impairment of mating behavior, but severing the vomeronasal nerves significantly reduces copulatory behavior in more than one third of treated males; and vomeronasal deafferentation in addition to $ZnSO_4$ treatment eliminates copulation in all male hamsters (93). Powers and Winans (93) concluded that bilateral olfactory bulbectomy results in suppression of copulation as a result of destruction of both the main and the accessory olfactory bulbs. Vomeronasal function is of primary importance for male hamster copulatory performance because main olfactory processes without vomeronasal input can mediate reproductive behavior in only a subpopulation of animals (94).

The advantages of emitting a chemosignal at or shortly before the fertile period allows females to ensure the availability of the male at a time when conception is possible. Males of some species also emit chemosignals related to reproduction. These are apparently more related to social than reproductive status and may provide the female information important to her selection of a mate.

The preputial gland of intact male rats and mice is responsible for odors in the urine that are attractive to females (95,96). This gland is very sensitive to androgens and enlarges greatly with approaching maturity (57). Lipid fractions of the preputial gland contain alkyl acetates, which are attractive to female rats, though DMDS and dimethyl sulfite, also present in preputial tissue extracts, are components attractive only to male rats (95). Urine from dominant male mice is more attractive to estrous females than that from subordinate males (97). Estrous female lemmings exhibit preferences for odors of dominant males over subordinates, whereas diestrous females are more attracted to odors of submissive males (75). The female hamster exhibits similar olfactory discriminations and preferences (98). Metabolic changes correlating with social experience or status or changes in testosterone levels may be reflected in the male's urine. Estrous female hamsters are able to distinguish by odor alone between socially experienced and naive males. They demonstrated preferences for males that would later become dominant after fighting (99,100).

Among mice, a mechanism of female discrimination may also be mediated through a chemosignal. A nonvolatile excitatory factor exists in female mouse urine that elicits ultrasounds in male mice (101,102); it is believed that production of and response to ultrasounds are related to gonadal hormone levels and possibly to reproductive fitness (102).

Olfactory bulbectomy, used extensively in studying rodent reproductive behavior in both males and females (103,104), produces a variety of effects on behavior that may not be attributable to anosmia alone (66,105). Olfactory bulbectomy results in destruction of the accessory olfactory bulb (12) and may interfere directly with a hormone-sensitive response pattern, mounting behavior (105). Total bulbectomy results in an increased sensitivity to estradiol benzoate, as assessed by the female's enhanced tendency to display lordosis (105). It is unlikely that anosmia alone would produce such effects. Because a variety of behavioral changes, including attenuation of copulatory behavior, are associated with bulbectomy, it is difficult to demonstrate the role of olfaction in mating behavior using this technique alone. In addition, an observed effect may be caused by damage to more than the chemoreceptor system targeted. For example, the septal organ and the nervus terminalis would be denervated by olfactory bulbectomy, and the trigeminal could be partly affected, thus leaving open the possibility that a change after olfactory bulbectomy resulted from interruption of their function and not that of the main olfactory system (12).

Although higher primates have a relatively undeveloped olfactory system and are considered to be microsomatic, evidence exists for at least some involvement of chemosignals as sex attractants in rhesus monkeys (106). The early work of Michael and Keverne (107) showed that male monkeys with an intact sense of smell would press a bar repeatedly to gain access to ovariectomized females that had been treated intravaginally with estradiol, but would not bar-press for stimulus females when their nasal passages had been plugged. Although later studies have shown that anosmia fails to interfere with male copulatory behavior in pairing tests (108), it is still likely that anosmia interferes with the males' ability to recognize receptive females.

Further work by Michael and coworkers has shown that when under estrogen stimulation, the vagina of the female rhesus monkey is the source of odor cues that are attractive to males (109), and that a complex mixture of fatty acids in the vaginal secretion is the active signal (110). Questions have arisen whether the fatty acids are the active agent and whether sufficient subjects were used to permit generalization (111). Six male monkeys were used in the early studies by Michael and coworkers, and these were tested repeatedly. Using a larger number of males tested in a somewhat different manner, Goldfoot et al. (112) did not find increases in sexual behavior of males toward ovariectomized females treated with natural or artificial vaginal odorants. Differences between these two sets of studies may relate to methodological differences in the way the experiments were designed and executed, as well as to differences in approach. Michael and Keverne focused almost entirely on responses to changes in attractivity, whereas Goldfoot and coworkers examined actual sexual performance (i.e., ejaculation by the male). The rhesus monkey is a highly adaptable primate that uses all of its senses and capitalizes on prior experience. The critical event in a mating is insemination, and the male monkey can succeed in this in the absence of its olfactory sense (112). Nevertheless, olfactory communication may be involved in the sequence of events leading to insemination if the sense is available.

Individual Identification and Mate Selection

Recognition of individual adults, especially as potential mating partners, is important to maintain species isolation, ensure genetic diversity, and maintain established mating relationships. Only with such individual recognition can successful and efficient reproduction occur.

Species recognition may be mediated by chemical signals. Animals are thought to learn at an early age the olfactory characteristics of their species in a manner analogous to visual imprinting (102). Reciprocally cross-fostered pygmy mice and house mice (113) and cross-fostered hamsters of different species (80) exhibit increased preferences for odors of the foster parent species later as adults. Male black-tailed deer raised with a surrogate mother scented with the rump gland secretion of a female pronghorn antelope later preferred to associate with pronghorn rather than with a female of their own species (114). For a review of the ontogeny of olfactory influences on vertebrate behavior, see Cheal (115).

Once the correct species is chosen as a mate, it is advantageous for mammals to chose mating partners that minimize inbreeding (116). One of the more remarkable findings regarding chemical signals is that laboratory-born, inbred strains of mice can discriminate odors of mice that differ genetically only at a chromosomal region occupied by genes that also code for the immune system (117). This complex of genes, known as the major histocompatibility complex (MHC), not only imparts important immunological functions and cell surface recognition elements of development (118) but is the source of chemosensory information in the urine that enables mice to identify one another as individuals (119,120). Yamazaki et al. (119,120) observed that inbred mice display mating preferences for mice of the opposite MHC type. Heterozygosity produced by such matings could increase the immune system's vigor and range of operation, so such matings could be of considerable adaptive significance (118). Potts et al. (121) have determined the nature and strength of selection operating on MHC genes in seminatural populations of house mice. Their results show that reproductive mechanisms, primarily mating preferences, result in 27% fewer MHC-homozygous offspring than expected from random mating.

Hamsters clearly may recognize each other on the basis of olfactory cues (86). Male hamsters may discriminate between odors of prior mates and novel females. They prefer the odors of novel females, either mated or unmated (122). The female's flank gland was found to be of importance for this discrimination. Male rats with multiple mate experience can identify individual females by body odor and exhibit preferences for the odor of their original partners only before the first ejaculation of the copulatory sequence, after which a novel female's odors become more attractive (123). Female rats prefer to approach the odor of a previous mating partner rather than a new male (124). Male and female rats can discriminate between partners that have and have not copulated on the day of testing (114).

PRIMING PHEROMONES

In mammals, a complex series of physiological and behavioral events must be orchestrated to ensure reproduction. Priming pheromones play an important role in reproduction by influencing the onset of puberty and the timing of the estrus cycle, and by their ability to interfere with pregnancy. The house mouse has been the animal model of choice in this field, but many of the effects of pheromones have been identified in other species as well.

Acceleration of Puberty

Puberty must occur at an appropriate time in an individual's development to promote maximum reproductive efficiency. In several species, priming pheromones have been shown to play an important role in regulating the onset of puberty. The physiological events occurring at the time of puberty have been reviewed extensively in this volume (139).

The onset of puberty in a variety of species has been shown to result from male stimulation (Table 2). In most, urine seems to convey the signal, but in others, such as the tamarin, specialized glands may be involved (138), or skin secretions deposited on the fur, as in sheep (133).

Evidence that chemical communication could be involved in the regulation of puberty onset was presented by Vandenbergh (125,126), who found that exposing juvenile female mice to bedding materials soiled by an adult male induced them to attain puberty approximately 10 days sooner than females exposed to clean bedding. Male urine was quickly identified as the active ingredient in soiled bedding that induced earlier sexual maturation (140,141).

Females require only 3 successive days of exposure any time between the ages of 21 and 29 days to respond to male urine (142). This finding clearly demonstrates the "priming" aspect of this pheromonal stimulation. The presence of the stimulus, male urine, is not required at the time of the response. Three days of exposure as much as 10 to 12 days earlier primed the advanced onset of puberty. Drickamer (143) showed that exposure of a juvenile female to male stimuli for brief periods once daily induced accelerated

TABLE 2. *Some mammals in which the onset of puberty in females has been accelerated by stimuli from adult males*

Species	Male stimulus	Female response measured	References
Mus musculus (house mouse)	Urine	10- to 20-Day advance in first estrus, double uterine weight	(125,126)
Peromyscus maniculatus (prairie deermouse)	Urine	Double uterine weight	(127)
Microtus pennsylvanicus (meadow vole)	Urine	16-Day advance in first estrus	(128)
Microtus ochrogaster (prairie vole)	Urine	14-Day advance in first estrus, quadruple uterine weight	(45,129)
Microtus pinetorum (pine vole)	Urine	Advance first estrus, increase uterine weight	(130)
Bos taurus (cow)	Urine	Increased frequency of puberty among heifers	(131)
Ovis aries (sheep)	Wool and wax	Termination of seasonal anestrus	(132,133)
Notomys alexis (hopping mouse)	Presence	8-Day advance in first estrus	(134)
Dicronstonyx groenlandicus (collared lemming)	Presence	14-Day advance in first estrus	(135)
Sus scrofa (pig)	Presence	27-Day advance in puberty	(136)
Capra aegagrus (goat)	Presence	Peaks in conceptions	(137)
Saguinus fuscilollis (tamarin)	Presence	Advanced age at first conception	(138)

sexual maturation. Females placed in cages recently inhabited by males and in cages containing a resident male for lengths of time varying from 10 minutes to 24 hours showed accelerated puberty. The presence of the male induced an earlier onset of puberty than just the chemical signal alone, and a longer exposure to the chemical signal is required for the effect. This confirms earlier findings (144,145) that a male pheromone explains only a portion of the male effect on female sexual development.

Male urine, at a dose of 0.03 mL/day for 8 days, applied to the nose of the juvenile female doubled uterine weight (146). Male urine diluted with water retains activity to remarkably low levels (147,148). In a systematic experiment using serial dilutions of urine, Drickamer (148) showed that a daily dose of 0.001 mL male urine induced the complete acceleratory effect and that the effect of the urine was not completely lost until a dilution of 1:10,000 was reached. Thus, the priming pheromone is active at very low levels.

The presence of the puberty-accelerating pheromone is androgen dependent. The urine of juvenile male mice failed to accelerate puberty (149), and the adult male loses its capacity to accelerate puberty in females within 10 to 15 days after castration. The puberty-accelerating pheromone is restored to the urine of the castrate male within 60 hours after a single 1-mg dose of testosterone propionate. A graded series of testosterone propionate doses restores pheromonal activity in a dose-dependent fashion (149). The urine of female mice, which is normally inactive, gains activity after the females are injected with testosterone propionate (149).

The association between androgen levels and the social status of males (67,150) is reflected by the finding that the puberty-accelerating pheromone of the male was influenced by his social status (151). Urine from dominant male mice accelerated the onset of puberty in female mice, whereas the urine of subordinate mice did not differ from water in its effect on puberty (152). By observing groups of four mice, Drickamer (143) was able to classify males as dominant, ranking second or third, or ranking lowest. Urine from these males was collected and painted on the noses of juvenile females. Urine from the dominant male resulted in the most pronounced acceleration of puberty; middle-ranking males had an intermediate effect; and the urine of the lowest-ranking male had no effect on puberty onset compared with juvenile females with water painted on their noses.

The puberty-accelerating pheromone from male urine has been identified by Novotny and colleagues as several low–molecular-weight urinary components bound to the major urinary proteins (MUP) (153,154). These compounds are 2-sec-butyldihydrothiazole (SBT), 3,4-dehydro-exo-brevicomin (DHB), 6-hydroxy-6-methyl-3-heptanone, and α- and β-farnesenes.

The bioassay used for testing urinary fractions is based on uterine weight changes in response to exposure of test substances on the oronasal groove. Such exposure results in an approximate doubling of uterine weight (146). The uterine weight change increase occurs as a consequence of estrogen production by the ovary in response to luteinizing hormone (LH) released by the pituitary. The release of LH can occur as rapidly as 20 minutes after exposure to an adult male or his urine (155,156).

The association of the puberty accelerating pheromones with proteins is particularly interesting. It has long been known that the urine of mice and some other small rodents contains a high concentration of proteins (157) and that the presence of these proteins is androgen dependent (158). These proteins, especially the MUP complex, have been used in a variety of genetic marker studies (159). The MUP complex has a molecular weight of approximately 17,000 Da. A subfraction of this protein has been suggested as the puberty-accelerating pheromone (160), but this has not been confirmed (153). Novotny and associates (154) have shown that the puberty-accelerating compounds, SBT and DHB, display strong affinity to MUPs. The MUPs then serve as carrier molecules (161).

Adult females can also accelerate the onset of puberty, but only when in estrus, pregnant, or lactating (162,163). Female chemosignals typically inhibit the development of other females (discussed later), so it is somewhat surprising that they can also accelerate sexual development. Perhaps, as Drickamer and Hoover (164) suggest, the release of acceleratory chemosignals by actively breeding females can signal other females, presumably kin, that social and environmental conditions are appropriate for successful reproduction. Support for this interpretation was provided when Drickamer (143) showed that grouping pregnant or lactating females at various densities in a variety of combinations resulted in the disappearance of the puberty-accelerating pheromone from these females.

Inhibition of Puberty

The onset of puberty occurs later in female mice reared in groups than in those reared singly (165,166), suggesting that interaction among the females results in suppression of sexual maturation. Drickamer (145) confirmed this by showing that puberty onset of isolated females could be delayed by exposure to bedding material soiled by grouped females. Urine from grouped females was shown to contain the suppressive pheromone by McIntosh and Drickamer (167). Urine from grouped females results in a 4- to 5-day delay in the onset of first estrus compared with juvenile females exposed to the urine of isolated females.

Although the bulk of the work on pheromonal inhibition of puberty has used the house mouse, a variety of other rodents and two species of callitrichid primates have been shown to demonstrate this phenomenon as well (Table 3). In cottontop tamarin (*Saguinus oedipus oedipus*) females born into breeding groups, time to first ovulation is delayed, and chemosignals are thought to play a role in the delay (168).

Pheromones often act at extremely low doses, and the puberty-inhibiting pheromone is a good example. Drickamer (148) showed that a daily dose of 0.001 mL urine from grouped female mice retained full activity, 0.0001 mL gave an intermediate response, and not until urine was diluted 1:100,000 did it lose potency to delay puberty. Although the puberty-inhibiting pheromone is potent at very low levels, it does not seem to be very stable, at least in comparison with the puberty-accelerating pheromone (169). Urine from grouped females placed on a glass plate and held at room temperature lost its potency to delay puberty within 7 days (170).

The mechanism underlying the puberty-inhibiting effect involves the vomeronasal organ (171). Adult female mice were grouped after being subjected to vomeronasalectomy or sham surgery. Grouped females not able to receive information through the vomeronasal organ failed to produce the puberty-inhibiting pheromone in their urine. Ovariectomy has no effect on the production of the puberty-inhibiting pheromone (149), but adrenalectomy, like vomeronasalectomy, abolishes the inhibitory effect of excreted urine from grouped females (172). Novotny et al. (173) demonstrated that a complex of six adrenal-dependent steroid metabolites in urine was likely to be responsible for inhibiting puberty in juvenile females, confirming a role for the adrenal gland. Additional work from this laboratory identified the adrenal-mediated metabolite 2,5-dimethylpyarzine as the pheromone responsible for the inhibition of puberty (174).

Understanding the mechanisms regulating the puberty-inhibiting pheromone is important because of evidence that this pheromone may contribute to the regulation of wild populations of house mice. Two field experiments have been conducted to determine the possible role of puberty-regulating pheromones on populations of house mice living on segments of highway cloverleaves. These areas, termed "highway islands" (184), are excellent sites to study small mammal populations because of their insularity and manageable size.

In the initial study (184), first-generation offspring of wild caught mice were released onto two highway islands. One population grew, and the other remained stable for 1 year. Resident mice were trapped at intervals to monitor population changes and to provide urine that was later tested for its potency to accelerate or inhibit puberty in laboratory females. Urine collected from adult males accelerated puberty in females equally when collected from the two populations, regardless of their density (185). Urine collected from adult females, on the other hand, was effective in delaying puberty only when collected from females residing in the dense population when it was at its peak. These results suggest that the acceleratory pheromone produced by the male is uninfluenced by population density, at least within the limits of the populations studied. Production of the female inhibitory pheromone, however, was directly dependent on density and thus could serve as an internal regulating mechanism modulating population growth through increasing generation time.

The study of freely growing populations on highway islands was followed by an experimental test of whether an acute increase in the density of the population could trigger the production of a puberty-inhibiting pheromone by the resident females in the population (186). Populations of mice were established on four similar highway islands. Using a counterbalanced design, alien females were released into each population to inflate the population abruptly. Populations were monitored using standard trapping procedures. Urine was collected from resident females before and after population "explosions" and tested for its ability to alter the onset of puberty in juvenile females in the laboratory. Urine collected 3 weeks after the population explosions delayed the onset of puberty by almost a week and was still effective, but less so, when collected 7 weeks after the population explosions. These results verify, in field experiments, that high population density can induce females to produce a puberty-inhibiting pheromone. This pheromone can then serve to reduce the rate of further population growth by increasing generation time. The theoretical implications of this work have been explored by Vandenbergh and Coppola (146).

Because regulation of the puberty-inhibiting pheromone may have important implications for our understanding of mouse population dynamics, several studies clarifying the mechanisms involved have been pursued. Studying groups of different density in the laboratory revealed that the more densely crowded the females were, the more rapidly they began to produce the inhibitory pheromone. When isolated from a dense group, females lose the ability to produce the inhibitory pheromone in their urine within 10 days (170). At least 4 to 7 days of exposure to the pheromone for a minimum of 1 hour per day is necessary to produce the delay in puberty (142,187).

Drickamer and Mikesic (188) established populations of house mice in large outdoor enclosures and then exposed the populations to either male urine, grouped female urine, or water over a 1-year period. Frequent trapping to monitor population density revealed that the populations exposed to male urine reached the highest density and those exposed to grouped female urine attained the lowest density; water-exposed populations were intermediate. These data provide additional experimental evidence that pheromones can play an important role in controlling population dynamics in house mice.

Keverne and coworkers (189) have shown that, in addition to elevating LH concentrations (155,156), primer pheromones can effect changes in prolactin secretion from the anterior pituitary. Exposure of female mice to males results in a rapid decrease of serum prolactin. The drop in prolactin concentrations is probably the result of dopamine release from the hypothalamic tuberoinfundibular dopaminergic neurons into the hypophyseal portal circulation. Injections of a dopamine agonist, bromocriptine, can reproduce the actions of known primer pheromones in female mice. Thus, neuroendocrine modulation of pituitary secretions appears to be the basic mechanism involved in priming pheromone effects on rodent reproduction.

Ovarian Cyclicity

Once puberty has been reached, chemosignals continue to play a role in regulating ovarian function in several species. When female rats live together, the onset of estrus becomes synchronous after approximately three cycles and is maintained as long as 12 cycles (186,187). Not all females are perfectly synchronized; 50% to 75% are in synchrony at any one time, with the others shifting in or out of synchrony.

The synchrony of ovarian cycles in grouped rats was shown to result from pheromonal communication when McClintock (190) showed that females sharing a recirculated air supply show ovarian synchrony similar to grouped females. McClintock (191) proposed that grouped female rats produce two chemosignals, one that phase-advances ovulation and thus shortens the cycle, and one that delays ovulation and thereby lengthens the cycle. To test this, airborne odors were presented continuously for 18 to 24 days to recipient isolated females. The donors of the odors were either in the follicular, the ovulatory, or the luteal phase of the of the rat's ovarian cycle. New donors were used each day to ensure a constancy of production of odors from a specific portion of the ovarian cycles (190). Constant follicular odors shortened and regularized the cycle. All females had 4-day cycles instead of the usual range of 3 to 6 days. Constant ovulatory odors increased the length and variability of the cycle, and odors from the luteal phase had no significant effect on the cycle. McClintock (190) points out that the presence of opposing signals suggests a coupled-oscillator mechanism that can be an effective generator of synchronized cycles. Later, she showed that the estrous cycle of female rats was shortened by constant exposure to odors from pregnant rats and was lengthened by odors from lactating rats and their pups (192). The fact that the signal is carried by an airborne substance suggests that the main olfactory system is involved, yet many other chemosignals influencing reproduction are received by the vomeronasal system (12). Resolution of this issue awaits experimentation.

In a provocative earlier study, McClintock (193) observed that menstrual cycles of women living in a dormitory became synchronized as the school year progressed. The degree of synchrony was directly related to the amount of time the women spent together. This and other evidence led McClintock to suggest that pheromones may be involved in the synchronization. For a more recent evaluation of human menstrual synchrony, see Weller and Weller (194). The source of the putative pheromone is the axilla and the likely active compounds are C6 to C11 unsaturated acids (195).

A survey of species in which the ovarian cycle is synchronized by pheromonal stimulation (Table 3) reveals that the male can also be a source of the synchronizing pheromone. The male synchronizing influence may act in concert with female stimuli, as

TABLE 3. *Species showing ovarian synchrony as a result of pheromonal stimulation from the same or opposite sex*

Species	Interaction	Source	References
Mesocricetus auratus (golden hamster)	F–F	Airborne	(199)
Microtus sp. (vole)	M–F	Urine	(128,200)
Rattus norvegicus (Norway rat)	F–F	Airborne	(190)
Capra hircus (goat)	M–F	Pen and urine	(203,204)
Bos taurus (cow)	F–F	Cervical mucus	(131)
Ovis aries (sheep)	M–F	Wool and wax	(132,133)
Homo sapiens (human)	M–F	Sweat	(194,195)

F, female; M, male.

shown by the early work of Whitten (196,197). Introduction of a male into a group of female house mice induces a synchronous wave of estrus 3 days later. The synchrony is markedly attenuated if the females are paired individually with males. Strong synchrony can be induced among isolated females if they are exposed to urine from grouped females before being paired with a male (198). In many species, the male pheromonal stimulation may synchronize estrus at the beginning of the mating season, as in sheep (45), or by enhancing reflex ovulation, as in voles (128). In fact, ovarian synchrony in repeated cycles is probably a rare event in mammalian species. Only humans and domestic species housed to regulate reproduction have the opportunity to show this phenomenon. Issues related to the adaptive significance of ovarian synchrony and suppression have been examined by McClintock (190) and Knowlton (201).

Ovarian suppression in mice as a result of grouping was first reported by Whitten (196,197). It was soon shown that the suppression was related to a chemical signal (202), which was later confirmed by Champlin (165). Females can be released from this inhibition by being isolated or by exposing them to an adult male or his urine (196,202). Removal from a group and pairing with a male results in a high proportion of conceptions occurring on the third night after separation from the group (196,202), a procedure that can be a useful husbandry technique to schedule litters. The active pheromone(s) are DHB and SBT, the same compounds responsible for puberty acceleration (205). In rats, there is no clear effect of grouping females on the regularity of ovarian cycles; if anything, cycles shortened from 5 to 4 days (202) when females were grouped.

Pregnancy Block

Even when pregnant, the female mouse is still subject to pheromonal stimulation from a male. Pioneering work by Bruce (206) showed that a high proportion of newly mated female mice separated from stud males and exposed to strange males terminate their pregnancy and promptly return to estrus. Further studies by Bruce (211) and Dominic (212) confirmed and extended this finding. This block to pregnancy, or the "Bruce effect," is effective in 20% to 80% of females, depending on the strangeness of the male. Males of a different genetic strain induce the greatest block to pregnancy. The stud male fails to block pregnancy presumably because the female has a memory of his odor (207). Additional studies suggest that N-methyl-D-aspartate may be involved in this memory trace (208).

An intact vomeronasal organ is necessary for the pregnancy block effect (43). Neuroendocrine and endocrine changes associated with pregnancy block have been reviewed by Marchlewska-Koj (209). She summarizes evidence showing that pheromonal stimuli from an alien male decrease prolactin release, which prevents corpus luteum formation and consequent pheromone release. Evidence is strong that a decreased concentration of progesterone is the final link in the hormonal chain of events resulting in failure of implantation. Administration of exogenous progesterone prevents the male-induced blockage of pregnancy (210).

Marchlewska-Koj (209) has shown that an active fraction can be separated from the protein fraction of male urine by column chromatography. It elutes at approximately the same location as the puberty-accelerating pheromone (169). This suggests that the two effects may be the result of the same pheromone, although further work is necessary to confirm this notion.

In voles, as in many mammals, ovulation occurs in response to male stimulation rather than on a cyclical basis. Such reflex ovulation is partly under pheromonal control, as discussed by Carter et al. (213). Estrus induction in the prairie vole (*Microtus ochrogaster*) depends on the presence of stimuli provided by the male (45,214). Females isolated from males show no signs of ovarian cyclicity. Brief contact with a male or the direct application of male urine initiates a prompt increase in uterine weight (45). A similar response occurs in the pine vole (*Microtus pinetorum*) (215). Interestingly, airborne chemical cues are ineffective; direct contact with urine is necessary, suggesting vomeronasal involvement. A single drop of male urine applied to the nose of a female vole sets in motion a sequence of neuroendocrine and endocrine events (216). Within 60 minutes, levels of LHRH rise in the posterior olfactory bulb, the terminus of fibers from the vomeronasal organ. LHRH remains unaffected in the anterior portion of the olfactory bulb. Concomitant with the rise in LHRH in the posterior olfactory bulb, norepinephrine levels fall, and no changes are found in dopamine concentration. Serum levels of LH were higher in the urine-treated females than in water-treated controls. These results suggest that LHRH and norepinephrine in the olfactory bulbs may mediate the LH release in response to pheromonal cues from the male vole.

In contrast to the activational effects of the male, female voles emit suppressive chemosignals that can delay the onset of puberty in juvenile females. This suppressive effect was detected in prairie voles when the increase in uterine weight that occurs after exposure to an unfamiliar male was prevented when male-exposed females were immediately housed with

TABLE 4. *A selection of mammals in which the onset of reproductive activity in females has been inhibited by stimuli from other females or the family*

Species	Stimulus	Female response measured	References
Mus musculus (house mouse)	Grouped females or their urine	5- to 7-day delay in first estrus	(166,167,170)
Peromyscus maniculatusr (deer mouse)	Female urine	Vaginal opening delayed	(175,176)
Microtus ochrogaster (prairie vole)	Female presence or urine	Blocked male activation of ovarian function	(177)
Microtus californicus (California vole)	Air supply or soiled bedding from family	Blocked male activation of ovarian function	(178,179)
Microtus pinetorum (pine vole)	Soiled bedding from family	Blocked male activation of ovarian function	(130)
Notomys alexis (hopping mouse)	Female presence	25-day delay in first estrus	(134)
Meriones unguiculatus (Mongolian gerbil)	Presence of lactating mother	Undeveloped scent glands and failure to mate	(180)
Sus scrofa (pig)	Urine from grouped gilts	First estrus and ovulation delayed	(181)
Callithrix jacchus (marmoset)	Presence of dominant females	Ovulation suppressed	(182)
Saguinus fusicollis (tamarin)	Presence of mother	Ovulation suppressed	(183)

another female or treated with female urine (177). In pine voles, a similar suppression of uterine growth occurs after exposure to bedding material soiled by the vole's family (130). Characterization of the source of the chemosignal or its chemical identity awaits further research.

CONCLUSION

Considerable progress has been made in understanding how pheromones influence reproduction in mammals in the decades that this topic has come under intense scientific scrutiny. It is clear that many mammals use chemical signals to help identify members of the opposite sex, kinship, and the reproductive status of their potential mates. The response is not automatic, as may be the case in nonmammalian species, but rather is molded by a blend of sensory stimuli in which pheromones play an important role. From even casual observations it was clear that wild and domestic animals were using chemical senses to make choices and to direct their behavior. What was perhaps less expected was that chemosignals could also drive reproductive functions in a physiological sense. Such chemosignals, or priming pheromones, have come under intense investigation. Major advances have been made in identifying the chemicals involved and in clarifying the molecular mechanism(s) involved in pheromone reception and transmission; in several cases, specific genes and their products have been identified.

In several species, the onset of puberty can be accelerated or inhibited by priming pheromones, and once adulthood is reached, pheromones can influence the characteristics of the female's ovarian cycle and induce a gonadotropin surge in the male. The identity of such priming pheromones has now been determined. Evidence is strong that the priming pheromones are relatively small molecules that are bound by urinary proteins. Identification of the vomeronasal organ as the receptor of priming pheromones ties these external chemical signals to the rich world of neurotransmitters in the central nervous system, and thus to endocrine changes in circulation that affect reproduction.

Major challenges remain in the field of chemosignals and reproduction. The interaction among various sensory signals remains to be understood. Further chemical characterization of pheromones and how they operate in a physiological sense is needed to understand how signaling and priming pheromones interact to permit adaptive behavioral and physiological functioning of individuals and populations.

ACKNOWLEDGMENTS

The author thanks M. D. Engell and B. C. Ryan for providing helpful criticism, and the members of the W. M. Keck Center for Behavioral Biology at North Carolina State University for stimulating discussions. All errors and omissions remain the responsibility of the author.

REFERENCES

1. Karlson, P., and Butenandt, A. (1959). Pheromones (ectohormones) in insects. *Annu. Rev. Entomol.* 4, 39–58.
2. Wilson, E. D., and Bossert, W. A. (1963). Chemical communication among animals. *Recent Prog. Horm. Res.* 19, 673–710.
3. Bronson, F. H. (1968). Pheromonal influences on mammalian reproduction. In *Perspectives in Reproduction and Sexual Behavior* (M. Diamond, Ed.), pp. 341–361. Indiana University Press, Bloomington, IN.
4. Johnston, R. E. (2003). Chemical communication in rodents: from pheromones to individual recognition. *J. Mammal.* 84, 1141–1162.

5. Brown, R. E., and MacDonald, D. W. (1985). *Social Odours in Mammals*. Oxford University Press, Oxford.
6. Albone, E. S. (1984). *Mammalian Semiochemistry: The Investigation of Chemical Signals Between Mammals*. John Wiley & Sons, Chichester, UK.
7. Doty, R. L. (1976). *Mammalian Olfaction: Reproductive Processes and Behavior*. Academic Press, New York.
8. Leon, M. (1983). Chemical communication in mother–young interactions. In *Pheromones and Reproduction in Mammals* (J. G. Vandenbergh, Ed.), pp. 39–78. Academic Press, New York.
9. Muller-Schwarze, D., and Mozell, M. M., Eds. (1982). *Chemical Signals in Vertebrates*. Plenum Press, New York.
10. Madison, D. M. (1976).Chemical communication in amphibians and reptiles. In *Chemical Signals in Vertebrates* (D. Muller-Schwarze and M. M. Mozell, Eds.), pp. 135–168. Plenum Press, New York.
11. Halpern, M., and Martinez-Marcos, A. (2003). Structure and function of the vomeronasal system: an update. *Prog. Neurobiol.* 70, 245–318.
11a. Sorenson, P. W., and Stacey, N. E. (1999). Evolution and specialization of fish hormonal pheromones. In *Advances in Chemical Signals in Vertebrates* (R. E. Johnston, D. Muller-Schwarze, and P. W. Sorensen, Eds.), pp. 2–47. Kluwer Academic/Plenum, New York.
12. Meredith, M. (1983). Sensory physiology of pheromone communication. In *Pheromones and Reproduction in Mammals* (J. G. Vandenbergh, Ed.), pp. 200–252. Academic Press, New York.
13. Dulac, C. (2000). Sensory coding of pheromone signals in mammals. *Curr. Opin. Neurobiol.* 10, 511–518.
14. Mombaerts, P. (1999). Molecular biology of odorant receptors in vertebrates. *Annu. Rev. Neurosci.* 22, 487–509.
15. Bhatnagar, K., and Reid, K. H. (1996). The human vomeronasal organ—historical perspectives. A study of Ruysch's and Jacobson's (1811) reports on the vomeronasal organ, with comparative comments in English translations. *Biomed. Res.* 7, 219–229.
16. Evans, C. (2003). *Vomeronasal Chemoreception in Vertebrates.* Imperial College Press, London.
17. Mendoza, A. S. (1993). Morphological studies on the rodent main and accessory olfactory systems: the regio olfactoria and vomeronasal organ. *Anat. Rec.* 175, 425–446.
18. Sugai, T., Sugitani, M., and Onoda, N. (2000). Novel subdivisions of the rat accessory olfactory bulb revealed by the combined method with lectin histochemistry, electrophysiological and optical recordings. *Neuroscience* 95, 23–32.
19. Krieger, J., Schmitt, A., Lobel, D., Gudermann, T., Schultz, G. Breer, H., and Boekhoff, I. (1999). Selective activation of G-protein subtypes in the vomeronasal organ upon stimulation with urine-derived compounds. *J. Biol. Chem.* 274, 4655–4662.
20. Meredith, M., Marques, O. M., O'Connell, R. J., and Stem, F. L. (1980). Vomeronasal pump: significance for male hamster sexual behavior. *Science* 207, 1224–1226.
21. Dulac, C., and Axel, R. (1995). A novel family of genes encoding putative pheromone receptors in mammals. *Cell* 83, 195–206.
22. Dulac, C., and Torello, T. A. (2003). Molecular detection of pheromone signals in mammals: from genes to behaviour. *Nat. Rev. Neurosci.* 4, 551–562.
23. Crosby, E. C., and Humphrey, T. (1938). Studies of the vertebrate telencephalon: I. The nuclear configuration of the olfactory and accessory olfactory formation and of the nucleus olfactorius anterior of certain reptiles, birds, and mammals. *J. Comp. Neurol.* 71, 121–213.
24. Moran, D. T., Jafek, B. W., and Rowley, J. C. III. (1991). The vomeronasal (Jacobson's) organ in man: ultrastructure and frequency of occurrence. *J. Steroid Biochem. Mol. Biol.* 39, 545–552.
25. Garcia-Velasco, J., and Mondragon, M. (1991). The incidence of the vomeronasal organ in 1000 human subjects and its possible clinical significance. *J. Steroid Biochem. Mol. Biol.* 39, 561–563.
26. Stensaas, L. J., Lavker, R. M., Monti-Bloch, L., Grosser, B. I., and Berliner, P. L. (1991). Ultrastructure of the human vomeronasal organ. *J. Steroid Biochem. Mol. Biol.* 39, 553–560.
27. McClintock, M. K. (1998). On the nature of mammalian and human pheromones. *Ann. N Y Acad. Sci.* 855, 390–392.
28. Trotier, D., Eliot, C., Wassef, M., Talmain, G., Bensimon, J. L., Doving, K. B., and Ferrand, J. (2000). The vomeronasal cavity in adult humans. *Chem. Senses* 25, 369–380.
29. Meredith, M. (2001). Human vomeronasal function: a critical review of best and worst cases. *Chem. Senses* 23, 433–445.
30. Wysocki, C. J. (1979). Neurobehavioral evidence for the involvement of the vomeronasal system in mammalian reproduction. *Neurosci. Biobehav. Rev.* 3, 301–341.
31. Estes, R. D. (1972). The role of the vomeronasal organ in mammalian reproduction. *Mammalia* 36, 315–341.
32. Ladewig, J., and Hart, B. L. (1980). Flehman and vomeronasal organ function in male goats. *Physiol. Behav.* 24, 1067–1071.
33. Salazar, I., and Quinteiro, P. (1998). Supporting tissue and vasculature of the vomeronasal organ: the rat at a model. *Microsc. Res. Tech.* 41, 492–505.
34. Meredith, M. (1991). Sensory processing in the main and accessory olfactory systems: comparisons and contrasts. *J. Steroid Biochem. Mol. Biol.* 39, 601–614.
35. Brown-Grant, K., Davidson, J. M., and Grieg, F. (1973). Induced ovulation in albino rats exposed to constant light. *J. Endocrinol.* 57, 7–22.
36. Johns, M. A. (1980). The role of the vomeronasal system in mammalian reproductive physiology. In *Chemical Signals* (D. Muller-Schwarze and R. M. Silverstein, Eds.), pp. 314–364. Plenum Press, New York.
37. Johns, M. A, Feder, H. H., Komisaruk, B. R., and Mayer, A. D. (1978). Urine induced reflex ovulation in anovulatory rats may be vomeronasal effect. *Nature* 272, 446–447.
38. Lomas, D. E, and Keverne, E. B. (1982). Role of the vomeronasal organ and prolactin in the acceleration of puberty in female mice. *J. Reprod. Fertil.* 66, 101–107.
39. Kaneko, N., Debski, E. A., Wilson, M. C., and Whitten, W. K. (1980). Puberty acceleration in mice: II. Evidence that the vomeronasal organ is a receptor for the primer pheromone in male mouse urine. *Biol. Reprod.* 22, 873–878.
40. Sanchez-Criado, J. E, and Gallego, A. (1979). Male induced precocious puberty in the female rat: role of the vomeronasal system. *Acta Endocrinol.* 91, Suppl. 255, 225–255.
41. Wysocki, C. J., Katz, Y., and Bernhard, R. (1983). Male vomeronasal organ mediates female-induced testosterone surges in mice. *Biol. Reprod.* 28, 917–922.
42. Sanchez-Criado, J. E. (1979). Bloqueo de las efectas feromonales en la rata par deaferentacion central del sistema olfatario accesario. *Rev. Esp. Fisiol.* 35, 137–142.
43. Bellringer, J. F, Pratt, H. P. M., and Keverne, E. B. (1980). Involvement of the vomeronasal organ and prolactin in pheromonal induction of delayed implantation in mice. *J. Reprod. Fertil.* 59, 223–228.
44. Reynolds, J., and Keverne, E. B. (1979). Accessory olfactory system and its role in pheromonally mediated suppression of oestrus in grouped mice. *J. Reprod. Fertil.* 57, 31–35.
45. Carter, C. S., Getz, L. L., Gavish, L., McDermott, J. L., and Arnold, P. (1980). Male-related pheromones and the activation of female reproduction in the prairie vole *Microtus ochrogaster*. *Biol. Reprod.* 23, 1038–1045.
46. Levai, O., and Strotmann, L. (2003). Projection pattern of nerve fibers from the septal organ: DiI-tracing studies with transgenic OMP mice. *Histochem. Cell Biol.* 120, 283–492.

47. Ma, M. H., Grosmaitre, X., and Iwemal, C. L., Baker, H., Greer, C. A., and Shepherd, G. M. (2003). Olfactory signal transduction in the mouse septal organ. *J. Neurosci.* 23, 317–324.
48. Weiler, E., and Farbman, A. I. (2003). The septal organ of the rat during postnatal development. *Chem. Senses* 28, 581–593.
49. Graziadei, P. P. C. (1977). Functional anatomy of the mammalian chemoreception system. In *Chemical Signals in Vertebrates* (D. Muller-Schwarze and M. M. Mozell, Eds.), pp. 435–454. Plenum Press, New York.
50. Schwanzel-Fukuda, M., and Pfaff, D. W. (1991). Migration of LHRH-immunoreactive neurons from the olfactory placode rationalizes olfactohormonal relationships. *J. Steroid Biochem. Mol. Biol*, 39, 565–572.
51. Schwanzel-Fukuda, M. (1999). Origin and migration of luteinizing hormone-releasing hormone neurons in mammals. *Microsc. Res. Tech.* 44, 2–10.
52. Wirsig-Wiechmann, C. R., and Wiechmann, A. E. (2001). The prairie vole vomeronasal organ is a target for gonadotropin-releasing hormone. *Chem. Senses* 26, 1193–1202.
53. Stoddart, D. M. (1990). *The Scented Ape: The Biology and Culture of Human Odour.* Cambridge University Press, Cambridge.
54. Signoret, J. P. (1991). Sexual pheromones in the domestic sheep: importance and limits in the regulation of reproductive physiology. *J. Steroid Biochem. Mol. Biol.* 39, 639–645.
55. Smith, B. A., and Block, M. L. (1991). Male saliva cues and female social choice in Mongolian gerbils. *Physiol. Behav.* 50, 379–384.
56. Muller-Schwarze, D. (1983). Scent glands in mammals and their functions. In *Advances in the Study of Mammalian Behavior* (J. F. Eisenberg and D. G. Kleiman, Eds.), pp. 150–197. American Society of Mammalogists, Shippensburg, PA.
57. Ebling, F. J. (1977). Hormonal control of mammalian skin glands. In *Chemical Signals in Vertebrates* (D. Muller-Schwarze and M. M. Mozell, Eds.), pp. 17–33. Plenum Press, New York.
58. Hurst, J. L., and Rich, T. J. (1999). Scent marks as competitive signals of mate quality. In *Advances in Chemical Signals in Vertebrates* (R. E. Johnston, D. Muller-Schwarze, and P. W. Sorensen, Eds.), pp. 209–225. Kluwer Academic/Plenum, New York.
59. Ferkin, M. H. (2001). The response of individuals to overmarks of conspecifics differs between two species of microtine rodents. In *Chemical Communication in Vertebrates* (A. Marchlewska-Koj, J. J. Lepri, and D. Muller-Schwarze, Eds.), Vol. 9, pp. 343–346. Kluwer Academic/Plenum, New York.
60. Thomas, S. A., and Wolff, J. O. (2002). Scent marking in voles: a reassessment of over marking, counter marking and self-advertisement. *Ethology* 108, 51–62.
61. LeMagnen, J. (1952). Les phenomenes olfactosexuels chez l'homme. *Arch. Sci. Physiol.* 6, 125–135.
62. Stern, J. J. (1970). Response of male rats to sex odors. *Physiol. Behav.* 5, 519–514.
63. Carr, W. J., and Caul, W. E. (1962). The effect of castration in the rat upon discrimination of sex odors. *Anim. Behav.* 10, 20–27.
64. Pfaff, D. W., and Pfaffmann, C. (1969). Olfactory and hormonal influences on the basal forebrain of the male rat. *Brain Res.* 15, 137–156.
65. Pfaff, D., and Gregory, E. (1971). Olfactory coding in olfactory bulb and medial forebrain bundle of normal and castrated male rats. *J. Neurophysiol.* 34, 208–216.
66. Alberts, J. R. (1974). Producing and interpreting olfactory deficits. *Physiol. Behav.* 12, 657–670.
67. Barkley, M. S., and Goldman, B. D. (1977). The effects of castration and silastic implants of testosterone on intermale aggression in the mouse. *Horm. Behav.* 9, 32–48.
68. Carr, W. J., Loeb, L. S., and Dissinger, M. L. (1965). Responses of rats to sex odors. *J. Comp. Physiol. Psychol.* 59, 370–377.
69. Lydell, K., and Doty, R. L. (1972). Male rat odor preferences for female urine as a function of sexual experience, urine age and urine sources. *Horm. Behav.* 3, 205–212.
70. Hayashi, S., and Kimura, T. (1974). Sex attractant emitted by female mice. *Physiol. Behav.* 13, 563–567.
71. Dixon, A. K., and Macintosh, J. M. (1975).The relationship between the physiological condition of female mice and the effects of their urine on the social behavior of adult males. *Anim. Behav.* 23, 513–520.
72. Johnston, R. E., and Bronson, F. H. (1982). Endocrine control of male mouse odors that elicit LH surges and attraction in males. *Biol. Reprod.* 27, 1174–1180.
73. Richmond, M., and Stehn, R. (1976). Olfaction and reproductive behavior in microtine rodents. In *Mammalian Olfaction* (R. L. Doty, Ed.), pp. 197–217. Academic Press, New York.
74. Huck, U. W., Banks, E. M., and Wang, S. C. (1981). Olfactory discrimination of social status in the brown lemming. *Behav. Neural Biol.* 33, 364–371.
75. Huck, U. W., and Banks, E. M. (1984). Social olfaction in male brown lemmings *(Lemmus sibiricus = trimucronatus)* and collared lemmings *(Dicrostonyx groenlandicus)*: I. Discrimination of species, sex and estrous condition. *J. Comp. Psychol.* 98, 54–59.
76. Sachs, B. D. (1997). Erection evoked in male rats by airborne scent from estrous females. *Physiol. Behav.* 62, 921–924.
77. Sachs, B. D. (1999). Airborne aphrodisiac odor from estrous rats: implication for pheromonal classification. In *Advances in Chemical Signals in Vertebrates* (R. E. Johnston, D. Muller-Schwarze, and P. W. Sorensen, Eds.), pp. 333–242. Kluwer Academic/Plenum, New York.
78. Lisk, R. D., Zeiss, J., and Ciaccio, L. A. (1972). The influence of olfaction on sexual behavior in male golden hamster *(Mesocricetus auratus)*. *J. Exp. Zool.* 181, 69–78.
79. Johnston, R. E. (1980). The role of vaginal secretion in hamster sexual behavior: males' responses to normal and vaginectomized females and their odors. *J. Comp. Physiol. Psychol.* 94, 905–913.
80. Murphy, M. R. (1980). Sexual preferences of male hamsters: importance of preweaning and adult experience, vaginal secretion and olfactory or vomeronasal sensation. *Behav. Neural Biol.* 30, 323–340.
81. Darby, E. M., Devore M., and Chorover, S. L. (1975). A presumptive sex pheromone in the hamster: some behavioral effects. *J. Comp. Physiol. Psychol.* 88, 496–502.
82. Singer, A. G., Clancy, A. N., Macrides, F., and Agosta, W. C. (1984). Chemical studies of hamster vaginal discharge: effects of endocrine ablation and protein digestion on behaviorally active macromolecular fractions. *Physiol. Behav.* 33, 639–643.
83. O'Connell, R. J., Singe, A. G., Pfaffmann, C., and Agosta, W. C. (1979). Pheromones: attraction to femtogram amounts of dimethyl disulfide and to mixtures of volatile compounds. *J. Chem. Ecol.* 5, 575–585.
84. Petrulis, A., and Johnston, R. E. (1995). A reevaluation of dimethyl disulfide as a sex attractant in golden hamsters. *Physiol. Behav.* 57, 779–784.
85. Laudauer, M. R., Liu, S., and Goldberg, N. (1980). Responses of male hamsters to the ear gland secretions of conspecifics. *Physiol. Behav.* 24, 1023–1026.
86. Johnston, R. E. (1985). Chemical communication: scent signals and scent marking. In *The Hamster* (H. Siegal, Ed.), pp. 133–154. Plenum Press, New York.
87. Johnston, R. E. (1980). Responses of male hamsters to odors of females in different reproductive states. *J. Comp. Physiol. Psychol.* 94, 894–904.
88. O'Connell, R. J., Singer, A. G., Stem, F. L., Jesmajian, S., and Agosta, W. C. (1981). Cyclic variations in the concentration of sex attractant pheromone in hamster vaginal discharge. *Behav. Neural Biol.* 31, 457–464.

89. Gregory, E., and Pritchard, W. S. (1983). The effects of neonatal androgenization of female hamsters on adult preferences for female hamster vaginal discharge. *Physiol. Behav.* 31, 861–864.
90. Johnston, R. E., Zahorik, D. M., Immler, K., and Zakon, H. (1978). Attractions of male sexual behavior by learned aversions to hamster vaginal secretions. *J. Comp. Physiol. Psychol.* 92, 85–93.
91. Kwan, M., and Johnston, R. E. (1980). The role of vaginal secretion in hamster sexual behavior: male's responses to normal and vaginectomized females or their odors. *J. Comp. Physiol. Psychol.* 94, 905–913.
92. Johnston, R. E. (1977). The causation of two scent marking behaviors in female golden hamsters (*Mesocricetus auratus*). *Anim. Behav.* 25, 317–327.
93. Powers, J. B., and Winans, S. S. (1975). Vomeronasal organ: critical role in mediating sexual behavior of the male hamster. *Science* 187, 961–963.
94. Winans, S. S., and Powers, J. B. (1977). Olfactory and vomeronasal deafferentation of male hamsters: histological and behavioral analysis. *Brain Res.* 126, 325–344.
95. Bronson F. H., and Caroom, D. (1971). Preputial gland of the male mouse: attractant function. *J. Reprod. Fertil.* 25, 279–282.
96. Gawienowski, A. (1977). Chemical attractants of the rat preputial gland. In *Chemical Signals in Vertebrates* (D. Muller-Schwarze and M. M. Mozell, Eds.), pp. 45–60. Plenum Press, New York.
97. Gawienowski, A. M., Orsulak, P. J., Stacewicz-Sapuntzakis, M., and Joseph, B. M. (1975). Presence of sex pheromone in preputial glands of male rats. *J. Endocrinol.* 67, 283–288.
98. Johnston, R. E. (1983). Chemical signals and reproductive behavior. In *Pheromones and Reproduction in Mammals* (J. G. Vandenbergh, Ed.), pp. 3–38. Academic Press, New York.
99. White, P. J., Fischer, R., and Meunier, G. (1984). The ability of females to predict male status via urinary odors. *Horm. Behav.* 18, 491–494.
100. White, P. J., Fischer, R., and Meunier, G. (1984). Discrimination of male social status by female hamsters. *Psychol. Rep.* 55, 487–492.
101. Nyby, J., Wysocki, C. J., Whitney, G., Sinzinno, G., and Schneider, J. (1973). Elicitation of male mouse ultrasonic vocalizations: I. Urinary cues. *J. Comp. Physiol. Psychol.* 93, 957–975.
102. Nyby, J., and Whitney, G. (1979). Experience affects behavioral responses to sex odors. In *Chemical Signals* (D. Muller-Schwarze and R. M. Silverstein, Eds.), pp. 173–192. Plenum Press New York.
103. Aron, C., Ross, J., and Asch, G. (1970). Effect of removal of olfactory bulb on mating behavior and ovulation in the rat. *Neuroendocrinology* 6, 109–117.
104. Larsson, K. (1969). Failure of gonadal and gonadotrophic hormones to compensate for an impaired sexual function in anosmic male rats. *Physiol. Behav.* 4, 733–737.
105. Meisel, R. L., Sachs, B. D. and Lumia, A. R. (1984). Olfactory bulb control of sexual function. In *Early Brain Damage* (S. Finger and C. R. Almli, Eds.), pp. 253–268. Academic Press, New York.
106. Keverne, E. B. (1983). Chemical communication in primate reproduction. In *Pheromones and Reproduction in Mammals* (J. G. Vandenbergh, Ed.), pp. 79–94. Academic Press, New York.
107. Michael, R. P, and Keverne, E. B. (1968). Pheromones in the communication of the sexual status in primates. *Nature* 218, 746–749.
108. Goldfoot, D. A., Essock-Vitale, S. M., Asa, C. S., Thornton, J. E, and Leshner, A. I. (1978). Anosmia in male rhesus monkeys does not alter copulatory activity with cycling females. *Science* 199, 1095–1096.
109. Michael, R. P., and Keverne, E. B. (1970). Primate sex pheromones of vaginal origin. *Nature* 225, 84–85.
110. Curtis, R. F., Ballantine, J. A., Keverne, E. B., Bonsall, R. W., and Michael, R. P. (1971). Identification of primate sexual pheromones and the properties of synthetic attractants. *Nature* 232, 396–398.
111. Goldfoot, D. A., Goy, R. W., Kravetz, M. A., and Freeman, S. K. (1976). Reply to Michael, Bonsall and Zumpe; reply to Keverne. *Horm. Behav.* 7, 373–378.
112. Goldfoot, D. A., Kravetz, M. A., Goy, R. W., and Freeman, S. K. (1976). Lack of effect of vaginal lavages and aliphatic acids on ejaculatory responses in rhesus monkeys. *Horm. Behav.* 7, 1–27.
113. Quadagno, D. M., and Banks, E. M. (1970). The effect of reciprocal cross fostering on the behavior of two species of rodents. *Mus musculus* and *Baiomys taylori ater. Anim. Behav.* 18, 379–390.
114. Muller-Schwarze, D., and Muller-Schwarze, C. (1971). Olfactory imprinting in a precocial mammal. *Nature* 229, 55–56.
115. Cheal, M. (1975). Social olfaction: a review of the ontogeny of olfactory influences on vertebrate behavior. *Behav. Biol.* 15, 1–25.
116. Hoogland, J. (1982). Prairie dogs avoid extreme inbreeding. *Science* 215, 1639–1641.
117. Yamazaki, K., Singer, A., Curan, M., and Beauchamp, G. K. (1999). Origins, functions, and chemistry of H-2 regulated odorants. In *Advances in Chemical Signals in Vertebrates* (R. E. Johnston, D. Muller-Schwarze, and P. W. Sorensen, Eds.), pp. 173–180. Kluwer Academic/Plenum, New York.
118. Goldstein, N. I., and Cagen, R. H. (1981). The major histocompatibility complex and olfactory receptors. In *Biochemistry of Taste and Olfaction* (R. H. Cagen and M. R. Kare, Eds.), pp. 93–107. Academic Press, New York.
119. Yamazaki, K., Yamaguchi, M., Beauchamp, G. K., Bard, J., Boyce, E. A., and Thomas, L. (1981). Chemosensation: an aspect of the uniqueness of the individual. In *Biochemistry of Taste and Olfaction* (R. H. Cagen and M. R. Kare, Eds.), pp. 85–92. Academic Press, New York.
120. Yamazaki, K, Yamaguchi, M., Boyse, E. A., and Thomas, L. (1980). The major histocompatibility complex as a source of odors imparting individuality among mice. In *Chemical Signals* (D. Muller-Schwarze and R. M. Silverstein, Eds.), pp. 267–274. Plenum Press, New York.
121. Potts, W. K., Manning, C. J., and Wakeland, E. K. (1991). Mating patterns in seminatural populations of mice influenced by MHC genotype. *Nature* 352, 619–621.
122. Johnston, R. E. (1982), Mechanisms of individual discrimination in hamsters. *Chem. Signals* 3, 245.
123. Krames, L., and Mastromatteo, L. A. (1973). Role of olfactory stimuli during copulation in male and female rats. *J. Comp. Physiol. Psychol.* 85, 528–535.
124. Carr, W. J., Krames, L., and Costonzo, D. J. (1970). Previous sexual experience and olfactory preference for novel versus original sex partners in rats. *J. Comp. Physiol.* 71, 216–222.
125. Vandenbergh, J. G. (1969). Male odor accelerates female sexual maturation in mice. *Endocrinology* 84, 658–660.
126. Vandenbergh, J. G. (1987). Regulation of puberty and its consequences on population dynamics of mice. *Am. Zool.* 27, 891–898.
127. Lombardo, D. L., and Terman, C. R. (1980). The influence of the social environment on sexual maturation of female deermice *Pheromyscus maniculatus bairdii. Res. Popul. Ecol.* 22, 93–100.
128. Baddaloo, E. G. Y., and Chulow, F. V. (1981). Effects of the male on growth, sexual maturation, and ovulation of young

female meadow voles, *Microtus pennsylvanicus*. *Can. J. Zool.* 59, 415–421.

129. Hasler, M. J., and Nalbandov, A. V. (1974). The effect of weanling and adult males on sexual maturation in female voles (*Microtus ochrogaster*). *Gen. Comp. Endocrinol.* 23, 237–238.
130. Lepri, J. J., and Vandenbergh, J. G. (1986). Puberty in pine voles, *Microtus pinetorum*, and the influence of chemosignals on female reproduction. *Biol. Reprod.* 34, 370–377.
131. Izard, M. K., and Vandenbergh, J. G. (1982). Priming pheromones from oestrous cows increase synchronization of oestrus in dairy heifers after PGF-2 injection. *J. Reprod. Fertil.* 66, 189–196.
132. Knight, T. W., Tervit, H. R., and Lynch, P. R. (1984). Effects of boar pheromones, ram's wool and presence of bucks on ovarian activity in anovular ewes early in the breeding season. *Anim. Reprod. Sci.* 6, 129–134.
133. Knight, T. W., and Lynch, P. R. (1980). Source of ram pheromones that stimulate ovulation in the ewe. *Anim. Reprod. Sci.* 3, 133–136.
134. Breed, W. G. (1976). Effect of environment on ovarian activity of wild hopping mice *Notomys alexis*. *J. Reprod. Fertil.* 47, 395–397.
135. Hasler, J. F., and Banks, E. M. (1975). The influence of mature males on sexual maturation in female collared lemmings (*Dicrostonyx groenlandicus*). *J. Reprod. Fertil.* 42, 583–586.
136. Brooks, P. H., and Cole, D. J. A. (1970). The effect of the presence of a boar on the attainment of puberty in gilts. *J. Reprod. Fertil.* 23, 435–440.
137. Chemineau, P. (1987). Possibilities for using bucks to stimulate ovarian and oestrous cycles in anovulatory goats – a review. *Livestock Prod. Sci.* 17, 135–147.
138. Epple, G., and Katz, Y. (1980). Social influences on first reproductive success and related behaviors in the saddle-back tamarin *Saguinus fuscicollis*, Callitrichidae. *Int. J. Primatol.* 1, 171–175.
139. Ojeda, S. R., and Skinner, M. K. (2006). Puberty in the rat. In *Knobil and Neill's Physiology of Reproduction* (J. D. Neill, Ed.), pp. 2061–2126. Academic Press, St. Louis.
140. Cowley, J. J., and Wise, O. R. (1972). Some effects of mouse urine on neonatal growth and reproduction. *Anim. Behav.* 20, 499–506.
141. Drickamer, L. C., and Murphy, R. X. (1978). Female mouse maturation: effects of excreted and bladder urine from juvenile and adult males. *Dev. Psychobiol.* 11, 63–72.
142. Colby, D. R., and Vandenbergh, J. G. (1979). Regulatory effects of urinary pheromones on puberty in the mouse. *Biol. Reprod.* 11, 268–279.
143. Drickamer, L. C. (1983). Effect of period of grouping of donors and duration of stimulus exposure on delay of puberty in female mice by a urinary chemosignals from grouped females. *J. Reprod. Fertil.* 69, 723–727.
144. Bronson, F. H., and Maruniak, J. A. (1975). Male-induced puberty in female mice: evidence for a synergistic action of social cues. *Biol. Reprod.* 13, 94–98.
145. Vandenbergh, J. G., Whitsett, J. M., and Lombardi, J. R. (1975). Partial isolation of a pheromone accelerating puberty in female mice. *J. Reprod. Fertil.* 43, 515–523.
146. Vandenbergh, J. G., and Coppola, D. (1986). The physiology and ecology of puberty modulation by primer pheromones. In *Advances in the Study of Behavior* (J. S. Rosenblatt, C. Beer, and M. Busnel, Eds.), pp. 71–107. Academic Press, New York.
147. Drickamer, L. C. (1982). Acceleration and delay of first vaginal oestrus in female mice by urinary chemosignals: dose levels and mixing urine treatment sources. *Anim. Behav.* 30, 456–460.
148. Drickamer, L. C. (1984). Effects of very small doses of urine on acceleration and delay of sexual maturation in female house mice. *J. Reprod. Fertil.* 71, 475–477.
149. Lombardi, J. R., Vandenbergh, J. G., and Whitsett, J. M. (1976). Androgen control of the sexual maturation pheromone in house mouse urine. *Biol. Reprod.* 15, 179–186.
150. Tollman, J., and King, J. A. (1956). The effects of testosterone propionate on aggression in male and female C57BL/10 mice. *Anim. Behav.* 4, 147–149.
151. Lombardi , J. R., and Vandenbergh, J. G. (1977). Pheromonally induced sexual maturation in females: regulation by the social environment of the male. *Science* 196, 545–546.
152. Drickamer, L. C. (1983). Male acceleration of puberty in female mice (*Mus musculus*). *J. Comp. Psychol.* 97, 191–200.
153. Novotny, M. V., Jemiolo, B., Wiesler, D., Ma, W., Harvey, S., Xu, F., Xie, T.-M., and Carmack, M. (1999). A unique urinary constituent, 6-hydroxy-6-methyl-3-heptanone, is a pheromone that accelerates puberty in female mice. *Chem. Biol.* 6, 377–383.
154. Novotny, M. V., Ma, W., Wiesler, D., and Zidek, L. (1999). Positive identification of the puberty-accelerating pheromone of the house mouse: the volatile ligands associating with the major urinary protein. *Proc. R. Soc. Lond. B Biol. Sci.* 266, 2017–2022.
155. Bronson, F. H., and Macmillan, B. (1983). Hormonal responses to primer pheromones. In *Pheromones and Reproduction in Mammals* (J. G. Vandenbergh, Ed.), pp. 175–199. Academic Press, New York.
156. Bronson, F. H., and Desjardens, C. (1974). Circulating concentrations of FSH, LH, estradiol and progesterone associated with acute, male induced puberty in female mice. *Endocrinology* 94, 1658–1668.
157. Parfentjev, I. A. (1932). Calcium and nitrogen content in urine of normal and cancer mice. *Proc. Soc. Exp. Biol. Med.* 29, 1285–1286.
158. Thung, P. J. (1962). Physiological proteinuria in mice. *Acta Physiol. Pharmacol. Neerl.* 10, 248–261.
159. Finlayson, J. S., Asofsky, R., Potter, M., and Runner, C. C. (1965). Major urinary protein complex of normal mice: origin. *Science* 149, 981–982.
160. Mucignat-Caretta, C., Caretta, A., and Cavaggioni, A. (1995). Acceleration of puberty onset in female mice by male urinary proteins. *J. Physiol. (Lond).* 486, 517–522.
161. Novotny, M. V. (2003). Pheromones, binding proteins and receptor responses in rodents. *Biochem. Soc. Trans.* 31, 117–122.
162. Drickamer, L. C. (1982). Chemosignal effects on puberty in young female mice: urine from pregnant and lactating females. *Dev. Psychobiol.* 16, 207–217.
163. Drickamer, L. C., McIntosh, T. K., and Rose, E. A. (1978). Effects of ovariectomy on the presence of maturation-delaying pheromone in the urine of female mice. *Horm. Behav.* 11, 131–137.
164. Drickamer, L. C., and Hoover, J. E. (1979). Effects of urine from pregnant and lactating female house mice on sexual maturation of juvenile females. *Dev. Psychobiol.* 12, 545–551.
165. Champlin, A. K. (1971). Suppression of oestrus in grouped mice: the effects of various densities and the possible nature of the stimulus. *J. Reprod. Fertil.* 27, 233–241.
166. Vandenbergh, J. G., Drickamer, L. C., and Colby, D. R. (1972). Social and dietary factors in the sexual maturation of female mice. *J. Reprod. Fertil.* 28, 397–405.
167. McIntosh, T. K., and Drickamer, L. C. (1977). Excreted urine, bladder urine, and the delay of sexual maturation in female house mice. *Anim. Behav.* 25, 999–1004.
168. Savage, A., Ziegler, T. E., and Snowdon, C. T. (1988). Sociosexual development, pair bond formation, and

mechanisms of fertility suppression in female cotton-top tamarins (*Saguinus oedipus*). *Am. J. Primatol.* 14, 345–359.
169. Vandenbergh, J. G., Finlayson, J. S., Dobrogosz, W. J., Dills, S. S., and Kost, T. A. (1976). Chromatographic separation of puberty accelerating pheromone from male mouse urine. *Biol. Reprod.* 15, 260–265.
170. Coppola, D. M., and Vandenbergh, J. G. (1985). Effect of density, duration of grouping and age of urine stimulus on the puberty delay pheromone in female mice. *J. Reprod. Fertil.* 73, 517–522.
171. Lepri, J. J., Wysocki, C. J., and Vandenbergh, J. G. (1985). Mouse vomeronasal organ: effects on chemosignal production and maternal behavior. *Physiol. Behav.* 35, 809–814.
172. Drickamer, L. C., and McIntosh, T. K. (1980). Effects of adrenalectomy on production and excretion of a pheromone delaying sexual maturation in grouped female mice. *Horm. Behav.* 14, 146–152.
173. Novotny, M., Jemiolo, B., Harvey, S., Wiesler, D., and Marchlewska-Koj, A. (1986). Adrenal-mediated endogenous metabolites inhibit puberty in female mice. *Science* 231, 722–725.
174. Jemiolo, B., and Novotny, M. (1994). Inhibition of sexual maturation in juvenile female and male mice by a chemosignal of female origin. *Physiol. Behav.* 55, 519–522.
175. Lombardi, J. R., and Whitsett, J. M. (1980). Effects of urine from conspecifics on sexual maturation in female prairie deermice, *Peromyscus maniculatus bairdii*. *J. Mammal.* 61, 766–768:
176. Haigh, G. R., Lounsburg, D. M., and Gordon, T. A. (1985). Pheromone induced reproductive inhibition in young female *Peromyscus leucopus*. *Biol. Reprod.* 33, 271–276.
177. Getz, L. L., Dluzen, D., and McDermott, J. L. (1983). Suppression of female reproductive maturation in *Microtus* by a female urine pheromone. *Behav. Proc.* 8, 59–64.
178. Batzli, G. Q., Getz, L. L., and Hurley, S. S. (1977). Suppression of growth and reproduction in microtine rodents by social factors. *J. Mammal.* 58, 583–591.
179. Rissman, E. F., and Johnston, R. E. (1985). Female reproductive development is not activated by male California voles exposed to family cues. *Biol. Reprod.* 32, 352–360.
180. Payman, B. C., and Swanson, H. H. (1980). Social influence on sexual maturation and breeding in the female Mongolian gerbil *Meriones unguiculatus*. *Anim. Behav.* 28, 528–535.
181. Clark, J. R., Faillace, L. S., Tribble, L. F., Orr, J. D. E., and Bell, R. W. (1985). Effects of composition and density of the group and "pheromone" treatment on puberty in gilts reared in confinement. In *2nd International Conference on Pig Reproduction* Columbia, MO, p. 57.
182. Abbott, D. H., and Hearn, J. P. (1976). Physical, hormonal and behavioral aspects of sexual development in the marmoset monkey, *Callithrix jaccus*. *J. Reprod. Fertil.* 53, 155–166.
183. Katz, Y., and Epple, G. (1980). Social influences on urinary estradiol cyclicity of female *Saguinus fuscicollis* (Callitrichidae). *Int. Cong. Primatol. Abstr.* 8, 219.
184. Massey, A., and Vandenbergh, J. G. (1980). Puberty delay by a urinary cue from female house mice in feral populations. *Science* 209, 821–822.
185. Massey, A., and Vandenbergh, J. G. (1981). Puberty acceleration by a urinary cue from male mice in feral population. *Biol. Reprod.* 24, 523–527.
186. Coppola, D. M., and Vandenbergh, J. G. (1986). Induction of a puberty regulating chemosignal in wild mice populations. *J. Mammal.* 68, 517–522.
187. Drickamer, L. C. (1977). Delay of sexual maturation in female house mice by exposure to grouped females or urine from grouped females. *J. Reprod. Fertil.* 51, 77–81.
188. Drickamer, L. C., and Mikesic, D. G. (1990). Urinary chemosignals, reproduction, and population size of house mice (*Mus domesticus*) living in field enclosures. *J. Chem. Ecol.* 16, 2955–2968.
189. Keverne, E. B. (1983). Pheromonal influences on the endocrine regulation of reproduction. *Trends Neurosci.* 6, 381–384.
190. McClintock, M. K. (1978). Estrous synchrony in the rat and its mediation by airborne chemical communication (*Rattus norvegicus*). *Horm. Behav.* 10, 264–276.
191. McClintock, M. K. (1983). Pheromonal regulation of the ovarian cycle: enhancement, suppression, and synchrony. In *Pheromones and Reproduction in Mammals* (J. G. Vandenbergh, Ed.), pp. 113–150. Academic Press, New York.
192. McClintock, M. K. (1983). Modulation of the estrous cycle by pheromones from pregnant and lactating rats. *Biol. Reprod.* 28, 823–829.
193. McClintock, M. K. (1971). Menstrual synchrony and suppression. *Nature* 229, 244–245.
194. Weller, L., and Weller, A. (1993). Human menstrual synchrony: a critical assessment. *Neurosci. Biobehav. Rev.* 17, 427–439.
195. Preti, G., and Wysocki, C. J. (1999). Human pheromones: releasers or primers. In *Advances in Chemical Signals in Vertebrates* (R. E. Johnston, D. Muller-Schwarze, and P. W. Sorensen, Eds.), pp. 315–332. Kluwer Academic/ Plenum, New York.
196. Whitten, W. K. (1956). Modification of the oestrus cycle of the mouse by external stimuli associated with the male. *J. Endocrinol.* 13, 399–404.
197. Whitten, W. K. (1958). Modification of the oestrus cycle of the mouse by external stimuli associated with the male: changes in the oestrus cycle determined by vaginal smears. *J. Endocrinol.* 17, 307–313.
198. Marsden, H. M., and Bronson, F. H. (1964). Estrous synchrony in mice: alteration by exposure to male urine. *Science* 144, 1469.
199. Handelmann, G., Ravizza, R., and Ray, W. J. (1980). Social dominance determines estrous entrainment among female hamsters. *Horm. Behav.* 14, 107–115.
200. Milligan, S. R. (1974). Social environment and ovulation in the vole, *Microtus agrestis*. *J. Reprod. Fertil.* 41, 35–47.
201. Knowlton, N. (1979). Reproductive synchrony, parental investment, and the evolutionary dynamics of sexual behavior. *Anim. Behav.* 27, 1022–1033.
202. Aron, C. L. (1979). Mechanisms of control of the reproduction function by olfactory stimuli in female mammals. *Physiol. Rev.* 59, 229–284.
203. Coblentz, B. (1976). Functions of scent-urination in ungulates with special reference to feral goats (*Capra hircus. L.*). *Am. Naturalist* 110, 549–557.
204. Shelton, M. (1980). Goats: influence of various exteroceptive factors in initiation of estrus and ovulation. *Int. Goat Sheep Res.* 1, 156.
205. Jemiolo, B., Harvey, S. and Novotny, M. (1986). Promotion of the Whitten effect in female mice by synthetic analogs of male urinary constituents. *Proc. Natl. Acad. Sci. U S A* 83, 4576–4579.
206. Bruce, H. M. (1979). An exteroceptive block to pregnancy in the mouse. *Nature* 184, 105.
207. Keverne, E. B., and de la Riva, C. (1982). Pheromones in mice: reciprocal interaction between the nose and brain. *Nature* 296, 148–150.
208. Brennan, P. A., and Keverne, E. B. (1997). Neural mechanisms of mammalian olfactory learning. *Prog. Neurobiol.* 51, 457–481.

209. Marchlewska-Koj, A. (1981). Pregnancy block elicited by male urinary peptides in mice. *J. Reprod. Fertil.* 61, 221–224.
210. Dominic, C. J. (1966). Observations on the reproductive pheromones in mice: II. Neuroendocrine mechanisms involved in the olfactory block to pregnancy. *J. Reprod. Fertil.* 11, 415–421.
211. Bruce, H. M. (1965). The effects of castration on the reproductive pheromones of male mice. *J. Reprod. Fertil.* 10, 141–143.
212. Dominic, C. J. (1964). Source of the male odor causing pregnancy block in mice. *J. Reprod. Fertil.* 8, 266–267.
213. Carter, C. S., Getz, L. L., and Cohen-Parsons, M. (1986). Relationships between social organization and behavioral endocrinology in a monogamous mammal. In *Advances in Animal Behavior* (J. Rosenblatt and C. Beer, Eds.), pp. 110–142. Academic Press, New York.
214. Richmond, M. E., and Conaway, C. H. (1969). Induced ovulation and oestrus in *Microtus ochrogaster*. *J. Reprod. Fertil. Suppl.* 6, 357–376.
215. Lepri, J. J., and Wysocki, C. J. (1987). Removal of the vomeronasal organ disrupts the activation of reproduction in female voles. *Physiol. Behav.* 40, 349–355.
216. Dluzen, D. E., Ramirez, V. D., Carter, C. S., and Getz L. L. (1981). Male vole urine changes luteinizing hormone-releasing hormone and norepinephrine in female olfactory bulb. *Science* 212, 573–575.

Reproductive Processes and Their Control

Knobil and Neill's Physiology of Reproduction, Third Edition
edited by Jimmy D. Neill,

CHAPTER 38

Puberty in the Rat

Sergio R. Ojeda[1] and Michael K. Skinner[2]

Historical Aspects, 2061
The Rat as a Model, 2062
The Female, 2063
Fetal Development, 2063
Postnatal Development, 2064
The Male, 2084
Fetal Development, 2084
Postnatal Development, 2085
Modulatory Influences Regulating the Timing of Puberty, 2089
Adrenal Gland, 2089
Somatic Growth, 2090
Pineal Gland, 2091
Pheromones, 2092
Environmental Factors and Toxins, 2092
Mechanism of the Onset of Puberty, 2093
"Gonadostat" Hypothesis, 2093
Gonad-Independent Increase in Central Drive, 2093
Loss of a Central Restraint, 2094
Activation of Excitatory Inputs, 2095
Glutamate–GABA Interactions and the Onset of Puberty, 2095
Conclusions, 2100
The Female, 2100
The Male, 2102
Acknowledgments, 2103
References, 2103

HISTORICAL ASPECTS

The first recorded use of the rat for experimental purposes was published in 1856 (1). Since its domestication in the early part of this century, the rat has probably contributed more substantially to the advancement of the biological sciences than any other laboratory species. Clearly, the story of the rat, as portrayed by Lindsey (2), is one of "ascendancy from the gutter to a place of nobility."

The use of the rat for the study of sexual development and puberty can be traced to the beginning of the 20th century, when investigators initiated studies to test the hypothesis that the ovaries contain a substance(s) responsible for the advent of sexual maturation. Essential to these undertakings was the pioneering work of Long and Evans, who in 1922 published a classic monograph (3) describing in detail the physiology of the rat estrous cycle and the initiation of female reproductive capacity. Long and Evans (3) demonstrated, for the first time, that the ovaries of immature rats develop rapidly when transplanted into adult rats. It was, however, the work of Allen and Doisy (4), published in 1924, that showed conclusively that extracts of follicular fluid injected into immature rats were capable of advancing vaginal opening. A year later, Frank et al. (5) presented evidence that a "lipoid extract" of placental tissue was not only able to induce precocious vaginal opening, but, more importantly, also advanced the first ovulation. These authors concluded that "puberty results from the elaboration, in sufficient amount, of the female hormone, and that the advent of puberty is not due to the removal of an inhibitory influence..."

That immature ovaries can modify the secretion of gonadotropin hormones from the anterior pituitary and that, in turn, these hormones can hasten the initiation of puberty by accelerating ovarian maturation were first demonstrated by Kallas in 1929 (6). This author found that when two immature female rats were joined parabiotically and one of them was castrated, the other one underwent precocious puberty. These experiments indicated that a dynamic relationship between the anterior pituitary and the ovaries is

[1]Division of Neuroscience, Oregon Regional Primate Research Center, Beaverton, Oregon;
[2]Center for Reproductive Biology, Washington State University, Pullman, Washington.

operative well before puberty; 3 years later, Hohlweg and Junkmann postulated that the brain is part of this axis (7). Surprisingly, however, their concept found little immediate support. This was perhaps due to the widespread acceptance of a hypothesis formulated by Moore and Price (8), also in 1932, which stated that a pituitary–gonadal interrelationship was sufficient to explain the maturation and initiation of reproductive capacity.

At this same time during the late 1920s and 1930s, McCullagh and colleagues used the rat model to study testis endocrinology (9,10). In a similar experiment as above, testes were isolated and extracts injected into castrated rats to maintain male reproductive tract morphology. The lipid substance was termed "androtin" (9); when further characterized it was found to correspond to testosterone (T) and precursors to the androgen family of steroids. Similar experiments also demonstrated the ability of a nonsteroidal testis extract to influence pituitary morphology; this testis hormone was termed "inhibin" (10). These early studies using a rat model helped establish our current understanding of testis endocrinology and the testis–hypothalamic–pituitary axis. The rat model was then used extensively in subsequent years to elucidate male reproduction.

It was not until 20 years later that the work of Harris and Jacobson (11) provided the basis for our current understanding of the pivotal role that the central nervous system (CNS) plays in the control of reproductive maturation. These authors showed that pituitaries from immature rats transplanted into the subarachnoid space under the median eminence of hypophysectomized adult animals were able to sustain estrous cyclicity. Subsequent experiments performed by Donovan and van der Werff ten Bosch (12) in the mid-1950s led to the concept of certain hypothalamic areas that exert inhibitory influences on gonadotropin secretion and that removal of these influences results in the initiation of puberty. A few years later, Elwers and Critchlow (13) found that lesions of the amygdala resulted in precocious ovarian activation, thereby providing evidence that extrahypothalamic structures are also involved in modulating gonadotropin secretion in developing rats. Although these early research efforts provided much insight into the neuroendocrinology of puberty, further significant progress was hampered by the lack of sensitive and accurate methods to measure changes in serum hormone levels. The 1970s witnessed an explosion of reports that, using the then-novel technique of radioimmunoassay (RIA), probed the neuroendocrine system from different angles in search of more definitive answers. In the 1990s, the availability of recombinant DNA technology provided a new set of powerful tools to explore the mechanisms underlying the initiation of puberty (14); in the current decade, the completion of the human genome, followed by that of the mouse, and the impending completion of the rat genome are beginning to open the doors to an entirely new world of scientific discovery that will permit the identification of functional networks rather than isolated molecules involved in controlling the pubertal process.

Beginning with the comprehensive review of Donovan and van der Werff ten Bosch (15), published in 1965, the physiology of rat puberty has been systematically reviewed over the years (16–23). This chapter provides an update of developments in this field. Similarities and differences with sexual maturation in other species will become apparent when this chapter is considered in conjunction with Chapters 39 and 40, which deal with puberty in the sheep and monkey, respectively.

THE RAT AS A MODEL

The rat is a convenient animal to use for the study of sexual development. It grows rapidly, reproduces frequently, is relatively inexpensive and easy to handle, and external signs of sexual maturity, though scanty, are readily detectable. In working with this animal, one may also assume that most of the basic mechanisms underlying the process of sexual maturation are conserved across species. If such an assumption is correct, then many of the results obtained in the rat may be extrapolated to other species, including the human. Among such mechanisms, the control of luteinizing hormone-releasing hormone (LHRH) by neurotransmitters, the initiation of gonadotropin secretion, the cellular components of steroid-positive and -negative feedback, and the control of ovarian and testis development represent only a few examples. Perhaps the most striking difference between the rat and the primate is the relative absence in the former of the juvenile hiatus of gonadotropin secretion, which is characteristic of the latter (24). This difference severely handicaps the rat as an animal model in which to investigate gonad-independent CNS-originated events that might be responsible for the decline in hypothalamic activity during the human juvenile period. Species such as the rhesus monkey appear to be the logical choice to examine this outstanding issue. However, the demonstration that the rat, like the primate, exhibits a diurnal change in the mode of LH secretion at the end of juvenile development (25) reiterates the possibility of using the rat as a model for the analysis of at least some aspects of this early event, which appears to be the primary hormonal manifestation of puberty onset.

From a developmental point of view, the rat is born at the stage comparable with 150 days of human gestational life (26). The gestational period of the rat lasts for 22 to 23 days. The first ovulation in most laboratory stocks occurs 35 to 45 days after birth. In males the first spermatozoa are seen in the lumen of seminiferous tubules by 40–45 days of age (27), and they reach the vas deferens 13 to 14 days later (28). Testicular descent occurs after day 15. Externally, the progression of puberty in the male rat can be followed by the growth of the testis and more precisely defined by the separation of the foreskin of the penis from the glans, known as balanopreputial separation (29). In females the only signal that puberty has occurred is canalization of the vagina, which normally is imperforated before puberty and later becomes patent as a consequence of estrogenic stimulation. Vaginal opening usually occurs on the day after the first preovulatory surge of gonadotropins has taken place (16,17,30,31). In most cases, vaginal lavages at opening show a majority of cornified cells (estrus), a condition that is followed within 1 to 2 days by the appearance of leukocytes, which soon become the predominant cell type (first diestrous phase of puberty).

Male rats, in general, do not display a postnatal period of testicular quiescence analogous to that observed in humans but, rather, show initiation and progression of testicular development at a very early age, postnatal days 8–10. Thus, terms such as "infantile" and "juvenile," when applied to male rats, should be treated with caution because they are not directly analogous to these phases in humans. Whereas developmental phases in the female rat have usually been defined in relation to the maturational stages of the ovary (16,32–34), sexual development of male rats has generally not been divided into specific maturational stages for the purpose of its study. Ramirez (17) suggested classification of the phases of rat puberty based on physiological parameters such as the changes in circulating gonadotropin levels and the alterations in steroid feedback mechanisms occurring at different postnatal ages. This classification was subsequently modified to include morphological and physiological parameters for both males and females (19). Accordingly, sexual development in the male rat can be divided into four phases: a *neonatal* period that comprises the first week after birth; an *infantile* period that extends from days 8 to 21; a *juvenile* period that ends around day 35; and the *peripubertal* period that ends at about 55 to 60 days of age [i.e., with the appearance of mature sperm in the vas deferens (28)]. Details of the functional and morphological parameters used to define these developmental periods are provided in (19).

Postnatal development of the female rat can also be divided into four phases, which have been described in detail earlier (19): a *neonatal* period that is initiated at birth and ends on postnatal day 7; an *infantile* period that extends from days 8 to 21; a *juvenile* period that ends around days 30 to 32; and a *peripubertal* period that has a variable duration but culminates with the occurrence of first ovulation (around days 38–40 for most laboratory stocks). Defining the end of juvenile development has been difficult; morphologically, the appearance of uterine fluid signals the beginning of the peripubertal period (19) but tells us little about the hormonal events responsible for the change. The end of the juvenile period can be more precisely defined as the time when morning–afternoon differences in pulsatile LH release become established (25).

An additional developmental phase that needs to be included (35) is a fetal or embryonic period, which for mechanistic purposes can be considered as starting at gestational day 12. It is around this time that sex determination is initiated and LHRH can first be detected in the fetal brain (36).

THE FEMALE

Fetal Development

Initiation of LHRH and Gonadotropin Secretion

Although LHRH neurons of adult rats are located primarily in the septal medial preoptic and hypothalamic areas of the brain (37,38), they are embryonically derived from cells in the olfactory placode (39–41). Most of these subsequently migrate caudally during fetal development toward their final destination in the forebrain. LHRH can be detected in the hypothalamus earlier than gonadotropins in the anterior pituitary. Immunoreactive LHRH can be detected in the brain of rat fetuses as early as day 12 (36), at which time LHRH-binding sites can also be found in the primordial anterior pituitary. If anterior pituitary anlagen are collected after day 13 and cultured in vitro, gonadotrophs differentiate spontaneously (42,43). However, if the glands are removed at an earlier age, the gonadotrophs fail to differentiate unless they are exposed to exogenous LHRH (43,44). These findings and the observation that the pituitary is vascularly connected to the brain between fetal days 12 to 14 (45,46) strongly suggest that hypothalamic LHRH plays a trophic function essential for the differentiation of pituitary gonadotrophs.

How the differentiation of LHRH neurons themselves is controlled remains unknown (47), but it appears to be controlled by transcription factor(s) different from those involved in determining the differentiation of the olfactory and respiratory epithelia from the nasal placode [for additional details see (47) and Chapter 28]. Hypothalamic LHRH content remains low until about gestational days 17–18; at this time it begins to increase, becoming substantially higher by the day of birth (36,42,48). Pituitary LH, as well as a LH release in response to LHRH, can also be detected around fetal day 17 (36,42,49). Pituitary follicle-stimulating hormone (FSH), on the other hand, becomes detectable much later, by days 19 to 21 (36,42,50). It would appear, however, that little LHRH reaches the adenohypophysis even at this time, because circulating gonadotropin levels remain at low values until the day of birth (36,42,48,49,51–53).

Embryonic Development of the Ovary

The indifferent gonad derives from the urogenital ridge under the control of transcription factors such as steroidogenic factor 1 (SF1), Wnt-1, Lim-1, and LHX (54). Upon sex determination around embryonic days 11–12 (E11-12), ovarian development requires the early expression of DAX-1 and Wnt-4 [reviewed in (54,55)]. The embryonic ovary does not have any major morphological changes in comparison with the testis. After sex determination (gestation days 12–14), the germ cells proliferate and form "nests" with aggregated germ cells surrounded by layers of squamous epithelial cells. These nests break down into primordial follicles after birth but are maintained throughout embryonic development in the rat. The embryonic ovary is not steroidogenic, and no hormones are produced. Therefore, the embryonic ovary primarily grows in size and germ cells proliferate into nests of aggregated germ cells.

Postnatal Development

Prepubertal Period

In this section we discuss the maturational events that occur at the hypothalamic–pituitary level and within the ovary during the neonatal, infantile, and juvenile periods and the developmental changes in functional relationship between these three basic components of the neuroendocrine reproductive axis.

Initiation of Ovarian Function: Extrapituitary Regulation. It is now clear that plasma gonadotropins are not involved in the development of the fetoneonatal ovary. Neither LH nor FSH receptors can be detected in ovaries of rats younger than 4 to 5 days of age (56–60). That initiation of follicular growth is gonadotropin independent was suggested by the findings that neither administration of pregnant mare serum gonadotropin (PMSG) or FSH (61,62) nor immunoneutralization of endogenous gonadotropins (63) alters the number of follicles that begin to grow during the first few days of postnatal life (64,65). More recently, deletion of the FSH (66) and LH receptor gene (67) by gene targeting unambiguously demonstrated the validity of these findings and firmly established the concept that early ovarian development is a gonadotropin-independent event.

The initiation of follicular formation in the rat is a dramatic event that takes place shortly after birth (68,69). Although very few, if any, primordial follicles are seen within the first 24 hours of birth, a marked increase occurs within the next 24 hours (Fig. 1). Before the first primordial follicles are formed after birth, the ovary consists of three main components: nests of germ cells, mesenchymal cells, and epithelial cells (70). As germ cells migrate into the primitive gonad, they form nests that are surrounded by epithelial cells of the ovarian rete (a group of convoluted tubules derived from the mesonephric kidney; see Chapter 7). It has been suggested that these cells, and not those derived from the coelomic epithelium, are destined to form the granulosa cells (71). Formation of primordial follicles after birth is preceded in the embryo by migration of mesenchymal cells that form stromal "pockets" containing groups of presumptive granulosa cells and clusters/nests of oocytes (72). Oocytes undergo extensive apoptosis while they are still within the nests. During fetal life, oocytes are protected by the high levels of progesterone (P) associated with gestation (Figs. 2 and 3), but after birth

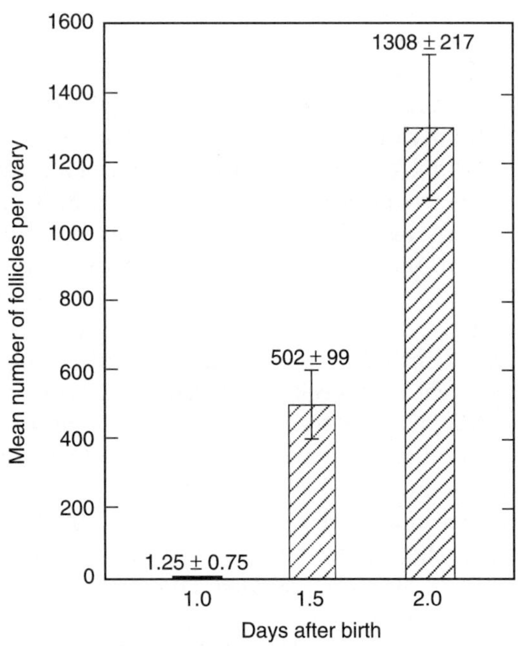

FIG. 1. Development of primordial follicles in the neonatal rat ovary. *Columns* represent mean values. *Vertical lines* indicate standard errors of the means (SEM). $N = 4$ for each column. [From (69), with permission.]

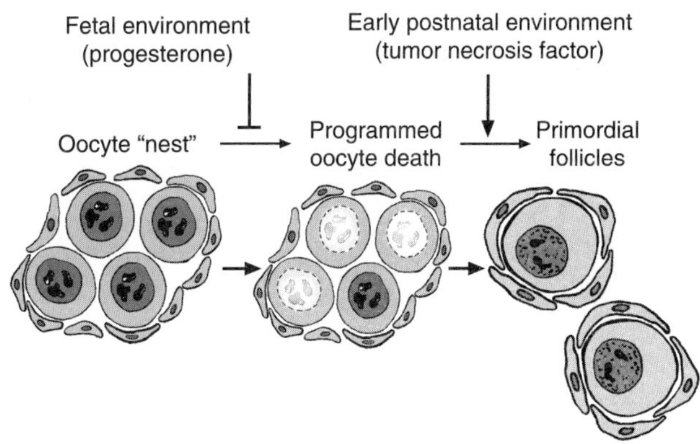

FIG. 2. Current model for the apoptosis of oocytes that occurs within 72 hours after birth in the rat ovary. Nests of oocytes maintained throughout embryonic ovarian development undergo a rapid loss of oocytes by apoptosis in response to TNF-α. The fetal environment of high progesterone concentrations blocks this apoptosis, which is initiated after birth as progesterone concentrations decline. Surviving oocytes form primordial follicles that consist of an oocyte surrounded by a single (complete or incomplete) layer of squamous pregranulosa cells.

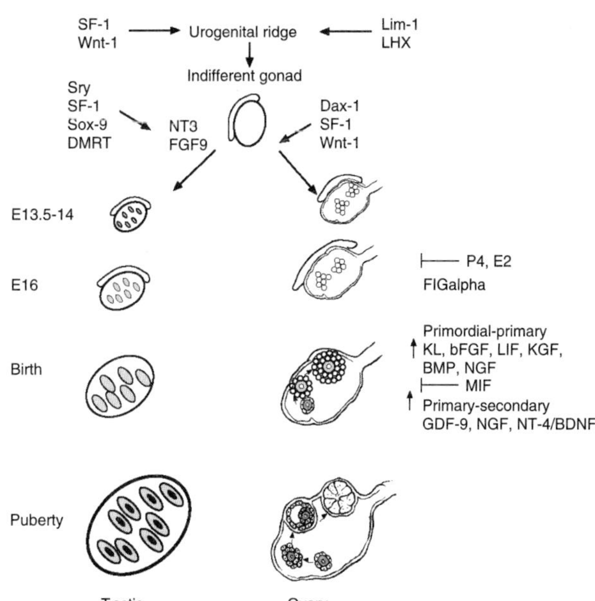

FIG. 3. Schematic of embryonic gonad development in the rat. The indifferent gonad is derived from the urogenital ridge under the control of transcription factors such as steroidogenic factor 1 (SF1), Wnt-1, Lim-1, and LHX. Upon sex determination around embryonic days 11–12 (E11-12), testis determining factor SRY, in combination with the transcription factors SOX-9 and DMRT, promote testis development. Neurotropin-3 (NT3) produced by Sertoli cells acts as a chemotactic factor for mesonephros cell migration into the gonad to promote cord formation at E13-E14. The testis subsequently grows and upon the onset of puberty seminiferous tubules are formed. Ovarian development requires the early expression of DAX-1 and Wnt-4; once nests of germ cells are formed they are maintained until birth. High progesterone (P4) and estradiol (E2) levels in the fetal ovary prevent the breakdown of the nests. Formation of primordial follicles requires the presence of FIGα. In the absence of this transcription factor, oocytes die and primordial follicles fail to form. Primordial follicles assemble after birth and subsequently develop into primary follicles via a process controlled by Nobox, an oocyte-specific homeodomain-encoding gene. Downstream factors required for the differentiation of primordial into primary follicles include growth factors such as kit ligand (KL), bFGF, nerve growth factor (NGF), leukemia inhibitory factor (LIF), keratinocyte growth factor (KGF), bone morphogenic proteins (BMP), and insulin. Anti-müllerian hormone (AMH) plays an inhibitory role in this process. Primary follicles become secondary (i.e., with two layers of granulosa cells under the influence of GDF-9 and the neurotrophins NT-4 and brain-derived neurotrophin factor). The small oval next to the gonad after sex determination represents the mesonephros. ↑, stimulation; ⊢, inhibition.

when P levels drop, oocytes become exposed to tumor necrosis factor α (73,74) (Fig. 2) and die in large numbers; those oocytes that survive form primordial follicles (75,76). This happens when somatic cells encircle a single oocyte and form a single (complete or incomplete) layer of flattened pregranulosa cells (Fig. 2). This organizational event is, at least in part, controlled by FIG alpha (77), a germ cell–specific basic-helix-loop-helix transcription factor; in the absence of FIG alpha, primordial follicles fail to form, oocytes are depleted and infertility ensues. Subsequent differentiation and development of primordial follicles is regulated by several factors produced locally by either granulosa cells or the oocyte itself (Fig. 3), including KL (the c-Kit ligand) (78,79), basic fibroblast growth factor (bFGF) (80), leukemia inhibitory factor (75), and the neurotrophins nerve growth factor, neurotrophin-3, and brain-derived neurotrophin factor (81,82).

Insulin growth factor (IGF) and keratinocyte growth factor have also been shown to facilitate the growth of primordial follicles [for references and discussion see (74)]. Anti-müllerian hormone, on the other hand, inhibits the process (83). In addition to these proteins, the adjacent stromal cells produce bone morphogenic protein-4, which promotes primordial follicle development (84). In the absence of GDF-9 (which is produced by primary follicle oocytes and not primordial) the oocytes continue to grow, but follicular development is arrested at the primary stage (78,85). In the absence of the trkB receptor (which recognizes neurotrophin-4 and brain-derived neurotrophin factor and is preferentially expressed in oocytes) the oocytes die, and this is followed by loss of follicular organization (82). Recently, Rajkovic et al. (86) demonstrated that *Nobox* (newborn ovary homeobox-encoding gene), an oocyte-specific homeobox gene (87), is critical for postnatal oocyte survival and for the differentiation of primordial into primary follicles (86). Thus, *Nobox* appears to be an upper-echelon gene that expressed in oocytes plays a fundamental role in ensuring both oocyte survival and early follicular growth. Its discovery gives considerable credence to the view that the oocyte is essential for both the assembly of new follicles and their subsequent development [for reviews see (74,88,89)].

Like follicular assembly and early follicle development, the initiation of ovarian steroidogenesis is also gonadotropin independent. Fetal ovaries cultured in vitro fail to respond to either FSH or LH with increases in aromatase activity (90). Nevertheless, exposure to exogenous cyclic AMP (cAMP) or activation of their adenylate cyclase system with forskolin results in increased aromatase activity, suggesting that first messengers other than FSH or LH operate within the ovary before follicular growth is initiated. These messengers signaling via cAMP-dependent pathways would be expected to be present in the ovary before the appearance of gonadotropin receptors. The neuromodulator vasoactive intestinal polypeptide (VIP) (91) and the neurotransmitter norepinephrine (NE) meet these criteria, because both reach the ovary via the extrinsic innervation before the appearance of primary follicles (69,92), act through cAMP, and are able to stimulate ovarian steroidogenesis [reviewed in (89)]. In fact, VIP can stimulate cAMP production and induce aromatase activity in fetal ovaries (90) well before acquisition of responsiveness to FSH. Thus, the initiation of ovarian steroidogenic activity might be at least in part determined by neurotransmitters reaching the gland via the extrinsic innervation, which in the rat precedes the assembly of primordial follicles (69).

The Hypothalamic–Pituitary Unit: Changes in Gonadotropin Secretion and Their Relationship to LHRH Release. Beginning shortly after birth, plasma FSH levels start to increase, reaching maximum values by day 12 (93,94). Thereafter, levels decline gradually, so that by the end of the juvenile period they are about one-fifth of the values present at day 12 (93,95). Plasma LH is also more elevated in neonatal–infantile rats than in juvenile animals (93,95,96), but the elevation is less evident than that of FSH. Release of LH in infantile rats appears to conform to a pattern of moderately elevated levels of the hormone interrupted by sporadic surges of release (95,97). Such bursts appear to become less evident as the animal grows and disappear completely with the advent of juvenile development (98).

How do these patterns of gonadotropin release relate to developmental changes in LHRH secretion? The hypothalamic content of LHRH increases markedly between the day of birth and the end of juvenile development (36,42,99). Although LHRH content is substantially lower during the neonatal–infantile period than during juvenile development, in vitro experiments have shown that the LHRH-releasing system is more responsive to a depolarizing stimulus around the second week of postnatal life than during the juvenile period (100). On the other hand, in vitro experiments using hypothalamic explants have shown that LHRH is secreted in a pulsatile fashion throughout prepubertal development (101,102). This suggests that the infantile LHRH neuronal system in the rat already has the capacity to generate LHRH secretory episodes in the absence of extrahypothalamic inputs. This pulsatility appears to be, at least in part, an intrinsic property of LHRH neurons because isolated clonal LHRH neurons in culture release LHRH in a pulsatile fashion in the absence of any neural connections (103).

An earlier study (101) showed that the frequency of LHRH pulses increases during prepubertal maturation of male rats. The existence of a similar increase in LHRH pulse frequency in females was recently demonstrated by Matagne et al. (104), thus offering an explanation as to why the secretion of FSH during the infantile phase of development differs so markedly from that of LH. The aforementioned findings suggest that as the reproductive hypothalamus matures during the first 2 weeks of postnatal life, LHRH is released as infrequent discharges that are sufficient to sustain a high level of FSH secretion but only generate transient bursts of LH release. Such a differential pattern of gonadotropin release has been observed in ovariectomized rhesus monkeys bearing hypothalamic lesions that were given low-frequency pulses of exogenous LHRH (105).

By the same reasoning, the decrease in FSH levels that follows the peak of secretion on day 12 would be caused by an increase frequency of LHRH discharges. Support for this concept is provided by two observations: (a) subcutaneous infusion of LHRH as 1-hour pulses every other hour to infantile ovariectomized rats via osmotic minipumps decreases plasma levels of FSH (Fig. 4A) but not LH (not shown), and (b) pulses of LHRH administered every 120–480 minutes increase the mRNA levels of FSH-β but not LH-β. On the other hand, LHRH pulses administered at a high frequency (e.g., every 8 minutes) produce an increase in the mRNA levels of LH-β but not FSH-β (106). Taken together, these results suggest that the frequency of LHRH discharge may play an important role in determining the pattern of FSH release during the infantile period. Contrary to an earlier belief, the decrease in FSH secretion that occurs during the infantile–juvenile transition period in the female rat is not exclusively due to activation of steroid estrogen–negative feedback mechanisms. Ovariectomy on postnatal day 6 results in increased plasma FSH levels that peak 10 days later; thereafter FSH secretion decreases markedly as the animals enter, and progress through, the juvenile period of development (Fig. 4B). Thus—like in humans and monkeys—FSH secretion in developing female rats is subjected to a central gonad-independent regulation that operates at the time when steroid-negative feedback control is becoming established (107) (see below). Unexpectedly, this similarity does not extend to the control of LH secretion. After rats are ovariectomized

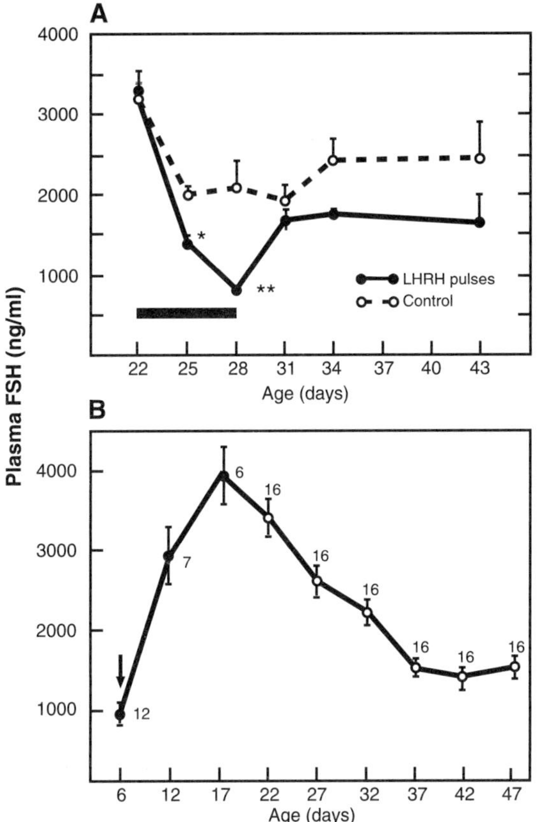

FIG. 4. A: Effect of exogenous LHRH on plasma FSH (top) and LH (bottom) levels in female rats. All the animals were ovariectomized on day 10. Between days 22 and 28 (*horizontal bar*) animals received either a 1-hour subcutaneous infusion of LHRH (1 μg) every other hour (*solid circles*) or served as untreated controls (*open circles*). The procedure for intermittent infusion of LHRH via osmotic minipump was based on that described by Lynch et al. (798). Each point represents the mean of five to seven animals, and the standard errors of the mean are depicted as vertical lines. *$P<.05$; **$P<.01$. **B:** Ovary-independent changes in FSH secretion in the prepubertal female rat. The animals were ovariectomized on postnatal day 6, and trunk blood was collected at the ages indicated in the abscissa. Notice that serum FSH levels increase first and then decrease at the time when FSH secretion normally decreases in ovary-intact rats. (From H. F. Urbanski and S. R. Ojeda, unpublished data.)

on postnatal day 6 LH levels increase, remaining maximally elevated throughout juvenile development (data not shown). A similar lack of centrally driven decrease in LH output has been reported in guinea pigs ovariectomized on day 2 of life (108).

A major factor contributing to the enhanced output of FSH in infantile rats is the relative ineffectiveness of estradiol (E_2)-negative feedback. Administration of E_2 has been repeatedly shown to be less effective in depressing gonadotropin levels in ovariectomized infantile rats than in juvenile animals (109–111). Of particular interest in this regard is the finding (112) that immunoneutralization of serum E_2 in infantile rats fails to induce a rise in circulating gonadotropin levels, which would be expected if E_2-negative feedback were operative at this time. The demonstration that RU2858, a synthetic estrogen that does not bind to α-fetoprotein (AFP, see below), strikingly inhibits the postovariectomy rise of serum gonadotropins in infantile rats (113) firmly established the concept that E_2-negative feedback fails to operate at this age because of the presence of AFP and not because of a lack of specific E_2 receptors. By day 16, AFP levels have decreased sufficiently, allowing free E_2 levels to be detected for the first time (114–116). Coinciding with this shift, the high serum FSH levels begin to decrease. Thus, it appears that the infantile increase in serum FSH levels may depend on at least three factors: a slow frequency of LHRH release, a facilitatory effect of 5α-reduced androgens produced in the anterior pituitary on LHRH-induced gonadotropin secretion, and the failure of E_2-negative feedback to operate at full capacity.

An additional factor contributing to the pattern of gonadotropin release observed in infantile female rats as compared with juvenile animals is a change in pituitary responsiveness to LHRH. It is well established that pituitary responsiveness to LHRH is much greater in infantile rats (117–119) than in juvenile animals. This increased responsiveness may be related to the existence of a greater percentage of gonadotrophs in the pituitaries of younger rats (120), and it may also be related to a direct facilitatory effect exerted by nonaromatizable androgens and P on the pituitary gland. Ovariectomy has been found to reduce the FSH response to LHRH, whereas dihydrotestosterone (DHT) or P treatment restores it (119). A stimulatory effect of DHT on FSH secretion has also been observed within 24 hours of its administration to immature female rats (121). A physiological role for DHT is further indicated by the fact that although its plasma levels are low in female rats, 5α-reductase activity is elevated in the anterior pituitary of infantile rats (122). Moreover, the developmental pattern of pituitary 5α-reductase activity closely follows that of serum FSH (122).

Other developmental changes include a shift in the molecular forms of pituitary FSH, from forms with low biological activity to those with high ones (123). However, both pituitary and serum FSH from infantile rats are biologically active, as they can induce ovarian growth and follicular development (124). As with FSH, there is some evidence to suggest that LH also changes in nature during puberty (125). With regard to LH secretion, infantile rats show sporadic bursts of LH secretion, which appear to result from activation of noradrenergic neurons, because the NE turnover in the preoptic area (POA) increases at the time of the LH secretory discharges (126). These bursts of LH release do not appear to be caused by an early expression of E_2-positive feedback (127) because

(a) E_2 is unable to induce an LH surge before postnatal day 15 (116,128), (b) passive immunization against E_2 fails to inhibit the LH surges (112), and (c) environmental disturbance inhibits their occurrence (129).

By the end of the infantile period, plasma FSH levels have decreased substantially and the bursts of LH secretion have disappeared. Throughout juvenile development, plasma FSH continues to decrease, though much less noticeably, and plasma LH levels remain at low values (93–98). The mode of LH release, however, is clearly pulsatile (25,130,131), with an interpulse interval of about 30 minutes (25).

LHRH Neurons and Their Transsynaptic/ Glial Control

LHRH Neurons. The low circulating gonadotropin levels seen in juvenile rats do not faithfully reflect the changes in hypothalamic activity that appear to occur during this period of development. This may be, at least in part, due to a low pituitary responsiveness to LHRH (117,119,132), which does not increase until the day of the first proestrus (133,134) (see below). During the juvenile period, LHRH gene expression is activated, with pro-LHRH mRNA levels increasing almost threefold (135). As LHRH gene transcription increases, the hypothalamus undergoes plastic rearrangements that may be important for the pubertal activation of LHRH release. One of these plastic changes involves LHRH neurons themselves (136). The other more general—but also poorly understood—plastic change occurs in neuronal subsets located in regions of the hypothalamus involved in the control of reproductive function, such as the arcuate nucleus (137). During the neonatal–infantile periods, the soma of most LHRH neurons has a smooth surface. In contrast, LHRH cells with an irregular surface ("spiny cells") become predominant between weeks 4 and 5 of postnatal life, a change that takes place even in the absence of the ovaries (138). Because in the prepubertal rat the total number of LHRH cells does not change, despite an increase in the number of spiny cells, it is evident that smooth LHRH cells have developed into spiny ones. These morphological changes may reflect an increase in the number of synaptic contact between controlling neuronal networks and LHRH neurons (139,140).

Although not formally demonstrated, it would be reasonable to assume that qualitative and/or quantitative changes in synaptic input to LHRH neurons might contribute to at least two events taking place during juvenile development: the prepubertal increase in responsiveness of the hypothalamic–pituitary unit to the positive feedback effect of E_2 (116) (see below) and the increase capacity of LHRH neurons to release LHRH in response to either excitatory amino acids (141) or prostaglandin E_2 (PGE_2) (132). The latter is, at least partially, maintained by circulating E_2 levels (142). During juvenile development, the cell–cell communication mechanisms that operate in the mature hypothalamus become firmly established and the LHRH neuronal network begins to operate under a dual inhibitory–facilitatory transsynaptic control and a glial growth factor–dependent regulatory input.

Excitatory Neurotransmission

Glutamatergic Neurons. Among the neuronal systems involved in the excitatory control of LHRH secretion, glutamatergic neurons stand out as major players. The facilitatory control they exert on LHRH neurons increases gradually during the infantile–juvenile periods as evidenced by the increasing ability of glutamate receptor stimulation to increase LHRH release during this phase of development (141). Hypothalamic glutamate content increases during prepubertal and peripubertal development (143,144), reaching a maximum after the onset of puberty (144). In addition, glutamate release from POA/medial basal hypothalamus explants (145,146) and the in vivo release of glutamate from the POA of female rats, measured in push–pull perfusates, increases during the juvenile–peripubertal periods (147), becoming markedly elevated before and during the preovulatory LH surge induced by either estrogen alone (148) or by the sequential treatment of estrogen plus P (149).

Metastin-producing Neurons. Metastin (kisspeptin), a peptide product of the *KiSS-1* gene (150,151), was shown to be a physiological ligand for the orphan receptor GPR-54 (151,152). Mutations of the *GPR-54* gene result in hypothalamic hypogonadism in humans (153,154) and impaired sexual development in mice (154,155), making it clear that the metastin–GPR-54 system provides a critical facilitatory transsynaptic input to LHRH neurons during sexual development. Recent findings have shown that metastin potently stimulates LHRH release in both rodents (156) and monkeys (157), further emphasizing the importance of this ligand–receptor complex in the developmental control of LHRH secretion.

Noradrenergic Neurons. Noradrenergic neurons have been extensively implicated in the control of pulsatile LHRH and LH release and as mediators of the ovarian steroid-stimulated preovulatory gonadotropin surge in rodents (158–160). NE turnover rate, which may more faithfully reflect changes in neurotransmitter activity than alterations in catecholamine content, increases in the medial preoptic area (MPOA) during prepubertal development of the female rat, reaching maximal values during the days preceding the first preovulatory surge of gonadotropins (161–164). PGE_2 mediates NE-induced LHRH release (165) via a Ca^{2+}-dependent (166,167) mechanism. Exposure of median eminence fragments from late juvenile female rats to NE resulted in a dose-related increase in PGE_2 and LHRH release (142), and in vivo treatment with E_2

increased the subsequent in vitro response of both PGE_2 and LHRH to NE. Because LHRH release from the median eminence of female rats increases from early juvenile (postnatal day 22) to peripubertal development (postnatal day 34) in response to PGE_2 and because the LHRH response to PGE_2 stimulation was blocked by ovariectomy and restored by E_2, it appears to be clear that the stimulatory action of NE on LHRH release is mediated by an E_2-sensitive PGE_2-dependent mechanism, which is further enhanced as puberty progresses (142).

The turnover rate of dopamine has been also shown to increase during prepubertal development (162,164). However, the ability of dopamine to stimulate adenylate cyclase activity decreases at this time, suggesting a loss in receptor sensitivity to the catecholamine (168). Although this change may be functionally related to the increasing prolactin (PRL) levels observed during development (95,98,169), it may also reflect a loss of an inhibitory dopamine tone on LHRH release (168). Perhaps these phenomena are related to a change in the influence that monoamines exert on gonadotropin secretion during infantile–juvenile development (170,171). Blockade of catecholamine synthesis in 16-day-old rats resulted in elevation in plasma LH levels, but the same treatment reduced LH levels in late juvenile 30-day-old rats (171), suggesting that although catecholamines inhibit LHRH secretion in young animals, they have a stimulatory effect as the animals approach the initiation of puberty.

Neuropeptide Y–producing Neurons. A neuropeptide implicated in the facilitatory transsynaptic control of rat puberty is neuropeptide Y (NPY) [for review see (23)]. For instance, the hypothalamic content of NPY increases during postnatal development of the rat, reaching maximal levels near the time of puberty (172,173). An increase in NPY mRNA abundance and NPY peptide levels, which parallels similar changes in LHRH mRNA and LHRH peptide content, occurs in the medial basal hypothalamus and medial POA of late (29-day-old) juvenile rats treated with estrogen and P to induce a preovulatory LH surge (174,175). Intraventricular infusion of an NPY antiserum delays the age at puberty in female rats (176), whereas a single intraventricular injection of NPY advances vaginal opening and the first ovulation in these animals (176). Moreover, NPY injected during a 5-day period into the lateral ventricle of chicks induces precocious puberty and an increase in food intake (177).

Inhibitory Neurotransmission

γ-Aminobutyric Acid. γ-Aminobutyric acid (GABA) is the dominant inhibitory neurotransmitter in the hypothalamus (178). GABA is synthesized from glutamate via a decarboxylation reaction catalyzed by two forms of the enzyme glutamic acid decarboxylase (GAD). Both the abundance of the mRNAs encoding these two forms (GAD-65 and GAD-67) and GABA concentrations themselves increase in the rat POA, first between neonatal and infantile development (143,179,180) and then again during the juvenile period (144,181). The developmental changes in GAD mRNA abundance and/or GABA content within the medial basal hypothalamus are less clear. Some show an increase in GAD mRNA content in the arcuate nucleus (179); others indicate that GABA concentrations in the rat medial basal hypothalamus do not change during the juvenile period (144). Push–pull perfusion experiments, on the other hand, have shown that GABA release from the POA decreases during the juvenile–peripubertal period (postnatal days 26 to 35) in female rats (147), thus suggesting that, like in the female monkey (182), the initiation of puberty in the female rat is preceded by a gradual loss of GABA-mediated inhibition.

Opioid Peptides. Because opioid peptides suppress pulsatile LH release (183–185) and because the opioid receptor antagonist naloxone counteracts the negative feedback effect of estrogen on gonadotropins in adult animals (186–192), a role for endogenous opioids in puberty has been proposed. In both immature rats and sheep, naloxone stimulates LH release (193–195), and administration of naloxone during early postnatal life elicits precocious puberty in female rats (196). Although studies carried out in sheep and monkey suggest that opioid peptides do not play a role in inhibiting gonadotropin secretion before puberty [see for instance (197); for additional references see (198,199)], the relative contribution of different opioid peptide systems to the regulation of pubertal LHRH release has not been functionally dissected. Recent experiments using mice in which *TTF-1*, a homeodomain gene involved in the transcriptional inhibition of proenkephalin expression (200), was conditionally deleted (see Potential Upstream Controlling Genes, below) suggest that enkephalin peptides produced by neurons of the lateral ventromedial nucleus of the hypothalamus do, in fact, exert an inhibitory control on female sexual development (201).

Glia-to-Neuron Communication. Glial cells regulate LHRH secretion via two related mechanisms that become established well before puberty: plastic rearrangements and direct cell–cell communication. Both mechanisms require the participation of growth factors; some of them, like transforming growth factor (TGF) β1 and β2, act via serine-threonine kinase receptors. Others, such as IGF-I, bFGF, and the family of epidermal growth factor (EGF)-related polypeptides, exert their effect by activating tyrosine kinase receptors. Additional messengers include nonpeptidergic molecules such as calcium, PGE_2, adenosine triphosphate, and glutamate. Figure 5 illustrates some of these glia-to-neuron signaling pathways.

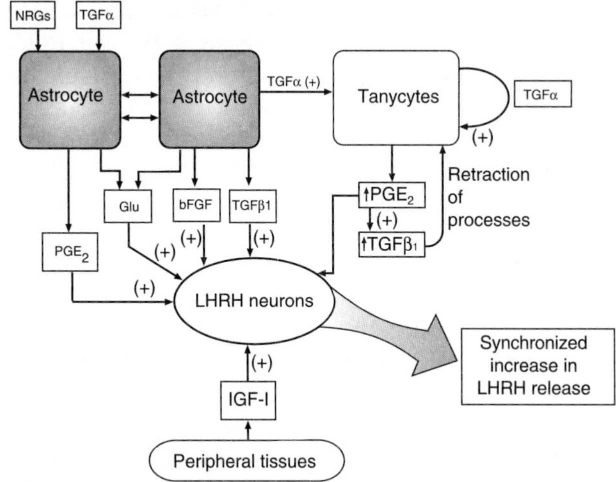

FIG. 5. Glial growth factors involved in the control of LHRH secretion. Although TGF-β1, IGF-I, and bFGF act directly on LHRH neurons via specific receptors expressed on these cells, TGF-α and neuregulins (NRGs) activate erbB receptors located on glial cells (astrocytes and tanycytes). As a result, glial cells release substances such as PGE_2 that stimulate LHRH release.

Growth Factors Acting via Serine-Threonine Kinase Receptors. A role for TGF-βs in the control of LHRH secretion was suggested by experiments showing that TGF-β1 secreted from astrocytes stimulates LHRH release from immortalized gonadotropin-releasing hormone (GnRH) neurons (202). Additional studies demonstrated that E_2 enhances TGF-β1 release from hypothalamic astrocytes (203), suggesting that astrocytic TGF-β1 may contribute to the intrahypothalamic mechanism by which E_2 induces the first preovulatory surge of LHRH. The presence of TGF-β receptors in adult LHRH neurons (204) indicates that, like immortalized GnRH neurons, GnRH neurons in situ may also be responsive to TGF-β1 stimulation. Although TGF-β1 can modify LHRH gene expression in vitro (202) and the hypothalamic expression of its encoding mRNA increases at the time of the preovulatory surge of gonadotropins (205), TGF-β1 does not stimulate LHRH release from the median eminence [(206), V. Prevot and S. Bouret, unpublished data], suggesting that its effects on LHRH neuronal function are limited to regulation of LHRH gene expression. Supporting this notion, a recent study showed that LHRH nerve terminals of the median eminence lack TGF-β receptors but that TGF-β1 can act on GnRH cell bodies in situ to decrease LHRH gene expression (207). Another study (208) showed that TGF-β1 acts at the median eminence to induce plastic rearrangements of glioependymal cells (tanycytes). Tanycytes line the ventral surface of the third ventricle and send their processes toward the external region of the median eminence (209), where they prevent the direct contact of GnRH axons with the perivascular space of the portal vasculature, via morphological specializations known as "end-feet." At the time of the preovulatory surge of gonadotropins, the end-feet retract and/or the basal lamina evaginates [reviewed in (210)], allowing direct contact of the LHRH terminals with the pericapillary space of the portal vessels. Treatment of tanycytes in culture with TGF-β1 reduces the migratory activity of the cells with 8 hours of treatment and induces retraction of the migrating cells, suggesting that the withdrawal of tanycyte end-feet seen in vivo may, at least in part, be due to TGF-β1 (Fig. 5).

Growth Factors Acting via Receptors with Tyrosine Kinase Activity. One of these growth factors is bFGF. Studies in immortalized GT1-7 LHRH-producing cells and native LHRH neurons showed that bFGF promotes survival, proliferation, and differentiation of these cells via activation of specific receptors (211,212). Other studies in GT1-7 cells demonstrated that bFGF promotes processing of the LHRH prohormone and elicits the release of LHRH precursor forms (213). bFGF also acts on astroglial cells to stimulate the release of molecules that promote GT1-7 neuronal differentiation (214). Its involvement in the mechanism underlying the preovulatory surge of gonadotropins is suggested by the increase in hypothalamic bFGF mRNA that occurs in the evening of proestrus (205), apparently as a result of estrogen action (215). Estrogen exerts this effect indirectly via stimulation of TGF-α production by hypothalamic astrocytes (215). Thus, both TGF-βs and bFGF contribute to the astroglial control of LHRH neuronal function (Fig. 5); however, the role that these growth factors may play in the control of puberty remains to be established.

In contrast, it is well established that other growth factors signaling via receptor tyrosine kinases are physiologically involved in the control of puberty. IGF-I is one of these factors. In both rats and primates (including humans), circulating IGF-I levels increase at puberty (216–218). Studies in female rats showed that although IGF-I is produced in astrocytes, the bulk of IGF-I present in the median eminence either at puberty or during the estrous cycle derives from peripheral sources such as the liver, instead of hypothalamic astrocytes (219). Rat tanycytes take up IGF-I from the bloodstream and presumably deliver the peptide to its sites of action within the hypothalamus (220). This uptake increases strikingly at the time of the preovulatory surge of gonadotropins (221), presumably due to a region-specific up-regulation of IGF-I receptors (219). A facilitatory role of IGF-I in the initiation of puberty has been shown by studies carried out in rats and nonhuman primates. On the one hand, central (219) or peripheral (222) administration of IGF-I advances puberty in both species. On the other, sexual development is delayed in mutant mice that, lacking growth hormone (GH) receptors, also have a deficiency in IGF-I production (223). Because the IGF-I involved in stimulating LHRH release is mostly of peripheral origin, IGF-I appears to be one of the peripheral metabolic signals postulated to facilitate the advent of the pubertal process (Fig. 5).

The EGF family of growth factors is another group of trophic factors acting via receptor tyrosine kinases with a well-documented role in the control of puberty. The EGF relatives, TGF-α and neuregulins (NRGs), are produced in hypothalamic astrocytes and stimulate LHRH secretion indirectly, via a cell–cell signaling mechanism that requires the intermediacy of astrocytes [reviewed in (224)]. To elicit LHRH release, these molecules bind to receptors of the erbB (erythroblastosis B) family located on the cell membrane of glial cells (instead of LHRH neurons) and elicit the release of chemical messengers that act directly on LHRH neurons to stimulate LHRH secretion (Fig. 6). Both hypothalamic astrocytes and tanycytes express erbB-1 receptors (208,225) (which bind TGF-α), but only astrocytes contain erbB-4 receptors (226) (which recognize NRGs). Nonetheless, upon ligand binding both receptors recruit the auxiliary receptor erbB-2 (208,226) (Fig. 6). In vitro studies identified PGE_2 as one of these messengers (226,227) and showed that the prostaglandin stimulates LHRH release by binding to prostaglandin receptor subtypes coupled to

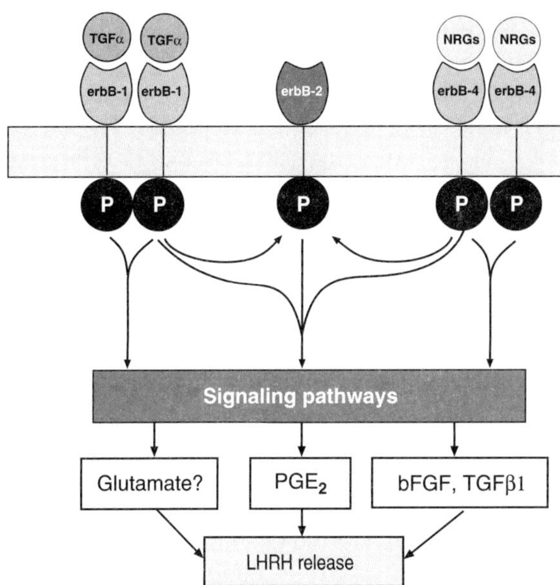

FIG. 6. Proposed activation of erbB-mediated signaling in hypothalamic astroglial cells during the initiation of female puberty in the rat. Binding of either TGF-α to erbB-1 receptors or NRGs to erbB-4 receptors results in phosphorylation (P) of the individual receptor and recruitment of the auxiliary subunit erbB-2. Formation of such heteromeric receptor complexes amplifies the capacity of the complex to activate intracellular signaling pathways, leading to the release of molecules, such as PGE_2, able to directly stimulate LHRH neurons. [Modified from (799) with permission.]

calcium mobilization and cAMP formation that are expressed on LHRH neurons (228). Other substances produced by astrocytes in response to erbB receptor stimulation are glutamate and the growth factors bFGF and TGF-β1 [reviewed in (23,229)] (Fig. 6).

The physiological contribution of astrocytic erbB receptors to the control of female sexual development has been demonstrated by a variety of approaches. For instance, both the pharmacological blockade of erbB-1 receptors in the median eminence (230) and a point mutation of the erbB-1 gene (231) result in delayed puberty. Conversely, female sexual development is hastened in transgenic mice expressing the human TGF-α gene under the control of an inducible promoter (232) and in rats carrying intrahypothalamic grafts of cells genetically modified to secrete TGF-α (233). These experiments showed that puberty was advanced only when the TGF-α–producing cells were grafted near LHRH neurons or their nerve terminals, suggesting that activation of a discrete subset of astrocytes near LHRH neurons may suffice to induce a pattern of LHRH secretion able to set in motion the pubertal process.

As indicated before, hypothalamic astrocytes also produce NRGs. Although NRGs can bind to both erbB-3 and erbB-4 receptors, hypothalamic astrocytes only express the erbB-4 gene (226), indicating that NRG signaling in these cells is exclusively mediated by an erbB-4/erbB-2 complex. The hypothalamic content of erbB-2 and erbB-4 mRNA begin to increase by the end of the juvenile period, when circulating sex steroid levels are still low, suggesting that the increase is independent of changes in sex steroid secretion. A further increase occurs near the first preovulatory surge of gonadotropins, a change that can be reproduced by treating immature rats with a sequential combination of estrogen and P. Thus, it appears that hypothalamic erbB-2 and erbB-4 expression increases first during juvenile development, driven by gonad-independent centrally originated mechanisms, and then at the time of puberty stimulated by gonadal steroids (226).

As in the case of erbB-1, a combination of neuroendocrine, molecular, and genetic approaches has defined the importance of erbB-2 and erbB-4 receptors in the control of female puberty. The contribution of erbB-2 was demonstrated by the striking delay in puberty caused by the in vivo disruption of hypothalamic erbB-2 receptor synthesis via intraventricular infusion of an antisense oligodeoxynucleotide directed against erbB-2 mRNA. Because erbB-2 is recruited by ligand-induced activation of both erbB-4 and erbB-1 receptors, additional experiments were carried out to tease both systems apart and to define the physiological importance of the NRG-dependent signaling system in the initiation of puberty. Mutant mice were generated to overexpress, in an astrocyte-specific fashion, a truncated erbB-4 protein that, lacking the intracellular domain, acts as a dominant negative receptor to block signaling via the intact receptor (234). Astrocytic release of PGE_2 was impaired and the LHRH response to NRGs was reduced in these animals, demonstrating that the mutated receptor disrupts the ability of hypothalamic astrocytes to respond to NRGs with the production of substances able to stimulate LHRH release. In keeping with these findings, pituitary gonadotropin release was reduced during female sexual development and puberty was delayed (234). Breeding these mutant animals to mice carrying defective erbB-1 receptors further delayed the onset of puberty and seriously compromised fertility (235), indicating that the astrocytic erbB-1 and erbB-4 signaling systems work in a coordinated fashion to facilitate LHRH release during sexual development and thus control the initiation of female puberty.

Maturation of the Hypothalamic–Pituitary–Ovarian Interrelationship

Development of Steroid-negative Feedback. A predominant feature in the maturation of steroid feedback mechanisms in the female rat is the relative inability of E_2-negative feedback to operate during neonatal–infantile development (109–113). During the first few postnatal days, this feedback relationship is absent, as ovariectomy of neonatal rats fails to activate gonadotropin release (236). When ovariectomy is performed after the first week of postnatal life (infantile period), serum gonadotropin levels increase, but—as already mentioned—E_2 has a reduced capacity to suppress this elevation (109–113). This defect is not due to a lack of hypothalamic–pituitary estrogen receptors but to the presence in serum and tissues of extremely high levels of AFP, which binds estrogen avidly (237,238).

It seems that during the infantile period of female development, aromatizable androgens play a leading role in the steroid-negative feedback control of gonadotropin secretion. Both T and androstenedione can be detected in the blood of infantile rats (98,107, 239). When T was administered via Silastic capsules to ovariectomized rats to mimic precastration levels of the androgen, it was found that physiological levels of T were effective in preventing the postcastration rise in serum gonadotropins (107). Whether the effect of T is due to its androgenic capacity per se or to prior local aromatization to estrogens is unknown. E_2, if given at sufficiently high doses that presumably overcome binding to AFP, can effectively inhibit gonadotropin release in infantile pubertal rats (110,111,113,116,240).

As levels of AFP decline, the capacity of E_2 to suppress gonadotropin release increases (109–111), so that it becomes maximally effective during the juvenile period (107,109–111). This enhanced effectiveness is maintained throughout juvenile development (107,240). In contrast to these changes in E_2-negative feedback, the capacity of aromatizable androgens to inhibit gonadotropin release remains relatively constant throughout infantile and juvenile development (107). It can therefore be concluded that the steroid-negative feedback regulation of gonadotropin release changes from a predominantly androgenic control during infantile days to a dual estrogenic–androgenic control during the juvenile period.

Development of E_2-Positive Feedback. Although administration of sufficiently high doses of E_2 to infantile rats suppresses circulating gonadotropin levels, a stimulatory effect of E_2 on LH release cannot be demonstrated before the third week of postnatal life (116,241). Early experiments found that injections of E_2 could not evoke a surge of LH in rats younger than 21 to 24 days of age (241,242). A more detailed study in which E_2 was administered via Silastic capsules and the resulting serum E_2 levels measured by RIA demonstrated that before day 15, E_2 levels as high as 400 pg/ml were unable to stimulate LH release (116) (Fig. 7). Between days 16 and 20, E_2 was effective in inducing an LH surge, but the serum levels needed to be twice as high as those observed on proestrus. After day 20, however, proestrous E_2 levels were sufficient. At all these ages the LH surge occurred 54 hours after implantation of the E_2 capsules, but when E_2 was administered to animals older than 28 days of age, an LH surge occurred within 30 hours. It is therefore evident that as the animal matures, the hypothalamic–pituitary unit becomes more sensitive to the stimulatory effect of E_2. This increased sensitivity may be, at least in part, the result of exposure to E_2, because pretreatment of juvenile rats with E_2 before an E_2 challenge advances the age at which LH responds to the steroid with a surge of secretion (242). In addition to these temporal changes, the magnitude of the LH response to E_2 also increases during the juvenile period (116,243), so that by days 30 to 32 the surge of LH induced by E_2 is indistinguishable from that normally seen at first proestrus. As discussed later, by the end of the juvenile period the LH-releasing system is so sensitive to E_2 that serum levels of the steroid even less than 50% of those seen at proestrus suffice to elicit large preovulatory-like LH discharges.

There is no doubt that in the rat, the stimulatory effect of E_2 on LH release involves the activation of LHRH secretion from the hypothalamus (134,244). Although more information is now known about the mechanism(s) underlying E_2-induced LHRH release (see below), much remains to be learned regarding the inability of E_2 to induce LH release in infantile rats.

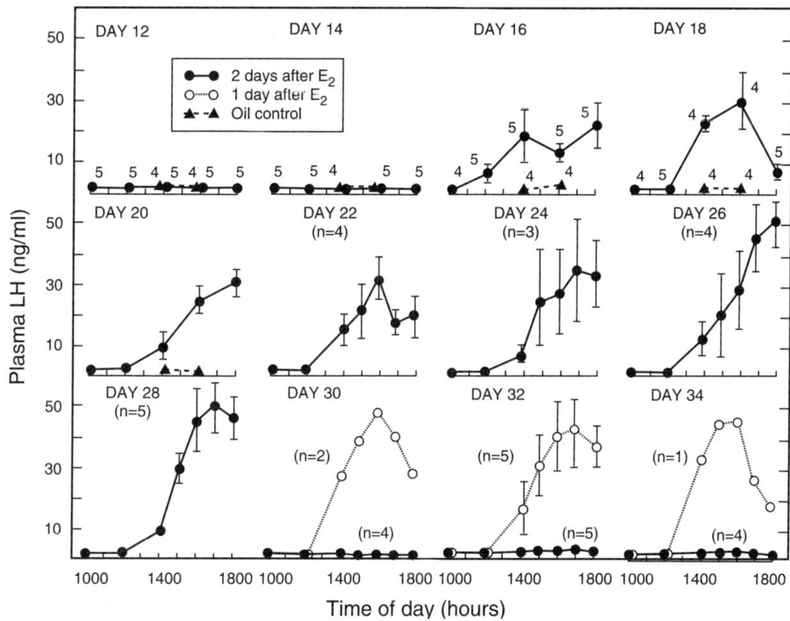

FIG. 7. Development of estrogen-positive feedback. Female rats of different prepubertal ages received a subcutaneous Silastic capsule (20 mm/100 g body weight) containing E_2 dissolved in corn oil (400 μg/ml). Animals older than 20 days of age were bled every 2 hours via an indwelling jugular cannula. Trunk blood was obtained from the younger animals at the times indicated in the abscissa. The animals received the E_2-containing capsule one (*open circle*) or two days (*closed circle*) before blood collection. [From (116) with permission.]

It seems clear, however, that even though the capacity of E_2 to stimulate LH release becomes established quite early, a precocious preovulatory surge of gonadotropins fails to occur because the ovary has not been yet subjected to the proper pulsatile pattern of LH stimulation necessary for the gland to produce E_2 in sufficient amounts, and for a sufficient period of time, to stimulate a surge of LH release.

The Ovary

Hormonal Control. Evaluation of parameters such as E_2 production (245), cAMP formation (246), and the number of gonadotropin receptors (56–60) has revealed that the ovary is relatively insensitive to gonadotropin stimulation during the first week of postnatal life. Nevertheless, primary follicles can already be observed by day 4 (64,65), and although initiation of follicular growth is gonadotropin independent (61–67), there is ample evidence that maintenance of follicular development depends on the continuous presence of gonadotropins (61–67,247–249).

The initial synthesis of ovarian FSH receptors in the infantile ovary is a process induced by molecules other than gonadotropins themselves. Thus far, two such molecules have been identified: the neuromodulator VIP (that reaches the ovary via the extrinsic innervation) and the neurotrophic nerve growth factor (that is produced by ovarian mesenchymal cells) (250). Because the ability of VIP to induce FSH receptors in fetal–neonatal ovaries involves stimulation of adenylate cyclase activity (90) and cAMP formation (250) whereas nerve growth factor increases FSH receptors via a cAMP-independent pathway (250), it can be concluded that the synthesis of FSH receptors in infantile ovaries is initiated by factors operating via both cAMP-dependent and cAMP-independent mechanisms.

Substantial evidence now exists that ovarian follicles become subjected to strong gonadotropin control during the second week of postnatal life (63–65). Deletion of the FSH receptor gene leads to arrest of follicular development at the secondary stage (251), whereas the absence of LH receptors only impairs the growth of antral follicles (67), indicating that FSH is the primary gonadotropin required for preantral follicular development. Deletion of the gene encoding FSH, instead of the FSH receptor, resulted in a similar phenotype, but the arrest of follicular growth occurred earlier, at the primary stage (249). Almost twice as many small follicles begin to move into a more advanced developmental stage during the second postnatal week than at later ages (252). The level of plasma FSH necessary for maintenance and/or formation of FSH receptors during the infantile period (days 10–20) is no greater than that observed during the late juvenile period (days 20 to 30–32) (~200 ng rat LH-RP-1/ml); suppression of serum FSH from the high infantile levels to the low juvenile values, using DHTP, failed to affect ovarian FSH receptor content even when serum LH was reduced to undetectable levels (253). During the infantile period, ovarian production of estrogen from exogenous precursors increases markedly (254,255) and FSH becomes able to induce aromatase activity (90). In addition, the ovary responds to endogenous increases in serum gonadotropins with steroid release (256).

Although it is clear that the lack of FSH prevents female sexual development and results in female infertility (249), it has been difficult to determine the impact that *high* infantile serum FSH levels have on subsequent sexual maturation (257,258). Because early follicular growth is enhanced at this time and completion of follicular growth takes 15 to 19 days (252), it appears that many of the follicles that begin to grow during this period represent the first cohort of follicles destined to ovulate at puberty.

The development of the neonatal–infantile ovary appears to be regulated by an additional modulatory mechanism, of maternal origin, which becomes established shortly after birth. Rat milk, like that of several other species, contains an LHRH-like substance biochemically indistinguishable from hypothalamic LHRH (259,260). After suckling, LHRH-like immunoreactivity is detected in the stomach content of the pups and an increase in LHRH levels is observed in plasma (260). Moreover, available LHRH receptors in the pup's ovaries decrease, an effect that can be prevented by prior intravenous administration of an antiserum to LHRH (260). Thus, LHRH of maternal origin is transferred to the pup via the milk; it crosses the gastrointestinal epithelium and reaches the ovary via the bloodstream, where it binds to specific receptors (260).

Milk LHRH behaves like hypothalamic LHRH in that it is able to both stimulate gonadotropin release from the anterior pituitary in vitro (260,261) and inhibit gonadotropin-induced E_2 and P release from granulosa cells in culture (260). Because rat pups suckle frequently throughout the entire day, milk LHRH is almost continuously available for binding to the infant ovary. Chronic exposure to continuous levels of LHRH depresses ovarian function [for a review see (262)], indicating that milk LHRH may be physiologically involved in restraining neonatal–infantile development of the pup ovary. This phenomenon may represent an evolutionary mechanism by which the mother rat regulates gonadal development of its offspring beyond intrauterine life. Indeed, available ovarian LHRH receptor content increases after postnatal day 15 (263,264), that is, at the time when pups begin to eat pelleted food and when nursing episodes become less frequent or cease altogether.

During juvenile development the ovary grows under the influence of low serum levels of LH and FSH (Fig. 8). Several earlier articles reviewed the morphology (65,71,252,258) and the hormonal control (265,266) of the immature ovary. The interested reader can find more detailed information in those articles. It is important to emphasize that throughout the juvenile period the ovary undergoes waves of follicular development and atresia (252,258,267), though in no instance does a crop of follicles reach the ovulatory stage.

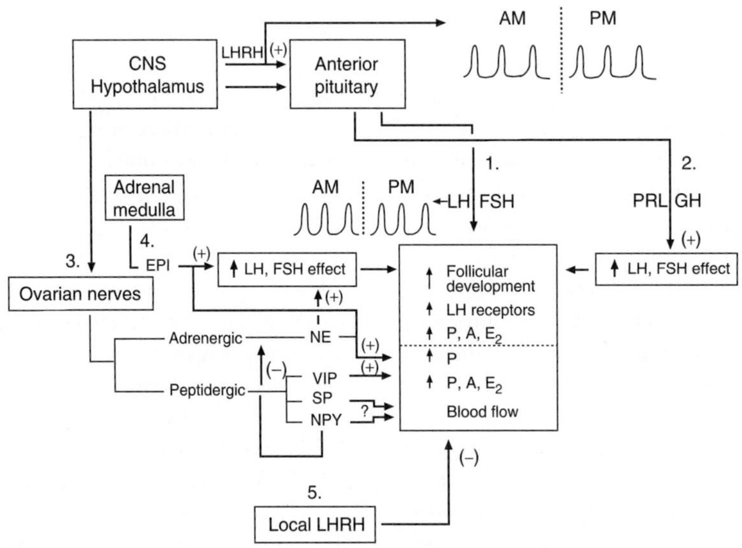

FIG. 8. Hormonal and neurogenic factors controlling ovarian development during the juvenile period of the female rat. A similar pattern of LH release in the mornings and afternoons is assumed to reflect similar patterns of LHRH release. Numbers indicate the different control mechanisms involved in regulating ovarian function. +, facilitatory; –, inhibitory; ?, effect not known; AM, morning; PM, afternoon; EPI, epinephrine; SP, substance P; A, androgens. [Modified from (35) with permission.]

This is probably due to the relative low levels of circulating gonadotropins, because the juvenile ovary can ovulate if challenged with sufficiently large amounts of exogenous gonadotropins (268,269).

Under normal conditions there appears to be a multiplicity of factors that, in addition to gonadotropins, contributes to regulate the gradual orderly maturation of the ovary (Fig. 8). The responsiveness of the ovary to gonadotropins is negligible during the early neonatal period, rises during infantile days, and becomes most prominent during the juvenile period (270,271). This change in responsiveness appears to be related, at least in part, to an increase in the number of gonadotropin receptors (59,60). Although the most pronounced change in FSH receptors appears to occur between postnatal days 4 and 16, human chorionic gonadotropin (hCG) (LH) receptors increase more rapidly during juvenile days (59).

The hypothetical inhibitory control exerted by milk LHRH ends at the initiation of juvenile development with the separation of the pups from their mother. Whether an LHRH-like peptide, produced locally or transported to the ovary via the nerves, contributes to the regulation of juvenile ovarian function is not clear at present. However, its presence is suggested by the changes in LHRH receptors observed at this time. The highest content of LHRH receptors occurs around day 25 (260,263,264). Thereafter, the content declines gradually toward the first proestrus, with the sharpest decrease being observed during early proestrus I (272) [i.e., the phase of puberty when uterine fluid becomes apparent for the first time (273)]. Although the prepubertal decrease in receptor content may reflect a declining LHRH inhibitory influence on the ovary, the validity of this hypothesis remains to be established.

In addition to gonadotropins, two anterior pituitary hormones appear to be involved in the regulation of prepubertal ovarian function. Both PRL and GH have been shown to support ovarian maturation by facilitating the effects of gonadotropins (Fig. 6). The secretion of both somatomammotropins is low at the beginning of the juvenile period and increases gradually thereafter (98,169,274,275). An adult-like pattern of GH release becomes established around the time of puberty (275), whereas a quotidian pattern of PRL release, characterized by PRL discharges occurring approximately every 3 hours, has been reported to occur by the beginning of the juvenile period (276). The most prominent PRL secretory episodes occur at mid-afternoon and during the early morning hours (277). As the animal approaches the end of the juvenile period, the nocturnal increases in PRL levels disappear, whereas the afternoon surge becomes even more pronounced (277). The afternoon increase in PRL levels can already be detected by the third week of postnatal life, that is, during the infantile period (278).

It has been known for several years that PRL accelerates the onset of puberty in female rats (279). This effect is observed after the systemic (127,279) or intrahypothalamic (280,281) administration of PRL. One of the mechanisms by which PRL exerts this effect is by enhancing ovarian responsiveness to gonadotropins (270). PRL may also act directly to enhance aromatase P-450 gene expression, as shown in ovaries from pregnant rats (282). Chronic stimulation of endogenous PRL release by blockade of dopaminergic receptors advances the onset of puberty and increases the P and E_2 response of the ovary to hCG and FSH (270,283). The ovaries from hyperprolactinemic rats exhibit an increased number of LH receptors (283), suggesting that PRL facilitates prepubertal ovarian development, at least in part, by facilitating LH actions. In contrast to these results, some authors have failed to find an advancing effect of PRL on the timing of puberty in female rats (284).

Other studies have shown that pharmacological inhibition of PRL secretion by administration of bromoergocryptine (CB-154), an ergot alkaloid that activates dopamine receptors, delays the onset of puberty and reduces the steroidal response to the ovary to gonadotropins (285). Hypoprolactinemic rats also show a reduced number of ovarian hCG (LH) receptors. Although concomitant administration of PRL can reestablish the inhibitory effect of CB-154, the possibility that subtle changes in LH release may play a role in these effects cannot be completely ruled out.

Experiments involving removal of the adrenal gland have revealed that PRL requires the presence of this gland to fully facilitate the ovarian steroidogenic response to gonadotropins (283,286). Adrenalectomy of hyperprolactinemic rats blunts the enhanced P and E_2 response of the ovaries to hCG, whereas corticosterone restores the P, but not the E_2, response (283,286). Because hyperprolactinemic-ovariectomized rats do not show increased serum levels of adrenal P or aromatizable androgens, it does not appear that these adrenal steroids mediate the facilitatory effects of PRL on the ovary. Rather, PRL and adrenal products may act at different steps in the sequence of events, leading to formation of preovulatory follicles (286).

Like PRL, GH has been shown to exert a facilitatory influence on sexual maturation of the female rat. Suppression of GH release by implantation of GH into the medial basal hypothalamus (287), by inoculation with the somatomedin-producing worm *Spirometra mansonoides* (288), or caused by the genetic ablation of the GH gene (289) delays the onset of puberty, as determined by the age at vaginal opening and at first ovulation. Part of the facilitatory effect of GH on sexual maturation is exerted at the level of the ovary because GH treatment in vivo increases the ovarian P response to gonadotropins in vitro (287). GH also facilitates FSH-induced formation of LH receptors and FSH-stimulated P secretion from cultured granulosa cells of immature hypophysectomized rats (290). Moreover, GH facilitates the stimulatory effect of cAMP and forskolin on P secretion, indicating that the hormone acts at more than one biochemical step to positively regulate granulosa cell function (290). At least some of these effects are directly mediated by GH, and not by IGF-I, because the ovaries of mice lacking GH show an IGF-I–independent deficiency in antral follicle growth (291). IGF-I is an important regulator of ovarian development; it is produced by granulosa cells and facilitates a number of FSH-dependent effects such as induction of aromatase activity, P secretion, and formation of LH receptors [for a review see (292)]. In the absence of IGF-1, follicular growth is interrupted at the late preantral/early antral stage (293).

Neural Control. Ovarian function is regulated not only by hormones but also by direct neural influences (Fig. 8) [for a review see (89,294)] that might provide a fine minute-to-minute control. The extrinsic innervation of the ovary reaches the gland via two main routes: the ovarian plexus nerve, which travels along the ovarian artery, and the superior ovarian nerve, which is associated with the suspensory ligament of the ovary (295–297). These extrinsic nerves are mainly sympathetic and sensory, but some are also parasympathetic (295). The sympathetic and sensory nerves originate from neurons located either in specific brain or spinal cord nuclei or in peripheral ganglia (298–300). The parasympathetic innervation, on the other hand, is provided by postganglionic cholinergic neurons. Their axons presumably reach the ovary via vagal efferents (301) and/or branches of the hypogastric plexus (299,301) that contribute to the ovarian plexus nerve. The sympathetic nerves of the gland contain catecholamines and NPY (302–305). The sensory nerves contain substance P and calcitonin gene-related peptide (298,304,306,307). VIP-containing fibers of both sympathetic and sensory nature (298,308) also innervate the ovary (91,304). The ovary becomes innervated before the initiation of folliculogenesis (69). Some of the early fibers are catecholaminergic; before birth their location in the ovary is limited to the innermost region of the gland, the ovarian medulla. This initial distribution suggests that follicular assembly, which also begins in the ovarian medulla (71,248,309), might occur under the influence of these developing nerves.

Because neurotransmitters contained in ovarian nerves potently stimulate ovarian steroidogenesis (see below), it has been postulated that ovarian nerves may facilitate prepubertal ovarian development [for a review, see (89)]. Consistent with this view, surgical or pharmacological removal of the extrinsic innervation—though sometimes producing small effects—has in general resulted in a reduction in the number of antral follicles [for a review see (302)]. Both passive and active immunosympathectomy during neonatal life result in permanent loss of the sympathetic innervation results and cause delayed follicular development, reduced ovarian steroid output, and marked irregularities of the rat estrous cycle (310,311).

Intrinsic Ovarian Neurons. In addition to the extrinsic innervation, the ovary also contains an intrinsic innervation. This innervation was first identified in the monkey ovary (312) and was shown to derive from small neurons containing low affinity nerve growth factor receptors (now known as p75 neurotrophin receptor, p75-NTR, because it recognizes all members of the nerve growth factor family [neurotrophins] with similar affinity). Some of these p75-NTR–positive cells were also immunoreactive to antibodies to tyrosine hydroxylase, demonstrating that they are catecholaminergic (312). A subsequent study revealed that neurons of the monkey ovary also express the gene encoding dopamine β-hydroxylase (313), the enzyme that metabolizes dopamine to NE. Surprisingly, the presence of neurons in the rat ovary appears to be strain specific; the neurons can only be seen in ovaries from Wistar rats but not Sprague-Dawley or Long-Evans rats (314,315). The reason for this difference is unclear, especially because the neurons have been detected not only in monkeys, but also in humans (316). The function of intrinsic ovarian neurons is not yet known; however, several lines of evidence suggest they may play a role(s) in controlling intraovarian physiological events. For instance, some of them produce catecholamines (312,315,316), suggesting they may influence steroidogenesis (317,318). Others also produce NPY (315), a peptide that extensively innervates the ovarian vasculature (305), suggesting a role in control of blood flow. It is important to note that only a small percentage of the ovarian neuronal population is catecholaminergic or contains NPY; hence, other neuronal phenotypes must also exist. Interestingly, the selective presence of neurons in the ovary of Wistar rats correlates with the longer reproductive life span in this strain than in other strains of rats (315). It is possible that ovarian neurons function as an independent intrinsic regulatory component capable of acting together with the extrinsic innervation to modify ovarian function.

Ovarian Nerves and Follicular Steroidogenesis. Ovarian steroidogenesis is stimulated by catecholamines and VIP (Fig. 8). Nerves containing NE reach the ovary via both the superior ovarian nerve and ovarian plexus nerve (297). Although the ovarian plexus nerve most innervates the ovarian vasculature, the majority of fibers derived from the superior ovarian nerve innervate the secretory compartments of the gland (i.e., interstitial cells), the thecal layer of follicles, and the corpus luteum (297,302,319,320). NE stimulates ovarian steroidogenesis via activation of β_2-adrenergic receptors (317,318). Activation of these receptors results in release of P from granulosa and luteal cells (317,321,322) and androgens from thecal cells (323,324). Of physiological significance is the fact that catecholamines facilitate the steroidogenic response of ovarian cells to low concentrations of gonadotropins, suggesting that the catecholaminergic sympathetic nerves contribute to amplify the effects of circulating gonadotropins on ovarian steroidogenesis [for reviews see (325,326)].

The ovarian content of β_2-adrenergic receptors decreases abruptly in the afternoon of the first proestrus and remains at low levels during estrus (318), but neither LH nor FSH can evoke a similar decrease in vitro (327). This indicates that catecholamines, rather than gonadotropins, are responsible for the decrease in receptor content seen in the afternoon of first proestrus. Interestingly, corticosterone at physiological levels can depress the β-adrenergic receptor content of granulosa cells in vitro by 60% (327), suggesting that ovarian β-adrenoreceptors may be tonically inhibited by corticosterone during prepubertal development.

That an activation of noradrenergic neurons projecting to the ovary indeed occurs on the day of proestrus is indicated by the finding of a rise in ovarian NE release at the time of the preovulatory surge of gonadotropins detected in push–pull perfusates of the ovary of freely moving animals (328). Furthermore, other experiments demonstrated that electrical stimulation of ovaries in vitro causes a greater increase in ^{3}H-NE release from the ovary of animals in proestrus and estrus than at other phases of the estrous cycle (329), indicating that the activity of ovarian sympathetic nerves is enhanced during the hours encompassing ovulation.

Regarding VIP, this neuropeptide is contained in nerve fibers that reach the ovary via the superior ovarian nerve (298,330). As in the case of catecholaminergic nerves, VIPergic fibers innervate the interstitial tissue, follicles, and the ovarian vasculature. VIP is a potent stimulus of P, E_2, and androgen secretion (91,331). The molecular mechanisms underlying these effects involve an enhanced synthesis of all three components of the side-chain cleavage–enzyme complex, the rate-limiting enzyme in steroid biosynthesis (332,333), and stimulation of aromatase activity (90,331). It appears that VIP targets a population of immature granulosa cells that is unresponsive to FSH (334), implying a role for the neuropeptide in early granulosa cell function. That one such role may be to induce granulosa cell differentiation is suggested by the facts that the actions of VIP are exerted via activation of cAMP synthesis and that cAMP itself induces granulosa cell differentiation (335). As indicated earlier, VIP may also participate in the control of early ovarian development.

In addition to this role, VIPergic nerves may contribute to facilitating the stimulatory effect of gonadotropins on ovarian steroid secretion at proestrus. This is suggested by the rapid drop in E_2 and P secretion that results from transection of the superior ovarian nerve in the afternoon of proestrus (321), at the time of elevated plasma gonadotropin levels, and in the absence of measurable changes in blood flow. Although part of this effect may be caused by the loss of NE inputs, the drop in E_2 secretion may be best attributed to the transection of VIPergic fibers because VIP, not catecholamines, stimulates estradiol secretion (91,317,318,331,336).

An additional piece of evidence supporting an involvement of ovarian nerves in the developmental regulation of ovarian function was provided by the demonstration that transplanted ovaries fully recover their ability to keep gonadotropin secretion in check around the time when their reinnervation is completed (337). These experiments also showed that if the reinnervation is prevented by neonatal immunosympathectomy, the negative feedback loop controlling gonadotropin secretion is disrupted because of the inability of the transplanted ovary to maintain FSH secretion at basal levels.

In summary, the nervous system influences the maturation of the ovary via two main routes, a hormonal and a neural one (Fig. 8). Whereas the former involves the secretion of hypothalamic factors that control the secretion of LH, FSH, PRL, and GH from the adenohypophysis, the latter directly links the CNS to the ovary via peptidergic and adrenergic nerves. Moreover, epinephrine of adrenal medullary origin may facilitate ovarian function after reaching the gland via the bloodstream. It appears that most of these regulatory mechanisms become firmly established during the juvenile period.

Onset of Puberty

The Initial Gonad-independent Activation of Pituitary Gonadotropin Secretion. The onset of puberty in rats is determined by a multiplicity of interrelated events, some of which originate during the infantile period. Nevertheless, the first hormonal manifestation that puberty is underway occurs only after the fourth postnatal week of development and is expressed as a diurnal change in the mode of LH release. The experiments of Meijs-Roelofs et al. (338) were the first to demonstrate an unequivocal prepubertal increase in mean LH levels, which became apparent 8 to 9 days before the expected day of first proestrus and appeared to be greater in the afternoons than in the mornings. An initial characterization of the mode of LH release in peripubertal rats (130) showed that starting around day 30 (vaginal opening in these rats occurred around day 38), a diurnal pattern of release developed in the female rat. This pattern was characterized by an afternoon increase in LH pulse amplitude. In subsequent experiments the mode of LH release was more precisely characterized in blood samples obtained every 5 minutes (25) by means of an automated blood sampling technique (339). The results clearly indicated that although in late 27- to 29-day-old juvenile rats afternoon plasma LH profiles were similar to morning levels (25), peripubertal 30- to 38-day-old rats had significantly greater basal LH levels and LH pulse amplitude in the afternoons than the younger animals (Fig. 9A,B). That this diurnal change in the mode of LH release is physiologically relevant to the functional development of the ovary was demonstrated by experiments in which peripubertal ovaries were perifused with LH regimens designed to mimic either morning or afternoon pulses of LH secretion (340). The results demonstrated that LH pulses of amplitudes similar to that seen in the afternoon of the peripubertal period elicited significantly more E_2 and P release than did morning-type LH pulses. In addition to showing an afternoon change in LH pulse amplitude, some peripubertal animals also exhibited a more sustained mid-afternoon episode of LH secretion (Fig. 9C), termed the LH "minisurge" (25). Mimicking such a secretory episode in vitro also led to enhanced E_2 and P secretion from the ovary (340), thereby suggesting that LH minisurges may also be important for the peripubertal activation of ovarian function.

It is clear that the afternoon increase in LH pulse amplitude is not ovarian dependent (341) and, more specifically, is not E_2 induced (342). On the other hand, the minisurges of LH secretion appear to be caused by subtle increases in serum E_2 levels (342). These conclusions derived from experiments in which short-term (48-hour) ovariectomy of prepubertal rats was found to result in greater plasma LH levels in the afternoon than in the morning (341). Detailed examination of the LH release pattern at this time, using an automated 5-minute bleeding paradigm (339), revealed that these high LH values were not due to a sustained surge of LH release but were, instead, the consequence of an increased amplitude of the LH pulses. When mean precastration serum E_2 levels were produced in short-term ovariectomized juvenile rats via subcutaneous Silastic capsules, pulsatile LH release was inhibited rather than enhanced. The use of larger doses of exogenous E_2 consistently failed to induce an

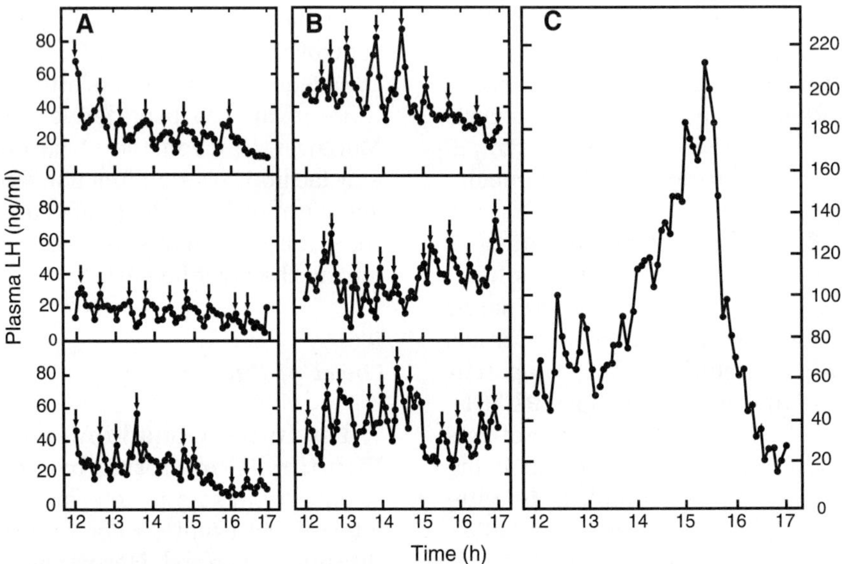

FIG. 9. Representative afternoon plasma LH profiles from (**A**) juvenile (27- to 29-day-old) and (**B**) peripubertal (30- to 38-day-old) female rats bled continuously for 5 hours. Six individual profiles from a total of 16 are depicted. Pulses of LH secretion are indicated by *arrows*. [From (25).] **C:** A minisurge of LH secretion occurring before the first preovulatory LH surge in the female rat. It is though that these minisurges are induced by small rises in estrogen levels resulting from the increase in pulsatile LH observed at the initiation of puberty. [From (25) with permission.]

increase in LH pulse amplitude but instead resulted in the appearance of minisurges and preovulatory-like surges of LH secretion (342).

An additional indication that an ovary-independent diurnal change in neuroendocrine central activity occurs during prepubertal development of the female rat came from studies examining PRL release. It is known that PRL secretion occurs episodically (276) and that during development of the female rat, PRL levels become more elevated in the afternoons than in the mornings (30,277,278). Ovariectomy of early juvenile rats does not abolish the afternoon surges of PRL (343), indicating that these secretory episodes are ovary independent. Other studies showed that when neonatal female rats were ovariectomized and their plasma patterns of PRL subsequently examined at different ages, at least 50% of the animals exhibited a mid-afternoon secretory episode of PRL release, even as late as 40 days after ovariectomy (344). Restoration of juvenile serum E_2 levels via subcutaneous E_2-containing Silastic capsules resulted in amplification of the surge, which still occurred at the same time of the day. Thus, these results (343,344) demonstrate that the neural mechanism responsible for the afternoon appearance of a PRL surge can develop in the absence of ovarian influences and that ovarian E_2 amplifies the magnitude of the PRL surge.

A quite different conclusion has been drawn from studies performed to elucidate the mechanisms determining the peripubertal "minisurges" of LH secretion (342). Such sustained episodes of release could not be detected in either short- or long-term ovariectomized rats. Only when circulating E_2 levels were slightly increased over juvenile values (via subcutaneous E_2-containing Silastic capsules) did a minisurge of LH secretion occur, indicating they are E_2 dependent.

Another factor that may play a role in accelerating ovarian maturation at the end of juvenile development is a change in the biological activity of circulating gonadotropins. As already mentioned, evidence exists that as the female rat matures, the biological activity of pituitary FSH increases (123). With respect to LH it appears that the hormone exists as different molecular forms in the bloodstream of prepubertal rats; when PMSG was administered to 27-day-old females, two different patterns of serum LH emerged, depending on which of two antisera were used in the RIA (345). Interestingly, this divergence was observed more clearly in the lighter animals (<60 g body weight), which also failed to ovulate. The same laboratory reported that LH released in PMSG-treated rats lighter than 60 g was measurable by RIA but was inactive in a cytochemical bioassay (125). In contrast, LH released in PMSG-treated rats weighing more than 60 g was active in both the RIA and the cytochemical bioassay used. Surprisingly, however, both types of LH were equally effective in stimulating T secretion from testicular interstitial cells.

There is little doubt that the diurnal changes in pulsatile LH release detected at the initiation of puberty result from the activation of LHRH secretion. The potential mechanisms underlying this activation are discussed later. It would appear to be clear, however, that the resulting changes in LH secretion are important for the continuation of puberty, because under the influence of these LH secretory episodes, the ovary is stimulated to produce more E_2 (340). In turn, subtle increases in E_2 levels appear to be able to evoke minisurges of LH secretion (342), which can induce further ovarian activation (340).

Phenotypic Changes at Puberty. Once the diurnal pattern of LH release becomes established, a new cascade of events develops that culminates with the first preovulatory surge of gonadotropin and the first ovulation. To examine the many components of this cascade, we have divided the process of puberty in the female rat into five different phases, mainly based on morphological criteria (30,273):

1. *Late juvenile*, previously (22) termed *anestrus:* This phase corresponds to the stage during which the changes in the mode of LH release begin to occur (25). Animals in this phase are about 30 days of age, their uteri are small (wet weight less than 100 mg), and, importantly, no intrauterine fluid can be detected. The vagina is always closed.

2. *Early proestrus:* Animals in this phase have larger uteri with intraluminal fluid; their vagina is closed. A further division into early proestrus-1 and early proestrus-2 subphases has been proposed, based on the amount of uterine fluid and the uterine weight observed (273).

3. *Late proestrus:* This phase corresponds to the day of *first proestrus*. Animals have large "ballooned" uteri full of fluid, with a wet weight greater than 200 mg. Their ovaries have large follicles. Most animals in this phase show closed vaginae.

4. *Estrus:* This is the day of first ovulation, when uterine fluid is no longer present, fresh corpora lutea can be readily discerned, the vagina is open, and vaginal cytology shows a predominance of cornified cells.

5. *First diestrus:* This phase of puberty is characterized by a vaginal cytology showing a predominance of leukocytes and by the presence of mature corpora lutea within the ovaries.

Activation of the Ovary. With the onset of puberty already determined by the activation of pulsatile LH release, the single most important event that remains to be defined is the timing of the first preovulatory surge of gonadotropins, which, in itself, represents the climax of female neuroendocrine reproductive maturation. Because the hypothalamic–pituitary unit is already fully competent to respond to preovulatory levels of estrogen with an LH surge, the occurrence and timing of this final event depend on the completion of ovarian maturation. Only when the ovary becomes capable of producing E_2 levels of sufficient magnitude, and for a sufficiently long period of time, will the preovulatory LH surge occur.

A multitude of maturational changes appears to be involved in hastening the acquisition of preovulatory competence by the ovary (Fig. 10). Whereas FSH receptor content is already maximal by the end of juvenile development, the number of LH receptors in granulosa cells increases dramatically between the anestrus and late proestrus phases of puberty (59,60). Concomitant with this increase, a decline in LHRH receptor content occurs, with the magnitude of the decrease being more pronounced between anestrus and early proestrus phases than at later times (272). The impact these two divergent changes in hormone receptor may have on ovarian function has already been discussed. It is noteworthy that during the days preceding the preovulatory gonadotropin surge, the steroidal responsiveness of the ovary to gonadotropins increases dramatically (271,273), most likely reflecting the progressive development of the follicles destined to ovulate at the first estrus.

The neurogenic component of the ovary also undergoes noticeable changes. The content of β-adrenergic receptors increases between anestrus and late proestrus phases and then declines abruptly at the time of the proestrous surge (318). Paralleling the increase in receptor content, the release of P in response to β-adrenergic stimulation becomes more prominent at late proestrus. However, the greatest increase in response occurs after ovulation, more specifically during the first estrus. Surprisingly, at this time the receptor content is low.

The concentration of VIP in the ovary, which remains almost unchanged between the second postnatal day and the end of juvenile development, increases significantly during the early part of the peripubertal period (days 30–35) (92). Moreover, the steroidogenic response to VIP undergoes profound changes at the time of puberty (91). The E_2 response to VIP, already distinct in juvenile rats, increases noticeably during the early proestrus and late proestrus phases of puberty. The P response to the peptide increases only moderately at this time and then strikingly after ovulation. Radioimmunoassayable substance P content in the ovary also increases between anestrus and late proestrus (346). Although the function of these changes in substance P content is unknown, it would not be unreasonable to suspect that they may be implicated in the edematization of the ovary that occurs at puberty (347) and/or in the changes in blood flow that occur during the estrous cycle (348,349).

Presumably as a consequence of this marked enhancement in facilitatory inputs to the ovary, the pattern of steroid production changes dramatically. Serum E_2 levels increase markedly between anestrus and late proestrus (31,273,350,351), reaching about 80 pg/ml during the morning of late proestrus (351). Serum P increases moderately before the LH surge, but serum T levels do so more prominently (350,351). This increase in T (or, in more general terms,

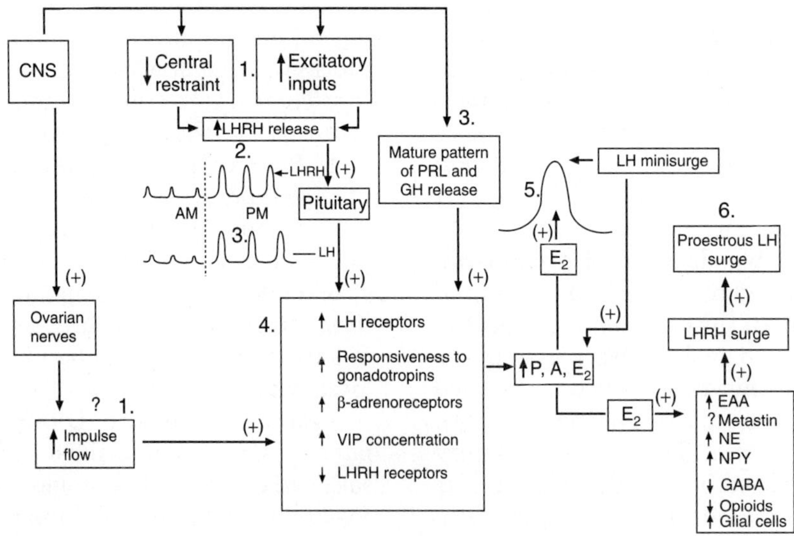

FIG. 10. Postulated cascade of events leading to the preovulatory activation of ovarian estradiol secretion at the onset of puberty in the female rat. It is proposed that the ovary is activated by both hormonal signals coming from the anterior pituitary and neuronal inputs arriving at the gland via the extrinsic innervation. For abbreviations, see Fig. 6. [Modified from (35) with permission.]

aromatizable androgens) appears to be relevant to the mechanism of vaginal opening. It has been shown that the production of early proestrous plasma levels of T, via T-containing Silastic capsules, in late juvenile rats results in precocious vaginal opening but does not advance the first ovulation (239). Examination of serum E_2 levels in animals treated with T shows that E_2 is not increased by the exposure to elevated T levels. Nevertheless, the vaginal epithelium is able to metabolize androgens via an aromatase-like reaction (352), suggesting that the hastening effect of physiological levels of T on vaginal opening may be due, at least in part, to local estrogen production by aromatization.

Although the secretion of E_2, P, and T increases, the secretion of 3a-androstanediol diminishes (353,354), a change that becomes much more pronounced during the hours encompassing the first preovulatory LH surge (355,356) and that seems to be elicited, at least in part, by rising PRL levels (355). Based on the findings that ovarian 5α-reductase activity decreases markedly at puberty (357) and that administration of 3α-diol delays the timing of first ovulation (358), it was proposed (357,358) that 3αa-diol is involved in restraining the onset of puberty. A decrease in serum 3α-diol, however, does not appear to have major consequences on gonadotropin release, because neither blockade of 5α-reductase activity nor administration of physiological levels of 3α-diol affects the time of puberty in female rats (356). Moreover, the prepubertal decrease in ovarian production of 3α-diol is not maintained because levels of the steroid increase again on the second proestrus (354).

First Preovulatory Surge of Gonadotropins. The acquisition by the ovary of the capacity to secrete sufficient E_2 for an adequate period of time represents the key event that determines the timing of puberty in the female rat. Passive immunoneutralization of circulating E_2 levels prevents the LH discharge and ovulation (359), underscoring the importance of the steroid for the occurrence of the first gonadotropin surge. Estrogen acts on both the anterior pituitary and the hypothalamus to bring about the proestrous surge. In the hypothalamus, it evokes a discharge of LHRH release (134); in the pituitary, it sensitizes the gonadotrophs to the stimulatory effect of LHRH (360). A direct stimulatory effect of P on LHRH release in immature rats has also been demonstrated (361). This finding suggests that the two- to threefold increase in serum P observed before proestrus (31,273,350,351) may have a role in facilitating the stimulatory effect of E_2 in LHRH release.

The stimulatory effect of P on LHRH release may be mediated, at least in part, by an increase in LHRH gene expression because P markedly increases steady-state LHRH mRNA levels within 4–6 hours of its administration to immature rats (230,362). In the female rat, an increase in pituitary responsiveness to LHRH is only observed on the day of proestrus (133,134), indicating that some elevation in basal LHRH output may be necessary for the responsiveness of the pituitary to increase in the presence of elevated E_2 levels. As a consequence of the release of LHRH and/or LHRH-related peptides (363), available pituitary LHRH receptors decline in the afternoon of proestrus (364,365). This phenomenon can be reversed by preventing the expression of neural events leading to LHRH release (364,365). The decrease in LHRH receptors appears to reflect ligand-induced unavailability of receptor for binding rather than representing a true loss of receptors (365).

Despite substantial efforts, there is still much to be learned about the hypothalamic mechanism by which E_2 activates an LHRH surge at puberty. Although LHRH neurons do not contain estrogen receptor-α (366,367), they do express estrogen receptor-β (368,369), indicating that, contrary to an earlier belief, LHRH neurons are subjected to a direct estrogen regulatory input (see Chapter 28). Based on these and other observations, the concept has been proposed that estrogen acts directly on LHRH neurons via estrogen receptor-β to suppress LHRH gene expression and indirectly via estrogen receptor-α (and/or estrogen receptor-β) on neurons synaptically connected to the LHRH neuronal network to activate LHRH neuronal activity and generate a preovulatory surge of LHRH secretion (370).

The neuronal networks most critically involved in the control of the preovulatory LHRH surge appear to be those that use GABA and excitatory amino acids as neurotransmitters (23). Nevertheless, there are several other neurotransmitter systems contributing to the process, either as downstream components of the EAA/GABA controlling system or as complementary transsynaptic regulatory pathways. These systems include stimulatory neurons such as serotoninergic (170,371–376), catecholaminergic (161,168, 377–380), and NPYergic neurons (158,381,382), and inhibitory neurons using opioid peptides as neuromodulators (383).

Excitatory Neurons

Glutamatergic Neurons. Glutamate and the glutamate receptor agonists NMDA and kainate all stimulate LHRH/LH release in immature rats (145,384–387). The importance of this stimulatory effect for the preovulatory surge of gonadotropins is demonstrated by the ability of NMDA receptor activation to induce precocious puberty in female rats (384,388) and male monkeys (389). Conversely, the administration of the NMDA receptor blockers, MK-801 or 2-amino-5-phosphonovaleric acid, delays the timing of puberty in rats (390–393), an effect not reproduced by the non-NMDA glutamate receptor antagonist, 6,7-dinitro-quinoxaline–2,3-dione (387). Furthermore, treatment with E_2 and P, which by itself increases plasma gonadotropin levels, has been shown to potentiate the stimulatory effect of NMDA on LH and FSH release (145). These findings have led to the conclusion that activation of NMDA receptors by endogenous glutamate is a physiological component of the central mechanism underlying the first preovulatory surge of gonadotropins in the rat (390–397).

Noradrenergic Neurons. In addition to amino acid neurotransmitters, a strong body of evidence implicates noradrenergic neurons in the preovulatory activation of LHRH release. For instance, the selective destruction of NE terminals by intraventricular injection of 6-hydroxydopamine blunts the LHRH surge induced by PMSG in immature rats (380). Also, microinjection of methyldopa, which results in the formation of "false" catecholamines, blocks PMSG or hCG-induced LH release when injected during the critical period of proestrus (378). NE turnover in the medial basal hypothalamus increases before the preovulatory surge of LH (161), a change due, at least partially, to E_2 (160,398). E_2 promotes NE release from hypothalamic slices in vitro (399), an effect that appears to be exerted directly on catecholaminergic neurons because they contain E_2 receptors (400). An involvement of NE-containing neurons in the area of the nucleus tractus solitarius (A2 cells), which project to the median eminence and rostral hypothalamus, is suggested by the finding that they respond to E_2 with genomic activation, as shown by an increase in c-*fos* expression (401).

NPY-producing Neurons. That NPY is also involved in the genesis of the first preovulatory surge of gonadotropins was demonstrated by the findings that the secretion of NPY into the portal blood increases in the afternoon of the first proestrus (172) and that immunoneutralization of NPY on the day of first proestrus inhibited the LHRH surge and attenuated the preovulatory increase in plasma LH (176). The presence of E_2 receptors in NPY neurons of the arcuate nucleus (402) suggests that E_2 acts directly on NPY neurons to affect their secretory activity (403). The NPYergic neurons involved in stimulating LHRH secretion may be those located in the hypothalamus rather than catecholaminergic neurons of the brainstem that coexpress the peptide with NE (158,382,404). It thus appears that estradiol activates more than one neurotransmitter system functionally coupled to LHRH neurons to elicit the first preovulatory surge of gonadotropins. Whether activation of these systems is a synchronized interdependent phenomenon or simply represents a high degree of redundancy aimed at ensuring the occurrence of the surge is not known.

Metastin Neurons. Although it is clear that activation of GPR-54 receptors by metastin is required for the initiation of puberty (153–155), it is not known whether this system is directly involved in mediating the stimulatory effect of estrogen on LHRH secretion at the time of the preovulatory surge of gonadotropins.

Inhibitory Neurons

GABAergic Neurons. Numerous studies have shown that GABA is inhibitory to LHRH/LH release at the time of the preovulatory surge of gonadotropins. For instance, GABA release (148,405) and tissue accumulation (406), as well as GAD-67 and GAD-65 mRNA levels (407,408), decrease in the MPOA before the initiation of an estrogen- or estrogen/P-induced LH surge in ovariectomized female rats. That similar changes may also occur during the natural preovulatory surge of LH secretion is suggested by the decrease in GAD-67 mRNA content (407,409) and GABA content (407) observed in neurons of the MPOA/septal region during the afternoon of proestrous in adult rats. GABA turnover rate also decreases before the estrogen-induced LH surge (406). Infusion of GABA or the $GABA_A$ receptor agonist muscimol into the third ventricle or MPOA blocks the preovulatory LH surge in intact female rats (410,411) and reduces the size of the estrogen-induced LH surge in ovariectomized rats (406). In contrast, both intraventricular and intravenous infusions of bicuculline, a $GABA_A$ receptor blocker, advance the timing of the LH surge in estrogen-primed ovariectomized rats and proestrous rats (412,413).

Opiatergic Neurons. Opiatergic neurons may directly influence LHRH neuronal activity because they synapse with LHRH neurons (414). In females, the inhibitory effect of opiates on gonadotropin release (193) decreases during the days preceding the first ovulation (383). This decreased effectiveness in opioid tone has been interpreted as being indicative of an involvement of the opiate system in mediating the pubertal "resetting of the gonadostat" (383), a conclusion based on the earlier suggestions that opiates mediate the negative feedback effect of gonadal steroids on LH release (415,416). A transient diurnal loss of opiate inhibitory tone may, in fact, contribute to the initiation of the peripubertal afternoon increase of LH release in females. This is suggested by the finding that the effectiveness of naloxone to enhance LH release decreases in the afternoon of the peripubertal period (417). For a detailed discussion of the "gonadostat resetting" hypothesis, see Mechanism of the Onset of Puberty, below.

Contribution of Glial Cells. It is now clear that activation of erbB-1 and erbB-4–mediated signaling in hypothalamic glial cells is required for the correct timing of the first preovulatory surge of gonadotropins [reviewed in (229)]. Although activation of these receptors in astrocytes results in PGE_2 release, TGF-α–induced activation of erbB-1 receptors on tanycytes results in PGE_2 and TGF-β1 release and causes plastic changes in tanycyte morphology that mimic those observed at the time of the preovulatory surge of gonadotropins (208). Initially, TGF-α promotes the growth of tanycytic processes and then—as TGF-β1 release increases as the result of PGE_2 stimulation—the processes retract. These in vitro results suggested that a similar mechanism operates at the time of the LH surge and that the sequential increase in hypothalamic TGF-α and TGF-β1 synthesis observed on the day of proestrus (205,230) ultimately lead to the retraction of tanycytic processes, which then allows LHRH nerve terminals to establish direct contact with the endothelial cells of the portal system (208). As part of the same cell–cell signaling process, E_2 induces astrocytic TGF-α synthesis (230,418), and TGF-α then mediates the stimulatory effect that E_2 exert on hypothalamic bFGF synthesis in the evening of proestrus (205,215).

Intracellular Mechanisms Underlying the First Preovulatory Discharge of LHRH. Earlier studies have suggested that E_2 may alter different steps in the sequence of biochemical events leading to LHRH release. The pathway best documented is perhaps that involving NE and PGE_2. The capacity of the hypothalamus to synthesize PGE_2 from arachidonic acid increases during the days preceding the preovulatory LH surge (419). This increase can be mimicked by the administration of E_2 to juvenile rats at a dose that evokes a premature surge of LH. Incubation of median eminences from animals exposed in vivo to proestrous levels of E_2 with various concentrations of NE or PGE_2 indicated that E_2 facilitates LHRH release by acting at two different biochemical steps (142). On one hand, it permits PGE_2 formation to be stimulated by lower doses of NE. On the other hand, it increases both the sensitivity and the responsiveness of the LHRH terminals to PGE_2.

These results suggest that E_2 exerts part of its stimulatory action on LHRH release by activating an NE-PGE_2-dependent pathway. It has also been shown that cAMP is involved in the process of LHRH secretion and that formation of cAMP occurs at a step subsequent to NE-induced synthesis of PGE_2 (420,421). Thus, PGE_2 increases cAMP production in hypothalamic fragments in vitro and methoxamine, an α_1-adrenergic agonist, enhances both cAMP formation and LHRH release, an effect that is blocked by indomethacin, an inhibitor of prostaglandin synthesis (420). Indomethacin also blocks NE-induced LHRH release without altering the stimulatory effect of PGE_2 on LHRH output (165). When adenylate cyclase activity is stimulated by forskolin, cholera toxin, or pertussis toxin, LHRH release is enhanced without an increase in PGE_2 formation (421). Because simultaneous exposure of median eminence nerve terminals to PGE_2 and forskolin does not result in a greater LHRH response than that elicited by either agent individually, PGE_2 and cAMP might act along a common pathway (420).

In addition to this PGE_2-cAMP pathway, the first preovulatory surge of LHRH appears to require another complementary pathway mediated by protein kinase C, a Ca^{2+}-activated phospholipid-dependent kinase [for a review see (422)]. Activation of protein kinase C, by either a synthetic diacylglycerol or a phorbol ester, results in LHRH release (423). Moreover, phospholipase C, which in intact cells catalyzes the hydrolysis of membrane polyphosphoinositides to generate diacylglycerol, also induces LHRH release. Activation of this protein kinase C–dependent pathway induces release of LHRH independently from the PGE_2-cAMP pathway. Thus, blockade of prostaglandin synthesis fails to suppress diacylglycerol-induced LHRH release and a phorbol ester–induced LHRH release without affecting PGE_2 formation. Coexistence of both pathways is indicated by the finding that exposure of median eminences to maximally effective doses of diacylglycerol or phorbol ester together with NE, PGE_2, or forskolin results in an additive effect on LHRH release (423). Activation of both cAMP and protein kinase C–dependent intracellular pathways results in up-regulation of LHRH gene expression (424) in addition to the changes in LHRH output.

Thus, simultaneous activation of both pathways may be required for the proestrous surge of LHRH to occur. The neurotransmitters involved in activating these pathways include NE, NPY, and excitatory amino acids. NPY is considered as a candidate, because it stimulates LHRH release in an intracellular Ca^{2+}-dependent prostaglandin-independent manner, via activation of Y-1 receptors (425,426). Glutamate is also a candidate because it uses NMDA and AMPA/kainate receptors to activate protein kinase C- and extracellular Ca^{2+}-dependent release of LHRH. Importantly, LHRH neurons respond to both NMDA and AMPA with increased firing activity (427), indicating they are directly innervated by glutamatergic neurons. An additional intracellular event triggered by preovulatory levels of E_2 in LHRH neurons is an increased expression of the early immediate genes c-*fos* and c-*jun* [(428) and references therein]. Although well documented, it is unclear whether such a change is required for the LHRH surge to occur. The ability of NMDA to increase LHRH release without affecting c-*fos* expression in LHRH neurons (429) indicates that glutamate-induced activation of LHRH release does not require c-*fos* as an intermediate.

It was earlier assumed that that both PGE_2 and cAMP operate within the LHRH neuron itself. It is now clear that PGE_2 is produced by glial cells and that, upon release from these cells, it acts on LHRH neurons to activate the subsequent intracellular steps, leading to LHRH release (211,227,228,430) (Fig. 5). Indeed, the neuronal LHRH producing cell line GT1-7 responds to PGE_2 with LHRH release but is unable to produce the prostaglandin in response to several secretagogues (431).

THE MALE

Fetal Development

Initiation of LHRH and Gonadotropin Secretion

As mentioned earlier, Aubert et al. (36) detected traces of LHRH in whole-brain extracts as early as gestational day 12. However, in this study, male and female tissue was pooled and so it is unknown whether sex differences exist during this early period of development. Likewise, even though it is clear that the hypothalamus contains radioimmunoassayable LHRH by gestational day 15 (48), it is unknown whether any sex difference exists at this time. Significantly, LHRH receptors are already detectable in the anterior pituitary of male rats at gestational day 16 (36), suggesting an active involvement of LHRH in the control of fetal pituitary function. The number of receptors then increases in parallel with hypothalamic LHRH content but with a phase delay of a few days. Studies using pooled pituitary homogenates from males and females have found radioimmunoassayable LH and FSH as early as gestational days 15 and 19, respectively (42). However, studies in which the sexes were segregated before use have yielded conflicting results with regard to sex differences. In one report, pituitary LH could be detected in male rats from gestational day 17 onward but not in females (51). The bulk of the evidence, however, indicates that pituitary LH can be detected in both sexes at this age and that pituitary LH content tends to be greater in the female (36,42,49).

Initiation of Testicular Function

Early in embryonic development, the gonad is bipotential (432) and can initiate male sex determination with the expression of the testis-determining factor (SRY) by Sertoli cells (433). In the rat this occurs at embryonic days 11–12 (Fig. 3). Other transcription factors involved include Sox9, SF1, DMRT, and Wnt-1 [reviewed in (54)].

Testis development starts with the formation of seminiferous cords. Germ cells migrating down the genital ridge eventually aggregate with Sertoli cells during formation of the cords. Cells of the mesonephros migrate into the early gonad and form a pre-peritubular cell that promotes seminiferous cord formation (434). Neurotrophin-3 production by the developing Sertoli cell acts as a chemotactic factor for the migrating mesonephros cells (435) as well as FGF9. Therefore, peritubular–Sertoli interactions appear to be present very early in testicular development. Neonatally, seminiferous cords exist and at the onset of puberty develop into seminiferous tubules. The stages of testis development can be categorized as (a) the initiation of testis determination and seminiferous cord formation, (b) the organization of the seminiferous cords and growth neonatally, (c) growth of the testis prepubertally, (d) onset of spermatogenesis and seminiferous tube formation at puberty, and (e) maintenance of optimal cellular functions and spermatogenesis in the adult (Fig. 3). The factors and cell–cell interactions that control this process remain to be elucidated.

Sexual differentiation of the male gonad begins very early in life, with the seminiferous cords formed at gestational day 13 in a hormone-independent manner [for a review see (436)]. Testicular LH receptors have been detected as early as gestational day 15.5 (437), and it is noteworthy that already at this time LH stimulation can cause an increase in cAMP and T production (438–440). The LH-receptor content increases further at gestational day 18.5, and maximum levels are attained around the time of birth. Coincident with this rise, the number of interstitial cells increases (441), as does testicular T content (437,442).

The adult population of interstitial cells is functionally different from the population found in the fetal testis (441,443). This fetal T plays an essential role in male sexual differentiation (444); its concentration declines soon after birth, due to the differentiation of more adult-stage Leydig cells. All embryonic cell types undergo rapid division toward the end of gestation, except germ cells that transiently arrest cell proliferation before birth. After birth, all testis cells proliferate as the testis grows. At the onset of puberty (around days 10–15 in the rat), Sertoli cells cease to divide and become postmitotic (445–448). Sertoli cells play a key role in the initiation of spermatogenesis and have been shown to possess FSH receptors from as early as gestational day 17.5 onward, rising markedly just before birth (437,449,450). Knockout mouse models for FSH, FSH receptor, and LH receptor show that embryonic gonad development, cell growth, and testis morphogenesis are hormone independent, such that neither gonadotropins nor steroids are required (66,67,249). Therefore, although the cells have the capacity to be hormone responsive, endocrine regulation of testis development is not critical.

Postnatal Development

In this section we discuss postnatal development of the hypothalamic–pituitary–testicular axis. Initially, a separate description is given of the changes that occur in each of the system's components. An attempt is then made to show how the basic components are integrated to produce a functional neuroendocrine unit and how this unit operates during the onset of puberty.

Hypothalamic–Pituitary Unit

In the male rat, hypothalamic LHRH levels continue to increase throughout postnatal development (444,451–453). However, unlike the situation in the female, where levels reach a maximum just before proestrus, in the male they increase even during adulthood. This ultimately results in a significant sex difference between the adult LHRH levels (48,99,454).

Similarly, the pituitary content of LH and FSH increases gradually with age, as does the responsiveness of the gland to LHRH stimulation (48,453, 455–457). In the male, the maximum FSH response occurs between 25 and 35 days of age, whereas the maximum LH response occurs between 35 and 45 days of age (118,458). This is several days later than the peak gonadotropin response observed in females (117,119). Examination of the ontogeny of pituitary LHRH receptors in the male rat has revealed a close correlation between the number of LHRH receptors and the pituitary content of LH. Both appear to stabilize when the animals enter the peripubertal phase of development (~30 days) (459,460). When expressed as concentration rather than as content, pituitary LHRH receptor levels show an increase during the first 4 weeks of life, reaching a peak at around 30 days, and then declining to the lower adult levels seen between 60 and 80 days of age (263,459). This decline during the latter part of sexual development is inversely correlated with rising serum T levels and therefore suggests an increased negative feedback action of testicular steroids on hypothalamic–pituitary function.

During the neonatal period, serum gonadotropin levels are high, though significantly lower than in females. Within a few days, however, the levels fall drastically (95,453). Numerous studies have attempted to characterize the subsequent developmental changes in gonadotropin secretion in the male rat, but a consistent pattern has not been forthcoming, especially for LH. Some reports suggest that serum LH concentrations increase as puberty approaches (95,453,461,462), whereas at the other extreme it has been suggested that they actually decrease (459,463). In most cases, however, no significantly consistent alterations have been observed in mean serum LH levels (48,93,98, 263,451,454). Such inconsistency most probably arises because of the pulsatile manner in which LH is released from the pituitary gland (464–466), even before puberty (131). In the female, this problem of fluctuating serum LH levels has been overcome either by using a very large number of animals in the study (338) or by examining an individual's pulsatile LH release pattern in detail (25,130). Whether or not definite puberty-related changes in the pulsatile LH release pattern occur in the male has yet to be convincingly established [for a comprehensive review of factors that regulate LH secretion in the rat, see (398)].

In contrast to the discrepant reports on LH secretion, there is a general consensus that sexual maturation in the male rat is associated with an increase in FSH secretion. Serum FSH levels rise during postnatal life and reach a maximum usually between 30 and 40 days of age. They then fall gradually as serum T concentrations increase and attain relatively low adult levels (48,93,95,98,451,453,459,461,462,467). Whether or not a peak of FSH secretion also occurs around day 12, as is the case for females, is uncertain but has been observed by some investigators (95,263,454,468).

In addition to changes in the amount of gonadotropin released or even possible changes in pulsatile pattern, some chemical characteristics of the hormones may also change during sexual development. This is suggested by a change in the isoelectric focusing pattern of FSH observed in male rats undergoing sexual maturation (453). Although these studies did not demonstrate whether such changes resulted in greater biological activity of the hormone, pleomorphism of pituitary FSH during prepubertal maturation in the female rat has been shown to be associated with the appearance of molecular forms of FSH that have greater biological activity (123).

The Testis

The somatic cells within the testis are hormone responsive and indirectly control the process of spermatogenesis. The Sertoli cells (469) form the basal and apical surface of the seminiferous tubule and provide the cytoarchitectural arrangements for the developing germinal cell (470). Tight junctional complexes between the Sertoli cells contribute to the maintenance of a blood–testis barrier (471) and create a unique environment within the tubule (472). The structure of the Sertoli cell has been reviewed by several investigators (470), and a three-dimensional reconstruction has increased appreciation for the complexity of the structural relationships within the seminiferous tubule (473). The biochemical analysis of the Sertoli cell has primarily focused on an examination of the components synthesized and secreted by the cell. The list of products is growing and includes steroids such as E_2 (474), metabolites such as lactate (475), and various proteins (476). Specific proteins that have been identified include androgen binding protein (477), plasminogen activator (478), testicular transferrin (479), testicular ceruloplasmin (480), inhibin (481), and several others (482). Most of these secretory products are hormonally regulated and provide useful markers of Sertoli cell differentiation. The differentiation of Sertoli cells throughout development is critical for the maintenance of testis function and spermatogenesis in the adult. A summary of this differentiation is shown in Fig. 11. This is initiated with the expression of SRY and cell fate determination and subsequent cell growth and expression of embryonic genes (i.e., MIS, aromatase). Before the onset of puberty the cell cycle is arrested and differentiated gene expression increases (i.e., transferrin and androgen binding protein). The control of this process is a combination of intrinsic transcriptional regulation and hormonal regulation.

Another somatic cell type in the seminiferous tubule is the peritubular myoid cell, which surrounds and forms the exterior wall of the tubule. Between peritubular cells and the basal surface of the Sertoli cells or spermatogonia is a basal lamina produced cooperatively by peritubular cells and Sertoli cells (483,484). Peritubular cells derive from the mesenchyma and can be isolated and maintained in cell culture (485,486). A biochemical analysis of peritubular cells has revealed a spectrum of secreted proteins (476,487,488). It is generally accepted that peritubular cells play an important structural role in the seminiferous tubule.

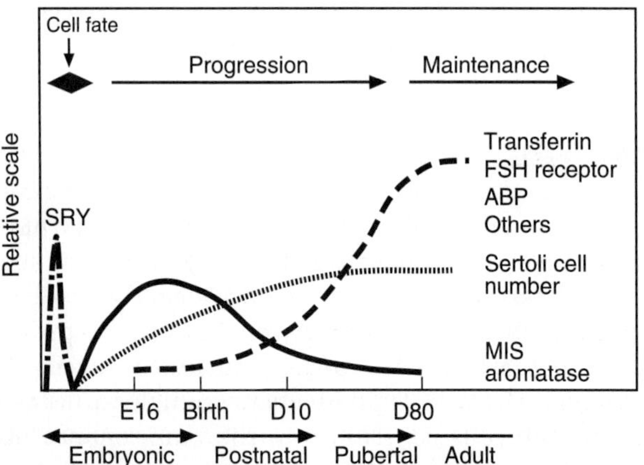

FIG. 11. Sertoli cell differentiation and developmental profile during embryonic and postnatal development. Cell fate is determined in the rat embryo between E11–12 through a transient increase in SRY expression and then Sertoli cells proliferate and numbers increase until the onset of puberty when the cell cycle is arrested and they become postmitotic. Relative gene expression is shown for genes involved in embryonic development of Sertoli cells (i.e., müllerian duct inhibitory substance [MIS] and aromatase) and genes required for maintaining differentiated Sertoli cell function (i.e., transferrin, FSH receptor, androgen binding protein [ABP]) and whose expression increases during pubertal development.

The importance of peritubular cells in the regulation of Sertoli cell function has also been considered (489).

The Leydig cell (490), which resides in the interstitial tissue, is another functionally important somatic cell in the testis and is responsible for the production of androgens under gonadotropin control. Leydig cells have a major influence on spermatogenesis through the actions of T on both the seminiferous tubule and the pituitary. Although the Leydig cell has numerous secretory products (489), the ability of the cell to produce androgens that act on the seminiferous tubules is one of the most significant secretory functions of the cell.

Analysis of the developmental changes in the somatic cells and germ cells within the rat testis has been reported (446,491). Details of such changes, as well as a review of physiological aspects of testicular development, are presented in Chapter 20. In the present section, discussion is focused on the two major routes by which testicular spermatogenesis and steroidogenesis in the rat are controlled by the hypothalamic–pituitary unit.

Hormonal Control. The endocrine control of spermatogenesis is the result of direct actions of hormones on somatic cells (492), with FSH acting directly on Sertoli cells and LH acting on Leydig cells. Androgens are known to have a dramatic influence on spermatogenesis, and high numbers of specific nuclear receptors have been found in Leydig, peritubular, and Sertoli cells (493). Sertoli cells can be influenced by a number of agents, including FSH, insulin, retinol, and T (494,495). Although the androgen receptor is present in Sertoli cells, purified populations of Sertoli cells have a negligible response to androgen in vitro with most cellular functions examined (494,495). A combination of direct and indirect actions of androgen via the peritubular cells are postulated (489). Recently, conditional deletion of the androgen receptor in Sertoli cells confirmed a critical role for androgens in the pubertal development of the testis, T production, and maintenance of normal spermatogenesis (496,497). Nonconditional knockout models had earlier yielded similar results (498).

FSH is an important regulator of Sertoli cell differentiation through the elevation of cellular cAMP levels. Sertoli cells are influenced by a number of different signal transduction systems, including cAMP, calcium mobilization, and protein kinase C (499). Nuclear receptors for agents such as retinoids also influence Sertoli cells (500,501). Although a number of regulatory agents have been shown to influence Sertoli cells through various signal transduction systems, androgens appear to be the most critical for the induction of pubertal development and maintenance of adult testis function. The FSH and FSH receptor knockout mouse models demonstrate FSH actions on

Sertoli cells are important for quantitative induction of pubertal development and maintenance of spermatogenesis (66,249). However, androgens alone can induce pubertal Sertoli development and induce and maintain spermatogenesis (66,249).

The most important androgen produced by the male gonad is T. It plays a pivotal role in several aspects of sexual maturation, including behavior, spermatogenesis, and differentiation and maintenance of accessory sex organs. T also exerts a tight control over gonadotropin secretion by a negative feedback loop.

During the prepubertal period, however, T is not the primary androgen produced by the rat testis. Marked prepubertal changes in testicular androgen production arise because of differences in the development of various enzymes (451,502–505). Most significantly, the activity of 5α-reductase develops during the infantile–juvenile period (502–505). Consequently, the primary androgens produced by the immature rat testes are androstenedione, 5α-androstanediol, and 5α-reduced steroids such as DHT. From around day 25 onward, the activity of other enzymes also becomes apparent, most notably 17α-hydroxylase, $C_{17\text{-}20}$ lyase, and 17β-hydroxysteroid dehydrogenase (506,507). Therefore, as the testes develop, more of the C_{21} precursors pass along the T-synthesizing pathway. The activity of 5α-reductase declines after about 40 days of age, leaving T as the major testicular androgen in the adult (503,508–510). Aromatase activity in the Sertoli cells declines during maturation (Fig. 11) but increases in the interstitial cells (511). Nevertheless, testicular production of estrogen is considerably lower than that of the androgens.

During early postnatal development, serum T levels are low. Although the levels begin to increase around the first week of life, the most pronounced rate of increase occurs much later between 50 and 60 days of age (98,462,510,512–514). Furthermore, it has been shown that the actual pattern of T secretion is highly correlated but phase delayed, with pulses of LH secretion (464). With regard to puberty, it is clear that changes in the secretion of the pituitary hormones precede the maturation of the testes. It is well established that FSH binds within the seminiferous tubules to facilitate spermatogenesis, whereas LH stimulates T secretion by a direct action on the interstitial cells. It is also clear that prepubertal increases in serum FSH levels promote the formation of gonadotropin receptors within the testes (18,443,462,515). Additionally, FSH has been shown to enhance the production of steroid biosynthetic enzymes (516). Overall, testicular growth proceeds in parallel with the FSH increase (93). The effect of LH on the release of steroids from immature testes can be blocked by hypophysectomy, but treatment of the hypophysectomized animals with FSH restores the effectiveness of LH (515). This is probably due to the capacity of FSH to increase testicular LH receptors (515). In view of these findings, it appears that a puberty-related increase in the production of T might theoretically occur irrespective of whether or not the tonic levels of serum LH actually change (assuming that the prepubertal increase in serum FSH levels has primed the gonad to the actions of LH). It remains to be established whether subtle changes in the pulsatile pattern of LH secretion play a role in modulating prepubertal steroidogenesis in the male as they do in the female (25,342). As shown in the FSH knockout model (249), FSH is not essential for testicular development, but it does facilitate and allow quantitative male reproductive function.

FSH is not the only hormone that induces testicular responsiveness to LH. This ability appears to be shared, to some extent, with GH and PRL (18,517,518), both of which show a progressive rise in their serum levels during sexual development (95,98,454,461). On the other hand, some reports have shown that the seminiferous tubules of GH-deficient rats have a qualitatively normal morphology during pubertal development; testicular endocrine function is also normal in these animals despite the small size of their testes (519). A correlation with the GH secretagogue receptor also correlates with pubertal development (520). GH likely facilitates the process but is not essential. The involvement of thyroid hormones in testicular development is also unclear. In support of its active role is the observation that thyroidectomy of immature rats severely inhibits gametogenesis and interstitial cell development (521,522). Interestingly, however, the induction of transient hypothyroidism during neonatal life, using a reversible goitrogen, ultimately results in a lasting enlargement of the testes and other reproductive organs when the animals become adults (523). The actions of thyroid hormones are in part due to optimizing the steroidogenic capacity of Leydig cells (524). LHRH has also been shown to exert some control over prepubertal steroidogenesis. Receptors to LHRH have been demonstrated in testicular interstitial tissue, and a finding suggests that their binding capacity increases between 30 and 40 days of age, subsequently falling to a stable level by day 60 (263,525). Moreover, it has been shown that LHRH can directly inhibit steroidogenesis in the rat testis (262,526), although the precise chemical identity of gonadal LHRH remains to be elucidated [for reviews see (262,527,528)]. A number of other cytokines and growth factors have also been correlated with the onset and progression of testis puberty, including antral natriuretic peptide (529), hepatocyte growth factor (530), antiapoptotic family member (bcl-w) (531), and interleukin (532). The physiological role and importance of most of these factors during puberty remains to be determined.

Testicular steroidogenesis is therefore influenced by at least these important factors: the secretory pattern of hypothalamohypophyseal hormones, the responsiveness of the testes to these hormones, and intrinsic testis control mechanisms. An analogous situation exists for the physiological actions of T during puberty, at which time these actions are exerted only on specific target tissues. Because steroid hormones circulate in the plasma predominantly in a protein-bound form, it is important to establish whether or not steroid-protein binding changes during prepubertal development. Apparently it does not (533,534), but the metabolic clearance rate of T might. For instance, when Smith et al. (535) gave male rats subcutaneous Silastic capsules containing T, they found a progressive age-related decrease in the clearance rate of T from the circulation. On the other hand, in an earlier study by Ulrich and Kent (536), the half-life for the disappearance of radiolabeled T was found to be similar in immature and adult animals. In addition to changes in the steroid production, the developing testes also show an alteration in the capacity to secrete certain proteins (476). One of the most important is androgen binding protein, which specifically binds T and 5α-DHT and is thought to be under the direct control of FSH. Serum levels of androgen binding protein rise sharply after birth, reaching a maximum at around 3 weeks of age (Fig. 11). Its decline in the circulation seems to be related to the formation of the blood–testis barrier, after which androgen binding protein is released primarily into the epididymis (436).

Neural Control. Earlier work suggested that testicular function in the rat is regulated not only by hormonal factors, but also directly by autonomic innervation (537,538). Testicular denervation in the immature rat has been shown to cause a significant reduction in testicular weight—especially if performed during the infantile period, when gonadotropic stimulation is still relatively low (539). Severe disruption of the seminiferous tubules, but not the interstitial tissue, was also found. Furthermore, essentially similar results were obtained when testicular sympathetic nerve terminals were destroyed with 6-hydroxydopamine (539). More recent studies firmly established this concept by identifying the neural pathways that connect the testis with the nervous system (540). These experiments, carried out by injecting the transganglionic retrograde tracer pseudorabies virus into the testis, showed that the virus reached the spinal cord, brainstem, and paraventricular nucleus of the hypothalamus. Additional experiments demonstrated the involvement of this neural pathway in the process by which corticotropin-releasing hormone and interleukin-1β injected intracerebrally inhibit the T response to hCG (540). The same authors also demonstrated that this neural pathway requires the intermediacy of the paraventricular nucleus of the hypothalamus (541).

Hypothalamic–Pituitary–Testicular Interrelationship

One of the classic examples of a negative feedback loop in endocrine systems is that of T and the hypothalamic–pituitary unit. The existence of such a loop in the rat has been demonstrated by surgical intervention. It is well known, for example, that orchidectomy leads to an immediate increase in the level of gonadotropin secretion, which can readily be suppressed by the administration of exogenous T. Interestingly, such a response is observed early in life, even during the neonatal period (236,542–546). The increase in gonadotropin secretion, immediately after castration, appears to be greatest in sexually mature animals (546), although a week later serum gonadotropin levels are the same irrespective of age (544). It should perhaps be emphasized that this lack of marked age-related response to castration contrasts notably with the observations made in agonadal humans (see Chapter 40).

The notion that the hypothalamic–pituitary unit becomes progressively less sensitive to testicular-negative feedback has been the basis for a popular hypothesis that tries to explain the initiation of sexual maturation in the male rat. The reasoning behind this hypothesis is that as feedback sensitivity decreases, the hypothalamic–pituitary unit becomes more effective at stimulating testicular development. Also, serum T levels are able to increase without completely suppressing the secretion of the gonadotropins. Initial support for this hypothesis was provided by Ramirez and McCann (542), who used the ovarian-ascorbic acid depletion bioassay to demonstrate an age-related decrease in T feedback sensitivity in orchidectomized rats. Numerous other studies using RIAs have now either confirmed or extended these findings (535,546–549). Interestingly, when the steroid treatment was delayed for 5 days after orchidectomy, an age-related alteration in feedback sensitivity was not found (18). It has subsequently been suggested that removal of gonadal feedback might elevate the sensitivity threshold of the hypothalamic–pituitary axis (546,550). To date, however, neither actual changes in sensitivity (e.g., at the receptor level) have been demonstrated nor is it clear whether such changes are relevant to the processes involved in the initiation of sexual development. Indeed, Nazian and Piacsek (551,552) have shown that the exposure of male rats to low ambient temperatures delays the onset of puberty, even though a corresponding change in negative feedback sensitivity was not detected (for further discussion of this subject, see "Gonadostat" Hypothesis, below).

A second negative feedback loop has been hypothesized for numerous years to account for the testicular control of FSH secretion. The need for such a hypothesis is quite evident when one considers the findings that as testicular development progresses, serum FSH levels fall, even though mean LH levels increase little or remain relatively unchanged. Evidence exists supporting the involvement of inhibin in the feedback regulation of FSH secretion in maturing rats (553,554). Interestingly, the distribution of inhibin appears to change from a predominantly extratubular to an intratubular pattern as the testes mature (554). Inhibin B is the predominant form expressed by Sertoli cells during pubertal development, whereas production of inhibin A is high neonatally (555,556). Circulating inhibin levels are especially high during the neonatal period, which correlates to low circulating FSH levels (557,558). Because serum inhibin levels decrease between 15 and 25 days of age (553), it is likely that this decline is involved in part in determining the increase in serum FSH levels that occur after day 25. An additional control point is provided by the age-related decline in pituitary binding sites for inhibin (553). Neither case, however, explains the subsequent fall in serum FSH that occurs after day 40. Clearly, the negative feedback of Sertoli cell–produced inhibin on FSH secretion is a critical testis–pituitary control point; however, the interplay with other regulatory mechanisms remains to be elucidated.

Pubertal Activation of the Hypothalamic–Pituitary–Testicular Axis

The current consensus is that the number of pituitary LHRH receptors reflects, at least on a short-term basis, hypothalamic secretion of LHRH (527,528). Because both the hypothalamic content of LHRH and the pituitary content of LHRH receptors begin to increase early in life, it is very probable that developmental changes within the LHRH-releasing centers provide one of the earliest stimuli for initiating pubertal. Several lines of evidence obtained from the male rat support this view. Similar to females, LHRH neurons in males undergo morphological changes as puberty approaches (136,138). As already discussed, these changes consist of a significant increase in the proportion of cells with spiny-like processes as opposed to cells with a smooth surface. Also, as in the case of the female, such morphological changes in the LHRH neurons might reflect an increase in puberty-related synaptic inputs to the cells and might be associated with an increase in central gonadal-independent drive. Likewise, this increase in central drive may be the primary factor responsible for the initiation of male puberty.

Another approach used to investigate the role of LHRH in the initiation of puberty has been to block the action of LHRH by passive or active immunization. If this blockade is initiated during the infantile period (before postnatal day 15), then testicular function is permanently impaired (559–561). Still another approach is that used by Bourguignon and colleagues (101,102), who examined the pulsatile release of LHRH from the medial basal hypothalami of male rats in vitro. A puberty-related increase in pulse frequency was detected, although a concomitant increase in pulse amplitude cannot be completely ruled out.

The possibility that the responsiveness of the pituitary gland to LHRH stimulation changes during maturation has already been mentioned, and it is particularly interesting that the peak gonadotropin responses occur during the peripubertal period of development (118,458). Furthermore, androgens have been shown to have direct effects on the pituitary gland; the studies of Nazian and Mahesh (562,563) imply that T can potentiate the pituitary response to LHRH in immature, but not adult, animals.

Another important developmental step in the pubertal activation of the hypothalamic–pituitary–testicular axis is the change in responsiveness of the testes to hypothalamic–pituitary stimulation (see above). The role of such changes in the initiation of sexual development is, however, more likely to be secondary to neuroendocrine changes first occurring at the central level. The same is probably also true for hypothetical alterations in the sensitivity of hypothalamic–pituitary sensitivity to testicular negative feedback.

MODULATORY INFLUENCES REGULATING THE TIMING OF PUBERTY

Adrenal Gland

A variety of experiments reported over the years has demonstrated that removal of the adrenal gland delays the age at which both vaginal opening and first ovulation occur [for reviews see (564,565)]. Most, if not all, studies on the involvement of the adrenal gland in the onset of puberty have been performed using female rats, perhaps because of the inherent difficulties encountered in trying to alter the timing of puberty in the male rat (16,17). Adrenalectomy may delay puberty by at least four different, but not necessarily exclusive, mechanisms: retardation of bodily growth, loss of circulating corticosterone levels, elimination of circulating epinephrine caused by the removal of the adrenal medulla, and increased secretion of adrenocorticotropic hormone resulting from the loss of corticosteroid-negative feedback.

Adrenalectomized rats do not grow as much as intact controls (565), probably because of the loss of corticosteroid support to intermediate metabolism. Although a low rate of bodily growth may by itself contribute to the delay in puberty (243,566), corticosterone replacement alone has been shown to normalize the time of vaginal opening and first ovulation (564). This effect of corticosterone appears to be exerted, at least in part, at the level of the ovary. Corticosterone facilitates the stimulatory effect of FSH on ovarian steroidogenesis (567) and supports the amplifying effect of PRL on the response of the ovary to gonadotropins (283,286).

Further evidence that the adrenal gland contributes to the regulation of ovarian function is provided by the findings that adrenalectomy decreases the ovarian response to PMSG (568), reduces the number of medium to large-size follicles (569), and inhibits ovarian compensatory hypertrophy (570). Earlier experiments in which adrenalectomy was followed by reimplantation of the adrenal glands indicated that the medulla did not have any influence on the timing of puberty (571). However, selective removal of the adrenal medulla, leaving the cortex intact, delays the age of vaginal opening and first ovulation without altering plasma corticosterone levels or body weight (572). Because medullectomized rats had reduced plasma epinephrine levels and epinephrine stimulated P secretion from granulosa cells at nanomolar concentrations, the conclusion was drawn that a loss of medullary products, presumably epinephrine, is responsible for the delay in puberty (572).

Reports suggesting that adrenocorticotropic hormone may exert inhibitory effects on the reproductive system have been published (573,574). Nevertheless, no agreement has been reached as to whether this effect of adrenocorticotropic hormone is direct or is mediated by the adrenal gland.

Somatic Growth

The influence of metabolic cues on the timing of puberty has been suspected for many years. The early work of Kennedy (575) indicated that vaginal opening was more related to body weight than to chronological age. Subsequently, Kennedy and Mitra (566) postulated that the state of somatic growth is reported to the hypothalamus by some metabolic cue that acts as a signal to initiate puberty. Years later, Frisch and Revelle (576) made a proposal, based on epidemiologic data collected on human females, that the attainment of a critical body weight is essential for puberty to occur. This hypothesis was subsequently modified to state that a particular ratio of lean to fat tissue and a minimum percentage of body fat are necessary for the occurrence of puberty (577). Experiments performed in the rat by Frisch and colleagues (578) showed that females fed a high-fat diet had their first estrus earlier than rats fed a low-fat diet but that the caloric intake at vaginal opening or first estrus was similar in both cases. These authors postulated that the constant caloric intake per unit of body weight at estrus may indicate that puberty occurs at a particular composition of fat/lean mass or fat/body weight. Further support for this hypothesis was provided by a subsequent report from the same laboratory (579), demonstrating that the relative percentages of body water, protein, and fat do not vary at the first estrus in female rats fed either a high-fat or a low-fat diet.

Though attractive, this hypothesis has been disputed by other investigators. Some have concluded that for the timing of puberty, growth rate is more important than attainment of a particular body weight (580). These authors have also presented evidence showing that the attainment of a certain percentage of body fat does not represent a signal for puberty onset and have contended that sexual maturation and increasing body fat are parallel, rather than causally related, phenomena (580). Others have further examined this issue and concluded that attainment of a specific percentage of body protein is more likely to be the metabolic signal for puberty (581).

In spite of these divergent conclusions, it is clear that metabolic disturbances associated with weight loss or decreased rate of body growth are inhibitory to the reproductive system [for a review see (582,583)]. This is in keeping with the hypothesis of Kennedy and Mitra (566), which states that chemical signals derived from the body's metabolic activity influence the reproductive hypothalamus and contribute to its activation at puberty. Some of these signals have now been identified and have been shown to include essential fatty acids, IGF-I, and leptin.

Essential Fatty Acids

Involvement of essential fatty acids in the control of puberty was indicated by studies in which animals fed a diet deficient in essential fatty acids during fetal life were found to have delayed vaginal opening and first ovulation (584). Formation of PGE_2 in response to appropriate stimuli was decreased in both hypothalamic and ovarian tissue. Moreover, the LH surge response to E_2 positive was delayed and the E_2 response of the ovary to hCG was diminished, indicating that essential fatty acids are required at all levels of the hypothalamic–pituitary-ovarian axis for normal function (584).

IGF-I

Plasma IGF-I levels increase strikingly during the onset of puberty in both rodents and primates (216,585,586). At least in rats, this increase occurs independently of the gonads (216). In vitro exposure of median eminences from juvenile rats to IGF-I resulted in a dose-related increase in LHRH release (587), suggesting that IGF-I may be physiologically involved in facilitating the changes in LHRH secretion that occur at puberty. Because synthesis of IGF-I in the postnatal hypothalamus is limited (588), the low levels of IGF-I mRNA detected in the medial basal hypothalamus do not appear to vary during puberty (219), and IGF-I receptors are highly concentrated in the median eminence (589–591), it has been proposed that IGF-I of peripheral origin constitutes the predominant source of IGF-I available to the median eminence during peripubertal sexual development (219,587). This hypothesis also states that the elevation in plasma IGF-I levels that accompanies the pubertal process may contribute to enhance LHRH release, and thus implicates IGF-I as one of the elusive metabolic "signals" involved in regulating the timing of mammalian puberty. In agreement with these observations, Pazos et al. (592) reported that the intracerebral administration of purified antibodies to IGF-I delayed sexual maturation in male rats, thereby lending additional support to the notion of a central stimulatory role for IGF-I in the developmental control of LHRH release.

Leptin

It has been postulated that leptin, a peptide synthesized in adipose tissue and secreted into the general circulation, is a key metabolic regulator of the pubertal process. Consistent with this view, circulating leptin levels increase (593–595) and leptin binding activity decreases (596) with puberty. Furthermore, leptin treatment advances puberty in mice (597,598) and prevents the delay in the onset of puberty in underfed female rats (599,600). Leptin may stimulate LHRH secretion via excitatory and inhibitory neuronal subsets synaptically connected to LHRH neurons (601–603). Whether leptin can act directly on LHRH neurons to stimulate LHRH release remains controversial; native LHRH neurons in the monkey hypothalamus do not contain leptin receptor mRNA (604), but the immortalized LHRH neuronal cell line GT1-7 has been shown to express the mRNA encoding the long (and biologically active) leptin receptor OB-Rb and to respond to low doses of leptin with LHRH release (605). In contrast to native LHRH neurons, the leptin receptor gene is expressed in both NPY and pro-opiomelanocortin neurons (604). Because leptin may also be produced in neurons of the rat hypothalamus (606), it is also possible that neuronally derived leptin contributes to regulating LHRH secretion via transsynaptic mechanisms.

Leptin can also act directly on the gonads to influence both ovary and testis functions. The actions on the ovarian follicle could directly affect steroidal feedback to the hypothalamus (607). Therefore, in addition to the central actions of leptin, gonadal actions also can influence cellular functions, hormone production, and development (608,609).

Despite the robustness of these findings, a wealth of information indicates that leptin is not the cue that times the initiation of puberty in rodents (599,610,611), humans (610,612–614), and monkeys (615–617) [for further discussion see (23,618)]. Instead, it is currently believed that leptin plays a permissive role important for the unfolding of the pubertal process (611).

Pineal Gland

At the present time there is little doubt that the pineal gland of mammals functions as an endocrine organ that transduces neuronal information about day length into endocrine secretions [for a review see (619–625)]. Melatonin is an indoleamine produced and released by the pineal gland (626). Alterations in the secretory pattern of this hormone mediate many, if not all, of the pineal effects. Perhaps one of the most impressive demonstrations of the pineal function, pertaining to the reproductive system, is found in seasonally breeding rodents such as hamsters. Normally, when sexually mature male Syrian hamsters (*Mesocricetus auratus*) are transferred from long to short photoperiodic conditions (i.e., <12.5 hours of light per day), their testes regress within 8 to 10 weeks and the animals become sexually quiescent (627–630). Surgical removal of the pineal gland or its sympathetic denervation completely blocks these photoperiod-induced changes (631). On the other hand, when exogenous melatonin is administered to such animals, in an appropriate mode, pinealectomy can be functionally reversed (632). In contrast, the influence of the pineal gland on the reproductive system of the laboratory rat appears to be much less dramatic. This is not unexpected because the rat does not have a well-defined breeding season, even in the feral condition. Under typical laboratory conditions of ad libitum feeding, constant ambient temperature, and constant long photoperiods, the pineal gland is functionally inert (620), and its contribution to the control of sexual maturation is negligible.

Although some researchers have been unable to show an affect of melatonin on sexual maturation in the rat (633), others have shown that daily melatonin injections given at an appropriate time of day can suppress the growth and functional activity of the ovary, delay vaginal opening, and inhibit PMS-induced ovulation in immature rats (634–644). In the male rat, pinealectomy advances sexual maturation, whereas daily afternoon injections of melatonin reduce the weight of the testes and seminal vesicles; decrease plasma levels of testosterone, LH, and FSH; and decrease the number of pituitary LHRH receptors. These inhibitory effects are observed only when melatonin is administered to animals during the juvenile–peripubertal period and not earlier or later (645–648).

Melatonin appears to act directly on the hypothalamus to exert these effects. Melatonin receptors are present in the hypothalamus (649–651), and these receptors develop before birth (652). Of the two cloned melatonin receptors, the Mel_{1a} receptor is the predominant subtype expressed in brain, and thus it accounts for most neurobiological actions of melatonin (651). Additionally, a few in vitro studies have suggested that melatonin may also affect the pituitary gland (653,654) and even the gonads directly (655,656). Although the above considerations indicate that pineal gland does influence the initiation of rat puberty, it does not appear that it plays a major role in the control of puberty in either humans or monkeys (23).

Pheromones

Pheromones are chemicals released from the body specifically to serve as a means of communication between members of the same species (657) (also see Chapter 37). Because the rat does not possess specialized external sweat glands, the most likely sources of its pheromones are the urine, the feces, and the skin that covers the mammary glands (658–660). A classic example of how pheromones can affect the reproductive system of rodents is the so-called Whitten effect (661). Exposure of female mice to male mice or to their excreta induces estrus in approximately half of the females three nights later. In the context of puberty, it is well established that sexual maturation in the female mouse can be accelerated by the presence of an adult male or even by introducing the female to a cage previously occupied by the male. On the other hand, if female mice are reared in groups, the onset of puberty is delayed compared with that of singly caged animals [for reviews see (662–664)]. Similar studies in the rat are less well documented but suggest that advancement or retardation of sexual maturation by male or female pheromones, respectively, is nowhere near as impressive as in the mouse and is, at best, only marginally significant (665,666).

Environmental Factors and Toxins

Environmental toxins are compounds the gestating mothers of developing embryos are exposed to that can have effects on gonadal development and puberty. The most critical class of these toxins regarding pubertal development is the endocrine disruptors. Endocrine disruptors alter an endocrine pathway through specific hormone (i.e., steroid) receptors to influence cellular function and developmental processes. A wide variety of compounds from plant extracts, plastics, pesticides, and fertilizers has been shown to have endocrine disruptor activity and influence sexual maturation and the pubertal process (667). The most sensitive exposure period is during embryonic and neonatal development, but exposure during pubertal development has effects as well. Neonatal exposure to estrogen can induce altered estrogen receptor expression in the rat testis (668). This indirectly influences the ability of Leydig cells to function and delays the onset and progression of puberty (669). The female is less sensitive to embryonic–neonatal exposure, so the male appears to have a more critical reliance on the androgen–estrogen balance for normal onset and progression of pubertal development (670). Estrogenic compounds such as methoxychlor and phthalates can mimic these actions of estrogen, and fetal or neonatal exposures can alter the onset and progression of puberty (671–673). The plant extract genistein (674) and commonly used compounds such as bisphenol A (675,676) and atrazine (677,678) also have similar effects. In the male, both estrogenic and antiandrogenic compounds (e.g., vinclozolin) have been shown to alter testis development and the onset of puberty. Actions during late fetal development on the testis can influence somatic cell and germ cell functions, as all the cells have the steroid receptors. Neonatal exposure could also influence the central hypothalamic–pituitary development, but because the male is more sensitive than the female, likely gonadal actions are more important. Female neonatal exposure has less dramatic effects, perhaps due to the later stage development of the postnatal ovary. The primary impact on puberty is a delay in the onset of puberty, as well as progression. More severe toxicology can permanently reduce gonadal function in the adult as well. Observations from these toxicology studies and actions of environmental toxins (i.e., endocrine disruptors) indicate that abnormal exposure and action of steroids during critical developmental periods can impact subsequent pubertal development. Further analysis of the molecular mechanism behind these endocrine disruptor actions will provide insight into the factors controlling puberty.

MECHANISM OF THE ONSET OF PUBERTY

"Gonadostat" Hypothesis

The "gonadostat resetting" hypothesis, formulated to explain the onset of puberty, is based on experiments originally performed in the rat (542,679). It originated in the early 1930s when Hohlweg and Dohrn (680) reported that formation of castration cells in the pituitary gland could be prevented in immature rats by a dose of E_2 that is approximately 1% of that required in the adult. Ramirez and McCann coined the phrase "resetting of the gonadostat" and, based on their results using castrated male rats and T replacement therapy, developed the hypothesis as it is known today (542,679). This hypothesis states that as the animal matures, the sensitivity of the hypothalamic–pituitary unit to steroid-negative feedback would decrease. The resulting gradual increase in gonadotropin levels would, in turn, stimulate further release of gonadal steroids.

Once RIA technology became available and permitted the measurement of circulating gonadotropin levels, these early findings were confirmed in both females (681,682) and males (535,546,552,553). Steele and Weisz (681) demonstrated that when prepubertal ovariectomized rats were infused with E_2 for several days, plasma LH levels were suppressed initially and "escaped" from E_2 inhibitory control at the time of vaginal opening. Further support for the gonadostat hypothesis was provided by the observation that E_2 implanted into the hypothalamus was more effective in suppressing the postovariectomy rise of gonadotropins in immature than in adult rats (683).

Subsequent studies brought the concept that this change in sensitivity is responsible for the onset of puberty into question (684). These studies examined the inhibitory effect of E_2 on LH secretion at several intervals before and after the first ovulation (684) and demonstrated that the loss in responsiveness to E_2-negative feedback occurs *after*, not before, the first ovulation. Consistent with these findings, other investigators subsequently reported that the effectiveness of E_2 in suppressing gonadotropin release changes little before puberty (685).

Results obtained in humans further supported this concept by demonstrating that in both males and females, the gonadotropin-secreting system is remarkably sensitive to the inhibitory effect of gonadal steroids during both early and midpuberty, that is, when gonadotropin secretion has already begun to increase (686,687). Thus, puberty in the rat is not initiated by changes in inhibitory feedback control; instead, it results from a gonad-independent centrally originating mechanism(s) that leads to the activation of LHRH release.

Gonad-Independent Increase in Central Drive

It is currently believed that LHRH secretion augments at puberty as a consequence of an increase in neural activity, commonly referred to as "central drive" (22,24,688). In the female rat, this increase in central drive is best recognized by the afternoon increase in pulsatile LH release that begins at the end of juvenile development (postnatal days 28–30) (25,130,341). Two (complementary) mechanisms have been postulated to initiate this event (Fig. 12). One mechanism, known as the central "restraint" of puberty (218,688), consists in the loss of a tonic transsynaptic inhibitory tone that maintains LHRH secretion in check during prepubertal days (688). This inhibition, shown to be mediated by GABA in the female monkey (689,690), has never been formally demonstrated in the rat [for further discussion and references see (23)]. The other mechanism is an increase in excitatory inputs facilitatory to LHRH secretion. According to this latter concept, puberty would not be initiated earlier because of a relative absence of excitatory inputs to LHRH neurons, and the pubertal activation of LHRH release would result from an increase in these excitatory influences (22,24). It would appear, however, that neither a pure increase in excitatory inputs nor an isolated loss

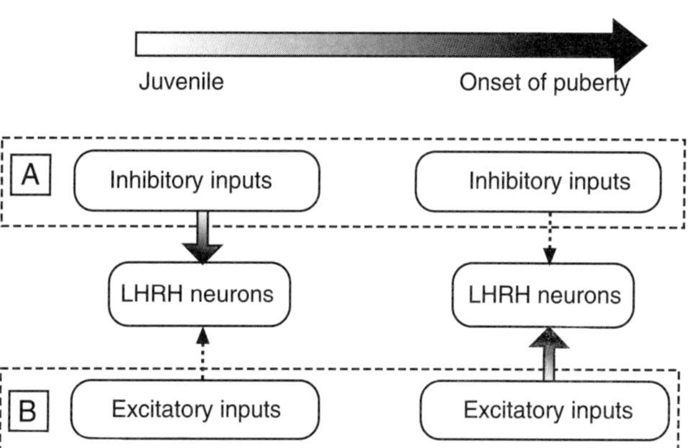

FIG. 12. Current hypotheses concerning the central mechanism(s) responsible for setting puberty in motion. **A:** Episodic LHRH secretion is postulated to be tonically restrained during adulthood by inhibitory transsynaptic inputs; puberty is thereby initiated because of a reduction in this central restraint. **B:** LHRH secretion is postulated to be low during juvenile days because of the lack of excitatory inputs (direct and/or indirect) to LHRH neurons; puberty is therefore initiated because there is an increase in these stimulatory influences. [From (800) with permission.]

of inhibitory transsynaptic inputs to LHRH neurons is the sole mechanism responsible for the activation of LHRH secretion at puberty. Instead, current evidence suggests that this activation is brought about by a combination of interrelated events. Two of these events are transsynaptic (a decrease in GABAergic inhibition and an increase in glutamatergic stimulation to LHRH neurons), and the third involves activation of a reciprocal glial–neuronal communication pathway. The demonstration that mutations of GPR-54, a receptor for the KiSS-1–derived peptide metastin (151,152), results in hypothalamic hypogonadism in humans (153,154) and delayed puberty in mice (154,155) has strongly implicated this gene and its ligand metastin in the transsynaptic excitatory control of puberty. Metastin is a peptide derived from a larger precursor encoded by *KiSS-1*, a gene originally shown to have metastasis antagonistic activity (150). In the hypothalamus, most KiSS-1 mRNA-containing neurons are found in the arcuate nucleus (156,157). Although it is not known yet if metastin-producing neurons receive glutamatergic or GABAergic input, they might innervate LHRH neurons directly because LHRH neurons express the GPR-54 receptor (691) (at least in cichlid fish, the only species thus far examined). Because of the major effect that inactivation of the genes encoding either the GPR-54 receptor or its ligand have on sexual development, it appears likely that the KiSS-1/GPR-54 signaling system represent a critical pathway required for the facilitatory transsynaptic control of LHRH release at puberty.

Although LHRH neurons undergo a series of maturational changes during juvenile development (see Postnatal Development, above), the neurons themselves do not constitute a limiting factor for puberty to occur, because their secretory activity can be prematurely enhanced by experimental manipulations. Examples of such manipulations are the electrical stimulation of the hypothalamus (692–694) and the administration of NMDA (384,392), both of which have been shown to be effective in accelerating the onset of puberty. It would then appear that puberty is initiated by the changing activity of neural pathways functionally and anatomically connected to LHRH neurons. The increase in LHRH mRNA level (135) and neuropeptide content (99) that occurs during the juvenile peripubertal transition period, the greater LHRH secretory response to secretagogues observed in late juvenile animals as compared with early juveniles (142), and the increased number of LHRH neurons with a more mature, "spiny," morphological appearance detected in juvenile–peripubertal animals as compared with younger rats (136) may be maturational changes that facilitate the secretory response of LHRH neurons to the pubertal increase in central drive.

Loss of a Central Restraint

The concept that certain areas of the brain exert a tonic inhibitory control of sexual development originated from the early work of Donovan and van der Werff ten Bosch (12). These authors demonstrated that electrolytic lesions of the anterior hypothalamic area resulted in precocious puberty in female rats and postulated that the lesions initiated puberty because they eliminated an important area for steroid inhibitory control. The ability of hypothalamic lesions to hasten the onset of rat puberty has been amply confirmed by other authors (695–698). Inhibitory influences originating within the limbic system and reaching the hypothalamus via the stria terminalis (699,700) have also been implicated in the restraint of female puberty (13,701,702).

As indicated above, the advancement of puberty brought about by hypothalamic lesions was attributed for many years to the destruction of a center inhibitory to gonadotropin secretion (12,703). The validity of this concept was later challenged by the findings that these lesions do not eliminate the ability of ovarian steroids to suppress gonadotropin release (704) and do not result in elevated basal levels of plasma gonadotropins (698). Subsequent studies demonstrated that lesions of the anterior hypothalamic area result in a striking increase in TGF-α production by reactive astrocytes surrounding the lesion site (705) and an increased expression of erbB-1 receptors in these cells (706). Pharmacological blockade of these receptors prevented the acceleration of puberty induced by the lesion (705), indicating that hypothalamic lesions advance puberty via activation of erbB-1–mediated signaling pathways (705,707). Because neither LHRH neurons in the normal hypothalamus nor in the lesioned hypothalamus express erbB-1 receptors (225,706), it is clear that TGF-α binds to erbB-1 receptors located on glial cells to induce the secretion of bioactive molecules capable of stimulating LHRH release. PGE_2 is one of these substances (206).

Other experiments provided additional insights into the sequence of events by which hypothalamic lesions enhance the secretory activity of LHRH neurons (707). Within 20 hours after injury there is a marked genomic activation of cells adjacent to the lesion site, including astrocytes, as revealed by an increase in c-*fos* expression (707). Although this activation does not occur in LHRH neurons, they respond to the injury with a loss in spininess. This change probably represents reversal to a more immature morphological type, because spiny LHRH neurons have been shown to be more densely innervated (139) and to be metabolically more active than smooth neurons (708). These initial changes are followed by an intermediate phase (24 hours to 4 days) during which the astrocytes near the lesion site become reactive and exhibit an increased expression of the genes encoding TGF-α and its receptor (705,706). At this time, the morphology of LHRH neurons returns to normal (707), and TGF-α may begin to stimulate glial production of PGE_2, which then acts on LHRH neurons to stimulate LHRH release (206). This glial-dependent stimulation of LHRH secretion requires neither activation of LHRH gene expression nor enhanced processing of the LHRH precursor peptide (707).

Although these findings demonstrate that hypothalamic lesions do not advance puberty because they remove an inhibitory system controlling the LHRH neuronal network, they do not challenge the concept that a loss of an inhibitory control is a contributing factor to the pubertal activation of LHRH secretion. As indicated before, this inhibitory tone is dually exerted by GABAergic and opiatergic neurons and serves to counteract the stimulatory effect of excitatory neurotransmitter systems on the LHRH neuronal network.

Activation of Excitatory Inputs

Just like inhibition of GABAergic neurotransmission advances puberty (at least in female monkeys) [reviewed in (23)], the selective manipulation of glutamatergic neurotransmission—the primary mode of excitatory transsynaptic communication used by hypothalamic neurons (709)—accelerates the initiation of puberty. Thus, activation of NMDA receptors via administration of NMDA advances puberty in both rats and monkeys (389,390) due to an increase in LHRH secretion (710). Conversely, blockade of NMDA neurotransmission by a noncompetitive use-dependent antagonist of NMDA receptors delayed the initiation of puberty in female rats (390), demonstrating the physiological importance of this excitatory transsynaptic mechanism in the control of the pubertal process. Other experiments have shown that the activity of hypothalamic NMDA receptors increases at the onset of rat puberty (141) and that this activation occurs in a gonad-independent manner (711).

Although LHRH neurons are directly innervated by glutamatergic processes (712), it does not appear that NMDA acts directly on LHRH neurons to elicit LHRH release, because only a small subset of these cells express the receptors during rat prepubertal development (713). Instead, glutamatergic neurons may use kainate receptors for direct regulation of LHRH secretion—as they do to activate many other neurosecretory neurons of the hypothalamus (714). Stimulation of these receptors increases LH release when injected systemically to rats (715) and increases LHRH release from median eminences in vitro (716) Furthermore, during prepubertal development of the female rat, nearly 50% of LHRH neurons express the mRNA encoding the kainate-preferring glutamate receptor subunit KA2 (717). LHRH neurons are able to assemble functional kainate receptors because they also express GluR5 mRNA (718), which encodes a receptor subunit required for the heterodimeric assembly of high-affinity functional kainate channels (719). Thus, LHRH secretion is under a tight glutamate-dependent facilitatory control mediated by two classes of ionotropic glutamate receptors, one (NMDA) located on interneurons synaptically connected to LHRH neurons and the other (kainate) on LHRH neurons themselves. The critical importance of this excitatory control is made evident by the loss of LHRH pulsatility that results from the blockade of endogenous glutamate synthesis in the developing rat hypothalamus (720). This finding led to the notion that an adequate availability of endogenous glutamate is essential for the peripubertal activation of changes in pulsatile LHRH release (720). In addition to glutamate, there is evidence that an adequate complement of NMDA, AMPA, and kainate receptors in the intact hypothalamus is required for normal pulsatile LH release (721).

Glutamate–GABA Interactions and the Onset of Puberty

Although experimental manipulations of the GABAergic or glutamatergic system result in opposite changes in gonadotropin output, it is now clear that both systems are tightly intertwined. For instance, an electron microscopy examination of the primate hypothalamus revealed an abundance of glutamatergic/GABAergic synaptic interactions (722). Moreover, glutamate is the natural precursor for GABA synthesis, so that any changes in glutamate availability directly affect GABA synthesis. Of unquestionable physiological importance are the findings that the synaptic activity of GABAergic neurons is extensively regulated by glutamatergic neurons, in an either inhibitory or stimulatory manner (723–729). Vice versa, changes in GABA synthesis cause immediate alterations in glutamate levels, as shown by the increase in glutamate release that occurs after the in vivo intrahypothalamic blockade of GABA synthesis (730).

Based on these considerations, it would appear to be clear that changes in glutamatergic and GABAergic neurotransmission are critical components of the central mechanism underlying the activation of LHRH release at puberty (Fig. 13). Which of the

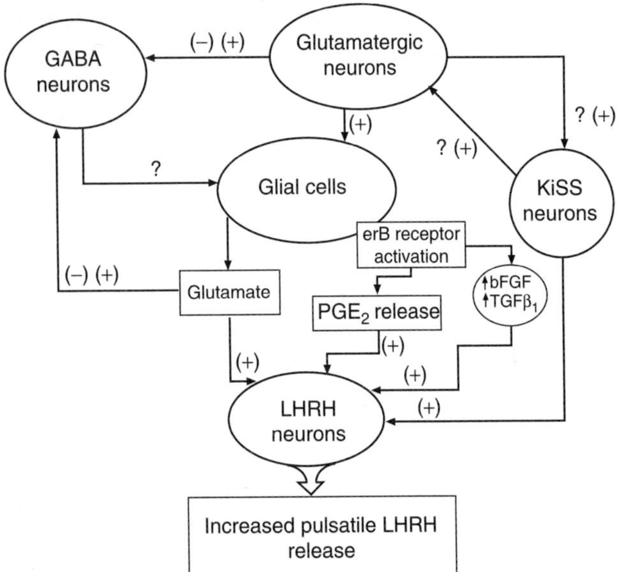

FIG. 13. Postulated cell–cell communication pathways that operating in the hypothalamus contribute to the pubertal activation of LHRH secretion.

two events occurs first is an issue not yet resolved. We favor the possibility that in the rat the activation of glutamatergic transmission precedes the loss in GABAergic inhibitory output. This concept is supported by findings showing that activation of either kainate receptors in the hippocampus or AMPA receptors in the cerebellum of the rat brain causes inhibition of GABA release (724–727). Assuming that similar interactions operate in the peripubertal hypothalamus, the pubertal decrease in GABA transmission can be envisioned as resulting from an increase in kainate/AMPA receptor-mediated excitatory amino acid inputs to hypothalamic GABAergic neurons. Although no morphological studies of the rat hypothalamus have been reported, an examination of the monkey hypothalamus revealed that GABAergic neurons receive a strong glutamatergic input (722) and that the hypothalamus as a whole contains twice as many glutamatergic than GABAergic synapses (722). The alternative possibility that a decrease in GABAergic inhibitory tone precedes the increase in glutamatergic output at the onset of puberty is supported by experiments conducted in female monkeys. One of these studies showed that glutamate release into the S-ME of peripubertal monkeys increases after experimentally inducing a decrease in GABA synthesis (730); another showed that GABA neurons innervate a fraction of glutamatergic neurons in the primate hypothalamus (722). An additional mechanism by which a decrease in GABA production could accentuate glutamatergic activity is by relieving an inhibitory clamp on NMDA receptor–mediated transmission. In the adult brain, GABAergic neurons normally inhibit NMDA receptors via indirect mechanisms, thus preventing the manifestation of NMDA receptor–dependent plastic events (731).

It would appear that a glutamate-to-GABA hierarchical arrangement may already be in place during the infantile–juvenile transition period, at the time when the initial postnatal increase in central drive is subsiding and FSH secretion is decreasing (Fig. 4). The existence of such an arrangement is suggested by the demonstration that glutamate, acting via metabotropic receptors, inhibits GABA excitatory activity in developing hypothalamic neurons at both presynaptic and postsynaptic sites (723). In considering the general principles governing the development of GABAergic neurons in brain (732) and the temporal sequence of organization of GABAergic and glutamatergic synapses in the rodent hypothalamus (733,734), it can be postulated that the neonatal–infantile activation of LHRH secretion is caused by the postnatal development of excitatory GABAergic synapses, an event that precedes the appearance of both mature inhibitory GABAergic synapses (732,733) and functional glutamatergic synapses (734,735). Thus, the postinfantile decrease in LHRH secretion would be determined, to a significant extent, both by the conversion of GABA neurotransmission from excitatory to inhibitory and by a glutamate-dependent metabotropic receptor-mediated inhibition of excitatory GABAergic synaptic activity. Whether the direct $GABA_A$ receptor–mediated input to LHRH neurons also switches from excitatory to inhibitory during infantile days or near the initiation of puberty remains controversial. Using a combination of molecular and electrophysiological approaches, DeFazio et al. (736) showed that regardless of age or reproductive condition, LHRH neurons always respond to $GABA_A$ receptor activation with excitation. In contrast, Han et al. (737) found that the response of LHRH neurons switches from depolarization to hyperpolarization at puberty. The same group subsequently reported that this change occurs in response to changes in endogenous GABA availability (738). The reasons for the discrepancy are not obvious, but one potential explanation might reside in the number of LHRH neurons expressing the NKCC2 transporter in different strains of mice. This transporter causes chloride extrusion, and thus it maintains low levels of intracellular chloride, a condition required for $GABA_A$ receptor stimulation to inhibit neuronal activity [for references and discussion see (736)]. Because LHRH neurons in mice are heterogenous for NKCC2 expression (736,739), the fraction of LHRH expressing the NKCC2 transported might be different in different strains of mice. Further investigation is required to resolve this issue.

If one accepts the idea that GABAergic and glutamatergic neurons provide the critical transsynaptic control of puberty, then one also has to accommodate the view that the neurotransmitter NE and the neuromodulator NPY play secondary roles (Fig. 13). Further discussion of this subject can be found in (23).

KiSS-1/GPR-54 System

Although it is still unknown if this peptidergic system is regulated by glutamatergic and/or GABAergic inputs, it appears to be clear that it plays a major role in the control of puberty and that it is a crucial component of the excitatory transsynaptic control of LHRH secretion (Fig. 13). Future experiments will undoubtedly identify the neuronal systems involved in the transsynaptic control of Kiss-1 production neurons.

Glial–Neuronal Interactions and the Onset of Puberty

The intimate relationship that exists between astrocytes and neurons in the control of LHRH secretion during sexual development is suggested by the mere presence of a large variety of neurotransmitter receptors in astroglial cells (740–743) and by the ability of astrocytes to produce both neurotransmitters and other cell–cell signaling molecules (744–746) similarly recognized by both neurons and astrocytes. Much information has been gathered in recent years demonstrating that astrocytes play a major role in the process by which excitatory amino acids affect neuronal function (747). Recently, the existence of a glutamate-dependent neuron-to-glia signaling pathway that may contribute to coordinating the transsynaptic and glial activation of LHRH secretion at puberty has been provided (748). This study showed that hypothalamic astrocytes contain at least one subtype of metabotropic receptor (mGluR5) and three subtypes of AMPA receptors (GluR1, 2, and 3). It also demonstrated that concomitant exposure of the astrocytes to AMPA and tACPD (a metabotropic receptor agonist), but not to each agonist separately, resulted in recruitment of erbB-1 and erbB-4 receptors and their respective ligands to the glial cell membrane, transactivation of the receptors via a mechanism requiring metalloproteinase activity, and increased expression of erbB-1 and erbB-2 receptors. These results led to the conclusion that a neuron-to-glia glutamatergic pathway represents a basic cell–cell communication mechanism used by the neuroendocrine brain to coordinate the facilitatory transsynaptic and astroglial input to LHRH neurons during sexual development. Previous studies had shown that simultaneous activation of AMPA and metabotropic receptors in astrocytes results in PGE_2 release and a PGE_2-dependent increase in glial glutamate release (749). Because astrocytic glutamate released in response to PGE_2 triggers calcium fluxes in a large number of neurons in the vicinity of the activated astrocytes, it has been postulated that this astrocytic PGE_2-glutamatergic-dependent signaling mechanism is an essential component of the process by which astrocytes regulate synaptic transmission in the CNS (749).

Lending further credence to this concept is the demonstration that astrocytes are required for the establishment of functional synaptic connections and for the efficacy of synaptic communication (750). Astrocytes appear to exert at least part of these actions by secreting protein factors required for neuronal membrane excitability (751) and by producing growth factors of the EGF family able to up-regulate the expression of specific glutamate receptors (752) and/or regulate synaptic plasticity via activation of specific erbB receptor subtypes, such as erbB-4 (753). Because astrocytes also potentiate GABAergic inhibitory synaptic transmission (754) and respond to GABA stimulation with depolarization (743,755), it is evident that a dynamic bidirectional communication exists between astroglial cells and neurons that use excitatory and inhibitory amino acids as neurotransmitters (Fig. 13). The critical importance of these two neuronal systems in the developmental control of LHRH release and the emerging evidence implicating astroglial cells in this process indicate that such a tripartite communication system is a fundamental mechanism used by the neuroendocrine brain to control the advent and progression of mammalian sexual maturation. An additional aspect of astrocyte physiology that requires consideration is the likely possibility of neuroendocrine astrocytes having specialized roles in different hypothalamic regions. Such a specialization is suggested by the brain region–specific expression of neuropeptides, neurotransmitter receptors, and growth factors in astrocytes (418,756,757). In addition, functional compartmentalization of astrocytic function may also occur in response to signals from the local microenvironment, as suggested by the disruption of gap-junctional communication that results from the overexpression of erbB-2 receptors (758). It is thus possible that growth factors and other locally produced signaling molecules may regulate the propagation of signals between different astrocytic subsets, thereby allowing astrocytes to regulate particular hypothalamic functions in both temporal- and region-specific manners. The ability of noradrenergic (759), glutamatergic (759,760), and GABAergic (743) neurotransmission to affect astrocytic morphology provides inferential support to this view. Additional credence to this view is provided by the demonstration of opposite effects of astrocytes from different brain regions on glutamatergic and GABAergic neurotransmission.

Although neocortical astrocytes increase neuronal activity via release of glutamate acting on NMDA receptors (761), hippocampal astrocytes facilitate GABA-mediated inhibitory synaptic transmission between interneurons and pyramidal cells (754).

The aforementioned considerations make clear that the increase in central drive underlying the initiation of mammalian puberty depends on a complex array of transsynaptic and glial–neuronal interactions, carried out by neurons utilizing amino acids of both excitatory and inhibitory nature as neurotransmitters, and astroglial cells producing cell–cell signaling molecules able to back-regulate neuronal function (Fig. 13). A fundamental question that then arises is that of the upstream controlling mechanisms that may integrate neuron-to-neuron and neuronal–glial communication during sexual development.

Potential Upstream Controlling Genes

Thus far, two genes of the homeobox family of transcriptional regulators have been implicated in the hypothalamic control of puberty: *Oct-2* (762), a POU homeodomain gene (763), and *TTF-1/Tebp/Nkx-2.1* (200), a member of the Nkx family of homeodomain genes (764). Both genes were identified in the hypothalamus of immature female rats subjected to puberty-inducing lesions of the hypothalamus.

Like other POU-domain genes, *Oct-2* is detected throughout the embryonic ventral forebrain (763). In contrast, *TTF-1* is selectively expressed in the developing diencephalon (765). In the postnatal brain, *Oct-2* expression persists in several hypothalamic nuclei, including the suprachiasmatic nucleus (SCN), the medial mammillary nucleus, and the olfactory bulb (766). However, *Oct-2* proteins are more abundant in cultured astrocytes than neurons (767,768), indicating that *Oct-2* gene products may preferentially control the transcription of astroglial genes. In keeping with this idea, lesions of the anterior hypothalamus that elicit sexual precocity in female rats led to a rapid activation of *Oct-2* expression in TGF-α-containing astrocytes surrounding the lesion site (762). This increased expression was accompanied by differential expression of Oct-2a, b, and c, three alternatively spliced forms of the *Oct-2* gene expressed in brain. Because similar changes were observed during normal pre- and peripubertal development, the conclusion was reached that activation of *Oct-2* gene expression in the neuroendocrine brain is part of the developmental program that unfolds during the initiation of female puberty. Importantly, *Oct-2* mRNA levels increased during juvenile development, before the pubertal increase in sex steroid secretion, suggesting that the initial activation of the *Oct-2* gene is a gonad-independent event. *Oct-2* transactivates the TGF-α gene via binding to octamer-like motifs present in the TGF-α promoter (762) that are similar to the octamer motif recognized by *Oct-2* in immunoglobin promoters (769).

The physiological importance of *Oct-2* of this transcriptional regulation in the control of puberty was demonstrated by experiments in which an antisense oligodeoxynucleotide against *Oct-2* mRNA depressed TGF-α mRNA levels in isolated hypothalamic astrocytes and delayed the age at first ovulation when infused into the third ventricle of the brain (762). Other POU-domain proteins that may contribute to the regulation of puberty process are Oct-1, required for the transcriptional activity of the neuron-specific LHRH gene enhancer (770), and SCIP/Tst-1, which acts as a repressor of LHRH gene activity (771). Because other homeobox-containing genes, such as those of the Nkx family (see below), may also be involved in the transcriptional control of hypothalamic genes at puberty, POU-domain proteins might represent only a fraction of a larger group of upstream genes controlling the integration of neuron–glia communication during female sexual development.

In addition to POU-domain genes (763), some members of the Nkx-2 family of homeodomain genes—in particular *TTF-1/Tebp/Nkx2.1*—have been shown to play an essential role in the embryonic development of the hypothalamus (765,772,773). Deletion of the *TTF-1* gene results in profound hypothalamic abnormalities, including underdevelopment of the ventromedial and dorsomedial nuclei, fusion of the walls of the third ventricle in its ventral aspect, and absence of the arcuate nucleus (773). *TTF-1* was found to be the Nkx gene preferentially activated in neurons after puberty-inducing hypothalamic lesions (200). In the intact prepubertal brain, *TTF-1* expression is for the most part restricted to the hypothalamus, and within this region, it is localized to discrete cellular subsets. These cells include ependymal/subependymal cells of the third ventricle/astroglial cells of the median eminence, preproenkephalinergic neurons of the lateroventromedial nucleus of the medial basal hypothalamus, and LHRH neurons of the POA. Hypothalamic *TTF-1* mRNA levels remained at low basal values during the first 3 weeks after birth, increasing transiently during juvenile development, before the initiation of puberty. Downstream genes controlled by *TTF-1* include erbB-2, LHRH, and proenkephalin (200). *TTF-1* colocalizes with erbB-2 in ependymal/subependymal cells and increases erbB-2 promoter activity in functional promoter assays (200), suggesting that the prepubertal increase in hypothalamic erbB-2 mRNA levels (226), which is temporally correlated with the changes in *TTF-1* gene expression, is—at least in part—brought about by a *TTF-1*–mediated transactivation of the erbB-2 gene promoter.

TTF-1 transactivates the LHRH promoter (200), suggesting that *TTF-1* plays a facilitatory role in LHRH biosynthesis and thus might be one of the factors responsible for the prepubertal increase in LHRH gene expression detected in prepubertal female rats (135). In addition to LHRH neurons, the *TTF-1* gene is abundantly expressed in preproenkephalinergic neurons of the lateroventral portion of the ventromedial nucleus, but in contrast to its stimulatory effect on LHRH gene transcription, *TTF-1* inhibits preproenkephalin promoter activity (200) suggesting that *TTF-1* functions in vivo as a regulator of preproenkephalin synthesis. A modulatory effect of *TTF-1* on preproenkephalin gene expression may have relevance to both the control of puberty and the pubertal activation of sexual behavior. Opioid peptides in general restrain the initiation of female puberty by inhibiting LHRH release (774,775), and preproenkephalinergic neurons of the lateroventromedial nucleus have been shown to mediate E_2-induced female sexual behavior (776). Recent experiments using the Cre-loxP conditional system for gene deletion demonstrated that loss of *TTF-1* expression in the hypothalamus does indeed result in delayed puberty (201) and that this loss is accompanied by increased expression of the proenkephalin gene in the hypothalamus. Moreover, *TTF-1*–deficient mice exhibited premature termination of their reproductive life span, suggesting that *TTF-1* is required not only for the normal time of female puberty, but also to maintain adult reproductive function (201). Thus, *TTF-1* may play a role in coordinating the activation of cellular subsets positively coupled to the pubertal process (such as LHRH neurons and erbB-2–containing ependymal/glial cells of the median eminence) and the repression of neuronal networks (such as preproenkephalinergic neurons) that are inhibitory to the initiation of puberty. Moreover, the finding that the expression of both *TTF-1* and *Oct-2* increases during the juvenile period, in the absence of changes in circulating sex steroid levels, suggests that both genes are components of the hierarchy of genes responsible for the centrally originated gonad-independent increase in central drive that sets in motion female puberty (Fig. 14).

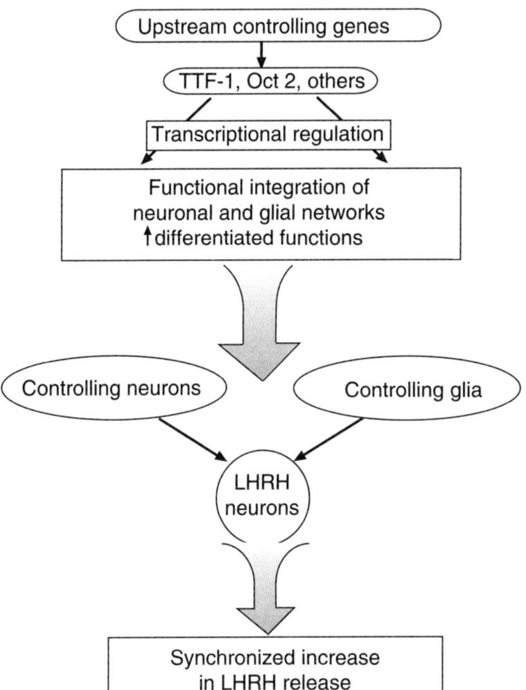

FIG. 14. Upstream genes postulated to participate in the hypothalamic control of puberty. *TTF-1* is expressed in both selected neuronal populations of the hypothalamus (LHRH, proenkephalin) and ependymoglial cells of the median eminence. In neurons *TTF-1* increases LHRH gene expression and represses proenkephalin gene expression; in ependymoglial cells it transactivates the erbB-2 promoter and facilitate erbB-1–dependent differentiated processes. Oct-2 is predominantly expressed in astroglial cells where it controls TGF-α gene expression.

Integration of a Clock Mechanism

As indicated previously, the earliest hormonal change associated with the initiation of rat puberty is an amplification of a diurnal pattern of LHRH/LH release (25). In addition to these changes, activity of the NE (159) and NPY (382) systems has been shown to be higher in the afternoon than in the morning, as opposed to activity of the opiate and GABAergic systems, which is lower in the afternoon than in the morning (405,417). Notwithstanding the importance of establishing a functionally appropriate integration between excitatory/inhibitory neurotransmitter systems and LHRH neurons, the synaptic interconnectivity of LHRH neurons themselves (777,778) may be another factor that ensures the synchronized activation of the LHRH neuronal network to secrete physiologically effective pulses of LHRH secretion and, then, a preovulatory discharge of the neuropeptide.

There is little doubt that this diurnal rhythm is controlled by a neural circadian clock located in the SCN of the hypothalamus (779,780). The identities of both the synaptic pathways conveying information from the circadian clock to the LHRH neuronal network and that of the transsynaptic/cell–cell signaling molecules involved in the process are, however, incompletely understood. In contrast, substantial progress has been made in recent years toward elucidating the molecular mechanisms governing the function of the master circadian clock entrained by the daily light–dark cycle (781,782). A critical, but still only partially answered, question (780,783) concerns the identity of the communication pathways and signaling molecules used by the SCN to entrain the diurnal fluctuations in LHRH output. Immunohistochemical and retrograde tracing studies have shown that LHRH neurons receive a substantial VIPergic input from the SCN (780,784). However, a more critical neuropeptide appears to be arginine-vasopressin, because coculture of SCN with POA tissues showed that diurnal fluctuations in LHRH release coincide with release of arginine-vasopressin instead of VIP (785). In addition to arginine-vasopressin and VIP, both excitatory and inhibitory amino acid neurotransmitters appear to be involved in generating a clock-dependent transsynaptic output from the SCN (786,787). A role for glial cells in this process is suggested by the finding that SCN astroglial cells are the cell type showing the most consistent oscillatory response to glutamate (786), the major neurotransmitter mediating the visual input to the SCN (788).

In addition to these conventional candidates, two novel molecules shown to be required for the maintenance of circadian rhythmicity (789) stand out as appealing candidates for a role in SCN–neuroendocrine communication. The photolyase-like proteins Cry1 and Cry2 are mammalian homologs of a family of plant blue-light receptors known as cryptochromes; they were found in the SCN, as well as in a variety of other tissues, and are thought to act as circadian photoreceptors able to detect light directly and thus photoentrain the biological clock of different tissues (789). Because in their absence there is a complete loss of free-running locomotor rhythmicity (789), they appear to be essential components of the SCN circadian clock.

It is then clear that much more research is needed to acquire a more complete understanding of the mechanisms by which the SCN circadian clock entrains the diurnal changes in activity of the LHRH pulse generator at the time of puberty. Such an understanding requires the identification of the transsynaptic and cell–cell communication pathways involved and the functional characterization of the key molecules underlying this communication.

Modulatory Role of Gonadal Steroids

E_2 may facilitate maturation of the neuroendocrine brain through its neurotrophic effects. E_2 enhances neurite growth (790) and has been shown to induce synaptogenesis in the hypothalamus of juvenile rats at doses that elicit a preovulatory surge of gonadotropins (791). The neurotrophic effects of E_2 appear to involve an increase in gene expression of the cytoskeletal proteins GAP-43 (792), class II, β-tubulin (793), and Tau (794). GAP-43 is a phosphoprotein predominantly found in developing axons, class II β-tubulin isotype is a microtubule protein predominantly expressed in developing and injured neurons, and Tau is also a microtubule-associated protein found in elongating axons. In addition to these neurotrophic effects, E_2 acts on the hypothalamus to increase TGF-α gene expression (418) and erbB-2 and erbB-4 receptor expression (226), which mediate the effects of NRGs on astroglial cells involved in the control of LHRH neurons.

CONCLUSIONS

The Female

The developmental process that leads to puberty in the female rat is based on an extraordinarily complex series of interrelated events (Fig. 15). The CNS plays a critical role by controlling both anterior pituitary function, through the secretion of hypothalamic factors, and the ovary via pituitary hormones and direct neural inputs. Gonadotropins play a decisive role at all stages of sexual development, with the exception of the fetal and neonatal periods, during which ovarian development is gonadotropin independent. Three major periods of activation of gonadotropin secretion can be identified during postnatal development of the rat.

The *first activational period* occurs during infantile development (postnatal days 7–21) and is manifested as an enhancement in FSH secretion, with sporadic elevation in LH levels. Although this strikingly different pattern of gonadotropin release may result, to a significant extent, from the interplay of steroidal influences, its primary determinant appears to be of central origin. It may, in fact, reflect a "disorganized" activity of LHRH neurons resulting from conflicting incoming information. On the one hand, the GABAergic excitatory tone that characterizes the immature brain before the second week of life switches to inhibition; on the other hand, glutamatergic synapses become firmly established, providing the necessary counterbalance to the inhibitory GABAergic tone. As this transsynaptic control develops, glial cells proliferate and establish their functional relationships among themselves and with different neuronal subsets via growth factor–initiated signaling. As a consequence of this maturational instability, LHRH neurons may fail to discharge synchronously. Infrequent randomly occurring episodes of synchronization may result in low-frequency LHRH discharges, which in turn would maintain high plasma FSH levels. A strong inhibitory (opiate) influence capable of slowing down the frequency of LHRH discharges may also contribute to the infantile pattern of gonadotropin release.

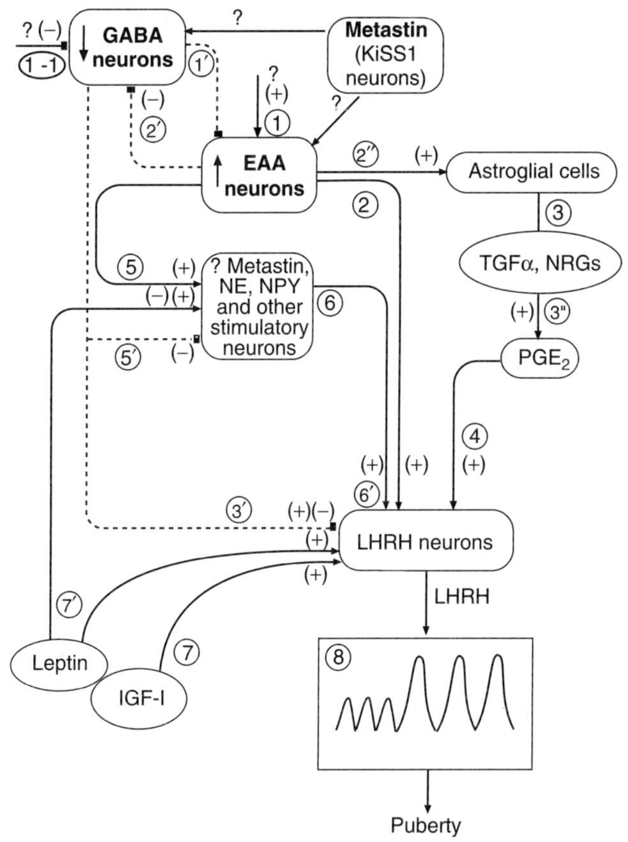

FIG. 15. The central events underlying the initiation of puberty. An increase in episodic LHRH secretion is activated by an integrative mechanism involving both neuron-to-neuron and glia-to-neuron communication processes. The neuron-to-neuron (transsynaptic) component involves the simultaneous activation of excitatory transsynaptic inputs and a reduction in inhibitory inputs to LHRH neurons. The main excitatory transsynaptic input affecting LHRH release is postulated to be provided by neurons that use excitatory amino acids (EAA) as neurotransmitters. NE and NPY-containing neurons provide a complementary facilitatory input to the LHRH neuronal network. The chief inhibitory neuronal input restraining LHRH release is provided by GABAergic neurons (GABA). The glia-to-neuron component of the process involves the production of the EGF-related peptides TGF-α and NRGs by astroglial cells and the stimulatory effect of these peptides on the glial production of neuroactive substances, such as PGE_2, which act directly on LHRH neurons to elicit LHRH secretion. Current evidence suggests that an increase in EAA neurotransmission (1) may be predominantly responsible for the pubertal activation of pulsatile LHRH secretion in the rat, whereas a decrease in an inhibitory GABAergic tone (1-1) may be a primary transsynaptic event underlying the process in the rhesus monkey. Evidence collected in rodents indicates that excitatory amino acids can act directly on LHRH neurons (2) to enhance LHRH release and on GABA neurons (2′) to inhibit GABA neurotransmission. It is also possible, however, that a primary decrease in GABA release (1′) triggers the pubertal increase in EAA neurotransmission. EAA acting on astroglial cells (2=>) appear to facilitate the glial production of TGF-α/NRGs (3). Although the decrease in GABA transmission begins to relieve LHRH neurons from GABAergic inhibitory control (3′), the increased production of EGF-related glial peptides stimulates the formation of PGE_2 (3′=), which acts directly on LHRH neurons to elicit LHRH release (4), amplifying the stimulatory effect of EAA (2) and magnifying the loss of GABAergic inhibitory control (3′). Upon the initiation of these changes, a further increase in EAA stimulation (5) and further decrease in GABAergic inhibitory tone (5′) lead to activation of NE and NPY neuronal systems (6). Both NE and NPY would then contribute to the progression of the pubertal process by stimulating LHRH secretion (6′). There is also evidence that an inhibitory NPY influence (6-1) on LHRH release is lifted at the end of juvenile development and thus contributes to the initiation of the pubertal process. The metabolic signals, leptin and IGF-I, further the process along by stimulating LHRH secretion either directly (7) or via functionally connected neuronal networks (7′). The resulting increase in pulsatile LHRH secretion (8) then results in the initiation of puberty. –, inhibition; +, stimulation. [Modified with permission from (800).]

During the second week of postnatal life, ovarian development comes under firm gonadotropin control. Driven by the increased FSH levels, a large crop of primordial follicles, some of which are destined to ovulate at puberty, is incorporated into a proliferative pool and begins to grow.

The *second activational period* denotes the end of juvenile development and results in the first overt hormonal manifestation of the onset of puberty. It is initiated by a coordinated loss of inhibitory transsynaptic control (GABA, opioids) coupled to an increase in excitatory inputs to the LHRH neuronal network. Both neurons and glial cells contribute to this facilitatory control. The primary neuronal networks involved are those that use glutamate for neurotransmission, but downstream networks such as noradrenergic and NPYergic neurons also participate. KiSS-1–producing neurons are critical components of this facilitatory control, but the nature of the transsynaptic input they receive has not been yet identified. Glial cells facilitate the process via cell–cell signaling pathways set in motion by growth factors acting via both serine-threonine kinase and tyrosine kinase receptors. A major glia-to-neuron signaling pathway required for the timely initiation of the pubertal process is initiated by TGF-α and NRGs, which when binding to erbB-1 and erbB-4 receptors located on glial cells elicit the release of bioactive substances that stimulate LHRH secretion. The net outcome of this changing neuronal–glial input to the LHRH neuronal network is a change in basal LH levels and pulsatile LH release. Both basal release of LH and the magnitude of LH secretory episodes increase in the afternoon. Establishment of this diurnal rhythm in LH release likely reflects coordination of the hypothalamic circadian oscillator that resides within the SCN with neuronal systems stimulatory to LHRH release. As this occurs, there is also an enhanced synchronization of LHRH discharges, perhaps resulting from completion of the synaptic and glial circuitry impinging on LHRH neurons, the interconnections between these neurons, and their morphological and functional maturation.

Recent studies are beginning to disclose the existence of a hierarchy of upstream regulatory genes that may coordinate the plethora of cell–cell interactions required for the pubertal activation of pulsatile LHRH release. These controlling genes may perform this task by establishing within each participating cell type the necessary conditions for productive cell–cell interactions of an either neuron-to-neuron or glia-to-neuron nature.

The overall process may be facilitated by the neurotrophic effects of E_2 but is not determined by a decrease in hypothalamic sensitivity to negative feedback. As a result of these neuroendocrine changes, an increased secretion of gonadotropins stimulates the ovary to produce more E_2, which then evokes minisurges of LH secretion. In turn, the surge further stimulates ovarian development and steroidogenesis.

The *third and final activational period* occurs more abruptly and is predominantly determined by an increased output of ovarian steroids, especially E_2. This period corresponds to the preovulatory discharge of LHRH, which evokes the first surge of gonadotropins. It is clear that this last activational phase is a result of the expression of E_2-positive feedback. E_2-positive feedback develops by the end of the infantile period. At this time the ovary is unable to secrete sufficient levels of E_2 to produce a surge of LH. Completion of ovarian maturation is therefore essential for the preovulatory gonadotropin discharge to occur. The ovaries grow under the influence of pituitary gonadotropins. During neonatal–infantile development, this influence appears to be modulated by milk LHRH; afterward, it is facilitated by GH and PRL. An additional facilitatory control mechanism seems to be provided by adrenergic and peptidergic nerves, neurotrophic factors produced within the ovary, and also epinephrine of adrenal medullary origin. The adrenergic inputs to the ovary facilitate P secretion either directly or by amplifying the effect of gonadotropins. VIPergic nerves appear to be involved in stimulating the secretion of all the main steroids (E_2, P, and aromatizable androgens). NPYergic nerves appear to modulate the availability of NE to its receptors and in conjunction with substance P may also be involved in regulation of blood flow.

Under the influence of all these factors, the steroid output of the ovary increases more rapidly as proestrus approaches. The pace of follicular development is accelerated by an increase in LH receptors and possibly by a decrease in local LHRH inhibitory influences. The increased serum E_2 levels then set in motion the central component of E_2-positive feedback by enhancing the transsynaptic activity of stimulatory neuronal systems associated to the LHRH neuronal network and the secretory activity of associated glia. The gonadotropin discharge induces the first ovulation, which occurs on the early morning of estrus. On this day, estrogenic stimulation of the vaginal epithelium results in canalization of the vagina, an event that provides the first somatic manifestation that reproductive competence has been attained.

The Male

Male sexual maturation initiates during embryonic development at the onset of testis determination and as gonadal androgens begin to be produced for reproductive tract development. Therefore, the testis becomes an endocrine organ earlier than the ovary, which initiates steroidogenesis postnatally. Although the testis has the capacity to respond to gonadotropins and steroids, testis formation and embryonic development are hormone independent. This has been shown using LHRH, FSH, FSH receptor, and LH receptor knockout models (66,67,249) that demonstrate normal testis development, cell growth, and cellular differentiation occurring in the absence of these endocrine agents.

Contrary to the female, the capacity for interrelations between the hypothalamic–pituitary unit and the male gonads are already functional before birth, but similar to the ovary a functional unit develops prepubertally to allow a gonadal axis to form. Also similar to the female, the primary events that set into motion the onset of male puberty originate within the CNS. Studies using male hamsters have shown that the major endocrine events associated with puberty are preceded by an increase in the size of LHRH neuronal perikarya (795). These changes may reflect changes in the synthesis and/or alteration in the secretory pattern of LHRH, leading to the initiation or enhancement of the pubertal rise of gonadotropin secretion. Such endocrine developments in turn influence the growth and maturation of the testes. One of the most significant testicular developments occurring at this time involves an increased steroidogenic capacity associated with an increased production of testosterone. In addition, as the animal matures, the testes become more sensitive to the stimulatory actions of the gonadotropins, primarily because of elevated FSH secretion, and so the level of T production becomes enhanced. Meanwhile, both T and the gonadotropins provide the basic stimulus for initiating spermatogenesis. Endocrine procedures and knockout mouse models have shown that T is essential to initiate and obtain adult testis function. FSH facilitates and allows a quantitative optimal testis function to be obtained. In the absence of FSH, the onset of puberty is delayed and progression is prolonged, and testis development can occur qualitatively but not quantitatively. Therefore, FSH actions on Sertoli cells help coordinate the onset and progression of puberty, but androgen is the primary driver of the process. The increased output of T also produces the physical alterations that are typically associated with puberty, namely the development and maintenance of the accessory sex organs. In addition, T and inhibin act as negative feedback agents to attenuate gonadotropin secretion. It has previously been hypothesized that the onset of male puberty results from a decrease in sensitivity of the hypothalamic–pituitary unit to negative feedback. This change in sensitivity would permit gonadotropin secretion to increase in spite of an enhanced production of androgens by the developing testes. The hypothesis, however, does not explain the divergent patterns of FSH and LH secretion observed in developing males.

In summary, the central neuroendocrine regulation of LHRH production and gonadotropin secretion initiates the onset of puberty through LH actions on the Leydig cells to increase androgen production and FSH actions on the Sertoli cell to enhance Leydig cell function and initiate Sertoli cell differentiation needed for the induction of spermatogenesis. The negative feedback of androgens and inhibin regulate gonadotropin production during the progression of puberty to the adult stage. The coordinated control of this hypothalamic–pituitary axis with the testis involves molecular events at the LHRH neuron and pituitary levels and at the testis somatic cell level (796,797). The resulting endocrine events control the onset and progression of puberty in the rat.

ACKNOWLEDGMENTS

We thank Janie Gliessman for excellent secretarial assistance. This research was supported by National Institutes of Health grants HD25123 and HD24870, National Institute of Child Health and Human Development NIH/NIH through cooperative Grant U54 HD18185 as part of the Specialized Cooperative Centers Program in Reproduction Research, NIH RR00163 for the operation of the Oregon National Primate Research Center (to S. R. O.), and the Washington State University and University of Idaho Center for Reproductive Biology (to M. K. S.).

REFERENCES

1. Philipeaux, J. M. (1856). Note sur l' extirpation des capsules surrenales chez les rats albinos (*Mus rattus*). *C. R. Hebd. Seances Acad. Sci.* 43, 904–906.
2. Lindsey, J. R. (1979). Historical foundations. In *The Laboratory Rat* (H. J. Baker, J.-R. Lindsey, and S. H. Weisbroth, Eds.), Vol. 1, pp. 1–36. Academic Press, New York.
3. Long, J. A., and Evans, H. M. (1922). *The Oestrous Cycle in the Rat and Its Associated Phenomena, Memoirs of the University of California*, Vol. 6. University of California Press, Berkeley.
4. Allen, E., and Doisy, E. A. (1924). The induction of a sexually mature condition in immature females by injection of the ovarian follicular hormone. *Am. J. Physiol.* 69, 577–588.
5. Frank, R. T., Kingery, H. M., and Gustavson, R. G. (1925). The female sex hormone II. An analysis of factors producing puberty. *JAMA* 85, 1558–1559.
6. Kallas, H. (1929). Puberté précose par parabiose. *C. R. Soc. Biol.* 100, 979–980.
7. Hohlweg, W., and Junkmann, K. (1932). Die hormonal-nervöse regulierung der function des hypophysenvorderlappens. *Klin. Woschenschr.* 11, 321–323.
8. Moore, C. R., and Price, D. (1932). Gonad hormone functions, and the reciprocal influence between gonads and hypophysis with its bearing on the problem of sex hormone antagonism. *Am. J. Anat.* 50, 13–71.

9. Walsh, E. L., Cuyler, W. K., and McCullagh, D. R. (1934). The physiologic maintenance of the male sex glands. *Am. J. Physiol.* 107, 508.
10. McCullagh, D. R. (1932). Dual endocrine activity of the testes. *Science* 76, 19–20.
11. Harris, G. W., and Jacobson, D. (1952). Functional grafts of the anterior pituitary gland. *Proc. R. Soc. Lond. Biol.* 139, 263–276.
12. Donovan, B. T., and van der Werff ten Bosch, J. J. (1956). Precocious puberty in rats with hypothalamic lesions. *Nature* 178, 745.
13. Elwers, M., and Critchlow, V. (1960). Precocious ovarian stimulation following hypothalamic and amygdaloid lesions in rats. *Am. J. Physiol.* 198, 381–385.
14. Ojeda, S. R., Prevot, V., and Heger, S. (2001). Regulation of puberty. *Curr. Opin. Endocrinol. Diabetes* 8, 154–160.
15. Donovan, B. T., and van der Werff ten Bosch, J. J. (1965). Physiology of puberty. In *Monographs of the Physiological Society, No. 15* (H. Barcroft, H. Davson, and W. D. M. Paton, Eds.), pp. 1–204. Williams & Wilkins, Baltimore.
16. Critchlow, V., and Bar-Sela, M. (1967). Control of the onset of puberty. In *Neuroendocrinology* (L. Martini and W. F. Ganong, Eds.), Vol. II, pp. 101–162. Academic Press, New York.
17. Ramirez, V. D. (1973). Endocrinology of puberty. In *Handbook of Physiology* (R. O. Greep and E. B. Astwood, Eds.), Vol. II, Section 7, pp. 1–28. American Physiological Society, Washington, DC.
18. Odell, W. D., and Swerdloff, R. S. (1976). Etiologies of sexual maturation: a model system based on the sexually maturing rat. *Rec. Prog. Horm. Res.* 32, 245–288.
19. Ojeda, S. R., Andrews, W. W., Advis, J. P., and Smith-White, S. (1980). Recent advances in the endocrinology of puberty. *Endocr. Rev.* 1, 228–257.
20. Ojeda, S. R., Smith (White), S. S., Urbanski, H. F., and Aguado, L. I. (1984). The onset of female puberty: underlying neuroendocrine mechanisms. In *Neuroendocrine Perspectives* (E. E. Müller and R. M. MacLeod, Eds.), pp. 225–278. Elsevier, New York.
21. Ojeda, S. R., and Urbanski, H. F. (1988). Puberty in the rat. In *The Physiology of Reproduction* (E. Knobil and J. D. Neill, Eds.), pp. 1699–1737. Raven Press, New York.
22. Ojeda, S. R., and Urbanski, H. F. (1994). Puberty in the rat. In *The Physiology of Reproduction* (E. Knobil and J. D. Neill, Eds.), pp. 363–409. Raven Press, New York.
23. Ojeda, S. R., and Terasawa, E. (2002). Neuroendocrine regulation of puberty. In *Hormones, Brain and Behavior* (D. Pfaff, A. Arnold, A. Etgen, S. Fahrbach, R. Moss, and R. Rubin, Eds.), Vol. 4, pp. 589–659. Elsevier, New York.
24. Plant, T. M. (1994). Puberty in primates. In *The Physiology of Reproduction* (E. Knobil and J. D. Neill, Eds.), pp. 453–485. Raven Press, New York.
25. Urbanski, H. F., and Ojeda, S. R. (1985). The juvenile-peripubertal transition period in the female rat: establishment of a diurnal pattern of pulsatile luteinizing hormone secretion. *Endocrinology* 117, 644–649.
26. Tanner, J. M. (1974). Sequence and tempo in the somatic changes in puberty. In *The Control of the Onset of Puberty* (M. M. Grumbach, G. D. Grave, and F. E. Mayer, Eds.), pp. 448–470. John Wiley & Sons, New York.
27. Clermont, L., and Perey, B. (1957). Quantitative study of the cell population of the seminiferous tubules in immature rats. *Am. J. Anat.* 100, 241–267.
28. Clegg, E. J. (1960). The age at which male rats become fertile. *J. Reprod. Fertil.* 1, 119–120.
29. Korenbrot, C. C., Huhtaniemi, I. T., and Weiner, R. L. (1977). Preputial separation as an external sign of pubertal development in the male rat. *Biol. Reprod.* 17, 298–303.
30. Ojeda, S. R., Wheaton, J. E., Jameson, H. E., and McCann, S. M. (1976). The onset of puberty in the female rat. I. Changes in plasma prolactin, gonadotropins, LHRH levels and hypothalamic LHRH content. *Endocrinology* 98, 630–638.
31. Meijs-Roelofs, H. M. A., Uilenbroek, J. Th. J., de Greef, W. J., de Jong, F. H., and Kramer, D. (1975). Gonadotropin and steroid levels around the time of first ovulation in the rat. *J. Endocrinol.* 67, 275–282.
32. Hisaw, F. L. (1947). Development of the Graaffian follicle and ovulation. *Physiol. Rev.* 27, 95–119.
33. Dawson, A. B., and McCabe, M. (1951). The interstitial tissue of the ovary in infantile and juvenile rats. *J. Morphol.* 88, 543–571.
34. Rennels, E. G. (1951). Influence of hormones on the histochemistry of ovarian interstitial tissue of the immature rat. *Am. J. Anat.* 88, 63–107.
35. Ojeda, S. R., Urbanski, H. F., and Ahmed, C. E. (1986). The onset of female puberty. Studies in the rat. *Rec. Prog. Horm. Res.* 42, 385–442.
36. Aubert, M. L., Begeot, M., Winiger, B. P., Morel, G., Sizonenko, P. C., and Dubois, P. M. (1985). Ontogeny of hypothalamic luteinizing hormone-releasing hormone (GnRH) and pituitary GnRH receptors in fetal and neonatal rats. *Endocrinology* 116, 1565–1576.
37. Silverman, A.-J., Krey, L. C., and Zimmerman, E. A. (1979). A comparative study of the luteinizing hormone releasing hormone (LHRH) neuronal networks in mammals. *Biol. Reprod.* 20, 98–110.
38. King, J. C., and Anthony, E. L. P. (1984). LHRH neurons and their projections in humans and other mammals: species comparisons. *Peptides* 5(Suppl 1), 195–207.
39. Schwanzel-Fukuda, M., Morrell, J. I., and Pfaff, D. W. (1985). Ontogenesis of neurons producing luteinizing hormone-releasing hormone (LHRH) in the nervus terminalis of the rat. *J. Comp. Neurol.* 238, 348–364.
40. Schwanzel-Fukuda, M., and Pfaff, D. W. (1990). The migration of luteinizing hormone-releasing hormone (LHRH) neurons from the medial olfactory placode into the medial basal forebrain. *Experientia* 46, 956–962.
41. Jennes, L. (1989). Prenatal development of the gonadotropin-releasing hormone-containing systems in rat brain. *Brain Res.* 482, 97–108.
42. Nemeskéri, A., Halász, B., and Kurcz, M. (1983). Ontogenesis of the rat hypothalamo-adenohypophyseal system and inherent capacity of the fetal pituitary to differentiate into hormone-synthesizing and releasing cells. In *The Anterior Pituitary Gland* (A. S. Bhatnagar, Ed.), pp. 341–354. Raven Press, New York.
43. Begeot, M., Dubois, M. P., and Dubois, P. M. (1983). Comparative study *in vivo* and *in vitro* of the differentiation of immunoreactive gonadotropic cells in fetal rat anterior pituitary. *Neuroendocrinology* 37, 52–58.
44. Begeot, M., Morel, G., Rivest, R. W., Aubert, M. L., Dubois, M. P., and Dubois, P. M. (1984). Influence of gonadoliberin (GnRH) on the differentiation of rat gonadotrophs: an *in vivo* and *in vitro* study. *Neuroendocrinology* 38, 217–225.
45. Szabo, K., and Csanyi, K. (1982). The vascular architecture of the developing pituitary-median eminence complex in the rat. *Cell. Tissue Res.* 224, 563–577.
46. Fink, G., and Smith, G. C. (1971). Ultrastructural features of the developing hypothalamo-hypophysial axis in the rat. *Z. Zellforsch.* 119, 208–226.
47. Wray, S. (2002). Development of gonadotropin-releasing hormone-1 neurons. *Front. Neuroendocrinol.* 23, 292–316.
48. Chiappa, S. A., and Fink, G. (1977). Releasing factor and hormonal changes in the hypothalamic-pituitary-gonadotrophin and adrenocorticotrophin systems before and after birth and

puberty in male, female and androgenized female rats. *J. Endocrinol.* 72, 211–224.

49. Salisbury, R. L., Dudley, S. D., and Weisz, J. (1982). Effect of gonadotropin-releasing hormone on circulating levels of immunoreactive luteinizing hormone in fetal rats. *Neuroendocrinology* 35, 265–269.
50. Watanabe, Y. G., and Daikoku, S. (1979). An immunohistochemical study on the cytogenesis of adenohypophysial cells in fetal rats. *Dev. Biol.* 68, 557–567.
51. Chowdhury, M., and Steinberger, E. (1976). Pituitary and plasma levels of gonadotrophins in foetal and newborn male and female rats. *J. Endocrinol.* 69, 381–384.
52. Slob, A. K., Ooms, M. P., and Vreeburg, J. T. M. (1980). Prenatal and early postnatal sex differences in plasma and gonadal testosterone and plasma luteinizing hormone in female and male rats. *J. Endocrinol.* 87, 81–87.
53. Picon, R., and Habert, R. (1981). A sensitive bioassay for luteinizing hormone-like activity applied to systemic plasma of foetal rats. *Acta Endocrinol. (Copenh).* 97, 176–180.
54. Parker, K. L., and Schimmer, B. P. (2006). Embryology and genetics of the mammalian gonads and ducts. In *Knobil and Neill's Physiology of Reproduction* (J. D. Neill, Ed.), pp. 313–336. Elsevier, St. Louis.
55. Burns, K. H., and Matzuk, M. M. (2006). Preimplantation embryogenesis. In *Knobil and Neill's Physiology of Reproduction* (J. D. Neill, Ed.), pp. 261–310. Elsevier, St. Louis.
56. Presl, J., Pospisil, J., Figarová, V., and Wagner, V. (1972). Developmental changes in uptake of radioactivity by the ovaries, pituitary and uterus after ^{125}I-labelled human chorionic gonadotrophin administration in rats. *J. Endocrinol.* 52, 585–586.
57. Siebers, J. W., Peters, F., Zenzes, M. T., Schmidtke, J., and Engel, W. (1977). Binding of human chorionic gonadotrophin to rat ovary during development. *J. Endocrinol.* 73, 491–496.
58. Peluso, J. J., Steger, R. W., and Hafez, E. S. E. (1976). Development of gonadotrophin-binding sites in the immature rat ovary. *J. Reprod. Fertil.* 47, 55–58.
59. Smith-White, S., and Ojeda, S. R. (1981). Changes in ovarian luteinizing hormone (LH) and follicle stimulating hormone (FSH) receptor content and in gonadotropin-induced ornithine decarboxylase activity during prepubertal and pubertal development of the rat. *Endocrinology* 109, 152–161.
60. Uilenbroek, J. Th. J., and van der Linden, R. (1983). Changes in gonadotrophin binding to rat ovaries during sexual maturation. *Acta Endocrinol. (Copenh).* 103, 413–419.
61. Peters, H., Byskov, A. G., Lintern-Moore, S., Faber, M., and Andersen, M. (1973). The effect of gonadotrophin on follicle growth initiation in the neonatal mouse ovary. *J. Reprod. Fertil.* 35, 139–141.
62. Ben-Or, S. (1963). Morphological and functional development of the ovary of the mouse. I. Morphology and histochemistry of the developing ovary in normal conditions and after FSH treatment. *J. Embryol. Exp. Morphol.* 11, 1–11.
63. Eshkol, A., Lunenfeld, B., and Peters, H. (1970). Ovarian development in infant mice. Dependence on gonadotrophic hormones. In *Gonadotrophins and Ovarian Development* (W. R. Butt, A. C. Crooke, and M. E. Ryle, Eds.), pp. 249–258. Livingstone, London.
64. Peters, H. (1979). Some aspects of early follicular development. In *Ovarian Follicular Development and Function* (A. R. Midgley and W. A. Sadler, Eds.), pp. 1–13. Raven Press, New York.
65. Lunenfeld, B., Kraiem, Z., and Eshkol, A. (1975). The function of the growing follicle. *J. Reprod. Fertil.* 45, 567–574.
66. Dierich, A., Sairam, M. R., Monaco, L., Fimia, G. M., Gansmuller, A., LeMeur, M., and Sassone-Corsi, P. (1998). Impairing follicle-stimulating hormone (FSH) signaling *in vivo*: targeted disruption of the FSH receptor leads to aberrant gametogenesis and hormonal imbalance. *Proc. Natl. Acad. Sci. U S A* 95, 13612–13617.
67. Zhang, F.-P., Poutanen, M., Wilbertz, J., and Huhtaniemi, I. (2001). Normal prenatal but arrested postnatal sexual development of luteinizing hormone receptor knockout (LuRKO) mice. *Mol. Endocrinol.* 15, 172–183.
68. Rajah, R., and Hirshfield, A. N. (1991). The changing architecture of the rat ovary during the immediate postpartum period: a three dimensional (3D) reconstruction. *Biol. Reprod.* 44, 152.
69. Malamed, S., Gibney, J. A., and Ojeda, S. R. (1992). Ovarian innervation develops before initiation of folliculogenesis in the rat. *Cell. Tissue Res.* 270, 87–93.
70. Byskov, A. G. (1986). Differentiation of mammalian embryonic gonad. *Physiol. Rev.* 66, 71–117.
71. Hirshfield, A. N. (1991). Development of follicles in the mammalian ovary. *Int. Rev. Cytol.* 124, 43–101.
72. Rajah, R., Glaser, E. M., and Hirshfield, A. N. (1992). The changing architecture of the neonatal rat ovary during histogenesis. *Dev. Dyn.* 194, 177–192.
73. Morrison, L. J., and Marcinkiewicz, J. L. (2002). Tumor necrosis factor á enhances oocyte/follicle apoptosis in the neonatal rat ovary. *Biol. Reprod.* 66, 450–457.
74. Kezel, P., Nilsson, E., and Skinner, M. K. (2002). Cell-cell interactions in primordial follicle assembly and development. *Front. Biosci.* 7, d1990–d1996.
75. Nilsson, E. E., Kezele, P., and Skinner, M. K. (2002). Leukemia inhibitory factor (LIF) promotes the primordial to primary follicle transition in rat ovaries. *Mol. Cell Endocrinol.* 188, 65–73.
76. Kezele, P., and Skinner, M. K. (2003). Regulation of ovarian primordial follicle assembly and development by estrogen and progesterone: endocrine model of follicle assembly. *Endocrinology* 144, 3329–3337.
77. Soyal, S. M., Amieh, A., and Dean, J. (2000). FIGa, a germ cell-specific transcription factor required for ovarian follicle formation. *Development* 127, 4645–4654.
78. Huang, E. J., Manova, K., Packer, A. I., Sanchez, S., Bachvarova, R. F., and Besmer, P. (1993). The murine steel panda mutation affects kit ligand expression and growth of early ovarian follicles. *Dev. Biol.* 157, 100–109.
79. Parrott, J. A., and Skinner, M. K. (1999). Kit-ligand/stem cell factor induces primordial follicle development and initiates folliculogenesis. *Endocrinology* 140, 4262–4271.
80. Nilsson, E., Parrott, J. A., and Skinner, M. K. (2001). Basic fibroblast growth factor induces primordial follicle development and initiates folliculogenesis. *Mol. Cell Endocrinol.* 175, 123–130.
81. Dissen, G. A., Romero, C., Newman Hirshfield, A., and Ojeda, S. R. (2001). Nerve growth factor is required for early follicular development in the mammalian ovary. *Endocrinology* 142, 2078–2086.
82. Paredes, A., Romero, C., Dissen, G. A., DeChiara, T. M., Reichardt, L., Cornea, A., Ojeda, S. R., and Xu, B. (2004). TrkB receptors are required for follicular growth and oocyte survival in the mammalian ovary. *Dev. Biol.* 267, 430–449.
83. Durlinger, A. L. L., Gruijters, M. J. G., Kramer, P., Karels, B., Ingraham, H. A., Nachtigal, M. W., Uilenbroek, J. Th. J., Grootegoed, J. A., and Themmen, A. P. N. (2002). Anti-müllerian hormone inhibits initiation of primordial follicle growth in the mouse ovary. *Endocrinology* 143, 1076–1084.
84. Nilsson, E. E., and Skinner, M. K. (2003). Bone morphogenetic protein-4 acts as an ovarian follicle survival factor and promotes primordial follicle development. *Biol. Reprod.* 69, 1265–1272.
85. Dong, J., Albertini, D. F., Nishimori, K., Kumar, T. R., Lu, N., and Matzuk, M. M. (1996). Growth differentiation factor-9

is required during early ovarian folliculogenesis. *Nature* 383, 531–535.

86. Rajkovic, A., Pangas, S. A., Ballow, D., Suzumori, N., and Matzuk, M. M. (2004). NOBOX deficiency disrupts early folliculogenesis and oocyte-specific gene expression. *Science* 305, 1157–1159.
87. Suzumori, N., Yan, C., Matzuk, M. M., and Rajkovic, A. (2002). Nobox is a homeobox-encoding gene preferentially expressed in primordial and growing oocytes. *Mech. Dev.* 111, 137–141.
88. Matzuk, M. M., Burns, K. H., Viveiros, M. M., and Eppig, J. J. (2002). Intercellular communication in the mammalian ovary: oocytes carry the conversation. *Science* 296, 2178–2180.
89. Dissen, G. A., Paredes, A., Romero, C., Dees, W. L., and Ojeda, S. R. (2004). Neural and neurotrophic control of ovarian development. In *The Ovary* (P. Leung and E. Adashi, Eds.), pp. 3–23. Academic Press, San Diego.
90. George, F. W., and Ojeda, S. R. (1987). Vasoactive intestinal peptide enhances aromatase activity in the neonatal rat ovary before development of primary follicles or responsiveness to follicle-stimulating hormone. *Proc. Natl. Acad. Sci. U S A* 84, 5803–5807.
91. Ahmed, C. E., Dees, W. L., and Ojeda, S. R. (1986). The immature rat ovary is innervated by vasoactive intestinal peptide (VIP)-containing fibers and responds to VIP with steroid secretion. *Endocrinology* 118, 1682–1689.
92. Advis, J. P., Ahmed, C. E., and Ojeda, S. R. (1989). Direct hypothalamic control of vasoactive intestinal peptide (VIP) levels in the developing rat ovary. *Brain Res. Bull.* 22, 605–610.
93. Ojeda, S. R., and Ramirez, V. D. (1972). Plasma levels of LH and FSH in maturing rats: response to hemigonadectomy. *Endocrinology* 90, 466–472.
94. Kragt, C. L., and Dahlgren, J. (1972). Development of neural regulation of follicle-stimulating hormone (FSH) secretion. *Neuroendocrinology* 9, 30–40.
95. Döhler, K. D., and Wuttke, W. (1974). Serum LH, FSH, prolactin and progesterone from birth to puberty in female and male rats. *Endocrinology* 94, 1003–1008.
96. Meijs-Roelofs, H. M. A., Uilenbroek, J. Th. J., Osman, P., and Welschen, R. (1973). Serum levels of gonadotropins and follicular growth in prepubertal rats. In *The Development and Maturation of the Ovary and Its Function* (H. Peters, Ed.), pp. 3–11. Excerpta Medica, Amsterdam.
97. McKinnon, P. C. B., Mattock, J. M., and ter Haar, M. B. (1976). Serum gonadotrophin levels during development in male, female and androgenized female rats and the effect of general disturbance on high luteinizing hormone levels. *J. Endocrinol.* 70, 361–371.
98. Döhler, K. D., and Wuttke, W. (1975). Changes with age in levels of serum gonadotropins, prolactin, and gonadal steroids in prepubertal male and female rats. *Endocrinology* 97, 898–907.
99. Araki, S., Toran-Allerand, C. D., Ferin, M., and Vande Wiele, R. L. (1975). Immunoreactive gonadotropin-releasing hormone (Gn-RH) during maturation in the rat: ontogeny of regional hypothalamic differences. *Endocrinology* 97, 693–697.
100. Hompes, P. G. A., Vermes, I., Tilders, F. J. H., and Schoemaker, J. (1982). *In vitro* release of LHRH from the hypothalamus of female rats during prepubertal development. *Neuroendocrinology* 35, 8–12.
101. Bourguignon, J.-P., and Franchimont, P. (1984). Puberty-related increase in episodic LHRH release from rat hypothalamus *in vitro*. *Endocrinology* 114, 1941–1943.
102. Bourguignon, J.-P., Gerard, A., Fawe, L., Alvarez-Gonzalez, M.-L., and Franchimont, P. (1991). Neuroendocrine control of the onset of puberty: secretion of gonadotrophin-releasing hormone from rat hypothalamic explants. *Acta Paediatr. Scand.* 372(Suppl), 19–25.
103. Melrose, P. (1987). Steroid effects on the secretory modalities of gonadotropin-releasing hormone. *Endocrinology* 121, 190–199.
104. Matagne, V., Rasier, G., Lebrethon, M.-C., Gérard, A., and Bourguignon, J.-P. (2004). Estradiol stimulation of pulsatile gonadotropin-releasing hormone secretion *in vitro*: correlation with prenatal exposure to sex steroids and induction of sexual precocity *in vivo*. *Endocrinology* 145, 2775–2783.
105. Wildt, L., Hausler, A., Marshall, G., Hutchinson, J. S., Plant, T. M., Belchtez, P. E., and Knobil, E. (1981). Frequency and amplitude of gonadotropin-releasing hormone stimulation and gonadotropin secretion in the rhesus monkey. *Endocrinology* 109, 376–385.
106. Dalkin, A. C., Haisenleder, D. J., and Ortolano, G. A. (1989). The frequency of gonadotropin-releasing-hormone stimulation differentially regulates gonadotropin subunit messenger ribonucleic acid expression. *Endocrinology* 125, 917–924.
107. Andrews, W. W., and Ojeda, S. R. (1981). A quantitative analysis of the maturation of steroid negative feedbacks controlling gonadotropin release in the female rat: transition from an androgenic to a predominantly estrogenic control. *Endocrinology* 108, 1313–1320.
108. Fraser, M. O., and Plant, T. M. (1989). Further studies on the role of the gonads in determining the ontogeny of gonadotropin secretion in the guinea pig (*Cavia porcelus*). *Endocrinology* 125, 906–911.
109. Ojeda, S. R., and Ramirez, V. D. (1973). Short-term steroid treatment on plasma LH and FSH in castrated rats from birth to puberty. *Neuroendocrinology* 13, 100–114.
110. Ojeda, S. R., Kalra, S. P., and McCann, S. M. (1975). Further studies on the maturation of the estrogen negative feedback on gonadotropin release in the prepubertal female rat. *Neuroendocrinology* 18, 242–255.
111. Meijs-Roelofs, H. M. A., and Kramer, P. (1979). Maturation of the inhibitory feedback action of oestrogen on follicle-stimulating hormone secretion in the immature female rat: a role for alphafoetoprotein. *J. Endocrinol.* 81, 199–208.
112. Frawley, L. S., and Henricks, D. M. (1979). Mode of gonadotropin secretion in infantile female rats and the role of estrogen in feedback regulation. *Endocrinology* 105, 1064–1072.
113. Andrews, W. W., and Ojeda, S. R. (1977). On the feedback actions of estrogen on gonadotropins and prolactin release in infantile female rats. *Endocrinology* 101, 1517–1523.
114. Puig-Duran, E., Greenstein, B. D., and MacKinnon, P. C. B. (1977). Follicle-stimulating hormone and the unbound oestradiol-17b fraction in plasma of immature female rats treated with a-foetoprotein-enriched extracts. *J. Endocrinol.* 75, 52P.
115. Germain, B. L., Campbell, P. S., and Anderson, J. N. (1978). Role of the serum estrogen-binding protein in the control of tissue estradiol levels during postnatal development of the female rat. *Endocrinology* 103, 1401–1410.
116. Andrews, W. W., Mizejewski, G. J., and Ojeda, S. R. (1981). Development of estradiol-positive feedback on luteinizing hormone release in the female rat: a quantitative study. *Endocrinology* 109, 1404–1413.
117. Debeljuk, L., Arimura, A., and Schally, A. V. (1972). Pituitary responsiveness to LH-releasing hormone in intact female rats of different ages. *Endocrinology* 90, 1499–1502.
118. Dullaart, J. (1977). Immature rat pituitary glands *in vitro*: age- and sex-related changes in luteinizing hormone releasing hormone-stimulated gonadotrophin release. *J. Endocrinol.* 73, 309–319.
119. Ojeda, S. R., Jameson, H. E., and McCann, S. M. (1977). Developmental changes in pituitary responsiveness to luteinizing hormone-releasing hormone (LHRH) in the

female rat: ovarian-adrenal influence during the infantile period. *Endocrinology* 100, 440–451.
120. Denef, C., Hautekeete, E., de Wolf, A., and Vanderschueren, B. (1978). Pituitary basophils from immature male and female rats: distribution of gonadotrophs and thyrotrophs as studied by unit gravity sedimentation. *Endocrinology* 103, 724–735.
121. Campbell, G. T., and Ramaley, J. A. (1978). Ovarian-dependent development of a stimulatory effect of a 5α-dihydrotestosterone on serum follicle stimulating hormone in the ovariectomized rat. *Biol. Reprod.* 19, 612–619.
122. Denef, C., Magnus, C., and McEwen, B. S. (1974). Sex dependent changes in pituitary 5α-dihydrotestosterone and 3α-androstanediol formation during postnatal development and puberty in the rat. *Endocrinology* 94, 1265–1274.
123. Chappel, S. C., Ulloa-Aguirre, A., and Ramaley, J. A. (1983). Sexual maturation in female rats: time course of the appearance of multiple species of anterior pituitary FSH. *Biol. Reprod.* 28, 196–205.
124. Ojeda, S. R., and Jameson, H. E. (1977). Studies on the biological properties and gel filtration behavior of pituitary and serum follicle stimulating hormone of infantile female rats. *Endocrinology* 101, 475–484.
125. Buckingham, J. C., and Wilson, C. A. (1985). Peripubertal changes in the nature of LH. *J. Endocrinol.* 104, 173–177.
126. Honna, K., Höhn, K.-G., and Wuttke, W. (1979). Involvement of catecholamines in eliciting LH peaks in 15-day-old female rats: effects of treatment with prolactin. *Brain Res.* 179, 271–279.
127. Wuttke, W., Döhler, K. D., and Gelato, M. (1976). Estrogens and prolactin as possible regulators of puberty. *J. Endocrinol.* 68, 391–396.
128. Puig-Duran, E., and MacKinnon, P. C. B. (1978). Ontogeny of prolactin and luteinizing hormone responses to oestrogen and progesterone in immature rats. *J. Endocrinol.* 76, 321–331.
129. MacKinnon, P. C. B., and Mattock, J. (1976). The effects of handling on episodic release of LH in immature rats. *Ann. Biol. Anim. Biochim. Biophys.* 16, 345–348.
130. Andrews, W. W., and Ojeda, S. R. (1981). A detailed analysis of the serum LH secretory profiles of conscious free-moving female rats during the time of puberty. *Endocrinology* 109, 2032–2039.
131. Kimura, F., and Kawakami, M. (1982). Episodic LH secretion in the immature male and female rat as assessed by sequential blood sampling. *Neuroendocrinology* 35, 128–132.
132. Andrews, W. W., and Ojeda, S. R. (1978). The control of LH release in prepubertal female rats: indirect evidence for an enhanced ability of the hypothalamus to release LHRH as the pituitary responsiveness to LHRH declines. *J. Endocrinol.* 78, 281–282.
133. Castro-Vazquez, A., and Ojeda, S. R. (1977). Changes in pituitary responsiveness to LHRH during puberty in the female rat: initiation of the priming effect. *Neuroendocrinology* 23, 88–98.
134. Sarkar, D. K., and Fink, G. (1979). Mechanism of the first spontaneous gonadotrophin surge and that induced by pregnant mare serum and effects of neonatal androgen in the rat. *J. Endocrinol.* 83, 339–354.
135. Jakubowski, M., Blum, M., and Roberts, J. L. (1991). Postnatal development of gonadotropin-releasing hormone and cyclophilin gene expression in the female and male rat brain. *Endocrinology* 128, 2702–2708.
136. Wray, S., and Hoffman, G. (1986). Postnatal morphological changes in rat LHRH neurons correlated with sexual maturation. *Neuroendocrinology* 43, 93–97.
137. Matsumoto, A., and Arai, Y. (1976). Developmental changes in synaptic formation in the hypothalamic arcuate nucleus of female rats. *Cell. Tissue Res.* 169, 143–146.
138. Wray, S., and Gainer, H. (1987). Effect of neonatal gonadectomy on the postnatal development of LHRH cell subtypes in male and female rats. *Neuroendocrinology* 45, 413–419.
139. Jennes, L., Stumpf, W. E., and Sheedy, M. E. (1985). Ultrastructural characterization of gonadotropin-releasing hormone (GnRH)-producing neurons. *J. Comp. Neurol.* 232, 534–547.
140. Wray, S., and Hoffman, G. (1986). Catecholamine innervation of LH-RH neurons: a developmental study. *Brain Res.* 399, 327–331.
141. Bourguignon, J.-P., Gerard, A., Mathieu, J., Mathieu, A., and Franchimont, P. (1990). Maturation of the hypothalamic control of pulsatile gonadotropin-releasing hormone secretion at onset of puberty. I. Increased activation of N-methyl-D-aspartate receptors. *Endocrinology* 127, 873–881.
142. Ojeda, S. R., Urbanski, H. F., Katz, K. H., and Costa, M. E. (1986). Activation of estradiol positive feedback at puberty: estradiol sensitizes the LHRH releasing system at two different biochemical steps. *Neuroendocrinology* 43, 259–265.
143. Flügge, G., Wuttke, W., and Fuchs, E. (1986). Postnatal development of transmitter systems: sexual differentiation of the GABAergic system and effects of muscimol. *Int. J. Dev. Neurosci.* 4, 319–326.
144. Goroll, D., Arias, P., and Wuttke, W. (1994). Ontogenic changes in the hypothalamic levels of amino acid neurotransmitters in the female rat. *Brain Res. Dev. Brain Res.* 77, 183–188.
145. Carbone, S., Szwarcfarb, B., Otero Losada, M. E., and Moguilevsky, J. A. (1992). Effects of ovarian steroids on the gonadotropin response to N-methyl-D-aspartate and on hypothalamic excitatory amino acid levels during sexual maturation in female rats. *Endocrinology* 130, 1365–1370.
146. Carbone, S., Szwarcfarb, B., Losada, M., and Moguilevsky, J. A. (1995). Effects of ovarian hormones on the hypothalamic excitatory amino acid system during sexual maturation in female rats. *Neuroendocrinology* 61, 235–242.
147. Goroll, D., Arias, P., and Wuttke, W. (1993). Preoptic release of amino acid neuro-transmitters evaluated in peripubertal and young adult female rats by push-pull perfusion. *Neuroendocrinology* 58, 11–15.
148. Jarry, H., Hirsch, B., Leonhardt, S., and Wuttke, W. (1992). Amino acid neurotransmitter release in the preoptic area of rats during the positive feedback actions of estradiol on LH release. *Neuroendocrinology* 56, 133–140.
149. Ping, L., Mahesh, V. B., Wiedmeier, V. T., and Brann, D. W. (1994). Release of glutamate and aspartate from the preoptic area during the progesterone-induced LH surge: in vivo microdialysis studies. *Neuroendocrinology* 59, 318–324.
150. Ohtaki, T., Shintani, Y., Honda, S., Matsumoto, H., Hori, A., Kanehashi, K., Terao, Y., Kumano, S., Takatsu, Y., Masuda, Y., Ishibashi, Y., Watanabe, T., Asada, M., Yamada, T., Suenaga, M., Kitada, C., Usuki, S., Kurokawa, T., Onda, H., Nishimura, O., and Fujino, M. (2001). Metastasis suppressor gene KiSS-1 encodes peptide ligand of a G-protein-coupled receptor. *Nature* 411, 613–617.
151. Kotani, M., Detheux, M., Vandenbogaerde, A., Communi, D., Vanderwinden, J. M., Le Poul, E., Brezillon, S., Tyldesley, R., Suarez-Huerta, N., Vandeput, F., Blanpain, C., Schiffmann, S. N., Vassart, G., and Parmentier, M. (2001). The metastasis suppressor gene KiSS-1 encodes kisspeptins, the natural ligands of the orphan G protein-coupled receptor GPR54. *J. Biol. Chem.* 276, 34631–34636.
152. Muir, A. I., Chamberlain, L., Elshourbagy, N. A., Michalovich, D., Moore, D. J., Calamari, A., Szekeres, P. G., Sarau, H. M., Chambers, J. K., Murdock, P., Steplewski, K., Shabon, U., Miller, J. E., Middleton, S. E., Darker, J. G., Larminie, C. G., Wilson, S., Bergsma, D. J., Emson, P., Faull, R.,

Philpott, K. L., and Harrison, D. C. (2001). AXOR12, a novel human G protein-coupled receptor, activated by the peptide KiSS-1. *J. Biol. Chem.* 276, 28969–28975.

153. de Roux, N., Genin, E., Carel, J.-C., Matsuda, F., Chaussain, J.-L., and Milgrom, E. (2003). Hypogonadotropic hypogonadism due to loss of function of the KiSS1-derived peptide receptor GPR54. *Proc. Natl. Acad. Sci. U S A* 100, 10972–10976.
154. Seminara, S. B., Messager, S., Chatzidaki, E. E., Thresher, R. R., Acierno, J. S., Jr., Shagoury, J. K., Bo-Abbas, Y., Kuohung, W., Schwinof, K. M., Hendrick, A. G., Zahn, D., Dixon, J., Kaiser, U. B., Slaugenhaupt, S. A., Gusella, J. F., O'Rahilly, S., Carlton, M. B., Crowley, W. F., Jr., Aparicio, S. A., and Colledge, W. H. (2003). The GPR54 gene as a regulator of puberty. *N. Engl. J. Med.* 349, 1614–1627.
155. Funes, S., Hedrick, J. A., Vassileva, G., Markowitz, L., Abbondanzo, S., Golovko, A., Yang, S., Monsma, F. J., and Gustafson, E. L. (2003). The KiSS-1 receptor GPR54 is essential for the development of the murine reproductive system. *Biochem. Biophys. Res. Commun.* 312, 1357–1363.
156. Gottsch, M. L., Smith, J. T., Popa, S. M., Cunningham, M. J., Acohido, B. V., Hohmann, J., Crowley, W. F., Jr., Seminara, S., Clifton, D. K., and Steiner, R. A. (2004). Stimulation of gonadotropin secretion by metastin in the mouse. *Prog. 86th Ann. Mtg. Endocrine Soc.* Abstr P2-210, p. 356.
157. Shahab, M., Mastronardi, C., Plant, T. M., Ojeda, S. R., Crowley, W. F., Jr., and Seminara, S. B. (2004). Hypothalamic GPR-54 expression and signaling during the peripubertal period in the rhesus monkey (*Macaca mulatta*). *Prog. 86th Ann. Mtg. Endocrine Soc.* Abstr P3-269, p. 529.
158. Kalra, S. P. (1986). Neural circuitry involved in the control of LHRH secretion: a model for preovulatory LH release. In *Frontiers in Neuroendocrinology* (W. F. Ganong and L. Martini, Eds.), pp. 31–75. Raven Press, New York.
159. Barraclough, C. A., and Wise, P. M. (1982). The role of catecholamines in the regulation of pituitary luteinizing hormone and follicle-stimulating hormone secretion. *Endocr. Rev.* 3, 91–119.
160. Ramirez, V. D., Feder, H. H., and Sawyer, C. H. (1984). The role of brain catecholamines in the regulation of LH secretion: a critical inquiry. In *Frontiers in Neuroendocrinology* (L. Martini and W. F. Ganong, Eds.), Vol. 8, pp. 27–84. Raven Press, New York.
161. Advis, J. P., Simpkins, J. W., Chen, H. T., and Meites, J. (1978). Relation of biogenic amines to onset of puberty in the female rat. *Endocrinology* 103, 11–16.
162. Höhn, K.-G., and Wuttke, W. (1979). Ontogeny of catecholamine turnover rates in limbic and hypothalamic structures in relation to serum prolactin and gonadotropin levels. *Brain Res.* 179, 281–293.
163. Donoso, A. O., and Banzan, A. M. (1975). Brain and hypothalamic catecholamine turnover in relation with puberty in the female rat. *Acta Physiol. Lat. Am.* 25, 37–44.
164. Raum, W. J., Glass, A. R., and Swerdloff, R. S. (1980). Changes in hypothalamic catecholamine neurotransmitters and pituitary gonadotropins in the immature female rat: relationships to the gonadostat theory of puberty onset. *Endocrinology* 106, 1253–1258.
165. Ojeda, S. R., Negro-Vilar, A., and McCann, S. M. (1979). Release of prostaglandin E (PGEs) by hypothalamic tissue: evidence for their involvement in catecholamine-induced LHRH release. *Endocrinology* 104, 617–624.
166. Ojeda, S. R., and Negro-Vilar, A. (1984). Release of prostaglandin E_2 from the hypothalamus depends on extracellular Ca^{+2} availability: relation to LHRH release. *Neuroendocrinology* 39, 442–447.
167. Ojeda, S. R., and Negro-Vilar, A. (1985). Prostaglandin E_2-induced LHRH release involves mobilization of intracellular Ca^{+2}. *Endocrinology* 116, 1763–1770.
168. Lamberts, R., and Wuttke, W. (1981). Puberty of female rats may in part be explained by decreased hypothalamic dopamine receptor sensitivity. *Brain Res.* 215, 375–381.
169. Voogt, J. L., Chen, C. L., and Meites, J. (1970). Serum and pituitary prolactin levels before, during, and after puberty in female rats. *Am. J. Physiol.* 218, 396–399.
170. Moguilevsky, J. A., Faigon, M. R., Scacchi, P., and Szwarcfarb, B. (1985). Effect of the serotoninergic system on luteinizing hormone secretion in prepubertal female rats. *Neuroendocrinology* 40, 135–138.
171. Moguilevsky, J. A., Arias, P., Szwarcfarb, B., Carbone, S., and Rondina, D. (1990). Sexual maturation modifies the catecholaminergic control of gonadotropin secretion and the effect of ovarian hormones on hypothalamic neurotransmitters in female rats. *Neuroendocrinology* 52, 393–398.
172. Sutton, S. W., Mitsugi, N., Plotsky, P. M., and Sarkar, D. K. (1988). Neuropeptide Y (NPY): a possible role in the initiation of puberty. *Endocrinology* 123, 2152–2154.
173. Corder, R., Walker, C. D., Gaillard, R. C., and Aubert, M. L. (1992). Inhibition of sexual maturation in male rats by melatonin: evidence linking the mechanism of action to changes in the regulation of hypothalamic neuropeptide Y. *J. Neuroendocrinol.* 4, 1–7.
174. O'Connor, J. L., Wade, M. F., Brann, D. W., and Mahesh, V. B. (1995). Evidence that progesterone modulates anterior pituitary neuropeptide Y levels during the progesterone-induced gonadotropin surge in the estrogen-primed intact immature female rat. *J. Steroid Biochem.* 52, 497–504.
175. Brann, D. W., McDonald, J. K., Putnam, C. D., and Mahesh, V. B. (1991). Regulation of hypothalamic gonadotropin-releasing hormone and neuropeptide Y concentrations by progesterone and corticosteroids in immature rats: correlation with luteinizing hormone and follicle-stimulating hormone release. *Neuroendocrinology* 54, 425–432.
176. Minami, S., Frautschy, S. A., Plotsky, P. M., Sutton, S. W., and Sarkar, D. K. (1990). Facilitatory role of neuropeptide Y on the onset of puberty: effect of immunoneutralization of neuropeptide Y on the release of luteinizing hormone and luteinizing-hormone-releasing hormone. *Neuroendocrinology* 52, 112–115.
177. Fraley, G. S., and Kuenzel, W. J. (1993). Precocious puberty in chicks (*Galls domesticus*) induced by central injection of neuropeptide Y. *Life Sci.* 52, 1649–1656.
178. Decavel, C., and van den Pol, A. N. (1990). GABA: A dominant neurotransmitter in the hypothalamus. *J. Comp. Neurol.* 302, 1019–1037.
179. Davis, A. M., Grattan, D. R., Selmanoff, M., and McCarthy, M. M. (1996). Sex differences in glutamic acid decarboxylase mRNA in neonatal rat brain: implications for sexual differentiation. *Horm. Behav.* 30, 538–552.
180. Davis, A. M., Ward, S. C., Selmanoff, M., Herbison, A. E., and McCarthy, M. M. (1999). Developmental sex differences in amino acid neurotransmitter levels in hypothalamic and limbic areas of rat brain. *Neuroscience* 90, 1471–1482.
181. Roth, C., Leonhardt, S., Theiling, K., Lakomek, M., Jarry, H., and Wuttke, W. (1998). Ontogeny of the GnRH-, glutaminase- and glutamate decarboxylase-gene expression in the hypothalamus of female rats. *Brain Res. Dev. Brain Res.* 110, 105–114.
182. Terasawa, E., and Fernandez, D. L. (2001). Neurobiological mechanisms of the onset of puberty in primates. *Endocr. Rev.* 22, 111–151.
183. Ferin, M., Wehrenberg, W. B., Lam, N. Y., Alston, E. J., and Vande Wiele, R. L. (1982). Effects and site action of morphine

on gonadotropin secretion in the female rhesus monkey. *Endocrinology* 111, 1652–1656.

184. Mallory, D. S., Bona-Gallo, A., and Gallo, R. V. (1989). Neurotransmitter involvement in naloxone-induced stimulation of pulsatile LH release on day 8 of pregnancy in the rat. *Brain Res. Bull.* 22, 1015–1021.
185. Kesner, J. S., Kaufman, J. M., Wilson, R. C., Kuroda, G., and Knobil, E. (1986). The effect of morphine on the electrophysiological activity of the hypothalamic luteinizing hormone-releasing hormone pulse generator in the rhesus monkey. *Neuroendocrinology* 43, 686–688.
186. Van Vugt, D. A., Lam, N. Y., and Ferin, M. (1984). Reduced frequency of pulsatile luteinizing hormone secretion in the luteal phase of the rhesus monkey. Involvement of endogenous opiates. *Endocrinology* 115, 1095–1101.
187. Orstead, K. M., Hess, D. L., and Spies, H. G. (1987). Opiatergic inhibition of pulsatile luteinizing hormone release during the menstrual cycle of rhesus macaques. *Proc. Soc. Exp. Biol. Med.* 184, 312–319.
188. Devorshak-Harvey, E., Bona-Gallo, A., and Gallo, R. V. (1987). Endogenous opioid peptide regulation of pulsatile luteinizing hormone secretion during pregnancy in the rat. *Neuroendocrinology* 46, 369–378.
189. Rossmanith, W. G., Mortola, J. F., and Yen, S. S. (1988). Role of endogenous opioid peptides in the initiation of the midcycle luteinizing hormone surge in normal cycling women. *J. Clin. Endocrinol. Metab.* 67, 695–700.
190. Ji, W.-Z., Kaynard, A. H., Pau, K.-Y. F., Hess, D. L., Baughman, W. L., and Spies, H. G. (1989). Endogenous opiates regulate the nocturnal reduction in luteinizing hormone pulse frequency during the luteal phase of the macaque menstrual cycle. *Biol. Reprod.* 41, 1024–1033.
191. Evans, W. S., Weltman, M. L., Johnson, M. L., Weltman, A., Veldhuis, J. D., and Rogol, A. D. (1992). Effects of opioid receptor blockade on luteinizing hormone (LH) pulses and interpulse LH concentrations in normal women during the early phase of the menstrual cycle. *J. Endocrinol. Invest.* 15, 525–531.
192. Grosser, P. M., O'Byrne, K. T., Williams, C. L., Thalabard, J. C., Hotchkiss, J., and Knobil, E. (1993). Effects of naloxone on estrogen-induced changes in hypothalamic gonadotropin-releasing hormone pulse generator activity in the rhesus monkey. *Neuroendocrinology* 57, 115–119.
193. Blank, M. S., Panerai, A. E., and Friesen, H. G. (1979). Opioid peptides modulate hormone secretion during sexual development. *Science* 203, 1129–1131.
194. Ebling, F. J., Schwartz, M. L., and Foster, D. L. (1989). Endogenous opioid regulation of pulsatile luteinizing hormone secretion during sexual maturation in the female sheep. *Endocrinology* 125, 369–383.
195. Cutler, L., Egli, C. A., Styne, D. M., Kaplan, S. L., and Grumbach, M. M. (1985). Hormone ontogeny in the ovine fetus. XVIII. The effect of an opioid antagonist on luteinizing hormone secretion. *Endocrinology* 116, 1997–2002.
196. Sirinathsinghji, D. J. S., Motta, M., and Martini, L. (1985). Induction of precocious puberty in the female rat after chronic naloxone administration during the neonatal period: the opiate "brake" on prepubertal gonadotrophin secretion. *J. Endocrinol.* 104, 299–307.
197. Medhamurthy, R., Gay, V. L., and Plant, T. M. (1990). The prepubertal hiatus in gonadotropin secretion in the male rhesus monkey (*Macaca mulatta*) does not appear to involve endogenous opioid peptide restraint of hypothalamic gonadotropin-releasing hormone release. *Endocrinology* 126, 1036–1042.
198. Foster, D. L., and Jackson, L. M. (2006). Puberty in the sheep. In *Knobil and Neill's Physiology of Reproduction* (J. D. Neill, Ed.), pp. 2127–2176. Elsevier, St. Louis.
199. Plant, T. M., and Witchel, S. F. (2006). Puberty in non-human primates and humans. In *Knobil and Neill's Physiology of Reproduction* (J. D. Neill, Ed.), pp. 2177–2230. Academic Press, St. Louis.
200. Lee, B. J., Cho, G. J., Norgren, R., Junier, M.-P., Hill, D. F., Tapia, V., Costa, M. E., and Ojeda, S. R. (2001). TTF-1, a homeodomain gene required for diencephalic morphogenesis, is postnatally expressed in the neuroendocrine brain in a developmentally regulated and cell-specific fashion. *Mol. Cell. Neurosci.* 17, 107–126.
201. Mastronardi, C., Smiley, G., Kuswakabe, T., Kawagushi, A., Cabrera, R., Mungenast, A., Kimura, S., and Ojeda, S. R. (2004). Neuronal deletion of the T/EBP gene delays female puberty and causes premature reproductive senescence. Program No. 143. 2, 2004 Abstract Viewer. Society for Neuroscience, Washington, DC, Online.
202. Galbiati, M., Zanisi, M., Messi, E., Cavarretta, I., Martini, L., and Melcangi, R. C. (1996). Transforming growth factor-b and astrocytic conditioned medium influence luteinizing hormone-releasing hormone gene expression in the hypothalamic cell line GT1. *Endocrinology* 137, 5605–5609.
203. Buchanan, C. D., Mahesh, V. B., and Brann, D. W. (2000). Estrogen-astrocyte-luteinizing hormone-releasing hormone signaling: a role for transforming growth factor-b. *Biol. Reprod.* 62, 1710–1721.
204. Prevot, V., Bouret, S., Croix, D., Takumi, T., Jennes, L., Mitchell, V., and Beauvillain, J. C. (2000). Evidence that members of the TGFb superfamily play a role in regulation of the GnRH neuroendocrine axis: expression of a type 1 serine-threonine kinase receptor for TGFb and activin in GnRH neurones and hypothalamic areas of the female rat. *J. Neuroendocrinol.* 12, 665–670.
205. Galbiati, M., Magnaghi, V., Martini, L., and Melcangi, R. C. (2001). Hypothalamic transforming growth factor b1 and basic fibroblast growth factor mRNA expression is modified during the rat oestrous cycle. *J. Neuroendocrinol.* 13, 483–489.
206. Ojeda, S. R., Urbanski, H. F., Costa, M. E., Hill, D. F., and Moholt-Siebert, M. (1990). Involvement of transforming growth factor a in the release of luteinizing hormone-releasing hormone from the developing female hypothalamus. *Proc. Natl. Acad. Sci. U S A* 87, 9698–9702.
207. Bouret, S., De Seranno, S., Beauvillain, J. C., and Prevot, V. (2004). Transforming growth factor b_1 may directly influence gonadotropin-releasing hormone gene expression in the rat hypothalamus. *Endocrinology* 145, 1794–1801.
208. Prevot, V., Cornea, A., Mungenast, A., Smiley, G., and Ojeda, S. R. (2003). Activation of erbB-1 signaling in tanycytes of the median eminence stimulates transforming growth factor b_1 release via prostaglandin E2 production and induces cell plasticity. *J. Neuroscience* 23, 10622–10632.
209. Kozlowski, G. P., and Coates, P. W. (1985). Ependymoneuronal specializations between LHRH fibers and cells of the cerebroventricular system. *Cell. Tissue Res.* 242, 301–311.
210. Prevot, V. (2002). Glial-neuronal-endothelial interactions are involved in the control of GnRH secretion. *J. Neuroendocrinol.* 14, 247–255.
211. Voigt, P., Ma, Y. J., Gonzalez, D., Fahrenbach, W. H., Wetsel, W. C., Berg-von der Emde, K., Hill, D. F., Taylor, K. G., Costa, M. E., Seidah, N. G., and Ojeda, S. R. (1996). Neural and glial-mediated effects of growth factors acting via tyrosine kinase receptors on LHRH neurons. *Endocrinology* 137, 2593–2605.
212. Tsai, P.-S., Werner, S., and Weiner, R. I. (1995). Basic fibroblast growth factor is a neurotropic factor in GT1 gonadotropin-releasing hormone neuronal cell lines. *Endocrinology* 136, 3831–3838.
213. Wetsel, W. C., Hill, D. F., and Ojeda, S. R. (1996). Basic fibroblast growth factor regulates the conversion of pro-luteinizing hormone-releasing hormone (LHRH) to LHRH

in immortalized hypothalamic neurons. *Endocrinology* 137, 2606–2616.

214. González-Manchón, C., Bilezikjian, L. M., Corrigan, A. Z., Mellon, P. L., and Vale, W. (1991). Activin-A modulates gonadotropin-releasing hormone secretion from a gonadotropin-releasing hormone-secreting neuronal cell line. *Neuroendocrinology* 54, 373–377.
215. Galbiati, M., Martini, L., and Melcangi, R. C. (2002). Oestrogens, via transforming growth factor a, modulate basic fibroblast growth factor synthesis in hypothalamic astrocytes: *in vitro* observations. *J. Neuroendocrinol.* 14, 829–835.
216. Handelsman, D. J., Spaliviero, J. A., Scott, C. D., and Baxter, R. C. (1987). Hormonal regulation of the peripubertal surge of insulin-like growth factor-I in the rat. *Endocrinology* 120, 491–496.
217. Copeland, K. C., Eichberg, J. W., Parker, C. R. Jr., and Bartke, A. (1985). Puberty in the chimpanzee: comatomedin-C and its relationship to somatic growth and steroid hormone concentrations. *J. Clin. Endocrinol. Metab.* 60, 1154–1160.
218. Grumbach, M. M., and Styne, D. M. (1998). Puberty: ontogeny, neuroendocrinology, physiology, and disorders. In *Williams Textbook of Endocrinology* (R. H. Williams, D. W. Foster, H. Kroenenberg, P. R. Larsen, and R. Zorab, Eds.), pp. 1509–1625. W. B. Saunders, Philadelphia.
219. Hiney, J. K., Srivastava, V., Nyberg, C. L., Ojeda, S. R., and Dees, W. L. (1996). Insulin-like growth factor I of peripheral origin acts centrally to accelerate the initiation of female puberty. *Endocrinology* 137, 3717–3728.
220. Garcia-Segura, L. M., Naftolin, F., Hutchison, J. B., Azcoitia, I., and Chowen, J. A. (1999). Role of astroglia in estrogen regulation of synaptic plasticity and brain repair. *J. Neurobiol.* 40, 574–584.
221. Dueñas, M., Luquin, S., Chowen, J. A., Torres-Aleman, I., Naftolin, F., and Garcia-Segura, L. M. (1994). Gonadal hormone regulation of insulin-like growth factor-I-like immunoreactivity in hypothalamic astroglia of developing and adult rats. *Neuroendocrinology* 59, 528–538.
222. Wilson, M. E. (1998). Premature elevation in serum insulin-like growth factor-I advances first ovulation in rhesus monkeys. *J. Endocrinol.* 158, 247–257.
223. Danilovich, N., Wernsing, D., Coschigano, K. T., Kopchick, J. J., and Bartke, A. (1999). Deficits in female reproductive function in GH-R-KO mice: role of IGF-I. *Endocrinology* 140, 2637–2640.
224. Ojeda, S. R., Ma, Y. J., Lee, B. J., and Prevot, V. (2000). Glia-to-neuron signaling and the neuroendocrine control of female puberty. *Rec. Prog. Horm. Res.* 55, 197–224.
225. Ma, Y. J., Hill, D. F., Junier, M.-P., Costa, M. E., Felder, S. E., and Ojeda, S. R. (1994). Expression of epidermal growth factor receptor changes in the hypothalamus during the onset of female puberty. *Mol. Cell. Neurosci.* 5, 246–262.
226. Ma, Y. J., Hill, D. F., Creswick, K. E., Costa, M. E., and Ojeda, S. R. (1999). Neuregulins signaling via a glial erbB2/erbB4 receptor complex contribute to the neuroendocrine control of mammalian sexual development. *J. Neurosci.* 19, 9913–9927.
227. Ma, Y. J., Berg-von der Emde, K., Rage, F., Wetsel, W. C., and Ojeda, S. R. (1997). Hypothalamic astrocytes respond to transforming growth factor alpha with secretion of neuroactive substances that stimulate the release of luteinizing hormone-releasing hormone. *Endocrinology* 138, 19–25.
228. Rage, F., Lee, B. J., Ma, Y. J., and Ojeda, S. R. (1997). Estradiol enhances prostaglandin E_2 receptor gene expression in luteinizing hormone-releasing hormone (LHRH) neurons and facilitates the LHRH response to PGE_2 by activating a glia-to-neuron signaling pathway. *J. Neurosci.* 17, 9145–9156.
229. Ojeda, S. R., Prevot, V., Heger, S., Lomniczi, A., Dziedzic, B., and Mungenast, A. (2003). Glia-to neuron signaling and the neuroendocrine control of female puberty. *Ann. Med.* 35, 244–255.
230. Ma, Y. J., Junier, M.-P., Costa, M. E., and Ojeda, S. R. (1992). Transforming growth factor alpha (TGFα) gene expression in the hypothalamus is developmentally regulated and linked to sexual maturation. *Neuron* 9, 657–670.
231. Apostolakis, E. M., Garai, J., Lohmann, J. E., Clark, J. H., and O'Malley, B. W. (2000). Epidermal growth factor activates reproductive behavior independent of ovarian steroids in female rodents. *Mol. Endocrinol.* 14, 1086–1098.
232. Ma, Y. J., Dissen, G. A., Merlino, G., Coquelin, A., and Ojeda, S. R. (1994). Overexpression of a human transforming growth factor alpha (TGFα) transgene reveals a dual antagonistic role of TGFα in female sexual development. *Endocrinology* 135, 1392–1400.
233. Rage, F., Hill, D. F., Sena-Esteves, M., Breakefield, X. O., Coffey, R. J., Costa, M. E., McCann, S. M., and Ojeda, S. R. (1997). Targeting transforming growth factor α expression to discrete loci of the neuroendocrine brain induces female sexual precocity. *Proc. Natl. Acad. Sci. U S A* 94, 2735–2740.
234. Prevot, V., Rio, C., Cho, G. J., Lomniczi, A., Heger, S., Neville, C. M., Rosenthal, N. A., Ojeda, S. R., and Corfas, G. (2003). Normal female sexual development requires neuregulin-erbB receptor signaling in hypothalamic astrocytes. *J. Neurosci.* 23, 230–239.
235. Prevot, V., Lomniczi, A., Corfas, G., and Ojeda, S. R. ErbB-1 and erbB-4 receptors act in concert to facilitate both female sexual development and mature reproductive function. *Endocrinology* (submitted).
236. Goldman, B. D., Grazia, Y. R., Kamberi, I. A., and Porter, J. C. (1971). Serum gonadotropin concentrations in intact and castrated neonatal rats. *Endocrinology* 88, 771–776.
237. Raymond, J. P., Mercier-Bodard, C., and Baulieu, E. E. (1971). Rat estradiol binding plasma protein (EBP). *Steroids* 18, 767–788.
238. Nunez, E., Engelmann, F., Bennassayag, C., and Joyle, M. F. (1971). Identification et purification préliminaire de la foeto protéine liant les oestrogènes dans le sérum de rats nouveaunes. *C. R. Acad. Sci. (Paris)* 273, 834–841.
239. Mathews, D., Andrews, W. W., Parker R., Jr., and Ojeda, S. R. (1987). A role for aromatizable androgens in female puberty. *Biol. Reprod.* 36, 836–843.
240. Steele, R. E. (1977). Role of the ovaries in maturation of the estradiol-luteinizing hormone negative feedback system of the pubertal rat. *Endocrinology* 101, 587–597.
241. Puig-Duran, E., and MacKinnon, P. C. B. (1976). Patterns of LH and prolactin release following steroid manipulations in the rat during development. *Ann. Biol. Anim. Biochem. Biophys.* 16, 373–375.
242. Caligaris, L., Astrada, J. J., and Taleisnik, S. (1972). Influence of age on the release of luteinizing hormone induced by oestrogen and progesterone in immature rats. *J. Endocrinol.* 55, 97–103.
243. Ronnekleiv, O. K., Ojeda, S. R., and McCann, S. M. (1978). Undernutrition, puberty and development of the estrogen positive feedback in the female rat. *Biol. Reprod.* 149, 274–277.
244. Sarkar, D. K., Chiappa, S. A., Fink, G., and Sherwood, N. M. (1976). Gonadotropin-releasing hormone surge in pro-oestrous rats. *Nature* 264, 461–463.
245. Lamprecht, S. A., Kohen, F., Ausher, J., Zor, U., and Lindner, H. R. (1976). Hormonal stimulation of oestradiol-17b release from the rat ovary during early postnatal development. *J. Endocrinol.* 68, 343–344.
246. Hunzicker-Dunn, M., and Birnbaumer, L. (1976). Adenylcyclase activities in ovarian tissues. III. Regulation of responsiveness to LH, FSH and PGE_1 in the prepubertal,

cycling, pregnant, and pseudopregnant rat. *Endocrinology* 99, 198–210.

247. Funkenstein, B., and Nimrod, A. (1982). Control of developmental processes in cultured neonatal ovaries. In *Development and Function of Reproductive Organs* (A. G. Byskov and H. Peters, Eds.), pp. 307–318. Excerpta Medica, Amsterdam.
248. Byskov, A. G., and Lintern-Moore, S. (1973). Follicle formation in the immature mouse ovary: the role of the rete ovarii. *J. Anat.* 116, 207–217.
249. Kumar, T. R., Wang, Y., Lu, N., and Matsuk, M. M. (1997). Follicle stimulating hormone is required for ovarian follicle maturation but not male fertility. *Nat. Genet.* 15, 201–204.
250. Romero, C., Paredes, A., Dissen, G. A., and Ojeda, S. R. (2002). Nerve growth factor induces the expression of functional FSH receptors in newly formed follicles of the rat ovary. *Endocrinology* 143, 1485–1494.
251. Oktay, K., Briggs, D., and Gosden, R. G. (1997). Ontogeny of follicle-stimulating hormone receptor gene expression in isolated human ovarian follicles. *J. Clin. Endocrinol. Metab.* 82, 3748–3751.
252. Hage, A. J., Groen-Klevant, A. C., and Welschen, R. (1978). Follicle growth in the immature rat ovary. *Acta Endocrinol. (Copenh).* 88, 375–382.
253. Smith, S. S., and Ojeda, S. R. (1986). Neonatal release of gonadotropin is essential for development of ovarian FSH receptors. *Biol. Reprod.* 34, 219–228.
254. Quatroppani, S. L., and Weisz, J. (1973). Conversion of progesterone to estrone and estradiol *in vitro* by the ovary of the infantile rat in relation to the development of its interstitial tissue. *Endocrinology* 53, 1269–1976.
255. Smeaton, T. C., Arcondoulis, D. E., and Steele, P. A. (1975). The synthesis of testosterone and estradiol-17b by the gonads of neonatal rats *in vitro*. *Steroids* 26, 181–182.
256. Andrews, W. W., Heiman, J. R., Porter, J. R., and Ojeda, S. R. (1981). The infantile female rat: *in vivo* ovarian and adrenal steroidogenic responses to exogenous administration or to endogenously induced elevations of gonadotropins and ACTH. *Biol. Reprod.* 24, 597–608.
257. Uilenbroek, J. Th. J., and de Wolff-Exalto, E. A. (1979). Prepubertal follicular development and ovarian activity in adulthood after suppression of gonadotropin levels in immature female rats. *Biol. Reprod.* 20, 384–389.
258. Schwartz, N. B. (1974). The role of FSH and LH and of their antibodies on follicular growth and on ovulation. *Biol. Reprod.* 10, 236–272.
259. Amarant, T., Fridkin, M., and Koch, Y. (1982). Luteinizing hormone-releasing hormone and thyrotropin-releasing hormone in human and bovine milk. *Eur. J. Biochem.* 127, 647–650.
260. Smith (White), S. S., and Ojeda, S. R. (1984). Maternal modulation of infantile ovarian development and available ovarian LHRH receptors via milk "LHRH." *Endocrinology* 115, 1973–1983.
261. Baram, T., Koch, Y., Hazum, E., and Fridkin, M. (1977). Gonadotropin-releasing hormone in milk. *Science* 198, 300–301.
262. Hsueh, A. J. W., and Jones, P. B. C. (1981). Extrapituitary actions of gonadotropin-releasing hormone. *Endocr. Rev.* 2, 437–461.
263. Dalkin, A. C., Bourne, G. A., Pieper, D. R., Regiani, S., and Marshall, J. C. (1981). Pituitary and gonadal gonadotropin-releasing hormone receptors during sexual maturation in the rat. *Endocrinology* 108, 1658–1664.
264. Smith White, S., and Ojeda, S. R. (1981). Changes in ovarian LHRH receptor content during the onset of puberty in the female rat. *Endocrinology* 108, 347–349.
265. Richards, J. S. (1994). Hormonal control of gene expression in the ovary. *Endocr. Rev.* 15, 725–751.
266. Richards, J. S. (2001). Perspective: the ovarian follicle. A perspective in 2001. *Endocrinology* 142, 2184–2193.
267. Ingram, D. L. (1962). Atresia. In *The Ovary* (S. Zuckerman, Ed.), Vol. 1, pp. 247–273. Academic Press, New York.
268. Zarrow, M. X., and Quinn, D. L. (1963). Superovulation in the immature rat following treatment with PMS alone and inhibition of PMS-induced ovulation. *J. Endocrinol.* 26, 181–188.
269. McCormack, C. E., and Meyer, R. K. (1964). Minimal age for induction of ovulation with progesterone in rats: evidence for neural control. *Endocrinology* 74, 793–799.
270. Advis, J. P., and Ojeda, S. R. (1978). Hyperprolactinemia-induced precocious puberty in the female rat: ovarian site of action. *Endocrinology* 103, 924–935.
271. Uilenbroek, J. Th. J., Wouterson, P. J. A., and van der Linden, R. (1983). Steroid production *in vitro* by rat ovaries during sexual maturation. *J. Endocrinol.* 99, 469–475.
272. Smith White, S., and Ojeda, S. R. (1984). The peripubertal decline in ovarian LHRH receptor content: characterization and distribution. *Neuroendocrinology* 36, 449–456.
273. Advis, J. P., Andrews, W. W., and Ojeda, S. R. (1979). Changes in ovarian steroidal and prostaglandin E responsiveness to gonadotropins during the onset of puberty in the female rat. *Endocrinology* 104, 653–658.
274. Ojeda, S. R., and Jameson, H. R. (1977). Developmental patterns of plasma and pituitary growth hormone (GH) in the female rat. *Endocrinology* 100, 881–889.
275. Eden, S. (1979). Age and sex-related differences in episodic growth hormone secretion in the rat. *Endocrinology* 105, 555–560.
276. Kimura, F., Tsai, C.-W., and Kawakami, M. (1983). Detailed profile of prolactin secretion in the immature female rat: evidence for the existence of an ultradian rhythm. *Endocrinol. Jpn.* 30, 297–303.
277. Kimura, F., and Kawakami, M. (1980). Two daily surges of prolactin secretion in the immature female rat. *Endocrinology* 107, 172–175.
278. Ojeda, S. R., Krulich, L., and Jameson, H. E. (1976). Developmental patterns of plasma and pituitary TSH and prolactin and hypothalamic TRH in the female rat. *Endocr. Res. Commun.* 3, 387–406.
279. Clemens, J. A., Minagushi, H., Storey, R., Voogt, J. L., and Meites, J. (1969). Induction of precocious puberty in female rats by prolactin. *Neuroendocrinology* 4, 150–156.
280. Voogt, J. L., Clemens, J. A., and Meites, J. (1969). Stimulation of pituitary FSH release in immature female rats by prolactin implant in median eminence. *Neuroendocrinology* 4, 157–163.
281. Advis, J. P., Andrews, W. W., and Ojeda, S. R. (1982). Studies on the central effect of prolactin in inducing precocious puberty in the female rat. *Brain Res. Bull.* 8, 449–458.
282. Hickey, G. J., Oonk, R. B., Hall, P. F., and Richards, J. S. (1989). Aromatase cytochrome P450 and cholesterol side-chain cleavage cytochrome P450 in corpora lutea of pregnant rats: diverse regulation by peptide and steroid hormones. *Endocrinology* 125, 1673–1682.
283. Advis, J. P., Richards, J. S., and Ojeda, S. R. (1981). Hyperprolactinemia-induced precocious puberty. I. Studies on the mechanisms by which prolactin (PRL) enhances ovarian progesterone responsiveness to gonadotropins in prepubertal rats. *Endocrinology* 108, 1333–1342.
284. Gonzalez, M. D., López, F., and Aguilar, E. (1984). Involvement of prolactin in the onset of puberty in female rats. *J. Endocrinol.* 101, 63–68.
285. Advis, J. P., Smith White, S., and Ojeda, S. R. (1981). Delayed puberty induced by chronic suppression of prolactin release in the female rat. *Endocrinology* 109, 1321–1330.

286. Advis, J. P., Aguado, L. I., and Ojeda, S. R. (1983). Hyperprolactinemia enhances ovarian estrogen responsiveness to gonadotropins in prepubertal rats: antagonistic effect of adrenalectomy. *Biol. Reprod.* 29, 181–184.
287. Advis, J. P., Smith White, S., and Ojeda, S. R. (1981). Activation of GH short loop negative feedback delays puberty in the female rat. *Endocrinology* 108, 1343–1352.
288. Ramaley, J. A., and Phares, C. K. (1980). Delay of puberty onset in females by suppression of growth hormone. *Endocrinology* 106, 1989–1993.
289. Carbone, S., Szwarcfarb, B., Otera Losada, M., and Moguilevsky, J. A. (1992). Effects of ovarian steroids on the gonadotropin response to N-methyl-D-aspartate and on hypothalamic excitatory amino acid levels during sexual maturation in female rats. *Endocrinology* 130, 1365–1370.
290. Jia, X. C., Kalmijn, J., and Hsueh, A. J. W. (1986). Growth hormone enhances follicle stimulating hormone-induced differentiation of cultured rat granulosa cells. *Endocrinology* 118, 1401–1409.
291. Bachelot, A., Monget, P., Imbert-Bolloré, P., Coshigano, K., Kopchick, J. J., Kelly, P. A., and Binart, N. (2002). Growth hormone is required for ovarian follicular growth. *Endocrinology* 143, 4104–4112.
292. Adashi, E. Y., Resnick, C. E., D'Ercole, A. J., Svoboda, M. E., and Van Wyk, J. J. (1985). Insulin-like growth factors as intraovarian regulators of granulosa cell growth function. *Endocr. Rev.* 6, 400–420.
293. Baker, J., Hardy, M. P., Zhou, J., Bondy, C., Lupu, F., Bellvé, A. R., and Efstratiadis, A. (1996). Effects of an *Igf1* gene null mutation on mouse reproduction. *Mol. Endocrinol.* 10, 903–918.
294. Dissen, G. A., Romero, C., Paredes, A., and Ojeda, S. R. (2002). Neurotrophic control of ovarian development. *Microsc. Res. Tech.* 59, 509–515.
295. Burden, H. W. (1978). Ovarian innervation. In *The Vertebrate Ovary: Comparative Biology and Evolution* (R. E. Jones, Ed.), pp. 615–638. Plenum Press, New York.
296. Baljet, B., and Drukker, J. (1980). The extrinsic innervation of the pelvic organs in the female rat. *Acta Anat.* 107, 241–267.
297. Lawrence, I. E., Jr., and Burden, H. W. (1980). The origin of the extrinsic adrenergic innervation to the rat ovary. *Anat. Rec.* 196, 51–59.
298. Klein, C. M., and Burden, H. W. (1988). Substance P- and vasoactive intestinal polypeptide (VIP)-immunoreactive nerve fibers in relation to ovarian postganglionic perikarya in para- and prevertebral ganglia: evidence from combined retrograde tracing and immunocytochemistry. *Cell. Tissue Res.* 252, 403–410.
299. Burden, H. W., Leonard, M., Smith, C. P., and Lawrence, I. E. J. (1983). The sensory innervation of the ovary: a horseradish peroxidase study in the rat. *Anat. Rec.* 207, 623–627.
300. Gerendai, I., Toth, I. E., Boldogkoi, Z., Medvecxky, I., and Halasz, B. (1998). Neuronal labeling in the rat brain and spinal cord from the ovary using viral transneuronal tracing technique. *Neuroendocrinology* 68, 244–256.
301. Burden, H. W., and Lawrence, I. E. J. (1978). Experimental studies on the acetylcholinesterase-positive nerves in the ovary of the rat. *Anat. Rec.* 190, 233–241.
302. Burden, H. W. (1985). The adrenergic innervation of mammalian ovaries. In *Catecholamines as Hormone Regulators* (N. Ben-Jonathan, J. M. Bahr, and R. I. Weiner, Eds.), pp. 261–278. Raven Press, New York.
303. McNeill, D. L., and Burden, H. W. (1986). Neuropeptide Y and somatostatin immunoreactive perikarya in preaortic ganglia projecting to the rat ovary. *J. Reprod. Fertil.* 78, 727–732.
304. Papka, R. E., Cotton, J. P., and Traurig, H. H. (1985). Comparative distribution of neuropeptide tyrosine-, vasoactive intestinal polypeptide-, substance P-immunoreactive, acetylcholinesterase-positive and noradrenergic nerves in the reproductive tract of the female rat. *Cell. Tissue Res.* 242, 475–490.
305. McDonald, J. K., Dees, W. L., Ahmed, C. E., Noe, B. D., and Ojeda, S. R. (1987). Biochemical and immunocytochemical characterization of neuropeptide Y in the immature rat ovary. *Endocrinology* 120, 1703–1710.
306. Dees, W. L., Kozlowski, G. P., Dey, R., and Ojeda, S. R. (1985). Evidence for the existence of substance P in the prepubertal rat ovary. II. Immunocytochemical localization. *Biol. Reprod.* 33, 471–476.
307. Kannisto, P., Ekblad, E., Helm, G., Owman, Ch., Sjöberg, N.-O., Stjernquist, M., Sundler, F., and Walles, B. (1986). Existence and coexistence of peptides in nerves of the mammalian ovary and oviduct demonstrated by immunocytochemistry. *Histochemistry* 86, 25–34.
308. McNeill, D. L., and Burden, H. W. (1987). Neuropeptides in sensory perikarya projecting to the rat ovary. *Am. J. Anat.* 179, 269–276.
309. Byskov, A. G., and Hoyer, P. E. (1988). Embryology of mammalian gonads and ducts. In *The Physiology of Reproduction* (E. Knobil and J. D. Neill, Eds.), pp. 265–302. Raven Press, New York.
310. Lara, H. E., McDonald, J. K., and Ojeda, S. R. (1990). Involvement of nerve growth factor in female sexual development. *Endocrinology* 126, 364–375.
311. Lara, H. E., McDonald, J. K., Ahmed, C. E., and Ojeda, S. R. (1990). Guanethidine-mediated destruction of ovarian sympathetic nerves disrupts ovarian development and function in rats. *Endocrinology* 127, 2199–2209.
312. Dees, W. L., Hiney, J. K., Schultea, T. D., Mayerhofer, A., Danilchik, M., Dissen, G. A., and Ojeda, S. R. (1995). The primate ovary contains a population of catecholaminergic neuron-like cells expressing nerve growth factor receptors. *Endocrinology* 136, 5760–5768.
313. Mayerhofer, A., Smith, G. D., Danilchik, M., Levine, J. E., Wolf, D. P., Dissen, G. A., and Ojeda, S. R. (1998). Oocytes are a source of catecholamines in the primate ovary: evidence for a cell-cell regulatory loop. *Proc. Natl. Acad. Sci. U S A* 95, 10990–10995.
314. D'Albora, H., and Barcia, J. J. (1996). Intrinsic neuronal cell bodies in the rat ovary. *Neurosci. Lett.* 205, 65–67.
315. D'Albora, H., Lombide, P., and Ojeda, S. R. (2000). Intrinsic neurons in the rat ovary: an immunohistochemical study. *Cell. Tissue Res.* 300, 47–56.
316. Anesetti, G., Lombide, P., D'Albora, H., and Ojeda, S. R. (2001). Intrinsic neurons in the human ovary. *Cell. Tissue Res.* 306, 231–237.
317. Adashi, E. Y., and Hsueh, A. J. W. (1981). Stimulation of β_2-adrenergic responsiveness by follicle-stimulating hormone in rat granulosa cells *in vitro* and *in vivo*. *Endocrinology* 108, 2170–2178.
318. Aguado, L. I., Petrovic, S. L., and Ojeda, S. R. (1982). Ovarian β-adrenergic receptors during the onset of puberty: characterization, distribution, and coupling to steroidogenic responses. *Endocrinology* 110, 1124–1132.
319. Flerko, B. (1977). Sympathetic nerve function in the ovary. In *Endocrinology, Vol. 1: Proceedings of the V International Congress of Endocrinology* (V. H. T. James, Ed.), pp. 210–214. Excerpta Medica, Amsterdam.
320. Jorgensen, J. C., Kannisto, P., Liedberg, F., Ottesen, B., Owman, C., and Schmidt, G. (1991). The influence of neuropeptide Y and norepinephrine on ovulation in the rat ovary. *Peptides* 12, 975–982.
321. Aguado, L. I., and Ojeda, S. R. (1984). Ovarian adrenergic nerves play a role in maintaining preovulatory steroid secretion. *Endocrinology* 114, 1944–1946.

322. Weiss, G. K., Dail, W. G., and Ratner, A. (1982). Evidence for direct neural control of ovarian steroidogenesis in rats. *J. Reprod. Fertil.* 65, 507–511.
323. Dyer, C. A., and Erickson, G. F. (1985). Norepinephrine amplifies human chorionic gonadotropin-stimulated androgen biosynthesis by ovarian theca-interstitial cells. *Endocrinology* 116, 1645–1652.
324. Hernandez, E. R., Jimenez, J. L., Payne, D. W., and Adashi, E. Y. (1988). Adrenergic regulation of ovarian androgen biosynthesis is mediated via b_2-adrenergic theca-interstitial cell recognition sites. *Endocrinology* 122, 1592–1602.
325. Hsueh, A. J. W., Adashi, E. Y., Jones, P. B. C., and Welsh, T. H. J. (1984). Hormonal regulation of the differentiation of cultured ovarian granulosa cells. *Endocr. Rev.* 5, 76–127.
326. Ojeda, S. R., and Lara, H. E. (1989). Role of the sympathetic nervous system in the regulation of ovarian function. In *The Menstrual Cycle and Its Disorders* (K. M. Pirke, W. Wuttke, and U. Schweiger, Eds.), pp. 26–32. Springer-Verlag, Berlin.
327. Aguado, L. I., and Ojeda, S. R. (1986). Prepubertal rat ovary: hormonal modulation of β-adrenergic receptors and of progesterone response to adrenergic stimulation. *Biol. Reprod.* 34, 45–50.
328. Wolf, R., Meier-Fleitmann, A., Duker, E.-M., and Wuttke, W. (1986). Intraovarian secretion of catecholamines, oxytocin, beta-endorphin, and gamma-amino-butyric-acid in freely moving rats: development of a push-pull tubing method. *Biol. Reprod.* 35, 599–607.
329. Ferruz, J., Barria, A., Galleguillos, X., and Lara, H. E. (1991). Release of norepinephrine from the rat ovary: local modulation by gonadotropins. *Biol. Reprod.* 45, 592–597.
330. Dees, W. L., Ahmed, C. E., and Ojeda, S. R. (1986). Substance P- and vasoactive intestinal peptide-containing fibers reach the ovary by independent routes. *Endocrinology* 119, 638–641.
331. Davoren, J. B., and Hsueh, A. J. W. (1985). Vasoactive intestinal peptide: a novel stimulator of steroidogenesis by cultured rat granulosa cells. *Biol. Reprod.* 33, 37–52.
332. Trzeciak, W. H., Ahmed, C. E., Simpson, E. R., and Ojeda, S. R. (1986). Vasoactive intestinal peptide induces the synthesis of the cholesterol side-chain cleavage enzyme complex in cultured rat ovarian granulosa cells. *Proc. Natl. Acad. Sci. U S A* 83, 7490–7494.
333. Trzeciak, W. H., Waterman, M. R., Simpson, E. R., and Ojeda, S. R. (1987). Vasoactive intestinal peptide regulates cholesterol side-chain cleavage cytochrome P-450 (P-450scc) gene expression in granulosa cells from immature rat ovaries. *Mol. Endocrinol.* 1, 500–504.
334. Kasson, B. G., Meidan, R., Davoren, J. B., and Hsueh, A. J. W. (1985). Identification of subpopulations of rat granulosa cells: sedimentation properties and hormonal responsiveness. *Endocrinology* 117, 1027–1034.
335. Ranta, T., Knecht, M., Darbon, J.-M., Baukal, A. J., and Catt, K. J. (1984). Induction of granulosa cell differentiation by forskolin: stimulation of adenosine 3′,5′-monophosphate production, progesterone synthesis, and luteinizing hormone receptor expression. *Endocrinology* 114, 845–850.
336. Aguado, L. I., and Ojeda, S. R. (1984). Prepubertal ovarian function is finely regulated by direct adrenergic influences. Role of the adrenergic innervation. *Endocrinology* 114, 1845–1853.
337. Lara, H. E., Dees, W. L., Hiney, J. K., Dissen, G. A., Rivier, C., and Ojeda, S. R. (1991). Functional recovery of the developing rat ovary after transplantation: contribution of the extrinsic innervation. *Endocrinology* 129, 1849–1860.
338. Meijs-Roelofs, H. M. A., Kramer, P., and Sander, H. J. (1983). Changes in serum concentrations of luteinizing hormone in the female rat approaching puberty. *J. Endocrinol.* 98, 241–249.
339. Urbanski, H. F., Urbanski, D., and Ojeda, S. R. (1984). An automated system for the study of pulsatile hormone secretion in the immature rat. *Neuroendocrinology* 38, 403–406.
340. Urbanski, H. F., and Ojeda, S. R. (1985). *In vitro* simulation of prepubertal changes in pulsatile luteinizing hormone release enhances progesterone and 17β-estradiol secretion from immature ovaries. *Endocrinology* 117, 638–643.
341. Urbanski, H. F., and Ojeda, S. R. (1987). Gonadal-independent activation of enhanced afternoon luteinizing hormone release during pubertal development in the female rat. *Endocrinology* 121, 907–913.
342. Urbanski, H. F., and Ojeda, S. R. (1986). Development of afternoon minisurges of luteinizing hormone secretion in prepubertal female rats is ovary dependent. *Endocrinology* 118, 1187–1193.
343. Kimura, F., and Kawakami, M. (1981). Two daily surges of prolactin secretion preserved after ovariectomy in the immature rat. *Endocrinol. Jpn.* 28, 647–652.
344. Urbanski, H. F., and Ojeda, S. R. (1987). Development and amplification of midafternoon surges of prolactin secretion in ovariectomized immature rats. *J. Endocrinol.* 110, 361–366.
345. ter Haar, M. B., and Wilson, C. A. (1978). Evidence for pleiomorphism of luteinizing hormone in peripubertal female rats. *J. Endocrinol.* 79, 133–134.
346. Ojeda, S. R., Costa, M., Katz, K. H., and Hersh, L. B. (1985). Evidence for the existence of substance P in the prepubertal rat ovary. I. Biochemical and physiological studies. *Biol. Reprod.* 33, 286–295.
347. Osman, P. (1975). Preovulatory changes in the ovaries during the first spontaneous pro-oestrus in the rat. *J. Endocrinol.* 67, 259–265.
348. Garris, D. R., and Curry, T. E., Jr. (1983). Ovarian blood flow in the rat: association with body weight, the estrous cycle, and pseudopregnancy. *Proc. Soc. Exp. Biol. Med.* 174, 198–204.
349. Varga, B., Horvath, E., Folly, G., and Stark, E. (1985). Study of the luteinizing hormone-induced increase of ovarian blood flow during the estrous cycle in the rat. *Biol. Reprod.* 32, 480–488.
350. Parker, C. R., Jr., and Mahesh, V. B. (1976). Hormonal events surrounding the natural onset of puberty in female rats. *Biol. Reprod.* 14, 347–353.
351. Andrews, W. W., Advis, J. P., and Ojeda, S. R. (1980). The first proestrus in the female rat: circulating steroid levels preceding and accompanying the preovulatory LH surge. *Proc. Soc. Exp. Biol. Med.* 163, 305–309.
352. Lephart, E. D., Mathews, D., Noble, J., and Ojeda, S. R. (1989). The vaginal epithelium of immature rats metabolizes androgens through an aromatase-like reaction: changes during the time of puberty. *Biol. Reprod.* 40, 259–267.
353. Ruf, K. B. (1983). On the significance of 5α-androstane-3α-17β-diol in the peripubertal female rat. *J. Steroid Biochem.* 19, 887–890.
354. Meijs-Roelofs, H. M. A., Kramer, P., Cappellen, W. A. V., and Woutersen, P. J. A. (1985). Changes in serum concentration and ovarian content of 5α-androstane-3α-17β-diol in the female rat approaching first ovulation. *Biol. Reprod.* 32, 301–308.
355. Advis, J. P., Wiener, S., and Ojeda, S. R. (1981). The changes in ovarian 3α-androstanediol response to gonadotropins during puberty in the rat. Modulatory role of prolactin (PRL). *Endocrinology* 109, 223–228.
356. Ojeda, S. R., Katz, K. H., Costa, M. E., and Advis, J. P. (1983). Effect of experimental alterations in serum levels of 5α-androstane-3α-17β-diol on the timing of puberty in the female rat. *Neuroendocrinology* 39, 19–24.
357. Eckstein, B., Mechoulam, R., and Burstein, S. H. (1970). Identification of 5α-androstane-3α-17β-diol as a principal

metabolite of pregnenolone in rat ovary at onset of puberty. *Nature* 228, 866–867.
358. Eckstein, B., Yehud, S., Shani, J., and Goldhaber, G. (1976). Suppression of luteinizing hormone release by 5α-androstane-3α-17β-diol and its 3β epimer in immature ovariectomized rats. *J. Endocrinol.* 70, 25–30.
359. Ferin, M., Zimmering, P. E., and Vande Wiele, L. (1969). Effects of antibodies to estradiol-17β on PMS-induced ovulation in immature rats. *Endocrinology* 84, 893–900.
360. Aiyer, M. S., and Fink, G. (1974). The role of sex steroid hormones in modulating the responsiveness of the anterior pituitary gland to luteinizing hormone releasing factor in the female rat. *J. Endocrinol.* 62, 553–572.
361. Kim, K., and Ramirez, V. D. (1982). *In vitro* progesterone stimulates the release of luteinizing hormone-releasing hormone from superfused hypothalamic tissue from ovariectomized estradiol-primed prepubertal rats. *Endocrinology* 111, 750–757.
362. Kim, K., Lee, B. J., Park, Y., and Cho, W. K. (1989). Progesterone increases messenger ribonucleic acid (mRNA) encoding luteinizing hormone releasing hormone (LHRH) level in the hypothalamus of ovariectomized estradiol-primed prepubertal rats. *Mol. Brain Res.* 6, 151–158.
363. Seeburg, P. H., and Adelman, J. P. (1984). Characterization of cDNA for precursor of human luteinizing hormone releasing hormone. *Nature* 311, 666–668.
364. Savoy-Moore, R. T., Schwartz, N. B., Duncan, J. A., and Marshall, J. C. (1981). Pituitary gonadotropin-releasing hormone receptors on proestrus: effect of pentobarbital blockade of ovulation in the rat. *Endocrinology* 109, 1360–1364.
365. Smith, S. S., and Ojeda, S. R. (1985). Estradiol-induced LH release in juvenile female rats: nembutal blocks the decline and the exposure of non-available pituitary LHRH receptor sites. *Neuroendocrinology* 41, 246–251.
366. Shivers, B. D., Harlan, R. E., Morrell, J. I., and Pfaff, D. W. (1983). Absence of oestradiol concentration in cell nuclei of LHRH-immunoreactive neurones. *Nature* 304, 345–347.
367. Herbison, A. E. (1998). Multimodal influence of estrogen upon gonadotropin-releasing hormone neurons. *Endocrine Rev.* 19, 302–330.
368. Hrabovszky, E., Shughrue, P. J., Merchenthaler, I., Hajszán, T., Carpenter, C. D., Liposits, Z., and Petersen, S. L. (2000). Detection of estrogen receptor-b messenger ribonucleic acid and ^{125}I-estrogen binding sites in luteinizing hormone-releasing hormone neurons of the rat brain. *Endocrinology* 141, 3506–3509.
369. Skynner, M. J., Sim, J. A., and Herbison, A. E. (1999). Detection of estrogen receptor α and β messenger ribonucleic acids in adult gonadotropin-releasing hormone neurons. *Endocrinology* 140, 5195–5201.
370. Petersen, S. L., Ottem, E. N., and Carpenter, C. D. (2003). Direct and indirect regulation of gonadotropin-releasing hormone neurons by estradiol. *Biol. Reprod.* 69, 1771–1778.
371. Brown, P. S. (1971). Pituitary FSH in immature female rats treated with drugs that inhibit or antagonize the actions of catecholamines or 5-HT. *Neuroendocrinology* 7, 183–192.
372. Aguilar, R., Pinilla, L., González, D., López, F., Bellido, C., and Aguilar, E. (1990). Participation of the serotoninergic system in the onset of puberty in female rats. *Biog. Amines* 7, 81–87.
373. Wilson, C. A., Andrews, M., Handley, J. C., Lemon, M., and Les, T. (1977). The role of hypothalamic serotonin (5-HT) before ovulation in immature rats treated with pregnant mare serum. *Psychoneuroendocrinology* 2, 267–274.
374. Chen, H. T., Sylvester, P. W., Ieire, T., and Meites, J. (1981). Potentiation of luteinizing hormone release by serotonin agonists in ovariectomized, steroid-primed rats. *Endocrinology* 108, 948–952.
375. Hery, M., LaPlante, E., and Kordon, C. (1976). Participation of serotonin in the phasic release of LH. I. Evidence from pharmacological experiments. *Endocrinology* 99, 496–503.
376. Walker, R. F. (1983). Serotonin-reinstatement of luteinizing hormone surges after loss of positive feedback in ovariectomized rats bearing subcutaneous capsules containing oestrogen. *J. Endocrinol.* 98, 7–17.
377. Müller, E. E., Cocchi, D., Villa, A., Zanbotti, F., and Fraschini, F. (1972). Involvement of brain catecholamines in the gonadotropin-releasing mechanism(s) before puberty. *Endocrinology* 90, 1267–1276.
378. Kordon, C., and Glowinski, J. (1969). Selective inhibition of superovulation by blockade of dopamine synthesis during the "critical period" in the immature rat. *Endocrinology* 85, 924–931.
379. Fukse, K., Hökfelt, T., Sundstedt, C. D., Ahren, K., and Hamberger, L. (1972). Amine turnover changes in the tubero-infundibular dopamine (DA) neurons in immature rats injected with PMS. *Neuroendocrinology* 10, 282–300.
380. Sarkar, D. K., Smith, G. C., and Fink, G. (1981). Effect of manipulating central catecholamines on puberty and the surge of luteinizing hormone and gonadotropin releasing hormone induced by pregnant mare serum gonadotropin in female rats. *Brain Res.* 213, 335–349.
381. McDonald, J. K. (1987). NPY and related substances. *Crit. Rev. Neurobiol.* 4, 97–135.
382. Kalra, S. P., and Crowley, W. R. (1992). Neuropeptide Y: a novel neuroendocrine peptide in the control of pituitary hormone secretion, and its relation to luteinizing hormone. In *Frontiers in Neuroendocrinology* (W. F. Ganong and L. Martini, Eds.), Vol. 13, pp. 1–46. Raven Press, New York.
383. Wilkinson, M., and Bhanot, R. (1982). A puberty-related attenuation of opiate peptide-induced inhibition of LH secretion. *Endocrinology* 110, 1046–1048.
384. Urbanski, H. F., and Ojeda, S. R. (1987). Activation of luteinizing hormone-releasing hormone release advances the onset of female puberty. *Neuroendocrinology* 46, 273–276.
385. Cicero, T. J., Meyer, E. R., and Bell, R. D. (1988). Characterization and possible opioid modulation of N-methyl-D-aspartic acid induced increases in serum luteinizing hormone levels in the developing male rat. *Life Sci.* 42, 1725–1732.
386. Bourguignon, J.-P., Gérard, A., and Frachimont, P. (1989). Direct activation of gonadotropin-releasing hormone secretion through different receptors to neuroexcitatory amino acids. *Neuroendocrinology* 49, 402–408.
387. Brann, D. W., Zamorano, P. L., Ping, L., and Mahesh, V. B. (1993). Role of excitatory amino acid neurotransmission during puberty in the female rat. *Mol. Cell. Neurosci.* 4, 107–112.
388. Smyth, C., and Wilkinson, M. (1994). A critical period for glutamate receptor-mediated induction of precocious puberty in female rats. *J. Neuroendocrinol.* 6, 275–284.
389. Plant, T. M., Gay, V. L., Marshall, G. R., and Arslan, M. (1989). Puberty in monkeys is triggered by chemical stimulation of the hypothalamus. *Proc. Natl. Acad. Sci. U S A* 86, 2506–2510.
390. Urbanski, H. F., and Ojeda, S. R. (1990). A role for N-methyl-D-aspartate (NMDA) receptors in the control of LH secretion and initiation of female puberty. *Endocrinology* 126, 1774–1776.
391. Meijs-Roelofs, H. M. A., Kramer, P., and van Leeuwen, E. C. M. (1991). The N-methyl-D-aspartate receptor antagonist MK-801 delays the onset of puberty and may acutely block the first spontaneous LH surge and ovulation in the rat. *J. Endocrinol.* 131, 435–441.
392. MacDonald, M. C., and Wilkinson, M. (1990). Peripubertal treatment with N-methyl-*D*-aspartic acid or neonatally with monosodium glutamate accelerates sexual maturation in

female rats, an effect reversed by MK-801. *Neuroendocrinology* 52, 143–149.

393. Wu, F. C. W., Howe, D. C., and Naylor, A. M. (1990). N-methyl-DL-aspartate receptor antagonism by D-2-amino-5-phosphonovaleric acid delays onset of puberty in the female rat. *J. Neuroendocrinol.* 2, 627–631.
394. Veneroni, O., Cocilovo, L., Müller, E. E., and Cocchi, D. (1990). Delay of puberty and impairment of growth in female rats given a non competitive antagonist of NMDA receptors. *Life Sci.* 47, 1253–1260.
395. Brann, D. W., and Mahesh, V. B. (1991). Endogenous excitatory amino acid involvement in the preovulatory and steroid-induced surge of gonadotropins in the female rat. *Endocrinology* 128, 1541–1547.
396. Brann, D. W., and Mahesh, V. B. (1991). Endogenous excitatory amino acid regulation of the progesterone-induced LH and FSH surge in estrogen-primed ovariectomized rats. *Neuroendocrinology* 53, 107–110.
397. Brann, D. W., and Mahesh, V. B. (1992). Excitatory amino acid neurotransmission: evidence for a role in neuroendocrine regulation. *Trends Endocrinol. Metab.* 3, 122–126.
398. Kalra, S. P., and Kalra, P. S. (1983). Neural regulation of luteinizing hormone secretion in the rat. *Endocr. Rev.* 4, 311–351.
399. Paul, S. M., Axelrod, J., Saavedra, J. M., and Skolnick, P. (1979). Estrogen-induced efflux of endogenous catecholamines from the hypothalamus *in vitro*. *Brain Res.* 178, 499–505.
400. Sar, M., and Stumpf, W. E. (1981). Central noradrenergic neurons concentrate ^{3}H-oestradiol. *Nature* 289, 500–502.
401. Jennes, L., Jennes, M. E., Purvis, C., and Nees, M. (1992). c-*fos* expression in noradrenergic A2 neurons of the rat during the estrous cycle and after steroid hormone treatments. *Brain Res.* 586, 171–175.
402. Sar, M., Sahu, A., Crowley, W. R., and Kalra, S. P. (1990). Localization of neuropeptide-Y immunoreactivity in estradiol-concentrating cells in the hypothalamus. *Endocrinology* 127, 2752–2756.
403. Sahu, A., Phelps, C. P., White, J. D., Crowley, W. R., Kalra, S. P., and Kalra, P. S. (1992). Steroidal regulation of hypothalamic neuropeptide Y release and gene expression. *Endocrinology* 130, 3331–3336.
404. Crowley, W. R., and Kalra, S. P. (1987). Neuropeptide Y stimulates the release of luteinizing hormone-releasing hormone from medial basal hypothalamus *in vitro*: modulation by ovarian hormones. *Neuroendocrinology* 46, 97–103.
405. Jarry, H., Perschl, A., and Wuttke, W. (1988). Further evidence that preoptic anterior hypothalamic GABAergic neurons are part of the GnRH pulse and surge generator. *Acta Endocrinol.* 118, 573–579.
406. Seltzer, A. M., and Donoso, A. O. (1992). Restraining action of GABA on estradiol-induced LH surge in the rat: GABA activity in brain nuclei and effects of GABA mimetics in the medial preoptic nucleus. *Neuroendocrinology* 55, 28–34.
407. Grattan, D. R., Rocca, M. S., Strauss, K. I., Sagrillo, C. A., Selmanoff, M., and McCarthy, M. M. (1996). GABAergic neuronal activity and mRNA levels for both forms of glutamic acid decarboxylase (GAD_{65} and GAD_{67}) are reduced in the diagonal band of Broca during the afternoon of proestrus. *Brain Res.* 733, 46–55.
408. Unda, R., Brann, D. W., and Mahesh, V. B. (1995). Progesterone suppression of glutamate acid decarboxylase (GAD_{67}) mRNA levels in the preoptic area: correlation to the luteinizing hormone surge. *Neuroendocrinology* 62, 562–570.
409. Herbison, A. E., Augood, S. J., and McGowan, E. M. (1992). Expression of glutamic acid decarboxylase messenger RNA in rat medial preoptic area neurones during the oestrous cycle and after ovariectomy. *Mol. Brain Res.* 14, 310–316.
410. Herbison, A. E., and Dyer, R. G. (1991). Effect on luteinizing hormone secretion of GABA receptor modulation in the medial preoptic area at the time of proestrous luteinizing hormone surge. *Neuroendocrinology* 53, 317–320.
411. Morello, H., Caligaris, L., Haymal, B., and Taleisnik, S. (1989). Inhibition of proestrous LH surge and ovulation in rats evoked by stimulation of the medial raphe nucleus involves a GABA-mediated mechanism. *Neuroendocrinology* 50, 81–87.
412. Mitsushima, D., Jinnai, K., and Kimura, F. (1997). Possible role of the g-aminobutyric acid-A receptor system in the timing of the proestrous luteinizing hormone surge in rats. *Endocrinology* 138, 1944–1948.
413. Kimura, F., and Jinnai, K. (1994). Bicuculline infusions advance the timing of luteinizing hormone surge in proestrous rats: comparisons with naloxone effects. *Horm. Behav.* 28, 424–430.
414. Thind, K. K., and Goldsmith, P. C. (1988). Infundibular gonadotropin-releasing hormone neurons are inhibited by direct opioid and autoregulatory synapses in juvenile monkeys. *Neuroendocrinology* 47, 203–216.
415. Cicero, T. J., Meyer, E. R., Gabriel, S. M., Bell, R. D., and Wilcox, C. E. (1980). Morphine exerts testosterone-like effects in the hypothalamus of the castrated male rat. *Brain Res.* 202, 151–164.
416. Sylvester, P. W., Van Vugt, D. A., Aylsworth, C. A., Hanson, E. A., and Meites, J. (1982). Effects of morphine and naloxone on inhibition by ovarian hormones of pulsatile release of LH in ovariectomized rats. *Neuroendocrinology* 34, 269–273.
417. Blank, M. S., and Mann, D. R. (1981). Diurnal influences on serum luteinizing hormone responses to opiate receptor blockade with naloxone or to luteinizing hormone-releasing hormone in the immature female rat. *Proc. Soc. Exp. Biol. Med.* 168, 338–343.
418. Ma, Y. J., Berg-von der Emde, K., Moholt-Siebert, M., Hill, D. F., and Ojeda, S. R. (1994). Region-specific regulation of transforming growth factor a (TGFa) gene expression in astrocytes of the neuroendocrine brain. *J. Neurosci.* 14, 5644–5651.
419. Ojeda, S. R., and Campbell, W. B. (1982). An increase in hypothalamic capacity to synthesize prostaglandin E_2 precedes the first preovulatory surge of gonadotropins. *Endocrinology* 111, 1031–1037.
420. Ramirez, V. D., Kim, K., and Dluzen, D. (1985). Progesterone action on the LHRH and the nigrostriatal dopamine neuronal systems: in vitro and in vivo studies. *Rec. Prog. Horm. Res.* 41, 421–472.
421. Ojeda, S. R., Urbanski, H. F., Katz, K. H., and Costa, M. E. (1985). Stimulation of cyclic adenosine 3′,5′-monophosphate production enhances hypothalamic luteinizing hormone-releasing hormone release without increasing prostaglandin E_2 synthesis: studies in prepubertal female rats. *Endocrinology* 117, 1175–1178.
422. Nishizuka, Y. (1983). Phospholipid degradation and signal translation for protein phosphorylation. *Trends Biochem. Sci.* 8, 13–16.
423. Ojeda, S. R., Urbanski, H. F., Katz, K. H., Costa, M. E., and Conn, P. M. (1986). Activation of two different but complementary biochemical pathways stimulates release of hypothalamic luteinizing hormone- releasing hormone. *Proc. Natl. Acad. Sci. U S A* 83, 4932–4936.
424. Lee, B. J., Kim, K., and Cho, W. K. (1990). Activation of intracellular pathways with forskolin and phorbol ester increases LHRH mRNA level in the rat hypothalamus superfused *in vitro*. *Mol. Brain Res.* 8, 185–191.
425. Kalra, S. P., Fuentes, M., Fournier, A., Parker, S. L., and Crowley, W. R. (1992). Involvement of the Y-1 receptor subtype in the regulation of luteinizing hormone secretion by neuropeptide Y in rats. *Endocrinology* 130, 3323–3330.

426. Sabatino, F. D., Collins, P., and McDonald, J. K. (1989). Neuropeptide-Y stimulation of luteinizing hormone-releasing hormone secretion from the median eminence *in vitro* by estrogen-dependent and extracellular Ca^{2+}-independent mechanisms. *Endocrinology* 124, 2089–2098.
427. Kuehl-Kovarik, M. C., Pouliot, W. A., Halterman, G. L., Handa, R. J., Dudek, F. E., and Partin, K. M. (2002). Episodic bursting activity and response to excitatory amino acids in acutely dissociated gonadotropin-releasing hormone neurons genetically targeted with green fluorescent protein. *J. Neurosci.* 22, 2313–2322.
428. Lee, W.-S., Abbud, R., Smith, M. S., and Hoffman, G. E. (1992). LHRH neurons express cJun protein during the proestrous surge of luteinizing hormone. *Endocrinology* 130, 3101–3103.
429. Lee, W. S., Abbud, R., Hoffman, G. E., and Smith, M. S. (1993). Effects of N-methyl-D-aspartate receptor activation on cFos expression in luteinizing hormone-releasing hormone nuerons in female rats. *Endocrinology* 133, 2248–2254.
430. Ojeda, S. R., Dissen, G. A., and Junier, M.-P. (1992). Neurotrophic factors and female sexual development. In *Frontiers in Neuroendocrinology* (W. F. Ganong and L. Martini, Eds.), Vol. 13, pp. 120–162. Raven Press, New York.
431. Negro-Vilar, A., Wetsel, W. C., Mellon, P. M., and Valenca, M. M. (1991). Secretagogue-stimulated secretion of LHRH and GAP elicited from a perifused immortalized LHRH neuronal cell line. *Prog. 73rd Ann. Mtg. Endocrine Soc.* p. 458.
432. Jost, A., Vigier, B., Prepin, J., and Perchellet, J. P. (1973). Studies on sex differentiation in mammals. *Rec. Prog. Horm. Res.* 29, 1–41.
433. Koopman, P., Gubbay, J., Vivian, N., Goodfellow, P., and Lovell-Badge, R. (1991). Male development of chromosomally female mice transgenic for Sry. *Nature* 351, 117–121.
434. Buehr, M., Gu, S., and McLaren, A. (1993). Mesonephric contribution to testis differentiation in the fetal mouse. *Development* 117, 273–281.
435. Cupp, A. S., Uzumcu, M., and Skinner, M. K. (2003). Chemotactic role of neurotropin 3 in the embryonic testis that facilitates male sex determination. *Biol. Reprod.* 68, 2033–2037.
436. Waites, G. M. H., Speight, A. C., and Jenkins, N. (1985). The functional maturation of the Sertoli cell and Leydig cell in the mammalian testis. *J. Reprod. Fertil.* 75, 317–326.
437. Warren, D. W., Huhtaniemi, I. T., Tapanainen, J., Dufau, M. L., and Catt, K. J. (1984). Ontogeny of gonadotropin receptors in the fetal and neonatal rat testis. *Endocrinology* 114, 470–476.
438. Warren, D. W., Haltmeyer, G. C., and Eik-Nes, K. B. (1975). The effect of gonadotrophins on the fetal and neonatal rat testis. *Endocrinology* 96, 1226–1229.
439. Feldman, S. C., and Bloch, E. (1978). Developmental pattern of testosterone synthesis by fetal rat testes in response to luteinizing hormone. *Endocrinology* 102, 999–1007.
440. Picon, R., and Gangnerau, M.-N. (1980). Acquisition of sensitivity to LH in relation to foetal development. Stimulation of cyclic AMP and testosterone production in the rat testis. *Mol. Cell. Endocrinol.* 18, 137–150.
441. Lording, D. W., and de Kretser, D. M. (1972). Comparative ultrastructural and histochemical studies of the interstitial cells of the rat testis during fetal and postnatal development. *J. Reprod. Fertil.* 29, 261–269.
442. Weisz, J., and Ward, I. L. (1980). Plasma testosterone and progesterone titers of pregnant rats, their male and female fetuses, and neonatal offspring. *Endocrinology* 106, 306–316.
443. Huhtaniemi, I. T., Nozu, K., Warren, D. W., Dufau, M. L., and Catt, K. J. (1982). Acquisition of regulatory mechanisms for gonadotropin receptors and steroidogenesis in the maturing rat testis. *Endocrinology* 111, 1711–1720.
444. Desjardin, C. (1981). Endocrine signaling and male reproduction. *Biol. Reprod.* 24, 1–21.
445. van den Dungen, H. M., van Dieten, J. A. M. J., van Rees, G. P., and Schoemaker, J. (1990). Testicular weight, tubular diameter and number of sertoli cells in rats are decreased after early prepubertal administration of an LHRH-antagonist; the quality of spermatozoa is not impaired. *Life Sci.* 46, 1081–1089.
446. Wang, Z. X., Wreford, N. G. M., and de Kretser, D. M. (1989). Determination of Sertoli cell numbers in the developing rat testis by stereological methods. *Int. J. Androl.* 12, 58–64.
447. Steinberger, A., and Steinberger, E. (1971). Replication pattern of Sertoli cells in maturing rat testis *in vivo* and in organ culture. *Biol. Reprod.* 4, 84–87.
448. Orth, J. (1982). Proliferation of Sertoli cells in fetal and postnatal rats: a quantitative autoradiographic study. *Anat. Rec.* 203, 485–492.
449. Hodgson, Y., Robertson, D. M., and de Kretser, D. M. (1983). The regulation of testicular function. *Int. Rev. Physiol.* 27, 275–327.
450. Steinberger, A., Thanki, K. H., and Siegal, B. (1974). FSH binding in rat testes during maturation and following hypophysectomy. Cellular localization of FSH receptors. In *Current Topics in Molecular Endocrinology* (M. L. Dufau and A. R. Means, Eds.), Vol. 1, pp. 177–192. Plenum Press, New York.
451. Payne, A. H., Kelch, R. P., Murono, E. P., and Kerlan, J. T. (1977). Hypothalamic, pituitary and testicular function during sexual maturation of the male rat. *J. Endocrinol.* 72, 17–26.
452. Brabant, G., Ray, A. K., Wagner, T. O. F., Krech, R., and von zur Mühlen, A. (1985). Peripubertal variation of GnRH-release from *in vitro* superfused hypothalami of male rats. *Neuroendocrinol. Lett.* 7, 197–202.
453. Chappel, S. C., and Ramaley, J. A. (1985). Changes in the isoelectric focusing profile of pituitary follicle-stimulating hormone in the developing male rat. *Biol. Reprod.* 32, 567–573.
454. Kamberi, I. A., de Vellis, J., Bacleon, E. S., and Inglish, D. (1980). Hormonal patterns of the hypothalamo-pituitary-gonadal axis in the rat during postnatal development and sexual maturation. *Endokrinologie* 75, 129–140.
455. Lisk, R. D. (1968). Luteinizing hormone in the pituitary gland of the albino rat: concentration and content as a function of sex and age. *Neuroendocrinology* 3, 18–24.
456. Kragt, C. L., and Ganong, W. F. (1968). Pituitary FSH content in male rats at various ages. *Proc. Soc. Exp. Biol. Med.* 128, 965–967.
457. Dupon, C., and Schwartz, N. B. (1971). Pituitary LH patterns in prepubertal normal and testosterone-sterilized rats. *Neuroendocrinology* 7, 236–248.
458. Debeljuk, L., Arimura, A., and Schally, A. V. (1972). Studies on the pituitary responsiveness to luteinizing hormone-releasing hormone (LH-RH) in intact male rats of different ages. *Endocrinology* 90, 585–588.
459. Chan, V., Clayton, R. N., Knox, G., and Catt, K. J. (1981). Ontogeny of pituitary GnRH receptors in the rat. *Endocrinology* 108, 2086–2092.
460. Duncan, J. A., Dalkin, A. C., Barkan, A., Regiani, S., and Marshall, J. C. (1983). Gonadal regulation of pituitary gonadotropin-releasing hormone receptors during sexual maturation in the rat. *Endocrinology* 113, 2238–2246.
461. Negro-Vilar, A., Krulich, L., and McCann, S. M. (1973). Changes in serum prolactin and gonadotropins during sexual development of the male rat. *Endocrinology* 93, 660–664.
462. Ketelslegers, J.-M., Hetzel, W. D., Sherins, R. J., and Catt, K. J. (1978). Developmental changes in testicular gonadotropin receptors: plasma gonadotropins and plasma testosterone in the rat. *Endocrinology* 103, 212–222.
463. Miyachi, Y., Nieschlag, E., and Lipsett, M. B. (1973). The secretion of gonadotropins and testosterone by the neonatal male rat. *Endocrinology* 92, 1–5.

464. Ellis, G. B., and Desjardin, C. (1982). Male rats secrete luteinizing hormone and testosterone episodically. *Endocrinology* 110, 1618–1627.
465. Ellis, G. B., Desjardin, C., and Fraser, H. M. (1983). Control of pulsatile LH release in male rats. *Neuroendocrinology* 37, 177–183.
466. Ellis, G. B., and Desjardin, C. (1984). Orchidectomy unleashes pulsatile luteinizing hormone secretion in the rat. *Biol. Reprod.* 30, 619–627.
467. Nazian, S. J., and Cameron, D. F. (1992). Termination of the peripubertal FSH increase in male rats. *Am. J. Physiol.* 262, E179–E184.
468. Tsutsui, K. (1991). Pituitary and gonadal hormone-dependent and -independent induction of follicle-stimulating hormone receptors in the developing testis. *Endocrinology* 128, 477–487.
469. Sertoli, E. (1865). On the existence of special branched cells in the seminiferous tubule of the human testes. *Morgangni* 7, 31–39.
470. Fawcett, D. W. (1975). The ultrastructure and functions of the Sertoli cell. In *Handbook of Physiology* (R. O. Greep and E. W. Hamilton, Eds.), pp. 22–55. American Physiological Society, Washington, DC.
471. Setchell, B. P., and Waites, G. M. H. (1975). In *Handbook of Physiology* (R. O. Greep and E. W. Hamilton, Eds.), pp. 143–172. American Physiological Society, Washington, DC.
472. Waites, G. M., and Gladwell, R. T. (1982). Physiological significance of fluid secretion in the testis and blood-testis barrier. *Physiol. Rev.* 62, 624–671.
473. Russell, L. D., Tallon-Doran, M., Weber, J. E., Wong, V., and Peterson, R. N. (1983). Three-dimensional reconstruction of a rat stage V Sertoli cell. III. A study of specific cellular relationships. *Am. J. Anat.* 167, 181–192.
474. Dorrington, J. H., Fritz, I. B., and Armstrong, D. T. (1978). Control of testicular estrogen synthesis. *Biol. Reprod.* 18, 55–64.
475. Robinson, R., and Fritz, I. B. (1981). Metabolism of glucose by Sertoli cells in culture. *Biol. Reprod.* 24, 1032–1041.
476. Kissinger, C., Skinner, M. K., and Griswold, M. D. (1982). Analysis of Sertoli cell-secreted proteins by two dimensional gel electrophoresis. *Biol. Reprod.* 27, 233–240.
477. Hansson, V., Weddington, S. C., Petrusz, P., Ritzen, E. M., Nayfeh, S. N., and French, F. S. (1975). FSH stimulation of testicular androgen binding protein (ABP): comparison of ABP response and ovarian augmentation. *Endocrinology* 97, 469–473.
478. Lacroix, M., Smith, F. E., and Fritz, I. B. (1977). Secretion of plasminogen activator by Sertoli cell enriched cultures. *Mol. Cell Endocrinol.* 9, 227–236.
479. Skinner, M. K., and Griswold, M. D. (1980). Sertoli cells synthesize and secrete transferrin-like protein. *J. Biol. Chem.* 255, 9523–9525.
480. Skinner, M. K., and Griswold, M. D. (1983). Sertoli cells synthesize and secrete a ceruloplasmin-like protein. *Biol. Reprod.* 28, 1225–1229.
481. de Jong, F. H., and Robertson, D. M. (1985). Inhibin: 1985 update on action and purification. *Mol. Cell Endocrinol.* 42, 95–103.
482. Griswold, M. D. (1988). Protein secretions of Sertoli cells. *Int. Rev. Cytol.* 110, 133–156.
483. Skinner, M. K., Cosand, W. L., and Griswold, M. D. (1984). Purification and characterization of testicular transferrin secreted by rat Sertoli cells. *Biochem. J.* 218, 313–320.
484. Dym, M., and Fawcett, D. W. (1970). The blood-testis barrier in the rat and the physiological compartmentation of the seminiferous epithelium. *Biol. Reprod.* 3, 308–326.
485. Tung, P. S., and Fritz, I. B. (1977). Isolation and culture of testis cells. A morphology at characterization. In *Techniques of Human Andrology* (E. S. E. Hafez, Ed.), pp. 125–146. Elsevier, Holland.
486. Anthony, C. T., and Skinner, M. K. (1989). Cytochemical and biochemical characterization of testicular peritubular myoid cells. *Biol. Reprod.* 40, 811–823.
487. Skinner, M. K., and Fritz, I. B. (1985). Testicular peritubular cells secrete a protein under androgen control that modulates Sertoli cell functions. *Proc. Natl. Acad. Sci. U S A* 82, 114–118.
488. Tung, P. S., Skinner, M. K., and Fritz, I. B. (1984). Fibronectin synthesis is a marker for peritubular cell contaminants in Sertoli cell-enriched cultures. *Biol. Reprod.* 30, 199–211.
489. Skinner, M. K. (1991). Cell-cell interactions in the testis. *Endocr. Rev.* 12, 45–77.
490. Ascoli, M. (1985). "Luteinizing Hormone Action and Receptors." CRC Press, Boca Raton, FL.
491. Zhengwei, Y., Wreford, N. G., and de Kretser, D. M. (1990). A quantitative study of spermatogenesis in the developing rat testis. *Biol. Reprod.* 43, 629–635.
492. Fritz, I. B. (1978). Sites of action of androgens and follicle stimulating hormones on cells in seminiferous tubule. In *Biochemical Actions of Hormones* (G. Litwack, Ed.), pp. 249–281. Academic Press, New York.
493. Verhoeven, G. (1980). Androgen receptor in cultured interstitital cells derived from immature rat testis. *J. Steroid Biochem.* 13, 469–474.
494. Skinner, M. K., and Griswold, M. D. (1982). Secretion of testicular transferrin by cultured Sertoli cells is regulated by hormones and retinoids. *Biol. Reprod.* 27, 211–221.
495. Perez-Infante, V., Bardin, C. W., Gunsalus, G. L., Musto, N. A., Rich, K. A., and Mather, J. P. (1986). Differential regulation of testicular transferrin and androgen-binding protein secretion in primary cultures of rat Sertoli cells. *Endocrinology* 118, 383–392.
496. Chang, C., Chen, Y.-T., Yeh, S.-D., Xu, Q., Wang, R.-S., Guillou, F., Lardy, H., and Yeh, S. (2004). Infertility with defective spermatogenesis and hypotestosteronemia in male mice lacking the androgen receptor in Sertoli cells. *Proc. Natl. Acad. Sci. U S A* 101, 6876–6881.
497. De Gendt, K., Swinnen, J. V., Saunders, P. T. K., Schoonjans, L., Dewerchin, M., Devos, A., Tan, K., Atanassova, N., Claessens, F., Lécureuil, C., Heyns, W., Carmeliet, P., Guillou, F., Sharpe, R. M., and Verhoeven, G. (2004). A Sertoli cell-selective knockout of the androgen receptor causes spermatogenic arrest in meiosis. *Proc. Natl. Acad. Sci. U S A* 101, 1327–1332.
498. Yeh, S., Tsai, M. Y., Xu, Q., Mu, X. M., Lardy, H., Huang, K. E., Lin, H., Yeh, S. D., Altuwaijri, S., Zhou, X., Xing, L., Boyce, B. F., Hung, M. C., Zhang, S., Gan, L., Chang, C., and Hung, M. C. (2002). Generation and characterization of androgen receptor knockout (ARKO) mice: an in vivo model for the study of androgen functions in selective tissues. [Erratum published in *Proc. Natl. Acad. Sci. U S A* 99(23):15245, 2002.] *Proc. Natl. Acad. Sci. U S A* 99, 13498–13503.
499. Heindel, J. J. (1990). cAMP response of cultured Sertoli cells from immature and adult hamsters. Effect of hypophysectomy and cryptorchidism. *J. Androl.* 11, 59–65.
500. Kim, K. H., and Griswold, M. D. (1990). The regulation of retinoic acid receptor mRNA levels during spermatogenesis. *Mol. Endocrinol.* 4, 1679–1688.
501. Gorczynska, E., and Handelsman, D. J. (1993). Requirement for transmembrane sodium flux in maintenance of cytosolic calcium levels in rat Sertoli cells. *Am. J. Physiol.* 264, E863–E867.
502. Rivarola, M. A., Podesta, E. J., and Chemes, H. E. (1972). *In vitro* testosterone-^{14}C metabolism by rat seminiferous tubules at different stages of development: formation of 5α- androstanediol at meiosis. *Endocrinology* 91, 537–542.

503. Podesta, E. J., and Rivarola, M. A. (1974). Concentration of androgens in whole testes, seminiferous tubules and interstitial tissue of rats at different stages of development. *Endocrinology* 95, 455–461.
504. Wiebe, J. P. (1976). Steroidogenesis in rat Leydig cells: changes in activity of 5-ane and 5-ene 3β hydroxysteroid dehydrogenases during sexual maturation. *Endocrinology* 98, 505–513.
505. Preslock, J. P. (1980). Steroidogenesis in the mammalian testis. *Endocr. Rev.* 1, 132–139.
506. Ivell, R., Balvers, M., Anand, R. J., Paust, H. J., McKinnell, C., and Sharpe, R. (2003). Differentiation-dependent expression of 17beta-hydroxysteroid dehydrogenase, type 10, in the rodent testis: effect of aging in Leydig cells. *Endocrinology* 144, 3130–3137.
507. Killian, J., Pratis, K., Clifton, R. J., Stanton, P. G., Robertson, D. M., and O'Donnell, L. (2003). 5alpha-Reductase isoenzymes 1 and 2 in the rat testis during postnatal development. *Biol. Reprod.* 68, 1711–1718.
508. Gupta, D., Zarzycki, J., and Rager, K. (1975). Plasma testosterone and dihydrotestosterone in male rats during sexual maturation and following orchidectomy and experimental bilateral cryptorchidism. *Steroids* 25, 33–42.
509. Moger, W. H. (1977). Serum 5α-androstane-3α,17β-diol androsterone and testosterone concentrations in the male rat. Influence of age and gonadotropin stimulation. *Endocrinology* 100, 1027–1032.
510. Piacsek, B. E., and Goodspeed, M. P. (1978). Maturation of the pituitary-gonadal system in the male rat. *J. Reprod. Fertil.* 52, 29–35.
511. Rommerts, F. F. G., de Jong, F. H., Brinkmann, A. O., and van der Molen, H. J. (1982). Development and cellular localization of rat testicular aromatase activity. *J. Reprod. Fertil.* 65, 281–288.
512. Resko, J. A., Feder, H. H., and Goy, R. W. (1968). Androgen concentrations in plasma and testis of developing rats. *J. Endocrinol.* 40, 485–491.
513. Knorr, D. W., Vanha-Pertulla, T., and Lipsett, M. B. (1970). Structure and function of rat testis through pubescence. *Endocrinology* 86, 1298–1304.
514. Grota, L. J. (1971). Effects of age and experience on plasma testosterone. *Neuroendocrinology* 8, 136–143.
515. Odell, W. D., Swerdloff, R. S., Jacobs, H. S., and Hescox, M. A. (1973). FSH induction of sensitivity to LH: one cause of sexual maturation in the male rat. *Endocrinology* 92, 160–165.
516. Murono, E. P., and Payne, A. H. (1979). Testicular maturation in the rat. *In vivo* effects of gonadotropins on steroidogenic enzymes in the hypophysectomized immature rat. *Biol. Reprod.* 20, 911–917.
517. Zipf, W. B., Payne, A. H., and Kelch, R. P. (1978). Prolactin, growth hormone and luteinizing hormone in the maintenance of testicular luteinizing hormone receptors. *Endocrinology* 103, 595–600.
518. Bartke, A. (1980). Role of prolactin in reproduction in male mammals. *Fed. Proc.* 39, 2577–2581.
519. Bartlett, J. M. S., Charlton, H. M., Robinson, I. C. A. F., and Nieschlag, E. (1990). Pubertal development and testicular function in the male growth hormone-deficient rat. *J. Endocrinol.* 126, 193–201.
520. Barreiro, M. L., Suominen, J. S., Gaytan, F., Pinilla, L., Chopin, L. K., Casanueva, F. F., Dieguez, C., Aguilar, E., Toppari, J., and Tena-Sempere, M. (2003). Developmental, stage-specific, and hormonally regulated expression of growth hormone secretagogue receptor messenger RNA in rat testis. *Biol. Reprod.* 68, 1631–1640.
521. Chowdury, A. R., Gautam, A. K., and Chatterjee, B. B. (1984). Thyroid-testis interrelationship during the development and sexual maturity of the rat. *Arch. Androl.* 13, 233–239.
522. Chowdury, A. R., and Arora, U. (1984). Role of thyroid in testicular development of immature rat. *Arch. Androl.* 12, 49–51.
523. Cooke, P. S., and Meisami, E. (1991). Early hypothyroidism in rats causes increased adult testis and reproductive organ size but does not change testosterone levels. *Endocrinology* 129, 237–243.
524. Ando, S., Sirianni, R., Forastieri, P., Casaburi, I., Lanzino, M., Rago, V., Giordano, F., Giordano, C., Carpino, A., and Pezzi, V. (2001). Aromatase expression in prepuberal Sertoli cells: effect of thyroid hormone. *Mol. Cell Endocrinol.* 178, 11–21.
525. Bourne, G. A., Regiani, S., Payne, A. H., and Marshall, J. C. (1980). Testicular GnRH receptors—characterization and localization of interstitial tissue. *J. Clin. Endocrinol. Metab.* 51, 407–409.
526. Hsueh, A. J. W., and Erickson, G. F. (1979). Extra-pituitary inhibition of testicular function by luteinizing hormone releasing hormone. *Nature* 281, 66–67.
527. Clayton, R. N., and Catt, K. J. (1981). Gonadotropin-releasing hormone receptors: Characterization, physiological regulation, and relationship to reproductive function. *Endocr. Rev.* 2, 186–209.
528. Clayton, R. N. (1985). Role of GnRH in the maturation of pituitary gonadotroph function. *J. Reprod. Fertil.* 75, 307–315.
529. Muller, D., Mukhopadhyay, A. K., Speth, R. C., Guidone, G., Potthast, R., Potter, L. R., and Middendorff, R. (2004). Spatiotemporal regulation of the two atrial natriuretic peptide receptors in testis. *Endocrinology* 145, 1392–1401.
530. Catizone, A., Ricci, G., Arista, V., Innocenzi, A., and Galdieri, M. (1999). Hepatocyte growth factor and c-MET are expressed in rat prepuberal testis. *Endocrinology* 140, 3106–3113.
531. Giannattasio, A., Angeletti, G., De Rosa, M., Zarrilli, S., Ambrosino, M., Cimmino, A., Coppola, C., Panza, G., Calafiore, R., Colao, A., Abete, O., and Lombardi, G. (2002). RNA expression bcl-w, a new related protein Bcl-2 family, and caspase-3 in isolated sertoli cells from pre-pubertal rat testes. *J. Endocrinol. Invest* 25, RC23–RC25.
532. Jonsson, C. K., Zetterstrom, R. H., Holst, M., Parvinen, M., and Soder, O. (1999). Constitutive expression of interleukin-1alpha messenger ribonucleic acid in rat Sertoli cells is dependent upon interaction with germ cells. *Endocrinology* 140, 3755–3761.
533. Murphy, B. E. P. (1968). Binding of testosterone and estradiol in plasma. *Can. J. Biochem.* 46, 299–302.
534. Ramalay, J. A. (1971). Steroid binding to serum proteins in maturing male and female rats. *Endocrinology* 89, 545–552.
535. Smith, E. R., Damassa, D. A., and Davidson, J. M. (1977). Feedback regulation and male puberty: testosterone-luteinizing hormone relationships in the developing rat. *Endocrinology* 101, 173–180.
536. Ulrich, R. S., and Kent, J. R. (1968). The disappearance rates of sex steroids in immature and mature rats. *Proc. Soc. Exp. Biol. Med.* 128, 1093–1096.
537. Frankel, A. I., and Ryan, E. L. (1981). Testicular innervation is necessary for the response of plasma testosterone levels to acute stress. *Biol. Reprod.* 24, 491–495.
538. Lamano Carvalho, T. L., Favaretto, A. L. V., Antunes-Rodrigues, J., and Ferreira, A. L. (1985). Evaluation of the frequency of stages VII-VIII and XIV of the tubular cycle of adult rats, 30 d after phenolization of the spermatic cord. Response to chronic administration of hCG. *Anat. Anz.* 159, 33–41.
539. Nagai, K., Murano, S., Minokoshi, Y., Okuda, H., and Kinutani, M. (1982). Effects of denervation and local 6-hydroxydopamine injection on testicular growth in rats. *Experientia* 38, 592–594.
540. Lee, S., Miselis, R., and Rivier, C. (2002). Anatomical and functional evidence for a neural hypothalamic-testicular pathway that is independent of the pituitary. *Endocrinology* 143, 4447–4454.

541. Selvage, D., and Rivier, C. (2003). Importance of the paraventricular nucleus of the hypothalamus as a component of a neural pathway between the brain and the testes that modulates testosterone secretion independently of the pituitary. *Endocrinology* 144, 594–598.
542. Ramirez, V. D., and McCann, S. M. (1965). Inhibitory effect of testosterone on luteinizing hormone secretion in immature and adult rats. *Endocrinology* 76, 412–417.
543. Yamayoto, M., Diebel, N. D., and Bogdanove, E. M. (1970). Analysis of initial and delayed effects of orchidectomy and ovariectomy on pituitary and serum LH levels in adult and immature rats. *Endocrinology* 86, 1102–1111.
544. Swerdloff, R. S., Walsh, P. C., Jacobs, H. S., and Odell, W. D. (1971). Serum LH and FSH during sexual maturation in the male rat. *Endocrinology* 88, 120–128.
545. Amatayakul, K., Ryan, R., Uozumi, T., and Albert, A. (1971). A reinvestigation of testicular-anterior pituitary relationships in the rat. I. Effects of castration and cryptorchidism. *Endocrinology* 88, 872–880.
546. Negro-Vilar, A., Ojeda, S. R., and McCann, S. M. (1973). Evidence for changes in sensitivity to testosterone negative feedback on gonadotropin release during sexual development in the male rat. *Endocrinology* 93, 729–735.
547. Smith, E. R., and Davidson, J. M. (1967). Differential responses to hypothalamic testosterone in relation to male puberty. *Am. J. Physiol.* 212, 1385–1390.
548. Bloch, G. J., Kragt, C. L., and Masken, J. F. (1971). Plasma LH levels in male rats of various ages: effects of castration and treatment with testosterone proprionate (TP). *Fed. Proc.* 30, 475a.
549. Fujii, T., Sakamoto, M., and Hayashi, K. (1980). Time course of the negative feedback effect of testosterone on serum gonadotropins in prepubertal male rats. *Endocrinol. Jpn.* 27, 753–758.
550. McCann, S. M., Ojeda, S. R., and Negro-Vilar, A. (1974). Sex steroid, pituitary and hypothalamic hormones during puberty in experimental animals. In *Control of the Onset of Puberty* (M. M. Grumbach, G. D. Grave, and F. E. Mayer, Eds.), pp. 1–31. John Wiley & Sons, New York.
551. Nazian, S. J., and Piacsek, B. E. (1977). Maturation of the reproductive system in male rats raised at a low ambient temperature. *Biol. Reprod.* 17, 668–675.
552. Nazian, S. J., and Piacsek, B. E. (1978). Sexual maturation of the cold-exposed male rat: alterations in secondary sexual organ sensitivity to testosterone. *Biol. Reprod.* 19, 256–260.
553. Sheth, A. R., Vanage, G. R., Vaze, A. Y., and Thakur, A. N. (1980). Negative feedback of secretion of follicle-stimulating hormone by the pituitary gland of developing male rats. *J. Endocrinol.* 87, 401–407.
554. Johnson, M. H., Thakur, A. N., and Sheth, A. R. (1981). Distribution of inhibin in testes of neonatal, pubertal, and adult rats. *Arch. Androl.* 7, 127–131.
555. de Kretser, D. M., Buzzard, J. J., Okuma, Y., O'Connor, A. E., Hayashi, T., Lin, S.-Y., Morrison, J. R., Loveland, K. L., and Hedger, M. P. (2004). The role of activin, follistatin and inhibin in testicular physiology. *Mol. Cell. Endocrinol.* 225, 57–64.
556. de Kretser, D. M., Loveland, K. L., Meehan, T., O'Bryan, M. K., Phillips, D. J., and Wreford, N. G. (2001). Inhibins, activins and follistatin: actions on the testis. *Mol. Cell. Endocrinol.* 180, 87–92.
557. Vitale, R., Fawcett, D. W., and Dym, M. (1973). The normal development of the blood-testis barrier and the effects of clomiphene and estrogen treatment. *Anat. Rec.* 176, 333–344.
558. Means, A. R., Fakunding, J. L., Huckins, C., Tindall, D. J., and Vitale, R. (1976). Follicle-stimulating hormone, the Sertoli cell, and spermatogenesis. *Rec. Prog. Horm. Res.* 32, 477–527.
559. Bercu, B. B., Jackson, I. M. D., Sawin, C. T., Safaii, P., and Rechlin, S. (1977). Permanent impairment of testicular development after transient immunological blockade of endogenous luteinizing hormone releasing hormone in the neonatal rat. *Endocrinology* 101, 1871–1879.
560. Bercu, B. B. (1982). Neutralization of gonadotropin-releasing hormone in neonatal rats with permanent impairment of the hypothalamic-pituitary-testicular axis. *Biol. Reprod.* 26, 612–619.
561. Vogel, D. L., Gunsalus, G. L., Bercu, B. B., Musto, N. A., and Bardin, C. W. (1983). Sertoli cell maturation is impaired by neonatal passive immunization with antiserum to luteinizing hormone-releasing hormone. *Endocrinology* 112, 1118–1121.
562. Nazian, S. J., and Mahesh, V. B. (1979). Sexual maturation of the male rat. Influence of androgens on the pituitary response to single or multiple injections of LHRH. *Neuroendocrinology* 29, 313–322.
563. Mahesh, V. B., and Nazian, S. J. (1979). Role of sex steroids in the initiation of puberty. *J. Steroid Biochem.* 11, 587–591.
564. Ramaley, J. A. (1978). The adrenal rhythm and puberty onset in the female rat. *Life Sci.* 23, 2079–2088.
565. Meijs-Roelofs, H. M. A., and Moll, J. (1978). Sexual maturation and the adrenal glands. *J. Reprod. Fertil.* 52, 413–418.
566. Kennedy, G. C., and Mitra, J. (1963). Body weight and food intake as initiating factors for puberty in the rat. *J. Physiol.* 166, 408–418.
567. Adashi, E. Y., Jones, P. B. C., and Hsueh, A. J. W. (1981). Synergistic effect of glucocorticoids on the stimulation of progesterone production by follicle-stimulating hormone in cultured rat granulosa cells. *Endocrinology* 109, 1888–1894.
568. Mandl, A. M. (1954). The sensitivity of adrenalectomized rats to gonadotrophins. *J. Endocrinol.* 11, 359–376.
569. Peppler, R. D., and Jacobs, J. J. (1976). The effect of adrenalectomy on ovulation and follicular development in the rat. *Biol. Reprod.* 15, 173–178.
570. Edgren, R. A., and Peterson, D. L. (1964). The influence of adrenalectomy and corticoid replacement on ovarian compensatory hypertrophy in rats. *Acta Endocrinol. (Copenh).* 47, 485–490.
571. Gorski, M. E., and Lawton, I. E. (1973). Adrenal involvement in determining the time of onset of puberty in the rat. *Endocrinology* 93, 1232–1234.
572. Aguado, L. I., and Ojeda, S. R. (1984). Effect of selective removal of the adrenal medulla on female sexual development. *Biol. Reprod.* 31, 605–618.
573. Christian, J. J. (1964). Actions of ACTH in intact and corticoid-maintained adrenalectomized female albino mice with emphasis on the reproductive tract. *Endocrinology* 75, 653–669.
574. MacFarland, L. A., and Mann, D. R. (1977). The inhibitory effects of ACTH and adrenalectomy on reproductive maturation in female rats. *Biol. Reprod.* 16, 306–314.
575. Kennedy, G. C. (1957). The development with age of hypothalamic restraint upon the appetite of the rat. *J. Endocrinol.* 16, 9–17.
576. Frisch, R. E., and Revelle, R. (1970). Height and weight at menarche and a hypothesis of critical body weight and adolescent events. *Science* 109, 397–399.
577. Frisch, R. E. (1980). Pubertal adipose tissue: is it necessary for normal sexual maturation? Evidence from the rat and human female. *Fed. Proc.* 39, 2395–2400.
578. Frisch, R. E., Hegsted, D. M., and Yoshinaga, K. (1975). Body weight and food intake at early estrus of rats on a high-fat diet. *Proc. Natl. Acad. Sci. U S A* 72, 4172–4176.
579. Frisch, R. E., Hegsted, D. M., and Yoshinaga, K. (1977). Carcass components at first estrus of rats on high-fat and low-fat diets: body water, protein and fat. *Proc. Natl. Acad. Sci. U S A* 74, 379–383.
580. Glass, A. R., Dahms, W. T., and Swerdloff, R. S. (1979). Body fat at puberty in rats: alteration by changes in diet. *Pediatr. Res.* 13, 7–9.

581. Wilen, R., and Naftolin, F. (1978). Pubertal food intake and body length, weight, and composition in the feed-restricted female rat: comparison with well fed animals. *Pediatr. Res.* 12, 263–267.
582. Steiner, R. A., Cameron, J. L., McNeill, T. H., Cliffton, D. K., and Bremmer, W. J. (1983). Metabolic signals for the onset of puberty. In *Neuroendocrine Aspects of Reproduction* (R. L. Norman, Ed.), pp. 183–228. Academic Press, New York.
583. Delemarre-van de Waal HA, van Coeverden, S. C., and Engelbregt, M. T. (2002). Factors affecting onset of puberty. *Horm. Res.* 57(Suppl 2), 15–18.
584. Smith White, S. S., and Ojeda, S. R. (1983). Essential fatty acid deficiency delays the onset of puberty by inhibition of hypothalamic and ovarian function. *Soc. Neurosci. Abstr.* 9, 1095.
585. Copeland, K. C., Kuehl, T. J., and Castracane, V. D. (1982). Pubertal endocrinology of the baboon: elevated somatomedin-C/insulin-like growth factor I at puberty. *J. Clin. Endocrinol. Metab.* 55, 1198–1201.
586. Luna, A. M., Wilson, D. M., Wibbelsman, C. J., Brown, R. C., Nagashima, R. J., Hintz, R. L., and Rosenfeld, R. G. (1983). Somatomedians in adolescence: a cross-sectional study of the effect of puberty on plasma insulin-like growth factor I and II levels. *J. Clin. Endocrinol. Metab.* 57, 268–271.
587. Hiney, J. K., Ojeda, S. R., and Dees, W. L. (1991). Insulin-like growth factor-I: a possible metabolic signal involved in the regulation of female puberty. *Neuroendocrinology* 54, 420–423.
588. Rotwein, P., Burgess, S. K., Milbrandt, J. D., and Krause, J. E. (1988). Differential expression of insulin-like growth factor genes in rat central nervous system. *Proc. Natl. Acad. Sci. U S A* 85, 265–269.
589. Matsuo, K., Niwa, M., Kurihara, M., Shigematsu, K., Yamashita, S., Ozaki, M., and Nagataki, S. (1989). Receptor autoradiographic analysis of insulin-like growth factor-I (IGF-I) binding sites in rat forebrain and pituitary gland. *Cell. Mol. Neurobiol.* 9, 357–367.
590. Bohannon, N. J., Corp, E. S., Wilcox, B. J., Figlewicz, D. P., Dorsa, D. M., and Baskin, D. G. (1988). Localization of binding sites for insulin-like growth factor-I (IGF-I) in the rat brain by quantitative autoradiography. *Brain Res.* 444, 205–213.
591. Werther, G. A., Hogg, A., Oldfield, B. J., McKinley, M. J., Figdor, R., and Mendelsohn, F. A. O. (1989). Localization and characterization of insulin-like growth factor-I receptors in rat brain and pituitary gland using *in vitro* autoradiography and computerized densitometry. A distinct distribution from insulin receptors. *J. Neuroendocrinol.* 1, 369–377.
592. Pazos, F., Sánchez-Franco, F., Balsa, J., López-Fernandez, J., Escalada, J., and Cacicedo, L. (1999). Regulation of gonadal and somatotropic axis by chronic intraventricular infusion of insulin-like growth factor 1 antibody at the initiation of puberty in male rats. *Neuroendocrinology* 69, 408–416.
593. Quanbeck, C., Sherwood, N. M., Millar, R. P., and Terasawa, E. (1997). Two populations of luteinizing hormone-releasing hormone neurons in the forebrain of the rhesus macaque during embryonic development. *J. Comp. Neurol.* 380, 293–309.
594. Ellinwood, W. E., and Resko, J. A. (1980). Sex differences in biologically active and immunoreactive gonadotropins in the fetal circulation of rhesus monkeys. *Endocrinology* 107, 902–907.
595. Terasawa, E., Quanbeck, C. D., Schultz, C. A., Burich, A. J., Luchansky, L. L., and Claude, P. (1993). A primary cell culture system of luteinizing hormone releasing hormone neurons derived from embryonic olfactory placode in the rhesus monkey. *Endocrinology* 133, 2379–2390.
596. Quinton, N. D., Smith, R. F., Clayton, P. E., Gill, M. S., Shalet, S., Justice, S. K., Simon, S. A., Walters, S., Postel-Vinay, M. C., Blakemore, A. I., and Ross, R. J. (1999). Leptin binding activity changes with age: the link between leptin and puberty. *J. Clin. Endocrinol. Metab.* 84, 2336–2341.
597. Ahima, R. S., Dushay, J., Flier, S. N., Prabakaran, D., and Flier, J. S. (1997). Leptin accelerates the onset of puberty in normal female mice. *J. Clin. Endocrinol. Metab.* 99, 391–395.
598. Chehab, F. F., Mounzih, K., Lu, R., and Lim, M. E. (1997). Early onset of reproductive function in normal female mice treated with leptin. *Science* 275, 88–90.
599. Cheung, C. C., Thornton, J. E., Kuijper, J. L., Weigle, D. S., Clifton, D. K., and Steiner, R. A. (1997). Leptin is a metabolic gate for the onset of puberty in the female rat. *Endocrinology* 138, 855–858.
600. Gruaz, N. M., Lalaoui, M., Pierroz, D. D., Englaro, P., Sizonenko, P. C., Blum, W. F., and Aubert, M. L. (1998). Chronic administration of leptin into the lateral ventricle induces sexual maturation in severely food-restricted female rats. *J. Neuroendocrinol.* 10, 627–633.
601. Seth, A., Stanley, S., Jethwa, P., Gardiner, J., Ghatei, M., and Bloom, S. (2004). Galanin-like peptide stimulates the release of gonadotropin-releasing hormone *in vitro* and may mediate the effects of leptin on the hypothalamo-pituitary-gonadal axis. *Endocrinology* 145, 743–750.
602. Reynoso, R., Ponzo, O. J., Szwarcfarb, B., Rondina, D., Carbone, S., Rimoldi, G., Scacchi, P., and Moguilevsky, J. A. (2003). Effect of leptin on hypothalamic release of GnRH and neurotransmitter amino acids during sexual maturation in female rats. *Exp. Clin. Endocrinol. Diabetes* 111, 274–277.
603. Watanobe, H. (2002). Leptin directly acts within the hypothalamus to stimulate gonadotropin-releasing hormone secretion *in vivo* in rats. *J. Physiol.* 545, 255–268.
604. Finn, P. D., Cunningham, M. J., Pau, K.-Y. F., Spies, H. G., Clifton, D. K., and Steiner, R. A. (1998). The stimulatory effect of leptin on the neuroendocrine reproductive axis of the monkey. *Endocrinology* 139, 4652–4662.
605. Magni, P., Vettor, R., Pagano, C., Calcagno, A., Beretta, E., Messi, E., Zanisi, M., Martini, L., and Motta, M. (1999). Expression of a leptin receptor in immortalized gonadotropin-releasing hormone-secreting neurons. *Endocrinology* 140, 1581–1585.
606. Morash, B., Li, A., Murphy, P. R., Wilkinson, M., and Ur, E. (1999). Leptin gene expression in the brain and pituitary gland. *Endocrinology* 140, 5995–5997.
607. Almog, B., Gold, R., Tajima, K., Dantes, A., Salim, K., Rubinstein, M., Barkan, D., Homburg, R., Lessing, J. B., Nevo, N., Gertler, A., and Amsterdam, A. (2001). Leptin attenuates follicular apoptosis and accelerates the onset of puberty in immature rats. *Mol. Cell Endocrinol.* 183, 179–191.
608. Urbanski, H. F. (2001). Leptin and puberty. *Trends Endocrinol. Metab.* 12, 428–429.
609. Mann, D. R., and Plant, T. M. (2002). Leptin and pubertal development. *Semin. Reprod. Med.* 20, 93–102.
610. Pierroz, D. D., Gruaz-Gumowski, N. M., Lalaoui, M., Blum, W. F., and Aubert, M. L. (2000). Leptin and NPY regulation of sexual maturation in the rat. In *The Onset of Puberty in Perspective* (J.-P. Bourguignon and T. M. Plant, Eds.), pp. 339–349. Elsevier Science B. V., Amsterdam.
611. Cheung, C. C., Thornton, J. E., Nurani, S. D., Clifton, D. K., and Steiner, R. A. (2001). A reassessment of leptin's role in triggering the onset of puberty in the rat and mouse. *Neuroendocrinology* 74, 12–21.
612. Clayton, P. E., Gill, M. S., Hall, C. M., Tillmann, V., Whatmore, A. J., and Price, D. A. (1997). Serum leptin through childhood and adolescence. *Clin. Endocrinol.* 46, 727–733.
613. Witchel, S. F., Arslanian, S., and Lee, P. A. (1999). Leptin concentrations in precocious puberty or untimely puberty with and without GnRH analogue therapy. *J. Pediatr. Endocrinol. Metab.* 12, 839–845.

614. Heger, S., Partsch, C. J., Peter, M., Blum, W. F., Kiess, W., and Sippell, W. G. (1999). Serum leptin levels in patients with progressive central precocious puberty. *Pediatr. Res.* 46, 71–75.
615. Plant, T. M., and Durrant, A. R. (1997). Circulating leptin does not appear to provide a signal for triggering the initiation of puberty in the male rhesus monkey (*Macaca mulatta*). *Endocrinology* 138, 4505–4508.
616. Urbanski, H. F., and Pau, K.-Y. F. (1998). A biphasic developmental pattern of circulating leptin in the male rhesus macaque (*Macaca mulatta*). *Endocrinology* 139, 2284–2286.
617. Mann, D. R., Akinbami, M. A., Gould, K. G., and Castracane, V. D. (2000). A longitudinal study of leptin during development in the male rhesus monkey: the effect of body composition and season on circulating leptin levels. *Biol. Reprod.* 62, 285–291.
618. Mann, D. R., and Plant, T. M. (2003). Leptin and pubertal development in higher primates. In *Leptin and Reproduction* (M. C. Henson and V. D. Castracane, Eds.), pp. 133–150. Kluwer Academic/Plenum Publishers, New York.
619. Wurtman, R. J., and Anton-Tay, F. (1969). The mammalian pineal gland as a neuroendocrine transducer. *Rec. Prog. Horm. Res.* 25, 493–522.
620. Reiter, R. J. (1980). The pineal and its hormones in the control of reproduction in mammals. *Endocr. Rev.* 1, 109–131.
621. Cardinali, D. P. (1981). Melatonin. A mammalian pineal hormone. *Endocr. Rev.* 2, 327–346.
622. Reiter, R. J. (1983). The pineal gland: An intermediary between the environment and the endocrine system. *Psychoneuroendocrinology* 8, 31–40.
623. Goldman, B. D., and Darrow, J. M. (1983). The pineal gland and mammalian photoperiodism. *Neuroendocrinology* 37, 386–396.
624. Darrow, J. M., and Goldman, B. D. (1986). Circadian regulation of pineal melatonin and reproduction in the Djungarian hamster. *J. Biol. Rhythms* 1, 39–54.
625. Reiter, R. J. (1991). Pineal melatonin: cell biology of its synthesis and of its physiological interactions. *Endocr. Rev.* 12, 151–180.
626. Lerner, A. B., Case, J. D., and Takahashi, Y. (1960). Isolation of melatonin and 5-methoxyindole-3-acetic acid from bovine pineal glands. *J. Biol. Chem.* 235, 1992–1997.
627. Gaston, S., and Menaker, M. (1967). Photoperiodic control of hamster's testis. *Science* 158, 925–928.
628. Reiter, R. J. (1973). Pineal control of a seasonal reproductive rhythm in male golden hamsters exposed to natural daylight and temperature. *Endocrinology* 92, 423–430.
629. Urbanski, H. F., Simpson, S. M., Ellis, D. E., and Follett, B. K. (1983). Secretion of follicle-stimulating hormone and luteinizing hormone in castrated golden hamsters during exposure to various photoperiods and to natural daylengths. *J. Endocrinol.* 99, 379–386.
630. Urbanski, H. F., Kim, S. O., and Connolly, M. L. (1990). The influence of photoperiod and 6-methoxybenzoxazolinone on the reproductive axis of inbred LSH/Ss Lak male hamsters. *J. Reprod. Fertil.* 90, 157–163.
631. Hoffman, R. A., and Reiter, R. J. (1965). Pineal gland: influence of gonads of male hamsters. *Science* 148, 1609–1610.
632. Goldman, B. D., Hall, V., Hollister, C., Roychoudhury, P., Tamarkin, L., and Westrom, W. (1979). Effects of melatonin on the reproductive system in intact and pinealectomized male hamsters maintained under various photoperiods. *Endocrinology* 104, 82–88.
633. Badawi, H., and Wilkinson, M. (1988). Lack of effect of melatonin on sexual maturation in female rats. *J. Reprod. Fertil.* 83, 273–278.
634. Wurtman, R. J., Axelrod, J., and Chu, E. W. (1963). Melatonin, a pineal substance: effect on the rat ovary. *Science* 141, 277–278.
635. Wurtman, R. J., Axelrod, J., and Phillips, L. S. (1963). Melatonin synthesis in the pineal gland: control by light. *Science* 142, 1071–1073.
636. Kincl, F. A., and Benagiano, G. (1967). The failure of the pineal gland removal in neonatal animals to influence reproduction. *Acta Endocrinol. (Copenh).* 54, 189–192.
637. Motta, M., Fraschini, F., and Martini, L. (1967). Endocrine effects of pineal gland and of melatonin. *Proc. Soc. Exp. Biol. Med.* 126, 431–435.
638. Dunaway, J. E. (1969). Alterations in the timing of PMS-induced ovulation following pinealectomy. *Neuroendocrinology* 5, 281–289.
639. Collu, R., Fraschini, F., and Martini, L. (1971). The effect of pineal methoxyindoles on rat vaginal opening time. *J. Endocrinol.* 50, 679–683.
640. Longnecker, D. E., and Gallo, D. G. (1971). The inhibition of PMSG-induced ovulation in immature rats. *Proc. Soc. Exp. Biol. Med.* 137, 623–625.
641. Relkin, R. (1971). Relative efficacy of pinealectomy, hypothalamic and amygdaloid lesions in advancing puberty. *Endocrinology* 88, 415–418.
642. Rivest, R. W., Lang, U., Aubert, M. L., and Sizonenko, P. C. (1985). Daily administration of melatonin delays rat vaginal opening and disrupts the first estrous cycles: evidence that these effects are synchronized by the onset of light. *Endocrinology* 116, 779–787.
643. Lang, U. (1986). Melatonin and puberty. *Pineal Res. Rev.* 4, 199–243.
644. Rivest, R. W., Aubert, M. L., Lang, U., and Sizonenko, P. C. (1986). Puberty in the rat: modulation by melatonin and light. *J. Neural Transm.* 21, 81–108.
645. Lang, U., Aubert, M. L., Conne, B. S., Bradtke, J. C., and Sizonenko, P. C. (1983). Influence of exogenous melatonin on melatonin secretion and the neuroendocrine reproductive axis of intact male rats during sexual maturation. *Endocrinology* 112, 1578–1584.
646. Lang, U., Rivest, R. W., Schlaepfer, L. V., Bradtke, J. C., Aubert, M. L., and Sizonenko, P. C. (1984). Diurnal rhythm of melatonin action on sexual maturation of male rats. *Neuroendocrinology* 38, 261–268.
647. Lang, U., Aubert, M. L., Rivest, R. W., Vinas-Bradtke, J. C., and Sizonenko, P. C. (1984). Daily afternoon administration of melatonin does not irreversibly inhibit sexual maturation in the male rat. *Endocrinology* 115, 2303–2310.
648. Aubert, M. L., Rivest, R. W., Lang, U., Winiger, B. P., and Sizonenko, P. C. (1988). Delayed sexual maturation induced by daily melatonin administration eliminates the LH response to naloxone despite normal responsiveness to GnRH in juvenile male rats. *Neuroendocrinology* 48, 72–80.
649. Fraschini, F., Mess, B. M., Piva, F., and Martini, L. (1968). Brain receptors senstive to indole compounds: function in control of luteinizing hormone secretion. *Science* 159, 1104–1105.
650. Reppert, S. M., Weaver, D. R., and Ebisawa, T. (1994). Cloning and characterization of a mammalian melatonin receptor that mediates reproductive and circadian responses. *Neuron* 13, 1177–1185.
651. Liu, C., Weaver, D. R., Jin, X., Shearman, L. P., Pieschi, R. L., Gribkoff, V. K., and Reppert, S. M. (1997). Molecular dissection of two distinct actions of melatonin on the suprachiasmatic circadian clock. *Neuron* 19, 91–102.
652. Williams, L. M., Martinoli, M. G., Titchener, L. T., and Pelletier, G. (1991). The ontogeny of central melatonin binding sites in the rat. *Endocrinology* 128, 2083–2090.
653. Martin, J. E., McKellar, S., and Klein, D. C. (1980). Melatonin inhibition of the *in vivo* pituitary response to luteinizing hormone-releasing hormone in the neonatal rat. *Neuroendocrinology* 31, 13–17.

654. Martin, J. E., and Sattler, C. (1982). Selectivity of melatonin pituitary inhibition for luteinizing hormone-releasing hormone. *Neuroendocrinology* 34, 112–116.
655. Ellis, L. C. (1969). The direct action of melatonin and serotonin on testicular androgen production *in vitro*. *J. Reprod. Fertil.* 18, 159.
656. Ellis, L. C. (1972). Inhibition of rat testicular androgen synthesis *in vitro* by melatonin and serotonin. *Endocrinology* 90, 17–28.
657. Rekwot, P. I., Ogwu, D., Oyedipe, E. O., and Sekoni, V. O. (2001). The role of pheromones and biostimulation in animal reproduction. *Anim. Reprod. Sci.* 65, 157–170.
658. Richards, D. B., and Stevens, A. (1974). Evidence for marking with urine by rats. *Behav. Biol.* 12, 517–523.
659. Leon, M. (1975). Dietary control of maternal pheromone in the lactating rat. *Physiol. Behav.* 14, 311–319.
660. Hofer, M. A., Shair, H., and Singh, P. (1976). Evidence that maternal ventral skin substances promote suckling in infant rats. *Physiol. Behav.* 17, 131–136.
661. Whitten, W. K. (1956). Modification of the oestrus cycle of the mouse by external stimuli associated with the male. *J. Endocrinol.* 13, 399–404.
662. Vandenbergh, J. G. (1983). Pheromonal regulation of puberty. In *Pheromones and Reproduction in Mammals* (J. G. Vandenbergh, Ed.), pp. 95–112. Academic Press, London.
663. Vandenbergh, J. G. (1983). Social factors controlling puberty in the female mouse. In *Hormones and Behavior in Higher Vertebrates* (J. Balthazart, E. Prove, and R. Gilles, Eds.), pp. 342–349. Springer, Berlin.
664. Dluzen, D., Guan, X., and Vandenbergh, J. G. (1992). Puberty acceleration in female mice induced with a partially purified male urine extract: effects on catecholamine release from the olfactory bulbs and hypothalamus. *Brain Res.* 585, 367–371.
665. Vandenbergh, J. G. (1976). Acceleration of sexual maturation in female rats by male stimulation. *J. Reprod. Fertil.* 46, 451–453.
666. Slob, A. K., van Es, G., and van der Werff ten Bosch, J. J. (1985). Social factors and puberty in female rats. *J. Endocrinol.* 104, 309–313.
667. Stoker, T. E., Parks, L. G., Gray, L. E., and Cooper, R. L. (2000). Endocrine-disrupting chemicals: prepubertal exposures and effects on sexual maturation and thyroid function in the male rat. A focus on the EDSTAC recommendations. Endocrine Disrupter Screening and Testing Advisory Committee. *Crit Rev. Toxicol.* 30, 197–252.
668. Tena-Sempere, M., Navarro, J., Pinilla, L., Gonzalez, L. C., Huhtaniemi, I., and Aguilar, E. (2000). Neonatal exposure to estrogen differentially alters estrogen receptor alpha and beta mRNA expression in rat testis during postnatal development. *J. Endocrinol.* 165, 345–357.
669. Sharpe, R. M., Rivas, A., Walker, M., McKinnell, C., and Fisher, J. S. (2003). Effect of neonatal treatment of rats with potent or weak (environmental) oestrogens, or with a GnRH antagonist, on Leydig cell development and function through puberty into adulthood. *Int. J. Androl* 26, 26–36.
670. Williams, K., McKinnell, C., Saunders, P. T., Walker, M., Fisher, J. S., Turner, K. J., Atanassova, N., and Sharpe, M. (2001). Neonatal exposure to potent and environmental oestrogens and abnormalities of the male reproductive system in the rat: evidence for importance of the androgen-oestrogen balance and assessment of the relevance to man. *Hum. Reprod. Update* 7, 236–247.
671. Atanassova, N., McKinnell, C., Turner, K. J., Walker, M., Fisher, J. S., Morley, M., Millar, M. R., Groome, N. P., and Sharpe, R. M. (2000). Comparative effects of neonatal exposure of male rats to potent and weak (environmental) estrogens on spermatogenesis at puberty and the relationship to adult testis size and fertility: evidence for stimulatory effects of low estrogen levels. *Endocrinology* 141, 3898–3907.
672. Kuwada, M., Kawashima, R., Nakamura, K., Kojima, H., Hasumi, H., Maki, J., and Sugano, S. (2002). Neonatal exposure to endocrine disruptors suppresses juvenile testis weight and steroidogenesis but spermatogenesis is considerably restored during puberty. *Biochem. Biophys. Res. Commun.* 295, 193–197.
673. Masutomi, N., Shibutani, M., Takagi, H., Uneyama, C., Takahashi, N., and Hirose, M. (2003). Impact of dietary exposure to methoxychlor, genistein, or diisononyl phthalate during the perinatal period on the development of the rat endocrine/reproductive systems in later life. *Toxicology* 192, 149–170.
674. Nagao, T., Yoshimura, S., Saito, Y., Nakagomi, M., Usumi, K., and Ono, H. (2001). Reproductive effects in male and female rats of neonatal exposure to genistein. *Reprod. Toxicol.* 15, 399–411.
675. Tan, B. L., Kassim, N. M., and Mohd, M. A. (2003). Assessment of pubertal development in juvenile male rats after sub-acute exposure to bisphenol A and nonylphenol. *Toxicol. Lett.* 143, 261–270.
676. Kwon, S., Stedman, D. B., Elswick, B. A., Cattley, R. C., and Welsch, F. (2000). Pubertal development and reproductive functions of Crl:CD BR Sprague-Dawley rats exposed to bisphenol A during prenatal and postnatal development. *Toxicol. Sci.* 55, 399–406.
677. Stoker, T. E., Laws, S. C., Guidici, D. L., and Cooper, R. L. (2000). The effect of atrazine on puberty in male wistar rats: an evaluation in the protocol for the assessment of pubertal development and thyroid function. *Toxicol. Sci.* 58, 50–59.
678. Stoker, T. E., Guidici, D. L., Laws, S. C., and Cooper, R. L. (2002). The effects of atrazine metabolites on puberty and thyroid function in the male Wistar rat. *Toxicol. Sci.* 67, 198–206.
679. Ramirez, V. D., and McCann, S. M. (1963). Comparison of the regulation of luteinizing hormone (LH) secretion in immature and adult rats. *Endocrinology* 72, 452–464.
680. Hohlweg, W., and Dohrn, M. (1932). Über die beziehungen zwischen hypophysenvorderlappen und keimdrüsen. *Klin. Wochenschr.* 11, 233–235.
681. Steele, R. E., and Weisz, J. (1974). Changes in sensitivity of the estradiol-LH feedback system with puberty in the female rat. *Endocrinology* 95, 513–520.
682. Eldridge, J. C., McPherson, J. C. I., and Mahesh, V. D. (1974). Maturation of the negative feedback control of gonadotropin secretion in the female rat. *Endocrinology* 94, 1536–1540.
683. Döcke, F., Rohde, W., Bao, T. N., Braun, W., and Dörner, G. (1978). Studies on the time and localization of the pubertal desensitization to oestrogen in female rats. *Endokrinologie* 71, 257–265.
684. Andrews, W. W., Advis, J. P., and Ojeda, S. R. (1981). The maturation of estradiol negative feedback in female rats: Evidence that the resetting of the hypothalamic "gonadostat" does not precede the first preovulatory surge of gonadotropins. *Endocrinology* 109, 2022–2031.
685. Kawagoe, S., and Hiroi, M. (1983). Maturation of negative and positive estrogen feedback in the prepubertal female rat. *Endocrinol. Jpn.* 30, 435–441.
686. Maruca, J., Kulin, H. E., and Santner, S. J. (1983). Perturbations of negative feedback sensitivity in agonadal patients undergoing estrogen replacement therapy. *J. Clin. Endocrinol. Metab.* 56, 53–59.
687. Kelch, R. P., Marshall, J. C., Saunder, S.-E., Hopwood, N. J., and Bremner, W. J. (1983). Gonadotropin regulation during human puberty. In *Neuroendocrine Aspects of Reproduction* (R. Norman, Ed.), pp. 229–256. Academic Press, New York.

688. Terasawa, E. (1999). Hypothalamic control of the onset of puberty. *Curr. Opin. Endocrinol. Diabetes* 6, 44–49.
689. Mitsushima, D., Hei, D. L., and Terasawa, E. (1994). Gamma-aminobutyric acid is an inhibitory neurotransmitter restricting the release of luteinizing hormone-releasing hormone before the onset of puberty. *Proc. Natl. Acad. Sci. U S A* 91, 395–399.
690. Keen, K. L., Burich, A. J., Mitsushima, D., Kasuya, E., and Terasawa, E. (1999). Effects of pulsatile infusion of the $GABA_A$ receptor blocker bicuculline on the onset of puberty in female rhesus monkeys. *Endocrinology* 140, 5257–5266.
691. Parhar, I. S., Ogawa, S., and Sakuma, Y. (2004). Laser captured single digoxigenin-labeled neurons of gonadotropin-releasing hormone types reveal a novel G protein-coupled receptor (GPR54) during maturation in cichlid fish. *Endocrinology* 145, 3613–3618.
692. Young-Lai, E. V., Holmes, M. J., and Ruf, K. B. (1976). Changes in the concentration of LH, FSH and estrogen in the immature female rat during precocious sexual maturation induced by electrochemical stimulation of the brain. *Horm. Res.* 7, 34–42.
693. Moll, J., Meijs-Roelofs, H. M. A., Kramer, P., and Dullart, J. (1976). Effects of electrical stimulation and of direct current and high frequency lesions of the hypothalamus on gonadotropin release and puberty in female rats. *Ann. Biol. Anim. Biochem. Biophys.* 16, 433–441.
694. Meijs-Roelofs, H. M. A. (1972). Effect of electrical stimulation of the hypothalamus on gonadotrophin release and the onset of puberty. *J. Endocrinol.* 54, 277–284.
695. Bogdanove, E. M., and Schoen, H. C. (1959). Precocious sexual development in female rats with hypothalamic lesions. *Proc. Soc. Exp. Biol. Med.* 100, 664–669.
696. Gellert, R. J., and Ganong, W. F. (1960). Precocious puberty in rats with hypothalamic lesions. *Acta Endocrinol. (Copenh).* 33, 569–576.
697. Schiavi, R. C. (1964). Effects of anterior and posterior hypothalamic lesions on precocious sexual maturation. *Am. J. Physiol.* 206, 805–810.
698. Advis, J. P., and Ramirez, V. D. (1977). Plasma levels of LH and FSH in female rats with precocious puberty induced by hypothalamic lesions. *Biol. Reprod.* 17, 313–320.
699. Velasco, M. E., and Taleisnik, S. (1969). Release of gonadotropins induced by amygdaloid stimulation in the rat. *Endocrinology* 84, 132–139.
700. Elwers, M., and Critchlow, V. (1961). Precocious ovarian stimulation following interruption of stria terminalis. *Am. J. Physiol.* 201, 281–284.
701. Elwers, M., and Critchlow, V. (1966). Delayed puberty following electrical stimulation of amygdala in female rats. *Am. J. Physiol.* 211, 1103–1107.
702. Döcke, F. (1977). Neural control factors in puberty. In *Endocrinology* (V. H. T. James, Ed.), Vol. 1, pp. 287–290. Excerpta Medica, Amsterdam.
703. Sherwood, N. M., and Timiras, P. S. (1974). Comparison of direct-current and radio-frequency-current lesions in the rostral hypothalamus with respect to sexual maturation in the female rat. *Endocrinology* 94, 1275–1285.
704. Ruf, K. B., and Sharpe, M. S. (1979). The relevance of animal models for the understanding of the process of puberty in humans. In *Clinical Neuroendocrinology: A Pathophysiological Approach* (G. Tolis, F. Labrie, J. B. Martin, and F. Naftolin, Eds.), pp. 239–245. Raven Press, New York.
705. Junier, M.-P., Ma, Y. J., Costa, M. E., Hoffman, G., Hill, D. F., and Ojeda, S. R. (1991). Transforming growth factor alpha contributes to the mechanism by which hypothalamic injury induces precocious puberty. *Proc. Natl. Acad. Sci. U S A* 88, 9743–9747.
706. Junier, M.-P., Hill, D. F., Costa, M. E., Felder, S., and Ojeda, S. R. (1993). Hypothalamic lesions that induce female precocious puberty activate glial expression of the epidermal growth factor receptor gene: Differential regulation of alternatively spliced transcripts. *J. Neurosci.* 13, 703–713.
707. Junier, M.-P., Wolff, A., Hoffman, G. E., Ma, Y. J., and Ojeda, S. R. (1992). Effect of hypothalamic lesions that induce precocious puberty on the morphological and functional maturation of the luteinizing hormone-releasing hormone neuronal system. *Endocrinology* 131, 787–798.
708. Witkin, J. W., and Demasio, K. A. (1989). Density of synaptic input is identical to smooth-contoured and thorny GnRH neurons in the male rat. *Soc. Neurosci. Abstr.* 15, 1082.
709. van den Pol, A. N., and Trombley, P. Q. (1993). Glutamate neurons in hypothalamus regulate excitatory transmission. *J. Neurosci.* 13, 2829–2836.
710. Claypool, L. E., Kasuya, E., Saitoh, Y., Marzban, F., and Terasawa, E. (2000). *N*-methyl D,L-aspartate induces the release of luteinizing hormone-releasing hormone in the prepubertal and pubertal female rhesus monkey as measured by *in vivo* push-pull perfusion in the stalk-median eminence. *Endocrinology* 141, 219–228.
711. Bourguignon, J.-P., Gérard, A., Alvarez-Gonzalez, M.-L., Fawe, L., and Franchimont, P. (1992). Gonadal-independent developmental changes in activation of N-methyl-*D*-aspartate receptors involved in gonadotropin-releasing hormone secretion. *Neuroendocrinology* 55, 634–641.
712. Goldsmith, P. C., Thind, K. K., Perera, A. D., and Plant, T. M. (1994). Glutamate-immunoreactive neurons and their gonadotropin-releasing hormone-neuronal interactions in the monkey hypothalamus. *Endocrinology* 134, 858–868.
713. Gore, A. C., Wu, T. J., Rosenberg, J. J., and Roberts, J. L. (1996). Gonadotropin-releasing hormone and NMDA receptor gene expression and colocalization change during puberty in female rats. *J. Neurosci.* 16, 5281–5289.
714. van den Pol, A. N., Wuarin, J.-P., and Dudek, F. E. (1990). Glutamate, the dominant excitatory transmitter in neuroendocrine regulation. *Science* 250, 1276–1278.
715. Price, M. T., Olney, J. W., and Cicero, T. J. (1978). Acute elevations of serum luteinizing hormone induced by kainic acid, *N*-methyl aspartic acid or homocystic acid. *Neuroendocrinology* 26, 352–358.
716. Donoso, A. O., López, F. J., and Negro-Vilar, A. (1990). Glutamate receptors of the non-*N*-methyl-D-aspartic acid type mediate the increase in luteinizing hormone releasing hormone release by excitatory amino acid *in vitro*. *Endocrinology* 126, 414–420.
717. Eyigor, O., and Jennes, L. (1997). Expression of glutamate receptor subunit mRNAs in gonadotropin-releasing hormone neurons during the sexual maturation of the female rat. *Neuroendocrinology* 66, 122–129.
718. Eyigor, O., and Jennes, L. (2000). Kainate receptor subunit-positive gonadotropin-releasing hormone neurons express c-Fos during the steroid-induced luteinizing hormone surge in the female rat. *Endocrinology* 141, 779–786.
719. Seeburg, P. H. (1993). The molecular biology of mammalian glutamate receptor channels. *Trends Neurosci.* 16, 359–365.
720. Bourguignon, J.-P., Gerard, A., Alvarez Gonzalez, M.-L., Purnelle, G., and Franchimont, P. (1995). Endogenous glutamate involvement in pulsatile secretion of gonadotropin-releasing hormone: evidence from effect of glutamine and developmental changes. *Endocrinology* 136, 911–916.
721. Brann, D. W. (1995). Glutamate: a major excitatory transmitter in neuroendocrine regulation. *Neuroendocrinology* 61, 213–225.
722. Thind, K. K., and Goldsmith, P. C. (1995). Glutamate and gabaergic neurointeractions in the monkey hypothalamus: a quantitative immunomorphological study. *Neuroendocrinology* 61, 471–485.

723. van den Pol, A. N., Gao, X.-B., Patrylo, P. R., Ghosh, P. K., and Obrietan, K. (1998). Glutamate inhibits GABA excitatory activity in developing neurons. *J. Neurosci.* 18, 10749–10761.
724. Min, M.-Y., Melyan, Z., and Kullmann, D. M. (1999). Synaptically released glutamate reduces γ-aminobutyric acid (GABA)ergic inhibition in the hippocampus via kainate receptors. *Proc. Natl. Acad. Sci. U S A* 96, 9932–9937.
725. Rodriguez-Moreno, A., Herreras, O., and Lerma, J. (1997). Kainate receptors presynaptically downregulate GABAergic inhibition in the rat hippocampus. *Neuron* 19, 893–901.
726. Rodriguez-Moreno, A., and Lerma, J. (1998). Kainate receptor modulation of GABA release involves a metabotropic function. *Neuron* 20, 1211–1218.
727. Satake, S., Saitow, F., Yamada, J., and Konishi, S. (2000). Synaptic activation of AMPA receptors inhibits GABA release from cerebellar interneurons. *Nat. Neurosci.* 3, 551–558.
728. Clarke, V. R. J., Ballyk, B. A., Hoo, K. H., Mandelzys, A., Pellizzari, A., Bath, C. P., Thomas, J., sharpe, E. F., Davies, C. H., Ornstein, P. L., Schoepp, D. D., Kamboj, R. K., Collingridge, G. L., Lodge, D., and Bleakman, D. (1997). A hippocampal GluR5 kainate receptor regulating inhibitory synaptic transmission. *Nature* 389, 599–603.
729. Memo, M., Bovolin, P., Costa, E., and Grayson, D. R. (1991). Regulation of γ-aminobutyric acid$_A$ receptor subunit expression by activation of *N*-methyl-D-aspartate-selective glutamate receptors. *Mol. Pharmacol.* 39, 599–603.
730. Terasawa, E., Luchansky, L. L., Kasuya, E., and Nyberg, C. L. (1999). An increase in glutamate release follows a decrease in gamma aminobutyric acid and the pubertal increase in luteinizing hormone releasing hormone in female rhesus monkeys. *J. Neuroendocrinol.* 11, 275–282.
731. Ben-Ari, Y., Khazipov, R., Leinekugel, X., Caillard, O., and Gaiarsa, J.-L. (1997). GABA$_A$, NMDA and AMPA receptors: a developmentally regulated "ménage à trois." *Trends Neurosci.* 20, 523–529.
732. Cherubini, E., Gaiarsa, J. L., and Ben-Ari, Y. (1991). GABA: an excitatory transmitter in early postnatal life. *Trends Neurosci.* 14, 515–519.
733. Obrietan, K., and van den Pol, A. N. (1995). GABA neurotransmission in the hypothalamus: developmental reversal from Ca^{2+} elevating to depressing. *J. Neurosci.* 15, 5065–5077.
734. Chen, G., Trombley, P. Q., and van den Pol, A. N. (1995). GABA receptors precede glutamate receptors in hypothalamic development: differential regulation by astrocytes. *J. Neurophysiol.* 74, 1473–1483.
735. Insel, T. R., Miller, L. P., and Gelhard, R. E. (1990). The ontogeny of excitatory amino acid receptors in rat forebrain. I. N-methyl-D-aspartate and quisqualate receptors. *Neuroscience* 35, 31–43.
736. DeFazio, R. A., Heger, S., Ojeda, S. R., and Moenter, S. M. (2002). Activation of A-type γ-aminobutyric acid receptors excites gonadotropin-releasing hormone neurons. *Mol. Endocrinol.* 16, 2872–2891.
737. Han, S. K., Abraham, I. M., and Herbison, A. E. (2002). Effect of GABA on GnRH neurons switches from depolzarization to hyperpolarizational puberty in the female mouse. *Endocrinology* 143, 1459–1466.
738. Han, S. K., Todman, M. G., and Herbison, A. E. (2004). Endogenous GABA release inhibits the firing of adult gonadotropin-releasing neurons. *Endocrinology* 145, 495–499.
739. Leupen, S. M., Tobet, S. A., Crowley, W. F., Jr., and Kaila, K. (2003). Heterogeneous expression of the potassium-chloride cotransporter KCC2 in gonadotropin-releasing hormone neurons of the adult mouse. *Endocrinology* 144, 3031–3036.
740. Gallo, V., and Ghiani, C. A. (2000). Glutamate receptors in glia: new cells, new inputs and new functions. *Trends Pharmacol. Sci.* 21, 252–258.
741. Murphy, S., and Pearce, B. (1987). Functional receptors for neurotransmitters on astroglial cells. *Neuroscience* 22, 381–394.
742. Bevan, S. (1990). Ion channels and neurotransmitter receptors in glia. *Semin. Neurosci.* 2, 467–481.
743. Verkhratsky, A., and Steinhauser, C. (2000). Ion channels in glial cells. *Brain Res. Rev.* 32, 380–412.
744. Martin, D. L. (1992). Synthesis and release of neuroactive substances by glial cells. *Glia* 5, 81–94.
745. Barres, B. A. (1991). New roles for glia. *J. Neurosci.* 11, 3685–3694.
746. Garcia-Segura, L. M., Chowen, J. A., and Naftolin, F. (1996). Endocrine glia: roles of glial cells in the brain actions of steroid and thyroid hormones and in the regulation of hormone secretion. *Front. Neuroendocrinol.* 17, 180–211.
747. Fields, R. D., and Stevens-Graham, B. (2002). New insights into neuron-glia communication. *Science* 298, 556–562.
748. Dziedzic, B., Prevot, V., Lomniczi, A., Jung, H., Cornea, A., and Ojeda, S. R. (2003). Neuron-to-glia signaling mediated by excitatory amino acid receptors regulates erbB receptor function in astroglial cells of the neuroendocrine brain. *J. Neurosci.* 23, 915–926.
749. Bezzi, P., Carmignoto, G., Pasti, L., Vesce, S., Rossi, D., Rizzini, B. L., Pozzan, T., and Volterra, A. (1998). Prostaglandins stimulate calcium-dependent glutamate release in astrocytes. *Nature* 391, 281–285.
750. Pfrieger, F. W., and Barres, B. A. (1997). Synaptic efficacy enhanced by glial cells in vitro. *Science* 277, 1684–1687.
751. Yuan, L.-L., and Ganetsky, B. (1999). A glial-neuronal signaling pathway revealed by mutations in a neurexin-related protein. *Science* 283, 1343–1345.
752. Ozaki, M., Sasner, M., Yano, R., Lu, H. S., and Buonanno, A. (1997). Neuregulin-β induces expression of an NMDA-receptor subunit. *Nature* 390, 691–694.
753. Huang, Y. Z., Won, S., Ali, D. W., Wang, Q., Tanowitz, M., Du, Q. S., Pelkey, K. A., Yang, D. J., Xiong, W. C., Salter, M. W., and Mei, L. (2000). Regulation of neuregulin signaling by PSD-95 interacting with ErbB4 at CNS synapses. *Neuron* 26, 443–455.
754. Kang, J., Jiang, L., Goldman, S. A., and Nedergaard, M. (1998). Astrocyte-mediated potentiation of inhibitory synaptic transmission. *Nat. Neurosci.* 1, 683–692.
755. Fraser, D. D., Mudrick-Donnon, L. A., and MacVicer, B. A. (1994). Astrocytic GABA receptors. *Glia* 11, 83–93.
756. Shinoda, H., Marini, A. M., Cosi, C., and Schwartz, J. P. (1989). Brain region and gene specificity of neuropeptide gene expression in cultured astrocytes. *Science* 245, 415–417.
757. Ernsberger, P., Iacovitta, L., and Reis, D. J. (1990). Astrocytes cultured from specific brain regions differ in their expression of adrenergic binding sites. *Brain Res.* 517, 202–208.
758. Hofer, A., Sáez, J. C., Chang, C. C., Trosko, J. E., Spray, D. C., and Dermietzel, R. (1996). C-erbB2/neu transfection induces gap junctional communication incompetence in glial cells. *J. Neurosci.* 16, 4311–4321.
759. Shao, Y., and McCarthy, K. D. (1994). Plasticity of astrocytes. *Glia* 11, 147–155.
760. Cornell-Bell, A. H., Thomas, P. G., and Smith, S. J. (1990). The excitatory neurotransmitter glutamate causes filopodia formation in cultured hippocampal astrocytes. *Glia* 3, 322–334.
761. Parpura, V., Basarsky, T. A., Liu, F., Jeftinija, K., Jeftinija, S., and Haydon, P. G. (1994). Glutamate-mediated astrocyte-neuron signalling. *Nature* 369, 744–747.
762. Ojeda, S. R., Hill, J., Hill, D. F., Costa, M. E., Tapia, V., Cornea, A., and Ma, Y. J. (1999). The Oct-2 POU-domain gene in the neuroendocrine brain: a transcriptional regulator of mammalian puberty. *Endocrinology* 140, 3774–3789.
763. Treacy, M. N., and Rosenfeld, M. G. (1992). Expression of a family of POU-domain protein regulatory genes during

development of the central nervous system. *Annu. Rev. Neurosci.* 15, 139–165.

764. Harvey, R. P. (1996). *NK-2* homeobox genes and heart development. *Dev. Biol.* 178, 203–216.
765. Lazzaro, D., Price, M., De Felice, M., and Di Lauro, R. (1991). The transcription factor TTF-1 is expressed at the onset of thyroid and lung morphogenesis and in restricted regions of the foetal brain. *Development* 113, 1093–1104.
766. Alvarez-Bolado, G., Rosenfeld, M. G., and Swanson, L. W. (1995). Model of forebrain regionalization based on spatiotemporal patterns of POU-III homeobox gene expression, birthdates, and morphological features. *J. Comp. Neurol.* 355, 237–295.
767. Hatzopoulos, A. K., Stoykova, A. S., Erselius, J. R., Goulding, M., Neuman, T., and Gruss, P. (1990). Structure and expression of the mouse Oct2a and Oct2b, two differentially spliced products of the same gene. *Development* 109, 349–362.
768. Schreiber, E., Harshman, K., Kemler, I., Malipiero, U., Schaffner, W., and Fontana, A. (1990). Astrocytes and glioblastoma cells express novel octamer-DNA binding proteins distinct from the ubiquitous Oct-1 and B cell type Oct-2 proteins. *Nucleic Acids Res.* 18, 5495–5503.
769. Ryan, A. K., and Rosenfeld, M. G. (1997). POU domain family values: flexibility, partnerships, and developmental codes. *Genes Dev.* 11, 1207–1225.
770. Clark, M. E., and Mellon, P. L. (1995). The POU homeodomain transcription factor Oct-1 is essential for activity of the gonadotropin-releasing hormone neuron-specific enhancer. *Mol. Cell. Biol.* 15, 6169–6177.
771. Wierman, M. E., Xiong, X., Kepa, J. K., Spaulding, A. J., Jacobsen, B. M., Fang, Z., Nilaver, G., and Ojeda, S. R. (1997). Repression of gonadotropin-releasing hormone (GnRH) promoter activity by the POU homeodomain transcription factor, SCIP/Oct-6/Tst-1: a regulatory mechanism of phenotype expression? *Mol. Cell. Biol.* 17, 1652–1665.
772. Price, M., Lazzaro, D., Pohl, T., Mattei, M.-G., Rüther, U., Olivo, J.-C., Duboule, D., and Di Lauro, R. (1992). Regional expression of the homeobox gene *Nkx-2.2* in the developing mammalian forebrain. *Neuron* 8, 241–255.
773. Kimura, S., Hara, Y., Pineau, T., Fernandez-Salguero, P., Fox, C. H., Ward, J. M., and Gonzalez, F. J. (1996). The *T/ebp* null mouse: thyroid-specific enhancer-binding protein is essential for the organogenesis of the thyroid, lung, ventral forebrain, and pituitary. *Genes Dev.* 10, 60–69.
774. Bhanot, R., and Wilkinson, M. (1983). Opiatergic control of gonadotropin secretion during puberty in the rat. *Endocrinology* 113, 596–602.
775. Brann, D. W., Zamorano, P. L., Putnam-Roberts, C. D., and Mahesh, V. B. (1992). Gamma-aminobutyric acid-opioid interactions in the regulation of gonadotropin secretion in the immature female rat. *Neuroendocrinology* 56, 445–452.
776. Lauber, A. H., Romano, G. J., Mobbs, C. V., Howells, R. D., and Pfaff, D. W. (1990). Estradiol induction of proenkephalin messenger RNA in hypothalamus: dose-response and relation to reproductive behavior in the female rat. *Mol. Brain Res.* 8, 47–54.
777. Leranth, C., Segura, L. M. G., Palkovits, M., MacLusky, N. J., Shanabrough, M., and Naftolin, F. (1985). The LH-RH-containing neuronal network in the preoptic area of the rat: demonstration of LH-RH-containing nerve terminals in synaptic contact with LH-RH neurons. *Brain Res.* 345, 332–336.
778. Pelletier, G. (1987). Demonstration of contacts between neurons staining for LHRH in the preoptic area of the rat brain. *Neuroendocrinology* 46, 457–459.
779. Turek, F. W., and Van Cauter, E. (1994). Rhythms in reproduction. In *The Physiology of Reproduction* (E. Knobil and J. D. Neill, Eds.), pp. 487–540. Raven Press, New York.
780. Van der Beek, E. M. (1996). Circadian control of reproduction in the female rat. *Prog. Brain Res.* 111, 295–320.
781. Reppert, S. M. (1998). A clockwork explosion! *Neuron* 1, 4.
782. Hall, J. C. (1995). Tripping along the trail to the molecular mechanisms of biological clocks. *Trends Neurosci.* 18, 230–240.
783. Watts, A. G. (1991). The efferent projections of the suprachiasmatic nucleus: anatomical insights into the control of circadian rhythms. In *Suprachiasmatic Nucleus: The Mind's Clock* (D. C. Klein, R. Y. Moore, and S. M. Reppert, Eds.), pp. 77–106. Oxford University Press, New York.
784. Reppert, S. M., and Uhl, G. R. (1987). Vasopressin messenger ribonucleic acid in supraoptic and suprachiasmatic nuclei: appearance and circadian regulation during development. *Endocrinology* 120, 2483–2487.
785. Funabashi, T., Shinohara, K., Mitsushima, D., and Kimura, F. (2000). Gonadotropin-releasing hormone exhibits circadian rhythm in phase with arginine-vasopressin in co-cultures of the female rat preoptic area and suprachiasmatic nucleus. *J. Neuroendocrinol.* 12, 521–528.
786. van den Pol, A. N., Finkbeiner, S. M., and Cornell-Bell, A. H. (1992). Calcium excitability and oscillations in suprachiasmatic nucleus neurons and glia in vitro. *J. Neurosci.* 12, 2648–2664.
787. Cagampang, F. R. A., Rattray, M., Powell, J. F., Campbell, I. C., and Coen, C. W. (1996). Circadian changes of glutamate decarboxylase 65 and 67 mRNA in the rat suprachiasmatic nuclei. *NeuroReport* 7, 1925–1928.
788. Shibata, S., Liou, S. Y., and Ueki, S. (1986). Influence of excitatory amino acid receptor antagonists and of baclofen on synaptic transmission in the optic nerve to the suprachiasmatic nucleus in slices of rat hypothalamus. *Neuropharmacology* 25, 403–409.
789. van der Horst, G. T. J., Muijtjens, M., Kobayashi, K., Takano, R., Kanno, S.-I., Takao, M., de Wit, J., Verkerk, A., Eker, A. P. M., van Leenen, D., Buijs, R., Bootsma, D., Hoeijmakers, J. H. J., and Yasui, A. (1999). Mammalian Cry1 and Cry2 are essential for maintenance of circadian rhythms. *Nature* 398, 627–630.
790. Toran-Allerand, C. D., Ellis, L., and Pfenninger, K. H. (1988). Estrogen and insulin synergism in neurite growth enhancement *in vitro*: mediation of steroid effects by interaction with growth factors? *Dev. Brain Res.* 41, 87–100.
791. Clough, R. W., and Rodriguez-Sierra, J. F. (1983). Synaptic changes in the hypothalamus of the prepubertal female rat administered estrogen. *Am. J. Anat.* 167, 205–214.
792. Lustig, R. H., Sudol, M., Pfaff, D. W., and Federoff, H. J. (1991). Estrogenic regulation and sex dimorphism of growth-associated protein 43 kDA (GAP-43) messenger RNA in the rat. *Mol. Brain Res.* 11, 125–132.
793. Rogers, L. C., Junier, M.-P., Farmer, S. R., and Ojeda, S. R. (1991). A sex-related difference in the developmental expression of class II b-tubulin messenger RNA in rat hypothalamus. *Mol. Cell. Neurosci.* 2, 130–138.
794. Ferreira, A., and Caceres, A. (1991). Estrogen-enhanced neurite growth: evidence for a selective induction of Tau and stable microtubules. *J. Neurosci.* 11, 392–400.
795. Urbanski, H. F., Doan, A., Pierce, M., Fahrenbach, W. H., and Collins, P. M. (1992). Maturation of the hypothalamo-pituitary-gonadal axis of male Syrian hamsters. *Biol. Reprod.* 46, 991–996.
796. Frojdman, K., Harley, V. R., and Pelliniemi, L. J. (2000). Sox9 protein in rat sertoli cells is age and stage dependent. *Histochem. Cell Biol.* 113, 31–36.
797. Beau, C., Rauch, M., Joulin, V., Jegou, B., and Guerrier, D. (2000). GATA-1 is a potential repressor of anti-Müllerian hormone expression during the establishment of puberty in the mouse. *Mol. Reprod. Dev.* 56, 124–138.

798. Lynch, H. J., Rivest, R. W., and Wurtman, R. J. (1980). Artificial induction of melatonin rhythms by programmed microinfusion. *Neuroendocrinology* 31, 106–111.

799. Ojeda, S. R., and Ma, Y. J. (1999). Glial-neuronal interactions in the neuroendocrine control of mammalian puberty: facilitatory effects of gonadal steroids. *J. Neurobiol.* 40, 528–540.

800. Ojeda, S. R., and Bilger, M. (1999). Neuroendocrine regulation of puberty. In *Neuroendocrinology in Physiology and Medicine* (P. M. Conn and M. Freeman, Eds.), pp. 197–224. Humana Press, Totowa, NJ.

Knobil and Neill's Physiology of Reproduction, Third Edition
edited by Jimmy D. Neill,
Elsevier © 2006

CHAPTER 39

Puberty in the Sheep

Douglas L. Foster and Leslie M. Jackson

General Considerations about Puberty in the Sheep, 2127
Overview, 2127
Definitions of "Puberty" in the Female, 2128
Neuroendocrine Sequence during the Transition into Adulthood, 2129
Onset of High Frequency GnRH Secretion Initiates the Transition into Adulthood, 2129
Events after the First LH Surge, 2138
Determinants Timing the Transition into Adulthood, 2140
Internal Determinants, 2141
External Determinants, 2148
Sex Differences in Timing the Transition into Adulthood, 2160
Sex Differences in the Pubertal Increase in GnRH Pulse Frequency, 2160
Programming of Sex Differences in the Timing of the Pubertal GnRH Rise by Prenatal Steroids, 2161
Programming of Sex Differences in Other GnRH Controls by Prenatal Steroids, 2162
Malprogramming of GnRH Controls: A Case for Fetal Origin of Reproductive Dysfunctions, 2163
GnRH Neuroanatomy in Males and Females, 2164
Sex Differences in Response to Internal and External Factors Timing Puberty, 2165
Integrated Model for Puberty, 2167
Acknowledgments, 2168
References, 2168

GENERAL CONSIDERATIONS ABOUT PUBERTY IN THE SHEEP

Overview

In the temporal sequence of domestication, the sheep is preceded only by the dog. People of an agricultural Mesolithic culture began the gradual domestication process of the wild Urial sheep (*Ovis orientalis*) probably in southwest Asia about 9000 B.C.; the Argali (*O. ammon*) and Mouflon (*O. musimon*) are now considered to have also contributed toward the Asiatic and European breeds of domestic sheep (*O. aries*) [for review see (1)]. The sheep, prized for its production of wool, meat, and milk, has been a successful species adapting to divergent climatological conditions ranging from desert to Arctic areas and often living in marginal areas unsuitable for cultivation. As pointed out by Hammond (2) over 60 years ago, this "universal provider" is the most widely used domestic animal species for studies of female reproductive physiology. This intensive use of the sheep by biomedical and agricultural scientists continues as model systems are developed for studies of the ovarian cycle (see Chapter 44), seasonal breeding (see Chapter 41), and parturition (see Chapter 55).

From an agricultural standpoint, the sheep is an economically important species worldwide. Understanding how puberty is timed ultimately leads to practical strategies for decreasing the generation interval in farm animals, thereby increasing the overall production of food and fiber. From the physician's viewpoint, using the sheep as an experimental model to study the pubertal process can lead to novel treatment of

Department of Obstetrics and Gynecology, Reproductive Sciences Program, University of Michigan, Ann Arbor, Michigan.

disorders of sexual maturation in children. Finally, from a broad vantage point, mechanisms underlying initiation of fertility in the sheep can be compared with those of other well-studied species to better understand the conservation and evolution of developmental strategies in differing ecologies. In this regard, the sheep is precocious in its development compared with the laboratory rat and rhesus monkey, both of which are altricial species. The lamb is well developed at birth, and many stages of maturation that are attained postnatally in species such as the rat (see Chapter 38) and primate (see Chapter 40) are achieved prenatally in the sheep. This rapid maturation is illustrated in Fig. 1, which shows the relative growth curves for the sheep and for the human, another altricial species. Both are born at approximately the same body weight, but the lamb grows much more rapidly. Whereas by 30 weeks of age the weight of the human infant has only doubled, that of the lamb has increased 10-fold, and reproductive cycles have begun. Therefore, the degree of sexual development attained by the lamb in slightly greater than 6 months requires 13 to 15 years in the child. Thus, the compressed developmental period of the lamb facilitates studies of sexual maturation of large long-lived mammalian females.

Because its manifestation of puberty is subject to modification by external factors, the sheep is representative of many mammalian species living in natural conditions. Most mammals reproduce seasonally, and as a consequence their sexual development is influenced by environmental changes (food availability, photoperiod, temperature, rainfall, etc.). Which of these cues predominates depends on several factors, including physical size of the species (larger

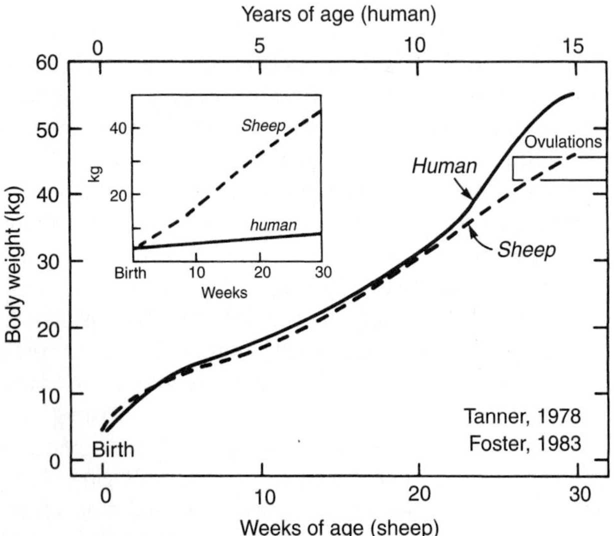

FIG. 1. Growth based on body weight from birth through the initiation of ovulation for sheep (mean) and humans (50th percentile). Absolute growth during the first 30 weeks is shown (*inset*). [From (268) with permission.]

animals have lower metabolic rates and more capacity for energy storage), longevity, and seasonal variation in local environment (see Chapter 41). The modern human deviates from the general rule that breeding in mammals is seasonal. Because of this, it could be argued that the expression of sexual maturity in most mammals is more complex than that of the human. As is the case for all species, the young of seasonal breeders must grow to become sexually mature; however, they must also coordinate the timing of puberty with environmental changes such that their first matings occur during the adult breeding season. Thus, the lamb provides an excellent model for investigations of how internal and external cues are channeled through multiple inputs into the brain to influence mechanisms that are timing onset of fertility. Finally, we are becoming increasingly aware that sex differences exist in certain mechanisms timing puberty and that such differences are organized prenatally. The long gestation of the sheep, approximately 5 months, coupled with the large size of the offspring, permit detailed study of how and when fetal hormones exert their organizing action on the brain before birth to cause sex differences in the expression of pulsatile hormone secretion timing puberty.

In view of the foregoing considerations, this chapter is organized into four major themes: (a) neuroendocrine sequence during the transition into adulthood, (b) internal determinants timing puberty, (c) external determinants timing puberty, and (d) sex differences in the timing of puberty. The emphasis is on the female because, as evident in the last section, the basic developmental plan is female. The female reproductive system of the sheep, as for other species, is more complex than that of the male because the production of periodic ovulations is a more complex process than the continuous production of spermatozoa. In the presence of the testes, testosterone and its androgenic and estrogenic metabolites masculinize and defeminize the developing somatic and neural tissue to result in a male and its simpler reproductive system.

Definitions of "Puberty" in the Female

The first mating of the lamb can result in a pregnancy, and therefore many investigators use "first estrus" as the definition of puberty in the sheep. Others equate the initiation of ovulation with "puberty." Because first ovulation and first estrus typically occur within 2 to 3 weeks of each other, distinction between the two to define the onset of reproductive cycles normally has little practical usefulness, considering that the interval between birth and either of these events is at least 6 months. Several patterns have been noted with respect to the temporal relationship between first ovulation and first estrus (Fig. 2). First estrus, when mating becomes possible, requires an increase in circulating progesterone before the preovulatory increase in estradiol to facilitate sexual behavior. Thus, first estrus is usually preceded one full cycle length (16 days) by the well-known "silent ovulation," an ovulation *not* accompanied by estrous behavior (3). This silent ovulation can be preceded by another silent ovulation and a cycle only 6 days long (Fig. 2, pattern I) (4–6). This short cycle is initiated by the first preovulatory surge of luteinizing hormone (LH), and during the resultant luteal phase, the rise in circulating progesterone is not sufficient to induce estrous behavior at the time of the next preovulatory increase in circulating estradiol (the final silent ovulation). Variations on this sequence can occur. For example, a short luteal phase may be followed by first estrus at the next LH surge if progesterone secretion is low but sufficient (Fig. 2, pattern II). However, the short luteal phase is not required for initiation of cyclic ovarian function, because some lambs exhibit one silent ovulation followed by a normal luteal phase such that estrus accompanies the next cycle (Fig. 2, pattern III). As noted later in the chapter, other definitions of puberty are applicable to reproductive development. For example, "neuroendocrine puberty" refers to maturation of the brain drive to the gonads, and this definition is used in studies of steroid negative feedback control. In addition, because the discussion encompasses the broad aspects of seasonal influences on the timing of first ovulation and estrus, differences between "onset of breeding season" and puberty are considered. The reader is invited to deliberate the nuances of these related terms and concepts that apply to the transition to adulthood in the seasonally breeding sheep.

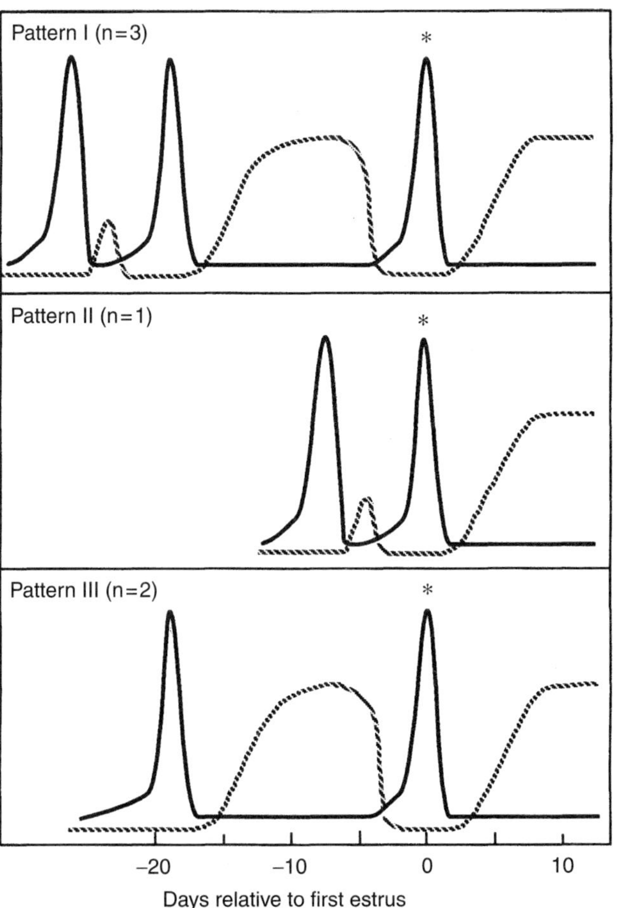

FIG. 2. Summary diagram of various patterns of progesterone at puberty in the sheep; time of first estrus (*asterisks*), changes in LH secretion (*solid line*), and progesterone patterns (*dashed line*). [From (6) with permission.]

NEUROENDOCRINE SEQUENCE DURING THE TRANSITION INTO ADULTHOOD

Onset of High Frequency GnRH Secretion Initiates the Transition into Adulthood

Because ovulation requires a preovulatory surge of LH and follicle-stimulating hormone (FSH), an explanation for the onset of the surge mode of gonadotropin secretion is central to any hypothesis for puberty in the female. The hormonal events that lead to the preovulatory gonadotropin surge during the follicular phase of the adult estrous cycle have been particularly well characterized for the sheep (see Chapter 44). The major neuroendocrine features of the periovulatory period phase are depicted in Fig. 3 (right). During the 48- to 72-hour follicular phase, gonadotropin-releasing hormone (GnRH) pulse frequency becomes more rapid, and a sustained rise in basal LH secretion occurs as the frequency of LH pulses increases to greater than one per hour. The increased gonadotropin stimulus drives one or more ovarian follicles to the preovulatory stage; estradiol production increases to high levels to activate the gonadotropin surge mechanism. Ovulation occurs, and one or more corpora lutea that secrete progesterone are formed.

In the immature female lamb, preovulatory gonadotropin surges do not occur spontaneously (Fig. 3). Although the surge system is capable of function from a very early age, it remains dormant until puberty. According to the hypothetical scheme illustrated, the reason is related to the activity of the mechanism controlling the tonic (pulsatile) mode of LH secretion, which along with FSH is ultimately responsible for the production of estradiol by the ovarian follicle. In the prepubertal female, LH pulses do occur, and their pattern reflects that of their trophic stimulus, GnRH. The amplitudes of LH pulses before puberty are as

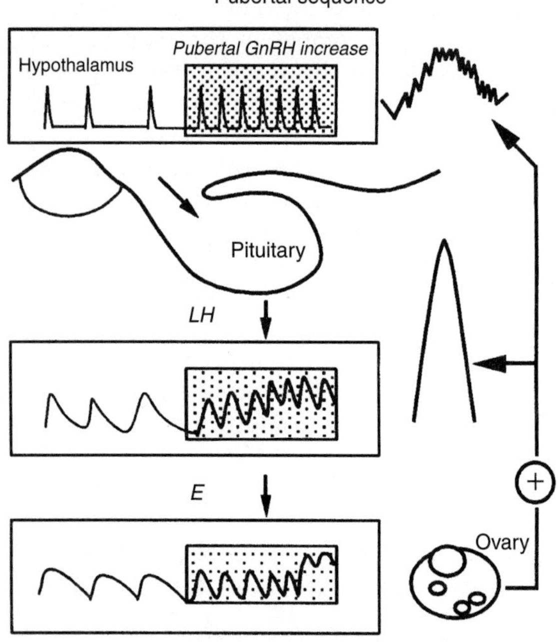

FIG. 3. Schematized relationships among gonadotropin-releasing hormone (GnRH), luteinizing hormone (LH), and estradiol (E) during the pubertal transition in the female sheep (*shaded areas*).

great as or greater than those of the adult. The frequency, however, is much lower than that during the follicular phase, with the LH discharge rate in the immature female being no faster than once every 90 to 120 minutes for any prolonged period. During low frequency secretion, concentrations of circulating LH between each pulse return to low levels, and only transient rises in circulating estradiol are produced by small ovarian follicles. Thus, before puberty, these low frequency LH pulses do not provide the necessary stimulus for follicular development, and therefore the sustained rise in estradiol cannot be produced to activate the preovulatory surge mechanism. Although the young lamb is capable of generating the requisite high frequency GnRH/LH pulses, this pattern is not produced because sensitivity to estradiol feedback inhibition is high. During puberty, a pronounced reduction in sensitivity to estradiol inhibition of GnRH secretion occurs, and the high frequency rhythm of GnRH secretion is first expressed. As a result, the follicular phase begins and culminates in the first GnRH surge and ovulation. Subsequent events that occur during the pubertal transition, which include the short luteal phase and first expression of sexual receptivity, are considered at the end of this section. Important to the consideration of the feedback control of GnRH secretion is that with estradiol losing its effectiveness as a potent negative feedback agent during the pubertal transition, once ovulations begin this feedback regulation is assumed by progesterone, which makes its first appearance with the emergence of corpora lutea.

The foregoing working hypothesis focuses on the mechanism governing the frequency of GnRH secretion, which is the rate-limiting system in the onset of reproductive function in the female lamb. This is based on several considerations: (a) the *potential* for the gonadotropin surge mechanism to respond to the stimulatory feedback action of estradiol exists long before puberty, (b) the *potential* for the ovary to produce follicular phase levels of estradiol in response to hourly pulses of LH exists long before puberty, (c) the *potential* to produce high frequency GnRH pulses exists long before puberty, and (d) the response to estradiol inhibition of GnRH secretion decreases during the pubertal period. Evidence to support each of these contentions is presented below.

Competency of the Preovulatory Gonadotropin Surge Mechanism before Puberty

Numerous studies have demonstrated unambiguously that exogenous estradiol can induce an LH surge in the sexually immature female sheep (7–14). As illustrated in Fig. 4, the potential to respond to the stimulatory feedback action of estradiol is attained within a few weeks after birth. In the study shown, the amplitude of the LH surge increased progressively with age. By 12 to 20 weeks, the magnitudes of the induced LH surges were well within the range of those capable of producing ovulation. The sensitivity of the preovulatory gonadotropin surge system has been tested in the prepubertal female about 10 weeks before the first spontaneous ovulation (11). Various lengths of Silastic capsules containing crystalline

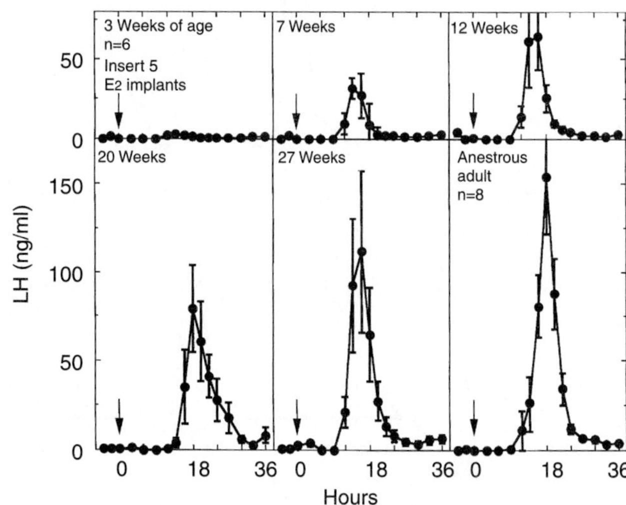

FIG. 4. Development of the response to the stimulatory feedback action of estradiol on LH secretion in the female sheep. Estradiol was administered (*arrows*) subcutaneously by Silastic capsules for 96 hours at each age in the same lambs. [From (9) with permission.]

estradiol were used to produce physiological rises of estradiol over a range of 2–10 pg/ml after endogenous estradiol had been removed by ovariectomy. The sensitivity in the immature female was found to be equal to that of the adult, and, remarkably, an estradiol increment as small as 2 pg/ml induced an LH surge. Moreover, administration of progesterone before estradiol in the immature female effectively prevented the induced surge (9), indicating that the potential for this feedback control becomes established well before the formation of the source of progesterone secretion, the corpus luteum. Taken together, these findings demonstrate that the mechanism governing the preovulatory surge system is exquisitely sensitive to the stimulatory feedback action of estradiol long before puberty. Therefore, the failure of the preovulatory LH surge to occur spontaneously in the sexually immature lamb must be due to the lack of a sustained rise in endogenous estradiol.

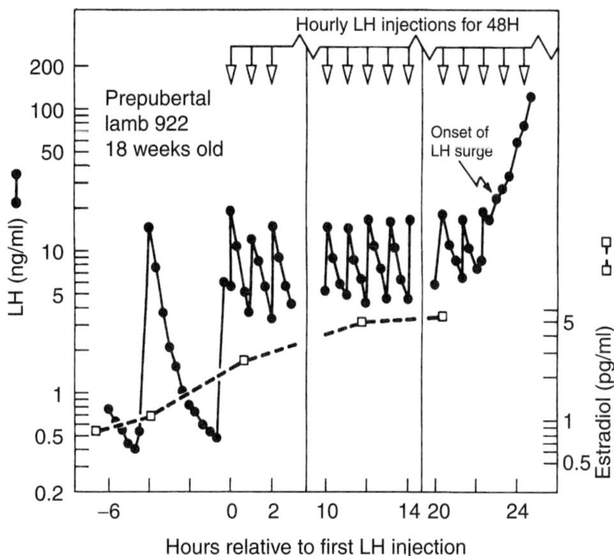

FIG. 5. Rapid administration of LH induces ovulation in the immature sheep. Circulating LH and estradiol concentrations are shown at various times before and during hourly injections (*arrows*) of purified ovine LH in an individual female. High LH values between hours 22 and 24 reflect the initial portion of a preovulatory LH surge that was followed by an 11-day luteal phase with three corpora lutea. [From (269) with permission.]

Competency of the Ovary before Puberty

The ovary matures rapidly in the sheep. Antral follicles are evident before birth, but they do not respond to exogenous gonadotropins until after 2 to 4 weeks of age (15,16), when granulosa and thecal layers become well developed (17). At a slightly greater age (5–6 weeks), exogenous gonadotropins (pregnant mare serum gonadotropin, human chorionic gonadotropin) can induce ovulation and formation of corpora lutea (18). These exogenous gonadotropins can act by stimulating follicular growth and increasing estradiol production to activate the preovulatory gonadotropin surge mechanism (19), much the same as occurs during the normal follicular phase (see Chapter 44). Moreover, the capacity of gonadotropin-primed follicles from 10- to 16-week-old lambs to secrete sex steroids in vitro was found to be similar to that of follicles from adults (20). The precocious development of the lamb ovary has been used to practical advantage. Induction of ovulation by exogenous gonadotropins before the normal time of puberty (21,22) permits early mating to reduce the generation interval. Although in principle this could increase the overall lifetime agricultural production of the female sheep, fertility is low in such females as is the case for young females of any species.

The competency of the lamb ovary to behave in an adult-like manner before puberty has been demonstrated in a more physiologic framework (14,23). In the study shown in Fig. 5, purified ovine LH was injected (intravenously) hourly for 48 hours into prepubertal lambs to simulate a key aspect of the follicular phase of the estrous cycle—*high frequency LH pulses*. The dose of LH used was chosen to replicate the amplitude of the pulses that are characteristic of the immature lamb (Fig. 5, 6-hour period before treatment). In response to hourly LH pulses, circulating estradiol rose to follicular phase concentrations and a preovulatory surge occurred that resulted in ovulation and the formation of progesterone-secreting corpora lutea. Remarkably, in some prepubertal females of the study, only 22 hourly injections were required to initiate the sequence of ovulatory events. By contrast, high amplitude low frequency pulses of exogenous LH are ineffectual in this regard (14,23). Therefore, the absence of large sustained estradiol rises in the prepubertal lamb is not due to immaturity of the ovary but rather to the lack of a sustained high frequency gonadotropic stimulus.

Parenthetically, it should be mentioned that induction of ovulation in the immature female sheep by administration of gonadotropins does not advance the time of spontaneous puberty. For example, prepubertal females in which an ovulatory cycle is induced by hourly LH injections fail to exhibit multiple consecutive reproductive cycles after the corpora lutea of the induced ovulation regress (14,23). Such lambs spontaneously initiate their first estrous cycles at the normal age (~30 weeks) (23). According to the working hypothesis, the immature females return to the anovulatory condition after luteal regression because the frequency of endogenous LH pulses remains low, and hence a preovulatory follicle cannot be developed.

Competency of the Pituitary to Secrete Gonadotropins before Puberty

The foregoing considerations about the ovary and surge mechanism infer that simple hypogonadotropism underlies the anovulatory condition of the immature sheep. More specifically, the deficiency appears to be in LH, rather than FSH, secretion because administration of LH *alone* induces the follicular phase sequence (23) (Fig. 5). A three- to fourfold increase in LH pulse frequency was reported by Huffman et al. (24) to occur gradually over the last week before first ovulation in the lamb (Fig. 6). Before that period, the frequency is variable, although it is typically low, with the interval between pulses being 2 to 3 hours or more (24–28).

In contrast to LH, circulating FSH in the lamb is well within the adult range by 10 to 12 weeks of age (5,25). A subtle rise in peripheral FSH concentrations, however, has been detected during the peripubertal period in samples of blood collected every 4 hours (6). However, this increase was not detected in another study when circulating FSH was monitored frequently (12-minute intervals for 6 hours) at various times before puberty (24). Thus, a role for FSH in the transition into adulthood in the sheep cannot be easily resolved. In the mature ewe, FSH may be only permissive with respect to estradiol secretion, although it plays an important role in regulating follicular growth and the timing of follicular maturation (see Chapter 44).

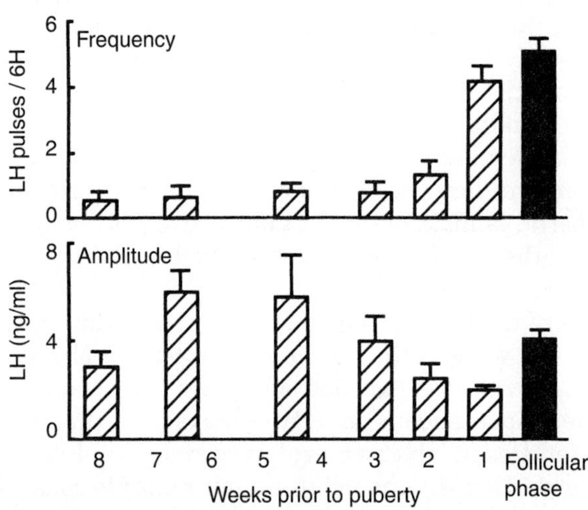

FIG. 6. Mean (± SEM) LH pulse frequency (*top panel*) and amplitude (*bottom panel*) during the prepubertal period (*striped bars*) and the first postpubertal follicular phase (*solid bars*). Values for weeks 4–5 and 6–7 were combined to increase the number of observations contributing to the means. Each mean frequency was based on values from seven or eight lambs. Because of the absence of LH pulses during some collection periods, mean pulse amplitudes were based on values from four (weeks 3, 6–7, and 8), five (weeks 2 and 4–5), or seven (week 1, follicular phase) lambs. [Redrawn from (24) with permission.]

Perhaps qualitative changes, rather than quantitative changes, in FSH may be important to the transition into adulthood. This possibility was tested in female lambs during experimental induction of puberty (29), as shown in Fig. 7. The first follicular phase was induced by pulsatile administration of GnRH (2 ng/kg every 2 hours for 24 hours and then hourly for the next 48 hours) in a manner previously found to synchronize the onset of the preovulatory gonadotropin surge (30). Serum concentrations of immunoreactive LH and FSH, as measured by standard radioimmunoassays, bioactive circulating FSH

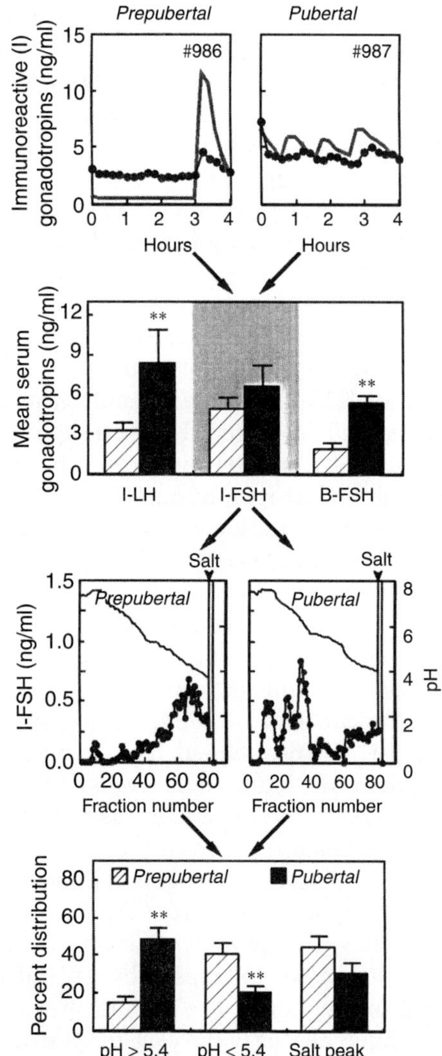

FIG. 7. Top panel: Circulating immunoreactive follicle stimulating hormone (I-FSH) (*closed circles*) and LH (*solid line*) in a prepubertal lamb (*left*) and in a lamb in which the first (pubertal) follicular phase was induced by administration of GnRH (*right*). **Second panel:** Mean (± SEM) concentrations of immunoreactive (I) and bioactive (B) gonadotropins in prepubertal (*striped bars*) and pubertal (*solid bars*) lambs. **Third panel:** Distribution pattern of I-FSH (*solid circles*) in serum of a prepubertal (*left*) and pubertal (*right*) lamb. **Bottom panel:** Mean (± SEM) percent distribution of serum I-FSH isoforms in three pH pools. [Redrawn from (29).]

as measured by the Sertoli cell aromatase assay, and the distribution pattern of FSH isoforms after chromatofocusing separation of circulating FSH from six GnRH-treated female lambs were compared with those of six untreated prepubertal lambs (Fig. 7). The LH pattern in prepubertal lambs, as expected, was characterized by low frequency and high amplitude LH pulses; the lambs in which puberty was induced had high frequency low amplitude LH pulses as dictated by the pattern and level of GnRH administration (Fig. 7, top panel). Although immunoreactive FSH concentrations were similar in the prepubertal and pubertal lambs, serum bioactive FSH concentrations were greater in the pubertal than in the prepubertal lambs, suggesting qualitative changes (Fig. 7, second panel). This increase in bioactivity was also associated with a change in the distribution pattern of circulating FSH isoforms; the pubertal lambs had a higher proportion of the less acidic serum FSH isoforms as compared with the prepubertal lambs (Fig. 7, bottom two panels). Because various FSH isoforms differ in their immunopotency, biopotency, receptor binding ability, and clearance, a shift in the predominant molecular form of FSH could occur during the pubertal process to increase biological activity of the gonadotropin. Interestingly, however, the increase in biological potency of FSH may be secondary to the increase in ovarian steroids that occurs during the pubertal transition because pulsatile administration of GnRH does not alter the isoform distribution patterns of circulating FSH in the absence of the ovaries (31). Therefore, it is likely that the increase in GnRH frequency leads to an increase in LH pulse frequency, which in turn augments steroidogenesis by the ovarian follicle. These increases in circulating ovarian steroids then lead to the increased biological potency of FSH during puberty.

At the outset, the simplest explanation for the low frequency LH pulses before puberty is that the neuroendocrine system is not sufficiently developed to produce high frequency GnRH pulses. This is not the case, however, as can be demonstrated by simply removing the ovaries. In the absence of steroid feedback inhibition during the early postnatal period, hourly LH pulses are readily produced (Fig. 8, bottom), much the same as those that initiate the follicular phase (Fig. 5). Therefore, the potential to produce high frequency LH pulses exists from within a few weeks after birth in the sheep. As shown below, they reflect high frequency GnRH secretion (Fig. 9). Finally, because high frequency LH pulses can be produced before puberty (in the absence of steroid feedback), we can also conclude that the pituitary is responsive to GnRH. Moreover, administration of GnRH readily produces LH pulses, and if the administration is prolonged for 3 days (30) or more (32), this can produce sufficient stimulation of the pituitary–ovarian axis to induce ovulation.

Competency of the Hypothalamus to Produce GnRH before Puberty

Determination of the pattern of GnRH to which the pituitary is exposed has been difficult for any species largely because of the inaccessibility of the circulation between the hypothalamus and pituitary (portal vasculature) and because of the minute amounts of the neuropeptide actually secreted into the pituitary circulation. As reviewed by Caraty et al. (33)

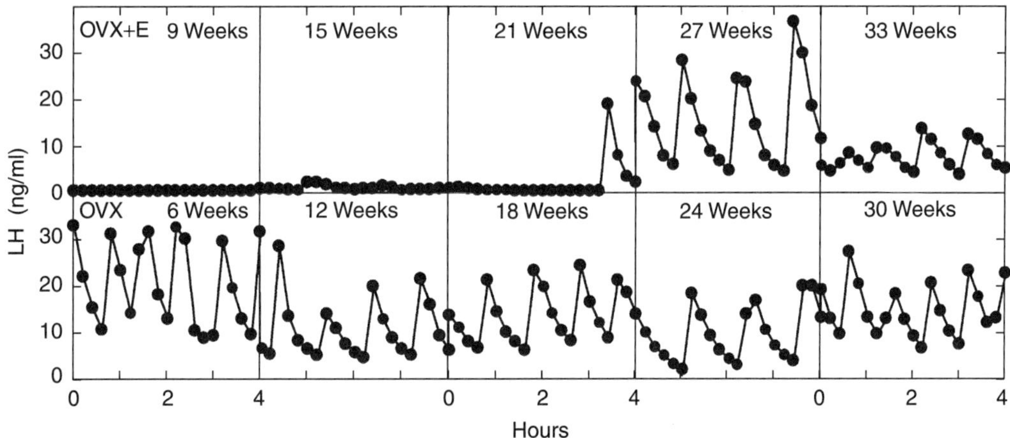

FIG. 8. Patterns of circulating LH in the same ovariectomized (OVX) lamb in the presence (weeks 9, 15, 21, 27, and 33) and absence (weeks 6, 12, 18, 24, and 30) of exogenous estradiol (E, 2 pg/ml by implant). Samples were collected 3 weeks after insertion (*top*) or removal (*bottom*) of implant. Ovariectomy was at 3 weeks of age. [From (270) with permission.]

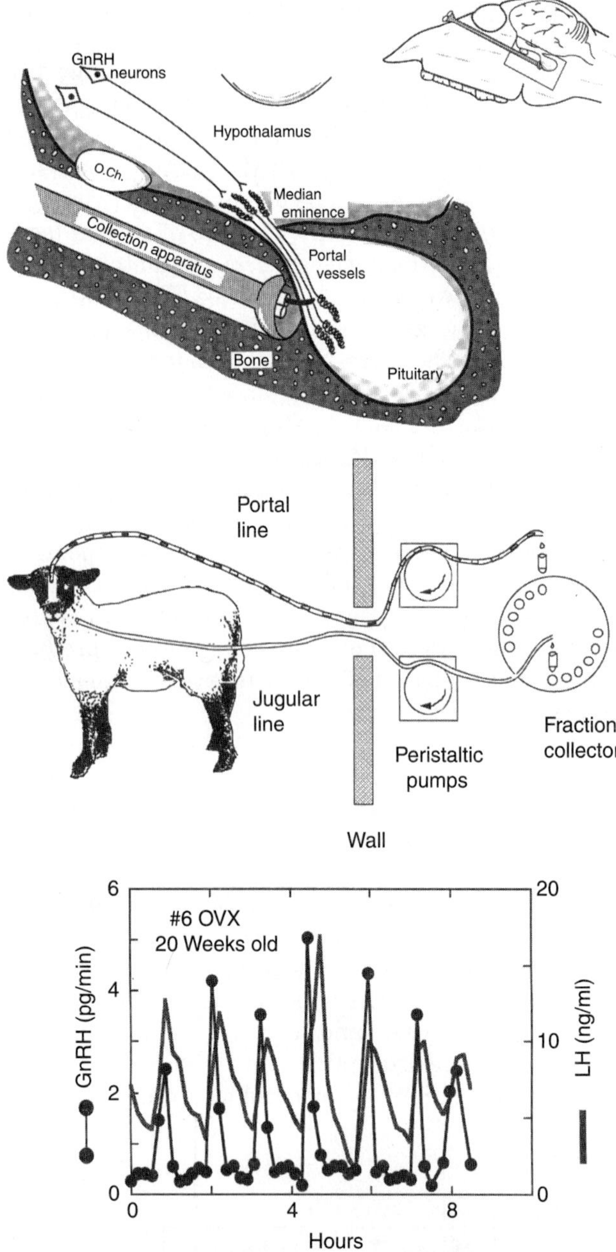

FIG. 9. Technique for remote collection of pituitary portal blood. One week after surgery, portal vessels are lesioned and blood collects into the artificial sinus where it is withdrawn through the collection apparatus and transferred into sample vials in a remote location. The jugular blood for assay of LH is collected simultaneously. The bottom panel shows the patterns of GnRH (*solid circles*) and LH (*solid line*) secretion in a prepubertal ovariectomized female. Note the high concordance of GnRH pulses and LH pulses. [Redrawn from (271) and data from (42) with permission.]

in 1994, Clarke and Cummins in Australia developed a technique to collect blood from the portal vasculature of fully conscious sheep. This technique was subsequently refined by Caraty and Locatelli in France for collection of samples for up to 24 hours by means of an implantable device. Finally, continuous improvement of the technique by Karsch in the United States led to remote sampling. Typically, in such studies the device is surgically implanted. Then, a few days later, samples of portal blood (for measurement of GnRH) and jugular blood (for measurement of LH) are collected at 5- to 10-minute intervals, although more frequent collections (up to 30 seconds) have been performed. Such an approach offers the advantages of being able to determine the pattern of GnRH and LH simultaneously in unanesthetized and undisturbed individuals. This technique has been miniaturized and adapted successfully to the lamb (Fig. 9). When examined in the prepubertal lamb (19 weeks of age) without ovaries, the pattern of LH appears to faithfully reflect the pattern of GnRH secretion (Fig. 9, bottom). Each LH pulse was preceded by a GnRH pulse; moreover, each obvious GnRH pulse resulted in an LH pulse. Such findings provide the necessary direct evidence that the developing female sheep has the ability to produce high frequency GnRH pulses long before puberty.

Decrease in Sensitivity to Estradiol Feedback Inhibition of GnRH Secretion during Puberty

The "gonadostat" hypothesis offers a conceptual explanation for why high frequency GnRH pulses are not expressed in the immature female lamb. The hypothesis, in the case of the female, is that concentrations of circulating gonadotropins are low in the immature individual because the system governing the secretion of GnRH is very sensitive to the inhibitory feedback action of ovarian steroids such as estradiol. The sensitivity then decreases as the time of puberty approaches, thereby allowing gonadotropin secretion to increase, and, consequently, the ovary is stimulated to function in an adult manner. This hypothesis, the origins of which may be traced back nearly half a century [for a brief history, see (34,35)], has been refined over the years and now includes the regulation of pulsatile GnRH secretion. Interestingly, although the original hypothesis was based on work conducted in the female rat, controversy exists about whether it is applicable to this rodent (35,36) (see Chapter 38). Moreover, in the rhesus monkey, another well-studied species, the hypothesis may only apply to the final stages of sexual maturation (see discussion at the end of this section).

Generally, sensitivity implies a given response to a given dose, and when describing the changing suppressive effects of estradiol on the secretion of GnRH/LH in the developing sheep, "sensitivity" has often been used interchangeably with "responsiveness." In this respect, a dose-response to estradiol inhibition reveals that very low doses of estradiol (<1 pg/ml) delivered by subcutaneous implant reduce the frequency of LH pulses to 50% within 24 hours in the prepubertal ovariectomized sheep, producing hourly pulses of LH. A slightly greater dose (1–2 pg/ml) completely suppresses LH pulses (37). Because of this highly suppressive effect of extremely low doses of estradiol in the ovariectomized prepubertal sheep model, it is of interest that the "response" of LH secretion to inhibition by a fixed, chronic, low dose of estradiol decreases markedly during the pubertal period (38). Whereas low physiological concentrations of exogenous circulating estradiol were able to prevent high-level LH secretion in ovariectomized lambs before the typical age at puberty, the same estradiol levels were ineffective in this regard afterward (Fig. 10). Circulating LH increased in these chronically estradiol-treated agonadal lambs when ovulations were beginning in ovarian-intact females. This leads to the conclusion that the decrease in feedback sensitivity to estradiol inhibition allows the follicular phase pattern of high frequency LH pulses to become expressed for the first time in the presence of the ovaries. This would rectify the typically hypogonadotropic condition of the developing lamb, and the ovarian follicle would be driven through the follicular phase of the first successful reproductive cycle, one that results in ovulation.

Detailed studies of the model used to examine changes in set point to steroid negative feedback have revealed that estradiol administration (via Silastic implant) reduces LH pulse frequency, but not amplitude, in the immature ovariectomized lamb (37,39). Accordingly, the subsequent increase in basal LH in such estradiol-treated females during the pubertal period reflects an increase in LH pulse frequency (40). This phenomenon, whereby high frequency LH pulses become expressed during the pubertal period as a consequence of a reduction in sensitivity to estradiol inhibition, can readily be demonstrated by periodically opening and closing the inhibitory feedback loop between estradiol and LH secretion within the same individual. For example, as shown in Fig. 8, removal of the ovaries at 3 weeks of age results in high frequency LH secretion by 6 weeks of age; insertion of an estradiol implant then suppresses LH secretion as evidenced by the pattern at 9 weeks of age. Subsequent removal of the estradiol again reveals the potential for high frequency LH secretion. This relationship was maintained upon insertion and removal of the estradiol implant with the same lamb until approximately 27 weeks of age when insertion of the same estradiol implant had no apparent effect. This is about the age at first ovulation in intact lambs. Thus, before the age at puberty, estradiol is extremely effective in suppressing pulsatile LH secretion, and it loses its effectiveness as the time of puberty is approached. Progesterone then

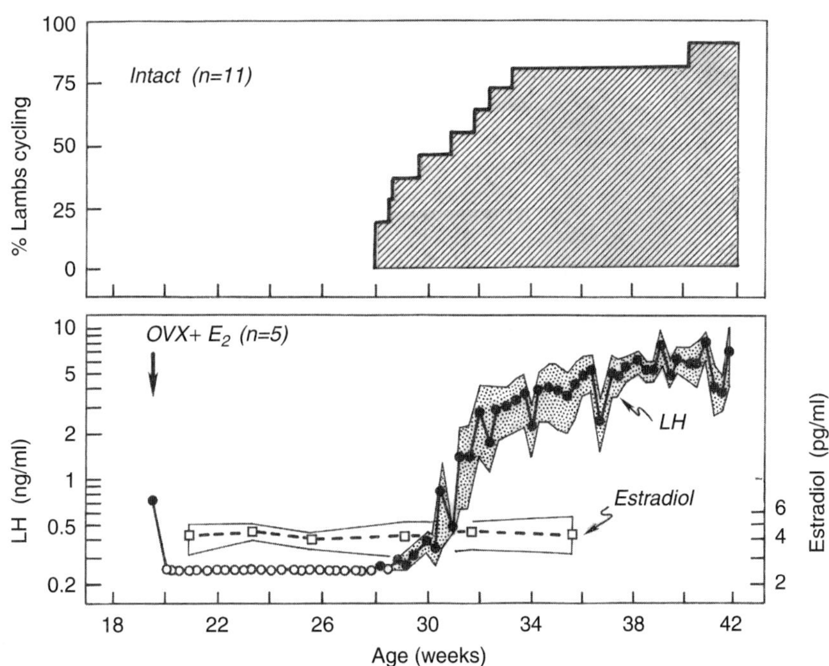

FIG. 10. Response to estradiol inhibition of tonic LH secretion decreases during puberty. **Top:** Onset of reproductive cycles in untreated lambs based on the appearance of luteal-phase concentrations of circulating progesterone. **Bottom:** Mean (± SEM) concentrations of circulating LH and estradiol (E_2) in ovariectomized lambs treated chronically with Silastic implants of estradiol from the time of ovariectomy (*arrow*); undetectable LH values (*open circles*). [From (200) with permission.]

assumes the role of the major inhibitory feedback control of GnRH secretion. Although the prepubertal female sheep is responsive to the negative feedback action of progesterone on LH secretion (41), this has no physiological relevance until the source of its secretion appears after first ovulation.

Using the portal blood collection technique shown in Fig. 9, the relationship between GnRH and LH in ovariectomized lambs has been examined during the prepubertal (~20 weeks of age), peripubertal (~25 weeks of age), and postpubertal (>30 weeks of age) period (Fig. 11) (42). The patterns of secretion were studied in the absence of estradiol (Fig. 11, left) and in its presence (Fig. 11, right); the quantity of estradiol administered by subcutaneous implant was in the low physiological range to permit the characterization of patterns of GnRH secretion even before puberty. In the absence of estradiol replacement, high frequency GnRH pulses occur. The frequency of pulses appears to increase during the peripubertal period. It is evident that the system governing GnRH secretion of the prepubertal female is more responsive to the inhibitory feedback action of estradiol than that of the postpubertal lamb. Whereas low physiological levels of estradiol were effective in reducing GnRH pulse frequency during the prepubertal and peripubertal periods, they eventually became ineffective in this regard, as indicated for the postpubertal female. When viewed in the context of the gonadostat hypothesis and pubertal process, these results offer additional strong evidence that in the female sheep, the mechanism governing GnRH secretion is highly responsive to estradiol inhibitory feedback before puberty and that a decrease in response of the mechanism to estradiol feedback is responsible for the increase in GnRH pulse frequency that underlies the transition into adulthood. The apparent increase in response of the peripubertal female to estrogen negative feedback is currently without explanation (Fig. 11, right panels). In this study, increasing amounts of exogenous estradiol were delivered by subcutaneous implant (prepubertal, 3 mm; peripubertal, 5 mm; postpubertal, 7.5 mm), but serum concentrations of estradiol in all groups were near or below assay sensitivity of 0.5–1.0 pg/ml. Perhaps the hypersensitive peripubertal female produced a robust response to a small, albeit undetectable, increase in estradiol (from 3- to 5-mm implant). Finally, it is unlikely that the apparent increase in the steroid-independent drive alone (Fig. 11, left panels) can account for the mechanism of feedback sensitivity because even twice the dose of chronic estradiol used in prepubertal females was ineffective in suppressing GnRH/LH secretion in postpubertal females.

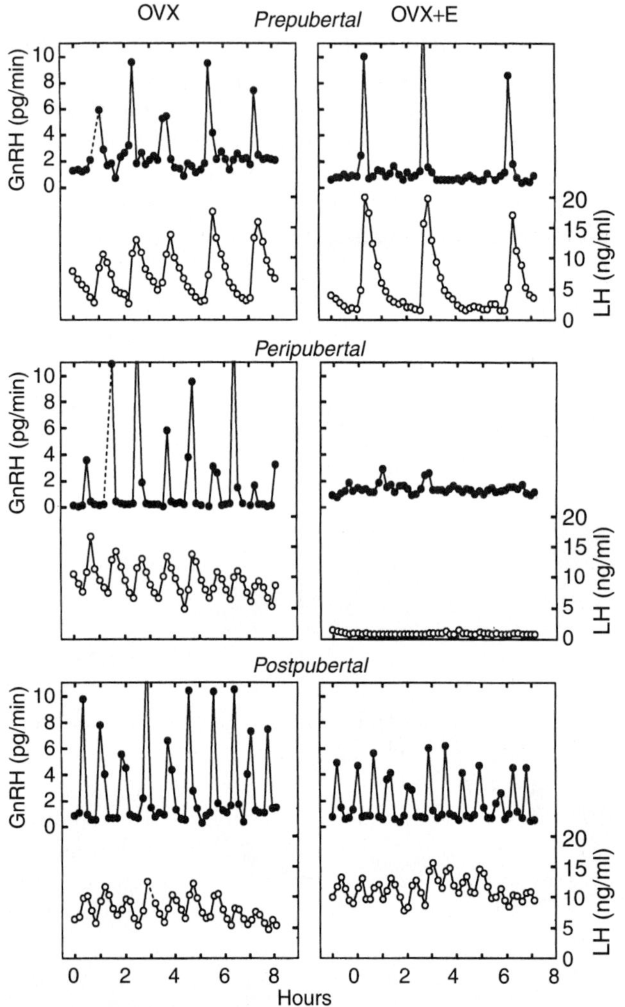

FIG. 11. Patterns of GnRH in portal blood (*solid circles*) and LH (*open circles*) in peripheral blood from representative ovariectomized lambs at three ages: prepubertal, ~20 weeks (*top*); peripubertal, ~25 weeks (*center*); and postpubertal, ~30 weeks (*bottom*). Study represents lambs with no estradiol treatment (*left*) and those with estradiol treatment (*right*). [From (42) with permission.]

Ultimately, the changing sensitivity to inhibitory steroid feedback control of GnRH secretion must be explained in electrophysiological/neurobiochemical terms. In the adult of several species, there is strong inferential evidence for the changing electrophysiological activity of neurons that periodically secrete GnRH into the pituitary portal vasculature. LH pulses have been found to occur closely in time with volleys of electrical activity recorded from multiple neurons in the adult hypothalamus [sheep (43), rat (44), monkey (45), and goat (46)]. A limited study has been performed in the developing sheep. In anesthetized male lambs, Pelletier and colleagues (47) recorded multiunit activity in an area between the optic chiasma and median eminence that was found to be related to pulsatile release of LH in adult ovariectomized ewes. Rhythmic changes in firing rates were recorded at 40- to 60-minute intervals, but because of the anesthesia, LH secretion was reduced such that pulses could not be detected. Hence, no correlation between electrical activity and LH secretion could be determined in this study.

In searching for a biochemical explanation for the observed decrease in sensitivity to estradiol during the pubertal period, Wilkinson and Landymore (48) suggested that the opioid system may play a role. According to their hypothesis, endogenous opioids mediate inhibitory steroid feedback and a reduction in opioid inhibition provides a neurochemical explanation for resetting the gonadostat (i.e., an opiostat). This hypothesis has been investigated in the developing sheep by administering an inhibitor of endogenous opioids (naloxone) to female lambs at various ages (49). Some were reared to undergo normal puberty (Fig. 12, top panel, for representative females), whereas others were exposed to a photoperiod treatment that delayed puberty (Fig. 12, bottom panels). In both groups, naloxone had no effect on LH secretion at the earliest age tested (9 weeks, data not shown), but at all later developmental stages, the opioid antagonist increased LH pulse frequency. Importantly, the incremental change in LH pulse frequency after naloxone treatment did not differ with age. Even with photoperiodically delayed puberty, robust responses were induced after opioid receptor blockade. This leads to the conclusion that endogenous opioid mechanisms provide important inhibitory control of pulsatile LH secretion in the developing sheep. However, the ubiquity of response to opioid blockade in pre- and postpubertal females does not support the hypothesis that a reduction in opioid inhibition underlies puberty. A similar conclusion was reached in a comparative study of opioid blockade of LH secretion in developing male and female lambs; the response to naloxone administration was not sexually differentiated (50) despite the fact that the timing of puberty was (51,52). This provides additional evidence that the response to naloxone develops very early in life and is not related to time of puberty. Finally, the basic premise that endogenous opioids mediate inhibitory steroid feedback must be rejected because opioid receptor blockade in the absence of the ovaries further increases LH pulse frequency in the lamb (49).

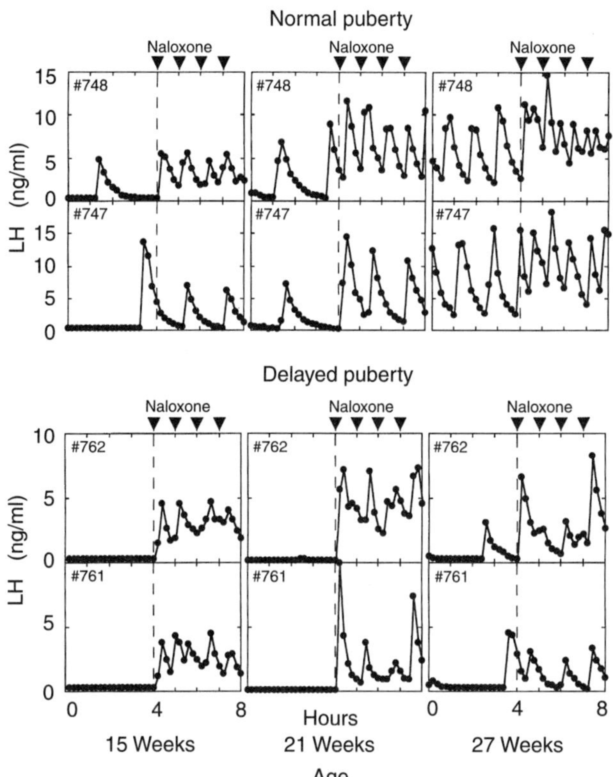

FIG. 12. Effects of naloxone on serum LH concentrations in developing estradiol-treated ovariectomized lambs at 15, 21, and 27 weeks of age. Data from four representative individuals are shown. During each naloxone test, the first 4 hours served as a control period; four intravenous injections of naloxone (1 mg/kg body weight) were administered hourly for 4 hours (*arrows*). **Top:** Two individuals (748 and 747) were raised on a permissive photoperiod; note the age-related increase in LH pulse frequency during the control periods. **Bottom:** Two individuals (762 and 761) were raised on an inhibitory photoperiod; note that LH secretion remains low in the control periods throughout the experiment. [Redrawn from (49) with permission.]

Workers using the sheep, ferret, and heifer as experimental models have embraced the gonadostat hypothesis as an explanation for the ultimate timing of the pubertal rise in GnRH/LH secretion [for review see (53)], but those working with the rhesus monkey, a subhuman primate, have differing perspectives (see Chapter 40). Studies (54,55) have been conducted testing the gonadostat hypothesis in the rhesus monkey using the same protocol as in the sheep, namely measurement of circulating LH in the absence of the ovaries but in the presence of a fixed concentration of circulating estradiol (by implant). The results reveal that low physiological concentrations of the steroid produce a potent inhibition of LH secretion for a prolonged period in the ovariectomized postmenarchial rhesus monkey; however, a rise in circulating LH eventually occurs, and when it does it is over the same age range as for first ovulation. Similar observations have been made by Wilson (55) for the control of FSH secretion by estradiol in the developing rhesus monkey. Taken together, such findings lead to the deduction that the sensitivity to estradiol negative inhibition decreases in the postmenarchial monkey to allow the potential for high GnRH secretion to be expressed in much the same way as occurs in the sheep. Thus, ovulation cannot be initiated without the reduction in response to estradiol inhibitory feedback. Transient returns to periods of hypersensitivity to estradiol negative feedback and hypogonadotropism do occur in the young monkey, and they perhaps explain the long anovulatory menstrual cycles, which are characteristic of early adulthood and its periods of irregular adolescent sterility (56).

The gonadostat hypothesis provides an adequate explanation for the increase in GnRH secretion during the final stages of sexual maturation in the post-menarchial primate, but the hypothesis cannot account for the prolonged hiatus in GnRH secretion that dominates the prepubertal period of the primate (see Chapter 40). During this extended time, removal of the ovaries does not result in high level GnRH/LH secretion; therefore, considerations of sensitivity to steroid inhibitory feedback control of GnRH secretion are irrelevant. The mechanism for this "steroid-independent" hiatus in GnRH secretion is unknown, although it is the subject of intense investigation, particularly in the primate. Explanations for the regulatory mechanisms responsible for the prolonged hypogonadotropism in the premenarchial primate are of great interest, but they should neither overshadow nor be overshadowed by the gonadostat hypothesis because the latter mechanism relates only to the *final* stages of sexual maturation. At this stage, if the primate were like the sheep, steroid-independent drive to GnRH secretion would be highly developed but would not be fully expressed in the presence of low circulating concentrations of estradiol. Because this high sensitivity to estradiol negative feedback must decrease before the pubertal rise in GnRH can occur, steroid-dependent regulation would be considered the final mechanism that dictates the timing of the first successful follicular phase culminating in ovulation. Viewed from this perspective, both control systems would be used in the timing of first ovulations with the steroid-independent system providing coarse control and the steroid-dependent system providing fine control. The steroid-independent control of GnRH secretion would determine the year when first ovulation is possible, whereas the steroid-dependent control mechanism would determine the week in which first ovulation occurs.

In comparison with the altricial rhesus monkey, the precocial sheep has developed many components of the hypothalamic–hypophyseal–ovarian axis necessary for puberty (stage III) within a few weeks after birth (Fig. 13, bottom). Thus, in the postnatal sheep, the mechanism underlying estradiol hyperresponsiveness dominates the vast majority of the developmental period until first ovulation at 7 months of age. For the monkey, the analogous period when sensitivity to steroid feedback inhibition is the primary regulator of LH secretion comprises only the final 25–30% of its postnatal development (~10 months between menarche at 30 months of age and first ovulation at 40 months). It is noteworthy that the developing sheep may have the analogous mechanism for nonsteroidal depression of GnRH secretion that characterizes much

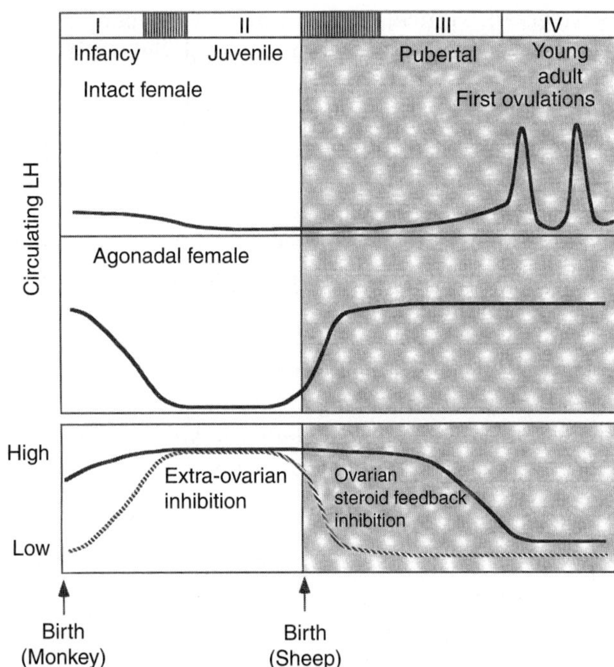

FIG. 13. Analogous phases of sexual maturation based on the control of LH secretion in the monkey and sheep. The times of birth for each within the maturational scheme are indicated. Central to the hypothesis is that the sheep is born at a relatively advanced stage of sexual maturation, one that is comparable with that developed by the monkey shortly before menarche. [From (53) with permission.]

of development in the primate (Fig. 13, stage II, juvenile). However, because the sheep is a precocial species and is born in an advanced stage of reproductive development, this nonsteroidal inhibitory system is proposed to function before birth and during the neonatal period [for review see (53,57)]. Although this unifying hypothesis is attractive, it remains untested. Circulating LH in both sexes diminishes during the final stages of gestation in the absence of the gonad (58), but the source of the "nongonadal" inhibition cannot be determined given the fetus remains exposed to many exogenous substances, including maternal and placental steroids. The postnatal period of hypogonadotropism, which is demonstrably gonadal steroid independent, can be prolonged in the sheep if growth is retarded (see Energetics of Sexual Maturation). In such slowly growing lambs, removal of the gonads does not result in the high frequency GnRH pulses typical of rapidly growing lambs. The issue of sensitivity to steroid feedback becomes important only after such animals have sufficient energy available to support the production of high frequency GnRH secretion in the absence of any steroid feedback inhibition.

Events after the First LH Surge

An understanding of the first gonadotropin surge and the changes that precede this event is clearly important to unraveling the initiation of the pubertal process. Nevertheless, later changes occur in the sheep that are worthy of mention to provide the broad picture of the transition into adulthood. Sexual maturation continues after the first ovulation. There are changes in the control of pulsatile LH secretion, normal luteal function and sexual receptivity become established, and a seasonal pattern of reproduction becomes manifest, with the young female returning to the anovulatory condition, much the same as before puberty.

Postpubertal Shift in Steroid Control of LH Secretion

During the pubertal period, the hypothalamohypophyseal system controlling pulsatile GnRH secretion decreases its ability to be inhibited by estradiol. High frequency GnRH pulses do not occur indefinitely thereafter because a qualitative shift occurs in their steroid feedback control. The new inhibitory steroid is progesterone, and the new timing mechanism that dictates the intervals between ovulatory cycles is its source of secretion—the corpus luteum. According to the hypothesis developed for neuroendocrine control of the estrous cycle, rising luteal phase progesterone concentrations decrease GnRH pulse frequency (see Chapter 44). The resultant low frequency LH pulses do not sufficiently stimulate the ovary to develop a preovulatory follicle. After regression of the corpus luteum and the decline in circulating progesterone, LH pulse frequency again increases, and another 2- to 3-day follicular phase begins to set in motion the next reproductive cycle. This sequence of progesterone dominance and withdrawal is repeated at 15- to 17-day intervals (length of estrous cycle). The inverse relationship between circulating progesterone and LH pulse frequency during the first estrous cycles of the postpubertal female has been characterized (25). As with the many other control systems, the potential for this feedback system to function exists well before puberty; experimentally, exogenous progesterone can decrease tonic LH secretion in the immature ovariectomized lamb (41).

Short Luteal Phase

The normally high luteal phase progesterone concentrations that suppress GnRH secretion to low levels may not appear during the first reproductive cycle. Rather, a small rise in circulating progesterone often occurs after the first preovulatory LH surge; another LH surge occurs 6 to 7 days later, which is then followed by a normal luteal phase in most lambs (Fig. 2, patterns I and II). The mechanism underlying the short luteal phase in the first reproductive cycle is incompletely understood. Removal of the uterus prevents regression of such corpora lutea (59), suggesting that a prostaglandin may be involved in this phenomenon, by virtue of either unusually high initial secretion of the uterine hormone or a high susceptibility of the first corpora lutea to the luteolytic action of prostaglandins. Based on studies of induced corpora lutea with a short life span in the anestrous adult ewe, some workers postulate that the luteal phase defect results from an asynchrony between follicular development and the LH surge (60–62). Most investigators believe that the presence of the brief progesterone rise ensures that subsequent luteal function is normal. Some contend that the critical synchronizing action of progesterone is at the level of the neuroendocrine system (62,63), whereas others suggest that it is directly on the ovary (60,61). Regardless of the cause of the short luteal phase, it is not a prerequisite for initiation of normal repetitive ovulatory cycles because it does not occur in all lambs (Fig. 2, pattern III) (6,14).

Sexual Receptivity

In the adult female sheep, progesterone plays an important "priming" role for estrogen induction of sexual behavior (64–67) (see Chapter 44). As determined by treatment of the ovariectomized ewe with physiological patterns and levels of steroids, the sequence for induction of estrus is an increase in circulating progesterone, withdrawal of progesterone, and an increase in circulating estradiol (68). The requirement for this hormonal sequence may be used to explain why the first and even second ovulation of the lamb during the pubertal period is not accompanied by estrous behavior (3,6) (Fig. 2). Before first ovulation, there is no corpus luteum, and hence there is no major increase in circulating progesterone to prime the neural centers that regulate sexual behavior to respond to the sustained estradiol rise that induces the first LH surge. The resultant short luteal phase generates only small amounts of progesterone, and therefore the increase in circulating estradiol before the second LH surge does not usually result in the expression of sexual behavior. Finally, after exposure of the brain to high concentrations of progesterone for several days during the first normal luteal phase, estrus first occurs in response to the rise in circulating estradiol typically during the third follicular phase. Therefore, as many as two silent ovulations (ovulations without estrus) may go unnoticed by the ram. Interestingly, estrus without ovulation has been reported in female lambs (69), and this raises the question of whether the adult might have a more stringent requirement for progesterone priming.

Other age-related differences have been noted in that the behavioral signs of sexual receptivity in the lamb, when they finally occur, are much weaker than in the fully mature female (70). Moreover, the duration of estrus is shorter in lambs (70–72). A fertile mating can occur during the first reproductive cycles, and a pregnancy can be carried to term (25,73). Compared with mature females, however, fertility of postpubertal lambs is lower [for reviews see (21,22)]. There is greater fertilization failure and embryo mortality (70), and this has been attributed to developmental abnormalities, including the timing of ovulation relative to estrus and a lower viability of the ovum (22).

Postpubertal Sterility: Seasonal Anestrus

Unequivocally, the greatest contributor to postpubertal infertility in the female sheep is seasonal anestrus (74). Under natural conditions, puberty in spring-born lambs is attained between 25 and 35 weeks of age, and estrous cycles occur at 15- to 17-day intervals. However, after only a few reproductive cycles, the young female becomes anovulatory and no longer exhibits estrous behavior (Fig. 14, top). As determined through use of an experimental model, the ovariectomized estradiol-treated lamb, the female once again becomes hypersensitive to estradiol feedback inhibition of LH secretion (Fig. 14, bottom). Thus, even before 1 year of age, the postpubertal lamb reverses its sensitivity to estradiol negative feedback and reverts to the anovulatory condition. This terminates the first breeding season.

The mechanism underlying seasonal cessation of estrous cycles in the sheep is hypogonadotropism due to insufficient GnRH secretion. The last cycle of the season occurs when the female attempts to drive the follicular phase, under increasing sensitivity to estradiol feedback inhibition that begins in late winter and early spring (see Seasonal Reproduction in Adult Sheep). After luteal regression and the decline in progesterone, GnRH pulses start to increase and begin the next follicular phase, but the GnRH secretory mechanism again becomes locked in a tight negative feedback loop with estradiol. Under increasing estradiol hypersensitivity, the follicular phase cannot be completed because the high frequency LH pulses required for development of the follicle to the preovulatory stage cannot be produced. Consequently, not enough estradiol is secreted to activate the preovulatory gonadotropin surge system or induce estrous behavior. There is a marked seasonal change in the inhibitory feedback effect of estradiol on the pulsatile secretion of GnRH in the adult ovariectomized ewe (75). During the breeding season, low doses of estradiol have no discernible inhibitory effect on GnRH secretion, whereas during the anestrous season, they produce a profound depression of GnRH pulse frequency.

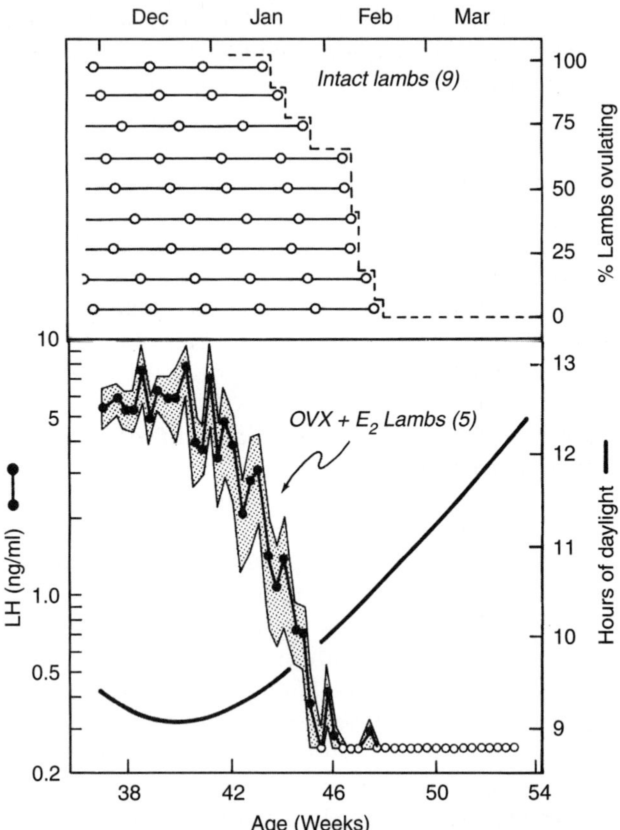

FIG. 14. Postpubertal infertility in lambs due to seasonal anestrus caused by an increase in sensitivity to estradiol inhibition of LH secretion. **Top:** Cessation of estrous cycles in individual lambs outdoors during late winter; estrous behavior (*open circles*). **Bottom:** Mean (± SEM) concentrations of circulating LH in ovariectomized estradiol-treated lambs raised under the same conditions as the intact females. Undetectable LH values (<0.25 ng/ml) (*open circles*); hours of natural light (*continuous solid line*). [Redrawn from (74) with permission.]

DETERMINANTS TIMING THE TRANSITION INTO ADULTHOOD

It is clear that the production of sustained high frequency GnRH pulses initiates the sequence of events that activate the ovary. Consideration of the details of brain mechanisms involved in the production of these high GnRH pulse frequencies can be found in Chapter 28 and Chapter 38. The fundamental question underlying the biology of puberty is what mechanism controls when this all-important increase in GnRH pulse frequency will occur. Is the timing by growth cues? How important are nongrowth cues? Is there a developmental clock regulated by genes? Why does puberty occur at different times in females and males? To begin to explore such issues, it is convenient to consider the sources of timing as internal and external.

Internal Determinants

In virtually all living organisms, sufficient growth is required to begin reproduction (Fig. 15). However, we remain fundamentally ignorant of how growth is perceived by the central nervous system in any vertebrate species. Whatever explanation ultimately evolves for mammals, it must include how somatic signals communicate with the hypothalamus to influence control of steroid-independent GnRH secretion as well as sensitivity of GnRH secretion to inhibitory gonadal steroid feedback.

Energetics of Sexual Maturation

Kennedy and Mitra (76) hypothesized over 40 years ago that changes in energy availability are important cues for puberty. They argued that sexual maturation is completed during an individual's transition from a state of reduced energy availability because of the demands of thermoregulation and growth to one of energy surplus when sufficient energy becomes available to support reproductive function. Whereas it is pointed out by Bronson and Manning (77) that the energetic cost of the physiological processes involved in ovulation are minuscule, the energetic consequences of pregnancy and lactation are expensive. A similar argument could be made for the male in which little energy is required for spermatogenesis, but sexual reproduction ultimately results in great energy expenditure during intraspecific competition for females.

Against this background, the energetics of the developing sheep can be considered. At birth, the lamb has almost no body fat (78,79) and therefore relies on milk from the mother for a continuous supply of energy. Compounding the lack of stored energy is the high energy expenditure of the newborn. In the very young lamb, the ratio of large surface area to relatively small body mass results in rapid heat loss, and metabolic rate must be correspondingly high to maintain core body temperature (78,79). The greatest energetic cost of the young lamb is thermoregulation and cellular maintenance (Fig. 16), and most of the remaining energy intake is allocated to muscular activity and growth. There is no excess energy as evidenced by the virtual absence of fat, and, consequently, according to the energetic hypothesis being developed, the priority for nonessential processes such as reproduction is low. The lamb remains reproductively inactive due to hypogonadotropism. As food consumption increases and the young lamb grows, body mass becomes larger and the relative surface area decreases, resulting in a corresponding decrease in metabolic rate. The eventual decline in metabolic rate allows more internal energy to become available for other physiologic demands,

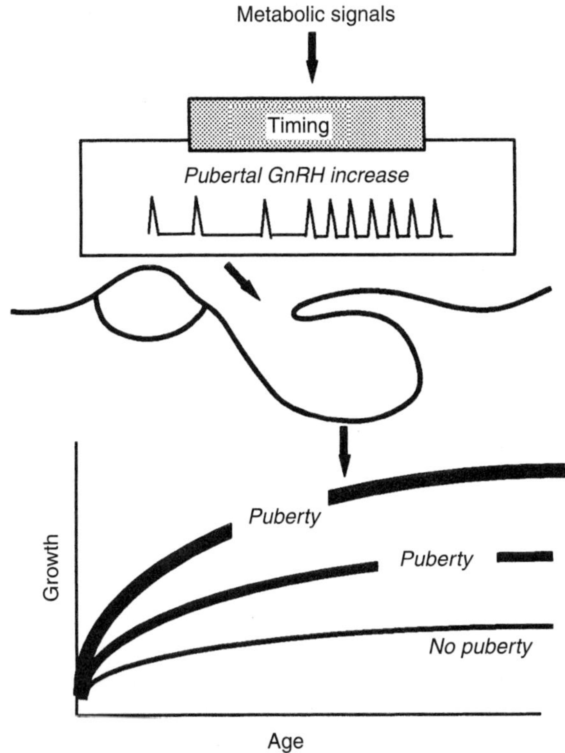

FIG. 15. Growth and the timing of puberty.

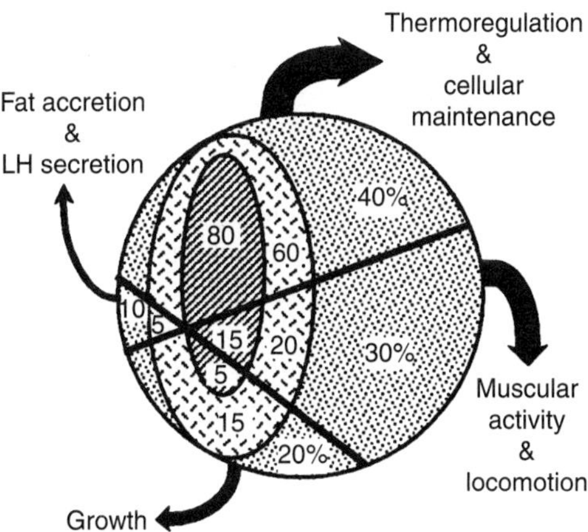

FIG. 16. Partitioning of metabolic fuel during development in the sheep. The circles represent increasing body mass from birth (*inner circle*) to maturity (*outer circle*). Each circle is divided into four quadrants to represent the estimated relative proportion of energy allocated at each stage of development. (Modified from D. C. Bucholtz, Ph.D. Thesis, University of Michigan.)

such as fat storage and reproduction. Central to this hypothesis is how the brain monitors energy metabolism both on a moment-to-moment basis and over the long term.

One potential approach to unravel how somatic signals in brain pathways might govern GnRH secretion during sexual maturation is to study the influence of slow or restricted growth. With reduced energy intake, most available energy is used for cell maintenance and thermoregulation and very little remains for growth. Thus, there is a delay in reaching the transition point at which metabolic rate declines and excess energy becomes available for fat deposition and reproduction (i.e., puberty is delayed). Eventually, if growth is sufficient, some energy can be expended for the eventual consequences of reproduction (pregnancy/lactation or male–male competition). Although one could argue that studying the slowly growing or growth-retarded individual may not reveal mechanisms timing normal puberty, it is axiomatic that the rate of growth depends on the available nutrition. Thus, because of vast differences in food supply, a wide range of "normal" growth rates occurs in nature (outside the laboratory) for each species. Studying individuals at the lower end of the growth rate spectrum can provide practical information about the timing of puberty in this species in agricultural operations, particularly in semiarid regions where food supply is unpredictable. Of greater importance is the fact that such studies could lead to some elementary understanding of how the brain normally monitors growth to determine when it is sufficient to increase GnRH secretory activity.

The influence of growth on GnRH secretion is readily apparent when the *ontogeny* of pulsatile LH secretion is examined in a population of developing lambs. Although the frequency of LH pulses increases in the neonatal period, the correlation is much greater with body weight (growth) than with age. This relationship is demonstrable in the presence (80) or absence (D. C. Bucholtz and D. L. Foster, unpublished observation) of the gonads. It is likely that such results reflect the differences in energy supplies available to individuals before weaning because the rates of lactation vary considerably between mothers. Experimentally, the relationship between LH pulse frequency and energy availability can be readily demonstrated by carefully controlling energy intake in lambs removed from their mothers and fed an artificial milk substitute (81). Reducing the daily intake of calories to such suckling animals greatly retards the increase in LH pulse frequency. Eventually, such animals begin high frequency LH secretion; however, they remain energetically sensitive because they have minimal energy (fat) reserves. Compared with rapidly growing peers

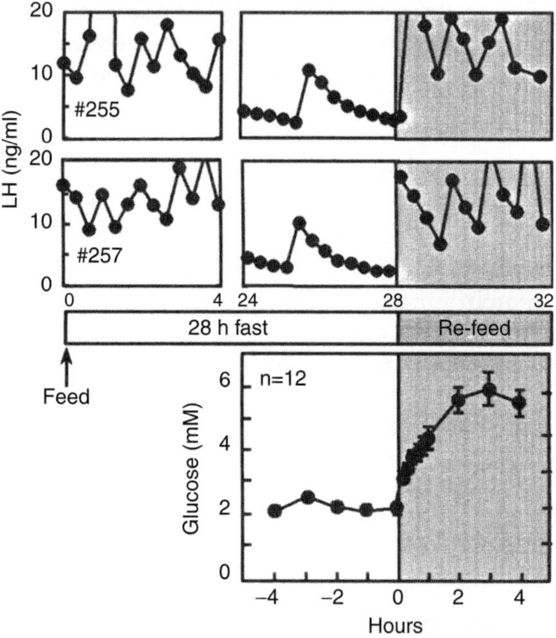

FIG. 17. Top: Influence of fasting and refeeding in two (255 and 257) slowly growing male lambs raised on artificial milk. LH pulse frequency diminishes upon missing normally scheduled meals *(left)* and increases upon refeeding *(right, shaded area)*. **Bottom:** Rapid increase in circulating glucose upon refeeding *(shaded area)* with artificial milk. [From (81).]

that have accumulated sufficient energy reserves to withstand short-term (24-hour) fasting, such slowly growing lambs exhibit pronounced moment-to-moment regulation of LH secretion (Fig. 17). Short-term fasting of such lambs gradually depresses LH secretion. Within 4 hours after realimentation on an artificial milk diet, LH pulse frequency increases rapidly, in some animals as quickly as within 20 minutes of refeeding. These findings indicate that in the absence of large energy reserves (fat) during development, LH secretion is highly sensitive to external sources of metabolic fuel (e.g., milk) on a moment-to-moment basis. From a broader perspective, growth may influence the time course of sexual maturation via the accumulation of somatic fuel stores capable of sustaining GnRH secretion between meals.

A long-held popular hypothesis, largely emanating from clinical literature and reviewed by Frisch (82), is that a certain level of fatness must be achieved to begin reproductive cycles in the female. However, until the recent discovery of the adipocyte-derived hormone leptin (83), there has been no mechanistic hypothesis about how fatness per se could ultimately affect the activity of GnRH neurons. Even after numerous studies examining the importance of leptin in the process of sexual maturation, it remains unclear whether leptin is the long-awaited answer to how the brain knows that the body is big enough to initiate reproductive activity or just one of many permissive factors that provide information about energy balance and regulate activation of the reproductive axis.

Leptin is a 16-kDa protein secreted primarily by adipocytes that acts within the hypothalamus to suppress appetite (84,85). The first link between leptin and reproductive function was made after the observation that administering leptin to obese, hypogonadal, leptin-deficient *ob/ob* mice suppresses appetite, reduces body weight, and establishes fertility (86). In adult sheep, peripheral concentrations of leptin are associated with measures of body fat (87–89), and circulating leptin levels typically decrease during periods of reduced energy intake or availability (88–90) that are associated with reduced gonadotropin secretion. For leptin to serve as an important blood-borne metabolic signal timing the pubertal increase in GnRH secretion, two criteria must be satisfied: (a) puberty is associated with an altered pattern of peripheral leptin secretion, and (b) exogenous leptin stimulates reproductive neuroendocrine function or advances the onset of puberty in otherwise sexually immature individuals (91).

A developmental increase in peripheral concentrations of leptin has been reported in the human (92–95), monkey (96–98), and rat (99–102), although the pattern differs between males and females and may not always correspond temporally with indices of sexual maturation. In general, leptin levels are biphasic in developing males, with the second peak occurring before puberty followed by a gradual decline in circulating leptin. Females also exhibit a peripubertal increase in leptin, but concentrations of leptin continue to increase after puberty. Although the contrasting effects of gonadal steroids on leptin secretion help explain sex differences in leptin levels after puberty, the differences in leptin levels with respect to the timing of puberty are less clear. In both sexes, the developmental changes in circulating concentrations of leptin typically precede, rather than coincide with the onset of high frequency GnRH and LH secretion. From these data on patterns of leptin, it is difficult to make a case that leptin triggers or initiates puberty. Nonetheless, leptin could be involved in timing the onset of puberty by providing information about growth and somatic maturation that is integrated with other developmental or environmental cues.

In the sheep, the pattern of leptin seems to depend on the sex, gonadal status, and nutritional conditions. In the neonatal lamb, serum concentrations of leptin are low (~2 ng/ml), as would be expected in view of its low amount of body fat (103). An initial increase in circulating leptin occurs during the first week of life in both females and castrated males, and this is thought to result from the transfer of leptin from the mother to offspring during early lactation. Leptin then gradually declines between 1 and 7 weeks of age in both sexes. Throughout this preruminant period (weaning typically occurs at 8–10 weeks of age), plasma leptin levels are regulated principally by level of nutrition (104). In males reared from birth on an ad libitum milk substitute diet, plasma concentrations of leptin continuously increased as lambs grew to a body weight of 20 kg at approximately 8 weeks of age, and leptin was positively correlated with body fat as a percent of empty body weight. However, when lambs were fed milk substitute to provide a low plane of nutrition, plasma leptin did not increase and leptin was not correlated with the amount of body fat, even though the percentage of empty body weight attributable to fat increased with age. In a study of the patterns of LH secretion in males fed a solid diet of moderate energy, leptin concentrations increased between 6 and 12 weeks of age (L. M. Jackson, R. A. Ehrhardt, Y. R. Boisclair, and D. L. Foster, unpublished data). Leptin levels were positively correlated with body weight at 6 weeks of age (prepubertal), but the correlation was not significant at older ages. Plasma concentrations of leptin also correlated with the frequency of LH pulses in prepubertal males, which suggests that the growth-dependent increase in LH pulse frequency could be associated with an increase in plasma concentrations of leptin. Whether a causal relationship exists between peripheral leptin levels and the onset of high frequency LH secretion in male sheep will require additional studies. Furthermore, peripheral leptin levels throughout development have not been characterized in female sheep; therefore, whether there are sex differences in the importance of leptin as a pubertal signal in this species is not known.

With regard to the second criterion for leptin as a pubertal signal, there is no consensus on the ability of exogenous leptin to stimulate gonadotropin secretion or advance puberty in sexually immature individuals. This could be due, in part, to the anorexigenic effects of leptin and the importance of adequate nutrition for activation of reproduction. In some studies, repeated injections of leptin advance puberty in female mice (105,106) and rats (107). Other studies failed to corroborate this finding (108,109), although exogenous leptin can prevent a delay in puberty caused by undernutrition (99). In agonadal juvenile male rhesus monkeys (16–20 months of age), short-term infusion of recombinant human leptin (16 days) did not increase LH secretion (110). However, daily injections of another recombinant human leptin from 12 to 30 months of age in prepubertal female rhesus monkeys increased gonadotropin secretion and advanced menarche (98). These contrasting results in the rhesus monkey could be related to the differing experimental approaches, differing gonadal steroid milieu, or sex differences in responsiveness to leptin as a metabolic signal. In addition, other developmental signals could regulate the prepubertal individual's response to leptin.

In developing sheep, the effect of exogenous leptin on reproductive neuroendocrine function has been studied in both sexes. In prepubertal females, LH secretion was not stimulated after 8–10 days of recombinant ovine leptin treatment delivered either intracerebroventricularly (111) or peripherally by intravenous infusion (112). With peripheral leptin treatment, feed intake was not suppressed, but the intracerebroventricular (ICV) treatment dramatically suppressed feed intake to the extent that one would predict a decrease in LH secretion. One could speculate that because the expected reduction in LH pulses did not occur, the centrally delivered exogenous leptin actually preserved pulsatile LH secretion. Thus, there is no clear evidence that GnRH or LH secretion is stimulated by artificially increasing leptin levels in young female sheep.

In the developing male sheep, the ability of leptin to stimulate LH secretion appears to depend on the stage of development and energy availability. The treatment paradigm used was previously shown to increase pulsatile LH secretion in a nutritionally hypogonadotropic adult model (gonadectomized, steroid-replaced, short-term fasted) (113). In this fasting model, subcutaneous administration of recombinant human leptin increases LH pulse frequency. Furthermore, leptin treatment is as effective at restoring reproductive neuroendocrine function as is refeeding. When the same dose of leptin that restored LH secretion in fasting adults was administered to prepubertal male lambs, LH pulse frequency increased, and the magnitude of the response depended on individual pretreatment LH pulse frequencies (113). These different patterns of gonadotropin secretion reflect the natural variability in levels of maturity at the same age. Leptin treatment did not increase the frequency of LH pulses in lambs not yet producing LH pulses or in those individuals already producing hourly (or faster) pulses. However, in some lambs that were not producing pulses before treatment and those lambs producing low frequency pulses (one to three pulses in 4 hours), LH pulse frequency increased after leptin treatment (Fig. 18). The ability of leptin to stimulate LH secretion in the prepubertal male contrasts with its effects in the female, which raises the possibility of sex differences in responses to growth cues timing the pubertal increase in GnRH secretion. However, as discussed below, female sheep are highly photoperiodic with respect to the timing of puberty, so it remains to be determined whether the sex difference

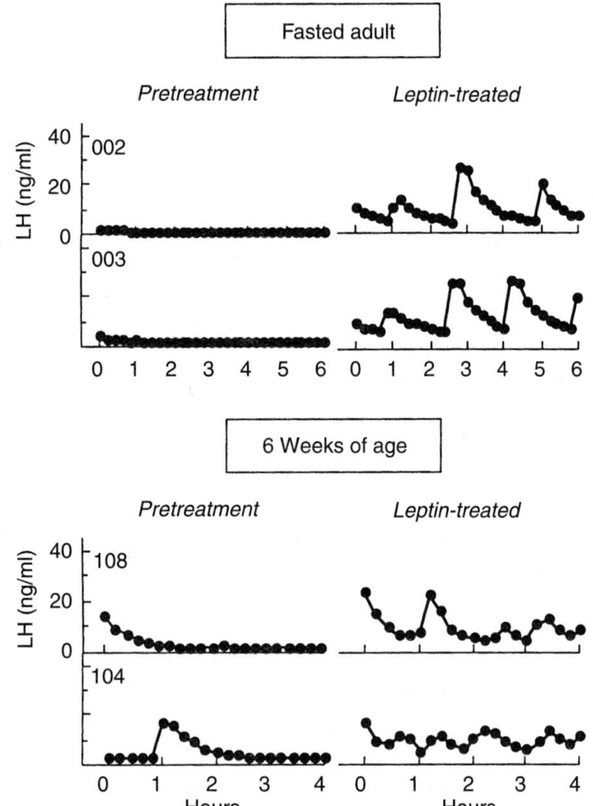

FIG. 18. Patterns of LH secretion in representative gonadectomized estradiol-treated male sheep before and after treatment with recombinant human leptin (12.5 μg/kg subcutaneous every 4 hours). **Top:** LH pulse frequency is reduced in adults fasted for 48 hours (*left*); 48 hours of leptin treatment while continuing the fast increases LH secretion (*right*). **Bottom:** LH pulse frequency is low in nonfasted growing lambs at 6 weeks of age (*left*); pulsatile LH secretion is increased after 72 hours of leptin treatment (*right*), although responses were variable (see text for discussion). [Data from (113).]

in the response to leptin in prepubertal individuals is due to differences in the responses to growth cues, photoperiod, or both.

The response in prepubertal male sheep to leptin differed from that in fasted adults. In all the nutritionally hypogonadotropic adults, leptin increased LH secretion, but this was not the case in all the developmentally hypogonadotropic lambs. Such a finding suggests additional developmental or metabolic signals that control the hypothalamic–pituitary response to leptin before puberty, and hypogonadotropism in the prepubertal individual experiencing limited energy availability due to rapid growth may not be functionally equivalent to nutritionally induced infertility in the adult. In the growth-retarded male lamb, responsiveness to leptin treatment does not improve with increasing age (Fig. 19) (114). In fact, continued growth restriction reduced pulsatile LH secretion in the absence of leptin treatment and eliminated a neuroendocrine response to leptin in individuals who had been responsive at an earlier age when growth

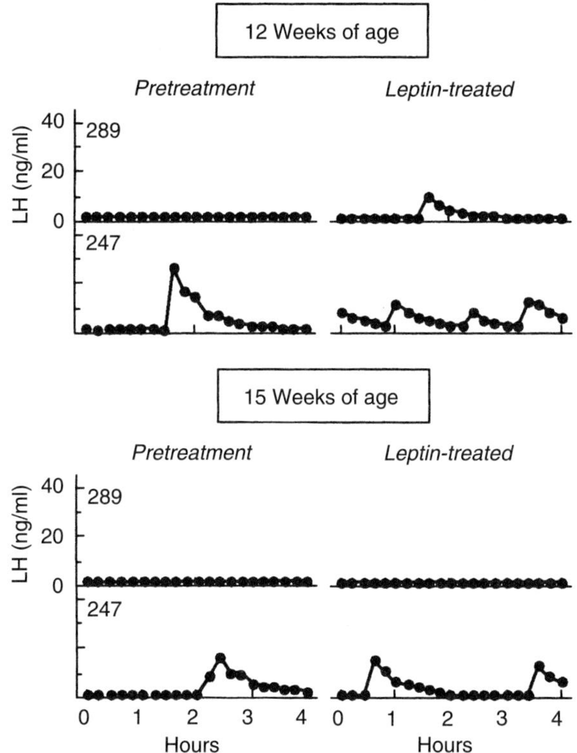

FIG. 19. Patterns of LH secretion in representative gonadectomized estradiol-treated male lambs fed a restricted postweaning diet to maintain body weight at 20 kg and treated with recombinant human leptin (12.5 μg/kg subcutaneous every 4 hours) for 72 hours. **Top:** Exogenous leptin stimulated LH secretion at 12 weeks of age. **Bottom:** Restricted growth prevented a developmental increase in pretreatment pulsatile LH secretion at 15 weeks of age, and leptin treatment was less effective at increasing LH secretion. [Data from (114).]

was normal or less severely restricted. These findings indicate that the mechanism regulating the onset of high frequency GnRH secretion relies on one or more metabolic cues for information regarding growth and energy balance. Although leptin can stimulate gonadotropin secretion in the prepubertal sheep, the response appears to be regulated by other developmental or metabolic signals, and thus this fat-derived hormone must be considered as a permissive signal timing puberty.

Before the discovery of leptin as a potential candidate metabolic cue for puberty, the correlation of fatness with onset of fertility prompted other studies of the role of energy metabolism in timing puberty. Evidence for energy control of central mechanisms involved in the regulation of LH secretion stems from studies using competitive inhibitors of glucose (2-deoxyglucose, 2DG) and free fatty acids (methylpalmoxorate). As reviewed by Wade and Schneider (115), historically such blockers have been used in studies of appetite control, but their usefulness in the study of reproduction was later realized. For example, concurrent treatment with the inhibitors of both fatty acid and glucose oxidation in the hamster disrupts estrous cyclicity (116). When such an approach was applied to the developing sheep, pulsatile secretion of LH was suppressed within a few hours; interestingly, these treatments suppressed frequency in some lambs, whereas in others they suppressed amplitude (117). Moreover, because the fatty acid blocker was administered after the glucose blocker, it is uncertain whether the major suppression of LH occurred in response to inhibition of glucose metabolism or fatty acid metabolism. In another study, when 2DG was administered alone, clear pulses of LH in the ovariectomized lamb were abolished (118). Presumably, the 2DG blocked the release of endogenous GnRH because both exogenous GnRH and a GnRH secretagogue (*N*-methyl-D,L-aspartate) were able to maintain pulsatile LH secretion when they were administered within 1 hour of 2DG treatment. Such studies provide evidence for a role of energy balance in the moment-to-moment regulation of LH secretion. Interestingly, leptin may modulate glucose availability to provide information about energy balance because administration of 2DG blocks the ability of leptin to stimulate LH secretion during fasting (119). Another possibility is that leptin, glucose, and perhaps other substances serve as independent metabolic signals that are integrated within the brain to provide information about overall energy balance. If this is the case, perhaps a hierarchy of metabolic signals exists and the action of one signal is dependent on the relative input provided by the remaining constellation of signals.

The metabolic hormone insulin-like growth factor I (IGF-1) has been proposed to be an important signal for puberty because circulating levels of the hormone increase during puberty in the rat (120,121), monkey (97,122), and sheep (123,124). In rodents, disruption of the growth hormone (GH)–IGF-1 axis by deletion of the gene for the GH receptor (125) or passive immunization against GH-releasing hormone (126) reduces growth and delays sexual maturation. Furthermore, administration of IGF-1 increases peripheral concentrations of LH and advances puberty in the female rat (127); in female monkeys, IGF-1 accelerates the decline in sensitivity to the negative feedback effects of estradiol (128) and advances the age at first ovulation (129).

In female lambs, the age at puberty is correlated with body weight and concentrations of circulating IGF-1, and the GH response to a GH-releasing hormone challenge tend to be greater in those individuals reaching puberty at younger ages (130). Similarly, plasma concentrations of IGF-1 in developing male lambs are positively correlated with testis size, plasma testosterone, and body weight, and IGF-1 levels are reduced in juvenile sheep during feed restriction and delayed puberty (123). A direct effect of IGF-1 on the timing of puberty in sheep has not been studied, but chronic treatment with GH increases plasma concentrations of IGF-1 in prepubertal females and advances the age at which high frequency (hourly) pulses of LH are produced (131). In adult sheep, peripheral IGF-1 acutely increases LH secretion (123), and the magnitude of the response is amplified in estradiol-treated or undernourished individuals. However, contrary to evidence in the rat (132), the ovine hypothalamus appears to contain few IGF-1 receptors (133), although IGF-1 could act through hypothalamic insulin or estradiol receptors to regulate GnRH secretion. IGF-1 receptors and binding proteins are abundant in the ovine pituitary gland (134), and IGF-1 stimulates LH secretion in vitro from cultured ovine pituitary cells in a dose-dependent manner (133). Furthermore, IGF-1 increases in peripheral circulation but not in cerebrospinal fluid after an increase in nutrients that stimulates gonadotropin secretion (135), and ICV administration of IGF-1 does not alter pulsatile LH secretion (136). Thus, the major impact of IGF-1 on gonadotropin secretion in the sheep appears to be at the level of the pituitary gland.

Insulin is another metabolic hormone that has been proposed to regulate LH (GnRH) secretion [for review see (137)]. Specifically, low nutrition would result in low insulin concentrations that, in turn, would provide information to the GnRH neurosecretory system to reduce its function. This hypothesis has been tested in the growth-retarded lamb by acute injection of insulin into the lateral ventricles in an attempt to replicate improved energy balance and thereby to provide a positive signal for GnRH secretion (138). Central administration of insulin minimized possible peripheral effects on glucose, amino acids, free fatty acids, and other metabolites that may affect neuroendocrine function. Acute injections of insulin failed to stimulate LH secretion, which indicates that the increased circulating concentrations of insulin associated with increased feed intake after growth restriction may not be causal to the accompanying increase in LH secretion. Such findings do not support the role of peripheral insulin alone as an important signal mediating the effects of metabolism on GnRH secretion in the lamb. However, in another study, mean LH concentrations increased after 2 and 4 days of ICV infusion of insulin to diet-restricted pubertal female sheep (139), despite the hypophagia induced by the insulin treatment. The difference between the responses to acute and chronic insulin treatment raises the possibility that central levels of insulin must be integrated over time or with other metabolic factors to provide information about nutrient status to the GnRH neuronal system.

Although not studied in the developing sheep, the influence of the other side of the energy equation, namely energy expenditure, has been studied in rodents. For example, without an increase in energy intake, either cold exposure (140) or exercise (141) delays puberty in the female rat. Presumably, this is through the suppression of GnRH secretion because pulsatile infusions of GnRH reinstate normal sexual development (142).

Signals and Pathways of Energy Metabolism in Modulating GnRH Secretion

Where energy metabolism might be transduced into signals affecting GnRH secretion and what neurotransmitter pathways are involved are yet unknown. Only limited information has accumulated on central mechanisms responsible for steroid-independent hypogonadotropism during slow growth. Various lines of evidence indicate that the reduced gonadotropin secretion in the growth-retarded lamb reflects reduced GnRH secretion, rather than inadequate pituitary function (Fig. 20). *First*, the frequency of GnRH pulses, as measured in the pituitary portal vasculature, is markedly reduced in growth-retarded lambs compared with rapidly growing lambs (143). *Second*, GnRH

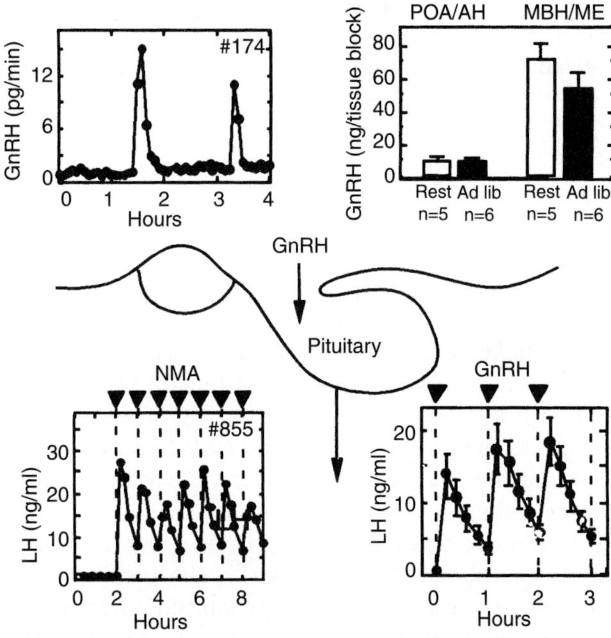

FIG. 20. Central mechanisms controlling pulsatile LH secretion in the nutritionally growth-restricted female lamb. See text for explanation of insets. **Top:** Pattern of GnRH in pituitary portal blood of a representative growth-retarded lamb (*left*). Mean (± SEM) hypothalamic GnRH content in lambs fed restricted (Rest) and ad libitum (Ad lib) diets. POA/AH, preoptic area/anterior hypothalamus; MBH/ME, mediobasal hypothalamus/median eminence (*right*). **Bottom:** Patterns of LH secretion in a representative growth-restricted lamb injected with *N*-methyl-D,L-aspartate (5 mg/kg body weight) (*left*). Mean (± SEM) LH concentrations in lambs injected with GnRH (5 ng/kg body weight, *n* = 15) (*right*). [From (272); portal GnRH data from (143).]

content in the median eminence (terminals) and preoptic area (cell bodies) is not diminished in growth-retarded lambs (144). *Third*, hourly intravenous administration of the putative neurotransmitter glutamate/aspartate (*N*-methyl-D,L-aspartate) to nutritionally growth-retarded lambs produces corresponding LH pulses that are similar in size to those induced by a physiological dose of GnRH (144). *Fourth*, the anterior pituitary gland contains adequate concentrations of LH and FSH (145), and physiological doses of GnRH readily induce LH and FSH secretion (144). These findings, considered together, infer that the central mechanisms controlling GnRH release, rather than synthesis of GnRH or LH, are limiting LH secretion when sexual maturation is delayed by growth retardation.

There is an increasing amount of information regarding the identity of neurotransmitter systems that regulate GnRH secretion and that could be altered during nutritionally induced growth retardation and hypogonadotropism. Increased inhibition by endogenous opioids was one of the first mechanisms postulated whereby metabolic changes influence the pulsatile secretion of GnRH. In the rat, there is limited evidence that fasting impairs LH secretion by activating an endogenous opioid pathway (146,147). In the growth-restricted lamb, however, such evidence has not been obtained. Although endogenous opioids appear to be involved tonically in the steroid-independent inhibition of LH secretion in the lamb, as discussed in the section on Decrease in Sensitivity to Estradiol Feedback Inhibition of GnRH Secretion during Puberty, whether this inhibition is heightened during growth retardation is uncertain. Blockade of endogenous opioid receptors with naloxone in the hypogonadotropic gonadectomized lamb does not increase LH pulse frequency at doses that are effective in this regard in the rapidly growing agonadal female (Fig. 21). Even greater doses of naloxone (144) or the long-acting antagonist, naltrexone (H. I'Anson and D. L. Foster, unpublished observation), have not been able to disinhibit LH secretion. One explanation is that other inhibitory neuronal pathways are also activated during periods of reduced energy metabolism. Thus, opioidergic inhibition of GnRH secretion may occur, but it would be masked by one or more other inhibitory neurotransmitter pathways, and blockade of only opioid receptors by an antagonist would be without an effect on GnRH secretion. The use of multiple pharmacologic agents, in addition to the opioid antagonist, would be required to test this possibility. Nevertheless, based on the existing evidence, we conclude that a massive increase in opioid tone cannot be the primary cause of hypogonadotropism in the growth-retarded lamb.

Neuropeptide Y (NPY) is another candidate neurotransmitter for modulation of GnRH secretion during

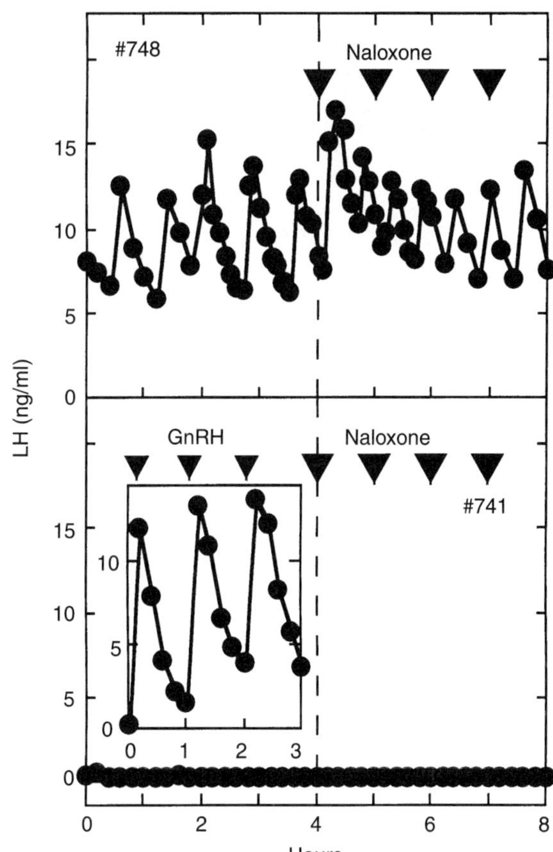

FIG. 21. Top: Circulating LH in a rapidly growing ovariectomized lamb before and after treatment with naloxone (1 mg/kg body weight). **Bottom:** Circulating LH in a growth-retarded ovariectomized lamb treated with the same dose of naloxone. Response to three GnRH injections (5 ng/kg body weight) on the previous day is shown (*inset*). [Redrawn from (270).]

inadequate growth during low nutrition. Hypothalamic NPY neurons are responsive to metabolic signals as evidenced by colocalization of leptin receptors on NPY neurons in rodents (148) and nonhuman primates (149) and expression of leptin receptor mRNA in the arcuate nucleus in the sheep, an area that also contains many NPY neurons (150). Furthermore, hypothalamic NPY mRNA expression (151,152) and secretion of NPY (152) are reduced in response to leptin treatment. Concentrations of NPY in the cerebrospinal fluid are higher in food-restricted, ovariectomized, estradiol-replaced lambs than in well-nourished lambs (153). In ovariectomized sheep, ICV infusion of NPY decreases LH secretion (153–155). As shown in Fig. 22, high frequency LH pulses are produced in the absence of steroid negative feedback, and a single ICV injection of porcine NPY resulted in an immediate but transient suppression of LH secretion (Fig. 22, bottom) whereas a lower dose was without effect (Fig. 22, top).

Because ICV injections of NPY also stimulate feeding in sheep (156,157) as in other species [for review

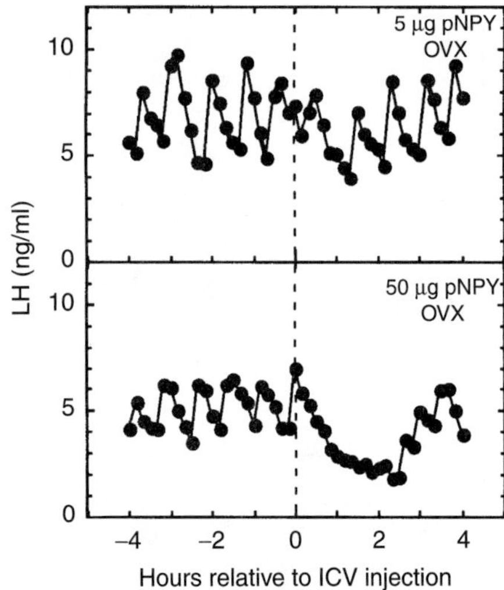

FIG. 22. Circulating LH in representative ovariectomized sheep before and after intracerebroventricular injection of 5 μg porcine NPY *(top)* and 50 μg porcine NPY *(bottom)*. [From (153) with permission.]

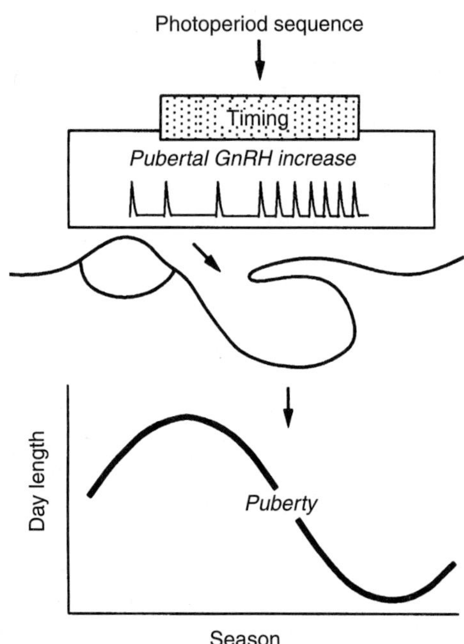

FIG. 23. Photoperiod and the timing of puberty.

see (158)], the neuropeptide may serve a dual role in the pathway of signal transduction that conveys information about metabolic cues to the neuronal systems controlling GnRH secretion and ingestive behavior. In a prepubertal individual, elevated NPY levels would suppress GnRH secretion until circulating metabolic signals (e.g., leptin) reflect somatic maturation at which point the inhibitory NPY input would diminish and GnRH secretion would increase.

It is clear that energy balance is important to the ontogeny and maintenance of pulsatile GnRH secretion, and our understanding of the metabolic signals that potentially provide information to the brain about energy availability is expanding. However, a number of questions remain unanswered about the signals, sensors, and pathways by which growth cues time puberty. For example, what specific aspect of metabolism or growth provides information to the brain that sufficient energy is available to begin reproductive activities that are induced by GnRH secretion? Is there one ultimate signal or is some combination of metabolic cues required? Which specific brain regions are involved in transforming signals of energy availability to signals modulating GnRH secretion? Do growth-related signals about energy metabolism that regulate steroid-independent GnRH secretion also modulate sensitivity to steroid feedback inhibition of GnRH secretion? Finally, how are these internal cues integrated with external cues that provide information about the environment to the developing individual?

External Determinants

Bearing in mind that the sheep is a seasonally breeding species, it is logical to consider that external cues might well become integrated with internal growth-related cues to influence the activity of the GnRH pulse generator (Fig. 23). Accordingly, if environmental information is received indicating that the season for puberty is inappropriate, the expression of high frequency GnRH pulses will not be permitted, despite the fact that growth-related information indicates that an adequate size or sufficient energy balance for puberty has been attained. Similarly, sexual maturity will not occur if seasonal cues indicate that the appropriate time of the year for puberty has arrived but metabolic cues indicate that the lamb cannot bear the energetic consequences of increasing GnRH pulse frequency and initiating reproductive function. The sustained increase in GnRH pulse frequency occurs only when internal and external cues indicate that both size and season are favorable for initiating a pregnancy. To appreciate mechanisms of how seasonal cues might impact neuroendocrine mechanisms regulating the onset of fertility in the female lamb, the seasonal nature of reproduction in the adult female sheep must be considered (see below and Chapter 41) [for review see (159,160)].

Seasonal Reproduction in Adult Sheep

Although the sheep has long been associated with humans, it retains a seasonal pattern of reproduction (Fig. 24). Conception is restricted to a certain periods of the year—the *breeding season*. During the remaining *anestrous season*, the female is anovulatory and does not exhibit estrous behavior. As with other seasonal breeders, the ability of the sheep to turn the

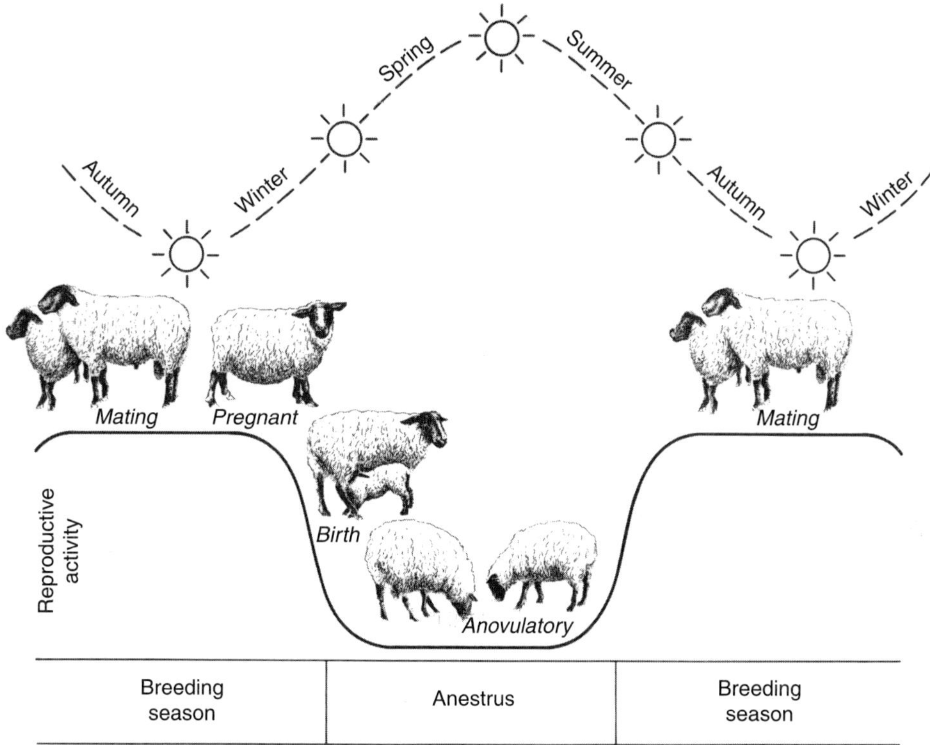

FIG. 24. Seasonal reproduction in female sheep of the Suffolk breed. During the breeding season (~6 months, autumn and winter), when reproductive activity is high, ovulations occur at 16- to 17-day intervals and mating is possible; during the anestrous season (~6 months, spring and summer), mating does not occur because the female is anovulatory. [From (200) with permission.]

reproductive system off and on, a process of reversible fertility, may be viewed as "nature's own contraceptive" (161). Because the length of pregnancy in the ewe is about 5 months, the breeding season occurs in autumn and winter so that the births of lambs are clustered in spring, the "season of plenty." This reproductive strategy allows access to high-quality forage, both to meet the demands of lactation for the ewe and to meet the subsequent growth requirements of the postweaning lamb. As pointed out later, birth in spring is advantageous in another regard. It allows most postnatal development to occur during summer so that puberty can occur during the same year, in the first autumn after birth. Artificial selection and better nutrition have extended the breeding season, and now a gradation exists from the monoestrous state of some wild species to the polyestrous condition of domestic breeds (72). Various breeds are capable of mating for as many as 8 to 10 months of the year (162,163), and some individuals have been reported to exhibit estrous cycles at 15- to 17-day intervals throughout the year (73,164). This suggests that the potential for continuous reproduction exists in the sheep, and artificial selection could make this possible in the future. Presently, assisted reproductive technologies, including hormonal manipulations, facilitate conception and birth of lambs at any time of the year if so desired.

Numerous types of experiments have demonstrated that day length is the primary environmental factor timing the annual reproductive cycle of the sheep. Originally, Marshall (165) determined that transferring female sheep across the equator to a locale where environmental variables were 6 months out of phase eventually shifted the breeding season 6 months. More definitively, when artificial photoperiods are used to reverse the annual changes in day length that occur under natural conditions, the breeding and anestrous seasons are reversed (166–168). When changes in annual photoperiod are accelerated to 90-day cycles of long days and short days, the reproductive cycle is also accelerated to produce two breeding seasons and two anestrous seasons per year (169). Finally, manipulation of the circadian pattern of melatonin, the pineal hormone involved in transducing photoperiodic information into a neuroendocrine signal, results in changes in timing of when reproductive cycles occur [for review see (160)]. These types of investigations in which photoperiod, or its pathway of transmission into neuroendocrine signals, is changed offer compelling evidence for a deterministic role for day length in synchronizing breeding activity among mature female sheep.

As discussed below, this pathway is also used by the developing female to synchronize its first reproductive cycles to the autumnal breeding season of the adult.

Photoperiod and Puberty

The month of birth influences the age at puberty, and this is most evident in those breeds that exhibit the most pronounced seasonal reproduction. The classic study by Hammond (2) in 1944 led to the general conclusion that regardless of season of birth, lambs would reach puberty only during the breeding season. Moreover, it has been noted in several breeds of sheep that winter-born and spring-born lambs reach puberty at about the same time of the year, but because the latter are born several months later, they are much younger at the first mating (2,72,73,163,170). Births later in the year (summer) may even further reduce the age at puberty (5,163,171). However, births extremely late in the year (autumn) are not advantageous because the age at puberty is no longer reduced. By contrast, such lambs delay initiation of reproductive cycles until the following year to coincide with the subsequent breeding season (2,73,163,170,172).

Long-standing evidence that photoperiod is the major environmental factor phasing the annual rhythm of reproduction in the adult female led to the logical consideration that seasonal changes in day length might also serve as the outdoor cue to synchronize the time of puberty to the appropriate season for first conception. This is summarized in Fig. 25. According to this scheme, the developing sheep maintains a photoperiod history from a very early age using the same photoneuroendocrine pathway as does the adult to transduce seasonal changes in day length into humoral signals that eventually modulate neuroendocrine function. The long days of spring and summer, followed by the decreasing day length of autumn, provide the stimulatory (permissive) photoperiod sequence to initiate puberty. As reviewed elsewhere for the adult (159), information about length of day (hours of light versus hours of darkness) is relayed from the retina through a retinohypothalamic tract via the suprachiasmatic nuclei, which provide circadian input; the information then proceeds through the paraventricular nuclei and then via a multisynaptic pathway involving the superior cervical ganglia to the pineal gland. Thus, the environmental light–dark cycle entrains the circadian-based melatonin rhythm such that melatonin is only secreted for the duration of the dark phase of the photoperiod. Long days produce a short duration of high melatonin secretion, whereas short days induce a long duration. These patterns of melatonin form the basis of a neuroendocrine record of photoperiod experience that modulates the frequency of GnRH secretion. Through this pathway, many classes of rhythms become involved in information transfer about season to the young sheep. The *annual* cycle of photoperiod changes influence the *circadian* pineal rhythm to influence the *ultradian* brain rhythms that modulate the *circhoral* production of the neuropeptide GnRH that gives rise to the pattern of LH to which the ovary is exposed.

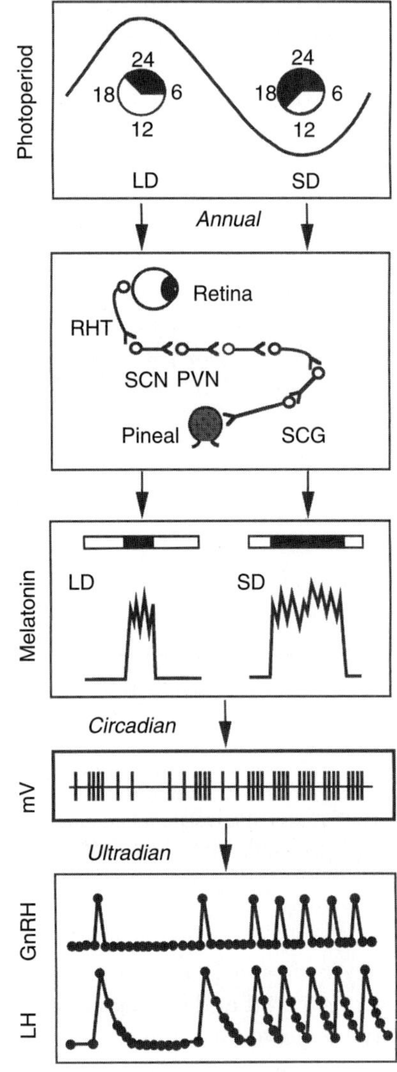

FIG. 25. Interplay among rhythms timing puberty involving day length, the pineal gland, electrical activity of GnRH neurons, and GnRH/LH secretion. LD, long day; SD, short day; RHT, retinohypothalamic tract; SCN, suprachiasmatic nucleus; PVN, paraventricular nucleus; SCG, superior cervical ganglion. [Redrawn from (183).]

Evidence for key elements of the foregoing integrated mechanism include the following: (a) a seasonally reversed artificial photoperiod sequence can reverse seasonally delayed puberty, (b) normal puberty requires that long days must be experienced before short days, (c) slow LH pulse frequency due to prolonged hypersensitivity to estradiol negative feedback is associated with photoperiodically delayed puberty, and (d) removal of pineal gland function delays puberty, and timed infusions of melatonin restore puberty to normal.

Spring-born Versus Autumn-born Lamb: A Case of Delayed Puberty. A study of the age at puberty in lambs (Suffolk breed) born in spring or in autumn was one of the first to implicate photoperiod as an important environmental timer of puberty (172). In that study, shown in Fig. 26, onset of cycles was determined by estrous behavior using mature vasectomized males housed continuously with the lambs; the increase in circulating progesterone to luteal phase concentrations was used to verify that reproductive cycles had begun. In a natural photoperiod, spring-born lambs attained puberty at about 30 weeks of age during the breeding season (Fig. 26, second panel).

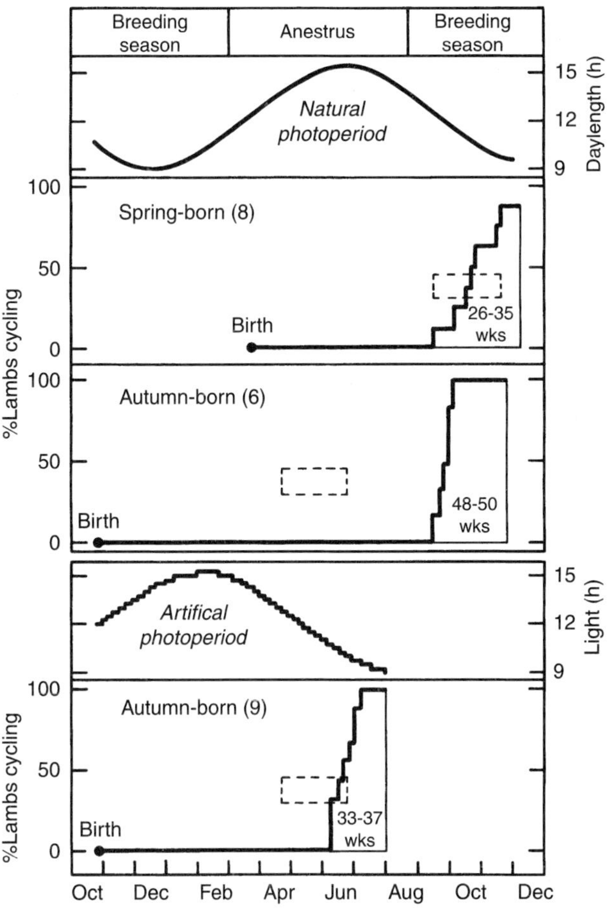

FIG. 26. Delay in puberty in autumn-born lambs due to inappropriate photoperiod during development. Shown are the ages and months at first ovulation in spring-born and autumn-born females raised in natural photoperiod (*first three panels*) and in autumn-born females reared in an artificial seasonally reversed photoperiod (*last two panels*). Ovulations were initiated in spring-born females during the age range of 26–35 weeks (*dashed rectangle*). [From (19) with permission.]

By contrast, when autumn-born lambs raised in a natural photoperiod reached the pubertal age of 30 weeks during the anestrous season, reproductive cycles did not occur despite their having sufficient somatic development for initiation of ovulations (Fig. 26, third panel). Rather, puberty was delayed until the following breeding season, when the lambs were nearly 1 year old. This seasonally induced delay in onset of reproductive cycles can be attributed to day length. Evidence supporting this inference is that the age at puberty was restored to normal in autumn-born females when they were raised in an artificial annually reversed photoperiod simulating a spring birth (Fig. 26, bottom panels).

Although the foregoing study provides evidence that lambs of the Suffolk breed are highly photoperiodic with respect to timing of puberty, this may not be the case in all breeds. Figure 27 illustrates a study of the effect of season on timing of puberty in the D'man breed (173). The D'man is a subtropical sheep that has evolved under a highly intensified production system in the oases of pre-Saharan regions with a pattern of relatively continuous reproduction (90% conception rate in both spring and autumn) (1,174). Thus, it provides an exception to the rule that reproduction in the sheep is highly seasonal. As a comparison with the D'man breed in the same area (Morocco),

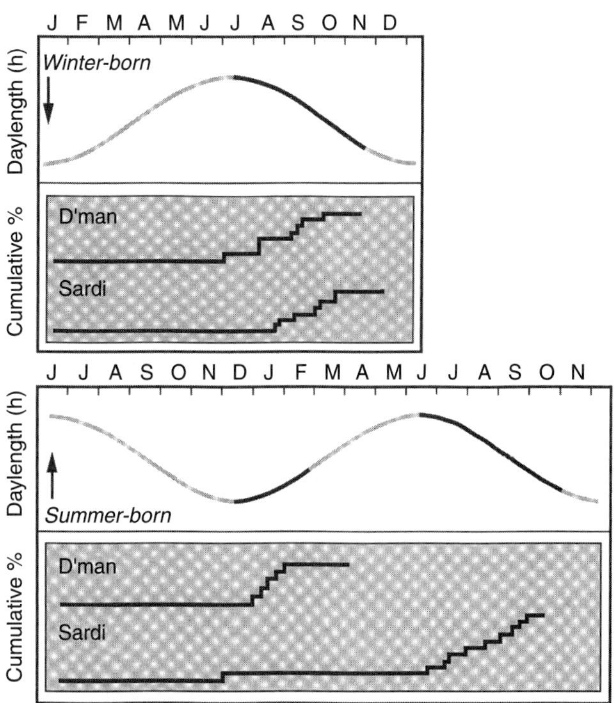

FIG. 27. Timing of puberty, as evidenced by the first rise in circulating progesterone to luteal phase concentrations in lambs of the D'man breed and of the Sardi breed. Lambs were born either in winter (*top*) or summer (*bottom*) and raised in natural environmental conditions in Morocco. [Data from (173), with permission.]

the Sardi breed is raised under much more open conditions and is seasonal in its pattern of reproduction. In this study conducted in Morocco under a natural photoperiod, conception was regulated so that lambs of both breeds were born in two different periods of the year, January and June. Winter-born lambs of both breeds attained puberty as evidenced by initiation of luteal function in the autumn, much the same as the Suffolk lamb as shown in Fig. 26. However, in the strongly seasonal Sardi breed, lambs born in the summer delayed puberty, again much like the autumn-born Suffolk female. By contrast, summer-born lambs of the D'man breed initiated reproductive cycles at the same age as when they were born in the winter. Taken together with information about seasonality of reproduction in the adults of these breeds, one can conclude that seasonal cues play a role in timing of puberty only if reproduction in the adult is seasonal.

Photoperiod is used to time puberty through the timing of the decrease in sensitivity to estradiol inhibition of LH secretion. This is shown in Fig. 28 for the ovariectomized model in which the potential to produce high frequency LH pulses typical of the follicular phase is revealed at an early age (e.g., by 12 weeks of age). In the presence of physiological levels of circulating estradiol (by subcutaneous implant), only a low frequency is maintained; the time when high frequency pulses are expressed depends on photoperiod history. The lamb raised in a stimulatory (i.e., permissive) photoperiod does not express the high frequency LH pulses until 25–30 weeks of age in the presence of estradiol (Fig. 28, top panels, OVX+E), despite its having achieved this potential at a much earlier age (Fig. 28, top panels, OVX). However, the lamb raised in an inhibitory photoperiod does not reduce its sensitivity to estrogen feedback inhibition at the same time as the lamb raised under the stimulatory photoperiod, and LH pulse frequency remains suppressed for much longer (Fig. 28, bottom panels, OVX+E).

In the context of season of birth and the timing of puberty (Fig. 26), the delay in initiation of reproductive cycles in the autumn-born lamb reared outdoors

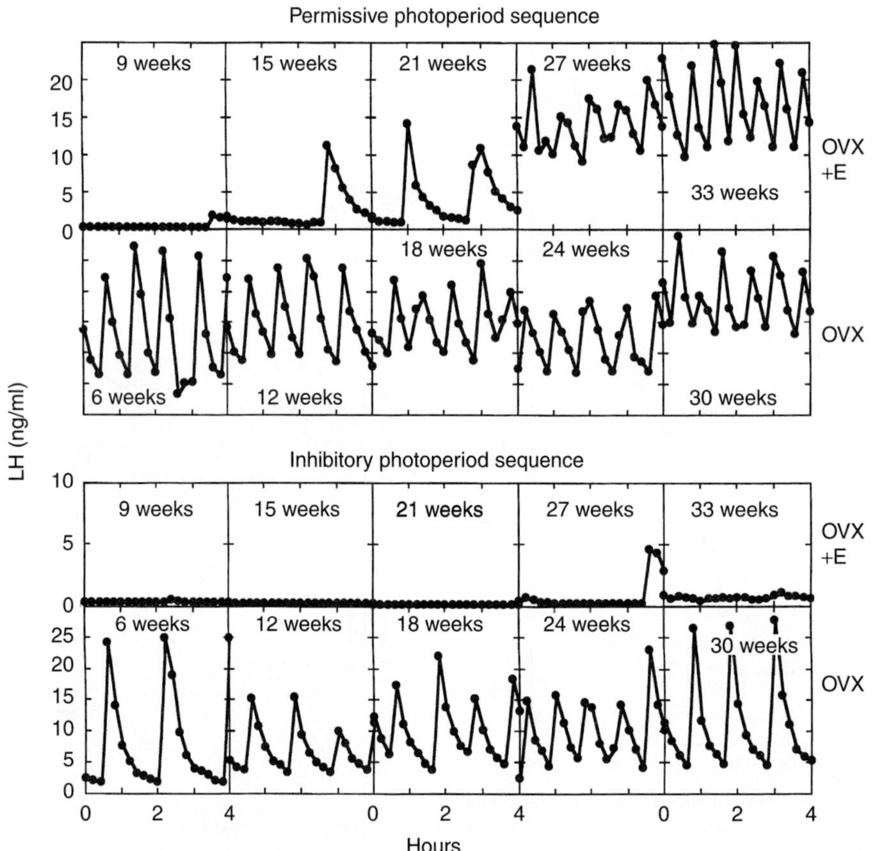

FIG. 28. LH patterns at different ages in a lamb maintained in a permissive photoperiod sequence for puberty (*top two panels*) and in a lamb maintained in an inhibitory photoperiod sequence for puberty (*bottom two panels*). For each lamb, upper panels depict LH patterns 3 weeks after replacement of a constant release estradiol capsule; lower panels depict LH patterns 3 weeks after removal of estradiol. [From (40) with permission.]

arises from prolonged hypersensitivity to estradiol negative feedback (172). Whereas the response to estradiol feedback inhibition of LH secretion diminishes in the spring-born lamb at 30 weeks of age during decreasing day length, the autumn-born female remains hypersensitive to estradiol at this age because of increasing day length. This maintains tonic LH secretion at low levels, thereby preventing the initiation of preovulatory follicular development (and subsequent pregnancy) during the long days of summer in these females that were born in the "wrong" season. The reduction in feedback sensitivity then occurs during the decreasing day length of autumn to synchronize puberty to the breeding season. Finally, this delay in reduction of sensitivity to estradiol negative feedback can be prevented experimentally. When the autumn-born female is raised indoors in an artificial annually reversed photocycle, the decrease in feedback sensitivity is restored to the normal age (~30 weeks) (172). Therefore, photoperiod cues, as well as growth-related cues, are used to time the initiation of ovulations in the developing sheep.

Studies of neuroendocrine timing of puberty in nonseasonal breeds of sheep have only begun. In this regard, the pattern of LH secretion during puberty in Sardi (strong seasonal breeder) and in D'man (aseasonal) accords with the timing of puberty, as shown in Fig. 27. In the D'man breed, the increase in LH secretion was the same regardless of whether the lambs were born in winter or summer. By contrast, in the summer-born Sardi female, much like the Suffolk female, the increase in LH pulse frequency did not occur at the usual age and was delayed until the next breeding season. Thus, in some breeds of sheep, only growth cues may be used to time puberty, whereas in others, both growth cues and photoperiod cues are important [L. Derqaoui, unpublished observation and (173)]. Studies of female Soay sheep using photoperiodic and growth manipulations suggested that within the body weight range examined, photoperiod is the dominant cue timing puberty in this less domesticated breed (175). The finding that photoperiod can induce puberty in Soay females with very limited growth while the same degree of growth restriction delays puberty in Suffolk females could simply reflect an extremely low threshold for the energy balance required to initiate high frequency GnRH secretion in the more feral breed. However, it could also represent an adaptation to harsh environmental conditions in which adequate forage during lactation and weaning are critical for offspring survival. Strict adherence to photoperiodic cues for timing puberty would optimize reproductive success by minimizing the number of births that would occur during late winter or early spring when food availability is unpredictable.

Photoperiod Sequence for Puberty. The simplest hypothesis for photoperiodic timing of puberty in the female sheep is that the lamb can begin reproductive cycles during short days but not during long days. Thus, the spring-born lamb could monitor photoperiod, and when day length becomes sufficiently short in the autumn, puberty occurs. Indeed, females raised from birth under long daily artificial photoperiods do not begin reproductive cycles during the normal age (25–35 weeks) (176). Eventually, ovulations began during the second year, although in many long-day-reared lambs they occur at irregular intervals, and many result in short luteal phases. Whereas the latter finding importantly indicates that long days block puberty, the eventual onset of reproductive cycles implies that the lamb is able to reject photoperiod information (become photorefractory) so that puberty cannot be forestalled indefinitely in an unfavorable light environment. Apparently, after growth requirements have been satisfied, the window of opportunity to achieve fertility can be shifted but not closed entirely. This underscores the timing role, rather than driving role played by photoperiod in puberty in the female sheep.

An unexpected finding was that exposure of the female sheep to short days alone delays puberty, much the same as long days. When lambs are raised from birth under continuous short days, repetitive ovulations do not begin during the first year after birth (Fig. 29, top). It appears that lambs must experience alternate photoperiods to begin reproductive cycles at the appropriate age. Specifically, long days must precede the short days during which puberty is attained. Systematically reducing the duration of long-day exposure in the study shown in Fig. 29 revealed that as little as 1 week would induce reproductive cycles in lambs otherwise raised in short days. Clearly, the photoperiodic timing of puberty in the female lamb is much more complex than the simple notion that the mere experience of short days alone initiates reproductive cycles. The finding of a long-day requirement to begin ovulation in this young short-day breeder raises the possibility that the lamb maintains a *photoperiod history* and that it uses the long days of spring and summer as a reference to time puberty to the short days of autumn.

Pineal Gland and Melatonin. Presently, important sites for melatonin action appear to be in the hypothalamus, but uncertainty exists about the mechanisms that translate information about the pattern of melatonin to alter the pattern of GnRH secretion (see Chapter 41). Using microimplants of melatonin into various areas of the brain, targets for melatonin action that affect LH secretion in the sheep have been identified in the mediobasal hypothalamus (177) and pre-mammillary region (178). Interestingly, several

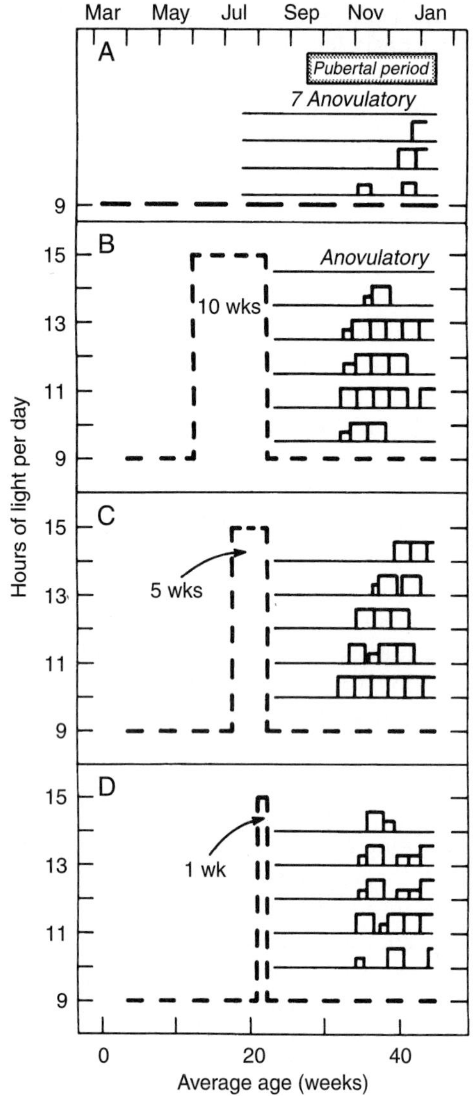

FIG. 29. Alternate photoperiods initiate reproductive cycles. Spring-born lambs were raised entirely in artificial short days of 9 hours light and 15 hours dark (9L:15D) or with 10 weeks, 5 weeks, or 1 week of long days of 15 hours light and 9 hours dark (15L:9D). Reproductive cycles with normal luteal phases (*large blocks*); cycles with short luteal phases (*small blocks*); photoperiod (*broken lines*). Each horizontal line shows data for an individual lamb, except in the top panel, which indicates that 7 of 10 lambs remained anovulatory. [From (200) with permission.]

studies (179–182) have reported binding sites of melatonin in the pars tuberalis of the pituitary, but it is now thought that this site is important for the action of melatonin on prolactin secretion rather than GnRH secretion. The issue of a timekeeping relay is essential to our understanding of melatonin action because of the long lag time (several weeks) between changes in the pattern of melatonin secretion and the appropriate LH response (159,183). Moreover, studies in the adult ewe have found that melatonin does not have an acute action (within minutes) in the modulation of LH secretion (184). In view of the foregoing considerations about the long time constant, it is possible that melatonin modulates GnRH secretion through inducing changes in morphology by remodeling yet unknown elements within the brain. Some potential targets for this remodeling are suggested by the finding in female sheep that glial ensheathment of GnRH neurons and the number of afferent inputs to GnRH neurons are greater during the breeding season than during seasonal anestrus (185). The lack of exact understanding about how melatonin interacts with mechanisms controlling GnRH secretion, however, should not detract from the important concept that the pattern of circulating melatonin encodes day length and this information is then used to increase or decrease the sensitivity of the GnRH neurosecretory system to estradiol negative feedback to alter the frequency of LH secretion.

The sheep is exposed to the photoperiod-linked melatonin rhythm very early in life. Melatonin has been detected in the circulation of the fetus, and the pattern exhibits a well-defined rhythm with the highest levels occurring during the hours of darkness (186,187). The pattern in the fetus mimics that of the mother, although its amplitude is reduced. This is for good reason because the mother generates the rhythm as evidenced by the finding that removal of the maternal pineal abolishes the nocturnal increases in melatonin in both the maternal and fetal circulation (188,189). This leads to the conclusion that the fetal pineal gland may not produce melatonin and that the melatonin in the fetal circulation is derived from the maternal pineal gland. Such a conclusion is reinforced by the finding that exogenous melatonin can cross the placenta from the maternal to the fetal side (188–190). However, melatonin has been detected in the fetal pineal gland (186), raising the possibility that melatonin could be secreted in very small quantities before birth because the pineal gland does not store melatonin appreciably (191). In view of the depressed melatonin in fetuses of pinealectomized mothers, however, the secretion of melatonin by the fetal pineal may be insignificant. Moreover, if the fetal pineal were to secrete melatonin, then the circadian pattern of secretion would not accurately reflect day length because it is not obvious how the fetus would be able to detect changes in ambient photoperiod to phase its secretion with the daily light–dark cycle.

Perhaps distinct nocturnal melatonin secretion begins only after birth when the newborn is first exposed to retinal light changes. Indeed, some but not all lambs exhibit a clear nocturnal melatonin rise by 1 week of age. For example, in studies of four sets of newborn twins (male/female) of the same age and weight, a circadian rhythm of melatonin was clearly evident in both individuals of two pairs, but in the other two pairs only ambiguous patterns were found (192). Because the differences in ontogeny of melatonin were greater between twin pairs than between siblings within a pair, there may be some genctic basis for the differential rate of development of circadian melatonin rhythmicity. By at least 6 weeks of age, melatonin rhythm is well established (193) (Fig. 30). Moreover, the young lamb can shift its melatonin rhythm to accord with a new photoperiod based on the patterns of circulating melatonin in lambs and their mothers shifted from long days (short durations of daily melatonin secretion) to short days (long durations of daily melatonin secretion). As shown in Fig. 30 (left), the daily period of increased melatonin, which is of short duration during long days, begins to expand during the first long night. By 2 weeks of age, the melatonin rhythm was entrained completely in all lambs; of importance, the rate of entrainment for the lambs was the same as for the mothers (Fig. 30, right). Arguments that either melatonin may be transferred from the lactating mother through the milk or that the melatonin rhythm of the lamb is entrained through some other maternal mechanism are not

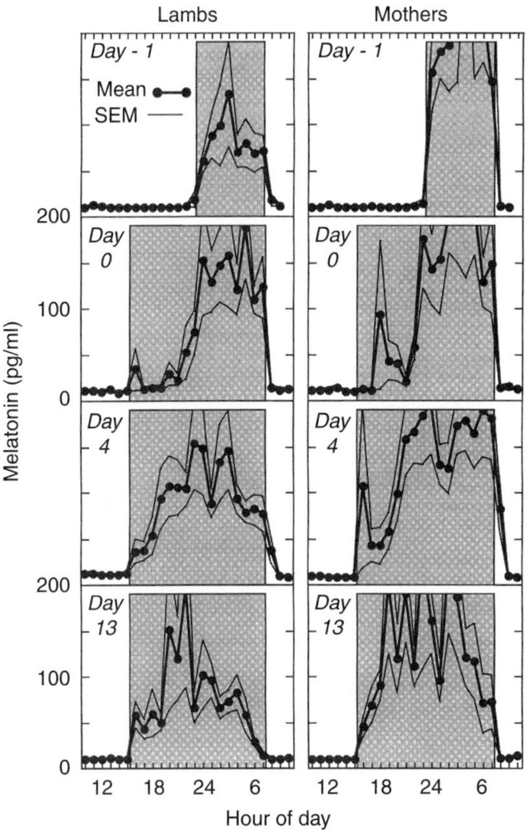

FIG. 30. Change in melatonin pattern after shift from long days to short days. Circulating melatonin (mean ± SEM, $n = 8$) is shown during a 24-hour period for 7-week-old lambs (*left*) and their mothers (*right*) on long days (day −1), on the day of the shift to short days (day 0), and on days 4 and 13; periods of darkness (*shaded areas*). [From (193) with permission.]

likely to be tenable based on considerations reviewed elsewhere (193). Thus, in view of the foregoing data from the fetus and early postnatal lamb, the developing sheep receives information about ambient day length via the mother before birth and then via its own secretory profile of melatonin during the early postnatal period.

Surgical removal of the adult pineal gland abolishes the inductive reproductive response of the female to short days and the inhibitory response to long days (159). However, an initial study of the role of the pineal gland in the onset of puberty in the female sheep was unable to detect any abnormal timing of first matings in pinealectomized lambs outdoors (194). This surprising finding was attributed to the possible photoperiod insensitivity of the Merino, a breed of sheep with a relatively long annual breeding season and an indistinct anestrous period. A subsequent study from the same laboratory found that pinealectomy delayed puberty in crossbred Merino lambs (× Border Leicester × Dorset breeds) (195); clearly, breed differences in the photoperiod modulation of puberty remain to be clarified for the sheep. Another approach to disruption of the pathway for transmission of photoperiod information is ablation of the sympathetic innervation of the pineal gland by bilateral removal of the superior cervical ganglia (see Fig. 25 for pathway). After this procedure, melatonin secretion may continue, albeit at very low levels, and any modest increases are not consistently related to any particular time of the photocycle (Fig. 31, bottom inset). Superior cervical ganglionectomy of lambs of the Suffolk breed at an early age (6 weeks) delays initiation of reproductive cycles, suggesting that normal pineal function is necessary for timing puberty to the appropriate season (Fig. 31) (196). Cycles do begin in lambs in which pineal function is absent [pinealectomized (195)] or impaired [ganglionectomized (196)], but they are unusually late; such females may even attain puberty at the time when untreated lambs end their first breeding season and enter seasonal anestrus. From these studies and those using continuous long days or short days, the concept is reinforced that photoperiod cues can readily modify the timing of puberty, but they cannot forestall or advance puberty outside of certain limits. Finally, melatonin is involved in regulating the timing of puberty by photoperiod in the female sheep (195–198) because the pattern of melatonin encodes information about day length, as in the adult.

An antigonadotropic role for melatonin has been proposed for the timing of puberty in children [for review see (199)]. This hypothesis is largely based on a decline with age in the amplitude of nighttime melatonin levels in children; thus, it has been argued that the inverse relationship between melatonin secretion

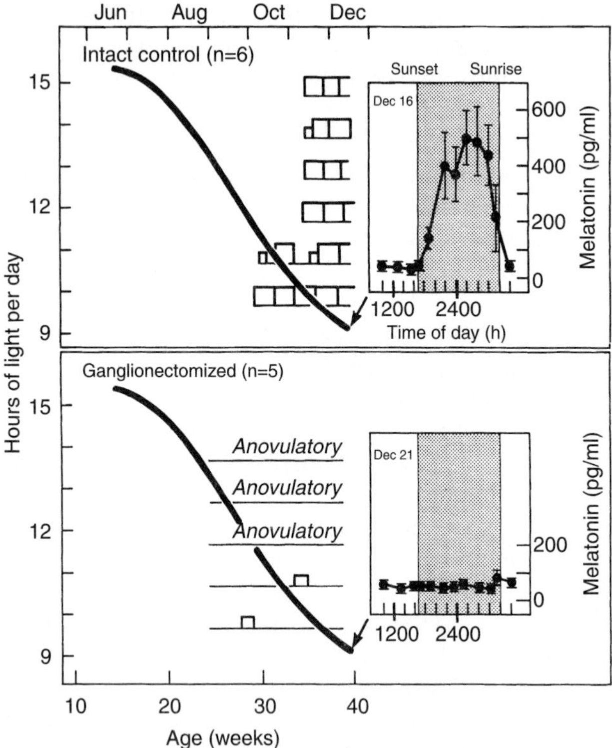

FIG. 31. Denervation of the pineal gland of the lamb prevents onset of repetitive reproductive cycles at the normal age in natural environment (*bottom*). The 24-hour pattern of circulating melatonin (mean ± SEM; *insets*) is shown at 40 weeks in lambs in which the superior cervical ganglia were removed bilaterally at 6 weeks of age (*bottom*); also shown is the melatonin pattern in unoperated postpubertal lambs that were exhibiting normal reproductive cycles (*top*) (see Fig. 29 for details of coding of luteal phases). [From (269) with permission.]

and the well-documented pubertal rise in nocturnal LH concentrations is a cause and effect relationship. The discordance among other reports of the temporal patterns of circulating melatonin in children has been briefly discussed elsewhere (200–202). A study in children (203) concedes that until manipulative research strategies are used to test the antigonadotropic hypothesis in children, we cannot discount that the declining melatonin concentrations during human development may only represent coincidental phenomena, rather than a cause and effect relationship. Studies conducted in the subhuman primate fail to support the notion of a simple antigonadotropic role for melatonin (see Chapter 40 for discussion). The prepubertal hiatus in gonadotropin secretion in the agonadal male rhesus monkey cannot be prevented by removal of the pineal (204). Moreover, providing a short-day melatonin pattern by increasing the duration of the nocturnal melatonin rise in female rhesus monkeys actually results in precocious menarche rather than its delay (205). Such an observation also reinforces the understanding that the initiation of cyclic ovarian function in the subhuman primate, like the sheep, can be influenced by external cues.

Evidence against a possible role for melatonin in retarding puberty in the female lamb is compelling for several reasons. First, based on an antigonadotropic role for melatonin, one would expect early puberty without normal pineal gland function. As demonstrated above, in the absence of melatonin, the initiation of ovulation is delayed; furthermore, brief replacement of a long-day pattern of melatonin prevents the delay. Second, even with a normal functioning pineal gland, puberty should occur when there is the least secretion of melatonin if the hormone was antigonadotropic. Under natural conditions, puberty occurs in the female lamb in the autumn when the length of night is increasing; hence, the duration of melatonin secretion is increasing by several hours and the overall amount is increasing. Third, the amplitude of the melatonin rhythm in the maturing lamb increases rather than decreases. Claypool et al. (192) found within a few weeks after birth, substantial quantities of melatonin are secreted nightly; a progressive increase in amplitude of nocturnal melatonin levels occurs with age such that during the peripubertal period, the hormone is secreted in high concentrations. Rodway et al. (206) failed to find any decrease in maximum nighttime concentrations of circulating melatonin as the lamb approaches puberty. In view of current information, extensive consideration of the amplitude is not warranted. Studies of augmentation of endogenous melatonin secretion by means of timed nightly infusions of the hormone (207) provide no evidence that the amplitude of the nocturnal melatonin rise carries any information.

Interestingly, chronic administration of melatonin by implants, which is a relatively easy method to deliver the indolamine, has been used in an attempt to alter the onset of ovarian cycles in the lamb. Constant-release devices do not produce a circadian pattern of hormone, but as in the adult (208,209) chronic melatonin treatment may be interpreted as a series of short days. In this regard, melatonin implants have been reported to have no effect or to advance or delay puberty in pineal-intact lambs raised outdoors (196–198). The varied effect appears to be related to the age when chronic treatment began; delayed puberty was associated with early initiation of melatonin administration, and precocious puberty was associated with a later insertion of implants. Similar results have been obtained for the timing of puberty and age at exposure to short days (176). A great deal remains to be learned from such studies that attempt to use principles about photoperiod to time onset of reproductive cycles through practical approaches that do not require circadian delivery of melatonin [for discussion see (183)].

Ontogeny of Photoperiodic Responses. It is clear that the young lamb can receive photoperiod information even before birth via the maternal melatonin rhythm (187–189). Can such information be used to influence the neuroendocrine control of hormone secretion? Thus far, studies have been largely conducted with the regulation of prolactin because it is well known that the secretion of this hormone is increased by long days and decreased by short days. It is tempting to include that photoperiod can influence fetal neuroendocrine function from the observation that the level of fetal prolactin in late gestation differs according to time of year (210). More direct evidence that photoperiod is the key seasonal factor regulating early prolactin secretion has been derived from a study using artificial photoperiod (211). In that study, the results of which are shown in Fig. 32, pregnant ewes were maintained in either short days (8:16-hour light:dark cycle) or long days (16:8-hour light:dark cycle) for the final third of gestation. Then, on the day of birth all lambs and their mothers were transferred to an intermediate photoperiod (12:12-hour light:dark cycle), and prolactin was monitored daily. In lambs gestated in long days, circulating prolactin at birth was high, which most likely reflects a carryover of this pattern of secretion from before birth. By contrast, lambs gestated in short days had low concentrations of circulating prolactin at birth. During the neonatal period, the high levels of prolactin in long-day gestated lambs declined in the equatorial photoperiod, whereas in short-day gestated lambs circulating prolactin increased in response to the 12:12-hour light:dark photoperiod. It is noteworthy that no sex differences occurred in patterns of circulating prolactin, a finding that is of general interest for the later discussion of differences in the timing of puberty in male and female lambs. The results of this study demonstrate two aspects of the fetal response to photoperiod cues received by the mother. First, photoperiod can modulate neuroendocrine functions before birth, thus accounting for the differential prolactin secretion at birth. Second, the lamb begins to construct a photoperiod history before birth with which it can interpret photoperiod signals postnatally. Therefore, in the study illustrated, long-day gestated lambs respond to the 12:12-hour light:dark photoperiod as a short day because it is a decrease relative to the photoperiodic signals received in utero; the short-day gestated lambs react as if this postnatal photoperiod were a long day.

Growth during Different Seasons and Timing of Puberty. Seasonal cues in the form of photoneuroendocrine signals are used by the developing female sheep to time the decrease in sensitivity to steroid feedback and express the high frequency GnRH pulses that drive the pubertal follicular phase. However, this can only occur if growth cues related to energy metabolism indicate that physiologic size is appropriate to support the consequences of pregnancy and lactation.

This interaction between growth, as influenced by nutrition, and environment is illustrated in Fig. 33, which presents the time of initiation of reproductive cycles in groups of lambs that are born at the same time of year (spring) but for which the growth rates differ. The study (171) was conducted in a natural photoperiod, and the rate of growth was altered by changing the availability of food. Some females, well fed from birth, grew rapidly and puberty occurred at the usual age (group A, ~30 weeks). Other lambs (groups B, C, and D) were placed on a restricted diet for varying periods after weaning and before they were fed ad libitum. The onset of reproductive cycles was delayed by various amounts in such females.

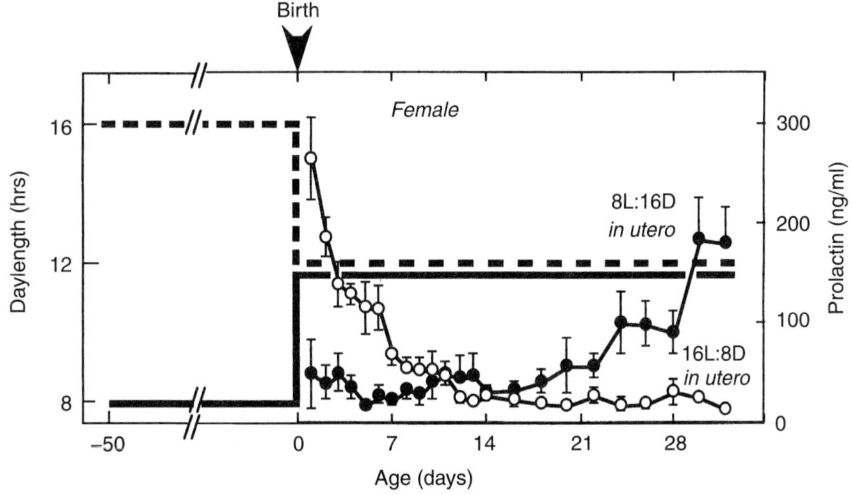

FIG. 32. Circulating prolactin concentrations in female lambs in a 12:12-h light:dark cycle, born to mothers maintained in a 16:8-hour light:dark cycle (*open circles*) or 8:16-hour light:dark cycle (*solid circles*) during late gestation. Values are means ± SEM, $n = 7$. [From (273) with permission.]

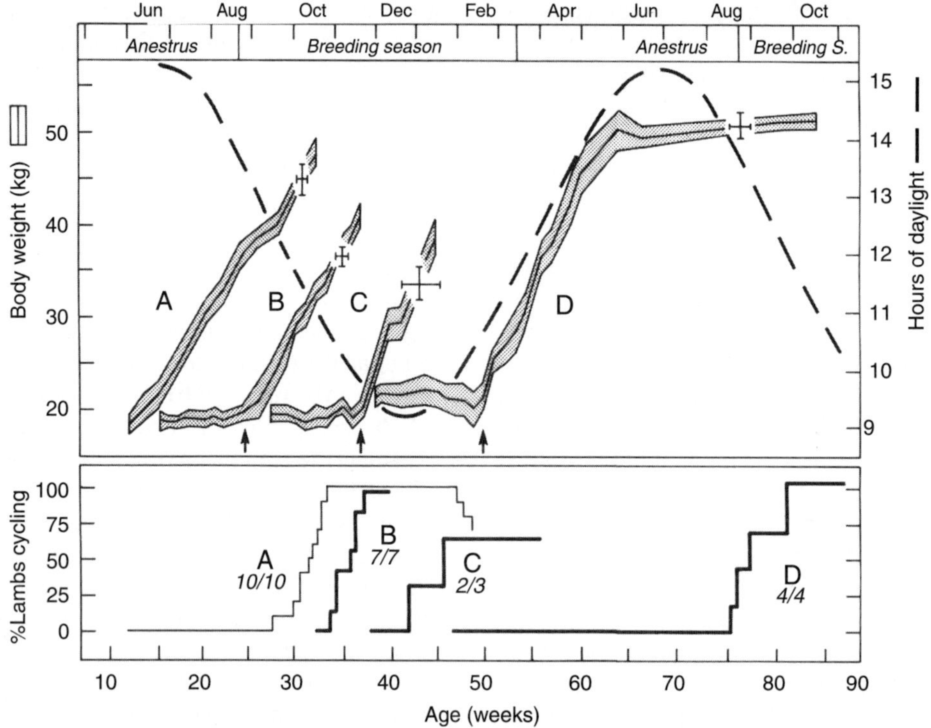

FIG. 33. Season and growth influence timing of onset of reproductive cycles in female lambs (top, mean ± SEM body weight and age at first luteal phase; bottom, data for individuals). Lambs were born in spring (March) and were raised outdoors (*broken line*, photoperiod). They were either fed ad libitum after weaning at 10 weeks of age (group A) or were placed on a restricted diet of similar composition (groups B–D); at various ages (*arrows*), feeding ad libitum was begun in food-restricted lambs. [From (171) with permission.]

Despite their having experienced an appropriate photoperiod for initiation of ovulations, puberty did not occur when they were growth retarded, presumably because sufficient energy was not available to support the consequences of activation of the reproductive system. This underscores the importance of somatic growth. Photoperiod time measurement is maintained in growth-retarded lambs (212), and when they were then induced to grow during the autumn and winter breeding season (groups B and C), cycles began at a smaller size (i.e., 35 kg) than for controls (i.e., 45 kg). This raises the possibility that the normally growing females (group A) had grown sufficiently to have the potential to attain puberty earlier in the year (August), but day lengths were too long at the younger age (i.e., 25 weeks) to reduce hypersensitivity to estradiol feedback inhibition. Later in the year (October), when day length became much shorter, sensitivity to estradiol negative feedback could be reduced to initiate reproductive cycles (i.e., 30 weeks of age). Finally, when the phase of rapid growth was induced during the anestrous season (group D), the lambs grew well beyond the normal size required to initiate reproductive cycles, but they remained anovulatory. These females continued to be hypersensitive to estradiol negative feedback because of the long days of summer, and it was not until the decreasing day lengths of autumn that the cycles could begin. Thus, both season and size serve as determinants timing the initiation of reproductive cycles in the lamb through photoperiod and growth-related cues. This same conclusion was reached earlier when season of birth, rather than growth, was experimentally altered (Fig. 26).

It is of interest that photoperiod itself can modify growth rate [for review see (213)]. In lambs, increasing daily exposure to light increases both food intake and efficiency in terms of amount of feed consumed to amount of weight gained. The hormonal mechanisms for the photoperiod-induced increases in growth remain to be established. The anabolic effect of long days does not appear to be associated with alterations in GH, insulin, or thyroxine because these hormones are unaffected by photoperiod. Although pinealectomy prevents photoperiod-induced changes in prolactin and photoperiod-induced increases in growth rate in sheep (214), there is no definite proof that prolactin is the hormonal mediator of increased growth under long days.

Sexual Maturity Versus Puberty Versus Onset of Breeding Season

It may be seemingly difficult to distinguish among the terms "sexual maturity," "puberty," and "onset of breeding season" because of the seasonal nature of reproduction in the adult. Puberty in the young female and onset of breeding activity in the adult after seasonal anestrus are both transitions from anovulatory conditions to those of cyclic ovarian function. An increase in GnRH pulse frequency is considered to underlie each transition (42,75). Because of this, the hypothalamus can be viewed as the ultimate common final pathway governing the two states of reproductive activity [for reviews see (215,216)]. The rhythm of the GnRH neurosecretory system is modified by estradiol such that during periods of high feedback sensitivity, its activity is slowed. The female lamb is born exquisitely sensitive to estradiol negative feedback; the reduction in sensitivity first occurs at puberty (Fig. 10) and then occurs at each transition from the anestrous to the breeding season. Thus, the ability to alter feedback sensitivity, first used by the lamb as a mechanism to regulate the tempo of sexual maturation, is retained during adulthood to regulate seasonal reproduction. In this broad sense, the onset of the breeding season in the sheep could be considered as an *annual puberty*, a notion proposed two decades ago by Geschwind (217). Accordingly, the notion of recurrent puberty could be applied as well to other physiological situations in which fertility resumes after a period of anovulation, such as resumption of reproductive cycles after lactational anestrus or after severe undernutrition. On the other hand, the definition of *puberty* in a seasonal breeder, as in a nonseasonal breeder, must be reserved to describe the consequences of *developmental cues* that lead to the initiation of fertility for the first time. In the fully mature female, where growth and development have ceased, the *onset of the breeding season* could be considered to reflect the consequences of *environmental cues* that are used to phase the preexisting annual rhythm of reproduction.

Because the developing sheep uses environmental cues to time the first cycles (puberty) to the appropriate season for conception (breeding season), a key question becomes, "When does *sexual maturity* occur?" To provide an appropriate setting to begin to answer this question, it is instructive to return to the case of the autumn-born lamb and the timing of its first reproductive activity (Fig. 26). Recall that such lambs remain anovulatory during the summer anestrous season despite their having attained the appropriate age and size for initiation of reproductive activity. Then, several months later at the onset of the adult breeding season, they initiate ovulations without further growth. There are two conceptually different explanations to account for this delay in the autumn-born lamb: (a) development is retarded by inhibitory photoperiod (long days), and the delayed onset of cyclicity until permissive photoperiods (short days) reflects delayed sexual maturity, and (b) development is not retarded; instead, long-day photoperiod induces seasonal anestrous at sexual maturity, and in this case, the onset of cyclicity simply reflects the onset of a breeding season when day length decreases sufficiently. Therefore, in such females the transition to adulthood is actually "masked." If the latter argument were valid, then even in the normal spring-born lamb sexual maturity is masked. Such an argument has been intimated from the results found in Fig. 33 where rapidly growing lambs (group A) initiated reproductive cycles at a size (~45 kg) much larger than that required for this phenomenon (~30–35 kg). According to this explanation, they continued to grow and remained anovulatory until day length became shorter. Support for this has been obtained by simply exposing lambs earlier in life to a permissive (stimulatory) photoperiod that markedly advances initiation of reproductive activity to a younger age and smaller size (218) (Fig. 34). In the example shown, exposure of females to a gradually decreasing artificial day length beginning at birth allowed reproductive activity to be initiated at 20 weeks of age. Typically, puberty occurs at 30 weeks of age outdoors when day length begins to decrease at approximately 15 weeks of age. Thus, advancing the decrease in day length by 15 weeks progressed puberty by 10 weeks. By using such approaches, the minimal age for the onset of reproductive activity as timed by growth cues should eventually be revealed. It is absolutely essential that this knowledge be obtained to establish when growth cues serve as important internal determinants of somatic size timing onset of reproductive activity

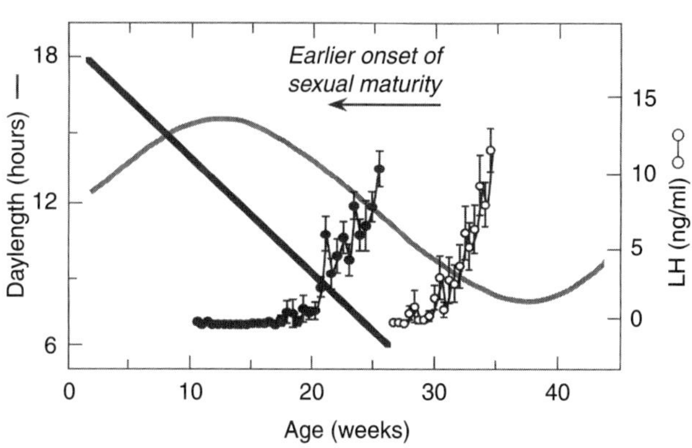

FIG. 34. Increase in mean (± SEM) circulating LH in rapidly growing, estradiol-treated, ovariectomized lambs exposed to either natural photoperiod (*open circles, n* = 7) or to decreasing artificial photoperiod (*solid circles, n* = 8) from birth. [Artificial photoperiod data from (218).]

and when photoperiod cues serve as an important external determinant synchronizing the expression of sexual maturity with the appropriate season. Only then will we be able to rid ourselves of the confusing terminologies presently used and be able to explain how reproductive activity is timed in more conceptual terms. For now, it would appear that "sexual maturity" of the female (potential to ovulate) is attained well before "puberty" (onset of ovulations).

SEX DIFFERENCES IN TIMING THE TRANSITION INTO ADULTHOOD

There is a profound sex difference in the timing of puberty in many species, including the sheep (Fig. 35). Spring-born male lambs begin reproductive development at about 10 weeks of age in midsummer, as evidenced by the onset of spermatogenesis (52,219), whereas females remain prepubertal until 25–35 weeks in the autumn when they first ovulate (Fig. 36). This strategy maximizes the reproductive opportunity for each sex. It is reasonable that the young male sheep initiates the pubertal process earlier. A longer time is required for spermatogenesis (5–10 weeks) than for the development of an early antral follicle to the preovulatory stage (few days). More importantly,

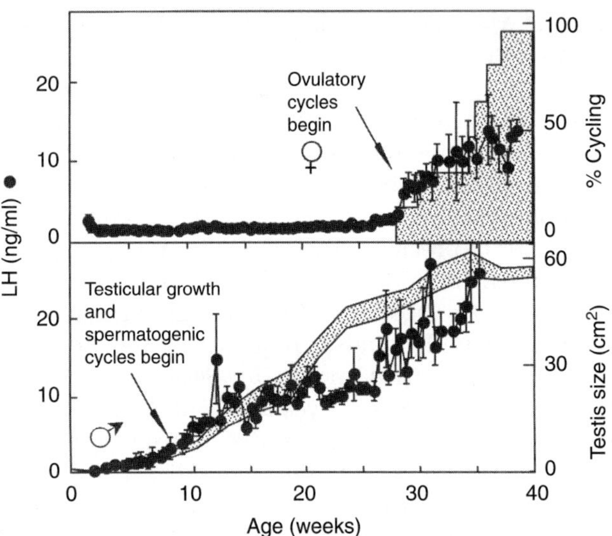

FIG. 36. Sex differences in the timing of the pubertal increase in GnRH secretion and gonadal activation. Mean (± SEM) LH concentrations (*solid circles*) in gonadectomized female (*top*) and male (*bottom*) sheep treated chronically with a constant level of estradiol by subcutaneous Silastic implant. The time of the increase in LH secretion is associated with the onset of ovulatory cycles (*shaded histogram*) in intact females and testicular growth and spermatogenesis (*shaded line*) in intact males. [Redrawn from (274).]

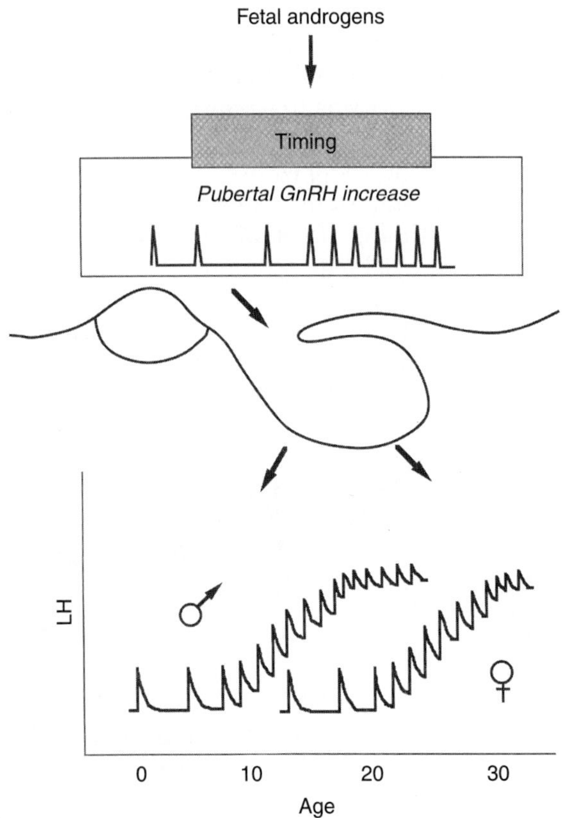

FIG. 35. Sex differences and the timing of puberty in sheep.

the prolonged exposure to progressively increasing levels of testosterone is necessary to continue to program and reinforce behaviors necessary for competition with older males and mating with females. Such a reproductive strategy would be particularly relevant in species where the first mating and conception occur in the first autumn after birth. By contrast, very fast maturing short-lived species (i.e., rodents) that first mate within weeks after birth in spring and summer may not be expected to exhibit such pronounced differences in the age at puberty. Similarly, in long-lived species that mature over an extended time (years), early puberty in the male would have no obvious advantage. In fact, males of higher primate species, such as the rhesus monkey and the human, mature at an even slower rate than females.

Sex Differences in the Pubertal Increase in GnRH Pulse Frequency

Studies of the underlying cause of the sex difference in initiation of puberty face the obvious problem of how to compare reproductive development between the two sexes. In most species, the outward signs of puberty in males and females are different and are therefore difficult to equate. These physical and behavioral changes during puberty are produced by sex steroid hormones, but differences in quality, quantity, and pattern of steroids arise from differences in the production of the male and female gametes (continuous liberation of spermatozoa verses the cyclic release of ova). Common to both sexes, however, is the antecedent rise in LH (GnRH) secretion during puberty when the sensitivity to gonadal steroid feedback inhibition decreases. This phenomenon has been demonstrated in both sexes (Fig. 36). Therefore, studies on sexual differentiation of the timing of puberty can use the pubertal LH rise as an index of "neuroendocrine sexual maturity" because it allows comparison of sex differences in the development of control systems within the brain unencumbered by differences in the gonad (ovary versus testis). In this model (neonatally gonadectomized lamb replaced with a constant feedback of exogenous steroids), the pubertal increase in LH is robust in both sexes. It positively correlates with indices of sexual maturation (initiation of testicular maturation and ovarian cycles) in gonad-intact males and females. The sustained increase in LH secretion beginning shortly after birth in the male indicates that it decreases its sensitivity to steroid negative feedback much earlier than does the female. The model uses estradiol as the steroid negative feedback hormone in both females (to replace estradiol secreted by the ovary) and males (to replace estradiol produced by aromatization of testosterone in the brain) (220–222).

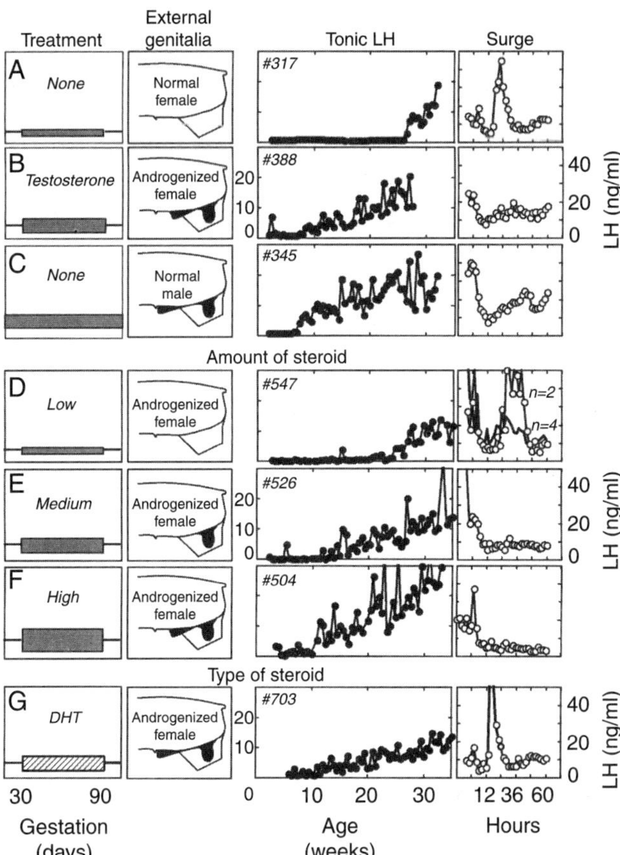

FIG. 37. Prenatal exposure to testosterone (**A–F**) or dihydrotestosterone (**G**) and its effect on external genitalia (*left*), the developmental pattern of tonic LH secretion (*middle*), and the LH surge (*right*). The lambs were gonadectomized within 3 weeks after birth and implanted with a Silastic capsule containing estradiol. [From (274).]

Programming of Sex Differences in the Timing of the Pubertal GnRH Rise by Prenatal Steroids

The pubertal increase in LH secretion occurs much earlier in the male than in the female because of the organizing action of testicular steroids on the brain before birth to control the sensitivity to the feedback action of estrogen after birth. This inference was made because prenatal exposure of the female fetus to exogenous testosterone advances the age when high frequency LH pulses are expressed (neuroendocrine sexual maturity, model system, estradiol implant, no ovaries). In the study illustrated in Fig. 37A–C using this model, testosterone was administered to pregnant ewes, and as expected this treatment masculinized the external genitalia of female lambs (52). Importantly, the pubertal increase in circulating LH in these developing females occurred at the same time as that in males, which was several months earlier than normal females. In this same study, more frequent measurements provide evidence that this precocious increase in LH is due to an increase in LH pulse frequency, not amplitude. The *time of testosterone exposure* is important. The critical period is early in development. Exposure of the unborn female between days 30 and 90 of gestation (147 days term) produces the profound change in the timing of the pubertal GnRH rise, whereas this same treatment from days 89–135 of pregnancy is without effect (data not shown) (52). The *duration of testosterone exposure* is important. Treatment for a short duration (e.g., days 30–60 or days 60–90) is effective, but the timing of the pubertal LH rise is advanced, albeit less so (data not shown) (223,224). The *amount of testosterone exposure* is important as determined in a dose-response study (Fig. 37D–F) (225). The patterns of LH secretion were related to the amount of testosterone experienced as assessed by the external genitalia. Those treatments producing the greatest external masculinization also produced the earliest decreases in response to steroid feedback and hence the earliest increase in LH secretion. The *type of steroid* is important. Because testosterone can be converted to a more potent androgen or an estrogen by the brain, the type of steroid action could not be determined in studies in which testosterone is administered. Administration of dihydrotestosterone (Fig. 37G), the androgenic metabolite that cannot be converted to an estrogen, also advanced the time of the pubertal rise in LH, suggesting that androgenic action is involved in this programming (226). Although the pure androgen produced much the same results as testosterone, they were not identical, leaving some possibility for a role of estrogens in organizing the timing of the pubertal increase in GnRH secretion.

Masculinization of the timing of puberty most likely occurs in the neural control of GnRH secretion because the effect of prenatal androgens was to increase the frequency rather than the amplitude of LH pulses. Androgen effects on pituitary function are less likely as the pituitary response to GnRH challenge is neither affected by androgenization nor is it sexually differentiated in the sheep (227,228).

Programming of Sex Differences in Other GnRH Controls by Prenatal Steroids

In addition to organizing the timing of the pubertal increase in GnRH secretion by modifying the sensitivity to estradiol negative feedback, prenatal steroids also program sex differences in the function of other steroid feedback controls. This would be expected because the regulation of reproductive function is more complex in the female than in the male because the liberation of ova is discontinuous, whereas spermatozoa production is continuous. Such differences in regulation would be manifest during puberty when gametes are first produced. Indeed, in the sheep the neuroendocrine regulation of ovulatory cycles requires four steroid feedback mechanisms to control GnRH secretion. Two feedback mechanisms regulate the surge mode of GnRH secretion—the positive feedback action of estrogens to induce a GnRH surge and the ability of progesterone to block this positive feedback action, whereas two feedback mechanisms control the pulsatile mode—the negative feedback of estrogens and the negative feedback of progesterone. In the male, three of these mechanisms are not necessary, and for central regulation of the control of spermatogenesis, only the simple estrogen negative feedback is required. To explain how these differences arise, one must consider whether the female adds the extra feedback controls or the male removes them. The latter is the most salient explanation. According to this hypothesis, multiple feedback mechanisms are inherent in both sexes and require no further developmental input for normal function during adulthood. However, if the individual is a male, the irrelevant steroid feedback controls of GnRH secretion are either abolished or become functionally inoperative through programmed reductions in sensitivity. This central programming in the male occurs though exposure of the developing brain to his own testicular steroids that act as androgens and estrogens to organize the brain during a critical period of development. In the sheep, neuroendocrine programming begins before birth and continues well into postnatal life. Evidence for this scheme is accumulating [for reviews see (229,230)].

GnRH Surge System

The GnRH surge system is not functional in the male. Prenatal treatment with testosterone at high doses perturbs the ability to respond to the positive feedback action of estrogen in young adults (high doses abolish the response; low doses prolong the latent period). In a direct test of the hypothesis that the failure of the LH surge to occur in the normal male or in females prenatally treated with testosterone, GnRH was measured in the pituitary portal system during an attempt to induce a preovulatory surge with estradiol (Fig. 38) (231). A prolonged discharge

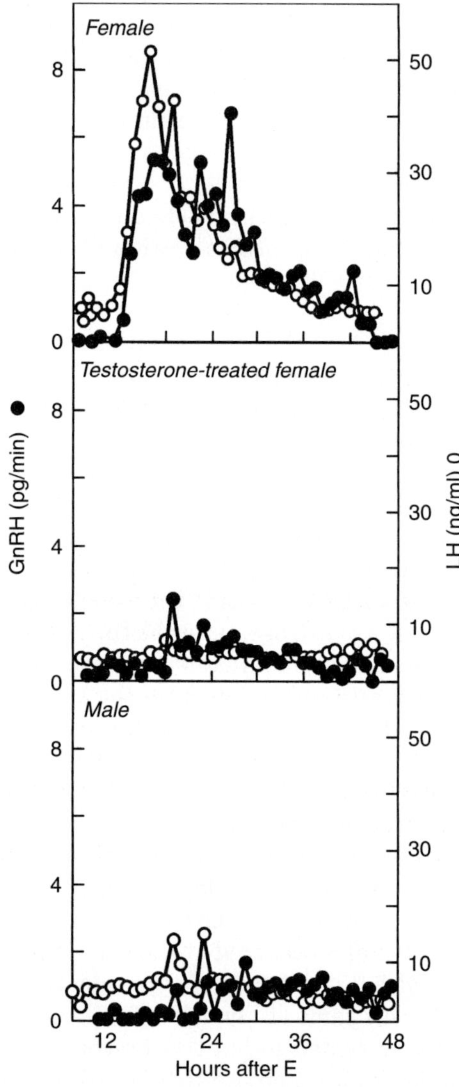

FIG. 38. Sex differences in the positive feedback effects of estradiol. Treatment with estradiol produces a surge of GnRH (*solid circles*) and LH (*open circles*) in a female (*top*) sheep, but no surge response is evident in the female treated prenatally with testosterone (*middle*) or the male (*bottom*). [Redrawn from (231).]

of GnRH was not detected in individuals prenatally exposed to endogenous or exogenous testosterone, indicating that indeed the early programming of the GnRH surge mechanism is at the level of the brain. Whereas exposure to high doses of testosterone abolishes the surge mechanism (Fig. 37, E and F, right), low doses do not (Fig. 37D, right) (225). In low dose females, the surge system is retained during adulthood, but the latent period from the administration of estradiol to the LH surge is prolonged. Prenatal administration of dihydrotestosterone, the potent nonaromatizable androgen, has no effect on the GnRH surge system (Fig. 37G, right) (226), suggesting that the programming of the GnRH surge mechanism is due to the estrogenic action of testosterone. This is of interest because prenatal treatment of the female with dihydrotestosterone reduces the sensitivity to the negative feedback action of estrogen early in life and results in a precocious pubertal rise in LH secretion (Fig. 38G). Thus, the tonic and surge modes of GnRH secretion may well be organized by the two different metabolites of testosterone. Perhaps prenatal androgens organize the level of sensitivity to the negative feedback action of steroids and the timing of the pubertal GnRH rise, whereas prenatal estrogens organize the mechanisms involved with the positive feedback action of estrogen and production of the GnRH surge.

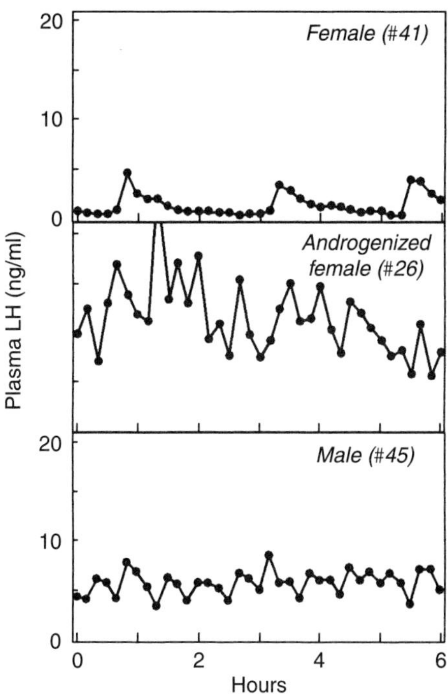

FIG. 39. Tonic LH secretion in representative, gonadectomized, progesterone-treated female (*top*), androgenized female (*middle*), and male (*bottom*) sheep. Patterns of pulsatile LH secretion are shown after 2 days of chronic progesterone treatment by subcutaneous implant. Progesterone reduced LH pulse frequency only in control females. [Redrawn from (232).]

Progesterone Feedback Systems

Two progesterone feedback controls of GnRH exist in the female sheep. One regulates the surge mode of secretion, and during the ovulatory cycle progesterone can block the positive feedback action of estrogen to prevent the GnRH surge. How this regulatory system is sexually differentiated is not known. It may be an associated part of the surge mechanism that is differentiated by estrogen. The alternative possibility that it is differentiated by androgen would not be expected, but, if so, it would be of interest in understanding the overall rules for sexual differentiation of GnRH feedback controls. The other feedback action of progesterone is clearly sexually differentiated, namely the response to progesterone inhibition of tonic GnRH secretion that prevents high frequency GnRH secretion from occurring during the luteal phase. When this negative feedback action was tested in both sexes in the absence of endogenous gonadal steroids, the male sheep was found to be much less responsive than the female; prenatal treatment of females with testosterone reduced their response to feedback inhibition of LH secretion by progesterone (Fig. 39) (232,233).

It will be of interest to understand how independent the development of the tonic mode of GnRH secretion is from the surge mode. As originally conceived from studies of the differentiation of the surge mode, the "critical period for sexual differentiation" denotes the broad period of sensitivity to organization by gonadal steroids [for review see (234)]. Perhaps separate but overlapping critical periods exist for masculinization of tonic GnRH secretion and defeminization of the surge mode of GnRH secretion. An emerging concept is that the "critical period" is not one single entity. Thus far in the sheep, discrete critical periods for the control of anatomy, behavior, and neuroendocrine function have already been noted (235–237). Each sexually dimorphic trait may have unique requirements for its sexual orientation with regard to the duration or type (androgen versus estrogen) of steroid exposure. We already have evidence to support this possibility based on the different organizational actions of androgens and estrogens on the feedback mechanisms controlling the tonic and surge modes of GnRH secretion.

Malprogramming of GnRH Controls: A Case for Fetal Origin of Reproductive Dysfunctions

In view of the foregoing understanding of sexual differentiation of GnRH feedback controls, malprogramming of these systems would be more problematic in the female than in the male. It is not clear why the male must alter GnRH feedback controls that it never uses. The surge system serves no purpose and could simply lie dormant because there is no signal for its activation (sustained estradiol rise). This appears to be the case in some higher primates in which administration of estradiol produces an LH surge (238,239). In fact, there is evidence, albeit limited, that cyclic activity of subcutaneous ovarian grafts can be supported in adult castrated male rhesus monkeys (one of four exhibited multiple ovarian cycles) (240). However, in other species, such as the rat, classic studies by several investigators, most notably Roger Gorski, revealed that the LH surge system cannot be induced to function in the male [for reviews see (241,242)], much the same as in the sheep. The same logic applies to the progesterone-regulated GnRH feedback controls that serve no obvious function in the male sheep where the sensitivity to progesterone negative feedback is profoundly diminished (232). Whether the second function of progesterone (blockade of the positive feedback action of estrogen) is altered in the male is yet unknown because this sex has no testable surge mechanism. Perhaps defeminization of the GnRH surge system is a collateral consequence of testosterone exposure during the critical period that is necessary for masculinization of tonic LH secretion (via androgenic actions) and reproductive behavior (via estrogenic actions).

If all the female-specific feedback controls were potentially functional in the male but never activated, there would be no adverse consequences and fertility would be normal. The opposite would not be true in the female, and programming of GnRH feedback controls toward those of the male would result in impaired fertility or infertility. The specific kind and severity of these defects depend on which of the feedback mechanisms are affected. This in turn depends on the type and quantity of hormone(s) to which the female fetus is exposed as well as the time and duration of exposure. It is clear that abolishing the surge mechanism would produce anovulation. More insidious would be incomplete programming by exposure to excess testosterone, and this type of damaged system could delay the GnRH surge to a later stage of follicular development. The consequences of ovulating in a postmature follicle would be a defective corpus luteum and a short or inadequate luteal phase. At best, cycles would be irregular. Similarly, programming of the progesterone feedback mechanisms would disrupt ovarian cyclicity. For example, a reduction in the efficacy of progesterone to regulate pulsatile GnRH secretion during the luteal phase, when the steroid serves as the primary negative feedback hormone, would disrupt the normal timing of the onset of the next follicular phase. Similarly, programming the second action of progesterone may result in the failure of progesterone to block the surge system.

The normal source of the programming steroids is within the fetus, as normally occurs in the male. Those from extrauterine sources, arising either from within or from outside the mother, may be problematic. Of contemporary practical concern are the natural and synthetic environmental substances that could act as endocrine disruptors through their androgenic or estrogenic actions to malprogram the developing GnRH control systems of the female, thereby impairing fertility during adulthood. Compounds with antiandrogenic actions are of equal concern to the developing male.

GnRH Neuroanatomy in Males and Females

It is not yet known where androgens and estrogens act in the brain to exert their organizing effects to sexually differentiate the control of GnRH secretion. GnRH neurons in the sheep brain have not been detected earlier than 43 days of gestation (term, ~150 days) (243). In the sheep (244), like the mouse (245,246), rat (247), and monkey (248), GnRH neurons arise from cells in the nasal placode and migrate along the nervus terminalis to their destination in the hypothalamus and preoptic area. Perhaps during the critical period of sexual differentiation, androgens exert their action on the GnRH neurosecretory system by altering GnRH number, morphology, or location. This explanation is unlikely in view of results obtained from examining the anatomy of GnRH-immunoreactive neurons in the preoptic area/hypothalamus of male and female lambs during the latter part of the critical period for sexual differentiation, at 85 days of gestation (249). The total number of GnRH neurons and their relative distribution were remarkably similar in males and females. Similarly, the percentages of different morphological subtypes (unipolar, bipolar, and multipolar) were the same in both sexes. Furthermore, these parameters in the fetal male and female were not different from those in the adult ewe (250,251), suggesting that substantial postnatal maturation of GnRH neurons is unlikely. These findings are not unique to the sheep because sex differences in GnRH-immunolabeled neurons have not been found in the developing rat (252) or in the fetal rhesus monkey (253). Although these types of studies do not provide an anatomical basis for the sex difference in the control of GnRH secretion in the lamb, ultrastructural work has revealed differences. At the electron microscopic level, there are more synaptic contacts on GnRH neurons of adult females than males, and prenatal exposure of females to testosterone decreases the number of synapses in females (Fig. 40, top) (254). Whether this is due to the organizing action of androgen or estrogen on the surge or tonic mode is not yet known. GnRH neuronal activity in

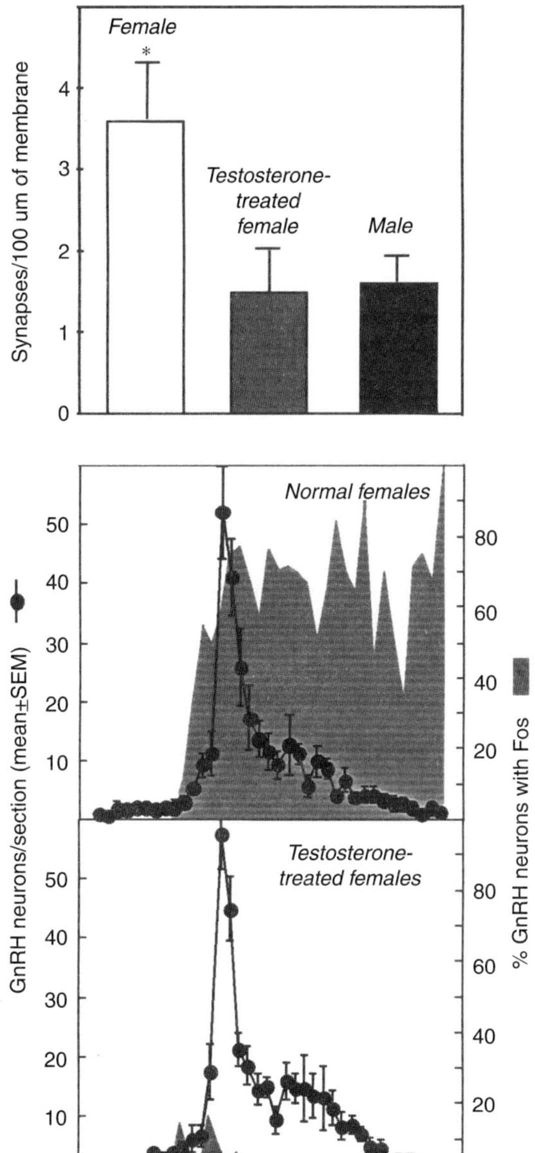

FIG. 40. Sex differences in GnRH neuronal anatomy and function. **Top:** Number of synapses on GnRH neurons in 12-month-old lambs is decreased by prenatal steroid exposure in testosterone-treated females and normal males. n = 5 lambs per group; 5–7 GnRH neurons/brain). **Middle and bottom:** Rostrocaudal distribution of GnRH neurons (*solid circles*) and Fos colocalization with GnRH neurons (*shaded area*) during the stimulatory feedback action of estrogen in normal (*middle*) and testosterone-treated (*bottom*) females. [Data for synaptic input from (249); middle and bottom panels redrawn from (255).]

response to estradiol is programmed prenatally by testosterone (255) as determined by comparison of Fos colocalization with GnRH in neurons located between the tenia tecta and the mediobasal hypothalamus in normal and androgenized females during a surge-inducing dose of estradiol at 45 weeks of age (Fig. 40, middle and bottom). As expected, a surge of LH was induced in normal females but not in females exposed to testosterone in utero. The number and distribution of GnRH neurons was not influenced by prenatal testosterone exposure, a finding similar to that in the fetal lamb (249). However, Fos colocalization with GnRH was reduced by prenatal testosterone. In normal females 66% of GnRH neurons within the preoptic area and anterior and mediobasal hypothalamus expressed Fos, whereas only 1.7% of these neurons were Fos positive in females exposed to testosterone before birth.

Sex Differences in Response to Internal and External Factors Timing Puberty

Two major physiological determinants timing the pubertal LH (GnRH) rise in the female sheep are growth and photoperiod. The response to one or both of these must be sexually differentiated because male lambs attain puberty at a smaller body size and under longer photoperiods (spring/summer) compared with female lambs (autumn). Moreover, the prenatally androgenized female lamb also exhibits puberty at a smaller size and longer day length, leading to the hypothesis that testosterone emanating from the fetal testes acts within the brain during the critical period of development before birth to alter the reproductive response either to growth cues or to photoperiod. As indicated in the next two sections, the evidence obtained thus far implicates a sexual differentiation of the photoperiod response mechanism rather than those systems responding to growth-related cues.

Sex Differences in Response to Growth Cues

Perhaps growth requirements for the female should be more critical than for the male because the ultimate metabolic demands for reproduction (i.e., pregnancy and lactation) are greater in the female. According to this hypothesis by Bronson (256), the neuroendocrine axis of the female should be more sensitive than the male to alterations in metabolic signals, such as those that must occur during growth restriction. To test this, one study determined the effect of chronic growth retardation on LH secretion in both sexes and, in addition, determined whether the gonadotropin response to acute increases and decreases in nutrition is sexually differentiated (257). The gonads were removed to avoid the confounding influence of gonadal steroids. No sex difference in gonadotropin secretion under chronic growth restriction was apparent, and LH pulse frequency became depressed equally in males and females during long-term malnutrition. For the acute responses a slight sex difference occurred, one in which males were somewhat more responsive than females to increases and decreases in level of nutrition on gonadotropin secretion. The combined results indicate that the reproductive neuroendocrine system in the developing male is at least as sensitive, and perhaps even more so, than that of the female to changes in level of nutrition. Such findings would not support the notion that the growth requirements for the neuroendocrine control of puberty as studied by modulating energy intake are markedly sexually differentiated.

Sex Differences in Response to Photoperiod

As discussed earlier, in the female sheep puberty is not normally timed by growth cues, but, instead, reproductive maturation is expressed in response to photoperiod cues. Much less is known about the role of photoperiod and maturation in the male. Studies conducted in several breeds (Ile de France, Merino-Border Leicester, and Suffolk) lead to the conclusion that the developing male sheep may be relatively insensitive to photoperiod (197,258,259). This is illustrated by raising males and females in an artificial photoperiod simulating that which occurs naturally after birth in the spring (Fig. 41, top) or in an artificially reversed photoperiod simulating birth in the autumn (Fig. 41, bottom) (260,261). As expected in the natural simulate photoperiod, the pubertal LH rise in females began between the usual ages of 25 and 30 weeks, but in the reversed photoperiod this was delayed beyond the end of the study. In males, the pubertal LH rise began at the same time in either photoperiod. In androgenized females, prenatal exposure to testosterone not only prevented the delay in the pubertal LH rise that occurred in the female in an inappropriate photoperiod, but advanced it to a much earlier age, one similar to that of the male. In the context of the hypothesis for sexual differentiation of the timing of puberty by photoperiod, these findings indicate that the default mechanism is to delay expression of sexual maturation during increasing or long photoperiods of spring and summer. In the absence of prenatal testosterone, as in the case of the female, this delays the initiation of pregnancy until the decreasing day lengths of autumn. Recall that this strategy results in spring-born young. The presence of fetal testosterone during a critical period of sexual differentiation alters the photoperiodic timing mechanism such that the male lamb will begin reproductive activity, regardless of day length, provided that growth cues are appropriate. Interestingly, the male does not remain unresponsive to photoperiod. By the second year or earlier, the control of GnRH begins to respond to this seasonal cue as an adult and reproductive activity waxes and wanes, becoming locked into seasons of increasing and decreasing fertility. In some primitive breeds, such as the Soay sheep, the testes regress markedly during the nonbreeding season to become aspermic. Thus, the male is programmed for a temporary hiatus in photoperiod control of its reproductive system during in its formative stages in the transition into adulthood.

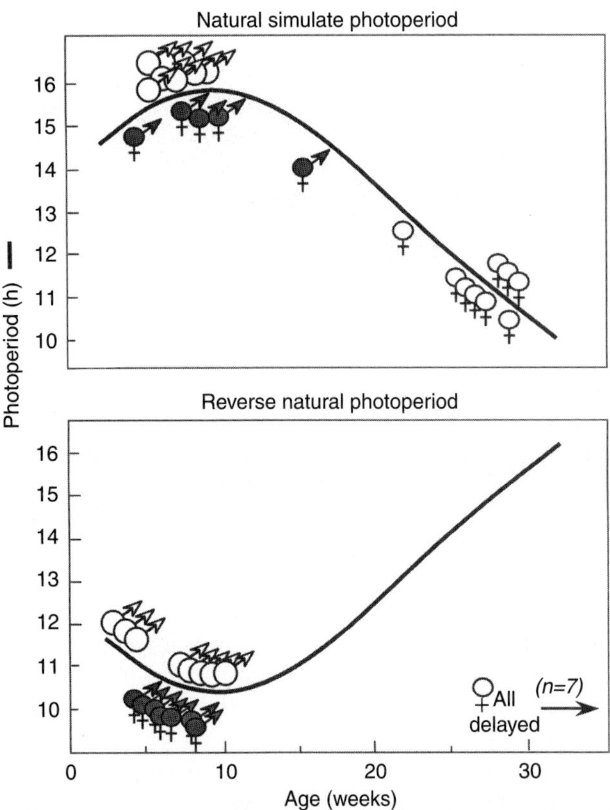

FIG. 41. Photoperiodic regulation of the timing of the puberty in male, female, and prenatally testosterone-treated female sheep. Symbols indicate timing of pubertal LH rise. **Top:** Animals raised indoors in a simulated natural photoperiod. **Bottom:** Animals raised indoors in a reverse natural photoperiod. Puberty was delayed in females in the reverse natural photoperiod beyond the end of the study. [Redrawn from (274).]

The sex difference in response to photoperiod is not due to a difference in the pattern of melatonin secretion. The pattern of melatonin is similar among males and females during development (192), indicating that the marked difference in timing of puberty in males and females does not originate from the pineal gland. Rather, it arises from a postpineal difference in the mechanism by which the melatonin signal is subsequently used to govern the expression of sexual maturation. This type of reproductive development in which the timing of puberty in the male is not influenced by photoperiod differs from the developmental strategy known thus far for several long-day breeding species. In the short-lived grasshopper mouse (262) and field vole (263), both sexes are responsive to photoperiod before puberty such that reproductive development is advanced by long days and retarded by short days. In the golden hamster, photoperiod does not influence the timing of puberty in either sex; apparently, the reproductive neuroendocrine system becomes responsive to photoperiod during adulthood (264–267).

INTEGRATED MODEL FOR PUBERTY

Figure 42 summarizes and synthesizes an integrated model for puberty in the lamb by focusing on the GnRH neurosecretory system and what type of information is integrated to increase the frequency of GnRH pulses that drives the transition into adulthood. This model accounts for both the sex difference in when this transition occurs and the different signals that time it in males and females. The core pattern of GnRH secretion in the sheep, like in other long-lived species, is an increase in activity beginning before birth and then a decline of varying durations followed by the pubertal rise. The response to the early increase in GnRH secretion occurring before birth programs the type of information that is used to produce the sex-specific pattern of GnRH secretion after birth. If no testes are present in utero, the multiple neuroendocrine feedback controls remain operative, and the (female) lamb becomes photoperiodic with respect to the timing of puberty using heightened sensitivity to steroid negative feedback. Although metabolic cues providc information that sufficient growth has occurred to begin reproduction, the young female cannot express her sexual maturity because the photoperiod is not appropriate. Sexual maturity is masked, the ovary remains quiescent, and she immediately enters seasonal anestrus. Once she has experienced the requisite photoperiod, the long days of summer followed by the short days of autumn, the neuroendocrine sensitivity to steroid negative feedback decreases and high frequency GnRH pulses are expressed. The other feedback controls remain operative, and ovulatory cycles are initiated to begin the breeding season. This strategy allows young females to synchronize the beginning of reproductive activity with those of mature seasonally breeding females.

By contrast, the male lamb tracks photoperiod in utero much like the female lamb but does not use photoperiod to time the expression of high frequency GnRH pulses. In response to the first increase in GnRH secretion and the resultant gonadotropin drive to the testes before birth, testosterone is secreted to masculinize the photoneuroendocrine system; this effectively reduces the reproductive response to day length.

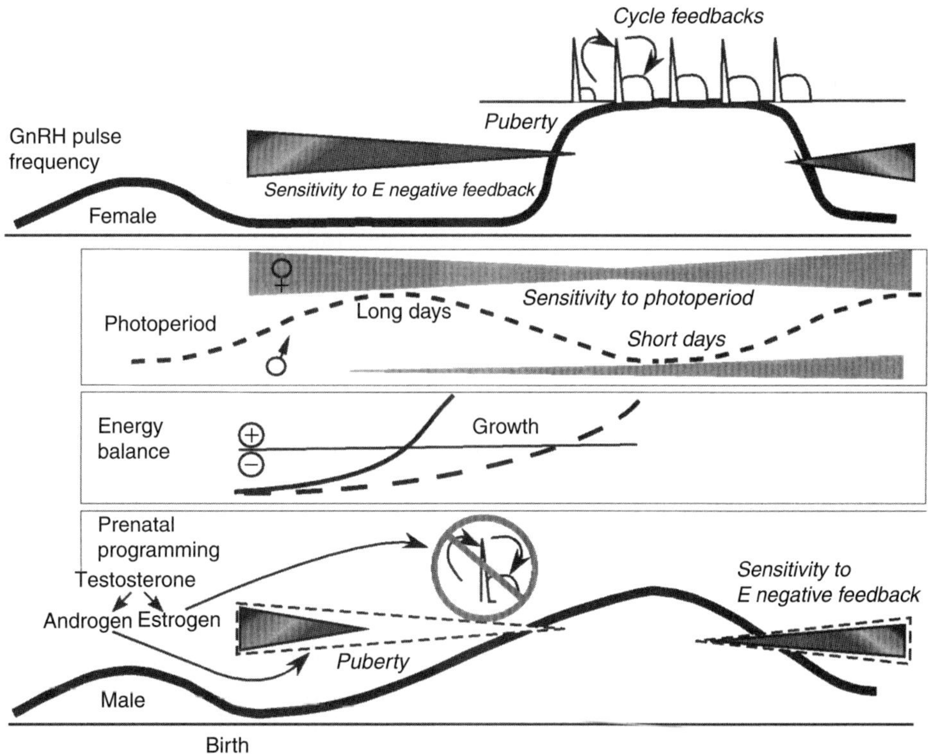

FIG. 42. Integrated model for internal and external determinants timing the onset of puberty in the sheep.

Rather than masculinization, this programming could be considered defeminization because the testicular steroid action modifies an innate function, much like the defeminization of the unnecessary feedback controls in the male that underlie ovulatory cycles. By removing the photoperiod requirement for the expression of high frequency GnRH pulses, puberty in the male begins once metabolic signals provide information that energy balance is sufficient to begin reproductive activity. Most interestingly, after further maturation, in the second year the reproductive system of the young male also becomes regulated by photoperiodic cues, much like that for the female to result in seasonal changes in sensitivity to negative feedback of GnRH secretion. Thus, whereas metabolic cues time hypothalamic sexual maturity in both sexes, the presence or absence of prenatal programming by testosterone and its metabolites determines whether photoperiod serves as a permissive cue to begin reproductive activity. Viewed in a broad sense, different signals time "puberty" in the two sexes. Metabolic signals time puberty in the male, but photoperiod cues time puberty in the female. Perhaps the hiatus in photoperiodic control in the male during the first year is necessary to increase the GnRH/LH drive to the testes early to develop the behavior necessary to begin competition with older males for females. This strategy for sex differences in timing of puberty is a prime example of how fundamental mechanisms have been conserved, whereas speciation has led to adaptive modification of the control of these mechanisms to allow for optimal timing of the onset of fertility.

ACKNOWLEDGMENTS

We are grateful to the undergraduate and graduate students, postdoctoral fellows, and visiting scientists in our laboratory and to the numerous other investigators who have generated the data described in this chapter; to the technical staff for work in animal experimentation, especially Mr. Douglas D. Doop; to the many colleagues of the Reproductive Sciences Program, especially Drs. Vasantha Padmanabhan and Fred J. Karsch and their laboratories; to the National Institutes of Health, the National Science Foundation, and the U.S. Department of Agriculture for research support; and to Ms. Juanita Pelt and Mr. Paul W. Slotten for technical assistance in the preparation of the manuscript.

REFERENCES

1. Ryder, M. L., Ed. (1983). *Sheep and Man*. Duckworth, London.
2. Hammond, J. Jr. (1944). On the breeding season in the sheep. *J. Agric. Sci.* 34, 97–105.
3. Foote, W. C., Sefidbakht, N., and Madsen, M. A. (1970). Pubertal estrus and ovulation and subsequent estrous cycle patterns in the ewe. *J. Anim. Sci.* 30, 86–90.
4. Berardinelli, J. G., Dailey, R. A., Butcher, R. L., and Inskeep, E. K. (1980). Source of circulating progesterone in prepubertal ewes. *Biol. Reprod.* 22, 233–236.
5. Fitzgerald, J., and Butler, W. R. (1982). Seasonal effects and hormonal patterns related to puberty in ewe lambs. *Biol. Reprod.* 27, 853–863.
6. Ryan, K. D., Goodman, R. L., Karsch, F. J., Legan, S. J., and Foster, D. L. (1991). Patterns of circulating gonadotropins and ovarian steroids during the first periovulatory period in the developing sheep. *Biol. Reprod.* 45, 471–477.
7. Land, R. B., Thimonier, J., and Pelletier, J. (1970). Possibility of the induction of LH secretion by injection of estrogen in the female lamb as a function of age. *C. R. Acad. Sci. Hebd. Seances Acad. Sci. D* 271, 1549–1551.
8. Squires, E. L., Scaramuzzi, R. J., Caldwell, B. V., and Inskeep, E. K. (1972). LH release and ovulation in the prepubertal lamb. *J. Anim. Sci.* 34, 612–614.
9. Foster, D. L., and Karsch, F. J. (1975). Development of the mechanism regulating the preovulatory surge of luteinizing hormone in sheep. *Endocrinology* 97, 1205–1209.
10. Tran, C. T., Edey, T. N., and Findlay, J. K. (1979). Pituitary response of prepubertal lambs to oestradiol-17β. *Aust. J. Biol. Sci.* 32, 463–467.
11. Foster, D. L. (1984). Preovulatory gonadotropin surge system of prepubertal female sheep is exquisitely sensitive to the stimulatory feedback action of estradiol. *Endocrinology* 115, 1186–1189.
12. Adams, T. E., Horton, M. B., Watson, J. G., and Adams, B. M. (1984). Bioactive luteinizing hormone (LH) during prepubertal development in ewe lambs. In *Proceedings of the 10th International Congress on Animal Reproduction and Artificial Insemination*, June 1984, (Abst. 1), Urbana-Champaign, IL.
13. Foster, D. L., and Olster, D. H. (1985). Effect of restricted nutrition on puberty in the lamb: patterns of tonic luteinizing hormone (LH) secretion and competency of the LH surge system. *Endocrinology* 116, 375–381.
14. Keisler, D. H., Inskeep, E. K., and Dailey, R. A. (1985). Roles of pattern of secretion of luteinizing hormone and the ovary in attainment of puberty in ewe lambs. *Domest. Anim. Endocrinol.* 2, 123–132.
15. Mansour, A. M. (1959). The hormonal control of ovulation in the immature lamb. *J. Agric. Sci.* 52, 87–94.
16. Mauleon, P. (1969). Oogenesis and folliculogenesis. In *Reproduction in Domestic Animals* (H. H. Cole and P. T. Cupps, Eds.), pp. 187–215. Academic Press, New York.
17. Kennedy, J. P., Worthington, C. A., and Cole, E. R. (1974). The postnatal development of the ovary and uterus of the Merino lamb. *J. Reprod. Fertil.* 36, 275–282.
18. Worthington, C. A., and Kennedy, J. P. (1979). Ovarian response to exogenous hormones in six-week-old lambs. *Aust. J. Biol. Sci.* 32, 91–95.
19. Foster, D. L., and Ryan, K. D. (1981). Endocrine mechanisms governing transition into adulthood in the female sheep. *J. Reprod. Fert.* 30 (Suppl.), 75–90.
20. Trounson, A. O., Willadsen, S. M., and Moor, R. M. (1977). Reproductive function in prepubertal lambs: ovulation, embryo development and ovarian steroidogenesis. *J. Reprod. Fertil.* 49, 69–75.
21. Dyrmundsson, O. R. (1983). The influence of environmental factors on the attainment of puberty in ewe lambs. In *Sheep Production* (W. Haresign, Ed.), pp. 393–408. Butterworths, London.
22. Quirke, J. F., Adams, T. E., and Hanrahan, J. P. (1983). Artificial induction of puberty in ewe lambs. In *Sheep Production* (W. Haresign, Ed.), pp. 409–429. Butterworths, London.

23. Foster, D. L., Ryan, K. D., and Papkoff, H. (1984). Hourly administration of luteinizing hormone induces ovulation in prepubertal female sheep. *Endocrinology* 115, 1179–1185.

24. Huffman, L. J., Inskeep, E. K., and Goodman, R. L. (1987). Changes in episodic luteinizing hormone secretion leading to puberty in the lamb. *Biol. Reprod.* 37, 755–761.

25. Foster, D. L., Lemons, J. A., Jaffe, R. B., and Niswender, G. D. (1975). Sequential patterns of circulating luteinizing hormone and follicle-stimulating hormone in female sheep from early postnatal life through the first estrous cycles. *Endocrinology* 97, 985–994.

26. Bindon, B. M., and Turner, H. N. (1974). Plasma LH of the prepubertal lamb: a possible early indicator of fertility. *J. Reprod. Fertil.* 39, 85–88.

27. Foster, D. L., Jaffe, R. B., and Niswender, G. D. (1975). Sequential patterns of circulating LH and FSH in female sheep during the early postnatal period: effect of gonadectomy. *Endocrinology* 96, 15–22.

28. Echternkamp, S. E., and Laster, D. B. (1976). Plasma LH concentrations for prepubertal, postpubertal anestrous, and cyclic ewes of varying fecundity. *J. Anim. Sci.* 42, 444–447.

29. Padmanabhan, V., Mieher, C. D., Borondy, M., I'Anson, H., Wood, R. I., Landefeld, T. D., Foster, D. L., and Beitins, I. Z. (1992). Circulating bioactive follicle-stimulating hormone and less acidic follicle-stimulating hormone isoforms increase during experimental induction of puberty in the female lamb. *Endocrinology* 131, 213–220.

30. I'Anson, H., Herbosa, C. G., Ebling, F. J. P., Wood, R. I., Bucholtz, D. C., Mieher, C. D., Foster, D. L., and Padmanabhan, V. (1993). Hypothalamic versus pituitary stimulation of luteinizing hormone secretion in the prepubertal female lamb. *Neuroendocrinology* 57, 467–475.

31. Padmanabhan, V., Reno, K. M., Borondy, M., Landefeld, T. D., Ebling, F. J. P., Foster, D. L., and Beitins, I. Z. (1992). Effect of nutritional repletion on pituitary and serum FSH isoform distribution in growth-retarded lambs. *Biol. Reprod.* 46, 964–971.

32. Pirl, K. G., and Adams, T. E. (1987). Induction of precocious puberty in ewe lambs by pulsatile administration of GnRH. *J. Reprod. Fertil.* 80, 355–359.

33. Caraty, A., Locatelli, A., Moenter, S. M., and Karsch, F. J. (1994). Sampling of hypophyseal portal blood of conscious sheep for direct monitoring of hypothalamic neurosecretory substances. In *Methods in Neuroscience, Pulsatility in Neuroendocrine Systems* (J. E. Levine and P. M. Conn, Eds.), pp. 162–183. Academic Press, New York.

34. Ramirez, V. D. (1973). Endocrinology of puberty. In *Handbook of Physiology, Endocrinology* (R. O. Greep and E. B. Astwood, Eds.), Vol. 2, Part 1, pp. 1–28. American Physiological Society, Washington, DC.

35. Andrews, W. W., Advis, J. P., and Ojeda, S. R. (1981). The maturation of estradiol-negative feedback in female rats: evidence that resetting of the hypothalamic "gonadostat" does not precede the first preovulatory surge of gonadotropins. *Endocrinology* 109, 2022–2039.

36. Döcke, F., Rhode, W., Gerber, P., and Dörner, G. (1984). Evidence that desensitization to the negative estrogen feedback is a prepubertal and not a postpubertal event in female rats. *Exp. Clin. Endocrinol.* 84, 1–6.

37. Friedman, C. R., I'Anson, H., and Manning, J. M. (1992). Acute and chronic inhibitory feedback of physiological concentrations of estradiol on LH pulse frequency in the prepubertal female sheep. *Biol. Reprod.* 46 (Suppl. 1), p. 80, Abst. 117.

38. Foster, D. L., and Ryan, K. D. (1979). Endocrine mechanisms governing transition into adulthood: a marked decrease in inhibitory feedback action of estradiol on tonic secretion of luteinizing hormone in the lamb during puberty. *Endocrinology* 105, 896–904.

39. Foster, D. L., Ryan K. D., Goodman, R. L., Legan S. L., Karsch, F. J., and Yellon, S. M. (1986). Delayed puberty in lambs chronically treated with oestradiol. *J. Reprod. Fert.* 78, 111–117.

40. Ebling, F. J. P., Kushler, R. H., and Foster, D. L. (1990). Pulsatile LH secretion during sexual maturation in the female sheep: photoperiodic regulation in the presence and absence of ovarian steroid feedback as determined in the same individual. *Neuroendocrinology* 52, 229–237.

41. Foster, D. L., and Karsch, F. J. (1976). Inhibition of tonic secretion of luteinizing hormone by progesterone in immature female sheep. *Endocrinology* 99, 1–6.

42. Manning, J. M., Herbosa, C. G., Pelt, J., Karsch, F. J., and Foster, D. L. (1993). Pattern of GnRH secretion in the pituitary portal circulation of the female sheep during puberty. In *Twenty-Third Annual Meeting of the Society for Neuroscience*, Washington, DC, Abst. 258.2.

43. Thiery, J. C., and Pelletier, J. (1981). Multiunit activity in the anterior median eminence and adjacent areas of the hypothalamus of the ewe in relation to LH secretion. *Neuroendocrinology* 32, 217–224.

44. Kawakami, M., Uemura, T., and Hayashi, R. (1982). Electrophysiological correlates of pulsatile gonadotropin secretion in rats. *Neuroendocrinology* 35, 63–67.

45. Wilson, R. C., Kesner, J. S., Kaufman, J. M., Uemura, T., Akema, T., and Knobil, E. (1984). Central electrophysiologic correlates of pulsatile luteinizing hormone secretion in the rhesus monkey. *Neuroendocrinology* 39, 256–260.

46. Tanaka, T., Mori, Y., and Hoshino, K. (1992). Hypothalamic GnRH pulse generator activity during the estradiol-induced LH surge in ovariectomized goats. *Neuroendocrinology* 56, 641–645.

47. Pelletier, J., Carrez-Camous, S., and Thiery, J. C. (1981). Basic neuroendocrine events before puberty in cattle, sheep and pigs. *J. Reprod. Fertil.* 30 (Suppl.), 91–102.

48. Wilkinson, M., and Landymore, K. M. (1989). Do brain opioid peptides regulate the onset of puberty? In *Brain Opioid Systems in Reproduction* (R. G. Dyer and R. J. Bicknell, Eds.), pp. 70–91. Oxford University Press, New York.

49. Ebling, F. J. P., Schwartz, M. L., and Foster, D. L. (1989). Endogenous opioid regulation of pulsatile luteinizing hormone secretion during sexual maturation in the female sheep. *Endocrinology* 125, 369–383.

50. Wood, R. I., I'Anson, H., Ebling, F. J. P., and Foster, D. L. (1993). Opioid inhibition of luteinizing hormone secretion compared in developing male and female sheep. *Neuroendocrinology* 56, 822–830.

51. Claypool, L. E., and Foster, D. L. (1990). Sexual differentiation of the mechanism controlling pulsatile secretion of luteinizing hormone contributes to sexual differences in the timing of puberty in sheep. *Endocrinology* 126, 1206–1215.

52. Wood, R. I., Ebling, F. J. P., I'Anson, H., Yellon, S. M., and Foster, D. L. (1991). Prenatal androgens time neuroendocrine sexual maturation. *Endocrinology* 128, 2457–2468.

53. Foster, D. L., Ebling, F. J. P., Ryan, K. D., and Yellon, S. M. (1989). Mechanisms timing puberty: a comparative approach. In *Control of the Onset of Puberty III* (H. A. Delemarre-van de Waal, T. M. Plant, G. P. van Rees, and J. Schoemaker, Eds.), pp. 227–249. Elsevier, Amsterdam.

54. Rapisarda, J. J., Bergman, K. S., Steiner R. A., and Foster, D. L. (1983). Response to estradiol inhibition of tonic luteinizing hormone secretion decreases during the final stage of puberty in the rhesus monkey. *Endocrinology* 112, 1172–1179.

55. Wilson, M. E. (1989). Relationship between growth and puberty in the rhesus monkey. In *Control of the Onset of Puberty III* (H. A. Delemarre-van de Waal, T. M. Plant, G. P. van Rees, and J. Schoemaker, Eds.), pp. 137–149, Elsevier, Amsterdam.

56. Foster, D. L. (1977). Luteinizing hormone and progesterone secretion during sexual maturation of the rhesus monkey.

Short luteal phases during the initial menstrual cycles. *Biol. Reprod.* 17, 584–590.
57. Foster, D. L. (1980). Comparative development of mammalian females: proposed analogies among patterns of LH secretion in various species. In *Problems in Pediatric Endocrinology* (C. LaCauza and A. W. Root, Eds.), pp. 193–210. Academic Press, New York.
58. Mesiano, S., Hart, C. S., Heyer, B. W., Kaplan, S. L., and Grumbach, M. M. (1991). Hormone ontogeny in the ovine fetus XXVI. A sex difference in the effect of castration on the hypothalamic-pituitary gonadotropin unit in the ovine fetus. *Endocrinology* 129, 3073–3079.
59. Keisler, D. H., Inskeep, E. K., and Dailey, R. A. (1983). First luteal tissue in ewe lambs: influence on subsequent ovarian activity and response to hysterectomy. *J. Anim. Sci.* 57, 150–156.
60. McLeod, B. J., and Haresign, W. (1984). Evidence that progesterone may influence subsequent luteal function in the ewe by modulating preovulatory follicle development. *J. Reprod. Fertil.* 70, 381–386.
61. Legan, S. J., I'Anson, H., Fitzgerald, B. P., and Akaydin, M. S. Jr. (1985). Importance of short luteal phases in the endocrine mechanism controlling initiation of estrous cycles in anestrous ewes. *Endocrinology* 117, 1530–1536.
62. Pearce, D. T., Martin, G. B., and Oldham, C. M. (1985). Corpora lutea with a short life-span induced by rams in seasonally anovular ewes are prevented by progesterone delaying the preovulatory surge of LH. *J. Reprod. Fertil.* 75, 79–84.
63. Ryan, K. D., and Foster, D. L. (1978). Two LH surges at puberty in the female lamb: possible role of progesterone. *Biol. Reprod.* 18 (Suppl. 1), p. 80, Abst. 118.
64. Dutt, R. H. (1953). Induction of estrus and ovulation in anestrual ewes by use of progesterone and pregnant mare serum. *J. Anim. Sci.* 12, 515–523.
65. Robinson, T. J. (1954). The necessity for progesterone with estrogen for the induction of recurrent estrus in the ovariectomized ewe. *Endocrinology* 55, 403–408.
66. Scaramuzzi, R. J., Tillson, S. A., Thorneycroft, I. H., and Caldwell, B. V. (1971). Action of exogenous progesterone and estrogen on behavioral estrus and luteinizing hormone levels in the ovariectomized ewe. *Endocrinology* 88, 1184–1189.
67. Domanski, E., Przekop, F., and Skubiszewski, B. (1972). Interaction of progesterone and estrogens on the hypothalamic center controlling estrous behavior in sheep. *Acta Neurobiol. Exp.* 32, 763–766.
68. Karsch, F. J., Legan, S. J., Ryan, K. D., and Foster, D. L. (1980). Importance of estradiol and progesterone in regulating LH secretion and estrous behavior during the sheep estrous cycle. *Biol. Reprod.* 23, 404–413.
69. Edey, T. N., Chu, T. T., Kilgour, R., Smith, J. F., and Tervit, H. R. (1977). Estrus without ovulation in pubertal ewes. *Theriogenology* 7, 11–15.
70. Edey, T. N., Kilgour, R., and Bremner, K. (1978). Sexual behavior and reproductive performance of ewe lambs at and after puberty. *J. Agric. Sci.* 90, 83–91.
71. Dyrmundsson, O. R. (1978). Studies on the breeding season of Icelandic ewes and ewe lambs. *J. Agric. Sci.* 90, 275–281.
72. Hafez, E. S. E. (1952). Studies on the breeding season and reproduction of the ewe. *J. Agric. Sci.* 42, 189–265.
73. Watson, R. H., and Gamble, L. C. (1961). Puberty in the Merino ewe with special reference to the influence of season of birth upon its occurrence. *Aust. J. Agric. Res.* 12, 124–138.
74. Foster, D. L., and Ryan, K. D. (1981). Premature seasonal inhibition of tonic LH secretion by oestradiol in the female lamb and its consequences. *J. Reprod. Fertil.* 63, 289–294.
75. Karsch, F. J., Dahl, G. E., Evans, N. P., Manning, J. M., Mayfield, K., Moenter, S. M., and Foster, D. L. (1993). Seasonal changes in gonadotropin-releasing hormone secretion in the ewe: alteration in response to the negative feedback action of estradiol. *Biol. Reprod.* 49, 1377–1383.
76. Kennedy, G. C., and Mitra, J. (1963). Body weight and food intake as initiating factors for puberty in the rat. *J. Physiol.* 166, 408–418.
77. Bronson, F. H., and Manning, J. M. (1991). Minireview—the energetic regulation of ovulation: a realistic role for body fat. *Biol. Reprod.* 44, 945–950.
78. Webster, A. J. F. (1979). Energy metabolism and requirements. In *Digestive Physiology and Nutrition of Ruminants* (D. C. Church, Ed.), "Nutrition," Vol. 2, pp. 210–229. Oxford Press, Oregon.
79. Walker, D. M. (1979). Nutrition of preruminants. In *Digestive Physiology and Nutrition of Ruminants* (D. C. Church, Ed.), "Nutrition," Vol. 2, pp. 258–280. Oxford Press, Oregon.
80. Foster, D. L., Mickelson, I. H., Ryan, K. D., Coon, G. A., Drongowski, R. A., and Holt, J. A. (1978). Ontogeny of pulsatile luteinizing hormone and testosterone secretion in male lambs. *Endocrinology* 102, 1137–1146.
81. Bucholtz, D. C., Manning, J. M., Herbosa, C. G., Schillo, K. K., and Foster, D. L. (1993). The energetics of LH secretion: a temporally-focused view of sexual maturation. In *Twenty-third Annual Meeting of the Society for Neuroscience*, Washington, DC, Abst. 349.10.
82. Frisch, R. E. (1991). Body weight, body fat and ovulation. *Trends Endocrinol. Metab.* 2, 191–197.
83. Zhang, Y., Proenca, R., Maffei, M., Barone, M., Leopold, L., and Friedman, J. M. (1994). Positional cloning of the mouse obese gene and its human homologue. *Nature* 372, 425–432.
84. Halaas, J. L., Gajiwala, S., Maffei, M., Cohen, S. L., Chait, B. T., Rabinowitz, D., Lallone, R. L., Burley, S. K., and Friedman, J. M. (1995). Weight-reducing effects of the plasma protein encoded by the obese gene. *Science* 269, 543–546.
85. Weigle, D., Bukowski, T., Foster, D., Holderman, S., Kramer, J., Lasser, G., Lofton-Day, C., Prunkard, D., Raymond, C., and Kuijper, J. (1995). Recombinant ob protein reduces feeding and body weight in the ob/ob mouse. *J. Clin. Invest.* 96, 2065–2070.
86. Barash, I. A., Cheung, C. C., Weigle, D. S., Ren, H., Kabigting, E. B., Kuijper, J. L., Clifton, D. K., and Steiner, R. A. (1996). Leptin is a metabolic signal to the reproductive system. *Endocrinology* 137, 3144–3147.
87. Delevaud, C., Bocquier, F., Chilliard, Y., Keisler, D. H., Gertler, A., and Kann, G. (2000). Plasma leptin determination in ruminants: effect of nutritional status and body fatness on plasma leptin concentration assessed by a specific RIA in sheep. *J. Endocrinol.* 164, 519–526.
88. Blache, D., Tellam, R. L., Chagas, L. M., Blackberry, M. A., Vercoe, P. E., and Martin, G. B. (2000). Level of nutrition affects leptin concentrations in plasma and cerebrospinal fluid in sheep. *J. Endocrinol.* 165, 625–637.
89. Daniel, J. A., Whitlock, B. K., Baker, J. A., Steele, B., Morrison, C. D., Keisler, D. H., and Sartin, J. L. (2002). Effect of body fat mass and nutritional status on 24-hour leptin profiles in ewes. *J. Anim. Sci.* 80, 1083–1089.
90. Nagatani, S., Zeng, Y., Keisler, D. H., Foster, D. L., and Jaffe, C. A. (2000). Leptin regulates pulsatile luteinizing hormone and growth hormone secretion in the sheep. *Endocrinology* 141, 3965–3975.
91. Foster, D. L., and Nagatani, S. (1999). Physiological perspectives on leptin as a regulator of reproduction: role in timing puberty. *Biol. Reprod.* 60, 205–215.
92. Mantzoros, C. S., Flier, J. S, and Rogol, A. D. (1997). A longitudinal assessment of hormonal and physical alterations during normal puberty in boys. V. Rising leptin levels may signal the onset of puberty. *J. Clin. Endocrinol. Metab.* 82, 1066–1070.
93. Ahmed, M. L., Ong, K. K., Morrell, D. J., Cox, L., Drayer, N., Perry, L., Preece, M. A., and Dunger, D. B. (1999).

Longitudinal study of leptin concentrations during puberty: sex differences and relationship to changes in body composition. *J. Clin. Endocrinol. Metab.* 84, 899–905.

94. Kratzsch, J., Lammert, A., Bottner, A., Seidel, B., Mueller, G., Thiery, J., Hebebrand, J., and Kiess, W. (2002). Circulating soluble leptin receptor and free leptin index during childhood, puberty, and adolescence. *J. Clin. Endocrinol. Metab.* 87, 4587–4594.
95. Mann, D. R., Johnson, A. O., Gimpel, T., and Castracane, V. D. (2003). Changes in circulating leptin, leptin receptor, and gonadal hormones from infancy until advanced age in humans. *J. Clin. Endocrinol. Metab.* 88, 3339–3345.
96. Urbanski, H. F., and Pau, K. Y. (1998). A biphasic developmental pattern of circulating leptin in the male rhesus macaque (*Macaca mulatta*). *Endocrinology* 139, 2284–2286.
97. Suter, K. J., Pohl, C. R., and Wilson, M. E. (2000). Circulating concentrations of nocturnal leptin, growth hormone, and insulin-like growth factor-I increase before the onset of puberty in agonadal male monkeys: potential signals for the initiation of puberty. *J. Clin. Endocrinol. Metab.* 85, 808–814.
98. Wilson, M. E., Fisher, J., Chikazawa, K., Yoda, R., Legendre, A., Mook, D., and Gould, K. G. (2003). Leptin administration increases nocturnal concentrations of luteinizing hormone and growth hormone in juvenile female rhesus monkeys. *J. Clin. Endocrinol. Metab.* 88, 4874–4883.
99. Gruaz, N. M., Lalaoui, M., Pierroz, D. D., Englaro, P., Sizonenko, P. C., Blum, W. F., and Aubert, M. L. (1998). Chronic administration of leptin into the lateral ventricle induces sexual maturation in severely food-restricted female rats. *J. Neuroendocrinol.* 10, 627–633.
100. Nagatani, S., Guthikonda, P., and Foster, D. L. (2000). Appearance of a nocturnal peak of leptin secretion in the pubertal rat. *Horm. Behav.* 37, 345–352.
101. Watanobe, H., and Schioth, H. B. (2002). Postnatal profile of plasma leptin concentrations in male and female rats: relation with the maturation of the pituitary-gonadal axis. *Regul. Pept.* 105, 23–28.
102. Smith, J. T., and Waddell, B. J. (2003). Developmental changes in plasma leptin and hypothalamic leptin receptor expression in the rat: peripubertal changes and the emergence of sex differences. *J. Endocrinol.* 176, 313–319.
103. McFadin, E. L., Morrison, C. D., Buff, P. R., Whitley, N. C., and Keisler, D. H. (2002). Leptin concentrations in periparturient ewes and their subsequent offspring. *J. Anim. Sci.* 80, 738–743.
104. Ehrhardt, R. A., Greenwood, P. L., Bell, A. W., and Boisclair, Y. R. (2003). Plasma leptin is regulated predominantly by nutrition in preruminant lambs. *J. Nutr.* 133, 4196–4201.
105. Chehab, F. F., Mounzih, K., Ronghua, L., and Lim, M. E. (1997). Early onset of reproductive function in normal female mice treated with leptin. *Science* 275, 88–90.
106. Ahima, R. S., Dushay, J., Flier, S. N., Prabakaran, D., and Flier, J. S. (1997). Leptin accelerates the onset of puberty in normal female mice. *J. Clin. Invest.* 99, 391–395.
107. Almog, B., Gold, R., Tajima, K., Dantes, A., Salim, K., Rubinstein, M., Barkan, D., Homburg, R., Lessing, J. B., Nevo, N., Gertler, A., and Amsterdam, A. (2001). Leptin attenuates follicular apoptosis and accelerates the onset of puberty in immature rats. *Mol. Cell Endocrinol.* 183, 179–191.
108. Cheung, C. C., Thornton, J. E., Kuijper, J. L., Weigle, D. S., Clifton, D. K., and Steiner, R. A. (1997). Leptin is a metabolic gate for the onset of puberty in the female rat. *Endocrinology* 138, 855–858.
109. Cheung, C. C., Thornton, J. E., Nurani, S. D., Clifton, D. K., and Steiner, R. A. (2001). A reassessment of leptin's role in triggering the onset of puberty in the rat and mouse. *Neuroendocrinology* 74, 12–21.
110. Barker-Gibb, M. L., Sahu, A., Pohl, C. R., and Plant, T. M. (2002). Elevating circulating leptin in prepubertal male rhesus monkeys (*Macaca mulatta*) does not elicit precocious gonadotropin-releasing hormone release, assessed indirectly. *J. Clin. Endocrinol. Metab.* 87, 4976–4983.
111. Morrison, C. D., Daniel, J. A., Holmberg, B. J., Djiane, J., Raver, N., Gertler, A., and Keisler, D. H. (2001). Central infusion of leptin into well-fed and undernourished ewe lambs: effects on feed intake and serum concentrations of growth hormone and luteinizing hormone. *J. Endocrinol.* 168, 317–324.
112. Morrison, C. D., Wood, R., McFadin, E. L., Whitley, N. C., and Keisler, D. H. (2002). Effect of intravenous infusion of recombinant ovine leptin on feed intake and serum concentrations of GH, LH, insulin, IGF-1, cortisol, and thyroxine in growing prepubertal ewe lambs. *Domest. Anim. Endocrinol.* 22, 103–112.
113. Jackson, L. M., Ichimaru, T., Jaffe, C. A., and Foster, D. L. (2002). Effects of leptin on LH secretion during developmental and nutritional hypogonadotropism. In *Program of the 84th Meeting of the Endocrine Society*, San Francisco, Abst. OR45-1.
114. Jackson, L. M., Ichimaru, T., Farhat, J. S., Malcolm, K. D., Mulla, C. M., Jaffe, C. A., and Foster, D. L. (2003). Leptin stimulation of gonadotropin secretion in prepubertal, growth-restricted lambs. In *33rd Annual Meeting of the Society for Neuroscience*, New Orleans, Abst. 237.3.
115. Wade, G. N., and Schneider, J. E. (1992). Metabolic fuels and reproduction in female mammals. *Neurosci. Biobehav. Rev.* 16, 235–272.
116. Schneider, J. E., and Wade, G. N. (1989). Availabilty of metabolic fuels controls estrous cyclicity of Syrian hamsters. *Science* 244, 1326–1328.
117. Bucholtz, D. C., Manning, J. M., Schillo, K. K., and Foster, D. L. (1992). LH secretion is acutely sensitive to metabolic fuel deprivation. *Biol. Reprod.* 46 (Suppl. 1), p. 131, Abst. 324.
118. Bucholtz, D. C., Vidwans, N. M., Herbosa, C. G., Schillo, K. K., and Foster, D. L. (1996). Metabolic interfaces between growth and reproduction. V. Pulsatile luteinizing hormone secretion is dependent on glucose availability. *Endocrinology* 137, 601–607.
119. Nagatani, S., Thompson, R. C., and Foster, D. L. (2001). Prevention of glucoprivic stimulation of corticosterone secretion by leptin does not restore high frequency luteinizing hormone pulses in rats. *J. Neuroendocrinol.* 13, 371–377.
120. Handelsman, D. J., Spaliviero, J. A., Scott, C. D., and Baxter, R. C. (1987). Hormonal regulation of the peripubertal surge of insulin-like growth factor-1 in the rat. *Endocrinology* 120, 491–496.
121. Gabriel, S. M., Roncancio, J. R., and Ruiz, N. S. (1992). Growth hormone pulsatility and the endocrine milieu during sexual maturation in male and female rats. *Neuroendocrinology* 56, 619–628.
122. Wilson, M. E. (2001). The impact of the GH-IGF-I axis on gonadotropin secretion: inferences from animal models. *J. Pediatr. Endocrinol. Metab.* 14, 115–140.
123. Adam, C. L., Findlay, P. A., and Moore, A. H. (1997). Effects of insulin-like growth factor-1 on luteinizing hormone secretion in sheep. *Anim. Reprod. Sci.* 50, 45–56.
124. Roberts, C. A., McCutcheon, S. N., Blair, H. T., Gluckman, P. D., and Breier, B. H. (1990). Developmental patterns of plasma insulin-like growth factor-1 concentrations in sheep. *Domest. Anim. Endocrinol.* 7, 457–463.
125. Danilovich, N., Wernsing, D., Coschigano, K. T., Kopchick, J. J., and Bartke, A. (1999). Deficits in female reproductive function in GH-R-KO mice; role of IGF-I. *Endocrinology* 140, 2637–2640.
126. Arsenijevic, Y., Wehrenberg, W. B., Conz, A., Eshkol, A., Sizonenko, P. C., and Aubert, M. L. (1989). Growth hormone

(GH) deprivation induced by passive immunization against rat GH-releasing factor delays sexual maturation in the male rat. *Endocrinology* 124, 3050–3059.

127. Hiney, J. K., Srivastava, V., Nyberg, C. L., Ojeda, S. R., and Dees, W. L. (1996). Insulin-like growth factor I of peripheral origin acts centrally to accelerate the initiation of female puberty. *Endocrinology* 137, 3717–3728.
128. Wilson, M. E. (1995). IGF-I administration advances the decrease in hypersensitivity to oestradiol negative feedback inhibition of serum LH in adolescent female rhesus monkeys. *J. Endocrinol.* 145, 121–130.
129. Wilson, M. E. (1998). Premature elevation in serum insulin-like growth factor-I advances first ovulation in rhesus monkeys. *J. Endocrinol.* 158, 247–257.
130. Shirley, K. L., Hernandez, J. A., Hallford, D. M., and Thomas, M. G. (2001). Serum luteinizing hormone, growth hormone and insulin-like growth factor-I after releasing hormone challenge in prepubertal ewe lambs selected for twinning. *Theriogenology* 56, 867–877.
131. Suttie, J. M., Kostyo, J. L., Ebling, F. J., Wood, R. I., Bucholtz, D. C., Skottner, A., Adel, T. E., Towns, R. J., and Foster, D. L. (1991). Metabolic interfaces between growth and reproduction. IV. Chronic pulsatile administration of growth hormone and the timing of puberty in the female sheep. *Endocrinology* 129, 2024–2032.
132. Daftary, S. S., and Gore, A. C. (2004). The hypothalamic insulin-like growth factor-1 receptor and its relationship to gonadotropin-releasing hormone neurones during postnatal development. *J. Neuroendocrinol.* 16, 160–169.
133. Adam, C. L., Gadd, T. S., Findlay, P. A., and Wathes, D. C. (2000). IGF-I stimulation of luteinizing hormone secretion, IGF-binding proteins (IGFBPs) and expression of mRNAs for IGFs, IGF receptors and IGFBPs in the ovine pituitary gland. *J. Endocrinol.* 166, 247–254.
134. Williams, L. M., Kelly, D., Hannah, L. T., and Morgan, P. J. (1995). Localization of ^{125}I-IGF-I binding on the ovine pars tuberalis. *J. Neuroendocrinol.* 7, 931–938.
135. Miller, D. W., Blache, D., Boukhliq, R., Curlewis, J. D., and Martin, G. B. (1998). Central metabolic messengers and the effects of nutrition on gonadotrophin secretion in sheep. *J. Reprod. Fertil.* 112, 347–356.
136. Blache, D., Chagas, L. M., Blackberry, M. A., Vercoe, P. E., and Martin G. B. (2000). Metabolic factors affecting the reproductive axis in male sheep. *J. Reprod. Fertil.* 120, 1–11.
137. Cameron, J. L., Hansen, P. D., McNeill, T. H., Koerker, D. J., Clifton, D. J., Rogers, K. V., Bremner, W. J., and Steiner, R. A. (1985). Metabolic cues for the onset of puberty in primate species. In *Adolescence in Females* (C. Flamigni, S. Venturoli, and J. R. Givens, Eds.), pp. 59–78. Year Book Medical Publishers, Chicago.
138. Hileman, S. M., Schillo, K. K., and Hall, J. B. (1993). Effects of acute, intracerebroventricular administration of insulin on serum concentrations of luteinizing hormone, insulin, and glucose in ovariectomized lambs during restricted and ad libitum feed intake. *Biol. Reprod.* 48, 117–124.
139. Daniel, J. A., Thomas, M. G., Hale, C. S., Simmons, J. M., and Keisler, D. H. (2000). Effect of cerebroventricular infusion of insulin and (or) glucose on hypothalamic expression of leptin receptor and pituitary secretion of LH in diet-restricted ewes. *Domest. Anim. Endocrinol.* 18, 177–185.
140. Manning, J. M., and Bronson, F. H. (1990). The effects of low temperature and food intake on ovulation in domestic mice. *Phys. Zool.* 63, 938–948.
141. Manning, J. M., and Bronson, F. H. (1989). Effects of prolonged exercise on puberty and luteinizing hormone secretion in female rats. *Am. J. Phys.* 257, R1359–R1364.
142. Manning, J. M., and Bronson, F. H. (1991). Suppression of puberty in rats by exercise: effects on hormone levels and reversal with GnRH infusion. *Am. J. Phys.* 260, R717–R723.
143. I'Anson, H., Manning, J. M., Herbosa, C. G., Pelt, J., Friedman, C. R., Wood R. I., Bucholtz, D. C., and Foster, D. L. (2000). Central inhibition of gonadotropin-releasing hormone secretion in growth-restricted hypogonadotropic female sheep. *Endocrinology* 141, 520–527.
144. Ebling, F. J. P., Wood, R. I., Karsch, F. J., Vannerson, L. A., Suttie, J. M., Bucholtz, D. C., Schall, R. E., and Foster, D. L. (1990). Metabolic interfaces between growth and reproduction: III. Central mechanisms controlling pulsatile LH secretion in the nutritionally growth-limited female lamb. *Endocrinology* 126, 2719–2727.
145. Landefeld, T. D., Ebling, F. J. P., Suttie, J. M., Vannerson, L. A., Padmanabhan, V., Beitins, I. Z., and Foster, D. L. (1989). Metabolic interfaces between growth and reproduction. II. Characterization of changes in messenger ribonucleic acid concentrations of gonadotropin subunits, growth hormone, and prolactin in nutritionally growth-limited lambs and the differential effects of increased nutrition. *Endocrinology* 125, 351–356.
146. Briski, K. P., Quigley, K., and Meites, J. (1984). Endogenous opiate involvement in acute and chronic stress-induced changes in plasma LH concentrations in the male rat. *Life Sci.* 34, 2485–2493.
147. Dyer, R. G., Mansfield, S., Corbet, H., and Dean, A. D. P. (1985). Fasting impairs LH secretion in female rats by activating an inhibitory opioid pathway. *J. Endocrinol.* 105, 91–97.
148. Håkasson, M.-L., Brown, H., Ghilardi, N., Skoda, R. C., and Meister, B. (1998). Leptin receptor immunoreactivity in chemically defined target neurons of the hypothalamus. *J. Neurosci.* 18, 559–572.
149. Finn, P. D., Cunningham, M. J., Pau, K.-Y. F., Spies, H. G., Clifton, D. K., and Steiner, R. A. (1998). The stimulatory effect of leptin on the nueroendocrine reproductive axis of the monkey. *Endocrinology* 139, 4652–4662.
150. Dyer, C. J., Simmons, J. M., Matteri, R. L., and Keisler, D. H. (1997). Leptin receptor mRNA is expressed in ewe anterior pituitary and adipose tissues and is differentially expressed in hypothalamic regions of well-fed and feed-restricted ewes. *Domest. Anim. Endocrinol.* 14, 119–128.
151. Schwartz, M. W., Baskin, D. G., Bukowski, T. R., Kuijper, J. L., Foster, D., Lasser, G., Prunkard, D. E., Porte Jr., D., Woods, S. C., Seely, R. J., and Weigle, D. S. (1996). Specificity of leptin action on elevated blood glucose levels and hypothalamic neuropeptide gene expression in *ob/ob* mice. *Diabetes* 45, 531–535.
152. Wang, Q., Bing, C., Al-Barazanji, K., Mossakowaska, D. E., Wang, X. M., McBay, D. L., Neville, W. A., Taddayon, M., Pickavance, L., Dryden, S., Thomas, M. E., McHale, M. T., Gloyer, I. S., Wilson, S., Buckingham, R., Arch, J. R., Trayhurn, P., and Williams, G. (1997). Interactions between leptin and hypothalamic neuropeptide Y neurons in the control of food intake and energy homeostasis in the rat. *Diabetes* 46, 335–341.
153. McShane, T. M., May, T., Miner, J. L., and Keisler, D. H. (1992). Central actions of neuropeptide-Y may provide a neuromodulatory link between nutrition and reproduction. *Biol. Reprod.* 46, 1151–1157.
154. Malven, P. V., Haglof, S. A., and DeGroot, H. (1992). Effects of intracerebral administration of neuropeptide-Y on secretion of luteinizing hormone in ovariectomized sheep. *Brain Res. Bull.* 28, 871–875.
155. Morrison, C. D., Daniel, J. A., Hampton, J. H., Buff, P. R., McShane, T. M., Thomas, M. G., and Keisler, D. H. (2003). Luteinizing hormone and growth hormone secretion in ewes infused intracerebroventricularly with neuropeptide Y. *Domest. Anim. Endocrinol.* 24, 69–80.

156. Miner, J. L., Della-Fera, M. A., Paterson, J. A., and Baile, C. A. (1989). Lateral cerebroventricular injection of neuropeptide Y stimulates feeding in sheep. *Am. J. Phys.* 257, R383–R387.
157. Miner, J. L., Della-Fera, M. A., and Paterson, J. A. (1990). Blockade of satiety factors by central injection of neuropeptide Y in sheep. *J. Anim. Sci.* 68, 3805–3811.
158. Berelowitz, M., Bruno, J. F., and White, J. D. (1992). Regulation of hypothalamic neuropeptide expression by peripheral metabolism. *Trends Endocrinol. Metab.* 3, 127–133.
159. Karsch, F. J., Bittman, E. L., Foster, D. L., Goodman, R. L., Legan, S. J. and Robinson, J. E. (1984). Neuroendocrine basis of seasonal reproduction. *Rec. Prog. Horm. Res.* 40, 185–232.
160. Karsch, F. J., Woodfill, C. J. I., Malpaux, B., Robinson, J. E., and Wayne, N. L. (1991). Melatonin and mammalian photoperiodism: synchronization of annual reproductive cycles. In *Suprachiasmatic Nucleus: The Mind's Clock* (R. Moore, S. Reppert, and D. Klein, Eds.), pp. 217–232. Oxford University Press, New York.
161. Lincoln, G. A., and Short, R. V. (1980). Seasonal breeding: nature's contraceptive. *Rec. Prog. Horm. Res.* 36, 1–52.
162. Fletcher, I. C., and Geytenbeek, P. E. (1970). Seasonal variation in the ovarian activity of Merino ewes. *Aust. J. Exp. Agric. Anim. Husb.* 10, 267–270.
163. Mallampati, R. S., Pope, A. L., and Casida, L. E. (1971). Breeding pattern in Targhee ewes and ewe lambs throughout the year. *J. Anim. Sci.* 33, 1278–1281.
164. Wiggins, E. L., Barker, H. B., and Miller, W. W. (1970). Estrual activity in open Rambouillet ewes. *J. Anim. Sci.* 30, 405–408.
165. Marshall, F. H. A. (1937). On change over in oestrus cycle in animals after transference across equator, with further observations on incidence of breeding season and factors controlling sexual periodicity. *Proc. R. Soc. Lond. B* 122, 413–428.
166. Yeates, N. T. M. (1949). The breeding season of the sheep with particular reference to its modification by artificial means using light. *J. Agric. Sci.* 39, 1–43.
167. Thwaites, C. J. (1965). Photoperiodic control of breeding activity in the Southdown ewe with particular reference to the effects of an equatorial light regime. *J. Agric. Sci.* 65, 57–64.
168. Wodzicka-Tomaszewska, M., Hutchinson, J. C. D., and Bennett, J. W. (1967). Control of the annual rhythm of breeding in ewes: effect of an equatorial daylength with reversed thermal seasons. *J. Agric. Sci.* 68, 61–67.
169. Legan, S. J., and Karsch, F. J. (1980). Photoperiodic control of seasonal breeding in ewes: modulation of the negative feedback action of estradiol. *Biol. Reprod.* 23, 1061–1068.
170. Dyrmundsson, O. R., and Lees, J. L. (1982). Attainment of puberty and reproductive performance in Clun Forest ewe lambs. *J. Agric. Sci.* 78, 39–45.
171. Foster, D. L., and Ryan, K. D. (1990). Puberty in the lamb: sexual maturation of a seasonal breeder in a changing environment. In *Control of the Onset of Puberty II* (P. C. Sizonenko, M. L. Aubert, and M. M. Grumbach, Eds.), pp. 108–142. William & Wilkins, Baltimore.
172. Foster, D. L. (1981). Mechanism for delay of first ovulation in lambs born in the wrong season (Fall). *Biol. Reprod.* 25, 85–92.
173. Derqaoui, L. (1993). Onset of puberty and development of reproductive capacity in D'man and Sardi breeds of sheep and their crosses. Ph.D. Thesis, Agronomy and Veterinary Medicine Institute, Rabat.
174. Lahlou-Kassi, A., and Marie, M. (1985). Sexual and ovarian function of the D'man ewe. In *Genetics of Reproduction in Sheep* (R. B. Land and D. W. Robinson, Eds.), pp. 245–260. Butterworths, London.
175. Adam, C. L., Findlay, P. A., Kyle, C. E., and Young, P. (1998). Effect of restricted nutrition on timing of puberty in female Soay sheep. *J. Reprod. Fertil.* 112, 31–37.
176. Yellon, S. M., and Foster, D. L. (1985). Alternate photoperiods time puberty in the female lamb. *Endocrinology* 116, 2090–2097.
177. Malpaux, B., Daveau, A., Maurice, F., Gayrard, V., and Thiery, J.-C. (1993). Short-day effects of melatonin on luteinizing hormone secretion in the ewe: evidence for central sites of action in the mediobasal hypothalamus. *Biol. Reprod.* 48, 752–760.
178. Malpaux, B., Daveau, A., Maurice-Mandon, F., Duarte, G., and Chemineau, P. (1998). Evidence that melatonin acts in the premammillary hpothalamic area to control reproduction in the ewe: presence of binding sites and stimulation of luteinizing hormone secretion by *in situ* mmicroimplant delivery. *Endocrinology* 139, 1508–1516.
179. Bittman, E. L., and Weaver, D. R. (1990). The distribution of melatonin binding sites in neuroendocrine tissues of the ewe. *Biol. Reprod.* 43, 986–993.
180. de Reviers M. M., Ravault, J. P., Tillet, Y., and Pelletier, J. (1989). Melatonin binding sites in the sheep pars tuberalis. *Neurosci. Lett.* 100, 89–93.
181. Morgan, P. J., Williams, L. M., Davidson, G., Lawson, W., and Howell., H. (1989). Melatonin receptors on ovine pars tuberalis: characterization and autoradiographical localization. *J. Neuroendocrinol.* 1, 1–4.
182. Stankoc, B., Cozzi, B., Lucini, V., Scaglione, F., and Fraschini, F. (1991). Characterization and mapping of melatonin receptors in the brain of three mammalian species: rabbit, horse and sheep. *Neuroendocrinology* 53, 214–221.
183. Foster, D. L., Ebling, F. J. P., Claypool, L. E., and Wood, R. I. (1989). Photoperiodic timing of puberty in sheep. In *Research in Perinatal Medicine, Vol. IX: Development of Circadian Rhythmicity and Photoperiodism in Mammals* (S. M. Reppert, Ed.), pp. 103–153. Perinatology Press, Ithaca, NY.
184. Robinson, J. E., Kaynard, A. H., and Karsch, F. J. (1986). Does melatonin alter pituitary responsiveness to GnRH in the ewe? *Neuroendocrinology* 43, 635–640.
185. Jansen, H. T., Cutter, C. T., Hardy, S., Lehman, M. N., and Goodman, R. L. (2003). Seasonal plasticity within the gonadotropin-releasing hormone (GnRH) system of the ewe: change in identified GnRH inputs and glial association. *Endocrinology* 144, 3663–3676.
186. Kennaway, D. J., Matthews, C. D., Seamark, R. F., Philipou, G., and Schilthuis, M. (1977). On the presence of melatonin in pineal glands and plasma of foetal sheep. *J. Steroid Biochem.* 8, 559–563.
187. Yellon, S. M., and Longo, L. D. (1987). Melatonin rhythms in fetal and maternal circulation during pregnancy in sheep. *Am. J. Physiol.* 252, E799–E802.
188. McMillen, I. C., and Nowak, R. (1989). Maternal pinealectomy abolishes the diurnal rhythm in plasma melatonin concentrations in the fetal sheep and pregnant ewe during late gestation. *J. Endocrinol.* 120, 459–464.
189. Yellon, S. M., and Longo, L. D. (1988). Effect of maternal pinealectomy and reverse photoperiod on the circadian melatonin rhythm in the sheep and fetus during the last trimester of pregnancy. *Biol. Reprod.* 39, 1093–1099.
190. Reppert, S. M., Shea, R. A., Anderson, A., and Klein, D. C. (1979). Maternal-fetal transfer of melatonin in a non-human primate. *Pediatr. Res.* 13, 788–791.
191. Reiter, R. J. (1991). Pineal gland—interface between the photoperiodic environment and the endocrine system. *Trends Endocrinol. Metab.* 2, 13–19.
192. Claypool, L. E., Wood, R. I., Yellon, S. M., and Foster, D. L. (1989). The ontogeny of melatonin secretion in the lamb. *Endocrinology* 124, 2135–2143.
193. Wood, R. I., Claypool, L. E., Ebling, F. J. P., and Foster, D. L. (1989). Entrainment of the melatonin rhythms in early postnatal lambs and their mothers. *J. Biol. Rhythms* 4, 457–465.

194. Matthews, C. D., Kennaway, D. J., and Seamark, R. F. (1981). Reproductive studies in pinealectomized sheep. In *Pineal Function* (C. D. Matthews and R. F. Seamark, Eds.), pp. 137–142. Elsevier/North-Holland, Amsterdam.
195. Kennaway, D. J., Gilmore, T. A., and Dunstan, E. A. (1985). Pinealectomy delays puberty in ewe lambs. *J. Reprod. Fertil.* 74, 119–125.
196. Foster, D. L., Ebling, F. J. P., Claypool, L. E., and Woodfill, C. J. I. (1988). Cessation of long-day melatonin rhythms time puberty in a short day breeder. *Endocrinology* 123, 1636–1641.
197. Kennaway, D. J., and Gilmore, T. A. (1984). Effects of melatonin implants in ewe lambs. *J. Reprod. Fertil.* 70, 39–45.
198. Nowak, R., and Rodway, R. G. (1985). Effect of intravaginal implants on the onset of ovarian activity in adult and prepubertal ewes. *J. Reprod. Fertil.* 74, 287–293.
199. Waldhauser, F., and Steger, H. (1986). Changes in melatonin secretion with age and pubescence. *J. Neural Transm.* 21(Suppl.), 183–197.
200. Foster, D. L., Karsch, F. J., Olster, D. H., Ryan, K. D. and Yellon, S. M. (1986). Determinants of puberty in a seasonal breeder. *Rec. Prog. Horm. Res.* 42, 331–384.
201. Klein, D. C. (1984). Melatonin and puberty. *Science* 224, 6.
202. Tamarkin, L., Baird, C. J., and Almeida, O. F. X. (1985). Melatonin: a coordinating system for mammalian reproduction? *Science* 227, 714–720.
203. Cavallo, A. (1992). Plasma melatonin rhythm in normal puberty: interactions of age and pubertal stages. *Neuroendocrinology* 55, 372–379.
204. Plant, T. M., and Zorub, D. S. (1986). Pinealectomy in agonadal infantile male rhesus monkeys (*Macaca mulatta*) does not interrupt initiation of the prepubertal hiatus in gonadotropin secretion. *Endocrinology* 118, 227–232.
205. Wilson, M. E., and Gordon, T. P. (1989). Short-day melatonin pattern advances puberty in seasonally breeding rhesus monkeys (*Macaca mulatta*). *J. Reprod. Fertil.* 86, 435–444.
206. Rodway, R. G., Swift, A. D., Nowak, R., Smith, J. A., and Padwick, J. A. (1985). Plasma concentrations of melatonin and the onset of puberty in the female lamb. *Anim. Reprod. Sci.* 8, 241–246.
207. Foster, D. L., Ebling, F. J. P., Claypool, L. E., Wood, R. I., Adel, T. E., and Schramm, W. (1989). Amplitude modulation of the nightly melatonin rise in the neonatal lamb and the subsequent timing of puberty. *Biol. Reprod.* 40, 920–928.
208. Lincoln, G. A., and Ebling, F. J. P. (1985). Effect of constant-release implants of melatonin on seasonal cycles in reproduction, prolactin secretion, and moulting in rams. *J. Reprod. Fertil.* 73, 241–253.
209. O'Callaghan, D., Karsch, F. J., Boland, M. P., and Roche, J. F. (1992). What photoperiodic signal is provided by a continuous release melatonin implant? *Biol. Reprod.* 45, 927–933.
210. Bassett, J. M., Bomford, J., and Mott., J. C. (1988). Photoperiod: an important regulator of plasma prolactin concentration in fetal lambs during late gestation. *Q. J. Exp. Physiol.* 73, 241–244.
211. Ebling, F. J. P., Wood, R. I., Suttie, J. M., Adel, T. E., and Foster, D. L. (1989). Prenatal photoperiod influences neonatal prolactin secretion in the sheep. *Endocrinology* 125, 384–391.
212. Foster, D. L., and Yellon, S. M. (1985). Photoperiodic time measurement is maintained in undernourished lambs with delayed puberty. *J. Reprod. Fertil.* 75, 203–208.
213. Tucker, H. A., Petitclerc, D., and Zinn, S. A. (1984). The influence of photoperiod on body weight gain, body composition, nutrient intake and hormone secretion. *J. Anim. Sci.* 59, 1610–1620.
214. Brown, E. B., and Forbes, J. M. (1980). Diurnal variations of plasma prolactin in growing sheep under two lighting regimes and the effect of pinealectomy. *J. Endocrinol.* 84, 91–99.
215. Karsch, F. J., and Foster, D. L. (1981). Environmental control of seasonal breeding: a common final mechanism governing seasonal breeding and sexual maturation. In *Environmental Factors in Mammal Reproduction* (D. P. Gilmore, and B. Cook, Eds.), pp. 30–53. MacMillan Press, London.
216. Ebling, F. J. P., and Foster, D. L. (1989). Seasonal breeding-a model for puberty? In *Control of the Onset of Puberty III* (H. A. Delemarre-van de Waal, T. M. Plant, G. P. van Rees, and J. Schoemaker, Eds.), pp. 253–264. Elsevier Science Publishers B.V., Amsterdam.
217. Geschwind, I. I. (1974). Discussion of McCann, S. M., Ojeda, S., and Negro-Vilar, A.: Sex steroid, pituitary and hypothalamic hormones during puberty in experimental animals. In *Control of the Onset of Puberty* (M. M. Grumbach, G. D. Grave, and F. E. Mayer, Eds.), pp. 27–28. John Wiley & Sons, New York.
218. Herbosa, C. G., Wood, R. I., I'Anson, H., and Foster, D. L. (1993). Prenatal photoperiod and the timing of puberty in the female lamb. *Biol. Reprod.* 50, 1367–1376.
219. Ortavant, R. (1958). Le cycle spermatogenetique chez le belier. Ph.D. Thesis, Paris University, Paris.
220. Olster, D. H., and Foster, D. L. (1986). Control of gonadotropin secretion in the male during puberty: aA decrease in response to steroid inhibitory feedback in the absence of an increase in steroid-independent drive in the sheep. *Endocrinology* 118, 2225–2234.
221. Schanbacher, B. D., and Ford, J. J. (1976). Seasonal profiles of plasma luteinizing hormone, testosterone and estradiol in the ram. *Endocrinology* 99, 752–757.
222. Schanbacher, B. D. (1984). Regulation of luteinizing hormone secretion in male sheep by endogenous estrogen. *Endocrinology* 115, 944–950.
223. Wood, R. I., Mehta, V., Herbosa, C. G., and Foster, D. L. (1995). Prenatal testosterone differentially masculinizes tonic and surge modes of LH secretion in the developing sheep. *Neuroendocrinology* 56, 822–830.
224. Herbosa, C. G., and Foster, D. L. (1996). Defeminization of the reproductive response to photoperiod occurs early in prenatal development in the sheep. *Biol. Reprod.* 54, 420–428.
225. Kosut, S. S., Wood, R. I., Herbosa-Encarnacion, C. G., and Foster, D. L. (1997). Prenatal androgens time neuroendocrine puberty in sheep: effects of dose. *Endocrinology* 38, 1072–1077.
226. Masek, K. S., Wood, R. I., and Foster, D. L. (1999). Prenatal dihydrotestosterone differentially masculinizes tonic and surge modes of luteinizing hormone secretion in sheep. *Endocrinology* 140, 3459–3466.
227. Clarke, I. J., Scaramuzzi, R. J., and Short, R. V. (1977). Ovulation in prenatally androgenized ewes. *J. Endocrinol.* 73, 385–389.
228. Wilson, P. R., and Tarttelin, M. F. (1978). Studies of sexual differentiation in sheep. II. Evidence of postnatal hypothalamic hypophysiotropic depression of basal LH secretion following prenatal androgenization. *Acta Endocrinol. (Copenh).* 89, 190–195.
229. Wood, R. I., Robinson, J. E., Forsdike, R. A., Padmanabhan, V., and Foster, D. L. (2000). Sexual differentiation of the timing of sexual maturation in sheep. In *The Onset of Puberty in Perpective* (J.-P. Bourguignon, and T. M. Plant, Eds.), pp. 269–287. Elsevier Science Publishers, Amsterdam.
230. Foster, D. L., Padmanabhan, V., Wood, R. I., and Robinson, J. E. (2002). Sexual differentiation of the neuroendocrine control of gonadotrophin secretion: concepts derived from sheep models. *Reproduction* 59 (Suppl.), 83–99.
231. Herbosa, C. G., Dahl, G. E., Evans, N. P., Pelt, J., Wood, R. I., and Foster, D. L. (1996). Sexual differentiation of the surge mode of gonadotropin secretion: prenatal androgens abolish the gonadotropin-releasing hormone surge in the sheep. *J. Neuroendocrinol.* 8, 627–633.

232. Robinson, J. E., Forsdike, R. A., and Taylor, J. A. (1999). In utero exposure of female lambs to testosterone reduces the sensitivity of the gonadotropin-releasing hormone neuronal network to inhibition by progesterone. *Endocrinology* 140, 5797–5805.
233. Robinson, J. E., Birch, R. A., Foster, D. L., and Padmanabhan, V. (2002). Prenatal exposure of the ovine fetus to androgens sexually differentiates the steroid feedback mechanisms that control gonadotropin releasing hormone secretion and disrupts ovarian cycles. *Arch. Sex. Behav.* 31, 35–41.
234. Goy, R. W., and McEwen, B. S., Eds. (1980). *Sexual Differentiation of the Brain*. MIT Press, Cambridge, MA.
235. Clarke, I. J., Scaramuzzi, R. J., and Short, R. V. (1976). Effects of testosterone implants in pregnant ewes on their female offspring. *J. Embryol. Exp. Morph.* 36, 87–99.
236. Short, R. V. (1974). Sexual differentiation of the brain of the sheep. In *Endocinologie Sexuelle de la Periode Perinatale* (M. G. Forest and J. Bertrand, Eds.), Vol. 32, pp. 121–142. INSERM, Paris.
237. Clarke, I. J. (1977). The sexual behaviour of prenatally androgenized ewes observed in the field. *J. Reprod. Fertil.* 49, 311–315.
238. Karsch, F. J., Dierschke, D. J., and Knobil, E. (1973). Sexual differentiation of pituitary function: apparent difference between primates and rodents. *Science* 179, 484–486.
239. Steiner, R. A., Clifton, D. K., Spies, H. G., and Resko, J. A. (1976). Sexual differentiation and feedback control of luteinizing hormone secretion in the rhesus monkey. *Biol. Reprod.* 15, 206–212.
240. Norman, R. L., and Spies, H. G. (1986). Cyclic ovarian function in a male macaque: additional evidence for a lack of sexual differentiation in the physiological mechanisms that regulate the cyclic release of gonadotropins in primates. *Endocrinology* 118, 2608–2610.
241. Gorski, R. A. (1991). Sexual differentiation of the endocrine brain and its control. In *Brain Endocrinology* (M. Motta, Ed.), pp. 71–104. Raven Press, New York.
242. Gorski, R. A. (2000). Sexual differentiation of the nervous system. In *Principles of Neural Science* (E. R. Kandel, J. Schwartz, and T. Jessel, Eds.), pp. 1130–1148. McGraw-Hill, New York.
243. Caldani, M. (1986). Mise en évidence immunohistochimique des neurones a LHRH du mouton (*Ovis aries*): systématisation et ontogenese. Thesis, Université de Paris-VI, Paris.
244. Caldani, M., Antoine, M., Batailler, M., and Duittoz, A. (1995). Ontogeny of GnRH systems. *J. Reprod. Fertil.* 49 (Suppl.), 147–162.
245. Wray, S., Nieburges, A., and Elkabes, S. (1989). Spatiotemporal cell expression of luteinizing hormone-releasing hormone in the prenatal mouse: evidence for an embryonic origin in the olfactory placode. *Brain Res. Dev.* 46, 309–318.
246. Schwanzel-Fukuda, M., and Pfaff, D. W. (1989). Origin of luteinizing hormone-releasing hormone neurons. *Nature* 338, 161–164.
247. Daikoku-Ishido, H., Okamura, Y., Yanihara, N., and Daikoku, S. (1990). Development of the hypothalamic luteinizing hormone-releasing hormone-containing neuron system in the rat: in vivo and in vitro transplanation studies. *Dev. Biol.* 140, 374–387.
248. Ronnekleiv, O. K., and Resko, J. A. (1990). Ontogeny of GnRH-containing neurons in early fetal development of rhesus macaques. *Endocrinology* 126, 498–511.
249. Wood, R. I., Newman, S. W., Lehman, M. N., and Foster, D. L. (1992). GnRH neurons in the fetal lamb hypothalamus are similar in males and females. *Neuroendocrinology* 55, 427–433.
250. Lehman, M. N., Robinson, J. E., Karsch, F. J., and Silverman, A.-J. (1986). Immunocytochemical localization of luteinizing hormone-releasing hormone (LHRH) pathways in the sheep brain during anestrous and the mid-luteal phase. *J. Comp. Neurol.* 244, 19–35.
251. Caldani, M., Batailler, M., Thiery, J. C., and Dubois, M. P. (1988). LHRH-immunoreactive structures in the sheep brain. *Histochemistry* 89, 129–139.
252. Wray, S., and Hoffman, G. E. (1986). Postnatal morphological changes in rat LH-RH neurons correlated with sexual maturation. *Neuroendocrinology* 43, 93–97.
253. Goldsmith, P. C., and Song, T. (1987). The gonadotropin-releasing hormone containing ventral hypothalamic tract in the fetal rhesus monkey. *J. Comp. Neurol.* 257, 130–139.
254. Kim, S.-J., Foster, D. L., and Wood, R. I. (1999). Prenatal testosterone masculinizes synaptic input to GnRH neurons in sheep. *Biol. Reprod.* 61, 599–605.
255. Wood, R. I., Kim, S.-J., and Foster, D. L. (1996). Prenatal androgens defeminize activation of GnRH neurons in response to estradiol stimulation. *J. Neuroendocrinol.* 8, 617–625.
256. Bronson, F. H. (1989). *Mammalian Reproductive Biology*. The University of Chicago Press, Chicago.
257. Wood, R. I., Ebling, F. J. P., and Foster, D. L. (1991). Sex differences in nutritional modulation of gonadotropin secretion during development: studies in the growth-retarded lamb. *Biol. Reprod.* 44, 632–639.
258. Courot, M., de Reviers, M., and Pelletier, J. (1975). Variations in pituitary and blood LH during puberty in the male lamb. Relation to time of birth. *Ann. Biol. Bioch. Biophys.* 15, 509–516.
259. Claypool, L. E. (1985). Sexual maturation of sheep: photoperiod that delays puberty in the female does not retard prepubertal testicular growth. *Biol. Reprod.* 32 (Suppl. 1), 176 (Abst. 278).
260. Wood, R. I., Ebling, F. J. P., I'Anson, H., and Foster, D. L. (1991). The timing of neuroendocrine sexual maturity in the male lamb by photoperiod. *Biol. Reprod.* 45, 82–88.
261. Herbosa, C. G., Wood, R. I., and Foster, D. L. (1995). Prenatal androgens modify the reproductive response to photoperiod in developing sheep. *Biol. Reprod.* 52, 163–169.
262. Frost, D., and Zucker, I. (1983). Photoperiod and melatonin influence seasonal gonadal cycles in the grasshopper mouse (*Onychomys leucogaster*). *J. Reprod. Fertil.* 69, 237–244.
263. Spears, N., and Clarke, J. R. (1987). Effect of nutrition, temperature and photoperiod on the rate of sexual maturation of the field vole (*Microtus agrestis*). *J. Reprod. Fertil.* 80, 175–181.
264. Darrow, J. M., Davis, F. C., Elliott, J. A., Stetson, M. H., Turek, F. W., and Menaker, M. (1980). Influence of photoperiod on reproductive development in the golden hamster. *Biol. Reprod.* 22, 443–450.
265. Rollag, M. D., Dipinto, M. N., and Stetson, M. H. (1982). Ontogeny of the gonadal response of golden hamsters to short photoperiod, blinding, and melatoin. *Biol. Reprod.* 27, 898–902.
266. Sisk, C. L., and Turek, F. W. (1983). Gonadal growth and gonadal hormones do not participate in the development of responsiveness to photoperiod in the golden hamster. *Biol. Reprod.* 29, 439–445.
267. Sisk, C. L., and Turek, F. W. (1987). Reproductive responsiveness to short photoperiod develops postnatally in male golden hamsters. *J. Androl.* 8, 91–96.
268. Foster, D. L., Olster, D. H., and Yellon, S. M. (1985). Neuroendocrine regulation of puberty by nutrition and photoperiod. In *Adolescence in Females* (C. Flamigni, S. Venturoli, and J. R. Givens, Eds.), pp. 1–21. Year Book Medical Publishers, Chicago.

269. Foster, D. L., Yellon, S. M., and Olster, D. H. (1985). Internal and external determinants of the timing of puberty in the female. *J. Reprod. Fertil.* 75, 327–344.

270. Foster, D. L., Ebling, F. J. P., Vannerson, L. A., Wood, R. I., and Fenner, D. E. (1988). Regulation of puberty in the lamb: internal and external cues. In *Progress in Endocrinology* (H. Imura, K. Shizume, and S. Yoshida, Eds.), Vol. 2, pp. 861–866. Elsevier, Amsterdam.

271. Moenter, S. M., Caraty, A., and Karsch, F. J. (1990). The estradiol-induced surge of gonadotropin-releasing hormone in the ewe. *Endocrinology* 127, 1375–1384.

272. I'Anson, H., Foster, D. L., Foxcroft G. R., and Booth, P. J. (1991). Nutrition and reproduction. In *Oxford Reviews of Reproductive Biology* (S. R. Milligan, Ed.), Vol. 13, pp. 239–311. Oxford University Press, Oxford.

273. Ebling, F. J. P., and Foster, D. L. (1989). Pineal melatonin rhythms and the timing of puberty in mammals. *Experientia* 45, 946–954.

274. Wood, R. I., and Foster, D. L. (1998). Sexual differentiation of reproductive neuroendocrine function. *Rev. Reprod.* 3, 130–140.

Knobil and Neill's Physiology of Reproduction, Third Edition
edited by Jimmy D. Neill,
Elsevier © 2006

CHAPTER 40

Puberty in Nonhuman Primates* and Humans

Tony M. Plant[1] and Selma F. Witchel[2]

Definition of Puberty, 2177
Recognition of Puberty, 2178
Historical Perspectives, 2178
Gonadal and Adrenal Function during Puberty, 2179
The Testis, 2179
The Ovary, 2180
The Adrenal, 2181
Pubertal Staging, 2182
Secular Trends, 2183
Pituitary Gonadotrope Function during Puberty, 2183
The Neuroendocrine Control System Governing Gonadotropin Secretion, 2184
Overview, 2184
Ontogeny, 2185
Nongonadal Determinant of the Hypogonadotropic State of Juvenile Development, 2189
Neuroendocrine Basis of the Juvenile Hiatus in Gonadotropin Secretion, 2191
Neurobiological Bases of the Juvenile Brake on Pulsatile GnRH Release, 2194
Theoretical Physiological Controls Governing the Duration of the Prepubertal Hiatus in GnRH Secretion, 2201
Somatometer, 2201
Pubertal Clock, 2207
Genetics and Puberty Genes, 2208
Fetal Programming and the Onset of Puberty, 2209
Exteroceptive Cues and the Timing of Puberty, 2209
Season, 2209
Social Factors, 2210
Food Availability, 2211
Disorders of Pubertal Development, 2211
Precocious Puberty, 2211
Delayed Puberty, 2212
Conclusion, 2213
Acknowledgments, 2213
References, 2213

DEFINITION OF PUBERTY

Attempts by biologists to define puberty have involved considerable debate and discussion that have, on occasion, led to the notion that the process of puberty does not readily lend itself to definition (2,3). On the other hand, Webster's view of the process, namely, as "the period of becoming first capable of reproducing sexually, marked by maturation of the genital organs, development of secondary sex characteristics, and, in the human and in higher primates, by the first occurrence of menstruation in the female" (4), requires little clarification. It is succinct, conceptually straightforward, and therefore acceptable to both lay and scientific communities. Of keynote importance in Webster's definition is the phrase "the period of becoming," because it correctly implies that puberty is a temporal process involving transition

*For the purpose of the present discussion, Romer's classification of the chordates was used (1). Romer divides the order of primates into the suborders *Lemuroidea* (tree shrews and lemurs), *Tarsioidea* (*Tarsius*), and *Anthropoidea* (monkeys, apes, and humans). Anthropoidea is further subdivided into the infraorders *Platyrrhini* (New World monkeys and marmosets) and *Catarrhina*. Three families of higher primate, namely *Cercopithecidae* (Old World monkeys), *Simiidae* (humans-like apes), and *Hominidae* (human), comprise the infraorder *Catarrhina*.

[1]Department of Cell Biology and Physiology, University of Pittsburth School of Medicine;
[2]Department of Pediatrics, Children's Hospital of Pittsburgh, Pittsburgh, Pennsylvania.

and development. Development may be viewed as the summation of differentiation and growth, and in higher primates normal puberty is observed during the terminal phase of growth. The transition into primate puberty is driven to a variable extent by two physiologic processes: gonadarche and adrenarche (5–7). Gonadarche comprises growth and maturation of the gonads and is associated with increased secretion of sex steroids and with the initiation of folliculogenesis and ovulation in the female and spermatogenesis in the male. Adrenarche comprises maturation of the adrenal cortex and is associated with increased secretion of adrenal androgens. Whereas gonadarche is obviously critical for puberty in all species, including nonprimates, adrenarche appears to be peculiar to humans and the Great Apes (8–11). In humans, the physical manifestations of adrenarche are known as pubarche and include the appearance of pubic and axillary hair, acne, and adult apocrine odor. The absence of adrenarche in humans neither prevents fertility nor dramatically influences the timing and tempo of gonadarche (7,12,13), and thus from a biological perspective adrenarche can be viewed as a temporal corollary of the pubertal process rather than as an integral and fundamental component of this important developmental event.

Perhaps the most striking feature of sexual development in primates is the pronounced interval that exists from birth until the initiation of the pubertal process. For example, although the sexually mature common marmoset[2] has a body weight similar to that of an adult rat (300 g), puberty in this small New World monkey is not observed until at least 6 to 12 months later than that in the laboratory rodent (14). The delay in the onset of puberty is most conspicuous in our own species, where initiation of spermatogenesis and ovarian cyclicity does not normally occur until the second decade of life. Thus, puberty in primates is separated in a most definite fashion from perinatal ontogeny. In these species, therefore, identification of the initiation of the pubertal process and its physiological and behavioral concomitants is a relatively simple task. At the other end of the spectrum of mammalian development are the rapidly maturing species, such as laboratory rodents, in which the processes of differentiation and puberty are much more difficult to separate, a state of affairs that contributes to the difficulty in defining their puberty.

In humans, the prolonged period of development from birth to adulthood may be divided into five life history stages, namely infancy, childhood, juvenile, adolescence (pubertal), and adulthood (15). Infancy is defined as the period when nourishment of the offspring is obtained via suckling or from formula and ends at the time of weaning. Childhood is the period during which the offspring remain dependent on their parents (or surrogate elders) for food and protection. In contrast to children, juveniles are no longer dependent on parents (elders) for survival. The juvenile phase of development is terminated by the onset of puberty, and at this time the individual enters adolescence. Because childhood is only observed in hominids (15), for the purpose of this chapter the term juvenile is used to describe the phase of development between infancy and puberty in both human and nonhuman primates.

RECOGNITION OF PUBERTY

In primates, as in other mammalian species, the state of having entered the development stage defined as puberty, which in catarrhines may span a time frame of several years, is recognized by the cascade of morphological, physiological, and behavioral sequela of increased gonadal and, in some cases, adrenal activity. Some of these biological changes are relatively discrete and therefore provide quantitative markers of this transitional phase in development. Many others, however, are less obvious and more difficult to monitor. It has not been possible to pinpoint the beginning or the ending of puberty in primates with any degree of precision, and it has not been possible to provide for any of these species a complete chronicle of events that unfold as an individual progresses through puberty. It is important to appreciate at the outset that the foregoing hiatus of descriptive information has minimal impact on our ability to construct and to test physiological models that have been, or may in the future be, proposed to account for the timing of puberty in primates. The purpose of this chapter is to examine the control systems that underlie the transition into puberty rather than to provide a comprehensive review of the morphological, physiological, and behavioral concomitants associated with puberty.

HISTORICAL PERSPECTIVES

The conceptual framework, which serves as the basis for our classical understanding of puberty, has been elegantly reviewed by Donovan and Van der

[2]The following trivial names have been used: African green monkey, *Cercopithecus aethiops*; baboon, *Papio*; bonnet monkey, *Macaca radiata*; capuchin monkey, *Cebus apelle*; chimpanzee, *Pan*; common marmoset, *Callithrix jacchus*; cotton-top tamarin, *Sanguinous oedipus Oedipus*; crab-eating or cynomolgus monkey, *Macaca fascicularis*; gorilla, *Gorilla*; hanuman langur, *Presbytis entellus*; mandrill, *Mandrillus sphinx*; owl monkey, *Aotus trivirgatus griseimembra*; pigtail monkey, *Macaca nemestrina*; rhesus monkey, *Macaca mulatta*; saddle-back tamarin, *Saquinus fuscicollis*; sooty mangabey, *Cercocebus atys*; squirrel monkey, *Saimiri sciureus*; and talapoin monkey, *Miopithecus talapoin*.

Werff ten Bosch (2). Only the major steps in the progression of thinking on this subject need to be reiterated here. That a major component of puberty involves a change in gonadal activity became apparent before the time of Christ, as knowledge of the effects of castration during prepubertal development in our own species first began to accumulate (16). Similarly, the idea that the pubertal change in gonadal activity is of a humoral nature is not of modern origin and, as pointed out by Donovan (17), dates back to at least the 17th century. The signal finding of Foa (18) at the turn of the present century, that transplantation of ovaries from an immature animal into an adult host led to a premature onset of pubertal activity in the transplanted gonad, provided the foundation for the view that although maturational processes may occur within the gonad of the immature animal, these are terminated long before the onset of puberty. Confirmation of Foa's original finding by several other groups led Lipschutz and his colleagues (19) to formulate in 1929 the "law of puberty," which stated that the onset of pubertal changes in the gonad is determined by maturation of a somatic component rather than by the gonad itself.

That the somatic component of the control system governing puberty might be the pituitary gland was suggested by the early finding of Crowe et al. (20) that partial ablation of the adenohypophysis in the prepubertal canine prevented the onset of sexual maturation. Definitive proof for the pituitary's role in puberty was provided 17 years later by Smith and Engle (21), who convincingly demonstrated that implantation of pituitary fragments in immature rats and mice led to premature sexual maturation in these species. However, with the subsequent forging of contemporary understanding of the neural regulation of anterior pituitary function by Geoffrey Harris and others, the role of the pituitary gland in puberty was relegated to that of a slave to the brain (22). In a fashion analogous with Foa's earlier ovarian transplantation experiments, Harris and Jacobsohn (23) found that when pituitary glands taken from immature rats were transplanted to a site beneath the median eminence of their hypophysectomized mothers, estrous cyclicity was reinitiated in the recipient animals well before puberty would have been expected to occur in the donors if the latter had remained intact. This result provides compelling evidence in support of our current concept that an intrinsic immaturity of the pituitary is not the proximate cause of ovarian and testicular quiescence during prepubertal development in many mammalian species.

The finding from one of our laboratories that chronic intermittent neurochemical stimulation of the hypothalamus of the prepubertal monkey imposes a precocious adult pattern of activity in the hypothalamus–pituitary–testicular axis indicates that the network of gonadotropin-releasing hormone (GnRH) neurons, which provide the major drive to this axis, must also be viewed as a nonlimiting component of the control system that governs the timing of puberty (24). Instead, the onset of puberty appears to depend on the ontogeny of central neural mechanisms that are upstream of the GnRH neuron and are able to dictate the postnatal secretion of the releasing factor expressed by these neurons. The critical importance of the central nervous system (CNS) in the pubertal process in primates is graphically reinforced by an extensive clinical literature documenting an association between lesions of the human brain, particularly those of the hypothalamus and adjacent areas, and disorders of puberty, such as GnRH-dependent precocious and delayed gonadarche (25–27).

GONADAL AND ADRENAL FUNCTION DURING PUBERTY

The Testis

The growth of the human testis during the first decade of postnatal life is unremarkable (28–30). A dramatic increase in testicular volume between 9 and 13 years from 2 ml to the adult volume ranging from 12 to 25 ml (28–30) marks male gonadarche. The chimpanzee exhibits a pattern of testicular growth similar to that of humans, although the phase of pubertal enlargement is initiated at approximately 6 years of age (31–34). In rhesus, crab-eating, and bonnet monkeys and in baboons, there is also little growth of the testis during prepubertal development (35–38). At around 3 years of age in these species of Old World monkey, testicular size increases markedly within a period of 1.5 years, from a volume of approximately 2.5 ml to an adult capacity of 25 ml. In the mandrill, pubertal testicular enlargement is observed between 4 and 8 years of age (39). A similar developmental pattern of testicular growth has been reported for the common marmoset, although, in the case of this New World primate, initiation and completion of the pubertal phase of rapid growth take place at a much earlier age (8 and 21 months, respectively) than in higher primates (14). In two other species of platyrrhine, the owl monkey and the capuchin monkey, a marked pubertal acceleration in testicular growth was only noted in the latter species (40,41).

The testes of postinfantile prepubertal macaques and humans contain few, if any, recognizable Leydig cells (42–46), and testicular testosterone secretion during this phase of development in these and other species of primate is minimal (29,34,36–39,46–57). The precise temporal relationship among the acceleration in testicular growth, the development of Leydig cells, and the increase in testosterone secretion during early puberty has not been systematically studied in any species. The development of Leydig cells and a rise in intratesticular androgen concentration, however, likely precede both the acceleration in testicular growth and the rise in circulating androgen levels. In humans, daytime levels of plasma testosterone begin to rise above prepubertal concentrations of 0.2 ng/ml during the tenth year of life and continue to increase progressively before reaching, at 14 to 15 years of age, a value (6 ng/ml) typical of that observed in adult men (29,47,48). The initiation of the pubertal rise in circulating daytime testosterone levels has been observed to occur between the ages of 6 and 8 years in the chimpanzee (34,49); 4 and 8 years in the mandrill (39); 2.5 and 4 years in the rhesus, crab-eating, and African green monkeys, the sooty mangabey, and the baboon (50,52–57); 2.5 and 3 years in the squirrel monkey (51); and approximately 1 year in the common marmoset and the owl monkey (14,40,58). Studies of humans and the rhesus monkey show that initial activation of testicular testosterone secretion during early puberty occurs nocturnally (54,59–61). In the rhesus macaque, a nighttime increment in circulating testosterone concentrations may be observed as early as 2 to 3 years of age, several months before the onset of the pubertal rise in daytime concentrations of this steroid (54). Although longitudinal studies of the peripubertal time course in testosterone secretion are scant, the available data for humans (62,63) and the rhesus monkey (54) (T. M. Plant, unpublished observations) suggest that the transition from a prepubertal to an adult pattern of secretion may occur relatively briskly.[3]

The age at which spermatogenesis is initiated has been less precisely defined than that for the pubertal onset of steroidogenic activity by the Leydig cell because of the difficulty in obtaining serial ejaculates and testicular biopsies. Limited data are available for macaques and humans only, and no attempt has been made to systematically relate the maturation of seminiferous tubule morphology to the pubertal initiation of Leydig cell function. In contrast to the steroidogenic component of the primate testis, which appears to be quiescent during the postinfantile prepubertal stage of development, the seminiferous cords lengthen during this phase of development due to proliferation of Sertoli cells and undifferentiated type A spermatogonia (42–44,64–67) (Simorangkir and Plant, unpublished observations). The acceleration in testicular growth during early puberty results largely from an increase in diameter and tortuosity of the seminiferous tubules associated with the appearance of mature Sertoli cells and proliferation of germ cells. Although spermatogenesis has been generally recognized to become established between the 12th and 16th year of life in humans (42,43,64), studies of the incidence of spermaturia in boys suggest that the onset of sperm production may occur in association with the initiation of accelerated testicular growth (68,69). In a qualitative sense, the postnatal ontogeny of the germinal epithelium in macaques appears to be comparable with that described for humans. In these Old World monkeys, sperm are first observed in the seminiferous tubules between 3 and 4 years of age (38,44,65). In New World monkeys, spermatogenesis is probably initiated during the first year of life; spermatocytes have been reported in the marmoset testis at 35 weeks of age (58), and cotton-top tamarin males as young as 22 months of age may impregnate female (70).

[3]Longitudinal studies of sexual development in humans also reveal a striking increase in testicular growth during puberty (2).

The Ovary

Knowledge of the ontogeny of the primate ovary has been derived primarily from studies of the human and rhesus monkey (71–75). In contrast to the postnatal pattern of testicular growth, the size of the ovary increases in a rectilinear fashion from infancy to adulthood. Moreover, growth and atresia of ovarian follicles begin in utero and continue during the prepubertal years. From a peak of approximately 6–7 million oocytes in the human female fetus at mid-gestation, some 350,000 remain at birth due to losses secondary to apoptosis (76,77). Enlargement of the ovary before puberty is the result of age-related increases in the number and size of antral follicles and in the quantity of medullary stroma. Sonography shows small follicles in the ovaries of the majority of prepubertal girls with increasing variation in follicular size postmenarche. Although the size attained by a growing antral follicle before atresia intercedes is variable, Graafian (preovulatory) follicles are not observed during prepubertal development. Whereas primordial and preantral follicles predominate during the prepubertal years, small antral follicles can develop before puberty. With further maturation, the adolescent ovary may appear on ultrasound images to be multi- or polycystic (78).

The availability of sensitive and specific methodology to measure the low serum estradiol concentrations in prepubertal girls and nonhuman primate females is limited. Two observations in prepubertal female monkeys, however, demonstrate that the prepubertal ovary actively secretes estradiol. First, ovarian vein concentrations of estradiol are three- to fourfold greater than peripheral levels (79) and, second, circulating estradiol concentrations decline after ovariectomy (80). That the prepubertal human ovary also secretes estradiol is supported by the finding, obtained with an ultrasensitive estradiol bioassay, that circulating levels of this steroid are significantly greater in prepubertal girls than in boys (81,82). Cross-sectional studies of girls have demonstrated that between 8 and 10 years of age, estradiol concentrations in serum samples collected during the morning begin to increase to values typically seen in adult women during the early follicular phase of the menstrual cycle (83–85). Concomitant with the rising estradiol concentrations, the early clinical feature of gonadarche, namely the initiation of breast development, occurs. Uterine volume increases throughout puberty with the greatest increase occurring during mid-puberty (86). In the female rhesus monkey, circulating estradiol concentrations rise between 2.5 and 3 years of age in association with nipple growth and an increase in perineal swelling and coloration (87,88).

In the human female, further development of secondary sexual characteristics occurs as puberty progresses and culminates in menarche, which generally occurs 2–3 years after the onset of breast development (30). The Third National Health and Nutrition Examination Survey, conducted between 1988 and 1994, reported the average age at menarche to be 12.54 years in North American girls, with an average age estimated to be 12.6 years for whites and 12.14 years for blacks (89). Menarche presumably represents acute withdrawal of hormonal support of estrogen-induced endometrial proliferation. Typically, menstrual cycles during the first year after menarche are often irregular and anovulatory (90–95), with most ranging in duration from 21 to 45 days. Within 5 years of menarche, most cycles are regular and range from 21 to 35 days. Fertility during the peripubertal years appears to be compromised by a relatively high incidence of menstrual cycles with short luteal phases (92,94,95).

Although menarche occurs in the great apes and Old World monkeys at a younger age than that in the human female (Table 1), the postmenarcheal phase of development in these species is also characterized by a variable phase of adolescent infertility (90,96,99,102) that, in the case of the rhesus monkey, is the result of a high incidence of anovulatory and short luteal phase cycles (108,109).

In the common marmoset and squirrel monkey, species of New World primate in which menstruation is not observed, ovulation as indicated by elevated plasma progesterone levels and conception may occur as early as 2.5 to 3 years of age (14,51). In other species of platyrrhine monkey—for example, the saddle-back tamarin, in which conception has been reported as young as 210 to 230 days of age (110)—the transition into puberty occurs within the first year of life under appropriate environmental and social conditions (see below).

TABLE 1. *Age at menarche in Old World monkeys and great apes*

Species	Age (yr)	References
Chimpanzee	7–11	(96–98)
Gorilla	6–8	(99)
Baboon	3–4	(100, 101)
Talapoin monkey	3–4	(102)
Pigtail monkey	2–4	(103)
Rhesus monkey	2–3	(104,105)
Crab-eating monkey	2–3	(106)
Hanuman langur	2–3	(107)

The Adrenal

The adrenal cortex consists of three histologically and functionally distinct zones: the glomerulosa, producing mineralocorticoids (aldosterone); the fasciculata, producing glucocorticoids (cortisol); and the reticularis, producing the androgens, dehydroepiandrosterone (DHEA), DHEA-sulfate (DHEAS), and androstenedione. The pituitary regulates circulating cortisol concentrations through adrenocorticotropin (ACTH) secretion. ACTH is a peptide derived following proteolytic cleavages of pro-opiomelanocortin. In the adrenal cortex, ACTH binds to its cognate receptor, a G-protein–coupled 7-transmembrane domain receptor that signals via cyclic AMP and protein kinase A. ACTH has both acute and chronic effects on the adrenal cortex. Acutely, it promotes uptake of low density lipoprotein, stimulates cholesterol esterase activity, enhances synthesis and phosphorylation of steroid acute regulatory protein, and promotes cortisol secretion. The chronic effects of ACTH involve stimulation of transcription and translation of steroidogenic enzyme genes. ACTH secretion is controlled by hypothalamic releasing factors, such as corticotropin-releasing hormone and vasopressin.

Adrenarche refers to the altered pattern of adrenal steroidogenesis that begins at 6–7 years of age in humans and is characterized by increased DHEAS secretion in association with enlargement of the zona reticularis. Circulating DHEAS concentrations peak during the second decade of human life and then progressively decline (111). The onset of adrenarche is associated with increased 17,20-lyase activity and decreased 3β-hydroxysteroid dehydrogenase activity, leading to increased secretion of DHEA, DHEAS, and androstenedione (112). Although designated as adrenal androgens, these steroid hormones do not activate the androgen receptor. Rather, they serve as precursors for testosterone and dihydrotesterone in target tissues.

In contrast to the role of hypothalamic GnRH in the initiation of gonadarche (see below), corticotropin-releasing hormone and vasopressin do not appear to trigger the onset of adrenarche. Adrenarche occurs independently of developmental changes in the hypothalamic–pituitary–gonadal axis. For example, children with gonadal dysgenesis experience normal adrenarche and pubarche (growth of pubic hair), whereas children with primary adrenal insufficiency may have normal gonadarche (7,12,113). No significant changes in ACTH concentrations have been noted during adrenarche and no specific adrenal androgen stimulating factors have been isolated from the pituitary. Nevertheless, decreased DHEAS concentrations in patients with ACTH resistance due, in some cases, to mutations in the ACTH receptor gene indicate that ACTH plays a permissive role in adrenarche (114).

Available data suggest that undefined gradual developmental changes within the adrenal cortex underlie adrenarche. Many of the enzymes responsible for adrenal steroidogenesis are members of the cytochrome P-450 family. One such enzyme, P450c17, encoded by the *CYP17* gene has two distinct activities, 17α-hydroxylase and 17-20-lyase. The relative rates of these two enzymatic steps modulate the pattern of adrenal steroidogenesis. Similar to other microsomal cytochrome P-450 enzymes, cytochrome P-450 oxidoreductase is required by P450c17 to shuttle electrons from nicotinamide-adenine dinucleotide phosphate to the central heme. Cytochrome b_5 is a heme protein. Increased DHEA and DHEAS secretion at adrenarche has been associated with a relative increase in the 17,20-lyase activity of P450c17 and a relative decrease in the activity of the 3β-hydroxysteroid dehydrogenase. At this time, increased expression of cytochrome P-450 oxidoreductase, cytochrome b_5, DHEA sulfotransferase, and decreased expression of 3β-hydroxysteroid dehydrogenase in the zona reticularis have been detected (115). In vitro studies have indicated that when the molar ratio of cytochrome P-450 oxidoreductase to P450c17 is increased, 17,20-lyase activity increases (116). Addition of purified cytochrome b_5 to a P450c17 assay system increased 17,20-lyase activity (117,118). Although increased serine/ threonine phosphorylation of P450c17 increases 17,20-lyase activity (119), it remains to be determined whether the phosphorylation status of P450c17 is the critical factor regulating the onset of adrenarche. Recent data suggest that 3β-hydroxysteroid dehydrogenase expression in the adrenal cortex is regulated by members of the NR4A family of orphan nuclear receptors (120). GATA-6, a zinc finger transcription factor, is preferentially expressed in the zona reticularis where it appears to interact with steroidogenic factor-1 to promote adrenal androgen biosynthesis (121).

Although findings have been inconsistent, clinical studies suggest that insulin, insulin-like growth factor-1 (IGF-1), and growth hormone (GH) concentrations also influence the timing and tempo of adrenarche (122). Clinical observations regarding associations between adrenarche and body size/ fatness have been inconsistent (7,123). Comparison of IGF-1 concentrations among prepubertal children has shown higher concentrations in African-American children (124). Whether these ethnic differences contribute to the earlier onset of adrenarche or to increased incidence of premature pubarche in African-American girls is unknown.

PUBERTAL STAGING

The physical changes associated with the onset and progression of puberty in higher primates have only been systematically defined in humans. Specifically, the genital and pubic hair changes that emerge in association with the onset of puberty in boys and girls are classified into five stages: Stage 1 is prepubertal and stage 5 is adult (27). These physical changes may be the result of either gonadarche, as in the case of breast or testicular enlargement, or adrenarche, as in the case of pubic hair development. Although the physical sequelae of gonadarche and adrenarche generally occur concomitantly, a discordance of the two processes may also be observed in normal development.

Increased estrogen secretion promotes breast development in girls. The development of breast buds with increased areolar diameter is considered to be stage 2, greater enlargement of the breasts occurs in stage 3, accompanied by increased pigmentation of the areolae and nipples. During stage 4, the areolae are mounded above the breast tissue. Recession of the areola to the general breast contour represents breast stage 5. Additional effects of estrogen at this stage of development include cornification of the vaginal mucosa, uterine growth, and morphogenesis of an adult female body habitus.

In girls, increased adrenal androgen secretion is considered to be responsible for the development of darker hair usually along the labia, which is classified as pubic hair stage 2. During pubic hair stage 3 the hair is darker and coarser, spreading over the pubic symphysis with gradual progression to a full female escutcheon. Apocrine odor may precede or accompany the development of pubic hair. Associated findings include acne and oiliness of skin and hair.

For boys, an increase in testicular volume is considered genital stage 2. At stage 2, the testes are approximately 4–8 ml in volume with the longest axis being approximately 2.5 cm. With increasing testicular volume, the scrotum enlarges and is accompanied by decreased rugation of the skin. The volume of the mature human testis is approximately 20–25 ml and represents increased growth of the seminiferous tubule due to Sertoli cell proliferation and differentiation and initiation of spermatogenesis. At genital stage 3, further growth of the testes has occurred, and the length and diameter of the penis has increased. At genital stage 4, penile size has increased with darkening of the scrotal skin.

Male pubic hair stage 2 consists of downy hair at the base of the penis. For pubic hair stage 3, the hair is longer, darker, and extends over the junction of the pubic bones. For pubic hair stage 4, the extent of hair has increased but has not yet achieved the adult male escutcheon. Other secondary sexual characteristics in boys include increased size of the larynx, deepening of the voice, increased bone mass, and increased muscle strength. Approximately 3 years after the appearance of pubic hair, terminal hair appears in androgen dependent regions on the face and trunk, where it may develop for years thereafter. The distribution and density of beard, chest, abdominal, and back hair varies considerably, presumably reflecting genetic differences. The appearance of spermatozoa in early morning urine specimens (spermaturia) occurs during genital stage 3. Transient gynecomastia (breast development) is observed in approximately 50% of boys (125). Typically, this is most prominent in mid-puberty when the ratio of circulating concentrations of estradiol to testosterone is relatively high.

The pubertal growth spurt in girls occurs concurrently with the onset of breast development. Usually only 4–6 cm of growth occurs after menarche. In boys, the pubertal growth spurt, with an average height velocity of 9.5 cm per year, occurs around genital stages 3–4. In general, the age at peak height velocity shows an inverse relationship with the magnitude of the growth spurt.

SECULAR TRENDS

Over the last century and a half, the age at menarche in North America, Europe, and other industrialized areas of the world appears to have decreased dramatically from 16 to 17 years of age in the middle of the 19th century to 12 to 13 years of age in the late 20th century (13,89,126–129). This well-established trend toward an earlier age at menarche, which in the early 21st century may be coming to a halt (130–135), is generally attributed to a corresponding secular improvement in the standard of living, including better nutrition and health care. Although analogous data for puberty in the human male are scant, in a study of American boys who were taller and heavier than those in previous cohorts, pubic hair and testicular development was reported to occur earlier than previously recognized (136). Interestingly, a secular trend toward an earlier age at menarche has also been reported for the rhesus monkey, although in this species the weight at menarche appears to increase with the age of menarche (104).

PITUITARY GONADOTROPE FUNCTION DURING PUBERTY

Our understanding of the pattern of basal or tonic[4] gonadotropin secretion during puberty in primates has been obtained primarily from studies of humans. With the advent in the late 1960s of radioimmunoassay (RIA) for the determination of luteinizing hormone (LH) and follicle-stimulating hormone (FSH) in small volumes of serum, a large number of research groups examined, in comprehensive cross-sectional studies, the time courses of circulating LH and FSH concentrations in normal boys and girls during pubertal development (17,85, 137–143). The general statement may be made that the transition into puberty in both boys and girls is tightly coupled with an initiation of a progressive increase in the overall secretory activity of the pituitary gonadotropes, as reflected by the time courses in circulating LH and FSH concentrations during this phase of human development (144,145). With the advent of the more sensitive immunoradiometric assay or immunofluorometric assay (IFMA) for the measurement of blood gonadotropin concentrations, it is now clear that FSH and LH levels rise concomitantly at puberty in both boys and girls and that the increase in LH at this stage of development is particularly robust (146–151). In the case of LH, where bioassay has also been applied, the pubertal rise in

[4]Conceptually, it is sometimes useful to define two modes of gonadotropin secretion. The basal or tonic mode accounts for the relatively low circulating LH and FSH concentrations observed in adult males and in females during the greater part of the menstrual cycle. The surge mode of gonadotropin secretion, which is elicited by a stimulatory action of estradiol, underlies the striking increase in circulating LH and FSH concentrations observed before ovulation. This issue is further discussed in Chapter 45.

this gonadotropin is revealed to be an immediate, rapid, and striking event (152,153).

Although studies of the time courses of basal gonadotropin secretion during juvenile and pubertal development in nonhuman primates, specifically the chimpanzee (34,154), the macaques (36–38,54,57, 88,155), and the sooty mangabey (53), have in general been collectively less comprehensive than those of humans, the coupling between a rise in circulating LH and FSH concentrations and the transition into puberty is, presumably, a general characteristic of sexual maturation in higher primates. Indeed, the switch to an LH dominated pattern of gonadotropin secretion at puberty is reflected in the male rhesus monkey by a greater proportion of LH immunoactive gonadotrophs in the pituitary of the adult compared with that of the juvenile (156).

It is also during puberty that the gonadotropes first acquire the capacity to respond to the stimulatory action of estradiol on LH and FSH secretion.[5] In the rhesus monkey, and probably in humans, this takes place after menarche (87,157–159).

The fundamental role of increased gonadotropin secretion to initiate gonadal development at puberty is well established. Compelling evidence in support of this idea is provided by findings that, in humans, spermatogenesis and ovulation may occur at very young ages as a result of premature gonadotropic stimulation of the gonad (27,141,160–162). Conversely, a delay in the onset of puberty in humans is associated with low circulating gonadotropin concentrations and hypogonadism (27,141,152,163). Inhibition of gonadotropin secretion in individuals with GnRH dependent precocious gonadarche results in a return to testicular and ovarian quiescence (27,161,162, 164), whereas chronic stimulation of gonadotropin secretion in subjects with delayed puberty secondary to hypogonadotropic hypogonadism results in the pubertal-like activation of the testis and ovary and in fertility (165–168).

The idea that the relative quiescence of the primate gonad from late infancy to puberty cannot be accounted for by an intrinsic immaturity of the testis and ovary is further reinforced by the findings that stimulation of the gonad in the prepubertal rhesus monkey with either exogenous or endogenous gonadotropin leads to the premature onset of ovulation and spermatogenesis (24,169–173). Taking the foregoing considerations together, it is clear that the hypogonadotropic state that characterizes the juvenile primate is the proximate cause underlying the protracted phase of relative gonadal quiescence from late infancy to puberty in these species.

[5]It is to be noted that the ability of estradiol to elicit gonadotropin surges during infantile development has not been tested.

It is also apparent that an understanding of the physiological mechanisms that regulate the timing of puberty requires an appreciation of the control system that governs gonadotropin secretion.

THE NEUROENDOCRINE CONTROL SYSTEM GOVERNING GONADOTROPIN SECRETION

Overview

Before embarking on a discussion of the ontogenesis of the control system that governs gonadotropin secretion, we provide perspective by presenting a brief overview of the essential features of the regulation of gonadotropin release in the adult. For a detailed examination of this subject, the reader is referred to Chapter 45. The primary stimulus to pituitary gonadotropin secretion is generated within the CNS and is transmitted to the gonadotropes of the adenohypophysis via the hypophyseal portal circulation by an intermittent hormonal signal in the form of the hypothalamic releasing factor, GnRH (mammalian GnRH, GnRH-I). Parenthetically, it should be noted that a second GnRH (GnRH-II) is also found in the primate hypothalamus (174,175), and the gene encoding the GnRH-II receptor is expressed in the anterior pituitary (176,177). Although intravenous administration of a bolus of GnRH-II to the rhesus monkey is able to elicit LH release (178) and GnRH-II is developmentally regulated in this species (175), the role of this peptide in regulating the pituitary axis is unknown. Throughout this chapter we use GnRH to describe GnRH-I. Although the distribution of GnRH perikarya and their projections within the primate hypothalamus has received considerable attention (see Chapter 28), the neural mechanisms and their attendant anatomical substrates that underlie the intermittent pattern of GnRH release remain to be fully understood. In the present review, the neural mechanisms responsible for generating the hypophysiotropic drive to the pituitary gonadotropes is viewed, for some purposes, as a "black box" resident within the hypothalamus and termed, in accordance with classical terminology, the GnRH pulse generator. The hypothalamic GnRH pulse generator, together with the pituitary gonadotropes, thus comprises the basic neuroendocrine unit responsible for LH and FSH secretion.

Gonadotropin release in the adult is regulated, in part, by negative feedback actions of the gonadal hormones on LH and FSH secretion that may be exerted either directly on the gonadotropes to suppress their responsivity to GnRH stimulation or indirectly at a suprapituitary locus to modulate the frequency and/or amplitude of the GnRH pulse generator. Removal of such negative feedback influences results in a chronic hypersecretion of both LH and FSH, which occasions a sustained elevation in circulating gonadotropin concentrations an order of magnitude higher than those observed in intact subjects. In the absence of feedback by gonadal hormones, the hypothalamic GnRH pulse generator of sexually mature male and female primates appears to exhibit an operational frequency of approximately 1 cycle per hour. This "adult" pattern of intermittent GnRH release, as reflected in the circhoral discharge of pituitary LH release, has been observed in male and female rhesus monkeys castrated postpubertally, in ovariectomized and postmenopausal women, and in hypogonadal men (179).

In the female, the characteristic time course of circulating gonadotropin concentrations throughout the primate menstrual cycle, namely, relatively stable basal concentrations throughout the greater part of the follicular and luteal phases of the cycle, interrupted at mid-cycle by a massive and abrupt peak in LH and FSH concentrations, is generally attributed to the tonic and surge modes of secretion, respectively. The regulation of these two secretory modes is achieved by the interplay of the feedback actions of ovarian hormones on LH and FSH secretion. The control of tonic gonadotropin secretion during the follicular and luteal phases is the result of a negative feedback action of ovarian hormones (primarily estradiol) on LH and FSH release. In addition to the inhibitory action of estradiol on LH and FSH secretion, this ovarian steroid also exerts a stimulatory action on gonadotropin release. The so-called positive feedback action of estradiol, which is the principal trigger for the preovulatory gonadotropin surge, is elicited, at least in the human and macaque female, when circulating levels of this steroid are maintained for approximately 36 hours at a threshold concentration of 200 to 300 pg/ml.

The neuroendocrine component of the control system that underlies the preovulatory gonadotropin surge in higher primates is presumed to involve the same elements that are involved in the regulation of basal gonadotropin secretion, namely, the hypothalamic GnRH pulse generator and the pituitary gonadotropes. The concept of a distinct control system (surge center) governing the preovulatory gonadotropin discharge, which was originally proposed to account for the generation of the preovulatory LH surge in the rat (180), is generally omitted from current models of the primate menstrual cycle (see Chapter 45). Thus, any impairment in hypothalamic GnRH pulse generator activity that results in a reduction in the hypophysiotropic drive to the gonadotropes, and therefore in tonic LH and FSH secretion, would be expected to be associated with concomitant deficits in the positive feedback action of estradiol.

Feedback control of gonadotropin secretion in the male involves only negative loops, comprised primarily of testosterone and inhibin B that regulate LH and FSH, respectively (179,181). It should be noted that the absence of positive feedback control in the male is the consequence of hormonal status rather than of fundamental sex differentiation of the primate GnRH pulse generator and pituitary gonadotrophs (182–185).

In both the male and the female, a pulsatile GnRH drive is critical for sustained gonadotropin secretion, and if this is compromised, for whatever reason, as occurs, for example, in severe undernutrition and during strenuous exercise training, individuals become hypogonadotropic (see Chapter 47). Thus, the pituitary gonadotrophs may be viewed as a slave to the hypothalamic GnRH pulse generator, and this analogy should be borne in mind as the ontogeny of the neuroendocrine unit responsible for gonadotropin secretion is considered.

Ontogeny

Although descriptions of the time courses of circulating LH and FSH concentrations during fetal development in primates are incomplete, it has been established for humans and rhesus monkeys that secretion of these glycoprotein hormones by the fetus increases considerably during mid- to late pregnancy, resulting in peak plasma concentrations between 100 and 150 days of gestation (13,186–188). Interestingly, at this stage of pregnancy, circulating LH and FSH concentrations in the fetus may attain values comparable with those observed in the open-loop or castrate situation in adults. The magnitude of the fetal rise in circulating FSH and LH concentrations is most striking in the female (186–188[6]). An inhibitory action of testicular hormones on LH and FSH secretion at the level of the hypothalamic–pituitary unit provides the most likely explanation for the relatively lower circulating LH and FSH concentrations in the male fetus. Compelling evidence for this view has been provided by Resko and Ellinwood (189), who demonstrated that orchidectomy between 98 and 104 days of fetal development in the rhesus monkey elicits a dramatic postcastration hypersecretion of gonadotropin that, within 3 weeks, results in a rise in circulating LH and FSH to concentrations observed in intact female fetuses of comparable age. Ovariectomy at this stage of fetal development

[6]Although Kaplan et al. (186) did not observe a sex difference in fetal LH concentration in blood, the latter workers did observe that fetal pituitary LH content is greatest in females.

does not appear to influence gonadotropin secretion. Although the FSH/LH ratio in plasma from human female fetuses (187) and late-gestation female rhesus fetuses (189) appears to be higher than that reported for their male counterparts, the significance of this sex difference, which persists into postnatal development, is not fully understood (see below).

During the latter half of human gestation, FSH and LH concentrations in the fetal circulation decline, and, at the time of birth, low gonadotropin levels are consistently observed in cord blood (190). A similar suppression of gonadotropin secretion during late fetal development is also observed in the rhesus monkey (188), although in the case of this species, elevated LH and FSH concentrations are sustained for a relatively greater proportion of fetal development before declining abruptly at the end of gestation.

The neuroendocrine mechanisms that underlie the characteristic ontogeny of gonadotropin secretion throughout fetal development in higher primates are poorly understood. Circumstantial evidence, however, suggests that the principal stimulus to the fetal gonadotropes is provided, as in the case of the adult, by an intermittent pattern of GnRH discharge by the hypothalamus. In humans, the hypothalamic–hypophyseal portal circulation and the major hypothalamic nuclei appear to differentiate by mid-gestation (186). Interestingly, the cells that synthesize and secrete GnRH, in contrast to other neurons in the CNS, originate outside the brain in the olfactory area and migrate to the hypothalamus during fetal development (191–194). In the rhesus monkey, there are about 2,000 GnRH neurons (195), and they reach the adult pattern of distribution within the septopreopticoinfundibular pathway between days 50 and 135 of gestation (193). Immunoactive GnRH has been extracted from human fetal hypothalamus early in gestation (196), and the location of GnRH perikarya and their projections in the fetal brain of humans, as revealed by immunocytochemistry, appears to be similar to that in adults (197,198). The gonadotropes in the fetal pituitary gland of humans and rhesus monkeys are able to respond to stimulation with synthetic GnRH (199,200). Studies using chronically catheterized ovine fetuses, a species in which the ontogeny of gonadotropin secretion during fetal development appears to resemble that in higher primates (201), have revealed the presence of distinct LH pulses during this phase of development (202). Such discharges of LH are presumably occasioned, as in the adult, by the episodic release of GnRH by the hypothalamic pulse generator of the fetal brain. This view is supported by the findings that chemical stimulation of the fetal ovine hypothalamus with a glutamate receptor agonist (*N*-methyl-D-aspartate [NMDA]) elicits GnRH release (203) and that daily administration of a GnRH agonist desensitizes the fetal gonadotrope (204).

The factors responsible for the characteristic bell-shaped time course in gonadotropin secretion during fetal development in humans are unclear. The best explanation for this phenomenon is that proposed by Grumbach and colleagues (13,186). These investigators suggested that differentiation of pituitary gonadotropes and the initiation of a hypophysiotropic drive to these cells are established early in fetal development and precede the ability of this neuroendocrine complex to respond to the inhibitory actions of gonadal and placental steroids on LH and FSH release. Thus, during the first half of fetal development, gonadotropin secretion is unrestrained and increases to a level reminiscent of that observed in the open-loop situation in adults. Thereafter, the fetal hypothalamic–pituitary unit presumably acquires the capacity to respond to the feedback actions of sex steroids, and circulating LH and FSH concentrations decline throughout the second half of fetal development as the placental production of estrogen and progesterone rises.

That gonadotropin secretion in the primate fetus is restrained during late gestation by the inhibitory action of placental steroids is further suggested by the observation that, after parturition and the resulting removal of placental influences, circulating LH and FSH rise to concentrations markedly greater than those observed later in prepubertal development (36,151,190,205–212). In human, chimpanzee, macaque, sooty mangabey, and marmoset males, this phase of heightened gonadotropin secretion during neonatal development is associated with an elevation in testicular testosterone secretion that produces circulating levels of this steroid comparable with those observed in adults (36,205,206,210,211,213). Continuous treatment of infantile male rhesus and marmoset monkeys with a GnRH receptor agonist results, as in adults, in a complete suppression of the pituitary–testicular axis (58,214), confirming that the drive to gonadotropin secretion at this stage of development is provided by hypothalamic GnRH secretion. Further study of the rhesus monkey has led to the notion that the hypothalamic component of the control system that governs testicular function is fully mature by the neonatal stage of development. Bilateral orchidectomy at 1 week of age elicits a postcastration rise in LH and FSH secretion seemingly indistinguishable from that observed after castration in sexually mature animals (208), demonstrating the integrity of the negative feedback control of pituitary gonadotropin secretion during infantile development. Moreover, in such neonatal castrates, LH secretion is episodic and exhibits a pulse frequency typical of open-loop LH secretion in animals castrated postpubertally, namely, 1 pulse an hour (215). The latter finding suggests that the hypothalamic GnRH pulse generator of the infantile male primate, like that of the postpubertal animal, is capable of generating the adult circhoral pattern of intermittent GnRH secretion. Pulsatile LH secretion has also been reported in normal human infants (216,217). In addition, infantile male rhesus monkeys exhibit a diurnal variation in testicular testosterone secretion that is quantitatively similar to that typically observed in pubertal and fully adult animals and is correlated, as in adults, with a corresponding diurnal variation in LH release (210,218). The latter observation suggests that the hypothalamic GnRH pulse generator of the infant is probably modulated by the same exteroceptive factors that regulate testicular function in the adult. Thus, in higher primates and perhaps in the marmoset, the hypothalamic–pituitary unit that governs testicular function differentiates to full maturity during fetal development. Testicular testosterone secretion during fetal development and infancy may, in turn, play a role in determining the tempo of the pubertal activation of the hypothalamic–pituitary–testicular axis (58,219–221) and impact on the development of affective behaviors (222,223), the growth of the male external genitalia (224,225), and the immune system (226).

The hypothalamic–pituitary unit in female *catarrhine* primates (humans, chimpanzees, and rhesus monkeys) also exhibits heightened activity during early neonatal development, resulting in an elevation in circulating LH and FSH concentrations, which are maintained for a variable period of time, depending on the identity of the species, before declining to the low levels characteristic of the juvenile state (151,190,207–209,211,212,227). Although the foregoing findings support, at first sight, the view that the female hypothalamic–pituitary unit is also fully mature by birth, this question is worthy of further consideration. In infantile girls, circulating FSH concentrations are sustained at elevated levels for several years and often exceed, in dramatic fashion, those observed in adults during the follicular phase of the menstrual cycle, rising on occasion into the adult castrate range (190,207,212). Circulating LH, on the other hand, exhibits a developmental pattern similar to that observed in the infantile boy, with a rise to low adult levels during the first few months of postnatal life (151,190,207,211,212). This sex difference in gonadotropin secretion during infantile human development, which has been detected as early as the first day of postnatal life (217,228), results in an elevated ratio of FSH to LH in the peripheral circulation of the female. A similar sexual dimorphism has also been observed in the chimpanzee and the rhesus monkey (209,218,229).

Studies of the rhesus monkey suggest that differential feedback signals cannot account for this sex difference in gonadotropin secretion: Castration at 1 week of age does not abolish the sex difference in the time courses of gonadotropin secretion during the subsequent 40 weeks of life (227). Similarly, in anorchic infantile boys, circulating FSH concentrations are lower than those in age-matched patients with Turner syndrome (230). Moreover, in agonadal infantile monkeys the relatively high FSH/LH ratio in the female is associated with a frequency of pulsatile LH secretion (one pulse every 4 hours) that is markedly slower than that in age-matched castrated males (one pulse every hour) (Fig. 1). The latter finding, in turn, suggests that the hypothalamic GnRH pulse generator of the infantile female, in contrast to that of the neonatal male, is not capable of operating at the adult

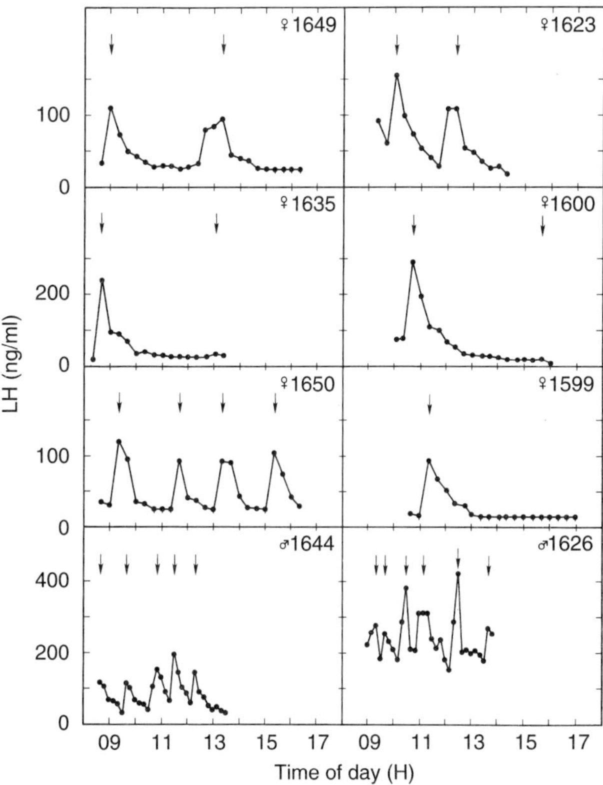

FIG. 1. Moment-to-moment changes in circulating LH concentrations at 4–7 weeks of age in six infantile female (*top three panels*) and two infantile male (*lower panel*) rhesus monkeys that were bilaterally gonadectomized at 1 week of age. *Arrows* indicate increments in LH identified as pulses by PULSAR. Note that the frequency of LH pulses, and presumably therefore of GnRH discharges, in agonadal infantile females is markedly slower than that in agonadal males of similar age. [Reprinted with permission from (227).]

circhoral frequency. In earlier editions of this book, an early influence of testicular androgen on the developing fetal brain was speculated to be the likely mechanism to account for this sex difference in hypothalamic function during infantile development. Although we still hold to this view, a more recent clinical study indicated that circulating concentrations of LH and testosterone are very low during the first 3 months of postnatal life of infants with complete androgen insensitivity due to mutations of the androgen receptor (231). This finding is in direct contrast to the intuitive expectation that gonadotropin secretion in nonandrogenized infantile males would exhibit an exaggerated female pattern of release in association with elevated testosterone levels, as is seen in patients with partial androgen insensitivity (231). Resolution of this intriguing paradox is likely to be particularly informative.

The foregoing suggestion that the hypothalamic GnRH pulse generator in the infantile female monkey operates at less than the adult circhoral frequency may be of considerable conceptual importance: If subsequent study establishes that the slow frequency of hypothalamic GnRH discharge in the infantile female reflects an "immature" pulse generator that fails to develop the capacity to operate at the adult circhoral frequency during fetal development, then a postnatal maturation stage of the GnRH pulse generator will have to be added to the schemata of primate puberty.

Although the potential of the neuroendocrine control system that governs gonadal function is fully realized during infantile development in male primates and partially in females, an immediate transition from this early adult-like phase of hormonal activity to a state of true sexual maturity does not occur. Instead, the activity of the hypothalamic–pituitary unit is arrested in late infancy, gonadotropin secretion declines to the low levels characteristic of the juvenile primate, and the protracted prepubertal period of relative gonadal quiescence is initiated. Several years later, the hypothalamic–pituitary unit is reawakened, and puberty is initiated.

It is to be noted that during infantile development, in contrast to the later juvenile phase of prepubertal development, the gametogenic functions of the ovary and testis are not operative in spite of the adult-like gonadotropin drive to the gonad. In the infantile testis, Sertoli cells and undifferentiated type A spermatogonia proliferate, but the production of differentiated type B spermatogonia, the initial step in spermatogenesis, does not occur (66,67,232–234). This deficit in seminiferous tubular function is likely to be accounted for by "immaturity" of the Sertoli cell. The precise age at which the human or nonhuman primate Sertoli cell acquires the capacity to respond to FSH and testosterone has not been established.

In humans and the rhesus monkey, the pubertal reawakening of the hypothalamic–pituitary–Leydig cell axis is characterized by nocturnal elevations of both LH and testosterone secretion (54,59,60, 149,235–239). Nighttime elevations in LH and testosterone secretion, however, are not unique to this phase of development and are observed in infantile and fully adult male rhesus monkeys (210,218,240) and in juvenile boys (149,237,238) when sensitive IFMAs are used to measure circulating LH. Nocturnal secretion of gonadotropin in juvenile boys is also reflected in urinary LH and FSH excretion (241). The failure to obtain evidence for diurnal LH secretion in the juvenile male monkey probably reflects the relatively poor sensitivity of the current assays available for this purpose. Thus, it seems that the neuroendocrine mechanisms that underlie the nocturnal elevation in LH and testicular testosterone secretion during puberty are the same as those that are operative in infantile and juvenile primates. The neuroendocrine basis of the pubertal pattern of LH release in men, however, may differ from that in boys. In this regard, pulsatile LH secretion during the early hours of sleep in pubertal boys appears to be resistant to the feedback action of testosterone (242,243), and at this stage of development, in contrast to the situation in adults, the suppressive effect of testosterone on LH release appears independent of the endogenous opioids (244,245).

Increased nocturnal activity in the pituitary–Leydig cell axis is sleep related in boys (246) and presumably reflects a sleep-associated augmentation in GnRH release by the hypothalamus. It is not known whether the nocturnal elevation in LH secretion in the monkey is sleep related. The pubertal activation of the hypothalamic–pituitary axis in the female is also characterized by nocturnal elevations in LH secretion (155,247–250), which, in the case of the human female, are also sleep related (247,248).

Studies of the agonadal female monkey indicate that, as in the male, the mechanisms for diurnal modulation of hypothalamic GnRH release are also extant during infantile and juvenile development (229,251). As is the case with the human male, the relationship between sleep and GnRH release appears to change with pubertal development, with sleep inducing a deceleration in pulsatile LH release during the follicular phase of the human menstrual cycle (252). An analogous situation appears to exist in the monkey, because the GnRH pulse generator of the adult female slows during the night (253).

It is also during the pubertal phase of development that spontaneous preovulatory gonadotropin surges are first observed in the female.[7] This mode of gonadotropin secretion, which is elicited by the positive or stimulatory action of estradiol on LH and FSH release, is responsible for ovulation, including the first pubertal ovulation. The absence of preovulatory gonadotropin surges during early puberty cannot be attributed solely to the failure of the ovaries to generate an estrogen stimulus of sufficient magnitude or duration to evoke a positive feedback action on gonadotropin release. In the female rhesus monkey, exogenous estrogen challenges during the immediate postmenarcheal period (4–8 months) fail, as do similar stimuli in prepubertal girls, to elicit LH discharges (157,158,254). Moreover, in postmenarcheal female monkeys, spontaneous elevations in circulating estradiol concentrations comparable with preovulatory levels in adult animals have been observed in the absence of a subsequent gonadotropin surge and ovulation (254). A discussion of the mechanisms that underlie the inability of the hypothalamic–pituitary unit to respond to the positive feedback action of estrogen until relatively late in puberty must be postponed until consideration has been given to the nature of the hiatus of tonic gonadotropin secretion between infancy and puberty.

Nongonadal Determinant of the Hypogonadotropic State of Juvenile Development

As described in the section above, the postnatal pattern of gonadotropin secretion in higher primates is characterized by elevated LH and FSH release during infancy and puberty (continuing into adulthood) separated by a relatively hypogonadotropic state during juvenile development. An important feature of this characteristic "primate" pattern of gonadotropin secretion postnatally is that its fundamental temporal aspects are preserved in the absence of the gonad. As shown in Fig. 2, this feature of primate development is exemplified by the rhesus monkey rendered agonadal at birth by either ovariectomy or orchidectomy (54,229). The nongonadal reduction in LH and FSH secretion during juvenile development is also readily apparent from data obtained with agonadal boys and girls (141,255–261). In the latter group of individuals, which are comprised primarily of phenotypically female subjects with Turner syndrome, circulating LH concentrations that exist between 4 and 10 years of age are often in the range reported for normal children and are always well below values reported for postmenopausal subjects. In both humans and the rhesus monkey, the degree and duration of the juvenile reduction in gonadotropin secretion is noticeably more marked in the male (Fig. 2). The relatively weaker agonadal "brake" or restraint on gonadotropin release during juvenile development in the female is most clearly reflected by FSH secretion. In agonadal female monkeys and girls, plasma concentrations of this gonadotropin during the juvenile phase of development are consistently greater than those in normal age-matched subjects. As in infancy, the relatively high circulating FSH/LH ratio in the agonadal juvenile female is associated with a low LH pulse frequency (251,258,259).

The sex difference in the degree to which gonadotropin secretion is restrained by the nongonadal brake during juvenile development is reinforced by studies in juvenile monkeys of the effect

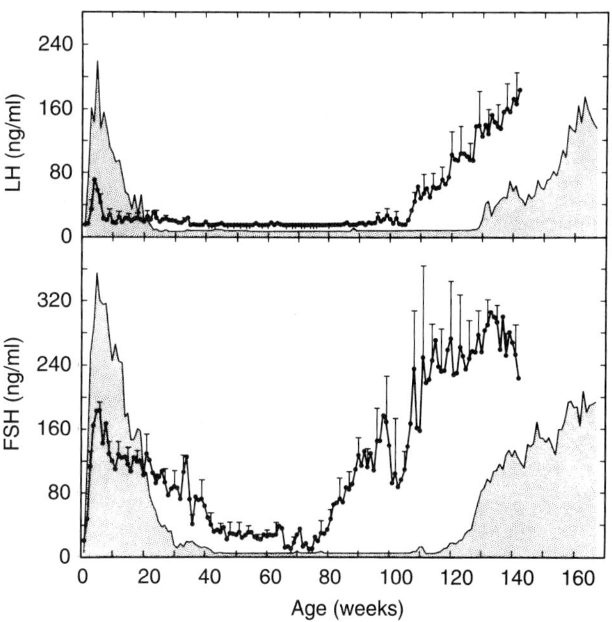

FIG. 2. Time courses of circulating mean LH (*top panel*) and FSH (*bottom panel*) concentrations determined in blood samples collected in the morning from birth until 142–166 weeks of age in rhesus monkeys ovariectomized (•, $n = 6$) and orchidectomized (*stippled area*, $n = 4$) at 1 week of age. Note that the prepubertal hiatus in the secretion of FSH, and LH to a lesser extent, in agonadal females is truncated in comparison with that in castrated males. This difference between agonadal males and females is further exaggerated when nighttime concentrations of LH and FSH are examined (not shown). *Vertical bars* above datum points indicate SEMs. [The data for males are redrawn with permission from (54).]

[7] Although spontaneous surge secretion of gonadotropin is only observed in the female, the ability to generate this mode of LH and FSH release in primates is not peculiar to the female. In the male, exposure of the hypothalamic–pituitary axis to an appropriate estrogen stimulus elicits a surge of gonadotropin (182–185).

of castration on the subsequent pattern of LH and FSH secretion. Orchidectomy at this stage of development fails to elicit a postcastration rise in either LH or FSH (80,254,262). Instead, plasma levels of these gonadotropins are maintained for many months at values that are indistinguishable from those of intact animals. Although the results of analogous studies in the prepubertal female are somewhat inconsistent (80,229,251,254,263,264), it is now generally accepted that, in contrast to the male, ovariectomy of juvenile rhesus monkeys between 12 and 15 months of age results in a postcastration increase in gonadotropin secretion, which is most marked at night and most noticeable for FSH (229,251,264).

Taking the foregoing considerations together it may be concluded that the low levels of circulating gonadotropins during juvenile development, which guarantee the relative quiescence of the gonad at this stage of primate development, are governed by a nongonadal brake, the effects of which are amplified, particularly in the female by inhibitory gonadal feedback. The site of action of the nongonadal brake is exerted at level of the hypothalamus via GnRH release, and this issue is explored in the next section.

When the nongonadal brake is waxing and waning during infancy and at the termination of the juvenile phase of development, respectively, inhibitory feedback actions of gonadal hormones on gonadotropin secretion, in both sexes, play major roles in setting the quantitative level of FSH and LH secretion and in modulating the precise duration of the hypogonadotropic juvenile state. Such feedback control during removal of the nongonadal brake at the termination of the juvenile phase of development affects the tempo of gonadal maturation during puberty. For example, the negative feedback action of estradiol on gonadotropin release has been proposed as an important component of the neuroendocrine mechanisms that account for the temporal delay between menarche and first ovulation (263,265,266). In this regard, it has been convincingly demonstrated for the rhesus monkey that the circulating estradiol concentrations required to prevent an elevation in gonadotropin after ovariectomy during the immediate postmenarcheal period are lower than those required to inhibit a postcastration hypersecretion of LH and FSH in adult females (263,265). Similarly, in teenaged girls with gonadal dysgenesis, the dose of conjugated estrogen required to suppress urinary excretion of LH and FSH into the normal range is less than that required for postmenopausal women (267). Such steroid-imposed restraint on tonic gonadotropin secretion, which, according to studies of the pubertal monkey, is mediated at least in part by a hypothalamic site of action of estrogen to inhibit GnRH release (268,269), is associated with, and is probably responsible for, the reduced efficacy of estrogen to exert its positive feedback action on gonadotropin secretion during this phase of pubertal development (109,157). This latter view is consistent with the finding that estrogen-induced LH surges may be elicited in agonadal female monkeys at an earlier age than in intact animals (270). Moreover, the onset of positive feedback in castrated monkeys is observed in association with the initiation of the pubertal rise in tonic gonadotropin secretion (270). Thus, it is reasonable to infer that ovulation does not generally take place immediately after menarche because the reawakening hypothalamic–pituitary unit is held in check by the relatively small amounts of estradiol produced by the ovary at this stage of development. A detailed discussion of these ideas has been presented by Foster and colleagues (265). The operation of a similar feedback mechanism may also be invoked to account for the observation in the male rhesus monkey that the pubertal postcastration rise in LH and FSH secretion at 2 years of age precedes, by approximately 6 months, a detectable increase in testicular testosterone secretion in intact monkeys (54).

In the previous edition of the present chapter (271), attempts were made to place the foregoing information on the role of gonadal steroid feedback on LH and FSH release in determining the primate pattern in gonadotropin secretion from birth to puberty into the context of the so-called gonadostat hypothesis, a model originally created to account for the timing of puberty in the rat (272–276) and later applied to the human (277,278). We now consider that such a discussion has little but historical interest, for several reasons. First, the terms used to describe developmental changes in the operation of the gonadostat, such as “sensitivity” and “resetting,” have never been rigorously defined by those that have found, or still find, the concept useful. It is therefore not possible to provide a critical evaluation of the gonadostat hypothesis. Second, the notion of a gonadostat is of little value to investigators applying contemporary approaches to better understand either the neurobiology of the nongonadal brake on gonadotropin secretion in the juvenile or the molecular biology underlying steroid feedback on LH and FSH release. Third, the gonadostat hypothesis as it relates to puberty in the rat is now questioned (see Chapter 38). For these reasons a detailed discussion of the gonadostat hypothesis has been omitted from this chapter.

Before leaving the role of sex steroids in the developmental control of gonadotropin release, it is

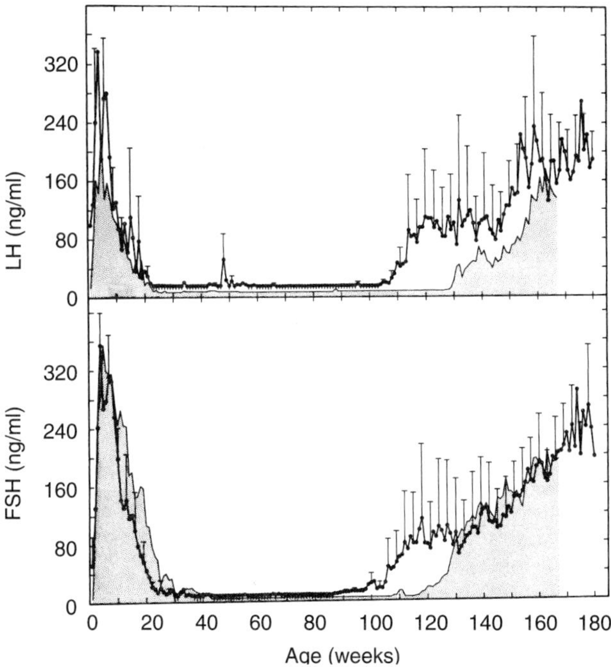

FIG. 3. Time courses of circulating mean LH (*top panel*) and FSH (*bottom panel*) concentrations from birth until 166–180 weeks of age in a group of rhesus monkeys orchidectomized at 1 week of age (stippled area, $n = 4$) and in a group of similar agonadal males adrenalectomized at 4 weeks of age (•, $n = 3$, birth to 40 weeks; $n = 5$, 41–180 weeks). The apparent earlier initiation of the pubertal mode of open-loop gonadotropin secretion in adrenalectomized animals is attributable to one animal. In the remaining four adrenalectomized animals, the pubertal rise in FSH secretion was not observed until 136 weeks of age or later. Vertical lines above datum points indicate SEMs. [The data for castrated males are redrawn with permission from (54) and those for castrated/adrenalectomized males are, in part, redrawn with permission from (218).]

worth noting that the possibility that adrenal sex steroids exert an inhibitory action on LH and FSH release in the juvenile primate is, at best, remote. Bilateral adrenalectomy at 1 month of age in agonadal infantile male rhesus monkeys does not interrupt the prepubertal hiatus in gonadotropin secretion (279) (Fig. 3), and human subjects with primary adrenal insufficiency do not exhibit precocious puberty (280).

Neuroendocrine Basis of the Juvenile Hiatus in Gonadotropin Secretion

Direct and indirect evidence underlines the now universally accepted view that the brake on gonadotropin secretion during juvenile development in primates is the result of a reduction in pulsatile GnRH release from the hypothalamus. Here, it is important to note that the notion of a brake is conceptual and, as such, the term should not necessarily be equated with inhibitory inputs. In the premenarcheal rhesus monkey, spontaneous ovulatory menstrual cycles may be preciously induced by a chronic intermittent infusion of exogenous GnRH (172) (Fig. 4), and in cognate studies of the male macaque, premature initiation of testosterone secretion and spermatogenesis has been reported (173,281). In agonadal male rhesus monkeys between 15 and 18 months of age, which is the phase of development characterized by undetectable circulating LH and FSH concentrations (Fig. 2), the chronic intravenous intermittent infusion of GnRH elicits a prompt hypersecretion of both gonadotropins (218, 282,283). These results demonstrating that the pituitary is not limiting in the initiation of puberty in macaques are entirely consistent with clinical findings that pulsatile GnRH treatment results in the full activation of the pituitary–gonadal axis in human subjects with delayed puberty occasioned by defects in endogenous GnRH release (165–167). Direct evidence for a pubertal elevation in the hypophysiotropic drive to the gonadotroph has been provided by Terasawa and her colleagues (105,284), who monitored GnRH release in perfusates of hypothalamus in the female monkey and documented an increase in the release of this neuropeptide at the time of puberty.

Although the LH response of the gonadotropes to a single-bolus stimulation with exogenous GnRH is markedly greater in humans and monkeys during the infantile and pubertal phases of development than during the intervening years (36,55,285–288), there is no reason to suspect that this ontogenic pattern in pituitary behavior is independent of the temporally related developmental changes in endogenous hypophysiotropic drive to the gonadotrope.

The finding that in children aged 2–10 years urinary excretion of the gonadotropic hormones increases at night provided the first evidence to support the notion that the hypophysiotropic drive to the pituitary–gonadal axis is not completely obliterated during the prepubertal phase of development (241). The most convincing evidence for this view is provided, not surprisingly, from studies of the female, where the juvenile brake on gonadotropin secretion is less intense than in the male. In agonadal female monkeys and in girls with Turner syndrome, robust pulsatile LH discharges and dramatic nocturnal elevations in the secretion of this gonadotropin have been demonstrated at ages corresponding to the phase of juvenile development in normal girls and female monkeys (229,251,258,259). Although analogous studies of males are less extensive, the finding that in agonadal juvenile boys circulating gonadotropins may be elevated to concentrations above those seen in age-matched intact

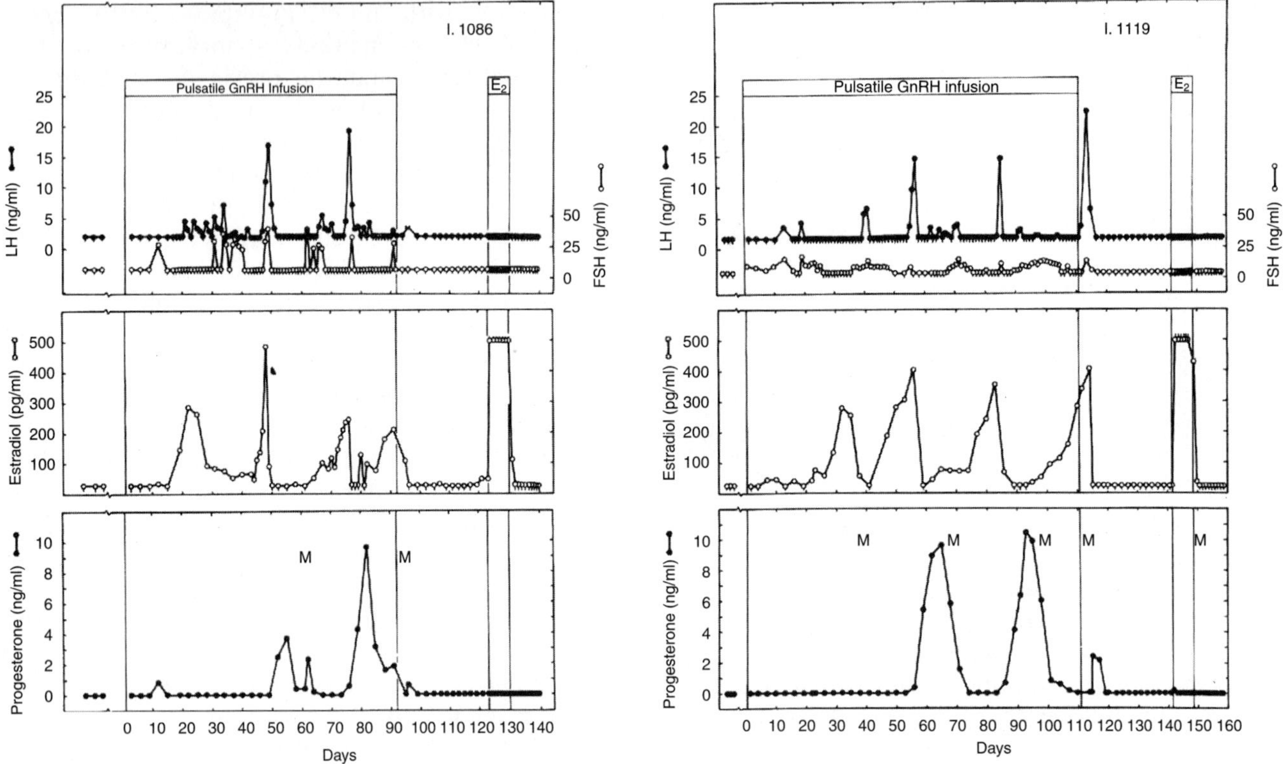

FIG. 4. Ovulatory ovarian cycles in two premenarcheal rhesus monkeys induced by a chronic intermittent intravenous infusion of GnRH (1 pulse an hour) initiated on day 0. Note that the pituitary–ovarian axis reverted to a prepubertal state after termination of GnRH treatment on days 92 and 111, and subsequent administration of estradiol (indicated by the *unshaded bar* labeled estradiol) failed to induce a gonadotropin surge. M, occurrence of menstruation. In this study, the rhesus LH pituitary preparation used to quantitate circulating LH concentrations was WDP-X-47BC. The immuno-potency of WDP-X-47BC is 3.6 times greater than that of WP-XV-20, which was used as standard in the RIA used to obtain data shown in Figures 1, 2, 3, and 7. [Redrawn with permission from (172).]

controls and exhibit pulsatility (255) indicates the presence of a residual hypophysiotropic drive in the human male during this phase of development. The prepubertal restraint on the GnRH pulse generator appears very intense in the male rhesus monkey, where differences in LH and FSH concentrations in the open- and closed-loop situation have not been observed (54,289). Although this may reflect assay sensitivity, LH pulses have not been observed in agonadal juvenile male rhesus monkeys in which pituitary responsiveness to GnRH had been greatly heightened by prior treatment with a pulsatile regimen of exogenous GnRH (251).

Whether the activity of the GnRH pulse generator is detectable in the ovarian intact closed-loop situation during prepubertal development is less clear. Whereas several studies have described LH pulsatility in prepubertal girls (as defined by Tanner stages), in many instances the subjects studied were approaching an age when the onset of puberty was to be anticipated (250,290–292). Thus, the possibility that the brake on pulsatile GnRH release was already being withdrawn cannot be excluded. This uncertainty, however, was eliminated in the study by Wu et al. (293), who, using IFMA, described LH and FSH pulses in the circulation of girls between 4 and 8 years of age. It should be noted that IFMA is at least an order of magnitude more sensitive than RIA, and thus much of the pulsatility in gonadotropin release identified during juvenile development by the later technology is undetectable by RIA. Moreover, the neuroendocrine basis of the small-amplitude secretory episodes of LH identified by IFMA during juvenile development, and specifically whether this mode of secretion is GnRH dependent, remains to be established. Although low levels of GnRH have been recovered in hypothalamic perfusates from ovarian intact juvenile monkeys (284), the functional significance of small fluctuations in GnRH content observed in the later study remains unclear.

When studies of pulsatile LH secretion in juvenile boys are evaluated, discrepancies between results obtained with various assay systems are, again, apparent. Studies using RIA or IFMA have identified pulsatile modes of LH secretion in prepubertal boys (255,290,293,294), some of whom ranged from 4 to 8 years of age (237,238,293). Episodic variation in circulating LH concentrations has also been observed in young (<36 months of age) male macaques (36,37), and in this species, passive immunization of males less than 30 months of age with anti-GnRH γ-globulin is reported to suppress circulating LH concentrations (295). In evaluating these observations in macaques, however, it is to be noted that in this genus transition into puberty, which is heralded by the onset of nocturnal testicular testosterone secretion, may occur as early as 2 years of age (54).

It seems reasonable to conclude that GnRH pulse generator activity during juvenile development is not abolished but rather is heavily diminished to a variable degree depending on sex and, perhaps, species. Conceptually, the issue of whether activity of the GnRH neuronal network of the juvenile is abolished or partially restrained is less interesting than elucidating the nature and control of the neurobiological brake that restrains GnRH release during juvenile development.

During pubertal development, descriptions of robust patterns of pulsatile LH release have been obtained. Gonadotropin secretion during early human puberty is characteristically low during wakefulness and augmented in association with sleep, when large amplitude LH pulses are observed (59,149,235,237,238,246–248,250,290,296–298). Because marked circadian changes in pituitary responsivity to GnRH stimulation are unlikely, the large nocturnal LH pulses probably reflect high amplitude GnRH discharges. The frequency of these discharges, originally observed only at night but later during the day, appears to increase noticeably from early to mid-puberty in both boys and girls (61,147,294,296,297). A decrease in LH pulse frequency later in pubertal development has been reported in both boys and girls (296,297,299). In boys, this phenomenon has been attributed (297) to the action of rising testosterone concentrations to decelerate the frequency of the GnRH pulse generator (179). However, not all studies of LH pulse frequency during puberty in boys and girls have revealed the foregoing changes in the frequency of intermittent LH secretion (290). Additionally, a direct estimate of hypothalamic GnRH release during puberty in the female monkey failed to detect an increase in frequency from early to mid-puberty (284). Longitudinal studies of agonadal male rhesus monkeys, in which pituitary responsiveness to endogenous hypothalamic GnRH discharges was heightened with a "priming" infusion of the synthetic decapeptide, have provided insight into the time course of the pubertal resurgence in pulsatile GnRH release in the absence of testicular steroid feedback (283). As reflected by moment to moment changes in LH secretion in such animals, nocturnal GnRH pulse generator activity accelerates explosively at the termination of the juvenile phase of development from less than one pulse in 7 hours to approximately four pulses in 7 hours over a period of less than 6 weeks (Fig. 5). Although this rapid tempo of the pubertal reawakening of the GnRH pulse generator in agonadal animals is presumably decreased in the normal situation by testicular inputs, it seems reasonable to propose that in male primates, the acceleration of the GnRH pulse generator during puberty is an early and rapidly completed neurobiological event in the initiation of this phase of development (283).

Indirect estimates of GnRH pulse amplitude during puberty have been made by comparing the amplitude of endogenous LH pulses with those elicited in the same pubertal subjects by bolus infusions of GnRH (previously shown to provide the pituitary with a hypophysiotropic stimulus comparable with that generated by the hypothalamus after puberty). Using this approach, two groups have concluded that GnRH pulse amplitude is similar in

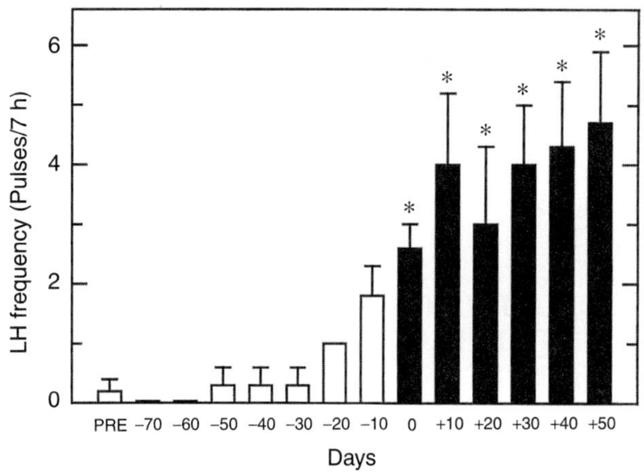

FIG. 5. Peripubertal developmental changes in GnRH pulse generator activity, as reflected by nocturnal frequency of LH secretory episodes, obtained by aligning data for five agonadal monkeys to the onset of the prepubertal resurgence in pulsatile GnRH release (day 0). Heightened pituitary responsiveness to GnRH stimulation was maintained by a "priming" infusion of pulsatile GnRH between assessment windows. Ages of animals at day 0 ranged from 24–29 months. PRE, Values for the earliest assessment window examined in each monkey, which ranged from 14–22 months of age; *asterisk*, significantly different from PRE (P <0.05). N for each time point varies from 3 to 5. [Reprinted with permission from (283).]

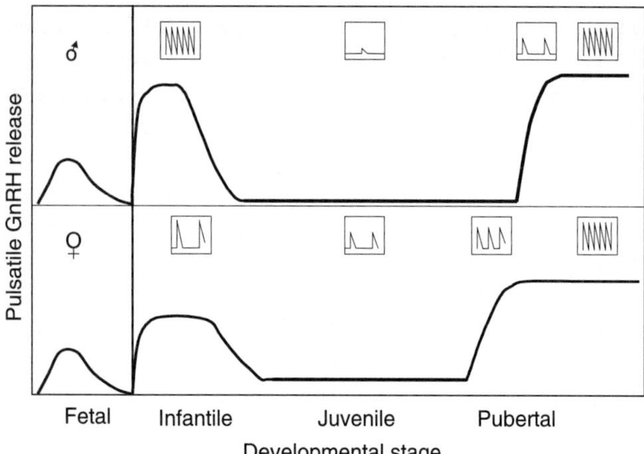

FIG. 6. A model of the impact of the neurobiological brake on pulsatile GnRH release during juvenile development in males (*above*) and females (*below*). Insets indicate frequency of pulsatile GnRH release at respective stages of development. [Reprinted with permission from Chapterhouse Codex Ltd., Plant, T. M. (2002). Control of onset of puberty in primates. *Top. Endocrinol.* 20, 2–5.]

adult men and pubertal boys during sleep (61,300). From the foregoing considerations it seems reasonable to conclude that the earliest detectable pubertal mode of operation of the GnRH pulse generator is composed of a high amplitude intermediate frequency discharge at night that probably reflects an amplification of a comparable preexisting prepubertal mode of activity. As puberty begins to progress, intermittent GnRH discharge occurs during the day and the integrated 24-hour frequency of the GnRH pulse generator accelerates. The time course of GnRH pulse generator activity during the mid- to postpubertal phase of development that is associated with increasing secretion of gonadal hormones is less clear. A schematic representation of GnRH pulse generator activity from fetal development until the entry into adulthood in a representative male and female higher primate is shown in Fig. 6.

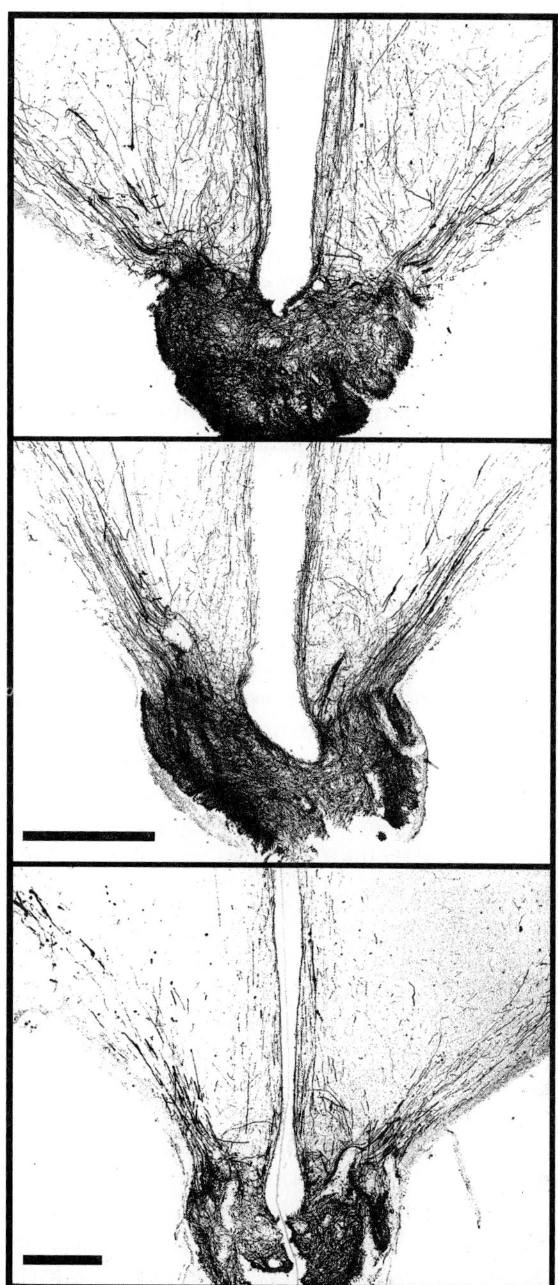

FIG. 7. Distribution of the GnRH neuronal network, as revealed by immunocytochemistry, in the mid-tuberal level of the mediobasal hypothalamus of an infant (*top*), juvenile (*center*), and adult (*bottom*) male rhesus monkey. Bar, 500 μm. [Redrawn with permission from (309).]

NEUROBIOLOGICAL BASES OF THE JUVENILE BRAKE ON PULSATILE GnRH RELEASE

In contrast to what might be anticipated, there is no evidence to suggest that expression of the gene encoding GnRH is repressed to any major extent between infancy and puberty (301). In the agonadal male macaque, hypothalamic GnRH content remains unchanged throughout postnatal life (262,302), and GnRH mRNA levels at the time of the pubertal resurgence of GnRH release show only a modest increase (262), the magnitude of which is markedly lessened in intact animals (303,304). Moreover, in the rhesus monkey and most other species, the distribution of GnRH neurons and their peptide content, as reflected by immunocytochemical analysis, are comparable in infant, juvenile, and adult animals (295, 305–309) (Fig. 7).

Not only is GnRH expressed at high levels before the pubertal resurgence of pulsatile GnRH release, but trains of intermittent GnRH discharges with a periodicity and amplitude similar to that generated by the hypothalamus of the adult may be provoked from the brain of juvenile male monkeys in response to hourly stimulation with NMDA (24,282,310) (Fig. 8). NMDA is an analogue of the acidic amino acid glutamate, a major excitatory neurotransmitter of the mammalian CNS (311,312). Although studies of NMDA action on the hypothalamus of the prepubertal female monkey have used only bolus administration of the glutamate receptor agonist (313,314), repetitive stimulation of the stalk median eminence region of the juvenile female monkey electrically or by direct application of an α_1-adrenergic receptor agonist also elicits a train of GnRH discharges (315,316). Pulsatile GnRH release from the prepubertal male monkey has also been observed in response to repetitive intravenous injections of the gut peptide, cholecystokinin (317). These findings suggest that GnRH neurons in the juvenile hypothalamus are endowed with the molecular machinery required for generating an intermittent discharge of the peptide. Moreover, this molecular machinery may be prematurely activated with apparent ease either experimentally (24) or pathophysiologically, as in GnRH dependent precocious gonardarche (27) (see below). The biosynthetic properties of hypothalamic GnRH neurons in the juvenile may be contrasted with those of the gonadotroph at a comparable stage of development. In most instances the magnitude of biosynthesis of a given secretory product is strongly coupled to the degree of cellular activation (318). Thus, as expected in the absence of a GnRH drive during the juvenile phase of development, LH content of the gonadotroph and circulating levels of LH are at least an order of magnitude lower than those observed in the infant or adult (302). In addition, the mRNA levels encoding the LH-β and FSH-β gonadotropin subunits are low (Winters and Plant, unpublished observations), and the pituitary requires weeks of intermittent GnRH stimulation before an adult like pattern of

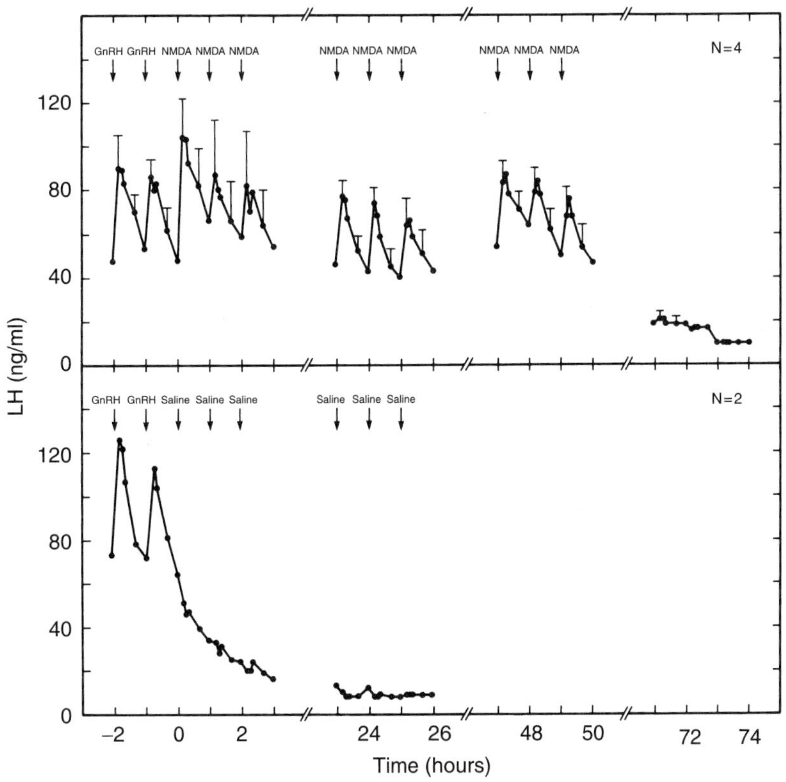

FIG. 8. The top panel shows pituitary LH responses in agonadal GnRH-primed juvenile male rhesus monkeys (n = 4) during the last two priming infusions of GnRH, during hourly intravenous infusions of NMDA, initiated at time 0 hours and maintained for 52 hours, and during a 3-hour window on the day after termination of intermittent NMDA administration. The bottom panel shows the expected decline in plasma LH concentrations that occurred on a separate occasion after initiation of an intermittent infusion of vehicle (saline) after a similar period of GnRH priming in two of the four animals. Vertical arrows indicate infusion of GnRH, NMDA, or saline. Note that NMDA and saline injections were administered without interruption for 50 and 25 hours, respectively; only those injections to which the LH response was monitored, however, are indicated. Vertical lines above data points indicate SEM. [Redrawn with permission from (310).]

gonadotropin secretion is elicited (283). From these data, it must be concluded that the GnRH neuron in the hypothalamus of the juvenile monkey is probably receiving stimulatory afferent input (thereby maintaining the cell's biosynthetic potential) and therefore that some other mechanism accounts for the prepubertal decline in the hypophysiotropic drive to the gonadotrope. Two attractive possibilities are suppression of release by an inhibitory synaptic input on the GnRH axon terminal and the attenuation of a pulse-generating signal on either the cell body or terminal.

With regard to the first possibility, the opioid peptides were the first to attract attention in this regard because of their established inhibitory action on pulsatile GnRH release in sexually mature monkeys and humans (319–323) and because endogenous opioid peptides are known to exert presynaptic inhibition in the magnocellular system (324). Administration of opioid antagonists such as naloxone and naltrexone, however, has consistently failed to release the prepubertal brake on the GnRH pulse generator as reflected by the failure of these drugs to elicit LH secretion in prepubertal children (325–328), chimpanzees (329), and rhesus monkeys (313, 330,331). The failure to detect naloxone-induced GnRH release during the prepubertal period cannot be accounted for by the decreased sensitivity of the gonadotrope during this phase of development because in studies of the monkey, the responsivity of these pituitary cells to GnRH was greatly heightened before the studies (329,330). Additionally, puberty in the male monkey appears to be associated with an up-regulation of the pro-opiomelanocortin gene (303).

γ-Aminobutyric acid (GABA) is a major inhibitory neurotransmitter in the CNS (332), and an elegant series of studies by Terasawa and Fernandez (105) provided unequivocal evidence that this amino acid plays an important role in the onset of puberty in the female. Using push–pull perfusion to quantitate neuropeptide and neurotransmitter release in the hypothalamus of the female rhesus monkey, Terasawa and her colleagues (333) demonstrated that the pubertal resurgence in GnRH secretion is temporally coupled to a decrease in GABA. In addition, acute inactivation of the $GABA_A$ receptor or reduction of GABA tone in this region of the hypothalamus of the prepubertal female monkey with bicuculline, a $GABA_A$ receptor antagonist, or antisense oligodeoxynucleotide for the mRNA encoding the GABA synthesizing enzyme, glutamic acid decarboxylase 67, respectively, elicited an immediate discharge of GnRH (333,334). Interestingly, the pubertal decrease in hypothalamic GABA tone in the female monkey is associated with, and may be the trigger for, an elevation in glutamate release at this stage of development (335) (Fig. 9). This inverse relationship between pubertal changes in GABA and glutamate would eliminate the need to posit a direct inhibitory action of GABA on the GnRH neuron. This may become an important consideration when the current controversy regarding the question of whether GABA action at the membrane of the GnRH neuron in the postnatal hypothalamus leads to depolarization or hyperpolarization is resolved (301). Finally, Terasawa's group (336) showed that chronic repetitive administration of bicuculline into the third cerebroventricle in the juvenile female monkey leads to precocious puberty,

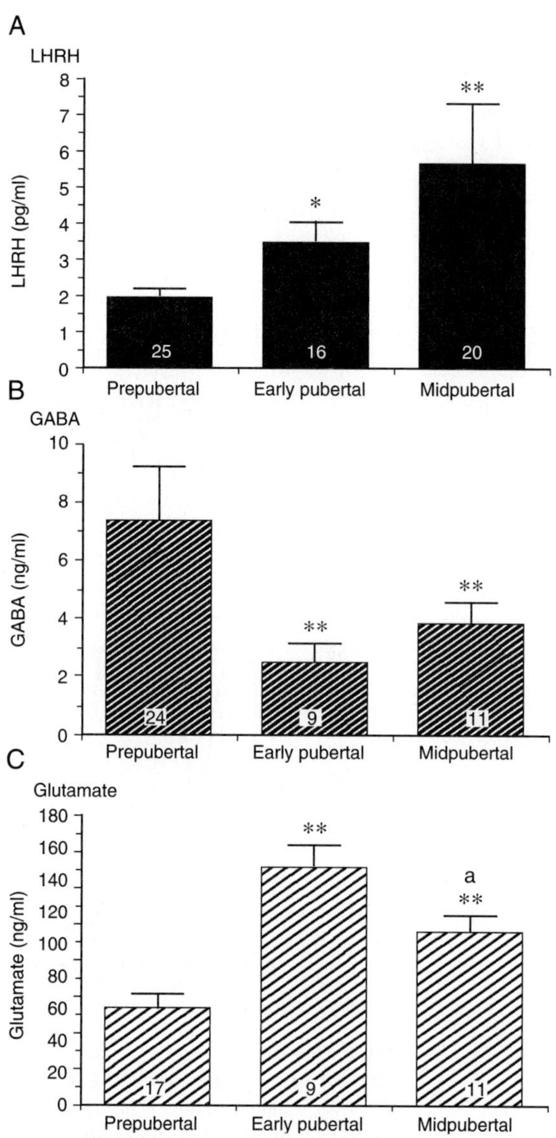

FIG. 9. Release of GnRH (**A**), GABA (**B**), and glutamate (**C**) in the stalk-median eminence of ovariectomized pre-, early-, and mid-pubertal female rhesus monkeys, as estimated in hypothalamic perfusates. *$P < 0.05$ vs. prepubertal; **$P < 0.01$ vs. prepubertal; [a]$P < 0.05$ vs. early pubertal. $n = 9–25$ (see individual bars). [Reprinted with permission from (335).]

as reflected by premature menarche and first ovulation. The foregoing studies were conducted in ovarian intact animals, and therefore it is unclear whether GABAergic neurons represent a component of the fundamental agonadal brake or the ovarian feedback mechanism that amplifies the central nongonadal brake. In this regard, there is no evidence for peripubertal changes in the expression of glutamic acid decarboxylase 65 and 67 in the male monkey (262,337), in which nongonadal restraint is most profound. Developmental changes in the expression of these genes have not been studied in the female.

Additional support for a role of GABA comes from clinical observations in which administration of loreclezole, a GABA agonist, to an 11-month-old girl with epilepsy and precocious gonadarche led to decreased LH secretion and regression in pubertal development (338). Pubertal delay has also been reported in epileptic boys treated with valproic acid, a drug with GABAergic activity (339). It should be noted, however, that the pharmacology of valproic acid is complex, and others have reported that this drug, like other antiepileptic agents, does not invariable delay sexual maturation (340).

There is also evidence obtained primarily with the male rhesus monkey that neuropeptide Y (NPY) is an important component of the neurobiological brake restraining GnRH release during juvenile development (262). First, changes in hypothalamic *NPY* expression and NPY content during the transition from the juvenile to pubertal phase of development in agonadal male monkeys are inversely related to those in pulsatile GnRH release; namely *NPY* expression is elevated during the juvenile phase of development and reduced in association with the pubertal resurgence in GnRH pulse generator activity (Fig. 10). Second, a marked inhibition of pulsatile GnRH release is observed in agonadal postpubertal male and female monkeys after central administration of NPY (341–343). Although the inhibition of GnRH release in the postpubertal situation is mediated, at least in part, by the NPY Y1 receptor, the pharmacological data supporting a role for this receptor subtype in the prepubertal brake on pulsatile GnRH release are tenuous (341). Also, the search for inactivating mutations of the NPY Y1 receptor gene in children with GnRH dependent precocious gonadarche (see below) has not been successful (344,345).

In addition to the posited inhibitory action of NPY on GnRH release during juvenile development (discussed above), this neuropeptide has also been implicated in the stimulation of GnRH release at the time of puberty and in adulthood. In both pubertal and postpubertal female monkeys, but not in premenarcheal subjects, NPY infusion in the region of the median eminence stimulates GnRH release (343,346,347) and infusion of NPY antibody or antisense oligodeoxynucleotide for NPY mRNA in this hypothalamic area suppresses GnRH release (346,348). The pubertal emergence of this stimulatory action of NPY on GnRH release is associated with an amplification of pulsatile release of NPY in the median eminence that is temporally coupled with that of GnRH (346,349). The most parsimonious explanation to account for the paradox of the inhibitory and excitatory actions of NPY on GnRH release is to posit two subsets of NPY neurons (350). One subset comprises a component of the brake that holds the GnRH pulse generator in check during juvenile development. The other subset comprises a component of the GnRH pulse generating neuronal network that is only manifest after the juvenile brake on the GnRH pulse generator has been lifted. Whether these two subsets of

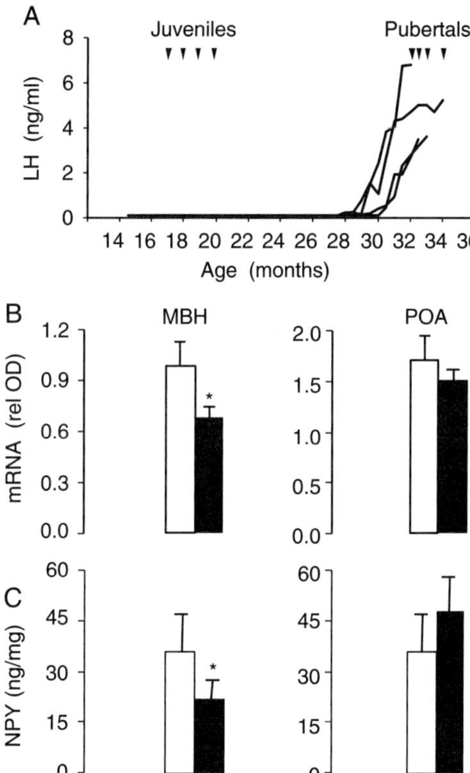

FIG. 10. Relationship between the pubertal resurgence in GnRH pulse generator activity, as reflected by circulating LH concentrations (**A**) and NPY mRNA and peptide levels in the mediobasal hypothalamus (MBH, *left*, **B** and **C**) and preoptic area (POA, *right*, B and C) in agonadal male monkeys. (A) *Arrowheads* indicate ages of animals when hypothalamic tissue was extracted. (B) Relative OD (means ± SD) of the NPY mRNAs in juvenile (*open bar*, $n = 4$) and pubertal monkeys (*filled bar*, $n = 4$). Cyclo mRNA was used for normalization. (C) Hypothalamic NPY peptide content (ng/mg protein, means ± SD, $n = 4$ in each group). $^*P < 0.05$. [Redrawn with permission from (262).]

NPY perikarya are anatomically distinct has not been examined.

Turning now to the possibility that the brake on pulsatile GnRH release during juvenile development is occasioned by the attenuation of a pulse-generating signal, glutamate and kisspeptin-1 must be viewed as the most likely candidates. As described above, hypothalamic glutamate release increases at the time of puberty in the female rhesus monkey (335) (Fig. 9) and in the juvenile male monkey repetitive activation of hypothalamic NMDA receptors using an intermittent intravenous infusion of NMDA elicits a sustained pulsatile discharge of GnRH (310; Fig. 8). In the gonadally intact animal this results in a precocious and complete pubertal activation of the pituitary–testicular axis (24). Because a similar mode of administration of the non-NMDA receptor agonist, kainic acid, at the same stage of development produces a progressively decremental train of GnRH discharges (351), it may be argued that glutamate, acting at the NMDA receptor subtype, represents a missing neurochemical component underlying the decreased pulse-generating signal responsible for the juvenile phase of development. Furthermore, a preliminary study indicates that the competitive NMDA receptor antagonist, DL-2-amino-5-phosphono-pentanoic acid, can interrupt GnRH pulse generator activity in the adult monkey (352). Whether the action of glutamate is exerted directly on GnRH neurons remains unclear. Axosomatic Glu-GnRH contacts have not been observed at either the light or ultrastructural level in the hypothalamus of adult male monkeys, and axodendritic contacts are rare (353). Moreover, the predominant glutamate receptor used by GnRH neurons in the rat appears to be that for kainite (see Chapter 38). Thus, at the present time, it seems reasonable to propose that the action of glutamate to drive intermittent pubertal GnRH release may be of an indirect nature (see below).

A novel signaling mechanism regulating GnRH release has recently emerged from studies of the genetic basis of normosmic hypogonadotropic hypogonadal patients (354,355). This signaling system involves GPR54, a G-protein–coupled receptor (356), and its ligand, kisspeptin-1, the 145 amino acid prohormone of *KiSS-1* (357). Three inactivating mutations of *GPR54* were reported: a homozygous deletion single nucleotide variant in one (exon 3) of the 5 exons of the gene (355), a compound heterozygous mutation in exon 5 (355), and a 155 nucleotide homozygous deletion at the junction of intron 4 and exon 5 (354). All affected individuals exhibited hypogonadotropic hypogonadism, with absent or incomplete pubertal development and without evidence of anosmia or hypothalamic and pituitary lesions. These clinical observations have been complemented by findings in the juvenile male monkey that intracerebroventricular or intravenous injections of kisspeptin-10 (112–121) elicit precocious GnRH release (Fig. 11) and that in both male and female monkeys, expression of *KiSS-1* is increased in association with the pubertal resurgence of pulsatile GnRH release (Figs. 12 and 13) (358). Additionally, hybridization histochemistry revealed robust *KiSS-1* and *GPR54* expression in the medial basal hypothalamus (MBH). Taken together, these findings indicate that increased GPR54 signaling contributes to the pubertal resurgence in GnRH release. As has been reported for cichlid fish (359), GnRH neurons of the hypothalamus of the juvenile monkey are known to express GPR54 (360). Therefore, if prolonged stimulation with kisspeptin-1 is shown in the future to elicit sustained pulsatile GnRH release, it would then be tempting to speculate that kisspeptin-1 provides the proximate pubertal drive to the GnRH neuron, which is imposed by an upstream glutamate signal. In this regard, the noncompetitive NMDA antagonist, MK 801, does not appear to attenuate kisspeptin induced LH release in the rat (361).

Peripheral intravenous administration of cholecystokinin, like that of kisspeptin-1 derived peptides,

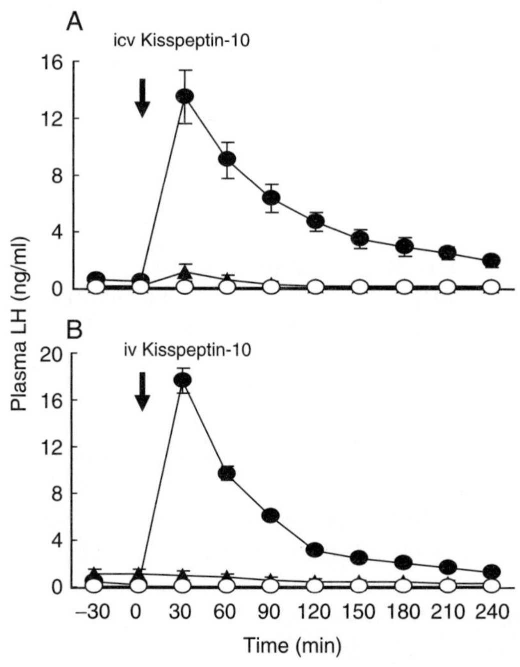

FIG. 11. Central (**A**, $n = 3–4$) or peripheral (**B**, $n = 2$) administration of a 100 μg of a fragment of kisspeptin [kisspeptin-1 (112–121)] ●–● to agonadal juvenile male monkeys elicits a precocious GnRH discharge, as reflected by plasma LH levels (means ± SEM). The LH rise is absent in vehicle treated animals (▲–▲) and abolished by prior administration of a GnRH receptor antagonist, acyline (○–○). [Reprinted with permission from (358).]

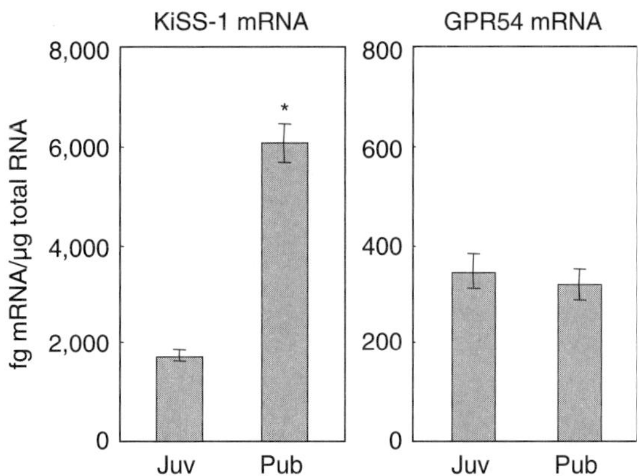

FIG. 12. Changes in KiSS-1 and GPR54 mRNA content (means ± SEM) determined by real-time polymerase chain reaction in medial basal hypothalamus of agonadal male monkeys before (Juv) or after (Pub) the pubertal resurgence of GnRH release. Hypothalamic RNA obtained from animals shown in Fig. 10. KiSS-1 mRNA content in the MBH was significantly greater (*P <0.005) after the pubertal resurgence of GnRH, whereas changes in GPR54 mRNA content during these developmental stages were unremarkable. Three animals were examined per group. [Reprinted with permission from (358).]

is also able to elicit precocious GnRH release in the agonadal juvenile male monkey (317). In the context of the timing of the pubertal resurgence in GnRH release, this action of cholecystokinin, which is associated with a generalized activation of the neuroendocrine hypothalamus, has not been examined further. Finally, acute central administration of methoxamine, an α_1-adrenergic receptor agonist, can elicit a GnRH discharge in the premenarcheal rhesus monkey (316), whereas intravenous injection

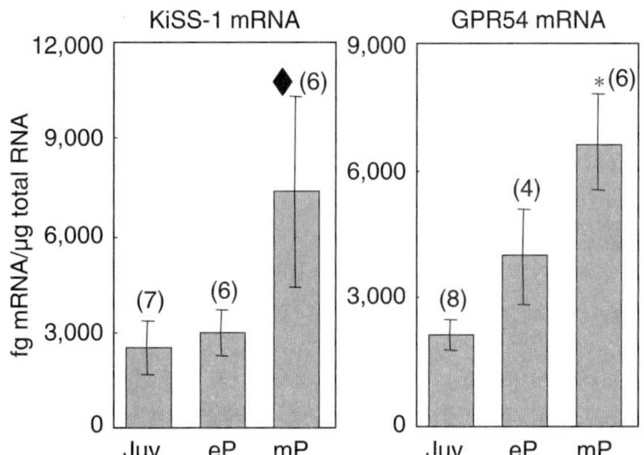

FIG. 13. Changes in KiSS-1 and GPR54 mRNA content (means ± SEM) determined by real-time polymerase chain reaction in medial basal hypothalamus (MBH) of intact female monkeys during pubertal development. KiSS-1 and GPR54 mRNA content increased in the MBH with pubertal development. Numbers in parentheses indicate the number of animals per group. *P <0.005 (analysis of variance), ♦P <0.05 (chi-square with threshold for bins set at 6,000 fg KiSS-1 mRNA/µg total RNA). Juv, Juvenile; eP, Early pubertal; mP, Mid-pubertal. [Reprinted with permission from (358).]

of the α-adrenergic receptor antagonist phentolamine blocks GnRH pulsatility in the adult female (362).

If the pubertal resurgence of pulsatile GnRH release is produced by a transsynaptic signal, it might be expected that the identity of such a signal would be reflected in the afferent neural input to the network of GnRH neurons and in the profile of receptor genes expressed in GnRH neurons. Although studies of the cell biology of GnRH neurons have been restricted primarily to nonprimate species (see Chapter 28), work from Goldsmith's laboratory indicates that GnRH neurons in the hypothalamus of the juvenile monkey receive synaptic input from other GnRH cells (363), β-endorphin neurons (363), arginine vasopressin neurons (364), and glutamate neurons (353). Our own studies indicate that GnRH neurons express GPR54 (360) and are innervated by NPY (350). Although the role of glutamate, β-endorphin, kisspeptin-1, and NPY in the control of the onset of puberty has been probed using pharmacological approaches (see above), that of arginine vasopressin has not been considered.

Glial-derived growth factors, such as transforming growth factor-α (TGF-α), neuroregulins, and their cognate erbB transmembrane tyrosine kinase receptors, play a role in the timing of puberty in the rat (see Chapter 38). The actions of these growth factors on GnRH neurons are stimulatory and indirect, involving autocrine/paracrine actions of factors such as prostaglandin E_2 released from neighboring astrocytes. Studies of the monkey in this regard are limited: TGF-α in the hypothalamus of the female rhesus monkey is increased dramatically at the time of puberty (304), although in the agonadal male expression of this gene is only marginally increased at a corresponding stage of hypothalamic development (262). Further support for the view that glial factors may play a role in the pubertal resurgence in GnRH release in higher primates is provided by findings that hypothalamic lesions result, on the one hand, in an astrocytic response that includes activation of TGF-α and expression of erbB receptors (365,366), and, on the other hand, may be associated with precocious gonadarche in children (367–369) and in juvenile monkeys (370). Thus, precocity in primates associated with hypothalamic lesions and in children who have been subjected to chemotherapy or cranial irradiation or who have sustained head trauma (367–369) may result from increased erbB signaling rather than by interruption of inhibitory inputs to the GnRH network, as was previously proposed (371). Additionally, hamartomas removed from two patients with precocious gonadarche have been found to be rich in astrocytes containing TGF-α but devoid of GnRH neurons (372), again indicating a role of increased erbB signaling in the etiology of this precocity. These arguments, which have been primarily articulated by Ojeda and his colleagues (371), are consistent with the finding in the rat that infusion of an erbB receptor antagonist at the site of a hypothalamic lesion prevents the puberty advancing action of the lesion (365).

Regardless of the identity and relative importance of the transsynaptic and astroglial systems that appears to underlie the pubertal resurgence of pulsatile GnRH release, it is not unreasonable to propose that this developmental event must involve a reorganization of the neural substrate that governs the control of the neurons that release GnRH. Because the capacity for structural plasticity within the mammalian brain appears to be associated with the expression of the "embryonic" or polysialic acid neural cell adhesion molecule (373), it might be predicted that areas of the hypothalamus containing GnRH neurons might express the polysialic acid neural cell adhesion molecule. A study of the rhesus monkey using immunocytochemical andimmunoblot analysis has demonstrated that this cell surface glycoprotein is expressed in the region of the arcuate nucleus and median eminence (374), an area of the hypothalamus that contains neuroendocrine GnRH neurons (375) and encompasses the entire terminal field of all GnRH projections to the primary plexus of the hypophyseal portal circulation. This latter finding thus reinforces the notion that puberty may be occasioned by a structural reorganization within a region of the hypothalamus that retains the capacity for plasticity into adulthood and that encompasses a component of the neuronal network responsible for regulating GnRH release. Indeed, quantitative ultrastructural studies have demonstrated that synaptic input to GnRH perikarya in the MBH of the agonadal male monkey declines in association with the pubertal resurgence in pulsatile GnRH release (Fig. 14). Interestingly, this structural remodeling during peripubertal development in the monkey has only been observed in the MBH (376,377). In contrast, the degree of glial ensheathment of GnRH neurons in the MBH of agonadal monkeys did not appear to be developmentally regulated (376), although peripubertal changes in glial ensheathment have been reported in the region of the organum vasculosum of the lamina terminalis in the female rhesus monkey (377).

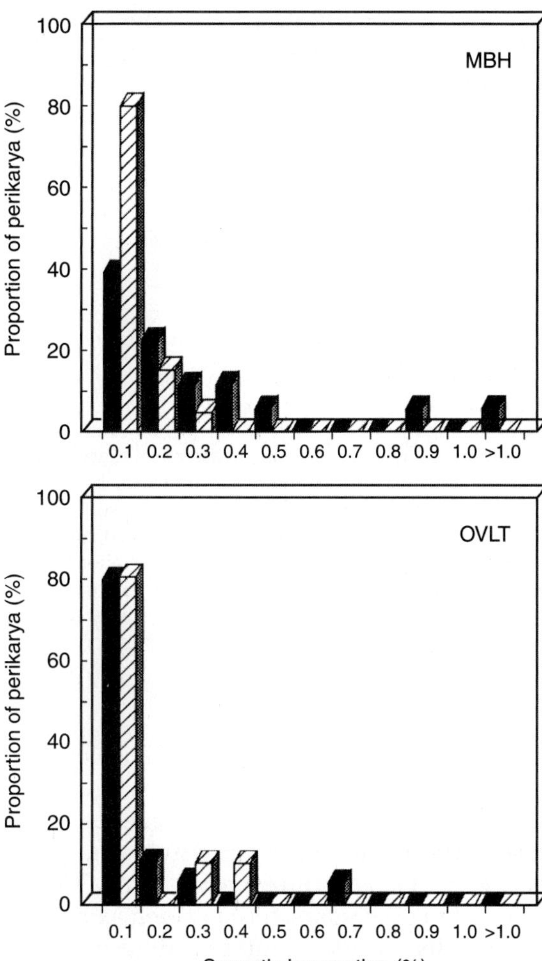

FIG. 14. Differences in synaptic input onto GnRH perikarya in juvenile (*solid bars*) and adult (*hatched bars*) male monkeys as reflected by the frequency distribution of GnRH cell bodies in the medial basal hypothalamus (**top**) and organun vasculosum of the lamina terminal (**bottom**) with increasing percentages of their perikaryal membrane occupied by synaptic densities. [Reprinted with permission from (376).]

In studying the neurobiological basis of the brake that holds pulsatile GnRH release in check from infancy to puberty, attention has focused almost exclusively on the resurgence in activity at the time of puberty. Of equal importance, however, are the mechanisms responsible for the reduction in pulsatile GnRH release during late infancy. The idea that the neurobiology underlying the imposition of the postnatal brake on pulsatile GnRH release is thrown into reverse at the termination of the juvenile phase of development is the most parsimonious explanation to account for the time course of GnRH release from birth to puberty in higher primates. In the case of NPY receptor and GPR54 signaling, there appears to be some merit to this notion. NPY gene and peptide expression in the MBH declines during this developmental transition (262), and low gonadotropin levels were recently described in an infantile boy with an inactivating mutation of GPR54 (378). On the other hand, several additional molecular concomitants associated with bringing the GnRH pulse generator into check during late infancy do not mirror those seen at the time this neuronal network is reactivated at the end of the juvenile phase of development (379). It has also been hypothesized that the reduction in pulsatile

GnRH release in late infancy is occasioned by a switch in GABA activity from excitatory to inhibitory (380) at this stage of development. This view is based on the finding that a developmental switch in the electrophysiological action of GABA on neurons in the rat brain occurs around postnatal days 7–14 (381). In primates, however, it is to be expected that this electrophysiological developmental milestone would take place prenatally and is therefore unlikely to be involved in the restraint of the pulse generator during infancy in these species.

The anatomical location of the neurobiological components of the control system that dictates the postnatal ontogeny of GnRH has been examined using classical techniques. In this regard, attempts have been made to deafferentate the MBH from the rest of the brain in immature monkeys and to determine the consequences of this perturbation on the timing of the onset of puberty (382,383). However, this surgical intervention in prepubertal animals, in contrast to its impact in adults (384), failed to abolish the diurnal variation in the secretion of cortisol, suggesting that isolation of the GnRH network from extrahypothalamic inputs may have been incomplete. Thus, although major alterations in the timing of menarche and other indices of puberty were not observed after deafferentation of the MBH, it is premature to conclude that the entire neural component of the puberty control system resides within this area of the hypothalamus.

We close this section with a brief comment on the value of experimental models using agonadal monkeys to investigate the neurobiological bases of the brake imposed on pulsatile GnRH release during juvenile development. Because the postnatal hypothalamus retains the capacity for both structural and molecular plasticity, in the intact situation it becomes impossible to distinguish whether a hypothalamic correlate of puberty, for example, a change in expression of a particular gene or the appearance or loss of a specific synaptic interaction, represents the primary event underlying the resurgence of GnRH pulsatility or is simply a secondary consequence due to increased gonadal steroid secretion that has been triggered by the fundamental hypothalamic event being sought. These potential confounding influences are eliminated when agonadal models are used.

THEORETICAL PHYSIOLOGICAL CONTROLS GOVERNING THE DURATION OF THE PREPUBERTAL HIATUS IN GnRH SECRETION

Two theoretical physiological control systems may be invoked to account for the prepubertal hiatus in GnRH pulse generator activity in higher primates. One involves the concept of a central neural timekeeping mechanism that is able to measure age and that is programmed to operate a gating system that would interrupt GnRH pulse generator activity from infancy to puberty. The other model requires positing a central neural mechanism that is capable of tracking one or more parameters of growth. Such a somatometer would, in turn, govern the postnatal ontogeny of GnRH release in much the same fashion proposed for the first model of a pubertal clock.

Somatometer

In considering a somatometer, it has been long viewed as axiomatic by some (385) that growth to maturity is required before an animal can breed. Thus, the notion that the attainment of a particular body size or composition provides the trigger for initiation of puberty has attracted considerable attention. Examination of the relationship between physical growth and the onset of puberty in primates has been restricted primarily to the human female, with menarche providing the usual end point. Attention was first drawn to a close correlation between the age at menarche and the rate of skeletal maturation (386,387). Later, Frisch and Revelle (388), stimulated by studies of the relationship between body weight and the onset of puberty in rats (389), drew attention to their observation of a relationship between menarche and the attainment of a mean body weight of approximately 47 kg. This association led the latter workers (388) to propose the hypothesis "that attainment of a body weight in the critical range causes a change in metabolic rate, which, in turn, reduces the sensitivity of the hypothalamus to estrogen, thus altering the ovarian-hypothalamus 'feedback'." It may, of course, be argued equally well that such changes in metabolic rate could lead, instead, to an increase in drive to or within the hypothalamic GnRH pulse generator.

Shortly after the formulation of this so-called critical weight hypothesis, Frisch and her colleagues (390) modified their views to include the concept of threshold or critical body composition for puberty. Specifically, menarche was argued to be most closely correlated with attainment of a particular proportion of body fat. According to such hypotheses, the secular trend toward an earlier age at menarche (see above) is attributed to attainment of a threshold body weight or composition at a progressively younger age, as a result of a corresponding secular improvement in the standard of living, including better nutrition and health care. Although the specific hypotheses of Frisch and her colleagues have been criticized on the basis of methodological and statistical grounds (391–395), the isolation in 1994 of leptin, a protein that is secreted by adipocytes and regulates feeding behavior and body weight by providing the hypothalamus with information on fat mass (396, 397), rekindled interest in the Frisch hypotheses. Indeed, an enormous effort has been expended over the last decade to examine the role of this adipocyte hormone in the timing of gonadarche.

Leptin

If circulating leptin is to serve as the signal initiating the pubertal resurgence of pulsatile GnRH release, two conditions must be satisfied. First, an elevation in plasma leptin concentration should precede the pubertal increase in GnRH release, and, second, a premature elevation of circulating leptin levels would result in precocious GnRH release and an early onset of gonadarche. With regard to the first condition, cross-sectional and longitudinal studies of both boys and girls indicate that, in general, circulating leptin concentrations increase during pubertal development (398–402). Moreover, the pubertal increases in circulating leptin concentrations in humans are inversely related to changes in levels of the soluble leptin receptor (403,404), indicating that the rise in bioavailable leptin at this stage of development may be more robust than that of total immunoactive leptin. The grouping of leptin data with respect to Tanner stages of pubertal development in the foregoing studies and the paucity of detailed longitudinal information on both leptin levels and LH pulsatility during transition from Tanner stage 1 to stage 2, however, do not allow the question of whether circulating leptin increases immediately before the initiation of gonadarche to be addressed. One exception to this situation has been provided by the longitudinal study of eight boys in whom increments in circulating leptin concentrations were reported to occur before nocturnal elevations in testosterone secretion (405). As previously noted (406), however, the leptin levels reported in the latter study were extremely variable, and the seemingly tantalizing increments of circulating levels of leptin were actually rather trivial in many of the subjects. In contrast to normal puberty, circulating leptin concentrations were unchanged at the time delayed gonadarche was initiated in boys with constitutional delay in growth and puberty (407).

Studies of the relationship between circulating leptin levels and the onset of gonadarche in other species of higher primate have been restricted to the rhesus monkey. In the female, circulating leptin levels at the initiation of gonadarche (approximately 24 months of age), as identified by increased nocturnal LH secretion, have been reported to be greater than those observed at 14 months of age (408). In the male monkey, however, the longitudinal relationship between the onset of nocturnal testosterone secretion and plasma leptin levels is unremarkable (409). Results from cross-sectional studies of the male monkey (410,411) are also consistent with the view that gonadarche in the male is triggered in the absence of a rise in circulating leptin. Thus, the report that in the agonadal rhesus monkey, plasma leptin concentrations increase in association with the pubertal resurgence of pulsatile GnRH release (412) is surprising.

With regard to the second condition that needs to be met if leptin is to be considered as a candidate for the trigger of gonadarche, daily administration of recombinant human leptin in intact female rhesus monkeys, starting at 1 year of age, was associated with increased circulating LH concentrations at 14 months of age, when compared with those in the control group (408). Nocturnal LH secretion was also detected earlier in the leptin-treated group, and this was associated with menarche advanced by approximately 2 months. The mechanisms underlying the leptin associated amplification of endocrine activity in the pituitary–ovarian axis in the premenarcheal monkey is unclear, but a precocious pubertal resurgence in pulsatile GnRH release is unlikely to be involved, as reflected by the absence of premature ovulatory menstrual cycles, which are readily produced when an adult hypophysiotropic drive is prematurely provided to the gonadotrophs (172). In a simpler experiment, albeit limited by a relative acute duration, a continuous 16-day intravenous infusion of recombinant human leptin to agonadal juvenile males, equivalent in age to 4- to 5-year-old boys, did not trigger precocious activation of GnRH pulse generator activity (413). Also of interest in this regard is the report that LH, FSH, and testosterone concentrations remained prepubertal in a 5-year-old leptin-deficient boy during 12 months of replacement with recombinant human leptin calculated to achieve circulating leptin levels that would not exceed 150% of that for age-matched normal boys (414). Finally, although circulating leptin levels have been reported in both boys and girls with GnRH-dependent precocious gonadarche (415,416), understandably, leptin measurements before manifestation of advanced sexual development were not reported.

Despite the lack of compelling data to support the hypothesis that increased leptin signaling provides the pivotal cue for the pubertal resurgence of pulsatile GnRH release, several clinical observations support the idea that this hormone is nevertheless obligatory, in a permissive sense, for gonadarche to be initiated. Primary hypogonadotropic hypogonadism at 23–34 years of age has been reported in association with morbid obesity in one male and two female subjects with a missense mutation in the leptin gene (417,418). Failure to undergo pubertal development between 14 and 19 years of age has been described in three sisters homozygous for a mutation in the leptin receptor gene (401). Primary amenorrhea at 15–17 years of age was reported in three patients with congenital generalized lipodystrophy associated with Seip-Berardinelli syndrome (418). Although one female subject with leptin deficiency subsequently entered puberty spontaneously (418), the foregoing clinical observations suggest that leptin signaling is required for normal gonadarche. On the other hand, normal pubertal development was reported in two female subjects with severe lipodystrophy (419) and in one girl with congenital contractual arachnodactyly (420), all of whom were markedly hypoleptinemic. Interestingly, although the lipodystrophic subjects with normal pubertal development had measurable concentrations of circulating immunoactive leptin (0.7–1.6 ng/ml), they were no greater than those in the subjects with Seip-Berardinelli syndrome (0.8–1.0 ng/ml) who failed to enter puberty. The most likely explanation for these inconsistent findings is that immunoactive leptin concentrations in blood may not accurately reflect the bioavailability of the circulating hormone.

Additional support for the notion that a permissive action of leptin is necessary for gonadarche to occur is provided by studies of the effects of hormone replacement in a small number of leptin-deficient human subjects of pubertal or postpubertal age who exhibit absent or incomplete gonadarche. In a girl with a frameshift mutation in the leptin gene and with a chronological and bone age of 9 and 12.5 years, respectively, a pulsatile pattern of LH secretion, indicative of a pubertal mode of GnRH pulse generator activity, was observed after 12 months of replacement with recombinant leptin to correct her morbid obesity (421). Continued leptin replacement to this girl was associated with menarche at 12.1 years of age (bone age, 13.4 years), and 1.5 years later Tanner stage 3 was reported (414). It should be noted that in this subject, bone age, which is considered a better predictor of gonadarche than chronological age (27,422–424), was pubertal at the start of leptin replacement. Therefore, the action of leptin in this subject may be viewed as permissive. Leptin replacement was also reported to facilitate menstruation in the women with primary or secondary amenorrhea in association with congenital general lipodystrophy (425).

These clinical observations suggest that a low threshold concentration of plasma leptin is needed for gonadarche to unfold (406): a notion that is consistent with the finding in underweight postpubertal women that regular menstruation was associated with leptin levels in the range of 1.9 ng/ml (426). This permissive action of leptin on hypothalamic GnRH release at puberty is also necessary for maintenance of gonadal function throughout postpubertal development. Leptin also appears to serve a similar role in rodents (427). Although this action of the hormone may be viewed as obligatory for gonadarche to unfold, leptin does not provide the signal that times the resurgence of pulsatile GnRH release and therefore is not the cue that terminates the prepubertal state. Recent studies of the male rhesus monkey also suggest that the restraint of GnRH pulse generator activity occurring during late infancy is unrelated to changes in circulating leptin concentrations at this earlier but equally critical stage of development (428).

Growth Hormone

GH plays a major role in the regulation of somatic growth, and therefore the finding that puberty is markedly delayed in children with isolated GH deficiency (429) raises the possibility that GH may be involved in the interface between somatic development and the timing of the onset of puberty. Although this is an attractive notion, it has been generally recognized that changes in GH secretion during puberty, at least in humans, are the consequence rather than the cause of the underlying neurobiological events triggering puberty (430). Thus, the pubertal rise in GH secretion in humans (431–436) and the concomitant increase in circulating somatomedin-C (IGF-1) concentrations (437–440), which has also been reported for the rhesus monkey (441,442), baboon (52), and chimpanzee (32), are absent in agonadal subjects of pubertal age (443) and may be mimicked in sexually infantile children or in subjects with delayed puberty by treatment with sex steroids (444–447). Moreover, the enhanced GH secretion that is observed in precocious puberty is suppressed when the drive to the pituitary–gonadotrope axis is interrupted with GnRH analogue treatment (439,448).

The notion of a causal relationship between the neuroendocrine axis governing growth and that regulating gonadarche has been championed by Wilson and his colleagues at the Yerkes National Primate Research Center based primarily on their studies of the female rhesus monkey. This was prompted by the finding that in the perimenarcheal female rhesus monkey, the onset of the pubertal rise in LH secretion was preceded by a marked daytime rise in the circulating GH concentration (88).

Moreover, the same temporal relationship between the pubertal LH increase and GH secretion was observed in animals that were ovariectomized premenarcheally and subsequently treated with estradiol (88), suggesting that, in contrast to humans, the increase in GH secretion at this stage of development may not be triggered by the rise in estrogen secretion. This view was supported by analyses of a series of blood samples collected during a longitudinal study of agonadal male rhesus monkeys suggesting GH secretion may increase in association with the pubertal resurgence in pulsatile GnRH release in the absence of the testis (412,449). The effects of perturbing GH status on the timing and tempo of puberty in the female rhesus monkey have been reported on several occasions. Treatment of premenarcheal monkeys with human GH was associated with a premature increase in pubertal LH secretion, menarche, and first ovulation (450). On the other hand, IGF-1 administration to premenarcheal female rhesus monkeys advanced only first ovulation (451) and attempts to suppress GH

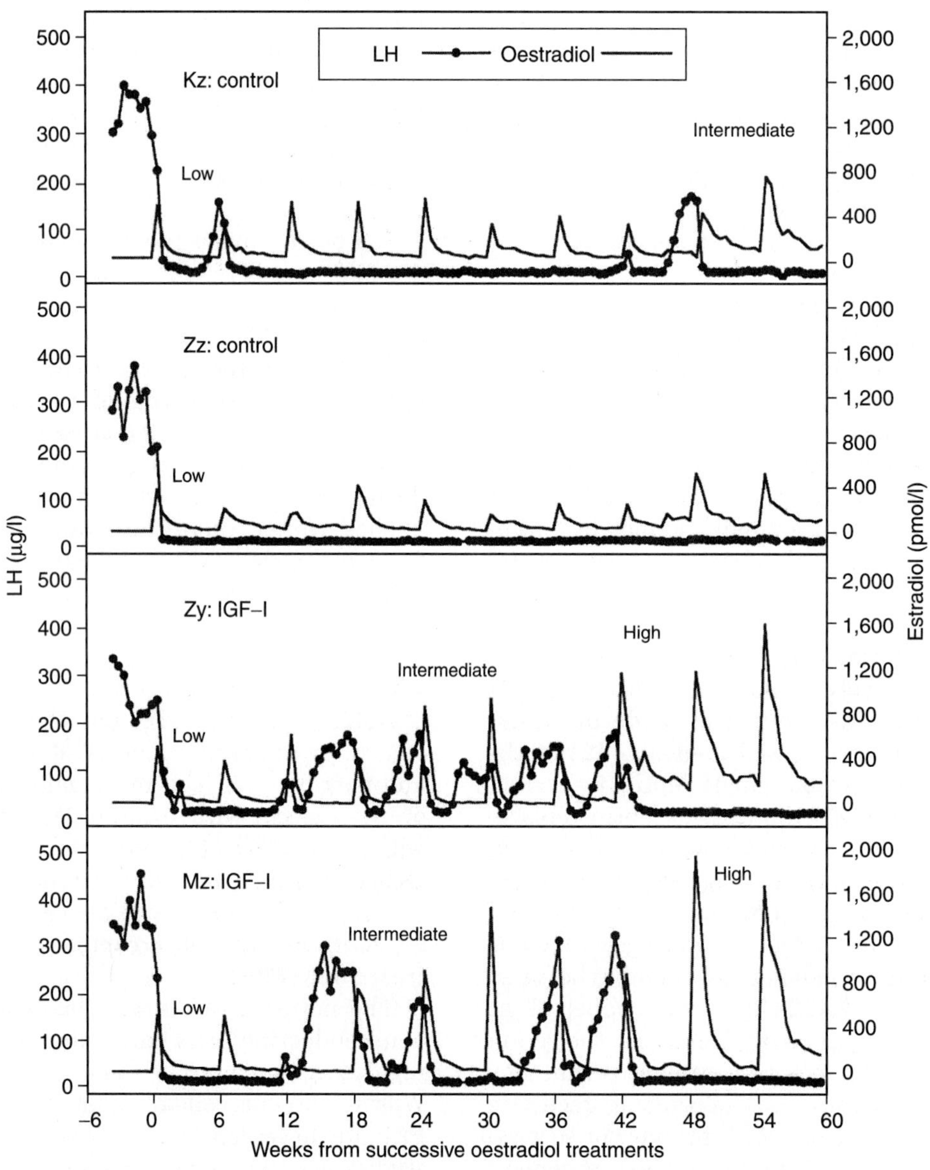

FIG. 15. Developmental changes in serum LH concentrations (*solid circles*) in four female monkeys ovariectomized at approximately 13 months of age, treated with IGF-1 (60–300 µg/day) or control from 16 to 18 months of age and replaced with estradiol starting at approximately 25 months of age (week 0). Circulating estradiol concentrations are shown by the continuous line. The IGF-1 treated animals are shown in the bottom two panels. [Reproduced with permission from (453).]

and IGF-1 in premenarcheal females by treatment with somatostatin delayed only first ovulation (442, 452). Moreover, the timing of the pubertal LH increase in ovariectomized estrogen-treated monkeys was not influenced by GH treatment (450), and IGF-1 administration to untreated agonadal females was without affect on this parameter of pubertal development (453). The current view of the Yerkes group (452) may be summarized as follows: GH plays no physiological role in the timing of the pubertal resurgence in pulsatile GnRH release but does govern the tempo of pubertal development in the female monkey after activation of the critical neurobiological event triggering increased GnRH release. One mechanism whereby GH accelerates the tempo of pubertal development in the female appears to involve an action of IGF-1 at the hypothalamic and/or pituitary level to impair the negative feedback action of estradiol secreted by the pubertal ovary on gonadotropin secretion (453) (Fig. 15).

Melatonin

Several lines of indirect evidence have been used to support the idea that melatonin may provide a signal that is read by the putative somatometer. Circulating concentrations of this indolamine in prepubertal children and female monkeys are higher than those in adults, and a decline in this plasma hormone occurs during juvenile and pubertal development (454–459). The production rate of melatonin, as reflected by studies of the urinary excretion of 6-hydroxymelatonin, the principal metabolite of this pineal hormone, either declines or remains constant during childhood and puberty (460–463). It has been known since the beginning of the century that pineal tumors may be associated with sexual precocity in humans (464), and the argument is made that those pineal tumors that destroy the parenchyma of the gland suppress the release of melatonin, which in turn leads to a premature reactivation of pulsatile GnRH release (465). It should be noted, however, that pineal tumors may themselves secrete gonadotropic substances (466,467) and may impinge on adjacent areas of the hypothalamus and thus induce premature sexual development by the same mechanism that leads to other forms of lesion-induced precocious gonadarche (see above). In patients with idiopathic precocious gonadarche, circulating melatonin concentrations are lower than those in age-matched individuals and similar to those of normal pubertal subjects (468). Moreover, the suppression of circulating melatonin concentrations in patients with this form of precocity is not the consequence of elevated gonadal hormones, because down-regulation of the pituitary–gonadal axis after treatment with GnRH analogues fails to restore melatonin to normal levels (468,469). The latter result is consistent with the finding that the prepubertal decline in circulating melatonin concentration in the female rhesus monkey is independent of the changes in estradiol secretion at this stage of development (470).

If elevated circulating melatonin levels during prepubertal development are indeed monitored by a somatometer, which in turn holds pulsatile GnRH release in check, then removal of the pineal gland during this phase of development should lead to a premature reactivation of the GnRH pulse generator. Studies in the monkey, however, suggest that this is not the case (218,471). Pinealectomy in agonadal males during the prepubertal hiatus in pulsatile GnRH release does not appear to influence the timing of the pubertal reactivation of the GnRH pulse generator (Fig. 16), and the decline in GnRH pulse generator activity during infancy in pinealectomized monkeys is indistinguishable from that in sham-castrated animals (471). Moreover, increasing the duration of the nocturnal elevation in circulating melatonin in the female rhesus monkey during the third year of life advances the age of menarche and ovulation (472). This latter finding probably reflects an action of the control system governing seasonal influences on the hypothalamic–pituitary–ovarian axis rather than a perturbation of the neurobiological brake that dictates the duration of the prepubertal hiatus in pulsatile GnRH release. Thus, it must be concluded that there is no compelling evidence for the view that the signal for the pubertal resurgence of pulsatile GnRH release is provided by a decline in circulating melatonin concentrations.

Metabolic Hormones and Substrates

Steiner and colleagues (473,474) proposed that the reduction in GnRH secretion from infancy to puberty may be the consequence of decreased availability of certain metabolic hormones or substrates. These investigators proposed that in the juvenile primate, the GnRH pulse generator is subjected to a metabolic milieu similar to that associated with fasting states in adults, such as in anorexia nervosa, in which hypothalamic GnRH release also appears to be compromised (475,476). In a somewhat similar vein, Winterer et al. (477) proposed that recognition of a critical level of available calories by the primate brain leads to the reawakening of the dormant GnRH pulse generator and therefore to the initiation of gonadarche. According to the latter hypothesis, a reduction in available calories below the critical level in adults could lead to impairment in GnRH release.

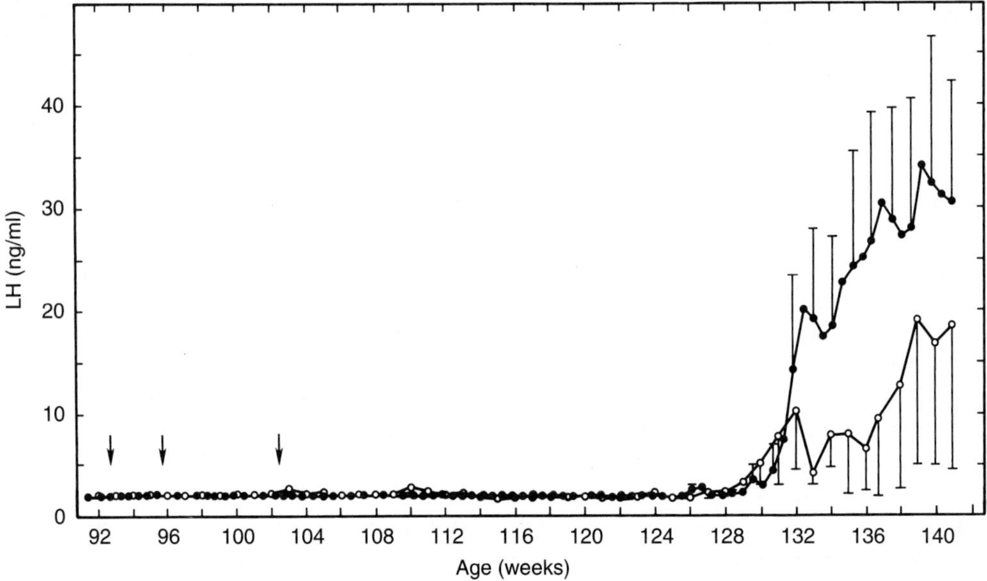

FIG. 16. Failure of pinealectomy to induce a precocious pubertal reactivation of the GnRH pulse generator (as reflected by circulating LH concentrations) in agonadal male rhesus monkeys. Data for pinealectomized animals (●–●, $n = 3$) are compared with those from a similar group of pineal-intact monkeys (○–○, $n = 3$). *Arrows* indicate the age at pinealectomy. LH concentrations are expressed in terms of WDP-X-47BC (see legend to Fig. 4 for further details). The time courses of FSH secretion in the two groups of animals (not shown) were similar to that of LH. [Redrawn with permission from (218).]

Several lines of circumstantial evidence, which have been reviewed by others (473,474,477), may be used to support such metabolic theories for the onset of puberty. The inhibition of gonadotropin secretion in adult monkeys and humans that result from restricted food intake may be observed, as in the case of the prepubertal hiatus in LH and FSH release, in the presence of gonadal steroids (478,479). Moreover, the pattern of hypothalamic GnRH pulse generator activity, as reflected by pulsatile patterns of LH secretion in female anorectics, appears to resemble closely those described for prepubertal and pubertal girls (475). Anorexia nervosa and its related disorder, bulimia nervosa, however, are complex diseases characterized by psychosocial features, including difficulties with emotional regulation and adult autonomy and tendencies toward obsessive-compulsive personality disorder in addition to weight loss. Moreover, dysregulation of the hypothalamic–pituitary–adrenal axis, characterized by elevated cortisol concentrations in the face of normal ACTH concentrations, reduced peripheral metabolism of cortisol, and lack of suppressibility of ACTH by dexamethasone, is common (480). Overactivity of the serotonergic system is also present (481). Although no specific genes have been identified, family and twin studies support the hypothesis that genetic factors influence the risk for both these disorders (482). Thus, anorexia nervosa represents the outcome of complex interaction between multiple genetic factors, early life experiences, and environment during the peripubertal years.

Although many of the metabolic sequelae of short-term fasting (such as the decline in circulating levels of glucose, amino acids, and insulin as well as the increase in plasma concentrations of free fatty acids and β-hydroxybutyrate) tend to be more pronounced in juveniles than in adult primates (483–485), these developmental differences do not, in general, appear to emerge until 12 hours or more into a fast. Nevertheless, the suggestion that fasting concentrations of circulating insulin in infants and adults are higher than those in children and juvenile monkeys (483–486) and therefore parallel the developmental pattern of hypothalamic GnRH pulse generator activity is of particular interest. The notion that insulin may serve as a metabolic trigger for puberty is reinforced by the proposal that in the context of the regulation of appetite, the primate hypothalamus has the capacity to monitor circulating insulin levels (487,488). Further credibility would be given to the foregoing speculation by the demonstration that fasting insulin levels also rise in agonadal children and monkeys concomitantly with the reawakening of the GnRH pulse generator.

With the exception of Steiner and colleagues, few investigators have attempted to address directly the foregoing metabolic hypotheses of puberty. In technically difficult studies, Steiner and colleagues (473,474) examined the gonadotropin response to a sustained intravenous infusion of glucose and amino acids administered to castrated male crab-eating monkeys between 18.5 and 31 months of age. This hyperalimentation, which resulted in a 75% increase in daily caloric intake and presumably stimulated insulin secretion, was associated with an elevation in circulating LH concentrations in approximately 50% of the animals studied. In evaluating this tantalizing observation, it should be noted that in the closely related rhesus macaque, gonadotropin secretion in prepubertally castrated males may be first detected as early as 2 years of age (54). It is therefore not inconceivable that spontaneous reawakening of the hypothalamic GnRH pulse generator may have been initiated during hyperalimentation in the older crab-eating macaques. The latter possibility would be excluded, of course, by a return of circulating LH to preinfusion control levels after termination of hyperalimentation, but such data were only reported for one animal. Unfortunately, these observations made more than 20 years ago have not been pursued.

Other Factors

Although adrenarche typically precedes puberty (6), the possibility that in humans a putative somatometer responds either directly or indirectly to maturation of the adrenal appears to be very unlikely. The most compelling evidence that may be cited for the latter view is the finding that the onset of gonadarche does not appear to be noticeably delayed in children with primary adrenal insufficiency present before the age at adrenarche (280). Conversely, premature adrenarche is not associated with any apparent signs of precocious gonadarche (280), and idiopathic GnRH dependent precocious gonadarche has been observed before adrenarche (7). From a comparative perspective, it is also difficult to argue that a rise in adrenal androgen secretion is a critical event in the pubertal reawakening of the hypothalamic GnRH pulse generator in higher primates, because adrenarche is not observed in baboons, macaques, mangabeys (10,11,53,56,489), and, presumably, other species of Old World monkeys.

Although exposure during prepubertal development to elevated androgens occasioned by adrenarche does not appear to underlie the pubertal reactivation of hypothalamic GnRH secretion, exposure to supraphysiological concentrations of plasma androgens during this phase of development may lead to premature reawakening of this neuroendocrine system. Young children with congenital adrenal hyperplasia may exhibit signs of true precocious puberty [e.g., sleep-augmented LH release (490)] or, more frequently, may progress into a state of true sexual maturity shortly after the adrenal pathology is diagnosed and excessive adrenal androgen production is prevented by corticosteroid replacement treatment (280, 491,492). Interestingly, long-term testosterone treatment of rhesus monkeys during the first year of life mimics the effects of adrenal hyperplasia in humans. In the female monkey, van Wagenen (493) found that the age of menarche was advanced by 1 year, and in the agonadal male monkey, we recently reported that early testosterone treatment led to a remarkably precocious pubertal increase in LH secretion (424). The mechanisms whereby supraphysiological and early exposure of androgens appear to result in the premature reawakening of the hypothalamic GnRH pulse generator are not known. Because treatment of agonadal male monkeys during the first year of life with high levels of estradiol or dihydrotestosterone (424) did not accelerate the onset of the pubertal resurgence in LH release, it seems reasonable to postulate that the effect of testosterone may be mediated by local aromatization within the specific hypothalamic areas. Alternatively, the influence of excess androgen on the timing of the pubertal process may be mediated indirectly by a metabolic signal induced by steroid action on peripheral structures (422). Thus, the appeal for the notion that attainment by whatever mechanism of a critical level of somatic maturation provides a trigger for the initiation of puberty is again reinforced.

Pubertal Clock

Only scant attention has been paid to the possibility that the timing of the onset of puberty is governed by an endogenous biological clock (309,494). Although there is no doubt that the mammalian CNS is endowed with timekeeping systems capable of tracking both circadian and circannual periods (495,496), the role, if any, of such mechanisms in dictating the ontogeny of the hypothalamic GnRH pulse generator has not been established. It is to be expected, however, that if such endogenous clocks are involved in the regulation of hypothalamic GnRH secretion throughout development, then selective damage to the timing mechanism would lead to a major alteration in the age at which puberty is initiated. In this regard, there is a considerable body of evidence, albeit obtained primarily from the study of nonprimate species, indicating that important timekeeping functions of the mammalian brain are subserved by the suprachiasmatic area of the hypothalamus (496). Attempts to dissociate the hypothalamic GnRH pulse generator in the prepubertal female rhesus monkey from the influence of more rostral brain regions, including the suprachiasmatic nucleus, by surgically transecting neural pathways between these two areas has not consistently resulted in a major alteration in the timing of menarche and other indices of puberty (382,383). However, as stated earlier, this surgical intervention in juvenile animals failed to abolish the diurnal rhythm in cortisol secretion.

Placement of lesions in the anterior hypothalamus of premenarcheal female rhesus monkeys has also been reported to be without effect on the age of menarche (497). In a subsequent study of the same problem, however, a significant number of prepubertal female monkeys with anterior hypothalamic lesions failed to thrive, and when data from these animals are excluded from numerical analysis, a lesion-induced advancement in menarche and first ovulation emerges (370). In humans, on the other hand, lesions of the anterior hypothalamus have generally been observed in association with delayed sexual maturation or hypogonadism (498). Moreover, the specificity of the neurosurgical procedure used for the placement of anterior lesions in the monkey is unclear, because similar lesions in the posterior hypothalamus also advance menarche and first ovulation in intact animals and the pubertal reinitiation of open-loop LH secretion in ovariectomized monkeys (370,499). The latter finding is consistent with the generally accepted view that posterior hypothalamic lesions that involve the tuber cinereum and mammillary bodies result in precocious puberty in humans (25,26,500). As discussed above, the puberty advancing action of some lesions may result from increased expression in the vicinity of the lesion of glial-derived growth factors, such as TGF-α and neuroregulins. Although further study of the sequelae of destructive hypothalamic lesions in the primate brain should not necessarily be discouraged, such an approach to date has not provided significant insight into the nature of the control system that times the pubertal reawakening of the hypothalamic GnRH pulse generator.

GENETICS AND PUBERTY GENES

As with other complex developmental processes, multiple genes are likely to modulate the onset and tempo of puberty (501). In this regard, the recognition that genetics plays a role in regulating the timing of gonadarche has been highlighted (502). Greater concordance in age of menarche and other indicators of the progression of gonadarche, such as peak height velocity and plasma LH concentrations, have generally been found in monozygotic twins when compared with dizygotic twins (503–510). The impact of genetics on the timing of gonadarche is further indicated by the findings of a high degree of heritability in the age of menarche (507,509) and, in the United States, of a significantly earlier age of menarche in African-American girls than their white counterparts (130,511–513). Similarly, girls in London of Afro-Caribbean and Indo-Pakistani parentage have an earlier age at menarche than white girls of British parentage (514). In addition, precocious gonadarche in girls may be familial and transmitted in an autosomal dominant mode (515). Attempts to identify genetic mutations or polymorphisms in such children, however, have not been fruitful (344,345). Although similar analyses of patients with delayed gonadarche due to constitutional delay of growth and puberty have not been reported to date, genotyping of individuals with this "normal" variant of pubertal development may be rewarding. Indeed, studies of a relatively large number of subjects with constitutional delay of growth and puberty indicated a genetic predisposition, as reflected by the finding that parents of children with constitutional delay of growth and puberty recall a later age of menarche and the adolescent growth spurt than those of a control group of children (516,517).

It should be noted, however, that when a genetic influence is established, it remains unclear whether the genotype is directly determining the timing of the pubertal resurgence of GnRH release [perhaps a clock gene in analogy to circadian timekeeping (518)] or simply governing permissive factors that are required for the resurgence of pulsatile GnRH release to be manifest once the "alarm" of a putative pubertal clock has sounded. Similarly, mutations of several genes involved in the differentiation and early development of the hypothalamic–pituitary– gonadal axis have been reported (519), and such mutations may have profound effects on pubertal development. However, as previously argued (301), such mutations may contribute little to our understanding of the control system responsible for the timing of gonadarche. For example, mutations of the *KAL-1* gene, which encodes anosmin, an extracellular glycoprotein that is necessary for the early embryonic migration of GnRH neurons into the hypothalamus (520), may lead to hypogonadotropic hypogonadism and absence of puberty. Although such mutations have provided significant insight into the early development of the reproductive axis, they have contributed little to our understanding of the control system that times the pubertal resurgence of pulsatile GnRH release.

We suggest therefore that the term "puberty" gene should be restricted to those genes that are specifically involved in the timing of either gonadarche or adrenarche. In the case of gonadarche, the proteins encoded by these genes would determine the age of the pubertal resurgence of pulsatile GnRH release by regulating the timing of the application or withdrawal of transsynaptic and/or glial inputs to the GnRH neuronal network of the juvenile hypothalamus. Conceivably, such putative puberty genes could time the resurgence of pulsatile GnRH release not only by triggering a hypothalamic signal at puberty but also by determining the duration of the prepubertal brake on pulsatile GnRH release or by timing the "turn off" of the GnRH pulse generator during infancy. The most likely, although theoretically not only, site for expression of such puberty genes would be the brain and probably the hypothalamus. In the case of adrenarche, the developmental increase in adrenal androgen secretion would be timed by genes that are expressed in the adrenal and unrelated to any brain genes controlling the onset of gonadarche. Despite the accelerating pace of elucidation into the genetic causes of disorders of pubertal development using genome-wide linkage scans, association studies, and haplotype-block approaches (521), specific puberty genes have yet to be identified. A gene that has recently attracted attention in this regard is *GPR54* (see above). However, the one case report of hypogonadotropism in an infantile boy with mutations of *GPR54* (378) suggests that the receptor encoded by this gene is probably critical for pulsatile GnRH release at all stages of development.

FETAL PROGRAMMING AND THE ONSET OF PUBERTY

Epidemiological observations and subsequent clinical and experimental data have demonstrated an association between intrauterine growth retardation and risk for several disorders in adulthood (522). In the case of pubertal development, adrenarche has been reported to be precocious and/or exaggerated in both boys and girls born small for gestational age. Some, but not all, girls with precocious adrenarche have an increased risk for ovarian hyperinsulinemic hyperandrogenism (early polycystic ovarian syndrome) at the time of gonadarche (522–524). The condition of polycystic ovarian syndrome refers to a heterogeneous familial disorder characterized by hyperandrogenism, chronic anovulation, hyperinsulinemia, and insulin resistance. The elevated androgen concentrations are due to excessive ovarian and/or adrenal androgen secretion. For some girls, premature pubarche is the first manifestation of early polycystic ovarian syndrome. In addition to elevated testosterone and/or androstenedione concentrations, data suggest that a notable neuroendocrine manifestation of polycystic ovarian syndrome, LH hypersecretion, is secondary to hyperandrogenism and develops several years postmenarche (525,526).

In contrast to adrenarche, the timing of menarche in girls born small for gestational age does not appear to be markedly different from girls born at an appropriate size for gestational age (527–529). In boys born small for gestational age, the timing of puberty appears to be normal, although subfertility has been reported postpubertally (529,530).

The role of fetal androgen in programming the neural component of the hypothalamic–pituitary–gonadal axis, and therefore potentially in timing the onset of gonadarche, is currently receiving considerable attention. Testosterone secretion by the fetal testis is presumable responsible for the greater frequency of pulsatile GnRH release in infant male monkeys compared with that in age-matched females (Fig. 1) and for the subsequent dimorphism in the developmental pattern of GnRH release as reflected by gonadotropin levels in agonadal monkeys and humans (Figs. 2 and 6). Prenatal exposure of female monkeys to elevated levels of androgens administered in large quantities to their mothers has been associated with increased LH pulse frequency, postnatal hyperandrogenism, insulin resistance, and decreased pancreatic beta cell function (531,532). These data suggest that the prenatal androgen milieu may influence the risk to develop polycystic ovarian syndrome (533). Additionally, androgen exposure early during gestation in the monkey has been reported to impair oocyte developmental competence (534). Children with congenital adrenal hyperplasia are exposed to elevated levels of androgens during fetal development, and although postnatally excessive androgen secretion may be controlled by glucocorticoid replacement, the timing of gonadarche and/or adrenarche in these patients may be perturbed. Moreover, although young women with congenital adrenal hyperplasia may exhibit regular menstrual cyclicity, fertility is frequently subnormal (535).

EXTEROCEPTIVE CUES AND THE TIMING OF PUBERTY

Season

Reproduction in many species of primates involves a seasonal pattern in breeding (536), which is presumably dictated, as in nonprimate species, by photoperiod and possibly other environmental cues (537) (see Chapter 41). The rhesus monkey living outdoors provides a striking example of a seasonal breeding primate: Ovulation generally occurs during the fall, and births are usually restricted to the spring (538). Thus, as previously noted by Rowell (102) and graphically demonstrated by Wilson et al. (441), the typical feral female experiences her first and second breeding season as an infant and juvenile of approximately 6 and 18 months of age, respectively; menarche occurs during the third breeding season, at approximately 30 months of age; and puberty is completed with first ovulation and conception by approximately 42 months of age during the fourth breeding season. Spring-born females that exhibit unusually early or late first ovulation do so 1 year before or after the age at which ovulation typically occurs (441). Moreover, when females that are born in the fall, because of contrived laboratory conditions, are reared from birth in an outdoor environment, first ovulation and conception typically occur in the fall at the end of the third year of life rather than at 3.5 years of age as in spring-born animals (441). Thus, under natural conditions, season imposes a quantum effect on the timing of puberty in the rhesus monkey, and, as previously remarked by Rowell (102), the onset of menarche or first ovulation is better predicted by counting time in breeding seasons than in months. An analogous quantal relationship between breeding season and the initiation of puberty has also been reported for male rhesus and bonnet macaques (50,539) and for male squirrel monkeys (51).

That the hypothalamic GnRH pulse generator is an important relay station in mediating the influence of season on the timing of puberty is indicated by the finding that testicular quiescence during the nonbreeding season may be reversed in the monkey by intermittent stimulation of the pituitary–gonadal axis with exogenous GnRH (540). The finding that menarche and first ovulation are advanced in spring-born female monkeys by imposing a short-day pattern of circulating melatonin during the long days of the summer preceding the third breeding season (472) suggests that the pineal gland is an important component of the interoceptive mechanism mediating the seasonal influence on the timing of the onset of puberty in this species.

Although the foregoing discussion underlines the robust effect seasonally related exteroceptive cues may have on the timing of the onset of puberty, it is important to remember that there is no evidence to indicate that such cues actually regulate the fundamental control system governing the onset of puberty, namely, the neurobiological brake that switches the GnRH pulse generator off in infancy and on at the termination of prepubertal development. It appears that irrespective of season, the latter control is independently programmed and the capacity for puberty is achieved independently of season. However, in a seasonal breeding species, puberty may only be manifest when the capacity for puberty and the breeding season coincide.

Social Factors

In nonprimate species, it is well established that social cues influence the timing of the onset of puberty (541). Cognate mechanisms appear extant in certain species of primates, and this is exemplified by the South American marmosets and tamarins. These small equatorial primates live in family groups consisting of a monogamous pair of adults and their offspring (110). Interestingly, the breeding female appears to suppress the ability of her adult daughters to reproduce (110,542), possibly by inhibiting ovulation (14). In Old World monkeys, social dominance also appears to be associated with earlier pubertal development in both males and females (221,543,544). The behavioral and neuroendocrine mechanisms responsible for this intriguing relationship between social dominance and the neuroendocrine axis governing reproductive function in callitrichids are unknown, but it is to be anticipated that modulation of the activity of the hypothalamic GnRH pulse generator is involved. Moreover, according to contemporary thinking, social subordination before puberty represents one of several "pediatric" stresses [for a comprehensive review of this subject see (545)] and as such any modulation on the GnRH pulse generator would be the result of activation of the hypothalamic–pituitary–adrenal axis (546). In humans, the role of psychosocial or socioemotional stressors generated by family dynamics and impacting on children, particularly girls, in timing the onset and tempo of puberty is an area of sociology receiving increasing attention (545,546). Interestingly, in this regard some presumed socioemotional stressors, for example, absence of the father, appear to be associated with an acceleration in the timing of puberty (545,547,548). Another interesting association between behavior and the onset of human puberty is the precocious gonadarche that has been reported in children, particularly girls, adopted by European parents from developing world regions, such as Asia and South America (549). Ethnic background does not appear to be involved, and a similar phenomenon has also been observed in immigrant girls arriving with their parents (550). The mechanism responsible for the association between adoption or migration and puberty in girls is unclear. The implication that endocrine disruptors may be involved in this form of precocity (550), however, suggests that the phenomenon may be unrelated to social factors.

Food Availability

In mammalian species in general, food availability (the balance between energy intake and expenditure) is another major environmental factor modulating the timing of sexual maturation (126,545,551) (see Chapter 47 and Chapter 39). In the human female, the increased metabolic demand that may be imposed by strenuous exercise such as long-distance running, gymnastics, and ballet is often associated with delayed or interrupted pubertal development (552,553). In such individuals, in whom deviation from ideal body weight may be minimal, the frequency of pulsatile LH secretion is less than that in sedentary women during the follicular phase of the menstrual cycle (554–557). Male wrestlers lose body weight, fat mass, and lean body mass during the competitive season, and disruption of the neuroendocrine axes governing testicular function and growth may be observed at this time (558). These changes reverse within months after the conclusion of the wrestling season. Together these findings suggest that metabolic signals, at least those associated with restricted food intake and/or exercise, have the capacity to impair the activity of the hypothalamic GnRH pulse generator.

Chronic disease is often associated with inadequate nutrition leading to impaired GnRH secretion and delayed or interrupted puberty. Children with sickle cell anemia, perinatally acquired human immunodeficiency virus infection, diabetes mellitus, chronic renal disease, and inflammatory bowel disease tend to have delayed puberty (559–561). Among girls with cystic fibrosis, age at menarche was delayed but correlated with nutritional status and maternal age at menarche (562). Effective treatment of the underlying disorder is generally associated with restored pubertal progression. For example, after successful renal transplantation, children with chronic renal disease manifest resumption of gonadotropin secretion and progressive pubertal development.

Undernutrition, or restricted nutrition with intense exercise, may delay the onset of puberty in primates in one of two ways. First, growth is impaired, and if, as proposed earlier, a somatometer governs the postnatal ontogeny of the hypothalamic GnRH pulse generator, a retardation in growth will be perceived by this control system, and the reawakening of the pulse generator will be delayed. Alternatively, the manifestation of puberty in the undernourished juvenile may be occasioned by the same mechanisms that underlie restoration of reproductive function in malnourished adults. Restricted food intake, imposed experimentally in the rhesus monkey or voluntarily in human subjects with anorexia nervosa, appears to inhibit gonadotropin secretion primarily by an action at a suprapituitary level that is mediated by suppression of intermittent hypothalamic GnRH secretion (479,563). Thus, in the undernourished juvenile, the same inhibitory mechanisms that arrest the hypothalamic GnRH pulse generator in food-restricted adults are presumably activated, and therefore any waning of the neurobiological brake on intermittent GnRH secretion in the juvenile would not become manifest because of the additional and independent check imposed by restricted food intake. Thus, gonadarche—the consequence of the reawakening of the GnRH pulse generator—is further delayed.

DISORDERS OF PUBERTAL DEVELOPMENT

To date, with the exception of one report in the chimpanzee (564), disorders of puberty in primates have only been described for humans, and these have been recently reviewed in detail (27). In the present chapter, we restrict our discussion to those particular disorders that, when their etiology is fully established, are most likely to add major advances to the understanding of the physiology of primate puberty. Because the key event in the onset of puberty is the resurgence of pulsatile GnRH release at gonadarche and because the neurobiology underlying this phenomenon remains a fascinating mystery, examination of the genetic bases of GnRH dependent disorders of human development will be the most revealing. This is not to say that GnRH independent disorders of pubertal development are neither of physiological interest nor of clinical importance. In fact, disorders of gonadal and adrenal steroidogenesis or the failure of the pituitary gland to undergo normal development may have profound impacts on growth and sexual development.

Disorders of human pubertal development fall into the broad categories of precocious and delayed puberty and both may be either familial or sporadic. In the former, patterns of inheritance include autosomal recessive, autosomal dominant, and X-linked recessive.

Precocious Puberty

Precocious puberty may be classified as being either GnRH dependent or GnRH independent.

GnRH dependent precocious puberty, or more accurately precocious gonadarche, is the result of a premature resurgence or incomplete suppression of the hypothalamic GnRH pulse generator. GnRH independent forms represent a group of disorders associated with increased sex steroid secretion that occurs in the absence of GnRH stimulated gonadotropin secretion.

In GnRH dependent precocious gonadarche, the normal pubertal sequence of hypothalamic events appears to be recapitulated, but the process starts at an inappropriately early age. This disorder is diagnosed more often in girls than in boys, perhaps because suppression of the GnRH pulse generator during childhood and juvenile development in girls is less robust than in boys (see above). Although often considered to be idiopathic in girls, an organic etiology can usually be identified in boys. Despite the relatively mature physical appearance of children with precocity, their cognitive and emotional development is normal for chronological age. Thus, the guilelessness and naïveté of such children exposes them to an increased risk to suffer sexual abuse with affected girls at risk of becoming pregnant.

Hypothalamic hamartomas, congenital malformations comprised of heterotropic neural tissue usually located on the floor of the third ventricle or attached to the tuber cinereum, are a common etiology of precocious gonadarche (500,565). The tumors can be classified as parahypothalamic when attached or suspended from the floor of the third ventricle or as intrahypothalamic, in which the mass is enveloped by the hypothalamus and distorts the third ventricle. Hamartomas do not grow over time and do not metastasize. Although gelastic or laughing seizures can be associated with precocious puberty due to hypothalamic hamartomas, most patients do not exhibit neurological symptoms. Interestingly, histological examination of hypothalamic hamartoma tissue has shown immunoactive GnRH perikarya in association with capillary networks (565–568) and/or astroglial-derived growth factors such as TGF-α (372). The surgical removal of such small peduncled hamartomas in infants with precocious puberty has been reported to result in a complete arrest of sexual development (565,568). The intriguing possibility that such hamartomas contain an ancillary GnRH pulse generator that is emancipated from normal developmental controls must therefore be considered. On the other hand, astroglial factors such as TGF-α may provide an ectopic drive to GnRH neurons with a normal distribution in the hypothalamus.

Other CNS lesions, hydrocephalus, meningomyelocele and certain neurodevelopmental disorders, previous head trauma, CNS infections, and CNS radiation may also be associated with GnRH dependent precocious gonadarche (569,570). CNS radiation used to treat intracranial tumors or used prophylactically for malignancies can induce precocious GnRH dependent gonadarche, perhaps as a result of an astroglial response with increased TGF-α production (371).

Precocious GnRH dependent gonadarche may also be observed in children with virilizing disorders such as congenital adrenal hyperplasia and familial male-limited precocious puberty (testotoxicosis) in whom skeletal maturation is usually significantly advanced. This particular pathophysiological condition may be replicated experimentally in the monkey by exogenous testosterone administration (424,493). Whether the action of testosterone to prematurely reawaken the GnRH pulse generator is exerted on the CNS or on peripheral tissue, such as, for example, the skeleton, is unclear.

The treatment of choice for children with GnRH dependent precocious gonadarche is a GnRH receptor agonist (571). GnRH receptor agonists are modifications of GnRH having greater resistance to degradation and increased affinity for the GnRH receptor than the native decapeptide. They are therefore perceived by the pituitary as a continuous GnRH stimulation, which induces down-regulation of GnRH receptor function and leads to decreased gonadotropin secretion. GnRH receptor agonists are available as daily injections or depot forms that are currently administered every 28 days. The major goals of treatment are to prevent further pubertal progression until appropriate for chronological age and to attain normal adult height. In situations of coexisting GH deficiency, combined treatment with recombinant human GH may be helpful to preserve height potential.

Delayed Puberty

In the context of understanding the timing of normal puberty, constitutional delay of growth and puberty is of most interest, because this condition, which is currently considered an extreme of the normal variation of pubertal timing, aggregates in families and therefore must have a genetic basis (see above). Typically, both gonadarche and adrenarche are delayed.

Delayed or absent pubertal development secondary to hypogonadotropic hypogonadism is reported to affect 1 in 7,500 males and 1 in 70,000 females (572), and investigations of families with this disorder have identified several underlying monogenic mutations, specifically in the *KAL-1*, *FGFR1*, *GPR54*, and *GnRHR* genes (354,355,520,573,574). Because the onset of the hypogonadotropic state resulting from mutations of *KAL-1*, *FGFR1*, *GnRHR*, and probably

GPR54 (378) occurs early in fetal development, these genes have not been posited to be specifically involved in the resurgence in pulsatile GnRH release at the end of the juvenile phase of development.

CONCLUSION

The study of puberty in primates has been restricted almost exclusively to the infraorder *Catarrhina*. In such higher primates, puberty may be viewed as manifesting itself as a result of the cascade of physiological and behavioral processes that are set in motion by the reawakening of a restrained but functional hypothalamic GnRH pulse generator. In males, maturation of the neural timing mechanism that governs pulsatile GnRH secretion appears to be completed during fetal development, but in the female the ontogeny of this system may not be completed until after birth. The protracted brake on GnRH secretion from late infancy until puberty that is a developmental hallmark of higher primates is imposed primarily by extragonadal mechanisms. The rising circulating concentrations of sex steroids produced as a result of gonadarche, however, retard the pubertal resurgence in pulsatile GnRH release that, in the absence of gonadal steroids, appears to occur explosively. Thus, the tempo of puberty is dictated by an interplay of central and gonadal factors. Although physiological models[8] with a central neural timekeeping system or growth-tracking device have been proposed to account for the timing of puberty, the identity of the control systems that dictate the arrest and reawakening of the hypothalamic GnRH pulse generator during postnatal life remains a fundamental problem in developmental biology. Similarly, the neurobiological components that underlie the resurgence of pulsatile GnRH release late in juvenile development and therefore that trigger the onset of puberty are only partially understood. Contemporary conceptualization of this event, however, involves a component of structural and molecular remodeling (plasticity) within the hypothalamic area containing the GnRH neurons. From a comparative perspective, studies of New World monkeys and marmosets, *Tarsius*, and tree shrews and lemurs are scant, and therefore it is not presently possible to speculate whether the timing of puberty in the lower primates is subserved by the same mechanisms that govern this process in Old World monkeys, apes, and humans.

[8]Attempts have also been made to construct mathematical models to account for the timing of the onset of puberty in primates (575,576).

ACKNOWLEDGMENTS

Work from the authors' laboratories described in this chapter was supported by the National Institutes of Health grants U54-HD08610 (to T. M. P.), R01 HD13254 (to T. M. P.), R01 HD16851 (to T. M. P.), R29-HD34808 (to S. F. W.), and 5M01-RR-00084 (General Clinical Research Center).

REFERENCES

1. Romer, A. S. (1959). *The Vertebrate Story*. University of Chicago Press, Chicago.
2. Donovan, B. T., and Van der Werff ten Bosch, J. J. (1965). *Physiology of Puberty*. Edward Arnold, London.
3. Ojeda, S. R., Andrews, W. W., Advis, J. P., and Smith White, S. (1980). Recent advances in the endocrinology of puberty. *Endocr. Rev*. 1, 228–257.
4. Gove, P. B., Ed. (1961). *Webster's Third New International Dictionary of the English Language Unabridged*. G. & C. Merriam, Springfield, IL.
5. Talbot, N. B., Sobel, E. H., McArthur, J. W., and Crawford, J. D. (1952). *Funtional Endocrinology From Birth through Adolescence*. Harvard University Press, Cambridge.
6. Reiter, E. O., Fuldauer, V. G., and Root, A. W. (1977). Secretion of the adrenal androgen, dehydroepiandrosterone sulfate, during normal infancy, childhood, and adolescence, in sick infants, and in children with endocrinologic abnormalities. *J. Pediatr*. 90, 766–770.
7. Sklar, C. A., Kaplan, S. L., and Grumbach, M. M. (1980). Evidence for dissociation between adrenarche and gonadarche: studies in patients with idiopathic precocious puberty, gonadal dysgenesis, isolated gonadotropin deficiency, and constitutionally delayed growth and adolescence. *J. Clin. Endocrinol. Metab*. 51, 548–556.
8. Korth-Schutz, S., Levine, L. S., and New, M. I. (1976). Serum androgens in normal prepubertal and pubertal children and in children with precocious adrenarche. *J. Clin. Endocrinol. Metab*. 42, 117–124.
9. Parker, L. N., Sack, J., Fisher, D. A., and Odell, W. D. (1978). The adrenarche: prolactin, gonadotropins, adrenal androgens, and cortisol. *J. Clin. Endocrinol. Metab*. 416, 396–401.
10. Cutler, G. B. Jr., Glenn, M., Bush, M., Hodgen, G. D., Graham, C. E., and Loriaux, D. L. (1978). Adrenarche: a survey of rodents, domestic animals, and primates. *Endocrinology* 103, 2112–2118.
11. Smail, P. J., Faiman, C., Hobson, W. C., Fuller, G. B., and Winter, J. S. D. (1982). Further studies on adrenarche in nonhuman primates. *Endocrinology* 111, 844–848.
12. Counts, D. R., Pescovitz, O. H., Barnes, K. M., Hench, K. D., Chrousos, G. P., Sherins, R. J., Comite, F., Loriaux, D. L., and Cutler, G. B. Jr. (1987). Dissociation of adrenarche and gonadarche in precocious puberty and in isolated hypogonadotropic hypogonadism. *J. Clin. Endocrinol. Metab*. 64, 1174–1178.
13. Grumbach, M. M., and Styne, D. M. (2003). Puberty: Ontogeny, neuroendocrinology, physiology, and disorders. In *Williams Textbook of Endocrinology* (P. R. Larsen, H. M. Kronenberg, S. Melmed, and K. S. Polonsky, Eds.), pp. 1115–1286. W. B. Saunders, Philadelphia.
14. Abbott, D. H., and Hearn, J. P. (1978). Physical, hormonal and behavioural aspects of sexual development in the marmoset monkey, *Callithrix jacchus*. *J. Reprod. Fertil*. 53, 155–166.

15. Bogin, B. (1995). Growth and development: recent evolutionary and biocultural research. In *Biological Anthropology: The State of the Science* (N. T. Boaz , and L. D. Wolfe, Eds.), pp. 49–70. International Institute for Human Evolutionary Research, Bend, OR.
16. Mack, W. S. (1964). Ruminations on the testis. *Proc. Soc. Med.* 57, 47–51.
17. Donovan, B. T. (1963). The timing of puberty. In *The Scientific Basis of Medicine, Annual Reviews 1963* (British Postgraduate Medical Foundation, Ed.), pp. 53–75. The Athlone Press, University of London, London.
18. Foa, C. (1900). La greffe des ovaries, en relation avec quelques questions de biologic generate. *Arch. Ital. Biol.* 34, 43–73.
19. Lipschutz, A., Kallas, H., and Paez, R. (1929). Hypophyse und Gesetz der Pubertat. *Pflugers Arch.* 221, 695–712.
20. Crowe, S. J., Gushing, H., and Homans, J. (1910). Experimental hypophysectomy. *Bull. Johns Hopkins Hosp.* 21, 127–169.
21. Smith, P. E., and Engle, E. T. (1927). Experimental evidence regarding the role of the anterior pituitary in the development and regulation of the genital system. *Am. J. Anat.* 40, 159–217.
22. Harris, G. W. (1955). *Neural Control of the Pituitary Gland.* Edward Arnold, London.
23. Harris, G. W., and Jacobsohn, D. (1952). Functional grafts of the anterior pituitary gland. *Proc. R. Soc. Lond. Biol.* 139, 263–276.
24. Plant, T. M., Gay, V. L., Marshall, G. R., and Arslan, M. (1989). Puberty in monkeys is triggered by chemical stimulation of the hypothalamus. *Proc. Natl. Acad. Sci. U S A* 86, 2506–2510.
25. Weinberger, L. M., and Grant, F. C. (1941). Precocious puberty and tumors of the hypothalamus. *Arch. Intern. Med.* 67, 762–792.
26. Barnes, N. D., Cloutier, M. D., and Hayles, A. B. (1972). The central nervous system and precocious puberty. In *Control of the Onset of Puberty* (M. M. Grumbach, G. D. Grave, and F. E. Mayer, Eds.), pp. 213–228. John Wiley & Sons, New York.
27. Witchel, S. F, and Plant, T. M. (2004). Puberty: gonadarche and adrenarche. In *Yen and Jaffe's Reproductive Endocrinology* (J. F. Strauss III and R. L. Barbieri, Eds.), pp. 493–535. Elesevier, Saunders, Philadelphia.
28. Schonfeld, W. A. (1943). Primary and secondary sexual characteristics. Study of their development in males from birth through maturity with biometric study of penis and testis. *Am. J. Dis. Child.* 65, 535–549.
29. Winter, J. S. D., and Faiman, C. (1972). Pituitary-gonadal relations in male children and adolescents. *Pediatr. Res.* 6, 126–135.
30. Marshall, W. A. (1978). Puberty. In *Human Growth, Vol. 2: Postnatal Growth* (F. Falkner and J. M. Tanner, Eds.), pp. 141–181. Plenum Press, New York.
31. Kraemer, H. C., Horvat, J. R., Doering, C., and McGinnis, R. R. (1982). Male chimpanzee development focusing on adolescence: integration of behavioral with physiological changes. *Primates* 23, 393–405.
32 . Copeland, K. C., Eichberg, J. W., Parker, C. R. Jr., and Bartke, A. (1985). Puberty in the chimpanzee: somatomedin-C and its relationship to somatic growth and steroid hormone concentrations. *J. Clin. Endocrinol. Metab.* 60, 1154–1160.
33. Marson, J., Meuris, S., Cooper, R. W., and Jouannet, P. (1991). Puberty in the male chimpanzee: progressive maturation of semen characteristics. *Biol. Reprod.* 44, 448–455.
34. Marson, J., Meuris, S., Cooper, R. W., and Jouannet, P. (1991). Puberty in the male chimpanzee: time-related variations in luteinizing hormone, follicle-stimulating hormone, and testosterone. *Biol. Reprod.* 44, 456–460.
35. Copeland, K. C., Kuehl, T. J., Reyes, P., and Castracane, V. D. (1981). The baboon as a model for puberty: growth, testis size, plasma, testosterone and somatomedin-C. *Pediatr. Res.* 15, 1547 (abst. 56).
36. Steiner, R. A., and Bremner, W. J. (1981). Endocrine correlates of sexual development in the male monkey, *Macaca fascicularis. Endocrinology* 109, 914–919.
37. Bercu, B. B., Lee, B. C., Pineda, J. L., Spiliotis, B. E., Denman, D. W. III, Hoffman, H. J., Brown, T. J., and Sachs, H. C. (1983). Male sexual development in the monkey. I. Cross-sectional analysis of pulsatile hypothalamic-pituitary-testicular function. *J. Clin. Endocrinol. Metab.* 56, 1214–1226.
38. Aravindan, G. R., Ravindranath, N., Gopalakrishnan, K., and Moudgal, N. R. (1990). DNA flow-cytometric analysis of testicular germ cell populations of the bonnet monkey (*Macaca radiata*) as a function of sexual maturity. *J. Reprod. Fertil.* 89, 397–406.
39. Wickings, E. J, and Dixson, A. F. (1992). Testicular function, secondary sexual development, and social status in male mandrills (*Mandrillus sphinx*). *Physiol. Behav.* 52, 909–916.
40. Dixson, A. F., Gardner, J. S., and Bonney, R. C. (1980). Puberty in the male owl monkey (*Aotus trivirgatus griseimembrd*): a study of physical and hormonal development. *Int. J. Primatol.* 1, 129–139.
41. Rey, R. A., Campo, S. M., Bedecarrás, P., Nagle, C. A., and Chemes, H. E. (1993). Is infancy a quiescent period of testicular development? Histological, morphometric, and functional study of the seminiferous tubules of the Cebus monkey from birth to the end of puberty. *J. Clin. Endocrinol. Metab.* 76, 1325–1331.
42. Sniffen, R. C. (1950). The testis I. The normal testis. *Arch. Pathol.* 50, 259–284.
43. Albert, A., Underdahl, L. O., Greene, L. F., and Lorenz, N. (1953). Male hypogonadism. I. The normal testis. *Proc. Mayo Clin.* 28, 409–422.
44. Van Wagenen, G., and Simpson, M. E. (1954). Testicular development in the rhesus monkey. *Anat. Rec.* 118, 231–251.
45. Mancini, R. E., Vilar, O., Lavieri, J. C., Andrada, J. A., and Heinrich, J. J. (1963). Development of Leydig cells in the normal testis. A cytological, cytochemical and quantitative study. *Am. J. Anat.* 112, 203–214.
46. Fouquet, J. P., Meusy-Dessolle, N., and Dang, D. C. (1984). Relationship between Leydig cell morphometry and plasma testosterone during postnatal development of the monkey *Macaca fascicularis. Reprod. Nutr. Dev.* 24, 281–296.
47. Frasier, S. D., Gafford, F., and Horton, R. (1969). Plasma androgens in childhood and adolescence. *J. Clin. Endocrinol. Metab.* 29, 1404–1408.
48. August, G. P., Grumbach, M. M., and Kaplan, S. L. (1972). Hormonal changes in puberty. III. Correlation of plasma testosterone, LH, FSH, testicular size, and bone age with male pubertal development. *J. Clin. Endocrinol. Metab.* 34, 319–326.
49. Martin, D. E., Swenson, R. B., and Collins, D. C. (1977). Correlation of serum testosterone levels with age in male chimpanzees. *Steroids* 29, 471–481.
50. Rose, R. M., Bernstein, I, S., Gordon, T. P., and Lindsley, J. G. (1978). Changes in testosterone and behavior during adolescence in the male rhesus monkey. *Psychosom. Med.* 40, 60–70.
51. Coe, C. L., Chen, J., Lowe, E. L., Davidson, J. M., and Levine, S. (1981). Hormonal and behavioral changes at puberty in the squirrel monkey. *Horm. Behav* . 15, 36–53.
52. Copeland, K. C., Kuehl, T. J., and Castracane, V. D. (1982). Pubertal endocrinology of the baboon: elevated somatomedin-C/insulin-like growth factor I at puberty. *J. Clin. Endocrinol. Metab.* 55, 1198–1201.
53. Mann, D. R., Castracane, V., McLaughlin, F., Gould, K. G., and Collins, D. C. (1983). Developmental patterns of serum luteinizing hormone, gonadal and adrenal steroids in the sooty mangabey (*Cercocebus atys*). *Biol. Reprod.* 28, 279–284.

54. Plant, T. M. (1985). A study of the role of the postnatal testes in determining the ontogeny of gonadotropin secretion in the male rhesus monkey (*Macaca mulatta*). *Endocrinology* 116, 1341–1350.
55. Brady, A. G., Koritnik, D. R., and Stevens, S. W. (1985). Hormonal responses to gonadotropin releasing hormone during testicular development in male African green monkeys. *J. Med. Primatol.* 14, 293–304.
56. Meusy-Dessolle, N., and Dang, D. C. (1985). Plasma concentrations of testosterone, dihydrotestosterone, Δ_4-androstenedione, dehydroepiandrosterone and estradiol-17β in the crab-eating monkey (*Macaca fascicularis*) from birth to adulthood. *J. Reprod. Fertil.* 74, 347–359.
57. Arslan, M., Mahmood, S., Khurshid, S., Naqvi, S. M. S., Afzal, M. A., and Baig, S. M. (1986). Changes in circulating levels of immunoreactive follicle stimulating hormone, luteinizing hormone, and testosterone during sexual development in the rhesus monkey, *Macaca mulatta*. *J. Med. Primatol.* 15, 351–359.
58. Lunn, S. F., Recio, R., Morris, K., and Fraser, H. M. (1994). Blockade of the neonatal rise in testosterone by a gonadotrophin-releasing hormone antagonist: effects on timing of puberty and sexual behaviour in the male marmoset monkey. *J. Endocrinol.* 141, 439–447.
59. Boyar, R. M., Rosenfeld, R. S., Kapen, S., Finkelstein, J. W., Roffwarg, H. P., Weitzman, E. D., and Hellman, L. (1974). Human puberty. Simultaneous augmented secretion of luteinizing hormone and testosterone during sleep. *Clin. Invest.* 54, 609–618.
60. Judd, H. L., Parker, D. C., Siler, T. M., and Yen, S. S. C. (1974). The nocturnal rise of plasma testosterone in pubertal boys. *J. Clin. Endocrinol. Metab.* 38, 710–713.
61. Wu, F. C. W., Borrow, S. M., Nicol, K., Elton, R., and Hunter, W. M. (1989). Ontogeny of pulsatile gonadotrophin secretion and pituitary responsiveness in male puberty in man: a mixed longitudinal and cross-sectional study. *J. Endocrinol.* 123, 347–359.
62. Knorr, D., Bidlingmaier, F., Butenandt, O., Fendel, H., and Ehrt-Wehle, R. (1974). Plasma testosterone in male puberty. *Acta Endocrinol.* 75, 181–194.
63. Lee, P. A., Jaffe, R. B., and Midgley, A. R. Jr. (1974). Serum gonadotropin, testosterone and prolactin concentrations throughout puberty in boys: a longitudinal study. *J. Clin. Endocrinol. Metab.* 39, 664–672.
64. Mancini, R. E., Narbaitz, R., and Lavieri, J. C. (1960). Origin and development of the germinative epithelium and Sertoli cells in the human testis: cytological, cytochemical, and quantitative study. *Anat. Rec.* 136, 477–483.
65. Dang, D. C., and Meusy-Dessolle, N. (1984). Quantitative study of testis histology and plasma androgens at onset of spermatogenesis in the prepubertal laboratory-born macaque (*Macaca fascicularis*). *Arch. Androl.* 12, 43–51.
66. Simorangkir, D. R., Marshall, G. R., and Plant T. M. (2003). Sertoli cell proliferation during prepubertal development in the rhesus monkey (*Macaca mulatta*) is maximal during infancy when gonadotropin secretion is robust. *J. Clin. Endocrinol. Metab.* 88, 4984–4989.
67. Sharpe, R. M., McKinnell, C., Kivlin, C., and Fisher, J. S. (2003). Proliferation and functional maturation of Sertoli cells, and their relevance to disorders of testis function in adulthood. *Reproduction* 125, 769–784.
68. Richardson, D. W., and Short, R. V. (1978). Fertility in adolescence. Time of onset of sperm production in boys. *J. Biosoc. Sci.* 5, 15–25.
69. Nielsen, C. T., Skakkebaek, N. E., Richardson, D. W., Darling, J. A., Hunter, W. M., Jorgensen, M., Nielsen, A., Ingerslev, O., Keiding, N., and Muller, J. (1986). Onset of the release of spermatozoa (spermarche) in boys in relation to age, testicular growth, pubic hair and height. *J. Clin. Endocrinol. Metab.* 62, 532–535.
70. Ginther, A. J., Carlson, A. A., Ziegler, T. E., and Snowdon, C. T. (2002). Neonatal and pubertal development in males of a cooperatively breeding primate, the cotton-top tamarin (*Saguius Oedipus Oedipus*). *Biol. Reprod.* 66, 282–290.
71. Van Wagenen, G., and Simpson, M. E. (1973). *Postnatal Development of the Ovary in* Homo sapiens *and* Macaca mulatta. Yale University Press, New Haven, CT.
72. Ross, G. T. (1976). Gonadotropins and preantral follicular maturation in women. *Fertil. Steril.* 25, 522–543.
73. Peters, H., Himelstein-Braw, R., and Faber, M. (1976). The normal development of the ovary in childhood. *Acta Endocrinol.* 82, 617–630.
74. Herter, L. D., Golendziner, E., Flores, J. A., Becker, E. Jr., and Spritzer, P. M. (2002). Ovarian and uterine sonography in healthy girls between 1 and 13 years old: correlation of findings with age and pubertal status. *Am. J. Roentgenol.* 178, 1531–1536.
75. Sforza, C., Ranzi, A., Ferrario, V. F., and Forabosco, A. (2004). Growth patterns of human ovarian volume during intrauterine and postnatal organogenesis. *Early Hum. Dev.* 80, 7–17.
76. Forabosco, A., Sforza, C., De Pol, A., Vizzotto, L., Marzona, L., and Ferrario, V. F. (1991). Morphometric study of the human neonatal ovary. *Anat. Rec.* 231, 201–208.
77. De Pol, A., Vaccina, F., Forabosco, A., Cavazzuti E., and Marzona, L. (1997). Apoptosis of germ cells during human prenatal oogenesis. *Hum. Reprod.* 12, 2235–2241.
78. Giorlandino, C., Gleicher, N., Taramanni, C., Vizzone, A., Gentill, P., Mancuso, S., and Forleo, R. (1989). Ovarian development of the female child and adolescent: I Morphology. *Int. J. Gynaecol. Obstet.* 29, 57–63.
79. Williams, R. F., Turner, C. K., and Hodgen, G. D. (1982). The late pubertal cascade in perimenarcheal monkeys: onset of asymmetrical ovarian estradiol secretion and bioassayable luteinizing hormone release. *J. Clin. Endocrinol. Metab.* 55, 660–665.
80. Winter, J. S. D., Ellsworth, L., Fuller, G., Hobson, W. C., Reyes, F. I., and Faiman, C. (1987). The role of gonadal steroids in feedback regulation of gonadotropin secretion at different stages of primate development. *Acta Endocrinol.* 14, 257–268.
81. Klein, K. O., Baron, J., Colli, M. J., McDonnell, D. P., and Cutler, G. B. Jr. (1994). Estrogen levels in childhood determined by an ultrasensitive recombinant cell bioassay. *J. Clin. Invest.* 94, 2475–2480.
82. Paris, F., Servant, N., Térouanne, B., Balaguer, P., Nicolas, J. C., and Sultan, C. (2002). A new recombinant cell bioassay for ultrasensitive determination of serum estrogenic bioactivity in children. *J. Clin. Endocrinol. Metab.* 87, 791–797.
83. Jenner, M. R., Kelch, R. P., Kaplan, S. L., and Grumbach, M. M. (1972). Hormonal changes in puberty. IV. Plasma estradiol, LH, and FSH in prepubertal children, pubertal females, and in precocious puberty, premature thelarche, hypogonadism and in a child with a feminizing ovarian tumor. *J. Clin. Endocrinol. Metab.* 34, 521–530.
84. Bidlingmaier, F., Wagner-Barnach, M., Butenandt, O., and Kron, D. (1973). Plasma estrogen in childhood and puberty under physiologic and pathologic conditions. *Pediatr. Res.* 7, 901–907.
85. Winter, J. S. D., and Faiman, C. (1973). Pituitary-gonadal relations in female children and adolescents. *Pediatr. Res.* 7, 948–953.
86. Holm, K., Laursen, E. M., Brocks, V., and Müller, J. (1995). Pubertal maturation of the internal genitalia: an ultrasound evaluation of 166 healthy girls. *Ultrasound Obstet. Gynecol.* 6, 175–181.

87. Terasawa, E., Nass, T. E., Yeoman, R. R., Loose, M. D., and Schultz, N. J. (1983). Hypothalamic control of puberty in the female rhesus macaque. In *Neuroendocrine Aspects of Reproduction* (R. L. Norman, Ed.), pp. 149–182. Academic Press, New York.
88. Wilson, M. E., Gordon, T. P., and Collins, D. C. (1986). Ontogeny of luteinizing hormone secretion and first ovulation in seasonal breeding rhesus monkeys. *Endocrinology* 118, 293–301.
89. Anderson, S. E., Dallal, G. E., and Must, A. (2003). Relative weight and race influence average age at menarche: results from two nationally representative surveys of US girls studied 25 years apart. *Pediatrics* 111, 844–850.
90. Hartman, C. G. (1931). On the relative sterility of the adolescent organism. *Science* 14, 226–221.
91. Treloar, A. E., Boynton, R. E., Behn, B. G., and Brown, B. W. (1967). Variation of the human menstrual cycle through reproductive life. *Int. J. Fertil.* 12, 77–126.
92. Doring, G. K. (1969). The incidence of anovular cycles in women. *J. Reprod. Fertil.* 6, 77–81.
93. Vollman, R. F. (1977). *The Menstrual Cycle*. W. B. Saunders, Philadelphia.
94. Apter, D., Viinikka, L., and Vihko, R. (1978). Hormonal pattern of adolescent menstrual cycles. *J. Clin. Endocrinol. Metab.* 47, 944–954.
95. Metcalf, M. G., Skidmore, D. S., Lowry, G. F., and Mackenzie, J. A. (1983). Incidence of ovulation in the years after the menarche. *J. Endocrinol.* 97, 213–219.
96. Young, W. C., and Yerkes, R. M. (1943). Factors influencing the reproductive cycle in the chimpanzee: the period of adolescent sterility and related problems. *Endocrinology* 33, 121–154.
97. Tutin, C. E. G. (1980). Reproductive behaviour of wild chimpanzees in the Gombe National Park, Tanzania. *J. Reprod. Fertil.* 28, 43–57.
98. Watts, E. S., and Gavan, J. A. (1982). Postnatal growth of nonhuman primates: the problem of the adolescent spurt. *Hum. Biol.* 54, 53–70.
99. Harcourt, A. H., Fossey, D., Stewart, K. J., and Watts, D. P. (1980). Reproduction in wild gorillas and some comparison with chimpanzees. *J. Reprod. Fertil.* 28, 59–70.
100. Gillman, J., and Gilbert, C. (1946). The reproductive cycle of the chacma baboon (*Papio ursinus*) with special reference to the problems of menstrual irregularities as assessed by the behaviour of the sex skin. *S. Afr. J. Med. Sci.* 11, 1–54.
101. Glassman, D. M., Coelho, A. M. Jr., Casey, K. D., and Bramblett, C. A. (1948). Weight growth in savannah baboons: a longitudinal study from birth to adulthood. *Growth* 48, 425–433.
102. Rowell, T. E. (1977). Variation in age at puberty in monkeys. *Folia. Primatol.* 27, 284–290.
103. Erwin, N., and Erwin, J. (1976). Age of menarche in pig-tail monkeys (*Macaca nemestrina*): a cross-sectional survey of perineal tumescence. *Theiriogenology* 5, 261–266.
104. Wilen, R., and Naftolin, F. (1976). Age, weight and weight gain in the individual pubertal female rhesus monkey (*Macca mulatta*). *Biol. Reprod.* 15, 356–360.
105. Terasawa, E., and Fernandez, D. L. (2001). Neurobiological mechanisms of the onset of puberty in primates. *Endocrinol. Rev.* 22, 111–151.
106. Honjo, S., Cho, F., and Terao, K. (1984). Establishing the cynomolgus monkey as a laboratory animal. *Adv. Vet. Sci. Comp. Med.* 28, 51–80.
107. Winkler, P., Loch, H., and Vogel, C. (1948). Life history of hanuman langurs (*Presbytis entellus*). *Folia. Primatol.* 43, l–23.
108. Foster, D. L. (1977). Luteinizing hormone and progesterone secretion during sexual maturation of the rhesus monkey: short luteal phases during the initial menstrual cycles. *Biol. Reprod.* 17, 584–590.
109. Resko, J. A., Goy, R. W., Robinson, J. A., and Norman, R. L. (1982). The pubescent rhesus monkey: some characteristics of the menstrual cycle. *Biol. Reprod.* 27, 354–361.
110. Epple, G., and Katz, Y. (1980). Social influences on first reproductive success and related behaviors in the saddle-back tamarin (*Saquinus fuscicollis, Callitrichidae*). *Int. J. Primatol.* 1, 171–183.
111. Labrie, F., Belanger, A., Cusan, L., Gomez, J. L., and Candas, B. (1997). Marked decline in serum concentrations of adrenal C19 sex steroid precursors and conjugated androgen metabolites during aging. *J. Clin. Endocrinol. Metab.* 82, 2396–2402.
112. Gell, J. S., Carr, B. R., Sasano, H., Atkins, B., Margraf, L., Mason, J. I., and Rainey, W. E. (1998). Adrenarche results from development of a 3β-hydroxysteroid dehydrogenase-deficient adrenal reticularis. *J. Clin. Endocrinol. Metab.* 83, 3695–3701.
113. Palmert, M. R., Hayden, D. L., Mansfield, M. J., Crigler, J. F. Jr., Crowley, W. F. Jr., Chandler, D. W., and Boepple, P. A. (2001). The longitudinal study of adrenal maturation during gonadal suppression: evidence that adrenarche is a gradual process. *J. Clin. Endocrinol. Metab.* 86, 4536–4542.
114. Weber, A., Clark, A. J., Perry, L. A., Honour, J. W., and Savage, M. O. (1997). Diminished adrenal androgen secretion in familial glucocorticoid deficiency implicates a significant role for ACTH in the induction of adrenarche. *Clin. Endocrinol. (Oxf.)* 46, 431–437.
115. Rainey, W. E., Carr, B. R., Sasano, H., Suzuki, T., and Mason, J. I. (2002). Dissecting human adrenal androgen production. *Trends Endocrinol. Metab.* 13, 234–239.
116. Yanagibashi, K., and Hall, P. F. (1986). Role of electron transport in the regulation of the lyase activity of C21 side-chain cleavage P-450 from porcine adenal and testicular microsomes. *J. Biol. Chem.* 261, 8429–8433.
117. Onoda, M., and Hall, P. F. (1982). Cytochrome b5 stimulates purified testicular microsomal cytochrome P-450 (C21 side-chain cleavage). *Biochem. Biophys. Res. Commun.* 108, 454–460.
118. Auchus, R. J., Lee, T. C., and Miller, W. L. (1998). Cytochrome b5 augments the 17,20-lyase activity of human P450c17 without direct electron transfer. *J. Biol. Chem.* 273, 3158–3165.
119. Zhang, L. H., Rodriguez, H., Ohno, S., and Miller, W. L. (1995). Serine phosphorylation of human P450c17 increases 17,20-lyase activity: implications for adrenarche and the polycystic ovary syndrome. *Proc. Natl. Acad. Sci. USA* 92, 10619–10623.
120. Bassett, M. H., Suzuki, T., Sasano, H., De Vries, C. J. M., Jimenez, P. T., Carr, B. R., and Rainey, W. E. (2004). The orphan nuclear receptor NGFIB regulates transcription of 3β-hydroxysteroid dehydrogenase. Implications for the control of adrenal functional zonation. *J. Biol. Chem.* 279, 37622–37630.
121. Jimenez, P., Saner, K., Mayhew, B., and Rainey, W. E. (2003). GATA-6 is expressed in the human adrenal and regulates transcription of genes required for adrenal androgen biosynthesis. *Endocrinology* 144, 4285–4288.
122. Guercio, G., Rivarola, M. A., Chaler, E., Maceiras, M., and Belgorosky, A. (2003). Relationship between the growth hormone/insulin-like growth factor-I axis, insulin sensitivity, and adrenal androgens in normal prepubertal and pubertal girls. *J. Clin. Endocrinol. Metab.* 88, 1389–1393.
123. Remer, T., and Manz, F. (1999). Role of nutritional status in the regulation of adrenarche. *J. Clin. Endocrinol. Metab.* 84, 3936–3944.

124. Yanovski, J. A., Sovik, K. N., Nguyen, T. T., and Sebring, N. G. (2000). Insulin-like growth factors and bone mineral density in African American and white girls. *J. Pediatr.* 137, 826–832.
125. Biro, F. M., Lucky, A. W., Huster, G. A., and Morrison, J. A. (1990). Hormonal studies and physical maturation in adolescent gynecomastia. *J. Pediatr.* 116, 450–455.
126. Tanner, J. M. (1962). *Growth at Adolescence*. Blackwell Scientific Publications, Oxford.
127. Frisch, R. E. (1972). Weight at menarche: similarity for well nourished and undernourished girls at differing ages, and evidence for historical constancy. *Pediatrics* 50, 445–450.
128. Wyshak, G., and Frisch, R. E. (1982). Evidence for a secular trend in age of menarche. *N. Engl. J. Med.* 306, 1033–1035.
129. Macmahon, B. (1973). *Age at Menarche: United States*. DHW Publication. Rockville, MD, National Center for Health Statistics, Report No. HRA 74-1615.
130. Herman-Giddens, M. E., Slora, E. J., Wasserman, R. C., Bourdony, C. J., Bhapkar, M. V., Koch, G. G., and Hasemeler, C. M. (1997). Secondary sexual characteristics and menses in young girls seen in office practice: a study from the Pediatric Research in Office Settings Network. *Pediatrics* 99, 505–512.
131. Hauspie, R. C., Vercauteren, M., and Susanne, C. (1996). Secular changes in growth. *Horm. Res.* 45, 8–17.
132. Huen, K. F., Leung, S. S., Lau, J. T., Cheung, A. Y., Leung, N. K., and Chiu, M. C. (1997). Secular trend in the sexual maturation of southern Chinese girls. *Acta Paediatr.* 86, 1121–1124.
133. Fredriks, A. M., van Buuren, S., Burgmeijer, R. J., Meulmeester, J. F., Beuker, R. J., Brugman, E., Roede, M. J., Verloove-Vanhorick, S. P., and Wit, J. M. (2000). Continuing positive secular growth change in The Netherlands 1955–1997. *Pediatr. Res.* 47, 316–323.
134. Adams Hillard, P. J. (2002). Menstruation in young girls: a clinical perspective. *Obstet. Gynecol.* 99, 655–662.
135. Styne, D. M. (2004). Puberty, obesity and ethnicity. *Trends Endocrinol. Metab.* 15, 472–478.
136. Herman-Giddens, M. E., Wang, L., and Koch, G. (2001). Secondary sexual characteristics in boys: estimates from the national health and nutrition examination survey III, 1988–1994. *Arch. Pediatr. Adolesc. Med.* 155, 1022–1028.
137. Yen, S. S. C., Vicic, W. J., and Kearchner, D. V. (1969). Gonadotropin levels in puberty. I. Serum luteinizing hormone. *J. Clin. Endocrinol. Metab.* 29, 382–385.
138. Yen, S. S. C., and Vicic, W. J. (1970). Serum follicle-stimulating hormone levels in puberty. *Am. J. Obstet. Gynecol.* 106, 134–137.
139. Burr, I. M., Sizonenko, P. C., Kaplan, S. L., and Grumbach, M. M. (1970). Hormonal changes in puberty. I. Correlation of serum luteinizing hormone and follicle stimulating hormone with stages of puberty, testicular size of bone age in normal boys. *Pediatr. Res.* 4, 25–35.
140. Lee, P. A., Midgley, A. R. Jr., and Jaffe, R. B. (1970). Regulation of human gonadotropins. VI. Serum follicle stimulating and luteinizing hormone determinations in children. *J. Clin. Endocrinol. Metab.* 31, 248–253.
141. Penny, R., Guyda, H. J., Baghdassarian, A., Johanson, A. J., and Blizzard, R. M. (1970). Correlation of serum follicular stimulating hormone (FSH) and luteinizing hormone (LH) as measured by radioimmunoassay in disorders of sexual development. *J. Clin. Invest.* 49, 1847–1852.
142. Sizonenko, P. C., Burr, I. M., Kaplan, S. L., and Grumbach, M. M. (1970). Hormonal changes in puberty. II. Correlation of serum luteinizing hormone and follicle stimulating hormone with stages of puberty and bone age in normal girls. *Pediatr. Res.* 4, 36–45.
143. Lee, P. A., Xenakis, T., Winer, J., and Matsenbaugh, S. (1976). Puberty in girls: correlation of serum levels of gonadotropins, prolactin, androgens, estrogens, and progestins with physical changes. *J. Clin. Endocrinol. Metab.* 43, 775–784.
144. McArthur, J. W. (1972). Gonadotropin in relation to sexual maturity. In *Gonadotropins* (B. B. Saxena, C. G. Beling, and H. M. Gandy, Eds.), pp. 487–501. John Wiley & Sons, New York.
145. Root, A. W. (1973). Endocrinology of puberty. I. Normal sexual maturation. *J. Pediatr.* 83, 1–19.
146. Apter, D., Cacciatore, B., Alfthan, H., and Stenman, U-H. (1989). Serum luteinizing hormone concentrations increase 100-fold in females from 7 years of age to adulthood, as measured by time-resolved immunofluorometric assay. *J. Clin. Endocrinol. Metab.* 68, 53–57.
147. Dunkel, L., Alfthan, H., Steraman, U.-H., Selstam, G., Roseberg, S., and Albertsson-Wikland, K. (1992). Developmental changes in 24-hour profiles of luteinizing hormone and follicle-stimulating hormone from puberty to midstages of puberty in boys. *J. Clin. Endocrinol. Metab.* 74, 890–897.
148. Andersson, A.-M., Juul, A., Peterson, J. H., Müller, J., Groome, N. P., and Skakkebæk, N. E. (1997). Serum inhibin B in healthy pubertal and adolescent boys: relation to age, stage of puberty, and follicle-stimulating hormone, luteinizing hormone, testosterone, and estradiol levels. *J. Clin. Endocrinol. Metab.* 82, 3976–3981.
149. Albertsson-Wikland, K., Rosberg, S., Lannering, B., Dunkel, L., Selstam, G., and Norjavaara, E. (1997). Twenty-four-hour profiles of luteinizing hormone, follicular-stimulating hormone, testosterone, and estradiol levels: a semilongitudinal study throughout puberty in healthy boys. *J. Clin. Endocrinol. Metab.* 82, 541–549.
150. Sehested, A., Juul, A. A., Andersson, A. M., Petersen, J. H., Jensen, T. K., Müller, J., and Skakkebæk, N. E. (2000). Serum inhibin A and inhibin B in healthy prepubertal, pubertal, and adolescent girls and adult women: relation to age, stage of puberty, menstrual cycle, follicle-stimulating hormone, luteinizing hormone, and estradiol levels. *J. Clin. Endocrinol. Metab.* 85, 1634–1640.
151. Chada, M., Prusa, R., Bronsky, J., Pechová, M., Kotaska, K., and Lisá, L. (2003). Inhibin B, follicle stimulating hormone, luteinizing hormone, and estradiol and their relationship to the regulation of follicle development in girls during childhood and puberty. *Physiol. Res.* 52, 341–346.
152. Lucky, A. W., Rich, B. H., Rosenfield, R. L., Fang, V. S., and Roche-Bender, N. (1980). LH bioactivity increases more than immunoreactivity during puberty. *J. Pediatr.* 97, 205–213.
153. Reiter, E. O., Beitins, I. Z., Ostrea, T., and Gutai, J. P. (1982). Bioassayable luteinizing hormone during childhood and adolescence and in patients with delayed pubertal development. *J. Clin. Endocrinol. Metab.* 54, 155–161.
154. Winter, J. S. D., Faiman, C., Hobson, W. C., and Reyes, F. I. (1980). The endocrine basis of sexual development in the chimpanzee. *J. Reprod. Fertil.* 28, 131–138.
155. Terasawa, E., Bridson, W. E., Nass, T. E., Noonan, J. J., and Dierschke, D. J. (1984). Developmental changes in the luteinizing hormone secretory pattern in peripubertal female rhesus monkeys: comparison between gonadally intact and ovariectomized animals. *Endocrinology* 115, 2233–2240.
156. Meeran, D., Urbanski, H. F., Gregory, S. J., Townsend, J., and Tortonese, D. J. (2003). Developmental changes in the hormonal identity of gonadotroph cells in the rhesus monkey pituitary gland. *J. Clin. Endocrinol. Metab.* 88, 2934–2942.
157. Dierschke, D. J., Weiss, G., and Knobil, E. (1974). Sexual maturation in the female rhesus monkey and the development of

estrogen-induced gonadotropic hormone release. *Endocrinology* 94, 198–206.
158. Reiter, E. O., Kulin, H. E., and Hamwood, S. M. (1974). The absence of positive feedback between estrogen and luteinizing hormone in sexually immature girls. *Pediatr. Res.* 8, 740–745.
159. Presl, J., Horejsi, J., Stroufova, A., and Herzmann, J. (1976). Sexual maturation in girls and the development of estrogen-induced gonadotropic hormone release. *Ann. Biol. Anim. Biochem. Biophys.* 16, 377–383.
160. Wilkins, L. (1965). *The Diagnosis and Treatment of Endocrine Disorders in Childhood Adolescence.* Charles C. Thomas, Springfield, IL.
161. Luder, A. S., Holland, F. J., Costigan, D. C., Jenner, M. R., Wielgosz, G., and Fazekas, A. T. A. (1984). Intranasal and subcutaneous treatment of central precocious puberty in both sexes with a long-acting analog of luteinizing hormone-releasing hormone. *J. Clin. Endocrinol. Metab.* 58, 966–972.
162. Styne, D. M., Harris, D. A., Egli, C. A., Conte, F. A., Kaplan, S. L., Rivier, J., Vale, W., and Grumbach, M. M. (1985). Treatment of true precocious puberty with a potent luteinizing hormone-releasing factor agonist: effect on growth, sexual maturation, pelvic sonography and the hypothalamic-pituitary-gonadal axis. *J. Clin. Endocrinol. Metab.* 61, 142–151.
163. Styne, D. M., and Grumbach, M. M. (1978). Puberty in the male and female: its physiology and disorders. In *Reproductive Endocrinology, Physiology, Pathophysiology and Clinical Management* (S. S. C. Yen and R. B. Jaffe, Eds.), pp. 189–239. W. B. Saunders, Philadelphia.
164. Mansfield, M. J., Beardsworth, D. E., Loughlin, J. S., Crawford, J. D., Bode, H. H., Rivier, J., Vale, W., Kushner, D. C., Crigler, J. F. Jr., and Crowley W. F. Jr. (1983). Long-term treatment of central precocious puberty with a long-acting analogue of luteinizing hormone releasing hormone. *N. Engl. J. Med.* 309, 1286–1290.
165. Crowley, W. F. Jr., and McArthur, J. W. (1980). Simulation of the normal menstrual cycle in Kallman's syndrome by pulsatile administration of luteinizing hormone-releasing hormone (LHRH). *J. Clin. Endocrinol. Metab.* 51, 173–175.
166. Hoffman, A. R., and Crowley, W. F. Jr. (1982). Induction of puberty in men by long-term pulsatile administration of low dose gonadotropin-releasing hormone. *N. Engl. J. Med.* 307, 1237–1241.
167. Delemarre-van de Waal, H. A., and Schoemaker, J. (1983). Induction of puberty by prolonged pulsatile LRH administration. *Acta Endocrinol.* 102, 603–609.
168. Seibel, M. M., Claman, P., Oskowitz, S. P., McArdle, C., and Weinstein, F. G. (1985). Events surrounding the initiation of puberty with long-term subcutaneous pulsatile gonadotropin-releasing hormone in a female patient with Kallman's syndrome. *J. Clin. Endocrinol. Metab.* 61, 575–579.
169. Alien, E. (1928). Precocious sexual development from anterior hypophysis implants in a monkey. *Anat. Rec.* 39, 315–323.
170. Bennett, W. I., Dufau, M. L., Catt, K. J., and Tullner, W. W. (1973). Effect of human menopausal gonadotropin upon spermatogenesis and testosterone production in juvenile rhesus monkeys. *Endocrinology* 92, 813–821.
171. Weiss, F., Rifkin, I., and Atkinson, L. E. (1976). Induction of ovulation in pre-menarchial rhesus monkeys with human gonadotropins. *Biol. Reprod.* 14, 401–404.
172. Wildt, L., Marshall, G., and Knobil, E. (1980). Experimental induction of puberty in the infantile female rhesus monkey. *Science* 207, 1373–1375.
173. Marshall, G. R., Wickings, E. J., and Nieschlag, E. (1985). Pulsatile GnRH administration in immature male monkeys. *Acta Endocrinol.* 108, 267 (abst. 199).
174. Chen, A., Yahalom, D., Ben-Aroya, N., Kaganovsky, E., Okon, E., and Koch, Y. (1998). A second isoform of gonadotropin-releasing hormone is present in the brain of human and rodents. *FEBS Lett.* 435, 199–203.
175. Latimer, V. S., Kohama, S. G., Garyfallou, V. T., and Urbanski, H. F. (2001). A developmental increase in the expression of messenger ribonucleic acid encoding a second form of gonadotropin-releasing hormone in the rhesus macaque hypothalamus. *J. Clin. Endocrinol. Metab.* 86, 324–329.
176. Neill, J. D., Duck, L. W., Sellers, J. C., and Musgrove, L. C. (2001). A gonadotropin-releasing hormone (GnRH) receptor specific for GnRH II in primates. *Biochem. Biophys. Res. Comm.* 282, 1012–1018.
177. Millar, R., Lowe, S., Conklin, D., Pawson, A., Maudsley, S., Troskie, B., Ott, T., Millar, M., Lincoln, G., Sellar, R., Faurholm, B., Scobie, G., Kuestner, R., Terasawa, E., and Katz, A. (2001). A novel mammalian receptor for the evolutionarily conserved type II GnRH. *Proc. Natl. Acad. Sci. USA* 98, 9636–9641.
178. Densmore, V. S., and Urbanski, H. F. (2003). Relative effect of gonadotropin-releasing hormone (GnRH)-I and GnRH-II on gonadotropin release. *J. Clin. Endocrinol. Metab.* 88, 2126–2134.
179. Plant, T. M. (1986). Gonadal regulation of hypothalamic gonadotropin-releasing hormone release in primates. *Endocr. Rev.* 7, 75–88.
180. Schwartz, N. B. (1969). A model for the regulation of ovulation in the rat. *Rec. Prog. Horm. Res.* 25, 1–43.
181. Ramaswamy, S., and Plant, T. M. (2001). Operation of the follicle-stimulating hormone (FSH)-inhibin B feedback loop in the control of primate spermatogenesis. *Mol. Cell. Endocrinol.* 180, 93–101.
182. Karsch, F. J., Dierschke, D. J., and Knobil, E. (1973). Sexual differentiation of pituitary function: apparent differences between primates and rodents. *Science* 179, 484–486.
183. Kulin, H. E., and Reiter, E. O. (1976). Gonadotropin and testosterone measurements after estrogen administration to adult men, prepubertal and pubertal boys, and men with hypogonadotropism: evidence for maturation of positive feedback in the male. *Pediatr. Res.* 10, 46–51.
184. Barbarino, A., and DeMarinis, L. (1980). Estrogen induction of luteinizing hormone release in castrated adult human males. *J. Clin. Endocrinol. Metab.* 51, 280–286.
185. Norman, R. L., and Spies, H. G. (1986). Cyclic ovarian function in a male macaque: additional evidence for a lack of sexual differentiation in the physiological mechanisms that regulate the cyclic release of gonadotropins in primates. *Endocrinology* 118, 2608–2610.
186. Kaplan, S. L., Grumbach, M. M., and Aubert, M. L. (1976). The ontogenesis of pituitary hormones and hypothalamic factors in the human fetus: maturation of central nervous system regulation of anterior pituitary function. *Rec. Prog. Horm. Res.* 32, 161 –234.
187. Clements, J. A., Reyes, F. I., Winter, J. S. D., and Faiman, C. (1976). Studies on human sexual development. III. Fetal pituitary and serum, and amniotic fluid concentrations of LH, CG, and FSH. *J. Clin. Endocrinol. Metab.* 42, 9–19.
188. Ellinwood, W. E., and Resko, J. A. (1980). Sex differences in biologically active and immunoreactive gonadotropins in the fetal circulation of rhesus monkeys. *Endocrinology* 107, 902–907.
189. Resko, J. A., and Ellinwood, W. E. (1985). Negative feedback regulation of gonadotropin secretion by androgens in fetal rhesus macaques. *Biol. Reprod.* 33, 346–352.
190. Winter, J. S. D., Faiman, C., Hobson, W. C., Prasad, A. V., and Reyes, F. I. (1975). Pituitary-gonadal relations in infancy. I. Patterns of serum gonadotropin concentrations

from birth to four years of age in man and chimpanzee. *J. Clin. Endocrinol. Metab.* 40, 545–551.
191. Schwanzel-Fukuda, M., and Pfaff, D. W. (1989). Origins of luteinizing hormone-releasing hormone neurons. *Nature* 338, 161–164.
192. Wray, S., Grant, P., and Gainer, H. (1989). Evidence that cells expressing luteinizing hormone-releasing hormone mRNA in the mouse are derived from progenitor cells in the olfactory placode. *Proc. Natl. Acad. Sci. USA* 86, 8132–8136.
193. Ronnekleiv, O. K., and Resko, J. A. (1990). Ontogeny of gonadotropin-releasing hormone-containing neurons in early fetal development of rhesus macaques. *Endocrinology* 126, 498–511.
194. Quanbeck, C., Sherwood, N. M., Millar, R. P., and Terasawa, E. (1997). Two populations of luteinizing hormone-releasing hormone neurons in the forebrain of the rhesus macaque during embryonic development. *J. Comp. Neurol.* 380, 293–309.
195. Silverman, A.-J., Antunes, J. L., Abrams, G. M., Nilaver, G., Thau, R., Robinson, J. A., Ferin, M., and Krey, L. C. (1982). The luteinizing hormone-releasing hormone pathways in rhesus (*Macaca mulatta*) and pigtailed (*Macaca nemestrina*) monkeys: new observations on thick, unembedded sections. *J. Comp. Neurol.* 211, 309–317.
196. Aubert, M. L., Grumbach, M. M., and Kaplan, S. L. (1977). The ontogenesis of human fetal hormones. IV. Somatostatin, luteinizing hormone releasing factor, and thyrotropin releasing factor in hypothalamus and cerebral cortex of human fetuses 10–22 weeks of age. *J. Clin. Endocrinol. Metab.* 44, 1130–1141.
197. Bugnon, C., Bloch, B., and Fellmann, D. (1977). Cytoimmunological study of the ontogenesis of the gonadotropic hypothalamic-pituitary axis in the human fetus. *J. Steroid. Biochem.* 8, 565–574.
198. Paulin, C., Dubois, M. P., Barry, J., and Dubois, P. M. (1977). Immunofluorescence study of LH-RH producing cells in the human fetal hypothalamus. *Cell. Tissue Res.* 182, 341–346.
199. Groom, G. V., and Boyns, A. R. (1973). Effect of hypothalamic releasing factors and steroids on release of gonadotropins by organ cultures of human fetal pituitaries. *J. Endocrinol.* 59, 511–522.
200. Huhtaniemi, I. T., Koritnik, D. R., Korenbrot, C. C., Mennin, S., Foster, D. B., and Jaffe, R. B. (1979). Stimulation of pituitary-testicular function with gonadotropin-releasing hormone in fetal and infant monkeys. *Endocrinology* 105, 109–114.
201. Sklar, C. A., Mueller, P. L., Gluckman, P., Kaplan, S. L., Rudolph, A. M., and Grumbach, M. M. (1981). Hormone ontogeny in the ovine fetus. VII. Circulating luteinizing hormone and follicle-stimulating hormone in mid- and late gestation. *Endocrinology* 108, 874–880.
202. Clark, S. J., Ellis, N., Styne, D. M., Gluckman, P. D., Kaplan, S. L., and Grumbach, M. M. (1984). Hormone ontogeny in the ovine fetus. XVII. Demonstration of pulsatile luteinizing hormone secretion by the fetal pituitary gland. *Endocrinology* 115, 1774–1779.
203. Bettendorf, M., Albers, N., de Zegher, F., Hart, C. S., Kaplan, S. L., and Grumbach, M. M. (1988). A neuroexcitatory amino acid analogue, N-methyl-D,L-aspartate (NMDA), elicits LH and FSH release in the ovine fetus by a central mechanism. 70th Annual Meeting of the Endocrine Society, New Orleans, Abstract No. 1071.
204. Clark, S. J., Hauffa, B. P., Rodens, K. P., Styne, D. L., Kaplan, S. L., and Grumbach, M. M. (1989). Hormone ontogeny in the ovine fetus. XIX. The effect of a potent luteinizing hormone-releasing factor agonist on gonadotropin and testosterone release in the fetus and neonate. *Pediatr. Res.* 25, 347–352.
205. Forest, M. G., Sizonenko, P. C., Cathiard, A. M., and Betrand, J. (1974). Hypophysogonadal function in humans during the first year of life. I. Evidence for testicular activity in early infancy. *J. Clin. Invest.* 53, 819–828.
206. Frawley, L. S., and Neill, J. D. (1979). Age related changes in serum levels of gonadotropins and testosterone in infantile male rhesus monkeys. *Biol. Reprod.* 20, 1147–1151.
207. Hammond, G. L., Koivisto, M., Kouvalainen, K., and Vihko, R. (1979). Serum steroids and pituitary hormones in infants with particular reference to testicular activity. *J. Clin. Endocrinol. Metab.* 49, 40–45.
208. Plant, T. M. (1980). The effects of neonatal orchidectomy on the developmental pattern of gonadotropin secretion in the male rhesus monkey (*Macaca mulatta*). *Endocrinology* 106, 1451–1454.
209. Fuller, G. B., Faiman, C., Winter, J. S. D., Reyes, F. I., and Hobson, W. C. (1982). Sex-dependent gonadotropin concentrations in infant chimpanzees and rhesus monkeys. *Proc. Soc. Exp. Biol. Med.* 169, 494–500.
210. Plant, T. M. (1982). A striking diurnal variation in plasma testosterone concentrations in infantile male rhesus monkeys (*Macaca mulatta*). *Neuroendocrinology* 35, 370–373.
211. Schmidt, H., and Schwarz, H. P. (2000). Serum concentrations of LH and FSH in the healthy newborn. *Eur. J. Endocrinol.* 143, 213–215.
212. Chellakooty, M., Schmidt, I. M., Haavisto, A. M., Boisen, K. A., Damgaard, I. N., Mau, C., Petersen, J. H., Juul, A., Skakkebæk, N. E., and Main, K. M. (2003). Inhibin A, inhibin B, follicle-stimulating hormone, luteinizing hormone, estradiol, and sex hormone-binding globulin levels in 473 healthy infant girls. *J. Clin. Endocrinol. Metab.* 88, 3515–3520.
213. Winter, J. S. D., Hughes, I. A., Reyes, F. I., and Faiman, C. (1976). Pituitary-gonadal relations in infancy. 2. Patterns of serum gonadal steroid concentrations in man from birth to two years of age. *J. Clin. Endocrinol. Metab.* 42, 679–686.
214. Mann, D. R., Davis-DaSilva, M., Wallen, K., Coan, P., Evans, D. E., and Collins, D. C. (1984). Blockade of neonatal activation of the pituitary-testicular axis with continuous administration of a gonadotropin-releasing hormone agonist in male rhesus monkeys. *J. Clin. Endocrinol. Metab.* 59, 207–211.
215. Plant, T. M. (1982). Pulsatile luteinizing hormone secretion in the neonatal male rhesus monkey (*Macaca mulatta*). *J. Endocrinol.* 93, 71–74.
216. Waldhauser, F., Weissenbacher, G., Frisch, H., and Pollak, A. (1981). Pulsatile secretion of gonadotropins in early infancy. *Eur. J. Pediatr.* 137, 71–74.
217. de Zegher, F., Devlieger, H., and Veldhuis, J. D. (1992). Pulsatile and sexually dimorphic secretion of luteinizing hormone in the human infant on the day of birth. *Pediatr. Res.* 32, 605–607.
218. Plant, T. M., and Zorub, D. S. (1982). The role of nongonadal restraint of gonadotropin secretion in the delay of onset of puberty in the rhesus monkey (*Macaca mulatta*). *J. Anim. Sci.* 55, 43–55.
219. Mann, D. R., Gould, K. G., Collins, D. C., and Wallen, K. (1989). Blockade of neonatal activation of the pituitary-testicular axis: effect on peripubertal luteinizing hormone and testosterone secretion and on testicular development in male monkeys. *J. Clin. Endocrinol. Metab.* 68, 600–607.
220. Mann, D. R., Akinbami, M. A., Gould, K. G., Tanner, J. M., and Wallen, K. (1993). Neonatal treatment of male monkeys with a gonadotropin-releasing hormone agonist alters differentiation of central nervous system centers that regulate sexual and skeletal development. *J. Clin. Endocrinol. Metab.* 76, 1319–1324.

221. Mann, D. R., Akinbami, M. A., Gould, K. G., Paul, K., and Wallen, K. (1998). Sexual maturation in male rhesus monkeys: importance of neonatal testosterone exposure and social rank. *J. Endocrinol.* 156, 493–501.

222. Lunn, S. F., Cowen, G. M., and Fraser, H. M. (1997). Blockade of the neonatal increase in testosterone by a GnRH antagonist: the free androgen index, reproductive capacity and postmortem findings in the male marmoset monkey. *J. Endocrinol.* 154, 125–131.

223. Nevison, C. M., Brown, G. R., and Dixson, A. F. (1997). Effects of altering testosterone in early infancy on social behaviour in captive yearling rhesus monkeys. *Physiol. Behav.* 62, 1397–1403.

224. Brown, G. R., Nevison, C. M., Fraser, H. M., and Dixson, A. F. (1999). Manipulation of postnatal testosterone levels affects phallic and clitoral development in infant rhesus monkeys. *Int. J. Androl.* 22, 119–128.

225. Main, K. M., Schmidt, I. M., and Skakkebæk, N. E. (2000). A possible role for reproductive hormones in newborn boys: progressive hypogonadism without the postnatal testosterone peak. *J. Clin. Endocrinol. Metab.* 85, 4905–4907.

226. Mann, D. R., Akinbami, M. A., Lunn, S. F., Fraser, H. M, Gould, K. G., and Ansari, A. A. (2000). Endocrine-immune interaction: alterations in immune function resulting from neonatal treatment with a GnRH antagonist and seasonality in male primates. *Am. J. Reprod. Immunol.* 44, 30–40.

227. Plant, T. M. (1986). A striking sex difference in the gonadotropin response to gonadectomy during infantile development in the rhesus monkey (*Macaca mulatta*). *Endocrinology* 119, 539–545.

228. Massa, G., de Zegher, M. G., and Vanderschueren-Lodeweyckx, M. (1992). Serum levels of immunoreactive inhibin, FSH, and LH in human infants at preterm and term birth. *Biol. Neonate.* 61, 150–155.

229. Pohl, C. R., deRidder, C. M., and Plant, T. M. (1995). Gonadal and nongonadal mechanisms contribute to the prepubertal hiatus in gonadotropin secretion in the female rhesus monkey (*Macaca mulatta*). *J. Clin. Endocrinol. Metab.* 80, 2094–2101.

230. Lustig, R. H., Conte, F. A., Kogan, B. A., and Grumbach, M. M. (1987). Ontogeny of gonadotropin secretion in congenital anorchism: sexual dimorphism versus syndrome of gonadal dysgenesis and diagnostic considerations. *J. Urol.* 138, 587–591.

231. Bouvattier, C., Carel, J.-C., Lecointre, C., David, A., Sultan, C., Bertrand, A.-M., Morel, Y., and Chaussain, J.-L. (2002). Postnatal changes of T, LH, and FSH in 46,XY infants with mutations in the AR gene. *J. Clin. Endocrinol. Metab.* 87, 29–32.

232. Marshall, G. R., and Plant, T. M. (1996). Puberty occurring either spontaneously or induced precociously in rhesus monkey (*Macaca mulatta*) is associated with a marked proliferation of Sertoli cells. *Biol. Reprod.* 54, 1192–1199.

233. Simorangkir, D. R., Marshall, G. R., and Plant, T. M. (2003). Discordancy between proliferation of type A spermatogonia and Sertoli cells during prepubertal development in the monkey. XVII North American Testis Workshop, Phoenix, AR, March, Abstract No. 76.

234. Simorangkir, D. R., Marshall, G. R., Schlatt, S., and Plant, T. M. (2004). Proliferation of undifferentiated type A spermatogonia (spermatogonial stem cells) during infancy and prior to puberty in the rhesus monkey (*Macaca mulatta*) as revealed by BrdU-labeling. 13th European Testis Workshop, Dunblane, Scotland, April.

235. Parker, D. C., Judd, H. L., Rossman, L. G., and Yen, S. S. C. (1975). Pubertal sleep-wake patterns of episodic LH, FSH and testosterone release in twin boys. *J. Clin. Endocrinol. Metab.* 40, 1099–1109.

236. Kelch, R. P., Hopwood, N. J., Sauder, S., and Marshall, J. C. (1985). Evidence for decreased secretion of gonadotropin-releasing hormone in pubertal boys during short-term testosterone treatment. *Pediatr. Res.* 19, 112–117.

237. Wu, F. C. W., Butler, G. E., Kelnar, C. J., Huhtaniemi, I., and Veldhuis, J. D. (1996). Ontogeny of pulsatile gonadotropin releasing hormone secretion from midchildhood, through puberty, to adulthood in the human male: a study using deconvolution analysis and an ultrasensitive immunofluorometric assay. *J. Clin. Endocrinol. Metab.* 81, 1798–1805.

238. Mitamura, R., Yano, K., Suzuki, N., Ito, Y., Makita, Y., and Okuno, A. (1999). Diurnal rhythms of luteinizing hormone, follicle-stimulating hormone, and testosterone secretion before the onset of male puberty. *J. Clin. Endocrinol. Metab.* 84, 29–37.

239. Crofton, P. M., Evans, A. E. M., Wallace, A. M., Groome, N. P., and Kelnar, C. J. H. (2004). Nocturnal secretory dynamics of inhibin B and testosterone in pre- and peripubertal boys. *J. Clin. Endocrinol. Metab.* 89, 867–874.

240. Plant, T. M. (1981). Time courses of concentrations of circulating gonadotropin, prolactin, testosterone, and cortisol in adult male rhesus monkeys (*Macaca mulatta*) throughout the 24 h light-dark cycle. *Biol. Reprod.* 25, 244–252.

241. Kulin, H. E., Moore, R. G. Jr., and Santner, S. J. (1976). Circadian rhythms in gonadotropin excretion in prepubertal and pubertal children. *J. Clin. Endocrinol. Metab.* 42, 770–773.

242. Hale, P. M., Khoury, S., Foster, C. M., Beitins, I. Z., Hopwood, N. J., Marshall, J. C., and Kelch, R. P. (1988). Increased luteinizing hormone pulse frequency during sleep in early to midpubertal boys: effects of testosterone infusion. *J. Clin. Endocrinol. Metab.* 66, 785–791.

243. Foster, C. M., Hassing, J. M., Mendes, T. M., Hale, P. M., Padmanabhan, V., Hopwood, N. J., Beitins, I. Z., Marshall, J. C., and Kelch, R. P. (1989). Testosterone infusion reduces nocturnal luteinizing hormone pulse frequency in pubertal boys. *J. Clin. Endocrinol. Metab.* 69, 1213–1220.

244. Kletter, G. B., Foster, C. M., Brown, M. B., Padmanabhan, V., Beitins, I. Z., Marshall, J. C., and Kelch, R. P. (1991). Naloxone does not reverse the suppressive effects of testosterone infusion on luteinizing hormone secretion in pubertal boys. *J. Clin. Endocrinol. Metab.* 73, 1241–1247.

245. Kletter, G. B., Foster, C. M., Beitins, I. Z., Marshall, J. C., and Kelch, R. P. (1992). Acute effects of testosterone infusion and naloxone on luteinizing hormone secretion in normal men. *J. Clin. Endocrinol. Metab.* 75, 1215–1219.

246. Kapen, S., Boyar, R. M., Finkelstein, J. W., Hellman, L., and Weitzman, E. D. (1974). Effect of sleep-wake cycle reversal on luteinizing hormone secretory pattern in puberty. *J. Clin. Endocrinol. Metab.* 39, 293–299.

247. Boyar, R. M., Wu, R. H. K., Roffwarg, H., Kapen, S., Weitzman, E. D., Hellman, L., and Finkelstein, J. W. (1976). Human puberty: 24-hour estradiol patterns in pubertal girls. *J. Clin. Endocrinol. Metab.* 43, 1418–1421.

248. Boyar, R. M. (1978). Control of the onset of puberty. *Annu. Rev. Med.* 29, 509–520.

249. Beck, W., and Wuttke, W. (1980). Diurnal variations of plasma luteinizing hormone, follicle-stimulating hormone, and prolactin in boys and girls from birth to puberty. *J. Clin. Endocrinol. Metab.* 50, 635–639.

250. Cemeroglu, A. P., Foster, C. M., Warner, R., Kletter, G. B., Marshall, J. C., and Kelch, R. P. (1996). Comparison of the neuroendocrine control of pubertal maturation in girls and boys with spontaneous puberty and in hypogonadal girls. *J. Clin. Endocrinol. Metab.* 81, 4352–4357.

251. Suter, K. J., Pohl, C. R., and Plant, T. M. (1999). Indirect assessment of pulsatile gonadotropin-releasing hormone

release in agonadal prepubertal rhesus monkeys (*Macaca mulatta*). *J Endocrinol.* 160, 35–41, 1999.

252. Rossmanith, W. G. (1998). The impact of sleep on gonadotropin secretion. *Gynecol. Endocrinol.* 12, 381–389.
253. O'Byrne, K. T., Thalabard, J.-C., Chiappini, S. E., Chen, M.-D., Hotchkiss, J., and Knobil, E. (1993). Ambient light modifies gonadotropin-releasing hormone pulse generator frequency in the rhesus monkey. *Endocrinology* 133, 1520–1524.
254. Dierschke, D. J., Karsch, F. J., Weick, R. F., Weiss, G., Hotchkiss, J., and Knobil, E. (1974). Hypothalamic-pituitary regulation of puberty: feedback control of gonadotropin secretion in the rhesus monkey. In *Control of the Onset of Puberty* (M. M. Grumbach, G. D. Grave, and F. E. Mayer, Eds.), pp. 104–114. John Wiley & Sons, New York.
255. Dunkel, L., Alfthan, H., Stenman, U. H., Perheenentupa, J. (1990). Gonadal control of pulsatile secretion of luteinizing hormone and follicle-stimulating hormone in prepubertal boys evaluated by ultrasensitive time-resolved immunofluorometric assays. *J. Clin. Endocrinol. Metab.* 70, 107–114.
256. Winter, J. S. D., and Faiman, C. (1972). Serum gonadotropin concentrations in agonadal children and adults. *J. Clin. Endocrinol. Metab.* 35, 561–564.
257. Conte, F. A., Grumbach, M. M., and Kaplan, S. L. (1975). A diphasic pattern of gonadotropin secretion in patients with the syndrome of gonadal dysgenesis. *J. Clin. Endocrinol. Metab.* 40, 670–674.
258. Levine Ross, J., Loriaux, D. L., and Cutler, G. B. Jr. (1983). Developmental changes in neuroendocrine regulation of gonadotropin secretion in gonadal dysgenesis. *J. Clin. Endocrinol. Metab.* 57, 288–293.
259. Hosoda, A., Fujieda, K., Matsuura, N., Okuno, A., and Yuri, K. (1991). Age-related change of pulsatile gonadotropin secretion in Turner syndrome. *Pediatr. Res.* 29, 196–200.
260. Nathwani, N. C., Hindmarsh, P. C., Massarano, A. A., and Brook, C. G. D. (1998). Gonadotrophin pulsatility in girls with the Turner syndrome: modulation by exogenous sex steroids. *Clin. Endocrinol.* 49, 107–113.
261. Ropelato, M. G., Escobar, M. E., Gottlieb, S., and Bergadá, C. (1997). Gonadotropin secretion in prepubertal normal and agonadal children evaluated by ultrasensitive time-resolved immunofluorometric assays. *Horm. Res.* 48, 164–172.
262. El Majdoubi, M., Sahu, A., Ramaswamy, S., and Plant, T. M. (2000). Neuropeptide Y: a hypothalamic brake restraining the onset of puberty in primates. *Proc. Natl. Acad. Sci. USA* 97, 6179–6184.
263. Wilson, M. E. (1995). IGF-I administration advances the decrease in hypersensitivity to oestradiol negative feedback inhibition of serum LH in adolescent female rhesus monkeys. *J. Endocrinol.* 145, 121–30.
264. Wilson, M. E., Fisher, J., and Chikazawa, K. (2004). Estradiol negative feedback regulates nocturnal luteinizing hormone and follicle-stimulating hormone secretion in prepubertal female rhesus monkeys. *J. Clin. Endocrinol. Metab.* 89, 3973–3978.
265. Rapisarda, J. J., Bergman, K. S., Steiner, R. A., and Foster, D. L. (1983). Response to estradiol inhibition of tonic luteinizing hormone secretion decreases during the final stage of puberty in the rhesus monkey. *Endocrinology* 112, 1172–1179.
266. Kulin, H. E. (1985). Developmental aspects of sex steroid-gonadotropin interactions in man. In *Adolescence in Females* (J. Givens, C. Flamingi, and S. Venturoli, Eds.), pp. 179–196. Year Book Medical Publishers, Chicago.
267. Maruca, J., Kulin, H. E., and Santner, S. J. (1983). Perturbations of negative feedback sensitivity in agonadal patients undergoing estrogen replacement therapy. *J. Clin. Endocrinol. Metab.* 56, 53–59.
268. Chongthammakun, S., Claypool, L. E., and Terasawa, E. (1993). Ovariectomy increases *in vivo* luteinizing hormone-releasing hormone release in pubertal, but not prepubertal, female rhesus monkeys. *J. Neuroendocrinol.* 5, 41–50.
269. Chongthammakun, S., and Terasawa, E. (1993). Negative feedback effects of estrogen on luteinizing hormone-releasing hormone release occur in pubertal, but not prepubertal, ovariectomized female rhesus monkeys. *Endocrinology* 132, 735–743.
270. Terasawa, E. (1985). Developmental changes in the positive feedback effect of estrogen on luteinizing hormone release in ovariectomized female rhesus monkeys. *Endocrinology* 117, 2490–2497.
271. Plant, T. M. (1994). Puberty in primates. In *The Physiology of Reproduction* (E. Knobil and J. D. Neill, Eds.), pp. 453–485. Raven Press, New York.
272. Dohrn, M., and Hohlweg, W. (1931). Hormonale Beziehungen zwischen Hypophysenvorder-lappen und Keimdriisen. In *Proceedings of the Second International Congress on Sexual Research*, pp. 436–442. Oliver and Boyd, Edinburgh.
273. Hoogstra, M. J., and deJongh, S. E. (1955). The problem of the occurrence of castration cells in the hypophysis in very young animals. *Acta Physiol. Pharmacol. Neerl.* 4, 145–152.
274. Byrnes, W. W., and Meyer, R. K. (1951). Effect of physiological amounts of estrogen in the secretion of follicle stimulating and luteinizing hormone. *Endocrinology* 49, 449–460.
275. Ramirez, D. V., and McCann, S. M. (1963). Comparison of the regulation of luteinizing hormone (LH) secretion in immature and adult rats. *Endocrinology* 72, 452–473.
276. Smith, E. R., Damassa, D. A., and Davidson, J. M. (1977). Feedback regulation and male puberty: testosterone-luteinizing hormone relationships in the developing rat. *Endocrinology* 101, 173–180.
277. Kulin, H. E., Grumbach, M. M., and Kaplan, S. L. (1969). Changing sensitivity of the pubertal gonadal hypothalamic feedback mechanism in man. *Science* 166, 1012–1013.
278. Kelch, R P., Kaplan, S. L., and Grumbach, M. M. (1973). Suppression of urinary and plasma follicle-stimulating hormone by exogenous estrogens in prepubertal and pubertal children. *J. Clin. Invest.* 52, 1122–1128.
279. Plant, T. M., and Zorub, D. S. (1984). A study of the role of the adrenal glands in the initiation of the hiatus in gonadotropin secretion during prepubertal development in the male rhesus monkey (*Macaca mulatta*). *Endocrinology* 114, 560–565.
280. Grumbach, M. M., Richards, G. E., Conte, F. A., and Kaplan, S. L. (1978). Clinical disorders of adrenal function and puberty: an assessment of the role of the adrenal cortex in normal and abnormal puberty in man and evidence for an ACTH-like pituitary adrenal androgen stimulating hormone. In *The Endocrine Function of the Human Adrenal Cortex* (V. H. T. James, M. Serio, G. Giusti, and L. Martini, Eds.), pp. 583–612. Academic Press, London.
281. Plant, T. M. (1988). Neuroendocrine basis of puberty in the rhesus monkey (*Macaca mulatta*). In *Frontiers in Neuroendocrinology* (L. Martini and W. F. Ganong, Eds.), Vol. 10, pp. 215–238. Raven Press, New York.
282. Gay, V. L., and Plant, T. M. (1987). N-Methyl-DL-aspartate (NMA) elicits hypothalamic GnRH release in prepubertal male rhesus monkeys (*Macaca mulatta*). *Endocrinology* 120, 2289–2296.
283. Suter, K. J., Pohl, C. R., and Plant, T. M. (1998). The pattern and tempo of the pubertal reaugmentation of open-loop pulsatile gonadotropin-releasing hormone release assessed indirectly in the male rhesus monkey (*Macaca mulatta*). *Endocrinology* 139, 2774–2783.
284. Watanabe, G., and Terasawa, E. (1989). *In vivo* release of luteinizing hormone releasing hormone increases with

puberty in the female rhesus monkey. *Endocrinology* 125, 92–99.
285. Monroe, S. E., Yamamoto, M., and Jaffe, R. B. (1983). Changes in gonadotrope responsitivity to gonadotropin releasing hormone during development of the rhesus monkey. *Biol. Reprod.* 29, 422–431.
286. Grumbach, M. M., Roth, J. C., Kaplan, S. L., and Kelch, R. P. (1974). Hypothalamic-pituitary regulation of puberty in man: evidence and concepts derived from clinical research. In *Control of the Onset of Puberty* (M. M. Grumbach, G. D. Grave, and F. E. Mayer, Eds.), pp. 115–166. John Wiley & Sons, New York.
287. Gamier, P. E., Chaussain, J. L., Binet, E., Schlumberger, A., and Job, J. C. (1974). Effect of synthetic luteinizing hormone-releasing hormone (LH-RH) on the release of gonadotropins in children and adolescents. VI. Relation to age, sex and puberty. *Acta Endocrinol.* 77, 422–434.
288. Conte, F. A., Grumbach, M. M., Kaplan, S. L., and Reiter, E. O. (1980). Correlation or luteinizing hormone-releasing factor-induced luteinizing hormone and follicle-stimulating hormone release from infancy to 19 years with the changing pattern of gonadotropin secretion in agonadal patients: relation to the restraint of puberty. *J Clin. Endocrinol. Metab.* 50, 163–188.
289. Bercu, B. B., Lee, B. C., Spiliotis, B. E., Pineda, J. L., Denman, D. W. III, Hoffman, H. J., and Brown, T. J. (1983). Male sexual development in the monkey. II. Cross-sectional analysis of pulsatile hypothalamic-pituitary secretion in castrated males. *J. Clin. Endocrinol. Metab.* 56, 1227–1235.
290. Oerter, K. E., Uriarte, M. M., Rose, S. R., Barnes, K. M., and Cutler, G. B. Jr. (1990). Gonadotropin secretory dynamics during puberty in normal girls and boys. *J. Clin. Endocrinol. Metab.* 71,1251–1258.
291. Clark, P. A., Iranmanesh, A., Veldhuis, J. D., and Rogol, A. D. (1997). Comparison of pulsatile luteinizing hormone secretion between prepubertal children and young adults: evidence for a mass/amplitude-dependent difference without gender or day/night contrasts. *J. Clin. Endocrinol. Metab.* 82, 2950–2955.
292. Manasco, P. K., Umach, D. M., Muly, S. M., Godwin, D. C., Negro-Vilar, A., Culler, M. D., and Underwood, L. E. (1997). Ontogeny of gonadotrophin and inhibin secretion in normal girls through puberty based on overnight serial sampling and a comparison with normal boys. *Hum. Reprod.* 12, 2108–2114.
293. Wu, F. C. W., Butler, G. E., Kelnar, C. J. H., Stirling, H. F., and Huhtaniemi, I. (1991). Patterns of pulsatile luteinizing hormone and follicle-stimulating hormone secretion in prepubertal (midchildhood) boys and girls and patients with idiopathic hypogonadotropic hypogonadism (Kallmann's syndrome): a study using an ultrasensitive time-resolved immunofluorometric assay. *J. Clin. Endocrinol. Metab.* 72, 1229–1237.
294. Dunkel, L., Alfthan, H., Stenman, U. H., Tapanainen, P., and Perheentupa, J. (1990). Pulsatile secretion of LH and FSH in prepubertal and early pubertal boys revealed by ultrasensitive time-resolved immunofluorometric assays. *Pediatr. Res.* 27, 215–219.
295. Cameron, J. L, McNeill, T. H., Fraser, H. M., Bremner, W. J., Clifton, D. K., and Steiner, R. A. (1985). The role of endogenous gonadotropin-releasing hormone in the control of luteinizing hormone and testosterone secretion in the juvenile male monkey, *Macaca fascicularis*. *Biol. Reprod.* 33, 147–156.
296. Wennink, J. M. B., Delemarre-van de Waal, H. A., Schoemaker, R., Schoemaker, H., and Schoemaker, J. (1990). Luteinizing hormone and follicle stimulating hormone secretion patterns in girls throughout puberty measured using highly sensitive immunoradiometric assays. *Clin. Endocrinol.* 33, 333–344.
297. Wennink, J. M. B., Delemarre-van de Waal, H. A., Schoemaker, R., Schoemaker, H., and Schoemaker, J. (1989). Luteinizing hormone and follicle stimulating hormone secretion patterns in boys throughout puberty measured using highly sensitive immunoradiometric assays. *Clin. Endocrinol.* 31, 551–564.
298. Wu, F. C. W., Butler, G. E., Kelnar, C. J. H., and Sellar, R. E. (1990). Patterns of pulsatile luteinizing hormone secretion before and during the onset of puberty in boys: a study using an immunoradiometric assay. *J. Clin. Endocrinol. Metab.* 70, 629–637.
299. Dunger, D. B., Matthews, D. R., Edge, J., Jones, J., and Preece, M. A. (1991). Evidence for temporal coupling of growth hormone, prolactin, LH and FSH pulsatility overnight during normal puberty. *J. Endocrinol.* 130, 141–149.
300. Corley, K. P., Valk, T. W., Kelch, R. P., and Marshall, J. C. (1981). Estimation of GnRH pulse amplitude during pubertal development. *Pediatr. Res.* 15, 157–162.
301. Plant, T. M., and Barker-Gibb, M. L. (2004). Neurobiological mechanisms of puberty in higher primates. *Hum. Reprod. Update* 10, 1–11.
302. Fraser, M. O., Pohl, C. R., and Plant, T. M. (1989). The hypogonadotropic state of the prepubertal male rhesus monkey (*Macaca mulatta*) is not associated with a decrease in hypothalamic gonadotropin releasing hormone content. *Biol. Reprod.* 40, 972–980.
303. Vician, L., Adams, L. A., Clifton, D. K., and Steiner, R. A. (1991). Pubertal changes in proopiomelanocortin and gonadotropin releasing hormone gene expression in the brain of the male monkey. *Mol. Cell. Neurosci.* 2, 31–38.
304. Ma, Y. J., Costa, M. E., and Ojeda, S. R. (1994). Developmental expression of the genes encoding transforming growth factor alpha and its receptor in the hypothalamus of female rhesus macaques. *Neuroendocrinology* 60, 346–359.
305. Barry, J. (1977). Immunofluorescence study of LRF neurons in man. *Cell. Tissue Res.* 181, 1–14.
306. Marshall, P. E., and Goldsmith, P. C. (1980). Neuroregulatory and neuroendocrine GnRH pathways in the hypothalamus and forebrain of the baboon. *Brain Res.* 193, 353–372.
307. Goldsmith, P. C., Lamberts, R., and Brezina, L. R. (1983). Gonadotropin-releasing hormone neurons and pathways in the primate hypothalamus and forebrain. In *Neuroendocrine Aspects of Reproduction* (R. L. Norman, Ed.), pp. 7–45. Academic Press, New York.
308. Silverman, A. J., Antunes, J. L., Ferin, M., and Zimmerman, E. A. (1977). The distribution of luteinizing hormone-releasing hormone in the hypothalamus of the rhesus monkey. *Endocrinology* 101, 134–142.
309. Perera, A. D., and Plant, T. M. (1992). The neurobiology of primate puberty. In *Functional Anatomy of the Neuroendocrine Hypothalamus*. Proceedings of the Ciba Foundation Symposium No. 168, pp. 252–267. John Wiley & Sons, Chichester.
310. Gay, V. L., and Plant, T. M. (1988). Sustained intermittent release of gonadotropin-releasing hormone in the prepubertal male rhesus monkey induced by N-methyl-*DL*-aspartic acid. *Neuroendocrinology* 48, 147–152.
311. Watkins, J. C., and Evans, R. H. (1981). Excitatory amino acid transmitters. *Annu. Rev. Pharmacol. Toxicol.* 21, 165–204.
312. van den Pol, A. N., Wuarin, J. P., and Dudek, F. E. (1990). Glutamate, the dominant excitatory transmitter in neuroendocrine regulation. *Science* 250, 1276–1278.
313. Plant, T. M., and Medhamurthy, R. (1990). Recent studies of the neuroendocrine basis of the prepubertal restraint of

pulsatile GnRH release in primates. In *Neuroendocrine Regulation of Reproduction* (S. S. C. Yen, and W. W. Vale, Eds.), pp. 19–28. Serono Symposia, Norwell.

314. Claypool, L. E., Kasuya, E., Saitoh, Y., Marzban, F., and Terasawa, E. (2000). *N*-methyl D,L-aspartate induces the release of luteinizing hormone-releasing hormone in the prepubertal and pubertal female rhesus monkey as measured by *in vivo* push-pull perfusion in the stalk-median eminence. *Endocrinology* 141, 219–228.
315. Claypool, L. E., Watanabe, G., and Terasawa, E. (1990). Effects of electrical stimulation of the medial basal hypothalamus on the *in vivo* release of luteinizing hormone-releasing hormone in the prepubertal and peripubertal female monkey. *Endocrinology* 127, 3014–3022.
316. Gore, A. C., and Terasawa, E. (1991). A role for norepinephrine in the control of puberty in the female rhesus monkey, *Macaca mulatta*. *Endocrinology* 129, 3009–3017.
317. Perera, A. D., Verbalis, J. G., Mikuma, N., Majumdar, S. S., and Plant, T. M. (1993). Cholecystokinin stimulates gonadotropin-releasing hormone release in the monkey (*Macaca mulatta*). *Endocrinology* 132, 1723–1728.
318. Uhl, G. R., and Nishimori, T. (1990). Neuropeptide gene expression and neural activity: assessing a working hypothesis in nucleus caudalis and dorsal horn neurons expressing preproenkephalin and preprodynorphin. *Cell. Mol. Neurobiol.* 10, 73–98.
319. Ropert, J. F., Quigley, M. E., and Yen, S. S. C. (1981). Endogenous opiates modulate pulsatile luteinizing hormone release in humans. *J. Clin Endocrinol. Metab.* 52, 583–585.
320. Veldhius, J. D., Rogol, A. D., Samojlik, E., and Ertel, N. H. (1984). Role of endogenous opiates in the expression of negative feedback action of androgens and estrogens on pulsatile properties of luteinizing hormone secretion in man. *J. Clin. Invest.* 74, 47–55.
321. Van Vugt, D. A., Lam, N. Y., and Fenn, M. (1984). Reduced frequency of pulsatile luteinizing hormone secretion in the luteal phase of the rhesus monkey. Involvement of endogenous opiates. *Endocrinology* 115, 1095–1101.
322. Fraioli, F., Cappa, M., Fabbri. A., Gnessi, L., Moretti, C., Borrelli, P., and Isidori, A. (1984). Lack of endogenous opioid inhibitory tone on LH secretion in early puberty. *Clin. Endocrinol.* 20, 299–305.
323. Kesner, J. S., Kaufman, J-M., Wilson, R. C., Kuroda, G., and Knobil, E. (1986). The effect of morphine on the electrophysiological activity of the hypothalamic luteinizing hormone-releasing hormone pulse generator activity in the rhesus monkey. *Neuroendocrinology* 43, 686–688.
324. Bicknell, R. J. (1985). Endogenous opioid peptides and hypothalamic neuroendocrine neurons. *J. Endocrinol.* 107, 437–446.
325. Sauder, S. E., Case, G. D., Hopwood, N. J., Kelch, R. P., and Marshall, J. C. (1984). The effects of opiate antagonism on gonadotropin secretion in children and in women with hypothalamic amenorrhea. *Pediatr. Res.* 18, 322–328.
326. Mauras, N., Veldhuis, J. D., and Rogol, A. D. (1986). Role of endogenous opiates in pubertal maturation: opposing actions of naltrexone in prepubertal and late pubertal boys. *J. Clin. Endocrinol. Metab.* 62, 1256–1263.
327. Ulloa-Aguirre, A., Mendez, J. P., Gonzalez-Castillo, A., Carranza-Lira, S., Garza-Flores, J., and Peres-Palacios, G. (1988). Changes in the responsiveness of luteinizing hormone secretion to infusion of the opioid antagonist naloxone throughout male sexual maturation. *Clin. Endocrinol.* 29, 17–28.
328. Petraglia, F., Larizza, D., Maghnie, M., Facchinetti, F., Volpe, A., Bernasconi, S., Genazzani, A. R., and Severi, F. (1988). Impairment of the opioidergic control of luteinizing hormone secretion in Turner's syndrome: lack of effect of gonadal steroid therapy. *J. Clin. Endocrinol. Metab.* 66, 1024–1028.
329. Blank, M. S., and Murphy, J. R. (1991). Luteinizing hormone sensitivity to naloxone in maturing male chimpanzees. *Brain Res. Bull.* 27, 241–245.
330. Medhamurthy, R., Gay, V. L., and Plant, T. M. (1990). The prepubertal hiatus in gonadotropin secretion in the male rhesus monkey (*Macaca mulatta*) does not appear to involve endogenous opioid peptide restraint of hypothalamic gonadotropin-releasing hormone release. *Endocrinology* 126, 1036–1042.
331. Osborn, R. G., Fuller, G. B., Hobson, W. C., Reyes, F. I., and Faiman, C. (1989). Role of endogenous opiates in the modulation of gonadotropin secretion in infant monkeys (*Macaca mulatta*): effects of naltrexone. *J. Med. Primatol.* 18, 405–414.
332. Decavel, D., and van den Pol, A. N. (1990). GABA: a dominant neurotransmitter in the hypothalamus. *J. Comp. Neurol.* 302, 1019–1037.
333. Mitsushima, D., Hei, D. L., and Terasawa, E. (1994). γ-Aminobutyric acid is an inhibitory neurotransmitter restricting the release of luteinizing hormone-releasing hormone before the onset of puberty. *Proc. Natl. Acad. Sci. USA* 91, 395–399.
334. Mitsushima, D., Marzban, F., Luchansky, L. L., Burich, A. J., Keen, K. L., Durning, M., Golos, T. G., and Terasawa, E. (1996). Role of glutamic acid decarboxylase in the prepubertal inhibition of the luteinizing hormone releasing hormone release in female rhesus monkeys. *J. Neuroscience* 16, 2563–2573.
335. Terasawa, E., Luchansky, L. L., Kasuya, E., and Nyberg, C. L. (1999). An increase in glutamate release follows a decrease in gamma aminobutyric acid and the pubertal increase in luteinizing hormone releasing hormone release in the female rhesus monkeys. *J. Neuroendocrinol.* 11, 275–282.
336. Keen, K. L., Burich, A. J., Mitsushima, D., Kasuya, E., and Terasawa, E. (1999). Effects of pulsatile infusion of the $GABA_A$ receptor blocker bicuculline on the onset of puberty in female rhesus monkeys. *Endocrinology* 140, 5257–5266.
337. Urbanski, H. F., Rodrigues, S. M., Garyfallou, V. T., and Kohama, S. G. (1998). Regional distribution of glutamic acid decarboxylase (GAD_{65} and GAD_{67}) mRNA in the hypothalamus of male rhesus before and after pubety. *Mol. Brain Res.* 57, 86–91.
338. Bourguignon, J.-P., Jaeken, J., Gerard, A., and de Zegher, F. (1997). Amino acid neurotransmission and initiation of puberty: evidence from nonketotic hyperglycinemia in a female infant and gonadotropin-releasing hormone secretion by rat hypothalamic explants. *J. Clin. Endocrinol. Metab.* 82, 1899–1903.
339. El-Khayat, H. A., Shatla, H. M., Ali, G. K. H., Abdulgani, M. O., Tomoum, H. U., and Attya, H. A. (2003). Physical and hormonal profile of male sexual development in epilepsy. *Epilepsia* 44, 447–452.
340. Rättyä, J., Vainionpää, L., Knip, M., Lanning, P., and Isojärvi, I. T. (1999). The effects of valproate, carbamazepine, and oxcarbazepine on growth and sexual maturation in girls with epilepsy. *Pediatrics* 103, 588–593.
341. Shahab, M., Balasubramaniam, A., Sahu, A., and Plant, T. M. (2003). Central nervous system receptors involved in mediating the inhibitory action of neuropeptide Y on luteinizing hormone secretion in the male rhesus monkey (*Macaca mulatta*). *J. Neuroendocrinol.* 15, 965–970.
342. Kaynard, A. H., Pau, K.-Y. F., Hess, D. L., and Spies, H. G. (1990). Third-ventricular infusion of neuropeptide Y suppresses luteinizing hormone secretion in ovariectomized rhesus macaques. *Endocrinology* 127, 2437–2444.

343. Pau, K.-Y. F., Berria, M., Hess, D. L., and Spies, H. G. (1995). Hypothalamic site-dependent effects of neuropeptide Y on gonadotropin-releasing hormone secretion in rhesus macaques. *J. Neuroendocrinol.* 7, 63–67.
344. Barker-Gibb, M., Plant, T. M., White, C., Lee, P, A., and Witchel, S. F. (2004). Genotype analysis of the neuropeptide Y (NPY) Y1 and NPY Y5 receptor genes in gonadotropin releasing hormone (GnRH) dependent precocious gonadarche. *Fert. Steril.* 82, 491–494.
345. Freitas, K. C., Brito, V. N., Arnhold. I. J., Mendonca, B. B., and Latronico, A. C. (2003). Molecular analysis of two distinct candidate genes (GABRA1 and NPY-Y1) that could be implicated in the human puberty timing. 85th Annual Meeting of The Endocrine Society, Philadelphia, p. 70, Abstract No. OR8-5.
346. Gore, A. C., Mitsushima, D., and Terasawa, E. (1993). A possible role of neuropeptide Y in the control of the onset of puberty in female rhesus monkeys. *Neuroendocrinology* 58, 23–34.
347. Woller, M. J., and Terasawa, E. (1991). Infusion of neuropeptide Y into the stalk-median eminence stimulates *in vivo* release of luteinizing hormone-releasing hormone in gonadectomized rhesus monkeys. *Endocrinology* 128, 1144–1150.
348. Kasuya, E., Mizuno, M., Watanabe, G., and Terasawa, E. (1998). Effects of an antisense oligodeoxynucleotide for neuropeptide Y mRNA on *in vivo* luteinizing hormone-releasing hormone release in ovariectomized female rhesus monkeys. *Regul. Pept.* 75–76, 319–325.
349. Woller, M. J., McDonald, J. K., Reboussin, D. M., and Terasawa, E. (1992). Neuropeptide Y is a neuromodulator of pulsatile luteinizing hormone-releasing hormone release in the gonadectomized rhesus monkey. *Endocrinology* 130, 2333–2342.
350. Plant, T. M. (2001). Neurobiological bases underlying the control of the onset of puberty in the rhesus monkey: a representative higher primate. *Front. Neuroendocrinol.* 22, 107–139.
351. Medhamurthy, R., Gay, V. L., and Plant, T. M. (1992). Repetitive injections of *L*-glutamic acid, in contrast to those of N-methyl-*D,L*-aspartic acid, fail to elicit sustained hypothalamic GnRH release in the prepubertal male rhesus monkey (*Macaca mulatta*). *Neuroendocrinology* 55, 660–666.
352. Grosser, P. M., Williams, C. L., Thalabard, J. C., Nishihara, M., Brody, S., and Knobil, E. (1989). N-methyl-D,L-aspartate (NMDA) receptor blockade of the hypothalamic gonadotropin-releasing hormone (GnRH) pulse generator activity in the rhesus monkey. 71st Annual Meeting of The Endocrine Society, Abstract No. 942.
353. Goldsmith, P. C., Thind, K. K., Perera, A. D., and Plant, T. M. (1994). Glutamate-immunoreactive neurons and their gonadotropin-releasing hormone-neuronal interactions in the monkey hypothalamus. *Endocrinology* 134, 858–868.
354. de Roux, N., Genin, E., Carel, J-C., Matsuda, F., Chaussain, J-L., and Milgrom, E., (2003). Hypogonadotropic hypogonadism due to loss of function of the KiSS1-derived peptide receptor GPR54. *Proc. Natl. Acad. Sci.* 100, 10972–10976.
355. Seminara, S. B., Messager, S., Chatzidaki, E. E., Thresher, R. R., Acierno, J. S. Jr., Shagoury, J. K., Bo-Abbas, Y., Kuohung, W., Schwinof, K. M., Hendrick, A. G., Zahn, D., Dixon, J., Kaiser, U. B., Slaugenhaupt, S. A., Gusella, J. F., O'Rahilly, S., Carlton, M. B. L., Crowley, W. F. Jr., Aparicio, S. A. J. R., and Colledge, W. H. (2004). The *GPR54* gene as a regulator of puberty. *N. Engl. J. Med.* 349, 1614–1627.
356. Lee, D. K., Nguyen, T., O'Neill, G. P., Cheng, R., Liu, Y., Howard, A. D., Coulombe, N., Tan, C. P., Tang-Nguyen, A.-T., George, S. R., and O'Dowd, B. F. (1999). Discovery of a receptor related to galanin receptors. *FEBS Lett.* 446, 103–107.
357. Harms, J. F., Welch, D. R., and Miele, M. E. (2003). KiSS1 metastasis suppression and emergent pathways. *Clin. Exp. Metastasis* 20, 11–18.
358. Shahab, M., Mastronardi, C., Seminara, S. B., Crowley, W. F., Ojeda, S. R., and Plant, T. M. (2005). Increased hypothalamic GPR54 signaling: a potential mechanism for initiation of puberty in primates. *Proc. Natl. Acad. Sci. USA* 102, 2129–2134.
359. Parhar, I. S., Ogawa, S., and Sakuma, Y. (2004). Laser-captured single digoxigenin-labeled neurons of gonadotropin-releasing hormone types reveal a novel G protein-coupled receptor (Gpr54) during maturation in cichlid fish. *Endocrinology* 145, 3613–3618.
360. Shibata, M., Gibbs, R. B., Shahab, M., and Plant, T. M. (2005). GnRH neurons in the peripubertal male rhesus monkey (*Macaca mulatta*) express GPR54: Implication for the control of primate puberty. 87th Annual Meeting of The Endocrine Society, June 2005, San Diego, Abstract. No. p1–98.
361. Navarro, V. M., Castellano, J. M., Fernàndez- Fernàndez, R., Tovar, S., Roa, J., Mayen, A., Nogueiras, R., Vazquez, M. J., Barreiro, M. L., Magni, P., Aguilar, E., Dieguez, C., Pinilla, L., and Tena-Sempere, M. (2005). Characterization of the potent luteinizing hormone-releasing activity of KiSS–1 peptide, the natural ligand of GPR54. *Endocrinology* 146, 156–163.
362. Kaufman, J.-M., Kesner, J. S., Wilson, R. C., and Knobil, E. (1985). Electrophysiological manifestation of luteinizing hormone-releasing hormone pulse generator activity in the rhesus monkey: influence of α-adrenergic and dopaminergic blocking agents. *Endocrinology* 116, 1327–1333.
363. Thind, K. K., and Goldsmith, P. C. (1988). Infundibular gonadotropin-releasing hormone neurons are inhibited by direct opioid and autoregulatory synapses in juvenile monkeys. *Neuroendocrinology* 47, 203–216.
364. Thind, K. K., Boggan, J. E., and Goldsmith, P. C. (1991). Interactions between vasopressin- and gonadotropin-releasing-hormone-containing neuroendocrine neurons in the monkey supraoptic nucleus. *Neuroendocrinology* 53, 287–297.
365. Junier, M.-P., Ma, Y. J., Costa, M. E., Hoffman, G., Hill, D. F., and Ojeda, S. R. (1991). Transforming growth factor α contributes to the mechanism by which hypothalamic injury induces precocious puberty. *Proc. Natl. Acad. Sci. USA* 88, 9743–9747.
366. Junier, M.-P., Hill, D. F., Costa, M. E., Felder, S., and Ojeda, S. R. (1993). Hypothalamic lesions that induce female precocious puberty activate glial expression of the epidermal growth factor receptor gene: differential regulation of alternatively spliced transcripts. *J. Neurosci.* 13, 703–713.
367. Brauner, R., Rappaport, R., Nicod, C., Malandry, F., Thibaud, E., Pierre-Kahn, A., Renier, D., Sainte-Rose, C., and Hirsch, J. F. (1987). Pubertes precoces vraies au cours de hydrocephalie non tumonale. *Arch. Fr. Pediatr.* 44, 433–436.
368. Sockalosky, J. J., Kriel, R. L., Krach, L. E., and Sheehan, M. (1987). Precocious puberty after traumatic brain injury. *J. Pediatr.* 110, 373–377.
369. Brauner, R., and Rappaport, R. (1985). Precocious puberty secondary to cranial irradiation for tumors distant from the hypothalamo-pituitary area. *Horm. Res.* 22, 78–82.
370. Terasawa, E., Noonan, J. J., Nass, T. E., and Loose, M. D. (1984). Posterior hypothalamic lesions advance the onset of puberty in the female rhesus monkey. *Endocrinology* 115, 2241–2250.
371. Ojeda, S. R., and Heger, S. (2001). New thoughts on female precocious puberty. *J. Pediatr. Endocrinol. Metab.* 14, 245–56.
372. Jung, H., Carmel, P., Schwartz, M. S., Witkin, J. W., Bentele, K. H. P., Westphal, M., Piatt, J. H., Costa, M. E., Cornea, A., Ma, Y. J., and Ojeda, S. R. (1999). Some hypothalamic

hamartomas contain transforming growth factor α, a puberty-inducing growth factor, but not luteinizing hormone-releasing hormone neurons. *J. Clin. Endocrinol. Metab.* 84, 4695–4701.

373. Theodosis, D. T., Rougon, G., and Poulain, D. A. (1991). Retention of embryonic features by an adult neuronal system capable of plasticity: polysialated neural cell adhesion molecule in the hypothalamc-neurohypophysial system. *Proc. Natl. Acad. Sci. USA* 88, 5494–5498.
374. Perera, A. D., Langenaur, C. F., and Plant, T. M. (1993). Postnatal expression of polysialic acid- neural cell adhesion molecule in the hypothalamus of the male rhesus monkey (*Macaca mulatta*). *Endocrinology* 133, 2729–2735.
375. Goldsmith, P. C., Thind, K. K., Song, T., Kim, E, J., Boggan, J. E. (1990). Location of the neuroendocrine gonadotropin-releasing hormone neurons in the monkey hypothalamus by retrograde tracing and immunostaining. *J. Neuroendocrinol.* 2, 157–168.
376. Perera, A. D., and Plant, T. M. (1997). Ultrastructural studies of neuronal correlates of the pubertal reaugmentation of hypothalamic gonadotropin-releasing hormone (GnRH) release in the rhesus monkey (*Macaca mulatta*). *J. Comp. Neurol.* 385, 71–82.
377. Witkin, J. W., O'Sullivan, H., and Ferin, M. (1995). Glial ensheathment of GnRH neurons in pubertal female rhesus macaques. *J. Neuroendocrinol.* 7, 665–671.
378. Semple, R. K., Achermann, J. C., Ellery, J., Farooqi, I. S., Karet, F. E., Stanhope, R. G., O'Rahilly, S. O., and Aparicio, S. A. (2005). Two novel missense mutations in GPR54 in a patient with hypogonadotropic hypogonadism. *J. Clin. Endocrin. Metab.* 90, 1849–1855.
379. El Majdoubi, M., Sahu, A., and Plant, T. M. (2000). Changes in hypothalamic gene expression associated with the arrest of pulsatile gonadotropin-releasing hormone release during infancy in the agonadal male rhesus monkey (*Macaca mulatta*). *Endocrinology* 141, 3273–3277.
380. Ojeda, S. R., and Terasawa, E. (2002). Neuroendocrine regulation of puberty. In *Hormones, Brain and Behavior* (D. W. Pfaff, A. P. Arnold, A. M. Etgen, S. E. Fahrbach, and R. T. Rubin, Eds.), Vol. 4, pp. 589–659. Elsevier Science, San Diego.
381. Cherubini, E., Gaiarsa, J. L., and Ben-Ari, Y. (1991). GABA: an excitatory transmitter in early postnatal life. *Trends Neurosci.* 14, 515–519.
382. Norman, R. L., and Spies, H. G. (1981). Brain lesions in infant female rhesus monkeys: effects on menarche and first ovulation and on diurnal rhythms of prolactin and cortisol. *Endocrinology* 108, 1723–1729.
383. Krey, L. C., Hess, D. L., Butler, W. R., Espinosa-Campos, J., Lu, K. H., Piva, F., Plant, T. M., and Knobil, E. (1981). Medial basal hypothalamic disconnection and the onset of puberty in the female rhesus monkey. *Endocrinology* 108, 1944–1948.
384. Krey, L. C., Lu, K. H., Butler, W. R., Hotchkiss, J., Piva, F., and Knobil, E. (1975). Surgical disconnection of the medial basal hypothalamus and pituitary function in the rhesus monkey. II. GH and cortisol secretion. *Endocrinology* 96, 1088–1093.
385. Bullough, W. S. (1951). *Vertebrate Sexual Cycles*. Methuen, London.
386. Simmons, K., and Greulich, W. W. (1943). Menarcheal age and the height, weight, and skeletal age of girls age 7 to 17 years. *J. Pediatr.* 22, 518–548.
387. Marshall, W. A. (1974). Interrelationships of skeletal maturation, sexual development and somatic growth in man. *Ann. Hum. Biol.* 1, 29–40.
388. Frisch, R. E., and Revelle, R. (1970). Height and weight at menarche and a hynothesis of critical body weights and adolescent events. *Science* 169, 397–399.
389. Kennedy, G. C., and Mitra, J. (1963). Body weight and food intake as initiating factors for puberty in the rat. *J. Physiol.* 166, 408–418.
390. Frisch, R. E., Revelle, R., and Cook, S. (1973). Components of weight at menarche and the initiation of the adolescent growth spurt in girls: estimated total water, lean body weight and fat. *Hum. Biol.* 45, 469–483.
391. Johnston, F. E., Malina, R. M., Galbraith, M. A., Frisch, R. E., Revelle, R., and Cook S. (1971). Height, weight and age at menarche and the "critical weight" hypothesis. *Science* 174, 1148–1149.
392. Johnston, F. E., Roche, A. F., Schell, L. M., and Wettenhall, N. B. (1975). Critical weight at menarche. *Am. J. Dis. Child.* 129, 19–23.
393. Trussell, J. (1978). Menarche and fatness: reexamination of the critical body composition hypothesis. *Science* 200, 1506–1509.
394. Reeves, J. (1979). Estimating fatness. *Science* 204, 881.
395. Ellison, R. T. (1981). Threshold hypotheses, developmental age, and menstrual function. *Am. J. Phys. Anthropol.* 54, 337–340.
396. Friedman, J. M., and Halaas, J. L. (1998). Leptin and the regulation of body weight in mammals. *Nature* 395, 763–770.
397. Schwartz, M. W., Woods, S. C., Porte, D. Jr., Seeley, R. J., and Baskin, D. G. (2000). Central nervous system control of food intake. *Nature* 404, 661–671.
398. Ahmed, M. L., Ong, K. K., Morell, D. J., Cox, L., Drayer, M. Perry, L., Preece, M. A., and Dunger, D. B. (1999). Longitudinal study of leptin concentrations during puberty: sex differences and relationship to changes in body composition. *J. Clin. Endocrinol. Metab.* 84, 899–905.
399. Ankarberg-Lindgren, C., Dahlgren, J., Carlsson, B., Rosberg, S., Carlsson, L., Wikland, K. A., and Norjavaara, E. (2001). Leptin levels show diurnal variation throughout puberty in healthy children, and follow a gender-specific pattern. *Eur. J. Endocrinol.* 145, 43–51.
400. Carlsson, B., Ankarberg, C., Rosberg, S., Norjavaara, E., Albertsson-Wikland, K., and Carlsson, L. M. (1997). Serum leptin concentrations in relation to pubertal development. *Arch. Dis. Child.* 77, 396–340.
401. Clément, K., Vaisse, C., Lahlou, N., Cabrol, S., Pelloux, V., Cassuto, D., Gourmelen, M., Dina, C., Chambaz, J., Lacorte, J.-M., Basdevant, A., Bougnères, P., Lebouc, Y., Froguel, P., and Guy-Grand, B. (1998). A mutation in the human leptin receptor gene causes obesity and pituitary dysfunction. *Nature* 392, 398–401.
402. Garcia-Mayor, R. V., Andrade, M. A., Rios, M., Lage, M., Dieguez, C., and Casanueva, F. F. (1997). Serum leptin levels in normal children: relationship to age, gender, body mass index, pituitary-gonadal hormones, and pubertal stage. *J. Clin. Endocrinol. Metab.* 82, 2849–2855.
403. Kratzsch, J., Lammert, A., Bottner, A., Seidel, B., Mueller, G., Thiery, J., Hebebrand, J., and Kiess, W. (2002). Circulating soluble leptin receptor and free leptin index during childhood, puberty, and adolescence. *J. Clin. Endocrinol. Metab.* 87, 4587–4594.
404. Mann, D. R., Johnson, A. O. K., Gimpel, T., and Castracane, V. D. (2003). Changes in circulating leptin, leptin receptor, and gonadal hormones from infancy until advanced age in humans. *J. Clin. Endocrinol. Metab.* 88, 3339–3345.
405. Mantzoros, C. S., Flier, J. S., and Rogol, A. D. (1997). A longitudinal assessment of hormonal and physical alterations during normal puberty in boys. V. Rising leptin levels may signal the onset of puberty. *J. Clin. Endocrinol. Metab.* 82, 1066–1070.
406. Mann, D. R., and Plant, T. M. (2002). Leptin and pubertal development. In *Seminars in Reproductive Medicine: Leptin and Reproduction* (V. D. Castracane and M. C. Henson, Eds.), pp. 93–102. Thieme Medical Publishers, New York.

407. Gill, M. S., Hall, C. M., Tillmann, V., and Clayton, P. E. (1999). Constitutional delay in growth and puberty (CDGP) is associated with hypoleptinaemia. *Clin. Endocrinol.* 50, 721–726.
408. Wilson, M. E., Fisher, J., Chikazawa, K., Yoda, R., Legendre, A., Mook, D., and Gould, K. G. (2003). Leptin administration increases nocturnal concentrations of luteinizing hormone and growth hormone in juvenile female rhesus monkeys. *J. Clin. Endocrinol. Metab.* 88, 4874–4883.
409. Plant, T. M., and Durrant, A. R. (1997). Circulating leptin does not appear to provide a signal for triggering the initiation of puberty in the male rhesus monkey (*Macaca mulatta*). *Endocrinology* 138, 4505–4508.
410. Mann, D. R., Akinbami, M. A., Gould, K. G., and Castracane, V. D. (2000). A longitudinal study of leptin during development in the male rhesus monkey: the effect of body composition and season on circulating leptin levels. *Biol. Reprod.* 62, 285–291.
411. Urbanski, H. F., and Pau, K.-Y. F. (1998). A biphasic developmental pattern of circulating leptin in the male rhesus macaque (*Macaca mulatta*). *Endocrinology* 139, 2284–2286.
412. Suter, K. J., Pohl, C. R., and Wilson, M. E. (2000). Circulating concentrations of nocturnal leptin, growth hormone, and insulin-like growth factor-I increase before the onset of puberty in agonadal male monkeys: potential signals for the initiation of puberty. *J. Clin. Endocrinol. Metab.* 85, 808–814.
413. Barker-Gibb, M., Sahu, A., Pohl, C. R., and Plant, T. M. (2002). Elevating circulating leptin in prepubertal male rhesus monkeys (*Macaca mulatta*) does not elicit precocious gonadotropin-releasing hormone release, assessed indirectly. *J. Clin. Endocrinol. Metab.* 87, 4976–4983.
414. Farooqi, I. S., Matarese, G., Lord, G. M., Keogh, J. M., Lawrence, E., Agwu, C., Sanna, V., Jebb, S. A., Perna, F., Fontana, S., Lechler, R. I., DePaoli, A. M., and O'Rahilly, S. (2002). Beneficial effects of leptin on obesity, T cell hyporesponsiveness, and neuroendocrine/metabolic dysfunction of human congenital leptin deficiency. *J. Clin. Invest.* 110, 1093–1103.
415. Palmert, M. R., Radovick, S., and Boepple, P. A. (1998). Leptin levels in children with central precocious puberty. *J. Clin. Endocrinol. Metab.* 83, 2260–2265.
416. Heger, S., Partsch, C.-J., Peter, M., Blum, W. F., Kiess, W., and Sippell, W. G. (1999). Serum leptin levels in patients with progressive central precocious puberty. *Pediatr. Res.* 46, 71–75.
417. Strobel, A., Issad, T., Camion, L., Ozata, M., and Strosberg, A. D. (1998). A leptin missense mutation associated with hypogonadism and morbid obesity. *Nat. Genet.* 18, 213–215.
418. Ozata, M., Ozdemir, I. C., Caglayan, I., and Licinio, J. (1999). Human leptin deficiency caused by a missense mutation: multiple endocrine defects, decreased sympathetic tone, and immune system dysfunction indicate new targets for leptin action, greater central than peripheral resistance to the effects of leptin, and spontaneous correction of leptin-mediated defects. *J. Clin. Endocrinol. Metab.* 84, 3686–3695.
419. Andreelli, F., Hanaire-Broutin, H., Laville, M., Tauber, J. P., Riou, J. P., and Thivolet, C. (2000). Normal reproductive function in leptin-deficient patients with lipoatropic diabetes. *J. Clin. Endocrinol. Metab.* 85, 715–719.
420. Galler, A., Schuster, V., and Kiess, W. (2001). Pubertal adipose tissue: is it really necessary for normal sexual maturation? *Eur. J. Endocrinol.* 145, 807–808.
421. Farooqi, I. S., Jebb, S. A., Langmack, G., Lawrence, E., Cheetham, C. H., Prentice, A. M., Hughes, I. A., McCamish, M. A., and O'Rahilly, S. (1999). Effects of recombinant leptin therapy in a child with congenital leptin deficiency. *N. Engl. J. Med.* 341, 879–884.
422. Plant, T. M., Fraser, M. O., Medhamurthy, R., and Gay, V. L. (1989). Somatogenic control of GnRH neuronal synchronization during development in primates: a speculation. In *Control of the Onset of Puberty III* (H. A. Delemarre-van de Waal, T. M. Plant, G. P. van Rees, and J. Schoemaker, Eds.), pp. 111–121. Elsevier, Amsterdam.
423. Flor-Cisneros, A., Leschek, E. W., Merke, D. P., Barnes, K. M., Coco, M., Cutler, G. B. Jr., and Baron, J. (2004). In boys with abnormal developmental tempo, maturation of the skeleton and the hypothalamic-pituitary- gonadal axis remains synchronous. *J. Clin. Endocrinol. Metab.* 89, 236–241.
424. Fraser, M. O., Arslan, M., and Plant, T. M. (2005). Androgen and estrogen treatment, alone, or in combination differentially influence bone maturation and hypothalamic mechanisms timing puberty in the male rhesus monkey (*Macaca mulatta*). *Ped. Res.* 57, 141–148.
425. Oral, E. A., Ruiz, E., Andewelt, A., Sebring, N., Wagner, A. J., DePaoli, A. M., and Gorden, P. (2002). Effect of leptin replacement on pituitary hormone regulation in patients with severe lipodystrophy. *J. Clin. Endocrinol. Metab.* 87, 3110–3117.
426. Köpp, W., Blum, W. F., von Prittwitz, S., Ziegler, A., Lübbert, H., Emons, G., Herzog, W., Herpertz, S., Deter, H. C., Remschmidt, H., and Hebebrand, J. (1997). Low leptin levels predict amenorrhea in underweight and eating disordered females. *Mol. Psychiat.* 2, 335–340.
427. Cheung, C. C., Thornton, J. E., Nurani, S. D., Clifton, D. K., and Steiner, R. A. (2001). A reassessment of leptin's role in triggering the onset of puberty in the rat and mouse. *Neuroendocrinology* 74, 12–21.
428. Bhat, G. K., Plant, T. M., and Mann, D. R. (2005). Relationship between serum concentrations of leptin, soluble leptin receptor, testosterone and IGF-1, and growth during the first year of postnatal life in the male rhesus monkey (*Macaca mulatta*). *Eur J. Endocrinol* 153, 153–158.
429. Tanner, J. M., and Whitehouse, R. H. (1975). A note on the bone age at which patients with true isolated growth hormone deficiency enter puberty. *J. Clin. Endocrinol. Metab.* 41, 788–790.
430. Giustina, A., and Veldhuis, J. D. (1998). Pathophysiology of the neuroregulation of growth hormone secretion in experimental animals and the human. *Endocr. Rev.* 19, 717–797.
431. Finkelstein, J. W., Roffwarg, H. P., Boyar, R. M., Kream, J., and Hellman, L. (1972). Age-related change in the twenty-four-hour spontaneous secretion of growth hormone. *J. Clin. Endocrinol. Metab.* 35, 665–670.
432. Zadik, Z., Chalew, S. A., McCarter, R. J. Jr., Meistas, M., and Kowarski, A. A. (1985). The influence of age on the 24-hour integrated concentration of growth hormone in normal individuals. *J. Clin. Endocrinol. Metab.* 60, 513–516.
433. Wennink, J. M. B., Delemarre-van de Waal, H. A., Schoemaker, R., Blaauw, G., van den Braken, C., and Schoemaker, J. (1990). Growth hormone secretion patterns in relation to LH and testosterone secretion throughout normal male puberty. *Acta Endocrinol.* 123, 263–270.
434. Wennick, J. M. B., Delemarre-van de Waal, H. A., Schoemaker, R., Blaauw, G., van den Braken, C., and Schoemaker, J. (1991). Growth hormone secretion patterns in relation to LH and estradiol secretion throughout normal female puberty. *Acta Endocrinol.* 124, 129–135.
435. Martha, P. M. Jr., Rogol, A. D., Veldhuis, J. D., Kerrigan, J. R., Goodman, D. W., and Blizzard, R. M. (1989). Alterations in the pulsatile properties of circulating growth hormone concentrations during puberty in boys. *J. Clin. Endocrinol. Metab.* 69, 563–570.
436. Martha, P. M. Jr., German, K. M., Blizzard, R. M., Rogol, A. D., and Veldhuis, J. D. (1992). Endogenous growth hormone secretion and clearance rates in normal boys, as determined by deconvolution analysis: relationship to age, pubertal status, and body mass. *J. Clin. Endocrinol. Metab.* 74, 336–344.

437. Rosenfield, R. I., Furlanetto, R., and Bock, D. (1983). Relationship of somatomedin-C concentrations to pubertal changes. *J. Pediatr.* 103, 723–728.
438. Luna, A. M., Wilson, D. M., Wibbelsman, C. J., Brown, R. C., Nagashima, R. J., Hintz, R. L., and Rosenfeld, R. G. (1983). Somatomedins in adolescence: a cross-sectional study of the effect of puberty on plasma insulin-like growth factor I and II levels. *J. Clin. Endocrinol. Metab.* 57, 268–271.
439. Harris, D. A., van Vliet, G., Egli, C. A., Grumbach, M. M., Kaplan, S. L., Styne, D. M., and Vainsel, M. (1985). Somatomedin-C in normal puberty and in true precocious puberty before and after treatment with potent luteinizing hormone-releasing hormone agonist. *J. Clin. Endocrinol. Metab.* 61, 152–159.
440. Johansen, J. S., Giwercman, A., Hartwell, D., Nielsen, C. T., Price, P. A., Christiansen, C., and Skakkebaek, N. E. (1988). Serum bone Gla-protein as a marker of bone growth in children and adolescents: correlation with age, height, serum insulin-like growth factor I, and serum testosterone. *J. Clin. Endocrinol. Metab.* 67, 273–278.
441. Wilson, M. E., Gordon, T. P., Blank. M. S., and Collins, D. C. (1984). Timing of sexual maturity in female rhesus monkeys (*Macaca mulatta*) housed outdoors. *J. Reprod. Fertil.* 70, 625–633.
442. Wilson, M. E., and Tanner, J. M. (1994). Somatostatin analog treatment slows growth and the tempo of reproductive maturation in female rhesus monkeys. *J. Clin. Endocrinol. Metab.* 79, 495–501.
443. Levine Ross, L., Myerson Long, L., Loriaux, D. L., and Cutler, G. B. Jr. (1985). Growth hormone secretory dynamics in Turner syndrome. *J. Pediatr.* 106, 202–206.
444. Parker, M. W., Johanson, A. J., Rogol, A. D., Kaiser, D. L., and Blizzard, R. M. (1984). Effect of testosterone on somatomedin-C concentrations in pre-pubertal boys. *J. Clin. Endocrinol. Metab.* 58, 87–90.
445. Rosenfeld, R. L., and Furlanetto, R. (1985). Physiologic testosterone or estradiol induction of puberty increases plasma somatomedin-C. *J. Pediatr.* 107, 415–417.
446. Link, K., Blizzard, R. M., Evans, W. S., Kaiser, D. L., Parker, M. W., and Rogol, A. D. (1986). The effect of androgens on the pulsatile release and the twenty-four-hour mean concentration of growth hormone in pre-pubertal males. *J. Clin. Endocrinol. Metab.* 62, 159–164.
447. Mauras, N., Rogol, A. D., and Veldhuis, J. D. (1989). Specific, time-dependent actions of low-dose ethinyl estradiol administration on the episodic release of growth hormone, follicle-stimulating hormone, and luteinizing hormone in prepubertal girls with Turner's syndrome. *J. Clin. Endocrinol. Metab.* 69, 1053–1058.
448. Mansfield, M. J., Rudlin, C. R., Crigler, J. F. Jr., Karol, K. A., Crawford, J. D., Boepple, P. A., and Crowley, W. F. Jr. (1988). Changes in growth and serum growth hormone and plasma somatomedin-C levels during suppression of gonadal sex steroid secretion in girls with central precocious puberty. *J. Clin. Endocrinol. Metab.* 66, 3–9.
449. Suter, K. J. (2004). The ontogeny of pulsatile growth hormone secretion and its temporal relationship to the onset of puberty in the agonadal male rhesus monkey (*Macaca mulatta*). *J. Clin. Endocrinol. Metab.* 89, 2275–2280.
450. Wilson, M. E., Gordon, T. P., Rudman, C. G., and Tanner, J. M. (1989). Effects of growth hormone on the tempo of sexual maturation in female rhesus monkeys. *J. Clin. Endocrinol. Metab.* 68, 29–38.
451. Wilson, M. E. (1998). Premature elevation in serum insulin-like growth factor-I advances first ovulation in rhesus monkeys. *J. Endocrinol.* 158, 247–257.
452. Wilson, M. E., Chikazwa, K., Fisher, J., Mook, D., and Gould, K. G. (2004). Reduced growth hormone secretion prolongs puberty but does not delay the developmental increase in luteinizing hormone in the absence of gonadal negative feedback. *Biol. Reprod.* 71, 588–597.
453. Wilson, M. E. (1995). IGF-I administration advances the decrease in hypersensitivity to oestradiol negative feedback inhibition of serum LH in adolescent female rhesus monkeys. *J. Endocrinol.* 145, 121–130.
454. Silman, R. E., Leone, R. M., Hooper, R. J. L., and Preece, M. A. (1979). Melatonin, the pineal gland and human puberty. *Nature* 282, 301–303.
455. Waldhauser, F., Weiszenbacher, G., Frisch, H., Zeitlhuber, U., Waldhauser, M., and Wurtman, R. J. (1984). Fall in nocturnal serum melatonin during prepuberty and pubescence. *Lancet* 1, 362–365.
456. Waldhauser, F., Weiszenbacher, G., Tatzer, E., Gisinger, B., Waldhauser, M., Schemper, M., and Frisch, H. (1988). Alterations in nocturnal serum melatonin levels in humans with growth and aging. *J. Clin. Endocrinol. Metab.* 66, 648–652.
457. Attanasio, A., Borrelli, P., and Gupta, D. (1985). Circadian rhythms in serum melatonin from infancy to adolescence. *J. Clin. Endocrinol. Metab.* 61, 388–390.
458. Wilson, M. E., and Gordon, T. P. (1988). Nocturnal changes in serum melatonin during female puberty in rhesus monkeys: a longitudinal study. *J. Endocrinol.* 121, 553–562.
459. Salti, R., Galluzzi, F., Bindi, G., Perfetto, F., Tarquini, R., Halberg, F., and Cornélissen, G. (2000). Nocturnal melatonin patterns in children. *J. Clin. Endocrinol. Metab.* 85, 2137–2144.
460. Tetsuo, M., Poth, M., and Markey, S. P. (1982). Melatonin metabolite excretion during childhood and puberty. *J. Clin. Endocrinol. Metab.* 55, 311–313.
461. Young, I. M., Francis, P. L., Leone, A. M., Stovell, P., and Silman, R. E. (1988). Constant pineal output and increasing body mass account for declining melatonin levels during human growth and sexual maturation. *J. Pineal. Res.* 5, 71–85.
462. Cavallo, C., and Dolan, L. M. (1996). 6-Hydroxymelatonin sulfate excretion in human puberty. *J. Pineal Res.* 21, 225–230.
463. Commentz, J. C., Uhlig, H., Henke, A., Hellwege, H. H., and Willig, R. P. (1997). Melatonin and 6-hydroxymelatonin sulfate excretion is inversely correlated with gonadal development in children. *Hormone Res.* 47, 97–101.
464. Ariës Kappers, J. (1981). A survey of advances in pineal research. In *The Pineal Gland, Vol. 1*, (R. J. Reiter, Ed.), pp. 1–25. CRC Press, Boca Raton, FL.
465. Kitay, J. I., and Altschule, M. D. (1954). 'The Pineal Gland. A Review of the Physiological Literature." Harvard University Press, Cambridge.
466. Reichlin, S. (1981). Neuroendocrinology. In *Textbook of Endocrinology* (R. H. Williams, Ed.), pp. 588–645. W. B. Saunders, Philadelphia.
467. Brown, G. M., and Niles, L. P. (1982). Studies on melatonin and other pineal factors. In *Clinical Neuroendocrinology* (G. M. Besser, and L. Martini, Eds.), Vol. 2, pp. 205–264. Academic Press, New York.
468. Waldhauser, F., Boepple, P. A., Schemper, M., Mansfield, M. J., and Crowley, W. F. Jr. (1991). Serum melatonin in central precocious puberty is lower than in age-matched prepubertal children. *J. Clin. Endocrinol. Metab.* 73, 793–796.
469. Berga, S. L., Jones, K. L., Kaufmann, S., Yen, S. S. C. (1989). Nocturnal melatonin levels are unaltered by ovarian suppression in girls with central precocious puberty. *Fertil. Steril.* 52, 936–941.
470. Wilson, M. E., Lackey, S., Chikazawa, K, Gordon, T. (1993). The amplitude of nocturnal melatonin concentrations is not decreased by oestradiol and does not alter reproductive

functions in adolescent or adult female rhesus monkeys. *J. Endocrinol.* 137, 299–309.
471. Plant, T. M., and Zorub, D. S. (1986). Pinealectomy in agonadal infantile male rhesus monkeys (*Macaca mulatta*) does not interrupt initiation of the prepubertal hiatus in gonadotropin secretion. *Endocrinology* 118, 221–232.
472. Wilson, M. E., and Gordon, T. P. (1989). Short-day melatonin pattern advances puberty in seasonally breeding rhesus monkeys (*Macaca mulatta*). *J. Reprod. Fertil.* 86, 435–444.
473. Steiner, R. A., Cameron, J. L., McNeill, T. H., Clifton, D. K., and Bremner, W. J. (1983). Metabolic signals for the onset of puberty. In *Neuroendocrine Aspects of Reproduction* (R. L. Norman, Ed.), pp. 183–227. Academic Press, New York:
474. Cameron, J. L., Hansen, P. P., McNeill, T. H., Koerker, D. J., Clifton, D. K., Rogers, K. V., Bremner, W. J., and Steiner, R. A. (1985). Metabolic cues for the onset of puberty in primate species. In *Adolescence in Females* (C. Flagmini, S. Venturoli, and J. R. Givens, Eds.), pp. 59–78. Yearbook Medical Publishers, Chicago.
475. Boyar, R. M., Katz, J., Finkelstein, J. W., Kapen, S., Weiner, H., Weitzman, E. D., and Hellman, L. (1974). Anorexia nervosa. Immaturity of the 24-hour luteinizing hormone secretory pattern. *N. Engl. J. Med.* 291, 861–865.
476. Marshall, J. C., and Kelch, R. P. (1979). Low dose pulsatile gonadotropin-releasing hormone in anorexia nervosa: a model for human pubertal development. *J. Clin. Endocrinol. Metab.* 49, 712–718.
477. Winterer, J., Cutler, G. B. Jr., and Loriaux, D. L. (1984). Caloric balance, brain to body ratio, and the timing of menarche. *Med. Hypotheses* 15, 87–91.
478. Kauli, R., Gusewitz, R., Galazer, A., Prager-Lewin, R., Gil-Ad, I., and Laron, Z. (1982). Effect of anorexia nervosa on gonadotropin secretion in a patient with gonadal dysgenesis. *Ada. Endocrinol.* 100, 363–368.
479. Dubey, A. K., Cameron, J. L., Steiner, R. A., and Plant, T. M. (1986). Inhibition of gonadotropin secretion in castrated male rhesus monkeys (*Macaca mulatta*) induced by dietary restriction: analogy with the prepubertal hiatus of gonadotropin release. *Endocrinology* 118, 518–525.
480. Connan, F., Campbell, I. C., Katzman, M., Lightman, S. L., and Treasure, J. (2003). A neurodevelopmental model for anorexia nervosa. *Physiol. Behav.* 79, 13–24.
481. Kaye, W. H. (1997). Anorexa nervosa, obsessional behavior, and serotonin. *Psychopharmacol. Bull.* 33, 335–344.
482. Gorwood, P., Kipman, A., and Foulon, C. (2003). The human genetics of anorexia nervosa. *Eur. J. Pharmacol.* 480, 163–170.
483. Haymond, M. W, Karl, I. E., Clarke, W. L., Pagliara, A. S., and Santiago, J. V. (1982). Differences in circulating gluconeogenic substrates during short-term fasting in men, women, and children. *Metabolism* 31, 33–42.
484. Kerr, D. S., Hansen, I. L., and Levy, M. M. (1983). Metabolic and hormonal responses of children and adolescents to fasting and 2-deoxyglucose. *Metabolism* 32, 951–959.
485. Cameron, J. L., Koerker, D. J., and Steiner, R. A. (1985). Metabolic changes during maturation of male monkeys: Possible signals for onset of puberty. *Am. J. Physiol.* 249, E385–E391.
486. Slone, D., Soeldner, J. S., Steinke, J., and Crigler, J. F. Jr. (1966). Serum insulin measurements in children with idiopathic spontaneous hypoglycemia and in normal infants, children and adults. *N. Engl. J. Med.* 274, 820–826.
487. Woods, S. C., Lotter, E. C., McKay, L. D., and Porter, D. Jr. (1979). Chronic intracerebroventricular infusion of insulin reduces food intake and body weight of baboons. *Nature* 282, 503–505.
488. Schwartz, M. W., Figlewicz, D. P., Baskin, D. G., Woods, S. C., and Porte, D. Jr. (1992). Insulin in the brain: a hormonal regulator of energy balance. *Endo. Rev.* 13, 387–414.
489. Castracane, V. D., Cutler, G. B. Jr., and Loriaux, D. L. (1981). Pubertal endocrinology of the baboon: adrenarche. *Am. J. Physiol.* 241, E305–E309.
490. Boyar, R. M., Finkelstein, J. W., David, R., Roffwarg, H., Kapen, S., Weitzman, E. D., and Hellman, L. (1973). Twenty-four hour patterns of plasma luteinizing hormone and follicle-stimulating hormone in sexual precocity. *N. Engl. J. Med.* 289, 282–286.
491. Penny, R., Olambiwonnu, N. O., and Frasier, S. D. (1973). Precocious puberty following treatment in a 6-year-old male with congenital adrenal hyperplasia: studies of serum luteinizing hormone (LH), serum follicle stimulating hormone (FSH) and plasma testosterone. *J. Clin. Endocrinol. Metab.* 36, 920–924.
492. Pescovitz, O. H., Comite, F., Cassorla, F., Dwyer, A. J., Poth, M. A., Sperling, M. A., Hench, K., McNemar, A., Skerda, M., and Loriaux, D. L. (1984). True precocious puberty complicating congenital adrenal hyperplasia: treatment with luteinizing hormone-releasing hormone analog. *J. Clin. Endocrinol. Metab.* 58, 857–861.
493. van Wagenen, G. (1949). Accelerated growth with sexual precocity in female monkeys receiving testosterone propionate. *Endocrinology* 45, 544–546.
494. Ramaley, J. A. (1980). Biological clocks and puberty onset. *Fed. Proc.* 39, 2355–2359.
495. Pengelley, E. T., Ed. (1974). *Circannual Clocks*. Academic Press, New York.
496. Turek, F. W. (1985). Circadian neural rhythms in mammals. *Annu. Rev. Physiol.* 47, 49–64.
497. van der Werff ten Bosch, J. J., Dierschke, D. J., Terasawa, E., and Slob, A. K. (1983). Anterior hypothalamic lesions and pubertal development in female rhesus monkeys. *Behav. Brain Res.* 17, 321–330.
498. Bauer, H. G. (1954). Endocrine and other clinical manifestations of hypothalamic disease. *J. Clin. Endocrinol. Metab.* 14, 13–31.
499. Schultz, N. J., and Terasawa, E. (1988). Posterior hypothalamic lesions advance the time of the pubertal changes in luteinizing hormone release in ovariectomized female rhesus monkeys. *Endocrinology* 123, 445–455.
500. Grumbach, M. M., and Styne, D. M. (1992). Puberty: ontogeny, neuroendocrinology, physiology, and disorders. In *Williams Textbook of Endocrinology* (J. D. Wilson and D. W. Foster, Eds.), pp. 1139–1221. W. B. Saunders, Philadelphia.
501. Lander, E. S., and Schork, N. J. (1994). Genetic dissection of complex traits. *Science* 265, 2037–2048.
502. Palmert, M. R., and Boepple, P. A. (2001). Variation in the timing of puberty: clinical spectrum and genetic investigation. *J. Clin. Endocrinol. Metab.* 86, 2364–2368.
503. Wasada, T., Akamine, Y., Oma, H., Nawata, H., Motomatsu, T., Kato, K., and Ibayashi, H. (1978). Studies on the changes of serum FSH, LH and prolactin concentrations in pubertal twins. *Endocrinol. Japan* 25, 575–582.
504. Treloar, S. A., and Martin, N. G. (1990). Age at menarche as a fitness trait: nonadditive genetic variance detected in a large twin sample. *Am. J. Hum. Genet.* 47, 137–148.
505. Kaprio, J., Rimpelä, A., Winter, T., Viken, R. J., Rimpelä, M., and Rose, R. J. (1995). Common genetic influences on BMI and age at menarche. *Hum. Biol.* 67, 739–753.
506. Sharma, J. C. (1983). The genetic contribution to pubertal growth and development studied by longitudinal growth data on twins. *Ann. Hum. Biol.* 10, 163–171.
507. Fischbein, S. (1977). Onset of puberty in MX and DZ twins. *Acta Genet. Med. Gemellol. (Roma)* 26, 151–158.
508. Fischbein, S. (1977). Intra-pair similarity in physical growth of monozygotic and of dizygotic twins during puberty. *Ann. Hum. Biol.* 4, 417–430.

509. Gedda, L., and Brenci, G. (1975). Twins as a natural test of chronogenetics. *Acta Genet. Med. Gemellol. (Roma)* 24, 15–30.
510 Eaves, L., Silberg, J., Foley, D., Bulik, C., Maes, H., Erkanli, A., Angold, A., Costello, E. J., and Worthman, C. (2004). Genetic and environmental influences on the relative timing of pubertal change. *Twin Res.* 7, 471–481.
511. Chumlea, W. C., Schubert, C. M., Roche, A. F., Kulin, H. E., Lee, P. A., Himes, J. H., and Sun, S. S. (2003). Age at menarche and racial comparisons in US girls. *Pediatrics* 111, 110–113.
512. Sun, S. S., Schubert, C. M., Chumlea, W. C., Roche, A. F., Kulin, H. E., Lee, P. A., Himes, J. H., and Ryan, A. S. (2002). National estimates of the timing of sexual maturation and racial differences among US children. *Pediatrics* 110, 911–919.
513. Biro, F. M., McMahon, R. P., Striegel-Moore, R., Crawford, P. B., Obarzanek, E., Morrison, J. A., Barton, B. A., and Falkner, F. (2001). Impact of timing of pubertal maturation on growth in black and white female adolescents: the National Heart, Lung, and Blood Institute Growth and Health Study. *J. Pediatr.* 138, 636–643.
514. Ulijaszek, S. J., Evans, E., and Miller, D. S. (1991). Age at menarche of European, Afro-Caribbean and Indo-Pakistani schoolgirls living in London. *Ann. Hum. Genet.* 18, 167–175.
515. de Vries, L., Kauschansky, A., Shohat, M., and Phillip, M. (2004). Familial central precocious puberty suggests autosomal dominant inheritance. *J. Clin. Endocrinol. Metab.* 89, 1794–1800.
516. Sedlmeyer, I. L., and Palmert, M. R. (2002). Delayed puberty: analysis of a large case series from an academic center. *J. Clin. Endocrinol. Metab.* 87, 1613–1620.
517. Sedlmeyer, I. L., Hirschhorn, J. N., and Palmert, M. R. (2002). Pedigree analysis of constitutional delay of growth and maturation: determination of familial aggregation and inheritance patterns. *J. Clin. Endocrinol. Metab.* 87, 5581–5586.
518. Takahashi, J. S. (2004). Finding new clock components: past and future. *J. Biol. Rhythms.* 19, 339–347.
519. Kalantaridou, S. M., and Chrousos, G. P. (2002). Monogenic disorders of puberty. *J. Clin. Endocrinol. Metab.* 87, 2481–2494.
520. Hardelin, J. P. (2001). Kallmann syndrome: towards molecular pathogenesis. *Mol. Cell. Endocrinol.* 179, 75–81.
521. Palmert, M. R., and Hirschhorn, J. N. (2003). Genetic approaches to stature, pubertal timining, and other complex traits. *Mol. Genet. Metab.* 80, 1–10.
522. Barker, D. J. P. (2004). The developmental origins of well-being. *Philos. Trans. R. Soc. Lond. B Biol. Sci.* 359, 1359–1366.
523. Ibanez, L., Potau, N., Marcos, M. V., and de Zegher, F. (1999). Exaggerated adrenarche and hyperinsulinism in adolescent girls born small for gestational age. *J. Clin. Endocrinol. Metab.* 84, 4739–4741.
524. Francois, I., and de Zegher, F. (1997). Adrenarche and fetal growth. *Pediatr. Res.* 41, 440–442.
525. Adams, J. M., Taylor, A. E., Crowley, W. F. Jr., and Hall, J. E. (2004). Polycystic ovarian morphology with regular ovulatory cycles: insights into the pathophysiology of polycystic ovarian syndrome. *J. Clin. Endocrinol. Metab.* 89, 4343–4350.
526. Ibanez, L., de Zegher, F., and Potau, N. (1999). Anovulation after precocious pubarche: early markers and time course in adolescence. *J. Clin. Endocrinol. Metab.* 84, 2691–2695.
527. Hokken-Koèlega, A. C. S. (2002). Timing of puberty and fetal growth. *Best Pract. Res. Clin. Endocrinol. Metab.* 16, 65–71.
528. Persson, I., Ahlsson, F., Ewald, U., Tuvemo, T., Qingyuan, M., von Rosen, D., and Proos, L. (1999). Influence of perinatal factors on the onset of puberty in boys and girls: implication for interpretation of link with risk of long term diseases. *Am. J. Epidemiol.* 150, 747–755.
529. Leger, J., Levy-Marchal, C., Bloch, J., Pinet, A., Chevenne, D., Porquet, D., Collin, D., and Czernichow, P. (1997). Reduced final height and indications for insulin resistance in 20 year olds born small for gestational age: regional cohort study. *Br. Med. J.* 315, 341–347.
530. Francois, I., deZegher, F., Spiessens, C., D'Hooghe, T., and Vanderschueren, D. (1997). Low birth weight and subsequent male subfertility. *Pediatr. Res.* 42, 899–901.
531. Eisner, J. R., Dumesic, D. A., Kemnitz, J. W., and Abbott, D. H. (2000). Timing of prenatal androgen excess determines differential impairment in insulin secretion and action in adult female rhesus monkeys. *J. Clin. Endocrinol. Metab.* 85, 1206–1210.
532. Dumesic, D. A., Abbott, D. H., Eisner, J. R., and Goy, R. W. (1997). Prenatal exposure of female rhesus monkeys to testosterone propionate increases serum luteinizing hormone levels in adulthood. *Fertil. Steril.* 67, 155–163.
533. Eisner, J. R., Barnett, M. A., Dumesic, D. A., and Abbott, D. H. (2002). Ovarian hyperandrogenism in adult female rhesus monkeys exposed to prenatral androgen excess. *Fertil. Steril.* 77, 167–172.
534. Dumesic, D. A., Schramm, R. D., Peterson, E., Paprocki, A. M., Zhou, R., and Abbott, D. H. (2002). Impaired developmental competence of oocytes in adult prenatally androgenized female rhesus monkeys undergoing gonadotropin stimulation for *in vitro* fertilization. *J. Clin. Endocrinol. Metab.* 87, 1111–1119.
535. Barnes, R. B., Rosenfield, R. L., Ehrmann, D. A., Cara, J. F., Cuttler, L., Levitsky, L. L., and Rosenthal, I. M. (1994). Ovarian hyperandrogynism as a result of congenital adrenal virilizing disorders: evidence for perinatal masculinization of neuroendocrine function in women. *J. Clin. Endocrinol. Metab.* 79, 1328–1333.
536. Lancaster, J. B., and Lee, R. B. (1965). The annual reproductive cycle in monkeys and apes. In *Primate Behavior* (I. Devore, Ed.), pp. 486–513. Holt, Rinehart and Winston, New York.
537. Karsch, F. J., Bittman, E. L., Foster, D. L., Goodman, R. L., Legan, S. J., and Robinson, J. E. (1984). Neuroendocrine basis of seasonal reproduction. *Rec. Prog. Horm. Res.* 40, 185–225.
538. Walker, M. L., Wilson, M. E., and Gordon, T. P. (1984). Endocrine control of the seasonal occurrence of ovulation in rhesus monkeys housed outdoodrs. *Endocrinology* 114, 1074–1081.
539. Glick, B. B. (1979). Testicular size, testosterone level, and body weight in male *Macaca radiata*. *Folia. Primatol.* 32, 268–289.
540. Wickings, E. J., Zaidi, P., Brabant, G., and Nieschlag, E. (1981). Stimulation of pituitary and testicular functions with LH-RH agonist or pulsatile LH-RH treatment in the rhesus monkey during the non-breeding season. *J. Reprod. Fertil.* 63, 129–136.
541. Vandenbergh, J. G. (1983). Pheromonal regulation of puberty. In *Pheromones and Reproduction in Mammals* (J. Vandenbergh, Ed.), pp. 95–112. Academic Press, New York.
542. French, J. A. (1997). Proximate regulation of singular breeding in *Callitrichid* primates. In *Cooperative Breeding in Mammals* (N. G. Solomon and J. A. French, Eds.), pp. 34–75. Cambridge University Press, Cambridge.
543. Dixson, A. F., and Nevison, C. M. (1997). The socioendocrinology of adolescent development in male rhesus monkeys (*Macaca mulatta*). *Horm. Behav.* 31, 126–135.
544. Bercovitch, F. B. (1993). Dominance rank and reproductive maturation in male rhesus macaques (*Macaca mulatta*). *J. Reprod. Fertil.* 99, 113 120.
545. Ellis, B. J. (2004). Timing of pubertal maturation in girls: an integrated life history approach. *Psychol. Bull.* 130, 920–958.

546. Charmandari, E., Kino, T., Souvatzoglou, E., and Chrousos, G. P. (2003). Pediatric stress: hormonal mediators and human development. *Horm. Res.* 59, 161–179.
547. Graber, J. A., Brooks-Gunn, J., and Warren, M. P. (1995). The antecedents of menarcheal age: heredity, family environment, and stressful life events. *Child Dev.* 66, 346–359.
548. Belsky, J., Steinberg, L., and Draper, P. (1991). Childhood experience, interpersonal development, and reproductive strategy: an evolutionary theory of socialization. *Child Dev.* 62, 647–670.
549. Mul, D., Oostdijk, W., and Drop, S. L. (2002). Early puberty in adopted children. *Horm. Res.* 57, 1–9.
550. Krstevska-Konstantinova, M., Charlier, C., Craen, M., Du Caju, M., Heinrichs, C., de Beaufort, C., Plomteux, G., and Bourguignon, J.-P. (2001). Sexual precocity after immigration from develoing countries to Belgium: evidence of previous exposire to organochlorine pesticides. *Hum. Reprod.* 16, 1020–1026.
551. Marshall, F. H. A. (1980). *The Physiology of Reproduction*. Longmans, Green, London.
552. Feicht, B., Johnson, T. S., Martin, B. J., Sparkes, K. E., and Wagner, W. W. Jr. (1978). Secondary amenorrhoea in athletes. *Lancet* 2, 1145–1146.
553. Warren, M. P. (1980). The effects of exercise on pubertal progression and reproductive function in girls. *J. Clin. Endocrinol. Metab.* 51, 1150–1157.
554. Cumming, D. C., Vickovic, M. M., Wall, S. R., and Fluker, M. R. (1985). Defects in pulsatile LH release in normally menstruating runners. *J. Clin. Endocrinol. Metab.* 60, 810–812.
555. Veldhuis, J. D., Evans, W. S., Demers, L. M., Thorner, M. O., Wakat, D., and Rogol, A. D. (1985). Altered neuroendocrine regulation of gonadotropin secretion in women distance runners. *J. Clin. Endocrinol. Metab.* 61, 557–563.
556. Reame, N. E., Sauder, S. E., Case, G. D., Kelch, R. P., and Marshall, J. C. (1985). Pulsatile gonadotropin secretion in women with hypothalamic amenorrhea: evidence that reduced frequency of gonadotropin-releasing hormone secretion is the mechanism of persistent anovulation. *J. Clin. Endocrinol. Metab.* 61, 851–858.
557. Loucks, A. B., Mortola, J. F., Girton, L., and Yen, S. S. C. (1989). Alterations in the hypothalamic-pituitary-ovarian and the hypothalamic-pituitary-adrenal axes in athletic women. *J. Clin. Endocrinol. Metab.* 68, 402–411.
558. Rogol, A. D. (2002). Androgens and puberty. *Mol. Cell. Endocrinol.* 198, 25–29.
559. Serjeant, G. R., Singhal, A., and Hambleton, I. R. (2001). Sickle cell disease and age at menarche in Jamaican girls: observations from a cohort study. *Arch. Dis. Child.* 85, 375–378.
560. Buchacz, K., Rogol, A. D., Lindsey, J. D., Wilson, C. M., Hughes, M. D., Seage, G. R. III, Oleske, J. M., Rogers, A. S., and Pediatric AIDS Clinical Trials Group 219 Study Team. (2003). Delayed onset of pubertal development in children and adolescents with perinatally acquired HIV infection. *J. Acquir. Immune Defic. Syndr.* 33, 56–65.
561. Ballinger, A. B., Savage, M. O., and Sanderson, I. R. (2003). Delayed puberty associated with inflammatory bowel disease. *Pediatr. Res.* 53, 205–210.
562. Arrigo, T., De Luca, F., Lucanto, C., Lombardo, M., Rulli, I., Salzano, G., and Lombardo, F. (2004). Nutritional, glycometabolic and genetic factors affecting menarcheal age in cystic fibrosis. *Diabetes Nutr. Metab.* 17, 114–119.
563. Cameron, J. L. (1996). Regulation of reproductive hormone secretion in primates by short-term changes in nutrition. *Rev. Reprod.* 1, 117–126.
564. Winterer, J., Merriam, G. R., Gross, E., Sly, D. L., Faiman, C., Loriaux, D. L., and Cutler, G. B. Jr. (1984). Idiopathic precocious puberty in the chimpanzee: a case report. *J. Med. Primatol.* 13, 73–79.
565. Wolman, L., and Balmforth, G. V. (1963). Precocious puberty due to a hypothalamic hamartoma in a patient surviving to late middle age. *J. Neurol. Neurosurg. Psychiatry* 26, 275–280.
566. Judge, D. M., Kulin, H. E., Page, R., Santen, R., and Trapukdi, S. (1977). Hypothalamic hamartoma. A source of luteinizing-hormone-releasing factor in precocious puberty. *N. Engl. J. Med.* 296, 7–10.
567. Hochman, H. I., Judge, D. M., and Reichlin, S. (1980). Precocious puberty and hypothalamic hamartoma. *Pediatrics* 67, 236–244.
568. Price, R. A., Lee, P. A., Albright, A. L., Ronnekleiv, O. K., and Gutai, J. P. (1984). Treatment of sexual precocity by removal of a luteinizing-releasing hormone secreting hamartoma. *J. Am. Med. Assoc.* 251, 2247–2249.
569. Adan, L., Bussieres, L., Dinand, V., Zerah, M., Pierre-Kahn, A., and Brauner, R. (2000). Growth, puberty, and hypothalamic-pituitary function in children with suprasellar arachnoid cyst. *Eur. J. Pediatr.* 159, 348–355.
570. Siddiqi, S. U., Van Dyke, D. C., Donohoue, P., and McBrien, D. M. (1999). Premature sexual development in individuals with neurodevelopmental disabilities. *Dev. Med. Child Neurol.* 41, 392–395.
571. Partsch, C. J., Heger, S., and Sippell, W. G. (2002). Management and outcome of central precocious puberty. *Clin. Endocrinol. (Oxf.)* 56, 129–148.
572. Hay, C., and Wu, F. (2002). Genetics and hypogonadotropic hypogonadism. *Curr. Opin. Obstet. Gynecol.* 14, 303–308.
573. de Roux, N., Young, J., Misrahi, M., Schaison, G., and Milgrom, E. (1999). Loss of function mutations of the GnRH receptor: a new cause of hypogonadotropic hypogonadism. *J. Pediatr. Endocrinol. Metab.* 12, 267–275.
574. Dodé, C., Levilliers, J., Dupont, J.-M., De Paepe, A., Le Dû, N., Soussi-Yanicostas, N., Coimbra, R. S., Delmaghani, S., Compain-Nouaille, S., Baverel, F., Pêcheux, C., Le Tessier, D., Cruaud, C., Delpech, M., Speleman, F., Vermeulen, S., Amalfitano, A., Bachelot, Y., Bouchard, P., Cabrol, S., Carel, J.-C., Delemarre-van de Waal, H., Goulet-Salmon, B., Kottler, M.-L., Richard, O., Sanchez-Franco, F., Saura, R., Young, J., Petit, C., and Hardelin, J.-P. (2003). Loss-of-function mutations in $FGFR_1$ cause autosomal dominant Kallmann syndrome. *Nat. Genet.* 33, 463–465.
575. Smith, W. R. (1980). Hypothalamic regulation of pituitary secretion of luteinizing hormone. 11. Feedback control of gonadotropin secretion. *Bull. Math. Biol.* 42, 57–78.
576. Lehrer, S. (1983). Puberty and resonance: a hypothesis. *Mt. Sinai J. Med.* 50, 39–43.

Knobil and Neill's Physiology of Reproduction, Third Edition
edited by Jimmy D. Neill,
Elsevier © 2006

CHAPTER 41

Seasonal Regulation of Reproduction in Mammals

Benoît Malpaux

Introduction, 2231
Seasonal Rhythms of Reproduction, 2231
Description of Seasonal Rhythm of Reproduction, 2231
Ultimate Versus Proximate Environmental Factors, 2233
Photoperiod as the Main Proximate Factor, 2234
Circannual Versus Noncircannual Species, 2234
Importance of Photoperiodic History, 2236
Transduction of Light–Dark Information into a Rhythm of Melatonin Secretion, 2237
Retinohypothalamopineal Pathways, 2237
The Pineal Gland, 2239
Melatonin Biosynthesis and Its Regulation, 2240
Light–Dark Control of Melatonin Synthesis, 2243
Melatonin Transport and Catabolism, 2245
Melatonin Regulation of the Hypothalamic–Pituitary–Gonadal Axis, 2245
Role of the Pineal Gland and Melatonin in Transducing Photoperiodic Information, 2245
Sites of Action of Melatonin, 2248
Melatonin Receptors and Seasonal Breeding, 2250
The PT: A Model for Analyzing Melatonin Duration Decoding Mechanisms, 2253
Melatonin Control of Reproduction, 2253
Interneurons Connecting to GnRH, 2256
Thyroid Hormones as Potential Mediators of Seasonal Plasticity, 2260
Genetic Control of Seasonality, 2264
Evidence for a Genetic Control, 2264
Individual Genetic Variation and Photoneuroendocrine Pathways, 2264
Human Reproductive Seasonality, 2266
Conclusion, 2266
Acknowledgments, 2267
References, 2267

INTRODUCTION

Mammals occur in nearly every habitat from the equator to the poles. Most of these habitats experience at least some degree of seasonal fluctuation in climate and food availability becoming more pronounced with increasing latitude (1). To cope with these changes, most long-lived species that spend at least part of the year in regions where annual differences in environmental conditions are very pronounced exhibit seasonal cycles of physiological functions, behavior, and morphology. These may include, depending on the species, migration, pelage, metabolic rate, and growth changes. However, the main adaptive mechanism widely observed in the wild is the ability to restrict breeding activity to a time of year when the most energetically challenging stages of reproduction (late pregnancy, birth, lactation, weaning of the young) coincide with annual peaks in food availability, the most propitious conditions for survival of the neonate. Gametogenesis, sexual behavior, and pregnancy are timed accordingly (2). Under cold and temperate climates, these conditions are met in the spring and early summer. In tropical and arid areas, the rainy season is often the limiting factor, so the birth season is shifted accordingly (3).

UMR INRA-CNRS-Université de Tours, Physiologie de la Reproduction et des Comportements, Centre INRA de Tours, Nouzilly, France.

The adaptive importance of this trait has been reviewed in detail elsewhere (2) and can be illustrated from two examples. To reintroduce European ibex into Czechoslovakia, native ibex were crossbred with the Turkish and Nubian subspecies. However, these subspecies had different breeding seasons, and the offspring bred earlier than the native ibex. Their progeny were born in winter when food was scarce and the climate harsh and consequently died (4). Also, an epidemiological study of rural human populations of Gambia indicates that season of birth strongly influences the adult survival prospects. Individuals born during the annual "hungry season" are up to 10 times more likely to die prematurely. No effect of season of birth on survival is observed before the age of 15, but this effect is manifested first around puberty and amplified with increasing age. A permanent effect of malnutrition on the development of the immune system during fetal growth seems to be a likely explanation (5).

The general objective of this chapter is to provide an overview of the physiological processes that allow mammals to breed at the optimal time of year given the annual changes in environmental factors of their habitat. The first part provides descriptive information about seasonal rhythms of reproduction and its control by environmental factors. With photoperiod being the main environmental parameter controlling reproduction, the mechanisms of action of photoperiod on seasonal breeding are then analyzed. This integration of photoperiod is a multistep process involving two major components analyzed in two parts. First, the light–dark cycle is transduced into an endocrine signal, namely the circadian secretion of melatonin produced by the pineal gland. Second, the rhythm of melatonin secretion is decoded to regulate the activity of the hypothalamic–pituitary–gonadal axis. Finally, the genetic control of seasonal breeding is discussed with some references to seasonal variations in human fertility.

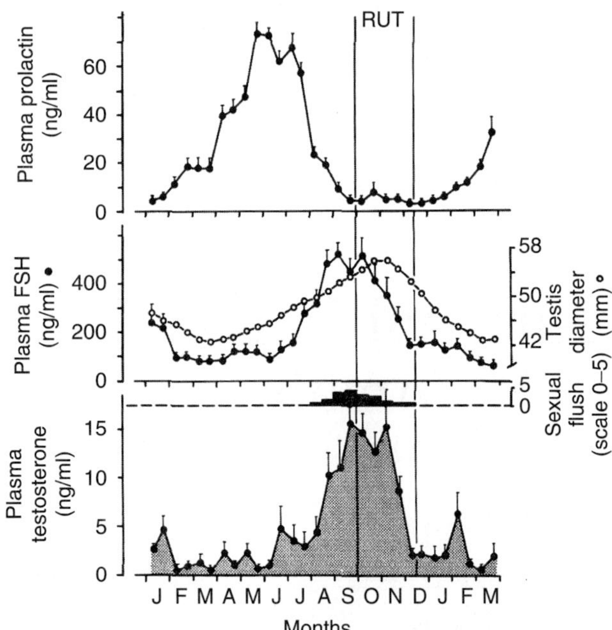

FIG. 1. Seasonal change in the concentration of prolactin, FSH, and testosterone and testicular diameter and sexual skin flush in adult Soay rams living outdoors near Edinburgh. [Adapted from (328).]

SEASONAL RHYTHMS OF REPRODUCTION

Description of Seasonal Rhythm of Reproduction

Reproductive activity in many mammalian populations shows dramatic seasonal variations with certain species oscillating between periods of reproductive activity and quiescence (Fig. 1). The time of the year when animals are reproductively active varies from species to species and depends mainly on the length of gestation. The period of mating is observed either in spring and early summer in animals with either a very short gestational period (e.g., the ferret, hamsters, voles) or a gestational period of about a year (e.g., the horse). Male Syrian hamsters undergo gonadal regression in the fall and remain reproductively quiescent throughout winter while in their burrows. Some time before the vernal equinox, and before emerging from their burrows, the males undergo spontaneous testicular recrudescence and the females begin cycling again (Fig. 2) (6). In contrast, mating occurs in autumn in animals with a gestational period of around 6 months (e.g., sheep, goats, deer) (7,8). The Suffolk ewe becomes sexually active in the late summer to early autumn. Unless she becomes pregnant, repeated 16-day estrous cycles persist until late winter when a 6-month anestrous period begins (7). In male sheep, sperm production varies greatly throughout the year. In rams of the Ile-de-France breed, daily sperm production is 50% to 80% as high in autumn as in spring (Fig. 3) (8). In some mammals, in addition to timing fertilization, environmental cues also time the implantation of the embryo after an obligatory diapause of the blastocysts (e.g., European badger, mink, roe deer). For example, in roe deer ovulation and fertilization take place in July, and after 5 months of embryonic diapause, implantation occurs in December and birth in May (9).

The characteristics of the seasonal rhythm of reproduction vary greatly among species and among latitudes. As pointed out by Bronson (1), from an analysis of three groups of animals in different locations of the northern hemisphere (Peromyscus,

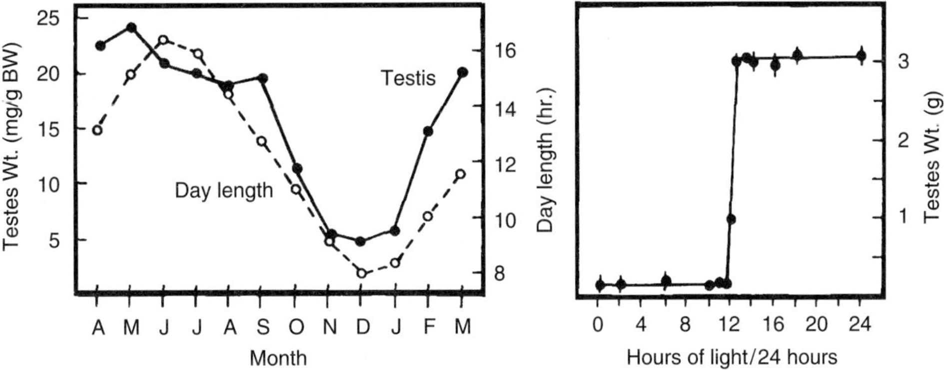

FIG. 2. Changes in the weight of the testes of Syrian hamsters maintained outdoors at a latitude of 49 degrees (*left panel*) or under conditions of controlled manipulation of photoperiod in the laboratory (*right panels*). [Adapted from (2,28,491).]

lagomorphs, deer), the annual pattern of reproduction has three main characteristics. First, all species show a short well-defined breeding season in the northern part of their ranges, and all become year around breeders in the southern part. This can be illustrated with intermediate-sized rabbits and hares of the order Lagomorpha that range from near the equator to over 80°N (Fig. 4). The large arctic hare that lives at a latitude of about 80 degrees in Canada produces only one litter per year. In contrast, the cottontail rabbit shows continuous breeding in southern Texas, Florida, and Venezuela (2). Second, the latitude at which the shift from restricted to continuous breeding occurs varies with body size; the shift occurs at higher latitudes in the smaller *Peromyscus*, lower in the lagomorphs, and lowest of all in the deer. Third, the degree of year-to-year variation as well as locale-to-locale variation decreases as the size of the animal increases (2).

Ultimate Versus Proximate Environmental Factors

Environmental factors that act on seasonal rhythms operate at two levels, the ultimate and the proximate (1,10,11). Ultimate factors are variables that in the course of evolution exert selection pressure to restrict breeding activity to a particular time of the year when it is most likely to be successful. Reproductive timing mechanisms are under high selection pressures because temporal errors can result in wasted effort and can jeopardize survival. The main ultimate factors include dietary and climatic factors,

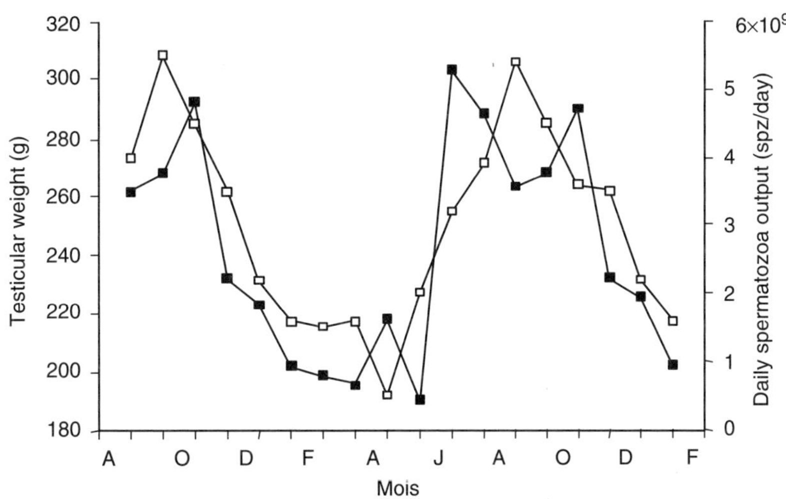

FIG. 3. Seasonal changes in daily spermatozoa output (*closed symbols*) and testicular weight (*open symbols*) in adult Ile-de-France rams living at a latitude of 45°N. [Adapted from (492).]

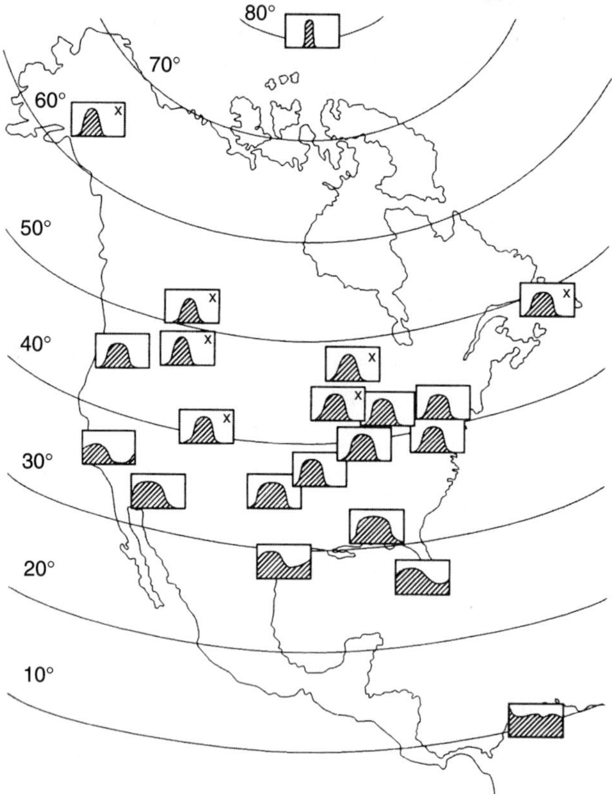

FIG. 4. Breeding seasons of three lagomorphs (cotton-tail rabbit, snowshoe hare, and arctic hare). The left axis of each rectangle represents January, the right axis represents December, and the stippled area between indicates the annual breeding pattern observed over a 12-month period (percentage of females pregnant in monthly kill samples). Those patterns observed for the snowshoe hare are marked with an X. [Reproduced from (2).]

particularly food availability, rainfall, and temperature. Some of the other ultimate cues of seasonal breeding are predator pressure and competition between species.

Proximate factors provide immediate cueing for the control of annual rhythms by regulating physiological processes. Many seasonal functions including reproduction, migration, and hibernation require long periods of development and preparation. This preparation must often be initiated when the conditions are far from optimal, and the fitness of an animal is improved if it is capable of obtaining predictive information concerning future environmental conditions. Thus, mechanisms have evolved for the use of reliable predictive information so that the individual is prepared for the times at which the ultimate factors are optimal. For a proximate factor to be a reliable predictor of seasonal events, it must be consistent and stable year after year. The annual cycle of photoperiod, the light fraction of the 24-hour day, is the most noise-free environmental variable and consequently is the major source of environmental information in the control of a variety of seasonal activities.

Photoperiod as the Main Proximate Factor

The use of photoperiod as a predictive cue was first demonstrated in a mammal by Baker and Ransom (12). They observed that in a colony of field voles held on 15 hours of light, reproduction was continuous. However, when exposed to 9 hours of light per day, reproduction was blocked. Since then, photoperiodism has been shown to influence reproduction in most mammalian orders but has been most thoroughly studied in hamsters and sheep. In Siberian hamsters, exposure to short days induces reproductive inhibition, winter molt, onset of hibernation, and increase in body mass, whereas maintenance of animals in long days prevents these changes (13). When sexually mature male Syrian hamsters are exposed to photoperiods shorter than 12.5 hours of light per day, their circulating luteinizing hormone (LH), follicle-stimulating hormone (FSH), and testosterone concentrations decline markedly. Within 8–12 weeks their testes regress and spermatogenetic activity ceases (Fig. 2) (14). In sheep, the reversal of the photoperiodic cycle, without any modification of other environmental factors, causes the breeding season to phase shift by 6 months, that is, decreasing day length is associated with reproductive activation and increasing day length with reproductive arrest (15). The reduction of the photoperiodic cycle to 6 months induces the appearance of two periods of activity every year (7,8). Photoperiod is also critical in these species for the attainment of puberty (see Chapter 39).

A clear feature of photoperiodic responses observed in all these treatments is that most photoperiod-induced changes are expressed after a latency of several weeks. Stimulation of ovulation after short-day exposure takes about 6 weeks in sheep and induction of testicular regression by short days in hamsters about 8 weeks (7,14). This feature of the response is consistent with the predictive nature of photoperiodic information.

Circannual Versus Noncircannual Species

Photoperiodic species are divided in two main categories depending on how photoperiod controls the annual rhythms and on the endogenous nature of the annual changes. Circannual species display endogenously generated rhythms, and photoperiod synchronizes these internally generated rhythms (referred to as type II rhythms). Noncircannual species do not display endogenous rhythms and are at least partially dependent on the environment for the generation of the seasonal cycle.

Circannual Species

A circannual rhythm is composed of an endogenously generated sequence of events that takes approximately 1 year to complete. Three conditions must be met for an annual rhythm to be considered as circannual (16). First, the cycle must persist for at least 2 years in experimental conditions that provide no external information about its period (e.g., constant photoperiod, temperature). Second, its period in such conditions must differ from 365 days, demonstrating that the cycle is not synchronized by an annually recurring environmental cue. Third, desynchrony must be observed among individuals in such conditions, also demonstrating that the cycle is not being driven by an environmental cue.

Circannual cycles have been described in animals exposed to a constant photoperiod, instead of constant darkness, which is sufficient to prevent the animal from receiving any information about time of the year. Circannual rhythms are expressed by many mammals, including bats, ferrets, squirrels, marmots, sheep, deer, and primates (16,17). One of the best studied examples is the change in body weight, reproduction, and hibernation in the Golden-mantled ground squirrel. During summer, body weight increases as a result of fat storage and reaches a plateau in the autumn. At this time, animals enter their underground burrows where they spend the winter. During winter they enter deep torpor with occasional arousal every 1 to 14 days. The final arousal in the spring leads to the resumption of activity above the ground. Upon emergence, ground squirrels are in reproductive condition and soon mate; females bear a single litter each year. The cycle begins again with the completion of the reproductive phase. When held on a fixed photoperiodic regimen of 10 hours of light per day, they manifest annual rhythms of body mass and reproduction for several years; the period of these free-running rhythms is less than 12 months.

How photoperiod entrains a circannual rhythm has been well studied in the case of reproductive activity in the ewe. This animal normally enters the breeding season in the autumn and, in the absence of pregnancy, has regular ovulatory cycles until mid-winter. Ovulations then cease, and the ewe remains anovulatory throughout the summer. When these animals are deprived of information about time of year, either by being kept in constant photoperiod or by blinding, they continue in most cases to show changes in ovulatory activity, if they are intact, or in LH secretion if they are ovariectomized, bearing an estradiol implant (18,19). If ewes are kept for 5 years in constant short photoperiod (8 hours of light), a cycle of LH secretion is observed with a period shorter than a year and cycles are desynchronous among individuals and out of phase with respect to the cycles of sheep maintained outdoors (Fig. 5) (19). The expression of this circannual rhythm is responsible for the timing of the onset and the end of the breeding season (20,21).

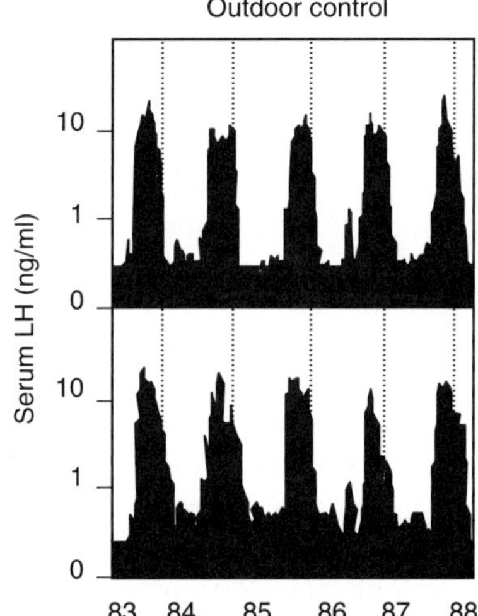

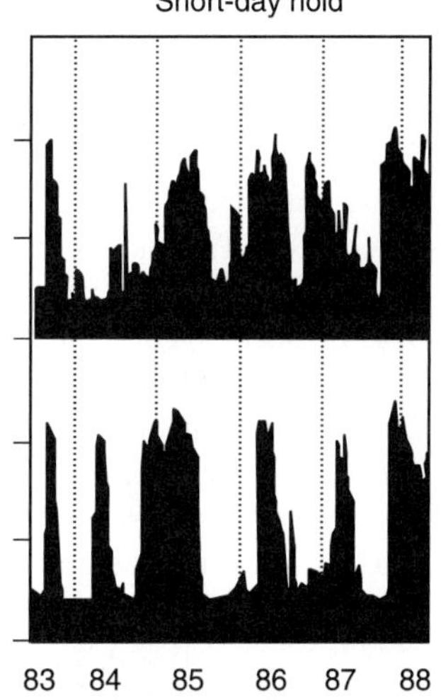

FIG. 5. Changes in LH concentrations for two ewes kept in constant short days (*right panels*) and two control ewes kept in natural photoperiod (*left panels*) throughout a 5-year study (1983–1988). All ewes were ovariectomized and treated with a subcutaneous implant of oetradiol. Dotted vertical lines indicate limits between years. [Adapted from (19).]

The adaptive advantage of circannual rhythms is relatively obvious in hibernating animals. The endogenous cyclicity allows them to measure the passage of time and to prepare for future events while they are deprived of external information about time of year. However, why are other animals equipped with circannual rhythms when external cues are present all the time? The seasonal variations in physiological functions could be directly driven by seasonal cues. One possible answer is that circannual rhythms, like other self-sustained rhythms, show a certain amount of inertia toward zeitgeber stimuli. The endogenous rhythms do not react immediately to changes in environmental cues and therefore are relatively nonsusceptible to irregular fluctuations in environmental conditions. This characteristic is probably important in migrating animals by allowing them to keep an internal timing of the year when they are exposed to conflicting or disturbing information about time of year as they travel across latitudes and environmental influences (16).

Noncircannual Species

Many mammalian species fail to show endogenous cycles of seasonal functions when housed under a fixed photoperiod. The best examples of this lack of circannual rhythm are many short-lived rodents that remain reproductively active as long as they are maintained in a long photoperiod. The Syrian hamster requires short days to induce gonadal regression at the end of the breeding season (Fig. 2). Subsequent gonadal recrudescence that occurs in mid-winter does not require an active photoperiodic drive; the transition into breeding season is generated by the development of refractoriness to the short days of winter, which allows this process to occur when days are still short. For the appropriate response to the decreasing day lengths of the following autumn to occur, the state of photoresponsiveness must be restored (14). This is accomplished only if the individual experiences several weeks of exposure to long days, as occurs each summer in the wild. Although the development of refractoriness is a form of endogenous timekeeping, the requirement for short days to inhibit reproductive activity and that for long days to restore photoresponsiveness makes hamsters unable to exhibit circannual cycle of reproductive activity whether they are exposed to short or to long days. It was proposed that the interval timer that underlies the development of refractoriness to short days (after about 20 weeks of exposure to this day length in Siberian hamsters) and the generation of circannual rhythms may be the same and represents the ancestral form of long-term timekeeping (22). Interestingly, a small proportion of Siberian hamsters (4 of 57) displays circannual rhythms of reproductive activity when exposed to short days continuously for 2 years (23). It is proposed that these animals may display a circannual rhythm in sensitivity to short-day inhibition. Such a rhythm has been shown in a related short-lived species, the European hamster (24).

Importance of Photoperiodic History

The response to photoperiod in mammals is not based solely on absolute day length. Animals respond differently to a given day length if it increases or decreases. In the ewe, 13 hours of light per day causes a stimulation of LH secretion if animals have been exposed previously to 16 hours of light. In contrast, 13 hours of light per day inhibits LH secretion if animals have been exposed previously to 10 hours of light (25). Similarly, transfer of Siberian hamsters from 16 hours to 13.5 hours of light results in gonadal regression, whereas transfer from 10 hours to 13.5 hours of light induces gonadal growth (26). Exposure to the same photoperiod produces opposite effects depending on the photoperiodic history of the animals. Similar results have been obtained in several other species, and the importance of photoperiodic history appears to be a general characteristic of photoperiodic responses [for review see (27)]. Photoperiodic history effects force us to reconsider the popular notion of "critical photoperiod" that was developed from a consideration of the effect of photoperiod in Syrian hamsters (27). In hamsters raised in 14 hours of light per day and transferred to various shorter day lengths, gonadal regression was observed for animals moved to 12 hours of light or shorter photoperiods, whereas reproductive activity was maintained in photoperiods of 12.5 hours of light or longer. In hamsters exposed to 6 hours of light with regressed testis and transferred to various longer day lengths, testicular recrudescence was seen only for animals moved to 12.5 hours of light or longer photoperiods (Fig. 2). Thus, 12.5 hours of light per day was said to be the critical day length for maintenance of reproductive activity in this species (28). This term "critical photoperiod" may be useful in some cases in the context of a given photoperiodic history.

In the ewe, progressively changing photoperiod throughout the year is also important to determine the length of the breeding season. Ewes submitted to a single-step decrease in photoperiod from 16 hours to either 8 or 12 hours of light per day show a stimulation of LH secretion that lasts for only 50 to 60 days. In contrast, if day length is decreased first from 16 to 12 hours and then at the time of LH increase to 8 hours of light, the period of elevated LH levels lasts almost twice as long (29). It should be noted that photoperiodic history is also important prenatally.

Postnatal testicular development of Siberian hamsters in 14 hours of light is rapid if mothers were exposed during gestation to a photoperiod shorter than 14 hours of light. In contrast, if the photoperiod of gestation is longer than 14 hours of light per day, postnatal testicular development occurs slowly (30). Because each day length occurs twice a year, once in summer/autumn and once in winter/spring, it is the ability of integrating photoperiodic history and the change in photoperiod that allow animals to determine precisely the time of year.

This ability to use photoperiodic history does not discount any importance of absolute day length. It should be noted that experiments in the Siberian hamster using progressive increases and decreases in photoperiod have shown that not only is the pattern of gradual change in day length important, but the absolute length of the day also communicates photoperiodic information (31).

TRANSDUCTION OF LIGHT–DARK INFORMATION INTO A RHYTHM OF MELATONIN SECRETION

Retinohypothalamopineal Pathways

Photoreceptors

Photic information is relayed to the pineal gland through a multisynaptic neural pathway including the central clock, the suprachiasmatic nuclei (SCN) (32), where it is then transduced into an endocrine signal that takes the form of a diurnally modulated rhythm of melatonin production (Fig. 6). Light is perceived by the retina because enucleation abolishes the light control of pineal melatonin production (33). Retinal rods and cones with their opsin-based visual pigments are necessary for the conscious perception of light. They were also thought to be the only ocular photoreceptors involved in forwarding light–dark information. However, the regulation of pineal melatonin production as well as the entrainment of circadian rhythms by light remains intact in a transgenic mouse model (*rd/rd cl*) that lacks all functional rods and cones. Enucleation abolishes these light effects in these null mice. Ocular photoreception is therefore not confined to rods and cones, and novel ocular photoreceptors are involved (34). The photopigment, melanopsin (*Opn4*), originally isolated from photosensitive melanophores of *Xenopus laevis* (35), has emerged as a strong candidate to mediate these effects of light. A subset of retinal ganglion cells expresses melanopsin (36) and projects to the SCN (37–40). These cells are intrinsically light responsive as their light-evoked depolarizations persist after anatomical and pharmacological isolation (41,42). In melanopsin knockout mice, light-dependent entrainment is attenuated. Bright light can still entrain circadian rhythms but not as much as in mice expressing melanopsin (43–46). The intrinsic photosensitivity of melanopsin retinal ganglion cells is abolished in these knockout mice, indicating the indispensable role of melanopsin in the cellular response (47). In mice lacking melanopsin and functional rods and cones, entrainment to light–dark cycles and masking response to light are completely abolished (44). The visual (rod–cone) and nonvisual (melanopsin retinal ganglion cells) systems together therefore seem to provide all of the photic input for circadian responses to light, and the nature of the interaction between these two systems remains to be elucidated.

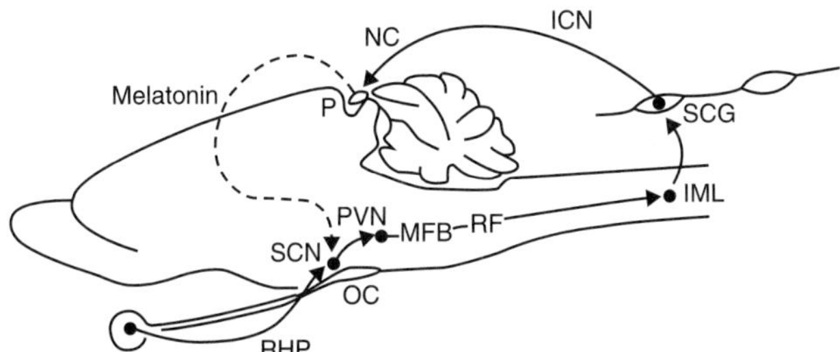

FIG. 6. Mammalian melatonin rhythm generating system. P, pineal gland; SCN, suprachiasmatic nuclei; RHP, retinohypothalamic projection; PVN: paraventricular nucleus, IML, intermediolateral cell column; ICN, inferior carotid nerve; NC, nervi conarii; OC, Optic chiasm, MFB, medial forebrain bundle, RF, reticular formation. [Reproduced from (114).]

Retinohypothalamic Tract

From retinal photoreceptors, photic information is conveyed to the SCN via the retinohypothalamic tract that generally terminates within the ventral and lateral SCN. The retinohypothalamic tract neurotransmitters are mainly glutamate (48,49) and pituitary adenylate cyclase-activating peptide (50). In response to light, they are released from the retinohypothalamic tract terminal and stimulate their receptors expressed in the SCN. Photic and nonphotic information also reaches the SCN indirectly through the thalamic intergeniculate leaflet (neuropeptide Y [NPY], enkephalin, and γ-aminobutyric acid) and the raphe nucleus (serotonin) (51–54).

Suprachiasmatic Nuclei

The SCN are localized in the anteroventral hypothalamus, and a large body of evidence has established that the central pacemaker resides in these hypothalamic nuclei [for review see (55)]. Lesioning of the SCN causes a disappearance of most circadian rhythms, including the secretory rhythm of melatonin (56), and makes the animal arrhythmic, whereas transplantation of fetal SCN restores circadian rhythmicity in such lesioned animals with a periodicity characteristic of the donor animal (57,58). It is considered that the ventrolateral part (core) of the SCN is the clock input area for the synchronizing events, whereas the dorsomedian part (shell) contains the oscillator and the output of the timing information (59). SCN neurons are mainly vasoactive intestinal peptide, vasopressin, gastrin-releasing peptide, somatostatin, and γ-aminobutyric acid (60). The basic mechanism responsible for the generation of the circadian rhythms is intrinsic to individual cells in the SCN, and specific contacts between cells are not necessary for operation of these autonomous single-cell circadian oscillators (61).

In response to light, immediate early genes (*c-fos*, *junB*, *NGF1A*, *NGF1B*) are expressed in the SCN. Numerous pieces of evidence have demonstrated that the induction of fos is functionally related to the synchronization process. Among them, light-induced fos induction occurs only at those times of the circadian cycle when light is able to phase shift the rhythm (62–65), fos induction is reduced by drugs that inhibit the phase-shifting effect of light (66–69), and antisense oligonucleotides against *c-fos* block the phase shifts induced by light (70).

Important progress has been made recently in the understanding of the molecular basis of circadian timing in mammals [for reviews see (71,72)] (Fig. 7). Circadian clocks arise from the expression and temporally regulated activity of specific genes and gene products. The intracellular clock mechanism in the mouse involves interacting positive and negative transcriptional feedback loops that drive recurrent rhythms in the RNA and protein levels of key clock components (71). Basic helix-loop-helix-PAS transcription factors CLOCK and BMAL1 act as positive regulators, and three PERIOD proteins (PER1, PER2, and PER3) and two CRYPTOCHROME proteins (CRY1 and CRY2) operate as negative regulators (71,73). CLOCK and BMAL1 form heterodimers that bind to E-box enhancers to activate the transcription of *Per* and *Cry* genes. As the levels of PER proteins increase, they complex with CRY proteins. The complexes are transported to the

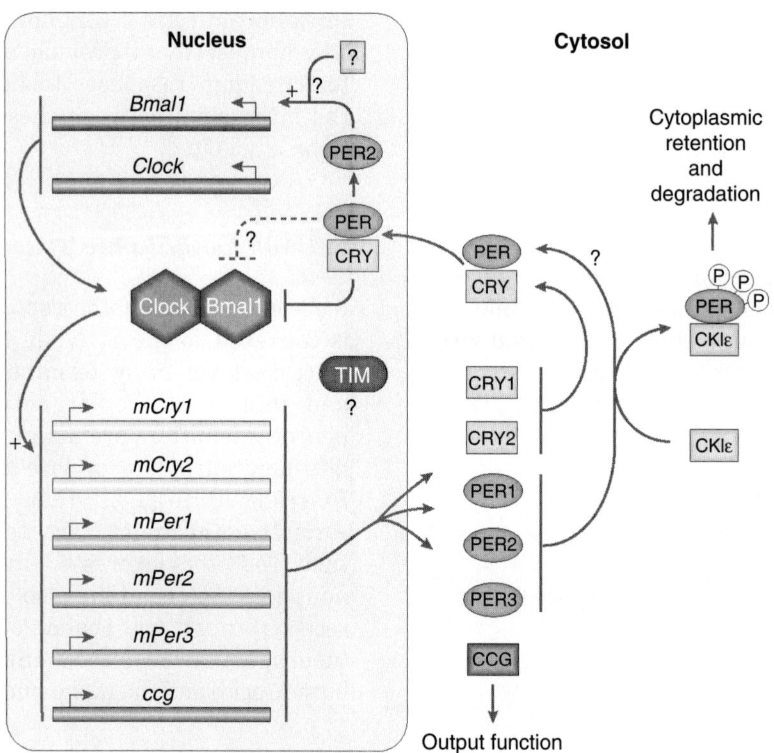

FIG. 7. Model of the molecular mechanisms involved in the mammalian circadian clock. At least two interlocked feedback loops are involved in generating rhythms: a negative autoregulatory loop of CRY and possibly PER proteins on their own genes (through inhibition of the transcriptional activator CLOCK-BMAL1) and a positive effect of PER2 on the expression of BMAL1. Casein kinase Iε (CKIε) regulates the stability and the nuclear translocation of PER proteins. CLOCK-BMAL1 also regulates clock-controlled genes (*ccg*) and hence rhythmic output. [Reproduced from (493).]

nucleus and inhibit the CLOCK-BMAL1–dependent transcriptional activation, resulting in a decrease in PER and CRY transcripts; this forms the negative feedback loop. On the other hand, CLOCK-BMAL1 heterodimer also activates the expression of an orphan nuclear receptor REV-ERBα that acts through Rev-Erb/ROR response elements in the *Bmal1* promoter to repress the transcription of positive regulator gene (74). As a result, Bmal1 RNA levels fall, whereas *Per* and *Cry* RNA levels rise. When the mCRY proteins enter the nucleus to inhibit *mPer* and *mCry* transcription, they also inhibit *rev-Erbα* transcription resulting in a de-repression (activation) of Bmal transcription (71,75); this forms the positive feedback loop. Expression of clock genes therefore oscillates because of the delay in the feedback loops, regulated in part by phosphorylation of the clock proteins that control their stability, nuclear reentry, and transcription complex formation. This model involving gene clocks to generate circadian rhythms represents a considerable advance in the understanding of the clock, but it has to be determined how these mechanisms result in a long period oscillation close to 24 hours.

Among clock genes, *Per1* and *Per2* appear to play a major role in photic entrainment in the SCN. Light exposure induces acute expression of *Per1* and *Per2* 30 and 120 min after the onset of light stimulation (76–78). Importantly, these genes are induced by light only at night, that is, when light causes a phase shift of circadian rhythms. Antisense oligonucleotides against *Per1* inhibit phase-dependent phase shifts of the clock by light (phase delay in early night and phase advance in late night) (79,80), suggesting an important role of *Per1* in clock synchronization. Similarly, *Per2* was suggested to be involved in phase delay of the clock (81). Consistent with these observations, light-dependent phase shifts of the clock are impaired in mice deficient in *Per1* or *Per2* (82).

The Pineal Gland

The pathway connecting the SCN to the pineal gland comprises several steps. The signal is first conveyed to the paraventricular nucleus of the hypothalamus, then down to the intermediolateral cell column of the thoracic part of the spinal cord, and back up to the superior cervical ganglion. Finally, the information is transmitted via the postganglionic sympathetic neurons, the nervi conarii, to the pineal gland (83). Although there is some evidence for parasympathetic (84) and central innervation (85) of the pineal gland, the function of this neural input remains unclear.

The structure and ultrastructure of the mammalian pineal have been described in several reviews (86,87). The pineal gland develops as an evagination of the diencephalic roof. Although the pineal bodies of different mammals possess this common developmental origin, there are variations with respect to pineal shape and position (86). The sheep pineal is a pea-like organ, forming a solid mass between the habenular and posterior commissures and belonging to the proximal type. The bulk of the pineal tissue lies closely related to the third ventricle and is supported by the stalk attaching the pineal body, which is separated by the pineal recess (86). The human pineal gland is a piriform organ and presents the same anatomical organization as in sheep. In contrast, the pineal gland of rodents is basically a long rod-like structure extending from the third ventricle to immediately beneath the skull (86). For instance, in hamster, the pineal organ consists of two separated parts connected by the pineal stalk, a major part named "superficial pineal" and a minor part named "deep pineal," closely related to the third ventricle. Although it is much smaller, the deeper part is made of functional pinealocytes (88). In lower vertebrates, the pineal gland is directly photosensitive, but in mammals it consists of neuroendocrine cells, the pinealocytes, that although insensitive to direct light (89) still exhibit expression of various photoreceptor markers, including rhodopsin kinase (90), opsin (91), and S-antigen (92). The mammalian pineal gland is a rather homogenous tissue containing mainly pinealocytes (>90%), a few glial and phagocytic cells, and rare neurons.

The pineal gland is highly vascularized (93) and has a blood flow estimated to be 4 mL/min/g, second only to that of the kidney (94) (Fig. 8). Blood is supplied to the pineal gland by vessels that originate in the choroids plexus, and venous drainage from its capillary network ultimately empties into the great cerebral vein of Galen (93). The density of the capillary network in and around the pineal gland explains why changes in pineal secretory activity are rapidly reflected in the peripheral circulation.

In addition to secretion into the blood, melatonin can also be released directly in the cerebrospinal fluid (CSF) (95,96) (Fig. 9). The contact area between the pineal gland and the third ventricle is the pineal recess, an evagination of the dorsocaudal part of the ventricular lumen of the third ventricle. In rodents, only the deep pineal gland is in contact with the pineal recess (97). In other species, like humans or sheep, the large unique pineal is located directly on the dorsal part of the brainstem (98). The ventricular lumen is normally bordered by an ependyma consisting of cubical or cylindrical epithelial cells, the lateral membranes of which are connected by intermediate and gap junctions. In the areas covering the circumventricular organs, the classic cylindrical

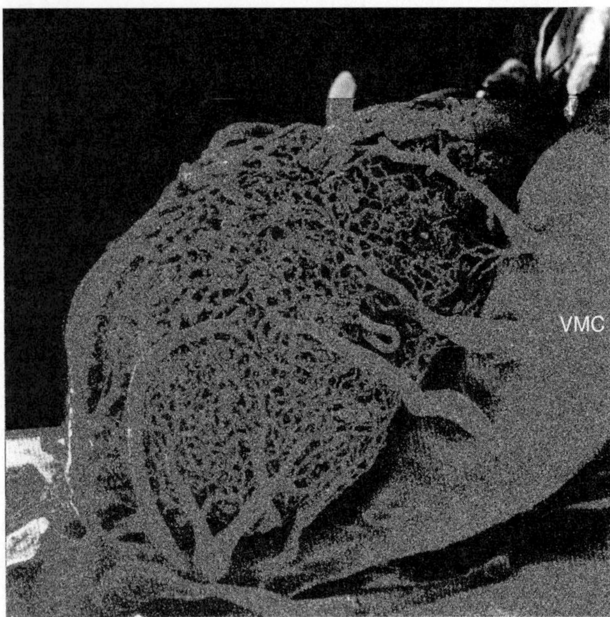

FIG. 8. Latex cast of the venous drainage of the rat pineal. Twelve to 16 veins usually drain the gland in this species; these vessels empty into the great cerebral vein (*right*). VMC, vena magna cerebri. [Reproduced from (494).]

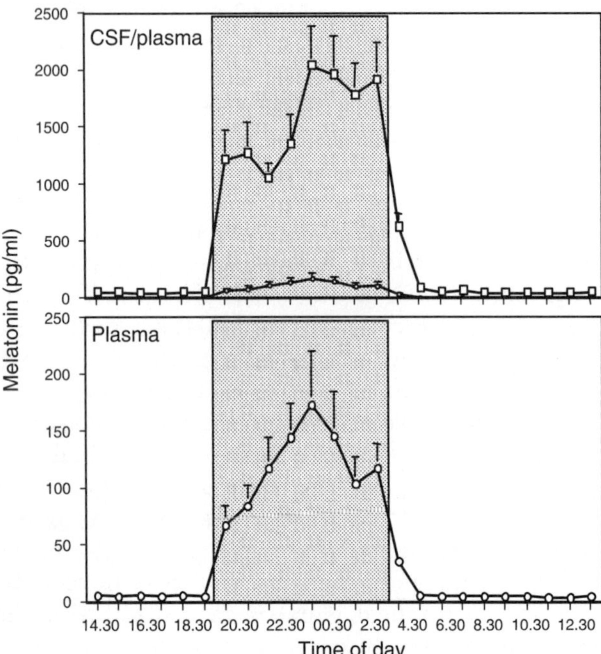

FIG. 9. Profiles of melatonin concentrations in the jugular blood (*top and bottom panels, open circle*) or in the third ventricle CSF (*top panel, open square*). Scale is expanded on the bottom panel to visualize the shape of the melatonin profile in plasma (same curve as above). Note that levels are 20-fold higher in the CSF and coding for duration of melatonin presence is similar in the two compartments. Dark period is represented by shaded area. [Adapted from (95).]

ependymal cells often change their morphology and the lateral membranes of the ependymal cells are endowed with tight junctions, which prevents hydrophilic molecules from being exchanged passively across the brain–CSF interface of these areas (99,100). Importantly, the classic ependymal lining of the third ventricle disappears in most parts of the pineal recess apposing the basal part of the gland and pineal cells.

A similar morphology of the pineal recess has been described in various species regardless of whether the pineal is divided in two parts or is unique [hamster (101), vole (102), brush-tailed possum (103), Mongolian gerbil (104), opossum (105), and sheep (106)]. In sheep, it was shown that the pineal cells bulging into the CSF were pinealocytes on the basis of their serotonin immunoreactivity (106). In addition, melatonin is released into the pineal recess because melatonin levels within the pineal recess are at least tenfold higher than in the ventral part of the third ventricle, and surgical obliteration of the pineal recess leads to a marked reduction of CSF melatonin levels without affecting blood melatonin levels (107). Also, the pineal gland–CSF interface is highly permeable to various molecules. Molecules newly synthesized within the pinealocytes, for example, melatonin, could reach the CSF by two nonexclusive mechanisms: direct release from cells that protrude into the CSF and release into the pineal interstitial fluid that subsequently drains into the ventricular lumen (106). Therefore, although the pineal forms a unique mass in species like sheep or humans, melatonin in CSF originates from pinealocytes located in the basal part of the pineal gland (i.e., next to the CSF), whereas melatonin that is found in the peripheral circulation comes from pinealocytes neighboring blood vessels throughout the pineal gland. This functional division of the pineal gland (basal part producing CSF melatonin, remainder and major part of the pineal producing blood melatonin) in species such as the sheep in which the pineal forms a unique mass may be compared with the situation in rodents where there is division into a superficial and a deep pineal; only thc decp pineal is in contact with the pineal recess. It is possible that in rodents the deep pineal gland releases its secretory products mainly in the CSF, whereas the superficial pineal gland secretes melatonin into the blood.

Melatonin Biosynthesis and Its Regulation

Melatonin was discovered in 1958 by Lerner et al. (108) because of its exquisitely potent effects on melanophore pigment dispersion. It had been known for many years (109) that pineal gland extracts could lighten the skin color of tadpoles, which is an aggregating action of melanosomes. More than 250,000 bovine pineal glands were processed chromatographically to identify the bioactive compound, *N*-acetyl-5-methoxytryptamine (108). This novel compound was named melatonin because it could affect melanin pigmentation in frog skin and was chemically related to serotonin (110). Melatonin has no effect on melanophores in mammals. The biosynthetic pathways of melatonin have been reviewed (111–115) (Fig. 10). The precursor of melatonin is the essential amino acid L-tryptophan, which is taken up from the bloodstream and converted to serotonin in two steps: Tryptophan is first metabolized into 5-hydroxy-tryptophan by tryptophan hydroxylase in the pineal mitochondria and is then transformed into serotonin in the pineal cytosol by an aromatic amino acid decarboxylase. This transformation of tryptophan into serotonin occurs during both day and night. Serotonin concentrations in the pineal gland are higher during the day than at night as a result of the nocturnal transformation of serotonin into melatonin. Serotonin is first acetylated by arylalkylamine-*N*-acetyl-transferase (AANAT) into *N*-acetyl-serotonin that is then O-methylated by hydroxyindole-O-methyltransferase (HIOMT) to form melatonin. The activity of AANAT displays marked day/night variations [>100-fold nocturnal increase (115–117)], whereas that of HIOMT only increases by about 50% at night (118–120). AANAT is therefore considered the "rate-limiting" enzyme for melatonin synthesis in the pineal gland. Because melatonin is a lipophilic molecule, it is not stored within the pinealocytes and is released into pineal interstitial space immediately.

Norepinephrine (NE) is the primary neurotransmitter of the sympathetic fibers, and it is released in large quantities at the onset of darkness; the nocturnal release is about 100-fold as high as that in daytime (121). Numerous approaches have shown the critical role of this increase in NE release to stimulate melatonin synthesis [for review see (115)].

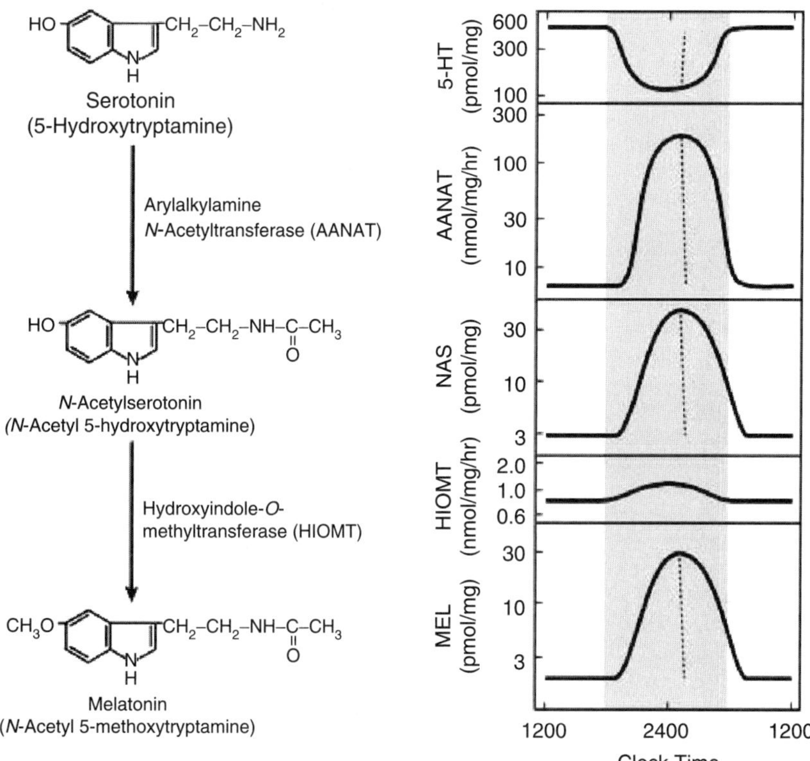

FIG. 10. Daily rhythm in pineal indoles. Serotonin (5-hydroxytryptamine; 5-HT) levels are high during the day and decrease at night (*dark panel*) because of a large increase in arylalkylamine *N*-acetyltransferase (AANAT) protein and activity. AANAT transfers an acetyl group from acetyl CoA to 5-HT. The increase in AANAT activity results in an increase in the intracellular concentration of *N*-acetylserotonin (*N*-acetyl 5-hydroxytryptamine; NAS), which is converted to melatonin (*N*-acetyl 5-methoxytryptamine; MEL) by hydroxyindole-*O*-methyltransferase (HIOMT). Exposure to light at night causes a return to the day time pattern (*dotted line*). [Reproduced from (114).]

NE acts upon α_1- and β-adrenergic receptors in the pinealocyte membrane (122). Activation of the β-adrenergic receptors appears as the major way by which NE stimulates melatonin biosynthesis during darkness. In vivo injections of the β-adrenergic receptor agonist, isoproterenol, during the day stimulates AANAT activity to levels observed at nighttime, whereas an in vivo injection of the antagonist, propranolol, inhibits the nocturnal increase in AANAT (115,123,124). The β-adrenergic effects of NE are mediated via the GTP-binding protein G_S and lead to an activation of the adenylate cyclase, which leads to a 10-fold increase in cyclic AMP (cAMP) levels (125). The maximal increase in cAMP is actually reached when the α_1-adrenergic receptors are activated at the same time (126). Stimulation of α_1-adrenergic receptors alone has no effect on cAMP accumulation but causes a large increase in

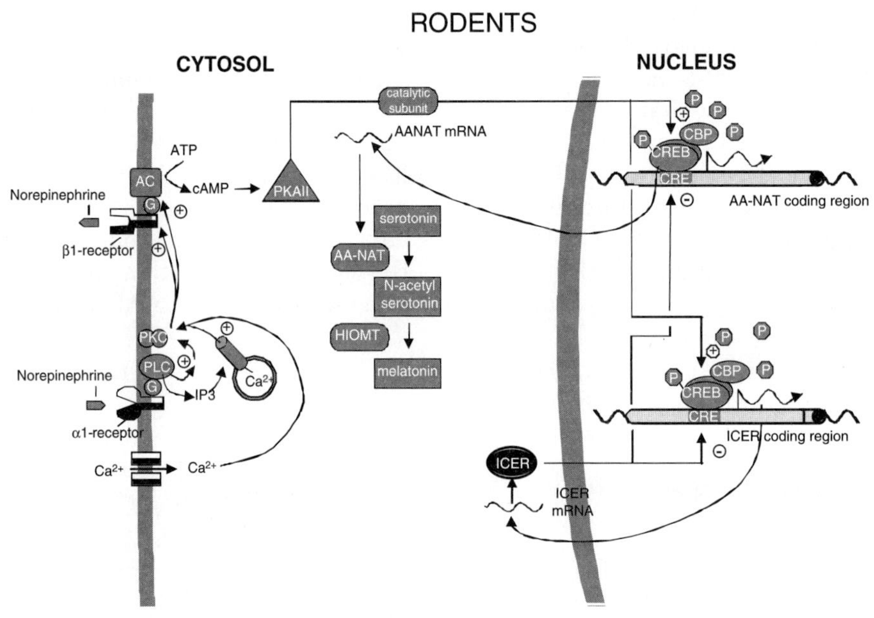

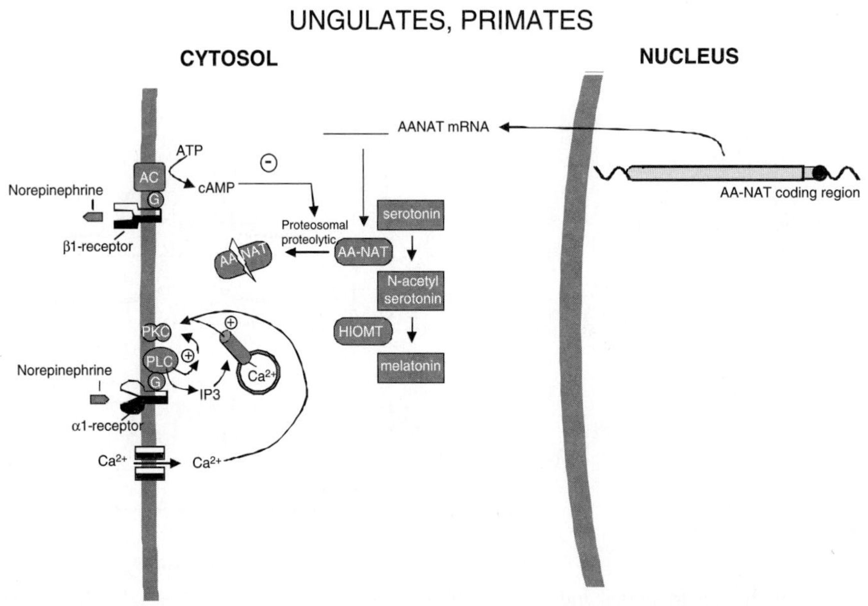

FIG. 11. Schematic diagram of the action of second and third messengers in rodent (*above*) and ungulate and primate (*bottom*) pinealocytes. AA-NAT, arylalkylamine *N*-acetyltransferase; AC, adenylate cyclase; CBP, CREB binding protein; CRE, cyclic AMP response element; G, GTP-binding protein; HIOMT, hydroxyindole-*O*-methyltransferase; ICER, inducible cyclic AMP early repressor; IP3, inositoltrisphosphate; P, phosphate groups; PKAII, cyclic AMP-dependent protein kinase type II; PKC, protein kinase C; PLC, phospholipase C. [Reproduced from (139).]

the intracellular concentration of free calcium ions, Ca^{2+}, and activation of protein kinase C (127,128).

The regulation of AANAT activity by day/night changes in cAMP levels differs greatly between two groups of species: (a) rodents and (b) ungulates and humans (129) (Fig. 11). In rodents, cAMP activates cAMP-dependent protein kinase, which phosphorylates cAMP response element binding protein (130), and in turn binds to cAMP response element sites in the *AANAT* promoter to activate transcription (131), thereby inducing a large increase in *AANAT* mRNA expression (117,132,133). A negative feedback mechanism is responsible for the fall in *AANAT* mRNA expression at the end of the night: Phosphorylated cAMP response element binding protein stimulates synthesis of the inducible cAMP early repressor that reaches a peak in the pineal gland during the second part of the night (134,135) and shuts down transcription of *AANAT* (136). This transcriptional control is supplemented by posttranscriptional mechanisms leading to a degradation of the AANAT protein by cAMP-dependent changes in proteasomal proteolysis (137). This proteolysis allows for a more rapid switch-off in AANAT protein than transcriptional processes. As a consequence of light exposure, melatonin synthesis decreases very rapidly because the enzyme is degraded with a half-time of 3.5 minutes (138,139).

In ungulates and primates, posttranscriptional mechanisms are more important to explain the rhythmic melatonin biosynthesis as the mRNA for AANAT fluctuates only marginally. In sheep and primates, the nocturnal increase in melatonin and AANAT activity is accompanied by a minimal increase in AANAT mRNA expression (140–143). In addition, in bovine pinealocytes, treatment with NE does not change constitutively elevated *AANAT* mRNA levels but does increase AANAT protein and activity. Furthermore, the NE-induced rise in protein and activity is blocked by a protein synthesis inhibitor but not by a transcriptional inhibitor (144). Also, in ovine pinealocytes, stimulation of cAMP by forskolin increases AANAT levels (145). It is thus proposed that in these species AANAT protein is constantly synthesized but instantly degraded by proteasomal proteolysis during the day (absence of cAMP); at night, increased levels of cAMP protects the protein from degradation and thereby enhances AANAT activity (144). This model, which emphasizes proteasomal proteolysis as the dominant switch-on mechanism of AANAT activity, may explain the rapid increase in melatonin secretion at the onset of the night in ungulates and primates by a continuous availability of the AANAT mRNA.

HIOMT activity is also regulated. However, in contrast to AANAT, which displays rapid changes in activity in response to both light signals and NE stimulation thereby providing an ON or OFF switch for melatonin secretion, HIOMT appears to be regulated on a longer term basis to modulate the level of nocturnal melatonin secretion [for review see (115)]. HIOMT activity is regulated over several days or weeks by the nocturnal noradrenergic stimulation of the pineal gland. Exposure of rat to continuous light or superior cervical ganglionectomy induces a large decrease (two- to threefold) in HIOMT activity, an amount comparable with animals kept in DD. This effect is reversed by daily injections of a noradrenergic agonist (146–148). This regulation of HIOMT activity on a long-term basis is due in part to the high stability of the protein [half-life > 24 hours (149)]. The regulation of HIOMT activity may be relevant to the seasonal regulation of the amplitude of the melatonin secretory rhythm in several species (115). For instance, in the Siberian hamster, pineal HIOMT activity is twice as high in short days as in long days, and this increase parallels a twofold increase in the amplitude of the nocturnal melatonin peak (150). These changes in HIOMT activity could result from changes in NE [as suggested in the rat (151)] or NPY [as suggested in the European hamster (152) input to the pineal]. In addition to melatonin, the pineal gland synthesizes other 5-methoxyindoles from serotonin, such as 5-methoxytryptophol and 5-methoxytryptamine, in a changing way according to photoperiod and season (153–155).

Light–Dark Control of Melatonin Synthesis

In virtually all species studied, melatonin is synthesized and secreted during the dark phase of the day (Figs. 9 and 10). This is true whether species are active at night or during the day. The lack of storage of melatonin in pinealocytes on the one hand and the rapid hepatic destruction of melatonin on the other cause a very close relationship between melatonin synthesis and presence in biological fluids.

The 24-hour profile of melatonin secretion varies greatly between species, and a classification into three groups has been proposed (156). In type A, melatonin levels remain low during the first part of the night; they increase only in the second half of the night and decrease to daytime levels before light onset. This pattern is the least common and has been described in Syrian hamster, Mongolian gerbil, and house mouse. In type B, soon after night onset, melatonin levels increase slowly to reach a peak at mid-darkness; during the second part of the night, melatonin falls to reach daytime levels near the time of light onset; this pattern has been described, among other species, in rat and humans. In type C, melatonin levels reach a plateau soon after the onset of the night [10–30 minutes in sheep (157)], remain elevated throughout the entire night, and decrease soon before or around the time of light onset; this pattern has been described, among other species, in sheep and Siberian hamsters. It is important to note that, regardless of the type of pattern, the duration of elevated melatonin secretion increases when the night is lengthened.

The magnitude of the nighttime rise in melatonin varies greatly between species. In the same individual, it is highly reproducible in amplitude, details of the profile, and in timing in a synchronized environment, almost like a hormonal fingerprint (158). Between individuals there are large differences in amplitude (159,160). For example, in sheep kept in the same environment, circulating blood nighttime levels may vary from 50 to 1,000 pg/ml between individuals of similar age and origin (161). This variability is under strong genetic control because the heritability coefficient (heritable part of the phenotypic variance) was found to be 0.53 in humans and 0.45 in sheep (161,162). From a physiological point of view, this variability does not result from differences in melatonin catabolism but from melatonin secretion (163). This variability in secretion is related to the size of the pineal and the number of pinealocytes that it contains rather than to the activity of enzymes of the biosynthetic pathways; this difference in size is established as early as 1 week of age in sheep (164,165). The genes involved in the control of such variability probably act during development in utero or very early postnatal life.

The light–dark cycle controls the 24-hour melatonin secretory in two ways: entrainment of the circadian rhythm of melatonin secretion and inhibition of melatonin secretion by light. The diurnal cycle of melatonin release is the expression of a circadian rhythm as it free-runs with a period close to 24 hours in constant darkness (166–169). This rhythm is entrained with melatonin secretion occurring at night by light–dark cycle characterized by a 24-hour period; this entrainment is the primary effect of light in natural conditions. The SCN is involved in generating and transmitting to the pineal gland circadian information, but it now clearly appears that the SCN is also a photoperiodic structure and is involved in the integration of seasonal information that it relays to the pineal gland [for reviews see (170–172)]. Several approaches have allowed investigators to demonstrate that the functional state of the SCN depends on photoperiod. When sheep are transferred from natural photoperiod to continuous darkness, melatonin secretion reflects the state of the clock in the absence of any direct influence of light. Using this paradigm, the duration of detectable melatonin in the plasma is shorter in summer than in winter, indicating that seasonal information is stored in the SCN (173). Also, the circadian rhythm of c-*fos* induction by light exposure in the ventral SCN of rats and hamster varies with photoperiod, the duration of photosensitivity being shorter in animals entrained to long photoperiods than to short photoperiods (174,175). When animals are changed from long to short days, decompression of the phase of c-*fos* induction occurs gradually (about 3 weeks). Importantly, this photoperiodic dependence of c-*fos* induction is not pineal dependent because it persists in pinealectomized animals (176,177). Similarly, in the hamster, the duration of high daytime *Per1* mRNA expands on long days and contracts on short days in a pineal-independent manner (178,179). This photoperiodic effect persists when animals are maintained in continuous dim red light (180). Also, multiunit neuronal activity rhythm in slices of the Syrian hamster SCN depends on photoperiod, and this capacity persists in vitro (181). This encoding of photoperiodic information in the SCN is responsible for driving the pineal and shaping differentially the pattern of melatonin secretion between long and short days.

However, light can also influence melatonin secretion by overriding the signal transmitted by the clock and exerting an acute inhibitory effect on melatonin synthesis and release irrespective of circadian time; this second effect of light causes remarkably rapid changes by inhibition of AANAT activity [for review see (114)]. In addition to entrainment, it is the second possible mechanism for light to determine the characteristics of the melatonin secretory by inhibiting melatonin secretion at the beginning or at the end of the period when the endogenous clock dictates secretion by the pineal (159). The relative importance of these two mechanisms, entrainment and inhibition, in shaping the melatonin rhythm according to photoperiod appears to differ among species. In rodents, the entraining effect is dominant as illustrated by the slow and gradual lengthening of the duration of melatonin secretion when animals are changed from long to short days; 3 to 4 weeks are necessary for a full adjustment (182). In contrast, in ungulates, the inhibitory effect of light is important to gate the melatonin peak at both ends of the melatonin peak. When day length is changed abruptly, the duration of melatonin secretion changes as soon as the first night and is fully adapted after 3 or 4 days (183). Also, when animals are exposed to constant darkness after being in long days, the duration of melatonin secretion expands within the first day of transfer to constant darkness (184).

Melatonin Transport and Catabolism

A good proportion (60–80%) of circulating melatonin is bound to plasma proteins, albumin, and α_1-acid glycoprotein, but the physiological significance of this binding has not been seriously considered at this stage, mainly because of its low affinity (on the order of 1 mmol/L) (185,186). Melatonin is rapidly metabolized: The half-life of circulating melatonin after exogenous administration is 10 to 60 minutes in humans (187), about 15 minutes in the rabbit (188), and 20 minutes in sheep (163). The main enzymatic pathway involves hydroxylation at the 6-position. Melatonin 6-hydroxylase activity is localized in hepatic microsomes, and CYP1A2 appears to be the main catalyst of the hydroxylation (189). The 6-hydroxymelatonin is then excreted in the urine after sulfate or glucuronide conjugation (159). In the brain, melatonin is also metabolized into *N*-acetyl-5-methoxykynurenamine or *N*-acetyl-*N*-formyl-5-methoxykynurenamine (190).

MELATONIN REGULATION OF THE HYPOTHALAMIC–PITUITARY–GONADAL AXIS

Role of the Pineal Gland and Melatonin in Transducing Photoperiodic Information

Role of the Pineal Gland

The importance of the pineal gland to transduce photoperiodic information was demonstrated by numerous experiments showing that the effect of photoperiod on seasonal reproduction is profoundly altered after removal or denervation of the pineal gland [for review see (83,191,192)]. In Syrian hamsters, pinealectomy prevents the gonadal regression normally brought about by experimental or natural short photoperiod (193,194). In mink, a short-day breeder, a remarkable effect of pinealectomy was observed in a 3-year experiment: After removal of the gland, animals display sexual rest and remain permanently reproductively inactive (195). In sheep, the demonstration of the role of the pineal was made more difficult because the early pinealectomy experiments failed to show a significant effect of the removal of the gland on the seasonal pattern of reproduction (196–198). The reasons for this failure to show a role of the pineal gland in the timing of the breeding season are now better understood. First, because there is an endogenous rhythm of reproduction in this species, the persistence of alternations between breeding and anestrus is expected even though the external signal is removed. Second, this endogenous cycle can be indirectly synchronized in pinealectomized ewes through social signals transmitted by their pineal-intact conspecifics (male or female) (199). A clear demonstration of the role of the pineal gland came from studies in which animals were tested for their ability to respond to controlled photoperiodic changes. Soay rams in which the innervation of the pineal was removed by superior cervical ganglionectomy were neither sensitive to the inhibitory effects of long days nor responsive to the stimulatory effects of short days (200). Similarly, pinealectomy makes rams or ewes unable to display synchronized changes in reproductive activity or in gonadotropin secretion in response to alternations between long and short days (183,201,202).

Role of Melatonin and Critical Features of the Rhythm of Secretion

The critical secretory product of the pineal gland involved in transducing photoperiodic information is melatonin. Timed injections of melatonin reverse the effect of pinealectomy on gonadotropin secretion and cause gonadal regression in hamsters (203). This early work on the role of the pineal gland and melatonin in long-day breeders, mainly in hamster, led to the notion of the antigonadal action of melatonin. However, although this term still appears occasionally in the literature, more recent data, particularly those obtained in short-day breeders, have led to a more generalized concept of the role of melatonin in the control of seasonal rhythms. Specifically, pinealectomy suppresses responses to both short and long photoperiod, and melatonin, depending on its specific pattern, reinstates both these responses [for review see (204)]. Thus, the role of melatonin is to provide an endocrine code for day length. This role was well illustrated by replacement studies in pinealectomized animals. In the short-day breeding sheep, melatonin delivered into the peripheral circulation to mimic short-day or long-day profiles can reproduce the stimulatory and inhibitory effects on LH secretion of short and long days, respectively (205,206). In the long-day breeding hamster, on the other hand, short-duration melatonin infusions are inductive of reproductive activity, whereas long-duration infusions are inhibitory (207,208).

The characteristic of the melatonin secretory rhythm that conveys the photoperiodic information to affect seasonal rhythms has been a matter of controversy. The rhythm of melatonin secretion is characterized by three features that could code for day length: amplitude, duration (length of time with elevated levels), and phase (presence of melatonin at a given time of day to coincide with a period of sensitivity to melatonin). Amplitude received some attention because in some wild species, such as European hamster, red deer (*Cervus elaphus*), reindeer (*Rangifer tarandus*), and muskoxen (*Ovibos moschatus*), strong seasonal changes in the amplitude of nocturnal secretion of melatonin have been reported (209–212). However, amplitude is unlikely to be important because it does not consistently differ between short and long days in most species and it is highly variable among individuals (160). In addition, in sheep, the characteristics of the breeding season do not differ between high and low melatonin secretors (213), and in hamsters, experimental manipulation of the amplitude of the nighttime rise does not alter the reproductive response to melatonin (204). Although, we cannot exclude that mean concentration of nocturnal melatonin may modulate reproductive response to photoperiod, it is not important in the discrimination between long and short days. The amplitude of the nocturnal elevation of melatonin must simply exceed a minimum level that is necessary for a nocturnal rise to be registered.

Evidence for the phase hypothesis was obtained from studies in which single injections of pharmacological doses of melatonin administered daily in late afternoon to Syrian hamsters caused gonadal regression. Injections at other times of the day, except for a brief interval immediately before lights-on, were ineffective (204,214). A severe limitation of these studies using daily injections of melatonin was that it is virtually impossible to control either the duration or the amplitude of the elevation of melatonin in the blood. In addition, in pineal-intact animals, the exogenous hormone given at the end of the light phase could have been effective because it added to the endogenous secretion to produce a single daily signal of long duration. When physiological doses of melatonin are infused into pinealectomized Siberian hamsters or sheep for fixed durations but at different times relative to the light–dark cycle, the response varies only with the duration of the infusion, not with time of the day (204,206–208,215). For instance, in pinealectomized sheep exposed to long days, melatonin is equally effective in causing an inhibition of gonadotrophic activity whether it is infused for 8 hours every night or in the middle of the 16-hour day (215). Furthermore, when melatonin signals are delivered at experimental periodicities of less than 24 hours and therefore scan across the prevailing light–dark cycle, short-day like responses still occur in Syrian hamsters (216). However, all these studies supporting the duration hypothesis present a limitation in that infusions are given on a regular and predictable basis. It therefore remains possible that a rhythm of sensitivity to melatonin is established by the melatonin signals themselves. One of the most convincing pieces of evidence in favor of duration being critical is therefore the observation that long melatonin pulses infused randomly at one of three different phases of the day induce the short-day gonadal response in pinealectomized Syrian hamster (217). These results and others obtained in several species led to the concept that the nocturnal duration of the melatonin elevation is the critical feature of the melatonin signal for its action on photoperiodic physiological functions. This is also in keeping with the fact that the duration of melatonin secretion is always positively correlated to the length of the night. However, this issue remains controversial, and two observations are not consistent with the duration hypothesis. First, two 2.5-hour infusions of melatonin given 3 hours apart produce a reproductive inhibition in Syrian hamsters, suggesting that melatonin does not need to be present continuously to be read as a short day. One interpretation is that the first infusion of melatonin synchronizes a rhythm of sensitivity and that the second infusion occurs during this period of sensitivity (218). These data would be fully conclusive once it is demonstrated that melatonin disappears from the target tissue during the 3-hour interinfusion interval. Second, a 1-hour infusion of melatonin produces a short-day response in pinealectomized Siberian hamsters but only if it is given immediately after lights-off and not at other times of the daily cycle (219). This suggests that responsiveness to melatonin varies throughout the day, but the physiological significance of the presence of melatonin for 1 hour has to be evaluated.

The definition of duration of melatonin is relative, not absolute. As discussed above, the interpretation of the duration of a given melatonin pattern depends on the photoperiodic history of the animal. A change in day length is always accompanied by a similar change in the duration of melatonin secretion (25). Therefore, the response to a given melatonin duration depends on the previous pattern of change in duration, and the ability to perceive gradual changes in photoperiod is the reflection of mechanisms allowing the discrimination of similar changes in the duration of melatonin secretion. This was tested directly in Siberian hamsters in which melatonin was infused with gradual duration changes; the pattern of change in melatonin duration determined the testicular response. The central nervous system appears to compare the duration of a given melatonin signal with those that preceded it (220). Animals must therefore encode a representation of prior durations of melatonin signals. In Siberian hamsters, the memory of prior photoperiods persists for at least 6.5 weeks and is absent after 20 weeks, suggesting that only information about recent melatonin signals is stored (221).

Melatonin as a Time-Giver Messenger

The annual changes in the duration of the melatonin signal provide the animal with an index of time, and, in this respect, melatonin can be considered as a "time-giver" messenger. This role of melatonin is particularly well illustrated in circannual species. The pineal by means of its secretion of melatonin is necessary for the entrainment of circannual rhythms. Pinealectomized female golden-mantled ground squirrels kept for 3 years in simulated natural photoperiod generate circannual rhythms of reproduction that are no longer entrained by photoperiod changes; they become asynchronous among individuals and present a period longer than 1 year (222). Similarly, pinealectomized ewes exposed to natural changes in day length display circannual cycles of LH secretion that are similar to those observed when pineal-intact animals are kept in constant photoperiod (19,223). It is possible to infuse melatonin daily with various durations to restore "photoperiodic information" corresponding to different times of year (224). Using this model, the importance of a specific portion of the annual photoperiodic cycle for the entrainment of the circannual rhythm of LH secretion can be tested by replacing melatonin in pinealectomized ewes only at this portion. Provision of "photoperiodic information," by means of a circadian melatonin pattern, for as little as 70 days each year can maintain entrainment of the circannual rhythm of LH secretion (223). Nevertheless, not all seasonal photoperiodic information is equally effective (Fig. 12). Replacement with melatonin patterns mimicking

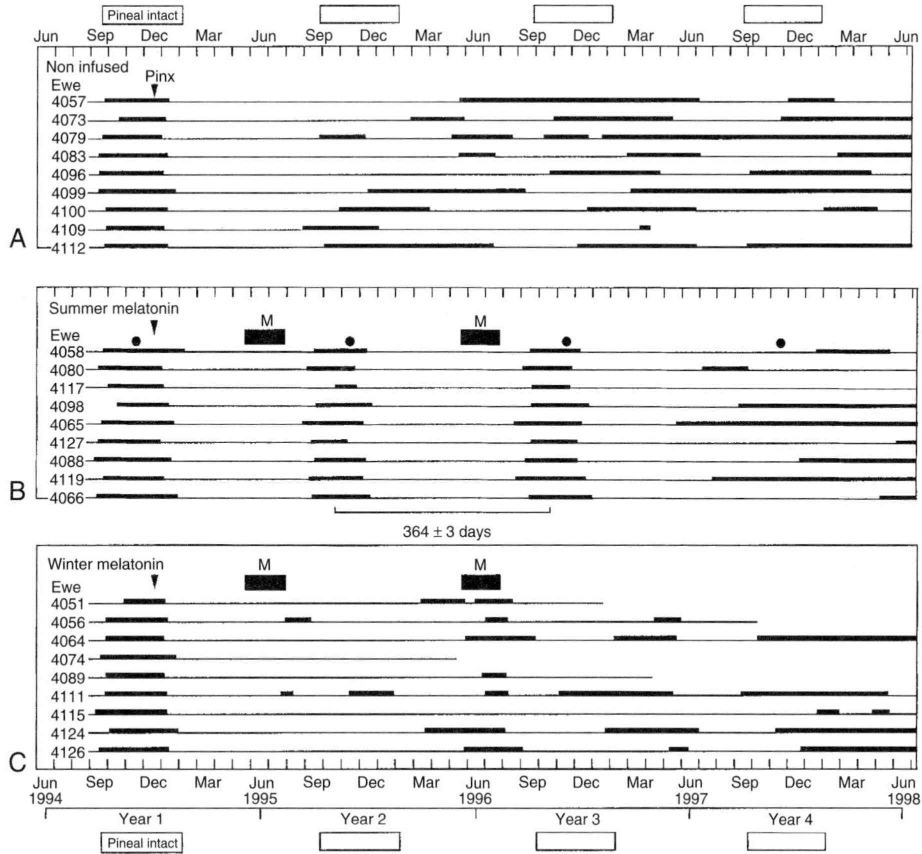

FIG. 12. Timing of high and low stages of circannual LH cycles in three groups of pinealectomized ewes: controls noninfused with melatonin (**A**), infused with a summer melatonin pattern (**B**), or a winter-melatonin pattern (**C**). *Filled circles* represent mean date of mid-point of LH rise in the summer-melatonin group before pinealectomy, included for reference. *Thick and thin* portions of horizontal lines depict respective high and low stages of LH cycles; discontinued lines in some ewes during the latter half of the study indicate observations were terminated due to development of health problems. *Solid boxes* indicate times of melatonin (M) infusion; *open boxes* at top and bottom depict timing of high stages of LH cycles in pineal-intact ewes. [Reproduced from (495).]

those secreted in spring/summer is more effective than an autumnal pattern, and replacement with the winter pattern is ineffective. Not all cues are equally effective at synchronization, and the quality of the photoperiodic signal, a long day, is critical for entrainment of this circannual rhythm (225,226). A similar conclusion was reached in some hibernating animals, ground squirrels, and woodchucks in which photoperiodic exposure in spring/summer could maintain precisely timed seasonal cycles despite the disruption of photoperiodic signals during periods of hibernation in autumn/winter (227,228). Results of all these studies indicate that seasonal changes in the duration of nocturnal melatonin secretion are critical for the entrainment of the circannual rhythms to a period of 12 months, and the entire annual cycle of duration changes is not required for an effective synchronization.

Observations in Siberian hamsters suggest interesting similarities between circannual and noncircannual species. In Siberian hamsters, gonadal recrudescence in late winter/early spring occurs as a result of an interval timer mechanism that renders animals refractory to the short inhibitory day lengths. This internal process is triggered by the first 6 to 12 weeks of exposure to short day lengths and is manifested by gonadal recrudescence many weeks later (229). When young are born late in the breeding season (August and later), reproductive maturation is delayed until the next breeding season and its timing is determined by this interval timer mechanism. Interestingly, gonadal development occurs at the same date in males born in early August or in late September (i.e., at different ages) (230). The duration measured by the interval timer is therefore flexible so that cohorts born at different times of the previous breeding season are rendered synchronous in terms of gonadal development. This adjustment results mainly from the action of photoperiod perceived between 3 and 9 weeks of age that phase advance the internal process leading to gonadal development in animals born in late September (231). In circannual species, annual changes in the duration of the melatonin signal synchronize the endogenously generated rhythm. Similarly, in noncircannual species, these changes in the melatonin signal can therefore alter the timing of an endogenous process that mediates the winter/spring seasonal transition (231).

Sites of Action of Melatonin

Hypothalamic Target Sites

The identification of the sites of action of melatonin is made difficult because melatonin influences many physiological functions (159). Melatonin could either act at a single site that would then be implicated in the regulation of many seasonal functions or act at multiple sites, each involved in regulating one seasonal function. This difficulty is expanded by the localization of high-affinity melatonin receptors in a wide variety of tissues in the body [see above and for review (232)].

The first demonstration that melatonin acted in the brain was provided by studies in the white-footed mouse using small beeswax pellets impregnated with melatonin (233). More evidence for a brain site was obtained with a rhythmic pattern of intracerebral melatonin delivery in pinealectomized white-footed mice (234) or in Syrian hamsters and gerbils (235,236). Also, central passive immunization against melatonin blocked photoperiodic responses in the Syrian hamster (237). However, these studies could not define a precise target within the brain.

The development of the melatonin probe, 2-(^{125}I)-iodomelatonin (^{125}I-melatonin), allowed the identification of putative target sites of melatonin within the hypothalamic–hypophyseal system by receptor autoradiography. Binding with this ligand was described in more than 100 cerebral structures [for review see (232,238,239)]. However, the number and the nature of labeled structures vary considerably between species, and only two structures, the pars tuberalis (PT) and, to a lesser degree, the SCN, appear to display melatonin receptors in almost all species studied. The PT drew most attention because the density of binding was much higher that in any other hypothalamic or pituitary site, a consistent feature among species investigated to date. Several studies have also found a seasonal variation in the melatonin binding site density and in the cytology of the PT (240). Finally, the PT is the most dorsal region of the anterior pituitary and cuffs around the infundibular stalk. All the portal vessels carrying factors (e.g., gonadotropin-releasing hormone [GnRH]) from the hypothalamus to the pars distalis of the pituitary pass through the PT. Thus, the PT is in a prime anatomical position for interacting with both the median eminence and the pars distalis of the pituitary and could therefore influence hypothalamic–hypophyseal function. It was proposed that melatonin could act on its receptors in the PT to alter the release of LH. LH would then diffuse into the overlying median eminence to alter the release of GnRH from the axons of the GnRH neurons (241). Indeed, in vitro, melatonin incubated with median eminence–PT tissue from male rats reduces the release of LH and causes a release of GnRH; LH antiserum also increases the release of GnRH, whereas addition of LH inhibits it. Nevertheless, the physiological significance of this observation is questionable given the lack of evidence for a short-loop feedback of LH on GnRH release in vivo (242).

Most importantly and surprisingly, functional studies in sheep and hamsters have led to the conclusion that the PT does not mediate the action of melatonin on the neuroendocrine reproductive axis. Indeed, in the ewe, melatonin delivered directly to the PT does not appear to modify the secretion of LH: Neither the placement of a melatonin microimplant directly against the anterior face of the PT nor the discrete insertion of a microimplant into the PT modify LH secretion (243,244). In contrast, microimplants placed in the medial basal hypothalamus or the third ventricle stimulated LH release (244,245). The placement of microimplants in the PT or pars distalis of Soay rams led to similar conclusions (246). These studies provide definitive evidence that the hypothalamus, not the PT, is the important melatonin target for transducing the effects of this indoleamine on the reproductive neuroendocrine axis.

In sheep, melatonin binding was found in the premammillary hypothalamic (PMH) area: the PMH is located at the base of the brain and limited dorsally by the fornix, extends 3 mm on either side of the third ventricle, is posterior to the infundibular recess, and is delimited caudally by the mammillary bodies (247,248) (Fig. 13). A clear relationship was observed between the proximity of melatonin microimplants to the area of binding and the effectiveness of these microimplants to stimulate LH, strongly suggesting that melatonin targets are located in the PMH (248). Consistent with this, lesions placed in the premammillary–caudal arcuate region of the hypothalamus of Soay rams altered both the amplitude and the timing of the photoperiod-induced gonadotrophin cycle (249). In addition, mRNA for the oMT1 melatonin receptor, thought to mediate the effect of melatonin on reproduction, is present in the PMH with a higher transcript expression at the end of the night than at the end of the day (250).

In hamsters, a hypothalamic localization of the target sites of melatonin for acting on reproduction was also strongly suggested. In Syrian hamsters, lesions of the dorsomedial hypothalamus that encompass melatonin binding sites block the gonadotropic response of male Syrian hamsters to short photoperiod or short-day-like infusions of melatonin, and lesions of the ventromedial hypothalamus cause premature testicular recrudescence in inhibitory photoperiod (251,252). Furthermore, an overlap between melatonin binding and androgen receptor immunoreactivity is found in the dorsomedial nucleus. Because a modulation of negative steroid feedback is a key mechanism mediating the action of melatonin on gonadotropin secretion (see below),

Anterior hypothalamus

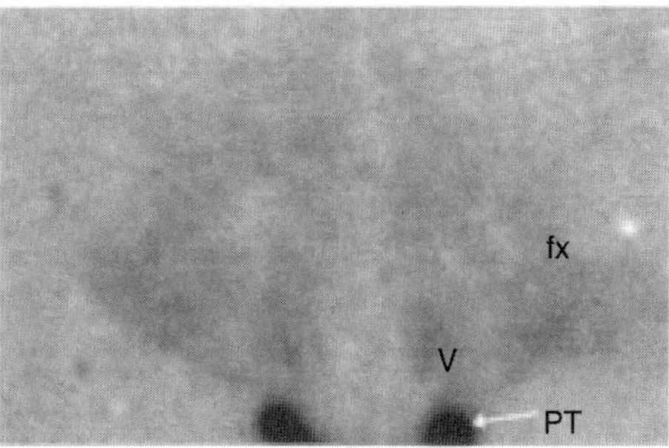

Posterior hypothalamus

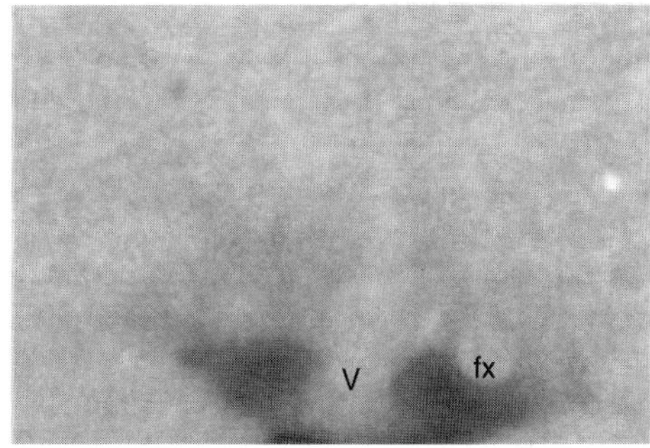

FIG. 13. Coronal sections of sheep hypothalamus illustrating distribution of ^{125}I-melatonin binding sites. **Top:** Section in the anterior hypothalamus. Note the high level of binding in the pars tuberalis (PT). **Bottom:** Section in the posterior hypothalamus. Note the presence of binding in the premammillary area (base of the brain, limited dorsally by the fornix [fx] and on either side of the third ventricle [v]). [Adapted from (248).]

it was suggested that this interaction between melatonin and steroids could take place in this area in Syrian hamsters (253). In Siberian hamsters, inhibition of testis growth was observed after localized infusions of very small amounts of melatonin into the SCN, thalamic reuniens nucleus, or thalamic paraventricular nucleus, three sites that exhibit ^{125}I-melatonin binding in this species and express MT1 mRNA (254,255). The implication of these three structures in Siberian hamsters was further suggested by a study that investigated both testicular regression and the development of photorefractoriness (256). Microimplants of melatonin were left in place for 40 weeks in one of these three nuclei and, in agreement with the previous study, induced testicular regression within 6 weeks. Twelve weeks later, this treatment no longer suppressed reproduction and gonadal recrudescence occurred, implying that each of the three brain sites can become refractory. Hamsters were then given a systemic melatonin infusion subcutaneously, which immediately initiated a second gonadal regression, suggesting that neurons at each site become refractory to melatonin without compromising responsiveness of the other sites (256). Surprisingly, given the proposed implication of these three structures, lesion of only the SCN prevents long-duration melatonin infusions from causing gonadal regression in Siberian hamsters (257). This observation also reveals an interesting species difference because lesions of the SCN do not prevent reproductive responses to melatonin in the Syrian hamster, the spotted skunk, or the mink (258–260). This species difference could be related to the higher binding and the different regional distribution of ^{125}I-melatonin in Siberian than in the Syrian hamster in these nuclei (261). The above-mentioned results of functional studies and the observed diversity of the distribution of melatonin binding sites in the brain suggest that melatonin may act in several sites to exert its various reproductive effects (gonadotropin secretion, gonadal activities, sexual behavior, etc.) and to store information about previous melatonin signal duration with interspecies differences.

The PT and the Control of Prolactin Secretion

The PT mediates, at least in part, the action of melatonin on prolactin secretion. Prolactin secretion is increased in summer and decreased in winter in a wide range of mammals. These seasonal variations are controlled by the annual changes in photoperiod, and as in the case of the reproductive axis, this photoperiodic effect is mediated by the circadian rhythm of melatonin (262). In rams with a surgical hypothalamic–pituitary disconnection in which the central control of prolactin secretion is abolished, photoperiodic regulation of prolactin secretion persists and the ability of a short-day-like melatonin signal to inhibit the secretion of prolactin is maintained (263). Also, in the ewe, melatonin implants inserted within the PT have an inhibitory effect on prolactin release (243). This suggests that melatonin acts directly at the level of the pituitary gland to regulate prolactin secretion, in contrast to its control of reproductive activity at the hypothalamic level. This dual site of action of melatonin for controlling gonadotropin and prolactin is consistent with the observations that lesions of the dorsomedial hypothalamus in Syrian hamsters disrupt the gonadotropin response to melatonin and leave intact the prolactin response (251). In the adult pituitary, expression of melatonin receptors is restricted to the PT. Evidence for the presence of melatonin receptors on lactotrophs has not been obtained, and melatonin is unable to regulate directly either prolactin gene expression or secretion in the pars distalis (264–266), which suggests that the target of melatonin to drive changes in prolactin secretion is the PT. It has been hypothesized that melatonin regulates changes in prolactin secretion by controlling the release by the PT of a prolactin secretagogue that would act in the pars distalis by a paracrine effect (267). In favor of this hypothesis, the PT from sheep, cattle, and Syrian hamster secrete a peptide, termed tuberalin, that stimulates prolactin secretion from primary cultures of pars distalis (266,268,269). Melatonin receptors appear to be expressed only in PT-specific thyrotroph in adult rodents (270,271). A large number of secretory molecules are expressed in the PT and the identity of tuberalin is not known yet (272).

The dual sites of action of melatonin for controlling gonadotropin and prolactin secretion, respectively, are interesting in relation to the photoperiodic regulation of these two functions. Indeed, the action of photoperiod and melatonin on reproductive activity is characterized by long latency and the importance of photoperiodic history that is consistent with an action at the level of the brain for the storage of information. In contrast, photoperiodic history is not critical to the regulation of prolactin secretion (273), and the action of melatonin is relatively rapid [only a few days (244)], which is consistent with a more "classical" endocrine regulation at the level of the pituitary.

Melatonin Receptors and Seasonal Breeding

Melatonin Receptors

The high-affinity agonist radioligand 2-(^{125}I)-iodomelatonin (^{125}I-melatonin) has been widely used to examine the characteristics of melatonin receptors. Initially, ^{125}I-melatonin binding sites were classified on the basis of kinetic and pharmacological properties into ML-1 and ML-2 classes (274,275). The ML-1 binding site represents high-affinity melatonin receptors ($K_d < 200$ pM) and has been identified as a membrane-associated receptor sensitive to guanine nucleotides (G protein–coupled receptors) and coupled to the inhibition of adenylyl cyclase via a pertussis toxin sensitive G protein (239). The ML-2 binding site was initially detected in the hamster brain (276) and displays a lower affinity for melatonin (in the nanomolar range) (274). It was renamed the MT3 melatonin receptor and has been described in hamster as the human homologue of the cytoplasmic protein quinone reductase 2 (277). The sequence of the isolated product revealed that MT3 is not a G protein–coupled melatonin receptor but rather the enzyme quinone reductase (277,278).

Three melatonin receptor subtypes have now been cloned in the ML-1 class. The first of these was cloned by using a cDNA library constructed from *Xenopus* dermal melanophores and a mammalian cell expression cloning strategy (279); it was named Mel1c. Subsequently, two further subtypes, MT1 and MT2 [previously called Mel1a and Mel1b (280)], were isolated from different species, including human and mouse (281–283). Mel1c was detected only in birds, fish, and *Xenopus* (284–287). Of the family of the three melatonin receptor subtypes of the ML-1 class in vertebrates, only two, MT1 and MT2, are therefore present in mammals.

The coding region of MT1 and MT2 is made of two exons separated by an intron of about 10 kb (281,282). Exon 1 codes for the N-terminal region until the first intracellular loop of the receptors, and exon 2 codes for the rest of the protein. MT1 receptor gene was localized on chromosome 4 in humans and chromosome 8 in mice (282). MT2 receptor gene was localized on chromosome 11 in humans and chromosome 9 in mice (281). MT1 and MT2 melatonin receptor subtypes show relatively high identity at the amino acid level (about 55% overall and 70% within transmembrane domains in humans) (281). The receptors are made of about 300 amino acids with a molecular weight of 40 kDa and present similar pharmacological properties [e.g., picomolar range affinity (288)].

Selective ligands for MT1 and MT2 melatonin receptor subtypes have been identified and should constitute valuable pharmacological tools to study the respective functional role of these receptor subtypes (289,290). Structurally, high affinity melatonin receptor subtypes define a distinct receptor family within the superfamily of G protein–coupled receptors because they have been shown to be functionally coupled to both pertussis toxin sensitive and pertussis toxin insensitive G proteins (291,292). Melatonin, through activation of MT1 receptors, elicits multiple cellular responses described in Figure 14 [for review see (293–295)]. MT1 and MT2 can exist as constitutive homo- and hetero-oligomers in living cells. When expressed at physiological levels in HEK 293 cells, most MT1 and MT2 receptors are organized into specific receptor dimers (296). The engagement of MT1 receptors into MT1 homodimers or MT1/MT2 heterodimers appears to be governed exclusively by the relative expression levels of MT1 and MT2 receptors. In contrast, MT2 receptors have a higher tendency to form heterodimers with MT1 receptors than to form homodimers, suggesting that MT2 receptors may be preferentially engaged into heterodimers in cells coexpressing both receptors. The documented coexpression of MT1 and MT2 receptors in many melatonin sensitive tissues, such as the SCN (297), suggests that heterodimerization could occur in native mammalian tissues assuming simultaneous expression of both receptors in the same cells (298).

An orphan receptor with no known ligand or function but structurally related to the melatonin receptors has been classified as a member of the melatonin receptor subfamily and has been cloned in humans and sheep (299,300). It is called melatonin-related receptor and its identity of sequence is 45% with MT1 and MT2 receptors. This receptor does not bind ^{125}I-melatonin or ^{3}H-melatonin but is widely distributed in neuroendocrine tissues (dorsal medial hypothalamus, lateral hypothalamus, arcuate nucleus, epithelial layer lining the third ventricle, etc.) with expression conservation apparent in sheep, mouse, rat, and hamster, suggesting a conserved biological function (301).

Several factors are known to influence the regulation of the melatonin receptor, including developmental stage, presence of melatonin itself, and time of year. Concerning development, in Syrian hamsters, the density of MT1 receptors in the SCN is drastically reduced after birth (302). MT1 melatonin receptors and MT1 gene are expressed in the pars distalis of newborn rats, but their density declines drastically thereafter and they are no longer expressed by day 21 (303). In contrast, MT1 receptors remain expressed in the PT (304). This down-regulation of MT1 receptors in the pars distalis accounts for the transient sensitivity of rat gonadotroph to melatonin in the neonatal period (303). The establishment of a mature hypothalamic GnRH input appears to drive this postnatal decline in pituitary MT1 receptors (305).

In rat SCN and PT, MT1 receptor mRNA levels and binding capacities follow a diurnal rhythm with elevated values during the day, suggesting that the expression of the receptor is regulated by melatonin itself (306,307). A similar variation was observed in the PT of sheep (250,308). In primary cultures of PT, exposure to melatonin for 24 hours reduces ^{125}I-melatonin specific binding (309). This effect of melatonin would involve protein kinase C–dependent pathways (310). The expression of melatonin binding sites in the PT also varies with time if year. In the male mink, the density of melatonin receptor is higher during winter than during summer (311). In contrast, in the hedgehog, the European hamster, and the sheep, density of binding is higher in summer (312–314).

Melatonin Receptor Subtype Involved in the Seasonal Responses

In mice the targeted disruption of MT1 showed that this receptor subtype is involved in the regulation of circadian rhythms (315). Unfortunately, this species does not display seasonal variations in reproduction, and therefore no conclusion can be reached

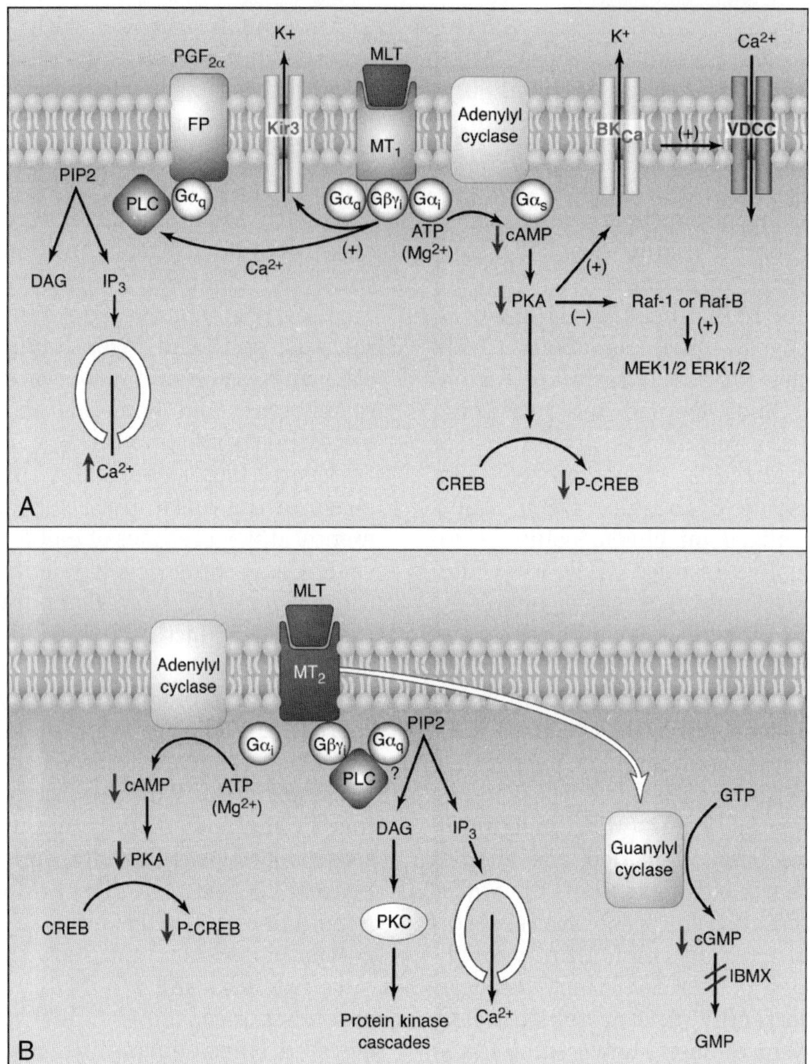

FIG. 14. Putative signaling pathways activated by MT1 and MT2 melatonin receptors. **A:** Multiple signaling pathways for MT1 melatonin receptors coupled to Gαi and Gαq. **B:** Signaling pathways coupled to MT2 melatonin receptor activation. No direct evidence for MT2 receptors coupling to Gq has been reported, so the pathway leading to PKC activation remains putative. PIP2, phosphatidylinositol 4,5-bisphosphate; PLC, phospholipase C; DAG, diacylglycerol; PKA, protein kinase A; CREB, cAMP-responsive element binding protein; ER, endoplasmic reticulum;VDCC, voltage-dependent Ca^{2+} channel; BKCa, calcium-activated potassium channel; FP, receptor for prostaglandin F2α; PGF2α, prostaglandin F2α; IBMX, isobutylmethylxantine; ATP, adenosine triphosphate; MLT, melatonin; GTP, guanosine triphosphate; GMP, guanosine monophosphate. [Reproduced from (293).]

concerning the role of MT1 in mediating seasonal effects. The MT2 receptor does not seem to play a major role in regulating seasonal functions. Indeed, in Siberian hamster, only the MT1 receptor is expressed because the MT2 receptor gene cannot encode a functional receptor due to nonsense mutations in the coding region of the receptor cDNA. This natural knockout of the MT2 receptor gene does not alter seasonal reproductive responses (316). Similarly, sheep, a seasonal species, seem to express only the MT1 receptor subtype among high-affinity melatonin receptors (250,317). This subtype was called MT1 from its higher identity (i.e., 79% of homology in amino acid sequence with human [h] MT1 compared with 56% for the hMT2 receptor). In addition, in comparison with human melatonin receptors expressed in CHO cells, ovine (o) MT1 presents a different binding profile to hMT2 but similar to hMT1 (318). These data gathered in hamsters and in sheep suggest that MT1 mediates the seasonal effect of melatonin on reproduction. This conclusion is substantiated by the presence of MT1 mRNA in putative melatonin target tissues that regulate reproduction [SCN in Siberian hamsters and PMH in sheep (250,255)].

The PT: A Model for Analyzing Melatonin Duration Decoding Mechanisms

Because of its high density of melatonin receptors and its involvement in the photoperiodic regulation of prolactin secretion, the PT has been extensively used to analyze the mechanisms of action of melatonin at the cellular and molecular levels [for review see (319)]. It has allowed investigators to identify the major role of the cAMP-dependent signal transduction (see above) (319). More recently, it has allowed the assimilation of information on putative molecular mechanisms involved in decoding of the duration of melatonin and therefore particularly relevant to the action of photoperiod.

The first mechanism potentially relevant to melatonin duration decoding is the phenomenon of sensitization of adenylate cyclase by melatonin (320). Interestingly, prolonged incubation of PT cell cultures with melatonin results in an enhanced stimulation of cAMP synthesis in response to forskolin (309). The key feature of this sensitization is the time-dependent effect of melatonin. With less than 4 hours of exposure to melatonin, the enhanced cAMP accumulation in response to forskolin was absent or minimal. As the time of exposure increased, cAMP accumulation also increased up to a maximum of 16 hours of exposure. Over a range of melatonin durations that mimic changes from long to short photoperiods, this sensitization of PT cells therefore increases in proportion to the duration of melatonin incubation. This sensitization probably results from changes in tyrosine phosphorylation of adenylyl cyclase (321). It was further suggested that melatonin signal duration could modulate the induction of cAMP by stimulatory factors, such as pituitary adenylate cyclase-activating peptide or adenosine (322,323). This time-dependent sensitization could result in long-term up- or down-regulation of gene transcription in the target cells in vivo according to the duration of melatonin presence and could thus be part of the mechanism allowing changes in melatonin duration to be decoded (319).

The second mechanism derives from the analysis of clock gene expression. The 24-hour expression profiles of several clock genes change with photoperiod. In Syrian and Siberian hamsters, *Per1* gene expression in the PT increases in the early light phase (first 3 hours of light) in both short and long days, but the amplitude of these peaks is greatly attenuated under short days (178,179). These changes in the expression of *Per1* in the PT are pineal dependent (324). In contrast, in the SCN, the daily changes in *Per1* expression are pineal independent and their amplitude does not vary with photoperiod (324). In sheep, the analysis of the diurnal profiles of expression of several clock genes in the PT revealed interesting phase relationships among them (325). The peak expression of the period genes (*Per1* and *Per2*) is similar to that of rodents and occurs in the early light phase (3–7 hours after lights-on). In contrast, the peak expression of the cryptochrome genes (*Cry1* and *Cry2*) occurs in the early night in both short and long days (325). When animals are submitted to an abrupt change from short days to long days, the expression profile of *Per* and *Cry* genes changes to a long-day profile rapidly (within 24 hours); this change is melatonin dependent (326). The activation of *Per* genes is locked to dawn and the activation of *Cry* genes is locked to dusk, and thus the phase relationship between the expression profiles of the two sets of genes varies with photoperiod. As demonstrated in the SCN, dimerization of the PER and CRY proteins in the cytoplasm is required for translocation in the nucleus [see above and for review see (71)]. The interval between expression of *Per* and *Cry* may affect opportunities for dimerization of their gene products in cells of the PT and thus the transcriptional control of downstream genes specific to the function of melatonin in the PT (327). This could constitute the basis for the photoperiodic control of prolactin secretion in the PT and possibly contribute to the decoding of melatonin duration in other target tissues of melatonin.

Melatonin Control of Reproduction

Steroid-Dependent and -Independent Effects of Photoperiod on Gonadotropin Secretion

Photoperiod induces changes in reproductive activity through modifications in the secretion of LH and FSH. In the male Syrian hamster, short photoperiod-induced testicular atrophy is accompanied by decreases in pituitary and serum LH and FSH, leading to testicular regression and lowered levels of circulating testosterone (14). Similar changes are observed in Soay rams transferred from short to long days (328). In intact females, seasonal variations in gonadotropin secretion are not so clear because of the changes occurring during the ovarian cycle at the time of the breeding season. Photoperiod regulates the secretion of gonadotropins by means of two complementary mechanisms: one independent of and the other dependent on gonadal steroids (7,14). The steroid-independent effects of photoperiod are observed in gonadectomized animals. In castrated rams and ovariectomized ewes, the secretion of LH is lower during inhibitory long days or the anestrus season than during short days or the breeding season, being one pulse every hour or every 30 minutes, respectively, in ovariectomized ewes (i.e., a twofold difference) (329–332). This difference between LH secretion during long days and short days is greatly increased in the presence of estradiol or testosterone, which demonstrates a steroid-dependent effect of photoperiod. In ovariectomized ewes treated with an estradiol implant mimicking mid-follicular phase levels, an LH pulse is observed either every 12 to 24 hours or every 30 minutes depending on whether animals are experiencing long or short days (i.e., at least a 20-fold difference) (330,333). A similar dual action of photoperiod on gonadotropin secretion is demonstrated in other species (14). For example, short-day exposed hamsters exhibit a considerable increase in sensitivity to the negative feedback actions of testosterone, and the levels of gonadotropins are also higher in long-day castrates than in short-day castrates (334–338). The steroid-independent effect of photoperiod on gonadotropin secretion is not dependent on adrenal steroids as the difference between short days and long days persist after both ovariectomy and adrenalectomy (339,340). Melatonin mediates both the steroid-dependent and -independent effects of photoperiod on gonadotropin secretion (335,341). How seasonal variations in the mechanisms controlling pulsatile LH secretion account for the seasonal variations in ovulatory activity has been particularly well studied in the ewe and is described in Chapter 44.

GnRH Neurons as the Final Nervous Pathway of Melatonin Action

Melatonin can act at three classical levels of the hypothalamic–hypophyseal–gonadal axis to modulate reproductive activity. Melatonin acts at the level of the gonads, where it modulates steroidogenesis by both Leydig cells (342,343) and ovarian theca cell (344). However, these effects of melatonin do not appear to be related to the seasonal regulation of reproduction. In the neonatal rat pituitary gland, melatonin inhibits the GnRH-stimulated secretion of LH and FSH from the pars distalis region (345). Over the initial weeks of postnatal life, this response to melatonin declines in parallel with a loss of melatonin receptors (346), and the physiological significance of this effect is unknown. In adults, no study has shown a clear effect of melatonin on the responsiveness to GnRH. In the ewe, the amount of LH released after a massive injection of GnRH (100 μg) was not influenced by melatonin administered either 3 hours before or simultaneously with the GnRH challenge (347). A similar conclusion was reached by studying the response to an injection of GnRH (1 μg/kg) given 30 minutes after an injection of melatonin in prepubertal or adult ewes (348). In another study, the LH response to hourly pulses of physiological doses of GnRH (2 ng/kg of body weight) was not different during the day (low melatonin) from that at night (high melatonin). Also, the characteristics of endogenously generated LH secretion (amplitude and frequency of pulses) were not different between day and night (349). Finally, in ovariectomized goats bearing an estradiol implant, the amplitude of endogenously generated LH pulses was not different between animals receiving a 2-mg injection of melatonin every day and controls (350).

Although an action of melatonin at the level of the pituitary and the gonads may contribute marginally to the photoperiodic control of reproductive activity, the critical step of this regulation is the neural processing of the duration of melatonin secretion to regulate the secretion of GnRH.

Melatonin treatment causes an increase in the immunoreactivity of GnRH in cell bodies located in the hypothalamus and increased density of GnRH-containing elements in the median eminence, in addition to its suppressive effects on gonadal function in white-footed mice (351,352). In the jerboa, a desert long-day hibernating breeder, the number of GnRH immunoreactive cells and fibers decrease in the autumn when animals are sexually quiescent (353). In male Syrian hamsters, immunoreactive GnRH in the hypothalamus is detectable in long days and in short days and the number of GnRH neurons and their pattern of distribution does not change in short days despite a large reduction of LH and FSH pituitary content and plasma concentration in this situation; some studies suggested a tendency for an increase in the number of GnRH neurons in short days (354–356). GnRH mRNA levels and the number of cells expressing GnRH mRNA are not different between animals exposed to short days for 16 weeks and controls maintained in long photoperiods, despite the marked difference in reproductive status between the two groups (357). In male Siberian hamsters, a rapid and transient increase in the distribution of detectable GnRH mRNA-containing cells is observed 2 days after transfer to stimulatory long days. This increase may be an early step in the stimulation of FSH secretion and gonadal growth that occur days later (358). All these data based on the assessment of GnRH mRNA or immunoreactivity suggest that a regulation of GnRH secretion rather than synthesis is more likely to explain the effect of melatonin on gonadotropin secretion.

Indeed, in vitro, the release of GnRH can be modulated by melatonin in the male mink (359). Mediobasal hypothalamic explants including the PT were incubated with melatonin for 8 or 16 hours, mimicking the profile of melatonin during long days and short days, respectively. Explants obtained at three times of the year (July, September, and November) released more GnRH when exposed to melatonin for 16 hours compared with 8 hours or with controls. A stimulation of synthesis by melatonin was also suggested by the persistence of the increase in GnRH in the medium for about 45 minutes after the withdrawal of melatonin. In vitro, a long exposure to melatonin, mimicking a single long night, therefore stimulates the release and possibly the synthesis of GnRH, and this effect depends heavily on the duration of exposure to melatonin (359).

This effect of season and melatonin on GnRH release was verified in vivo in sheep in which it is possible to measure directly and reliably the release of GnRH in the hypophyseal portal system (360). The frequency of GnRH pulses in seasonally anovulatory ewes is low, on the order of one pulse in 12 hours; this is comparable with that during the luteal phase of the estrous cycle when pulsatile GnRH secretion is at its minimal rate for the breeding season (361). Moreover, after withdrawal of progesterone implants, the frequency of GnRH pulses remains low in seasonally anestrous ewes, whereas it increases dramatically in breeding season ewes (eight pulses in 6 hours) (361). In ovariectomized ewes bearing an estradiol implant (i.e., sex steroid levels are constant), melatonin, given as a short-day profile, causes an increase in the frequency of pulsatile GnRH release (362) (Fig. 15). Interestingly, in this model, lengthening the daily duration of melatonin presence by exposure to short days or by a short-day-like melatonin treatment causes a stimulation of LH pulsatile secretion after a delay of 40–60 days, a lag time also necessary for ovulation to appear in intact ewes in response to the same treatments (7). GnRH and LH pulse frequency increase simultaneously, indicating that the functional cause of the long latency for melatonin to change LH pulse frequency and induce ovulation therefore lies in the control of GnRH release (362).

In the murine immortalized GnRH-secreting GT1-7 cell line (363), melatonin has been shown to down-regulate GnRH gene expression in a 24-hour cyclic pattern. GnRH mRNA levels were reduced to approximately 40% of basal after 12 hours and 36 hours of incubation with 1 nM melatonin and returned to normal after 24 or 48 hours of this treatment (364). GT1-7 cells express both MT1 and MT2 receptors, and the effect of melatonin on these cells is blocked by the melatonin receptor antagonist

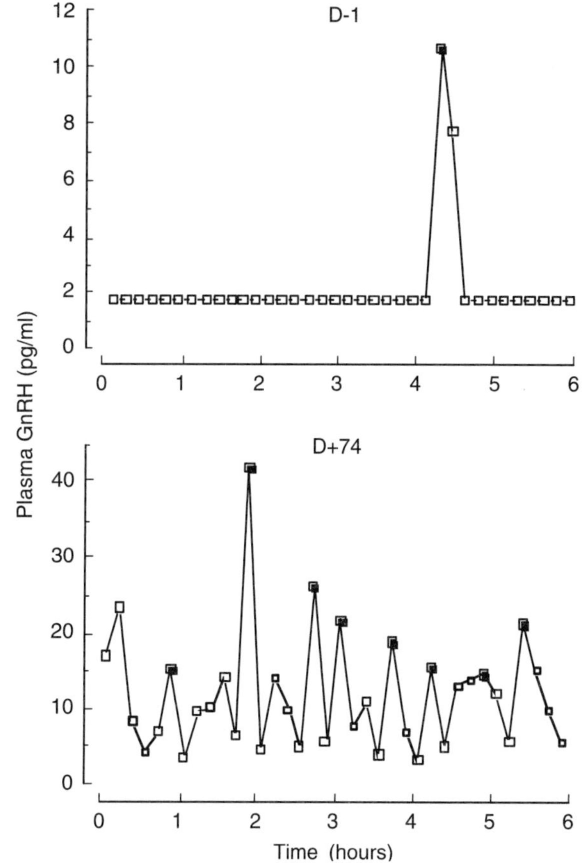

FIG. 15. Effect of melatonin on pulsatile GnRH secretion measured in pituitary portal blood. Examples of individual profiles of GnRH secretions before (D-1, **top**) or 74 days after (**bottom**) the insertion of a melatonin implant providing a short-day-like melatonin signal. Ewes were ovariectomized and treated with a subcutaneous estradiol implant and were exposed to long days (16 hours of light). Blood samples were obtained every 10 minutes for 6 hours on each occasion. The closed symbols depict detected pulses. [Adapted from (362).]

luzindole (364). Melatonin inhibits forskolin-induced cAMP accumulation through a PTX-sensitive mechanism and increases protein kinase C activity in GT1-7 cells (365). Melatonin appears to mediate its effect on GnRH mRNA expression at the transcriptional level (364). The effect of melatonin on GnRH gene transcription has been localized to five specific regions within the 300-bp neuron specific enhancer of the GnRH gene. One sequence of this region (–1736 to –1723) contains putative ROR responsive elements. This is interesting because the orphan nuclear receptors, RORα and RZRβ, both members of the ROR/RZR retinoid-related receptor family, have been suggested to bind melatonin (366). However, studies using electrophoretic mobility shift assay suggest that this region does not bind RORα but instead appears to engage the binding of C/EBPβ and COUP-TFI to regulate GnRH gene expression (367). This direct action of melatonin on GT1-7 neurons suggests that melatonin may have a direct action on GnRH neurons, and this may constitute an element of the physiological effect of melatonin on GnRH secretion. However, GT1-7 neurons are a transformed clonal population of cells outside of their natural environment, and these results should be validated in situ (364).

Interneurons Connecting to GnRH

As previously described in this chapter, the final action of melatonin at the level of the central nervous system is a control of GnRH secretion, mainly through modulation of steroid-negative feedback. However, the photoperiodic effects of melatonin on GnRH, at least for the most part, are not direct, and it is generally accepted that they are mediated by hypothalamic networks that impinge on GnRH neurons for at least two reasons. First, GnRH immunoreactivity does not overlap with melatonin binding sites in the melatonin target mediating seasonal reproduction [dorsomedial hypothalamus or SCN in hamster, PMH in sheep (248,253,368)]. Second, several neurotransmitter systems have been shown to be involved in the regulation of LH secretion by season or by melatonin. This neuronal network connecting melatonin target cells to GnRH is poorly understood. In addition, the main effect of melatonin is a modification of steroid feedback onto GnRH secretion, and therefore melatonin-dependent and steroid-dependent pathways are related.

Target Cells of Melatonin

The phenotype of the cells bearing melatonin receptors in the target nuclei mediating the seasonal effect of melatonin is poorly decrypted, mainly because of the unavailability of an antibody raised against the MT1 receptor of seasonal species. In Siberian hamster, many SCN cells that express MT1 receptor mRNA coexpress AVP mRNA (Fig. 16). This suggests that this peptide may participate in

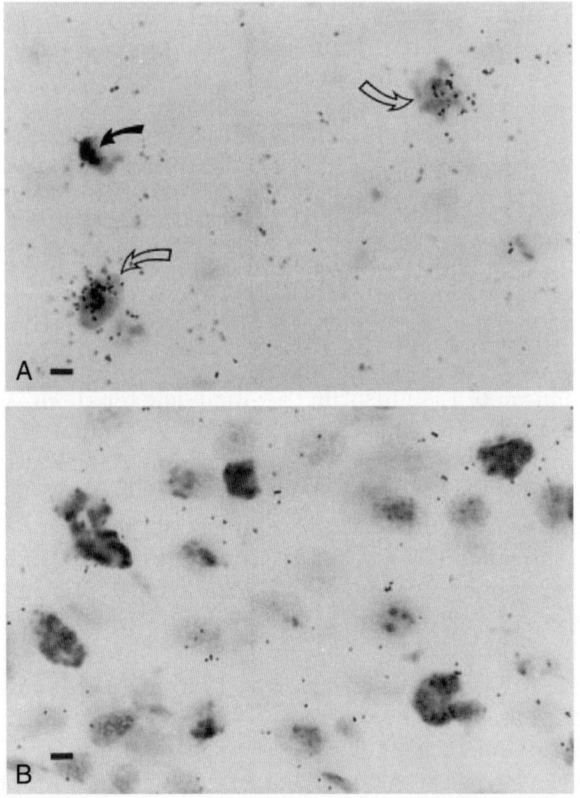

FIG. 16. Photomicrograph illustrating double label in situ hybridization in Siberian hamster SCN. **A:** SCN cells that expressed MEL1a receptor mRNA also expressed AVP mRNA (*open curved arrows*). A SCN cell that was single labeled for AVP[15] mRNA (*filled curved arrow*) is also shown. **B:** SCN section hybridized with a combination of anti-sense dig-labeled probe for AVP mRNA with sense-control ^{35}S-labeled probe for MEL1a receptor mRNA. Labeling for AVP mRNA, but not for MEL1a receptor mRNA, indicates specificity and lack of chemography. Toluidine blue counterstain. Bar=6 μM. [Reproduced from (255).]

the response to seasonal photoperiodic signals. These cells may respond to melatonin directly and possibly relay the melatonin signal to other regions of the brain that regulate seasonal reproduction (255). In sheep, oMT1 mRNA has also been described in the premammillary area, the target of melatonin, but, thus far, the low level of expression of the transcripts has not allowed colocalization studies (250). This area of binding is a heterogeneous and complex region made of three nuclei, the caudal arcuate nucleus, the ventral division of the premammillary nucleus, and the ventral tuberomammillary nucleus. The caudal arcuate nucleus contains neurons immunoreactive for tyrosine hydroxylase, dynorphin, estrogen receptor α, cocaine- and amphetamine-regulated transcript peptide, and nitric oxide synthetase. The ventral division of the premammillary nucleus also contains nitric oxide synthetase and cocaine- and amphetamine-regulated transcript peptide cells. In the ventral tuberomammillary nucleus, only fibers for orexin were identified (369). The description of these various neuronal phenotypes should now help determine the functional contribution of each of them to the action of melatonin. Downstream from the target cells of melatonin, some neural substrates mediating the effects of melatonin on GnRH have been identified by a combination of anatomical, pharmacological, and lesion studies conducted primarily in sheep.

Catecholamines

The role of dopamine (DA) in the regulation of GnRH secretion is well documented, and numerous pharmacological and anatomical data as well as lesions of dopaminergic structures (370,371) suggest that DA and NE neurons are involved in transducing the inhibitory effects of estradiol on GnRH release during the anestrous season in the ewe (372). This effect involves a dopaminergic cell group (A15) that is stimulated by estradiol and is responsible for the inhibition of GnRH secretion during seasonal anestrus (371,372). This implication of DA as mediator of estradiol-negative feedback is presented in Chapter 44.

Independently of estradiol, photoperiod, via melatonin, modulates the activity of tyrosine hydroxylase (TH) in the DA terminals present in the median eminence in the ewe. Specifically, exposure to stimulatory short days results in decreased dopaminergic activity in the median eminence as assessed by a reduction in both DA content and TH activity (373,374). No effect of short-day exposure on NE content in this structure or on TH activity in the other hypothalamic areas investigated was found (374). The stimulation of LH secretion by a melatonin implant causes a parallel reduction in TH activity that suggests strongly that the effect of photoperiod on TH activity is mediated by melatonin (375). The inhibition of median eminence TH activity by short days or by treatment with a melatonin implant is expressed at a time when the inhibition of prolactin secretion is already maximal, suggesting that these photoperiod-induced changes in TH activity are independent of the regulation of prolactin secretion (375) (Fig. 17). Rather, they are temporally related to the photoperiodic regulation of LH secretion. This change is important in relation to the regulation of GnRH. Indeed, the pharmacological blockade of TH by local perfusion of α-methyl-para-tyrosine in the median eminence of long-day photoinhibited ewes led to an increase in LH secretion (376). This finding is consistent with the effect of pimozide implants in the median eminence of anestrous ewes (377). These data suggest, therefore, that an increase in TH activity in the median eminence is an important component of the inhibitory effect of melatonin on GnRH output. Interestingly, the pharmacological blockade of TH activity no longer causes a stimulation of LH secretion when it is applied after a more protracted exposure to long days (71 days) (378), suggesting that increased TH activity is critical in the initial phase of the inhibitory effect of long days and is no longer the limiting factor for the maintenance of the long-day inhibition (378). This observation is consistent with the hypothesis that different neural pathways underlie photostimulation and photoinhibition or, at minimum, specific stages of these conditions (379). For instance, lesions in the anterior hypothalamic area of the ewe delay, but do not suppress, the inhibitory effect of a shift from long to short days (379). The photoperiodic modulation of TH activity appears to be estradiol independent because the photoperiod-induced changes in TH activity are similar in OVX and OVX+E ewes (374). Thus, in contrast to the A15 dopaminergic nuclei involved in the modulation of the estradiol-negative feedback, the dopaminergic neurones of the median eminence appear to be involved upstream relative to the integration of the estradiol signal. The median eminence in sheep is a structure rich in dopaminergic terminals but contains no TH-immunoreactive perikarya (380). DA fibers form synaptic contacts with neurons secreting GnRH and DA inhibits the release of GnRH from median eminence explants in vitro (381), providing anatomical support for ME as the terminal step for the pathway regulating GnRH secretion. This effect of DA at the median eminence is mediated by D2-like dopaminergic receptors (382). The localization of the cell bodies with efferent axonal projections to

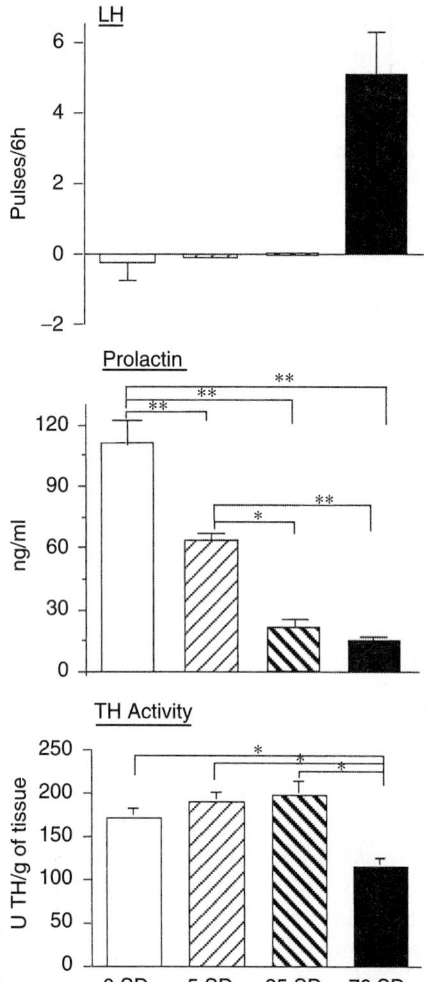

FIG. 17. Mean (± SEM) difference in the number of LH pulses per 6 hours between a control period in long days (LD, 16 hours of light) and the end of the experiment after various lengths of exposure to short days (8 hours of light, SD) (**top**), mean (± SEM) plasma prolactin concentrations at the end of the experiment (**middle**), and mean (± SEM) tyrosine hydroxylase (TH) activity in the median eminence of ewes. All ewes were ovariectomized and treated with a subcutaneous implant of estradiol. They were pretreated with LD and then allocated to one of four groups exposed to 0, 5, 25, or 76 SD. LH secretory profiles were determined from LH concentration measured in blood samples obtained every 10 minutes for 6 hours at two different periods of the photoperiodic treatment: (a) 7 days before the end of the LD period and (b) 3 to 4 days before the end of the experiment (i.e., in LD or on average after 3, 23, or 74 SD). Prolactin concentrations were determined in every third sample during the last 3 hours of the second sampling period. TH activity was measured individually in the stalk-median eminence of the same ovariectomized ewes ($n = 7$ per group). *$P < 0.05$; **$P < 0.01$. [Adapted from (375).]

the median eminence and, more generally, the anatomical and functional relationships between the A15 nuclei and the median eminence have yet to be determined.

In the male sheep, the peripheral administration of a D2 dopaminergic receptor antagonist (sulpiride) induces an immediate increase in LH secretion in long days but not in short days (383). Also, the placement of microimplants of sulpiride (D2 antagonist) to bromocriptine (D2 agonist) in the mediobasal hypothalamus modifies the stimulatory effects of microimplants of melatonin on the gonadotrophic axis in sexually inactive rams exposed to long days (384). Altogether, results in sheep indicate that a modulation of hypothalamic dopaminergic pathways is a mechanism by which melatonin mediates the effects of photoperiod on reproductive function in this species.

In male Syrian hamsters, exposure to inhibitory short days, presumably through melatonin secretion, reduces DA and NE turnover in the median eminence (385). This change is associated with a decrease in L-aromatic amino acid decarboxylase positive cells in the arcuate nucleus without changes in TH-immunoreactive cell number or TH activity (385). Suppression of LH and FSH release in this species is presumably related to reduced NE activity, whereas reduced DA turnover may represent a consequence of suppression of prolactin levels by short days (385).

Serotonin

Serotonin is involved in the seasonal inhibition of LH secretion in the ewe. The injection of an antagonist (cyproheptadine) of this indolamine causes stimulation in LH secretion during seasonal anestrus, not during the breeding season (386,387). This effect is observed in ovariectomized ewes both treated and untreated with estradiol. Serotonin plays a critical role in one particular case of inhibition of LH secretion, that is, refractoriness to short days (388,389). Refractoriness to short days is expressed in ewes that have been exposed continuously to short days for at least 150 days and display a decrease in LH secretion, although they are still exposed to the initially stimulatory day length. This phenomenon is probably the initial event of the expression of the circannual rhythm of reproduction and is critical to timing the cessation of the breeding season in natural photoperiod (390). In this situation, systemic administration of serotonin antagonists can temporarily reverse the suppression of pulsatile LH secretion (391). The use of specific antagonists, and especially ketanserin, led to the suggestion that $5HT2_A$ receptors are probably involved in this inhibitory effect of serotonin (391). The inhibition of pulsatile LH secretion during the establishment of refractoriness to short days is associated with changes in ^{3}H-ketanserin binding that suggests a change in $5HT2_A$ receptor binding capacities in the PMH (392). A parallel change in $5HT2_A$ gene expression was observed by in situ hybridization (393). To be consistent with the inhibitory effect of serotonin, an increase in $5HT2_A$

receptor binding capacities and gene expression were expected during this period of low LH secretion. Surprisingly, the opposite was found, that is, decreased ^{3}H-ketanserin binding and mRNA density. During the expression of the circannual rhythm of reproduction in pinealectomized ewes, similar changes were observed. Both $5HT2_A$ receptor gene expression and $5HT2_A$ receptor density were greater in ewes expressing a high pulsatile LH secretion compared with those expressing a low pulsatile LH secretion at the time of sacrifice (394). These results suggest an implication of the $5HT2_A$ receptor system of the PMH in the generation of endogenous circannual rhythms. This provides novel evidence for a putative site that may contribute to the generation of these rhythms. This possibility is further strengthened by the evidence that the PMH is thought to be the target site of melatonin mediating seasonal effects and therefore synchronization of the circannual rhythm of reproduction (see above) (248).

Opioid

Endogenous opioid peptides are involved in the regulation of pulsatile GnRH secretion (see Chapter 44). The injection of a μ-opioid receptor antagonist, naloxone, transitorily activates LH secretion in a variety of species, including seasonal species such as rams (395), ewes (396), mares (397), stallions (398), and male Syrian hamsters (399). This response is dependent on the presence of gonadal steroids and varies during the photoperiod-induced reproductive cycle. In horses, the stimulation of LH secretion by naloxone is larger during the season of sexual rest than during the breeding season, which is consistent with a modulation of the inhibitory tone of endogenous opioids by photoperiod. Along the same lines, in hamsters, the regression of the testis induced by exposure to short days can be delayed with repeated naloxone injections (399). In sheep, the role of opioids appears to be more complex. A stimulation of LH secretion is observed in sexually active males under short days, whereas little or no effect is obtained in sexually inactive animals exposed to long days (395). The activity of the inhibitory endogenous opioid systems therefore appears to be increased under short days. This hypothesis is consistent with the observation that in the arcuate nucleus of the ram, the expression of the mRNA encoding pro-opiomelanocortin, the precursor of β-endorphin, is higher in short days than in long days (400). It was proposed that this apparently paradoxical stimulation of an inhibitory system when animals are photostimulated results from an interaction between dopaminergic and opioid systems in which dopaminergic pathways are inhibitory not only to the GnRH neurons (see above) but also to opioid networks. According to this hypothesis, photoperiod-induced changes in dopaminergic tone would induce reversed changes in opioid tone. In favor of this hypothesis, activation of dopaminergic receptors with the dopaminergic agonist bromocriptine blocks the stimulatory effect of the opioid antagonist naloxone on pulsatile LH secretion observed in short days. Conversely, blockade of dopaminergic receptors with the specific antagonist sulpiride enhances the stimulatory effect of naloxone in long days (401). Photoperiod-induced changes in opioid tone would therefore be secondary to the photoperiodic regulation of dopaminergic activity.

NPY

Several studies in sheep suggest that season or photoperiod, presumably through melatonin secretion, induces changes in NPY mRNA and/or protein levels in the arcuate nucleus that are consistent with a reinforced inhibitory action of NPY in long days. In the ram, NPY mRNA levels are higher in long-day than in short-day photoperiods (402). In castrated rams, NPY mRNA levels are not different between long days and short days. However, in these animals, testosterone stimulates mRNA levels in long days but not in short days (403). The long-day increase in NPY gene expression is therefore testosterone dependent. In the ewe, the number of NPY immunoreactive neurons is increased during the midluteal phase of the estrous cycle of short-day treated animals as compared with the anestrous condition of long-day treated animals (404). In ovariectomized ewes, the number of NPY immunoreactive neurons was higher during the breeding season that during the anestrous season. Importantly, within seasons, estrogen did not affect NPY cell number (405). The origin and the significance of this sexual dichotomy in the regulation of NPY by photoperiod and steroids are unclear. Although NPY containing neurons synapse directly onto GnRH neurons (406), it is difficult to determine at this stage whether photoperiod-induced changes in NPY are involved in regulating seasonal fluctuations in LH secretion and reproductive activity or in voluntary food intake.

Excitatory Amino Acids

The role of excitatory amino acids, glutamate and aspartate, in regulating GnRH release has attracted attention because of the ability of their agonists to elicit GnRH and LH release in a variety of species (407).

Their implication in the photoperiodic regulation of LH secretion was suggested by experiments involving administration of *N*-methyl-D,L-aspartic acid (NMDA), an agonist at the NMDA receptor and a major subtype of excitatory amino acid receptor. Such an administration acutely stimulates GnRH and LH secretion in seasonal species such as sheep and hamsters (408–410). In rams exposed to inhibitory long-day photoperiod with regressed testis, NMDA rapidly induces a prolonged period of LH secretion. In contrast, during the phase of testicular recrudescence after exposure to stimulatory short days, sensitivity to NMDA was markedly reduced, and only a transient response could be elicited (411). In male Syrian hamsters, a higher dose of NMDA is required to stimulate LH secretion in reproductively active long-day treated animals than in short-day treated animals with regressed testis (410). Also, in adult Siberian hamsters, chronic treatment with NMDA, injected once daily, maintains reproductive activity in hamsters transferred from a stimulatory long-day photoperiod to an inhibitory short-day photoperiod (412). Although glutamate is involved in the photic regulation of circadian rhythmicity at the level of the retinohypothalamic tract, this action of NMDA cannot be explained by an effect on the melatonin secretory rhythm. Indeed, the differential effect of NMDA in photoinhibited and photostimulated ewes is maintained if stimulation is induced by melatonin administration instead of short-day exposure (413). Furthermore, in Siberian hamsters, the chronic effect of NMDA is observed on the photoperiod-induced changes in reproduction but not on other functions (pelage color, body weight) (414). These experiments suggest that a change in excitatory amino acid input may be part of the mechanism mediating the action of melatonin on GnRH secretion. Consistent with this hypothesis, photoperiod modifies the density of NMDA receptors in the preoptic area of hamsters (415).

Besides the neurotransmitters or receptors mentioned above, future research will probably reveal the involvement of other neuronal systems in mediating the action of melatonin on GnRH neurons. Regardless of the nature of these neurons, it is clear that the action of melatonin on GnRH neurons most likely involves a complex network of interacting neurons.

Thyroid Hormones as Potential Mediators of Seasonal Plasticity

Thyroid Hormones and Seasonality

Role of Thyroid Hormones. The presence of thyroid hormones is obligatory for expression of the seasonal reproductive cycle in a broad spectrum of species. The first demonstration of a role of thyroid hormones in seasonal reproduction was made in starlings in which thyroidectomy prevented the seasonal decline in testicular activity and sustained the breeding season indefinitely (416). It was later demonstrated that thyroidectomy prevented the development of photorefractoriness, a spontaneous change responsible for the termination of the breeding season (417). Since then, multiple studies have indicated that thyroid hormones are essential for the maintenance of seasonal reproductive changes in a variety of species of birds (418,419) and mammals (golden hamsters, edible dormouse, mink, wild rabbit, red deer, and sheep) (420–427). In adult male Siberian hamsters, inhibition of thyroid gland function by addition of thiourea to drinking water does not interfere with gonadal regression upon exposure to short days but does cause an earlier gonadal recrudescence; thyroid hormones are therefore critical for the prolonged maintenance of reproductive inhibition in short days (428). In sheep, ablation of the thyroid maintains ovarian cyclicity without interruption (423). Thyroidectomy blocks neuroendocrine changes that lead to the transition from the breeding season to anestrus, whether the transition is induced by exposure to inhibitory photoperiod or driven by the endogenous rhythm that underlies the seasonal changes of reproductive activity in this species (429,430). This effect of thyroidectomy is fully prevented by replacement of thyroxine (T_4) at a dose that restores a physiological serum concentration of that hormone (431). In males, thyroid hormones are also critical for the expression of the seasonal reproductive cycle, but two sex differences are noticeable. First, in contrast to females in which seasonality is fully abrogated, thyroidectomized males appear to express some residual seasonality with a decline in scrotal circumference and FSH secretion during the nonbreeding season (422,424,425,427). Second, thyroidectomy of rams during seasonal sexual rest results in rapid sustained testis growth and increased FSH secretion; in contrast, in ewes the surgery has no effect if conducted during seasonal anestrus, allowing a breeding season to begin at the normal time. Once breeding is initiated, however, it does not stop (425,432,433). In contrast to the main central neural effect of thyroid hormones (see below), this short-term effect of thyroidectomy is due to a reduction in the negative feedback efficacy of endogenous inhibin at the anterior pituitary that increases FSH secretion, which in turn stimulates testicular growth (434).

Although peripheral concentrations of T_4 display seasonal variations that appear to be related to the reproductive cycle (highest concentrations at the time of the end of the breeding season) in most species (426), results in sheep suggest that the change in circulating T_4 is not needed to drive the transition to anestrus; rather, thyroid hormones need only to be present for occurrence of this transition and their role is permissive rather than dose dependent (435). Furthermore, the presence of T_4 is required only during a discrete period late in the breeding season to develop anestrus at the usual time. This time frame of thyroid hormone action lasts 60–90 days, and thyroid hormones do not need to be present the rest of the year for the expression of the full seasonal cycle of reproductive activity (436). Sensitivity to T_4 (tetraiodothyronine) is, however, maintained beyond the end of the breeding season. If T_4 is given in April, neuroendocrine activity still falls to baseline. By June and July, the responsiveness to T_4 is lost, indicating that the ewes gradually lose responsiveness to T_4 during the latter half of the anestrous season (437) (Fig. 18). Therefore, a window exists of sensitivity to melatonin that is much longer (November to June) than the critical period (November to January) when thyroid hormones are essential to obtain a correct timing of the end of the breeding season. The existence of normal seasonal prolactin and melatonin patterns in thyroidectomized animals indicates that perception of photoperiodic changes by other neuroendocrine systems can occur without thyroid hormones (437,438). Thyroid hormones are therefore required for the transduction of the melatonin signal to the GnRH neurosecretory system, and the site of action of thyroid hormone in this regard is proximal to the GnRH network.

Mechanisms for Thyroid Hormone Action. The primary targets of the thyroid hormones are located within the brain where they ultimately impact central mechanisms controlling seasonal GnRH inhibition (439). A central site of action is indicated

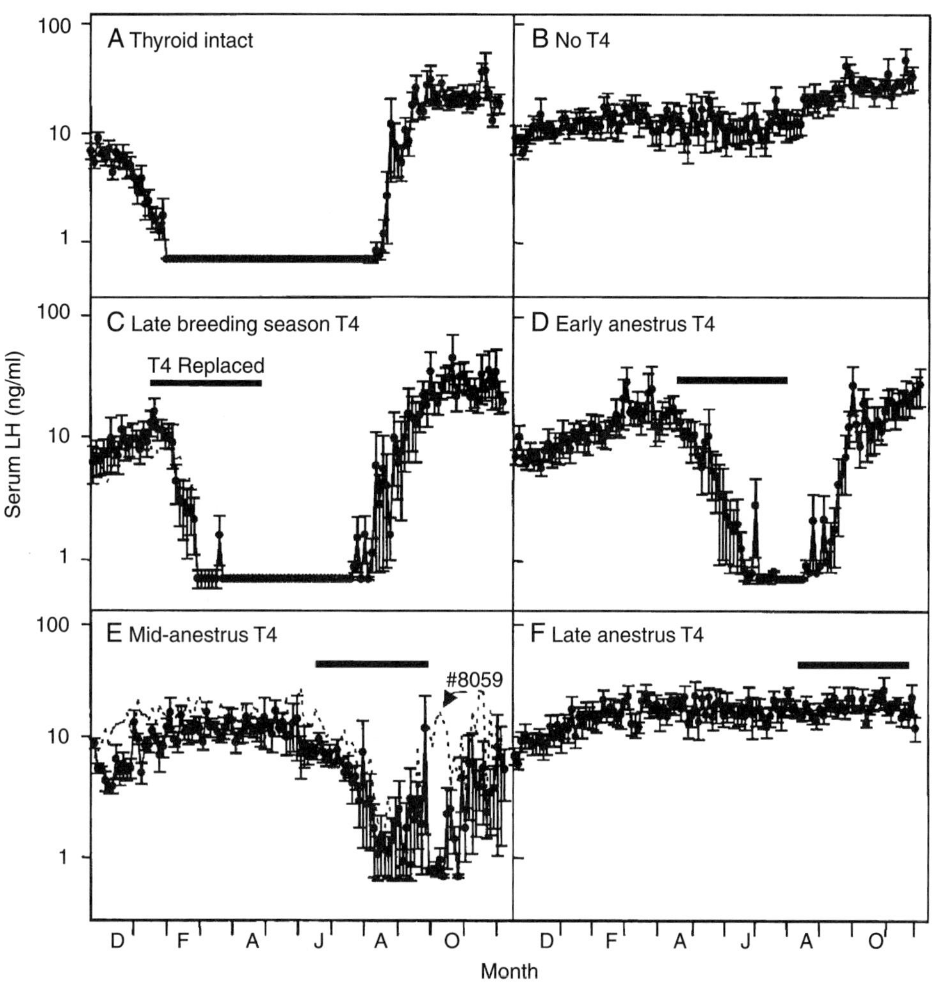

FIG. 18. Changes (mean ± SEM) in serum LH concentrations in thyroid-intact controls (**A**), thyroidectomized controls (**B**), or thyroidectomized ewes replaced with T_4 beginning in the late breeding season (**C**), early anestrus (**D**), mid-anestrus (**E**), or late anestrus (**F**). Shaded bars depict the time of T_4 replacement. LH concentrations are plotted on a logarithmic scale. [Adapted from (437).]

by the finding that low doses of T_4 provided during the period of sensitivity to thyroid hormone restores appropriately timed seasonal LH inhibition in thyroidectomized ewes only when delivered intracerebroventricularly (440). This central effect was further substantiated by the description of two potential specific hypothalamic regions, the premammillary region and the ventromedial preoptic area, in which thyroid hormones act to allow the seasonal inhibition of LH secretion (432). Both of these sites are densely populated with thyroid hormone receptors in the sheep (441). The ventromedial preoptic area is interesting because it is thought to contain estradiol receptors mediating the negative feedback of this steroid during the anestrous season and not during the breeding season (442). The premammillary region is interesting because it is a target site of melatonin for the control of seasonal reproduction in this species (248), suggesting that this area requires thyroid hormones to respond to melatonin; in contrast, the PT does not require thyroid hormones to respond to melatonin for the control of prolactin secretion because seasonal oscillations in prolactin are not disrupted by thyroidectomy in the ewe. The premammillary region is also the area where $5HT2_A$ binding capacities and gene expression vary between the active and inactive stages of the circannual rhythm of reproduction in the ewe, and these changes could constitute a key element of the generation of this circannual rhythm (see above for details). In this respect, it is worth noting that ^{3}H-ketanserin binding capacity is decreased in thyroidectomized rats and that this effect of thyroidectomy can be reversed with low doses of T_4 (443). This raises the hypothesis that the permissive action of T_4 for the expression of the circannual rhythm is mediated by changes in the $5HT2_A$ system in the premammillary area.

Although T_4 plasma levels play a permissive role, hypothalamic T_4 concentrations may serve a functional role in seasonal timing. Using cDNA microarrays, Prendergast et al. (428) identified a class of genes encoding T_4 binding proteins (transthyretin, T_4 binding globulin, and albumin) whose expression is reduced in the hypothalamus of Siberian hamsters that are reproductively refractory to inhibitory short day length compared with animals photosensitive to either short or long days. This reduced expression of these genes is correlated with a reduced uptake of T_4 in the hypothalamus because hypothalamic concentrations of T_4 are reduced in photorefractory animals, although peripheral concentrations do not differ between photorefractory and photosensitive animals. Moreover, resensitizing photorefractory hamsters to short days by 12 weeks of exposure to long days increased hypothalamic T_4 uptake to values similar to those of photosensitive hamsters that had never been exposed to short days (Fig. 19). Prendergast et al. (428) hypothesized that reproductive photoresponsiveness in Siberian hamsters is T_4 independent for most of the annual cycle, as previously demonstrated in the ewe, but that the sustained reproductive inhibition becomes T_4 dependent after prolonged exposure to short days. If a spontaneous down-regulation of hypothalamic thyroid hormone binding protein expression coincided with such a T_4-dependent phase, the resulting reduction in T_4 availability to hypothalamic target tissues could be responsible for the manifestation of photorefractoriness (428). Exposure to long days in summer could restore a level of thyroid hormone in the hypothalamus sufficient to permit sustained reproductive inhibition during the following short days of autumn (428).

Recently, it was shown that photoperiod can also influence the expression of type 2 iodothyronine deiodinase (*dio2*), the enzyme that converts prohormone T_4 to bioactive T_3 (triiodothyronine). Expression of

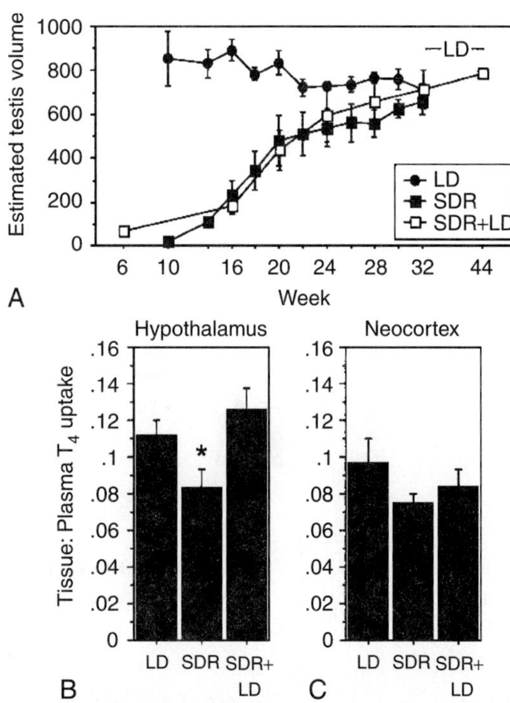

FIG. 19. Diminished T_4 uptake in the hypothalamus of photorefractory hamsters. **A:** Mean estimated testis volume (± SEM) over the 44-week experiment for hamsters that had been exposed to short or long day lengths for 32 weeks (SDR and LD, respectively) and in hamsters exposed to short day lengths for 32 weeks followed by long day lengths for 12 weeks (SDR + LD; a photoperiod treatment that restores photosensitivity in photorefractory hamsters). **B, C:** Tissue-to-plasma ratios (± SEM) of ^{125}I-L-T_4 uptake in the hypothalamus and neocortex of photosensitive (LD and SDR + LD) and photorefractory (SDR) hamsters. SDR hamsters exhibited lower T_4 uptake relative to LD hamsters; breaking of refractoriness by exposure to LD restored elevated hypothalamic T_4 uptake (*$P < 0.05$ vs. all other groups). [Reproduced from (428).]

dio2 was found in the ependymal cell layer lining the infralateral walls of the third ventricle, the cell-clear zone overlying the tuberoinfundibular sulcus and surrounding the blood vessels in the arcuate nucleus in Siberian hamsters. In these structures, the expression of *dio2* was lower in short-day than in long-day photosensitive animals. The inhibitory effect of short days was mimicked by daily injections of melatonin to animals exposed to long days (444). Clearly, it can be hypothesized that these two mechanisms, uptake of T_4 and conversion of T_4 to T_3, could act synergistically to regulate locally the concentrations of bioactive T_3 concentrations in the hypothalamic target tissues. The comparison of *dio2* activity between photorefractory and photosensitive hamsters would provide valuable information in this respect. It would also be of interest to determine whether T_4 uptake and conversion of T_4 to T_3 could participate in seasonal timing in species, like sheep or ground squirrels, in which the role of photoperiod is to entrain a self-sustaining circannual rhythm of reproduction. Of particular interest would be to determine whether the window of sensitivity to T_4 demonstrated in the ewe between November and June (437) corresponds to a time of increased transport of T_4 to the brain or increased conversion of T_4 to T_3.

Seasonal Plasticity

Thyroid hormones may be an important factor necessary for seasonal morphological changes that could be involved in the regulation of GnRH secretion. Thyroid hormones are essential for morphological changes in the nervous system, for instance, they are critical to normal neonatal brain development and peripheral nerve regeneration (445–447). It is possible that they play an analogous role in seasonal animals by causing the prerequired morphological rearrangements for the seasonal changes in GnRH secretion. Evidence for such rearrangements was obtained at the level of the GnRH neurons of the ewe. At the light microscopic level, immunostained GnRH neurons in the preoptic area have longer and more numerous dendrites during anestrus than during the breeding season. However, these changes could reflect alterations in the content or localization of immunodetectable GnRH rather than actual morphological changes (448). At the electron microscopic level, GnRH neurons in the preoptic area receive more than twice the mean number of synaptic inputs per unit of plasma membrane during the breeding season as during anestrus; this difference was seen in both axodendritic and axosomatic inputs, although GnRH dendrites received more synaptic input than GnRH somas (449). Interestingly, these changes are not dependent on changes in ovarian steroid levels because they are found both in intact animals and in ovariectomized ewes bearing estradiol implants (449); this differs markedly from other examples of seasonal plasticity, such as those in the forebrain nuclei controlling song in canaries, which are steroid controlled (450). This seasonal change in neural input to GnRH neuron, identified based on the proximity of terminal immunoreactivity to the presumptive GnRH cell membrane, was supported by a recent study using immunofluorescent visualization of GnRH combined with confocal microscopy and optical sectioning techniques (451). This study further demonstrated that specific GnRH afferents undergo seasonal remodeling and contribute to this overall change: NPY and γ-aminobutyric acid inputs increased during the breeding season and β-endorphin inputs were greater during either anestrus (GnRH somas) or breeding season (GnRH dendrites) (451). In another study, an increase in the percentage of GnRH neurons receiving γ-aminobutyric acid inputs was lower in midluteal phase breeding season ewes than in anestrous ewes, also suggesting a seasonal remodeling. However, an effect of progesterone cannot be excluded (452).

In addition, GnRH proximal dendrites are much more heavily associated with glia during the breeding season than during anestrus. In contrast, somas of rostral GnRH neurons exhibit more glial association during anestrus, but the difference was not seen with caudal GnRH neurons and was smaller than the effect seen on dendrites (451). Neural cell adhesion molecule (NCAM) and its polysialylated form, PSA-NCAM, which promotes plasticity by modifying the stability of cell–cell contacts, are also thought to be functionally significant to seasonal plasticity of the GnRH neurosecretory system. Indeed, in the Siberian hamster, photoperiod-induced changes in gonadotropin secretion are associated with variations in the relative amounts of PSA and other NCAM isoforms in the basal hypothalamus (453). In sheep, the expression of PSA-NCAM is intimately associated with preoptic GnRH neurons, and this expression increases during the breeding season (454). These combined increases in PSA-NCAM and glial association may help to ensure the long-term stability of the increase in net stimulatory inputs to GnRH neurons during the breeding season (451). It is interesting to note that in the Japanese quail, a retraction of glial cell end-feet associated with GnRH terminals in the median eminence is related to increased GnRH release in long days and could result from a photoinduced increase in T_3 in glial cells, suggesting a similar influence of thyroid hormones on seasonal morphological changes critical to the regulation of GnRH release (455,456).

GENETIC CONTROL OF SEASONALITY

Evidence for a Genetic Control

The expression of the seasonal variations in reproductive activity differs greatly between species but also between populations of the same species and even between individuals of the same geographical populations for wild species or the same breed for domesticated species [for reviews see (1,457)]. Within a species, different breeding populations may inhabit different types of environment, and these may exert different selective pressure leading to various seasonal patterns of reproductive activity. A well-studied example of mammalian species exhibiting different seasonal profiles according to its location is the white-footed mouse, *Peromyscus leucopus*. Mice captured in Connecticut exhibit photoperiodic control of reproductive activity when they are tested in the laboratory. In contrast, mice of the same species captured in Georgia fail to exhibit responses to photoperiodic changes (458). When bred in the laboratory, animals from these two populations continue to be photoperiodic and nonphotoperiodic, respectively, suggesting this trait is heritable (458). In some natural breeding populations, a large variability in photoperiodic responsiveness can be observed among individuals. For example, white-footed mice captured in Michigan vary greatly in their reproductive response to photoperiod (459). When they were subjected to mass selection in the laboratory for or against photoresponsiveness, the first generation yielded two lines of mice that were 80% photoresponsive or 80% nonresponsive, respectively (Fig. 20). Three more generations of selection failed to notably change these proportions, suggesting that a single variable locus may be responsible for much of the heritable variance present in the population (459). Similar observations were made in deer mice, field voles, and Siberian hamsters (460–462).

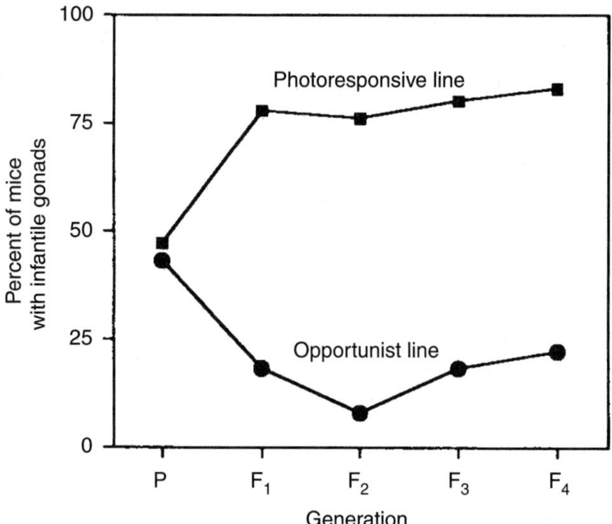

FIG. 20. Percentage of individuals, males and females combined, that were classified as reproductively photoresponsive in two lines of white-footed mice, one selected for reproductive photoresponsiveness and the other in which selection was against that trait, over four generations of selection. [Adapted from (459).]

Quantitative genetic analyses have addressed the genetic basis for differences in seasonality and/or photoresponsiveness. Breeds of sheep present various profiles of reproductive seasonality according to the latitude they originate from, ranging from several months of inactivity in high latitudes to continuous breeding in tropical latitudes (463). In Mediterranean latitudes, great variability exists within breeds in terms of the presence and duration of anestrus. Some ewes completely cease ovulatory activity, whereas others show isolated ovulation during anestrus or continue to cycle throughout the year (464,465). For instance, in Merinos d'Arles ewes, about 30% of the females display spontaneous ovulation in April (466). This trait is repeatable and heritable (R^2 and h^2 equal to 0.3 and 0.2, respectively), suggesting that alleles of specific genes control the presence or absence of spontaneous spring ovulations (466). In another study, heritability was calculated on ewes selected for or against fertility in the spring (467). They were joined with rams in November (spring in the south hemisphere), and two divergent lines different by the date of lambing were established (earlier date in the less seasonal line). This trait, date of lambing, is also repeatable and heritable (R^2 and h^2 equal to 0.36 and 0.23, respectively). In Siberian hamsters, h^2 estimates for photoresponsiveness range from 0.20 to 0.52 (468). A study determined the genetic parameters of the development of photorefractoriness in Djungarian hamsters. All individuals that undergo short-day induced gonadal regression eventually become refractory and exhibit gonadal recrudescence, but not all individuals do so at precisely the same time (221). Parent–offspring regression analyses revealed a heritability estimate, h^2, of 0.91 ± 0.40 for female offspring and 0.35 ± 0.20 for male offspring (469). This difference of heritability between males and females in this interval timer governing photorefractoriness and therefore timing spring breeding suggests a greater contribution of environmental factors to expression of this trait in males (469).

Individual Genetic Variation and Photoneuroendocrine Pathways

Individual genetic variation in photoresponsiveness can occur at one or several stages of the photoneuroendocrine pathways, linking the eyes to the gonads. In Djungarian hamsters, differences in the regulation of melatonin secretion were found between photoresponsive and nonresponsive animals. The duration of the melatonin does not extend upon transfer from a long-day to a short-day photoperiod (470). Along the same lines, melatonin concentrations are lower in ewes selected for high fertility in the spring (low seasonality) compared with control animals of the same breed (143 vs. 184 pg/ml), suggesting the existence of a genetic or physiological link between the amplitude of the melatonin profile and the expression of seasonality (471). However, this link was not confirmed in a reverse experiment in which ewes chosen for their high or low plasma concentrations of melatonin did not differ in the timing of their breeding season (213). In white-footed mice, a comparison of physiological mechanisms between two lines of animals selected for or against responsiveness to the inhibitory effects of short days revealed three differences relevant to the control of reproduction (472). First, the density of ^{125}I-melatonin binding was lower in responsive than in nonresponsive animals kept in short days in two areas, the medial preoptic area and the bed nucleus of the stria terminalis (473). The medial preoptic area has been previously identified as a potential site of melatonin action in this species (233), suggesting that this difference in binding may contribute to differences in photoresponsiveness. Second, the inhibition of LH secretion by estradiol was stronger in responsive than in nonresponsive mice in short days, suggesting that variation in sensitivity to estradiol-negative feedback is reduced in nonresponsive animals (472). Third, in short days, responsive mice had less immunoreactive GnRH neurons than did nonresponsive animals (472).

Studies in sheep have indicated that polymorphisms in the gene encoding MT1 receptor could explain a part of the variability in the expression of reproductive seasonality. Relative to the sequence for oMT1 published by Reppert et al. (282), a variant form was first described with eight base changes, three of which resulted in amino acids substitutions in the receptor (474). Further studies of the sheep receptor showed an even higher degree of polymorphism in exon II of this gene (475,476). Exon II has two restriction fragment length polymorphism (RFLP) sites, one for *Mnl* I and the second for *Rsa* I enzymes (475). The *Mnl* I RFLP site is characterized by a mutation leading to the absence of the specific *MnlI* cleavage site at position 605 of the coding sequence. Interestingly, the incidence of the *MnlI* RFLP was found to be variable in breeds with difference in reproductive seasonality (475,477). In addition, an association between genotype and seasonal ovulatory activity was described (476). Genotypic frequency for the *MnlI* polymorphism were compared in Merinos d'Arles ewes with a history of spontaneous ovulatory activity in April and ewes of the same breed with a history of anovulation in this period. The distributions of the two groups of ewes between genotypes were different. Interestingly, genotype –/– (i.e., animals carrying the mutation on both chromosomes) was never found in ewes with a history of spontaneous ovulatory activity in April (476) (Fig. 21). Also, the density of ^{125}I-melatonin binding sites in the PT was about 50% as low in ewes of genotype +/+ as in those of genotype –/– (476).

Along the same lines, fertility for autumn lambing (i.e., fertilization in the spring) is reduced in animals genotyped –/– in Virginia Tech out-of-season mating line (crossbreed line of 50% Dorset, 25% Rambouillet, and 25% Finnsheep) compared with ewes with at least one copy of the + allele (478). Additional evidence for a relationship between the – allele and anovulation in the spring was provided by a study in small-tailed Han sheep in China (479). The ovulatory activity of this breed is not seasonal, and in a sample of 150 ewes, genotype –/– was never found despite a frequency of heterozygotes of 50%. Altogether, these data collected from isolated physiological events (spring spontaneous ovulatory activity and fertility after mating in spring) suggest that a genotypic effect associated with the *Mnl* I RFLP site could account for the variability observed in the expression of seasonality both between and within breeds. The mutation leading to the suppression of *Mnl* I site is silent and cannot explain the associated

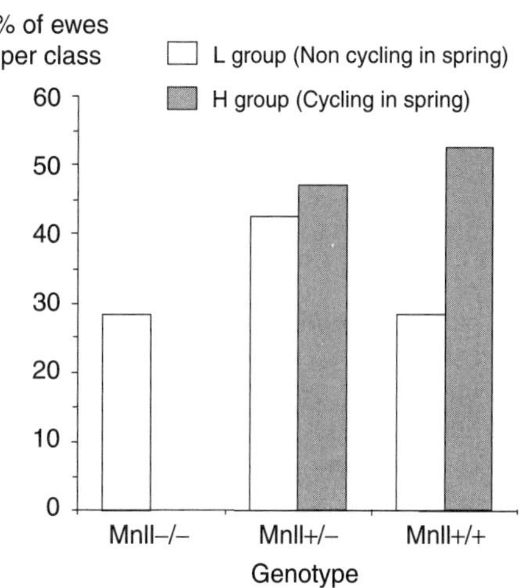

FIG. 21. Distribution of Merinos d'Arles ewes according to their genotype (presence or absence of the polymorphic site *Mnl* I of the melatonin receptor MT1). Ewes were selected according to their ability to display ovulatory cycle in the spring (group H, weak seasonal variations) or not (group L, strong seasonal variations). The frequency distribution of genotypes is different for the two groups. [Adapted from (476).]

phenotype differences. Consequently, it has to be viewed as a genetic marker for quantitative trait loci responsible for seasonal variability. Interestingly, the silent mutation linked to the RFPL site is always found to be associated with a nonconservative mutation at position 706, leading to the substitution of a valine by an isoleucine in the fifth transmembrane domain, which could be a causal mutation responsible for variability in seasonal reproduction (476). Altogether, these data therefore indicate that detectable genetic variation is present at multiple sites in this complex neuroendocrine pathway, and this suggests the implication of several genes to explain phenotypic variability.

Human Reproductive Seasonality

Concerning genetic differences among species, the question of reproductive seasonality in humans has been raised many times. Clearly, human reproduction is not strictly seasonal, in the sense that conceptions are not restricted to a part of the year. However, most studies suggest a seasonality in successful conceptions [for review see (480–484)]. A thorough analysis of correlations between conception rates and many environmental factors in 166 regions of the earth suggested that photoperiod and, to a lesser degree, temperature are the main environmental factors contributing to the annual conception rhythm (485). A spring increase in conception rate is observed in many countries of the northern and southern hemispheres, and at higher latitude, where changes in day length are more pronounced, this increase is steeper. In some parts of the world, like North America and Eastern Europe, strong bimodal waveforms with a second peak of conceptions in winter are found. In the United States, the winter peak has become dominant since the 1930s. It has been proposed that this winter conception peak is the consequence of an overriding effect of hot summer temperatures in continental climates of Eastern Europe and the northern United States (480,481). Correlation studies indeed indicate that extreme temperatures (above 20°C and below 5°C) decrease the probability of conception. Industrialization is related to a decrease in the amplitude of the human conception rhythms, most likely as a consequence of shifting work from outside to inside and thus reducing the daily and annual changes of light and temperature (481). Obviously, a limitation of these conclusions is that they come from epidemiological studies and correlation analyses. Experimental test of the role of photoperiod in humans is not conceivable. However, it is worth noting that seasonal breeding is common in nonhuman primates, and in some species of prosimians, Old World and New World monkeys, photoperiod has been demonstrated experimentally to be an important environmental factor controlling reproduction (486). For example, in the rhesus monkey, females display seasonal changes in the occurrence of ovulation and males in testicular function with a peak of activity during the short days of autumn and winter (487,488). Alternations of long days and short days with a 32-week period entrain the changes in reproductive function in the male (487). In humans, the pathway from the retina to the pineal is functional, and melatonin is secreted at night as in other mammals (482). Whether the duration of melatonin secretion varies with season is more controversial, and conflicting results have been produced. It seems, however, that the duration of melatonin secretion is longer in winter than in summer at high-temperate, subpolar, and polar latitudes (482). Importantly, exposure of humans to artificial long or short photoperiods leads to a difference in the duration of secretion, indicating the retina–pineal axis is capable of detecting changes in photoperiod (489). Altogether, these studies suggest that humans may be seasonally photoperiodic. The impact of environmental factors is greatly reduced in industrialized countries, and this reduces drastically the expression of this biological property. It is also likely that large individual variations in photoresponsiveness exist, and this may be the cause of many inconsistencies observed in the studies of human sensitivity to photoperiod (484).

CONCLUSION

The integration of photoperiod by seasonal breeders involves complex pathways that can be divided into two major, pre- and postpineal, steps. The prepineal circuitry is relatively well understood and allows the transduction of day/night information into a rhythm of melatonin secretion. This process allows organisms to be informed about time of year by means of the annual changes in the duration of melatonin signal. This melatonin signal is then decrypted by target sites to regulate the seasonal changes in various physiological functions. These postpineal pathways most likely differ between seasonal functions as already shown for the seasonal control of prolactin and gonadotropin secretion. Concerning seasonal breeding, the integration of the melatonin signal involves more than reversible changes between active and inactive states by responding to opposite photoperiodic signals. Phenomena such as memory of prior melatonin exposure, refractoriness to a melatonin signal, and self-sustained circannual rhythms play a critical role in determining the seasonal variations in reproduction. They allow a very fine-tuning of the breeding season to environmental constraints and lead to considerable variations in the timing of breeding among species and populations. Despite the knowledge accumulated over the years on the basis of melatonin signal integration, basic questions remain unanswered, such as how does the same melatonin signal produce opposite effects in long-day and short-day breeders? How does the same melatonin signal produce opposite effects depending on the prior melatonin exposure of the individual? Significant advances have been made to identify sites of action and receptors of melatonin, but the implication of progress in these areas to seasonal breeding is still unclear. Similarly, the existence of a neuronal network involving several neurotransmitters connecting melatonin targets to GnRH neurons has been repeatedly substantiated, but the precise organization of this network and the significance of its seasonal plasticity are only marginally understood.

Cloning of melatonin receptor subtypes, development of specific melatonin analogues, knowledge of the molecular basis of the clock and the role of clock genes in melatonin signal decoding, and description of the neuronal network controlled by melatonin should give insights into the long-term mechanisms underlying the control of GnRH secretion by melatonin. It is also worth noting that overriding periods of reproductive quiescence in seasonal species has attracted much interest from breeders of farm species in their quest to meet the constant demand of consumers for meat or milk. The current knowledge on photoperiodism and seasonal breeding has already led to the development of practical methods to control the season of reproduction in domesticated mammalian species such as sheep, goats, horses, and mink. The timely use of melatonin, preceded or not by exposure to long days to reset the circannual rhythm of reproduction, allows ovulation and spermatogenesis to be induced during the rest season and high fertility after mating to be obtained in both sheep and goats (490). Identification of polymorphisms in specific genes related to the expression of seasonality can also open the way for genetic selection of aseasonal individuals by increasing the frequency of favorable alleles for reduced seasonality.

Although the emphasis in this chapter has been on photoperiod because this parameter is the most widely used, the best described and the best understood factor from a neuroendocrine viewpoint, other parameters such as diet, temperature, humidity, and stress interact with photoperiod to fine-tune seasonal variations in reproductive activity. A challenge for future research will be to unravel the neuroendocrine pathways that allow these factors to interact with photoperiod and melatonin signal.

ACKNOWLEDGMENTS

I am grateful to Philippe Chemineau for valuable comments on this manuscript and to Patricia Volland-Nail and Odile Moulin for their help with the preparation of the manuscript.

REFERENCES

1. Bronson, F. H. (1989). "Mammalian Reproductive Biology." The University of Chicago Press, London.
2. Bronson, F. H., and Heideman, P. D. (1994). Seasonal regulation of reproduction in mammals. In *The Physiology of Reproduction* (E. Knobil, and J. D. Neill, Eds.), pp. 541–83. Raven Press, New York.
3. Zucker, I., Johnston, P. G., and Frost, D. (1980). Comparative physiological and biochronometric analyses of rodent seasonal reproductive cycles. *Prog. Reprod. Biol.* 5, 102–133.
4. Turcek, F. J., and Stiavnica, B. (1951). Effect of introductions on two game populations in czechoslovakia. *J. Wildlife Manage.* 15, 113–114.
5. Moore, S. E., Cole, T. J., Poskitt, E. M., Sonko, B. J., Whitehead, R. G., McGregor, I. A., and Prentice, A. M. (1997). Season of birth predicts mortality in rural Gambia. *Nature* 388, 434.
6. Reiter, R. J. (1975). Exogenous and endogenous control of the annual reproductive cycle in the male golden hamster: participation of the pineal gland. *J. Exp. Zool.* 191, 111–120.
7. Karsch, F. J., Bittman, E. L., Foster, D. L., Goodman, R. L., Legan, S. J., and Robinson, J. E. (1984). Neuroendocrine basis of seasonal reproduction. *Rec. Prog. Horm. Res.* 40, 185–232.
8. Ortavant, R., Pelletier, J., Ravault, J. P., Thimonier, J., and Volland-Nail, P. (1985). Photoperiod: main proximal and distal factor of the circannual cycle of reproduction in farm mammals. *Oxf. Rev. Reprod. Biol.* 7, 305–345.
9. Aitken, R. J. (1974). Delayed implantation in roe deer (*Capreolus capreolus*). *J. Reprod. Fertil.* 39, 225–233.
10. Baker, J. R. (1938). The evolution of breeding seasons. In *Evolution: Essays on Aspects of Evolutionnary Biology* (G. R. de Beer, Ed.), pp. 161–177. Oxford University Press, Oxford.
11. Thompson, A. L. (1950). Factors determining the breeding season of birds: an introductory review. *Ibis* 92, 173–184.
12. Baker, J. R., and Ramsom, R. M. (1932). Factors affecting the breeding of the field mouse (*Microtus agrestis*). *Proc. R. Soc. Lond.* 110, 313–322.
13. Hoffmann, K. (1973). The influence of photoperiod and melatonin on testis size, body weight, and pelage colour in the Djungurian hamster (*Phodopus sungorus*). *J. Comp. Physiol.* 95, 267.
14. Steger, R. W., Matt, K., and Bartke, A. (1985). Neuroendocrine regulation of seasonal reproductive activity in the male golden hamster. *Neurosci. Biobehav. Rev.* 9, 191–201.
15. Thwaites, C. J. (1965). Photoperiodic control of breeding activity in the Southdown ewewith particular reference to the effects of an equatorial light regime. *J. Agric. Sci. Cambr.* 65, 57–65.
16. Gwinner, E. (1986). *Circannual Rhythms.* Springer-Verlag, Berlin.
17. Zucker, I. (2001). Circannual rhyhms: mammals. In *Handbook of Behavioral Neurobiology: Circadian Clocks* (J. Takahashi, F. W. Turek, and R. Y. Moore, Eds.), pp. 511–530. Plenum-Kluwer, New York.
18. Jansen, H. T., and Jackson, G. L. (1993). Circannual rhythms in the ewe: patterns of ovarian cycles and prolactin secretion under two different constant photoperiods. *Biol. Reprod.* 49, 627–634.

19. Karsch, F. J., Robinson, J. E., Woodfill, C. J., and Brown, M. B. (1989). Circannual cycles of luteinizing hormone and prolactin secretion in ewes during prolonged exposure to a fixed photoperiod: evidence for an endogenous reproductive rhythm. *Biol. Reprod.* 41, 1034–1046.
20. Robinson, J. E., and Karsch, F. J. (1988). Timing the breeding season of the ewe: what is the role of daylength? *Reprod. Nutr. Dev.* 28, 365–374.
21. Malpaux, B., Robinson, J. E., Wayne, N. L., and Karsch, F. J. (1989). Regulation of the onset of the breeding season of the ewe: importance of long days and of an endogenous reproductive rhythm. *J. Endocrinol.* 122, 269–278.
22. Prendergast, B. J., Nelson, R. J., and Zucker, I. (2002). Mammalian seasonal rhythms: behavior and neuroendocrine substrates. In *Hormones, Brain, and Behavior* (D. W. Pfaff, Ed.), pp. 93–156. Academic Press, San Diego.
23. Anchordoquy, H. C., and Lynch, G. R. (2000). Evidence of an annual rhythm in a small proportion of Siberian hamsters exposed to chronic short days. *J. Biol. Rhythms* 15, 122–125.
24. Saboureau, M., Masson-Pevet, M., Canguilhem, B., and Pevet, P. (1999). Circannual reproductive rhythm in the European hamster (*Cricetus cricetus*): demonstration of the existence of an annual phase of sensitivity to short photoperiod. *J. Pineal Res.* 26, 9–16.
25. Robinson, J. E., and Karsch, F. J. (1987). Photoperiodic history and a changing melatonin pattern can determine the neuroendocrine response of the ewe to daylength. *J. Reprod. Fertil.* 80, 159–165.
26. Hoffmann, K., Illnerova, H., and Vanecek, J. (1986). Change in duration of the nighttime melatonin peak may be a signal driving photoperiodic responses in the Djungarian hamster (*Phodopus sungorus*). *Neurosci. Lett.* 67, 68–72.
27. Goldman, B. D. (2001). Mammalian photoperiodic system: formal properties and neuroendocrine mechanisms of photoperiodic time measurement. *J. Biol. Rhythms* 16, 283–301.
28. Elliott, J. A. (1976). Circadian rhythms and photoperiodic time measurement in mammals. *Fed. Proc.* 35, 2339–2346.
29. Malpaux, B., Robinson, J. E., Brown, M. B., and Karsch, F. J. (1988). Importance of changing photoperiod and melatonin secretory pattern in determining the length of the breeding season in the Suffolk ewe. *J. Reprod. Fertil.* 83, 461–470.
30. Horton, T. H., and Stetson, M. H. (1990). Modification of pubertal photoperiodic responses in rodents by factors acting during fetal development. *Prog. Clin. Biol. Res.* 342, 639–647.
31. Gorman, M. R., and Zucker, I. (1998). Mammalian photoperiodism: new perspectives from the use of simulated natural photoperiod. In *Biological Clocks. Mechanisms and Applications* (Y. Touitou, Ed.), pp. 195–204. Elsevier, Amsterdam.
32. Moore, R. Y. (1996). Entrainment pathways and the functional organization of the circadian system. *Prog. Brain Res.* 111, 103–119.
33. Legan, S. J., and Karsch, F. J. (1983). Importance of retinal photoreceptors to the photoperiodic control of seasonal breeding in the ewe. *Biol. Reprod.* 29, 316–325.
34. Lucas, R. J., Freedman, M. S., Munoz, M., Garcia-Fernandez, J. M., and Foster, R. G. (1999). Regulation of the mammalian pineal by non-rod, non-cone, ocular photoreceptors. *Science* 284, 505–507.
35. Provencio, I., Jiang, G., De Grip, W. J., Hayes, W. P., and Rollag, M. D. (1998). Melanopsin: an opsin in melanophores, brain, and eye. *Proc. Natl. Acad. Sci. U S A* 95, 340–345.
36. Provencio, I., Rodriguez, I. R., Jiang, G., Hayes, W. P., Moreira, E. F., and Rollag, M. D. (2000). A novel human opsin in the inner retina. *J. Neurosci.* 20, 600–605.
37. Moore, R. Y., Speh, J. C., and Card, J. P. (1995). The retinohypothalamic tract originates from a distinct subset of retinal ganglion cells. *J. Comp. Neurol.* 352, 351–366.
38. Provencio, I., Cooper, H. M., and Foster, R. G. (1998). Retinal projections in mice with inherited retinal degeneration: implications for circadian photoentrainment. *J. Comp. Neurol.* 395, 417–439.
39. Gooley, J. J., Lu, J., Chou, T. C., Scammell, T. E., and Saper, C. B. (2001). Melanopsin in cells of origin of the retinohypothalamic tract. *Nat. Neurosci.* 4, 1165.
40. Hannibal, J., Hindersson, P., Knudsen, S. M., Georg, B., and Fahrenkrug, J. (2002). The photopigment melanopsin is exclusively present in pituitary adenylate cyclase-activating polypeptide-containing retinal ganglion cells of the retinohypothalamic tract. *J. Neurosci.* 22, RC191.
41. Berson, D. M., Dunn, F. A., and Takao, M. (2002). Phototransduction by retinal ganglion cells that set the circadian clock. *Science* 295, 1070–1073.
42. Sekaran, S., Foster, R. G., Lucas, R. J., and Hankins, M. W. (2003). Calcium imaging reveals a network of intrinsically light-sensitive inner-retinal neurons. *Curr. Biol.* 13, 1290–1298.
43. Ruby, N. F., Brennan, T. J., Xie, X., Cao, V., Franken, P., Heller, H. C., and O'Hara, B. F. (2002). Role of melanopsin in circadian responses to light. *Science* 298, 2211–2213.
44. Hattar, S., Lucas, R. J., Mrosovsky, N., Thompson, S., Douglas, R. H., Hankins, M. W., Lem, J., Biel, M., Hofmann, F., Foster, R. G., and Yau, K. W. (2003). Melanopsin and rod-cone photoreceptive systems account for all major accessory visual functions in mice. *Nature* 424, 76–81.
45. Panda, S., Provencio, I., Tu, D. C., Pires, S. S., Rollag, M. D., Castrucci, A. M., Pletcher, M. T., Sato, T. K., Wiltshire, T., Andahazy, M., Kay, S. A., Van Gelder, R. N., and Hogenesch, J. B. (2003). Melanopsin is required for non-image-forming photic responses in blind mice. *Science* 301, 525–527.
46. Mrosovsky, N., and Hattar, S. (2003). Impaired masking responses to light in melanopsin-knockout mice. *Chronobiol. Int.* 20, 989–999.
47. Lucas, R. J., Hattar, S., Takao, M., Berson, D. M., Foster, R. G., and Yau, K. W. (2003). Diminished pupillary light reflex at high irradiances in melanopsin-knockout mice. *Science* 299, 245–247.
48. van den Pol, A. N. (1991). Glutamate and aspartate immunoreactivity in hypothalamic presynaptic axons. *J. Neurosci.* 11, 2087–2101.
49. Ding, J. M., Faiman, L. E., Hurst, W. J., Kuriashkina, L. R., and Gillette, M. U. (1997). Resetting the biological clock: mediation of nocturnal CREB phosphorylation via light, glutamate, and nitric oxide. *J. Neurosci.* 17, 667–675.
50. Hannibal, J., Ding, J. M., Chen, D., Fahrenkrug, J., Larsen, P. J., Gillette, M. U., and Mikkelsen, J. D. (1997). Pituitary adenylate cyclase-activating peptide (PACAP) in the retinohypothalamic tract: a potential daytime regulator of the biological clock. *J. Neurosci.* 17, 2637–2644.
51. Moore, R. Y. (1978). Neural control of pineal function in mammals and birds. *J. Neural Transm.* Suppl., 47–58.
52. Card, J. P., and Moore, R. Y. (1982). Ventral lateral geniculate nucleus efferents to the rat suprachiasmatic nucleus exhibit avian pancreatic polypeptide-like immunoreactivity. *J. Comp. Neurol.* 206, 390–396.
53. Moore, R. Y., and Speh, J. C. (1993). GABA is the principal neurotransmitter of the circadian system. *Neurosci. Lett.* 150, 112–116.
54. Morin, L. P., and Blanchard, J. H. (2001). Neuromodulator content of hamster intergeniculate leaflet neurons and their projection to the suprachiasmatic nucleus or visual midbrain. *J. Comp. Neurol.* 437, 79–90.
55. Klein, D. C., Moore, R. Y., and Reppert, S. M., Eds. (1991). *Suprachiasmatic Nucleus: The Mind's Clock*. Oxford University Press, New York.

56. Tessonneaud, A., Locatelli, A., Caldani, M., and Viguier-Martinez, M. C. (1995). Bilateral lesions of the suprachiasmatic nuclei alter the nocturnal melatonin secretion in sheep. *J. Neuroendocrinol.* 7, 145–152.
57. Turek, F. W., Swann, J., and Earnest, D. J. (1984). Role of the circadian system in reproductive phenomena. *Rec. Prog. Horm. Res.* 40, 143–183.
58. LeSauter, J., and Silver, R. (1998). Output signals of the SCN. *Chronobiol. Int.* 15, 535–550.
59. Moore, R. Y., and Silver, R. (1998). Suprachiasmatic nucleus organization. *Chronobiol. Int.* 15, 475–487.
60. van Esseveldt, K. E., Lehman, M. N., and Boer, G. J. (2000). The suprachiasmatic nucleus and the circadian time-keeping system revisited. *Brain Res. Rev.* 33, 34–77.
61. Welsh, D. K., Logothetis, D. E., Meister, M., and Reppert, S. M. (1995). Individual neurons dissociated from rat suprachiasmatic nucleus express independently phased circadian firing rhythms. *Neuron* 14, 697–706.
62. Kornhauser, J. M., Nelson, D. E., Mayo, K. E., and Takahashi, J. S. (1990). Photic and circadian regulation of c-fos gene expression in the hamster suprachiasmatic nucleus. *Neuron* 5, 127–134.
63. Ginty, D. D., Kornhauser, J. M., Thompson, M. A., Bading, H., Mayo, K. E., Takahashi, J. S., and Greenberg, M. E. (1993). Regulation of CREB phosphorylation in the suprachiasmatic nucleus by light and a circadian clock. *Science* 260, 238–241.
64. Rea, M. A. (1992). Different populations of cells in the suprachiasmatic nuclei express c-fos in association with light-induced phase delays and advances of the free-running activity rhythm in hamsters. *Brain Res.* 579, 107–112.
65. Sutin, E. L., and Kilduff, T. S. (1992). Circadian and light-induced expression of immediate early gene mRNAs in the rat suprachiasmatic nucleus. *Brain Res. Mol. Brain Res.* 15, 281–290.
66. Abe, H., Rusak, B., and Robertson, H. A. (1991). Photic induction of Fos protein in the suprachiasmatic nucleus is inhibited by the NMDA receptor antagonist MK-801. *Neurosci. Lett.* 127, 9–12.
67. Abe, H., Rusak, B., and Robertson, H. A. (1992). NMDA and non-NMDA receptor antagonists inhibit photic induction of Fos protein in the hamster suprachiasmatic nucleus. *Brain Res. Bull.* 28, 831–835.
68. Rea, M. A., Buckley, B., and Lutton, L. M. (1993). Local administration of EAA antagonists blocks light-induced phase shifts and c-fos expression in hamster SCN. *Am. J. Physiol.* 265, R1191–R1198.
69. Zhang, Y., Zee, P. C., Kirby, J. D., Takahashi, J. S., and Turek, F. W. (1993). A cholinergic antagonist, mecamylamine, blocks light-induced fos immunoreactivity in specific regions of the hamster suprachiasmatic nucleus. *Brain Res.* 615, 107–112.
70. Wollnik, F., Brysch, W., Uhlmann, E., Gillardon, F., Bravo, R., Zimmermann, M., Schlingensiepen, K. H., and Herdegen, T. (1995). Block of c-Fos and JunB expression by antisense oligonucleotides inhibits light-induced phase shifts of the mammalian circadian clock. *Eur. J. Neurosci.* 7, 388–393.
71. Reppert, S. M., and Weaver, D. R. (2002). Coordination of circadian timing in mammals. *Nature* 418, 935–941.
72. Hirota, T., and Fukada, Y. (2004). Resetting mechanism of central and peripheral circadian clocks in mammals. *Zoolog. Sci.* 21, 359–368.
73. King, D. P., and Takahashi, J. S. (2000). Molecular genetics of circadian rhythms in mammals. *Annu. Rev. Neurosci.* 23, 713–742.
74. Ueda, H. R., Chen, W., Adachi, A., Wakamatsu, H., Hayashi, S., Takasugi, T., Nagano, M., Nakahama, K., Suzuki, Y., Sugano, S., Iino, M., Shigeyoshi, Y., and Hashimoto, S. (2002). A transcription factor response element for gene expression during circadian night. *Nature* 418, 534–539.
75. Yu, W., Nomura, M., and Ikeda, M. (2002). Interactivating feedback loops within the mammalian clock: BMAL1 is negatively autoregulated and upregulated by CRY1, CRY2, and PER2. *Biochem. Biophys. Res. Commun.* 290, 933–941.
76. Albrecht, U., Sun, Z. S., Eichele, G., and Lee, C. C. (1997). A differential response of two putative mammalian circadian regulators, mper1 and mper2, to light. *Cell* 91, 1055–1064.
77. Shearman, L. P., Zylka, M. J., Weaver, D. R., Kolakowski, L. F. Jr., and Reppert, S. M. (1997). Two period homologs: circadian expression and photic regulation in the suprachiasmatic nuclei. *Neuron* 19, 1261–1269.
78. Shigeyoshi, Y., Taguchi, K., Yamamoto, S., Takekida, S., Yan, L., Tei, H., Moriya, T., Shibata, S., Loros, J. J., Dunlap, J. C., and Okamura, H. (1997). Light-induced resetting of a mammalian circadian clock is associated with rapid induction of the mPer1 transcript. *Cell* 91, 1043–1053.
79. Akiyama, M., Kouzu, Y., Takahashi, S., Wakamatsu, H., Moriya, T., Maetani, M., Watanabe, S., Tei, H., Sakaki, Y., and Shibata, S. (1999). Inhibition of light- or glutamate-induced mPer1 expression represses the phase shifts into the mouse circadian locomotor and suprachiasmatic firing rhythms. *J. Neurosci.* 19, 1115–1121.
80. Tischkau, S. A., Mitchell, J. W., Tyan, S. H., Buchanan, G. F., and Gillette, M. U. (2003). Ca^{2+}/cAMP response element-binding protein (CREB)-dependent activation of Per1 is required for light-induced signaling in the suprachiasmatic nucleus circadian clock. *J. Biol. Chem.* 278, 718–723.
81. Wakamatsu, H., Takahashi, S., Moriya, T., Inouye, S. T., Okamura, H., Akiyama, M., and Shibata, S. (2001). Additive effect of mPer1 and mPer2 antisense oligonucleotides on light-induced phase shift. *Neuroreport* 12, 127–131.
82. Albrecht, U., Zheng, B., Larkin, D., Sun, Z. S., and Lee, C. C. (2001). MPer1 and Mper2 are essential for normal resetting of the circadian clock. *J. Biol. Rhythms* 16, 100–104.
83. Reiter, R. J. (1981). The mammalian pineal gland: structure and function. *Am. J. Anat.* 162, 287–313.
84. Phansuwan-Pujito, P., Moller, M., and Govitrapong, P. (1999). Cholinergic innervation and function in the mammalian pineal gland. *Microsc. Res. Tech.* 46, 281–295.
85. Moller, M. (1999). Introduction to mammalian pineal innervation. *Microsc. Res. Tech.* 46, 235–238.
86. Vollrath, L. (1981). The pineal organ. In *Handbuch der mikroskopischen anatomie des menchen* (A. Oksche and L. Vollrath, Eds.). Springer-Verlag, Berlin.
87. Pevet, P. (1983). Anatomy of the pineal gland of mammals. In *Current Endocrinology: The Pineal Gland* (R. Relkin, Ed.), pp. 1–75. Elsevier Biomedical, Amsterdam.
88. Garidou, M. L., Bartol, I., Calgari, C., Pevet, P., and Simonneaux, V. (2001). In vivo observation of a non-noradrenergic regulation of arylalkylamine N-acetyltransferase gene expression in the rat pineal complex. *Neuroscience* 105, 721–729.
89. Collin, J. (1971). Differentiation and regression of the cells of the sensory line in the epiphysis cerebri. In *The Pineal Gland* (G. E. W. Wolstenholme and J. Knight, Eds.), pp. 79–125. Churchill-Livingston, Edinburgh.
90. Zhao, X., Yokoyama, K., Whitten, M. E., Huang, J., Gelb, M. H., and Palczewski, K. (1999). A novel form of rhodopsin kinase from chicken retina and pineal gland. *FEBS Lett.* 454, 115–121.
91. Foster, R. G., Provencio, I., Bovee-Geurts, P. H., and DeGrip, W. J. (2003). The photoreceptive capacity of the developing pineal gland and eye of the golden hamster (*Mesocricetus auratus*). *J. Neuroendocrinol.* 15, 355–363.
92. Mirshahi, M., Faure, J. P., Brisson, P., Falcon, J., Guerlotte, J., and Collin, J. (1984). S-antigen immunoreactivity in retinal rods and cones and pineal photosensitive cells. *Biol. Cell.* 52, 195–198.

93. Duvernoy, H. M., Parratte, B., Tatu, L., and Vuillier, F. (2000). The human pineal gland: relationships with surrounding structures and blood supply. *Neurol. Res.* 22, 747–790.
94. Arendt, J. (1986). Role of the pineal gland and melatonin in seasonal reproductive function in mammals. *Oxf. Rev. Reprod. Biol.* 8, 266–320.
95. Skinner, D. C., and Malpaux, B. (1999). High melatonin concentrations in third ventricular cerebrospinal fluid are not due to Galen vein blood recirculating through the choroid plexus. *Endocrinology* 140, 4399–4405.
96. Tricoire, H., Moller, M., Chemineau, P., and Malpaux, B. (2003). Origin of cerebrospinal fluid melatonin and possible function in the integration of photoperiod. *Reprod. Suppl.* 61, 311–321.
97. Moller, M. (1981). The ultrastructure of the deep pineal gland of the Mongolian gerbil and mouse: granular vesicle localisation and innervation. In *Pineal Function* (C. D. Matthews and R. F. Seamark, Eds.), pp. 257–266. Elsevier, Amsterdam.
98. Moller, M. (1992). Fine structure of the pinealopetal innervation of the mammalian pineal gland. *Microsc. Res. Tech.* 21, 188–204.
99. Krisch, B., Leonhardt, H., and Buchheim, W. (1978). The functional and structural border between the CSF- and blood-milieu in the circumventricular organs (organum vasculosum laminae terminalis, subfornical organ, area postrema) of the rat. *Cell Tissue Res.* 195, 485–497.
100. Leonhardt, v. H. (1980). Ependym und Circumventriculäre Organe. In *Handbuch der mikroskopischen Anatomie des Menschen* (A. Oksche, and L. Vollrath, Eds.), part 10 (Neuroglia). Springer-Verlag, Heidelberg.
101. Hewing, M. (1978). A liquor contacting area in the pineal recess of the golden hamster (*Mesocricetus auratus*). *Anat. Embryol. (Berl).* 153, 295–304.
102. Hewing, M. (1980). Cerebrospinal fluid-contacting area in the pineal recess of the vole (*Microtus agrestis*), guinea pig (*Cavia cobaya*), and rhesus monkey (*Macaca mulatta*). *Cell Tissue Res.* 209, 473–484.
103. Tulsi, R. S. (1979). A scanning electron microscope study of the pineal recess of the adult brush-tailed possum (*Trichosurus vulpecula*). *J. Anat.* 129, 521–530.
104. Welsh, M. G. (1983). CSF-contacting pinealocytes in the pineal recess of the Mongolian gerbil: a correlative scanning and transmission electron microscope study. *Am. J. Anat.* 166, 483–493.
105. Cosenza, R. M., Sousa Neto, J. A., and Machado, A. B. (1990). The pineal recess of the opossum: a scanning and transmission electron microscope study. *Microsc. Electron. Biol. Cell.* 14, 101–113.
106. Tricoire, H., Malpaux, B., and Moller, M. (2003). Cellular lining of the sheep pineal recess studied by light-, transmission-, and scanning electron microscopy: morphologic indications for a direct secretion of melatonin from the pineal gland to the cerebrospinal fluid. *J. Comp. Neurol.* 456, 39–47.
107. Tricoire, H., Locatelli, A., Chemineau, P., and Malpaux, B. (2002). Melatonin enters the cerebrospinal fluid through the pineal recess. *Endocrinology* 143, 84–90.
108. Lerner, A. B., Case, J. D., Takahashi, Y., Lee, T. H., and Mori, W. (1958). Isolation of melatonin, the pineal gland factor that lightens melanocytes. *Adv. Exp. Med. Biol.* 80, 2587–2592.
109. McCord, C. P., and Allen, F. B. (1917). Evidence associating pineal gland function with alterations in pigmentation. *J. Exp. Zool.* 23, 207–224.
110. Lerner, A. B. (1999). Melatonin—without the hype. *Adv. Exp. Med. Biol.* 460, 1–3.
111. Wurtman, R. J., and Axelrod, J. (1968). The formation, metabolism and physiologic effects of melatonin. *Adv. Pharmacol.* 6, 141–151.
112. Klein, D. C., Auerbach, D. A., Namboodiri, M. A., and Wheler, G. H. T. (1981). Indole metabolism in the mammalian pineal gland. In *The Pineal Gland* (R. J. Reiter, Ed.), pp. 199–227. CRC Press, Boca Raton, FL.
113. Sugden, D. (1989). Melatonin biosynthesis in the mammalian pineal gland. *Experientia* 45, 922–932.
114. Ganguly, S., Coon, S. L., and Klein, D. C. (2002). Control of melatonin synthesis in the mammalian pineal gland: the critical role of serotonin acetylation. *Cell Tissue Res.* 309, 127–137.
115. Simonneaux, V., and Ribelayga, C. (2003). Generation of the melatonin endocrine message in mammals: a review of the complex regulation of melatonin synthesis by norepinephrine, peptides, and other pineal transmitters. *Pharmacol. Rev.* 55, 325–395.
116. Klein, D. C., and Weller, J. L. (1970). Indole metabolism in the pineal gland: a circadian rhythm in N-acetyltransferase. *Science* 169, 1093–1095.
117. Roseboom, P. H., Coon, S. L., Baler, R., McCune, S. K., Weller, J. L., and Klein, D. C. (1996). Melatonin synthesis: analysis of the more than 150-fold nocturnal increase in serotonin N-acetyltransferase messenger ribonucleic acid in the rat pineal gland. *Endocrinology* 137, 3033–3045.
118. Axelrod, J., Wurtman, R. J., and Snyder, S. H. (1965). Control of hydroxyindole O-methyltransferase activity in the rat pineal gland by environmental lighting. *J. Biol. Chem.* 240, 949–954.
119. Quay, W. B. (1967). Lack of day-night rhythm and effect of darkness in rat pineal content of N-acetylserotonin-O-methyltransferase. *Physiologist* 10, 286.
120. Ribelayga, C., Gauer, F., Calgari, C., Pevet, P., and Simonneaux, V. (1999). Photoneural regulation of rat pineal hydroxyindole-O-methyltransferase (HIOMT) messenger ribonucleic acid expression: an analysis of its complex relationship with HIOMT activity. *Endocrinology* 140, 1375–1384.
121. Drijfhout, W. J., van der Linde, A. G., Kooi, S. E., Grol, C. J., and Westerink, B. H. (1996). Norepinephrine release in the rat pineal gland: the input from the biological clock measured by in vivo microdialysis. *J. Neurochem.* 66, 748–755.
122. Klein, D. C. (1985). *Photoneural Regulation of the Mammalian Pineal Gland*, pp. 38–56. CIBA Foundation Symposia.
123. Deguchi, T., and Axelrod, J. (1972). Induction and superinduction of serotonin N-acetyltransferase by adrenergic drugs and denervation in rat pineal organ. *Proc. Natl. Acad. Sci. U S A* 69, 2208–2211.
124. Vanecek, J., and Illnerova, H. (1983). The effect of daily evening isoproterenol administration on reproductive organ growth in male rats treated neonatally with testosterone propionate. *Experientia* 39, 332–333.
125. Strada, S. J., Klein, D. C., Weller, J., and Weiss, B. (1972). Effect of norepinephrine on the concentration of adenosine 3′,5′-monophosphate of rat pineal gland in organ culture. *Endocrinology* 90, 1470–1475.
126. Vanecek, J., Sugden, D., Weller, J., and Klein, D. C. (1985). Atypical synergistic alpha 1- and beta-adrenergic regulation of adenosine 3′,5′-monophosphate and guanosine 3′,5′-monophosphate in rat pinealocytes. *Endocrinology* 116, 2167–2173.
127. Sugden, D., and Klein, D. C. (1987). A cholera toxin substrate regulates cyclic GMP content of rat pinealocytes. *J. Biol. Chem.* 262, 7447–7450.
128. Schomerus, C., Laedtke, E., and Korf, H. W. (1995). Calcium responses of isolated, immunocytochemically identified rat pinealocytes to noradrenergic, cholinergic and vasopressinergic stimulations. *Neurochem. Int.* 27, 163–175.

129. Stehle, J. H., von Gall, C., Schomerus, C., and Korf, H. W. (2001). Of rodents and ungulates and melatonin: creating a uniform code for darkness by different signaling mechanisms. *J. Biol. Rhythms* 16, 312–325.
130. Maronde, E., Pfeffer, M., Olcese, J., Molina, C. A., Schlotter, F., Dehghani, F., Korf, H. W., and Stehle, J. H. (1999). Transcription factors in neuroendocrine regulation: rhythmic changes in pCREB and ICER levels frame melatonin synthesis. *J. Neurosci.* 19, 3326–3336.
131. Baler, R., Covington, S., and Klein, D. C. (1999). Rat arylalkylamine N-acetyltransferase gene: upstream and intronic components of a bipartite promoter. *Biol. Cell.* 91, 699–705.
132. Borjigin, J., Wang, M. M., and Snyder, S. H. (1995). Diurnal variation in mRNA encoding serotonin N-acetyltransferase in pineal gland. *Nature* 378, 783–785.
133. Gauer, F., Poirel, V. J., Garidou, M. L., Simonneaux, V., and Pevet, P. (1999). Molecular cloning of the arylalkylamine-N-acetyltransferase and daily variations of its mRNA expression in the Syrian hamster pineal gland. *Brain Res. Mol. Brain Res.* 71, 87–95.
134. Stehle, J. H., Foulkes, N. S., Molina, C. A., Simonneaux, V., Pevet, P., and Sassone-Corsi, P. (1993). Adrenergic signals direct rhythmic expression of transcriptional repressor CREM in the pineal gland. *Nature* 365, 314–320.
135. Diaz, E., Garidou, M. L., Dardente, H., Salingre, A., Pevet, P., and Simonneaux, V. (2003). Expression and regulation of Icer mRNA in the Syrian hamster pineal gland. *Brain Res. Mol. Brain Res.* 112, 163–169.
136. Foulkes, N. S., Borjigin, J., Snyder, S. H., and Sassone-Corsi, P. (1996). Transcriptional control of circadian hormone synthesis via the CREM feedback loop. *Proc. Natl. Acad. Sci. U S A* 93, 14140–14145.
137. Gastel, J. A., Roseboom, P. H., Rinaldi, P. A., Weller, J. L., and Klein, D. C. (1998). Melatonin production: proteasomal proteolysis in serotonin N-acetyltransferase regulation. *Science* 279, 1358–1360.
138. Klein, D. C., and Weller, J. L. (1972). Rapid light-induced decrease in pineal serotonin N-acetyltransferase activity. *Science* 177, 532–533.
139. Stehle, J. H., von Gall, C., Schomerus, C., and Korf, H. W. (2001). Of rodents and ungulates and melatonin: creating a uniform code for darkness by different signaling mechanisms. *J. Biol. Rhythms* 16, 312–325.
140. Coon, S. L., Roseboom, P. H., Baler, R., Weller, J. L., Namboodiri, M. A., Koonin, E. V., and Klein, D. C. (1995). Pineal serotonin N-acetyltransferase: expression cloning and molecular analysis. *Science* 270, 1681–1683.
141. Klein, D. C., Coon, S. L., Roseboom, P. H., Weller, J. L., Bernard, M., Gastel, J. A., Zatz, M., Iuvone, P. M., Rodriguez, I. R., Begay, V., Falcon, J., Cahill, G. M., Cassone, V. M., and Baler, R. (1997). The melatonin rhythm-generating enzyme: molecular regulation of serotonin N-acetyltransferase in the pineal gland. *Rec. Prog. Horm. Res.* 52, 307–357; discussion 357–358.
142. Privat, K., Ravault, J. P., Chesneau, D., and Fevre-Montange, M. (1999). Day/night variation of tryptophan hydroxylase and serotonin N-acetyltransferase mRNA levels in the ovine pineal gland and retina. *J. Pineal Res.* 26, 193–203.
143. Johnston, J. D., Bashforth, R., Diack, A., Andersson, H., Lincoln, G. A., and Hazlerigg, D. G. (2004). Rhythmic melatonin secretion does not correlate with the expression of arylalkylamine N-acetyltransferase, inducible cyclic amp early repressor, period1 or cryptochrome1 mRNA in the sheep pineal. *Neuroscience* 124, 789–795.
144. Schomerus, C., Korf, H. W., Laedtke, E., Weller, J. L., and Klein, D. C. (2000). Selective adrenergic/cyclic AMP-dependent switch-off of proteasomal proteolysis alone switches on neural signal transduction: an example from the pineal gland. *J. Neurochem.* 75, 2123–2132.
145. Fleming, J. V., Barrett, P., Coon, S. L., Klein, D. C., and Morgan, P. J. (1999). Ovine arylalkylamine N-acetyltransferase in the pineal and pituitary glands: differences in function and regulation. *Endocrinology* 140, 972–978.
146. Yang, H. Y., and Neff, N. H. (1976). Hydroxyindole O-methyltransferase: an immunochemical study of the neuronal regulation of the pineal enzyme. *Mol. Pharmacol.* 12, 433–439.
147. Sugden, D., and Klein, D. C. (1983). Beta-adrenergic receptor control of rat pineal hydroxyindole-O-methyltransferase. *Endocrinology* 113, 348–353.
148. Ribelayga, C., Pevet, P., and Simonneaux, V. (1997). Adrenergic and peptidergic regulations of hydroxyindole-O-methyltransferase activity in rat pineal gland. *Brain Res.* 777, 247–250.
149. Sugden, D., Cena, V., and Klein, D. C. (1987). Hydroxyindole O-methyltransferase. *Methods Enzymol.* 142, 590–596.
150. Ribelayga, C., Pevet, P., and Simonneaux, V. (2000). HIOMT drives the photoperiodic changes in the amplitude of the melatonin peak of the Siberian hamster. *Am. J. Physiol. Regul. Integr. Comp. Physiol.* 278, R1339–R1345.
151. Ribelayga, C., Garidou, M. L., Malan, A., Gauer, F., Calgari, C., Pevet, P., and Simonneaux, V. (1999). Photoperiodic control of the rat pineal arylalkylamine-N-acetyltransferase and hydroxyindole-O-methyltransferase gene expression and its effect on melatonin synthesis. *J. Biol. Rhythms* 14, 105–115.
152. Moller, M., Masson-Pevet, M., and Pevet, P. (1998). Annual variations of the NPYergic innervation of the pineal gland of the European hamster (*Cricetus cricetus*). A quantitative immunohistochemical study. *Cell Tissue Res.* 291, 423–431.
153. Skene, D. J., Pevet, P., Vivien-Roels, B., Masson-Pevet, M., and Arendt, J. (1987). Effect of different photoperiods on concentrations of 5-methoxytryptophol and melatonin in the pineal gland of the Syrian hamster. *J. Endocrinol.* 114, 301–9.
154. Skene, D. J., Vivien-Roels, B., and Pevet, P. (1991). Day and nighttime concentrations of 5-methoxytryptophol and melatonin in the retina and pineal gland from different classes of vertebrates. *Gen. Comp. Endocrinol.* 84, 405–411.
155. Raynaud, F., and Pevet, P. (1991). Determination of 5-methoxyindoles in pineal gland and plasma samples by high-performance liquid chromatography with electrochemical detection. *J. Chromatogr.* 564, 103–113.
156. Reiter, R. J. (1991). Pineal melatonin: cell biology of its synthesis and of its physiological interactions. *Endocr. Rev.* 12, 151–180.
157. Ravault, J. P., and Chesneau, D. (1999). The onset of increased melatonin secretion after the onset of darkness in sheep depends on the photoperiod. *J. Pineal Res.* 27, 1–8.
158. Arendt, J. (1998). Melatonin and the pineal gland: influence on mammalian seasonal and circadian physiology. *Rev. Reprod.* 3, 13–22.
159. Arendt, J. (1995). *Melatonin and the Mammalian Pineal Gland*. Chapman & Hall, London.
160. Chemineau, P., Daveau, A., Bodin, L., Zarazaga, L., Gomez-Brunet, A., and Malpaux, B. (2002). Sheep as a mammalian model of genetic variability in melatonin. *Reprod. Suppl.* 59, 181–190.
161. Zarazaga, L. A., Malpaux, B., Bodin, L., and Chemineau, P. (1998). The large variability in melatonin blood levels in ewes is under strong genetic influence. *Am. J. Physiol.* 274, E607–E610.
162. Wetterberg, L., Iselius, L., and Lindsten, J. (1983). Genetic regulation of melatonin excretion in urine. A preliminary report. *Clin. Genet.* 24, 399–402.

163. Zarazaga, L. A., Malpaux, B., Guillaume, D., Bodin, L., and Chemineau, P. (1998). Genetic variability in melatonin concentrations in ewes originates in its synthesis, not in its catabolism. *Am. J. Physiol.* 274, E1086–E1090.
164. Coon, S. L., Zarazaga, L. A., Malpaux, B., Ravault, J. P., Bodin, L., Voisin, P., Weller, J. L., Klein, D. C., and Chemineau, P. (1999). Genetic variability in plasma melatonin in sheep is due to pineal weight, not to variations in enzyme activities. *Am. J. Physiol.* 277, E792–E797.
165. Gomez Brunet, A., Malpaux, B., Daveau, A., Taragnat, C., and Chemineau, P. (2002). Genetic variability in melatonin secretion originates in the number of pinealocytes in sheep. *J. Endocrinol.* 172, 397–404.
166. Yellon, S. M., Tamarkin, L., Pratt, B. L., and Goldman, B. D. (1982). Pineal melatonin in the Djungarian hamster: photoperiodic regulation of a circadian rhythm. *Endocrinology* 111, 488–492.
167. Ebling, F. J., Lincoln, G. A., Wollnik, F., and Anderson, N. (1988). Effects of constant darkness and constant light on circadian organization and reproductive responses in the ram. *J. Biol. Rhythms* 3, 365–384.
168. Ravault, J. P., Arendt, J., Tobler, I., Chesneau, D., and Maulin, O. (1989). Entrainment of melatonin rhythms in rams by symmetrical light-dark cycles of different period length. *Chronobiol. Int.* 6, 329–339.
169. Maywood, E. S., Hastings, M. H., Max, M., Ampleford, E., Menaker, M., and Loudon, A. S. (1993). Circadian and daily rhythms of melatonin in the blood and pineal gland of free-running and entrained Syrian hamsters. *J. Endocrinol.* 136, 65–73.
170. Schwartz, W. J., de la Iglesia, H. O., Zlomanczuk, P., and Illnerova, H. (2001). Encoding le quattro stagioni within the mammalian brain: photoperiodic orchestration through the suprachiasmatic nucleus. *J. Biol. Rhythms* 16, 302–311.
171. Hofman, M. A., and Swaab, D. F. (2002). A brain for all seasons: cellular and molecular mechanisms of photoperiodic plasticity. *Prog. Brain Res.* 138, 255–280.
172. Hofman, M. A. (2004). The brain's calendar: neural mechanisms of seasonal timing. *Biol. Rev. Camb. Philos. Soc.* 79, 61–77.
173. Matthews, C. D., Seamark, R. F., and Guerin, M. V. (1992). Plasma melatonin profiles of Romney Marsh sheep in natural photoperiod and in acutely extended darkness. *J. Reprod. Fertil.* 95, 869–875.
174. Sumova, A., Travnickova, Z., Peters, R., Schwartz, W. J., and Illnerova, H. (1995). The rat suprachiasmatic nucleus is a clock for all seasons. *Proc. Natl. Acad. Sci. U S A* 92, 7754–7758.
175. Vuillez, P., Jacob, N., Teclemariam-Mesbah, R., and Pevet, P. (1996). In Syrian and European hamsters, the duration of sensitive phase to light of the suprachiasmatic nuclei depends on the photoperiod. *Neurosci. Lett.* 208, 37–40.
176. Sumova, A., and Illnerova, H. (1996). Endogenous melatonin signal does not mediate the effect of photoperiod on the rat suprachiasmatic nucleus. *Brain Res.* 725, 281–283.
177. Jacob, N., Vuillez, P., and Pevet, P. (1997). Photoperiod does not act on the suprachiasmatic nucleus photosensitive phase through the endogenous melatonin, in the Syrian hamster. *Neurosci. Lett.* 229, 117–120.
178. Messager, S., Hazlerigg, D. G., Mercer, J. G., and Morgan, P. J. (2000). Photoperiod differentially regulates the expression of Per1 and ICER in the pars tuberalis and the suprachiasmatic nucleus of the Siberian hamster. *Eur. J. Neurosci.* 12, 2865–2870.
179. Messager, S., Ross, A. W., Barrett, P., and Morgan, P. J. (1999). Decoding photoperiodic time through Per1 and ICER gene amplitude. *Proc. Natl. Acad. Sci. U S A* 96, 9938–9943.
180. Nuesslein-Hildesheim, B., O'Brien, J. A., Ebling, F. J., Maywood, E. S., and Hastings, M. H. (2000). The circadian cycle of mPER clock gene products in the suprachiasmatic nucleus of the siberian hamster encodes both daily and seasonal time. *Eur. J. Neurosci.* 12, 2856–2864.
181. Mrugala, M., Zlomanczuk, P., Jagota, A., and Schwartz, W. J. (2000). Rhythmic multiunit neural activity in slices of hamster suprachiasmatic nucleus reflect prior photoperiod. *Am. J. Physiol. Regul. Integr. Comp. Physiol.* 278, R987–R994.
182. Illnerova, H., and Vanecek, J. (1988). Entrainment of the rat pineal rhythm in melatonin production by light. *Reprod. Nutr. Dev.* 28, 515–526.
183. Bittman, E. L., Karsch, F. J., and Hopkins, J. W. (1983). Role of the pineal gland in ovine photoperiodism: regulation of seasonal breeding and negative feedback effects of estradiol upon luteinizing hormone secretion. *Endocrinology* 113, 329–336.
184. Picazo, R. A., and Lincoln, G. A. (1995). Light control of the duration of the daily melatonin signal under long and short days in the Soay ram. Role of inhibition and entrainment. *J. Biol. Rhythms* 10, 55–63.
185. Morin, D., Simon, N., Depres-Brummer, P., Levi, F., Tillement, J. P., and Urien, S. (1997). Melatonin high-affinity binding to alpha-1-acid glycoprotein in human serum. *Pharmacology* 54, 271–275.
186. Kennaway, D. J., and Voultsios, A. (1998). Circadian rhythm of free melatonin in human plasma. *J. Clin. Endocrinol. Metab.* 83, 1013–1015.
187. Waldhauser, F., Waldhauser, M., Lieberman, H. R., Deng, M. H., Lynch, H. J., and Wurtman, R. J. (1984). Bioavailability of oral melatonin in humans. *Neuroendocrinology* 39, 307–313.
188. Bechgaard, E., Lindhardt, K., and Martinsen, L. (1999). Intranasal absorption of melatonin in vivo bioavailability study. *Int. J. Pharm.* 182, 1–5.
189. Skene, D. J., Papagiannidou, E., Hashemi, E., Snelling, J., Lewis, D. F., Fernandez, M., and Ioannides, C. (2001). Contribution of CYP1A2 in the hepatic metabolism of melatonin: studies with isolated microsomal preparations and liver slices. *J. Pineal Res.* 31, 333–342.
190. Hirata, F., Hayaishi, O., Tokuyama, T., and Seno, S. (1974). In vitro and in vivo formation of two new metabolites of melatonin. *J. Biol. Chem.* 249, 1311–1313.
191. Karsch, F. J., Bittman, E. L., Foster, D. L., Goodman, R. L., Legan, S. J., and Robinson, J. E. (1984). Neuroendocrine basis of seasonal reproduction. *Rec. Prog. Horm. Res.* 40, 185–232.
192. Goldman, B. D. (2001). Mammalian photoperiodic system: formal properties and neuroendocrine mechanisms of photoperiodic time measurement. *J. Biol. Rhythms* 16, 283–301.
193. Czyba, J. C., Girod, C., and Durand, M. (1964). Sur l'antagonisme épiphyso-hypophysaire et les variations saisonnières de la spermatogénése chez le hamster doré (*Mesocricetus auratus*). *C. R. Soc. Biol. (Paris)* 188, 742.
194. Hoffman, R. A., and Reiter, R. J. (1965). Pineal gland: influence on gonads of male hamsters. *Science* 148, 1609–1611.
195. Boissin-Agasse, L., Jacquet, J. M., Lacroix, A., and Boissin, J. (1988). Long-term effects of pinealectomy on testicular function, luteinizing hormone-releasing hormone hypothalamic system, and plasma prolactin levels in the mink, a short-day breeder. *J. Pineal Res.* 5, 385–396.
196. Roche, J. F., Karsch, F. J., Foster, D. L., Takagi, S., and Dziuk, P. J. (1970). Effect of pinealectomy on estrus, ovulation and luteinizing hormone in ewes. *Biol. Reprod.* 2, 251–254.
197. Seamark, R. F., Kennaway, D. J., Matthews, C. D., Fellenberg, A. J., Phillipou, G., Kotaras, P., McIntosh, J. E., Dunstan, E., and Obst, J. M. (1981). The role of the pineal gland in seasonality. *J. Reprod. Fertil.* 30 (Suppl.), 15–21.

198. Kennaway, D. J., Dunstan, E. A., Gilmore, T. A., and Seamark, R. F. (1984). Effects of pinealectomy, oestradiol and melatonin on plasma prolactin and LH secretion in ovariectomized sheep. *J. Endocrinol.* 102, 199–207.
199. Wayne, N. L., Malpaux, B., and Karsch, F. J. (1989). Social cues can play a role in timing onset of the breeding season of the ewe. *J. Reprod. Fertil.* 87, 707–713.
200. Lincoln, G. A. (1979). Photoperiodic control of seasonal breeding in the ram: participation of the cranial sympathetic nervous system. *J. Endocrinol.* 82, 135–147.
201. Barrell, G. K., and Lapwood, K. R. (1979). Effects of pinealectomy on the secretion of luteinizing hormone, testosterone and prolactin in rams exposed to various lighting regimes. *J. Endocrinol.* 80, 397–405.
202. Brinklow, B. R., and Forbes, J. M. (1984). Effect of pinealectomy on the plasma concentrations of prolactin, cortisol and testosterone in sheep in short and skeleton long photoperiods. *J. Endocrinol.* 100, 287–294.
203. Goldman, B., Hall, V., Hollister, C., Roychoudhury, P., Tamarkin, L., and Westrom, W. (1979). Effects of melatonin on the reproductive system in intact and pinealectomized male hamster maintained under various photoperiods. *Endocrinology* 104, 82–88.
204. Bartness, T. J., Powers, J. B., Hastings, M. H., Bittman, E. L., and Goldman, B. D. (1993). The timed infusion paradigm for melatonin delivery: what has it taught us about the melatonin signal, its reception, and the photoperiodic control of seasonal responses? *J. Pineal Res.* 15, 161–190.
205. Bittman, E. L., Dempsey, R. J., and Karsch, F. J. (1983). Pineal melatonin secretion drives the reproductive response to daylength in the ewe. *Endocrinology* 113, 2276–2283.
206. Bittman, E. L., and Karsch, F. J. (1984). Nightly duration of pineal melatonin secretion determines the reproductive response to inhibitory day length in the ewe. *Biol. Reprod.* 30, 585–593.
207. Carter, D. S., and Goldman, B. D. (1983). Antigonadal effects of timed melatonin infusion in pinealectomized male Djungarian hamsters (*Phodopus sungorus sungorus*): duration is the critical parameter. *Endocrinology* 113, 1261–1267.
208. Carter, D. S., and Goldman, B. D. (1983). Progonadal role of the pineal in the Djungarian hamster (*Phodopus sungorus sungorus*): mediation by melatonin. *Endocrinology* 113, 1268–1273.
209. Vivien-Roels, B., Pitrosky, B., Zitouni, M., Malan, A., Canguilhem, B., Bonn, D., and Pevet, P. (1997). Environmental control of the seasonal variations in the daily pattern of melatonin synthesis in the European hamster, *Cricetus cricetus. Gen. Comp. Endocrinol.* 106, 85–94.
210. Eloranta, E., Timisjarvi, J., Nieminen, M., Ojutkangas, V., Leppaluoto, J., and Vakkuri, O. (1992). Seasonal and daily patterns in melatonin secretion in female reindeer and their calves. *Endocrinology* 130, 1645–1652.
211. Tedesco, S. C., Flood, P. F., Morton, D. J., and Reiter, R. J. (1992). Seasonal melatonin and luteinizing hormone rhythms in muskosen at 52°N. *Rangifer* 12, 197–201.
212. Garcia, A., Landete-Castillejos, T., Zarazaga, L., Garde, J., and Gallego, L. (2003). Seasonal changes in melatonin concentrations in female Iberian red deer (*Cervus elaphus hispanicus*). *J. Pineal Res.* 34, 161–166.
213. Zarazaga, L. A., Malpaux, B., and Chemineau, P. (2003). Amplitude of the plasma melatonin nycthemeral rhythms is not associated with the dates of onset and offset of the seasonal ovulatory activity in the Ile-de-France ewe. *Reprod. Nutr. Dev.* 43, 167–177.
214. Watson-Whitmyre, M., and Stetson, M. H. (1983). Simulation of peak pineal melatonin release restores sensitivity to evening melatonin injections in pinealectomized hamsters. *Endocrinology* 112, 763–765.
215. Wayne, N. L., Malpaux, B., and Karsch, F. J. (1988). How does melatonin code for day length in the ewe: duration of nocturnal melatonin release or coincidence of melatonin with a light-entrained sensitive period? *Biol. Reprod.* 39, 66–75.
216. Maywood, E. S., Buttery, R. C., Vance, G. H., Herbert, J., and Hastings, M. H. (1990). Gonadal responses of the male Syrian hamster to programmed infusions of melatonin are sensitive to signal duration and frequency but not to signal phase nor to lesions of the suprachiasmatic nuclei. *Biol. Reprod.* 43, 174–182.
217. Grosse, J., Maywood, E. S., Ebling, F. J., and Hastings, M. H. (1993). Testicular regression in pinealectomized Syrian hamsters following infusions of melatonin delivered on non-circadian schedules. *Biol. Reprod.* 49, 666–674.
218. Pitrosky, B., Kirsch, R., Vivien-Roels, B., Georg-Bentz, I., Canguilhem, B., and Pevet, P. (1995). The photoperiodic response in Syrian hamster depends upon a melatonin-driven circadian rhythm of sensitivity to melatonin. *J. Neuroendocrinol.* 7, 889–895.
219. Gunduz, B., and Stetson, M. H. (2001). A test of the coincidence and duration models of melatonin action in Siberian hamsters: the effects of 1-hr melatonin infusions on testicular development in intact and pinealectomized prepubertal Phodopus sungorus. *J. Pineal Res.* 30, 97–107.
220. Gorman, M. R., and Zucker, I. (1997). Pattern of change in melatonin duration determines testicular responses in Siberian hamsters, *Phodopus sungorus. Biol. Reprod.* 56, 668–673.
221. Prendergast, B. J., Gorman, M. R., and Zucker, I. (2000). Establishment and persistence of photoperiodic memory in hamsters. *Proc. Natl. Acad. Sci. U S A* 97, 5586–5591.
222. Hiebert, S. M., Thomas, E. M., Lee, T. M., Pelz, K. M., Yellon, S. M., and Zucker, I. (2000). Photic entrainment of circannual rhythms in golden-mantled ground squirrels: role of the pineal gland. *J. Biol. Rhythms* 15, 126–134.
223. Woodfill, C. J., Robinson, J. E., Malpaux, B., and Karsch, F. J. (1991). Synchronization of the circannual reproductive rhythm of the ewe by discrete photoperiodic signals. *Biol. Reprod.* 45, 110–121.
224. Bittman, E. L., Dempsey, R. J., and Karsch, F. J. (1983). Pineal melatonin secretion drives the reproductive response to daylength in the ewe. *Endocrinology* 113, 2276–2283.
225. Woodfill, C. J., Wayne, N. L., Moenter, S. M., and Karsch, F. J. (1994). Photoperiodic synchronization of a circannual reproductive rhythm in sheep: identification of season-specific time cues. *Biol. Reprod.* 50, 965–976.
226. Barrell, G. K., Thrun, L. A., Brown, M. E., Viguie, C., and Karsch, F. J. (2000). Importance of photoperiodic signal quality to entrainment of the circannual reproductive rhythm of the ewe. *Biol. Reprod.* 63, 769–774.
227. Concannon, P., Roberts, P., Baldwin, B., and Tennant, B. (1997). Long-term entrainment of circannual reproductive and metabolic cycles by Northern and Southern Hemisphere photoperiods in woodchucks (*Marmota monax*). *Biol. Reprod.* 57, 1008–1015.
228. Zucker, I. (1985). Pineal gland influences period of circannual rhythms of ground squirrels. *Am. J. Physiol.* 249, R111–R115.
229. Prendergast, B. J., Flynn, A. K., and Zucker, I. (2000). Triggering of neuroendocrine refractoriness to short-day patterns of melatonin in Siberian hamsters. *J. Neuroendocrinol.* 12, 303–310.
230. Gorman, M. R. (2001). A plastic interval timer synchronizes pubertal development of summer- and fall-born hamsters. *Am. J. Physiol. Regul. Integr. Comp. Physiol.* 281, R1613–R1623.

231. Gorman, M. R. (2003). Melatonin implants disrupt developmental synchrony regulated by flexible interval timers. *J. Neuroendocrinol.* 15, 1084–1094.
232. Bittman, E. L. (1993). The sites and consequences of melatonin binding in mammals. *Am. Zool.* 33, 200–211.
233. Glass, J. D., and Lynch, G. R. (1982). Evidence for a brain site of melatonin action in the white-footed mouse, *Peromyscus leucopus*. *Neuroendocrinology* 34, 1–6.
234. Dowell, S. F., and Lynch, G. R. (1987). Duration of the melatonin pulse in the hypothalamus controls testicular function in pinealectomized mice (*Peromyscus leucopus*). *Biol. Reprod.* 36, 1095–1101.
235. Hastings, M. H., Walker, A. P., Roberts, A. C., and Herbert, J. (1988). Intra-hypothalamic melatonin blocks photoperiodic responsiveness in the male Syrian hamster. *Neuroscience* 24, 987–991.
236. Devries, M. J., Ferreira, S. A., and Glass, J. D. (1989). Evidence that short photoperiod-induced gonadal regression in the Mongolian gerbil is mediated by the action of melatonin in the medial hypothalamus. *Brain Res.* 494, 241–246.
237. Bonnefond, C., Walker, A. P., Stutz, J. A., Maywood, E., Juss, T. S., Herbert, J., and Hastings, M. H. (1989). The hypothalamus and photoperiodic control of FSH secretion by melatonin in the male Syrian hamster. *J. Endocrinol.* 122, 247–254.
238. Masson-Pevet, M., George, D., Kalsbeek, A., Saboureau, M., Lakhdar-Ghazal, N., and Pevet, P. (1994). An attempt to correlate brain areas containing melatonin-binding sites with rhythmic functions: a study in five hibernator species. *Cell Tissue Res.* 278, 97–106.
239. Morgan, P. J., Barrett, P., Howell, H. E., and Helliwell, R. (1994). Melatonin receptors: localization, molecular pharmacology and physiological significance. *Neurochem. Int.* 24, 101–146.
240. Rutten, A., Hewing, M., and Wittkowski, W. (1988). Seasonal ultrastructural changes of the hypophyseal pars tuberalis in the hedgehog (*Erinaceus europaeus* L.). *Acta Anat. (Basel).* 133, 217–223.
241. Nakazawa, K., Marubayashi, U., and McCann, S. M. (1991). Mediation of the short-loop negative feedback of luteinizing hormone (LH) on LH-releasing hormone release by melatonin-induced inhibition of LH release from the pars tuberalis. *Proc. Natl. Acad. Sci. U S A* 88, 7576–7579.
242. Skinner, D. C., Evans, N. P., and Malpaux, B. (1997). Does a short loop feedback mechanism for the control of luteinizing hormone secretion exist in the ewe? *Endocrinology* 138, 4220–4226.
243. Malpaux, B., Skinner, D. C., and Maurice, F. (1995). The ovine pars tuberalis does not appear to be targeted by melatonin to modulate luteinizing hormone secretion, but may be important for prolactin release. *J. Neuroendocrinol.* 7, 199–206.
244. Malpaux, B., Viguie, C., Skinner, D. C., Thiery, J. C., and Chemineau, P. (1997). Control of the circannual rhythm of reproduction by melatonin in the ewe. *Brain Res. Bull.* 44, 431–438.
245. Malpaux, B., Daveau, A., Maurice, F., Gayrard, V., and Thiéry, J.-C. (1993). Short-day effects of melatonin on luteinizing hormone secretion in the ewe: evidence for central sites of action in the mediobasal hypothalamus. *Biol. Reprod.* 48, 752–760.
246. Lincoln, G. A., and Maeda, K. I. (1992). Reproductive effects of placing micro-implants of melatonin in the mediobasal hypothalamus and preoptic area in rams. *J. Endocrinol.* 132, 201–215.
247. Chabot, V., Caldani, M., de Reviers, M. M., and Pelletier, J. (1998). Localization and quantification of melatonin receptors in the diencephalon and posterior telencephalon of the sheep brain. *J. Pineal Res.* 24, 50–57.
248. Malpaux, B., Daveau, A., Maurice-Mandon, F., Duarte, G., and Chemineau, P. (1998). Evidence that melatonin acts in the premammillary hypothalamic area to control reproduction in the ewe: presence of binding sites and stimulation of luteinizing hormone secretion by in situ microimplant delivery. *Endocrinology* 139, 1508–1516.
249. Lincoln, G. A. (2002). Neuroendocrine regulation of seasonal gonadotrophin and prolactin rhythms: lessons from the Soay ram model. *Reprod. Suppl.* 59, 131–147.
250. Migaud, M., Daveau, A., and Malpaux, B. (2004). MTRN1A melatonin receptors in the ovine pre-mammillary hypothalamus: day-night variation in the expression of the transcripts. *Biol. Reprod.*
251. Maywood, E. S., and Hastings, M. H. (1995). Lesions of the iodomelatonin-binding sites of the mediobasal hypothalamus spare the lactotropic, but block the gonadotropic response of male Syrian hamsters to short photoperiod and to melatonin. *Endocrinology* 136, 144–153.
252. Bae, H. H., Mangels, R. A., Cho, B. S., Dark, J., Yellon, S. M., and Zucker, I. (1999). Ventromedial hypothalamic mediation of photoperiodic gonadal responses in male Syrian hamsters. *J. Biol. Rhythms* 14, 391–401.
253. Maywood, E. S., Bittman, E. L., and Hastings, M. H. (1996). Lesions of the melatonin- and androgen-responsive tissue of the dorsomedial nucleus of the hypothalamus block the gonadal response of male Syrian hamsters to programmed infusions of melatonin. *Biol. Reprod.* 54, 470–477.
254. Badura, L. L., and Goldman, B. D. (1992). Central sites mediating reproductive responses to melatonin in juvenile male Siberian hamsters. *Brain Res.* 598, 98–106.
255. Song, C. K., Bartness, T. J., Petersen, S. L., and Bittman, E. L. (2000). Co-expression of melatonin (MEL1a) receptor and arginine vasopressin mRNAs in the Siberian hamster suprachiasmatic nucleus. *J. Neuroendocrinol.* 12, 627–634.
256. Freeman, D. A., and Zucker, I. (2001). Refractoriness to melatonin occurs independently at multiple brain sites in Siberian hamsters. *Proc. Natl. Acad. Sci. U S A* 98, 6447–6452.
257. Bartness, T. J., Goldman, B. D., and Bittman, E. L. (1991). SCN lesions block responses to systemic melatonin infusions in Siberian hamsters. *Am. J. Physiol.* 260, R102–R112.
258. Bittman, E. L., Crandell, R. G., and Lehman, M. N. (1989). Influences of the paraventricular and suprachiasmatic nuclei and olfactory bulbs on melatonin responses in the golden hamster. *Biol. Reprod.* 40, 118–126.
259. Berria, M., DeSantis, M., and Mead, R. A. (1988). Effects of suprachiasmatic nuclear ablation and melatonin on delayed implantation in the spotted skunk. *Neuroendocrinology* 48, 371–375.
260. Maurel, D. L., Ben Saad, M. M., Roch, G., and Siaud, P. (2002). Testicular activity is restored by melatonin replacement after suprachiasmatic nucleus lesion or superior cervical ganglionectomy in mink. *J. Pineal Res.* 32, 15–20.
261. Maywood, E. S., Bittman, E. L., Ebling, F. J., Barrett, P., Morgan, P., and Hastings, M. H. (1995). Regional distribution of iodomelatonin binding sites within the suprachiasmatic nucleus of the Syrian hamster and the Siberian hamster. *J. Neuroendocrinol.* 7, 215–223.
262. Lincoln, G. A., and Clarke, I. J. (2002). Noradrenaline and dopamine regulation of prolactin secretion in sheep: role in prolactin homeostasis but not photoperiodism. *J. Neuroendocrinol.* 14, 36–44.
263. Lincoln, G. A., and Clarke, I. J. (1994). Photoperiodically-induced cycles in the secretion of prolactin in hypothalamo-pituitary disconnected rams: evidence for translation of the

melatonin signal in the pituitary gland. *J. Neuroendocrinol.* 6, 251–260.

264. Helliwell, R. J., and Williams, L. M. (1994). The development of melatonin-binding sites in the ovine fetus. *J. Endocrinol.* 142, 475–484.
265. Skinner, D. C., and Robinson, J. E. (1995). Melatonin-binding sites in the gonadotroph-enriched zona tuberalis of ewes. *J. Reprod. Fertil.* 104, 243–250.
266. Stirland, J. A., Johnston, J. D., Cagampang, F. R., Morgan, P. J., Castro, M. G., White, M. R., Davis, J. R., and Loudon, A. S. (2001). Photoperiodic regulation of prolactin gene expression in the Syrian hamster by a pars tuberalis-derived factor. *J. Neuroendocrinol.* 13, 147–157.
267. Morgan, P. J., and Williams, L. M. (1996). The pars tuberalis of the pituitary: a gateway for neuroendocrine output. *Rev. Reprod.* 1, 153–161.
268. Morgan, P. J., Webster, C. A., Mercer, J. G., Ross, A. W., Hazlerigg, D. G., MacLean, A., and Barrett, P. (1996). The ovine pars tuberalis secretes a factor(s) that regulates gene expression in both lactotropic and nonlactotropic pituitary cells. *Endocrinology* 137, 4018–4026.
269. Lafarque, M., Oliveros, L., and Aguado, L. (1998). (Effect of adenohypophyseal pars tuberalis secretions on pars distalis prolactin liberation). *Medicina (B. Aires)* 58, 36–40.
270. Klosen, P., Bienvenu, C., Demarteau, O., Dardente, H., Guerrero, H., Pevet, P., and Masson-Pevet, M. (2002). The Mt1 melatonin receptor and RORbeta receptor are co-localized in specific TSH-immunoreactive cells in the pars tuberalis of the rat pituitary. *J. Histochem. Cytochem.* 50, 1647–1657.
271. Dardente, H., Klosen, P., Pevet, P., and Masson-Pevet, M. (2003). MT1 melatonin receptor mRNA expressing cells in the pars tuberalis of the European hamster: effect of photoperiod. *J. Neuroendocrinol.* 15, 778–786.
272. Wittkowski, W., Bockmann, J., Kreutz, M. R., and Bockers, T. M. (1999). Cell and molecular biology of the pars tuberalis of the pituitary. *Int. Rev. Cytol.* 185, 157–194.
273. Hastings, M. H., Walker, A. P., Powers, J. B., Hutchison, J., Steel, E. A., and Herbert, J. (1989). Differential effects of photoperiodic history on the responses of gonadotrophins and prolactin to intermediate daylengths in the male Syrian hamster. *J. Biol. Rhythms* 4, 335–350.
274. Dubocovich, M. L. (1988). Pharmacology and function of melatonin receptors. *FASEB J.* 2, 2765–2773.
275. Dubocovich, M. L. (1995). Melatonin receptors: are there multiple subtypes? *Trends Pharmacol. Sci.* 16, 50–56.
276. Duncan, M. J., Takahashi, J. S., and Dubocovich, M. L. (1989). Characteristics and autoradiographic localization of 2-(125I)iodomelatonin binding sites in Djungarian hamster brain. *Endocrinology* 125, 1011–108.
277. Nosjean, O., Ferro, M., Coge, F., Beauverger, P., Henlin, J. M., Lefoulon, F., Fauchere, J. L., Delagrange, P., Canet, E., and Boutin, J. A. (2000). Identification of the melatonin-binding site MT3 as the quinone reductase 2. *J. Biol. Chem.* 275, 31311–31317.
278. Nosjean, O., Nicolas, J. P., Klupsch, F., Delagrange, P., Canet, E., and Boutin, J. A. (2001). Comparative pharmacological studies of melatonin receptors: MT1, MT2 and MT3/QR2. Tissue distribution of MT3/QR2. *Biochem. Pharmacol.* 61, 1369–1379.
279. Ebisawa, T., Karne, S., Lerner, M. R., and Reppert, S. M. (1994). Expression cloning of a high-affinity melatonin receptor from Xenopus dermal melanophores. *Proc. Natl. Acad. Sci. U S A* 91, 6133–6137.
280. Dubocovich, M. L., Cardinali, D. P., Delagrange, P., Krause, D. N., Strosberg, D., Sugden, D., and Yocca, F. D. (2000). Melatonin receptors. In *The IUPHAR Compendium of Receptor Characterization and Classification*, pp. 270–277. IUPHAR Media, London.
281. Reppert, S. M., Godson, C., Mahle, C. D., Weaver, D. R., Slaugenhaupt, S. A., and Gusella, J. F. (1995). Molecular characterization of a second melatonin receptor expressed in human retina and brain: the Mel1b melatonin receptor. *Proc. Natl. Acad. Sci. U S A* 92, 8734–738.
282. Reppert, S. M., Weaver, D. R., and Ebisawa, T. (1994). Cloning and characterization of a mammalian melatonin receptor that mediates reproductive and circadian responses. *Neuron* 13, 1177–1185.
283. Roca, A. L., Godson, C., Weaver, D. R., and Reppert, S. M. (1996). Structure, characterization, and expression of the gene encoding the mouse Mel1a melatonin receptor. *Endocrinology* 137, 3469–3477.
284. Liu, F., Yuan, H., Sugamori, K. S., Hamadanizadeh, A., Lee, F. J., Pang, S. F., Brown, G. M., Pristupa, Z. B., and Niznik, H. B. (1995). Molecular and functional characterization of a partial cDNA encoding a novel chicken brain melatonin receptor. *FEBS Lett.* 374, 273–278.
285. Reppert, S. M., Weaver, D. R., Cassone, V. M., Godson, C., and Kolakowski, L. F. Jr. (1995). Melatonin receptors are for the birds: molecular analysis of two receptor subtypes differentially expressed in chick brain. *Neuron* 15, 1003–1015.
286. Jockers, R., Petit, L., Lacroix, I., de Coppet, P., Barrett, P., Morgan, P. J., Guardiola, B., Delagrange, P., Marullo, S., and Strosberg, A. D. (1997). Novel isoforms of Mel1c melatonin receptors modulating intracellular cyclic guanosine 3′,5′-monophosphate levels. *Mol. Endocrinol.* 11, 1070–1081.
287. Mazurais, D., Brierley, I., Anglade, I., Drew, J., Randall, C., Bromage, N., Michel, D., Kah, O., and Williams, L. M. (1999). Central melatonin receptors in the rainbow trout: comparative distribution of ligand binding and gene expression. *J. Comp. Neurol.* 409, 313–324.
288. Sugden, D., Pickering, H., Teh, M. T., and Garratt, P. J. (1997). Melatonin receptor pharmacology: toward subtype specificity. *Biol. Cell.* 89, 531–537.
289. Dubocovich, M. L., Masana, M. I., Iacob, S., and Sauri, D. M. (1997). Melatonin receptor antagonists that differentiate between the human Mel1a and Mel1b recombinant subtypes are used to assess the pharmacological profile of the rabbit retina ML1 presynaptic heteroreceptor. *Naunyn. Schmiedebergs Arch. Pharmacol.* 355, 365–375.
290. Audinot, V., Mailliet, F., Lahaye-Brasseur, C., Bonnaud, A., Le Gall, A., Amosse, C., Dromaint, S., Rodriguez, M., Nagel, N., Galizzi, J. P., Malpaux, B., Guillaumet, G., Lesieur, D., Lefoulon, F., Renard, P., Delagrange, P., and Boutin, J. A. (2003). New selective ligands of human cloned melatonin MT1 and MT2 receptors. *Naunyn. Schmiedebergs Arch. Pharmacol.* 367, 553–561.
291. Brydon, L., Roka, F., Petit, L., de Coppet, P., Tissot, M., Barrett, P., Morgan, P. J., Nanoff, C., Strosberg, A. D., and Jockers, R. (1999). Dual signaling of human Mel1a melatonin receptors via G(i2), G(i3), and G(q/11) proteins. *Mol. Endocrinol.* 13, 2025–2038.
292. Drew, J. E., Barrett, P., Conway, S., Delagrange, P., and Morgan, P. J. (2002). Differential coupling of the extreme C-terminus of G protein alpha subunits to the G protein-coupled melatonin receptors. *Biochim. Biophys. Acta* 1592, 185–192.
293. Masana, M. I., and Dubocovich, M. L. (2001). Melatonin receptor signaling: finding the path through the dark. *Sci. STKE* 2001, E39.
294. Barrett, P., Conway, S., and Morgan, P. J. (2003). Digging deep—structure-function relationships in the melatonin receptor family. *J. Pineal Res.* 35, 221–230.

295. von Gall, C., Stehle, J. H., and Weaver, D. R. (2002). Mammalian melatonin receptors: molecular biology and signal transduction. *Cell Tissue Res.* 309, 151–162.
296. Ayoub, M. A., Couturier, C., Lucas-Meunier, E., Angers, S., Fossier, P., Bouvier, M., and Jockers, R. (2002). Monitoring of ligand-independent dimerization and ligand-induced conformational changes of melatonin receptors in living cells by bioluminescence resonance energy transfer. *J. Biol. Chem.* 277, 21522–21528.
297. Dubocovich, M. L., Rivera-Bermudez, M. A., Gerdin, M. J., and Masana, M. I. (2003). Molecular pharmacology, regulation and function of mammalian melatonin receptors. *Front. Biosci.* 8, d1093–d1108.
298. Ayoub, M. A., Levoye, A., Delagrange, P., and Jockers, R. (2004). Preferential formation of MT1/MT2 melatonin receptor heterodimers with distinct ligand interaction properties compared with MT2 homodimers. *Mol. Pharmacol.* 66, 312–321.
299. Reppert, S. M., Weaver, D. R., Ebisawa, T., Mahle, C. D., and Kolakowski, L. F. Jr. (1996). Cloning of a melatonin-related receptor from human pituitary. *FEBS Lett.* 386, 219–224.
300. Drew, J. E., Barrett, P., Williams, L. M., Conway, S., and Morgan, P. J. (1998). The ovine melatonin-related receptor: cloning and preliminary distribution and binding studies. *J. Neuroendocrinol.* 10, 651–661.
301. Drew, J. E., Barrett, P., Mercer, J. G., Moar, K. M., Canet, E., Delagrange, P., and Morgan, P. J. (2001). Localization of the melatonin-related receptor in the rodent brain and peripheral tissues. *J. Neuroendocrinol.* 13, 453–458.
302. Duncan, M. J., and Davis, F. C. (1993). Developmental appearance and age related changes in specific 2-(125I)iodomelatonin binding sites in the suprachiasmatic nuclei of female Syrian hamsters. *Brain Res. Dev.* 73, 205–212.
303. Vanecek, J. (1988). The melatonin receptors in rat ontogenesis. *Neuroendocrinology* 48, 201–203.
304. Vanecek, J. (1998). Cellular mechanisms of melatonin action. *Physiol. Rev.* 78, 687–721.
305. Johnston, J. D., Messager, S., Ebling, F. J., Williams, L. M., Barrett, P., and Hazlerigg, D. G. (2003). Gonadotrophin-releasing hormone drives melatonin receptor down-regulation in the developing pituitary gland. *Proc. Natl. Acad. Sci. U S A* 100, 2831–2835.
306. Gauer, F., Masson-Pevet, M., and Pevet, P. (1993). Melatonin receptor density is regulated in rat pars tuberalis and suprachiasmatic nuclei by melatonin itself. *Brain Res.* 602, 153–156.
307. Barrett, P., MacLean, A., Davidson, G., and Morgan, P. J. (1996). Regulation of the Mel 1a melatonin receptor mRNA and protein levels in the ovine pars tuberalis: evidence for a cyclic adenosine 3′,5′-monophosphate-independent Mel 1a receptor coupling and an autoregulatory mechanism of expression. *Mol. Endocrinol.* 10, 892–902.
308. Piketty, V., and Pelletier, J. (1993). Melatonin receptors in the lamb pars tuberalis/median eminence throughout the day. *Neuroendocrinology* 58, 359–365.
309. Hazlerigg, D. G., Gonzalez-Brito, A., Lawson, W., Hastings, M. H., and Morgan, P. J. (1993). Prolonged exposure to melatonin leads to time-dependent sensitization of adenylate cyclase and down-regulates melatonin receptors in pars tuberalis cells from ovine pituitary. *Endocrinology* 132, 285–292.
310. Barrett, P., Davidson, G., Hazlerigg, D. G., Morris, M. A., Ross, A. W., and Morgan, P. J. (1998). Mel 1a melatonin receptor expression is regulated by protein kinase C and an additional pathway addressed by the protein kinase C inhibitor Ro 31-8220 in ovine pars tuberalis cells. *Endocrinology* 139, 163–171.
311. Messager, S., Caillol, M., George, D., and Martinet, L. (1997). Seasonal variation of melatonin binding sites in the pars tuberalis of the male mink (*Mustela vison*). *J. Neuroendocrinol.* 9, 523–528.
312. Gauer, F., Masson-Pevet, M., Saboureau, M., George, D., and Pevet, P. (1993). Differential seasonal regulation of melatonin receptor density in the pars tuberalis and the suprachiasmatic nuclei: a study in the hedgehog (*Erinaceus europaeus*, L.). *J. Neuroendocrinol.* 5, 685–690.
313. Skene, D. J., Masson-Pevet, M., and Pevet, P. (1993). Seasonal changes in melatonin binding sites in the pars tuberalis of male European hamsters and the effect of testosterone manipulation. *Endocrinology* 132, 1682–1686.
314. Piketty, V. (2001). Absence of sexual dimorphism in pars tuberalis (125I)-melatonin binding sites of lambs slaughtered in June and in October. *J. Pineal Res.* 30, 50–55.
315. Liu, C., Weaver, D. R., Jin, X., Shearman, L. P., Pieschl, R. L., Gribkoff, V. K., and Reppert, S. M. (1997). Molecular dissection of two distinct actions of melatonin on the suprachiasmatic circadian clock. *Neuron* 19, 91–102.
316. Weaver, D. R., Liu, C., and Reppert, S. M. (1996). Nature's knockout: the Mel1b receptor is not necessary for reproductive and circadian responses to melatonin in Siberian hamsters. *Mol. Endocrinol.* 10, 1478–1487.
317. Drew, J. E., Barrett, P., Williams, L. M., Conway, S., and Morgan, P. J. (1998). The ovine melatonin-related receptor: cloning and preliminary distribution and binding studies. *J. Neuroendocrinol.* 10, 651–661.
318. Mailliet, F., Audinot, V., Malpaux, B., Bonnaud, A., Delagrange, P., Migaud, M., Barrett, P., Viaud-Massuard, M. C., Lesieur, D., Lefoulon, F., Renard, P., and Boutin, J. A. (2004). Molecular pharmacology of the ovine melatonin receptor: comparison with recombinant human MT1 and MT2 receptors. *Biochem. Pharmacol.* 67, 667–677.
319. Hazlerigg, D. G., Morgan, P. J., and Messager, S. (2001). Decoding photoperiodic time and melatonin in mammals: what can we learn from the pars tuberalis? *J. Biol. Rhythms* 16, 326–335.
320. Barrett, P., Schuster, C., Mercer, J., and Morgan, P. J. (2003). Sensitization: a mechanism for melatonin action in the pars tuberalis. *J. Neuroendocrinol.* 15, 415–421.
321. Barrett, P., Choi, W. S., Morris, M., and Morgan, P. (2000). A role for tyrosine phosphorylation in the regulation and sensitization of adenylate cyclase by melatonin. *FASEB J.* 14, 1619–1628.
322. Barrett, P., Messager, S., Schuster, C., Moar, K. M., Mercer, J. G., and Morgan, P. J. (2002). Pituitary adenylate cyclase-activating polypeptide acts as a paracrine regulator of melatonin-responsive cells of the ovine pars tuberalis. *Endocrinology* 143, 2366–2375.
323. von Gall, C., Garabette, M. L., Kell, C. A., Frenzel, S., Dehghani, F., Schumm-Draeger, P. M., Weaver, D. R., Korf, H. W., Hastings, M. H., and Stehle, J. H. (2002). Rhythmic gene expression in pituitary depends on heterologous sensitization by the neurohormone melatonin. *Nat. Neurosci.* 5, 234–238.
324. Messager, S., Garabette, M. L., Hastings, M. H., and Hazlerigg, D. G. (2001). Tissue-specific abolition of Per1 expression in the pars tuberalis by pinealectomy in the Syrian hamster. *Neuroreport* 12, 579–582.
325. Lincoln, G., Messager, S., Andersson, H., and Hazlerigg, D. (2002). Temporal expression of seven clock genes in the suprachiasmatic nucleus and the pars tuberalis of the sheep: evidence for an internal coincidence timer. *Proc. Natl. Acad. Sci. U S A* 99, 13890–13895.
326. Hazlerigg, D. G., Andersson, H., Johnston, J. D., and Lincoln, G. (2004). Molecular characterization of the

long-day response in the Soay sheep, a seasonal mammal. *Curr. Biol.* 14, 334–339.
327. Lincoln, G. A., Andersson, H., and Hazlerigg, D. (2003). Clock genes and the long-term regulation of prolactin secretion: evidence for a photoperiod/circannual timer in the pars tuberalis. *J. Neuroendocrinol.* 15, 390–397.
328. Lincoln, G. A., and Short, R. V. (1980). Seasonal breeding: nature's contraceptive. *Rec. Prog. Horm. Res.* 36, 1–52.
329. Pelletier, J., and Ortavant, R. (1975). Photoperiodic control of LH release in the ram. I. Influence of increasing and decreasing light photoperiods. *Acta Endocrinol. (Copenh).* 78, 435–441.
330. Goodman, R. L., Bittman, E. L., Foster, D. L., and Karsch, F. J. (1982). Alterations in the control of luteinizing hormone pulse frequency underlie the seasonal variation in estradiol negative feedback in the ewe. *Biol. Reprod.* 27, 580–589.
331. Robinson, J. E., Radford, H. M., and Karsch, F. J. (1985). Seasonal changes in pulsatile luteinizing hormone (LH) secretion in the ewe: relationship of frequency of LH pulses to day length and response to estradiol negative feedback. *Biol. Reprod.* 33, 324–334.
332. Montgomery, G. W., Martin, G. B., and Pelletier, J. (1985). Changes in pulsatile LH secretion after ovariectomy in Ile-de-France ewes in two seasons. *J. Reprod. Fertil.* 73, 173–183.
333. Pelletier, J., and Ortavant, R. (1975). Photoperiodic control of LH release in the ram. II. Light-androgens interaction. *Acta Endocrinol. (Copenh).* 78, 442–450.
334. Tamarkin, L., Hutchison, J. S., and Goldman, B. D. (1976). Regulation of serum gonadotropins by photoperiod and testicular hormone in the Syrian hamster. *Endocrinology* 99, 1528–1533.
335. Glass, J. D., and Dolan, P. L. (1988). Melatonin acts in the brain to mediate seasonal steroid inhibition of luteinizing hormone secretion in the white-footed mouse (*Peromyscus leucopus*). *Proc. Soc. Exp. Biol. Med.* 188, 375–380.
336. Ellis, G. B., and Turek, F. W. (1980). Photoperiodic regulation of serum luteinizing hormone and follicle-stimulating hormone in castrated and castrated-adrenalectomized male hamsters. *Endocrinology* 106, 1338–1344.
337. Ellis, G. B., and Turek, F. W. (1980). Photoperiod-induced change in responsiveness of the hypothalamic-pituitary axis to exogenous 5 alpha-dihydrotestosterone and 17 beta-estradiol in castrated male hamsters. *Neuroendocrinology* 31, 205–209.
338. Turek, F. W., Elliott, J. A., Alvis, J. D., and Menaker, M. (1975). The interaction of castration and photoperiod in the regulation of hypophyseal and serum gonadotropin levels in male golden hamsters. *Endocrinology* 96, 854–860.
339. Bittman, E. L., and Goldman, B. D. (1979). Serum levels of gonadotrophins in hamsters exposed to short photoperiods: effects of adrenalectomy and ovariectomy. *J. Endocrinol.* 83, 113–118.
340. Montgomery, G. W., Martin, G. B., Locatelli, A., and Pelletier, J. (1987). Effect of adrenalectomy on LH release in sheep during the anoestrous season. *J. Endocrinol.* 114, 437–442.
341. Bittman, E. L., Kaynard, A. H., Olster, D. H., Robinson, J. E., Yellon, S. M., and Karsch, F. J. (1985). Pineal melatonin mediates photoperiodic control of pulsatile luteinizing hormone secretion in the ewe. *Neuroendocrinology* 40, 409–418.
342. Valenti, S., Giusti, M., Guido, R., and Giordano, G. (1997). Melatonin receptors are present in adult rat Leydig cells and are coupled through a pertussis toxin-sensitive G-protein. *Eur. J. Endocrinol.* 136, 633–639.
343. Li, L., Xu, J. N., Wong, Y. H., Wong, J. T., Pang, S. F., and Shiu, S. Y. (1998). Molecular and cellular analyses of melatonin receptor-mediated cAMP signaling in rat corpus epididymis. *J. Pineal Res.* 25, 219–228.
344. Tanavde, V. S., and Maitra, A. (2003). In vitro modulation of steroidogenesis and gene expression by melatonin: a study with porcine antral follicles. *Endocr. Res.* 29, 399–410.
345. Martin, J. E., and Klein, D. C. (1976). Melatonin inhibition of the neonatal pituitary response to luteinizing hormone-releasing factor. *Science* 191, 301–302.
346. Vanecek, J. (1999). Inhibitory effect of melatonin on GnRH-induced LH release. *Rev. Reprod.* 4, 67–72.
347. Symons, A. M., and Arendt, J. (1982). Lack of effect of melatonin on the pituitary response to LH-RH in the ewe. *J. Reprod. Fertil.* 64, 103–106.
348. Rodway, R. G., and Swift, A. D. (1983). Melatonin and the pituitary response to GnRH in prepubertal and adult ewes. *Horm. Metab. Res.* 15, 349–351.
349. Robinson, J. E., Kaynard, A. H., and Karsch, F. J. (1986). Does melatonin alter pituitary responsiveness to gonadotropin-releasing hormone in the ewe? *Neuroendocrinology* 43, 635–640.
350. Chemineau, P., Martin, G. B., Saumande, J., and Normant, E. (1988). Seasonal and hormonal control of pulsatile LH secretion in the dairy goat (*Capra hircus*). *J. Reprod. Fertil.* 83, 91–98.
351. Glass, J. D. (1986). Short photoperiod-induced gonadal regression: effects on the gonadotropin-releasing hormone (GnRH) neuronal system of the white-footed mouse, *Peromyscus leucopus*. *Biol. Reprod.* 35, 733–743.
352. Glass, J. D., and Knotts, L. K. (1987). A brain site for the antigonadal action of melatonin in the white-footed mouse (*Peromyscus leucopus*): involvement of the immunoreactive GnRH neuronal system. *Neuroendocrinology* 46, 48–55.
353. El Ouezzani, S., Tramu, G., and Magoul, R. (2000). The gonadotropin-releasing hormone neurosecretory system of the jerboa (Jaculus orientalis) and its seasonal variations. *J. Neuroendocrinol.* 12, 1205–1212.
354. Pickard, G. E., and Silverman, A. J. (1979). Effects of photoperiod on hypothalamic luteinizing hormone releasing hormone in the male hamster. *J. Endocrinol.* 83, 421–428.
355. Urbanski, H. F., Doan, A., and Pierce, M. (1991). Immunocytochemical investigation of luteinizing hormone-releasing hormone neurons in Syrian hamsters maintained under long or short days. *Biol. Reprod.* 44, 687–692.
356. Ronchi, E., Aoki, C., Krey, L. C., and Pfaff, D. W. (1992). Immunocytochemical study of GnRH and GnRH-associated peptide in male Syrian hamsters as a function of photoperiod and gonadal alterations. *Neuroendocrinology* 55, 134–145.
357. Brown, D. I., Garyfallou, V. T., and Urbanski, H. F. (2001). Photoperiodic modulation of GnRH mRNA in the male Syrian hamster. *Brain Res. Mol. Brain Res.* 89, 119–125.
358. Bernard, D. J., Abuav-Nussbaum, R., Horton, T. H., and Turek, F. W. (1999). Photoperiodic effects on gonadotropin-releasing hormone (GnRH) content and the GnRH-immunoreactive neuronal system of male Siberian hamsters. *Biol. Reprod.* 60, 272–276.
359. Messager, S., Caillol, M., and Martinet, L. (1999). Long-term exposure of hypothalamic explants to melatonin alters the release of gonadotrophin releasing hormone and the density of melatonin binding sites in the pars tuberalis of the male mink (*Mustela vison*). *J. Pineal Res.* 26, 17–27.
360. Caraty, A., Evans, N. P., Fabre-Nys, C. J., and Karsch, E. J. (1995). The preovulatory gonadotrophin-releasing hormone surge: a neuroendocrine signal for ovulation. *J. Reprod. Fertil.* 49 (Suppl.), 245–255.
361. Barrell, G. K., Moenter, S. M., Caraty, A., and Karsch, F. J. (1992). Seasonal changes of gonadotropin-releasing hormone secretion in the ewe. *Biol. Reprod.* 46, 1130–1135.
362. Viguie, C., Caraty, A., Locatelli, A., and Malpaux, B. (1995). Regulation of luteinizing hormone-releasing hormone (LHRH) secretion by melatonin in the ewe. I. Simultaneous

delayed increase in LHRH and luteinizing hormone pulsatile secretion. *Biol. Reprod.* 52, 1114–1120.

363. Mellon, P. L., Windle, J. J., Goldsmith, P. C., Padula, C. A., Roberts, J. L., and Weiner, R. I. (1990). Immortalization of hypothalamic GnRH neurons by genetically targeted tumorigenesis. *Neuron* 5, 1–10.
364. Roy, D., Angelini, N. L., Fujieda, H., Brown, G. M., and Belsham, D. D. (2001). Cyclical regulation of GnRH gene expression in GT1-7 GnRH-secreting neurons by melatonin. *Endocrinology* 142, 4711–4720.
365. Roy, D., and Belsham, D. D. (2002). Melatonin receptor activation regulates GnRH gene expression and secretion in GT1-7 GnRH neurons. Signal transduction mechanisms. *J. Biol. Chem.* 277, 251–258.
366. Carlberg, C., and Wiesenberg, I. (1995). The orphan receptor family RZR/ROR, melatonin and 5-lipoxygenase: an unexpected relationship. *J. Pineal Res.* 18, 171–178.
367. Gillespie, J. M., Roy, D., Cui, H., and Belsham, D. D. (2004). Repression of gonadotropin-releasing hormone (GnRH) gene expression by melatonin may involve transcription factors COUP-TFI and C/EBP beta binding at the GnRH enhancer. *Neuroendocrinology* 79, 63–72.
368. Lehman, M. N., Goodman, R. L., Karsch, F. J., Jackson, G. L., Berriman, S. J., and Jansen, H. T. (1997). The GnRH system of seasonal breeders: anatomy and plasticity. *Brain Res. Bull.* 44, 445–457.
369. Sliwowska, J. H., Billings, H. J., Goodman, R. L., Coolen, L. M., and Lehman, M. N. (2004). The premammillary hypothalamic area of the ewe: anatomical characterization of a melatonin target area mediating seasonal reproduction. *Biol. Reprod.* 70, 1768–1775.
370. Thiery, J. C., Martin, G. B., Tillet, Y., Caldani, M., Quentin, M., Jamain, C., and Ravault, J. P. (1989). Role of hypothalamic catecholamines in the regulation of luteinizing hormone and prolactin secretion in the ewe during seasonal anestrus. *Neuroendocrinology* 49, 80–87.
371. Goodman, R. L. (1994). Neuroendocrine control of the ovine estrous cycle. In *The Physiology of Reproduction* (E. Knobil, and J. D. Neill, Eds.), pp. 659–709. Raven Press, New York.
372. Thiery, J. C., Gayrard, V., Le Corre, S., Viguie, C., Martin, G. B., Chemineau, P., and Malpaux, B. (1995). Dopaminergic control of LH secretion by the A15 nucleus in anoestrous ewes. *J. Reprod. Fertil.* 49 (Suppl.), 285–296.
373. Thiery, J. C. (1991). Monoamine content of the stalk-median eminence and hypothalamus in adult female sheep as affected by daylength. *J. Neuroendocrinol.* 3, 407–411.
374. Viguie, C., Thibault, J., Thiery, J. C., Tillet, Y., and Malpaux, B. (1996). Photoperiodic modulation of monoamines and amino-acids involved in the control of prolactin and LH secretion in the ewe: evidence for a regulation of tyrosine hydroxylase activity. *J. Neuroendocrinol.* 8, 465–474.
375. Viguie, C., Thibault, J., Thiery, J. C., Tillet, Y., and Malpaux, B. (1997). Characterization of the short day-induced decrease in median eminence tyrosine hydroxylase activity in the ewe: temporal relationship to the changes in luteinizing hormone and prolactin secretion and short day-like effect of melatonin. *Endocrinology* 138, 499–506.
376. Viguie, C., Picard, S., Thiery, J. C., and Malpaux, B. (1998). Blockade of tyrosine hydroxylase activity in the median eminence partially reverses the long day-induced inhibition of pulsatile LH secretion in the ewe. *J. Neuroendocrinol.* 10, 551–558.
377. Havern, R. L., Whisnant, C. S., and Goodman, R. L. (1991). Hypothalamic sites of catecholamine inhibition of luteinizing hormone in the anestrous ewe. *Biol. Reprod.* 44, 476–482.
378. Bertrand, F., Viguie, C., Picard, S., and Malpaux, B. (1998). Median eminence dopaminergic activation is critical for the early long-day inhibition of luteinizing hormone secretion in the ewe. *Endocrinology* 139, 5094–5102.
379. Hileman, S. M., Kuehl, D. E., and Jackson, G. L. (1994). Effect of anterior hypothalamic area lesions on photoperiod-induced shifts in reproductive activity of the ewe. *Endocrinology* 135, 1816–1823.
380. Tillet, Y. (1995). Distribution of neurotransmitters in the sheep brain. *J. Reprod. Fertil.* 49 (Suppl.), 199–220.
381. Kuljis, R. O., and Advis, J. P. (1989). Immunocytochemical and physiological evidence of a synapse between dopamine- and luteinizing hormone releasing hormone-containing neurons in the ewe median eminence. *Endocrinology* 124, 1579–1581.
382. Bertrand, F., Thiery, J., Picard, S., and Malpaux, B. (1999). Implication of D2-like dopaminergic receptors in the median eminence during the establishment of long-day inhibition of LH secretion in the ewe. *J. Endocrinol.* 163, 243–254.
383. Tortonese, D. J., and Lincoln, G. A. (1994). Photoperiodic modulation of the dopaminergic control of pulsatile LH secretion in sheep. *J. Endocrinol.* 143, 25–32.
384. Tortonese, D. J., and Lincoln, G. A. (1995). Effects of melatonin in the mediobasal hypothalamus on the secretion of gonadotrophins in sheep: role of dopaminergic pathways. *J. Endocrinol.* 146, 543–552.
385. Steger, R. W., Juszczak, M., Fadden, C., and Bartke, A. (1995). Photoperiod effects on neurohypophyseal and tuberoinfundibular dopamine metabolism in the male hamster. *Endocrinology* 136, 3000–3006.
386. Le Corre, S., and Chemineau, P. (1993). Control of photoperiodic inhibition of luteinizing hormone secretion by dopaminergic and serotonergic systems in ovariectomized Ile-de-France ewes supplemented with oestradiol. *J. Reprod. Fertil.* 97, 367–373.
387. Meyer, S. L., and Goodman, R. L. (1986). Separate neural systems mediate the steroid-dependent and steroid-independent suppression of tonic luteinizing hormone secretion in the anestrous ewe. *Biol. Reprod.* 35, 562–571.
388. Whisnant, C. S., and Goodman, R. L. (1990). Further evidence that serotonin mediates the steroid-independent inhibition of luteinizing hormone secretion in anestrous ewes. *Biol. Reprod.* 42, 656–661.
389. Kao, C., Schaeffer, D. J., and Jackson, G. L. (1992). Different neuroendocrine systems modulate pulsatile luteinizing hormone secretion in photosuppressed and photorefractory ewes. *Biol. Reprod.* 46, 425–434.
390. Robinson, J. E., and Karsch, F. J. (1984). Refractoriness to inductive day lengths terminates the breeding season of the Suffolk ewe. *Biol. Reprod.* 31, 656–663.
391. Le Corre, S., and Chemineau, P. (1993). Serotonergic 5HT2 receptors mediate the inhibitory action of serotonin on luteinizing hormone secretion in ovariectomized, estradiol-treated ewes that are refractory to short days. *Biol. Reprod.* 49, 140–147.
392. Le Corre, S., Segu, L., Caldani, M., and Chemineau, P. (1994). Differences in ketanserin binding in the ventromedial hypothalamus of ewes responsive or refractory to short days. *Neuroendocrinology* 60, 589–600.
393. Pelletier, J., Auzan, C., Daveau, A., Clauser, E., and Chemineau, P. (1999). Sheep 5HT2A receptors: partial cloning of the coding sequence and mRNA localization by in situ hybridization in the ewe hypothalamus. *Cell Tissue Res.* 295, 231–239.
394. Chemineau, P., Daveau, A., Pelletier, J., Malpaux, B., Karsch, F. J., and Viguie, C. (2003). Changes in the 5-HT2A receptor system in the pre-mammillary hypothalamus of the ewe are related to regulation of LH pulsatile secretion by an endogenous circannual rhythm. *BMC Neurosci.* 4, 1.

395. Lincoln, G. A., Ebling, F. J., and Martin, G. B. (1987). Endogenous opioid control of pulsatile LH secretion in rams: modulation by photoperiod and gonadal steroids. *J. Endocrinol.* 115, 425–438.
396. Brooks, A. N., Lamming, G. E., Lees, P. D., and Haynes, N. B. (1986). Opioid modulation of LH secretion in the ewe. *J. Reprod. Fertil.* 76, 693–708.
397. Aurich, A., Schlote, S., Hoppen, H. O., Klug, E., Hoppe, H., and Aurich, J. E. (1994). Effects of the opioid antagonist naloxone on release of luteinizing hormone in mares during the anovulatory season. *J. Endocrinol.* 142, 139–144.
398. Aurich, C., Sieme, H., Hoppe, H., and Schlote, S. (1994). Involvment of endogenous opioids in the regulation of LH and testosterone release in the male horse. *J. Reprod. Fertil.* 102, 327–336.
399. Chen, H. J., Targovnik, J., McMillan, L., and Randall, S. (1984). Age difference in endogenous opiate modulation of short photoperiod-induced testicular regression in golden hamsters. *J. Endocrinol.* 101, 1–6.
400. Hileman, S. M., Kuehl, D. E., and Jackson, G. L. (1998). Photoperiod affects the ability of testosterone to alter proopiomelanocortin mRNA, but not luteinizing hormone-releasing hormone mRNA, levels in male sheep. *J. Neuroendocrinol.* 10, 587–592.
401. Tortonese, D. J. (1999). Interaction between hypothalamic dopaminergic and opioidergic systems in the photoperiodic regulation of pulsatile luteinizing hormone secretion in sheep. *Endocrinology* 140, 750–757.
402. Clarke, I. J., Rao, A., Chilliard, Y., Delavaud, C., and Lincoln, G. A. (2003). Photoperiod effects on gene expression for hypothalamic appetite-regulating peptides and food intake in the ram. *Am. J. Physiol. Regul. Integr. Comp. Physiol.* 284, R101–R115.
403. Dobbins, A., Lubbers, L. S., Jackson, G. L., Kuehl, D. E., and Hileman, S. M. (2004). Neuropeptide Y gene expression in male sheep: influence of photoperiod and testosterone. *Neuroendocrinology* 79, 82–89.
404. Skinner, D. C., and Herbison, A. E. (1997). Effects of photoperiod on estrogen receptor, tyrosine hydroxylase, neuropeptide Y, and beta-endorphin immunoreactivity in the ewe hypothalamus. *Endocrinology* 138, 2585–2595.
405. Barker-Gibb, M. L., and Clarke, I. J. (2000). Effect of season on neuropeptide Y and galanin within the hypothalamus of the ewe in relation to plasma luteinizing hormone concentrations and the breeding season: an immunohistochemical analysis. *J. Neuroendocrinol.* 12, 618–626.
406. Tillet, Y., Caldani, M., and Batailler, M. (1989). Anatomical relationships of monoaminergic and neuropeptide Y-containing fibres with luteinizing hormone-releasing hormone systems in the preoptic area of the sheep brain: immunohistochemical studies. *J. Chem. Neuroanat.* 2, 319–326.
407. Brann, D. W., and Mahesh, V. B. (1994). Excitatory amino acids: function and significance in reproduction and neuroendocrine regulation. *Front. Neuroendocrinol.* 15, 3–49.
408. Estienne, M. J., Schillo, K. K., Hileman, S. M., Green, M. A., and Hayes, S. H. (1990). Effect of N-methyl-D,L-aspartate on luteinizing hormone secretion in ovariectomized ewes in the absence and presence of estradiol. *Biol. Reprod.* 42, 126–130.
409. Jansen, H. T., Khalid, M., and Jackson, G. L. (1991). N-methyl-D,L-aspartate induces a transient increase in LH secretion in the seasonally anestrous ewe. *Domest. Anim. Endocrinol.* 8, 55–62.
410. Hui, Y., Hastings, M. H., Maywood, E. S., and Ebling, F. J. (1992). Photoperiodic regulation of glutamatergic stimulation of secretion of luteinizing hormone in male Syrian hamsters. *J. Reprod. Fertil.* 95, 935–946.
411. Lincoln, G. A., and Wu, F. C. W. (1991). Luteinizing hormone response to N-methyl-D,L-aspartate during a photoperiodically-induced reproductive cycle in the ram. *J. Neuroendocrinol.* 3, 309–317.
412. Urbanski, H. F. (1990). A role for N-methyl-D-aspartate receptors in the control of seasonal breeding. *Endocrinology* 127, 2223–2228.
413. Viguie, C., Caraty, A., Locatelli, A., and Malpaux, B. (1995). Regulation of luteinizing hormone-releasing hormone (LHRH) secretion by melatonin in the ewe. II. Changes in N-methyl-D,L-aspartic acid-induced LHRH release during the stimulation of luteinizing hormone secretion by melatonin. *Biol. Reprod.* 52, 1156–1161.
414. Ebling, F. J., Alexander, I. H., Urbanski, H. F., and Hastings, M. H. (1995). Effects of N-methyl-D-aspartate (NMDA) on seasonal cycles of reproduction, body weight and pelage colour in the male Siberian hamster. *J. Neuroendocrinol.* 7, 555–566.
415. Urbanski, H. F., and Pierce, M. (1991). Photoperiodic control of seasonal breeding in Syrian hamsters: involvment of excitatory amino acid receptors. *Neuroendocrinol. Lett.* 14, 33–37.
416. Woitkewitsch, A. A. (1940). Dependence of seasonal periodicity in gonadal changes on the thyroid gland in *Sturnus vulgaris* L. *C. R. Acad. Sci. Moscow* 28, 741–745.
417. Wieselthier, A. S., and Van tienhoven, A. (1972). The effect of thyroidectomy on testicular size and on the photorefractory period in the starling (*Sturnus vulgaris* L.). *J. Exp. Zool.* 179, 331–338.
418. Jallageas, M., and Assenmacher, I. (1979). Further evidence for reciprocal interactions between the annual sexual and thyroid cycles in male Peking ducks. *Gen. Comp. Endocrinol.* 37, 44–51.
419. Follett, B. K., and Nicholls, T. J. (1985). Influences of thyroidectomy and thyroxine replacement on photoperiodically controlled reproduction in quail. *J. Endocrinol.* 107, 211–221.
420. Vriend, J. (1985). Effects of melatonin and thyroxine replacement on thyrotropin, luteinizing hormone, and prolactin in male hypothyroid hamsters. *Endocrinology* 117, 2402–2407.
421. Jallageas, M., and Assenmacher, I. (1986). Effects of castration and thyroidectomy on the annual biological cycles of the edible dormouse *Glis glis*. *Gen. Comp. Endocrinol.* 63, 301–308.
422. Jacquet, J. M., Coutant, C., Maurel, D., Boissin-Agasse, L., and Boissin, J. (1986). Effect of thyroidectomy on variations during the spring and summer of the testicular activity and blood prolactin in the mink. *C. R. Acad. Sci. III* 303, 367–370.
423. Nicholls, T. J., Follett, B. K., Goldsmith, A. R., and Pearson, H. (1988). Possible homologies between photorefractoriness in sheep and birds: the effect of thyroidectomy on the length of the ewe's breeding season. *Reprod. Nutr. Dev.* 28, 375–385.
424. Shi, Z. D., and Barrell, G. K. (1992). Requirement of thyroid function for the expression of seasonal reproductive and related changes in red deer (*Cervus elaphus*) stags. *J. Reprod. Fertil.* 94, 251–259.
425. Parkinson, T. J., and Follett, B. K. (1994). Effect of thyroidectomy upon seasonality in rams. *J. Reprod. Fertil.* 101, 51–58.
426. Karsch, F. J., Dahl, G. E., Hachigian, T. M., and Thrun, L. A. (1995). Involvement of thyroid hormones in seasonal reproduction. *J. Reprod. Fertil.* 49 (Suppl.), 409–422.
427. Ben Saad, M. M., and Maurel, D. L. (2004). Reciprocal interaction between seasonal testis and thyroid activity in Zembra Island wild rabbits (*Oryctolagus cuniculus*): effects of castration, thyroidectomy, temperature, and photoperiod. *Biol. Reprod.* 70, 1001–1009.

428. Prendergast, B. J., Mosinger, B., Jr., Kolattukudy, P. E., and Nelson, R. J. (2002). Hypothalamic gene expression in reproductively photoresponsive and photorefractory Siberian hamsters. *Proc. Natl. Acad. Sci. U S A* 99, 16291–16296.
429. Moenter, S. M., Woodfill, C. J., and Karsch, F. J. (1991). Role of the thyroid gland in seasonal reproduction: thyroidectomy blocks seasonal suppression of reproductive neuroendocrine activity in ewes. *Endocrinology* 128, 1337–1344.
430. Dahl, G. E., Evans, N. P., Moenter, S. M., and Karsch, F. J. (1994). The thyroid gland is required for reproductive neuroendocrine responses to photoperiod in the ewe. *Endocrinology* 135, 10–15.
431. Webster, J. R., Moenter, S. M., Woodfill, C. J., and Karsch, F. J. (1991). Role of the thyroid gland in seasonal reproduction. II. Thyroxine allows a season-specific suppression of gonadotropin secretion in sheep. *Endocrinology* 129, 176–183.
432. Anderson, G. M., Hardy, S. L., Valent, M., Billings, H. J., Connors, J. M., and Goodman, R. L. (2003). Evidence that thyroid hormones act in the ventromedial preoptic area and the premammillary region of the brain to allow the termination of the breeding season in the ewe. *Endocrinology* 144, 2892–2901.
433. Thrun, L. A., Dahl, G. E., Evans, N. P., and Karsch, F. J. (1997). Effect of thyroidectomy on maintenance of seasonal reproductive suppression in the ewe. *Biol. Reprod.* 56, 1035–1040.
434. Anderson, G. M., Lapwood, K. R., Knight, P. G., and Parkinson, T. J. (2003). The reproductive response of rams to thyroidectomy: mediation by impaired inhibin feedback rather than a change in LH pulsatility. *Reproduction* 126, 353–364.
435. Dahl, G. E., Evans, N. P., Thrun, L. A., and Karsch, F. J. (1995). Thyroxine is permissive to seasonal transitions in reproductive neuroendocrine activity in the ewe. *Biol. Reprod.* 52, 690–696.
436. Thrun, L. A., Dahl, G. E., Evans, N. P., and Karsch, F. J. (1997). A critical period for thyroid hormone action on seasonal changes in reproductive neuroendocrine function in the ewe. *Endocrinology* 138, 3402–3409.
437. Billings, H. J., Viguie, C., Karsch, F. J., Goodman, R. L., Connors, J. M., and Anderson, G. M. (2002). Temporal requirements of thyroid hormones for seasonal changes in LH secretion. *Endocrinology* 143, 2618–2625.
438. Anderson, G. M., and Barrell, G. K. (1998). Effects of thyroidectomy and thyroxine replacement on seasonal reproduction in the red deer hind. *J. Reprod. Fertil.* 113, 239–250.
439. Webster, J. R., Moenter, S. M., Barrell, G. K., Lehman, M. N., and Karsch, F. J. (1991). Role of the thyroid gland in seasonal reproduction. III. Thyroidectomy blocks seasonal suppression of gonadotropin-releasing hormone secretion in sheep. *Endocrinology* 129, 1635–1643.
440. Viguie, C., Battaglia, D. F., Krasa, H. B., Thrun, L. A., and Karsch, F. J. (1999). Thyroid hormones act primarily within the brain to promote the seasonal inhibition of luteinizing hormone secretion in the ewe. *Endocrinology* 140, 1111–1117.
441. Jansen, H. T., Lubbers, L. S., Macchia, E., DeGroot, L. J., and Lehman, M. N. (1997). Thyroid hormone receptor (alpha) distribution in hamster and sheep brain: colocalization in gonadotropin-releasing hormone and other identified neurons. *Endocrinology* 138, 5039–5047.
442. Anderson, G. M., Connors, J. M., Hardy, S. L., Valent, M., and Goodman, R. L. (2001). Oestradiol microimplants in the ventromedial preoptic area inhibit secretion of luteinizing hormone via dopamine neurones in anoestrous ewes. *J. Neuroendocrinol.* 13, 1051–1058.
443. Kulikov, A., Moreau, X., and Jeanningros, R. (1999). Effects of experimental hypothyroidism on 5-HT1A, 5-HT2A receptors, 5-HT uptake sites and tryptophan hydroxylase activity in mature rat brain1. *Neuroendocrinology* 69, 453–459.
444. Watanabe, M., Yasuo, S., Watanabe, T., Yamamura, T., Nakao, N., Ebihara, S., and Yoshimura, T. (2004). Photoperiodic regulation of type 2 deiodinase gene in Djungarian hamster: possible homologies between avian and mammalian photoperiodic regulation of reproduction. *Endocrinology* 145, 1546–1549.
445. Koibuchi, N., and Chin, W. W. (2000). Thyroid hormone action and brain development. *Trends Endocrinol. Metab.* 11, 123–128.
446. Barakat-Walter, I. (1999). Role of thyroid hormones and their receptors in peripheral nerve regeneration. *J. Neurobiol.* 40, 541–559.
447. Bernal, J. (2002). Action of thyroid hormone in brain. *J. Endocrinol. Invest.* 25, 268–288.
448. Lehman, M. N., Karsch, F. J., Robinson, J. E., and Silverman, A. J. (1988). Ultrastructure and synaptic organization of luteinizing hormone-releasing hormone (LHRH) neurons in the anestrous ewe. *J. Comp. Neurol.* 273, 447–458.
449. Xiong, J. J., Karsch, F. J., and Lehman, M. N. (1997). Evidence for seasonal plasticity in the gonadotropin-releasing hormone (GnRH) system of the ewe: changes in synaptic inputs onto GnRH neurons. *Endocrinology* 138, 1240–1250.
450. Nottebohm, F., Nottebohm, M. E., and Crane, L. (1986). Developmental and seasonal changes in canary song and their relation to changes in the anatomy of song-control nuclei. *Behav. Neural Biol.* 46, 445–471.
451. Jansen, H. T., Cutter, C., Hardy, S., Lehman, M. N., and Goodman, R. L. (2003). Seasonal plasticity within the gonadotropin-releasing hormone (GnRH) system of the ewe: changes in identified GnRH inputs and glial association. *Endocrinology* 144, 3663–3676.
452. Pompolo, S., Pereira, A., Kaneko, T., and Clarke, I. J. (2003). Seasonal changes in the inputs to gonadotropin-releasing hormone neurones in the ewe brain: an assessment by conventional fluorescence and confocal microscopy. *J. Neuroendocrinol.* 15, 538–545.
453. Lee, W., Watanabe, M., and Glass, J. D. (1995). Photoperiod affects the expression of neural cell adhesion molecule and polysialic acid in the hypothalamus of the Siberian hamster. *Brain Res.* 690, 64–72.
454. Viguie, C., Jansen, H. T., Glass, J. D., Watanabe, M., Billings, H. J., Coolen, L., Lehman, M. N., and Karsch, F. J. (2001). Potential for polysialylated form of neural cell adhesion molecule-mediated neuroplasticity within the gonadotropin-releasing hormone neurosecretory system of the ewe. *Endocrinology* 142, 1317–1324.
455. Yamamura, T., Hirunagi, K., Ebihara, S., and Yoshimura, T. (2004). Seasonal morphological changes in the neuro-glial interaction between gonadotropin-releasing hormone nerve terminals and glial endfeet in Japanese quail. *Endocrinology* 145, 4264–4267.
456. Yoshimura, T., Yasuo, S., Watanabe, M., Iigo, M., Yamamura, T., Hirunagi, K., and Ebihara, S. (2003). Light-induced hormone conversion of T_4 to T_3 regulates photoperiodic response of gonads in birds. *Nature* 426, 178–181.
457. Goldman, B. D., and Nelson, R. J. (1993). Melatonin and seasonality in mammals. In *Melatonin: Biosynthesis, Physiological Effects, and Clinical Applications* (H. S. Yu and R. J. Reiter, Eds.), pp. 225–252. CRC Press, Boca Raton, FL.
458. Lynch, G. R., Heath, H. W., and Johnston, C. M. (1981). Effect of geographical origin on the photoperiodic control of reproduction in the white-footed mouse, *Peromyscus leucopus*. *Biol. Reprod.* 25, 475–480.

459. Heideman, P. D., and Bronson, F. H. (1991). Characteristics of a genetic polymorphism for reproductive photoresponsiveness in the white-footed mouse (*Peromyscus leucopus*). *Biol. Reprod.* 44, 1189–1196.
460. Desjardins, C., Bronson, F. H., and Blank, J. L. (1986). Genetic selection for reproductive photoresponsiveness in deer mice. *Nature* 322, 172–173.
461. Spears, N., and Clarke, J. R. (1988). Selection in field voles (*Microtus agrestis*) for gonadal growth under short photoperiods. *J. Anim. Ecol.* 57, 61–70.
462. Lynch, G. R., Lynch, C. B., and Kliman, R. M. (1989). Genetic analyses of photoresponsiveness in the Djungarian hamster, *Phodopus sungorus*. *J. Comp. Physiol. A* 164, 475–481.
463. Chemineau, P., Daveau, A., Cognie, Y., Aumont, G., and Chesneau, D. (2004). Seasonal ovulatory activity exists in tropical Creole female goats and Black Belly ewes subjected to a temperate photoperiod. *BMC Physiol.* 4, 12.
464. Thimonier, J., and Mauleon, P. (1969). Variations saisonnières du comportement d'oestrus et des activités ovarienne et hypophysaire chez les ovins. *Ann. Biol. Anim. Biochim. Biophys.* 9, 233–250.
465. Avdi, M., Banos, G., Stefos, K., and Chemineau, P. (2004). Seasonal variation in testicular volume and sexual behavior of Chios and Serres rams. *Theriogenology* 62, 275–282.
466. Hanocq, E., Bodin, L., Thimonier, J., Teyssier, J., Malpaux, B., and Chemineau, P. (1999). Genetic parameters of spontaneous spring ovulatory activity in Merinos d'Arles sheep. *Genet. Sel. Evol.* 31, 77–90.
467. Smith, J. F., and Jiohnson, D. L., et al. (1992). Genetic parameters and performance of flocks selected for advanced lambing date. *Proc. N. Z. Soc. Anim. Product.* 52, 129–131.
468. Kliman, R. M., and Lynch, G. R. (1992). Evidence for genetic variation in the occurrence of the photoresponse of the Djungarian hamster, *Phodopus sungorus*. *J. Biol. Rhythms* 7, 161–173.
469. Prendergast, B. J., Renstrom, R. A., and Nelson, R. J. (2004). Genetic analyses of a seasonal interval timer. *J. Biol. Rhythms* 19, 298–311.
470. Puchalski, W., and Lynch, G. R. (1986). Evidence for differences in the circadian organization of hamsters exposed to short day photoperiod. *J. Comp. Physiol. A* 159, 7–11.
471. Notter, D. R., and Chemineau, P. (2001). Nocturnal melatonin and prolactin plasma concentrations in sheep selected for fertility in autumn lambing. *J. Anim. Sci.* 79, 2895–2901.
472. Heideman, P. D. (2004). Top-down approaches to the study of natural variation in complex physiological pathways using the white-footed mouse (*Peromyscus leucopus*) as a model. *Ilar J.* 45, 4–13.
473. Heideman, P. D., Kane, S. L., and Goodnight, A. L. (1999). Differences in hypothalamic 2-(125I)iodomelatonin binding in photoresponsive and non-photoresponsive white-footed mice, *Peromyscus leucopus*. *Brain Res.* 840, 56–64.
474. Barrett, P., Conway, S., Jockers, R., Strosberg, A. D., Guardiola-Lemaitre, B., Delagrange, P., and Morgan, P. J. (1997). Cloning and functional analysis of a polymorphic variant of the ovine Mel 1a melatonin receptor. *Biochim. Biophys. Acta* 1356, 299–307.
475. Messer, L. A., Wang, L., Tuggle, C. K., Yerle, M., Chardon, P., Pomp, D., Womack, J. E., Barendse, W., Crawford, A. M., Notter, D. R., and Rothschild, M. F. (1997). Mapping of the melatonin receptor 1a (MTNR1A) gene in pigs, sheep, and cattle. *Mamm. Genome* 8, 368–370.
476. Pelletier, J., Bodin, L., Hanocq, E., Malpaux, B., Teyssier, J., Thimonier, J., and Chemineau, P. (2000). Association between expression of reproductive seasonality and alleles of the gene for Mel(1a) receptor in the ewe. *Biol. Reprod.* 62, 1096–1101.
477. Notter, D. R., and Cockett, N. E. (2005). Opportunities for detection and use of QTL influencing seasonal reproduction in sheep: a review. *Genet. Sel. Evol.* 37 (Suppl. 1), S39–S53.
478. Notter, D. R., Cockett, N. E., and Hadfield, T. S. (2003). Evaluation of melatonin receptor 1a as a candidate gene influencing reproduction in an autumn-lambing sheep flock. *J. Anim. Sci.* 81, 912–917.
479. Chu, M. X., Ji, C. L., and Chen, G. H. (2003). Association between PCR-RFLP of melatonin receptor 1a gene and high prolificacy in small tail Han sheep. *Asian-Aust. J. Anim. Sci.* 16, 1701–1704.
480. Roenneberg, T., and Aschoff, J. (1990). Annual rhythm of human reproduction. I. Biology, sociology, or both? *J. Biol. Rhythms* 5, 195–216.
481. Roenneberg, T. (2004). The decline in human seasonality. *J. Biol. Rhythms* 19, 193–195; discussion 196–197.
482. Wehr, T. A. (2001). Photoperiodism in humans and other primates: evidence and implications. *J. Biol. Rhythms* 16, 348–364.
483. Lam, D. A., and Miron, J. A. (1991). Seasonality of births in human populations. *Soc. Biol.* 38, 51–78.
484. Bronson, F. H. (2004). Are humans seasonally photoperiodic? *J. Biol. Rhythms* 19, 180–192.
485. Roenneberg, T., and Aschoff, J. (1990). Annual rhythm of human reproduction. II. Environmental correlations. *J. Biol. Rhythms* 5, 217–239.
486. Lindburg, D. G. (1987). Seasonality of reproduction in primates. In *Comparative Primate Biology, Vol. 2B: Behavior, Cognition and Motivation* (G. Mitchell and J. Erwin, Eds.), pp. 167–218. Alan R. Liss, New York.
487. Chik, C. L., Almeida, O. F., Libre, E. A., Booth, J. D., Renquist, D., and Merriam, G. R. (1992). Photoperiod-driven changes in reproductive function in male rhesus monkeys. *J. Clin. Endocrinol. Metab.* 74, 1068–1074.
488. Walker, M. L., Wilson, M. E., and Gordon, T. P. (1984). Endocrine control of the seasonal occurrence of ovulation in rhesus monkeys housed outdoors. *Endocrinology* 114, 1074–1081.
489. Wehr, T. A., Moul, D. E., Barbato, G., Giesen, H. A., Seidel, J. A., Barker, C., and Bender, C. (1993). Conservation of photoperiod-responsive mechanisms in humans. *Am. J. Physiol.* 265, R846–R857.
490. Chemineau, P., Malpaux, B., Delgadillo, J. A., Guerin, Y., Ravault, J. P., Thimonier, J., and Pelletier, J. (1992). Control of sheep and goat reproduction: use of light and melatonin. *Anim. Reprod. Sci.* 30, 157–184.
491. Vendreley, E., Guerillot, C., Basseville, C., and Delage, C. (1971). Poids testiculaire et spermatogénése du hamster doré au cours du cycle saisonnier. *C. R. Soc. Biol. (Paris)* 165, 1562–1565.
492. Dacheux, J. L., Pisselet, C., Blanc, M. R., Hochereau-de-Reviers, M. T., and Courot, M. (1981). Seasonal variations in rete testis fluid secretion and sperm production in different breeds of ram. *J. Reprod. Fertil.* 61, 363–371.
493. Cermakian, N., and Sassone-Corsi, P. (2000). Multilevel regulation of the circadian clock. *Nat. Rev. Mol. Cell Biol.* 1, 59–67.
494. Hodde, K. C. (1979). The vascularization of the rat pineal organ. *Prog. Brain Res.* 52, 39–44.
495. Barrell, G. K., Thrun, L. A., Brown, M. E., Viguie, C., and Karsch, F. J. (2000). Importance of photoperiodic signal quality to entrainment of the circannual reproductive rhythm of the ewe. *Biol. Reprod.* 63, 769–774.

Knobil and Neill's Physiology of Reproduction, Third Edition
edited by Jimmy D. Neill,
Elsevier © 2006

CHAPTER 42

Neuroendocrine Control of Mating-Induced Ovulation

Alexander S. Kauffman and Emilie F. Rissman

Abstract, 2283
Introduction, 2284
Evolution, Adaptive Function, and Phylogenetic Distribution of Mating-Induced Ovulation, 2285
Stimuli Requirements for Inducing the LH Surge and Ovulation in Reflex Ovulators, 2289
Behavioral Estrus of Reflex Ovulators, 2289
Exogenous Stimuli Requirements for Inducing the LH Surge and Ovulation, 2291
Nontactile Cues Influencing Ovulation in Reflex Ovulators, 2293
Endocrine Correlates of Coitus-Induced Ovulation: Hypothalamic, Pituitary, and Ovarian Hormone Secretion, 2294
Gonadotropin-Releasing Hormone (GnRH), 2296
Luteinizing Hormone (LH) and Follicle-Stimulating Hormone (FSH), 2297
Prolactin, 2300
Ovarian Progestins, 2301
Neural Pathways and Neurotransmitters Mediating the Control of Reflex Ovulation, 2302
The Hypothalamic Component and GnRH Neurons, 2303
The Brainstem Component and the Role of Norepinephrine, 2305
Regulatory Actions of Neuropeptide Y, 2307
Acetylcholine, Opioids, and Other Putative Neuroregulatory Factors, 2308
The Fine-Line between Induced and Spontaneous Ovulators: Male-Facilitated Ovulation and Puberty, 2309
Ovulation Advance in Seasonally Breeding Ungulates, 2309
Male-Facilitated Ovulation in Rodents that Normally Exhibit Spontaneous Cycles, 2310
Section Summary, 2312
Exceptions to the Rules: How Spontaneous Begets Induced Ovulation, 2312
Anesthetics and Estrous Cycles, 2312
Circadian Clocks and Induced Ovulation in Rats, 2312
Aging and Persistent Estrous in Rats, 2313
Brain Grafts Restore Fertility and Induced Ovulation in Mouse Mutants Lacking GnRH, 2314
Section Summary, 2314
Conclusions and Future Directions, 2314
Acknowledgments, 2316
References, 2316

ABSTRACT

Ovulation is a complex process whose timing is critical for reproductive success. Typically, mammalian species are categorized as having one of two types of ovulatory patterns. The most commonly studied laboratory species, and humans, exhibit spontaneous ovulation, in which the maturation of ovarian follicles and the spontaneous release of mature ova occur at a specific time during a regularly recurring ovarian cycle. Another pattern for ovulation, mating-induced ovulation, is found in a wide variety of mammals and may be the more primitive condition. Mating-induced ovulation (also called reflex ovulation) includes a

Department of Biochemistry and Molecular Genetics, University of Virginia School of Medicine, Charlottesville, Virginia.

variable amount of spontaneous follicular development along with an absolute requirement of mating to evoke ovulation. In this chapter we review evolutionary aspects of this process, including a comprehensive review of the data on individual species that show this pattern. We describe the behavioral requirements for inducing ovulation from a comparative perspective, and discuss the hormonal correlates of induced ovulation, including the secretion of hypothalamic, pituitary, and ovarian hormones. These sections are followed by a detailed description of the neural inputs and neurotransmitters involved in mediating reflex ovulation. In the final section we consider the question of whether spontaneous and induced ovulation are distinct entities or individual parts of a single continuum by reviewing the literature on male-facilitated estrous and ovulation. Here we also describe "exceptions to the rules," several cases in which rats and mice (normally spontaneous ovulators) can shift into an induced ovulatory mode. It is our hope that this chapter will stimulate new research on this process that has enormous practical and translational implications for the field of reproduction.

INTRODUCTION

The first observations of mating-induced ovulation were made over a century ago in ferrets and rabbits. In these pioneering studies, ovaries were compared in unmated and mated females and ovulation was noted only in the latter case (1,2). These findings were in stark contrast to those in humans and rats in which ovulation could be detected in virgin individuals. Thus, two categories of ovulation were formulated: spontaneous versus mating-induced ovulation (also known as reflex ovulation). Spontaneous ovulation encompasses the natural maturation of ovarian follicles and the spontaneous release of mature ova at a specific time during a regularly recurring ovarian cycle. In contrast, mating-induced ovulation includes a variable amount of spontaneous follicular development along with an absolute requirement of mating to evoke ovulation. One hundred years later, the distinction between these two ovulatory patterns has proven to be very useful. Although we know more about the control of ovulation in rats and mice than in many other mammals, the number of species that are confirmed as being mating-induced ovulators has grown substantially over the years, and much of the basic physiology and endocrinology that accompanies this ovulatory pattern have also been elucidated.

In the first part of this chapter (Evolution, Adaptive Function, and Phylogenetic Distribution of Mating-Induced Ovulation), we review the latest evidence on the occurrence of induced ovulation and present an exhaustive list of mammalian species that display this trait. In these studies the dependent measures vary from counting backwards from birth to determine the time of conception, to quantification of luteinizing hormone (LH) pre-ovulatory surges, to histological examination of ovulated ova in oviducts and of corpora lutea. Moreover, it is important to note that the current list of species that are induced ovulators is very likely to be incomplete since the requirements for studying and identifying this process include housing and mating animals under strictly controlled laboratory conditions. While smaller and domestic species (such as rodents or agricultural livestock) can usually meet this requirement, many other animals which are not easily housed or studied cannot; breeding programs for endangered and/or zoo populations may augment our basic knowledge in this area. As the list of induced ovulating species grows, our ability to test hypotheses on the evolution of this trait is also enhanced; we present several of these adaptive hypotheses in the section on Evolution, Adaptive Function, and Phylogenetic Distribution of Mating-Induced Ovulation and discuss the supporting evidence (or lack thereof) for each.

In the section on Stimuli Requirements for Inducing the LH Surge and Ovulation in Reflex Ovulators we consider the various stimuli that are required to evoke ovulation in the subset of reflex-ovulating species in which data are currently available. The amounts and timing of copulatory stimulation needed to promote either the pre-ovulatory LH surge or ovulation per se show tremendous variability, both within and between species. Likewise, the types of mating-related stimuli and cues that play a role in promoting ovulating also show species-to-species variability. In addition, a theme first introduced here, and which we elaborate on later in the section on The Fine-Line Between Induced and Spontaneous Ovulators: Male-Facilitated Ovulation and Puberty; Exceptions to the Rules: How Spontaneous Begets Induced Ovulation, is one that other authors (3) have noted as well: the issue of male-facilitated versus mating-induced ovulation, and the degree of similarity between them. As discussed later, many animals that spontaneously ovulate often require male-related cues, including in some cases coitus, to maximize their fertility and reproductive success. It is tempting to speculate that male-facilitated ovulation is merely a lesser degree of mating induced ovulation and, perhaps, a derived physiological remnant of an ancestral induced ovulation trait. Though this hypothesis has yet to be tested, it is supported by the fact that several spontaneous ovulators can, under specific conditions, exhibit reflex ovulation, a phenomenon discussed in detail in the section on Exceptions to the Rules: How Spontaneous Begets Induced Ovulation.

Hormone patterns that are associated with induced ovulation are considered next (in the section on Endocrine Correlates of Coitus-Induced Ovulation: Hypothalamic, Pituitary, and Ovarian Hormone Secretion). The timing of ovarian steroid and LH secretion in spontaneous and induced ovulating species can be very different with the LH surge sometimes occurring before peak circulating levels of steroids in some reflex ovulators. In this section we review the neuroendocrine responses to mating and artificial vagino-cervical stimulation in numerous reflex ovulators, including shrews, rabbits, mustelids, cats, and several species of voles and camels. Specific attention is given to the precise temporal pattern of hormone release after mating, with particular focus on gonadotropin-releasing hormone (GnRH), LH, follicle-stimulating hormone (FSH), prolactin, and progesterone. One exciting aspect of these data is the elegant coordination exhibited between behavior and hormone responses.

Work on the neural mechanisms underlying reflex ovulation is still in its infancy, though rapid advances in this area are currently being made. Many of the regulatory brain sites and neural pathways are presently being determined, as are many of the critical regulatory neurotransmitters involved in mediating this ovulatory process. This work, described in detail in the section on Neural Pathways and Neurotransmitters Mediating the Control of Reflex Ovulation, has great potential for clinical populations (as does the hormonal data described in the section on Endocrine Correlates of Coitus-Induced Ovulation), as anovulation is one of the major causes of infertility in women (4,5). Thus, there is an urgent need for new technologies and medical applications in this area. Because of the possible mechanistic similarities between spontaneous and induced ovulation, a better understanding of the neuroendocrinology of reflex ovulation will undoubtedly lead to new health-related treatments and technologies for humans.

In the final two sections we discuss male-facilitated ovulation and unique cases in which rats and mice can shift from being spontaneous ovulators to exhibiting mating induced ovulation. While the genetics of reflex ovulation is completely unknown, the fact that this trait can be revealed in these common laboratory animals suggests that very few gene changes are needed to shift from one mode to the other. These data are discussed in regards to the relationship between the two ovulatory patterns, and the likelihood that they are opposing, yet related, degrees of a single ovulatory continuum. Again, the clinical implications and future impact of studies of these models will be very valuable.

EVOLUTION, ADAPTIVE FUNCTION, AND PHYLOGENETIC DISTRIBUTION OF MATING-INDUCED OVULATION

Humans and the overwhelming majority of laboratory species used in biological research are spontaneous ovulators, giving the impression that this mode of ovulation is the most prominent type among mammals. However, increasing data from a diverse range of mammalian species indicate that induced ovulation may be equally as common as spontaneous ovulation. Indeed, several authors have speculated that reflex ovulation may in fact be more widespread and the prevailing mechanism used by mammals whereas spontaneous ovulation may be a more restricted and specialized mechanism (6,7). Unfortunately, as noted by Conaway (7), because most of the common domestic and laboratory species (i.e., nonhuman primates, rats, mice, hamsters, and sheep) are spontaneous ovulators, this point is often overlooked. However, with the exception of primates, induced ovulation is present in most mammalian orders that have been examined (Table 1) (6). In some orders, such as Insectivora, Lagomorpha, and Carnivora, induced ovulation appears to be the prevailing pattern whereas in other orders, such as Rodentia and Artiodactyla, both spontaneous and induced ovulation exist (Fig. 1).

Several authors have proposed that induced ovulation is the more ancestral (early evolved) trait from which spontaneous ovulation eventually evolved (7–9). In support of this, Conaway (7) noted that induced ovulation is observed in all members of the primitive order Insectivora and, within rodents, primarily in the older families Scuiridae (squirrels) and Microtinae (voles and lemmings). More recent species of rodents tend to be spontaneous ovulators (7), as do most members of more recently evolved mammalian orders such as ungulates (excluding several families of Artiodactyla). Furthermore, multiple species that are generally regarded as spontaneous ovulators, such as rats, mice, and humans, can become induced ovulators under some environmental or physiological conditions (a condition referred to as "facultative induced ovulation") (9–15). This raises the likelihood that spontaneous ovulators possess the underlying neural and/or hormonal pathways that mediate coitus-induced ovulation, lending support to the hypothesis that spontaneous ovulation evolved from the more primitive induced ovulation. The idea that induced and spontaneous ovulation are merely

TABLE 1. *Social, seasonal, and behavioral parameters affecting the induction of ovulation in reflex ovulators*

Species	Social or solitary	Seasonal breeder	Estrus and female receptivity	Mating stimuli needed to induce ovulation	Refs.
Short-tailed shrew (*Blarina brevicauda*)	Solitary	Yes	Not cyclic Estrus duration up to 33 d in absence of male	Multiple ejaculatory series	(18,388,389)
Least shrew (*Cryptotis parva*)	Social	No	Not cyclic Estrus induced by male Estrus duration 1–2 d	Multiple ejaculatory series	(3,389)
White-toothed shrew (*Crocidura russula monacha*)	Weakly social		Not cyclic Estrus induced by male		(3,389)
Musk shrew (*Suncus murinus*)	Solitary	Weakly seasonal	Not cyclic Estrus induced by male	Multiple ejaculatory series (often over several days)	(67,87, 390–392)
Hedgehog (*Erinaceus europeaus*)	Solitary	Yes	Prolonged estrus during breeding seasons		(18,389)
Collared lemmings (*Dicrostonyx groenlandicus*)	Solitary	Yes	No regular cycles Estrus lasts 6–30 d Possibly male-induced		(33,389,393)
Meadow voles (*Microtus pennsylvanicus*)	Solitary	Yes	No regular cycles Prolonged receptivity Likely not male-induced	Multiple ejaculatory series	(34,36,38,50,116)
Montane vole (*Microtus montanus*)	Solitary	Yes	Not cyclic Constantly receptive or induced rapidly by male	Multiple ejaculatory series	(38,50,112,178)
Bank vole (*Clythriomomys glareolus*)		Yes	Not cyclic Estrus induced by male	Single copulatory series	(46,47,394)
European pine vole (*Pitymys subterraneus*)	Solitary	Yes	Not cyclic Likely induced by male		(131,395)
California vole (*Microtus californicus*)	Solitary	No	Irregular cycles Estrus duration 5–21 d	Single copulatory series	(32,76)
Japanese field vole (*Microtus montebelli*)					(396)
Townsend's vole (*Microtus townsendii*)	Solitary	Yes	Not cyclic Estrus duration 1–19 d		(35)
Pine voles (*Microtus pinetorum*)	Social	Yes	Estrus induced by male		(48–50)
Russian vole (*Microtus rossiaemerionalis*)			Estrus induced by male		(51)
Prairie voles (*Microtus ochrogaster*)	Social	No	Not cyclic Estrus induced by male	Multiple ejaculatory series	(41–44,84, 397,398)
Short-tailed field vole (*Microtus agrestis*)	Solitary	Yes	Not cyclic constantly receptive	1 complete intromission: 92% 1 ejaculatory series: 100%	(38,50,75,106, 115,399–401)
Tree mouse (*Phenacomys longicaudus*)		Yes	Estrus duration ~6 d, but can last 3 wk if unmated		(3)
13-lined ground squirrel (*Citellus tridecemlineatus*)	Mostly solitary	Yes	Irregular cycles Estrus duration 2–4 wks followed sporadically by briefer periods	Single copulatory series	(18,402)
5-striped Indian palm squirrel (*Funambulus pennanti*)	Social	Yes	Estrus duration 3 d		(18)
Woodchuck (*Marmota monax*)	Solitary	Yes	Not cyclic Prolonged receptivity for 2–4 weeks		(192)
Rabbit (*Oryctolagus cuniculus*)	Social	No (weakly seasonal in some cases)	Irregular cycles, Estrus duration 2–15 d repeating every 4–7 d	Single copulatory series	(6,18,65,72,132, 229,389, 403–405)
Common Hare (*Lepus europaeus*)	Solitary	Yes	Estrus duration 14 d	Single copulatory series	(3,389)
Mink (*Mustela vison*)	Solitary	Yes	Not cyclic 3–4 wk of estrus	Single intromission	(68,96,100, 185,406)

TABLE 1. *Social, seasonal, and behavioral parameters affecting the induction of ovulation in reflex ovulators—cont'd*

Species	Social or solitary	Seasonal breeder	Estrus and female receptivity	Mating stimuli needed to induce ovulation	Refs.
Ferret (*Mustela furo*)	Solitary	Yes	Not cyclic estrus may last 4 mo if unmated	Single intromission	(1,3,69,70, 99,389)
Domestic cat (*Felis catus*)	Solitary (social in a few cases)	Yes	Cyclic Estrus duration 4–8 d	Multiple ejaculatory series (but single mating may work depending on day of estrus)	(88–90, 92–94,407)
Bears (family: *Ursidae*)	Solitary	Yes	Prolonged estrus duration (up to 1–2 mo)		(389,408)
Raccoons (*Procyon lotor*)	Usually solitary (social in some cases)	Yes	Estrus duration 2–3 d	1 copulatory series	(3,18,389)
Skunks (*Mephitius mephitis*)	Solitary	Yes	Prolonged estrus for up to 2 mo		(389)
Canada otter (*Lutra Canadensis*)	Solitary	Yes	Prolonged estrus duration up to 6 wk in absence of mating		(18,389)
Bactrian camels (*Camel bactrianus*)	Solitary or social	Yes	Not cyclic Estrus duration 1–7 d Intervals between estrus variable: 10–40 d		(104,389,409)
Dromedary camels (*Camelus dromedaries*)	Social	Yes	Irregular cycles Estrus duration 4–6 d May last 2–3 wk in absence of male	Single copulatory series	(77,78,103,389, 409,410)
Llama (*Lama glama*)		Yes	Not cyclic Prolonged periods of receptivity	Single copulatory series	(79,111,173, 411–413)
Alpaca (*Lama pacos*)	Social	Yes	Not cyclic Variable estrus duration 7–36 days	Single copulatory series	(79–81,409, 413,414)

extremes of a single physiological continuum will be discussed later in the section on Exceptions to the Rules: How Spontaneous Begets Induced Ovulation.

If induced ovulation is a primitive trait from which spontaneous ovulation later evolved, the former trait should be more prevalent in "primitive" (early evolved) mammals such as monotremes or marsupials, as well as in nonmammalian vertebrates related to mammalian ancestors (amphibians, reptiles, birds). In general, this hypothesis is not supported: review of the literature indicates that very few nonmammalian species, if any, have been shown to be induced ovulators, the possible exceptions being the red-sided garter snake and the whiptail lizard, the latter of which may be a facultative-induced ovulator (16,17). This suggests that induced ovulation evolved primarily in mammals after their evolutionary divergence from other vertebrates. While no studies addressing ovulation have been performed in Monotremata, in the primitive mammalian order Marsupialia, the majority of species studied have long been assumed to be spontaneous ovulators (18). However, this assumption is likely untrue, as recent data has implicated several marsupial species, most notably koalas and brush-tailed bettongs, as being reflex ovulators (19–23). Furthermore, previous reports that several other marsupials such as brush-tailed possums and the grey short-tailed possum were spontaneous ovulators have more recently been called into question in light of data suggesting that these species are induced ovulators (24–30). In view of this recent evidence, it may be necessary to re-examine the presence of induced ovulation in other marsupial species. One possibility may be that induced ovulation evolved in a subset of marsupial species, and the trait was maintained in subsequent eutherian species. Clearly, the occurrence of induced ovulation needs to be assessed in other "primitive" mammalian orders, such as Monotremata, Edentata, Philodata, and Dermoptera (Fig. 1). Currently, only a single species from these four orders has been studied (the nine-banded armadillo), and it appears to spontaneously ovulate (18,31). The dearth of information regarding ovulation in these primitive

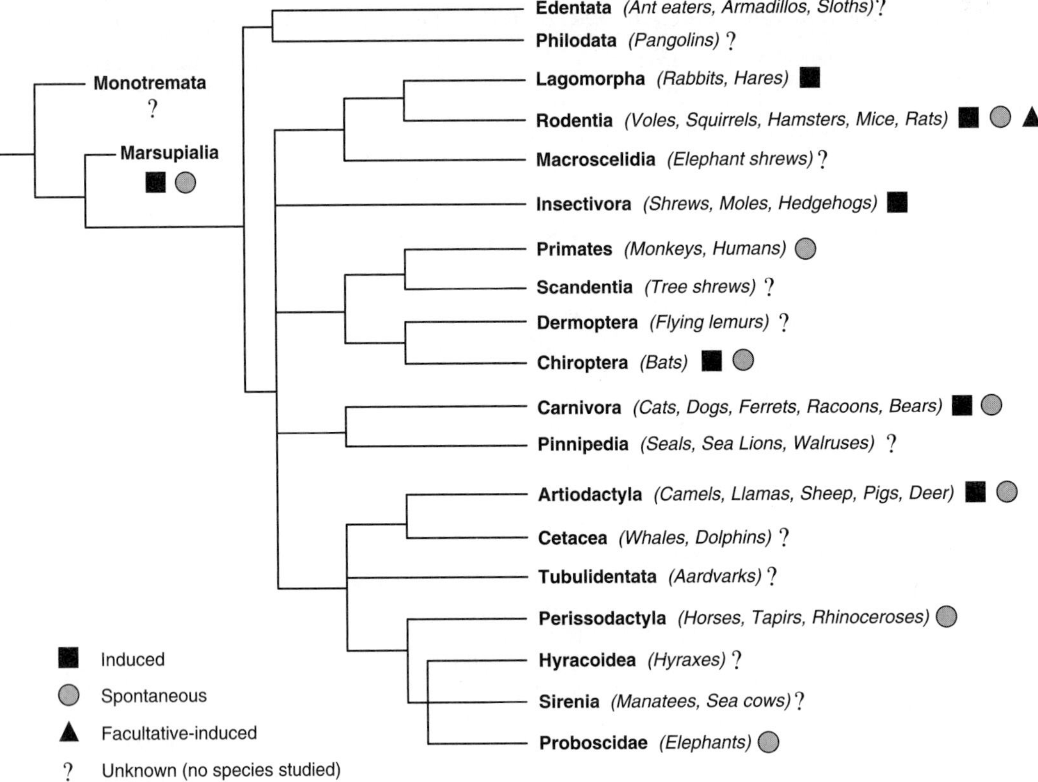

FIG. 1. Phylogenetic representation of the occurrence of mating-induced (■), facultative-induced (▲), and spontaneous (●) ovulation in mammalian orders. In several orders, species with either induced or spontaneous ovulation patterns have been found (Marsupialia, Rodentia, Chiroptera, Carnivora, and Artiodactyla). Two orders so far appear to contain only species with induced ovulation (Insectivora and Lagomorpha). Induced ovulation is absent from Primates, Perissodactyla and Proboscidae. In nearly half of all orders, it is not presently clear which ovulation subtypes exist due to lack of experimental evidence (represented by **?**): Monotremata, Edentata, Philodata, Macroscelidia, Scandentia, Dermoptera, Pinnipedia, Tubulidentata, Hyracoidea, and Sirenia. See text and Tables 1 and 2 for additional detail and references. [Figure adapted from (3).]

orders precludes any definitive conclusions regarding the evolution of the trait from being made. Once more data are available, the phylogenetic pattern of induced ovulation in mammals can be determined, thereby allowing the assessment of whether induced ovulation is a primitive trait. However, based on the current data, there does not appear to be any discernible phylogenetic pattern of distribution of ovulation type.

To further complicate the issue, both induced and spontaneous ovulation are present in many mammalian orders, including marsupials, rodents, carnivores, and artiodactyls, making it difficult to determine which is the more primitive ancestral trait. It has been suggested, and is likely, that induced ovulation evolved multiple times within mammals; the presence of this trait in various species may therefore reflect one or more unique environmental or social selection pressures. Greenwald (32) was one of the first to suggest that induced ovulation might be most valuable to species that regularly undergo marked population fluctuations, such as California voles (*Microtus californicus*). For such animals, reflex ovulation would greatly enhance the possibility that when a male encounters a female, she will be ready to immediately ovulate and become pregnant, thereby promoting a rapid recovery from low population levels. In contrast to spontaneous ovulators, the mating behavior and rate of successful fertilization for induced ovulators would therefore not be dictated by events intrinsic to the female (i.e., hormonal and behavioral estrous cycles) that could potentially render females nonreceptive for long periods of time. This population cycles hypothesis for explaining the adaptive value of reflex ovulation was echoed by several other researchers studying mating-induced ovulation in other rodents (33–35).

An additional corollary hypothesis presented by Zarrow and Clark (8) suggested that induced ovulation would be particularly useful in solitary, nongregarious animals in which encounters with members of the opposite sex were infrequent and unpredictable. The adaptive value of induced ovulation would therefore be related to population density, not just for species that normally exhibit cyclic fluctuations in population size, but also for asocial or dispersed animals that are solitary for most of the time. Thus, for animals with solitary social systems, such as the Townsend's vole (*Microtus townsendii*) and the meadow vole (*Microtus pennsylvanicus*), mating-induced ovulation would ensure that ovulation and subsequent fertilization always occur when a male encounters a female, which happens infrequently and unpredictably (35,36). While this hypothesis applies to many induced ovulators, it is not universal and does not generalize to all species. Most notably, there are several species of reflex ovulators commonly known to live in groups in which sexual partners are probably available most of the time (6). For example, pine voles and prairie voles are reportedly group-living species which display induced ovulation, as are rabbits. Furthermore, solitary species exist that are spontaneous ovulators (for example, porcupines). Even so, review of the literature yields an interesting pattern in which a large majority of induced ovulators (~70% of species for which data is available) are primarily solitary animals with minimal group-living (Table 1).

Focusing specifically on carnivores, Lariviere and Ferguson (37) performed a meta-analysis to test several hypotheses related to the evolution of induced ovulation. In addition to a trend for induced ovulators to exhibit exaggerated sexual dimorphism and larger home ranges, this study found strong correlations between induced ovulation and several life-history traits, including seasonality and mating system; carnivores that were induced ovulators were more likely to be seasonal breeders and to exhibit primarily multi-male (polygynous) mating systems whereas spontaneous ovulators were less seasonal and were either monogamous, polygynous, or polygamous, in similar proportions (37). The authors concluded that induced ovulation in carnivores evolved through sexual selection as a reproductive strategy that allows males the assurance of egg fertilization during short periods of pair bonding and benefits females via post-copulatory mate choice (based on the degree of copulatory stimulation necessary to induce ovulation and a functional corpora luteum) (37). It is unclear whether similar correlations between ovulation type and life-history traits exists in other mammalian orders, though it is likely they are not universal: rabbits, least shrews, and prairie voles are not seasonal breeders (interestingly, they are also not solitary). Even so, review of the literature indicates that at least 75% of induced ovulators are seasonal breeders (Table 1). It appears likely that induced ovulation evolved multiple times, not necessarily under the same selection pressures (ecological, social, or sexual). Thus, for different reasons, the same trait (induced ovulation) may have developed independently several times as an adaptive solution to different challenges or environmental pressures. If so, it is possible that the regulatory mechanisms underlying induced ovulation may have evolved differently in various species that exhibit the trait; for this reason, detailed species comparisons are a requisite for studying any mechanistic aspect of this biological phenomenon.

STIMULI REQUIREMENTS FOR INDUCING THE LH SURGE AND OVULATION IN REFLEX OVULATORS

As Sawrey and Dewsbury (38) noted, labeling a species an induced ovulator tells very little of the total story regarding its reproductive physiology and behavior; for example, even in closely related species, variability exists in the nature and timing of the stimuli able to induce ovulation. In all reflex ovulators, ovulation induction is primarily contingent upon tactile cues, specifically vagino-cervical stimulation derived from coitus; however, as discussed in this section, there is considerable species-to-species variation in how much stimulation is actually needed to evoke ovulation (9,38). Similarly, there is wide variation among species in the latency for ovulation to occur after mating, ranging from several hours in bank voles to almost 2 days in camels and mink (Fig. 2). Lastly, nontactile stimuli associated with mating, such as olfactory, pheromonal, or auditory cues, may facilitate the occurrence of mating-induced ovulation in some reflex ovulators but not others.

Behavioral Estrus of Reflex Ovulators

The stimuli for mating-induced ovulation must be studied in context, i.e., the female must copulate in order to ovulate. Whether or not a female engages in sexual activity is dependent, in almost all cases, on her state of estrus and behavioral receptivity. Therefore, it is necessary to discuss the cues, both internal and external, that underlie behavioral estrus in female reflex ovulators.

There is considerable variability in the duration and periodicity of behavioral estrus in reflex ovulators (Table 1) (3). Some species, such as cats and Bactrian camels, exhibit well-defined periodic cycles of female receptivity lasting several days, while other species, such as ferrets and California voles, will stay constantly receptive for weeks or months, often displaying "spontaneous" periods of estrus lasting

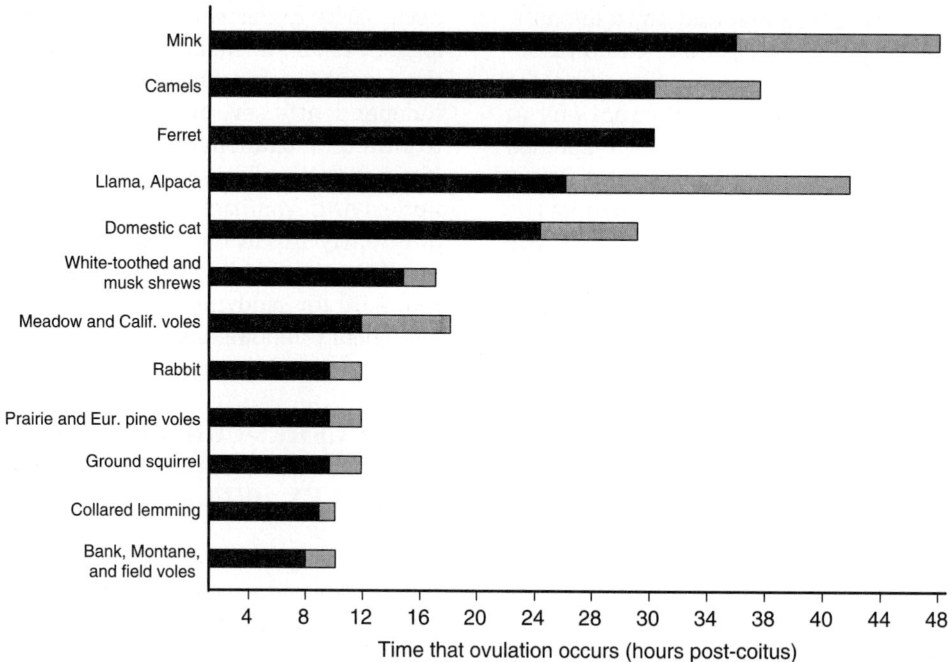

FIG. 2. The timing of ovulation relative to the onset of mating in several species that exhibit reflex ovulation. The *black bars* represent the total time from mating to the earliest reported onset of ovulation. The *gray bars* represent the reported variability in the timing of ovulation. The species are ordered by latency to ovulate after mating, from most rapid (**bottom**) to slowest (**top**). See Table 1 for references.

long durations and recurring at irregular intervals. In addition, other species such as musk shrews or pine voles will never display sexual receptivity unless they are first exposed to signals from a conspecific male; in the absence of male cues, females of these species are permanently anestrus and remain sexually unreceptive.

The duration and occurrence of estrus periods are strongly influenced by social and environmental cues (3). Many induced ovulators are seasonal breeders; thus, seasonal changes in photoperiod dictate the display of female receptivity, restricting it to specific times of the year when reproduction would be most favorable (based on environmental availability of energy and resources). For example, both the 13-lined ground squirrel (*Citellus tridecemlineatus*) and the ferret (*Mustela furo*) are seasonal breeders, curtailing reproduction during the short days of fall and winter and re-activating it in the spring and summer when food is most plentiful. As discussed in the section on Evolution, Adaptive Function, and Phylogenetic Distribution of Mating-Induced Ovulation, seasonality appears to be one of the strongest correlates of induced ovulation. Lariviere and Ferguson (37) found strong correlations between induced ovulation and seasonality in the order Carnivora, with species that are induced ovulators more likely to be seasonal than those that are spontaneous ovulators. This pattern extends to other mammalian orders as well; while not all reflex ovulators are seasonal breeders, more than 75% of the species for which data is available display some degree of seasonality in their reproduction (Table 1).

In some species of reflex ovulators, social cues affect a female's receptive state. The ability of social stimuli, most notably pheromonal, olfactory, and tactile cues of conspecific males, to decrease a female's latency to display estrus, or more often, to actually induce estrus, has been documented in many species of insectivores and rodents. For example, female musk shrews (*Suncus murinus*) do not exhibit regularly defined periods of sexual receptivity; instead, direct exposure to a male for 30 to 60 minutes triggers a sudden onset of receptivity at which point the female stops responding aggressively to the male and begins to display receptive behaviors (39,40). Similar to musk shrews, prairie voles (*Microtus ochrogaster*) have estrus which is induced by males (41,42). In reproductively naive female prairie voles, direct contact with male urine or male-soiled bedding, in the absence of physical contact, results in increased uterine weights, though it does not reliably elicit behavioral estrus (defined by lordosis) (42). However, when female voles are physically exposed to a male for 18 hours and then tested 6 hours later, 70% of females show lordosis, suggesting that direct contact with a male may be necessary to induce and maintain behavioral estrus in this species (42). Additional studies have determined that ovarian estradiol levels, uterine weights, uterine protein levels, and high-affinity estrogen binding sites in the brain all increase in virgin prairie voles after direct exposure to a male (43–45). Other vole species exhibiting male-induced estrus include pine voles, bank voles, and Russian voles (46–51) (Table 1).

In most spontaneously ovulating species, such as rats, mice, and primates, behavioral receptivity is promoted by the actions of endogenous estrogens and progesterone (Pg) (see Chapter 34). For example, ovariectomized mice and rats will not display receptive sexual behaviors even in the presence of an actively mounting male; however, these females will become receptive and engage in the full complement of copulatory behaviors following a timed regimen of exogenous estrogen and Pg administration (usually given 48 hr and 4 hr, respectively, prior to male exposure). In most cases, Pg is either necessary or strongly facilitory in inducing this behavioral receptivity. In contrast, though binding sites for Pg have been reported in several brain regions of induced ovulators (3,52,53), there is little evidence that Pg is needed for the display of sexual receptivity in induced ovulators. Administration of exogenous estradiol to ovariectomized ferrets, cats, rabbits, and reflexively-ovulating rodents induces maximal sexual receptivity, even in the absence of concurrent Pg treatment (3,54–58). Furthermore, Pg administered to estrogen-primed induced ovulators either had no stimulatory effect on female receptivity or in some cases significantly decreased sexual behaviors (52,58–61). Likewise, in some reflex ovulators such as musk shrews, female receptivity occurs in the presence of low circulating gonadal hormones, precluding a necessary role for Pg in promoting estrus (61–63). Collectively, these findings suggest that many induced ovulators do not require Pg to be sexually receptive; however, as noted by Bakker and Baum (3), this lack of facilitory effect by Pg is not exclusive to reflex ovulators, as female rhesus monkeys do not show increased sexual behavior following Pg treatment (64). Furthermore, many induced ovulators have not been tested, and the role of Pg should be examined in additional species before any definitive, and more generalizable, conclusions can be made.

Exogenous Stimuli Requirements for Inducing the LH Surge and Ovulation

A female that copulates is subject to stimuli in several modalities, any number or combination of which might potentially facilitate or induce ovulation. For example, in the process of mating, a female is exposed to not only physical (tactile) stimuli from the male (both genital and somatic), but also to olfactory, pheromonal, visual, and auditory cues, many of which are received concurrently. Despite the abundance of cues, it appears that tactile stimulation of the vagina/cervix during coitus is the primary inducer of ovulation, though as discussed later, other cues often augment the ovulatory response of the female to coitus.

Although genital stimulation is the main exogenous cue that evokes ovulation, there is wide variation among induced ovulators in the minimal copulatory stimuli necessary to achieve this response (Table 1). The degree of stimulation ranges from a single, rapid intromission in rabbits to multiple matings and ejaculatory series over multiple days in musk shrews (6,65–67). Though the amount of copulatory stimuli needed to induce ovulation may depend on a combination of physiological, ecological, social, or phylogenetic factors, there is currently no clear understanding as to why some species require more than others. For example, neither body size nor social system is consistently predictive of how much coital stimulation a given species requires: short-tailed field voles and prairie voles are two closely related species of similar body size and yet the former ovulates after a single intromission whereas the latter ovulates only after repeated ejaculatory series. Likewise, ferrets, musk shrews, and meadow voles are all solitary (non-gregarious) species but the latter two species ovulate only after multiple ejaculatory series whereas ferrets can ovulate after a single, partial intromission.

In general, induced ovulators can be divided into three loose categories of required coital stimuli. The first category includes species which require minimal copulatory stimulation and in which ovulation can be induced rapidly (within minutes or less). In these animals, multiple intromissions and an ejaculation are oftentimes not required: interrupted coitus or even a single intromission can usually elicit maximal ovulatory rates. Animals falling into this category include mustelids, lagomorphs, and the short-tailed field vole (*Microtus agrestis*) (Table 1). Both minks and ferrets normally copulate for an hour or two, but interrupted coitus lasting only 5 minutes is enough to induce ovulation (68,69). In fact, Carroll et al. (70) reported that a single intromission lasting just 1 minute could induce an ovulatory LH surge in female ferrets; the same effect was not observed when females were mounted and neck-gripped without intromission, indicating that the activation of the LH-releasing pathway was specific to vagino-cervical stimulation. The fastest copulatory time-course to promote ovulation has been reported for rabbits, in which a typical ovulation-inducing copulatory series can last less than a minute (oftentimes, a single intromission followed by immediate ejaculation) and has been observed in several cases to occur as fast as

1 second or less (6,65,71–74). In the short-tailed field vole, an interrupted intromission lasting less than 10 seconds is sufficient to evoke ovulation in only 25% of the animals whereas a complete intromission lasting 22 seconds induces ovulation in 92% of females (75).

The second category of stimuli requirements for inducing ovulation is comprised of species that require, at minimum, a single complete copulatory series, usually consisting of multiple intromissions followed by an ejaculation. This group includes several rodent species, raccoons, camels, and hares (Table 1). In bank voles (*Clythriomomys glareolus*), limited matings of 1 to 3 intromissions results in ovulation in ~83% of animals; a complete copulatory series induces ovulation in all females (46). Likewise, in California voles (*Microtus californicus*), a single ejaculation series lasts 6 minutes on average and results in 100% ovulation rates (76). Similarly, most members of the Camildae ovulate after a single complete mating. In Dromedary camels (*Camelus dromedaries*) 70% or more release LH and ovulate after a single copulatory series lasting 10 to 15 minutes (77,78). Similarly, 100% of llamas (*Lama glama*) ovulate after a single mating, which can range between 10 to 50 minutes (79,80). In alpacas (*Lama pacos*), a single mating normally lasting 25 to 35 minutes results in 82% to 100% of females ovulating, whereas interrupted coitus lasting only 5 minutes results in 60% or less ovulating (79,81). Similar findings for a single, short-duration copulatory series inducing ovulation have been reported for 13-lined ground squirrels (*Citellis tridecemlinatus*) and raccoons (*Procyon lotor*) (3,18,82).

The third category of stimuli requirements for ovulation induction includes species that require multiple ejaculatory series, occurring either consecutively on the same day or over the course of several days. This group includes several species of vole, all shrews, and most probably cats (Table 1). In meadow voles (*Microtus Pennsylvanicus*), Montane voles (*M. montanus*), and prairie voles (*M. ochrogaster*), increasing copulatory stimuli associated with increasing numbers of ejaculatory series results in elevated rates of ovulation. For example, 75% of meadow voles ovulate after one complete copulatory series, whereas 100% ovulate after multiple series (83). For Montane voles, ovulation rates in females increase from 25% after one ejaculatory series to 75% after two coital series; only with three or more consecutive matings do ovulations rate approach 100%. Though Gray et al. (84) reported that most (~90%) female prairie voles ovulate after only one ejaculatory series, Carter and colleagues (43,85) consistently found that two ejaculatory series (lasting ~30 minutes) were not sufficient to induce ovulation in prairie voles; in contrast, more than 80% of females allowed *ad-libitum* mating for 24 hours ovulated. Carter noted that pairs in this species normally mate intermittently for 25 to 30 hours, suggesting that longer durations of copulation are naturally required to achieve ovulation (43). This conjecture was supported by Roberts et al. (86) who found that only 38% of female prairie voles mating for less than 10 hours ovulate, whereas 100% of females mating for more than 12 hours ovulated.

Most, if not all, species of shrews require multiple copulatory series to induce ovulation. While females of both short-tailed shrews (*Blarina brevicauda*) and least shrews (*Cryptotis parva*) have been reported to require multiple ejaculations to ovulate (3,18), the most data on this topic exist for the Asian musk shrew (*Suncus murinus*). For this species, not only do females often require multiple copulatory series to ovulate, but in most cases, females (especially virgins) need to experience multiple copulations on multiple days. While one ejaculatory series can sometimes induce ovulation in virgin females, the percentage of females who show this is fairly low (ranging from 9%–39%) (66,67,87). Likewise, a single 4-hour mating duration (encompassing three or fewer ejaculation series) resulted in only 25% to 50% of virgin female musk shrews ovulating, whereas 75% to 80% ovulated after 3 or more consecutive days of mating (66,67). Interestingly, multiple copulatory bouts separated by 24 days also enhanced ovulation in shrews (66). These findings indicate that prior sexual experience increases ovulation rates, and that initial matings prime and/or permanently change the neuroendocrine and neural mechanisms underlying reflex ovulation for future sexual encounters. These priming effects of virgin matings require coital stimulation and are not simply due to increased male exposure: only 20% to 29% of virgin females given nonsexual male contact or exposure to male urine ovulated when later paired with a male (in contrast to 82% of sexually experienced females) (66).

The majority of data suggest that cats, like shrews, require multiple ejaculatory series to ovulate, although this is open to some debate. Wildt et al. (88) reported that 25% to 33% of female cats ovulate after a single mating, whereas 83% ovulated when allowed to mate 3 times on a single day. Similarly, Concannon et al. (89) found that a single copulation induced ovulation in 50% of female cats whereas multiple copulatory series resulted in 100% ovulation rate. Furthermore, multiple matings on a single day resulted in higher LH release than a single mating (90,91), with more than 80% of the multiply-mated females exhibiting a pre-ovulatory LH surge (92). However, both Johnson and Gay (90) and Tsutsui and Stabenfeldt (93) reported high ovulatory rates (~100%) in cats receiving only a single mating series.

These authors, along with Banks and Stabenfeldt (94), speculated that female cats, which display behavioral estrus for a duration of 4 to 6 days, are more likely to ovulate in response to a single copulation if it occurs on day 3 or 4 of estrus. This may be due, in part, to the level of estrogen secretion, which peaks at this time. Though not tested, increased estrogen exposure over the course of estrus could increase a female's sensitivity to somatic tactile stimuli as well as pituitary sensitivity to GnRH, either of which could facilitate ovulation induction (94). Regardless of the underlying mechanism, it appears that the day of estrus influences the ability of a single copulatory series to induce ovulation in cats, but that ovulation can occur on any estrus day provided there is "extra" stimulation via multiple copulations. Whether or not this extra copulatory stimulation acts to increase estrogen secretion (thereby advancing the stage of estrus) is unknown and remains to be tested.

Several intriguing questions arise from the study of required copulatory stimuli in reflex ovulators, all of which have yet to be addressed. In species requiring multiple copulatory series, how do the secretion and binding of regulatory hormones and neurotransmitters change with increasing stimulation and successive mating bouts? For example, how does the copulatory-induced release of GnRH, prolactin, or Pg (see Endocrine Correlates of Coitus-Induced Ovulation: Hypothalamic, Pituitary, and Ovarian Hormone Secretion) vary between the first and last ejaculatory series in Montane voles or musk shrews? Furthermore, what is the nature of the copulatory signal that finally triggers ovulation in such animals requiring multiple copulatory series? Determining answers to these issues may help elucidate the regulatory pathways and physiological and molecular mechanisms underling this pattern of ovulation.

Interestingly, most of the species in the category 3 (multiple copulatory series) exhibit behavioral estrus which is not spontaneous or cyclical, but rather which is induced by the male. Conversely, most of the species that ovulate after just one copulatory series or after a single intromission (categories 1 and 2) tend to be acyclic (or weakly cyclic) with variable durations of "spontaneous" estrus that is not contingent upon male exposure (Table 1). Thus, the animals that require the most coital stimuli to ovulate (shrews, most voles) are those that also require males to activate their sexual receptivity (i.e., induce estrus behavior), whereas those species that ovulate with the least amount of copulatory stimulation (rabbits, ferrets) are those that are already in estrus before interacting with a male. While there are certainly exceptions to this observation (cats, short-tailed shrews), it is an interesting trend that could benefit from additional data and future study.

There is currently no information available on ovulation-inducing mating requirements for many reflex ovulators, including lemmings, woodchucks, numerous vole and insectivore species, and several of the larger carnivores (bears, skunks, lions, etc). Additionally, there is limited data for many species, making their placement into one of the three categories tentative. For example, it is unknown whether species that typically ovulate after one copulatory series, such as California voles or camels, will still ovulate after interrupted copulations (single intromissions) as do ferrets and minks. Furthermore, it is unclear if species shown to ovulate after multiple ejaculatory series will still do so after a single series depending on the day of mating in relation to estrus (as in cats). Future studies should resolve these issues to better determine how different degrees of coital stimulation initiate the induction of ovulation and how the neuroendocrine pathways controlling ovulation are themselves regulated by the act of copulation.

Nontactile Cues Influencing Ovulation in Reflex Ovulators

Though tactile stimulation received during coitus appears to be the driving force behind reflex ovulation, several lines of evidence indicate that other factors are often required. Using artificial vaginal-cervical stimulation (AVS) with glass rods or vibrating probes, ovulation has been induced in several reflex ovulators including minks, ferrets, bank voles, rabbits, and cats (6,95–99). However, in most cases excluding cats, this form of tactile stimulation resulted in lower rates of ovulation and LH release than normally achieved with natural copulation (18,46,99,100). For, example, only 44% of female bank voles ovulated when given ten 1-minute pulses of AVS designed to mimic a full copulatory series (46); in contrast, similar stimulation paired afterwards with natural copulation resulted in 90% ovulatory rate. In rabbits, the effect of AVS is usually only seen in estrogen-primed females or intact females that are in high estrus conditions (6,101,102); additionally, Ramirez and Soufi (6) reported that only 60% of females given AVS responded with increased GnRH release and of these, none ovulated. Other studies in camels (78,103,104), musk shrews (E. F. Rissman, unpublished observations), and several species of voles (32,75,105,106) have found no ability of AVS to induce ovulation.

Of the 11 induced ovulating species in which AVS data is available, 63% did not ovulate when AVS was given alone. Collectively, these findings indicate that although vagino-cervical stimulation is critical to inducing ovulation, additional stimuli/cues are often also necessary to facilitate the ovulatory response.

Such stimuli normally experienced during copulation include olfactory and pheromonal cues, visual or auditory stimuli, or tactile stimulation of nongenital regions (such as neck- or flank-gripping). For example, in ferrets, ovulation was only induced by AVS when the female was simultaneously neck-gripped by a conspecific male (99). Of additional consideration is the possibility that females may be selectively responsive to certain parameters of genital stimulation and that the AVS is not of the proper frequency, duration, or pressure to induce ovulation.

Olfactory or pheromonal cues may be important in facilitating ovulation in some induced ovulators. Female musk shrews require an intact olfactory system to ovulate in response to mating; female shrews subjected to olfactory bulbectomy ovulated far less than females receiving a sham surgery (10% vs 64%), even though both groups exhibited comparable female receptivity and mating behavior. This indicates that olfactory detection, in addition to coital stimuli, is critical for mating-induced ovulation in the musk shrew (107). This is not the case, however, for all induced ovulators. Meadow voles that sustain lesions of the vomeronasal organ still mate and ovulate, while similar findings have been reported for ferrets and rabbits (36,108,109); intriguingly, none of these three species exhibit male-induced estrus whereas the female musk shrew's sexual receptivity is activated by the male. The role of the vomeronasal system in affecting mating-induced ovulation may therefore be related to the degree to which activation of a female's reproductive system (estrus and receptivity) is dependent on the presence of a male. This issue should be addressed by examining other species of reflex ovulators in which the role of vomeronasal and olfactory systems have yet to be studied. In addition, *cfos* studies examining the post-coital activation of various sensory regions of the brain, such as the main and accessory olfactory bulbs, would be useful in elucidating the facilitory role of nontactile cues in ovulation.

Several studies have reported high ovulation rates in several members of the Camelidae following artificial insemination (80,104,110,111). The authors of these studies conclude that semen itself, in the absence of coital stimulation, can induce ovulation in these animals. In Bactrian camels, 87% of females ovulated when inseminated with semen and 75% ovulated when given just seminal fluid (104). Another study found that 100% of Bactrian females ovulated and exhibited LH surges following semen administration into the vagina (110). Although it is unclear how much incidental AVS these females received while being inseminated, this is unlikely to account for the ovulations as "control" inseminations (milk) were ineffective and Bactrian camels do not normally ovulate in response to AVS (104). Similar findings of semen-induced ovulation have been reported for llamas (80). However, other studies have reported no ovulatory effect of semen in the closely related alpaca or Dromedary camel (78,81). If ovulations in llamas and Bactrian camels are truly induced by semen, it raises several questions: (a) What is the "ovulating factor" in the semen? (b) How is the chemical signal present in semen transduced into a neural and/or neuroendocrine signal that eventually triggers pre-ovulatory GnRH release? (c) Do other species typically classified as "induced ovulators" also ovulate in response to semen per se rather than, or in addition to, copulatory stimulation?

Interestingly, infrequent *spontaneous* ovulation has been reported for several induced ovulators, including mink, meadow voles, Montane voles, short-tailed field voles, llamas, alpacas, and cats. The percentage of females reportedly exhibiting spontaneous ovulations typically ranges from 5% to 16% (81, 111–116), and in the case of cats, as high as 67% (91,117). However, the validity of such "spontaneous" ovulations is questionable. In cats and minks, ovulations have been reported for group-housed females or females given nonsexual contact with males (68,91, 117,118). Similarly, "spontaneous" ovulations were observed in several vole species in response to exposure to male bedding or nonsexual contact with males (112,113,115,119). In each case, it is unclear how much stimulation (tactile, olfactory, pheromonal, or otherwise) each female received from the males or co-housed females. In the case of the cat, it should be noted that ovulation was assessed indirectly based on faecal Pg concentrations (91,117). However, if the females in each of the above cases really did not engage in copulation, it suggests that noncoital cues can, in rare instances, induce ovulation by themselves. The mechanism underlying such nontactile induction of ovulation is unclear but could possibly be related to increased estrogen secretion in already-primed females. Future studies are required to tease apart the modulatory roles of tactile, olfactory, pheromonal, and other cues and to determine how such cues are physiologically transduced and subsequently integrated in the brain.

ENDOCRINE CORRELATES OF COITUS-INDUCED OVULATION: HYPOTHALAMIC, PITUITARY, AND OVARIAN HORMONE SECRETION

Mechanistically, the fundamental difference between induced and spontaneous ovulators is the ability of the latter to repeatedly generate a surge in LH release from the pituitary in the absence of mating or environmental cues, hence "spontaneous."

TABLE 2. *Hormonal profiles and latencies to ovulation after coitus in induced ovulators*

Species	Onset of ovulation	GnRH profile	LH profile	Progesterone (PG) and prolactin (PRL)	Refs.
Short-tailed shrew (*Blarina brevicauda*)	55–71 h				(3)
White-toothed shrew (*Crocidura russula monacha*)	15–17 h				(3)
Musk shrew (*Suncus murinus*)	14–17 h	Increased within1 h		PG: no change for 18 h then begin to increase	(415) (390) (416) (201) (417) (151)
Collared lemmings (*Dicrostonyx groenlandicus*)	9 h				(33)
Meadow voles (*M. Pennsylvanicus*)	12–18 h		Increased by 30 min Elevated for 1 h	PRL: increased at 24 h and 48 h	(418) (177)
Montane vole (*M. montanus*)	8–10 h		Increased at 60 min	PG: increased at 1 h	(419) (178)
Bank vole (*Clthriomomys glareolus*)	6–14 h				(394)
European pine vole (*Pitymys subterraneus*)	10 h				(131)
California vole (*M. californicus*)	15–18 h				(32)
Japanese field vole (*M. montebelli*)	9 h				(396)
Prairie voles (*M. ochrogaster*)	10–11 h 10–12 h			PG: low at 24 h, increased by 3 d	(41) (86) (43)
Short-tailed field vole (*M. agrestis*)	8–10 h 9–12	Increased within 10 min	Increased 5–10 min Peaks 20–40 min Elevated for 90 min		(399) (3) (145) (126) (174) (3) (82)
13-lined ground squirrel (*Citellus tridecemlineatus*)	10–12 h				
5-striped Indian palm squirrel (*Funambulus pennanti*)	24 h				(3)
Woodchuck (*Marmota monax*)	20–32 h			PG: increased at 20–36 h	(192)
Rabbit (*Oryctolagus cuniculus*)	10–12 h	Increased 7.5 min Peaks 10–30 min Elevated for 2–3 h	Increased within 3–10 min Peaks 60–120 min Elevated for 4–6 h	PG: increased within 5–60 min. Elevated for 5 h PRL: immediate decrease after mating? Rise 30–90 min Elevated for 5 h	(2,241) (142) (143) (140) (133) (157) (158) (159) (160) (162) (261) (199) (155,195)
Hare (*Lepus europaeus*)	<24 h				(162) (180) (3)
Mink (*Mustela vison*)	36–52 h		Increased by 16 min Elevated for 12+ h	PRL: increases 5 d	(96) (68, 169,170, 406,420) (185) (187) (118, 186)
Ferret (*Mustela furo*)	30 h	Massive "immediate" release in vitro	Increased by 15–60min Peaks 4–6 h Elevated for 12+ h	PG: increased beginning at 5 d	(3) (70) (123) (421) (144) (188)
Domestic cat (*Felis catus*)	24–29 h		Increased by 5–10 min Peaks 20–120 min (variable) Elevated for 4–10 h	PG: increased 56–72 h	(167) (94) (88) (89) (90) (92)
Bactrian camels (*Camel bactrianus*)	30–36 h		Increased at 4 h Elevated for 8 h	PG: increased at 3–5 d	(110)
Dromedary (*Camelus dromedaries*)	24–48 h		Increased by 60 min Peaks 2–4 h Elevated 9–13 h		(3) (103) (77)
Llama (*lama glama*)	26–42 h		Increased by 30–60 min Peaks 2–3 h Elevated for 5–6 h		(79) (173) (413) (114)
Alpaca (*lama pacos*)	26–30 h		Increased by 60–90 min Peaks 3–4 h Elevated for 5–6 h		(79) (80) (173,414)

However, while it is not cyclical or endogenously cued, the critical signal in inducing ovulation in reflex ovulators is also a massive "ovulatory" surge in LH. This LH surge is itself assumed to be induced by a sudden and massive release of GnRH, which, in turn, is triggered via neural and hormonal pathways activated by mating. The precise temporal patterns of hormonal secretions activated by mating have been studied in detail for a wide range of induced ovulators from almost all representative orders. This section reviews and compares some of the findings on this topic from multiple species of reflex ovulators.

Gonadotropin-Releasing Hormone (GnRH)

The pattern and timing of GnRH release and/or neuronal responses to mating has been studied in only a few induced ovulators including rabbits, ferrets, musk shrews, and the short-tailed field vole. The relatively low number of species for which GnRH data are available owes primarily to the technical difficulty in directly measuring plasma GnRH in the hypophyseal portal system. Unlike LH or ovarian steroids, GnRH is not secreted into the systemic circulation and cannot easily be determined by sampling peripheral plasma. In fact, to date, direct sampling of plasma GnRH from the portal bloodstream has not been accomplished for any induced ovulator. Rather, all the data existing for GnRH release has come from immunocytochemical, push-pull perfusion (PPP), and radio-immuno assay (RIA) analyses of forebrain regions containing GnRH neurons. Though there is no direct experimental evidence that GnRH is released into the portal blood system upon mating, the likelihood of its occurrence is well supported by circumstantial data (6). Synthetic GnRH peptides administered intracerebroventricularly (icv) or intraperitoneally can stimulate the release of LH in several reflex ovulators (78,79,120–126) as well as induce ovulation (122,127–131). Furthermore, one study in rabbits showed that inhibitory antagonists of GnRH given 30 minutes prior to mating reduce both the coitally induced LH surge and ovulation (125).

Much of the data on coitally induced GnRH release exists for rabbits, in which the development and use of the PPP technique has allowed investigators to accurately and precisely measure hypothalamic GnRH secretion in vivo in conscious, freely moving females (6,132). The research laboratories of Ramirez and Spies have both used PPP in rabbits to demonstrate an in vivo pulsatile baseline GnRH release, similar to that observed in spontaneous ovulating species such as the rat, monkey, and sheep (132–139). Furthermore, both Spies' and Ramirez's groups have examined the pattern of GnRH release in female rabbits following copulation. Kaynard et al. (140) determined that GnRH secretion from the medial basal hypothalamus (MBH) and anterior hypothalamus (AH) was increased 10 to 30 minutes following copulation, peaking after 30 to 90 minutes. The increase in GnRH release induced by coitus was approximately 50 to 100 times that of the basal levels and did not return to baseline levels for more than 2 to 3 hours (140). Greater increases in GnRH release were found when the push-pull cannula was placed in the tuberal region of the median eminence than in the basal AH (140). Lin and Ramirez (141) reported similar effects of mating on GnRH release, with hypothalamic secretion of GnRH from the median eminence increased within 10 minutes of copulation. More recently, Yang et al. (142) used microdialysis to show that GnRH is released from the arcuate nucleus-median eminence as quickly as 7.5 minutes after mating and peaks approximately 20 minutes post-coitus; as in other studies (140,143), GnRH levels remained high (30x) for several hours before returning to low baseline levels.

Data regarding GnRH release in induced ovulators other than rabbits is not as abundant, though several studies in ferrets and shrews have attempted to analyze the time-course of changes in GnRH-neuron number and content following mating. Investigators in Baum's laboratory have studied coitus-induced changes in GnRH release from ferrets using in vitro perifusion analysis. Lambert et al. (144) collected individual slices of the MBH from mated female ferrets and determined that the in vitro release and content of GnRH from these slices was significantly reduced 15 minutes after a single intromission. These MBH GnRH levels remained low for several hours post-coitus and then increased again (144). This finding indicates that a single intromission causes a release of GnRH from the female ferret's MBH which thereby rapidly depletes neuronal terminals of the peptide (3). A similar outcome was reported for short-tailed field voles in which a 56% decrease in hypothalamic GnRH peptide content was observed just 5 to 10 minutes after initiation of mating (145). Further evidence supporting a rapid time course of GnRH release post-coitus comes from a study in ferrets demonstrating a 50% decrease in the number of hypothalamic GnRH-immunoreactive neurons 75 minutes after AVS given in the presence of a neck-gripping male (99).

Additional studies have used expression of the immediate early gene *cfos* as a marker of neuronal activation in GnRH neurons after mating. In ferrets, a single intromission lasting 5 minutes caused an increase in the percentage of GnRH cells in regions of the MBH (including arcuate nucleus) and posterior hypothalamus that express Fos protein when analyzed 60 to 90 minutes post-coitus (108,146,147); the percentage of Fos-expressing GnRH cells returned to basal levels less than 3 hours post-coitus, indicating an acute, short-term increase of Fos activity. Because translation of the Fos protein often takes 70 to 90 minutes to reach maximal levels (148,149), the presence of high Fos content 90 minutes after mating indicates that this transcription factor, and thus GnRH neurons, are likely activated immediately (within minutes) after a single intromission. It has been speculated that increased Fos activity in these neurons may serve to increase GnRH gene transcription, thereby replenishing intracellular MBH GnRH stores within the first hour after intromission when GnRH is released (3). Similar findings have been reported for rabbits, in which *cfos* mRNA and Fos protein expression in hypothalamic GnRH neurons are enhanced after mating within 30 and 90 minutes, respectively (150).

Studies in musk shrews have yielded contrasting results to those in ferrets and rabbits. The present data suggest that in shrews, a single ejaculatory series comprising mounts, intromissions, and ejaculation (taking a duration of 30–75 minutes) does not result in a rapid, sustained increase in GnRH release. Specifically, females sacrificed 1 hour after mating show no change in GnRH content in the ventral forebrain (including the preoptic area [POA]) or olfactory regions (including the tenia tecta); however, 8 to 15 hours after receiving an ejaculation, females have greatly increased GnRH content in both regions (151). Furthermore, females have increased proGnRH-immunoreactive (ir) cell numbers in the ventral forebrain 1 hour after mating, indicating increased synthesis of the GnRH precursor (152). One possible explanation for these findings is that if GnRH stores are depleted quickly within the first several minutes after mating (as in other reflex ovulators), the "lack of change" in GnRH peptide content observed at 1 hour post-coitus actually represents a restoration back to pre-mating baseline levels. Thus, the changes in GnRH content over time may be due to a rapid, long-term replenishing and "re-stocking" of the peptide, as evidenced by the increase in pro-GnRH peptide synthesis. The higher than baseline levels observed at 15 hours post-mating could represent a priming mechanism for additional matings and may explain why second and third copulatory series often result in a higher incidence of ovulation in musk shrews: their GnRH system is augmented by previous matings so that the GnRH surge is larger and more likely to evoke ovulation. Alternatively, the GnRH ovulatory surge may occur much later in shrews than it does in other reflex ovulators. Shrews may therefore increase their GnRH content slowly over the first several hours post-coitus in order to subsequently provide a massive GnRH output for eventually inducing ovulations. Unfortunately, the exact time of the ovulatory GnRH and LH surges in musk shrews is not known since there is currently no LH assay available for this species (151) and GnRH-microdialysis studies have not been conducted. Although GnRH-ir cell numbers increase within minutes of meeting a male, it is unclear if female shrews, like other induced ovulators, exhibit a GnRH surge within minutes of mating, or whether this surge's latency is prolonged by several hours. It would not be surprising if the latency to the ovulatory surge was greater in shrews, as these animals require much more copulatory stimulation than most reflex ovulators to initiate ovulation; thus, the activation of their reproductive system appears to be less rapid and immediate than that of other species. Eventual development of a reliable LH assay for this species, as well as data sampling over more time points post-coitus, will provide a clearer and more accurate picture of the time-course of mating-induced GnRH and LH release in shrews.

Luteinizing Hormone (LH) and Follicle-Stimulating Hormone (FSH)

In contrast to GnRH patterns following mating, there is an abundance of data from a wide range of reflex-ovulating species regarding the pattern and time-course of gonadotropin release, primarily LH, after copulation. As it is much easier to measure LH in the peripheral bloodstream than to measure GnRH in the portal system, much is known about the relationship between mating and LH secretion. In fact, data regarding LH release is available from more than a third of the currently known induced ovulating species. To date, only in Insectivores has the use of an RIA to measure plasma LH been unsuccessful. Though the general pattern is the same, several parameters of copulation-induced LH release, including latency to increase, latency to peak, and total duration of elevated LH, exhibit species-differences (Fig. 3).

Currently, the most data on the coitus-induced LH surge exists for rabbits. Early experiments by Fee et al. (153) and Westman and Jacobsohn (154) determined that hypophysectomy performed 1 hour after mating did not prevent the occurrence of ovulation, indicating that sufficient gonadotropin was released in the first hour to support ovulation. More than 25 years later, additional studies based on bioassays and pituitary analyses attempted to better define the time-course of LH release. Hilliard et al. (155) measured LH indirectly by assessing the in vivo progestin release from ovaries infused with plasma from recently mated female rabbits. The authors determined that

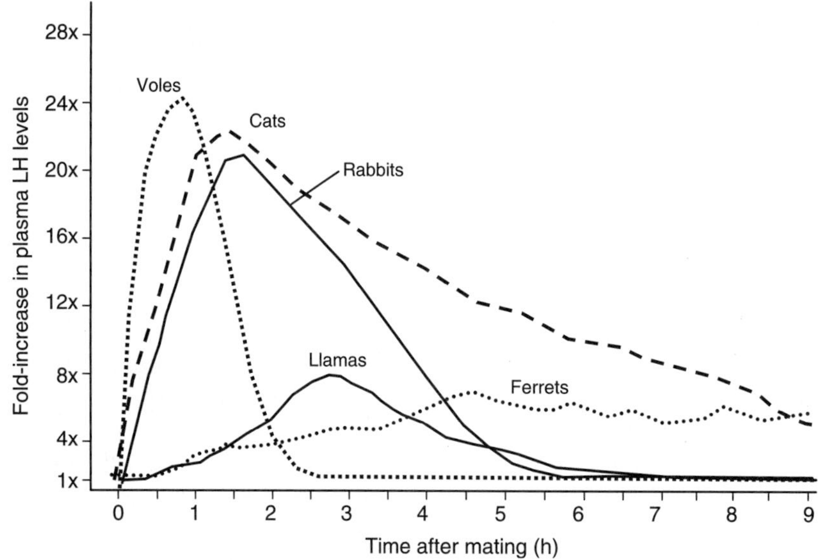

FIG. 3. A schematic representation of the timing of ovulatory LH surges after a single copulation in five animals that display induced ovulation. Because the values of LH levels (as measured in ng/ml) vary depending on the laboratory and assay used, the data are plotted as fold-increase above baseline levels and represent a generalized representation averaged from several studies per species. The timing of the latency to peak and the duration of elevated LH levels for each animal is representative of typical species-specific time courses; of note is the variability between species in how rapidly (and for how long) LH is secreted after a single mating. Vole = Montane and short-tailed field voles which are nearly identical in the timings of their LH pattern (70,79,88,89,123,126,140,160,161,163,173,174,176,178,422).

gonadotropin output increased rapidly within 30 minutes, peaking approximately 2 hours after mating, and remaining elevated above baseline levels for 4 to 6 hours (155). Because the donor blood included all pituitary hormones, it was impossible to separate the effects of LH in this study from those that might have been caused by FSH. Brown-Grant et al. (156) measured anterior pituitary content of LH of female rabbits and found that it was greatly reduced 4 hours after mating whereas Desjardins et al. (157), using a similar technique, reported a significant decrease in pituitary LH content as early as 15 minutes post-coitus. In the latter study, the decreased LH content was maximal 7 hours after mating (157).

More accurate LH data were obtained in the early 1970s with the use of RIA. Using a guinea pig anti-ovine antibody to LH, Scaramuzzi et al. (158) first reported an increase in LH beginning 30 minutes after mating, with peak levels as high as 100-fold attained after 60 to 90 minutes. LH remained elevated significantly above baseline levels for approximately 5 hours post-coitus (158). Dufy-Barbe et al. (159) used an RIA employing a more sensitive rabbit anti-ovine LH antibody and determined that LH began to increase 10 to 20 minutes following mating, peaked around 90 minutes, and remained elevated for approximately 4 hours. Similar findings in rabbits have since been reported by numerous authors with a consistent pattern emerging: LH increases within the first 10 to 15 minutes or less, peaks around 1 to 2 hours, and remains elevated for a total of 4 to 5 hours before returning to baseline levels (140,142,143,160–163) (Figs. 4 and 5).

Unlike LH, FSH does not tend to show major increases in plasma levels immediately after mating in the rabbit (159) (Fig. 4). Desjardins et al. (157) reported no significant decrease in anterior pituitary FSH content in the first several hours after mating. However, other studies have shown a sluggish, minor increase in circulating FSH that peaks 3 to 5 hours after coitus and reaches maximal levels that are only 3 to 6 times higher than pre-coital levels (140,159,161–164). This rise in FSH returns to baseline levels within a few hours after peaking (Fig. 4). However, a secondary rise in FSH, equally small in amplitude, begins 12 to 24 hours after mating and lasts 16 to 24 hours (6,161,162,164). This secondary increase in FSH levels is often called the post-ovulatory FSH rise, as ovulation occurs 12 hours after mating. The post-ovulatory FSH increase, unlike the first FSH increase, is GnRH independent, as GnRH antagonists do not prevent its occurrence (165). Instead, FSH likely rises following ovulation in response to decreased negative feedback on the pituitary by ovarian inhibin (which preferentially inhibits FSH secretion from pituitary cells in vitro) (164–166). The decrease in

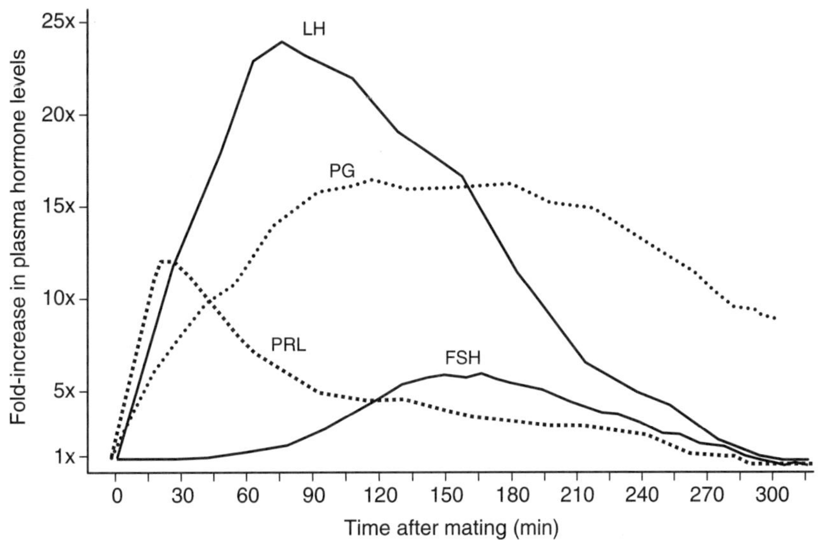

FIG. 4. A composite schematic showing typical changes in plasma luteinizing hormone (LH), prolactin (PRL), follicle-stimulating hormone (FSH), and progesterone (PG) after a single copulation in a female rabbit. In most cases, plasma was taken by venipuncture of the marginal vein of the ear or the jugular vein. Data are plotted as fold-increase in hormone levels above pre-mating baseline levels (140,160–163, 165,179,180,199,200,422).

inhibin secretion may be a direct result of follicular rupture and ovulation, though it is unclear why FSH levels return to baseline again approximately 1 day later.

After rabbits, cats have been the induced ovulators of choice to study coitally induced gonadotropin release. Though cats were known to be reflex ovulators for over 60 years and the interval from mating to ovulation was well established, studies examining LH release after mating were not reported until the early 1980s (Fig. 3). Concannon et al. (89), using RIA, first demonstrated an increase in LH 10 minutes post-coitus in cats mated on estrus day 3. The LH levels in this study peaked after approximately 1 hour (>20x) and stayed elevated for more than 8 hours. Interestingly, animals permitted four copulatory series on the same day had higher levels of LH release and longer durations of elevated LH (approximately twice as high in both cases) than females mated just once (89). Mounting without an intromission resulted in a minor increase in LH (2x) that lasted less than an hour (89). In the same year, Wildt et al. (88) reported that mating caused an increase in LH secretion at 1 hour post-coitus that showed a wide range in latency to peak (1–6 hours post coitus). LH levels in this study remained elevated for at least 9 to 12 hours (88). As in Concannon et al.'s (89) study, LH levels were higher (though often peaked later) in cats mated multiple times on 1 day versus singly-mated animals (88). Wildt et al. (88) did not measure LH levels at any time point less than 1 hour after mating; however, Johnson and Gay (90) subsequently measured LH at 6 different time-points in the first hour and found that LH began to increase as soon as 5 minutes after mating and often peaked after 30 to 60 minutes. Similar findings regarding the pattern of LH release in cats have since been reported in naturally-estrus and estrogen-primed females: LH tends to increase quickly within the first 5 to 10 minutes, peaks in approximately 1 hour, and remains elevated above baseline levels for about 8 hours (wide range from 4–14 hours) (Fig. 3) (90,92,94,167,168). In most cases, multiple matings and longer copulatory stimulation in cats results in higher LH levels and greater durations of elevated LH secretion. Surprisingly, studies in cats have not looked at FSH release in response to mating or GnRH injections, and thus the time-course of FSH secretion, if any, is unknown for this species.

There are good data on LH induction in mustelids, primarily from work performed with ferrets by Baum and colleagues. The first published report of LH release in mated ferrets was in 1985 by Carroll et al. (70). This study observed a rise in LH 1 to 2 hours after a single intromission, peaking at 6 to 8 hours, and staying elevated for approximately 12 hours. There was no effect of neck-gripping, mounting, or pelvic thrusting on LH release in the absence of intromissions (70). Though single intromissions in ferrets can vary drastically in duration from 3 to 150 minutes, there was no correlation between intromission duration and the height or duration of the LH peak (70).

This differs from cats in which prolonged mating enhances amplitude and duration of LH secretion. An additional difference is that the LH rise peaks later in ferrets than in rabbits, and remains high for a longer duration (Fig. 3). A later study by Carroll et al. (123) reported similar findings: LH peaked 2 to 4 hours after mating and stayed elevated for approximately 12 hours or more. Interestingly, this study found that baseline LH secretion in estrus females is not strongly pulsatile, but that pulsatility increases in frequency and amplitude following mating (123). Wersinger et al. (147) corroborated the earlier findings of Carroll et al. (70) by showing that LH was not elevated in mated female ferrets during the first several hours after mating but was higher 6 to 12 hours post-coitus. Only two studies have examined LH secretion in minks. In the first, LH was elevated in less than 6 hours after mating and remained high for almost a day (169). However, plasma was only assayed 2 times per day, prohibiting a detailed analysis of the time course. A more recent study looking at more time points determined that LH was significantly elevated an average of 16 minutes after mating, remained high at 12 hours post-coitus, and was decreased to baseline by 24 hours (170).

As is the case with cats, FSH has not been measured in minks or ferrets following intromissions or prolonged copulatory series. However, two studies in estrous ferrets have determined that FSH secretion does not change from basal levels in response to GnRH injections or hypothalamic electrical stimulation; in contrast, LH did increase in response to these stimuli, suggesting that FSH may not be induced by mating in mustelids (171,172). However, FSH was only monitored during the first 3 to 4 hours after mating; it is possible that FSH increases slowly in ferrets as it does in rabbits, and that appreciable FSH increases are not visible until 4 to 6 hours or later. Future studies in minks and ferrets addressing the specific time-course of FSH release in response to mating should resolve this issue.

There is only minimal information from the Camalidae on gonadotropin induction after mating. In female Dromedarian camels, plasma LH was analyzed via RIA every 10 minutes for up to 12 hours post-coitus. LH levels began to rise 60 minutes after mating, peaked at 2 to 4 hours, and remained elevated 9 to 12 hours before declining to baseline (77). The magnitude of LH increase in this study was lower than that seen in other reduced ovulators, with maximal levels in the camels ranging from 3 to 9 times that of baseline, pre-mating levels. In the Bactrian camel, LH was elevated 4 hours after mating with a peak that was almost 3 times that of basal levels; LH levels began to drop after 8 hours and were no longer elevated when assayed after 24 hours (110). This study measured LH only 3 times each day, thus the picture of the LH pattern is limited. Neither camel study measured FSH secretion.

Several studies by Bravo et al. (79,173) analyzed the time-course of LH secretion in llamas and alpacas. In both species, LH increased 30 to 60 minutes after mating, peaked at 2 to 4 hours, and remained elevated until to 6 hours post-coitus (Fig. 3). Unlike in rabbits, there was no change in FSH secretion in either llamas or alpacas during the first 6 hours after mating, though no additional measurements were taken after this time (173).

In contrast to other induced ovulators, various vole species tend to show a more acute LH response to mating (Fig. 3). In the short-tailed field vole, increases in LH have been detected as soon as 5 to 10 minutes after mating, with LH levels peaking within 20 to 40 minutes (126,145,174–176). In most cases, LH had returned to baseline levels by 2 hours post-coitus (126, 174). Interestingly, Charlton et al. (174) reported that a second LH surge could be obtained in females mated 90 minutes after the initial mating; the timing and magnitude of this second LH rise was not different than the first one that occurred 90 minutes earlier. In the meadow vole, increases in LH have been reported beginning 30 minutes after mating and peaking at 30 to 60 minutes before returning to baseline after 90 minutes (177). LH secretion has been studied in the Montane vole, but plasma was only assayed at one time-point after mating; in this case, LH was significantly increased 60 minutes after mating, reaching levels that were almost 40 times that of unmated females (178). Similar to cats and ferrets, no data currently exist for FSH secretion in female voles after mating.

Prolactin

Information on prolactin is minimal and only exists for a select few species including rabbits, meadow voles, and short-tailed field voles. However, the data currently available are interesting enough to warrant future studies in other species looking at post-coital prolactin release. In rabbits, Desjardins et al. (157) measured a significant decrease in anterior pituitary prolactin content in the first 15 to 45 minutes post-coitus which remained low for several hours. Similarly, Kaynard et al. (140) reported a large and immediate surge in plasma prolactin induced by coitus in female rabbits. Specifically, prolactin levels rose almost tenfold 20 minutes after mating and returned to baseline levels after 6 to 7 hours (Fig. 4). Similar rises in prolactin secretion were not observed in unmated does. Pau et al. (179) also reported post-coital induction of prolactin surges in both intact and ovariectomized, estrogen-treated rabbits. However, these data are in contrast to other rabbit studies by Fuchs et al. (180) and Kaufmann et al. (162) who reported an immediate drop in plasma prolactin levels after mating followed by a rapid restoration to pre-mating values within 60 to 90 minutes. It is not clear what caused the discrepancies in results, although Kaynard et al. (140) noted that the pre-coital "baseline" levels of prolactin in the Fuchs study were considerably elevated, and thus the rabbits in the Fuchs and Kaufmann studies may have been hyper-prolactinemic due to experimental stress (181,182). If so, the sudden drop in prolactin immediately upon mating may have been the result of a temporary "relief of stress" due to copulatory activity and interacting with a male.

It is unclear what role increased prolactin would serve in the several hours after mating, though one hypothesis is that it is important for the maintenance of steroidogenesis in the rabbit ovary and corpus luteum (140). In the short-tailed field vole, bank vole, and meadow vole, development of a functional corpus luteum after ovulation is dependent on a similar post-mating surge of prolactin (176,177, 183,184). In these species, plasma levels of prolactin significantly increase 10 to 30 minutes after mating (176,183) and are also elevated 1 to 2 days later (177). Fertility and successful pregnancy in reflex ovulators may therefore be contingent, to some degree, on a post-coital increase in prolactin. In support of this, exogenous administration of prolactin after mating significantly increases the percentage of voles with functional corpora lutea as well as the frequency of successful pregnancies (176,183,184). Though the data appear promising, the hypothesized role of prolactin as a luteotrophic factor during the first several hours and days following mating needs to be tested in other induced ovulating species before any conclusions can be made.

Ovarian Progestins

In most reflex ovulators, progesterone (Pg) levels begin to rise several days after mating in association with increased activity of the corpus luteum and initiation of pregnancy. In these cases, plasma progestin levels begin to increase 3 to 15 days post-coitus, depending on the species. Reflex-ovulating species that have been shown to exhibit such gradual post-coital increases in progestin secretion include minks, ferrets, camels, prairie voles, cats, rabbits, woodchucks, and musk shrews (43,87,92,94,110,118,167,185–192). However, whether or not the release of Pg is affected acutely, in the first several hours after mating, has received little formal attention apart from work done in rabbits and two species of voles and shrews.

The rabbit ovary produces several steroids including estrogen, Pg, 20-α-hydroxyprogesterone (20-α-OHP), testosterone, and androstenedione (6,193). Notably, both of the progestins are secreted into the blood in quantities more than 1,000-fold higher than estrogen and androgens (ng/ml versus pg/ml), with 20-a-OHP being secreted in quantities approximately tenfold that of Pg (6). Progestin secretion in rabbits has typically been measured via two routes: ovarian output and plasma levels. The former method has been used to determine the amount of hormone produced by the ovary by direct measurement from the ovarian vein (typically expressed as ovarian output per unit time (μg hormone/ovary/hr). The latter method has analyzed peripheral plasma levels of progestin hormone by assaying blood obtained from either the jugular vein or marginal ear vein (typically expressed in ng/ml).

Ovarian output (μg/ovary/hr) of both progestins increases immediately following mating in the rabbit. Ovarian secretion of Pg is elevated above pre-mating levels within 5 to 30 minutes post-coitus, peaks around 90 to 120 minutes, and remains high for at least 6 hours (194,195). Ovarian output of 20-a-OHP also increases rapidly after mating, beginning 10 to 30 minutes post-coitus, peaking at 2 to 3 hours, and remaining elevated for about 5 to 8 hours (155,195–197). The increase in 20-a-OHP ovarian secretion rate tends to be of a much larger magnitude than that of Pg (ten- to 20-fold vs. three- to fivefold, respectively) (6). Similarly, the peripheral plasma levels of both Pg and 20-a-OHP also increase after mating, but as expected, both show a slight lag behind the increase in ovarian output of the hormones. Based on data from several studies, a general pattern for the time-course of progestin levels in the blood after coitus emerges: plasma concentrations (ng/ml) of both progestins begin to increase within 5 to 60 minutes after mating, peak between 6 to 180 minutes, and remain elevated for more than 5 hours (Fig. 4) (162, 163,198–200). Similar effects on plasma progestin levels, both in the timing and magnitude of their secretion, have been reported for rabbits in response to either hypothalamic electrical stimulation or ovulatory-inducing LH infusions (196,200). Interestingly, both the ovarian output and peripheral plasma concentrations of both progestins are low (at baseline levels) 10 hours after mating, just before the time of ovulation (6,155,195). Whether or not low progestin levels at the time of ovulation are a necessary prerequisite for follicular rupture and ova expulsion is unknown and remains to be tested.

While numerous studies have measured Pg secretion after mating in additional species of reflex ovulators, the vast majority of these studies have looked primarily on a daily (long-term) basis. In contrast, there are few data regarding short-term Pg secretion in the first several hours post-coitus. This is unfortunate given the abundant data in rabbits showing a rapid rise in Pg in the first few hours after mating. It is possible that researchers studying other species have missed the initial post-coital Pg surge by assaying blood only once or twice a day at long-term intervals (i.e., once every 12–24 hours).

Hourly, short-term Pg levels after mating have been measured in a few species. Similar to rabbits, Montane voles show a two- to threefold increase in plasma Pg levels within 1 hour of mating (178); this was the only time point reported, so it is unknown whether hormone levels were still elevated at subsequent hourly time-points or if the levels had returned to baseline prior to ovulation. Likewise, musk shrews exhibit a slight increase in plasma Pg (twofold) immediately after mating (within minutes), with hormone levels no longer elevated later at 5 hours post-coitus (201). Though the sample size of this particular study was very low (only 1–4 shrews tested per time point), these results were confirmed by a later study using larger sample sizes (10–11 per group) in which plasma Pg levels of female shrews increased two- to fourfold at both 7 minutes and 2 hours after sexual interactions with a male (202). In contrast to these findings in shrews and Montane voles, no significant increase in Pg was observed in prairie voles 1 hour after mating (43), or in cats or camels 4 or 8 hours post-coitus (92,110). Future studies need to analyze additional (earlier) time points in these species, as well as measure hourly Pg secretion in other reflex ovulators, to address the possibility that acute progestin secretion plays a critical role in the induced ovulatory process.

The functional role of increased Pg and 20-a-OHP immediately after mating has been discussed by several authors. Hilliard et al. (195,203) found that injections of 20-a-OHP and Pg both stimulated LH release in females rabbits and postulated that progestins act as a positive feedback hormone to prolong and heighten post-coital LH release from the pituitary after the initial LH surge. In support of this feedback hypothesis are findings that either Pg or 20-a-OHP, infused as pulses directly into the hypothalamus, stimulates the release of GnRH (137,139, 204,205). However, this hypothesis was challenged by Goodman and Neill (163) who reported that 60% of female rabbits ovariectomized 15 minutes after mating exhibit similar coitus-induced LH patterns indistinguishable from that of intact females. This study suggests that progestins are not required after the first 15 minutes post-coitus to induce an ovulatory LH surge. However, as several studies have reported increased Pg levels as early as 5 to 10 minutes after mating, it remains possible that elevated progestin secretion in the first 15 minutes (prior to ovulation) may be critical for inducing LH surges and/or ovulation. Adding to the complexity of this issue, evidence exists showing that continual, nonpulsatile administration of Pg (given via Silastic capsules) actually decreases GnRH release and prevents an increase in LH secretion (139). Kawakami and Sawyer (206) also determined that subcutaneous administration of Pg facilitated ovulation in response to vagino-cervical stimulation after several hours, but inhibited it over the long term (after 24 hours). Thus, Pg may have both positive and negative feedback effects on the hypothalamic–pituitary–gonadal (HPG) axis, depending on the duration, timing, and pulsatility of its secretion.

Other functions may exist for post-coital, pre-ovulatory rises in progestins. Gray et al. (1976) and Hilliard and Eaton (195) both noted that Pg has many postulated effects on mammalian female reproduction, including stimulation of enzymes involved in promoting follicular rupture (207), promoting contractions in the oviduct to aid in egg transport (208,209), and modulation of female sexual receptivity. Any of these regulatory actions might be promoted by acute, post-coital surges in progestins. Future work is needed to elucidate the exact function(s) of mating-induced progestin release, as well as to determine which reflex-ovulating species exhibit such short-term secretions. In addition, it will also be important to tease apart the respective roles (and mechanisms of action) of post-coital Pg and 20-α-OHP, especially considering that the latter hormone is secreted at a significantly greater magnitude (ten- to 20-fold higher than Pg).

NEURAL PATHWAYS AND NEUROTRANSMITTERS MEDIATING THE CONTROL OF REFLEX OVULATION

The neural substrates and pathways underlying the activation and regulation of induced ovulation are not entirely understood, though recent technical advances in molecular and neurobiology have greatly improved our knowledge of this topic. In this section we summarize the current knowledge available regarding which brain regions and neurotransmitters/neuropeptides are involved in the neural arc of reflex-ovulation that begins in the periphery and ends ultimately at the anterior pituitary. The most detailed information regarding this topic is available for rabbits, although additional data exists to a lesser degree for several other species as well, including ferrets and cats.

The Hypothalamic Component and GnRH Neurons

There is minimal information for rabbits or other reflex ovulators on which peripheral neural pathways activated by coitus transduce the tactile signals of mating from the somatic periphery to the brain. During copulation, contact with a male potentially activates many tactile-sensing areas of the female central nervous system which receive input from numerous nerves including the perineal nerve, pudendal nerve, pelvic nerve, and several somatic nerves from the skin and body (6). It is presently unclear which of these nerves are critical to relaying the coital signal to the brain to induce ovulation. In the female cat, lesion of the pelvic plexus or the hypogastric nerves inhibited ovulation in the majority of animals in response to artificial vaginal stimulation, though a small percentage of females still ovulated after bilateral section of the pelvic nerve (210). At present, more research is needed in reflex ovulating species to fully address this issue.

The initial pioneering studies of the neural substrates in the central nervous system underlying reflex ovulation were performed in the late 1930s through the 1950s by the laboratories of Harris, Sawyer, and Faure. These early studies, using primarily rabbits, clearly established the role of the hypothalamic-preoptic region in regulating ovulation. Harris (211) first reported that brief electrical stimulation of the hypothalamus induced ovulation in female rabbits. Later, Sawyer (212) showed that lesions of various regions of the hypothalamus, such as the basal tuberal region, the arcuate nucleus, the ventro-medial hypothalamus, and the median eminence, blocked mating-induced ovulation. In contrast, lesions of the mammillary region, the AH, the organum vasculosum lamina terminalis (OVLT), or the medial amygdala did not prevent coitus-induced ovulation (212,213). Vincent et al. (214) subsequently reported increased electrical activity (EEG) in the arcuate and lateral hypothalamic regions following mechanical vagino-cervical stimulation in rabbits. Similar increases in hypothalamic electrical activity, both single-unit and multi-unit, were observed in estrous cats experiencing vaginal stimulation, though no responses were evoked in anestrous cats (97,215–217).

Several studies of the topographical localization of GnRH neurons in rabbits indicated that GnRH cells and fibers are widely distributed in the hypothalamus (218, 219). Functional support of this finding was provided by Ramirez et al. (220) who used the PPP to show that GnRH is basally released in a pulsatile manner from an extensive area of the MBH, including the rostral, medial, and posterior regions. These data are consistent with previous studies by Faure (221), Vincent et al. (214), and Kawakami and Sawyer (222) which demonstrated a diffuse distribution of hypothalamic neurons in rabbits that are electrically activated after vagino-cervical stimulation or natural copulation, including the arcuate nucleus, premammillary region, and posterolateral hypothalamic region. Similar to the rabbit, the feline brain also contains a wide anatomical distribution of GnRH cells and fibers, with the latter present both in the hypothalamus and in extra-hypothalamic regions including the thalamus and brainstem (223,224). Within the cat hypothalamus, GnRH-neurons are diffusely distributed in, but not limited to, the AH, the preoptic area, the arcuate nucleus, the periventricular nucleus, and the supraoptic nucleus (223,225). Minks and ferrets also exhibit a wide distribution of GnRH cells in the brain and hypothalamus, though a major concentration of GnRH perikarya (80% in minks) are present specifically in the MBH (226,227). Electrical stimulation of the ferret hypothalamus induced LH release and ovulation in both estrous and anestrous females (171,211), with the greatest responses observed when the electrode was in the preoptic area, MBH, or ventromedial hypothalamus. Vagino-cervical stimulation of estrogen-primed ferrets in the presence of a neck-gripping male resulted in a 50% decrease in neurons expressing GnRH immunoreactivity throughout the hypothalamus, and was accompanied by an LH surge (99).

A wide distribution of GnRH neurons in reflex ovulators, along with the fact that hypothalamic lesions in a wide range of nuclei prevent ovulation, suggests that multiple sensory and regulatory pathways may exist to control the output of GnRH. In support of this hypothesis, Voloschin and Gallardo (228) reported that copulatory information in the rabbit was relayed to the MBH via numerous, widely-distributed pathways. This conjecture was based on the finding that in order to block coitus-induced ovulation, it was necessary to make a complete, 360° circular deafferentation of the hypothalamus. In contrast, semi-circular cuts removing neural input from only several sides of the hypothalamus were ineffective in preventing mating from activating ovulation (228). Multiple inputs to the MBH likely correspond to multi-sensory stimuli received by reflex ovulators during copulation, including visual, audio, tactile, and pheromonal cues (229).

Though diffuse throughout the hypothalamus, GnRH neurons that control the ovulatory LH surge appear to be distinct and experimentally distinguishable from those that regulate behavioral estrus and receptivity. Sawyer (212) showed that small lesions in either the basal tuberal region of the hypothalamus or the arcuate nucleus specifically block coitus-induced ovulation in female rabbits without causing ovarian atrophy or decreasing behavioral estrus. In contrast, lesions located in the mammillary region of the hypothalamus failed to prevent coitus- or drug-induced ovulation but did result in a permanent behavioral anestrus (212). Related to these findings, Hayward et al. (196) determined that electrical stimulation of the posterior median eminence and the basal tuberal hypothalamus induced ovulation and gonadotropin release whereas similar stimulation in other hypothalamic regions were less successful. Thus, while both processes rely on the actions of GnRH, either directly or indirectly, the neuroendocrine control of ovulation and estrus behavior in rabbits can be dissociated and are likely regulated by distinct populations of GnRH cells and/or projections (6). This paradigm is also true for cats, in which hypothalamic lesions of the MBH prevented ovulation but left estrus mating behavior intact whereas lesions of the AH destroyed female receptive behavior but did not prevent ovulatory responses (230). Likewise, electrical stimulation of the MBH, but not the AH, induced LH release and ovulation in estrogen-primed female cats, indicating a functional difference in these two GnRH-containing areas of felines (98,230,231).

In vitro GnRH release from hypothalamic tissue has been studied in several reflex ovulators. In ferrets, Lambert et al. (144) studied the effect of copulation on in vitro release of GnRH from MBH fragments collected from mated females. In this experiment, both GnRH release and content were reduced in the MBH of estrous females 15 minutes post-coitus, indicating a massive release of GnRH immediately after mating which depletes releasable stores of the peptide (144). In rabbits, both the labs of Spies and Ramirez have utilized in vivo PPP to study GnRH release in unrestrained animals. In addition to determining the pattern and time-course of GnRH secretion after mating, as discussed above in the section on Endocrine Correlates of Coitus-Induced Ovulation, PPP allowed these investigators to determine which hypothalamic regions were involved in inducing ovulation. Kaynard et al. (140) reported a massive 100-fold increase in GnRH release from the rabbit MBH within 1 hour after mating; similar increases were not observed when females were sham-mated. Histology revealed that the majority of cannula tips were located in the arcuate nucleus, and tuberal region of the median eminence. Much smaller (and slower) increases in GnRH release on the order of sevenfold were observed in mated females when the PPP cannula were located in the AH (encompassing the suprachiasmatic and paraventricular nuclei) (140). Independently, Lin and Ramirez (141) showed that a rise in GnRH secretion immediately after mating was only detected when the PPP cannula was located in the tuberal portion of the rabbit MBH as opposed to the dorsal region of the hypothalamus. Similar increases in post-coital GnRH release from the rabbit MBH, concurrent with ovulatory LH surges, have since been measured using microdialysis (142). Together, these PPP data indicate that the major terminal field of GnRH cells is in the arcuate-median eminence regions of the MBH.

The expression of the proto-oncogene *cfos*, indexed by either *cfos* mRNA or Fos protein, has been used by several investigators as a morphological indicator of genomic activation and cellular activity. Reyna-Neyra et al. (232) found a twofold increase in *cfos* expression, using reverse-transcription PCR, in the hypothalamus of female rabbits immediately after mating. Similar increases in *cfos* expression post-coitus were not observed in the hippocampus or preoptic area. Caba et al. (150) also characterized the expression of *cfos* mRNA by using in situ hybridization in several forebrain regions of mated rabbits. Hypothalamic *cfos* mRNA expression was enhanced 30 minutes after mating, especially in the anteroventral periventricular nucleus (AVPV, projects to the medial preoptic area which contains GnRH neurons) and ventrolateral hypothalamus (VLH, adjacent to the arcuate nucleus and rich in estrogen receptors), as well as in the bed nucleus of the stria terminalis (BNST, which receives neural input from olfactory bulb and projects to the AVPV) (150). At 60 minutes after coitus, *cfos* expression was still increased in the AVPV and VLH, but no longer in the BNST. In rabbits and ferrets, mating was also followed by increased expression of Fos protein in numerous regions including the medial preoptic area, BNST, MBH, AVPV, and the VLH (146,147,150,233,234). Additional Fos activation in mated female ferrets was also observed in the amygdala (AMG), main olfactory bulb, and the thalamic central tegmental field (CTF) (147,233,234). In addition, in both rabbits and ferrets, double-labeled Fos-GnRH neurons, as determined with immuno-cytochemistry, were observed after mating in several forebrain areas including the POA, the anterior hypothalamus, the BNST, the MBH, and the VLH (108,146,147,150,233). Thus, copulation in these reflex ovulators activates numerous hypothalamic GnRH populations in several discrete loci.

The wide distribution of *cfos* activation in the forebrain suggests that coital signals reach the hypothalamus from multiple, separate neural pathways, including those arising in the olfactory bulb and the brainstem, as formerly proposed (228,229). Bakker et al. (108) differentiated between neural sites activated by genital stimulation versus olfactory cues by looking at post-coital Fos activation in nares-occluded and intact female ferrets. Intact females showed increased Fos production in multiple brain sites including the main olfactory bulb (MOB), AMG, POA, BNST, and VLH. However, when olfactory input was blocked via nares-occlusion, the same sites still showed Fos activation except for the MOB and the AMG (108,234). This suggests that male odor cues from the olfactory forebrain are transmitted to the hypothalamus via the AMG, whereas tactile cues from stimulation of the genitals are not. In support of this, females exposed to male odor cues without copulation had increased Fos protein in the AMG and MOB, but not in the preoptic area, BNST, AH, or MBH-GnRH neurons (108,233); these findings reinforce the dogma that, while odors and pheromones may be facilitory or modulatory to inducing ovulation, copulation itself (i.e., vagino-cervical stimulation) is the primary cue that activates ovulation. However, in ferrets, as in several other reflex ovulators, vagino-cervical stimulation alone is unsuccessful in inducing ovulation but must be concurrently paired with other male cues (such as male neck gripping and olfactory cues). Thus, while genital stimulation may be the principal stimulus, females may require concurrent modulatory input of other signals as well. The Fos data of Bakker et al. (108) suggest that the same number of GnRH cells in the hypothalamus are activated by coitus with or without olfactory input. However, ovulation or LH surges were not measured in these studies. Thus, it is unknown whether these increases in GnRH cell activity, as measured by presence of Fos, in nares-occluded females result in high ovulatory rates. Future Fos studies that measure LH and ovulation, as well as implement MOB or AMG lesions, will better elucidate how mating stimuli are processed and transmitted to the MBH.

The Brainstem Component and the Role of Norepinephrine

Evidence linking the hypothalamic GnRH neurons which regulate the LH surge with hindbrain and brainstem regions was first gathered indirectly in pioneering work done by Sawyer and colleagues in the late 1940s and early 1950s. These investigators determined early on that catecholamines, specifically norepinephrine (NE), played a critical role in mating-induced ovulation in rabbits. As discussed later, NE-containing cells are located primarily in the midbrain and brainstem and project to numerous forebrain regions including the hypothalamus (235). Experiments in Sawyer's lab showed that an intravenous injection of epinephrine or icv infusions of NE into the third ventricle induced LH release and ovulation in estrogen-primed rabbits (236,237). Furthermore, treatment of female rabbits with the α-adrenergic blocking agent dibenamine prevented mating-induced ovulation (71). Donovan and Harris (238) failed to induce ovulation with infusions of catecholamines directly into the pituitary, indicating that NE's actions were likely upstream in hypothalamic regions.

Numerous studies in rabbits have subsequently confirmed a role of NE in regulating GnRH cells and the ovulatory LH surge. Histological studies have determined that NE fibers and terminals are located diffusely throughout the hypothalamus in a primarily bimodal distribution rostro-caudally, with peaks in the medial preoptic area and the ventromedial region (239–241). Infusions of NE into the third ventricle stimulated an LH surge and subsequent ovulation of estrogen-primed females, and the effectiveness of NE to induce GnRH release or ovulation was inhibited with various α-adrenergic blocking agents, such as SKF-501 or prozosin (109,143,242). Ramirez and Spies' labs independently used the PPP technique to observe in vivo release of GnRH in response to infusions of NE directly into the rabbit hypothalamus (Fig. 5) (138,220). Interestingly, this effect was less potent in females that were not treated with estrogen (138), suggesting that estrogen primes GnRH neurons or affects NE synthesis or release (6). An additional PPP study by Kaynard et al. (140) demonstrated a 200% rise in NE levels in the MBH following mating along with concurrent rises in GnRH and LH. Using microdialysis and HPLC, Yang et al. (142) extended these findings by elegantly showing that the post-coital rise in NE in the MBH preceded the rise in GnRH by 2.5 to 7.5 minutes (Fig. 6). Because it is released minutes earlier than GnRH, NE is likely an upstream signal of GnRH release and may serve as the critical trigger for activating GnRH terminals to release the peptide (132).

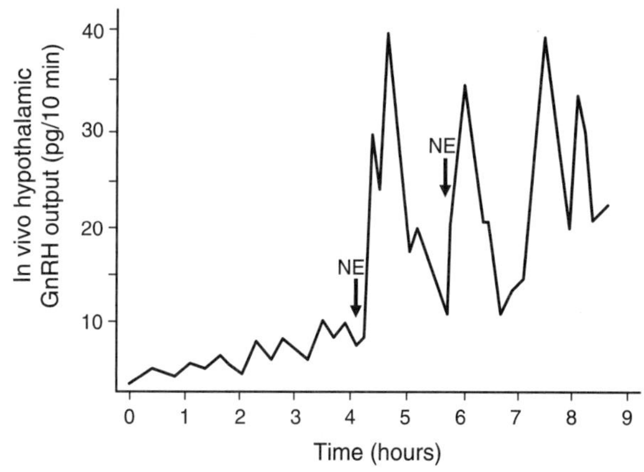

FIG. 5. This figure shows the stimulatory effect of norepinephrine (NE) on GnRH release from the hypothalamus of an intact female rabbit. Hypothalamic GnRH output was measured in vivo with push-pull perfusion before and after two localized central infusions of NE (10–6 M). The times of infusion are indicated on the graph by the two *arrows*. [Figure adapted from (6).]

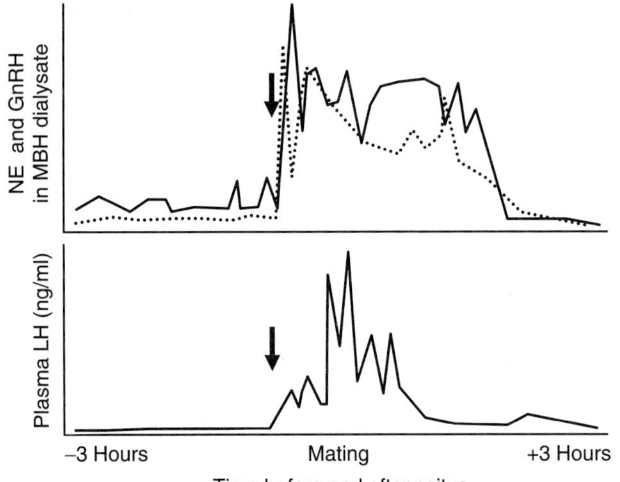

FIG. 6. The temporal relationship between coitus and the hypothalamic release of norepinephrine (NE) and gonadotropin releasing hormone (GnRH) in female rabbits (**top panel**). The concurrent pattern of pituitary luteinizing hormone (LH) release is also shown (**bottom panel**). NE (*dashed line*) and GnRH (*solid line*) were measured from microdialysate, with the probe located in the arcuate nucleus-median eminence of the MBH. Note that after mating NE peaks minutes before GnRH surge peaks. LH (ng/ml) measurements were taken from plasma of same animals. The *arrow* represents time of mating. [Figure adapted from (142).]

Acute intravenous cocaine administration, which blocks re-uptake of NE in neural synapses and thus increases its bioavailability (162), increased the ovulation rate of female rabbits when given immediately after mating (243). Similar findings were recently reported by Pau et al. (244) studying the role of presynaptic NE transport proteins (NET) which regulate NE uptake and, thus, its concentrations at the synapse. Microdialysis infusion of desipramine, a specific NET blocker, into the median eminence-arcuate nucleus region of female rabbits significantly elevated NE levels concurrent with increased pulses in GnRH release. Thus, increased NE availability results in increased GnRH release and higher ovulation rates, further supporting the role for NE in regulating GnRH secretion in induced ovulators.

Though Przekop and Domanski (245) found that acute infusion of the neurotoxin 6-hydroxydopamine (6-OHDA) caused temporary blockade of ovulation, Rabii et al. (246) subsequently reported that repeated infusions of the 6-OHDA into the third ventricle, in dosages shown to chronically depress hypothalamic NE by more than 80%, failed to block the copulation-induced LH surge in estrogen-primed rabbits (246). Thus, the coitus-induced ovulating mechanism is still functional in chronically hypothalamic-NE deficient animals, suggesting that NE may normally play only a modulatory role in the induction of LH release in reflex ovulators. It is possible that in the absence of NE, alternative regulatory systems are recruited to maintain the reproductive mechanisms (246); such additional signals could include NPY, as discussed below. Alternatively, it is possible that only a small fraction of the available NE pool is needed to regulate GnRH release in response to mating. As 6-OHDA only depletes an estimated 80% of the hypothalamic NE, there is still a small amount (~20%) available which may be sufficient to preserve its physiological function. In addition to affecting NE-producing neurons, 6-OHDA also depletes dopaminergic neurons; however, these studies did not address whether the observed effects were related to lack of dopamine.

The location of NE-producing neurons in the brains of reflex ovulators has been well-studied by Blessing and others (247–255). While NE-containing terminals exist in the hypothalamus (253,256), NE-producing neurons are located in the brainstem in several loci that include the medulla oblongata (the A1 noradrenergic group), the nucleus of the solitary tract (A2 group), the lateral tegmentum in the pons (A5 group), and the locus coeruleus (LC; A6 group) (235,247,253–255,257). Studies in the 1970s by Rose demonstrated increased electrical activity in both the medulla and the pons of female cats after artificial vagino-cervical stimulation (258–260), though ovulation and/or GnRH release were not assessed. In rabbits, the expression of the *cfos* gene was markedly increased 30 minutes after mating in several brainstem regions, including the A1 and A2 areas, and double labeling of Fos protein and dopamine-β-hydroxylase (DH; the enzyme that converts dopamine to NE) occurred in the same regions when observed at 90 minutes post-coitus (235). These results suggest that coital neural signals are transported via the spinal cord to the caudal brainstem where they activate (directly or indirectly) NE-containing neurons in specific loci (A1 and A2 regions) (235) (Fig. 7). In support of this, expression of tyrosine hydroxylase (TH; the rate limiting enzyme in NE and dopamine synthesis) mRNA increased within 15 to 30 minutes following mating in several rabbit brainstem nuclei, including the A1 and A2 areas (235), and the LC (A6 area) (261,262). In female ferrets, mating induced a rapid increase in the number of co-labeled Fos-TH cells in both the A2 and A6 regions, as well as in the MOB (255). Enhanced TH activity in rabbit and ferret brainstem areas may be critical for rapid recovery of hypothalamic-NE depletion following a massive post-coital release (261).

The data above suggest that several brainstem sites (A1, A2, and A6) may be the direct neural links between the periphery and the hypothalamus (Fig. 7). As previously noted (3), in female rats, lesions of the LC blocked pulsatile LH release as well as NE input into the preoptic area and MBH, indicating that this nucleus plays a significant role in regulating NE release and pituitary function (263,264). Retrograde and anterograde tracing studies performed in rats and cats have shown that the hypothalamus receives NE projections from the A1 (medullary lateral tegmentum) and A6 (the LC) regions (265–271). Neural connections between the pons and A2 area (nucleus of solitary tract) and the hypothalamus have also been suggested (251,266,270–272). Thus, these sites may act independently or in concert as a "relay station" for the peripheral coital signals transmitted to the hypothalamus (251). However, as Pau et al. (251) noted, it is unlikely that the LC (A6) receives direct coital signals from the spinal cord, as retrograde tracing studies have established that the LC receives caudal afferent input from only two nuclei in the medulla. Thus, it is possible that spinal coital signals from the periphery may first project to the medulla, which then projects to the hypothalamus via other secondary brainstem relay areas such as the LC or A1 and A2 regions (235,251) (Fig. 7). Additional studies implementing retrograde and anterograde tracers will be helpful in defining the neural pathways and connections in the coital-reflex arc of induced ovulators. Intriguingly, bidirectional innervation exists between many brainstem regions and the hypothalamus. For example, Wersinger et al. (255) noted that the LC, which sends projections to the hypothalamus, also receives forebrain input from the POA, BNST, and other hypothalamic nuclei. This bidirectionality in neural connections may represent feedback loops and/or modulatory pathways by which the brain regulates the activity of forebrain and brainstem areas in mediating the coitus-induced reflex.

Regulatory Actions of Neuropeptide Y

In addition to NE, there is sufficient evidence that neuropeptide Y (NPY) is involved in regulating GnRH secretion in induced ovulators. NPY is a 36 amino acid peptide that is widely distributed in neural tissue and released in the hypothalamus in a pulsatile fashion (273–275). Studies have shown hypothalamic NPY to be present in, or proximal to, GnRH neurons (132,275–277). The effects of NPY on GnRH secretion in rabbits have been well documented by Spies' lab. Exogenous infusions of NPY directly into the hypothalamus increased both GnRH and LH release in intact and estrogen-primed, ovariectomized females (Fig. 8) (274,276,278). Furthermore, immuno-neutralization of NPY using specific NPY antibodies injected icv suppressed plasma LH

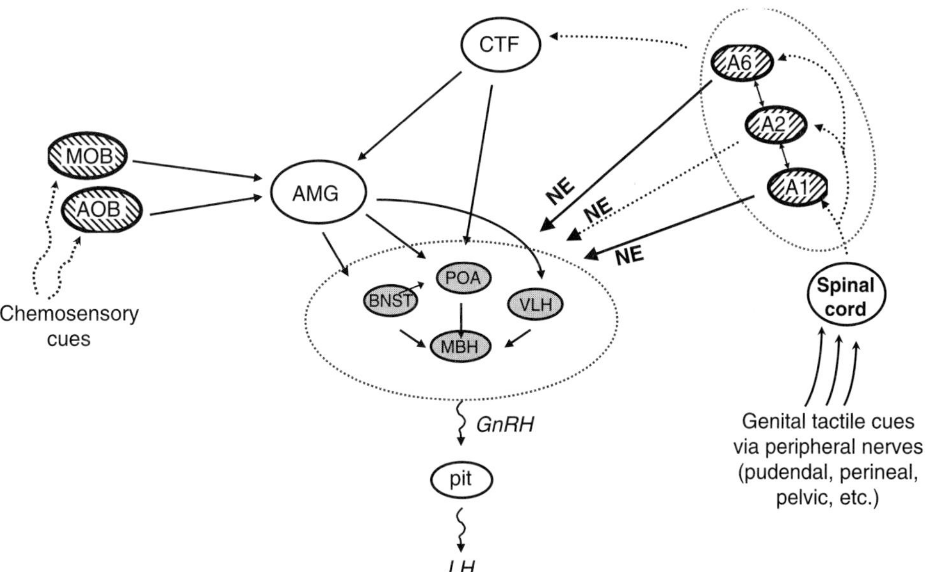

FIG. 7. A schematic diagram of the putative neural pathways from the periphery, brainstem, and olfactory forebrain leading to the hypothalamus and culminating in the ovulatory surge of luteinizing hormone (LH). Two pathways are depicted, one that transduces olfactory and/or pheromonal cues and the other tactile cues associated with copulation. Chemosensory cues are received in the main (MOB) and accessory (AOB) olfactory bulbs and these regions project to the amygdala (AMG). The amygdala projects to several preoptic (POA) and hypothalamic sites in which gonadotropin releasing hormone (GnRH) is ultimately released, in turn stimulating pituitary (pit) secretion of LH. The spinal cord projects (directly and/or indirectly) to several brainstem-noradrenergic nuclei (A1, A2, A6) which contain norepinephrine (NE) producing neurons. These brainstem sites project to various regions of the POA/hypothalamus where convergence with olfactory input is likely. In addition it is likely that as yet to be clarified brainstem projections to the central tegmental field (CTF) can influence ovulation via direct projections to the AMG and the POA. BNST, Bed nucleus of the stria terminalis; VLH, ventrolateral hypothalamus; MBH, mediobasal hypothalamus. *Dotted lines* represent putative (direct or indirect) pathways.

concentrations and LH pulsatility in intact female rabbits (279). Interestingly, NPY infusions into the MBH of ovariectomized, nonestrogen-primed rabbits resulted in *decreased* (or unchanged) GnRH and LH release (276). Thus, NPY stimulates the GnRH axis when rabbits are intact or ovariectomized and primed with estrogen, but does not activate GnRH release when estrogen is absent (274). Though these data suggest that estrogen plays a role in modulating NPY's regulatory actions, the precise mechanism through which this occurs is currently unknown. Likewise, the mechanism by which NPY stimulates GnRH secretion is unclear, some evidence indicates that α-adrenergic receptors are one component. Ovariectomized, estrogen-treated rabbits infused with prazosin an (α-1 adrenergic receptor antagonist) did not release GnRH in response to hypothalamic NPY infusions (Fig. 8) (278). In contrast, similar rabbits treated with NPY alone or NPY and yohimbine (an α-2 receptor blocker) did release GnRH (Fig. 8) (278). In relation to this, coitus-induced GnRH release in rabbits is disrupted by hypothalamic prazosin infusions, providing further evidence for a role of α-1 adrenergic receptors (143). Collectively, these data suggest that NPY's effects on GnRH release are mediated by NE acting on α-1 receptors. Further studies are needed to determine if NPY directly causes the release of NE in the hypothalamus of induced ovulators and by what mechanism.

Interestingly, NPY mRNA- and peptide-containing neurons have been localized in the brainstem of cats and rabbits in several regions also shown to contain NE, TH, and dopamine-β-hydroxylase, including the lateral tegmentum (A1) and the nucleus of the solitary tract (A2) (251,280). In rabbits, little to no NPY mRNA was found in the A6 (LC) or A5 (pons lateral tegmental nucleus) regions. Thus, NPY–NE interactions in the control of GnRH release may exist in both the hypothalamus and the brainstem (251,281,282); future work is needed to differentiate the roles of NPY and NE in these distinct areas of the brain.

Acetylcholine, Opioids, and Other Putative Neuroregulatory Factors

There is some evidence that acetylcholine (Ach) may be involved in the control of reflex ovulation. Sawyer (283,284) found that atropine (a selective cholinergic muscarinic receptor antagonist) given intravenously immediately after mating prevented ovulation in female rabbits. Similarly, another anticholinergic drug, methantheline (Banthine), strongly inhibited copulation-induced ovulation when given within a few seconds post-mating (285). Interestingly, the timing of these drugs' actions suggested that the cholinergic component of the reflex arc was upstream from the adrenergic (NE) component: when atropine treatment was given after the first 15 seconds post-coitus, most rabbits still ovulated, whereas the majority of rabbits given dibenamine (an α-adrenergic blocker) in the same time-frame did not ovulate (6,283). Thus, the effects of Ach are acute and occur in the first few seconds post-coitus whereas the actions of NE, while still occurring soon after mating, are likely not initiated as soon. Whether or not Ach plays a direct role in subsequently releasing NE from hypothalamic terminals is unknown. Since these early studies performed more than 50 years, little additional research has been done characterizing the role

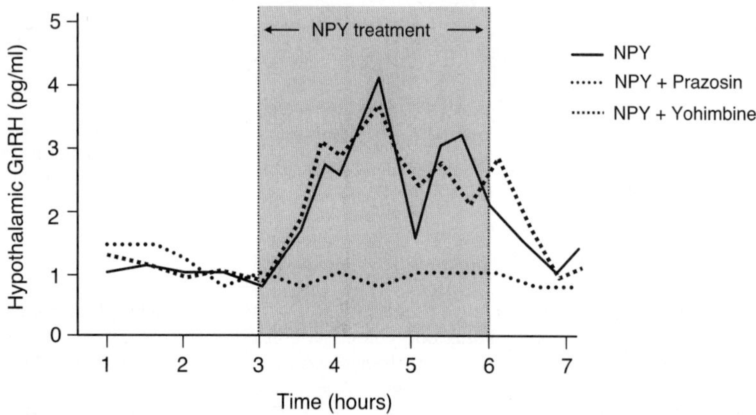

FIG. 8. Pattern of mediobasal hypothalamic GnRH release in female rabbits before and after central administration of NPY, as determined using push-pull perfusion. During the treatment period, NPY was given alone or in combination with either prazosin (an α1-adrenergic antagonist) or yohimbine (an α2 adrenergic antagonist). NPY caused a significant increase in GnRH secretion that was prevented with concurrent administration of prazosin but not yohimbine. These findings suggest that NPY's stimulatory effects on GnRH are mediated by, or dependent upon, the α1- but not the α2-adrenergic receptor. [Adapted from (278).]

of Ach and cholinergic receptors in reflex ovulation. Centeno et al. (262) recently reported preliminary data that the coitus-induced LH surge is attenuated with central administration of α-BTX and that mRNA expression of the rate-limiting enzyme for Ach synthesis, choline acetyltransferase, increases after mating in several brainstem regions. These authors postulate that the brainstem is where coital stimuli from the periphery are transformed into noradrenergic and cholinergic signals that stimulate GnRH release (262). In support of this, nicotinic cholinergic receptors have been found in several brainstem sites within or adjacent to NE cells in the A1, A2, and A6 areas; the number of neurons expressing these receptors increased 1 hour after mating, indicating their activation soon after mating (262). The cholinergic receptors in the A1, A2, and A6 areas may mediate effects of Ach on coitus-induced increases in TH expression and NE release.

The role of endogenous opioid peptides in the control of ovulation has been studied extensively in spontaneous ovulators [reviewed in (3,286–288)] but only to a minor extent in induced ovulators. In rabbits, naloxone (an opioid antagonist) infused either intravenously or intrahypothalamically stimulates GnRH and LH pulses in ovariectomized females but not in intact, estrous females (289). In another study, morphine-sulfate given subcutaneously to intact female rabbits inhibited LH secretion but this inhibitory effect could be prevented with naloxone (290). Naloxone increased LH pulse amplitude in ovariectomized rabbits and treatment with estrogen blocked this action (291,292). Similar findings were reported for ferrets and cats, in which naloxone given to block opioid receptors increased LH pulse frequency and promoted mating in ovariectomized females (293–295); in the case of ferrets, the stimulatory effects of naloxone were inhibited by estrogen replacement in the ovariectomized females. These data indicate that endogenous opioids might exert inhibitory actions on the GnRH-LH system, and that the hormonal milieu, in particular estrogen, may be critical for such opioid functions. However, as noted by Bakker and Baum (3), there is likely no major role of endogenous opioids in regulating the coitus-induced LH surge as naloxone treatment is only stimulatory in the absence of circulating estrogens, a physiological state which does not exist in the majority reflex ovulators at the time of mating (musk shrews being one exception).

Additional neurotransmitters and regulatory signals which have been suggested to influence the GnRH and LH surges in induced ovulators include dopamine (242), serotonin (296,297), GABA (298), and substance P (299). Except for substance P, which augmented LH release when infused in rabbits (299), each of these factors seems to play an inhibitory or modulatory role in the control of mating-induced ovulation, though to date only minimal data exists to support this assertion. Furthermore, as noted by Ramirez and Soufi (6), neither dopamine turnover rate nor serotonin levels in the hypothalamus change after mating (300), suggesting these neurotransmitters play only a modulatory role, if any, in the endogenous coital-reflex arc. Additional research is necessary to determine the roles of these putative regulators in the process of induced ovulation.

THE FINE-LINE BETWEEN INDUCED AND SPONTANEOUS OVULATORS: MALE-FACILITATED OVULATION AND PUBERTY

Other authors have commented (3), as have we in previous sections of this chapter, on cases where species assumed to exclusively display spontaneous ovulation have the capacity to display some malleability in their cycles as influenced by social cues. In the majority of these examples, exposure to males, or some social cue(s) derived from males, advances the onset of the spontaneous ovulation during the estrous cycle. Although exposure to male cues with or without physical contact cannot induce ovulation in these animals it can hasten its onset. In this section we review the literature on this topic and explore the differences and similarities between this phenomenon and copulation-induced ovulation. In a later section we discuss special instances when reflex ovulation can occur in spontaneous ovulators.

Ovulation Advance in Seasonally Breeding Ungulates

The presence of a male, or his pheromones, or in some cases even his vocalizations, can advance the onset of breeding in several species of female ungulates (301). In all cases, these effects are found in highly social animals with well-defined breeding seasons. Probably the best-studied species in this regard is the domestic sheep (*Ovis aries*). The so-called "ram-effect" has been used by farmers to accelerate the onset of the mating season by bringing female ewes into estrous as early as possible (301,302). Introduction of rams to anestrous groups of ewes results in LH surges within 3 hours in the females, followed by ovulation about nine days later (303,304). Female ewes that are pinealectomized come into breeding condition later in the season than pineal-intact females; however, when pinealectomized females are housed with rams they begin to show LH increases at the same time as control (unoperated) ewes (305). A ram, his wool, or his urine, can be used to facilitate the onset of LH surges, which culminate in ovulation in anoestrous ewes [reviewed by (301,306)]. Perhaps paradoxically, the accessory olfactory system (i.e., vomeronasal organ) is not required for this effect (307) and one report suggests that complete olfactory bulbectomy also fails to eliminate the effect of rams on LH-surge onset in multiparous ewes (308). However, because wool is a potent stimulus, current work is focused on purifying the pheromone(s) from rams' wool (303,309).

A similar male-effect on female LH secretion is noted in goats. When anestrous goats are exposed to hair from a gonadally intact male, the frequency of LH pulses increases (302). One method used to study responses to pheromones in female goats is chronic implantation of electrodes in the medial basal hypothalamus. Before, during, and after various stimuli are presented, multiple unit activity recordings are collected. Male goat hair stimulates these electrical responses within minutes, and, in addition, preovulatory-like LH surges occur directly after each bout of electrical activity (310–312). These data suggest that male pheromones have direct activational effects on GnRH release in spontaneous ovulators. An additional example is pigs (313) in which mating per se enhances the onset of ovulation and shortens its duration.

An example of male-facilitated ovulation which may not be mediated by olfaction has been reported in free-ranging red deer, which live in female herds containing a single territorial male. Prior to and during the breeding season, the territorial male engages in roaring, a vocalization that may function to repel other males and attract females to the group (314). An additional function of roaring was demonstrated by a study employing audio-playbacks from stags. Conception dates were used to measure onset of breeding. Interestingly, females exposed to calling in advance of the breeding season, via audio playback or authentic stag calling, conceived significantly earlier than control females. Whether or not ovulation actually occurred earlier was not directly assessed, but the findings are intriguing and warrant additional study.

Male-Facilitated Ovulation in Rodents that Normally Exhibit Spontaneous Cycles

There are well-documented cases in both rats and mice in which exposure to males affects the temporal onset of ovulation, and hastens its occurrence during the cycle (additional circumstances in both species when copulation can directly induce ovulation is covered in the section on Exceptions to the Rules). In one paradigm, rats with regular 5-day cycles were studied to determine whether pheromones can shorten cycle lengths (315–317). About 50% of 5-day females had 4-day cycles following exposure to male urine. Progesterone levels in ovarian venous blood were examined in females to determine the ovarian response to male urine. Females that responded to male pheromones had lower Pg levels on diestrous days than did 5-day cycling females (316). Combined with data on follicular development these data suggest that male pheromones enhance follicular development and thus shorten cycle durations from 5 to 4 days. In another paradigm, male-facilitated LH release was noted in ovariectomized adult rats treated with a modest dose (5 μg/100 g body weight) of estradiol 3 days prior to LH monitoring (318). Females were fitted with indwelling jugular cannula and several hours later exposed to a social or pheromonal stimulus. Between 3 to 4 hours later the control rats that received only estradiol experienced a mild increase in LH. However, this increase began one hour earlier in females exposed to intact males, their soiled-bedding, or castrated males. Moreover, the maximally elevated levels of LH were at least twofold higher in the female rats exposed to males (318). This effect was blocked when the procedure was repeated with females that underwent vomeronasal organ, accessory olfactory bulb, or ventral premammillary nuclei lesions (318,319).

A final paradigm developed to examine effects of pheromones on estrous cycles in rats employs estrogenized, cycling females (320). The main difference between this paradigm and the previously described method is that ovaries remain intact. A large dose of estradiol (2 mg) produces persistent vaginal estrus (PVE) which can be reversed by exposure to male urine. Interestingly, several neuroendocrine features of the cycles prior to estrogen treatment (such as plasma concentrations of estradiol and Pg) are disrupted after estrogen treatment but are subsequently reversed by urine treatment (321). When females undergo vomeronasal organ ablation prior to the estradiol treatment, PVE is still induced, but the condition is no longer reversed by male urine (322). This model is thought to reflect some aspects of persistent estrous exhibited by aging female rats.

Studies conducted in mice have focused on the role of males and male-related cues in the pubertal transition between anestrous and first ovulation. It can be argued from the perspective of mouse ecology that this is the most important period for reproduction. In the wild, few female mice have a chance to reproduce at all and still fewer can re-breed after their initial pregnancy. Moreover, given the life expectancy of wild mice the advancement of reproductive capabilities by as few as 3 days may make a long-term difference in an individual female's reproductive success

(323). Work conducted in the early 1970s identified some intriguing strain differences in the ability of females, primed with sub-threshold levels of either human chorionic gonadotropin (hCG) or pregnant mare serum gonadotropin (PMSG), to respond to male pheromones with increased incidence of ovulation and/or numbers of ova shed (324,325). The basic paradigm used prepubertal females housed across a wire-barrier from adult males for 2 days and then treated with a low dose of either hCG or PMSG; ovulation was determined 24 hours later. The percentage of mice ovulating and numbers of ova were increased by the pre-exposure of males in 3 of 5 mouse strains (325). In another study, the stimulus quality of males was examined. Strains of females that did (SWR/J) or did not (C57BL/6J) respond to male priming were used, and a cross-strain exposure of males was used. Data demonstrated that females of both strains can respond to priming when a male SWR/J or hybrid is provided but not when C57BL/6J males were used. Thus males of the C57BL/6J strain have a relatively poor capacity to induce ovulation (324). The nature of the male stimuli causing these changes was not investigated in these studies.

The classic work on puberty acceleration induced by male contact was initiated by an observation that group-housed female mice became acyclic, but that cyclicity could be restored, and often synchronized, by exposure to male urine (326). Studies of puberty acceleration (327,328) used females at critical age and weight intervals. Prepubertal female mice can experience accelerated puberty after contact with urine from a testes-intact male, an androgenized female, or an adult male (327,329). Examination of a series of male-related cues revealed that pheromones and tactile cues are critical for this effect. Mixtures of cues, such as cohabitation with a castrate and exposure to urine from another testes-intact male accelerated puberty (330) whereas exposure to just urine or urine along with a male on the other side of a screen barrier were ineffective. The best hormonal description of male-induced puberty illustrates a small surge in LH about 1 hour after contact with a male (330,331) (Fig. 9). This is followed by a surge in estradiol 12 hours later, at the time of lights off, and a second smaller estradiol surge 24 hours later. A large, ovulatory LH surge then occurs at lights off 2.5 days after the female is introduced to the male (331). Estradiol injections, in place of male exposure, can also induce the ovulatory LH surge in these females (332). The timing of these events is reminiscent of the data presented earlier in this section in ewes, rats, and other mammals that display earlier gonadotropin release in response to males. Voles are close relatives of mice, and although they display copulation-induced ovulation, many vole species also require priming with a male prior to exhibiting behavioral estrous (Table 1) (43). Unfortunately, few hormone data exist in any species of voles that map the changes in plasma levels of steroids and gonadotropins for several days over the course of estrous and ovulation induction. We speculate that if such a profile were developed, it would likely be strikingly similar to the pattern seen in mice. The only major difference would be that the beginning of behavioral estrous in the vole would start around hour 40, with mating and the induction of ovulation occurring at lights off on the

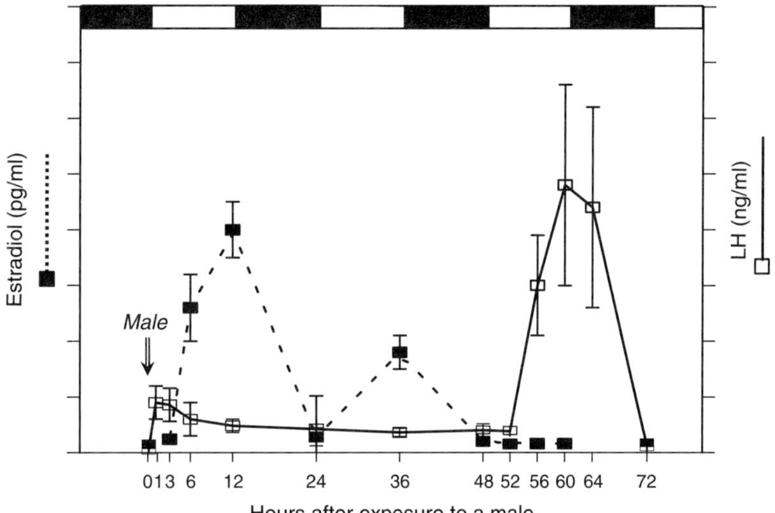

FIG. 9. Patterns of estradiol and LH in plasma of a peri-pubertal female mouse (a spontaneous ovulator in adulthood). The sampling begins when an adult male is placed with the female (at time 0) and continues for 72 hours. LH (*solid line*) increases slightly within an hour after the male is introduced and an ovulatory LH surge is noted 56 hours later. Estradiol concentrations (*dashed line*) are elevated in the first 12 hours of contact with a male and a secondary peak is seen 24 hours later (i.e., 36 hours after male exposure). At the top of the graph, the daily light-dark cycle is depicted (12 hours light, 12 hours dark per day). [Adapted from (331).]

third day of pairing (hour 56). In fact, it is very likely that at the time that *Microtus* and *Mus* diverged evolutionarily, the common ancestor had a reproductive pattern similar to the contemporary vole and the mouse subsequently evolved spontaneous estrous cycles (but with the initiation of the female's cycles still dependent on male social cues).

Section Summary

In summary, we have shown that while ovulation may not be physically induced by mating, it is certainly *facilitated* by contact with an appropriate male mating partner in a number of spontaneously ovulating species. Though it is unclear if the animals that show male-facilitated ovulation arose from species with induced ovulation, it is interesting that all of the animals discussed here are in mammalian orders that include both induced and spontaneously ovulators (Rodentia and Artiodactyla). Another common pattern is that, in the vast majority of cases, the male-related cues that facilitate ovulation are chemical cues. Whether the vomeronasal and/or main olfactory pathways are involved is not yet clear and may vary between species. However, the role of olfactory cues is reminiscent of the origin of the GnRH neurons in the olfactory placode. In many adult mammals, GnRH neurons and fibers reside in olfactory-related portions of the brain. Thus, a transduction pathway from nose to ovulation may be an important aspect of the underlying mechanisms of induced ovulation.

EXCEPTIONS TO THE RULES: HOW SPONTANEOUS BEGETS INDUCED OVULATION

Several experimental conditions (3 in rats and 1 in mice) exist which cause a shift from spontaneous ovulation to an acyclic condition typified by persistent vaginal and behavioral estrous. Under these conditions mating can induce ovulation. Much of this work was inspired by pioneering studies by Everett and Sawyer who began their investigations of induced ovulation in the late 1940s. Here we will speculate about the mechanisms that may have evolved along with the development of spontaneous ovulation which normally mask mating-induced ovulation in mice and rats.

Anesthetics and Estrous Cycles

The first study to demonstrate mechanisms underlying mating-induced ovulation in rats was reported by Harrington and his colleagues (333). This group and others (including Everett) had observed that treating females with various barbiturates or tranquilizers commonly used as anesthetics for surgery caused a temporary anovulatory condition. Chlorpromazine is a potent anti-psychotic drug that dampens activity of dopaminergic neurons by acting as an antagonist at the D2 receptor. In addition it has antagonistic actions on α-1 adrenergic receptors. Regularly cycling female rats injected with chlorpromazine became anovulatory (but this could be reversed by treatment with hCG). Of more interest, an overnight co-habitation with a sexually active, but not with a sexually inactive, adult male stimulated ovulation in more than half of the chlorpromazine treated females. A similar study used pentobarbital or phenobarbital to block ovulation in normally cycling rats (13). The mode of action of these compounds is via activation of the GABA-A receptor which affects calcium channels. Drugs were given on two consecutive days and on the third day (proestrous) females were paired with males overnight. Several rats became pregnant or pseudopregnant (implying ovulation had occurred) and those that did not showed little indication that mating had occurred. The unpaired control rats did not ovulate. Finally, when immature rats were first treated with PMS to induce follicular development and then given chlorpromazine, ovulation was blocked (8). In this study, three methods were used to induce ovulation in these acyclic rats and all were successful; mating (77% ovulated), mechanical vagino-cervical stimulation (82%), and electrical stimulation of the cervix (50% ovulated). This study provided clear-cut evidence for the display of reflex ovulation in a species that normally would only ovulate spontaneously during its estrous cycle.

Circadian Clocks and Induced Ovulation in Rats

Constant light (LL) has many effects on endogenous circadian rhythms including influencing the temporal control of corticosterone secretion and locomotor activity (334,335). In addition, when adult female rats are maintained in LL their estrous cycles cease and persistent estrous ensues, as judged by vaginal smears and behavioral receptivity (10). The vast majority of these females that experience mating, including intromissions, exhibit induced ovulation. This phenomenon cannot be explained by a contribution of adrenal steroids since adrenalectomized females housed in LL also display ovulation after mating (10). Another study by Johns et al. (336) showed that male-soiled bedding could also induce ovulation in 50% of female rats housed on constant light.

While exposure to urine alone had little effect, the action of soiled bedding was blocked by vomeronasal organ occlusion. This study has yet to be replicated, but challenges the notion that vagino-cervical stimulation is responsible for mating-induced ovulation under LL conditions.

The ability of rats housed in constant light to shift to reflex ovulation might be attributed to a disinhibition of circadian control of the LH surge. Other evidence that supports this hypothesis comes from studies that lesion the suprachiasmatic nuclei (SCN; the mammalian circadian clock) of female rats. After electrolytic lesions that remove at least 75% of the nucleus, acyclicity characterized by persistent anovulation and daily vaginal cornification was the predominant pattern. All but one of the female rats with persistent estrous displayed sexual receptivity when tested with males, and 58% ovulated (337). These lesions also disrupt daily corticosterone rhythms, pineal serotonin N-acetyl transferase activity, and locomotor activity rhythms confirming that the circadian clock was not functional (337). The effects of SCN lesions on mating-induced ovulation were replicated in a knife cut model (338). Female rats received precise neural cuts which were ventral to the SCN and extended dorsal to block passage of fibers between the preoptic area and the medial basal hypothalamus. One week after surgery, 78% of the females had persistent estrous and 62% ovulated after an overnight pairing with males that included sexual activity (338). Because the knife cuts disrupt both SCN afferents and efferents, as well as GnRH fibers en route to the median eminence, it is not clear what the underlying mechanisms may be in this case.

To differentiate between the disruptive effects of lesions on the clock versus their effects on the amount of GnRH available for release, studies have examined the effects of very small lesions in the SCN or POA on subsequent feedback regulation of LH secretion (339,340). Adult female rats were subjected to electrolytic lesions and estrous cycles were monitored. Lesions in either the SCN or the medial preoptic/anterior hypothalamus (now recognized as the anteroventral periventricular nucleus or AVPV) produced anovulation and persistent vaginal estrous. Lesions nearby these structures produced irregular estrous cycles. However only lesions in the medial preoptic, which extended a bit caudal of the nucleus, or included some of the anterior SCN, disrupted positive feedback. In an earlier study SCN-lesioned ovariectomized females had elevated levels of LH as did control rats (341). Thus, SCN lesions may disrupt estrous cyclicity and produce persistent estrous by alterations in circadian regulation of the ovulatory GnRH surge.

Aging and Persistent Estrous in Rats

It is well known that in rats, and other animals, ovarian cycles become irregular as reproductive senescence approaches (342). In middle-aged rats (about 10–12 months of age) constant lordosis can be exhibited on a daily basis for a period of several months before cycles cease and females enter permanent diestrous (342). The majority of middle-aged rats in persistent estrus (which no longer spontaneously ovulate) displayed lordosis in response to a sexually active male. These female rats also had post-mating LH surges that triggered ovulation (343,344). Interestingly, the majority of females with taped vaginas that spent the evening receiving mounts only from sexually active males also ovulated (344). Perhaps under these special conditions aged females are hypersensitive to ano-genital stimulation and thus mounting and flank stimulation promote ovulation. When treated with a dose of estradiol that produced positive feedback in young rats, the middle-aged rats did not respond with a LH release (343). Thus, in middle-aged rats, the neuroendocrine mechanism that responds to mating is intact while LH responses to steroids are not operational. Perhaps as the HPG axis ages, steroidal regulation of GnRH is disrupted, and the ability of mating to induce LH is enhanced to the point of an ovulation-producing surge. This discrepancy may be caused in part by reductions of Pg receptor mRNA present in several hypothalamic nuclei in persistent estrus middle-aged females (345).

While it is unclear why middle-aged rats display anovulatory cycles and induced ovulation, several different factors have been identified that will restore estrous cyclicity in these animals. Implants of L-dopa into the medial preoptic area, but not nearby sites temporarily restored cycles in middle-aged females in persistent estrous (346). Systemic treatment with naltrexone was also able to reinstate cycles in rats with persistent estrous (347). Perhaps dopamine and/or opioids regulate the LH surge differently in induced versus spontaneously ovulating animals. Few data are available to address this question which could be the basis for elegant comparative studies.

A final set of studies in rats asked if the same pheromonal cues that can enhance ovulation in young females can also reinstate cycles in aged rats. Aged (7–13 months of age) females were smeared daily and those with persistent vaginal cornification where subjected to daily contact with male urine or saline delivered directly on their nares (322,348). While receiving male urine, females displayed a drop in the numbers of estrous vaginal smears. Plasma LH, FSH, estradiol, Pg, and prolactin quantified two hours after male urine treatment revealed that estradiol, Pg, and FSH were all elevated by contact with male urine (348). Hypothalamic content of GnRH was also assessed and this peptide was depleted in brain after urine treatment. It is interesting that LH was not affected despite the large change in GnRH content; the reason for this outcome is not clear.

Brain Grafts Restore Fertility and Induced Ovulation in Mouse Mutants Lacking GnRH

Mice lacking functional GnRH gene were identified in the late 1970s (349). These spontaneous mutants, commonly called hypogonadal mice (*hpg*), are reproductively immature throughout life. Females have delayed or absent vaginal opening, tiny ovaries with no follicular development, and very low plasma levels of gonadotropins (i.e., no spontaneous ovulations). However, as adults reproductive function can be restored in these mice by implanting fetal grafts from the hypothalamic/preoptic into the third ventricle (350,351). Indicative of reflex ovulation, the mice with fetal grafts display constant vaginal estrus, a pronounced LH surge 10 minutes after mating, and they normally conceive on the day they are first paired with males (350–353). When grafted females were tested for positive feedback responses to Pg the majority failed to display an LH surge. However, the fact that *hpg* mice with grafts have an unusual hormonal milieu, both when gonad-intact and after ovariectomy, make this a difficult experiment to interpret (354). The *hpg* mice also exhibit a good deal of variability in their reproductive responses to the grafts, likely caused by variability in graft location and survival.

One measure of GnRH neuron activation is *cfos* expression within GnRH cells. This paradigm was used to determine if mating and/or steroids activated GnRH neurons in grafted *hpg* females (355). Ovary-intact females were primed with steroid hormones and received no mating, intromissions only, or intromissions and an ejaculation. Females primed with Pg and experiencing ejaculations or just intromissions had maximal (40%–47%) co-expression of *cfos* in GnRH neurons. This is truly a synergistic effect since females that did not receive Pg but did mate had low levels of co-expression, as did females that received Pg but no copulation. These results suggest that grafts receive afferent connections from the host tissue. In fact successful grafts that produce mice with ovarian development have more synaptophysin, a synaptic vesicle protein, containing appositions than do nonfunctional grafts (356). Where the afferents project from is unclear, but at least two neurotransmitters, β-endorphin (357) and galanin (358), have been identified in afferents that contact grafted GnRH neurons.

Section Summary

One take-home message from these studies is that the relationship between the GnRH pulse-generator and the circadian clock may be an important, yet under-studied, feature of induced ovulation. In the cases described here for rats, a loss in circadian function is coupled with the expression of induced ovulation in these normally spontaneously ovulating animals. Tranquilizers and barbiturates can reset the circadian clock (359). Both SCN lesions and constant light disrupt or alter many circadian rhythms (360). One well-characterized effect of aging is a disruption in sleep/wake cycles and dampening of other circadian rhythms (361–363). How the circadian clock regulates GnRH pulsatility is an area of active study but the primary models in use are rats, hamsters, and mice (364), all animal with spontaneous ovulation. Clearly, work on the relationship between ovulation and the SCN in an animal with induced ovulation would be extremely useful.

Unlike the rat examples, there is no known relationship between induced ovulation in *hpg* mice and circadian rhythms. Interestingly, we are aware of no studies that have documented whether *hpg* mice have normal circadian rhythms, either before or after they receive fetal grafts. In addition, several clock-gene mutant mice have been generated (365,366). Recently estrous cycles and pregnancy was monitored in the *Clock* mutant (367). Mutant females had abnormal vaginal smears with more cornified smears than did wild-type females, but no data on the timing or quality of their mating behaviors was given. We predict that these mutants may have the ability to display mating-induced ovulation. Interestingly, in Drosophila, mating behavior has a circadian rhythm, which is disrupted in period- and timeless-null mutant females (368). Certainly many interesting questions on the origins of induced ovulation could be studied in mice, including how many genetic changes are needed to shift from spontaneous to induced ovulation or vice versa.

CONCLUSIONS AND FUTURE DIRECTIONS

Mating-induced ovulation is wide-spread among different mammalian orders and species therein. As more data are made available, we predict that the actual prevalence of this pattern will prove to be even greater than is now suspected. Furthermore, when individual variability within members of the same species is studied, facultative-induced ovulation may turnout to be a common pattern in spontaneous ovulators. A number of species have male-facilitated ovulation, and thus possess the ability to fine-tune their reproductive efforts to the presence of mating partners. Many adaptive arguments can be advanced to suggest that this would be in the best interest of the female's fitness and reproductive success. It is also worth noting that even in the domestic laboratory species, estrous cycles in females are entrained to both environmental cues (light cycles) and also to social cues (326,369). Given the selective breeding that laboratory and domestic farm animals have undergone, it is safe to assume that the mechanisms that underlie these responses to males are important because they have been evolutionarily preserved.

Because induced and spontaneous ovulating species often reside in the same phylogenetic orders and families, comparative work on the mechanisms and genetics that differentiate these two systems would be extremely worthwhile. Patterns of hormone release in peri-pubertal female mice are extremely similar to what is known about male-induced estrous that proceeds mating and ovulation in voles. A related category of steroidal regulation that should be examined is the contribution of adrenal steroids in regards to puberty and/or ovulation induction. Prairie voles have very high levels of corticosterone at the time that they mate (370,371) and cortisol facilitates receptivity in female musk shrews (372). Furthermore, there are many newly discovered neurotransmitters (e.g., urocortin and orexin) that interact with the hypothalamic–pituitary–adrenal axis and the role of these factors, as well as the role of adrenal steroids, in induced ovulation is currently unknown.

Another avenue for future studies is the relationship between peripheral and neural peptides that influence metabolic signaling and induced ovulation. Peptides such as NPY, cholecystokinin (CCK), leptin, corticotrophin-releasing hormone (CRH), and GnRH II may play dual roles in, or act as intermediate steps between, the regulation of energetic balance and ovulation. The regulatory roles of some of these candidates have been well-studied in animals with spontaneous ovulation (373–377) but data in species with induced ovulation are lacking. One meritorious candidate is leptin, which is produced by fat cells but has central effects such as advancing puberty onset and stimulating LH secretion (377–379). In humans, low leptin levels are associated with hypothalamic amenorrhea (374) and recent work shows that leptin treatment can elevate LH levels in women with this condition (380). However, to our knowledge, no information exists on the role of leptin in any species that exhibit induced ovulation. In our work with musk shrews, leptin levels are undetectable which correlates nicely with very low levels of body fat in these animals (381). However, another candidate neuropeptide, GnRH II, has been shown to affect both food intake and sexual behavior in female musk shrews (382–384). While the presence of this peptide in brains of other mammals is controversial, recently we have found that central GnRH II infusions stimulate mating behavior in underfed female mice as they do in female musk shrews (382,384,385). We hypothesize that this hormone, along with other regulatory peptides, may provide an important link between energetics and ovulation in many induced ovulating species.

Activation of GnRH neurons by steroids or neurotransmitters has received little attention in induced ovulating species. The discovery of estrogen receptors (ER)-β in GnRH cells (386) opens up a new avenue to investigate in mice and rats, but no one has even mapped the neural distribution of this receptor in an induced ovulator. While *cfos* expression may be a good indicator of neural activity in mice and rats, it may not have the same role in GnRH neurons in mating-induced ovulators. It is also possible that other transcription factors, such as *c-jun*, *jun-B*, or *fra-1* (387), may be involved and these may be expressed at different times relative to mating and/or ovulation. Moreover, the GnRH promoter has yet to be sequenced in any species of induced ovulator, and thus the genetic and molecular control of GnRH production is also unknown. In fact, one of the exciting new fields in behavior genetics is how behavior affects genetic and epigenetic function. Induced ovulation could be an excellent system to study these types of effects.

The neural circuitry for induced ovulation has been well examined in rabbit and ferrets, but far less is known about other species. The sheer number of stimuli that converge on the GnRH neurons to regulate hormone release is large in both induced and spontaneous species. In ferrets, several studies have tried to parcel out the olfactory, tactile, and vagino-cervical pathways, but much more can still be done, including further defining critical brain regions and their projections as well as the numerous regulatory neurotransmitters that, in addition NE and NPY, are likely involved. The huge advantage to doing this type of work in an induced ovulator is that the timing of the cues can be synchronized (or not) by the experimenter and their impact on LH, GnRH, and/or ovulation can be assessed. In fact, induced ovulators could provide an elegant system to study the physiology of GnRH neurons relative to hormone release and mating. While some work like this has been conducted more could be done and effects of sensory inputs on different aspects of GnRH integration could be determined.

One temporary draw back to studies in species exhibiting induced ovulation is that these are not model genetic organisms. Within mammals, the human and mouse genomes have received top priority; the rat and chimpanzee are also on the list of mammals whose genomes will be mapped soon. Given the improving technology and homologies between organisms, it is likely that the genome of at least one of the commonly used species with mating-induced ovulation will be characterized in the next 5 years. This accomplishment will open up new molecular and genetic techniques that can be used to answer some basic and esoteric questions. For example, once we have identified critical genes that influence the trait we will be able to finally determine if induced ovulation is in fact the primitive, ancestral condition in mammals.

ACKNOWLEDGMENTS

The authors' work on this project has been funded by NIH grants K02 MH01349 and R01 MH068729, and by NIH grant NRSA F32 MH070084.

REFERENCES

1. Marshall, F. H. A. (1904). The estrous cycle in the common ferret. *Q. J. Microsc. Sci.* 48, 323–345.
2. Heape, W. (1905). Ovulation and degeneration of ova in the rabbit. *Proc. Roy. Soc. Lond. (Biol).* 76, 260–268.
3. Bakker, J., and Baum, M. J. (2000). Neuroendocrine regulation of GnRH release in induced ovulators. *Front. Neuroendocrinol.* 21, 220–262.
4. Hamilton-Fairley, D., and Taylor, A. (2003). Anovulation. *BMJ* 327, 546–549.
5. Ballen, A. (1997). Anovulatory infertility and ovulation induction. Policy and Practice Subcommittee of the British Fertility Society. *Hum. Reprod.* 12, 83–87.
6. Ramirez, V. D., and Soufi, W. L. (1994). The neuroendocrine control of the rabbit ovarian cycle. In *The Physiology of Reproduction* (E. Knobil and J. D. Neill, Eds.), pp. 585–611. Raven Press, New York.
7. Conaway, C. H. (1971). Ecological adaptation and mammalian reproduction. *Biol. Reprod.* 4, 239–247.
8. Zarrow, M. X., and Clark, J. H. (1968). Ovulation following vaginal stimulation in a spontaneous ovulator and its implications. *J. Endocrinol.* 40, 343–352.
9. Jochle, W. (1975). Current research in coitus-induced ovulation: a review. *J. Reprod. Fertil.* Suppl 22, 165–207.
10. Brown-Grant, K., Davidson J. M., and Greig F. (1973). Induced ovulation in albino rats exposed to constant light. *J. Endocrinol.* 57, 7–22.
11. Brown-Grant, K. (1977). The induction of pseudopregnancy and pregnancy by mating in albino rats exposed to constant light. *Horm. Behav.* 8, 62–76.
12. Brown-Grant, K. (1969). The induction of ovulation during pregnancy in the rat. *J. Endocrinol.* 43, 529–538.
13. Everett, J. W. (1967). Provoked ovulation or long-delayed pseudopregnancy from coital stimuli in barbiturate-blocked rats. *Endocrinology* 80, 145–154.
14. Weir, B. J. (1969). The induction of ovulation in the chinchilla. *J. Endocrinol.* 43, 55–60
15. Weir, B. J. (1973). The induction of ovulation and oestrus in the chinchilla. *J. Reprod. Fertil.* 33, 61–68.
16. Mendonca, M. T., and Crews, D. (1989). Effect of fall mating on ovarian development in the red-sided garter snake. *Am. J. Physiol.* 257, R1548–R1550.
17. Crews, D., Grassman, M., and Lindzey, J. (1986). Behavioral facilitation of reproduction in sexual and unisexual whiptail lizards. *Proc. Natl. Acad. Sci. U S A* 83, 9547–9550.
18. Asdell, S. A. (1964). *Patterns of Mammalian Reproduction.* Comstock Publishing, Ithaca, NY.
19. Smith, M. J. (1994). Male-induced oestrus and ovulation in female brush-tailed bettongs (Bettongia penicillata) suckling a young in the pouch. *Reprod. Fertil. Dev.* 6, 445–449.
20. Smith, M. J. (1992). Evidence from the oestrous cycle for male-induced ovulation in Bettongia penicillata (Marsupialia). *J. Reprod. Fertil.* 95, 283–289.
21. Hinds, L. A., and Smith, M. J. (1992). Evidence from plasma progesterone concentrations for male-induced ovulation in the brush-tailed bettong, Bettongia penicillata. *J. Reprod. Fertil.* 95, 291–302.
22. Johnston, S. D., McGowan, M. R., O'Callaghan, P., Cox R., and Nicolson, V. (2000). Natural and artificial methods for inducing the luteal phase in the koala (Phascolarctos cinereus). *J. Reprod. Fertil.* 120, 59–64.
23. Johnston, S. D., McGowan, M. R., O'Callaghan, P., Cox, R., and Nicolson, V. (2000). Studies of the oestrous cycle, oestrus and pregnancy in the koala (Phascolarctos cinereus). *J. Reprod. Fertil.* 120, 49–57.
24. Mate, K. E., Robinson, E. S., Vandeberg, J. L., and Pedersen, R. A. (1994). Timetable of in vivo embryonic development in the grey short-tailed opossum (Monodelphis domestica). *Mol. Reprod. Dev.* 39, 365–374.
25. Fadem, B. H. (1985). Evidence for the activation of female reproduction by males in a marsupial, the gray short-tailed opossum (Monodelphis domestica). *Biol. Reprod.* 33, 112–116.
26. Fadem, B. H., and Rayve, R. S. (1985). Characteristics of the oestrous cycle and influence of social factors in grey short-tailed opossums (Monodelphis domestica). *J. Reprod. Fertil.* 73, 337–342.
27. Fadem, B. H. (1987). Activation of estrus by pheromones in a marsupial: stimulus control and endocrine factors. *Biol. Reprod.* 36, 328–332.
28. Taggart, D. A., Shimmin, G. A., McCloud, P., and Temple-Smith, P. D. (1999). Timing of mating, sperm dynamics, and ovulation in a wild population of agile Antechinus (Marsupialia: dasyuridae). *Biol. Reprod.* 60, 283–289.
29. Crawford, J. L., McLeod, B. J., Thompson, E. G., and Hurst, P. R. (1998). Presence of males affects the incidence of ovulation after pouch young removal in brushtail possums (Trichosurus vulpecula). *Anim. Reprod. Sci.* 51, 45–55.
30. Hinds, L. A., Reader, M., Wernberg-Moller, S., and Saunders, N. R. (1992). Hormonal evidence for induced ovulation in Monodelphis domestica. *J. Reprod. Fertil.* 95, 303–312.
31. Peppler, R. D., Hossler, F. E., and Stone, S. C. (1986). Determination of reproductive maturity in the female nine-banded armadillo (Dasypus novemcinctus). *J. Reprod. Fertil.* 76, 141–146.
32. Greenwald, G. S. (1956). The reproductive cycle of the field mouse, *Mictrotus californicus*. *J. Mammalogy* 37, 213–222.
33. Hasler, J. F., Dziuk, P. J., and Banks, E. M. (1974). Ovulation and related phenomena in the collared lemming (Dicrostonyx groenlandicus). *J. Reprod. Fertil.* 38, 21–28.
34. Clulow, F. V., and Mallory, F. F. (1970). Oestrus and induced ovulation in the meadow vole, Microtus pennsylvanicus. *J. Reprod. Fertil.* 23, 341–343.
35. MacFarlane, J. D., and Taylor, J. M. (1982). Nature of estrus and ovulation in *Microtus townsendii* (Bachman). *J. Mammalogy* 63, 104–109.
36. Meek, L. R., Lee, T. M., Rogers, E. A., and Hernandez, R. G. (1994). Effect of vomeronasal organ removal on behavioral estrus and mating latency in female meadow voles (Microtus pennsylvanicus). *Biol. Reprod.* 51, 400–404.

37. Lariviere, S., and Feruson, S. H. (2003). Evolution of induced ovulation in North American carnivores. *J. Mammalogy* 84, 937–947.
38. Sawrey, D. K., and Dewsbury, D. A. (1985). Control of ovulation, vaginal estrus, and behavioral receptivity in voles (Microtus). *Neurosci. Biobehav. Rev.* 9, 563–571.
39. Rissman, E. F. (1996). Behavioral regulation of gonadotropin-releasing hormone. *Biol. Reprod.* 54, 413–419.
40. Rissman, E. F. (1989). Male-related chemical cues promote sexual receptivity in the female musk shrew. *Behav. Neural. Biol.* 51, 114–120.
41. Richmond, M., and Conaway, C. H. (1969). Induced ovulation and oestrus in *Microtus ochrogaster*. *J. Reprod. Fertil. Suppl.* 6, 357–376.
42. Carter, C. S., Witt, D. M., Schneider, J., Harris, Z. L., and Volkening, D. (1987). Male stimuli are necessary for female sexual behavior and uterine growth in prairie voles (Microtus ochrogaster). *Horm. Behav.* 21, 74–82.
43. Carter, C. S., Witt, D. M., Manock, S. R., Adams, K. A., Bahr, J. M., and Carlstead, K. (1989). Hormonal correlates of sexual behavior and ovulation in male-induced and postpartum estrus in female prairie voles. *Physiol. Behav.* 46, 941–948.
44. Cohen-Parsons, M., and Carter, C. S. (1987). Males increase serum estrogen and estrogen receptor binding in brain of female voles. *Physiol. Behav.* 39, 309–314.
45. Cohen-Parsons, M., and Roy, E. J. (1989). Social stimuli augment estrogen receptor binding in preoptic area of female prairie voles. *Brain Res.* 476, 363–366.
46. Andersson, C. B., and Gustafsson, T. O. (1982). Effect of limited and complete mating on ovaries and adrenals in bank voles, Clethrionomys glareolus. *J. Reprod. Fertil.* 64, 431–435.
47. Clarke, J. R., and Hellwing, S. (1983). Fertility of the post-partum bank vole (Clethrionomys glareolus). *J. Reprod. Fertil.* 68, 241–246.
48. Schadler, M. H., and Butterstein, G. M. (1979). Reproduction in the pine vole, *Microtus pinetorum*. *J. Mammalogy* 60, 841–844.
49. Lepri, J. J., and Vandenbergh, J. G. (1986). Puberty in pine voles, Microtus pinetorum, and the influence of chemosignals on female reproduction. *Biol. Reprod.* 34, 370–377.
50. Taylor, S. A., Salo, A. L., and Dewsbury, D. A. (1992). Estrus induction in four species of voles (Microtus). *J. Comp. Psychol.* 106, 366–373.
51. Widayati, D. T., Mekada, K., Oda, S., Zholnerovskaya, E., Zakiyan, S. M., and Fukuta, K. (2003). Reproductive features of the Russian vole in laboratory breeding. *Exp. Anim.* 52, 329–334.
52. Baum, M. J., Gerlach, J. L., Krey, L. C., and McEwen, B. S. (1986). Biochemical and radioautographic analysis of estrogen-inducible progestin receptors in female ferret brain and pituitary: correlations with effects of progesterone on sexual behavior and gonadotropin-releasing hormone-stimulated secretion of luteinizing hormone. *Brain Res.* 368, 296–309.
53. Cohen-Parsons, M., and Carter, C. S. (1988). Males increase progestin receptor binding in brain of female voles. *Physiol. Behav.* 42, 191–197.
54. Baum, M. J., and Schretlen, P. J. (1978). Oestrogenic induction of sexual behaviour in ovariectomized ferrets housed under short or long photoperiods. *J. Endocrinol.* 78, 295–296.
55. Beyer, C., and McDonald, P. (1973). Hormonal control of sexual behaviour in the female rabbit. *Adv. Reprod. Physiol.* 6, 185–219.
56. Whalen, R. E., and Hardy, D. F. (1970). Induction of receptivity in female rats and cats with estrogen and testosterone. *Physiol. Behav.* 5, 529–533.
57. Hasler, J. F., and Banks, E. M. (1976). The behavioral and somatic effects of ovariectomy and replacement therapy in female collared lemmings (Dicrostonyx groenlandicus). *Horm. Behav.* 7, 59–74.
58. Dluzen, D. E., and Carter, C. S. (1979). Ovarian hormones regulating sexual and social behaviors in female prairie voles, Microtus ochrogaster. *Physiol. Behav.* 23, 597–600.
59. Huck, U. W., Carter, C. S., and Banks, E. M. (1979). Estrogen and progesterone interactions influencing sexual and social behavior in the brown lemming, Lemmus trimucronatus. *Horm. Behav.* 12, 40–49.
60. Huck, U. W., Carter, C. S., and Banks, E. M. (1982). Natural or hormone-induced sexual and social behaviors in the female brown lemming (Lemmus trimucronatus). *Horm. Behav.* 16, 199–207.
61. Rissman, E. F., and Bronson, F. H. (1987). Role of the ovary and adrenal gland in the sexual behavior of the musk shrew, Suncus murinus. *Biol. Reprod.* 36, 664–668.
62. Fortman, M., Dellovade, T. L., and Rissman, E. F. (1992). Adrenal contribution to the induction of sexual behavior in the female musk shrew. *Horm. Behav.* 26, 76–86.
63. Rissman, E .F., Clendenon, A. L., and Krohmer, R. W. (1990). Role of androgens in the regulation of sexual behavior in the female musk shrew. *Neuroendocrinology* 51, 468–473.
64. Baum, M. J., Keverne, E. B., Everitt, B. J., Herbert, J., and De Greef, W. J. (1977). Effects of progesterone and estradiol on sexual attractivity of female rhesus monkeys. *Physiol. Behav.* 18, 659–670.
65. Beyer, C., Velazquez, J., Larsson, K., and Contreras, J. L. (1980). Androgen regulation of the motor copulatory pattern in the male New Zealand white rabbit. *Horm. Behav.* 14, 179–190.
66. Rissman, E. F. (1992). Mating induces puberty in the female musk shrew. *Biol. Reprod.* 47, 473–477.
67. Clendenon, A. L., and Rissman, E. F. (1990). Prolonged copulatory behavior facilitates pregnancy success in the musk shrew. *Physiol. Behav.* 47, 831–835.
68. Adams, C. E. (1981). Observations on the induction of ovulation and expulsion of uterine eggs in the mink, Mustela vison. *J. Reprod. Fertil.* 63, 241–248.
69. Miller, B. J., Anderson, S. H. (1989). Failure of fertilization following abbreviated copulation in the ferret (Mustela putorius furo). *J. Exp. Zool.* 249, 85–89.
70. Carroll, R. S., Erskine, M. S., Doherty, P. C., Lundell L. A., and Baum M. J. (1985). Coital stimuli controlling luteinizing hormone secretion and ovulation in the female ferret. *Biol. Reprod.* 32, 925–933.
71. Sawyer, C. H., Markee, J. E., and Hollinshead, W. H. (1947). Inhibition of ovulation in the rabbit by the adrenergic-blocking agent dibenamine. *Endocrinology* 41, 395–402.
72. Beyer, C., and Rivaud, N. (1969). Sexual behavior in pregnant and lactating domestic rabbits. *Physiol. Behav.* 4, 753–757.
73. Contreras, J. L., and Beyer, C. (1979). A polygraphic analysis of mounting and ejaculation in the New Zealand white rabbit. *Physiol. Behav.* 23, 939–943.
74. Beyer, C., Contreras, J. L., Morali, G., and Larsson, K. (1981). Effects of castration and sex steroid treatment on the motor copulatory pattern of the rat. *Physiol. Behav.* 27, 727–730.
75. Milligan, S. R. (1975). Mating, ovulation and corpus luteum function in the vole, Microtus agrestis. *J. Reprod. Fertil.* 42, 35–44.
76. Kenney, A. M., Hartung, T. G., and Dewsbury, D. A. (1979). Copulatory behavior and the initiation of pregnancy in California voles (Microtus californicus). *Brain Behav. Evol.* 16, 176–191.
77. Marie, M., and Anouassi, A. (1986). Mating-induced luteinizing hormone surge and ovulation in the female camel (Camelus dromedarius). *Biol. Reprod.* 35, 792–798.
78. Skidmore, J. A., Billah, M., and Allen, W. R. (1996). The ovarian follicular wave pattern and induction of ovulation in the mated and non-mated one-humped camel (Camelus dromedarius). *J. Reprod. Fertil.* 106, 185–192.
79. Bravo, P. W., Stabenfeldt, G. H., Fowler, M. E., and Lasley, B. L. (1992). Pituitary response to repeated copulation and/or gonadotropin-releasing hormone administration in llamas and alpacas. *Biol. Reprod.* 47, 884–888.

80. Sumar, J. B. (1999). Reproduction in female South American domestic camelids. *J. Reprod. Fertil. Suppl.* 54, 169–178.
81. Fernandez-Baca, S., Madden, D. H., and Novoa, C. (1970). Effect of different mating stimuli on induction of ovulation in the alpaca. *J. Reprod. Fertil.* 22, 261–267.
82. Foster, M. A. (1934). The reproductive cycle in the female ground squirrel, *Citellus tridecemlineatus*. *Am. J. Anat.* 54, 487–511.
83. Sawrey, D. K., and Dewsbury, D. A. (1985). Control of ovulation, vaginal estrus, and behavioral receptivity in voles (Microtus). *Neurosci. Biobehav. Rev.* 9, 563–571.
84. Gray, G. D., Zerylnick, M., Davis, H. N., and Dewsbury, D. A. (1974). Effects of variations in the male copulatory behavior on ovulation and implantation in prairie voles, Microtus ochrogaster. *Horm. Behav.* 5, 389–396.
85. Carter, C. S., and Getz, L. L. (1985). Social and hormonal determinants of reproductive patterns in the prairie vole. In *Comparative Neurobiology* (R. Gilles and J. Balthazart, Eds.), pp 18–39. Springer-Verlag, Berlin.
86. Roberts, R. L., Wolf, K. N., Sprangel, M. E., Rall, W. F, and Wildt, D. E. (1999). Prolonged mating in prairie voles (Microtus ochrogaster) increases likelihood of ovulation and embryo number. *Biol. Reprod.* 60, 756–762.
87. Fortune, J. E., Eppig, J. J., and Rissman, E. F. (1992). Mating stimulates estradiol production by ovaries of the musk shrew (Suncus murinus). *Biol. Reprod.* 46, 885–891.
88. Wildt, D. E., Seager, S. W., and Chakraborty, P. K. (1980). Effect of copulatory stimuli on incidence of ovulation and on serum luteinizing hormone in the cat. *Endocrinology* 107, 1212–1217.
89. Concannon, P., Hodgson, B., and Lein, D. (1980). Reflex LH release in estrous cats following single and multiple copulations. *Biol. Reprod.* 23, 111–117.
90. Johnson, L. M., and Gay, V. L. (1981). Luteinizing hormone in the cat. II. Mating-induced secretion. *Endocrinology* 109, 247–252.
91. Gudermuth, D. F., Newton, L., Daels, P., and Concannon, P. (1997). Incidence of spontaneous ovulation in young, group-housed cats based on serum and faecal concentrations of progesterone. *J. Reprod. Fertil. Suppl.* 51, 177–184.
92. Wildt, D. E., Chan, S. Y., Seager, S. W., and Chakraborty, P. K. (1981). Ovarian activity, circulating hormones, and sexual behavior in the cat. I. Relationships during the coitus-induced luteal phase and the estrous period without mating. *Biol. Reprod.* 25, 15–28.
93. Tsutsui, T., and Stabenfeldt, G. H. (1993). Biology of ovarian cycles, pregnancy and pseudopregnancy in the domestic cat. *J. Reprod. Fertil. Suppl.* 47, 29–35.
94. Banks, D. H., and Stabenfeldt, G. (1982). Luteinizing hormone release in the cat in response to coitus on consecutive days of estrus. *Biol. Reprod.* 26, 603–611.
95. Greulich, W. W. (1934). Artificially induced ovulation in the cat (*Felis domestica*). *Anat. Rec.* 48, 217–224.
96. Enders, R. K. (1952). Reproduction in the mink (*Mustela vison*). *Proc. Am. Phil. Soc.* 96, 691–741.
97. Ratner, A., Koenig, J. Q., and Frazier, D. T. (1971). Hypothalamic unit activity in the cat: effects of estrogen and vaginal stimulation. *Proc. Soc. Exp. Biol. Med.* 137, 321–326.
98. Sawyer, C. H. (1970). Electrophysiological correlates of release of pituitary ovulating hormones. *Fed. Proc.* 29, 1895–1899.
99. Bibeau, C. E., Tobet, S. A., Anthony, E. L. P., Carroll, R. S., Baum, M. J., and King, J. C. (1991). Vaginocervical stimulation of ferrets induces release of luteinizing hormone-releasing hormone. *J. Neuroendocrinol.* 3, 29–36.
100. Sundqvist, C., Ellis, L. C., Bartke, A. (1988). Reproductive endocrinology of the mink (Mustela vison). *Endocr. Rev.* 9, 247–266.
101. Carlyle, A., and Williams, T. D. (1961). Artificially induced ovulation in the rabbit. *J. Physiol. (Lond).* 157, 42P–43P.
102. Sawyer, C. H. (1949). Reflex induction of ovulation in the estrogen-treated rabbit by artificial vaginal stimulation. *Anat. Rec.* 103, 502.
103. Al Eknah, M. M. (2000). Reproduction in Old World camels. *Anim. Reprod. Sci.* 60–61, 583–592.
104. Chen, B. X., Yuen, Z. X., and Pan, G. W. (1985). Semen-induced ovulation in the bactrian camel (Camelus bactrianus). *J. Reprod. Fertil.* 74, 335–339.
105. Kenney, A. M., Lanier, D. L., and Dewsbury, D. A. (1977). Effects of vaginal-cervical stimulation in seven species of muroid rodents. *J. Reprod. Fertil.* 49, 305–309.
106. Breed, W. G. (1967). Ovulation in the genus Microtus. *Nature* 214, 826.
107. Rissman, E. F., and Li, X. (2000). Olfactory bulbectomy blocks mating-induced ovulation in musk shrews (Suncus murinus). *Biol. Reprod.* 62, 1052–1058.
108. Bakker, J., Kelliher, K. R., and Baum, M. J. (2001). Mating induces gonadotropin-releasing hormone neuronal activation in anosmic female ferrets. *Biol. Reprod.* 64, 1100–1105.
109. Sawyer, C. H., and Radford, H. M. (1978). Effects of intraventricular injections of norepinephrine on brain-pituitary-ovarian function in the rabbit. *Brain Res.* 146, 83–93.
110. Xu, Y. S., Wang, H. Y., Zeng, G. Q., Jiang, G. T., and Gao, Y. H. (1985). Hormone concentrations before and after semen-induced ovulation in the bactrian camel (Camelus bactrianus). *J. Reprod. Fertil.* 74, 341–346.
111. England, B. G., Foote, W. C., Matthews, D. H., Cardozo, A. G., and Riera, S. (1969). Ovulation and corpus luteum function in the llama (Lama glama). *J. Endocrinol.* 45, 505–513.
112. Gray G. D., Davis H. N., Zerylnick M., and Dewsbury D. A. (1974). Oestrus and induced ovulation in montane voles, *Microtus montanus*. *J. Reprod. Fertil.* 38, 193–196.
113. Milligan, S. R. (1975). Further observations on the influence of the social environment on ovulation in the vole, Microtus agrestis. *J. Reprod. Fertil.* 44, 543–544.
114. Adams, G. P., Griffin, P. G., and Ginther, O. J. (1989). In situ morphologic dynamics of ovaries, uterus, and cervix in llamas. *Biol. Reprod.* 41, 551–558.
115. Milligan, S. R. (1974). Social environment and ovulation in the vole, Microtus agrestis. *J. Reprod. Fertil.* 41, 35–47.
116. Baddaloo, E. G. Y., and Clulow, F. V. (1981). Effects of the male on growth, sexual maturation, and ovulation of young, female meadow voles, *Microtus pennsylvanicus*. *Can. J. Zool.* 59, 415–421.
117. Lawler, D. F., Johnston, S. D., Hegstad, R. L., Keltner, D. G., and Owens, S. F. (1993). Ovulation without cervical stimulation in domestic cats. *J. Reprod. Fertil. Suppl.* 47, 57–61.
118. Moller, O. M. (1974). Plasma progesterone before and after ovariectomy in unmated and pregnant mink, Mustela vison. *J. Reprod. Fertil.* 37, 367–372.
119. Baddaloo, E. G. Y., and Clulow, F. V. (1981). Effects of the male on growth, sexual maturation, and ovulation of young female meadow voles, *Microtus pennsylvanicus*. *Can. J. Zool.* 59, 415–421.
120. Kanematsu, S., Scaramuzzi, R. J., Hilliard, J., and Sawyer, C. H. (1974). Patterns of ovulation-inducing LH release following coitus, electrical stimulation and exogenous LH-RH in the rabbit. *Endocrinology* 95, 247–252.
121. Humphrey, R. R., Dermody, W. C., Brink, H. O., Bousley, F. G., Schottin, N. H., Sakowski, R., Vaitkus, J. W., Veloso, H. T., and Reel, J. R. (1973). Induction of luteinizing hormone (LH) release and ovulation in rats, hamsters, and rabbits by synthetic luteinizing hormone-releasing factor (LRF). *Endocrinology* 92, 1515–1526.

122. Chakraborty, P. K., Wildt, D. E., and Seager, S. W. (1979). Serum luteinizing hormone and ovulatory response to luteinizing hormone-releasing hormone in the estrous and anestrous domestic cat. *Lab. Anim. Sci.* 29, 338–344.
123. Carroll, R. S., Erskine, M. S., and Baum, M. J. (1987). Sex difference in the effect of mating on the pulsatile secretion of luteinizing hormone in a reflex ovulator, the ferret. *Endocrinology* 121, 1349–1359.
124. Dluzen, D. E., Ramirez, V. D., Carter, C. S., and Getz, L. L. (1981). Male vole urine changes luteinizing hormone-releasing hormone and norepinephrine in female olfactory bulb. *Science* 212, 573–575.
125. Phelps, C. P., Coy, D. H., Schally, A. V., and Sawyer, C. H. (1977). Blockade of LH release and ovulation in the rabbit with inhibitory analogues of luteinizing hormone releasing hormone. *Endocrinology* 100, 1526–1532.
126. Milligan, S. R. (1981). Analysis of the LH surges induced by mating and LH-RH in the vole, Microtus agrestis. *J. Reprod. Fertil.* 63, 39–45.
127. Yanaihara, T., Sakurai, Y., Okinaga, S., and Arai, K. (1973). Induction of ovulation in rabbits by synthetic luteinizing hormone-releasing hormone. *Am. J. Obstet. Gynecol.* 115, 855–856.
128. Hilliard, J., Schally, A. V., and Sawyer, C. H. (1971). Progesterone blockage of the ovulatory response to intrapituitary infusion of LHRH in rabbits. *Endocrinology* 88, 730–736.
129. Bruni, J. E., Montemurro, D. G., and Clattenburg, R. E. (1982). Ovulation in the rabbit following intraventricular and intravenous administration of synthetic Gn-RH. *Acta Endocrinol. (Copenh.).* 100, 285–289.
130. Rissman, E. F., Alones, V. E., Craig-Veit, C. B., and Millam, J. R. (1995). Distribution of chicken-II gonadotropin-releasing hormone in mammalian brain. *J. Comp. Neurol.* 357, 524–531.
131. Jemiolo, B. (1983). Ovulation and fertilization in the vole, Pitymys subterraneus. *Biol. Reprod.* 28, 523–527.
132. Spies, H. G., Pau, K. Y., and Yang, S. P. (1997). Coital and estrogen signals: a contrast in the preovulatory neuroendocrine networks of rabbits and rhesus monkeys. *Biol. Reprod.* 56, 310–319.
133. Ramirez, V. D., Pickle, R. L., and Lin, W. W. (1991). In vivo models for the study of gonadotropin and LHRH secretion. *J. Steroid Biochem. Mol. Biol.* 40, 143–154.
134. Pau, K. Y., Gliessman, P. M., Hess, D. L., Ronnekleiv, O. K., Levine, J. E., and Spies, H. G. (1990). Acute administration of estrogen suppresses LH secretion without altering GnRH release in ovariectomized rhesus macaques. *Brain Res.* 517, 229–235.
135. Levine, J. E., and Ramirez, V. D. (1982). Luteinizing hormone-releasing hormone release during the rat estrous cycle and after ovariectomy, as estimated with push-pull cannulae. *Endocrinology* 111, 1439–1448.
136. Levine, J. E., Pau, K. Y., Ramirez, V. D., and Jackson, G. L. (1982). Simultaneous measurement of luteinizing hormone-releasing hormone and luteinizing hormone release in unanesthetized, ovariectomized sheep. *Endocrinology* 111, 1449–1455.
137. Ramirez, V. D., Chen, J. C., Nduka, E., Lin, W., and Ramirez, A. D. (1986). Push-pull perfusion of the hypothalamus and the caudate nucleus in conscious, unrestrained animals. *Ann. N Y Acad. Sci.* 473, 434–448.
138. Pau, K. Y., and Spies, H. G. (1986). Estrogen-dependent effects of norepinephrine on hypothalamic gonadotropin-releasing hormone release in the rabbit. *Brain Res.* 399, 15–23.
139. Pau, K. Y., Orstead, K. M., Hess, D. L., and Spies, H. G. (1986). Feedback effects of ovarian steroids on the hypothalamic-hypophyseal axis in the rabbit. *Biol. Reprod.* 35, 1009–1023.
140. Kaynard, A. H., Pau, K. Y., Hess, D. L., and Spies, H. G. (1990). Gonadotropin-releasing hormone and norepinephrine release from the rabbit mediobasal and anterior hypothalamus during the mating-induced luteinizing hormone surge. *Endocrinology* 127, 1176–1185.
141. Lin, W. W., and Ramirez, V. D. (1991). Effect of mating behavior on luteinizing hormone-releasing hormone release in female rabbits as monitored with push-pull cannulae. *Neuroendocrinology* 53, 229–235.
142. Yang, S., Pau, K. F., Hess, D. L., and Spies, H. G. (1996). Sexual dimorphism in secretion of hypothalamic gonadotropin-releasing hormone and norepinephrine after coitus in rabbits. *Endocrinology* 137, 2683–2693.
143. Yang, S. P., Pau, K. Y., Airhart, N., and Spies, H. G. (1998). Attenuation of gonadotropin-releasing hormone reflex to coitus by alpha1-adrenergic receptor blockade in the rabbit. *Proc. Soc. Exp. Biol. Med.* 218, 204–209.
144. Lambert, G. M., Rubin, B. S., and Baum, M. J. (1992). Sexual dimorphism in the effects of mating on the in vitro release of LHRH from the ferret mediobasal hypothalamus. *Physiol. Behav.* 52, 809–813.
145. Versi, E., Chiappa, S. A., Fink, G., and Charlton, H. M. (1982). Effect of copulation on the hypothalamic content of gonadotrophic hormone-releasing hormone in the vole, Microtus agrestis. *J. Reprod. Fertil.* 64, 491–494.
146. Lambert, G. M., Rubin, B. S., and Baum, M. J. (1992). Sex difference in the effect of mating on c-fos expression in luteinizing hormone-releasing hormone neurons of the ferret forebrain. *Endocrinology* 131, 1473–1480.
147. Wersinger, S. R., and Baum, M. J. (1996). The temporal pattern of mating-induced immediate-early gene product immunoreactivity in LHRH and non-LHRH neurons of the estrous ferret forebrain. *J. Neuroendocrinol.* 8, 345–359.
148. Sagar, S. M., Sharp, F. R., and Curran, T. (1988). Expression of c-fos protein in brain: metabolic mapping at the cellular level. *Science* 240, 1328–1331.
149. Dragunow, M., and Faull, R. (1989). The use of c-fos as a metabolic marker in neuronal pathway tracing. *J. Neurosci. Methods* 29, 261–265.
150. Caba, M., Pau, K. Y., Beyer, C., Gonzalez, A., Silver, R., and Spies, H. G. (2000). Coitus-induced activation of c-fos and gonadotropin-releasing hormone in hypothalamic neurons in female rabbits. *Brain Res. Mol. Brain Res.* 78, 69–79.
151. Dellovade, T. L., Ottinger, M. A., and Rissman, E. F. (1995). Mating alters gonadotropin-releasing hormone cell number and content. *Endocrinology* 136, 1648–1657.
152. Tai, V. C., Schiml, P. A., Li, X., and Rissman, E. F. (1997). Behavioral interactions have rapid effects on immunoreactivity of prohormone and gonadotropin-releasing hormone peptide. *Brain Res.* 772, 87–94.
153. Fee, A. R., and Parkes, A. S. (1929). Studies on ovulation. I. The relation of the anterior pituitary body to ovulation in the rabbit. *J. Physiol. (Lond).* 67, 383–388.
154. Westman, A., and Jacobsohn, D. (1940). *Acta Obst. Gynecol. Scand.* 20, 392.
155. Hilliard, J., Hayward, J. N., and Sawyer, C. H. (1964). Postcoital patterns of secretion of pituitary gonadotropin and ovarian progestin in the rabbit. *Endocrinology* 75, 957–963.
156. Brown-Grant, K., El Kabir, D. J., and Fink, G. (1968). The effect of mating on pituitary luteinizing hormone and thyrotrophic hormone content in the female rabbit. *J. Endocrinol.* 41, 91–94.
157. Desjardins, C., Kirton, K. T., and Hafs, H. D. (1967). Anterior pituitary levels of FSH, LH, ACTH and prolactin after mating in female rabbits. *Proc. Soc. Exp. Biol. Med.* 126, 23–26.
158. Scaramuzzi, R. J., Blake, C. A., Papkoff, H., Hilliard, J., and Sawyer, C. H. (1972). Radioimmunoassay of rabbit luteinizing

hormone: serum levels during various reproductive states. *Endocrinology* 90, 1285–1291.
159. Dufy-Barbe, L., Franchimont, P., and Faure, J. M. (1973). Time-courses of LH and FSH release after mating in the female rabbit. *Endocrinology* 92, 1318–1321.
160. Jones, E. E., Bain, J. B., and Odell, W. D. (1976). Postcoital luteinizing hormone release in male and female rabbits as determined by radioimmunoassay. *Fertil. Steril.* 27, 848–852.
161. Osteen, K. G., and Mills, T. M. (1979). Serum LH and FSH levels in the pregnant rabbit. *Proc. Soc. Exp. Biol. Med.* 162, 454–457.
162. Kaufmann, R. A., Savoy-Moore, R. T., Subramanian, M. G., and Moghissi, K. S. (1992). Cocaine inhibits mating-induced, but not human chorionic gonadotropin-stimulated, ovulation in the rabbit. *Biol. Reprod.* 46, 641–647.
163. Goodman, A. L., and Neill, J. D. (1976). Ovarian regulation of postcoital gonadotropin release in the rabbit: reexamination of a functional role for 20 alphadihydroprogesterone. *Endocrinology* 99, 852–860.
164. Mills, T., Copland, A., and Osteen, K. (1981). Factors affecting the postovulatory surge of FSH in the rabbit. *Biol. Reprod.* 25, 530–535.
165. Mills, T. M., Copland, J. A., Coy, D. H., and Schally, A. V. (1983). Is the postovulatory release of follicle-stimulating hormone in the rabbit mediated by luteinizing hormone-releasing hormone? *Endocrinology* 113, 1020–1024.
166. Goodman, A. L. (1984). In vitro evidence of a role for inhibin in female rabbits. *Am. J. Physiol.* 246, E243–E248.
167. Shille, V. M., Munro, C., Farmer, S. W., Papkoff, H., and Stabenfeldt, G. H. (1983). Ovarian and endocrine responses in the cat after coitus. *J. Reprod. Fertil.* 69, 29–39.
168. Concannon, P. W., Lein, D. H., and Hodgson, B. G. (1989). Self-limiting reflex luteinizing hormone release and sexual behavior during extended periods of unrestricted copulatory activity in estrous domestic cats. *Biol. Reprod.* 40, 1179–1187.
169. Tauson, A. H. (1993). Effect of body condition and dietary energy supply on reproductive processes in the female mink (Mustela vison). *J. Reprod. Fertil.* Suppl 47, 37–45.
170. Tauson, A. H., Fink, R., Forsberg, M., Lagerkvist, C. G., and Wamberg, S. (2000). LH release in mink (Mustela vison). Pattern of the LH surge and effect of metabolic status. *Reprod. Nutr. Dev.* 40, 229–247.
171. Donovan, B. T., and ter Haar, M. B. (1977). Stimulation of the Hypothalamus and FSH and LH secretion in the ferret. *Neuroendocrinology* 23, 268–278.
172. Donovan, B. T., and ter Haar, M. B. (1977). Effects of luteinizing hormone releasing hormone on plasma follicle-stimulating hormone and luteinizing hormone levels in the ferret. *J. Endocrinol.* 73, 37–52.
173. Bravo, P. W., Stabenfeldt, G. H., Lasley, B. L., and Fowler, M. E. (1991). The effect of ovarian follicle size on pituitary and ovarian responses to copulation in domesticated South American camelids. *Biol. Reprod.* 45, 553–559.
174. Charlton, H. M., Naftolin, F., Sood, M. C., and Worth, R. W. (1975). The effect of mating upon LH release in make and female boles of the species *Mictrotus agrestis*. *J. Reprod. Fertil.* 42, 167–170.
175. Sutherland, R. C., Fink, G., and Charlton, H. M. (1984). Effect of mating on the metabolic activity of the brain and pituitary gland assessed by [14C]2-deoxyglucose in a reflex ovulator, the vole (Microtus agrestis). *Brain Res.* 311, 317–322.
176. Milligan, S. R., and MacKinnon, P. C. (1976). Correlation of plasma LH and prolactin levels with the fate of the corpus luteum in the vole, Microtus agrestis. *J. Reprod. Fertil.* 47, 111–113.
177. Meek, L. R., and Lee, T. M. (1994). Luteinizing hormone and prolactin in mated female meadow voles housed in long and short day lengths. *Biol. Reprod.* 51, 725–730.
178. Gray, G. D., Davis, H. N., Kenney, A. M., and Dewsbury, D. A. (1976). Effect of mating on plasma levels of LH and progesterone in montane voles (Microtus montanus). *J. Reprod. Fertil.* 47, 89–91.
179. Pau, C. Y., Pau, K. Y., Berria, M., and Spies, H. G. (2000). Ovarian influence on gonadotropin and prolactin release in mated rabbits. *Endocrine* 13, 25–35.
180. Fuchs, A. R., Cubile, L., and Dawood, M. Y. (1981). Effects of mating on levels of oxytocin and prolactin in the plasma of male and female rabbits. *J. Endocrinol.* 90, 245–253.
181. McNeilly, A. S., and Friesen, H. G. (1978). Heterologous radioimmunoassay for rabbit prolactin. *Endocrinology* 102, 1539–1547.
182. McNeilly, A. S., and Friesen, H. G. (1978). Prolactin during pregnancy and lactation in the rabbit. *Endocrinology* 102, 1548–1554.
183. Charlton, H. M., Milligan, S. R., and Versi, E. (1978). Studies on the control of the corpus luteum in the vole, Microtus agrestis. *J. Reprod. Fertil.* 52, 283–288.
184. Westlin, L. M. (1982). Increased fertility in young primiparous female bank voles, Clethrionomys glareolus, treated with prolactin or progesterone after mating. *J. Reprod. Fertil.* 66, 113–115.
185. Lagerkvist, G., Einarsson, E. J., Forsberg, M., and Gustafsson, H. (1992). Profiles of oestradiol-17 beta and progesterone and follicular development during the reproductive season in mink (Mustela vison). *J. Reprod. Fertil.* 94, 11–21.
186. Moller, O. M. (1973). The progesterone concentrations in the peripheral plasma of the mink (Mustela vison) during pregnancy. *J. Endocrinol.* 56, 121–132.
187. Martinet, L., Allais, C., and Allain, D. (1981). The role of prolactin and LH in luteal function and blastocyst growth in mink (Mustela vison). *J. Reprod. Fertil. Suppl.* 29, 119–130.
188. Villars, T. A., Erskine, M. S., Lambert, G. M., Jacobson, D., Weaver, C. E., and Baum, M. J. (1990). Endocrine correlates of mating-induced reductions in estrous behaviour in an induced ovulator, the ferret. *Horm. Behav.* 24, 198–214.
189. Elias, E., Bedrak, E., and Cohen, D. (1985). Induction of oestrus in the camel (Camelus dromedarius) during seasonal anoestrus. *J. Reprod. Fertil.* 74, 519–525.
190. Schmidt, P. M., Chakraborty, P. K., and Wildt, D. E. (1983). Ovarian activity, circulating hormones and sexual behavior in the cat. II. Relationships during pregnancy, parturition, lactation and the postpartum estrus. *Biol. Reprod.* 28, 657–671.
191. Furumura, K., Ota, K., Yokoyama, A., and Oda, S. (1983). Mammary growth and plasma progesterone level during pregnancy in the house musk shrew, Suncus murinus Linnaeus. *Endocrinol. Jpn.* 30, 621–630.
192. Hikim, A. P., Woolf, A., Bartke, A., and Amador A. G. (1992). Further observations on estrus and ovulation in woodchucks (Marmota monax) in captivity. *Biol. Reprod.* 46, 10–16.
193. Hilliard, J., Pang, C. N., Scaramuzzi, R. J., Penardi, R., and Sawyer, C. H. (1974). Secretion rates of estradiol, testosterone, and progesterone from right and left rabbit ovaries cannulated concurrently or successively. *Biol. Reprod.* 10, 364–369.
194. Bahr, J. M. (1978). Simultaneous measurement of steroids in follicular fluid and ovarian venous blood in the rabbit. *Biol. Reprod.* 18, 193–197.
195. Hilliard, J., and Eaton, L. W. Jr. (1971). Estradiol-17 beta, progesterone and 20-alpha-hydroxypregn-4-en-3-one in rabbit ovarian venous plasma. II. From mating through implantation. *Endocrinology* 89, 522–527.

196. Hayward, J. N., Hilliard, J., and Sawyer, C. H. (1964). Time of release of pituitary gonadotropin induced by electrical stimulation of the rabbit brain. *Endocrinology* 74, 108–113.
197. Hilliard, J., Spies, H. G., Lucas, L., and Sawyer, C. H. (1968). Effect of prolactin on progestin release and cholesterol storage by rabbit ovarian interstitium. *Endocrinology* 82, 122–131.
198. Bahr, J., Shahabi, N., Waldron, E., and Nalbandov, A. V. (1976). Plasma steroid concentrations in conscious and anesthetized rabbits. *Proc. Soc. Exp. Biol. Med.* 152, 210–212.
199. Fuchs, A. R., and Beling, C. (1974). Evidence for early ovarian recognition of blastocysts in rabbits. *Endocrinology* 95, 1054–1058.
200. Eiler, H., Bahr, J., and Nalbandov, A. V. (1977). Ovarian steroids in the uterine lumena. I. Effect of LH injection and mating in rabbits. *Biol. Reprod.* 17, 459–464.
201. Hasler, M. J., and Nalbandov, A. V. (1980). Ovulation, ovum maturation and changes in plasma and adrenal progesterone concentrations in the musk shrew (Suncus murinus). *Biol. Reprod.* 22, 377–381.
202. Rissman, E. F., and Crews, D. (1988). Hormonal correlates of sexual behavior in the female musk shrew: the role of estradiol. *Physiol. Behav.* 44, 1–7.
203. Hilliard, J., Penardi, R., and Sawyer, C. H. (1967). A functional role for 20-alpha-hydroxypregn-4-en-3-one in the rabbit. *Endocrinology* 80, 901–909.
204. Lin, W. W., and Ramirez, V. D. (1988). Effect of pulsatile infusion of progesterone on the in vivo activity of the luteinizing hormone-releasing hormone neural apparatus of awake unrestrained female and male rabbits. *Endocrinology* 122, 868–876.
205. Lin, W. W., and Ramirez, V. D. (1990). Infusion of progestins into the hypothalamus of female New Zealand white rabbits: effect on in vivo luteinizing hormone-releasing hormone release as determined with push-pull perfusion. *Endocrinology* 126, 261–272.
206. Kawakami, M., and Sawyer, C. H. (1959). Neuroendocrine correlates of changes in brain activity thresholds by sex steroids and pituitary hormones. *Endocrinology* 65, 652–668.
207. Rondell, P. (1970). Follicular processes in ovulation. *Fed. Proc.* 29, 1875–1879.
208. Boling, J. L., and Blandau, R. J. (1971). Egg transport through the ampullae of the oviducts of rabbits under various experimental conditions. *Biol. Reprod.* 4, 174–184.
209. Blandau, R. J., Boling, J. L., Halbert, S., and Verdugo, P. (1975). Methods for studying oviductal physiology. *Gynecol. Invest.* 6, 123–145.
210. Diakow, C. (1971). Effects of genital desensitization on mating behavior and ovulation in the female cat. *Physiol. Behav.* 7, 47–54.
211. Harris, G. W. (1937). The induction of ovulation in the rabbit by electrical stimulation of the hypothalamo-hypophysial mechanism. *Proc. Roy. Soc. Lond. (Biol).* 122, 374–394.
212. Sawyer, C. H. (1959). Effect of brain lesions on estrous behavior and reflexogenous ovulation in the rabbit. *J. Exp. Zool.* 142, 227–246.
213. Lescure, H., Dufy, B., Leonardelli, J., and Bensch, C. L. (1978). Organum vasculosum laminae terminalis and reflex ovulation in the rabbit. *Brain Res.* 154, 209–213.
214. Vincent, J. D., Dufy, B., and Faure, J. M. (1970). Effects of vaginal stimulation on hypothalamic single units in unanesthetized rabbits. *Experientia* 26, 1266–1267.
215. Porter, R. W., Cavanaugh, E. B., Critchlow, B. V., and Sawyer, C. H. (1957). Localized changes in the electrical activity of the hypothalamus in estrous cats following vaginal stimulation. *Am. J. Physiol.* 189, 145–151.
216. Kawakami, M., and Saito, H. (1967). Unit activity in the hypothalamus of the cat: effect of genital stimuli, luteinizing hormone and oxytocin. *Jpn. J. Physiol.* 17, 466–486.
217. Alcaraz, M., Guzman-Flores, C., Salas, M., and Beyer, C. (1969). Effect of estrogen on the responsivity of hypothalamic and mesencephalic neurons in the female cat. *Brain Res.* 15, 439–446.
218. Barry, J. (1976). Characterization and topography of LHRH neurons in the rabbit. *Neurosci. Lett.* 2, 201–205.
219. Foster, W. G., and Younglai, E. V. (1991). An immunohistochemical study of the GnRH neuron morphology and topography in the adult female rabbit hypothalamus. *Am. J. Anat.* 191, 293–300.
220. Ramirez, V. D., Ramirez, A. D., Slamet, W., and Nduka, E. (1986). Functional characteristics of the luteinizing hormone-releasing hormone pulse generator in conscious unrestrained female rabbits: activation by norepinephrine. *Endocrinology* 118, 2331–2339.
221. Faure, J. M., Vincent, J. D., Bensch, C., Favarel-Garrigues, B., and Dufy, B. (1967). [Elementary activity in the hypothalamus during ovulation in chronically ovulating rabbits]. *J. Physiol. (Paris)* 59, 405.
222. Kawakami, M., and Sawyer, C. H. (1959). Induction of behavioral and electroencephalographic changes in the rabbit by hormone administration or brain stimulation. *Endocrinology* 65, 631–643.
223. Belda, M., Covenas, R., Narvaez, J. A., Aguirre, J. A., and Tramu, G. (2000). Distribution of luteinizing hormone-releasing hormone in the upper brainstem and diencephalon of the cat: an immunocytochemical study. *Brain Res. Bull.* 51, 281–291.
224. Hubbard, J. I., Hyland, B. I., and Sirett, N. E. (1990). Luteinizing hormone release in the anesthetized cat following stimulation in the diagonal band of Broca, dorsal septum and fornix. *Neuroendocrinology* 52, 434–440.
225. Barry, J., and Dubois, M. P. (1975). Immunofluorescence study of LRF-producing neurons in the cat and the dog. *Neuroendocrinology* 18, 290–298.
226. Toumi, F. N., Martinet, L., and Peytevin, J. (1992). Gonadotropin-releasing hormone neurons and pathways in the brain of the female mink (Mustela vison). *Cell. Tissue Res.* 270, 383–393.
227. King, J. C., and Anthony, E. L. (1984). LHRH neurons and their projections in humans and other mammals: species comparisons. *Peptides* 5, Suppl 1, 195–207.
228. Voloschin, L. M., and Gallardo, E. A. (1976). Effect of surgical disconnection of the medial basal hypothalamus on postcoital reflex ovulation in the rabbit. *Endocrinology* 99, 959–962.
229. Brooks, C. M. (1937). The role of the cerbral cortex andof various sense organs in the excitation of mating activity in the rabbit. *Am. J. Physiol.* 120, 544–553.
230. Robison, B. L., and Sawyer, C. H. (1987). Hypothalamic control of ovulation and behavioral estrus in the cat. *Brain Res.* 418, 41–51.
231. Sirett, N. E., Hyland, B. I., Hubbard, J. I., Lapwood, K. R., and Elgar, H. J. (1986). Luteinizing hormone release in the anaesthetised cat following electrical stimulation of limbic structures. *Neuroendocrinology* 42, 128–136.
232. Reyna-Neyra, A., Camacho-Arroyo, I., Cerbon, M. A., and Gonzalez-Mariscal, G. (2000). Mating modifies c-fos expression in the brain of male and female rabbits. *Neurosci. Lett.* 284, 1–4.
233. Wersinger, S. R., and Baum, M. J. (1997). Sexually dimorphic processing of somatosensory and chemosensory inputs to forebrain luteinizing hormone-releasing hormone neurons in mated ferrets. *Endocrinology* 138, 1121–1129.
234. Bakker, J., Woodley, S. K., Kelliher, K. R, and Baum, M. J. (2002). Sexually dimorphic activation of galanin neurones in

the ferret's dorsomedial preoptic area/anterior hypothalamus after mating. *J. Neuroendocrinol.* 14, 116–125.

235. Caba, M., Bao, J., Pau, K. Y., and Spies, H. G. (2000). Molecular activation of noradrenergic neurons in the rabbit brainstem after coitus. *Brain Res. Mol. Brain Res.* 77, 222–231.
236. Sawyer, C. H., Markee, J. E., and Everett, J. W. (1950). Activation of the adenohypophesis by intravenous injection of epinephrine in the atropinized rabbit. *Endocrinology* 46, 536–543.
237. Sawyer, C. H. (1952). Stimulation of ovulation in the rabbit by the intraventricular injection of epinephrine or norepinephrine. *Anat. Rec.* 112, 385.
238. Donovan, B. T., and Harris, G. W. (1956). Adrenergic agents and the release of gonadotrophic hormone in the rabbit. *J. Physiol. (Lond).* 132, 577–585.
239. Fink, G., Smith, G. C., McMaster, R., Osborne, L. W., and Chiappa, S. A. (1975). The catecholamine-containing tubero-infundibular system and the control of luteinizing hormone release in the rabbit. *Brain Res.* 89, 71–80.
240. Metcalf, G. (1974). The use of stereotactic dissection followed by fluorimetric assay, to determine the distribution of noradrenaline, dopamine and 5-hydroxytryptamine in the preoptic hypothalamic area of rabbit brain: an alternative approach to histochemistry. *Br. J. Pharmacol.* 50, 189–195.
241. Bensch, C., Lescure, H., Robert, J., and Faure, J. M. (1975). Catecholamine histofluorescence in the median eminence of female rabbits activated by mating. *J. Neural Transm.* 36, 1–18.
242. Sawyer, C. H., Hilliard, J., Kanematsu, S., Scaramuzzi, R., and Blake, C. A. (1974). Effects of intraventricular infusions of norepinephrine and dopamine on LH release and ovulation in the rabbit. *Neuroendocrinology* 15, 328–337.
243. Atlas, S. J., and Wallach, E. E. (1991). Effects of intravenous cociaine on reproductive function in the mated rabbit. *Am. J. Obstet. Gynecol.* 165, 1785–1790.
244. Pau, K. Y., Lee, C. J., Cowles, A., Yang, S. P., Hess, D. L., and Spies, H. G. (1998). Possible involvement of norepinephrine transporter activity in the pulsatility of hypothalamic gonadotropin-releasing hormone release: influence of the gonad. *J. Neuroendocrinol.* 10, 21–29.
245. Przekop, F., and Domanski, E. (1976). Role of catecholamines in release of gonadotrophic hormones in the rabbit. *Acta Physiol. Pol.* 27, 163–168.
246. Rabii, J., Ehlers, C., Clifton, D., and Sawyer, C. H. (1980). Effects of intraventricular infusions of 6-hydroxydopamine (6-OHDA) on pituitary LH release and ovulation in the rabbit. *Neuroendocrinology* 30, 362–368.
247. Blessing, W. W., Chalmers, J. P., and Howe, P. R. (1978). Distribution of catecholamine-containing cell bodies in the rabbit central nervous system. *J. Comp. Neurol.* 179, 407–423.
248. Blessing, W. W., Goodchild, A. K., Dampney, R. A., and Chalmers, J. P. (1981). Cell groups in the lower brain stem of the rabbit projecting to the spinal cord, with special reference to catecholamine-containing neurons. *Brain Res.* 221, 35–55.
249. Blessing, W. W., Jaeger, C. B., Ruggiero, D. A., and Reis, D. J. (1982). Hypothalamic projections of medullary catecholamine neurons in the rabbit: a combined catecholamine fluorescence and HRP transport study. *Brain Res. Bull.* 9, 279–286.
250. McAllen, R. M., and Blessing, W. W. (1987). Neurons (presumably A1-cells) projecting from the caudal ventrolateral medulla to the region of the supraoptic nucleus respond to baroreceptor inputs in the rabbit. *Neurosci. Lett.* 73, 247–252.
251. Pau, K. Y., Ma, Y. J., Yu, J. H., Yang, S. P., Airhart, N., and Spies, H. G. (1997). Topographic comparison of the expression of norepinephrine transporter, tyrosine hydroxylase and neuropeptide Y mRNA in association with dopamine beta-hydroxylase neurons in the rabbit brainstem. *Brain Res. Mol. Brain Res.* 48, 367–381.
252. Jones, B. E., and Friedman, L. (1983). Atlas of catecholamine perikarya, varicosities and pathways in the brainstem of the cat. *J. Comp. Neurol.* 215, 382–396.
253. Sakai, K., Yoshimoto, Y., Luppi, P. H., Fort P., el Mansari, M., Salvert, D., and Jouvet, M. (1990). Lower brainstem afferents to the cat posterior hypothalamus: a double-labeling study. *Brain Res. Bull.* 24, 437–455.
254. Sakai, K. (1991). Physiological properties and afferent connections of the locus coeruleus and adjacent tegmental neurons involved in the generation of paradoxical sleep in the cat. *Prog. Brain Res.* 88, 31–45.
255. Wersinger, S. R., and Baum, M. J. (1997). Sexually dimorphic activation of midbrain tyrosine hydroxylase neurons after mating or exposure to chemosensory cues in the ferret. *Biol. Reprod.* 56, 1407–1414.
256. Palkovits, M. (1981). Catecholamines in the hypothalamus: an anatomical review. *Neuroendocrinology* 33, 123–128.
257. Dahlstrom, A., and Fuxe, K. (1964). Evidence for the existence of monoamine-containing neurons in the central nervous system. I. Demonstration of monoamines in the cell bodies of brainstem neurons. *Acta Physiol. Scand. Suppl.* 232, 1–55.
258. Rose, J. D., and Sutin, J. (1973). Responses of single units in the medulla to genital stimulation in estrous and anestrous cats. *Brain Res.* 50, 87–99.
259. Rose, J. D. (1975). Response properties and anatomical organization of pontine and medullary units responsive to vaginal stimulation in the cat. *Brain Res.* 97, 79–93.
260. Rose, J. D. (1973). Pontine unit responses to genital stimulation in estrous and anestrous cats. *Brain Res.* 62, 291–296.
261. Yang, S. P., Pau, K. Y., and Spies, H. G. (1997). Tyrosine hydroxylase and norepinephrine transporter mRNA levels increase in locus coeruleus after coitus in rabbits. *J. Mol. Endocrinol.* 19, 311–319.
262. Centeno, M. L., Luo, J., Lindstrom, J. M., Caba, M., and Pau, K. Y. (2004). Expression of alpha4 and alpha7 nicotinic receptors in the brainstem of female rabbits after coitus. *Brain Res.* 1012, 1–12.
263. Anselmo-Franci, J. A., Franci, C. R., Krulich, L., Antunes-Rodrigues, J., and McCann, S. M. (1997). Locus coeruleus lesions decrease norepinephrine input into the medial preoptic area and medial basal hypothalamus and block the LH, FSH and prolactin preovulatory surge. *Brain Res.* 767, 289–296.
264. Anselmo-Franci, J. A., Rocha-Barros, V. M., Franci, C. R., and McCann, S. M. (1999). Locus ceruleus lesions block pulsatile LH release in ovariectomized rats. *Brain Res.* 833, 86–92.
265. Loughlin, S. E., Foote, S. L., and Bloom, F. E. (1986). Efferent projections of nucleus locus coeruleus: topographic organization of cells of origin demonstrated by three-dimensional reconstruction. *Neuroscience* 18, 291–306.
266. Cunningham, E. T., Jr., and Sawchenko, P. E. (1988). Anatomical specificity of noradrenergic inputs to the paraventricular and supraoptic nuclei of the rat hypothalamus. *J. Comp. Neurol.* 274, 60–76.
267. Fallon, J. H., and Loughlin, S. E. (1982). Monoamine innervation of the forebrain: collateralization. *Brain Res. Bull.* 9, 295–307.
268. Mason, S. T., and Fibiger, H. C. (1979). Regional topography within noradrenergic locus coeruleus as revealed by retrograde transport of horseradish peroxidase. *J. Comp. Neurol.* 187, 703–724.

269. Palkovits, M., and Jacobowitz, D. M. (1974). Topographic atlas of catecholamine and acetylcholinesterase-containing neurons in the rat brain. II. Hindbrain (mesencephalon, rhombencephalon). *J. Comp. Neurol.* 157, 29–42.
270. Sawchenko, P. E., and Swanson, L. W. (1982). The organization of noradrenergic pathways from the brainstem to the paraventricular and supraoptic nuclei in the rat. *Brain Res.* 257, 275–325.
271. McBride, R. L., and Sutin, J. (1976). Projections of the locus coeruleus and adjacent pontine tegmentum in the cat. *J. Comp. Neurol.* 165, 265–284.
272. McBride, R. L., and Sutin, J. (1977). Amygdaloid and pontine projections to the ventromedial nucleus of the hypothalamus. *J. Comp. Neurol.* 174, 377–396.
273. Pau, K. Y., Khorram, O., Kaynard, A. H., and Spies, H. G. (1989). Simultaneous induction of neuropeptide Y and gonadotropin-releasing hormone release in the rabbit hypothalamus. *Neuroendocrinology* 49, 197–201.
274. Khorram, O., Pau, K. Y., and Spies, H. G. (1988). Release of hypothalamic neuropeptide Y and effects of exogenous NPY on the release of hypothalamic GnRH and pituitary gonadotropins in intact and ovariectomized does in vitro. *Peptides* 9, 411–417.
275. Hu, H., Rao, J. K., Prasad, C., and Jayaraman, A. (1987). Localization of neuropeptide Y-like immunoreactivity in the cat hypothalamus. *Peptides* 8, 569–573.
276. Khorram, O., Pau, K. Y., and Spies, H. G. (1987). Bimodal effects of neuropeptide Y on hypothalamic release of gonadotropin-releasing hormone in conscious rabbits. *Neuroendocrinology* 45, 290–297.
277. Leger, L., Charnay, Y., Danger, J. M., Vaudry, H., Pelletier, G., Dubois, P. M., and Jouvet, M. (1987). Mapping of neuropeptide Y-like immunoreactivity in the feline hypothalamus and hypophysis. *J. Comp. Neurol.* 255, 283–292.
278. Berria, M., Pau, K. Y., and Spies, H. G. (1991). Evidence for alpha 1-adrenergic involvement in neuropeptide Y-stimulated GnRH release in female rabbits. *Neuroendocrinology* 53, 480–486.
279. Kaynard, A. H., and Spies, H. G. (1991). Immunoneutralization of neuropeptide Y suppresses luteinizing hormone secretion in rabbits. *Endocrinology* 128, 2769–2775.
280. Massari, V. J., Hornby, P. J., Friedman, E. K., Milner, T. A., Gillis, R. A., and Gatti, P. J. (1990). Distribution of neuropeptide Y-like immunoreactive perikarya and processes in the medulla of the cat. *Neurosci. Lett.* 115, 37–42.
281. Edwards, G. L., Cunningham, J. T., Beltz, T. G., and Johnson, A. K. (1989). Neuropeptide Y-immunoreactive cells in the caudal medulla project to the median preoptic nucleus. *Neurosci. Lett.* 105, 19–26.
282. Yang, S. N., Tang, Y. H., Yang, X. L., and Wang, S. (1992). Effects of neuropeptide Y injected into A1 noradrenergic nucleus on blood pressure and catecholamines in the plasma of cats. *Zhongguo Yao Li Xue Bao* 13, 400–402.
283. Sawyer, C. H. (1964). Control of secretion of gonadotropins. In *Gonadotropins* (H. H. Cole, Ed.), pp. 113–159. W. H. Freeman, San Francisco.
284. Sawyer, C. H., Markee, J. E., and Townsend, B. F. (1949). Cholinergic and adrenergic components in the neurohumoral control of LH release in the rabbit. *Endocrinology* 44, 18–37.
285. Sawyer, C. H., Markee, J. E., and Everett, J. W. (1951). Blockade of neurogenic stimulation of the rabbit adenohypophysis by banthine. *Am. J. Physiol.* 166, 223–228.
286. Ferin, M., Van Vugt, D., and Wardlaw, S. (1984). The hypothalamic control of the menstrual cycle and the role of endogenous opioid peptides. *Recent Prog. Horm. Res.* 40, 441–485.
287. Yen, S. S., Quigley, M. E., Reid, R. L., Ropert, J. F., and Cetel, N. S. (1985). *Neuroendocrinology* of opioid peptides and their role in the control of gonadotropin and prolactin secretion. *Am. J. Obstet. Gynecol.* 152, 485–493.
288. Piva, F., Limonta, P., Dondi, D., Pimpinelli, F., Martini, L., and Maggi, R. (1995). Effects of steroids on the brain opioid system. *J. Steroid Biochem. Mol. Biol.* 53, 343–348.
289. Orstead, K. M., and Spies, H. G. (1987). Inhibition of hypothalamic gonadotropin-releasing hormone release by endogenous opioid peptides in the female rabbit. *Neuroendocrinology* 46, 14–23.
290. YoungLai, E. V., and Byrne, A. (1989). Opioidergic control of gonadotropin secretion in the female rabbit: divergent effects of morphine on secretion of follicle-stimulating hormone and luteinizing hormone. *Can. J. Physiol. Pharmacol.* 67, 1486–1492.
291. YoungLai, E. V., Wilkinson, M., and Thompson, N. (1988). Effects of ovariectomy and estradiol replacement on naloxone induced LH secretion in the female rabbit. *J. Steroid Biochem.* 29, 347–351.
292. Younglai, E. V., Thompson, N., and Wilkinson, M. (1988). Opioidergic control of luteinizing hormone release in the female rabbit: influence of ovariectomy and steroid replacement on pulsatile secretion. *Biol. Reprod.* 39, 630–637.
293. Lambert, G. M., Erskine, M. S., and Baum, M. J. (1990). Effect of naloxone on the pulsatile secretion of luteinizing hormone in gonadectomized male and female ferrets before and after oestradiol replacement. *J. Neuroendocrinol.* 2, 701–705.
294. Hubbard, J. I., Hyland, B. I., and Sirett, N. E. (1990). Stimulation of the LH release by naloxone in anaesthetized cats after ovariectomy. *Neurosci. Lett.* 108, 295–302.
295. Aiudi, G., Cinone, M., Sciorsci, R. L., and Minoia, P. (2001). Induction of fertile oestrus in cats by administration of hCG and calcium-naloxone. *J. Reprod. Fertil.* Suppl 57, 335–337.
296. Przekop, F., Skubiszewski, B., and Domanski, E. (1975). The effects of indoleamines (serotonin and melatonin) on induction of ovulation in rabbits. *Acta Physiol. Pol.* 26, 395–398.
297. Mishra, N., Tangri, K. K., Bhargava, K. P., and Gupta M. L. (1990). Evidence for the inhibitory effect of 5-hydroxytryptamine at central and peripheral sites on ovulation in rabbits. *Clin. Exp. Pharmacol. Physiol.* 17, 595–599.
298. Marshall, F. H. A., Verney, E. B., and Vogt, M. (1939). The occurrance of ovulation in the rabbit as a result of stimulation of the central nervous system by drugs. *J. Physiol. (Lond).* 97, 128–132.
299. Traczyk, W. Z., Pau, K. Y., Kaynard, A. H., and Spies, H. G. (1992). Modulatory role of substance P on gonadotropin and prolactin secretion in the rabbit. *J. Physiol. Pharmacol.* 43, 279–297.
300. Endroczi, E. (1978). Monoaminergic control of pituitary gonadotropin secretion in refletory ovulatory animals. In *Clinical Psychoneuroendocrinology in Reproduction* (L. Carenza, P. Pancheri, and L. Zichella Eds.), pp. 177–188. Academic Press, London.
301. Rekwot, P. I., Ogwu, D., Oyedipe, E. O., and Sekoni, V. O. (2001). The role of pheromones and biostimulation in animal reproduction. *Anim. Reprod. Sci.* 65, 157–170.
302. Martin, G., Oldham, C. M., Cognie, Y., and Pearce, D. T. (1986). The physiological responses of anovulatory ewes to the introduction of rams-a review. *Livestock Prod. Sci.* 15, 219–247.
303. Over, R., Cohen-Tannoudji, J., Dehnhard, M., Claus, R., and Signoret, J. P. (1990). Effect of pheromones from male goats on LH-secretion in anoestrous ewes. *Physiol. Behav.* 48, 665–668.
304. Rosa, H. J., Juniper, D. T., and Bryant, M. J. (2000). Effects of recent sexual experience and melatonin treatment of rams on plasma testosterone concentration, sexual behaviour and ability to induce ovulation in seasonally anoestrous ewes. *J. Reprod. Fertil.* 120, 169–176.

305. Wayne, N. L., Malpaux, B., and Karsch, F. J. (1989). Social cues can play a role in timing onset of the breeding season of the ewe. *J. Reprod. Fertil.* 87, 707–713.
306. Signoret, J. P. (1991). Sexual pheromones in the domestic sheep: importance and limits in the regulation of reproductive physiology. *J. Steroid Biochem. Mol. Biol.* 39, 639–645.
307. Cohen-Tannoudji, J., Lavenet, C., Locatelli, A., Tillet, Y., and Signoret, J. P. (1989). Non-involvement of the accessory olfactory system in the LH response of anoestrous ewes to male odour. *J. Reprod. Fertil.* 86, 135–144.
308. Cohen-Tannoudji, J., Locatelli, A., and Signoret, J. P. (1986). Non-pheromonal stimulation by the male of LH release in the anoestrous ewe. *Physiol. Behav.* 36, 921–924.
309. Cohen-Tannoudji, J., Einhorn, J., and Signoret, J. P. (1994). Ram sexual pheromone: first approach of chemical identification. *Physiol. Behav.* 56, 955–961.
310. Iwata, E., Wakabayashi, Y., Matsuse, S., Kikusui, T., Takeuchi ,Y., and Mori, Y. (2001). Induction of primer pheromone production by dihydrotestosterone in the male goat. *J. Vet. Med. Sci.* 63, 347–348.
311. Hamada, T., Nakajima, M., Takeuchi, Y., and Mori, Y. (1996). Pheromone-induced stimulation of hypothalamic gonadotropin-releasing hormone pulse generator in ovariectomized, estrogen-primed goats. *Neuroendocrinology* 64, 313–319.
312. Iwata, E., Wakabayashi, Y., Kakuma, Y., Kikusui, T., Takeuchi, Y., and Mori, Y. (2000). Testosterone-dependent primer pheromone production in the sebaceous gland of male goat. *Biol. Reprod.* 62, 806–810.
313. Signoret, J. P., Du Mesnil du Buisson, F., and Mauleon, P. (1972). Effect of mating on the onset and duration of ovulation in the sow. *J. Reprod. Fertil.* 31, 327–330.
314. McComb, K. (1987). Roaring by red deer stags advances the date of oestrus in hinds. *Nature* 330, 648–649.
315. Aron, C. (1971). Presumed involvement of pheromones in mating behavior in the rat. *Horm. Behav.* 2, 315–323.
316. Chateau, D., Roos, J., Plas-Roser, S., Roos, M., and Aron, C. (1976). Hormonal mechanisms involved in the control of oestrous cycle duration by the odour of urine in the rat. *Acta Endocrinol. (Copenh).* 82, 426–435.
317. Aron, C. (1979). Mechanisms of control of the reproductive function by olfactory stimuli in female mammals. *Physiol. Rev.* 59, 229–284.
318. Beltramino, C., and Taleisnik, S. (1983). Release of LH in the female rat by olfactory stimuli. Effect of the removal of the vomeronasal organs or lesioning of the accessory olfactory bulbs. *Neuroendocrinology* 36, 53–58.
319. Beltramino, C., and Taleisnik, S. (1985). Ventral premammillary nuclei mediate pheromonal-induced LH release stimuli in the rat. *Neuroendocrinology* 41, 119–124.
320. Mora, O. A., and Cabrera, M. M. (1994). Pheromonal male-induced diestrus and cyclicity in aging intact and young estrogenized female rats. *Biol. Reprod.* 50, 603–606.
321. Mora, A., and Sanchez-Criado, J. E. (2002). Effect of adrenergic blockade on the pheromonal restoration of cyclic activity in young oestrogen-primed persistent oestrous female rats. *J. Neuroendocrinol.* 14, 200–206.
322. Mora, O. A., and Cabrera, M. M. (1997). The pheromonal restoration of cyclic activity in young estrogenized persistent estrus female rats is a vomeronasal effect. *Life Sci.* 60, 493–498.
323. Bronson, F. H. (1979). The reproductive ecology of the house mouse. *Q. Rev. Biol.* 54, 265–299.
324. Eleftheriou, B. E., Bailey, D. W., and Zarrow, M. X. (1972). A gene controlling male pheromonal facilitation of PMSG-induced ovulation in mice. *J. Reprod. Fertil.* 31, 155–158.
325. Zarrow, M. X., Eleftheriou, B. E., and Denenberg, V. H. (1972). Pheromonal facilitation of HCG-induced ovulation in different strains of immature mice. *Biol. Reprod.* 6, 277–280.
326. Whitten, W. K. (1959). Occurrence of anoestrus in mice caged in groups. *J. Endocrinol.* 18, 102–107.
327. Vandenbergh, J. G. (1967). Effect of the presence of a male on the sexual maturation of female mice. *Endocrinology* 81, 345–349.
328. Bronson, F. H. (1983). Hormonal responses to primer pheromones. In *Pheromones and Reproduction in Mammals* (J. G. Vandenbergh, Ed.). Academic Press, New York.
329. Bronson, F. H., and Whitten, W. K. (1968). Oestrus-accelerating pheromone of mice: assay, androgen-dependency and presence in bladder urine. *J. Reprod. Fertil.* 15, 131–134.
330. Bronson, F. H., and Maruniak, J. A. (1975). Male-induced puberty in female mice: evidence for a synergistic action of social cues. *Biol. Reprod.* 13, 94–98.
331. Bronson, F. H., and Desjardins, C. (1974). Circulating concentrations of FSH, LH, estradiol, and progesterone associated with acute, male-induced puberty in female mice. *Endocrinology* 94, 1658–1668.
332. Bronson, F. H. (1975). Male-induced precocial puberty in female mice: confirmation of the role of estrogen. *Endocrinology* 96, 511–514.
333. Harrington, F. E., Eggert, R. G., Wilbur, R. D., and Linkenheimer, W. H. (1966). Effect of coitus on chlorpromazine inhibition of ovulation in the rat. *Endocrinology* 79, 1130–1134.
334. Morimoto, Y., Oishi, T., Arisue, K., Ogawa, Z., and Tanaka, F. (1975). Circadian rhythm of plasma corticosteroid in adult female rats: chronological shifts in abnormal lighting regimens and connection with oestrous cycle. *Acta Endocrinol. (Copenh).* 80, 527–541.
335. Depres-Brummer, P., Levi, F., Metzger, G., and Touitou, Y. (1995). Light-induced suppression of the rat circadian system. *Am. J. Physiol.* 268, R1111–R1116.
336. Johns, M. A., Feder, H. H., Komisaruk, B. R., and Mayer, A. D. (1978). Urine-induced reflex ovulation in anovulatory rats may be a vomeronasal effect. *Nature* 272, 446–448.
337. Brown-Grant, K., and Raisman, G. (1977). Abnormalities in reproductive function associated with the destruction of the suprachiasmatic nuclei in female rats. *Proc. R. Soc. Lond. B. Biol. Sci.* 198, 279–296.
338. Taleisnik, S., Sherwood, M. R., and Raisman, G. (1979). Dissociation of spontaneous and mating induced ovulation by frontal hypothalamic deafferentations in the rat. *Brain Res.* 169, 155–162.
339. Wiegand, S. J., Terasawa, E., Bridson, W. E., and Goy, R. W. (1980). Effects of discrete lesions of preoptic and suprachiasmatic structures in the female rat. Alterations in the feedback regulation of gonadotropin secretion. *Neuroendocrinology* 31, 147–157.
340. Wiegand, S. J., and Terasawa, E. (1982). Discrete lesions reveal functional heterogeneity of suprachiasmatic structures in regulation of gonadotropin secretion in the female rat. *Neuroendocrinology* 34, 395–404.
341. Antunes-Rodrigues, J., and McCann, S. M. (1967). Effect of suprachiasmatic lesions on the regulation of luteinizing hormone secretion in the female rat. *Endocrinology* 81, 666–670.
342. LeFevre, J., and McClintock, M. K. (1988). Reproductive senescence in female rats: a longitudinal study of individual differences in estrous cycles and behavior. *Biol. Reprod.* 38, 780–789.
343. Matt, D. W., Coquelin, A., and Lu, J. K. (1987). Neuroendocrine control of luteinizing hormone secretion

and reproductive function in spontaneously persistent-estrous aging rats. *Biol. Reprod.* 37, 1198–1206.

344. Day, J. R., Morales, T. H., and Lu, J. K. (1988). Male stimulation of luteinizing hormone surge, progesterone secretion and ovulation in spontaneously persistent-estrous, aging rats. *Biol. Reprod.* 38, 1019–1026.
345. Mills, R. H., Romeo, H. E., Lu, J. K., and Micevych, P. E. (2002). Site-specific decrease of progesterone receptor mRNA expression in the hypothalamus of middle-aged persistently estrus rats. *Brain Res.* 955, 200–206.
346. Cooper, R. L., Brandt, S. J., Linnoila, M., and Walker, R. F. (1979). Induced ovulation in aged female rats by L-dopa implants into the medial preoptic area. *Neuroendocrinology* 28, 234–240.
347. Field, E. A., and Kuhn, C. M. (1989). Opiate antagonist treatment reinstates estrous cycles in middle-aged persistent-estrous rats. *Biol. Reprod.* 40, 714–719.
348. Mora, O. A., Cabrera, M. M., and Sanchez-Criado, J. E. (1994). Hormonal pattern of the pheromonal restoration of cyclic activity in aging irregularly cycling and persistent-estrus female rats. *Biol. Reprod.* 51, 920–925.
349. Cattanach, B. M., Iddon, C. A., Charlton, H. M., Chiappa, S. A., and Fink, G. (1977). Gonadotrophin-releasing hormone deficiency in a mutant mouse with hypogonadism. *Nature* 269, 338–340.
350. Gibson, M. J., Moscovitz, H. C., Kokoris, G. J., and Silverman, A. J. (1987). Plasma LH rises rapidly following mating in hypogonadal female mice with preoptic area (POA) brain grafts. *Brain Res.* 424, 133–138.
351. Gibson, M. J., Moscovitz, H. C., Kokoris, G. J., and Silverman, A. J. (1987). Female sexual behavior in hypogonadal mice with GnRH-containing brain grafts. *Horm. Behav.* 21, 211–222.
352. Gibson, M. J., Krieger, D. T., Charlton, H. M., Zimmerman, E. A., Silverman, A. J., and Perlow, M. J. (1984). Mating and pregnancy can occur in genetically hypogonadal mice with preoptic area brain grafts. *Science* 225, 949–951.
353. Gibson, M. J., Charlton, H. M., Perlow, M. J., Zimmerman, E. A., Davies, T. F., and Krieger, D. T. (1984). Preoptic area brain grafts in hypogonadal (hpg) female mice abolish effects of congenital hypothalamic gonadotropin-releasing hormone (GnRH) deficiency. *Endocrinology* 114, 1938–1940.
354. Gibson, M. J., Kokoris, G. J., and Silverman, A. J. (1988). Positive feedback in hypogonadal female mice with preoptic area brain transplants. *Neuroendocrinology* 48, 112–119.
355. Wu, T. J., Silverman, A. J., and Gibson, M. J. (1996). FOS expression in grafted gonadotropin-releasing hormone neurons in hypogonadal mouse: mating and steroid induction. *J. Neurobiol.* 31, 67–76.
356. Rajendren, G., and Gibson, M. J. (2001). Increased synaptic input to gonadotropin releasing hormone cells in preoptic area grafts that support reproductive development in female hypogonadal mice. *J. Neuroendocrinol.* 13, 270–274.
357. Miller, G. M., Silverman, A. J., Rogers, M. C., and Gibson, M. J. (1995). Neuromodulation of transplanted gonadotropin-releasing hormone neurons in male and female hypogonadal mice with preoptic area brain grafts. *Biol. Reprod.* 52, 572–583.
358. Rajendren, G. (2002). Increased galanin synapses onto activated gonadotropin-releasing hormone neuronal cell bodies in normal female mice and in functional preoptic area grafts in hypogonadal mice. *J. Neuroendocrinol.* 14, 435–441.
359. Turek, F. W., and Losee-Olson, S. H. (1987). Dose response curve for the phase-shifting effect of triazolam on the mammalian circadian clock. *Life Sci.* 40, 1033–1038.
360. Weaver, D. R. (1998). The suprachiasmatic nucleus: a 25-year retrospective. *J. Biol. Rhythms* 13, 100–112.
361. Pandi-Perumal, S. R., Seils, L. K., Kayumov, L., Ralph, M. R., Lowe, A., Moller, H., and Swaab, D. F. (2002). Senescence, sleep, and circadian rhythms. *Ageing Res. Rev.* 1, 559–604.
362. Weitzman, E. D., Moline, M. L., Czeisler, C. A., and Zimmerman, J. C. (1982). Chronobiology of aging: temperature, sleep-wake rhythms and entrainment. *Neurobiol. Aging* 3, 299–309.
363. Welsh, D. K., Richardson, G. S., and Dement, W. C. (1986). Effect of age on the circadian pattern of sleep and wakefulness in the mouse. *J. Gerontol.* 41, 579–586.
364. Barbacka-Surowiak, G., Surowiak, J., and Stoklosowa, S. (2003). The involvement of suprachiasmatic nuclei in the regulation of estrous cycles in rodents. *Reprod. Biol.* 3, 99–129.
365. King, D. P., and Takahashi, J. S. (2000). Molecular genetics of circadian rhythms in mammals. *Annu. Rev. Neurosci.* 23, 713–742.
366. Allada, R., Emery, P., Takahashi, J. S., and Rosbash, M. (2001). Stopping time: the genetics of fly and mouse circadian clocks. *Annu. Rev. Neurosci.* 24, 1091–1119.
367. Miller, B., Olson, S. L., Turek, F. W., Levine, J. E., Horotn, T. H., and Takshashi, J. S. (2004). Circadian clock mutant disrupts estrous cyclicity and maintenance of pregnancy. *Curr. Biol.* 10, 1367–1373.
368. Sakai T., and Ishida, N. (2001). Circadian rhythms of female mating activity governed by clock genes in Drosophila. *Proc. Natl. Acad. Sci. U S A* 98, 9221–9225.
369. McClintock, M. K. (1984). Estrous synchrony: modulation of ovarian cycle length by female pheromones. *Physiol. Behav.* 32, 701–705.
370. DeVries, A. C., DeVries, M. B., Taymans, S. E., and Carter, C. S. (1996). The effects of stress on social preferences are sexually dimorphic in prairie voles. *Proc. Natl. Acad. Sci. U S A* 93, 11980–11984.
371. Taymans, S. E., DeVries, A. C., DeVries, M. B., Nelson, R. J., Friedman, T. C., Castro, M., Detera-Wadleigh, S., Carter, C. S., and Chrousos, G. P. (1997). The hypothalamic-pituitary-adrenal axis of prairie voles (Microtus ochrogaster): evidence for target tissue glucocorticoid resistance. *Gen. Comp. Endocrinol.* 106, 48–61.
372. Schiml, P. A., and Rissman, E. F. (1999). Cortisol facilitates induction of sexual behavior in the female musk shrew (Suncus murinus). *Behav. Neurosci.* 113, 166–175.
373. Catzeflis, C., Pierroz, D. D., Rohner-Jeanrenaud, F., Rivier, J. E., Sizonenko, P. C., and Aubert, M. L. (1993). Neuropeptide Y administered chronically into the lateral ventricle profoundly inhibits both the gonadotropic and the somatotropic axis in intact adult female rats. *Endocrinology* 132, 224–234.
374. Miller, K. K., Parulekar, M. S., Schoenfeld, E., Anderson, E., Hubbard, J., Klibanski, A., and Grinspoon, S. K. (1998). Decreased leptin levels in normal weight women with hypothalamic amenorrhea: the effects of body composition and nutritional intake. *J. Clin. Endocrinol. Metab.* 83, 2309–2312.
375. Geary, N. (2001). Estradiol, CCK and satiation. *Peptides* 22, 1251–1263.
376. Aubert, M. L., Pierroz, D. D., Gruaz, N. M., d'Alleves, V., Vuagnat, B. A., Pralong, F. P., Blum, W. F., and Sizonenko, P. C. (1998). Metabolic control of sexual function and growth: role of neuropeptide Y and leptin. *Mol. Cell Endocrinol.* 140, 107–113.
377. Schneider, J. E. (2004). Energy balance and reproduction. *Physiol. Behav.* 81, 289–317.
378. Dearth, R. K., Hiney, J. K., and Dees, W. L. (2000). Leptin acts centrally to induce the prepubertal secretion of luteinizing hormone in the female rat. *Peptides* 21, 387–392.
379. Mann, D. R., and Plant, T. M. (2002). Leptin and pubertal development. *Semin. Reprod. Med.* 20, 93–102.
380. Welt, C. K., Chan, J. L., Bullen, J., Murphy, R., Smith, P., DePaoli, A. M., Karalis, A., and Mantzoros, C. S. (2004).

Recombinant human leptin in women with hypothalamic amenorrhea. *N. Engl. J. Med.* 351, 987–997.
381. Temple, J. L., Schneider, J. E., Scott, D. K., Korutz, A., and Rissman, E. F. (2002). Mating behavior is controlled by acute changes in metabolic fuels. *Am. J. Physiol. Regul. Integr. Comp. Physiol.* 282, R782–R790.
382. Kauffman, A. S., and Rissman, E. F. (2004). A critical role for the evolutionarily conserved gonadotropin-releasing hormone II: mediation of energy status and female sexual behavior. *Endocrinology* 145, 3639–3646.
383. Kauffman, A. S., and Rissman, E. F. (2004). The evolutionarily conserved gonadotropin-releasing hormone II modifies food intake. *Endocrinology* 145, 686–691.
384. Kauffman, A. S. (2004). Emerging functions of gonadotropin-releasing hormone ii in mammalian physiology and behaviour. *J. Neuroendocrinol.* 16, 794–806.
385. Temple, J. L., Millar, R. P., and Rissman, E. F. (2003). An evolutionarily conserved form of gonadotropin-releasing hormone coordinates energy and reproductive behavior. *Endocrinology* 144, 13–19.
386. Legan, S. J., and Tsai, H. W. (2003). Oestrogen receptor-alpha and -beta immunoreactivity in gonadotropin-releasing hormone neurones after ovariectomy and chronic exposure to oestradiol. *J. Neuroendocrinol.* 15, 1164–1170.
387. Robertson, H. A. (1992). Immediate-early genes, neuronal plasticity, and memory. *Biochem. Cell Biol.* 70, 729–737.
388. Pearson, O. P. (1944). Reproduction in the shrew (Blarina brevicauda). *Am. J. Anat.* 75, 39.
389. Nowak, R. N., and Paradiso, J. L. (1983). *Walker's Mammals of the World*. Johns Hopkins University Press, Baltimore.
390. Dryden, G. L. (1969). Reproduction in *Suncus murinus*. *J. Reprod. Fertil. Suppl.* 6, 377–396.
391. Rissman, E. F., Silveira, J., and Bronson, F. H. (1988). Patterns of sexual receptivity in the female musk shrew (Suncus murinus). *Horm. Behav.* 22, 186–193.
392. Rissman, E. F., Nelson, R. J., Blank, J. L., and Bronson, F. H. (1987). Reproductive response of a tropical mammal, the musk shrew (Suncus murinus), to photoperiod. *J. Reprod. Fertil.* 81, 563–566.
393. Hasler, J. F. (1975). A review of reproduction and sexual maturation in the microtine rodents. *Biologist* 57, 52–86.
394. Clarke, J. R., Clulow, F. V., and Grieg, F. (1970). Ovulation in the bank vole, Clethrionomys glareolus. *J. Reprod. Fertil.* 23, 531.
395. Frankiewicz, J., and Marchlewska-Koj, A. (1985). Effects of conspecifics on sexual maturation in female European pine voles (*Pitymys subterraneus*). *J. Reprod. Fertil.* 74, 153–156.
396. Goto, N., and Hashizume, R. (1978). Pattern of ovulation in *Microtus montebelli*. *J. Mammal Soc. Japn.* 4, 181–188.
397. Carter, C. S., Getz, L. L., Gavish, L., McDermott, J. L., and Arnold, P. (1980). Male-related pheromones and the activation of female reproduction in the prairie vole (Microtus ochrogaster). *Biol. Reprod.* 23, 1038–1045.
398. Hnatczuk, O. C., and Morrell, J. I. (1995). Interaction of male sensory cues and estradiol in the induction of estrus in the prairie vole. *Physiol. Behav.* 58, 785–790.
399. Breed, W. G., and Clarke, J. R. (1970). Ovulation and associated histological changes in the ovary following coitus in the vole (Microtus agrestis). *J. Reprod. Fertil.* 22, 173–175.
400. Breed, W. G. (1969). Oestrus and ovarian histology in the lactating vole (Microtus agrestis). *J. Reprod. Fertil.* 18, 33–42.
401. Breed, W. G., and Clarke, J. R. (1970). Effect of photoperiod on ovarian function in the vole, Microtus agrestis. *J. Reprod. Fertil.* 23, 189–192.
402. Landau, I. T., and Holmes, W. G. (1988). Mating of captive thirteen-lined ground squirrels and the annual timing of estrus. *Horm. Behav.* 22, 474-487.
403. Beyer, C., Rivaud, N., and Cruz, M. L. (1970). Initiation of sexual behavior in prepuberally ovariectomized rabbits. *Endocrinology* 86, 171–174.
404. Fee, A. R., and Parkes, A. S. (1930). Studies on ovulation. III Effect of vaginal anesthesia on ovulation in the rabbit. *Am. J. Physiol.* 70, 385–388.
405. Brooks, C. M. (1935). Studies in the neural basis of ovulation in the rabbit. *Am. J. Physiol.* 113, 18–19.
406. Sundqvist, C., Amador, A. G., and Bartke, A. (1989). Reproduction and fertility in the mink (Mustela vision). *J. Reprod. Fertil.* 85, 413–441.
407. Wildt, D. E., Guthrie, S. C., and Seager, S. W. (1978). Ovarian and behavioral cyclicity of the laboratory maintained cat. *Horm. Behav.* 10, 251–257.
408. Boone, W. R., Keck, B. B., Catlin, J. C., Casey, K. J., Boone, E. T., Dye, P. S., Schuett, R. J., Tsubota, T., and Bahr, J. C. (2004). Evidence that bears are induced ovulators. *Theriogenology* 61, 1163–1169.
409. Novoa, C. (1970). Reproduction in Camelidae. *J. Reprod. Fertil.* 22, 3–20.
410. Skidmore, J. A., Billah, M., and Allen, W. R. (1995). The ovarian follicular wave pattern in the mated and non-mated dromedary camel (Camelus dromedarius). *J. Reprod. Fertil. Suppl.* 49, 545–548.
411. Bravo, P. W., Fowler, M. E., Stabenfeldt, G. H., and Lasley, B. L. (1990). Ovarian follicular dynamics in the llama. *Biol. Reprod.* 43, 579–585.
412. England, B. G., Foote, W. C., Cardozo, A. G., Matthews, D. H., and Riera, S. (1971). Oestrous and mating behaviour in the llama (Llama glama). *Anim. Behav.* 19, 722–726.
413. Brown, B. W. (2000). A review on reproduction in South American camelids. *Anim. Reprod. Sci.* 58, 169–195.
414. San-Martin, M., Copaira, M., Zuniga, J., Rodreguez, R., Bustinz,a G., and Acosta, L. (1968). Aspects of reproduction in the alpaca. *J. Reprod. Fertil.* 16, 395–399.
415. Mortia, S. (1968). On the copulation and ovulation of the Riukiu musk shrew, *Suncus murinus riukiuanus Kuroda*. *Sci. Bull. Fac. Lib. Arts Edu.*, Nagasaki Univ. 19, 85–95.
416. Bedford, J. M., Mori, T., and Oda, S. (1997). Ovulation induction and gamete transport in the female tract of the musk shrew, Suncus murinus. *J. Reprod. Fertil.* 110, 115–125.
417. Matsuzaki, T., Matsuzaki, K., Yokoyama, M., and Saito, M. (1997). The period of ovulation and presence of the first polar body of ova ovulated in the house musk shrew (Suncus murinus). *Exp. Anim.* 46, 291–296.
418. Lee, C., Horvath, D. J., Metcalfe, R. W., and Inskeep, E. K. (1970). Ovulation in *Microtus pennsylvanicus* in a laboratory environment. *Lab. Anim. Care* 20, 1098.
419. Cross, P. C. (1972). Observations on the induction of ovulation in *Microtus montanus*. *J. Mammalogy* 53, 210.
420. Hansson, A. (1947). The physiology of reproduction in mink, *Mustela vison* (Shreb) with special reference to delayed implantation. *Acta Zool.* 28, 1–136.
421. Bakker, J., Rubin, B. S., and Baum, M. J. (1999). Changes in mediobasal hypothalamic gonadotropin-releasing hormone messenger ribonucleic acid levels induced by mating or ovariectomy in a reflex ovulator, the ferret. *Endocrinology* 140, 595–602.
422. Orstead, K. M., Hess, D. L., and Spies, H. G. (1988). Pulsatile patterns of gonadotropins and ovarian steroids during estrus and pseudopregnancy in the rabbit. *Biol. Reprod.* 38, 733–743.

Knobil and Neill's Physiology of Reproduction, Third Edition
edited by Jimmy D. Neill,
Elsevier © 2006

CHAPTER 43

Neuroendocrine Control of the Ovarian Cycle of the Rat

Marc E. Freeman

Introduction, 2328
Ovarian Events during the Estrous Cycle, 2329
Oogenesis, 2329
Folliculogenesis, 2330
Ovulation, 2332
Corpus Luteum Formation and Activation, 2332
Secretion of Ovarian Steroids during the Estrous Cycle, 2333
Patterns of Pituitary Hormone Secretion during the Estrous Cycle, 2335
Historical Perspective, 2335
Pattern of Hormone Secretion, 2336
Neural Events during the Estrous Cycle, 2339
The Circadian Control of the Ovarian Cycle, 2339
Role of the Hypothalamus, 2340
Peptidergic Neurohormones Controlling the Estrous Cycle, 2341
Biogenic Amine Neurotransmitters Controlling the Estrous Cycle, 2345
Peptide Transmitters as Possible Regulators of the Estrous Cycle, 2347
Gases as Possible Regulators of the Estrous Cycle, 2348
Growth Factors as Possible Regulators of the Estrous Cycle, 2349
Control of Pituitary Hormone Secretion during the Estrous Cycle of the Rat, 2350
Control of LH Secretion during the Estrous Cycle, 2350
Control of Follicle-Stimulating Hormone Secretion during the Estrous Cycle, 2357
Control of Prolactin Secretion during the Estrous Cycle, 2360
The Active Luteal Phase during the Ovarian Cycle of the Rat, 2363
"Rescue" of the Corpus Luteum, 2363
The Unique Pattern of Prolactin Secretion, 2364
Control of Mating-Induced Prolactin Secretion, 2364
Significance of the Short Luteal Phase, 2367
References, 2367

ABSTRACT

Since the early 20th century, the laboratory rat has generously provided us with information about its reproductive biology that we could use to study other mammals. As a result, we not only have a great understanding of the complexities of regulation of ovarian cyclicity in the rat, but we have taken the rat's clues and followed them in studying the controls of the ovarian cycle in domesticated food- and clothing-producing mammals, our pets and even ourselves. In addition, the rat has allowed us to "pick his brains" so that we may understand the neuroendocrine control of the ovarian cycle. This chapter takes the historical approach in describing our understanding and, in some cases, our misunderstanding of the control of the ovarian cycle by the brain and its interaction with the anterior pituitary gland and ovaries. Thus, it is with great fondness that I dedicate this chapter to rats of the varied strains for

Department of Biological Science, Biomedical Research Facility, Florida State University, Tallahassee, Florida.

their generous contribution to our understanding of mammalian reproductive biology.

INTRODUCTION

An understanding of the controls of the ovarian cycle represents a focal point at which man can have an influence on his socioeconomic status. He can regulate the fertility of estrual mammals which he depends on for food and clothing or those which compete with him for common resources. Much of what we know of the control of the ovarian cycle of many spontaneous ovulating, nonseasonal estrual mammals is based on our knowledge of the controls of the estrous cycle of the rat. However, unlike the ovarian cycles of other spontaneous ovulating mammals described in this section (sheep and monkey), the rat is unique in that the cycle is characterized by a brief luteal phase. Moreover, the events of the cycle are largely under photoperiodic control. That is, the lighting periodicity plays a dominant role in the incidence and duration of the stages of the ovarian cycle. This chapter will present a detailed description of the estrous cycle of the rat, with emphasis on the neuroendocrine controls. This chapter was intended as an historical approach to this topic when the first edition of this book appeared in 1988. The same chapter in the second edition continued this strategy. This current edition again emphasizes that approach but updates with significant contributions to our knowledge of the controls of the ovarian cycle made up to early 2004.

The word *estrus* (oestrus or oestrum, the British spelling) is a Latin adaptation of the Greek word *oistros* meaning gadfly, sting, or frenzy. This term was first used by Heape (1) to describe the "special period of sexual desire of the female" and distinguish it from *rut* in the male. Heape further described distinct stages of the cycle as it applied to mammals during the breeding season. He used the term *anestrus* to describe the nonbreeding season or period of rest in the female mammal when the ovaries and accessory reproductive organs are relatively quiescent and attempts of mating by the male are resisted. Heape also used the prefixes *pro-*, *di-*, and *met-* along with the suffix *-estrus* to describe the stages of the cycle between the periods of *estrus* during the sexual season. The first part of the cycle he termed *proestrus*, and characterized as the time when an animal is "coming on heat." The next period, *estrus*, Heape described as "only at that time that the female is willing to receive the male, and fruitful coition rendered possible in most, if not all, mammals." In the absence of conception, estrus is succeeded by a short recovery period called *metestrus* during which the estrus changes in the reproductive tract subside. The following period, *diestrus*, is of variable duration in different species. During this time, the ovarian secretions prepare the reproductive tract for receipt of the ovum, newly fertilized shortly after mating on estrus. If fertilization has not taken place, the mammal returns to proestrus and the cycle begins anew. This description of the various reproductive stages, though originally applied to the estrous cycle of seasonal breeders (1), with time, was applied to the estrous cycle of all mammals; seasonal or nonseasonal, polyestrous or monoestrous. However, the timing of the various stages and the behavioral and morphological characteristics of the stages may vary between estrual mammals.

The laboratory rat is a nonseasonal, spontaneously ovulating, polyestrous mammal. That is, the ovarian cycle continues throughout the year (as opposed to ovarian cycles restricted to one season as in the sheep). Ovulation is not dependent on overt nervous stimulation (as opposed to the requirement of the mating stimulus in rabbits). Ovulation occurs every 4 to 5 days throughout the year (the dog ovulates once in the spring and once in the fall). The mean length of the 1,999 cycles in the rat was described by Long and Evans (2) to be 5.4 days with a range of 3 to 38 days. If cycles of greater than 8 days are excluded, the mean is 4.8 days which compare favorably with 4.4 days of Blandau, Boling, and Young (3) and 4.5 days of Astwood (4). In addition, the length of the various phases of the estrous cycle based on vaginal smear pattern has also been described (2,4–7). Proestrus lasts for 12 to 14 hours; estrus, 25 to 27 hours; metestrus, 6 to 8 hours; and diestrus, 55 to 57 hours. The stages of the estrous cycle can be conveniently determined by monitoring, under low magnification, the cell types which appear in the vaginal smear. These cell types are shown in Fig. 1. Proestrus is characterized by a predominance of nucleated epithelial cells. These cells are distinctly round, visibly nucleated and appear in clusters at the height of proestrus. Occasionally, a few cornified squamous epithelial cells may appear as well. On the next day, estrus, the dominant cell is the cornified squamous epithelial cell. This cell is without an observable nucleus, contains a highly granular cytoplasm and is irregularly shaped. These also appear in large groups. During metestrus (often called diestrus-1) the dominant cell type is the leukocyte which appears along with a significant number of nucleated epithelial cells. The leukocytes are small with granular cytoplasm and, if examined under higher magnification, usually contain a vesiculated nucleus. Once again, the following day, proestrus, is characterized by the complete absence of leukocytes in the vaginal smear and the predominance of nucleated epithelial cells. The 5-day cycle may appear as an additional day of cornification (estrus) or an additional day of leukocyte infiltration (diestrus).

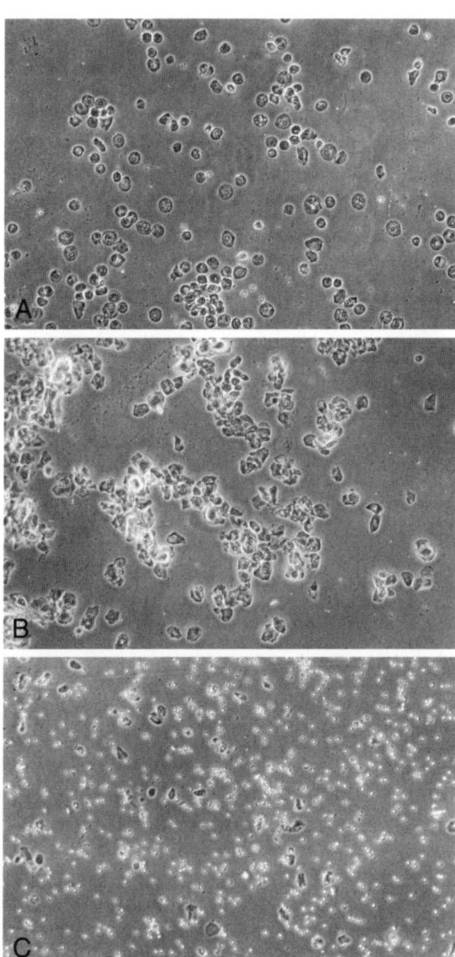

FIG. 1. Cell types appearing in the vaginal smear of rats during the stages of the estrous cycle of the rat (magnification × 100). **A**: *Proestrus* is characterized by a predominance of nucleated epithelial cells which are round and bear an easily visible nucleus. These may appear in clusters or individually. Occasionally a few cornified squamous epithelial cells may appear in the field. Note the complete absence of leukocytes in the smear. **B**: *Estrus* is distinctively characterized by large numbers of cornified squamous epithelial cells which occur in clusters. These cells are without visible nuclei, contain a granular cytoplasm and are irregularly shaped. Note the complete absence of leukocytes and nucleated epithelial cells. **C**: *Diestrus-1* and *-2* is characterized by a predominance of small leukocytes interspersed by a few nucleated epithelial or cornified squamous epithelial cells. The vaginal smears obtained during diestrus-1 and -2 are virtually indistinguishable.

OVARIAN EVENTS DURING THE ESTROUS CYCLE

Oogenesis

The intrafollicular ovum has progressed during fetal life through prophase of meiosis I to the dictyate or "resting stage" by 4 days after birth (8). This stage is characterized by the presence of a distinct germinal vesicle or nucleus, the faint appearance of the diploid number of chromosomes (n = 42) and a distinct nucleolus (Fig. 2A). The oocyte remains at this stage until "selected" for maturation on proestrus in the adult. The exact mechanism by which the oocytes are "selected" on proestrus has not been determined. The timing of the resumption of the maturation process has been reported with respect to the onset of sexual receptivity (9), the phases of "natural" or artificial illumination (8), and the timing of the ovulation inducing release of luteinizing hormone (LH) (10). In the adult, the final maturational step of the oocyte begins within the follicle and is known as the first meiotic division. This process is a reduction division and is controlled by the preovulatory surge of LH (11). That is, the chromosome number is reduced from the diploid content of 42 to the haploid number of 21. The first signal that meiosis I has been re-initiated is breakdown of the germinal vesicle, the disappearance of the nuclear membrane and chromosome thickening (Fig. 2B). This occurs between 17:00 and 18:00 hours on proestrus (8) or approximately 2 to 3 hours after the beginning of the ovulation-inducing release of LH and 7 to 9 hours before the expected time of ovulation (10). During the first metaphase the paired chromosomes are aligned along a metaphase plate (Fig. 2C). They are attached to their respective centromeres by spindle fibers. At anaphase I (Fig. 2D) the chromosomes begin separating with the result that half the chromosomal material (n = 21) is lost by polar body extrusion during telophase I (Fig. 2E,F). Extrusion of the first polar body has been variously reported to be complete by 4 to 8 hours after the beginning of sexual receptivity (9,12,13), by 21:00 hours proestrus (8), or by 22:00 hours on proestrus to 03:00 hours estrus (10). Once the first polar body is extruded, ovulation occurs. This occurs by 8 to 10 hours after the beginning of sexual receptivity (9,13–15), 9 to 10 hours after the resumption of meiosis I (8) or by 03:00 to 05:00 hours on estrus (10). The newly ovulated egg found in the upper third of the oviduct is now haploid with one polar body extruded and the remaining chromosomes aligned along a new metaphase plate. The tubal ovum, now in metaphase II, is available for fertilization by

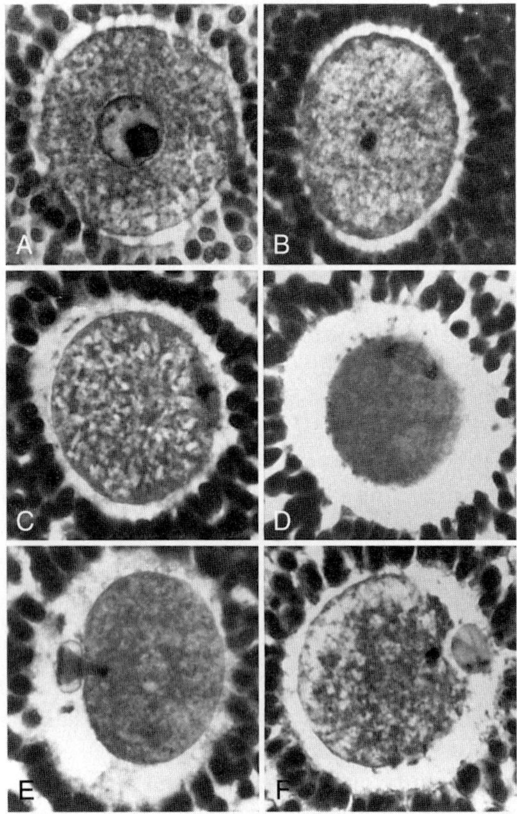

FIG. 2. Stages of oocyte meiotic development on the afternoon of proestrus through the morning of estrus in the rat (magnification × 938). **A**: *Dictyate or resting stage of prophase I.* The oocyte remains in this stage through 18:00 hours (24-h system) proestrus. Note the intact germinal vesicle and darkly stained nucleolus. **B**: *Late prophase I.* The oocyte begins to show these changes from 18:00 to 21:00 hours on proestrus initiated in response to a surge quantity of LH secretion earlier that day. Note the breakdown of the germinal vesicle and the appearance of the thickened chromatin mass. **C**: *Metaphase I.* The oocyte presents this stage from 21:00 to 24:00 hours on proestrus. Note the alignment of chromosomes along the metaphase plate and their attachment to the centrosome by the spindle fibers. **D**: *Anaphase I.* The chromosomes begin to separate from 21:00 to 24:00 hours on proestrus perpendicular to the axis of the metaphase plate. **E**: *Early telophase I.* Polar body extrusion begins between 24:00 hours proestrus and 03:00 hours estrus. Note the complete pinching off of the polar body membrane from the oocyte membrane as well as the intact spindle fibers. **F**: Late telophase I. Polar body extrusion complete by 03:00 hours estrus. The oocyte is now ready for extrusion from the rupturing follicle and eventual fertilization. Ovulation occurs 03:00 to 06:00 hours estrus.

the haploid sperm. The second meiotic division proceeds if the ovum is fertilized. In the absence of fertilization, the oocyte breaks down and a new group of intrafollicular oocytes begin maturation at the next proestrus, 4 days later.

Folliculogenesis

The maturational process of the oocyte described above occurs within the ovarian follicle. An egg is considered to be in a follicle when a definite layer of cells become arranged around the egg in a uniform capsule of follicular epithelium. This occurs during fetal development (16). The follicular epithelial cells are derived from embryonic stroma. Such follicles, with a single layer of simple squamous epithelium around the egg, are called primordial follicles. The squamous epithelium, in turn, enlarges to form a low columnar epithelium in the neonate. The primordial follicles are therefore known as primary follicles (Fig. 3). This simple columnar epithelium multiplies, first forming a double, and finally a stratified cuboidal follicular epithelium. At this stage the epithelial cells are clearly recognized as granulosa cells while the follicles are identified as secondary follicles. Multilayered secondary follicles transform into vesicular follicles with the formation of multiple fluid-filled intracellular spaces. These spaces eventually enlarge and coalesce into a single larger cavity, the antrum. The structure is now termed a tertiary follicle, which upon preovulatory growth is known as a Graafian follicle. The oocyte together with its attached granulosa cells which once occupied the center of the secondary follicle now becomes eccentric forming a hillock, the cumulus oophorus. The granulosa cells surrounding the encapsulated oocyte form a ring known as the corona radiata. The fluid contents of the antrum, the liquor folliculi, increase in volume as ovulation approaches. The enlargement of the granulosa cell layer is accompanied by the development of an outer encapsulating sheath derived from the stroma. This constitutes the theca which can be divided into a theca interna and theca externa. This division is first noted in the secondary follicle. The theca interna cells rapidly enlarge and assume a polygonal shape with vacuolated cytoplasm and vesicular nuclei. These cells are enmeshed in a reticular and fibrous network and contain a plexus of capillaries and lymphatics. This layer hypertrophies as proestrus approaches and is considered one of the major sites of estrogen production during the cycle (17,18). Indeed, owing to its glandular appearance, it has been referred to as the thecal gland (19). The outermost follicular layer, the theca externa, contains contractile tissue which is thought to play a role in ovulation (20,21). Not all follicles present on the ovary at the time of birth are destined to ovulate. Indeed, in the rat, Arai (22) calculated a decline in the total number of oocytes from 35,000 at the time of birth to 11,000 20 days later. This is an attrition rate of over 1,000 oocytes per day. If continued, the ovaries would be totally depleted by the time of sexual maturity. Obviously, this attrition rate does not continue. This attrition, known as atresia, is a normal circumstance throughout the life of the rat.

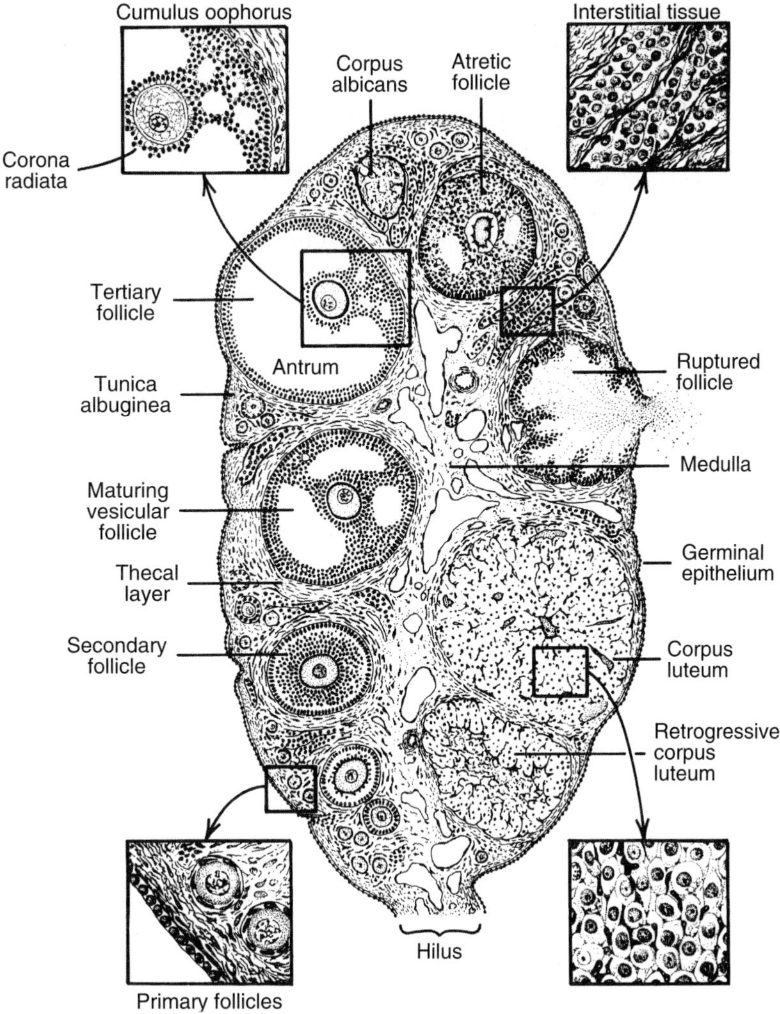

FIG. 3. Composite stages of ovarian follicular development through ovulation, corpus luteum formation, and regression.

Though some atresia of primordial and primary follicles occurs, atresia of secondary and vesicular follicles is most conspicuous during each cycle. The appearance of the atretic follicles differ depending upon the stage of the follicle at the time degeneration begins and the extent of the retrogressive changes at that time (16). The questions of what determines which follicles ovulate and whether atresia begins in the oocyte or follicular epithelium have remained largely unanswered (23). Also, the mechanism for maintenance of the viable follicle pool is not clear. However, recent work has suggested that that juvenile and adult mouse ovaries possess mitotically active germ cells that, based on rates of oocyte degeneration (atresia) and clearance, are needed to continuously replenish the follicle pool (24).

The first recognizable atretic changes are found in the epithelial cells lining the antral boundary of vesicular follicles and in the cells just external to the corona radiata of secondary follicles. Under conventional staining the first signs of the process is a shrinking of the nucleus and condensation of the chromatin mass of the granulosa cells. This is known as *pyknosis*. In addition, the granulosa cell nucleus may rupture and the chromatin disintegrates into formless granules which are extruded from the cell. This is known as *karyorrhexis*. In vesicular follicles, these nuclear fragments and pyknotic cells tend to slough off and float into the liquor folliculi and are readily noticeable. Such sloughing results in shrinkage of the cumulus oophorus and smoothing of its antral surfaces, a process which allows the ovum to eventually enter the liquor folliculi (23). In the rat, the zona pellucida undergoes degeneration followed rapidly by the oocyte. The granulosa cells also degenerate completely. However, the theca interna persists as interstitial masses which eventually break up into small groups of cells scattered within the medulla.

The selection of follicles in the cycling rat ovary for preovulatory growth and maturation begins as a presumptive random event. Using a classification system for developing mouse follicles proposed by Pedersen and Peters (25), Richards and Midgley (26) have developed a model for follicular selection. For descriptive purposes follicles are classified as (a) small follicles just entering the pool of committed, growing follicles, which is distinguished from the large pool of quiescent, nongrowing follicles; (b) large preantral follicles; and (c) large antral preovulatory follicles. Some follicles begin to grow during each day of the cycle. This growth continues until either atresia or ovulation occurs. Those follicles that do ovulate on the morning of estrus actually began to grow 19 days earlier as a member of a larger group of growing follicles. These follicles continue to grow to large preantral stages. At this stage, the follicles either become atretic or, during the proestrus preceding the cycle in which they will ultimately rupture, are "committed" to the ovulatory pool. This model proposes that the follicles must arrive at the "pre-committed" stage in synchrony with the anticipated surge of gonadotrophic hormones from the pituitary gland. Those asynchronous follicles degenerate if gonadotrophin stimulation does not occur within a critical time.

By mapping the size distribution of large antral follicles (>390 μm) on each day of the rat estrous cycle one can observe a distinct variation on each day (27,28). On late proestrus, preovulatory follicles are 500 to 750 μm just prior to rupture. On estrus, after ovulation and during metestrus, only follicles between 390 and 500 μm are present. During diestrus, follicles greater than 500 μm were just noted. The absence of 390 to 500 μm follicles on proestrus suggests that follicles of this size on late estrus and early metestrus are the ones that were selected to grow through diestrus and early proestrus and ultimately destined to ovulate (27). Moreover, the absence of 390 to 500 μm follicles on proestrus followed by their dramatic appearance on estrus further suggests that an event on proestrus selects these follicles to begin growth for the next estrous cycle. The control of this selection process will be described in another section.

Ovulation

The follicular events immediately preceding rupture and expulsion of the egg are easily manipulated and have been best described in the rabbit (20,21). It is believed that the ovulatory event and its controls are generalizable to most mammals. The ovulatory process of the rat has also been described but not in as great detail (2,29). Rather than describe the changes from proestrus through ovulation, Blandau (29) treated immature rats with gonadotrophic preparations and observed the process of ovulation in over 300 follicles. The ovulatory follicles bulge from the ovarian surface and appear turgid and highly vascular. Within minutes after stimulation by gonadotrophic hormone the macula pellucida or stigma forms at the apex of the preovulatory follicle. This is the first sign that the ovulatory process has begun. Beyond the periphery of the stigma a dilated ring of blood vessels forms. As follicular rupture approaches, the superficial germinal epithelial cells covering the stigma are breaking into small clumps. The underlying stroma thins and a delicate vesicle that may be a bleb of viscous follicular fluid forms. The actual escape of the ovum following breakdown of the stigma was observed in 166 cases. If the unattached cumulus oophorus and enclosed oocyte lay close to the ruptured stigma, they act as a plug. The extrusion of the cumulus mass and egg took 11 seconds to 12 minutes and 40 seconds. First the granulosa cells begin to ooze through the plugged orifice followed very quickly by the remaining egg mass. Once the egg is extruded, a great burst of follicular fluid follows. In the majority of cases the cumulus and egg do not lie next to the rupturing stigma. Under these circumstances, the follicular fluid bursts through as if under pressure. The cumulus and oocyte would be extruded subsequently in a stream. The oocyte rapidly finds its way to the ciliated fimbriated end of the oviduct. Within an hour after ovulation eggs surrounded by cumulus masses can be recognized in the upper third of the oviduct. All ovulations are complete in 1.5 hours.

Corpus Luteum Formation and Activation

During the estrous cycle of the rat, three or more generations of corpora lutea may be present on the ovary from the immediately preceding ovulatory cycles (2). It appears that the cells of the corpus luteum have a dual origin. That is, based on anatomical and histochemical observations (30), the corpus luteum arises from both the granulosa and thecal cells of the preovulatory follicle. Each set of corpora persist morphologically for 12 to 14 days. In the rat, each cyclic set of corpora lutea can be distinguished on the basis of size (2,31), vascularity (32) and stain characteristics (17,33,34). By diestrus, the newly formed corpora have attained their maximal size which is maintained through metestrus of the following cycle. By diestrus of that second cycle the corpora lutea abruptly regress. This regression coincides with closure of blood vessels, the appearance of areas of degeneration, leukocytic infiltration (32), increase content of 20-α hydroxysteroid dehydrogenase (34) and increase cholesterol content (33). Such corpora lutea, if present in an unmated animal, are referred to as "nonfunctional" corpora lutea. That is, the corpora lutea do not secrete sufficient progesterone to support a decidual reaction induced by mechanical or chemical trauma of the uterine endometrium. Actually, the corpora are not completely "nonfunctional" since progesterone is secreted by the newly formed corpora lutea from metestrus through diestrus (35). In the absence of luteotrophic support they do, however, secrete copious amounts of another progestin which is a metabolite of progesterone, 20α-hydroxyprogesterone (Fig. 4). This steroid will not support a decidual reaction nor hold the hypothalamo–pituitary machinery controlling ovulation in check. As the secretion of luteal progesterone declines through diestrus, a new ovulation takes place shortly thereafter. Thus, the "luteal phase" of rats (as well as hamsters and mice) is atypical. Other mammals presented relatively long luteal phases (11–14 days) with extensive release of high levels of progesterone, while the corpora lutea of unmated rats secrete lower levels of progesterone for only 1 to 2 days. This serves as a basis for the short cycles in rodents. If, on the other hand, the animals are allowed to mate and become pregnant or if the animals mate with infertile males, or if the uterine cervix is stimulated artificially, the pituitary gland secretes sufficient quantities of a "luteotrophin" to "rescue" the corpus luteum and allow it to persist for a number of days approximating a typical mammalian luteal phase (35). If the mating is fertile this luteal phase persists for the entire period of pregnancy, 20 to 22 days. If the mating is infertile or artificial, the corpus luteum persists for 12 to 14 days. This period is known as pseudopregnancy. The luteotrophin released by the mating stimulus is pituitary prolactin. The physiological basis for this extended luteal phase of the rat brought about by the mating stimulus will be described in a later section of this chapter. In addition to support for luteal function provided by the pituitary gland, there is an abundant literature to suggest that the placenta secretes a luteotrophin which maintains the corpus luteum of the latter part of pregnancy. Moreover, compelling evidence leads us to believe that the decidual tissue may contain activity, independent of the well-established pituitary and varying placental activities (36–40), which can serve as a luteotrophin (41). Finally, though corpus luteum regression or luteolysis could be merely the consequence of withdrawal of luteotrophic support, there is ample evidence that the uterus may secrete "luteolysins" which directly or indirectly hasten luteal death (42).

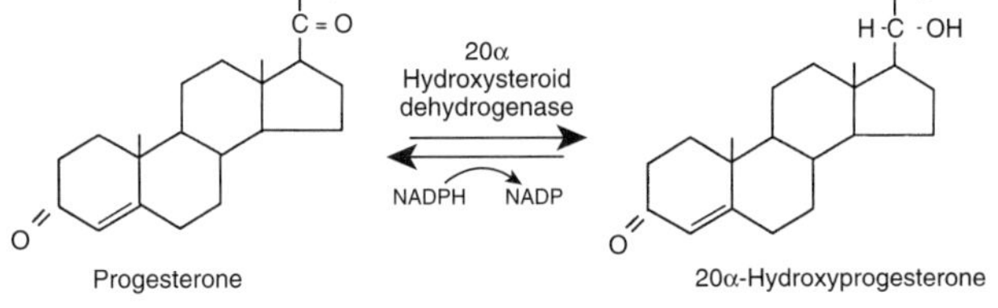

CH_3 C=O
20α Hydroxysteroid dehydrogenase
NADPH NADP
CH_3 H-C-OH
Progesterone
20α-Hydroxyprogesterone

FIG. 4. Fate of progesterone synthesized by newly formed corpora lutea in the rat. During the estrous cycle the newly formed corpus luteum contains high levels of 20α hydroxysteroid dehydrogenase. This enzyme converts progesterone to 20α hydroxyprogesterone. This latter hormone is not active in most *progestational assays* and will not support a pregnancy or a decidual reaction in the uterus. Prolactin, secreted during pregnancy or pseudopregnancy is responsible for suppressing the activity of this enzyme, hence maintaining mainly progesterone secretion from the corpora lutea.

Secretion of Ovarian Steroids during the Estrous Cycle

The preovulatory period of the estrous cycle is characterized by a growth of ovarian follicles and a concomitant enhanced secretion of estrogen. The secretion rate of estradiol into ovarian venous plasma is low on estrus, begins to rise significantly by late on metestrus through the morning of diestrus and reaches peak concentrations by the afternoon of proestrus (43,44). The mean peripheral plasma concentration of estradiol (Fig. 5) reflects the pattern found in ovarian vein blood (35,45–48). That is, in the 4-day cycling rat, peripheral plasma levels of estradiol are basal through estrus. Late on metestrus through early diestrus, plasma levels begin to rise. This increase continues through diestrus and early proestrus to reach peak values and plateau by mid-proestrus. During the early evening, shortly before the dark interval in the colony, estradiol levels fall rapidly reaching basal values by the early morning hours of estrus. The cell types responsible for the secretion of estradiol throughout the cycle have been described in an experiment by Falck (49). Co-transplants of estradiol-sensitive tissue, the vagina, and various ovarian cellular compartments were made to the anterior chamber of the eye. Only transplantation of both theca interna and granulosa cells caused typical "estrogenic" changes in the cotransplanted vaginal tissue.

In rats with 4-day estrous cycles, peripheral plasma levels of testosterone and androstenedione are similar in pattern to estradiol (50,51). The similar

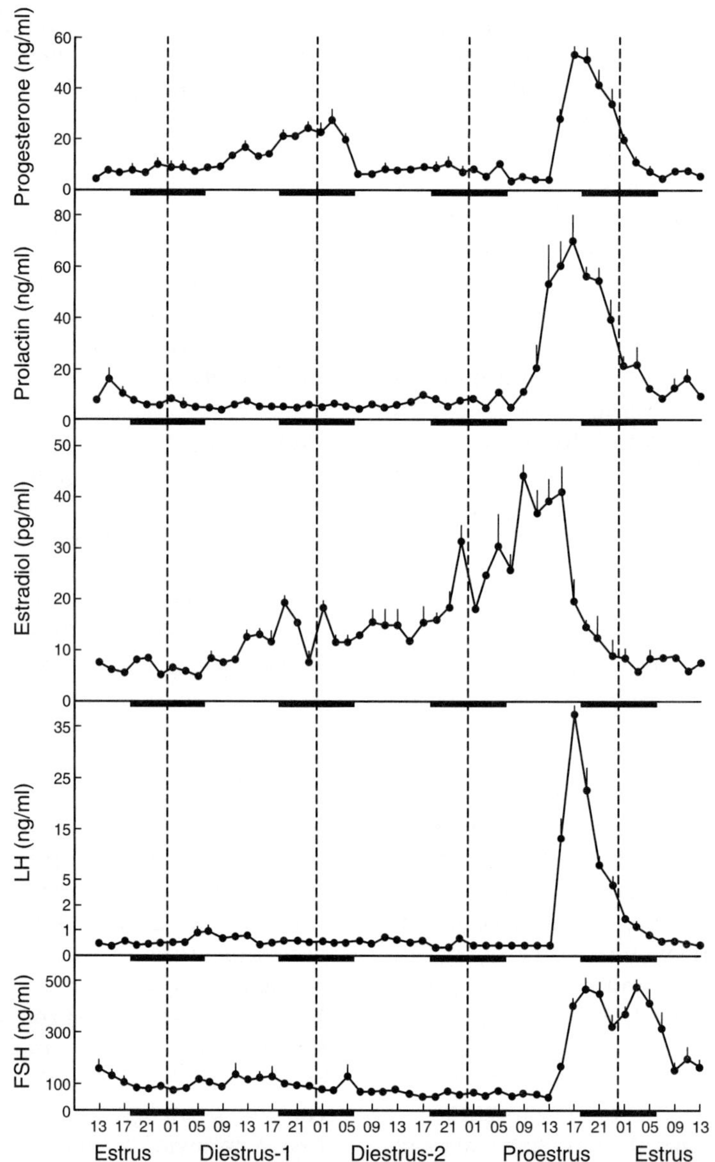

FIG. 5. Concentrations of progesterone, prolactin, estradiol, LH, and FSH in peripheral plasma obtained at 2-hour intervals throughout each day of the 4-day estrous cycle of the rat. Each point represents the mean hormone concentration (±SE) of 5 to 6 rats. *Black bars* represent the dark interval in the animal room (06:00–18:00) and the numbers below them represent the time-of-day (10–2) in terms of the 24-hour clock. [From (34).]

pattern of secretion of the androgens and estrogens may reflect the fact that the synthesis, secretion and controls of the two classes of hormones are interrelated.

In adult rats, the dominant progestin secreted into ovarian vein blood during the estrous cycle is 20 α-hydroxypregn-4-en-3-one (52–56) also known as 20 α-hydroxyprogesterone (20 α-OH-P). This progestin (Fig. 4) is a metabolite of progesterone and its synthesis is catalyzed by the enzyme 20 α-hydroxysteroid dehydrogenase (57). 20 α-hydroxyprogesterone is relatively inactive in classic progesterone bioassays. That is, it will not support the development of deciduomata in pseudopregnant rats (58,59). There are 2 peaks of both 20 α-hydroxyprogesterone and progesterone secretion into ovarian vein blood during the cycle. The first peak occurs during the afternoon of metestrus and both steroids probably arise from the newly formed corpora lutea (52,53). The second peak occurs during the late afternoon of proestrus (52,53). The 20 α-OH-progesterone peak at this time probably comes from the corpora lutea while the progesterone arises from the granulosa cells of the preovulatory follicle (60). The secretion of these progestins into ovarian vein blood is mostly mirrored in their pattern of secretion in peripheral plasma (35,46,48,61–63). In 4-day cycling rats, the secretion of 20 α-hydroxyprogesterone into peripheral plasma appears quite variable. The lowest levels are observed during the morning of diestrus. However, peripheral levels rise during the evening of diestrus and the morning of proestrus to reach peak levels by 20:00 hours on proestrus. The secretion of progesterone is more regular (Fig. 5). A large increase of follicular origin takes place during the afternoon and evening of proestrus. This increase occurs nearly simultaneously with the major ovulation-inducing increase of LH secretion. It reaches peak levels around the time of the LH peak in the early evening and returns to basal levels by the morning of estrus. A second major peak of luteal origin begins about mid-day on metestrus, is protracted through the early morning of diestrus and falls to basal levels shortly after the onset of illumination on diestrus. This rise and fall represents the unsuccessful attempt of the newly formed corpus luteum to survive. The control of these patterns of progestin secretion during the estrous cycle as well as their physiological significance will be discussed in subsequent sections.

PATTERNS OF PITUITARY HORMONE SECRETION DURING THE ESTROUS CYCLE

Historical Perspective

As stated so eloquently by Roy Greep in his "History of Research on Anterior Hypophysial Hormones" (64): "the elucidation of the relation of the anterior pituitary to gonadal function had been developing ever so gradually from dim beginnings much like the breaking of day, an event that terminates with sunrise and casting light over the scene." It is most difficult to pinpoint the earliest evidence that the anterior pituitary gland plays a role in reproductive cyclicity. Indeed, in their classic 1922 description of the estrous cycle of the rat (2), Long and Evans did not allude to a potential role for the pituitary gland in control of the reproductive cycle. However, in a series of studies published at the same time, they noted that growth-promoting pituitary extracts also caused ovarian enlargement and excessive luteinization while estrous cycles were either abolished or lengthened (65–68). However, the most striking observations, made independently by P. E. Smith (69,70) as well as Zondek and Aschheim (71,72), were that hypophysectomy caused ovarian atrophy in rats (69,70). In addition, implants of fresh anterior lobe tissue caused immediate growth and maturation of ovaries and uterus in hypophysectomized rats (69,70) and precocious sexual maturation in immature rats (71,72). Based on these studies and similar observations by others (64), by 1927 it had been largely accepted that the anterior pituitary gland secretes substances into the bloodstream which stimulates the gonads. Based on studies of gonad-stimulating activity in blood and urine of pregnant, ovariectomized and postmenopausal women, Zondek (73,74) suggested that the pituitary gland possessed two gonad-stimulating activities which he termed prolan A for follicle-stimulating activity and prolan B for luteinizing activity. These separate activities were also described by Evans and Simpson (75) who recognized a synergistic action of pituitary implants or extracts with prolan B activity caused ovarian stimulation in immature rats. However, their efforts to separate the activities were only marginally successful. We now know that the activity in urine of postmenopausal women, designated human menopausal gonadotrophin (hMG), is comprised of a mixture of follicle-stimulating hormone (FSH) and LH, both of pituitary origin. The first report of a separation of these biological activities is generally credited to Fevold and co-workers (76) with later confirmation by this same group (77,78). A "two-hormone" hypothesis for gonadotrophic function was now accepted (64). Such elucidation of separate biological activities took a "mere" 10 years to crystallize. It took almost another 20 years for the chemical identity of one of these, LH, to be described (79). Though other workers have made contributions leading to final isolation (64), the isolation and characterization of FSH is credited to Papkoff and co-workers (80,81). We now know that both FSH and LH are each composed of two peptide subunits in virtually all mammals (82). One subunit, designated α is common to FSH, LH, and thyroid-stimulating hormone (TSH). The other subunit, β, is structurally specific for each gonadotrophin. This specificity is responsible for the characteristic biologic activity of each. Finally, each gonadotrophin has characteristic carbohydrate moieties. The carbohydrate components may play a role in separate biologic activities (82).

The other hormone of the anterior pituitary gland of significance to the ovarian cycle of the rat is prolactin. Evans and Simpson (83) reported that the mammary glands of adult virgin rats were developed and milk synthesis initiated by extracts of bovine anterior pituitary glands. Riddle and collaborators (84,85) showed milk secretion by the crop sac of the pigeon was under the control of the anterior pituitary gland and that, on purification, the hormone producing crop milk in pigeons was the same as that which produced milk secretion in mammals.

They named the hormone prolactin. In addition to its myriad activities in nonmammalian vertebrates (64), prolactin was shown to "lengthen the estrous cycle" in rats by Lahr and Riddle (86). Though they noted the increased size of the corpora lutea in prolactin-treated rats, they did not appreciate that cycle lengthening was due to the luteotrophic activity of prolactin. The observation that pituitary prolactin plays a role in maintaining progestational corpora lutea was made independently by Astwood (87) and Evans and collaborators (88,89). Thus, beside the "pro-lactational" role shared by all mammals, pituitary prolactin participates in maintenance of corpora lutea in rats, mice, and hamsters. Early attempts at isolating biologically active prolactin met with equivocal results. Indeed, using differing bioassays, sources of active preparations, and purification procedures, it was variously claimed that prolactin and growth hormone preparations shared biological properties. It wasn't until the mid-1960s that separability of nonprimate growth hormone and prolactin was firmly established. In addition, it was also claimed that the pituitary gland contained a class of hormones separate from prolactin which stimulated mammary development without lactation in rats, rabbits, and guinea pigs. These putative "hormones" were known as mammogens. There are serious doubts at this time that such activities exist. The interested reader is referred to Greep (64) for review of what were two raging controversies. Though a vast amount of work has been contributed by others, the primary structure of ovine (90) and human (91) prolactin has been determined.

Control of the ovarian cycle of the rat by anterior pituitary hormones has been elucidated by various tools used to measure spontaneous release of the hormones during the estrous cycle. Indeed, LH was measured by ovarian ascorbic acid depletion bioassay (92) and FSH by direct augmentation of the ovarian weight response to hCG (93) or indirectly by uterine weight enhancement (94). Prolactin activity was quantitated by the growth of the pigeon crop sac (85). These assays all have the distinct advantage of outstanding hormonal and tissue specificity. They all share the inherent disadvantage of lack of sensitivity such that cyclic pattern of pituitary hormone secretion into blood must be inferred on the basis of changing content in the gland. Blood hormone levels were largely undetectable in samples obtained from individual animals. The concentration of hormone in the gland is the result of a balance between synthesis and release. Since these processes can vary independently of each other, declining levels of hormone in the gland may reflect decreased synthesis and release or release overtaking synthesis. Conversely, elevated levels may be the result of decreased release in the face of unchanging synthesis or acceleration of synthesis with or without compensatory increased release. For these reasons, the only definitive endpoint for hormone release is the concentration in blood. Sensitive radioimunoassays are now available for rat FSH and LH (95) as well as rat prolactin (96), which will detect serum levels of these hormones. The development and availability of materials for these assays through the National Institutes of Health has represented a major milestone on our road to acquiring knowledge of the control of ovarian cycles. The following descriptions of the pattern of pituitary hormone secretion are syntheses of the work of several laboratories (35,46,48,61,62) using these materials. All rats were under 14:10 or 12:12 light-dark cycle with noon being the midpoint of the light period (97,98). Under these conditions the majority of the cycles are 4-days in length (97). This description will be confined to animals with 4-day ovulatory cycles. The actual secretory patterns of these pituitary hormones are shown in Fig. 5 along with the circulating patterns of estradiol and progesterone. The interested reader is referred to the work of Nequin, Alvarez, and Schwartz (48) for a detailed description of differences in 4- and 5-day cycles.

Pattern of Hormone Secretion

Pattern of LH Secretion throughout the Estrous Cycle

The serum levels of LH are at the lowest from early on the morning of estrus, shortly after ovulation, through metestrus, diestrus, and mid-day on proestrus (Fig. 5). Though most studies report basal, unchanging levels of LH over this time (35,46,48, 61,62), one study reports slight but significant diurnal variation throughout the cycle (99). This circadian pattern from estrus through diestrus consisted of a small elevation of LH each day at the midpoint of the light period and the lowest levels by midnight of each day. On the afternoon of proestrus, about 14:00 to 15:00 hours, the circulating levels of LH begin to increase rapidly and ultimately reach peak levels by 17:00 to 19:00 hours on that same evening (Fig. 5). This rapid "surge" of LH induces follicular rupture and ovulation. Thereafter, these blood levels begin to decline and reach basal levels by early on the morning of estrus. When blood samples are obtained at 6- to 10-minute intervals during each day of the estrous cycle, a pulsatile or ultradian pattern of LH secretion is observed (100–103). Specifically, the pulses occur at periodicities of around 55 to 60 minutes and amplitudes, expressed in terms of the NIAMDD-LH-RP-1 standard, of 15 to 40 ng/ml whole blood from metestrus through the morning

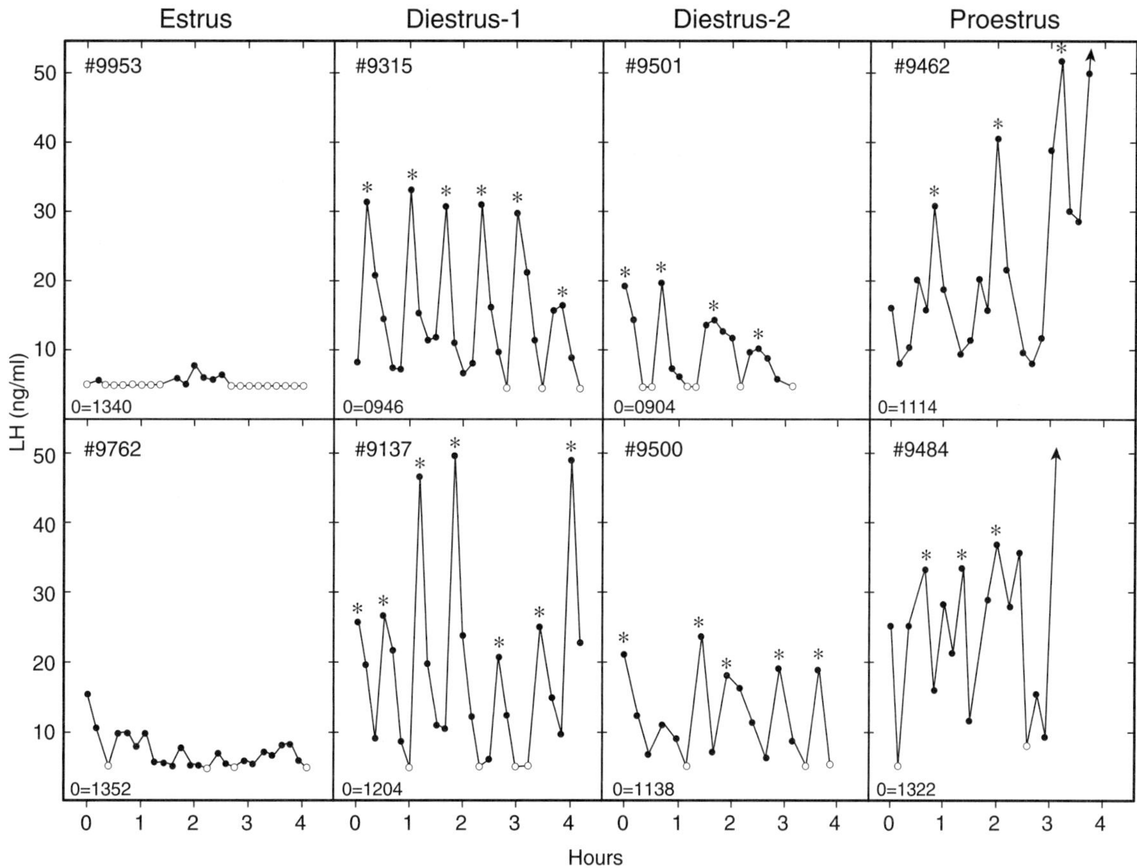

FIG. 6. Concentrations of LH in peripheral plasma obtained at 10-minute intervals for 4 hours from each of two representative animals during each day of the 4-day estrous cycle of the rat. The animal numbers are presented in the upper left corner of each block and the time of sampling initiation is given as zero-time in the lower left. Statistically defined pulses are indicated with *asterisks* and the *open circles* represent values below the detection limits of the assay. [From (99).]

of proestrus (Fig. 6). During the afternoon of proestrus the ovulation-inducing surge of LH has been characterized as composed of either declining pulse frequency with an amplitude of 600 ng/ml (101) or a single constant almost unvarying linear pulse of LH secretion with an amplitude of nearly 200-times baseline (Fig. 7) in some animals (99, 100,104). This single pulse declines through the early morning of estrus. By the late morning and throughout the day of estrus pulses of LH are infrequent (102) or absent (100). The circhoral pulses return by the morning of metestrus and continue through proestrus as before. The control of this unique pattern of LH secretion will be described in later sections.

Pattern of FSH Secretion throughout the Estrous Cycle

The pattern of FSH secretion throughout the estrous cycle is similar to that of LH (35,46,95,99). Specifically, basal levels of FSH are secreted from late on estrus through metestrus, diestrus and mid-day on proestrus (Fig. 5). From mid-afternoon onward, proestrus FSH secretion increases simultaneous with LH. That is, by 14:00 to 15:00 hours the circulating levels of FSH begin to increase rapidly and ultimately reach peak levels by 17:00 to 19:00 hours that same evening. Though both FSH and LH levels begin to decline to baseline after this time, during the early morning of estrus a secondary rise of FSH begins and peaks shortly thereafter. Circulating levels of FSH then begin to decline and reach baseline by the early evening of estrus. This early initial phase of FSH secretion shares some controls with the contemporaneous LH "surge" while the estrous phase of FSH secretion is under different controls. Circadian and ultradian pulses of FSH similar to LH have not been described during the estrous cycle (35,46,95,99). That is not to indicate that they do not exist, but merely that the available FSH radioimmunoassays are not sensitive enough to distinguish low level pulses of FSH secretion during the estrous cycle. On the other

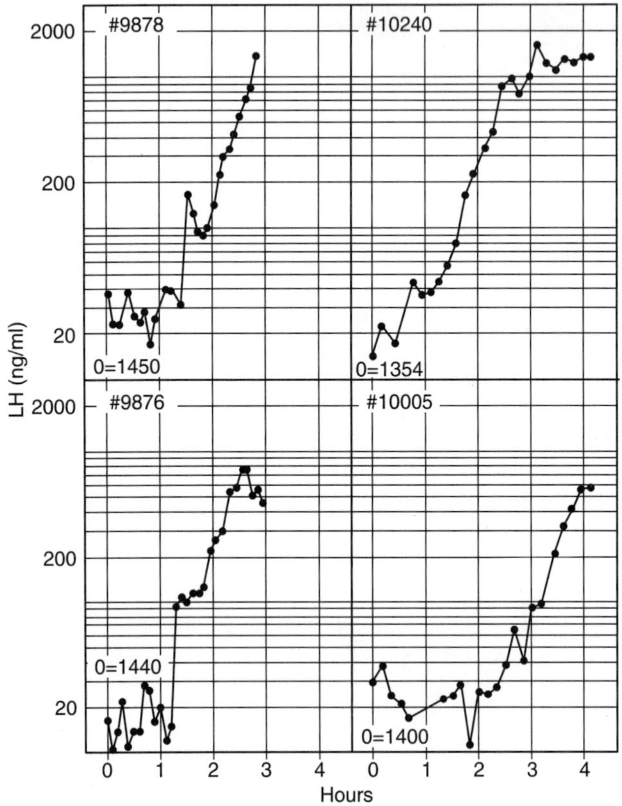

FIG. 7. A semilog plot of the LH concentration in blood samples obtained at 6-minute *(left panels)* or 10-minute *(right panels)* intervals during the afternoon of proestrus. The details are described in Fig. 6. Note the absence of discrete multiple pulses of LH during the preovulatory surge during proestrus. [(From (99).]

hand, pulsatile LH and FSH secretion does occur in the gonadectomized animal but the interpulse interval of LH is significantly shorter than that of FSH (105). The similarities and differences in control of FSH and LH secretion during the estrous cycle will be described in a later section.

Pattern of Prolactin Secretion throughout the Estrous Cycle

The pattern of prolactin secretion throughout the estrous cycle is also similar to that of LH (35,46,95,99). That is, basal levels of prolactin are secreted from the evening of estrus through the early morning of proestrus (Fig. 5). During the afternoon of proestrus a "surge" of prolactin, similar in timing to that of LH, is observed (35,99). While most laboratories have described a single surge of prolactin on proestrus (35,99,106), other laboratories have reported a secondary increase on estrus (46) or continuously elevated prolactin levels on proestrus, estrus and metestrus (107). Since these latter patterns may have been due to method and frequency of blood collection it is generally accepted that the afternoon of proestrus is the only time that a major "surge" of prolactin secretion occurs.

Gonadotrophin Subunit and Prolactin mRNAs in the Pituitary Gland throughout the Estrous Cycle

Gene expression for the α subunit (108), the β subunit of LH (108), the FSH-β subunit (109) and prolactin (110,111) have been described in the anterior pituitary gland throughout the estrous cycle of the rat (Fig. 8). Alpha mRNA is heightened by 08:00 hours diestrus-2 reaching maximal values by 20:00 hours and declined abruptly by 24:00 hours (108). The content of α-mRNA remained low during proestrus, estrus and diestrus-1. In contrast, LH-β mRNA was low on estrus and diestrus-1, increased at 08:00 hours diestrus-2, and then fell rapidly and remained low throughout the morning of proestrus (108). By 14:00 hours on proestrus a second increase began which peaked at 17:00 hours and then declined rapidly. This peak preceded the preovulatory surge

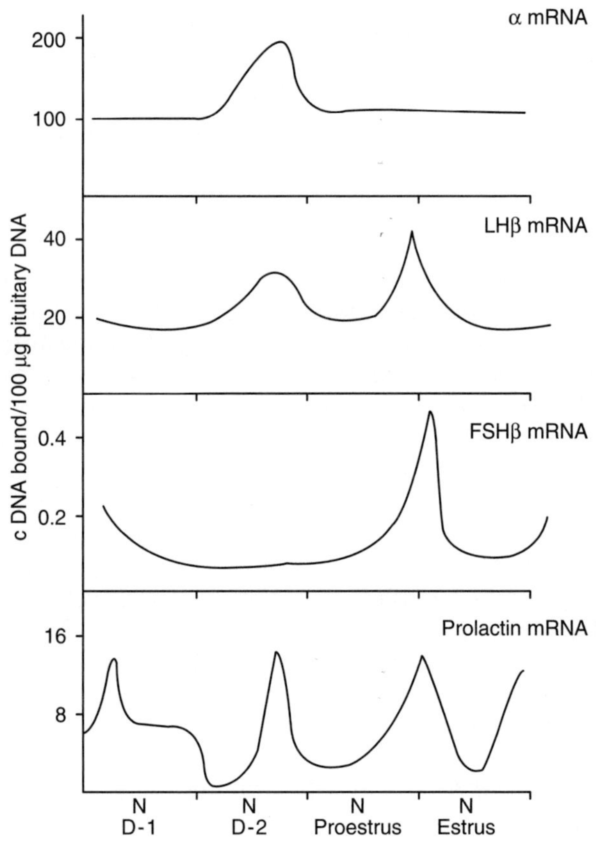

FIG. 8. Pituitary content of mRNA, LH-β mRNA, FSH-β mRNA, and prolactin mRNA throughout the estrous cycle of the rat. [Redrawn from (107–110).]

of LH secretion. By 22:00 hours proestrus the levels were equivalent to estrus and diestrus-1. FSH-β mRNA increased from 20:00 hours on proestrus to a peak value by 02:00 hours estrus, returned to basal levels by 08:00 hours and remained low through 20:00 hours estrus (109). This peak corresponded to the surge of FSH secretion on proestrus. Another smaller increase began by 23:00 hours estrus which remained elevated throughout the afternoon of diestrus-1 though FSH secretion was basal over that time. The pattern of prolactin mRNA during the estrous cycle was distinctly different than that of the gonadotrophin subunit messages (110,111). There was a striking increase of prolactin mRNA content in the anterior pituitary gland at specific intervals during each day of the estrous cycle (110,111). During proestrus and estrus prolactin mRNA was higher at 23:00 hours than during any other time on each day. In contrast, prolactin mRNA levels were highest at 08:00 hours on diestrus-1. During diestrus-2, prolactin mRNA values peaked between 17:00 and 20:00 hours. The only clear correlation between pituitary or plasma prolactin and prolactin mRNA was found on the afternoon of proestrus when a morning rise in prolactin mRNA preceded a morning increase in pituitary prolactin content (110).

NEURAL EVENTS DURING THE ESTROUS CYCLE

The Circadian Control of the Ovarian Cycle

We have earlier described the secretion of LH as an ultradian pattern. That is, frequent blood sampling reveals that LH secretion is largely pulsatile with frequencies of less than an hour (100–103). The pulses originate as the result of pulsatile release of luteinizing hormone-releasing hormone (LHRH) presumably into the portal vasculature (112). This is not the only patterned rhythmic control of the rat's ovarian cycle. Indeed, Everett and Sawyer (113) showed that the LH-release mechanism is under a 24-hour photoperiodic control and thus can fall under the category of a circadian rhythm. These elegant studies (Fig. 9) showed that administration of pentobarbital between 14:00 and 15:45 hours on the afternoon of proestrus to animals under daily illumination from 05:00 to 19:00 hours resulted in blockade of ovulation on estrus for 1 day (Fig. 9C). Treatment with the same drug again the next day, estrus, blocked ovulation a further 24 hours (Fig. 9E). Treatment at times other than 14:00 to 15:45 hours (Fig. 9B,D) was much less effective in blocking ovulation on each day. These data led to the concept of a daily "critical period": that interval during

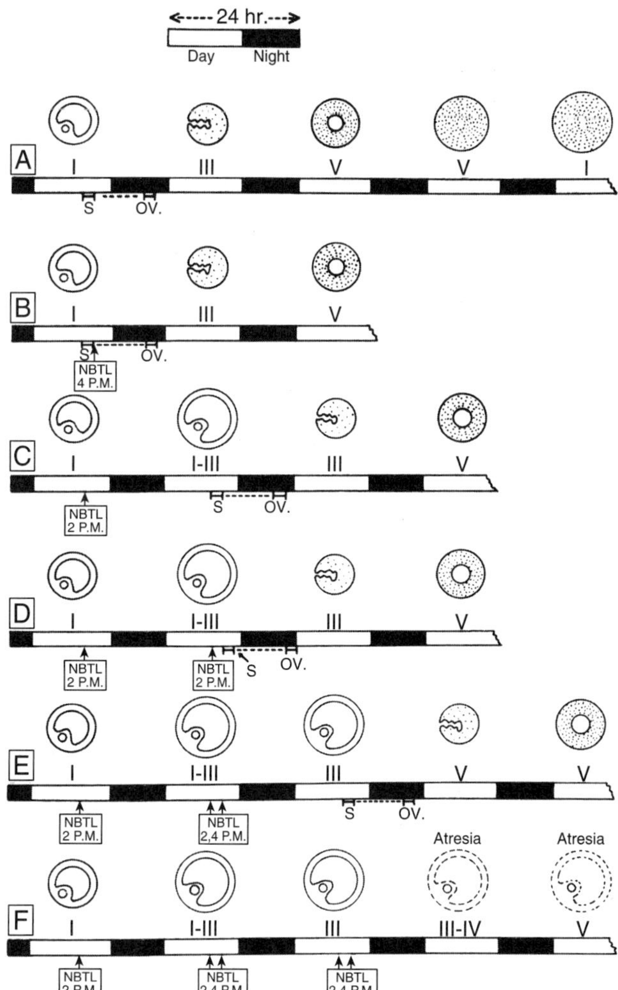

FIG. 9. The classic experiments of Everett and Sawyer (112) illustrating the experimental basis for the 24-hour periodicity in the LH-release mechanism during the estrous cycle of the rat. The *light* and *dark bars* represent a 24-hour continuum; s defines the *critical period* and *ov* the *time of ovulation*. The *Roman numerals* represent vaginal smear stages: in each group (A-F) I represents the afternoon of proestrus. Group A represents the timing of the events of the normal cycle from ovulation through corpus luteum formation. Administration of the central anesthetic Nembutal (NBTL, sodium pentobarbital) at 4 p.m. on proestrus does not disturb the timing of the normal cyclic events (Group B). However, a single injection of NBTL at 2 p.m. on proestrus postpones the release of a surge quantity of LH and ovulation 24 hours (Group C). Repetition of the treatment at the same time the following day, estrus, was insufficient to block LH release and ovulation and ovulation an additional 24 hours (Group D). On the other hand, administration of NBTL at 2 p.m. on proestrus and at both 2 and 4 p.m. on estrus blocks the surge of LH and ovulation 48 hours (Group E). The follicles cannot remain competent to respond to a delayed LH surge beyond 72 hours. Consequently, atresia results from such a delay (Group F). These experiments have indicated that a hypothalamic signal occurs daily between 2 and 4 p.m., the *critical period*, which leads to LH release. However, the ovarian steroid milieu is appropriate for such release only on proestrus. [From (112).]

which a daily neural event leads to LH release and subsequent ovulation. The timing of the "critical period" can be shifted by altering the "Zeitgeber" or time cue. That is, shifting of the phase but not the duration of the lighting period will shift the "critical period" by a corresponding time (98). Thus, environmental light-dark cycles act to regulate ovarian cycles via entrainment of their circadian system. In fact, with prolonged exposure to the lack of a daily photoperiod, such as constant light, the rat will gradually become acyclic as manifested by constant vaginal cornification (114). A number of studies have implicated the paired suprachiasmatic nuclei (SCN) of the hypothalamus as the transducers of the signal provided by the light-dark rhythm. Specifically there are neural connections between the eye and the SCN (115–117). Lesioning or isolation of the SCN presumably prevents transduction of the lighting periodicity and subsequently terminates the estrous cycle in rats (118,119). Thus, one of the signals triggering LH release and subsequent ovulation is provided by the circadian system so that there is only a limited "window" during each day when LH release could potentially be initiated. This daily potential for LH release is determined by transduction of the lighting periodicity, the Zeitgeiber, by the SCN. The interaction of ovarian hormones with the lighting signals eventuating in LH, FSH and prolactin release will be described in the next sections.

Role of the Hypothalamus

The roots of the young science of *neuroendocrinology* can be traced historically through the endocrinology of reproduction (120,121). Indeed, experiments of Fee and Parkes (122,123) suggested that a neural stimulus was involved in coitally induced ovulation in rabbits. However, the classic, albeit crude, experiments proving participation of the brain in ovulation reflex to coitus was the demonstration that passage of an electrical current through the head induced ovulation (124). This was quickly followed by the demonstration that local hypothalamic stimulation influenced ovulation (125) and that transection of the pituitary stalk with barrier placement blocked postcoital ovulation (126). In the rat, Hohlweg and collaborators (127,128) demonstrated that transplantation of the pituitary to areas distant from the brain resulted in failure of "castration cells" to appear in response to gonadectomy. In addition, Westman and Jacobsohn (129,130) showed that effects of estrogen on the pituitary of prepubertal rats could be prevented by transection of the pituitary stalk and placement of a barrier to prevent re-growth. Unfortunately, at this time it was incorrectly concluded that the hypothalamo–pituitary connections were exclusively neural and these deficit effects were merely due to transection of nerves. A role for the vascular connections between the hypothalamus and anterior pituitary gland were not yet fully appreciated.

The first description of a hypothalamo–hypophysial portal vasculature is credited to Popa and Fielding (131) who made the observation in rabbits. They observed that the blood flowed from the anterior pituitary toward the hypothalamus. The flow from hypothalamus to pituitary was correctly described in the frog by Houssay and collaborators in 1935 (132) and in the cat by Wislocki and King (133). The view was later confirmed in many species including the rat (134,135). The generalized configuration of the portal vasculature is that of a primary capillary bed lying within the median eminence and infundibular stem [Fig. 10; and (136)]. Long portal vessels arise from the upper infundibulum and drain toward sinusoids within the anterior pituitary. In the rat the main supply to the anterior pituitary is through these long portal vessels. Short portal vessels derived from capillary loops of the lower infundibulum adjacent to the neural lobe pass through the intermediate lobe and terminate in the anterior pituitary gland. It is interesting to note that Popa and Fielding's original suggestion of the direction of blood flow may not have been entirely incorrect. Using an elegant approach (137) and cannulation of single portal vessels (138), Porter and his collaborators (139) have proposed retrograde transport of pituitary hormones from the pituitary gland to the hypothalamus through "retrograde vascular channels." Blood flow from the pituitary gland to the brain has also been shown in sheep (140), monkeys (141,142), but not in pigs (143). The physiological significance of such vascular arrangements to the ovarian cycle of the rat will be discussed in a later section.

The description of the hypothalamo-hypophysial portal system led Harris to question his earlier (125) assumption that the ovulatory response to electrical stimulation of the hypothalamus was exclusively due to excitation of neural pathways to the pituitary gland. In fact, Green and Harris (135) suggested that the variation in response to hypophysial stalk section was due to regeneration of the transected hypophysial portal vessels. Harris proved this by demonstrating that restoration of reproductive function in stalk-transected rats was proportional to the degree of portal vessel regeneration (144,145). Indeed, this was also noted earlier by Greep (146) who showed that autotransplant or homotransplant of pituitary grafts back to the sella turcica of

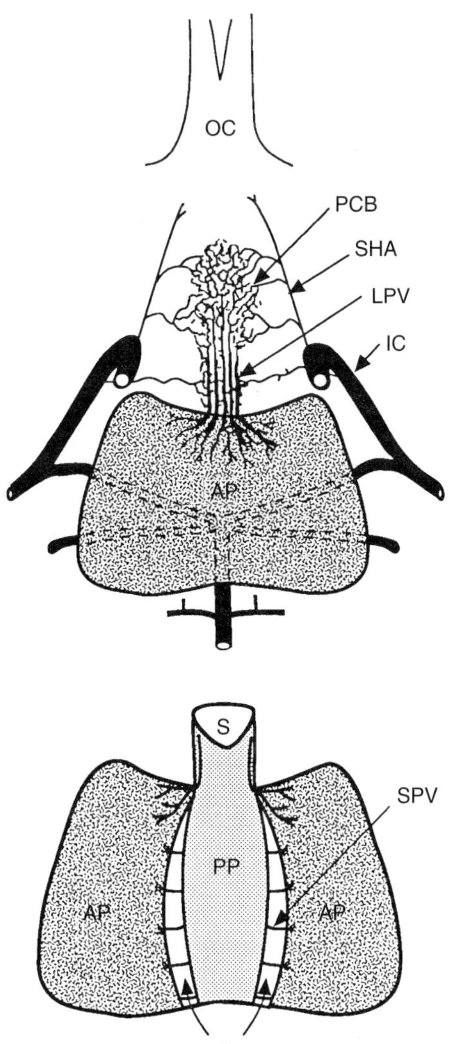

FIG. 10. Diagrammatic representation of the blood supply to the anterior pituitary (AP) gland through the long and short portal vessels. *Top* is a ventral view and *bottom* is a dorsal view. The superior hypophysial arteries (SHA) arise from the internal carotid artery (IC). The SHA, in turn, supplies blood to the primary capillary bed (PCB) in the median eminence. The long portal vessels (LPV) arise from the PCB and pass down the stalk (S) to the AP. The short portal vessels (SPV) arise from capillaries in the posterior pituitary (PP), pass through the intermediate lobe (IL) to the AP. Thus neurohormones can arrive at sites of action in the AP from axons which terminate in the median eminence or the PP.

hypophysectomized rats restored full pituitary function. The significance of this operation was not appreciated until Harris and Jacobsohn (147) transplanted pituitary glands from neonatal donor rats to various neural and nonneural sites in hypophysectomized recipient rats. Only when the gland was transplanted to just beneath the median eminence and vascularized from the median eminence were full reproductive and other pituitary functions restored. With these and other observations it was finally appreciated that hypothalamic control of reproductive function was due to a neurovascular link between the hypothalamus and pituitary gland. The chemical nature of the molecules synthesized by neurons and transported by this specialized vascular link are described in the following sections.

Peptidergic Neurohormones Controlling the Estrous Cycle

An activity which is capable of releasing LH was found in extracts of median eminence by both McCann (148) and Harris (149) working independently. McCann later suggested that a luteinizing hormone releasing *hormone* (LHRH) existed in the hypothalamus which activated the release of LH (150). In 1971 Schally's laboratory successfully isolated the decapeptide and identified its chemical structure (151,152). With the synthesis of the decapeptide, antibodies were produced for purposes of radioimmunoassay (153), and immunohistochemical localization of LHRH-neurons within the brain (154,155). The peptide fulfills all the expectations as a neurohormone: identification in neurons, release into portal blood and activity correlated with release of LH into peripheral plasma.

LHRH has been localized in the rodent brain using a number of imaginative approaches. Wheaton and co-workers (156) measured LHRH by radioimmunoassay in frozen sections of rat brain and found the majority of the activity in the infundibular stem and median eminence, the site at which axon terminals and the primary capillary bed of the hypophysial portal system meet (Fig. 11). In addition, smaller but significant quantities of LHRH were measured in the anterior hypothalamus, the suprachiasmatic nucleus and rostrally in the preoptic area. Similar observations were made by Palkovits and collaborators (157–159) and Selmanoff and collaborators (160) using an innovative technique of punching tissue from specific brain regions. In addition, LHRH is found in regions of the circumventricular organs particularly the organum vasculosum of the lamina terminalis (OVLT), the vascular structure at the tip of the third ventricle within the preoptic area. Using various immunocytochemical approaches LHRH neurons have been shown to be small and distributed rostro-caudally from the diagonal band of Broca (DBB) to the premamillary region and extend up to 2 mm on either side of the midline in the preoptic region and medial forebrain bundle (161–168). LHRH-positive cells are concentrated in the DBB, stria terminalis, medial nucleus of the septum, medial preoptic area (MPOA) and the anterior hypothalamic area (AHA). LHRH-containing perikarya have also

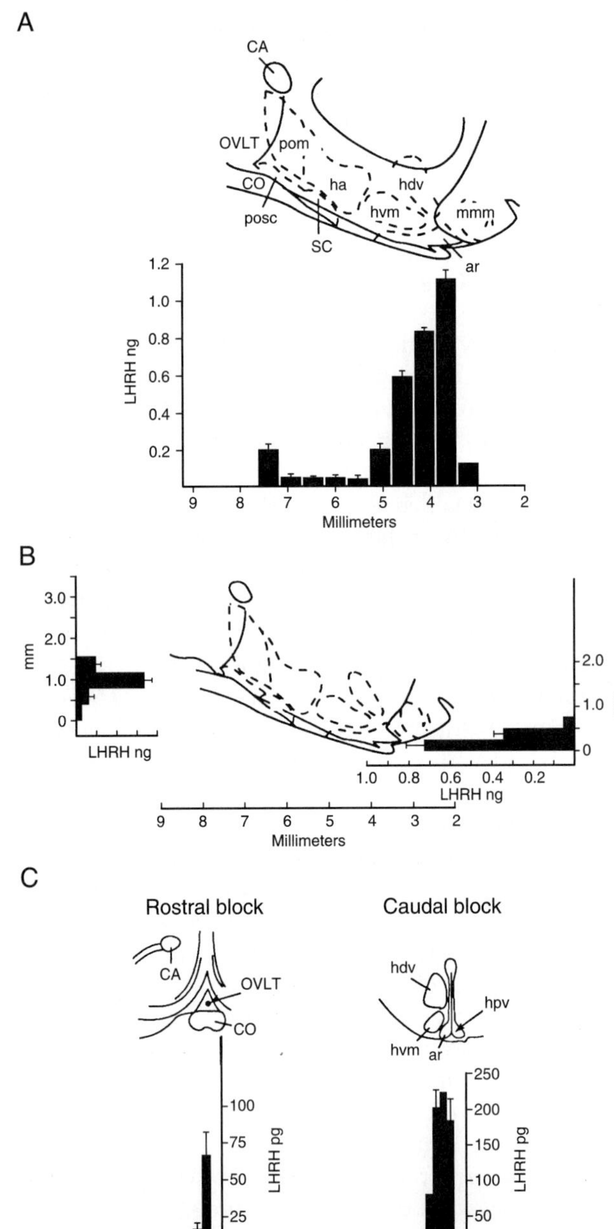

FIG. 11. Immunoassayable LHRH in extracts of frontal (panel *A*), horizontal (panel *B*), and sagittal (panel *C*) sections of the rat brain. The *bars* of each histogram represent the mean (+SE) LHRH content in sections projecting through parasagittal (panels *A* and *B*) or frontal (panel *C*) images of the hypothalamus of the rat. *ar*, arcuate nucleus; *ha*, anterior hypothalamic nucleus; *hdv*, dorsomedial hypothalamic nucleus; *hpv*, periventricular hypothalamic nucleus; *hvm*, ventromedial hypothalamic nucleus; *pom*, medial preoptic nucleus; *posc*, pars suprachiasmatic-preoptic nucleus; *mmm*, pars medialis-medial mamillary nucleus; *sc*, suprachiasmatic nucleus; *CA*, anterior commissure; *CO*, optic chiasm; *I*, infundibulum; *OVLT*, organum vasculosum of the lamina terminalis. Note that the greatest content of LHRH was in a broad band corresponding to the ar with lesser but significant amounts in the OVLT. [From (155).]

been found in the retrochiasmatic area of the medial basal hypothalamus (167–169). LHRH neurons which project into the median eminence (ME) are concentrated basally in the midline regions from the MPOA rostrally and through the retrochiasmatic area caudally. These fibers terminate on capillaries of the primary plexus of the hypothalamo-hypophysial portal vasculature (170).

There is reasonably good correlation between the location of LHRH cell bodies and axons and experimental manipulations at these sites which affect the ovulatory discharge of LH. Using an experimental model in which the proestrous preovulatory "surge" of LH has been blocked, Everett (171) has mapped a funnel-shaped septal-preoptico-tuberal pathway which responds to electrochemical stimulation by releasing an ovulatory amount of LH. These fibers are fan-shaped in the septal complex, narrow through the MPOA-AHA and converge as a compact band in the ME (Fig. 12). Electrochemical stimulation along this band excited LH release (172). However, despite the presence of LHRH fibers, stimulation of the lateral preoptic area, the median forebrain bundle and dorsal or ventral medial hypothalamus failed to excite LH release (171,172). Moreover, a large electrolytic lesion placed within the MPOA produced a series of consecutive pseudopregnancies bounded by ovulation (173,174). In fact, treatment with dopamine agonists which abolishes the secretion of prolactin in these animals induces the return of normal ovulatory cycles (173) suggesting that the long anovulatory periods may be associated with and due to the secretion of prolactin (175), not to damage of the MPOA. These studies suggest that a large group of neurons within the DBB-septal-MPOA may not regulate cyclic

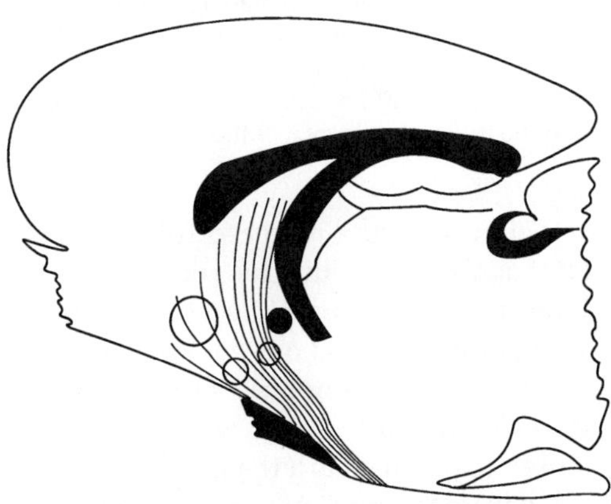

FIG. 12. Sagittal view of the rat brain diagramming the course of fibers which, when lesioned, affect ovulation. Widely dispersed origins are in the septal complex where relatively large irritative lesions are required to activate the number of fibers that can be affected by small lesions in the preoptic area. [From (170).]

LH release. However, the lesion of discrete midline sites through this area such as the OVLT (176), the medial preoptic nucleus (MPN) (177), or SCN (173) produced anovulation and anestrus or constant estrus. Furthermore, lesion of the OVLT results in a significant decline of LHRH within the ME (178). It is interesting to note that transection of caudal efferent projections from these midline structures by means of small frontal cuts failed to block the ovulatory discharge of LH while extended frontal cuts encompassing midline as well as all anterolateral input to the basal hypothalamus completely blocked the ovulation-inducing release of LH (179). Thus, the exact role of rostral inputs to the medial basal hypothalamus governing the ovulation-inducing surge on proestrus has not been unequivocally resolved.

Regardless of the site of origin of the LHRH neurons causing release of LH from the pituitary gland, it is now fully recognized that a neurogenic stimulus on the afternoon of proestrus triggers a cascade of events within the hypothalamus eventuating in LHRH release from nerve terminals within the ME. This, in turn, is the stimulus for the ovulation-inducing surge of LH in the rat. Just as the secretion of the LH surge on proestrus is reflected in altered pituitary levels of LH (114), so too is a surge of LHRH release into portal blood on proestrus reflected by changing levels of the decapeptide in the hypothalamus. Using indirect estimates of bioactive LHRH activity in hypothalamic extract by assaying their LH-releasing capabilities from pituitaries in vitro, there appeared to be enhanced LH-releasing activity in hypothalami around the time of the LH surge on proestrus and a rapid decline in hypothalamic content associated with the surge (180–182). LHRH concentrations have also been measured in the brain by immunohistochemistry or radioimmunoassay. In confirmation of the earlier studies, it seems that LHRH activity of the medial basal hypothalamus declines as the proestrous surge of LH begins (183,184) or as peak peripheral levels of LH are approached (185–187). There does not appear to be general agreement on the increase in hypothalamic LHRH-content prior to the surge (188–190). This may be due merely to variation in dissection technique of the various hypothalamic areas. In one study (191) six areas along the preoptic-suprachiasmatic-tuberoinfundibular tract were dissected by the Palkovits micropunch technique prior to and through the ovulation-inducing surge of LH on proestrus. Between 09:00 and 12:00 hours on proestrus there was a large increase in the concentration of LHRH in the median eminence as well as the suprachiasmatic, medial preoptic and anterior hypothalamic nuclei. Smaller, but significant, increases were observed in the retrochiasmatic area as well as the suprachiasmatic preoptic nucleus. As with other studies, a large dramatic fall in LHRH concentration occurred in all areas between 12:00 and 15:00 hours when the surge of LH begins. Presumably, these changes are a reflection of release of LHRH into portal blood (Fig. 13). Indeed, using a sophisticated surgical approach coupled with an anesthetic which does not appear to block the ovulatory surge of LH secretion on proestrus, Sarkar and collaborators (192); and later Ching (193,194) described an increase in radioimmunoassayable LHRH concentration in hypophysial blood which shortly precedes the preovulatory surge of LH release on proestrus. Unfortunately, current methodologies are such that it is difficult to obtain a reasonable amount of portal plasma frequently enough to determine if the pulsatile nature of LH release on proestrus in the rat (100,101) is a reflection of changes in LHRH pulse frequency and amplitude or merely the consequence of steroid-induced alteration in pituitary sensitivity without a variation in LHRH frequency or amplitude. In a larger mammal, the sheep, LHRH has been described as pulsatile in portal blood with pulse-amplitude increasing at the time that the ovulation-inducing surge of LH is released (195). On the other hand, in the rat, by frequently perfusing the medial basal hypothalamus with artificial cerebrospinal fluid and then sampling that

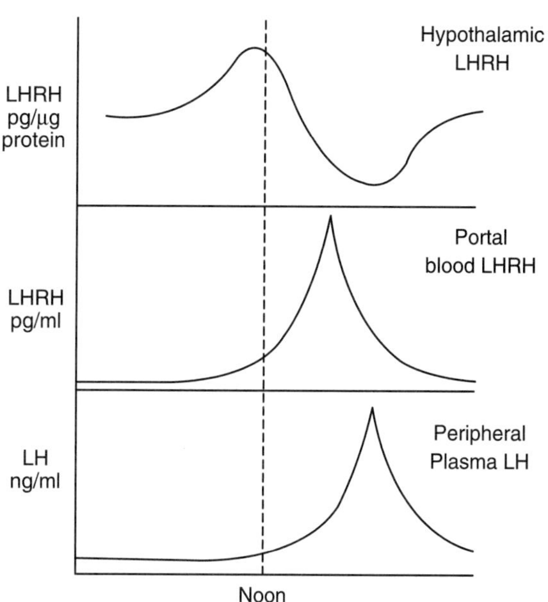

FIG. 13. Diagrammatic representation of the changes in hypothalamic concentration of LHRH (439), portal blood concentration of LHRH (191), and peripheral plasma concentration of LH (34) throughout proestrus in the rat. Note that preceding the release of the ovulation inducing surge of LH there is a dramatic decline in hypothalamic concentration of LHRH and a corresponding increase in portal blood levels of LHRH.

fluid for assay of LHRH content (push-pull perfusion), Levine and Ramirez (112) have described the neurosecretory "activity" of LHRH neurons during the estrous cycle. Presumably, these are the neurons from which axon terminals LHRH is secreted into hypophysial portal blood. Under these conditions pulsatile LHRH release can be evaluated on proestrus. During proestrus, the average LHRH interpulse interval was 48 minutes. Superimposed upon this is a biphasic pattern of LHRH secretion. The first phase consisted of a pulse of greater amplitude (1.2–2.0 pg/12 min) than the basal pulses (<1.2 pg/12 min). This was followed by a single large pulse (2.0–7.0 pg/12 min) approximately 2 to 3 hours later. It is this pulse which presumably stimulates the release of the LH surge (112). These data, coupled with the observation that immunoneutralization of LHRH on proestrus blocks the surge of LH (196) firmly establishes the decapeptide as the physiological luteinizing hormone-releasing hormone. The multiple pulses of LHRH serve a physiological role in the economy of LH release. Indeed it has been shown that small pulses of LHRH "prime" the gonadotroph cell to be maximally response to subsequent release of a larger LHRH pulse on proestrus (197,198). Though the decapeptide has been unequivocally identified as *the* LH-releasing hormone in mammals, it is not clear that the termination of the LH surge is due exclusively to termination of LHRH secretion into hypophysial portal blood. Indeed, a peptide of hypothalamic origin which inhibits LHRH-stimulated LH secretion has been partially characterized (199). Administration of antisera to the peptide results in heightened release of LH on proestrus or in response to a challenge with estradiol (200).

Though LHRH occupies the primacy as the *LH-releasing hormone*, others often arise as alternative or supporting candidates. The current crop of candidates is headed by oxytocin, neuropeptide Y and galanin. There is abundant evidence that oxytocin acts at the anterior pituitary gland to augment the stimulatory effects of LHRH on LH secretion (172,201–212). Oxytocin is found in portal blood (213,214) in highest amounts on the afternoon of proestrus (214). Intravenous administration of an inhibitor to oxytocin on the afternoon of proestrus blocks the preovulatory release of LH (215). These data suggest a role for oxytocin as a neurohumoral LH-releasing hormone. Similar dual neuromodulatory and neurohumoral roles have been suggested for neuropeptide Y in control of LH secretion (216). There is abundant evidence that neuropeptide Y acts a neurohormone to modulate LHRH-stimulated LH secretion (217,218). Neuropeptide Y is found in the hypophysial portal circulation in high concentrations (219,220). It regulates the binding of LHRH to its receptor on the gonadotroph cell membrane (221,222), potentiates LHRH-stimulated LH surges (223), and modulates calcium-transients in gonadotrophs (224). These effects are probably mediated through the Y-1 receptor (225). Hypothalamic neuropeptide Y gene expression increases prior to the onset of the LH surge (226,227). This suggests a role for neuropeptide Y as a neurohormone participating with LHRH to regulate LH secretion at the gonadotroph. Finally, there is evidence for a neuromodulatory, neuroendocrine and paracrine/autocrine role for galanin in the control of pituitary LH secretion. Galanin will stimulate LH secretion in vivo (228). It may act as a neurohormone since it is colocalized with LHRH (229) and has been found in portal blood (230). Whether these three sites of galanin localization are all significant for the control of LH secretion is unknown at this time.

Though LHRH will release both FSH and LH (231) there is still some uncertainty if one or possibly two releasing hormones exist. In the latter case, LHRH could release both LH and FSH while a distinct FSH-releasing factor (FSH-RF) may exist. Indeed, a radiofrequency lesion placed in the paraventricular nucleus-dorsal anterior hypothalamic area (PVN-DAHA) lowered the plasma FSH levels of ovariectomized rats without disturbing the blood levels of LH or the concentration of LHRH in the median eminence (232). Similarly, electrochemical stimulation of the DAHA caused FSH but no LH release (233). Moreover, the posterior median eminence has greater FSH-releasing activity than can be accounted for by its content of LHRH (234). Finally, a bioactive FSH-RF has been isolated by size exclusion and ion exchange chromatography (235,236). Fractions with greater FSH-releasing activity than could be accounted for by immunoreactive LHRH was detected prior to the immunoreactive LHRH peak. Thus, not only have hypothalamic areas been identified which preferentially release FSH, but FSH-releasing activity has been localized and extracted from the hypothalamus. If indeed a distinct FSH-RF does exist, it may play an important role in the biphasic pattern of FSH secretion on proestrus through estrus. Administration of antiserum to LHRH (237,238) or an LHRH antagonist (239) blocks the preovulatory surge of LH but the secondary surge of FSH on estrus is undisturbed by treatment with the antiserum or antagonist. These data, taken together, argue for a distinct FSH-RF which may control the secondary surge of FSH on estrus. However, other studies described later, argue that the secondary surge is independent of hypothalamic control.

The physiological control by peptidergic neurohormones of the secretion of prolactin during the estrous cycle is less clear than that of LH and FSH. A number of neuropeptides have been demonstrated to stimulate prolactin release. However, the most likely candidates as neurohumoral prolactin-releasing hormones (PRHs) during the estrous cycle are thyrotropin-releasing hormone (TRH), vasoactive intestinal polypeptide (VIP), and oxytocin (OT). All three peptides are found in hypophysial portal blood (213,214,240,241) and stimulate prolactin-release in vitro (242,243) and in vivo (243,244). The indication that TRH plays a physiological role in the release of prolactin during the estrous cycle is the observation that antiserum to TRH attenuates the basal secretion of prolactin on diestrus and the surge on proestrus (245). The notion that VIP may be involved in the regulation of prolactin secretion during the estrous cycle is supported by the fact that an antiserum to VIP administered on the afternoon of proestrus attenuates the peak phase of the prolactin surge later that day (246). However, the levels of VIP in the hypothalamus and anterior pituitary do not change during the surge of prolactin on proestrus (247). Therefore, a role for VIP in the cyclic release of prolactin remains uncertain. The concentration of oxytocin in hypophysial portal blood increases on the afternoon of proestrus just prior to the surge of prolactin release into peripheral plasma (214). Peripheral administration of an oxytocin antagonist on the afternoon of proestrus completely blocked the proestrous surge of prolactin (215). Such data provide compelling evidence for a role of oxytocin in the release of prolactin on the afternoon of proestrus.

Biogenic Amine Neurotransmitters Controlling the Estrous Cycle

The first indication that the catecholamines may be involved in ovarian cyclicity and ovulation was the demonstration that the α-adrenergic antagonists dibenamine and SKF-501 blocked ovulation presumably by preventing the release of an ovulation-inducing surge of LH (248–250). Furthermore, blockade of the α adrenergic postsynaptic receptor with chlorpromazine or depletion of catecholamines from presynaptic nerve terminals with reserpine also blocks ovulation (251). Reserpine acts by inducing a slow release of catecholamines from the storage granules in the nerve terminals and inhibits Mg^{+2}/ATP-dependent reuptake. Under these conditions intraneuronal monoamine oxidase deaminates the transmitters and thus causes a long lasting depletion of norepinephrine and dopamine (252). The inhibitory effects of depletors on ovulation can be prevented by inhibitors of monoamine oxidase such as pargyline or nialamide (253). These drugs prevent metabolism of the catecholamines and thus stimulate gonadotrophin release by presumably regulating LHRH release into portal blood.

Beside the pharmacologic studies, significant information could be gained if release of the biogenic amine could be quantitated directly at the synaptic cleft. Since this is technically impractical and tissue concentrations are an insensitive index of neuronal activity (254), changes in the turnover rate of the amine pool is used as an index of catecholamine neuronal activity. Increased turnover rates have suggested enhanced neuronal activity and conversely decreased neuronal activity is revealed by declining turnover rates. These have been used to correlate catecholamine neuronal activity and gonadotropin release. Two basic approaches have been taken. In one, a steady state method, labeled tyrosine or catecholamine precursors are injected at doses which do not disturb the endogenous precursor pool. The appearance of labeled catecholamines in the brain after systemic or intraventricular administration of precursor indicates biosynthesis while the disappearance of labeled catecholamine from the brain indicates turnover. This procedure has the inherent disadvantage of being relatively insensitive and thus requiring that large areas of the brain be dissected. Nevertheless, useful descriptions of cyclic catecholamine synthesis in the brain during the estrous cycle of the rat have been presented (255). On the other hand, there are several non-steady state methods that are in use today. Of these, the most widely used is the change in catecholamine content after treatment with α-methylparatyrosine, an inhibitor of tyrosine hydroxylase. This method has been used to describe changes in catecholamine dynamics during the estrous cycle (256–258). Other methods used to characterize both catecholamine, indolamine and peptide dynamics in vivo in specific areas of the brain include push-pull perfusion (25–263), microdialysis (264–271), and in vivo voltammetry or electrochemistry (272,273).

The methods used to measure the catecholamines include microfluorometric assays (257), radioenzymatic assays (274), and high-performance liquid chromatography coupled with electrochemical detection (275). Under conditions of synthesis inhibition with α-methylparatyrosine the index of neural activity is the "turnover rate" which provides an estimate of the amount of neurotransmitter released per microgram protein per unit time. For example, with the semiquantitative microfluorometric assay Lofstrom (257) observed that NE turnover is high in the subependymal layer of the rat during proestrus and low on all other days while DA turnover rates in the lateral palisade zone are low on proestrus and high on other days of the cycle. While measuring the catecholamines by the more sensitive and specific radioenzymatic assay, Rance and co-workers (258) had been able to thoroughly describe turnover rates of catecholamines in microdissected brain areas during the afternoon and early evening in proestrous rats. These were correlated with changes in LHRH concentrations in the median eminence and plasma concentrations of LH, FSH, prolactin, estradiol, and progesterone. The concentrations of LHRH in the median eminence rose during the morning of proestrus. At this time plasma estradiol levels were elevated but both progesterone and FSH and LH remained basal. During this same time both norepinephrine (NE) and dopamine (DA) turnover rates in the MPN, SCN, arcuate nucleus (ARN), and the ME were low. Correlated with the beginning of the FSH and LH surges on proestrus, the ME concentrations of LHRH declined. This decline was probably a reflection of LHRH release into portal blood (192) and is correlated with the release of LH and FSH into peripheral plasma. During this time of enhanced, LHRH, LH, and FSH release, turnover rates of NE in the median eminence remain elevated while those of DA decline. DA turnover rates also decline in the ARN but not the MPN. This decline in turnover of DA in the ME is correlated with enhanced release of pituitary prolactin and is a reflection of the physiological inhibitory control the hypothalamus exerts over the secretion of prolactin from the pituitary gland. Norepinephrine turnover rates remain elevated in the MPOA, SCN, and ARN and may be a reflection of the dynamic stimulatory control that NE exerts over the activity of LHRH neurons and subsequently the LH surge on proestrus. Finally, the turnover rate of epinephrine (E) in the hypothalamus from morning to afternoon of proestrus has not been unequivocally characterized (258,276). On this basis, a role in LH release on proestrus has not been overwhelmingly accepted. However, pharmacological evidence, described later, suggests a role for E in regulating LH secretion during the estrous cycle.

Though less compelling, there is evidence developed later in this chapter that the indoleamine, serotonin (5-hydroxytryptamine, 5-HT), may play a neurotransmitter role in the release of LH during the estrous cycle. Unfortunately, an exact role is less clear cut since serotonin has been implicated as both an inhibitory (277–279) and stimulatory (280) neurotransmitter involved in control of LH secretion. Additionally, though not directly shown to play a role in the regulation of the estrous cycle as yet, other stimulatory neurotransmitter candidates are emerging. Specifically, there is a class of excitatory amino acids within the central nervous system. These are represented by L-glutamic acid (L-Glu) and L-aspartic acid. They are present in high concentrations in axon terminals and are probably released by an exocytotic mechanism in the synaptic cleft. Both glutamate and aspartate are released from the medial preoptic area during a progesterone-induced LH surge (281). The central actions of L-Glu probably involve two main postsynaptic receptors which have been characterized by radioligand studies using the highly active agonist N-methyl-D-aspartate (NMDA) and kainic (KA) or quisqualic (QA) acid (non-NMDA). In general, NMDA treatment increases circulating LH (282). There is little doubt that it does so by acting within the hypothalamus (283). Both NMDA and KA will release LHRH from medial basal hypothalamic explants in vitro (284). Thus, both NMDA and non-NMDA receptors are involved in stimulating LHRH and subsequently LH release. Indeed, antagonism of the NMDA receptor will result in blunting of pulsatile LHRH release (285). Furthermore, endogenous amino acid neurotransmission involving both NMDA and non-NMDA receptors are also involved in the estradiol-induced release of an LH surge in ovariectomized rats (286). Central administration of AMPA (alpha-amino-3-hydroxy-5-methylisoxazole-4-propionic acid) a non-NMDA–type glutamate receptor agonist resulted in a stimulation of LH release in the estrogen-primed ovariectomized adult rat (287). AMPA was also found to potently stimulate LHRH release in vitro from mediobasal hypothalamic (MBH) fragments obtained from estrogen-primed ovariectomized adult rats, and this effect was blocked by the selective AMPA receptor antagonist, NBQX (287). The mechanism of action of AMPA appears to differ from that of NMDA as AMPA, in contrast to NMDA, failed to elevate nitric oxide synthase activity in the hypothalamus. The effect of AMPA on LH secretion is steroid dependent, as central administration of AMPA stimulates LH release in estrogen-primed ovariectomized rats, but is inhibitory to LH release in nonestrogen-primed ovariectomized rats. In contrast, AMPA stimulated LHRH release equally well from MBH fragments obtained from estrogen-primed or nonestrogen-primed ovariectomized rats (287). Though no data are available to directly show that the excitatory amino acids play a role in cyclic release of LH in the rat, these data indicate that it is indeed possible. Finally, it has been appreciated for some time that gamma amino butyric acid (GABA) in the central nervous system is a potent inhibitor of phasic LH secretion (288,289). Though the mechanism and site within the brain for this inhibition is not known, there is some suggestion that GABA may interfere with noradrenergic transmission within the locus coeruleus (290,291). Thus, one would expect GABAergic neurons to be active at all times during the estrous cycle except the afternoon of proestrous. Indirect evidence supports such a relationship (292). However, as with many areas of science, this too is not without controversy. Though there is little doubt that this inhibitory effect is modulated by the $GABA_A$ receptor (292). Indeed, endogenous GABA release inhibits the firing of adult LHRH neurons (293). On the other hand, activation of $GABA_A$ receptors excites the electrical activity of adult LHRH neurons (294). As of this time, the resolution of the controversy is not in hand.

Peptide Transmitters as Possible Regulators of the Estrous Cycle

Peptides such as LHRH qualify as neurohormones on the basis of their presence in hypophysial portal blood and their action directly on cells in the anterior pituitary gland. However, there are numerous neuropeptides and neuroactive amino acids which may play a neurotransmitter or neuromodulatory role directly within the hypothalamus.

Perhaps the most widely studied peptides which may be affecting cyclic LH, FSH, and prolactin secretion are those which belong to the family of endogenous opioid peptides. One of these, β-endorphin, inhibits both pulsatile (295–297) and preovulatory (298,299) release of the gonadotrophins. Beta-endorphin is found in portal blood and its concentration is inversely correlated with preovulatory LH release on proestrus and postcastration secretory patterns of LH (300–302). However, the opioid may not qualify as a neurohormone since it does not suppress LH or prolactin secretion at the pituitary gland (303). It is assumed that these suppressive effects are exerted by β-endorphin neurons projecting into the ARN-ME (304) and suppressing LHRH release into portal blood (298,305). It has been suggested that β-endorphin may mediate the suppressive effects of neuropeptide Y (NPY) (306) and orexin (307) administered centrally on LH secretion. Though the endorphins may qualify as inhibitory neurotransmitters controlling cyclic LH release, the role of the enkephalins in the physiological regulation of gonadotrophin secretion is less clear. Systemic injection of methionine-enkephalin decreases LH release in ovariectomized rats (295), whereas it is without effect on pulsatile LH secretion when given in the third ventricle (308). Neutralization of endogenous dynorphin peptides, the endogenous ligands for the kappa-opioid receptors, by push-pull perfusion of the MPOA with antibodies specific for dynorphin A1-17 or A1-8 from 10:30 to 13:55 hours on proestrus, prematurely advance the increase in plasma LH levels normally occurring on this day of the estrous cycle (309). These results suggest that a reduction in prodynorphin gene expression on the afternoon of proestrus may be one event involved in a possible decrease in dynorphin inhibitory tone on the ovulatory LH surge-generating signal. On the other hand, leucine-enkephalin *stimulated* LH release in ovariectomized rats. Under these conditions, we cannot assign an unambiguous physiological role for the enkephalins in control of cyclic hormone release.

We earlier established a neurohumoral role for neuropeptide Y (NPY), oxytocin and galanin in the control of LH secretion. The data are equally compelling for a neuromodulatory role for these as well.

NPY stimulates the release of LHRH from the medial basal hypothalamus in vitro (310), and synaptic contacts have been identified between LHRH and NPY-containing axonal fibers in the septo-preoptic area (311). Most of the literature shows that NPY enhances the LH and FSH-secretory responses to LHRH in vivo (217,219,223,312–315). LH surges are attenuated in NPY-knockout mice (316). These neuromodulatory effects are probably mediated by Y1 receptors (317). Estradiol induces NPY1 receptor expression through nonligand-dependent activation of progesterone receptors (318,319). Though the bulk of the evidence suggests that neuropeptide Y exerts stimulatory effects on LHRH neurosecretion, there are also data to suggest that neuropeptide Y may inhibit LHRH neurosecretion by acting centrally at a Y-5 receptor (320). Thus, it appears quite likely that NPY enhances the surge mode of LHRH neuronal activity through Y-1 receptors yet inhibits basal LHRH neuronal activity through a Y-5 receptor. On the other hand, stimulation of basal LHRH release may be attributed to a Y-4 receptor (321).Taken together, it seems that summoning the well-known link between undernutrition and the loss of fertility, the stimulation of NPY activity by negative energy balance may also lead to an inhibitory effect on LHRH neurons (322,323). Oxytocin may also act as a neuromodulator. Central but not peripheral administration of an antiserum to oxytocin blocks the proestrous surge of LH (324) suggesting a neuromodulatory role for oxytocin in control of LHRH secretion. Excitatory effects of oxytocin may occur at LHRH axon terminals (325) stimulating the activity of nitric oxide synthase via norepinephrine, resulting in an increased release of nitric oxide, which increases prostaglandin-E2 release that in turn induces LHRH release (326,327). Galanin may also act as a neuromodulator since it stimulates norepinephrine release from hypothalamic slices (328). Since galanin has been found to be an estrogen-inducible secretory product of the anterior pituitary (329,330), it could subserve a paracrine/autocrine role in that gland (331). Moreover, since galanin genes are expressed in hypothalamic LHRH-containing neurons, there is a possibility of autocrine regulation at the hypothalamic level (332).

Other neuropeptides have been shown to affect release of LH under a variety of experimental conditions. However, to date, none have been shown to play a truly physiological role in control of cyclic release of LH. The neuropeptides shown to inhibit LH release include arginine-vasotocin (333), cholecystokinin (334), neurotensin (335), and gastrin (336). Those neuropeptides which have been shown to stimulate LH release include substance P (335). In contrast, angiotensin II (337,338), vasoactive intestinal peptide (339–341) by a hypothalamic site of action on LHRH neurons (342), and human pancreatic peptide (343) have been shown to have dual effects on LH secretion: the peptides are inhibitory in untreated ovariectomized animals and stimulatory in steroid-treated ovariectomized rats.

A new class of peptides which were originally described for their orexigenic or satiety properties have more recently been implicated in control of reproductive events such as puberty and the estrous cycle. It is not surprising that these peptides may have a critical role in reproduction since it has been well known that reproduction is optimally tied to feeding state. Although the adipocyte-derived satiety hormone, leptin, is not the primary signal that initiates the onset of puberty, it acts in a permissive fashion, as a metabolic gate, to allow pubertal maturation to proceed if and when metabolic resources are deemed adequate (344,345). In the adult, leptin stimulates LH, FSH, and prolactin release in vitro (346) and in vivo (347). Leptin also stimulates LHRH secretion in vivo (347) and in vitro (348). This is not terribly surprising since leptin receptors have been found in the anterior pituitary gland (349), in areas of the hypothalamus known to control secretion of these hormones (350,351), on hypothalamic neurons containing estrogen receptors (352) and on immortalized LHRH-secreting neurons (353). However, the physiological significance of such observations should be tempered with caution since acute immunoneutralization of endogenous leptin in circulation is without effect on the proestrous as well as the estrogen/progesterone-induced surges of LH and prolactin (354). On the other hand, cycling rats treated with leptin antiserum for 8 days become anestrus (355).

Other members of orexigenic neuropeptides also affect reproductive cyclicity. The orexins (356), also known as hypocretins (357), are cleaved from a common precursor, prepro-orexin (130 aa residues) to form orexin A (residues 33-36; hypocretin 1 28-66) and orexin B (residues 69-96; hypocretin 2 residues 69-97). Orexin receptors are widely distributed throughout the hypothalamus. A subtype of the receptor, orexin 1, is found on LHRH neuronal cell bodies in the hypothalamus but not on LHRH terminals in the median eminence (358). Orexin A fibers are also found in close apposition to 75% to 85% of these LHRH cell bodies (358). Orexin A peptide is a potent stimulator of LHRH release from hypothalamic explants obtained from proestrous rats with diminished effects on explants obtained at other stages of the estrous cycle (359). Similarly, intracerebroventricular injection of orexin A into estrogen/progesterone-treated ovariectomized rats, stimulates LH release in a dose-dependent manner (360). However, the actions of orexin A on LH release may be site-specific. Indeed, injection of the peptide into the rostral preoptic area stimulates LH secretion; whereas, injection into the medial preoptic area or the arcuate nucleus/median eminence area inhibits LH secretion in ovariectomized estrogen/progesterone-treated rats (361). Finally, hypothalamic content of orexin A is greater on proestrus than any other day of the estrous cycle in rats (362). These data, taken together, make a much more compelling case for a role of the orexins than that of leptin in regulating the ovarian cycle of the rat.

To add to the multiplicity of controls, a new ligand and its receptor which may play a role in LHRH neuronal development and activity has recently been described. A metastasis-suppressor gene, KiSS-1, encodes a peptide, kisspeptin-54 (metastin), which binds to a G-protein–coupled receptor, GPR54 (363,364). A mutation in this gene has been associated with idiopathic hypogonadotropic hypogonadism in man (365). The receptor is found on LHRH-like neurons of cichlid fish (366). Both KiSS-1 and GPR54 mRNAs are found in the rat hypothalamus (367). Intracerebroventricular or peripheral administration of metastin or its active fragment, kisspeptin-10, stimulates LH as well as FSH secretion (368–371). These data suggest that metastin may act both at the hypothalamus and anterior pituitary gland to regulate gonadotropin secretion. It is to be anticipated that this will be a highly active area of research in the near future.

Gases as Possible Regulators of the Estrous Cycle

Though the role of peptide and amine neurotransmitters in the regulation of the estrous cycle have been appreciated for some time, it is only recently that the neuroendocrine community has found that the gases carbon monoxide (CO) and nitric oxide (NO) play roles as neurotransmitters regulating the hypothalamo–pituitary axis. These gases are unlike classical peptide and amine neurotransmitters in that they are not released from synaptic vesicles, do not interact with discrete receptor proteins and do not function exclusively at synapses.

CO is produced in the brain as a product of heme metabolism catalyzed by the enzyme heme oxygenase (HO). The byproduct is biliverdin (372–375). HO was originally described as existing as two isoenzymes: HO-1, the inducible form and HO-2, the constituitive form (376,377). A third form, HO-3, has also been found in rat brain (378). One of the highest areas of HO production in the brain is the hypothalamus (379). Hematin, a substrate for CO production is capable of stimulating LHRH release from hypothalamic fragments (380). Zinc protoporphyrin IX, an HO inhibitor, or hemoglobin, a gas scavenging molecule which has a far greater affinity for CO than oxygen, will block the stimulatory effect of hematin on LHRH secretion (372). There is also evidence that CO may mediate the stimulatory effect of NPY on LHRH production (261). Not only does CO act as a neurotransmitter controlling LHRH secretion, but the presence of HO-1 and HO-2 in the pituitary suggests a paracrine role as well (381). Moreover, in ovariectomized rats, the release of an estrogen-induced surge of LH secretion is advanced by 2 hours by prolonged treatment with chromium mesoporphyrin (CrMP), an inhibitor of HO (381). This effect was reversed by administration of hematin (381). Hematin alone has no influence on the estrous cycle of the rat; whereas, CrMP reduced the occurrence of the estrus phase of the rat's estrous cycle (382). Taken together, these data hint at a role for CO as either a neurotransmitter or a paracrine factor that plays a role in regulating the estrous cycle of the rat.

NO is synthesized from the substrate L-arginine catalyzed by the enzyme nitric oxide synthase (NOS). There are three major isoforms of NOS that have been identified at this time: neuronal NOS (n-NOS or type 1), macrophage NOS (m-NOS or type 2), and endothelial NOS (e-NOS or type 3) (383). n-NOS and e-NOS are constitutive, calcium- and calmodulin-dependent enzymes; whereas, m-NOS is an inducible calcium- and calmodulin-dependent enzyme. For full activity all three isoforms require FMN, FAD, NADPH and tetrahydrobiopterin as co-factors (383). Both regulatory and binding proteins have been described for n-NOS. A 10-kd protein designated *p*rotein *i*nhibitor of *n*-NOS (PIN) has been found in the brain (384). As the name implies, this protein physically interacts with and inhibits the activity of n-NOS specifically. There is little work as yet describing a physiological role for PIN in neuroendocrine function. At least two postsynaptic binding proteins, designated PSD-93 and PSD-95, bear PDZ domains that recognize and interact with similar motifs on n-NOS (385–388). Once again, the precise role of these binding proteins to neuroendocrine function is unknown.

NOS has been found in the PVN, SON, OVLT, MPOA, SCN, ME, and ARN (389–393). Immortalized LHRH neurons contain both n-NOS mRNA and protein (394,395). In these cell lines NO has been reported to play either an inhibitory role (396) or a cGMP-dependent stimulatory role (394,397) in LHRH secretion. However, n-NOS is not co-localized in LHRH neurons obtained from rats (394,398), but is found in neurons bearing NMDA receptors (389).

At the current time it is believed that NO modulates LHRH release indirectly by influencing the LHRH secretory response to a host of stimulatory factors. These include the mediation of the LHRH secretory response to estradiol (399–404), NPY (261,405,406) and leptin (407,408). Finally, NO has been found in gonadotrophs of the anterior pituitary gland (409) and is stimulated (410) or inhibited (411) by estradiol and stimulated by LHRH (412). A precise role for NO of pituitary origin has yet to be described.

Growth Factors as Possible Regulators of the Estrous Cycle

Among the growth factors, epidermal growth factor (EGF), insulin-like growth factors (IGF), and nerve growth factor (NGF) may be implicated in the neuroendocrine control of the ovarian cycle of the rat (413). Much of the data suggest a supportive role in the facilitation of LH and FSH secretion.

Prior to the preovulatory release of LH during the late afternoon of proestrus there is a fourfold increase in the proportion of gonadotrophs in the anterior lobe of the pituitary gland (414–417). During this same period, there is an increased expression of EGF and EGF receptor on gonadotrophs to reach a peak on the afternoon of proestrus (418,419). Both EGF itself and LHRH independently increased the expression of EGF receptor on gonadotrophs (413). In addition to its mitogenic effect (420), EGF may also be responsible for differentiation of gonadotrophs bearing LHRH receptors (421,422). EGF release from gonadotrophs appears to be controlled by estradiol (423) as well as LHRH (424). Taken together, it seems that EGF subserves an important autocrine or paracrine role by increasing the proportion of gonadotrophs responsive to LHRH which will produce the ovulation-inducing quantum of LH.

The case for a physiological role for IGF in the neuroendocrine control of the ovarian cycle of the rat is less compelling. Evidence suggests that IGF-I may act at both the pituitary and hypothalamic levels. Indeed, IGF-I and IGF-II both enhance basal and LHRH-stimulated LH release in vitro (425–427). During the estrous cycle of the rat, the number of synaptic inputs to arcuate neuronal somas decreases between the morning of proestrus and the morning of estrus. This decline in synaptic inputs and the accompanying increase in glial ensheathing of neuronal somas are blocked by an IGF-I receptor antagonist. In contrast, the IGF-I receptor antagonist did not affect the basal number of synapses or the morphology of synaptic terminals or length of the synaptic contacts. These findings indicate that IGF-I receptor activation may be involved in the phasic remodeling of arcuate nucleus synapses during the estrous cycle (428).

Nerve growth factor is a 26-kd protein found co-localized in virtually every cell of the anterior pituitary gland (429–432) and secreted (429). Many of these same cells also contain the high affinity form of the NGF receptor (430). NGF is also found in the ovary (433) as well as in the central nervous system (434). However, the levels of the protein or the message do not appear to differ throughout the estrous cycle of the rat (434). Though there are no data to suggest a direct neuroendocrine control of the ovarian cycle, its importance may lie in a potential role in differentiation of somatomammotrophs into mammotrophs (431,435,436).

CONTROL OF PITUITARY HORMONE SECRETION DURING THE ESTROUS CYCLE OF THE RAT

The cyclic release of FSH, LH, and prolactin is the result of positive (stimulatory) and negative (inhibitory) feedback control by the ovarian steroids acting at both the hypothalamus and pituitary gland. The following is the description of this physiological interplay which eventuates in both basal secretion from late estrus through early proestrus and the preovulatory "surge" patterns of these hormones on late proestrus (Fig. 5).

Control of LH Secretion during the Estrous Cycle

Role of Ovarian Steroids Secreted throughout the Estrous Cycle

Inhibition of LH Secretion. In the intact female rat the secretion of LH is low from late estrus to early proestrus due to the negative feedback provided by estrogen and progesterone secreted by the ovaries. This negative feedback was first recognized by Moore and Price in 1932 (437). Such inhibitory effects on LH secretion can be most easily demonstrated by removing the ovaries. Early observations using bioassay endpoints (114) have shown that both pituitary and plasma levels of LH rise following gonadectomy of the rat. These observations were extended using the more sensitive radioimmunoassays. Plasma levels of LH rise slowly over a number of days following ovariectomy (438–440). The earliest significant rise occurs by 18 to 24 hours after ovariectomy with diestrus being the most responsive day for removal of the inhibitory feedback response (439,440). As pointed out by Schwartz and McCormack (114), the hundreds of attempts at restoring LH levels to pregonadectomy levels with estrogen and progesterone have presented varying experimental approaches often without reference to the intact animal. It is now generally accepted that the most potent ovarian steroid for inhibiting LH secretion is estrogen (114). In long-term ovariectomized rats LH levels are significantly lowered by a single injection of as little as 0.2 micrograms of estradiol benzoate (441). Progesterone administration alone is ineffective in lowering the elevated LH levels in ovariectomized rats (442,443,439). However, administration of progesterone to estrogen-treated ovariectomized rats lowers LH levels below those of animals treated with estrogen alone (442). Moreover, simulation of the basal levels of estradiol and progesterone secreted from metestrus through early diestrus prevents the postcastration rise of LH in animals ovariectomized on metestrus (444). From these data it can be inferred that the levels of estradiol and progesterone secreted from metestrus through diestrus are capable of restraining the hypothalamo–pituitary axis and leading to basal secretion of LH during this time. However, now that we appreciate that LH is secreted in a pulsatile manner during the rat's estrous cycle (100–102), it is now apparent that estrogen and progesterone differentially modify LH secretion by influencing the frequency and amplitude of the pulses. It is also apparent that these steroids affect pulse amplitude and frequency differently on the various days of the cycle as well as in long-term ovariectomized rats. Indeed, in long-term ovariectomized rats estradiol suppresses mean LH levels by decreasing pulse frequency (445). Using slightly different experimental approaches, both progesterone (446) and estradiol (447) have been assigned the role of inhibiting mean LH levels from metestrus through diestrus by suppressing pulse amplitude. During the interval from estrus to metestrus, however, mean LH increases due to increased pulse amplitude and frequency which is independent of ovarian steroid negative feedback control (448). It seems likely that the steroids exert their negative feedback on LH secretion by modifying the amplitude and frequency of the LH pulses. This effect is probably exerted at the hypothalamic level by directly or indirectly modifying the pulsatile release of LHRH into the hypophysial portal blood supply to the pituitary gland.

Stimulation of LH Secretion. There is a voluminous literature which indicates that ovarian estradiol, secreted on diestrus, is the positive feedback stimulus for the ovulation-inducing surge of LH. There are several experimental approaches which have attested to this fact. Specifically, under either a 14:10 or 12:12 light-dark cycle with noon as the midpoint of the light interval (97,98,449) ovariectomized rats will respond to a dose of exogenous estradiol by presenting a proestrous-like LH surge the following day with the same timing and duration as the intact rat (450–458). However the magnitude of this induced surge differs from the spontaneous surge in proestrus. That is, baseline values are not as low and peak values are not as high as those on proestrus. In addition, rather than secreting a surge of LH on the single day after estradiol administration, ovariectomized animals respond with LH surges of equivalent timing on each of at least the following 3 days (450–458). Thus, a daily neural event must be coupled with the presence of estrogen to eventuate in daily LH surges in the ovariectomized rat. This confirms Everett and Sawyer's (113) original proposal of a 24-hour periodicity in the LH release apparatus. Other approaches have demonstrated that ovarian estradiol secreted on diestrus is required to activate the LH-surge response on proestrus. Specifically administration of estrogen inhibitors (459), estradiol antiserum (460,461) or ovariectomy (462) on diestrus blocks the proestrous surge of LH. On the other hand, administration of these agents or removal of the ovaries early on proestrus is ineffective in preventing this response later that same day. These data, taken together, suggest that the rising titer of estradiol secreted on diestrus is the major stimulus for the increased secretion of LH on proestrus.

The primary stimulus for the increased secretion of estrogen is the tonic pattern of LH secretion from metestrus through diestrus. Administration of sufficient antiserum to LH to neutralize tonic LH levels on diestrus-2 blocks the subsequent LH surge on proestrus (463,464). Administration of that same dose of LH antiserum on early proestrus failed to block the LH surge late that afternoon (463). Administration of the synthetic estrogen diethylstilbestrol, or LH antiserum preincubated with LH, on diestrus reversed the inhibitory effects of the LH antiserum. Thus, the tonic levels of LH secreted on diestrus contribute to the stimulation of estradiol secretion by the ovarian follicle which in turn stimulates LH secretion.

Unlike the long-term ovariectomized rat which responds to a brief exposure to estradiol with daily LH surges, the intact rat presents LH surges only every 4 to 5 days in response to ovarian estradiol. Moreover, unlike the long-term ovariectomized rat, the acutely ovariectomized rat requires *chronic* exposure to estradiol to secrete daily LH surges (456,457). Thus, some ovarian event associated with the LH surge on proestrus prevents expression of the daily neural signal eventuating in daily LH release. Indeed, it has been shown that expression of the daily LH surge mechanism is limited to proestrus by the heightened secretion of progesterone on proestrus (450,454). Not only is the proestrous surge of progesterone inhibitory to the secretion of a surge quantity of LH on the days following proestrus, but there is every indication that it also contributes to the secretion of the surge of LH on proestrus. This idea first arose from the observation of Everett's in 1948 (465) that injection of progesterone on the day prior to proestrus of a 5-day cycling rat would advance LH release and ovulation 24 hours. In fact, progesterone administered early on the morning of proestrus will advance LH release by several hours (466). Similarly, progesterone administered to ovariectomized, estradiol benzoate-pretreated rats will induce an LH surge during the afternoon of that same day (442). In addition, progesterone will enhance the magnitude of the estrogen-induced LH surge (442). However, regardless of when injected, progesterone will not advance the LH surge beyond the early afternoon. Thus, LH release induced by progesterone is keyed to the time of day, not to the time the steroid was injected. Taken together, it seems that the proestrous surge of progesterone plays a dual role in controlling LH release during the cycle. Specifically, an initial early response which insures the timeliness and increases the magnitude of the coincident LH surge induced by estrogen secretion during diestrus followed by a later extinction of the expression of the estrogen-induced 24-hours periodic release of LH. At this time there is no described role for the other major increase in progesterone secretion which occurs from late metestrus through early diestrus.

The stimulus for the increase in progesterone secretion on proestrus appears to be the proestrous surge of LH itself. Hypophysectomy on proestrus (56) or administration of pentobarbital to block the proestrous surge of LH (61,467) prevents the major rise in progesterone which occurs later that day. Administration of LH reversed the effect of pentobarbital blockade of progesterone secretion (61,467). While the progesterone secreted on proestrus arises from the preovulatory ovarian follicle and is controlled by the proestrous surge of LH, the increase in progesterone from metestrus through diestrus is secreted ephemerally from the newly formed corpora lutea and appears to be autonomous of pituitary support. Indeed, Uchida and co-workers (56) found that hypophysectomy on estrus, shortly after corpus luteum formation, did not prevent the increase of progesterone on metestrus. Thus, this increase occurs independently of continued luteotrophic support.

Interestingly, there appears to be a nonligand-dependent role for the progesterone receptor (PR) in induction of a proestrus-like LH surge by estradiol (468). Pharmacologic blockade of the PR blocks the LH and primary FSH surge on proestrus in rats (469–471). Ovariectomized PR knockout mice will not respond to an estradiol challenge or a self-priming paradigm of LHRH administration with significant LH secretion (472). Peripheral administration of the PR antagonist blocks the estrogen-induced increase in LHRH levels in the median eminence and the proestrous surge of LH in plasma (473). Moreover, implantation of an antisense PR oligonucleotide into the third ventricle prevented PR expression in the anteroventral periventricular region of the hypothalamus and blocked estrogen-induced LH surges (473). Such PR activation requires cAMP (474). Taken together, these data suggest expression of estradiol-induced surges of LH secretion requires ligand-*independent* PR activation. If this is the case, the precise role of the ligand, progesterone, must be more thoroughly evaluated. One possibility is that progesterone down-regulates the activated PR (475,476).

Role of Catechol Estrogens. There is now considerable evidence that the brain converts estrone and estradiol to 2- and 4-hydroxylated derivatives known as catechol estrogens (477). The 2- or 4-phenolic hydroxyl group is located on carbon 2 or 4 ortho to the existing hydroxyl group on carbon 3 (Fig. 14). These catechol estrogens may be the active form in which the steroid exerts its physiological effect on the cyclic secretion of LH and perhaps even FSH and prolactin. Catechol estradiol competes with estradiol for estrogen-binding sites in the anterior pituitary gland and hypothalamus (478). In addition, the catechol estrogens compete for dopamine binding sites on anterior pituitary membranes (479). With a great deal of variation in experimental design and animal model, the catechol estrogens have been implicated in the stimulatory and inhibitory feedback control of LH secretion. It appears that there is a considerable difference in physiological responses to the 2- and 4-hydroxylated estrogens (480,481). In the adult, castrated rat, 4-hydroxyestradiol induces a dramatic decrease

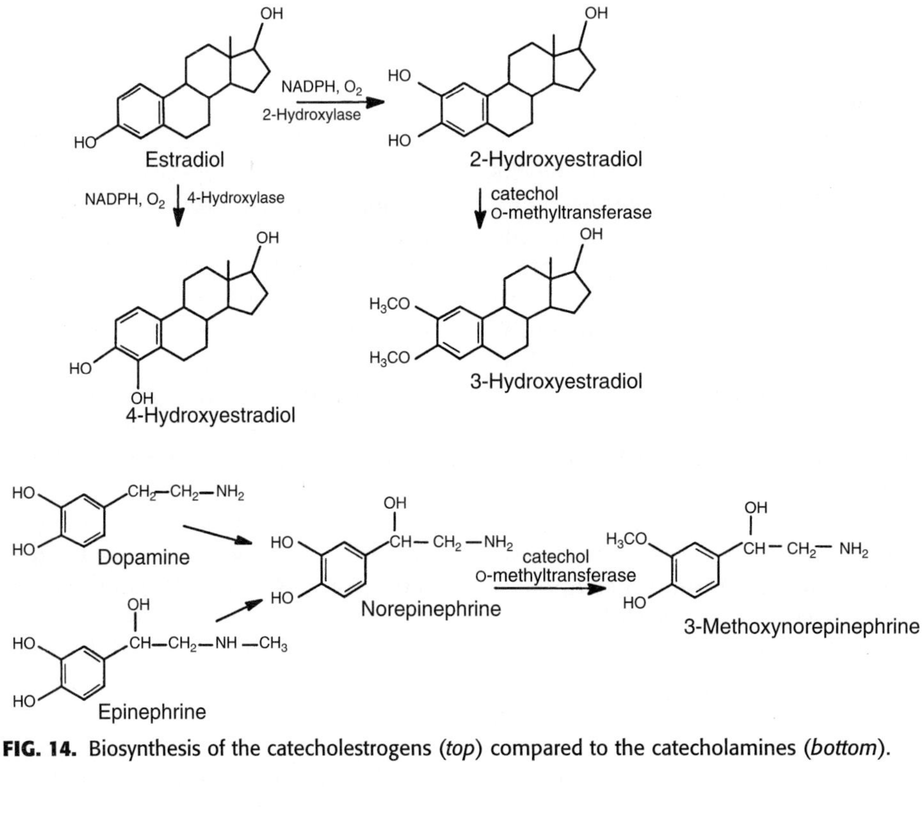

FIG. 14. Biosynthesis of the catecholestrogens (*top*) compared to the catecholamines (*bottom*).

in plasma LH concentration while 2-hydroxyestradiol exerts no effect even at higher doses (482,483). The positive feedback of 2-hydroxyestradiol can be demonstrated in the immature rat only if a massive dose was administered or the steroid injected on 2 consecutive days (484). On the other hand, in adult cycling rats 2-hydroxyestradiol did not affect a preovulatory-like surge of LH (485). However, a sharp rise in the activity of estradiol-2-hydroxylase in subcellular fractions of rat brain occurs just prior to the LH surge on proestrus (486), suggesting an enhanced synthetic capability and a role of 2-hydroxyestradiol at this time. But, such data are not conclusive since it has been shown that estradiol secreted on diestrus, but not proestrus, is responsible for stimulating the LH surge on proestrus (460,461). Nevertheless, though the data are fraught with inconsistencies, it appears that the catechol estrogens may play a significant role in the control of LH secretion during the estrous cycle of the rat.

Sites at Which the Ovarian Steroids Act in Controlling LH Secretion during the Estrous Cycle

The Anterior Pituitary as a Site for Negative Feedback of LH Secretion by Ovarian Steroids. Whether the pituitary is a target for the inhibitory feedback effect of ovarian steroids on LH secretion has been largely unresolved (487,488). That ovariectomy increases (489,490) and estradiol decreases (490–493) the magnitude of LH released in response to LHRH administration, suggests that the pituitary is indeed a sight of negative feedback by estradiol. However, the inhibitory effects of estradiol on pituitary responsiveness to LHRH lasts only 3 to 8 hours even though basal LH release is suppressed (490–493). Moreover, heightened LH responsiveness is only noticeable shortly after ovariectomy before basal LH secretion increases (490). These data suggest that the basal secretion of LH from estrus through early diestrus is due to a direct inhibitory effect of ovarian steroids, particularly estradiol, on the pituitary gland. However, the initial response to ovariectomy may be increased LHRH pulse frequency leading to an increase in pituitary receptors to LHRH (494). The heightened response of the pituitary gland to exogenous LHRH occurs much later (495). Thus, the major site of negative feedback may not be at the pituitary gland but may be at the LH pulse generator within the hypothalamus. Under these conditions, removal of the inhibitory input at the hypothalamus in ovariectomized animals would result in enhanced LHRH pulse frequency and apparent heightened sensitivity of the pituitary gland as a consequence of the enhanced LHRH frequency, rather than removal of inhibitory ovarian steroids.

Neural Sites at Which Ovarian Steroids Act to Inhibit LH Secretion. A great deal of work has focused on the diencephalon as the neural site at which ovarian steroids exert their negative feedback effects. Most older studies have eliminated rostral diencephalic sites such as the diagonal band of Broca, septal complex, preoptic area, and periventricular nucleus as sites at which ovarian steroids act to inhibit LH release (496). The tonic center in the medial basal hypothalamus has been described as the site at which estradiol acts to inhibit LH secretion (497,498). However, these conclusions were based on sampling at daily intervals. We now appreciate, on the basis of 6 minute sampling intervals, that the pulse generator in the CNS regulates the pulsatile patterns of LH secretion which vary throughout the estrous cycle (100,101). Indeed, removal of the ovaries on metestrus results in an increase in both the amplitude and frequency of the LH pulses through diestrus (250). The increase in pulse amplitude is due to the absence of ovarian progesterone while the increase in pulse frequency is due to the removal of both estradiol and progesterone (446). While the pituitary gland may have been eliminated as a site at which the steroids act to modify the LH secretory response to the LH pulse generator (495), the exact site(s) in the CNS potentially responsive to the inhibitory effects of estradiol on pulsatile LHRH secretion have not been widely described. Contrary to the studies which eliminated forebrain areas as sites through which estrogen might act to inhibit LH release, other studies based on localized implantation of estradiol benzoate have implicated the preoptic suprachiasmatic area, the nucleus of the tractus diagonalis and the diagonal band of Broca as sites at which estradiol diminishes mean LH concentration and LH pulse frequency (499). LH pulse amplitude was unaffected in these studies. Implants within the bed nucleus of the stria terminalis, the medial preoptic area, the medial septal nucleus, the anterior hypothalamic area or the 3rd ventricle were ineffective in influencing LH secretory pulses (499). In addition to steroid implantation studies, classical lesion and transection studies have revealed that the LH pulse generation must be located in the arcuate nucleus-median eminence area of the hypothalamus (500). In light of these studies, it seems that the ovarian steroids must maintain the basal LH secretion characteristic of estrus through early proestrus by acting upon neurons in rostral areas of the brain. These neurons, in turn, must synapse with cells of the pulse generator perhaps lying within the arcuate nucleus median eminence area. The ovarian steroids may act upon these rostral cells to result in inhibition of the frequency or amplitude of LH-releasing signals provided to the pituitary gland by the pulse generator. It is not known if the steroids are exciting inhibitory neurons or inhibiting excitatory neurons in the rostral areas. Perhaps the steroids are acting at rostral aminergic neurons which regulate peptidergic neurons in the arcuate nucleus-median eminence area.

The Anterior Pituitary Gland as a Site at Which Ovarian Steroids Stimulate LH Secretion. There is little doubt that the rising titers of estradiol through diestrus and the surge quantities of progesterone secreted on proestrus act at the gonadotroph cell to enhance its responsiveness to LHRH. Indeed, both basal and LHRH-stimulated LH release from cultured anterior pituitary cells are enhanced by the addition of estradiol to the culture medium (501). Similarly, pituitary cells obtained during estrogen-dominated phases of the cycle are most responsive to LHRH (502). In addition, within the first 3 to 8 hours of introduction to the media, progesterone magnifies the estradiol-induced enhancement of the LH-secretory response to LHRH (503). However, by 48 hours after introduction, progesterone significantly depresses the estrogen-induced enhanced LH secretory response to LHRH (504). This effect of addition of steroids to pituitary cells in vitro mirrors the LH-secretory response to LHRH in vivo (197). That is, from diestrus through early proestrus, rats secrete increasing amounts of estradiol and also respond to a challenge of LHRH with increasing maximal increments in plasma concentration of LH. Coinciding with the proestrous surge of progesterone the pituitary gland presents a larger acute increment in LH secretory response to LHRH which is only present for approximately 2 hours. After this time the LH response to LHRH declines rapidly; reaching a nadir by the early morning of metestrus. These effects can be mimicked in vivo by appropriately timed administration of steroids to ovariectomized rats (505). These data, and others, convincingly demonstrate that ovarian estradiol secreted from metestrus through early proestrus and ovarian progesterone secreted during the afternoon of proestrus sensitize the pituitary gland to respond to LHRH. The later effects of the proestrous surge of progesterone are to blunt or completely inhibit the response to LHRH. There is some evidence that the sole role of estradiol may be to modify the response of the pituitary gland to a circadian signal from the hypothalamus. Indeed, in both the cycling (189) and ovariectomized (506) rat, LHRH content of the preoptic area of the diencephalon declines at the same time each day, perhaps a reflection of release of the decapeptide into portal blood. However, on estrus, metestrus and diestrus (189) or in the absence of any substitution with ovarian steroids (506) such decline does not result in release of an LH surge. On the other hand, during the estrogen-dominated phase of the estrous cycle, proestrus, such decline in LHRH content is accompanied by release of a surge quantity of LH (189). Similarly, administration of estradiol to ovariectomized rats does not modify the daily decline in LHRH content but results in release of a surge-like quantity of LH each day (506). It has been suggested that the proestrous surge of progesterone acts by extinguishing the response of the gonadotroph cell to this circadian stimulus (450), thus limiting the expression of the circadian rhythm to the next estrogen-dominated phase of the cycle, proestrus. Taken together, these data suggest that the proestrous surge of LH is the consequence of enhanced sensitivity of the pituitary gland brought about by the rising titers of estradiol from diestrus through proestrus. A further increment in pituitary sensitivity is the result of the initial phases of the proestrous surge of progesterone. The relative insensitivity of the gland on estrus through diestrus may be the late response to the proestrous surge of progesterone.

Just how estrogen and progesterone modify cyclic response of the pituitary gland to LHRH is not known at this time. It had been rather widely suggested that the ovarian steroids enhance the pituitary response to LHRH by increasing LHRH receptor number (507–509) while the diminished response induced by progesterone was a function of declining receptor number (510). However, it now appears that the estrogen-induced heightened or progesterone-regulated diminished responsiveness may occur independent of an increase in LHRH receptor number and may involve a postreceptor mechanism (511–513).

Control of LH Subunit Message. There is little doubt that the gonads influence the expression of the LH subunit mRNAs. Ovariectomy dramatically enhances the expression of the α subunit mRNA and, to a lesser extent, LH-β subunit mRNA in the pituitary gland (514). Moreover, treatment of ovariectomized rats with an estrogen and progesterone regimen which lowers serum levels of LH also lowers α subunit mRNA and LH-β subunit mRNA levels in the pituitary gland (515). Ovariectomized rats treated with estradiol to induce a preovulatory-like surge of LH secretion presented a concomitant increase in α subunit mRNA without change in LH-β mRNA (516). However, when rat pituitary fragments were incubated with estradiol for 2 or 6 hours, LH-β subunit mRNA synthesis was increased (517). These data suggest an additional subunit message control mechanism in cyclic female rats. Indeed, in female rats low amplitude LHRH pulses elevate only LH-β subunit mRNA, whereas higher amplitude pulses enhance both LH-β subunit and α subunit mRNAs (518).

Neural Sites at Which Ovarian Steroids Act to Induce Release of LH. At the moment, the evidence is overwhelming that estrogen acts at the diencephalon to stimulate release of an ovulation-inducing surge quantity of LH on proestrus. Most of the evidence points to rostral hypothalamic sites as the location for these stimulatory effects on LH secretion. Specifically, estradiol implanted directly into the preoptic-anterior hypothalamic area elicits LH release (519,520). Progesterone implanted in this same area also exerts a stimulatory effect on LH secretion in estrogen-primed animals (520). Implantation of steroids in other regions of the brain are largely without stimulatory effect on LH secretion. In addition, destruction of rostral inputs to the medial basal hypothalamus, while having no influence on the LH secretory response to ovariectomy or the negative feedback of ovarian steroids on LH secretion (488), prevents the stimulatory effects of ovarian steroids on release of a surge-like pattern of LH secretion (190,521–523). Also, direct implantation of progesterone within the medial preoptic area blocked the expression of the 24-hour periodic stimulus for release of a surge quantity of LH (451). Placement in other neural areas was ineffective. These and other data have served the argument that the LH "surge center" responsive to the positive feedback of estradiol and progesterone as well as the inhibitory effect of progesterone lies within the rostral hypothalamus. Moreover, the centers controlling the tonic or basal secretion of LH lie within the medial basal hypothalamus. Functional connections between the preoptic-anterior hypothalamus-suprachiasmatic area and the medial basal hypothalamus are required to elicit the positive feedback response to estradiol and progesterone (121).

Interrelationship Among Steroid Hormones, Neural Peptides, and Monoamines in Controlling Cyclic Release of LH

Steroid Effects on LHRH Secretion. As noted earlier, LHRH output by the medial basal hypothalamus, as judged by push-pull perfusion, is pulsatile in nature with enhanced output prior to the expected surge of LH on proestrus (112). Moreover, the concentrations of LHRH in hypophysial portal blood also increases dramatically just prior to the surge of LH on proestrus (192). There is little doubt that the increased output of LHRH by the medial basal hypothalamus is due to the enhanced secretion of estradiol on diestrus exciting peptidergic neurons in the preoptic-anterior hypothalamic area either directly or indirectly, to release a surge quantity of LHRH into portal blood. That is, estradiol benzoate stimulates release of LHRH into hypophysial portal blood in a pattern equivalent to that secreted spontaneously on proestrus (524). The LHRH secretory response to electrical stimulation of the preoptic area was also enhanced by prior estradiol administration (525). However, stimulation of the median eminence, though resulting in a greater LHRH secretory response than preoptic stimulation, did not respond to estradiol with further increased release of LHRH (525). Thus, estradiol must act within the preoptic area to induce LHRH secretion from axon terminals in the median eminence. Estradiol receptors were originally reported to be absent on the cell bodies of LHRH neurons (526). Its action may involve stimulation of the synthesis of LHRH mRNA in areas of the brain which correspond to those which have been identified as "estrogen sensitive" (527–530). However, this would require the presence of estradiol receptors on either LHRH neurons or nearby interneurons. Recent evidence suggests that estrogen receptor (ER) α and β mRNA are found in adult LHRH neurons (531) and ER-α in perikarya of neurons projecting to the vicinity of LHRH perikarya in the rostral preoptic area (532). Indeed, it is possible that estradiol acts on ER-β–containing neurons to stimulate LHRH release and ER-α neurons to inhibit LHRH release (533–536). Further details of the various isoforms of the ER may be found in the Chapter 28.

Though the proestrous surge of progesterone enhances the LH secretory response on proestrus, it may not enhance LHRH secretion into portal blood (524). This would suggest that progesterone enhances LH release on proestrus only by sensitizing the pituitary to receipt of an estrogen-induced surge of LHRH. However, in vitro perifusion of fragments of the preoptic area-medial basal hypothalamus complex of adult ovariectomized rats treated with estrogen and progesterone in vivo resulted in enhanced release of LHRH relative to hypothalami of rats treated with estradiol alone (537). Using the presence of the marker protein, cFos, which signifies neuronal activity, it has been shown that LHRH neurons in the hypothalamus are "activated" shortly after the proestrous surge of progesterone (538). Moreover, using the same approach, it has been shown that progesterone enhances the surge of LH in estradiol-treated ovariectomized rats by further activating LHRH neurons (539). However, it is not known if such "activation" eventuates in LHRH release into portal blood. Since each approach (portal blood vs.

in vitro perifusion vs LHRH neuronal activation) is fraught with uncertainties, it cannot be concluded unequivocally that the proestrous surge of progesterone enhances LH release indirectly by acting on LHRH secreting neurons in the preoptic-medial basal hypothalamic complex. On the other hand, it seems certain that the rising titers of estradiol on diestrus through proestrus act at the preoptic-anterior hypothalamic level to either directly or indirectly enhance LHRH release into hypophysial portal blood. Estradiol may act indirectly by stimulating norepinephrine activity (540–542) which in turn stimulates LHRH release (258). The role of monoamines is discussed in the following section.

Role of Central Monoamines in Cyclic Release of LHRH. Most of the early data suggesting a role of the catecholamines in directly controlling LHRH release is inferential. That is, neurophysiological and neuropharmacologic manipulations within the diencephalon release LH, whereas application of pharmacologic agents directly to the pituitary gland, in most cases, does not affect LH release. These data, along with a few direct measurements of LHRH in vitro (543) and in vivo (544) have served to support the voluminous literature indicating that the catecholamines affect LH release by modifying the activity of LHRH neurons (496). However, the issue of which catecholamine(s) is (are) involved in cyclic release of LHRH by ovarian steroids is far from resolved.

Norepinephrine appears to be the physiological neurotransmitter controlling LH release. However, this catecholamine has a dual effect on LH release. Specifically, in the untreated ovariectomized rat, infusion of norepinephrine or adrenergic agonists into the third cerebral ventricle inhibits pulsatile LH release (545,546). In addition, stimulation of the noradrenergic cell group in the locus coeruleus blocks the release of LH on proestrus (547). Presumably, norepinephrine inhibits neurons in the medial preoptic area responsible for pulsatile LHRH release into portal blood (548). This inhibition is mediated by excitation of β-adrenergic receptors (549,550) since the effects can be blocked with propranolol, a β-blocker, but not phenoxybenzamine, an α-blocker. On the other hand, in ovariectomized rats in which serum LH was maintained at basal levels by ovarian steroids, norepinephrine or α- (but not β-) agonists induced LHRH (551) and LH release (546,552,553). Moreover, treatment with α-antagonists blocked the release of the preovulatory "surge" of LH on proestrus (554). The noradrenergic fibers responsible for stimulation of LHRH and subsequently LH secretion also appear to arise from cell bodies in the locus coeruleus (555–559). These data, taken together, suggest that LH release may be inhibited by a β-adrenergic mechanism in the preoptic area and stimulated via excitation of an α-adrenergic receptor on LHRH neurons. The latter conclusions are also supported by norepinephrine turnover rates on the afternoon of proestrus (258).

Just as norepinephrine has been implicated in the control of the LH surge on the afternoon of proestrus, there are equally convincing data suggesting that central epinephrine plays a similar role (496). For example, blockade of the proestrous surge of LH with pentobarbital (560) or LY78335, an inhibitor of phenylethanolamine-N-methyl-transferase, the enzyme which converts norepinephrine to epinephrine (561) could only be reversed by administration of epinephrine in the cerebral ventricles. Administration of dopamine or norepinephrine was ineffective. Moreover, selective pharmacological suppression of epinephrine synthesis delayed the proestrus-like afternoon accumulation of LHRH in the median eminence and the resultant proestrus-like LH surge in ovariectomized estrogen-progesterone–treated rats (562). These experiments, together with others, suggest a role for epinephrine in stimulating LHRH and subsequently LH release on the afternoon of proestrus. The exact relationship between the function of norepinephrine and epinephrine in controlling LHRH and subsequent LH release is not clear at this time. One suggestion (561) is that norepinephrine stimulates accumulation of LHRH in the median eminence and that epinephrine stimulates release into the portal circulation. The stimulation of these aminergic neurons, in turn, might be a function of ovarian estradiol secreted on diestrus and perhaps even progesterone secreted on proestrus (562).

Most of the current literature provides an uncertain role for dopamine in control of the cyclic release of LH in the rat. Dopamine turnover in the median eminence declines prior to the LH surge on proestrus (258). Moreover, activation of central dopamine receptors inhibits the pulsatile release of LH (563–565). These studies imply that dopamine may be a physiological inhibitor of LHRH and consequently LH release. However, dopamine has also been shown to stimulate LHRH release from median eminence fragments in vitro (566). This suggests a stimulatory role for dopamine in cyclic LH release. Though McCann's laboratory has shown that intraventricular injection of dopamine will cause LH release in ovariectomized steroid-treated rats (567,568), this has not been confirmed in other laboratories (546,552). Therefore, at the moment a role for dopamine in control of the cyclic release of LH has not been unequivocally established.

There are a large number of studies implicating 5-hydroxytryptamine, or serotonin, derived from cell bodies in the dorsal and median raphe nuclei as both inhibitory or stimulatory to LHRH and consequent

LH release. Enhancement of serotonin activity in regions of LHRH neurons along the preoptic-tuberal pathway depresses episodic LH release (277–279). On the other hand, a large number of studies also indicate a stimulatory role for serotonin in the LH "surge" mechanism on proestrus. Indeed, the timing role of the SCN in the circadian nature of the LH release apparatus may be a function of its serotoninergic innervation from the raphe (569–571). There appears to be a diurnal rhythm of serotonin metabolism within the SCN (569). Moreover, pharmacologic blockade of this rhythm disrupts the proestrous surge of LH (280) as well as the estrogen-induced daily surge of LH in ovariectomized rats (571). The usual questions of specificity in pharmacologic experiments such as these obtain. However, lesions of the midbrain raphe, which reduce hypothalamic levels of serotonin, also diminish the estrogen-induced surge of LH in ovariectomized rats (570,571). Taken together, these data imply that serotonin plays a dual role in release of LH: inhibitory to the ultradian rhythm of the LH pulse generator and stimulatory to the daily periodicity of the circadian oscillator. The physiological significance of this dual control remains to be clarified.

Role of Endogenous Opioids in Cyclic Release of LHRH. Of all the neurotransmitter-like peptides implicated in the control of LH secretion, only the endogenous opioids have been shown to play a physiological role in the release of LH during the estrous cycle. Much of our understanding arises from the use of opioid receptor agonists such as morphine sulfate, analogs of the opiates, or opioid receptor antagonists such as naloxone or naltrexone.

Morphine and the endogenous opioids block the proestrous surge of LHRH and subsequently LH (298) as well as the LH surge induced by ovarian steroids in ovariectomized rats (572,573). These effects may be due to the suppression of norepinephrine and epinephrine release from neurons which control LHRH secretion (572,574). Through use of the opiate receptor blocker, naloxone, it is apparent that endogenous opioid tone is greatest on the days of metestrus, diestrus and estrus and diminished immediately before and during the LH surge on proestrus (575,576). In addition, the opiate tone is also reduced around the time of LH release in ovariectomized (577) or ovariectomized, steroid-treated rats (576). Interestingly, β-endorphin levels in portal blood are inversely correlated with dynamic states of LH secretion on proestrus and after castration (301). These data, taken together, suggest that the endogenous opioids may play a physiologically significant restrictive role in control of LH secretion. The changes in opioid tone may be controlled, in turn, by the ovarian steroids.

Control of Follicle-Stimulating Hormone Secretion during the Estrous Cycle

It is apparent from the earlier discussion (Pattern of FSH Secretion Throughout the Estrous Cycle) that the pattern of secretion of FSH bears some basic similarities to that of LH. The basic differences are the bimodal pattern of FSH secretion on proestrus through early estrus (35,46,95,99) and the longer interpulse interval of FSH than LH in gonadectomized animals (105). It will be apparent from the following discussion that these differences may be attributable to differential control of FSH and LH secretion by the hypothalamus as well as by ovarian steroidal and nonsteroidal hormones. Given the inherent limitations of the measuring tools for FSH, it is difficult to make statements regarding the nature of the pulse generator for FSH. Consequently, we do not know if the pulse generator regulating LH secretion and that regulating FSH secretion are one in the same.

Role of Ovarian Steroid and Nonsteroidal Hormones Secreted Throughout the Estrous Cycle

The similarities in control of FSH and LH secretion by ovarian steroids abound. For example, just as the most potent negative feedback for LH is estradiol secreted from late estrus through early proestrus, the same relationship obtains for inhibition of FSH secretion (114). Moreover, secretion of progesterone during this same interval potentiates the inhibitory effect of estrogen on secretion of FSH. However, not all of the inhibition of FSH secretion can be attributed to the ovarian steroids. Much of the physiological literature favors a major role for a previously unidentified nonsteroidal ovarian factor variously called inhibin in the male (578), and originally designated inhibin-F (540) and folliculostatin (579) in the female or the genderless term, gonadostatin (580) for its possible role in suppressing LH as well as FSH secretion. This activity is synthesized by the granulosa cells of the ovarian follicle (581) and appears in high concentration in follicular fluid. In fact, porcine follicular fluid has served as a reliable source of inhibin bioactivity. In addition, the bioactivity or immunoreactivity is found in ovarian venous plasma (582–584), the corpus luteum of pseudopregnancy (585) and the anterior pituitary gland within the gonadotroph (586) of the rat.

The initial evidence for the existence of a nonsteroidal ovarian factor regulating cyclic FSH secretion was indirect. Simulation of the patterns of estradiol and progesterone circulating throughout the estrous cycle are sufficient to maintain basal and induce surge quantities of LH in ovariectomized rats (587). However, both basal levels and surge concentrations of FSH are 2 to 3 times higher in ovariectomized rats than comparable secretory states in intact cycling animals (587). Such observations argued for another ovarian factor which inhibits excessive secretion of FSH and thus prevents uncontrolled follicular maturation. When administration of inhibin-containing porcine follicular fluid was superimposed upon the steroid treatments, both basal and peak concentrations of FSH were lowered to levels similar to those seen in intact cycling rats (587). Interestingly, there was an inverse relationship between the amount of inhibin-like bioactivity in ovarian venous plasma and the secretion of FSH throughout the estrous cycle (582). That is during periods of basal secretion of FSH, inhibin-like activity in ovarian venous plasma is high while during the proestrous-estrous surge of FSH the activity declines.

The activity of porcine follicular fluid (pFF) attributed to inhibin was subsequently isolated and partially characterized as a 32-kd protein of 2 chains of 18 (designated the α-subunit) and 14 kd (the β-subunit), respectively (588). Two forms of 32-kd inhibin, designated inhibin A and B (Fig. 15), were actually isolated (588). The two forms share the same α-subunit but their β-subunits, though similar, are distinguishable by N-terminal microsequencing (Fig. 15). The cDNA clones of the α- and β subunits of inhibins A and B have been isolated from a porcine cDNA library and the N-terminal sequences of α-, βa-, and βb subunits used to design oligonucleotide probes (589). The three rat inhibin subunits were subsequently isolated using the porcine cDNA clones (590,591). The inhibins of the various species are highly conserved. In addition to the varying isoforms of inhibin, two forms of an inhibin binding protein, designated InhBP-L and InhBP-S, have been characterized in the anterior pituitary of the rat (592). Like circulating inhibin A and inhibin B levels, pituitary InhBP-L and InhBP-S mRNA levels are dynamically regulated across the cycle and are negatively correlated with serum FSH levels. Expression of both forms of InhBP is also positively correlated with serum inhibin B, but not inhibin A, levels.

During the course of purification of inhibin, two separate and distinct FSH-regulating activities were also found (588). One of these, designated activin (Fig. 15), showed potent FSH-releasing properties in the inhibin bioassay (588). When the amino acid sequence of the inhibins was deduced, it was found that they bore striking similarities to TGF-β (589). Though TGF-β will enhance FSH release in vitro (593), the FSH-releasing activities in the preparation were found to be dimers of the β subunits of the inhibins (Fig. 15). One, designated activin is βaβb (594) while the other, designated activin A, is βaβa (595,596). A third, activin B (βbβb), has been proposed (597). A third FSH-inhibitory activity distinct from the inhibins was also isolated from pFF (Fig. 15) during the original search for inhibin (588). This single-chain glycosylated polypeptide has a MW of 35-kd and has been designated follistatin (598). It was subsequently shown that follistatin exists as 32, 35, and 39-kd glycoproteins (599). The possibility exists that the varying forms

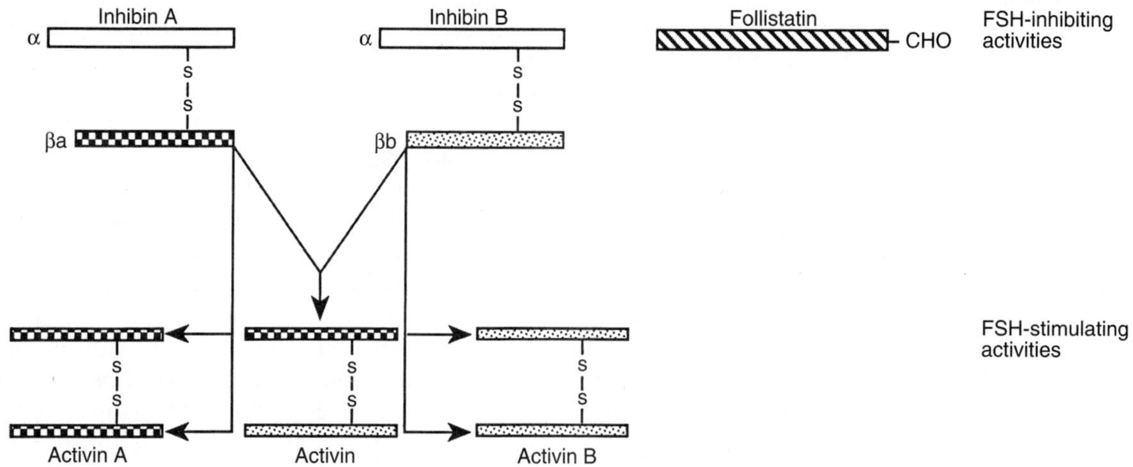

FIG. 15. Diagrammatic representation of the three FSH-inhibiting proteins (inhibin A, inhibin B, follistatin) and three FSH-stimulating proteins (activin, activin A, activin B) found in the gonads of the rat. Similar *patterned bars* represent similar amino acid sequences in the respective subunits. *Arrows* point to proposed structural relationships between the inhibins and activins. See text for details.

may be accounted for by varying degrees of glycosylation. Follistatin will specifically inhibit FSH release from rat pituitary cells in vitro (600). However, there is also evidence that follistatin may be an activin-binding protein (601) or may exert a direct effect on granulosa cells via an autocrine or paracrine mechanism (602).

Much of what we have learned about inhibin, activin, and follistatin (603) is due to the purification of reasonable quantities of these materials as well as the advent of recombinant technology. Using recombinant technologies, the peptides can be synthesized and used for biological studies as well as antibodies produced for immunoneutralization or radioimmunoassay. Recombinant inhibin will suppress FSH but not LH release in ovariectomized rats (604) but suppress both LH and FSH release on proestrus without an impact on ovulation rate 5 days later (605). Similarly, recombinant activin A (606) or purified activin A (607) stimulates FSH synthesis and secretion in vivo (606), stabilizes (608) and enhances (606) FSH-β mRNA in vitro and enhances estradiol secretion in hypophysectomized rats (607). Thus, activin A may not only stimulate FSH secretion, but may also play an autocrine or paracrine role in controlling ovarian estradiol secretion. Administration of antisera to the inhibin preparations not only enhances FSH release during the estrous cycle but also increases ovulation rate during the next estrous cycle (609,610). Measurements of immunoreactive inhibin in peripheral plasma throughout the estrous cycle of the rat (610–612) confirms the measurements of bioactive inhibin in ovarian venous plasma (582). Serum concentrations of inhibin were high during the morning of proestrus and then declined sharply to the lowest levels by 24:00 hours on proestrus after the preovulatory LH and FSH surges. Peripheral plasma levels of inhibin then returned to peak levels by 15:00 hours on the afternoon of estrus from which they again rapidly declined to another low level by 24:00 hours on estrus. Thus, serum levels of immunoreactive inhibin are inversely correlated with those of FSH (35).

The estradiol-induced proestrous surge of LHRH is responsible for increased pituitary follistatin gene expression that is associated with the primary FSH surge on proestrus (613,614). At the time of the trough between the primary and secondary surge of FSH on proestrus (Fig. 5), there is decline in pituitary follistatin, an increase in free circulating activin A and pituitary content of FSH-β mRNA (586). These three events lead to the subsequent secondary surge of FSH secretion.

Much of the in vitro evidence points to a stimulatory effect of ovarian steroids on FSH secretion directly. Estradiol appears to stimulate LH and FSH secretion by acting on the pituitary directly (504). Moreover, progesterone potentiates this stimulatory effect of estradiol on gonadotrophin secretion (504). However, unlike its direct stimulatory effect on LH secretion, the apparent effect of progesterone on FSH secretion may be secondary to its stimulatory effect on FSH synthesis (504). The other ovarian steroid which appears to preferentially stimulate cyclic FSH release perhaps by acting directly on the pituitary gland is testosterone (615). This is predicted by the fact that neutralization of circulating levels of androgens (50,51) partially blocks the early estrous secondary surge of FSH without disturbing the proestrous surge of LH or FSH. Androgens and progesterone may act within the dorsomedial nucleus-anterior hypothalamic area (232) as well as the pituitary gland (504,615) selectively to stimulate FSH secretion. Thus, it appears that estrogen and progesterone share a stimulatory role in controlling the release of FSH and LH on proestrus. It appears that androgens may play a further role in contributing to the secondary rise of FSH on estrus.

Interrelationship of Steroid and Nonsteroidal Hormones, Neural Peptides, and Monoamines Controlling the Cyclic Release of FSH

As previously described, the rising titers of estradiol by early diestrus have excited specific peptidergic neurons within the hypothalamus to release a larger pulse of LHRH into the portal vasculature during the afternoon of proestrus (192,616). While the early (proestrous) phase of the FSH surge and the entire proestrous surge of LH are dependent on this release of LHRH into portal blood, several lines of evidence suggest that the secondary (estrus) phase of FSH secretion may be independent of hypothalamic support. When spontaneous FSH release on proestrus is blocked with phenobarbital, infusion of LHRH reinstitutes the proestrous phase of FSH release (617). Moreover, administration of an antiserum to (237) or an antagonist of LHRH on the early afternoon of proestrus (239,618) was effective in blocking the proestrous phase of FSH release. However, these same treatments were ineffective in preventing the estrous phase of FSH release when given on late proestrus. In addition, complete hypothalamic deafferentation on late proestrus does not prevent the estrus phase of FSH release (619). Such data emphasize the fact that the proestrous phase of FSH secretion is dependent on direct hypothalamic stimulation while the estrous phase may be independent of direct support.

While it is clear that inhibin is the ovarian peptide and follistatin is the pituitary peptide that inhibit

FSH-β subunit message, it is also apparent that LHRH enhances FSH-β mRNA in the female rat that is dependent upon pulse amplitude (518). Low amplitude LHRH pulses are more effective as stimulators of FSH-β and LH-β subunit mRNAs while higher amplitude LHRH pulses are required for α subunit mRNA synthesis (518).

Though LHRH is effective as a pharmacologic agent in reinstating FSH release (617), it is not yet completely certain that it is in fact the physiological releasing hormone for all phases of FSH release. The regulation of the hypothalamic-independent estrous phase of FSH secretion remains the subject of informed speculation from which we can paint a picture. As we have detailed above, ovarian estradiol secreted through diestrus stimulates specialized hypothalamic neurons to secrete LHRH into the portal vasculature on the early afternoon of proestrus. This LHRH in turn stimulates the pituitary gland to release LH and perhaps also FSH (620). Follicular progesterone released in response to the proestrous surge of LH acts on estrogen-sensitized cells in the pituitary gland to enhance the secretion of LH and the synthesis as well as secretion of FSH on proestrus. The secondary rise of FSH on early estrus may be a consequence of the stimulation by the primary FSH surge secreted on proestrus (621,622). This primary FSH surge may either act directly on the pituitary gland or perhaps even on the hypothalamus (623) via retrograde blood flow (139). Alternatively, the secondary phase of FSH secretion on estrus may merely be a delayed or protracted response to the initial release of LHRH from the hypothalamus on proestrus. This is suggested by the fact that pituitary fragments from proestrous rats superfused in vitro with a 10-minute pulse of LHRH respond with an immediate pulse of LH lasting approximately one hours from baseline to baseline. On the other hand, FSH secretion in response to this 10 min pulse of LHRH is biphasic. The first phase lasting approximately 1.5 hours and the second phase persisting over 4 hours (624). Thus, the secondary phase of FSH secretion may be the delayed response of androgen-sensitized gonadotrophs (51,615) to LHRH subsequent to inhibin-withdrawal (582,625). The decline of inhibin secretion into ovarian venous plasma (582) is due to the proestrous surges of LH as well as FSH (583,612). The decline of inhibin secretion into ovarian venous blood leads to enhanced pulsatile FSH secretion (105) presumably at the pituitary level, thus allowing the secondary FSH surge to occur. The following resumption of inhibin secretion into ovarian venous blood by late estrus (582) corresponds to the recruitment of a new crop of ovarian follicles initiated in turn by this secondary surge of FSH on estrus (626,627). Thus, inhibin secretion is inhibited by the surges of LH and FSH on proestrus and stimulated by the secondary surge of FSH secretion on estrus.

In general, the neurotransmitter system regulating release of FSH during the estrous cycle is unknown. This is due largely to the uncertainty of the nature of the hypophysiotropic stimulus arising from the hypothalamus. As summarized above, one school of thought proposes that differential FSH and LH secretion during the estrous cycle is merely due to changing sensitivity of the FSH-secreting cell to LHRH brought about by inhibin of ovarian origin. Alternatively, differential secretion of FSH and LH may also be due to release of a peptidergic FSHRH distinct from LHRH (628). As noted earlier, the argument has arisen as a result of the demonstration of hypothalamic areas controlling solely FSH (233) as well as extracts of hypothalamic tissue containing mainly FSH—with virtually no LH-releasing activity (236). If LHRH is also the physiological FSH-releasing hormone (620), one might expect that depending on experimental design, some manipulations of central monoamines affect FSH as well as LH secretion. Thus, in situations in which the sensitivity of the FSH-secreting cell is not modified by inhibin, one would expect that pharmacologic stimulation of the central dopaminergic or noradrenergic system might affect LH and FSH secretion. However, microinjection of dopamine, norepinephrine or epinephrine into the third brain ventricle of steroid-primed ovariectomized rats induced release of LH but not FSH (568). As noted earlier, these amines will induce release of LHRH from the hypothalamus. It seems that their inability to release FSH in the face of LH release argues that LHRH may not be a physiological releaser of FSH and that the dopaminergic or noradrenergic system may also not be involved. Obviously, an understanding of the hypothalamic control of FSH secretion is one of the great voids in completing our understanding of the estrous cycle and its controls.

Control of Prolactin Secretion during the Estrous Cycle

As noted earlier, the pattern of secretion of prolactin during the estrous cycle is quite similar to that of LH. That is, it is secreted in low basal amounts for the major portion of the cycle except for a "surge" quantity on the afternoon of proestrus. This surge begins approximately 2 hours prior to the preovulatory surge of LH on proestrus but reaches peak values coincident with that of LH on proestrus. Enumeration of the similarities and differences in controls of the 2 surges follow.

Role of Ovarian Steroids in Control of the Cyclic Secretion of Prolactin

There is an overwhelming literature which clearly states that the dominant and perhaps sole stimulus for the proestrous surge of prolactin is the rising titer of estrogen on diestrus through early proestrus. Indeed, just as neutralization of the circulating levels of estradiol with an antiserum blocks the proestrous surge of LH, it also blocks the proestrous surge of prolactin (461). Similarly, just as estrogen will induce daily afternoon surges of LH in ovariectomized rats, it will also stimulate daily surges of prolactin (458). These data, taken together, clearly show that the proestrous surges of LH and prolactin are both linked to a 24-hour clock wherein expression is closely linked to the presence of estradiol. Moreover, progesterone administered on diestrus-3 of a 5-day cycle will advance the proestrous LH (465) and prolactin (629) surges 24 hours. However, unlike the same effects on LH, progesterone neither enhances magnitude of estrogen-induced prolactin surges nor extinguishes the estrogen-induced diurnal rhythm of prolactin release (M. E. Freeman, unpublished). Though estrogen administration will elevate basal levels of plasma prolactin in male rats, surge levels of prolactin do not result from such administration (458). Such a sexually differential response results from androgenization of the hypothalamo–pituitary axis of the males during neonatal life (458).

Sites and Mechanisms Whereby Ovarian Steroids Control Prolactin Secretion during the Estrous Cycle

There is little doubt that one locus for the stimulatory action of estradiol is the lactotroph cell of the pituitary gland. Indeed, incubation of enzymatically dissociated pituitary cells with estradiol enhances both the basal- and TRH-stimulated release of prolactin (630). Moreover, coincubation with progesterone prevents the stimulatory effect of estradiol while progesterone alone exerts no effect (630). Estradiol probably exerts its excitatory effect by blocking the inhibitory influence of dopamine. The dopamine agonist, CB-154, is less effective in inhibiting basal or TRH-stimulated prolactin secretion in the presence of estradiol (630). This antidopaminergic action of estradiol is probably not due to antagonism of the dopamine receptor (630) but may act at a postreceptor calcium- or cyclic AMP-dependent step (630,631). In a sense, estradiol functions as a prolactin responsiveness factor; decreasing responsiveness to dopamine and allowing some spontaneous prolactin release as well as increasing responsiveness to TRH (632).

Estradiol also stimulates the proestrous surge of prolactin by acting at the hypothalamic level. Earlier studies revealed that a cut placed immediately behind the optic chiasm blocked the induction of a proestrous-like surge of prolactin by exogenous estradiol (633). Since this retrochiasmatic cut transected caudal efferents from the MPOA as well as the SCN, and the latter is a timing regulator (115) one cannot be absolutely certain, on the basis of transection studies, of the exact site of estrogen action in inducing the proestrous surge of prolactin. It is more certain however, on the basis of focal lesion as well as localized implantation studies that both the MPOA and the SCN play a role. For example, lesions of the MPOA prevented the estrogen-dependent proestrus surge of prolactin (634) and the estrogen-induced afternoon surge of prolactin in ovariectomized rats (635). In addition, lesions of the SCN blocked the estrogen-induced afternoon surge of prolactin in ovariectomized rats (636–638). These data suggest that both the MPOA and the SCN are involved in the cyclic prolactin secretory response to ovarian estradiol. They do not indicate if one or both sites respond directly to estrogen or indirectly through synaptic influence with other estrogen-responsive areas. On the other hand, implantation of cannulae containing sufficiently dilute crystalline estradiol into hypothalami of ovariectomized rats revealed sites which responded to such implants with proestrous-like surges of prolactin (638). Systemic leakage was monitored by vaginal smears and uterine weight patterns. This study revealed that the MPOA and the ventromedial hypothalamus (VMH) were most sensitive to estradiol implantation while placement in other sites involved in the estradiol-response, the SCN and anterior pituitary, did not induce proestrus-like surges of prolactin (638). Though this study suggested that the SCN is probably not estrogen-responsive directly, more recent studies have suggested that estradiol acts within the SCN ensuring the fidelity of each prolactin surge. Indeed, administration of estradiol increased the expression of the mRNA for the clock gene Cry 2 within the SCN but was without affect on the expression of Cry 1 mRNA (639).

More recent data implicates cell bodies in the locus coeruleus (presumably noradrenergic) in the elaboration of the estradiol-induced surge of prolactin on proestrus (640). Indeed, lesion of the locus coeruleus blocked both the proestrous surge and the estradiol induced surge of prolactin without influencing basal secretion. Such an observation suggests the involvement of norepinephrine in the surge of prolactin.

Dopamine and Candidate Prolactin-Releasing Hormones of Hypothalamic Origin

The chemical identity of the neurotransmitters and neurohormones regulating this estrogen-dependent proestrous surge of prolactin is partially known. We now appreciate the fact that though the dominant hypothalamic input regulating pituitary prolactin release is inhibitory, the release of the proestrous surge of prolactin probably results in a delicate balance of acute removal of inhibitory input coupled with simultaneous stimulatory input from the hypothalamus (641). It has been shown, though not without contradiction, that dopaminergic binding sites (642), the turnover of dopamine (643), the activity of dopaminergic neurons (644), and dopamine levels in portal blood (645) are depressed on the day of proestrus corresponding to an increased secretion of prolactin into peripheral plasma. Dopamine receptors in the anterior pituitary gland have been reported to increase (642) and decrease (646) during the afternoon of proestrus. The source of dopamine controlling prolactin secretion is believed to be tuberoinfundibular neurons (TIDA) with cell bodies in the arcuate nucleus of the hypothalamus and axons terminating upon capillary tufts in the external layer of the median eminence. Long portal vessels emanating from these tufts deliver dopamine to the anterior lobe of the pituitary gland (647,648). There is also ample evidence to suggest that dopamine arrives at the anterior lobe of the pituitary gland from the posterior lobe that contains axon terminals of tuberohypophysial dopaminergic (THDA) neurons where cell bodies are found in the rostral part of the arcuate nucleus (649,650). Finally, dopamine may arrive from the intermediate lobe that contains axon terminals of periventricular hypophysial dopaminergic (PHDA) neurons where cell bodies are found in the periventricular nucleus (651–653). The short portal vessels connecting the three lobes of the pituitary gland may serve as the vascular pathway through which dopamine is transported to the anterior lobe from the neural and intermediate lobes. Indeed, the fact that posterior lobectomy (654) or neurointermediate lobe denervation (650) results in chronic elevation of prolactin secretion implicates both THDA and PHDA neurons in the control of prolactin secretion.

The issue of the control of the activity of these dopaminergic neurons on proestrus is far from settled. The content of dopamine in the median eminence and portal blood declines prior to the onset of the surge of prolactin secretion on proestrus (645–647). The proestrus surge of progesterone has been implicated in the decline in dopaminergic tone in vivo (657), while estradiol has been implicated in the short-term decline in the in vitro activity (658) and gene expression (659) of the rate-limiting enzyme in dopamine synthesis, tyrosine hydroxylase. Progesterone reverses the estradiol-induced decrease in tyrosine hydroxylase mRNA levels in the arcuate nucleus (660). Not only does the proestrous surge of progesterone suppress the expression of tyrosine hydroxylase mRNA in the arcuate nucleus (661), but it inactivates the enzyme by inducing dephosphorylation (662). It is quite likely that both steroids play a role in the prolactin-releasing decrease in dopamine synthesis and neurosecretion: estradiol initiating (658) and progesterone prolonging (663) the decline in tyrosine hydroxylase gene expression on proestrus.

The signal for the reactivation of the dopaminergic neurons and the resultant decline in prolactin secretion at the end of proestrus are the peak levels of prolactin on proestrus (664). Indeed, immunoneutralization of circulating prolactin on proestrus prevents the elevation of dopamine concentration (665) and activation (666) of all three populations of hypothalamic dopaminergic neurons.

The induction of an estrogen-stimulated proestrus-like surge of prolactin is associated with a decline of dopamine levels and increase in TRH levels in portal blood (667). Such evidence suggests that dopamine is the physiological prolactin inhibitor prior to the afternoon of proestrus and that a diminution of the levels of this neurohormone in portal blood is requisite for a surge of prolactin secretion during the afternoon. Based on pharmacologic approaches (632), candidates for the converse stimulatory hormones are many. The earliest physiological evidence pointed to the tripeptide TRH as the stimulatory neurohormone on the afternoon of proestrus. Indeed, TRH is a potent stimulator of prolactin release (632). The concentration of TRH in hypophysial portal plasma increases on the afternoon of proestrus corresponding to the times of the prolactin surge (668). However, a concomitant surge of TSH does not accompany the proestrous surge of prolactin (669). This is most surprising since suckling, a potent stimulus for prolactin release, also releases TSH (669). It is possible that during the estrous cycle, receptors on lactotrophs are more abundant or have a higher affinity for the peptide than receptors on thyrotrophs thus allowing the lactotroph the most sensitive secretory response to minute changes in levels of TRH in portal plasma. Immunoneutralization of these changing blood levels during proestrus have resulted in attenuation (245) or merely altered the timing (670) of the proestrous surge of prolactin. Nevertheless, it is now inferred from studies in ovariectomized, estrogen-treated rats (667) and lactating rats (671–673) as well as in vitro studies (674) that optimal prolactin release on proestrus in response to estrogen and progesterone or during lactation in response to suckling requires a diminution in dopamine release in concert with an increase in a PRH release into portal blood.

Though not as thoroughly documented, there is abundant evidence that OT (215,675) and VIP may play a physiological role as a PRH on the afternoon of proestrus [see (676) for references]. On the other hand, somatostatin, GABA, atrial natriuretic hormone, vasopressin, calcitonin, NPY, AII, Galanin, substance P, bombesin-like peptides, neurotensin, and even dopamine itself have been implicated as PRHs, but their physiological significance has not been satisfactorily unmasked at this time [see (676) for references].

Peptides isolated from bovine hypothalamus through the use of an orphaned receptor extracted from human pituitary glands (677), when synthesized (678), specifically stimulated the release of prolactin from pituitary cells of lactating rats. Two peptide sequences of 20 and 31 amino acids were isolated and quickly named *prolactin-releasing peptide* (PrRP). The PrRP mRNA for each is most abundant in the medulla oblongata (679–681) whereas only moderate amounts are found in the hypothalamus and the anterior lobe of the pituitary gland (680,682,683). Neither the 20 or 31-aa peptide will stimulate prolactin secretion from pituitary cells obtained from male rats; whereas, those obtained from random cycling female rats respond only to the highest dose of either peptide (684). When administered intravenously, the 31-aa peptide stimulated prolactin secretion in male as well as female rats in a dose-dependent manner with females responding in a more sensitive manner, particularly during proestrus (685). Immunocytochemical approaches have revealed PrRP-positive fibers in the paraventricular and supraoptic nuclei of the hypothalamus and in the neural lobe of the pituitary gland with cell bodies found in the dorsomedial and ventromedial nuclei of the hypothalamus (686,683). Interestingly, the presence of PrRP-positive fibers in the neural lobe and the ability of the peptide to release oxytocin (687) suggests that prolactin-releasing properties of OT may be mediated through PrRP. Alternatively, the absence of PrRP-positive fibers in the external layer of the median eminence and presence in the posterior lobe of the pituitary gland (683,686,688) suggest the PrRP may reach the anterior lobe through the short rather than the long portal vessels. Though meager, some studies suggest that PrRP may play a role in the preovulatory release of prolactin and even LH on proestrus (689–692). In contrast, there is emerging a literature which illustrates the role of PrRP in autonomic homeostasis (693). Indeed, PrRP has been implicated in regulation of cardiovascular function (694–696), stress responses (697–700), ingestive behavior and energy balance (701–703) and sleep (704–706). Taken together, it is clear that the rush to (mis-) name this peptide has not led to an unambiguous conclusion of a physiological role.

Possible Roles for Prolactin Secreted during the Estrous Cycle

While it is quite clear that prolactin is luteotrophic during pseudopregnancy and early pregnancy, an indispensable role for prolactin in control of the hypothalamo–pituitary–ovarian events of the estrous cycle has not yet been firmly established. It was initially suggested that the proestrous surge of prolactin plays a luteolytic role during the estrous cycle (707,708). However, the corpora lutea are already secreting relatively small amounts of progesterone by proestrus so it seems unlikely that such input is necessary. On the other hand, a role for the proestrous surge of prolactin in control of female sexual behavior has been suggested (709). Inhibition of the proestrous surge of prolactin with a dopamine agonist attenuates the lordosis quotient (LQ) on proestrus while replacement with ovine prolactin restores the LQ to normal. The enhanced LQ is also associated with the estrogen-induced prolactin surge in ovariectomized rats. However, it is not certain at this point if prolactin has a direct physiological role in enhancing female sexual behavior or if it may act by stimulating adrenal progesterone secretion (709). Finally, there are both pituitary and ovarian roles which have recently been attributed to prolactin secreted during the estrous cycle. Specifically, the proestrous surge of prolactin may attenuate the pituitary LHRH-receptor level (710). This would explain the decline in LHRH receptor which occurs on proestrus. Secondly, within the granulosa cells of the rapidly developing ovarian follicle on proestrus, the surge of prolactin may be responsible for inhibiting estrogen-synthesizing aromatase enzyme activity (711) as reflected in the rapid decline in blood estrogen levels on the early afternoon of proestrus after the prolactin surge begins (35). However, interference with either of these activities does not appear to have a major impact on the regular expression of the normal estrous cycle in the rat.

THE ACTIVE LUTEAL PHASE DURING THE OVARIAN CYCLE OF THE RAT

"Rescue" of the Corpus Luteum

Earlier in this chapter it was noted that the corpora lutea formed during the estrous cycle of the rat were only ephemerally active structures. That is, except for a brief interval shortly after ovulation, the corpora lutea do not secrete sufficient progesterone to maintain a decidual reaction in the uterus. Thus, the luteal phase is relatively brief unlike those species which have spontaneous prolonged functional luteal phases of 14 to 16 days duration such as guinea pigs, sheep, cows, pigs, monkeys, and humans. However, the corpora lutea can be "rescued" from the quiescent state and transformed to the active state (35,712) by injection of prolactin, or transplantation of pituitary glands beneath the kidney capsule which secrete copious amounts of prolactin when removed from hypothalamic restraint (713). In addition, corpora lutea may be rescued on diestrus by mating with a male rat on the previous proestrus. The corpora lutea of pregnancy persist for 20 to 22 days. Finally, mating with a sterile male or artificial stimulation of the uterine cervix results in a period of extended luteal function lasting 12 to 14 days. This interval is known as pseudopregnancy and is characterized by a period of heightened prolactin secretion from the pituitary gland in response to the mating stimulus (714,715). It is this prolactin secretion which initially "rescues" the corpus luteum on diestrus (35) and allows it to respond to other luteotrophic factors (41).

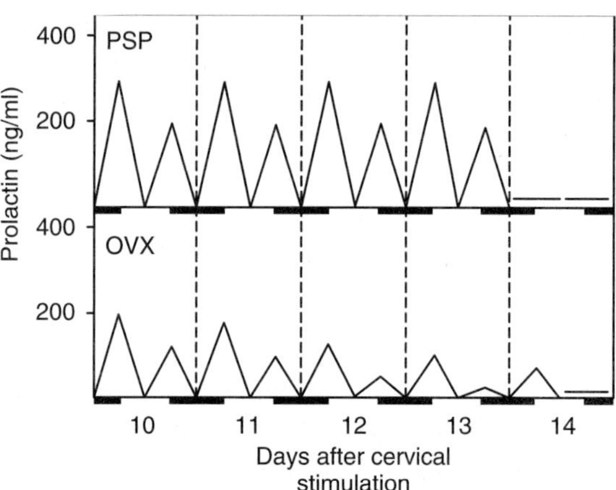

FIG. 16. Diagrammatic representation of the nocturnal and diurnal surges of PRL on days 10 to 14 after cervical stimulation in animals with intact ovaries (PSP, *upper panels*) and long-term ovariectomized rats (OVX, *bottom panels*). Values on the ordinate are approximate serum hormone concentrations (ng/ml NIAMDD-PRL-RP-1). *Black bars* on abscissa are the dark intervals in the animal quarters (18:00–06:00 h). During PSP the larger nocturnal and smaller diurnal surges persist for 12 to 13 days and end abruptly. The surges of PRL in OVX rats are smaller than those of PSP rats. Moreover, as the days following cervical stimulation elapse, the magnitude of the surges gradually wanes. However, the surges themselves persist beyond a normal PSP.

The Unique Pattern of Prolactin Secretion

Rather than being constantly elevated, the pattern of prolactin secretion in response to mating is characterized by two daily surges (716,714,715). One of these surges, called *diurnal* begins in the afternoon (13:00–15:00 hours), reaches peak values during the early evening (17:00–19:00 hours) and returns to basal levels by 24:00 hours. The other surge, called *nocturnal*, begins by 01:00 hours, peaks by 03:00 to 07:00 hours, and approaches baseline values by 11:00 hours. This recurrent pattern (Fig. 16, *top*) of prolactin secretion continues throughout pseudopregnancy but ends abruptly by day 12 or 13 (717,718). This does not represent the typical neuroendocrine response to an acute stimulus. On the contrary, the prolactin secretory response to suckling exemplifies the typical neuroendocrine reflex. Temporally, the response is tightly coupled to the stimulus. Within minutes of introduction of hungry pups, the nursing mother's blood levels of prolactin increase. In addition, the magnitude of the increase is related to the intensity of the stimulus. That is, larger litters induce release of more prolactin from lactating mothers than do smaller litters. The response is tightly stimulus-bound in that each subsequent secretory episode requires reinforcement by reapplication of the suckling stimulus provided by the pups. In contrast, the prolactin secretory response to mating has some unique features not shared with other neuroendocrine reflexes. Unlike the constantly elevated prolactin secretion induced by suckling, mating-induced prolactin secretion is surge-like, coupled to the time of day (719) and is repetitive even without reapplication of the stimulus (714,715).

Control of Mating-Induced Prolactin Secretion

Ovarian and Uterine Controls

The surges are somewhat independent of the ovaries in that they continue in an attenuated form if the ovaries are removed within hours of cervical stimulation or can be initiated by and continue for several days after brief cervical stimulation of long-term ovariectomized rats (717,720). Thus, the basic neuroendocrine reflex remains in the absence of ovarian steroids. However, ovarian steroids do play a role in the control of mating-induced prolactin surges. Progesterone magnifies the nocturnal surge and prolongs the recurrent nature of both the nocturnal and diurnal surges of prolactin (717,720). Moreover, estradiol magnifies the diurnal surge and abolishes the nocturnal surge. Though the steroids are not required for initiation or maintenance, they seem to play a role in the termination of these surges on a timely basis. The surges end abruptly by day 12 to 13 in pseudopregnant rats

(Fig. 16) but wane gradually in ovariectomized, cervically stimulated rats through day 14 (717). This result implies an ovarian role in their abrupt termination. Indeed, the diurnal surge is terminated by the gradual withdrawal of progesterone as the corpus luteum wanes at the end of pseudopregnancy, whereas the nocturnal surge is terminated by the rising level of estradiol secreted by the newly developing follicle at this time (717).

Just as the disappearance of prolactin surges at the end of pseudopregnancy is associated with a decline in luteal activity (717,721,722), the presence of the surges beyond day 13 is associated with lengthening of the functional lifespan of the corpus luteum. Specifically, in hysterectomized rats, regression of the corpus luteum is prevented (42) and the surges of prolactin continue for several days beyond a normal pseudopregnancy (629,723). Though it is possible that the prolactin surges are terminated indirectly because of regression of the corpus luteum induced by a uterine luteolytic agent (42), the uterus may also secrete a substance that acts directly at the hypothalamo–pituitary axis to terminate prolactin surges. Indeed, the nocturnal surges of prolactin are still present 16 days after cervical stimulation in long-term ovariectomized rats in the absence of the uterus but not when the uterus is in place (723). Because there is no luteal tissue to be maintained in these animals, the continuation of the surge must be due, not to persistent progesterone secretion, but to direct restraint provided by the uterus. In fact, the uterus must be present for the ovarian steroids to effectively inhibit the surges of prolactin secretion (724). This implies that the nonpregnant uterus may secrete an activity that acts at the hypothalamo–pituitary axis to modify prolactin secretion. Such prolactin inhibitory activity has been extracted from the uterus and shown to be specific for its tissue of origin and hormone on which it acts (725). Moreover, it has been demonstrated that uterine epithelial cells secrete the activity while its presence in peripheral blood is correlated with the presence of the uterus (726). Though there are many questions remaining unanswered regarding the prolactin-inhibitory activity of the uterus, these data suggest that the nonpregnant uterus may function as an endocrine organ regulating prolactin secretion.

Neural Sites Mediating the Prolactin Secretory Response to Cervical Stimulation

The pelvic nerve is the primary sensory afferent from the cervix (727). The message provided by cervical stimulation ascends to the brainstem through the anterolateral columns of the spinal cord.

The central neural areas responsive to cervical stimulation include the lateral reticular nucleus, the central gray, the gigantocellular reticular nucleus and the dorsal raphe nucleus (728–730). The induction of nocturnal surges of prolactin can be prevented by transections in the ventromedial and lateral part of the midbrain (731).

Within the hypothalamus, three areas have been implicated in the control of prolactin release in response to cervical stimulation. Lesions of the dorsal MPOA cause repetitive pseudopregnancies (173) characterized by nocturnal but no diurnal surges of prolactin (174). Moreover, the lesion will block the diurnal surge induced by cervical stimulation (174). These studies suggest that the MPOA possesses two types of neurons: those tonically inhibitory to nocturnal surges and those potentially stimulatory to diurnal surges. Cervical stimulation acts by inhibiting the former and stimulating the latter (732). Since destruction of the MPOA leads to consecutive spontaneous luteal phases approximating those of other mammals, one can speculate on the evolutionary significance of the primacy of the MPOA. The activation of repetitive luteal phases by destruction of the MPOA suggests that the rat has a short estrous cycle characterized by a quiescent corpus luteum due to the tonic inhibitory activity provided by the MPOA over secretion of the luteotrophic hormone, prolactin. Perhaps phylogenetically, an active MPOA which stimulates ovulation-inducing LH surges and inhibits luteal-activating prolactin surges in rodents is the more primitive form. The more advanced form may be mammals not dependent on the MPOA for LH secretion, ovulation, and subsequent luteal activation. Indeed, rhesus monkeys secrete perfectly normal LH surges in response to estrogen when only their medial basal hypothalami are intact (733). The second area implicated in mating induced prolactin secretion is the dorsomedial-ventromedial area of the hypothalamus (DMN-VMN). Electrical stimulation of the DMN-VMN induces a pseudopregnancy characterized by nocturnal and diurnal surges of prolactin (174,734). However, lesion of this area does not interfere with estrous cyclicity, the ability to mate and support a pregnancy to term, or secrete a nocturnal surge of prolactin (734–736). Such a lesion only interferes with the diurnal surge of prolactin (735). In fact, the intact DMN-VMN is required for the stimulatory but not the inhibitory control of the prolactin surges by the MPOA (737). The third area of the hypothalamus shown to be important in mating-induced prolactin secretion is the SCN. When lesioned, both surges of prolactin that had been induced by mating are completely abolished (738). Since this area has been characterized as a master pacemaker for prolactin release (739), it is probably the timer involved in transducing the time-of-day signal for these regular prolactin surges (115). At the moment, the functional interrelationships between these areas is not yet fully appreciated.

Neuroendocrine Control of the Prolactin Secretory Response to Cervical Stimulation

Though there is a reasonably expansive literature describing the neuroendocrine control of suckling-induced prolactin secretion (632), the same is not true describing control of mating-induced prolactin release. A single study describes the relationship between dopamine levels in hypophysial portal blood and the release of prolactin surges in cervically stimulated rats (740). As expected, there was an inverse relationship between dopamine levels in portal blood and prolactin surges in peripheral blood. That is, during the surges, dopamine levels were 36% lower in cervically stimulated than control ovariectomized animals, while during intersurge intervals there was no difference between dopamine levels in portal blood of control and cervically stimulated rats.

The reason for this periodic inhibition of dopamine release into portal blood has been studied using the accumulation of L-dopa in the median eminence as an index of tuberoinfundibular dopaminergic nerve activity (741). In this assay system, the amount of L-dopa accumulated in the median eminence is inversely related to the activity of dopaminergic neurons. The activity of these dopaminergic neurons, in turn, is inversely correlated with the prolactin surges induced by mating and may indeed be controlled by them (741). Thus, as the prolactin surge occurs in response to diminution of dopamine secretion into portal blood, these high levels of prolactin may feedback to stimulate the activity of these tuberoinfundibular dopaminergic neurons which results in dopamine release into portal blood and consequent inhibition of prolactin secretion (742). Such feedback loops result in phasic rather than enhanced basal secretion and may be the basis for the two daily surges of prolactin secretion in mated animals. This issue, coupled with the possible involvement of a mating-induced prolactin releasing factor, has not been clarified.

To determine if the decrease in dopamine levels is solely responsible for release of prolactin in cervically stimulated rats, the synthesis of dopamine was blocked with α-methyl para tyrosine (AMPT) in ovariectomized noncervically stimulated rats. Dopamine was then infused peripherally at rates which resulted in portal blood levels simulating those of cervically stimulated rats (740). Dopamine concentration in physiologic ranges inhibited prolactin secretion but pharmacologic quantities were required to suppress prolactin to basal levels characteristic of the intersurge intervals. Moreover, a 36% decrease of infused dopamine resulted in only a 1.5-fold increase in prolactin levels while a 36% spontaneous decrease in dopamine secreted into portal blood in the cervically stimulated rat was associated with a four- to fivefold increase in prolactin secretion (740). Such observations may argue for the involvement of another additional inhibiting factor or a heretofore unidentified prolactin releasing factor activated by cervical stimulation.

The description of an endogenous stimulatory rhythm regulating prolactin secretion (743) would be compatible with this notion. The fact that the prolactin secretory response of ovariectomized rats to pharmacologic dopamine inhibition is greatest at times of day corresponding to the nocturnal and diurnal surges of prolactin (743) argues for active *stimulatory* input to the lactotroph by the hypothalamus at those times. This input is normally not expressed in unmated animals since dopamine is tonically present to mask the response to the endogenous stimulatory rhythm. Quite possibly, mating merely acts to lower dopaminergic tone sufficiently to allow this prolactin releasing rhythm to be expressed. The components of this endogenous stimulatory rhythm have been shown to be oxytocin, VIP, and serotonin (744). Indeed, pharmacologic blockade of oxytocin receptors will interfere with both the nocturnal and diurnal surges of prolactin secretion in cervically stimulated animals while VIP antagonism blocks only the nocturnal surge and serotonin inhibition blocks the diurnal surge (745). This has led to the hypothesis (745) that oxytocin is the neurohormone which stimulates the lactotroph to release prolactin at these two times of day and that VIP is a neurotransmitter which excites oxytocin release into portal blood in the early morning and serotonin is the neurotransmitter that stimulates the early evening surge (Fig. 16). In ovariectomized rats, oxytocinergic neurons express the marker of neuronal activity, cFos, in greatest abundance during the early morning and early evening (746). In addition, VIP content of the paraventricular nucleus is highest in the early morning and serotonin highest in the early evening (747). These data suggest that the two surges of prolactin secretion are the consequence of the unmasking of an endogenous stimulatory rhythm by the dopamine lowering effects of the mating stimulus and further suggest that *any* exteroceptive stimulus known to release prolactin may be subject to the same rhythm (748). More recent studies have emphasized the complexity of the control of mating-induced prolactin secretion (749). Oxytocin appears to stimulate prolactin secretion by a calcium-dependent mechanism. In mated animals, the activity of dopaminergic neurons during the early morning just prior to the initiation of the nocturnal surge of prolactin decreases (750). This diminution appears to be due to the influence of VIP originating originating in SCN neurons terminating upon the three populations of neuroendocrine hypothalamic dopaminergic neurons (717,720). These neurons have VIP receptors and placement of VIP antisense oligonucleotides into the SCN prevents the fall in dopamine that precedes the release of prolactin. Similarly, placement of VIP antisense oligonucleotides into the SCN abolished the afternoon increase in oxytocin and prolactin in cervically stimulated animals (749). These data, taken together, imply that mating induces prolactin secretion through a twofold mechanism: removal of the inhibitory influence of dopamine and initiation of the stimulatory effects of oxytocin. Both of these events are regulated by VIP.

Significance of the Short Luteal Phase

As noted earlier, the corpora lutea of rats, mice and hamsters secrete progesterone for only a brief period after ovulation. In the absence of spontaneous luteotrophic support the waning corpora lutea marshal in new waves of ovarian follicles and hence new ovulation. Due to the brevity of the luteal phase in rodents, ovulation recurs every 4 to 5 days. One can speculate on the selective advantage of short cycles. In those cycles in which an ovulation is not associated with a fertile mating and conception, it would not be advantageous to have a protracted period, the luteal phase, in preparation for continuation of a pregnancy which did not take place. On the other hand, it seems most efficient if an active luteal phase which prepares the uterus for a conceptus is only associated with a mating stimulus. In that regard, a luteotrophic mechanism activated by a mating stimulus is an example of such an efficient system.

REFERENCES

1. Heape, W. (1900). The "sexual season" of mammals and the relation of the "pro-oestrum" to menstruation. *Quart. J. Micr. Sci.* 44, 1–70.
2. Long, J. A., and Evans, H. M. (1922). The estrous cycle of the rat and its associated phenomena. *Mem. Univ. Calif.* 6, 1–148.
3. Blandau, R. J., Boling, J. L., and Young, W. C. (1941). The length of heat in the albino rat as determined by the copulatory response. *Anat. Rec.* 79, 453–463.
4. Astwood, E. B. (1939). Changes in weight and water content of the uterus of the normal adult rat. *Am. J. Physiol.* 126, 162–170.
5. Hartman, C. G. (1944). Some new observations on the vaginal smear of the rat. *Yale J. Biol. Med.* 17, 99–112.
6. Mandl, A. M. (1951). The phases of the oestrous cycle in the adult white rat. *J. Exp. Biol.* 28, 576–584.
7. Mandl, A. M. (1951). Cyclical changes in the vaginal smear of adult ovariectomized rats. *J. Exp. Biol.* 28, 585–592.
8. Mandl, A. M. (1963). Pre-ovulatory changes in the oocyte of the adult rat. *Proc. Roy. Soc. B.* 158, 105–118.
9. Odor, D. L. (1955). The temporal relationship of the first maturation division of rat ova to the onset of heat. *Am. J. Anat.* 97, 461–491.
10. Tsafriri, A., and Kraicer, P. F. (1972). The time sequence of ovum maturation in the rat. *J. Reprod. Fert.* 29, 387–393.
11. Ayalon, D., Tsafriri, A., Lindner, H. R., Cordova, T., and Harell, A. (1972). Serum gonadotrophin levels in pro-oestrous rats in relation to the resumption of meiosis by the oocytes. *J. Reprod. Fert.* 31, 51–58.
12. Blandau, R. J. (1945). The first maturation division of the rat ovum. *Anat. Rec.* 92, 449–457.
13. Boling, J. L., Blandau, R. J., Soderwall, A. L., and Young, W. C. (1941). Growth of the Graafian follicle and the time of ovulation in the albino rat. *Anat. Rec.* 79, 313–331.
14. Young, W. C. (1961). The hormones and mating behavior. In *Sex and Internal Secretions* (W. C. Young, Ed.), pp. 1173–1239. Williams & Wilkins, Baltimore.
15. Young, W. C., Boling, J. L., and Blandau, R. J. (1941). The vaginal smear picture, sexual receptivity and time of ovulation in the albino rat. *Anat. Rec.* 80, 37–45.
16. Mossman, H. W., and Duke, K. L. (1973). Comparative Morphology of the Mammalian Ovary.
17. Deane, H. W. (1952). Histochemical observations on the ovary and oviduct of the albino rat during the estrous cycle. *Am. J. Anat.* 91, 363–413.
18. Dempsey, E. W., and Bassett, D. L. (1943). Observations on the fluorescence, birefringence and histochemistry of the rat ovary during the reproductive cycle. *Endocrinology* 33, 384–401.
19. Harris, R. J. (1962). The structure of the ovary. In *The Ovary* (S. Zuckerman, Ed.), pp. 143–187. Academic Press, New York.
20. Espey, L. L. (1978). Ovulation. In *The Vertebrate Ovary. Comparative Biology and Evolution* (R. E. Jones, Ed.), pp. 503–532. Plenum Press, New York.
21. Lipner, H. (1973). Mechanism of mammalian ovulation. In *Handbook of Physiology, Endocrinology* (R. Greep, Ed.), pp. 409–437. American Physiological Society, Washington, DC.
22. Arai, H. (1920). On the post-natal development of the ovary (albino rat), with special reference to the number of ova. *Am. J. Anat.* 27, 405–462.
23. Ingram, D. L. (1962). Atresia. In *The Ovary* (S. Zuckerman, Ed.), pp. 247–274. Academic Press, New York.
24. Johnson, J., Canning, J., Kaneko, T., Pru, J. K., and Tilly, J. L. (2004). Germline stem cells and follicular renewal in the postnatal mammalian ovary. *Nature* 428, 145–150.
25. Pedersen, T., and Peters, H. (1968). Proposal for a classification of oocytes and follicles in the mouse ovary. *J. Reprod. Fert.* 17, 555–557.
26. Richards, J. S., and Midgley, A. R., Jr. (1976). Protein hormone action: a key to understanding follicular and luteal cell development. *Biol. Reprod.* 14, 82–94.
27. Hirshfield, A. N., and Midgley, A. R., Jr. (1978). Morphometric analysis of follicular development in the rat. *Biol. Reprod.* 19, 597–605.
28. Peppler, R. D., and Greenwald, G. S. (1970). Influence of unilateral ovariectomy on follicular development in cycling rats. *Am. J. Anat.* 127, 9–14.
29. Blandau, R. J. (1955). Ovulation in the living albino rat. *Fertil. Steril.* 6, 391–404.

30. Greenwald, G. S., and Rothchild, I. (1968). Formation and maintenance of corpora lutea in laboratory animals. *J .Anim. Sci.* 27, Suppl. I, 139–162.
31. Boling, J. L. (1942). Growth and regression of corpora lutea during the normal estrous cycle of the rat. *Anat. Rec.* 82, 131–145.
32. Bassett, D. L. (1943). The changes in the vascular pattern of the ovary of the albino rat during the estrous cycle. *Am. J. Anat.* 73, 251–291.
33. Everett, J. W. (1945). The microscopically demonstrable lipids of cyclic corpora lutea in the rat. *Am. J. Anat.* 77, 293–323.
34. Pupkin, M., Bratt, H., Weisz, J., Lloyd, C. W., and Balogh, K., Jr. (1966). Dehydrogenases in the rat ovary. I. A histochemical study of 5- 3 b-and 20 a-hydroxysteroid dehydrogenases and enzymes of carbohydrate oxidation during the estrous cycle. *Endocrinology* 79, 316–327.
35. Smith, M. S., Freeman, M. E., and Neill, J. D. (1975). The control of progesterone secretion during the estrous cycle and early pseudopregnancy in the rat: prolactin, gonadotropin and steroid levels associated with rescue of the corpus luteum of pseudopregnancy. *Endocrinology* 96, 219–226.
36. Deb, S., Hashizume, K., Boone, K., Southard, J. N., Talamantes, F., Rawitch, A., and Soares, M. J. (1989). Antipeptide antibodies reveal structural and functional characteristics of rat placental lactogen-II. *Mol. Cell. Endocr.* 63, 45–56.
37. Campbell, W. J., Deb, S., Kwok, S. C. M., Joslin, J. A., and Soares, M. J. (1989). Differential expression of placental lactogen-II and prolactin-like protein-A in the rat chorioallantoic placenta. *Endocrinology* 125, 1565–1574.
38. Robertson, M. C., Croze, F., Schroedter, I. C., and Friesen, H. G. (1990). Molecular cloning and expression of rat placental lactogen-I complementary deoxyribonucleic acid. *Endocrinology* 127, 702–710.
39. Faria, T. N., Deb, S., Kwok, S. C. M., Talamantes, F., and Soares, M. J. (1990). Ontogeny of placental lactogen-I and placental lactogen-II expression in the developing rat placenta. *Dev. Biol.* 141, 279–291.
40. Robertson, M. C., Schroedter, I. C., and Friesen, H. G. (1991). Molecular cloning and expression of rat placental lactogen-I_v, a variant of rPL-I present in late pregnant rat placenta. *Endocrinology* 129, 2746–2756.
41. Rothchild, I. (1981). The regulation of the mammalian corpus luteum. *Rec. Progr. Horm. Res.* 37, 183–298.
42. Melampy, R. M., and Anderson, L. L. (1968). Role of the uterus in corpus luteum function. *J. Anim. Sci.* 27, Suppl. I, 77–96.
43. Shaikh, A. A. (1971). Estrone and estradiol levels in the ovarian venous blood from rats during the estrous cycle and pregnancy. *Biol. Reprod.* 5, 297–307.
44. Yoshinaga, K., Hawkins, R. A., and Stocker, J. F. (1969). Estrogen secretion by the rat ovary in vivo during the estrous cycle and pregnancy. *Endocrinology* 85, 103–112.
45. Brown-Grant, K., Exley, D., and Naftolin, F. (1970). Peripheral plasma estradiol and luteinizing hormone concentrations during the estrous cycle of the rat. *J. Endocrinol.* 48, 295–296.
46. Butcher, R. L., Collins, W. E., and Fugo, N. W. (1974). Plasma concentration of LH, FSH, prolactin, progesterone and estradiol-17b throughout the 4-day estrous cycle of the rat. *Endocrinology* 94, 1704–1708.
47. Naftolin, F., Brown-Grant, K., and Corker, C. S. (1972). Plasma and pituitary luteinizing hormone and peripheral plasma oestradiol concentrations in the normal oestrous cycle of the rat and after experimental manipulation of the cycle. *J. Endocrinol.* 53, 17–30.
48. Nequin, L. G., Alvarez, J., and Schwartz, N. B. (1979). Measurement of serum steroid and gonadotropin levels and uterine and ovarian variables throughout 4 day and 5 day estrous cycles in the rat. *Biol. Reprod.* 20, 659–670.
49. Falck, B. (1959). Site of production of oestrogen in rat ovary as studied in microtransplants. *Acta Physiol. Scand. Suppl.* 47, 94–101.
50. Dupon, C., and Kim, M. H. (1973). Peripheral plasma levels of testosterone, androstenedione and oestradiol during the rat estrous cycle. *J. Endocrinol.* 59, 653–654.
51. Gay, V. L., and Tomacari, R. L. (1974). Follicle stimulating hormone in the female rat: cyclic release is dependent on circulating androgen. *Science* 184, 75–77.
52. Eto, T., Masuda, H., Suzuki, Y., and Hosi, T. (1962). Progesterone and pregn-4-en-20 a-ol-3-one in rat ovarian venous blood at different stages in reproductive cycle. *Jpn. J. Anim. Reprod.* 8, 34–40.
53. Hashimoto, I., Hendricks, D. M., Anderson, L. L., and Melampy, R. M. (1968). Progesterone and pregn-4-en-20 a-ol-3-one in ovarian venous blood during various reproductive states in the rat. *Endocrinology* 82, 333–341.
54. Piacsek, B. E., Schneider, T. C., and Gay, V. L. (1971). Sequential study of luteinizing hormone (LH) and "progestin" secretion on the afternoon of proestrus in the rat. *Endocrinology* 89, 39–45.
55. Telegdy, G., and Endroczi, E. (1963). The ovarian secretion of progesterone and 20 a-hydroxpregn-4-en-3-one in rats during the estrous cycle. *Steroids* 2, 119–123.
56. Uchida, K., Kadowaki, M., and Miyake, T. (1969). Ovarian secretion of progesterone and 20 a-hydroxypregn-4-en-3-one during rat estrous cycle in chronological relation to pituitary release of luteinizing hormone. *Endocrinol. Jpn.* 16, 227–237.
57. Wiest, W. G., Kidwell, W. R., and Kirschbaum, T. H. G. (1963). Induction of rat ovarian 2-a-hydroxysteroid dehydrogenase activity by gonadotropic hormone administration. *Steroids* 2, 617–629.
58. Allen, W. C. (1931). I. Cyclical alterations in the endometrium of the rat during the normal cycle, pseudo-pregnancy and pregnancy. II. Production of deciduomata during pregnancy. *Anat. Rec.* 48, 65–103.
59. Wilcox, R. B., and Wiest, R. G. (1960). Comparative effectiveness of progesterone and 4-pregnene-20 a-ol-3-one in the development of deciduomata. *Endocrinology* 67, 281–282.
60. Hashimoto, I., and Wiest, W. G. (1969). Correlation of the secretion of ovarian steroids with function of a single generation of corpora lutea in the immature rat. *Endocrinology* 84, 873–885.
61. Barraclough, C. A., Collu, R., Massa, R., and Martini, L. (1971). Temporal interrelationships between plasma LH, ovarian secretion rates and peripheral plasma progestin concentrations in the rat: effects of nembutal and exogenous gonadotropins. *Endocrinology* 88, 1437–1447.
62. Goldman, B. D., Kamberi, I. A., Sitteri, P. K., and Porter, J. C. (1969). Temporal relationship of progestin secretion, LH release and ovulation in rats. *Endocrinology* 85, 1137–1143.
63. Shaikh, A. A., and Shaikh, S. A. (1975). Adrenal and ovarian steroid secretion in the rat estrous cycle temporally related to gonadotropins and steroid levels found in peripheral plasma. *Endocrinology* 96, 37–44.
64. Greep, R. O. (1974). History of research on anterior hypophysial hormones. In *Handbook of Physiology* (R. O. Greep, E. B. Astwood, E. Knobil, and W. H. Sawyer, Eds.), Section 7, Endocrinology, pp. 1–27. American Physiological Society, Washington, DC.
65. Evans, H. M., and Long, J. A. (1921). The effect of feeding the anterior lobe of the hypophysis on the oestrous cycle of the rat. *Anat. Rec.* 21, 62–62.
66. Evans, H. M., and Long, J. A. (1921). The effect of the anterior lobe of the hypophysis administered intraperitoneally upon growth and maturity and oestrous cycles of the rat. *Anat. Rec.* 21, 62–63.
67. Evans, H. M., and Long, J. A. (1922). Characteristic effects upon growth, oestrus and ovulation induced by the intraperitoneally

administration of fresh anterior hypophyseal substance. *Anat. Rec.* 23, 19–19.
68. Evans, H. M., and Long, J. A. (1922). Characteristic effects upon growth, oestrus and ovulation induced by the intraperitoneally administration of fresh anterior hypophyseal substance. *Proc. Natl. Acad. Sci. U S A* 8, 38–39.
69. Smith, P. E. (1926). Ablation and transplantation of the hypophyses in the rat. *Anat. Rec.* 32, 221–221.
70. Smith, P. E. (1926). Hastening development of female genital system by daily homoplastic pituitary transplants. *Proc. Soc. Exp. Biol. Med.* 24, 131–132.
71. Aschheim, S. (1926). Uber die funktion des ovariums. *Z. Geburtshille Gynaekol.* 90, 387–392.
72. Zondek, B., and Aschheim, S. (1926). Uber die funktion des ovariums. *Z. Geburtshilfe Gynaekol.* 90, 372–376.
73. Zondek, B. (1930). Uber die hormone des hypophysenvorderlappens. I. Wachstums hormon, follikelreifungshormon (Prolan A, Luteinisierungshormon) Prolan B, Stoffwechselhormon. *Klin. Wochschr.* 9, 245–248.
74. Zondek, B. (1930). Uber die hormone des hypophysenvorderlappens. II. Follikelreifungshormon Prolan A-Klimakteriumkastration. *Klin. Wochschr.* 9, 393–396.
75. Evans, H. M., and Simpson, M. E. (1929). A comparison of the ovarian changes produced in immature animals by implants of hypophyseal tissue and hormone from the urine of pregnant women. *Am. J. Physiol.* 89, 381–387.
76. Fevold, H. L., Hisaw, F. L., and Leonard, S. L. (1931). The gonad-stimulating and the luteinizing hormones of the anterior lobe of the hypophysis. *Am. J. Physiol.* 97, 291–301.
77. Greep, R. O., and Fevold, H. L. (1937). The spermatogenic and secretory function of the gonads of hypophysectomized adult rats treated with pituitary FSH and LH. *Endocrinology* 21, 611–618.
78. Greep, R. O., Fevold, H. L., and Hisaw, F. L. (1936). Effects of two hypophyseal gonadotropic hormones on the reproductive system of the male rat. *Anat. Rec.* 65, 261–272.
79. Squire, P. G., and Li, C. H. (1959). Purification and properties of interstitial cell-stimulating hormone from sheep pituitary gland. *J. Biol. Chem.* 234, 520–525.
80. Papkoff, H., Candotti, A., and Li, C. H. (1964). Purification of ovine follicle stimulating hormone (FSH). *Fed. Proc.* 23, 410–410.
81. Papkoff, H., Gospodarowicz, D., and Li, C. H. (1967). Purification and properties of ovine follicle stimulating hormone. *Arch. Biochem. Biophys.* 120, 434–439.
82. Sairam, M. R., and Papkoff, H. (1974). Chemistry of pituitary gonadotrophins. In *Handbook of Physiology* (R. O. Greep, E. B. Astwood, E. Knobil, and W. H. Sawyer, Eds.), Section 7, Endocrinology, Volume 4, Part 2, pp. 111–131. American Physiological Society, Washington, DC.
83. Evans, H. M., and Simpson, M. E. (1929). Hyperplasia of the mammary apparatus of adult virgin females induced by anterior hypophyseal hormones. *Proc. Soc. Exp. Biol. Med.* 26, 598–598.
84. Riddle, O., Bates, R. W., and Dykshorn, S. W. (1932). A new hormone of the anterior pituitary. *Proc. Soc. Exp. Biol. Med.* 29, 1211–1215.
85. Riddle, O., Bates, R. W., and Dykshorn, S. W. (1933). The preparation, identification and assay of prolactin—a hormone of the anterior pituitary. *Am .J. Physiol.* 105, 191–216.
86. Lahr, L., and Riddle, O. (1936). Temporary suppression of estrous cycles in the rat by prolactin. *Proc. Soc. Exp. Biol. Med.* 34, 880–893.
87. Astwood, E. B. (1941). The regulation of corpus luteum function by hypophysial luteotropin. *Endocrinology* 28, 309–320.
88. Evans, H. M., Simpson, M. E., and Lyons, W. R. (1941). Influence of lactogenic preparations on production of traumatic placentoma in rat. *Proc. Soc. Exptl. Biol. Med.* 46, 586–590.
89. Evans, H. M., Simpson, M. E., Lyons, W. R., and Turpeinen, K. (1941). Anterior pituitary hormones which favor production of traumatic uterine placentoma. *Endocrinology* 28, 933–945.
90. Li, C. H., Dixon, J. S., Lo, T.B., Pankov, Y. A., and Schmidt, K. D. (1969). Amino acid sequence of ovine lactogenic hormone. *Nature* 224, 695–696.
91. Niall, H. D., Hogan, M. L., Tregear, G. W., Segre, G. V., Hwang, P., and Friesen, H. (1973). The chemistry of growth hormone and lactogenic hormones. *Rec. Progr. Horm. Res.* 29, 387–416.
92. Parlow, A. F. (1961). Bioassay of pituitary luteinizing hormone by depletion of ovarian ascorbic acid. In *Human Pituitary Gonadotropins* (A. Albert, Ed.), pp. 300–312. Charles C. Thomas, Springfield, IL.
93. Steelman, S. L., and Pohley, F. M. (1953). Assay of the follicle stimulating hormone based on the augmentation with human chorionic gonadotropin. *Endocrinology* 53, 604–616.
94. Igarashi, M., and McCann, S. M. (1964). A new sensitive bioassay for follicle stimulating hormone (FSH). *Endocrinology* 74, 440–445.
95. Daane, T. A., and Parlow, A. F. (1971). Periovulatory patterns of rat serum follicle stimulating hormone and luteinizing hormone during the normal estrous cycle: effects of pentobarbital. *Endocrinology* 88, 653–663.
96. Neill, J. D., and Reichert, L. E. (1971). Development of a radioimmunoassay for rat prolactin and evaluation of the NIAMDD rat prolactin radioimmunoassay. *Endocrinology* 88, 548–555.
97. Hoffman, J. C. (1968). Effect of photoperiod on estrous cycle length in the rat. *Endocrinology* 83, 1355–1358.
98. Hoffman, J. C. (1969). Light and reproduction in the rat: effect of lighting schedule on ovulation blockade. *Biol. Reprod.* 1, 185–188.
99. Gay, V. L., Midgley, A. R., Jr., and Niswender, G. D. (1970). Patterns of gonadotrophin secretion associated with ovulation. *Fed. Proc.* 29, 1880–1887.
100. Fox, S. R., and Smith, M. S. (1985). Changes in the pulsatile pattern of luteinizing hormone secretion during the rat estrous cycle. *Endocrinology* 116, 1485–1492.
101. Gallo, R. V. (1981). Pulsatile LH release during the ovulatory LH surge on proestrus in the rat. *Biol. Reprod.* 24, 100–104.
102. Gallo, R. V. (1981). Pulsatile LH release during periods of low level LH secretion in the rat estrous cycle. *Biol. Reprod.* 24, 771–777.
103. Higuchi, T., and Kawakami, M. (1982). Changes in the characteristics of pulsatile luteinizing hormone secretion during the oestrous cycle and after ovariectomy and oestrogen treatment in female rats. *J. Endocrinol.* 94, 177–182.
104. Blake, C. A. (1976). A detailed characterization of the proestrous luteinizing hormone surge. *Endocrinology* 98, 445–450.
105. Lumpkin, M. D., DePaolo, L. V., and Negro-Vilar, A. (1984). Pulsatile release of follicle-stimulating hormone in ovariectomized rats is inhibited by porcine follicular fluid (Inhibin). *Endocrinology* 114, 201–206.
106. Neill, J. D. (1970). Effect of "stress" on serum prolactin and luteinizing hormone levels during the estrous cycle of the rat. *Endocrinology* 87, 1192–1197.
107. Amenomori, Y., Chen, C. L., and Meites, J. (1970). Serum prolactin levels in rats during different reproductive states. *Endocrinology* 86, 506–510.
108. Zmeili, S. M., Papavasiliou, S. S., Thorner, M. O., Evans, W. S., Marshall, J. C., and Landefeld, T. D. (1986). Alpha and luteinizing hormone b subunit messenger ribonucleic acids during the rat estrous cycle. *Endocrinology* 119, 1867–1869.
109. Ortolano, G. A., Haisenleder, D. J., Dalkin, A. C., Iliff-Sizemore, S. A., Landefeld, T. D., Maurer, R. A., and Marshall, J. C. (1988). Follicle-stimulating hormone b subunit messenger

ribonucleic acid concentrations during the rat estrous cycle. *Endocrinology* 123, 2946–2949.

110. Haisenleder, D. J., Ortolano, G. A., Landefeld, T. D., Zmeili, S. M., and Marshall, J. C. (1989). Prolactin messenger ribonucleic acid concentrations in 4-day cycling rats and during the prolactin surge. *Endocrinology* 124, 2023–2028.
111. Shimokawa, N., Kato, Y., Imai, K., and Wakabayashi, K. (1990). Changes in content of prolactin mRNA during the rat estrous cycle. *Exp. Clin. Endocrinol.* 96, 8–14.
112. Levine, J. E., and Ramirez, V. D. (1982). Luteinizing hormone–releasing hormone release during the rat estrous cycle and after ovariectomy, as estimated with push–pull cannulae. *Endocrinology* 111, 1439–1448.
113. Everett, J. W., and Sawyer, C. H. (1950). A 24-hour periodicity in the "LH-release apparatus" of female rats, disclosed by barbiturate sedation. *Endocrinology* 47, 198–218.
114. Schwartz, N. B., and McCormack, C. E. (1972). Reproduction: gonadal function and its regulation. *Ann. Rev. Physiol.* 34, 425–472.
115. Moore, R. Y. (1978). Central neural control of circadian rhythms. In *Frontiers in Neuroendocrinology* (W. F. Ganong, and L. Martini, Eds.), pp. 185–206. Raven Press, New York.
116. Berk, M. L., and Finkelstein, J. A. (1981). An autoradiographic determination of the efferent projections of the suprachiasmatic nucleus of the hypothalamus. *Brain Res.* 226, 1–13.
117. Stephan, F. S., Berkley, K. S., and Moss, R. L. (1981). Efferent connections of the rat suprachiasmatic nucleus. *Neuroscience* 6, 2625–2641.
118. Nunez, A. A., and Stephan, F. K. (1977). The effects of hypothalamic knife cuts on drinking rhythms and the estrous cycle of the rat. *Behav. Biol.* 20, 224–234.
119. Raisman, G., and Brown-Grant, K. (1977). The "suprachiasmatic syndrome": endocrine and behavioral abnormalities following lesions of the suprachiasmatic nuclei in the female rat. *Proc. R. Soc. Lond. (Biol.)* 198, 297–314.
120. Bogdanove, E. M. (1964). Role of brain in regulation of pituitary gonadotropin secretion. *Vit. Horm.* 22, 205–260.
121. Everett, J. W. (1978). The mammalian hypothalamo–pituitary axis. In *The Endocrine Hypothalamus* (S. L. Jeffcoate and J. S. M. Hutchinson, Eds.), pp. 1–33. Academic Press, New York.
122. Fee, A. R., and Parkes, A. S. (1929). Studies on ovulation. I. The relation of the anterior pituitary body to ovulation in the rabbit. *J. Physiol.* 67, 383–388.
123. Fee, A. R., and Parkes, A. S. (1930). Studies on ovulation. III. Effect of vaginal anesthesia on ovulation in the rabbit. *J. Physiol.* 70, 385–388.
124. Marshall, F. H. A., and Verney, E. B. (1936). The occurence of ovulation and pseudopregnancy in the rabbit as a result of central nervous system stimulation. *J. Physiol* 86, 327–336.
125. Harris, G. W. (1937). Induction of ovulation in the rabbit by electrical stimulation of the hypothalamo-hypophysial mechanism. *Proc. Roy. Soc. (Lond.) Ser. (Biol)* 122, 374–394.
126. Westman, A., and Jacobsohn, D. (1937). Experimentelle untersuchungen uber die bedeutung des hypophysen-zwischenhirnsystems fur die produktion gonadotropen hormone des hypophysenvorderlappens. *Acta Obst. Gynec. Scand.* 17, 235–265.
127. Hohlweg, W., and Dohrn, M. (1932). Uber die beziehungen zwishchen hypophysen vorder-lappen und keimdrusen. *Klin. Uschr.* 11, 233–235.
128. Hohlweg, W., and Junkmann, K. (1932). Die hormonal-nervose regulierung der funktion des hypophysenvorder lappens. *Klin. Uschr.* 11, 321–323.
129. Westman, A., and Jacobsohn, D. (1938). Endokrinologische untersuchungen an ratten mit durchtrennten hypophysenstiel. VI. Produktion und abgabe der gonadotropen hormone. *Acta Path. Microbiol. Scand.* 15, 445–453.
130. Westman, A., and Jacobsohn, D. (1938). Endokrinologische untersuchungen an ratten mit hypophysenstiel. III. Uber die luteinizierende wirkung des follikel hormons. *Acta Obst. Gynec. Scand.* 18, 115–123.
131. Popa, G. T., and Fielding, V. (1930). A portal circulation from the pituitary to the hypothalamic region. *J. Anat.* 65, 88–91.
132. Houssay, B. A., Biassotti, A., and Sammartino, R. (1935). Modifications fonctionelles de l'hypophyse apres les lesions infundibulotuberiennes chez le crapaud. *C. R. Seances Soc. Biol. Fil.* 120, 725–727.
133. Wislocki, G. B., and King, L. S. (1936). The permeability of the hypophysis and hypothalamus to vital dyes, with a study of the hypophysial vascular supply. *Am. J. Anat.* 58, 421–472.
134. Green, J. D., and Harris, G. W. (1947). The neurovascular link between the neurohypophysis and adenohypophysis. *J. Endocrinol.* 5, 136–146.
135. Green, J. D., and Harris, G. W. (1949). Observations of the hypophysial portal vessels of the living rat. *J. Physiol.* 108, 359–361.
136. Ambach, G., and Palkovits, M. (1979). The blood supply of the hypothalamus in the rat. In *Handbook of the Hypothalamus* (P. K. Morgane, and J. Panksepp, Eds.), Vol. 1, Anatomy of the Hypothalamus, pp. 267–377. Marcel Dekker, New York.
137. Porter, J. C., and Smith, K. R. (1967). Collection of hypophysial stalk blood in rats. *Endocrinology* 81, 1182–1185.
138. Porter, J. C., Mical, R. S., Kamberi, I. A., and Grazia, Y. R. (1970). A procedure for the cannulation of a pituitary stalk portal vessel and perfusion of the pars distalis in the rat. *Endocrinology* 87, 197–201.
139. Oliver, C., Mical, R. S., and Porter, J. C. (1977). Hypothalamo–pituitary vasculature: evidence for retrograde blood flow in the pituitary stalk. *Endocrinology* 101, 598–604.
140. Bergland, R. M., Davis, S. L., and Page, R. B. (1977). Pituitary secretes to brain. Experiments in sheep. *Lancet* 2, 276–278.
141. Bergland, R. M., and Page, R. B. (1978). Can the pituitary secrete directly to the brain? (Affirmative anatomical evidence). *Endocrinology* 102, 1325–1338.
142. Bergland, R. M., and Page, R. B. (1979). Pituitary–brain vascular relations: a new paradigm. *Science* 204, 18–24.
143. Page, R. B. (1983). Directional pituitary flow: a microcinephotographic study. *Endocrinology* 112, 157–165.
144. Harris, G. W. (1949). Regeneration of the hypophysio-portal vessels. *Nature* 163, 70–70.
145. Harris, G. W. (1950). Oestrous rhythm, pseudopregnancy and the pituitary stalk in the rat. *J. Physiol.* 111, 347–360.
146. Greep, R. O. (1936). Functional pituitary grafts in rats. *Proc. Soc. Exp. Biol. Med.* 34, 754–755.
147. Harris, G. W., and Jacobsohn, D. (1952). Functional grafts of the anterior pituitary gland. *Proc. Roy. Soc. (Lond.) Ser. (Biol.)* 139, 263–276.
148. McCann, S. M., Taleisnik, S., and Friedman, H. M. (1960). LH-releasing activity in hypothalamic extracts. *Proc. Soc. Exp. Biol. Med.* 104, 432–434.
149. Harris, G. W. (1961). The pituitary stalk and ovulation. In *Control of Ovulation* (C. A. Villee, Ed.), pp. 56–74. Pergamon Press, London.
150. McCann, S. M. (1962). A hypothalamic luteinizing hormone–releasing factor. *Am. J. Physiol.* 202, 395–400.
151. Matsuo, H., Baba, Y., Nair, R. M. G., Arimura, A., and Schally, A. V. (1971). Structure of the porcine LH- and FSH-releasing hormone. I. The proposed amino acid sequence. *Biochem. Biophys. Res. Commun.* 43, 1334–1339.
152. Schally, A. V., Arimura, A., Baba, Y., Nair, R. M. G., Matsuo, H., Redding, T. W., Debeljuk, L., and White, W. F. (1971).

Isolation and properties of the FSH and LH-releasing hormone. *Biochem. Biophys. Res. Commun.* 43, 393–399.

153. Nett, T., Akbar, A., Niswender, G., Hedlund, M., and White, W. (1973). A radioimmunoassay for gonadotropin-releasing hormone in serum. *J. Clin. Endocrinol. Metab.* 36, 880–885.
154. Barry, J., Dubois, M., and Poulain, P. (1973). LRF producing cells of the mammalian hypothalamus: a fluorescent antibody study. *Z. Zellforsch.* 146, 351–366.
155. Liposits, Z. S., Setalo, G. Y., and Flerko, B. (1984). Application of the silver-gold intensified 3, 3'-diaminobenzidine chromogen to the light and electron microscopic detection of the luteinizing hormone-releasing hormone system of the rat brain. *Neuroscience* 13, 513–525.
156. Wheaton, J. E., Krulich, L., and McCann, S. M. (1975). Localization of luteinizing hormone releasing hormone in the preoptic area and hypothalamus of the rat using radioimmunoassay. *Endocrinology* 97, 30–38.
157. Kizer, J. S., Palkovits, M., and Brownstein, M. (1976). Releasing factors in the circumventricular organs of the rat brain. *Endocrinology* 98, 311–317.
158. Palkovits, M. (1973). Isolated removal of hypothalamic or other brain nuclei of the rat. *Brain Res.* 59, 449–450.
159. Palkovits, M., Arimura, A., Brownstein, M., Schally, A. V., and Saavedra, J. M. (1974). Luteinizing hormone-releasing hormone (LH-RH) content of the hypothalamic nuclei in rat. *Endocrinology* 95, 554–558.
160. Selmanoff, M., Wise, P. M., and Barraclough, C. A. (1980). Regional distribution of luteinizing hormone-releasing hormone (LH-RH) in rat brain determined by microdissection and radioimmunoassay. *Brain Res.* 192, 421–432.
161. Baker, B. L., Dermody, W. C., and Reel, J. R. (1975). Distribution of gonadotropin-releasing hormone in the rat brain as observed with immunocytochemistry. *Endocrinology* 97, 125–135.
162. Bennet-Clark, C., and Joseph, S. A. (1982). Immunocytochemical distribution of LHRH neurons and processes in the rat hypothalamic and extrahypothalamic location. *Cell Tissue Res.* 221, 493–504.
163. King, J. C., Parsons, J., Erlandsen, S. L., and Williams, T. H. (1974). Luteinizing hormone-releasing hormone pathway of the rat hypothalamus revealed by the unlabeled antibody-antiperoxidase method. *Cell Tissue Res.* 153, 211–217.
164. King, J., Tobet, S., Snavely, F., and Arimura, A. (1982). LHRH immunopositive cells and their projections to the median eminence and organum vasculosum of the lamina-terminalis. *J. Comp. Neurol.* 209, 287–300.
165. Naik, D. V. (1975). Immunoreactive LHRH neurons in the hypothalamus identified by light and histo fluorescence microscopy. *Cell Tissue Res.* 157, 423–436.
166. Setalo, G., Vigh, S., Schally, A. V., Arimura, A., and Flerko, B. (1975). LHRH-containing neural elements in the rat hypothalamus. *Endocrinology* 96, 135–142.
167. Shivers, B. D., Harlan, R., Morrell, J., and Pfaff, D. W. (1983). Immunocytochemical localization of luteinizing hormone-releasing hormone in the male and female rat brains. *Neuroendocrinology* 36, 1–12.
168. Witkin, J., Paden, C. M., and Silverman, A. (1982). The luteinizing hormone-releasing hormone (LHRH) systems in the rat brain. *Neuroendocrinology* 35, 429–438.
169. Kelly, M. J., Ronnekleiv, O., and Eskay, R. L. (1982). Immunocytochemical localization of luteinizing hormone-releasing hormone in neurons in the medial basal hypothalamus of the female rat. *Exp. Brain Res.* 48, 97–106.
170. Hokfelt, T., Fuxe, K., Goldstein, M., Johansson, O., Fraser, H., and Jeffcoate, S. (1975). Immunofluorescent mapping of central monoamines and releasing hormone (LRH) systems. In *Anatomical Neuroendocrinology* (W. E. Stumpf, and L. D. Grant, Eds.), pp. 381–392. Karger, Basel.
171. Everett, J. W., Radford, H. M., and Holsinger, J. (1964). Electrolytic irritative lesions in the hypothalamus and other forebrain areas: effects on luteinizing hormone release and the ovarian cycle. In *Hormonal Steroids* (L. Martini, and A. Pecile, Eds.), pp. 235–246. Academic Press, New York.
172. Kalra, S. P., Ajika, K., Krulich, L., Fawcett, C. P., Quijada, M., and McCann, S. M. (1971). Effects of hypothalamic and preoptic electrochemical stimulation on gonadotropins and prolactin release in proestrous rats. *Endocrinology* 88, 1150–1158.
173. Clemens, J. A., Smalstig, E. B., and Sawyer, B. D. (1976). Studies on the role of the preoptic area in the control of reproductive function in the rat. *Endocrinology* 99, 728–735.
174. Freeman, M. E., and Banks, J. A. (1980). Hypothalamic sites which control the surges of prolactin secretion induced by cervical stimulation. *Endocrinology* 106, 668–673.
175. Smith, M. S. (1980). Role of prolactin in mammalian reproduction. In *Reproductive Physiology III, International Review of Physiology* (R. O. Greep, Ed.), pp. 249–276. University Park Press, Baltimore.
176. Piva, F., Limonta, P., and Martini, L. (1982). Role of the organum vasculosum laminae terminalis in the control of gonadotropin secretion in the rat. *J. Endocrinol.* 93, 355–364.
177. Wiegand, S. J., Terasawa, E., and Bridson, W. E. (1978). Persistent estrus and blockade of progesterone-induced LH release following lesions which do not damage the suprachiasmatic nucleus. *Endocrinology* 102, 1645–1648.
178. Samson, W. K., and McCann, S. M. (1979). Effects of lesions in the organum vasculosum lamina terminalis on the hypothalamic distribution of luteinizing hormone-releasing hormone and gonadotropin secretion in the ovariectomized rat. *Endocrinology* 105, 939–946.
179. Halasz, B., and Gorski, R. A. (1967). Gonadotrophic hormone secretion in female rats after partial or total interruption of neural afferents to the medial basal hypothalamus. *Endocrinology* 80, 608–622.
180. Chowers, I., and McCann, S. M. (1965). Content of luteinizing hormone releasing factor and luteinizing hormone during the estrous cycle and after changes in gonadal steroid titers. *Endocrinology* 76, 700–708.
181. Kalra, S. P., Krulich, L., and McCann, S. M. (1973). Changes in gonadotropin-releasing factor content in the rat hypothalamus following electrochemical stimulation of anterior hypothalamic area and during the estrous cycle. *Neuroendocrinology* 12, 321–333.
182. Ramirez, V. D., and Sawyer, C. H. (1965). Fluctuations in hypothalamic LH-RF (luteinizing hormone-releasing factor) during the rat estrous cycle. *Endocrinology* 76, 282–289.
183. Asai, T., and Wakabayashi, K. (1975). Changes in hypothalamic LH-RF content during the rat estrous cycle. *Endocrinol. Jpn.* 22, 319–326.
184. Kalra, P. S., and Kalra, S. P. (1977). Temporal changes in the hypothalamic and serum luteinizing hormone-releasing hormone (LHRH) levels and the circulating ovarian steroids during the rat estrous cycle. *Acta Endocrinol.* 85, 449–455.
185. Barr, G. D., and Barraclough, C. A. (1978). Temporal changes in medial basal hypothalamic LH-RH correlated with plasma LH during the rat estrous cycle and following electrochemical stimulation of the medial preoptic area in pentobarbital treated proestrous rats. *Brain Res.* 148, 413–423.
186. Oshima, I., Morishita, H., Omura, K., and Saito, S. (1978). Changes in hypothalamic LH-RH content and blood levels of LH-RH, gonadotropin and estradiol during the preovulatory stage of rat estrous cycle. *Endocrinol. Jpn.* 25, 607–611.

187. Polkowska, J., and Jutisz, M. (1979). Local changes in immunoreactive gonadotropin releasing hormone in the rat median eminence during the estrous cycle. Correlation with the pituitary luteinizing hormone. *Neuroendocrinology* 28, 281–288.
188. Araki, S., Ferin, M., Zimmerman, E. A., and Vande Wiele, R. L. (1975). Ovarian modulation of immunoreactive gonadotropin releasing hormone (Gn-RH) in the rat brain: evidence for a differential effect on the anterior and midhypothalamus. *Endocrinology* 96, 644–650.
189. Kalra, S. P. (1976). Circadian rhythm in luteinizing hormone–releasing hormone (LH-RH) content of preoptic area during the rat estrous cycle. *Brain Res.* 104, 354–358.
190. Kalra, S. P. (1976). Tissue levels of luteinizing hormone–releasing hormone in the preoptic area and hypothalamus, and serum concentrations of gonadotropins following anterior hypothalamic deafferentation and estrogen treatment of the female rat. *Endocrinology* 99, 101–107.
191. Wise, P. M., Rance, N., Selmanoff, M. K., and Barraclough, C. A. (1981). Changes in radioimmunoassayable luteinizing hormone–releasing hormone in discrete brain areas of the rat at various times on proestrus, diestrous day 1, and after phenobarbital administration. *Endocrinology* 108, 2179–2185.
192. Sarkar, D. K., Chiappa, S. A., Fink, G., and Sherwood, N. M. (1976). Gonadotropin-releasing hormone surge in pro-oestrous rats. *Nature* 264, 461–463.
193. Ching, M. (1982). Correlative surges of LHRH, LH and FSH in pituitary stalk plasma and systemic plasma of rat during proestrus. Effect of anesthetics. *Neuroendocrinology* 34, 279–285.
194. Ching, M. (1982). Differential effect of althesin versus chloralose-urethane anesthesia on in vivo LHRH and LH release in the proestrous rat. *Endocr. Res. Commun.* 9, 37–46.
195. Moenter, S. M., Caraty, A., Locatelli, A., and Karsch, F. J. (1991). Pattern of gonadotropin-releasing hormone (GnRH) secretion leading up to ovulation in the ewe: existence of a preovulatory GnRH surge. *Endocrinology* 129, 1175–1182.
196. Koch, Y., Chobsieng, P., Zor, U., Fridkin, M., and Lindner, H. R. (1973). Suppression of gonadotropin secretion and prevention of ovulation in the rat by antiserum to synthetic gonadotropin-releasing hormone. *Biochem. Biophys. Res. Commun.* 55, 623–626.
197. Aiyer, M. S., Fink, G., and Greig, F. (1974). Changes in the sensitivity of the pituitary gland to luteinizing hormone releasing factor during the oestrous cycle of the rat. *J. Endocrinol.* 60, 47–64.
198. Aiyer, M. S., Chiappa, S. A., and Fink, G. (1974). A priming effect of luteinizing hormone releasing factor on the anterior pituitary gland in the female rat. *J. Endocrinol.* 62, 573–588.
199. Hwan, J. C., and Freeman, M. E. (1987). Partial purification of a hypothalamic factor that inhibits gonadotropin-releasing hormone-stimulated luteinizing hormone release. *Endocrinology* 120, 483–490.
200. Hwan, J. C., and Freeman, M. E. (1987). A physiological role for luteinizing hormone release-inhibiting factor of hypothalamic origin. *Endocrinology* 121, 1099–1103.
201. Robinson, G., and Evans, J. J. (1990). Oxytocin has a role in gonadotrophin regulation in rats. *J. Endocrinol.* 125, 425–432.
202. Robinson, G., and Evans, J. J. (1991). Oxytocin influences preovulatory follicular development and advances ovulation in rats. *Acta Endocrinol. (Copenh).* 125, 109–112.
203. Robinson, G., Evans, J. J., and Catt, K. J. (1992). Oxytocin stimulates LH production by the anterior pituitary gland of the rat. *J. Endocrinol.* 132, 277–283.
204. Evans, J. J., Catt, K. J., Dobbs, B. R., Moorhouse, A. R., and Robinson, G. (1993). Oxytocin modulation of gonadotrophin release. *Regul. Pept.* 45, 243–246.
205. Evans, J. J., and Tulloch, S. (1995). Effects of administration of oxytocin in association with gonadotropin-releasing hormone on luteinizing hormone levels in rats in vivo. *Peptides* 16, 145–150.
206. Evans, J. J., Hurd, S. J., and Mason, D. R. (1995). Oxytocin modulates the luteinizing hormone response of the rat anterior pituitary to gonadotrophin-releasing hormone *in vitro*. *J. Endocrinol.* 145, 113–119.
207. Hurd, S. J., Evans, J. J., and Mason, D. R. (1996). Oxytocin augmentation of gonadotropin-releasing hormone-stimulated luteinizing hormone release is modulated by cyclic AMP. *Life Sci.* 59, 713–717.
208. Evans, J. J. (1996). Oxytocin and the control of LH. *J. Endocrinol.* 151, 169–174.
209. Evans, J. J., Janmohamed, S., and Forsling, M. L. (1999). Gonadotrophin-releasing hormone and oxytocin secretion from the hypothalamus *in vitro* during prooestrus: the effects of time of day and melatonin. *Brain Res. Bull.* 48, 93–97.
210. Abbas, M. M., and Evans, J. J. (2000). Regulation of c-fos protein in gonadotrope cells by oxytocin and gonadotropin-releasing hormone. *Neuroendocrinology* 71, 292–300.
211. Evans, J. J., Youssef, A. H., Abbas, M. M., and Schwartz, J. (1999). GnRH and oxytocin have nonidentical effects on the cellular LH response by gonadotrophs at pro-oestrus. *J. Endocrinol.* 163, 345–351.
212. Evans, J. J., Pragg, F. L., and Mason, D. R. (2001). Release of luteinizing hormone from the anterior pituitary gland in vitro can be concurrently regulated by at least three peptides: gonadotropin-releasing hormone, oxytocin and neuropeptide Y. *Neuroendocrinology* 73, 408–416.
213. Gibbs, D. M. (1984). High concentrations of oxytocin in hypophysial portal plasma. *Endocrinology* 114, 1216–1218.
214. Sarkar, D. K., and Gibbs, D. M. (1984). Cylcic variation of oxytocin in the blood of pituitary portal vessels of rats. *Neuroendocrinology* 39, 481–483.
215. Johnston, C. A., and Negro-Vilar, A. (1988). Role of oxytocin on prolactin secretion during proestrus and in different physiological or pharmacological paradigms. *Endocrinology* 122, 341–350.
216. Kalra, S. P., and Crowley, W. R. (1992). Neuropeptide Y: a novel neuroendocrine peptide in the control of pituitary hormone secretion, and its relation to luteinizing hormone. In *Frontiers in Neuroendocrinology* (W. F. Ganong, and L. Martini, Eds.), pp. 1–46. Raven Press, New York.
217. Bauer-Dantoin, A. C., Knox, K. L., Schwartz, N. B., and Levine, J. E. (1993). Estrous cycle stage-dependent effects of neuropeptide-Y on luteinizing hormone (LH)-releasing hormone-stimulated LH and follicle-stimulating hormone secretion from anterior pituitary fragments in vitro. *Endocrinology* 133, 2413–2417.
218. Freeman, M. E. (1993). Neuropeptide Y: a unique member of the constellation of gonadotropin-releasing hormones. *Endocrinology* 133, 2411–2412.
219. Sutton, S. W., Toyama, T. T., Otto, S., and Plotsky, P. M. (1988). Evidence that neuropeptide Y (NPY) released into the hypophysial-portal circulation participates in priming gonadotropes to the effects of gonadotropin releasing hormone (GnRH). *Endocrinology* 123, 1208–1210.
220. McDonald, J. K., Koenig, J. I., Gibbs, D. M., Collins, P., and Noe, B. D. (1987). High concentrations of neuropeptide Y in pituitary portal blood of rats. *Neuroendocrinology* 46, 538–541.
221. Parker, S. L., Kalra, S. P., and Crowley, W. R. (1991). Neuropeptide Y modulates the binding of a gonadotropin-releasing hormone (GnRH) analog to anterior pituitary GnRH receptor sites. *Endocrinology* 128, 2309–2316.
222. Leblanc, P., L'Héritier, A., Rasolonjanahary, R., and Kordon, C. (1994). Neuropeptide Y enhances LHRH binding

to rat gonadotrophs in primary culture. *Neuropeptides* 26, 87–92.

223. Bauer-Dantoin, A. C., McDonald, J. K., and Levine, J. E. (1991). Neuropeptide Y potentiates luteinizing hormone (LH)-releasing hormone-stimulated LH surges in pentobarbital-blocked proestrous rats. *Endocrinology* 129, 402–408.
224. Shangold, G. A., and Miller, R. J. (1990). Direct neuropeptide Y-induced modulation of gonadotrope intracellular calcium transients and gonadotropin secretion. *Endocrinology* 126, 2336–2342.
225. Leupen, S. M., Besecke, L. M., and Levine, J. E. (1997). Neuropeptide Y Y1-receptor stimulation is required for physiological amplification of preovulatory luteinizing hormone surges. *Endocrinology* 138, 2735–2739.
226. Sahu, A., Crowley, W. R., and Kalra, S. P. (1994). Hypothalamic neuropeptide-Y gene expression increases before the onset of the ovarian steroid-induced luteinizing hormone surge. *Endocrinology* 134, 1018–1022.
227. Sahu, A., Crowley, W. R., and Kalra, S. P. (1995). Evidence that hypothalamic neuropeptide Y gene expression increases before the onset of the preovulatory LH surge. *J. Neuroendocrinol.* 7, 291–296.
228. Sahu, A., Crowley, W. R., Tatemoto, K., Balasubramaniam, A., and Kalra, S. P. (1987). Effects of neuropeptide Y, NPY, analog (norleucine4-NPY), galanin and neuropeptide K on LH release in ovariectomized (ovx) and ovx estrogen, progesterone-treated rats. *Peptides* 8, 921–926.
229. Coen, C. W., Montagnese, C., and Opacka-Juffry, J. (1990). Coexistence of gonadotrophin-releasing hormone and galanin: immunohistochemical and functional studies. *J. Neuroendocrinol.* 2, 107–111.
230. López, F. J., Meade, E. H., Jr., and Negro-Vilar, A. (1990). Development and characterization of a specific and sensitive radioimmunoassay for rat galanin: measurement in brain tissue, hypophyseal portal and peripheral serum. *Brain Res. Bull.* 24, 395–399.
231. Schally, A. V., Arimura, A., Kastin, A. J., Matsuo, H., Baba, Y., Redding, T. W., Nair, R. M. G., Debeljuk, L., and White, W. F. (1971). The gonadotropin-releasing hormone: a single hypothalamic peptide regulates the secretion of both LH and FSH. *Science* 173, 1036–1037.
232. Lumpkin, M. D., and McCann, S. M. (1984). Effect of destruction of the dorsal anterior hypothalamus on follicle-stimulating hormone secretion in the rat. *Endocrinology* 115, 2473–2480.
233. Chappel, S. C., and Barraclough, C. A. (1976). Hypothalamic regulation of pituitary FSH secretion. *Endocrinology* 98, 927–935.
234. Mizunuma, H., Samson, W. K., Lumpkin, M. D., and McCann, S. M. (1983). Evidence for an FSH-releasing factor in the posterior portion of the median eminence. *Life Sci.* 33, 2003–2009.
235. Dhariwal, A. P. S., Nallar, R., Batt, M., and McCann, S. M. (1965). Separation of follicle-stimulating hormone-releasing factor from luteinizing hormone-releasing factor. *Endocrinology* 76, 290–294.
236. Mizunuma, H., Samson, W. K., Lumpkin, M. D., Moltz, J. H., Fawcett, C. P., and McCann, S. M. (1983). Purification of a bioactive FSH-releasing factor (FSHRF). *Brain Res. Bull.* 10, 623–629.
237. Blake, C. A., and Kelch, R. P. (1981). Administration of antiluteinizing hormone-releasing hormone serum to rats: effects of periovulatory secretion of luteinizing hormone and follicle-stimulating hormone. *Endocrinology* 109, 2175–2179.
238. Hasegawa, Y., Miyamoto, K., Yazaki, C., and Igarashi, M. (1981). Regulation of the second surge of follicle-stimulating hormone. Effects of antiluteinizing hormone-releasing hormone serum and pentobarbital. *Endocrinology* 109, 130–135.
239. Condon, T. P., Heber, D., Stewart, J. M., Sawyer, C. H., and Whitmoyer, D. I. (1984). Differential gonadotropin secretion: blockade of periovulatory LH but not FSH secretion by a potent LHRH antagonist. *Neuroendocrinology* 38, 357–361.
240. Eskay, R. L., Oliver, C., Ben-Jonathan, N., and Porter, J. C. (1975). Hypothalamic hormones in portal and systemic blood. In *Hypothalamic Hormones: Chemistry, Physiology, Pharmacology and Clinical Uses* (M. Motta, P. G. Crosignani, and L. Martini, Eds.), pp. 125–137. Academic Press, New York.
241. Shimatsu, A., Kato, Y., Matsushita, N., Katakami, H., Yanaihara, N., and Imura, H. (1981). Immunoreactive vasoactive intestinal polypeptide in rat hypophysial portal blood. *Endocrinology* 108, 395–398.
242. Clemens, J. A., and Shaar, C. J. (1980). Control of prolactin secretion in mammals. *Fed. Proc.* 39, 2588–2592.
243. Kato, Y., Iwasaki, Y., Iwasaki, J., Abe, H., Yanaihara, N., and Imura, H. (1978). Prolactin release by vasoactive intestinal polypeptide in rats. *Endocrinology* 103, 554–558.
244. Muller, G. P., Chen, J. H., and Meites, J. (1973). In vivo stimulation of prolactin release in the rat by synthetic TRH. *Proc. Soc. Exp. Biol. Med.* 144, 613–615.
245. Koch, Y., Goldhaber, G., Fireman, I., Zor, U., Shani, J., and Tal, E. (1977). Suppression of prolactin and thyrotropin secretion in the rat by antiserum to thyrotropin-releasing hormone. *Endocrinology* 100, 1476–1478.
246. Murai, I., Reichlin, S., and Ben-Jonathan, N. (1989). The peak phase of the proestrous prolactin surge is blocked by either posterior pituitary lobectomy or antisera to vasoactive intestinal peptide. *Endocrinology* 124, 1050–1055.
247. Watanobe, H., Sasaki, S., and Takebe, K. (1991). Vasoactive intestinal peptide levels in the hypothalamus and anterior pituitary do not change in association with the pre-ovulatory prolactin surge in the rat. *Neuropeptides* 20, 159–162.
248. Everett, J. W., Sawyer, C. H., and Markee, J. E. (1949). A neurogenic timing factor in control of the ovulatory discharge of luteinizing hormone in the cyclic rat. *Endocrinology* 44, 234–250.
249. Sawyer, C. H., Everett, J. W., and Markee, J. E. (1949). A neural factor in the mechanism by which estrogen induces the release of luteinizing hormone in the rat. *Endocrinology* 44, 218–233.
250. Sawyer, C. H., Markee, J. E., and Everett, J. W. (1950). Further experiments on blocking pituitary activation in the rabbit and the rat. *J. Exp. Zool.* 113, 659–682.
251. Barraclough, C. A., and Sawyer, C. H. (1957). Blockade of release of pituitary ovulating hormone in the rat by chlorpromiazine and reserpine: possible mechanisms of action. *Endocrinology* 61, 341–351.
252. Glowinski, J., and Axelrod, J. (1965). Effect of drugs on the uptake, release and metabolism of 3H-norepinephrine in the rat brain. *J. Pharm. Exp. Therap.* 149, 43–49.
253. Meyerson, B. J., and Sawyer, C. H. (1968). Monoamines and ovulation in the rat. *Endocrinology* 83, 170–176.
254. Carlsson, A. (1964). Functional significance of drug-induced changes in brain monoamine levels. *Prog. Brain Res.* 8, 9–27.
255. Zschaeck, L. L., and Wurtman, R. J. (1973). Brain 3H-catechol synthesis and the vaginal estrous cycle. *Neuroendocrinology* 11, 144–149.
256. Demarest, K. T., Johnston, C. A., and Moore, K. E. (1981). Biochemical indices of catecholaminergic neuronal activity in the median eminence during the estrous cycle of the rat. *Neuroendocrinology* 32, 24–27.
257. Lofstrom, A. (1977). Catecholamine turnover alterations in discrete areas of the median eminence of the 4- and 5-day cyclic rat. *Brain Res.* 120, 113–131.
258. Rance, N., Wise, P. M., Selmanoff, M. K., and Barraclough, C. A. (1981). Catecholamine turnover rates in discrete

hypothalamic areas and associated with changes in median eminence luteinizing hormone-releasing hormone and serum gonadotropins on proestrus and diestrous day 1. *Endocrinology* 108, 1795–1802.

259. Demling, J., Fuchs, E., Braumert, M., and Wuttke, W. (1985). Preoptic catecholamine, GABA, and glutamate release in ovariectomized and ovariectomized estrogen-primed rats utilizing a push–pull cannula technique. *Neuroendocrinology* 41, 212–218.
260. Grattan, D. R., Park, S. K., and Selmanoff, M. (1995). Orchidectomy and NMDA increase GnRH secretion as measured by push–pull perfusion of rat anterior pituitary. *Am. J. Physiol. Endocrinol. Metab.* 268, E685–E692.
261. Kohsaka, A., Watanobe, H., Kakizaki, Y., and Suda, T. (1999). A comparative study of the effects of nitric oxide and carbon monoxide on the in vivo release of gonadotropin-releasing hormone and neuropeptide Y from rat hypothalamus during the estradiol-induced luteinizing hormone surge: estimation by push–pull perfusion. *Neuroendocrinology* 69, 245–253.
262. Grattan, D. R., Park, S. K., and Selmanoff, M. (1995). Orchidectomy and NMDA increase GnRH secretion as measured by push–pull perfusion of rat anterior pituitary. *Am. J. Physiol. Endocrinol. Metab.* 268, E685–E692.
263. Cramer, O. M., and Barraclough, C. A. (1978). The actions of serotonin, norepinephrine, and epinephrine on hypothalamic processes leading to adenohypophyseal luteinizing hormone release. *Endocrinology* 103, 694–703.
264. Egawa, M., Hoebel, B. G., and Stone, E. A. (1988). Use of microdialysis to measure brain noradrenergic receptor function in vivo. *Brain Res.* 458, 303–308.
265. Kendrick, K. M. (1990). Microdialysis measurement of in vivo neuropeptide release. *J. Neurosci. Methods* 34, 35–46.
266. Justice, J. B., Jr. (1993). Quantitative microdialysis of neurotransmitters. *J. Neurosci. Methods* 48, 263–276.
267. Mas, M., Gonzalez-Mora, J. L., and Hernandez, L. (1996). In vivo monitoring of brain neurotransmitter release for the assessment of neuroendocrine interactions. *Cell. Mol. Neurobiol.* 16, 383–396.
268. Gerin, C., and Privat, A. (1996). Evaluation of the function of microdialysis probes permanently implanted into the rat CNS and coupled to an on-line HPLC system of analysis. *J. Neurosci. Methods* 66, 81–92.
269. Westerink, B. H. C., and Timmerman, W. (1999). Do neurotransmitters sampled by brain microdialysis reflect functional release? *Anal. Chim. Acta* 379, 263–274.
270. Zimmer, L., Kodas, E., Guilloteau, D., Garreau, L., Besnard, J. C., and Chalon, S. (2000). Microdialysis as a tool for in vivo study of dopamine transporter function in rat brains. *J. Neurosci. Methods* 103, 137–144.
271. Westerink, B. H. C., and De Vries, J. B. (2001). A method to evaluate the diffusion rate of drugs from a microdialysis probe through brain tissue. *J. Neurosci. Methods* 109, 53–58.
272. Mulchahey, J. J., and Neill, J. D. (1986). Dopamine levels in the anterior pituitary gland montitored by in vivo electrochemistry. *Brain Res.* 386, 332–340.
273. Thompson, T. L., and Moss, R. L. (1997). Modulation of mesolimbic dopaminergic activity over the rat estrous cycle. *Neurosci. Lett.* 229, 145–148.
274. Ben-Jonathan, N., and Porter, J. C. (1976). A sensitive radioenzymatic assay for dopamine, norepinephrine, and epinephrine in plasma and tissue. *Endocrinology* 98, 1497–1507.
275. Plotsky, P. M., Gibbs, D. M., and Neill, J. D. (1978). Liquid chromatographic-electrochemical measurement of dopamine in hypophysial stalk blood of rats. *Endocrinology* 102, 1887–1900.
276. Coombs, M. C., and Coen, C. W. (1983). Adrenaline turnover rates in the medial preoptic area and medial basal hypothalamus in relation to the release of luteinizing hormone in female rat. *Neuroscience* 10, 207–210.
277. Arendash, G. W., and Gallo, R. V. (1978). Serotonin involvement in the inhibition of episodic luteinizing hormone release during electrical stimulation of the midbrain dorsal raphe in ovariectomized rats. *Endocrinology* 102, 1199–1206.
278. Gallo, R. V., and Moberg, G. D. (1977). Serotonin mediated inhibition of episodic luteinizing hormone release during electrical stimulation of the arcuate nucleus in ovariectomized rats. *Endocrinology* 100, 945–954.
279. Rasmussen, D. D., Jacobs, W., Kissinger, P. T., and Malven, P. V. (1981). Plasma luteinizing hormone in ovariectomized rats following pharmacologic manipulation of endogenous brain serotonin. *Brain Res.* 229, 230–235.
280. Walker, R. F. (1983). Quantitative and temporal aspects of serotonin's facilitatory action on phasic secretion of luteinizing hormone in female rats. *Neuroendocrinology* 36, 468–474.
281. Ping, L., Mahesh, V. B., Wiedmeier, V. T., and Brann, D. W. (1994). Release of glutamate and aspartate from the preoptic area during the progesterone-induced LH surge: in vivo microdialysis studies. *Neuroendocrinology* 59, 318–324.
282. Farah, J. M., Jr., Rao, T. S., Mick, S. J., Coyne, K. E., and Iyengar, S. (1991). *N*-methyl-D-aspartate treatment increases circulating adrenocorticotropin and luteinizing hormone in the rat. *Endocrinology* 128, 1875–1880.
283. Ondo, J. G., Wheeler, D. D., and Dom, R. M. (1988). Hypothalamic site of action for N-methyl-D-aspartate (NMDA) on LH secretion. *Life Sci.* 43, 2283–2286.
284. Bourguignon, J. P., Gérard, A., and Franchimont, P. (1989). Direct activation of gonadotropin-releasing hormone secretion through different receptors to neuroexcitatory amino acids. *Neuroendocrinology* 49, 402–408.
285. Bourguignon, J. P., Gerard, A., Mathieu, J., Simons, J., and Franchimont, P. (1989). Pulsatile release of gonadotropin-releasing hormone from hypothalamic explants is restrained by blockade of N-Methyl-D,L-aspartate receptors. *Endocrinology* 125, 1090–1096.
286. Lopez, F. J., Donoso, A. O., and Negro-Vilar, A. (1990). Endogenous excitatory amino acid neurotransmission regulates the estradiol-induced LH surge in ovariectomized rats. *Endocrinology* 126, 1771–1773.
287. Ping, L., Mahesh, V. B., Bhat, G. K., and Brann, D. W. (1997). Regulation of gonadotropin-releasing hormone and luteinizing hormone secretion by AMPA receptors. Evidence for a physiological role of AMPA receptors in the steroid-induced luteinizing hormone surge. *Neuroendocrinology* 66, 246–253.
288. Donoso, A. O., and Banzan, A. M. (1984). Effects of increase of brain GABA levels on the hypothalamic-pituitary-luteinizing hormone axis in rats. *Acta Endocr.* 106, 298–304.
289. Donoso, A. O., and Banzan, A. M. (1986). Blockade of the LH surge and ovulation by GABA-T inhibitory drugs that increase brain GABA levels in rats. *Psychoneuroendocrinology* 11, 429–436.
290. Landa, A. I., and Donoso, A. O. (1987). Blockade of pro-oestrus LH surge and ovulation by GABA increase in the rat locus coeruleus. *Acta Endocr.* 115, 490–496.
291. Akema, T., Chiba, A., and Kimura, F. (1990). On the relationship between noradrenergic stimulatory and GABAergic inhibitory systems in the control of luteinizing hormone secretion in female rats. *Neuroendocrinology* 52, 566–572.
292. Herbison, A. E., and Dyer, R. G. (1991). Effect on luteinizing hormone secretion of GABA receptor modulation in the medial preoptic area at the time of proestrous luteinizing hormone surge. *Neuroendocrinology* 53, 317–320.

293. Han, S. K., Todman, M. G., and Herbison, A. E. (2004). Endogenous GABA release inhibits the firing of adult gonadotropin-releasing hormone neurons. *Endocrinology* 145, 495–499.
294. Defazio, R. A., Heger, S., Ojeda, S. R., and Moenter, S. M. (2002). Activation of A-type gamma-aminobutyric acid receptors excites gonadotropin-releasing hormone neurons. *Mol. Endocrinol.* 16, 2872–2891.
295. Bruni, J. F., Van Vugt, D., Marshall, S., and Meites, J. (1977). Effects of naloxone, morphine and methionine enkephalin on serum prolactin, luteinizing hormone, follicle-stimulating hormone, thyroid stimulating hormone and growth hormone. *Life Sci.* 21, 461–466.
296. Fox, S. R., Hofer, M. T., and Smith, M. S. (1984). Naloxone disinhibits pulsatile LH secretion during estrus of the rat estrous cycle but does not stimulate LH release during lactation. *Soc. Neurosci. Abstr.* 10, 1211–1211.
297. Kinoshita, F., Nakai, Y., Katakami, H., Kato, Y., Yajima, H., and Imura, H. (1980). Effect of b endorphin on pulsatile luteinizing hormone release in conscious castrated rats. *Life Sci.* 27, 843–846.
298. Ching, M. (1983). Morphine suppresses the proestrous surge of GnRH in pituitary portal plasma of rats. *Endocrinology* 112, 2209–2211.
299. Ieiri, T., Chen, H. T., Campbell, G. A., and Meites, J. (1980). Effects of naloxone and morphine on the proestrous surge of prolactin and gonadotropins in the rat. *Endocrinology* 106, 1568–1570.
300. Hulse, G. K., Coleman, G. J., Copolov, D. L., and Clements, J. A. (1984). Relationship between endogenous opioids and the oestrous cycle in the rat. *J. Endocrinol.* 100, 271–275.
301. Sarkar, D. K., and Yen, S. S. C. (1985). Changes in b-endorphin–like immunoreactivity in pituitary portal blood during the estrous cycle and after ovariectomy in rats. *Endocrinology* 116, 2075–2079.
302. Sarkar, D. K., and Minami, S. (1995). Diurnal variation in luteinizing hormone-releasing hormone and b-endorphin release in pituitary portal plasma during the rat estrous cycle. *Biol. Reprod.* 53, 38–45.
303. Meites, J., Bruni, J. F., Van Vugt, D. A., and Smith, A. F. (1979). Relation of endogenous opioid peptides and morphine to neuroendocrine functions. *Life Sci.* 24, 1325–1336.
304. Bloom, F., Battenberg, E., Rossier, J., Ling, N., and Guillemin, R. (1978). Neurons containing b-endorphin in rat brain exist separately from those containing enkephalin: immunocytochemical studies. *Proc. Natl. Acad. Sci. U S A* 75, 1591–1595.
305. Drouva, S. V., Epelbaum, J., Tapia-Arancibia, T., Laplante, E., and Kordon, C. (1981). Opiate receptors modulate LHRH and SRIF release from mediobasal hypothalamic neurons. *Neuroendocrinology* 32, 163–167.
306. Xu, B., Sahu, A., Crowley, W. R., Leranth, C., Horvath, T., and Kalra, S. P. (1993). Role of neuropeptide-Y in episodic luteinizing hormone release in ovariectomized rats: an excitatory component and opioid involvement. *Endocrinology* 133, 747–754.
307. Irahara, M., Tamura, T., Matuzaki, T., Saito, S., Yasui, T., Yamano, S., Kamada, M., and Aono, T. (2001). Orexin-A suppresses the pulsatile secretion of luteinizing hormone via b-endorphin. *Biochem. Biophys. Res. Commun.* 281, 232–236.
308. Leadem, C .A., and Kalra, S. P. (1985). Effects of endogenous opioid peptides and opiates on luteinizing hormone and prolactin secretion in ovariectomized rats. *Neuroendocrinology* 41, 342–352.
309. Zhang, Q., McCoy, J. M., and Gallo, R. V. (2002). Further studies on possible dynorphin involvement in the ovulatory luteinizing hormone surge in the proestrous rat. *Endocrine* 18, 231–238.
310. Crowley, W. R., and Kalra, S. P. (1987). Neuropeptide Y stimulates the release of luteinizing hormone-releasing hormone from medial basal hypothalamus in vitro: modulation by ovarian hormones. *Neuroendocrinology* 46, 97–103.
311. Tsuruo, Y., Kawano, H., Kagotani, Y., Hisano, S., Daikoku, S., Chihara, K., Zhang, T., and Yanaihara, N. (1990). Morphological evidence for neuronal regulation of luteinizing hormone-releasing hormone-containing neurons by neuropeptide Y in the rat septo-preoptic area. *Neurosci. Lett.* 110, 261–266.
312. Guy, J., Li, S., and Pelletier, G. (1988). Studies on the physiological role and mechanism of action of neuropeptide Y in the regulation of luteinizing hormone secretion in the rat. *Regul. Pept.* 23, 209–216.
313. Crowley, W. R., Hassid, A., and Kalra, S. P. (1987). Neuropeptide Y enhances the release of luteinizing hormone (LH) induced by LH-releasing hormone. *Endocrinology* 120, 941–945.
314. Bauer-Dantoin, A. C., McDonald, J. K., and Levine, J. E. (1990). Neuropeptide Y enhances LHRH-stimulated LH surges in pentobarbital-blocked proestrous rats. *Abstr. Soc. Neurosci.* 16,
315. Bauer-Dantoin, A. C., McDonald, J. K., and Levine, J. E. (1992). Neuropeptide Y potentiates luteinizing hormone (LH)-releasing hormone-induced LH secretion only under conditions leading to preovulatory LH surges. *Endocrinology* 131, 2946–2952.
316. Xu, M., Hill, J. W., and Levine, J. E. (2000). Attenuation of luteinizing hormone surges in neuropeptide Y knockout mice. *Neuroendocrinology* 72, 263–271.
317. Li, C., Chen, P. L., and Smith, M. S. (1999). Morphological evidence for direct interaction between arcuate nucleus neuropeptide Y (NPY) neurons and gonadotropin-releasing hormone neurons and the possible involvement of NPYY1 receptors. *Endocrinology* 140, 5382–5390.
318. Xu, M., Urban, J. H., Hill, J. W., and Levine, J. E. (2000). Regulation of hypothalamic neuropeptide Y Y1 receptor gene expression during the estrous cycle: role of progesterone receptors. *Endocrinology* 141, 3319–3327.
319. Hill, J. W., Urban, J. H., Xu, M., and Levine, J. E. (2004). Estrogen induces neuropeptide Y (NPY) Y1 receptor gene expression and responsiveness to NPY in gonadotrope-enriched pituitary cell cultures. *Endocrinology* 145, 2283–2290.
320. Raposinho, P. D., Broqua, P., Pierroz, D. D., Hayward, A., Dumont, Y., Quirion, R., Junien, J. L., and Aubert, M. L. (1999). Evidence that the inhibition of luteinizing hormone secretion exerted by central administration of neuropeptide Y (NPY) in the rat is predominantly mediated by the NPY-Y5 receptor subtype. *Endocrinology* 140, 4046–4055.
321. Jain, M. R., Pu, S. Y., Kalra, P. S., and Kalra, S. P. (1999). Evidence that stimulation of two modalities of pituitary luteinizing hormone release in ovarian steroid–primed ovariectomized rats may involve neuropeptide YY1 and Y4 receptors. *Endocrinology* 140, 5171–5177.
322. Kalra, S. P., and Horvath, T. L. (1998). Neuroendocrine interactions between galanin, opioids, and neuropeptide Y in the control of reproduction and appetite. *Ann. N Y Acad. Sci.* 863, 236–240.
323. Kalra, S. P., and Kalra, P. S. (2004). NPY—an endearing journey in search of a neurochemical on/off switch for appetite, sex and reproduction. *Peptides* 25, 465–471.
324. Johnston, C. A., Lopez, F., Samson, W. K., and Negro-Vilar, A. (1990). Physiologically important role for central oxytocin in the preovulatory release of luteinizing hormone. *Neurosci. Lett.* 120, 256–258.

325. Selvage, D., and Johnston, C. A. (2001). Central stimulatory influence of oxytocin on preovulatory gonadotropin-releasing hormone requires more than the median eminence. *Neuroendocrinology* 74, 129–134.
326. Rettori, V., Canteros, G., and McCann, S. M. (1997). Interaction between NO and oxytocin: influence on LHRH release. *Braz. J. Med. Biol. Res.* 30, 453–457.
327. Rettori, V., Canteros, G., Renoso, R., Gimeno, M., and McCann, S. M. (1997). Oxytocin stimulates the release of luteinizing hormone-releasing hormone from medial basal hypothalamic explants by releasing nitric oxide. *Proc. Natl. Acad. Sci. U S A* 94, 2741–2744.
328. Tsuda, K., Yokoo, H., and Goldstein, M. (1989). Neuropeptide Y and galanin in norepinephrine release in hypothalamic slices. *Hypertension* 14, 81–86.
329. Kaplan, L. M., Gabriel, S. M., Koenig, J. I., Sunday, M. E., Spindel, E. R., Martin, J. B., and Chin, W. W. (1988). Galanin is and estrogen-inducible, secretory product of the rat anterior pituitary. *Proc. Natl. Acad. Sci. U S A* 85, 7408–7412.
330. O'Halloran, D. J., Jones, P. M., Steel, J. H., Gon, G., Giaid, A., Ghatei, M. A., Polak, J. M., and Bloom, S. R. (1990). Effect of endocrine manipulation on anterior pituitary galanin in the rat. *Endocrinology* 127, 467–475.
331. Todd, J. F., Small, C. J., Akinsanya, K. O., Stanley, S. A., Smith, D. M., and Bloom, S. R. (1998). Galanin is a paracrine inhibitor of gonadotroph function in the female rat. *Endocrinology* 139, 4222–4229.
332. Rossmanith, W. G., Clifton, D. K., and Steiner, R. A. (1996). Galanin gene expression in hypothalamic GnRH-containing neurons of the rat: a model for autocrine regulation. *Horm. Metab. Res.* 28, 257–266.
333. Chessman, D. W., Osland, R. B., and Forsham, P. H. (1977). Suppression of preovulatory surge of luteinizing hormone and subsequent ovulation in the rat by arginine vasotocin. *Endocrinology* 101, 1194–1202.
334. Vijayan, E., Samson, W. K., and McCann, S. M. (1979). In vivo and in vitro effects of cholecystokinin on gonadotropin, prolactin, growth hormone and thyrotropin release in the rat. *Brain Res.* 17, 295–302.
335. Vijayan, E., and McCann, S. M. (1979). In vivo and in vitro effects of substance P and neurotensin on gonadotropin and prolactin release. *Endocrinology* 105, 64–68.
336. Vijayan, E., Samson, W. K., and McCann, S. M. (1978). Effects of intraventricular injection of gastrin on release of LH, prolactin, TSH and GH in conscious ovariectomized rats. *Life Sci.* 23, 2225–2232.
337. Steele, M. K., Gallo, R. V., and Ganong, W. F. (1985). Stimulatory or inhibitory effects of angiotensin II upon LH secretion in ovariectomized rats: a function of gonadal steroids. *Neuroendocrinology* 40, 210–216.
338. Ganong, W. F. (1995). Reproduction and the renin–angiotensin system. *Neurosci. Biobehav. Rev.* 19, 241–250.
339. Vijayan, E., Samson, W. K., Said, S. I., and McCann, S. M. (1979). Vasoactive intestinal peptide (VIP): evidence for a hypothalamic site of action to release growth hormone, luteinizing hormone and prolactin in conscious ovariectomized rats. *Endocrinology* 104, 53–57.
340. Stobie, K. M., and Weick, R. F. (1989). Vasoactive intestinal peptide inhibits luteinizing hormone secretion: the inhibition is not mediated by dopamine. *Neuroendocrinology* 49, 597–603.
341. Weick, R. F., and Stobie, K. M. (1995). Role of VIP in the regulation of LH secretion in the female rat. *Neurosci. Biobehav. Rev.* 19, 251–259.
342. Van der Beek, E. M., Wiegant, V. M., Van der Donk, H. A., Van den Hurk, R., and Buijs, R. M. (1993). Lesions of the suprachiasmatic nucleus indicate the presence of a direct vasoactive intestinal polypeptide-containing projection to gonadotrophin-releasing hormone neurons in the female rat. *J. Neuroendocrinol.* 5, 137–144.
343. Kalra, S. P., and Crowley, W. R. (1984). Differential effects of human pancreatic peptide on LH release in rat. *Neuroendocrinology* 38, 511–513.
344. Cunningham, M. J., Clifton, D. K., and Steiner, R. A. (1999). Leptin's actions on the reproductive axis: perspectives and mechanisms. *Biol. Reprod.* 60, 216–222.
345. Cheung, C. C., Thornton, J. E., Kuijper, J. L., Weigle, D. S., Clifton, D. K., and Steiner, R. A. (1997). Leptin is a metabolic gate for the onset of puberty in the female rat. *Endocrinology* 138, 855–858.
346. Yu, W. H., Kimura, M., Walczewska, A., Karanth, S., and McCann, S. M. (1997). Role of leptin in hypothalamic–pituitary function. *Proc. Natl. Acad. Sci. U S A* 94, 1023–1028.
347. Watanobe, H. (2002). Leptin directly acts within the hypothalamus to stimulate gonadotropin-releasing hormone secretion in vivo in rats. *J. Physiol.* 545.1, 255–268.
348. Woller, M., Tessmer, S., Neff, D., Nguema, A. A., Van Roo, B., and Waechter-Brulla, D. (2001). Leptin stimulates gonadotropin releasing hormone release from cultured intact hemihypothalami and enzymatically dispersed neurons. *Proc. Soc. Exp. Biol. Med.* 226, 591–596.
349. Sone, M., Nagata, H., Takekoshi, S., and Osamura, R. Y. (2001). Expression and localization of leptin receptor in the normal rat pituitary gland. *Cell Tissue Res.* 305, 351–356.
350. Zamorano, P. L., Mahesh, V. B., De Sevilla, L. M., Chorich, L. P., Bhat, G. K., and Brann, D. W. (1997). Expression and localization of the leptin receptor in endocrine and neuroendocrine tissues of the rat. *Neuroendocrinology* 65, 223–228.
351. Brann, D. W., Wade, M. F., Dhandapani, K. M., Mahesh, V. B., and Buchanan ,C. D. (2002). Leptin and reproduction. *Steroids* 67, 95–104.
352. Diano, S., Kalra, S. P., Sakamoto, H., and Horvath, T. L. (1998). Leptin receptors in estrogen receptor-containing neurons of the female rat hypothalamus. *Brain Res.* 812, 256–259.
353. Magni, P., Vettor, R., Pagano, C., Calcagno, A., Beretta, E., Messi, E., Zanisi, M., Martini, L., and Motta, M. (1999). Expression of a leptin receptor in immortalized gonadotropin-releasing hormone-secreting neurons. *Endocrinology* 140, 1581–1585.
354. Gonzalez, L. C., Pinilla, L., Tena-Sempere, M., Dieguez, C., Casanueva, F. F., and Aguilar, E. (2000). Effect of acute immunoneutralization of endogenous leptin on prolactin and LH secretion during the afternoon of pro-oestrus or in steroid-treated ovariectomized female rats. *J. Reprod. Fertil.* 118, 39–45.
355. Carro, E., Pinilla, L., Seoane, L. M., Considine, R. V., Aguilar, E., Casanueva, F. F., and Dieguez, C. (1997). Influence of endogenous leptin tone on the estrous cycle and luteinizing hormone pulsatility in female rats. *Neuroendocrinology* 66, 375–377.
356. Sakurai, T., Amemiya, A., Ishii, M., Matsuzaki, I., Chemelli, R. M., Tanaka, H., Williams, S. C., Richardson, J. A., Kozlowski, G. P., Wilson, S., Arch, J. R., Buckingham, R. E., Haynes, A. C., Carr, S. A., Annan, R. S., McNulty, D. E., Liu, W. S., Terrett, J. A., Elshourbagy, N. A., Bergsma, D. J., and Yanagisawa, M. (1998). Orexins and orexin receptors: a family of hypothalamic neuropeptides and G protein-coupled receptors that regulate feeding behavior. *Cell* 92, 573–585.
357. De Lecea, L., Kilduff, T. S., Peyron, C., Gao, X. B., Foye, P. E., Danielson, P. E., Fukuhara, C., Battenberg, E. L., Gautvik, V. T., Bartlett, F. S., II, Frankel, W. N., Van den Pol, A. N., Bloom, F. E., Gautvik, K. M., and Sutcliffe, J. G. (1998). The hypocretins: hypothalamus-specific peptides with neuroexcitatory activity. *Proc. Natl. Acad. Sci. U S A* 95, 322–327.

358. Campbell, R. E., Grove, K. L., and Smith, M. S. (2003). Gonadotropin-releasing hormone neurons coexpress orexin 1 receptor immunoreactivity and receive direct contacts by orexin fibers. *Endocrinology* 144, 1542–1548.
359. Russell, S. H., Small, C. J., Kennedy, A. R., Stanley, S. A., Seth, A., Murphy, K. G., Taheri, S., Ghatei, M. A., and Bloom, S. R. (2001). Orexin A interactions in the hypothalamo–pituitary gonadal axis. *Endocrinology* 142, 5294–5302.
360. Pu, S., Jain, M. R., Kalra, P. S., and Kalra, S. P. (1998). Orexins, a novel family of hypothalamic neuropeptides, modulate pituitary luteinizing hormone secretion in an ovarian steroid-dependent manner. *Regul. Pept.* 78, 133–136.
361. Small, C. J., Goubillon, M. L., Murray, J. F., Siddiqui, A., Grimshaw, S. E., Young, H., Sivanesan, V., Kalamatianos, T., Kennedy, A. R., Coen, C. W., Bloom, S. R., and Wilson, C. A. (2003). Central orexin A has site-specific effects on luteinizing hormone release in female rats. *Endocrinology* 144, 3225–3236.
362. Porkka-Heiskanen, T., Kalinchuk, A., Alanko, L., Huhtaniemi, I., and Stenberg, D. (2004). Orexin A and B levels in the hypothalamus of female rats: the effects of the estrous cycle and age. *Eur. J. Endocrinol.* 150, 737–742.
363. Kotani, M., Detheux, M., Vandenbogaerde, A., Communi, D., Vanderwinden, J. M., Le Poul, E., Brezillon, S., Tyldesley, R., Suarez-Huerta, N., Vandeput, F., Blanpain, C., Schiffmann, S. N., Vassart, G., and Parmentier, M. (2001). The metastasis suppressor gene KiSS-1 encodes kisspeptins, the natural ligands of the orphan G protein-coupled receptor GPR54. *J. Biol. Chem.* 276, 34631–34636.
364. Ohtaki, T., Shintani, Y., Honda, S., Matsumoto, H., Hori, A., Kanehashi, K., Terao, Y., Kumano, S., Takatsu, Y., Masuda, Y., Ishibashi, Y., Watanabe, T., Asada, M., Yamada, T., Suenaga, M., Kitada, C., Usuki, S., Kurokawa, T., Onda, H., Nishimura, O., and Fujino, M. (2001). Metastasis suppressor gene KiSS-1 encodes peptide ligand of a G-protein–coupled receptor. *Nature* 411, 613–617.
365. de Roux, N., Genin, E., Carel, J. C., Matsuda, F., Chaussain, J. L., and Milgrom, E. (2003). Hypogonadotropic hypogonadism due to loss of function of the KiSS1-derived peptide receptor GPR54. *Proc. Natl. Acad. Sci. U S A* 100, 10972–10976.
366. Parhar, I. S., Ogawa, S., and Sakuma, Y. (2004). Laser-captured single digoxigenin-labeled neurons of gonadotropin-releasing hormone types reveal a novel G protein–coupled receptor (Gpr54) during maturation in cichlid fish. *Endocrinology* 145, 3613–3618.
367. Navarro, V. M., Castellano, J. M., Fernandez-Fernandez, R., Barreiro, M. L., Roa, J., Sanchez-Criado, J. E., Aguilar, E., Dieguez, C., Pinilla, L., and Tena-Sempere, M. (2004). Developmental and hormonally regulated messenger ribonucleic acid expression of KiSS-1 and its putative receptor, GPR54, in rat hypothalamus and potent luteinizing hormone-releasing activity of KiSS-1 peptide. *Endocrinology* 145, 4565–4574.
368. Gottsch, M. L., Cunningham, M. J., Smith, J. T., Popa, S. M., Acohido, B. V., Crowley, W. F., Seminara, S., Clifton, D. K., and Steiner, R. A. (2004). A role for kisspeptins in the regulation of gonadotropin secretion in the mouse. *Endocrinology* 145, 4073–4077.
369. Navarro, V. M., Castellano, J. M., Fernandez-Fernandez, R., Tovar, S., Roa, J., Mayen, A., Barreiro, M. L., Casanueva, F. F., Aguilar, E., Dieguez, C., Pinilla, L., and Tena-Sempere, M. (2005). Effects of KiSS-1 peptide, the natural ligand of GPR54, on follicle-stimulating hormone secretion in the rat. *Endocrinology* 146, 1689–1697.
370. Navarro, V. M., Castellano, J. M., Fernandez-Fernandez, R., Tovar, S., Roa, J., Mayen, A., Nogueiras, R., Vazquez, M. J., Barreiro, M. L., Magni, P., Aguilar, E., Dieguez, C., Pinilla, L., and Tena-Sempere, M. (2005). Characterization of the potent luteinizing hormone-releasing activity of KiSS-1 peptide, the natural ligand of GPR54. *Endocrinology* 146, 156–163.
371. Thompson, E. L., Patterson, M., Murphy, K. G., Smith, K. L., Dhillo, W. S., Todd, J. F., Ghatei, M. A., and Bloom, S. R. (2004). Central and peripheral administration of kisspeptin-10 stimulates the hypothalamic–pituitary–gonadal axis. *J. Neuroendocrinol.* 16, 850–858.
372. Brann, D. W., Bhat, G. K., Lamar, C. A., and Mahesh, V .B. (1997). Gaseous transmitters and neuroendocrine regulation. *Neuroendocrinology* 65, 385–395.
373. Grossman, A., Costa, A., Forsling, M. L., Jacobs, R., Kostoglou-Athanassiou, I., Nappi, G., Navarra, P., and Satta, M. A. (1997). Gaseous neurotransmitters in the hypothalamus. The roles of nitric oxide and carbon monoxide in neuroendocrinology. *Horm. Metabol. Res.* 29, 477–482.
374. Mancuso, C., Preziosi, P., Grossman, A. B., and Navarra, P. (1997). The role of carbon monoxide in the regulation of neuroendocrine function. *Neuroimmunomodulation* 4, 225–229.
375. Dawson, T. M., and Snyder, S. H. (1994). Gases as biological messengers: nitric oxide and carbon monoxide in the brain. *J. Neurosci.* 14, 5147–5159.
376. Maines, M. D., Trakshel, G. M., and Kutty, R. K. (1986). Characterization of two constitutive forms of rat liver microsomal heme oxygenase. Only one molecular species of the enzyme is inducible. *J. Biol. Chem.* 261, 411–419.
377. Trakshel, G. M., and Maines, M. D. (1989). Multiplicity of heme oxygenase isozymes. HO-1 and HO-2 are different molecular species in rat and rabbit. *J. Biol. Chem.* 264, 1323–1328.
378. McCoubrey, W. K., Jr., Huang, T. J., and Maines, M. D. (1997). Isolation and characterization of a cDNA from the rat brain that encodes hemoprotein heme oxygenase-3. *Eur. J. Biochem.* 247, 725–732.
379. Laitinen, J. T., and Juvonen, R. O. (1995). A sensitive microassay reveals marked regional differences in the capacity of rat brain to generate carbon monoxide. *Brain Res.* 694, 246–252.
380. Lamar, C. A., Mahesh, V. B., and Brann, D. W. (1996). Regulation of gonadotropin-releasing hormone (GnRH) secretion by heme molecules: a regulatory role for carbon monoxide? *Endocrinology* 137, 790–793.
381. Alexandreanu, I. C., and Lawson, D. M. (2003). Heme oxygenase in the rat anterior pituitary: immunohistochemical localization and possible role in gonadotropin and prolactin secretion. *Exp. Biol. Med. (Maywood.)* 228, 64–69.
382. Alexandreanu, I. C., and Lawson, D. M. (2002). Effects of chronic administration of a heme oxygenase substrate or inhibitor on progression of the estrous cycle, pregnancy and lactation of Sprague-Dawley rats. *Life Sci.* 72, 153–162.
383. Arzumanian, V., Stankevicius, E., Laukeviciene, A., and Kevelaitis, E. (2003). Mechanisms of nitric oxide synthesis and action in cells. *Medicina (Kaunas.)* 39, 535–541.
384. Jaffrey, S. R., and Snyder, S. H. (1996). PIN: an associated protein inhibitor of neuronal nitric oxide synthase. *Science* 274, 774–777.
385. Brenman, J. E., Christopherson, K. S., Craven, S. E., McGee, A. W., and Bredt, D. S. (1996). Cloning and characterization of postsynaptic density 93, a nitric oxide synthase interacting protein. *J. Neurosci.* 16, 7407–7415.
386. Brenman, J. E., Chao, D. S., Gee, S. H., McGee, A. W., Craven, S. E., Santillano, D. R., Wu, Z., Huang, F., Xia, H., Peters, M. F., Froehner, S. C., and Bredt, D. S. (1996). Interaction of nitric oxide synthase with the postsynaptic density protein PSD-95 and alpha1-syntrophin mediated by PDZ domains. *Cell* 84, 757–767.
387. Gomperts, S. N. (1996). Clustering membrane proteins: it's all coming together with the PSD-95/SAP90 protein family. *Cell* 84, 659–662.

388. Brenman, J. E., and Bredt, D. S. (1996). Nitric oxide signaling in the nervous system. *Methods Enzymol.* 269, 119–129.
389. Bhat, G. K., Mahesh, V. B., Lamar, C. A., Ping, L., Aguan, K., and Brann, D. W. (1995). Histochemical localization of nitric oxide neurons in the hypothalamus: association with gonadotropin-releasing hormone neurons and co-localization with N-methyl-*D*-aspartate receptors. *Neuroendocrinology* 62, 187–197.
390. Prevot, V., Bouret, S., Stefano, G. B., and Beauvillain, J. C. (2000). Median eminence nitric oxide signaling. *Brain Res. Rev.* 34, 27–41.
391. Mitome, M., Shirakawa, T., Oshima, S., Nakamura, W., and Oguchi, H. (2001). Circadian rhythm of nitric oxide production in the dorsal region of the suprachiasmatic nucleus in rats. *Neurosci. Lett.* 303, 161–164.
392. Knauf, C., Ferreira, S., Hamdane, M., Mailliot, C., Prevot, V., Beauvillain, J. C., and Croix, D. (2001). Variation of endothelial nitric oxide synthase synthesis in the median eminence during the rat estrous cycle: an additional argument for the implication of vascular blood vessel in the control of GnRH release. *Endocrinology* 142, 4288–4294.
393. Knauf, C., Prevot, V., Stefano, G. B., Mortreux, G., Beauvillain, J. C., and Croix, D. (2001). Evidence for a spontaneous nitric oxide release from the rat mėdian eminence: influence on gonadotropin-releasing hormone release. *Endocrinology* 142, 2343–2350.
394. Mahachoklertwattana, P., Sanchez, J., Kaplan, S. L., and Grumbach, M. M. (1994). *N*-methyl-D-aspartate (NMDA) receptors mediate the release of gonadotropin-releasing hormone (GnRH) by NMDA in a hypothalamic GnRH neuronal cell line (GT1-1). *Endocrinology* 134, 1023–1030.
395. Belsham, D. D., Wetsel, W. C., and Mellon, P. L. (1996). NMDA and nitric oxide act through the cGMP signal transduction pathway to repress hypothalamic gonadotropin-releasing hormone gene expression. *EMBO J.* 15, 538–547.
396. Sortino, M. A., Aleppo, G., Scapagnini, U., and Canonico, P. L. (1994). Involvement of nitric oxide in the regulation of gonadotropin-releasing hormone release from the GT1–1 neuronal cell line. *Endocrinology* 134, 1782–1787.
397. Moretto, M., Lopez, F. J., and Negro-Vilar, A. (1993). Nitric oxide regulates luteinizing hormone-releasing hormone secretion. *Endocrinology* 133, 2399–2402.
398. Herbison, A. E., Simonian, S. X., Norris, P. J., and Emson, P. C. (1996). Relationship of neuronal nitric oxide synthase immunoreactivity to GnRH neurons in the ovariectomized and intact female rat. *J. Neuroendocrinol.* 8, 73–82.
399. Bonavera, J. J., Sahu, A., Kalra, P. S., and Kalra, S. P. (1993). Evidence that nitric oxide may mediate the ovarian steroid-induced luteinizing hormone surge: involvement of excitatory amino acids. *Endocrinology* 133, 2481–2487.
400. Bonavera, J. J., Sahu, A., Kalra, P. S., and Kalra, S. P. (1994). Evidence in support of nitric oxide (NO) involvement in the cyclic release of prolactin and LH surges. *Brain Res.* 660, 175–179.
401. Ceccatelli, S., Grandison, L., Scott, R. E. M., Pfaff, D. W., and Kow, L. M. (1996). Estradiol regulation of nitric oxide synthase mRNAs in rat hypothalamus. *Neuroendocrinology* 64, 357–363.
402. Prevot, V., Croix, D., Rialas, C. M., Poulain, P., Fricchione, G. L., Stefano, G. B., and Beauvillain, J. C. (1999). Estradiol coupling to endothelial nitric oxide stimulates gonadotropin-releasing hormone release from rat median eminence via a membrane receptor. *Endocrinology* 140, 652–659.
403. Moreno, A., and Franci, C. R. (2004). Estrogen modulates the action of nitric oxide in the medial preoptic area on luteinizing hormone and prolactin secretion. *Life Sci.* 74, 2049–2059.
404. Aguan, K., Mahesh, V. B., Ping, L., Bhat, G., and Brann, D. W. (1996). Evidence for a physiological role for nitric oxide in the regulation of the LH surge: effect of central administration of antisense oligonucleotides to nitric oxide synthase. *Neuroendocrinology* 64, 449–455.
405. Kalra, P. S., Bonavera, J. J., and Kalra, S. P. (1995). Central administration of antisense oligodeoxynucleotides to neuropeptide Y (NPY) mRNA reveals the critical role of newly synthesized NPY in regulation of LHRH release. *Regul. Pept.* 59, 215–220.
406. Bonavera, J. J., Kalra, P. S., and Kalra, S. P. (1996). L-arginine nitric oxide amplifies the magnitude and duration of the luteinizing hormone surge induced by estrogen: involvement of neuropeptide Y. *Endocrinology* 137, 1956–1962.
407. McCann, S. M., Karanth, S., Kimura, M., Yu, W. H., and Rettori, V. (1996). The role of nitric oxide (NO) in control of hypothalamic–pituitary function. *Rev. Bras. Biol.* 56 Su 1 Pt 1, 105–112.
408. Watanobe, H., and Schiöth, H. B. (2001). Nitric oxide mediates leptin-induced preovulatory luteinizing hormone and prolactin surges in rats. *Brain Res.* 923, 193–197.
409. Ceccatelli, S., Hulting, A.-L., Zhang, X., Gustafsson, L., Villar, M., and Hökfelt, T. (1993). Nitric oxide synthase in the rat anterior pituitary gland and the role of nitric oxide in regulation of luteinizing hormone secretion. *Proc. Natl. Acad. Sci. U S A* 90, 11292–11296.
410. Tsumori, M., Murakami, Y., Koshimura, K., and Kato, Y. (2002). Growth hormone-releasing hormone and gonadotropin-releasing hormone stimulate nitric oxide production in 17b-estradiol-primed rat anterior pituitary cells. *Endocrine* 17, 215–218.
411. Qian, X., Jin, L., and Lloyd, R. V. (1999). Estrogen down-regulates neuronal nitric oxide synthase in rat anterior pituitary cells and GH_3 tumors. *Endocrine* 11, 123–130.
412. Lozach, A., Garrel, G., Lerrant, Y., Bérault, A., and Counis, R. (1998). GnRH-dependent up-regulation of nitric oxide synthase I level in pituitary gonadotrophs mediates cGMP elevation during rat proestrus. *Mol. Cell. Endocrinol.* 143, 43–51.
413. Childs, G. V., and Armstrong, J. (2001). Sites of epidermal growth factor synthesis and action in the pituitary: paracrine and autocrine interactions. *Clin. Exp. Pharmacol. Physiol.* 28, 249–252.
414. Childs, G. V., Unabia, G., and Rougeau, D. (1994). Cells that express luteinizing hormone (LH) and follicle-stimulating hormone (FSH) beta-subunit messenger ribonucleic acids during the estrous cycle: the major contributors contain LH beta, FSH beta, and/or growth hormone. *Endocrinology* 134, 990–997.
415. Childs, G. V., Unabia, G., Lee, B. L., and Lloyd, J. (1992). Maturation of follicle-stimulating hormone gonadotropes during the rat estrous cycle. *Endocrinology* 131, 29–36.
416. Childs, G. V., Unabia, G., and Lloyd, J. (1992). Recruitment and maturation of small subsets of luteinizing hormone gonadotropes during the estrous cycle. *Endocrinology* 130, 335–344.
417. Childs, G. V., Unabia, G., Tibolt, R., and Lloyd, J. M. (1987). Cytological factors that support nonparallel secretion of luteinizing hormone and follicle-stimulating hormone during the estrous cycle. *Endocrinology* 121, 1801–1813.
418. Armstrong, J., and Childs, G. V. (1997). Changes in expression of epidermal growth factor receptors by anterior pituitary cells during the estrous cycle: cyclic expression by gonadotropes. *Endocrinology* 138, 1903–1908.
419. Armstrong, J. L., and Childs, G. V. (1997). Regulation of expression of epidermal growth factor receptors in gonadotropes by epidermal growth factor and estradiol: studies in cycling female rats. *Endocrinology* 138, 5434–5441.

420. Childs, G. V., and Unabia, G. (2001). Epidermal growth factor and gonadotropin-releasing hormone stimulate proliferation of enriched population of gonadotropes. *Endocrinology* 142, 847–853.
421. Childs, G. V., Unabia, G., and Miller, B. T. (1994). Cytochemical detection of gonadotropin-releasing hormone-binding sites on rat pituitary cells with luteinizing hormone, follicle-stimulating hormone, and growth hormone antigens during diestrous up-regulation. *Endocrinology* 134, 1943–1951.
422. Leblanc, P., L'Héritier, A., and Kordon, C. (1997). Cryptic gonadotropin-releasing hormone receptors of rat pituitary cells in culture are unmasked by epidermal growth factor. *Endocrinology* 138, 574–579.
423. Mouihate, A., and Lestage, J. (1995). Estrogen increases the release of epidermal growth factor from individual pituitary cells in female rats. *J. Endocrinol.* 146, 495–500.
424. Mouihate, A., Verrier, D., and Lestage, J. (1995). EGF release by rat gonadotroph cells: characteristics and effects of LHRH. *Life Sci.* 58, 107–114.
425. Kanematsu, T., Irahara, M., Miyake, T., Shitsukawa, K., and Aono, T. (1991). Effect of insulin-like growth factor I on gonadotropin release from the hypothalamus–pituitary axis in vitro. *Acta Endocr.* 125, 227–233.
426. Soldani, R., Cagnacci, A., and Yen, S. S. C. (1994). Insulin, insulin-like growth factor I (IGF-I) and IGF-II enhance basal and gonadotrophin-releasing hormone-stimulated luteinizing hormone release from rat anterior pituitary cells in vitro. *Eur. J. Endocrinol.* 131, 641–645.
427. Xia, Y. X., Weiss, J. M., Polack, S., Diedrich, K., and Ortmann, O. (2001). Interactions of insulin-like growth factor-I, insulin and estradiol with GnRH-stimulated luteinizing hormone release from female rat gonadotrophs. *Eur. J. Endocrinol.* 144, 73–79.
428. Fernandez-Galaz, M. C., Naftolin, F., and Garcia-Segura, L. M. (1999). Phasic synaptic remodeling of the rat arcuate nucleus during the estrous cycle depends on insulin-like growth factor-I receptor activation. *J. Neurosci. Res.* 55, 286–292.
429. Patterson, J. C., and Childs, G. V. (1994). Nerve growth factor in the anterior pituitary: regulation of secretion. *Endocrinology* 135, 1697–1704.
430. Patterson, J. C., and Childs, G. V. (1994). Nerve growth factor and its receptor in the anterior pituitary. *Endocrinology* 135, 1689–1696.
431. Missale, C., Boroni, F., Sigala, S., Buriani, A., Fabris, M., Leon, A., Dal Toso, R., and Spano, P. F. (1996). Nerve growth factor in the anterior pituitary: localization in mammotroph-cells and cosecretion with prolactin by a dopamine-regulated mechanism. *Proc. Natl. Acad. Sci. U S A* 93, 4240–4245.
432. Burnham, P., Conner, J. M., and Varon, S. (1995). Colocalization of NGF and TSH-like immunoreactivity in cultures of adult rat anterior pituitary cells. *J. Neurosci. Res.* 41, 73–78.
433. Dissen, G. A., Lara, H. E., Leyton, V., Paredes, A., Hill, D. F., Costa, M. E., Martinez-Serrano, A., and Ojeda, S. R. (2000). Intraovarian excess of nerve growth factor increases androgen secretion and disrupts estrous cyclicity in the rat. *Endocrinology* 141, 1073–1082.
434. Gibbs, R. B. (1998). Levels of trkA and BDNF mRNA, but not NGF mRNA, fluctuate across the estrous cycle and increase in response to acute hormone replacement. *Brain Res.* 787, 259–268.
435. Missale, C., Boroni, F., Frassine, M., Caruso, A., and Spano, P. (1995). Nerve growth factor promotes the differentiation of pituitary mammotroph cells *in vitro*. *Endocrinology* 136, 1205–1213.
436. Missale, C. (1999). Nerve growth factor in the neuroendocrine system. *Neurosci. Res. Commun.* 24, 61–70.
437. Moore, C. R., and Price, D. (1932). Gonadal hormone functions and the reciprocal influence between gonads and hypophysis, with its bearing on sex hormone antagonism. *Am. J. Anat.* 50, 13–71.
438. Gay, V. L., and Midgley, A. R. J. (1969). Response of the adult rat to orchidectomy and ovariectomy as determined by LH radioimmunoassay. *Endocrinology* 84, 1359–1364.
439. Tapper, C. M., Naftolin, F., and Brown-Grant, K. (1972). Influence of the reproductive state at the time of operation on the early response to ovariectomy in the rat. *J. Endocrinol.* 53, 47–57.
440. Yamamoto, M., Diebel, N. D., and Bogdanove, E. M. (1970). Analysis of initial and delayed effects of orchidectomy and ovariectomy on pituitary and serum LH levels in adult and immature rats. *Endocrinology* 86, 1102–1111.
441. Ajika, K., Krulich, L., Fawcett, C. P., and McCann, S. M. (1972). Effects of estrogen on plasma and pituitary gonadotropins and prolactin and on hypothalamic releasing and inhibitory factors. *Neuroendocrinology* 9, 304–315.
442. Caligaris, L., Astrada, J. J., and Taleisnik, S. (1971). Biphasic effect of progesterone on the release of gonadotropin in rats. *Endocrinology* 89, 331–337.
443. Swerdloff, R. S., Jacobs, H. S., and Odell, W. D. (1972). Synergistic role of progestogens in estrogen induction of LH and FSH surge. *Endocrinology* 90, 1529–1536.
444. Goodman, R. L. (1978). A quantitative analysis of the physiological role of estradiol and progesterone in the control of tonic and surge secretion of luteinizing hormone in the rat. *Endocrinology* 102, 142–150.
445. Weick, R. F., and Noh, K. A. (1984). Inhibitory effects of estrogen and progesterone on several parameters of pulsatile LH release in the ovariectomized rat. *Neuroendocrinology* 38, 351–356.
446. Leipheimer, R. E., Bona-Gallo, A., and Gallo, R. V. (1984). The influence of progesterone and estradiol on the acute changes in pulsatile luteinizing hormone release induced by ovariectomy on diestrus day 1 in the rat. *Endocrinology* 114, 1605–1612.
447. Goodman, R. L., and Daniel, K. (1985). Modulation of pulsatile luteinizing hormone secretion by ovarian steroids in the rat. *Biol. Reprod.* 32, 217–225.
448. Gallo, R. V., and Bona-Gallo, A. (1985). Lack of ovarian steroid negative feedback on pulsatile luteinizing hormone release between estrus and diestrous day 1 in the rat estrous cycle. *Endocrinology* 116, 1525–1528.
449. Hoffman, J. C. (1970). Light and reproduction in the rat: effects of photoperiod length on albino rats from two different breeders. *Biol. Reprod.* 2, 255–261.
450. Banks, J. A., and Freeman, M. E. (1978). The temporal requirement of progesterone on proestrus for extinction of the estrogen-induced daily signal controlling luteinizing hormone release in the rat. *Endocrinology* 102, 426–432.
451. Banks, J. A., and Freeman, M. E. (1980). Inhibition of the daily LH release mechanism by progesterone acting at the hypothalamus. *Biol. Reprod.* 22, 217–222.
452. Banks, J. A., Mick, C., and Freeman, M. E. (1980). A possible cause for the differing responses of the luteinizing hormone surge mechanism of ovariectomized rats to short term exposure to estradiol. *Endocrinology* 106, 1677–1681.
453. Caligaris, L., Astrada, J. J., and Taleisnik, S. (1971). Release of luteinizing hormone induced by estrogen injection into ovariectomized rats. *Endocrinology* 88, 810–815.
454. Freeman, M. E., Dupke, K. C., and Croteau, C. M. (1976). Extinction of the estrogen-induced daily signal for LH release in the rat: a role for the proestrous surge of progesterone. *Endocrinology* 99, 223–229.

455. Jackson, G. L. (1972). Effect of actinomycin D on estrogen-induced release of luteinizing hormone in ovariectomized rats. *Endocrinology* 91, 1284–1287.
456. Legan, S. J., Coon, G. A., and Karsch, F. J. (1975). Role of estrogen as initiator of daily LH surges in the ovariectomized rat. *Endocrinology* 96, 50–56.
457. Legan, S. J., and Karsch, F. J. (1975). A daily signal for the LH surge in the rat. *Endocrinology* 96, 57–62.
458. Neill, J. D. (1972). Sexual differences in the hypothalamic regulation of prolactin secretion. *Endocrinology* 90, 1154–1159.
459. Shirley, B., Wolinsky, J., and Schwartz, N. B. (1968). Effect of a single injection of an estrogen antagonist on the estrous cycle of the rat. *Endocrinology* 82, 959–968.
460. Ferin, M., Tempone, A., Zimmering, P. E., and Vande Wiele, R. L. (1969). Effects of antibodies to 17b-estradiol and progesterone on the estrous cycle of the rat. *Endocrinology* 85, 1070–1083.
461. Neill, J. D., Freeman, M. E., and Tillson, S. A. (1971). Control of the proestrus surge of prolactin and luteinizing hormone secretion by estrogens in the rat. *Endocrinology* 89, 1448–1453.
462. Schwartz, N. B. (1964). Acute effects of ovariectomy on pituitary LH, uterine weight, and vaginal cornification. *Am. J. Physiol.* 207, 1251–1259.
463. Ely, C. A., and Schwartz, N. B. (1971). Elucidation of the role of luteinizing hormone in estrogen secretion and ovulation by use of antigonadotropic sera. *Endocrinology* 89, 1103–1108.
464. Freeman, M. E., Reichert, L. E. J., and Neill, J. D. (1972). Regulation of the proestrus surge of prolactin secretion by gonadotropin and estrogens in the rat. *Endocrinology* 90, 232–238.
465. Everett, J. W. (1948). Progesterone and estrogen in the experimental control of ovulation time and other features of the estrous cycle in the rat. *Endocrinology* 43, 389–405.
466. Brown-Grant, K., and Naftolin, F. (1972). Facilitation of luteinizing hormone secretion in the female rat by progesterone. *J. Endocrinol.* 53, 37–46.
467. Ichikawa, S., Morioka, H., and Sawada, T. (1972). Acute effect of gonadotrophins on the secretion of progestins by the rat ovary. *Endocrinology* 90, 1356–1362.
468. Levine, J. E., Chappell, P. E., Schneider, J. S., Sleiter, N. C., and Szabo, M. (2001). Progesterone receptors as neuroendocrine integrators. *Front. Neuroendocrinol.* 22, 69–106.
469. Ringstrom, S. J., Szabo, M., Kilen, S. M., Saberi, S., Knox, K. L., and Schwartz, N. B. (1997). The antiprogestins RU486 and ZK98299 affect follicle-stimulating hormone secretion differentially on estrus, but not on proestrus. *Endocrinology* 138, 2286–2290.
470. Szabo, M., Knox, K. L., Ringstrom, S. J., Perlyn, C. A., Sutandi, S., and Schwartz, N. B. (1996). Mechanism of the inhibitory action of RU486 on the secondary follicle-stimulating hormone surge. *Endocrinology* 137, 85–89.
471. Rao, I. M., and Mahesh, V. B. (1986). Role of progesterone in the modulation of the preovulatory surge of gonadotropins and ovulation in the pregnant mare's serum gonadotropin-primed immature rat and the adult rat. *Biol. Reprod.* 35, 1154–1161.
472. Chappell, P. E., Schneider, J. S., Kim, P., Xu, M., Lydon, J. P., O'Malley, B. W., and Levine, J. E. (1999). Absence of gonadotropin surges and gonadotropin-releasing hormone self-priming in ovariectomized (OVX), estrogen (E_2)-treated, progesterone receptor knockout (PRKO) mice. *Endocrinology* 140, 3653–3658.
473. Chappell, P. E., and Levine, J. E. (2000). Stimulation of gonadotropin-releasing hormone surges by estrogen. I. Role of hypothalamic progesterone receptors. *Endocrinology* 141, 1477–1485.
474. Chappell, P. E., Lee, J., and Levine, J. E. (2000). Stimulation of gonadotropin-releasing hormone surges by estrogen. II. Role of cyclic adenosine 3′5′-monophosphate. *Endocrinology* 141, 1486–1492.
475. Auger, A. P., LaRiccia, L. M., Moffatt, C. A., and Blaustein, J. D. (2000). Progesterone, but not progesterone-independent activation of progestin receptors by a mating stimulus, rapidly decreases progestin receptor immunoreactivity in female rat brain. *Horm. Behav.* 37, 135–144.
476. Guerra-Araiza, C., Villamar-Cruz, O., González-Arenas, A., Chavira, R., and Camacho-Arroyo, I. (2003). Changes in progesterone receptor isoforms content in the rat brain during the oestrous cycle and after oestradiol and progesterone treatments. *J. Neuroendocrinol.* 15, 984–990.
477. Fishman, J., and Norton, B. (1975). Catechol-estrogen formation in the central nervous system of the rat. *Endocrinology* 96, 1054–1059.
478. Davies, I. J., Naftolin, F., Ryan, K. J., Fishman, J., and Siu, J. (1975). The affinity of catechol estrogens for estrogen receptors in the pituitary and anterior hypothalamus of the rat. *Endocrinology* 97, 554–557.
479. Schaeffer, J. M., and Hsueh, A. J. W. (1979). 2-hydroxyestradiol interaction with dopamine receptor binding in rat anterior pituitary. *J. Biol. Chem.* 254, 5606–5608.
480. Fishman, J. (1976). The catechol estrogens. *Neuroendocrinology* 22, 363–374.
481. Fishman, J. (1981). Biological action of catechol oestrogens. *J. Endocrinol.* 89, 59P–65P.
482. Foreman, M. (1980). Effect of exogenous 2-hydroxyestradiol (2-OH-E2) on serum LH, FSH and prolactin in orchidectomised rats. *Endocrinology* 106, 144–144.
483. Parvizi, N., Ellendorf, F., and Wuttke, W. (1981). Catecholestrogens in the brain: action on pituitary hormone secretion and catecholamine turnover. In *Steroid Hormone Regulation of the Brain* (K. Fuxe, J.-A. Gustaffsson, and L. Wetterberg, Eds.), pp. 107–107. Pergamon Press, Oxford.
484. Rodriguez-Sierra, J. F., and Blake, C. A. (1982). Catecholestrogens and release of anterior pituitary gland hormones. I. Luteinizing hormone. *Endocrinology* 110, 318–324.
485. Negro-Vilar, A. (1979). Estradiol-induced LH and prolactin surges in cycling rats are not duplicated by 2-hydroxyestradiol. *IRSC Med. Sci.* 7, 114–114.
486. Fishman, J., Norton, B. I., and Krey, L. (1980). 2-hydroxylation of estrogen in the brain participates in the initiation of the preovulatory LH surge in the rat. *Biochem. Biophys. Res. Commun.* 93, 471–477.
487. Goodman, R. L., and Knobil, E. (1981). The sites of action of ovarian steroids in regulation of LH secretion. *Neuroendocrinology* 32, 57–63.
488. McCann, S. M. (1974). Regulation of secretion of follicle stimulating hormone and luteinizing hormone. In *Handbook of Physiology* (E. Knobil, and W. H. Sawyer, Eds.), pp. 489–517. American Physiological Society, Washington, DC.
489. Cooper, K. J., Fawcett, C. P., and McCann, S. M. (1974). Augmentation of pituitary responsiveness to luteinizing hormone/follicle stimulating hormone-releasing factor (LH-RF) as a result of acute ovariectomy in the four-day cyclic rat. *Endocrinology* 96, 1123–1129.
490. Henderson, S. R., Baker, C., and Fink, G. (1977). Oestradiol-17b and pituitary responsiveness to luteinizing hormone releasing factor in the rat: a study using rectangular pulses of oestradiol-17b monitored by non-chromatographic radioimmunoassay. *J. Endocrinol.* 73, 441–453.
491. DeKoning, J., van Dieten, J. A. M. J., and van Rees, G. P. (1976). Inhibitory and augmentative effects of oestradiol on LH-RH–induced release of LH by anterior pituitary glands

from intact female rats in vitro. *Mol. Cell. Endocrinol.* 5, 321–326.
492. Libertun, C., Orias, R., and McCann, S. M. (1974). Biphasic effect of estrogen on the sensitivity of the pituitary to luteinizing hormone-releasing factor (LRF). *Endocrinology* 94, 1094–1100.
493. Vilchez-Martinez, J. A., Arimura, A., Debeljuk, L., and Schally, A. V. (1974). Biphasic effect of estradiol benzoate on the pituitary responsiveness to LH-RH. *Endocrinology* 94, 1300–1303.
494. Katt, J. A., Duncan, J. A., Herbon, L., Barkan, A., and Marshall, J. C. (1985). The frequency of gonadotropin-releasing hormone stimulation determines the number of pituitary gonadotropin-releasing hormone receptors. *Endocrinology* 116, 2113–2115.
495. Leipheimer, R. E., and Gallo, R. V. (1983). Acute and long-term changes in central and pituitary mechanisms regulating pulsatile luteinizing hormone secretion after ovariectomy in the rat. *Neuroendocrinology* 37, 421–426.
496. Kalra, S. P., and Kalra, P. S. (1983). Neural regulation of luteinizing hormone secretion in the rat. *Endocr. Rev.* 4, 311–351.
497. Blake, C. A. (1977). A medial basal hypothalamic site of synergistic action of estrogen and progesterone on the inhibition of pituitary luteinizing hormone release. *Endocrinology* 101, 1130–1134.
498. Smith, E. R., and Davidson, J. M. (1974). Location of feedback receptors; effects of intracranially implanted steroids on plasma LH and LRF response. *Endocrinology* 95, 1566–1573.
499. Akema, T., Tadokoro, Y., and Kimura, F. (1984). Regional specificity in the effect of estrogen implantation within the forebrain on the frequency of pulsatile luteinizing hormone secretion in the ovariectomized rat. *Neuroendocrinology* 39, 517–523.
500. Soper, B. D. (1980). Hypothalamic and extrahypothalamic mediation of pulsatile discharges of luteinizing hormone in the ovariectomized rat. *Endocrinology* 106, 348–355.
501. Drouin, J., Lagace, L., and Labrie, F. (1976). Estradiol-induced increase of the LH responsiveness to LH releasing hormone (LHRH) in rat anterior pituitary cells in culture. *Endocrinology* 99, 1477–1481.
502. Baldwin, D. M., and Downs, T. R. (1980). Release of LH and FSH by anterior pituitary cell suspensions from female rats during the estrous cycle and from estrogen-treated ovariectomized rats. *Biol. Reprod.* 24, 581–590.
503. Lagace, L., Massicotte, J., and Labrie, F. (1980). Acute stimulatory effects of progesterone on luteinizing hormone and follicle-stimulating hormone release in rat anterior pituitary cells in culture. *Endocrinology* 106, 684–689.
504. Drouin, J., and Labrie, F. (1981). Interactions between 17 beta-estradiol and progesterone in the control of luteinizing hormone and follicle-stimulating hormone release in rat anterior pituitary cells in culture. *Endocrinology* 108, 52–57.
505. Aiyer, M. S., and Fink, G. (1974). The role of sex steroid hormones in modulating the responsiveness of the anterior pituitary gland to luteinizing hormone releasing factor in the female rat. *J. Endocrinol.* 62, 553–572.
506. Snabes, M. C., Kelch, R. P., and Karsch, F. J. (1977). A daily neural signal for luteinizing hormone release in the untreated ovariectomized rat: changes in gonadotropin-releasing hormone content of the preoptic area and hypothalamus throughout the day. *Endocrinology* 100, 1521–1525.
507. Clayton, R. N., and Catt, K. (1981). Gonadotrophin-releasing hormone receptors: characterization physiological regulation and relationship to reproductive function. *Endocr. Rev.* 2, 186–206.
508. Clayton, R. N., Solano, A. R., Garcia-Vela, A., Dufau, M. L., and Catt, K. J. (1980). Regulation of pituitary receptors for gonadotropin-releasing hormone during the rat estrous cycle. *Endocrinology* 107, 699–706.
509. Savoy-Moore, R. T., Schwartz, N. B., Duncan, J. A., and Marshall, J. C. (1980). Pituitary gonadotrophin-releasing hormone receptors during the estrous cycle. *Science* 209, 942–944.
510. Witcher, J. A., Nearhoof, K. F., and Freeman, M. E. (1984). Secretion of luteinizing hormone (LH) and pituitary receptors for LH-releasing hormone as modified by the proestrous surge of progesterone. *Endocrinology* 115, 2189–2194.
511. Baldwin, D. M., Bourne, G. A., and Marshall, J. C. (1984). Pituitary LH responsiveness to GnRH in vitro as related to GnRH receptor number. *Am. J. Physiol.* 247, E651–E656.
512. Barkan, A. L., Regiani, S. R., Duncan, J. A., and Marshall, J. C. (1983). Pituitary gonadotropin-releasing hormone receptors during gonadotropin surges in ovariectomized estradiol-treated rats. *Endocrinology* 112, 1042–1048.
513. Pieper, D. R., Gala, R. R., Schiff, M. A., Regiani, S. R., and Marshall, J. C. (1984). Pituitary gonadotropin releasing hormone (GnRH) receptor responses to GnRH in hypothalamus-lesioned rats: inhibition of responses by hyperprolactinemia and evidence that testosterone and estradiol modulate gonadotropin secretion at postreceptor sites. *Endocrinology* 115, 1190–1196.
514. Corbani, M., Counis, R., Starzec, A., and Jutisz, M. (1984). Effect of gonadectomy on pituitary levels of mRNA encoding gonadotropin subunits and secretion of luteinizing hormone. *Mol. Cell. Endocr.* 35, 83–87.
515. Corbani, M., Counis, R., Wolinska-Witort, E., D'Angelo-Bernard, G., Moumni, M., and Jutisz, M. (1990). Synergistic effects of progesterone and oestradiol on rat LH subunit mRNA. *J Mol. Endocrinol.* 4, 119–125.
516. Haisenleder, D. J., Barkan, A. L., Papavasiliou, S., Zmeili, S. M., Dee, C., Jameel, M. L., Ortolano, G. A., El-Gewely, M. R., and Marshall, J. C. (1988). LH subunit mRNA concentrations during LH surge in ovariectomized estradiol-replaced rats. *Am. J. Physiol.* 254, E99–E99.
517. Shupnik, M. A., Gharib, S. D., and Chin, W. W. (1989). Divergent effects of estradiol on gonadotropin gene transcription in pituitary fragments. *Mol. Endocrinol.* 3, 474–480.
518. Haisenleder, D. J., Ortolano, G. A., Dalkin, A. C., Ellis, T. R., Paul, S. J., and Marshall, J. C. (1990). Differential regulation of gonadotropin subunit gene expression by gonadotropin-releasing hormone pulse amplitude in female rats. *Endocrinology* 127, 2869–2875.
519. Goodman, R. L. (1978). The site of the positive feedback action of estradiol in the rat. *Endocrinology* 102, 151–159.
520. Kalra, S. P., and McCann, S. M. (1975). The stimulatory effect on gonadotropin release of implants of estradiol or progesterone in certain sites in the central nervous system. *Neuroendocrinology* 19, 289–302.
521. Bishop, W., Kalra, P. S., Fawcett, C. P., Krulich, L., and McCann, S. M. (1972). The effects of hypothalamic lesions on the release of gonadotropins and prolactin in response to estrogen and progesterone treatment in female rats. *Endocrinology* 91, 1404–1410.
522. Halasz, B., and Pupp, L. (1965). Hormone secretion of the anterior pituitary gland after physical interruption of all nervous pathways to the hypophysiotropic area. *Endocrinology* 77, 553–562.
523. Taleisnik, S., Caligaris, L., Beltramino, C., and Caceres, A. (1978). Stimulatory and inhibitory effects of ovarian steroids on gonadotrophin secretion in ovariectomized rats after anterior hypothalamic deafferentation. *J. Endocrinol.* 77, 11–23.

524. Sarkar, D. K., and Fink, G. (1979). Effects of gonadal steroids on output of luteinizing hormone releasing factor into pituitary stalk blood in the female rat. *J. Endocrinol.* 80, 303–313.
525. Sherwood, N. M., Chiappa, S. A., and Fink, G. (1976). Immunoreactive luteinizing hormone releasing factor in pituitary stalk blood from female rats: sex steroid modulation of response to electrical stimulation of preoptic area or median eminence. *J. Endocrinol.* 70, 501–511.
526. Shivers, B. D., Harlan, R. E., Morrell, J. I., and Pfaff, D. W. (1983). Absence of oestradiol concentration in cell nuclei of LHRH-immunoreactive neurones. *Nature* 304, 345–347.
527. Shivers, B. D., Harlan, R. E., Hejmancik, J. F., Conn, P. M., and Pfaff, D. W. (1986). Localization of cells containing LHRH-like mRNA in rat forebrain using in situ hybridization. *Endocrinology* 118, 883–886.
528. Zoeller, R. T., Seeburg, P. H., and Young WS, 3rd. (1988). In situ hybridization histochemistry for messenger ribonucleic acid (mRNA) encoding gonadotropin-releasing hormone (GnRH): effect of estrogen on cellular levels of GnRH mRNA in female rat brain. *Endocrinology* 122, 2570–2578.
529. Rothfeld, J., Hejtmancik, J. F., Conn, P. M., and Pfaff, D. W. (1989). In situ hybridization for LHRH mRNA following estrogen treatment. *Mol. Brain Res.* 6, 121–125.
530. Rosie, R., Thomson, E., and Fink, G. (1990). Oestrogen positive feedback stimulates the synthesis of LHRH mRNA in neurones of the rostral diencephalon of the rat. *J. Endocrinol.* 124, 285–289.
531. Skynner, M. J., Sim, J. A., and Herbison, A. E. (1999). Detection of estrogen receptor a and b messenger ribonucleic acids in adult gonadotropin-releasing hormone neurons. *Endocrinology* 140, 5195–5201.
532. Simonian, S. X., Spratt, D. P., and Herbison, A. E. (1999). Identification and characterization of estrogen receptor a-containing neurons projecting to the vicinity of the gonadotropin-releasing hormone perikarya in the rostral preoptic area of the rat. *J. Comp. Neurol.* 411, 346–358.
533. Herbison, A. E. (1998). Multimodal influence of estrogen upon gonadotropin-releasing hormone neurons. *Endocr. Rev.* 19, 302–330.
534. Herbison, A. E., and Pape, J. R. (2001). New evidence for estrogen receptors in gonadotropin-releasing hormone neurons. *Front. Neuroendocrinol.* 22, 292–308.
535. Abrahám, I. M., Han, S. K., Todman, M. G., Korach, K. S., and Herbison, A. E. (2003). Estrogen receptor b mediates rapid estrogen actions on gonadotropin-releasing hormone neurons *in vivo*. *J. Neurosci.* 23, 5771–5777.
536. Dorling, A. A., Todman, M. G., Korach, K. S., and Herbison, A. E. (2003). Critical role for estrogen receptor alpha in negative feedback regulation of gonadotropin-releasing hormone mRNA expression in the female mouse. *Neuroendocrinology* 78, 204–209.
537. Leadem, C. A., and Kalra, S. P. (1984). Stimulation with estrogen and progesterone of luteinizing hormone (LH)-releasing hormone release from adult female rat hypothalami: correlation with the LH surge. *Endocrinology* 114, 51–56.
538. Lee, W.-S., Smith, M. S., and Hoffman, G. E. (1990). Luteinizing hormone-releasing hormone neurons express Fos protein during the proestrous surge of luteinizing hormone. *Proc. Natl. Acad. Sci. U S A* 87, 5163–5167.
539. Lee, W.-S., Smith, M. S., and Hoffman, G. E. (1990). Progesterone enhances the surge of luteinizing hormone by increasing the activation of luteinizing hormone-releasing hormone neurons. *Endocrinology* 127, 2604–2606.
540. Wise, P. M., Rance, N., and Barraclough, C. A. (1981). Effects of estradiol and progesterone on catecholamine turnover rates in discrete hypothalamic regions in ovariectomized rats. *Endocrinology* 108, 2186–2193.
541. Condé, G. L., Bicknell, R. J., and Herbison, A. E. (1995). Changing patterns of Fos expression in brainstem catecholaminergic neurons during the rat oestrous cycle. *Brain Res.* 672, 68–76.
542. Condé, G. L., Herbison, A. E., Fernandez-Galaz, C., and Bicknell, R. J. (1996). Estrogen uncouples noradrenergic activation of Fos expression in the female rat preoptic area. *Brain Res.* 735, 197–207.
543. Miyake, A., Tasaka, K., Sakumoto, T., Kawamura, Y., Aono, T., and Kurachi, K. (1983). Norepinephrine induces releases of both H-RH from the hypothalamus and LH from the rat pituitary in vitro. *Endocrinol. Jpn.* 30, 509–512.
544. Sarkar, D. K., and Fink, G. (1981). Gonadotropin releasing hormone surge: possible modulation through postsynaptic a-adrenoreceptors and two pharmacologically distinct dopamine receptors. *Endocrinology* 108, 862–867.
545. Gallo, R. V. (1984). Further studies on norepinephrine-induced suppression of pulsatile luteinizing hormone release in ovariectomized rats. *Neuroendocrinology* 39, 120–125.
546. Gallo, R. V., and Drouva, S. V. (1979). Effect of intraventricular infusion of catecholamines on luteinizing hormone release in ovariectomized and ovariectomized, steroid-primed rats. *Neuroendocrinology* 29, 149–162.
547. Dotti, C., and Taleisnik, S. (1982). Inhibition of the release of LH and ovulation by activation of the noradrenergic system. Effect of interrupting the ascending pathways. *Brain Res.* 249, 281–290.
548. Leipheimer, R. E., and Gallo, R. V. (1985). Medial preoptic area involvement in norepinephrine-induced suppression of pulsatile luteinizing hormone release in ovariectomized rats. *Neuroendocrinology* 40, 345–351.
549. Caceres, A., and Taleisnik, S. (1980). Inhibition of secretion of luteinizing hormone induced by electrochemical stimulation of the anterior cingulate cortex mediated by a beta-adrenergic mechanism. *J. Endocrinol.* 87, 419–429.
550. Dotti, C., and Taleisnik, S. (1984). Beta-adrenergic receptors in the premammillary nucleus mediate the inhibition of LH release evoked by locus ceruleus stimulation. *Neuroendocrinology* 38, 6–11.
551. Krieg, R. J., and Ching, M. C. H. (1982). Stimulation of luteinizing hormone releasing hormone (LHRH) secretion by norepinephrine (NE) in steroid-primed or unprimed ovariectomized (OVX) rats. *Fed. Proc.* 42, 976–976.
552. Krieg, R. J., and Sawyer, C. H. (1976). Effects of intraventricular catecholamines on LH release in ovariectomized steroid-primed rats. *Endocrinology* 99, 411–419.
553. Leung, P. C. K., Arendash, G. W., Whitmoyer, D. I., Gorski, R. A., and Sawyer, C. H. (1982). Differential effects of central adrenoceptor agonists on luteinizing hormone release. *Neuroendocrinology* 34, 207–214.
554. Kalra, S. P., and McCann, S. M. (1974). Effects of drugs modifying catecholamine synthesis on plasma LH and ovulation in the rat. *Neuroendocrinology* 15, 79–91.
555. Anselmo-Franci, J. A., Franci, C. R., Krulich, L., Antunes-Rodrigues, J., and McCann, S. M. (1997). Locus coeruleus lesions decrease norepinephrine input into the medial preoptic area and medial basal hypothalamus and block the LH, FSH and prolactin preovulatory surge. *Brain Res.* 767, 289–296.
556. Anselmo-Franci, J. A., Rocha-Barros, V. M., Franci, C. R., and McCann, S. M. (1999). Locus ceruleus lesions block pulsatile LH release in ovariectomized rats. *Brain Res.* 833, 86–92.
557. Helena, C. V., Franci, C. R., and Anselmo-Franci, J. A. (2002). Luteinizing hormone and luteinizing hormone-releasing

hormone secretion is under locus coeruleus control in female rats. *Brain Res.* 955, 245–252.

558. Martins-Afférri, M. P., Ferreira-Silva, I. A., Franci, C. R., and Anselmo-Franci, J. A. (2003). LHRH release depends on Locus Coeruleus noradrenergic inputs to the medial preoptic area and median eminence. *Brain Res. Bull.* 61, 521–527.
559. Helena, C. V. V., Franci, C. R., and Anselmo-Franci, J. A. (2002). Luteinizing hormone and luteinizing hormone-releasing hormone secretion is under locus coeruleus control in female rats. *Brain Res.* 955, 245–252.
560. Rubinstein, L., and Sawyer, C. H. (1970). Role of catecholamines in the release of pituitary ovulating hormone(s) in the rat. *Endocrinology* 86, 988–995.
561. Kalra, S. P. (1985). Catecholamine involvement in preovulatory LH release: reassessment of the role of epinephrine. *Neuroendocrinology* 40, 139–144.
562. Adler, B. A., Johnson, M. D., Lynch, C. O., and Crowley, W. R. (1983). Evidence that norepinephrine and epinephrine systems mediate the stimulatory effects of ovarian hormones on luteinizing hormone and luteinizing hormone-releasing hormone. *Endocrinology* 113, 1431–1438.
563. Drouva, S. V., and Gallo, R. V. (1977). Further evidence for inhibition of episodic luteinizing hormone release in ovariectomized rats by stimulation of dopamine receptors. *Endocrinology* 100, 792–298.
564. Gallo, R. V. (1978). The effect of blockade of dopamine receptors on the inhibition of episodic luteinizing hormone release during electrical stimulation of the arcuate nucleus in ovariectomized rats. *Endocrinology* 102, 1026–1035.
565. Gallo, R. V. (1980). Neuroendocrine regulation of pulsatile luteinizing hormone release in the rat. *Neuroendocrinology* 30, 122–131.
566. Negro-Vilar, A., Ojeda, S. R., and McCann, S. M. (1979). Catecholaminergic modulation of luteinizing hormone-releasing hormone release by median eminence terminals in vitro. *Endocrinology* 104, 1749–1757.
567. Schneider, H. P. G., and McCann, S. M. (1969). Possible role of dopamine as transmitter to promote discharge of LH-releasing factor. *Endocrinology* 85, 121–132.
568. Vijayan, E., and McCann, S. M. (1978). Re-evaluation of the role of catecholamines in control of gonadotropin and prolactin release. *Neuroendocrinology* 25, 150–165.
569. Hery, M., Faudon, G., and Hery, F. (1982). Daily variations in serotonin metabolism in the suprachiasmatic nucleus of the rat: influence of estradiol impregnation. *J. Endocrinol.* 94, 157–166.
570. Hery, M., Laplante, E., and Kordon, C. (1978). Participation of serotonin in the phasic release of luteinizing hormone. II. Effects of lesions of serotonin containing pathways in the central nervous system. *Endocrinology* 102, 1019–1025.
571. Hery, M., Laplante, E., and Kordon, C. (1976). Participation of serotonin in the phasic release of LH. I. Evidence from pharmacological experiments. *Endocrinology* 99, 496–503.
572. Adler, B. A., and Crowley, W. R. (1984). Modulation of luteinizing hormone release and catecholamine activity by opiates in the female rat. *Neuroendocrinology* 38, 248–253.
573. Kalra, S. P., and Simpkins, J. W. (1981). Evidence for noradrenergic mediation of opioid effects on luteinizing hormone secretion. *Endocrinology* 109, 776–782.
574. Kalra, S. P., and Gallo, R. V. (1983). Effects of intraventricular catecholamines on luteinizing hormone release in morphine-treated rats. *Endocrinology* 113, 23–28.
575. Allen, L. G., and Kalra, S. P. (1986). Evidence that decrease in opioid tone may evoke preovulatory luteinizing hormone release in the rat. *Endocrinology* 118, 2375–2381.
576. Gabriel, S. M., Simpkins, J. W., and Kalra, S. P. (1983). Modulation of endogenous opioid influence on luteinizing hormone secretion by progesterone and estradiol. *Endocrinology* 113, 1806–1811.
577. Bhanot, R., and Wilkinson, M. (1983). Opiatergic control of LH secretion is eliminated by gonadectomy. *Endocrinology* 112, 399–40l.
578. McCullagh, D. R. (1932). Dual endocrine activity of the testis. *Science* 76, 19–20.
579. Hoffman, J. C., Lorenzen, J. R., Neil, T., and Schwartz, N. B. (1979). Selective suppression of the primary surge of follicle-stimulating hormone in the rat: further evidence for folliculostatin in porcine follicular fluid. *Endocrinology* 105, 200–203.
580. Williams, A. T., and Lipner, H. (1981). Negative feedback control of gonadotropin secretion by chronically administered estradiol and porcine follicular fluid (gonadostatin) in ovariectomized rats. *Endocrinology* 109, 1496–1501.
581. Erickson, G. F., and Hsueh, A. J. W. (1978). Secretion of "inhibin" by rat granulosa cells in vitro. *Endocrinology* 103, 1960–1963.
582. De Paolo, L. V., Shander, D., Wise, P. M., Barraclough, C. A., and Channing, C. P. (1979). Identification of inhibin-like activity in ovarian venous plasma of rats during the estrous cycle. *Endocrinology* 105, 647–654.
583. Shander, D., Anderson, L. D., and Barraclough, C. A. (1980). FSH and LH affect the endogenous release of pituitary FSH and the ovarian secretion of inhibin in rats. *Endocrinology* 106, 1047–1053.
584. Shander, D., Anderson, L. D., Barraclough, C. A., and Channing, C. P. (1980). Interactions of porcine follicular fluid with ovarian steroids and luteinizing hormone-releasing hormone on the secretion of luteinizing hormone and follicle stimulating hormone by cultured pituitary cells. *Endocrinology* l06, 237–242.
585. Foreman, D., and Wilder, M. (1995). Inhibin is present in the corpus luteum of pseudopregnant rats and can interact with progesterone to inhibit resumption of estrous cycles. *Endocr. Res.* 21, 595–614.
586. Besecke, L. M., Guendner, M. J., Sluss, P. A., Polak, A. G., Woodruff, T. K., Jameson, J. L., Bauer-Dantoin, A. C., and Weiss, J. (1997). Pituitary follistatin regulates activin-mediated production of follicle-stimulating hormone during the rat estrous cycle. *Endocrinology* 138, 2841–2848.
587. Williams, A. T., and Lipner, H. (1982). The contribution of gonadostatin (inhibin-F) to the control of gonadotropin secretion in a simulated estrous cycle in steroid-treated ovariectomized rats. *Endocrinology* 111, 231–237.
588. Ling, N., Ying, S.-Y., Veno, N., Esch, F., Denoroy, L., and Guillemin, R. P. (1985). Isolation and partial characterization of a Mr 32,000 protein with inhibin activity from porcine follicular fluid. *Proc. Natl. Acad. Sci. U S A* 82, 7217–7221.
589. Mason, A. J., Hayflick, J. S., Ling, N., Esch, F., Ueno, N., Ying, S.-Y., Guillemin, R., Niall, H., and Seeburg, P. H. (1986). Complementary DNA sequences of ovarian follicular fluid inhibin show precursor structure and homology with transforming growth factor-b. *Nature* 318, 659–663.
590. Esch, F. S., Shimasaki, S., Cooksey, K., Mercado, M., Mason, A. J., Ying, S. Y., Ueno, N., and Ling, N. (1987). Complementary Deoxyribonucleic Acid (cDNA) cloning and DNA sequence analysis of rat Ovarian Inhibins. *Mol. Endocrinol.* 1, 388–388.
591. Woodruff, T. K., Meunier, H., Jones, P. B., Hsueh, A. J., and Mayo, K. E. (1987). Rat inhibin:molecular cloning of a and b-subunit complementary deoxyribonucleic acids and expression in the ovary. *Mol. Endocrinol.* 1, 561–568.
592. Bernard, D. J., and Woodruff, T. K. (2001). Inhibin binding protein in rats: alternative transcripts and regulation in the

pituitary across the estrous cycle. *Mol. Endocrinol.* 15, 654–667.

593. Ying, S.-Y., Becker, A., Ling, N., Ueno, N., and Guillemin, R. (1986). Type beta transforming growth factor (TGFb) is a potent stimulator of the basal secretion of follicle stimulating hormone (FSH) in a pituitary monolayer system. *Biochem. Biophys. Res. Commun.* 135, 950–956.
594. Ling, N., Ying, S. Y., Ueno, N., Shimasaki, S., Esch, F., Hotta, M., and Guillemin, R. (1986). Pituitary gland FSH is released by a heterodimer of the b-subunits from the two forms of inhibin. *Nature* 321, 779–782.
595. Vale, W., Rivier, J., Ying, S. Y., Vaughan, J., McClintock, R., Corrigan, A., Woo, W., Karr, D., and Spiess, J. (1986). Purification and characterization of an FSH releasing protein from porcine ovarian follicular fluid. *Nature* 321, 776–779.
596. Ling, N., Ying, S. Y., Ueno, N., Shimasaki, S., Esch, F., Hotta, M., and Guillemin, R. (1986). A homodimer of the b-subunits of inhibin A stimulates the secretion of pituitary follicle stimulating hormone. *Biochem. Biophys. Res. Commun.* 138, 1129–1137.
597. Burger, H. G., and Igarashi, M. (1988). Inhibin: definition and nomenclature, including related substances. *Mol. Endocrinol.* 2, 391–391.
598. Ueno, N., Ling, N., Ying, S.-Y., Esch, F., Shimasaki, S., and Guillemin, R. (1987). Isolation and partial characterization of follistatin, a novel Mr 35,000 monomeric protein which inhibits the release of follicle stimulating hormone. *Proc. Natl. Acad. Sci. U S A* 84, 8282–8286.
599. Shimasaki, S., Koga, M., Esch, F., Mercado, M., Cooksey, K., Koba, A., and Ling, N. (1988). Porcine Follistatin gene structure supports two forms of mature Follistatin produced by alternative splicing. *Biochem. Biophys. Res. Commun.* 152, 717–723.
600. Ying, S.-Y., Becker, A., Swanson, G., Tan, P., Ling, N., Esch, F., Ueno, N., Shimasaki, S., and Guillemin, R. (1987). Follistatin specifically inhibits Pituitary Follicle Stimulating Hormone release in vitro. *Biochem. Biophys. Res. Commun.* 149, 133–139.
601. Nakamura, T., Takio, K., Eto, Y., Shibai, H., Titani, K., and Sugino, H. (1990). Activin-binding protein from rat ovary is follistatin. *Science* 247, 836–838.
602. Xiao, S., Findlay, J. K., and Robertson, D. M. (1990). The effect of bovine activin and follicle-stimulating hormone (FSH) suppressing protein/follistatin on FSH-induced differentiation of rat granulosa cells in vitro. *Mol. Cell. Endocrinol.* 69, 1–8.
603. Bilezikjian, L. M., Blount, A. L., Corrigan, A. Z., Leal, A., Chen, Y., and Vale, W. W. (2001). Actions of activins, inhibins and follistatins: implications in anterior pituitary function. *Clin. Exp. Pharmacol. Physiol* 28, 244–248.
604. Rivier, C., Schwall, R., Mason, A., Burton, L., Vaughan, J., and Vale, W. (1991). Effect of recombinant inhibin on luteinizing hormone and follicle-stimulating hormone secretion in the rat. *Endocrinology* 128, 1548–1554.
605. Rivier, C., Schwall, R., Mason, A., Burton, L., and Vale, W. (1991). Effect of recombinant inhibin on gonadotropin secretion during proestrus and estrus in the rat. *Endocrinology* 128, 2223–2228.
606. Carroll, R. S., Kowash, P. M., Lofgren, J. A., Schwall, R. H., and Chin, W. W. (1991). *In vivo* regulation of FSH synthesis by inhibin and activin. *Endocrinology* 129, 3299–3304.
607. Doi, M., Igarashi, M., Hasegawa, Y., Eto, Y., Shibai, H., Miura, T., and Ibuki, Y. (1992). In vivo action of activin-A on pituitary-gonadal system. *Endocrinology* 130, 139–144.
608. Carroll, R. S., Corrigan, A. Z., Vale, W., and Chin, W. W. (1991). Activin stabilizes follicle-stimulating hormone-beta messenger ribonucleic acid levels. *Endocrinology* 129, 1721–1726.
609. Sander, H. J., Kramer, P., Van Leeuwen, E. C. M., Van Cappellen, W. A., Meijs-Roelofs, H. M. A., and de Jong, F. H. (1991). Ovulation rate, follicle population and FSH levels in cyclic rats after administration of an inhibin-neutralizing antiserum. *J. Endocrinol.* 130, 297–303.
610. Rivier, C., and Vale, W. (1989). Immunoneutralization of endogenous inhibin modifies hormone secretion and ovulation rate in the rat. *Endocrinology* 125, 152–157.
611. Hasegawa, Y., Miyamoto, K., and Igarashi, M. (1989). Changes in serum concentration of immunoreactive inhibin during the estrous cycle of the rat. *J. Endocrinol.* 121, 91–100.
612. Rivier, C., Roberts, V., and Vale, W. (1989). Possible role of luteinizing hormone and follicle-stimulating hormone in modulating inhibin secretion and expression during the estrous cycle of the rat. *Endocrinology* 125, 876–882.
613. Halvorson, L. M., Weiss, J., Bauer-Dantoin, A. C., and Jameson, J. L. (1994). Dynamic regulation of pituitary follistatin messenger ribonucleic acids during the rat estrous cycle. *Endocrinology* 134, 1247–1253.
614. Bauer-Dantoin, A. C., Weiss, J., and Jameson, J. L. (1996). Gonadotropin-releasing hormone regulation of pituitary follistatin gene expression during the primary follicle-stimulating hormone surge. *Endocrinology* 137, 1634–1639.
615. Drouin, J., and Labrie, F. (1976). Selective effect of androgens on LH and FSH release in anterior pituitary cells in culture. *Endocrinology* 98, 1528–1538.
616. Sarkar, D. K., and Fink, G. (1980). Luteinizing hormone releasing factor in pituitary stalk plasma from long-term ovariectomized rats: effects of steroids. *J. Endocrinol.* 86, 511–524.
617. Blake, C. A. (1976). Stimulation of the early phase of the proestrous follicle-stimulating hormone rise after infusion of luteinizing hormone-releasing hormone in phenobarbital-blocked rats. *Endocrinology* 98, 461–467.
618. Schwartz, N. B., Rivier, C., Rivier, J., and Vale, W. W. (1985). Effect of gonadotropin-releasing hormone antagonists on serum follicle-stimulating hormone and luteinizing hormone under conditions of singular follicle-stimulating hormone secretion. *Biol. Reprod.* 32, 391–398.
619. Rush, M. E., Ashiru, O. A., and Blake, C. A. (1982). Effects of complete hypothalamic deafferentation on the estrous phase of FSH release in the cyclic rat. *Biol. Reprod.* 26, 399–403.
620. Wise, P. M., Rance, N., Barr, G. D., and Barraclough, C. A. (1979). Further evidence that LHRH also is FSH-RH. *Endocrinology* 104, 940–947.
621. Ashiru, O. A., and Blake, C. A. (1979). Stimulation of endogenous FSH release during estrus by exogenous FSH or LH at proestrus in the phenobarbital-blocked rat. *Endocrinology* 105, 1162–1167.
622. Ashiru, O. A., and Blake, C. A. (1980). Variations in the effectiveness with which rat FSH can stimulate its own secretion during the rat estrous cycle. *Endocrinology* 106, 476–480.
623. Chappel, S. C., Ulloa-Aguirre, A., and Coutifaris, C. (1983). Biosynthesis and secretion of follicle-stimulating hormone. *Endocr. Rev.* 4, 179–211.
624. Turgeon, J. L., and Waring, D. W. (1982). Differential changes in the rate and pattern of follicle-stimulating hormone secretion from pituitaries of cyclic rats superfused in vitro. *Endocrinology* 111, 66–73.
625. Schwartz, N. B., and Channing, C. P. (1977). Evidence for ovarian "inhibin" suppression of the secondary rise in serum follicle stimulating hormone levels in proestrous rats by injection of porcine follicular fluid. *Proc. Natl. Acad. Sci. U S A* 74, 5721–5724.
626. Hirshfield, A. N. (1979). The role of FSH in the recruitment of large follicles. In *Ovarian Follicular Development and*

Function (A. R. Midgley, and W. A. Sadler, Eds.), pp. 19–22. Raven Press, New York.
627. Hirshfield, A. N. (1981). Role of FSH in the selection of ovulatory follicles. In *Dynamics of Ovarian Function* (N. B. Schwartz, and M. Hunzicker-Dunn, Eds.), pp. 79–82. Raven Press, New York.
628. McCann, S. M., Mizunuma, H., Samson, W. K., and Lumpkin, M. D. (1983). Differential hypothalamic control of FSH secretion: a review. *Psychoneuroendocrinology* 8, 299–308.
629. Neill, J. D., and Smith, M. S. (1974). Pituitary–ovarian interrelationships in the rat. In *Current Topics in Experimental Endocrinology* (V. H. T. James and L. Martini, Eds.), pp. 73–106. Academic Press, New York.
630. Labrie, F., Ferland, L., DiPaolo, T., and Veilleux, R. (1980). Modulation of prolactin secretion by sex steroids and thyroid hormones. In *Central and Peripheral Regulation of Prolactin* (R. M. MacLeod, and U. Scapagnini, Eds.), pp. 97–113. Raven Press, New York.
631. Dannies, P. S., and Tam, S. W. (1982). Mechanism of regulation of prolactin release. In *Cellular Regulation of Secretion and Release* (P. M. Conn, Ed.), pp. 529–568. Academic Press, New York.
632. Leong, D. A., Frawley, S., and Neill, J. D. (1983). Neuroendocrine control of prolactin secretion. *Ann. Rev. Physiol.* 45, 109–127.
633. Neill, J. D. (1974). Prolactin:Its Secretion and Control. In *Handbook of Physiology* (R. O. Greep, E. B. Astwood, E. Knobil, and W. H. Sawyer, Eds.), Section 7, Endocrinology, Vol. IV, Part 2, pp. 469–488. American Physiological Society, Washington, DC.
634. Kimura, F., and Kawakami, M. (1978). Reanalysis of the preoptic afferents involved in the surge of LH, FSH and prolactin release in the proestrous rat. *Neuroendocrinology* 27, 74–85.
635. Pan, J.-T., and Gala, R. R. (1985). Central nervous system regions involved in the estrogen-induced afternoon prolactin surge. I. Lesion studies. *Endocrinology* 117, 382–387.
636. Kawakami, M., and Arita, J. (1981). Effects of lesions in the medial basal part of the suprachiasmatic area on prolactin and gonadotropin surges induced by estrogen and progesterone treatment in ovariectomized rats. *Neuroendocrinology* 32, 242–247.
637. Kawakami, M., Arita, J., and Yoshioka, E. (1980). Loss of estrogen-induced daily surges of prolactin and gonadotropins by suprachiasmatic nucleus lesions in ovariectomized rats. *Endocrinology* 106, 1087–1092.
638. Pan, J.-T., and Gala, R. R. (1985). Central nervous system regions involved in the estrogen-induced afternoon prolacin surge. II. Implantation studies. *Endocrinology* 117, 388–395.
639. Nakamura, T. J., Shinohara, K., Funabashi, T., and Kimura, F. (2001). Effect of estrogen on the expression of Cry1 and Cry2 mRNAs in the suprachiasmatic nucleus of female rats. *Neurosci. Res.* 41, 251–255.
640. Poletini, M. O., Szawka, R. E., Franci, C. R., and Anselmo-Franci, J. A. (2004). Role of the locus coeruleus in the prolactin secretion of female rats. *Brain Res. Bull.* 63, 331–338.
641. Samson, W. K., Taylor, M. M., and Baker, J. R. (2003). Prolactin-releasing peptides. *Regul. Pept.* 114, 1–5.
642. Pasqualini, C., Lenoir, V., El Abed, A., and Kerdelhue, B. (1984). Anterior pituitary dopamine receptors during the rat estrous cycle. *Neuroendocrinology* 38, 39–44.
643. DeMaria, J. E., Livingstone, J. D., and Freeman, M. E. (1998). Characterization of the dopaminergic input to the pituitary gland throughout the estrous cycle of the rat. *Neuroendocrinology* 67, 377–383.
644. Lerant, A., and Freeman, M. E. (1997). Dopaminergic neurons in periventricular and arcuate nuclei of proestrous and ovariectomized rats: endogenous diurnal rhythm of Fos-related antigens expression. *Neuroendocrinology* 65, 436–445.
645. Ben-Jonathan, N., Oliver, C., Weiner, H. J., Mical, R. S., and Porter, J. C. (1977). Dopamine in hypophysial portal plasma of the rat during the estrous cycle and throughout pregnancy. *Endocrinology* 100, 452–458.
646. Heiman, M. L., and Ben-Jonathan, N. (1982). Dopaminergic receptors in the rat anterior pituitary change during the estrous cycle. *Endocrinology* 111, 37–41.
647. Ben-Jonathan, N. (1985). Dopamine: a prolactin-inhibiting hormone. *Endocr. Rev.* 6, 564–589.
648. Ben-Jonathan, N., and Hnasko, R. (2001). Dopamine as a prolactin (PRL) inhibitor. *Endocr. Rev.* 22, 724–763.
649. Arbogast, L. A., Murai, I., and Ben-Jonathan, N. (1989). Differential alterations in dopamine turnover rates in the stalk-median eminence and posterior pituitary during the preovulatory prolactin surge. *Neuroendocrinology* 49, 525–530.
650. DeMaria, J. E., Zelena, D., Vecsernyés, M., Nagy, G. M., and Freeman, M. E. (1998). The effect of neurointermediate lobe denervation on hypothalamic neuroendocrine dopaminergic neurons. *Brain Res.* 806, 89–94.
651. Goudreau, J. L., Lindley, S. E., Lookingland, K. J., and Moore, K. E. (1990). Functional and neurochemical evidence that tuberohypophysial dopaminergic neurons in the rat project from the periventricular nucleus to the intermediate lobe of the pituitary. *Abstr. Soc. Neurosci.* 16,
652. Goudreau, J. L., Lindley, S. E., Lookingland, K. J., and Moore, K. E. (1992). Evidence that hypothalamic periventricular dopamine neurons innervate the intermediate lobe of the rat pituitary. *Neuroendocrinology* 56, 100–105.
653. Goudreau, J. L., Falls, W. M., Lookingland, K. J., and Moore, K. E. (1995). Periventricular-hypophysial dopaminergic neurons innervate the intermediate but not the neural lobe of the rat pituitary gland. *Neuroendocrinology* 62, 147–154.
654. Murai, I., and Ben-Jonathan, N. (1986). Chronic posterior pituitary lobectomy: prolonged elevation of plasma prolactin and interruption of cyclicity. *Neuroendocrinology* 43, 453–458.
655. Arbogast, L. A., and Ben-Jonathan, N. (1988). The preovulatory prolactin surge: an evaluation of the role of dopamine. *Endocrinology* 123, 2690–2695.
656. Ben-Jonathan, N., Arbogast, L. A., and Hyde, J. F. (1989). Neuroendrocrine regulation of prolactin release. *Prog. Neurobiol.* 33, 399–447.
657. Arbogast, L. A., and Ben-Jonathan, N. (1990). The preovulatory prolactin surge is prolonged by a progesterone-dependent dopaminergic mechanism. *Endocrinology* 126, 246–252.
658. Pasqualini, C., Guibert, B., and Leviel, V. (1993). Short-term inhibitory effect of estradiol on tyrosine hydroxylase activity in tuberoinfundibular dopaminergic neurons in vitro. *J. Neurochem.* 60, 1707–1713.
659. Blum, M., McEwen, B. S., and Roberts, J. L. (1987). Transcriptional analysis of tyrosine hydroxylase gene expression in the tuberoinfundibular dopaminergic neurons of the rat arcuate nucleus after estrogen treatment. *J. Biol. Chem.* 262, 817–821.
660. Arbogast, L. A., and Voogt, J. L. (1993). Progesterone reverses the estradiol-induced decrease in tyrosine hydroxylase mRNA levels in the arcuate nucleus. *Neuroendocrinology* 58, 501–510.
661. Arbogast, L. A., and Voogt, J. L. (1994). Progesterone suppresses tyrosine hydroxylase messenger ribonucleic acid levels in the arcuate nucleus on proestrus. *Endocrinology* 135, 343–350.
662. Arbogast, L. A., and Voogt, J. L. (2002). Progesterone induces dephosphorylation and inactivation of tyrosine

hydroxylase in rat hypothalamic dopaminergic neurons. *Neuroendocrinology* 75, 273–281.

663. Arbogast, L. A., and Ben-Jonathan, N. (1989). Tyrosine hydroxylase in the stalk-median eminence and posterior pituitary is inactivated only during the plateau phase of the preovulatory prolactin surge. *Endocrinology* 125, 667–674.
664. DeMaria, J. E., Lerant, A. A., and Freeman, M. E. (1999). Prolactin activates all three populations of hypothalamic neuroendocrine dopaminergic neurons in ovariectomized rats. *Brain Res.* 837, 236–241.
665. DeMaria, J. E., Nagy, G. M., and Freeman, M. E. (2000). Immunoneutralization of prolactin prevents stimulatory feedback of prolactin on hypothalamic neuroendocrine dopaminergic neurons. *Endocrine* 12, 333–337.
666. Lerant, A. A., DeMaria, J. E., and Freeman, M. E. (2001). Decreased expression of fos-related antigens (FRAs) in the hypothalamic dopaminergic neurons after immunoneutralization of endogenous prolactin. *Endocrine* 16, 181–187.
667. deGreef, W. J., Klootwijk, W., Karels, B., and Visser, T. J. (1985). Levels of dopamine and thyrotrophin-releasing hormone in hypophysial stalk blood during an estrogen-stimulated surge of prolactin in the ovariectomized rat. *J. Endocrinol.* 105, 107–112.
668. Fink, G., Koch, Y., and Ben Aroya, N. (1982). Release of the thyrotropin releasing hormone into hypophysial portal blood is high relative to other neuropeptides and may be related to prolactin secretion. *Brain Res.* 243, 186–189.
669. Blake, C. A. (1974). Stimulation of pituitary prolactin and TSH release in lactating and proestrous rats. *Endocrinology* 94, 503–508.
670. Horn, A. M., Fraser, H. M., and Fink, G. (1985). Effects of antiserum to thyrotrophin-releasing hormone on the concentrations of plasma prolactin, thyrotrophin and LH in the pro-estrous rat. *J. Endocrinol.* 104, 205–209.
671. Ben-Jonathan, N., Neill, M. A., Arbogast, L. A., Peters, L. L., and Hoefer, M. T. (1980). Dopamine in hypophysial portal blood: relationship to circulating prolactin in pregnant and lactating rats. *Endocrinology* 106, 690–696.
672. deGreef, W. J., and Visser, T. J. (1981). Evidence for the involvement of hypothalamic dopamine and thyrotropin-releasing hormone in suckling-induced release of prolactin. *J. Endocrinol* 91, 213–223.
673. Plotsky, P. M., and Neill, J. D. (1982). Interactions of dopamine and thyrotropin-releasing hormone in the regulation of prolactin release in lactating rats. *Endocrinology* 111, 168–173.
674. Fagin, K. D., and Neill, J. D. (1981). The effect of dopamine on thyrotropin-releasing hormone-induced prolactin secretion in vitro. *Endocrinology* 109, 1835–1840.
675. Bale, T. L., Dorsa, D. M., and Johnston, C. A. (1995). Oxytocin receptor mRNA expression in the ventromedial hypothalamus during the estrous cycle. *J. Neurosci.* 15, 5058–5064.
676. Freeman, M. E., Kanyicska, B., Lerant, A., and Nagy, G. M. (2000). Prolactin: structure, function, and regulation of secretion. *Physiol. Rev.* 80, 1523–1631.
677. Hinuma, S., Habata, Y., Fujii, R., Kawamata, Y., Hosoya, N., Fukusumi, S., Kitada, C., Masuo, Y., Asano, T., Matsumoto, H., Sekiguchi, M., Kurokawa, T., Nishimura, O., Onda, H., and Fujino, M. (1998). A prolactin-releasing peptide in the brain. *Nature* 393, 272–276.
678. Nishimura, O., Moriya, T., Suenaga, M., Tanaka, Y., Itoh, T., Koyama, N., Fujii, R., Hinuma, S., Kitada, C., and Fujino, M. (1998). Synthesis of new peptides with prolactin-releasing activity by a combination of recombinant DNA technology and a cysteine-specific cyanylation reaction. *Chem. Pharm. Bull. (Tokyo)* 46, 1490–1492.
679. Minami, S., Nakata, T., Tokita, R., Onodera, H., and Imaki, J. (1999). Cellular localization of prolactin-releasing peptide messenger RNA in the rat brain. *Neurosci. Lett.* 266, 73–75.
680. Fujii, R., Fukusumi, S., Hosoya, M., Kawamata, Y., Habata, Y., Hinuma, S., Sekiguchi, M., Kitada, C., Kurokawa, T., Nishimura, O., Onda, H., Sumino, Y., and Fujino, M. (1999). Tissue distribution of prolactin-releasing peptide (PrRP) and its receptor. *Regul. Pept.* 83, 1–10.
681. Tokita, R., Nakata, T., Katsumata, H., Konishi, S., Onodera, H., Imaki, J., and Minami, S. (1999). Prolactin secretion in response to prolactin-releasing peptide and the expression of the prolactin-releasing peptide gene in the medulla oblongata are estrogen dependent in rats. *Neurosci. Lett.* 276, 103–106.
682. Matsumoto, H., Murakami, Y., Horikoshi, Y., Noguchi, J., Habata, Y., Kitada, C., Hinuma, S., Onda, H., and Fujino, M. (1999). Distribution and characterization of immunoreactive prolactin-releasing peptide (PrRP) in rat tissue and plasma. *Biochem. Biophys. Res. Commun.* 257, 264–268.
683. Maruyama, M., Matsumoto, H., Fujiwara, K., Kitada, C., Hinuma, S., Onda, H., Fujino, M., and Inoue, K. (1999). Immunocytochemical localization of prolactin-releasing peptide in the rat brain. *Endocrinology* 140, 2326–2333.
684. Samson, W. K., Resch, Z. T., Murphy, T. C., and Chang, J.-K. (1998). Gender-biased activity of the novel prolactin releasing peptides: comparison with thyrotropin releasing hormone reveals only pharmacological effects. *Endocrine* 9, 289–291.
685. Matsumoto, H., Noguchi, J., Horikoshi, Y., Kawamata, Y., Kitada, C., Hinuma, S., Onda, H., Nishimura, O., and Fujino, M. (1999). Stimulation of prolactin release by prolactin-releasing peptide in rats. *Biochem. Biophys. Res. Commun.* 259, 321–324.
686. Iijima, N., Kataoka, Y., Kakihara, K., Bamba, H., Tamada, Y., Hayashi, S., Matsuda, T., Tanaka, M., Honjyo, H., Hosoya, W., Hinuma, S., and Ibata, Y. (1999). Cytochemical study of prolactin-releasing peptide (PrRP) in the rat brain. *Neuroreport* 10, 1713–1716.
687. Maruyama, M., Matsumoto, H., Fujiwara, K., Noguchi, J., Kitada, C., Hinuma, S., Onda, H., Nishimura, O., Fujino, M., Higuchi, T., and Inoue, K. (1999). Central administration of prolactin-releasing peptide stimulates oxytocin release in rats. *Neurosci. Lett.* 276, 193–196.
688. Yamakawa, K., Kudo, K., Kanba, S., and Arita, J. (1999). Distribution of prolactin-releasing peptide-immunoreactive neurons in the rat hypothalamus. *Neurosci. Lett.* 267, 113–116.
689. Hizume, T., Watanobe, H., Yoneda, M., Suda, T., and Schiöth, H. B. (2000). Involvement of prolactin-releasing peptide in the preovulatory luteinizing hormone and prolactin surges in the rat. *Biochem. Biophys. Res. Commun.* 279, 35–39.
690. Watanobe, H. (2001). In vivo release of prolactin-releasing peptide in rat hypothalamus in association with luteinizing hormone and prolactin surges. *Neuroendocrinology* 74, 359–366.
691. Yuan, Z. F., and Pan, J. T. (2002). Involvement of angiotensin II, TRH and prolactin-releasing peptide in the estrogen-induced afternoon prolactin surge in female rats: studies using antisense technology. *Life Sci.* 71, 899–910.
692. Anderson, S. T., Kokay, I. C., Lang, T., Grattan, D. R., and Curlewis, J. D. (2003). Quantification of prolactin-releasing peptide (PrRP) mRNA expression in specific brain regions of the rat during the oestrous cycle and in lactation. *Brain Res.* 973, 64–73.
693. Roland, B. L., Sutton, S. W., Wilson, S. J., Luo, L., Pyati, J., Huvar, R., Erlander, M. G., and Lovenberg, T. W. (1999). Anatomical distribution of prolactin-releasing peptide

and its receptor suggests additional functions in the central nervous system and periphery. *Endocrinology* 140, 5736–5745.
694. Samson, W. K., Resch, Z. T., and Murphy, T. C. (2000). A novel action of the newly described prolactin-releasing peptides: cardiovascular regulation. *Brain Res.* 858, 19–25.
695. Satoh, F., Smith, D. M., Gardiner, J. V., Mahmoodi, M., Murphy, K. G., Ghatei, M. A., and Bloom, S. R. (2000). Characterization and distribution of prolactin releasing peptide (PrRP) binding sites in the rat—evidence for a novel binding site subtype in cardiac and skeletal muscle. *Brit. J. Pharmacol.* 129, 1787–1793.
696. Horiuchi, J., Saigusa, T., Sugiyama, N., Kanba, S., Nishida, Y., Sato, Y., Hinuma, S., and Arita, J. (2002). Effects of prolactin-releasing peptide microinjection into the ventrolateral medulla on arterial pressure and sympathetic activity in rats. *Brain Res.* 958, 201–209.
697. Watanobe, H., Schiöth, H. B., Wikberg, J. E. S., and Suda, T. (2000). Evaluation of the role for prolactin-releasing peptide in prolactin secretion induced by ether stress and suckling in the rat: comparison with vasoactive intestinal peptide. *Brain Res.* 865, 91–96.
698. Lin, S. H. S., Leslie, F. M., and Civelli, O. (2002). Neurochemical properties of the prolactin releasing peptide (PrRP) receptor expressing neurons: evidence for a role of PrRP as a regulator of stress and nociception. *Brain Res.* 952, 15–30.
699. Samson, W. K., Keown, C., Samson, C. K., Samson, H. W., Lane, B., Baker, J. R., and Taylor, M. M. (2003). Prolactin-releasing peptide and its homolog RFRP-1 act in hypothalamus but not in anterior pituitary gland to stimulate stress hormone secretion. *Endocrine* 20, 59–66.
700. Morales, T., and Sawchenko, P. E. (2003). Brainstem prolactin-releasing peptide neurons are sensitive to stress and lactation. *Neuroscience* 121, 771–778.
701. Lawrence, C. B., Celsi, F., Brennand, J., and Luckman, S. M. (2000). Alternative role for prolactin-releasing peptide in the regulation of food intake. *Nat. Neurosci.* 3, 645–646.
702. Vergoni, A. V., Watanobe, H., Guidetti, G., Savino, G., Bertolini, A., and Schiöth, H. B. (2002). Effect of repeated administration of prolactin releasing peptide on feeding behavior in rats. *Brain Res.* 955, 207–213.
703. Ellacott, K. L. J., Lawrence, C. B., Pritchard, L. E., and Luckman, S. M. (2003). Repeated administration of the anorectic factor prolactin-releasing peptide leads to tolerance to its effects on energy homeostasis. *Am. J. Physiol. Regul. Integr. Comp. Physiol.* 285, R1005–R1010.
704. Zhang, S. Q., Kimura, M., and Inoue, S. (2000). Effects of prolactin-releasing peptide (PrRP) on sleep regulation in rats. *Psychiatry Clin. Neurosci.* 54, 262–264.
705. Zhang, S. Q., Inoué, S., and Kimura, M. (2001). Sleep-promoting activity of prolactin-releasing peptide (PrRP) in the rat. *Neuroreport* 12, 3173–3176.
706. Lin, S. H. S., Arai, A. C., España, R. A., Berridge, C. W., Leslie, F. M., Huguenard, J. R., Vergnes, M., and Civelli, O. (2002). Prolactin-releasing peptide (PrRP) promotes awakening and suppresses absence seizures. *Neuroscience* 114, 229–238.
707. Malven, P. V., and Sawyer, C. H. (1966). A luteolytic action of prolactin in hypophysectomized rats. *Endocrinology* 79, 268–274.
708. Wuttke, W., and Meites, J. (1971). Luteolytic role of prolactin during the estrous cycle of the rat. *Proc. Soc. Exp. Biol. Med.* 137, 988–991.
709. Witcher, J. A., and Freeman, M. E. (1985). The proestrous surge of prolactin enhances sexual receptivity in the rat. *Biol. Reprod.* 32, 834–839.
710. Duncan, J., Herbon, L., and Marshall, J. C. (1985). Estradiol and prolactin modulate GnRH regulation of pituitary GnRH receptors and gonadotropin secretion in female rats. *Endocrinology* 116, 272–272.
711. Wang, C., Hsueh, A. J. W., and Erickson, G. F. (1981). Direct inhibition of estrogen production in rat granulosa cells by prolactin. In *Dynamics of Ovarian Function* (N. B. Schwartz, and M. Hunzicker-Dunn, Eds.), pp. 61–66. Raven Press, New York.
712. Smith, M. S., McLean, B., and Neill, J. D. (1976). Prolactin: the initial luteotropic stimulus of pseudopregnancy in the rat. *Endocrinology* 98, 1370–1377.
713. Everett, J. W. (1961). The mammalian female reproductive cycle and its controlling mechanisms. In *Sex and Internal Secretions* (W. C. Young, Ed.), pp. 497–555. Williams & Wilkins, Baltimore.
714. Freeman, M. E., and Neill, J. D. (1972). The pattern of prolactin secretion during pseudopregnancy in the rat: a daily nocturnal surge. *Endocrinology* 90, 1292–1294.
715. Freeman, M. E., Smith, M. S., Nazian, S. J., and Neill, J. D. (1974). Ovarian and hypothalamic control of the daily surges of prolactin secretion during pseudopregnancy in the rat. *Endocrinology* 94, 875–882.
716. Butcher, R. L., Fugo, N. W., and Collins, W. C. (1972). Semicircadian rhythm in plasma levels of prolactin during early gestation in the rat. *Endocrinology* 90, 1125–1127.
717. Gorospe, W. C., and Freeman, M. E. (1981). An ovarian role in prolonging and terminating the two surges of prolactin in pseudopregnant rats. *Endocrinology* 108, 1293–1298.
718. Smith, M. S., and Neill, J. D. (1976). Termination at midpregnancy of the two daily surges of plasma prolactin initiated by mating in the rat. *Endocrinology* 98, 696–706.
719. Smith, M. S., and Neill, J. D. (1976). A critical period for cervically-stimulated prolactin release. *Endocrinology* 98, 324–328.
720. Freeman, M. E., and Sterman, J. R. (1978). Ovarian steroid modulation of prolactin surges in cervically stimulated ovariectomized rats. *Endocrinology* 102, 1915–1920.
721. deGreef, W. J., and Zeilmaker, G. H. (1978). Regulation of prolactin secretion during the luteal phase of the rat. *Endocrinology* 102, 1190–1198.
722. Welschen, R., Osman, P., Dullart, J., de Greef, W. J., Uilenbroek, J., and de Jong, F. H. (1975). Levels of follicle-stimulating hormone, estradiol-17b and progesterone, and follicular growth in the pseudopregnant rat. *J. Endocrinol.* 64, 37–47.
723. Freeman, M. E. (1979). A direct effect of the uterus on the surges of prolactin induced by cervical stimulation in the rat. *Endocrinology* 105, 387–390.
724. Gorospe, W. C., Mick, C., and Freeman, M. E. (1981). The obligatory role of the uterus for termination of prolactin surges initiated by cervical stimulation. *Endocrinology* 109, 1–4.
725. Gorospe, W. C., and Freeman, M. E. (1984). Uterus of the rat contains prolactin inhibitory activity. *Am. J. Physiol.* 247, E251–E257.
726. Gorospe, W. C., and Freeman, M. E. (1985). Detection of prolactin inhibitory activity in uterine epithelial cell secretions and rat serum. *Endocrinology* 116, 1559–1564.
727. Reiner, P., Woolsey, J., Adler, N., and Morrison, A. (1981). A gross anatomical study of the peripheral nerves associated with reproductive function in the female albino rat. In *Neuroendocrinology of Reproduction* (N. T. Adler, Ed.), pp. 545–549. Plenum Press, New York.
728. Allen, T. O., Adler, N. T., Greenberg, J. H., and Reivich, M. (1981). Vaginocervical stimulation selectively increases metabolic activity in the rat brain. *Science* 211, 1068–1072.

729. Hornby, J. B., and Rose, J. D. (1976). Responses of caudal brain stem neurons to vaginal and somatosensory stimulation in the rat and evidence of genital-nociceptive interactions. *Exp. Neurol.* 51, 363–376.
730. Kawakami, M., and Kubo, K. (1971). Neuro-correlate of limbic-hypothalamo-pituitary-gonadal axis in the rat: change in limbic-hypothalamic unit activity induced by vaginal and electrical stimulation. *Neuroendocrinology* 7, 65–89.
731. Kawakami, M., and Arita, J. (1982). Midbrain pathways for the initiation and maintenance of the nocturnal prolactin surge in pseudopregnant rats. *Endocrinology* 110, 1977–1982.
732. Gunnett, J. W., and Freeman, M. E. (1984). Hypothalamic regulation of mating induced prolactin release: effect of electrical stimulation of the medial preoptic area in conscious female rats. *Neuroendocrinology* 38, 12–16.
733. Hess, D. L., Wilkins, R. H., Moossy, V., Chang, J. L., Plant, T. M., McCormack, J. T., Nakai, Y., and Knobil, E. (1977). Estrogen-induced gonadotropin surges in decerebrated female rhesus monkeys with medial basal hypothalamic peninsulae. *Endocrinology* 101, 1264–1271.
734. Gunnett, J. W., Mick, C., and Freeman, M. E. (1981). The role of the dorsomedial-ventromedial area of the hypothalamus in the control of prolactin secretion induced by cervical stimulation. *Endocrinology* 109, 1846–1850.
735. Chateau, D., Plas-Roser, S., and Aron, C. L. (1981). Follicular growth, ovulatory phenomena and ventromedial nucleus lesions in the rat. *Endokrinologie* 77, 257–268.
736. Yokoyama, A., and Ota, K. (1959). Effect of oxytocin replacement on lactation in rats bearing hypothalamic lesions. *Endocrinol. Jpn.* 6, 268–276.
737. Gunnett, J. W., and Freeman, M. E. (1985). The interaction of the medial preoptic area and the dorsomedial-ventromedial nuclei of the hypothalamus in the regulation of the mating-induced release of prolactin. *Neuroendocrinology* 40, 232–237.
738. Bethea, C. L., and Neill, J. D. (1980). Lesions of the suprachiasmatic nuclei abolish the cervically stimulated prolactin surges in the rat. *Endocrinology* 107, 1–5.
739. Gunnett, J. W., and Freeman, M. E. (1983). The mating-induced release of prolactin: a unique neuroendocrine response. *Endocr. Rev.* 4, 44–61.
740. deGreef, W. J., and Neill, J. D. (1979). Dopamine levels in hypophysial stalk plasma of the rat during surges of prolactin secretion induced by cervical stimulation. *Endocrinology* 105, 1093–1099.
741. Demarest, K. T., Moore, K. E., and Riegle, G. D. (1983). Role of prolactin feedback in the semicircadian rhythm of tuberoinfundibular dopaminergic neuronal activity during early pregnancy in the rat. *Neuroendocrinology* 36, 371–375.
742. Demarest, K. T., Riegle, G. D., and Moore, K. E. (1984). Prolactin-induced activation of tuberoinfundibular dopaminergic neurons:evidence for both a rapid 'tonic' and a delayed 'induction' component. *Neuroendocrinology* 38, 467–475.
743. Arey, B. J., Averill, R. L., and Freeman, M. E. (1989). A sex-specific endogenous stimulatory rhythm regulating prolactin secretion. *Endocrinology* 124, 119–123.
744. Arey, B. J., and Freeman, M. E. (1989). Hypothalamic factors involved in the endogenous stimulatory rhythm regulating prolactin secretion. *Endocrinology* 124, 878–883.
745. Arey, B. J., and Freeman, M. E. (1990). Oxytocin, vasoactive intestinal peptide and serotonin regulate the mating-induced surges of prolactin secretion in the rat. *Endocrinology* 126, 279–284.
746. Arey, B. J., and Freeman, M. E. (1992). Activity of oxytocinergic neurons in the paraventricular nucleus mirrors the periodicity of the endogenous stimulatory rhythm regulating prolactin secretion. *Endocrinology* 130, 126–132.
747. Arey, B. J., and Freeman, M. E. (1992). Activity of vasoactive intestinal peptide and serotonin in the paraventricular nucleus reflects the periodicity of the endogenous stimulatory rhythm regulating prolactin secretion. *Endocrinology* 131, 736–742.
748. Arey, B. J., Kanyicska, B., and Freeman, M. E. (1991). The endogenous stimulatory rhythm regulating prolactin secretion is present in the lactating rat. *Neuroendocrinology* 53, 35–40.
749. Egli, M., Bertram, R., Sellix, M. T., and Freeman, M. E. (2004). Rhythmic secretion of prolactin in rats: action of oxytocin coordinated by vasoactive intestinal polypeptide of suprachiasmatic nucleus origin. *Endocrinology* 145, 3386–3394.
750. Lerant, A., Herman, M. E., and Freeman, M. E. (1996). Dopaminergic neurons of periventricular and arcuate nuclei of pseudopregnant rats: semicircadian rhythm in fos-related antigens immunoreactivities and in dopamine concentration. *Endocrinology* 137, 3621–3628.

Knobil and Neill's Physiology of Reproduction, Third Edition
edited by Jimmy D. Neill,
Elsevier © 2006

CHAPTER 44

Neuroendocrine Control of the Ovarian Cycle of the Sheep

Robert L. Goodman[1] and E. Keith Inskeep[2]

Introduction, 2389
Method of Analysis, 2390
Description of the Ovine Estrous Cycle, 2390
Estrous Behavior, 2390
Hypothalamo–Hypophyseal Unit, 2391
Ovary, 2391
Uterine Prostaglandins, 2396
Analysis of the Estrous Cycle, 2397
Control of Estrous Behavior, 2397
Control of Hypothalamo–Hypophyseal Function, 2398
Control of Ovarian Function, 2408
Regulation of Secretion of $PGF_{2}\alpha$ by the Uterus and Transport to the Ovary, 2416
Resynthesis, a Model for the Ovine Estrous Cycle, 2418
Environmental Control of Ovarian Cyclicity in the Ewe, 2421
Photoperiodic Control of Seasonal Breeding, 2421
Effects of Stressors, 2423
Effects of Nutrition, 2425
Effects of Pheromones and Social Interactions, 2426
Conclusion, 2428
Acknowledgments, 2428
References, 2428

INTRODUCTION

Considerable progress has been made over the last decade in describing the mechanistic basis of the events responsible for the ovine estrous cycle. Because of this wealth of new information, we have modified the overall approach of this review to focus on these new developments and to identify gaps in our current knowledge. We also have added new sections on follicular dynamics and ovulation that were not considered in the previous version. These changes in approach, and space constraints, required that we be more selective in citations than in previous versions of this chapter, but we have endeavored to maintain a balance between current work and important earlier observations.

Patterns of reproductive function in the ewe are dominated by two distinct rhythms (1). The first of these is a 16- to 17-day estrous cycle. The duration of this cycle is remarkably constant, with an estimated 95% of ewes showing estrous cycles within the normal range of 14 to 19 days (2). There are some differences in cycle lengths among breeds and with age (2,3), but these differences are relatively small (= 1 day). The other major rhythm in ovine reproduction is an annual cycle of ovarian activity. In most breeds of sheep (1–3), normal estrous cycles occur in the fall and winter (breeding season), but ovarian cycles cease in the spring and summer (anestrus). From an adaptive point of view, this limited duration of fertility ensures that lambs are born in the spring, when environmental conditions are favorable for their survival. In contrast to the constancy of estrous cycle length, seasonal reproductive patterns vary dramatically; some primitive mountain sheep have only one estrous cycle during the breeding season, while Merino ewes

[1]Department of Physiology;
[2]Division of Animal and Veterinary Sciences, West Virginia University, Morgantown, West Virginia.

show estrous cycles throughout most of the year (1). This variability appears to reflect the environment in which the breeds were developed, with breeds originating in harsher conditions having shorter breeding seasons (3). It should be kept in mind that these breed differences in seasonality reflect differences in the degree of ovarian activity during spring and summer. The presence or absence of ovulation represents a somewhat arbitrary dividing line in what is, in fact, a continuum of ovarian function. Thus, seasonal variations in fertility are evident even in ewes capable of breeding throughout the year and, in seasonal breeders, considerable ovarian activity (e.g., follicular development) occurs during anestrus.

The estrous cycle of the ewe is essentially analogous to the ovarian cycles of other spontaneously ovulating species. It is thus not surprising that the neuroendocrine mechanisms controlling the ovine ovarian cycle are similar in some respects to those in primates (Chapter 45) and rats (Chapter 43). However, the seasonal aspect of reproductive function in the ewe represents an additional challenge for any model of these control systems. Namely, it must be able to account for both the events occurring within the estrous cycle and for the annual cessation and re-initiation of ovarian cycles. In addition to these theoretical issues, the occurrence of a seasonal anovulatory period provides a natural experimental model for examining the control of ovulation. Consequently, experiments examining the control of the estrous cycle and the annual reproductive cycle of sheep became intertwined over the years so that our understanding of each has developed in parallel. Nevertheless, it is useful for didactic reasons to consider these two cycles separately and we will first consider the control of the estrous cycle. Moreover, considerable recent work has focused on the effects of environmental stimuli (i.e., stressors), which can acutely influence ovarian function within a season. Therefore, we have expanded the final portion of this chapter to consider both seasonal breeding and the effects of other environmental stimuli on ovine reproduction.

METHOD OF ANALYSIS

The ovine estrous cycle results from the coordinated interaction of four tissues: brain, pituitary, ovary, and uterus. To understand the control of this cycle, we must analyze the communication among these tissues, which occurs largely via seven hormones: gonadotropin-releasing hormone (GnRH) from the hypothalamus, luteinizing hormone (LH), and follicle-stimulating hormone (FSH) from the pituitary, estradiol, inhibin and progesterone from the ovary, and prostaglandin $F_2\alpha$ ($PGF_2\alpha$) from the uterus and corpora lutea. In analyzing the physiological interactions of these hormones, two factors are of paramount importance. First, for the action of a particular hormone to be important in controlling ovarian cyclicity, it must be evident at physiological concentrations of the hormone. Thus, particular emphasis will be placed on those studies that have avoided pharmacological hormone treatments or artificial manipulations with little relevance to physiological events. Second, changes in circulating concentrations of these hormones occur in a specific sequence at very predictable times during the estrous cycle. As a result, examination of the temporal relationship between two specific hormones provides one useful indication of the causal linkage between the two.

Because of the importance of these two factors, this chapter will begin with a detailed description of the behavioral and hormonal patterns that occur throughout the estrous cycle. We then dissect the cycle into its functional components and consider the control of each component individually. The third step will be synthesis of these components into a model that can account for the events of the estrous cycle. Finally, we will briefly consider how these mechanisms are modified by a number of environmental stimuli to produce alterations in reproductive function.

DESCRIPTION OF THE OVINE ESTROUS CYCLE

Estrous Behavior

When approached by a ram, a ewe in estrus will stand still with her head lowered. She often wags her tail and looks over her shoulder as the ram nudges her flank and/or anogenital region. At the peak of estrus, the ewe may actively seek a ram and initiate the sexual episode by nudging his forequarters (4,5). Courtship behavior is usually followed by mounting and copulation with several copulations occurring during a single estrous period. In general, behavioral estrus lasts 24 to 36 hours in the ewe, although the duration of estrus is approximately 50% longer in breeds with high ovulation rates (6,7). The duration of estrous behavior is also influenced by pheromones from the ram; if ewes are continuously penned with a ram, estrus ends much earlier than when the two sexes are in contact intermittently (5,8). Despite this variability in estrous duration, the onset of estrus is coupled with the pre-ovulatory LH surge so that ovulation occurs on the average 24 to 30 hours after the onset of estrous behavior (2,9). Because of this correlation, the onset of estrus is usually used as a marker to time the events of the ovine estrous cycle and, by convention, the first day of estrus (and the LH surge) is designated day 0 of the estrous cycle.

Hypothalamo–Hypophyseal Unit

LH and GnRH Patterns

The development of a technique for monitoring GnRH secretion in the hypophyseal portal blood of unanesthetized sheep (10,11), has produced considerable information on the pattern of endogenous GnRH secretion during the ovine estrous cycle that is not available in any other species. These data have demonstrated a very close correlation between GnRH and LH secretion. Therefore, we have chosen to describe secretion of these two hormones in tandem, rather than sequentially. Historically, two functionally distinct modes of LH secretion were identified in the ewe (12), as in the rat and primate (see Chapters 43 and 45), each controlling different aspects of ovarian function. Tonic LH secretion occurs in a pulsatile pattern at a low level throughout the cycle and is important for ovarian steroidogenesis; the LH surge occurs around estrus and induces ovulation and formation of the corpus luteum. Tonic and surge secretion are controlled by different feedback mechanisms and, after the identification of GnRH, the concept of two different modes of secretion under separate control was applied to this neurohormone.

Tonic LH secretion during the cycle (Fig. 1) is inversely correlated to luteal function (13) so that LH concentrations decline from days 1 to 9 as progesterone secretion from the corpus luteum increases, remains low until day 14, and then increases during the follicular phase following luteolysis (13–16). In contrast, when averaged over several hours, tonic GnRH secretion does not vary markedly throughout the cycle; in fact, there is no significant difference in mean GnRH secretion rates between luteal and follicular phases (Fig. 1) (17,18). However, this invariant average GnRH secretion rate masks a marked variation in the pattern of pulsatile GnRH secretion (Fig. 2). During the mid-luteal phase low frequency, high amplitude GnRH pulses are observed; following luteolysis the frequency increases and the amplitude decreases (17,18). Because GnRH pulses drive LH pulses, LH pulse frequency follows the same pattern (Fig. 2): it is high in the early part of the cycle, decreases to a nadir late in the luteal phase and then increases progressively throughout the follicular phase (15,19,20). LH pulse amplitude either does not differ between mid-luteal and follicular phases (15,20), or decreases in parallel with GnRH pulse amplitude during the follicular phase (19). Thus, mean LH concentrations closely correlate with LH pulse frequencies, but not amplitudes, throughout most of the estrous cycle.

The LH surge represents a brief, massive outpouring of LH from the pituitary (Fig. 1). LH concentrations rise rapidly within 4 to 8 hours to a peak of 100 to 200 ng/ml (50–100 times basal levels) and then decline just as rapidly so that the surge lasts approximately 12 hours. The onset of the LH surge is usually closely coupled to the onset of estrous behavior (7,21) and the interval between the LH surge and ovulation (22–26 h) is remarkably constant (22). At the same time, there is a dramatic 20- to 40-fold increase in GnRH secretion rate (18) (Fig. 1), which reflects a change in the pattern of GnRH release from episodic to continuous (23). This change appears to occur gradually (24) starting with increased GnRH secretion between pulses, followed by an increase in both interpulse GnRH secretion and GnRH pulse amplitude, and then continuous release during peak secretion (23,24). Although the initial phases of the pre-ovulatory surges of GnRH and LH are relatively synchronous, elevated GnRH secretion is maintained for many hours (18) after the termination of the LH surge (Fig. 1).

FSH Secretion

Ovine FSH has been much more difficult to measure than ovine LH, but there is now considerable agreement (15,25–29) on the general features of FSH release throughout the estrous cycle (Fig. 1). During the follicular phase, FSH concentrations decline to a nadir on the day before the pre-ovulatory LH surge (15,25–29). There is an FSH surge coincident with the LH surge and FSH concentrations fall in parallel with LH to a nadir at 8 to 12 hours after the pre-ovulatory peak. FSH then increases to a second peak 20 to 28 hours after the first followed by a second nadir between 36 to 48 hours. A gradual increase in FSH concentrations is often observed sometime between days 4 and 6 of the luteal phase (26,28–30), that correlates with the second "wave" of follicular development (see below). Small increments in FSH, lasting about a day, occur at 3- to 6-day intervals throughout the luteal phase (30–33), each preceding a wave of follicular development (30,34).

Ovary

Follicular Dynamics

Formation of primordial follicles begins in the ovigerous cords near the interface of the cortex and

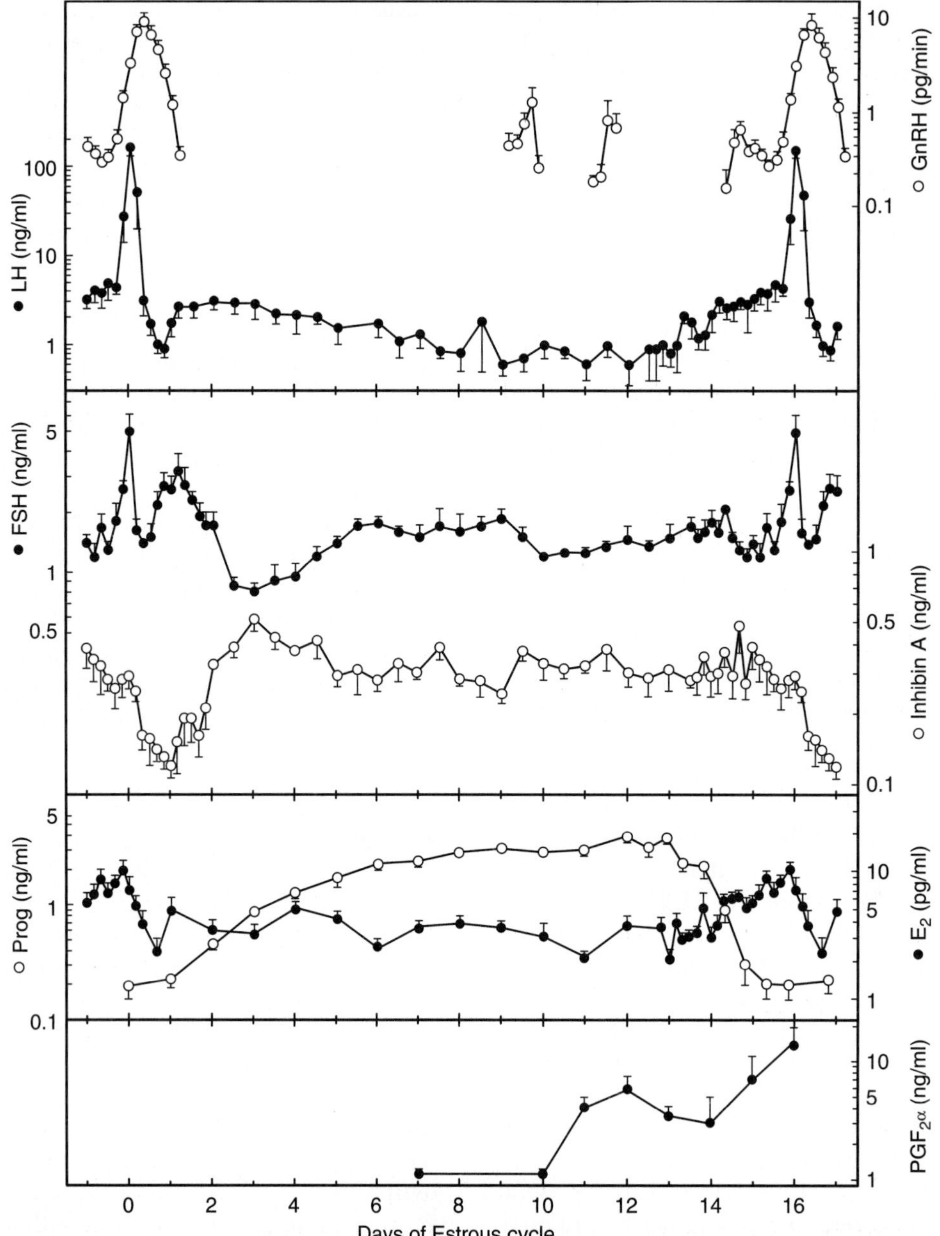

FIG. 1. Changes in mean (± SEM) secretion rate of GnRH into hypophyseal portal blood, peripheral concentrations of gonadotropins and major ovarian hormones, and uterine vein concentrations of $PGF_2\alpha$ throughout the ovine estrous cycle. Data for all hormones (except $PGF_2\alpha$) during the periovulatory period (days −1 to +2 and 13–17) are normalized to the peak of the LH surge (day 0 and day 16) and represent values monitored every 4 hours. During the remainder of the cycle, data collected every 12 or 24 hours are presented. [GnRH values were calculated from data in (18), LH and ovarian steroid concentrations are from (27). (Copyright 1981, The Endocrine Society.) FSH and inhibin A from (29), and $PGF_2\alpha$ values are derived from data in (103).]

medulla of the fetal ovary around day 75 of gestation (35), a time when the maximum number of germ cells is found within the fetal ovary (36). Initially, each primordial follicle consists of an oocyte, surrounded by a single layer of squamous granulosal cells, although in some cases, one or more of these cells is cuboidal. The number of granulosal cells per primordial follicle is highly variable, but correlates with the diameter of the oocyte (37). Sawyer et al. (35) concluded that the majority of pregranulosal cells arose from the surface epithelium of the developing ovary, because the ovigerous cords were open to the surface epithelium, but not to the ovarian stroma. The pregranulosal cells segregate the oocytes within

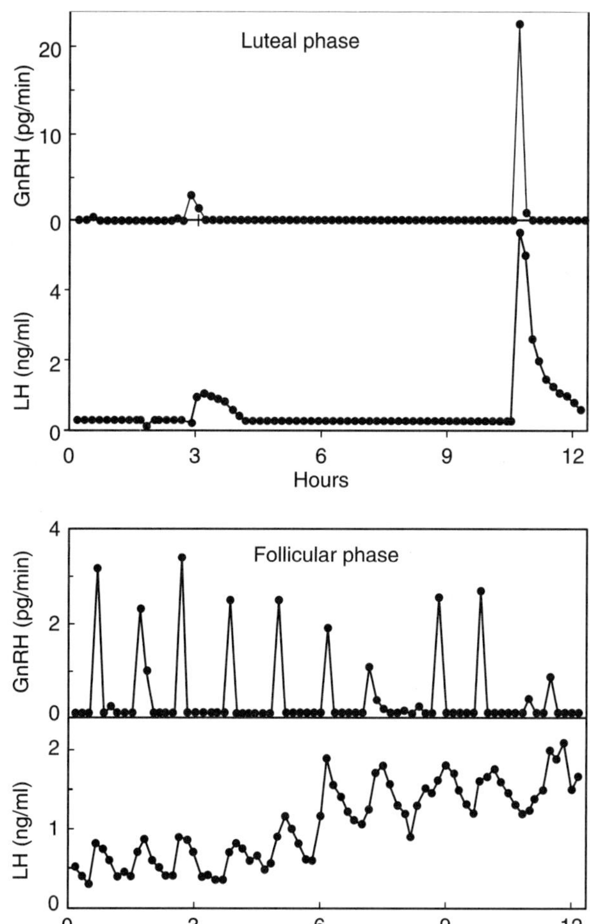

FIG. 2. Pulsatile patterns of hypophyseal portal GnRH secretion and jugular concentrations of LH during the late luteal phase (days 12 to 13) and late follicular phase (39 to 51 h after progesterone withdrawal). Note difference in the scale of the y-axis between luteal and follicular phases. [Drawn from data provided by F. J. Karsch work described in (18).]

the ovigerous cords before these cords regress, and develop into the granulosal cells as follicles form. More than 75% of the germ cells are lost by day 90 of gestation (36), presumably by apoptosis (38). The ovaries of young ewes contain from 40,000 to 300,000 primordial follicles (39,40). Numbers of antral follicles in ewe lambs are high at birth and decrease and increase at various times during the prepubertal period (41,42), with no clear pattern of follicular waves (42). In 10-day observation periods, there were an average of 2.4 peaks of FSH concentrations, but an average of only 0.8 follicles grew to as large as 5 mm.

In the adult ewe, early workers, making observations at slaughter or laparotomy, showed greater follicular activity in association with a functional corpus luteum [reviewed by Dailey et al. (43)]. Numbers of total follicles and small follicles varied with day of the estrous cycle, with two periods of enhanced development, interpreted as apparent waves (43,44). In contrast, Smeaton and Robertson (45) found three to four apparent waves of follicular activity during an estrous cycle. Importantly, the ovulatory follicle could not be identified on the basis of the largest diameter until 2 days before estrus (46). More recently, repeated ultrasonography (47,48) and daily laparoscopy for 18 days (49) have been used to assess follicular dynamics. The latter workers found an organized pattern of three follicular waves in Suffolk ewes (49), whereas the former did not find clear wave patterns in ewes of other mixed breeding (47,48). Evans (50) summarized the findings of 15 laparoscopic and ultrasonographic studies and concluded that waves can be demonstrated if one defines a wave on the basis of follicles that arise at 3 mm in diameter and grow to at least 5 mm in diameter (34). However, the number of waves detected ranged from two to six in the various studies, waves overlapped in some cases, and ovulatory follicles arose from both the penultimate and ultimate waves (51) and as early as day 9 or as late as day 14 of the cycle (52). Detailed hormonal analysis (30,34,51,53–56) indicated that each of the follicular waves is preceded by a transitory increase in FSH concentrations so that FSH was inversely related to follicular diameter (Fig. 3). Bartlewski et al. (54) and Souza et al. (57) agree that follicular waves continue during anestrus, however, Souza et al. did not find a direct association of the waves with FSH in anestrous Finn-Merino ewes.

Landau et al. (58) found that ovulatory follicles arose over a greater range of days in ewes with twin than in those with single ovulations, and that twin ovulations appeared to result from reduced atresia of both smaller and larger follicles. That is consistent with ovulations from both the ultimate and penultimate waves, i.e., larger follicles from the penultimate wave were retained and more small ones in the ultimate wave grew to ovulatory size and competence. The most rapid apparent turnover of follicles was reported in the Merino del Pais, a monovulatory breed in Spain, in which seven different follicles became the largest during an estrous cycle (59). The fact that ovulatory follicles arose at different times and from different apparent waves of development (*vide supra*) is consonant with the conclusion by Driancourt et al. (60) that active follicular "dominance is probably not operative" in the ewe. Dominance has been difficult to demonstrate, except during the follicular phase (48,50,59). Using detailed high resolution ultrasonography, Duggavathi et al. (61) found no significant changes in numbers of small or medium follicles around the days of wave emergence; thus they concluded that there is little evidence for selection of recruited follicles and follicular dominance in the ewe. Gonzalez-Bulnes et al. (62)

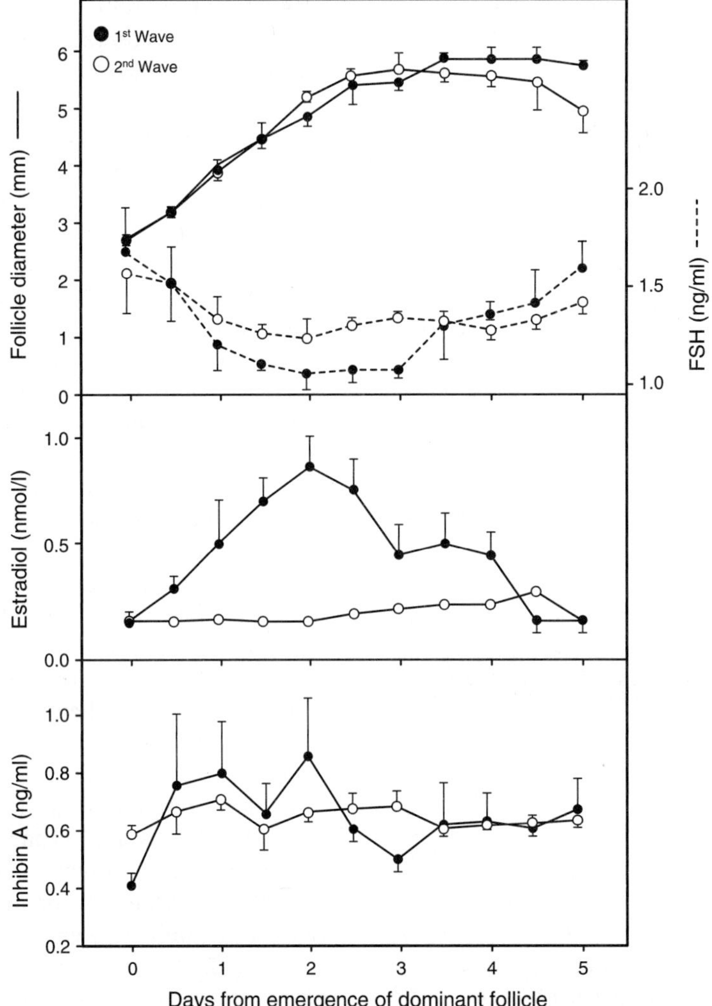

FIG. 3. Relationships among diameter of the largest follicle (**top panel,** *solid lines*), peripheral FSH concentrations (**top panel,** *dashed lines*) and concentrations of estradiol (**middle panel**) and inhibin A (**lower panel**) in ovarian venous plasma during the first and second follicular waves of the estrous cycle. Values (mean ±SEM) are normalized to the day of first emergence (diameter >2.5 mm) of the largest follicle, which occurred on days 1 to 2 and 6 to 7 for waves 1 and 2. [Data replotted from Souza et al. (30); reproduced by permission of the Society of Reproduction.]

reached a different conclusion; they followed follicular development by daily transrectal ultrasonography in Scottish Blackface ewes treated with exogenous LH pulses every 90 minutes over 14 days. Large follicles reached a mean maximum diameter of 7.2 ± 0.5 mm on day 5.5 ± 2.6 after first detection. In the presence of a persistent dominant follicle, no other follicle grew larger than 4 mm in diameter and numbers of new growing follicles and smaller follicles decreased. Concentrations of FSH decreased as dominant follicles grew, in relation to the decline in the number of remaining follicles. The decrease in the number of growing follicles was greatest in the ovary ipsilateral to the dominant follicle. Thus, they concluded that a large follicle depressed the recruitment and growth of other follicles by both systemic and local inhibitors, but it must be recognized that the frequent LH pulses made conditions in their study comparable to the follicular phase.

Ovulation (Follicular Rupture)

Knowledge of ovulation in mammals has been increased greatly by the extensive studies stimulated by reviews of the subject by Lawrence Espey in 1978 and 1980 and JoAnne Richards in 1980 (63–65). The process in the ewe does not differ particularly from that described in other species (66,67), but recent work in ewes has contributed to greater understanding of the role of the ovarian surface epithelium (68) and of the involvement of tumor necrosis factor-α (TNF-α) (69,70) in rupture of the follicle.

The ovulatory cascade of physiological changes in follicles includes changes in steroidogenesis (71), secretion of prostaglandins (72), the microcirculation and histology (73,74), as well as changes in numerous cytokines and enzymes (75,76). An initial effect of the LH surge is to increase 3′5′ cyclic adenosine monophosphate (c-AMP) by 4 through 12 hours (77), which blocks aromatase and thus the synthesis of estradiol-17β from progesterone via testosterone. The resulting increase in progesterone in the follicular tissue increases secretion of prostaglandins. In the ovulatory follicle wall, concentrations of PGE_2 and $PGF_2\alpha$ increased at 8 hours, remained elevated through 12 hours, and PGE_2, but not $PGF_2\alpha$, then decreased by 16 to 20 hours (72,77,78). Intrafollicular treatments with LH or FSH increased PGE_2 in granulosal cells, and both PGE_2 and $PGF_2\alpha$ in the theca (79). Conversion of PGE_2 to $PGF_2\alpha$ in the theca, by the enzyme 9-keto-PGE_2-reductase is dependent on follicular progesterone secretion (77,78,80–82). Isoxazol, an inhibitor of progesterone synthesis, blocked ovulation in sheep, and it was restored by progesterone or $PGF_2\alpha$, but not by PGE_2 (77,78,80,83).

Based on studies cited in the previous paragraph, increased secretion of prostaglandins has been considered obligatory for follicular rupture. However, Murdoch and Lund (70) showed that the dosage of indomethacin required to prevent rupture of pre-ovulatory follicles was much greater than the amount required to suppress secretion of $PGF_2\alpha$. The higher dosage blocked apoptosis in both ovarian surface epithelium and granulosal cells in the apex of the pre-ovulatory follicle. These workers had found earlier that TNF-α was required for collagenolysis in the pre-ovulatory follicle (84) and ovulation (69) and was released from apical, but not basal, thecal endothelial cells as ovulation approached. In the later study (70), indomethacin did not prevent the release of TNF-α from thecal endothelial cells in the apex of the follicle, but prevented its action, an effect that was confirmed by showing that indomethacin inhibited TNF-α–induced apoptosis in ovarian surface epithelial cells in vitro.

Murdoch and McDonnel (68) proposed the following sequence of events in the ovulatory follicle in response to the LH surge. LH stimulates the secretion of urokinase-type plasminogen activator (uPA) from the ovarian surface epithelium. The uPA stimulates conversion of interstitial plasminogen to plasmin, which activates latent collagenases in the apical follicular wall and releases TNF-α from the thecal endothelium. Collagenolysis, potentiated by increases in transcription of matrix metalloproteinases in response to TNF-α, weakens and thins the apical follicular wall, leading to stigma formation and rupture. Because progesterone and $PGF_2\alpha$ can each overcome the inhibitory effect of isoxazol, it may be logical to conclude that stimulation of synthesis of $PGF_2\alpha$ by progesterone and the action of $PGF_2\alpha$ provide an alternative or redundant mechanism for stimulating collagenolysis and apoptosis, or for assuring responsiveness of the latter processes to TNF-α.

Ovarian Inhibin Secretion

Studies evaluating mRNA (85,86) and correlating inhibin secretion rate with the presence of follicles or corpora lutea (85,87) demonstrated that this hormone is synthesized only by granulosal cells of antral follicles in the ewe. Although mRNA for both β_A and β_B can be found in ovine follicles, inhibin-A appears to be the primary form of inhibin secreted by the ovary (29); in fact, inhibin-B could not be detected by ELISA in ovine plasma or follicular fluid (87a). Large follicles are an important source of inhibin, but smaller follicles also appear to contribute significantly to ovarian inhibin secretion (85,88). Because the latter are not an important source of estradiol (89), the ovarian secretion of these two follicular hormones is not always coupled (28,29).

Many of the early reports describing inhibin secretion in ewes relied on immunoassays that detected primarily the alpha subunit of inhibin to avoid possible cross-reaction with the beta subunits of activin. However, these assays may have measured both inhibin and the free alpha subunit. The development of two-site ELISA assays has overcome these technical problems and provided relatively consistent estimates of inhibin A secretion throughout the ovine estrous cycle (Fig. 1). In contrast to estradiol secretion, there is no change in either peripheral inhibin concentrations (28) or inhibin secretion rate from the ovary (29) during the follicular phase. Inhibin begins to decline within a few hours after the peak of the LH surge, reaching a nadir about 24 hours later that coincides with the peak of the second FSH surge. There is then an increase in inhibin secretion between days 1 and 3 of the luteal phase, followed by a slight decline over the next 1 to 2 days (28,29). No further consistent changes in inhibin are seen throughout the rest of the luteal phase, although transitory decreases occasionally correlate with increases in FSH secretion and initiation of follicular waves.

The early assays detected episodic secretion of inhibin in ovarian venous samples, but inhibin pulses did not vary with stage of the cycle and were not correlated with those of estradiol or with any changes in LH or FSH concentrations (90). These inhibin pulses are not evident in the peripheral circulation, because the episodic fluctuations in secretion are damped by the 50-minute long half-life (87) of this hormone. This episodic release has yet to be confirmed with more specific assays.

Ovarian Steroid Secretion

Studies on the secretion of ovarian steroids have focused on three steroids: progesterone, estradiol, and androstenedione. In vivo, progesterone appears to be secreted into the blood exclusively by the corpus luteum (89), whereas estradiol is secreted only by follicles (91). Although androstenedione originates from follicles, corpora lutea, and stroma (91), the secretion pattern of androstenedione correlates closely with estradiol (92,93), so that it most likely derives primarily from follicular tissue.

Progesterone. Progesterone concentrations are virtually undetectable early in the cycle (13) and then rise gradually from days 2 to 8 to reach a maximal concentration ranging from 1.5 to 3 ng/ml, depending on the breed (6,7). Progesterone concentrations remain relatively constant from days 8 to 14 and then rapidly fall to undetectable levels over the next 1 to 2 days (Fig. 1). In contrast to the rat and primate (Chapters 43 and 45), progesterone secretion does not increase in response to the pre-ovulatory LH surge in the sheep (89), but it does increase within the pre-ovulatory follicle(s) and plays an important role in the ovulatory process (77).

This pattern of progesterone concentrations is remarkably constant across breeds. Breeds with high ovulation rates do have slightly higher progesterone concentrations at mid-cycle than those with low ovulation rates (6,7), but the increment in progesterone secretion is considerably less than the increase in number of corpora lutea. Maximal progesterone concentrations also vary within the breeding season, with higher concentrations occurring in the middle than at the beginning or end of the season (94).

Estradiol and Androstenedione. Regardless of whether estradiol is measured in the ovarian venous (15,92) or peripheral circulation (13,16,95), there is general agreement that the principal estradiol rise occurs during the 2- to 3-day follicular phase of the estrous cycle (Fig. 1). Estradiol secretion begins to rise after the first drop in progesterone concentrations at luteolysis (16,92), and increases five- to tenfold over the next few days. This estradiol increment peaks at the start of the pre-ovulatory LH surge (Fig. 1) and then rapidly declines to basal concentrations (15,16). Estradiol secretion during this period appears to come largely from the follicle destined to ovulate, which is growing rapidly at this time (89).

There is less agreement on whether significant increments in estradiol secretion occur at other times of the cycle. Most (13,30,92), but not all (95), studies observed a peak in estradiol secretion around day 4 of the cycle that correlates with the first follicular wave (Fig. 3). Estradiol secretion does occur throughout the rest of the luteal phase (30,93), but mean concentrations remain relatively low and show no consistent pattern during this period (Figs. 1, 3). The pattern of androstenedione secretion closely parallels that for estradiol (15,92,93), except for the latter portion of the follicular phase when androstenedione release plateaus while estradiol output continues to rise (15).

When measured at frequent intervals in individual sheep, both progesterone (92,93,96,97) and estradiol (15,19,93) show marked fluctuations, indicative of an intermittent secretory pattern analogous to tonic GnRH and LH secretion. The episodic secretion of estradiol is thought to be driven by LH pulses because increases in secretion of this steroid are always preceded by LH pulses (19,93). Both estradiol pulse frequency and amplitude increase during the sustained estradiol rise of the follicular phase (15,19). In contrast, progesterone pulses are not correlated with the occurrence of LH pulses (92,93,97).

Uterine Prostaglandins

With the discovery in 1970 that prostaglandin $F_2\alpha$ ($PGF_2\alpha$) is luteolytic in the sheep (98), interest immediately arose as to the pattern of $PGF_2\alpha$ during the cycle. However, assessment of uterine $PGF_2\alpha$ secretion in vivo is complicated by two factors. First, it must be measured in the utero-ovarian or uterine vein because prostaglandins are metabolized almost completely in one passage through the lungs (99) and because some metabolism occurs locally in the uterus and/or ovary (100,101). Second, $PGF_2\alpha$ is secreted episodically (92,102–105), so that blood samples must be collected relatively frequently to determine the actual secretory pattern. When these technical issues are taken into account, a consistent picture of $PGF_2\alpha$ secretion emerges. From days 2 to 10 of the estrous cycle, there is little secretion of $PGF_2\alpha$ from the uterus (92,104,106,107). Episodes of $PGF_2\alpha$ secretion first appear between days 11 and 13 of the cycle and the frequency of these episodes then increases over the next 2 to 3 days (92,100,102–107). These changes in episodic prostaglandin release are reflected in mean concentrations of $PGF_2\alpha$ in the uterine vein (92,103,106,107) which increase late in the luteal phase, plateau during the early follicular phase, and then increase again to a peak after progesterone declines (Fig. 1).

ANALYSIS OF THE ESTROUS CYCLE

In this section, we will consider the hormonal control of each component of the ovine estrous cycle. This is not intended to be a complete review of each component, but instead will focus on those aspects of control directly relevant to events of the estrous cycle and recent mechanistic observations.

Control of Estrous Behavior

Hormonal Control

Under physiological conditions, both progesterone and estradiol are necessary for the induction of estrous behavior in the ewe. Although androgens can induce estrous behavior, they do not appear to be physiologically important (108). Estradiol treatment alone can produce estrus, but in the absence of progesterone pretreatment, pharmacological doses are required (95,109–111). Progesterone pretreatment increases the sensitivity to estrogen (109) and the intensity of behavioral estrus (112). This action of progesterone is augmented by concurrent treatment with estrogen (5) and generally requires prolonged (= 6 days) exposure (110), although treatments as short as five days are effective in anestrous ewes (113). Estradiol is also essential for estrous behavior (95,114), with low concentrations of estradiol (3 pg/ml) sufficient to induce estrus in virtually all ewes (115). Higher estradiol concentrations decrease the latency to estrus (Fig. 4) and increase the ram seeking activity of the ewe (5). Both high concentrations of progesterone prior to estradiol (116) and maximal secretion of estradiol are required for the synchronization of estrus with the pre-ovulatory LH surge. Progesterone decreases the latency to estrus (109), but increases that of the LH surge (95,117), moving the two events closer together, and the full pre-ovulatory estradiol rise completes this synchronization (Fig. 4). Finally, the fall in estradiol at the time of the LH surge is important for termination of estrus, because prolonged estrogen treatment doubles its duration (118).

The temporal relationship between progesterone and estradiol during the estrous cycle is critical for initiation of estrus. In contrast to the facilitory effects of pretreatment with progesterone, when

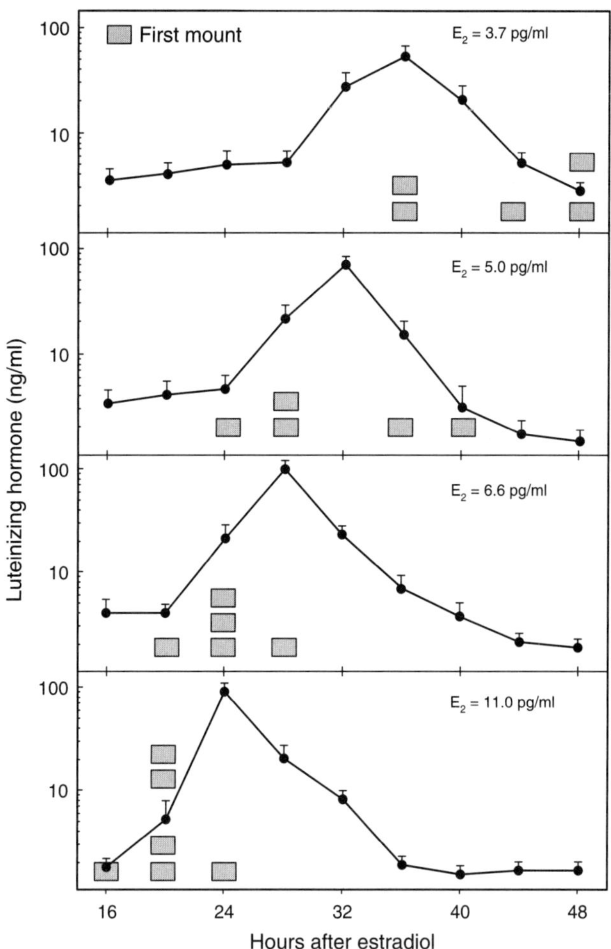

FIG. 4. Effects of different estradiol (E_2) concentrations on the synchronization of the LH surge and the onset of estrous behavior. Breeding season ewes, pretreated with steroid-containing Silastic implants to produce luteal-phase concentrations of progesterone and E_2, were ovariectomized. Progesterone implants were removed to simulate luteolysis and, 16 hours later, additional E_2 capsules were inserted to produce four different concentrations of this steroid, ranging from 3.7 to 11 pg/ml. Blood samples were collected and onset of estrous behavior was monitored every 4 hrs for the next 2 days. Mean (±SEM) LH concentrations (*solid circles*) and onset of estrous behavior in each ewe (*shaded boxes*) are shown. [Details of experimental protocol can be found in Goodman et al. (115).]

progesterone is given simultaneously with estrogen it blocks estrous behavior (95,111,119). Estrogen treatment is highly effective if given 1 to 2 days after cessation of progesterone administration (95,112), but its potency gradually declines with time after progesterone treatment (111,119). Thus, the ideal steroid pattern for induction of estrus in the ewe is a prolonged period of elevated progesterone followed by a fall in progesterone and a rise in estradiol over the next 1 to 2 days, a pattern very similar to that observed during the ovine follicular phase (Fig. 1).

The physiological importance of both steroids is further emphasized by the absence of estrus at the transitions between the anestrous and breeding seasons. At the start of the breeding season, the first ovulation is not accompanied by estrous behavior (94,109), because of the lack of progesterone priming. At the end of the breeding season, estrus does not occur after luteolysis (94), because estradiol concentrations fail to rise (120).

Neural Control

In contrast to the wealth of information on the hormonal control of estrous behavior in the ewe, the neural substrates for this behavior remain largely unknown, although the ventromedial nucleus (VMN) appears to play an important role. Knife cuts that extend into this nucleus blocked the ability of steroids to induce estrus (121) and microimplants of estradiol into it produced estrous behavior in progesterone-pretreated ewes (122). Moreover progesterone increased expression of estrogen receptor alpha (ER-α) in this region (123) which might explain the need for progesterone priming. Estrogen microimplants in the preoptic area (POA) failed to induce estrus (122), but progesterone microimplants blocked the ability of systemic estrogen treatment to induce sexual receptivity when placed in either the POA or VMN (124). To date two neurotransmitters have been implicated in the control of estrous behavior in the ewe: dopamine (DA) and GnRH. DA release in the VMN was stimulated by prolonged progesterone treatment, doubled in response to progesterone withdrawal, and then fell rapidly if estradiol was administered to induce estrus (125). This pattern led to the proposal that after progesterone withdrawal, DA has first a stimulatory and then an inhibitory effect on the subsequent expression of estrus. This hypothesis was recently supported by the observation that a DA D2-receptor agonist increased the intensity of estrous behavior when given into the VMN before systemic estradiol treatment, and decreased it when given to this area after estradiol (126). Interestingly, these treatments had no effect on the onset or duration of estrous behavior, suggesting that other neurotransmitters are involved in triggering and terminating this behavior. In fact, the pre-ovulatory GnRH surge appears to be involved in the latter. During the surge, GnRH is released into both the portal circulation and the cerebrospinal fluid (CSF) in the third ventricle (127). Administration of a GnRH antagonist into CSF during the latter part of the GnRH surge prematurely terminated estrous behavior (128), providing strong evidence that GnRH in CSF helps ensure that estrous behavior lasts for 36 hours in the ewe.

Control of Hypothalamo–Hypophyseal Function

Neuroanatomy

As in many other species (Chapter 28), GnRH neurons are scattered throughout the ovine diencephalon (129,130), with about 50% of them located in the POA (Fig. 5) and the rest found in the diagonal band of Broca (dBB), anterior hypothalamic area (AHA), and mediobasal hypothalamus (MBH), which extends from the posterior border of the optic chiasm to the rostral edge of the mammillary body and includes the arcuate nucleus (ARC) and VMN. Most of these cells are multipolar with extensive dendritic processes, but some bipolar cells can be seen.

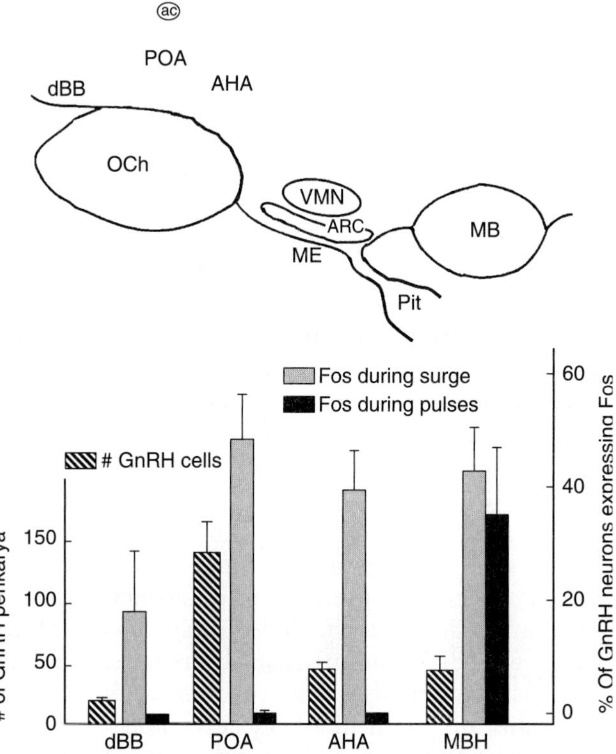

FIG. 5. Diencephalic areas important for reproductive function in the ewe. **Top panel:** Schematic drawing of a parasagittal section through the ovine preoptic area and hypothalamus illustrating the major areas involved in the regulation of estrous behavior and GnRH secretion. ac, anterior commissure; AHA, anterior hypothalamic area; ARC, arcuate nucleus; dBB: diagonal band of Broca; MB, mammillary body; ME, median eminence; OCh, optic chiasm; POA, preoptic area; Pit, pituitary; VMN, ventromedial nucleus. **Bottom panel:** Distribution of GnRH perikarya (*cross-hatched bars*) and GnRH neurons activated (percentage containing Fos) during the preovulatory GnRH surge (*shaded bars*) and during pulsatile GnRH secretion (*solid bars*). Data are mean ±SEM; note that scale for percentage of GnRH neurons expressing Fos is on the right y-axis. [Values for GnRH perikarya and Fos expression during pulsatile secretion (copyright 1999, The Endocrine Society) are replotted from Boukhliq et al. (185); those for Fos expression during the surge plotted from data presented in Moenter et al. (186).]

Although there is good evidence for anatomical and functional subpopulations of GnRH neurons (see below), there are no regional differences in the percentage of GnRH neurons projecting to the median eminence (131). Thus, GnRH fibers arising from cells in the dBB, POA, and AHA, that travel either periventricularly or laterally at the base of the brain (129,130), are responsible for most of the GnRH terminals in the external zone of the median eminence (132).

Immunocytochemical studies over the last decade have defined many of the steroid-responsive neural systems that are likely to mediate the feedback actions of estradiol and progesterone. The most well-characterized are ER-α–containing cells (133–139), which are concentrated in a number of areas including the lateral septum (LS), bed nucleus of the stria terminalis (bNST), POA, AHA, ventrolateral portions of the VMN, arcuate nucleus (ARC), and brain stem areas containing the A1 and A2 noradrenergic (NE) cell groups. Progesterone receptors (PR) are found in many ER-α–positive neurons (140) so that they show a similar, but somewhat more restricted distribution, with high numbers of cells in the POA, VMN, ARC, and AHA (141–143). Cells containing ER-β mRNA are found in many of these same nuclei, but with different densities than ER-α–positive cells, except for the ARC, which does not contain ER-β (138).

Steroids are unlikely to exert their effects directly on GnRH neurons because these cells do not contain ER-α (144,145) or PR (142); it is not known if ovine GnRH neurons contain ER-β, but there is no evidence that this receptor plays an important role in control of gonadotropin secretion in the ewe. Because of the apparent importance of afferent inputs, considerable effort has been made to determine steroid-responsive neural systems that project to GnRH neurons. To date, nine neurotransmitters (146–150) have been identified in boutons that make close contacts at the light microscopic level with GnRH perikarya (Table 1): four classical neurotransmitters (DA, γ-aminobutyric acid [GABA], glutamate, and norepinephrine [NE]) and five neuropeptides (dynorphin, β-endorphin, neurokinin B (NKB), neuropeptide Y [NPY], and somatostatin). Although there are regional differences among GnRH perikarya in the level of innervation by specific neurotransmitters (148,149), in general the neuropeptides provide more input than the classical neurotransmitters, except for glutamate. Synaptic contacts containing three of these (dynorphin, β-endorphin, NPY) have been identified with electron microscopy (149,151). Some of these signaling molecules are found in the same synaptic terminals; there is a 50%–60% overlap of NPY and NE (147), glutamate is found in 15% of the NE terminals (152), and dynorphin and neurokinin B may represent the same set of ARC neurons (153). These nine neurotransmitters probably account for almost all the synaptic inputs onto GnRH soma, but other inputs to their dendrites have yet to be identified (148). It should be kept in mind that ER-α–containing neurons in the ARC (but not elsewhere) also project to the median eminence (154), where they may contact GnRH terminals; DA synapses on GnRH terminals in the median eminence have been identified with electron microscopy (155).

These nine neurotransmitters are co-localized with ERα (and/or PR) in neuronal cell bodies (134–137,139,143–146,152), most of which are in either the POA/AHA and/or ARC nuclei (Table 1).

TABLE 1. *Steroid-responsive neurons that may project to POA GnRH perikarya*

Neurotransmitter or neuropeptide	% GnRH cells contacted	Area of cell bodies	% Cells containing ERα	Neurons project to POA	Effect on GnRH
Dopamine	12%–18%	POA/AHA (A14)	3%–17%[a]	No	Inhibit and
		ARC (A12)	3%–13%[a]	No	Stimulate
GABA	20%	POA	44%	?	Inhibit
Glutamate	73%	bNST	35%–70%[c]	No	Stimulate
		POA	41%–50%[c]	No	
		ARC	48%–77%[c]	Yes	
Norepinephrine	19%	Brain stem A1	20%–83%[c]	Yes	Stimulate
		Brain stem A2	16%–60%[c]	No	and Inhibit
Dynorphin	35%	POA/AHA	90%[b]	?	Inhibit
		ARC	90%[b]	?	
β-Endorphin	65%	ARC	20%	Yes	Inhibit
Neurokinin B	24%–39%	ARC	97%	Yes	Unknown
Neuropeptide Y	25%	ARC	4%–12%[c]	?	Inhibit and
		Brain stem[d]	?		Stimulate
Somatostatin	80%	VMN	30%	?	Inhibit

[a]Variability based on different reports.
[b]Co-localization of PR (100% of PR cells in ARC contain ER-α).
[c]Variability based on rostral-caudal differences with nucleus.
[d]Based on co-localization of NPY with DBH in boutons contacting GnRH perikarya.

There are no NE perikarya in the ovine diencephalon; NE neurons in the brain stem (A1 and A2 groups) contain ER-α, but not PR (141). This correlation between steroid-containing perikarya and synaptic contacts on to GnRH neurons does not demonstrate monosynaptic connections; more direct tract-tracing studies are needed to support such a conclusion. Based on injections of retrograde tract tracers into the medial POA, NE neurons from both the ventrolateral medulla (A1 group) and nucleus of the solitary tract (A2 group) project to the medial POA. However, only ER-α–positive cells in the A1 project to this area (137,156,157). Similar data have demonstrated that high numbers of ER-α–positive cells in the LS, POA, VMN, and ARC project to the POA (157). The phenotype of the estrogen-responsive GnRH afferents in most of these areas has not been determined, but there is tract-tracing evidence that ARC neurons containing glutamate (152), but not those containing DA (156), project to the POA. While retrograde tract tracing can be used to demonstrate neurons projecting to the POA, this approach cannot establish direct connections to GnRH cells. Injections of anterograde tracers are needed to address this issue; two such studies have been done with injections into the VMN/ARC area, but one observed direct projections to POA GnRH perikarya (158) and the other did not (159). Finally, it is important to keep in mind that other neurotransmitters are very likely involved in mediating steroid negative feedback. For example, nitric oxide synthase (NOS) is co-localized with steroid receptor-containing cells in VMN neurons (160), but NOS-positive synaptic contacts on GnRH neurons have yet to be identified.

There is general agreement that dynorphin (149), β-endorphin (161,162), and GABA (163,164) inhibit and glutamate (165) stimulates GnRH release in the ewe (Table 1). Dopamine and NPY can have both inhibitory (166,167) and stimulatory (168,169) effects depending on the site of administration, and NE has either inhibitory or stimulatory effects depending on the endocrine condition of the ewe (170). There is little known about the effects of NKB and somatostatin, although there is one report that intracerebroventricular (icv) administration of somatostatin inhibited LH secretion in OVX ewes (171).

Control of Tonic GnRH and LH Secretion by Ovarian Steroids

Three ovarian steroids have been implicated in the control of tonic LH secretion: androstenedione, estradiol, and progesterone. Although antibodies against androstenedione increase tonic LH secretion in ovary-intact ewes (172), most evidence indicates this is an indirect effect and that this androgen plays no role in the control of tonic LH secretion during the estrous cycle (173,174). Perhaps the strongest evidence against a physiological negative feedback role for androstenedione is the effect of reproducing normal progesterone and estradiol concentrations in acutely ovariectomized (OVX) ewes (95). As illustrated in Fig. 6, such an "artificial estrous cycle" produced tonic LH concentrations in steroid-treated OVX animals that were virtually identical to those seen during the normal estrous cycle. This contrasts with tonic FSH concentrations, which remained elevated above those in ovary-intact ewes, an observation implicating a physiological role for inhibin in ewes (see below). When this pattern of progesterone or estradiol was produced in the absence of the other steroid, some inhibition of tonic LH secretion was observed, but neither steroid was able to maintain normal tonic LH concentrations (95). Thus, estradiol and progesterone are both necessary and sufficient to account for the pattern and mean concentrations of tonic LH, but not FSH, during the ovine estrous cycle.

Quantitatively, progesterone is more important than estradiol in determining the pattern of tonic LH secretion during the estrous cycle. Studies in acutely OVX ewes have demonstrated that progesterone produces a strong inhibition of pulsatile LH secretion (95,175,176) that is due to suppression of LH (173,177) and GnRH (178,179) pulse frequency. In contrast, during the breeding season, estradiol produces only a weak inhibition of mean LH concentrations (175,180). Estradiol inhibits GnRH (181) and LH (20,173,177) pulse amplitude, but it also stimulates pulse frequency (20,181). Each of these individual actions is dose-dependent (181), but because they tend to offset each other, basal estradiol concentrations are as effective as peak concentrations in suppressing mean LH (176). Estradiol enhances the inhibitory actions of progesterone (173,176,182), but this action is also not dose-dependent (176). Consequently, the inhibitory effects of estradiol are relatively constant throughout the cycle. In contrast, progesterone exerts a dose-dependent suppression of tonic LH secretion (95,176) so that changes in progesterone concentrations are largely responsible for the pattern of LH secretion. During the early luteal phase, tonic LH secretion decreases because the increasing concentrations of progesterone suppress GnRH and LH pulse frequency. Tonic LH concentrations then remain low as long as progesterone concentrations are elevated and hold GnRH pulse frequency in check. The fall in progesterone at luteolysis allows the frequency of the pulse generator to increase, an effect that is further enhanced by estradiol. Estradiol does inhibit pulse amplitude during the follicular phase, but this does not completely compensate for the increase in pulse frequency so that tonic LH secretion increases as the follicular phase progress.

The GnRH Pulse Generator. A key first step in understanding the mechanisms of steroid negative

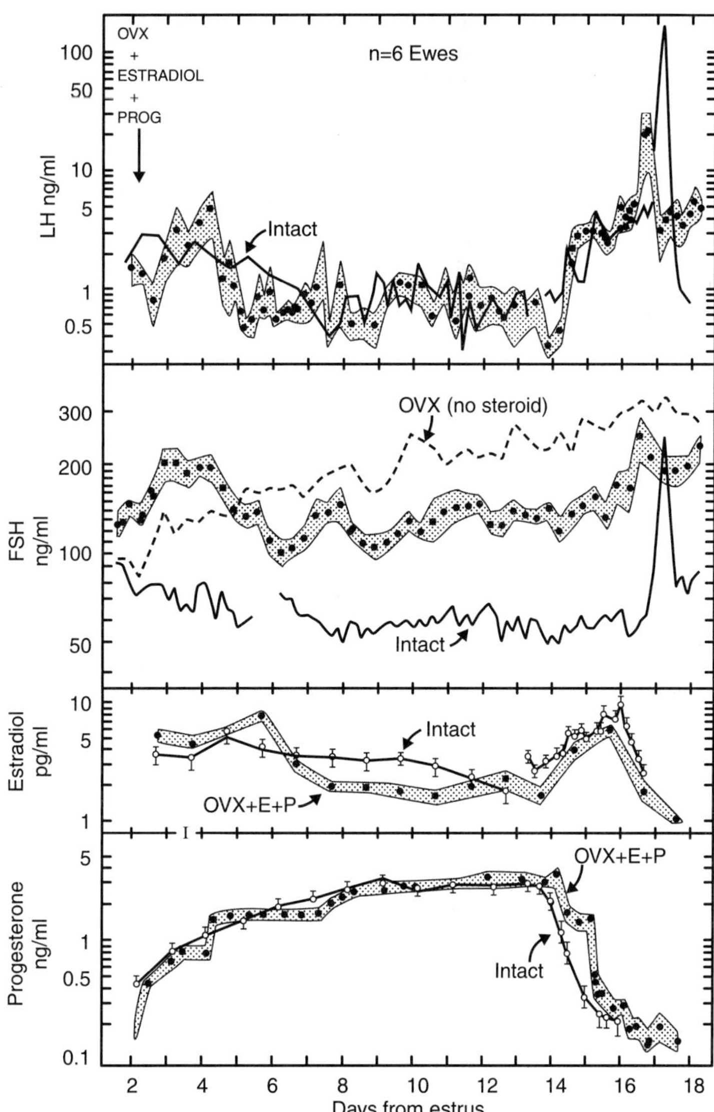

FIG. 6. Effects of physiological concentrations of estradiol (E) and progesterone (P) concentrations on secretion of LH and FSH in ovariectomized (OVX) ewes. Animals were OVX on day 2 after estrus and given E and P implants to simulate the pattern of these two steroids during the ovine estrous cycle. In the bottom two panels, mean concentrations (*solid circles*) ±SEM (*shaded area*) of E and P in these steroid-treated OVX ewes are compared with those in ovary-intact animals (*open circles*). In the top two panels, mean concentrations (*solid circles*) ±SEM (*shaded area*) of LH and FSH in these steroid-treated OVX ewes are compared with mean concentrations of LH and FSH in ovary-intact animals (*solid lines*) and mean FSH concentrations in untreated OVX ewes (*dashed line*). [Data reprinted from Karsch et al. (95) with permission of the Society for Study of Reproduction and Goodman et al. (27) (copyright 1981, The Endocrine Society).]

feedback is identifying the neural systems responsible for the generation of GnRH pulses. While these systems remain largely unknown, three lines of evidence implicate GnRH neurons within the MBH. First, knife-cut studies demonstrated that neither anterior nor posterior inputs are necessary for normal pulsatile LH secretion in OVX ewes (121,183,184). Similar cuts in the ARC completely disrupted episodic LH release, while cuts between the AHA and ARC decreased LH pulse amplitude without altering LH pulse frequency (121,183). Second, studies using the early immediate gene product, Fos, as an index of neural activation demonstrated that GnRH neurons in the MBH, but not elsewhere, are activated by stimuli that induce episodic LH secretion (185). This contrasts with the LH surge (186), during which GnRH neurons in all areas are activated (Fig. 5). Finally, there is one report (187) that bursts in multi-unit electrical activity temporally correlated with LH pulses are observed in the MBH, although it is not clear that these recordings reflect activity of GnRH neurons.

The episodic nature of GnRH release implies that the GnRH neurons controlling tonic LH secretion must function in synchrony, an inference supported by detailed analysis of pulsatile GnRH secretion rates (188). These studies showed that episodic GnRH secretion occurs essentially as a square wave, with very rapid onsets and offsets and no secretion between pulses, indicating tight coupling of the responsible GnRH neurons. Immortalized GnRH cells can show episodic secretion, but comparison of the pattern of hormone release from GnRH cells in vitro (189) and in vivo (188,190,191) indicates that their activity is much more tightly synchronized when they are in situ. Thus, it appears likely that other neural elements normally contribute to their synchronization and two of these have been implicated in the sheep: GnRH and endogenous opioid peptides (EOP).

Systemic administration of a GnRH antagonist to ewes can increase GnRH pulse frequency, pulse amplitude, and GnRH secretion between pulses, with the latter being the most consistent effect (178). These data raise the possibility that ultra-short loop negative feedback may contribute to the generation of GnRH pulses, but other data in sheep are not consistent with this hypothesis (192,193). If GnRH does inhibit its own secretion, this is most likely an indirect effect, because no GnRH synapses have been observed on GnRH perikarya in ewes (149,194). The possibility that an EOP system plays a part in the GnRH pulse generator arises from data (191) that the EOP-receptor antagonist, naloxone, increased both the duration of GnRH release during a pulse and GnRH secretion between pulses. Thus, inhibition by EOP appears to help terminate each GnRH pulse and prevent random firing of GnRH neurons between pulses. Because these effects of naloxone were observed in chronically OVX ewes, they are unlikely to represent blockade of any residual negative feedback system. The data on the potential roles of GnRH and EOP fit well with a model for generation of episodic GnRH secretion (195). According to this model, the initial firing of GnRH neurons stimulates an inhibitory system (EOP?) which would then (after a delay of a few minutes) produce sufficient inhibitory input back onto the GnRH neurons to terminate the pulse. Test of this hypothesis awaits identification of the actual EOP neurons involved and their innervation by GnRH synapses.

Mechanisms Mediating Progesterone Negative Feedback. Progesterone has one primary negative feedback action: inhibition of GnRH pulse frequency. Although progesterone can have inhibitory effects directly on the pituitary (196–200), the physiological relevance of these remains unclear, because pituitary effects are not always observed in vivo (177,201,202) and this steroid specifically inhibits LH pulse frequency (173,177), an action that reflects inhibition of GnRH pulse frequency. While there is some evidence that progesterone increases GABA (203) and decreases glutamate (204) release in the POA, most work has pointed to EOP neurons as the primary mediators of progesterone negative feedback. EOP agonists act in the brain to inhibit LH secretion (162,205) and EOP antagonists increase LH pulse frequency (206–209) and GnRH secretion (210) in luteal phase and progesterone-treated OVX ewes, but usually not in OVX or estradiol-treated OVX (OVX+E) animals (206–209,211). Four recent lines of evidence support the hypothesis that ARC dynorphin neurons, acting via κ-EOP receptors, may be the EOP mediating the negative feedback actions of progesterone. First, microinfusions of progesterone in the MBH, but not in the POA, inhibited LH pulse frequency (212). Second, virtually all dynorphin neurons in the ARC contain PR (143), the receptor mediating progesterone negative feedback (179). Third, local administration of a specific κ-antagonist to the MBH increased LH pulse frequency in luteal phase ewes, whereas μ- and δ-antagonists were ineffective when placed in this area (149). Fourth, light microscopic analysis indicated that dynorphin-positive boutons make close contacts on almost all GnRH perikarya in the MBH, but contact fewer than half of the GnRH cell bodies in the POA (149). In contrast, β-endorphin-containing varicosities make close contacts on approximately the same percentage of GnRH perikarya in all diencephalic areas. Both dynorphin and β-endorphin also act in the POA to inhibit LH pulse frequency during the luteal phase (149,161,213,214). However, knife cuts between the POA and MBH did not disrupt the normal luteal phase (215) or the EOP systems mediating progesterone negative feedback (216). These data led to the hypothesis that progesterone acts on dynorphin neurons in the ARC to stimulate release of this EOP, which then acts directly on GnRH neurons in the MBH to inhibit GnRH pulse frequency.

Mechanisms of Estradiol Negative Feedback. These are much more complex than those of progesterone because estradiol produces four different effects on episodic GnRH/LH secretion. During the luteal phase it increases the ability of progesterone to inhibit GnRH pulse frequency (179,182). During the follicular phase, estradiol: (a) decreases GnRH and LH pulse amplitude (181), (b) increases pulse frequency (181,190), and (c) alters the shape of GnRH pulses and increases GnRH release between pulses (190). The mechanisms underlying these actions are not well understood, in part because direct assessments of GnRH secretion are necessary to monitor most of them.

The primary negative feedback action of estradiol is to inhibit LH pulse amplitude. Most evidence indicates that this reflects estradiol actions at both the pituitary and hypothalamus. Acutely (within a few hours), estradiol treatment decreases the response of the pituitary to GnRH, both in vivo (196,217) and in vitro (218). In fact, under some circumstances, estrogen treatment completely suppresses LH pulses without disrupting GnRH secretion into the portal circulation (217). The cellular mechanisms of this inhibitory action have not been completely elucidated. One important negative feedback action of estradiol at the pituitary is to acutely decrease expression of mRNA for the α and LH-β subunits (219–221), but most other actions of estradiol on the gonadotrope (i.e., GnRH-R expression) are stimulatory (see below). Estradiol probably has other inhibitory effects because the rapidity with which it suppresses LH secretion (217) is unlikely to be accounted for solely by a decrease in LH synthesis. Estradiol-induced increases in GnRH pulse frequency may indirectly decrease LH pulse amplitude, because increasing the frequency of GnRH administration decreases LH pulse amplitude in response to each GnRH injection (201,223,224).

Estradiol also decreases the amount of GnRH released per pulse (181,190,225), an action that may occur in both the POA and MBH (226). Of the nine neural systems identified to contain ERα and project to GnRH neurons (Table 1), possible roles of EOP, GABA and NE in estrogen negative feedback have been examined. Although most early work monitoring LH pulse amplitude indicated that EOP mediate this action of estradiol (206,208,209), the effects of naloxone on GnRH pulse amplitude in OVX and OVX+E ewes (191) were not consistent with this hypothesis. It is also unlikely that GABA mediates this action of estrogen because microinjections of GABA receptor antagonists into the POA failed to block estrogen negative feedback in breeding season ewes (163). On the other hand, there are three reports indicating that NE mediates this action of estradiol: (a) estradiol rapidly induced Fos in A1 NE neurons (227); (b) microimplants of α-adrenergic antagonists into the POA increased LH pulse amplitude during the follicular phase and in OVX+E ewes, but not in OVX animals (228); and (c) microinjections of NE into the POA inhibited LH secretion in OVX+E ewes (170). Interestingly, these inhibitory effects of NE microinjections were not observed in the absence of estradiol (170). Thus, most data are consistent with the hypothesis that estrogen inhibits GnRH pulse amplitude in breeding season ewes by increasing the release of NE from A1 neurons projecting to the POA and/or increasing the response of GnRH neurons to this neurotransmitter.

There is little known about how estrogen exerts its other three actions on pulsatile GnRH secretion in breeding season ewes. The ability of estradiol to increase progesterone inhibition of GnRH pulse frequency most likely involves the dynorphin neurons postulated to mediate progesterone negative feedback. They contain both ER-α and PR (140,143) so this action of estradiol may reflect the well-established ability of this steroid to increase expression of the PR (141). The mechanisms by which estradiol increases GnRH pulse frequency are completely unknown. It is unlikely that they involve EOP or NE because antagonists to these neurotransmitters either have no effect on pulse frequency in OVX+E ewes (191,228) or increase frequency during the follicular phase (208).

One interesting recent observation is that estradiol dramatically alters GnRH pulse shape (Fig. 7) and increases the release of GnRH between pulses (190). The mechanisms of these actions of estradiol have not yet been examined, but it is interesting to note that the EOP antagonist, naloxone, produces the opposite effect (Fig. 7) on GnRH pulse shape (191). This raises the possibility that the increase in β-endorphin release that occurs in the follicular phase (230,231) may mediate effects of estradiol on

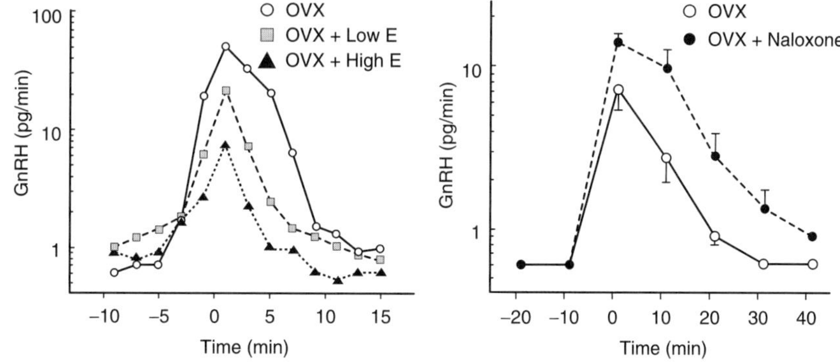

FIG. 7. Effects of estradiol (E) and an EOP antagonist (naloxone) on the shape of GnRH pulses in OVX ewes. **Left panel:** Mean GnRH secretion rate, normalized to the peak of each pulse, in samples collected every 2 minutes from acutely OVX ewes treated with either no steroid, or estradiol implants to produce luteal-phase (low E) or peak follicular phase (high E) concentrations of E. **Right panel:** Mean GnRH secretion rate, normalized to the peak of each pulse, in samples collected every 10 minutes from ewes OVX 2 weeks earlier, before (OVX) and during (OVX+naloxone) iv infusion of naloxone. Note difference in axes between two panels. [Reprinted from Goodman et al. (229).]

GnRH pulse shape. In contrast, both estradiol and naloxone increase the release of GnRH between pulses, so that this action of estradiol cannot be due to an increase in EOP. Finally, these changes in GnRH pulse shape are greatly magnified at the start of the pre-ovulatory GnRH surge (24) so that these effects of estradiol may reflect positive, rather than negative, feedback actions of this steroid.

Control of the GnRH and LH Surges

It is now well established that the follicular phase estradiol rise is the ovarian signal that induces the pre-ovulatory surges of GnRH and LH [evidence reviewed by Goodman (232)]. In contrast to the rat (Chapter 43), neither progesterone nor the PR augments the LH surge because (a) circulating progesterone concentrations do not increase during the periovulatory period of the ovine estrous cycle (92) and (b) the PR antagonist, RU486, does not block the estradiol-induced surges of GnRH and LH in the ewe (233). Like estrous behavior, induction of the LH surge does not require the full pre-ovulatory estradiol rise; an increase to 3 to 4 pg/ml is sufficient (115). However, greater amounts of estradiol do increase the magnitude of the LH surge (115,234) and synchronize it with the onset of estrus (Fig. 4); peak estradiol concentrations of 7 to 10 pg/ml are required to mimic events during the estrous cycle. The effects of progesterone are more complex. Progesterone pretreatment and withdrawal delay the onset of the GnRH/LH surge, contributing to its synchronization with estrus (95,117), and augment the amplitude of the GnRH surge (235). However, if progesterone is elevated during an estradiol stimulus, it blocks the LH/GnRH surge (95,119,236).

Mechanisms of Action of Estradiol. The sites of estradiol action in triggering the LH surge are still under investigation, but it is clear that stimulatory effects of estradiol occur at both the hypothalamus and pituitary. The amount of LH released in response to exogenous GnRH is maximal late in the follicular phase of the estrous cycle (237), but this could be due to either the follicular-phase increase in estradiol or GnRH secretion because both the former (238–240) and the latter (241) can increase pituitary responsiveness to exogenous GnRH. This conundrum has been resolved by studies in which endogenous GnRH secretion was blocked and replaced with exogenous GnRH pulses (196,242,243). In such "hypophysiotropically clamped" ewes, estrogen administration stimulates an increment in LH that resembles an LH surge temporally but has a magnitude that is only 10% to 20% of the normal surge.

The stimulatory effects of estradiol on the gonadotrope appear to be multifaceted. It increases expression of GnRH-receptor, an action that is important for the maintenance of the elevated GnRH receptor number that first occurs in response to increased GnRH pulse frequency following luteolysis [reviewed by Turzillo and Nett (244)]. Prolonged estradiol treatment also increases mRNA for the α- and LH-β subunits (219–221,245), enhances the movement of LH-containing granules toward the cell membrane of gonadotropes (246,247) and induces formation of new calcium channels (248), all of which would enhance pituitary responsiveness to GnRH. Interestingly, these actions of estradiol at the pituitary require longer exposure to estradiol than does the pre-ovulatory GnRH surge (249).

Direct measurements of GnRH have clearly demonstrated that estradiol also acts in the brain. As discussed above, estradiol not only causes a dramatic increase in mean GnRH release, but also alters the pattern of GnRH secretion so that it moves progressively from episodic to continuous (23,24,181). As illustrated in Fig. 8, GnRH secretion between pulses begins to increase just prior to the start of the LH surge; there is then a dramatic increase in both GnRH pulse amplitude and interpulse GnRH secretion during the ascending phase which eventually leads to sustained erratic GnRH release during peak secretion. The mechanisms by which estradiol produces these effects are still poorly understood, but two questions have been resolved. First, studies using local administration of estradiol have demonstrated that it acts in the VMN to induce the surge (122,226). Second, there is not an anatomically distinct subpopulation of GnRH neurons involved; approximately 40% of GnRH neurons in all hypothalamic areas (Fig. 5) express Fos during the surge (186). These data are consistent with the observation that knife cuts between the POA-AHA, where most of the GnRH perikarya are found, and the MBH decrease the amplitude, but do not completely block, the estrogen-induced LH surge (183,184). The percentage of GnRH neurons expressing Fos increases with time after the onset of the surge (186,250) so that changes in the pattern of GnRH secretion (Fig. 8) may reflect the progressive recruitment of more GnRH neurons during the ascending phase of GnRH release.

At least eight different neurotransmitters have been implicated in the control of the GnRH surge, but for most of them the data are limited to a few studies (Table 2). In analyzing this information, we have made use of a conceptual model (251) that divides the surge-induction process into three stages: (a) activation, a period lasting between 4 to 14 hours during which estradiol must be present, (b) transmission,

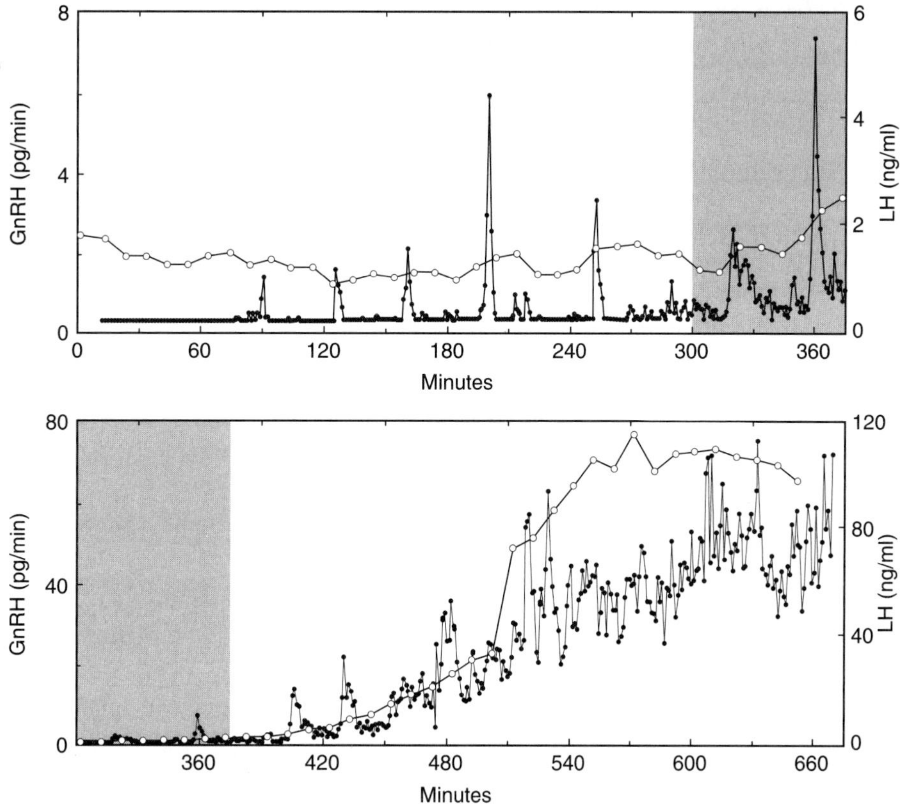

FIG. 8. Time course of changes in the pattern of GnRH secretion leading up to and during the estradiol-induced preovulatory surges of GnRH and LH. GnRH secretion rates (*solid circles*) monitored every min and peripheral LH concentrations (*open circles*) monitored every 10 minutes in an individual ewe are presented. Note the differences in the y-axis between the top and bottom panels. The *shaded area* depicts the time of overlap of values from top and bottom graphs. [Reproduced from Evans et al. (24) (copyright 1995, The Endocrine Society).]

TABLE 2. *Effects of estradiol on transmitters implicated in induction of GnRH surge*

Transmitter	Activation phase (E_2 required)	Transmission phase (E_2 not required)	GnRH release (E_2 not required)	Caveats
Norepinephrine	Fos in A1 cells	Increased NE pulses and DBH[b] release in POA	Increased in ME	Antagonist doesn't always block
β-Endorphin	Decreased mRNA[a] No change in Fos	Increased in ME Agonists delay surge No change in mRNA[a]	Fall in ME No change in mRNA[a]	Antagonist doesn't advance
ARC NPY	No change in Fos or mRNA[a]	Increased in ME	Increased in ME Antagonist delays	Inconsistent data
ARC Glutamate	Increased Fos			Limited data
POA GABA	Increased release	Gradual decline		Limited data
ARC NKB	Decreased mRNA[a]	No change in Fos	No change in Fos	Limited data
Dopamine	No change in Fos		DA turnover increased in VMN Antagonist blocks Increased Fos	Agonist doesn't advance
VMN	Increased mRNA[a]			Inhibits GnRH
Somatostatin	No change in Fos			Limited data

[a]mRNA for precursor of neurotransmitter.
[b]DBH (dopamine-β-hydroxylase) which is stored in NE-containing secretory vesicles.

which lasts for 6 to 8 hours and does not require the presence of estradiol, and (c) GnRH release, which also does not require the presence of estradiol. Of the eight transmitters, two have received considerable attention: stimulation by NE and inhibition by β-endorphin. Estrogen stimulates A1 NE neurons (227) and NE concentrations in the ME increase during the GnRH surge (230). NE input to the POA may well be involved in the transmission phase because there is both direct (252) and indirect (253) evidence for NE release in this area prior to the surge. On the other hand, microinjections of NE to the POA had largely inhibitory effects (170) and an α-adrenergic antagonist did not consistently block the estrogen-induced LH surge (254). There is an increase in ME β-endorphin concentrations during the follicular phase (230,231), followed by a fall during the GnRH surge (230). Less consistent results have been obtained with measurements of the corresponding mRNA, with one group finding a decrease 4 hours after the start of estrogen treatment (255), and another reporting no change during induction of the surge (256). Nevertheless, the pattern of β-endorphin raises the possibility that withdrawal of an EOP inhibition contributes to the onset of the surge and EOP agonists can delay the surge (257,258). However, EOP antagonists do not advance the surge (259,260) so an EOP brake is unlikely to be the last limiting factor before the onset of GnRH release. Several groups have examined a possible role for NPY, but no consistent picture has emerged. Advis et al. (169) presented evidence that increased NPY release at the level of the ME triggers the LH surge, but icv administration of NPY antisera has been reported to delay (261) or advance (262) the surge. Administration of NPY to ewes has been reported to increase (169), decrease (167,263,264,) or have no effect (261) on LH secretion.

Of the other five neurotransmitters, three of them [glutamate (152), GABA (252), and NKB(265)] have been implicated in only one report (Table 2). The other two (DA and somatostatin) may be involved in estrogen actions in the VMN. Dopaminergic activity increased in the VMN coincident with the surge and blockade of DA (and NE) synthesis in this area delayed the surge (168). These observations are consistent with early data that a DA antagonist could block the surge (254), but administration of a DA agonist to the VMN during the transmission phase did not advanced the surge (126). Because VMN somatostatin neurons contain ER-α (139), they are a potential target for the positive feedback effects of estrogen in this nucleus and estrogen treatment increases somatostatin mRNA during the activation phase (171). Estrogen did not induce Fos in these neurons acutely (266), but 81% of them contained Fos during the GnRH surge (139). One group (171) observed only inhibitory effects of somatostatin on LH release, but this was in OVX ewes, so it is important to determine its effects following estrogen treatment. In summary, one can propose the following model (admittedly speculative) for the estrogen-induced GnRH surge. Estradiol acts on VMN somatostatin neurons, and probably A1 NE neurons, to activate the surge mechanism. During the transmission phase there is an increase in stimulatory NE (and NPY?) input to GnRH neurons, but this is balanced by an increase in inhibition by β-endorphin. Following removal of the EOP brake, GnRH release is driven by DA stimulation of VMN (somatostatin or glutamate?) neurons, augmented by NE and possibly NPY release in the ME.

Mechanism of Action of Progesterone. Much less is known about the mechanisms of the two actions of progesterone. The stimulatory effect of progesterone pretreatment on the GnRH surge was discovered only recently (235) because it is not reflected in the pre-ovulatory LH surge (12,119,235). This action of progesterone accounts for discrepancies in initial estimates of the size of estrogen-induced GnRH surge (18,225,267,268) and may underlie the well-established role for this steroid in initiation of estrous behavior discussed above. Progesterone pretreatment did not increase Fos expression in GnRH neurons during the estradiol-induced GnRH surge (269), but did increase Fos in non-GnRH POA neurons. This raises the possibility that the latter neurons may provide stimulatory input that augments GnRH release, although they also could be involved in behavioral estrus.

The ability of progesterone to block the LH surge is due to inhibition of the GnRH surge (236), not inhibitory effects of this steroid at the pituitary (196,201). It is mediated by the classical PR (233) and prevents both the activation (270) and transmission (271) phases of the estradiol-induced surge. Progesterone treatment throughout both phases blocked Fos expression in GnRH neurons, but did not consistently affect Fos expression in other POA neurons (269). This is consistent with the report that progesterone acts in the VMN, not the POA, to block the LH surge (124), but the effects of progesterone on Fos expression in the VMN have not been examined. Because progesterone negative feedback is mediated by EOP, these transmitters could also be responsible for progesterone blockade of the GnRH surge. However studies with the EOP antagonist, naloxone (272,273), have demonstrated that EOP are not involved in progesterone blockade of activation and play only a minor role in the disruption of transmission. Progesterone also decreases NE release in the POA (203), but the potential role of this action in blockade of the GnRH surge has not been investigated.

Control of FSH Secretion. The pre-ovulatory FSH surge is controlled by the same neural mechanisms that produce the coincident LH surge. The pre-ovulatory rise in estradiol secretion triggers the FSH surge (27,274,275) and progesterone blocks the estrogen-induced FSH surge (276). In the case of the FSH surge, only actions of estradiol in the brain appear to be important; estrogen either does not increase (221,222), or decreases (245,277) amounts of FSH-β mRNA in the pituitary, and estrogen injection in stalk-sectioned ewes receiving a constant GnRH stimulus failed to produce any change in FSH secretion (196).

The control of tonic FSH secretion throughout the remainder of the estrous cycle is more complicated because of the involvement of factors at three different levels: ovary, pituitary, and hypothalamus [reviewed by Padmanabhan and colleagues (278,279)]. With regard to ovarian factors, work in the late 1980s and early 1990s, led to the consensus that the negative feedback actions of both estradiol and inhibin control tonic FSH secretion (232,280,281). That inhibin (or another inhibitory ovarian hormone) was involved in the control of tonic FSH secretion was established by replacement studies with physiological concentrations of progesterone and estradiol in OVX ewes (27,95), which produced normal tonic LH secretion, but failed to suppress FSH to physiological concentrations (Fig. 6). This was followed by a careful dose-response study (280) that provided strong evidence that inhibin alone could not prevent the postcastration rise in FSH secretion, but the combination of inhibin and estradiol was very effective in this regard. Both hormones act at the pituitary to decrease FSH-β mRNA and thus FSH synthesis and secretion (245,277,282–285). The ability of estradiol to increase GnRH pulse frequency probably also contributes to its inhibitory effects on FSH secretion, because low frequency pulses favor FSH release (223,286).

While there has been considerable progress in elucidating the mechanisms of action of estradiol and inhibin in the last decade, the relative importance of these two hormones in the normal control of FSH secretion remains unresolved. Based on the patterns of estradiol and inhibin during the cycle (Fig. 1), one can infer that the follicular phase estradiol rise is responsible for the fall in FSH concentrations at this time, while the post-ovulatory fall, and then rebound, in inhibin concentrations probably accounts for the "second FSH surge" on day 1 on the cycle (29). It is less clear how ovarian hormones control the FSH fluctuations associated with follicular waves during the remainder of the cycle; as illustrated in Fig. 3, estradiol and inhibin can account for changes in FSH associated with the first wave, but not with subsequent ones (30). Similarly, in anestrous ewes, transient increases in FSH concentrations associated with follicular waves cannot be accounted for by changes in secretion of estradiol or inhibin (54). This raises the possibility that other ovarian hormones may be involved in the control of FSH secretion and there is recent evidence for free circulating activin in the ewe (287).

Interest in pituitary factors that could act in an autocrine and/or a paracrine fashion arose from observations of GnRH-independent secretion of FSH (288,289), which represents a significant fraction of FSH secretion from the ovine pituitary (290,291). Because of work in other species, this interest quickly focused on the roles of activin, inhibin and follistatin. Activin is a member of the transforming growth factor-β (TGF-β) family that consists of a β-β dimer. It is a potent stimulator of FSH synthesis and secretion. Inhibin (also in the TGF-β family) is an α-β dimer and inhibits FSH by disrupting the actions of activin, while follistatin inhibits FSH release by binding activin. All these activities have been demonstrated in the ewe (282,285,287,292,293) and the ovine pituitary produces activin (285), inhibin and follistatin (292). To further complicate this story, recent work has raised the possibility that two bone morphogenetic proteins (BMP-6 and BMP-7) that are also members of the TGF-β superfamily stimulate FSH synthesis and secretion (294). There is strong evidence that local production of activin contributes to the maintenance of FSH secretion (285,287). Although there are currently few data on the regulation of these proteins and their physiological function in sheep, there is one report that estradiol inhibits activin (β_B) mRNA (285), and another that activin (but not steroids or hypothalamic hormones) stimulates follistatin secretion (292). Thus, any inferences on the intra-pituitary control of FSH in the ewe must depend largely on data from other species.

One of the more intriguing recent developments in the control of FSH secretion has been data from ewes for the existence of a distinct hypothalamic FSH-releasing factor (FSH-RF) (278). The possibility of such a factor in sheep initially arose from reports that pituitary stalk-section completely blocks FSH secretion (223), but chronic immunoneutralization of GnRH does not (289). More recently, Padmanabhan et al. (290) have made use of hypophyseal portal blood collections to monitor FSH release directly from the pituitary to identify clear episodes of FSH secretion that are not always evident in jugular samples. These data demonstrated that each GnRH pulse produces an episode of FSH secretion, but that there are also numerous FSH pulses that occur independently of GnRH release (e.g., see Fig. 9). The independence of the latter from GnRH was clearly demonstrated by

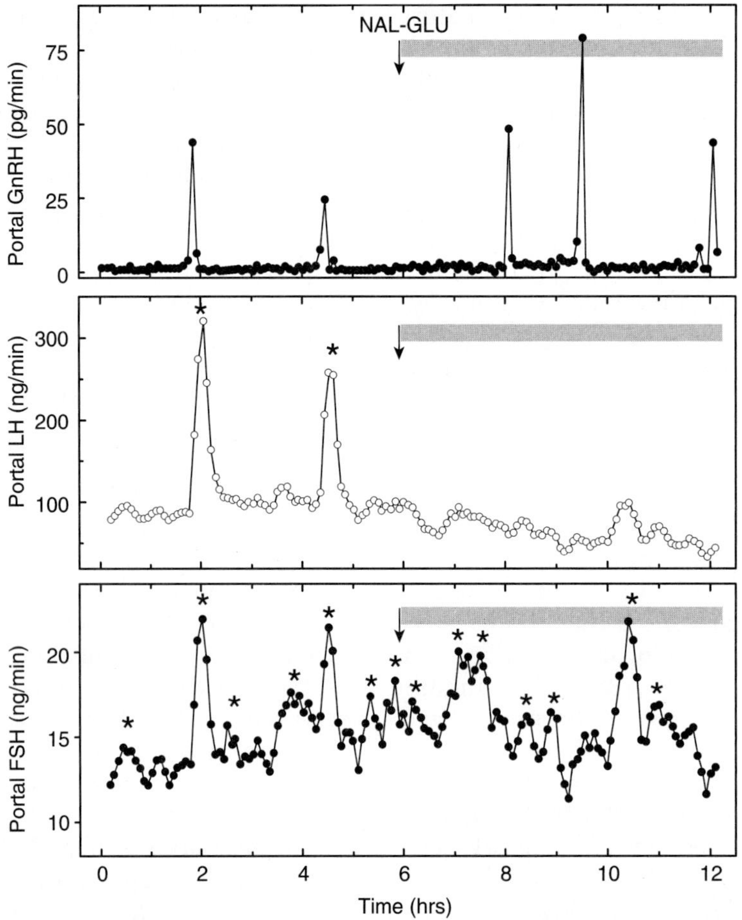

FIG. 9. Secretion rates of GnRH, LH, and FSH estimated in samples collected every 5 minutes from the surface of the anterior pituitary. The GnRH receptor antagonist (Nal-Glu) was injected iv at 6 hours to block the effects of endogenous GnRH pulses on the gonadotropes. *Peaks of statistically identified pulses of LH and FSH. Detailed descriptions of the experimental protocol and analyses can be found in Padmanabhan et al. (291) (copyright 2003, The Endocrine Society), from which these data were taken.

the administration of a GnRH receptor antagonist (291) that completely blocked LH pulses without significantly affecting FSH pulses (Fig. 9). One can draw the strong inference from these results that the hypothalamus produces a FSH-RF that is distinct from GnRH, but acceptance of this hypothesis awaits the isolation and identification of this factor. Interestingly, in some, but not all ewes, GnRH-independent pulses in FSH were associated with small increments in LH detectable in portal samples (Fig. 9), which raises the possibility that this putative FSH-RF has some LH-RH activity. This could reflect an action on gonadotropes that contain both LH and FSH (286).

Control of Ovarian Function

A complete review of the control of ovarian function in the ewe is beyond the scope of this chapter. Instead, this section will focus on factors responsible for follicular development, luteal development and regression, and hormone secretion throughout the ovine estrous cycle.

Control of Follicular Development

In 1993, a group of workers from Australia, New Zealand, Scotland, and England summarized the literature and proposed a model for folliculogenesis and follicular selection in the determination of ovulation rate in the ewe (40). In that model, atresia was considered minimal in primordial follicles, but once a follicle began to grow, it became subject to atresia. Development of an antrum and growth to as much as 2 mm in diameter could occur in the absence of pituitary gonadotropic support (295,296). Proliferation of granulosal cells continued until they reached over 100,000 in a committed follicle. Such a follicle could not return to a quiescent state, but exhibited a low rate of atresia. Under conditions of low concentrations of

FSH and LH, thecal cells differentiated and expressed LH receptors. The now gonadotropin-responsive follicle was subject to some atresia. With continued growth and increases in LH and FSH, the gonadotropin-responsive follicle acquired aromatase and became gonadotropin-dependent, acquiring LH receptors on granulosal cells. The gonadotropin-dependent follicle had greater aromatase activity and required support of both LH and low concentrations of FSH; thus, it was subject to a high rate of atresia, especially during the luteal phase when progesterone limited LH pulse frequency (173,177). In this model, the ovulatory-sized follicle became atretic if an LH surge did not occur within about 72 hours.

Based on recent reviews (297–299), the model presented by Scaramuzzi et al. (40) has gained general acceptance, with two modifications. First, as illustrated in Fig. 10, replacement studies with FSH and LH demonstrated that a sustained increment in FSH alone was sufficient for development of follicles capable of ovulating in response to an LH surge (300–302); withdrawal of FSH in this model induced atresia unless there was a simultaneous increase in LH pulse frequency (302). This transfer of follicular support from FSH to LH occurs normally during the follicular phase and accounts for the continued growth of the pre-ovulatory follicle(s) in the face of falling FSH concentrations. It can also be demonstrated in

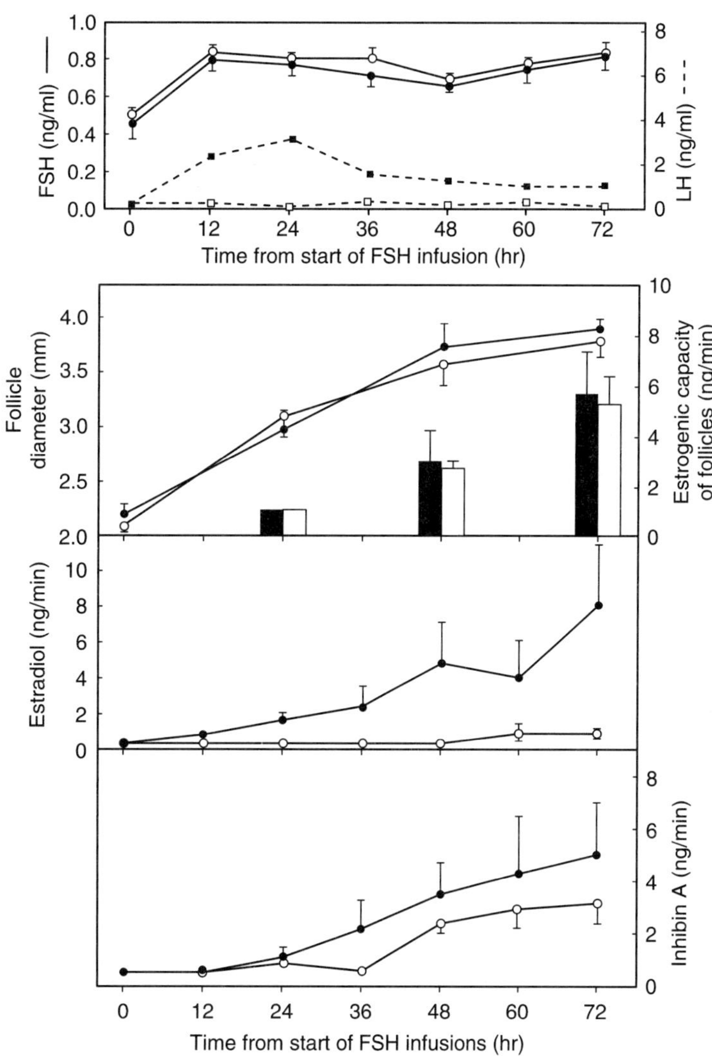

FIG. 10. Effects of FSH, with and without LH, on the development of, and hormone secretion from, follicles in ewes with an ovary autotransplanted in the neck. Gonadotropin secretion was suppressed by long-term treatment with a GnRH antagonist. FSH was infused iv for 72 hours, with (*closed symbols*) and without (*open symbols*) episodic LH injections (every 4 h). **Top panel** depicts FSH (*circles*) and LH (*squares*), concentrations (top panel); **bottom panels** present follicular diameter and secretion rates of estradiol and inhibin. Estrogenic capacity of the follicles (peak estradiol secretion rate in response to a single iv injection of LH) was determined daily. [Data redrawn from Campbell et al. (301); reproduced by permission of the Society for Endocrinology).]

the luteal phase, if a rapid LH pulse pattern is induced artificially by exogenous administration of LH (62) or by production of low progesterone concentrations that do not strongly inhibit LH pulse frequency (52,303). Thus the low LH pulse frequency in luteal-phase ewes is permissive for the follicular turnover necessary to prevent ovulation of aged ova after luteolysis occurs (52,304). Second, Campbell et al. (305), using grafts of ovarian tissue, found that a normal follicular hierarchy was established two months earlier in hypergonadotropic than in hypogonadotropic ewes, and concluded that gonadotropins can affect the rate of pre-antral follicular development. Recent work has focused on intraovarian factors involved in follicular development, other than steroids and inhibin, including components of the insulin-like growth factor (IGF) system, and endothelial cell mitogens such as vascular endothelial growth factor (VEGF) and fibroblast growth factor-2. The reader is referred to two recent reviews (297,299) and the reports by Gonzalez-Bulnes et al. (62,306) for discussions of possible roles of intraovarian factors in follicular development.

Control of Ovulation Rate

Ovulation rate is subject to control by hormonal, environmental, and genetic factors. The seasonal influence, in parallel with the incidence of ovarian cycles (307) and nutritional effects (308), has been well documented and will be considered later. Evidence has accumulated that low progesterone concentrations in peripheral blood can increase ovulation rate, independent of nutritional management. In studies of effects of low progesterone on persistence of follicles, there was a trend towards an increased ovulation rate (52) and a prolific breed of ewes had lower concentrations of progesterone than a nonprolific breed (51). Anestrous ewes pretreated to produce progesterone levels lower than those in the luteal phase before ram introduction had a greater ovulation rate than ewes with ram introduction alone (309). Thus, treatment at progesterone withdrawal and ram introduction with low dosages of FSH, that had been shown to increase ovulation rate in anestrous ewes induced to ovulate with ram introduction, did not increase ovulation rate beyond that obtained in progesterone pretreated ewes (309).

Several instances have now been identified in which a single gene has a major influence on ovulation rate [reviewed by Montgomery et al. (310)]. These genes vary in the amount by which ovulation rate is increased by the effective allele, and in physiological characteristics that have been identified as under the influence of the gene. For example, the Booroola gene ($FecB^B$) increases secretion of FSH. Either one or two copies of $FecB^B$, or a single copy of the Inverdale gene ($FecX^I$) leads to follicles that have fewer granulosal cells, acquire LH receptors earlier, and mature at a smaller size than in noncarriers. The effect of age on reproductive function was examined in ewes with or without the $FecB^B$ allele (306). Older ewes with $FecB^B$ maintained higher ovulation rates than noncarriers and both ovulatory follicles and corpora lutea were smaller, as observed before (310). However, numbers of developing follicles did not differ between genotypes, and concentrations of FSH increased and those of inhibin A decreased in older ewes, regardless of genotype. Thus, the authors suggested that action of the $FecB^B$ allele is more likely at the ovary than the pituitary and that it reduces the rate of follicular atresia.

The Inverdale and Hanna ($FecX^H$) genes are both found on the X chromosome and exhibit a mutation in bone morphogenetic protein 15 (BMP-15) (311) which is expressed in the oocyte. Both this protein and growth differentiation factor-9 (GDF-9; also expressed in the oocyte) are required for normal follicular development and ovulation (312). While a single copy increases ovulation rate by approximately one, two copies lead to a streak ovary and no ovulation, because follicular development stops at the primary follicle stage. The Booroola gene is on chromosome 6 and represents a point mutation in the intracellular threonine kinase signaling domain of the bone morphogenetic protein-1B receptor (BMPR-1B), which is expressed in oocytes, granulosal cells and luteal cells [reviewed by McNatty et al. (311,312a)]. A single copy increases ovulation rate by one or two oocytes and two copies by three to ten oocytes more than noncarriers. The Lacaune gene, found in the major French breed on chromosome 11, increases ovulation rate by about one oocyte in the heterozygote and produces five to seven ovulations in putative homozygotes (312). The Woodlands gene, characterized as a maternally-imprinted gene on the X chromosome, increased ovulation rate by 0.4 oocytes in daughters of carrier rams (313).

The mechanisms of action of the bone morphogenetic proteins and growth differentiation factors have only recently begun to be investigated in sheep (311). In a study using immunization against either BMP-15 or GDF-9 (314), ovulation rate was increased in both cases, with fewer single ovulations and more cases of three or greater ovulations.

Control of Estradiol Secretion

Most investigations on ovarian estradiol secretion have focused on the role of LH and FSH in controlling the follicular-phase estradiol rise because of its established role in induction of ovulation and estrous behavior. The positive correlation between mean tonic LH secretion and estradiol concentrations during the follicular phase (13–16) and the close temporal correlation between endogenous pulses of LH and estradiol (15,19,93) led to the hypothesis that the increase in episodic LH secretion that occurs following luteolysis drives the pre-ovulatory estradiol rise (13–16). This hypothesis was supported by observations that blockade of episodic LH secretion during the follicular phase with progesterone (14,16) or GnRH antisera (315,316) disrupted the pre-ovulatory estradiol rise and ovulation. Conversely, mimicking the follicular phase LH increment by either infusion (317,318) or intermittent injections of LH (318,319) in anestrous ewes produced a pre-ovulatory–like estradiol rise (317,319), and almost invariably induced an LH surge and formation of a corpus luteum (317–319). FSH either remained unchanged (317) or declined slightly (319) during the LH-induced estradiol rise.

The hypothesis that tonic LH was the sole driver of the pre-ovulatory estradiol rise was challenged by a series of interesting studies in which gonadotropin secretion was suppressed with a long-acting GnRH agonist (320–322). In this model, FSH administration in vivo, even in the absence of LH pulses, stimulated follicular development and increased estradiol production by these follicles in vitro (320–322). These observations, together with the evidence that FSH infusion during the normal follicular phase increased estradiol secretion (323), while suppression of FSH release decreased aromatase activity (324) and suppressed estradiol secretion (325), raised the possibility that FSH plays a preeminent role in stimulating pre-ovulatory follicular development and estradiol secretion (321).

This conflict was resolved by the development of a model using chronic treatment with a GnRH antagonist that markedly suppressed both LH and FSH secretion in breeding season ewes (300–302). FSH infusion alone stimulated growth of ovulatory follicles, but little or no estradiol secretion occurred (Fig. 10). When the follicular phase pattern of episodic LH secretion was added to FSH, it had no effect on follicular growth but markedly stimulated steroid secretion. Thus, although FSH secretion clearly plays an important role in controlling follicular development and ovulation rate, it appears to be largely permissive for the pre-ovulatory estradiol rise. With the hormonal regulation of the pre-ovulatory estradiol rise largely resolved, more recent work has turned toward intra-follicular factors. In vivo studies have demonstrated inhibition of estradiol secretion by TGF-α, EGF, bFGF, and inhibin A (300,326,327) and stimulation by IGF-1 (300,328). However these data must be interpreted cautiously because different effects are sometimes observed in vitro [e.g., inhibin A stimulated estradiol secretion, (329)] and some of these effects may reflect an action on follicular development (297).

Most evidence is consistent with the view that tonic LH is also the primary hormonal stimulus for estradiol secretion during the luteal phase. Exogenous LH (330,331), but not FSH (330), stimulates estradiol secretion and each endogenous LH pulse induces an estradiol pulse (93) during the luteal phase. This critical role for LH is illustrated by the different estradiol secretion patterns during the first and second follicular waves (30); during the first wave when LH pulse frequency is high (because of low progesterone) there is a progressive increase in estradiol secretion, but in the second wave when LH pulse frequency is suppressed there is no increase in estradiol secretion, despite the same rate of follicular growth (Fig. 3).

In contrast to the stimulatory effects of tonic LH secretion on estradiol release, the LH surge terminates estradiol secretion (15,317,332). The LH surge appears to directly inhibit aromatase activity (77,333) so that estradiol concentrations fall before any changes in androgen production are evident (15,334). In addition to FSH and LH, progesterone can modulate estradiol secretion, and it does so by two mechanisms. First, by inhibiting tonic LH secretion, progesterone indirectly suppresses estradiol production (14,16). Second, progesterone directly inhibits the response of the follicle to exogenous LH (317). Such a direct inhibitory effect of progesterone may account, in part, for the observation that LH pulses during the luteal phase produce smaller estradiol increments than do comparable LH pulses during the follicular phase (15) or anestrous season (335,336), but the mechanism of this effect has yet to be established.

Control of Inhibin Secretion

In the previous edition of this chapter (232), we concluded that neither LH (316,337) nor FSH (338) directly stimulates inhibin secretion. While chronic treatment with FSH increases inhibin production (Fig. 10), this probably reflects a stimulation of follicular development (323,339). More recent work has supported these conclusions. Inhibin secretion does increase with infusion of FSH and falls with FSH withdrawal, but these responses are delayed more than 24 hours and correlate with follicular growth and atresia, respectively (301,302). When episodic LH was added to the FSH infusion, it had no effect on inhibin secretion (301,302). The pre-ovulatory gonadotropin surge appears to suppress ovarian inhibin secretion because inhibin concentrations fall to their lowest levels during the estrous cycle within a few hours of the surge (28,29). This has been attributed to the induction of ovulation and atresia in antral follicles by the LH surge (28), but artificial induction of ovulation using only LH did not dramatically inhibit inhibin secretion (302). The latter observation raises the interesting possibility that the pre-ovulatory FSH surge, or some other hormone released at this time, causes the rapid decline in inhibin on day 1 of the estrous cycle.

Control of Luteal Progesterone Secretion

As reviewed in detail in the previous version of this chapter (232), the functional activity and demise of the corpus luteum during the ovine estrous cycle are controlled by luteotropic factors from the pituitary (LH) and luteolytic factors from the uterus ($PGF_2\alpha$) (340–342), both of which, in turn, are regulated by progesterone. Actions of these hormones on corpora lutea on a comparative basis have been reviewed recently (343,344); these authors have pointed out several interesting new findings, including evidence that the uterus produces luteotropic factors early in the estrous cycle (345,346), while it was already known that the caruncles may produce factors that limit progesterone secretion (347). Here we will focus on other new information raised in the last decade.

Luteal Development. A key factor in luteal development is angiogenesis, because of the extensive vascular supply required for this active gland. One cytokine involved is VEGF, a potent mitogen for vascular endothelial cells that also stimulates migration of endothelial cells (348). In ovine corpora lutea, VEGF mRNA was greatest early in the estrous cycle and least during luteal regression. Redmer et al. (349) subsequently localized VEGF protein to capillary pericytes and vascular smooth muscle cells in the early corpus luteum. Pericytes were plentiful in the early corpus luteum, but rapidly decreased as the gland matured. They proposed that the capillary pericytes play a critical role in angiogenesis during the follicular-luteal transformation. Another growth factor that has received recent study is stem cell factor. It is secreted by both large and small steroidogenic luteal cells (350) and (based on the location of its receptor, c-kit) acts on small luteal cells and endothelial cells, but not on large luteal cells (351). Stem cell factor was released at relatively constant rates throughout the cycle, but c-kit mRNA was low on day 3, increased as the corpus luteum developed through days 7 to 13, and then decreased somewhat by day 16 (350). However, no specific functional role for stem cell factor in luteal development or luteolysis has been established.

Estrogen, acting via ER-β (352), may play a role in activating secretion of progesterone by the early corpus luteum, in contrast to the fact that tonic concentrations of estradiol are required for luteolysis (353,354). This hypothesis is consistent with the finding that blockade of aromatase delayed the initial rise in secretion of progesterone early in the estrous cycle (355). In those ewes, the duration of secretion of progesterone by the corpus luteum was normal; it simply began and ended later in relation to ovulation, a result that, in turn, is consonant with the effect of progesterone in timing luteolysis [discussed in Goodman (232), reviewed in detail by Goff (356), and now generally accepted].

Oxytocin is generally known for its potential involvement in luteal regression, but an oxytocin antagonist that inhibited luteal regression in goats was also shown to inhibit luteal development and function in that species (357). We mention this finding because that action interfered with secretion of not only progesterone, but also the luteotropic prostaglandin, prostacyclin (PGI_2), and was overcome by treatment with PGI_2. The source of PGI_2 was not identified in that study, but, as discussed later in this section, it can be secreted by the corpus luteum (358). Such an effect of oxytocin adds to the controversy relative to the real role of luteal oxytocin in secretion of prostaglandins in ruminants, which will be discussed in the section on control of uterine secretion of $PGF_2\alpha$.

Luteal Support. Although LH is the primary luteotropic hormone during the ovine estrous cycle, it appears to play largely a permissive or supportive role in progesterone secretion. Concentrations of LH during the cycle are correlated inversely with concentrations of progesterone (13,14), so that progesterone secretion is maximal when LH concentrations are low (Fig. 1). Most progesterone secretion derives from large luteal cells that are unresponsive to LH, because they normally function at maximal capacity. In addition, growth hormone (GH), perhaps acting through IGF-I, has been implicated as a supplementary luteotropin in ruminants, based mainly on data in the cow (359). The ovine corpus luteum expresses GH receptor and IGF-I and its receptor (360,361), and treatment with a combination of LH and GH restored luteal weight to near normal in hypophysectomized ewes (361). Treatment with LH restored concentrations of mRNA for GH receptor in corpora lutea of the hypophysectomized ewes, while treatment with GH restored mRNA for IGF-I. However, mRNA for IGF-I or GH receptor did not change after a luteolytic dose of $PGF_2\alpha$. Perhaps, regulation of the action of GH through IGF-I is achieved by changes in the IGF binding protein-1, which is increased in response to $PGF_2\alpha$ in cattle (362).

More recent work has been aimed at understanding the biochemical sites of stimulation of progesterone synthesis by LH in luteal cells, the details of which have been reviewed (343,363). These include activation of protein kinase A, transport of cholesterol through the cytoplasm, and transport of cholesterol from outer to inner mitochondrial membrane. Enhanced cholesterol transport was associated with phosphorylation of steroidogenic acute regulatory protein (StAR) by protein kinase A. In a recent study (364), large luteal cells had sixfold more StAR than small luteal cells in ovine corpora lutea and concentrations of StAR were correlated with serum concentrations of progesterone throughout the estrous cycle. Similarly, endozepine, another protein involved in transport of cholesterol across the mitochondrial membrane, was 2.4-fold greater in large cells, leading those authors to propose that these proteins played a constitutive role in large luteal cells, which, as mentioned above, secrete most of the progesterone, but are unresponsive to LH.

Luteal Regression. Luteolysis is defined as structural demise of the corpus luteum. During luteolysis, loss of the capacity to synthesize and secrete progesterone (365) is followed by loss of the cells that comprise the corpus luteum [reviewed by Pate (366)]. In ruminants, natural luteolysis is dependent upon the presence of the uterus. The existence of a uterine luteolytic factor was first inferred from the prolongation of luteal function following hysterectomy (367). It was soon realized that this factor was acting locally, at least in part, because unilateral hysterectomy affected luteal function on the ipsilateral, but not contralateral, ovary (368–370). The evidence that $PGF_{2}\alpha$ is the uterine luteolysin in the ewe has been reviewed several times (342,344,371,372) and was summarized in the previous version of this chapter (232). $PGF_{2}\alpha$ also inhibits progesterone secretion, and does so at dosages that do not induce structural luteolysis (373).

It has been well documented that $PGF_{2}\alpha$ has at least four major actions in the corpus luteum (343). The temporal sequence of these major, and many other, effects of a luteolytic dose of $PGF_{2}\alpha$ is illustrated in Fig. 11. First, $PGF_{2}\alpha$ stimulates luteal secretion of oxytocin, within 2 to 5 minutes after

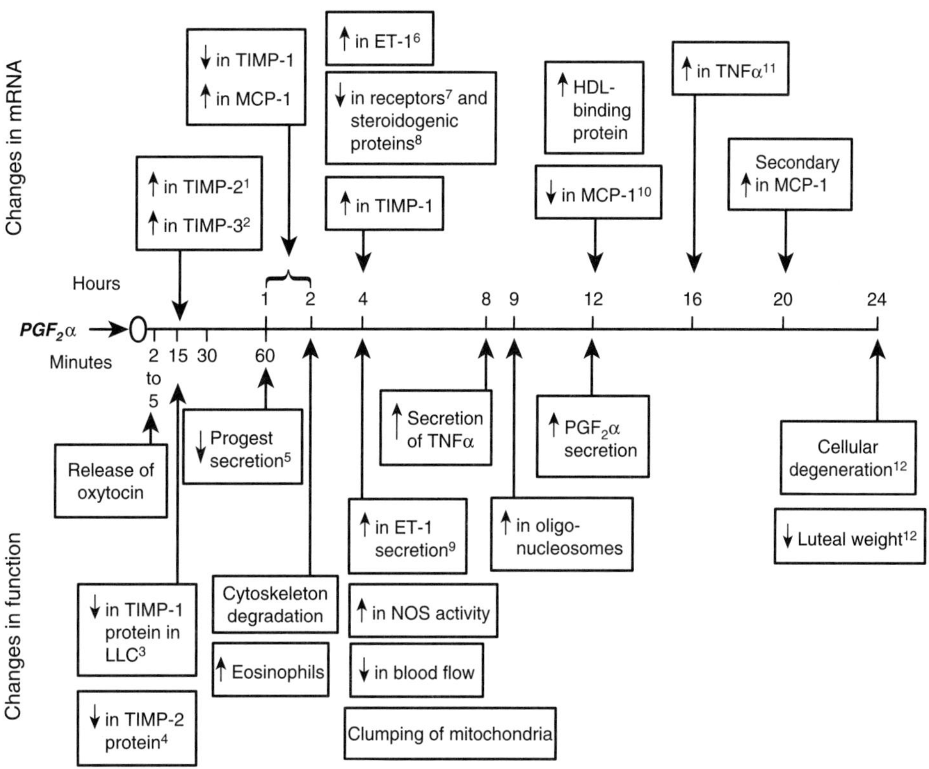

FIG. 11. Time course of intraluteal events in response to $PGF_{2}\alpha$. See text for details of how these changes take place. Data on messenger RNA are presented above the timeline and data on functional and structural changes are presented below the timeline. [Data were adapted from (342,373,374,377–379,384–386,389,401–405).]

[1]Continues through 48 hours. [2]Continues through 6 hours. [3]Continues through 8 hours. [4]Continues through 24 hours. [5]More often detected by 2 to 4 hours because of sampling at same times as for blood flow. [6]Continues through 12 hours. [7]Low density lipoprotein, LH and $PGF_{2}\alpha$. [8]StAR, 3βHSD, P450 scc. [9]Continues through 20 hours. [10]Continues through 16 hours. [11]Occurs in luteal macrophages. [12]Through 36 hours and beyond.

an i.m. injection (374) or within 15 minutes after beginning infusion into a uterine lymphatic vessel (375). Because the stimulation of oxytocin release by $PGF_2\alpha$ has recently been reviewed by Stormshak (376), we will focus here on the other three actions of this prostaglandin. Second is a rapid decrease in luteal blood flow (detected by 4 h), concomitant with the third action of $PGF_2\alpha$, a decrease in the secretion of progesterone (detected as early as 1–2 h and usually by 4 h) due to activation of the protein kinase C (PKC) second messenger pathway. Fourth, $PGF_2\alpha$-induced increases in intracellular calcium appear to cause cell death via apoptotic mechanisms and structural luteolysis. Much of the research in the ruminant corpus luteum in the last decade has been directed to the mechanism by which $PGF_2\alpha$ induces these effects, including the possible involvement of a fifth action, stimulation of luteal production of the vasoactive factors, prostaglandins, endothelin, angiotensin II and nitric oxide, as intermediates, or collaborators in a luteolytic cascade.

While functional (i.e., decrease in progesterone secretion) and structural luteolysis have traditionally been viewed as distinct processes, there is recent evidence that they may be coupled. Injection of $PGF_2\alpha$ induced depletion of tissue inhibitor of metalloproteinase-1 (TIMP-1) protein in large luteal cells within 15 minutes (377) (Fig. 11). Messenger RNA for TIMP-1 decreased at 1 and 2 hours after $PGF_2\alpha$, but increased beyond control values at 4 and 6 hours. In contrast, mRNA for TIMP-2 and TIMP-3 increased following $PGF_2\alpha$. Furthermore, Murdoch (378) has shown degradation of the cytoskeleton in ovine large luteal cells within 2 hours after treatment with $PGF_2\alpha$. Clumping of mitochondria in one side of the cell was observed at 4 hours (378). These changes were followed by an increase in matrix metalloproteinase 2 (MMP-2) activity at 8 hours (379) and an increase in oligonucleosomes at 9 hours (373). Thus, changes expected to alter the structure of the extracellular matrix, and to be associated with cell death and structural luteolysis, preceded the nadir in progesterone concentrations at about 8 hours, and may play a role in functional as well as structural luteolysis. This possibility is supported by the observation that sub-luteolytic infusion of $PGF_2\alpha$ induced precipitous declines in TIMP-1 and -2 protein lasting 8 or 24 hours, respectively (379). Thus, rapid structural changes may influence steroidogenic capacity of luteal cells by inducing changes in conformation or localization of LH receptors in the plasma membrane (344). In support of this hypothesis, mRNA for TIMP-2 declined in corpora lutea during the late luteal phase (380). In contrast, mRNA in ovine corpora lutea increased for MMP-14 on day 4 and was greatest for MMP-2 on day 10, associated with increases in gelatinolytic activity (381). MMP-2 and -9 were localized primarily in large luteal cells, but MMP-2 appeared to be present in endothelial cells on day 15. These authors concluded that remodeling of the extracellular matrix occurs throughout the luteal phase and may "facilitate processes important to luteal development, maintenance and regression."

A component of luteal regression appears to be the influx of immune cells into the tissue, which is important for structural luteolysis [see reviews by Pate (366,382) and Penny (383)]. Monocyte chemoattractant protein-1 (MCP-1) may be responsible for the influx of macrophages into the corpus luteum because increases in mRNA for MCP-1 were detected with 1 to 2 hours after injection of $PGF_2\alpha$ (384,385). This increase was followed by a decrease at 12 to 16 hours, and a secondary increase by 24 hours. A similar, but more prolonged response followed stimulation of protein kinase C, whereas treatment with luteotropin (LH) or stimulation of cAMP/protein kinase A reduced mRNA for MCP-1 (385). In addition, eosinophils increased within regressing ovine corpora lutea, and in response to treatment with $PGF_2\alpha$, such an increase was observed within 2 hours (386) (Fig. 11), particularly in association with connective tissue trabeculae and blood vessels.

The fact that corpora lutea of domestic animals can secrete prostaglandins has been known for some time (387,388), but was largely ignored in development of concepts of the regulation of luteolysis and luteal maintenance. To test whether luteal secretion of $PGF_2\alpha$ is required for luteolysis, Griffeth et al. (389) placed implants containing two dosages of indomethacin in the corpus luteum of ewes on day 11 of the estrous cycle. Serum progesterone declined by day 17 and ewes returned to estrus, but luteal weights on day 18 were heavier with either dosage of indomethacin than in control ewes. In a second experiment, peaks of the inactive metabolite of $PGF_2\alpha$, 15-keto, 13,14-dihydro-$PGF_2\alpha$ (PGFM), observed during 12 to 36 hours after initiation of luteolysis, did not differ between intact and hysterectomized ewes, indicating that the $PGF_2\alpha$ secreted in response to treatment was not of uterine origin.

Factors modifying luteal secretion of luteotropic (PGI_2, PGE_2) and luteolytic ($PGF_2\alpha$) prostaglandins have been examined in the cow (358), but have received little attention in the ewe. For that reason, the bovine data will be summarized briefly here. Synthesis of PGI_2 (measured as its metabolite 6-keto-$PGF_1\alpha$) by bovine luteal cells in vitro was greater early in the estrous cycle (390), but was not affected by LH (391). Content of $PGF_2\alpha$ was lower than that of 6-keto-$PGF_1\alpha$ at all stages of the cycle, but luteal production of $PGF_2\alpha$ was greater in cells from day-5 corpora lutea than from corpora lutea collected later in the cycle. Progesterone decreased synthesis of both prostaglandins by bovine luteal cells from days 6 and 10 of the cycle in culture, resulting in an increased ratio of 6-keto-$PGF_1\alpha$ (luteotropic) to $PGF_2\alpha$ (luteolytic), but had no effect on cultured cells from days 4 or 18 (392). Thus loss in the ability of progesterone to suppress luteal secretion of $PGF_2\alpha$ might contribute to luteolysis. PGI_2 suppressed basal secretion of PGE_2 and $PGF_2\alpha$, and $PGF_2\alpha$ suppressed secretion of PGI_2; PGE_2 had no effect on synthesis of either PGI_2 or $PGF_2\alpha$ (393). Thus, during luteolysis, $PGF_2\alpha$ might suppress production of a luteotropic agent (PGI_2) that inhibits local production of $PGF_2\alpha$. Pate's group further examined effects of cytokines on synthesis of PGI_2, PGE_2, and $PGF_2\alpha$. Interleukin-1β stimulated synthesis of all three prostaglandins; interferon-γ initially decreased, but then stimulated secretion of PGI_2 and $PGF_2\alpha$ (394) and TNF-α increased synthesis of both PGI_2 and $PGF_2\alpha$, but was ultimately cytotoxic when combined with interferon γ (395), perhaps by upregulating interferon receptors (J. L. Pate, personal communication). These cytokines are produced by immune cells (discussed above) and their potential functional significance in luteolysis has been addressed in some detail (366,382).

Three other vasoactive substances have been shown to increase in ruminant corpora lutea in response to $PGF_2\alpha$: endothelin 1 (ET-1) [reviewed by Meidan et al. (396)], angiotensin II (397–399), and nitric oxide (400). Only endothelin has received any detailed study in the ewe, but it was recently reported that activity of endothelial nitric oxide synthase increased at 4 hours after treatment with $PGF_2\alpha$ at mid-cycle in the ewe (401).

A luteolytic dose of $PGF_2\alpha$ stimulated gene expression for ET-1 in ovine corpora lutea (402) and in a study using microdialysis in vivo (403), perfusion of $PGF_2\alpha$ increased secretion of ET-1 over a period of 12 hours (Fig. 11). ET-1 inhibited basal and LH-stimulated secretion of progesterone in dispersed ovine luteal cells and the effect was blocked by an antagonist specific for the receptor for endothelin, ET_A (402). The ET_A receptor antagonist also delayed the decrease in plasma progesterone in response to $PGF_2\alpha$. Injection of 100 μg ET-1 alone did not regress ovine corpora lutea, but regression was observed in ewes treated with 100 μg ET-1 in combination with a subluteolytic dose of $PGF_2\alpha$ (402). Likewise, $PGF_2\alpha$ followed by ET-1, simultaneous perfusion of $PGF_2\alpha$ and ET-1, two successive perfusions of ET-1, or ET-1 followed by TNFα decreased progesterone release and caused a massive release of oxytocin in the study using microdialysis of ovine corpora lutea (403). Finally TNF-α, which is secreted by luteal cells and macrophages and was discussed earlier in relation to its role in follicular rupture, was elevated in regressing ovine corpora lutea after progesterone secretion had decreased (404) (Fig. 11), and has been proposed as a component of luteal regression in cattle and pigs (396). Considering all the observations in sheep and cattle, it appears that intraluteal secretion of $PGF_2\alpha$, ET-1, and angiotensin II in response to uterine $PGF_2\alpha$ is integral to luteal regression during the estrous cycle in ruminants, and that these substances interact in the luteolytic cascade.

Factors Affecting Luteal Sensitivity to $PGF_2\alpha$. An interesting research area in recent years has been the effort to understand differences in luteal sensitivity to regression by $PGF_2\alpha$. These studies have been conducted in cattle, sheep and pigs, and there appears to be a great deal of commonality among the species. In the ewe, a single injection of $PGF_2\alpha$ usually does not cause luteolysis before day 5 (reviewed in 232,343,406). However, longer acting analogues of $PGF_2\alpha$ in sheep (406) or repeated treatments in cows (362) were effective as early as day 3 of the cycle. Luteal sensitivity could not be accounted for by changes in concentrations of receptor for $PGF_2\alpha$, which did not differ between days 2 and 15 or 4, 10, and 15 in the ewe (407,408).

In studies with single treatments, changes in response to $PGF_2\alpha$ between early and mid-luteal phases have been observed in mRNA for cyclooxygenase-2 (prostaglandinG/H synthase-2) (409), MCP-1 mRNA (385), and TIMP-1 protein (377). The protein for cyclooxygenase-2 increased in response to $PGF_2\alpha$ in mid-cycle ovine corpora lutea (410). Similarly, Arosh et al. (411) have shown that mRNA's for PGE synthase, PG transporter and receptors for PGE_2 and $PGF_2\alpha$ vary with stage of cycle in the bovine corpus luteum. Preferential expression during periods of luteal resistance to $PGF_2\alpha$ favored PGE_2 synthesis, transport and signaling; in contrast, during periods of luteal sensitivity to $PGF_2\alpha$, PGF receptor and $PGF_2\alpha$ synthesis (via aldoketoreductase 1B5, which also can inactivate progesterone to 20α-OH-pregn-4-ene-3-one) were favored. These differences might act in concert with the decrease later in the cycle in mRNA for and activity of prostaglandin dehydrogenase (PGDH), which inactivates $PGF_2\alpha$ to PGFM, in the corpus luteum (101). Activity of that enzyme was greater in pregnant than in nonpregnant ewes on day 14 post-estrus, despite a lack of differences in mRNA on day 12.5 (412). However, there were no changes in mRNA for PGDH with stage of cycle in the cow (411).

Another source of differences in sensitivity to $PGF_2\alpha$ was identified by Juengel et al. (408). They found that mRNAs for two inhibitors of protein kinase C, the second messenger system for $PGF_2\alpha$, were greater on day 4 than on days 10 and 15 of the cycle. The mRNAs for these inhibitors did not change in response to $PGF_2\alpha$; thus, they might confer resistance to the luteolysin in young corpora lutea. Overall, it appears that components of prostaglandin synthesis, metabolism and action can all vary within the corpus luteum, and the relative importance of each of these changes in sensitivity to $PGF_2\alpha$ remains to be determined.

In the pig, treatment with $PGF_2\alpha$ on day 17, when the corpora lutea are sensitive, but not on day 9, when they are not sensitive, decreased mRNA for ER-α at 10 hours after treatment (413). ERα increased from days 6 to 9, a period of increasing sensitivity to $PGF_2\alpha$, in the ovine corpus luteum (414). Combining these facts with the observation, discussed above, that ER-β was greatest in the early corpus luteum in the ewe (352), one could propose that the differing actions of estrogen at different stages of the cycle are mediated by different receptors. Many years ago, Piper and Foote (415) observed that daily treatment with estradiol, beginning early in the cycle, maintained ovine corpora lutea. However, a single treatment with estradiol at mid-cycle or later is well known to be luteolytic, by virtue of causing increased uterine secretion of $PGF_2\alpha$ (416).

Regulation of Secretion of $PGF_2\alpha$ by the Uterus and Transport to the Ovary

There is still considerable controversy as to the mechanism by which the increase in $PGF_2\alpha$ secretion that causes luteolysis is initiated and maintained. Three ovarian hormones have been implicated in regulation of uterine $PGF_2\alpha$: progesterone and oxytocin, which are produced by the ovine corpus luteum, and estradiol (417–419). In the previous version of this chapter (232), and more recently (344,356), the evidence that each of these hormones can affect secretion of $PGF_2\alpha$ and the physiological significance of each hormone was reviewed in detail.

Progesterone clearly plays a major role in regulating the timing of the increases in episodic uterine secretion of $PGF_2\alpha$ (107) and modulates the magnitude of those increases (420,421). As illustrated in Fig. 12, injection of progesterone at estrus advanced the occurrence of the initial increase in $PGF_2\alpha$ from 12.3 ± 1.0 to 8.0 ± 0.3 days after estrus (107). Conversely, the progesterone antagonist, RU486, delayed the onset of $PGF_2\alpha$ release and luteolysis,

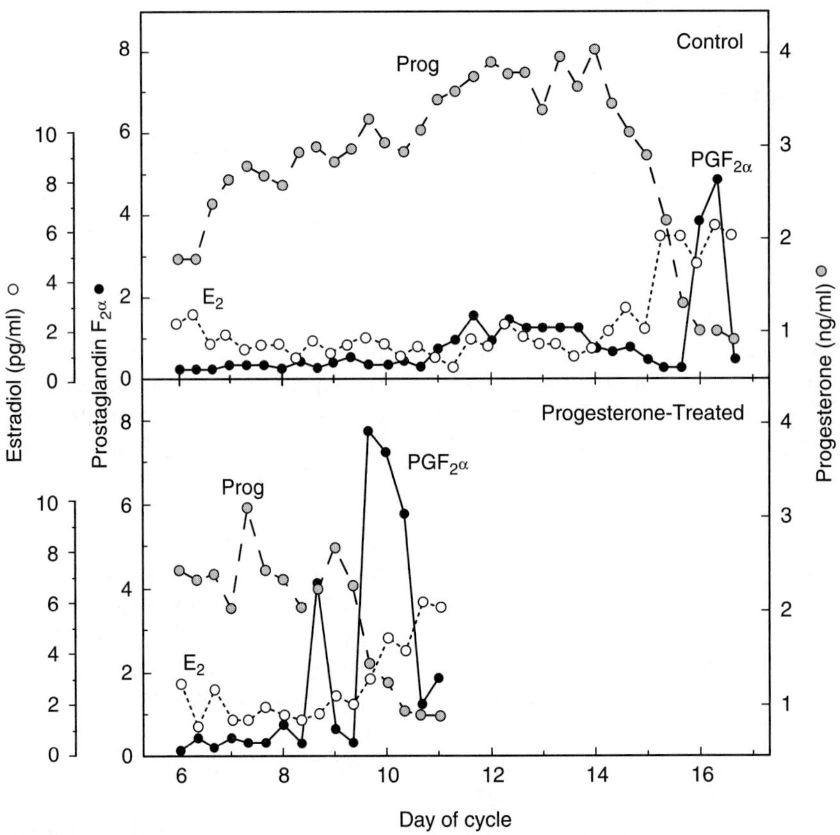

FIG. 12. Effects of progesterone treatment early in the estrous cycle (days 0 and 1) on mean utero-ovarian vein concentrations of $PGF_2\alpha$ (*solid circles* and *line*) and mean peripheral concentrations of progesterone (*shaded symbols* and *dashed line*) and estradiol (*open symbols* and *dotted line*) in samples collected every 8 hours from days 6 to 17 of the estrous cycle. $PGF_2\alpha$ was measured in samples via an indwelling catheter and a rolling 3-sample average was plotted. [Replotted from (107), with permission from the Society for the Study of Reproduction.]

if given early during the luteal phase (422). The ability of progesterone to properly time luteolysis is dependent upon previous exposure of the uterus to progesterone (423) followed by estrogen (424). Data in cattle indicated that the key effect is the establishment of increased numbers of progesterone receptors and decreased numbers of oxytocin receptors in the uterus (425). Thus, as demonstrated in both ewes (426,427) and cows (424,428), progesterone from one cycle, and the subsequent rise in estrogen at estrus prevent premature release of $PGF_2\alpha$ in the next cycle. The modulating effect of progesterone on secretion of $PGF_2\alpha$ in the ewe is illustrated by the fact that progesterone withdrawal on days 12 to 13 of the estrous cycle enhanced secretion of $PGF_2\alpha$ over the next few days (420,423,429), whereas midrange secretion patterns (sufficient to induce luteolysis) were maintained by continued progesterone (420,421).

In addition to the role of estradiol in development of progesterone receptors discussed above, tonic secretion of follicular estradiol is necessary for normal luteolysis, as demonstrated in studies using follicle ablation, x-irradiation, ovariectomy and hysterectomy (353,354,430–432). However, no increase in secretion of estradiol is required to support increases in secretion of $PGF_2\alpha$ preceding normal luteolysis (420,421); the increase in estradiol begins after the initiation of luteolysis (92), and blocking it with exogenous progestogen had no effect on luteolysis (429,432). Nevertheless, exogenous estrogen causes luteolysis in mid- to late luteal phase, an effect that requires previous exposure of the uterus to progesterone (102,416,434,435), and is mediated by increases in secretion of $PGF_2\alpha$ (416). Treatment with estradiol increased the secretion of $PGF_2\alpha$ in response to oxytocin, in anestrous ewes, early in the estrous cycle, and during the late luteal phase (reviewed by 232,426,436). This might be the mechanism by which treatment with a dosage of estradiol-17β (80 μg twice daily) that was not luteolytic alone, was able to extend the occurrence of luteolysis in response to the presence of an intrauterine device from day 3 until at least day 5 of the luteal phase (437).

As documented in previous reviews (232, 344,438), oxytocin will stimulate release of $PGF_2\alpha$ from the uterus. Thus, it has been used in many studies as a test of the ability of the uterus to secrete $PGF_2\alpha$ under different physiological conditions. However, its significance in controlling secretion of $PGF_2\alpha$ responsible for luteolysis was questioned as early as 1986, when it was observed that the frequency of pulses of oxytocin did not differ between non-pregnant and pregnant ewes (439), despite the fact that luteolysis was prevented in the latter. Nevertheless, the concept that became accepted to a large extent, as delineated in the reviews cited above was as follows:

Progesterone down-regulates its own receptors as the luteal phase progresses, allowing up-regulation of oxytocin receptors in the endometrium. Then pulses of oxytocin, from the corpus luteum, stimulate episodes of uterine secretion of $PGF_2\alpha$, which in turn, stimulate secretion of luteal oxytocin, initiating a positive feedback loop between uterine $PGF_2\alpha$ and luteal oxytocin. This loop was proposed to play a central role in the generation of episodic secretion of these two hormones (422,440).

Recent data have cast doubt on the requirement for luteal oxytocin in luteal regression in ruminants; exogenous oxytocin was ineffective at any stage of the cycle in the sheep (441). Depletion of oxytocin from the bovine corpus luteum (by treatment with noradrenaline) before the expected time of luteolysis did not affect duration of the estrous cycle (442) and an oxytocin receptor antagonist did not alter luteolysis in cattle (443). McCracken et al. (444) confirmed this finding after depleting oxytocin from the ovine corpus luteum (with infusions of $PGF_2\alpha$). Finally, Arnold et al. (445) found that oxytocin did not stimulate secretion of $PGF_2\alpha$ by endometrial explants from either cyclic or pregnant cows on day 17 post-estrus.

As pointed out by both Goodman (232) and McCracken (344), episodes of secretion of $PGF_2\alpha$ and oxytocin occur independently of each other as well as together (420,421,446,447), and more data support the importance of $PGF_2\alpha$ in causing release of oxytocin than vice versa (374,439,440,448). In addition, the increase in uterine responsiveness to oxytocin late in the estrous cycle (449,450) may occur after the start of luteolysis (451).

Obviously, there are many questions about the validity of the concept that pulses of luteal oxytocin generate episodes of uterine secretion of $PGF_2\alpha$. Therefore, Chami et al. (452) examined the possibility that uterine platelet-activating factor (PAF), which was known to induce uterine release of $PGF_2\alpha$ in the same manner as oxytocin, could act as a pulse generator for $PGF_2\alpha$. They found that treatment of ovariectomized ewes with progesterone and estrogen, mimicking the normal luteal phase, caused release of PAF into the uterine lumen. Further, exogenous PAF stimulated secretion of $PGF_2\alpha$, as measured by PGFM, and interferon τ, which reduces pulses of PGFM in pregnant ewes, inhibited the effect of PAF. Finally, oxytocin and PAF together produced a synergistic effect on PGFM. They concluded that PAF fulfilled the criteria of an endogenous initiator of release of $PGF_2\alpha$, and that the synergism of PAF and oxytocin could account for the high amplitude of secretion of $PGF_2\alpha$ during luteolysis in the ewe.

Based on the effects following partial hysterectomy (368,370) and vascular anastomosis (370), initiation of luteolysis by $PGF_2\alpha$ appears to be a local effect between each uterine horn and its ipsilateral ovary [reviewed by Ginther (370)]. It was postulated that $PGF_2\alpha$ enters the ovarian artery from the utero-ovarian vein, via a countercurrent exchange mechanism (370) that allows $PGF_2\alpha$ to reach the ovary without entering the pulmonary circulation, where it would be enzymatically inactivated in the lungs (453). More recently, Bonnin et al. (454) have presented strong evidence for a re-interpretation that the process involves both rapid systemic transport (12% of $PGF_2\alpha$ was not oxidized by the lungs) and a slower local diffusion pathway that utilizes the uterovarian veins and lymph vessels.

Knowledge of the involvement of the lymph vessels has come from the work of Brian Heap and colleagues, who first reported that $PGF_2\alpha$ from the uterine lumen was transferred into uterine lymph and that more $PGF_2\alpha$ was transferred from uterine lymph vessels to the adjacent ovary, corpus luteum and ovarian vein than from the utero-ovarian vein (455). In subsequent studies (456,457), they demonstrated that $PGF_2\alpha$ infused into a uterine lymphatic vessel caused a rapid release of oxytocin from both ovaries containing corpora lutea, whereas infusion into a single ovarian lymphatic vessel caused release of oxytocin only from the adjacent ovary containing a corpus luteum.

Utilizing extensive data from the pig and Heap's observations in the sheep, Krzymowski and Stefanczyk-Krzymowska (458) have proposed a new theory to explain how luteolysis is initiated. They pointed to the fact that uterine weight increases 40% to 60% under the influence of progesterone and that blood flow to the uterus decreased by 60% to 70% by day 12 of the luteal phase in pigs (when increased secretion of $PGF_2\alpha$ begins in that species). Thus, a decrease in oxygen and resultant apoptosis of endometrial cells lead to secretion of $PGF_2\alpha$, in their model. They argued further that episodic increases in $PGF_2\alpha$ in uterine venous blood "do not confirm pulsatile secretion of $PGF_2\alpha$," but rather may result from rhythmic uterine contractions in response to oxytocin. It will be interesting to see how these new concepts, relative to roles of PAF and uterine blood supply, withstand the scrutiny of future investigations.

RESYNTHESIS, A MODEL FOR THE OVINE ESTROUS CYCLE

In this section, we will use the previous analysis on the control of the individual components to describe how these components interact to produce the ovine estrous cycle. This systems analysis approach, and space constraints, preclude inclusion of some of the detailed mechanisms described in each of the previous sections. In this synthesis, it is useful to make a distinction between ovarian events controlled by LH from those controlled by FSH because they function largely independently of each other.

The estrous cycle can be viewed as a sequence of causally-related endocrine events, with each step in the sequence initiating the subsequent hormonal change. Because the estrous cycle is a recurring chain of events, the description of these events can begin, in theory, at any point in the cycle. However, the pre-ovulatory LH surge is a particularly appropriate starting point because it initiates formation of a new ovarian structure and occurs on what is conventionally taken as day 0 of the ovine estrous cycle.

The LH surge produces ovulation and luteinization of the follicular remnants, and as the corpus luteum develops, progesterone concentrations begin to rise (Fig. 13). Because the luteotropic actions of LH are permissive in the ewe, the pattern of progesterone throughout the luteal phase (from ovulation to just before luteolysis) reflects internal factors within the corpus luteum and its secretory capacity; it is thus an indirect consequence of the luteinizing actions of the LH surge. The elevated progesterone concentrations during the luteal phase have several major actions critical to the estrous cycle. Progesterone primes the centers in the brain controlling estrous behavior and the GnRH surge so that the follicular-phase estradiol rise induces estrus and a maximal GnRH surge. The early increase in progesterone (days 3–5) inhibits uterine $PGF_2\alpha$ secretion, but also programs the uterus to release $PGF_2\alpha$ 7 to 8 days later. It also modifies uterine function so that no premature release of $PGF_2\alpha$ occurs early in the next estrous cycle, ensuring a normal luteal phase in that cycle. Finally, progesterone inhibits tonic LH secretion by suppressing GnRH pulse frequency, so that circulating LH concentrations decrease to a nadir at mid-cycle and remain low until luteolysis. The low concentrations of LH are necessary for follicular waves and the minimal support for luteal function allows the increase in $PGF_2\alpha$ that occurs between days 11 and 13 to induce luteolysis, and consequently progesterone concentrations fall precipitously. The fall in progesterone at luteolysis releases three hypothalamic neural systems from the inhibitory effects of this steroid. It allows the centers controlling the GnRH/LH surge and estrous behavior to respond to estradiol and allows the GnRH pulse generator to increase in frequency. The resulting rise in tonic LH concentrations maintains follicular growth and stimulates estradiol secretion from the ovary. The sustained follicular-phase elevation in circulating

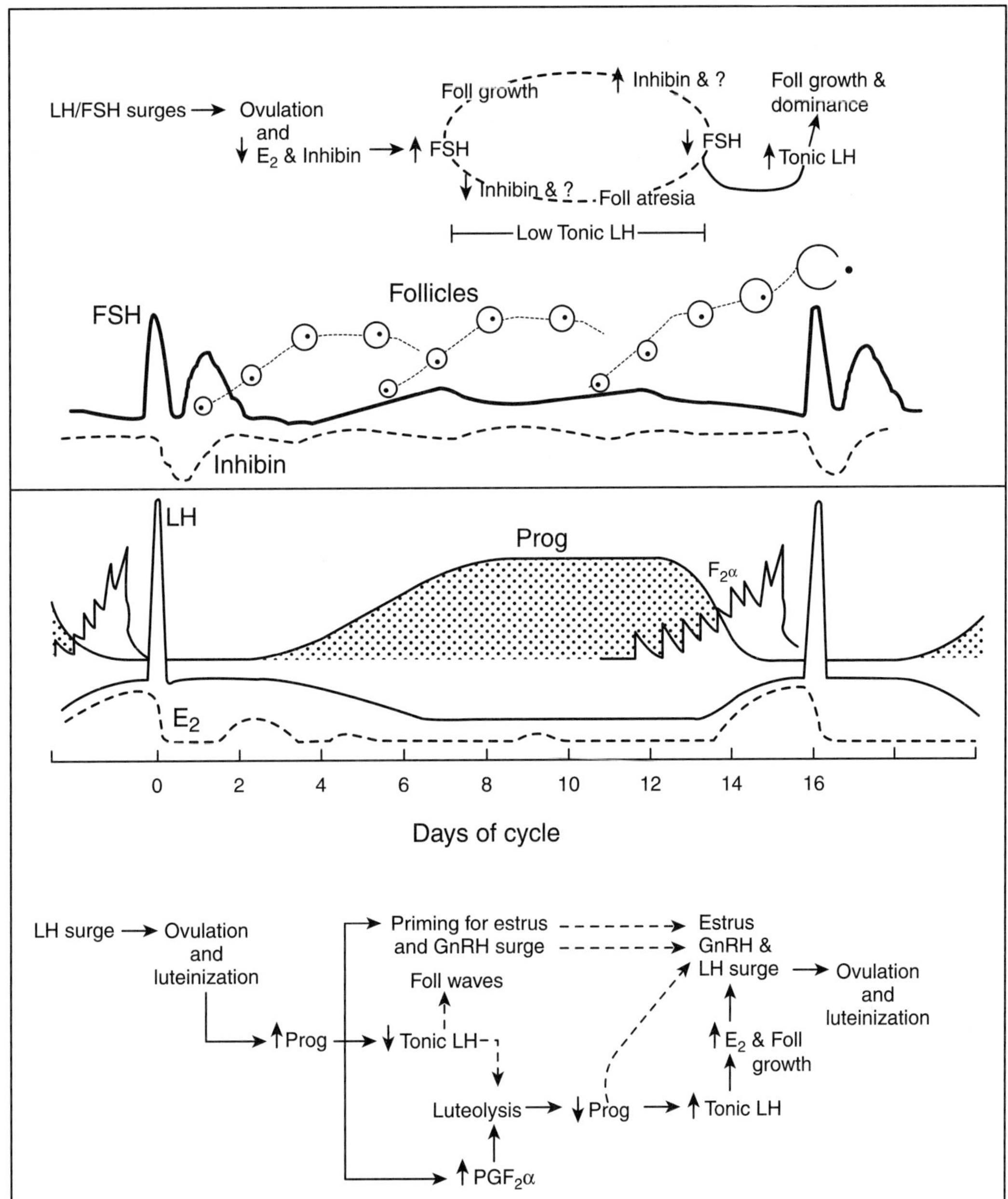

FIG. 13. Model for the two major feedback systems controlling the ovine estrous cycle. The top half of the figure presents a schematic depiction of circulating concentrations of FSH and inhibin, the feedback cycle thought to be responsible for waves of follicular development during the luteal phase, and the role of tonic LH secretion in the final stages of follicular growth in the follicular phase. The bottom half of the figure depicts a schematic pattern of progesterone (Prog), LH, estradiol (E_2), and $PGF_{2}\alpha$ secretion, and the sequence of causal relationships that determines the timing of major events during the estrous cycle. In the latter, *solid arrows* represent direct effects, and *dashed arrows* represent permissive actions. See text for further details.

estradiol triggers estrous behavior and the preovulatory GnRH and LH surges, and is also important for programming the uterus for normal $PGF_{2}\alpha$ secretion during the next estrous cycle. The LH surge, in turn, terminates estradiol secretion and induces ovulation and luteinization so the next cycle begins.

At the same time these events are occurring, a feedback loop between developing follicles and FSH secretion determines follicular dynamics. The preovulatory gonadotropin surges terminate secretion of both estradiol (caused by the LH surge) and inhibin (caused by the FSH surge?) from the ovary, and the fall in these two hormones allows FSH concentrations to increase on day 1. This increase in FSH stimulates follicular development and, in the presence of elevated tonic LH levels, both estradiol and inhibin concentrations increase as a cohort of follicles develop (Fig. 3). These (and possibly a third) ovarian hormones, in turn inhibit FSH secretion, which eventually causes atresia in these follicles if tonic LH secretion is also low. As follicular function wanes, the fall in their negative feedback signal allows FSH concentrations to increase again and a second wave of follicles begins to develop. This temporal sequence is postulated to recur throughout the luteal phase (except there is little, if any, increase in peripheral estradiol concentrations), producing several waves of follicular development. It should be noted, however, that the limited data now available do not support a role for inhibin in controlling the changes in FSH secretion associated with follicular waves so that an unidentified follicular hormone may be critical for this feedback loop. Regardless of the follicular signals involved, the characteristics of the feedback loop between FSH and follicular hormones, and the low concentrations of LH, prevent the development of active follicular dominance in the ewe. This, in turn, ensures that the ovaries always contain follicles ready to begin the final phase of pre-ovulatory development if luteolysis occurs, which accounts for the short duration of the follicular phase (459). When luteolysis occurs, the resulting increasing frequency of LH pulses maintains follicular growth in the face of falling FSH concentrations and promotes follicular dominance so that in most breeds only 1 to 2 ova are ovulated.

The feedback loop between follicles and FSH secretion functions in parallel with, and relatively independent of, the progesterone feedback loops controlling luteolysis and the events of the follicular phase. Consequently changes in FSH concentrations play little role in timing the events of the estrous cycle. It is thus not surprising that changes in ovulation rate have little influence on the control of ovarian cycles; estrous cycles in multiple ovulators are very similar to those in ewes ovulating a single ovum (6,7).

One important characteristic of this model is the major role of ovarian steroids in control of the estrous cycle. Perhaps the three critical events timing the estrous cycle are luteolysis, estrous behavior, and ovulation-luteinization, and the occurrence of all three are ultimately controlled by ovarian steroids. Progesterone, by ensuring that $PGF_2\alpha$ secretion increases late in the luteal phase, controls luteolysis, while the follicular-phase estradiol rise is responsible for estrous behavior and, by triggering the LH surge, ovulation. It is interesting to note that the secretory products of the corpus luteum and pre-ovulatory follicle initiate events leading to the destruction of each structure. The function of both tissues is thus self-limiting, a characteristic that allows for the cyclic nature of ovarian activity. While the pre-ovulatory estradiol rise is critical for events during the follicular phase, it is in turn controlled by progesterone. If progesterone concentrations are high, tonic LH secretion is suppressed, and consequently the estradiol rise cannot occur; if progesterone falls, tonic LH secretion increases and stimulates the pre-ovulatory estradiol increment. One can thus infer that in the ewe, as in the rhesus monkey (460), a "pelvic clock" controls the estrous cycle, but in the case of the sheep the corpus luteum represents a larger component of this clock than in primates.

That the corpus luteum plays a primary role in timing the events of the ovine estrous cycle is not a new concept. Indeed it was first suggested in the 1930s following the experimental observation that lutectomy at mid-cycle resulted in estrous behavior and ovulation within 2 to 3 days (2), and a critical role for progesterone was incorporated into the first model for control of estrus and ovulation in the ewe (9). This model, developed by Robertson prior to the general availability of radioimmunoassays, accurately predicted most of the critical events leading from the demise of the corpus luteum to the induction of ovulation (9); its only inaccurate prediction was that an increase in FSH was the stimulus for the pre-ovulatory estradiol rise. The importance of progesterone to the ovine estrous cycle was further confirmed by the first measurements of circulating hormonal concentrations throughout the cycle, so that Hansel and Echternkamp concluded in their 1972 review (461) that "the dominant event in the cycle... seems to be the rapid regression of the corpus luteum at the appropriate time." The emphasis on the importance of luteolysis to the control of the estrous cycle sparked interest in the mechanisms controlling luteal lifespan and led to the discovery of the luteolytic effects of $PGF_2\alpha$ (98). This discovery, in turn, enabled Goding and associates to develop a model encompassing the complete estrous cycle (331,462,463), a model that set the groundwork for many of our current ideas. Subsequent work tended to emphasize the role of estradiol (464), so it was not until the discovery of the negative feedback actions of progesterone (175) and the importance of these actions to control of the follicular phase (13,14) that investigators began to attribute a major role to progesterone in models for the estrous cycle (334,465). This in turn led to the model presented in our previous chapter that accounted for the timing of major events in the ovine estrous cycle, but did not include FSH and follicular dynamics (232). Thus, the current model (Fig. 13), which incorporates recent data on the latter, is the most complete representation of the ovine estrous cycle developed to date.

There is one other important characteristic of this model that is relevant to the control of seasonal reproductive patterns in the ewe. Because the estrous cycle is a sequence of hormonal events, with each step depending on the previous one, disruption of a single step will prevent all subsequent events and thereby cause anovulation. A good example of this phenomenon is pregnancy. The presence of a conceptus in utero blocks the luteolytic actions of $PGF_{2}\alpha$ (103,466), probably by stimulating uterine secretion of the luteotropin, PGE_2 (103,467). Consequently, progesterone does not fall, tonic LH secretion and estradiol do not increase, and ovulation does not occur. This anovulatory condition will then continue until removal of the embryo/fetus, by death or parturition, allows progesterone concentrations to fall. Similarly, disruption of a single endocrine event accounts for the anovulatory condition of anestrus, although in this case the blockade occurs at a different step in the chain of events comprising the ovine estrous cycle.

ENVIRONMENTAL CONTROL OF OVARIAN CYCLICITY IN THE EWE

So far, this model has considered only the role of endogenous hormonal signals in the control of the ovine estrous cycle. However, a variety of external environmental cues can also influence ovarian function in the ewe. The most obvious example of environmental influences is photoperiod, which determines when the breeding and anestrous seasons occur (468; Chapter 41). However, a number of other environmental factors, including stressors (469, Chapter 48), nutrition (308,470), and pheromones (471) influence reproductive function in the ewe. As there are excellent recent reviews on these topics (cited above), here we will briefly consider each, emphasizing the underlying changes in hypothalamic function and their effects on events within the estrous cycle.

Photoperiodic Control of Seasonal Breeding

That the annual reproductive cycle of the ewe is controlled by photoperiod was first proposed by Marshall (472) and verified experimentally by Yeates (473). The effects of photoperiod in the ewe reflect a complex interaction between day length (hours of light/day) and an endogenous circannual rhythm that are described in detail in Chapter 41. Thus, we will focus on the physiological mechanisms by which changes in day length alter ovarian function. The key observation that laid the foundation for our understanding of these events was the demonstration of a dramatic seasonal variation in the negative feedback actions of estradiol (180) that is due to seasonal changes in inhibition of GnRH and LH pulse frequency (173,474,475). During anestrus, estradiol suppresses GnRH and LH pulse frequency and thereby strongly inhibits tonic LH secretion, whereas during the breeding season this steroid cannot inhibit pulse frequency and thus produces only a modest inhibition in mean LH concentrations. There is also a modest decrease in LH pulse frequency evident during anestrus in ovariectomized ewes (474,475), but the physiological significance of this "steroid-independent" effect of photoperiod is unclear.

Neuroendocrine Mechanisms Mediating Seasonal Change in Response to Estradiol Negative Feedback

The mechanisms by which a change in photoperiod is transformed into an alteration in estradiol negative feedback can be divided into three major steps (476–478). First, photoperiodic information is received by the retina and a neural signal is transmitted via the suprachiasmatic nucleus, paraventricular nucleus and superior cervical ganglion to the pineal gland. Second, the pineal gland transduces the neural signal into an endocrine signal: the serum and CSF melatonin patterns. The pineal secretes melatonin only at night, so the duration of the daily increment in melatonin concentrations is a hormonal analogue of the external photoperiod. Third, the melatonin pattern is translated in the hypothalamus to produce a change in its response to estradiol negative feedback.

The exact mechanisms by which melatonin alters hypothalamic function to produce changes in estradiol negative feedback are still under investigation. There is strong evidence that melatonin acts in an area just rostral to the mammillary bodies (the premammillary region) to control changes in estradiol negative feedback at the transition from anestrus to the breeding season (479,480); whether it acts in this area to time the transition to anestrus has not been examined for technical reasons. The premammillary region contains three major nuclei (481), but their roles in mediating the actions of melatonin have yet to be determined. There is also sufficient information to construct a model for the changes in hypothalamic function responsible for the increased response to estradiol negative feedback in anestrus. This model grew out of the observation that pentobarbital anesthesia increased LH pulse frequency during anestrus, but not in the breeding season (482). Because pentobarbital decreases neuronal activity (483), these data imply that during anestrus, LH pulse frequency is held in check by a set of inhibitory neurons; pentobarbital decreases the activity of these inhibitory neurons, allowing pulsatile LH secretion to increase. This inference led to the proposal that the long-day

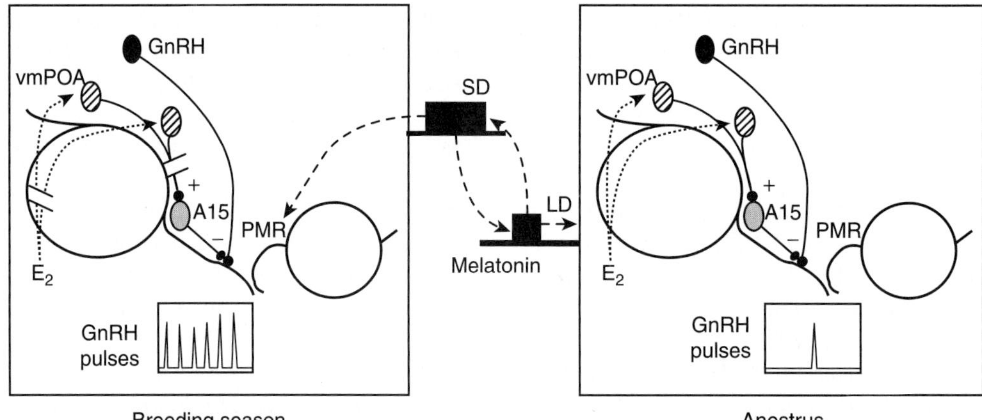

FIG. 14. Model for the seasonal changes in response to estradiol (E_2) negative feedback that postulates an E_2-responsive inhibitory neural system is active only during anestrus. During the breeding season (*left*), this system is inactive so that E_2 cannot inhibit GnRH pulse frequency and progesterone is the primary regulator of tonic GnRH secretion. During anestrus (*right*), this system is active, allowing E_2 to act on ER-α–containing neurons in the vmPOA and retrochiasmatic area, which in turn stimulate activity of A15 DA neurons. These DA neurons act in the MBH to inhibit GnRH pulse frequency, thus inhibiting LH secretion. The long-day (LD) melatonin pattern of anestrus activates this neural circuit by increasing the responsiveness of the ER-α–containing neurons and/or in their ability to stimulate A15 DA neurons. The short-day (SD) melatonin pattern acts in the premammillary region (PMR) to inactivate this system during the transition to the breeding season.

photoperiod of anestrus activates a set of estradiol-sensitive neurons that inhibit the GnRH pulse generator (Fig. 14). Estradiol suppresses GnRH pulse frequency by increasing the activity of these neurons, but can do so only in anestrus because these neurons are active only at this time of year.

Work over the last decade has identified some of the critical components of this photoperiod-controlled inhibitory neural system (Fig. 14). At its center is a group of DA neurons clustered at the base of the brain just posterior to the optic chiasm, known as the A15 DA cell group (484–487). These neurons are stimulated by estradiol in anestrus (187,488,489), but not in the breeding season (489), and are necessary for estradiol negative feedback during anestrus (485,490). There is both anatomical (468,491) and pharmacological (492,493) evidence that these A15 neurons project caudally to the MBH and median eminence (494), where they inhibit GnRH pulse frequency. Although estradiol stimulates the A15 DA neurons during anestrus, these neurons do not contain ER-α (135,144) or ER-β (495), observations that led to a search for estradiol-responsive afferents. Early work (484,496,497) raised the possibility that NE neurons in the brain stem could provide this input, but other data (498,499) are not consistent with this possibility. More recent work has implicated estrogen-responsive neurons in the ventromedial POA (500) and retrochiasmatic area (501). ER-α–containing neurons in these areas project to the A15 (468), and local administration of estradiol to either area inhibits LH pulse frequency via a DA system in anestrus, but not in the breeding season (501–503). The phenotype of these neurons has yet to be identified, but there are preliminary data implicating both nitric oxide (504) and GABA (505).

How the activity of this neural circuit is turned on and off by melatonin remains a complete mystery. There are neural projections from the premammillary region to different components of this neural circuit (506) and interesting seasonal changes in the morphology of both A15 and GnRH neurons have been observed. There is a decrease in synaptic contacts onto POA GnRH perikarya in anestrus (148,507), which appears to reflect a shift from stimulatory inputs in the breeding season to inhibitory inputs in anestrus, but no change in DA close contacts was observed (147). There is an increase in synaptic contacts onto A15 DA neurons in anestrus, and a corresponding increase in their dendritic trees (508), but the functional significance of these morphological changes has not been tested. It is also interesting that the transition from breeding season to anestrus is dependent on permissive actions of thyroid hormones (509) (see Chapter 41) in the vmPOA and premammillary region (510).

How Does the Annual Variation in Estradiol Negative Feedback Account for Seasonal Breeding?

As noted in the section on Resynthesis, a Model for the Ovine Estrous Cycle, because the estrous cycle is a chain of causally related events (Fig. 13), disruption of any step will break the chain and produce anovulation. In terms of seasonal breeding, the critical step is the follicular phase increase in tonic LH secretion or, more specifically, LH pulse frequency, which drives the pre-ovulatory estradiol rise (511). During the breeding season, progesterone is the primary inhibitor of GnRH pulse frequency. Hence, when progesterone concentrations fall at luteolysis, GnRH and LH pulse frequencies increase and the latter stimulates ovarian estradiol secretion; the estradiol rise then triggers the LH surge and ovulation occurs (Fig. 15, *left panel*). At the transition into anestrus, however, the inhibitory neural systems activated by long-day photoperiod allow estradiol to become the primary negative-feedback steroid controlling GnRH pulse frequency. Consequently, following regression of the last corpus luteum of the breeding season, GnRH and LH pulse frequency do not increase. Without the increase in tonic LH secretion, there is no pre-ovulatory estradiol rise to induce an LH surge and estrous behavior, and anovulation and anestrus result (Fig. 15, *middle panels*). Ewes remain in anestrus as long as the photoperiodically controlled inhibitory neurons can hold LH pulse frequency in check. In the autumn, however, the activity of these neural systems declines. Thus, in the absence of progesterone, pulsatile LH secretion increases and stimulates an estradiol rise, which triggers an LH surge, and the first ovulation of the breeding season occurs (Fig. 15, *right panel*).

It should be noted that the mechanism controlling tonic LH secretion is not the only hormonal system

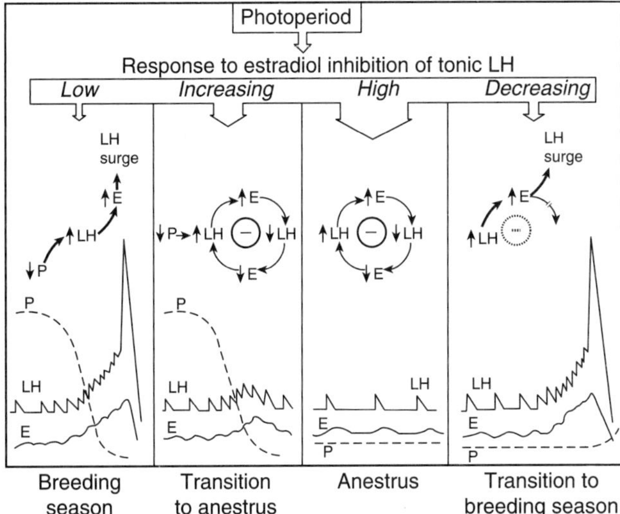

FIG. 15. Model for the control of seasonal breeding in the ewe. **Left to right:** endocrine events: (a) during the follicular phase in the breeding season; (b) at luteolysis during the transition to anestrus; (c) throughout anestrus; (d) during the first follicular phase at the transition to the breeding season. [Reproduced from Karsch et al. (511).]

to show seasonal variations in sheep. A variety of other seasonally related changes has been observed in anestrous ewes, including a decrease in the ability of estradiol to induce estrous behavior (115,512,513), a slight reduction in ovarian responsiveness to LH (120), and a marked increase in circulating prolactin concentrations (514,515). Although these changes may modulate ovarian function and thereby contribute to the maintenance of anestrus, none is sufficient to account for seasonal breeding in the ewe (115,511,516). In contrast, the photoperiod-controlled alterations in tonic LH secretion can explain both the cessation of ovarian cycles in the spring and their reinitiation in the fall (511).

Effects of Stressors

Over the last decade there has been considerable new information developed on effects of stressors on ovine reproduction. Synthesis of this new information is complicated because a variety of different stressors has been used, and it is often difficult for the investigator to assess the degree of stress on the animal. To overcome the latter problem, most workers have used the magnitude of the increase in cortisol secretion as an index of the degree of stress perceived by the animal. The effects of a large variety of stressors have been examined in ewes (469,517) and all of them produce inhibitory effects, although the degree of inhibition depends on the nature of the stressor. Of these, there is significant information on four stressors, which can be ranked (from least to greatest) based on the increase in cortisol concentrations they induce (Fig. 16): isolation/restraint (518–521), transport (522–525), insulin-induced hypoglycemia (518,526–530), and endotoxin (531–538).

In assessing the effects of these stressors, it should be kept in mind that there is considerable inter-animal variation in response to any stressor, which has been attributed to a number of factors including hormonal environment (518,520,525,537), habituation to the stressor (519,524,533), and gender differences (517,518). Nevertheless, some useful conclusions can be drawn on the effects of stressors on GnRH secretion. All four of these stressors inhibit LH pulse frequency in breeding season ewes (520,525–527,531) and endotoxin (531,535,536) and hypoglycemia (526) also consistently inhibit LH pulse amplitude (Fig. 16). Unlike the effects of photoperiod, the actions of stressors appear to be largely independent of steroid negative feedback. The inhibition of LH pulse frequency is clearly evident in the absence of ovarian steroids (520,525–527,531,535) and the magnitude of this response is not markedly altered by estradiol and/or progesterone (520,525). The inhibition of LH pulse

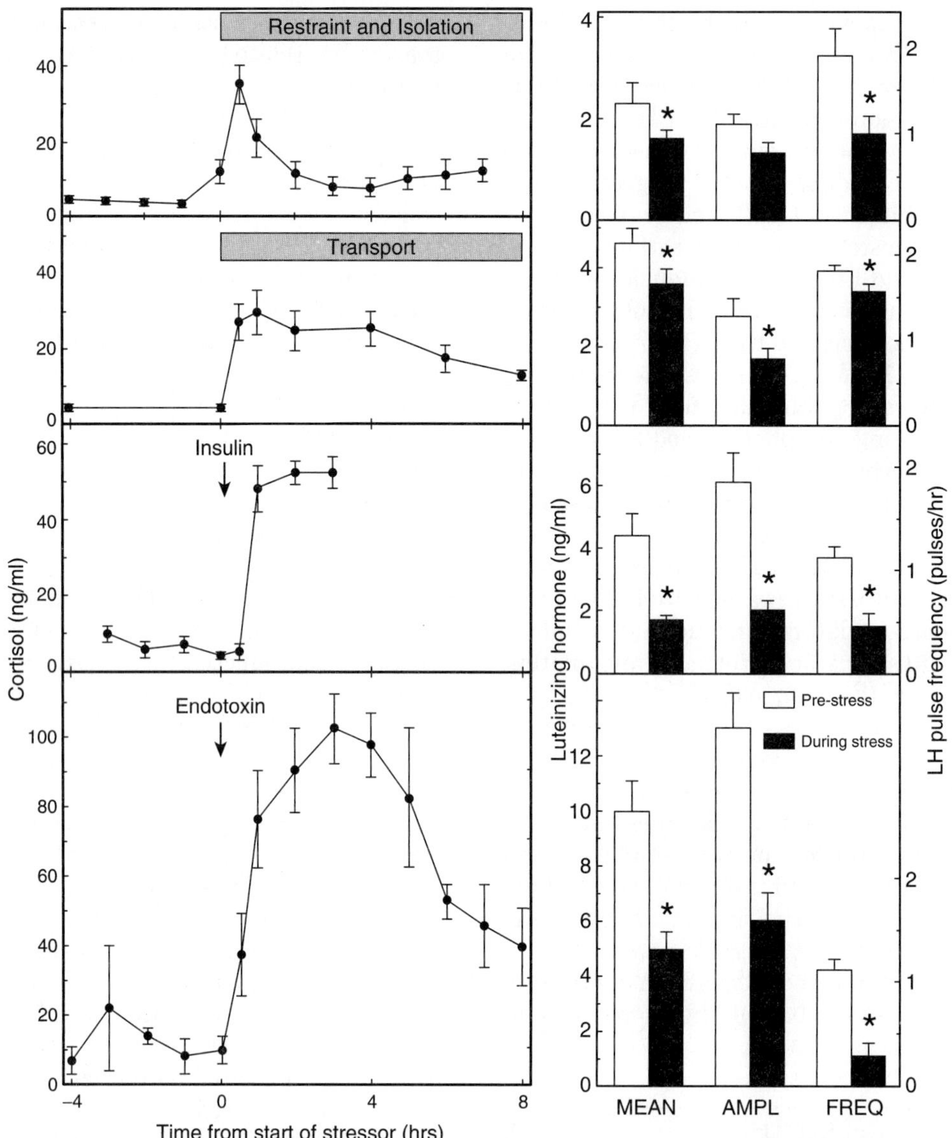

FIG. 16. Effect of four different stressors on mean (±SEM) concentrations of cortisol (**left panels**) and mean (+SEM) LH concentrations, LH pulse amplitudes (AMPL), and frequencies (FREQ) (**right panels**). [Data on cortisol concentrations were replotted from: Turner et al. (518) for restraint and insulin (reproduced by permission of the Society for Endocrinology); Phogat et al. (524) for transport, Williams et al. (535) for endotoxin (copyright 2001, The Endocrine Society).] LH values are for the 4 hours before (*open bars*) and during (*closed bars*) stress and are taken from: Tilbrook et al. (520) for restraint, Dobson et al. (525) for transport, Clarke et al. (526) for insulin, and Battaglia et al. (531) for endotoxin (copyright 1997, The Endocrine Society).

amplitude by isolation/restraint requires the presence of estradiol (520), but this is not true for the other three stressors that have been examined (526,531,535,536). Interestingly, transport and endotoxin failed to inhibit LH pulse frequency in ovariectomized anestrous ewes (525,538) so that there may be some overlap in the pathways by which photoperiod and these stressors inhibit GnRH pulse frequency. There is strong evidence that endotoxin and transport decrease LH pulse amplitude by inhibiting the response of the pituitary to GnRH (524,535); no pituitary effects of isolation/restrain stress were observed in OVX ewes (520).

In addition to inhibiting tonic GnRH secretion, the stress of transport, endotoxin, or hypoglycemia disrupts the ability of estradiol to induce an LH surge; the effects of isolation/restraint stress on the LH surge have not been tested. Interestingly, these three stressors disrupt estradiol positive feedback by different mechanisms. Transport will sometimes delay the LH surge (for about 2 h), but only if it occurs during the period just before surge onset (522–524) so that it may slow down the transmission phase of estradiol action. Endotoxin specifically blocks the initial activation phase without affecting subsequent events (533). Insulin-induced hypoglycemia blocks both the activation phase and the actual release of GnRH during the surge (527), and is the only agent known to affect the latter. Thus, stressors can act at multiple steps between luteolysis and ovulation, and the results of studies during the follicular phase (523,534) are consistent with more than one site of action.

There is much less information available about the neural pathways by which stressors inhibit reproductive function in ewes. Because stressors activate the HPA axis, the different components of this system have all been examined for possible inhibitory effects on GnRH or LH secretion. Although all four stressors increase release of the two primary stimulators of ACTH, corticotrophin-releasing hormone (CRH) and arginine vasopressin (AVP) (469,529,530,532), these hormones are unlikely to be the primary inhibitors of GnRH because icv administration of CRH or AVP either had no effect (526) or actually stimulated (539,540) LH secretion. ACTH administration can delay the LH surge (541,542), whereas administration of cortisol or dexamethasone does not (543,544). However, ACTH also stimulates progesterone secretion from the adrenal cortex, and there is some evidence that this steroid mediates the inhibitory effects of ACTH on the LH surge (469,542). On the other hand, progesterone does not mediate the effects of insulin-induced hypoglycemia on the LH surge (545). Cortisol inhibits LH pulse amplitude (538,546,547) by decreasing the pituitary response to GnRH via type II glucocorticoid receptors (548,549), but blockade of cortisol synthesis does not prevent the endotoxin-induced inhibition of either LH or GnRH pulses (538). Recent preliminary data indicate that this action of cortisol is dependent upon an intact connection between hypothalamus and pituitary (550), which raises the possibility that it stimulates release of a hypothalamic hormone that decreases the response of the pituitary to GnRH.

Other neural systems activated by these stressors may well play critical roles in suppression of GnRH secretion. There is some evidence implicating EOP in the inhibition of episodic GnRH secretion (526) and the GnRH surge (545), but EOP do not appear to mediate the effects of transport stress (522). Prostaglandins mediate the effects of endotoxin on temperature and have been implicated in the inhibition of LH pulses by endotoxin (536), but are not involved in the inhibition of the LH surge by this stressor (537). Recent indirect data implicated stress-induced activation of the adrenal medulla in inhibition of LH secretion (521), but this was a fairly transitory effect. Thus, it appears likely that stress-specific pathways mediate the inhibitory actions of different stressors on the reproductive axis.

Effects of Nutrition

Most studies on the interactions of nutrition and reproduction over the last decade have focused on the mechanisms by which restricted nutrition inhibits ovarian cyclicity (Chapter 47). This topic has received little attention in ewes because adult animals, in contrast to prepubertal lambs (Chapter 39), are remarkably resistant to nutritional deprivation. Undernutrition can induce anestrus in ewes, but this requires extremely severe restrictions (551). Similarly, food restriction that produced a 25% reduction in body weight over several months did not alter LH or FSH secretion in OVX ewes (552). More severe weight loss (~33%) produced a modest inhibition of LH pulse amplitude (553) or frequency (554,555) in OVX ewes, but did not always suppress mean LH and FSH concentrations (553,554). Extreme food restriction did markedly suppress gonadotropin secretion in OVX ewes (556) by inhibiting GnRH secretion (557), but this could reflect the stressful effects of such treatments, particularly in light of the recent report that several days of fasting increased adrenal activity in OVX ewes (558).

In contrast, more modest changes in nutrition can affect ovulation rate in adult ewes (308,559,560). This effect of nutrition has been recognized for over a century and has been the focus of considerable work over the last few decades (reviewed in 308). Nutritional influences can be both long-term, manifested by body condition [static (561)] and short-term in response to increased food intake [dynamic (58,561)]. Increased body condition from thin to moderate (559,562,563), or from moderate to obese (564), increased ovulation rate. Increased food intake, in the absence of any differences in body condition, also increased ovulation rate (560,565,566). The most dramatic example of the latter is the increase in ovulation rate observed after a 5-day food supplement late in the luteal phase (567), an effect that can be produced by a 5-day infusion of either glucose (568) or branch-chained amino acids (569). The effects of nutrition are influenced by breed (570) and an abrupt increase in food intake produces stronger effects in thin ewes (559), probably because these animals have a higher potential to increase ovulation rate. Interestingly, an increase in body condition was correlated with greater numbers of large follicles (563,571), while increased food intake produced a larger number of medium-sized follicles (560,572). Most workers believe that increases in both medium and large follicles are due to a decrease in the rate of atresia (58,308,560,572).

The effects of nutrition on ovulation rate most likely reflect changes in gonadotropin secretion, because they do not influence the ovarian responsiveness to FSH (573). Furthermore, the direct infusion of insulin and glucose to the ovary decreased secretion of androstenedione and estradiol (574). There are reports that increased nutrition stimulated LH (564,565,570,575) or FSH (563,568,570) secretion at specific times of the estrous cycle, but numerous studies have also observed no changes in LH (563,566–569,574,576) or FSH (565–567,569,569, 576,577). This inability to observe consistent changes in LH or FSH secretion may reflect the dynamic nature of the negative feedback loop between ovarian follicles and gonadotropin secretion; an increase in number of ovulatory follicles (which presumably precedes the increase in ovulation rate) would be expected to produce a stronger inhibitory signal that brings LH and FSH secretion back down to concentrations seen in the absence of increased nutritional input. In support of this possibility, increases in both body condition (576) and food intake (566) that did not alter LH and FSH concentrations in ovary-intact ewes, increased them in acutely OVX animals (566,576) and altered the response to exogenous ovarian hormones in acutely OVX ewes (578). Food intake increased metabolism of progesterone (579) so that a reduction in steroid negative feedback could be one basis for increased ovulation rate in response to short-term increases in nutrient intake. Thomas et al. (580) modeled these changes with injections of pentobarbital and showed equivalent changes in liver metabolism and ovulation rate. However, one would expect a more consistent increase in LH pulse frequency if this change were sufficient to account for the effects of food intake on ovulation rate.

There is essentially no information on the neural mechanisms by which nutrition influences ovulation rate in ewes. It is tempting to speculate that leptin may be involved because it is increased by both increased food intake and body condition (470,581,582) and leptin can stimulate gonadotropin secretion in ewes under some circumstances (553). However, there are no reports on the effects of leptin on ovulation rate. A role for insulin has also been proposed because secretion of this hormone increases in response to acute increases in food intake (567–569), but direct administration of insulin to the ovary inhibited steroid production (574). There is some information implicating orexin neurons in the stimulatory effects of food intake on gonadotropin secretion in males (583,584), and orexin neurons innervate GnRH perikarya in ewes (585), but no study has directly tested their role in controlling ovulation rate.

Effects of Pheromones and Social Interactions

While the actions of stress and nutrition are most important during the breeding season, the effects of social and pheromonal cues are most evident during anestrus. The most striking example of this phenomenon is the induction of ovulation in anestrous ewes by exposure to rams (471,586,587), which is often referred to as the "ram effect." It requires introduction of a ram that is new to the anestrous ewes (588,589) and is mediated primarily by pheromonal cues (590,591), although other sensory stimuli may be involved (588). The pheromones mediating this effect appear to be produced by the sudoriferous glands in the skin (590) and their production is controlled by androgens (592,593). This effect of androgens may account for the observations that rams with higher reproductive activity are more effective

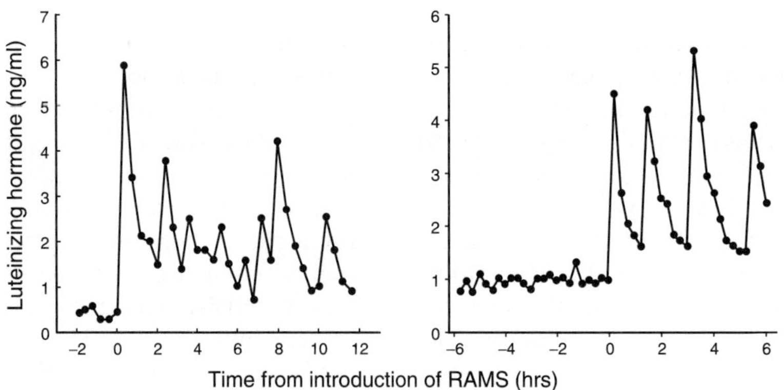

FIG. 17. Effects of introduction of a ram on pulsatile LH secretion in an ovary-intact anestrous ewe (*left panel*) and an estradiol-treated OVX ewe during anestrus (*right panel*). Rams introduced at time 0. [Redrawn from Martin (586).]

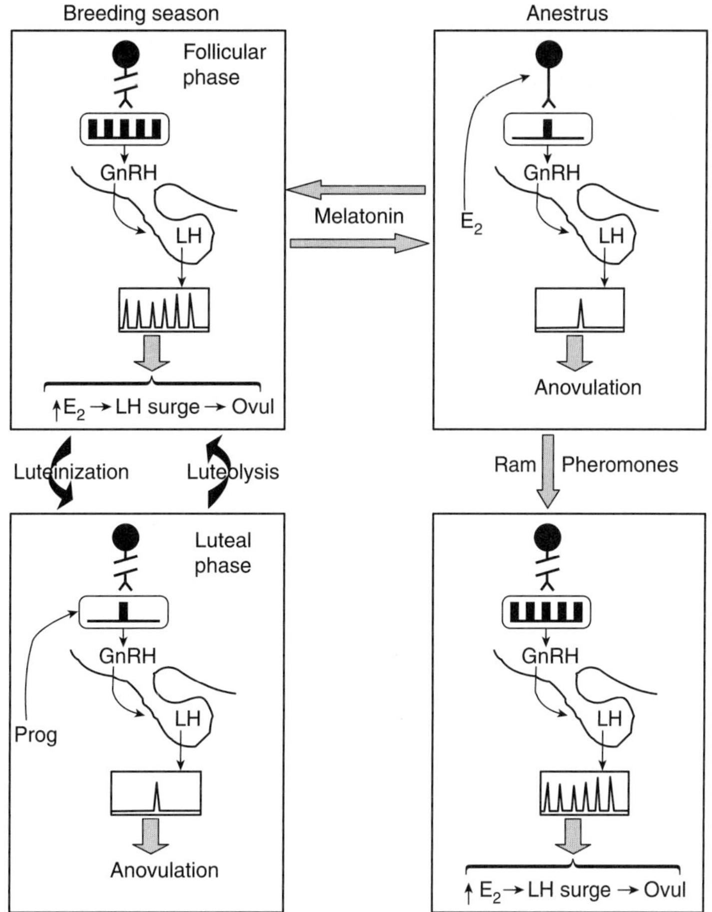

FIG. 18. Control of ovulation in the ewe. The frequency of pulsatile LH secretion, by controlling the preovulatory estradiol (E_2) rise, determines whether or not ovulation occurs. **Left panels:** During the breeding season, progesterone (Prog) is the primary regulator of LH pulse frequency. **Bottom left:** During the luteal phase, Prog inhibits the hypothalamic pulse generator, producing anovulation. **Top left:** Following luteolysis, LH pulse frequency increases and initiates the follicular phase events leading to ovulation (not shown: stressors can inhibit these events by suppressing pulsatile LH secretion and disrupting the E_2-induced LH surge). **Right panels:** During anestrus, the long-day melatonin pattern activates an E_2-sensitive inhibitory neural system. **Top right:** In anestrous ewes isolated from rams, E_2 stimulates the inhibitory neural system which inhibits GnRH and LH pulse frequency and ensures anovulation. **Bottom right:** Pheromones from a novel ram shut down the inhibitory system, allowing LH pulse frequency to increase and produce ovulation during anestrus.

in inducing ovulation (593–595), and that prior exposure of rams to estrous females [which stimulates the male reproductive system (596)] enhances their ability to induce ovulation in anestrous ewes (471,597). The effectiveness of rams also is influenced by the responsiveness of females. Some of this variation appears to reflect the "depth" of anestrus (471,586). Breeds with limited anestrous periods respond more readily to rams (598,599) than do breeds with more prolonged anestrus (599,600). Ewes are generally more responsive late in anestrus than they are in the middle of the anovulatory period (599,601).

As illustrated in Fig. 17, the exposure of anestrous ewes to a novel ram produces an immediate increase in LH pulse frequency (587,600,602–604), which is followed by follicular growth (605) and a pre-ovulatory LH surge (587,606); concentrations of FSH tend to fall during this induced follicular phase (605,606). Ovulations usually occur within 2 to 3 days of ram introduction, but can sometimes be delayed so that they occur 4 to 6 days later (607). These male pheromones have been postulated to disrupt estradiol negative feedback because introduction of rams readily increases LH pulse frequency in estradiol-treated OVX ewes (602), but has little (598), or no (602), effect in untreated OVX animals. However, the neural mechanisms responsible for this change in episodic GnRH secretion remain completely unknown. In fact, there has been surprisingly little work on the basic mechanisms underlying this phenomenon during the last decade, although there have been a number of studies on practical aspects of ram introduction [reviewed by Ungerfeld et al. (471)] and on the stimulatory effects of females on rams [reviewed by Sanford et al. (596)].

CONCLUSION

The ovine estrous cycle is controlled by a complex interaction between cues from the external environment and internal hormonal signals. Environmental factors influence ovulation rate and determine whether or not estrous cycles occur, but are not directly involved in the events of the estrous cycle. Instead, these endocrine events reflect a coordinated hormonal communication among the brain, pituitary, ovaries, and uterus. The primary coordinator of this communication within the ewe is progesterone, which controls the timing of the major endocrine events of the estrous cycle (Fig. 13). Superimposed on these internal events are four major environmental cues. First, changes in nutrition influence the number of healthy follicles and thus ovulation rate. Second, stressors inhibit follicular phase events, delaying the occurrence of ovulation. Finally, photoperiod and pheromones control when estrous cycles occur. Long-day photoperiods, by activating inhibitory neural systems, actively suppress ovarian cycles, but these inhibitory effects of photoperiod can be disrupted by introduction of pheromones from the ram. These two external cues appear to influence ovarian cycles largely by controlling the negative feedback action of estradiol. Inhibitory photoperiods induce anovulation by increasing estradiol negative feedback, and pheromones trigger ovulation by blocking estradiol negative feedback.

Although the overall regulation of the ovine estrous cycle involves a large number of hormonal interactions, if one focuses on just the control of ovulation, an important unifying principle emerges. Namely, it is the frequency of pulsatile LH release that determines whether or not ovulation occurs. LH pulse frequency is critical because a relatively fast frequency is required for the sustained estradiol rise that triggers the pre-ovulatory LH surge. Consequently, the activity of the hypothalamic GnRH pulse generator is a pivotal control point under a variety of conditions (Fig. 18). During the breeding season, estradiol does not inhibit pulse frequency, but progesterone does. Thus, at this time of year, the pattern of progesterone determines the timing of ovulation. When progesterone is high, pulse frequency is low, and anovulation results; when progesterone falls at luteolysis, pulse frequency increases, and ovulation occurs within 2 to 3 days, unless stressors inhibit LH pulse frequency. During anestrus, estradiol gains the capacity to suppress the hypothalamic pulse generator, so that LH pulse frequency remains low in the absence of progesterone (Fig. 18). As a result, the animals remain anovulatory as long as estradiol is able to hold episodic LH pulses in check. Exposure of ewes to pheromones from rams is one means of disrupting estradiol negative feedback on the GnRH pulse generator. This stimulus, in turn, allows LH pulse frequency to increase, so that introduction of rams induces ovulation in anestrous ewes (Fig. 18). Alternatively, the negative feedback action of estradiol is lost as the effects of long-day photoperiod wane in autumn. Thus, LH pulse frequency increases, which initiates ovulation and the start of the breeding season. With the return of ovulatory cycles, progesterone concentrations increase, and this steroid resumes control of the hypothalamic pulse generator and hence the occurrence of ovulation.

ACKNOWLEDGMENTS

Supported by grants from NIH (HD17864 and HD39916), Hatch 421 (Multistate Project Northeast 1007), and a grant from the West Virginia Legislature for economic development of the sheep industry.

REFERENCES

1. Marshall, F. H. A. (1904). The oestrous cycle and the formation of the corpus luteum in the sheep. *Phil. Trans. Roy. Soc. London, Ser. B.* 196, 47–97.
2. McKenzie, F. F., and Terrill, C. E. (1937). Estrus, ovulation, and related phenomena in the ewe. *Missouri Agr. Exper. Sta. Res. Bull.* 264, 5–88.
3. Hafez, E. S. E. (1952). Studies on the breeding season and reproduction of the ewe. *J. Agric. Sci.* 42, 189–265.
4. Banks, E. M. (1964). Some aspects of sexual behavior in domestic sheep, *Ovis aries*. *Behavior* 23, 249–278.
5. Fabre-Nys, C., and Martin, G. B. (1991). Hormonal control of proceptive and receptive sexual behavior and the preovulatory LH surge in the ewe: Reassessment of the respective roles of estradiol, testosterone, and progesterone. *Horm. Behav.* 25, 295–312.
6. Bindon, B. M., Blanc, M. R., Pelletier, J., Terqui, M., and Thimonier, J. (1979). Periovulatory gonadotrophin and ovarian steroid patterns in sheep of breeds with differing fecundity. *J. Reprod. Fertil.* 55, 15–25.
7. Quirke, J. F., Hanrahan, J. P., and Gosling, J. P. (1979). Plasma progesterone levels throughout the oestrous cycle and release of LH at oestrus in sheep with different ovulation rates. *J. Reprod. Fertil.* 55, 37–44.
8. Fletcher, I. C., and Lindsay, D. R. (1971). Effect of rams on the duration of oestrous behavior in ewes. *J. Reprod. Fertil.* 25, 253–259.
9. Robertson, H. A. (1969). The endogenous control of estrus and ovulation in sheep, cattle, and swine. *Vit. Horm.* 27, 91–130.
10. Clarke, I. J., and Cummins, J. T. (1982). The temporal relationship between gonadotropin releasing hormone (GnRH) and luteinizing hormone (LH) secretion in ovariectomized ewes. *Endocrinology* 111, 1737–1739.

11. Caraty, A., and Locatelli, A. (1988). Effect of time after castration on secretion of LHRH and LH in the ram. *J. Reprod. Fertil.* 82, 263–269.
12. Goding, J. R., Blockey, MA. deB., Brown, J. M., Catt, K. J., and Cumming, I. A. (1970). The role of oestrogen in the control of the oestrous cycle in the ewe. *J. Reprod. Fertil.* 21, 368–369.
13. Hauger, R. L., Karsch, F. J., and Foster, D. L. (1977). A new concept for control of the estrous cycle of the ewe based on the temporal relationships between luteinizing hormone, estradiol and progesterone in peripheral serum and evidence that progesterone inhibits tonic LH secretion. *Endocrinology* 101, 807–817.
14. Baird, D. T., and Scaramuzzi, R. J. (1976). Changes in the secretion of ovarian steroids and pituitary luteinizing hormone in the peri-ovulatory period in the ewe: the effect of progesterone. *J. Endocrinol.* 70, 237–245.
15. Baird, D. T., Swanston, I. A., and McNeilly, A. S. (1981). Relationship between LH, FSH, and prolactin concentration and the secretion of androgens and estrogens by the preovulatory follicle in the ewe. *Biol. Reprod.* 24, 1013–1025.
16. Karsch, F. J., Foster, D. L., Legan, S. J., Ryan, K. D., and Peter, G. K. (1979). Control of the preovulatory endocrine events in the ewe: interrelationship of estradiol, progesterone, and luteinizing hormone. *Endocrinology* 105, 421–426.
17. Clarke, I. J., Thomas, G. B., Yao, B., and Cummings, J. T. (1987). GnRH secretion throughout the ovine estrous cycle. *Neuroendocrinology* 46, 82–88.
18. Moenter, S. M., Caraty, A., Locatelli, A., and Karsch, F. J. (1991). Pattern of gonadotropin-releasing hormone (GnRH) secretion leading up to ovulation in the ewe: Existence of a preovulatory GnRH surge. *Endocrinology* 129, 1175–1182.
19. Baird, D. T. (1978). Pulsatile secretion of LH and ovarian estradiol during the follicular phase of the sheep estrous cycle. *Biol. Reprod.* 18, 359–364.
20. Karsch, F. J., Foster, D. L., Bittman, E. L., and Goodman, R. L. (1983). A role for estradiol in enhancing luteinizing hormone pulse frequency during the follicular phase of the estrous cycle of sheep. *Endocrinology* 113, 1333–1339.
21. Mauer, R. E., Revenal, P., Johnson, E. S., Moyer, R. H., Hirata, A., and White, W. F. (1972). Levels of luteinizing hormone in sera of ewes near the time of estrus as determined by radioimmunoassay. *J. Anim. Sci.* 34, 88–92.
22. Cumming, I. A., Buckmaster, J. M., Blockey, M. A. DeB., Goding, J. R., Winfield, C. G., and Baxter, R. W. (1973). Constancy of interval between luteinizing hormone release and ovulation in the ewe. *Biol. Reprod.* 9, 24–29.
23. Moenter, S. M., Brand, R. C., and Karsch, F. J. (1992). Dynamics of gonadotropin-releasing hormone (GnRH) secretion during the GnRH surge: insights into the mechanism of GnRH surge induction. *Endocrinology* 130, 2978–2984.
24. Evans, N. P., Dahl, G. E., Mauger, D. T., Padmanabhan, V., Thrun, L. A., and Karsch, F. J. (1995). Does estradiol induce the preovulatory gonadotropin-releasing hormone (GnRH) surge in the ewe by inducing a progressive change in the mode of operation of the GnRH neurosecretory system? *Endocrinology* 136, 5511–5519.
25. L' Hermite, M., Niswender, G. D., Reichert, L. E. Jr., and Midgley, A. R. Jr. (1972). Serum follicle-stimulating hormone in sheep as measured by radioimmunoassay. *Biol. Reprod.* 6, 325–332.
26. Pant, H. C., Hopkinson, C. R. N., and Fitzpatrick, R. J. (1977). Concentration of oestradiol, progesterone, luteinizing hormone and follicle-stimulating hormone in the jugular venous plasma of ewes during the oestrous cycle. *J. Endocrinol.* 73, 247–255.
27. Goodman, R. L., Pickover, S. M., and Karsch, F. J. (1981). Ovarian feedback control of follicle-stimulating hormone in the ewe: evidence for selective suppression. *Endocrinology* 108, 772–777.
28. Souza, C. J. H., Campbell, B. K., Webb, R., and Baird, D. T. (1997). Secretion of inhibin A and follicular dynamics throughout the estrous cycle in the sheep with and without the Booroola gene (*Fec*^{B}). *Endocrinology* 138, 5333–5340.
29. Knight, P. G., Feist, S. A., Tannetta, D. S., Bleach, E. C. L., Fowler, P. A., O'Brien, M., and Groome, N. P. (1998). Measurement of inhibin-A (α βA dimer) during the oestrous cycle, after manipulation of ovarian activity and during pregnancy in ewes. *J. Reprod. Fertil.* 113, 159–166.
30. Souza, C. J. H., Campbell, B. K., and Baird, D. T. (1998). Follicular waves and concentrations of steroids and inhibin A in ovarian venous blood during the luteal phase of the oestrous cycle in ewes with an ovarian autotransplant. *J. Endocrinol.* 156, 563–572.
31. Souza, C. J. H., Campbell, B. K., and Baird, D. T. (1997). Follicular dynamics and ovarian steroid secretion in sheep during the follicular and early luteal phases of the estrous cycle. *Biol. Reprod.* 56, 483–488.
32. Miller, K. F., Nordheim, E. V., and Ginther, O. J. (1981). Periodic fluctuations in FSH concentrations during the ovine estrous cycle. *Theriogenology* 16, 669–679.
33. Bister, J. L., and Paquay, R. (1983). Fluctuations in the plasma levels of follicle-stimulating hormone during estrous cycle, anestrus, gestation and lactation in the ewe: evidence for an endogenous rhythm of FSH release. *Theriogenology* 19, 565–582.
34. Ginther, O. J., Kot, K., and Wiltbank, M. C. (1995). Associations between emergence of follicular waves and fluctuations in FSH concentrations during the estrous cycle in ewes. *Theriogenology* 43, 689–703.
35. Sawyer, H. R., Smith, P., Heath, D. A., Juengel, J. L., Wakefield, S. J., and McNatty, K. P. (2002). Formation of ovarian follicles during fetal development in sheep. *Biol. Reprod.* 66, 1134–1150.
36. Smith, P., O, W. S., Hudson, N. L., Shaw, L., Heath, D. A., Condell, L., Phillips, D. J., and McNatty, K. P. (1993). Effects of the Booroola gene (Fec^B) on body weight, ovarian development and hormone concentrations during fetal life. *J. Reprod. Fertil.* 98, 41–54.
37. Lundy, T., Smith, P., O'Connell, A., Hudson, N. L., and McNatty, K. P. (1999). Populations of granulosa cells in small follicles of the sheep ovary. *J. Reprod. Fertil.* 115, 251–262.
38. Reynaud, K., and Driancourt, M. A. (2000). Oocyte attrition. *Mol. Cell. Endocrinol.* 163, 101–108.
39. Land, R. B. (1970). Number of oocytes present at birth in the ovaries of pure and Finnish Landrace cross Blackface and Welsh sheep. *J. Reprod. Fertil.* 21, 517–521.
40. Scaramuzzi, R. J., Adams, N. R., Baird, D. T., Campbell, B. K., Downing, J. A., Findlay, J. K., Henderson, K. M., Martin, G. B., McNatty, K. P., McNeilly, A. S., and Tsonis, C. G. (1993). A model for follicle selection and the determination of ovulation rate in the ewe. *Reprod. Fertil. Dev.* 5, 459–478.
41. Bartlewski, P. M., Beard, A. P., Cook, S. J., and Rawlings, N. C. (2002). Ovarian activity during sexual maturation and following introduction of the ram to ewe lambs. *Small Rum. Res.* 43, 37–44.
42. Rawlings, N. C., Evans, A. C. O., Honaramooz, A., and Bartlewski, P. M. (2003). Antral follicle growth and endocrine changes in prepubertal cattle, sheep and goats. *Anim. Reprod. Sci.* 78, 259–270.
43. Dailey, R. A., Fogwell, R. L., and Thayne, W. V. (1982). Distribution of visible follicles on the ovarian surface in ewes. *J. Anim. Sci.* 54, 1196–1204.
44. Brand, A., and de Jong, W. H. (1973). Qualitative and quantitative micromorphological investigations of the tertiary follicle population during the oestrous cycle in sheep. *J. Reprod. Fertil.* 33, 431–439.

45. Smeaton, T. C., and Robertson, H. A. (1971). Studies on the growth and atresia of graafian follicles in the ovary of the sheep. *J. Reprod. Fertil.* 25, 243–252
46. Bherer, J., Matton, P., and Dufour, J. J. (1977). Fate of the two largest follicles in the ewe after injection of gonadotropins at two stages of the estrous cycle. *Proc. Soc. Exp. Biol. Med.* 154, 412–414.
47. Schrick, F. N., Surface, R. A., Pritchard, J. Y., Dailey, R. A., Townsend, E. C., and Inskeep, E. K. (1993). Ovarian structures during the estrous cycle and early pregnancy in ewes. *Biol. Reprod.* 49, 1133–1140.
48. Ravindra, J. P., Rawlings, N. C., Evans, A. C. O., and Adams, G. P. (1994). Ultrasonographic study of ovarian follicular dynamics in ewes during the oestrous cycle. *J. Reprod. Fertil.* 101, 501–509.
49. Noel, B., Bister, J. L., and Paquay, R. (1993). Ovarian follicular dynamics in Suffolk ewes at different periods of the year. *J. Reprod. Fertil.* 99, 695–700.
50. Evans, A. C. O. (2003). Ovarian follicle growth and consequences for fertility in sheep. *Anim. Reprod. Sci.* 78, 289–306.
51. Bartlewski, P. M., Beard, A. P., Cook, S. J., Chandolia, R. K., Honaramooz, A., and Rawlings, N. C. (1999). Ovarian antral follicular dynamics and their relationships with endocrine variables throughout the oestrous cycle in breeds of sheep differing in prolificacy. *J. Reprod. Fertil.* 115, 111–124.
52. Johnson, S. K., Dailey, R. A., Inskeep, E. K., and Lewis, P. E. (1996). Effect of peripheral concentrations of progesterone on follicular growth and fertility in ewes. *Dom. Anim. Endocrinol.* 13, 69–79.
53. Bartlewski, P. M., Beard, A. P., and Rawlings, N. C. (2000). An ultrasound-aided study of temporal relationships between the patterns of LH/FSH secretion, development of ovulatory-sized antral follicles and formation of corpora lutea in ewes. *Theriogenology* 54, 229–245.
54. Bartlewski, P. M., Vanderpol, J., Beard, A. P., Cook, S. J., and Rawlings, N. C. (2000). Ovarian antral follicular dynamics and their associations with peripheral concentrations of gonadotrophins and ovarian steroids in anoestrous Finnish Landrace ewes. *Anim. Reprod. Sci.* 58, 273–291.
55. Evans, A. C. O., Duffy, P., Quinn, K. M., Knight, P. G., and Boland, M. P. (2001). Follicular waves are associated with transient fluctuations in FSH but not oestradiol or inhibin-A concentrations in anoestrous ewes. *J. Anim. Sci.* 72, 547–554.
56. Gibbons, J. R., Kot, K., Thomas, D. L., Wiltbank, M. C., and Ginther, O. J. (1999). Follicular and FSH dynamics in ewes with a history of high and low ovulation rates. *Theriogenology* 52, 1005–1020.
57. Souza, C. J. H., Campbell, B. K., and Baird D. T. (1996). Follicular dynamics and ovarian steroid secretion in sheep during anoestrus. *J. Reprod. Fertil.* 108, 101–106.
58. Landau, S., Houghton, J. A. S., Mawhinney, J. R., and Inskeep, E. K. (1996). Protein sources affect follicular dynamics in ewes near the onset of the breeding season. *Reprod. Fertil. Dev.* 8, 1021–1028.
59. Lopez-Sebastian, A., Gonzalez de Bulnes, A., Santiago-Moreno, J., Gomez-Brunet, A., Townsend, E. C., and Inskeep, E. K. (1997). Patterns of follicular development during the estrous cycle in monovular Merino del Pais ewes. *Anim. Reprod. Sci.* 48, 279–291.
60. Driancourt, M. A., Webb, R., and Fry, R. C. (1991). Does follicular dominance occur in ewes? *J. Reprod. Fertil.* 93, 63–70.
61. Duggavathi, R., Bartlewski, P. M., Barrett, D. M. W., and Rawlings, N. C. (2003). Use of high resolution transrectal ultrasonography to assess changes in numbers of small ovarian antral follicles and their relationships to the emergence of follicular waves in cyclic ewes. *Theriogenology* 60, 495–510.
62. Gonzalez-Bulnes, A., Souza, C. J., Campbell, B. K., and Baird, D. T. (2004). Systemic and intraovarian effects of dominant follicles on ovine follicular growth. *Anim. Reprod. Sci.* 84, 107–119.
63. Espey, L. L. (1978). Ovarian contractility and its relationship to ovulation: a review. *Biol. Reprod.* 19, 540–551.
64. Espey, L. L. (1980). Ovulation as an inflammatory reaction—a hypothesis. *Biol. Reprod.* 22, 73–106.
65. Richards, J. S. (1980). Maturation of ovarian follicles: actions and interactions of pituitary and ovarian hormones. *Physiol. Rev.* 60, 51–89.
66. Espey, L. L. (1994). Current status of the 1980 hypothesis that mammalian ovulation is comparable to an inflammatory reaction. *Biol. Reprod.* 50, 233–238.
67. Espey, L. L., and Richards, J. S. (2002). Temporal and spatial patterns of ovarian gene transcription following an ovulatory dose of gonadotropin in the rat. *Biol. Reprod.* 67, 1662–1670.
68. Murdoch, W. J., and McDonnel, A. C. (2002). Roles of the ovarian surface epithelium in ovulation and carcinogenesis. *Reproduction* 123, 743–750.
69. Murdoch, W. J., Colgin, D. C., and Ellis, J. A. (1997). Role of tumor necrosis factor-α in the ovulatory mechanism of ewes. *J. Anim. Sci.* 75, 1601–1605.
70. Murdoch, W. J., and Lund, S. A. (1999). Prostaglandin-independent anovulatory mechanism of indomethacin action: Inhibition of tumor necrosis factor alpha-induced sheep ovarian cell apoptosis. *Biol. Reprod.* 61, 1655–1659.
71. Murdoch, W. J., and Dunn, T. G. (1982). Alterations in follicular steroid hormones during the preovulatory period in the ewe. *Biol. Reprod.* 27, 300–307.
72. Murdoch, W. J., Dailey, R. A., and Inskeep, E. K. (1981). Preovulatory changes in prostaglandins E_2 and $F_2\alpha$ in ovine follicles. *J. Anim. Sci.* 53, 192–205.
73. Murdoch, W. J., Nix, K. J., and Dunn, T. G. (1983). Dynamics of ovarian blood supply to periovulatory follicles of the ewe. *Biol. Reprod.* 28, 1001–1006.
74. Cavender, J. L., and Murdoch, W. J. (1988). Morphological studies on the microcirculatory system of periovulatory ovine follicles. *Biol. Reprod.* 39, 989–997.
75. Murdoch, W. J., and McCormick, R. J. (1993). Mechanisms and physiological implications of leukocyte chemoattraction into periovulatory ovine follicles. *J. Reprod. Fertil.* 97, 375–380.
76. Gottsch, M. L., Van Kirk, E. A., and Murdoch, W. J. (2000). Tumour necrosis factor alpha up-regulates matrix metalloproteinase-2 activity in periovulatory follicles: Metamorphic and endocrine implications. *Reprod. Fertil. Dev.* 12, 75–80.
77. Murdoch, W. J., Peterson, T. A., Van Kirk, E. A., Vincent, D. L., and Inskeep, E. K. (1986). Interactive roles of progesterone, prostaglandins and collagenase in the ovulatory mechanisms of the ewe. *Biol. Reprod.* 35, 1187–1194.
78 Murdoch, W. J., Hansen, T. R., and McPherson, L. A. (1993). Role of eicosanoids in vertebrate ovulation. *Prostaglandins* 46, 85–115.
79. Murdoch, W. J., and McCormick, R. J. (1991). Dose-dependent effects of indomethacin on ovulation in sheep: Relationship to follicular prostaglandin production, steroidogenesis, collagenolysis and leukocyte chemotaxis. *Biol. Reprod.* 45, 907–911.
80. Murdoch, W. J., Slaughter, R. G., and Ji, T. H. (1991). In situ hybridization analysis of ovarian endoperoxide synthase mRNA throughout the periovulatory period of the ewe. *Dom. Anim. Endocrinol.* 8, 455–457.
81. Murdoch, W. J., and Ferris, M. L. (1988). Prostaglandin E_2-9-keto-reductase activity of preovulatory ovine follicles. *J. Anim. Sci.* 66, 2924–2929.

82. Murdoch, W. J., and Cavender, J. L. (1989). Effect of indomethacin on the vascular architecture of preovulatory ovine follicles: Possible implication in the luteinized unruptured follicle syndrome. *Fertil. Steril.* 51, 153–155.
83. Murdoch, W. J., and Cavender, J. L. (1987). Mechanisms of ovulation. *Adv. Contracept. Deliv. Syst.* 3, 353–367.
84. Johnson, M. L., Murdoch, J., Van Kirk, E. A., Kaltenbach, J. E., and Murdoch, W. J. (1999). Tumor necrosis factor α regulates collagenolytic activity in preovulatory ovine follicles: relationship to cytokine secretion by the oocyte-cumulus cell complex. *Biol. Reprod.* 61, 1581–1585.
85. Engelhardt, H., Smith, K. B., McNeilly, A. S., and Baird, D. T. (1993). Expression of messenger ribonucleic acid for inhibin subunits and ovarian secretion of inhibin and estradiol at various stages of the sheep estrous cycle. *Biol. Reprod.* 49, 281–294.
86. Tisdall, D. J., Hudson, N., Smith, P., and McNatty, K. P. (1994). Localization of ovine follistatin and alpha and beta A inhibin mRNA in the sheep ovary during the oestrous cycle. *J. Mol. Endocrinol.* 12, 181–193.
87. Findlay, J. K., Tsonis, C. G., Staples, L. D., and Cahill, R. N. P. (1986). Inhibin secretion by the sheep ovary. *J. Reprod Fertil.* 76, 751–761.
87a. McNeilly, A. S., Souza, C. J. H., Baird, D. T., Swanston, I. A., McVerry, J., Crawford, J., Cranfield, M., and Lincoln, G. A. (2002). Production of Inhibin A not B in rams: changes in plasma inhibin A during testis growth, and expression of inhibin/activin subunit mRNA in adult testis. *Reproduction* 123, 827–835.
88. Campbell, B. K., McNeilly, A. S., Mann, G. E., and Baird, D. T. (1991). The effect of stage of estrous cycle and follicular maturation on ovarian inhibin production in sheep. *Biol. Reprod.* 44, 483–490.
89. Bjersing, L., Hay, M. F., Kann, G., Moor, R. M., Naftolin, F., Scaramuzzi, R. J., Short, R. V., and Younglai, E. V. (1972). Changes in gonadotrophins, ovarian steroids and follicular morphology in sheep at oestrus. *J. Endocrinol.* 52, 465–479.
90. Campbell, B. K., Mann, G. E., McNeilly, A. S., and Baird, D. T. (1990). Pulsatile secretion of inhibin, oestradiol and androstenedione by the ovary of the sheep during the oestrous cycle. *J. Endocrinol.* 126, 385–393.
91. Baird, D. T., and Scaramuzzi, R. J. (1976). The source of ovarian oestradiol and androstenedione in the sheep during the luteal phase. *Acta Endocrinol. (Copenh.)* 83, 402–409.
92. Baird, D. T., Land, R. B., Scaramuzzi, R. J., and Wheeler, A. G. (1976). Endocrine changes associated with luteal regression in the ewe: the secretion of ovarian oestradiol, progesterone and androstenedione and uterine prostaglandin F2α throughout the oestrous cycle. *J. Endocrinol.* 69, 275–286.
93. Baird, D. T., Swanston, I., and Scaramuzzi, R. J. (1976). Release of LH and secretion of ovarian steroids in sheep during the luteal phase of the estrous cycle. *Endocrinology* 98, 1490–1496.
94. Wheeler, A. G., and Land, R. B. (1977). Seasonal variation in oestrus and ovarian activity of Finnish Landrace, Tasmanian Merino and Scottish Blackface ewes. *Anim. Prod.* 24, 363–376.
95. Karsch, F. J., Legan, S. J., Ryan, K. D., and Foster, D. L. (1980). Importance of estradiol and progesterone in regulating LH secretion and estrous behavior during the sheep estrous cycle. *Biol. Reprod.* 23, 404–413.
96. McNeilly, A. S., and Fraser, H. M. (1987). Effect of gonadotropin-releasing hormone agonist-induced suppression of LH and FSH on follicle growth and corpus luteum function in the ewe. *J. Endocrinol.* 115, 273–282.
97. Alecozay, A. A., Selcer, K. W., Clark, J. R., Burns, J. M., Norman, R. L., Niswender, G. D., and Leavitt, W. W. (1988). Pattern of ovarian progesterone secretion during the luteal phase of the ovine estrous cycle. *Biol. Reprod.* 39, 287–294.
98. McCracken, J. A., Glew, M. E., and Scaramuzzi, R. J. (1970). Corpus luteum regression induced by prostaglandin F2α. *J. Clin. Endocrinol. Metabol.* 30, 544–546.
99. Samuelsson, B., Granstrom, E., Green, K., and Hambert, M. (1971). Metabolism of prostaglandins. *Ann. N. Y. Acad. Sci.* 180, 138–163.
100. Lewis, G. S., Wilson, L. Jr., Wilks, J. W., Pexton, J. E., Fogwell, R. L., Ford, S. P., Butcher, R. L., Thayne, W. V., and Inskeep, E. K. (1977). PGF2α and its metabolites in uterine and jugular venous plasma and endometrium of ewes during early pregnancy. *J. Anim. Sci.* 45, 320–327.
101. Silva, P. J., Juengel, J. J., Rollyson, M. K., and Niswender, G. D. (2000). Prostaglandin metabolism in the ovine corpus luteum: Catabolism of prostaglandin $F_{2\alpha}$ ($PGF_{2\alpha}$) coincides with resistance of the corpus luteum to $PGF_{2\alpha}$. *Biol. Reprod.* 63, 1229–1236.
102. Barcikowski, B., Carlson, J. C., Wilson, L., and McCracken, J. A. (1974). The effect of endogenous and exogenous estradiol-17β on the release of prostaglandin F2α from the ovine uterus. *Endocrinology* 95, 1340–1349.
103. Inskeep, E. K., and Murdoch, W. J. (1980). Relation of ovarian functions to uterine and ovarian secretion of prostaglandins during the estrous cycle and early pregnancy in the ewe and cow. In *Reproductive Physiology* (R. O. Greep, Ed.), III. International Review of Physiology, Vol. 22, pp. 325–356. University Park Press, Baltimore.
104. Bland, K. P., Horton, E. W., and Poyser, N. L. (1971). Levels of prostaglandin F2α in the uterine venous blood of sheep during the oestrous cycle. *Life Sci.* 10, 509–517.
105. Pexton, J. E., Weems, C. W., and Inskeep, E. K. (1975). Prostaglandins F in uterine and ovarian venous plasma from nonpregnant and pregnant ewes collected by cannulation. *Prostaglandins* 9, 501–509.
106. Nett, T. M., Staigmiller, R. B., Akbar, A. M., Diekman, M. A., Ellinwood, W. E., and Niswender, G. D. (1976). Secretion of prostaglandin F2α in cycling and pregnant ewes. *J. Anim. Sci.* 42, 876–880.
107. Ottobre, J. S., Lewis, G. S., Thayne, W. V., and Inskeep, E. K. (1980). Mechanism by which progesterone shortens the estrous cycle of the ewe. *Biol. Reprod.* 23, 1046–1053.
108. Martensz, N. D., and Scaramuzzi, R. J. (1979). Plasma concentrations of luteinizing hormone, follicle-stimulating hormone and progesterone during the breeding season in ewes immunized against androstenedione or testosterone. *J. Endocrinol.* 81, 249–259.
109. Robinson, T. J. (1954). The necessity for progesterone with estrogen for the induction of recurrent estrus in the ovariectomized ewe. *Endocrinology* 55, 403–408.
110. Robinson, T. J., Moore, N. W., and Binet, F. E. (1956). The effect of the duration of progesterone pretreatment on the response of the spayed ewe to oestrogen. *J. Endocrinol.* 14, 1–7.
111. Moore, N. W., and Robinson, T. J. (1957). The behavioural and vaginal response of the spayed ewe to oestrogen injected at various times relative to the injection of progesterone. *J. Endocrinol.* 15, 360–365.
112. Fabre-Nys, C., and Martin, G. B. (1991). Roles of progesterone and oestradiol in determining the temporal sequence and quantitative expression of sexual receptivity and the preovulatory LH surge in the ewe. *J. Endocrinol.* 130, 367–379.
113. Knights, M., Hoehn, T., Lewis, P. E., and Inskeep, E. K. (2001). Effectiveness of intravaginal progesterone inserts and FSH for inducing synchronized estrus and increasing lambing rate in anestrous ewes. *J. Anim. Sci.* 79, 1120–1131.
114. Fairclough, R. J., Smith, J. F., and Peterson, A. J. (1976). Passive immunization against oestradiol-17β and its effect on luteolysis, oestrus and ovulation in the ewe. *J. Reprod. Fertil.* 48, 169–171.

115. Goodman, R. L., Legan, S. J., Ryan, K. D., Foster, D. L., and Karsch, F. J. (1981). Importance of variations in behavioural and feedback actions of oestradiol to the control of seasonal breeding in the ewe. *J. Endocrinol.* 89, 229–240.
116. Lewis, P. E., Bolt, D. J., and Inskeep, E. K. (1974). Pattern of luteinizing hormone release in progestin-treated ewes. *J. Anim. Sci.* 38, 1204–1209.
117. Skinner, D. C., Harris, T. G., and Evans, N. P. (2000). Duration and amplitude of the luteal phase progesterone increment times the estradiol-induced luteinizing hormone surge in ewes. *Biol. Reprod.* 63, 1135–1142.
118. Fabre-Nys, C., Martin, G. B., and Venier, G. (1993). Analysis of the hormonal control of female sexual behavior and the preovulatory LH surge in the ewe: Roles of quantity of estradiol and duration of its presence. *Horm. Behav.* 27, 108–121.
119. Scaramuzzi, R. J., Tillson, S. A., Thorneycroft, I. H., and Caldwell, B. V. (1971). Action of exogenous progesterone and estrogen on behavioral estrus and luteinizing hormone levels in the ovariectomized ewe. *Endocrinology* 88, 1184–1189.
120. Legan, S. J., Goodman, R. L., Ryan, K. D., Foster, D. L., and Karsch, F. J. (1985). Can the transition into anoestrus in the ewe be accounted for solely by insufficient tonic LH secretion? *J. Endocrinol.* 106, 55–60.
121. Thiery, J. C., Pelletier, J., and Signoret, J. P. (1978). Effect of hypothalamic deafferentation on LH and sexual behaviour in ovariectomized ewes under hormonally induced oestrous cycle. *Ann. Biol. Anim. Bioch. Biophys.* 186, 1413–1426.
122. Blache, D., Fabre-Nys, C. J., and Venier, G. (1991). Ventromedial hypothalamus as a target for oestradiol action on proceptivity, receptivity and luteinizing hormone surge of the ewe. *Brain Res.* 546, 241–249.
123. Blache, D., Batailler, M., and Fabre-Nys, C. (1994). Oestrogen receptors in the preoptico-hypothalamic continuum: Immunohistochemical study of the distribution and cell density during oestrous cycle in ovariectomized ewes. *J. Neuroendocrinol.* 6, 329–339.
124. Blache, D., Fabre-Nys, C., and Venier, G. (1996). Inhibition of sexual behavior and the luteinizing hormone surge by intracerebral progesterone implants in the female sheep. *Brain Res.* 741, 117–122.
125. Fabre-Nys, C., Blache, D., Hinton, M. R., Goode, J. A., and Kendrick, K. M. (1994). Microdialysis measurement of neurochemical changes in the mediobasal hypothalamus of ovariectomized ewes during oestrus. *Brain Res.* 649, 282–296.
126. Fabre-Nys, C., Chesneau, D., De La Riva, C., Hinton, M. R., Locatelli, A., Ohkura, S., and Kendrick, K. M. (2003). Biphasic role of dopamine on female sexual behaviour via D2 receptors in the mediobasal hypothalamus. *Neuropharmacology* 44, 354–366.
127. Skinner, D. C., Malpaux, B., Delaleu, B., and Caraty, A. (1995). Luteinizing hormone (LH)–releasing hormone in third ventricular cerebrospinal fluid of the ewe: correlation with LH pulses and LH surge. *Endocrinology* 136, 3230–3237.
128. Caraty, A., Delaleu, B., Chesneau, D., and Fabre-Nys, C. (2002). Sequential role of E2 and GnRH for the expression of estrous behavior in ewes. *Endocrinology* 143, 139–145.
129. Lehman, M. N., Robinson, J. E., Karsch, F. J., and Silverman, A. J. (1986). Immunocytochemical localization of luteinizing hormone-releasing hormone (LHRH) in the sheep brain during anestrus and the mid-luteal phase of the estrous cycle *J. Comp. Neurol.* 244, 19–35.
130. Caldani, Batailler, M., Thiery, J. C., and Dubois, M. P. (1988). LHRH-immunoreactive structures in the sheep brain. *Histochemistry* 89, 129–139.
131. Jansen, H. T., Hileman, S. M., Lubbers, L. S., Kuehl, D. E., Jackson, G. L., and Lehman, M. N. (1997). Identification and distribution of neuroendocrine gonadotropin-releasing hormone neurons in the ewe. *Biol. Reprod.* 56, 655–662.
132. Polkowska, J. (1981). Immunocytochemistry of luteinizing hormone releasing hormone (LHRH) and gonadotropic hormones in the sheep after anterior deafferentations of the hypothalamus. *Cell Tissue Res.* 220, 637–649.
133. Lehman, M. N., Ebling, F. J. P., Moenter, S. M., and Karsch, F. J. (1993). Distribution of estrogen receptor-immunoreactive cells in the sheep brain. *Endocrinology* 133, 876–886.
134. Herbison, A. E. (1995). Neurochemical identity of neurones expressing oestrogen and androgen receptors in sheep hypothalamus. *J. Reprod. Fertil. Suppl.* 49, 271–283.
135. Skinner, D. C., and Herbison, A. E. (1997). Effects of photoperiod on estrogen receptor, tyrosine hydroxylase, neuropeptide Y, and β-endorphin immunoreactivity in the ewe hypothalamus. *Endocrinology* 138, 2585–2595.
136. Simonian, S. X., Delaleu, B., Caraty, A., and Herbison, A. E. (1998). Estrogen receptor expression in brainstem noradrenergic neurons of the sheep. *Neuroendocrinology* 67, 392–402.
137. Scott, C. J., Rawson, J. A., Pereira, A. M., and Clarke, I. J. (1999). Oestrogen receptors in the brainstem of the female sheep: Relationship to noradrenergic cells and cells projecting to the medial preoptic area. *J. Neuroendocrinol.* 11, 745–755.
138. Scott, C. J., Tilbrook, A. J., Simmons, D. M., Rawson, J. A., Chu, S., Fuller, P. J., Ing, N. H., and Clarke, I. J. (2000). The distribution of cells containing estrogen receptor-a (ERa) and ERβ messenger ribonucleic acid in the preoptic area and hypothalamus of the sheep: Comparison of males and females. *Endocrinology* 141, 2951–2962.
139. Scanlan, N., Dufourny, L., and Skinner, D. C. (2003). Somatostatin-14 neurons in the ovine hypothalamus: colocalization with estrogen receptor α and somatostatin-28 (1–12) immunoreactivity, and activation in response to estradiol. *Biol. Reprod.* 69, 1319–1324.
140. Dufourny, L., and Skinner, D. C. (2002). Progesterone receptor, estrogen receptor α, and the type II glucocorticoid receptor are coexpressed in the same neurons of the ovine preoptic area and arcuate nucleus: A triple immunolabeling study. *Biol. Reprod.* 67, 1605–1612.
141. Scott, C. J., Pereira, A. M., Rawson, J. A., Simmons, D. M., Rossmanith, W. G., Ing, N. H., and Clarke, I. J. (2000). The distribution of progesterone receptor immunoreactivity and mRNA in the preoptic area and hypothalamus of the ewe: Upregulation of progesterone receptor mRNA in the mediobasal hypothalamus by oestrogen. *J. Neuroendocrinol.* 12, 565–575.
142. Skinner, D. C., Caraty, A., and Allingham, R. (2001). Unmasking the progesterone receptor in the preoptic area and hypothalamus of the ewe: No colocalization with gonadotropin-releasing neurons. *Endocrinology* 142, 573–579.
143. Foradori, C. D., Coolen, L. M., Fitzgerald, M. E., Skinner, D. C., Goodman, R. L., and Lehman, M. N. (2002). Colocalization of progesterone receptors in the parvicellular dynorphin neurons of the ovine preoptic area and hypothalamus. *Endocrinology* 143, 4366–4374.
144. Lehman, M. N., and Karsch, F. J. (1993). Do gonadotropin-releasing hormone, tyrosine hydroxylase-, and β-endorphin-immunoreactive neurons contain estrogen receptors? A double-label immunocytochemical study in the Suffolk ewe. *Endocrinology* 133, 887–895.
145. Herbison, A. E., Robinson, J. E., and Skinner, D. C. (1993). Distribution of estrogen receptor-immunoreactive cells in the preoptic area of the ewe: Co-localization with glutamic acid decarboxylase but not luteinizing hormone-releasing hormone. *Neuroendocrinology* 57, 751–759.
146. Goubillon, M. L., Forsdike, R. A., Robinson, J. E., Ciofi, P., Caraty, A., and Herbison, A. E. (2000). Identification of neurokinin B-expressing neurons as a highly estrogen-receptive, sexually dimorphic cell group in the ovine arcuate nucleus. *Endocrinology* 141, 4218–4225.

147. Pompolo, S., Pereira, A., Kaneko, T., and Clarke, I. J. (2003). Seasonal changes in the inputs to gonadotropin-releasing hormone neurones in the ewe brain: An assessment by conventional fluorescence and confocal microscopy. *J. Neuroendocrinol.* 15, 538–545.
148. Jansen, H. T., Cutter, C. T., Hardy, S., Lehman, N. M., and Goodman, R. L. (2003). Seasonal plasticity in the GnRH system of the ewe: changes in identified GnRH inputs and in glial association. *Endocrinology* 144, 3663–3676.
149. Goodman, R. L., Coolen, L. M., Anderson, G. M., Hardy, S. L., Valent, M., Connors, J., Fitzgerald, M. E., and Lehman, M. N. (2004). Evidence that dynorphin plays a major role in mediating progesterone negative feedback on gonadotropin-releasing hormone neurons in sheep. *Endocrinology* 145, 2959–2967.
150. Robinson, J. E., Birch, R. A., Grindrod, J. A. E., Taylor, J. A., and Unsworth, W. P. (2003). Sexually differentiated regulation of GnRH release by gonadal steroid hormones in sheep. *Reprod. Suppl.* 61, 299–310.
151. Norgren, R. B., and Lehman, M. N. (1989). A double-label pre-embedding immunoperoxidase technique for electron microscopy using diaminobenzidine and tetramethylbenzidine as markers. *J. Histochem. Cytochem.* 37, 1283–1289.
152. Pompolo, S., Pereira, A., Scott, J. C., Fujiyma, F., and Clarke, I. J. (2003). Evidence for estrogenic regulation of gonadotropin-releasing hormone neurons by glutamatergic neurons in the ewe brain: An immunohistochemical study using an antibody against vesicular glutamate transporter-2. *J. Comp. Neurol.* 465, 136–144.
153. Lehman, M. N., Goodman, R. L., and Foradori, C. D. (2003). Colocalization of dynorphin A and neurokinin B immunoreactivity in the ovine arcuate nucleus. *Prog. Ann. Meet. Soc. Neurosci.* Abstr. 611.18.
154. Jansen, H. T., Hileman, S. M., Lubbers, L. S., Jackson, G. L., and Lehman, M. N. (1996). A subset of estrogen receptor-containing neurons project to the median eminence in the ewe. *J. Neuroendocrinol.* 8, 921–927.
155. Kuljis, R. O., and Advis, J. P. (1989). Immunocytochemical and physiological evidence of a synapse between dopamine- and luteinizing hormone releasing hormone-containing neurons in the ewe median eminence. *Endocrinology* 124, 1579–1581.
156. Tillet, Y., Batailler, M., and Thibault, J. (1993). Neuronal projections to the medial preoptic area in the sheep with special reference to monoaminergic afferents: Immunohistochemical and retrograde tract tracing studies. *J. Comp. Neurol.* 330, 195–220.
157. Goubillion, M. L., Delaleu, B., Tillet, Y., Caraty, A., and Herbison, A. E. (1999). Localization of estrogen-receptive neurons projecting to the GnRH neuron-containing rostral preoptic area of the ewe. *Neuroendocrinology* 70, 228–236.
158. Goubillion, M. L., Caraty, A., and Herbison, A. E. (2002). Evidence in favour of a direct input from the ventromedial nucleus to gonadotropin-releasing hormone neurones in the ewe: An anterograde tracing study. *J. Neuroendocrinol.* 14, 95–100.
159. Pompolo, S., Rawson, J. A., and Clarke, I. J. (2001). Projections from the arcuate/ventromedial region of the hypothalamus to the preoptic area and bed nucleus of the stria terminalis in the brain of the ewe: lack of direct input to gonadotropin-releasing hormone neurons. *Brain Res.* 904, 1–12.
160. Dufourny, L., and Skinner, D. C. (2002). Influence of estradiol on NADPH diaphorase/neuronal nitric oxide synthase activity and colocalization with progesterone or type II glucocorticoid receptors in ovine hypothalamus. *Biol. Reprod.* 67, 829–836.
161. Weesner, G. D., and Malven, P. V. (1990). Intracerebral immunoneutralization of beta-endorphin and met-enkephalin disinhibits release of pituitary luteinizing hormone in sheep. *Neuroendocrionology* 52, 382–388.
162. Curlewis, J. D., Naylor, A. M., Rhind, S. M., and McNeilly, A. S. (1991). Effects of beta-endorphin on pulsatile luteinizing hormone and prolactin secretion during the follicular phase in the ewe. *J. Neuroendol.* 3, 123–126.
163. Scott, C. J., and Clarke, I. J. (1993). Inhibition of luteinizing hormone secretion in ovariectomized ewes during the breeding season by γ-aminobutyric acid (GABA) is mediated by GABA-A receptors, but not GABA-B receptors. *Endocrinology* 132, 1789–1796.
164. Scott, C. J., and Clarke, I. J. (1993). Evidence that changes in the function of the subtypes of the receptors for γ-aminobutyric acid may be involved in the seasonal changes in the negative feedback effects of estrogen on gonadotropin-releasing hormone secretion and plasma luteinizing hormone levels in the ewe. *Endocrinology* 133, 2904–2912.
165. Estienne, M. J., Schillo, K. K., Hileman, S. M., Green, M. A., and Hayes, S. H. (1990). Effect of N-methyl-d,l-aspartate on luteinizing hormone secretion in ovariectomized ewes in the absence and presence of estradiol. *Biol. Reprod.* 42, 126–130.
166. Meyer, S. L., and Goodman, R. L. (1986). Separate neural systems mediate the steroid-dependent and steroid-independent suppression of tonic luteinizing hormone secretion in the anestrous ewe. *Biol. Reprod.* 35, 562–571.
167. Barker-Gibb, M. L., Scott, C. J., Boublik, J. H., and Clarke, I. J. (1995). The role of neuropeptide-Y (NPY) in the control of LH secretion in the ewe with respect to season, NPY receptor subtype and the site of action in the hypothalamus. *J. Endocrinol.* 147, 565–579.
168. Anderson, S. T., Walsh, J. P., Tillet, Y., Clarke, I. J., and Curlewis, J. D. (2001). Dopaminergic input to the ventromedial hypothalamus facilitates the oestrogen-induced luteinizing hormone surge in ewes. *Neuroendocrinology* 73, 91–101.
169. Advis, J. P., Klein, J., Kulis, R. O., Sarkar, D. K., McDonald, J. M., and Conover, C. A. (2003). Regulation of gonadotropin releasing hormone release by neuropeptide Y at the median eminence during the preovulatory period in ewes. *Neuroendocrinology* 77, 246–257.
170. Scott, C. J., Cummins, J. T., and Clarke, I. J. (1992). Effects on plasma luteinizing hormone levels of microinjection of noradrenaline and adrenaline into the septo-preoptic area of the brain of the ovariectomized ewe: Changes with season and chronic oestrogen treatment Journal of *Neuroendocrinology* 4, 131–141.
171. Pillon, D., Caraty, A., Fabre-Nys, C., Lomet, D., Cateau, M., and Bruneau, G. (2004). Regulation by estradiol of hypothalamic somatostatin gene expression: possible involvement of somatostatin in the control of LH secretion in the ewe. *Biol. Reprod.* 71, 38–44.
172. Pathiraja, N., Carr, W. R., Fordyce, M., Forster, J., Land, R. B., and Morris, B. A. (1984). Concentration of gonadotrophins in the plasma of sheep given gonadal steroid antisera to raise ovulation rate. *J. Reprod. Fertil.* 72, 93–100.
173. Martin, G. B., Scaramuzzi, R. J., and Henstridge, J. D. (1983). Effects of oestradiol, progesterone and androstenedione on the pulsatile secretion of luteinizing hormone in ovariectomized ewes during spring and autumn. *J. Endocrinol.* 96, 181–193.
174. Scaramuzzi, R. J., Baird, D. T., Clark, I. J., Martensz, N. D., and Van Look, P. F. A. (1980). Ovarian morphology and the concentration of steroids during the oestrous cycle of sheep actively immunized against androstenedione. *J. Reprod. Fertil.* 58, 27–35.
175. Karsch, F. J., Legan, S. J., Hauger, R. L., and Foster, D. L. (1977). Negative feedback action of progesterone on tonic luteinizing hormone secretion in the ewe: dependence on the ovaries. *Endocrinology* 101, 800–806.

176. Goodman, R. L., Legan, S. J., Ryan, K. D., Foster, D. L., and Karsch, F. J. (1980). Two effects of estradiol that normally contribute to the control of tonic LH secretion in the ewe. *Biol. Reprod.* 23, 415–422.
177. Goodman, R. L., and Karsch, F. J. (1980). Pulsatile secretion of luteinizing hormone: differential suppression by ovarian steroids. *Endocrinology* 107, 1286–1290.
178. Padmanabhan, V., Evans, N. P., Dahl, G. E., McFadden, K. L., Mauger, D. T., and Karsch, F. J. (1995). Evidence for short or ultrashort loop negative feedback of gonadotropin-releasing hormone secretion. *Neuroendocrinology* 62, 248–258.
179. Skinner, D. C., Evans, N. P., Delaleu, B., Goodman, R. L., Bouchard, P., and Caraty, A. (1998). The negative feedback actions of progesterone on gonadotropin-releasing hormone secretion are transduced by the classical progesterone receptor. *Proc. Nat. Acad. Sci. U S A* 95, 10978–10983.
180. Legan, S. J., Karsch, F. J., and Foster, D. L. (1977). The endocrine control of seasonal reproductive function in the ewe: a marked change in response to the negative feedback action of estradiol on luteinizing hormone secretion. *Endocrinology* 101, 818–824.
181. Evans, N. P., Dahl, G. E., Glover, B. H., and Karsch, F. J. (1994). Central regulation of pulsatile gonadotropin-releasing hormone (GnRH) secretion by estradiol during the period leading up to the preovulatory GnRH surge in the ewe. *Endocrinology* 134, 1806–1811.
182. Goodman, R. L., Bittman, E. L., Foster, D. L., and Karsch, F. J. (1981). The endocrine basis of the synergistic suppression of luteinizing hormone by estradiol and progesterone. *Endocrinology* 109, 1414–1417.
183. Jackson, G. L., Kuehl, D., McDowell, K., and Zaleski, A. (1978). Effect of hypothalamic deafferentation on secretion of luteinizing hormone in the ewe. *Biol. Reprod.* 17, 808–819.
184. Pau K.-Y., F., Kuehl, D. E., and Jackson, G. L. (1982). Effect of frontal hypothalamic deafferentation on luteinizing hormone secretion and seasonal breeding in the ewe. *Biol. Reprod.* 27, 999–1009.
185. Boukhliq, R., Goodman, R. L., Berriman, S. J., Adrian, B., and Lehman, M. N. (1999). A subset of gonadotropin-releasing hormone neurons in the ovine medial basal hypothalamus is activated during increased pulsatile luteinizing hormone secretion. *Endocrinology* 140, 5929–5936.
186. Moenter, S. M., Karsch, F. J., and Lehman, M. N. (1993). Fos expression during the estradiol-induced gonadotropin-releasing hormone (GnRH) surge of the ewe: Induction in GnRH and other neurons. *Endocrinology* 133, 896–903.
187. Goodman, R. L., Thiery, J. C., Delaleu, B., and Malpaux, B. (2000). Estradiol increases multi-unit electrical activity in the A15 area of ewes exposed to inhibitory photoperiods. *Biol. Reprod.* 63, 1352–1357.
188. Moenter, S. M., Brand, R. M., Midgley, A. R., and Karsch, F. J. (1992). Dynamics of gonadotropin-releasing hormone release during a pulse. *Endocrinology* 130, 503–510.
189. Wetsel, W. C., Valenca, M. M., Merchenthaler, I., Liposits, Z., Lopez, F. J., Weiner, R. I., Mellon, P. L., and Negro-Vilar, A. (1992). Intrinsic pulsatile secretory activity of immortalized luteinizing hormone-releasing hormone-secreting neurons. *Proc. Nat. Acad. Sci. U S A* 89, 4149–4153.
190. Evans, N. P., Dahl, G. E., Mauger, D. T., and Karsch, F. J. (1995). Estradiol induces both qualitative and quantitative changes in the pattern of gonadotropin-releasing hormone secretion during the presurge period in the ewe. *Endocrinology* 136, 1603–1609.
191. Goodman, R. L., Parfitt, D. B., Evans, N. P., Dahl, G. E., and Karsch, F. J. (1995). Endogenous opioid peptides control the amplitude and shape of gonadotropin-releasing hormone pulses in the ewe. *Endocrinology* 136, 2412–2420.
192. Naylor, A. M., Porter, D. W. F., and Lincoln, D. W. (1989). Inhibitory effect of central LHRH on LH secretion in the ovariectomized ewe. *Neuroendocrinology* 49, 531–536.
193. Caraty, A., Locatelli, A., Delaleu, B., Spitz, I. M., Schatz, B., and Bouchard, P. (1990). Gonadotropin-releasing hormone (GnRH) agonists and GnRH antagonists do not alter endogenous GnRH secretion in short-term castrated rams. *Endocrinology* 127, 2519–1523.
194. Lehman, M. N., Karsch, F. J., Robinson, J. E., and Silverman, A. J. (1988). Ultrastructure and synaptic organization of luteinizing hormone-releasing hormone (LHRH) neurons in the anestrous ewe. *J. Comp. Neurol.* 273, 447–458.
195. Brown, D., Herbison, A., Robinson, J. E., Marrs, R. W., and Leng, G. (1994). Modeling the luteinizing hormone-releasing hormone pulse generator. *Neuroscience* 63, 869–879.
196. Clarke, I. J., and Cummins, J. T. (1984). Direct pituitary effects of estrogen and progesterone on gonadotropin secretion in the ovariectomized ewe. *Neuroendocrinology* 39, 267–274.
197. Batra, S. K., and Miller, W. L. (1985). Progesterone decreases the responsiveness of ovine pituitary cultures to luteinizing hormone-releasing hormone. *Endocrinology* 117, 1436–1440.
198. Girmus, R. L., and Wise, M. E. (1991). Direct pituitary effects of estradiol and progesterone on luteinizing hormone release, stores, and subunit messenger ribonucleic acids. *Biol. Reprod.* 45, 128–134.
199. Girmus, R. L., and Wise, M. E. (1992). Progesterone directly inhibits pituitary luteinizing hormone secretion in an estradiol-dependent manner. *Biol. Reprod.* 46, 710–714.
200. Clarke, I. J., Cummins, J. T., Crowder, M. E., and Nett, T. M. (1989). Long-term negative feedback effects of oestrogen and progesterone on the pituitary gland of the long-term ovariectomized ewe. *J. Endocrinol.* 120, 207–214.
201. Kaynard, A. H., and Karsch, F. J. (1988). Hypophyseal actions of pulsatile gonadotropin-releasing hormone in the ewe: development and application of a new experimental model. *Neuroendocrinology* 48, 287–295.
202. Wheaton, J. E., and Mullett, M. A. (1982). Effects of progesterone treatment on basal and LH-RH–induced plasma LH concentrations in anoestrous and ovariectomized sheep. *J. Reprod. Fertil.* 64, 325–328.
203. Robinson, J. E., and Kendrick, K. M. (1992). Inhibition of luteinizing hormone secretion in the ewe by progesterone: associated changes in the release of gamma-aminobutyric acid and noradrenaline in the preoptic area as measured by intracranial microdialysis. *J. Neuroendol.* 4, 231–236.
204. Caraty, A., Skinner, D. C., Huyge, B., Bertault, T., Picard, S., Delaleu, B., Malpaux, B., and Thiery, J. C. (1998). The GnRH increase following progesterone withdrawal is associated with an increased glutamatergic tone in the preoptic area of the ovariectomized ewe. *Ann. N. Y. Acad. Sci.* 15, 363–364.
205. Horton, R. J. E., Francis, H., and Clarke, I. J. (1989). Seasonal and steroid-dependent effects on the modulation of LH secretion in the ewe by the intracerebroventricularly administered β-endorphin or naloxone. *J. Endocrinol.* 122, 509–517.
206. Trout, W. E., and Malven, P. V. (1987). Effects of exogenous estradiol-17β and progesterone on naloxone-reversible inhibition of the release of luteinizing hormone in ewes. *J. Anim. Sci.* 65, 1602–1609.
207. Brooks, A. N., Lamming, G. E., Lees, P. D., and Haynes, N. B. (1986). Opioid modulation of LH secretion in the ewe. *J. Reprod. Fertil.* 76, 693–708.
208. Yang, K., Haynes, N. B., Lamming, G. E., and Brooks, A. N. (1988). Ovarian steroid hormone involvement in endogenous opioid modulation of LH secretion in mature ewes during the breeding and non-breeding seasons. *J. Reprod. Fertil.* 83, 129–139.

209. Whisnant, C. S., and Goodman, R. L. (1988). Effects of an opioid antagonist on pulsatile luteinizing hormone secretion in the ewe vary with changes in steroid negative feedback. *Biol. Reprod.* 39, 1032–1038.
210. Horton, R. J. E., Cummins, J. T., and Clarke, I. J. (1987). Naloxone evokes large-amplitude GnRH pulses in luteal-phase ewes. *J. Reprod. Fertil.* 81, 277–286.
211. Schillo, K. K., Kuehl, D., and Jackson, G. L. (1985). Do endogenous opioid peptides mediate the effects of photoperiod on release of luteinizing hormone and prolactin in ovariectomized ewes? *Biol. Reprod.* 32, 779–787.
212. Skinner, D. C., Delaleu, B., and Caraty, A. (2000). Where in the brain does progesterone inhibit GnRH secretion? *Prog. Ann. Meet. Soc. Neurosci.* Abstr. 582.7.
213. Malven, P. V., Stanisiewski, E. P., and Haglof, S. A. (1990). Ovine brain areas sensitive to naloxone-induced stimulation of luteinizing hormone release. *Neuroendocrinology* 52, 373–381.
214. Whisnant, C. S., Havern, R. L., and Goodman, R. L. (1991). Endogenous opioid suppression of luteinizing hormone pulse frequency and amplitude in the ewe: hypothalamic sites of action. *Neuroendocrinology* 54, 587–593.
215. Jackson, G. L., Leshin, L. S., and Schillo, K. K. (1986). Effect of frontal hypothalamic deafferentation on duration of breeding season and melatonin secretion in the ewe. *Biol. Reprod.* 35, 1277–1288.
216. Whisnant, C. S., and Goodman, R. L. (1994). Effect of anterior hypothalamic deafferentation on the negative feedback of gonadal steroids on luteinizing hormone pulse frequency in the ewe. *Dom. Anim. Endocrinol.* 11, 151–159.
217. Caraty, A., Locatelli, A., and Martin, G. B. (1989). Biphasic response in the secretion of gonadotropin-releasing hormone in ovariectomized ewes injected with estradiol. *J. Endocrinol.* 123, 375–382.
218. Miller, W. L., Knight, M. M., Grimek, H. J., and Gorski, J. (1977). Estrogen regulation of follicle stimulating hormone in cell cultures of sheep pituitaries. *Endocrinology* 100, 1306–1316.
219. Landefeld, T., Kepa, J., and Karsch, F. J. (1984). Estradiol feedback effects on the alpha-subunit mRNA in the sheep pituitary gland: correlation with serum and pituitary luteinizing hormone concentrations. *Proc. Nat. Acad. Sci. U S A* 81, 1322–1326.
220. Landefeld, T., and Kepa, J. (1984). Regulation of LH beta subunit mRNA in the sheep pituitary gland during different feedback states of estradiol. *Biochem. Biophys. Res. Comm.*122, 1307–1313.
221. Herring, R. D., Hamernik, D. L., Kile, J. P., Sousa, M. E., and Nett, T. M. (1991). Chronic administration of estradiol produces a triphasic effect on serum concentrations of gonadotropins and messenger ribonucleic acid for gonadotropin subunits, but not on pituitary content of gonadotropins, in ovariectomized ewes. *Biol. Reprod.* 45, 151–156.
222. Di Gregorio, G. B., and Nett, T. M. (1995). Estradiol and progesterone influence the synthesis of gonadotropins in the absence of gonadotropin-releasing hormone in the ewe. *Biol. Reprod.* 53, 166–172.
223. Clarke, I. J., Cummins, J. T., Findlay, J. K., Burman, K. J., and Doughton, B. W. (1984). Effects on plasma luteinizing hormone and follicle stimulating hormone of varying the frequency and amplitude of gonadotropin-releasing hormone pulses in ovariectomized ewes with hypothalamo-pituitary disconnection. *Neuroendocrinology* 39, 214 221.
224. Clarke, I. J., and Cummins, J. T. (1985). GnRH pulse frequency determines LH pulse amplitude by altering the amount of releasable LH in the pituitary glands of ewes. *J. Reprod. Fertil.* 73, 425–431.
225. Moenter, S. E., Caraty, A., and Karsch, F. J. (1990). The estradiol-induced surge of gonadotropin-releasing hormone in the ewe. *Endocrinology* 127, 1375–1384.
226. Caraty, A., Fabre-Nys, C., Delaleu, B., Locatelli, A., Bruneau, G., Karsch, F. J., and Herbison, A. (1998). Evidence that the mediobasal hypothalamus is the primary site of action of estradiol in inducing the preovulatory gonadotropin releasing hormone surge in the ewe. *Endocrinology* 139, 1752–1760.
227. Rawson, J. A., Scott, C. J., Pereira, A. M., Jakubowska, A., and Clarke, I. J. (2001). Noradrenergic projections from the A1 field to the preoptic area in the brain of the ewe and fos responses to oestrogen in the A1 cells. *J. Neuroendol.* 13, 129–138.
228. Goodman, R. L., Havern, R. L., and Whisnant, C. S. (1996). Alpha-adrenergic neurons inhibit luteinizing hormone pulse amplitude in breeding season ewes. *Biol. Reprod.* 54, 380–386.
229. Goodman, R. L., Gibson, M., Skinner, D. C., and Lehman, M. N. (2002). Neuroendocrine control of pulsatile GnRH secretion during the ovarian cycle: evidence from the ewe. *Reprod, Suppl.* 59, 41–56.
230. Domanski, E., Chomicka, L. K., Ostrowska, A., Gajewska, A., and Mateusiak, K. (1991). Release of luteinizing hormone-releasing hormone, β-endorphin and noradrenaline by the nucleus infundibularis/median eminence during periovulatory period in sheep. *Neuroendocrinology* 54, 151–158.
231. Conover, C. D., Kuljis, R. O., Rabii, J., and Advis, J. P. (1993). Beta-endorphin regulation of luteinizing hormone-releasing hormone release at the median eminence in ewes: immunocytochemical and physiological evidence *Neuroendocrinology* 57, 1182–1195.
232. Goodman, R. L. (1994). Neuroendocrine control of the ovine estrous cycle. In *The Physiology of Reproduction* (E. Knobil and J. D. Neill, Eds.), Vol. 2, pp. 659–710. Raven Press, New York.
233. Skinner, D. C., Bouchard, P., and Caraty, A. (1999). The progesterone blockade of the luteinizing hormone surge is overcome by RU486. *J. Neuroendocrinol.* 11, 637–641.
234. Beck, T. W., and Reeves, J. J. (1973). Serum luteinizing hormone (LH) in ewes treated with various dosages of 17β-estradiol at three stages of the anestrous season. *J. Anim. Sci.* 36, 566–570.
235. Caraty, A., and Skinner, D. C. (1999). Progesterone priming is essential for the full expression of the positive feedback effect of estradiol in inducing the preovulatory gonadotropin-releasing hormone surge in the ewe. *Endocrinology* 140, 165–170.
236. Kasa-Vubu, J. Z., Dahl, G. E., Evans, N. P., Thrun, L. A., Moenter, S. M., Padmanabhan, V., and Karsch, F. J. (1992). Progesterone blocks the estradiol-induced gonadotropin discharge in the ewe by inhibiting the surge of gonadotropin-releasing hormone. *Endocrinology* 131, 208–212.
237. Hooley, R. D., Baxter, R. W., Chamley, W. A., Cumming, I. A., Jonas, H. A., and Findlay, J. D. (1974). FSH and LH response to gonadotropin-releasing hormone during the ovine estrous cycle and following progesterone administration. *Endocrinology* 95, 937–942.
238. Coppings, R. J., and Malven, P. V. (1976). Biphasic effect of estradiol on mechanisms regulating LH release in ovariectomized sheep. *Neuroendocrinology* 21, 146–156.
239. Moss, G. E., and Nett, T. M. (1980). GnRH interaction with anterior pituitary. IV. Effect of estradiol-17β on GnRH-mediated release of LH from ovine pituitary cells obtained during the breeding season, anestrous season, and period of transition into or out of the breeding season. *Biol. Reprod.* 23, 398–403.

240. Huang, E. S.-R., and Miller, W. L. (1980). Effects of estradiol-17β on basal and luteinizing hormone releasing hormone-induced secretion of luteinizing hormone and follicle stimulating hormone by ovine pituitary cell culture. *Biol. Reprod.* 23, 124–134.
241. Crighton, D. B., and Foster, J. P. (1977). Luteinizing hormone release after two injections of synthetic luteinizing hormone releasing hormone in the ewe. *J. Endocrinol.* 72, 59–67.
242. Herman, M. E., and Adams, T. E. (1990). Gonadotropin secretion in ovariectomized ewes: effect of passive immunization against gonadotropin-releasing hormone GnRH and infusion of a GnRH agonist and estradiol. *Biol. Reprod.* 42, 273–280.
243. Kaynard, A. H., Malpaux, B., Robinson, J. E., Wayne, N. L., and Karsch, F. J. (1988). Importance of pituitary and neural actions of estradiol in induction of the luteinizing hormone surge in the ewe. *Neuroendocrinology* 48, 296–303.
244. Turzillo, A. M., and Nett, T. M. (1999). Regulation of GnRH receptor gene expression in sheep and cattle. *J. Reprod. Fertil. Suppl.* 54, 75–86.
245. Landefeld, T. D., Bagnell, T., and Levitan, I. (1989). Effects of estradiol on gonadotropin subunit messenger ribonucleic acid amounts during an induced gonadotropin surge in anestrous ewes. *Mol. Endocrinol.* 3, 10–14.
246. Thomas, S. G., and Clarke, I. J. (1997). The positive feedback action of estrogen mobilizes LH-containing, but not FSH-containing secretory granules in ovine gonadotropes. *Endocrinology* 138, 1347–1350.
247. Thomas, S. G., Takahashi, M., Neill, J. D., and Clarke, I. J. (1998). Components of the neuronal exocytotic machinery in the anterior pituitary of the ovariectomized ewe and the effects of oestrogen in gonadotropes as studied with confocal microscopy. *Neuroendocrinology* 67, 244–259.
248. Heyward, P. M., and Clarke, I. J. (1995). A transient effect of estrogen on calcium currents and electrophysiological response to gonadotropin-releasing hormone in ovine gonadotropes. *Neuroendocrinology* 62, 543–552.
249. Evans, N. P., Dahl, G. E., Padmanabhan, V., Thrun, L. A, and Karsch, F. J. (1997). Estradiol requirements for induction and maintenance of the gonadotropin-releasing hormone surge: Implications for neuroendocrine processing of the estradiol signal. *Endocrinology* 138, 5408–5414.
250. Lehman, M. N., Berriman, S. J., Jansen, H. T., Parfitt, D., Dahl, G. E., Evans, N. P., and Karsch, F. J. (1994). Is Fos expression in GnRH neurons during the estradiol-induced surge of LH in the ewe associated with neurosecretory or behavioral events? *Biol. Reprod.* 50 Suppl., Abstr. 521.
251. Evans, N. P., Richter, T. A., Skinner, D. C., and Robinson, J. E. (2002). Neuroendocrine mechanisms underlying the effects of progesterone on the oestradiol-induced GnRH/LH surge. *Reprod. Suppl.* 59, 47–66.
252. Robinson, J. E., Kendrick, K. M., and Lambart, C. E. (1991). Changes in the release of gamma-aminobutyric acid and catecholamines in the preoptic/septal area prior to and during the preovulatory surge of luteinizing hormone in the ewe. *J. Neuroendocrinol.* 3, 393–399.
253. Clarke, I. J., Scott, C. J., Pereira, A., and Rawson, J. (1999). Levels of dopamine beta hydroxylase immunoreactivity in the preoptic hypothalamus of the ovariectomized ewe following injection of oestrogen: evidence for increased noradrenergic release around the time of the oestrogen-induced surge in luteinizing hormone. *J. Neuroendocrinol.* 11, 503–512.
254. Jackson, G. L. (1977). Effect of adrenergic blocking drugs on secretion of luteinizing hormone in the ovariectomized ewe. *Biol. Reprod.* 16, 543–548.
255. Pillon, D., Caraty, A., Fabre-Nys, C., and Bruneau, G. (2003). Early decrease of proopiomelanocortin but not neuropeptide Y mRNA expression in the mediobasal hypothalamus of the ewe, during the estradiol-induced preovulatory LH surge. *Gen. Comp. Endocrinol.* 134, 264–272.
256. Walsh, J. P., Rao, A., Simmons, D. M., and Clarke, I. J. (1998). Proopiomelanocortin mRNA levels in ovine hypothalamus are not reduced at the time of the preovulatory luteinizing hormone surge. *J. Neuroendocrinol.* 10, 803–808.
257. Walsh, J. P., and Clarke, I. J. (1996). Effects of central administration of highly selective opioid μ-, σ-, and κ-receptor agonists on plasma luteinizing hormone (LH), prolactin, and the estrogen-induced LH surge in ovariectomized ewes. *Endocrinology* 137, 3540–3648.
258. Walsh, J. P., and Clarke, I. J. (1998). Blockade of the oestrogen-induced luteinizing hormone surge in ovariectomized ewes by a highly selective opioid mu-receptor agonist: evidence for site of action. *Neuroendocrinology* 67, 164–170.
259. Horton, R. J. E., and Clarke, I. J. (1988). Lack of an effect of morphine or naloxone on the oestrogen-induced LH surge in anoestrous ewes. *J. Endocrinol.* 119, 89–93.
260. Currie, W. D., Joseph, I. B. J. K., and Rawlings, N. C. (1991). Morphine, naloxone and the gonadotropin surge in ewes. *J. Reprod. Fertil.* 92, 407–414.
261. Porter, D. W. F., Naylor, A. M., McNeilly, A. S., and Lincoln, D. W. (1993). Endocrine actions of central neuropeptide Y in the ewe: activation of the hypothalamo-pituitary-adrenal axis by exogenous neuropeptide Y and role of endogenous neuropeptide Y in the secretion of luteinizing hormone during the oestrous cycle. *J. Neuroendocrinol.* 5, 153–174.
262. Malven, P. V., Haglof, S. A., and Jiang, H. (1995). Serum concentrations of luteinizing hormone, growth hormone, and prolactin in untreated and estradiol-treated ovariectomized ewes after immunoneutralization of hypothalamic neuropeptide Y. *J. Anim. Sci.* 73, 2105–2112.
263. McShane, T. M., May, T., Miner, J. L., and Keisler, D. H. (1992). Central actions of neuropeptide-Y may provide a neuromodulatory link between nutrition and reproduction. *Biol. Reprod.* 46, 1151–1157.
264. Malven, P. V., Haglof, S. A., and DeGroot, H. (1992). Effects of intracerebral administration of neuropeptide-Y on secretion of luteinizing hormone in ovariectomized sheep. *Brian Res. Bull.* 28, 871–875.
265. Pillon, D., Caraty, A., Fabre-Nys, C., and Bruneau, G. (2003). Short-term effect of oestradiol on neurokinin B mRNA expression in the infundibular nucleus of ewes. *J. Neuroendocrinol.* 15, 749–753.
266. Clarke, I. J., Pompolo, S., Scott, C. J., Rawson, J. A., Caddy, D., Jakubowska, A. E., and Pereira, A. M. (2001). Cells in the arcuate nucleus and ventromedial nucleus of the ovariectomized ewe that respond to oestrogen: A study using fos immunohistochemistry. *J. Neuroendocrinol.* 13, 934–941.
267. Cummins, J. T., and Clarke, I. J. (1985). Increased gonadotropin-releasing hormone pulse frequency associated with estrogen-induced luteinizing hormone surges in ovariectomized ewes. *Endocrinology* 116, 2376–2383.
268. Clarke, I. J. (1988). Gonadotropin-releasing hormone secretion (GnRH) in anoestrous ewes and the induction of GnRH surges by oestrogen. *J. Endocrinol.* 117, 355–360.
269. Richter, T. A., Robinson, J. E., and Evans, N. P. (2001). Progesterone treatment that either blocks or augments the estradiol-induced gonadotropin-releasing hormone surge is associated with different patterns of hypothalamic neural activation. *Neuroendocrinology* 73, 378–386.
270. Richter, T. A., Robinson, J. E., and Evans, N. P. (2002). Progesterone blocks the estradiol-stimulated luteinizing hormone surge by disrupting activation in response to a stimulatory estradiol signal in the ewe. *Biol. Reprod.* 67, 119–125.

271. Harris, T. G., Dye, S., Robinson, J. E., Skinner, D. C., and Evans, N. P. (1999). Progesterone can block transmission of the estradiol-induced signal for luteinizing hormone surge generation during a specific period of time immediately after activation of the gonadotropin-releasing hormone surge-generating system. *Endocrinology* 140, 827–834.
272. Currie, W. D., Ravindra, J., Kingsbury, D. L., and Rawlings, N. C. (1992). Independence of progesterone blockade of the luteinizing hormone surge in ewes from opioid activity at naloxone-sensitive receptors. *J. Reprod. Fertil.* 95, 249–255.
273. Richter, T. A., Spackman, D. S., Robinson, J. E., Dye, S., Harris, T. G., Skinner, D. C., and Evans, N. P. (2001). Role of endogenous opioid peptides in mediating progesterone-induced disruption of the activation and transmission stages of the GnRH surge induction process. *Endocrinology* 142, 5212–5219.
274. Jonas, H. A., Salamonsen, L. A., Burger, H. G., Chamley, W. A., Cumming, I. A., Findlay, J. K., and Goding, J. R. (1973). Release of FSH after administration of gonadotrophin-releasing hormone or estradiol to the anestrous ewe. *Endocrinology* 92, 862–865.
275. Reeves, J. J., Beck, T. W., and Nett, T. M. (1974). Serum FSH in anestrous ewes treated with 17β-estradiol. *J. Anim. Sci.* 38, 374–377.
276. Cumming, I. A., Baxter, R. W., Buckmaster, J. M., Jonas, H. A., Findlay, J. K., and Goding, J. R. (1974). Control of FSH secretion in the ewe. *J. Reprod. Fertil.* 36, 447–448.
277. Mercer, J. E., Clements, J. A., Funder, J. W., and Clarke, I. J. (1989). Regulation of follicle-stimulating hormone β and common α-subunit messenger ribonucleic acid by gonadotropin-releasing hormone and estrogen in the sheep pituitary. *Neuroendocrinology* 50, 321–326.
278. Padmanabhan, V., and McNeilly, A. S. (2001). Is there an FSH-releasing factor? *Reproduction* 121, 21–30.
279. Padmanabhan, V., Karsch, F. J., and Lee, J. S. (2002). Hypothalamic, pituitary and gonadal regulation of FSH. *Reprod. Suppl.* 59, 67–82.
280. Martin, G., Price, C. A., Thiery, J.-C., and Webb, R. (1988). Interactions between inhibin, oestradiol and progesterone in the control of gonadotrophin secretion in the ewe. *J. Reprod. Fertil.* 82, 319–328.
281. Baird, D. T., Campbell, B. K., Mann, G. E., and McNeilly, A. S. (1991). Inhibin and oestradiol in the control of FSH secretion in the sheep. *J. Reprod. Fertil. Suppl.* 43, 125–138.
282. Gregg, D. W., Schwall, R. H., and Nett, T. M. (1991). Regulation of gonadotropin secretion and number of gonadotropin-releasing hormone receptors by inhibin, activin-A, and estradiol. *Biol. Reprod.* 44, 725–732.
283. Clarke, I. J., Rao, A., Fallest, P. C., and Shupnik, M. A. (1993). Transcription rate of the follicle stimulating hormone (FSH) β subunit gene is reduced by inhibin in sheep but does not fully explain the decrease in mRNA. *Mol. Cell. Endocrinol.* 91, 211–216.
284. Turzillo, A. M., and Nett, T. M. (1997). Effects of bovine follicular fluid and passive immunization against gonadotropin-releasing hormone (GnRH) on messenger ribonucleic acid for GnRH receptor and gonadotropin subunits in ovariectomized ewes. *Biol. Reprod.* 56, 1537–1543.
285. Baratta, M., West, L. A., Turzillo, A. M., and Nett, T. M. (2001). Activin modulates differential effects of estradiol on synthesis and secretion of follicle-stimulating hormone in ovine pituitary cells. *Biol. Reprod.* 64, 714–719.
286. Molter-Gerard, C., Fontaine, J., Guerin, S., and Taragnat, C. (1999). Differential regulation of the gonadotropin storage pattern by gonadotropin-releasing hormone pulse frequency in the ewe. *Biol. Reprod.* 60, 1224–1230.
287. Padmanabhan, V., Battaglia, D., Brown, M. B., Karsch, F. J., Lee, J. S., Pan, W., Phillips, D. J., and Cleeff, J. V. (2002). Neuroendocrine control of follicle-stimulating hormone (FSH) secretion: II Is follistatin-induced suppression of FSH secretion mediated via changes in activin availability and does it involve changes in gonadotropin-releasing hormone secretion? *Biol. Reprod.* 66, 1395–1402.
288. Fraser, H. M., and McNeilly, A. S. (1983). Differential effects of LH-RH immunoneutralization on LH and FSH secretion in the ewe. *J. Reprod. Fertil.* 69, 569–577.
289. Fraser, H. M., Clarke, I. J., and McNeilly, A. S. (1981). Effect of oestrogen and an LH-RH agonist on the release of gonadotrophins in ovariectomized ewes deprived of LH-RH. *J. Reprod. Fertil.* 62, 131–140.
290. Padmanabhan, V., McFadden, K., Mauger, D. T., Karsch, F. J., and Midgley, A. R. Jr. (1997). Neuroendocrine control of follicle-stimulating hormone (FSH) secretion. I. Direct evidence for separate episodic and basal components of FSH secretion. *Endocrinology* 138, 424–432.
291. Padmanabhan, V., Brown, M. B., Dahl, G. E., Evans, N. P., Karsch, F. J., Mauger, D. T., Neill, J. D., and Cleeff, J. V. (2003). Neuroendocrine control of follicle-stimulating hormone (FSH) secretion. III. Is there a gonadotropin-releasing hormone-independent component of episodic FSH secretion in ovariectomized and luteal phase ewes? *Endocrinology* 144, 1380–1392.
292. Farnworth, P. G., Thean, E., Robertson, D. M., and Schwartz, J. (1995). Ovine anterior pituitary production of follistatin *in vitro*. *Endocrinology* 136, 4397–4406.
293. Huang, H.-J., Sebastian, J., Strahl, B. D., Wu, J. C., and Miller, W. L. (2001). Transcriptional regulation of the ovine follicle-stimulating hormone-β gene by activin and gonadotropin-releasing hormone (GnRH): involvement of two proximal activator protein-1 sites for GnRH stimulation. *Endocrinology* 142, 2267–2274.
294. Huang, H.-J., Wu, J. C., Su, P., Zhirnov, O., and Miller, W. L. (2001). A novel role for bone morphogenetic proteins in the synthesis of follicle-stimulating hormone. *Endocrinology* 142, 2275–2283.
295. Dufour, J., Cahill, L. P., and Mauleon, P. (1979). Short- and long-term effects of hypophysectomy and unilateral ovariectomy on ovarian follicular populations. *J. Reprod. Fertil.* 57, 301–309
296. McNatty, K. P., Heath, D. A., Hudson, N., and Clarke, I. J. (1990). Effect of long-term hypophysectomy on ovarian follicular populations and gonadotrophin-induced adenosine cyclic 3′,5′ monophosphate output by follicles from Booroola ewes with or without the F gene. *J. Reprod. Fertil.* 90, 515–522.
297. Webb, R., Nicholas, B., Gong, J. G., Campbell, B. K., Gutierrez, C. G., Garverick, H. A., and Armstrong, D. G. (2003). Mechanisms regulating follicular development and selection of the dominant follicle. *Reprod. Suppl.* 61, 71–90.
298. Campbell, B. K., Souza, C., Gong, J., Webb, R., Kendall, N., Marsters, P., Robinson, G., Mitchell, A., Telfer, E. E., and Baird, D. T. (2003). Domestic ruminants as models for the elucidation of the mechanisms controlling ovarian follicle development in humans. *Reprod. Suppl.* 61, 429–443.
299. Hunter, M. G., Robinson, R. S., Mann, G. E., and Webb, R. (2004). Endocrine and paracrine control of follicular development and ovulation rate in farm species. *Anim. Reprod. Sci.* 82–83, 461–477.
300. Campbell, B. K., Scaramuzzi, R. J., and Webb, R. (1995). Control of antral follicular development and selection in sheep and cattle. *J. Reprod. Fertil. Suppl.* 49, 335–350.
301. Campbell, B. K., Dobson, H., and Scaramuzzi, R. J. (1998). Ovarian function in ewes made hypogonadal with GnRH antagonist and stimulated with FSH in the presence or absence of low amplitude LH pulses. *J. Endocrinol.* 156, 213–222.

302. Campbell, B. K., Dobson, H., Baird, D. T., and Scaramuzzi, R. J. (1999). Examination of the relative role of FSH and LH in the mechanism of ovulatory follicle selection in sheep. *J. Reprod. Fertil.* 117, 355–367.
303. Viñoles, C., Meikle, A., Forsberg, M., and Rubianes, E. (2001). The effect of subluteal levels of exogenous progesterone on follicular dynamics and endocrine patterns during early luteal phase if the ewe. *Theriogenology* 51, 1351–1361.
304. Viñoles, C., Forsberg, M., Banchero, G., and Rubianes, E. (2001). Effect of long-term and short-term progestogen treatment on follicular development and pregnancy rate in cyclic ewes. *Theriogenology* 55, 993–1004.
305. Campbell, B. K., Telfer, E. E., Webb. R., and Baird, D. T. (2004). Evidence of a role for follicle-stimulating hormone in controlling the rate of preantral follicle development in sheep. *Endocrinology* 145, 1870–1879.
306. Gonzalez-Bulnes, A., Souza, C. J. H., Campbell, B. K., and Baird, D. T. (2004). Effect of ageing on hormone secretion and follicular dynamics in sheep with and without the Booroola gene. *Endocrinology* 145, 2858–2864.
307. Hulet, C. V., Shelton, M., Gallagher, J. R., and Price, D. A. (1974). Effects of origin and environment on reproductive phenomena in Rambouillet ewes. I. Breeding season and ovulation. *J. Anim. Sci.* 38, 1210–1217.
308. Downing, J. A., and Scaramuzzi, R. J. (1991). Nutrient effects on ovulation rate, ovarian function and the secretion of gonadotrophic and metabolic hormones in sheep. *J. Reprod. Fertil. Suppl.* 43: 209–227
309. Knights, M., Baptiste, Q. S., Dixon, A. B., Pate, J. L., Marsh, D. J., Inskeep, E. K., and Lewis, P. E. (2003). Effects of dosage of FSH, vehicle and time of treatment on ovulation rate and prolificacy in ewes during the anestrous season. *Small Rum. Res.* 50, 1–9.
310. Montgomery, G. W., Galloway, S. M., Davis, G. H., and McNatty, K. P. (2001). Genes controlling ovulation rate in sheep. *Reproduction* 121, 843–852.
311. McNatty, K. P., Juengel, J. L., Wilson, T., Galloway, S. M., Davis, G. H., Hudson, N. L., Moeller, C. L., Cranfield, M., Reader, K. L., Laitinen, M. P. E., Groome, N. P., Sawyer, H. R., and Ritvos, O. (2003). Oocyte-derived growth factors and ovulation rate in sheep. *Reprod. Suppl.* 61, 339–351.
312. Mulsant, P., Lecerf, F., Fabre, S., Bodin, L., Thimonier, J., Monget, P., Lanneluc, L., Monniaux, D., Teyssier, J., and Elsen, J.-M. (2003). Prolificacy genes in sheep: the French genetic programmes. *Reprod. Suppl.* 61, 353–359.
312a. McNatty, K. P., Moore, L. G., Hudson, N. L., Quirke, L. D., Lawrence, S. B., Reader, K., Hanrahan, J. P., Smith, P., Groome, N. P., Laitinen, M. Ritvos, O., and Juengel, J. L. (2004). The oocyte and its role in regulating ovulation rate: a new paradigm in reproductive biology. *Reproduction* 128, 379–386.
313. Davis, G. H., Dodds, K. G., Wheeler, R., and Jay, N. P. (2001). Evidence that an imprinted gene on the X-chromosome increases ovulation rate in sheep. *Biol. Reprod.* 64, 216–221.
314. Juengel, J. L., Hudson, N. L., Whiting, L., and McNatty, K. P. (2004). Effects of immunization against bone morphogenetic protein 15 and growth differentiation factor 9 on ovulation rate, fertilization, and pregnancy in ewes. *Biol. Reprod.* 70, 557–561.
315. McNeilly, A. S., Fraser, H. M., and Baird, D. T. (1984). Effect of immunoneutralization of LH-releasing hormone on LH, FSH and ovarian steroid secretion in the preovulatory phase of the oestrous cycle in the ewe. *J. Endocrinol.* 101, 213–219.
316. Campbell, B. K., McNeilly, A. S., Picton, H. M., and Baird, D. T. (1990). The effect of a potent gonadotrophin-releasing hormone antagonist on ovarian secretion of oestradiol, inhibin and androstenedione and the concentration of LH and FSH during the follicular phase of the sheep oestrous cycle. *J. Endocrinol.* 126, 377–384.
317. Goodman, R. L., Reichert, L. E. Jr., Legan, S. J., Ryan, K. D., Foster, D. L., and Karsch, F. J. (1981). Role of gonadotropins and progesterone in determining the preovulatory estradiol rise in the ewe. *Biol. Reprod.* 25, 134–142.
318. McNatty, K. P., Gibb, M., Dobson, C., and Thurley, D. C. (1981). Evidence that changes in luteinizing hormone secretion regulate the growth of the preovulatory follicle in the ewe. *J. Endocrinol.* 90, 375–389.
319. McNeilly, A. S., O'Connell, M., and Baird, D. T. (1982). Induction of ovulation and normal luteal function by pulsed injections of luteinizing hormone in anestrous ewes. *Endocrinology* 110, 1292–1299.
320. Picton, H. M., Tsonis, C. G., and McNeilly, A. S. (1990). FSH causes a time-dependent stimulation of preovulatory follicle growth in the absence of pulsatile LH secretion in ewes chronically treated with gonadotrophin-releasing hormone agonist. *J. Endocrinol.* 126, 297–307.
321. Picton, H. M., Tsonis, C. G., and McNeilly, A. S. (1990). The antagonistic effect of exogenous LH pulses on FSH-stimulated preovulatory follicle growth in ewes chronically treated with a gonadotrophin-releasing hormone agonist. *J. Endocrinol.* 127, 273–283.
322. Picton, H. M., and McNeilly, A. S. (1991). Effect of basal and pulsatile LH release on FSH-stimulated follicle growth in ewes chronically treated with gonadotrophin-releasing hormone agonist. *J. Endocrinol.* 128, 449–456.
323. Tsonis, C. G., McNeilly, A. S., and Baird, D. T. (1988). Inhibin secretion by the sheep ovary during the luteal and follicular phases of the oestrous cycle and following stimulation with FSH. *J. Endocrinol.* 117, 283–291.
324. McNatty, K. P., Hudson, N., Gibb, M., Ball, K., Henderson, K. M., Heath, D. A., Lun, S., and Kieboom, L. E. (1985). FSH influences follicle viability, oestradiol biosynthesis and ovulation rate in Romney ewes. *J. Reprod. Fertil.* 75, 121–131.
325. Baird, D. T., Campbell, B. K., and McNeilly, A. S. (1990). Ovine follicular fluid suppresses the ovarian secretion of androgens, oestradiol and inhibin. *J. Endocrinol.* 127, 23–32.
326. Campbell, B. K., and Scaramuzzi, R. J. (1996). The effect of ovarian arterial infusion of human recombinant inhibin and bovine follicular fluid on ovarian hormone secretion by ewes with an autotransplanted ovary. *J. Endocrinol.* 149, 531–540.
327. Campbell, B. K. (1999). The modulation of gonadotrophic hormone action on the ovary by paracrine and autocrine factors. *Anat. Histol. Embryol.* 28, 247–251.
328. Scaramuzzi, R. J., Murray, J. F., Downing, J. A., and Campbell, B. K. (1999). The effects of exogenous growth hormone on follicular steroid secretion and ovulation rate in sheep. *Dom. Anim. Endocrinol.* 17, 269–277.
329. Campbell, B. K., and Baird, D. T. (2001). Inhibin A is a follicle stimulating hormone-responsive marker of granulosa cell differentiation, which has both autocrine and paracrine actions in sheep. *J. Endocrinol.* 169, 333–345.
330. McCracken, J. A., Uno, A., Goding, J. R., Ichikawa, Y., and Baird, D. T. (1969). The *in vivo* effects of sheep pituitary gonadotrophins on the secretion of steroids by the autotransplanted ovary of the ewe. *J. Endocrinol.* 45, 425–440.
331. McCracken, J. A., Baird, D. T., and Goding, J. R. (1971). Factors affecting the secretion of steroids from the transplanted ovary in the sheep. *Recent Prog. Horm. Res.* 27, 537–582.
332. Moor, R. M. (1974). The ovarian follicle of the sheep: inhibition of oestrogen secretion by luteinizing hormone. *J. Endocrinol.* 61, 455–463.
333. Rado, A., McCracken, J. A., and Baird, D. T. (1970). The formation of oestrogens by the autotransplanted ovary of the ewe perfused *in vivo* with C_{19} steroids. *Acta Endocrin. (Copenh.)* 65, 244–260.

334. Baird, D. T., and McNeilly, A. S. (1981). Gonadotrophic control of follicular development and function during the oestrous cycle of the ewe. *J. Reprod. Fertil. Suppl.* 30, 119–133.
335. Scaramuzzi, R. J., and Baird, D. T. (1977). Pulsatile release of luteinizing hormone and the secretion of ovarian steroids in sheep during anestrus. *Endocrinology* 101, 1801–1806.
336. Legan, S. J., and Karsch, F. J. (1979). Neuroendocrine regulation of the estrous cycle and seasonal breeding in the ewe. *Biol. Reprod.* 20, 74–85.
337. Campbell, B. K., McNeilly, A. S., and Baird, D. T. (1989). Episodic ovarian inhibin secretion is not due to LH pulses in anoestrous ewes. *J. Endocrinol.* 123, 173–179.
338. Campbell, B. K., Picton, H. M., McNeilly, A. S., and Baird, D. T. (1991). Effect of FSH on ovarian inhibin secretion in anoestrous ewes. *J. Reprod. Fertil.* 91, 501–509.
339. McNeilly, A. S., Swanston, I. A., Crow, W., Tsonis, C. G., and Baird, D. T. (1989). Changes in the plasma concentrations of inhibin throughout the normal sheep oestrous cycle and after the infusion of FSH. *J. Endocrinol.* 120, 295–305.
340. Niswender, G. D., Schwall, R. H., Fitz, T. A., Farin, C. E., and Sawyer, H. R. (1985). Regulation of luteal function in domestic ruminants: new concepts. *Recent Prog. Horm. Res.* 41, 101–142.
341. Denamur, R. (1974). Luteotrophic factors in the sheep. *J. Reprod. Fertil.* 38, 251–259.
342. Goding, J. R. (1974). The demonstration that PGF2α is the uterine luteolysin in the ewe. *J. Reprod. Fertil.* 38, 261–271.
343. Niswender, G. D., Juengel, J. L., Silva, P. J., Rollyson, M. K., and McIntush, E. W. (2000). Mechanisms controlling the function and life span of the corpus luteum. *Physiol. Rev.* 80, 1–29.
344. McCracken, J. A. (2004). Prostaglandin $F_2\alpha$: the luteolytic hormone. In *The Eicosanoids* (P. Curtis-Prior, Ed.), pp. 525–545. John Wiley & Sons, Hoboken, NJ.
345. Weems, C. W., Weems, Y. S., and Randel, R. D. (2005). Prostaglandins and reproduction in female farm animals. *Vet. J.* 74, 45–59.
346. Senkot, A. P., and Murdoch, W. J. (1991). Luteotropic influence of the ovine uterus. *Dom. Anim. Endocrinol.* 8, 369–374.
347. Robinson, J. S., Challis, J. R. G., Furr, B. J. A., Louis, T. M., and Thorburn, G. D. (1976). Is the sheep corpus luteum subject to tonic inhibition during the luteal phase of the estrous cycle? *Eur. J. Gynec. Reprod. Biol.* 6, 191–199.
348. Reynolds, L. P., and Redmer, D. A. (1998). Expression of the angiogenic factors, basic fibroblast growth factor and vascular endothelial growth factor, in the ovary. *J. Anim. Sci.* 76, 1671–1681.
349. Redmer, D. A., Doraiswamy, V., Bortnem, B. J., Fisher, K., Jablonska-Shariff, A., Grazul-Bilska, A. T., and Reynolds, L. P. (2001). Evidence for a role of capillary pericytes in vascular growth of the developing ovine corpus luteum. *Biol. Reprod.* 65, 879–889
350. Gentry, P. C., Smith, G. W., Anthony, R. V., Zhang, Z., Long, D. K., and Smith, M. F. (1996). Characterization of ovine stem cell factor mRNA and protein throughout the luteal phase. *Biol. Reprod.* 54, 970–979.
351. Gentry, P. C., Smith, G. W., Leighr, D. R., Bao, B., and Smith, M. F. (1998). Ontogeny of stem cell factor receptor (c-kit) messenger ribonucleic acid in the ovine corpus luteum. *Biol. Reprod.* 59, 983–990.
352. Cardenas, H., Burke, K. A, Bigsby, R. M., Pope, W. F., and Nephew, K. P. (2001). Estrogen receptor β in the sheep ovary during the estrous cycle and early pregnancy. *Biol. Reprod.* 65, 128–134.
353. Hixon, J. E., Gengenbach, D. R., and Hansel, W. (1975). Failure of prostaglandin F2α to cause luteal regression in ewes after destruction of ovarian follicles by x-irradiation. *Biol. Reprod.* 13, 126–135.
354. Gengenbach, D. R., Hixon, J. E., and Hansel, W. (1977). A luteolytic interaction between estradiol and prostaglandin F2α in hysterectomized ewes. *Biol. Reprod.* 16, 571–579.
355. Benoit, A. M., Inskeep, E. K., and Dailey, R. A. (1992). Effect of a nonsteroidal aromatase inhibitor on *in vitro* and *in vivo* secretion of estradiol and on the estrous cycle in ewes. *Dom. Anim. Endocrinol.* 9, 313–327.
356. Goff, A. K. (2004). Steroid hormone modulation of prostaglandin secretion in the ruminant endometrium during the estrous cycle. *Biol. Reprod.* 71, 11–16
357. Homeida, A. M., and El Aknah, M. M. (1992). Inhibition of luteal function by oxytocin antagonist in goats (*Capra hircus*). *J. Reprod. Fertil.* 94, 279–285.
358. Milvae, R. A. (1986). Role of luteal prostaglandins in the control of bovine corpus luteum functions. *J. Anim. Sci.* 62, Suppl. 2, 72–78.
359. Lucy, M. C., Bilby, C. R., Kirby, C. J., Yuan, W., and Boyd, C. K. (1999). Role of growth hormone in development and maintenance of follicles and corpora lutea. *J. Reprod. Fertil.* 54, 49–59.
360. Perks, C. M., Denning-Kendall, P. A., Gilmour, R. S., and Wathes, D. C. (1995). Localization of messenger ribonucleic acid for insulin-like growth factor I (IGF-I), IGF-II and the type 1 IGF receptor in the ovine ovary throughout the estrous cycle. *Endocrinology* 136, 5266–5273.
361. Juengel, J. L., Nett, T. M., Anthony, R. V., and Niswender, G. D. (1997). Effects of luteotropic and luteolytic hormones on expression of mRNA encoding insulin-like growth factor-1 and growth hormone receptor in the ovine corpus luteum. *J. Reprod. Fertil.* 110, 291–298.
362. Sayre, B. L., Taft, R., Inskeep, E. K., and Killefer, J. (2000). Increased expression of insulin-like growth factor binding protein-1 during induced regression of bovine corpora lutea. *Biol. Reprod.* 63:21–29.
363. Juengel, J. L., and Niswender, G. D. (1999). Molecular regulation of luteal progesterone synthesis in domestic ruminants. *J. Reprod. Fertil. Suppl.* 54, 193–205.
364. Bogan, R., Pickering, M., Escudero, J., Ku, C.-Y., and Niswender, G. D. (2004). Quantification of endozepine and StAR protein in the ovine corpus luteum throughout the estrous cycle. *Biol. Reprod.* 71, Suppl. 1, Abstr. 767.
365. McGuire, W. J., Juengel, J. L., and Niswender, G. D. (1994). Protein kinase C second messenger system mediates the antisteroidogenic effects of prostaglandin $F_{2\alpha}$ in the ovine corpus luteum in vivo. *Biol. Reprod.* 51, 800–806.
366. Pate, J. L. (1995). Involvement of immune cells in regulation of ovarian function. *J. Reprod. Fertil. Suppl.* 49, 365–377.
367. Wiltbank, J. N., and Casida, L. E. (1956). Alteration of ovarian activity by hysterectomy. *J. Anim. Sci.* 15, 134–140.
368. Inskeep, E. K., and Butcher, R. L. (1966). Local component of utero-ovarian relationships in the ewe. *J. Anim. Sci.* 25, 1164–1168.
369. Moor, R. M., and Rowson, L. E. A. (1966). Local uterine mechanisms affecting luteal function in the sheep. *J. Reprod. Fertil.* 11, 307–310.
370. Ginther, O. J. (1974). Internal regulation of physiological processes through local venoarterial pathways: a review. *J. Anim. Sci.* 39, 550–564.
371. Baird, D. T., and Scaramuzzi, R. J. (1975). Prostaglandin F2α and luteal regression in the ewe: comparison with 16-aryloxyprostaglandin (ICI 80,996). *Ann. Biol. Anim. Bioch. Biophys.* 15, 161–174.
372. McCracken, J. A., Carlson, J. C., Glew, M. E., Goding, J. R., Baird, D. T., Green, K., and Samuelsson, B. (1972). Prostaglandin F2α identified as a luteolytic hormone in sheep. *Nature* 238, 129–134.
373. Juengel, J. J., Haworth, J. D., Rollyson, M. K., Silva, P. J., Sawyer, H. R., and Niswender, G. D. (2000). Effect of dose

of prostaglandin $F_2\alpha$ on steroidogenic components and oligonucleosomes in ovine luteal tissue. *Biol. Reprod.* 62, 1047–1051.

374. Lamsa, J. C., Kot, S. J., Eldering, J. A., Nay, M. G., and McCracken, J. A. (1989). Prostaglandin $F_{2\alpha}$-stimulated release of ovarian oxytocin in the sheep *in vivo*: threshold and dose dependency. *Biol. Reprod.* 40, 1215–1223.
375. Heap, R. B., Fleet, I. R., Davis, A. J., Goode, J. A., Hamon, M. H., Walters, D. E., and Flint, A. P. (1989). Neurotransmitters and lymphatic-vascular transfer of prostaglandin$F_{2\alpha}$ stimulate ovarian oxytocin output in sheep. *J. Endocrinol.* 122, 147–159.
376. Stormshak, F. (2003). Biochemical and endocrine aspects of oxytocin production by the mammalian corpus luteum. *Reprod. Biol. Endocrinol.* 1, 92–97. (Available at: *http://www.rbej.com/content/1/1/92*).
377. Ricke, W. A., Smith, G. W., McIntush, E. W., and Smith, M. F. (2002). Analysis of luteal tissue inhibitor of metalloproteinase-1 -2 and -3 during prostaglandin $F_{2\alpha}$-induced luteolysis. *Biol. Reprod.* 66, 1387–1394.
378. Murdoch, W. J. (1996). Microtubular dynamics in granulosa cells of periovulatory follicles and granulosa-derived (large) lutein cells of sheep: relationships to the steroidogenic folliculo-luteal shift and functional luteolysis. *Biol. Reprod.* 54, 1135–1140.
379. Towle, T. A., Tsang, P. C. W., Milvae, R. A., Newbury, M. K., and McCracken J. A. (2002). Dynamic in vivo changes in tissue inhibitors of metalloproteinases 1 and 2, and matrix metalloproteinases 2 and 9, during prostaglandin $F_{2\alpha}$-induced luteolysis in sheep. *Biol. Reprod.* 66, 1515–1521.
380. Smith, G. W., McCrone, S., Petersen, S. L., and Smith, M. F. (1995). Expression of messenger ribonucleic acid encoding tissue inhibitor of metalloproteinases-2 within ovine follicles and corpora lutea. *Endocrinology* 136, 570–576.
381. Ricke, W. A., Smith, G. W., Reynolds, L. P., Redmer, D. A., and Smith, M. F. (2002). Matrix metalloproteinase (2, 9, and 14) expression, localization and activity in ovine corpora lutea throughout the estrous cycle. *Biol. Reprod.* 66, 1083–1094.
382. Pate, J. L. (2003). Lives in the balance: responsiveness of the corpus luteum to uterine and embryonic signals. *Reprod. Suppl.* 61, 207–217.
383. Penny, L. A. (2000). Monocyte chemoattractant protein in luteolysis. *Rev. Reprod.* 5, 63–66.
384. Tsai, S.-J., Juengel, J. L., and Wiltbank, M. C. (1997). Hormonal regulation of monocyte chemoattractant protein-1 messenger ribonucleic acid expression in corpora lutea. *Endocrinology* 138, 4517–4520.
385. Haworth, J. D., Rollyson, M. K., Silva, P., McIntush, E. W., and Niswender, G. D. (1998). Messenger ribonucleic acid encoding monocyte chemoattractant protein-1 is expressed by ovine corpus luteum in response to prostaglandin $F_2\alpha$. *Biol. Reprod.* 58, 169–174.
386. Murdoch, W. J. (1987). Treatment of sheep with prostaglandin $F_2\alpha$ enhances production of a luteal chemoattractant for eosinophils. *Am. J. Reprod. Immunol. Microbiol.* 15, 52–56.
387. Rexroad, C. E. Jr., and Guthrie, H. D. (1979). Prostaglandin $F_2\alpha$ and progesterone release in vitro by ovine luteal tissue during induced luteolysis. *Adv. Exp. Med. Biol.* 112, 639–644.
388. Wade, D. E., and Lewis, G. S. (1996). Exogenous prostaglandin $F_2\alpha$ stimulates utero-ovarian release of prostaglandin $F_2\alpha$ in sheep: a possible component of the luteolytic mechanism of action of exogenous prostaglandin $F_2\alpha$. *Dom. Anim. Endocrinol.* 13, 383–398.
389. Griffeth, R. J., Nett, T. M., Burns, P. D., Escudero, J. M., Inskeep, E. K., and Niswender, G. D. (2002). Is luteal production of $PGF_2\alpha$ required for luteolysis? *Biol. Reprod.* 66, Suppl. 1, Abstr. 465.
390. Milvae, R. A., and Hansel, W. (1983). Prostacyclin, prostaglandin $F_2\alpha$ and progesterone production by bovine luteal cells during the estrous cycle. *Biol. Reprod.* 29, 1063–1068.
391. Pate, J. L., and Condon, W. A. (1984). Effects of prostaglandin F_2a on agonist-induced progesterone production in cultured bovine luteal cells. *Biol. Reprod.* 31, 427–435.
392. Pate, J. L. (1988). Regulation of prostaglandin synthesis by progesterone in the bovine corpus luteum. *Prostaglandins* 36, 303–315.
393. Nothnick, W. B., and Pate, J. L. (1990). Interleukin-1β is a potent stimulator of prostaglandin synthesis in bovine luteal cells. *Biol. Reprod.* 43, 898–903.
394. Fairchild, D. L., and Pate, J. L. (1991). Modulation of bovine luteal cell synthetic capacity by interferon-gamma. *Biol. Reprod.* 44, 357–363.
395. Fairchild-Benyo, D. L., and Pate, J. L. (1992). Tumor Necrosis factor-α alters bovine luteal cell synthetic capacity and viability. *Endocrinology* 130, 854–860.
396. Meidan, R., Milvae, R. A., Weiss, S., Levy, N., and Friedman, A. (1999). Intraovarian regulation of luteolysis. *J. Reprod. Fertil. Suppl.* 54, 217–228.
397. Hayashi, K., and Miyamoto, A. (1999). Angiotensin II interacts with prostaglandin $F_2\alpha$ and endothelin-1 as a local luteolytic factor in the bovine corpus luteum in vitro. *Biol. Reprod.* 60, 1104–1109.
398. Hayashi, K., Miyamoto, A., Berisha, B., Kosmann, M. R., Okuda, K., and Schams, D. (2000). Regulation of angiotensin II production and angiotensin receptors in microvascular endothelial cells from bovine corpus luteum. *Biol. Reprod.* 62, 162–167.
399. Schams, D., Berisha, B., Neuvians, T., Amselgruber, W., and Kraetzl, W. D. (2003). Real-time changes of the local vasoactive peptide systems (angiotensin, endothelin) in the bovine corpus luteum after induced luteal regression. *Mol. Reprod. Dev.* 65, 57–66.
400. Skarzynski, D. J., Jaroszewski, J. J., Bah, M. M., Deptula, K. M., Barszczewska, B., Gawronska, B., and Hansel, W. (2003). Administration of nitric oxide synthase inhibitor counteracts prostaglandin $F_2\alpha$-induced luteolysis in cattle. *Biol. Reprod.* 68, 1674–1681.
401. Reynolds, L., Vonnahme, K., Borowczyk, E., Johnson, M. L., Redmer, D. Borowicz, P., and Grazul-Bilska, A. (2004). Profile of angiogenic factors during luteal regression induced by PGF in sheep. *Biol. Reprod.* 71 (Special Issue), Abstr. 182.
402. Hinckley, S. T., and Milvae, R. A. (2001). Endothelin-1 mediates prostaglandin $F_2\alpha$-induced luteal regression in the ewe. *Biol. Reprod.* 64, 1619–1623.
403. Ohtani, M., Takase, S., Wijayagunawardane, M. B. P., Tetsuka, M., and Miyamoto, A. (2004). Local interaction of prostaglandin $F_{2\alpha}$ with endothelin-1 and tumor necrosis factor-α on the release of progesterone and oxytocin in ovine corpora lutea in vivo: a possible implication for a luteolytic cascade. *Reproduction* 127, 117–124.
404. Ji, I., Slaughter, R. G., Ellis, J. A., Ji, T. H., and Murdoch, W. J. (1991). Analyses of ovine corpora lutea for tumor necrosis factor mRNA and bioactivity during prostaglandin-induced luteolysis. *Mol. Cell. Endocrinol.* 81, 77–80.
405. Inskeep, E. K., Smutny, W. J., Butcher, R. L., and Pexton, J. E. (1975). Effects of intrafollicular injections of prostaglandins in nonpregnant and pregnant ewes. *J. Animal Sci.* 41, 1098–1104.
406. Rubianes, E., Menchaca, A., and Carbajal, B. (2003). Response of the 1–5 day–aged ovine corpus luteum to prostaglandin $F_2\alpha$. *Anim. Reprod. Sci.* 78, 47–55.
407. Juengel, J. L., Wiltbank, M. C., Meberg, B. M., and Niswender, G. D. (1996). Regulation of steady-state

concentrations of messenger ribonucleic acid encoding prostaglandin $F_{2\alpha}$ receptor in ovine corpus luteum. *Biol. Reprod.* 54, 1096–1102.
408. Juengel, J. L., Melner, M. H., Clapper, J. A., Turzillo, A. M., Moss, G. E., Nett, T. M., and Niswender, G. D. (1998). Steady-state concentrations of mRNA encoding two inhibitors of protein kinase C in ovine luteal tissue. *J. Reprod. Fertil.* 113, 299–305.
409. Tsai, S.-J., and Wiltbank, M. C. (1997). Prostaglandin $F_2\alpha$ induces expression of prostaglandin G/H synthase-2 in the ovine corpus luteum: a potential positive feedback loop during luteolysis. *Biol. Reprod.* 57, 1016–1022.
410. Allen, J. F., Tsang, P. C. W., Milvae, R. A., Keator, C., and McCracken, J. A. (2003). Dynamic in vivo changes in protein levels of cyclooxygenases-1 and -2 and steroidogenic acute regulatory protein during prostaglandin $F_2\alpha$-induced luteolysis in sheep. *Biol. Reprod.* 68, Suppl. 1, Abstr. 72.
411. Arosh, J. A., Banu, S. K., Chapdelaine, P., Madore, E., Sirois, J., and Fortier, M. A. (2004). Prostaglandin biosynthesis, transport and signaling in corpus luteum: A basis for autoregulation of luteal function. *Endocrinology* 145, 2551–2560.
412. Costine, B. A., Inskeep, K., and Wilson, M. E. (2004). Downregulation of luteal prostaglandin F synthase in pregnant, but not in nonpregnant ewes after treatment with prostaglandin $F_2\alpha$ on day 12. *Biol. Reprod.* 71 (Special Issue), Abstr. 214.
413. Diaz, F. J., Crenshaw, T. D., and Wiltbank, M. C. (2000). Prostaglandin $F_2\alpha$ induces distinct physiological responses in porcine corpora lutea after acquisition of luteolytic capacity. *Biol. Reprod.* 63, 1504–1512.
414. Zieba, D., Bilinska, B., Schmalz-Fraczek, B., and Murawski, M. (2000). Immunohistochemical localization of estrogen receptors in the ovine corpus luteum throughout the estrous cycle. *Folia Histochem. Cytobiol.* 38, 111–117.
415. Piper, E. L., and Foote, W. C. (1968). Ovulation and corpus luteum maintenance in ewes treated with 17–β oestradiol. *J. Reprod. Fertil.* 16, 253–259.
416. Ford, S. P., Weems, C. W., Pitts, R. E., Pexton, J. E., Butcher, R. L., and Inskeep, E. K. (1975). Effects of estradiol-17β and progesterone on prostaglandins F in sheep uteri and venous plasma. *J. Anim. Sci.* 41, 1407–1413.
417. Wathes, D. C., and Swann, R. W. (1982). Is oxytocin an ovarian hormone? *Nature* 297, 225–227.
418. Schams, D., Lahlou-Kassi, A., and Glatzel, P. (1982). Oxytocin concentrations in peripheral blood during the oestrous cycle and after ovariectomy in two breeds of sheep with low and high fecundity. *J. Endocrinol.* 92, 9–13.
419. Watkins, W. B., Moore, L. G., Flint, A. P. F., and Sheldrick, E. L. (1984). Secretion of neurophysins by the ovary in sheep. *Peptides* 5, 61–64.
420. Ottobre, J. S., Vincent, D. L., Silvia, W. J., and Inskeep, E. K. (1984). Aspects of regulation of uterine secretion of prostaglandins during the oestrous cycle and early pregnancy. *Anim. Reprod. Sci.* 7, 75–100.
421. Vincent, D. L., and Inskeep, E. K. (1986). Role of progesterone in regulating uterovarian venous concentrations of PGF_{2a} and PGE_2 during the estrous cycle and early pregnancy in ewes. *Prostaglandins* 31, 715–733.
422. Morgan, G. L., Geisert, R. D., McCann, J. P., Bazer, F. W., Ott, T. L., Mirando, M. A., and Stewart, M. (1993). Failure of luteolysis and extension of the interoestrous interval in sheep treated with the progesterone antagonist mifepristone (RU 486). *J. Reprod. Fertil.* 98, 451–457.
423. Vallet, J. L., Lamming, G. E., and Batten, M. (1990). Control of endometrial oxytocin receptor and uterine response to oxytocin by progesterone and oestradiol in the ewe. *J. Reprod. Fertil.* 90, 625–634.
424. Kieborz-Loos, K. R., Garverick, H. A., Keisler, D. H., Hamilton, S. A., Salfen, B. E., Youngquist, R. S., and Smith, M. F. (2003). Oxytocin-induced secretion of prostaglandin $F_{2\alpha}$ in postpartum beef cows: effects of progesterone and estradiol-17β treatment. *J. Anim. Sci.* 81, 1830–1836.
425. Zollers, W. G., Garverick, H. A., Smith, M. F., Salfen, B. E., and Youngquist, R. S. (1993). Concentrations of progesterone and oxytocin receptors in endometrium of postpartum cows expected to have a short or normal oestrous cycle. *J. Reprod. Fertil.* 97, 329–337.
426. Hunter, M. G. (1991). Characteristics and causes of the inadequate corpus luteum. *J. Reprod. Fertil. Suppl.* 43, 91–99
427. Beard, A. P., and Hunter, M. G. (1994). Effects of progesterone pretreatment on the oxytocin receptor concentration and the response to oxytocin during the simulated early luteal phase in the ovariectomized ewe. *J. Reprod. Fertil.* 102, 57–63.
428. Cooper, D. A., Carver, D. A., Villeneuve, P., Silvia, W. J., and Inskeep, E. K. (1991). Effects of progestogen treatment on concentrations of prostaglandins and oxytocin in plasma from the posterior vena cava of post-partum beef cows. *J. Reprod. Fertil.* 91, 411–421.
429. Fairclough, R. J., Moore, L. G., McGowan, L. T., Smith, J. F., and Watkins, W. B. (1983). Effect of exogenous progestogens on plasma concentrations of the oxytocin-associated neurophysin and 13, 14-dihydro-15-keto-prostaglandin F in intact and ovariectomized ewes over the time of expected luteal regression. *Biol. Reprod.* 29, 271–277.
430. Karsch, F. J., Noveroske, J. W., Roche, J. F., Norton, H. W., and Nalbandov, A. V. (1970). Maintenance of ovine corpora lutea in the absence of ovarian follicles. *Endocrinology* 87, 1228–1236.
431. Ginther, O. J. (1971). Response of corpora lutea to cauterization of follicles in sheep. *Amer. J. Veterin. Res.* 32, 59–62.
432. Roche, J. F., Karsch, F. J., Foster, D. L., and Dziuk, P. J. (1974). Serum LH in ewes following sequential removal of ovarian follicles, corpora lutea and stroma. *J. Reprod. Fertil.* 40, 215–218.
433. Echternkamp, S. E., Bolt, D. J., and Hawk, H. W. (1976). Ovarian and pituitary hormones in blood of progestogen-treated ewes. *J. Anim. Sci.* 42, 893–900.
434. Caldwell, B. V., Tillson, S. A., Brock, W. A., and Speroff, L. (1972). The effects of exogenous progesterone and estradiol on prostaglandin F levels in ovariectomized ewes. *Prostaglandins* 1, 217–228.
435. Scaramuzzi, R. J., Baird, D. T., Boyle, H. P., Land, R. B., and Wheeler, A. G. (1977). The secretion of prostaglandin F from the autotransplanted uterus of the ewe. *J. Reprod. Fertil.* 49, 157–160.
436. Beard, A. P., and Hunter, M. G. (1994). Effects of bovine follicular fluid and exogenous oestradiol on the GnRH-induced short luteal phase in anoestrous ewes. *J. Reprod. Fertil.* 100, 211–217.
437. Warren, J. E. Jr., and Hawk, H. W. (1971). Enhancement by exogenous estradiol of IUD-induced luteal regression in the ewe. *Biol. Reprod.* 4, 48–51.
438. Silvia, W. J., Lewis, G. S., McCracken, J. A., Thatcher, W. W., and Wilson, L. Jr. (1991). Hormonal regulation of uterine secretion of prostaglandin $F_{2\alpha}$ during luteolysis in ruminants. *Biol. Reprod.* 45, 655–663.
439. Hooper, S. B., Watkins, W. B., and Thorburn, G. D. (1986). Oxytocin, oxytocin-associated neurophysin, and prostaglandin $F_{2\alpha}$ concentrations in the utero-ovarian vein of pregnant and nonpregnant sheep. *Endocrinology* 119, 2590–2597.

440. Fairclough, R. J., Moore, L. G., McGowan, L. T., Peterson, A. J., Smith, J. F., Tervit, H. R., and Watkins, W. B. (1980). Temporal relationship between plasma concentrations of 13,14-dihydro-15-keto-prostaglandin F and neurophysin I/II around luteolysis in sheep. *Prostaglandins* 20, 199–208.
441. Milvae, R. A., Duby, R. T., Tritschler, J. P. II, Pekala, R. F., Gnatek, G. G., Bushmich, S. L., and Schreiber, D. T. Jr. (1991). Function and lifespan of corpora lutea in ewes treated with exogenous oxytocin. *J. Reprod. Fertil.* 92, 133–138.
442. Kotwica, J., and Skarzynski, D. (1993). Influence of oxytocin removal from the corpus luteum on secretory function and duration of the oestrous cycle in cattle. *J. Reprod. Fertil.* 97, 411–417.
443. Kotwica, J., Skarzynski, D., Bogacki, D., Melin, P., and Starostka, B. (1997). The use of an oxytocin antagonist to study the function of ovarian oxytocin during luteolysis in cattle. *Theriogenology* 48, 1287–1299.
444. McCracken, J. A., Amico, J. A., Newbury, M. K., and Milvae, R. A. (2000). Depletion of oxytocin from ovine corpus luteum does not extend cycle length. *Biol. Reprod.* 62, Suppl. 1, Abstr. 92.
445. Arnold, D. R., Binelli, M., Yonk, J., Alexenko, A. P., Drost, M., Wilcox, C. J., and. Thatcher, W. W. (2000). Intracellular regulation of endometrial $PGF_2\alpha$ and PGE_2 production in dairy cows during early pregnancy and following treatment with recombinant interferon-τ. *Dom. Anim. Endocrinol.* 18, 199–216.
446. Gilbert, C. L., Jenkins, K., and Wathes, D. C. (1991). Pulsatile release of oxytocin into the circulation of the ewe during oestrus, mating and the early luteal phase. *J. Reprod. Fertil.* 91, 337–346.
447. Mitchell, M. D., Kraemer, D. L., Brennecke, S. P., and Webb, R. (1982). Pulsatile release of oxytocin during the estrous cycle, pregnancy and parturition in sheep. *Biol. Reprod.* 27, 1169–1173.
448. Moore, L. G., Choy, V. J., Elliot, R. L., and Watkins, W. B. (1986). Evidence for the pulsatile release of PGF-2α inducing the release of ovarian oxytocin during luteolysis in the ewe. *J. Reprod. Fertil.* 76, 159–166.
449. Fairclough, R. J., Moore, L. G., Peterson, A. J., and Watkins, W. B. (1984). Effect of oxytocin on plasma concentrations of 13,14-dihydro-15-keto prostaglandin F and the oxytocin-associated neurophysin during the estrous cycle and early pregnancy in the ewe. *Biol. Reprod.* 31, 36–43.
450. Flint, A. P. F., Sheldrick, E. L., McCann, T. J., and Jones, D. S. C. (1990). Luteal oxytocin: characteristics and the control of synchronous episodes of oxytocin and $PGF_{2\alpha}$ secretion at luteolysis in ruminants. *Dom. Anim. Endocrinol.* 7, 111–124.
451. Sheldrick, E. L., and Flint, A. P. F. (1985). Endocrine control of uterine oxytocin receptors in the ewe. *J. Endocrinol.* 106, 249–258.
452. Chami, O., Megevand, A., Ott, T., Bazer, F., and O'Neill, C. (1999). Platelet-activating factor may act as an endogenous pulse generator for sheep of luteolytic $PGF_{2\alpha}$ release. *Am. J. Physiol.* 276 (*Endocrinol. Metabol.* 39), E783–E792.
453. Piper, P. J., Vane, J. R., and Wyllie, A. H. (1970). Inactivation of prostaglandins by the lungs. *Nature* 225, 600–604.
454. Bonnin, P., Huynh, L., L'Haridon, R., Chene, N., and Martal, J. (1999). Transport of uterine $PGF_2\alpha$ to the ovaries by systemic circulation and local lymphovenous-arterial diffusion during luteolysis in sheep. *J. Reprod. Fertil.* 116, 199–210.
455. Heap, R. B., Fleet, I. R., and Hamon, M. (1985). Prostaglandin F-2α is transferred from the uterus to the ovary in the sheep by lymphatic and blood vascular pathways. *J. Reprod. Fertil.* 74, 645–656.
456. Heap, R. B., Fleet, I. R., Davis, A. J., Goode, J. A., Hamon, M. H., Walters, D. E., and Flint, A. P. F. (1989). Neurotransmitters and lymphatic-vascular transfer of prostaglandin $F_{2\alpha}$ stimulate ovarian oxytocin output in sheep. *J. Endocrinol.* 122, 147–159.
457. Fleet, I. R., Davis, A. J., Goode, J. A., Hamon, M., Collier, R. J., and Heap, R. B. (1994). Unilateral control of ovarian oxytocin release and the facilitatory effects of insulin-like growth factor-I in sheep. *J. Reprod. Fertil.* 100, 623–628.
458. Krzymowski, T., and Stefanczyk-Krzymowska, S. (2002). Uterine blood supply as a main factor in the regulation of the estrous cycle – a new theory. *Reprod. Biol.* 2, 93–114.
459. Baird, D. T., Baker, T. G., McNatty, K. P., and Neal, P. (1975). Relationship between the secretion of the corpus luteum and the length of the follicular phase of the ovarian cycle. *J. Reprod. Fertil.* 45, 611–619.
460. Knobil, E. (1974). On the control of gonadotropin secretion in the Rhesus monkey. *Recent Prog. Horm. Res.* 30, 1–46.
461. Hansel, W., and Echternkamp, S. E. (1972). Control of ovarian function in domestic animals. *Am. Zoologist* 12, 225–243.
462. Cumming, I. A., Brown, J. M., Blockey, M. A. deB., and Goding, J. R. (1971). Regulation of the oestrous cycle in the ewe. *J. Reprod. Fertil.* 24, 148–149.
463. Cumming, I. A. (1975). The ovine and bovine oestrous cycle. *J. Reprod. Fertil.* 43, 583–596.
464. Goding, J. R., Buckmaster, J. M., Cerini, J. C., Cerini, M. E. D., Chamley, W. A., Cumming, I. A., Fell, L. R., Findlay, J. K., and Jonas, H. (1973). Gonadotrophins in the ovine oestrous cycle. *J. Reprod. Fertil. Suppl.* 18, 31–37.
465. Karsch, F. J. (1980). Seasonal reproduction: a saga of reversible fertility. *Physiologist* 23, 29–38.
466. Pexton, J. E., Weens, C. W., and Inskeep, E. K. (1975). Prostaglandins F in uterine venous plasma, ovarian arterial and venous plasma and in ovarian and luteal tissue of pregnant and non-pregnant ewes. *J. Anim. Sci.* 41, 154–159.
467. Lewis, G. S., Jenkins, P. E., Foywell, R. L., and Inskeep, E. K. (1978). Concentrations of prostaglandins E_2 and $F_{2\alpha}$ and their relationship to luteal function in early pregnant ewes. *J. Anim. Sci.* 47, 1314–1323.
468. Lehman, M. N., Coolen, L. M., Goodman, R. L., Viguie, C., Billings, H. J., and Karsch, F. J. (2002). Seasonal plasticity in the brain: The use of large animal models for neuroanatomical research. *Reprod. Suppl.* 59, 149–165
469. Smith, R. F., Ghuman, S. P. S., Evans, N. P., Karsch, F. J., and Dobson, H. (2003). Stress and the control of LH secretion in the ewe. *Reprod. Suppl.* 61, 267–282.
470. Adams, C. L., Aracher, Z. A., and Miller, D. W. (2003). Leptin actions on the reproductive neuroendocrine axis in sheep. *Reprod. Suppl.* 61, 283–297.
471. Ungerfield, R., Forsberg, M., and Rubianes, E. (2004). Overview of the response of anoestrous ewes to the ram effect. *Reprod. Fertil. Dev.* 16, 479–490.
472. Marshall, F. H. A. (1937). On change over in oestrous cycle in animals after transference across equator, with further observations on incidence of breeding season and factors controlling sexual periodicity. *Proc. R. Soc. Lond. Ser. B.* 122, 413–428.
473. Yeates, N. T. M. (1949). The breeding season of the sheep with particular reference to its modification by artificial means using light. *J. Agric. Sci.* 39, 1–43.
474. Goodman, R. L., Bittman, E. L., Foster, D. L., and Karsch, F. J. (1982). Alterations in the control of luteinizing hormone pulse frequency underlie the seasonal variation in estradiol negative feedback in the ewe. *Biol. Reprod.* 27, 580–589.
475. Barrel, G. K., Moenter, S. M., Caraty, A., and Karsch, F. J. (1992). Seasonal changes in gonadotropin-releasing hormone secretion in the ewe. *Biol. Reprod.* 46, 1130–1135.

476. Legan, S. J., and Winans, S. S. (1981). The photoneuroendocrine control of seasonal breeding in the ewe. *Gen. Comp. Endocrinol.* 45, 317–328.
477. Karsch, F. J., Bittman, E. L., Foster, D. L, Goodman, R. L., Legan, S. J., and Robinson, J. E. (1984). Neuroendocrine basis of seasonal reproduction. *Recent Prog. Horm. Res.* 40, 185–232.
478. Turek F. W., and Campbell, C. S. (1979). Photoperiodic regulation of neuroendocrine-gonadal activity. *Biol. Reprod.* 20:32–50.
479. Malpaux, B., Daveau, A., Maurice, F., Gayrard, V., and Thiéry, J. C. (1993). Short day effects of melatonin on luteinizing hormone secretion in the ewe: evidence for central sites of action in the mediobasal hypothalamus. *Biol. Reprod.* 48, 752–760.
480. Malpaux, B., Daveau, A., Maurice-Mandon, F., Duarte, G., and Chemineau, P. (1998). Evidence that melatonin acts in the premammillary hypothalamic area to control reproduction in the ewe: presence of binding sites and stimulation of luteinizing hormone secretion by in situ microimplant delivery. *Endocrinology* 139, 1508–1516.
481. Sliwowska, J. H., Billings, H. J., Goodman, R. L., Coolen, L. M., and Lehman, M. N. (2004). The premammillary hypothalamic area of the ewe: anatomical characterization of a melatonin target area mediating seasonal reproduction. *Biol. Reprod.* 70, 1768–1775.
482. Goodman, R. L., and Meyer, S. L. (1984). Effects of pentobarbital anesthesia on tonic luteinizing hormone secretion in the ewe: evidence for active inhibition of luteinizing hormone in anestrus. *Biol. Reprod.* 30, 374–381.
483. Barker, J. L., and Ranson, B. R. (1978). Pentobarbitone pharmacology of mammalian central neurons grown in tissue culture. *J. Physiol.* 280, 355–372.
484. Meyer, S. L., and Goodman, R. L. (1985). Neurotransmitters involved in mediating the steroid-dependent suppression of pulsatile luteinizing hormone secretion in anestrous ewes: effects of receptor antagonists. *Endocrinology* 116, 2054–2061.
485. Thiéry, J.-C., Martin, G. B., Caldani, M., Quentin, M., Jamain, C., and Ravault, J.-P. (1989). Role of hypothalamic catecholamines in the regulation of luteinizing hormone and prolactin secretion in the ewe during seasonal anestrus. *Neuroendocrinology* 49, 80–87.
486. Tillet, Y., Batailler, M., Kreiger-Poullet, M., and Thibault, J. (1990). Presence of dopamine-immunoreactive cell bodies in the catecholaminergic group A15 of the sheep brain. *Histochemistry* 93, 327–333.
487. Tillet, J. C., Gayrard, V., Le Corre, S., Viguie, C., Martin, G. B., Chemineau, P., and Malpaux, B. (1995). Dopaminergic control of LH secretion by the A15 nucleus in anoestrous ewes. *J. Reprod. Fertil. Suppl.* 49, 285–296.
488. Gayrard, V., Malpaux, B., Tillet, Y., and Thiery, J. C. (1994). Estradiol increases tyrosine hydroxylase activity of the A15 nucleus dopaminergic neurons during long days in the ewe. *Biol. Reprod.* 50, 1168–1177.
489. Lehman, M. N., Durham, D. M., Jansen, H. T., Adrian, B. A., and Goodman, R. L. (1996). Dopaminergic A14/A15 neurons are activated during estradiol negative feedback in the anestrous ewe. *Endocrinology* 137, 4443–4450.
490. Havern, R. L., Whisnant, C. S., and Goodman, R. L. (1994). Dopaminergic structures in the ovine hypothalamus mediating estradiol negative feedback in anestrous ewes. *Endocrinology* 134, 1905–1914.
491. Gayrard, V., Thiery, J. C., Thibault, J., and Tillet, Y. (1995). Efferent projections from the retrochiasmatic area to the median eminence and to the pars nervosa of the hypophysis with special reference to the A15 dopaminergic cell group in the sheep. *Cell Tiss. Res.* 281, 561–567.
492. Havern, R. L., Whisnant, C. S., and Goodman, R. L. (1991). Hypothalamic sites of catecholamine inhibition of luteinizing hormone in the anestrous ewe. *Biol. Reprod.* 44, 476–482.
493. Bertrand, F., Viguie, C., Picard, S., and Malpaux, B. (1998). Median eminence dopaminergic activation is critical for the early long-day inhibition of luteinizing hormone secretion in the ewe. *Endocrinology* 139, 5094–5102.
494. Thiéry, J. C. (1991). Monoamine content of the stalk-median eminence and hypothalamus in adult female sheep as affected by daylength. *J. Neuroendocrinol.* 3, 407–411.
495. Lubbers, L. S., Greco, B., Hileman, S. M., Schwartz, P. E., and Blaustein, J. D. (1999). Localization of tyrosine hydroxylase (TH) and estrogen receptor (ER) in the retrochiasmatic region of the hypothalamus of female sheep. *Prog. Ann. Meet. Soc. Neurosci.* Abstr. 572.11.
496. Goodman, R. L. (1989). Functional organization of the catecholaminergic neural systems inhibiting luteinizing hormone secretion in anestrous ewes. *Neuroendocrinology* 50, 406–412.
497. Tillet, Y., and Thibault, J. (1993). Morphological relationships between tyrosine hydroxylase-immunoreactive neurons and dopamine-β-hydroxylase-immunoreactive fibers in dopamine cell group A15 of the sheep. *J. Chem. Neuroanat.* 6, 69–78.
498. Goodman, R. L., Robinson, J. E., Kendrick, K. M., and Dyer, R. D. (1995). Is the inhibitory action of estradiol on luteinizing hormone pulse frequency in anestrous ewes mediated by noradrenergic neurons in the preoptic area? *Neuroendocrinology* 61, 284–292.
499. Anderson, G. M., Jansen, H. T., and Goodman, R. L. (1999). Noradrenergic A1 neurons are not activated by estradiol negative feedback on LH secretion in anestrous ewes. *Prog. Ann. Meet. Soc. Neurosci.* Abstr. 582.4.
500. Stefanovic, I., Adrian, B., Jansen, H. T., Lehman, M. N., and Goodman, R. L. (2000). The ability of estradiol to induce Fos expression in a subset of estrogen receptor-α containing neurons in the preoptic area of the ewe depends on reproductive status. *Endocrinology* 141, 190–196.
501. Gallegos-Sanchez, J., Delaleu, B., Caraty, A., Malpaux, B., and Thiery, J. C. (1997). Estradiol acts locally within the retrochiasmatic area to inhibit pulsatile luteinizing hormone release in the female sheep during anestrus. *Biol. Reprod.* 56, 1544–1549.
502. Anderson, G. M., Connors, J. M., Hardy, S. L., Valent, M., and Goodman, R. L. (2001). Oestradiol microimplants in the ventromedial preoptic area inhibit secretion of luteinising hormone via dopaminergic neurons in anoestrous ewes. *J. Neuroendocrinol.*13, 1051–1058.
503. Hardy, S. L., Anderson, G. M., Valent, M., Connors, J. M., and Goodman, R. L. (2003). Evidence that estrogen receptor alpha, but not beta, mediates seasonal changes in the response of the ovine retrochiasmatic area to estradiol. *Biol. Reprod.* 68, 846–852.
504. McManus, C. J., Hardy, S. L., Valent, M., Connors, J. M., and Goodman, R. L. (2004). Time of year alters response to nitric oxide synthase inhibition in seasonally anestrous ewes. *Prog. Ann. Meet. Soc. Neurosci.* Abstr. 756.23.
505. Goodman, R. L., McManus, C. J., Connors, J. M., Hileman, S. M., and Lehman, M. N. (2004). Evidence that γ-aminobutyric acid (GABA) is part of the neural circuit mediating estradiol negative feedback in anestrous ewes. *Prog. Ann. Meet. Soc. Neurosci.* Abstr. 143.5.
506. Sliwowska, J., Foradori, C., Coolen, L. M., Goodman, R. L., Hardy, S. L., Anderson, G. M., and Lehman, M. N. (2002). The premammillary hypothalamus area of the ewe: anatomical characterization and connections with circuitry mediating seasonal reproduction. *Prog. Ann. Meet. Soc. Neurosci.* Abstr. 572.6.

507. Xiong, J.-J., Karsch, F. J., and Lehman, M. N. (1997). Evidence for seasonal plasticity in the gonadotropin-releasing hormone (GnRH) system of the ewe: changes in synaptic inputs onto GnRH neurons. *Endocrinology* 138, 1240–1250
508. Adams, V. L., Salm, A., Connors, J. M., Coolen, L. M., and Goodman, R. L. (2001). Seasonal plasticity of A15 dopaminergic neurons in the ewe. *Prog. Ann. Meet. Soc. Neurosci.* Abstr. 631.13.
509. Karsch, F. J., Dahl, G. E., Hachigian, T. M., and Thrun, L. A. (1995). Involvement of thyroid hormones in seasonal reproduction. *J. Reprod. Fertil. Suppl.* 49, 409–422.
510. Anderson, G. M., Connors, J. M., Hardy, S. L., Valent, M., and Goodman, R. L. (2003). Evidence that thyroid hormones act in the ventromedial preoptic area and the premammillary region of the brain to allow the termination of the breeding season in the ewe. *Endocrinology* 144, 2892–2901.
511. Karsch, F. J., Goodman, R. L., and Legan, S. J. (1980). Feedback basis of seasonal breeding: test of an hypothesis. *J. Reprod. Fertil.* 58, 521–535.
512. Raeside, J. I., and McDonald, M. F. (1959). Seasonal changes in the oestrous response by the ovariectomized ewe to progesterone and estrogen. *Nature* 184, 458–459.
513. Gibson, W. R., and Robinson, T. J. (1971). The seasonal nature of reproductive phenomena in the sheep. *J Reprod. Fertil.* 24, 9–18.
514. Ducker, M. J., and Bowman, J. C. (1970). Photoperiodism in the ewe. 3. The effects of various patterns of increasing daylength on the onset of anoestrus in Clun Forest ewes. *Anim. Prod.* 12, 467–471.
515. Ortavant, R., Pelletier, J., Ravault, J. P., and Thimonier, J. (1978). Annual cyclic variations in prolactin in sheep. In *Environmental Endocrinology* (I. Assenmacher, and D. S. Farner, Eds.), pp. 75–78. Springer-Verlag, New York.
516. McNeilly, A. S., and Land, R. B. (1979). Effect of suppression of plasma prolactin on ovulation, plasma gonadotrophins and corpus luteum function in LH-RH-treated anestrous ewes. *J. Reprod. Fertil.* 56, 601–609.
517. Tilbrook, A. J., Turner, A. I., and Clarke, I. J. (2000). Effects of stress on reproduction in non-rodent mammals: the role of glucocorticoids and sex differences. *Rev. Reprod.* 5, 105–113.
518. Turner, A. I., Canny, B. J., Hobbs, R. J., Bond, J. D., Clarke, I. J., and Tilbrook, A. J. (2002). Influence of sex and gonadal status of sheep on cortisol secretion in response to ACTH and on cortisol and LH secretion in response to stress: importance of different stressors. *J. Endocrinol.* 173, 113–122.
519. Rasmussen, D. D., and Malven, P. V. (1983). Effects of confinement stress on episodic secretion of LH in ovariectomized sheep. *Neuroendocrinology* 36, 392–396.
520. Tilbrook, A. J., Canny, B. J., Serapiglia, M. D., Ambrose, T. J., and Clarke, I. J. (1999). Suppression of the secretion of luteinizing hormone due to isolation/restraint stress in gonadectomized rams and ewes is influenced by sex steroids. *J. Endocrinol.* 160, 469–481.
521. Stackpole, C. A., Turner, A. I., Clarke, I. J., Lambert, G. W., and Tilbrook, A. J. (2003). Seasonal differences in the effect of isolation and restraint stress on the luteinizing hormone response to gonadotropin-releasing hormone in hypothalamo-pituitary disconnected, gonadectomized rams and ewes. *Biol. Reprod.* 69, 1158–1164.
522. Smart, D., Forhead, A. J., Smith, R. F., and Dobson, H. (1994). Transport stress delays the oestradiol-induced LH surge by a non-opioidergic mechanism in the early postpartum ewe. *J. Endocrinol.* 142, 447–451.
523. Dobson, H., Tebble, J. E., Phogat, J. B., and Smith, R. F. (1999). Effect of transport on pulsatile and surge secretion of LH in ewes in the breeding season. *J. Reprod. Fertil.* 116, 1–8.
524. Phogat, J. B., Smith, R. F., and Dobson, H. (1999). Effect of transport on pituitary responsiveness to exogenous GnRH and oestradiol-induced LH release in intact ewes. *J. Reprod. Fertil.* 116, 9–18.
525. Dobson, H., Tebble, J. E., Ozturk, M., and Smith, R. F. (1999). Effect of transport on pulsatile LH release in ovariectomized ewes with or without prior steroid exposure at different times of year. *J. Reprod. Fertil.* 117, 213–222.
526. Clarke, I. J., Horton, R. J. E., and Doughton, B. W. (1990). Investigation of the mechanism by which insulin-induced hypoglycemia decreases luteinizing hormone secretion in ovariectomized ewes. *Endocrinology* 127, 1470–1476.
527. Medina, C. L., Nagatani, S., Darling, T. A., Bucholtz, D. C., Tsukamura, H., Maeda, K., and Foster, D. L. (1998). Glucose availability modulates the timing of the luteinizing hormone surge in the ewe. *J. Neuroendocinol.* 10, 785–792.
528. Engler, D., Pham, T., Fullerton, M. J., Funder, J. W., and Clarke, I. J. (1988). Studies of the regulation of the hypothalamic–pituitary–adrenal axis in sheep with hypothalamic–pituitary disconnection I. Effect of an audiovisual stimulus and insulin-induced hypoglycemia. *Neuroendocrinology* 48, 551–560.
529. Engler, D., Pham, T., Fullerton, M. J., Ooi, G., Funder, J. W., and Clarke, I. J. (1989). Studies of the secretion of corticotrophin-releasing factor and arginine vasopressin into the hypophysial-portal circulation of the conscious sheep I. Effect of an audiovisual stimulus and insulin-induced hypoglycemia. *Neuroendocrinology* 49, 367–381.
530. Canny, B. J., Funder, J. W., and Clarke, I. J. (1989). Glucocorticoids regulate ovine hypophysial portal levels of corticotrophin-releasing factor and arginine vasopressin in a stress-specific manner. *Endocrinology* 125, 2532–2539.
531. Battaglia, D. F., Bowen, J. M., Krasa, H. B., Thrun, L. A., Viguie, C., and Karsch, F. J. (1997). Endotoxin inhibits the reproductive neuroendocrine axis while stimulating adrenal steroids: A simultaneous view from hypophyseal portal and peripheral blood. *Endocrinology* 138, 4273–4281.
532. Battaglia, D. F., Brown, M. E., Krasa, H. B., Thrun, L. A., Viguie, C., and Karsch, F. J. (1998). Systemic challenge with endotoxin stimulates corticotropin-releasing hormone and arginine vasopressin secretion into hypophyseal portal blood: Coincidence with gonadotropin-releasing hormone suppression. *Endocrinology* 139, 4175–4181.
533. Battaglia, D. F., Beaver, A. B., Harris, T. G., Tanhehco, E., Viguie, C., and Karsch, F. J. (1999). Endotoxin disrupts the estradiol-induced luteinizing hormone surge: Interference with estradiol signal reading, not surge release. *Endocrinology* 140, 2471–2479.
534. Battaglia, D. F., Krasa, H. B., Padmanabhan, V., Viguie, C., and Karsch, F. J. (2000). Endocrine alterations that underlie endotoxin-induced disruption of the follicular phase in ewes. *Biol. Reprod.* 62, 45–53.
535. Williams, C. Y., Harris, T. G., Battaglia, D. F., Viguie, C., and Karsch, F. J. (2001). Endotoxin inhibits pituitary responsiveness to gonadotropin-releasing hormone. *Endocrinology* 142, 1915–1922.
536. Harris, T. G., Battaglia, D. F., Brown, M. E., Brown, M. B., Carlson, N. E., Viguie, C., Williams, C. Y., and Karsch, F. J. (2000). Prostaglandins mediate the endotoxin-induced suppression of pulsatile gonadotropin-releasing hormone and luteinizing hormone secretion in the ewe. *Endocrinology* 141, 1050–1058.
537. Breen, K. M., Billings, H. J., Debus, N., and Karsch, F. J. (2004). Endotoxin inhibits the surge secretion of gonadotropin-releasing hormone via a prostaglandin-independent pathway. *Endocrinology* 145, 221–227.

538. Debus, N., Breen, K. M., Barrell, G. K., Billings, H. J., Brown, M., Young, E. A., and Karsch, F. J. (2002). Does cortisol mediate endotoxin-induced inhibition of pulsatile luteinizing hormone and gonadotropin-releasing hormone secretion. *Endocrinology* 143, 3748–3758.
539. Naylor, A. M., Porter, D. W. F., and Lincoln, D. W. (1990). Central administration of corticotrophin-releasing factor in the sheep: effects on secretion of gonadotropins, prolactin, and cortisol. *J. Endocrinol.* 124, 117–125.
540. Caraty, A., Miller, D. W., Delaleu, B., and Martin, G. B. (1997). Stimulation of LH secretion in sheep by central administration of corticotrophin-releasing hormone. *J. Reprod. Fertil.* 111, 249–257.
541. Dobson, H., Essawy, S. A., and Alam, M. G. S. (1988). Suppression of LH response to gonadotrophin-releasing hormone or oestradiol by ACTH(1–24) treatment in anoestrous ewes. *J. Endocrinol.* 118, 193–197.
542. Phogat, J. N., Smith, R. F., and Dobson, H. (1999). Effect of adrenocorticotrophic hormone ($ACTH^{1-24}$) on ovine pituitary gland responsiveness to exogenous pulsatile GnRH and oestradiol-induced LH release in vivo. *Anim. Reprod. Sci.* 55, 193–203.
543. Morberg, G. P., Watson, J. G., Stoebel, D. P., and Cook, R. (1981). Effect of cortisol and dexamethasone on the oestrogen-induced release of luteinizing hormone in the anoestrous ewe. *J. Endocrinol.* 90, 221–225.
544. Phillips, D. J., and Clarke, I. J. (1990). Effects of the synthetic glucocorticoid, dexamethasone, on reproductive function in the ewe. *J. Endocrinol.* 126, 289–295.
545. Dobson, H., and Smith, R. F. (2000). What is stress and how does it affect reproduction? *Anim. Reprod. Sci.* 60, 743–752.
546. Porter, D. W. F., Lincoln, D. W., and Naylor, A. M. (1990). Plasma cortisol is increased during the inhibition of LH secretion by central LHRH in the ewe. *Neuroendocrinology* 51, 705–712.
547. McFarlane, M. S., Breen, K. M., Sakurai, H., Adams, B. M., and Adams, T. E. (2000). Effect of duration of infusion of stress-like concentrations of cortisol on follicular development and the preovulatory surge of LH in sheep. *Anim. Reprod. Sci.* 63, 167–175.
548. Breen, K. M., and Karsch, F. J. (2004). Does cortisol inhibit pulsatile luteinizing hormone secretion at the hypothalamic or pituitary level? *Endocrinology* 145, 692–698.
549. Breen, K. M., Stackpole, C. A., Clarke, I. J., Pytiak, A. V., Tilbrooke, A. J., Wagenmaker, E. R., Young, E. A., and Karsch, F. J. (2004). Does the type II glucocorticoid receptor mediate cortisol-induced suppression in pituitary responsiveness to gonadotropin-releasing hormone? *Endocrinology* 45, 2739–2746.
550. Stackpole, C. A., Turner, A. J., Clarke, I. J., Lambert, G. W., and Tilbrook, A. J. (2003). Cortisol does not suppress the luteinizing hormone (LH) response to gonadotropin-releasing hormone (GnRH) in hypothalamo-pituitary disconnected rams and ewes. *Biol. Reprod.* 68, Suppl. 1, Abstr. 429.
551. Allen, D. M., and Lamming, G. E. (1961). Nutrition and reproduction in the ewe. *J. Agric. Sci., Cambr.* 56, 67–69.
552. Barker-Gibb, M. L., and Clarke, I. J. (1996). Increase galanin and neuropeptide-Y immunoreactivity within the hypothalamus of ovariectomized ewes following a prolonged period of reduced body weight is associated with changes in plasma growth hormone but not gonadotropin levels. *Neuroendocrinology* 64, 194–207.
553. Henry, B. A., Goding, J. W., Tilbrook, A. J., Dunshea, F. R., and Clarke, I. J. (2001). Intracerebroventricular infusion of leptin elevates the secretion of luteinising hormone without affecting food intake in long-term food-restricted sheep, but increases growth hormone irrespective of bodyweight. *J. Endocrinol.* 168, 67–77.
554. Henry, B. A., Tilbrook, A. J., Dunshea, F. R., Rao, A., Blache, D., Martin, G. B., and Clarke, I. J. (2000). Long-term alterations in adiposity affect the expression of melanin-concentrating hormone and enkephalin but not proopiomelanocortin in the hypothalamus of ovariectomized ewes. *Endocrinology* 141, 1506–1514.
555. Thomas, G. B., Mercer, J. E., Karalix, T., Rao, A., Cummins, J. T., and Clarke, I. J. (1990). Effect of restricted feeding on the concentrations of growth hormone (GH), gonadotropins, and prolactin (PRL) in plasma, and on the amounts of messenger ribonucleic acid for GH, gonadotropin subunits, and PRL in the pituitary glands of adult ovariectomized ewes. *Endocrinology* 126, 1361–1367.
556. Tatman, W. R., Judkins, M. B., Dunn, T. G., and Moss, G. E. (1990). Luteinizing hormone in nutrient-restricted ovariectomized ewes. *J. Anim. Sci.* 68, 1097–1102.
557. Kile, J. P., Alexander, B. M., Moss, G. E., Hallford, D. M., and Nett, T. M. (1991). Gonadotrophin-releasing hormone overrides the negative feedback effect of reduced dietary energy on gonadotropin synthesis and secretion in ewes. *Endocrinology* 128, 843–849.
558. Kiyma, Z., Alexander, B. M., Van Kirk, E. A., Murdoch, W. J., Hallford, D. M., and Moss, G. E. (2004). Effects of feed restriction on reproductive and metabolic hormones in ewes. *J. Anim. Sci.* 82, 2548–2557
559. Gunn, R. G., Doney, J. M., and Russel, A. J. F. (1969). Fertility in Scottish Blackface ewes as influenced by nutrition and body condition at mating. *J. Agric. Sci., Cambr.* 73, 289–294.
560. Haresign, W. (1981). The influence of nutrition on reproduction in the ewe. 1. Effects on ovulation rate, follicle development, and luteinizing hormone release. *Anim. Prod.* 32, 197–202.
561. Coop, I. E. (1966). The response of ewes to flushing. *World Rev. Anim. Prod.* 4, 69–78.
562. Gunn, R. G., and Doney, J. M. (1975). The interaction of nutrition and body condition at mating on ovulation rate and early embryo mortality in Scottish Blackface ewes. *J. Agric. Sci., Cambr.* 85, 465–470.
563. Rhind, S. M., and McNeilly, A. S. (1986). Follicle populations, ovulation rates and plasma profiles of LH, FSH and prolactin in Scottish Blackface ewes in high and low levels of body condition. *Anim. Prod. Sci.* 10, 105–115.
564. Rhind, S. M., Leslie, I. D., Gunn, R. G., and Doney, J. M. (1986). Effects of high levels of body condition and food intake on plasma follicle stimulating hormone, luteinizing hormone, prolactin and progesterone profiles around mating in Greyface ewes. *Anim. Prod.* 43, 101–107.
565. Rhind, S. M., Leslie, I. D., Gunn, R. G., and Doney, J. M. (1985). Plasma levels of FSH, LH, prolactin and progesterone profiles of Cheviot ewes with different levels of intake before and after mating, and associated effects on reproductive performance. *Anim. Prod. Sci.* 8, 301–313.
566. Rhind, S. M., Martin, G. B., McMillen, S., Tsonis, C. G., and McNeilly, A. S. (1989). Effect of level of food intake of ewes on the secretion of LH and FSH and the pituitary response to gonadotrophin-releasing hormone in ovariectomized ewes. *J. Endocrinol.* 121, 325–330.
567. Downing, J. A., Joss, J., Connell, P., and Scaramuzzi, R. J. (1995). Ovulation rate and the concentrations of gonadotrophic and metabolic hormones in ewes fed lupin grain. *J. Reprod. Fertil.* 103, 137–145.
568. Downing, J. A., Joss, J., and Scaramuzzi, R. J. (1995). Ovulation rate and the concentrations of gonadotrophins and metabolic hormones in ewes infused with glucose during the late luteal phase of the oestrous cycle. *J. Endocrinol.* 146, 403–410.

569. Downing, J. A., Joss, J., and Scaramuzzi, R. J. (1995). A mixture of the branched chain amino acids leucine, isoleucine and valine increases ovulation rate in ewes when infused during the late luteal phase of the oestrous cycle: an effect that may be mediated by insulin. *J. Endocrinol.* 145, 315–323.
570. Rhind, S. M., and Schanbacher, B. D. (1991). Ovarian follicle populations and ovulation rates of Finnish Landrace cross ewes in different nutritional states and associated profiles of gonadotrophins, inhibin, growth hormone (GH) and insulin-like growth factor-1. *Dom. Anim. Endocrinol.* 8, 281–291.
571. McNeilly, A. S., Jonassen, J. A., and Rhind, S. M. (1987). Reduced ovarian follicular development as a consequence of low body condition in ewes. *Acta Endocrinol. (Copenh.)* 115, 75–83.
572. Rhind, S. M., and McNeilly, A. S. (1998). Effects of level of food intake on ovarian follicle number, size and steroidogenic capacity in the ewe. *Anim. Reprod. Sci.* 52, 131–138.
573. Rhind, S. M., Goddard, P. J., McMillen, S., and McNeilly, A. S. (1993). FSH-stimulated follicle development in ewes in high and low body condition and chronically treated with gonadotrophin-releasing agonist. *J. Reprod. Fertil.* 97, 451–456.
574. Downing, J. A., Joss, J., and Scaramuzzi, R. J. (1999). The effect of a direct arterial infusion of insulin and glucose on the ovarian secretion rates of androstenedione and oestradiol in ewes with an autotransplanted ovary. *J. Endocrinol.* 163, 531–541.
575. Haresign, W. (1981). The influence of nutrition on reproduction in the ewe. 2. Effects of undernutrition on pituitary responsiveness to luteinizing hormone-releasing hormone stimulation. *Anim. Prod.* 32, 257–260.
576. Rhind, S. M., McMillen, S., McKelvey, W. A. C., Rodriguez-Herrejon, F. F., and McNeilly, A. S. (1989). Effect of the body condition of ewes on the secretion of LH and FSH and the pituitary response to gonadotrophin-releasing hormone. *J. Endocrinol.* 120, 497–502.
577. Findlay, J. K., and Cumming, I. A. (1976). FSH in the ewe: effects of season, live weight and plane of nutrition on plasma FSH and ovulation rate. *Biol. Reprod.* 15, 335–342.
578. Boukhliq, R., Adams, N. R., and Martin, G. B. (1996). Effect of nutrition on the balance of production of ovarian and pituitary hormones in ewes. *Anim. Reprod. Sci.* 45, 59–70.
579. Parr, R. A., Davis, I. F., Miles, M. A., and Squires, T. J. (1993). Liver blood flow and metabolic clearance rate of progesterone in sheep. *Res. Vet. Sci.* 55, 311–316.
580. Thomas, D. L., Thomford, P. J., Crickman, J. G., Cobb, A. R., and Dziuk, P. J. (1987). Effects of plane of nutrition and phenobarbital during the pre-mating period on reproduction in ewes fed differentially during the summer and mated in the fall. *J. Anim. Sci.* 64, 1144–1152.
581. Daniel, J. A., Whitlock, B. K., Baker, J. A., Steele, B., Morrison, C. D., Keisler, D. H., and Sartin, J. L. (2002). Effect of body fat mass or nutritional status on 24-hour leptin profiles in ewes. *J. Anim. Sci.* 80, 1083–1089.
582. Munoz-Gutierrez, M., Blache, D., Martin, G. B., and Scaramuzzi, R. J. (2002). Folliculogenesis and ovarian expression of mRNA encoding aromatase in anoestrous sheep after 5 days of glucose or glucosamine infusion or supplementary lupin feeding. *Reproduction* 124, 721–731.
583. Blache, D., Adam, C. L., and Martin, G. B. (2002). The mature male sheep: a model to study the effects of nutrition on the reproductive axis. *Reprod. Suppl.* 59, 219–233.
584. Blache, D., Zhang, S., and Martin, G. B. (2003). Fertility in male sheep: modulators of the acute effects of nutrition on the reproductive axis of male sheep. *Reprod. Suppl.* 61, 387–402.
585. Iqbal, J., Pompolo, S., Sakurai, T., and Clarke, I. J. (2001). Evidence that orexin-containing neurons provide direct input to gonadotropin-releasing hormone neurons in the ovine hypothalamus. *J. Neuroendocrinol.* 13, 1033–1041.
586. Martin, G. B. (1984). Factors affecting the secretion of luteinizing hormone in the ewe. *Biol. Rev.* 59, 1–87.
587. Martin, G. B., Oldham, C. M., Cogie, Y., and Pearce, D. (1986). The physiological responses of anovulatory ewes to the introduction of rams—a review. *Livest. Prod. Sci.* 15, 219–247.
588. Pearce, G. P., and Oldham, C. M. (1988). Importance of non-olfactory ram stimuli in mediating ram-induced ovulation in the ewe. *J. Reprod. Fertil.* 84, 333–339.
589. Cushwa, W. T., Bradford, G. E., Stabenfeldt, G. H., Berger, Y. M., and Dally, M. R. (1992). Ram influence on ovarian and sexual activity in anestrous ewes: effects of isolation of ewes from rams before joining and date of ram introduction. *J. Anim. Sci.* 70, 1195–1200.
590. Knight, T. W., and Lynch, P. R. (1980). Source of ram pheromones that stimulate ovulation in the ewe. *Anim. Reprod. Sci.* 3, 133–136.
591. Knight, T. W., Tervit, H. R., and Lynch, P. R. (1983). Effect of boar pheromones, rams wool, and the presence of bucks on ovarian activity in anovular ewes early in the breeding season. *Anim. Reprod. Sci.* 6, 129–134.
592. Fulkerson, W. J., Adams, N. R., and Gherardi, P. B. (1981). Ability of castrate male sheep treated with oestrogen or testosterone to induce and detect oestrus in ewes. *App. Anim. Ethol.* 7, 57–66.
593. Signoret, J. P., Fulkerson, W. J., and Lindsay, D. R. (1982). Effectiveness of testosterone-treated wethers and ewes as teasers. *Appl. Anim. Ethol.* 9, 37–45.
594. Nugent, R. A. III, Notter, D. R., and McClure, W. H. (1988). Effects of ram preexposure and ram breed on fertility of ewes in summer breeding. *J. Anim. Sci.* 66, 1622–1626.
595. Iglesias, R. M. R., Ciccioli, N. H., Irazoqui, H., and Rodriguez, B. T. (1991). Importance of behavioral stimuli in ram-induced ovulation in seasonally anovular Corriedale ewes. *Appl. Anim. Behav. Sci.* 30, 323–332.
596. Sanford, L. M., Palmer, W. M., and Howland, B. E. (1974). Influence of sexual activity on serum levels of LH and testosterone in the ram. *Can. J. Anim. Sci.* 54, 579–585.
597. Walkden-Brown, S. W., Martin, G. B., and Restall, B. J. (1999). Role of male-female interaction in regulating reproduction in sheep and goats. *J. Reprod. Fertil.* 52, 243–257.
598. Martin, G. B., and Scaramuzzi, R. J. (1983). The induction of oestrus and ovulation in seasonally anovular ewes by exposure to rams. *J. Steroid Biochem.* 19, 869–875.
599. Nugent, R. A. III, Notter, D. R., and Beal, W. E. (1988). Effects of ewe breed and ram exposure on estrous behavior in May and June. *J. Anim. Sci.* 66, 1363–1370.
600. Chesworth, J. M., and Tait, A. (1974). A note on the effect of the presence of rams upon the amount of luteinizing hormone in the blood of ewes. *Anim. Prod.* 19, 107–110.
601. Edgar, D. G., and Bilkey, D. A. (1963). The influence of rams on the onset of the breeding season in ewes. *Proc. Soc. Anim. Prod.* 23, 79–87.
602. Martin, G. B., Scaramuzzi, R. J., and Lindsay, D. R. (1983). Effect of the introduction of rams during the anoestrous season on the pulsatile secretion of LH in ovariectomized ewes. *J. Reprod. Fertil.* 67, 47–55.
603. Poindron, P., Cognie, Y., Gayerie, F., Orgeur, P., Oldham, C. M., and Ravault, J.-P. (1980). Changes in gonadotrophins and prolactin levels in isolated (seasonally or lactationally) anovular ewes associated with ovulation caused by the introduction of rams. *Physiol. Behav.* 25, 227–236.

604. Martin, G. A., Oldham, C. M., and Lindsay, D. R. (1980). Increased plasma LH levels in seasonally anovular Merino ewes following introduction of rams. *Anim Reprod. Sci.* 3, 125–132.
605. Atkinson, S., and Williamson, P. (1985). Ram-induced growth of ovarian follicles and gonadotrophin inhibition in anoestrous ewes. *J. Reprod. Fertil.* 73, 185–189.
606. Oldham, C. M., Martin, G. B., and Knight, T. W. (1978). Stimulation of seasonally anovular Merino ewes by rams. I. Time from introduction of the rams to the preovulatory LH surge and ovulation. *Anim. Reprod. Sci.* 1, 283–290.
607. Ungerfeld, R., Pinczak, A. Fosberg, M., and Rubianes, E. (2003). Ovarian and endocrine responses of Corriedale ewes to "ram effect" in the non-breeding season. *Can. J. Anim. Sci.* 82, 599–602.

Knobil and Neill's Physiology of Reproduction, Third Edition
edited by Jimmy D. Neill,
Elsevier © 2006

CHAPTER 45

Control of Follicular Development, Corpus Luteum Function, the Maternal Recognition of Pregnancy, and the Neuroendocrine Regulation of the Menstrual Cycle in Higher Primates

Anthony J. Zeleznik[1] and Clifford R. Pohl[2]

Follicular Development and Its Control, 2451
Structural and Functional Characteristics of Follicular Development, 2451
Control of Follicular Estrogen Biosynthesis, 2453
Gonadotropic Control of Preovulatory Follicular Development, 2455
Selection of the Preovulatory Follicle, 2456
Corpus Luteum Formation and Function, 2461
Development of the Corpus Luteum, 2461
Endocrine Secretions of the Corpus Luteum, 2463
Luteotropic Hormones, 2463
Autocrine and Paracrine Regulation of Luteal Function, 2465
Cellular Regulation of Luteal Function, 2466
Regression of the Primate Corpus Luteum, 2468
Maternal Recognition of Pregnancy, 2470
Neuroendocrine Control of the Menstrual Cycle, 2475
Gonadotropin Secretion during the Menstrual Cycle, 2475
The Control of Gonadotropin Secretion, 2476
Frequency Changes of the GnRH Pulse Generator during the Menstrual Cycle, 2485
Summary of the Control System and Timing of the Menstrual Cycle, 2486
Acknowledgments, 2487
References, 2487

The history of the ovary as a reproductive and endocrine gland is summarized in contributions by Short and Corner (1–3). The anatomical presence of the ovary has been recognized since the time of Aristotle (384–322 B.C.), and the great anatomists of the 16th and early 17th centuries, Vesalius, Fallopius, and Fabricius, described the vesicular nature of the ovary. In his text, *De mulierum organis generationi inservietibus*, de Graaf not only described the ovarian (Graafian) follicle but also surmised that the corpus luteum, previously named by Malphigi, participated in reproduction. De Graaf noted that the presence of a corpus luteum within the ovary was associated with the presence of a fetus in the uterus and that there was a corresponding number of corpora lutea and fetuses in species that bore more than one young. Having deduced the role of the ovarian follicle in the production of the egg (although he concluded that the follicle was the egg itself), de Graaf proposed that the function of the corpus luteum was to

[1]Department of Cell Biology and Physiology, University of Pittsburgh School of Medicine, Pittsburgh;
[2]School of Health Sciences, Duquesne University, Pittsburgh, Pennsylvania.

squeeze the egg out of the ovary (3). The correct anatomical relationship between the ovarian follicle and the oocyte was identified by von Baer in 1827 (1).

The endocrine functions of the ovary began to be recognized in the mid-19th century as the result of clinical descriptions on the effects of ovariectomy in humans. In his 1856 text, Carl Ludwig noted that removal of the ovaries in humans led not only to the cessation of menstruation, but also to the shrinkage of the uterus (3). As presented by Corner (3), the initial thoughts regarding the atrophy of the uterus after castration were that either the blood or nerve supply to the uterus was disrupted by the surgery. The finding by Knauer in 1896 that castration in guinea pigs followed by grafting ovaries at different sites within the body prevented uterine atrophy indicated that bloodborne chemicals were produced by the ovaries that acted upon the uterus (3). The description by Edgar Allen in 1922 of the cyclic changes in the vagina of rodents and the association of these changes with the presence of mature follicles led to the hypothesis that the uterine development associated with estrus was due to secretions of the maturing follicle, which was subsequently confirmed by the findings that injections of small amounts of follicular fluid into castrated mice and rats led to changes in the vaginal epithelium typical of estrus (4–6). Armed with this sensitive bioassay for the hormone of estrus, Edward Doisy and colleagues purified estrone from human pregnancy urine and porcine follicular fluid (7,8).

During the early 1900s, research on ovarian secretions progressed on two fronts. In addition to the knowledge that removal of the ovary led to atrophy of the uterus, the role of the corpus luteum in the maintenance of pregnancy was documented when Fraenkel demonstrated that ablation of the corpus luteum in rabbits after mating resulted in the failure of the animals to establish pregnancy (9,10). The link between the secretions of the corpus luteum and the endometrium was made in 1910 by Ancel and Bouin when they described the histological changes of the rabbit endometrium associated with pregnancy and demonstrated that the removal of corpora lutea prevented the development of the gestational endometrium (11). In 1928, George Corner, after a decade of research on the anatomy and physiology of the corpus luteum, invited Willard M. Allen, a medical student skilled in organic chemistry, to take a year of sabbatical from his studies to work with him on the corpus luteum. This collaboration led to the isolation of a progestational agent from pig corpora lutea and the demonstration that this material maintained pregnancy in castrated rabbits (12,13). Research on the chemical nature of the progestational hormone was continued by Allen and others, and progesterone was obtained in crystalline form in 1934 (14,15).

The first half of the 20th century was also the period in which the role of the pituitary gland on ovarian function was defined. As elegantly chronicled by Greep (16), these years were a period of intense efforts to identify the hypophyseal hormones responsible for follicle growth and luteinization. The clinical findings that disorders of the pituitary glands in humans were sometimes associated with sexual immaturity suggested that the pituitary secreted a gonad-stimulating substance, observations that were confirmed when P. E. Smith demonstrated that the effects of hypophysectomy on the reproductive organs could be counteracted by extracts of pituitary tissue (17). In the 1920s, Ascheim and Zondek extracted urine from pregnant and menopausal women and identified the presence of two distinct gonadotropic activities: Urine from postmenopausal women stimulated follicular development, whereas urine from pregnant women stimulated luteinization (18,19). Isolation and biological characterization of two distinct pituitary gonadotropic hormones, follicle-stimulating hormone (FSH) and luteinizing hormone (LH), was accomplished in the mid-1940s (20).

The first direct evidence for adenohypophyseal control of ovarian function in primates was the independent studies of Allen in 1928 (21) and Hartman in 1930 (22) that demonstrated that treatment of immature and adult monkeys with pituitary extracts resulted in stimulation of ovarian function. Thereafter, investigations by others demonstrated that follicular development in monkeys could be initiated by pituitary extracts but not by extracts obtained from urine of pregnant women (23–25). The proof of two distinct anterior pituitary gonadotropic hormones in the 1940s led to rational approaches in the stimulation of ovulation in primates, and the work of van Wagenen and Simpson in the early 1950s demonstrated that follicular growth could be stimulated in monkeys with purified preparations of macaque gonadotropins (26). Knobil et al. (27) were the first to report that follicular development and ovulation could be elicited in hypophysectomized monkeys with purified gonadotropins, findings that were subsequently confirmed in humans (28,29).

As reviewed by Donovan (30), the notion that the central nervous system plays a role in reproduction evolved from early and scattered observations, including the enhancement of egg laying by swallows after the removal of eggs from their nests, the triggering of ovulation in the rabbit by coitus, and the restriction of breeding to specific seasons in several species (31). In humans, the recognition by clinicians that menstrual cycles can be altered by environmental and social stresses, such as those associated with new jobs and failed relationships, also contributed to the idea that the nervous system is more than just a bystander to the reproductive system. The previously described work that established the anterior pituitary gland as the source of the gonadotropic hormones pointed to the neighboring hypothalamus as a very enticing candidate for neural involvement in the regulation of FSH and LH secretion, and thus the discovery in animals that electrical stimulation of the hypothalamus, but not of the pituitary, caused the release of both hormones was both anticipated and gratifying (32,33). These observations coincided with the first clear statement, formulated by Green and Harris in 1947 (34), of the concept for neurovascular control of the anterior pituitary. Coupled with the initial demonstrations that neuroactive drugs such as dibenamine and atropine could block ovulation (35–37), proposals quickly emerged for adrenergic and cholinergic components in the central mechanisms that control pituitary release of LH (36,37). Although early efforts were directed toward establishing a pituitary site of action for adrenergic substances (38), these experiments were soon challenged on the basis of technical problems and inconsistencies (39). Subsequently, it was shown that the neurohumoral substance acting at the pituitary to release LH was not a monoamine but rather a small polypeptide extractable from stalk-median eminence (40,41). The structure of this substance was determined (42,43), and the decapeptide, now called gonadotropin-releasing hormone (GnRH), was synthesized (44–46). Later analyses of GnRH distribution performed by radioimmunoassay (47–49) and immunocytochemistry (50–55) localized it mainly to the mediobasal hypothalamic and preoptic-septal regions, a finding in general agreement with anatomical–physiological models of central components involved in the reproductive axis (56,57).

In attempts to produce follicular development in humans and subhuman primates with exogenous gonadotropins, the maturation of a single ovulatory follicle was a rare event, the typical response to exogenous gonadotropins being superovulation. The physiological mechanisms whereby a single follicle is selected to mature during the primate menstrual cycle continue to be an important problem to this day. Similarly, the mechanisms responsible for the regression of the primate corpus luteum at the termination of nonfertile menstrual cycles as well as its "rescue" during early pregnancy have not been completely resolved. The goal of this chapter is to review the extant literature in a manner that integrates the information gained from classical organ system physiological approaches with that obtained from studies at the tissue and cellular levels in reference to the mechanisms involved in the selection of a single preovulatory follicle as well as the regulation of the function and life span of the ephemeral corpus luteum in higher primates.

FOLLICULAR DEVELOPMENT AND ITS CONTROL

Structural and Functional Characteristics of Follicular Development

As illustrated in Fig. 1, the early follicular phase of the menstrual cycle (days 1–7 after menses) is characterized by low concentrations of estrogen; on about day 8 of the follicular phase, serum estrogen concentrations begin to rise and exponentially increase thereafter until the mid-cycle gonadotropin surge is triggered and ovulation occurs. (The neuroendocrine control of the menstrual cycle is discussed in detail in the section on Neuroendocrine Control of the Menstrual Cycle.) The rise in estrogen secretion during the mid-follicular phase reflects the emergence of the follicle destined to ovulate. At this time in the menstrual cycle, asymmetrical estrogen secretion, as assessed by direct sampling of ovarian venous blood, becomes evident and mechanical destruction of the largest visible follicle within the ovaries delays ovulation by the 2 weeks required for the maturation of a new preovulatory follicle (59,60). Although the

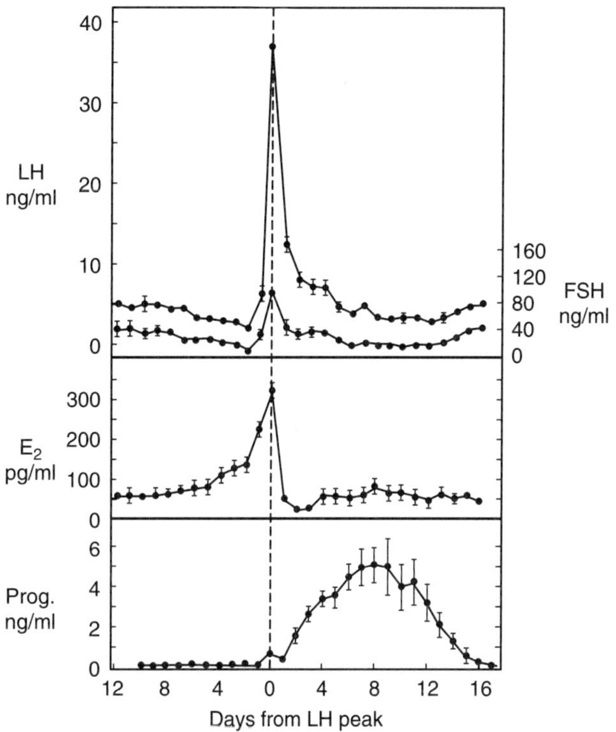

FIG. 1. Profiles of the concentrations of FSH, LH, estradiol, and progesterone throughout the rhesus monkey menstrual cycle. [Reproduced from (58), copyright 1980, The Endocrine Society.]

presence of the follicle destined to ovulate does not become apparent until the mid-follicular phase of the cycle, the complete maturation of the preovulatory follicle is a lengthy process that begins many menstrual cycles before the one during which it ovulates.

Preantral Follicular Development

Figure 2 illustrates the stages of follicular development from the primordial to the preovulatory stage. Folliculogenesis begins with the transition of primordial follicles, which are formed during fetal life and may remain dormant within the ovaries for as long as 40 years, to a primary follicle. This transition is characterized by the expansion of the epithelial-like granulosa cells to a cuboidal appearance, the initiation of cellular replication of the granulosa cells, and the initiation of the growth of the oocyte. After the preantral follicle acquires six to seven layers of granulosa cells, the antral cavity begins to form and the theca cell layer begins to develop. Based on measurements of mitotic indices and granulosa cell doubling times, Gougeon (61) estimated that the growth of a follicle from the primordial to the early antral stage in humans takes at least 85 days, whereas the final stages of follicular growth occur during of the follicular phase of the menstrual cycle, which lasts approximately 14 days.

Histological studies have demonstrated the presence of preantral and early antral follicles in ovaries removed from hypophysectomized and prepubertal monkeys, indicating that the early stages of follicular development up to and including the formation of early antral follicles are independent of the secretion of the pituitary gonadotropins FSH and LH (27,62–64). Similarly, morphological studies in humans and macaques identified all stages of preantral follicles throughout the entire menstrual cycle (65,66). Use of (^{3}H)-thymidine to label dividing granulosa cells in monkeys revealed that all stages of preantral follicles undergo DNA synthesis during the luteal phase of the menstrual cycle as well as during

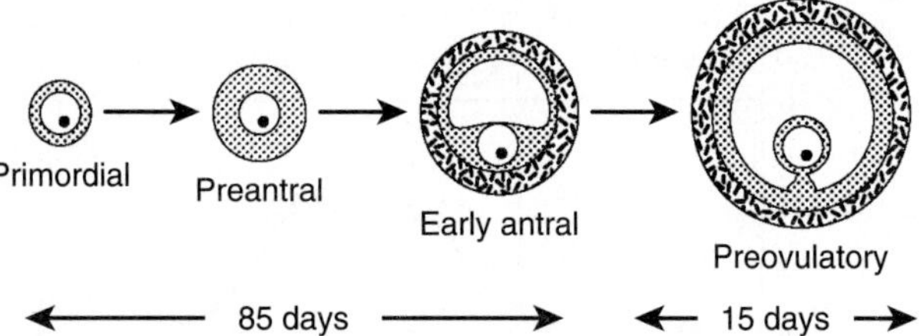

FIG. 2. Pattern of mammalian follicular development. Growth of follicles from the primordial to the early antral stages, characterized principally by proliferation of granulosa cells, takes approximately 85 days to complete. The maturation of an early antral follicle to a preovulatory follicle takes approximately 15 days.

the mid through late follicular phases after the selection of the dominant follicle (67).

Granulosa cells from early antral follicles isolated during the human menstrual cycle do not produce substantial amounts of estrogen under basal conditions, but the secretion of estrogen can be stimulated by prolonged incubation of the cells with FSH and androgen substrate, whereas they do not respond to LH, either acutely or after prolonged incubation (68). The responsiveness of developing follicles to the gonadotropins parallels the distribution of cell surface receptors for FSH and LH because it has been demonstrated that, like that of other subprimate species, granulosa cells of preantral and early antral follicles posses cell surface receptors for FSH but not for LH (69). Because early antral follicles are responsive to FSH and are present throughout the menstrual cycle as a result of the continuous entry of primordial follicles into the pool of developing follicles, it is generally accepted that the process of preantral follicular development serves to provide a constantly available source of maturing follicles for final development to the preovulatory stage when provided with appropriate gonadotropic support. This is best evidenced by the observations that preovulatory follicular development is rapidly stimulated in hypophysectomized monkeys and humans by the administration of exogenous FSH and LH (27–29).

Antral Follicular Development

The maturation of follicles beyond the early antral stage is under obligatory gonadotropic control. During this phase of development, striking structural and functional changes occur within the follicle that result in its ability to produce steroid hormones and the ability to undergo ovulation and luteinization in response to the mid-cycle gonadotropin surge. Because preantral follicles possess only FSH receptors, it is this gonadotropin that is responsible for the initiation of preovulatory follicular development and the associated biochemical changes within the follicle that accompany this process. As summarized by Hillier (70) and as largely documented in rats, stimulation of granulosa cells by FSH results in the coordinated activation of numerous genes whose products are responsible for the acquisition of steroidogenic capacity and the ability to respond to LH. Included among the mRNAs induced by FSH are cholesterol side chain cleavage cytochrome P-450 (P450scc), aromatase cytochrome P-450 (P450arom), and the LH receptor (71–75). The coordinated expression of P450arom and its encoded protein results in the ability of the follicle to produce estrogen, whereas the acquisition of the LH receptors on granulosa cells confers LH responsiveness to the follicle that results in its capacity to ovulate and luteinize in response to the mid-cycle gonadotropin surge (76).

Control of Follicular Estrogen Biosynthesis

Gonadotropic Hormones

The classical studies of Greep et al. (20) demonstrated in hypophysectomized rats that although highly purified FSH was able to stimulate follicular development, estrogen secretion, as measured by increases in uterine weights, was not stimulated unless LH was also administered. Administration of LH alone to hypophysectomized rats stimulated the repair of interstitial tissue but did not stimulate follicular development or estrogen secretion. These results, published over 60 years ago, provide the framework for virtually all of our fundamental knowledge regarding the hormonal control mechanisms that govern preovulatory follicular development in primates and subprimate species.

The production of estrogen by the maturing follicle requires the interaction between the theca cells and the granulosa cells (77). As revealed from studies conducted initially in subprimate species, theca cells under LH stimulation produce C19 androgens (testosterone and androstenedione). Because theca cells lack significant quantities of the aromatase enzyme, theca cells cannot metabolize androgens to estrogens (78). By contrast, granulosa cells lack the 17α-hydroxylase, C17,20 lyase activities that are required to metabolize C21 steroids to C19 androgens but under FSH stimulation acquire the aromatase enzyme that converts C19 androgens to estrogens (68). Thus, Greep's original observations that both FSH and LH are required for estrogen production can be explained by the requirements for theca cell production of androgens under the influence of LH as well as granulosa cell aromatization of androgens to estrogen under the influence of FSH.

This "two cell, two gonadotropin" model for estrogen biosynthesis is also operative in primates. Autoradiographical studies have demonstrated that theca cells of early antral and antral follicles possess LH receptors, whereas granulosa cells contain FSH receptors (69). Immunocytochemical analysis has demonstrated that theca cells, but not granulosa cells, possess 17α-hydroxylase, C17,29 lyase activity, whereas granulosa cells from preovulatory follicles, but not theca cells, stain positive for the aromatase enzyme (79). In vitro studies using radiolabeled steroids demonstrated minimal production of estrogen when theca cells or granulosa cells were incubated alone, whereas combining the two cell types resulted in significant production of estrogen (80). The requirement for androgen as a substrate for estrogen production by granulosa cells was demonstrated by the findings that addition of testosterone or androstenedione to isolated human granulosa cells resulted in estrogen secretion, whereas granulosa cells incubated in the absence of androgen produced minimal amounts of estrogen (81,82).

In humans, the FSH-mediated induction of aromatase is the rate-limiting step in follicular estrogen secretion. Granulosa cells collected from immature follicles throughout the follicular and luteal phases produced negligible quantities of estrogen when incubated in the presence of androgen substrate unless they were exposed to FSH for 48 hours (68). In contrast, granulosa cells collected from mature preovulatory follicles were able to metabolize androgens to estrogens immediately upon removal from the follicle in the absence of added FSH (83). In vivo, thecal androgen production does not appear to be rate limiting for the secretion of estrogen because androgen concentrations in the follicular fluids of antral follicles whose granulosa cells lack aromatase are similar to those of mature follicles whose granulosa cells possess aromatase activity (84–86).

Autocrine and Paracrine Factors

Although FSH and LH are the primary regulators of ovarian cellular function, it has become increasingly apparent that the ovarian cellular responses to the gonadotropins can be modified either in a positive or a negative manner by factors that are produced and have actions within the follicle. Those factors that have received the most attention with respect to the primate ovary include steroid hormones (androgens and estrogens), insulin-like growth factor (IGF)-I, and the inhibins and activins.

Steroid Hormones. The ability of estrogen to augment the actions of FSH on granulosa cells is well recognized, and it has been postulated that elevated estrogen concentrations present within the maturing follicle may contribute to the selection of the preovulatory follicle. This thesis is based largely on studies in rodents which demonstrated that treatment of hypophysectomized rats with estradiol or diethylstilbestrol greatly enhanced the responses of the ovary to exogenously administered FSH (87–89). However, rigorous studies in macaques failed to demonstrate a synergistic effect of diethylstilbestrol on FSH-stimulated follicular development in vivo (90). Messenger RNA for the estrogen receptor and immunoreactive estrogen receptor protein have been identified in baboon granulosa cells (91). Using specific antibodies that distinguish between the alpha and beta forms of the estrogen receptor, Saunders et al. (92) demonstrated that that the beta, but not the alpha, form of the estrogen receptor was present in granulosa cells of human and marmoset ovaries. Despite the apparent presence of estrogen receptors in the follicle, it does not appear that estrogen is absolutely required for gonadotropin-dependent antral follicular development. Women with inactivating mutations of aromatase cannot convert androgens into estrogen, and as a result, serum FSH levels are elevated due to a lack of estrogen-negative feedback (93,94). Studies on the morphology of ovaries of aromatase-deficient girls have revealed the presence of numerous large antral follicles with intact oocytes and normal-appearing granulosa cells that regressed after the suppression of FSH secretion by the administration of low doses of estrogen (95). These findings indicate that lack of estrogen production does not inhibit FSH-stimulated antral follicular development. However, it remains to be determined whether the large follicles present in the ovaries of aromatase-deficient humans are functionally normal and can ovulate in response to exogenous LH or human chorionic gonadotropin (hCG).

As mentioned previously, androgens are produced by theca cells under LH stimulation and are metabolized to estrogen by granulosa cells. In addition to serving as the obligatory substrate for aromatase, androgens potentiate FSH-stimulated induction of aromatase activity and inhibin production by granulosa cells in vitro (96,97). However, it is doubtful that selective production of androgens by individual follicles could influence which follicle subsequently responds to FSH because, as mentioned previously, androgen concentrations in antral fluids do not differ between immature and preovulatory follicles. Androgen receptors have been identified by immunocytochemistry in granulosa cells of preantral and antral follicles, but it is not known whether there are quantitative differences in receptor content among different sized follicles (98).

Although androgens augment the actions of FSH in vitro (99), paradoxically, treatment of rats in vivo with testosterone or dihydrotestosterone leads to follicular atresia and alterations in the cell cycle of granulosa cells (100,101). Moreover, administration of an androgen antagonist amplifies the stimulatory effects of FSH–hCG treatment in hypophysectomized rats, indicating that endogenous androgens are antagonistic to gonadotropin-stimulated ovarian function as well (102). Thus, depending on the experimental model, androgens have been shown to exert both positive and negative actions on ovarian function.

Studies in macaques have suggested that androgens appear to be positive regulators of follicular development because treatment of rhesus monkeys with testosterone or dihydrotestosterone increases the abundance of mRNA for the FSH receptor in granulosa cells and increases the growth of small follicles (103,104). However, when the actions of androgens on gonadotropin-stimulated ovarian function were tested in vivo in rhesus monkeys, no apparent stimulation of ovarian function was observed. Rather, dihydrotestosterone treatment suppressed the stimulatory actions of FSH and LH on ovarian steroid production and gonadotropin-stimulated increases in ovarian weights (105). In humans, it does not appear that androgens are absolutely essential for follicular growth because successful ovarian stimulation by FSH was achieved in a woman with 17α-hydroxylase deficiency, and granulosa cells isolated from this woman were able to produce estrogen when provided with an androgen substrate (106,107).

Insulin-like Growth Factors I and II. IGF-I synergizes with FSH in the induction of aromatase activity in primate granulosa cells as well as with LH in the production of androgens by theca cells (108–111). As such, IGF-I could contribute to the regulation of estrogen biosynthesis by the developing follicle. Although IGF-I is produced by rat granulosa cells (112), evidence to date in primates suggests that IGF-I is produced in ovarian interstitial cells but not in granulosa cells (113). However, IGF-I is present in follicular fluid, presumably derived from ultrafiltration of plasma, and could thus serve as a regulator of gonadotropin action on the maturing follicle (114). Messenger RNA for IGF-II is present in human granulosa cells and human granulosa cells secrete IGF-II in vitro (115,116), suggesting that this peptide could fulfill the role of an autocrine/paracrine agent. However, the biological actions of IGF-II differ from those of IGF-I as it has been demonstrated that although high concentrations of IGF-II increased basal aromatase activity by human granulosa cells, no synergistic effects between IGF-II and FSH were observed (117).

If IGF-I is present in follicular fluid as a result of filtration from plasma, it would be expected that all follicles with comparable blood flow would be exposed to similar IGF-I concentrations. However, the biological availability of IGF-I within individual follicles may be modulated by IGF binding proteins (118). Based on data obtained by Western analysis and ligand blotting studies, human follicular fluid was shown to contain at least five proteins that bound radiolabeled IGF-I with high affinity (119). The major species of binding protein, based on size and recognition by specific antibodies, was IGF binding protein 3. Culture of human granulosa cells with FSH reduced the secretion of IGF binding protein 3, as did the culture of human theca cells with LH (119). Because the sequestration of IGF-I by binding proteins decreases the concentration of free IGF-I, stimulation of a follicle by FSH and/or LH could suppress IGF binding protein synthesis, thereby reducing intrafollicular IGF binding protein concentrations, which in turn could increase the availability of free IGF-I to the follicle.

The hypothesis that IGF-I augments the actions of FSH and/or LH on the maturing follicle was tested in a nonhuman primate animal model. Rhesus moneys received an intravenous infusion of IGF-1 for 5 days before and continuing throughout a 15-day regimen of pulsatile infusion of FSH and LH (120). Serum estrogen and serum androstenedione concentrations were measured as an index of FSH and LH actions, and ovarian weights were measured at the end of the FSH/LH infusion. Results demonstrated that there was no augmentation of FSH/LH-stimulated estrogen or androstenedione production by IGF-1 and there was no increase in ovarian weights compared with FSH/LH treatment alone. The lack of effect of IGF-1 in this in vivo setting is consistent with results from an IGF-1–deficient Laron dwarf which indicated that these subjects respond normally to ovulation induction with exogenous gonadotropins (121).

Inhibin/Activin. The inhibins and activins are proteins composed of two of three peptide subunits, an α subunit, a βA subunit, and a βB subunit. Assembly of an α subunit with a βA subunit results in the formation of inhibin A, and assembly of an α subunit with a βB subunit results in inhibin B. Assembly of two identical β subunits results in the formation of the homodimeric activins (activin A and activin B), whereas assembly of a βA subunit and a βB subunit results in activin AB (122). Inhibins and activins were first identified on the basis of their actions on FSH secretion by the pituitary where it was shown that inhibins suppress FSH secretion and activins stimulate FSH secretion (123). Subsequently, it has been shown that the inhibins and activins may function as locally acting autocrine and paracrine factors in other organ systems, including the hematopoietic system, and during embryonic development (124,125).

The expression of inhibin and activin in the primate ovary is developmentally regulated. In cynomolgus monkeys, neither the α subunit nor the β subunits are present in preantral follicles, whereas granulosa cells in early antral follicles intensely express mRNA for the βB subunit. As early antral follicles mature to the preovulatory stage, the expression of the βB subunit is extinguished, whereas the expression of the α and the βA subunits is stimulated (126). These observations suggest that the early stages of antral follicular development may be influenced by activin, whereas the final stages of preovulatory follicle development may be influenced by inhibin. Similar findings have been reported in humans (127). One potential role for the inhibin/activin system in the primate ovary may be the coordination of the functional activities of the theca and the granulosa cells in the production of estrogen. As mentioned previously, the production of estrogen requires the synthesis of androgens by the theca cells under LH control and the metabolism of androgens to estrogen by granulosa cells in which aromatase activity is induced by FSH. Studies with isolated human theca cells have revealed that activin antagonizes the stimulatory effects of LH on androgen production, whereas inhibin potentiates the ability of LH to stimulate androgen production (110,128). These actions may serve to ensure that appropriate amounts of androgen are produced by the theca cells relative to the extant aromatase activity. Before stimulation by FSH and before the induction of aromatase, granulosa cells may produce activin, which would suppress theca androgen production and prevent excessive androgen secretion. As the follicle is stimulated by FSH, aromatase is induced as is the ability of the follicle to produce inhibin. The increased demand for androgens by granulosa cells of the maturing follicle for subsequent aromatization to estrogen could thus be met by the actions of inhibin, which augment LH-sensitive theca androgen production.

Gonadotropic Control of Preovulatory Follicular Development

Follicle-Stimulating Hormone

The essential role for FSH in preovulatory follicular development in primates was demonstrated unequivocally by Rabin et al. (129) in their investigations of a woman with an isolated deficiency of FSH. This "experiment of nature," a 22-year-old with primary amenorrhea, had undetectable concentrations of FSH (<3 mIU/ml), elevated levels of LH (50–90 mIU/ml), estrogen levels less than 10 pg/ml, and no evidence of antral follicular development assessed by laparoscopy and ovarian biopsy. Although initial attempts to restore ovulation in this individual with exogenous human menopausal gonadotropins were unsuccessful due to the production of anti-FSH antibodies in response to the "foreign" antigen, subsequent studies in which increasing amounts of FSH were administered to overcome the binding of the hormone to anti-FSH antibodies resulted in follicular development, estrogen production, ovulation, and successful pregnancy (130). The low estradiol concentrations and lack of follicular development in the presence of elevated LH supports the findings of Greep et al. (20) that LH, in the absence of FSH, is unable to initiate preovulatory follicular development and estrogen secretion. The failure to observe follicular development until FSH concentrations were restored despite the presence of LH in the circulation confirmed the absolute requirement of FSH for preovulatory follicular development and estrogen secretion. Further confirmation of the absolute role for FSH in preovulatory follicular growth was revealed by the identification of inactivating mutations of the FSH receptor in humans. This rare disorder is associated with the failure of ovarian follicles to advance beyond the early antral stages of maturation (131,132).

Luteinizing Hormone

The availability of pure FSH obtained by recombinant DNA technology (133) and its use in patients with gonadotropin deficiency has permitted investigations of actions of FSH in vivo in the apparent absence (or very low circulating levels) of LH. As would be predicted from Greep's observations, in these FSH- and LH-deficient patients, administration of pure FSH promoted large antral follicular development but did not result in the stimulation of estrogen secretion (134). In addition, inactivating mutations of the LH receptor have been identified in humans. The ovarian phenotype in these individuals is exactly what would have been predicted from the extant knowledge of the effects of LH on the maturing follicle. Thus, women with inactivating mutations of the LH receptor are hypoestrogenic and anovulatory (135). Interestingly, ovaries in one subject with a nonfunctional LH receptor were reported to be enlarged and polycystic (135). The likely reason for the ovarian enlargement is that serum FSH concentrations are above normal because of the lack of estrogen-mediated negative feedback.

Selection of the Preovulatory Follicle

Theoretical Considerations

The fact that numerous follicles can be stimulated to mature in humans and subhuman primates when exogenous gonadotropins are administered eliminates the possibility that only a single follicle is available for maturation during the follicular phase of the menstrual cycle. Although it is indisputable that the pituitary gonadotropins are indispensable for preovulatory follicular growth, how is it that a single follicle usually matures in view of the fact that FSH and LH are provided to all follicles by the circulatory system? The process of follicular selection must require the operation of a highly stringent regulatory process, the function of which is to ensure that only one of the many follicles with the potential to mature to the preovulatory state actually does so. In simple terms, follicle selection must involve a mechanism by which the presence of a maturing follicle in some manner inhibits the growth of other follicles without compromising its own development.

Initiation of Preovulatory Follicular Development

During the luteal phase of the menstrual cycle, preantral follicles continue to develop but the maturation of follicles beyond the early antral stage does not occur (65–67,82). Upon the regression of the corpus luteum, antral follicle growth resumes and ovulation occurs approximately 14 days later. That the hiatus in preovulatory follicular development during the luteal phase of the menstrual cycle is due to the presence of the corpus luteum has been demonstrated by the observations that removal of the corpus luteum before the expected time of luteal regression resulted in the prompt resumption of preovulatory follicular development and ovulation approximately 14 days later (136–138). The nature of inhibitory influences imposed by the corpus luteum upon preovulatory follicular development has been the subject of debate. In monkeys, because restoration of preovulatory follicle development after luteectomy was not necessarily followed by a statistical increase in FSH and LH secretion, it was proposed that the corpus luteum directly inhibited follicular development by reducing the ability of follicles to respond to gonadotropins (136). In humans, however, resumption of follicular development after the removal of the corpus luteum was preceded by a clear elevation of FSH concentrations (137,138). Although the hypothesis that the corpus luteum directly inhibits follicular growth has not been disproved, the available evidence to date suggests that the corpus luteum indirectly inhibits preovulatory follicular development by way of its secretions of estrogen, progesterone, and possibly inhibin, which suppress FSH secretion. Thus, as mentioned previously, preantral follicular development continues throughout the luteal phase, and hence the inhibitory influences imposed by the corpus luteum must be confined to the final stages of preovulatory folliculogenesis. The observations in monkeys that preovulatory follicular development can be readily stimulated during the luteal phase of the menstrual cycle either by direct administration of exogenous gonadotropins or by interfering with the feedback inhibition estrogen and progesterone upon gonadotropin secretion is consistent with the hypothesis that the failure of follicles to mature beyond the early antral stages during the luteal phase is due to insufficient gonadotropin concentrations (139,140). Finally, although systemic administration of progesterone, with its resultant feedback effects on gonadotropin secretion, after luteectomy in monkeys prolonged the interval until the next ovulation, local elevation of intraovarian progesterone concentrations by progesterone-imbedded Silastic wafers, which presumably was insufficient to restrain gonadotropin secretion, was without effect (141).

Selection of the Preovulatory Follicle

Primates are unique with respect to the long duration required for the maturation of the preovulatory follicle when compared with other species, such as sheep and cows. Baird et al. (142) proposed that the lengthy follicular phase of primates is because the primate corpus luteum, unlike that of sheep and cows, produces estrogen that suppresses FSH secretion below the level necessary to advance the development of early antral follicles. Both sheep and cows display waves of follicular development during the luteal phase in which antral follicles develop but do not ovulate, presumably because the secretion of progesterone by the corpus luteum decreases the frequency of LH pulses and blocks estrogen-induced LH surges (142–144). Upon the regression of the corpus luteum and the removal of the feedback inhibition of progesterone on LH secretion, follicles rapidly produce estrogen, which triggers an LH surge and ovulation (145). By contrast, because of the suppression of FSH secretion during the luteal phase of the primate menstrual cycle, follicular development does not progress beyond the very early antral stages and waves of antral follicular development do not occur; hence, more time is required for follicles to undergo FSH-dependent differentiation. Upon the demise of the corpus luteum and the elevation in plasma FSH and LH concentrations, preovulatory follicular development is initiated. Although the perimenstrual rise in serum FSH concentrations was recognized by Ross et al. (146) in their original descriptions of the hormonal profiles of the human menstrual cycle, the extent to which such a slight (30%–50%) rise in FSH concentrations during the early follicular phase participated in follicular development was not fully appreciated until Brown (147) demonstrated that changes in FSH concentrations on the order of 10%–30% are sufficient to initiate follicular development in anovulatory women.

Figure 3 illustrates the profile of FSH and estrogen concentrations during the follicular phase of the macaque menstrual cycle. Before the mid-cycle gonadotropin surge, there is a reciprocal relationship between the plasma concentrations of FSH and estradiol. During the early follicular phase before the emergence of a stimulated follicle, serum FSH concentrations are elevated, whereas estradiol concentrations are low. Approximately 5 days before the mid-cycle gonadotropin surge, serum estrogen concentrations begin to rise as the result of the emergence of a maturing follicle. Associated with the gradual increase in estradiol concentrations is a progressive fall in FSH concentrations due to the feedback actions of estrogen on gonadotropin secretion (148,149). This classical negative feedback relationship between estradiol and FSH is an essential component in the process of follicular selection. Because of the steady exit of follicles from the primordial pool, there is always a maturational distinct population of early antral follicles within the ovaries ready for development to the preovulatory stage under the influence of FSH. Because a growing follicle acquires sufficient aromatase activity as a result of FSH stimulation, its production of estrogen suppresses FSH secretion below that necessary to sustain the development of less mature follicles, which then undergo atresia.

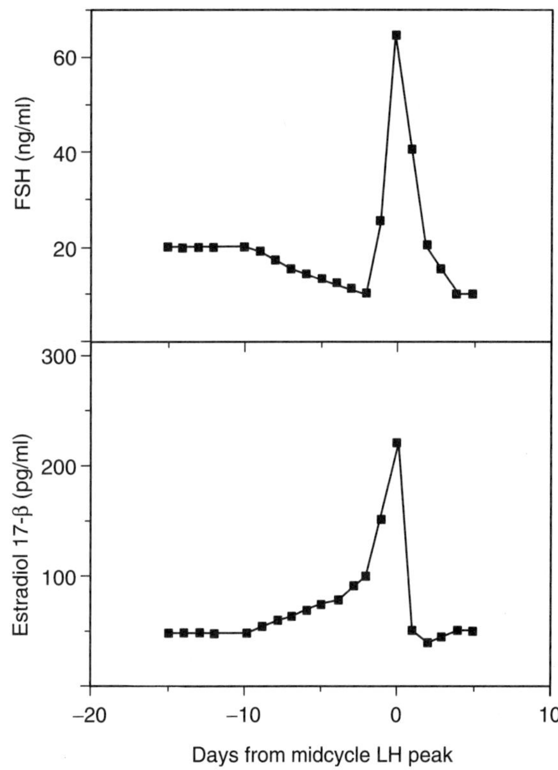

FIG. 3. Pattern of plasma FSH and estrogen concentrations during the follicular phase of the menstrual cycle. As plasma estrogen concentrations rise during the mid through late follicular phase of the menstrual cycle, there is a corresponding fall in FSH concentrations.

The hypothesis that the maturing follicle inhibits the development of less mature follicles by suppressing FSH secretion by way of its production of estrogen has been tested directly in humans and subhuman primates by manipulating the pattern of estrogen concentrations during the follicular phase of the menstrual cycle. In rhesus monkeys, subcutaneous insertion of estrogen-containing capsules on days 3 to 6 of the follicular phase of the menstrual cycle prematurely elevated estrogen concentrations by 50–80 pg/ml and resulted in a slight but significant fall in the plasma concentration of FSH and an interruption of spontaneous follicular development. When the estrogen capsules were removed, FSH concentrations once again became elevated, whereupon follicular development began anew and ovulation occurred 14 days later (150). Analogous studies in humans demonstrated that oral administration of ethinyl estradiol during the early follicular phase of the menstrual cycle resulted in a fall in FSH concentrations and a lengthening of the follicular phase, an effect that was negated by concurrent administration of an antiestrogen (151). This feedback model for follicle selection would also predict that negating the gonadotropin suppressing effects of estrogen during the mid through late follicular phases of the menstrual cycle should prevent the fall in FSH concentrations and result in the maturation of more than one preovulatory follicle. Indeed, passive immunization of rhesus monkeys with anti-estradiol antibodies during the mid through late follicular phases of the menstrual cycle prevented the fall in FSH concentrations and caused the maturation of more than one preovulatory follicle (152). In humans, it is well known that antagonizing the biological actions of estrogen with the antiestrogen clomiphene or inhibition of aromatase results in an augmentation of FSH secretion and maturation of more than one preovulatory follicle (153,154). Although these studies are consistent with the hypothesis that feedback inhibition of FSH secretion is the principal mechanism by which the maturing follicle suppresses the development of less mature follicle, it should be noted that the maturing follicle also produces inhibin and extracts of porcine follicular fluid suppress FSH secretion and follicular development in primates (155). Because both estradiol and inhibin are produced by the preovulatory follicle, it has been difficult to establish the relative importance of estrogen versus inhibin in the control of FSH secretion (156). In the passive immunization studies described previously, neutralization of estrogen resulted in the maturation of more than one preovulatory follicle, presumably in the presence of elevated inhibin concentrations that would have resulted from the increased number of maturing follicles (157). These observations therefore suggest that in the absence of estrogen-mediated suppression of FSH secretion, inhibin alone was not able to govern the development of a single preovulatory follicle Likewise, as mentioned previously, humans with aromatase deficiency have elevated levels of FSH in the presence of macroscopically normal large antral follicles, which are the primary source of circulating inhibin A (94,95), yet despite the presumed presence of elevated inhibin A, FSH levels in these subjects are elevated. However, until passive immunization studies are conducted in primates with anti-inhibin antibodies, a definitive conclusion regarding the relative contributions of estrogen and inhibin in the control of FSH secretion cannot be made. Regardless of whether inhibin or estrogen is the primary regulator of FSH secretion during the follicular phase of the menstrual cycle, the important fact is that the maturing follicle inhibits the maturation of other follicles by depriving them of gonadotropic support.

Given that FSH is absolutely required for preovulatory follicle growth and that the maturing follicle inhibits the development of other follicles by suppressing FSH secretion, how is it that the maturing follicle continues to develop in the presence of FSH concentrations that are unable to maintain the development of less mature follicles? The only explanation for this paradox is that as the follicle matures, it becomes less dependent on FSH such that the concentration of FSH necessary to initiate preovulatory follicular development is greater than the concentration of FSH necessary to maintain preovulatory follicular growth. This hypothesis was tested directly by infusion of highly purified human FSH and human LH into cynomolgus monkeys whose endogenous gonadotropin secretion was blocked by a GnRH antagonist (158). As shown in Fig. 4, when plasma FSH levels were maintained at approximately 10 mIU/ml, there was no evidence of estrogen secretion. Elevation of plasma FSH concentrations to approximately 20 mIU/ml resulted in the initiation of preovulatory follicular development as reflected by increasing concentrations of estrogen. Once preovulatory follicular growth was stimulated by FSH, reduction of plasma FSH concentrations to 8–10 mIU/ml over a 5-day period was associated with a continued rise in estrogen production, and histological examination of the ovary revealed the presence of a mature preovulatory follicle. That estrogen secretion continued to rise despite the progressive fall in FSH concentrations demonstrates that the maturing follicle, as a consequence of FSH simulation, exhibits a diminished requirement for FSH such that it continues to mature in the presence of FSH concentrations that are unable to initiate the development of less mature follicles. Similar findings have been observed in humans (159,160). These data support the hypothesis developed by Brown (147) of a threshold concentration of FSH that must be achieved to initiate preovulatory

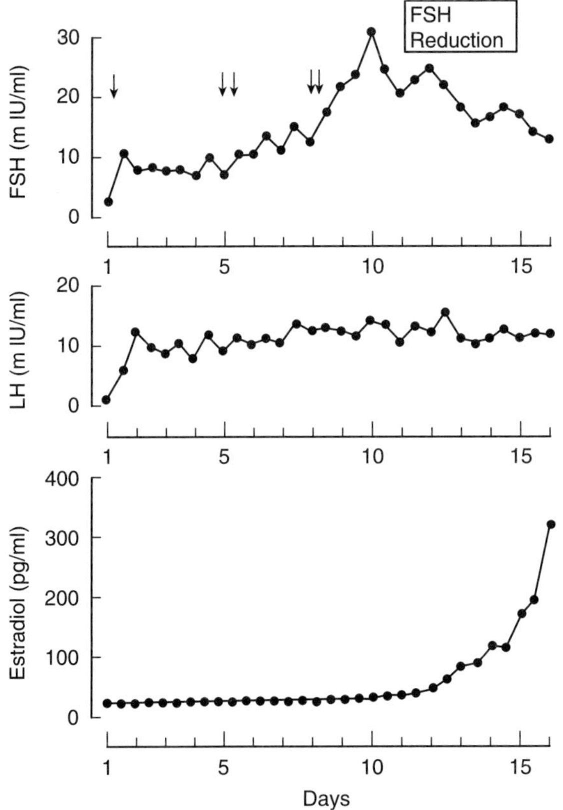

FIG. 4. Ovarian responses in a cynomolgus monkey that received hourly pulses of purified human FSH and human LH. The concentration of FSH in the infusate was increased every 4–5 days until estrogen secretion became evident; thereafter the concentration of FSH in the infusate was reduced by 12.5% per day for 5 days. Estrogen secretion, reflecting the presence of a maturing follicle, was stimulated when plasma concentrations reached ≈20 mIU/ml. When plasma FSH concentrations were reduced to levels that were unable to initiate estrogen secretion (10–15 mIU/ml), the production of estrogen by the stimulated follicle continued to increase in an exponential manner. [Reproduced from (114), with permission, copyright 1986, The Endocrine Society.]

follicle development. Not surprisingly, this experimentally determined threshold concentration of FSH is very close to the actual concentration of FSH measured during the early follicular phase of the human menstrual cycle when the initiation of preovulatory follicular development occurs (146). Once this threshold concentration of FSH is reached, follicles readily enter the final stages of preovulatory development. The number of follicles that develop to the preovulatory stage is dependent on the duration that FSH concentrations are maintained above threshold (158). During spontaneous menstrual cycles, the operation of the negative feedback loop between estrogen and FSH secretion provides moment-to-moment communication between the ovary and the hypothalamic–pituitary axis that results in the suppression of FSH secretion and the cessation of secondary follicular development upon the emergence of a maturing follicle (159,160).

Mechanisms by Which the Maturing Follicle Becomes Less Dependent on FSH

The final solution to the question of follicle selection requires the elucidation of the mechanism(s) by which the maturing follicle escapes its requirement for elevated FSH concentrations. During preovulatory follicular development, there are changes in the cellular functions of the FSH-stimulated follicle that may contribute. These development-dependent changes may occur at different levels, and the advantages that each may provide to the follicle are not mutually exclusive but rather likely provide a variety of "fail safe" mechanisms that ensure the maturing follicle is spared from the reduction in FSH concentrations.

LH and FSH Receptors. Increases in the density of cell surface receptors for FSH and LH increase the sensitivity of target cells to the hormone, as predicted by the law of mass action. In rats, the density of FSH receptors on granulosa cells increases during preovulatory follicular development (161). Although comparable studies have not been reported in primates because of the difficulty in obtaining sufficient amounts of granulosa cells from unstimulated follicles, analysis of the dose-response relationship between FSH concentrations and estrogen production by marmoset granulosa cells revealed that cells collected from preovulatory follicles were nearly four times more sensitive to FSH when compared with granulosa cells harvested from less mature follicles (162). A fourfold increase in responsiveness to FSH would more than compensate for a 50% decline in plasma FSH concentrations.

In addition to changes in the number of FSH receptors on granulosa cells, a major effect of FSH is to induce cell surface receptors for LH (163). The presence of LH receptors on granulosa cells of developing follicles could also serve to protect the follicle from the fall in plasma FSH concentrations by providing the follicle with the ability to respond to LH. Because preovulatory follicles are responsive to both FSH and LH whereas early antral follicles are responsive to only FSH (68,83), a reduction in FSH concentrations would reduce the only source of gonadotropic support to early antral follicles. By contrast, the FSH-mediated induction of LH receptors on granulosa cells could protect the mature follicle from reduced FSH concentrations by providing it with the ability to respond to both FSH and LH. The hypothesis that the selected follicle has reduced dependence on FSH because of its acquisition of LH receptors has been experimentally substantiated in humans (164,165) and in sheep (166).

Angiogenesis. In addition to increases in gonadotropin responsiveness, preferential delivery of gonadotropins to the maturing follicle could protect it from a fall in FSH concentrations. In rhesus monkeys, the density of the capillary network that supplies the preovulatory follicle is at least three times greater than that supplying other less mature follicles, and this increased density of capillaries results in a greater delivery of radioiodinated gonadotropin to the selected follicle (69). The findings that extracts of follicle cells or conditioned tissue culture media from granulosa cells were able to stimulate the proliferation of endothelial cells in vitro suggested that the maturing follicle may produce diffusible angiogenic factors that cause selective vascularization of the theca layer (167). A new family of heparin-binding endothelial growth factors has been identified. The first members of the family, vascular endothelial growth factor (VEGF) and vascular permeability factor, were isolated, respectively, from bovine pituitary cell–conditioned medium and a guinea pig tumor cell line (168,169). Human VEGF is a glycoprotein of 165 amino acids, whereas the structure of human vascular permeability factor is nearly identical to that of VEGF except it contains an additional internal amino acid sequence (169). Additional forms of VEGF have now been identified and appear to be derived from alternate exon splicing (170). VEGFs are secreted proteins that specifically stimulate endothelial cell proliferation, and receptors for VEGF are ubiquitous on capillary endothelial cells (171). In situ hybridization studies on ovaries from cynomolgus monkeys have demonstrated that mRNA for VEGF was expressed strongly by the maturing follicle but not by other antral follicles, indicating this protein could be responsible for the selective vascularization that accompanies preovulatory follicular development (172). Studies by Zimmermann et al. (173) tested the hypothesis that VEGF plays a role in follicular selection by treating rhesus monkeys with an antibody directed against the VEGF receptor and observed that administration of anti-VEGF receptor antibody, but not a control antibody, inhibited preovulatory follicular development.

In summary, the selection of the preovulatory follicle in primates is best explained by the concept of a critical "threshold" concentration of FSH that must be achieved to stimulate the development of early antral follicles to the preovulatory stage. As depicted in Fig. 5, during the luteal phase of the menstrual cycle the corpus luteum restrains FSH secretion, plasma FSH concentrations are held below that necessary to stimulate preovulatory follicular development, and follicular growth does not progress beyond the early antral stage. Upon the regression of the corpus luteum, the feedback inhibition of FSH secretion is relieved, plasma FSH concentrations rise, and early antral follicles undergo FSH-dependent growth and differentiation. The number of follicles that develop to the preovulatory stage depends on the duration that FSH concentrations are maintained above threshold values. By way of the FSH-dependent induction of aromatase activity, the leading follicle produces estrogen that suppresses FSH secretion below the threshold and further maturation of follicles is prevented, whereas the maturing follicle continues to develop because the FSH-mediated induction of LH receptors on the maturing follicle protects it from the its self-imposed reduction of FSH secretion. Thus, the FSH-mediated inductions of aromatase and LH receptors on granulosa cells are the principal components of the process of follicle selection. By acquiring aromatase, the maturing follicle produces estrogen that inhibits FSH secretion and terminates the maturation of less mature follicles while at the

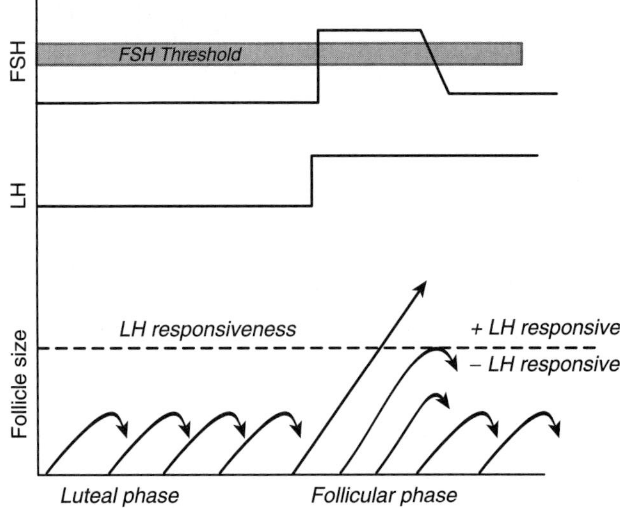

FIG. 5. The FSH threshold model for the selection of the preovulatory follicle. During the luteal phase of the menstrual cycle, circulating FSH concentrations are held below the FSH threshold by secretions of estrogen, progesterone, and inhibin by the corpus luteum and growing follicles do not advance beyond the preantral stage and undergo atresia. Upon the regression of the corpus luteum at the end of the menstrual cycle, the negative feedback restraint on FSH secretion is released and FSH concentrations rise above threshold levels. One (or occasionally more) of the maturing preantral follicles is stimulated in response to the elevation of FSH and develops both the aromatase enzyme and LH receptors. The acquisition of aromatase produces a rise in systemic levels of estradiol that results in the suppression of FSH secretion, which, in turn, prevents the maturation of less mature follicles. The FSH-stimulated induction of LH receptors and the acquisition of LH responsiveness of granulosa cells of the stimulated follicle permit the FSH-stimulated follicle to mature in the presence of FSH concentrations that are insufficient to stimulate the maturation of other less mature follicles that undergo atresia due to the lack of appropriate FSH stimulation.

same time the concomitant induction of LH receptors enables that follicle to continue to develop despite the fall in FSH concentrations

CORPUS LUTEUM FORMATION AND FUNCTION

After years of investigation of the function and regulation of the corpus luteum, George W. Corner commented, "These yellow bodies of the ovary have been puzzling to scientists ever since they were first described by Regner de Graaf" (3). In retrospect, this statement is particularly accurate regarding the quest for the understanding of the physiological regulation of this gland in mammals because the mechanisms that individual species have evolved to control the function of the corpus luteum exhibit marked variations (174). It appears, at least on the surface, that higher primates have developed a more simple solution to the regulation of the corpus luteum than most other species. As in other mammals, the primate corpus luteum plays a pivotal role in the control of reproduction and, in doing so, its functions must be regulated accordingly. In nonfertile menstrual cycles, the corpus luteum has a finite 14- to 16-day life span after which its secretions of steroid and protein hormones cease and the tissue is eventually removed from the ovary. The regression of the corpus luteum in the absence of conception is obligatory for the initiation of a new menstrual cycle because its secretions of estrogen, progesterone, and possibly inhibin exert negative feedback effects at the hypothalamic–pituitary axis that prevents the rises in FSH and LH secretion that are essential for the development of a new preovulatory follicle (139,142). On the other hand, during menstrual cycles in which conception occurs and implantation ensues, the prolongation of luteal function beyond its usual life span is obligatory for the maintenance of pregnancy until the placenta assumes the primary responsibility for the production of steroid hormones (175,176). An explanation of luteal function must therefore account for the mechanisms by which the functions of the corpus luteum are governed to accommodate its dual responsibilities.

Development of the Corpus Luteum

Preovulatory Determinants of Luteal Function

The corpus luteum is formed upon the culmination of preovulatory follicular development and the rupture of the mature Graafian follicle. As shown in Fig. 1, the mid-cycle gonadotropin surge marks the transition point at which the secretion of follicular estrogen is replaced as the major ovarian steroid by the production of progesterone by the corpus luteum. Although the corpus luteum is not formed until after ovulation, developmental events that transpire during the follicular phase under the control of FSH are essential for the appropriate functioning of the corpus luteum. Granulosa cells isolated from immature follicles are steroidogenically quiescent and require stimulation by FSH in vitro before the secretion of estrogen and progesterone becomes evident. By contrast, granulosa cells collected from preovulatory follicles of humans and subhuman primates are capable of producing both estrogen and progesterone immediately after isolation (68,177). This FSH-dependent differentiation of the granulosa cell is responsible not only for the production of estrogen by the maturing follicle, but also for the preparation of the granulosa–lutein cells to rapidly initiate progesterone secretion after ovulation. Studies in subprimate species have shown that FSH stimulation of granulosa cells during preovulatory follicular development is associated with increases in the cellular content of mRNAs and/or enzyme activities of P450scc, 3β-hydroxysteroid dehydrogenase, $\Delta^{5\text{-}4}$ isomerase (3β-HSD), P450arom, and the LH receptor, all of which are required for the secretion of steroid hormones by the corpus luteum (71–74). The importance of FSH on luteal function by way of its effects on the differentiation of the follicular granulosa cell was shown by in vivo studies in subhuman primates that demonstrated a disturbance in the normal pattern of FSH secretion during the follicular phase resulting in abnormal luteal phases characterized by reduced progesterone production (178,179).

Luteinization

Although the granulosa cells of preovulatory follicles manifest LH-responsive estrogen and progesterone secretion that is characteristic of luteal cells, the transformation of the granulosa cell into a luteal cell by the mid-cycle LH surge is associated with a further accretion of the steroidogenic capacity of the cell and a further increase in the steady-state levels of mRNAs for enzymes involved in progesterone production (75). This increased steroidogenic capacity occurs in parallel with changes in the morphology of the granulosa–lutein cells. In the rhesus monkey, granulosa cells collected from preovulatory follicles during the late follicular phase have a diameter of approximately 10 μm, contain sparse amounts of endoplasmic reticulum, and possess mitochondria with lamelliform cristae. By contrast, luteal cells of newly formed corpora lutea measure approximately 25 μm in diameter and possess abundant smooth and rough endoplasmic reticulum and mitochondria with tubular cristae (180). The structural remodeling of granulosa cells during luteinization is dependent on LH secretion because granulosa cells collected from rhesus monkey follicles before the LH surge fail to produce substantial amounts of progesterone in vitro and fail to exhibit morphological signs of luteinization (181). In addition to alterations in cellular morphology, the process of luteinization is associated with terminal cellular differentiation. Studies in rats have convincingly demonstrated that corpora lutea do not incorporate (^{3}H)thymidine into nuclei, a measure of DNA synthesis, whereas such incorporation is readily apparent in follicular granulosa cells (182). Although comparable in vivo labeling studies have not been performed in primates, analysis of rhesus monkey ovaries has indicated that the expression of markers of proliferation such as Ki67 and proliferating cell nuclear antigen are absent in the parenchymal cells of the corpus luteum (183,184). The molecular mechanisms by which cellular proliferation is curtailed in the corpus luteum is not completely understood, although it is now known that luteinization is associated with the reduced expression of cellular proteins thought to be involved in the regulation of cell cycle dynamics (185). The luteinizing stimulus generated by the LH surge thus results in a coordinated series of responses of granulosa cells that include intracellular remodeling, enhanced gene expression, and modulation of the cell cycle. Although the precise cellular mechanisms by which LH initiates the coordinated processes of luteinization is not known, it is likely that the large increase in cAMP or other intracellular signaling molecules triggered by the ovulatory LH surge either activates or represses signaling pathways that control cellular differentiation and proliferation (186). The morphological events associated with luteinization are not completed until 5–6 days after ovulation; thereafter, the luteal cells exhibit all the cellular characteristics associated with steroid-producing cells (187).

Angiogenesis

The granulosa cell layer of preovulatory follicles is not vascularized; the capillary network that supplies the follicle abruptly terminates at the basement membrane that separates the granulosa cells from the theca cells. As described by Corner (3), after ovulation, as the luteinizing granulosa layer grows thicker, blood vessels creep in from the theca layer and form a network that supplies every one of the large luteal cells. Within a few days after ovulation, the corpus luteum becomes highly vascularized such that on a tissue weight basis, blood flow to this gland is among the greatest of any tissue in the body (188). This increased vascularity, in addition to providing a conduit for the delivery of luteal steroids to the general circulation, is also necessary for the provision of cholesterol substrate in the form of low-density lipoproteins (LDLs) for progesterone biosynthesis (189).

The incisive experiments of Gospodarowicz and Thakral (190) demonstrated that the newly ovulated follicle produces diffusible angiogenic substances that direct capillary proliferation into the luteinizing tissue. As described previously, mRNA for the angiogenic factor VEGF is expressed intensely by the maturing follicle. As shown in Fig. 6, this intense expression continues after ovulation such that the newly formed corpus luteum also expresses mRNA for VEGF (172). The expression of mRNA for VEGF in the corpus luteum, as revealed by Northern analysis, correlates with luteal function in that it is extinguished during spontaneous luteal regression and declines prematurely after suppression of LH secretion by a GnRH antagonist (172). Stimulation of luteal function by administration of chorionic gonadotropin (CG) extends the expression of mRNA for VEGF beyond the expected time of luteal regression (172). The localization of mRNA for VEGF to structures within the primate ovary in which angiogenesis plays an important role in development is consistent with the hypothesis that the VEGF family of angiogenic factors may be instrumental in the formation of capillaries that accompanies both the development of the preovulatory follicle and the neovascularization that is essential for the functioning of the corpus luteum. Indeed, studies in nonhuman primates have shown that administration of VEGF antagonists at the time of ovulation or during the luteal phase of the menstrual cycle attenuates progesterone production by the corpus luteum (191,192).

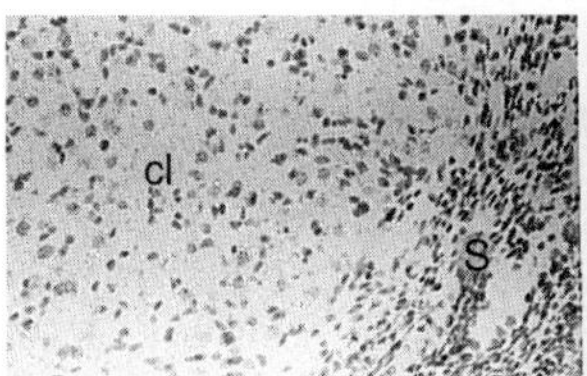

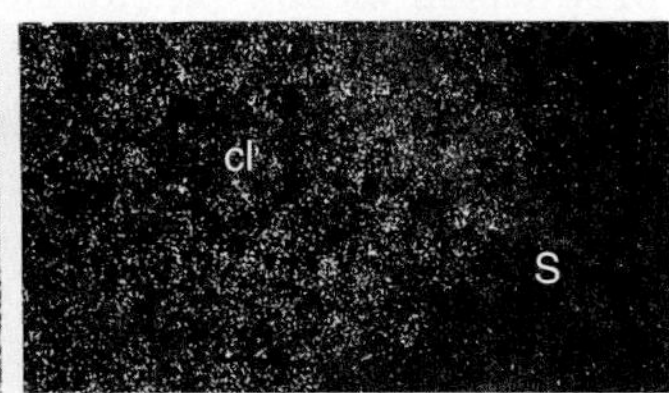

FIG. 6. In situ hybridization of mRNA for VEGF in the primate corpus luteum. The left panel illustrates a hematoxylin-stained section of a corpus luteum during the early luteal phase of the menstrual cycle. The right panel illustrates a dark-field photomicrograph of the same section incubated with a radiolabeled antisense RNA for VEGF and subjected to autoradiography. Note the presence of silver grains (*white*) in the corpus luteum but not the stromal tissue. [Reproduced from (172), copyright 1984, The Endocrine Society.]

Endocrine Secretions of the Corpus Luteum

Steroid Hormones

The description of the corpus luteum as an endocrine gland is based on its capacity to secrete progesterone and estrogen (3,193). The production of progesterone and estrogen by the primate corpus luteum likely is regulated by separate steroidogenic pathways. As described previously, the secretion of estrogen by the developing follicle requires the participation of both the theca cells for the production of androgen substrates and the granulosa cells for the aromatization of androgens to estrogens. Although the basement membrane that separates theca cells from granulosa cells is dissolved after ovulation, the primate corpus luteum retains some of the structural organization that was present in the preovulatory follicle. Using specific antibodies to steroidogenic enzymes, it has been shown that P45017α, a theca cell marker, is confined to the outer borders of the corpus luteum, whereas P450arom, a granulosa cell marker, is present within the central area of the corpus luteum (79). Assuming that the "two-cell" model for estrogen secretion by the maturing follicle can be extended to the corpus luteum, the production of estrogen therefore would require the coordinated activity of both theca lutein and granulosa lutein cells.

Progesterone production by the corpus luteum is dependent on cholesterol as a substrate. Although luteal cells can manufacture cholesterol de novo from acetate, as mentioned previously, the corpus luteum relies extensively on serum LDL-associated cholesterol for use as a precursor for the synthesis of progesterone (194,195). The dependence of normal luteal secretion of progesterone on LDL cholesterol is reinforced by the finding that progesterone concentrations during the luteal phase of the menstrual cycle in a woman with hypobetalipoproteinemia, a genetic disorder characterized by low plasma concentrations of LDL, were nearly tenfold lower than those of normal subjects (196).

Protein Hormones

In addition to steroid hormones, the primate corpus luteum also synthesizes and releases a variety of protein hormones, including relaxin (197,198) oxytocin (199–201), and inhibin (126,202–204). Enucleation of the corpus luteum of cynomolgus monkeys resulted in a rapid decline in serum inhibin concentrations, indicating that most inhibin in blood is of luteal cell origin (205). The secretion of inhibin throughout the luteal phase appears to reflect a unique function of the primate corpus luteum because inhibin does not appear to be produced to any great extent by corpora lutea of rodents and domestic animals (206–209).

Luteotropic Hormones

Luteinizing Hormone

As discussed in the section on follicular maturation, the early to mid-1900s was a period in which the nature and biological actions of the pituitary gonadotropins were revealed. The parallel studies on the identification and purification of prolactin and the demonstration that this hormone had stimulatory effects on the corpus luteum of rodents (210,211) confounded (as it still does today!) the identification of the anterior hypophyseal hormone responsible for the regulation of the primate corpus luteum. Hisaw (212), in his pioneering studies on the regulation of the rhesus corpus luteum, proposed that prolactin was the primary luteotropin responsible for the maintenance of the corpus luteum of the menstrual cycle, whereas CG was required to extend the functional life span of the corpus luteum of pregnancy. Although it was well established by the early 1960s that LH stimulated progesterone production by human corpora lutea in vitro (213,214), not until the early 1970s was it was shown that LH acutely increased progesterone production by the corpus luteum in vivo (215,216).

The requirement for LH in the control of progesterone secretion and the functional life span of the corpus luteum has been the subject of numerous debates. Gemzell (28) noted that induction of ovulation by hCG in hypophysectomized women who had been treated with FSH to induce follicular growth was followed by luteal phases of normal duration, suggesting that once formed, the human corpus luteum functioned independently of additional pituitary gonadotropin support. Subsequent studies by Vande Wiele et al. (29) demonstrated that the long half-life of hCG used to induce ovulation provided extended gonadotropic support to the corpus luteum; when LH was used to induce ovulation, the corpus luteum regressed after 5–6 days unless additional LH was given throughout the luteal phase. An obligatory requirement for LH in the maintenance of the corpus luteum was indicated when Moudgal et al. (217) passively immunized rhesus monkeys with antibodies against LH and found that immunized animals exhibited premature luteal regression, an observation later confirmed by others (218).

The role of LH in the maintenance of the primate corpus luteum was subsequently questioned by the observations that neither hypophysectomy nor treatment with a GnRH antagonist to suppress LH secretion truncated the luteal phase of macaques (219,220). However, studies in rhesus monkeys in which pituitary gonadotropin secretion was controlled directly by pulsatile infusion of synthetic GnRH provided compelling evidence that pituitary secretion of gonadotropins absolutely is required for luteal function (221). As first described by Knobil (222) and shown in Fig. 7 (top), rhesus monkeys rendered anovulatory by the placement of radiofrequency lesions in the medial basal hypothalamus (MBH) in which menstrual cycles are restored by hourly intravenous pulses of GnRH have luteal phases of normal 14- to 16-day duration. As shown in Fig. 7 (bottom), when gonadotropin secretion was interrupted in this experimental model by terminating the GnRH infusions, a rapid and precipitous fall in serum progesterone concentrations occurred and premature menstruation ensued in all animals. The initial report (210) that GnRH antagonists fail to shorten the life span of the corpus luteum appears to be dose related because others have shown in both humans and macaques that GnRH antagonists rapidly suppress luteal

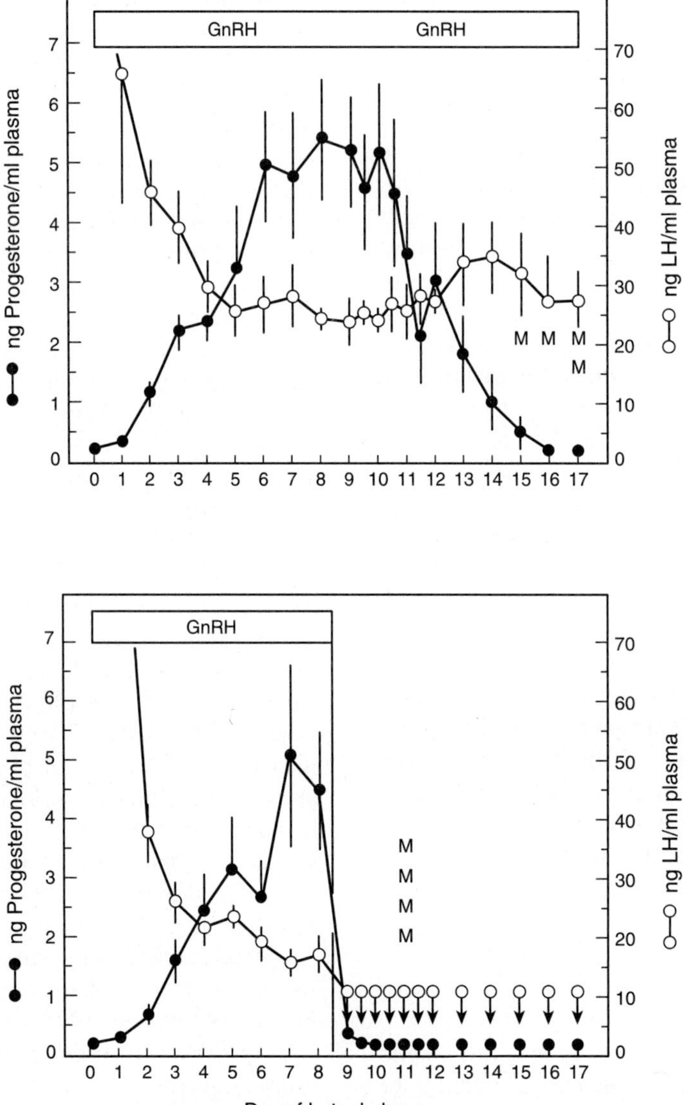

FIG. 7. Serum LH and progesterone concentrations in MBH-lesioned rhesus monkeys whose menstrual cycles were restored by a pulsatile infusion of synthetic GnRH. The top panel illustrates serum LH and progesterone concentrations of animals that received GnRH at a frequency of one pulse per hour throughout the menstrual cycle. Note that these animals have luteal phases of normal duration. The bottom panel illustrates data from animals in which the infusions of GnRH were terminated on day 8 of the luteal phase. Serum LH and progesterone concentrations fell rapidly upon the cessation of GnRH treatment and premature menses were observed in all animals. [Reproduced from (173), copyright 1984, The Endocrine Society.]

progesterone secretion and concomitantly evoke premature menses (223,224).

In addition to the secretion of progesterone, LH also is required for the production of estrogen and inhibin by the corpus luteum as plasma concentrations of both hormones decline after treatment with GnRH antagonists (225,226). Further, plasma concentrations of estrogen, inhibin, and relaxin are elevated during early pregnancy and in nonpregnant subjects in response to the administration of exogenous CG, suggesting that synthesis of these hormones is also gonadotropin regulated (227–231).

Follicle-Stimulating Hormone

Both FSH and LH secretion are reduced by GnRH antagonists, and the withdrawal of GnRH support to MBH lesioned animals curtails both FSH and LH secretion (232,233). Thus, experiments using these approaches to suppress LH secretion do not rule out the possibility that FSH may also participate in the regulation of the primate corpus luteum. FSH receptors have been identified in human corpora lutea (234), but FSH does not stimulate adenylate cyclase activity in membranes prepared from macaque corpora lutea (235) and does not stimulate progesterone secretion in either human or macaque corpora lutea in vitro (236,237). Although FSH has been shown to cause a slight (25%) increase in estrogen production by luteal cells isolated from human corpora lutea, this effect was observed in response to pharmacological concentrations of FSH (238), and the possibility that the stimulatory effects of supraphysiological concentrations of FSH on luteal cell steroidogenesis are not a consequence of slight LH contamination of the FSH preparation cannot be ruled out (236). Further, the observations that exogenous hCG or LH alone is able to maintain luteal function in hypophysectomized women indicate that FSH does not play an obligatory role in the regulation of the primate corpus luteum in vivo (28,29).

Prolactin

Prolactin is a principal luteotropin in rats and mice (239). Whether this hormone exerts luteotropic effects on the primate corpus luteum is less clear. Despite Hisaw's (212) failure to extend the functional life span of the rhesus monkey corpus luteum with prolactin, he nonetheless concluded that prolactin was the principal pituitary gonadotropin responsible for the maintenance of lutea function until CG production by the implanting trophoblast commences. Using the argument previously developed against FSH as a luteotropin, the fact that normal luteal phases can be established in hypophysectomized women after FSH and hCG treatment argues against an obligatory requirement for prolactin in the maintenance of luteal function in primates (28,29). Moreover, suppression of prolactin secretion during the luteal phase with a dopamine agonist does not interrupt normal luteal function in monkeys and humans (240,241).

Although prolactin may not be an essential luteotropin in primates, elevated levels of prolactin may sustain luteal function beyond its typical life span of 14–16 days. Some rhesus monkeys rendered anovulatory by the placement of lesions in the MBH have elevated serum prolactin concentrations, presumably due to the destruction of the dopaminergic pathways that inhibit prolactin secretion. Although these hyperprolactinemic animals undergo ovulatory menstrual cycles in response to administration of exogenous GnRH, in some animals progesterone concentrations failed to decline to undetectable levels at the end of the luteal phase but rather were maintained at approximately 1 ng/ml for extended periods beyond the expected time of luteal regression (242). That this extension of luteal function was due to prolactin was indicated by the findings that treatment of these hyperprolactinemic monkeys with the dopamine agonist bromocriptine resulted in a suppression in circulating prolactin concentrations and a prompt fall in serum progesterone levels. Similarly, lactation, which is associated with elevated prolactin concentrations, is known to stimulate progesterone secretion (243). Treatment of suckling monkeys with bromocriptine suppressed prolactin secretion and reduced progesterone to undetectable levels (242). Although the aforementioned results indicate that prolactin most likely does have stimulatory effects on the primate corpus luteum, it is unlikely that the corpus luteum is responsive to prolactin concentrations typically observed during the menstrual cycle and that the responses of the corpus luteum during suckling and surgically induced hyperprolactinemia reflects a response of the tissue to elevated concentrations of this hormone.

Autocrine and Paracrine Regulation of Luteal Function

Steroid Hormones

The macaque and human corpus luteum contains progesterone receptors, androgen receptors, and the β form of the estrogen receptor (98,244–246). In vivo studies in monkeys have shown that blocking the production of progesterone with a 3β-HSD inhibitor shortens the luteal phase, suggesting that intraovarian progesterone may have an essential role in corpus luteum function (247). However, to date the actual luteal processes that are regulated by progesterone have not been identified. As to be described elsewhere, it has been proposed that estrogen may directly act at the level of the corpus luteum to promote luteal regression. There have been no systematic studies undertaken to identify a role of androgens on corpus luteum function, other than to serve as a substrate for estrogen biosynthesis by the corpus luteum.

Proteins

Although a number of proteins, including inhibin, activin, and IGF-I, have been shown to influence steroid production by theca and granulosa cells of humans and subhuman primates (110,117), little information is available regarding the actions of these putative autocrine and paracrine agents on the primate corpus luteum. Studies have shown that activin reduced progesterone production by cultured macaque luteal cells, whereas inhibin had no effects (248). In vivo, peripheral infusion of activin A suppressed progesterone production and shortened the luteal phase of rhesus monkeys (249). However, in these studies, the infusion of activin also suppressed LH secretion. When activin was infused directly into the corpus luteum, neither progesterone production nor the length of the luteal phase was affected, indicating that the luteolytic effects of activin were most likely caused by a suppression of LH secretion rather than by a direct effect of activin on the corpus luteum.

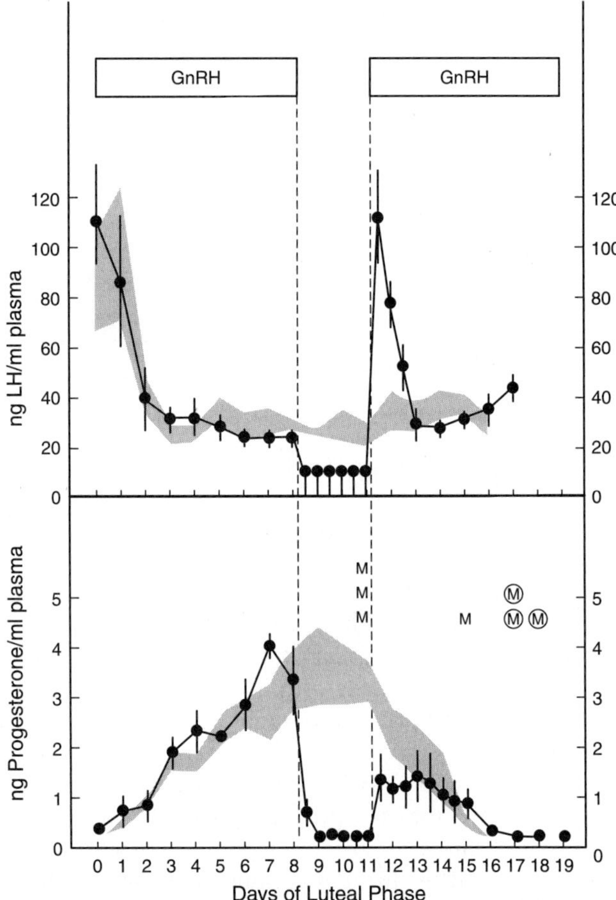

FIG. 8. The primate corpus luteum is capable of recovering from a transient withdrawal of pituitary gonadotropin support. MBH-lesioned rhesus monkeys whose menstrual cycles were driven by exogenous GnRH were deprived of the GnRH infusion from days 8 to 11 of the luteal phase of the menstrual cycle. During the interval of deprivation of gonadotropins, serum progesterone concentrations fell to nondetectable levels and premature menstruations occurred. However, when LH secretion was restored by reinstating the GnRH infusions, progesterone production recommenced and the corpus luteum regressed at its normal time, 16 days after ovulation. The shaded areas show data from animals in which GnRH was provided at a frequency of one pulse per hour throughout the luteal phase. [Reproduced from (198), copyright 1985, The Endocrine Society.]

Cellular Regulation of Luteal Function

The corpus luteum of nonfertile menstrual cycles has a finite 14- to 16-day life span during which time its functional presence is manifested by the secretion of progesterone and estrogen. Although serum progesterone concentrations are typically used to assess the functional status of the corpus luteum, studies in MBH-lesioned rhesus monkeys whose menstrual cycles are driven by pulsatile infusion of exogenous GnRH have shown that the gonadotropic regulation of progesterone secretion can be dissociated from the control of luteal life span (250). As shown in Fig. 8, the termination of GnRH infusions in MBH-lesioned animals resulted in a rapid fall in serum progesterone concentrations, demonstrating that pituitary gonadotropin secretion is required for progesterone production by the corpus luteum. However, when LH secretion was restored 3 days later in these animals, serum progesterone secretion resumed and persisted for a time that effectively completed the duration of a typical luteal phase. Although the withdrawal of LH for a 3-day period resulted in a suppression of progesterone secretion and premature menses, it did not compromise the functional life span of the corpus luteum. This suggests that although LH may be required for the acute stimulation of progesterone production, it may not be required for the long-term maintenance of the functional capacity of the corpus luteum. A complete investigation of the regulation of the primate corpus luteum must therefore address both the acute regulation of progesterone production and the long-term control of its steroidogenic capacity.

Acute Actions of Gonadotropins

Studies in both macaques and humans have revealed a close temporal relationship between secretory pulses of LH by the pituitary gland and episodes of progesterone and estrogen production by the corpus luteum (251,252). Elevations in plasma progesterone concentrations occur within minutes after individual pulses of LH. Because of the very rapid increases in steroid secretion in response to individual LH discharges, it is likely that the acute effects of LH on progesterone production by the corpus luteum reflect the channeling of cholesterol substrate through extant biosynthetic pathways rather than increases in the intracellular concentrations of enzymes responsible for progesterone synthesis. Addition of cAMP to dissociated luteal cells results in rapid increases in progesterone production (253), indicating that the acute stimulatory effects of LH on luteal steroidogenesis are mediated by cAMP. The precise intracellular mechanisms responsible for the acute mobilization of cholesterol are unknown but probably include the intracellular transport of cholesterol and its access to P450scc at the inner mitochondrial membrane mediated by the steroidogenic acute regulatory protein, StAR (254).

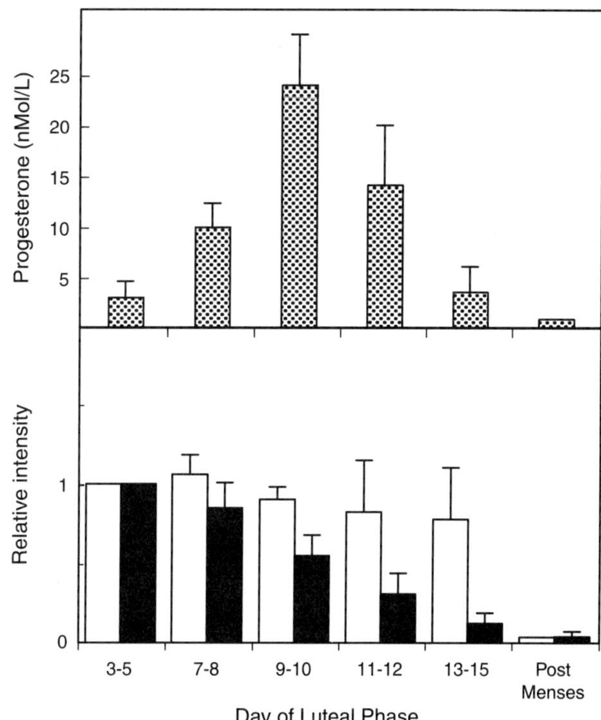

FIG. 9. Expression of mRNAs for P450scc and 3β-HSD throughout the luteal phase of cynomolgus monkeys. Corpora lutea were collected at defined stages of the luteal phase, as defined on the x-axis. The top panel illustrates serum progesterone concentrations of the animals immediately before removal of the corpora lutea. The bottom panel illustrates levels of mRNAs for P450scc (*open bars*) and 3β-HSD (*solid bars*). [Reproduced from (207), copyright 1991, The Endocrine Society.]

Long-Term Actions of Gonadotropins

Although it is well established that LH is obligatory for the acute production of progesterone by the corpus luteum, the extent to which LH is involved in the longer term regulation of the steroidogenic capacity of the corpus luteum has not been explored in great detail. The cloning of the genes for the enzymes involved in estrogen and progesterone biosynthesis has provided powerful tools for the investigation of luteal function at the cellular level (255–259). Figure 9 illustrates serum progesterone concentrations and the cellular expression of mRNAs for P450scc and 3β-HSD in corpora lutea at defined stages of the luteal phase of cynomolgus monkeys (260). Surprisingly, the highest concentrations of both mRNAs are seen shortly after ovulation. mRNA for P450scc remained relatively constant throughout the luteal phase until regression, whereas mRNA for 3β-HSD declined throughout the entire luteal phase. Similar findings were observed with human corpora lutea (75,261). The progressive fall in these mRNAs throughout the luteal phase parallels the steroidogenic capacity of isolated luteal cells because it has been shown that luteal cells collected from newly formed human corpora lutea have greater basal and gonadotropin stimulated progesterone production than cells obtained during the midluteal phase, when serum progesterone concentrations are maximal (236,237). The discrepancy between progesterone biosynthetic capacity in vitro and the in vivo secretion of progesterone by the newly formed corpus luteum likely reflects the fact that the newly formed corpus luteum has yet to establish an extensive microvascular network and therefore the delivery of precursor via LDL cholesterol to luteal cells is limited (189).

The progressive decline in mRNAs for enzymes involved in steroid secretion and steroidogenic capacity of isolated luteal cells in vitro throughout the luteal phase was interpreted as indicating that the absolute steroidogenic capacity of the corpus luteum is set by the mid-cycle gonadotropin surge and declines thereafter independently of the overall pattern of gonadotropin secretion (237,260). A similar conclusion was reached in the investigation in the rat corpus luteum, which demonstrated that after luteinization, the expression of mRNA for P450scc does not require continued gonadotropin support (262,263). However, the notion that LH does not directly participate in the long-term regulation of mRNAs for steroidogenic enzymes in the primate corpus luteum was dispelled when it was shown in cynomolgus monkeys that suppressing LH secretion during the luteal phase with a potent GnRH antagonist resulted in a decline in mRNA concentrations for P450scc and 3β-HSD to undetectable levels over a 3-day period (264). Thus, LH, in addition to its obligatory role in the acute stimulation of progesterone production, is also required for the long-term maintenance of steroidogenic capacity of the primate corpus luteum. These findings are consistent with numerous in vitro studies that demonstrated LH or cAMP stimulates the expression of mRNAs for enzymes involved in steroidogenesis in cultured human granulosa–lutein cells (265–267).

Regression of the Primate Corpus Luteum

Alteration of the Pattern of LH Secretion

Given that LH is required for both the acute and long-term regulation of luteal function, it would not be illogical to surmise that changes in LH secretion during the luteal phase could be responsible for the initiation of luteal regression. The most striking change in LH secretion during the luteal phase of both human and subhuman primates is the progesterone-mediated reduction of the frequency of LH pulses from approximately one pulse per hour during the early to midluteal phases to one pulse per 4–8 hours from the mid through late luteal phases (251,252). However, as originally documented by Knobil and colleagues (222) and later by others as shown in Fig. 8, maintenance of menstrual cycles in anovulatory rhesus monkeys and humans by unvarying GnRH pulse frequencies typical of the follicular and early luteal phases does not prolong the life span of the primate corpus luteum (268,269). Moreover, as shown in Fig. 10, when an LH pulse frequency of one pulse per 8 hours, which is typical of the mid through late luteal phases of the menstrual cycle when luteal regression occurs, is provided shortly after ovulation, serum progesterone concentrations increased progressively and normal 14- to 16-day luteal phases were observed in three of four animals studied (270).

Further evidence that luteal regression is not caused directly by a reduction of gonadotropic support to the corpus luteum has been obtained from MBH-lesioned monkeys in which plasma concentrations of LH were reduced during the luteal phase by limiting the amount of GnRH delivered per pulse (241). Reductions in plasma LH concentrations during the early luteal phase of the menstrual cycle by 50%–75% did not cause immediate luteal regression. Rather, progesterone secretion was maintained for 6–7 days after the reductions in plasma LH concentrations. These findings suggest that during the early through midluteal phases of the menstrual cycle there is a surfeit of LH, whereas during the mid through late luteal phases of the menstrual cycle, an age-dependent alteration of the sensitivity of the corpus luteum to LH occurs such that a more intense gonadotropic stimulus is required to sustain luteal function. Extension of this hypothesis would predict that luteal regression at the termination of nonfertile menstrual cycles is not due to a reduction in LH secretion per se but rather to the failure of the corpus luteum to respond to ambient concentrations of LH, a conclusion that completely supports Hisaw's

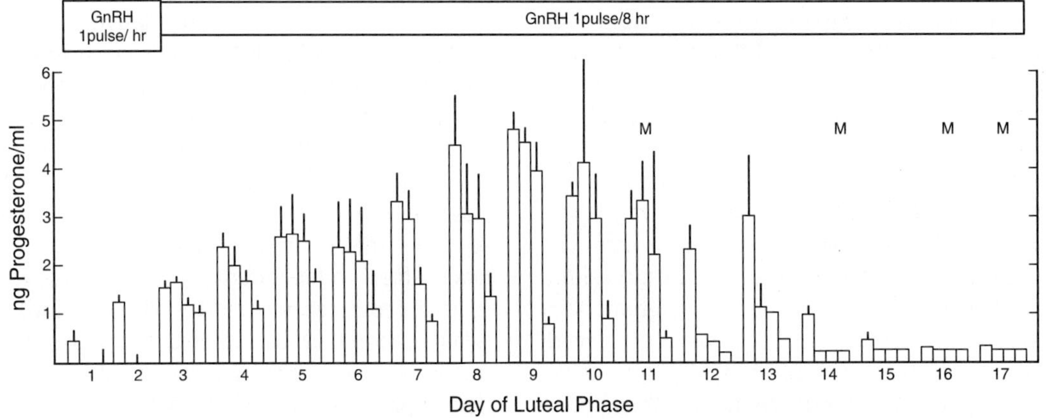

FIG. 10. Response of the macaque corpus luteum to a decrease in the frequency of LH pulses. MBH-lesioned rhesus monkeys received exogenous GnRH at a frequency of one pulse per hour to restore follicular development and ovulation. Shortly after ovulation, on days 2–3 of the luteal phase, the frequency of GnRH pulses was changed to one pulse every 8 hours, which is typical of the mid through late luteal phase. The individual bars over each day, beginning on day 3, reflect plasma progesterone concentrations obtained 30, 60, 120, and 240 minutes after a GnRH pulse. Despite the reduced LH pulse frequency, normal luteal function was observed in three of four animals. [Reproduced from (216), copyright 1986, The Endocrine Society.]

speculation, made in 1944, that "menstruation is not due necessarily to a lack or absence of pituitary gonadotropin but rather to a failure of the corpus luteum" (212).

Estrogen as a Luteolysin

Hoffman (271) noted in humans that direct injection of estrogen into the ovary containing the corpus luteum resulted in premature menstruation, whereas injection of estrogen either into the contralateral ovary or systemically did not shorten the menstrual cycle, an observation that has been confirmed by others in both humans and macaques (272,273). These observations led to the suggestion of a "self-destruct" mechanism by which internally produced estrogen leads to the regression of the corpus luteum (274), a hypothesis supported by the findings that concentrations of estradiol and estrone in macaque corpora lutea increase during the late luteal phase as progesterone concentrations decline (275). The luteolytic effect of estrogen was proposed to occur directly at the level of the corpus luteum because in many instances, premature regression of the corpus luteum did not appear to be accompanied by a reduction in plasma LH concentrations (272,276). The self-destruct hypothesis was supported further by the observation that the addition of estrogen to isolated macaque luteal cells inhibited both basal and gonadotropin-stimulated progesterone production (277,278).

This attractive hypothesis, which would account for the observations that luteal regression does not appear to due to a reduction in LH secretion, was questioned by subsequent observations in which sensitive bioassays were used to measure serum LH concentrations during estrogen-induced luteolysis. These studies revealed that estrogen-induced luteal regression was preceded by slight, but significant, declines in plasma LH concentrations during exposure to exogenous estrogen (279). The physiological relevance of estrogen-induced luteal regression was also questioned by studies in MBH-lesioned monkeys which demonstrated that exogenously administered estradiol, at dosages shown to be luteolytic in intact animals, failed to provoke premature luteal regression in monkeys in which hypophyseal drive was maintained constant by pulsatile infusion of GnRH (280). Unless there are totally different mechanisms responsible for luteolysis between GnRH-driven and spontaneously cycling animals, a possibility that is unlikely because of the remarkable similarities of their luteal phases, it would appear that because of its failure to induce luteal regression in GnRH driven animals, estrogen is not a physiologically important luteolysin during the spontaneous luteal phase. This conclusion is supported by the findings that neither the administration of aromatase inhibitors nor estrogen antagonists to spontaneously cycling nonhuman primates prolongs the duration of the luteal phase (281–283).

Prostaglandins

In many subprimate species, luteal regression is caused by the production of prostaglandins by the nongravid uterus. This is clearly not the case in primates, including humans, because hysterectomy does not result in prolonged luteal phases (284–286). However, numerous studies have shown that prostaglandins can affect luteal function. In vitro, prostaglandins E_2, I_2, and D_2 generally stimulate luteal function (287–292), whereas prostaglandins of the F series inhibit luteal functions (293–296). In vivo, infusions of prostaglandin $F_{2\alpha}$ directly into the corpus luteum have been shown to cause luteal regression (297–300). These findings, together with the observations that luteal cells produce prostaglandins in vitro (301–305), have led to the supposition that locally produced prostaglandins may exert direct luteolytic effects on the primate corpus luteum.

Although it is unequivocal that exogenously administered prostaglandins can induce premature luteal regression, it is uncertain whether the luteolytic effects of prostaglandins reflect physiological or pharmacological actions. Injections of indomethacin into rhesus monkeys at dosages that prevented the initiation of labor failed to prolong the life span of the corpus luteum (306). More recently it has been shown that direct infusion of the prostaglandin synthesis inhibitor meclofenamate into the rhesus monkey corpus luteum actually caused premature luteal regression, suggesting that prostaglandin synthesis may be required for normal luteal function as opposed to luteal regression (307). In this regard, it has been demonstrated that direct infusion of prostaglandin E_2 into the corpus luteum inhibits prostaglandin $F_{2\alpha}$-induced luteal regression in monkeys (308). However, infusion of prostaglandin E_2 alone during the time of luteal regression did not extend the life span of the corpus luteum, thus questioning the role of endogenous prostaglandin $F_{2\alpha}$ in the regression of the primate corpus luteum.

Cellular Correlates of Luteal Regression

Production of both estrogen and progesterone by isolated luteal cells under basal conditions and in response to tropic stimulation declines as the luteal phase progresses (236,237). The decline in steroidogenic capacity in aging corpora lutea is accompanied by a reduction in the number of LH receptors on luteal cells as well as a decline in LH-sensitive adenyl cyclase activity (309–311). However, it remains uncertain as to whether these changes in the transmembrane signaling system for LH are causes or consequences of luteal regression because reductions in serum progesterone concentrations may precede the decreases in LH receptors and LH-sensitive adenyl cyclase. In addition to changes in LH responsiveness, luteal cells collected from corpora lutea during the late luteal phase also exhibit reduced progesterone production in response to cAMP as well as in response to the addition of pregnenolone, the immediate precursor to progesterone (236). These data therefore suggest that the alterations in cellular function that accompany luteolysis are global in nature such that many aspects of luteal cell function are compromised.

The histological studies of Corner describe the morphological correlates of luteal regression in macaque and human corpora lutea (312). During the midluteal phase, approximately 9 days after ovulation when serum progesterone concentrations are at their zenith, the corpus luteum begins to exhibit signs of degeneration, including shrinkage of luteal cells and vacuolization of the cytoplasm. By day 11 of the luteal phase, 4–5 days before the expected time of menstruation, cell shrinkage becomes more pronounced and nuclear chromatin condensation and pyknosis are apparent. As mentioned previously, the process of luteinization of granulosa cells in vivo is not completed until 5–6 days after ovulation (187). On the basis of the morphological characteristics of luteal cells throughout the luteal phase, it would thus appear that luteolysis on a cellular level is a progressive phenomenon that begins shortly after the process of luteinization is completed.

Stouffer and colleagues (313) used flow cytometry to analyze the functional characteristics of luteal cells from rhesus monkey corpora lutea based on their sizes. At least two populations of luteal cells were distinguished, small luteal cells (15 μm diameter) and large luteal cells (>25 μm diameter). During the early luteal phase small luteal cells outnumbered large luteal cells by a ratio of 4:1, and although both small and large cells were responsive to hCG and cAMP, the absolute steroidogenic capacity of large cells exceeded that of small cells by ten- to 20-fold. As the luteal phase progressed, both the number of small and large cells declined, but the rate of decline of large cells exceeded that of small cells such that by the end of the luteal phase, the ratio of small to large cells was approximately 15:1 (314). In addition, from the midluteal phase onward, small cells were unresponsive to hCG and cAMP, whereas large cells retained responsiveness to both cAMP and hCG (249). These observations indicate a gradual shift in population of steroidogenic cells as the corpus luteum ages. If the "small" luteal cells isolated by flow cytometry correspond to the shrunken granulosa–lutein cells described by the Corners, it would then appear that luteal regression may be the result of a stochastic process in which individual luteal cells undergo a transformation from being gonadotropin responsive and steroidogenically competent to gonadotropin unresponsive cells and that at any given point in the luteal phase the functional status of the corpus luteum is determined by the relative proportions of each cell type. A progressive decrease in the percentage of gonadotropin-unresponsive cells during the mid through late luteal phases of the menstrual cycle could provide a mechanism by which the corpus luteum regresses without a change in gonadotropin secretion. In this regard, the morphological changes in luteal cells that occur during the mid through late luteal phases are typical of apoptotic cell death (315), and analyses of the human corpus luteum have revealed a progressive increase in the number of apoptotic cells during the mid through late luteal phases of the menstrual cycle (316). The determination of whether apoptosis is the cause or the result of a loss of responsiveness to gonadotropins is not known. Neither is it known whether the diminution of gonadotropin responsiveness is due solely to a consequence of luteal cell aging, such as a reduction of telomere length, or whether it is due to the actions of other factors that impinge on the luteal cell; this remains one of the major unanswered questions regarding luteal regression.

Maternal Recognition of Pregnancy

In the absence of a successful pregnancy, the primate corpus luteum has a finite 14- to 16-day life span, but its secretory activity is prolonged in menstrual cycles in which successful conception occurs. Indeed, the abrupt increase in serum progesterone concentrations that was seen in pregnant rhesus monkeys when spontaneous luteal regression normally would have occurred led Neill et al. (317) to conclude that the corpus luteum of the menstrual cycle is "rescued" during early pregnancy.

The findings by Ascheim and Zondek in 1927 (18) that urine of pregnant women contained a substance that stimulated luteinization in rats was the first evidence of a specific luteotropic factor associated with pregnancy, an observation that shortly thereafter resulted in the development of the first test for pregnancy. Because the biological activity of extracts from urine from pregnant women differed from that of extracts of urine from postmenopausal women, which also possessed the ability to stimulate follicular growth, Zondek (19) proposed the existence of two hormones, Prolan A, which stimulates follicle growth, and Prolan B, which stimulates luteinization. Although Zondek initially proposed that both Prolan A and B were of pituitary origin, the observations by Gey et al. (318) and Jones et al. (319) that cultured placenta trophoblast produced Prolan B provided definitive evidence for the placental production and secretion of this gonadotropin.

As mentioned previously, the insightful studies by Hisaw in 1944 demonstrated that injections of CG into rhesus monkeys during the luteal phase prolonged the functional life span of the corpus luteum, a seminal observation that was made without the benefit of chemical assays for either estrogen or progesterone. Based on careful monitoring of menstrual bleeding, coloration of sex skin, and characteristics of the vaginal mucosa, Hisaw correctly deduced that CG stimulated both estrogen and progesterone secretion by the corpus luteum. On the basis of these studies, Hisaw concluded that the newly formed corpus luteum is maintained by the pituitary gonadotropin, whereas the extension of the functional life span of the corpus luteum was a direct result of placental production of CG (212).

Role of CG in the Rescue of the Corpus Luteum

The requirement of CG for the prolongation of the life span of the corpus luteum was convincingly demonstrated by the findings that its neutralization by either active or passive immunization resulted in failure to support pregnancy (320–323). Although early studies in both humans and subhuman primates revealed a temporal association between rising concentrations of CG in peripheral plasma and progesterone and estrogen concentrations, a direct cause–effect relationship between the secretion of CG and the rescue of the corpus luteum was confounded by two observations. First, in some studies, increases in plasma concentrations of CG occurred after the rescue of the corpus luteum was apparent (324,325). Second, despite rising concentrations of CG, the heightened response of the corpus luteum is transient such that progesterone secretion declines in the presence of increasing concentrations of CG in blood (274).

The lack of a precise relationship between the initial detection of CG in plasma and luteal rescue was most likely due to the insensitivity of early bioassays for CG. Using sensitive radioimmunoassays for rhesus CG, Hodgen et al. (326) and Atkinson et al. (226) demonstrated that CG concentrations in plasma were detectable, on the average, 10–11 days after ovulation. However, this rise occurred 1 day after the initial rise in progesterone secretion, making it difficult to establish a causal role for the appearance of CG in the blood and the rescue of the corpus luteum. Using a very sensitive immunoassay, Baird et al. (327) demonstrated that in humans hCG in urine can be detected 7–8 days after ovulation when increases in urinary excretion of estrogen and progesterone metabolites also becomes apparent. It is well documented that human and nonhuman primate blastocysts can produce CG in vitro (328–330). Although it has been suggested that production of CG by preimplantation embryos and local delivery of the gonadotropin from the uterus to the ovary may contribute to the rescue of the corpus luteum (331,332), direct injections of small amounts of hCG into the uterine cavity of rhesus monkeys failed to stimulate progesterone secretion (333). As concluded by Ross (334), the most likely explanation for the slight temporal discrepancy between enhanced progesterone secretion by the corpus luteum and the initial detection of CG in blood is that the corpus luteum is more sensitive to low concentrations of CG than are existing assays.

More perplexing are the numerous observations that enhanced progesterone secretion by the corpus luteum is not sustained despite rising titers of CG in plasma (335,336), a phenomenon that also occurs in response to the administration of exogenous CG (227,337,338). Neill and Knobil (338) proposed that the decline in steroidogenesis after initial exposure to CG might be due to depletion of intracellular substrates required for steroidogenesis. More recently, it has been shown that prolonged stimulation of rhesus monkeys by hCG in vivo results in a progressive decrease in the ability of hCG to stimulate adenylate cyclase activity in membranes isolated from corpora lutea (339). This process of desensitization could account for the progressive decline in the secretion of progesterone by the corpus luteum in response to the progressive increases in plasma concentrations of CG that occur during early pregnancy. However, as deduced by Hisaw (212) and subsequently demonstrated later by Neill and Knobil (338), as shown in Fig. 11, the progressive decline in progesterone production in response to hCG is not accompanied by a parallel decline in estrogen secretion. These observations indicate that individual CG-dependent biosynthetic pathways of the corpus luteum may be differentially affected by luteal cell aging, a hypothesis that is supported by the observation of an age-dependent attenuation of the responsiveness of selected mRNAs to CG stimulation (340) As shown in Fig. 12, treatment of cynomolgus monkeys with hCG during the early luteal phase (days 5–10) uniformly increased mRNAs for $P450_{scc}$, 3β-HSD, $P450_{arom}$, and the LDL receptor. By contrast, when an identical regimen of hCG was provided during the late luteal phase (days 10–16), mRNAs for enzymes involved in progesterone

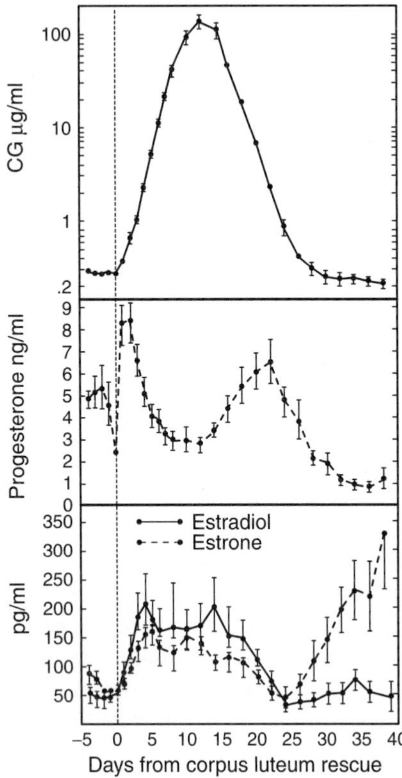

FIG. 11. Patterns of serum CG, progesterone, and estrogen concentrations during early pregnancy in the rhesus monkey. Results are standardized to the time of rescue of the corpus luteum (day 0) that occurs on the average 10–11 days after ovulation. Note the decline in serum progesterone that occurs after the first week of pregnancy despite rising concentrations of CG. By contrast, estrogen concentrations parallel the temporal pattern of CG. [Reproduced with permission from (226).]

production (P450scc and 3β-HSD) were not stimulated, whereas mRNAs for P450arom and the LDL receptor remained responsive to gonadotropin treatment. These findings indicate that luteal cell aging is associated with a selective loss of individual gonadotropin-sensitive biosynthetic pathways and agree well with the physiological data that luteal rescue (and exogenous hCG treatment) is associated with continuous stimulation of estrogen secretion but only transient stimulation of progesterone secretion and that the corpus luteum becomes less responsive to exogenous CG as the luteal phase progresses (227,337). Diminished progesterone production by the corpus luteum therefore does not appear to be due to an absolute refractoriness to CG but rather to a selective loss in specific luteal cell functions.

Mechanism of Action of CG

CG and LH share identical α subunits, whereas their β subunits differ in size and pattern of glycosylation (341). The additional 30 amino acid sequence at the carboxyl terminal of CG-β together with the different glycosylation pattern results in a plasma half-life of CG that is substantially longer (36 hours vs. approximately 20 minutes) than that of LH (342–344). Both CG and LH appear to occupy the same receptor (345) on luteal cells, both stimulate adenylate cyclase activity in luteal cell membranes (235), and both stimulate progesterone secretion by isolated luteal

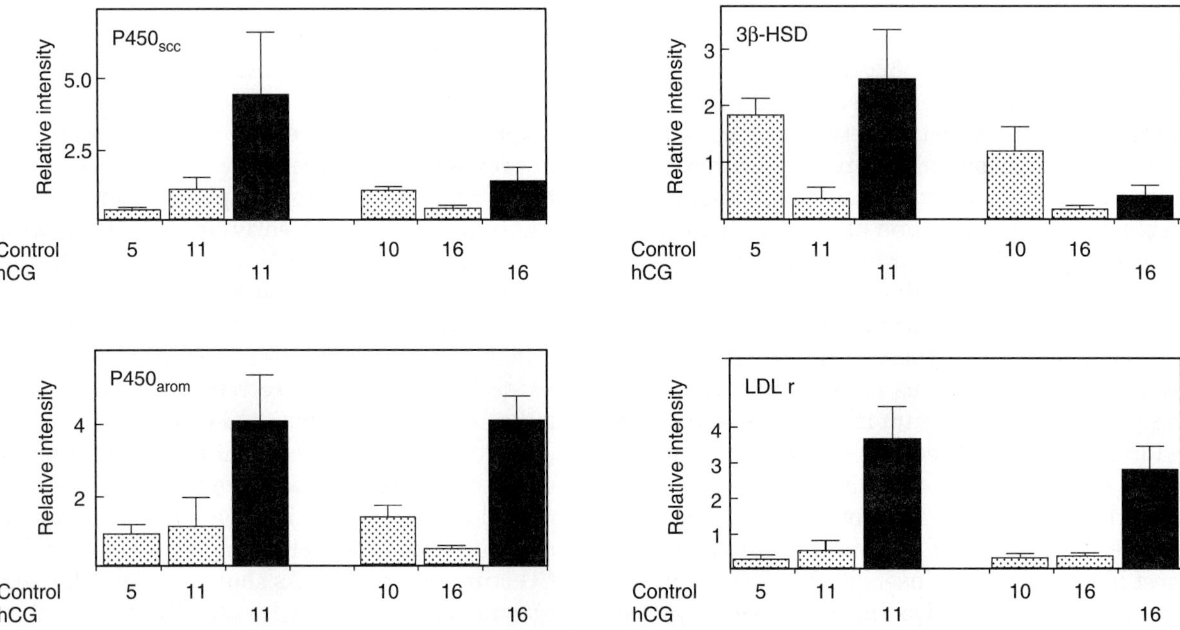

FIG. 12. Age-dependent expression of mRNAs in macaque corpora lutea in response to exogenous CG. Cynomolgus monkeys were treated with increasing amounts of hCG on either days 5–11 of the luteal phase or days 10–16 of the luteal phase. Corpora lutea were removed and RNA was extracted for Northern analysis. Treatment during the early luteal phase led to uniform increases in mRNAs for $P450_{scc}$, 3β-HSD, $P450_{arom}$, and the LDL receptor, whereas treatment with hCG during the mid through late luteal phase increased mRNAs for only $P450_{arom}$ and the LDL receptor. [Reproduced from (283), copyright 1992, The Endocrine Society.]

cells (236). The paradox regarding the rescue of the corpus luteum therefore is why the corpus luteum regresses in the presence of LH yet the similar molecule CG is able to prolong its functional life span.

Three hypotheses have been developed to account for the differential responses of the primate corpus luteum to LH and CG. One hypothesis is that the patterns of the secretory dynamics of LH and CG provide different gonadotropic stimuli to the corpus luteum (346). This hypothesis relates to the fact that whereas plasma concentrations of LH intermittently fall to very low values during the luteal phase due to the pulsatile nature of LH secretion by the pituitary gland, plasma concentrations of CG do not fall to these low levels due to both the continuous secretion of CG by the placenta and its longer circulatory half-life (347). It is thus possible that continuous exposure of the corpus luteum to CG provides a more intense gonadotropic stimulus to the corpus luteum than does the episodic stimuli provided by LH. A second hypothesis is that although CG and LH are structurally similar, the amino acid differences in the β chains of CG and LH as well as differences in the extent of glycosylation and sulfation (348) could result in an inherent difference in biological activity between the two hormones. This hypothesis is based largely on studies of rodent and ovine systems which demonstrated that the time course of steroid production in response to hCG was significantly prolonged when compared with that ovine LH, suggesting that the dynamics of the interaction of CG with the CG/LH receptor and/or the subsequent activation of intracellular signaling pathways may differ from that of LH (349,350). A third hypothesis is that the aging corpus luteum exhibits a diminished responsiveness to gonadotropins such that the ambient concentrations of LH present during the mid through late luteal phases of the menstrual cycle are unable to sustain its functions (115) and that functional life span of the corpus luteum is prolonged in early pregnancy because the exponentially increasing plasma concentrations of CG transiently overcome the diminishing responsiveness of the aging corpus luteum to LH.

These hypotheses were tested in cynomolgus monkeys. In this study, monkeys received intravenous infusions of either LH or CG in a continuous or an exponentially increasing manner beginning during the midluteal phase of the menstrual cycle and continuing beyond the expected time of luteal regression (351). As shown in the left panels of Fig. 13, a continuous infusion of either LH (top) or CG (bottom), which resulted in bioactive LH/CG concentrations 10 times greater than endogenous LH concentrations, did not prolong the functional life span of the corpus luteum These findings reinforce the notion that luteal regression at the end of nonfertile menstrual cycles is the result of a decrease in the responsiveness of the aging corpus luteum to LH because superimposition of exogenous LH or CG at tenfold greater concentrations than endogenous LH concentrations did not prolong the functional activity of the corpus luteum. Also, these results do not support the hypothesis that a simple switch from pulsatile LH secretion to the continuous secretion of CG is sufficient to rescue the corpus luteum because timely luteal regression occurred in the animals that received a continuous (i.e., nonpulsatile) infusion of LH.

In contrast, exponentially increasing concentrations of LH or CG (Fig. 13, right panels) prolonged progesterone production beyond the expected time of luteal regression. However, the responses to LH and CG were not identical because although the pattern of progesterone production in animals that received the exponentially increasing regimen of CG was similar to that of pregnant animals, the cumulative progesterone levels in response to the exponentially increasing LH infusion were 40% lower than those seen in the animals that received hCG, despite the fact that bioactive plasma concentrations of both gonadotropins were comparable throughout the duration of the infusions. Thus, it appears that luteal rescue requires CG rather than LH and that the corpus luteum be exposed to an exponential increase of this gonadotropin. The reason CG was more effective than LH in this experimental paradigm remains to be answered.

Interestingly, the duration of CG production by the placenta differs among primate species (352). In humans, hCG concentrations reach a maximum during the first trimester of pregnancy, decline thereafter, but remain detectable throughout the entire duration of pregnancy. A similar pattern of CG production is seen in apes, with the exception that the absolute plasma concentrations are an order of magnitude less than those seen in humans. In baboons and macaques, CG concentrations decline to near undetectable levels during mid-pregnancy. Whether there are functional requirements that underlie the different patterns of CG secretion, such as direct actions on the conceptus or uterus (353, 354), or whether these differences are merely consequences of evolution is not known.

The Luteal–Placental Shift

The sole function of the corpus luteum in the maintenance of pregnancy is the production of steroid hormones, because women with ovarian failure and ovariectomized monkeys can achieve successful term pregnancies after embryo transfer provided they receive exogenous estrogen and progesterone (355,356). In both humans and subhuman primates, the corpus luteum is essential until the placenta assumes the

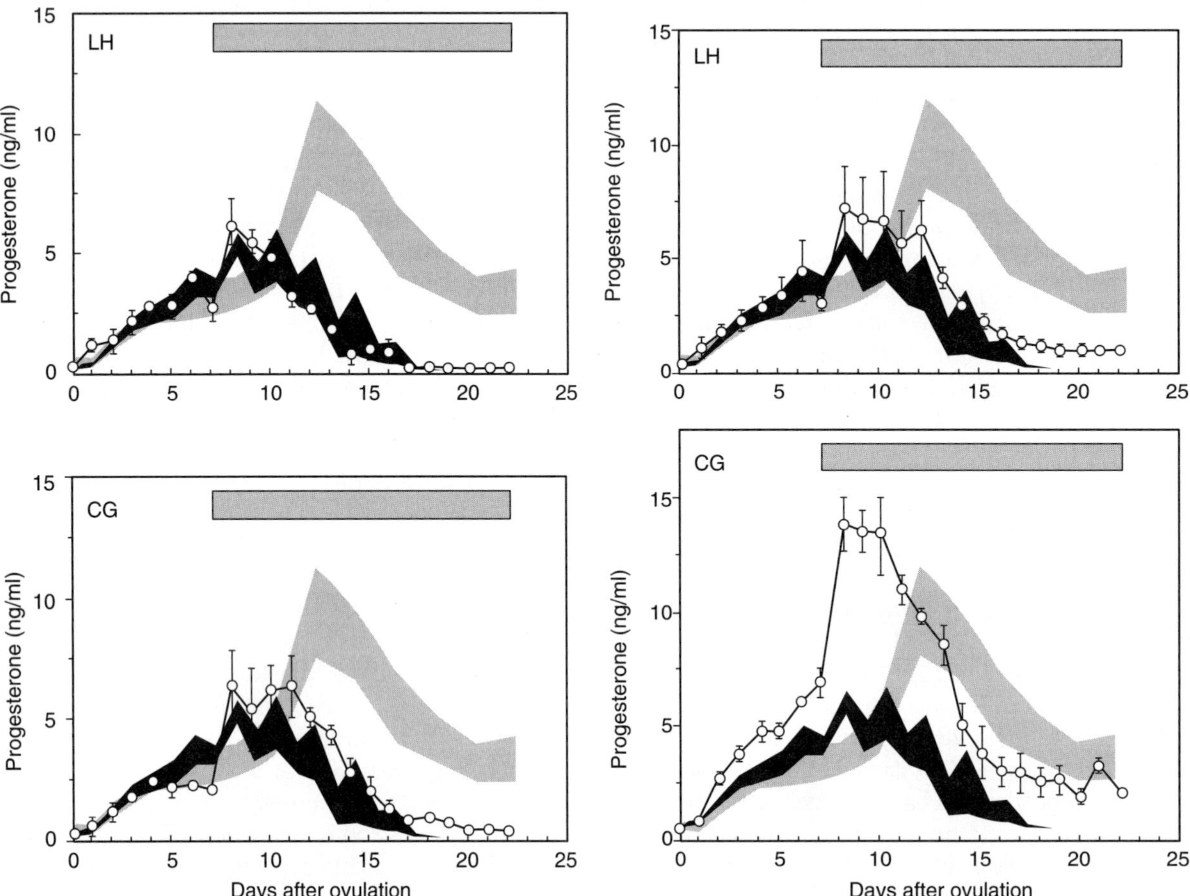

FIG. 13. Responses of the primate corpus luteum to exogenous LH and hCG. The duration of each gonadotropin treatment is shown by the shaded rectangle at the top of each panel. The left panels illustrate serum progesterone in animals that received intravenous infusions of LH and hCG at a constant rate. The right panels illustrate serum progesterone in animals that received intravenous infusions of LH and hCG at an exponentially increasing rate. The dark shaded area encompasses the means ± 1 standard error of the mean (SEM) of serum progesterone concentrations during luteal phases of three control monkeys, and the light shaded area encompasses the means ± 1 SEM of serum progesterone concentrations during luteal phases of three pregnant monkeys. [Redrawn from (351).]

major role of steroid secretion. Removal of the human corpus luteum before the fifth week after implantation resulted in a decline in serum progesterone concentrations and abortion (176). In rhesus monkeys it has been shown that successful pregnancies can be observed after the removal of the corpus luteum as early as 1 week after implantation despite a dramatic fall in serum progesterone and estrogen concentrations (357). That abortions can be prevented by replacement with progesterone alone in humans and subhuman primates after luteectomy indicates that this steroid, so aptly named by Corner and Allen for its progestational effects, is the sole luteal substance required for the maintenance of early pregnancy (358).

The fate of the corpus luteum during pregnancy has been followed in humans by the measurement of 17α-hydroxyprogesterone in blood. Because the 17α-hydroxylase enzyme is present in the corpus luteum but not in the placenta, presumably all 17α-hydroxyprogesterone in blood is derived from the corpus luteum. Plasma concentrations of 17α-hydroxyprogesterone peak on the fourth and fifth week of gestation and decline thereafter, whereas serum progesterone concentrations continue to rise after the fifth week of gestation, indicating that from week 5 onward, the major source of circulating progesterone is the placenta (336). Ovariectomy of rhesus monkeys on day 25 of pregnancy does not result in a dramatic fall in serum progesterone concentrations, indicating that by this time the placenta is the primary source of progesterone in this species (274).

Although steroid secretion by the corpus luteum wanes during early pregnancy, it does not appear that the corpus luteum becomes totally nonfunctional. It has been shown in humans at term pregnancy that venous blood draining the ovary bearing the corpus luteum contains greater concentrations of both progesterone and relaxin than that found in blood from the contralateral ovary and the peripheral circulation (198). In postpartum monkeys, progesterone concentrations in ovarian venous blood from the ovary that contains the corpus luteum are greater than those found in blood from either the contralateral ovary or the peripheral circulation (243). The fact that lactating animals have higher concentrations of progesterone in the peripheral circulation as well as in the blood draining the corpus luteum–bearing ovary suggests that prolactin may contribute to the maintenance of the low activity of luteal function observed during pregnancy and immediately after parturition. The physiological significance of the persistence of luteal activity throughout pregnancy and after parturition is not known in view of the observations that successful pregnancies occur when the corpus luteum is removed after the luteal–placental shift.

In summary, the formation of the corpus luteum represents the culmination of the ovarian events of the long process of gametogenesis and the beginning of the period in which the implanted embryo must be nourished and sheltered. The numerous cell divisions that granulosa cells undergo during preantral follicular development and their final differentiation into LH-responsive steroid-producing cells during the follicular phase of the menstrual cycle are not dissimilar to the morphogenesis of the elegant yellow monarch butterfly. The emergence of the corpus luteum from its follicular cocoon after ovulation begins the final episode of the ovarian cycle. Like a butterfly, the corpus luteum must fulfill a purpose that is essential for the preservation of the species, and when this function is completed, it dies. The data available to date entirely support Hisaw's conclusion, made more than 50 years ago, that the death of the primate corpus luteum is not due to a lack of pituitary gonadotropin secretion but rather to intrinsic changes within the corpus luteum. The simplest explanation is that there is an age-dependent decrease in the responsiveness of the primate corpus luteum to LH such that ambient concentrations of LH during the mid thorough late luteal phases of the menstrual cycle are no longer able to sustain its secretory activity. Rescue of the corpus luteum by CG is likely the result of the exponentially increasing concentrations of gonadotropic activity that overcome the diminished sensitivity of the aging gland to LH. The final challenge is to identify the cellular mechanisms that are responsible for the age-dependent loss in the responsiveness of the corpus luteum to LH.

NEUROENDOCRINE CONTROL OF THE MENSTRUAL CYCLE

Gonadotropin Secretion during the Menstrual Cycle

Primate species studied to date exhibit reproductive cycles that are qualitatively similar to those of all other mammals insofar as they share three common features: (a) a preovulatory phase of follicular development characterized by increasing levels of circulating estrogens, with gonadotropin levels remaining generally low; (b) a mid-cycle estrogen peak immediately preceding or coincident with the peak of a surge in gonadotropins that heralds incipient ovulation; and (c) a postovulatory phase characterized by high levels of progesterone, with gonadotropin levels again remaining low (Fig. 1).

Secretory Patterns across the Menstrual Cycle

As previously discussed, when the dominant follicle matures during the late follicular phase and produces increasing amounts of estradiol, the concentrations of this steroid rise in the circulation whereas those of the gonadotropins fall, especially FSH. Progesterone concentrations during this time are uniformly low due to the absence of a functional corpus luteum. Levels of dimeric inhibin A, which are low throughout most of the follicular phase, begin to rise near the end of this phase (359–362). Conversely, dimeric inhibin B rises in the early follicular phase and falls in the late follicular phase, roughly paralleling FSH but with a delay of 2–4 days (359–361).

As the follicular phase makes its transition into the mid-cycle periovulatory period, rising plasma estradiol is accompanied by the initiation of the preovulatory gonadotropin surge, which constitutes the central neuroendocrine event in the cycle and whose peak precedes ovulation by some 20–44 hours (363–373). As the gonadotropin surge develops, estradiol levels continue to rise briefly and then fall precipitously. These changes in estradiol concentrations during the surge are paralleled by a small rise and then a fall in progesterone and 17α-hydroxyprogesterone concentrations. Also at mid-cycle in the human, a peak of inhibin A coincides with the peak of gonadotropins, whereas a peak of inhibin B follows by 1–2 days (360,361). These mid-cycle elevations of the inhibins apparently do not occur, or are less apparent, in the macaque (359).

After ovulation, progesterone levels rise during the first 6 days of the luteal phase, remain elevated for approximately 3 days, and then begin to wane. In the rhesus monkey, menstrual flow, which defines the beginning of the next cycle, occurs approximately 3 days after plasma progesterone concentrations fall below 1 ng/ml (364,374). Estrogen production by the corpus luteum is also elevated during this phase, albeit to a greater degree in the human (29,202,375–377) and great apes (378–383) than in Old World macaques (58,384–390). The time course of circulating estradiol concentrations during the luteal phase is similar to that of progesterone. Inhibin A concentrations also rise and fall during the luteal phase in parallel with progesterone (359–362), whereas inhibin B levels are low (359–361). Circulating levels of the gonadotropins during most of the luteal phase are similar to or slightly lower than those of the follicular phase but begin to rise (especially FSH) near the end of the cycle during the luteal–follicular transition (29,202,372,376,391–400).

Pulsatile Secretion

A new dimension to the study of endocrine interactions and control was introduced with the discovery that moment-to-moment LH levels in the circulation are not constant but exhibit a characteristic oscillation that occurs with a period of approximately 1 hour (and was therefore termed "circhoral") in the ovariectomized rhesus monkey (401) [see (402) for Knobil's intriguing account of the discovery of this rhythm]. Ultradian rhythms in the plasma concentrations of gonadotropic hormones were soon discovered in many other mammalian species [see (403)], and today are recognized for all the major hormonal secretions of the anterior pituitary gland.

The rhythmic pattern of LH in the circulation appears as successive pulses of the hormone that quickly rise to a peak and then decay exponentially after a time course that approximates its half-life (401,404–412). The pulses are produced by the intermittent release of LH from pituitary gonadotropes without significant interpulse secretion (i.e., secretion between individual pulses) as determined by deconvolution analysis and other mathematical approaches (408–413). Pulses of both LH and FSH, the latter largely synchronous with the former, are clearly demonstrated in postmenopausal women and agonadal subjects of both sexes (404,414–419) but are more difficult to distinguish in eugonadal women during the follicular phase (375,415,416,420–424) due to the suppressive negative feedback actions of estrogen on pulse amplitude. As might be anticipated, ovarian responses to individual low-amplitude pulses of gonadotropins in the form of distinct estradiol pulses are difficult to demonstrate (420,425). Despite the rising mean levels of estradiol, the frequency of LH pulses during most of the follicular phase in women and monkeys is similar to that seen in postmenopausal women or ovariectomized monkeys (401,410,414,416,417,421–423,426,427).

In women, the pulsatile release of LH appears to be maintained during the preovulatory gonadotropin surge, with similar frequency but higher amplitude than that which occurs during the follicular phase (375,414,420–425,428–430). In the rhesus monkey, however, discrepant findings have been reported. Whereas some described pulses of high frequency and high amplitude during the surge (426,431,432), Knobil and colleagues (365,433) failed to detect such pulses during either the spontaneous mid-cycle surge or gonadotropin surges induced by exogenous estradiol. The explanation for these divergent findings in the rhesus monkey might include differences in pulse detection methods, the occurrence of assay errors that become amplified when ligand levels are high and changing rapidly, and the possibility that the physiological release mechanism for LH during the surge in these animals is sensitive to local uncontrolled variables in different laboratory settings that might acutely provoke or suppress pulsatile release. A similar discrepancy has been reported for the rat (434,435).

The luteal phase is distinguished by low-frequency high-amplitude LH pulses (251,252,375,411,414,420, 422,424,426,433,436,437) that, in contrast to gonadotropin pulses during the follicular phase, elicit distinct ovarian responses in the form of clearly demonstrable pulses of progesterone (251,252,420,437–440). As progesterone levels decline during the late luteal phase, the pulsatile pattern of LH release that characterizes the follicular phase is gradually reestablished (400,421,433). This highlights the general principle that pulse amplitude and pulse frequency often appear to be inversely related (401,409,423,433,441,442), resulting in similar daily production rates of LH during the follicular and luteal phases of the cycle (443).

A circadian component in the pulsatile release of LH has been observed in the form of a reduction in pulse frequency during sleep in women throughout the early follicular phase (412,421,423,444–449) and in rhesus monkeys throughout the follicular phase (433) and after ovariectomy (450). This contrasts with the higher frequency of nocturnal LH pulses observed in some men (451,452) and in adult male rhesus monkeys (453–455).

The Control of Gonadotropin Secretion

Whereas the ovarian cycle is directed by LH and FSH, the control of these gonadotropic hormones rests with the major secretory products of the hypophysiotropic hypothalamus and the ovary.

The Hypothalamus and the GnRH Pulse Generator

That the central nervous system plays an essential role in the control of gonadotropin secretion in the higher primates has been established beyond any doubt, based on experiments performed in the rhesus monkey and experiments of nature in the human (222,403,456–465). Much evidence has accrued to support the role of GnRH as the principal mediator for the central control of both LH and FSH. Thus, the administration of antisera to GnRH (466–468), GnRH antagonists (223,224,233,469–489), and chronically administered GnRH agonists (486,490–501) all inhibit gonadotropin secretion, whereas the acute administration of GnRH stimulates rapid increases in circulating levels of these hormones.

Alternative forms of GnRH (chicken GnRH-II and fish GnRH-III) and GnRH-II mRNA have been described in the hypothalamus of primates and other mammalian species (502–507). These isoforms of GnRH have LH-releasing bioactivity (504,507), and GnRH-II receptor cDNA has been cloned from primate RNA (508–510). Whereas GnRH-II receptors are expressed in gonadotropes (508), the neurons that synthesize GnRH-II appear to be different from those that produce GnRH (503). It remains to be determined whether these peptides, newly identified in mammals, play significant roles in the control of gonadotropin secretion in the higher vertebrates, including primates. Furthermore, there is some support for the view that FSH secretion is less dependent on GnRH than is LH (233,478,484,511,512), and some have suggested the existence of an independent FSH-releasing factor (511–513) that may be an isoform of GnRH (513). However, in the primate, it appears that the role of such a factor, if it exists, may be minor because most if not all of the neuroendocrine regulation of FSH, including its differential release relative to LH during the menstrual cycle, can be attributed to the pattern of episodic GnRH secretion in combination with the negative feedback actions of ovarian hormones (see section on Luteal–Follicular Transition).

The initial observations of circhoral LH release in the rhesus monkey (401) were followed up by the demonstration that systemic injections of neuroleptic agents such as chlorpromazine and haloperidol, or of the α-adrenergic blockers phentolamine and phenoxybenzamine, quickly resulted in an interruption of pulsatile LH release (514). These findings provided support for the view that in contrast to a pituitary site of origin, each LH pulse is a direct response to a bolus of GnRH released from nerve terminals in the median eminence and conveyed to the anterior pituitary by the pituitary portal vasculature. This conclusion provided the foundation for the concept of a central oscillator, or pulse generator, that is ultimately responsible for the intermittent discharge of GnRH from the terminals of GnRH neurons. In monkeys, pulses of GnRH have been reported in pituitary portal blood (515), in cerebrospinal fluid (516,517), and in hypothalamic push–pull perfusates (432,518–529). However, the most compelling demonstration that LH pulses are driven by GnRH pulses has been provided by studies in the ewe showing unambiguous correspondence between pulses of GnRH in pituitary stalk portal blood and pulses of LH in peripheral blood (530–532).

Physiological Significance of Pulsatile GnRH Secretion. The intermittent delivery of endogenous GnRH to the pituitary gland is essential for the maintenance and control of gonadotropin secretion. The recognition of this important principle arose from early attempts in ovariectomized rhesus monkeys to reestablish gonadotropin secretion using exogenous GnRH administration after the placement of radiofrequency lesions in the MBH that abolished endogenous GnRH secretion. Whereas a constant intravenous infusion of GnRH in such animals induced only a transitory release of LH and FSH (533), infusion of the decapeptide as 6-minute pulses once every hour reestablished normal gonadotropic function (232,490). Furthermore, continuous infusion of GnRH profoundly inhibited gonadotropin secretion previously established by the pulsatile administration of this hormone, and this inhibition was reversed when the pulsatile replacement regimen was reinstituted (490). The time required for continuous GnRH delivery to maximally suppress gonadotropin secretion was approximately 2 weeks (490). That this phenomenon is the consequence of the pattern of hypophysiotropic stimulation rather than of the total mass of GnRH delivered to the gonadotropes was suggested by the failure of a wide range of infusion rates of GnRH to restore gonadotropin secretion in monkeys with hypothalamic lesions when given in the continuous mode (490). Constant infusions of GnRH have been shown equally ineffective in eliciting sustained gonadotropin secretion in women (497,499,534). In ovariectomized rhesus monkeys with hypothalamic lesions, increasing the frequency of pulsatile GnRH replacement from the physiological rate of one pulse per hour to two, three, or five pulses per hour resulted in the development of pituitary refractoriness to GnRH stimulation with a decline in plasma LH and FSH to nearly undetectable levels (535). The cellular mechanisms by which pituitary gonadotropes become desensitized by continuous or excessively frequent exposure to GnRH or to its long-acting analogues are not fully understood but may involve down-regulation of GnRH receptors and activation of regulators of G protein signaling and/or G-protein–coupled receptor kinases (536–540). In any case, hypotheses that attempt to explain the refractoriness of the gonadotropes to continuous GnRH stimulation must take into account the observation that, in vivo, the phenomenon occurs over a period of days, as does its reversal with pulsatile GnRH administration (490,499,535). Deliberate pituitary gonadotrope desensitization for the purpose of inhibiting inappropriate gonadotropin secretion (e.g., in central precocious puberty) is an obvious clinical application (541–549).

In ovariectomized rhesus monkeys with hypothalamic lesions, decreasing the frequency of exogenous GnRH pulses from once per hour to once every 3 hours leads to an increase in circulating FSH and a decrease in plasma LH concentrations (535), thus lowering the ratio of LH to FSH in the circulation. Similar findings of a lower LH/FSH ratio with slow frequencies of exogenously administered GnRH pulses have been reported in men and women with hypothalamic hypogonadism (441,550–553), intact male monkeys (554), and ovary-intact female monkeys with hypothalamic lesions (555). When the gonad is present, the reduced LH/FSH ratio is due mainly to declining mean LH levels rather than to elevations in FSH. This shift in the observed LH/FSH ratio that is induced simply by reducing the frequency of pulsatile GnRH administration is most readily explained by an increased magnitude of pituitary LH and FSH release at each stimulated pulse (409,423,441,442,535), most likely due to continued biosynthesis in the face of a reduced discharge frequency that results in augmented releasable pools of these hormones (535). When considered along with their known half-lives in the circulation [47 minutes for LH (556), 240 minutes for FSH (556,557)], it becomes apparent that the larger pulses of LH produced with slow frequencies of GnRH delivery are completely cleared from the circulation during the longer interpulse intervals, whereas those of FSH are not entirely cleared due to its slower disappearance rate. It should be noted that in a physiological context, the ratio of LH to FSH is approximately the same during the follicular phase as during the luteal phase when hormone pulse frequencies are much reduced. The luteal phase LH/FSH ratio might be accounted for by elevated levels of inhibin A that occur at this time (359–362) and can suppress FSH secretion (249,558).

It is of interest that the reported insensitivity of the female macaque to GnRH (386,471,559–562) was also observed in the pulsatile replacement studies (490,535). The dose of GnRH per pulse required to sustain gonadotropin secretion at the circhoral frequency in the female rhesus monkey greatly exceeds that used to restore gonadotropin levels in humans with anorexia nervosa or Kallmann syndrome (269,563,564). Moreover, the dose-response characteristics of pulsatile GnRH infusion in the female monkey seem to permit only limited control of gonadotropin secretion through modulation of GnRH pulse amplitude. Increasing the minimum effective infusion rate tenfold failed to change LH concentrations and only slightly reduced FSH levels (535).

Localization of the GnRH Pulse Generator. In the primate, the MBH is the locus of both the cell bodies of hypophysiotropic GnRH neurons and the GnRH pulse generator, because complete neural disconnection of this region from the remainder of the central nervous system did not interfere with the circhoral pattern of gonadotropin secretion in ovariectomized monkeys (565). In agreement, immunocytochemical and in situ hybridization studies in primates have established that GnRH cell bodies reside in the MBH, among other areas (566–582). Within the MBH, discrete radiofrequency lesions that bilaterally destroyed the arcuate nuclei resulted in an abrupt fall in LH and FSH concentrations to undetectable levels without producing deficits in other anterior pituitary functions (533). In contrast, large lesions within the MBH that spared the arcuate nuclei had little or no effect on gonadotropin secretion (533), suggesting that this ventral portion of the periventricular zone in the MBH mediates the hypophysiotropic control of the gonadotropes in the rhesus monkey. Of interest in this regard are demonstrations that pulsatile GnRH release occurs in vitro from hypothalamic explants (583–586).

Electrophysiological Monitoring of GnRH Pulse Generator Activity. The ability of multiunit recording to detect electrical changes in a continuous fashion has prompted the use of this technique to observe the activity of the GnRH pulse generator in real time. Thus, as shown in Fig. 14, electrodes placed in the MBH have recorded abrupt increases in multiunit electrical activity (MUA) that are synchronous with the initiation of LH pulses in rhesus monkeys (433,442,587,588), rats (589–591), and goats (592,593). Because of the heterogeneous microarchitecture of the MBH, it has not been possible to determine whether the MUA volleys emanate from GnRH cell bodies, GnRH fibers, or other cellular elements (594). In the monkey, the tight one-to-one relationship between MUA volleys and LH pulses in the circulation is maintained when the frequency of these events is altered physiologically or with pharmacological agents (433,442,595–601), thus permitting the use of MUA as an independent measure of GnRH

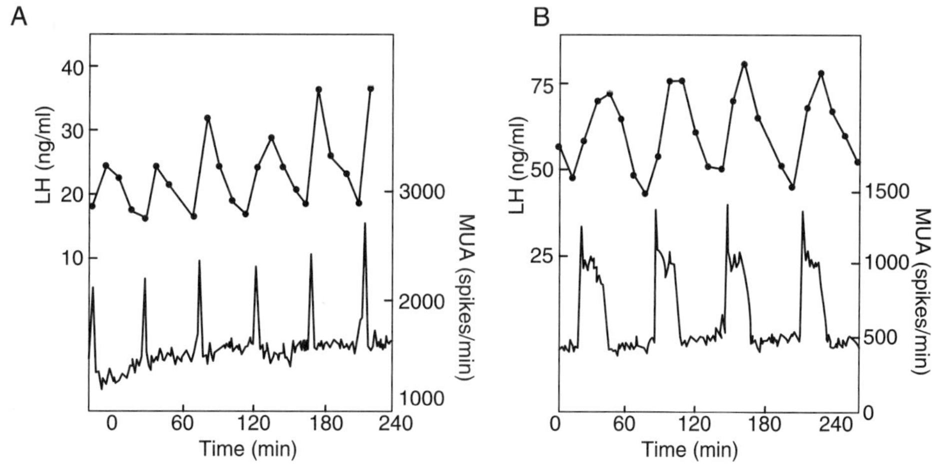

FIG. 14. MUA recorded from the mediobasal hypothalamus in a normal monkey on day 5 of the follicular phase of the menstrual cycle (A) and in an ovariectomized animal (B). Note the association of the MUA volleys with the initiation of LH pulses and the shorter duration of the volleys in the presence of the ovary. [Reproduced from (427) with permission, S. Karger AG, Basel.]

pulse generator activity. With the addition of radiotelemetry, MUA recording provides the means for continuous monitoring of the GnRH pulse generator in freely behaving animals (433).

In ovariectomized monkeys, the MUA signal consists of a rapid increase in spike frequency, followed by a plateau that lasts for approximately 20 minutes at a level somewhat lower than the initial peak, and then an abrupt return to baseline (442). The height of the peak in relation to the plateau that follows gives the former the appearance of an initial "overshoot."

In the presence of the ovary, the MUA signal is strikingly abbreviated in that its duration appears to be limited to the rapid "overshoot" phase and lasts only 2–3 minutes (Fig. 14) (427). The administration of estradiol to long-term ovariectomized monkeys reduces the duration of the MUA volleys to that of eugonadal animals within 3–5 hours (597). On the other hand, it takes 4–6 weeks to fully develop the converse response, namely, lengthening of the duration of MUA volleys after removal of the ovaries (427). This protracted time course, which may be associated with structural remodeling in the hypothalamus after ovariectomy [for discussion see (427)], is reminiscent of, but most likely unrelated to, the 3-week period required to fully develop the LH response to castration in this species (602). Using cluster analysis to extract single units from MUA signals, it was determined that the longer duration of MUA volleys in ovariectomized monkeys could not be attributed to sequential activation of different units but rather was due to longer bursts of individual cells (603). Despite the profound differences in the MUA signal between ovariectomized and estrogen-replaced animals, there appears to be no relation between MUA volley duration and the amplitude of associated LH pulses (604). This suggests that the full complement of GnRH that stimulates an LH pulse in the circulation is released at the very onset of the MUA volley (i.e., during the "overshoot" phase) without any further release of GnRH during the ensuing plateau. As previously suggested (604), perhaps the prolonged duration of increased hypothalamic electrical activity seen in the absence of the ovary is associated with the control of other systems, for example, the control of vasomotor activity in relation to hot flashes that coincide with LH pulses in postmenopausal women and that are readily reversed by estrogen administration (605–607).

Nature of the GnRH Pulse Generator. Whereas the GnRH pulse generator in primates is localized in the MBH, the cellular and molecular bases of pulse generating activity, including the mechanisms that ultimately determine the timing of secretory events at the terminals of GnRH neurons, remain unclear. One possible arrangement would involve nonsecretory components of the MBH that entrain the activities of appropriate GnRH neurons to orchestrate their synchronous discharge of neurosecretion. Another paradigm, and one that is gaining support from in vitro investigations (see below), assigns the pacemaking role to the GnRH neurons themselves. In either case, modulation by afferent inputs from other neurons is superimposed upon fundamental pulse generating activity to alter the frequency and amplitude of GnRH release.

The anatomical challenges for the operation of a GnRH pulse generator in primates are considerable. If this function resides in non-GnRH neurons, their somata and efferent projections to the GnRH neurons must exist within the boundaries of the MBH because pulse generator activity continues to function in monkeys with MBH–pituitary islands created with Halasz knife cuts (565). In such an arrangement where the pulse generator is external to the GnRH neurons, which part of the latter cell type receives the pulse generator signals? Because the cell bodies of GnRH neurons in the MBH are diffusely scattered (566–582) but their axons and terminals converge at the median eminence, one might reasonably suppose that these neurons would be coordinated more efficiently by a control that targets their terminals rather than their cell bodies or dendrites. Similar anatomical considerations pertain if the GnRH neurons themselves function as the pulse generator, because the diffuse arrangement of their cell bodies presents logistical challenges for their synchronization at the level of the soma.

As mentioned, the use of in vitro approaches has produced evidence in favor of the view that GnRH neurons themselves constitute the pulse generator. Although human MBH in vitro secretes GnRH in pulses (586), investigations using adult primate tissue are understandably rare, and studies on pulse-generating mechanisms have principally relied on the use of fetal cells and of tissue preparations from phylogenetically lower species. Experiments with the immortalized GnRH mouse cell line GT1 and its subclones (608–610) and with GnRH neurons obtained from embryonic monkey olfactory placode (611,612) have suggested that GnRH neurons may exist in a functional network whose properties derive from the coupled activities of the cells, many or all of which show spontaneous oscillations in membrane potentials, intracellular calcium concentrations, and perhaps other cellular behaviors. Whereas individual GnRH neurons have the capacity for episodic discharge and secretion, an important question is how are they synchronized to release the decapeptide into pituitary portal blood as discrete pulses? Synchronization among the neurons could be achieved by interneuronal contacts such as gap junctions, intercellular bridges, or synapses, although morphological studies in primates suggest that cytoplasmic bridges and synapses between GnRH neurons are rare occurrences (613,614). Alternatively, a diffusible substance such as nitric oxide might serve to synchronize hormone release by the GnRH neurons (610,612), and non-neuronal cells such as glia may also play a role (611,612,615). GnRH itself does not appear to act as an autocrine or ultrashort-loop feedback controller of its own secretion because intravenous or intracerebroventricular administration of a GnRH agonist to ovariectomized rhesus monkeys failed to alter the frequency of MUA volleys within the MBH (616,617).

Each of the in vitro models is substantially different from the mature genetically nontransformed condition of the intact adult GnRH pulse generator functioning in its native milieu with its integral connections and cellular associations. To fully understand the mechanisms that drive the rhythm of GnRH secretion therefore requires the collective insights provided by a variety of approaches, along with the development of mathematical models based on the performance of these systems (618,619). Pursuit of this problem is essential, given the obligatory role of the GnRH pulse generator in governing the reproductive system.

Central Modulation of GnRH Pulse Generator Activity. The spontaneous activity of the GnRH pulse generator is modulated by stimulatory and inhibitory afferent neural systems. Stimulatory neuromodulators include norepinephrine, glutamate, and neuropeptide Y (NPY); inhibitory neuromodulators include endogenous opioids and γ-aminobutyric acid (620). The neurosecretory peptide corticotropin-releasing hormone is also inhibitory to GnRH release (601,621). If inhibitory and stimulatory afferent neural systems simultaneously exert tones on the GnRH pulse generator, pulse generator activity might reflect the summation of their opposing actions, and pharmacological antagonism of one of the systems will tip the balance toward expression of the other, much like the behavior exhibited by the opposing autonomic innervations of the sinoatrial node in the heart.

Some neural systems may affect the GnRH pulse generator mechanism tonically, whereas others may exert their effects only after becoming activated under specific physiological conditions such as stress or changes in the hormonal milieu. Norepinephrine is possibly an example of the former because α-adrenergic blockers such as phentolamine, phenoxybenzamine, and prazosin reduce LH pulse frequency or completely abolish LH pulses in ovariectomized monkeys (514,520,596,622). In contrast, pulsatile LH release in the human does not appear to be affected by α-adrenergic blockers (404,415,623–625). It is unclear at present whether this discrepancy between the monkey and the human is a true difference or a product of the generally lower doses of antagonists used in human studies.

The endogenous opioids slow the frequency of the GnRH pulse generator, and the strength of opioidergic tone is related to the steroid milieu. Thus, administration of the opiate antagonist naloxone results in marked acceleration of pulsatile LH release in the luteal phase, mild acceleration in the late follicular phase, slight acceleration in the early follicular phase, and no effect after ovariectomy or in postmenopausal women (595,626–639). Responsiveness to naloxone is restored in postmenopausal women by pretreatment with estrogens and progestins (637,638,640). These findings suggest that opioidergic tone is minimal in the absence of circulating steroid hormones, an interpretation supported by the observation that β-endorphin levels in pituitary portal blood of rhesus monkeys are lowest near the beginning of the menstrual cycle or after ovariectomy and greatest during the luteal phase or in ovariectomized animals treated with estradiol and progesterone (641,642). Close contacts have been observed between GnRH neurons and opioidergic fibers in the human and monkey hypothalamus (571,572,643).

Although intraventricular infusion of NPY in the ovariectomized rhesus monkey inhibits LH release (644), NPY is apparently stimulatory to the GnRH pulse generator because its infusion directly into the stalk-median eminence stimulates GnRH release (645,646), and infusion of an antiserum against NPY into the same region suppresses pulsatile GnRH release measured in push–pull perfusates (525). Pulsatile release of NPY and GnRH also occurred in parallel, with NPY pulses leading GnRH pulses by approximately 5 minutes (525). NPY neurons are found in the hypothalamus (647–649) with fibers that show close associations with GnRH cell bodies and dendrites (576).

As an inhibitory modulator, γ-aminobutyric acid suppression of the GnRH pulse generator might be most significant during the prepubertal hiatus of gonadotropin secretion that characterizes the juvenile period of development in the primate (620,650). For the transition into adulthood, it has been suggested that stimulation by glutamate might accompany the withdrawal of γ-aminobutyric acid to facilitate the reactivation of the GnRH pulse generator that is responsible for the onset of puberty (620).

Ovarian Hormones

That ovarian hormones exert a suppressive effect on the gonadotropic secretions of the anterior pituitary gland was first articulated by Moore and Price (651) and is now understood as an example of negative feedback control, a central tenet of endocrine physiology. Whereas many of the classic studies concerning ovarian steroid hormone actions on gonadotropin secretion relied on mean plasma concentrations of LH and FSH as endpoints, it is now recognized that estradiol and progesterone exert feedback actions that can affect the frequency of the pulse generator and therefore the frequency of gonadotropin pulses. The additional information about feedback effects on pulsatile gonadotropin release is a useful lens through which to view and interpret data that are limited to mean hormone levels. Thus, progesterone dramatically reduces the frequency of the GnRH pulse generator during the luteal phase but has little effect on mean gonadotropin levels because the amplitude of the pulses is proportionately increased. In contrast, the feedback actions of estradiol on gonadotropin secretion throughout the follicular phase do not produce any major changes in pulse frequency (410,412,413,421,422,426,433).

Negative Feedback Actions of Ovarian Steroid Hormones. In normally cycling women and monkeys, estradiol restrains gonadotropin secretion (148,652,653) by diminishing the amplitude of gonadotropin pulses (654,655). In the rhesus monkey, opening the negative feedback loop by ovariectomy is followed within a day or two by detectable increases of plasma LH concentrations that continue rising to a maximum that exceeds follicular phase levels by an order of magnitude after approximately 3 weeks (602). Reclosing the negative feedback loop by administering estradiol at levels that approximate those of the follicular phase restores mean plasma gonadotropin concentrations to their previously low values (148, 149,652,654). The sensitivity of LH and FSH to negative feedback suppression by estradiol increases with time after ovariectomy (58,148).

As already noted, estradiol exerts little to no effect on pulse frequency during the follicular phase of the cycle. Furthermore, the frequency of the pulse generator in eugonadal monkeys during the follicular phase is comparable with that of ovariectomized animals (427,656). However, ovariectomized monkeys that receive exogenous estradiol to reestablish physiological (follicular phase) concentrations of the steroid show a slowing of the GnRH pulse generator (657) or even its complete arrest accompanied by a reduction in circulating LH to undetectable levels within a few hours (597). It was proposed that the presence of the ovary normally protects the hypothalamic pulse generator from steroid-induced suppression (597), although the identity of a putative ovarian factor that fulfills this role is presently unknown.

The observation that mean levels of circulating gonadotropins are roughly the same in the follicular phase, when progesterone is nearly absent, and in the luteal phase, when progesterone levels are maximally elevated, contributed to the early conclusion that estradiol and not progesterone is the principal ovarian negative feedback controller on gonadotropin secretion during the cycle (58). Nevertheless, progesterone is responsible for slowing the GnRH pulse generator during the luteal phase (Fig. 15) (414,422, 423,437,627,658), an action that might support the preferential secretion of FSH during the succeeding luteal-follicular transition (see section on Luteal-Follicular Transition). Although progesterone in the absence of estrogen has little effect on circulating

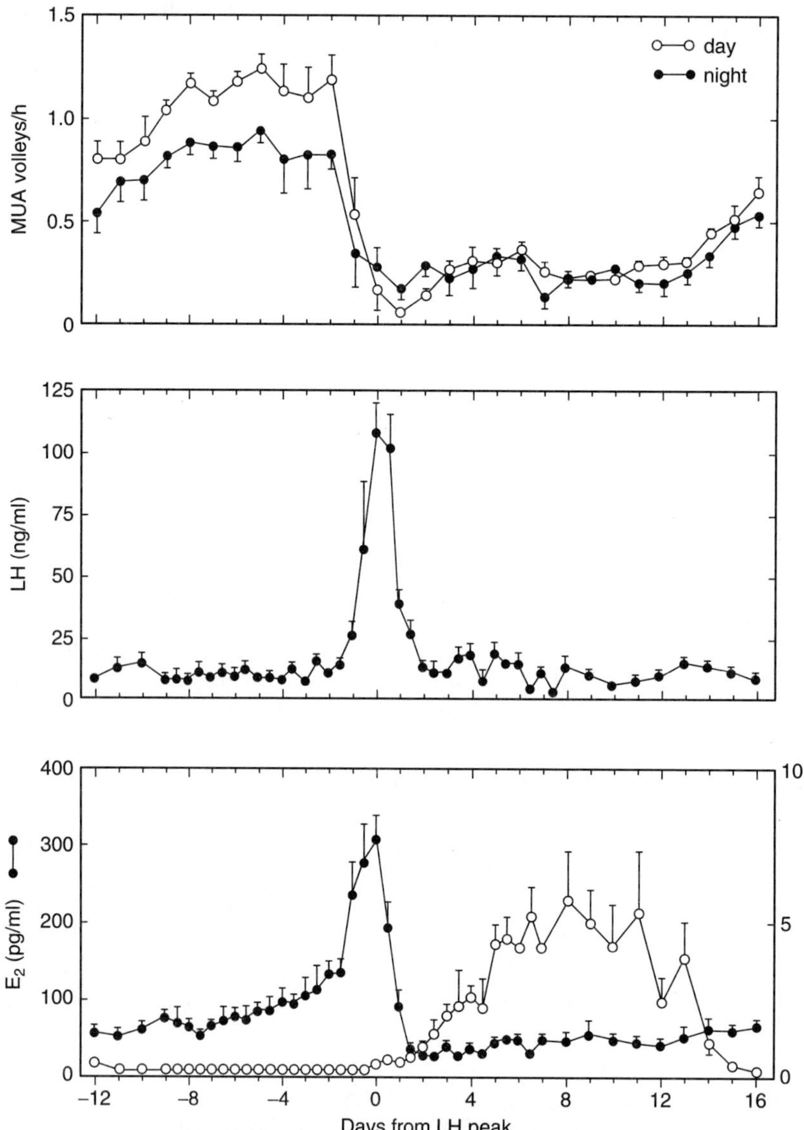

FIG. 15. Frequency of GnRH pulse generator activity (MUA volleys per hour) monitored by radiotelemetry during the menstrual cycle of unrestrained rhesus monkeys, with accompanying time courses of LH, estradiol, and progesterone. Data are normalized to the day of the mid-cycle LH peak (day 0). Each point represents the mean ± standard error of the mean (n = 5–12 observations). Note the slower frequency of the GnRH pulse generator at night during the follicular phase, the abrupt reduction in pulse generator activity with the onset of the LH surge, and the slower frequency of the pulse generator during the luteal phase in comparison with its activity during the follicular phase. [Reproduced from (433), copyright 1991, The Endocrine Society.]

levels of LH and FSH (654,659), progesterone can synergize with estradiol in the ovariectomized monkey to profoundly suppress the gonadotropins to undetectable levels (148), once again suggesting that the presence of the ovary normally protects the pulse generator from excessive negative feedback inhibition by steroid hormones.

Positive Feedback Actions of Ovarian Steroid Hormones. The initiation of the preovulatory surge of gonadotropins is the consequence of the stimulatory, or positive feedback, effect of estradiol. Circulating levels of this steroid in excess of 200 pg/ml, such as those which normally occur in the late follicular phase, when sustained for approximately 2 days (660,661), initiate gonadotropin surges in monkeys (652,660,662) and women (655,658,663–666). Whereas estradiol alone is sufficient to elicit fully developed surges of LH and FSH in the rhesus monkey (58,431,652,667–669), progesterone is additionally required for their full expression in the human. Thus, when estrogen alone is administered to postmenopausal women or to women during the follicular phase, the resulting surge of LH appears shortened and blunted (670) and that of FSH is minimal or nonexistent (671,672). When progesterone is administered 36–48 hours after estrogen administration, however, the resulting LH and FSH surges appear to be similar to normal preovulatory discharges (670,671). The small amount of progesterone produced before ovulation therefore seems to be necessary for the full development of the gonadotropin surge in the human. Although the precise timing for the initiation of the preovulatory progestin rise relative to the onset of the gonadotropin surge is controversial (202,363,365,366,372,373,384, 385,658,671,673–678), the initiating stimulus for the surge is considered to be the rising levels of estradiol produced by the maturing preovulatory follicle, with progesterone playing a facilitatory role in the human.

In addition to facilitating estrogen-induced surges, progesterone has the ability to induce LH discharges within several hours of its administration to estrogen-primed women (489,655,658,659,679–684) and monkeys (685–688). However, these discharges do not resemble normal preovulatory gonadotropin surges in that they have a short duration (usually less than 24 hours) and can be blocked by pentobarbital anesthesia (688). This differs sharply from the estrogen-induced release of LH, which is unaffected by pentobarbital (58). Furthermore, these acute releases triggered by progesterone demonstrate increased frequency and amplitude of LH pulses (680,686) and GnRH pulses, the latter associated with similar changes in pulsatile NPY release (527). Although this seemingly pharmacological effect of progesterone may be related to its action in facilitating the normal preovulatory surge, the functional relation between these mechanisms is unclear.

Surge-blocking Action of Progesterone. Whereas administration of exogenous estradiol during the follicular phase readily elicits a gonadotropin surge, attempts to induce surges during the luteal phase are unsuccessful due to a blocking action exerted by elevated levels of progesterone that exist at this time in the cycle (689). Blockade of surges has been demonstrated experimentally in eugonadal women (671,682) and monkeys (667,689–692) after administration of adequate doses of progesterone or related progestins, such as 17α-hydroxyprogesterone or 20α-dihydroxyprogesterone, at the same time as the estradiol stimulus. However, the blocking response requires the presence of the ovary, as progestin administration to ovariectomized monkeys produces only facilitating effects and not blockade of surges, even after chronic pretreatment with estradiol (667).

Sites of Action of Estradiol. Studies in the rhesus monkey have demonstrated intact positive and negative feedback actions of estradiol after the destruction of neural tissue in a variety of locations that spare the MBH, including (a) lesions in the rostral hypothalamus (222,693), (b) complete removal of the preoptic–anterior hypothalamus by aspiration (694), and (c) complete neural disconnection of the MBH from the remainder of the brain (565,695). The feedback actions of estradiol must therefore occur within the MBH-pituitary unit, a conclusion that prompted studies directed toward differentiating between the MBH and the pituitary as the locus of action. Early findings that estrogen administration acutely inhibits gonadotropin release in response to exogenously administered GnRH (386,696–698) suggested that the steroid's negative feedback effects are exerted at the pituitary. A further suggestion that the positive feedback site for estrogen action is not neural was provided by the observations that pentobarbital, phenobarbital, and morphine, all of these constituting agents that block the preovulatory surge in rats (699), cannot abolish estradiol-induced LH surges in the rhesus monkey (58). As already mentioned, estradiol does not appear to alter the frequency of the pulse generator under physiological conditions. Although microinjections of estradiol into the hypothalamus or third ventricle of rhesus monkeys were reported to decrease circulating LH levels (700–702) whereas estradiol implanted directly into the anterior pituitary produced only small effects (701,702), these results likely reflect Bogdanove's (703) implantation paradox whereby hypothalamic infusion permits efficient distribution of the steroid by the pituitary portal circulation to the entire anterior lobe.

To further distinguish between the MBH and pituitary as sites of action of estradiol, studies using the so-called hypophysiotropic clamp preparation were conducted. This consists of rhesus monkeys that effectively lack hypothalamic GnRH due to radiofrequency lesions in the arcuate region of the MBH that are administered chronic pulsatile infusions of GnRH at an unvarying frequency of one pulse per hour to reestablish pituitary gonadotropin secretion. In such monkeys that have also been ovariectomized, estradiol administration causes a prompt fall in circulating levels of LH and FSH (222,232,622), reflecting the steroid's negative feedback action, followed approximately 36 hours later by gonadotropin surges indistinguishable from those seen in nonlesioned control monkeys (222,232). In such animals with intact ovaries, not only are the biphasic actions of estradiol clearly evident, but also the monkeys exhibited normal ovulatory menstrual cycles (222,704). One monkey that received this unvarying pulsatile GnRH replacement regimen displayed 28-day ovulatory menstrual cycles for 9 months [see unpublished observations cited in (222)]. Similar results were obtained in monkeys administered pulsatile GnRH replacement after surgical transection of the pituitary stalk with insertion of a barrier to prevent reconnection of the pituitary portal vessels (222,705–707). Because women who lack endogenous GnRH also respond to unvarying pulsatile delivery of exogenous GnRH by initiating ovulatory cycles (268,269,552,707–712), the approach has significant clinical application for the treatment of infertility.

The hypophysiotropic clamp studies in the rhesus monkey led to the conclusion that estradiol exerts its negative and positive feedback actions on gonadotropin secretion at the pituitary level and, more specifically, on the gonadotrope itself. Because the pulsatile GnRH replacement regimen was constant throughout the experimental phase of these studies, it is evident that the role of GnRH in the context of the menstrual cycle, especially with respect to the initiation of the mid-cycle surge, is essentially permissive and not controlling.

Furthermore, although the positive feedback action of estradiol requires a suitably primed gonadotrope, that is, one that has had a recent history of exposure to GnRH, this releasing hormone support does not need to be present at the time the estrogen-induced surge is actually initiated, because gonadotropin surges can be elicited by estradiol administered up to 48 hours after the discontinuation of pulsatile GnRH replacement in MBH-lesioned monkeys (713). Indeed, in monkeys with an intact hypothalamus, the GnRH pulse generator is completely arrested during spontaneous or estrogen-induced gonadotropin surges (433,597). The conclusion that pulsatile GnRH stimulation does not need to be ongoing to initiate and sustain a gonadotropin surge is further supported by the findings of intact surges after estrogen administration when endogenous GnRH delivery or action has been compromised by (a) acute pituitary stalk section with barrier placement (714), (b) administration of α-adrenergic blocking agents (58), (c) neutralization of GnRH with an antiserum (466), and (d) receptor blockade with GnRH antagonists (479,715). In apparent contradiction with the latter finding, however, some have observed reduction, delay, or complete blockade of gonadotropin surges in the monkey after the administration of GnRH antagonists (716,717). Although these discrepancies are difficult to reconcile, the preponderance of evidence suggests that the gonadotropes in the monkey do not require acute GnRH stimulation during an estrogen-induced gonadotropin surge. Hence, the known estrogen-induced up-regulation of GnRH receptors and augmentation of pituitary responsiveness to GnRH that occur at this time (718) do not appear to be essential for successful production of the LH surge in the monkey.

In contrast, the gonadotropes in the human may have a greater dependency on continuing GnRH support during the surge because the mid-cycle surge in normally cycling women is consistently delayed after the administration of GnRH antagonists in the late follicular phase (481,482,487,719,720) and is interrupted if the antagonist is given during an ongoing surge (721). Estrogen augmentation of pituitary responsiveness to GnRH (661,722) may also play a role during the mid-cycle gonadotropin surge in the human.

Despite the indications cited above from studies in the rhesus monkey that GnRH may not be required to acutely drive the gonadotropin surge, the possibility remains that a surge of GnRH is released into the pituitary portal circulation at mid-cycle, possibly serving as a "fail-safe" mechanism to ensure the preovulatory discharge of gonadotropins. Estrogen-induced GnRH increases have been reported for the monkey in (a) pituitary portal blood of anesthetized animals with transected pituitary stalks (723,724), (b) push–pull perfusates of MBH (432,725), (c) cerebrospinal fluid collected from the third ventricle (517), and (d) culture media supporting hypothalamic explants (726). However, the GnRH surge in the monkey most likely does not involve synchronized release of the hormone from neurons of the GnRH pulse generator because activity of the latter is dramatically quiescent during the surge as monitored by MUA recording in the MBH (Fig. 15) (433,597). Although similar experimental approaches have not been taken in the human, a study using submaximal doses of a GnRH antagonist at different times of the cycle in normal women has suggested that the overall amount of GnRH released during the mid-cycle surge is reduced (727).

Sites of Action of Progesterone. Because progesterone slows the frequency of the GnRH pulse generator (414,422,423,437,627,658,728), this hormone's action must be exerted within the central nervous system. Its action, as well as the less robust decelerating effect of estradiol in ovariectomized subjects (597,657,729,730), appears to be mediated at least partly by endogenous opioids because decreases in pulse generator frequency triggered by these steroids can be reversed by opiate antagonists (436,626,627,629–631,731,736). As previously mentioned, the strength of opioid-induced slowing of the pulse generator is related to the steroid milieu (see section on Central Modulation of GnRH Pulse Generator Activity).

The facilitating action of progesterone on estrogen-induced gonadotropin surges may be exerted at the pituitary, as surges induced in MBH-lesioned monkeys receiving pulses of exogenous GnRH and bearing progesterone implants are advanced in time (737). In contrast, progesterone's ability to pharmacologically trigger brief episodes of gonadotropin release appears to be exerted centrally because the stimulated LH release occurs with increased frequency and amplitude of hormone pulses (680,686) and can be blocked by pentobarbital (688) and a GnRH antagonist (489).

The blockade of estrogen-induced gonadotropin surges exerted by progesterone during the luteal phase or experimentally is exerted in the central nervous system, because hypothalamic lesions abolish this blocking action (737). In contrast, progesterone blockade is preserved in monkeys with intact nervous systems receiving pulsatile GnRH administration (738). These results indicate that progesterone blockade of estrogen-induced gonadotropin surges is not the consequence of an interruption in GnRH release—a circumstance incapable of abolishing surges in any case (see previous section)—and suggest that progesterone may cause the production of an inhibitory factor from the central nervous system that acts at the level of the pituitary, thereby inhibiting the positive feedback action of estradiol. The inhibitor does not appear to be an endogenous opioid because an opiate antagonist was unable to reverse progesterone blockade of surges (739).

Negative Feedback Actions of Ovarian Inhibin. Although the inhibins were named for their selective suppression of FSH secretion, the negative feedback roles of inhibin A and inhibin B in a physiological context remain unclear in the female primate (740). Most interest lies in the possibilities that (a) withdrawal of circulating inhibin A might contribute to the perimenstrual rise in FSH associated with the luteal–follicular transition, and (b) increasing plasma concentrations of inhibin B during the mid-follicular phase might contribute to the falling levels of FSH that occur at this time. However, little direct support currently exists for either one of these feedback roles.

Administration of inhibin A to monkeys during the early follicular phase fails to suppress FSH concentrations (741) unless given in exceptionally high doses (558). Although infusions of inhibin A in monkeys during the mid-to-late luteal phases can suppress FSH levels without altering LH concentrations (249), similarly timed administration of antisera against either recombinant human inhibin A (741) or human inhibin α subunit (742) failed to elicit the expected rise in FSH concentrations.

The patterns of circulating concentrations of FSH and inhibin B during the follicular phase, in which FSH leads inhibin B by 2–4 days, suggest that the former drives the latter. Experimental support for this relation is provided by the findings that administration of a GnRH antagonist to monkeys during the late luteal phase suppresses inhibin B levels and that the suppression is readily reversed by FSH injections (359). In the mid-follicular phase when FSH levels were waning, administration of a pure antiestrogen to block the negative feedback action of estradiol on FSH secretion was followed by rises in the concentrations of both FSH and inhibin B (359). That the increased levels of inhibin B in this experiment might have exerted an inhibitory effect on FSH was suggested by the observation that FSH levels began to decline even while the antiestrogen treatment continued, whereas the rise in inhibin B levels persisted for 2 more days, but this interpretation is problematic because LH levels declined contemporaneously with FSH (359), thereby calling into question the specificity of the response. Thus, a negative feedback role for inhibin B during the normal ovarian cycle remains uncertain. However, an indication that such a role may exist is emerging from studies in normal older (premenopausal) women that show reduced circulating levels of inhibin B associated with elevated concentrations of FSH despite normal or slightly elevated concentrations of estradiol (743,744).

Frequency Changes of the GnRH Pulse Generator during the Menstrual Cycle

The importance of frequency of the GnRH pulse generator cannot be overstated for the production of normal ovulatory menstrual cycles. In particular, a modest slowing of the GnRH pulse generator has profound consequences for follicular development and ovulation, as determined in studies using hypophysiotropically clamped rhesus monkeys with intact ovaries. Whereas the administration of GnRH pulses at the follicular phase frequency of once every hour sustains normal ovulatory cycles, slowing the frequency of GnRH pulses to once every 2 or 3 hours disrupts or stops normal follicular development and prevents ovulation despite the maintenance of plasma FSH concentrations at normal levels or higher (555). A lesser reduction in frequency, achieved by administering GnRH pulses once every 90 minutes, reduces ovulation rate in the monkeys by 50% (555). Similar observations have been made in women with hypothalamic amenorrhea treated with pulsatile GnRH (552).

Although ovulatory menstrual cycles can be induced by a fixed frequency of exogenous GnRH pulses, the fact remains that pulse frequency does change during the natural menstrual cycle. However, as discussed below, the physiological significance of these changes in GnRH pulse frequency remains to be established.

Luteal Phase

Spontaneous menstrual cycles in monkeys and women include a period of marked reduction in the frequency of the GnRH pulse generator during the luteal phase. It has been reported in the rhesus monkey that exogenous administration of GnRH at a fast frequency typical of the follicular phase but imposed during the normal luteal phase leads to disturbances in subsequent cycles that are characterized by prolonged follicular phases and reduced progesterone levels during the luteal phases (745). On the basis of these observations, it was proposed that the normally reduced frequency of the pulse generator during the luteal phase of the menstrual cycle plays an important role for subsequent ovarian cyclicity. However, a similar study in normal women showed no such disruptive effects (746), and, as previously noted, studies using the hypophysiotropic clamp preparation in GnRH-deficient monkeys and women clearly demonstrate that regular menstrual cycles can be supported by a regimen of exogenous GnRH administration that is unvarying in frequency and amplitude (see section on Sites of Action of Estradiol).

It has been proposed that the slowing of the GnRH pulse generator may support differential changes in gonadotropin synthesis, storage, and secretion by the anterior pituitary. Pituitary contents of both gonadotropins rise during the follicular phase, plummet at mid-cycle as a result of the surge, and rise again during the luteal phase (747). However, the increase in gonadotropin content during the follicular phase is proportionately greater for LH, whereas the rise during the luteal phase is proportionately greater for FSH (747). These differential changes in pituitary content of LH and FSH might be the result of differences in the frequency of GnRH stimulation between the two phases of the cycle. As proposed by Marshall and colleagues (748,749), faster frequencies of GnRH appear to support preferential synthesis and release of LH, whereas slower frequencies favor FSH. Thus, during the luteal phase FSH might be synthesized at a faster rate than LH in the pituitary, where it accumulates due to the negative feedback action on release imposed by estradiol and possibly by inhibin A. As discussed in the section on FSH and LH Receptors, the maintenance of LH secretion in the presence of declining FSH secretion during the mid through late follicular phase of the menstrual cycle is likely key to the selection of the preovulatory follicle.

Luteal–Follicular Transition

Upon the regression of the corpus luteum, the frequency of the GnRH pulse generator increases due to the loss of the negative feedback action of progesterone (400,750) and circulating levels of both gonadotropins rise, with FSH concentrations rising to a greater degree than LH (202,391,392,397,398,400). In normally cycling monkeys and women, the differential release of the gonadotropins at the luteal–follicular transition might be the result of their differential accumulation in the gonadotropes during the preceding luteal phase (see previous section). However, it must be emphasized that the preferential rise in circulating FSH at this time of the cycle is clearly evident during ovarian cycles driven by an unvarying frequency of pulsatile GnRH replacement in monkeys with MBH lesions and in women with hypothalamic amenorrhea (268,269,452,704,708, 709,746). Thus, the explanation for the preferential rise in FSH levels during the luteal–follicular transition remains unclear.

Summary of the Control System and Timing of the Menstrual Cycle

The control of the menstrual cycle rests with the functional interactions among its three major components: the GnRH pulse generator, the pituitary gonadotropes, and the ovary. The GnRH pulse generator in the MBH is the central neural component of the control system. During the follicular phase of the cycle, this mechanism releases a bolus of GnRH into the pituitary portal circulation approximately once every hour, thereby providing the essential pulsatile stimulus to the gonadotropes. The latter respond to these pulsatile stimuli by releasing pulses of LH and FSH, which stimulate follicular development and the secretion of ovarian hormones. Estradiol levels rise in the circulation, and this hormone attenuates the amplitude of gonadotropin pulses through its negative feedback action on the gonadotropes without affecting pulse frequency. The induction of follicular maturation and the achievement of maximal circulating levels of estradiol require about 14 days in the human and the rhesus monkey.

At the end of the follicular phase, when estradiol concentrations have exceeded a threshold of 200 pg/ml for approximately 2 days, the negative feedback action of this steroid on the gonadotropes converts to a positive feedback effect, causing them to release the mid-cycle surge of LH and FSH. The small rise in progesterone levels that accompanies the onset of this event is required to facilitate the estrogen stimulus for the full development of the surge in the human. Although an increased release of GnRH into the pituitary portal vessels might occur at mid-cycle, it does not appear to be necessary for generating the gonadotropin surge. Pulsatile GnRH release continues throughout the mid-cycle surge in the human, whereas the GnRH pulse generator is arrested at this time in the monkey. Furthermore, the gonadotropin surge in the human is more dependent on ongoing GnRH secretion than that in the monkey. Estrogen augmentation of gonadotrope responsiveness to GnRH may also play a role in sustaining the gonadotropin surge in the human.

At 20–44 hours after its peak, the mid-cycle gonadotropin surge results in the rupture of the Graafian follicle, ovulation, and formation of the corpus luteum, which has a functional requirement for LH. Rising levels of progesterone, accompanied by estradiol, produce a slowing of pulsatile GnRH release, an effect that is mediated by opioidergic neurons that modulate the pulse generator. As a result, the release of gonadotropin pulses occurs with lower frequency but greater amplitude, such that mean levels of these hormones are similar to those during the follicular phase. If conception does not occur, the corpus luteum regresses approximately 14 days after its formation, estradiol and progesterone levels fall, and the inhibitory opioidergic tone on the GnRH pulse generator is removed, allowing it to resume its fast frequency mode of operation characteristic of the follicular phase.

Thus, the characteristic 28-day timing of the menstrual cycle is controlled by the ovary and represents the sum of the durations of follicular development and the functional life span of the corpus luteum. The system, however, has an absolute requirement for the activity of the GnRH pulse generator, which functions as a permissive mechanism that allows the functional interactions between the pituitary and the ovary to unfold.

ACKNOWLEDGMENTS

We acknowledge the late Ernst Knobil, Ph.D., for his mentorship in the development of our careers as researchers and teachers of physiology. Work from the Zeleznik laboratory described in this chapter was supported by the National Institutes of Health (HD 08610, HD12014, and HD16842).

REFERENCES

1. Corner, G. W. Sr. (1974). The early history of progesterone. *Gynecol. Invest.* 5, 106–112.
2. Short, R. V. (1979). The discovery of the ovaries. In *The Ovary* (S. Zuckerman and B. J. Weir, Eds.), pp. 1–39. Academic Press, London.
3. Corner G. W. (1943). *The Hormones in Human Reproduction*. Princeton University Press, Princeton.
4. Allen, E. (1922). The estrous cycle of the mouse. *Am. J. Anat.* 30, 297–371.
5. Allen, E., and Doisy, E. A. (1923). An ovarian hormone, preliminary report on its localization, extraction and partial purification and action in test animals. *J. Am. Med. Assoc.* 81, 819–821.
6. Allen, E., Francis, B. F., Robertson, L. L., Colgate, C. E., Johnston, C. G., Doisy, E. A., Kountz, W. B., and Gibson, H. (1924). The hormone of the ovarian follicle, its localization and action in test animals and additional points bearing upon the internal secretion of the ovary. *Am. J. Anat.* 34, 133–181.
7. Doisy, E. A., Veler, C. D., and Thayer, S. (1929). Folliculin from urine of pregnant women. *Am. J. Physiol.* 90, 329–330.
8. MacCorquodale, D. W., Thayer, S. A., and Doisy, E. A. (1936). The isolation of the principal estrogenic substance of liquor folliculi. *J. Biol. Chem.* 115, 435–448.
9. Fraenkel, L. (1903). Die funktion des corpus luteum. *Arch. Gynäk.* 67, 438–545.
10. Fraenkel, L. (1910). Neue experimente zur funktion des corpus luteum. *Arch. Gynäk.* 91, 705–761.
11. Bouin, P., and Ancel, P. (1910). Sur le déterminisme de la préparation de l'utérus à la fixation de l'oeuf. *J. Physiol. Path. Gen.* 12, 1–16.
12. Corner G. W., and Allen, M. W. (1929). Production of a special uterine reaction (progestational proliferation) by extracts of the corpus luteum. *Am. J. Physiol.* 88, 326–339.
13. Allen, M. W., and Corner G. W. (1930). Maintenance of pregnancy in rabbit after very early castration, by corpus luteum extracts. *Proc. Soc. Exp. Biol. Med.* 27, 403–405.
14. Fevold H. L., and Hisaw, F. L. (1930). Purification of a hormone of corpus luteum responsible for progestational development and other reactions. *Proc. Soc. Exp. Biol. Med.* 27, 606–608.
15. Wintersteiner, K., and Allen, W. M. (1934). Crystalline progesterone. *J. Biol. Chem.* 107, 321–336.
16. Greep, R. O. (1974). History of research on anterior hypophyseal hormones. In *Handbook of Physiology. Section 7. Endocrinology; Vol. 4, Part 2, The Pituitary Gland* (E. Knobil and W. H. Sawyer, Eds.), pp. 1–27. American Physiological Society, Washington, DC.
17. Smith, P. E. (1930). Hypophysectomy and replacement therapy in the rat. *Am. J. Anat.* 45, 205–274.
18. Ascheim, S., and Zondek, B. (1927). Hypophysenvorderlappenhormon und ovarialhormone im harn von schwangeren. *Klin. Wochschr.* 6, 1322.
19. Zondek, B. (1930). Über die hormone des hypophysenvorderlappens wachstumshormon, follikelreifungshormon (Prolan A, luteinisierungshormon) Prolan B, Stoffwechselhormon. *Klin. Wochschr.* 9, 393–396.
20. Greep, R. O., Van Dyke, H. B., and Chow, B. F. (1942). Gonadotropins of the swine pituitary: various biological effects of purified thylakentrin (FSH) and pure metakentrin (ICSH). *Endocrinology* 30, 635–649.
21. Allen, E. (1928). Precocious sexual development from anterior hypophysis implants in a monkey. *Anat. Rec.* 39, 315–323.
22. Hartman, C. G. (1930). Anterior lobe of the pig and the monkey ovary. *Proc. Soc. Exp. Biol. Med.* 27, 338–340.
23. Hisaw, F. L., Fevold, H. L., and Leonard, S. L. (1931). Effects of hypophyseal extracts on sexually immature monkeys. *Proc. Soc. Expl. Biol. Med.* 29, 204–206.
24. Engel, E. T. (1933). Biological differences in the response of the female macacus monkey to extracts of the anterior pituitary and of human pregnancy urine. *Am. J. Physiol.* 106, 145–155.
25. Hartman, C. G. (1942). Further attempts to cause ovulation by means of gonadotropes in the adult rhesus monkey. *Contrib. Embryol. Carnegie Inst. Wash.* 30, 113–126.
26. van Wagenen, G. (1968). Induction of ovulation in *Macaca mulatta*. *Fertil. Steril.* 19, 15–29.

27. Knobil, E., Kostyo, J. L., and Greep, R. O. (1959). Production of ovulation in the hypophysectomized rhesus monkey. *Endocrinology* 65, 487–493.
28. Gemzell, C. (1965). Induction of ovulation with human gonadotropins. *Rec. Prog. Horm. Res.* 21, 179–204.
29. Vande Wiele, R. L., Bogumil, J., Dyrenfurth, I., Ferin, M., Jewelewicz, R., Warren, M., Rizkallah, T., and Mikhail, G. (1970). Mechanisms regulating the menstrual cycle in women. *Rec. Prog. Horm. Res.* 26, 63–103.
30. Donovan, B. T. (1970). *Mammalian Neuroendocrinology.* McGraw-Hill, London.
31. Marshall, F. H. A. (1936). Sexual periodicity and the causes which determine it. *Phil. Trans. R. Soc. B* 226, 423–456.
32. Markee, J. E., Sawyer, C. H., and Hollinshead, W. H. (1946). Activation of the anterior hypophysis by electrical stimulation in the rabbit. *Endocrinology* 38, 345–357.
33. Harris, G. W. (1948). Electrical stimulation of the hypothalamus and the mechanism of neural control of the adenohypophysis. *J. Physiol. (Lond).* 107, 418–429.
34. Green, J. D., and Harris, G. W. (1947). The neurovascular link between the neurohypophysis and adenohypophysis. *J. Endocrinol.* 5, 136–146.
35. Sawyer, C. H., Markee, J. E., and Hollinshead, W. H. (1947). Inhibition of ovulation in the rabbit by the adrenergic-blocking agent dibenamine. *Endocrinology* 41, 395–402.
36. Sawyer, C. H., Everett, J. W., and Markee, J. E. (1949). A neural factor in the mechanism by which estrogen induces the release of luteinizing hormone in the rat. *Endocrinology* 44, 218–233.
37. Sawyer, C. H., Markee, J. E., and Townsend, B. F. (1949). Cholinergic and adrenergic components in the neurohumoral control of the release of LH in the rabbit. *Endocrinology* 44, 18–37.
38. Markee, J. E., Sawyer, C. H., and Hollinshead, W. H. (1948). Adrenergic control of the release of luteinizing hormone from the hypophysis of the rabbit. *Rec. Prog. Horm. Res.* 2, 117–131.
39. Harris, G. W. (1955). *Neural Control of the Pituitary Gland.* Edward Arnold, London.
40. McCann, S. M., Taleisnik, S., and Friedman, H. M. (1960). LH-releasing activity in hypothalamic extracts. *Proc. Soc. Exp. Biol. Med.* 104, 432–434.
41. McCann, S. M. (1962). A hypothalamic luteinizing-hormone-releasing factor. *Am. J. Physiol.* 202, 395–400.
42. Baba, Y., Matsuo, H., and Schally, A. V. (1971). Structure of the porcine LH- and FSH-releasing hormone. II. Confirmation of the proposed structure by conventional sequential analyses. *Biochem. Biophys. Res. Commun.* 44, 459–463.
43. Matsuo, H., Baba, Y., Nair, R. M. G., Arimura, A., and Schally, A. V. (1971). Structure of the porcine LH- and FSH-releasing hormone. I. The proposed amino acid sequence. *Biochem. Biophys. Res. Commun.* 43, 1334–1339.
44. Matsuo, H., Arimura, A., Nair, R. M. G., and Schally, A. V. (1971). Synthesis of the porcine LH- and FSH-releasing hormone by the solid-phase method. *Biochem. Biophys. Res. Commun.* 45, 822–827.
45. Monahan, M., Rivier, J., Burgus, R., Amoss, M., Blackwell, R., Vale, W., and Guillemin R. (1971). Synthèse totale par phase solide d'un décapeptide qui stimule la sécrétion des gonadotropines hypophysaires LH et FSH. *Comp. Rend. (Paris)* 273, 508–510.
46. Burgus, R., Butcher, M., Amoss, M., Ling, N., Monahan, M., Rivier, J., Fellows, R., Blackwell, R., Vale, W., and Guillemin, R. (1972). Primary structure of the ovine hypothalamic luteinizing hormone-releasing factor (LRF). *Proc. Natl. Acad. Sci. U S A* 69, 278–282.
47. Palkovits, M., Arimura, A., Brownstein, M., Schally, A. V., and Saavedra, J. M. (1974). Luteinizing hormone-releasing hormone (LH-RH) content of the hypothalamic nuclei in rat. *Endocrinology* 95, 554–558.
48. King, J. C., Williams, T. H., and Arimura, A. A. (1975). Localization of luteinizing hormone-releasing hormone in rat hypothalamus using radioimmunoassay. *J. Anat.* 120, 275–288.
49. Wheaton, J. E., Krulich, L., and McCann, S. M. (1975). Localization of luteinizing hormone-releasing hormone in the preoptic area and hypothalamus of the rat using radioimmunoassay. *Endocrinology* 97, 30–38.
50. Barry, J., Dubois, M. P., and Poulain, P. (1973). LRF producing cells of the mammalian hypothalamus. A fluorescent antibody study. *Z. Zellforsch.* 146, 351–366.
51. King, J. C., Parsons, J. A., Erlandsen, S. L., and Williams, T. H. (1974). Luteinizing hormone-releasing hormone (LH-RH) pathway of the rat hypothalamus revealed by the unlabeled antibody peroxidase-antiperoxidase method. *Cell Tissue Res.* 153, 211–217.
52. Zimmerman, E. A., Hsu, K. C., Ferin, M., and Kozlowski, G. P. (1974). Localization of gonadotropin-releasing hormone (Gn-RH) in the hypothalamus of the mouse by immunoperoxidase technique. *Endocrinology* 95, 1–8.
53. Baker, B. L., Dermody, W. C., and Reel, J. R. (1975). Distribution of gonadotropin-releasing hormone in the rat brain as observed with immunocytochemistry. *Endocrinology* 97, 125–135.
54. Sétáló, G., Vigh, S., Schally, A. V., Arimura, A., and Flerkó, B. (1975). LH-RH-containing neural elements in the rat hypothalamus. *Endocrinology* 67, 511–520.
55. Hoffman, G. E., Knigge, K. M., Moynihan, J. A., Melnyk, V., and Arimura, A. (1978). Neuronal fields containing luteinizing hormone releasing hormone (LHRH) in mouse brain. *Neuroscience* 3, 219–231.
56. Barraclough, C. A. (1966). Modifications in the CNS regulation of reproduction after exposure of prepubertal rats to steroid hormones. *Rec. Prog. Horm. Res.* 22, 503–529.
57. Halász, B. (1969). The endocrine effects of isolation of the hypothalamus from the rest of the brain. In *Frontiers in Neuroendocrinology, 1969* (W. F. Ganong and L. Martini, Eds.), pp. 307–342. Oxford University Press, New York.
58. Knobil, E. (1974). On the control of gonadotropin secretion in the rhesus monkey. *Rec. Prog. Horm. Res.* 30, 1–46.
59. Baird, D. T., and Fraser, I. S. (1975). Concentration of oestrone and oestradiol in follicular fluid and ovarian venous blood of women. *Clin. Endocrinol. (Oxf).* 4, 259–266.
60. Goodman, A. L., and Hodgen, G. D. (1983). The ovarian triad of the primate menstrual cycle. *Rec. Prog. Horm. Res.* 39, 1–73.
61. Gougeon, A. (1986). Dynamics of follicular growth in the human: a model from preliminary results. *Hum. Reprod.* 1, 81–87.
62. Peters, H., Byskov, A. G., Himelstein-Braw, R., and Faber, M. (1975). Follicular growth: the basic event in the mouse and human ovary. *J. Reprod. Fertil.* 45, 559–566.
63. van Wagenen, G., and Simpson, M. E. (1973). *Postnatal Development of the Ovary in* Homo sapiens *and* Macaca mulatta *and Experimental Induction of Ovulation in the Macaque*. Yale University Press, New Haven, CT.
64. Goldenberg, R. L., Powell, R. D., Rosen, S. W., and Ross, G. T. (1976). Ovarian morphology in women with anosmia and hypogonadotropic hypogonadism. *Am. J. Obstet. Gynecol.* 126, 91–94.
65. Block, E. (1951). Quantitative morphological investigations of the follicular system in women: variations in the different phases of the sexual cycle. *Acta Endocrinol. (Copenh).* 8, 33–54.

66. Koering, M. J. (1969). Cyclic changes in ovarian morphology during the menstrual cycle in *Macaca mulatta*. *Am. J. Anat.* 126, 73–101.
67. Zeleznik, A. J., Wildt, L., and Schuler, H. M. (1980). Characterization of ovarian folliculogenesis during the luteal phase of the menstrual cycle in rhesus monkeys using (3H) thymidine autoradiography. *Endocrinology* 107, 982–988.
68. Hillier, S. G., Reichert, L. E. Jr., and van Hall, E. V. (1981). Control of preovulatory follicular estrogen biosynthesis in the human ovary. *J. Clin. Endocrinol. Metab.* 52, 847–856.
69. Zeleznik, A. J., Schuler, H. M., and Reichert, L. E. Jr. (1981). Gonadotropin-binding sites in the rhesus monkey ovary: role of the vasculature in the selective distribution of human chorionic gonadotropin to the preovulatory follicle. *Endocrinology* 109, 356–362.
70. Hillier, S. G. (1991). Cellular basis of follicular endocrine function. In *Ovarian Endocrinology* (S. G. Hillier, Ed.), pp. 25–72. Blackwell Scientific Publications, Oxford.
71. Strauss, J. F., and Miller, W. L. (1991). Molecular basis of ovarian steroid synthesis. In *Ovarian Endocrinology* (S. G. Hillier, Ed.), pp. 25–72. Blackwell Scientific Publications, Oxford.
72. Richards, J. S., Jahnsen, T., Hedin, L., Lifka, J., Ratoosh, S. L., Durcia, J. M., and Goldring, N. B. (1987). Ovarian follicular development: from physiology to molecular biology. *Rec. Prog. Horm. Res.* 43, 231–270.
73. Hickey, G. J., Chen, S. A., Besman, M. J., Shively, J. E., Hall, P. F., Gaddy-Kurten, D., and Richards, J. S. (1988). Hormonal regulation, tissue distribution, and content of aromatase cytochrome P450 messenger ribonucleic acid and enzyme in rat ovarian follicles and corpora lutea: relationship to estradiol biosynthesis. *Endocrinology* 122, 1426–1436.
74. Segaloff, D. L., Wang, H. Y., and Richards, J. S. (1990). Hormonal regulation of luteinizing hormone/chorionic gonadotropin receptor mRNA in rat ovarian cells during follicular development and luteinization. *Mol. Endocrinol.* 4, 1856–1865.
75. Doody K. J., Lorence, M. C., Mason, J. I., and Simpson E. R. (1990). Expression of messenger ribonucleic acid species encoding steroidogenic enzymes in human follicles and corpora lutea throughout the menstrual cycle. *J. Clin. Endocrinol. Metab.* 70, 1041–1045.
76. Zeleznik, A. J., Keyes, P. L., Menon, K. M. J., Midgley, A. R. Jr., and Reichert, L. E. Jr. (1977). Development-dependent responses of ovarian follicles to FSH and hCG. *Am. J. Physiol.* 233, E229–E234.
77. Falk, B. (1959). Site of production of oestrogen in rat ovary as studied in microtransplants. *Acta Physiol. Scand.* 47 (Suppl 163), 1–101.
78. Hillier, S. G. (1981). Regulation of follicular oestrogen biosynthesis: a survey of current concepts. *J. Endocrinol.* 89, P3–P18.
79. Sasano, H., Okamoto, M., Mason, J. I., Simpson, E. R., Mendelson, C. R., Sasano, N., and Silverberg, S. G. (1989). Immunolocalization of aromatase, 17a-hydroxylase and side-chain cleavage cytochromes P-450 in the human ovary. *J. Reprod. Fertil.* 85, 163–169.
80. Ryan, K. J., Petro, Z., and Kaiser, J. (1968). Steroid formation by isolated and recombined ovarian granulosa and thecal cells. *J. Clin. Endocrinol.* 28, 355–358.
81. McNatty, K. P., Makris, A., Reinhold, V. N., De Grazia C., Osathanondh, R., and Ryan, K. J. (1979). Metabolism of androstenedione by human ovarian tissues in vitro with particular reference to reductase and aromatase activity. *Steroids* 34, 429–443.
82. McNatty, K. P., Hillier, S. G., van den Boogard, A. J. M., Trimbos-Kemper, T. C. M., Reichert, L. E. Jr., and van Hall, E. V. (1983). Follicular development during the luteal phase of the human menstrual cycle. *J. Clin. Endocrinol. Metab.* 56, 1022–1031.
83. Zeleznik, A. J., and Hillier, S. G. (1984). The role of gonadotropins in the selection of the preovulatory follicle. *Clin. Obstet. Gynecol.* 27, 927–940.
84. McNatty, K. P. (1978). Cyclic changes in antral fluid hormone concentrations in humans. *Clin. Endocrinol. Metab.* 7, 577–600.
85. Brailly, S. Gougeon, A., Milgrom, E., Bomsel-Helmreich, O., and Papiernik, E. (1981). Androgens and progestins in the human ovarian follicle: differences in the evolution of preovulatory, healthy, nonovulatory, and atretic follicles. *J. Clin. Endocrinol. Metab.* 53, 128–134.
86. Westergaard, L., Christensen, I. J., and McNatty, K. P. (1986). Steroid levels in ovarian follicular fluid related to follicle size and health status during the normal menstrual cycle in women. *Hum. Reprod.* 1, 227–232.
87. Goldenberg, R. L., Reiter, E. O., and Ross, G. T. (1973). Follicle response to exogenous gonadotropins: an estrogen mediated phenomenon. *Fertil. Steril.* 24, 121–125.
88. Richards, J. S. (1980). Maturation of ovarian follicles: actions and interactions of pituitary and ovarian hormones on follicular cell differentiation. *Physiol. Rev.* 60, 51–89.
89. Hillier, S. G., Zeleznik, A. J., Knazek, R. A., and Ross, G. T. (1980). Hormonal regulation of preovulatory follicle maturation in the rat. *J. Reprod. Fertil.* 60, 219–229.
90. Koering, M. J. (1987). Follicular maturation and atresia: morphological correlates. In *The Primate Ovary* (R. Stouffer, R. Brenner, and C. Phoenix, Eds.), pp. 3–23. Plenum Press, New York.
91. Billiar, R. B., Loukides, J. A., and Miller, M. M. (1992). Evidence for the presence of the estrogen receptor in the ovary of the baboon (*Papio anubis*). *J. Clin. Endocrinol. Metab.* 75, 1159–1165.
92. Saunders, P. T., Millar, M. R., Williams, K., Macpherson, S., Harkiss, D., Anderson, R. A., Orr, B., Groome, N. P., Scobie, G., and Fraser, H. M. (2000). Differential expression of estrogen receptor-alpha and -beta and androgen receptor in the ovaries of marmosets and humans. *Biol. Reprod.* 63, 1098–1105.
93. Ito, Y., Fisher C. R., Conte, F. A., Grumbach, M. M., and Simpson, E. R. (1993). Molecular basis of aromatase deficiency in an adult female with sexual infantilism and polycystic ovaries. *Proc. Natl. Acad. Sci. U S A* 90, 11673–11677.
94. Conte, F. A., Grumbach, M. M., Ito, Y., Fisher, C. R., and Simpson, E. R. (1994). A syndrome of female pseudohermaphrodism, hypergonadotropic hypogonadism, and multicystic ovaries associated with missense mutations in the gene encoding aromatase (P450arom). *J. Clin. Endocrinol. Metab.* 78, 1287–1292.
95. Mullis, P. E., Yoshimura, N., Kuhlmann, B., Lippuner, K., Jaeger, P., and Harada, H. (1997). Aromatase deficiency in a female who is compound heterozygote for two new point mutations in the P450arom gene: impact of estrogens on hypergonadotropic hypogonadism, multicystic ovaries, and bone densitometry in childhood. *J. Clin. Endocrinol. Metab.* 82, 1739–1745.
96. Harlow, C. R., Hillier, S. G., and Hodges, J. K. (1986). Androgen modulation of follicle stimulating hormone induced granulosa cell steroidogenesis in the primate ovary. *Endocrinology* 119, 1403–1405.
97. Hillier, S. G., Wickings, E. J., Illingworth, P. I., Yong, E. L., Reichert, L. E. Jr., Baird, D. T., and McNeilly, A. S. (1991). Control of immunoactive inhibin production by human granulosa cells. *Clin. Endocrinol. (Oxf).* 35, 71–78.
98. Hild-Petito, S., West, N. B., Brenner, R. M., and Stouffer, R. L. (1991). Localization of androgen receptor in the follicle and corpus luteum of the primate ovary during the menstrual cycle. *Biol. Reprod.* 44, 561–568.

99. El-Hefnawy, T., and Zeleznik, A. J. (2001). Synergism between FSH and activin in the regulation of proliferating cell nuclear antigen (PCNA) and cyclin D2 expression in rat granulosa cells. *Endocrinology* 142, 4357–4362.
100. Kaipia, A., and Hsueh, A. J. W. (1997). Regulation of ovarian follicle atresia. *Annu. Rev. Physiol.* 59, 349–363.
101. Pradeep P. K., Li, X., Peegel H., and Menon, K. M. J. (2002). Dihydrotestosterone inhibits granulosa cell proliferation by decreasing the cyclin D2 mRNA expression and cell cycle arrest at G1 phase. *Endocrinology* 143, 2930–2935.
102. Zeleznik, A. J., Hillier, S. G., and Ross, G. T. (1979). Follicle stimulating hormone-induced follicular development: an examination of the role of androgens. *Biol. Reprod.* 21, 673–681.
103. Weil, S., Vendola, K., Zhou, J., and Bondy, C. A. (1999). Androgen and follicle-stimulating hormone interactions in primate ovarian follicle development. *J. Clin. Endocrinol. Metab.* 84, 2951–2956.
104. Vendola, K. A., Zhou, J., Adesanya, O. O., Weil, S. J., and Bondy, C. A. (1998). Androgens stimulate early stages of follicular growth in the primate ovary. *J. Clin. Invest.* 101, 2622–2629.
105. Zeleznik, A. J., Little-Ihrig, L., and Ramasawamy, S. (2004). Administration of dihydrotestosterone to rhesus monkeys inhibits gonadotropin-stimulated ovarian steroidogenesis. *J. Clin. Endocrinol. Metab.* 89, 860–866.
106. Rabinovici, J., Blankstein, J., Goldman, B., Rudak, E., Dor, Y., Pariente, C., Geier, A., Lunenfeld, B., and Mashiach, S. (1989). In vitro fertilization and primary embryonic cleavage are possible in 17 alpha-hydroxylase deficiency despite extremely low intrafollicular 17 beta-estradiol. *J. Clin. Endocrinol. Metab.* 68, 693–697.
107. Pariente, C., Rabinovici, J., Lunenfeld, B., Rudak, E., Dor, J., Mashiach, S., Levran, D., Blankstein, J., and Geier, A. (1990). Steroid secretion by granulosa cells isolated from a woman with 17 alpha-hydroxylase deficiency. *J. Clin. Endocrinol. Metab.* 71, 984–987.
108. Erickson, G. F., Garzo, V. G., and Magoffin, D. A. (1989). Insulin-like growth factor I regulates aromatase activity in human granulosa and granulosa luteal cells. *J. Clin. Endocrinol. Metab.* 69, 716–724.
109. Bergh, C., Olsson, J. H., and Hillensjö, T. ((1991). Effect of insulin-like growth factor I on steroidogenesis in cultured human granulosa cells. *Acta Endocrinol. (Copenh).* 125, 177–185.
110. Hillier, S. G., Young, E. L., Illingworth, P. J., Baird, D. T., Schwall R. H., and Mason, A. J. (1991). Effect of recombinant inhibin on androgen synthesis in cultured human thecal cells. *Mol. Cell. Endocrinol.* 75, R1–R6.
111. Bergh, C., Carlsson, B., Olsson, J.-H., Selleskog, U., and Hillensjö, T. (1993). Regulation of androgen production in cultured human thecal cells by insulin-like growth factor I and insulin. *Fertil. Steril.* 59, 323–331.
112. Oliver, J. E., Aitman, T. J., Powell, J. F., Wiolson, C. A., and Clayton, R. N. (1989). Insulin-like growth factor I gene expression in the rat ovary is confined to the granulosa cells of developing follicles. *Endocrinology* 124, 2671–2679.
113. Hernandez, E. R., Hurwitz, A., Vera, A., Pellicer, A., Adashi, E., LeRoith, D., and Roberts, C. T. Jr. (1992). Expression of the genes encoding the insulin-like growth factors and their receptors in the human ovary. *J. Clin. Endocrinol. Metab.* 74, 419–425.
114. Eden, J. A., Jones, J., Carter, G. D., and Alaghband-Zadeh, J. (1988). A comparison of follicular fluid levels of insulin-like growth factor I in normal dominant and cohort follicles, polycystic and multicystic ovaries. *Clin. Endocrinol. (Oxf).* 29, 327–336.
115. Voutilainen, R., and Miller, W. L. (1987). Coordinate tropic hormone regulation of mRNAs for insulin-like growth factor II and the cholesterol side-chain-cleavage enzyme, P450scc, in human steroidogenic tissues. *Proc. Natl. Acad. Sci. U S A* 84, 1590–1594.
116. Ramasharma, K., and Li, C. H. (1987). Human pituitary and placental hormones control human insulin-like growth factor II secretion in human granulosa cells. *Proc. Natl. Acad. Sci. U S A* 84, 2643–2647.
117. Erickson, G. F., Magoffin, D. A., Cragun, J. R., and Chang, R. J. (1990). The effects of insulin and insulin-like growth factors I and II on estradiol production by granulosa cells of polycystic ovaries. *J. Clin. Endocrinol. Metab.* 70, 894–902.
118. Ooi, G. T. (1990). Insulin-like growth factor binding proteins (IGFBP'S): more than just 1, 2, 3. *Mol. Cell. Endocrinol.* 71, C39–C43.
119. San Roman, G. A., and Magoffin, D. A. (1992). Insulin-like growth factor binding proteins in ovarian follicles from women with polycystic ovarian disease: cellular source and levels in follicular fluid. *J. Clin. Endocrinol. Metab.* 75, 1010–1016.
120. Zeleznik, A. J., Little-Ihrig, L., and Ramasawamy, S. (2002). Administration of insulin-like growth factor I to rhesus monkeys does not augment gonadotropin-stimulated ovarian steroidogenesis. *J. Clin. Endocrinol. Metab.* 87, 5722–5729.
121. Dor, J., Ben-Shlomo, I., Lunenfeld, B., Pariente, C., Levran, D., Karasik, A., Seppala, M., and Mashiach, S. (1992). Insulin-like growth factor-I (IGF-I) may not be essential for ovarian follicular development: evidence from IGF-I deficiency. *J. Clin. Endocrinol. Metab.* 74, 539–542.
122. Ying, S.-Y. (1988). Inhibins, activins and follistatins: gonadal proteins modulating the secretion of follicle stimulating hormone. *Endocr. Rev.* 9, 267–293.
123. Tsonis, C. G., and Sharpe, R. M. (1986). Dual gonadal control of follicle stimulating hormone. *Nature* 321, 776–779.
124. Yu, J., Shao, L., Lemas, V., Yu, A. L., Vaughan, J., Rivier, J., and Vale, W. (1987). Importance of FSH-releasing protein and inhibin in erythroid differentiation. *Nature* 330, 724–725.
125. Slack, J. (1991). Molecule of the moment. *Nature* 349, 17–18.
126. Schwall, R. H., Mason, A. J., Wilcox, J. N., Bassett, S. G., and Zeleznik, A. J. (1990). Localization of inhibin/activin subunit mRNAs within the primate ovary. *Mol. Endocrinol.* 4, 75–79.
127. Yamoto, M., Minami, S., Nakano, R., and Kobayashi, M. (1992). Immunohistochemical localization of inhibin/activin subunits in human ovarian follicles during the menstrual cycle. *J. Clin. Endocrinol. Metab.* 74, 989–993.
128. Hillier, S. G. (1991). Paracrine control of follicular estrogen synthesis. *Sem. Reprod. Endocrinol.* 9, 332–340.
129. Rabin, D., Spitz, I., Bercovici, B., Bell, J., Laufer, A., Benvieniste, R., and Polishuk, W. (1972). Isolated deficiency of follicle stimulating hormone. *N. Engl. J. Med.* 287, 1313–1317.
130. Rabinowicz, D., Benveniste, R., Lindner, J., Lorber, D., and Daniell, J. (1979). Isolated follicle stimulating hormone deficiency revisited. *N. Engl. J. Med.* 300, 126–128.
131. Aittomaki, K., Lucena, J. L., Pakarinen, P., Sistonen, P., Tapanainen, J., Gromoll, J., Kaskikari, R., Sankila, E. M., Lehvaslaiho, H., Engel, A. R., Nieschlag E., Huhtaniemi I., and de la Chappelle, A. (1995). Mutation in the follicle-stimulating hormone receptor gene causes hereditary hypergonadotropic ovarian failure. *Cell* 82, 959–968.
132. Rannikko, A., Pakarinen, P., Manna, P. R., Beau, I., Misrahi, M., Aittomaki, K., and Huhtaniemi, I. (2002). Functional characterization of the human FSH receptor with an inactivating Ala189Val mutation. *Mol. Hum. Reprod.* 8, 311–317.

133. Mannaerts, B., de Leuw, R., Geelen, J., van Ravestein, A., van Wezenbeek, P., Schuurs, A., and Kloosterboer, H. (1991). Comparative in vitro and in vivo studies on the biological characteristics of recombinant follicle stimulating hormone. *Endocrinology* 129, 2623–2630.
134. Schoot, D. C., Coelingh Bennink, H. J. T., Mannaerts, B. M., Lamberts, S. W., Bouchard, P., and Fauser, B. C. (1992). Human recombinant follicle stimulating hormone induces growth of preovulatory follicles without concomitant increase in androgen and estrogen biosynthesis in a woman with isolated gonadotropin deficiency. *J. Clin. Endocrinol. Metab.* 74, 1471–1473.
135. Toledo, S. P., Brunner, H. G., Kraaij, R., Post, M., Dahia, P. L., Hayashida, C. Y., Kremer, H., and Themmen, A. P. (1996). An inactivating mutation of the luteinizing hormone receptor causes amenorrhea in a 46,XX female. *J. Clin. Endocrinol. Metab.* 81, 3850–3854.
136. Goodman, A. L., Nixon, W. E., Johnson, D. K., and Hodgen, G. D. (1977). Regulation of folliculogenesis in the cycling rhesus monkey: selection of the dominant follicle. *Endocrinology* 100, 155–161.
137. Nilsson, L., Wikland, M., and Hamberger, L. (1982). Recruitment of an ovulatory follicle in the human following follicle-ectomy and lute-ectomy. *Fertil. Steril.* 37, 30–34.
138. Baird, D. T., Bäckström, T., McNeilly, A. S., Smith, S. K., and Wathen, C. G. (1984). Effect of enucleation of the corpus luteum at different stages of the luteal phase of the human menstrual cycle on subsequent follicular development. *J. Reprod. Fertil.* 70, 615–624.
139. Zeleznik, A. J., and Resko, J. A. (1980). Progesterone does not inhibit gonadotropin-induced follicular maturation in the female rhesus monkey. *Endocrinology* 106, 1820–1826.
140. Zeleznik, A. J, Hutchison, J. S., and Schuler, H. M. (1987). Passive immunization with anti-estradiol antibodies during the luteal phase of the menstrual cycle potentiates the perimenstrual rise in serum gonadotropin concentrations and stimulates follicular growth in the cynomolgus monkey. *J. Reprod. Fertil.* 80, 403–410.
141. Goodman, A. L., and Hodgen, G. D. (1977). Systemic versus intraovarian progesterone replacement after luteectomy in rhesus monkeys: differential patterns of gonadotropins and follicle growth. *J. Clin. Endocrinol. Metab.* 45, 837–840.
142. Baird, D. T., Baker, T. G., McNatty, K. P., and Neal, P. (1975). Relationship between the secretion of the corpus luteum and the length of the follicular phase of the ovarian cycle. *J. Reprod. Fertil.* 45, 611–619.
143. Smeaton, T. C., and Robertson, H. A. (1971). Studies on the growth and atresia of Graafian follicles in the ovary of the sheep. *J. Reprod. Fertil.* 25, 243–252.
144. Siroos, J., and Fortune, J. E. (1990). Lengthening the bovine estrous cycle with low levels of exogenous progesterone: a model for studying ovarian follicular dominance. *Endocrinology* 127, 916–925.
145. Baird, D. T. (1983). Factors regulating the growth of the preovulatory follicle in the sheep and human. *J. Reprod. Fertil.* 69, 343–352.
146. Ross G. T., Cargille C. M., Lipsett M. B., Rayford P. L., Marshall J. R., Strott C. A., and Rodbard, D. (1970). Pituitary and gonadal hormones during spontaneous and induced menstrual cycles. *Rec. Prog. Horm. Res.* 26, 1–62.
147. Brown, J. B. (1978). Pituitary control of ovarian function-concepts derived from gonadotrophin therapy. *Aust. N. Z. J. Obstet. Gynaecol.* 18, 46–54.
148. Karsch, F. J., Weick, R. F., Hotchkiss, J., Dierschke, D. J., and Knobil, E. R. (1973). An analysis of the negative feedback control of gonadotropin secretion utilizing chronic implantation of ovarian steroids in ovariectomized rhesus monkeys. *Endocrinology* 93, 478–486.
149. Bassett, S. G., and Zeleznik, A. J. (1990). Acute suppression of FSH secretion by oestradiol in the ovariectomized rhesus monkey. *J. Reprod. Fertil.* 88, 441–446.
150. Zeleznik, A. J. (1981). Premature elevation of systemic estradiol reduces serum levels of follicle stimulating hormone and lengthens the follicular phase of the menstrual cycle in rhesus monkeys. *Endocrinology* 109, 352–355.
151. Vaitukaitis, J. L., Bermudez, J. A., Cargille, C. M., Lipsett, M. B., and Ross, G. T. (1971). New evidence for an anti-estrogenic action of clomiphene citrate in women. *J. Clin. Endocrinol. Metab.* 32, 503–508.
152. Zeleznik, J., Hutchison, J. S., and Schuler, H. M. (1985). Interference with the gonadotropin suppressing actions of estradiol in macaques overrides the selection of a single preovulatory follicle. *Endocrinology* 117, 991–999.
153. Speroff, L., Glass, R. H., and Kase, N. G. (1983). *Clinical Gynecologic Endocrinology and Infertility*, 3rd ed. Williams & Wilkins, Baltimore.
154. Mitwally, M. F., and Casper, R. F. (2004). Aromatase inhibitors in ovulation induction. *Sem. Reprod. Med.* 22, 61–78.
155. Channing, C. P., Anderson, L. D., Hoover, D. J., Gagliano, P., and Hodgen, G. D. (1981). Inhibitory effects of porcine follicular fluid on monkey serum FSH levels and follicular maturation. *Biol. Reprod.* 25, 267–272.
156. Lenton, E. A., deKretser, D. M., Woodward, A. J., and Robertson, D. M. (1991). Inhibin concentrations throughout the menstrual cycles of normal, infertile, and older women compared with those during spontaneous conception cycles. *J. Clin. Endocrinol. Metab.* 73, 1180–1190.
157. McLachlan, R. I., Roberrtson, D. M., Healy, D. L., deKretser, D. M., and Burger, H. G. (1986). Plasma inhibin levels during gonadotropin induced ovarian hyperstimulation for IVF: a new index of follicular function? *Lancet* 1, 1233–1234.
158. Zeleznik, A. J., and Kubik, C. J. (1986). Ovarian responses in macaques to pulsatile infusion of follicle stimulating hormone and luteinizing hormone: increased sensitivity of the maturing follicle to FSH. *Endocrinology* 119, 2025–2032.
159. Glasier, A. F., Baird, D. T., and Hillier, S. G. (1988). FSH and the control of follicular growth. *J. Steroid Biochem.* 32, 167–170.
160. Schoot, D. C., Pache, T. D., Hop, W. C., deJong, F. H., and Fauser, B. C. J. M. (1992). Growth patterns of ovarian follicles during induction of ovulation with decreasing doses of human menopausal gonadotropin following presumed selection in polycystic ovarian syndrome. *Fertil. Steril.* 57, 1117–1120.
161. LaPolt, P. S., Tilly, J. L., Aihara, T., Nishimori, K., and Hsueh, A. J. W. (1992). Gonadotropin-induced up- and down-regulation of ovarian follicle stimulating hormone receptor gene expression in immature rats: effects of pregnant mares serum gonadotropin, human chorionic gonadotropin and recombinant FSH. *Endocrinology* 130, 1289–1295.
162. Harlow, C. R., Shaw, H. J., Hillier, S. G., and Hodges, J. K. (1988). Factors influencing FSH-induced steroidogenesis in marmoset granulosa cells: effects of androgens at all stages of follicular development. *Endocrinology* 122, 2780–2787.
163. Zeleznik, A. J., Midgley, A. R. Jr., and Reichert, L. E. Jr. (1974). Granulosa cell maturation in the rat: increased binding of human chorionic gonadotropin following treatment with follicle stimulating hormone in vivo. *Endocrinology* 95, 818–825.
164. Sullivan, M. W., Stewart-Akers, A., Krasnow, J. S., Berga, S. L., and Zeleznik, A. J. (1999). Ovarian responses in women to

recombinant follicle-stimulating hormone and luteinizing hormone (LH): a role for LH in the final stages of follicular maturation. *J. Clin. Endocrinol. Metab.* 84, 228–232.
165. Filicori, M., Cognigni, G. E., Tabarelli, C., Pocognoli, P., Taraborrelli, S., Spettoli, D., and Ciampaglia, W. (2002). Stimulation and growth of antral ovarian follicles by selective LH activity administration in women. *J. Clin. Endocrinol. Metab.* 87, 1156–1161.
166. Campbell, B. K., Dobson, H., Baird, D. T., and Scaramuzzi, R. J. (1999). Examination of the relative role of FSH and LH in the mechanism of ovulatory follicle selection in sheep. *J. Reprod. Fertil.* 117, 355–367.
167. Koos, R. D. (1989). Potential relevance of angiogenic factors to ovarian physiology. *Sem. Reprod. Endocrinol.* 7, 29–40.
168. Leung, D. W., Cachianes G., Kuang, W. J., Goeddel, D. V., and Ferrara, N. (1989). Vascular endothelial growth factor is a secreted angiogenic mitogen. *Science* 246, 1306–1309.
169. Connolly, D. T., Heuvelman, D. M., Nelson, R., Olander, J. V., Eppley, B. L., Delfino, J. J., Siegel, N. R., Leimgruber R. M., and Feder, J. (1989). Tumor vascular permeability factor stimulates endothelial cell growth and angiogenesis. *J. Clin. Invest.* 84, 1470–1478.
170. Houck, K. A., Ferrara, N., Winer, J., Cachianes, G., Bing, L., and Leung, D. W. (1991). The vascular endothelial growth factor family: identification of a fourth molecular species and characterization of alternative splicing of RNA. *Mol. Endocrinol.* 5, 1806–1814.
171. Ferrara, N., Houck, K., Jakeman, L., and Leung, D. W. (1992). Molecular and biological properties of the vascular endothelial growth factor family of proteins. *Endocr. Rev.* 13, 18–32.
172. Ravindranath, N., Little-Ihrig, L. L., Phillips, H. S., Ferrara, N., and Zeleznik, A. J. (1992). Vascular endothelial growth factor messenger ribonucleic acid expression in the primate ovary. *Endocrinology* 131, 254–260.
173. Zimmermann, R. C., Xiao, E., Bohlen, P., and Ferin, M. (2002). Administration of antivascular endothelial growth factor receptor 2 antibody in the early follicular phase delays follicular selection and development in the rhesus monkey. *Endocrinology* 143, 2496–2502.
174. Rothchild, I. (1981). The regulation of the mammalian corpus luteum. *Rec. Prog. Horm. Res.* 37, 183–298.
175. Hartman C. G., and Corner, G. W. (1947). Removal of the corpus luteum and of the ovaries of the female monkey during pregnancy: observations and cautions. *Anat. Rec.* 98, 539–546.
176. Csapo, A. I., Pulkkinen M. O., and Wiest, W. G. (1973). Effects of luteectomy and progesterone replacement therapy in early pregnant patients. *Am. J. Obstet. Gynecol.* 115, 759–765.
177. Channing, C. P. (1980). Progesterone and estrogen secretion by cultured monkey ovarian cell types: influences of follicular size, serum luteinizing hormone levels, and follicular fluid estrogen levels. *Endocrinology* 107, 342–352.
178. Stouffer, R. L., and Hodgen, G. D. (1980). Induction of luteal phase defects in rhesus monkeys by follicular fluid administration at the onset of the menstrual cycle. *J. Clin. Endocrinol. Metab.* 51, 669–671.
179. Dierschke, D. J., Hutz, R. J., and Wolf, R. C. (1985). Induced follicular atresia in the rhesus monkey: strength-duration relationship of the estrogen stimulus. *Endocrinology* 117, 1397–1403.
180. Crisp, T. M., and Channing, C. P. (1972). Fine structural events correlated with progestin secretion during luteinization of rhesus monkey granulosa cells in culture. *Biol. Reprod.* 7, 55–72.
181. Channing, C. P. (1970). Influences of the in vivo and in vitro hormonal environment upon luteinization of granulosa cells in tissue culture. *Rec. Prog. Horm. Res.* 26, 589–622.
182. Hirschfield, A. N. (1984). Continuous (3H) thymidine infusion: a method for the study of follicular dynamics. *Biol. Reprod.* 30, 485–491.
183. Somers, J. P., Benyo, D. F., Little-Ihrig, L., and Zeleznik, A. J. (1995). Luteinization in primates is accompanied by loss of a 43-kilodalton adenosine 3′,5′-monophosphate response element-binding protein isoform. *Endocrinology* 136, 4762–4768.
184. Young, F. M., Rodger, F. E., Illingworth, P. J., and Fraser, H. M. (2000). Cell proliferation and vascular morphology in the marmoset corpus luteum. *Hum. Reprod.* 15, 557–566.
185. Chaffin, C. L., Schwinof, K. M., and Stouffer, R. L. (2001). Gonadotropin and steroid control of granulosa cell proliferation during the periovulatory interval in rhesus monkeys. *Biol. Reprod.* 65, 755–762.
186. Conti, M. (2002). Specificity of the cyclic adenosine 3′,5′-monophosphate signal in granulosa cell function. *Biol. Reprod.* 67, 1653–1661.
187. Adams, E. C., and Hertig, A. T. (1969). Studies on the human corpus luteum. I. Observations on the ulltrastructure of development and regression of the luteal cells during the menstrual cycle. *J. Cell Biol.* 41, 696–715.
188. Abdul-Karim R. W., and Bruce, N. (1973). Blood flow to the ovary and corpus luteum at different stages of gestation in the rabbit. *Fertil. Steril.* 24, 44–47.
189. Carr, B. R., MacDonald, P. C., and Simpson, E. R. (1982). The role of lipoproteins in the regulation of progesterone secretion by the human corpus luteum. *Fertil. Steril.* 38, 303–311.
190. Gospodarowicz, D., and Thakral, K. K. (1978). Production of a corpus luteum angiogenic factor responsible for proliferation of capillaries and neovascularization of the corpus luteum. *Proc. Natl. Acad. Sci. U S A* 75, 847–851.
191. Hazzard, T. M., Xu, F., and Stouffer, R. L. (2002). Injection of soluble vascular endothelial growth factor receptor 1 into the preovulatory follicle disrupts ovulation and subsequent luteal function in rhesus monkeys. *Biol. Reprod.* 67, 1305–1312.
192. Fraser, H. M., and Wulff, C. (2003). Angiogenesis in the corpus luteum. *Reprod. Biol. Endocrinol.* 1, 88.
193. Zander, J., Brendle, E., von Muenstermann, A. M., Diczfalusy, E., Martinsen, B., and Tillinger, K. G. (1959). Identification and estimation of oestradiol-17β and oestrone in human ovaries. *Acta. Obstet. Gynaecol. Scand.* 38, 724–736.
194. Tureck, R. W., and Strauss, J. F. III. (1982). Progesterone synthesis by luteinized human granulosa cells in culture: the role of de novo sterol synthesis and lipoprotein-carried sterol. *J. Clin. Endocrinol. Metab.* 54, 367–373.
195. Soto, E., Silavin, S. L., Tureck, R. W., and Strauss, J. F. III. (1984). Stimulation of progesterone synthesis in luteinized human granulosa cells by human chorionic gonadotropin and 8-bromo-adenosine 3′,5′- monophosphate: the effect of low density lipoproteins. *J. Clin. Endocrinol. Metab.* 58, 831–837.
196. Illingworth, D. R., Corbin, D. K., Kemp, E. D., and Keenan, E. J. (1982). Hormone changes during the menstrual cycle in a beta lipoproteinemia-reduced luteal phase progesterone in a patient with homozygous hypobetalipoproteinemia. *Proc. Natl. Acad. Sci. U S A* 79, 6685–6689.
197. Yki-Järvinen, H., Wahlström, T., and Seppälä, M. (1983). Immunohistochemical demonstration of relaxin in the genital tract of pregnant and nonpregnant women. *J. Clin. Endocrinol. Metab.* 57, 451–454.

198. Emmi, A. M., Skurnick, J., Goldsmith, L. T., Gagliardi, C. L., Schmidt, C. L. Kleinberg, D., and Weiss, G. (1991). Ovarian control of pituitary hormone secretion in early human pregnancy. *J. Clin. Endocrinol. Metab.* 72, 1359–1363.
199. Khan-Dawood, F. S., Marut, E. L., and Dawood, M. Y. (1984). Oxytocin in the corpus luteum of the cynomolgus monkey (*Macaca fascicularis*). *Endocrinology* 115, 570–574.
200. Wathes, D. C., Swann, R. W., Pickering, B. T., Porter, D. G., Hull, M. G. R., and Drife, J. O. (1982). Neurohypophyseal hormones in the human ovary. *Lancet* 2, 410–412.
201. Amico, J. A., and Zeleznik, A. J. (1990). Ovarian oxytocin and neurophysin concentrations in the cynomolgus monkey (*Macaca fascicularis*). *Endocr. Res.* 16, 17–30.
202. McLachlan, R. I., Robertson, D. M., Healy, D. L., Burger, H. G., and deKretser, D. M. (1987). Circulating immunoreactive inhibin levels during the normal human menstrual cycle. *J. Clin. Endocrinol. Metab.* 65, 954–961.
203. Davis, S. R., Krozowski, Z., McLachlan, R. I., and Burger, H. G. (1987). Inhibin gene expression in the human corpus luteum. *J. Endocrinol.* 115, R21–R23.
204. Tsonis, C. G., Hillier S. G., and Baird, D. T. (1987). Production of inhibin bioactivity by human granulosa lutein cells: stimulation by LH and testosterone in vitro. *J. Endocrinol.* 112, R11–R14.
205. Bassett, S. G., Winters, S. J., Keeping, H. S., and Zeleznik, A. J. (1990). Serum immunoreactive inhibin levels before and after luteectomy in the cynomolgus monkey (*Macaca fascicularis*). *J. Clin. Endocrinol. Metab.* 70, 590–594.
206. Woodruf, T. K., D'Agostino, J., Schwartz, N. B., and Mayo, K. E. (1988). Dynamic changes in inhibin messenger RNAs in rat ovarian follicles during the reproductive cycle. *Science* 239, 1296–1299.
207. Meunier, H., Cajander, S. B., Roberts, V. J., Rivier, C., Sawchenko, P. E., Hsueh, A. J. W., and Vale, W. (1988). Rapid changes in the expression of inhibin a, βA-, and βB-subunits in ovarian cell types during the rat estrous cycle. *Mol. Endocrinol.* 2, 1352–1363.
208. Torney, A. J., Hodgsason, Y. M., Forage, R., and deKretser, D. M. (1989). Cellular localization of inhibin mRNA in the bovine ovary by in situ hybridization. *J. Reprod. Fertil.* 86, 391–399.
209. Rodgers, R. J., Stuchbery, S. J., and Findlay, J. K. (1989). Inhibin mRNAs in ovine and bovine ovarian follicles and corpora lutea throughout the estrous cycle and gestation. *Mol. Cell. Endocrinol.* 62, 95–101.
210. Lahr L., and Riddle, O. (1936). Temporary suppression of estrous cycles in the rat by prolactin. *Proc. Soc. Exp. Biol. Med.* 34, 880–893.
211. Evans, H. M., Simpson, M. E., and Lyons, W. R. (1941). Influence of lactogenic preparations on production of traumatic placentoma in the rat. *Proc. Soc. Exp. Biol. Med.* 46, 586–590.
212. Hisaw, F. L. (1944). The placental gonadotropin and luteal function in monkeys (*Macaca mulatta*). *Yale J. Biol. Med.* 17, 121–137.
213. Rice, B. F., Hammerstein, J., and Savard, K. (1964). Steroid hormone formation in the human ovary. II. Actions of gonadotropin in vitro in the corpus luteum. *J. Clin. Endocrinol.* 24, 606–615.
214. Savard, K., Marsh, J. M., and Rice, B. F. (1965). Gonadotropins and ovarian steroidogenesis. *Rec. Prog. Horm. Res.* 21, 285–365.
215. Hanson, F. W., Powell, J. E., and Stevens, V. C. (1971). Effects of HCG and human pituitary LH on steroid secretion and functional lifespan of the human corpus luteum. *J. Clin. Endocrinol. Metab.* 32, 211–215.
216. MacDonald, G. J., and Greep, R. O. (1972). Ability of luteinizing hormone (LH) to acutely increase serum progesterone levels during the secretory phase of the rhesus menstrual cycle. *Fertil. Steril.* 23, 466–470.
217. Moudgal, N. R., MacDonald, G. J., and Greep, R. O. (1971). Effect of HCG antiserum on ovulation and corpus luteum formation in the monkey (*Macaca fascicularis*). *J. Clin. Endocrinol. Metab.* 32, 579–581.
218. Thau, R. B., Yamamoto, Y., Sundaram, K., and Spinola, P. G. (1983). Human chorionic gonadotropin stimulates luteal function in rhesus monkeys immunized against the β subunit of ovine luteinizing hormone. *Endocrinology* 112, 277–283.
219. Asch, R. H., Abou-Samara, M., Braunstein, G. D., and Pauerstein, C. J. (1982). Luteal function in hypophy sectomized rhesus monkeys. *J. Clin. Endocrinol. Metab.* 55, 154–161.
220. Balmaceda, J. P., Borghi, M. R., Coy, D. H., Schally, A. V., and Asch, R. H. (1983). Suppression of postovulatory gonadotropin levels does not affect corpus luteum function in rhesus monkeys. *J. Clin. Endocrinol. Metab.* 57, 866–868.
221. Hutchison, J. S., and Zeleznik, A. J. (1984). The rhesus monkey corpus luteum is dependent on pituitary gonadotropin secretion throughout the luteal phase of the menstrual cycle. *Endocrinology* 115, 1780–1786.
222. Knobil, E. (1980). The neuroendocrine control of the menstrual cycle. *Rec. Prog. Horm. Res.* 36, 53–88.
223. Fraser, H. M., Abbott, M., Laird, N. C., McNeilly, A. S., Nestor, J. J. Jr., and Vickery, B. H. (1986). Effects of an LH releasing hormone antagonist on the secretion of LH, FSH, prolactin and ovarian steroids at different stages of the luteal phase in the stumptailed macaque (*Macaca arctoides*). *J. Endocrinol.* 111, 83–90.
224. Mais, V., Kazer, R. R., Cetel, N. S., Rivier, J., Vale, W., and Yen, S. S. (1986). The dependency of folliculogenesis and corpus luteum function on pulsatile gonadotropin secretion in cycling women using a gonadotropin-releasing hormone antagonist as a probe. *J. Clin. Endocrinol. Metab.* 62, 1250–1255.
225. Fraser, H. M., Robertson, D. M., and deKretser, D. M. (1989). Immunoreactive inhibin concentrations in serum throughout the menstrual cycle of the macaque: suppression of inhibin during the luteal phase after treatment with an LHRH antagonist. *J. Endocrinol.* 121, R9–R12.
226. Atkinson, L. E., Hotchkiss, J., Fritz, G. R., Surve, A. H., Neill, J. D., and Knobil, E. (1975). Circulating levels of steroids and chorionic gonadotropin during pregnancy in the rhesus monkey with special attention to the rescue of the corpus luteum in early pregnancy. *Biol. Reprod.* 12, 335–345.
227. Ottobre, J. S., and Stouffer, R. L. (1984). Persistent versus transient stimulation of the macaque corpus luteum during prolonged exposure to human chorionic gonadotropin: a function of age of the corpus luteum. *Endocrinology* 114, 2175–2182.
228. Illingworth, P. J., Reddi, K., Smith, K., and Baird, D. T. (1990). Pharmacological "rescue" of the corpus luteum results in increased inhibin production. *Clin. Endocrinol. (Oxf).* 33, 323–332.
229. Quagliarello, J., Steinetz, B. G., and Weiss, G. (1979). Relaxin secretion in early pregnancy. *Obstet. Gynecol.* 53, 62–63.
230. Quagliarello, J., Goldsmith, L., Steinetz, B., Lustig, D. S., and Weiss, G. (1980). Induction of relaxin secretion in nonpregnant women by human chorionic gonadotropin. *J. Clin. Endocrinol. Metab.* 51, 74–77.
231. Ottobre, J. S., Nixon, W. E., and Stouffer, R. L. (1984). Induction of relaxin secretion in rhesus monkeys by human

chorionic gonadotropin: dependence on the age of the corpus luteum of the menstrual cycle. *Biol. Reprod.* 31, 1000–1006.

232. Nakai, Y., Plant, T. M., Hess, D. L., Keogh, E. J., and Knobil, E. (1978). On the sites of the negative and positive feedback actions of estradiol in the control of gonadotropin secretion in the rhesus monkey. *Endocrinology* 102, 1008–1014.
233. Hall, J. E., Whitcomb, R. W., Rivier, J., Vale, W., and Crowley, W. F. Jr. (1990). Differential regulation of luteinizing hormone, follicle stimulating hormone, and free a subunit secretion from the gonadotrope by gonadotropin-releasing hormone (GnRH): evidence from the use of two GnRH antagonists. *J. Clin. Endocrinol. Metab.* 70, 328–335.
234. McNeilly, A. S., Kerin, J., Swanston, I. A., Bramley, T. A., and Baird, D. T. (1980). Changes in the binding of human chorionic gonadotropin/luteinizing hormone, follicle stimulating hormone and prolactin to human corpora lutea during the menstrual cycle and pregnancy. *J. Endocrinol.* 87, 315–325.
235. Eyster, K. M., and Stouffer, R. L. (1985). Adenylate cyclase in the corpus luteum of the rhesus monkey. II. Sensitivity to nucleotides, gonadotropins, catecholamines and nonhormonal activators. *Endocrinology* 116, 1552–1558.
236. Stouffer, R. L., Nixon, W. E., Gulyas, B., and Hodgen, G. D. (1977). Gonadotropin-sensitive progesterone production by rhesus monkey luteal cells in vitro: a function of age of the corpus luteum during the menstrual cycle. *Endocrinology* 100, 506–512.
237. Fisch, B., Margara, R. A., Wintston, R. M., and Hillier, S. G. (1989). Cellular basis of luteal steroidogenesis in the human ovary. *J. Endocrinol.* 122, 303–311.
238. Ohara A., Mori, T., Taii, S., Ban, C., and Narimoto, K. (1987). Functional differentiation in steroidogenesis of two types of luteal cells isolated from mature corpora lutea of menstrual cycle. *J. Clin. Endocrinol. Metab.* 65, 1192–1200.
239. Hilliard, J. (1973). Corpus luteum function in guinea pigs, hamsters, rats, mice and rabbits. *Biol. Reprod.* 8, 203–221.
240. del Pozo, E., Goldstein, M., Friesen, H., Brun del Re, H., and Eppenberger, U. (1975). Lack of action of prolactin suppression in the regulation of the human menstrual cycle. *Am. J. Obstet. Gynecol.* 123, 719–723.
241. Zeleznik, A. J., and Little-Ihrig, L. L. (1990). Effect of reduced luteinizing hormone concentrations on corpus luteum function during the menstrual cycle of rhesus monkeys. *Endocrinology* 126, 2237–2244.
242. Richardson, D. W., Goldsmith, L., Pohl, C. R., Schallenberger, E., and Knobil, E. (1985). The role of prolactin in the regulation of the primate corpus luteum. *J. Clin. Endocrinol. Metab.* 60, 501–504.
243. Weiss, G., Dierschke, D. J., Karsch, F. J., Hotchkiss, J., Butler, W. R., and Knobil, E. (1973). The influence of lactation on luteal function in the rhesus monkey. *Endocrinology* 93, 954–959.
244. Duffy, D. M., Chaffin, C. L., and Stouffer, R. L. (2000). Expression of estrogen receptor alpha and beta in the rhesus monkey corpus luteum during the menstrual cycle: regulation by luteinizing hormone and progesterone. *Endocrinology* 141, 1711–1717.
245. Maybin, J. A., and Duncan, W. C. (2004). The human corpus luteum: which cells have progesterone receptors? *Reproduction* 128, 423–431.
246. Hillier, S. G., Tetsuka, M., and Fraser, H. M. (1997). Location and developmental regulation of androgen receptor in primate ovary. *Hum. Reprod.* 12, 107–111.
247. Duffy, D. M., Hess, D. L., and Stouffer, R. L. (1994). Acute administration of a 3 beta-hydroxysteroid dehydrogenase inhibitor to rhesus monkeys at the midluteal phase of the menstrual cycle: evidence for possible autocrine regulation of the primate corpus luteum by progesterone. *J. Clin. Endocrinol. Metab.* 79, 1587–1594.
248. Brannian, J. D., Woodruff, T. K., Mather, J. P., and Stouffer, R. L. (1992). Activin-A inhibits progesterone production by macaque luteal cells in culture. *J. Clin. Endocrinol. Metab.* 75, 756–761.
249. Stouffer, R. L., Dahl, K. D., Hess, D. L., Woodruff, T. K., Mather, J. P., and Molskness, T. A. (1994). Systemic and intraluteal infusion of inhibin A or activin A in rhesus monkeys during the luteal phase of the menstrual cycle. *Biol. Reprod.* 50, 888–895.
250. Hutchison, J. S., and Zeleznik, A. J. (1985). The corpus luteum of the primate menstrual cycle is capable of recovering from a transient withdrawal of pituitary gonadotropin support. *Endocrinology* 117, 1043–1049.
251. Ellinwood, W. E., Norman, R. L., and Spies, H. G. (1984). Changing frequency of pulsatile luteinizing hormone and progesterone secretion during the luteal phase of the menstrual cycle of rhesus monkeys. *Biol. Reprod.* 31, 714–722.
252. Filicori, M., Butler, J. P., and Crowley, W. F. Jr. (1984). Neuroendocrine regulation of the corpus luteum in the human. Evidence for pulsatile progesterone secretion. *J. Clin. Invest.*73, 1638–1647.
253. Marsh, J. M., and LeMaire W. J. (1974). Cyclic AMP accumulation and steroidogenesis in the human corpus luteum: effect of gonadotropins and prostaglandins. *J. Clin. Endocrinol. Metab.* 38, 99–106.
254. Stocco, D. M. (2001). StAR protein and the regulation of steroid hormone biosynthesis. *Annu. Rev. Physiol.* 63, 193–213.
255. Chung, B.-C., Matteson, K. J., Voutilainen, R., Mohandas, T. K., and Miller, W. L. (1986). Human cholesterol side-chain cleavage enzyme, P450SCC: cDNA cloning, assignment of the gene to chromosome 15, and expression in the placenta. *Proc. Natl. Acad. Sci. U S A* 83, 8962–8966.
256. Lorence, M. C., Murry, B. A., Trant, J. M., and Mason, J. I. (1990). Human 3b-hydroxysteroid dehydrogenase/D5-4 isomerase from placenta: expression in nonsteroidogenic cells of a protein that catalyzes the dehydrogenation/isomerization of C21 and C19 steroids. *Endocrinology* 126, 2493–2498.
257. Yamamoto, T., Davis, C. G., Brown, M. S., Schneider, W. J., Casey, M. L., Goldstein, J. L., and Russel, D. W. (1984). The human LDL receptor: a cysteine-rich protein with multiple Alu sequences in its mRNA. *Cell* 39, 27–38.
258. Simpson, E. R., Evans, C. T., Corbin, C. J., Powell, F. E., Ledesma, D. B., and Mendelson, C. R. (1987). Sequencing of cDNA inserts encoding aromatase cytochrome P-450 (P450AROM). *Mol. Cell. Endocrinol.* 52, 267–272.
259. Chung, B.-C., Picado-Leonard, J., Haniu, M., Bienkowski, V. M., Hall, P. F., Shively, J. E., and Miller, W. L. (1987). Cytochrome P450c17 (steroid 17a-hydroxylase/17,20 lyase): cloning of human adrenal and testis cDNAs indicates the same gene is expressed in both tissues. *Proc. Natl. Acad. Sci. U S A* 84, 407–411.
260. Bassett, S. G., Little-Ihrig, L. L., Mason, J. I., and Zeleznik, A. J. (1991). Expression of messenger ribonucleic acids that encode for 3b-hydroxysteroid dehydrogenase and cholesterol side-chain cleavage enzymes throughout the luteal phase of the macaque menstrual cycle. *J. Clin. Endocrinol. Metab.* 72, 362–366.
261. Duncan, W. C., Cowen, G. M., and Illingworth, P. J. (1999). Steroidogenic enzyme expression in human corpora lutea in the presence and absence of exogenous human chorionic gonadotrophin (HCG). *Mol. Hum. Reprod.* 5, 291–298.
262. Goldring, N. B., Durcia, J. M., Lifka, J., Hedin, L., Ratoosh, S. L., Miller, W. L., Orly, J., and Richards, J. S. (1987).

Cholesterol side-chain cleavage P450 messenger ribonucleic acid: evidence for hormonal regulation in rat ovarian follicles and constitutive expression in corpora lutea. *Endocrinology* 120, 1942–1950.

263. Oonk, R. B., Krasnow, J. S., Beattie, W. G., and Richards, J. S. (1989). Cyclic AMP-dependent and -independent regulation of cholesterol side chain cleavage cytochrome P-450 (P450scc) in rat ovarian granulosa cells and corpora lutea. *J. Biol. Chem.* 264, 21934–21942.

264. Ravindranath, N., Little-Ihrig, L., Benyo, D. F., and Zeleznik, A. J. (1992). Role of luteinizing hormone in the expression of cholesterol side chain cleavage cytochrome P450 and 3β-hydroxysteroid dehydrogenase, Δ5-4 isomerase messenger ribonucleic acids in the primate corpus luteum. *Endocrinology* 131, 2065–2070.

265. Voutilainen, R., Tapanainen, J., Chung, B. C., Matteson, K. J., and Miller, W. L. (1986). Hormonal regulation of P450scc (20,22 desmolase) and P450c17 (17a-hydroxylase/17,20 lyase) in cultured human granulosa cells. *J. Clin. Endocrinol. Metab.* 63, 202–207.

266. Golos, T. G., Strauss, J. F. III, and Miller, W. L. (1987). Regulation of low density lipoprotein receptor and cytochrome P450scc mRNA levels in human granulosa cells. *J. Steroid Biochem.* 27, 767–773.

267. McAllister, J. M., Mason, J. I., Byrd, W., Trant, J. M., Waterman, M. R., and Simpson, E. R. (1990). Proliferation human granulosa-lutein cells in long term monolayer culture: expression of aromatase, cholesterol side chain cleavage, and 3β-hydroxysteroid dehydrogenase. *J. Clin. Endocrinol. Metab.* 71, 26–33.

268. Leyendecker, G., Wildt, L., and Hansmann, M. (1980). Pregnancies following chorionic intermittent (pulsatile) administration of GnRH by means of a portable pump ("Zyklomat")—a new approach to infertility in hypothalamic amenorrhea. *J. Clin. Endocrinol. Metab.* 51, 1214–1216.

269. Crowley, W. F. Jr., and McArthur, J. W. (1980). Stimulation of the normal menstrual cycle in Kallman's syndrome by pulsatile administration of luteinizing hormone-releasing hormone (LHRH). *J. Clin. Endocrinol. Metab.* 51, 173–175.

270. Hutchison, J. S., Nelson, P. B., and Zeleznik, A. J (1986). Effects of different gonadotropin pulse frequencies on corpus luteum function during the menstrual cycle of rhesus monkeys. *Endocrinology* 119, 1964–1971.

271. Hoffman, F. (1960). Untersuchungen uber die hormonale beeinflussung der lebensdauer des corpus luteum in zyklus der frau. Geburtsh. *Frauenheilk.* 20, 1153–1159.

272. Karsch, F. J., Krey, L. C., Weick, R. F., Dierschke, D. J., and Knobil, E. (1973). Functional luteolysis in the rhesus monkey: the role of estrogen. *Endocrinology* 92, 1148–1152.

273. Gore, B. Z., Caldwell, B. V., and Speroff, L. (1973). Estrogen-induced human luteolysis. *J. Clin. Endocrinol. Metab.* 36, 615–617.

274. Knobil, E. (1973). On the regulation of the primate corpus luteum. *Biol. Reprod.* 8, 246–258.

275. Butler, W. R., Hotchkiss, J., and Knobil, E. (1975). Functional luteolysis in the rhesus monkey: ovarian estrogen and progesterone during the luteal phase of the menstrual cycle. *Endocrinology* 96, 1509–1512.

276. Karsch, F. J., and Sutton, G. P. (1976). An intra-ovarian site for the luteolytic action of estrogen in the rhesus monkey. *Endocrinology* 98, 553–561.

277. Stouffer, R. L., Nixon, W. E., and Hodgen, G. D. (1977). Estrogen inhibition of basal and gonadotropin-stimulated progesterone production by rhesus monkey luteal cells in vitro. *Endocrinology* 101, 1157–1163.

278. Depp, R., Cox, D. W., Pion, R. J., Conrad, J. H., and Heinrichs, W. L. (1973). Inhibition of the pregnenolone Δ5-3β-hydroxysteroid dehydrogenase-Δ5-4 isomerase system of human placenta and corpus luteum of pregnancy. *Gynecol. Invest.* 4, 106–120.

279. Schoonmaker, J. N., Bergman, K. S., Steiner, R. A., and Karsch, F. J. (1982). Estradiol-induced luteal regression in the rhesus monkey: evidence for an extraovarian site of action. *Endocrinology* 110, 1708–1715.

280. Hutchison, J. S., Kubik, C. J., Nelson, P. B., and Zeleznik, A. J. (1987). Estrogen induces premature luteal regression in rhesus monkeys during spontaneous menstrual cycles, but not in cycles driven by exogenous gonadotropin-releasing hormone. *Endocrinology* 121, 466–474.

281. Ellinwood, W. E., and Resko, J. A. (1983). Effect of inhibition of estrogen synthesis during the luteal phase on function of the corpus luteum in rhesus monkeys. *Biol. Reprod.* 28, 636–644.

282. Westfahl, P. K., and Resko, J. A. (1983). The effects of clomiphene on luteal function in the nonpregnant cynomolgus macaque. *Biol. Reprod.* 29, 963–969.

283. Albrecht, E. D., Haskins, A. L, Hodgen, G. D., and Pepe, G. J. (1981). Luteal function in baboons with administration of the antiestrogen ethamoxytriphetol (MER-25) throughout the luteal phase of the menstrual cycle. *Biol. Reprod.* 25, 451–457.

284. Neill, J. D., Johansson, E. D. B., and Knobil, E. (1969). Failure of hysterectomy to influence the normal pattern of cyclic progesterone secretion in the rhesus monkey. *Endocrinology* 84, 464–465.

285. Beling, C. G., Marcus, S. L., and Markham, S. M. (1970). Functional activity of the corpus luteum following hysterectomy. *J. Clin. Endocrinol. Metab.* 30, 30–39.

286. Castracane, V. D., Moore, G. T., and Shaikh, A. A. (1979). Ovarian function of hysterectomized Macaca fascicularis. *Biol. Reprod.* 20, 462–472.

287. Stouffer, R. L., Nixon, W. E., and Hodgen, G. D. (1979). Disparate effects of prostaglandins on basal and gonadotropin-stimulated progesterone production by luteal cells isolated from rhesus monkeys during the menstrual cycle and pregnancy. *Biol. Reprod.* 20, 897–903.

288. Hamberger, L., Hahlin, M., and Lindblom, B. (1987). The role of prostaglandins and catecholamines on human corpus luteum function. In *The Primate Ovary* (R. Stouffer, R. Brenner, and C. Phoenix, Eds.), pp. 191–205. Plenum Press, New York.

289. Molskness, T. A., VandeVoor, C. A., and Stouffer, R. L. (1987). Stimulatory and inhibitory effects of prostaglandins on the gonadotropin-sensitive adenylate cyclase in the monkey corpus luteum. *Prostaglandins* 34, 279–290.

290. Hahlin, M., Dennefors, B., Johanson, C., and Hamberger, L. (1988). Luteotropic effects of prostaglandin E2 on the human corpus luteum of the menstrual cycle and early pregnancy. *J. Clin. Endocrinol. Metab.* 66, 909–914.

291. Webley, G. E., Richardson, M. C., Summers P. M., Given, A., and Hearn, J. P. (1989). Changing responsiveness of luteal cells of the marmoset monkey (*Callithrix jacchus*) to luteotrophic and luteolytic agents during normal and conception cycles. *J. Reprod. Fertil.* 87, 301–310.

292. Bennegård, B., Hahlin, M., and Hamberger, L. (1990). Luteotropic effects of prostaglandins I2 and D2 on isolated human corpora lutea. *Fertil. Steril.* 54, 459–464.

293. Brannian, J. D., and Stouffer, R. L. (1991). Progesterone production by monkey luteal cell subpopulations at different stages of the menstrual cycle: changes in agonist responsiveness. *Biol. Reprod.* 44, 141–149.

294. Hamberger, L., Nilsson, L., Dennefors, B., Khan, and Sjögren, A. (1979). Cyclic AMP formation of isolated human corpora lutea in response to hCG-interference by PGF2α. *Prostaglandins* 17, 615–621.

295. Patwardhan, V. V., and Lanthier, A. (1984). Effect of prostaglandin F2α on the hCG-stimulated progesterone production by human corpora lutea. *Prostaglandins* 27, 465–473.
296. Houmard, B. S., Guan, Z., Stokes, B. T., and Ottobre, J. S. (1992). Activation of the phosphatidyl-inositol pathway in the primate corpus luteum by prostaglandin F2α. *Endocrinology* 131, 743–748.
297. Auletta, F. J., Speroff, L., and Caldwell, B. V. (1973). Prostaglandin F2-induced steroidogenesis and luteolysis in the primate corpus luteum. *J. Clin. Endocrinol. Metab.* 36, 405–407.
298. Wilks, J. W. (1980). Effect of (15S)-15-methyl-prostaglandin F2α methyl ester and estrogens upon the corpus luteum and conceptus of the rhesus monkey. *Prostaglandins* 20, 807–823.
299. Auletta, F. J., and Flint, A. P. (1988). Mechanisms controlling corpus luteum function in sheep, cows, nonhuman primates, and women especially in relation to the time of luteolysis. *Endocr. Rev.* 9, 88–105.
300. Auletta, F. J., Kamps, D. L., Pories, S., Bisset, J., and Gibson, M. (1984). An intra-corpus luteum site for the luteolytic action of prostaglandin F2α in the rhesus monkey. *Prostaglandins* 27, 285–298.
301. Challis, J. R. G., Calder, A. A., Dilley, S., Forster, C. S., Hillier, K., Hunter, D. J. S., MacKenzie, I. Z., and Thorburn, G. D. (1976). Production of prostaglandins E and F by corpora lutea, corpora albicantes and stroma from the human ovary. *J. Endocrinol.* 68, 401–408.
302. Vijayakumar, R., and Walters, W. A. W. (1983). Human luteal tissue prostaglandins, 17β-estradiol, and progesterone in relation to the growth and senescence of the corpus luteum. *Fertil. Steril.* 39, 298–303.
303. Patwardhan, V. V., and Lanthier, A. A. (1985). Luteal phase variations in endogenous concentrations of prostaglandins PGE and PGF and in the capacity for their in vitro formation in the human corpus luteum. *Prostaglandins* 30, 91–98.
304. Johnson, M. S. Ottobre, A. C., and Ottobre, J. S. (1988). Prostaglandin production by corpora lutea of rhesus monkeys: characterization of incubation conditions and examination of putative regulators. *Biol. Reprod.* 39, 839–846.
305. Houmard, B. S., and Ottobre, J. S. (1989). Progesterone and prostaglandin production by primate luteal cells collected at various stages of the luteal phase: modulation by calcium ionophore. *Biol. Reprod.* 41, 401–408.
306. Manaugh, L. C., and Novy , M. J. (1976). Effects of indomethicin on corpus luteum function and pregnancy in rhesus monkeys. *Fertil. Steril.* 27, 588–598.
307. Sargent, E. L., Baughman, W. L., Novy, M. J., and Stouffer, R. L. (1988). Intraluteal infusion of a prostaglandin synthesis inhibitor, sodium meclofenamate, causes premature luteolysis in rhesus monkeys. *Endocrinology* 123, 2261–2269.
308. Zelinski-Wooten, M. B., and Stouffer, R. L. (1990). Intraluteal infusions of prostaglandins of the E, D, I, and A series prevent PGF2α-induced, but not spontaneous, luteal regression in rhesus monkeys. *Biol. Reprod.* 43, 507–516.
309. Cameron, J. L., and Stouffer, R. L. (1982). Gonadotropin receptors of the primate corpus luteum. II. Changes in available luteinizing hormone and chorionic gonadotropin binding sites in macaque luteal membranes during the nonfertile menstrual cycle. *Endocrinology* 110, 2068–2073.
310. Eyster, K. M., Ottobre, J. S., and Stouffer, R. L .(1985). Adenylate cyclase in the corpus luteum of the rhesus monkey. III. Changes in basal and gonadotropin sensitive activities during the luteal phase of the menstrual cycle. *Endocrinology* 117, 1571–1577.
311. Madhra, M., Gay, E., Fraser, H. M., and Duncan, W. C. (2004). Alternative splicing of the human luteal LH receptor during luteolysis and maternal recognition of pregnancy. *Mol. Hum. Reprod.* 10, 599–603.
312. Corner, G. W. (1945). Development, organization, and breakdown of the corpus luteum in the rhesus monkey. *Carnegie Contrib. Embryol.* 31, 117–146.
313. Hild-Petito, S. A., Shiigi, S. M., and Stouffer, R. L. (1989). Isolation and characterization of cell subpopulations from the monkey corpus luteum of the menstrual cycle. *Biol. Reprod.* 40, 1075–1085.
314. Brannian, J. D., and Stouffer, R. L. (1991). Cellular approaches to understanding the function and regulation of the primate corpus luteum. *Sem. Reprod. Endocrinol.* 9, 341–351.
315. Kerr, J., Wyllie, A. H., and Currie, A. R. (1972). Apoptosis: a basic biological phenomenon with wide-ranging implications in tissue kinetics. *Br. J. Cancer* 26, 239–257.
316. Nakano, R. (1997). Control of the luteal function in humans. *Sem. Reprod. Endocrinol.* 15, 335–344.
317. Neill, J. D., Johansson, E. D. B., and Knobil, E. (1969). Patterns of circulating progesterone concentrations during the fertile menstrual cycle and the remainder of gestation in the rhesus monkey. *Endocrinology* 84, 45–48.
318. Gey, G. O., Seegar, G. E., and Hellman, L. M. (1938). The production of gonadotropic substance (Prolan) by placental cells in tissue culture. *Science* 88, 306–307.
319. Jones, G. E. S., Gey, G. D., and Gey, M. K. (1943). Hormone production by placental cells maintained in continuous culture. *Bull. Johns Hopkins Hosp.* 72, 26–38.
320. Hearn, J. P., Short, R. V., and Lunn, S. F. (1975). The effects of immunizing marmoset monkeys against the β subunit of hCG. In *Physiological Effects of Immunity Against Reproductive Hormones* (R. G. Edwards and M. H. Johnson, Eds.), pp. 229–247. Cambridge University Press, London.
321. Stevens, V. C. (1975). Antifertility effects from immunization with intact, subunits and fragments of hCG. In *Physiological Effects of Immunity Against Reproductive Hormones* (R. G. Edwards and M. H. Johnson, Eds.), pp. 249–274. Cambridge University Press, London.
322. Prahalada, S., Venkatramaiah, M., Roa, A. J., and Moudgal, N. R. (1975). Termination of pregnancy in macaques (*Macaca radiata*) using monkey antiserum to ovine LH. *Contraception* 12, 137–147.
323. Thau, R. B., and Sundram, K. (1980). The mechanisms of action of an antifertility vaccine in the rhesus monkey: reversal of the effects of antisera to the β subunit of ovine luteinizing hormone by medroxyprogesterone acetate. *Fertil. Steril.* 33, 317–320.
324. Tullner, W. W., and Hertz, R. O. (1966). Chorionic gonadotropin levels in the rhesus monkey during early pregnancy. *Endocrinology* 78, 204–207.
325. Arslan, M., Meyer, R. K., and Wolf, R. C. (1976). Chorionic gonadotropin in the blood and urine of pregnant rhesus monkeys. *Proc. Soc. Exp. Biol. Med.* 125, 349–352.
326. Hodgen, G. D., Niemanm, W. H., and Tullner, W. W. (1975). Duration of chorionic gonadotropin production by the placenta of the rhesus monkey. *Endocrinology* 96, 789–791.
327. Baird, D. D., Weinberg, C. R., Wilcox, A. J., McConnaughey, D. R., Musey, P. I., and Collins, D. C. (1991). Hormonal profiles of natural conception cycles ending in early, unrecognized pregnancy loss. *J. Clin. Endocrinol. Metab.* 72, 793–800.
328. Pope, C. E., Pope, V. Z., and Beck, L. R. (1982). Development of baboon preimplantation embryos to post-implantation stages in vitro. *Biol. Reprod.* 27, 915–923.

329. Fishel, S. B., Edwards, R. G., and Evans, C. J. (1984). Human chorionic gonadotropin secreted by preimplantation embryos cultured in vitro. *Science* 223, 816–818.
330. Seshagiri, P. B., and Hearn, J. P. (1993). In vitro development of in vivo produced rhesus monkey morulae and blastocysts to hatched, attached, and post-attached blastocyst stages: morphology and early secretion of chorionic gonadotropin. *Hum. Reprod.* 8, 279–287.
331. Meyer, R. K. (1972). Chorionic gonadotropin, corpus luteum function and embryo implantation in the rhesus monkey. *Acta Endocrinol. (Copenh).* 166 (Suppl.), 214–217.
332. Saxena, B. B., Kaali, S., and Landesman, R. (1977). The transport of chorionic gonadotropin through the reproductive tract. *Eur. J. Obstet. Gynaecol. Reprod. Biol.* 7, 1–4.
333. Wehrenberg, W. B., Dierschke, D. J., Wolf, R. C., and Vernon, M. W. (1980). Effect of intrauterine and intramuscular administration of human chorionic gonadotropin on corpus luteum function in cyclic rhesus monkeys. *Biol. Reprod.* 23, 10–14.
334. Ross, G. T. (1978). Human chorionic gonadotropin and maternal recognition of pregnancy. In *Maternal Recognition of Pregnancy," Ciba Foundation Symposium 64*, pp. 191–208. Excerpta Medica, Amsterdam.
335. Yoshimi, T., Strott, C. A., Marshall, J. R., and Lipsett, M. B. (1969). Corpus luteum function in early pregnancy. *J. Clin. Endocrinol. Metab.* 29, 225–230.
336. Tulchinsky, D., and Hobel, C. J. (1973). Plasma human chorionic gonadotropin, estrone, estradiol, progesterone and 17a hydroxyprogesterone in human pregnancy. III. Early normal pregnancy. *Am. J. Obstet. Gynecol.* 113, 751–758.
337. Wilks, J. W., and Noble, A. S. (1983). Steroidogenic responsiveness of the monkey corpus luteum to exogenous chorionic gonadotropin. *Endocrinology* 112, 1256–1266.
338. Neill, J. D., and Knobil, E. (1972). On the nature of the initial luteotropic stimulus of pregnancy in the rhesus monkey. *Endocrinology* 90, 34–38.
339. Stouffer, R. L. (1988). Perspectives on the corpus luteum of the menstrual cycle and early pregnancy. *Sem. Reprod. Endocrinol.* 6, 103–113.
340. Benyo, D. F., Little-Ihrig, L. L., and Zeleznik, A. J. (1993). Non-coordinated expression of luteal cell messenger ribonucleic acids during hCG stimulation of the primate corpus luteum. *Endocrinology* 124, 699–704.
341. Pierce, J. G., and Parsons, T. F. (1981). Glycoprotein hormones: structure and function. *Annu. Rev. Biochem.* 50, 465–495.
342. Parlow, A. F. (1968). Comparative bioassay of luteinizing hormone by three methods. In *Gonadotropins* (E. Rosemberg, Ed.), pp. 59–69. Geron-X, Los Altos, CA.
343. Rizkallah, T., Gurpide, E., and Vande Wiele, R. L. (1969). Metabolism of hCG in man. *J. Clin. Endocrinol. Metab.* 29, 92–100.
344. Morell, A. G., Gregoriadis, G., and Scheinberg, I. H. (1971). The role of sialic acid in determining the survival of glycoproteins in the circulation. *J. Biol. Chem.* 246, 1461–1467.
345. Cameron, J. L., and Stouffer, R. L. (1982). Gonadotropin receptors of the primate corpus luteum. I. Characterization of 125I-labeled human luteinizing hormone and human chorionic gonadotropin binding to luteal membranes from the rhesus monkey. *Endocrinology* 110, 2059–2067.
346. Stouffer, R. L. Ottobre, J. S., and VandeVoort, C. A. (1987). Regulation of the primate corpus luteum during early pregnancy. In *The Primate Ovary* (R. L. Stouffer, Ed.), pp. 207–220. Plenum Press, New York.
347. Monfort, S. L., Hess, D. L., Hendrickx, A. G., and Lasley, B. L. (1989). Absence of regular pulsatile gonadotropin secretion during implantation in the rhesus macaque. *Endocrinology* 125, 1766–1773.
348. Bousfield, G. R., Perry, W. M., and Ward, D. N. (1994). Gonadotropins: chemistry and biosynthesis. In *The Physiology of Reproduction* (E. Knobil and J. D. Neill, Eds.), pp. 1749–1792. Raven Press, New York.
349. Segaloff, D., Puett, D., and Ascoli, M. (1981). The dynamics of the steroidogenic response of perifused Leydig tumor cells to human chorionic gonadotropin, ovine luteinizing hormone, cholera toxin, and adenosine 3′,5′-cyclic monophosphate. *Endocrinology* 108, 632–638.
350. Niswender, G. D., Roess, D. A., Sawyer, H. R., Silvia, W. J., and Barisas, B. G. (1985). Differences in the lateral mobility of receptors for luteinizing hormone (LH) in the luteal cell plasma membrane when occupied by ovine LH versus human chorionic gonadotropin. *Endocrinology* 116, 164–169.
351. Zeleznik, A. J. (1999). In vivo responses of the primate corpus luteum to luteinizing hormone and chorionic gonadotropin. *Proc. Natl. Acad. Sci. U S A* 95, 11002–11007.
352. Hodgen, G. D. (1980). Patterns of secretion and antigenic similarities among primate chorionic gonadotropins: significance for fertility research. In *Chorionic Gonadotropin* (S. J. Segal, Ed.), pp. 53–63. Plenum Press, New York.
353. Hearn, J. P. (1986). The embryo-maternal dialogue during early pregnancy in primates. *J. Reprod. Fertil.* 76, 809–819.
354. Fazleabas, A. T., Donnelly, K. M., Srinivasan, S., Fortman, J. D., and Miller, J. B. (1999). Modulation of the baboon (*Papio anubis*) uterine endometrium by chorionic gonadotrophin during the period of uterine receptivity. *Proc. Natl. Acad. Sci. U S A* 96, 2543–2548.
355. Hodgen, G. D. (1983). Surrogate embryo transfer combined with estrogen-progesterone therapy in monkeys. Implantation, gestation and delivery without ovaries. *JAMA* 250, 2167–2172.
356. Lutjen, P., Trounson, A. O., Leeton, J., Findlay, J., Wood, C., and Renou, P. (1984). The establishment and maintenance of pregnancies using in vitro fertilization and embryo donation in a patient with primary ovarian failure. *Nature* 307, 174–175.
357. Goodman, A. L., and Hodgen, G. D. (1979). Corpus luteum-conceptus-follicle relationships during the fertile cycle in rhesus monkeys; pregnancy maintenance despite early luteal removal. *J. Clin. Endocrinol. Metab.* 49, 469–471.
358. Meyer, R. K., Wolf, R. C., and Arslan, M. (1969). Implantation and maintenance of pregnancy in progesterone-treated ovariectomized monkeys (*Macaca mulatta*). *2nd Int. Cong. Primatol.* 2, 30–35.
359. Fraser, H. M., Groome, N. P., and McNeilly, A. S. (1999). Follicle-stimulating hormone-inhibin B interactions during the follicular phase of the primate menstrual cycle revealed by gonadotropin-releasing hormone antagonist and antiestrogen treatment. *J. Clin. Endocrinol. Metab.* 84, 1365–1369.
360. Groome, N. P., Illingworth, P. J., O'Brien, M., Pai, R., Rodger, F. E., Mather, J. P., and McNeilly, A. S. (1996). Measurement of dimeric inhibin B throughout the human menstrual cycle. *J. Clin. Endocrinol. Metab.* 81, 1401–1405.
361. Sehested, A., Juul, A., Andersson, A. M., Petersen, J. H., Jensen, T. K., Müller, J., and Skakkebaek, N. E. (2000). Serum inhibin A and inhibin B in healthy prepubertal, pubertal, and adolescent girls and adult women: relation to age, stage of puberty, menstrual cycle, follicle-stimulating hormone, luteinizing hormone, and estradiol levels. *J. Clin. Endocrinol. Metab.* 85, 1634–1640.
362. Groome, N. P., Illingworth, P. J., O'Brien, M., Cooke, I., Ganesan, T. S., Baird, D. T., and McNeilly, A. S. (1994). Detection of dimeric inhibin throughout the human menstrual cycle by two-site enzyme immunoassay. *Clin. Endocrinol. (Oxf).* 40, 717–723.

363. Kirton, K. T., Niswender, G. D., Midgley, A. R. Jr., Jaffe, R. B., and Forbes, A. D. (1970). Serum luteinizing hormone and progesterone concentration during the menstrual cycle of the rhesus monkey. *J. Clin. Endocrinol. Metab.* 30, 105–110.
364. Monroe, S. E., Atkinson, L. E., and Knobil, E. (1970). Patterns of circulating luteinizing hormone and their relation to plasma progesterone levels during the menstrual cycle of the rhesus monkey. *Endocrinology* 87, 453–455.
365. Weick, R. F., Dierschke, D. J., Karsch, F. J., Butler, W. R., Hotchkiss, J., and Knobil, E. (1973). Periovulatory time courses of circulating gonadotropic and ovarian hormones in the rhesus monkey. *Endocrinology* 93, 1140–1147.
366. Pauerstein, C. J., Eddy, C. A., Croxatto, H. D., Hess, R., Siler-Khodr, T. M., and Croxatto, H. B. (1978). Temporal relationships of estrogen, progesterone, and luteinizing hormone levels to ovulation in women and infrahuman primates. *Am. J. Obstet. Gynecol.* 130, 876–884.
367. Croxatto, H. D., Ortiz, M. E., and Croxatto, H. B. (1980). Correlation between histologic dating of the human corpus luteum and the luteinizing hormone peak-biopsy interval. *Am. J. Obstet. Gynecol.* 136, 667–670.
368. World Health Organization Task Force Investigators. (1980). Temporal relationships between ovulation and defined changes in the concentration of plasma estradiol-17beta, luteinizing hormone, follicle-stimulating hormone, and progesterone: I. Probit analysis. *Am. J. Obstet. Gynecol.* 138, 383–390.
369. Testart, J., Frydman, R., Feinstein, M. C., Thebault, A., Roger, M., and Scholler, R. (1981). Interpretation of plasma luteinizing honnone assay for the collection of mature oocytes from women: definition of a luteinizing hormone surge-initiating rise. *Fertil. Steril.* 36, 50–54.
370. Testart, J., and Frydman, R. (1982). Minimum time lapse between luteinizing hormone surge or human chorionic gonadotropin administration and follicular rupture. *Fertil. Steril.* 37, 50–53.
371. Seibel, M. M. (1986). Luteinizing hormone and ovulation timing. *J. Reprod. Med.* 31 (Suppl 8), 754–759.
372. McLachlan, R. I., Cohen, N. L., Dahl, K. D., Bremner, W. J., and Soules, M. R. (1990). Serum inhibin levels during the periovulatory interval in normal women: relationships with sex steroid and gonadotropin levels. *Clin. Endocrinol. (Oxf).* 32, 39–48.
373. Fritz, M. A., McLachlan, R. I., Cohen, N. L., Dahl, K. D., Bremner, W. J., and Soules, M. R. (1992). Onset and characteristics of the midcycle surge in bioactive and immunoactive luteinizing hormone secretion in normal women: influence of physiological variations in periovulatory ovarian steroid honnone secretion. *J. Clin. Endocrinol. Metab.* 75, 489–493.
374. Goodman, A. L., and Hodgen, G. D. (1983). The ovarian triad of the primate menstrual cycle. *Rec. Prog. Horm. Res.* 39, 1–67.
375. Leyendecker, G., and Wildt, L. (1983). Control of gonadotropin secretion in women. In *Neuroendocrine Aspects of Reproduction* (R. L. Norman, Ed.), pp. 295–323. Academic Press, New York.
376. Thorneycroft, I. H., Mishell, D. R., Stone, S. C., Kharma, K. M., and Nakamura, R. M. (1971). The relation of serum 17-hydroxyprogesterone and estradiol-17beta levels during the human menstrual cycle. *Am. J. Obstet. Gynecol.* 111, 947–951.
377. Yen, S. S. C., Vandenberg, G., Tsai, C. C., and Siler, T. M. (1974). Causal relationship between hormonal variables in the menstrual cycle. In *Biorhythms and Human Reproduction* (M. Ferin, F. Halberg, R. M. Richart, and R. L. Vande Wiele, Eds.), pp. 219–238. Wiley, New York.
378. Reyes, F. I., Winter, J. S. D., Faiman, C., and Hobson, W. C. (1975). Serial serum levels of gonadotropins, prolactin and sex steroids in the non-pregnant and pregnant chimpanzee. *Endocrinology* 96, 1447–1455.
379. Graham, C. E. (1981). Menstrual cycle of the great apes. In *Reproductive Biology of the Great Apes* (C. E. Graham, Ed.), pp. 1–43. Academic Press, New York.
380. Steinetz, B. M., Randolph, C., and Mahoney, C. J. (1992). Serum concentrations of relaxin, chorionic gonadotropin, estradiol-17beta, and progesterone during the reproductive cycle of the chimpanzee (*Pan troglodytes*). *Endocrinology* 130, 3601–3607.
381. Nadler, R. D., Collins, D. C., and Blank, M. S. (1984). Luteinizing hormone and gonadal steroid levels during the menstrual cycle of orangutans. *J. Med. Primatol.* 13, 305–314.
382. Nadler, R. D., Graham, C. E., Collins, D. C., and Gould, K. G. (1979). Plasma gonadotropins, prolactin, gonadal steroids, and genital swelling during the menstrual cycle of lowland gorillas. *Endocrinology* 105, 290–296.
383. Nadler, R. D., Collins, D. C., Miller, L. C., and Graham, C. E. (1983). Menstrual cycle patterns of hormones and sexual behavior in gorillas. *Horm. Behav.* 17, 1–17.
384. Johansson, E. D. B., Wide, L., and Gemzell, C. (1971). Luteinizing hormone (LH) and progesterone in plasma and oestrogens in urine during 42 normal menstrual cycles. *Acta Endocrinol. (Copenh).* 68, 502–512.
385. Kling, O. R., and Westfahl, P. K. (1978). Steroid changes during the menstrual cycle of the baboon (*Papio cynocephalus*) and human. *Biol. Reprod.* 18, 392–400.
386. Krey, L. C., Butler, W. R., Weiss, G., Weick, R. F., Dierschke, D. J., and Knobil, E. (1973). Influences of endogenous and exogenous gonadal steroids on the actions of synthetic LRF in the rhesus monkey. In *Hypothalamic Hypophysiotropic Hormones. Physiological and Clinical Studies* (C. Gual and E. Rosemberg, Eds.), pp. 39–47. Excerpta Medica, Amsterdam.
387. Tillson, S. A., Swisher, D. A., Pharriss, B. B., Erickson, R. E., and Neill, J. D. (1976). Interrelationships between pituitary gonadotropins and ovarian steroids in baboons during continuous intrauterine progesterone treatment. *Biol. Reprod.* 15, 291–296.
388. Goodman, A. L., Descalzi, C. D., Johnson, D. K., and Hodgen, G. D. (1977). Composite pattern of circulating LH, FSH, estradiol, and progesterone during the menstrual cycle in cynomolgus monkeys. *Proc. Soc. Exp. Biol. Med.* 155, 479–481.
389. Shaikh, A. A., Naqvi, R. H., and Shaikh, S. A. (1978). Concentrations of oestradiol-17beta and progesterone in the peripheral plasma of the cynomologus monkey (*Macaca fascicularis*) in relation to the length of the menstrual cycle and its component phases. *J. Endocrinol.* 79, 1–7.
390. Thomson, J. A., Hess, D. L., Dahl, K. D., Iliff-Sizemore, S. A., Stouffer, R. L., and Wolf, D. P. (1992). The Sulawesi crested black macaque (*Macaca nigra*) menstrual cycle: changes in perineal tumescence and serum estradiol, progesterone, follicle-stimulating hormone, and luteinizing hormone levels. *Biol. Reprod.* 46, 879–884.
391. Reddi, K., Wickings, E. J., McNeilly, A. S., Baird, D. T., and Hillier, S. G. (1990). Circulating bioactive follicle stimulating hormone and immunoreactive inhibin levels during the normal menstrual cycle. *Clin. Endocrinol. (Oxf).* 33, 547–557.
392. Roseff, S. J., Bangah, M. L., Kettel, L. M., Vale, W., Rivier, J., Burger, H. G., and Yen, S. S. C. (1989). Dynamic changes in circulating inhibin levels during the luteal-follicular transition of the human menstrual cycle. *J. Clin. Endocrinol. Metab.* 69, 1033–1039.
393. Faiman, C., and Ryan, R. J. (1967). Serum follicle-stimulating hormone and luteinizing hormone concentrations during the

menstrual cycle as determined by radioimmunoassays. *J. Clin. Endocrinol. Metab.* 27, 1711–1716.
394. Midgley, A. R. Jr., and Jaffe, R. B. (1968). Regulation of human gonadotropins. IV. Correlation of serum concentrations of follicle-stimulating and luteinizing hormones during the menstrual cycle. *J. Clin. Endocrinol. Metab.* 28, 1699–1703.
395. Niswender, G. D., and Spies, H. G. (1973). Serum levels of luteinizing hormone, follicle-stimulating hormone and progesterone throughout the menstrual cycle of rhesus monkeys. *J. Clin. Endocrinol. Metab.* 37, 326–328.
396. Yamaji, T., Peckham, W. D., Atkinson, L. E., Dierschke, D. J., and Knobil, E. (1973). Radioimmunoassay of rhesus monkey follicle-stimulating hormone (RhFSH). *Endocrinology* 92, 1652–1659.
397. Resko, J. A., Norman, R. L., Niswender, G. D., and Spies, H. G. (1974). The relationship between progestins and gonadotropins during the late luteal phase of the menstrual cycle in rhesus monkeys. *Endocrinology* 94, 128–135.
398. Goodman, A. L., and Hodgen, G. D. (1982). Antifolliculogenic action of progesterone despite hypersecretion of FSH in monkeys. *Am. J. Physiol.* 243, E387–E397.
399. Jia, X. C., Kessel, B., Yen, S. S. C, Tucker, E. M., and Hsueh, A. J. W. (1986). Serum bioactive follicle stimulating hormone during the human menstrual cycle and in hyper- and hypogonadotropic states: application of a sensitive granulosa cell aromatase bioassay. *J. Clin. Endocrinol. Metab.* 62, 1243–1249.
400. Hall, J. E., Schoenfeld, D. A., Martin, K. A., and Crowley, W. F. Jr. (1992). Hypothalamic gonadotropin-releasing hormone secretion and follicle-stimulating hormone dynamics during the luteal-follicular transition. *J. Clin. Endocrinol. Metab.* 74, 600–607.
401. Dierschke, D. J., Bhattacharya, A. N., Atkinson, L. E., and Knobil, E. (1970). Circhoral oscillations of plasma LH in the ovariectomized rhesus monkey. *Endocrinology* 87, 850–853.
402. Knobil, E. (1992). Remembrance: the discovery of the hypothalamic gonadotropin-releasing hormone pulse generator and of its physiological significance. *Endocrinology* 131, 1005–1006.
403. Pohl, C. R., and Knobil, E. (1982). The role of the central nervous system in the control of ovarian function in higher primates. *Annu. Rev. Physiol.* 44, 583–593.
404. Santen, R. J., and Bardin, C. W. (1973). Episodic luteinizing hormone secretion in man: pulse analysis, clinical interpretation, physiological mechanisms. *J. Clin. Invest.* 52, 2617–2628.
405. Yen, S. S. C., Llerena, O., Little, B., and Pearson, O. H. (1968). Disappearance rates of endogenous luteinizing hormone and chorionic gonadotropin in man. *J. Clin. Endocrinol. Metab.* 28, 1763–1767.
406. Kohler, P. O., Ross, G. T., and Odell, W. D. (1968). Metabolic clearance and production rates of human luteinizing hormone in pre- and post-menopausal women. *J. Clin. Invest.* 47, 38–47.
407. Veldhuis, J. D., Fraioli, F., Rogol, A. D., and Dufau, M. L. (1986). Metabolic clearance of biologically active luteinizing hormone in man. *J. Clin. Invest.* 77, 1122–1128.
408. Veldhuis, J. D., and Johnson, M. L. (1988). In vivo dynamics of luteinizing hormone secretion and clearance in man: assessment by deconvolution mechanics. *J. Clin. Endocrinol. Metab.* 66, 1291–1300.
409. McIntosh, R. P., and McIntosh, J. E. A. (1985). Amplitude of episodic release of LH as a measure of pituitary function analysed from the timecourse of hormone levels in the blood: comparison of four menstrual cycles in an individual. *J. Endocrinol.* 107, 231–239.
410. Kazer, R. R., Liu, C. H., and Yen, S. S. C. (1987). Dependence of mean levels of circulating luteinizing hormone upon pulsatile amplitude and frequency. *J. Clin. Endocrinol. Metab.* 65, 796–800.
411. McIntosh, R. P., McIntosh, J. E. A., and Lazarus, L. (1988). A comparison of the dynamics of secretion of human growth hormone and LH pulses. *J. Endocrinol.* 118, 339–345.
412. Sollenberger, M. J., Carlsen, E. C., Johnson, M. L., Veldhuis, J. D., and Evans, W. S. (1990). Specific physiological regulation of luteinizing hormone secretory events throughout the human menstrual cycle: new insights into the pulsatile mode of gonadotropin release. *J. Neuroendocrinol.* 2, 845–852.
413. Veldhuis, J. D., Iranmanesh, A., Johnson, M. L., and Lizarralde, G. (1990). Twenty-four-hour rhythms in plasma concentrations of adenohypophyseal hormones are generated by distinct amplitude and/or frequency modulation of underlying pituitary secretory bursts. *J. Clin. Endocrinol. Metab.* 71, 1616–1623.
414. Yen, S. S. C., Tsai, C. C., Naftolin, F., Vandenberg, G., and Ajabor, L. (1972). Pulsatile patterns of gonadotropin release in subjects with and without ovarian function. *J. Clin. Endocrinol. Metab.* 34, 671–675.
415. Yen, S. S. C., Vandenberg, G., Tsai, C. C., and Parker, D. C. (1974). Ultradian fluctuations of gonadotropins. In *Biorhythms and Human Reproduction* (M. Ferin, R. M. Halberg, and R. L. Vande Wiele, Eds.), pp. 203–218. Wiley, New York.
416. Filicori, M., Marseguerra, M., Mimmi, P., Bolelli, G., Franceschetti, F., Possati, G., and Flamigni, C. (1982). The pattern of LH and FSH pulsatile release: physiological and clinical significance. In *The Gonadotropins: Basic Science and Clinical Aspects in Females* (C. Flamigni and J. R. Givens, Eds.), pp. 365–375. Academic Press, London.
417. Samuels, M. H., Veldhuis, J. D., Henry, P., and Ridgway, E. C. (1990). Pathophysiology of pulsatile and copulsatile release of thyroid-stimulating hormone, luteinizing hormone, follicle-stimulating hormone, and alpha-subunit. *J. Clin. Endocrinol. Metab.* 71, 425–432.
418. Veldhuis, J. D., Johnson, M. L., and Seneta, E. (1991). Analysis of the copulsatility of anterior pituitary hormones. *J. Clin. Endocrinol. Metab.* 73, 569–576.
419. Matsumoto, A. M., and Bremner, W. J. (1984). Modulation of pulsatile gonadotropin secretion by testosterone in man. *J. Clin. Endocrinol. Metab.* 58, 609–614.
420. Backstrom, C. T., McNeilly, A. S., Leask, R. M., and Baird, D. T. (1982). Pulsatile secretion of LH, FSH, prolactin, oestradiol and progesterone during the human menstrual cycle. *Clin. Endocrinol. (Oxf).* 17, 29–42.
421. Filicori, M., Santoro, N., Merriam, G. R., and Crowley, W. F. Jr. (1986). Characterization of the physiological pattern of episodic gonadotropin secretion throughout the human menstrual cycle. *J. Clin. Endocrinol. Metab.* 62, 1136–1144.
422. Filicori, M., and Crowley, W. F. Jr. (1983). Hypothalamic regulation of gonadotropin secretion in women. In *Neuroendocrine Aspects of Reproduction* (R. L. Norman, Ed.), pp. 285–294. Academic Press, New York.
423. Rossmanith, W. G., Liu, C. H., Laughlin, G. A., Mortola, J. F., Suh, B. Y., and Yen, S. S. C. (1990). Relative changes in LH pulsatility during the menstrual cycle: using data from hypogonadal women as a reference point. Clin *Endocrinol (Oxf).* 32, 647–660.
424. Reame, N., Sauder, S. E., Kelch, R. P., and Marshall, J. C. (1984). Pulsatile gonadotropin secretion during the human menstrual cycle: evidence for altered frequency of gonadotropin-releasing hormone secretion. J. *Clin. Endocrinol. Metab.* 59, 328–337.

425. Djahanbakhch, O., Warner, P., McNeilly, A. S., and Baird, D. T. (1984). Pulsatile release of LH and estradiol during the periovulatory period in women. *Clin. Endocrinol. (Oxf).* 20, 579–589.
426. Norman, R. L., Lindstrom, S. A., Bangsberg, D., Ellinwood, W. E., Gleissman, P., and Spies, H. G. (1984). Pulsatile secretion of luteinizing hormone during the menstrual cycle of rhesus macaques. *Endocrinology* 115, 261–266.
427. O'Byrne, K. T., Chen, M. D., Nishihara, M., Williams, C. L., Thalabard, J. C., Hotchkiss, J., and Knobil, E. (1993). Ovarian control of gonadotropin hormone-releasing hormone (GnRH) pulse generator activity in the rhesus monkey: duration of the associated hypothalamic signal. *Neuroendocrinology* 57, 588–592.
428. Steele, P. A., McDonnell, L. F., and Judd, S. J. (1986). Activity of gonadotropin-releasing hormone neurons during the preovulatory luteinizing hormone surge. *Fertil. Steril.* 45, 179–184.
429. Clayton, R. N., Royston, J. P., Chapman, J., Wilson, M., Obhrai, M., Sawers, R. S., and Lynch, S. S. (1987). Is changing hypothalamic activity important for control of ovulation? *Br. Med. J.* 295, 7–12.
430. Adams, J. M., Taylor, A. E., Schoenfeld, D. A., Crowley, W. F., Jr., and Hall, J. E. (1994). The midcycle gonadotropin surge in normal women occurs in the face of an unchanging gonadotropin-releasing hormone pulse frequency. *J. Clin. Endocrinol. Metab.* 79, 858–864.
431. Marut, E. J., Williams, R. F., Cowan, B. D., Lynch, A., Lerner, S. P., and Hodgen, G. D. (1981). Pulsatile pituitary gonadotropin secretion during maturation of the dominant follicle in monkeys: estrogen positive feedback enhances the biological activity of LH. *Endocrinology* 109, 2270–2272.
432. Levine, J. E., Norman, R. L., Gleissman, P. M., Oyama, T. T., Bangsberg, D. R., and Spies, H. G. (1985). In vivo gonadotropin-releasing hormone release and serum luteinizing hormone measurements in ovariectomized, estrogen-treated rhesus macaques. *Endocrinology* 117, 711–721.
433. O'Byrne, K. T., Thalabard, J. C., Grosser, P. M., Wilson, R. C., Williams, C. L., Chen, M. D., Ladendorf, D., Hotchkiss, J., and Knobil, E. (1991). Radiotelemetric monitoring of hypothalamic gonadotropin-releasing hormone activity throughout the menstrual cycle of the rhesus monkey. *Endocrinology* 129, 1207–1214.
434. Fox, S. R., and Smith, M. S. (1985). Changes in the pulsatile pattern of luteinizing hormone secretion during the rat estrus cycle. *Endocrinology* 116, 1485–1492.
435. Gallo, R. V. (1981). Pulsatile LH release during the ovulatory LH surge on proestrus in the rat. *Biol. Reprod.* 24, 100–104.
436. Van Vugt, D. A., Lam, N. Y., and Ferin, M. (1984). Reduced frequency of pulsatile luteinizing hormone secretion in the luteal phase of the rhesus monkey. Involvement of endogenous opioids. *Endocrinology* 115, 1095–1101.
437. Soules, M. R., Steiner, R. A., Clifton, D. K., Cohen, N. L., Aksel, S., and Bremner, W. J. (1984). Progesterone modulation of pulsatile luteinizing hormone secretion in normal women. *J. Clin. Endocrinol. Metab.* 58, 378–383.
438. Rossmanith, W. G., Laughlin, G. A., Mortola, J. F., and Yen, S. S. C. (1990). Secretory dynamics of oestradiol (E2) and progesterone (P4) during periods of relative pituitary LH quiescence in the midluteal phase of the menstrual cycle. *Clin. Endocrinol. (Oxf).* 32, 13–23.
439. Veldhuis, J. D., Christiansen, E., Evans, W. S., Kolp, L. A., Rogol, A. D., and Johnson, M. L. (1988). Physiological profiles of episodic progesterone release during the midluteal phase of the human menstrual cycle: analysis of circadian and ultradian rhythms, discrete pulse properties, and correlations with simultaneous luteinizing hormone release. *J. Clin. Endocrinol. Metab.* 66, 414–421.
440. Rossmanith, W. G., Laughlin, G. A., Mortola, J. F., Johnson, M. L., Veldhuis, J. D., and Yen, S. S. C. (1990). Pulsatile cosecretion of estradiol and progesterone by the midluteal phase corpus luteum: temporal link to luteinizing hormone pulses. *J. Clin. Endocrinol. Metab.* 70, 990–995.
441. Finkelstein, J. S., Badger, T. M., O'Dea, L. S. T. L., Spratt, D. I., and Crowley, W. F. (1988). Effects of decreasing the frequency of gonadotropin-releasing hormone stimulation on gonadotropin secretion in gonadotropin-releasing hormone-deficient men and perifused rat pituitary cells. *J. Clin. Invest.* 81, 1725–1733.
442. Wilson, R. C., Kesner, J. C., Kaufman, J. M., Uemura, T., Akema, T., and Knobil, E. (1984). Central electrophysiologic correlates of pulsatile luteinizing hormone secretion in the rhesus monkey. *Neuroendocrinology* 39, 256–260.
443. Veldhuis, J. D., and Johnson, M. L. (1990). A review and appraisal of deconvolution methods to evaluate in vivo neuroendocrine secretory events. *J. Neuroendocrinol.* 2, 755–771.
444. Kapen, S., Boyar, R., Perlow, M., Hellman, L., and Weitzman, E. D. (1973). Luteinizing hormone: changes in secretory pattern during sleep in adult women. *Life Sci.* 13, 693–701.
445. Kapen, S., Boyar, R., Hellman, L., and Weitzman, E. D. (1976). The relationship of luteinizing hormone secretion to sleep in women during the early follicular phase: effects of sleep reversal and a prolonged three-hour sleep-wake schedule. *J. Clin. Endocrinol. Metab.* 42, 1031–1040.
446. Soules, M. R., Steiner, R. A., Cohen, N. L., Bremner, W. J., and Clifton, D. K. (1985). Nocturnal slowing of pulsatile luteinizing hormone secretion in women during the follicular phase of the menstrual cycle. *J. Clin. Endocrinol. Metab.* 61, 43–49.
447. Rossmanith, W. G., and Yen, S. S. C. (1987). Sleep-associated decrease in luteinizing hormone pulse frequency during the early follicular phase of the menstrual cycle: evidence for an opioidergic mechanism. *J. Clin. Endocrinol. Metab.* 65, 715–718.
448. Rossmanith, W. G., Mortola, J. F., and Yen, S. S. C. (1988). Effects of dopaminergic blockade on the sleep-associated changes in luteinizing hormone pulsatility in early follicular phase women. *Neuroendocrinology* 48, 634–639.
449. Marshall, L. A., Martin, M. C., and Jaffe, R. B. (1988). Effect of preovulatory estradiol concentrations on luteinizing hormone diurnal secretory patterns: a hypothesis for the timing of the luteinizing hormone surge. *Am. J. Obstet. Gynecol.* 159, 1123–1127.
450. O'Byrne, K. T., Thalabard, J. C., Chiappini, S. E., Chen, M. D., Hotchkiss, J., and Knobil, E. (1993). Ambient light modifies gonadotropin-releasing hormone pulse generator frequency in the rhesus monkey. *Endocrinology* 133, 1520–1524.
451. Spratt, D. I., O'Dea, L. S., Schoenfeld, D., Butler, J., Rao, P. N., and Crowley, W. F. Jr. (1988). Neuroendocrine-gonadal axis in man: frequent sampling of LH, FSH, and testosterone. *Am. J. Physiol.* 254, E658–E666.
452. Crowley, W. F. Jr., Filicori, M., Spratt, D. I., and Santoro, N. F. (1985). The physiology of gonadotropin-releasing hormone (GnRH) secretion in men and women. *Rec. Prog. Horm. Res.* 41, 473–525.
453. Steiner, R. A., Peterson, A. P., Yu, J. Y. L, Conner, H., Gilbert, M., terPenning, B., and Bremner, W. J. (1980). Ultradian luteinizing hormone and testosterone rhythms

in the adult male monkey, *Macaca fascicularis*. *Endocrinology* 107, 1489–1493.

454. Plant, T. M. (1981). Time courses of concentrations of circulating gonadotropin, prolactin, testosterone, and cortisol in adult male rhesus monkeys (*Macaca mulatta*) throughout thc 24 h light-dark cyclc. *Biol. Reprod.* 25, 244–252.
455. Plant, T. M. (1986). Gonadal regulation of hypothalamic gonadotropin-releasing hormone release in primates. *Endocr. Rev.* 7, 75–88.
456. Pohl, C. R., and Hotchkiss, J. (1983). Neural control of reproductive functions in primates. *Int. Rev. Physiol.* 27, 147–175.
457. Spies, H. G., and Chappell, S. C. (1984). Mammals: nonhuman primates. In *Marshall's Physiology of Reproduction* (G. E. Lamming, Ed.), Vol. 1, pp. 659–712. Churchill Livingstone, New York.
458. Yen, S. S. C., and Lein, A. (1984). Mammals: man. In *Marshall's Physiology of Reproduction* (G. E. Lamming, Ed.), Vol. 1, pp. 713–788. Churchill Livingstone, New York.
459. Klinefelter, H. F. Jr., Albright, F., and Griswold, G. C. (1943). Experience with a quantitative test for normal or decreased amounts of follicle stimulating hormone in the urine in endocrinological diagnosis. *J. Clin. Endocrinol. Metab.* 3, 529–544.
460. Kallmann, F. J., Schoenfeld, W. A., and Barrera, S. E. (1944). The genetic aspects of primary eunuchoidism. *Am. J. Ment. Defic.* 48, 203–236.
461. Yen, S. S. C, Rebar, R., VandenBerg, G., and Judd, H. (1973). Hypothalamic amenorrhea and hypogonadotropinism: responses to synthetic LRF. *J. Clin. Endocrinol. Metab.* 36, 811–816.
462. Leyendecker, G. (1979). The pathophysiology of hypothalamic ovarian failure: diagnostic and therapeutical considerations. *Eur. J. Obstet. Gynaecol. Reprod. Biol.* 9, 175–186.
463. Spratt, D. I., Carr, D. B., Merriam, G. R., Scully, R. E., Rao, P. N., and Crowley, W. F. Jr. (1987). The spectrum of abnormal patterns of gonadotropin-releasing hormone secretion in men and idiopathic hypogonadotropic hypogonadism: clinical and laboratory correlations. *J. Clin. Endocrinol. Metab.* 64, 283–291.
464. Crowley, W. F. Jr., and Jameson, L. J. (1992). Gonadotropin-releasing hormone deficiency: perspectives from clinical investigation. *Endocr. Rev.* 13, 635–640.
465. Schwanzel-Fukuda, M., Jorgenson, K. L., Bergen, H. T., Weesner, G. D., and Pfaff, D. W. (1992). Biology of normal luteinizing hormone-releasing hormone neurons during and after their migration from olfactory placode. *Endocr. Rev.* 13, 623–634.
466. McCormack, J. T., Plant, T. M., Hess, D. L., and Knobil, E. (1977). The effect of luteinizing hormone releasing hormone (LHRH) antiserum administration on gonadotropin secretion in the rhesus monkey. *Endocrinology* 100, 663–667.
467. Fraser, H. M, McNeilly, A. S., and Popkin, R. M. (1984). Passive immunization against LH-RH: elucidation of the role of LH-RH in controlling LH and FSH secretion and LH-RH receptors. In *Immunological Aspects of Reproduction in Mammals* (D. B. Crighton, Ed.), pp. 399–418. Butterworths, London.
468. Fraser, H. M., McNeilly, A. S., Abbott, M., and Steiner, R. A. (1986). Effect of LHRH immunoneutralization on follicular development, the LH surge and luteal function in the stumptailed macaque monkey (*Macaca arctoides*). *J. Reprod. Fertil.* 76, 299–309.
469. Cetel, N. S., Rivier, J., Vale, W., and Yen, S. S. C. (1983). The dynamics of gonadotropin inhibition in women induced by an antagonistic analog of gonadotropin-releasing hormone. *J. Clin. Endocrinol. Metab.* 57, 62–65.
470. Kenigsberg, D., Littman, B. A., and Hodgen, G. D. (1984). Medical hypophysectomy. I. Dose-response using a gonadotropin-releasing hormone antagonist. *Fertil. Steril.* 42, 112–115.
471. Labrie, A., Belanger, A., and Dupont, A., Eds. (1984). *LHRH and Its Analogues: Basic and Clinical Aspects*. Excerpta Medica, Amsterdam.
472. Grady, R. R., Shin, L., Charlesworth, M. C., Cohen-Becker, I. R., Smith, M., Rivier, C., Rivier, J., Vale, W., and Schwartz, N. B. (1985). Differential suppression of follicle-stimulating hormone and luteinizing hormone secretion in vivo by a gonadotropin-releasing hormone antagonist. *Neuroendocrinology* 40, 246–252.
473. Fraser, H. M., Baird, D. T., McRae, G. I., Nestor, J. J. Jr., and Vickery, B. H. (1985). Suppression of luteal phase progesterone secretion in the stumptailed macaque by an antagonist analogue of luteinizing hormone releasing hormone. *J. Endocrinol.* 104, R1–R4.
474. Schmidt-Gollwitzer, M., and Schley R, Eds. (1985). *New Developments in Bioscience. I. LH-RH and Its Analogues: Fertility and Antifertility Aspects*. Walter de Gruyter, Berlin.
475. Kenigsberg, D., and Hodgen, G. D. (1986). Ovulation inhibition by administration of weekly gonadotropin-releasing hormone antagonist. *J. Clin. Endocrinol. Metab.* 62, 734–738.
476. Karten, M. J., and Rivier, J. E. (1986). Gonadotropin-releasing hormone analog design. Structure-function studies toward the development of agonists and antagonists: rationale and perspective. *Endocr. Rev.* 7, 44–66.
477. Vickery, B. H. (1986). Comparison of the potential for therapeutic utilities with gonadotropin-releasing honnone agonists and antagonists. *Endocr. Rev.* 7, 115–124.
478. Hall, J. E., Brodie, T. D., Badger, T. M., Rivier, J., Vale, W., Conn, P. M., Schoenfeld, D., and Crowley, W. F. Jr. (1988). Evidence of differential control of FSH and LH secretion by gonadotropin-releasing hormone (GnRH) from the use of a GnRH antagonist. *J. Clin. Endocrinol. Metab.* 67, 524–531.
479. Fraser, H. M. (1990). Inhibitory effects of treatment with an LHRH antagonist on the ovulatory cycle are reduced when administered during the late follicular phase. *Contraception* 41, 73–83.
480. Roseff, S. J., Kettel, L. M., Rivier, J., Burger, H. G., Baulieu, E., and Yen, S. S. C. (1990). Accelerated dissolution of luteal-endometrial integrity by the administration of antagonists to gonadotropin releasing hormone and progesterone to late-luteal phase women. *Fertil. Steril.* 54, 805–810.
481. Ditkoff, E. C., Cassidenti, D. L., Paulson, R. J., Sauer, M. V., Paul, W. L., Rivier, J., Yen, S. S. C, and Lobo, R. A. (1991). The gonadotropin-releasing hormone antagonist (Nal-Glu) acutely blocks the luteinizing hormone surge but allows for resumption of folliculogenesis in normal women. *Am. J. Obstet. Gynecol.* 165, 1811–1817.
482. Fluker, M. R., Marshall, L. A., Monroe, S. E., and Jaffe, R. B. (1991). Variable ovarian response to gonadotropin-releasing hormone antagonist-induced gonadotropin deprivation during different phases of the menstrual cycle. *J. Clin. Endocrinol. Metab.* 72, 912–919.
483. Gordon, K., Williams, R. F., Danforth, D. R., and Hodgen, G. D. (1991). Antide-induced suppression of pituitary gonadotropin and ovarian steroid secretion in cynomolgus monkeys: premature luteolysis and prolonged inhibition of folliculogenesis following a single treatment. *Biol. Reprod.* 44, 701–706.
484. Hall, J. E., Bhatta, N., Adams, J. M., Rivier, J. E., Vale, W. W., and Crowley, W. F. Jr. (1991). Variable tolerance of the

developing follicle and corpus luteum to gonadotropin-releasing hormone antagonist-induced gonadotropin withdrawal in the human. *J. Clin. Endocrinol. Metab.* 72, 993–1000.
485. Kettel, L. M., Roseff, S. J., Chiu, T. C., Bangah, M. L., Vale, W., Rivier, J., Burger, H. G., and Yen, S. S. C. (1991). Follicular arrest during the midfollicular phase of the menstrual cycle: a gonadotropin-releasing hormone antagonist imposed follicular-follicular transition. *J. Clin. Endocrinol. Metab.* 73, 644–649.
486. Couzinet, B., Lahlou, N., Thomas, G., Thalabard, J. C., Bouchard, P., Roger, M., and Schaison, G. (1991). Effects of gonadotropin releasing hormone antagonist and agonist on the pulsatile release of gonadotrophins and alpha-subunit in postmenopausal women. *Clin. Endocrinol. (Oxf).* 34, 477–483.
487. Frydman, R., Cornel, C., de Ziegler, D., Taieb, J., Spitz, I. M., and Bouchard, P. (1992). Spontaneous luteinizing hormone surges can be reliably prevented by the timely administration of a gonadotrophin releasing hormone antagonist (Nal-Glu) during the late follicular phase. *Hum. Reprod.* 7, 930–933.
488. Salameh, W., Bhasin, S., Steiner, B. S., Peterson, M., and Swerdloff, R. S. (1992). Effect of improved assay sensitivity on luteinizing hormone pulse detection after gonadotropin-releasing hormone antagonist treatment in man. *J. Clin. Endocrinol. Metab.* 75, 1479–1483.
489. Kolp, L. A., Pavlou, S. N., Urban, R. J., Rivier, J., Vale, W. W., and Veldhuis, J. D. (1992). Abrogation by a potent gonadotropin-releasing hormone antagonist of the estrogen/progesterone-stimulated surge-like release of luteinizing hormone and follicle-stimulating hormone in postmenopausal women. *J. Clin. Endocrinol. Metab.* 75, 993–997.
490. Belchetz, P. E., Plant, T. M., Nakai, Y., Keogh, E. J., and Knobil, E. (1978). Hypophysial responses to continuous and intermittent delivery of hypothalamic gonadotropin-releasing hormone. *Science* 202, 631–633.
491. Bergquist, C., Nillius, S. J., Bergh, T., Skarin, G., and Wide, L. (1979). Inhibitory effects of gonadotropin secretion and gonadal function in men during chronic luteinizing hormone-releasing hormone analogue. *Acta Endocrinol. (Copenh).* 91, 601–608.
492. Bergquist, C., Nillius, S. J., and Wide, L. (1979). Reduced gonadotropin secretion in postmenopausal women during treatment with a stimulatory LRH analogue. *J. Clin. Endocrinol. Metab.* 49, 472–474.
493. Linde, R., Doelle, G. C., Alexander, N., Kirshner, F., Vale, W., Rivier, J., and Rabin, D. (1981). Reversible inhibition of testicular steroidogenesis and spermatogenesis by a potent gonadotropin-releasing hormone agonist in normal men: an approach toward the development of a male contraceptive. *N. Engl. J. Med.* 305, 663–667.
494. Comite, F., Cutler, G. B., Rivier, J., Vale, W. W., Loriaux, D. L., and Crowley, W. F. (1981). Short-term treatment of idiopathic precocious puberty with a long-acting analog of luteinizing hormone-releasing hormone. *N. Engl. J. Med.* 305, 1546–1550.
495. Casper, R. F., and Yen, S. S. C. (1981). Menopausal flushes: effect of pituitary gonadotropin desensitization by a potent luteinizing hormone-releasing factor agonist. *J. Clin. Endocrinol. Metab.* 53, 1056–1058.
496. McNeil, L. W., McKenna, T. J., Lacroix, A., Benveniste, R., and Rabin, D. (1979). Seventy-two hour infusions of LHRH in normal men: gonadotropin and testicular steroid responses. *J. Clin. Endocrinol. Metab.* 49, 149–151.
497. Rabin, D., and McNeil, L. W. (1981). Pituitary and gonadal desensitization after continuous luteinizing hormone-releasing hormone infusion in normal females. *J. Clin. Endocrinol. Metab.* 51, 873–876.
498. Fraser, R. M., and Lunn, S. F. (1989). Rapid inhibitory effects of an LHRH agonist implant on the oestrogen-induced LH surge and the induction of a defective luteal phase after an agonist-induced ovulation in the macaque. *Hum. Reprod.* 4, 506–512.
499. Southworth, M. B., Matsumoto, A. M., Gross, K. M., Soules, M. R., and Bremner, W. J. (1991). The importance of signal pattern in the transmission of endocrine information: pituitary gonadotropin responses to continuous and pulsatile gonadotropin-releasing hormone. *J. Clin. Endocrinol. Metab.* 72, 1286–1289.
500. Matsumoto, A. M., Gross, K. M., and Bremner, W. J. (1991). The physiological significance of pulsatile LHRH secretion in man: gonadotropin responses to physiological doses of pulsatile versus continuous LHRH administration. *Int. J. Androl.* 14, 23–32.
501. Fraser, R. M., Haining, R., Cowen, G. M., Sandow, J., Smith, K. B., and Smith, S. K. (1992). Long-acting gonadotrophin releasing hormone agonist implant causes variable duration of suppression of ovarian steroid and inhibin secretion. *Clin. Endocrinol. (Oxf).* 36, 97–104.
502. Latimer, V. S., Kohama, S. G., Garyfallou, V. T., and Urbanski, H. F. (2001). A developmental increase in the expression of messenger ribonucleic acid encoding a second form of gonadotropin-releasing hormone in the rhesus macaque hypothalamus. *J. Clin. Endocrinol. Metab.* 86, 324–329.
503. Latimer, V. S., Rodrigues, S. M., Garyfallou, V. T., Kohama, S. G., White, R. B., Fernald, R. D., and Urbanski, H. F. (2000). Two molecular forms of gonadotropin-releasing hormone (GnRH-I and GnRH-II) are expressed by two separate populations of cells in the rhesus macaque hypothalamus. *Brain Res. Mol. Brain Res.* 75, 287–292.
504. Yahalom, D., Chen, A., Ben-Aroya, N., Rahimipour, S., Kaganovsky, E., Okon, E., Fridkin, M., and Koch, Y. (1999). The gonadotropin-releasing hormone family of neuropeptides in the brain of human, bovine and rat: identification of a third isoform. *FEBS Lett.* 463, 289–294.
505. Urbanski, H. F., White, R. B., Fernald, R. D., Kohama, S. G., Garyfallou, V. T., and Densmore, V. S. (1999). Regional expression of mRNA encoding a second form of gonadotropin-releasing hormone in the macaque brain. *Endocrinology* 140, 1945–1948.
506. Chen, A., Yahalom, D., Ben-Aroya, N., Kaganovsky, E., Okon, E., and Koch, Y. (1998). A second isoform of gonadotropin-releasing hormone is present in the brain of human and rodents. *FEBS Lett.* 435, 199–203.
507. Lescheid, D. W., Terasawa, E., Abler, L. A., Urbanski, H. F., Warby, C. M., Millar, R. P., and Sherwood, N. M. (1997). A second form of gonadotropin-releasing hormone (GnRH) with characteristics of chicken GnRH-II is present in the primate brain. *Endocrinology* 138, 5618–5629.
508. Millar, R., Lowe, S., Conklin, D., Pawson, A., Maudsley, S., Troskie, B., Ott, T., Millar, M., Lincoln, G., Sellar, R., Faurholm, B., Scobie, G., Kuestner, R., Terasawa, E., and Katz, A. (2001). A novel mammalian receptor for the evolutionarily conserved type II GnRH. *Proc. Natl. Acad. Sci. U S A* 98, 9636–9641.
509. Neill, J. D. (2002). GnRH and GnRH receptor genes in the human genome. *Endocrinology* 143, 737–743.
510. Neill, J. D., Duck, L. W., Sellers, J. C., and Musgrove, L. C. (2001). A gonadotropin-releasing hormone (GnRH) receptor specific for GnRH II in primates. *Biochem. Biophys. Res. Commun.* 282, 1012–1018.
511. Padmanabhan, V., and McNeilly, A. S. (2001). Is there an FSH-releasing factor? *Reproduction* 121, 21–30.
512. Padmanabhan, V., Brown, M. B., Dahl, G. E., Evans, N. P., Karsch, F. J., Mauger, D. T., Neill, J. D., and Van Cleeff, J. (2003).

Neuroendocrine control of follicle-stimulating hormone (FSH) secretion. III. Is there a gonadotropin-releasing hormone-independent component of episodic FSH secretion in ovariectomized and luteal phase ewes? *Endocrinology* 144, 1380–1392.

513. McCann, S. M., Karanth, S., Mastronardi, C. A., Dees, W. L., Childs, G., Miller, B., Sower, S., and Yu, W. H. (2001). Control of gonadotropin secretion by follicle-stimulating hormone-releasing factor, luteinizing hormone-releasing hormone, and leptin. *Arch. Med. Res.* 32, 476–485.
514. Bhattacharya, A. N., Dierschke, D. J., Yamaji, T., and Knobil, E. (1972). The pharmacologic blockade of the circhoral mode of LH secretion in the ovariectomized rhesus monkey. *Endocrinology* 90, 778–786.
515. Carmel, P. W., Araki, S., and Ferin, M. (1976). Pituitary stalk portal blood collection in rhesus monkeys: evidence for pulsatile release of GnRH. *Endocrinology* 99, 243–248.
516. Van Vugt, D. A., Diefenbach, W. D., Alston, E., and Ferin, M. (1985). Gonadotropin-releasing hormone pulses in third ventricular cerebrospinal fluid of ovariectomized rhesus monkeys: correlation with luteinizing hormone pulses. *Endocrinology* 117, 1550–1558.
517. Xia, L., Van Vugt, D., Alston, E. J., Luckhaus, J., and Ferin, M. (1992). A surge of gonadotropin-releasing hormone accompanies the estradiol-induced gonadotropin surge in the rhesus monkey. *Endocrinology* 131, 2812–2820.
518. Gearing, M., and Terasawa, E. (1988). Luteinizing hormone releasing hormone (LHRH) neuroterminals mapped using the push-pull perfusion method in the rhesus monkey. *Brain Res. Bull.* 21, 117–121.
519. Pau, K. Y. F, Gliessman, P. M., Hess, D. L., and Spies, H. G. (1988). Effects of estrogen on hypothalamic gonadotropin-releasing hormone release in castrated male rhesus macaques. *Brain Res.* 459, 70–75.
520. Pau, K. Y. F, Hess, D. L., Kaynard, A. H., Ji, W. Z., Gleissman, M., and Spies, H. G. (1989). Suppression of mediobasal hypothalamic gonadotropin-releasing hormone and plasma luteinizing hormone pulsatile patterns by phentolamine in ovariectomized rhesus macaques. *Endocrinology* 124, 891–898.
521. Pau, K. Y. F, Gleissman, P. M., Hess, D. L., Ronnekleiv, O. K., Levine, J. E., and Spies, H. G. (1990). Acute administration of estrogen suppresses LH secretion without altering GnRH release in ovariectomized rhesus monkeys. *Brain Res.* 517, 229–235.
522. Watanabe, G., and Terasawa, E. (1989). In vivo release of luteinizing hormone releasing hormone with puberty in the female rhesus monkey. *Endocrinology* 125, 92–99.
523. Claypool, L. E., Watanabe, G., and Terasawa, E. (1990). Effects of electrical stimulation of the medial basal hypothalamus on the in vivo release of luteinizing hormone-releasing hormone in the prepubertal and peripubertal female monkey. *Endocrinology* 127, 3014–3022.
524. Gore, A. C., and Terasawa, E. (1991). A role for norepinephrine in the control of puberty in the female rhesus monkey, Macaca mulatta. *Endocrinology* 129, 3009–3017.
525. Woller, M. J., McDonald, J. K., Reboussin, D. M., and Terasawa, E. (1992). Neuropeptide Y is a neuromodulator of pulsatile luteinizing hormone-releasing hormone release in the gonadectomized rhesus monkey. *Endocrinology* 130, 2333–2342.
526. Gore, A. C., Windsor-Engnell, B. M., and Terasawa, E. (2004). Menopausal increases in pulsatile gonadotropin-releasing hormone release in a nonhuman primate (*Macaca mulatta*). *Endocrinology* 145, 4653–4659.
527. Woller, M. J., and Terasawa E. (1994). Changes in pulsatile release of neuropeptide-Y and luteinizing hormone (LH)-releasing hormone during the progesterone-induced LH surge in rhesus monkeys. *Endocrinology* 135, 1679–1686.
528. Chongthammakun, S., Claypool, L. E., and Terasawa, E. (1993). Ovariectomy increases in vivo luteinizing hormone-releasing hormone release in pubertal, but not prepubertal, female rhesus monkeys. *J. Neuroendocrinol.* 5, 41–50.
529. Chongthammakun, S., and Terasawa, E. (1993). Negative feedback effects of estrogen on luteinizing hormone-releasing hormone release occur in pubertal, but not prepubertal, ovariectomized female rhesus monkeys. *Endocrinology* 132, 735–743.
530. Clarke, I. J., and Cummins, J. T. (1982). The temporal relationship between gonadotropin releasing hormone (GnRH) and luteinizing hormone (LH) secretion in ovariectomized ewes. *Endocrinology* 111, 1737–1739.
531. Moenter, S. M., Caraty, A., and Karsch, F. J. (1990). The estradiol-induced surge of gonadotropin-releasing hormone in the ewe. *Endocrinology* 127, 1375–1384.
532. Moenter, S. M., Brand, R. M., Midgley, A. R., and Karsch, F. J. (1992). Dynamics of gonadotropin-releasing hormone release during a pulse. *Endocrinology* 130, 503–510.
533. Plant, T. M., Krey, L. C., Moossy, J., McCormack, J. T., Hess, D. L., and Knobil, E. (1978). The arcuate nucleus and the control of gonadotropin and prolactin secretion in the female rhesus monkey (*Macaca mulatta*). *Endocrinology* 102, 52–62.
534. Heber, D., and Swerdloff, R. S. (1981). Down-regulation of pituitary gonadotropin secretion in postmenopausal females by continuous gonadotropin-releasing hormone administration. *J. Clin. Endocrinol. Metab.* 52, 171–172.
535. Wildt, L., Hausler, A., Marshall, G., Hutchison, J. S., Plant, T. M., Belchetz, P. E., and Knobil, E. (1981). Frequency and amplitude of gonadotropin-releasing hormone stimulation and gonadotropin secretion in the rhesus monkey. *Endocrinology* 109, 376–385.
536. Neill, J. D., Musgrove, L. C., Duck, L. W., and Sellers, J. C. (1999). High efficiency method for gene transfer in normal pituitary gonadotropes: adenoviral-mediated expression of G protein-coupled receptor kinase 2 suppresses luteinizing hormone secretion. *Endocrinology* 140, 2562–2569.
537. Cassina, M. P., Musgrove, L. C., Duck, L. W., Sellers, J. C., and Neill, J. D. (1999). Dual intracellular pathways in gonadotropin releasing hormone (GNRH) induced desensitization of luteinizing hormone (LH) secretion. *Life Sci.* 64, 2215–2223.
538. Neill, J. D., Duck, L. W., Musgrove, L. C., and Sellers, J. C. (1998). Potential regulatory roles for G protein-coupled receptor kinases and beta-arrestins in gonadotropin-releasing hormone receptor signaling. *Endocrinology* 139, 1781–1788.
539. Neill, J. D., Duck, L. W., Sellers, J. C., Musgrove, L. C., Scheschonka, A., Druey, K. M., and Kehrl, J. H. (1997). Potential role for a regulator of G protein signaling (RGS3) in gonadotropin-releasing hormone (GnRH) stimulated desensitization. *Endocrinology* 138, 843–846.
540. Castro-Fernandez, C., and Conn, P. M. (2002). Regulation of the gonadotropin-releasing hormone receptor (GnRHR) by RGS proteins: role of the GnRHR carboxyl-terminus. *Mol. Cell. Endocrinol.* 191, 149–156.
541. Crowley, W. F. Jr., Comite, F., Vale, W., Rivier, J., Loriaux, D. L., and Cutler, G. B. Jr. (1981). Therapeutic use of pituitary desensitization with a long acting LHRH agonist: a potential new treatment for idiopathic precocious puberty. *J. Clin. Endocrinol. Metab.* 52, 370–372.
542. Boepple, P. A., Mansfield, M. J., Wierman, M. E., Rudlin, C. R., Bode, II. II., Crigler, J. F. Jr., Crawford, J. D., and Crowley, W. F. Jr. (1986). Use of a potent, long acting

agonist of gonadotropin-releasing hormone in the treatment of precocious puberty. *Endocr. Rev.* 7, 24–33.

543. Manasco, P. K., Peskovitz, O. H., Feuillan, P. P., Hench, K. D., Barnes, K. M., Jones, J., Hill, S. C., Loriaux, D. L., and Cutler, G. B. Jr. (1988). Resumption of puberty after long term luteinizing hormone-releasing hormone agonist treatment of central precocious puberty. *J. Clin. Endocrinol. Metab.* 67, 368–372.
544. Kaplan, S. L., and Grumbach, M. M. (1989). True precocious puberty: treatment with GnRH-agonists. In *Control of the Onset of Puberty III* (H. Delemarre-van de Waal, T. M. Plant, G. P. van Rees, and J. Schoemaker, Eds.), pp. 357–373. Excerpta Medica, Amsterdam.
545. Boepple, P. A., Mansfield, M. J., Crawford, J. D., Crigler, J. F. Jr., Blizzard, R. M., and Crowley, W. F. Jr. (1991). Use of GnRH analogues as therapeutic probes in development: differential effects of sex steroids on linear growth vs. skeletal maturation in children with sexual precocity. In *Reproduction. Growth and Development* (A. Negro-Vilar and G. Perez-Palacios, Eds.), pp. 91–98. Raven Press, New York.
546. Vickery, B. H., and Lunenfeld, B., Eds. (1990). *GnRH Analogues in Cancer and Human Reproduction. Volume 2: GnRH Analogues in Reproduction and Gynecology*. Kluwer Academic Publishers, Dordrecht.
547. Vickery, B. H., Lunenfeld, B., Eds. (1990). *GnRH Analogues in Cancer and Human Reproduction. Volume 3: Benign and Malignant Tumors*. Kluwer Academic Publishers, Dordrecht.
548. Jay, N., Mansfield, M. J., Blizzard, R. M., Crowley, W. F. Jr., Schoenfeld, D., Rhubin, L., and Boepple, P. A. (1992). Ovulation and menstrual function of adolescent girls with central precocious puberty after therapy with gonadotropin-releasing hormone agonists. *J. Clin. Endocrinol. Metab.* 75, 890–894.
549. Conn, P. M., and Crowley, W. F. Jr. (1994). Gonadotropin-releasing hormone and its analogs. *Annu. Rev. Med.* 45, 391–405.
550. Gross, K. M., Matsumoto, A. M., and Bremner, W. J. (1987). Differential control of luteinizing hormone and follicle-stimulating hormone secretion by luteinizing hormone-releasing hormone pulse frequency in man. *J. Clin. Endocrinol. Metab.* 64, 675–680.
551. Spratt, D. I., Finkelstein, J. S., Butler, J. P., Badger, T. M., and Crowley, W. F. Jr. (1987). Effects of increasing the frequency of low doses of gonadotropin-releasing hormone (GnRH) on gonadotropin secretion in GnRH deficient men. *J. Clin. Endocrinol. Metab.* 64, 1179–1186.
552. Filicori, M., Flamigni, C., Campaniello, E., Ferrari, P., Meriggiola, M. C., Michelacci, L., Pareschi, A., and Valdiserri, A. (1989). Evidence for a specific role of GnRH pulse frequency in the control of the human menstrual cycle. *Am. J. Physiol.* 257, E930–E936.
553. Crowley, W. F. Jr., Whitcomb, R. W., Jameson, J. L., Weiss, J., Finkelstein, J. S., and O'Dea, L. St. L. (1991). Neuroendocrine control of human reproduction in the male. *Rec. Prog. Horm. Res.* 47, 27–62.
554. Adams, L. A., Clifton, D. K., Bremner, W. J., and Steiner, R. A. (1988). Testosterone modulates the differential release of luteinizing hormone and follicle-stimulating hormone that occurs in response to changing gonadotropin-releasing hormone pulse frequency in the male monkey, *Macaca fascicularis*. *Biol. Reprod.* 38, 156–162.
555. Pohl, C. R., Richardson, D. W., Hutchison, J. S., Germak, J. A., and Knobil, E. (1983). Hypophysiotropic signal frequency and the functioning of the pituitary-ovarian system in the rhesus monkey. *Endocrinology* 112, 2076–2080.
556. Veldhuis, J. D., Carlson, M. L., and Johnson, M. L. (1987). The pituitary gland secretes in bursts: appraising the nature of glandular secretory impulses by simultaneous multiple-parameter deconvolution of plasma hormone concentrations. Proc. *Natl. Acad. Sci. U S A* 84, 7686–7690.
557. Urban, R. J., Padmanabhan, V., Beitins, I., and Veldhuis, J. D. (1991). Metabolic clearance of human follicle-stimulating hormone assessed by radioimmunoassay, immunoradiometric assay, and in vitro Sertoli cell bioassay. *J. Clin. Endocrinol. Metab.* 73, 818–823.
558. Molskness, T. A., Woodruff, T. K., Hess, D. L., Dahl, K. D., and Stouffer, R. L. (1996). Recombinant human inhibin-A administered early in the menstrual cycle alters concurrent pituitary and follicular, plus subsequent luteal, function in rhesus monkeys. *J. Clin. Endocrinol. Metab.* 81, 4002–4006.
559. Ehara, Y., Ryan, K. J., and Yen, S. S. C. (1972). Insensitivity of synthetic LRF in LH-release of rhesus monkey. *Contraception* 6, 465–478.
560. Arimura, A., Spies, H. G., and Schally, A. V. (1973). Relative insensitivity of rhesus monkeys to the LH-releasing hormone (LH-RH). *J. Clin. Endocrinol. Metab.* 36, 372–374.
561. Ferin, M., Warren, M., Dyrenfurth, I., Vande Wiele, R. L., and White, W. F. (1974). Response of rhesus monkeys to LRH throughout the ovarian cycle. *J. Clin. Endocrinol. Metab.* 38, 231–237.
562. Frawley, L. S., Dailey, R. A., Tindall, G. T., and Neill, J. D. (1981). Increased LH secretory response to LHRH after hypophysial stalk transection of monkeys. *Neuroendocrinology* 32, 14–18.
563. Marshall, J. C., and Kelch, R. P. (1979). Low dose pulsatile gonadotropin-releasing hormone in anorexia nervosa: a model of human pubertal development. *J. Clin. Endocrinol. Metab.* 49, 712–718.
564. Valk, T. W., Corley, K. P., Kelch, R. P., and Marshall, J. C. (1980). Hypogonadotropic hypogonadism: hormonal responses to low dose pulsatile administration of gonadotropin-releasing hormone. *J. Clin. Endocrinol. Metab.* 51, 730–738.
565. Krey, L. C., Butler, W. R., and Knobil, E. (1975). Surgical disconnection of the medial basal hypothalamus and pituitary function in the rhesus monkey. I. Gonadotropin secretion. *Endocrinology* 96, 1073–1087.
566. Barry, J. (1977). Immunofluorescence study of LRF neurons in man. *Cell Tissue Res.* 181, 1–14.
567. Barry, J., and Carette, B. (1975). Immunofluorescence study of LRF neurons in primates. *Cell Tissue Res.* 164, 163–178.
568. Bugnon, C., Bloch, B., and Fellmann, D. (1977). Etude immuno-cytologique des neurones hypothalamiques a LH-RH chez le foetus humain. *Brain Res.* 128, 249–262.
569. Silverman, A. J., Antunes, J. L., Ferin, M., and Zimmerman, E. A. (1977). The distribution of luteinizing hormone-releasing hormone (LHRH) in the hypothalamus of the rhesus monkey. Light microscopic studies using immunoperoxidase technique. *Endocrinology* 101, 134–142.
570. Dudas, B., and Merchenthaler, I. (2004). Bi-directional associations between galanin and luteinizing hormone-releasing hormone neuronal systems in the human diencephalon. *Neuroscience* 127, 695–707.
571. Dudas, B., and Merchenthaler, I. (2004). Close anatomical associations between beta-endorphin and luteinizing hormone-releasing hormone neuronal systems in the human diencephalon. *Neuroscience* 124, 221–229.
572. Dudas, B., and Merchenthaler, I. (2003). Topography and associations of leu-enkephalin and luteinizing hormone-releasing hormone neuronal systems in the human diencephalon. *J. Clin. Endocrinol. Metab.* 88, 1842–1848.
573. Dudas, B., and Merchenthaler, I. (2002). Close juxtapositions between LHRH immunoreactive neurons and substance P immunoreactive axons in the human diencephalon. *J. Clin. Endocrinol. Metab.* 87, 2946–2953.

574. Dudas, B., and Merchenthaler, I. (2002). Close juxtapositions between luteinizing hormone-releasing hormone-immunoreactive neurons and corticotropin-releasing factor-immunoreactive axons in the human diencephalon. *J. Clin. Endocrinol. Metab.* 87, 5778–5784.
575. Dudas, B., and Merchenthaler, I. (2001). Catecholaminergic axons innervate LH-releasing hormone immunoreactive neurons of the human diencephalon. *J. Clin. Endocrinol. Metab.* 86, 5620–5626.
576. Dudas, B., Mihaly, A., and Merchenthaler, I. (2000). Topography and associations of luteinizing hormone-releasing hormone and neuropeptide Y-immunoreactive neuronal systems in the human diencephalon. *J. Comp. Neurol.* 427, 593–603.
577. Rance, N. E., Young, W. S. III, and McMullen, N. T. (1994). Topography of neurons expressing luteinizing hormone-releasing hormone gene transcripts in the human hypothalamus and basal forebrain. *J. Comp. Neurol.* 339, 573–586.
578. Bloch, B., Gaillard, R. C., Culler, M. D., and Negro-Vilar, A. (1992). Immunohistochemical detection of proluteinizing hormone-releasing hormone peptides in neurons in the human hypothalamus. *J. Clin. Endocrinol. Metab.* 74, 135–138.
579. Stopa, E. G., Koh, E. T., Svendsen, C. N., Rogers, W. T., Schwaber, J. S., and King, J. C. (1991). Computer-assisted mapping of immunoreactive mammalian gonadotropin-releasing hormone in adult human basal forebrain and amygdala. *Endocrinology* 128, 3199–3207.
580. Sung, W. L., Tsai, M. S., and Shen, C. L. (1988). The distribution of luteinizing hormone-releasing hormone (LHRH) neurons and fibers in the Formosan rock-monkey. *Proc. Natl. Sci. Coun. Repub. China-Pt. B Life Sci.* 12, 156–162.
581. Marshall, P. E., and Goldsmith, P. C. (1980). Neuroregulatory and neuroendocrine GnRH pathways in the hypothalamus and forebrain of the baboon. *Brain Res.* 193, 353–372.
582. Silverman, A. J., Antunes, J. L., Abrams, G. M., Nilaver, G., Thau, R., Robinson, J. A., Ferin, M., and Krey, L. (1982). The luteinizing hormone-releasing hormone pathways in rhesus (*Macaca mulatta*) and pigtailed (*Macaca nemestrina*) monkeys: new observations on thick, unembedded sections. *J. Comp. Neurol.* 211, 309–317.
583. Bourguignon, J. P., and Franchimont, P. (1984). Puberty-related increase in episodic LHRH release from rat hypothalamus in vitro. *Endocrinology* 114, 1941–1943.
584. Krsmanovic, L. Z., Stojilkovic, S. S., Merelli, F., Dufour, S. M., Virmani, M. A., and Catt, K. J. (1992). Calcium signalling and episodic secretion of gonadotropin-releasing hormone in hypothalamic neurons. *Proc. Natl. Acad. Sci. U S A* 89, 8462–8466.
585. McKibbin, P. E., and Belchetz, P. E. (1986). Prolonged pulsatile release of gonadotropin-releasing hormone from the guinea pig hypothalamus in vitro. *Life Sci.* 38, 2145–2150.
586. Rasmussen, D. D, Gambacciani, M., Swartz, W., Tueros, V. S., and Yen, S. S. C. (1989). Pulsatile gonadotropin-releasing hormone release from the human mediobasal hypothalamus in vitro: opiate receptor-mediated suppression. *Neuroendocrinology* 49, 150–156.
587. Knobil, E. (1981). Patterns of hypophysiotropic signals and gonadotropin secretion in the rhesus monkey. *Biol. Reprod.* 24, 44–49.
588. O'Byrne, K. T., and Knobil, E. (1993). Electrophysiological approaches to gonadotrophin releasing hormone pulse generator activity in the rhesus monkey. *Hum. Reprod.* 8 (Suppl 2), 37–40.
589. Kawakami, M., Uemura, T., and Hayashi, R. (1982). Electrophysiological correlates of pulsatile gonadotropin release in rats. *Neuroendocrinology* 35, 63–67.
590. Kimura, F., Nishihara, M., Hiruma, H., and Funabashi, T. (1991). Naloxone increases the frequency of the electrical activity of luteinizing hormone-releasing hormone pulse generator in long-term ovariectomized rats. *Neuroendocrinology* 53, 97–102.
591. Nishihara, M., Hiruma, H., and Kimura, F. (1991). Interactions between the noradrenergic and opioid peptidergic systems in controlling the electrical activity of luteinizing hormone-releasing hormone pulse generator in ovariectomized rats. *Neuroendocrinology* 54, 321–326.
592. Mori, Y., Nishihara, M., Tanaka, T., Shimizu, T., Yamaguchi, M., Takeuchi, T., and Hoshino, K. (1991). Chronic recording of electrophysiological manifestation of the hypothalamic gonadotropin-releasing hormone pulse generator activity in the goat. *Neuroendocrinology* 53, 392–395.
593. Tanaka, T., Ozawa, T., Hoshino, K., and Mori, Y. (1995). Changes in the gonadotropin-releasing hormone pulse generator activity during the estrous cycle in the goat. *Neuroendocrinology* 62, 553–561.
594. Silverman, A. J., Wilson, R. C., Kesner, J. S., and Knobil, E. (1986). Hypothalamic localization of multiunit electrical activity associated with pulsatile LH release in the rhesus monkey. *Neuroendocrinology* 44, 168–171.
595. Kesner, J. S., Kaufman, J. M., Wilson, R. C., Kuroda, G., and Knobil, E. (1986). The effect of morphine on the electrophysiological activity of the hypothalamic luteinizing hormone releasing hormone pulse generator in the rhesus monkey. *Neuroendocrinology* 43, 686–688.
596. Kaufman, J. M., Kesner, J. S., Wilson, R. C., and Knobil, E. (1985). Electrophysiological manifestation of luteinizing hormone-releasing hormone pulse generator activity in the rhesus monkey: influence of alpha-adrenergic and dopaminergic blocking agents. *Endocrinology* 116, 1327–1333.
597. Kesner, J. S., Wilson, R. C., Kaufman, J. M., Hotchkiss, J., Chen, Y., Yamamoto, H., Pardo, R. R., and Knobil, E. (1987). Unexpected responses of the hypothalamic gonadotropin-releasing hormone "pulse generator" to physiological estradiol inputs in the absence of the ovary. *Proc. Natl. Acad. Sci. U S A* 84, 8745–8749.
598. Williams, C. L., Nishihara, M., Thalabard, J. C., O'Byrne, K. T., Grosser, P. M., Hotchkiss, J., and Knobil, E. (1990). Duration and frequency of the multiunit electrical activity associated with the hypothalamic gonadotropin releasing hormone (GnRH) pulse generator in the rhesus monkey: differential effects of morphine. *Neuroendocrinology* 52, 225–228.
599. Williams, C. L., Nishihara, M., Thalabard, J. C., Grosser, P. M., Hotchkiss, J., and Knobil, E. (1990). Corticotropin-releasing factor and gonadotropin-releasing hormone pulse generator activity in the rhesus monkey: electrophysiological studies. *Neuroendocrinology* 52, 133–137.
600. Chen, M. D., O'Byrne, K. T., Chiappini, S. E., Hotchkiss, J., and Knobil, E. (1992). Hypoglycemic "stress" and gonadotropin releasing hormone pulse generator activity in the rhesus monkey: role of the ovary. *Neuroendocrinology* 56, 666–673.
601. Chen, M. D., Ordog, T., O'Byrne, K. T., Goldsmith, J. R., Connaughton, M. A., and Knobil, E. (1996). The insulin hypoglycemia-induced inhibition of gonadotropin-releasing hormone pulse generator activity in the rhesus monkey: roles of vasopressin and corticotropin-releasing factor. *Endocrinology* 137, 2012–2021.
602. Atkinson, L. E., Bhattacharya, A. N., Monroe, S. E., Dierschke, D. J., and Knobil, E. (1970). Effects of gonadectomy on plasma LH concentration in the rhesus monkey. *Endocrinology* 87, 847–849.

603. Cardenas, H., Ordog, T., O'Byrne, K. T., and Knobil, E. (1993). Single unit components of the hypothalamic multiunit electrical activity associated with the central signal generator that directs the pulsatile secretion of gonadotropic hormones. *Proc. Natl. Acad. Sci. U S A* 90, 9630–9634.
604. Williams, C. L., Thalabard, J. C., O'Byrne, K. T., Grosser, P. M., Nishihara, M., Hotchkiss, J., and Knobil, E. (1990). Duration of the phasic electrical activity of the hypothalamic gonadotropin-releasing hormone pulse generator and dynamics of luteinizing hormone pulses in the rhesus monkey. *Proc. Natl. Acad. Sci. U S A* 87, 8580–8582.
605. Caspar, R. F., Yen, S. S. C., and Wilkes, M. M. (1979). Menopausal flushes: a neuroendocrine link with pulsatile luteinizing hormone secretion. *Science* 205, 823–825.
606. Walsh, B., and Schiff, I. (1990). Vasomotor flushes. *Ann. N Y Acad. Sci.* 592, 346–356.
607. Kronenberg, F. (1990). Hot flashes: epidemiology and physiology. *Ann. N Y Acad. Sci.* 592, 52–86.
608. Moenter, S. M., DeFazio, A. R., Pitts, G. R., and Nunemaker, C. S. (2003). Mechanisms underlying episodic gonadotropin-releasing hormone secretion. *Front. Neuroendocrinol.* 24, 79–93.
609. Van Goor, F., Krsmanovic, L. Z., Catt, K. J., and Stojilkovic, S. S. (2000). Autocrine regulation of calcium influx and gonadotropin-releasing hormone secretion in hypothalamic neurons. *Biochem. Cell Biol.* 78, 359–370.
610. Lopez, F. J., Merchenthaler, I. J., Moretto, M., and Negro-Vilar, A. (1998). Modulating mechanisms of neuroendocrine cell activity: the LHRH pulse generator. *Cell. Mol. Neurobiol.* 18, 125–146.
611. Terasawa, E., Richter, T. A., and Keen, K. L. (2002). A role for non-neuronal cells in synchronization of intracellular calcium oscillations in primate LHRH neurons. *Prog. Brain Res.* 141, 283–291.
612. Terasawa, E. (1998). Cellular mechanism of pulsatile LHRH release. *Gen. Comp. Endocrinol.* 112, 283–295.
613. Witkin, J. W. (1999). Synchronized neuronal networks: the GnRH system. *Microsc. Res. Tech.* 44, 11–18.
614. Goldsmith, P. C., and Thind, K. K. (1995). Morphological basis for neuronal control of GnRH secretion in the monkey. In *The Neurobiology of Puberty* (T. M. Plant and P. A. Lee, Eds.), pp. 73–85. Journal of Endocrinology Limited, Bristol.
615. Richter, T. A., Keen, K. L., and Terasawa, E. (2002). Synchronization of Ca(2+) oscillations among primate LHRH neurons and nonneuronal cells in vitro. *J. Neurophysiol.* 88, 1559–1567.
616. Kesner, J. S., Kaufman, J. M., Wilson, R. C., Kuroda, G., and Knobil, E. (1986). On the short-loop feedback regulation of the hypothalamic luteinizing hormone releasing hormone "pulse generator" in the rhesus monkey. *Neuroendocrinology* 42, 109–111.
617. Ordog, T., Chen, M. D., Nishihara, M., Connaughton, M. A., Goldsmith, J. R., and Knobil, E. (1997). On the role of gonadotropin-releasing hormone (GnRH) in the operation of the GnRH pulse generator in the rhesus monkey. *Neuroendocrinology* 65, 307–313.
618. Gordan, J. D., Attardi, B. J., and Pfaff, D. W. (1998). Mathematical exploration of pulsatility in cultured gonadotropin-releasing hormone neurons. *Neuroendocrinology* 67, 2–17.
619. Brown, D., Herbison, A. E., Robinson, J. E., Marrs, R. W., and Leng, G. (1994). Modelling the luteinizing hormone-releasing hormone pulse generator. *Neuroscience* 63, 869–879.
620. Terasawa, E., and Fernandez, D. L. (2001). Neurobiological mechanisms of the onset of puberty in primates. *Endocr. Rev.* 22, 111–151.
621. Xiao, E., and Ferin, M. (1997). Stress-related disturbances of the menstrual cycle. *Ann. Med.* 29, 215–219.
622. Plant, T. M., Nakai, Y., Belchetz, P., Keogh, E. J., and Knobil, E. (1978). The sites of action of estradiol and phentolamine in the inhibition of the pulsatile, circhoral discharges of LH in the rhesus monkey (*Macaca mulatta*). *Endocrinology* 102, 1015–1018.
623. Veldhuis, J. D., Rogol, A. D., Williams, F. A., and Johnson, M. L. (1983). Do alpha-adrenergic mechanisms regulate spontaneous or opiate-modulated pulsatile luteinizing hormone secretion in man? *J. Clin. Endocrinol. Metab.* 57, 1292–1296.
624. Paradisi, R., Venturoli, S., Spada, M., Porcu, E., Magrini, O., Giambiasi, M. E., Capelli, M., and Flamigni, C. (1986). Alpha-noradrenergic mechanisms do not affect spontaneous pulsatile luteinizing hormone secretion in normal women. IRCS *J. Med. Sci.* 14, 235–236.
625. Paradisi, R., Venturoli, S., Capelli, M., Spada, M., Giambiasi, M. E., Magrini, O., Porcu, E., Fabbri, R., and Flamigni, C. (1987). Effects of alpha1-adrenergic blockade on pulsatile luteinizing hormone, follicle-stimulating hormone, and prolactin secretion in polycystic ovary syndrome. *J. Clin. Endocrinol. Metab.* 65, 841–846.
626. Ferin, M., Van Vugt, D., and Chernick, A. (1983). Central nervous system peptides and reproductive function in primates. In *Neuroendocrine Aspects of Reproduction* (R. L. Norman, Ed.), pp. 69–91. Academic Press, New York.
627. Ferin, M., Van Vugt, D., and Wardlaw, S. (1984). The hypothalamic control of the menstrual cycle and the role of endogenous opioid peptides. *Rec. Prog. Horm. Res.* 40, 441–480.
628. Reid, R. L., Quigley, M. E., and Yen, S. S. C. (1983). The disappearance of opioidergic regulation of gonadotropin secretion in post-menopausal women. *J. Clin. Endocrinol. Metab.* 57, 1107–1110.
629. Yen, S. S. C, Quigley, M. E., Reid, R. L., Ropert, J. F., and Cetel, N. S. (1985). Neuroendocrinology of opioid peptides and their role in the control of gonadotropin and prolactin secretion. *Am. J. Obstet. Gynecol.* 152, 485–493.
630. Ropert, J. F., Quigley, M. E., and Yen, S. S. C. (1981). Endogenous opiates modulate pulsatile luteinizing hormone release in humans. *J. Clin. Endocrinol. Metab.* 52, 583–585.
631. Van Vugt, D. A., Bakst, G., Dyrenfurth, I., and Ferin, M. (1983). Naloxone stimulation of luteinizing hormone secretion in the female monkey: influence of endocrine and experimental conditions. *Endocrinology* 113, 1858–1864.
632. Blankenstein, J., Reyes, F. I., Winter, J. S. D., and Faiman, C. (1981). Endorphins and the regulation of the human menstrual cycle. *Clin. Endocrinol. (Oxf).* 14, 287–294.
633. Gosselin, R. E., Blankstein, J., Dent, W. D., Hobson, W. C., Fuller, G. B., Reyes, F. I., Winter, J. S. D., and Faiman, C. (1983). Effects of naloxone and an enkephalin analogue on serum prolactin, cortisol and gonadotropins in the chimpanzee. *Endocrinology* 112, 2168–2173.
634. Rossmanith, W. G., Mortola, J. F., and Yen, S. S. C. (1988). Role of endogenous opioid peptides in the initiation of the midcycle luteinizing hormone surge in normal cycling women. *J. Clin. Endocrinol. Metab.* 67, 695–700.
635. Rossmanith, W. G., Wirth, U., Sterzik, K., and Yen, S. S. C. (1989). The effects of prolonged opioidergic blockade on LH pulsatile secretion during the menstrual cycle. *J. Endocrinol. Invest.* 12, 245–252.
636. Evans, W. S., Weltman, J. Y., Johnson, M. L., Weltman, A., Veldhuis, J. D., and Rogol, A. D. (1992). Effects of opioid

receptor blockade on luteinizing hormone (LH) pulses and interpulse LH concentrations in normal women during the early phase of the menstrual cycle. *J. Endocrinol. Invest.* 15, 525–531.
637. Shoupe, D., Montz, F. J., and Lobo, R. A. (1985). The effects of estrogen and progestin on endogenous opioid activity in oophorectomized women. *J. Clin. Endocrinol. Metab.* 60, 178–183.
638. DeFazio, J., Verheugen, C., Chetkowski, R., Nass, T., Judd, H. L., and Meldrum, D. R. (1984). The effects of naloxone on hot flashes and gonadotropin secretion in postmenopausal women. *J. Clin. Endocrinol. Metab.* 58, 578–580.
639. Quigley, M. E., and Yen, S. S. C. (1980). The role of endogenous opiates on LH secretion during the menstrual cycle. *J. Clin. Endocrinol. Metab.* 51, 179–181.
640. Casper, R. F., and Alapin-Rubillovitz, S. (1985). Progestins increase endogenous opioid peptide activity in postmenopausal women. *J. Clin. Endocrinol. Metab.* 60, 34–36.
641. Wehrenberg, W. B., Wardlaw, S. L., Frantz, A. G., and Ferin, M. (1982). Beta-endorphin in hypophyseal portal blood: variations throughout the menstrual cycle. *Endocrinology* 111, 879–881.
642. Wardlaw, S. L., Wehrenberg, W. B., Ferin, M., Antunes, J. L., and Frantz, A. (1982). Effect of sex steroids on beta-endorphin in hypophyseal portal blood. *J. Clin. Endocrinol. Metab.* 55, 877–881.
643. Thind, K. K., and Goldsmith, P. C. (1988). Infundibular GnRH neurons are inhibited by direct opioid and autoregulatory synapses in juvenile monkeys. *Neuroendocrinology* 47, 203–216.
644. Kaynard, A. H., Pau, K. Y. F., Hess, D. L., and Spies, H. G. (1990). Third-ventricular infusion of neuropeptide Y suppresses luteinizing hormone secretion in ovariectomized rhesus macaques. *Endocrinology* 127, 2437–2444.
645. Woller, M. J., and Terasawa, E. (1991). Infusion of neuropeptide Y into the stalk-median eminence stimulates in vivo release of luteinizing hormone-releasing hormone in gonadectomized rhesus monkeys. *Endocrinology* 128, 1144–1150.
646. Pau, K. Y., Berria, M., Hess, D. L., and Spies, H. G. (1995). Hypothalamic site-dependent effects of neuropeptide Y on gonadotropin-releasing hormone secretion in rhesus macaques. *J. Neuroendocrinol.* 7, 63–67.
647. McDonald, J. K. (1990). Role of neuropeptide Y in reproductive function. *Ann. N Y Acad. Sci.* 611, 258–272.
648. Khorram, O., Roselli, C. E., Ellinwood, W. E., and Spies, H. G. (1987). The measurement of neuropeptide Y in discrete hypothalamic and limbic regions of male rhesus macaques with a human NPY-directed anti-serum. *Peptides* 8, 159–163.
649. Thind, K. K., Boggan, J. E., and Goldsmith, P. C. (1993). Neuropeptide Y system of the female monkey hypothalamus: retrograde tracing and immunostaining. *Neuroendocrinology* 57, 289–298.
650. Plant, T. M., and Barker-Gibb, M. L. (2004). Neurobiological mechanisms of puberty in higher primates. *Hum. Reprod. Update* 10, 67–77.
651. Moore, G. R., and Price, D. (1932). Gonad hormone functions, and the reciprocal influence between gonads and hypophysis with its bearing on the problem of sex hormone antagonism. *Am. J. Anat.* 50, 13–71.
652. Karsch, F. J., Dierschke, D. J., Weick, R. F., Yamaji, T., Hotchkiss, J., and Knobil, E. (1973). Positive and negative feedback control by estrogen of luteinizing hormone secretion in the rhesus monkey. *Endocrinology* 92, 799–804.
653. Tsai, C. C., and Yen, S. S. (1971). Acute effects of intravenous infusion of 17-beta-estradiol on gonadotropin release in pre- and post-menopausal women. *J. Clin. Endocrinol. Metab.* 32, 766–771.
654. Yamaji, T., Dierschke, D. J., Bhattacharya, A. N., and Knobil, E. (1972). The negative feedback control by estradiol and progesterone of LH secretion in the ovariectomized rhesus monkey. *Endocrinology* 90, 771–777.
655. Yen, S. S. C, Tsai, C. C., Vandenberg, G., and Rebar, R. (1972). Gonadotropin dynamics in patients with gonadal dysgenesis: a model for the study of gonadotropin regulation. *J. Clin. Endocrinol. Metab.* 35, 897–904.
656. Hotchkiss, J., and Knobil, E. (1994). The menstrual cycle and its neuroendocrine control. In *The Physiology of Reproduction* (E. Knobil and J. D. Neill, Eds.), pp. 711–749. Raven Press, New York.
657. Grosser, P. M., O'Byrne, K. T., Williams, C. L., Thalabard, J. C., Hotchkiss, J., and Knobil E. (1993). Effects of naloxone on estrogen-induced changes in hypothalamic gonadotropin-releasing hormone pulse generator activity in the rhesus monkey. *Neuroendocrinology* 57, 115–119.
658. Leyendecker, G., Wardlaw, S., and Nocke, W. (1972). Experimental studies on the endocrine regulations during the periovulatory phase of the human menstrual cycle: the effects of exogenous 17beta-oestradiol and progesterone on the release of pituitary luteinizing and follicle stimulating hormones. *Acta Endocrinol. (Copenh).* 71, 160–178.
659. Nillius, S. J., and Wide, L. (1971). Effects of progesterone on the serum levels of FSH and LH in postmenopausal women treated with oestrogen. *Acta Endocrinol. (Copenh).* 67, 362–370.
660. Karsch, F. J., Weick, R. F., Butler, W. R., Dierschke, D. J., Krey, L. C., Weiss, G., Hotchkiss, J., Yamaji, T., and Knobil, E. (1973). Induced LH surges in the rhesus monkey: strength-duration characteristics of the estrogen stimulus. *Endocrinology* 92, 1740–1747.
661. Young, J. R., and Jaffe, R. B. (1976). Strength-duration characteristics of estrogen effects on gonadotropin response to gonadotropin-releasing hormone in women. II. Effects of varying concentrations of estradiol. *J. Clin. Endocrinol. Metab.* 42, 432–442.
662. Yamaji, J., Dierschke, D. J., Hotchkiss, J., Bhattacharya, A. N., Surve, A. H., and Knobil, E. (1971). Estrogen induction of LH release in the rhesus monkey. *Endocrinology* 89, 1034–1041.
663. Monroe, S. E., Jaffe, R. B., and Midgley, A. J. Jr. (1972). Regulation of human gonadotropins. XII. Increase in serum gonadotropins in response to estradiol. *J. Clin. Endocrinol. Metab.* 34, 342–347.
664. Nillius, S. J., and Wide, L. (1971). Induction of a midcycle-like peak of luteinizing hormone in young women by exogenous oestradiol-17beta. *J. Obstet. Gynaecol.* 78, 822–827.
665. Yen, S. S. C., and Tsai, C. C. (1971). The biphasic pattern in the feedback action of ethinyl estradiol on the release of pituitary FSH and LH. *J. Clin. Endocrinol. Metab.* 33, 882–887.
666. Yen, S. S. C., and Tsai, C. C. (1972). Acute gonadotropin release induced by exogenous estradiol during the mid-follicular phase of the menstrual cycle. *J. Clin. Endocrinol. Metab.* 34, 298–305.
667. Clifton, D. K., Steiner, R. A., Resko, J. A., and Spies, H. G. (1975). Estrogen-induced gonadotropin release in ovariectomized rhesus monkeys and its advancement by progesterone. *Biol. Reprod.* 13, 190–194.
668. Schenken, R. S., Williams, R. F., and Hodgen, G. D. (1990). Effect of 20alpha-dihydroprogesterone on the estrogen-induced bioactive luteinizing hormone surge in ovariectomized monkeys. *Fertil. Steril.* 54, 1146–1149.

669. Wolf, J. P., Hsiu, J. G., Anderson, T. L., Ulmann, A. S., Baulieu, E. E., and Hodgen, G. D. (1989). Noncompetitive antiestrogenic effect of RU 486 in blocking the estrogen-stimulated luteinizing hormone surge and the proliferative action of estradiol on endometrium in castrate monkeys. *Fertil. Steril.* 52, 1055–1060.
670. Liu, J. H., and Yen, S. S. C. (1983). Induction of midcycle gonadotropin surge by ovarian steroids in women: a critical evaluation. *J. Clin. Endocrinol. Metab.* 57, 797–802.
671. March, C. M., Goebelsmann, U., Nakamura, R. M., and Mishell, D. R. Jr. (1979). Roles of estradiol and progesterone in eliciting the midcycle luteinizing hormone and follicle-stimulating hormone surges. *J. Clin. Endocrinol. Metab.* 49, 507–513.
672. Lutjen, P. J., Findlay, J. K., Trounson, A. O., Leeton, J. F., and Chan, L. K. (1986). Effect on plasma gonadotropins of cyclic steroid replacement in women with premature ovarian failure. *J. Clin. Endocrinol. Metab.* 62, 419–423.
673. Bosu, W. T. K., Holmdahl, T. H., Johansson, E. D. B., and Gemzell, C. (1972). Peripheral plasma levels of oestrogens, progesterone and 17 alpha-OH progesterone during the menstrual cycle of the rhesus monkey. *Acta Endocrinol. (Copenh).* 71, 755–764.
674. Laborde, N., Carril, M., Cheviakoff, S., Croxatto, H. D., Pedroza, E., and Rosner, J. M. (1976). The secretion of progesterone during the periovulatory period in women with certified ovulations. *J. Clin. Endocrinol. Metab.* 43, 1157–1163.
675. Landgren, B. M., Aedo, A. R., Nunez, M., Cekan, S. Z., and Diczfalusy, E. (1977). Studies in the pattern of circulating steroids in the menstrual cycle. *Acta Endocrinol. (Copenh).* 84, 620–632.
676. Uemura, T., Suzuki, N., Tsuchihashi, S., and Shiojima, Y. (1978). Changes in serum 17 alpha-hydroxyprogesterone levels during periovulatory phase in women. *Endocrinol. Jpn.* 25, 295–298.
677. Hoff, J. D., Quigley, M. E., and Yen, S. S. C. (1983). Hormonal dynamics at midcycle: a reevaluation. *J. Clin. Endocrinol. Metab.* 57, 792–796.
678. Djahanbakhch, O., McNeilly, A. S., Warner, P. M., Swanston, I. A., and Baird D. T. (1984). Changes in plasma levels of prolactin, in relation to those of FSH, oestradiol, androstenedione and progesterone around the preovulatory surge of LH in women. *Clin. Endocrinol. (Oxf).* 20, 463–472.
679. Chang, R. J., and Jaffe, R. B. (1978). Progesterone effects on gonadotropin release in women pretreated with estradiol. *J. Clin. Endocrinol. Metab.* 47, 119–125.
680. Permazel, J. M., Lenton, E. A., Roberts, I., and Cooke, I. D. (1989). Acute effects of progesterone and the antiprogestin RU 486 on gonadotropin secretion in the follicular phase of the menstrual cycle. *J. Clin. Endocrinol. Metab.* 68, 960–965.
681. Odell, W. D., and Swerdloff, R. S. (1968). Progesterone-induced luteinizing and follicle-stimulating hormone surge in postmenopausal women: a simulated ovulatory peak. *Proc. Natl. Acad. Sci. U S A* 61, 529–536.
682. Leyendecker, G., Wildt, L., Gips, H., Nocke, W., and Plotz, E. J. (1976). Experimental studies on the positive feedback effect of progesterone, 17 alpha-hydroxyprogesterone and 20alpha-dihydroprogesterone on the pituitary release of LH and FSH in the human female: the estrogen priming of the progesterone feedback on pituitary gonadotropins in the eugonadal woman. *Arch. Gynak.* 221, 29–45.
683. Rakoff, J. S., and Yen, S. S. C. (1978). Progesterone induced acute release of prolactin in estrogen primed ovariectomized women. *J. Clin. Endocrinol. Metab.* 47, 918–921.
684. Steele, P. A., and Judd, S. J. (1988). Positive and negative feed-back effect of progesterone on luteinizing hormone secretion in post-menopausal women. *Clin. Endocrinol. (Oxf).* 29, 1–7.
685. Terasawa, E., Rodriguez-Sierra, J. F., Dierschke, D. J., Bridson, W. E., and Goy, R. W. (1980). Positive feedback effect of progesterone on luteinizing hormone (LH) release in cyclic female rhesus monkeys: LH response occurs in two phases. *J. Clin. Endocrinol. Metab.* 51, 1245–1250.
686. Terasawa, E., Krook, C., Eman, S., Watanabe, G., Bridson, W. E., Sholl, S. A., and Hei, D. L. (1987). Pulsatile luteinizing hormone (LH) release during the progesterone-induced LH surge in the female rhesus monkey. *Endocrinology* 120, 2265–2271.
687. Yeoman, R. R., and Terasawa, E. (1984). An increase in single-unit activity of the medial basal hypothalamus occurs during the progesterone-induced luteinizing hormone surge in the female rhesus monkey. *Endocrinology* 115, 2445–2452.
688. Terasawa, E., Noonan, J., and Bridson, W. E. (1982). Anaesthesia with pentobarbitone blocks the progesterone-induced luteinizing hormone surge in the ovariectomized rhesus monkey. *J. Endocrinol.* 92, 327–339.
689. Dierschke, D. J., Yamaji, T., Karsch, F. J., Weick, R. F., Weiss, G., and Knobil, E. (1973). Blockade by progesterone of estrogen-induced LH and FSH release in the rhesus monkey. *Endocrinology* 92, 1496–1501.
690. Spies, H. G., and Niswender, G. D. (1972). Effects of progesterone and estradiol on LH release and ovulation in rhesus monkeys. *Endocrinology* 90, 257–261.
691. Spies, H. G., and Niswender, G. D. (1971). Blockade of the surge of preovulatory serum luteinizing hormone and ovulation with exogenous progesterone in cycling rhesus (*Macaca mulatta*) monkeys. *J. Clin. Endocrinol. Metab.* 32, 309–316.
692. Helmond, F. A., Simons, P. A., and Hein, P. R. (1981). Strength and duration characteristics of the facilitatory and inhibitory effects of progesterone on the estrogen-induced gonadotropin surge in the female rhesus monkey. *Endocrinology* 108, 1837–1842.
693. Plant, T. M., Moossy, J., Hess, D. L., Nakai, Y., McCormack, J. T., and Knobil, E. (1979). Further studies on the effects of lesions in the rostral hypothalamus on gonadotropin secretion in the female rhesus monkey (*Macaca mulatta*). *Endocrinology* 105, 465–473.
694. Hess, D. L., Wilkins, R. H., Moossy, J., Chang, J. L., Plant, T. M., McCormack, J. T., Nakai, Y., and Knobil, E. (1977). Estrogen-induced surges in decerebrated female rhesus monkeys with medial basal hypothalamic peninsulae. *Endocrinology* 101, 1264–1271.
695. Ferin, M., Antunes, J. L., Zimmerman, E., Dyrenfurth, I., Frantz, A. G., Robinson, A., and Carmel, P. W. (1977). Endocrine function in female rhesus monkeys after hypothalamic disconnection. *Endocrinology* 101, 1611–1620.
696. Keye, W. R. Jr., and Jaffe, R. B. (1974). Modulation of pituitary gonadotropin response to gonadotropin-releasing hormone by estradiol. *J. Clin. Endocrinol. Metab.* 38, 805–810.
697. Thompson, I. E., Arfania, J., and Taymor, M. L. (1973). Effects of estrogen and progesterone on pituitary response to stimulation by luteinizing hormone-releasing factor. *J. Clin. Endocrinol. Metab.* 37, 152–155.
698. Yen, S. S., Vandenberg, G., and Siler, T. M. (1974). Modulation of pituitary responsiveness to LRF by estrogen. *J. Clin. Endocrinol. Metab.* 39, 170–177.
699. Blake, C. A. (1974). Differentiation between the "critical period," the "activation period" and the "potential activation

period" for neurohumoral stimulation of LH release in proestrus rats. *Endocrinology* 95, 572–578.

700. Ferin, M., Carmel, P. W., Zimmerman, E. A., Warren, M., Perez, R., and Vande Wiele, R. L. (1974). Location of intrahypothalamic estrogen-responsive sites influencing LH secretion in the female rhesus monkey. *Endocrinology* 95, 1059–1068.
701. Weick, R. F. (1983). Negative feedback and induction of the luteinizing hormone surge following implantation of estradiol into the hypothalamus and the pituitary in the ovariectomized rhesus monkey. In *The Anterior Pituitary Gland* (A. S. Bhatnagar, Ed.), pp. 127–137. Raven Press, New York.
702. Chappel, S. C., Resko, J. A., Norman, R. L., and Spies, H. G. (1981). Studies in rhesus monkeys on the site where estradiol inhibits gonadotropins: delivery of 17beta-estradiol to the hypothalamus and pituitary gland. *J. Clin. Endocrinol. Metab.* 52, 1–8.
703. Bogdanove, E. M. (1963). Direct gonad-pituitary feedback: an analysis of effects of intracranial estrogenic depots on gonadotropin secretion. *Endocrinology* 73, 696–712.
704. Knobil, E., Plant, T. M., Wildt, L., Belchetz, P. E., and Marshall, G. (1980). Neuroendocrine control of the rhesus monkey menstrual cycle: permissive role of the hypothalamic gonadotropin-releasing hormone (GnRH). *Science* 207, 1371–1373.
705. Richardson, D. W., Wildt, L., Hutchison, J. S., Pohl, C. R., Antunes, J. L., Ferin, M., and Knobil E. (1983). Induction of ovulatory menstrual cycles by an unvarying pulsatile GnRH infusion in pituitary stalk-sectioned monkeys. *Proceedings Endocrine Society 1983* (abstr 196).
706. Norman, R. L., Gliessman, P., Lindstrom, S. A., Hill, J., and Spies, H. G. (1982). Reinitiation of ovulatory cycles in pituitary stalk-sectioned rhesus monkeys: evidence for a specific hypothalamic message for the preovulatory release of luteinizing hormone. *Endocrinology* 111, 1874–1882.
707. Norman, R. L., Levine, J. E., and Spies, H. G. (1983). Control of gonadotropin secretion in primates: observations in stalk-sectioned rhesus macaques. In *Neuroendocrine Aspects of Reproduction* (R. L. Norman, Ed.), pp. 263–284. Academic Press, New York.
708. Hurley, D. M., Brian, R., Outch, K., Stockdale, J., Fry, A., Hackman, C., Clarke, I., Burger, H. G. (1984). Induction of ovulation and fertility in amenorrheic women by pulsatile low-dose gonadotropin-releasing hormone. *N. Engl. J. Med.* 310, 1069–1074.
709. Santoro, N., Filicori, M., and Crowley, W. F. Jr. (1986). Hypogonadotropic disorders in men and women: diagnosis and therapy with pulsatile gonadotropin-releasing hormone. *Endocr. Rev.* 7, 11–23.
710. Martin, K., Santoro, N., Hall, J., Filicori, M., Weirman, M., and Crowley, W. F. Jr. (1990). Management of ovulatory disorders with pulsatile gonadotropin-releasing hormone. *J. Clin. Endocrinol. Metab.* 71, 1081A–1081G.
711. Leyendecker, G., Waibel-Treber, S., and Wildt, L. (1990). The central control of follicular maturation and ovulation in the human. In *Oxford Review of Reproductive Biology* (S. R. Milligan, Ed.), Vol. 12, pp. 93–146. Oxford University Press, Oxford.
712. Filicori, M., Flamigni, C., Meriggiola, M. C., Ferrari, P., Michelacci, L., Campaniello, E., Valdiserri, A., and Cognigni, G. (1991). Endocrine response determines the clinical outcome of pulsatile gonadotropin-releasing hormone ovulation induction in different ovulatory disorders. *J. Clin. Endocrinol. Metab.* 72, 965–972.
713. Wildt, L., Hausler, A., Hutchison, J. S., Marshall, G., and Knobil, E. (1981). Estradiol as a gonadotropin releasing hormone in the rhesus monkey. *Endocrinology* 108, 2011–2013.
714. Ferin, M., Rosenblatt, H., Carmel, P. W., Antunes, J. L., and Vande Wiele, R. L. (1979). Estrogen-induced gonadotropin surges in female rhesus monkeys after pituitary stalk section. *Endocrinology* 104, 50–52.
715. Fraser, H. M., Lunn, S. F., Cowen, G. M., Smith, K. B., and Conn, P. M. (1991). Effect of late follicular phase administration of antide on ovulation and inhibin secretion in macaques. *Contraception* 44, 667–676.
716. Fraser, H. M., Lunn, S. F., Morris, K. D., and Deghenghi, R. (1997). Initiation of high dose gonadotrophin-releasing hormone antagonist treatment during the late follicular phase in the macaque abolishes luteal function irrespective of effects upon the luteinizing hormone surge. *Hum. Reprod.* 12, 430–435.
717. Norman, R. L., Rivier, J., Vale, W., and Spies, H. G. (1986). Inhibition of estradiol-induced gonadotropin release in ovariectomized rhesus macaques by a gonadotropin-releasing hormone antagonist. *Fertil. Steril.* 45, 288–291.
718. Adams, T. E., Norman, R. L., and Spies, H. G. (1981). Gonadotropin-releasing hormone receptor binding and pituitary responsiveness in estradiol-primed monkeys. *Science* 213, 1388–1390.
719. Dubourdieu, S., Charbonnel, B., d'Acremont, M.-F., Carreau, S., Spitz, I. M., and Bouchard, P. (1994). Effect of administration of a gonadotropin-releasing hormone (GnRH) antagonist (Nal-Glu) during the periovulatory period: the luteinizing hormone surge requires secretion of GnRH. *J. Clin. Endocrinol. Metab.* 78, 343–347.
720. Leroy, I., d'Acremont, M., Brailly-Tabard, S., Frydman, R., de Mouzon, J., and Bouchard, P. (1994). A single injection of a gonadotropin-releasing hormone (GnRH) antagonist (Cetrorelix) postpones the luteinizing hormone (LH) surge: further evidence for the role of GnRH during the LH surge. *Fertil. Steril.* 62, 461–467.
721. Christin-Maitre, S., Olivennes, F., Dubourdieu, S., Chabbert-Buffet, N., Charbonnel B, Frydman, R., and Bouchard, P. (2000). Effect of gonadotrophin-releasing hormone (GnRH) antagonist during the LH surge in normal women and during controlled ovarian hyperstimulation. *Clin. Endocrinol. (Oxf).* 52, 721–726.
722. Keye, W. R. Jr., and Jaffe, R. B. (1975). Strength-duration characteristics of estrogen effects on gonadotropin response to gonadotropin-releasing hormone in women. I. Effects of varying duration of estradiol administration. *J. Clin. Endocrinol. Metab.* 41, 1003–1008.
723. Neill, J. D., Patton, J. M., Daily, D. A., Tsou, R. C., and Tindall, G. T. (1977). Luteinizing hormone releasing hormone (LHRH) in pituitary portal blood of rhesus monkeys: relationship to level of LH release. *Endocrinology* 101, 430–434.
724. Neill, J. D., Dailey, R. A., Tsou, R. C., and Tindall, G. T. (1977). Secretion of luteinizing hormone releasing hormone (LHRH) in monkeys. *Adv. Exp. Med. Biol.* 87, 203–224.
725. Pau, K. Y., Berria, M., Hess, D. L., and Spies, H. G. (1993). Preovulatory gonadotropin-releasing hormone surge in ovarian-intact rhesus macaques. *Endocrinology* 133, 1650–1656.
726. Levine, J. E., Bethea, C. L., and Spies, H. G. (1985). In vitro gonadotropin-releasing hormone release from hypothalamic tissues of ovariectomized estrogen-treated cynomolgus macaques. *Endocrinology* 116, 431–438.
727. Hall, J. E., Taylor, A. E., Martin, K. A., Rivier, J., Schoenfeld, D. A., and Crowley, W. F. Jr. (1994). Decreased release of gonadotropin-releasing hormone during the preovulatory midcycle luteinizing hormone surge in normal women. *Proc. Natl. Acad. Sci. U S A* 91, 6894–6898.
728. Nippoldt, T. B., Reame, M. E., Kelch, R. P., and Marshall, J. C. (1989). The roles of estradiol and progesterone in decreasing

luteinizing hormone pulse frequency in the luteal phase of the menstrual cycle. *J. Clin. Endocrinol. Metab.* 69, 67–76.
729. Veldhuis, J. D., Evans, W. S., Rogol, A. D., Thorner, M. O., and Stumpf, P. (1987). Actions of estradiol on discrete attributes of the luteinizing hormone pulse signal in man: studies in postmenopausal women treated with pure estradiol. *J. Clin. Invest.* 79, 769–776.
730. Veldhuis, J. D., Rogol, A. D., Perez-Palacios, G., Stumpf, P., Kitchin, J. D., and Dufau, M. L. (1985). Endogenous opiates participate in the regulation of pulsatile luteinizing hormone release in an unopposed estrogen milieu: studies in estrogen-replaced, gonadectomized patients with testicular feminization. *J. Clin. Endocrinol. Metab.* 61, 790–793.
731. Veldhuis, J. D., Rogol, A. D., Samojlik, E., and Ertel, N. H. (1984). Role of endogenous opiates in the expression of negative feedback actions of androgen and estrogen on pulsatile properties of luteinizing hormone secretion in man. *J. Clin. Invest.* 74, 47–55.
732. Orstead, K. M., Hess, D. L., and Spies, H. G. (1987). Opiatergic inhibition of pulsatile luteinizing hormone release during the menstrual cycle of rhesus macaques. *Proc. Soc. Exp. Biol. Med.* 184, 312–319.
733. Vermesh, M., Silva, P. D., and Lobo, R. A. (1987). Endogenous opioids modulate the inhibitory effects of androgen on the hypothalamic-pituitary axis of normal women. *J. Clin. Endocrinol. Metab.* 65, 1183–1186.
734. Ellingboe, J., Veldhuis, J. D., Mendelson, J. H., Kuehnle, J. C., and Mello, N. K. (1982). Effect of endogenous opioid blockade on the amplitude and frequency of pulsatile LH secretion in normal men. *J. Clin. Endocrinol. Metab.* 54, 854–857.
735. Gindoff, P. R., Jewelewicz, R., Hembree, W., Wardlaw, S., and Ferin, M. (1988). Sustained effects of opioid antagonism during the normal human luteal phase. *J. Clin. Endocrinol. Metab.* 66, 1000–1004.
736. Steele, P. A., and Judd, S. J. (1986). Role of endogenous opioids in reducing the frequency of pulsatile luteinizing hormone secretion induced by progesterone in women. *Clin. Endocrinol. (Oxf).* 25, 669–674.
737. Wildt, L., Hutchison, J. S., Marshall, G., Pohl, C. R., and Knobil, E. (1981). On the site of action of progesterone in the blockade of the estradiol-induced gonadotropin discharge in the rhesus monkey. *Endocrinology* 109, 1293–1294.
738. Pohl, C. R., Richardson, D. W., Marshall, G., and Knobil, E. (1982). Mode of action of progesterone in the blockade of gonadotropin surges in the rhesus monkey. *Endocrinology* 110, 1454–1455.
739. Van Vugt, D. A., Heisler, L. E., and Reid, R. L. (1992). Progesterone inhibits the estrogen-induced gonadotropin surge in the rhesus monkey independent of endogenous opiates. *J. Clin. Endocrinol. Metab.* 74, 1312–1319.
740. Hayes, F. J., Hall, J. E., Boepple, P. A., and Crowley, W. F. Jr. (1998). Differential control of gonadotropin secretion in the human: endocrine role of inhibin. *J. Clin. Endocrinol. Metab.* 83, 1835–1841.
741. Fraser, H. M., and Tsonis, C. G. (1994). Manipulation of inhibin during the luteal-follicular phase transition of the primate menstrual cycle fails to affect FSH secretion. *J. Endocrinol.* 142, 181–186.
742. Fraser, H. M., Smith, K. B., Lunn, S. F., Cowen, G. M., Morris, K., and McNeilly, A. S. (1992). Immunoneutralization and immunocytochemical localization of inhibin alpha subunit during the mid-luteal phase in the stump-tailed macaque. *J. Endocrinol.* 133, 341–347.
743. Klein, N. A., Houmard, B. S., Hansen, K. R., Woodruff, T. K., Sluss, P. M., Bremner, W. J., and Soules, M. R. (2004). Age-related analysis of inhibin A, inhibin B, and activin a relative to the intercycle monotropic follicle-stimulating hormone rise in normal ovulatory women. *J. Clin. Endocrinol. Metab.* 89, 2977–2981.
744. Klein, N. A., Illingworth, P. J., Groome, N. P., McNeilly, A. S., Battaglia, D. E., and Soules, M. R. (1996). Decreased inhibin B secretion is associated with the monotropic FSH rise in older, ovulatory women: a study of serum and follicular fluid levels of dimeric inhibin A and B in spontaneous menstrual cycles. *J. Clin. Endocrinol. Metab.* 81, 2742–2745.
745. Lam, N. Y., and Ferin, M. (1987). Is the decrease in the hypophysiotropic signal frequency normally observed during the luteal phase important for menstrual cyclicity in the primate? *Endocrinology* 120, 2044–2049.
746. Soules, M. R., Steiner, R. A., Clifton, D. K., and Bremner, W. J. (1987). The effects of inducing a follicular phase gonadotropin secretory pattern in normal women during the luteal phase. *Fertil. Steril.* 47, 45–53.
747. Hotchkiss, J., Dierschke, D. J., Butler, W. R., Fritz, G. R., and Knobil, E. (1982). Relation between levels of circulating ovarian steroids and pituitary gonadotropin content during the menstrual cycle of the rhesus monkey. *Biol. Reprod.* 26, 241–248.
748. Marshall, J. C., and Griffin, M. L. (1993). The role of changing pulse frequency in the regulation of ovulation. *Hum. Reprod.* 8 (Suppl 2), 57–61.
749. Marshall, J. C., Dalkin, A. C., Haisenleder, D. J., Griffin, M. L., and Kelch, R. P. (1992). GnRH pulses—the regulators of human reproduction. *Trans. Am. Clin. Climatol. Assoc.* 104, 31–46.
750. McCartney, C. R., Gingrich, M. B., Hu, Y., Evans, W. S., and Marshall, J. C. (2002). Hypothalamic regulation of cyclic ovulation: evidence that the increase in gonadotropin-releasing hormone pulse frequency during the follicular phase reflects the gradual loss of the restraining effects of progesterone. *J. Clin. Endocrinol. Metab.* 87, 2194–2200.

Knobil and Neill's Physiology of Reproduction, Third Edition
edited by Jimmy D. Neill,
Elsevier © 2006

CHAPTER 46

Suckling and the Control of Gonadotropin Secretion

Alan S. McNeilly

The Rat, 2512
Estrus and Ovulation, 2512
Ovarian Activity, 2513
Pattern of Gonadotropin Secretion, 2514
Role of Ovarian Steroids and Opioids, 2515
Prolactin and Energy Balance, 2516
The Pig, 2516
Estrus and Ovulation, 2516
Ovarian Activity, 2517
Gonadotropins, 2517
Induction of Ovulation during Lactation, 2518
Prolactin, Opioids, and Nutritional Status, 2519
The Sheep, 2519
Estrus, Ovulation, and Gonadotropins, 2519
Prolactin, 2521
The Cow, 2522
Estrus and Ovulation, 2522
Ovarian Activity, 2522
Gonadotropins, 2523
Luteinizing Hormone, 2524
Response to Estradiol and Ovariectomy, 2524
Induction of Ovulation, 2525
Role of Prolactin, Opioids, and Negative Energy Balance, 2525
Nonhuman Primates, 2526
Ovarian Activity, 2526
Gonadotropins, 2526
Prolactin, 2528
Humans, 2528
Ovarian Activity, 2530
Gonadotropins, 2531
Luteinizing Hormone, 2531
Prolactin and Suckling, 2532
Influence of Nutritional Status, 2533
Mechanisms of Suckling-Induced Ovarian Inactivity, 2533
References, 2536

In all mammalian species the principal role of lactation is to supply the developing young with nutrition in the form of milk. However, in many (but not all) species, it also plays a major role in their reproductive strategy. In some species, such as the rat, ovulation occurs in the immediate postpartum period, but suckling causes a delay in implantation if conception occurs. In other species (e.g., the marmoset), lactation appears to have no influence on the resumption of ovulation; however, in Old World primates and humans, suckling can induce prolonged periods of ovulatory failure and infertility.

In general, in lactation the alteration in reproductive activity is caused by the suckling stimulus suppressing gonadotropin (principally luteinizing hormone) secretion and preventing normal follicular development and ovulation. However, the strategy adopted by different species is sufficiently diverse that a review of the role of suckling in controlling gonadotropin secretion can only be tackled initially

Medical Research Council Human Reproductive Sciences Unit, University of Edinburgh Centre for Reproductive Biology, Edinburgh, U.K.

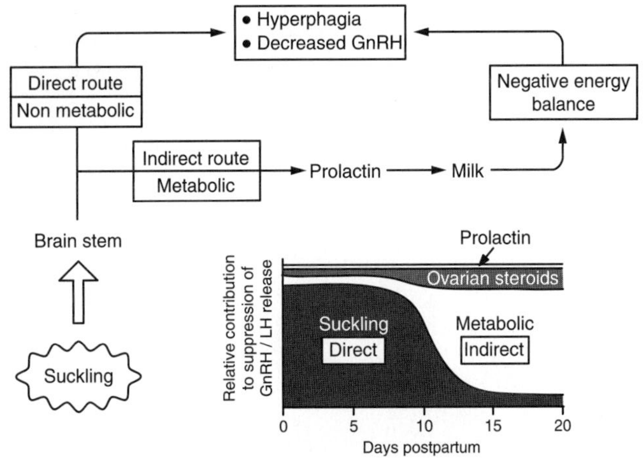

FIG. 1. Schematic of the effect of suckling on GnRH and gonadotrophin secretion, and hyperphagia, illustrating the direct effects of the neural suckling stimulus transmitted up the brain stem and the indirect effect through the influence of negative energy balance associated with the production of milk. The approximate timings of the changes in influence of these direct and indirect components throughout lactation are shown for the rat. [Adapted and modified from (72).]

by reviewing the changes during lactation in representatives of the different species. Lactation is associated with a suppression of fertility in a number of species other than those reviewed in detail. However, data on the changes in prolactin and gonadotropin levels and ovarian activity are almost absent and preclude any detailed discussion in this chapter, in which the principal aim has been to evaluate the effect of suckling on gonadotropin release. At that point it can be determined whether there are similarities in the mechanism of suckling-induced suppression of gonadotropin secretion that would suggest the involvement of a common underlying mechanism. In some species the energy burden of producing sufficient milk for growing offspring in later lactation results in a period of negative energy balance which influences gonadotrophin secretion and enhances the effects of the suckling stimulus (Fig. 1). In other species this effect seems to be very limited and suppression of gonadotrophin secretion appears to relate mainly if not solely to the direct effects of suckling alone. Whether, and at what point, in lactation these influences concur is variable across species, but it appears that the central mechanism through which the suppression of gonadotropin-releasing hormone (GnRH) pulsatile secretion occurs may be common, and will be discussed at the end of the chapter.

THE RAT

Estrus and Ovulation

The time course of events in the postpartum period in the rat is illustrated in Fig. 2. Whether lactation occurs or not, a postpartum estrus occurs within 24 hours of birth, with the exact timing of onset depending on the time of parturition but not of photoperiod (1–3). The characteristics of this estrus are similar to the estrous period of the cycle (1),

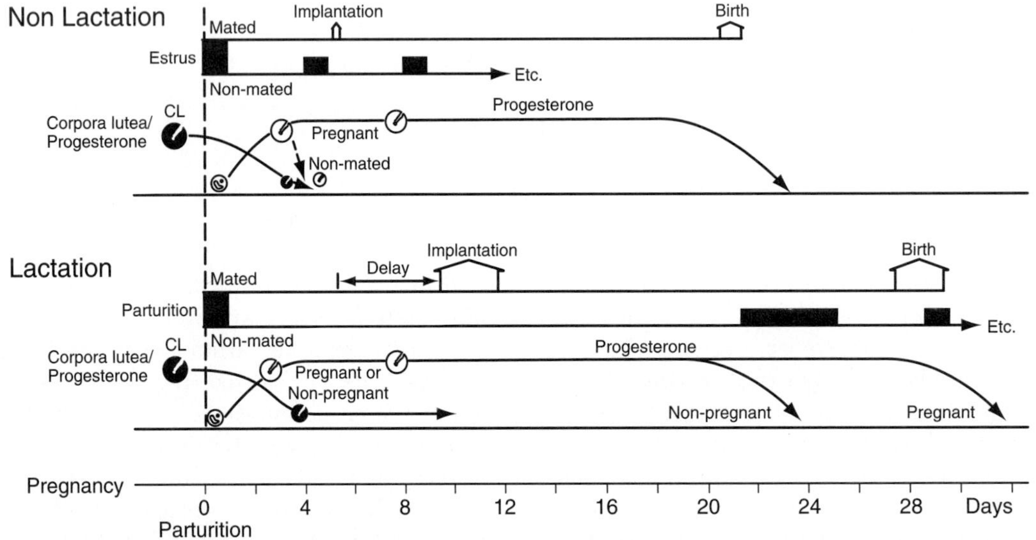

FIG. 2. Changes in the onset of estrous behavior (*black box*), implantation and birth (*open box*), the levels of progesterone secreted by the corpus luteum of pregnancy (*black circular* symbol), and the levels of progesterone formed at the postpartum estrus (*white circular* symbol) occurring within 24 hours of parturition in nonsuckled and lactating rats. Note that the corpus luteum of pregnancy regresses if the rat is not suckled but is maintained if lactation occurs.

although more ejaculations at mating appear to be necessary for pregnancy to occur postpartum than during the estrous cycle (4). It is probable that, in the natural habitat, postpartum estrus is the normal time of conception for female rats (5), thus allowing the female to be pregnant with the next litter while nurturing the first (6). Thus when one litter is weaned the second litter is delivered shortly thereafter and a third is conceived. However, to achieve this timing, implantation (which normally occurs on day 5 of pregnancy) is delayed until around day 10, a delay of 5 to 7 days during which the blastocysts enter a metabolically quiescent state (7). Pregnancy is thus extended during lactation such that weaning occurs around days 25 to 28 prior to parturition, with pregnancy having been extended from its normal 21 days to 28 to 30 days or longer.

If mating does not occur, then normal estrous cycles resume, with the second estrus occurring 4 to 5 days after the postpartum estrus. In contrast, estrus and ovulation are delayed during lactation for about 20 days in rats suckling 6 to 10 pups (8–11). The duration of this delay is dependent on the litter size (8,9), with the time interval between parturition and estrus varying between 13 and 36 days in rats with litters of between 1 and 24 pups (Fig. 2) (12). The effect of variations in litter size up to the normal maximum litter size of 13 is relatively small (8,9,12), and it is a common finding that ovulation will occur in late lactation even if attempts are made to keep the suckling stimulus at a maximum (12). However, with large litters (>12 pups), in which suckling avidity of individual pups is greater than normal or in which the avidity of suckling of a normal-sized litter (e.g., 12 pups) (Fig. 3) is maintained by providing the mother every day with a 24-hour food-and-water-deprived litter of the age corresponding to the lactational age of the mother, then the onset of estrus can be delayed for a considerably longer period (Fig. 3) (12). Thus the suppression of estrus during lactation in the rat is directly related to the strength or avidity of the suckling stimulus.

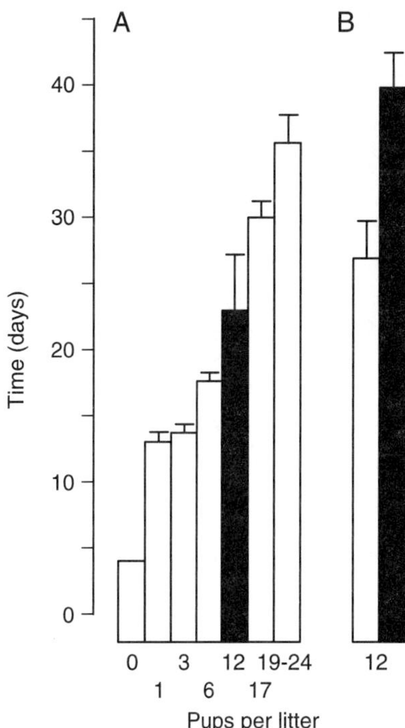

FIG. 3. The effects of litter size (**A**) and suckling activity (**B**) on the interval (mean ± SEM) between parturition and first estrus in the rat. Note that increasing the litter size extends this interval (**A**), and for a given litter size it can also be extended if suckling activity is maintained through lactation either when newborn litters of 12 pups are given to lactating rats every 7 days (**B**; *open bars*) or when litters of 12 pups of the age corresponding to the lactational age of the mother and deprived of food and water for 24 hours were given to her each day, starting on day 5 of lactation (*closed bars*). [Redrawn from (12).]

Ovarian Activity

Ovulation occurs within the first 24 to 36 hours after parturition (1,2), and the resulting corpora lutea (CL) are maintained throughout lactation (8,63). The weight of these increases to a plateau around day 10 of lactation, coinciding with an increase in the secretion of progesterone from these CL and a concomitant increase in plasma levels of progesterone to reach maximum levels of 60 to 150 ng/ml by days 6 to 8 lactation (7,9,13–17). The CL of pregnancy are also maintained through lactation, but both their weight and progesterone output decrease, and they are only about 20% as active as the CL of lactation (17). As in nonlactating pregnant or pseudopregnant rats, the CL, at least in early lactation, are dependent on prolactin for the maintenance of progesterone secretion (9,17–20). The amount of progesterone produced by the CL is also correlated with the litter size by the indirect effect of suckling intensity on the circulating levels of prolactin (15,17,21). During the second half of lactation progesterone production by the CL becomes more dependent on luteinizing hormone (LH) (22).

While the CL are maintained in lactation, all large antral follicles degenerate during the first half of lactation (7,9,10,17,23–27); also, the degree of suppression of follicle growth is related to the intensity of the suckling stimulus (17). Serum or plasma levels of estradiol (16,17,25,26) and inhibin (25–27) are consistently low during this time. However, a small increase to levels similar to those during the diestrous phase of the cycle occurs around day 10 to 12 (16), about the time when implantation of embryos, an estrogen-dependent event (28), occurs to end the lactation-induced diapause. After day 16 of lactation the number of healthy antral follicles starts to increase, although this is not necessarily accompanied by an increase in serum estradiol and inhibin levels (17,25,26). As will be described later, the absence of follicular development correlates with a low level of LH. In early lactation, no or very few follicles were present that were capable of ovulating in response to a single large injection of human chorionic gonadotropin (hCG), and those follicles present had low levels of follicle-stimulating hormone (FSH) receptors (24). Responsive follicles could be induced by pretreatment for 5 days with low levels of either hCG or LH (17,22,29). Thus, the failure of follicular development appears to be directly related to a lack of LH support, since a transient increase in FSH associated with reduced negative feedback following immunoneutralization of GnRH during lactation did not affect follicle growth (30).

Pattern of Gonadotropin Secretion

Follicle-Stimulating Hormone

In intact lactating rats, plasma or serum levels of FSH while low in the first 3 to 5 days of lactation (22,25) are equivalent to those in nonlactating rats throughout the rest of lactation (9,16,17,22,25,31,32). The pituitary content of FSH increases by day 5 of lactation to be equivalent to, and by day 10 onward generally exceeds, that of cycling female rats (17,33,34). The total amount of FSH released in vitro in response to GnRH stimulation of pituitaries from lactating and cycling females is also similar (33,34), although there was a reduced priming effect of GnRH on FSH release (34). These results suggest that the secretion of FSH is not suppressed during lactation in the rat. However, in the near absence of inhibitory products secreted by the inactive ovary, at least during the first half of lactation, it might be expected that FSH levels should be greater than those in intact cyclic rats. That this is not the case suggests a subtle deficiency in FSH release. An increase in FSH after ovariectomy or immunoneutralization of inhibin only occurred in the second half of lactation when follicle growth had resumed and plasma concentrations of inhibin were increasing (25). The increase in FSH could be blocked by treatment with inhibin but not steroids (25). Thus during early lactation, suckling alone, in the absence of inhibin from the ovary, causes suppression of FSH, presumably by inhibiting GnRH secretion (25,35). In late lactation, when GnRH secretion resumes, FSH is maintained at normal levels by the presence of inhibin secreted by follicles developing during lactation in response to the limited increase in LH secretion occurring during the second half of lactation.

Luteinizing Hormone

It is clear that until late lactation (i.e., beyond day 16) basal serum or plasma levels of LH are suppressed to around 25% of levels seen in the diestrous phase of the cycle (9,10,16,17,21–25,31–34,36–39). LH levels increase around day 20, coinciding with the decline in progesterone and increase in 20a-dihydroprogesterone levels at that time (16,17). The low levels of LH result from suppression of pulsatile secretion of LH at least to day 10 of lactation (Fig. 4) (40–43) in contrast to the normal pulsatile secretion of LH during the estrous cycle (44,45). Even the minimal suckling input of two pups is sufficient to suppress pulsatile LH secretion on day 10 of lactation (Fig. 4) (40). During early lactation the pituitary content of LH is decreased, associated with a decrease in LH-β and gonadotropin α-subunit mRNA abundance (46–48), increasing in mid-lactation to be similar to that of the diestrous female (16,17,21,31,37,49). Pituitary GnRH-receptor content (50,51) and mRNA (52) is decreased to only 10% of that in diestrous, in suckled females (50,53). The hypothalamic content of

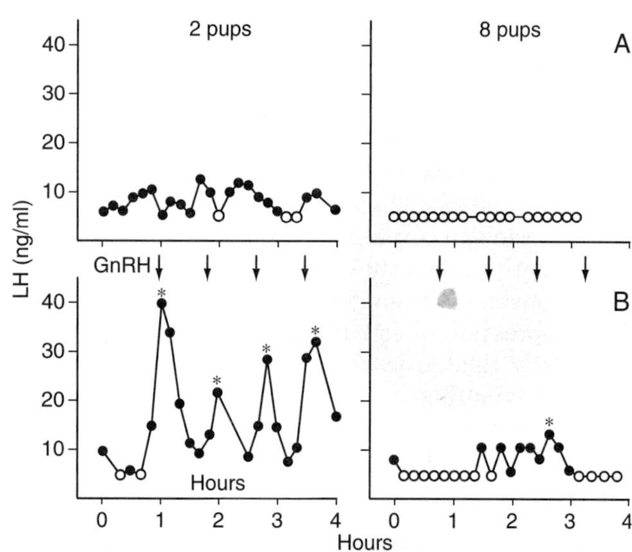

FIG. 4. The basal patterns of LH secretion **(A)** and response to 0.4 ng GnRH injections every 50 min (**B**; *arrows*) on day 10 of lactation in intact rats with two or eight pups. Blood samples were collected every 6 to 10 minutes for 3 to 4 hours. Note the absence of pulsatile secretion and only minimal basal LH secretion in the suckled intact rats. The adequate LH response to GnRH in the rats suckled by two pups indicates that the poor pulsatile LH pattern is not due to an inability of the pituitary to respond to GnRH. The very much reduced response to GnRH in the rats suckled by eight pups reflects a reduced pituitary content of LH caused by a greater suckling-induced suppression of hypothalamic GnRH release. [Redrawn from (40).]

GnRH may (31,51) or may not (54) be reduced in lactation but there is no release of GnRH, measured indirectly as LH, in response to the excitatory amino acids (55–57).

In spite of the reduction in GnRH receptors and pituitary content of LH, the pituitary remains responsive to exogenous GnRH, which causes the release of LH both in vivo (31,40,49,58,59) and in vitro (33,34), although in both situations the amount of LH released is reduced in comparison to diestrus. The decrease in sensitivity to GnRH stimulation is illustrated dramatically in vivo where repeated injections of "physiological" amounts of GnRH (0.4 or 0.5 ng) failed to induce normal pulsatile LH release on day 10 of lactation in mothers suckling eight pups but did induce a response in those suckling only two pups (Fig. 4) (40,59). Even with larger amounts of GnRH (20–500 ng) there is a decrease, in lactating rats (58,60), in the degree of potentiation of LH release compared with that normally seen in diestrous females (61). This decrease in response to GnRH, which occurs in vitro, appears to be related to a reduction in the post GnRH receptor changes in the gonadotrope (34), but this decrease in response can be overcome by longer treatment with GnRH (59). Thus, treatment of suckling rats with pulsatile GnRH for 24 hours from day 10 of lactation resulted in a dose-dependent increase in the pituitary GnRH receptor content to levels in nonsuckling rats and a concomitant dose-dependent increase in the amplitude of the pulses of LH released after each GnRH pulse (Fig. 4) (59,60). This response is similar to that seen after weaning (46–48,62), which results in a rapid increase in GnRH receptors to normal by 16 hours after weaning (46–48). In ovariectomized lactating rats the increase in GnRH receptors is accompanied by a return of pulsatile LH secretion within 12 hours after weaning, an increase in LH-β and gonadotropin α-subunit mRNA abundance by 24 hours, and a normal pattern of LH pulses after 48 to 72 hours (46,47,63). These changes are reversed if suckling resumes (63). However, in normal ovary-intact lactating rats the restoration of LH-β and gonadotropin α-subunit mRNA and content of LH in the pituitary and the resumption of normal pulsatile LH release was delayed by 24 to 48 hours due to the continued effect of progesterone from the CL of lactation on hypothalamic GnRH output after weaning (47,48). Nevertheless, follicle growth occurs and a preovulatory surge of LH and ovulation occurs within 72 hours of weaning at day 15 (24).

Role of Ovarian Steroids and Opioids

In the normal estrous cyclic female a preovulatory type LH surge can be induced by the injection of estrogen (64,65). In lactating females such a response still occurs, but the amount of LH released is significantly reduced (21,31), as is the response to progesterone after estrogen priming (58,66). The induction of daily surges of LH, which occurs when nonlactating females are continuously exposed to estradiol (64), is also abolished (21) or severely reduced (67). The magnitude of the response to estrogen also decreases as litter size increases (21), as does the LH response to GnRH injection. Thus in the lactating rat, the positive feedback response to estrogen remains intact even though it is considerably attenuated.

The relative contributions of progesterone, and other ovarian steroids and proteins, have been assessed by monitoring the responses of LH and FSH to ovariectomy during lactation. Suckling can prevent the post-castration rise in serum and pituitary levels of LH and, to a lesser extent, FSH (at least until mid-lactation, days 10–12) when ovariectomy is performed in early lactation (days 2–5) (25,27,30,32,63,68,69) and this suppressive effect of suckling is enhanced by increasing litter size (21). The inhibition of the LH response results from a maintained suppression of the pulsatile release of LH, at least to day 10 (40,46,48,69), associated with a failure of the normal ovariectomy-induced decrease in hypothalamic GnRH content in early lactation in the rat (70,71). When lactating mothers are ovariectomized in mid-lactation (days 10–15), suckling can reduce to 50%, but can no longer completely prevent, the increase in gonadotropins (68,69). Thus in early lactation, the suckling stimulus alone appears to be the major, if not only, component in the suppression of GnRH/LH pulsatile secretion (Fig. 4) (72). Around mid-lactation the influence of negative energy balance associated with the greatly increased nutrient, and hence milk, demand of the growing pups exerts an increasing influence, although maintained suckling remains essential (72,73).

The effects of steroids, particularly progesterone, on GnRH release are mediated in part by the actions of opioids within the hypothalamus. Thus in lactation in the rat where low levels of progesterone are maintained secreted by the CL of lactation (Fig. 1), opioids might be expected to play a role. Administration of β-endorphin caused a prolonged suppression of LH release (35) and adrenalectomy-induced increases in β-lipotropin, the precursor of β-endorphin, prevented the post-ovariectomy increase in LH and FSH in lactating rats (22), indicating that the opioid pathway involved in GnRH/LH can operate in lactation. However, blockade of opioid receptors with the broad spectrum opioid antagonist naloxone had only a marginal effect on LH secretion (74–76) suggesting that opioids are only marginally involved in suckling-induced suppression of GnRH/LH secretion.

Prolactin and Energy Balance

In common with most mammals, lactation in the rat is characterized by elevated levels of prolactin induced by suckling (77–79), and is inversely related to low levels of LH. Thus while the neural stimulus of suckling is obviously crucial for suppression of GnRH/LH pulsatile secretion, the humoral effects of the high levels of prolactin may also play a role in the suppression of GnRH output. This concept is supported by the fact that suppression of prolactin by CB-154 after ovariectomy in early lactation resulted in an increase in LH to 50% of the level seen in nonlactating ovariectomized females, and to 100% castration levels in mid-lactation (32,68). Administration of exogenous prolactin with the CB-154 reversed these effects. Thus, prolactin alone in the absence of ovarian steroids can augment the inhibitory effects of the neural suckling stimulus in preventing the postcastration rise in LH and FSH but is most effective in the later stages of lactation, when the suckling stimulus is declining (32,39,69).

This concept is further supported by the effects on estrus onset of altering prolactin at different stages of lactation. Suppression of prolactin by CB-154 in early lactation resulted in estrus occurring some 6 to 8 days later (14,36,39), whereas in mid-lactation it occurred very rapidly (39), again implying that the suppressive effects of prolactin became more important in later lactation. These responses appear to relate to the effect of CB-154 on LH, since treatment causes a rapid increase in LH in mid-lactation but has a minimal effect on LH in early lactation (36,69). In mid-lactation administration of prolactin causes a 50% reduction in the increase in GnRH receptors occurring after weaning (46), implying a direct effect of prolactin on hypothalamic release of GnRH. This appeared to be confirmed in a GnRH neuronal cell line developed after targeted oncogenesis of GnRH neurones in the mouse (80) but will require confirmation in normal GnRH neurones in the mouse and other species.

The suckling stimulus remains of key importance to suppression of gonadotropin and stimulation of prolactin secretion in lactation. Changes in both LH and prolactin appear to be closely linked, since basal levels of LH only begin to rise in intact lactating mothers as prolactin levels decline. In lactating mothers separated from all their pups for 8 hours (74) or from all but one pup for 23 hours (82) between days 6 and 10 of lactation, the reintroduction of the pups and the subsequent period of suckling was accompanied by a dramatic rise in prolactin, as expected [see (77)], as well as by a rapid decrease in LH (74,82), but not FSH (74), within 30 minutes of the onset of suckling. This decrease in LH occurred when the prolactin increase was prevented by CB-154.

However, during normal lactation in the rat the secretion of LH and prolactin is not as precisely interrelated as the pup-separation studies would suggest. The suckling activity of the pups declines throughout lactation, with a considerable variation in suckling activity during each 24-hour period (9,84). Although prolactin levels were higher and LH levels were lower over a 24-hour observation period in rats suckling 10 (compared with 5) pups (9), there was no precise relationship between suckling activity and the release of prolactin and LH (9,84). Indeed, maximum levels of prolactin could occur at a time when mothers were away from the pups for up to 60% of the time, whereas alterations in LH levels did not parallel the changes in either suckling activity or prolactin (84). Thus, the effect of the normal suckling pattern during lactation in the rat on the secretion of LH and prolactin may more readily be considered as a chronic alteration in hormone secretion rather than a series of acute changes occurring with every suckling episode. There is no doubt, however, that the suckling activity of the pups is the key component in the suppression of gonadotropin (principally LH) secretion responsible for the inhibition of ovarian activity. It is also clear that by mid-lactation the metabolic load of producing sufficient milk for the growing pups induces a state of negative energy balance. This leads to maternal hyperphagia, and the associated changes in the hypothalamus appear to play a major role in amplifying the effects of suckling on maintaining the suppression of GnRH/LH pulsatile secretion (72,73). These changes will be discussed in greater detail at the end of this chapter.

THE PIG

Estrus and Ovulation

The domestic pig generally remains anestrous during the first 4 to 6 weeks of lactation, and weaning the litter usually leads to estrus and ovulation within 3 to 10 days. The interval from weaning to estrus is negatively related to the number of piglets nursed (85); it also decreases with longer lactations, probably because of the reduction in suckling of older piglets (86–88). Reduction of the suckling intensity by separating piglets from sows for various periods (6–12 hours) each day from the second to third week of lactation (partial weaning) usually results in the occurrence of estrus during lactation in a high proportion of sows (89–94). Split weaning, where a proportion of the litter is weaned early, can also increase the percentage of sows exhibiting estrus by 5 to 10 days after weaning (92,94–97). These results clearly emphasize the important role of suckling intensity on the duration of anestrus, although anestrus can be maintained throughout early lactation with as few as one or two piglets suckling (98,99).

A proportion of the delay is due to the time taken for the hypothalamic–pituitary–ovarian axis to recover from the effects of pregnancy (100–102). In the absence of suckling, the first ovulation in the majority of sows occurs around 17 ± 7 days (103), although abnormal ovarian function often occurs up to 6 weeks after early weaning (104,105). In some sows removal of the piglets at birth results in immediate follicular development, estrogen secretion, and behavioral estrus (104).

Ovarian Activity

During the period of anestrus (up to 4 weeks postpartum), ovarian follicular development is characterized by a large population of small-sized follicles, and a small population of medium-sized follicles, but few large follicles (100,106–112). No CL other than those present throughout pregnancy are present, and plasma levels of both estradiol and progesterone are consistently low throughout the lactational period (113–119). During lactation the CL of pregnancy regress and rapidly lose the ability to secrete progesterone and relaxin (120), the latter related to a loss of relaxin mRNA expression (121).

The number of ovarian antral follicles does not change during lactational anestrus; however, there is a gradual increase in the size of follicles, with a concomitant decrease in atresia (109,110,122). During lactations of up to 8 weeks in duration, follicles rarely exceed 6 mm in diameter (109,110), with the largest preovulatory follicle in the normal estrous cycle being 7 to 10 mm [see (123)]. Weaning results in a rapid increase in the number of medium- and large-sized follicles, accompanied by a corresponding depletion of small-sized follicles (108,11,122).

In the pig, FSH appears to stimulate follicle development up to 5 to 6 mm, whereas LH is necessary for the final stages of follicle maturation and ovulation (124). The lack of follicular development during lactational anestrus appears to be due to a lack of LH stimulation since growth of large follicles occurs when pulsatile LH secretion is induced by pulsatile GnRH treatment (125).

Gonadotropins

Follicle-Stimulating Hormone

During lactational anestrus, pituitary levels of FSH return to normal soon after parturition (100,106,126), whereas plasma levels of FSH increase gradually throughout lactation (101,114,118,126). The release of FSH in response to GnRH remains constant and normal throughout lactation (114,116). Ovariectomy during lactation results in a normal increase in plasma levels of FSH (114). Since the low levels of plasma estrogen are unchanged by ovariectomy, it appears that FSH secretion is inhibited in lactation mainly by the secretion of inhibin, or other nonsteroidal products, by the ovary (114). However, suckling does reduce FSH output, even though this may be a marginal effect, because plasma levels increase significantly after weaning (Fig. 5) (117,118,122). Weaning of litters at 3 weeks (compared with 5 weeks) resulted in lower levels of FSH 2 to 3 days after weaning (118). The peak level of this post-weaning increase in FSH appears to correlate with the ovulation rate of the subsequent ovulation, but the level of FSH is not related to the interval between weaning and ovulation (117). There is an inverse relationship between FSH and LH after weaning in late lactation, with FSH levels declining as LH pulse frequency increases (127). This decrease in FSH probably relates to an increase in inhibin production by follicles developing in response to the increase in LH (128).

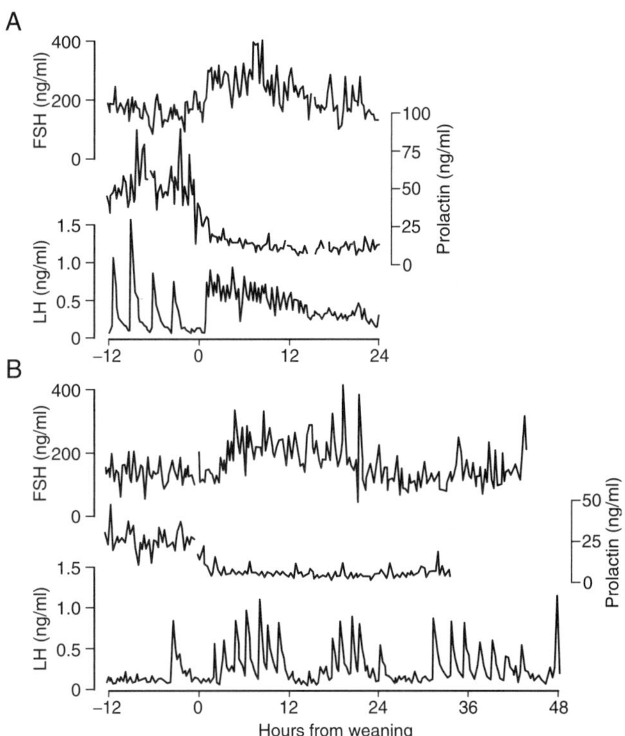

FIG. 5. Plasma levels of LH, FSH, and prolactin in two sows weaned around 21 days postpartum and with a subsequent weaning-to-estrus interval of (**A**) 4 days and (**B**) >11 days. Samples were obtained at 15-minute intervals. Note that before weaning pulsatile release of LH occurred with the frequency being greater in the sow with the earlier onset of estrus postweaning. After weaning, LH pulsatile release and FSH increase, while prolactin levels decline. [From (94).]

Luteinizing Hormone

Pulsatile LH secretion continues during pregnancy, acting to maintain CL function (98). At parturition in a number of sows high-amplitude pulsatile secretion of LH, equivalent to that occurring after ovariectomy, continues for up to 48 hours before declining and becoming completely suppressed as suckling intensity increases (101,129). Thereafter, in contrast to FSH, pituitary levels of LH remain suppressed throughout the first 3 to 4 weeks of lactational anestrus (100,106,122), basal levels of plasma LH remain low (98,112,114,118,119,122,130–132), and neither plasma nor pituitary levels of LH increase in response to ovariectomy performed before the third week of lactation (106,114). As with FSH, serum LH levels increase beyond the third week of lactation in both intact (114,133,134) and ovariectomized (114) sows and a similar increase in pituitary LH release occurs in response to GnRH (114,135,136). The suppression of serum levels of LH appears to be less in sows suckling small litters (2–4 piglets) as opposed to normal-sized litters (7–12 piglets), while there is greater suppression in sows with a long suckling duration (119); however, LH levels were still lower than normal (99).

The decrease in basal LH is directly related to a decrease in the pulsatile secretion of LH, which is either absent (98,131) or very infrequent (101,112,118, 119,127,137–140) during early lactation. Treatment with the excitatory amino acid N-methyl-D,L-aspartic acid increases GnRH-driven LH secretion in sows at this time, indicating that suckling was reducing the ability of the hypothalamus to release GnRH (100,126,141). Treatment of sows with low dose estradiol between days 24 and 27 of lactation abolished endogenous pulsatile LH secretion and caused a short-term inhibition of pulsatile GnRH-induced LH release (125) suggesting that estradiol also played an important part together with suckling, in maintaining the suppression of pulsatile GnRH/LH release. Pulsatile secretion with an interpulse interval of 2 to 5 hours was present by 21 to 24 days of lactation (117,127), and the frequency of these pulses was related to the interval from weaning to estrus, the higher frequency being correlated with an earlier estrus onset (Fig. 5), and the LH response to GnRH, but not directly to ovarian follicle activity (139).

In parallel with the restoration of basal and pulsatile release of LH, the ability of estradiol to stimulate a surge of LH (which in the normally cyclic sow occurs 50 to 55 hours after the endogenous increase in estradiol) also increases during lactation. No LH response occurs on day 5 of lactation; a limited response occurs up to day 15, with a normal response returning by around day 35 (133,142,143), although estrous behavior occurs in response to the estrogen throughout this time (143). This poor positive feedback response to estradiol in early lactation probably explains the reduced LH response and the pre-ovulatory surge to the natural and normal increase in endogenous estradiol secretion observed after the early weaning of sows around day 10 of lactation (18,144,145).

A significant increase in the basal secretion of LH occurs within 8 to 12 hours of weaning, from around day 21 of lactation onward (97,117,118,127). This is due to a dramatic increase in pulsatile secretion of LH, which increases over the next 2 to 3 days (117). This increase in LH release is related to a significant increase in the pituitary content of LH (106,122) and an increase in hypothalamic content of GnRH, the levels of which were suppressed during lactation (100,122,126).

The characteristics of the post-weaning pattern of pulsatile secretion of LH do not appear to correlate with the time from weaning to estrus onset, which can vary from 2 to 11 days (117). However, the increase in LH pulsatile secretion is more marked in sows with an early onset of estrus, which also correlated with a greater pulsatile secretion of LH immediately before weaning (117,127) and the LH response to GnRH before weaning (139). The occurrence of early estrus and ovulation suggests that follicles capable of ovulation were present in these sows around the time of weaning and that the higher LH pulsatile secretion may have been responsible for this enhanced follicle development, since there are no differences in FSH in early and late ovulators (117). The greater pulsatile secretion of LH in some sows during lactation may reflect a relative decrease in the suckling activity of the piglets.

Induction of Ovulation during Lactation

Ovulation can be induced by the administration of gonadotropins or GnRH during lactational anestrus, but the response is affected by the duration of lactation. Ovulation rate and fertility vary greatly among treatments; this is probably related to the degree of follicular development at the time of the treatment [see (123,124)]. In general, treatment with FSH or various doses of pregnant-mare serum gonadotropin (PMSG) alone or in combination with hCG fails to induce estrus and conception during the first 5 days of lactation, but the percentage of sows exhibiting estrus and conceiving increases after the third week of lactation.

The pulsatile administration, but not continuous infusion (107) of GnRH (1.5–2.0 μg/hr) for 60 (or more) hours will successfully induce the development of large estrogenic follicles (125), estrus and the induction of the preovulatory LH surge in most sows from day 24 of lactation onward; the percentage of sows responding has been found to decrease as treatment is started earlier in lactation (116,124). These results support the concept that failure of normal follicular development during lactation in the sow is due to insufficient pulsatile secretion of LH, which is dependent on the suckling stimulus. As this declines during lactation, the pulsatile LH secretion resumes and increases, thus allowing follicular development and, eventually, ovulation.

Prolactin, Opioids, and Nutritional Status

As in the other species, lactation in the pig is associated with increased levels of prolactin, with prolactin being released in response to the suckling stimulus (97,114,115,117,125,127,145–149). Treatment with estradiol does not affect prolactin secretion in late lactation (125), but an increase in ambient temperature to 30°C results in an increased release of prolactin in response to thyrotropin-releasing hormone (TRH) (138). Treatment of lactating sows with CB-154 to suppress prolactin resulted in a small increase in LH levels (significantly less than that occurring after weaning) no change (150,151), or a decrease in LH (152). Continual treatment throughout lactation only resulted in an increase in LH in late lactation (149). There appears to be no correlation between the plasma levels of prolactin throughout lactation in the sow and plasma levels of FSH or LH (114). In addition, the acute suckling stimulus, which releases prolactin, had no significant effect on endogenous levels of LH or the LH response to GnRH (114). This suggests that prolactin per se has either only a minimal suppressive or no effect on GnRH/LH secretion (149). However, it is the possible that prolactin may modify the ovarian response to gonadotropins (153) but this effect would appear to be of minimal significance overall.

Suckling appears to stimulate prolactin secretion via an opiate pathway, since treatment of lactating sows with naloxone reproducibly inhibits prolactin release (138,148,154) and morphine may increase prolactin acutely (155). However, a role for opiates in causing the suppression of GnRH and hence LH and FSH secretion is unclear. Treatment with naloxone may result in an increase in (148,154,156,157) or no effect on (138,158) pulsatile LH release and may increase FSH in late lactation (159). However, morphine decreased LH in half of the treated lactating sows, prevented the increase in pulsatile LH secretion, and delayed the onset of estrus after weaning (155). A direct inhibitory effect of β-endorphin on LH release from pig pituitary cells in vitro has also been reported (160). Thus the role of opiates in the suckling-induced suppression of GnRH in the sow remains unclear; any opiodergic or dopaminergic regulation of GnRH/LH secretion only appears to occur to any substantial effect in later lactation (149) possibly related to the resumption of some ovarian steroid secretion at this time.

Nutritional status is clearly important in sows suckling a number of piglets. In general ovarian function and litter performance after weaning was greater in sows that had a higher body mass and in those that had lost least protein during lactation (102,161,162). However, the effects of nutrition are variable and appear to depend on the ability of individual sows to utilize energy. For instance, sows with long duration suckling were heavier and lost less weight during lactation, but had a greater suppression of the GnRH/LH axis (119). The influence of nutrition in sows requires much further work, particularly since most studies relate to commercial rearing of pigs.

THE SHEEP

Estrus, Ovulation, and Gonadotropins

All breeds of sheep experience a period of anestrus during long days, with the breeding season beginning as day length declines in the autumn (see Chapter 41). The time of onset of estrous cycles in the autumn, and the number of cycles during the winter (i.e., duration of the breeding season before ewes become anestrus again in response to the increasing day length in the spring), are breed dependent. In many breeds, lambing occurs in the spring, so any period of lactational anestrus can coincide with the onset of seasonal anestrus. Thus to investigate the effect of suckling and lactation on ovarian function, ewes have to be mated at the end of the breeding season or during an induced ovulation in anestrus so that they will lamb during the breeding season. Only in this way can the true effect of suckling and lactation on the resumption of ovarian activity be studied in the absence of the confounding inhibitory influence of seasonal photoperiod induced anestrus.

In nonsuckled ewes, estrous cycles resume within 3 to 5 weeks postpartum (163–165) and are associated with a restoration to normal estrous cycle levels of pituitary LH (166–170), FSH (167–169), and GnRH receptor (169) content of the pituitary, a normal response to GnRH in vivo (168,169,171) and a restoration of normal levels of hypothalamic GnRH (168,169). Plasma levels of FSH (172,173) and LH (166) also return to normal, and estrus associated with ovulation and pregnancy occurs (165).

In ewes giving birth during the breeding season and suckling lambs, estrus may be delayed for at least 3 weeks (163–165,174–176), and some may fail to ovulate before the start of seasonal anestrus (177). This interval is variable between breeds of ewe (163–165,168,175,177). During the breeding season the pituitary content of GnRH receptors (169), LH, and abundance of mRNA for LH-β and gonadotropin α-subunits increase to normal cycle levels by day 22 postpartum in suckled ewes (Fig. 6) (176). Basal (174,176,178) and pulsatile (174,176) secretion of LH increase progressively during lactation, although the increase is slower in ewes suckling two or three lambs than a single lamb (174). The increase in LH pulse frequency is directly related to an increase in the frequency of pulsatile release of GnRH into the hypophyseal-portal vessels (176) and by day 21 postpartum both GnRH pulse frequency and amplitude were similar to that in the normal estrous cycle (179,180). A considerable number (62%) of GnRH pulses were not associated with pulses of LH on day 3 of lactation (176), probably due to the low pituitary content (181) and releasable pool of LH-containing granules essential for pulsatile LH secretion (182) at this time. As lactation progressed, most (62%–78%) GnRH pulses were associated with an LH pulse (Fig. 6) (176) as pituitary levels of LH, and presumably the number of LH-containing granules in gonadotropes (182), increased. Interestingly the increase in GnRH pulses at day 3 correlated with an increase in LH and gonadotropin α-subunit mRNA before any increase in pituitary LH content (176,181). In lactating Corriedale ewes lambing during seasonal anestrus, the pituitary content of LH, FSH, and prolactin, as well as the hypothalamic content of GnRH and plasma FSH, had increased

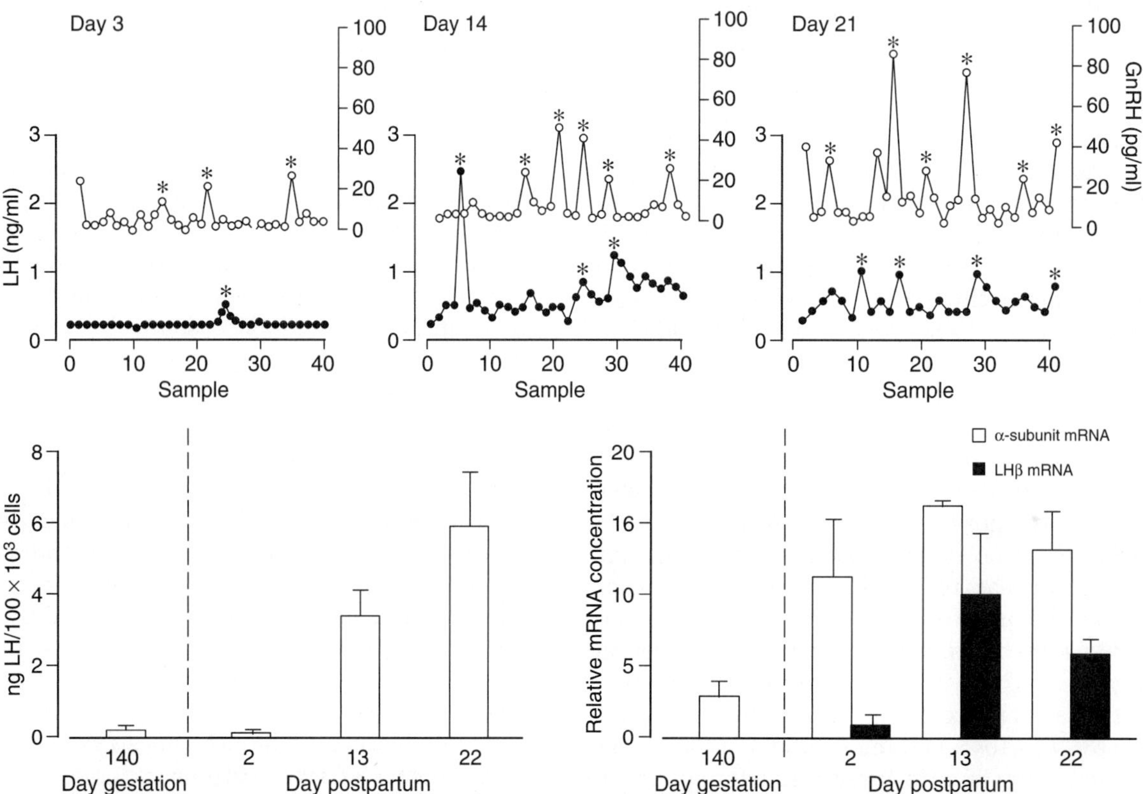

FIG. 6. Changes in the pituitary content of LH and the abundance of the mRNAs encoding the LH-β subunit and gonadotropin alpha-subunit in relation to the pulsatile release of GnRH (*open data points*) from the hypothalamus into the hypophyseal portal vessels and release of LH (*closed data points*) from the pituitary in suckled ewes from late pregnancy through to day 22 of lactation. Note that at term pituitary content of LH is very low related to undetectable LH-β mRNA. On day 3 postpartum pulsatile secretion of GnRH returns, but only a few GnRH pulses (*) result in the release of a pulse (*) of LH, related to the low levels of pituitary LH in releasable storage granules. The progressive increase in GnRH pulsatile release results in an increase in LH-β mRNA, LH pituitary content in the form of releasable granules, and hence release of pulses of LH from the pituitary. [From (176,181,182).]

by 11 to 17 days postpartum to those in seasonal anestrus (183). However, LH pulse frequency 8 to 14 days after parturition was greater in these ewes being suckled during anestrus than in nonsuckled anestrous ewes (183), possibly related to reduced ovarian steroid feedback.

The pituitary responsiveness to exogenous GnRH recovers to normal by 14 days postpartum in ewes lambing during the late breeding and early anestrous season (166,176,184). However, in spite of the normal LH response to exogenous GnRH and the increased pulse frequency of lactating ewes, seen by 14 to 20 days postpartum, only between 46% and 50% of ewes showed a positive feedback response of LH to estradiol by 30 days postpartum (177). This clearly implies an inadequacy of hypothalamic GnRH release, since the positive feedback response is dependent on GnRH (157,179,185). A similar decrease in the positive feedback response to estradiol was observed in ovariectomized lactating ewes also lambing during the breeding season (186). Indeed the pulse frequency of LH in these ewes was significantly less than that in ovariectomized nonlactating ewes (186). The depression of plasma LH following chronic estradiol treatment was also greater in ovariectomized lactating, as opposed to nonlactating, ewes; this indicates that there is a suckling-dependent enhanced sensitivity to the negative feedback effects of estradiol in the Merino ewe. Similar studies have yet to be carried out in other breeds of ewe.

Ovulation can be induced in lactating anestrous ewes by the continuous pulsatile administration of low doses of GnRH. In those ewes that ovulate (86%), inadequate CL function occurs if GnRH is delivered at a constant frequency (hourly) in the absence of progesterone pretreatment (187). However, if the frequency is progressively increased (from once every 3 hours to an hourly rate), ovulation with adequate luteal function can be induced in the majority (80%–90%) of ewes, with normal luteal function occurring in most ewes, especially if pretreated with progesterone before GnRH (188). Ovulation can be successfully induced in nonsuckling anestrous ewes with either LH (189,190) or GnRH with progesterone priming (191). This indicates that absence of normal follicular development during lactational and seasonal anestrus is primarily due to an inability to generate the increase in pulsatile GnRH and hence in LH pulsatile release for the final stages of follicular development before ovulation (192). Accelerated follicle growth related to a fast increase in plasma FSH after lambing may have a deleterious effect on subsequent ovarian function, with early ovulation resulting in the formation of an inadequate CL of short duration (175). In suckled ewes the delayed increase in LH pulses appears to delay development of follicles such that first ovulation is usually associated with a normal CL (175). There was no difference in the number of receptors for LH or prostaglandin F2-α (PGF), the sheep luteolysin, or the number of small or large luteal cells in short compared with normal lifespan CL postpartum (193). Thus, either excess PGF continues to be released from the involuting uterus or LH support is inadequate.

Prolactin

Plasma levels of prolactin are low during the breeding season, increase in response to increasing day length, and are high during the period of anestrus (194). Prolactin levels in lactating ewes, whether in the breeding season (183) or in anestrus (195,196), are greater than in nonlactating seasonally anestrous ewes, maintained by the suckling stimulus, which increases the secretion of β-endorphin (197). As in other species the prolactin response to suckling declines throughout lactation (198), as the pituitary–ovarian axis returns to normal in lactating ewes.

Although these changes in prolactin suggest a possible direct role for prolactin in suppressing LH secretion, there is little evidence to support this in the ewe. Treatment of lactating ewes with CB-154 to suppress prolactin may (163) or may not (173) advance the onset of estrus postpartum but does not appear to influence the basal or pulsatile release of LH in lactating ewes (199,200), even though the release of LH to exogenous GnRH may be increased (199). It is possible that high levels of prolactin may directly reduce ovarian estradiol (201) and progesterone (202) secretion and also reduce the positive feed back effect of estradiol on LH release (196). These effects may contribute to reduced fertility of lactating ewes following the induction of estrus (203,204), although inadequate LH secretion is a more probable explanation (205,206). Resumption of pulsatile GnRH and LH release and normal FSH secretion occurs while prolactin levels are several-fold greater than in the normal estrous cycle (175,176). Prolactin receptors have been identified on gonadotropes in the sheep (207) and interactions between dopamine and prolactin on the GnRH-induced release of LH between breeding season and anestrus ewes have been identified (208). How these observations relate to changes in LH secretion during the suckling-induced prolactin release in lactation in sheep remains to be explored. Thus, there is little evidence to support a central role for prolactin suppressing gonadotropin secretion postpartum, with the principal inhibition arising from the suckling stimulus itself (209).

THE COW

Estrus and Ovulation

In the lactating nonsuckled cow, estrous cyclicity resumes in the majority at around 30 to 70 days, but may occur as early as 20 days, postpartum (210–218). This first overt estrus is often preceded by ovulation with a lack of behavioral estrus, "silent heat," and the formation of an inadequate CL secreting reduced amounts of progesterone for a shorter than normal period (213,214,216,217,219). Suckling can delay this interval to estrus onset for a variable time up to 150 days postpartum (210,220–231). Thus typical times to the onset of estrus would be 25 days for milked cows and 57 days for suckled ones (232). Calf removal (either temporary or permanent) or the prevention of suckling by the use of nose plates, shortens the acyclic period (233–240). The interval to first ovulation may (241) or may not (224) be increased if dairy cows are milked more frequently. Similarly, the interval to first estrus may (241,242) or may not (243–246) be extended if the number of calves suckling is increased, although restricting suckling to a limited period each day may result in an earlier return to estrus (242,247,248).

Although it is clear that the interval to estrus onset in both milked and suckled lactating cows is longer than in nonlactating postpartum cows (221) and that suckling causes a longer delay than milking, there is considerable variation in the duration of this delay, an effect related to several environmental factors. Of these, photoperiod and nutrition appear to be the most important. Although cows do not display a period of seasonal anestrus, season does affect the duration of the interval to first estrus. Thus in both milked (213) and suckling (249–251) cows, those calving in late autumn and winter tend to have a longer period of acyclicity than those calving in the spring or summer (251–254), an effect that can be replicated by artificial alteration of the photoperiod (253). A parallel seasonal change in ovarian follicular development in nonlactating cows has also been reported (254), and the time from parturition to the emergence of the first dominant follicle was shorter in fall than spring calving dairy cows (217). This photoperiodic influence is also closely related to an effect of nutrition with heavier (and therefore presumed better-nourished with near normal energy balance) animals resuming estrous cyclicity early (244), with provision of increased dietary energy being associated with a reduction in the period of acyclicity (225,226,255–260). In contrast, poor nutrition resulting in a state of negative energy balance, is associated with a prolonged anestrous period (225,226,231, 259–266) and will be discussed later.

Ovarian Activity

Before first ovulation in both milked and suckling cows, plasma and milk levels of progesterone remain basal, thus confirming the absence of a functional CL (216–218,220,224,229,232,267). Plasma levels of estradiol also remain suppressed until near the time of first ovulation (217,224,229,230,268–270). This lack of steroid secretion is associated with a suppression of the final stages of follicular development prior to the endocrine changes preceding the first preovulatory LH surge (268,269,273–276). Healthy 10 mm follicles have been observed in weaned cows by 7 days, in milked cows by 10 days, and in suckling cows by 20 days postpartum (265,268,269). However, in milked cows the percentage of nonatretic follicles in animals before first ovulation decreased with time postpartum up to day 35 (275). However, in up to 20% of cows, the absence of ovulation of some of these large follicles may lead to the formation of follicle cysts associated with lower levels of progesterone (277). Ultrasound scans of the ovaries allowing tracking of follicle dynamics during lactation have shown that during the normal estrous cycle usually only one dominant follicle develops at any one time such that during the luteal phase between two and three dominant follicles grow, each regressing as a new dominant follicle develops (278–287). In milked cows a dominant follicle begins to develop within 7 to 20 days postpartum, the time being shorter in fall calving cows, and approximately 75% of these follicles ovulate (217,288). In the remainder, first ovulation is preceded by a variable number of waves of dominant follicle growth (217,283–285,288). Decreased fertility in dairy cows is associated with a prolonged time of follicle development related to reduced oocyte competence (283).

In suckled cows, in contrast, although the first dominant follicle also appears at the same time as in milked cows, around 10 days postpartum, only very few (11%) of these follicles ovulate (Fig. 7) (217,289–291). The interval from calving to first ovulation is characterized by the growth and regression of a variable number of dominant follicles (Fig. 7) (229,230,289–291). Following ovulation of the first dominant follicle in both milked and suckled cows the CL is almost always short lived (217,229,230,287,289). The dominant follicle destined to produce these inadequate CL may have reduced LH receptors on the theca and granulosa cells and decreased estradiol in follicular fluid (283,285,286,292,293). Low body condition at least over the first 9 weeks reduces the number of large follicles that are estrogenic (294). It is now clear that in both milked and suckled cows dominant follicles grow and regress throughout lactational anestrus, and any delay to first ovulation relates to a failure of adequate pulsatile LH secretion to generate sufficient

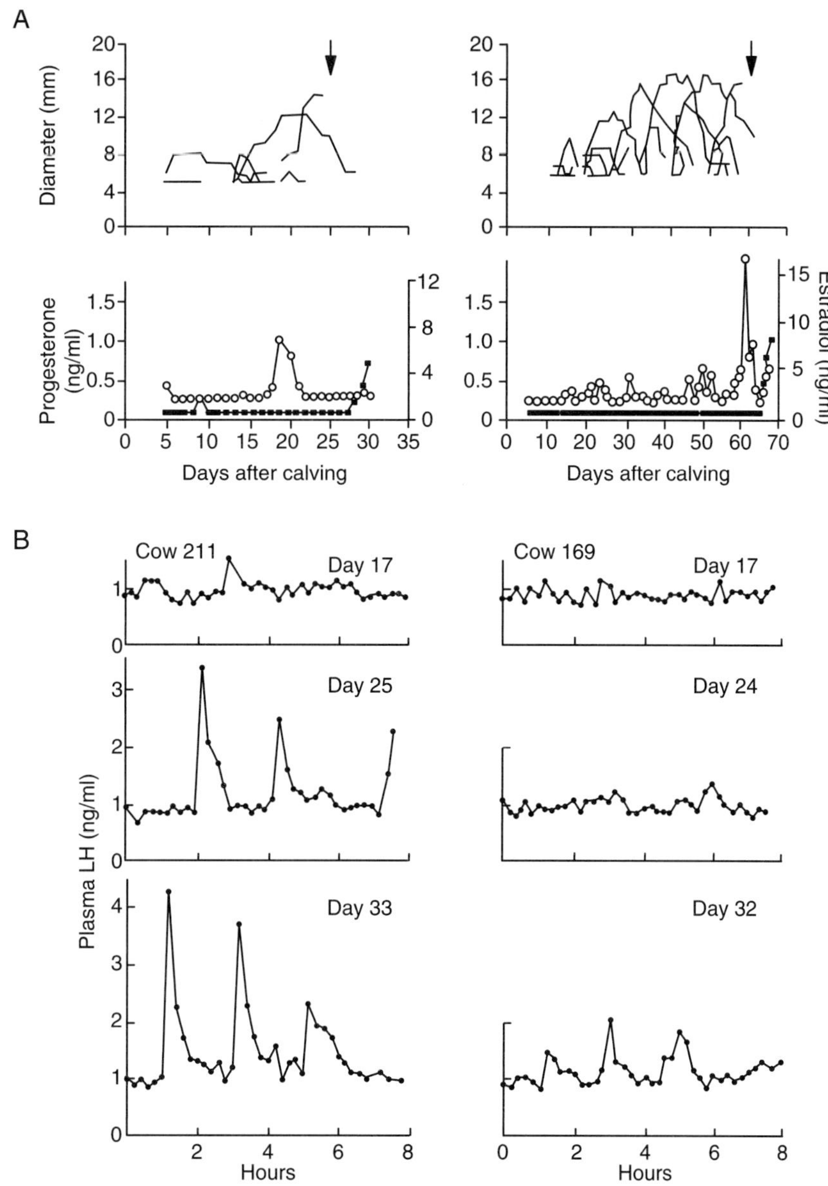

FIG. 7. Patterns of growth and regression of individual follicles monitored by ultrasound and levels of progesterone (*closed square*) and estradiol (*o*) in two suckled beef cows (**A**) and the appearance of the pulsatile release of LH during lactation in suckled beef cows (**B**). Note in **A** that the first dominant follicle only ovulated (indicated by the *arrow*) in 11% of cases and a highly variable number of dominant follicles grew and regressed in the remainder before ovulation occurred. Note in **B** that the earlier resumption of pulsatile LH secretion in cow 211 is associated with the earlier resumption of estrous cycles, related to ovulation 21 days earlier than cow 169. [Redrawn from (230,299).]

estradiol secretion from each dominant follicle to induce a preovulatory LH surge.

Gonadotropins

Follicle-Stimulating Hormone

In suckled cows, pituitary contents of FSH throughout lactation are not significantly different from those in the normal estrous cycle (260–263,267,276). After an initial increase after parturition, plasma levels of FSH remain equivalent to those in the normal estrous cycle throughout the period of lactational anestrus in both suckling and milked cows (214,216,227,245,297,299,300). In both beef and dairy cows plasma concentrations of FSH increase within 10 days of parturition and then fluctuate every 7 to 10 days in a regular wave pattern, with each increase preceding the emergence and growth of the next wave of dominant follicle growth (301–304) exactly as in the normal estrous cycle (283–285). The GnRH-induced release of FSH from pituitaries of suckling cows did not change throughout lactation up to 56 days postpartum (297). In contrast, the in vivo FSH response to both single large bolus injections (303,304) and small repeated injections (272) of GnRH declined with time postpartum. This probably reflects an increase in ovarian activity with the concomitant increase in secretion of inhibin and/or estradiol.

Luteinizing Hormone

The pituitary concentration of LH at the end of pregnancy is suppressed and increases throughout the postpartum period in suckling cows, reaching levels similar to those in cyclic animals by 40 to 50 days postpartum (267,276,295–298). Plasma LH concentrations also increase throughout the postpartum period (214,217,218,230,258,259,270,289,305–309), with this increase relating to a resumption of the pulsatile secretion of LH (Fig. 7) (215,217,218,224,230, 258,259,273,302,310,311), but no change in bioactivity of LH (312). This increase is paralleled by an increase in the release of LH from the pituitary in response to GnRH both in vivo (220,304,313,314) and in vitro (276,297). These changes in LH occur more rapidly in milked cows (around day 10 postpartum) than in suckled ones (up to day 50 postpartum) (217,230,236). The major increase in LH pulsatile secretion occurs in the few days preceding the first ovulation postpartum (29,230,232,251,273,315), with suckling prolonging the nonpulsatile period.

Since there is no change in either the affinity or the number of GnRH receptors in the pituitary (276,297), the absence of LH pulses is presumably due to a reduction in the hypothalamic release of GnRH. This has not been measured directly in vivo. Hypothalamic content of GnRH was greater in suckled than nonsuckled ovariectomized lactating cows (316), although in ovary-intact cows GnRH release from the stalk median eminence in vitro was not affected by weaning or level of nutrition (239). The delay in return to ovulation due to dietary restriction or negative energy balance is related to a delay in the resumption of pulsatile LH release (225,226,228,258,317). Increased dietary energy, which reduces the time to first ovulation, may not alter LH pulse frequency (258,259) but may increase LH pulse amplitude (232). The short-term infusion of insulin increased follicle estradiol production without altering pulsatile LH secretion (318) implying an effect of insulin directly on the ovary.

Response to Estradiol and Ovariectomy

Although the injection of estradiol into suckled cows may induce the positive feedback release of both LH and FSH by day 17 (319), this response may be reduced in comparison to that in nonsuckling cows during the post partum period (320). In milked cows, no response to estradiol was observed on day 5, but responsiveness had returned by day 15 postpartum (321). These results further suggest that suckling directly impairs GnRH secretion in vivo.

The increase in LH after ovariectomy on days 4 to 5 postpartum is delayed in cows suckled by one or two calves (315,321–323). Both basal levels and the pulsatile secretion of LH are decreased for 5 to 10 days after ovariectomy. Thereafter, both basal levels and pulse frequency of LH increase throughout the postpartum period, and levels are similar to those of nonsuckled ovariectomized cows by 40 to 60 days postpartum (315). This escape from the inhibitory effect of suckling can be further delayed if the ovariectomized suckled cows are treated with estradiol, when the increase in LH can be prevented for up to 72 days (Fig. 8) (322,323).

In contrast to the inhibitory effects of suckling on the post-ovariectomy rise in LH, the increase in plasma levels of FSH was similar in suckled and nonsuckled cows. These results imply that in the

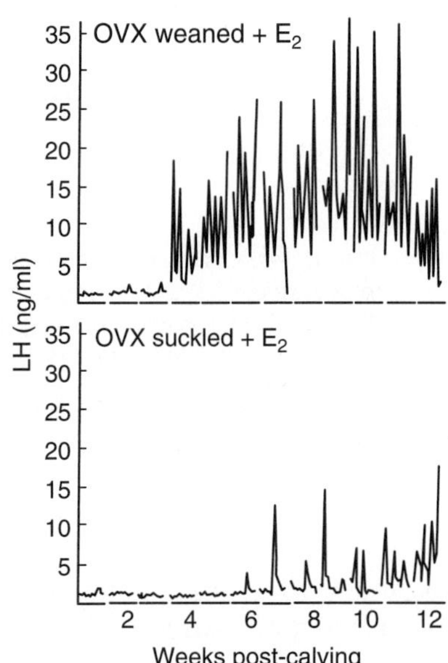

FIG. 8. Suckling-related increase in the sensitivity to the negative feedback effects of estradiol in beef cows postpartum. Cows were ovariectomized (OVX) at parturition and implanted with estradiol (+E_2). They were then either weaned or suckled. Note that E_2 delays, but does not prevent, the post-ovariectomy increase in pulsatile secretion of LH in weaned cows. This increase in LH was significantly reduced if the cows were suckled. [From (323).]

early postpartum period, when suckling is maximal, the suckling stimulus alone is sufficient to inhibit the pulsatile secretion of LH. Indeed treatment of suckled beef cows with an antiestrogen between days 20 and 30 postpartum did not affect the time to first estrus (324). However, as lactation proceeds, suckling maintains an increased hypothalamic sensitivity to the negative feedback effects of low levels of estradiol and further delays the onset of the preovulatory increase in the pulsatile secretion of LH.

Induction of Ovulation

Although ovulation can be induced in postpartum cows by single intramuscular injections of GnRH, the results are very variable due to variations in the stage of follicle development at the time of the GnRH injection [e.g., see (220,283,284,325–330)]. Even if ovulation does occur it may result in an inadequate CL. Greatest success occurs when follicular development is advanced and associated with increased levels of estradiol (232,328), which will occur earlier in the nonsuckled, as opposed to the suckled, lactating cow but cannot be induced by treatment with FSH (268).

A more successful treatment is the administration of frequent low doses of GnRH (311) or LH pulses alone (289), which mimics the normal increase in LH secretion in the preovulatory phase of the cow estrous cycle (283,284). Ovulation with normal luteal function can be induced in this way during the early postpartum period in suckled cows (289,299), but intermittent treatment of milked cows has yielded variable results (272,331). The results further support the concept that the failure of follicles to progress through the final stages of maturation to pre-ovulatory status and ovulation is due to a failure of the normal pulsatile secretion of LH as a result of decreased secretion of hypothalamic GnRH due to the suckling stimulus. However, in dairy cows in particular, the huge nutrient demands associated with producing high milk yields results in many instances in a negative energy balance which also has a major impact on the responsiveness of the ovaries to gonadotropins as discussed next.

Role of Prolactin, Opioids, and Negative Energy Balance

As in other species, lactation in the cow is associated with high levels of prolactin (224,332), in addition to the increase in prolactin secretion associated with increased day length and temperature (332,333). However, plasma levels of prolactin are similar in suckled and milked cows (224,334) and in cows suckling one or two calves (335), even though there is a considerable difference in the time to onset of estrus postpartum. Furthermore, suppression of plasma prolactin concentrations with CB-154 (243,314,336,337) does not affect the time to estrus onset, and infusion of prolactin failed to affect the pattern of LH secretion (310). It therefore seems improbable that prolactin plays any active role in the suppression of ovarian activity postpartum.

The role of opioids in the suckling-induced suppression of GnRH in the lactating cow remains unclear. Naloxone treatment in dairy cows did not affect LH (259) and in suckled cows had a minimal and variable effect to increase LH in early lactation (338,339), the response apparently depending on ovarian status (300). Levels of met-enkephalin were increased in the preoptic area (POA) of the hypothalamus in suckled cows with reduced energy intake regardless of suckling (239). However levels of β-endorphin in both the hypothalamus and POA were increased in suckled cows (239).

While the suckling stimulus, together with calf bonding (340) is clearly the major factor in suppressing GnRH/LH pulsatile secretion leading to failure of follicles, induced to grow by the normal levels of FSH, to reach the final stages of follicle maturation and secrete normal estradiol levels, many cows are in negative energy balance during lactation. In suckled beef cows negative energy balance associated with reduced body condition leads to a smaller size of pre-ovulatory follicle, which returns to normal with better feeding (341). The situation is extreme in dairy cows, particularly those selected for high milk yield (283–285) and is associated with decreased plasma levels of leptin (342–344), insulin (286,301,318) and IGF-1 (285,342,344). Both insulin and IGF-1 act locally within the ovary to increase responsiveness to gonadotropins (283,285,344). Increased dietary input, or infusion of insulin increases ovarian responsiveness to gonadotropins without any significant effect on gonadotropin secretion (301,318). This supports the concept that in cattle at least one significant component of negative energy balance is through direct effects at the level of the ovary, acting in concert with the indirect effects of negative energy balance within the hypothalamus to affect GnRH and gonadotropin secretion. These latter effects would operate through an endocrine loop, since, e.g., leptin levels are lower in lactating cows in negative energy balance and an effect via leptin at the hypothalamus, affecting GnRH secretion probably also operates.

NONHUMAN PRIMATES

Lactation-induced infertility has been recorded in a number of nonhuman primates, with the overt reasons for failure of conception being (a) inhibition of ovulation and (b) reduction or abolition of sexual activity. These primates include the great apes, the gorilla (345), and the chimpanzee (346–348); lesser Old World anthropoids such as the baboon (349); various species of macaque (346–348); and members of the genus Cercopithecus (349,359). Among the New World primates, an effect of lactation on reproduction has been observed in spider (360), howler (361), capuchin (362) and squirrel monkeys (363,364), a limited effect has been observed in the cotton-top tamarin (365), but none has been seen in the common marmoset (366,367).

The majority of studies on the influence of suckling on gonadotropin secretion and ovulation activity have been restricted to the rhesus monkey. Thus some caution must be exercised in assuming that these changes in the hypothalamus–pituitary–gonadal axis are truly representative of nonhuman primates in general. However, sufficient similarities exist between the rhesus and cynomolgus monkey during the period of lactational infertility such that they will be discussed together.

Ovarian Activity

An absence of follicular development associated with low plasma levels of estradiol (358,368,369) and inhibin (365) persists throughout most of the period of lactational infertility, up to around 150 days postpartum. Thereafter estradiol and inhibin increase as follicle growth resumes in response to increased LH secretion (358,365). In the nonsuckled monkey, normal follicle development with ovulation resumes by about 42 days postpartum, with first menses occurring about 14 days later (369). Earlier resumption of ovarian activity can be induced by appropriate gonadotropin treatment, suggesting that there is no intrinsic ovarian insensitivity to gonadotropins in the early postpartum period (370).

In nonsuckled animals, plasma progesterone levels decline rapidly after parturition (358,368,371–373); this is associated with the collapse of the CL of pregnancy (368). In contrast, in lactating suckled mothers, plasma progesterone levels, while decreasing over the first days postpartum, remain significantly higher than those in the normal follicular phase of the cycle for 2 to 3 months postpartum (368,369,372). This progesterone secretion is due to maintenance of the CL of pregnancy, as a result of the increased plasma levels of prolactin associated with lactation (372–374). Prolactin delays the functional luteolysis of the CL because it has been found that suppression of prolactin postpartum, either by weaning the infant (372) or by treating the mother with CB-154 (373,374), results in luteolysis and a decrease in progesterone secretion.

Gonadotropins

Plasma levels of FSH remain suppressed during early lactation but increase progressively, becoming equivalent to follicular-phase levels by 6 to 10 months postpartum (358,372). In contrast, plasma levels of LH are suppressed for up to 9 months postpartum and only increase to normal follicular-phase levels by around 12 months (372,375). In the first month postpartum, there is no response of either LH or FSH to GnRH (376). The suppression in basal levels of both LH and FSH is paralleled by a failure of LH response and a reduced FSH response to the positive feedback effects of estradiol benzoate for the first 3 to 6 months postpartum (Fig. 9) (372–374,376–378). The response then gradually increases, thus approaching a normal positive feedback response of FSH by 7 to 9 months and of LH by 10 months postpartum (372,374). In nonsuckled monkeys, a normal response returns by the end of the first month postpartum (372). Since in the rhesus monkey (379) and the stumptail macaque (380) the positive feedback effect of estrogen is predominantly due to a direct stimulatory action at the level of the pituitary gland, these results imply that suckling has inhibited the hypothalamic release of GnRH (372).

In accordance with this, the prompt rise in serum LH and FSH levels that normally follows ovariectomy is delayed for at least 30 days in postpartum females nursing infants (358,381). In suckled cynomolgus monkeys ovariectomized on day 14 postpartum, LH increased gradually but with considerable variation in both rate of increase and maximum LH concentrations reached (358).

A similar reduction in serum LH and FSH, as well as inhibition of the positive feedback effects of estradiol, occurs in non-postpartum monkeys fostering neonates (Fig. 9) (372). Thus, the suppression of gonadotropin secretion is primarily due to the suckling stimulus and is not dependent on antecedent events associated with gestation and parturition (374), but ovarian products are essential for maintaining suppression of LH and FSH throughout lactation (358). In contrast, a change in opioids does not appear to be involved in the suckling-induced

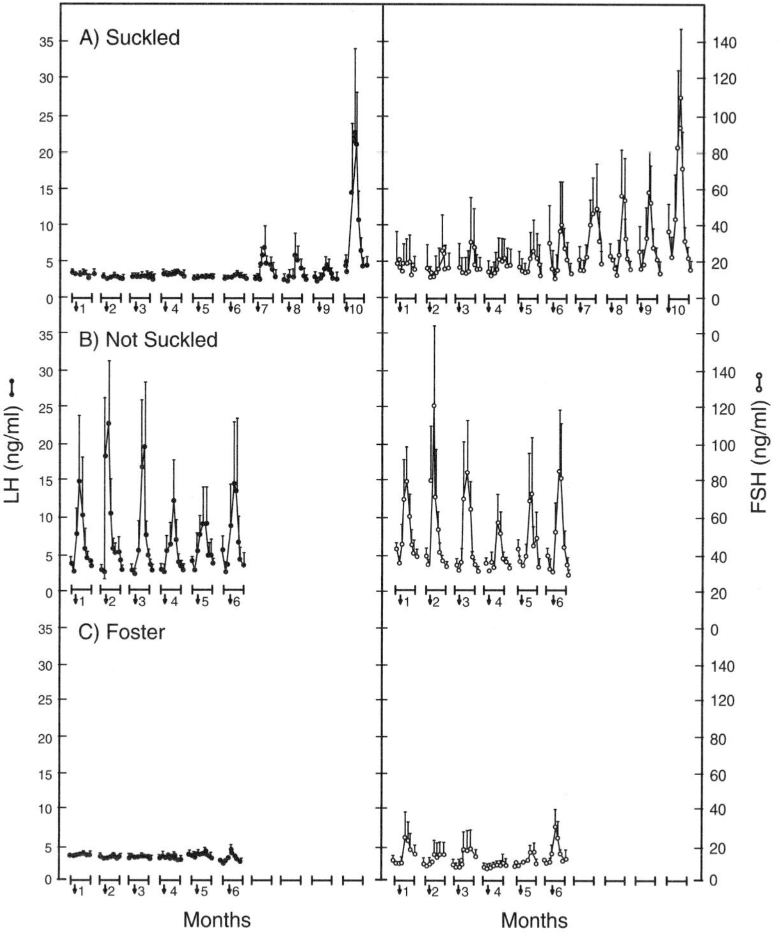

FIG. 9. Suckling-induced inhibition of the positive feedback effect of estradiol on LH and FSH release in the rhesus monkey. Estradiol benzoate was injected subcutaneously (*arrows*) each month postpartum in mothers suckled by their own infants (**A**), in mothers whose infants were weaned on the day of parturition (**B**), and in mothers suckled by foster infants from group B (**C**). Note the absence of gonadotropin responses during the initial months of nursing in both natural and foster mothers, but not in mothers weaned at the time of parturition. [From (372).]

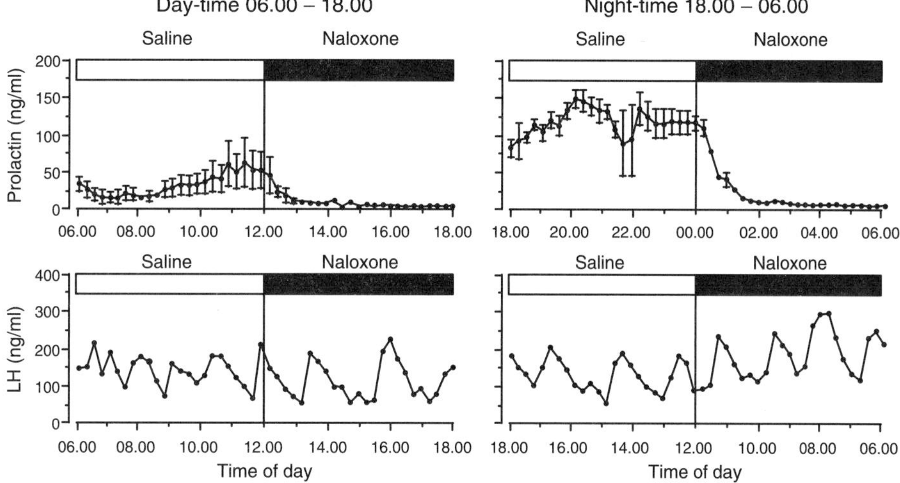

FIG. 10. Effects of naloxone and saline infusions during the daytime or nighttime on prolactin concentrations and pulsatile LH release on day 73 postpartum in nursing monkeys ovariectomized 58 days earlier. Note the rapid inhibition of prolactin secretion (mean, n = 3) during naloxone infusion but no effect on the pulsatile release of LH (individual profiles). [From (358).]

suppression of gonadotropins, since treatment with naloxone did not affect LH release in both ovary-intact and ovariectomized monkeys (Fig. 10) (358).

The suppression of pulsatile GnRH release induced by suckling has been confirmed indirectly in the rhesus monkey by the demonstration of suppression by suckling of multiunit electrical activity (MUA) recorded in the medio-basal hypothalamus, as a measure of the GnRH pulse generator and hence a surrogate for GnRH release. Each short burst of MUA is associated with the release of a pulse of LH supporting the concept that MUA relates directly to GnRH release (382). Suckling of nonpregnant ovary-intact normally cycling foster mothers by foster infants inhibited MUA activity and LH secretion, stopping menstrual cycles, but was without effect in ovariectomized mothers (Fig. 11) (375).

Prolactin

Plasma levels of prolactin are elevated throughout lactation (358,372,374). Although there is a clear increase in prolactin in response to suckling episodes when suckling frequency is low (383,384), no specific suckling-related episodes of prolactin release appear to occur during early lactation, when suckling frequency is relatively high (358,384,385). Suckling-induced prolactin release during both the day and night time is mediated by endogenous opioids, since prolactin secretion was completely blocked by naloxone treatment (Fig. 10) (358).

Injections of prolactin in bonnet monkeys have been reported to (a) reduce the LH response to GnRH (386) and (b) extend the period of lactation infertility (387). However, both basal levels of LH and FSH and the positive feedback effects of estradiol continued to be inhibited by suckling in rhesus monkeys treated with CB-154 to suppress prolactin throughout the postpartum period (374). Although the positive feedback response occurred somewhat earlier in the postpartum period (3 months) than in suckled monkeys with normally elevated prolactin levels (372), this may reflect a reduction in suckling activity of the neonates because milk secretion had been inhibited by the CB-154 treatment (374).

In the rhesus monkey suckling-induced suppression of MUA activity reflecting GnRH pulsatile secretion may be interrupted by suppression of prolactin secretion by CB-154 treatment, although this effect was not universal (Fig. 11) (375). Nevertheless it does show that prolactin under some circumstances can interfere with GnRH pulse generation during suckling, but the factors which determine whether there will or will not be an effect remain to be determined (375).

The prolactin-dependent maintenance of CL function and progesterone secretion does not appear to be important in the suppression of gonadotropin secretion. The plasma levels of progesterone, which are only maintained for 2 to 3 months postpartum, are well below those required for the inhibition of gonadotropin secretion (388). However, other ovarian factors, probably estradiol, do appear to be necessary (358,381).

Thus, the evidence to support a role for prolactin in the suppression of gonadotropin secretion and ovarian activity in the rhesus monkey is unclear. Indeed, lactation does not suppress fertility in the common marmoset even though prolactin levels are elevated during lactation (367). Thus, the suppression of fertility postpartum is mainly, if not exclusively, dependent on the suckling activity of the infant. Earlier conception postpartum occurs in rhesus monkey mothers, who actively encourage independent behavior of their infants by rejecting more of their infants' attempts to suckle (389). The period of anovulation is prolonged in primiparous mothers, which may have higher suckling frequencies than multiparous mothers (390–392), although there is also evidence for an increased sensitivity of primiparous mothers to the suckling stimulus (393). Nevertheless, with the direct evidence that prolactin can influence the effects of suckling on MUA activity in the hypothalamus of foster suckled rhesus monkeys, but only when the ovaries steroids are present, it is clear that prolactin may influence GnRH/LH secretion, but the effect is apparently unpredictable, even within the same animal (Fig. 11) (375).

HUMANS

Although there is no doubt that breastfeeding in women delays the resumption of ovarian cyclicity and menses, the duration of this delay is immensely variable both between and within different populations (394–424). Lactational amenorrhea may last from 2 to 3 months to 3 to 4 years, with the duration depending principally on the pattern of suckling: The more frequent or intense the suckling, the longer the delay and it is now clear from a number of studies that the lactational amenorrhoea method (LAM) encompassing full breastfeeding and amenorrhea over 6 months of lactation is a very reliable contraceptive [see (419,425)]. If women remain amenorrheic then this can be extended to 9 months of breastfeeding (423). Not only does this contraceptive effect continue to play a vital role in fertility control worldwide, breastfeeding is also associated with prevention of infant mortality and morbidity (426–429).

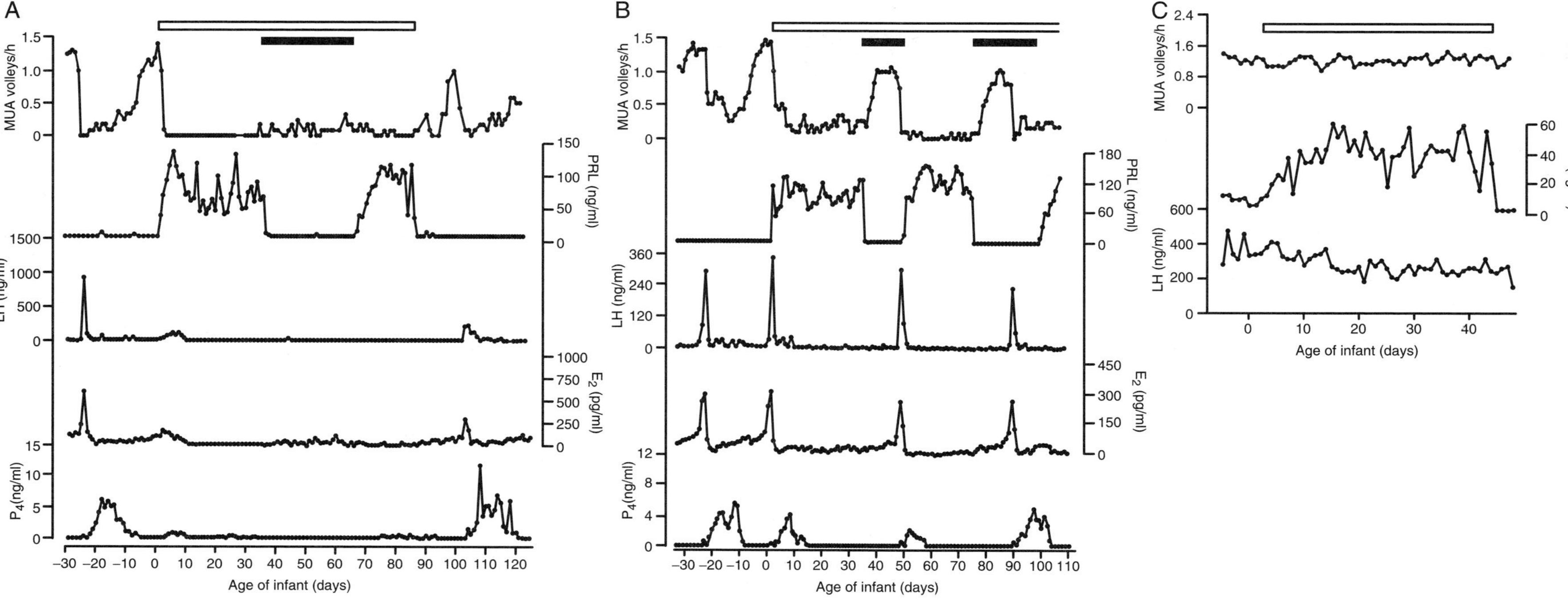

FIG. 11. Changes in multiunit electrical activity (MUA) recorded from the hypothalamus, as a surrogate for GnRH release, in relation to serum levels of prolactin (PRL), LH, estradiol (E2), and progesterone (P4) in ovary intact (**A, B**) or ovariectomized (**C**) adult female rhesus monkeys which were suckled by foster infants (*open bar*). During this suckling period the intact monkeys (**A, B**) were treated for different durations with bromocriptine (*closed bars*) to suppress the suckling-induced prolactin release. Note that in the absence of the ovaries (**C**) suckling did not affect MUA activity or LH secretion even though prolactin levels were elevated. In contrast, in ovary-intact monkeys (**A, B**) suckling suppressed MUA activity and inhibited LH secretion leading to a suppression of ovarian E2 and P4 secretion. These changes were associated with elevated levels of prolactin. When prolactin release was suppressed with bromocriptine during the suckling period there was either no effect (**A**), or a resumption (**B**) of MUA activity. This effect was variable and in the majority of cases there was no effect of suppression of prolactin on MUA activity. [Redrawn from (375).]

The postpartum period of amenorrhea is associated with raised levels of prolactin, and in most studies (394–397,399,400,406,407,422,430–433) the duration of lactational amenorrhea is closely correlated with this period of hyperprolactinemia. Suckling results in the release of prolactin (406,432,441), with the amount released being greater in the afternoon and night than in the morning (406,433,437,439). Both the amount of prolactin released in response to suckling and the basal levels of prolactin decrease with time postpartum and during lactation (397,399–401,406,407,43,437,439,440, 442–448) unrelated to any change in half-life (439). While the isoforms of the prolactin secreted may not change (384) the biological activity of prolactin may be increased during lactation in women (449,450) associated with prolonged lactational amenorrhea (450). This decline in prolactin levels is more noticeable in populations in which there is a relatively low frequency of breastfeeding episodes (3–5 per day) than when a high frequency (more than 10 breastfeeding episodes per day) is maintained throughout lactation. In the latter case the frequency of suckling and the inter-suckling intervals are key factors, and, although the response to suckling may be minimal, basal levels of prolactin remain elevated at all times (397,431,447,451).

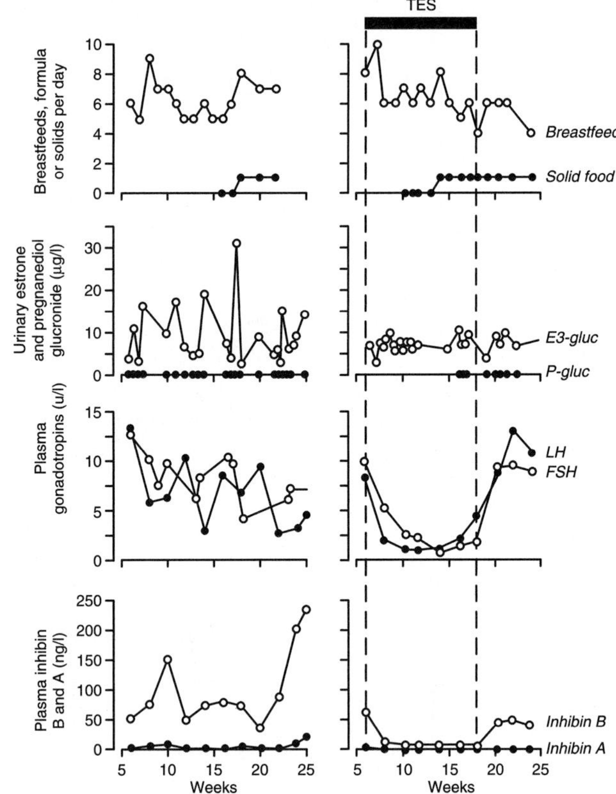

FIG. 12. Changes in the pattern of breastfeeding, urinary estrone and pregnanediol, and plasma LH and FSH, and inhibins A and B in representative breastfeeding women treated with a control (left-hand) or a low dose transdermal estradiol releasing (TES) patch (*closed box*) as used for HRT (right-hand). Note that while there was no effect of the TES on suckling activity, there was a dramatic suppression of both LH and FSH leading to suppression of ovarian secretion of steroids and inhibins, indicating inhibition of reduced follicle growth. These results indicate that there is an increase in sensitivity to the negative feedback effects of estradiol in early lactation in women. [Redrawn from (454).]

Ovarian Activity

Throughout the period of lactational amenorrhea in most women, ovarian follicular development, monitored by changes in plasma levels of estradiol (399,416,452–456) and inhibin A and B (454,455), and urinary estrogen excretion (396,408,409,454–463) as well as by ultrasonic visualization of the ovary (454–456,463), is inhibited (Fig. 12). Luteal structures never occur, and levels of urinary pregnanediol and plasma progesterone remain low throughout. Short-term elevations in estrogen secretion lasting a few days may occur, but the maximum levels achieved are reduced in comparison to those associated with the preovulatory follicle (396,402,408,454,455,464).

In contrast, in non-breastfeeding women, urinary estrogen excretion and plasma estradiol levels indicative of preovulatory follicle development resume within 20 to 30 days postpartum, and ovulation takes place within 4 to 8 weeks postpartum (396,402,457,458, 460,465,466). A proportion of women ovulate during breastfeeding, some ovulate before a menstrual period has occurred, and some will become pregnant without any intervening menses (396,402,418,467). In the majority of these cases, however, the luteal phase following ovulation is deficient but tends to improve with subsequent ovulatory cycles as lactation continues (396,399,402,409,417,418,451,461,467), a situation similar to the pattern of luteal activity during the resumption of menstrual cycles of non-breastfeeding women postpartum (459,461,465,466). The relatively high proportion of inadequate luteal phases is thought to be a principal cause of the reduced fertility of women who resume menstrual cycles while continuing to breastfeed (419).

Although gonadotropin stimulation immediately postpartum (7–15 days) fails to stimulate estrogen secretion (468,469), thereafter, exogenous gonadotropin (453,470) or treatment with pulsatile GnRH to induce increases in endogenous gonadotropins (463,417) will stimulate follicular development, estrogen secretion and ovulation. Thus, the ovary remains responsive to gonadotropin stimulation throughout the period of lactational amenorrhea, although induction of ovulation with GnRH pulses may lead initially to the formation of inadequate CL (463) with a return to normal luteal function if treatment is continued (471).

Gonadotropins

Follicle-Stimulating Hormone

Plasma levels of FSH increase within 3 weeks postpartum in breastfeeding women and remain within the normal range of menstrual cycles throughout the postpartum period (397,448,452–454,457,458,462, 472,473). The FSH response to GnRH also increases, and the response may be greater than that in the normal menstrual cycle (453,472,474–477). When ovarian feedback is minimal and GnRH/LH pulse frequency is very low in early lactation, FSH pulses coinciding with LH pulses may occur on rare occasions, but this association disappears as GnRH/LH pulse frequency increases (Fig. 13) (440).

Luteinizing Hormone

In contrast to FSH, LH levels only increase to the lower limit of normal by 20 days postpartum and remain suppressed in the majority of women throughout most of the period of lactational amenorrhea (397,399,407,440,452–458,462,463,472,478–480) (Figs. 12 and 13). The LH response to GnRH is also diminished during early lactation (453,474–478) but from around 12 weeks onward the LH responses to GnRH given as single bolus injections, or as a series of hourly pulses are within the normal nonlactating menstrual cycle range (463,471,476). The low LH levels are associated with very low-amplitude pulses of LH at a frequency similar to, or less than, that of the luteal phase (443,444,478). These pulses may occur either infrequently with normal amplitude (Fig. 13) (386) or at regular intervals with low-amplitude (480) throughout the 24 hours. There is no evidence for enhanced pulsatile LH secretion during sleep (440,480) as observed during puberty (481). In some cases, and at different times during the period of amenorrhea, basal levels of LH increase to within the normal menstrual-cycle range; also, normal pulsatile secretion of LH, equivalent to that in the early follicular phase, is observed (Fig. 13) (440,464,478,480). This enhanced LH secretion is often accompanied by an increase in urinary estrogen excretion, indicative of

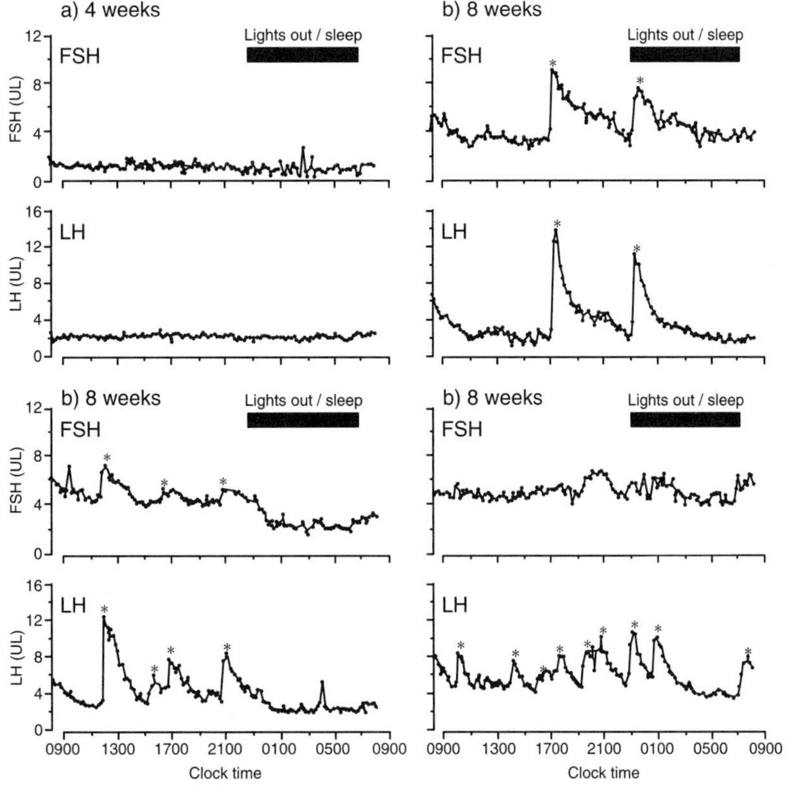

FIG. 13. Pattern of resumption of pulsatile secretion of LH and FSH over a 24-hour period in fully breastfeeding women at (a) 4 weeks and (b) 8 weeks postpartum. Note that FSH and LH pulses only coincide at low LH (GnRH) pulse frequencies when negative feedback signals from the ovary are minimal. Pulsatile LH (GnRH) secretion resumes at a variable frequency and at random over 24 hours, with pulses also occurring at nighttime when prolactin response to suckling was maximal. All women remained amenorrheic for at least 20 weeks after sampling. [From (440).]

follicular development, but levels rarely increase to be equivalent to those in the normal follicular phase of the cycle. If breastfeeding continues at the same suckling frequency, LH secretion declines again, and a period of ovarian inactivity resumes (419,464,478). If suckling frequency is reduced or is only minimal, the increase in LH secretion may be maintained, with a consequent increase in estradiol secretion related to follicle development, and ovulation may occur (408,409,458,462,478).

However, in fully breastfeeding women, in whom LH levels are suppressed, estradiol fails to induce a positive feedback response of either LH or FSH, and they are more sensitive to the negative feedback effects of estrogen (Fig. 12) (452,457,476,482,483) but not progesterone (455). This supports the concept that during breastfeeding in women any small increase in estradiol levels resulting from an increase in the pulsatile secretion of LH would be sufficient to switch off further secretion of LH, thus preventing normal follicular development (419,464,484). Treatment of non-breastfeeding women with the opioid antagonists naloxone (485) or naltrexone (479) did not affect LH between days 7 and 10 postpartum but transiently increased FSH and LH from day 15 onward, the response matching the increased response to GnRH (485) and the return of steroid secretion from the developing follicles. In breastfeeding women neither naloxone (476,477) nor naltrexone (479) affected LH or FSH.

Prolactin and Suckling

The close association between the plasma levels of prolactin and the duration of infertility has led to the suggestion that prolactin per se may be involved in the suppression of gonadotropin secretion in women (397,399,400,430,431). Indeed, a reduced prolactin response to suckling, particularly at night, is correlated with an earlier resumption of ovulatory cycles in some (406,432,450) but not other (440) studies. The increase in estradiol secretion associated with follicular development prior to first menses postpartum is negatively correlated with the plasma level of prolactin in some but not all studies (396,397,400,431,439). However, the suckling-related increase in prolactin is not acutely related to any inhibition of basal or pulsatile secretion of LH (433,436,440,458,478). It is clear, however, that the level of prolactin postpartum in women is related to the amount of suckling during a 24-hour period (396–398,406,430–433,439,440). Thus, the relationship between prolactin and resumption of ovarian activity may be indirect: The decline in prolactin levels reflects a decrease in suckling activity, while it is this decrease in suckling that allows the resumption of normal LH secretion (464). A difference in the sensitivity of the estradiol feedback system in terms of enhancing prolactin secretion and suppression of gonadotrophin secretion has been reported (486). Thus, women who had high prolactin levels associated with low estradiol levels at 38 weeks of pregnancy tended to have a prolonged duration of lactational amenorrhoea (486). Whether prolactin is involved directly is not clear, but an increase in sensitivity to estradiol in the prolactin secretion system could be mirrored by a similar increased sensitivity to estradiol in the gonadotropin axis, helping to maintain suppression of normal pulsatile GnRH secretion (454,476).

A continuation of breast-feeding at night is associated with a longer duration of lactational amenorrhea (396,431), although night time feeding tends to occur more often in women who breastfeed frequently over a 24-hour period. Nevertheless, a continuation of night time breastfeeding has been associated with the continued occurrence of inadequate luteal function when ovulatory cycles have resumed in women who continue to breastfeed (461,467). The prolactin response to suckling appears to be greater in the afternoon than in the morning, when suckling is infrequent (437). However, pulsatile release of LH occurs throughout the 24-hour period (433,440,480), and the presence or absence of LH pulses is not related to the pattern of prolactin secretion, particularly at night, when prolactin levels and responses to suckling are maximal and LH pulses occur (440). Because the start of the preovulatory surge during the normal menstrual cycle occurs at night in the majority of women (487,488), it may be that nighttime suckling is more likely to suppress the generation of a normal preovulatory LH surge, thus leading to the formation of an inadequate CL (464).

The possibility that high levels of prolactin might inhibit normal follicular development in women by a direct action at the ovary has been reviewed in detail previously (484,489). Although prolactin may (490) or may not (491,492) suppress estradiol secretion from human granulosa cells cultured in vitro, there is little evidence that a drug- or anesthetic-induced increase in prolactin around the preovulatory LH surge or during the luteal phase affects normal ovarian activity in any significant way (493). Indeed, an increase in prolactin secretion naturally occurs during the normal preovulatory surge of LH during the menstrual cycle (487). Drug-induced hyperprolactinemia during the follicular phase can block normal follicular development and ovulation (484); however, a direct effect of the drug on the release of LH, independent of its effect on prolactin, cannot be ruled out and has been reported (484). Thus, it is still not clear whether the elevated levels of prolactin during lactational amenorrhea in women play a significant role in the inhibition of LH release and ovarian activity. It is probable that, as in other species, prolactin plays only a minor role and that the principal effect is on direct suckling-induced inhibition of GnRH/LH release.

Influence of Nutritional Status

There is no clear relationship between maternal nutritional status and the duration of lactational amenorrhea (494–502) perhaps related to the fact that perceived nutritional requirements for breastfeeding women may be overestimated (497) due to an increase in metabolic efficiency during lactation in women (498). Nevertheless, it is well recognized that women experience a period of hyperphagia during breastfeeding, which is probably directly related to the suckling stimulus. At present there is no evidence of alterations in leptin levels during lactation in breastfeeding women (503,504), although these studies need to be extended to more long-term breastfeeding. Low body fat, which may reduce fertility in breastfeeding women once menstruation has resumed (407), is associated with a marginal increase in prolactin (395,398,411,413,431), probably related to an increase in suckling activity (499,500). However, body mass index (BMI) per se does not appear to affect the duration of lactational infertility since ovulatory cycles resumed at the same time in women with very different BMIs, even though well-nourished women resumed regular cycles earlier (502). This suggests that the resumption of GnRH/LH pulsatile secretion sufficient to stimulate follicle steroid secretion may occur earlier in better nourished women, but that the positive feedback response to estradiol allowing generation of the preovulatory LH surge may be delayed. Thus, it appears that lactational amenorrhoea occurs for at least 10 months in well-nourished women when in positive energy balance suggesting that the suckling stimulus alone is sufficient to suppress gonadotrophin secretion and ovarian activity. Thereafter with longer duration of lactational amenorrhoea the indirect influence of negative energy balance may become of increasing importance adding to the effect of suckling (505). In general there is no strong evidence for a major effect of nutrition in the duration of lactational amenorrhoea, at least during the early stages of lactation (501,506,507); but emerging evidence that when the demands of continued breastfeeding results in a negative energy balance in the mother, then this can prolong the duration, as seen in other species (Fig. 1).

MECHANISMS OF SUCKLING-INDUCED OVARIAN INACTIVITY

Several common features of the effects of suckling on gonadotropin release have emerged in the species for which adequate data are available. It is clear that the suckling stimulus alone at least in early lactation in most species, is sufficient to maintain an inhibition on the GnRH pulse generator limiting pulsatile LH secretion regardless of nutritional status, a direct route of action (Fig. 1). However, as lactation advances and the young demand more milk to supply their nutrient requirements, the effects of nutritional status becomes more important, an indirect route of action (Fig. 1) (72), but when this happens in lactation appears to differ between species, and will obviously depend on the requirements of the offspring. In species with large litter sizes such as rats, or where domestic animals have been bred for maximum milk yield, a state of negative energy balance will emerge fairly quickly, even though lactation is associated with a large increase in food intake, hyperphagia. In other species such as primates with one or two offspring this nutritional deficit may not appear to be so extreme and the role of negative energy balance may remain minimal. What is emerging is that there may be common pathways that link the effects of the direct neural suckling stimulus on the one hand, and negative energy balance on the other, in the mechanisms whereby pulsatile GnRH secretion is inhibited during lactation (Fig. 14) (72,73), or where in some cases ovarian responsiveness may be compromised in extreme cases (283,285,286,344). Before examining these potential pathways, the common features of suckling-induced suppression of gonadotropin secretion and consequent ovarian function will be summarized.

In most species, both pituitary and plasma FSH return to normal estrous or menstrual-cycle levels soon after delivery, with normal responses to GnRH paralleling these changes. There is also a rapid increase in FSH after ovariectomy, although this response may be somewhat suppressed when the suckling intensity is high, e.g., in the pig (114) or during early lactation in the monkey at a time when pituitary levels of FSH are also reduced (381). The secretion of FSH requires minimal GnRH stimulation, and a slow pulse frequency or low amplitude pulses of GnRH are sufficient to activate FSH-β mRNA transcription with subsequent synthesis of FSH which is then released mainly through a constitutive pathway with little storage in the pituitary gonadotropes (508–510). Thus, the amount of FSH released is mainly controlled by the rate of synthesis which in turn is controlled mainly by the negative feedback effects of gonadal steroids, particularly

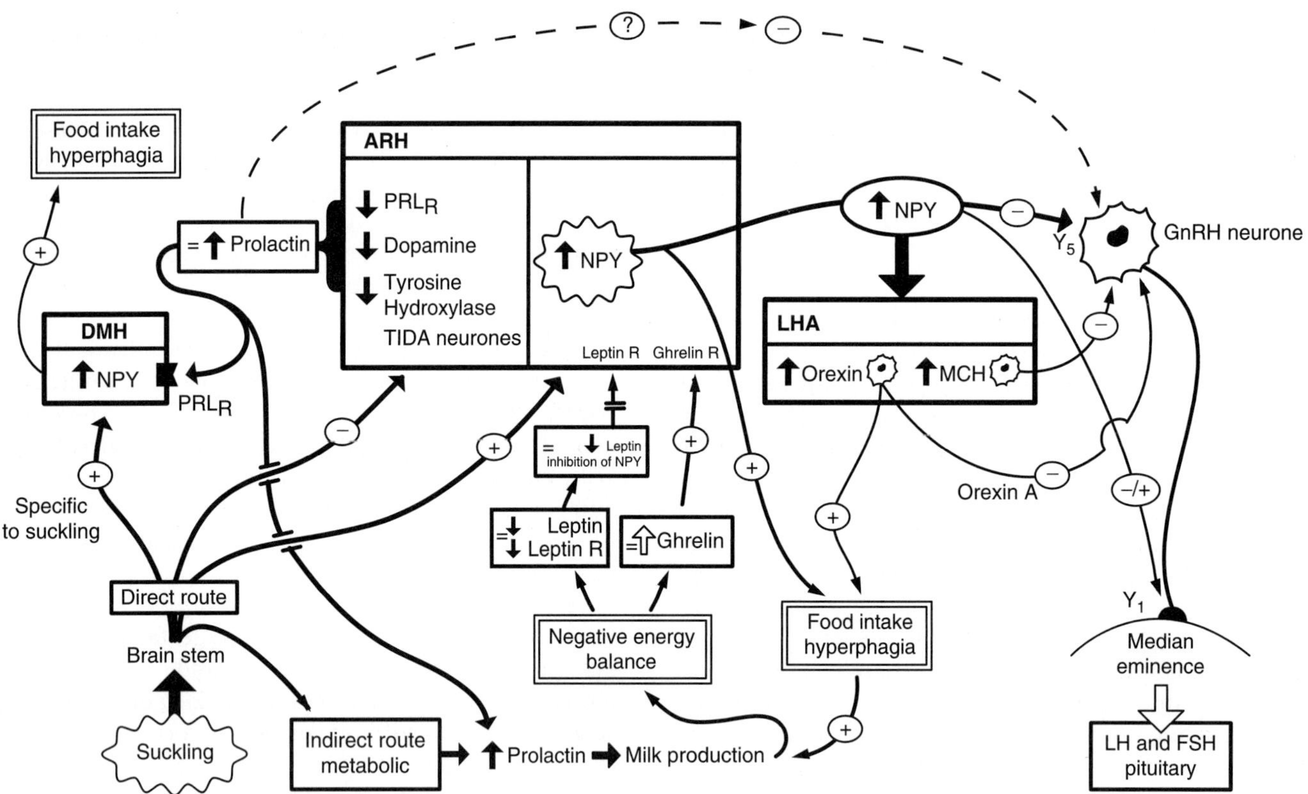

FIG. 14. Schematic diagram of the potential direct and indirect routes of the suckling stimulus within the hypothalamus that ultimately leads to the complete or partial suppression of pulsatile GnRH release resulting in inhibition of normal ovarian activity and the infertility associated with suckling. Note that the major final common pathway are the GnRH neurones with the principal immediate intermediate being neuropeptide Y (NPY) emanating from neurones in the arcuate nucleus (ARH) and signaling through Y-5 receptors in the cell body and Y-1 receptors at the terminals of GnRH neurones. These NPY neurones are directly stimulated by the suckling stimulus through the direct route, and by the indirect route through the effects of the negative energy balance induced by the metabolic load of producing milk. This latter route becomes more influential as lactation progresses, but if and when this additional effect of the indirect route occurs will depend on species (Fig. 1). Suckling also induces an increase in food intake (hyperphagia) though a direct effect on NPY-expressing neurons in the dorsomedial hypothalamus (DMH). The effects of negative energy balance through the indirect route are associated with changes orexin A expression in the lateral hypothalamic area (LHA) potentially directly affecting the GnRH neurons, and other factors such as increased ghrelin and decreased leptin which increase NPY expression through their respective receptors (Ghrelin R and Leptin R) in the ARH all resulting in hyperphagia. The increase in secretion of prolactin occurring through suckling-induced decreased dopamine release from tuberoinfundibular (TIDA) neurons in the ARH, appears to play an indirect role through maintenance of milk production. Most of these pathways have only been elucidated in the rat and will need to be confirmed in other species. The ultimate endpoint is a modulation of the GnRH pulse generator with suppression of normal pulsatile LH secretion and inhibition of ovarian function. *Thick lines* represent defined pathways elucidated in more than one species; *thin lines* indicate pathways clearly defined in the rat. The *dotted line* linking prolactin directly to the GnRH neurons indicates a potential, but unproven, pathway in vivo.

estradiol, and peptides, principally inhibin, through direct actions at the pituitary gland (419,508–510). The reasonably rapid return of apparently normal levels of FSH observed during lactation probably arises because suckling suppresses GnRH output but does not inhibit it completely, allowing a resumption of FSH-β mRNA transcription, and hence FSH synthesis and secretion. This may stimulate ovarian follicle development, but, because of the reduced GnRH input, LH pulsatile drive to the follicles to stimulate steroid production, particularly estradiol, is reduced resulting in a reduced negative feedback signal on FSH secretion. Thus, the major limiting factor with regard to continued follicular development is not FSH (419).

In contrast, in all species for which there are adequate data, suckling clearly inhibits the normal pulsatile release of LH and it is this absence of sufficient pulsatile LH secretion to drive the final stages of follicle development that limits ovarian activity. During early lactation in most species and throughout lactation when the suckling intensity is maintained, for example, by the multiple suckling of litters, LH pulses occur infrequently and the response to ovariectomy is suppressed. Since pulsatile LH release occurs in response to the pulsatile secretion of GnRH from the hypothalamus, releasing LH-containing granules from the gonadotropes [see (509)], the suckling stimulus appears to inhibit the normal pattern of hypothalamic release of GnRH with LH pulses occurring infrequently. As the suckling stimulus declines with advancing lactation (related to the growth of the young and their use of alternative food sources), the frequency of pulsatile release of LH increases. The failure of, or reduction in, positive feedback action of estradiol almost certainly relates to this decrease in the ability of the hypothalamus to increase GnRH output. Conversely, an enhanced negative feedback action of estradiol would be expected if the positive effect of GnRH was suppressed by the suckling stimulus. Only when suckling has declined to the point where GnRH release is no longer inhibited does normal pulsatile LH release occur, resulting in normal follicular development. Even then, inadequate CL may be formed, perhaps because of a reduction in the release of LH during and after the pre-ovulatory LH surge as a result of the continued suckling stimulus reducing the surge release of GnRH.

The close correlation between suckling and prolactin had led to the suggestion that the elevated levels of prolactin may be directly involved with the suppression of hypothalamic GnRH release but there is little evidence that prolactin has any major action directly on the ovary to reduce ovarian activity. Lactation demands the maintenance through the suckling stimulus of high levels of prolactin and this is achieved through a reduction in the negative feedback effect of prolactin which normally operates through the stimulation of dopamine release from tuberoinfundibular (TIDA) neurones in the arcuate nucleus (ARH) to directly inhibit prolactin release from the lactotropes. During lactation these TIDA neurones become less responsive to prolactin resulting in reduced release of dopamine, related to a reduction in tyrosine hydroxylase associated with a reduction in the expression of the long form of the prolactin receptor in these TIDA neurones (511–516). Opioid neurones acting through μ and κ, but not δ receptors are directly involved in this suckling-regulated decreased dopamine production by the TIDA neurones and explains why specific, or broad spectrum opioid receptor blockers or immunoneutralization of opioids suppress prolactin secretion during lactation in most species. These mechanisms that maintain prolactin secretion in lactation operate mostly through the TIDA neurons, and at present no link has been established between these TIDA neurones and the GnRH neuronal system (Fig. 14) (73).

Prolactin receptors are present on the dorsomedial (DMH) neurones in the hypothalamus and prolactin increases the direct action of suckling on stimulating neuropeptide Y (NPY) expression in these DMH neurones (Fig. 14) (517–519) which is related to increases in food intake [see (73)]. Again, there is no known link between DMH and GnRH neurones. Suckling also increases prolactin receptor expression in the ventrolateral and ventromedial preoptic (VLPO and VMPO, respectively) nuclei (520); and while there is no co-localization of prolactin receptors with any neurones, the VLPO contains a large number of GnRH neurones (521) and the VMPO has been implicated in the regulation of the preovulatory LH surge in the rat (522). Furthermore, while suckling increased GABA turnover in the VLPO in regions unconnected with GnRH neurons, an increase in GABA turnover has been linked with the reduction in the responsiveness of GnRH neurones to excitatory amino acid stimulation (523). Thus, while the sites at which, and mechanisms through which, prolactin may act in the hypothalamus are emerging, there is still no clear direct link between prolactin per se and suppression of the GnRH pulse generator.

A clearer picture is emerging in the link between suckling and the pathways involved in maintaining food intake (73) which appears to provide a common pathway for both the direct suckling stimulus, and the indirect nutritional energy balance input (Fig. 14). A major factor in regulating food intake is NPY particularly that expressed in the ARH neurones. Increases in NPY are associated with increased food intake, hyperphagia, a situation which occurs as a consequence of the suckling stimulus through a direct route (Fig. 14) (73). Within non-TIDA neurones in the ARH nucleus NPY expression is greatly increased in neurones that form direct links to the GnRH neurones either through Y5 receptors expressed on the GnRH neuronal cell bodies, or Y1 receptors on GnRH neurone fibers and terminals at the median eminence (517–519). This increase in NPY occurs independently of prolactin (517). Leptin, the feedback protein released from fat stores simulates LH secretion in the nonlactating animal and leptin receptors are present on NPY neurones in the ARH (524). During the stages of negative energy balance in lactation both leptin levels (524) and its receptors in the ARH (525) decrease. This would effectively reduce the positive effects of leptin on suppressing NPY expression in the ARH providing another route to suppression of GnRH neuronal activity (Fig. 14). Ghrelin, a GH releasing peptide produced in the stomach and gastrointestinal tract (526) is elevated in plasma in situations of energy deficit and stimulates food intake and reduces pulsatile LH secretion when given (527–529). Ghrelin receptors co-localize with NPY mRNA-expressing neurones in the ARH, again providing a link between negative energy balance and NPY (530). Finally NPY can increase the activity of orexin neurones in the lateral hypothalamic area which make direct contacts through orexin-1, orexin A-specific receptors, with GnRH neurones (531), and orexin inhibits LH secretion when steroid levels are low (528), as in early lactation at least (Fig. 14).

From all these observations a specific role for elevated NPY expression in neurones in the ARH in suppressing the GnRH pulse generator emerges. This increase in ARH NPY can be stimulated both through the direct effects of suckling alone, or through the indirect effects of reduced energy balance on the secretion of factors known to be altered in this state. The role of the suckling-induced increase in prolactin would appear to be mainly, if not exclusively, in the production of milk that leads to alterations in energy balance, enhancing the suppressive effects of the suckling stimulus in suppressing the GnRH pulse generator. The importance of the energy balance effects will depend on species, but would be without major effects, unless very extreme, in the absence of suckling. In most species where there is adequate data, there is no doubt that the strength of the suckling stimulus, be it the intensity of sucking within sucking bouts, the frequency or duration of suckling, or the inter-suckling interval, is the principal determinant of the duration of the suckling-induced delay in resumption of ovarian activity at least in early lactation. As lactation progresses the effects of suckling are amplified if a condition of negative energy balance related to continued milk production ensues. At what point, if ever, this transition from direct effects of suckling alone, to added indirect effects of negative energy balance occurs will depend on the metabolic load, and the ability to respond to this in individual mothers. It does now appear that a common pathway linking the direct suckling and indirect energy balance pathways has emerged in the potential role of NPY (Fig. 14). The final common pathway remains the regulation of pulsatile secretion of GnRH which ultimately controls pituitary gonadotropin secretion and hence ovarian function.

REFERENCES

1. Everett, J. W. (1961). The mammalian female reproductive cycle and its controlling mechanisms. In *Sex and Internal Secretions* (W. C. Young, Ed.), Vol. 1, pp. 497–555. Williams & Wilkins, Baltimore.
2. Connor, J. R., and Davis, H. N. (1980). Postpartum estrus in Norway rats: 1. Behaviour. *Biol. Reprod.* 23, 994–999.
3. Connor, J. R., and Davis, H. N. (1980). Postpartum estrus in Norway rats: II. Physiology. *Biol. Reprod.* 23, 1000–1006.
4. Davis, H. M., and Connor, J. R. (1980). Male modulation of female reproductive physiology during postpartum mating in Norway rats. *Behav. Neural. Biol.* 29, 128–131.
5. Bennett, J. P., and Vickery, B. H. (1970). Rats and mice. In *Reproduction and Breeding Techniques for Laboratory Animals* (E. S. E. Hafez, Ed.), pp. 299–315. Lea & Febiger, Philadelphia.
6. Conaway, C. H. (1971). Ecological adaptation and mammalian reproduction. *Biol. Reprod.* 1, 239–247.
7. Gosden, R. G., Russell, J. A., Clark, J., Piper, I., and McNeilly, A. S. (1981). Effects of inhibiting prolactin secretion on the maintenance of embryonic diapause in the suckling rat. *J. Endocrinol.* 88, 197–203.
8. Rothchild, I. (1960). The corpus luteum-pituitary relationship. The association between the cause of luteotrophin secretion and the cause of follicular quiescence during lactation: the basis for a tentative theory of the corpus luteum-pituitary relationship in the rat. *Endocrinology* 67, 9–41.
9. van der Shoot, P., Lankhorst, R. R., de Roo, J. A., and de Greef, W. J. (1978). Suckling stimulus, lactation and suppression of ovulation in the rat. *Endocrinology* 103, 949–956.
10. Sodersten, P., Hansen, S., and Eneroth, P. (1983). Inhibition of sexual behaviour in lactating rats. *J. Endocrinol.* 99, 189–197.
11. Maneckjee, R., and Moudgal, N. R. (1975). The onset of oestrus and ovulation in lactating rats. *J. Reprod. Fertil.* 43, 313–315.
12. Lindblom, C., Sodersten, P., and Eneroth, R. (1985). Effects of pup sucking behaviour on inhibition of sexual behaviour and ovulatory secretion of LH in lactating rats. *J. Endocrinol.* 104, 419–425.
13. Grota, U., and Eik-Nes, K. B. (1967). Plasma progesterone concentrations during pregnancy and lactation in the rat. *J. Reprod. Fertil.* 13, 83–91.
14. Tomogane, H., Ota, K., and Yokoyama, A. (1969). Progesterone and 2 hydroxypregn-4–en-3–one levels in ovarian vein blood of the rat throughout lactation. *Endocrinology* 44, 101–106.
15. Yoshinaga, K., Moudgal, N. R., and Greep, R. O. (1971). Progestin secretion by the ovary in lactating rats: effect of LH-antiserum, LH and prolactin. *Endocrinology* 88, 1125–1135.
16. Smith, M. S., and Neill, J. D. (1977). Inhibition of gonadotropin secretion during lactation in the rat: relative contribution of suckling and ovarian steroids. *Biol. Reprod.* 17, 255–261.
17. Taya, K., and Greenwald, G. S. (1982). Mechanisms of suppression of ovarian follicular development during lactation in the rat. *Biol. Reprod.* 27, 1090–1101.
18. Smith, M. S., McClean, B. K., and Neill, J. D. (1976). Prolactin: the initial luteotropic stimulus of pseudopregnancy in the rat. *Endocrinology* 98, 1370–1377.
19. Yoshinaga, K. (1974). Ovarian progestin secretion in lactating rats. Effects of intrabursal injection of prolactin antiserum, prolactin and LH. *Endocrinology* 94, 829–834.
20. Ford, J. J., and Yoshinaga, K. (1975). The role of prolactin in the luteotrophic process of lactating rats. *Endocrinology* 96, 335–339.
21. Coppings, R. J., and McCann, S. M. (1979). Stimulatory feedback action of estradiol in intact and ovariectomized lactating rats. *Biol. Reprod.* 21, 219–227.
22. Taya, K., and Sasamoto, S. (1990). Involvement of the adrenal gland in the suckling-induced decrease in LH and FSH secretion and the concomitant increase in prolactin secretion in the rat. *J. Endocrinol.* 125, 279–285.
23. Uilenbroek, J. T. J., van der Shoot, F., den Besten, D., and Lankhorst, R. R. (1982). A possible direct effect of prolactin on follicular activity. *Biol. Reprod.* 27, 119–125.

24. Lux-Lantos, V., Tesone, M., and Libertun, C. (1990). Ovarian follicle-stimulating hormone binding changes associated with the reinstatement of ovulatory cycles after lactation interruption in the rat. *Endocrinology* 126, 680–686.
25. Taya, K., and Sasamoto, S. (1991). Mechanisms responsible for suppression of FSH and LH during lactation in the rat. *J. Endocrinol.* 129, 119–130.
26. Yohkaichiya, T., O'Connor, A., and de-Kretser, D. M. (1991). Circulating immunoreactive inhibin, gonadotropin, and prolactin levels during pregnancy, lactation, and postweaning estrous cycle in the rat. *Biol. Reprod.* 44, 5–12.
27. Taya, K., Komura, H., Watanabe, G., and Sasamoto, S. (1989). Peripheral blood levels of immunoreactive inhibin during pseudopregnancy, pregnancy and lactation in the rat [erratum appears in *J. Endocrinol.* (1989). 122, 770]. *J. Endocrinol.* 121, 545–552.
28. Psychoyos, A. (1973). Hormonal control of ovoimplantation. *Vitam. Horm.* 31, 210–256.
29. Taya, K., and Sasamoto, S. (1988). Induced development of ovulatory follicles during the early stages of lactation by the administration of LH in the rat. *J. Endocrinol.* 116, 115–122.
30. Taya, K., and Sasamoto, S. (1988). Selective release of FSH in lactating rats during the period of follicular atresia induced by the administration of antiserum to LH-releasing hormone. *J. Endocrinol.* 118, 455–464.
31. Smith, M. S. (1978). A comparison of pituitary responsiveness to luteinizing hormone-releasing hormone during lactation and the estrous cycle of the rat. *Endocrinology* 102, 114–120.
32. Smith, M. S. (1978). The relative contribution of suckling and prolactin to the inhibition of gonadotropin secretion during lactation in the rat. *Biol. Reprod.* 19, 77–83.
33. Smith, M. S. (1982). Effect of pulsatile gonadotropin-releasing hormone on the release of luteinizing hormone and follicle-stimulating hormone in vitro by anterior pituitaries from lactating and cycling rats. *Endocrinology* 110, 882–891.
34. Smith, M. S. (1985). Release of luteinizing hormone follicle stimulating hormone from pituitaries of lactating rats by gonadotropin releasing hormone and high potassium concentration. *Endocrinology* 116, 1825–1834.
35. Taya, K., and Sasamoto, S. (1989). Inhibitory effects of corticotrophin-releasing factor and beta-endorphin on LH and FSH secretion in the lactating rat. *J. Endocrinol.* 120, 509–515.
36. Lu, K. H., Chen, H. T., Huang, H. H., Grandison, L., Marshall, S., and Meites, J. (1976). Relation between prolactin and gonadotrophin secretion in postpartum lactating rats. *J. Endocrinol.* 68, 241–250.
37. Ford, J. J., and Melampy, R. M. (1973). Gonadotropin levels in lactating rats: effect of ovariectomy. *Endocrinology* 93, 540–547.
38. Hammons, J., Velasco, M., and Rothchild, I. (1973). Effect of the sudden withdrawal or increase of suckling on serum LH levels in ovariectomized postparturient rats. *Endocrinology* 92, 205–211.
39. Hansen, S., Sodersten, P., and Eneroth, P. (1983). Mechanisms regulating hormone release and the duration of dioestrus in the lactating rat. *J. Endocrinol.* 99, 173–180.
40. Fox, S. R., and Smith, M. S. (1984). The suppression of pulsatile luteinizing hormone secretion during lactation. *Endocrinology* 115, 2045–2051.
41. Park, S. K., and Selmanoff, M. (1991). Dose-dependent suppression of post-castration luteinizing hormone secretion exerted by exogenous prolactin administration in male rats: a model for studying hyperprolactinemic hypogonadism. *Neuroendocrinology* 53, 404–410.
42. Brogan, R. S., Mitchell, S. E., Trayhurn, P., and Smith, M. S. (1999). Suppression of leptin during lactation: contribution of the suckling stimulus versus milk production. *Endocrinology* 140, 2621–2627.
43. Li, C., Chen, P., and Smith, M. S. (1998). The acute suckling induces the expression of neuropeptide Y (NPY) in cells in the dorsomedial hypothalamus and increases NPY expression in the arcuate nucleus. *Endocrinology* 139, 1645–1652.
44. Gallo, R. V. (1981). Pulsatile LH release during periods of low levels LH secretion in the rat estrous cycle. *Biol. Reprod.* 24, 771–776.
45. Higuchi, T., and Kawakami, M. (1982). Changes in the characteristics of pulsatile luteinizing hormone secretion during the oestrous cycle and after ovariectomy and oestrogen treatment in female rats. *J. Endocrinol.* 94, 177–182.
46. Smith, M. S., and Lee, L. R. (1989). Modulation of pituitary gonadotropin-releasing hormone receptors during lactation in the rat. *Endocrinology* 124, 1455–1461.
47. Lee, L. R., Haisenleder, D. J., Marshall, J. C., and Smith, M. S. (1989). Expression of alpha-subunit and luteinizing hormone (LH) beta messenger ribonucleic acid in the rat during lactation and after pup removal: relationship to pituitary gonadotropin-releasing hormone receptors and pulsatile LH secretion. *Endocrinology* 124, 775–782.
48. Lee, L. R., Haisenleder, D. J., Marshall, J. C., and Smith, M. S. (1989). Effects of progesterone on pulsatile luteinizing hormone (LH) secretion and LH subunit messenger ribonucleic acid during lactation in the rat. *Endocrinology* 124, 2128–2134.
49. Steger, R. W., and Peluso, J. J. (1978). Gonadotrophin regulation in the lactating rat. *Acta Endocrinol.* 88, 668–675.
50. Smith, M. S. (1984). Effects of the intensity of the suckling stimulus and ovarian steroids on pituitary gonadotropin-releasing hormone receptors during lactation. *Biol. Reprod.* 31, 548–555.
51. Minaguchi, H., and Meites, J. (1967). Effect of suckling on hypothalamic LH releasing factor and prolactin inhibiting factor, and on pituitary LH and prolactin. *Endocrinology* 80, 603–607.
52. Smith, M. S., and Rheinhart, J. (1993). Changes in gonadotropin-releasing hormone receptor mRNA content during lactation and after pup removal. *Endocrinology* 133, 2080–2084.
53. Clayton, R. N., Solano, A. R., Garcia-Vila, A., Dufau, M. L., and Catt, K. J. (1980). Regulation of pituitary receptors for gonadotropin-releasing hormone during the rat estrous cycle. *Endocrinology* 107, 699–705.
54. Marks, D. L., Smith, M. S., Clifton, D. K., and Steiner, R. A. (1993). Regulation of GnRH and galanin gene expression in GnRH neurons during lactation in the rat. *Endocrinology* 133, 1450–1455.
55. Pohl, C. R., Lee, L. R., and Smith, M. S. (1989). Qualitative changes in luteinizing hormone and prolactin responses to N-methyl-aspartic acid during lactation in the rat. *Endocrinology* 124, 1905–1911.
56. Abbud, R., Hoffman, G. E., and Smith, M. S. (1992). Lactation inhibits hippocampal and cortical expression of cFos in response to NMDA but not to kainite receptor agonists. *Mol. Cell. Neurosci.* 3, 244–250.
57. Abbud, R., Hoffman, G. E., and Smith, M. S. (1993). Cortical refractoriness to N-methyl-D, L-aspartic acid (NMA) stimulation in the lactating rat: recovery after pup removal and blockade of progesterone receptors. *Brain Res.* 604, 15–23.
58. Hodson, C. A., Simpkins, J. W., and Meites, J. (1978). Inhibition of luteinizing hormone release and luteinizing hormone releasing hormone action by the ovaries of postpartum lactating rats. *Endocrinology* 102, 832–836.
59. Lee, L. R., Paul, S. J., and Smith, M. S. (1989). Dose response effects of pulsatile GnRH administration on restoration of pituitary GnRH receptors and pulsatile LH secretion during lactation. *Neuroendocrinology* 49, 664–668.

60. Lu, K. H., Chen, H. T., Grandison, L., Huang, H. H., and Meites, J. (1976). Reduced luteinizing hormone release by synthetic luteinizing hormone releasing hormone (LHRH) in postpartum lactating rats. *Endocrinology* 98, 1235–1240.
61. Aiyer, M. S., Chiappa, S. A., and Fink, G. (1974). A priming effect of luteinizing hormone releasing factor on the anterior pituitary gland in the female rat. *J. Endocrinol.* 68, 573–588.
62. Ciofi, P., Fallon, J. H., Croix, D., Polak, J. M., and Tramu, G. (1991). Expression of neuropeptide Y precursor-immunoreactivity in the hypothalamic dopaminergic tubero-infundibular system during lactation in rodents. *Endocrinology* 128, 823–834.
63. Maeda, K. I., Tsukamura, H., Uchida, E., Ohkura, N., Ohkura, S., and Yokoyama, A. (1989). Changes in the pulsatile secretion of LH after the removal of and subsequent resuckling by pups in ovariectomized lactating rats. *J. Endocrinol.* 121, 277–283.
64. Legan, S. J., Coon, G. A., and Karsch, F. J. (1975). Role of estrogen as initiator of daily LH surges in the ovariectomized rat. *Endocrinology* 96, 50–56.
65. Blake, C. A. (1977). Effects of estrogen and progesterone on luteinizing hormone release in ovariectomized rats. *Endocrinology* 101, 1122–1129.
66. Lee, W.-S., Smith, M. S., and Hoffman, G. E. (1990). Progesterone enhances the surge of luteinizing hormone by increasing activation of luteinizing hormone-releasing neurons. *Endocrinology* 127, 2604–2611.
67. Tsukamura, H., Maeda, K. I., and Yokoyama, A. (1988). Effect of the suckling stimulus on daily LH surges induced by chronic oestrogen treatment in ovariectomized lactating rats. *J. Endocrinol.* 118, 311–316.
68. Smith, M. S. (1981). Site of action of prolactin in the suppression of gonadotropin secretion during lactation in the rat: effect on pituitary responsiveness to LHRH. *Biol. Reprod.* 124, 967–976.
69. Maeda, K., Uchida, E., Tsukamura, H., Ohkura, N., Ohkura, S., and Yokoyama, A. (1990). Prolactin does not mediate the suppressive effect of the suckling stimulus on luteinizing hormone secretion in ovariectomized lactating rats. *Endocrinol. Jpn.* 37, 405–411.
70. Culler, M. D., McArthur, N. H., Dees, W. L., Owens, R. E., and Harms, P. G. (1982). Immunocytochemical evidence that suckling inhibits the postovariectomy depletion of median eminence luteinizing hormone releasing hormone (LHRH). *Neuroendocrinology* 34, 258–264.
71. Culler, M. D., McArthur, N. H., Dees, W. L., Owens, R. E., and Harms, P. G. (1982). Inhibition of the postovariectomy depletion of hypothalamic luteinizing hormone releasing hormone (LHRH) by suckling. *Biol. Reprod.* 26, 633–639.
72. Tsukamura, H., and Maeda, K-I. (2001). Non-metabolic and metabolic factors causing lactational anestrus: rat models uncovering the neuroendocrine mechanisms underlying the suckling-induced changes in the mother. *Prog. Brain Res.* 133, 1–19.
73. Smith, M. S., and Grove, K. L. (2002). Integration of the regulation of reproductive function and energy balance: lactation as a model. *Front. Neuroendocrinol.* 23, 225–256.
74. Sirinanthsinghji, D. J. S., and Martini, L. (1984). Effects of bromocriptine and naloxone on plasma levels of prolactin, LH and FSH during suck ling in the female rat: response to gonadotrophin releasing hormone. *J. Endocrinol.* 100, 175–182.
75. Wu, T. J., McArthur, N. H., and Harms, P. G. (1992). A possible dual mechanism involved in the inhibitory influence of suckling on luteinizing hormone release in the ovariectomized rat. *Biol. Reprod.* 46, 653–657.
76. Andrews, Z. B., and Gratton, D. R. (2003). Opioid receptor subtypes involved in the regulation of prolactin secretion during pregnancy and lactation. *J. Neuroendocrinol.* 15, 227–236.
77. Neill, J. D., and Nagy, G. M. (1994). Prolactin secretion and its control. In *Physiology of Reproduction* (E. Knobil, and J. D. Neil, Eds.), pp. 1833–1860. Raven Press, New York.
78. Tucker, H. A. (1994). Lactation and its hormonal control. In *Physiology of Reproduction* (E. Knobil, and J. D. Neil, Eds.), pp. 1065–1098. Raven Press, New York.
79. Rothchild, I. (1981). The regulation of the mammalian corpus luteum. *Recent Prog. Horm. Res.* 37, 183–298.
80. Milenkovic, L., D'Angelo, G., Kelly, P. A., and Weiner, R. I. (1994). Inhibition of gonadotropin hormone-releasing hormone release by prolactin from GT1 neuronal cell lines through prolactin receptors. *Proc. Nat. Acad. Sci. U S A* 91, 1244–1247.
81. Selamoff, M. (1997). Commentary on the use of immortalised neuroendocrine cell lines for physiological research. *Endocrine* 6, 1–3.
82. lsherwood, K. M., and Cross, B. A. (1980). Effect of the suckling stimulus on secretion of prolactin and luteinizing hormone in conscious and anaesthetized rats. *J. Endocrinol.* 87, 437–444.
83. Dondi, D., Maggi, R., Panerai, A. E., Piva, F., and Limonta, P. (1991). Hypothalamic opiatergic tone during pregnancy, parturition and lactation in the rat. *Neuroendocrinology* 53, 460–466.
84. Sodersten, P., and Eneroth, P. (1984). Suckling and serum prolactin and LH concentrations in lactating rats. *J. Endocrinol.* 102, 251–256.
85. Fahmy, M. H., Holtman, W. B., and Baker, R. D. (1979). Failure to recycle after weaning and weaning to oestrus interval in cross-bred sows. *Anim. Prod.* 29, 193–202.
86. Self, H. L., and Grummer, R. H. (1958). The rate and economy of pig gains and the reproductive behaviour in sows when litters are weaned at 10, 21 or 56 days. *J. Anim. Sci.* 17, 862–868.
87. Moody, N. W., and Speer, V. C. (1971). Factors affecting sow farrowing interval. *J. Anim. Sci.* 32, 510–514.
88. Cole, D. J. A., Varley, M. A., and Hughes, P. E. (1975). Studies in sow reproduction. 2. The effect of lactation length on the subsequent reproductive performances of the sow. *Anim. Prod.* 20, 401–406.
89. Smith, D. M. (1961). The effect of daily separation of sows from their litters upon milk yield, creep intake, and energetic efficiency. *N. Z. J. Agric. Res.* 4, 232–245.
90. Walker, C., and England, D. (1978). Reproduction and litter performance of sows mated during lactation. *West Sect. Am. Soc. Anim. Sci.* 29, 115–116.
91. Thompson, L. H., Hanford, K. I., and Jensen, A. H. (1981). Estrus and fertility in lactating sows and piglet performance as influenced by limited nursing. *J. Anim. Sci.* 53, 1419–1423.
92. Stevenson, J. S., and Davis, D. L. (1984). Influence of reduced litter size and daily litter separation on fertility of sows at 2 to 5 weeks postpartum. *J. Anim. Sci.* 59, 284–293.
93. Kirkwood, R. N., Smith, W. C., and Lapwood, K. R. (1983). Effect of partial weaning of piglets on lactational oestrus and sow and piglet body weight changes. *N. Z. J. Exp. Agric.* 11, 231–233.
94. Henderson, R., and Hughes, F. E. (1984). The effects of partial weaning, movement and boar contact on the subsequent reproductive performance of lactating sows. *Anim. Prod.* 39, 131–135.
95. Stevenson, I. S., and Britt, J. H. (1981). Interval to estrus in sows and the performance of pigs after alteration of litter size during late lactation. *J. Anim. Sci.* 53, 177–181.

96. Cox, N. M., Britt, J. H., Armstrong, W. D., and Alhusen, H. D. (1983). Effect of feeding fat and altered weaning schedule on re-breeding in primiparous sows. *J. Anim. Sci.* 56, 21–29.
97. Rojkittikhun, T., Rojanasthien, S., Einarsson, S., Madej, A., and Lundeheim, N. (1991). Effect of fractionated weaning on hormonal patterns and weaning to oestrus interval in primiparous sows. *Acta Vet. Scand.* 32, 35–45.
98. Parvizi, N., Elsaesser, F., Smidt, D., and Ellendorff, F. (1976). Plasma luteinizing hormone and progesterone in the adult female pig during the oestrous cycle, late pregnancy and lactation and after ovariectomy and pertobarbitone treatment. *J. Endocrinol.* 169, 193–203.
99. Kurravongkrit, A. (1984). Clinical morphological and endocrinological studies in primiparous postpartum sows. Dissertation, Sveriges Lantbruksurivarsitet, Uppsala, Sweden.
100. Sesti, L. A. C., and Britt, J. H. (1994). Secretion of gonadotropins and estimated releasable pools of gonadotropin-releasing hormone and gonadotropins during establishment of suckling-induced inhibition of gonadotropin secretion in the sow. *Biol. Reprod.* 50, 1078–1086.
101. De Rensis, F., Hunter, M. G. and Foxcroft, G. R. (1993). Suckling-induced inhibition of luteinizing hormone secretion and follicular development in the early postpartum sow. *Biol. Reprod.* 48, 964–969.
102. Willis, H. J., Zak, L. J. and Foxcroft, G. R. (2003). Duration of lactation, endocrine and metabolic state, and fertility of primiparous sows. *J. Anim. Sci.* 81, 2088–2102.
103. Elliot, J. I., King, G. J., and Robertson, H. A. (1980). Reproductive performance of the sow subsequent to weaning piglets at birth. *Can. J. Anim. Sci.* 60, 65–71.
104. Kuravongkrit, A., Edquist, L. E., and Einarsson, S. (1983). Clinical and endocrinological studies in primiparous zero-weaned sows. 3. Hormonal patterns of ovarian disorders due to zero-weaning. *Zentrabl. Veterinarmed. A.* 30, 625–636.
105. Kuravongkrit, A., Kindahl, H., Einarsson, S., and Edquist, L. E. (1983). Clinical and endocrinological studies in primiparous zero-weaned sows. 1. Clinical and morphological findings with special reference to the effect of PGF treatment. *Zentrabl. Veterinarmed. A.* 30, 607–615.
106. Crighton, D. B., and Lamming, G. E. (1969). The lactational anoestnis of the sow: the status of the anterior pituitary-ovarian system during lactation and after weaning. *J. Endocrinol.* 43, 507–519.
107. Lauderdale, J. W., Kirkpatrick, R. L., First, N. L., Hauser, E. R., and Casida, L. E. (1965). Ovarian and pituitary gland changes in periparturient sows. *J. Anim. Sci.* 24, 1000–1103.
108. Palmer, W. M., Teague, H. S., and Venzke, W. G. (1965). Microscopic observations on the reproductive tract of the sow during lactation and early postweaning. *J. Anim. Sci.* 24, 1117–1125.
109. Palmer, W. M., Teague, H. S., and Venzke, W. G. (1965). Histological change in the reproductive tract of the sow during lactation and early post weaning. *J. Anim. Sci.* 24, 541–545.
110. Kunavongkrit, A., Einarsson, S., and Settergren, I. (1982). Follicular development in primiparous lactating sows. *Anim. Reprod. Sci.* 5, 47–56.
111. Dyck, G. W. (1983). Postweaning changes in the reproductive tract of the sow. *Can. J. Anim. Sci.* 63, 571–577.
112. Kermabon, A.-Y., Prunier, A., Djiane, J., and Salesse, R. (1995). Gonadotropins in lactating sows exposed to long or short days during pregnancy and lactation: serum concentrations and ovarian receptors. *Biol. Reprod.* 53, 1095–1102.
113. Ash, R. W., and Heap, R. B. (1975). Oestrogen progesterone and corticosteroid concentrations in peripheral plasma of sows during pregnancy, parturition, lactation and after weaning. *J. Endocrinol.* 64, 141–154.
114. Stevenson, J. S., Cox, N. M., and Britt, J. H. (1981). Role of the ovary in controlling luteinizing hormone, follicle stimulating hormone, and prolactin secretion during and after lactation in pigs. *Biol. Reprod.* 24, 341–353.
115. Duggan, R. T., Bryant, M. I., and Cunningham, F. J. (1982). Gonadotrophin, total oestrogen and progesterone concentrations in the plasma of lactating sows with particular reference to lactational oestrus. *J. Reprod. Fertil.* 64, 303–313.
116. Cox, N. M., and Britt, J. H. (1982). Pulsatile administration of gonadotrophin releasing hormone to lactating sows: endocrine changes associated with induction of fertile estrus. *Biol. Reprod.* 27, 1125–1137.
117. Shaw, H. J., and Foxcroft, G. R. (1985). Relationships between LH, FSH and prolactin secretion and reproductive activity in the weaned sow. *J. Reprod. Fertil.* 75, 17–28.
118. Edwards, S., and Foxcroft, G. R. (1983). Endocrine changes in sows weaned at two stages of lactation. *J. Reprod. Fertil.* 167, 161–172.
119. Hulten, F., Valros, A., Rundgren, M., and Einarson, S. (2002). Reproductive endocrinology and postweaning performance in the multiparous sow. Part 2. Influence of nursing behaviour. *Theriogenology* 58, 1519–1520.
120. Hunter, M. G., Denning-Kendall, P., Boulton, M. I., De-Rensis, F., Wild, M. L., and Foxcroft, G. R. (1992). Lack of stimulation of relaxin secretion in lactating sows by suckling in vivo or by oxytocin in vitro. *J. Reprod. Fertil.* 94, 121–128.
121. Bagnell, C. A., Tashima, L., Tsark, W., Ali, S. M., and McMurtry, I. P. (1990). Relaxin gene expression in the sow corpus luteum during the cycle, pregnancy, and lactation. *Endocrinology* 126, 2514–2520.
122. Cox, N. M., and Britt, J. H. (1982). Relationships between endogenous gonadotropin-releasing hormone, gonadotropins, and follicular develop ment after weaning in sows. *Biol. Reprod.* 27, 70–78.
123. Foxcroft, G. R., and Hunter, M. G. (1985). Basic physiology of follicular maturation in the pig. *J. Reprod. Fertil. Suppl.* 30, 1–19.
124. Britt, J. H., Armstrong, J. D., Cox, N. M., and Esbenshade, K. L. (1985). Control of follicular development during and after lactation in sows. *J. Reprod. Fert. Suppl.* 33, 37–54.
125. De Rensis, F., Hunter, M. G., Grant, S. A., Lancaster, R. T., and Foxcroft G. R. (1991). Effect of estrogen administration on endogenous and luteinizing hormone-releasing-hormone-induced luteinizing hormone secretion and follicular development in the lactating sow. *Biol. Reprod.* 44, 975–982.
126. Sesti, L. A. C., and Britt, J. H. (1993). Agonist-induced release of gonadotropin-releasing hormone, luteinizing hormone, and follicle-stimulating hormone and their associations with basal secretion of luteinizing hormone and follicle-stimulating hormone throughout lactation in sows. *Biol. Reprod.* 49, 332–339.
127. Foxcroft, G. R., Shaw, H. J., Hunter, M. G., Booth, P. J., and Lancaster, R. T. (1987). Relationships between luteinizing hormone, follicle stimulating hormone and prolactin secretion and ovarian follicular development in the weaned sow. *Biol. Reprod.* 36, 175–191.
128. Trout, W. E., Killen, J. H., Christenson, R. K., Schanbacher, B. D., and Ford, J. J. (1992). Effects of weaning on concentrations of inhibin in follicular fluid and plasma of sows. *J. Reprod. Fertil.* 94, 107–114.
129. Foxcroft, G. R. (1992). Nutritional and lactational regulation of fertility in the sow. *J. Reprod. Fertil. Suppl.* 45, 113–125.
130. Aherne, F. X., Christopherson, R. J., Thompson, J. R., and Hardin, R. T. (1976). Factors affecting the onset of puberty, postweaning estrus and blood hormone levels of Lacombe gilts. *Can. J. Anim. Sci.* 56, 681–692.

131. van der Wiel, D. F. M., van Landeghem, A. A. J., Willemse, A. H., and Bevers, M. M. (1979). Endocrine control of ovarian function after weaning in the domestic sow. *J. Endocrinol.* 80, 69P (Abstr.).
132. Dyck, G. W., Palmer, W. M., and Simaraks, S. (1979). Postweaning plasma concentrations of luteinizing hormone and estrogens in sows: effect of treatment with pregnant mares' serum gonadotropin or estradiol-17β plus progesterone. *Can. J. Anim. Sci.* 59, 159–166.
133. Elsaesser, F., and Parvizi, N. (1980). Partial recovery of the stimulatory oestrogen feedback action on LH release during late lactation in the pig. *J. Reprod. Fertil.* 59, 63–67.
134. Stevenson, J. S., and Britt, J. H. (1980). Luteinizing hormone, total estrogens and progesterone secretion during lactation and after weaning in sows. *Theriogenology* 14, 453–462.
135. Bevers, M. M., Willemse, A. H., Kruip, T. A. M., and van de Wiel, D. F. M. (1981). Prolactin levels and the LH response to synthetic LH-RH in the lactating sow. *Anim. Reprod. Sci.* 4, 155–163.
136. Rojansthien, S. A. M., Lundeheim, N., and Einarsson, S. (1987). Luteinizing hormone response to different doses of synthetic gonadotropin releasing hormone during early and late lactation in primiparous sows. *Anim. Reprod. Sci.* 13, 299–307.
137. Rojkittikhun, T., Einarsson, S., and Kindahl, H. (1991). A technique for continuously monitoring hormone levels in lactating sows and results obtained using it to study LH release. *Zentralbl. Veterinarmed. A.* 38, 344–349.
138. Barb, C. R., Estienne, M. J., and Kraeling, R. R. (1991). Endocrine changes in sows exposed to elevated ambient temperature during lactation. *Domest. Anim. Endocrinol.* 8, 117–127.
139. De Rensis, F., and Foxcroft, G. R. (1999). Correlation between LH response to challenges with GnRH and naloxone during lactation, and LH secretion and follicular development after weaning. *Anim. Reprod. Sci.* 28, 143–152.
140. De Rensis, F., Cosgrove, J. R., Willis, H. J., Hofacker, S., and Foxcroft, G. R. (1999). Ontogeny of the opioidergic regulation of LH and prolactin secretion in lactating sows, II: Interaction between suckling and morphine administration. *J. Reprod. Fertil.* 116, 243–252.
141. Sesti, L. A. C., and Britt. H. J. (1992). Elicitation of release of luteinizing hormone by N-methyl-D, L-aspartic acid during three paradigms of suppressed secretion of luteinizing hormone in the female pig. *Domest. Anim. Endocrinol.* 9, 105–114.
142. Ramirez, J. L., Bennett, W. A., and Cox, N. M. (1985). Effects of stage of lactation on estrous behaviour, LH and ovulation after administration of estradiol benzoate to sows. *J. Anim. Sci. (Suppl.).* 61, 44 (Abstr.).
143. Cox, N. M., Ramirez, J. L., Matamoros, I. A., and Bennett, W. A. (1988). Estrogen induces estrus unaccompanied by a preovulatory surge in luteinizing hormone in suckled sows. *Biol. Reprod.* 38, 592–596.
144. Kirkwood, R. M., Lapwood, K. R., Smith, W. C., Moller, K., and Garrick, D. J. (1984). Effects of estradiol benzoate treatment on the reproductive performance and endocrine status of sows after lactations of 10 or 35 days. *J. Reprod. Fertil.* 82, 329–337.
145. Kirkwood, R. M., Lapwood, K. R., Smith, W. C., and Anderson, I. L. (1984). Plasma concentrations of LH, prolactin, oestradiol-I7ß and progesterone in sows weaned after lactations of 10 or 35 days. *J. Reprod. Fertil.* 70, 95–102.
146. van Landeghem, A. A. J., and van de Wiel, D. F. M. (1978). Radioimmunoassay for porcine prolactin: plasma levels during lactation, suckling and weaning and after TRH administration. *Acta Endocrinol.* 88, 653–667.
147. Bevers, M. M., Willemse, A. H., and Kruip, T. A. M. (1978). Plasma prolactin levels in the sow during lactation and the postweaning period as measured by radioimmunoassay. *Biol. Reprod.* 19, 628–634.
148. Armstrong, J. D., Kraeling, R. R., and Britt, J. H. (1998). Effects of naloxone or transient weaning on secretion of LH and prolactin in lactating sows. *J. Reprod. Fertil.* 83, 301–308.
149. De Rensis, F., Quintavalla, F., and Foxcroft, G. R. (1998). Treatment of lactating sows with the dopamine agonist Cabergoline: effects on LH and prolactin secretion and responses to challenges with naloxone and morphine. *Anim. Reprod. Sci.* 51, 233–247.
150. Bevers, M. M., Willemse, A. H., and Kruip, T. A. M. (1983). The effect of bromocriptine on luteinizing hormone levels in the lactating sow: evidence for a suppressive action by prolactin and the suckling stimulus. *Acta Endocrinol.* 104, 261–265.
151. Mattioli, M., and Seren, E. (1985). Effect of bromocryptine treatment during lactational anestrus in pigs. In *Endocrine Causes of Seasonal and Lactational Anestrus in Animals* (F. Ellendorf, and F. Elsaesser, Eds.), pp. 165–178. Martinus Nijhoff, Dordecht.
152. Kraeling, R. R., Rampacek, G. B., Cox, N. M., and Kiser, T. E. (1982). Prolactin and luteinizing hormone secretion after bromocryptine (CB-154) treatment in lactating sows and ovariectomized gilts. *J. Anim. Sci.* 54, 1212–1220.
153. Mattioli, M., Conte, F., Galeati, G., and Seren, E. (1986). Effect of naloxone on plasma concentrations of prolactin and LH in lactating sows. *J. Reprod. Fertil.* 76, 167–173.
154. Armstrong, J. D., Kraeling, R. R., and Britt, J. H. (1988). Morphine suppresses luteinizing hormone concentrations in transiently weaned sows and delays onset of estrus after weaning. *J. Anim. Sci.* 66, 2215–2223.
155. Barb, C. R., Kraeling, R. R., Rampacek, G. B., and Whisnant, C. S. (1986). Opioid inhibition of luteinizing hormone secretion in the postpartum lactating sow. *Biol. Reprod.* 35, 368–371.
156. Kraeling, R. R., and Barb, C. R. (1990). Hypothalamic control of gonadotrophin and prolactin secretion in pigs. *J. Reprod. Fertil. Suppl.* 40, 3–17.
157. Armstrong, J. D., Esbenshade, K. L., Coffey, M. T., Heimer, E., and Campbell, R. (1990). Opioid control of growth hormone in the suckled sow is primarily mediated through growth hormone releasing factor. *Domest. Anim. Endocrinol.* 7, 191–198.
158. Barb, C. R., Kraeling, R. R., Rampacek, G. B., and Leshin, L. S. (1987). Opioid modulation of follicle stimulating hormone (FSH) and luteinizing hormone (LH) secretion in the postpartum sow. *Adv. Exp. Biol. Med.* 219, 647–652.
159. Barb, C. R., Barrett, J. B., Wright, J. T., Kraeling, R. R., and Rampacek, G. B. (1990). Opioid modulation of LH secretion by pig pituitary cells in vitro. *J. Reprod. Fertil.* 90, 213–219.
160. McNeilly, A. S. (1984). Prolactin and ovarian function. In *Neuroendocrine Perspectives 3* (E. E. Muller, and R. M. Macleod, Eds.), pp. 279–316. Elsevier Biomedical Press, Amsterdam
161. Clowes, E. J., Aherne, F. X., Foxcroft, G. R., and Baracos, V. E. (2003). Selective protein loss in lactating sows is associated with reduced litter growth and ovarian function. *J. Anim. Sci.* 81, 753–764.
162. Clowes, E. J., Aherne, F. X., Schefer, A. L., Foxcroft, G. R., and Baracos, V. E. (2003). Parturition body size and body protein loss during lactation influence performance during lactation and ovarian functon at weaning in first-parity sows. *J. Anim. Sci.* 81, 1517–1528.

163. Kann, G., Martinet, J., and Shirar, A. (1979). Hypothalamic-pituitary control during lactation in sheep. In *Control of Ovulation* (D. B. Crighton, N. B. Haynes, G. E. Foxcroft, and G. E. Lamming, Eds.), pp. 319–333. Butterworths, London.
164. Mauleon, P., and Dauzier, L. Variations de durée de l'anoestrus de lactation chez Ic brebis de race Ile-de-France. *Ann. Biol. Anim. Biochem. Biophys.* 5, 131–143.
165. Shevah, Y., Black, W. J. M., and Land, R. B. (1975). The effects of nutrition on the reproductive performance of Finn X Dorset ewes. 11. Postpartum ovarian activity, conception and the plasma concentration of progesterone and LH. *J. Reprod.* 45, 289–299.
166. Jenkin, G., Heap, R. B., and Symons, D. B. A. (1977). Pituitary responsiveness to synthetic LHRH and pituitary LH content at various reproductive stages in the sheep. *J. Reprod. Fertil.* 49, 207–214.
167. Chamley, W. A., Jones, H. A., and Parr, R. A. (1976). Content of LH, FSH, and growth hormone in pituitaries of pregnant and anestrus ewes. *Endocrinology* 98, 1535–1538.
168. Moss, G. E., Adams, T. E., Niswender, G. D., and Nett, T. M. (1980). Effects of pituitary and suckling on concentrations of pituitary gonadotropins, hypothalamic GnRH and pituitary responsiveness to GnRH in ewes. *J. Anim. Sci.* 50, 495–502.
169. Crowder, M. E., Glues, P. A., Tamanini, C., Moss, G. E., and Nett, T. M. (1982). Pituitary content of gonadotropins and GnRH-receptors in pregnant, postpartum and steroid-treated OVX ewes. *J. Anim. Sci.* 50, 495–502.
170. Restall, B. J., Kearins, R. D., and Starr, B. G. (1977). Studies of pituitary function in lactating ewes. *J. Reprod. Fertil.* 49, 291–296.
171. Wright, P. J., Jenkin, G., and Heap, R. B. (1981). Prolactin and LH release in response to LHRH and TRH in ewes during dioestrus, pregnancy and postpartum. *J. Reprod. Fertil.* 62, 447–453.
172. Wright, P. J., Jenkin, G., and Heap, R. B. (1981). Prolactin and LH release in response to LHRH and TRH in ewes during dioestrus, pregnancy and postpartum. *Biol. Reprod.* 15, 335–342.
173. Fitzgerald, B. P., and Cunningham, F. J. (1981). Effect of removal of lambs or treatment with bromocriptine on plasma concentrations of prolactin and FSH during the postpartum period in ewes lambing at different times during the breeding season. *J. Reprod. Fertil.* 56, 141–148.
174. Schirar, A., Cognie, Y., and Louault, F. (1990). Resumption of gonadotrophin release during the postpartum period in suckling and non-suckling ewes. *J. Reprod. Fertil.* 88, 593–604.
175. Schirar, A., Cognie, Y., Louault, F., Poulin, N., Levasseur, M. C., and Martinet, J. (1989). Resumption of oestrous behaviour and cyclic ovarian activity in suckling and non-suckling ewes. *J. Reprod. Fertil.* 87, 789–794.
176. Wise, M. E. (1990). Gonadotropin-releasing hormone secretion during the postpartum anestrous period of the ewe. *Biol. Reprod.* 43, 719–725.
177. Wright, P. J., Geytenbeek, P. E., Clarke, I .J., and Findlay, J. K. (1980). Pituitary responsiveness to LHRH, the occurrence of oestradiol-17fl-induced positive feedback and the resumption of oestrous cycles in ewes postpartum. *J. Reprod. Fertil.* 60, 171–176.
178. Restall, B. J., and Starr, B. G. (1977).The influence of season of lambing and lactation on reproductive activity and plasma LH concentrations in Memo ewes. *J. Reprod. Fertil.* 49, 297–303.
179. Clarke, I. J., Thomas, G. B., Yao, B., and Cummins, J. T. (1987). GnRH secretion throughout the ovine estrous cycle. *Neuroendocrinology* 46, 82–88.
180. Moenter, S. M., Caraty, A., Locatelli, A., and Karsch, F. J. (1991). Pattern of gonadotropin-releasing hormone (GnRH) secretion leading up to ovulation in the ewe: existence of a preovulatory GnRH surge. *Endocrinology* 129, 1175–1182.
181. Wise, M. E., Nilson, J. H., Nejedlik, M. T., and Nett, T. M. (1985). Measurement of messenger RNA for luteinizing hormone β-subunit and a-subunit during gestation and the postpartum period in ewes. *Biol. Reprod.* 33, 1009–1015.
182. Crawford, J., Currie, R. and McNeilly, A. S. (2000). Replenishment of LH stores of gonadotrophs in relation to gene expression, synthesis and secretion of LH after the preovulatory phase of the sheep oestrous cycle. *J. Endocrinol.* 167, 453–463.
183. Clarke, I. J., Wright, P. J., Chamley, W. A., and Burman, K. (1984). Differences in the reproductive endocrine status of ewes in the early postpartum period and during seasonal anoestrus. *J. Reprod. Fertil.* 70, 591–597.
184. Chamley, W. A., Findlay, J. K., Cumming, I. A., Buckmaster, J. M., and Goding, J. R. (1974). Effect of pregnancy on the LH response to synthetic gonadotropin releasing hormone in the ewe. *Endocrinology* 94, 291–293.
185. Fraser, H. M., and McNeilly, A. S. (1982). Effect of immunoneutralization of luteinizing hormone releasing hormone on the estrogen-induced luteinizing hormone and follicle-stimulating hormone surges in the ewe. *Biol. Reprod.* 27, 548–555.
186. Wright, P. J., Geytenbeek, P. E., Clarke, I. J., and Findlay, J. K. (1981). Evidence for a change in oestradiol negative feedback and LH pulse frequency in postpartum ewes. *J. Reprod. Fertil.* 61, 97–102.
187. Wright, P. J., Geytenbeek, P. E., Clarke, I. J., and Findlay, J. K. (1983). LH release and luteal function in postpartum acyclic ewes after the pulsatile ad ministration of LH-RH. *J. Reprod. Fertil.* 67, 257–262.
188. Wright, P. J., Geytenbeek, P. E., Clarke, I. J., and Findlay, J. K. (1984). Induction of plasma LH surges and normal luteal function in acyclic postpartum ewes by the pulsatile administration of LHRH. *J. Reprod. Fertil.* 71, 1–6.
189. McNeilly, A. S., O'Connell, M., and Baird, D. T. (1982). Induction of ovulation and normal luteal function by pulsed injections of luteinizing hormone in anestrous ewes. *Endocrinology* 110, 1292–1299.
190. McNatty, K. P., Hudson, N., and Gibb, M. (1984). Effects of long-term treatment with LH on induction of cyclic ovarian activity in seasonally anoestrous ewes. *J. Endocrinol.* 100, 67–73.
191. McLeod, B. J., Haresign, W., and Lamming, G. E. (1982). Response of GnRH with and without progesterone pretreatment. *J. Reprod. Fertil.* 65, 223–230.
192. Baird, D. T., and McNeilly, A. S. (1981). Gonadotrophin control of follicular development and secretion in the sheep oestrous cycle. *J. Reprod. Fertil. Suppl.* 30, 119–133.
193. Braden, T. D., Sawyer, H. R., and Niswender, G. D. (1989). Functional and morphological characteristics of the first corpus luteum formed after parturition in ewes. *J. Reprod. Fertil.* 86, 525–533.
194. Walton, J. S., McNeilly, J. R., McNeilly, A. S., and Cunningham F. J. (1977). Changes in blood levels of prolactin, LH, FSH and progesterone during anoestrus in the ewe. *J. Endocrinol.* 75, 127–136.
195. Lamming, G. E., Moseley, S. R., and McNeilly, J. R. (1974). Prolactin release in the sheep. *J. Reprod. Fertil.* 40, 151–168.
196. Kann, G., and Denamur, R. (1974). Possible role of prolactin during the oestrous cycle and gestation in the ewe. *J. Reprod. Fertil.* 39, 473–483.

197. Gordon, K., Renfree, M. B., Short, R. V., and Clarke, I. J. (1987). Hypothalamo pituitary portal blood concentration of during suckling in the ewe. *J. Reprod. Fertil.* 79, 397–408.
198. McNeilly, J. R. (1972). Observations on the pattern of prolactin release during suckling in the ewe. *J. Reprod. Fertil.* 31, 487–494.
199. Wright, P. J., Findlay, J. R., and Anderson, G. A. (1981). The failure of prolactin to enhance the inhibitory effect of oestradiol-17ß on LH synthesis and release in ewes. *J. Reprod. Fertil.* 62, 537–542.
200. Louw, B. P., Lishman, A. W., Botha, W. A., Arongie, P. A. R., Poultney, B. G., and Günter, M. J. (1976). The release of luteinizing hormone (LH) in ewes deprived of prolactin during lactations. *S. Afr. J. Anim. Sci.* 6, 87–91.
201. McNeilly, A. S., and Baird, D. T. (1983). Direct effect of prolactin induced by TRH injection, on ovarian oestradiol secretion in the ewe. *J. Reprod. Fertil.* 69, 559–568.
202. Rhind, S. M., Chesworth, J. M., and Robinson, J. J. (1978). A seasonal difference in ovine peripheral plasma prolactin and progesterone concentrations in early pregnancy and the relationship between the two hormones. *J. Reprod. Fertil.* 52, 79–81.
203. Restall, B. J., Kearins, R. D., Herdegen, J., and Carberry, P. (1978). The induction of reproductive activity in lactating ewes. *Aust. J. Agric. Res.* 29, 181–189.
204. Rhind, S. M., Robinson, J. J., Chesworth, J. M., and Crofts, R. M. J. (1980). Effects of season, lactation and plane of nutrition on prolactin concentrations in ovine plasma and the role of prolactin in the control of ewe fertility. *J. Reprod. Fertil.* 58, 145–152.
205. Kittock, R. I., Stellflug, J. N., and Lowry, S. R. (1983). Enhanced progesterone and pregnancy rate after gonadotropin administration in lactating ewes. *J. Anim. Sci.* 56, 652–655.
206. Pelletier, M., and Thimonier, J. (1973). Comparison of the induced preovulatory LH discharge in lactating and dry sheep during seasonal an oestrus. *J. Reprod. Fertil.* 33, 310–313.
207. Tortonese, D., Brooks, J., Ingleton, P. M., and McNeilly, A. S. (1998). Detection of prolactin receptor gene expression in the sheep pituitary gland and visualisation of the specific translation of the signal in gonadotrophs. *Endocrinology* 139, 5212–5223.
208. Gregory, S. J., Townsend, J., McNeilly, A. S., and Tortonese, D. J. (2004). Effects of prolactin on the luteinizing hormone response to gonadotropin releasing hormone in primary pituitary cell cultures during the ovine annual reproductive cycle. *Biol. Reprod.* 70, 1299–1305.
209. Kann, G., and Martinet, J. (1975). Prolactin levels and duration of postpartum anoestrus lactating ewes. *Nature* 257, 63–64.
210. Casida, L. E., Graves, W. E., Hauser, E. R., Lauderdale, J. W., Reisen, J. W., Saiduddin, S., and Tyler, W. J. (1968). Studies on the postpartum cow. *Bull. Wis. Agric. Exp. Stn.* 270, 1–54.
211. Hurnik, J. F., King, G. F., and Robertson, H. A. (1975). Estrus and related behaviour in postpartum Holstein cows. *Appl. An. Ethol.* 2, 55–62.
212. King, G. J., Hurnik, J. F., and Robertson, H. A. (1976). Ovarian function and estrus in dairy cows during early lactation. *J. Anim. Sci.* 42, 688–692.
213. Bulman, D. C., and Lamming, G. E. (1978). Milk progesterone levels in relation to conception, repeat breeding and factors influencing acyclicity in dairy cows. *J. Reprod. Fertil.* 54, 447–458.
214. Webb, R., Lamming, G. E., Haynes, N. B., and Foxcroft, G. R. (1980). Plasma progesterone and gonadotrophin concentrations and ovarian activity in postpartum dairy cows. *J. Reprod. Fertil.* 59, 133–143.
215. Stevenson, J. S., and Britt, J. H. (1978). Relationships among luteinizing hormone estradiol, progesterone, glucocorticoids, milk yield, body weight and postpartum ovarian activity in holstein cows. *J. Anim. Sci.* 48, 570–577.
216. Schams, D., Schallenberger, E., and Menzer, C. (1978). Profiles of LH, FSH and progesterone in postpartum dairy cows and their relationship to the commencement of cyclic function. *Theriogenology* 10, 453–468.
217. Savio, J. D., Boland, M. P., Hynes, N., and Roche, J. F. (1990). Resumption of follicular activity in the early post-partum period of dairy cows. *J. Reprod. Fertil.* 88, 569–579.
218. Schemm, S. R., Deaver, D. R., Griel, L. C. Jr., and Muller, L. D. (1990). Effects of recombinant bovine somatotropin on luteinizing hormone and ovarian function in lactating dairy cows. *Biol. Reprod.* 42, 815–821.
219. Spicer, U., Tucker, W. B., and Adams, G. D. (1990). Insulin-like growth factor-I in dairy cows: relationships among energy balance, body condition, ovarian activity, and estrous behavior. *J. Dairy Sci.* 73, 929–937.
220. Webb, R., Lamming, G. E., Haynes, N. B., Hafs, D., and Manns, J. G. (1977). Response of cyclic and postpartum suckled cows to injections of synthetic LHRH. *J. Reprod. Fertil.* 50, 203–210.
221. Short, R. E., Bellows, R. A., Moody, E. L., and Howland, B. E. (1972). Effects of suckling and mastectomy on bovine postpartum reproduction. *J. Anim. Sci.* 34, 70–74.
222. Oxenreider, S. L. (1968). Effects of suckling and ovarian function on postpartum reproductive activity in beef cows. *Am. J. Vet. Res.* 29, 2099–2102.
223. Wiltbank, J. N., and Cook, A. C. (1958). The comparative reproductive performance of nursed cows and milked cows. *J. Anim. Sci.* 17, 640–648.
224. Carruthers, T. D., and Hafs, H. D. (1980). Suckling and four times daily milking: influence on ovulation, estrus and serum luteinizing hormone, glucocorticoids and prolactin postpartum Holsteins. *J. Anim. Sci.* 50, 919–925.
225. Randel, R. D. (1990). Nutrition and postpartum rebreeding in cattle. *J. Anim. Sci.* 68, 853–862.
226. Williams, G. L. (1990). Suckling as a regulator of postpartum rebreeding in cattle: a review. *J. Anim. Sci.* 68, 831–852.
227. Bryner, R. W., Garcia-Winder, M., Lewis, P. E., Inskeep, E. K., and Butcher, R. L. (1990). Changes in hormonal profiles during the estrous cycle in old lactating beef cows. *Domest. Anim. Endocrinol.* 7, 181–189.
228. Schrick, F. N., Spitzer, J. C., Jenkins, T. C., Henricks, D. M., and Althen, T. G. (1990). Effect of dietary energy restriction on metabolic and endocrine responses during the estrous cycle of the suckled beef cow. *J. Anim. Sci.* 68, 3313–3321.
229. Perry, R. C., Corah, L. R., Kiracofe, G. H., Stevenson, J. S., and Beal, W. E. (1991). Endocrine changes and ultrasonography of ovaries in suckled beef cows during resumption of postpartum estrous cycles. *J. Anim. Sci.* 69, 2548–2555.
230. Murphy, M. G., Boland, M. P., and Roche, J. F. (1990). Pattern of follicular growth and resumption of ovarian activity in post-partum beef suckler cows. *J. Reprod. Fertil.* 90, 523–533.
231. Perry, R. C., Corah, L. R., and Cochran, R. C. (1991). Influence of dietary energy on follicular development, serum gonadotropins, and first postpartum ovulation in suckled beef cows. *J. Anim. Sci.* 69, 3762–3773.
232. Peters, A. R., and Lamming, G. E. (1984). Reproductive activity of the cow in the postpartum period. II. Endocrine, patterns and induction of ovulation. *Br. Vet. J.* 140, 269–280.

233. England, B. G., Hauser, E. R., and Casida, L. E. (1973). Some effects of unilateral ovariectomy in the postpartum beef cow. *J. Anim. Sci.* 36, 45–50.
234. Tribble, R. L., Sorensen, A. M., Woodward, T. L., Connor, J. S., Beverly, J. R., and Fleeger, J. L. (1973). Serum progestins and luteinizing hormone levels in non-suckled primiparous heifers. *Nature* 246, 494–495.
235. Connor, J. S., Tribble, R. L., Woodward, T. L., Beverly, J. R., Sorensen, A. M., Jr., and Fleeger, J. L. (1974). *J. Anim. Sci.* 39, 203 (Abstr.).
236. Randel, R. D., Short, R. E., and Bellows, R. A. (1976). Suckling effect on LH and progesterone in beef cows. *J. Anim. Sci.* 42, 267 (Abstr.).
237. Carter, M. L., Dierschke, D. J., Rutledge, J. J., and Hauser, E. R. (1980). Effect of gonadotropin releasing hormone and calf removal on pituitary ovarian function and reproductive performance in postpartum beef cows. *J. Anim. Sci.* 51, 903–909.
238. Shively, T. E., and Williams, G. L. (1989). Patterns of tonic luteinizing hormone release and ovulation frequency in suckled anestrous beef cows following varying intervals of temporary weaning. *Domest. Anim. Endocrinol.* 6, 379–387.
239. Connor, H. C., Houghton, P. L., Lemenager, R. P., Malven, P. V., Parfet, J. R., and Moss, G. E. (1990). Effect of dietary energy, body condition and calf removal on pituitary gonadotropins, gonadotropin-releasing hormone (GnRH) and hypothalamic opioids in beef cows. *Domest. Anim. Endocrinol.* 7, 403–411.
240. Mukasa-Mugerwa, E., Tegegne, A., and Franceschini, R. (1991). Influence of suckling and continuous cow-calf association on the resumption of post-partum ovarian function in Bos indicus cows monitored by plasma progesterone profiles. *Reprod. Nutr. Dev.* 31, 241–247.
241. Clapp, H. (1937). A factor in breeding efficiency of dairy cattle. *Proc. Am. Soc. Anim. Prod.* 37, 259–265.
242. Wetterman, R. P., Turman, E. J., Wyatt, R. D., and Totusek, R. (1978). Influence of suckling intensity in reproductive performance of range cows. *J. Anim. Sci.* 47, 342–346.
243. Montgomery, G. W. (1982). Influence of suckling frequency and bromocriptine treatment on the resumption of ovarian cycles in post partum beef cattle. *Theriogenology* 17, 551–563.
244. Peters, A. R., and Riley, G. M. (1982). Milk progesterone profiles and factors affecting postpartum ovarian activity in beef cows. *Anim. Prod.* 34, 145–153.
245. Lamming, G. E., Wathes, D. C., and Peters, A. R. (1981). Endocrine patterns of the postpartum cow. *J. Reprod. Fertil. Suppl.* 30, 155–170.
246. Peters, A. R., Lamming, G. E., and Fisher, M. W. (1981). A comparison of plasma LH concentrations in milked and suckling postpartum cows. *J. Reprod. Fertil.* 62, 567–573.
247. Suzuki, O., and Sato, M. (1984). Effects of once-daily suckling and twice-daily suckling on postpartum reproductive performance of beef cows and on growth and ingestive behaviour of their calves. *Jpn. Anim. Prod.* 30, 39–45.
248. Randel, R. O., and Welker, G. A. (1981). Effect of once-daily suckling on postpartum interval and cow-calf performance of first-calf Brahman X Hereford heifers. *J. Anim. Sci.* 53, 755–757.
249. Montgomery, G. W., Davis, G. H., and Hurrell, G. A. (1980). Interval from calving to first oestrus in autumn- and spring-calving herds in the same locality. *Proc. N. Z. Soc. Anim. Prod.* 40, 280–284.
250. Peters, A. R., and Riley, G. M. (1982). Is the cow a seasonal breeder? *Br. Vet. J.* 138, 533–537.
251. Hansen, P. J., and Hauser, E. R. (1983). Genotype X environmental interactions on reproductive traits of bovine females. III. Seasonal variation in postpartum reproduction as influenced by genotype, suckling and dietary regimen. *J. Anim. Sci.* 56, 1362–1369.
252. Hansen, P. J., and Hauser, E. R. (1984). Photoperiodic alteration of postpartum reproductive function in suckled cows. *Theriogenology* 22, 1–14.
253. Hansen, P. J. (1985). Seasonal modulation of puberty and the postpartum anestrus in cattle: a review. *Livestock Prod. Sci.* 12, 309–327.
254. McNatty, K. P., Hudson, N., Gibb, M., et al. (1984). Seasonal differences in ovarian activity in cows. *J. Endocrinol.* 102, 109–198.
255. Wiltbank, J. N., Rowden, W. W., Ingalls. J. E., and Zimmerman, D. R. (1964). Influence of post-partum energy level on reproductive performance of Hereford cows restricted in energy intake prior to calving. *J. Anim. Sci.* 23, 1049–1053.
256. Dunn, T. G., Ingalls, J. E., Zimmerman, D. R., and Wiltbank, J. N. (1969). Reproductive performance of 2–year-old Hereford and Angus heifers as influenced by pre- and post-calving energy intake. *J. Anim. Sci.* 29, 719–726.
257. Spicer, U., Tucker, W. B., and Adams, G. D. (1990). Insulin-like growth factor-I in dairy cows: relationships among energy balance, body condition, ovarian activity, and estrous behavior. *J. Dairy Sci.* 73, 929–937.
258. Lucy, M. C., Staples, C. R., Michel, F. M., Thatcher, W. W., and Bolt, D. J. (1991). Effect of feeding calcium soaps to early postpartum dairy cows on plasma prostaglandin F2 alpha, luteinizing hormone, and follicular growth. *J. Dairy Sci.* 74, 483–489.
259. Canfield, R. W., and Butler, W. R. (1991). Energy balance, first ovulation and the effects of naloxone on LH secretion in early postpartum dairy cows. *J. Anim. Sci.* 69, 740–746.
260. Butler, W. R., and Smith, R. D. (1989). Interrelationships between energy balance and postpartum reproductive function in dairy cattle. *J. Dairy Sci.* 72, 767–783.
261. Lamming, G. E. (1966). Nutrition and the endocrine system. *Nutr. Abstr.* 36, 1–13.
262. Laing, J. A., ed. (1970). Fertility and infertility in the domestic animals, 2nd ed. Bailliere, Tindall & Cassell, London.
263. Morrow, D. A. (1970). Diagnosis and prevention of infertility in cattle. *J. Dairy Sci.* 53, 961–969.
264. McClure, J. J. (1972). Blood glucose and female infertility. *Vet. Rec.* 91, 193.
265. Oxenreider, S. L., and Wagner, W. C. (1971). Effects of lactation and energy intake on postpartum ovarian activity in the cow. *J. Anim. Sci.* 33, 1025–1031.
266. Staples, C. R., Thatcher, W. W., and Clark, J. H. (1990). Relationship between ovarian activity and energy status during the early postpartum period of high producing dairy cows. *J. Dairy Sci.* 73, 938–947.
267. Labhsetwar, A. P., Collins, W. E., Tyler, W. J., and Casida, L. E. (1964). Some pituitary ovarian relationships in the periparturient cow. *J. Reprod. Fertil.* 38, 85–90.
268. Lishman, A. W., Allison, S. M. J., Fogwell, R. L., Butcher, R. L., and Inskeep, E. K. (1979). Follicular development and function of induced CL in underfed postpartum anestrous beef cows. *J. Anim. Sci.* 48, 867–875.
269. Hoffman, B., Schems, D., Gimenez, T., Ender, M. L., Hermann, C., and Karg, H. (1973). Change of progesterone, total oestrogens, corticosteroids, prolactin and LH in bovine peripheral plasma around parturition with special reference to the effect of exogenous corticoids and a prolactin inhibitor respectively. *Acta Endocrinol.* 73, 389–395.
270. Rawlings, N. C., Weir, L., Todd, B., Manns, J., and Hyland, J. H. (1980). Some endocrine changes associated

with the postpartum period of the suckling beef cow. *J. Reprod. Fertil.* 60, 301–308.

271. Walters, D., Schams, D., and Schallenberge, E. (1984). Pulsatile secretion of gonadotrophins, ovarian steroids and ovarian oxytocin during the luteal phase of the oestrous cycle in the cow. *J. Reprod. Fertil.* 71, 479–491.
272. Peters, A. R., Pimentel, M. G., and Lamming, G. E. (1985). Hormone responses to exogenous GnRH pulses in postpartum dairy cows. *J. Reprod. Fertil.* 75, 557–565.
273. Walters, D. L., Kaltenbach. C. C., Dunn, T. G., and Short, R. E. (1982). Pituitary and ovarian function in postpartum beef cows. I. Effect of suckling on serum and follicular fluid hormones and follicular gonadotropin receptors. *Biol. Reprod.* 26, 640–646.
274. Bellin, M. E., Hinshelwood, M. M., Hauser, E. R., and Ax, R. L. (1984). Influence of suckling and side of corpus luteum or pregnancy on folliculogenesis in postpartum cows. *Biol. Reprod.* 31, 849–855.
275. Dufour, J. I., and Roy, G. L. (1985). Distribution of ovarian follicular populations in the dairy cow within 35 days after parturition. *J. Reprod. Fertil.* 75, 85–90.
276. Moss, G. E., Parfet, J. R., Marvin, C. A., Allrich, R. D., and Diekman, M. A. (1985). Pituitary concentrations of gonadotropins and receptors for GnRH in suckled beef cows at various intervals and calving. *J. Anim. Sci.* 60, 285–293.
277. Halter, T. B., Hayes, S. H., Laranja da Fonseca, L. F., and Silva, W. J. (2003). Relationship between endogenous progesterone and follicular dynamics in lactating dairy cows with ovarian follicular cysts. *Biol. Reprod.* 69, 218–223.
278. Pierson, R. A., and Ginther, O. J. (1984). Ultrasonography of the bovine ovary. *Theriogenology* 21, 495–504.
279. Quirk, S. M., Hickey, G. J., and Fortune, J. E. (1986). Growth and regression of ovarian follicles during the follicular phase and during the oestrous cycle in heifers undergoing spontaneous and PGF-2a-induced luteolysis. *J. Reprod. Fertil.* 77, 211–219.
280. Savio, J. D., Keenan, L., Boland, M. P., and Roche, J. F. (1988). Pattern of growth of dominant follicles during the oestrous cycle of heifers. *J. Reprod. Fertil.* 83, 663–671.
281. Sirois, J., and Fortune, J. E. (1988). Ovarian follicular dynamics during the oestrous cycle in heifers monitored by real-time ultrasonography. *Biol. Reprod.* 39, 308–317.
282. Gong, J. G., Bramley, A., and Webb, R. (1991). The effect of recombinant somatotropin on ovarian function in heifers: follicular populations and peripheral hormones. *Biol. Reprod.* 45, 941–949.
283. Webb, R., Nicholas, B., Gong, J. G., Campbell, B. K., Gutierrez, C. G., Garverick, H. A., and Armstrong, D. G. (2003). Mechanisms regulating follicular development and selection of the dominant follicle. *Reprod. Suppl.* 61, 71–90.
284. Wiltbank, M. C., Gumen, A., and Sartori, R. (2001). Physiological classification of anovulatory conditions in cattle. *Theriogenology* 57, 21–51.
285. Webb, R., Garnsworthy, P. C., Gong, J.-G., and Armstrong, D. G. (2004). Control of follicular growth: local interactions and nutritional influences. *J. Anim. Sci.* 82, E63–E74.
286. Beam, S. W., and Butler, W. R. (1999). Effects of energy balance on follicular development and first ovulation in postpartum dairy cows. *J. Reprod. Fertil. Suppl.* 54, 4111–424.
287. Yavas, Y., and Walton, J. S. (2000). Postpartum acyclicity in suckled beef cows: a review. *Theriogenology* 54, 25–55.
288. Savio, J. D., Boland, M. P., and Roche, J. F. (1990). Development of dominant follicles and length of ovarian cycles in post-partum dairy cows. *J. Reprod. Fertil.* 88, 581–591.
289. Duffy, P., Crowe, M. A., Boland, M. P., and Roche, J. F. (2000). Effect of exogenous LH pulses on the fate of the first dominant follicle in postpartum beef cows nursing calves. *J. Reprod. Fertil.* 118, 9–17.
290. Lamb, G. C., Lynch, J. M., Greiger, D. M., Minton, J. E., and Stevenson, J. S. (1997). Ad libitum suckling by an unrelated calf in the presence or absence of a cow's own calf prolongs postpartum aovulation. *J. Anim. Sci.* 75, 2762–2769.
291. Stag, K., Spicer, L. J., Sreenan, J. M., Roche, J. F., and Diskin, M. G. (1998). Effect of calf isolation on follicular wave dynamics, gonadotropin and metabolic hormone changes, and interval to first ovulation in beef cows fed either of two energy levels postpartum. *Biol. Reprod.* 59, 777–783.
292. lnskeep, E. K., Braden, T. D., Lewis, P. E., Garcia-Winder, M., and Niswender, G. D. (1988). Receptors for luteinizing hormone and follicle-stimulating hormone in largest follicles of postpartum beef cows. *Biol. Reprod.* 38, 587–591.
293. Braden, T. D., King, M. E., Odde, K. G., and Niswender, G. D. (1989). Development of preovulatory follicles expected to form short-lived CL in beef cows. *J. Reprod. Fertil.* 85, 97–104.
294. Prado, R., Rhind, S. M., Wright, I. A., Russell, A. J. F., McMillen, S. M., Smith, A. J., and McNeilly, A. S. (1990). Ovarian follicle populations, steroidogenicity and micromorphology at five and nine weeks post partum in beef cows in two levels of body condition. *Anim. Prod.* 51, 103–108.
295. Cermak, D. L., Braden, T., Manns, J., Niswender, G. D., and Nett, T. M. Contents of hypothalamic GnRH, pituitary FSH and LH, and pituitary receptors for GnRH and estradiol in postpartum suckled beef cow. *J. Anim. Sci.* 57, 322.
296. Saiduddin, S., Riesen, W., Tyler, W. J., and Casida, L. E. (1968). Studies on the postpartum cow. 5.0.0. Relation of the postpartum interval to pituitary gonadotropins, ovarian follicular development and development and fertility in dairy cows. *Wis. Agr. Exp. Sin. Bull.* 270, 15.
297. Leung, K., Padmanabhan, V., Spicer, U., Tucker, H. A., Short, R. E., and Convey, E. M. (1986). Relationships between pituitary GnRH-binding sites and pituitary release of gonadotropins in postpartum beef cows. *J. Reprod. Fertil.* 76, 53–63.
298. Carruthers, T. D., Convey, E. M., Kesner, J. S., Hafs, H. D., and Cheng, K. W. (1980). The hypothalamo–pituitary–gonadotropin axis of suckled and non suckled dairy cows postpartum. *J. Anim. Sci.* 51, 949–957.
299. Riley, G. M., Peters, A. R., and Lamming, G. E. (1981). Induction of pulsatile LH release, FSH release and ovulation in postpartum acyclic beef cows by repeated small doses of GnRH. *J. Reprod. Fertil.* 63, 559–565.
300. Dobson, H. (1978). Radioimmunoassay of FSH in the plasma of postpartum dairy cows. *J. Reprod. Fertil.* 52, 45–49.
301. Gong, J.-G., Lee, W. J., Garnsworthy, P. C., and Webb, R. (2002). Effect of dietary-induced increases in circulating insulin concentrations during the early postpartum period on reproductive functon in dairy cows. *Reproduction* 123, 419–427.
302. Duffy, P., Crowe, M. A., Austin, E. J., Mihm, M., Boland, M. P., and Roche, J. F. (2004). The effect of eCG or estradiol at or after norgetomet removal on follicular dynamics, estrus and ovulation in early port-partum beef cows nursing calves. *Theriogenology* 61, 725–734.
303. Schallenberger, E., and Petersen, A. J. (1982). Effect of ovariectomy on tonic gonadotrophin secretion in cyclic and postpartum dairy cows. *J. Reprod. Fertil.* 64, 47–52.
304. Foster, J. P., Lamming, G. E., and Peters, A. R. (1980). Short-term relationship between plasma LH, FSH and progesterone concentrations in post partum dairy cows and the effect of GnRH injection. *J. Reprod. Fertil.* 59, 321–327.
305. Echternkamp, S. E., and Hansel, W. (1973). Concurrent changes in bovine plasma hormone levels prior to and during

the first postpartum estrous cycle. *J. Anim. Sci.* 37, 1362–1370.

306. Dickey, J. F., Herricks, D. M., and Bierly, S. (1975). Hormonal relationships in postpartum Holstein cows. *J. Anim. Sci.* 41, 348 (Abstr.).
307. Kesler, D. I., Weston, P. G., Pimentel, C. A., Troxel, T. R., Vincent, D. L., and Hixon, J. E. (1981), Diminution of the in vitro response to luteinizing hormone by CL induced by gonadotropin releasing hormone treatment of postpartum suckled beef cows. *J. Anim. Sci.* 53, 749–754.
308. Goodale, W. S., Garverick, H. A., Kesler, D. J., Bierschwal, C. J., Elmore, R. G., and Youngquist, R. S. (1978). Transitory change of hormones in plasma of postpartum dairy cows. *J. Dairy Sci.* 61, 740–746.
309. Troxel, T. R., Kesler, D. J., Noble, R. C., and Carlin, S. E. (1980). Ovulation and reproductive hormones following steroid pretreatment, calf removal and GnRH in postpartum suckled beef cows. *J. Anim. Sci.* 51, 652–659.
310. Forrest, D. W., Fleeger, J. L., Long, C. R., Sorensen, A. M. Jr., and Harms, P. G. (1980). Effects of exogenous prolactin on peripheral luteinizing hormone levels in ovariectomized cows. *Biol. Reprod.* 22, 197–201.
311. Rahe, H. M., Owens, R. E., Fleeger, J. L., Newton, H. J., and Harms, P. G. (1980). Pattern of plasma luteinizing hormone in the cyclic cow: dependence upon the period of the cycle. *Endocrinology* 107, 498–503.
312. Bastidas, P., Forrest, D. W., Del-Vecchio, R. P., and Randel, R. D. (1990). Biological and immunological luteinizing hormone activity and blood metabolites in postpartum Brahman cows. *J. Anim. Sci.* 68, 2771–2778.
313. Fernandes, L. C., Thatcher, W. W., Wilcox , C. J., and Call, E. P. (1978). LH release in response to GnRH during the postpartum period of dairy cows. *J. Anim. Sci.* 46, 443–448.
314. Schallenberger, E., Schams, D., and Zottmeier, K. (1978). Response of lutropin (LH) and follitropin (FSH) to the administration of gonadoliberin (GnRH) in pregnant and post-partum cattle including experiments with prolactin suppression. *Theriogenology* 10, 35–53.
315. Garcia-Winder, M., Imakawa, K., Day, M. L., Zalesky, D. D., Kittok, R. J., and Kinder, J. E. (1984). Effect of suckling and ovariectomy on the control of luteinizing hormone secretion during the postpartum period in beef cows. *Biol. Reprod.* 31, 771–778.
316. Zalesky, D. D., Forrest, D. W., McArthur, N. H., Wilson, J. M., Morris, D. L., and Harms, P. G. (1990). Suckling inhibits release of luteinizing hormone-releasing hormone from the bovine median eminence following ovariectomy. *J. Anim. Sci.* 68, 444–448.
317. Canfield, R. W., Sniffen, C. J., and Butler, W. R. (1990). Effects of excess degradable protein on postpartum reproduction and energy balance in dairy cattle. *J. Dairy. Sci.* 73, 2342–2349.
318. Butler, S. T., Pelton, S. H., and Butler, W. R. (2004). Insulin increases 17β-estradiol production by the dominant follicle of the first postpartum follicle wave in dairy cows. *Reproduction* 127, 537–545.
319. Peters, A. R. (1984). Effect of exogenous oestradiol-l7 on gonadotrophin secretion in postpartum beef cows. *J. Reprod. Fertil.* 72, 473–478.
320. Short, R. E., Randel, R. D., Staigmiller, R. B., and Bellows, R. A. (1979). Factors affecting estrogen-induced LH release in the cow. *Biol. Reprod.* 21, 683–689.
321. Schallenberger, E., Oerterer, U., and Hutterer, G. (1982). Neuroendocrine regulation of postpartum function. In *Factors Influencing Fertility in the Postpartum Cow* (H. Karg, and E. Schallenberger, Eds.), pp. 123–147. Martinus Nijhoff, The Hague.
322. Hinshelwood, M. M., Dierschke, D. J., and Hauser, E. R. (1985). Effect of suckling on the hypothalamic–pituitary axis in postpartum beef cows, independent of ovarian secretions. *Biol. Reprod.* 32, 290–300.
323. Acosta, B., Tarnavsky, G. K., Platt, T. E., et al. (1983). Nursing enhances the negative effect of estrogen on LH release in the cow. *J. Anim. Sci.* 57, 1530–1536.
324. Schramm, R. D., Roberge, S., and Reeves, J. J. (1991). Enclomiphene does not alter the postpartum interval of suckled beef cows. *J. Anim. Sci.* 69, 4112–4116.
325. Schams, D., Hoefer, F., Hoffman, B., Ender, M. L., and Karg, H. (1973). Effects of synthetic LH-RH treatment on bovine ovarian function during oestrous cycle and post partum period. *Acta. Endocrinol. (Suppl.)* 177, 296.
326. Britt, J. H., Kittock, R. J., and Harrison, D. S. (1974). Ovulation, estrus and endocrine responses after GnRH in early post partum cows. *J. Anim. Sci.* 39, 915–920.
327. Kesler, D. J., Weston, P. G., Pimentel, C. A., Troxel, T. R., Vincent, D. L., and Hixon, J. E. (1981). Diminution of the in vitro response to luteinizing hormone by CL induced by gonadotropin releasing hormone treatment of postpartum suckled beef cows. *J. Anim. Sci.* 53, 749–754.
328. Zaied, A. A., Garverick, H. A., Bierschwal, C. J., Elmore, R. G., Youngquist, R. S., and Sharp, A. J. (1980). Effect of ovarian activity and endogenous reproductive hormone on GnRH-induced ovarian cycles in post partum dairy cows. *J. Anim. Sci.* 50, 508–5l3.
329. Kesler, D. I., Garverick, H. A., Youngquist, R. S., Elmore, R. G., and Bierschwal, C. J. (1978). Ovarian and endocrine responses and reproductive performance following GnRH treatment in early postpartum dairy cows. *Theriogenology* 9, 363–369.
330. Lee, C. N., Maurice, E., Ax, R. L., Pennington, J. A., Hoffman, W. F., and Brown, M. D. (1983). Efficacy of gonadotropin-releasing hormone administered at the time of artificial insemination of heifers and post partum and repeat breeder dairy cows. *Am. J. Vet. Res.* 44, 2160–2163.
331. Vostermans, J. P. M., and Walton, J. S. (1985). Effect of intermittent injections of gonadotrophin releasing hormone at various frequencies on the release of luteinizing hormone and ovulation in dairy cows. *Anim. Reprod. Sci.* 8, 335–347.
332. Webb, R., and Lamming, G. E. (1981). Patterns of plasma prolactin in postpartum suckled cows. *J. Endocrinol.* 90, 391-396.
333. Peters, R. R., and Tucker, H. A. (1978). Prolactin and growth hormone responses to photoperiod. *Endocrinology* 103, 229–234.
334. Smith, J. F., Payne, E., Tervit, H. R., McGowan, L. T., Fairclough, R., Kilgour, R., and Goold, P. G. (1981). The effect of suckling upon the endocrine changes associated with anoestrus in identical twin dairy cows. *J. Reprod. Fertil. Suppl.* 30, 241–249.
335. Wheeler, M. B., Anderson, G. B., Munro, C. J., and Stabenfeldt, G. H. (1982). Prolactin response in beef cows and heifers suckling one or two calves. *J. Reprod. Fertil.* 64, 243–249.
336. Williams, G. L., and Ray, D. E. (1980). Hormonal and reproductive profiles of early postpartum beef heifers after prolactin suppression or steroid-induced luteal function. *J. Anim. Sci.* 50, 905–918.
337. Cummins, I. J., Findlay, J. K., and Lawson, R. A. S. (1977). An attempt at the use of CB154 (ergocryptine) to reduce postpartum anoestrus in Hereford heifers. *Theriogenology* 8, 193–198.
338. Rund, L. A., Leshin, L. S., Thompson, F. N., Rampacek, G. B., and Kiser, T. E. (1989). Influence of the ovary and suckling

on luteinizing hormone response to naloxone in postpartum beef cows. *J. Anim. Sci.* 67, 1527–1533.

339. Myers, T. R., Myers, D. A., Gregg, D. W., and Moss, G. E. (1989). Endogenous opioid suppression of release of luteinizing hormone during suckling in. postpartum anestrous beef cows. *Domest. Anim. Endocrinol.* 6, 183–190.
340. Griffith, M. K., and Williams, G. L. (1996). Roles of maternal vision and olfaction in suckling-mediated inhibition of luteinizing hormone secretion, expression of maternal selectivity, and lactational performance of beef cows. *Biol. Reprod.* 54, 761–768.
341. Rhodes, F. M., Fitzpatrick, L. A., Entwhistle, K. W., and De'ath, G. (1995). Sequential changes in ovarian follicular dynamics in Bos indicus heifers before and after nutritional anoestrous. *J. Reprod. Fertil.* 104, 41–49.
342. Meikle, A., Kulcsar, M, Chilliard, Y., Febel, H., Delavaud, C., Cavestany, D., and Chilibroste, P. (2004). Effects of parity and body condition at parturition on endocrine and reproductive parameters of the cow. *Reproduction* 127, 727–737.
343. Kadokawa, H., Blanche, D., Yamada, Y., and Martin, G. B. (2004). Relationships between changes in plasma concentrations of leptin before and after parturition and the timing of first post-partum ovulation in high-producing Holstein dairy cows. *Reprod. Fertil. Dev.* 12, 405–411.
344. Armstrong, D. G., Gong, J. G., and Webb, R. (2003). Interactions between nutrition and ovarian activity in cattle: physiological, cellular and molecular mechanisms. *Reprod. Suppl.* 61, 403–414.
345. Harcourt, A. H., Stewart, K. J., and Watts, D. P. (1980). Reproduction in wild gorillas and some comparisons with the chimpanzee. *J. Reprod. Fertil. Suppl.* 28, 59–70.
346. Douglas. J. D., and Butler T. M. (1970). Chimpanzee breeding at the 657 1st Aeromedical Research Laboratory. *Lab. Anim. Care* 20, 477–482.
347. Clegg, M. T., and Weaver, M. (1972). Chorionic gonadotrophin secretion during pregnancy in the chimpanzee (Pan troglodytes). *Proc. Soc. Exp. Biol. Med.* 139, 1170–1174.
348. Tutin, C. E. G. (1980). Reproductive behaviour of wild chimpanzees in the Gombe National Park, Tanzania. *J. Reprod. Fertil. Suppl.* 28, 43–57.
349. Rowell, T. E. (1974). Female reproduction cycles and social behaviour in primates. *Adv. Stud. Behav.* 4, 69–105.
350. Kaufman, J. C. (1965). A three-year study of mating behaviour in a free-ranging band of rhesus monkeys. *Ecology* 46, 500–512.
351. Fujiwara, J., Honjo, S., Imaizumi, K., and Imanichi, T. (1967). The postpartum menstruation of cynomolgus monkeys kept under laboratory conditions. *Jpn. Med. Sci. Biol.* 22, 181–185.
352. Vandenberg, J. G., and Vessey, S. (1968). Seasonal breeding of free-ranging rhesus monkeys and related ecological factors. *J. Reprod. Fertil.* 15, 71–79.
353. Weiss, G., Butler, W. R., Hotchkiss, J., Dierschke, D. J., and Knobil, E. (1976). Periparturitional serum concentrations of prolactin, the gonadotrophins and the gonadal hormones in the rhesus monkey. *Proc. Soc. Exp. Biol. Med.* 151, 113–116.
354. Drickamer, L. C. (1974). A ten-year summary of reproductive data for free-ranging Macaca mulatta. *Folia Primatol.* 21, 61–80.
355. Varley, M. A., and Vessey, S. J. (1977). Effects of geographic transfer on the timing of seasonal breeding of rhesus monkeys. *Folia Primatol.* 28, 52–59.
356. Nozaki, M., Watanabe, O., Taya, K., Katakai, Y., Wada, I., and Sasamoto, S. (1990). Changes in circulating inhibin levels during pregnancy and early lactation in the Japanese monkey. *Biol. Reprod.* 43, 444–449.
357. Maeda, K. I,. Tsukamura, H., Ohkura, S., Kanaizuka, T., and Suzuki, I. (1991). Suppression of ovarian activity during the breeding season in suckling Japanese monkey (Macaca muscata). *J. Reprod. Fertil.* 92, 371–375.
358. Gordon, K., Hodgen, G. D., and Richardson, D. W. (1992). Postpartum lactational anovulation in a nonhuman primate (Macaca fascicularis): endogenous opiate mediation of suckling-induced hyperprolactinaemia. *J. Clin. Endocrinol. Metab.* 75, 59–67.
359. Mallinson, J. J. C. (1971). Observations on the reproduction and development of ververt monkey with special reference to intersubspecific hybridization (Cercopithecus pygerythrus, Cuvier F, 1821). *Mammalia* 35, 598–609.
360. Wolf, R. H., Harrison, R. M., and Martin, T. W. (1975). A review of reproductive patterns in new world monkeys. *Lab. Anim. Sci.* 25, 814–821.
361. Carpenter, C. R. (1934). A field study of the behaviour and social relations of howling monkeys. *Comp. Psychol. Monogr.* 10, 1–68.
362. Recabarren, M. P., Vergara, M., Martinez, M. C., Gordon, K., and Serron-Ferre, M. (2000). Impact of lactation upon fertility in the New World primate capuchin monkey (Cebus apella). *J. Med. Primatol.* 29, 350–360.
363. Travis, J. C., and Holmes, W. N. (1974). Some physiological and behavioural changes associated with oestrus and pregnancy in the squirrel monkey (Saimiri sciureus). *J. Zool.* 174, 41–66.
364. Coe, C. L., and Rosenblum, L. A. (1978). Annual reproductive strategy of the squirrel monkey (Saimiri sciureus). *Folia Primatol.* 29, 19–42.
365. Ziegler, T. E., Widowski, T. M., Larson, M. L., and Snowdon, C. T. (1990). Nursing does affect the duration of the post-partum to ovulation interval in cotton-top tamarins (Saguinus oedipus). *J. Reprod. Fertil.* 90, 563–570.
366. Lunn, S. F., and McNeilly, A. S. (1982). Failure of lactation to have a consistent effect on interbirth interval in the common marmoset (Calithrix jacchus jacchus). *Folia Primatol.* 37, 99–105.
367. McNeilly, A. S., Abbott, D. H., Lunn, S. F., Chambers, P. C., and Hearn, J. P. (1981). Plasma prolactin concentrations during the ovarian cycle and lactation and their relationship to return of fertility postpartum in the common marmoset (Calithrix jacchus). *J. Reprod. Fertil.* 62, 353–360.
368. Weiss, G., Dierschke, D. L., Karsch, F. H., Hotchkiss, J., Butler, W. R., and Knobil, E. (1973). The influence of lactation on luteal function in the rhesus monkey. *Endocrinology* 93, 954–959.
369. Williams, R. F., Hodgen, G. D. (1983). Mechanisms of lactational anovulation in primates: nursing increases ovarian steroidogenesis. In *Factors Regulating Ovarian Function* (G. S. Greenwald, and P. F. Terranova, Eds.), pp. 17–21. Raven Press, New York.
370. di Zerega, G. S., Williams, R. F., Mann, M. L., and Hodgen G. D. (1979). Anovulation after pregnancy termination. Ovarian versus hypothalamic-pituitary factors. *J. Clin. Endocrinol. Metab.* 49, 594–599.
371. Goodman, A. L., and Hodgen, G. D. (1978). Postpartum patterns of circulating FSH, LH, prolactin, estradiol and progesterone in non suckling cynomolgus monkeys. *Steroids* 31, 731–744.
372. Plant, T. M., Schallenberger, E., Hess, D. L., McCormack, J. T., Dufy Barbe, L., and Knobil, E. (1980). The influence of suckling on estrogen- induced gonadotropin discharges in the female rhesus monkey (Macaca mulata). *Biol. Reprod.* 123, 760–766.
373. Richardson, D. W., Goldsmith, L. T., Pohi, C. R., Schallenberger, E., and Knobil, E. (1985). The role of prolactin

in the regulation of the primate corpus luteum. *J. Clin. Endocrinol. Metab.* 60, 501–504.

374. Schallenberger, E., Richardson, D. W., and Knobil, E. (1981). Role of prolactin in the lactational amenorrhoea of the rhesus monkey (Macaca mulatta). *Biol. Reprod.* 25, 370–374.
375. Ordog, T., Chen, M.-D., O'Byrne, K. T., Goldsmith, J. R., Connaughton, M. A., Hotchkiss, J, and Knobil, E. (1998). On the mechanism of lactational anovulation in the rhesus monkey. *Am. J. Physiol.* 274, E665–E676.
376. Ash, R. H., Koloulas, I. B., Smith, G., Eddy, C. A., and Balmaceda, J. P. (1984). Pituitary responsiveness to LHRH during gestation and puerperium in the rhesus monkey. *Acta Eur. Fertil.* 15, 15–23.
377. Yamaji, T., Dierschke, D. J., Hotchkiss, J., Bhatacharya, A. N., Surve, A. H., and Knobil E. (1971). Estrogen induction of LH release in the rhesus monkey. *Endocrinology* 89, 1034–1041.
378. Williams, R. F., Johnson, D. K., and Hodgen, G. D. (1979). Resumption of estrogen-induced gonadotropin surges in postpartum monkeys. *J. Clin. Endocrinol. Metab.* 49, 422–428.
379. Nakai, Y., Plant, T. M., Hess, D. L., Keogh, E. J., and Knobil, E. (1978). On the sites of negative and positive feedback action of estradiol in the control of gonadotropin secretion in the rhesus monkey. *Endocrinology* 102, 1008–1014.
380. Fraser, H. M., McNeilly, A. S., Abbott, M., and Steiner, R. A. (1986). Effect of LHRH immunoneutralization on follicular development, the LH surge and luteal function in the stump-tailed macaque monkey (Macaca arctiodes). *J. Reprod. Fertil.* 76, 299–309.
381. Weiss, G., Butler, W. R., Dierschke, D. J., and Knobil, E. (1976). Influence of suckling on gonadotropin secretion in the postpartum rhesus monkey. *Proc. Soc. Exp. Biol. Med.* 153, 330–331.
382. O'Byrne, K. T., Thalabard, J.-C., Grosser, P. M., Wilson, R. C., Williams, C. L., Chen, M.-D., Ladendorf, D., Hotchkiss, J., and Knobil, E. (1991). Radiotelemetric monitoring of hypothalamic gonadotropin-releasing hormone pulse generator activity throughout the menstrual cycle of the rhesus monkey. *Endocrinology* 129, 1207–1214.
383. Frawley, L. S., Mulchahey, J. J., and Neill, J. D. (1983). Nursing induces a biphasic release of prolactin in rhesus monkeys. *Endocrinology* 112, 558–561.
384. Aso, T., and Williams, R. F. (1985). Lactational amenorrhoea in monkeys: effects of suckling on prolactin secretion. *Endocrinology* 117, 1727–1734.
385. Gordon, K., Aso, T., and Williams, R. F. (1995). Lactational anovulation in non-human primates: restriction of nursing inhibits prolactin secretion without precipitating the return of ovulatory menstrual cyclicity in cynomolgus monkeys. *Contraception* 51, 265–272.
386. Maneckjee, R., Srinath, B. R., and Moudgal, N. R. (1976). Prolactin suppresses release of luteinizing hormone during lactation in the monkey. *Nature* 262, 507–508.
387. Maneckjee, R., and Moudgal, N. R. (1975). The onset of oestrus and ovulation in lactating rats. *J. Reprod. Fertil.* 44, 313–315.
388. Dierschke, D. J., Yamaji, T., Karsch, F. J., Weick, R. F., Weiss, G., and Knobil, E. (1973). Blockade by progesterone of estrogen-induced LH and FSH release in the rhesus monkey. *Endocrinology* 92, 1495–1501.
389. Simpson, M. J. A., Simpson, A. E., Hooley, J., and Zunz, M. (1981). Infant-related influences on birth intervals in rhesus monkey. *Nature* 290, 49–51.
390. Gomendio, M. (1989). Differences in fertility and suckling patterns between primiparous and multiparous rhesus mothers (Macaca mulata). *J. Reprod. Fertil.* 87, 529–542.
391. Pope, N. S., Gordon, T. P., and Wilson, M. E. (1986). Age, social rank, and lactational status influence ovulatory patterns in seasonally breeding rhesus monkeys. *Biol. Reprod.* 35, 353–359.
392. Wilson, M. E., Walker, M. L., and Gordon, T. P. (1983). Consequences of first pregnancy in rhesus monkeys. *Am. J. Phys. Anthropol.* 61, 103–110.
393. Wilson, M. E., Walker, M. L., Pope, N. S., and Gordon, T. P. (1988). Prolonged lactational infertility in adolescent rhesus monkeys. *Biol. Reprod.* 38, 163–174.
394. McNeilly, A. S. (1979). Effects of lactation on fertility. *Br. Med. Bull.* 35, 151–154.
395. Short, R. V. (1976). Lactation—the central control of reproduction. *Ciba Foundation Symposium* 45, 73–85.
396. Howie, P. W., and McNeilly, A. S. (1982). Effect of breast-feeding patterns on human birth intervals. *J. Reprod. Fertil.* 65, 545–557.
397. Delvoye, P., Badawi, M., Demaegd, M., and Robyn, C. (1978). Serum prolactin, gonadotrophins and estradiol in menstruating and amenorrhoeic women during two years of lactation. *Am. J. Obstet. Gynecol.* 130, 635–640.
398. Konner, M., and Worthman, C. (1990). Nursing frequency, gonadal function and birth spacing among Kung hunter gatherers. *Science* 207, 788–791.
399. Duchen, M. R., and McNeilly, A. S. (1980). Hyperprolactinaemia and long-term lactational amenorrhoea. *Clin. Endocrinol.* 12, 621–627.
400. Andersen, A. M., and Schioler, V. (1982). Influence of breast feeding pattern on pituitary–ovarian axis of women in an industrialized community. *Am. J. Obstet. Gynecol.* 143, 673–677.
401. Lunn, P. G., Prentice, A. M., Austin, S., and Whitehead, R. G. (1980). Influence of maternal diet on plasma-prolactin levels during lactation. *Lancet* 114, 623–625.
402. Gross, B. A., and Eastman, C. J. (1985). Prolactin and the return of ovulation in breast feeding women. *J. Biosoc. Sci. (Suppl.).* 9, 25–42.
403. Brown, J. B., Harrison, P., and Smith, M. A. (1985). A study of returning fertility after childbirth and during lactation by measurement of urinary oestrogen and pregnanediol excretion and cervical mucus production. *J. Biosoc. Sci. (Suppl.).* 9, 5–23.
404. Rivera, R., Ortiz, E., Barrera, M., Kennedy, K., and Bhiwandiwala, P. (1985). Preliminary observations on the return of ovarian function among breast feeding and post-partum non breast feeding women in a rural area of Mexico. *J. Biosoc. Sci. (Suppl.).* 9, 127–136.
405. Wood, J. W., Johnson, P. L., and Campbell, K. L. (1985). Demographic and endocrinological aspects of low natural fertility in highland New Guinea. *J. Biosoc. Sci.* 17, 57–79,
406. Diaz, S., Seron-Ferre, M., Cardenas, H., Schiapacasse, V., Brandeis, A., and Croxatto, H. B. (1989). Circadian variation of basal plasma prolactin, prolactin response to suckling, and length of amenorrhoea in nursing women. *J. Clin. Endocrinol. Metab.* 68, 945–955.
407. Fink, A. E., Fink, G., Wilson, H., Bennie, J., Carroll, S., and Dick, H. (1992). Lactation, nutrition and fertility and the secretion of prolactin and gonadotrophins in Mopan Mayan women. *J. Biosoc. Sci.* 24, 35–52.
408. Fraser, H. M., Dewart, P. J., Smith, S. K., Cowen, G. M., Sandow, J., and McNeilly, A. S. (1989). Luteinizing hormone releasing hormone agonist in post-partum contraception: endocrine effects on breastfeeding mothers. *J. Clin. Endocrinol. Metab.* 69, 995–1002.
409. Gray, R. H., Campbell, O. M., and Apelo, R. (1990). Risk of ovulation during lactation. *Lancet* 335, 25–29.
410. Israngkura, B., Kennedy, K. I., and Leelapatana, B. (1989). Breastfeeding and the return to ovulation in Bangkok. *Int. J. Gynecol. Obstet.* 30, 33.

411. Jones, R. E. (1989). Breast-feeding and post-partum amenorrhoea in Indonesia. *J. Biosoc. Sci.* 21, 83–95.
412. Khan, T., Kennedy, K. I., and Kazi, A. (1989). A study of breastfeeding and the return of menses and pregnancy in Karachi, Pakistan. *Contraception* 40, 365–378.
413. Rivera, R., Barrera, M., and Kennedy, K. I. (1988). Breastfeeding and the return to ovulation in Durango, Mexico. *Fertil. Steril.* 49, 780–786.
414. Rosner, A. E., and Schulman, S. K. (1990). Birth interval among breast-feeding women not using contraceptives. *Pediatrics* 86, 747.
415. Savina, G., and Kennedy K. I. (1989). The effect of a breast-feeding education program on Iactational amenorrhoea in the Philippines. *Stud. Fam. Plan.* 20, 203–207.
416. Shaaban, M. M., Sayed, G. H. and Ghaneimah, S. A. (1987). The recovery of ovarian function during breastfeeding. *J. Steroid Biochem.* 22, 19–25.
417. Shaaban, M. M., Kennedy, K. I., and Sayed, G. H. (1990). The recovery of fertility during breastfeeding in Assuit, Egypt. *J. Biosoc. Sci.* 22, 19–31.
418. Short, R. V., Lewis, P. R., and Renfree, M. B. (1991). Contraceptive effects of extended lactational amenorrhoea: beyond the Bellagio concensus. *Lancet* 337, 715–717.
419. McNeilly, A. S. (2001). Lactational control of Reproduction. *Reprod. Fertil. Dev.* 13, 583–590.
420. WHO. (1998). The World Health Organization multinational study of breastfeeding and lactational amenorrhoea. I. Description of infant feeding patterns and of the return of menses. *Fertil. Steril.* 70: 448–460.
421. WHO. (1998). The World Health Organization multinational study of breastfeeding and lactational amenorrhoea. II. Factors associated with the length of amenorrhoea. *Fertil. Steril.* 70, 461–471.
422. Stallings, J. Y., Worthman, C. M., and Pather-Brick, C. (1998). Biological and behavioural factors influence group differences in prolactin levels among breastfeeding Nepali women. *Am. J. Human Biol.* 10, 191–210.
423. Kennedy, K. I., Labbok, M. H., and Van Look, P. F. A. (1996). Consensus statement: lactational amenorrhoea method for family planning. *Int. J. Gynaecol.* 54, 55–57.
424. Visness. C. M., Kennedy, K. I., and Ramos, R. R. (1997). The duration and character of postpartum bleeding among breastfeeding women. *Obstet. Gynecol.* 89, 159–163.
425. Labbok, M. H., Pérez, A., Valdés, V., Sevilla, F., Wade, K. and Laukaran, V. H. (1994). The lactational amenorrhea method (LAM): a postpartum introductory family planning method with policy and program implications. *Adv. Contracept.* 10, 93–109.
426. Wilson, A. C., Fiorsyth, J. S., Greene, S. A., Irvine, L., Hau, C., and Howie, P. W. (1998). Relation of infant diet to childhood health: seven year follow up of cohort of children in Dundee infant feeding study. *Brit. Med. J.* 316, 21–25.
427. Kennedy, K. L., and Visness, C. M. (1992). Contraceptive efficacy of lactational amenorrhoea. *Lancet* 339, 227–229.
428. Kennedy, K. I., Rivera, R., and McNeilly, A. S. (1989). Consensus statement on the use of breastfeeding as a family planning method. *Contraception* 39, 477–496.
429. Thapa, S., Short, R. V. and Potts, M. (1988). Breast feeding, birth spacing and their effects on child survival. *Nature* 335, 679–682.
430. Gross, B. A., and Eastman, C. J. (1983). Effect of breast feeding status on prolactin secretion and resumption of menses. *Med. J. Aust.* 1, 313–320.
431. Wood, J. W., Lai, D., Johnson, F. L., Campbell, K. L., and Maslar, I. A. (1985). Lactation and birth spacing in highland New Guinea. *J. Biosoc. Sci. (Suppl.).* 9, 159–173.
432. Diaz, S., Cardenas, H., Brandeis, A., Miranda, P., Schiappacasse, V., Salvatierra, A. M., Herreros, C., Seron-Ferre, M. and Croxatto, H. B. (1991). Early differences in the endocrine profile of long and short lactational amenorrhoea. *J. Clin. Endocrinol. Metab.* 72, 195–201.
433. Tay, C. C. K., Glasier, A. F. and McNeilly, A. S. (1996). Twenty-four hour patterns of prolactin secretion during lactation and the relation to suckling and the resumption of fertility in breastfeeding women. *Hum. Reprod.* 11, 950–955.
434. Noel, G. L., Dimond, R. C., Wartofsky, L., Earl, J. M., and Frantz, A. G. (1974). Studies of prolactin and FSH secretion by continuous infusion of small amounts of thyrotropin releasing hormone (TRH). *J. Clin. Endocrinol. Metab.* 39, 5–17.
435. Tyson, J. E., Friesen, H. G., and Anderson, M. S. (1972). Human lactational and ovarian response to endogenous prolactin release. *Science* 177, 897–900.
436. Dawood, M. Y., Khan-Dawood, F. S., Wahi, R. S., and Fuchs, F. (1981). Oxytocin release and plasma anterior pituitary and gonadal hormones in women during lactation. *J. Clin. Endocrinol. Metab.* 52, 678–683.
437. Glasier, A., McNeilly, A. S., and Howie, P. W. (1984). The prolactin response to suckling. *Clin. Endocrinol.* 21, 109–116.
438. Liu, J. H., Lee, D. W., and Markoff, E. (1988). Differential release of prolactin variants in postpartum and early follicular phase women. *J. Clin. Endocrinol. Metab.* 71, 605–610.
439. Nunley, W. C., Urban, R. J., Kitchin, J. D., Bateman, B. G., Evans, W. S., and Veldhuis J. D. (191). Dynamics of pulsatile prolactin release during the postpartum lactational period. *J. Clin. Endocrinol. Metab.* 71, 287–293.
440. Tay, C. C. K., Glasier, A. F., and McNeilly, A. S. (1992). The twenty-four hour pattern of pulsatile luteinizing hormone, follicle stimulating hormone and prolactin release during the first eight weeks of lactational amennorhoea in breastfeeding women. *Hum. Reprod.* 7, 951–958.
441. Unvas-Moberg, K., Widstrom, A. M., Werner, S., Mattiesen, A. S., and Winberg, J. (1990). Oxytocin and prolactin levels in breast-feeding women. Correlation with milk yield and duration of breast-feeding. *Acta Obstet. Gynecol. Scand.* 69, 301–306.
442. Delvoye, P., Demaegd, M., Delogne-Desnoeck, J., and Brown, C. (1977). The influence of the frequency of nursing and of previous lactation experience on serum prolactin lactating mothers. *J. Biosoc. Sci.* 9, 447–451.
443. Madden, J. D., Boyar, R. M., MacDonald, P. C., and Porter, J. C. (1978). Analysis of secretory patterns of prolactin and gonadotropins during 24 hours in a lactating woman before and after resumption of menses. *Am. J. Obstet. Gynecol.* 132, 435–441.
444. Tyson, J. E., Carter, J. N., Andreassen, B., Huth, J., and Smith, B. (1978). Nursing-mediated prolactin and luteinizing hormone secretion during puerperal lactation. *Fertil. Steril.* 30, 154–162.
445. Hennart, P., Delogne-Desnoeck, J., Vis, H., and Robyn, C. (1981). Serum levels of prolactin and milk production in women during a lactation period of thirty months. *Clin. Endocrinol.* 14, 349–353.
446. McNeilly, A. S., Robinson, C. A. F., Houston, M. J., and Howie, P. W. (1983). Release of oxytocin and prolactin in response to suckling. *Br. Med. J.* 286, 257–259.
447. Delvoye, F., Delogne-Desnoeck, J., and Robyn, C. (1980). Hyperprolactinaemia during prolonged lactation: evidence for anovulatory cycles and inadequate corpus luteum. *Clin. Endocrinol.* 13, 243–247.
448. Bonnar, J., Franklin, M., Nott, P. N., and McNeilly, A. S. (1975). Effect of breastfeeding on pituitary-ovarian function after child-birth. *Br. Med. J.* 243, 82–84.

449. Ellis, L. A., and Picciano, M. F. (1995). Bioactive and immunoreactive prolactin variants in human milk. *Endocrinology* 136, 2711–2720.
450. Campino, C., Torres, C., Ampuero, S., Diaz, S., Gonzalez, G. B., and Seron-Ferre, M. (1999). Bioactivity of prolactin isoforms: lactation and recovery of menses in nursing women. *Human Reprod.* 14, 898–905.
451. Hennart, P., Hofvander, Y., Vis, H., and Robyn, C. (1985). Comparative study of nursing mothers in Africa (Zaire) and in Europe (Sweden): breast feeding behaviour, nutritional status, lactational hyperprolactinaemia and status of the menstrual cycle. *Clin. Endocrinol.* 22, 179–187.
452. Reyes F. I., Winter, J. S. D., and Faiman, C. (1972). Pituitary-ovarian interrelation ships during the puerperium. *Am. J. Obstet. Gynecol.* 14, 589–594.
453. Andreassen, B., and Tyson, J. E. (1976). Role of the hypothalamic–pituitary–ovarian axis in puerperal infertility. *J. Clin. Endocrinol. Metab.* 42, 1114–1122.
454. Peerhentupa, A., Chritchley, H. O. D., Illingworth, P. J. and McNeilly, A. S. (2000). Enhanced sensitivity to steroid negative feedback during breast-feeding: Low dose estradiol (transdermal estradiol supplementation) suppresses gonadotropins and ovarian activity assessed by inhibin B. *J. Clin. Endocrinol. Metab.* 85, 4280–4286.
455. Peerhentupa, A., Chritchley, H. O. D., Illingworth, P. J. and McNeilly, A. S. (2003). Effect of progestin-only pill on pituitary–ovarian axis activity during lactation. *Contraception* 67, 467–471.
456. Flynn, A., Docker, M., Brown, J. B., and Kennedy, K. I. (1991). Ultrasonographic patterns of ovarian activity during breastfeeding. *Amer. J. Obstet. Gynecol.* 165, 2027–31.
457. Baird, D. T., McNeilly, A. S., Sawers, R. S., and Sharpe, R. M. (1979). Failure of oestrogen-induced discharge of luteinizing hormone in lactating women. *J. Clin. Endocrinol. Metab.* 49, 500–509.
458. McNeilly, A. S., Howie, P. W., and Houston, M. J. (1980). Relationship of feeding patterns, prolactin and resumption of ovulation postpartum. In *Research Frontiers in Fertility Regulation* (G. I. Zatuchni, M. H. Labbok, and J. J. Sciarra, Eds.), pp. 102–116. Harper & Row, New York.
459. Howie, P. W., McNeilly, A. S., Houston, M. J., Cook, A., and Boyle, H. (1981). Effect of supplementary food on suckling patterns and ovarian activity during lactation. *Br. Med. J.* 283, 757–759.
460. Howie, P. W., McNeilly, A. S., Houston, M. J., Cook, A., and Boyle, H. (1982). Fertility after childbirth: postpartum ovulation and menstruation in bottle and breast feeding mothers. *Clin. Endocrinol.* 17, 323–332.
461. McNeilly, A. S., Howie, P. W., Houston, M. J., Cook, A., and Boyle, H. (1982). Fertility after childbirth: adequacy of postpartum luteal phases. *Clin. Endocrinol.* 17, 609–615.
462. Glasier, A., McNeilly, A. S., and Howie, P. W. (1983). Fertility after childbirth: changes in serum gonadotrophin levels in bottle and breast feeding women. *Clin. Endocrinol.* 19, 493–501.
463. Glasier, A., McNeilly, A. S., and Baird, D. T. (1986). Induction of ovarian activity by pulsatile infusion of LHRH in women with lactational amenorrhoea. *Clin. Endocrinol.* 24, 243–252.
464. McNeilly, A. S., Glasier, A., Howie, P. W. (1985). Endocrine control of lactational infertility. In *Maternal Nutrition and Lactational Infertility* (J. Dobbing, Ed.), pp. 1–16. Veney/Raven Press, New York.
465. Poindexter, A. N., Litter, M. B., and Besch, P. K. (1983). The recovery of normal plasma progesterone levels in the postpartum female. *Fertil. Steril.* 39, 494–498.
466. Gray, R. H., Campbell, O. M., and Zacur, H. A. (1987). Postpartum return of ovarian activity in nonbreastfeeding women monitored by urinary assays. *J. Clin. Endocrinol. Metab.* 64, 645–648.
467. McNeilly, A. S., Glasier, A. F., Howie, P. W., Houston, M. J., Cook, A., and Boyle, H. (1983). Fertility after childbirth: pregnancy associated with breastfeeding. *Clin. Endocrinol.* 18, 167–173.
468. Zarate, A., Canales, E. S., Soria, J., Ruiz, F., and MacGregor, C. (1972). Ovarian refractoriness during lactation in women: effect of gonadotropin stimulation. *Am. J. Obstet. Gynecol.* 112, 1130–1132.
469. Del Pozo, E., Goldstein, M., Friesen, H. G., Brun, R. E., and Eppenberger, U. (1975). Lack of action of prolactin suppression on the regulation of the human menstrual cycle. *Am. J. Obstet. Gynecol.* 123, 719–723.
470. Nakano, R., Mori, A., Kayashima, F., Washio, M., and Tojo, S. (1975). Ovarian response to exogenously administered human gonadotropins during the postpartum period. *Am. J. Obstet. Gynecol.* 121, 187–192.
471. Zinaman, M. J., Cartledge, T., Tomai, T., Tippett, P., and Merriam, G. R. (1995). Pulsatile GnRH stimulates normal cycle ovarian function in amenorrhoeic lactating postpartum women. *J. Clin. Endocrinol. Metab.* 80, 2088–2093.
472. Keye, W. R. Jr., and Jaffe, R. B. (1976). Changing patterns of FSH and LH response to gonadotropin-releasing hormone in the puerperium. *J. Clin. Endocrinol. Metab.* 42, 1133–1138.
473. Liu, J. H., and Park, K. H. (1988). Gonadotropin and prolactin secretion increases during sleep during the puerperium in nonlactating women. *J. Clin. Endocrinol. Metab.* 66, 839–845.
474. Jeppson, S., Rannevik, G., and Kullander, S. (1974). Studies on the decreased gonadotropin response after the administration of LH/FSH releasing hormone during pregnancy and puerperium. *Am. J. Obstet. Gynecol.* 120, 1029–1034.
475. Le Maire, W. J., Shapiro, A. G., Riggall, F., and Yang, N. S. T. (1974). Temporary pituitary insensitivity to stimulation by synthetic LRF during the postpartum period. *J. Clin. Endocrinol. Metab.* 38, 915–918.
476. Illingworth, P. J., Seaton, J. E. V., McKinlay, C., Reid-Thomas, V. and McNeilly, A. S. (1995). Low dose transdermal oestradiol suppresses gonadotrophin secretion in breastfeeding women. *Human Reprod.* 10, 1671–77.
477. Tay, C. C. K., Glasier, A. F., and McNeilly, A. S. (1993). Effect of antagonists of dopamine and opiates on the basal and GnRH-induced secretion of luteinizing hormone, follicle stimulating hormone and prolactin during lactational amenorrhoea in breastfeeding women. *Hum. Reprod.* 8, 532–539.
478. Glasier, A. F., McNeilly, A. S., and Howie, P. W. (1984). Pulsatile secretion of LH in relation to the resumption of ovarian activity post partum. *Clin. Endocrinol.* 20, 415–426.
479. Kremer, J. A., Borm, G., and Schellekens, L. A. (1991). Pulsatile secretion of luteinizing hormone and prolactin in lactating and nonlactating women and the response to naltrexone. *J. Clin. Endocrinol. Metab.* 72, 294–299.
480. Nunley, W. C., Urban, R. J., and Evans, W. S. (1991). Preservation of pulsatile luteinizing hormone release during postpartum lactational amennorhoea. *J. Clin. Endocrinol. Metab.* 73, 629–634.
481. Wu, F. C. W., Butler, G. E., Kelnar, C. H. J., Stirling, H. F., and Huhtaniemi, I. (1991). Patterns of pulsatile luteinising and follicle stimulating hormone secretion in prepubertal (midchildhood) boys and girls and patients with idiopathic hypogonadotrophic hypogonadism (Kallman's Syndrome): a study using an ultrasensitive time-resolved immunofluorescent assay *J. Clin. Endocrinol. Metab.* 72, 1229–1237.

482. Glass, M. R., Rudd, B. T., Lynch, S. S., and Butt, W. R. (1981). Oestrogen-gonadotrophin feedback mechanisms in the puerperium. *Clin. Endocrinol.* 14, 257–267.
483. Vermer, H. M., and Rolland, R. (1982). The influence of exogenous oestradiol benzoate on the pituitary responsiveness to LHRH during the puerperium in women. *Clin. Endocrinol.* 16, 251–258.
484. McNeilly, A. S. (1980). Prolactin and the control of gonadotrophin secretion in the female. *J. Reprod. Fertil.* 58, 537–549.
485. Ishizuka, B., Quigley, M. E., and Yen, S. S. C. (1984). Postpartum hypogonadotrophinism: evidence for increased opioid inhibition. *Clin. Endocrinol.* 20, 573–577.
486. Campino, C., Torres, C., Rioseco, A., Poblete, A., Pugin, E., Valdes, V., Catalan, S., Belmar, C., and Srron-Fere, M. (2001). Plasma prolactin/oestradiol ratio at 38 weeks gestation predicts the duration of lactational amenorrhoea *Hum. Reprod.* 16, 2540–2545.
487. Djahanbakhch, O., McNeilly, A. S., Warner, P. M., Swanston, I. A., and Baird, D. T. (1984). Changes in plasma levels of prolactin, in relation to those of FSH, oestradiol, androstenedione and progesterone around the preovulatory surge of LH in women. *Clin. Endocrinol.* 20, 463–472.
488. Edwards, R. G. (1981). Test-tube babies, 1981. *Nature* 293, 253–256.
489. McNeilly, A. S., Glasier, A., Jonassen, J., and Howie, P. W. (1982). Evidence for direct inhibition of ovarian function by prolactin. *J. Reprod. Fertil.* 65, 559–569.
490. McNatty, K. P., Sawers, R. S., and McNeilly, A. S. (1974). A possible role for prolactinin control of steroid secretion by the human Graafian follicle. *Nature* 250, 653–655.
491. McNeilly, A. S., Glasier, A., Swanston, I. A., and Djahanbakhch, O. (1983). Prolactin and the human ovary. In *Human Prolactin: Proceedings of the III International Congress* (G. Tolis, Ed.), pp. 173–178. Raven Press, New York.
492. Edwards, W. L., Haynes, S. P., and Wilson, J. D. (1982). Lack of suppressive effect of prolactin on progesterone production by human granulosa cells in vitro. *Proc. Aust. Soc. Endocrinol.* 25, 72–80.
493. McNeilly, A. S. (1985). Prolactin and the corpus luteum. In *The Luteal Phase* (S. L. Jeffcoate, Ed.), pp. 71–87. John Wiley & Sons, Chichester.
494. Hennart, F., Hofvander, Y., and Vis, H. (1985). Comparative study of nursing mothers in Africa (Zaire) and in Europe (Sweden): breastfeeding behaviour, nutritional status, lactational hyperprolactinaemia and status of the menstrual cycle. *Clin. Endocrinol.* 22, 179–185.
495. Lewis, P. R., Brown, J. B., Renfree, M. B. and Short, R. V. (1991). The resumption of ovulation and menstruation in a well-nourished population of women breastfeeding for an extended period of time. *Fertil. Steril.* 55, 529–36.
496. Lunn, P. G. (1985). Maternal nutrition and lactational infertility: the baby in the driving seat. In *Maternal Nutrition and Lactational Infertility* (J. Dobbing, Ed.), pp. 41–55. Veney/Raven Press, New York.
497. van-Raaij, J. M., Schonk, C. M., and Vermaat-Miedema, S. H. (1991). Energy cost of lactation, and energy balances of well-nourished Dutch lactating women: reappraisal of the extra energy requirements of lactation. *Am. J. Clin. Nutr.* 53, 612–619.
498. lllingworth, P. J., Jung, R. T., Howie, P. W. (1986). Diminution in energy expenditure during lactation. *Br. Med. J.* 292, 437–439.
499. Bowen-Jones, A., Thompson, C., and Drewett, R. F. (1982). Milk flow and suckling rates during breast feeding. *Dev. Med. Child Neurol.* 24, 625–630.
500. Delgado, H. L., Martorell, R., and Klein, R. E. (1982). Nutrition, lactation and birth interval components in rural Guatemala. *Am. J. Clin. Nutr.* 35, 1468–1473.
501. Ramachandran, P. (1992). Nutrition and its influence on the mother-child dyad. In *Nutrition and Population Link—Breastfeeding, Family Planning and Child Health* (United Nations Administrative Committee on Coordination, Subcommittee on Nutrition), Nutrition Policy Discussion Paper No. 11, pp. 47–54.
502. Wasalathanthri, S., and Tennekoon, K. H. (2001). Lactational amenorrhea/anovulation and some of their determinants: a comparison of well-nourished and undernourished women. *Fertil. Steril.* 76, 317–325.
503. Lage, M., Garcia-Mayor, R. V., Tome, M. A., Cordido, F., Valle-Inclan, F., Considine, R. V., Caro, J. F., Dieguez, C. and Casanueva, F. F. (1999). Serum leptin lives in women throughout pregnancy and the postpartum period and in women suffering spontaneous abortion. *Clin. Endocrinol.* 50, 211–216.
504. Sir-Petermann, T., Recabarren, S. E., Lobos, A., Maliqueo, M., and Wildt, L. (2001). Secretory pattern of leptin and LH during lactational amenorrhoea in breastfeeding normal and polycystic ovarian syndrome women. *Hum. Reprod.* 16, 244–249.
505. Valeggia, C., and Ellison, P. T. (2004). Lactational amenorrhoea in well-nourished women of Formosa, Argentina. *J. Biosoc. Sci.* 36, 573–595.
506. Meridith John, A. (1993). Statistical evidence of links between maternal nutrition and post-partum infertility. In *Biomedical and Demographic Determinants of Reproduction* (R. Gray, H. Lediron, and A. Spira, Eds.), pp. 372–390. Clarendon Press, Oxford.
507. Ford, K. and Huffman, S. L. (1993). Maternal nutrition, infant feeding, and post-partum amenorrhoea: recent evidence from Bangladesh. In *Biomedical and Demographic Determinants of Reproduction* (R. Gray, H. Lediron, and A. Spira, Eds.), pp. 383–390. Clarendon Press, Oxford.
508. Padmanabhan, V., and McNeilly, A. S. (2001). Is there an FSH-releasing factor? *Reproduction* 121, 21–30.
509. McNeilly, A. S., Crawford, J. L., Taragnat, C., Nicol, L., and McNeilly, J. R. (2003). The differential secretion of FSH and LH: regulation through genes, feedback and packaging *Reproduction Suppl.* 61, 463–476.
510. Nicol, L., McNeilly, J. R., Stridserg, M., and McNeilly, A. S. (2004). Differential release of gonadotropins: the role of secretogranin II in the regulated and constitutive release of LH and FSH. *J. Mol. Endocrinol.* 32, 467–480
511. Selmanoff, M., and Wise, P. M. (1981). Decreased dopamine turnover in the median eminence in response to suckling in the lactating rat. *Brain Res.* 212, 101–116.
512. Demarest, K. T., McKay, D. W., Riegle, G. D., and Moore, K. E. (1983). Biochemical indices of tuberoinfundibular dopaminergic neuronal activity during lactation: a lack of response to prolactin. *Neuroendocrinology* 36, 130–137.
513. Arbogast, L. A., and Voogt, J. L. (1998). Endogenous opioid peptides contribute to suckling-induced prolactin release by suppressing tyrosine hydroxylase activity and messenger ribonucleic acid levels in tuberoinfundibular dopaminergic neurons. *Endocrinology* 139, 2857–2862.
514. Gratton, D. R. (2001). The actions of prolactin in the brain during pregnancy and lactation. *Prog. Brain Res.* 133, 153–172.
515. Augustine, R. A., Kokay, I. C., Andrews, Z. B., Ladyman, S. R., and Gratton, D. R. (2003). Quantitation of prolactin receptor mRNA in the maternal brain during pregnancy and lactation. *J. Mol. Endocrinol.* 31, 221–232.

516. Callaghan, P., Klostermann, S., Prunty, D., and Tompkins, J. (2000). Immunoneutralization of endogenous opioid peptides prevents the suckling-induced prolactin increase and the inhibition of tuberoinfundibular dopaminergic neurons. *Neuroendocrinology* 71, 268–276.
517. Chen, P., and Smith, M. S. (2004). Regulation of hypothalamic neuropeptide Y messenger ribonucleic acid expression during lactation: role of prolactin. *Endocrinology* 145, 823–829.
518. Chen, P., and Smith, M. S. (2003). Suckling-induced activaton of neuronal input to the dorsomedial nucleus of the hypothalamus: candidates for mediating the activaton of DMH neuropeptide Y neurons during lactation. *Brain Res.* 984, 11–20.
519. Li, C., Chen, P., and Smith, M. S. (1999). Neuropeptide Y and tuberoinfundibular dopamine activities are altered during lactation: role of prolactin. *Endocrinology* 140, 118–123.
520. Pi, X., and Voogt, J. L. (2000). Effect of suckling on prolactin receptor immunoreactivity in the hypothalamus of the rat. *Neuroendocrinology* 71, 308–317.
521. Sargillo, C. A., Grattan, D. R., McCarthy, M. M., and Selmanoff, M. (1996). Hormonal and neurotransmitter regulation of GnRH gene expression and related reproductive behaviours. *Behav. Genet.* 26, 241–277.
522. Gu, G. B., and Simerly, R. B. (1997). Projections of the sexually dimorphic anteroventricular periventricular nucleus in the female rat. *J. Comp. Neurol.* 384, 142–164.
523. Kornblatt, J. J., and Grattan, D. R. (2001). Lactation alters gamma-aminobutyric acid neuronal activity in the hypothalamus and cerebral cortex in the rat. *Neuroendocrinology* 73, 175–184.
524. Keisler, D. H., Daniel, J. A., and Morrison, C. D. (1999). The role of leptin in nutritional status and reproductive function. *J. Reprod. Fertil. Suppl.* 54, 425–435.
525. Garcia, M. D., Casanueva, F. F., Dieguez, C., and Senaris, R. M. (2000). Gestational profile of leptin messenger ribonucleic acid (mRNA) content in the placenta and adipose tissue in the rat, and regulation of the mRNA levels of the leptin receptor subtypes in the hypothalamus during pregnancy and lactation. *Biol. Reprod.* 62, 698–703.
526. Date, Y., Kojima, M., Hosoda, H., Sawaguchi, A., Mondal, M. S., Suganuma, T., and Matsukura, S. (2000). Ghrelin, a novel growth hormone-releasing acetylated peptide, is synthesized in a distinct endocrine cell type in the gastrointestinal tracts of rats and humans. *Endocrinology* 141, 4255–4261.
527. Gnanapavan, S., Kola, B., Bustin, S. A., Morris, D. G., McGee, P., Fairclough, P., Bhattacharya, S., Carpenter, R., Grossman, A. B., and Korbonits, M. (2002). The tissue distribution of the mRNA of ghrelin and subtypes of its receptor, GHS-R, in humans. *J. Clin. Endocrinol. Metab.* 87, 2988–2991.
528. Futura, M., Funabashi, T., and Kimura, F. (2003). Intracerebroventricular administration of ghrelin suppresses pulsatile luteinizing hormone secretion in ovariectomized rats. *Biochem. Biophys. Res. Com.* 288, 780–785.
529. Seal, L. J., Cohen, M. A., Byrnes, A. E., Frost, G. E., Murphy, K. G., Dhillo, W. S., Ghatei, M. A., and Bloom, S. R. (2001). Ghrelin enhances appetite and increases food intake in humans. *J. Clin. Endocrinol. Metab.* 86, 5992–5995.
530. De Souza, M. J., Leidy, H. J., O'Donnell, E., Lasley, B., and Williams, N. I. (2004). Fasting ghrelin levels in physically active women: relationship with menstrual disturbances and metabolic hormones. *J. Clin. Endocrinol. Metab.* 89, 3535–3542.
531. Campbell, R. E., Grove, K. L., and Smith, M. S. (2003). Gonadotropin-releasing hormone neurons coexpress orexin 1 receptor immunoreactivity and receive direct contacts by orexin fibers. *Endocrinology* 144, 1542–1548.

Knobil and Neill's Physiology of Reproduction, Third Edition
edited by Jimmy D. Neill,
Elsevier © 2006

CHAPTER 47

Physiological Mechanisms Integrating Metabolism and Reproduction

Stephanie M. Krasnow and Robert A. Steiner

Energetics and Reproduction, 2553
Introduction, 2553
Environmental Influences on Reproduction, 2554
Role of Food Availability, 2554
Seasonal Influences, 2555
Factors that Influence Sensitivity to Metabolic Challenges, 2555
Early Observations, 2557
Metabolic Stressors that Impair Reproduction, 2558
Caloric Restriction, 2559
Exercise, 2562
Lactation, 2564
Cold Ambient Temperature, 2565
Signals Linking Nutrition, Growth, and Reproduction, 2566
Leptin, 2568
Insulin, 2575
Metabolic Fuels, 2579
Thyroid Hormones, 2582
Other Metabolic and Possible Neural Signals, 2583
Central Signaling Molecules that Integrate Metabolism and Reproduction, 2584
Routes of Entry into the CNS, 2584
Sites of Metabolic Cue Detection, 2584
Central Signaling Molecules, 2585
Metabolic Disorders and Reproductive Function, 2596
Obesity, 2596
Diabetes Mellitus, 2596
Lipodystrophies, 2597
Polycystic Ovary Syndrome, 2598
Acknowledgments, 2599
References, 2599

"... growth, nutrition, and reproduction are intimately allied processes..."

CHARLES DARWIN (1897)

ENERGETICS AND REPRODUCTION

Introduction

Reproduction is energetically expensive. An individual's reproductive capacity is gated by one's nutritional status, which reflects the homeostatic balance between energy intake and expenditure. Whereas maintaining sufficient energy availability is crucial for an individual's survival, preserving reproductive function is essential for the survival of the species. Thus, energetics and reproduction are inescapably linked, and animals have evolved complex physiological mechanisms that constrain reproduction as a function of metabolic state. Over the past several decades, we have come to learn that these mechanisms are extensively integrated, sharing a constellation of metabolic, endocrine, and neural components.

The goal of this chapter is to provide an overview of the physiological mechanisms integrating metabolism and reproduction. We begin by highlighting some of the ecological, evolutionary, and physiological principles that govern the codependency of energy regulation

Departments of Physiology and Biophysics, Obstetrics & Gynecology, and Biology, University of Washington, Seattle, Washington.

and reproductive success. Next, we illustrate these principles by describing ethological studies of wild animals in their natural habitats, experimental research on domesticated animals in the laboratory or field, and epidemiological studies of humans. The chapter is devoted to summarizing our current understanding of the physiological mechanisms that integrate energy homeostasis and reproduction, with emphasis given to signaling between metabolic hormones/fuels and the neuroendocrine reproductive axis. To underscore the clinical relevance of these complex relationships and to highlight opportunities for further investigation, we conclude by briefly describing several metabolic disorders that are often accompanied by impairment of reproductive function.

The relationship between nutrition and reproduction has been extensively investigated and documented across a wide spectrum of *Mammalia* [for an excellent review of this topic, see (1)]. In this chapter we limit our discussion of the metabolic control of reproduction to a few mammalian species, where there are extensive studies of particular species—either observational or experimental. Although generalizations that derive from such intensive studies of a few species may reveal important information about the mechanisms that link metabolism and reproduction, it is important to bear in mind that different species may have evolved unique adaptive mechanisms to solve the same physiological problem (beware of the pitfalls of such generalizations). For information about the link between energetics and reproduction in nonmammalian species, we refer the reader to (2–4).

Environmental Influences on Reproduction

An organism's reproductive potential is governed by the interactions between its genes and its environment. Although an organism's reproductive physiology and behavior are ultimately the products of its genetic makeup, a multitude of environmental factors can modulate reproductive function by acting on the animal's genetic substrate. A mammal's environment is not solely limited to its physical habitat (e.g., geographical location, climate, etc.), but also encompasses its dietary and social influences (1). Environmental factors can have bidirectional effects on reproductive function (i.e., promoting or prohibiting reproduction) and can influence reproductive capacity at all reproductive stages. Thus, an animal's environment can have a significant bearing on the timing of its sexual maturation, maintenance of reproductive function in adulthood, and (in some species) the onset of reproductive senescence.

Recent taxonomic estimates suggest that approximately 3,000 mammalian species currently inhabit the earth (5). These species live in extremely diverse environments extending from the equatorial jungles to the polar regions. Much of the environmental diversity on the earth can be attributed to differences in temperature, photoperiod, rainfall, humidity, and food sources. Even within a single habitat, environmental conditions often vary considerably, from season to season and from year to year (particularly within the temperate zone). By virtue of genetic variation and natural selection, mammals have acquired adaptations that have helped them cope with environmental constraints. As environmental conditions change over time, evolution forces either adaptation or demise. In this manner, animals have continually adapted their reproductive strategies to their changing environmental circumstances, and as a consequence, mammals have evolved a wide range of reproductive strategies.

Evolutionary success devolves from the reproductive success of individuals. Mammals have evolved many different strategies to optimize reproductive success. One common strategy for maximizing reproductive efficiency is to restrict reproductive efforts to times when environmental conditions are most conducive for bearing and nourishing offspring. Confining reproduction to favorable environmental conditions increases the probability of a successful outcome for each attempt to reproduce. When environmental conditions are unfavorable, sexual maturation may be delayed and reproductive efforts postponed until conditions improve. By precluding efforts at reproduction when environmental conditions are decidedly suboptimal, individuals can focus their energy on their own survival, instead of trying to produce offspring whose survival (and own reproductive success) is dubious.

Role of Food Availability

Of all the various environmental factors that shape reproductive potential, food availability has perhaps the greatest constraining influence, because food is the most proximate necessity for life and thus reproduction (1). All physiological processes consume energy. Available fuel is first diverted to processes that are essential for an organism's immediate survival, such as cellular maintenance, thermoregulation, sensory function, and the locomotor activity required to forage for food and flee from predators. Once these essential demands have been fulfilled, any extra energy can be allocated to processes that are less crucial for immediate survival, such as growth, storage of excess energy as fat, and reproduction. As long as energy intake is sufficient to meet all these demands, they can all continue unabated. However, if food becomes scarce or if a dramatic increase in whole body energy expenditure cannot be matched by increased food intake, the animal must order its priorities among the competing demands and often must make sacrifices. Growth and reproduction are usually accorded the lowest immediate priority, and thus they are often the first to be sacrificed. This can be viewed as an evolutionary adaptation, because it would be disadvantageous for individuals to waste valuable energy on growth or to increase the size of the population during times when food is in short supply and would be equally detrimental to the individual's own reproductive success. Indeed, spending energy on reproduction—if it requires ongoing expenditure for success—would certainly be disadvantageous to the species. Thus, it would make evolutionary sense that priority be given to the proximate needs for an individual's survival at the expense of longer term investments in reproductive activities.

Independent of its energetic (caloric) components, food also contains nutrients that mammals cannot synthesize themselves, including some vitamins, minerals, and essential amino acids. Under certain conditions, dietary supplementation with specific nutrients can enhance reproductive performance, and likewise, diets deficient in certain nutrients may compromise reproductive capability (1). Thus, an *adequate diet is more than adequate calories*. Unfortunately, the role of specific nutrients in regulating reproductive function has not been nearly as well characterized as that of the energetic component of food; thus, we do not consider this topic any further in our review. Interested readers can refer to (1) for more information about this subject.

Seasonal Influences

Many mammals reside in habitats that are characterized by pronounced seasonal fluctuations in environmental conditions, including temperature, light, and rainfall. Because all these climatic factors play a role in dictating food availability, most of these natural habitats are also susceptible to seasonal variations in the availability of energy resources. With notable exceptions (e.g., humans), most mammals that inhabit seasonally fluctuating environmental niches reproduce in a seasonal manner [for a detailed account of the extraordinarily diverse seasonal reproductive strategies used by mammals, see (1)]. Seasonal breeders exhibit annual cycles of reproductive activity, in which sexual development, mating, and conception are usually confined to a particular time of year. These animals spend the remainder of the year in a relative state of reproductive quiescence, which in some mammals entails a complete shutdown of the reproductive axis. In some species, only the females display seasonal changes in gonadal activity. In other species, both males and females exhibit reproductive quiescence during the nonbreeding season, albeit often to differing extents and with different time frames.

From an evolutionary standpoint, food availability is the principal selective pressure that shapes the reproductive strategies of seasonal breeders (1). Seasonal breeders maximize their reproductive success by timing their reproductive efforts so that the most energetically demanding stage of the females' reproductive cycle coincides with the time of year in which food is most abundant. In most female mammals, lactation is the most energetically expensive phase of their reproductive cycle. Therefore, in seasonally breeding mammals with relatively long gestational periods, this strategy often requires animals to mate during a time of year when environmental conditions are less than ideal. For example, sheep breed during the autumn and winter (when food is less readily available), so that ewes give birth and are lactating during the spring, when foraging conditions are optimal (6). Different species (as well as different populations within a single species) reproduce during different times of the year, depending in large part on where they live, what they eat, and when their food sources are most readily available. Although some species of seasonal breeders adjust their seasonal breeding patterns in direct response to alterations in food availability, mammals more commonly rely on a seasonal predictor to time their annual breeding cycles, such as the daily photoperiod. For a more detailed account of the seasonal regulation of reproduction in mammals, we refer the reader to Chapters 39 and 41.

Factors that Influence Sensitivity to Metabolic Challenges

The reproductive system of some mammals is more sensitive than others to reductions in energy availability. Of the various factors that dictate an animal's sensitivity to energetic challenges, three are worth describing in detail—sex, body size, and domestication.

Sex Differences

Reproduction imposes differential energetic demands on males and females. Gametogenesis is a relatively inexpensive process for both sexes. Except for the rare instances where the male provides significant parental care to its offspring, the male's relatively small energetic contribution usually ends after insemination. Gametogenesis has unique implications for the female. Although ovulation itself constitutes only a modest energetic burden, pregnancy imposes a large energetic drain. The pregnant mammal faces the colossal metabolic demands associated not only with gestation but also with lactation and further rearing her offspring to achieve their independence.

Because reproduction imposes different energetic requirements between the sexes, the impact of metabolic stress on reproduction is under different selection pressures between males and females. Thus, the male and female of a species are likely to display different adaptive responses to energetic challenges. If a female were to endeavor reproduction when food is scarce, the probability that she and her offspring would survive would be significantly reduced, because her ability to meet the large energetic demands associated with gestation and lactation would be compromised. In contrast, the gene pool of the male may be advantaged if the male remains reproductively competent and active, even during energetically unfavorable conditions. This is because spermatogenesis is energetically cheap. By investing a relatively small amount of energy in sustaining spermatogenesis, the evolutionary advantage to the male's own "selfish" genes would be enhanced, because the likelihood of their persistence in a subsequent generation would be increased. Although both sexes are sensitive to nutritional stress, females are usually—but not always—more susceptible than males to the detrimental effects of food deprivation; thus, the female reproductive axis generally has a lower threshold of inhibition in response to a nutritional challenge. Notwithstanding, just as one must exercise care about generalizing among species about the influence of nutrition on reproduction, one must also be aware that the mechanisms by which nutrition influences reproduction are often different between the sexes. Throughout this chapter, wherever it is relevant and known, we highlight these differences.

Size Differences

Body size influences the responsiveness of the reproductive system to metabolic stress. Small mammals are generally more sensitive than larger ones to nutritional challenges. First, small mammals have less energy reserves than large mammals and thus have reduced capacities to store energy when food is abundant and mobilize fuels when food is scarce. Second, small mammals possess a relatively large surface-to-volume ratio, which imposes a greater demand on thermogenesis. To maintain their body temperature, small mammals must maintain a high metabolic rate to meet their relatively large thermoregulatory demands. This requires small mammals to consume more food (per unit of body weight) than larger mammals. As a consequence of their large thermoregulatory demands and their relative paucity of energy reserves, small mammals frequently experience wide fluctuations in fat storage and mobilization. For example, laboratory mice that are only allowed access to food during the dark phase of the light–dark cycle mobilize approximately one-third of their entire body fat reserve during the subsequent light phase (7). So, with their limited energy stores and large thermogenic demands, small mammals are highly sensitive to even daily reductions in food availability and are likely to perish as a consequence of only relatively brief states of energy deficiency.

In contrast, large mammals have greater absolute energy reserves, mostly in the form of fat. Large mammals also have a relatively small surface-to-volume ratio and are more energetically efficient than smaller mammals. Compared with small mammals, larger mammals dissipate less heat per unit of body weight and consequently have lower thermogenic demands; thus, they function at a lower metabolic rate and consume less food per unit of body weight than smaller mammals. Larger mammals also tend to produce fewer offspring, and their offspring are smaller in relation to the size of the mother (compared with the offspring of small mammals). Consequently, the young of large mammals demand relatively less of their mother's metabolic resources during lactation than do small mammals. Prepubertal and adult animals of the same species also differ in their relative metabolic requirements (per unit of weight), which is attributable, in part, to differences in their body size, surface-to-volume ratios, and metabolic rate. This is reflected by the fact that the metabolic adaptation to fasting occurs more rapidly in young than adult animals, and thus the impact of fasting is usually greater on juvenile than adult animals (8).

Domestication

Most experiments investigating the link between nutrition and reproduction have been conducted in laboratory animals and livestock. The study of laboratory animals offers certain advantages over the study of animals in the wild, in terms of convenience, expense, and control over environmental circumstances; however, the laboratory environment represents an "unnatural" departure from "real" life in the wild. Laboratory animals are usually bred and housed under energetically favorable conditions; they are generally well fed and live in comfortable and relatively nonthermogenically demanding climates.

The same is usually true of economically valuable livestock, for which nutrition and living conditions are optimized to realize their full commercial value.

Laboratory animals and other domesticated species differ from their wild counterparts in important ways that erode the validity of inferences drawn from studies of one to the other. First, as a consequence of selective breeding, domesticated species and animals bred for laboratory investigation have considerably less genetic variability than their wild counterparts. As a consequence, what might be "true" in an entire population of laboratory animals may be valid for only a subset (or none) of an equivalent species in the wild. Second, in their optimized environment, domesticated and laboratory animals are generally less susceptible to adverse environmental conditions than their wild counterparts. From a nutritional standpoint, laboratory rodents live in a "utopian" milieu, where food is normally available, whereas wild animals are at the mercy of constantly fluctuating food sources. Wild mammals grow more slowly and often experience delayed puberty compared with their domesticated counterparts, attributable, no doubt, to the erratic availability of food (1). By concentrating studies of nutrition and reproduction on laboratory animals, scientists may have underestimated the diversity of reproductive strategies used by mammals in the wild, thus limiting our real understanding of the relationship between nutrition and reproduction under "natural" conditions.

The postponement of reproductive activities that may occur in response to a nutritional challenge should not be considered a pathophysiological event. Rather, it should be viewed as an adaptive strategy that promotes the survival of both the individual and the species. In the wild, few habitats offer a continuous abundance of food, and animals living under these "natural" circumstances face intermittent privation and hardship. Accordingly, it is not surprising that the interdependence of reproduction and nutrition applies to mammalian species ranging from musk shrews to whales to humans (9).

If these evolutionary, ecological, and physiological arguments hold true, one would expect to see evidence of the dependency of reproduction on nutrition and fuel stores in the natural environment. Indeed, overwhelming evidence supports this contention, with numerous reports attesting to the impact of nutrition on reproductive function across virtually all classes of vertebrates.

Early Observations

The notion that reproduction depends on adequate nutrition may be traced back thousands of years. One of the earliest examples of prehistoric art is a figurine of an obese woman known as the Venus of Willendorf. The sculpture was unearthed in Austria and has been dated to 25,000–23,000 B.C. Obesity was probably an unlikely characteristic of members of the hunter–gatherer societies that populated Europe during the Paleolithic period, given that food sources were often scarce and unreliable. Beyond its endowment of fat, the female idol has pronounced sexual characteristics, which has led some to speculate that the Venus of Willendorf was a fertility fetish (10). Thus, it is at least conceivable that even prehistoric humans were aware of the connection between fat and fertility. The ancient Greeks were most assuredly conscious of the relationship between nutrition and fertility. Aristotle made observations about the origins of female infertility in the 4th century B.C He posited that nutrition is the most important factor determining female fertility, noting that both underfed and overfed women displayed a reduced capacity to conceive and give birth (11).

In the late 19th century, Charles Darwin published seminal observations about the fertility of domesticated animals, which he noted were usually more prolific than their wild counterparts. Darwin suggested that the disparity between the fertility of domesticated and wild animals might be attributed, at least in part, to the more favorable environmental and nutritional conditions enjoyed by domesticated animals. Darwin speculated that even within the same individuals, fertility can be influenced by nutrition, noting that fertility improves when sheep living on sparse mountain meadows are brought down to graze on more lush lowland pastures (12).

Agrarian societies have long recognized the beneficial effects of proper nutrition on fertility, and various strategies have been adopted to maximize the reproductive success of livestock. For example, flushing is a commonly used agricultural practice, in which ewes that live on poor pastures are provided with supplemental nutrition during the weeks immediately preceding and during mating, which has consistently been shown to increase the frequency of twinning (13).

Over the past couple hundred years, biologists have noted a secular trend toward an earlier age of sexual maturation in human adolescents, which many have attributed to improvements in nutrition. In various populations of girls in Western Europe, the average age at menarche advanced considerably between 1830 and 1960, by approximately 4 months per decade (14). For example, the average age of menarche in Norwegian girls in 1844 was approximately 17 years, whereas by 1953 the mean age of menarche in Norway had decreased to 13.5 years (14) (Fig. 1). Similar trends have also been noted in the timing of male puberty, both with respect to the age

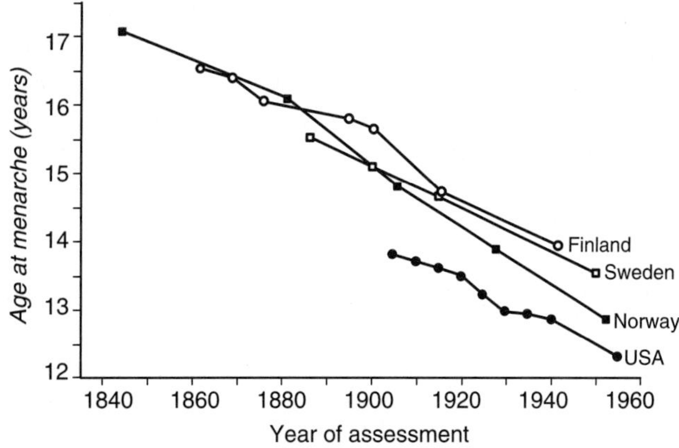

FIG. 1. Average age of menarche in girls in Finland, Sweden, Norway, and the United States from 1840 to 1960. [Redrawn from (14).]

of the adolescent growth spurt and the timing of voice change. In the middle of the 18th century, Johann Sebastian Bach led three choirs of schoolboys in Leipzig. Based on preserved records of the boys' ages and the presumed vocal compositions of the choirs, Daw estimated the average age of voice change in the choirboys (15). According to his extrapolations, there was a sharp increase in the average age of the altos (boys whose voices were in the process of changing) in the 1740s, with an average age of voice change of approximately 18 years. In comparison, the average age of voice change in schoolboys in London in 1959 was 13.3 years. The period of delayed timing of puberty in Leipzig during the 1740s coincided with the War of the Austrian Succession, which undoubtedly had adverse effects on the nutrition and general health of the choirboys (15).

Until the latter half of the 20th century, observations about the link between nutrition and reproduction were almost entirely anecdotal. Although the systems controlling body weight and reproduction had been studied well before this time, it was not until the 1950s that scientists began to investigate the interactions between these two control systems. In the 1950s and 1960s, Gordon Kennedy made pioneering efforts toward identifying the relationships between growth, nutrition, and reproduction when he methodically investigated the effects of growth and feeding on the timing of sexual maturation in female rats. In comparing groups of peripubertal female rats that were either optimally nourished or subjected to varying degrees of food and/or growth restriction, he observed that the timing of first estrus and first mating (hallmarks of puberty in female rats) were more closely related to body weight than to chronological age (16). Some years later, Frisch and Revelle (17) described similar findings in human females. Based on their studies of the association between body height and weight and the age at menarche in various populations of girls, they concluded that girls must attain a critical body weight as a physiological requirement for menarche (17). This hypothesis was later modified to allege that the attainment of a critical level of body fat (rather than body weight per se) more accurately predicts the onset of puberty, thus giving rise to the widely influential "critical fat hypothesis" (18). Since the seminal findings of Kennedy and Frisch, investigators have invested considerable effort toward elucidating the physiological mechanisms that couple energy homeostasis and reproduction. Such investigations have included observational studies of wild animals, laboratory animals, livestock, and humans. In addition, experimental work has been performed in a variety of species in which the metabolic status of animals has been purposely manipulated and the response of the reproductive axis studied in detail, with the goal of shedding some light on the mechanisms by which fat, metabolic fuels, and hormones might influence the reproductive axis.

METABOLIC STRESSORS THAT IMPAIR REPRODUCTION

Organisms maintain energy homeostasis through the coordinated regulation of fuel intake, utilization, and storage. Complex physiological systems and behaviors defend the sustained delivery of glucose and other fuels to the body's cells. Through the concerted actions of a host of peripheral and central mechanisms, animals maintain energy homeostasis by adjusting food intake to meet the body's energetic requirements and, conversely, by modulating energy expenditure to compensate for alterations in food intake. If food is readily available, animals match energy intake and expenditure with remarkable precision over long periods of time, maintaining energetic balance as well as a relatively constant mass of energy reserves.

Despite taut physiological controls, energy homeostasis is susceptible to disruption by a variety of metabolic stressors, particularly those that prevent consumption of sufficient fuel to meet energetic demands. In these situations, animals switch from an energy storing to an energy mobilizing state and rely on their energy reserves to provide substrates for metabolism. As an animal's energy stores are gradually depleted, energy availability becomes further reduced. Under circumstances in which energy availability is diminished because energy expenditure exceeds energy intake, the animal enters a state of negative energy balance.

Negative energy balance often stalls reproduction. The extent to which reproductive function may be inhibited by low energy availability is dependent, in part, on the nature of the metabolic challenge as well as on the severity of the energy deficit. Even subtle reductions in energy availability may result in diminished reproductive function, and severe metabolic challenges often result in a complete cessation of reproductive activity. The inhospitable effects of energy deficits on reproductive function can often be observed at all three levels of the hypothalamic–pituitary–gonadal (HPG) axis as well as in the quantity and quality of sexual and maternal behaviors. Although the consequences of metabolic stress on the reproductive axis may be manifest throughout the HPG axis, the brain (gonadotropin-releasing hormone [GnRH] neurons and their afferents) is thought to be the primary site of this integrative process.

In this section, we highlight several metabolic stressors that challenge energy homeostasis and inhibit reproductive function. We begin by describing situations in which energy balance is compromised as a result of reduced energy intake, such as would occur when animals are subjected to caloric restriction imposed by environmental constraints or experimental manipulations. We then present examples of circumstances in which negative energy balance results from the increased energetic demands imposed by strenuous physical activity, lactation, and prolonged exposure to cold ambient temperature.

Caloric Restriction

Our early human ancestors routinely encountered periods of time when food was scarce. This remains true today for many human beings who are impoverished or in war zones as well as in wild animals that exist at the mercy of seasonal or chance fluctuations in energy resources. In a state of negative energy balance, organisms must adjust the way they partition energy and sacrifice those processes that are nonessential for immediate survival. Consequently, reproduction is often among the first functions to be abandoned when energy availability is limited. The repercussions of inadequate caloric intake on reproductive function are manifest in both sexually immature and adult mammals. Although there is a considerable amount of overlap between the negative consequences of caloric restriction during these two distinct reproductive stages, for the purpose of this section, we consider the effects of food restriction on sexual maturation and adult reproductive function separately.

Sexual Maturation

During pubertal development, an animal experiences a series of physiological and behavioral events as it develops from the juvenile to the adult state (see Chapters 38–40). Although dramatic changes occur throughout the HPG axis with respect to neuroendocrine and gonadal function, puberty is ultimately a brain-limited phenomenon. Intermittent administration of GnRH or the excitatory amino acid agonist *N*-methyl-D-aspartate induces precocious puberty in juvenile monkeys (19,20), rats (21), and guinea pigs (22), demonstrating that the pituitary and gonads are functional in the prepubertal animal but are held in abeyance until the normal time of puberty onset. Puberty is triggered by the commencement of pulsatile GnRH secretion from the hypothalamus. Intermittent release of GnRH into the pituitary portal vasculature evokes adult patterns of gonadotropin (luteinizing hormone [LH] and follicle-stimulating hormone [FSH]) secretion from the anterior pituitary, which in turn act on the gonads to promote sex steroid synthesis, spermatogenesis, and ovulation. At puberty, the brain also initiates and orchestrates appropriate reproductive behaviors under the activational effects of gonadal steroids.

It is now considered axiomatic that a certain degree of somatic growth is requisite for sexual maturation (23). Somatic growth is influenced by both the quantity and quality of nutrition that an animal receives during prenatal and postnatal development. Depending on the severity of caloric restriction, growth occurs at a reduced rate or ceases altogether in chronically undernourished young mammals (16,24,25). Undernourished animals are generally smaller than well-fed animals as both groups approach the normal age for sexual maturation; thus, puberty typically occurs at a later chronological age in growth-restricted individuals. Domesticated animals usually attain sexual maturity at an earlier age than their wild counterparts, which is likely to be attributable to the accelerated growth rate of well-fed domesticated animals (1).

Chronic undernutrition during prenatal and/or postnatal development retards growth and delays sexual maturation in a wide range of species, including rats (16,26–29), mice (30), musk shrews (31), sheep (25), cattle (32), and pigs (33). Indications of prolonged sexual immaturity in food-restricted animals include immature gonadotropin secretory patterns, reduced plasma gonadal steroid concentrations, hypogonadism, impaired spermatogenesis, absent or shortened estrous cycles, anovulation, and absent or impaired mating behavior. The inhibitory effects of caloric restriction on the timing of puberty are also often manifest in gonadectomized animals, which indicates that reproductive disturbances can arise in food-restricted animals independently of alterations in gonadal steroid feedback (25,34–36), although food restriction may also alter (increase) an animal's sensitivity to the negative feedback effects of gonadal steroids (37,38). Depending on the species in question, males and females sometimes display differential sensitivities to the effects of chronic undernutrition on pubertal development. In prepubertal rats and mice, females are apparently more sensitive than males to the effects of altered nutritional intake on reproductive development (30,39). In contrast, male and female gonadectomized lambs appear to be equally sensitive to the suppressive effects of chronic undernutrition on gonadotropin secretion (36).

Children are also susceptible to the detrimental effects of poor nutrition on growth and reproductive development. The average age of sexual maturation (as determined by the onset of the adolescent growth spurt in both sexes and the occurrence of menarche in females) has occurred at progressively younger ages over the past couple hundred years (14) (Fig. 1). This phenomenon has widely been attributed to marked improvements in general health and nutrition during this period of time. There have been transient reversals in this trend during periods of war and/or economic hardship, when food was generally less available than in more prosperous times (14). In addition to this secular trend, differences in the timing of puberty have also been observed between contemporary well-nourished and undernourished children. A longitudinal study of chronically undernourished girls in Alabama revealed that these girls had reduced rates of skeletal growth and maturation compared with well-nourished girls as well as a delay in menarche by approximately 2 years (40). A similar effect of poor nutrition on the timing of puberty has been documented in children suffering from severe protein calorie malnutrition. In this study, none of the malnourished children who were of prepubertal age displayed any signs of impending pubertal development, which was in accordance with their reduced circulating levels of LH and FSH compared with age-matched healthy children (41).

The negative impact of chronic undernutrition on sexual maturation in humans has been documented most thoroughly in girls and women who suffer from anorexia nervosa. In these individuals, caloric deprivation is self-inflicted as a result of a psychological disturbance and distorted body image. If the onset of the syndrome occurs before puberty, women maintain age-inappropriate prepubertal patterns of gonadotropin secretion and display primary amenorrhea (42,43). Although anorexia nervosa is much less common in males, there are a few reports in the literature of delayed puberty in men who developed anorexia nervosa at a young age. Afflicted males are hypogonadal and have a reduced libido, prepubertal serum concentrations of gonadotropins and testosterone, and a juvenile LH response to a GnRH challenge (44,45).

Although severely undernourished people and other animals may linger indefinitely in a prepubertal state, the adverse effects of undernutrition on sexual maturation are usually reversible. Despite exhibiting a reduced rate of growth during early development, when undernourished animals are subsequently provided with free access to food, they usually experience a rapid "catch-up" growth. When chronically undernourished laboratory animals are allowed to feed freely, sexual maturation often resumes within hours to days (29,34,46,47) (Fig. 2). In contrast to the rapid restoration of reproductive development in chronically undernourished rodents and ruminants, the activational effects of improved nutrition on sexual maturation occur more gradually in undernourished humans. In anorexic patients, weight gain therapy often results in the gradual appearance of clinical and neuroendocrine signs of sexual development; however, this may not always be the case, even after these individuals achieve and maintain their ideal body weights for an extended period of time (44,45).

Delayed or suspended sexual maturation in chronically undernourished mammals is most likely attributable to a primary deficiency in the secretion of GnRH. The frequency and amplitude of pulsatile GnRH secretion are reduced by chronic food restriction in ovariectomized (OVX) ewe lambs (48), as is the secretion of GnRH from the median eminence of intact ewe lambs (49). In food-restricted female rats that fail to progress through puberty, pulsatile infusion of GnRH induces ovulation within 4–6 days after treatment commences (29). Intermittent administration of GnRH to chronically undernourished OVX lambs

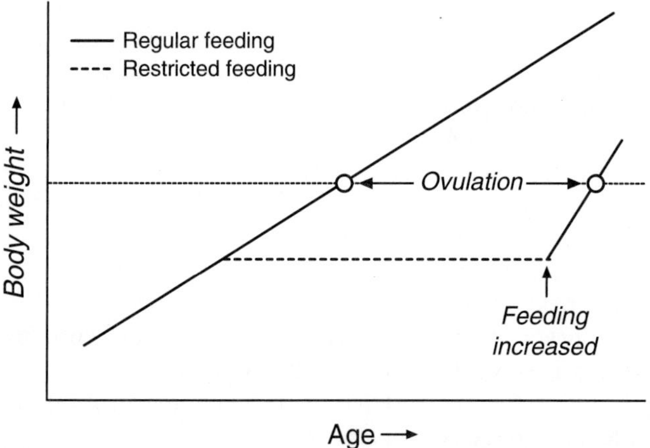

FIG. 2. Effect of chronic undernutrition on the timing of puberty in female rats. Although growth and sexual maturation (as assessed by first ovulation) are arrested in the food-restricted rats, ovulation occurs rapidly once these animals are allowed to feed regularly again. Notably, ovulation in the previously food-restricted rats occurs at an older age but at approximately the same body weight as in the rats that are fed normally throughout life. [Adapted from (29).]

elicits pulsatile LH secretion from the pituitary (34). These observations suggest that the gonadotropes of undernourished prepubertal rats and lambs are functional, despite the food restriction, and that they lack only sufficient stimulation by GnRH. Although growth-restricted ewe lambs have a normal number of GnRH neurons in the forebrain, their GnRH neurons are distributed in a different pattern than in normally fed lambs (50). Nutrition-induced alterations in GnRH secretion in prepubertal animals do not appear to be attributable to changes in GnRH synthesis, because the hypothalamic content of GnRH is not significantly altered by chronic undernutrition or refeeding (35,51). The observation that single or repeated injections of *N*-methyl-D,L-aspartate increase serum LH levels in undernourished lambs indicates that these animals are capable of releasing GnRH in response to chemical stimulation (51). In growth-restricted ewe lambs, administration of the nonspecific central nervous system (CNS) inhibitor pentobarbital increases mean LH levels and LH pulse frequency, which suggests that food restriction impairs LH release via a central inhibition of GnRH secretion (48).

Adult Reproductive Function

In addition to delaying the onset of sexual maturation, caloric restriction exerts inhibitory effects on adult reproductive function. As is the case in sexually immature animals, food restriction suppresses multiple facets of the reproductive axis in adult mammals. Both the duration and degree of undernutrition influence the extent of reproductive impairment, with prolonged underfeeding or total food deprivation producing the most profound reproductive deficiencies. Under some circumstances of extreme caloric restriction, adult mammals essentially revert to a prepubertal reproductive state.

There is a vast body of literature on the reproductive consequences of food restriction in adult female mammals, including rats (52,53), mice (54), hamsters (55–57), musk shrews (58), cattle (59), and ewes (60). Evidence of reproductive impairment in undernourished adult females includes reduced serum gonadotropin and gonadal steroid concentrations, impaired follicular maturation, disrupted estrous cyclicity, inhibition of the preovulatory LH surge, and decreased sexual receptivity. It remains unclear whether heightened ovarian steroid-negative feedback plays an important role in mediating the effects of caloric restriction on female reproductive function. In some experimental paradigms, food restriction suppresses LH secretion in intact and estrogen-treated OVX females but not in hormonally untreated OVX animals (61,62), whereas in other cases the effects of food restriction on the adult reproductive axis persist in the absence of gonadal steroids (60,63).

In adult males, severe undernutrition can result in reduced gonadotropin and testosterone secretion, atrophy of the testes and accessory organs, and impaired spermatogenesis (56,64–70). Both gonadal steroid-dependent and -independent mechanisms have been implicated in mediating the effects of underfeeding or fasting on male reproductive function (68,71,72). Although caloric restriction usually has a negative impact on reproductive function in males, the reproductive consequences of undernutrition are sometimes less severe in males than in females (1). From an evolutionary point of view, it makes sense that males would be less sensitive than females to the effects of malnutrition, because perpetuation of the male's genes may be optimized by continued opportunistic reproductive activity—even at the risk of an accelerating somatic decline in the face of starvation.

One striking feature of alterations in reproductive function caused by dietary manipulations is the rapidity with which the reproductive axis responds to reduced or improved caloric intake. Signs of reproductive impairment are often evident after 1 or 2 days of fasting, particularly in small mammals with their high metabolic rates and paucity of fat stores (1). The rapid inhibitory effects of food restriction on reproductive function are often rivaled by an equally rapid restoration of reproductive function after reinstitution of sufficient caloric intake (65,66,73). In fasted male rhesus macaques, pulsatile LH secretion is impaired within a few hours of missing a regularly scheduled meal but is rapidly reinstated (within minutes to hours) after the animals are provided with food (69) (Fig. 3). Furthermore, the ability of caloric intake to restore LH and testosterone secretion in fasted male monkeys is dependent on the size of the reinstated meal (i.e., LH and testosterone levels are positively correlated with meal size) (74).

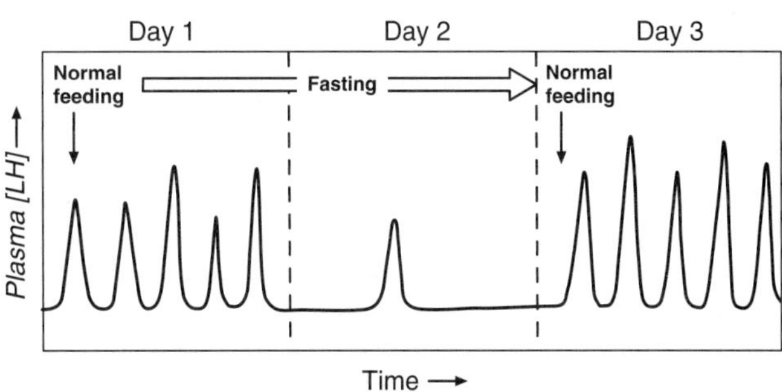

FIG. 3. Effects of fasting and subsequent refeeding on pulsatile LH secretion in male rhesus macaques. LH pulsatility is abolished when food is withheld but is rapidly reinstated when feeding resumes on the following day. [Adapted from (69).]

Poor nutrition also has a negative impact on the reproductive axis of the adult human. Brief periods of fasting are associated with reduced serum concentrations of gonadotropins and/or sex steroids in men (70,75,76) and women (77). Chronic undernourishment can elicit severe reproductive impairment. The observation that birth rates are reduced during and in the months after periods of war or famine is presumably due in part to diminished fertility in chronically undernourished women and men (78,79). From his studies of a population of !Kung San women in Botswana, van der Walt et al. (80) postulated that these women experience seasonal cycles of fertility in which ovulation is suppressed when food availability reaches a nadir. In this same population, the birth rate reaches a peak approximately 9 months after the presumed period of maximal food availability (80). Women who lose a significant amount of weight as a consequence of injudicious dieting often experience menstrual disturbances and infertility (81,82), both of which are usually reversed if adequate body weight is regained (81,83). Women with anorexia nervosa are commonly amenorrheic and have prepubertal or early pubertal LH secretory patterns (42). Although the reproductive abnormalities in adult women with anorexia nervosa are often corrected after weight gain (42,84), increased body weight alone is not always sufficient to restore reproductive function (85,86). Reproductive impairment is also evident in men with anorexia nervosa and is improved by weight gain (45,87–89).

Exogenous administration of GnRH generally attenuates or prevents the decline in pituitary and gonadal function caused by underfeeding or fasting in adult experimental animals (68,90,91) and humans (43,92,93). Although the expression of GnRH mRNA in the basal forebrain is normal in fasted adult male rats (94), GnRH content is elevated in nerve terminals in the median eminence (72), which suggests that GnRH secretion may be reduced, despite normal GnRH synthesis. Although many investigators have observed that undernourished humans and laboratory animals exhibit normal or even enhanced gonadotropin release in response to a GnRH challenge (66,69,93,95–97), others have observed alterations in the magnitude and/or timing of the LH response to GnRH treatment in food-restricted individuals (72,98–100). The apparently altered pituitary function in the latter experiments could conceivably be attributable to reduced levels of GnRH receptor in the pituitary (101), pituitary gonadotropin content (90), or gonadotropin subunit gene expression (90,91).

Exercise

An animal's daily energetic requirements can be markedly increased by prolonged or more frequent bouts of physical activity. Such physical activity can take many forms, ranging from food-seeking locomotor activity to recreational exercise. In contrast to the unlimited food availability that is enjoyed by most laboratory animals and by a large proportion of the human population, animals in the wild must continually forage for their food. If food sources are scarce, foraging can require an immense amount of energy and thus can significantly raise an animal's metabolic requirements. Humans also experience increased energetic demands during exercise, as do animals that are trained to exercise in the laboratory. Under these different circumstances, if the increased energetic demands are not compensated by an equivalent increase in caloric intake, negative energy balance ensues.

Negative energy balance that is imposed by strenuous physical activity commonly elicits disturbances in reproductive function. This relationship has been most extensively investigated in female athletes, particularly in women who participate in endurance sports that emphasize leanness for improved performance and/or aesthetic purposes (102). In addition to the dramatically increased energetic demands incurred by their intensive training schedules, many athletes also display pathological eating behaviors or else simply do not consume enough calories to compensate for their increased metabolic demands (102). Reproductive impairment is more commonly observed in these individuals than in individuals who exercise to a similar degree but consume enough calories to meet their increased energy requirements (103,104). These observations suggest that insufficient energy availability, rather than the physical and emotional stress associated with exercise, is the underlying cause of reproductive impairment in female athletes (105,106).

Intense exercise may delay the onset of puberty or impair ongoing reproductive function in female athletes. Delayed menarche has been reported in female athletes who began athletic training before the onset of puberty, including ballet dancers (107,108), gymnasts (109), distance runners (110), track and field athletes (111), and swimmers (110,112). Secondary amenorrhea and oligomenorrhea occur at a higher frequency in female athletes than in sedentary women (107,108,113). Studies of female runners have demonstrated that the frequency of amenorrhea is positively correlated with the number of miles the women run per week (113,114). Amenorrhea in female athletes is frequently accompanied by prepubertal patterns of gonadotropin secretion, anovulation, and reduced serum estradiol concentrations (108,113). When the intensity of exercise training is temporarily reduced due to injury, vacation, or the end of the training season, sexual development and/or menses commonly resume in ballet dancers (108), runners (113), and rowers (115). Thus, exercise-induced disruption of menstrual cyclicity is generally a reversible phenomenon.

Not all female athletes become amenorrheic during intense training. However, despite the regular occurrence of menses, less overt signs of neuroendocrine and ovarian impairment are commonly seen in eumenorrheic female athletes, such as reduced pulsatile LH secretion (116,117). Females do not need to be top-tiered competitive athletes to experience exercise-induced reproductive deficiencies, because even recreational moderately exercising athletes sometimes experience anovulatory cycles, shortened luteal phases, and reduced serum levels of estradiol and progesterone (118,119). When untrained college women were placed on an intense 2-month running schedule, most of the women developed irregular menstrual cycles that were characterized by abnormal luteal function, anovulation, and delayed menses. Regular menstrual cycles were reinstated in all the women within 6 months after ceasing the running routine (104).

Although not as widely documented as in female athletes, highly trained male athletes are also susceptible to exercise-related reproductive disturbances. In general, the consequences of strenuous exercise on the reproductive system of the male are more subtle than the effects on the female, and such effects in the male are confined to those who participate in sports in which a lean physique is desirable. A cross-sectional study of male distance runners revealed that compared with sedentary control subjects, distance runners have reduced serum concentrations of testosterone and prolactin, although these values are generally still within the low physiological range (120). However, compared with recreational athletes, male marathon runners exhibit reduced basal and GnRH-stimulated LH secretion (121). In amateur male wrestlers, the athletes' low body weight and body fat composition during the competitive season are associated with reduced serum concentrations of testosterone and prolactin (122).

Impaired reproductive function is also evident in laboratory animals that are subjected to prolonged periods of strenuous exercise. Impairment of reproduction with exercise is generally observed under circumstances in which animals are unable to compensate for the energetic cost of exercise by increasing their food intake. Although forced or voluntary wheel running has been shown to have an acceleratory effect on the timing of vaginal opening (with respect to body weight) and to increase the number of ovulatory cycles in peripubertal female rats and mice (123,124), prolonged exercise that is not matched by an appropriate increase in caloric intake generally stalls reproductive development. In an experiment that was designed to simulate the energetic cost-to-benefit ratios of foraging, peripubertal female mice were forced to run on a wheel to obtain food. As the foraging requirements increased (i.e., as the mice were required to complete a greater number of wheel revolutions to obtain a pellet of food), growth and reproductive development were delayed, but only in those animals that were unable to consume enough food to compensate for the increased cost of exercise (124). Reproductive development was blocked in peripubertal female rats that were placed on a similar foraging schedule but quickly resumed after the running requirement was relaxed; pulsatile LH secretion was initiated, and all the animals had estrous smears and ovulated within 4 or 5 days (125).

Exercise also has pronounced effects on reproductive function in adult female rodents. Female rats that are placed on an intensive running schedule for 2–3 months often have disrupted estrous cycles, as evidenced by an increased cycle length and reduced plasma estradiol concentrations (126,127). Basal LH secretion and pituitary LH protein content are also reduced in exercise-induced anestrous rats (127). Although female hamsters that voluntarily exercise on a running wheel display normal estrous cyclicity when they have unlimited access to food, exercising hamsters that are food deprived for 48 hours (and are thus unable to increase their energy intake to match their increased energy expenditure) show impairment of estrous cyclicity and lordosis behavior. Furthermore, food-deprived exercising hamsters display more pronounced reproductive deficiencies than food-deprived hamsters that are not provided with a running wheel, demonstrating that the additional energetic cost of exercise exaggerates the inhibitory effects of fasting on reproductive function (128).

Detrimental effects of intense exercise on ovarian cyclicity have also been observed in female monkeys. Williams et al. (129) demonstrated that adult female cynomolgus monkeys that are trained to run on a treadmill but not permitted to increase their food intake over the course of the exercise program become amenorrheic within 7–24 months after the onset of exercise. The transition to amenorrhea is fairly abrupt and is characterized by an increased cycle length (due to a prolonged follicular phase), anovulation, and progressively reduced serum concentrations of LH, FSH, estradiol, and progesterone (129). When the amenorrheic monkeys were subsequently allowed to consume more calories, menstrual cyclicity was reinstated, despite the fact that the monkeys were still maintained on the exercise program (Fig. 4). In these animals, serum concentrations of gonadotropins and gonadal steroids were elevated, and preovulatory LH surges occurred within 12–57 days after overfeeding commenced (130). Thus, as is the case in human female athletes, reduced energy availability

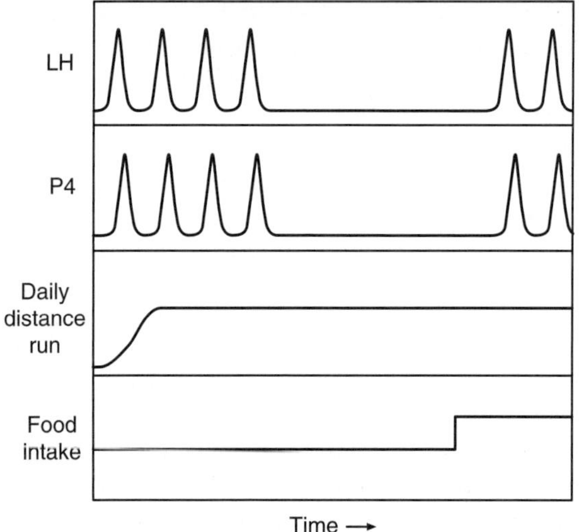

FIG. 4. Effects of prolonged exercise training and food consumption on menstrual cyclicity in female macaques. Menstrual cyclicity is eventually disrupted when monkeys are trained to run on a treadmill each day, as evidenced by the absence of LH surges and progesterone (P4) peaks. When the monkeys continue to exercise at the same level but are permitted to consume more food each day, menstrual cyclicity resumes. [Adapted from (130).]

appears to play a causal role in exercise-induced amenorrhea in female monkeys.

LH pulse frequency is reduced in exercising humans and rats, which implicates a reduction in pulsatile GnRH secretion as the underlying cause of exercise-induced reproductive impairment (116,117,131,132). Both amenorrheic and eumenorrheic female athletes exhibit normal or even enhanced LH release in response to exogenous GnRH treatment, which further suggests a deficit in hypothalamic GnRH secretion (116,131). Reduced GnRH secretion does not appear to be a consequence of a primary defect in GnRH synthesis, because hypothalamic GnRH content is elevated in exercising peripubertal female rats, despite the fact that they are acyclic and do not secrete LH in a pulsatile manner (132). Intermittent GnRH administration results in the initiation of estrous cyclicity and ovulation in exercising peripubertal rats within 3–6 days, even though the animals maintain a high level of physical activity during the GnRH infusion (132).

Lactation

Female mammals are routinely presented with an enormous metabolic challenge each time they give birth. The postpartum period is the most energetically demanding stage of the female reproductive cycle because of the substantial energetic cost of lactation. In addition to fulfilling her own energetic requirements, a lactating female must devote considerable energy and nutrients toward producing milk to nourish her offspring. The metabolic cost of milk production varies among species and even among individuals within a species, depending in part on the number of offspring per birth, duration of lactation, milk yield, and milk composition. In some mammals, the daily energetic cost of milk production exceeds the daily whole body energy requirements of nonlactating females (133). In general, the energetic cost of milk production increases over the course of lactation as a consequence of the amplifying nutritional needs of the continuously growing offspring.

Mammals use various strategies to meet the increased energy demands imposed by lactation. In part, females rely on the increased mass of adipose tissue deposited during gestation to provide energy for milk production (134). Stored fat plays a considerable role in supporting lactation in large mammals that possess sizable reserves but probably does not serve as a major energy source in smaller mammals with a paucity of adipose tissue. Because even large mammals with abundant fat reserves cannot rely solely on stored energy to meet the energy demands of lactation, lactating females must increase their energy intake by consuming more food (135,136). Hyperphagia is particularly pronounced during the later stages of lactation when the energetic cost of milk production is at its peak and when the female has already depleted a large portion of her energy reserves (1). For example, laboratory mice have been observed to double their food intake during the later stages of lactation, an effort that is probably even more pronounced in wild mice that must actively forage for food (1). In addition to increasing energy consumption and mobilizing energy reserves, some mammals (including humans) conserve energy during lactation by reducing energy expenditure, which they accomplish by increasing metabolic efficiency and/or decreasing physical activity (136). Despite these efforts to increase food intake and reduce energy expenditure, the immense energetic demands of the lactating female usually exceed energy intake.

From a purely energetic standpoint, it would be highly disadvantageous for a lactating animal to become pregnant again immediately after giving birth. If pregnancy were to occur, the female would be faced not only with the metabolic demands of nursing her existing offspring, but also with the sizable energetic cost associated with gestation. Given that lactating animals already exist in a state of negative energy balance, the added metabolic burden of pregnancy could severely jeopardize the survival of both females and their offspring. As an adaptive response, females of many (but not all) mammalian species experience a period of lactational infertility after parturition, thereby allowing females to concentrate their available energy on protecting and nourishing one offspring or litter at a time (137). Lactational infertility increases the likelihood of a manageable interval between successive births; for this reason, many women (particularly in developing countries) rely on breast-feeding as a contraceptive method (138). The duration and degree of lactational infertility is species dependent, and even within a species there is often a considerable degree of variability among individuals.

In some species, lactational infertility is attributable to an inhibition of ovarian function. In these species, lactational anestrus is characterized by inadequate follicular development, anovulation, and (in humans and nonhuman primates) amenorrhea. In other species (e.g., rats), lactational infertility results from a blockade of implantation. Regardless of the mechanism(s) by which lactating animals are rendered temporarily infertile, inadequate gonadotropin secretion from the anterior pituitary is typically observed during lactational infertility (137). Although serum levels of LH are reduced in lactating females compared with nonlactating females, circulating levels of FSH are often unaffected by lactation (139,140). Lactation prevents or delays the castration-induced rise in circulating LH levels in females that are OVX after parturition, demonstrating that LH secretion is compromised during lactation independently of the negative feedback effects of ovarian steroids (139,141–143). The positive feedback actions of estrogen on LH secretion are often altered by lactation, as evidenced by a reduced magnitude of the estrogen-induced LH surge in lactating rats (142,144), rhesus monkeys (145), and humans (146).

Evidence derived from studies of many species suggests that a reduction in GnRH release from the hypothalamus is a primary cause of insufficient LH secretion in lactating animals. Intermittent administration of GnRH to lactating animals can restore pulsatile LH secretion, follicular development, and ovarian cyclicity, thus indicating that the pituitary and ovary remain functional and responsive to trophic stimulation during lactational anestrus (147,148). GnRH synthesis does not appear to be compromised, because the expression of GnRH mRNA and peptide are not different between cycling and lactating rats (149). In female rhesus macaques suckling adopted infants, hypothalamic multiunit activity (an indirect measure of GnRH pulse generator activity) is reduced compared with multiunit activity during the follicular and luteal phases of the preceding menstrual cycle (150). Although lactating rats release LH in response to electrochemical stimulation of the medial preoptic area (MPOA) (151), these animals are insensitive to the LH-releasing effect of exogenous administration of *N*-methyl-D,L-aspartate (152). This implies that lactational infertility is attributable in part to an impairment of GnRH secretion as a result of refractoriness to afferent excitation.

Although negative energy balance and infertility occur in parallel during lactation, poor energy availability is only one of many factors that could contribute to the occurrence of lactational infertility (153). Food restriction or fasting commonly extends the length of lactational infertility (154,155). Conversely, nutritional supplementation has been shown to decrease the duration of lactational amenorrhea and infertility in malnourished women (156). Together, these observations suggest that the duration of lactational infertility is at least partly dictated by the female's nutritional status. However, the observation that suckling reduces LH secretion in lactating rats even when the females are not producing milk (and thus are not subject to the energetic drain of milk production) demonstrates that negative energy balance alone cannot fully account for lactational infertility (157). The mechanical action of suckling activates neural pathways that project to the brain, and the suckling stimulus itself is thought to influence energy homeostasis and reproductive function during lactation (153). Alterations in circulating levels of reproductive hormones, including estrogen, progesterone, and prolactin, also play important regulatory roles during lactation. Thus, the brain integrates both neural signals related to suckling and hormonal signals related to energy homeostasis during lactation and dictates reproductive output accordingly.

Cold Ambient Temperature

A significant proportion of a mammal's daily energy expenditure is devoted to maintaining thermal homeostasis. It has been estimated that average-sized mammals living at average ambient temperatures expend approximately 80%–90% of the energy that they obtain from food on thermoregulation (158). To sustain core temperature in cold environments, mammals must devote a considerable fraction of their energy resources toward heat production. Small mammals are particularly sensitive to the energetic demands imposed by exposure to cold ambient temperatures. First, compared with large mammals, the fat stores of small mammals usually comprise a relatively low percent of their body weight. Second, with large surface-to-volume ratios, small mammals lose heat to their surroundings more readily than larger mammals. As a result, small mammals must continuously generate heat just to maintain body temperature, which is accomplished by maintaining a high metabolic rate. Even a modest decrease in ambient temperature can constitute a considerable metabolic burden for small mammals, because it can markedly augment their already large thermoregulatory demands.

As one adaptive response to offset the increased thermal demands imposed by cold ambient temperatures, small mammals attempt to increase their food consumption. Although increasing food intake may be a viable strategy in freely fed laboratory rodents that are exposed to cold temperatures, small mammals living in the wild are rarely afforded this opportunity, especially during the winter when food is scarce and the energetic costs of foraging are high. As a result, mammals that are exposed to cold ambient temperatures for prolonged periods of time in their natural habitats are typically unable to consume and assimilate sufficient additional calories to compensate for their increased thermoregulatory demands, thus leading to a state of negative energy balance.

Mammals exposed to cold ambient temperatures often sacrifice reproductive function in favor of allocating available energy to more immediately pressing metabolic demands. For example, male and female laboratory rodents that are housed in a cold room before they reach puberty have delayed rates of growth and sexual maturation (159–162). Female mice that are born and reared in a cold environment also display signs of reproductive impairment during adulthood, including prolonged estrous cycles, reduced litter sizes, and increased interbirth intervals (160). Reproductive impairment also occurs in mammals that are not exposed to cold temperatures until they have reached adulthood. Adult female golden hamsters that are housed at 6°C for 25 or 50 days have reduced uterine weights, impaired follicular development, and a reduced frequency of ovulation compared with animals housed at 22°C (163). Low ambient temperature does not appear to have adverse effects on testicular weight or sperm count in male mice, which suggests that males are relatively resistant to the negative reproductive consequences of exposure to cold temperature (164).

The impact of cold environments on reproduction is mitigated by increased availability of food (165,166). Although not without exceptions, reproductive impairment in cold environments is generally limited to those individuals that are unable to increase their food intake to compensate for the increased thermal demands imposed by exposure to cold. In female Syrian hamsters housed at 5°C and fed diets that differ in caloric content or in ease of obtaining, the latency to cold-induced anestrus increases in parallel with caloric intake (167). Although ovarian steroid-primed OVX hamsters housed at 5°C for 3 days display normal estrous behavior if they are allowed to feed freely, food-restricted hamsters housed under the same environmental conditions exhibit impaired sexual receptivity (168). In mice that are food-restricted and housed for 4 weeks at 12°C, (a relatively cold environment), ovulation is suppressed; however, in cold-exposed mice that are permitted to increase their food intake, ovulation is unaffected until temperatures are lowered to –8°C, a temperature at which mice are physically incapable of processing greater quantities of food (169). In summary, the differential effects of cold ambient temperature on reproductive function in mammals that are either able or unable to increase their food consumption indicates that reproductive function is not impaired by cold temperatures per se but rather by the associated reduction in energy availability.

Unfortunately, there is only scant information in the literature concerning the mechanisms by which ambient temperature and thermoregulatory systems influence reproductive function. Thus, for now, we can only speculate about whether exposure to cold ambient temperature disrupts reproductive function by modulating GnRH secretion, although this would seem to be a plausible (and testable) hypothesis.

SIGNALS LINKING NUTRITION, GROWTH, AND REPRODUCTION

In the previous section, we described different metabolic challenges that may confront mammals in either a natural habitat or laboratory environment. Each of these metabolic perturbations results in diminished energy availability (as a consequence of insufficient energy intake and/or excessive energetic demands) and is accompanied by various signs of reproductive impairment, including a delayed onset of puberty in young animals and disrupted hormonal secretory patterns and reproductive behaviors in adults. Administration of GnRH often prevents or corrects these reproductive deficiencies, implying that under these circumstances, reduced hypothalamic drive to the pituitary is an underlying cause of impaired reproductive function. Thus, although signs of reproductive impairment are commonly observed at multiple sites along the HPG axis, compelling evidence suggests that disruptions in energy balance primarily impair reproductive function through a central mode of action.

To synchronize reproductive efforts with states of energy sufficiency (and, conversely, to prevent reproductive efforts during periods of energy deficiency), the neural circuits that control GnRH release (i.e., the GnRH pulse generator) must be provided with information about that status of peripheral energy metabolism and fuel reserves. For this to occur, the brain must be receptive to metabolic and/or neural cues that either directly or indirectly reflect the animal's nutritional status. Over the past several decades, investigators in the field have made great progress toward elucidating the identity and nature of the signals that convey metabolic information from the periphery to the GnRH pulse generator. Before providing examples of several hormones and metabolites that have been implicated as metabolic conduits to the neuroendocrine reproductive axis, we first describe the historical context of current research by recollecting a few pivotal experiments that inspired future investigations into the signaling mechanisms that link metabolism and reproduction.

Early inquiries into the afferent signaling mechanisms that relay information about energy metabolism to the CNS were focused on investigating the nature of the metabolic cues that influence the timing of the onset of puberty. Because animals must complete a certain degree of somatic growth before they reach sexual maturity (23), investigators initially sought evidence for a relationship between reproductive development and various growth-related signals. In the 1950s and 1960s, Kennedy and Mitra performed an elegant series of experiments in which they subjected young female rats to varying degrees of undernutrition by limiting their access to food before or after weaning and observed that these severely growth-restricted animals achieved various pubertal milestones (e.g., vaginal opening, first estrous smear, and first mating) at older chronological ages but at similar body weights as their optimally nourished siblings (16,170) (Fig. 5). The investigators concluded that the initiation of puberty is more closely related to body weight than to chronological age, and consequently they hypothesized that puberty is triggered by a signal that reflects the changes in energy balance (perhaps alterations in food intake and/or metabolic rate) that occur in young growing animals (16).

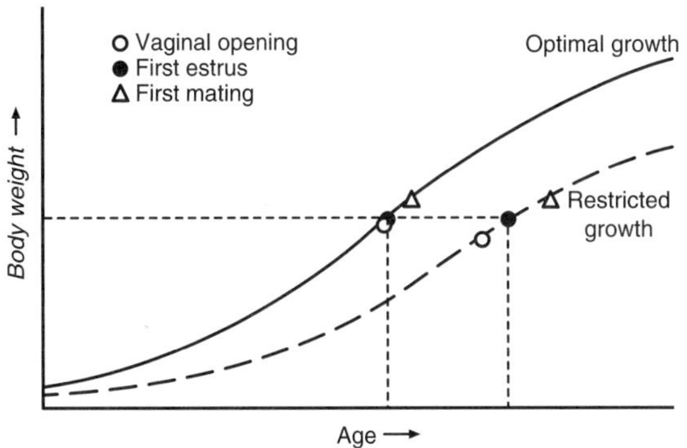

FIG. 5. Sexual maturation is delayed in growth-restricted female rats. Young female rats were either permitted to achieve optimal growth by maximizing their caloric intake during suckling (optimal growth) or were growth-restricted by limiting their caloric intake during suckling (restricted growth). Although the restricted growth rats experienced vaginal opening (*open circles*), first estrus (*closed circles*), and first mating (*triangles*) at significantly older ages than the optimal growth rats, the three pubertal milestones occurred at similar body weights in the two groups of animals (as illustrated for first estrus in the graph). [Adapted from (16) and (24).]

Less than a decade after Kennedy and Mitra reported their observations about puberty in female rats, Frisch and Revelle proposed that a similar relationship between body weight and the timing of puberty exists in human adolescents. Based on epidemiologic data interpolated from growth curves, these investigators concluded that early and late maturing girls experience menarche at different ages and heights but at similar body weights. Therefore, Frisch and Revelle hypothesized that menarche is triggered when girls attain a critical body weight (17). A few years later, the hypothesis was revised to contend that a minimum percentage of body fat is necessary for the initiation and maintenance of ovulation and menstrual cycles in women (18). In accord with the notion that puberty depends on the attainment of a critical level of fatness in human females, Frisch et al. (171) subsequently demonstrated that female rats fed either a low- or high-fat diet experience first estrus at different ages and body weights but at similar body compositions.

Although the "critical body weight" and "critical body fat" hypotheses have received considerable attention over the years, many investigators have expressed doubts about whether the apparent relationship between gross body composition and ovulatory cycles reflects a causal relationship between the phenomena or simply a close correlation that is not easily dissected from some other mechanistic link. On the one hand, the critical body fat hypothesis offers a relatively simplistic (and appealing) model for how the brain monitors the body's energetic status. Because adipose tissue is the principal form of stored energy in the body, the size of fat stores serves as a reasonably accurate reflection of energy availability in some mammals (particularly large mammals that possess significant amounts of body fat). However, Frisch's conclusions were based solely on correlational observations, which do not provide evidence of a direct causal relationship between the attainment of a certain body composition and the occurrence of ovulatory cycles. Although the notion of a minimum fat requirement for reproductive function has proven to be correlated in certain populations of amenorrheic women who possess very low levels of body fat (18), most evidence obtained from studies of women and experimental animals demonstrates that the attainment of a threshold body weight or level of adiposity is neither necessary nor sufficient for the onset or maintenance of ovulatory cycles (172). Thus, although the HPG axis appears to be sensitive to fluctuations in body fat content, reproductive function is not strictly dictated by the presence or absence of a minimum level of body fat.

Kennedy and Mitra speculated about the existence of a chemical signal(s) that conveys information about energy intake, energy metabolism, and/or energy storage to the neural circuits that regulate energy balance (16,173). These putative chemical signals could originate in adipose tissue or in other tissues that play a role in energy metabolism (e.g., pancreas, liver, thyroid, etc.) and could reflect short-term or long-term energy balance. Several different hypotheses have arisen regarding the identity of these putative chemical signals. According to the "adipostat" hypothesis, the brain receives information about peripheral energy stores by way of a chemical signal that carries information about the body's adipose reserves; the brain in turn alters food intake and metabolism to maintain a constant level of body fat (173,174). Although unproven when it was initially proposed more than a half century ago, there is now compelling evidence to support the idea that bloodborne signals from the periphery act on the brain to regulate body weight. Conceivably, the same metabolic signal(s) that regulates feeding and metabolism to maintain body weight homeostasis could also impinge on the neuroendocrine reproductive axis. Another mechanism by which the brain could receive information about nutritional status is by monitoring the availability of the metabolic fuels that serve as substrates for energy metabolism (9,175). In addition to metabolic fuels and hormones, there is also evidence suggesting that neural input from the periphery plays a role in transmitting metabolic information to the central pathways controlling energy homeostasis and reproduction (176,177).

In this section, we discuss several signaling molecules that have been implicated in reporting metabolic status to the HPG axis, focusing primarily on bloodborne metabolic cues. Our goal here is not to provide an exhaustive list of all the hormones and metabolites that may be involved in the regulation of energy balance and reproductive function, but rather to provide a detailed accounting of some of the most promising candidate molecules for linking metabolism and reproduction. Several caveats should be kept in mind while reading this section. First, like all of science, this should be viewed as a status report of work in progress, because the field is advancing at an exceedingly rapid pace. Second, it now seems highly improbable that a single metabolic signal is responsible for conveying metabolic information to the brain for the regulation of either body weight or reproduction. Rather, it would appear that the brain modulates energy homeostasis and reproductive output based on its ability to integrate multiple metabolic cues. Third, it can often be extremely difficult to distinguish between the actions of different metabolic signals, because changes in any one metabolic variable rarely occur in isolation of changes in other metabolic variables. Finally, given the diverse energy requirements and reproductive strategies of mammals, it is implausible that all mammalian species rely on the same suite of metabolic signals. Thus, great caution should be exercised about forming generalizations across species on this matter.

Leptin

Leptin as a Metabolic Signal to the Brain

The most direct means by which a bloodborne factor could relay information about the size of fat reserves to the brain would be via a hormonal signal that originates in adipose tissue itself. For more than 40 years, investigators had speculated about the existence of a fat-derived bloodborne signal that communicates nutritional information to the CNS, but their suppositions suffered for lack of any plausible candidate molecule to fulfill such a function (173). The discovery of the adipocyte-derived hormone leptin in 1994 generated hope that this might be the long-sought humoral link between body fat and the central pathways controlling energy homeostasis and reproductive processes.

Leptin is the protein product of the *ob* gene, which is highly conserved across vertebrate species (178). Leptin is primarily produced by adipocytes, where its synthesis and secretion are correlated with adipose mass and body mass index (178–180). Tissue *ob* mRNA expression and plasma levels of leptin are reduced when body fat stores are diminished by fasting or chronic food restriction. After a period of undernutrition, plasma leptin concentrations are restored by refeeding and subsequent body weight gain (179–181). Plasma leptin is typically elevated in obese individuals (179,180) and is reduced in women with anorexia nervosa (182) and in highly trained athletes with low percentages of body fat (183–185). In rodents and humans, leptin exhibits a diurnal pattern of secretion that is influenced by meal patterns (186–188). Because fluctuations in circulating levels of leptin reflect both short- and long-term alterations in energy availability, leptin is poised to relay information about prevailing metabolic status to various physiological systems.

Leptin's physiological effects are mediated by its interactions with leptin receptors (Ob-R), of which there are at least six isoforms (189–191). Although the functional roles of each of the six isoforms are only partially understood, Ob-Rb is believed to be the primary mediator of leptin's actions, because it is the only isoform that is capable of activating the full complement of leptin's downstream signaling pathways (192,193). Ob-R mRNA is expressed in several peripheral organs, including liver, brown and white adipose tissue, skeletal muscle, pituitary, ovary, testis, and uterus, and leptin exerts direct actions on many of these tissues in vitro (194,195). The presence of Ob-Rb mRNA in various regions of the brain indicates that the brain is also a leptin-sensitive tissue. Among the areas of the brain that express Ob-Rb are several hypothalamic nuclei that have been implicated in the control of body weight and reproduction (196–199). Although leptin is capable of crossing the blood–brain barrier via a saturable transport system (200), the presence of Ob-R immunoreactivity in several of the circumventricular organs implicates an alternate route by which leptin can modulate neuronal function (201).

Leptin and Energy Homeostasis

Leptin's effects on energy homeostasis are mediated by its combined actions on feeding, thermogenesis, glucose and lipid metabolism, and physical activity (202–204). Exogenous administration of leptin reduces body weight and adiposity both by suppressing food intake and enhancing energy expenditure (205–208). Leptin modulates energy intake and energy expenditure when it is administered directly into the brain, often at much lower doses than those that are required for leptin to alter feeding and metabolism when it is injected systemically, which implicates a central mode of action (203,205,209–211). Collectively, these observations demonstrate that leptin is ideally situated to serve as the afferent limb of a negative feedback circuit between fat reserves and the brain. When adipose mass increases, the attendant increase in plasma levels of leptin signals the brain to reduce food intake and increase energy expenditure, with the net result of reducing fat depots back to their initial size.

Leptin and Reproduction

In addition to leptin's ability to provide metabolic information to regions of the brain that control feeding behavior and autonomic function, leptin is also believed to relay nutritional information to multiple neuroendocrine systems, including the HPG axis (212). Over the past 10 years, considerable attention has been devoted to elucidating not only whether leptin is a critical component of the control systems that regulate reproductive processes, but also whether leptin functions as a metabolic cue synchronizing reproductive activity with nutritional status. In vivo, leptin stimulates the secretion of GnRH (213) and gonadotropins (214,215). Similar stimulatory effects of leptin have also been observed in hypothalamic and pituitary explants in vitro (214,216). In contrast to its stimulatory effects on hormone secretion at the levels of the hypothalamus and pituitary, leptin exerts an inhibitory effect on the synthesis and secretion of sex steroids in cultured ovarian and testicular tissue (217,218).

The extent to which leptin stimulates various indices of reproductive function often depends on the nutritional status of the animal in question. Although leptin's effects on HPG function are often subtle in well-fed animals, the effects of exogenous leptin administration are usually more pronounced in animals that display signs of reproductive impairment as a consequence of acute or chronic malnutrition. Because reduced caloric intake is associated with reductions in plasma leptin (179,181,212), low circulating levels of leptin could theoretically provide a signal of poor nutritional status to the HPG axis. In rats (63,219), mice (212), sheep (220), cattle (221), and monkeys (222) [but also see (223)], central or peripheral infusions of leptin to fasted animals attenuate or completely prevent the reductions in LH secretion that normally accompany acute food deprivation (Fig. 6). In female mice and hamsters, concomitant leptin replacement attenuates or prevents the fasting-induced inhibition of ovulation, estrous cyclicity, and sexual receptivity (212,224). In healthy men, administration of leptin during a 72-hour fast prevents the decline in testosterone secretion and LH pulsatility that are normally observed during a fast of this duration (225). In addition to the ability of exogenous leptin treatment to prevent impaired reproductive function resulting from acute food deprivation, leptin is also capable of partially restoring LH secretion in animals that have been subjected to chronic food restriction (226). In summary, these observations demonstrate that artificially maintaining plasma leptin concentrations at levels equal to or greater than those that occur in well-fed animals prevents many of the reproductive deficiencies that are commonly observed in undernourished animals. Thus, it would seem that the reproductive axis interprets high circulating levels of leptin to signify that the animal's metabolic status is conducive for reproductive success.

Rodent Models of Impaired Leptin Action

Various rodent models of leptin deficiency or impaired sensitivity to leptin have provided insight into the functional significance of leptin with respect to energy balance and HPG function. Early investigations of obese (*ob/ob*) mice provided a glimpse of

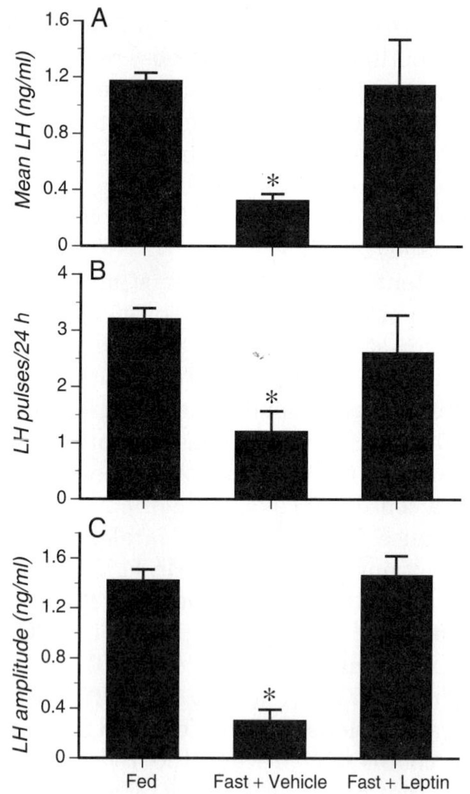

FIG. 6. Leptin reverses the inhibitory effect of food deprivation on pulsatile LH secretion. **(A)** Mean plasma LH concentrations, **(B)** LH pulse frequency, and **(C)** LH pulse amplitude in fed and fasted estradiol-treated OVX female rats. Serial blood samples were obtained while the rats fed normally (Fed) and again when the same animals were fasted for 48 hours while receiving peripheral injections of the vehicle (Fast + Vehicle) or leptin (3 μg/g; Fast + Leptin). *$P < 0.05$ vs. Fed and Fast + Leptin groups. [Redrawn from (63).]

the consequences of life without leptin, even before investigators learned the nature of these animals' genetic defect (227). *Ob/ob* mice are leptin deficient as a consequence of a nonsense mutation in the leptin gene, which results in the production of a truncated nonfunctional protein (178). In addition to their pronounced obesity, *ob/ob* mice exhibit many other metabolic, endocrine, and behavioral abnormalities, including hyperphagia, reduced metabolism, diabetes, and infertility (227–229).

Signs of reproductive impairment are evident at all three levels of the HPG axis in *ob/ob* mice. Adult male *ob/ob* mice are hypogonadal, have reduced basal plasma levels of gonadotropins and testosterone, and display increased sensitivity to the negative feedback effects of testosterone, all of which are characteristic of a prepubertal state (230). Female *ob/ob* mice share many of the same reproductive abnormalities as their male counterparts, in addition to their failure to ovulate or display normal estrous cycles (231). Low levels of circulating gonadotropins in the presence of reduced secretion of gonadal steroids implicate hypothalamic failure as the underlying cause of persistent sexual immaturity and sterility in *ob/ob* mice. The ovaries of female *ob/ob* mice are capable of being stimulated by exogenous gonadotropins, which provides further support for the notion that *ob/ob* mice lack adequate gonadotropin secretion from the pituitary (231). Studies that have examined whether the insufficient gonadotropin secretion in *ob/ob* mice is attributable to impaired GnRH secretion from the hypothalamus and/or reduced pituitary sensitivity to GnRH have yielded mixed results. Compared with lean mice, *ob/ob* mice have elevated hypothalamic GnRH content, and gonadotropin secretion is normal, enhanced, or reduced in response to exogenous GnRH stimulation (232–234). Although *ob/ob* mice on the C57BL/6 strain are sterile, approximately 40% of male *ob/ob* mice crossed onto a BALB/cJ background are reproductively competent, thus demonstrating the importance of modifier genes in determining the reproductive phenotypes of leptin-deficient mice (235).

Many of the metabolic and reproductive abnormalities of *ob/ob* mice are attenuated or reversed by exogenous leptin treatment. Peripheral or central injections of recombinant leptin significantly reduce food intake, body weight, and adiposity in *ob/ob* mice (205–208). Leptin-treated *ob/ob* mice lose more weight than *ob/ob* mice that are pair fed to the leptin-treated animals, suggesting that leptin reduces body weight in *ob/ob* mice not only by suppressing food intake but also by enhancing energy expenditure (206). Repeated injections of leptin also improve reproductive function in *ob/ob* mice. Fourteen days of peripheral leptin injections result in increased circulating gonadotropin concentrations and reproductive organ weights as well as in improved histological signs of gametogenesis and steroidogenesis in male and female *ob/ob* mice (236) (Fig. 7). Longer regimens of daily leptin treatment completely correct the sterility phenotypes of both sexes of *ob/ob* mice (237,238). Despite two reports in the literature that food restriction can sometimes induce fertility in male *ob/ob* mice (239,240), neither male nor female *ob/ob* mice that are pair fed to leptin-treated *ob/ob* mice exhibit any obvious signs of improved reproductive function (237,238). Notwithstanding the two early reports to the contrary, the fact that food restriction alone usually does not elicit activation of the reproductive system in *ob/ob* mice would seem to indicate that the improved reproductive function in leptin-treated *ob/ob* mice is a direct consequence of leptin treatment rather than a secondary effect of the weight loss and improved metabolic condition resulting from long-term administration of leptin.

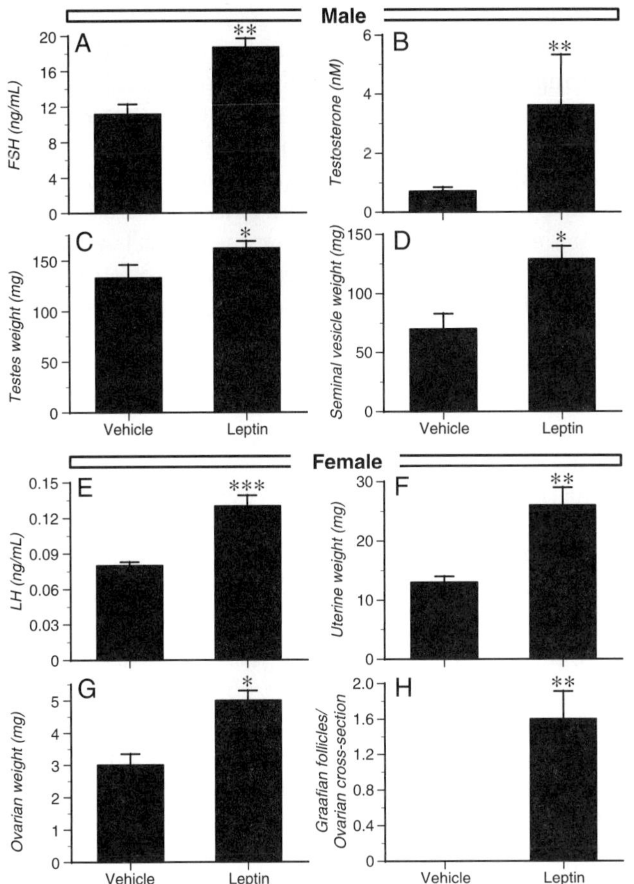

FIG. 7. Stimulatory effects of long-term leptin administration on (**A**) serum FSH concentrations, (**B**) serum testosterone concentrations, (**C**) testes weight, and (**D**) seminal vesicle weight in adult male *ob/ob* mice and (**E**) serum LH concentrations, (**F**) uterine weight, (**G**) ovarian weight, and (**H**) folliculogenesis in adult female *ob/ob* mice. Mice received two daily peripheral injections of leptin (50 μg) or the vehicle alone for 14 days. To avoid the confounding effects of reduced food intake in the leptin-treated animals, the vehicle-treated mice were pair fed with the leptin-treated mice (i.e., the vehicle-treated mice were food-restricted such that they ate an equivalent amount of food as the leptin-treated mice). *P <0.05, **P <0.005, and ***P <0.0005 vs. vehicle group. [Redrawn from (236) with permission. Copyright 1996, The Endocrine Society.]

Leptin's critical role in maintaining energy homeostasis and reproductive function is further highlighted by various rodent models of impaired leptin signaling. Diabetic (*db/db*) mice and fatty (*fa/fa*) rats have dramatically elevated plasma levels of leptin but are insensitive to leptin's actions due to inactivating mutations in the Ob-R gene (190,241,242). Mutant *db/db* mice and *fa/fa* rats possess many of the same metabolic, endocrine, and reproductive anomalies as *ob/ob* mice, including obesity and infertility (228,243–247). Several key observations underscore the importance of central leptin signaling for the support of both energy homeostasis and reproduction. First, *db/db* mice bearing a transgenic restoration of Ob-Rb expression targeted specifically to neurons are less obese than *db/db* mice, and second, the male transgenic mice (but curiously not the female transgenic mice) are fertile (248). Taking an opposite approach, Cohen and colleagues (249) generated transgenic mice that possess a neuron-specific disruption of Ob-R. Variable degrees of adiposity are observed in these animals, in an inverse correlation with the amount of residual Ob-R expression in the hypothalamus. Although fertility was not investigated in these mutant animals, male mice with low expression of hypothalamic Ob-R have normal testosterone levels, but female mice with similarly low hypothalamic Ob-R expression have reduced circulating estradiol (249). Finally, mice bearing a single point mutation in the intracellular loop of Ob-Rb that renders it incapable of activating signal transducer and activator of transcription 3 (STAT3) signaling pathways are obese and hyperphagic, yet 40% of the female mutants are fertile. This suggests that leptin's effects on body weight, but not reproduction, are mediated by STAT3 signal transduction pathways (250), indicating that the molecular signaling pathways for leptin's actions on energetics and reproduction are divergent proximal to STAT3 (at least in females).

Leptin and Ob-R Mutations in Humans

Although plasma leptin levels are elevated in most cases of human obesity, a few individuals with mutations in the leptin or Ob-R genes have been identified. Montague et al. (251) documented the clinical manifestations of leptin deficiency in two young children, both of whom were morbidly obese. When one of the children was treated with recombinant human leptin for 1 year, she experienced rapid and sustained weight loss, which persisted for the entire duration of leptin treatment. Although the girl was of prepubertal age when leptin treatment commenced, her serum levels of LH and FSH gradually increased over the course of leptin treatment and were exhibiting nocturnal patterns of pulsatile release after 12 months of treatment, consistent with an early stage of puberty (252). Strobel and colleagues (253) investigated the effects of leptin deficiency on HPG function in two adult leptin-deficient family members. Neither subject had progressed through puberty; the 34-year-old female had primary amenorrhea and the 22-year-old male was hypogonadal and had low basal testosterone levels. The male patient displayed normal gonadotropin release in response to exogenous GnRH, implicating hypothalamic hypogonadism as the primary defect (253). Daily leptin treatment for 18 months reversed the hypogonadism and triggered the onset of puberty in the male subject and induced regular ovulatory menstrual cycles in the female (254). Three morbidly obese sisters possessing a mutation in the Ob-R gene were also identified. Similar to the leptin-deficient adults, none of the women had spontaneously progressed through puberty. All three women were amenorrheic with low estradiol and low to normal levels of gonadotropins, again indicative of hypogonadotropic hypogonadism (255).

Leptin and Puberty

As mentioned in preceding sections, leptin-deficient mice and humans linger in a prepubertal state during adulthood, as do animals that are subjected to moderate or severe caloric restriction before the onset of puberty (25,29,230,253). In both of these circumstances, plasma leptin levels are markedly reduced. Elevating leptin levels (either exogenously, by administering leptin to leptin-deficient individuals, or endogenously, by refeeding food-restricted animals) rapidly induces the onset of sexual maturation (25,29, 237,238,252,254). Collectively, these observations suggest that attainment of some critical threshold of circulating leptin is a prerequisite for the initiation of puberty. Whether leptin serves as a trigger for sexual maturation or merely plays a permissive role in initiating the onset of puberty has been the subject of considerable debate. If leptin serves as the primary activational signal to the neuroendocrine centers that initiate maturation of the HPG axis, then one would expect to observe an increase in leptin secretion and/or enhanced sensitivity to leptin's actions before the onset of puberty. Although some reports in the literature demonstrate increased leptin secretion during the peripubertal period in rats (256,257), mice (258), rhesus monkeys (259), and humans (260–263), many others have not been able to detect an increase in leptin secretion immediately before the onset of sexual development in the same species (186,264–275). Ahima et al. (264) observed similar serum levels of leptin between mice of peripubertal age who exhibited vaginal opening and those who had not yet experienced this pubertal milestone. Similarly, serum leptin levels were not different between pubertal male macaques and macaques of similar age in which puberty was delayed by the neonatal administration of a GnRH antagonist (273). Furthermore, it is unlikely that the pubertal initiation of pulsatile gonadotropin secretion is a consequence of enhanced hypothalamic sensitivity to circulating leptin, because Ob-Rb mRNA expression in the hypothalamus of rats does not increase in temporal association with the onset of puberty (267,276).

If leptin does indeed serve as a trigger for sexual maturation, then one would also predict that administering leptin to prepubertal animals would advance the timing of puberty. Several groups have tested this hypothesis in a variety of species, but again these experiments yielded equivocal results. Two groups independently demonstrated that daily systemic leptin injections that commence at the time of weaning significantly advance the onset of puberty in female mice. In these experiments, the mean age at which vaginal opening, first estrus, and mating behavior occurred was advanced by 1–4 days compared with the ages at which each of these pubertal events occurred in vehicle-treated mice (258,264). In the female rat, a similar time course of leptin treatment results in an earlier age at vaginal opening and first ovulation as well as in earlier pubertal rises in plasma LH, FSH, and progesterone concentrations compared with vehicle-treated rats (277). The observation that transgenic mice that overexpress leptin have a mean age at vaginal opening that is approximately 2 days earlier than that observed in wild-type mice is consistent with the notion that high plasma levels of leptin trigger the onset of puberty in rodents (278). However, other investigators have observed that leptin treatment does not advance the timing of sexual maturation beyond when it would normally occur in well-fed mice and rats (267,279). On the other hand, leptin has been consistently shown to prevent the delay in puberty that is caused by food restriction in prepubertal female rats (256,279) and mice (267) when it is administered centrally or peripherally (Fig. 8). Discordant results have also been reported with respect to leptin's effects on the timing of puberty in rhesus macaques. In agonadal juvenile male monkeys, infusion of leptin for 16 days significantly elevated circulating leptin levels but did not induce precocious pulsatile GnRH secretion (which was assessed indirectly by measuring LH release) (280). However, in intact female juvenile macaques, a significantly longer period of leptin treatment (18 months) advanced the age at which increases in LH and estradiol secretion were observed as well as the average age at menarche, compared with untreated macaques (281).

Currently, it is difficult to draw any firm conclusions about the exact nature of leptin's role in determining the onset of sexual maturation. There are too many inconsistencies in the literature with respect to both the developmental patterns of endogenous leptin secretion and leptin's ability or inability to induce puberty before the time it would normally occur in well-fed animals. It seems likely that the differences in experimental findings and conclusions are attributable to differences between species, sex,

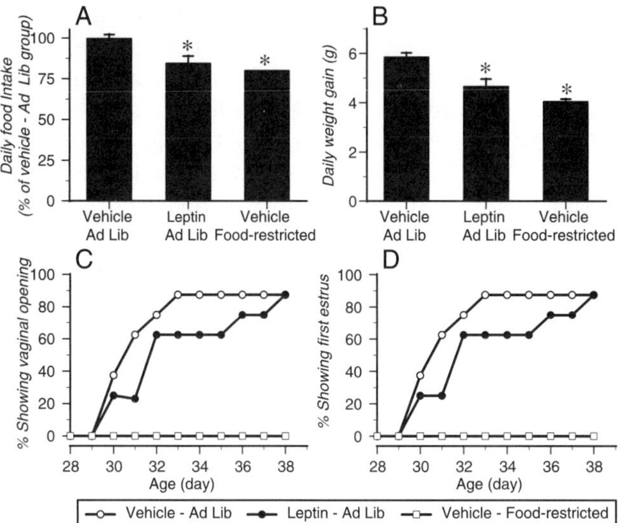

FIG. 8. Leptin treatment prevents the delay in puberty due to food restriction in female rats. (**A**) Daily food intake, (**B**) daily body weight gain, (**C**) timing of vaginal opening, and (**D**) timing of first estrus in female rats that were fed ad libitum and received twice daily peripheral injections of the vehicle (Vehicle-Ad Lib) or leptin (6.3 μg/g; Leptin-Ad Lib) beginning on postnatal day 18. A third group of vehicle-treated rats was food restricted (Vehicle-Food-restricted), such that they consumed the same amount of food as the leptin-treated animals. Despite the fact that food intake and body weight gain did not differ between the Leptin-Ad Lib and Vehicle-Food-restricted groups, the timing of vaginal opening and first estrus were significantly delayed only in the Vehicle-Food-restricted group. $^*P < 0.001$ vs. Vehicle-Ad Lib group. [Redrawn from (279) with permission. Copyright 1997, The Endocrine Society.]

and experimental designs that were used across the various studies. The outcomes of the several experiments in which leptin did not significantly advance the timing of puberty in well-fed animals argue against the hypothesis that leptin serves as a trigger for the onset of puberty. However, the fact that leptin reliably prevents the delay in puberty that normally occurs as a consequence of mild or moderate food restriction instead implicates a permissive role for leptin in controlling the onset of puberty, wherein leptin functions as a metabolic gate for the mechanisms that initiate puberty. When circulating leptin levels exceed a critical threshold, the gate opens and puberty occurs. However, if plasma leptin remains below this threshold (due to genetic leptin deficiency or an unfavorable nutritional plane), the gate remains closed and the HPG axis remains quiescent.

Leptin and Pregnancy/Lactation

In the female, leptin's role in the regulation of reproductive function extends to pregnancy and lactation. In rodents and humans, leptin levels increase during the later stages of pregnancy (282,283). Depending on the species in question, the increased circulating leptin concentrations during late pregnancy result from enhanced leptin secretion from a progressively increasing mass of adipose tissue and/or the synthesis and release of leptin from the fetus and placenta (284–286). Females exhibit a certain degree of leptin resistance during pregnancy (287,288). Although increases in circulating leptin levels would normally trigger a reduction in food intake, females instead display hyperphagia during pregnancy (289). Leptin resistance is advantageous in this physiological context, because it permits pregnant females to consume greater than normal amounts of food so that they can invest extra energy toward fetal growth and development as well as store extra calories in preparation for the large energetic demands of lactation. Although leptin-treated female *ob/ob* mice do not require leptin for the entire duration of pregnancy, there remains some uncertainty about whether leptin is essential during the early stages of gestation, including decidualization and implantation. Although one study has shown that leptin-treated female *ob/ob* mice cannot carry pups to term if leptin treatment is withdrawn before 6.5 days after copulation (290), another study has shown that withdrawing leptin immediately after conception (0.5 days after copulation) does not have deleterious effects on the ability of female *ob/ob* mice to bear young (288). Maternal plasma leptin concentrations decrease after birth as a consequence of the metabolic drain of milk production during lactation (157,283,286). However, the importance of leptin during lactation is underscored by the fact that female *ob/ob* mice usually fail to lactate, especially if leptin treatment is withdrawn before parturition (237,288,290). Leptin appears to be essential for mammary gland development during the later stages of pregnancy, because female *ob/ob* mice that are treated with leptin up through 14.5 days after copulation are unable to lactate, even if leptin treatment is reinstated after parturition (290).

Sites of Leptin's Actions

Leptin appears to be capable of exerting effects at all three levels of the HPG axis. Several lines of evidence support a role for central leptin signaling in the neuroendocrine control of reproduction. First, infusion of leptin directly into the brain stimulates various indices of reproductive function (214,226, 256,291,292). Second, the stimulatory effect of intracerebroventricular (ICV) leptin injection on LH secretion in female rats can be blocked by administration of GnRH antiserum (291). Third, female mice and rats with impaired leptin actions in the brain (either by a neuron-specific genetic disruption of Ob-Rb or by ICV injection of leptin antiserum) exhibit decreased pulsatile LH secretion, disrupted estrous cyclicity, and reduced plasma levels of estradiol (249,293). Fourth, low doses of leptin stimulate GnRH secretion from hypothalamic explants in vitro (214), and intrahypothalamic leptin perfusion induces GnRH release in food-deprived steroid-primed OVX female rats in vivo (213). Ob-R mRNA and protein are expressed in several hypothalamic nuclei, including those that play a role in the regulation of gonadotropin release and reproductive behaviors (196–198,222,294). Although GnRH-expressing GT1-7 cells express Ob-R mRNA (195,295), hypothalamic GnRH neurons do not appear to express Ob-R mRNA or immunoreactivity (222,294), thus implicating unidentified afferent input(s) to GnRH neurons in mediating leptin's effects on GnRH release.

Leptin also exerts direct effects on the pituitary and gonads. Ob-R mRNA and protein are expressed in the anterior pituitary (195,296) but have not been detected on gonadotropes (296). Leptin stimulates LH and FSH release from cultured rat pituitary cells in a dose-dependent manner, at levels comparable with those induced by GnRH (214). Leptin also has direct effects on the gonads, as evidenced by the presence of Ob-R mRNA in the ovary and testis (195,297), and the various effects of leptin on the synthesis and secretion of gonadal steroids in vitro (217,218).

Summary and Contradictions

Although leptin's actions on feeding and body weight regulation are well characterized, the precise nature of leptin's role in reproductive physiology is less understood. Scientists have devoted considerable efforts to ascertaining whether leptin is necessary and/or sufficient for normal reproductive function. Sexual maturation and fertility are severely compromised when leptin levels are markedly reduced (e.g., as a consequence of a sustained energy deficit) and when leptin is altogether absent (e.g., in *ob*/*ob* mice and in humans with leptin gene mutations), thus lending support to the supposition that leptin is obligatory for the normal maturation and functioning of the reproductive system. However, a number of recent reports have provided evidence to the contrary. For example, reproductive function is sometimes preserved in lipodystrophic patients who have very low plasma leptin levels (298). In female hamsters, pronounced fasting-induced reductions in serum leptin concentrations are not always accompanied by reproductive impairment (299). Although it is plausible that even very low levels of leptin are capable of providing a stimulatory signal to the reproductive axis under these circumstances, this explanation cannot account for the observation that a substantial percentage of leptin-deficient male *ob*/*ob* mice that are crossed onto a BALB/cJ background are fertile (235). This latter observation is inconsistent with the notion that normal reproductive function is dependent on achieving and maintaining a critical threshold level of leptin and consequently raises doubts as to whether leptin is truly essential for normal reproductive function.

The question of whether leptin is sufficient to activate the reproductive axis at puberty and maintain reproductive function in adulthood remains unsettled. On the one hand, treatment with exogenous leptin significantly improves reproductive function in leptin-deficient mice and humans and attenuates or reverses the inhibitory effects of fasting or moderate food restriction on sexual maturation and fertility in multiple species. However, treating animals with high doses of leptin typically does not advance the timing of puberty beyond when it would normally occur in well-fed animals. Although exogenously administered leptin can prevent the delay in the onset of puberty due to moderate food restriction, it cannot fully overcome the inhibitory effect of severe food restriction on the timing of puberty in female rats (279). Furthermore, leptin (administered either centrally or peripherally) is incapable of preventing fasting-induced anestrus in Syrian hamsters when intracellular glucose or fatty acid oxidation is blocked (224,300). Thus, elevated plasma or CNS concentrations of leptin are not always sufficient for normal reproductive function. Although there are still many unanswered questions about leptin's action on the reproductive axis, the weight of evidence to date suggests that leptin plays a permissive role in the regulation of reproductive function. According to this model, reproductive efforts are permitted when leptin levels are above a particular threshold but prohibited (usually) when leptin concentrations are below this threshold.

Leptin's complicated effects on peripheral metabolism have made it difficult to determine whether leptin's effects on reproduction are mediated directly by leptin itself or indirectly through leptin's influence on other metabolic factors (e.g., insulin and fuels), which in turn influence the reproductive axis. The presence of leptin receptors at all three levels of the HPG axis supports the contention that leptin has direct effects on reproductive function. However, the fact that leptin alters the intracellular availability and oxidation of glucose and fatty acids, at least in part through central mechanisms, raises the possibility that the observed effects of exogenously administered leptin on the reproductive axis are actually secondary to an increase in metabolic fuel availability and utilization (301). The latter hypothesis is bolstered by observations that peripherally or centrally administered leptin prevents the occurrence of anestrus in Syrian hamsters that are fasted for 48 or 60 hours but is incapable of preventing fasting-induced anestrus when metabolic fuel utilization is blocked by the administration of metabolic inhibitors (224,300). Based on our current knowledge, it seems probable that leptin influences reproduction through multiple signaling pathways and mechanisms, both directly and indirectly, although its physiological significance remains shrouded in uncertainty.

Insulin

Insulin as an Adiposity Signal to the Brain

Before the discovery of leptin, insulin was widely regarded as the most promising candidate for a humoral link between energy reserves and the reproductive axis, because insulin was the only identified metabolic hormone that fulfilled most or all the criteria for an adiposity signal. After leptin's discovery, insulin lost its status as "the" prime adiposity/satiety factor; however, today insulin is enjoying something of a renaissance as a key component in the constellation of metabolic hormones with signaling effects on the brain. Whereas insulin's effects on carbohydrate, fat, and protein metabolism in the periphery have been well established for many years, insulin's actions within the CNS have only recently begun to be elucidated. The results of several recent experiments have identified a role for central insulin signaling in the regulation of body weight and reproductive function, which has consequently sparked a renewed interest in insulin as a potential bloodborne link between nutrition and reproduction.

Insulin is synthesized and secreted by pancreatic beta cells. In the periphery, insulin functions as an anabolic hormone, promoting the cellular uptake and storage of metabolic fuels while concomitantly inhibiting the mobilization of energy reserves. In the presence of high circulating levels of insulin, energy that is derived from food is stored in skeletal muscle, liver, and adipose tissue. Although multiple neural, endocrine, and paracrine factors regulate insulin secretion, insulin is primarily released in response to the ingestion of nutrients (especially glucose). Consequently, insulin secretion varies in response to daily meal patterns (302). An elevation in the plasma level of insulin during and immediately after a meal permits an animal to store excess consumed energy for later use. Between meals or during extended periods of food deprivation or restricted caloric intake, plasma levels of insulin fall, triggering a shift from an energy-storing to an energy-mobilizing state. Despite the periodic fluctuations in circulating insulin levels that occur in response to nutrient intake, basal insulin secretion is positively correlated with adiposity (303). Because steady-state plasma insulin levels reflect the quantity of an animal's energy reserves, insulin has the potential to transmit information about long-term metabolic status to other physiological systems, including the reproductive axis.

Insulin and the CNS

The brain was initially believed to be an insulin-insensitive tissue, based on observations that the brain does not require insulin for glucose uptake or metabolism (304,305). However, many key discoveries over the past several decades have challenged this notion. Insulin (306), insulin receptors (307–309), and multiple insulin receptor signal transduction proteins (310–314) are expressed throughout the CNS. Insulin has been detected in the cerebrospinal fluid (CSF) of many species (315–317), where it is taken up by hypothalamic neurons lining the third cerebral ventricle (318). Although there is some evidence for local insulin synthesis within the CNS, most of the insulin present in the brain is probably of peripheral origin (317). Insulin is capable of entering the brain via two routes: receptor-mediated transport across the blood–brain barrier (316) and uptake through the circumventricular organs (319,320). Under most circumstances, concentrations of insulin in the CSF are positively correlated with basal plasma insulin levels (321,322). Fluctuations in plasma insulin concentrations that occur in response to nutrient intake, caloric deficiency, or glucose or insulin administration elicit parallel changes in the concentration of insulin in the CSF (321–324). However, because concentrations of insulin in the CSF fluctuate less dramatically than plasma insulin concentrations, insulin levels in the brain are thought to reflect the "integrated" insulin state of the animal and are therefore a more accurate index of an animal's steady-state nutritional plane (321).

Insulin and Energy Homeostasis

In addition to insulin's effects on energy storage and energy utilization in peripheral target tissues, insulin also influences energy balance via central mechanisms. Woods and colleagues (325) first revealed a role for insulin in the central regulation of body weight by demonstrating that infusions of insulin into the lateral cerebral ventricle of baboons resulted in rapid and sustained reductions in food intake and body weight. Likewise in the rat, infusions of insulin directly into the cerebral ventricles or into discrete hypothalamic nuclei inhibit feeding and reduce body weight (326–330). Central insulin administration also increases brown fat thermogenesis, which suggests that insulin reduces body weight through a combination of decreasing food intake and enhancing energy expenditure (331).

Experimental models and clinical disorders of compromised insulin action have provided insight into the physiological significance of central insulin signaling to energy homeostasis. Acute fasting or prolonged periods of caloric deprivation result in reductions in plasma and CSF insulin concentrations, which could contribute to the pronounced hyperphagia that occurs when refeeding is permitted (322–324). Humans and laboratory animals with uncontrolled diabetes mellitus also exhibit pronounced hyperphagia. In rats that are rendered diabetic by administration of streptozotocin (STZ, a toxin that selectively destroys pancreatic beta cells and thus depletes animals of endogenous insulin), central insulin infusion dramatically attenuates the hyperphagia that is normally associated with diabetes (332). Compared with their lean counterparts, obese Zucker rats are hyperphagic and have reduced insulin transport into the brain (333), impaired central insulin signaling (334), and reduced sensitivity to the effects of central insulin administration on feeding and body weight (328,334). Experimental manipulations that interfere with central insulin signaling pathways are also associated with increased food intake. In rats, injection of neutralizing insulin antibodies into the ventromedial hypothalamus increases nighttime feeding (335), and specific down-regulation of insulin receptors in the hypothalamus results in hyperphagia and increased fat mass (336). Hyperphagia and obesity also occur in mice bearing a neuron-specific disruption of the insulin receptor gene (337). Collectively, these observations demonstrate that insulin deficiency, impaired insulin transport into the brain, and/or disrupted insulin receptor signaling are associated with imbalances between energy intake and energy expenditure and are consistent with the notion that insulin signaling within the brain is essential for maintaining energy homeostasis.

Insulin and the Reproductive Axis

Insulin is not only poised to transmit information about the size of peripheral energy stores to the neural pathways that modulate energy intake and energy expenditure, it also has the potential to relay metabolic information to the neuroendocrine pathways that control reproductive function. Several studies have revealed an association between excursions in plasma levels of insulin and alterations in pulsatile LH secretion. Even relatively short periods of food deprivation elicit parallel reductions in plasma insulin levels and LH pulsatility (338,339). Conversely, refeeding after a fast is accompanied by increases in both insulin and LH secretion (340). However, in male rhesus macaques that are refed after a 24-hour fast, the reinstatement of pulsatile LH secretion is not dependent on insulin secretion, because LH pulsatility resumes even when plasma insulin is maintained at low levels by pharmacological intervention (340). Thus, it would appear that although reduced insulin levels might signal the hypothalamus and pituitary to inhibit LH secretion during periods of limited caloric intake, mechanisms other than meal-induced insulin secretion are capable of restoring LH secretion in response to refeeding.

Experiments in which animals are treated with insulin have yielded mixed results with respect to reproductive function. The interpretation of experiments in which insulin is injected peripherally is complicated by the hypoglycemia that results from administration of large doses of insulin. In many species (including humans), systemic insulin administration has inhibitory effects on several measures of reproductive function, including GnRH pulse generator activity (341,342), pulsatile LH secretion (343–348), estradiol secretion (349), timing of the estrogen-induced LH surge (344), estrous cyclicity (350,351), and fertility (351). In many instances, the adverse effects of systemic insulin treatment on reproductive function do not occur if euglycemia is maintained. In OVX ewes, the inhibitory effects of peripheral insulin administration on LH secretion and the timing of the estrogen-induced LH surge are reversed by the simultaneous infusion of glucose (343,344). In rats, the suppressive effect of insulin-induced hypoglycemia on LH pulse frequency is prevented by the infusion of glucose directly into the hypothalamus (345), and glucose infusions prevent insulin from reducing the electrical activity of the GnRH pulse generator in rats and goats (342,352). Notwithstanding, there is also evidence suggesting that systemic insulin administration alters reproductive function through glucose-independent mechanisms. For example, increasing peripheral insulin levels in mice results in increased LH secretion, regardless of circulating glucose concentrations (353).

Injecting insulin directly into the brain (at doses that do not alter peripheral insulin or glucose concentrations) is another way to circumvent insulin-induced hypoglycemia. However, experiments in which animals have received central insulin infusions have also yielded equivocal results. On the one hand, it has been reported that the central administration of insulin for 4 days increases LH pulse frequency in adult rams (354). However, another study of growth-restricted OVX ewe lambs showed that central infusions of insulin have no effect on any measure of LH secretion and actually decrease mean LH concentrations and LH pulse frequency when it is administered after 2 weeks of refeeding (355). It is difficult to reconcile such different experimental outcomes with a unifying hypothesis for the action of insulin on the neuroendocrine reproductive axis; however, it would appear that under some circumstances insulin can act centrally to influence pulsatile LH secretion, presumably by acting on the GnRH pulse generator, and that this action is confounded by the animal's state of glycemia.

Impaired Insulin Action and Reproduction

Experimental and clinical conditions of impaired insulin signaling have further reinforced an implicit role for insulin in the regulation of reproductive function, if only in a permissive sense. Reproductive dysfunction is a common symptom of uncontrolled diabetes mellitus in humans. Diabetic men are susceptible to impotence, reduced libido, low testosterone levels, and impaired sperm quality (356–360). Women with poorly controlled insulin-dependent diabetes mellitus often experience delayed menarche, secondary amenorrhea, and infertility (361). LH pulse frequency is reduced in diabetic women (362). Both hypothalamic and pituitary defects have been implicated as primary causes of diabetes-induced reproductive impairment, because gonadotropin secretion in response to exogenous GnRH administration is either normal (361) or reduced (363,364) in diabetic individuals.

Experimental animals with diabetes (whether spontaneously occurring or experimentally induced) provide a useful model of diabetes-induced reproductive dysfunction, because they share many of the same reproductive deficits as diabetic humans. STZ-treated diabetic male rats have reduced gonadotropin and testosterone levels, decreased activity of Leydig cell steroidogenic enzymes, disrupted spermatogenesis, and impaired sexual behaviors (365–368). Pituitary responsiveness to GnRH is unimpaired in STZ-treated diabetic male rats (369). GnRH levels are reduced in the medial basal hypothalamus of diabetic male rats (368). In vitro, basal GnRH secretion in diabetic male rats is normal, but KCl-evoked GnRH secretion is markedly reduced (370,371).

Female rats that are rendered diabetic by treatment with STZ or alloxan exhibit a constellation of reproductive deficits, including delayed sexual maturation (372), reduced ovulation (373–375), disrupted estrous cycles (372,375), absent or delayed LH surges (373,374,376), and impaired sexual behaviors (377,378). Although STZ-induced diabetes does not appear to alter the number or activity of GnRH neurons (376) or the hypothalamic GnRH content (379), GnRH content is reduced in the median eminence of diabetic female rats (380). Furthermore, the pituitary of diabetic female rats releases normal (374,379) or reduced (380,381) amounts of LH in response to GnRH, which could be indicative of central and/or pituitary defects.

Many of the reproductive abnormalities associated with diabetes are attenuated or normalized by peripheral insulin treatment. In diabetic ram lambs, the acute withdrawal and subsequent reinstatement of insulin treatment decreases and increases LH pulse frequency, respectively (382) (Fig. 9). Peripheral insulin administration also corrects many of the reproductive deficits observed in STZ-treated rats (365,374,378,383). The ability of insulin to reverse the reproductive defects associated with diabetes reflects, at least in part, insulin's action on the brain. Central insulin treatment prevents the blockade of the steroid-induced LH surge and impaired sexual behaviors that occur in untreated female diabetic rats (376,384), and LH pulse frequency is increased

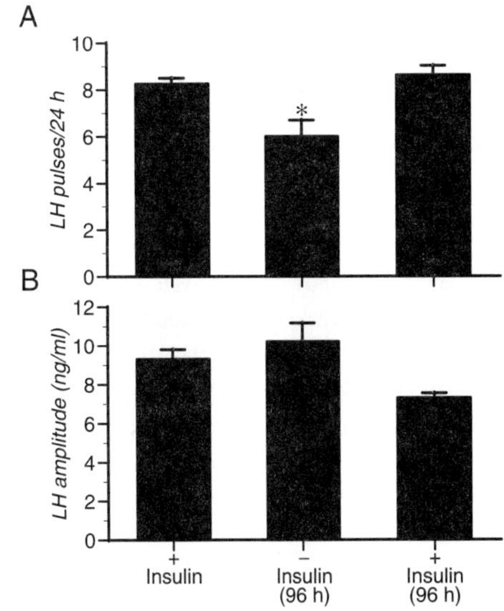

FIG. 9. Parameters of pulsatile LH secretion in insulin-deficient diabetic male lambs with or without systemic insulin supplementation. **(A)** LH pulse frequency and **(B)** LH pulse amplitude in male lambs that were rendered diabetic by STZ treatment at 12 weeks of age. At 26 weeks of age, serial blood samples were obtained from the animals during chronic insulin replacement (+ Insulin), after 96 hours of acute insulin withdrawal (– Insulin 96 h), and after 96 hours of acute insulin resupplementation after insulin withdrawal (+ Insulin 96 h). LH pulse frequency was suppressed by acute insulin withdrawal but was restored when insulin treatment was reinitiated. *P <0.005 vs. all other groups. [Redrawn from (382).]

in diabetic male sheep after 5 days of a continuous central infusion of insulin (385), indicating that insulin exerts a direct stimulatory effect on the GnRH pulse generator. Thus, based on studies of the effects of insulin on reproductive function in diabetic animals, we may infer that insulin acts permissively to sustain pulsatile GnRH and gonadotropin secretion in normal healthy animals; when this control system is disturbed, things run amok.

Impairment of reproductive function also occurs in genetic models of deficient insulin signaling. Female mice bearing targeted deletions of insulin receptor substrate-2 (IRS-2) have impaired fertility, which is accompanied by small ovaries with reduced numbers of primary and antral follicles. Most female IRS-2 null mutants are acyclic and lack corpora lutea. These mutants also have fewer numbers of gonadotropes in the pituitary as well as reduced plasma levels of LH, estrogen, and progesterone (386). Male and female mice carrying a neuron-specific deletion of the insulin receptor gene also display reduced fertility (337). Males show signs of impaired spermatogenesis, and females have reduced numbers of antral follicles and corpora lutea. Both sexes of these mutant animals have reduced circulating levels of LH and display either normal or enhanced LH release in response to administration of a GnRH receptor agonist, which suggests a central origin for their reproductive defects. Thus, an intact insulin signaling pathway in the brain is essential for the normal function of the neuroendocrine reproductive axis.

Sites and Mechanisms of Action

Many of insulin's effects on energy balance and reproductive function occur when insulin is injected directly into the brain. Likewise, disruptions in insulin signaling that are confined to the CNS result in abnormal body weight regulation and reproductive impairment. Together, these observations suggest that the brain is an important locus for insulin's actions on energy homeostasis and reproduction. Insulin receptors and insulin signal transduction proteins, such as IRS-1 and phosphatidylinositol-3 kinase, are expressed in the hypothalamus and notably in regions that have been implicated in the regulation of body weight and GnRH secretion (310,314). Although it is unknown whether GnRH neurons express insulin receptors, insulin receptors are apparently expressed in a GnRH-expressing cell line, GT1-7 cells (387). Insulin increases GnRH gene expression and stimulates its release from cultured fetal and adult rat hypothalamic neurons (353,388). In some circumstances of impaired insulin signaling and associated reproductive dysfunction, exogenous GnRH treatment can elicit normal or enhanced gonadotropin release, suggesting that the primary defect in these conditions is reduced GnRH secretion rather than impaired pituitary sensitivity to GnRH (337,362,369,374,379).

Although it seems likely that insulin exerts a direct modulatory effect on the neural circuits that govern pulsatile GnRH secretion, it is also conceivable that insulin indirectly influences GnRH neurosecretory function by altering the transport and availability of amino acid precursors for neurotransmitter synthesis (389). Monoamine neurotransmitters (e.g., serotonin and the catecholamines) regulate GnRH and gonadotropin secretion under a variety of circumstances [for a review see [390]). The synthesis of monoamine neurotransmitters is substrate limited and is therefore dependent on the delivery of amino acid precursors from the circulation into the brain (391). Tyrosine and tryptophan (the amino acid precursors for the catecholamines and serotonin, respectively) compete with the other large neutral amino acids for transport into the CNS via a common carrier system (391). By virtue of its ability to alter the ratios of amino acids in the circulation, insulin enhances the transport of tyrosine and tryptophan into the brain and increases brain serotonin concentrations and norepinephrine turnover rates (391). A potentially synergistic role for insulin and amino acids in the regulation of gonadotropin secretion was proposed over 20 years ago by Steiner and colleagues (389), who demonstrated that an intravenous infusion of dextrose (which stimulates endogenous insulin secretion) in combination with the essential amino acids (including tyrosine and tryptophan) elicited adult-like patterns of LH secretion in juvenile male macaques, whereas dextrose alone was without effect on gonadotropin secretion. Whether insulin's effects on neurotransmitter synthesis underlie or contribute to the actions of insulin on the neuroendocrine reproductive axis has yet to be directly tested.

Insulin has also been demonstrated to have direct effects on the pituitary and gonads. Insulin binding sites (309), IRS-1, and phosphatidylinositol-3 kinase are all expressed within the adenohypophysis (314), and insulin stimulates both basal and GnRH-induced LH and FSH release from primary pituitary cultures (392). In adult rats, insulin is present in testicular interstitial fluid, and insulin receptors are expressed by Leydig cells (393,394), from which insulin stimulates basal and human chorionic gonadotropin–induced testosterone production in vitro (395,396). In the ovary, insulin is present in follicular fluid, insulin receptors are expressed by granulosa cells, and insulin stimulates ovarian steroidogenesis (397).

Summary

Collectively, the available data suggest that insulin, like leptin, may serve as a humoral link between body fat stores and the central pathways that regulate and coordinate energy homeostasis and reproduction. Our comprehension of insulin's precise role in these processes has been hindered by the multiple complex actions of insulin. When reproductive impairment occurs as a result of manipulated insulin signaling, it is difficult to distinguish between the direct effects of insulin and those effects that are attributable to variables that were themselves altered by insulin. The literature is plagued by several inconsistencies regarding the effects of exogenous insulin administration on reproductive function. More recently, investigators have questioned how well insulin levels reflect body fat stores, given that changes in insulin concentrations often precede significant changes in adiposity (299). The metabolic and reproductive deficiencies that are observed when insulin is absent or its actions are impaired have been more consistent. At this point, the bulk of evidence obtained in various species suggests that insulin and intact insulin signaling pathways are necessary for full reproductive competence; however, insulin by itself is usually insufficient to restore reproductive function in the face of extreme metabolic challenges. Thus, the relative absence of insulin signaling might be a more critical signal to the pathways controlling energy homeostasis and reproductive function than is the presence of abundant insulin. A lack of insulin signaling communicates the perceived inadequacy of energy stores to the brain. However, when insulin is present above some threshold, it serves as a permissive signal to the reproductive axis, informing the brain that the animal has attained a sufficient plane of nutrition to allow reproduction to take place.

Metabolic Fuels

Metabolic Fuels Directly Reflect Energy Availability

Thus far, we have provided examples of metabolic cues that are poised to relay information to the reproductive axis about long-term fluctuations in energy availability. However, a growing body of evidence suggests that reproductive function is also influenced by short-term (i.e., minute-to-minute or hour-to-hour) alterations in metabolic status [for review see (398)]. The rapidity with which the HPG axis responds to an improved metabolic state is illustrated by multiple observations that pulsatile LH secretion commences or resumes within minutes to hours after food-restricted prepubertal or adult animals are given unlimited access to food (29,34,46,69). Under many circumstances of reduced or improved nutrition, reproductive function is altered in the absence of significant changes in body weight or body composition (69,299,399) and can occur independently of changes in circulating levels of leptin or insulin (299,400). These observations suggest that the reproductive axis cannot rely solely on adiposity signals to garner information about prevailing metabolic status but must also be attuned to signals that reflect more immediate shifts in energy availability, such as the availability of metabolic fuels [for review see (9)].

Mammals primarily rely on the intracellular oxidation of glucose and fatty acids to derive the energy necessary to support most physiological processes. Although peripheral tissues may derive energy from both glucose and fatty acid metabolism, the brain utilizes glucose as its principal source of energy. To be considered "available" for intracellular oxidation and thus energy production, metabolic fuels must not only be present in the body in sufficient quantities, they must also be capable of entering cells and feeding into their respective metabolic pathways. Healthy well-nourished animals generally possess sufficient quantities of oxidizable fuels as a consequence of ample intake of nutrients during meals and the presence of substantial energy reserves that can be readily mobilized in between bouts of feeding. Conversely, metabolic fuel availability can be severely compromised in food-restricted animals, depending on the duration and severity of caloric restriction. Regardless of an animal's long-term metabolic status, metabolic fuel availability varies as a function of acute alterations in energy intake, energy expenditure, and the body's ability to use metabolic fuels. The availability of oxidizable metabolic fuels therefore reflects the body's immediate capacity to supply substrates for energy generating pathways and thus has the potential to serve as a signal of short-term energetic status to various physiological systems, including those that regulate energy homeostasis and reproduction.

Metabolic Fuel Availability and Feeding

According to the glucostatic theory of food intake, alterations in intracellular glucose utilization are detected by glucoreceptors in the brain, which relay their information to the neural circuits that modulate feeding behavior. In this model, a reduction in intracellular glucose utilization (glucoprivation) indicates insufficient glucose availability and thus stimulates feeding, whereas an increase in intracellular glucose utilization elicits a perception of satiety (175). Although there are mixed reports in the literature regarding whether exogenous glucose administration restrains food intake (401–403), feeding is reduced by peripheral administration of glucagon, a hormone that increases circulating levels of glucose by promoting hepatic glucose output (404,405). On the other hand, reductions in glucose availability invariably elicit a compensatory drive to increase energy intake. Mammals that are hypoglycemic as a consequence of prolonged food restriction exhibit pronounced hyperphagia when they are subsequently allowed to freely refeed (406). Hunger and hyperphagia are also observed in humans and experimental animals that are rendered hypoglycemic by systemic injections of insulin and in diabetic animals having impaired cellular glucose uptake and oxidation as a function of insulin deficiency or resistance to insulin's actions (407).

Pharmacological blockade of intracellular glucose utilization can have profound effects on feeding behavior. Peripheral administration of metabolic inhibitors such as 2-deoxyglucose (2-DG) or 5-thioglucose, nonmetabolizable glucose analogues that cause glucoprivation by competitively inhibiting glycolysis, results in increased hunger and food intake in a variety of species, including rats, monkeys, and humans (408–410). Injection of 2-DG or 5-thioglucose directly into the cerebral ventricles stimulates feeding at lower doses than those that are required of peripheral injections, which suggests that glucoprivic feeding is mediated by central mechanisms (402,411–413).

Relatively few studies have addressed whether alterations in fatty acid availability modulate feeding behavior. Systemic administration of methyl palmoxirate or mercaptoacetate, molecules that induce a state of lipoprivation by impairing intracellular fatty acid metabolism, increases food intake in rats that are fed a medium- or high-fat diet but not in low-fat fed rats or high fat-fed hamsters (414–416). In rats, combined treatment with methyl palmoxirate and 2-DG elicits a greater stimulatory effect on feeding than that which would be expected from the additive effects of either metabolic inhibitor alone, which suggests that glucose and fatty acid availability exert a coordinated control over feeding behavior (417).

Metabolic Fuel Availability and Reproductive Function

Metabolic challenges that compromise fuel availability often have deleterious effects on reproductive function. Animals that are subjected to prolonged caloric restriction have reduced metabolic fuel availability and often display reproductive abnormalities; conversely, food-deprived animals that are subsequently given free access to food or are infused with nutrients show rapid and parallel improvements in fuel availability and reproductive function (29,34,47,418). Animals that are hypoglycemic as a consequence of systemic insulin treatment also exhibit signs of impaired reproductive function (341,343,345–351), as do humans and experimental animals with diabetes (419). Under some circumstances, coadministering glucose and insulin prevents both the hypoglycemia and the impaired LH secretion that normally result from systemic insulin treatment (343,344).

Although there are instances in which glucose availability and reproductive function decline or improve in parallel, there are also circumstances under which changes in reproductive function occur in the absence of significant changes in glucose availability. For example, Dubey and colleagues (68) observed that adult macaques that were food restricted for 3–5 weeks did not experience a significant reduction in plasma glucose concentrations yet still exhibited impaired LH secretion. Furthermore, the rapid restoration of pulsatile LH secretion that occurs in food-restricted monkeys after access to food is not dependent on a rise in glucose availability, because meals comprised of protein and fat restore pulsatile LH release as effectively as mixed nutrient or pure dextrose meals, despite the fact that ingestion of a protein/fat meal does not significantly elevate plasma glucose levels (418). Thus, at least in some primate species, reproductive function may be altered (impaired or improved) in the absence of corresponding changes in glucose availability, at least based simply on the state of glycemia, because glucose "availability" is also influenced by insulin and tissue sensitivity to insulin, among other factors. Experiments that have attempted to elucidate whether increased glucose availability is sufficient to stimulate the reproductive axis have also yielded mixed results. Chronic infusion of dextrose alone does not induce precocious pulsatile LH release in agonadal juvenile macaques (389) but has been shown to partially restore pulsatile LH secretion in a small number of fasted rhesus monkeys (420). Although female hamsters maintain normal estrous cyclicity when provided with only glucose during the first 2 days of their estrous cycle, consumption or injection of pure glucose after 48 hours of food restriction does not mimic the restorative effect of refeeding on mating behavior in female musk shrews (57,421).

In addition to their effects on food intake, pharmacological metabolic inhibitors also exert profound effects on the reproductive axis. In males, 2-DG–induced glucoprivation reduces the secretion of LH (422–426) and testosterone (422). In females, 2-DG

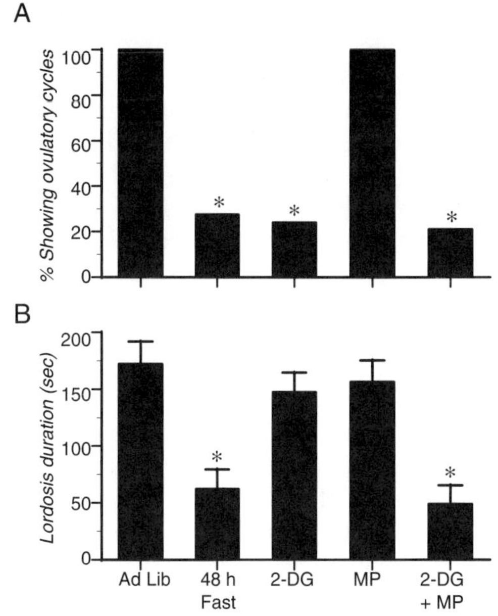

FIG. 10. Effects of glucoprivation and lipoprivation on ovarian cyclicity and sexual behavior in female Syrian hamsters. (**A**) Ovulatory cyclicity in intact female hamsters that were fed ad libitum, were fasted for 48 h, or were treated with 2-DG (2,000 mg/kg) or methyl palmoxirate (100 mg/kg) alone or in combination. (**B**) Estrous behavior in estrogen and progesterone-treated OVX female hamsters that were allowed to feed ad libitum, were fasted for 48 h, or were treated with 2-DG (750 mg/kg) and/or methyl palmoxirate (25 mg/kg). Whereas 2-DG administration mimicked the inhibitory effect of fasting on estrous cyclicity, only the combined treatment with 2-DG and methyl palmoxirate reproduced the suppressive effect of fasting on sexual behavior. $^*P < 0.05$ vs. Ad lib group. [Redrawn from (9).]

reduces GnRH pulse generator activity (342), suppresses pulsatile LH secretion (423,427,428), disrupts estrous cyclicity and ovulation (429,430) (Fig. 10), decreases the magnitude of the steroid-induced LH surge (431), and impairs mating behavior (421). Collectively, these observations demonstrate that the HPG axis is responsive to dramatic reductions in glucose availability and that reproductive function is often impaired when glucose availability is inadequate.

Blockade of fatty acid metabolism by administration of methyl palmoxirate induces anestrus in food-deprived fat hamsters but not in freely fed fat or lean hamsters (430,432). Treatment with mercaptoacetate prevents the rapid restoration of mating behavior that occurs after a brief period of refeeding in food-restricted female musk shrews (421). These observations could implicate an important role for fat metabolism in the metabolic regulation of reproductive function. Alternatively, the effects of methyl palmoxirate or mercaptoacetate on estrous cyclicity and behavior in food-restricted animals might reflect a disruption of the glucose-sparing effects of fat oxidation (433). The latter interpretation would fortify the notion that glucose availability is more important than fat availability in relaying energetic status to the neuroendocrine reproductive axis, which is congruent with the fact that the brain relies on glucose as its primary energy source.

There are also instances in which reproductive impairment occurs after simultaneous blockade of glucose and fat utilization but not after inhibition of either glucose or fat metabolism alone. For example, estrous behavior in steroid-primed OVX female Syrian hamsters is impaired by the coadministration of 2-DG and methyl palmoxirate but is unaffected when the same doses of either of the metabolic inhibitors are administered alone (432,434,435) (Fig. 10). Under these circumstances, as long as one fuel source is available in sufficient quantities, it can compensate for the body's temporary inability to oxidize the other fuel source. Based on these observations, one can infer that in some species, certain aspects of reproductive function are sensitive to the general availability of metabolic fuels rather than to the availability of a particular metabolic substrate.

Sites of Action

Although the brain's participation in sensing glucose and detecting glucoprivation is fairly well established (436–440), recent evidence suggests that the brain is also directly sensitive to elevations or reductions in fatty acid availability (441,442). Many of the effects of glucoprivation, but not lipoprivation, on reproductive function also appear to be mediated by central mechanisms. Injection of 2-DG or 5-thioglucose directly into the brain reduces the magnitude of the steroid-induced LH surge in OVX female rats (431), induces anestrus in female hamsters (433), and suppresses LH secretion in male rats (424,426) and gonadectomized male and female sheep (425,428). Injection of 2-DG into either the lateral or fourth cerebral ventricles of OVX ewes reduces LH pulsatility to a similar extent, which implies that glucose availability is sensed in the vicinity of the caudal hindbrain (428). GnRH neuronal activity in 2-DG–treated female rats and hamsters is reduced after 2-DG administration (as assessed by the reduced expression of the immediate early gene product Fos in GnRH neurons) (431,443). In castrated male lambs, the glucoprivation-induced reduction in LH pulse frequency does not appear to be attributable to a primary defect in GnRH secretion, considering that *N*-methyl-D,L-aspartate (a potent GnRH secretogogue) increases LH pulse frequency in 2-DG–treated animals (425). This last observation suggests that glucoprivation impairs LH secretion via its effect on a central locus that is upstream from GnRH neurons.

In contrast to the well-established effects of impaired glucose availability within the CNS, the pituitary does not appear to play a major role in modulating reproductive output as a function of intracellular glucose availability. Incubation of rat anterior pituitary cultures with 2-DG does not significantly alter either basal or GnRH-induced LH release (426). Furthermore, observations that exogenous GnRH treatment elicits normal or enhanced LH responses in 2-DG–treated gonadectomized female rats (422,423) and ewes (427) and that GnRH administration sustains pulsatile LH secretion in 2-DG–treated male lambs (425) demonstrate that pituitary function is not significantly impaired by glucoprivation. Thus, it would appear that the brain—not the pituitary—is the primary target for glucose sensing and orchestrating the appropriate response of the neuroendocrine reproductive axis to changes in glycemic state.

Summary

The notion that the reproductive axis directly responds to alterations in metabolic fuel availability is an attractive one, given that this would seem to be the most parsimonious mechanism by which the reproductive axis could be attuned to short-term shifts in energy availability. In addition to the putative effects of acute changes in metabolic fuel availability on reproductive function, alterations in metabolic fuel availability might also underlie some of the reproductive consequences of long-term changes in energy status. Leptin increases oxidative metabolism of both fatty acids and glucose (204,444), and leptin's ability to maintain or correct reproductive function in food-restricted animals can be blocked by the coadministration of inhibitors of glucose and/or fatty acid utilization (224,300,445). Insulin also has complex effects on carbohydrate and lipid metabolism in peripheral tissues. Although the brain has historically been viewed as one of the few organs that does not require insulin for glucose uptake, this notion has been brought into question by the identification of the insulin-regulated glucose transporter GLUT-4 in hypothalamic neurons (446). Thus, the availability of metabolic fuels might have an even broader impact on energy homeostasis and reproductive function than previously believed.

To date, the weight of evidence in support of the idea that glucose and/or fatty acid availability influences reproductive function is either based on the association between reproductive capacity and fuel availability or reflects the interpretation of results obtained after administration of antimetabolic agents, such as 2-DG. Although metabolic inhibitors provide a means by which to investigate the consequences of imposing a severe metabolic stress, the abrupt reductions in fuel availability that these agents induce do not mimic the more gradual changes in availability of metabolic fuels that would accompany the transition from a fed to a fasted state. Furthermore, it is not surprising that the reproductive axis is sensitive to glucoprivation, given that glucose is the brain's sole source of energy during nonfasting conditions. In this respect, it seems most plausible that the effects of glucoprivation are not specific to the neural pathways that regulate energy homeostasis and neuroendocrine function but might instead represent more global effects on neural function. Until more specific tools are developed that will allow us to isolate the effects of pharmacological manipulations of fuel availability to particular neuronal populations, further understanding of the role of metabolic fuels in regulating reproductive function will remain enigmatic.

Thyroid Hormones

The thyroid gland synthesizes and secretes hormones that play critical roles in the growth and development of many organ systems. Thyroid hormones also stimulate energy metabolism and are thus indirectly responsible for generating heat. The major secretory product of the thyroid, thyroxine (T_4), is converted to the more biologically potent triiodothyronine (T_3), which occurs in peripheral tissues. The secretion of T_4 from the thyroid is primarily driven by the release of thyroid-stimulating hormone from the anterior pituitary, which is in turn modulated by thyrotropin-releasing hormone neurons in the hypothalamus. Thyroid hormone receptors are members of the nuclear hormone receptor family, which exert direct effects on gene transcription when bound by their ligands. In accordance with its widespread role in regulating growth and metabolism, thyroid hormone binding sites and receptor mRNAs are distributed throughout the body, including various regions of the CNS (447,448).

Although hypothalamic–pituitary–thyroid function is commonly altered during states of negative energy balance, the directional shift depends on the nature of the metabolic stressor that compromises energy availability. Circulating levels of T_3 (and sometimes T_4) are diminished by fasting or food restriction (68,449–453). In humans, circulating T_3 concentrations are typically reduced in girls and women with anorexia nervosa and in other malnourished individuals (454–457). When food intake is insufficient to meet the body's energetic requirements, lessening the stimulatory actions of T_3 on metabolic rate reduces basal metabolic demands and thus permits animals to conserve their energy resources. In food-deprived animals, circulating levels of thyroid hormones are increased by refeeding (458,459) or by leptin treatment (212,460). In contrast to the inhibitory effects of reduced energy intake on thyroid function, small mammals that are exposed to cold ambient temperatures have elevated circulating levels of T_3—no doubt as a mechanism to increase thermogenesis and maintain core body temperature (461).

Much of what is currently known about the actions of thyroid hormones on the HPG axis is based on the observed consequences of thyroid hormone deficiency or excess on reproductive function. Hypothyroidism retards follicular development and impairs estrous cyclicity in adult female rats and hamsters (462,463) and reduces circulating levels of gonadotropins and gonadal steroids in male and female rats (462,464). These effects are generally reversed by treatment with either T_3 or T_4 (462–464). Female rats that are thyroidectomized before puberty show delayed sexual maturation (465). In contrast to the inhibitory effects of hypothyroidism on HPG function, sexual behavior is enhanced in thyroidectomized female rats (466). Thyroid hormones play critical roles in seasonal reproductive cyclicity in sheep, based on observations that thyroidectomized ewes continue to exhibit estrous cycles and display high frequency GnRH and LH pulses during the nonbreeding season (467–469), a phenomenon that is reversed by treatment with T_4 (470,471). Hypothyroid women commonly experience irregular menstrual cycles, anovulation, and infertility (472), and hypothyroid men often exhibit signs of hypogonadotropic hypogonadism (473). In contrast to the inhibitory effects of hypothyroidism on reproductive function in adults, precocious puberty has been reported in hypothyroid boys and girls (474,475).

Although reproductive impairment is frequently observed in association with thyroid hormone deficiency, hyperthyroidism can also have a negative impact on reproductive function. In adult female steroid-primed OVX rats and mice, the administration of high doses of T_3 or T_4 reduces lordosis behavior (466,476). Hyperthyroid prepubertal male lambs have reduced pulsatile LH secretion, decreased plasma testosterone levels, arrested testicular growth, and impaired spermatogenesis (477). The observation that hyperthyroid lambs retain a normal LH response to exogenous GnRH administration implies that a central defect is an underlying cause of arrested sexual maturation in these animals (478). Puberty is also delayed in female rats treated with T_4 during the first 5 days of postnatal life, as evidenced by a delay in the timing of vaginal opening and first estrus (479). Puberty is commonly delayed when hyperthyroidism occurs before puberty in girls, and hyperthyroid women often experience oligomenorrhea or amenorrhea (472,480).

Thyroid hormone receptors and/or binding sites are expressed at all three levels of the HPG axis; thus, thyroid hormones have the potential to alter reproductive function by acting at multiple loci (447,448,481–483). In sheep and hamsters, a subset of GnRH neurons express thyroid hormone receptors, which is consistent with the idea that thyroid hormones exert direct effects on GnRH synthesis and/or secretion (484). Thyroid hormones have been demonstrated to have direct effects on gonadal function in vitro (485–487).

Taken together, these observations suggest that any significant departure from a euthyroid state in either direction can have negative consequences on the reproductive axis. Although too little thyroid hormone might signify a need to conserve energy due to reduced energy intake, excessive thyroid hormone secretion might reflect the increased energetic demands imposed by thermogenesis. In either situation, energy availability is reduced. Hence, it is understandable that hypothyroidism and hyperthyroidism have partially overlapping effects on reproductive function. Whether alterations in thyroid hormone secretion and/or metabolism are involved in conveying metabolic information to the HPG axis when animals are subjected to metabolic challenges is less certain. Given the relatively low clearance rate of T_4 from the circulation, it seems an improbable candidate for providing information about metabolic status to the HPG axis during acute nutritional challenges. However, it remains possible that altered thyroid hormone secretion and/or action contributes to reproductive impairment brought about by more sustained disruptions in energy balance.

Other Metabolic and Possible Neural Signals

In addition to the metabolic factors that were mentioned above, a variety of other hormones has been implicated in relaying metabolic cues to the neuroendocrine reproductive axis, including growth hormone, insulin-like growth factor I, and glucocorticoids (488) (see Chapter 48). Several gastrointestinal peptides that are involved in modulating feeding behavior have also been demonstrated to influence reproductive function, such as cholecystokinin, glucagon-like peptide-1, and ghrelin (489–492).

Although we have solely focused on bloodborne metabolic signals, there is also evidence supporting the notion that neural cues convey information about metabolic status to the centers in the brain that control feeding and reproduction. Vagal afferents relay information to the CNS about meal size and constituency by transmitting signals from gastric mechanoreceptors and chemoreceptors. Meal-related signals that are transmitted by the vagus nerve are important in regulating feeding behavior, because transection of the vagus nerve abolishes the feeding effects of various nutrients and gut peptides (493). Vagal afferents have also been implicated in mediating the decline in reproductive function in rats during an acute fast. In estrogen-treated OVX female rats that are deprived of food for 48 hours, complete subdiaphragmatic vagotomy or selective sectioning of the gastric branch of the vagus nerve reverses the fasting-induced inhibition of pulsatile LH secretion, suggesting that vagal afferents are involved in inhibiting LH release during acute fasting (177). To our knowledge, it is unknown whether vagal afferents are involved in mediating the restorative effect of refeeding on pulsatile LH secretion in food-deprived animals or whether vagal input contributes to the reproductive impairment that arises during chronic food restriction.

CENTRAL SIGNALING MOLECULES THAT INTEGRATE METABOLISM AND REPRODUCTION

In the previous section, we discussed several metabolic signals that have been implicated in relaying metabolic information from the periphery to the neural circuits that govern energy homeostasis and reproductive function. In this section, we describe some of the central signaling pathways by which the brain detects, integrates, and relays metabolic information to the neural circuits that influence reproductive function. We first discuss the various routes by which circulating metabolic signals gain access to the brain as well as the principal sites in the brain where circulating metabolic cues are detected and integrated. The remainder of this section is devoted to describing several of the neurotransmitter and neuromodulator systems believed to participate in coupling nutrition and reproduction, with a focus on those signaling molecules that are anatomically and/or functionally linked to pathways mediating signaling from leptin, insulin, and metabolic fuels.

Routes of Entry into the CNS

The blood–brain barrier prevents many circulating metabolic substrates and peptide hormones from freely diffusing out of the cerebral blood vessels and into the brain. The blood–brain barrier, which is composed of nonfenestrated capillary endothelial cells joined by tight junctions and surrounded by astroglial processes, functionally isolates most of the brain parenchyma from large, hydrophilic, and/or charged bloodborne molecules, including peptide hormones (494). The choroid plexus forms a similar barrier between the blood and the CSF, thereby preventing many circulating molecules from gaining access to the brain parenchyma via the CSF in the ventricular system. To circumvent these anatomical barriers, many circulating metabolic signals that meet the exclusionary criteria (mentioned previously) are actively transported across the blood–brain barrier or blood–CSF barrier. Glucose, insulin, and leptin are all transported into the brain by carrier-mediated transport systems (200,304,495). Insulin receptors and leptin receptors are present in the cerebral vasculature and the choroid plexus and are believed to play important roles in their ligands' respective transport processes (189,496–498).

An alternate route by which circulating metabolic signals can gain access to the brain is by way of the circumventricular organs, which include the median eminence, the area postrema, the organum vasculosum lamina terminalis, and the subfornical organ. The circumventricular organs are regions of the brain that are unrestricted by the blood–brain barrier, because they are vascularized by fenestrated capillaries that allow molecules to freely diffuse into and out of the blood vessels (494). By nature of this anatomical arrangement, the circumventricular organs and their adjacent neural structures are poised to detect humoral signals from the periphery. Molecules that diffuse into the interstitium of the circumventricular organs can either act on local populations of neurons or can diffuse or be actively transported to adjacent brain structures or the CSF. The circumventricular organs contain insulin and leptin binding sites, which are particularly concentrated in the median eminence and the overlying hypothalamic arcuate nucleus (ARC), indicating that these neural structures have the capacity to detect leptin and insulin in the blood (200,201,319,320). In the caudal hindbrain, the area postrema plays a critical role in the detection of glucose availability (176).

Sites of Metabolic Cue Detection

GnRH Neurons

Based on their anatomical distribution pattern, GnRH neurons appear to be ideally situated to directly detect metabolic fuels and hormones that enter the brain from the systemic circulation. GnRH perikarya primarily reside within the anterior hypothalamus/POA and the mediobasal hypothalamus (499). Nerve terminals of hypophysiotropic GnRH neurons are located in the median eminence (500). These neuroanatomical regions surround the third cerebral ventricle, and hence GnRH neurons could theoretically receive direct input from peripheral metabolic signals having entered the CSF. Bloodborne metabolic signals could also gain access to GnRH cell bodies and nerve terminals by way of the circumventricular organs in the forebrain, namely the organum vasculosum lamina terminalis and the median eminence.

Despite the close proximity of GnRH neurons to the CSF and multiple circumventricular organs, there is scant evidence in favor of direct regulation of GnRH neurons by peripheral metabolic signals. GnRH neurons do not express detectable levels of leptin receptor mRNA or immunoreactivity, which argues against a direct modulatory effect of leptin on GnRH neuronal function (222,294). However, contrary to these anatomical findings, recent electrophysiological studies by Sullivan et al. (501) have demonstrated that leptin modulates the postsynaptic responses of GnRH neurons to GABAergic neurotransmission in vitro. Furthermore, leptin's postsynaptic effects on GABAergic neurotransmission in GnRH neurons are blocked by the intracellular application of an inhibitor of Janus kinase, a downstream component of central leptin signaling pathways (502). Together, these observations hint at the presence of functional leptin receptors on GnRH neurons and challenge the notion that GnRH neurons are not directly modulated by leptin. Although insulin binding sites are located in the hypothalamus and on GnRH-expressing GT1-7 cells (387,503), there are currently no reports of insulin receptor expression by GnRH neurons. Finally, it is unlikely that GnRH neurons play a major role in detecting alterations in intracellular glucose availability, because many of the adverse effects of glucoprivation on reproductive function are abolished in animals with aspiration lesions of the area postrema or by the immunotoxic destruction of hindbrain catecholaminergic neurons (504–506). These latter observations suggest that glucoprivic information is primarily relayed to GnRH neurons through transsynaptic pathways originating in the brainstem.

Afferent Inputs to GnRH Neurons

The meager evidence in favor of direct metabolic regulation of GnRH neuronal activity suggests that information about metabolic status is primarily transmitted to GnRH neurons by way of neuronal afferents. Over the years, a growing body of evidence has revealed that many of the afferent signaling pathways that relay metabolic information to the neural circuits controlling energy homeostasis are also responsible for conveying metabolic status to GnRH neurons. These neural pathways commonly originate in the hypothalamus or the brainstem, which have long been recognized as important loci in the regulation of feeding and energy expenditure (507). Many populations of neurons in discrete hypothalamic and brainstem nuclei are sensitive to alterations in energy balance. These circuits are modulated by circulating metabolic signals and relay this information to the effector circuits that control feeding and metabolism as well as gonadotropin secretion and reproductive behaviors. Collectively, these observations point toward a multitude of central signaling systems that could conceivably relay metabolic cues to the GnRH pulse generator. The following section focuses on several of the neurotransmitters and neuromodulators believed to be the most likely candidates for fulfilling this important function.

Central Signaling Molecules

Over the past several decades, dozens of signaling molecules in the CNS have been implicated in regulating energy balance and/or reproductive function, and the list is continually expanding. To add to this complexity, these signaling molecules do not act in isolation but instead interact with many other signaling pathways in the brain. Furthermore, it is improbable that a single neurotransmitter or neuromodulator is solely responsible for coupling reproductive function and energy homeostasis. Because reproduction is essential for the survival of a species, it is more likely that organisms possess several redundant signaling pathways and mechanisms that synchronize reproduction with prevailing energy availability. In the following section, we discuss several key neuropeptides that have been implicated in fulfilling this function. However, it should be noted that various classical neurotransmitter signaling systems have also been implicated in the regulation of reproductive function, including norepinephrine, dopamine, serotonin, GABA, and glutamate (390). For a more comprehensive account of the relationships between many of these neurotransmitter systems and GnRH neuronal function, we refer the reader to Chapter 28.

Neuropeptide Y

Neuropeptide Y (NPY) is a 36 amino acid member of the pancreatic polypeptide family (508). NPY is highly conserved across mammalian species and is one of the most abundantly expressed neuropeptides within the CNS (509,510). NPY mRNA and protein are expressed at high levels in the ARC and in brainstem catecholaminergic neurons (509,511,512). NPY-immunoreactive fibers are distributed throughout the forebrain, including multiple loci that are involved in regulating body weight and/or reproductive function (509). To date, six different subtypes of NPY receptors (Y1–Y6) have been identified (513). Over the past several years, many of NPY's numerous functions have been attributed to its interactions with a specific receptor subtype(s).

Since its discovery in 1982, a major focus of those who study the NPY signaling system has been to elucidate the role of NPY in the central regulation of energy homeostasis. Within the brain, NPY signaling has an impact on many aspects of energy balance, including feeding, thermogenesis, peripheral insulin secretion, and hepatic glucose output (514). Despite its diverse roles in modulating energy homeostasis, NPY is arguably best known for its ability to robustly stimulate food consumption in satiated animals. Injection of NPY into the lateral or third cerebral ventricles potently increases acute food intake in rats (515), mice (516), hamsters (517), sheep (518), and monkeys (519). In rodents, chronic infusion or injection of NPY into the cerebral ventricles or directly into the hypothalamic paraventricular nucleus (PVN) increases food intake, body weight gain, and fat mass (520–523). NPY neurons project to the PVN (524), where NPY is secreted in temporal association with increased appetite (525). A possible role for endogenous NPY in modulating energy balance is bolstered by several lines of evidence, including the observations that ICV injection of NPY antiserum decreases food intake and body weight gain (526) and that central infusion of antisense oligodeoxynucleotides to the mRNAs for either NPY (527) or various NPY receptor subtypes (528–530) reduces feeding and body weight gain in rats.

The synthesis and secretion of NPY in the hypothalamus are altered in response to a variety of metabolic stressors. Concordant with NPY's putative function as an orexigenic neuropeptide, hypothalamic levels of its mRNA and the peptide itself are increased by food deprivation (212,531–533) and chronic dietary restriction (534). Levels of NPY are elevated in the PVN of fasted rats (535), and the release of NPY into the PVN is increased by fasting and reduced by subsequent refeeding (525). The concentration of NPY in the CSF is elevated in chronically undernourished sheep (536) and in women with anorexia nervosa (537). In the latter group, concentrations of NPY in the CSF are normalized by weight gain (537). Levels of NPY mRNA and protein are also elevated during circumstances in which animals experience negative energy balance as a result of enhanced energy expenditure, including exercise (538), lactation (539,540), and exposure to cold temperature (533). In each of these cases, hypothalamic NPYergic tone is elevated in association with reduced energy availability, at a time when animals possess a powerful drive to consume food to relieve their energy deficits.

NPY neurons in the ARC are sensitive to changes in circulating leptin concentrations. In rats and mice, treatment with exogenous leptin during a prolonged fast prevents both the reduction in plasma leptin concentrations and the increase in the expression of NPY mRNA that normally occur in fasted animals (212,541). Obese rodents with a congenital deficiency in either leptin (*ob/ob* mice) or functional leptin receptors (*db/db* mice and *fa/fa* rats) are hyperphagic and have elevated levels of NPY mRNA in the hypothalamus (531,542,543). Enhanced NPYergic tone is partially responsible for the hyperphagia and obesity of *ob/ob* mice, because *ob/ob* mice in which the NPY gene has been disrupted eat less food and are leaner than normal *ob/ob* mice (544). Treatment with exogenous leptin reduces levels of NPY mRNA in *ob/ob* mice and does so independently of leptin's capacity to reduce food intake in these animals (545,546). Presumably, leptin exerts a direct modulatory effect on NPY neuronal function, because a substantial proportion of NPY neurons in the ARC express leptin receptors (222,547).

The expression of NPY in the hypothalamus is also regulated by insulin. Central injections of insulin during a 48-hour fast reverse the fasting-induced increases in NPY mRNA levels in the ARC and NPY immunoreactivity in the PVN (548). The expression of NPY in the hypothalamus is elevated in diabetic rodents (531,543,549), and central infusion of insulin reduces the expression of NPY mRNA in the ARC and the hyperphagia that accompanies diabetes (332). The selective down-regulation of hypothalamic insulin receptors by the central infusion of antisense oligodeoxynucleotides to insulin receptor mRNA elicits hyperphagia and increases NPY mRNA expression in the ARC (336). Although insulin receptors and insulin signal transduction proteins are expressed within the ARC (310,313,314), it is currently unknown whether NPY neurons express any of these signaling molecules.

NPY neurons are responsive to reductions in glucose availability and have been implicated in the feeding response to glucoprivation. Inhibition of intracellular glucose utilization by administration of the metabolic inhibitor 2-DG stimulates food intake and increases NPY mRNA in the ARC and in hindbrain catecholaminergic neurons (550,551). A causal role for NPY in glucoprivic feeding is suggested by the observation that food consumption in response to 2-DG treatment is significantly attenuated by the infusion of neutralizing antibodies to NPY into the PVN (552). Furthermore, ICV injection of 2-DG induces the expression of the immediate early gene *c-fos* in NPY neurons in the ARC, demonstrating that NPY neurons are transcriptionally activated by glucoprivic stimuli (553). Although NPY neurons in the ARC express the glucose-sensing enzyme glucokinase (554) and are sensitive to changes in extracellular glucose concentrations (555), synaptic input from the brainstem plays an important role in relaying glucoprivic information to NPY neurons in the ARC (556).

In addition to its well-established roles in the central regulation of energy intake and energy expenditure, NPY also exerts pronounced effects on the neuroendocrine reproductive axis (557). NPY fibers form synaptic contacts with GnRH cell bodies in the POA (558). In the mouse, approximately 50% of the NPY fibers that appose GnRH cell bodies originate in the ARC, whereas another 25% project from the brainstem (559). In the rat, Y5 receptors are expressed on GnRH cell bodies in the POA (560). Although Y1 receptors have not been detected on GnRH perikarya, Y1 receptors are expressed on fibers that are in close proximity to GnRH cell bodies in the POA (561). NPY fibers also project to the external zone of the median eminence (562), where they come in close contact with GnRH fibers (561). Y1 immunoreactivity has been detected on GnRH fibers and nerve terminals in the median eminence (561). Thus, NPY neurons are anatomically situated to modulate both the synthesis and secretion of GnRH. In addition to its putative direct effects on GnRH neurons, NPY is also secreted into the pituitary portal circulation and thus has the potential to act directly on the pituitary as well (562,563).

NPY exerts complex actions upon the reproductive axis. Investigations of the effects of NPY on GnRH and gonadotropin secretion have mostly focused on the female. In general, NPY's effects on GnRH/LH secretion are dependent on the gonadal steroid milieu. In gonad-intact animals or in OVX females that are treated with ovarian steroids, NPY generally has a stimulatory effect on GnRH and LH secretion. Acute ICV injection of NPY stimulates LH release in intact male and ovarian steroid-primed OVX female rats (564,565). Push–pull perfusion of NPY through the median eminence increases GnRH release in intact female rabbits and in estrogen-treated OVX monkeys (566,567), and NPY stimulates GnRH secretion from mediobasal hypothalamus or median eminence fragments from intact female rabbits and ovarian steroid-treated OVX rats in vitro (568–570). NPY has been postulated to play a critical role in the generation of the preovulatory LH surge, based on observations that NPY gene expression is increased in intact female rats during the afternoon and evening of proestrus (571) and in OVX rats before the ovarian steroid-induced LH surge (572). Furthermore, the secretion of NPY into the median eminence and pituitary portal circulation is increased before or during the LH surge in intact and steroid-primed OVX rats (563,573), and administration of NPY antiserum or NPY mRNA antisense oligodeoxynucleotides blocks the steroid-induced GnRH and LH surges in OVX rats and monkeys (563,574,575).

In marked contrast to its actions in animals with normal or elevated gonadal steroid concentrations, centrally administered NPY has predominantly inhibitory effects on LH secretion in hormonally untreated gonadectomized rats (564,576,577), mice (516), rabbits (566), sheep (536,578), cows (579), and rhesus monkeys (580). The inhibitory effect of centrally administered NPY on pulsatile LH release in OVX rats is reversed by exogenous administration of GnRH, which suggests that NPY impairs LH secretion by suppressing the hypothalamic drive to the pituitary (577). Indeed, perfusion of NPY into the median eminence of OVX rabbits reduces pulsatile GnRH secretion (566). However, in vitro perfusion of mediobasal hypothalamus fragments with NPY does not alter GnRH secretion in either OVX rats or rabbits, implying that NPY's ability to inhibit GnRH release in vivo requires neuronal input from additional brain sites (568,569).

Although NPY stimulates GnRH and LH secretion in intact animals when it is administered acutely, NPY inhibits reproductive function in intact animals when it is administered in a chronic fashion. Chronic ICV infusion of NPY delays sexual maturation in prepubertal female rats (581) and disrupts ovarian cyclicity in adult female rats (522). Infusion of NPY receptor agonists lengthens the estrous cycle in female rats and increases the duration of lactational diestrus (582). In intact male mice and rats, chronic ICV infusion of NPY reduces circulating levels of gonadotropins and testosterone, decreases testicular and prostate weights, and diminishes pituitary GnRH receptor content (523,583) (Fig. 11).

In addition to its effects on GnRH neurosecretory function, NPY also acts directly on the pituitary to modulate LH secretion. Although NPY has sustained stimulatory effects on LH and FSH secretion from pituitary fragments from intact female rabbits, NPY has only a transient stimulatory effect on gonadotropin secretion from OVX rabbit pituitary cells (569). Other investigators have observed no effect of NPY on LH secretion from cultured anterior pituitary cells from OVX female rats (584). However, NPY has been shown to enhance pituitary responsiveness

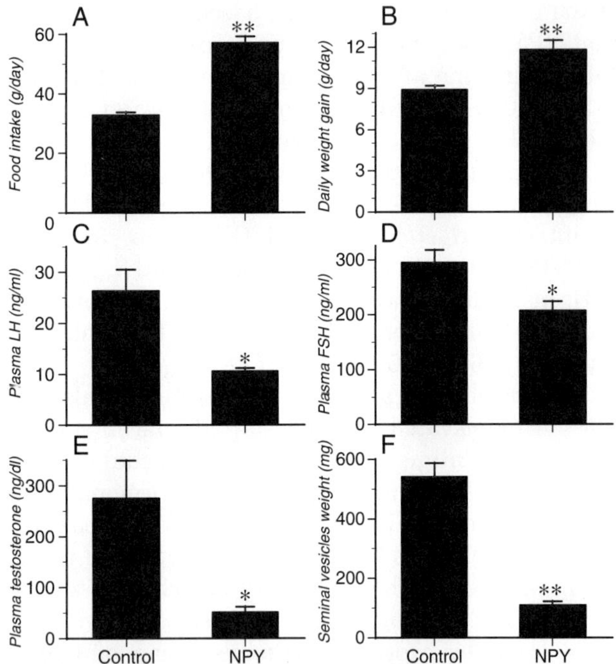

FIG. 11. Chronic central infusion of NPY increases food intake and suppresses reproductive function in intact male rats. **(A)** Daily food intake, **(B)** daily weight gain, **(C)** plasma LH concentrations, **(D)** plasma FSH concentrations, **(E)** plasma testosterone concentrations, and **(F)** seminal vesicle weight in rats that were infused with NPY (36 μg/day) into the lateral cerebral ventricle for 7 days. The control rats were sham-operated but did not receive an infusion. *P <0.05 and **P <0.001 vs. control rats. [Redrawn from (583).]

to GnRH, because NPY potentiates GnRH-induced LH secretion from anterior pituitary cell cultures and in pentobarbital-blocked proestrous female rats (584,585). Pituitary responsiveness to GnRH is impaired in the absence of NPY, as evidenced by the observation that NPY knockout mice exhibit reduced LH release in response to exogenous GnRH administration as well as steroid-induced LH surges that are reduced in magnitude (586).

NPY suppresses mating behavior in both sexes. Central administration of NPY decreases the occurrence of stereotyped sexual behaviors in male rats (587) and suppresses estrous behavior in steroid-primed OVX female rats and hamsters (587,588). Administration of NPY antiserum increases sexual receptivity in steroid-primed OVX female obese Zucker rats, which implicates a role for endogenous NPY in the suppression of estrous behaviors in these animals (526). NPY's effects on estrous behavior are most likely enacted within the forebrain, because lordosis duration is reduced when an NPY receptor agonist is injected into the lateral ventricle, MPOA, PVN, or anterior hypothalamus of female hamsters but not when the NPY analogue is injected into the fourth cerebral ventricle (589).

The generation and characterization of several strains of mutant mice have revealed an important role for NPY in mediating the effects of leptin deficiency on reproductive function. Compared with sterile *ob/ob* mice, fertility is improved in *ob/ob* mice with a targeted disruption of the NPY gene (544). Thirty percent of NPY-deficient male *ob/ob* mice are capable of siring offspring (compared with 5% of male *ob/ob* mice), and some female *ob/ob* mice lacking NPY are capable of giving birth and nursing litters (544). Fertility is also significantly improved in male and female *ob/ob* mice possessing a dysfunctional Y4 receptor gene (590), and seminal vesicle weights and pituitary LH content are normalized in Y1 receptor null male *ob/ob* mice (591). To our knowledge, there have not been any reports in the literature of similar experiments that have directly tested whether NPY plays a critical role in mediating the effects of insulin deficiency and/or low metabolic fuel availability on HPG function.

Does NPY play an essential role in the regulation of energy homeostasis and/or reproductive function? Surprisingly, NPY knockout mice are fertile and do not exhibit any overt abnormalities with respect to body weight, feeding behavior, or endocrine function (592,593). The seemingly normal body weight and reproductive phenotypes of NPY knockout mice do not preclude an important role for NPY in regulating body weight and reproduction in wild-type animals; rather, it seems likely that NPY knockout mice rely on either redundant or compensatory signaling pathways and molecules that compensate for the loss of NPY during development (592,593). Alternatively, it is conceivable that NPY signaling is dispensable under favorable metabolic conditions but becomes crucial only when animals are faced with an energetic challenge. This latter possibility is supported by the observation that fasting impairs LH secretion in intact and estrogen-treated OVX female wild-type mice but not in female NPY knockout mice (594). With the exception of the above-mentioned study in NPY-deficient mice, we are unaware of any other studies demonstrating that reducing endogenous NPYergic tone abolishes the adverse effects of food deprivation on HPG function. Such studies are necessary before one can draw any firm conclusions about whether enhanced NPY signaling plays a critical role in mediating the effects of reduced energy availability on the reproductive axis.

Taken together, these observations demonstrate that NPY modulates HPG function at the levels of the brain and the anterior pituitary. Under certain physiological conditions, NPY exerts a stimulatory effect on GnRH/LH release, most notably with respect to the generation and amplification of the preovulatory LH surge. However, NPY's inhibitory effects on reproductive function predominate when circulating levels of estrogen are significantly reduced and/or when NPY is chronically elevated. Both of these requirements are met during severe metabolic challenges, making it likely that NPY contributes to the reproductive impairment that commonly occurs during periods of reduced energy availability. From an adaptive standpoint, enhanced NPY signaling during a period of negative energy balance would elicit sensations of hunger and would thus provoke energy-deprived animals to search for food. At the same time, increased NPYergic tone would have negative repercussions on the reproductive axis, thus forcing energetically challenged animals to abandon their reproductive efforts until energetic conditions improve.

Pro-Opiomelanocortin–Derived Peptides

Pro-opiomelanocortin (POMC) is a precursor protein that is post-translationally cleaved into several smaller biologically active peptides, including the endogenous opioid β-endorphin and the melanocortin α-melanocyte-stimulating hormone (α-MSH) (595,596). Within the CNS, cell bodies expressing POMC mRNA and its peptide products reside within the ARC and in the nucleus of the solitary tract in the hindbrain (597–599). Fibers containing POMC-derived peptides are dispersed throughout the brain (597,598,600,601), as are receptors for melanocortin and opioid peptides (602–604). Of the five melanocortin receptors that have been cloned and characterized, only melanocortin-3 receptor (MC3-R) and melanocortin-4 receptor (MC4-R) are expressed within the brain at appreciable levels and are activated by α-MSH (602,603,605–607). Three major subfamilies of opioid receptors have been identified (mu, kappa, and delta); β-endorphin binds to both mu and delta receptors with high affinity (608).

Within the ARC, POMC neurons are transcriptionally regulated in response to alterations in metabolic status. POMC mRNA levels are reduced during states of negative energy balance, including lactation (540), fasting (94,609–611), and chronic food restriction (612). In contrast, POMC mRNA expression in the ARC is up-regulated by involuntary overfeeding (613). In humans, CSF concentrations of POMC-derived peptides are reduced in underweight individuals with anorexia nervosa but are normalized in anorexic women who have achieved a significant degree of weight gain (614).

POMC neurons in the ARC are sensitive to multiple peripheral metabolic cues, including leptin, insulin, and glucose. Leptin induces Fos expression in POMC neurons (615), increases the firing rate of POMC neurons in hypothalamic slices (616), and stimulates α-MSH secretion from hypothalamic explants in vitro (617). Levels of POMC mRNA are reduced in leptin-deficient *ob/ob* mice but are significantly increased when *ob/ob* mice are treated with leptin (609,610,618). Rodents with disrupted leptin signaling because of leptin receptor mutations also exhibit reductions in POMC mRNA levels and α-MSH peptide content (609,610,619). Fasting and STZ-induced diabetes are both associated with reduced circulating levels of insulin and leptin as well as decreased expression of POMC mRNA in the ARC (609,620). Peripheral administration of insulin normalizes POMC mRNA levels in STZ-treated diabetic rats (620), and exogenous administration of leptin or insulin increases POMC mRNA levels in fasted rats (609,621). POMC neurons in the ARC express leptin receptor mRNA (222,622) and protein (294) and are immunoreactive for the beta subunit of the insulin receptor (621), indicating that POMC neurons are likely to be direct hypothalamic targets for leptin and insulin. Demonstrations that POMC neurons express glucokinase (an enzyme that is necessary for glucose sensing) and exhibit reduced firing rates in response to reductions in extracellular glucose concentrations suggest that POMC neurons are also capable of directly detecting alterations in glucose availability (554,623). However, levels of POMC mRNA are unaltered by peripheral 2-DG treatment, which argues against a role for POMC peptide products in mediating the endocrine and behavioral responses to glucoprivation (550,624).

α-**MSH.** The central melanocortin system has received considerable attention over the past decade, ascribable to its prominent role in modulating energy intake and energy expenditure [for a review see (625)]. POMC neurons project to regions of the brain that receive and integrate feeding-related cues, including the ARC, PVN, the lateral hypothalamus, and the dorsal motor nucleus of the vagus (626). MC3-R and/or MC4-R are expressed in each of these regions (602,603). Although multiple melanocortin peptides have been shown to influence feeding behavior, α-MSH has been accorded a primary role in regulating food intake. The acute injection of α-MSH or melanocortin receptor agonists into the cerebral ventricles or directly into various hypothalamic nuclei reduces food intake and body weight (611,627–631). Conversely, food intake and body weight are increased after the central infusion of melanocortin receptor antagonists, implying that endogenous melanocortins normally exert a tonic restraint upon food consumption (628,629,632,633). The suppressive effects of leptin and insulin on food intake are at least partially mediated by central melanocortin signaling, because ICV administration of the nonspecific melanocortin receptor antagonist SHU9119 blocks or attenuates the ability of either hormone to reduce food intake (292,621,634). The functional relevance of an intact melanocortin signaling system to body weight regulation is underscored by observations that mice and humans with POMC gene mutations are obese (635,636), as are mice and humans with loss-of-function mutations in the MC3-R (637,638) or MC4-R genes (639–643).

In contrast to the relatively well-established roles of melanocortins (particularly α-MSH) in modulating energy intake, the relevance of central melanocortin signaling pathways to neuroendocrine reproductive function is less certain. POMC neurons project to the POA (644), where synaptic contacts have been observed between POMC-containing fibers and GnRH neurons (645,646). Although MC3-R and MC4-R mRNAs are expressed in the MPOA (602,603), it is currently unknown whether either receptor is expressed by GnRH neurons. However, it is worth noting that MC3-R and MC4-R transcripts are endogenously expressed by GnRH-producing neuronal cell lines, in which α-MSH stimulates cyclic AMP production and elicits GnRH secretion (647). Together, these observations hint at the existence of a neural pathway by which melanocortin signaling could impinge on GnRH neuronal function.

Several investigators have tested the effects of central or peripheral injections of α-MSH, melanocortin receptor analogues, and/or α-MSH antiserum and have reported a stimulatory (648–651), inhibitory (652–654), or no effect (292,655) of melanocortin signaling on GnRH and/or gonadotropin secretion. Thus, a clear picture of the physiological role (if any) of melanocortin signaling in the regulation of neuroendocrine reproductive function has yet to emerge. In contrast to its effects on hormone secretion, α-MSH has consistently been observed to stimulate lordosis behavior in female rats when it is injected into the hypothalamic ventromedial nucleus or the median eminence, thus implicating a potential role for melanocortin signaling in the generation of female-specific sexual behaviors (656,657).

The most compelling evidence against a critical role for melanocortin signaling in relaying nutritional signals to the reproductive axis is provided by repeated observations that reproductive function is preserved when central melanocortin signaling is chronically disrupted. In male rats, continuous ICV infusion of SHU9119 for 7 days does not affect either gonadotropin or testosterone secretion (633). Although the central administration of SHU9119 to *ob/ob* mice attenuates leptin's suppressive effects on feeding and body weight, the melanocortin receptor antagonist does not impair leptin's ability to improve various measures of reproductive function in the same animals (292). Finally, humans with MC4-R mutations advance through puberty at a normal age and are fertile as adults (641–643). Based on the observations noted above, it would appear that a functional melanocortin system is not essential for normal reproductive function. Although melanocortin signaling is critically involved in the neural control of food intake and body weight, central melanocortin pathways are unlikely to play a major role in conveying metabolic information to the neuroendocrine reproductive axis.

β-Endorphin. The endogenous opioid peptides (which include β-endorphin, the dynorphins, and the enkephalins) are involved in many physiological processes, including energy balance, neuroendocrine function, reward, nociception, and the body's complex responses to stress (658). Although β-endorphin and α-MSH are both cleaved from the POMC precursor protein and are secreted by the same populations of neurons, the two peptides exert several divergent effects within the CNS (595,659,660).

Although α-MSH has been shown to exert a tonic restraint upon food intake, β-endorphin stimulates food consumption in many species. β-Endorphin or opiate receptor agonists (e.g., morphine) enhance food intake, whereas β-endorphin neutralizing antibodies or opiate receptor antagonists (e.g., naloxone) inhibit feeding [for a review see (661)]. Chronic naloxone infusion in rats induces a sustained reduction in food intake and a transient decrease in body weight gain (662). Paradoxically, male mice that lack β-endorphin but retain normal expression of the other POMC-derived peptides are obese due to increased food intake, which suggests a previously unidentified anorectic role for endogenous β-endorphin signaling (663).

Endogenous opioid peptides exert an inhibitory effect on the HPG axis. Administration of β-endorphin or morphine suppresses LH secretion in males and females of several species (664–667) and blocks the preovulatory LH surge and ovulation in cycling female rats (668). Humans who undergo long-term morphine treatment for pain management experience hypogonadotropic hypogonadism, amenorrhea or irregular menstrual cycles, and reduced libido (669). Conversely, opiate receptor antagonists (e.g., naloxone and naltrexone) increase gonadotropin secretion in rats (664,670,671), rabbits (672), sheep (673), monkeys (667,674), and humans (675–677). Neither opiate receptor agonists nor antagonists alter the pituitary LH response to exogenous GnRH administration (667,670,677). Furthermore, treatment with a GnRH antagonist prevents naloxone from stimulating LH secretion (678). Taken together, these observations suggest that opioid signaling pathways influence gonadotropin secretion through a hypothalamic site of action. This supposition is further supported by observations that morphine reduces GnRH pulse generator activity in OVX rhesus monkeys (679), infusion of β-endorphin into the median eminence reduces GnRH secretion in lambs (49), and naloxone increases GnRH release both in vitro (680,681) and in vivo (672,682,683).

Although POMC neurons synapse upon GnRH neurons in the POA and ARC (645,646,684), GnRH neurons do not appear to express any of the classical opioid receptors (685,686). Notwithstanding, Lagrange et al. (687) observed that GnRH neurons are hyperpolarized by a mu opioid receptor agonist, even when synaptic transmission is abolished by the addition of the sodium channel blocker tetrodotoxin to the incubation medium. Thus, it is presently unclear whether β-endorphin suppresses GnRH secretion through direct actions on GnRH neurons or whether the inhibitory effects of β-endorphin are indirectly relayed to GnRH neurons by way of interneurons.

Endogenous opioid peptides are involved in mediating many of the endocrine, autonomic, and behavioral responses to a variety of stressors, including stress-induced inhibition of reproductive function. For example, rats that are subjected to various forms of acute or chronic stress exhibit reductions in LH release, but this effect is often reversed by the administration of opioid receptor antagonists or β-endorphin antiserum (688,689). Opioid signaling pathways have also been implicated in mediating the inhibitory effects of metabolic stressors on LH secretion. Opioid receptor antagonists have been reported to attenuate or reverse the suppression of LH secretion in fasted rats (688,690,691), lactating animals (692), and in some women with anorexia nervosa (693). However, naloxone does not significantly increase LH secretion in fasted monkeys (694) or in chronically food-restricted lambs (51) or pigs (695), which indicates that in these animals, endogenous opioids do not play a primary causal role in suppressing gonadotropin secretion in response to the nutritional stress of reduced energy availability.

Galanin-Like Peptide

Galanin-like peptide (GALP) is one of the most recent additions to the growing list of hypothalamic neuropeptides believed to relay information about nutritional status to the neuroendocrine reproductive axis. GALP is a 60 amino acid peptide that was originally isolated from the porcine hypothalamus and has since been cloned in the mouse, rat, pig, macaque, and human (696–698). In all species examined to date, the expression of GALP mRNA in the forebrain is confined to the ARC, median eminence, and the infundibular stalk (697–701). Takatsu et al. (702) confirmed GALP's distribution pattern at the protein level by demonstrating that GALP-immunoreactive cell bodies are confined to these same hypothalamic regions. Although a small percentage (~7%) of GALP neurons in the ARC coexpress α-MSH, most GALP neurons appear to be phenotypically distinct from other identified neuronal populations in the ARC (700,702–704).

The expression of GALP mRNA in the ARC is sensitive to negative energy balance and its attendant fluctuations in metabolic hormones and fuel availability (699,704,705). Both fasting and experimentally induced insulin-dependent diabetes mellitus are associated with decreased plasma concentrations of leptin and insulin as well as reduced GALP mRNA expression in the ARC (704,705). In fasted female rats, systemic leptin injections significantly increase the number of GALP mRNA-expressing cells in the ARC, suggesting that GALP neurons are positively regulated by leptin (699). This presumption is further supported by the observations that hypothalamic levels of GALP mRNA are dramatically reduced in leptin-deficient *ob/ob* mice and in rats *(fa/fa)* and mice (*db/db*) with deficient leptin receptor signaling (697,706,707) (Fig. 12). Daily injections of leptin into the lateral ventricle of *ob/ob* mice restore GALP mRNA to levels comparable with those seen in wild-type mice, thus demonstrating that leptin modulates the activity of GALP neurons through a central mechanism(s) (697). In addition to increasing cellular levels of GALP mRNA, leptin stimulates the secretion of GALP from hypothalamic explants in vitro (708). A direct modulatory effect of leptin on GALP neurons is inferred based on findings that the vast majority (88–98%) of GALP neurons in the ARC express leptin receptors (698,702).

GALP gene expression is also influenced by alterations in insulin secretion and/or sensitivity to

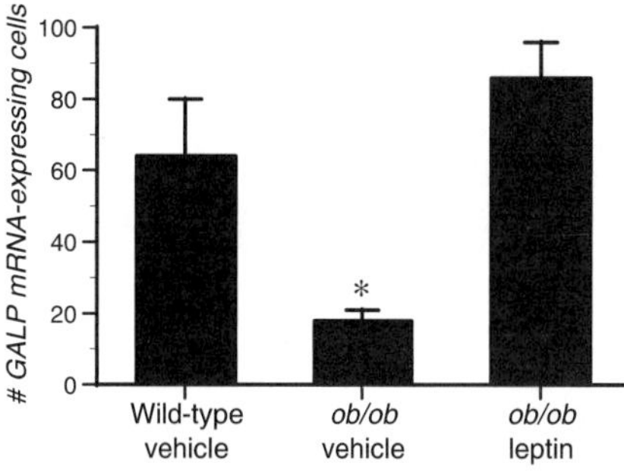

FIG. 12. GALP mRNA levels in the ARC of male wild-type and *ob/ob* mice that received daily ICV injections of leptin (0.02 nmol) or the vehicle alone for 7 days. The number of GALP mRNA-expressing cells was significantly reduced in the vehicle-treated *ob/ob* mice but was restored to wild-type levels by long-term leptin treatment. **P* <0.01 vs. all other groups. [Redrawn from (697) with permission. Copyright 2001, The Endocrine Society.]

insulin's actions. Compared with nondiabetic controls, diabetic rats that are rendered insulin deficient by STZ treatment show markedly reduced levels of GALP mRNA, which can be normalized by insulin replacement (705,707). The expression of GALP mRNA in the ARC is reduced in insulin-resistant obese Zucker rats but is significantly increased in these animals by treatment with exogenous insulin (707,709). ICV injections of insulin fully reverse the suppressive effect of a 48-hour fast on hypothalamic GALP mRNA levels, which implicates a role for central insulin signaling in the regulation of GALP neuronal function (705) (Fig. 13). The stimulatory effect of insulin on the expression of GALP mRNA is unlikely to be secondary to insulin-induced hypoglycemia, because central glucoprivation (induced by 2-DG) reduces GALP gene expression in the ARC in a manner similar to prolonged food deprivation (710).

Given that GALP gene expression in the ARC is reduced in energy-deficient states and is increased by hormones such as leptin and insulin, which are associated with energy repletion, one might predict that GALP neurons play an intermediary role in relaying metabolic signals to the neural circuits that control feeding and metabolism. GALP neurons project to hypothalamic nuclei that are involved in regulating feeding behavior (702), and ICV injection of GALP induces Fos expression in several feeding-related loci (711). Contrary to what might be expected of a neuropeptide whose expression is augmented by anorectic hormones such as insulin and leptin, ICV or intrahypothalamic injections of GALP acutely stimulate food consumption in rats (712–714). However, food intake and body weight gain are significantly

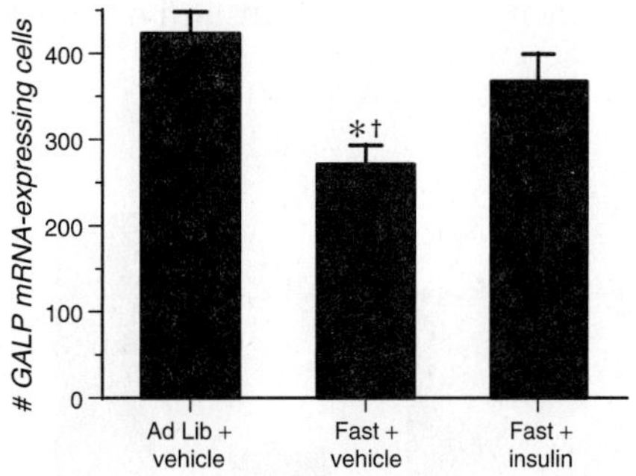

FIG. 13. Effects of food deprivation and central insulin treatment on GALP mRNA expression in the ARC. Male rats were allowed to feed ad libitum or were fasted for 48 hours, during which time the animals received five ICV injections of insulin (4 mU) or the vehicle alone. Compared with the ad libitum fed rats, GALP mRNA levels were significantly reduced by a 48-hour fast, an effect that was partially reversed by central insulin administration. *P <0.001 vs. Ad lib + Vehicle rats, †P <0.05 vs. Fast + Insulin rats. [Redrawn from (705). Copyright 2004, reprinted with permission from American Diabetes Association.]

reduced in rats and mice over the 24-hour period after an ICV injection of GALP (713,715). Long-term GALP treatment induces a transient suppression of food intake but a sustained reduction in body weight in *ob/ob* mice (716). Centrally administered GALP significantly increases core body temperature in rats via a prostaglandin-dependent mechanism (713). Long-term GALP treatment increases thermogenesis in *ob/ob* mice, as evidenced by elevated body temperatures and increased brown adipose tissue expression of uncoupling protein-1 mRNA and protein compared with vehicle-treated *ob/ob* mice (716). Collectively, these observations demonstrate that centrally administered GALP influences food intake and energy expenditure in rodents and implicate GALP neurons as likely components of the neural circuitry that regulates energy homeostasis.

A putative role for GALP in the regulation of HPG function was initially proposed based on neuroanatomical evidence that GALP neurons in the ARC send axonal projections to the MPOA and the bed nucleus of the stria terminalis, where GALP fibers closely appose GnRH-immunoreactive cell bodies and fibers (702). Furthermore, ICV injection of GALP induces Fos expression in a subset of GnRH neurons, indicating that these GnRH neurons are activated by exogenous GALP administration (717).

Injection of GALP into the lateral or third cerebral ventricles stimulates the release of LH and testosterone in male mice, rats, and rhesus macaques (704,715,717) (Fig. 14). GALP's stimulatory effect on LH secretion involves a GnRH-dependent pathway, because pretreatment with a GnRH antagonist completely abolishes GALP's stimulatory effect on LH release in rats and monkeys (704,717). GALP does not stimulate basal or GnRH-induced LH secretion from dispersed anterior pituitary cells, thus providing further evidence in support of a hypothalamic site of action (717). In vitro, GALP stimulates the secretion of GnRH from hypothalamic explants and from GT1-7 cells (708). A role for GALP in relaying leptin's stimulatory signal to the HPG axis is supported by the same group's finding that incubation of hypothalamic explants with GALP antiserum blocks leptin's ability to stimulate GnRH secretion (708).

Central injections of GALP activate transcription in a subset of neurons in the POA (717,718), a region that is important in the generation of male-specific sexual behaviors (719). In male rats, ICV administration of GALP robustly increases the frequency of mounting, intromission, and ejaculation and decreases the latency to all three behaviors, compared with treatment with the vehicle alone. The stimulatory effect of GALP on mating behavior persists (albeit with reduced magnitude) in castrated male rats, thus indicating that GALP modulates sexual

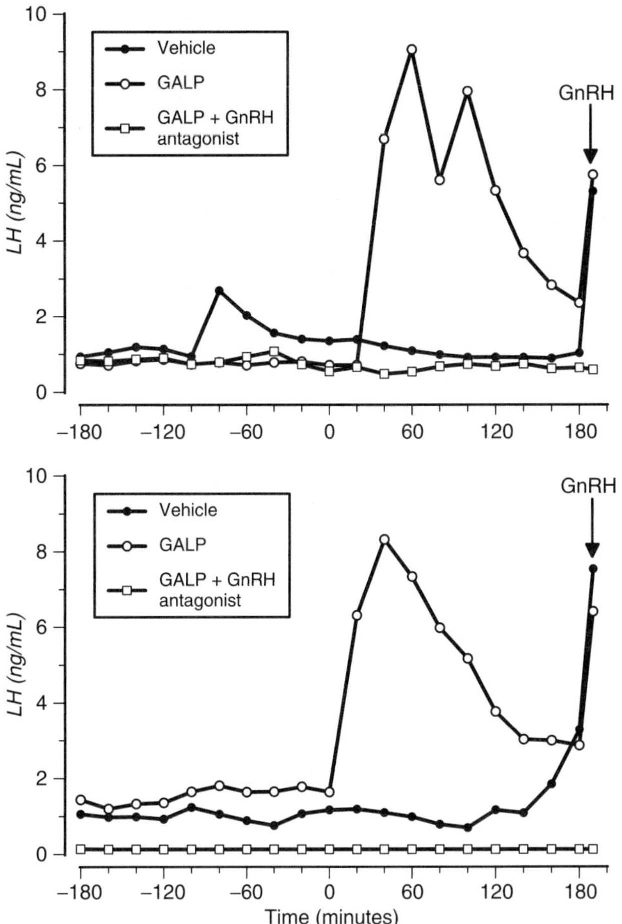

FIG. 14. Plasma LH profiles in two testosterone-implanted agonadal male rhesus macaques after an ICV injection of GALP (500 μg; *open circles*) or the vehicle alone (*filled circles*) at time zero. Pretreatment with the GnRH antagonist acyline (60 μg/kg body weight; *open squares*) blocked the stimulatory effect of GALP on LH secretion. Both monkeys received all three treatments on different days. The animals received an intravenous injection of GnRH (0.3 μg) 3 hours after the vehicle or GALP injection to confirm gonadotrope responsiveness. [Redrawn from (704) with permission. Copyright 2004, The Endocrine Society.]

behavior independently of testosterone and other testicular secretory products (720).

In summary, GALP neurons in the ARC are ideally situated to serve as a molecular link between energy metabolism and the HPG axis. GALP neuronal activity is modulated in response to various metabolic challenges, and GALP mRNA levels are altered by circulating metabolic signals. Presumably, GALP neurons integrate and relay metabolic information to the neural circuits that govern reproductive function at both the hormonal and behavioral levels. It is yet to be discovered whether GALP neurons interact directly with GnRH neurons to elicit GnRH secretion or GALP exerts its actions indirectly through interneurons. The endogenous receptor(s) for GALP has yet to be identified, which has impeded efforts to investigate the central effector mechanisms by which GALP exerts its effects on energy homeostasis and reproductive output. Multiple lines of evidence suggest that exogenous GALP administration mimics many of the central effects of leptin and insulin on energy balance and HPG function; however, it is currently unknown whether GALP neurons are downstream targets of leptin and/or insulin signaling pathways that are physiologically relevant to body weight regulation and reproduction or whether GALP modulates energy homeostasis and reproduction through parallel regulatory pathways. Hopefully, the advent of new molecular and pharmacological tools will aid in addressing these questions.

Corticotropin-Releasing Hormone

Corticotropin-releasing hormone (CRH) neurons in the hypothalamus play a pivotal role in initiating a host of endocrine, autonomic, and behavioral responses to a variety of physical and psychological stressors. CRH neurons are located in several regions of the brain, including the basal forebrain, hypothalamus, and brainstem (721). Within the hypothalamus, CRH-expressing neurons are primarily located in the parvicellular division of the PVN (721,722), an area that receives and processes information from various feeding-related hypothalamic and extrahypothalamic sites (723). CRH functions as a catabolic neuropeptide, reducing body weight through its combined effects of decreasing food consumption and increasing sympathetic output to brown adipose tissue (724,725). CRH neurons in the PVN express leptin receptors (294). CRH mRNA expression is reduced in food-restricted animals, an effect that is reversed by the central administration of leptin (541). Leptin also stimulates CRH release from hypothalamic slices in vitro (726). CRH neurons are believed to be downstream mediators of leptin's effects on energy homeostasis, because treatment with a CRH antagonist attenuates the inhibitory effects of leptin on food intake and body weight in rats and mice (727–729).

From an adaptive standpoint, it would be disadvantageous for an organism to devote significant amounts of energy toward reproduction during a potentially life-threatening, prolonged, stressful situation. Accordingly, one of the central actions of CRH may be to inhibit reproductive activity during stressful times. Injection of CRH into the cerebral ventricles suppresses sexual behavior in male rats (730), and infusion of CRH into the cerebral ventricles or directly into the ARC/ventromedial nucleus inhibits estrous behavior in steroid-primed female rats and hamsters (731,732). Steroid-primed female mice that overexpress CRH do not exhibit lordosis behavior or mate with male mice (733).

CRH also inhibits the secretion of gonadotropins (primarily LH but sometimes FSH too) in gonadectomized males and females of several species (689,734–737). In intact female rats, exogenous CRH administration blocks the preovulatory LH surge and inhibits ovulation (735). The observations that CRH does not alter LH release from dispersed anterior pituitary cells in vitro (734) and that CRH inhibits the secretion of GnRH both in vivo (737,738) and in vitro (739,740) imply a hypothalamic site of action for CRH. This inference is bolstered by the facts that CRH-containing nerve terminals form synaptic contacts with GnRH neurons in the MPOA (741) and that CRH alters GnRH transcription in a GnRH-expressing neuronal cell line (742). The inhibitory effects of CRH on reproductive function are believed to be at least partially mediated by endogenous opioid pathways, because treatment with opiate receptor antagonists or β-endorphin antiserum attenuates or reverses the inhibitory effects of CRH on GnRH and LH secretion (689,737,738,743,744), activity of the GnRH pulse generator (745), and sexual behavior (730,731).

CRH signaling pathways play a role in mediating the inhibitory effects of various stressors on reproductive function, including the nutritional stress associated with reduced energy availability. Several studies have revealed that interfering with central CRH signaling prevents or attenuates the impact of diminished energy availability on HPG function. CRH neurons have been proposed to mediate the inhibitory effect of glucoprivation on LH secretion, because pretreatment with a CRH antagonist prevents 2-DG from suppressing pulsatile LH release in estrogen-treated OVX rats (746). Administration of a CRH antagonist into the cerebral ventricles also prevents the fasting-induced inhibition of estrous behavior in female hamsters (732) as well as the fasting-induced reduction in pulsatile LH secretion in estrogen-treated OVX rats (747). However, the finding that fasting-induced reductions in LH and testosterone secretion persist in CRH knockout mice demonstrates that CRH neurons are not solely responsible for mediating the inhibitory effects of undernutrition on reproductive hormone secretion (748). For a more detailed account of the interactions between the hypothalamic–pituitary–adrenal and HPG axes, we refer the reader to Chapter 48.

Melanin-Concentrating Hormone

Melanin-concentrating hormone (MCH) is a 19 amino acid peptide that was isolated from the salmon pituitary and was originally noted for its role in altering skin coloration in fish (749). Within the mammalian brain, MCH-containing cell bodies are primarily located in the lateral hypothalamus, zona incerta, and the perifornical area (750,751). MCH-containing fibers are widely distributed throughout the CNS (751,752). MCH is the cognate ligand for the SLC-1 receptor (now termed the "MCH receptor"), which is expressed in many hypothalamic and brainstem nuclei (753–755).

MCH neurons in the lateral hypothalamus and zona incerta are regulated by nutritional cues. Levels of MCH mRNA are increased in leptin-deficient *ob/ob* mice and in fasted animals having reduced serum leptin concentrations but are reduced in these animals by leptin treatment (756,757). Leptin delivered intracerebroventricularly also reduces levels of hypothalamic MCH mRNA in nonfasted animals (758). MCH-expressing neurons in the lateral hypothalamus express leptin receptors, which would argue that leptin has direct effects on MCH neurons (294). Further evidence that MCH neurons are downstream targets of leptin include observations that MCH deficiency attenuates the obesity phenotype of *ob/ob* mice (759), and injection of MCH antiserum into the MPOA of estrogen-primed OVX rats prevents leptin from stimulating LH secretion when it is injected into the zona incerta (760).

Although the lateral hypothalamus contains a large population of glucose-sensitive neurons, there have been inconsistent findings about whether MCH neurons are sensitive to glucose availability. Although one group has reported that MCH mRNA expression in the lateral hypothalamus is up-regulated by glucoprivation (550), others have been unable to replicate this finding either in vivo (761) or in vitro (762). Because of these apparent discrepancies, it is presently unclear whether MCH neurons participate in mediating the feeding and/or reproductive responses to glucoprivation.

The lateral hypothalamus has long been recognized as a "feeding center" in the brain (763), and MCH neurons in the lateral hypothalamus have been identified as components of the neural circuitry that controls appetitive behavior. The acute injection of MCH into the cerebral ventricles stimulates food intake (756,764,765), whereas chronic central MCH infusion increases food consumption, body weight, and adipose tissue mass (765,766). Transgenic mice that overexpress MCH in the lateral hypothalamus develop obesity and insulin resistance (767). In contrast, mice bearing a targeted disruption of the MCH gene are hypophagic and lean, and to date, MCH is the only neuropeptide whose targeted deletion gives rise to a lean phenotype (768).

MCH has also been implicated in the regulation of reproductive function. In OVX/adrenalectomized rats that are primed with estrogen, injection of MCH into the MPOA or ventromedial nucleus enhances sexual receptivity (769). Plasma LH levels are increased when MCH is infused into the MPOA or median eminence, and LH release is reduced by injection of MCH antiserum into the MPOA of steroid-primed OVX/adrenalectomized rats (770). In contrast, Tsukamura et al. (771) observed that ICV MCH administration reduces circulating levels of LH and LH pulse frequency in adrenal-intact OVX rats that are treated with estrogen. This suggests that the effects of MCH on gonadotropin release might depend on circulating gonadal and adrenal steroid concentrations (771). MCH-containing fibers terminate in close proximity to pituitary portal blood vessels in the median eminence, and MCH stimulates LH and FSH release from pituitaries collected from female rats at proestrus (772). Notwithstanding these observations, other lines of evidence point toward a hypothalamic site of action for MCH. MCH-immunoreactive fibers are located in the POA (752), where they are found in close contact with GnRH neurons (133). Furthermore, MCH stimulates GnRH release from the median eminence of proestrous rats in vitro (772). Thus, MCH appears to modulate gonadotropin secretion both within the brain and at the level of the pituitary.

Orexins

The orexins (also known as hypocretins) are expressed by a distinct population of neurons in the lateral hypothalamus and are involved in the regulation of feeding behavior, neuroendocrine function, activity, and arousal [for review see (773)]. The orexins are derived from a larger precursor protein, prepro-orexin (preprohypocretin), which is cleaved into orexin-A (hypocretin 1) and orexin-B (hypocretin 2) (774,775). Orexin-containing cell bodies are located in the lateral hypothalamus and perifornical area and send fiber projections to virtually all regions of the CNS (774–777). Two orexin receptors have been identified, both of which are widely distributed throughout the hypothalamus and brainstem (778,779). According to van den Pol et al. (780), approximately one-third of lateral and medial hypothalamic neurons are responsive to orexins in vitro. Whereas the orexin receptor-1 selectively binds orexin-A, the orexin receptor-2 binds orexin-A and orexin-B with similar affinities (774).

Like MCH neurons, orexin neurons in the lateral hypothalamus are sensitive to alterations in energy balance. Although several investigators have reported that prepro-orexin mRNA is elevated in fasted animals (774,781), expression of the orexin precursor gene is apparently not up-regulated in chronically food-restricted animals (781). Orexin neurons express leptin receptors (782–784), and leptin treatment reverses the fasting-induced elevation of prepro-orexin mRNA in food-deprived rats (785). Prepro-orexin mRNA is increased by insulin-induced hypoglycemia (781) but is not altered by 2-DG–induced glucoprivation (550,781). In hypothalamic slices, a subset of orexin neurons is activated by decreases in extracellular glucose, suggesting that some orexin neurons in the lateral hypothalamus are glucose sensitive (786).

The orexins were named based on their orexigenic effects on food intake. Injection of orexin-A into the cerebral ventricles or directly into various hypothalamic nuclei stimulates food consumption (774,787–789). In contrast, the central injection of orexin-A antibodies decreases food intake (790). Although chronic ICV infusion of orexin-A increases daytime food intake, it does not alter either body weight or composition (789,791). The effects of orexin-B on feeding have been inconsistent, with several groups reporting either a transient stimulatory effect (774) or no effect at all (788,789). The orexins also increase metabolic rate and thermogenesis (792–794), and transgenic mice with ablated orexin neurons develop a late-onset obesity due to reduced energy expenditure (795).

Like many other hypothalamic neuropeptides, the orexins exert a steroid-dependent bimodal effect on LH secretion. In the presence of basal or elevated concentrations of gonadal steroids, the orexins stimulate gonadotropin secretion. In steroid-primed OVX rats, orexin-A and orexin-B stimulate LH release (796). Conversely, reductions in endogenous orexin signaling result in impaired LH secretion. Central administration of orexin-A neutralizing antibodies abolishes the LH surge in steroid-primed OVX rats (797). Pulsatile LH secretion is reduced in narcoleptic men with orexin deficiency (798). In contrast to the effects of orexins on LH secretion in the presence of basal or elevated gonadal steroid concentrations, ICV administration of orexin-A or orexin-B reduces mean plasma LH levels and LH pulse frequency in OVX rats and in OVX rats that are treated with a low dose of estrogen (796,799,800). Orexin-containing fibers are present in the POA and septal nuclei (775), where they closely appose GnRH neurons (783,801,802). Orexin receptor-1 is present on GnRH cell bodies in the POA but not on GnRH nerve terminals in the median eminence (801). In vitro, orexin-A stimulates GnRH release from hypothalamic explants in both sexes (802,803). Orexin-A does not alter basal LH release but inhibits GnRH-induced LH secretion from dispersed pituitary cells collected from female rats at proestrus (803). Although orexin neurons are anatomically and functionally linked to GnRH neurons, it is currently unknown whether orexin neurons play a role in coupling nutritional signals with the activity of the reproductive axis.

METABOLIC DISORDERS AND REPRODUCTIVE FUNCTION

The coupling between metabolism and reproduction likely involves complex interactions between numerous metabolic signaling pathways. In many cases, the functional significance of a particular metabolic signal to normal physiology becomes more apparent under circumstances in which its levels are altered and/or its actions are disrupted. In this section, we briefly describe four human disorders that are either characterized by or commonly associated with impaired metabolic regulation as well as varying degrees of reproductive dysfunction. The association (or lack thereof) between disrupted energy homeostasis and reproductive impairment sheds light on the relative importance of various metabolic signals to the pathogenesis and pathophysiology of metabolism-related reproductive disorders and provides insight into the role of these metabolic signals in normal reproductive physiology.

Obesity

Obesity has reached epidemic proportions in many developed and developing nations and is consequently well on its way to surpassing undernutrition and infectious diseases as the most significant contributor to ill health (804). Just as being underweight can have profound negative consequences on the reproductive axis, many obese individuals also experience variable degrees of reproductive impairment. Excess adiposity can have detrimental effects on both male and female reproductive function (805,806). Obese men sometimes experience impotence, reduced libido, and diminished plasma LH and testosterone concentrations (807,808). Amenorrhea and anovulation (809,810), infertility (811–813), and various complications during pregnancy and childbirth (814) occur with greater frequency in obese women than in age-matched women of normal body weight and/or body mass index. Some investigators have postulated that reproductive impairment in obese women is better predicted by body fat distribution than by body weight or body mass index, based on observations that anovulation and menstrual disturbances are more closely associated with upper body obesity than with lower body obesity or overall adiposity (815,816). In many cases, the adverse effects of obesity on reproductive function are reversible, because sustained weight loss often increases serum LH and testosterone levels in obese men (817) and improves menstrual cyclicity and fertility in obese women (818–820).

The factors that contribute to reproductive impairment in obese women are still poorly understood. Likewise, it is unclear why only a fraction of obese women experience reproductive dysfunction. Although there have been a few documented cases of leptin-deficient obese humans, most obese individuals have elevated circulating leptin concentrations, due to their excess adipose mass (179,180). Given their obese state despite having elevated plasma leptin concentrations, most obese individuals are believed to be at least partially resistant to leptin's weight- and appetite-reducing actions (179,180), perhaps as a consequence of a reduced capacity for leptin transport into the CNS (821,822). Thus, it is conceivable that obese women also possess a diminished sensitivity to leptin's stimulatory effects on the neuroendocrine reproductive axis. Although this hypothesis has not been directly tested in obese women, the reproductive deficiencies of mice with diet-induced obesity provide evidence that is consistent with this argument. Female mice that are fed a high fat diet ultimately become obese, with elevated plasma leptin concentrations and dramatically reduced hypothalamic leptin receptor expression, which collectively are suggestive of a leptin-resistant state. Compared with nonobese mice, leptin-resistant obese female mice have a significantly reduced pregnancy rate, which is most likely attributable to a central defect (823). Although leptin has a stimulatory effect on the reproductive axis at the hypothalamic and pituitary levels, leptin inhibits steroidogenesis when it is present at high concentrations (218). Therefore, the elevated levels of leptin in obese individuals might exert direct inhibitory effects on gonadal function.

Obese women with menstrual dysfunction commonly exhibit hyperandrogenism and insulin resistance, both of which are significantly improved by weight loss (819,824,825). The fact that reproductive function is also improved in many weight-reduced obese patients implicates elevated androgens and/or insulin resistance in the development of obesity-related menstrual disturbances. It has also been reported that plasma concentrations of NPY and β-endorphin are elevated in obese women (826,827). Given the complex pathophysiology of obesity, it is likely that obesity-associated reproductive impairment results from interactions between numerous endocrine, metabolic, and neural factors.

Diabetes Mellitus

Diabetes mellitus is a disease that is characterized by profound disturbances in carbohydrate, fat, and protein metabolism. The adverse metabolic consequences of diabetes are either directly or indirectly attributable to impaired insulin action, either as a consequence of insulin deficiency (type 1 diabetes) or of the development of tissue resistance to insulin's actions (type 2 diabetes). Reproductive dysfunction is but one of many diabetes-related complications and is particularly prevalent in type 1 diabetic patients due to the typical early onset of the disease during childhood or early adulthood. Women with poorly controlled type 1 diabetes often experience delayed menarche, reduced LH pulsatility, irregular menses, and infertility (361,362,397,419,828). Impotence and impaired sperm motility are common in men with type 1 diabetes (356–358). Many of the same metabolic and reproductive disturbances that occur in diabetic humans have also been observed in experimental models of type 1 diabetes, including rats that are rendered insulin-deficient by STZ or alloxan treatment. Diabetic male rats have reduced testosterone concentrations (367), diminished pulsatile LH secretion (365), and impaired copulatory behavior (368). STZ- or alloxan-treated diabetic female rats experience delayed puberty (372), reduced ovulation rates (374,375), loss of estrous cyclicity (372,375), and reduced sexual behavior (377,378).

In contrast to the well-established effects of poorly controlled type 1 diabetes on the HPG axis, much less has been reported about reproductive function in individuals with type 2 diabetes. The fact that type 2 diabetes typically occurs in conjunction with other metabolic and nonmetabolic diseases (e.g., obesity, hypertension, etc.) would make it difficult to isolate diabetes as the underlying cause of reproductive dysfunction in these individuals. Furthermore, type 2 diabetes usually has a late onset, often occurring at an age when fertility has already begun to decline.

Both metabolic function and reproductive capacity in type 1 diabetic patients are usually improved by insulin therapy. Insulin treatment reduces the incidence of menstrual dysfunction and improves fertility in women with type 1 diabetes (361,362,660,829). Similarly, peripheral insulin replacement reverses many of the adverse reproductive consequence of diabetes in STZ- or alloxan-treated rats (365,366,372, 374,378,383,830). Impaired reproductive function in diabetic individuals could conceivably be attributable to their reduced capacity for cellular glucose uptake and metabolism in the periphery, both of which are improved by insulin treatment. However, centrally administered insulin (at doses that do not affect peripheral glycemia) prevents the diabetes-induced blockade of the LH surge and impaired copulatory behavior in STZ-treated female rats (376,384) and increases LH pulse frequency in diabetic sheep (385). These observations imply that insulin acts directly on the brain to mediate the reversal of diabetes-related reproductive dysfunction. However, insulin also has potent effects on gonadal function (397); thus, the lack of insulin in type 1 diabetics could conceivably disrupt reproductive function at multiple levels of the HPG axis.

Leptin concentrations are rapidly reduced in STZ-treated diabetic rats but are normalized by insulin treatment (831). In children and adolescents with newly diagnosed type 1 diabetes (before the onset of insulin therapy), circulating leptin concentrations are significantly lower than in healthy control subjects, even after adjusting for differences in body mass index (832). After intensive insulin therapy, plasma leptin concentrations are higher in diabetic children and adults than in nondiabetic control subjects of similar age and body mass index (833,834), which raises the possibility that chronically elevated insulin levels in insulin-treated diabetic patients might directly stimulate leptin production and/or secretion. Whether alterations in circulating leptin levels play a role in the disruption of reproductive function in diabetic individuals and/or the improvement of reproductive function after insulin treatment remains to be explored.

Lipodystrophies

The lipodystrophies are a group of rare heterogeneous disorders characterized by a selective loss of adipose tissue (835). The degree of fat loss varies among the particular variants, ranging from localized fat loss (partial lipodystrophy) to a complete loss of visible adipose tissue (generalized lipodystrophy). Lipodystrophic patients typically exhibit a host of metabolic perturbations, including diabetes, insulin resistance, hypertriglyceridemia, and hyperphagia (835). These metabolic disturbances are also seen in several lines of transgenic mice with ablated adipose tissue (836–839). Serum concentrations of leptin are dramatically reduced in lipodystrophic humans and mice as a consequence of their diminished adipose tissue mass, thus their reduced capacity for leptin synthesis (837,839–841). Long-term leptin treatment significantly improves glycemic control in lipodystrophic women (842,843) and mice (844–846), which implicates relative leptin deficiency as an underlying cause of insulin resistance and diabetes in these individuals.

Humans and rodents with leptin or leptin receptor mutations exhibit hypogonadotropic hypogonadism and do not spontaneously progress through puberty (230,244–246,253,255). Therefore, one might predict that lipodystrophic individuals would have similar reproductive deficiencies due to their dramatically reduced circulating levels of leptin. Consistent with this hypothesis, Moitra and colleagues (837) reported that fertility was compromised in both sexes of their lipodystrophic mice. However, clinical investigations have yielded mixed results regarding reproductive function in lipodystrophic patients. Although menarche occurred at a normal age in one cohort of women with generalized lipodystrophy, amenorrhea and menstrual irregularities were frequently reported (840). In another clinical investigation, menstrual cyclicity was disrupted in 80% of the lipodystrophic female subjects but was restored by 4 months of leptin therapy (847). In contrast, Andreelli et al. (298) reported that reproductive function was preserved in two women with early-onset generalized lipodystrophy, as evidenced by their normal timing of menarche, regular menstrual cyclicity, serum gonadotropin and sex steroid concentrations within the normal range, and their normal LH response to exogenous GnRH administration (298). Additionally, several case studies have documented successful pregnancies in lipodystrophic women (848–850).

In stark contrast to the dramatic reproductive deficiencies that occur in humans who are unable to synthesize or respond to leptin, lipodystrophic girls progress through puberty at a normal pace, and at least some lipodystrophic women retain normal reproductive function throughout adulthood. These observations are inconsistent with the notion that achieving some critical threshold level of leptin is a prerequisite for triggering the onset and maintenance of reproductive function in human females but rather suggest that leptin plays neither an activational nor a permissive role in initiating the onset of puberty and maintaining adult reproductive function, at least in the human. Alternatively, because serum leptin concentrations are significantly reduced but are not altogether absent in lipodystrophic women, it remains at least conceivable that even very low levels of leptin provide a sufficient stimulatory signal to the neuroendocrine reproductive axis in these individuals, particularly because these individuals may be exquisitely sensitive to leptin.

Polycystic Ovary Syndrome

Polycystic ovary syndrome (PCOS) is believed to be the most common endocrine disorder in women of reproductive age (851), afflicting approximately 6–7% of premenopausal women (852,853). PCOS is a heterogeneous disorder of unknown etiology. Despite the varied nature of its clinical manifestations, there appears to be a general consensus that hyperandrogenism and chronic anovulation are the two hallmark features of PCOS (851,854). Many women with PCOS experience menstrual dysfunction dating back to menarche (855), display altered gonadotropin secretory profiles (856–858), and develop ovarian cysts as a consequence of arrested follicular development (859). Furthermore, PCOS is the most common cause of anovulatory infertility in women (860).

In addition to reproductive abnormalities, women with PCOS commonly experience disordered metabolism. Approximately 50% of women with PCOS are overweight or obese (861). Compared with nonobese women with PCOS, obese PCOS patients usually present with more severe endocrine and metabolic symptoms (861). Insulin resistance and compensatory hyperinsulinemia are also prevalent in PCOS patients, thus causing women with PCOS to have an increased risk of developing type 2 diabetes (862,863). Insulin resistance occurs independently of adiposity in women with PCOS, because both nonobese and obese PCOS patients display signs of reduced insulin sensitivity and obese PCOS patients are more insulin resistant than obese women without PCOS (864,865). Circulating insulin and androgen concentrations are positively correlated in both nonobese and obese PCOS patients (862,864), and hyperinsulinemia has been postulated to play a causal role in the development of hyperandrogenism and ovulatory dysfunction in women with PCOS (859,863).

Women with PCOS are commonly treated with insulin-sensitizing drugs (e.g., metformin, troglitazone, etc.), which not only improve insulin sensitivity and glucose homeostasis, but also reduce circulating levels of LH and androgens, increase ovulatory rates, and improve menstrual function and fertility (866–868). In obese PCOS patients, insulin sensitivity, menstrual cyclicity, and fertility are also improved by moderate weight loss (869). The observation that metformin and the antiandrogen spironolactone have similar beneficial effects on reproductive function in PCOS patients but that only metformin significantly improves insulin sensitivity argues against a role for insulin resistance as the primary cause of reproductive impairment in women with PCOS (867).

Insulin regulates leptin synthesis and secretion in rats and humans (870,871), and rodents with impaired leptin synthesis or signaling exhibit many of the same metabolic and reproductive defects as women with PCOS, including obesity, hyperinsulinemia, insulin resistance, and infertility (227,228,243, 245,246). Hence, it seems plausible that impaired leptin action contributes to the clinical manifestations of PCOS. Although a subset of patients with PCOS were initially reported to have disproportionately higher levels of leptin than would be predicted based on their body mass index alone (872), several subsequent reports have revealed that PCOS patients have normal circulating leptin concentrations when compared with healthy women of comparable body weight or body mass index (873–876). Leptin sensitivity in PCOS patients has yet to be examined. In conclusion, the contribution of metabolic factors to reproductive dysfunction in PCOS patients remains unresolved.

ACKNOWLEDGMENTS

We thank Patrick Clarke for generating the figures for this chapter. We are immensely grateful to Dr. Donald Clifton for many years of scientific collaboration, as well as to the countless students and postdoctoral fellows (past and present) in the Steiner-Clifton laboratory for their hard work, dedication, and significant scientific contributions. Finally, we are grateful to the National Institutes of Health and the National Science Foundation for their ongoing financial support of our research.

REFERENCES

1. Bronson, F. H. (1989). *Mammalian Reproductive Biology*. The University of Chicago Press, Chicago.
2. Girish, S., and Saidapur, S. K. (2000). Interrelationship between food availability, fat body, and ovarian cycles in the frog, *Rana tigrina*, with a discussion on the role of fat body in anuran reproduction. *J. Exp. Zool.* 286. 487–493.
3. Jorgensen, C. B. (1982). Factors controlling the ovarian cycle in a temperate zone anuran, the toad *Bufo bufo*: food uptake, nutritional state, and gonadotropin. *J. Exp. Zool.* 224, 437–443.
4. Aubret, F., Bonnet, X., Shine, R., and Lourdais, O. (2002). Fat is sexy for females but not males: the influence of body reserves on reproduction in snakes (*Vipera aspis*). *Horm. Behav.* 42, 135–147.
5. Novacek, M. J. (1992). Mammalian phylogeny: shaking the tree. *Nature* 356, 121–125.
6. Foster, D. L. (1994). Puberty in the sheep. In *The Physiology of Reproduction* (E. Knobil and J. D. Neill, Eds.), Vol. 2, pp. 411–451. Raven Press, New York.
7. Bronson, F. H. (1987). Susceptibility of the fat reserves of mice to natural challenges. *J. Comp. Physiol. (B)*. 157, 551–554.
8. Cameron, J. L., Koerker, D. J., and Steiner, R. A. (1985). Metabolic changes during maturation of male monkeys: possible signals for onset of puberty. *Am. J. Physiol.* 249, E385–E391.
9. Wade, G. N., Schneider, J. E., and Li, H. Y. (1996). Control of fertility by metabolic cues. *Am. J. Physiol.* 270, E1–E19.
10. Ehrenberg, M. R. (1989). *Women in Prehistory*. British Museum Publications, London.
11. Medvei, V. C. (1982). *A History of Endocrinology*. MTP Press, Lancaster.
12. Darwin, C. (1897). *The Variation of Animals and Plants Under Domestication*. Vol. 2. D. Appleton and Company, New York.
13. Coop, I. E. (1966). Effect of flushing on reproductive performance of ewes. *J. Agric. Sci. Cambr.* 67, 305–323.
14. Tanner, J. M. (1962). *Growth at Adolescence*, 2nd ed. Blackwell Scientific Publications, Oxford.
15. Daw, S. F. (1970). Age of boys' puberty in Leipzig, 1727–49, as indicated by voice breaking in J. S. Bach's choir members. *Hum. Biol.* 42, 87–89.
16. Kennedy, G. C., and Mitra, J. (1963). Body weight and food intake as initiating factors for puberty in the rat. *J. Physiol.* 166, 408–418.
17. Frisch, R. E., and Revelle, R. (1970). Height and weight at menarche and a hypothesis of critical body weights and adolescent events. *Science* 169, 397–399.
18. Frisch, R. E., and McArthur, J. W. (1974). Menstrual cycles: fatness as a determinant of minimum weight for height necessary for their maintenance or onset. *Science* 185, 949–951.
19. Wildt, L., Marshall, G., and Knobil, E. (1980). Experimental induction of puberty in the infantile female rhesus monkey. *Science* 207, 1373–1375.
20. Plant, T. M., Gay, V. L., Marshall, G. R., and Arslan, M. (1989). Puberty in monkeys is triggered by chemical stimulation of the hypothalamus. *Proc. Natl. Acad. Sci. U S A* 86, 2506–2510.
21. Urbanski, H. F., and Ojeda, S. R. (1987). Activation of luteinizing hormone-releasing hormone release advances the onset of female puberty. *Neuroendocrinology* 46, 273–276.
22. Loose, M. D., and Terasawa, E. (1985). Pulsatile infusion of luteinizing hormone-releasing hormone induces precocious puberty (vaginal opening and first ovulation) in the immature female guinea pig. *Biol. Reprod.* 33, 1084–1093.
23. Bullough, W. S. (1951). *Vertebrate Sexual Cycles*. Methuen, London.
24. Widdowson, E. M., and Kennedy, G. C. (1962). Rate of growth, mature weight and life-span. *Proc. R. Soc. Lond. B. Biol. Sci.* 156, 96–108.
25. Foster, D. L., and Olster, D. H. (1985). Effect of restricted nutrition on puberty in the lamb: patterns of tonic luteinizing hormone (LH) secretion and competency of the LH surge system. *Endocrinology* 116, 375–381.
26. Widdowson, E. M., Mavor, W. O., and McCance, R. A. (1964). The effect of undernutrition and rehabilitation on the development of the reproductive organs: rats. *J. Endocrinol.* 29, 119–126.
27. Larsson, K., Carlsson, S. G., Sourander, P., Forsstrom, B., Hansen, S., Henriksson, B., and Lindquist, A. (1974). Delayed onset of sexual activity of male rats subjected to pre- and postnatal undernutrition. *Physiol. Behav.* 13, 307–311.
28. Menendez-Patterson, A., Fernandez, S., Florez-Lozano, J., and Marin, B. (1982). Effect of early pre- and postnatal acquired malnutrition on development and sexual behavior in the rat. *Pharmacol. Biochem. Behav.* 17, 659–664.
29. Bronson, F. H. (1986). Food-restricted, prepubertal, female rats: rapid recovery of luteinizing hormone pulsing with excess food, and full recovery of pubertal development with gonadotropin-releasing hormone. *Endocrinology* 118, 2483–2487.
30. Hamilton, G. D., and Bronson, F. H. (1985). Food restriction and reproductive development in wild house mice. *Biol. Reprod.* 32, 773–778.
31. Wayne, N. L., Wade, G. N., and Rissman, E. F. (1991). Effects of food restriction and social cues on sexual maturation and growth in male musk shrews (*Suncus murinus*). *J. Reprod. Fertil.* 91, 385–392.
32. Day, M. L., Imakawa, K., Zalesky, D. D., Kittok, R. J., and Kinder, J. E. (1986). Effects of restriction of dietary energy intake during the prepubertal period on secretion of luteinizing hormone and responsiveness of the pituitary to luteinizing hormone-releasing hormone in heifers. *J. Anim. Sci.* 62, 1641–1648.
33. Dickerson, J. W., Gresham, G. A., and McCance, R. A. (1964). The effect of undernutrition and rehabilitation on the

development of the reproductive organs: pigs. *J. Endocrinol.* 29, 111–118.
34. Foster, D. L., Ebling, F. J., Micka, A. F., Vannerson, L. A., Bucholtz, D. C., Wood, R. I., Suttie, J. M., and Fenner, D. E. (1989). Metabolic interfaces between growth and reproduction. I. Nutritional modulation of gonadotropin, prolactin, and growth hormone secretion in the growth-limited female lamb. *Endocrinology* 125, 342–350.
35. Bronson, F. H. (1988). Effect of food manipulation on the GnRH-LH-estradiol axis of young female rats. *Am. J. Physiol.* 254, R616–R621.
36. Wood, R. I., Ebling, F. J., and Foster, D. L. (1991). Sex differences in nutritional modulation of gonadotropin secretion during development: studies in the growth-retarded lamb. *Biol. Reprod.* 44, 632–639.
37. Piacsek, B. E. (1985). Altered negative feedback response to ovariectomy and estrogen in prepubertal restricted-diet rats. *Biol. Reprod.* 32, 1062–1068.
38. Piacsek, B. E., Bonifer, T. M., and Tan, R. C. (1986). Altered testosterone feedback in pubertal male rats raised on reduced caloric intake. *J. Androl.* 7, 292–297.
39. Hamilton, G. D., and Bronson, F. H. (1986). Food restriction and reproductive development: male and female mice and male rats. *Am. J. Physiol.* 250, R370–R376.
40. Dreizen, S., Spirakis, C. N., and Stone, R. E. (1967). A comparison of skeletal growth and maturation in undernourished and well-nourished girls before and after menarche. *J. Pediatr.* 70, 256–263.
41. Chakravarty, I., Sreedhar, R., Ghosh, K. K., Card, D., and Bulusu, S. (1982). Circulating gonadotropin profile in severe cases of protein calorie malnutrition. *Fertil. Steril.* 37, 650–654.
42. Boyar, R. M., Katz, J., Finkelstein, J. W., Kapen, S., Weiner, H., Weitzman, E. D., and Hellman, L. (1974). Anorexia nervosa. Immaturity of the 24-hour luteinizing hormone secretory pattern. *N. Engl. J. Med.* 291, 861–865.
43. Marshall, J. C., and Kelch, R. P. (1979). Low dose pulsatile gonadotropin-releasing hormone in anorexia nervosa: a model of human pubertal development. *J. Clin. Endocrinol. Metab.* 49, 712–718.
44. Russ, M. J., Ackerman, S. H., Barakat, R., and Levy, B. (1986). Hypogonadotropic hypogonadism and delayed puberty in a man with anorexia nervosa. *Psychosomatics* 27, 737–739.
45. Hall, A., Delahunt, J. W., and Ellis, P. M. (1985). Anorexia nervosa in the male: clinical features and follow-up of nine patients. *J. Psychiatr. Res.* 19, 315–321.
46. Bronson, F. H., and Heideman, P. D. (1990). Short-term hormonal responses to food intake in peripubertal female rats. *Am. J. Physiol.* 259, R25–R31.
47. Sisk, C. L., and Bronson, F. H. (1986). Effects of food restriction and restoration on gonadotropin and growth hormone secretion in immature male rats. *Biol. Reprod.* 35, 554–561.
48. I'Anson, H., Manning, J. M., Herbosa, C. G., Pelt, J., Friedman, C. R., Wood, R. I., Bucholtz, D. C., and Foster, D. L. (2000). Central inhibition of gonadotropin-releasing hormone secretion in the growth-restricted hypogonadotropic female sheep. *Endocrinology* 141, 520–527.
49. Prasad, B. M., Conover, C. D., Sarkar, D. K., Rabii, J., and Advis, J. P. (1993). Feed restriction in prepubertal lambs: effect on puberty onset and on in vivo release of luteinizing-hormone-releasing hormone, neuropeptide Y and beta-endorphin from the posterior-lateral median eminence. *Neuroendocrinology* 57, 1171–1181.
50. I'Anson, H., Terry, S. K., Lehman, M. N., and Foster, D. L. (1997). Regional differences in the distribution of gonadotropin-releasing hormone cells between rapidly growing and growth-restricted prepubertal female sheep. *Endocrinology* 138, 230–236.
51. Ebling, F. J., Wood, R. I., Karsch, F. J., Vannerson, L. A., Suttie, J. M., Bucholtz, D. C., Schall, R. E., and Foster, D. L. (1990). Metabolic interfaces between growth and reproduction. III. Central mechanisms controlling pulsatile luteinizing hormone secretion in the nutritionally growth-limited female lamb. *Endocrinology* 126, 2719–2727.
52. Piacsek, B. E., and Meites, J. (1967). Reinitiation of gonadotropin release in underfed rats by constant light or epinephrine. *Endocrinology* 81, 535–541.
53. Howland, B. E. (1971). Gonadotrophin levels in female rats subjected to restricted feed intake. *J. Reprod. Fertil.* 27, 467–470.
54. Bronson, F. H., and Marsteller, F. A. (1985). Effect of short-term food deprivation on reproduction in female mice. *Biol. Reprod.* 33, 660–667.
55. Printz, R. H., and Greenwald, G. S. (1970). Effects of starvation on follicular development in the cyclic hamster. *Endocrinology* 86, 290–295.
56. Howland, B. E., and Skinner, K. R. (1973). Effect of starvation on LH levels in male and female hamsters. *J. Reprod. Fertil.* 32, 505–507.
57. Morin, L. P. (1986). Environment and hamster reproduction: responses to phase-specific starvation during estrous cycle. *Am. J. Physiol.* 251, R663–R669.
58. Gill, C. J., and Rissman, E. F. (1997). Female sexual behavior is inhibited by short- and long-term food restriction. *Physiol. Behav.* 61, 387–394.
59. Imakawa, K., Day, M. L., Garcia-Winder, M., Zalesky, D. D., Kittok, R. J., Schanbacher, B. D., and Kinder, J. E. (1986). Endocrine changes during restoration of estrous cycles following induction of anestrus by restricted nutrient intake in beef heifers. *J. Anim. Sci.* 63, 565–571.
60. Thomas, G. B., Mercer, J. E., Karalis, T., Rao, A., Cummins, J. T., and Clarke, I. J. (1990). Effect of restricted feeding on the concentrations of growth hormone (GH), gonadotropins, and prolactin (PRL) in plasma, and on the amounts of messenger ribonucleic acid for GH, gonadotropin subunits, and PRL in the pituitary glands of adult ovariectomized ewes. *Endocrinology* 126, 1361–1367.
61. Howland, B. E., and Ibrahim, E. A. (1973). Increased LH-suppressing effect of oestrogen in ovariectomized rate as a result of underfeeding. *J. Reprod. Fertil.* 35, 545–548.
62. Cagampang, F. R., Maeda, K., Yokoyama, A., and Ota, K. (1990). Effect of food deprivation on the pulsatile LH release in the cycling and ovariectomized female rat. *Horm. Metab. Res.* 22, 269–272.
63. Nagatani, S., Guthikonda, P., Thompson, R. C., Tsukamura, H., Maeda, K. I., and Foster, D. L. (1998). Evidence for GnRH regulation by leptin: leptin administration prevents reduced pulsatile LH secretion during fasting. *Neuroendocrinology* 67, 370–376.
64. Negro-Vilar, A., Dickerman, E., and Meites, J. (1971). Effects of starvation on hypothalamic FSH-RF and pituitary FSH in male rats. *Endocrinology* 88, 1246–1249.
65. Howland, B. E. (1975). The influence of feed restriction and subsequent re-feeding on gonadotrophin secretion and serum testosterone levels in male rats. *J. Reprod. Fertil.* 44, 429–436.
66. Campbell, G. A., Kurcz, M., Marshall, S., and Meites, J. (1977). Effects of starvation in rats on serum levels of follicle stimulating hormone, luteinizing hormone, thyrotropin, growth hormone and prolactin: response to LH-releasing hormone and thyrotropin-releasing hormone. *Endocrinology* 100, 580–587.
67. Blank, J. L., and Desjardins, C. (1985). Differential effects of food restriction on pituitary-testicular function in mice. *Am. J. Physiol.* 248, R181–R189.

68. Dubey, A. K., Cameron, J. L., Steiner, R. A., and Plant, T. M. (1986). Inhibition of gonadotropin secretion in castrated male rhesus monkeys (*Macaca mulatta*) induced by dietary restriction: analogy with the prepubertal hiatus of gonadotropin release. *Endocrinology* 118, 518–525.
69. Cameron, J. L., and Nosbisch, C. (1991). Suppression of pulsatile luteinizing hormone and testosterone secretion during short term food restriction in the adult male rhesus monkey (*Macaca mulatta*). *Endocrinology* 128, 1532–1540.
70. Cameron, J. L., Weltzin, T. E., McConaha, C., Helmreich, D. L., and Kaye, W. H. (1991). Slowing of pulsatile luteinizing hormone secretion in men after forty-eight hours of fasting. *J. Clin. Endocrinol. Metab.* 73, 35–41.
71. Howland, B. E., and Skinner, K. R. (1973). Effect of starvation on gonadotropin secretion in intact and castrated male rats. *Can. J. Physiol. Pharmacol.* 51, 759–762.
72. Pirke, K. M., and Spyra, B. (1981). Influence of starvation on testosterone-luteinizing hormone feedback in the rat. *Acta. Endocrinol. (Copenh).* 96, 413–421.
73. Temple, J. L., and Rissman, E. F. (2000). Brief refeeding restores reproductive readiness in food-restricted female musk shrews (*Suncus murinus*). *Horm. Behav.* 38, 21–28.
74. Parfitt, D. B., Church, K. R., and Cameron, J. L. (1991). Restoration of pulsatile luteinizing hormone secretion after fasting in rhesus monkeys (*Macaca mulatta*): dependence on size of the refeed meal. *Endocrinology* 129, 749–756.
75. Klibanski, A., Beitins, I. Z., Badger, T., Little, R., and McArthur, J. W. (1981). Reproductive function during fasting in men. *J. Clin. Endocrinol. Metab.* 53, 258–263.
76. Rojdmark, S. (1987). Influence of short-term fasting on the pituitary-testicular axis in normal men. *Horm. Res.* 25, 140–146.
77. Olson, B. R., Cartledge, T., Sebring, N., Defensor, R., and Nieman, L. (1995). Short-term fasting affects luteinizing hormone secretory dynamics but not reproductive function in normal-weight sedentary women. *J. Clin. Endocrinol. Metab.* 80, 1187–1193.
78. Stein, Z., and Susser, M. (1975). Fertility, fecundity, famine: food rations in the Dutch famine (1944/5) have a causal relation to fertility, and probably to fecundity. *Hum. Biol.* 47, 131–154.
79. Wynn, A., and Wynn, M. (1993). The effects of food shortage on human reproduction. *Nutr Health.* 9, 43–52.
80. van der Walt, L. A., Wilmsen, E. N., and Jenkins, T. (1978). Unusual sex hormone patterns among desert-dwelling hunter-gatherers. *J. Clin. Endocrinol. Metab.* 46, 658–663.
81. McArthur, J. W., O'Loughlin, K. M., Beitins, I. Z., Johnson, L., Hourihan, J., and Alonso, C. (1976). Endocrine studies during the refeeding of young women with nutritional amenorrhea and infertility. *Mayo Clin. Proc.* 51, 607–616.
82. Wentz, A. C. (1980). Body weight and amenorrhea. *Obstet. Gynecol.* 56, 482–487.
83. Bates, G. W., Bates, S. R., and Whitworth, N. S. (1982). Reproductive failure in women who practice weight control. *Fertil. Steril.* 37, 373–378.
84. Beumont, P. J., Carr, P. J., and Gelder, M G. (1973). Plasma levels of luteinizing hormone and of immunoreactive oestrogens (oestradiol) in anorexia nervosa: response to clomiphene citrate. *Psychol Med.* 3, 495–501.
85. Warren, M. P., and Vande Wiele, R. L. (1973). Clinical and metabolic features of anorexia nervosa. *Am J Obstet. Gynecol.* 117, 435–449.
86. Katz, J. L., Boyar, R., Roffwarg, H., Hellman, L., and Weiner, H. (1978). Weight and circadian luteinizing hormone secretory pattern in anorexia nervosa. *Psychosom. Med.* 40, 549–567.
87. Wheeler, M. J., Crisp, A. H., Hsu, L. K., and Chen, C. N. (1983). Reproductive hormone changes during weight gain in male anorectics. *Clin. Endocrinol. (Oxf).* 18, 423–429.
88. de Rosa, G., Corsello, S. M., de Rosa, E., Della Casa, S., Ruffilli, M. P., Grasso, P., and Pasargiklian, E. (1983). Endocrine study of anorexia nervosa. *Exp. Clin. Endocrinol.* 82, 160–172.
89. Tomova, A., and Kumanov, P. (1999). Sex differences and similarities of hormonal alterations in patients with anorexia nervosa. *Andrologia* 31, 143–147.
90. Kile, J. P., Alexander, B. M., Moss, G. E., Hallford, D. M., and Nett, T. M. (1991). Gonadotropin-releasing hormone overrides the negative effect of reduced dietary energy on gonadotropin synthesis and secretion in ewes. *Endocrinology* 128, 843–849.
91. Bergendahl, M., Perheentupa, A., and Huhtaniemi, I. (1991). Starvation-induced suppression of pituitary-testicular function in rats is reversed by pulsatile gonadotropin-releasing hormone substitution. *Biol. Reprod.* 44, 413–419.
92. Nillius, S. J., Fries, H., and Wide, L. (1975). Successful induction of follicular maturation and ovulation by prolonged treatment with LH-releasing hormone in women with anorexia nervosa. *Am J Obstet. Gynecol.* 122, 921–928.
93. Aloi, J. A., Bergendahl, M., Iranmanesh, A., and Veldhuis, J. D. (1997). Pulsatile intravenous gonadotropin-releasing hormone administration averts fasting-induced hypogonadotropism and hypoandrogenemia in healthy, normal weight men. *J. Clin. Endocrinol. Metab.* 82, 1543–1548.
94. Bergendahl, M., Wiemann, J. N., Clifton, D. K., Huhtaniemi, I., and Steiner, R. A. (1992). Short-term starvation decreases POMC mRNA but does not alter GnRH mRNA in the brain of adult male rats. *Neuroendocrinology* 56, 913–920.
95. Rojdmark, S. (1987). Increased gonadotropin responsiveness to gonadotropin-releasing hormone during fasting in normal subjects. *Metabolism* 36, 21–26.
96. Wiegelmann, W., and Solbach, H. G. (1972). Effects of LH-RH on plasma levels of LH and FSH in anorexia nervosa. *Horm. Metab. Res.* 4, 404.
97. Aono, T., Kinugasa, T., Yamamoto, T., Miyake, A., and Kurachi, K. (1975). Assessment of gonadotrophin secretion in women with anorexia nervosa. *Acta Endocrinol. (Copenh).* 80, 630–641.
98. Warren, M. P., Jewelewicz, R., Dyrenfurth, I., Ans, R., Khalaf, S., and Vande Wiele, R. L. (1975). The significance of weight loss in the evaluation of pituitary response to LH-RH in women with secondary amenorrhea. *J. Clin. Endocrinol. Metab.* 40, 601–611.
99. Vigersky, R. A., Andersen, A. E., Thompson, R. H., and Loriaux, D. L. (1977). Hypothalamic dysfunction in secondary amenorrhea associated with simple weight loss. *N. Engl. J. Med.* 297, 1141–1145.
100. Beumont, P. J., George, G. C., Pimstone, B. L., and Vinik, A. I. (1976). Body weight and the pituitary response to hypothalamic releasing hormones in patients with anorexia nervosa. *J. Clin. Endocrinol. Metab.* 43, 487–496.
101. Bergendahl, M., Perheentupa, A., and Huhtaniemi, I. (1989). Effect of short-term starvation on reproductive hormone gene expression, secretion and receptor levels in male rats. *J. Endocrinol.* 121, 409–417.
102. Warren, M. P., and Perlroth, N. E. (2001). The effects of intense exercise on the female reproductive system. *J. Endocrinol.* 170, 3–11.
103. Williams, N. I., Young, J. C., McArthur, J. W., Bullen, B., Skrinar, G. S., and Turnbull, B. (1995). Strenuous exercise with caloric restriction: effect on luteinizing hormone secretion. *Med. Sci. Sports Exerc.* 27, 1390–1398.
104. Bullen, B. A., Skrinar, G. S., Beitins, I. Z., von Mering, G., Turnbull, B. A., and McArthur, J. W. (1985). Induction of menstrual disorders by strenuous exercise in untrained women. *N. Engl. J. Med.* 312, 1349–1353.

105. Knobil, E. (1993). Inhibition of luteinizing hormone secretion by fasting and exercise: "stress" or specific metabolic signals? *Endocrinology* 132, 1879–1880.
106. Loucks, A. B., Verdun, M., and Heath, E. M. (1998). Low energy availability, not stress of exercise, alters LH pulsatility in exercising women. *J. Appl. Physiol.* 84, 37–46.
107. Frisch, R. E., Wyshak, G., and Vincent, L. (1980). Delayed menarche and amenorrhea in ballet dancers. *N. Engl. J. Med.* 303, 17–19.
108. Warren, M. P. (1980). The effects of exercise on pubertal progression and reproductive function in girls. *J. Clin. Endocrinol. Metab.* 51, 1150–1157.
109. Lindholm, C., Hagenfeldt, K., and Ringertz, B. M. (1994). Pubertal development in elite juvenile gymnasts. Effects of physical training. *Acta. Obstet. Gynaecol. Scand.* 73, 269–273.
110. Frisch, R. E., Gotz-Welbergen, A. V., McArthur, J. W., Albright, T., Witschi, J., Bullen, B., Birnholz, J., Reed, R. B., and Hermann, H. (1981). Delayed menarche and amenorrhea of college athletes in relation to age of onset of training. *JAMA* 246, 1559–1563.
111. Malina, R. M., Harper, A. B., Avent, H. H., and Campbell, D. E. (1973). Age at menarche in athletes and non-athletes. *Med. Sci. Sports.* 5, 11–13.
112. Stager, J. M., Robertshaw, D., and Miescher, E. (1984). Delayed menarche in swimmers in relation to age at onset of training and athletic performance. *Med. Sci. Sports Exerc.* 16, 550–555.
113. Dale, E., Gerlach, D. H., and Wilhite, A. L. (1979). Menstrual dysfunction in distance runners. *Obstet. Gynecol.* 54, 47–53.
114. Feicht, C. B., Johnson, T. S., Martin, B. J., Sparkes, K. E., and Wagner, W. W. Jr. (1978). Secondary amenorrhoea in athletes. *Lancet* 2, 1145–1146.
115. Erdelyi, G. J. (1962). Gynecological survey of female athletes. *J. Sports Med. Phys. Fit.* 2, 174–179.
116. Loucks, A. B., Mortola, J. F., Girton, L., and Yen, S. S. (1989). Alterations in the hypothalamic-pituitary-ovarian and the hypothalamic-pituitary-adrenal axes in athletic women. *J. Clin. Endocrinol. Metab.* 68, 402–411.
117. Cumming, D. C., Vickovic, M. M., Wall, S. R., and Fluker, M. R. (1985). Defects in pulsatile LH release in normally menstruating runners. *J. Clin. Endocrinol. Metab.* 60, 810–812.
118. De Souza, M. J., Miller, B. E., Loucks, A. B., Luciano, A. A., Pescatello, L. S., Campbell, C. G., and Lasley, B. L. (1998). High frequency of luteal phase deficiency and anovulation in recreational women runners: blunted elevation in follicle-stimulating hormone observed during luteal-follicular transition. *J. Clin. Endocrinol. Metab.* 83, 4220–4232.
119. Broocks, A., Pirke, K. M., Schweiger, U., Tuschl, R. J., Laessle, R. G., Strowitzki, T., Horl, E., Horl, T., Haas, W., and Jeschke, D. (1990). Cyclic ovarian function in recreational athletes. *J. Appl. Physiol.* 68, 2083–2086.
120. Wheeler, G. D., Wall, S. R., Belcastro, A. N., and Cumming, D. C. (1984). Reduced serum testosterone and prolactin levels in male distance runners. *JAMA* 252, 514–516.
121. MacConnie, S. E., Barkan, A., Lampman, R. M., Schork, M. A., and Beitins, I. Z. (1986). Decreased hypothalamic gonadotropin-releasing hormone secretion in male marathon runners. *N. Engl. J. Med.* 315, 411–417.
122. Strauss, R. H., Lanese, R. R., and Malarkey, W. B. (1985). Weight loss in amateur wrestlers and its effect on serum testosterone levels. *JAMA* 254, 3337–3338.
123. Bronson, F. H. (1987). Puberty in female rats: relative effect of exercise and food restriction. *Am. J. Physiol.* 252, R140–R144.
124. Perrigo, G., and Bronson, F. H. (1983). Foraging effort, food intake, fat deposition and puberty in female mice. *Biol. Reprod.* 29, 455–463.
125. Manning, J. M., and Bronson, F. H. (1989). Effects of prolonged exercise on puberty and luteinizing hormone secretion in female rats. *Am. J. Physiol.* 257, R1359–R1364.
126. Carlberg, K. A., and Fregly, M. J. (1985). Disruption of estrous cycles in exercise-trained rats. *Proc. Soc. Exp. Biol. Med.* 179, 21–24.
127. Caston, A. L., Farrell, P. A., and Deaver, D. R. (1995). Exercise training-induced changes in anterior pituitary gonadotrope of the female rat. *J. Appl. Physiol.* 79, 194–201.
128. Powers, J. B., Jetton, A. E., and Wade, G. N. (1994). Interactive effects of food deprivation and exercise on reproductive function in female hamsters. *Am. J. Physiol.* 267, R185–R190.
129. Williams, N. I., Caston-Balderrama, A. L., Helmreich, D. L., Parfitt, D. B., Nosbisch, C., and Cameron, J. L. (2001). Longitudinal changes in reproductive hormones and menstrual cyclicity in cynomolgus monkeys during strenuous exercise training: abrupt transition to exercise-induced amenorrhea. *Endocrinology* 142, 2381–2389.
130. Williams, N. I., Helmreich, D. L., Parfitt, D. B., Caston-Balderrama, A., and Cameron, J. L. (2001). Evidence for a causal role of low energy availability in the induction of menstrual cycle disturbances during strenuous exercise training. *J. Clin. Endocrinol. Metab.* 86, 5184–5193.
131. Veldhuis, J. D., Evans, W. S., Demers, L. M., Thorner, M. O., Wakat, D., and Rogol, A. D. (1985). Altered neuroendocrine regulation of gonadotropin secretion in women distance runners. *J. Clin. Endocrinol. Metab.* 61, 557–563.
132. Manning, J. M., and Bronson, F. H. (1991). Suppression of puberty in rats by exercise: effects on hormone levels and reversal with GnRH infusion. *Am. J. Physiol.* 260, R717–R723.
133. Smith, M. S., and Grove, K. L. (2002). Integration of the regulation of reproductive function and energy balance: lactation as a model. *Front Neuroendocrinol.* 23, 225–256.
134. Naismith, D. J., Richardson, D. P., and Pritchard, A. E. (1982). The utilization of protein and energy during lactation in the rat, with particular regard to the use of fat accumulated in pregnancy. *Br. J. Nutr.* 48, 433–441.
135. Ota, K., and Yokoyama, A. (1967). Body weight and food consumption of lactating rats: effects of ovariectomy and of arrest and resumption of suckling. *J. Endocrinol.* 38, 251–261.
136. Dufour, D. L., and Sauther, M. L. (2002). Comparative and evolutionary dimensions of the energetics of human pregnancy and lactation. *Am. J. Hum. Biol.* 14, 584–602.
137. McNeilly, A. S. (1994). Suckling and the control of gonadotropin secretion. In *The Physiology of Reproduction* (E. Knobil and J. D. Neill, Eds.), Vol. 2, pp. 1179–1212. Raven Press, New York.
138. Howie, P. W., and McNeilly, A. S. (1982). Effect of breast-feeding patterns on human birth intervals. *J. Reprod. Fertil.* 65, 545–557.
139. Smith, M. S., and Neill, J. D. (1977). Inhibition of gonadotropin secretion during lactation in the rat: relative contribution of suckling and ovarian steroids. *Biol. Reprod.* 17, 255–261.
140. Glasier, A., McNeilly, A. S., and Howie, P. W. (1983). Fertility after childbirth: changes in serum gonadotrophin levels in bottle and breast feeding women. *Clin. Endocrinol. (Oxf).* 19, 493–501.
141. Steger, R. W., and Peluso, J. J. (1978). Gonadotrophin regulation in the lactating rat. *Acta. Endocrinol. (Copenh).* 88, 668–675.

142. Coppings, R. J., and McCann, S. M. (1979). Stimulatory feedback action of estradiol in intact and ovariectomized lactating rats. *Biol. Reprod.* 21, 219–227.
143. Fox, S. R., and Smith, M. S. (1984). The suppression of pulsatile luteinizing hormone secretion during lactation in the rat. *Endocrinology* 115, 2045–2051.
144. Smith, M. S. (1978). Hypothalamic-pituitary responsiveness during lactation in the rat: estrogen-induced luteinizing hormone surges. *Endocrinology* 102, 121–127.
145. Plant, T. M., Schallenberger, E., Hess, D. L., McCormack, J. T., Dufy-Barbe, L., and Knobil, E. (1980). Influence of suckling on gonadotropin secretion in the female rhesus monkey (*Macaca mulatta*). *Biol. Reprod.* 23, 760–766.
146. Baird, D. T., McNeilly, A. S., Sawers, R. S., and Sharpe, R. M. (1979). Failure of estrogen-induced discharge of luteinizing hormone in lactating women. *J. Clin. Endocrinol. Metab.* 49, 500–506.
147. Lee, L. R., Paul, S. J., and Smith, M. S. (1989). Dose response effects of pulsatile GnRH administration on restoration of pituitary GnRH receptors and pulsatile LH secretion during lactation. *Neuroendocrinology* 49, 664–668.
148. Wright, P. J., Geytenbeek, P. E., Clarke, I. J., and Findlay, J. K. (1984). Induction of plasma LH surges and normal luteal function in acyclic post-partum ewes by the pulsatile administration of LH-RH. *J. Reprod. Fertil.* 71, 1–6.
149. Marks, D. L., Smith, M. S., Clifton, D. K., and Steiner, R. A. (1993). Regulation of gonadotropin-releasing hormone (GnRH) and galanin gene expression in GnRH neurons during lactation in the rat. *Endocrinology* 133, 1450–1458.
150. Ordog, T., Chen, M. D., O'Byrne, K. T., Goldsmith, J. R., Connaughton, M. A., Hotchkiss, J., and Knobil, E. (1998). On the mechanism of lactational anovulation in the rhesus monkey. *Am. J. Physiol.* 274, E665–E676.
151. Smith, M. S. (1982). Hypothalamic responsiveness during lactation in the rat: luteinizing hormone release after electrochemical stimulation. *Biol. Reprod.* 27, 125–131.
152. Pohl, C. R., Lee, L. R., and Smith, M. S. (1989). Qualitative changes in luteinizing hormone and prolactin responses to N-methyl-aspartic acid during lactation in the rat. *Endocrinology* 124, 1905–1911.
153. Tsukamura, H., and Maeda, K. (2001). Non-metabolic and metabolic factors causing lactational anestrus: rat models uncovering the neuroendocrine mechanism underlying the suckling-induced changes in the mother. *Prog. Brain Res.* 133, 187–205.
154. Woodside, B. (1991). Effects of food restriction on the length of lactational diestrus in rats. *Horm. Behav.* 25, 70–83.
155. Woodside, B., Abizaid, A., and Jafferali, S. (1998). Effect of acute food deprivation on lactational infertility in rats is reduced by leptin administration. *Am. J. Physiol.* 274, R1653–R1658.
156. Lunn, P., Watkinson, M., Prentice, A. M., Morrell, P., Austin, S., and Whitehead, R. G. (1981). Maternal nutrition and lactational amenorrhoea. *Lancet* 1, 1428–1429.
157. Brogan, R. S., Mitchell, S. E., Trayhurn, P., and Smith, M. S. (1999). Suppression of leptin during lactation: contribution of the suckling stimulus versus milk production. *Endocrinology* 140, 2621–2627.
158. Bartholomew, G. A. (1977). Energy metabolism. In *Animal Physiology: Principles and Adaptations* (M. S. Gordon, G. A. Bartholomew, A. D. Grinnel, C. B. Joregensen, and F. N. White, Eds.), pp. 57–110. Macmillan, New York.
159. Biggers, J. D., Ashoub, M. R., McLaren, A., and Michie, D. (1958). The growth and development of mice in three climatic environments. *J. Exp. Biol.* 35, 144–155.
160. Barnett, S. A. (1965). Adaptation of mice to cold. *Biol. Rev. Cambr. Philos. Soc.* 40, 5–51.
161. Nazian, S. J., and Piacsek, B. E. (1977). Maturation of the reproductive system in male rats raised at a low ambient temperature. *Biol. Reprod.* 17, 668–675.
162. Perrigo, G., and Bronson, F. H. (1985). Behavioral and physiological responses of female house mice to foraging variation. *Physiol. Behav.* 34, 437–440.
163. Reiter, R. J. (1968). Changes in the reproductive organs of cold-exposed and light-deprived female hamsters (*Mesocricetus auratus*). *J. Reprod. Fertil.* 16, 217–222.
164. Pryor, S., and Bronson, F. H. (1981). Relative and combined effects of low temperature, poor diet, and short daylength on the productivity of wild house mice. *Biol. Reprod.* 25, 734–743.
165. Marsteller, F. A., and Lynch, C. B. (1987). Reproductive responses to variation in temperature and food supply by house mice. II. Lactation. *Biol. Reprod.* 37, 844–850.
166. Marsteller, F. A., and Lynch, C. B. (1987). Reproductive responses to variation in temperature and food supply by house mice. I. Mating and pregnancy. *Biol. Reprod.* 37, 838–843.
167. Schneider, J. E., and Wade, G. N. (1990). Effects of diet and body fat content on cold-induced anestrus in Syrian hamsters. *Am. J. Physiol.* 259, R1198–R1204.
168. Early, A. H., Wade, G. N., and Lempicki, R. L. (1999). Effects of cold exposure on estrous behavior and neural estrogen receptor in Syrian hamsters. *Physiol. Behav.* 65, 763–768.
169. Manning, J. M., and Bronson, F. H. (1990). The effects of low temperature and food intake on ovulation in domestic mice. *Physiol Zool.* 63, 938–948.
170. Kennedy, G. C. (1957). The development with age of hypothalamic restraint upon the appetite of the rat. *J. Endocrinol.* 16, 9–17.
171. Frisch, R. E., Hegsted, D. M., and Yoshinaga, K. (1977). Carcass components at first estrus of rats on high-fat and low-fat diets: body water, protein, and fat. *Proc. Natl. Acad. Sci. U S A* 74, 379–383.
172. Bronson, F. H., and Manning, J. M. (1991). The energetic regulation of ovulation: a realistic role for body fat. *Biol. Reprod.* 44, 945–950.
173. Kennedy, G. C. (1953). The role of depot fat in the hypothalamic control of food intake in the rat. *Proc. R. Soc. Lond. B. Biol. Sci.* 140, 578–596.
174. Mayer, J. (1955). Regulation of energy intake and the body weight: the glucostatic theory and the lipostatic hypothesis. *Ann. N Y Acad. Sci.* 63, 15–43.
175. Mayer, J., Vitale, J. J., and Bates, M. W. (1951). Mechanism of the regulation of food intake. *Nature* 167, 562–563.
176. Ritter, S., and Taylor, J. S. (1990). Vagal sensory neurons are required for lipoprivic but not glucoprivic feeding in rats. *Am. J. Physiol.* 258, R1395–R1401.
177. Cagampang, F. R., Maeda, K., and Ota, K. (1992). Involvement of the gastric vagal nerve in the suppression of pulsatile luteinizing hormone release during acute fasting in rats. *Endocrinology* 130, 3003–3006.
178. Zhang, Y., Proenca, R., Maffei, M., Barone, M., Leopold, L., and Friedman, J. M. (1994). Positional cloning of the mouse obese gene and its human homologue. *Nature* 372, 425–432.
179. Maffei, M., Halaas, J., Ravussin, E., Pratley, R. E., Lee, G. H., Zhang, Y., Fei, H., Kim, S., Lallone, R., and Ranganathan, S., et al. (1995). Leptin levels in human and rodent: measurement of plasma leptin and ob RNA in obese and weight-reduced subjects. *Nat. Med.* 1, 1155–1161.
180. Considine, R. V., Sinha, M. K., Heiman, M. L., Kriauciunas, A., Stephens, T. W., Nyce, M. R., Ohannesian, J. P., Marco, C. C., McKee, L. J., and Bauer, T. L., et al. (1996). Serum immunoreactive-leptin concentrations in normal-weight and obese humans. *N. Engl. J. Med.* 334, 292–295.

181. Weigle, D. S., Duell, P. B., Connor, W. E., Steiner, R. A., Soules, M. R., and Kuijper, J. L. (1997). Effect of fasting, refeeding, and dietary fat restriction on plasma leptin levels. *J. Clin. Endocrinol. Metab.* 82, 561–565.
182. Grinspoon, S., Gulick, T., Askari, H., Landt, M., Lee, K., Anderson, E., Ma, Z., Vignati, L., Bowsher, R., Herzog, D., and Klibanski, A. (1996). Serum leptin levels in women with anorexia nervosa. *J. Clin. Endocrinol. Metab.* 81, 3861–3863.
183. Laughlin, G. A., and Yen, S. S. (1997). Hypoleptinemia in women athletes: absence of a diurnal rhythm with amenorrhea. *J. Clin. Endocrinol. Metab.* 82, 318–321.
184. Thong, F. S., McLean, C., and Graham, T. E. (2000). Plasma leptin in female athletes: relationship with body fat, reproductive, nutritional, and endocrine factors. *J. Appl. Physiol.* 88, 2037–2044.
185. Weimann, E., Blum, W. F., Witzel, C., Schwidergall, S., and Bohles, H. J. (1999). Hypoleptinemia in female and male elite gymnasts. *Eur. J. Clin. Invest.* 29, 853–860.
186. Ahima, R. S., Prabakaran, D., and Flier, J. S. (1998). Postnatal leptin surge and regulation of circadian rhythm of leptin by feeding. Implications for energy homeostasis and neuroendocrine function. *J. Clin. Invest.* 101, 1020–1027.
187. Sinha, M. K., Ohannesian, J. P., Heiman, M. L., Kriauciunas, A., Stephens, T. W., Magosin, S., Marco, C., and Caro, J. F. (1996). Nocturnal rise of leptin in lean, obese, and non-insulin-dependent diabetes mellitus subjects. *J. Clin. Invest.* 97, 1344–1347.
188. Schoeller, D. A., Cella, L. K., Sinha, M. K., and Caro, J. F. (1997). Entrainment of the diurnal rhythm of plasma leptin to meal timing. *J. Clin. Invest.* 100, 1882–1887.
189. Tartaglia, L. A., Dembski, M., Weng, X., Deng, N., Culpepper, J., Devos, R., Richards, G. J., Campfield, L. A., Clark, F. T., and Deeds, J., et al. (1995). Identification and expression cloning of a leptin receptor, OB-R. *Cell* 83, 1263–1271.
190. Lee, G. H., Proenca, R., Montez, J. M., Carroll, K. M., Darvishzadeh, J. G., Lee, J. I., and Friedman, J. M. (1996). Abnormal splicing of the leptin receptor in diabetic mice. *Nature* 379, 632–635.
191. Wang, M. Y., Zhou, Y. T., Newgard, C. B., and Unger, R. H. (1996). A novel leptin receptor isoform in rat. *FEBS Lett.* 392, 87–90.
192. Bjorbaek, C., Uotani, S., da Silva, B., and Flier, J. S. (1997). Divergent signaling capacities of the long and short isoforms of the leptin receptor. *J. Biol. Chem.* 272, 32686–32695.
193. Banks, A. S., Davis, S. M., Bates, S. H., and Myers, M. G. Jr. (2000). Activation of downstream signals by the long form of the leptin receptor. *J. Biol. Chem.* 275, 14563–14572.
194. Bjorbaek, C., and Kahn, B. B. (2004). Leptin signaling in the central nervous system and the periphery. *Rec. Prog. Horm. Res.* 59, 305–331.
195. Zamorano, P. L., Mahesh, V. B., De Sevilla, L. M., Chorich, L. P., Bhat, G. K., and Brann, D. W. (1997). Expression and localization of the leptin receptor in endocrine and neuroendocrine tissues of the rat. *Neuroendocrinology* 65, 223–228.
196. Mercer, J. G., Hoggard, N., Williams, L. M., Lawrence, C. B., Hannah, L. T., and Trayhurn, P. (1996). Localization of leptin receptor mRNA and the long form splice variant (Ob-Rb) in mouse hypothalamus and adjacent brain regions by in situ hybridization. *FEBS Lett.* 387, 113–116.
197. Fei, H., Okano, H. J., Li, C., Lee, G. H., Zhao, C., Darnell, R., and Friedman, J. M. (1997). Anatomic localization of alternatively spliced leptin receptors (Ob-R) in mouse brain and other tissues. *Proc. Natl. Acad. Sci. U S A* 94, 7001–7005.
198. Elmquist, J. K., Bjorbaek, C., Ahima, R. S., Flier, J. S., and Saper, C. B. (1998). Distributions of leptin receptor mRNA isoforms in the rat brain. *J. Comp. Neurol.* 395, 535–547.
199. Williams, L. M., Adam, C. L., Mercer, J. G., Moar, K. M., Slater, D., Hunter, L., Findlay, P. A., and Hoggard, N. (1999). Leptin receptor and neuropeptide Y gene expression in the sheep brain. *J. Neuroendocrinol.* 11, 165–169.
200. Banks, W. A., Kastin, A. J., Huang, W., Jaspan, J. B., and Maness, L. M. (1996). Leptin enters the brain by a saturable system independent of insulin. *Peptides* 17, 305–311.
201. Meister, B., and Hakansson, M. L. (2001). Leptin receptors in hypothalamus and circumventricular organs. *Clin. Exp. Pharmacol. Physiol.* 28, 610–617.
202. Friedman, J. M., and Halaas, J. L. (1998). Leptin and the regulation of body weight in mammals. *Nature* 395, 763–770.
203. van Dijk, G. (2001). The role of leptin in the regulation of energy balance and adiposity. *J. Neuroendocrinol.* 13, 913–921.
204. Kamohara, S., Burcelin, R., Halaas, J. L., Friedman, J. M., and Charron, M. J. (1997). Acute stimulation of glucose metabolism in mice by leptin treatment. *Nature* 389, 374–377.
205. Campfield, L. A., Smith, F. J., Guisez, Y., Devos, R., and Burn, P. (1995). Recombinant mouse OB protein: evidence for a peripheral signal linking adiposity and central neural networks. *Science* 269, 546–549.
206. Halaas, J. L., Gajiwala, K. S., Maffei, M., Cohen, S. L., Chait, B. T., Rabinowitz, D., Lallone, R. L., Burley, S. K., and Friedman, J. M. (1995). Weight-reducing effects of the plasma protein encoded by the obese gene. *Science* 269, 543–546.
207. Pelleymounter, M. A., Cullen, M. J., Baker, M. B., Hecht, R., Winters, D., Boone, T., and Collins, F. (1995). Effects of the obese gene product on body weight regulation in ob/ob mice. *Science* 269, 540–543.
208. Weigle, D. S., Bukowski, T. R., Foster, D. C., Holderman, S., Kramer, J. M., Lasser, G., Lofton-Day, C. E., Prunkard, D. E., Raymond, C., and Kuijper, J. L. (1995). Recombinant ob protein reduces feeding and body weight in the ob/ob mouse. *J. Clin. Invest.* 96, 2065–2070.
209. Halaas, J. L., Boozer, C., Blair-West, J., Fidahusein, N., Denton, D. A., and Friedman, J. M. (1997). Physiological response to long-term peripheral and central leptin infusion in lean and obese mice. *Proc. Natl. Acad. Sci. U S A* 94, 8878–8883.
210. Tang-Christensen, M., Havel, P. J., Jacobs, R. R., Larsen, P. J., and Cameron, J. L. (1999). Central administration of leptin inhibits food intake and activates the sympathetic nervous system in rhesus macaques. *J. Clin. Endocrinol. Metab.* 84, 711–717.
211. Henry, B. A., Goding, J. W., Alexander, W. S., Tilbrook, A. J., Canny, B. J., Dunshea, F., Rao, A., Mansell, A., and Clarke, I. J. (1999). Central administration of leptin to ovariectomized ewes inhibits food intake without affecting the secretion of hormones from the pituitary gland: evidence for a dissociation of effects on appetite and neuroendocrine function. *Endocrinology* 140, 1175–1182.
212. Ahima, R. S., Prabakaran, D., Mantzoros, C., Qu, D., Lowell, B., Maratos-Flier, E., and Flier, J. S. (1996). Role of leptin in the neuroendocrine response to fasting. *Nature* 382, 250–252.
213. Watanobe, H. (2002). Leptin directly acts within the hypothalamus to stimulate gonadotropin-releasing hormone secretion in vivo in rats. *J. Physiol.* 545, 255–268.
214. Yu, W. H., Kimura, M., Walczewska, A., Karanth, S., and McCann, S. M. (1997). Role of leptin in hypothalamic-pituitary function. *Proc. Natl. Acad. Sci. U S A* 94, 1023–1028.

215. Walczewska, A., Yu, W. H., Karanth, S., and McCann, S. M. (1999). Estrogen and leptin have differential effects on FSH and LH release in female rats. *Proc. Soc. Exp. Biol. Med.* 222, 170–177.
216. Ogura, K., Irahara, M., Kiyokawa, M., Tezuka, M., Matsuzaki, T., Yasui, T., Kamada, M., and Aono, T. (2001). Effects of leptin on secretion of LH and FSH from primary cultured female rat pituitary cells. *Eur. J. Endocrinol.* 144, 653–658.
217. Tena-Sempere, M., Pinilla, L., Gonzalez, L. C., Dieguez, C., Casanueva, F. F., and Aguilar, E. (1999). Leptin inhibits testosterone secretion from adult rat testis in vitro. *J. Endocrinol.* 161, 211–218.
218. Spicer, L. J. (2003). The effect of leptin on ovarian steroidogenesis. In *Leptin and Reproduction* (M. C. Henson and V. D. Castracane, Eds.), pp. 97–109. Kluwer Academic/Plenum Publishers, New York.
219. Kalra, S. P., Xu, B., Dube, M. G., Moldawer, L. L., Martin, D., and Kalra, P. S. (1998). Leptin and ciliary neurotropic factor (CNTF) inhibit fasting-induced suppression of luteinizing hormone release in rats: role of neuropeptide Y. *Neurosci. Lett.* 240, 45–49.
220. Nagatani, S., Zeng, Y., Keisler, D. H., Foster, D. L., and Jaffe, C. A. (2000). Leptin regulates pulsatile luteinizing hormone and growth hormone secretion in the sheep. *Endocrinology* 141, 3965–3975.
221. Maciel, M. N., Zieba, D. A., Amstalden, M., Keisler, D. H., Neves, J. P., and Williams, G. L. (2004). Leptin prevents fasting-mediated reductions in pulsatile secretion of luteinizing hormone and enhances its gonadotropin-releasing hormone-mediated release in heifers. *Biol. Reprod.* 70, 229–235.
222. Finn, P. D., Cunningham, M. J., Pau, K. Y., Spies, H. G., Clifton, D. K., and Steiner, R. A. (1998). The stimulatory effect of leptin on the neuroendocrine reproductive axis of the monkey. *Endocrinology* 139, 4652–4662.
223. Lado-Abeal, J., Hickox, J. R., Cheung, T. L., Veldhuis, J. D., Hardy, D. M., and Norman, R. L. (2000). Neuroendocrine consequences of fasting in adult male macaques, effects of recombinant rhesus macaque leptin infusion. *Neuroendocrinology* 71, 196–208.
224. Schneider, J. E., Goldman, M. D., Tang, S., Bean, B., Ji, H., and Friedman, M. I. (1998). Leptin indirectly affects estrous cycles by increasing metabolic fuel oxidation. *Horm. Behav.* 33, 217–228.
225. Chan, J. L., Heist, K., DePaoli, A. M., Veldhuis, J. D., and Mantzoros, C. S. (2003). The role of falling leptin levels in the neuroendocrine and metabolic adaptation to short-term starvation in healthy men. *J. Clin. Invest.* 111, 1409–1421.
226. Henry, B. A., Goding, J. W., Tilbrook, A. J., Dunshea, F. R., and Clarke, I. J. (2001). Intracerebroventricular infusion of leptin elevates the secretion of luteinising hormone without affecting food intake in long-term food-restricted sheep, but increases growth hormone irrespective of body weight. *J. Endocrinol.* 168, 67–77.
227. Ingalls, A. M., Dickie, M. M., and Snell, G. D. (1950). Obese, a new mutation in the house mouse. *J. Hered.* 41, 317–318.
228. Coleman, D. L. (1978). Obese and diabetes: two mutant genes causing diabetes-obesity syndromes in mice. *Diabetologia* 14, 141–148.
229. Bray, G. A., and York, D. A. (1979). Hypothalamic and genetic obesity in experimental animals: an autonomic and endocrine hypothesis. *Physiol. Rev.* 59, 719–809.
230. Swerdloff, R. S., Batt, R. A., and Bray, G. A. (1976). Reproductive hormonal function in the genetically obese (ob/ob) mouse. *Endocrinology* 98, 1359–1364.
231. Batt, R. A. (1972). The response of the reproductive system in the female mutant mouse, obese (genotype ob/b) to gonadotrophin-releasing hormones. *J. Reprod. Fertil.* 31, 496–497.
232. Swerdloff, R. S., Peterson, M., Vera, A., Batt, R. A., Heber, D., and Bray, G. A. (1978). The hypothalamic-pituitary axis in genetically obese (ob/ob) mice: response to luteinizing hormone-releasing hormone. *Endocrinology* 103, 542–547.
233. Wilkinson, M., and Moger, W. H. (1981). Transpubertal modulation of pituitary and testicular function in the ob/ob mouse. *Horm. Res.* 14, 95–103.
234. Batt, R. A., Everard, D. M., Gillies, G., Wilkinson, M., Wilson, C. A., and Yeo, T. A. (1982). Investigation into the hypogonadism of the obese mouse (genotype ob/ob). *J. Reprod. Fertil.* 64, 363–371.
235. Ewart-Toland, A., Mounzih, K., Qiu, J., and Chehab, F. F. (1999). Effect of the genetic background on the reproduction of leptin-deficient obese mice. *Endocrinology* 140, 732–738.
236. Barash, I. A., Cheung, C. C., Weigle, D. S., Ren, H., Kabigting, E. B., Kuijper, J. L., Clifton, D. K., and Steiner, R. A. (1996). Leptin is a metabolic signal to the reproductive system. *Endocrinology* 137, 3144–3147.
237. Chehab, F. F., Lim, M. E., and Lu, R. (1996). Correction of the sterility defect in homozygous obese female mice by treatment with the human recombinant leptin. *Nat. Genet.* 12, 318–320.
238. Mounzih, K., Lu, R., and Chehab, F. F. (1997). Leptin treatment rescues the sterility of genetically obese ob/ob males. *Endocrinology* 138, 1190–1193.
239. Lane, P. W., and Dickie, M. M. (1954). Fertile obese male mice. Relative sterility in obese males corrected by dietary restriction. *J. Hered.* 45, 56–58.
240. Eastcott, A. (1972). Influence of restricting the diet on breeding in genetically obese mice. *Lab Anim.* 6, 9–18.
241. Chen, H., Charlat, O., Tartaglia, L. A., Woolf, E. A., Weng, X., Ellis, S. J., Lakey, N. D., Culpepper, J., Moore, K. J., Breitbart, R. E., Duyk, G. M., Tepper, R. I., and Morgenstern, J. P. (1996). Evidence that the diabetes gene encodes the leptin receptor: identification of a mutation in the leptin receptor gene in db/db mice. *Cell* 84, 491–495.
242. Chua, S. C., Jr., Chung, W. K., Wu-Peng, X. S., Zhang, Y., Liu, S. M., Tartaglia, L., and Leibel, R. L. (1996). Phenotypes of mouse diabetes and rat fatty due to mutations in the OB (leptin) receptor. *Science* 271, 994–996.
243. Hummel, K. P., Dickie, M. M., and Coleman, D. L. (1966). Diabetes, a new mutation in the mouse. *Science* 153, 1127–1128.
244. Johnson, L. M., and Sidman, R. L. (1979). A reproductive endocrine profile in the diabetes (db) mutant mouse. *Biol. Reprod.* 20, 552–559.
245. Zucker, L. M., and Zucker, T. F. (1961). Fatty, a new mutation in the rat. *J. Hered.* 52, 275–278.
246. Saiduddin, S., Bray, G. A., York, D. A., and Swerdloff, R. S. (1973). Reproductive function in the genetically obese "fatty" rat. *Endocrinology* 93, 1251–1256.
247. Bray, G. A. (1977). The Zucker-fatty rat: a review. *Fed. Proc.* 36, 148–153.
248. Kowalski, T. J., Liu, S. M., Leibel, R. L., and Chua, S. C. Jr. (2001). Transgenic complementation of leptin-receptor deficiency. I. Rescue of the obesity/diabetes phenotype of LEPR-null mice expressing a LEPR-B transgene. *Diabetes* 50, 425–435.
249. Cohen, P., Zhao, C., Cai, X., Montez, J. M., Rohani, S. C., Feinstein, P., Mombaerts, P., and Friedman, J. M. (2001). Selective deletion of leptin receptor in neurons leads to obesity. *J. Clin. Invest.* 108, 1113–1121.

250. Bates, S. H., Stearns, W. H., Dundon, T. A., Schubert, M., Tso, A. W., Wang, Y., Banks, A. S., Lavery, H. J., Haq, A. K., Maratos-Flier, E., Neel, B. G., Schwartz, M. W., and Myers, M. G. Jr. (2003). STAT3 signalling is required for leptin regulation of energy balance but not reproduction. *Nature* 421, 856–859.
251. Montague, C. T., Farooqi, I. S., Whitehead, J. P., Soos, M. A., Rau, H., Wareham, N. J., Sewter, C. P., Digby, J. E., Mohammed, S. N., Hurst, J. A., Cheetham, C. H., Earley, A. R., Barnett, A. H., Prins, J. B., and O'Rahilly, S. (1997). Congenital leptin deficiency is associated with severe early-onset obesity in humans. *Nature* 387, 903–908.
252. Farooqi, I. S., Jebb, S. A., Langmack, G., Lawrence, E., Cheetham, C. H., Prentice, A. M., Hughes, I. A., McCamish, M. A., and O'Rahilly, S. (1999). Effects of recombinant leptin therapy in a child with congenital leptin deficiency. *N. Engl. J. Med.* 341, 879–884.
253. Strobel, A., Issad, T., Camoin, L., Ozata, M., and Strosberg, A. D. (1998). A leptin missense mutation associated with hypogonadism and morbid obesity. *Nat. Genet.* 18, 213–215.
254. Licinio, J., Caglayan, S., Ozata, M., Yildiz, B. O., de Miranda, P. B., O'Kirwan, F., Whitby, R., Liang, L., Cohen, P., Bhasin, S., Krauss, R. M., Veldhuis, J. D., Wagner, A. J., DePaoli, A. M., McCann, S. M., and Wong, M. L. (2004). Phenotypic effects of leptin replacement on morbid obesity, diabetes mellitus, hypogonadism, and behavior in leptin-deficient adults. *Proc. Natl. Acad. Sci. U S A* 101, 4531–4536.
255. Clement, K., Vaisse, C., Lahlou, N., Cabrol, S., Pelloux, V., Cassuto, D., Gourmelen, M., Dina, C., Chambaz, J., Lacorte, J. M., Basdevant, A., Bougneres, P., Lebouc, Y., Froguel, P., and Guy-Grand B. (1998). A mutation in the human leptin receptor gene causes obesity and pituitary dysfunction. *Nature* 392, 398–401.
256. Gruaz, N. M., Lalaoui, M., Pierroz, D. D., Englaro, P., Sizonenko, P. C., Blum, W. F., and Aubert, M. L. (1998). Chronic administration of leptin into the lateral ventricle induces sexual maturation in severely food-restricted female rats. *J. Neuroendocrinol.* 10, 627–633.
257. Nagatani, S., Guthikonda, P., and Foster, D. L. (2000). Appearance of a nocturnal peak of leptin secretion in the pubertal rat. *Horm. Behav.* 37, 345–352.
258. Chehab, F. F., Mounzih, K., Lu, R., and Lim, M. E. (1997). Early onset of reproductive function in normal female mice treated with leptin. *Science* 275, 88–90.
259. Suter, K. J., Pohl, C. R., and Wilson, M. E. (2000). Circulating concentrations of nocturnal leptin, growth hormone, and insulin-like growth factor-I increase before the onset of puberty in agonadal male monkeys: potential signals for the initiation of puberty. *J. Clin. Endocrinol. Metab.* 85, 808–814.
260. Mantzoros, C. S., Flier, J. S., and Rogol, A. D. (1997). A longitudinal assessment of hormonal and physical alterations during normal puberty in boys. V. Rising leptin levels may signal the onset of puberty. *J. Clin. Endocrinol. Metab.* 82, 1066–1070.
261. Carlsson, B., Ankarberg, C., Rosberg, S., Norjavaara, E., Albertsson-Wikland, K., and Carlsson, L. M. (1997). Serum leptin concentrations in relation to pubertal development. *Arch. Dis. Child.* 77, 396–400.
262. Mann, D. R., Johnson, A. O., Gimpel, T., and Castracane, V. D. (2003). Changes in circulating leptin, leptin receptor, and gonadal hormones from infancy until advanced age in humans. *J. Clin. Endocrinol. Metab.* 88, 3339–3345.
263. Ahmed, M. L., Ong, K. K., Morrell, D. J., Cox, L., Drayer, N., Perry, L., Preece, M. A., and Dunger, D. B. (1999). Longitudinal study of leptin concentrations during puberty: sex differences and relationship to changes in body composition. *J. Clin. Endocrinol. Metab.* 84, 899–905.
264. Ahima, R. S., Dushay, J., Flier, S. N., Prabakaran, D., and Flier, J. S. (1997). Leptin accelerates the onset of puberty in normal female mice. *J. Clin. Invest.* 99, 391–395.
265. Arslanian, S., Suprasongsin, C., Kalhan, S. C., Drash, A. L., Brna, R., and Janosky, J. E. (1998). Plasma leptin in children: relationship to puberty, gender, body composition, insulin sensitivity, and energy expenditure. *Metabolism* 47, 309–312.
266. Bronson, F. H. (2001). Puberty in female mice is not associated with increases in either body fat or leptin. *Endocrinology* 142, 4758–4761.
267. Cheung, C. C., Thornton, J. E., Nurani, S. D., Clifton, D. K., and Steiner, R. A. (2001). A reassessment of leptin's role in triggering the onset of puberty in the rat and mouse. *Neuroendocrinology* 74, 12–21.
268. Rayner, D. V., Dalgliesh, G. D., Duncan, J. S., Hardie, L. J., Hoggard, N., and Trayhurn, P. (1997). Postnatal development of the ob gene system: elevated leptin levels in suckling fa/fa rats. *Am. J. Physiol.* 273, R446–R450.
269. Devaskar, S. U., Ollesch, C., Rajakumar, R. A., and Rajakumar, P. A. (1997). Developmental changes in ob gene expression and circulating leptin peptide concentrations. *Biochem. Biophys. Res. Commun.* 238, 44–47.
270. Plant, T. M., and Durrant, A. R. (1997). Circulating leptin does not appear to provide a signal for triggering the initiation of puberty in the male rhesus monkey (*Macaca mulatta*). *Endocrinology* 138, 4505–4508.
271. Plant, T. M. (2001). Leptin, growth hormone, and the onset of primate puberty. *J. Clin. Endocrinol. Metab.* 86, 458–460.
272. Mann, D. R., Akinbami, M. A., Gould, K. G., and Castracane, V. D. (2000). A longitudinal study of leptin during development in the male rhesus monkey: the effect of body composition and season on circulating leptin levels. *Biol. Reprod.* 62, 285–291.
273. Mann, D. R., Akinbami, M. A., Gould, K. G., and Castracane, V. D. (2002). Leptin and thyroxine during sexual development in male monkeys: effect of neonatal gonadotropin-releasing hormone antagonist treatment and delayed puberty on the developmental pattern of leptin and thyroxine secretion. *Eur. J. Endocrinol.* 146, 891–898.
274. Urbanski, H. F., and Pau, K. Y. (1998). A biphasic developmental pattern of circulating leptin in the male rhesus macaque (*Macaca mulatta*). *Endocrinology* 139, 2284–2286.
275. Watanobe, H., and Schioth, H. B. (2002). Postnatal profile of plasma leptin concentrations in male and female rats: relation with the maturation of the pituitary-gonadal axis. *Regul. Pept.* 105, 23–28.
276. Smith, J. T., and Waddell, B. J. (2003). Developmental changes in plasma leptin and hypothalamic leptin receptor expression in the rat: peripubertal changes and the emergence of sex differences. *J. Endocrinol.* 176, 313–319.
277. Almog, B., Gold, R., Tajima, K., Dantes, A., Salim, K., Rubinstein, M., Barkan, D., Homburg, R., Lessing, J. B., Nevo, N., Gertler, A., and Amsterdam, A. (2001). Leptin attenuates follicular apoptosis and accelerates the onset of puberty in immature rats. *Mol. Cell. Endocrinol.* 183, 179–191.
278. Yura, S., Ogawa, Y., Sagawa, N., Masuzaki, H., Itoh, H., Ebihara, K., Aizawa-Abe, M., Fujii, S., and Nakao, K. (2000). Accelerated puberty and late-onset hypothalamic hypogonadism in female transgenic skinny mice overexpressing leptin. *J. Clin. Invest.* 105, 749–755.
279. Cheung, C. C., Thornton, J. E., Kuijper, J. L., Weigle, D. S., Clifton, D. K., and Steiner, R. A. (1997). Leptin is a metabolic

gate for the onset of puberty in the female rat. *Endocrinology* 138, 855–858.

280. Barker-Gibb, M. L., Sahu, A., Pohl, C. R., and Plant, T. M. (2002). Elevating circulating leptin in prepubertal male rhesus monkeys (*Macaca mulatta*) does not elicit precocious gonadotropin-releasing hormone release, assessed indirectly. *J. Clin. Endocrinol. Metab.* 87, 4976–4983.
281. Wilson, M. E., Fisher, J., Chikazawa, K., Yoda, R., Legendre, A., Mook, D., and Gould, K. G. (2003). Leptin administration increases nocturnal concentrations of luteinizing hormone and growth hormone in juvenile female rhesus monkeys. *J. Clin. Endocrinol. Metab.* 88, 4874–4883.
282. Kawai, M., Yamaguchi, M., Murakami, T., Shima, K., Murata, Y., and Kishi, K. (1997). The placenta is not the main source of leptin production in pregnant rat: gestational profile of leptin in plasma and adipose tissues. *Biochem. Biophys. Res. Commun.* 240, 798–802.
283. Hardie, L., Trayhurn, P., Abramovich, D., and Fowler, P. (1997). Circulating leptin in women: a longitudinal study in the menstrual cycle and during pregnancy. *Clin. Endocrinol. (Oxf).* 47, 101–106.
284. Hoggard, N., Hunter, L., Duncan, J. S., Williams, L. M., Trayhurn, P., and Mercer, J. G. (1997). Leptin and leptin receptor mRNA and protein expression in the murine fetus and placenta. *Proc. Natl. Acad. Sci. U S A* 94, 11073–11078.
285. Masuzaki, H., Ogawa, Y., Sagawa, N., Hosoda, K., Matsumoto, T., Mise, H., Nishimura, H., Yoshimasa, Y., Tanaka, I., Mori, T., and Nakao, K. (1997). Nonadipose tissue production of leptin: leptin as a novel placenta-derived hormone in humans. *Nat. Med.* 3, 1029–1033.
286. Tomimatsu, T., Yamaguchi, M., Murakami, T., Ogura, K., Sakata, M., Mitsuda, N., Kanzaki, T., Kurachi, H., Irahara, M., Miyake, A., Shima, K., Aono, T., and Murata, Y. (1997). Increase of mouse leptin production by adipose tissue after midpregnancy: gestational profile of serum leptin concentration. *Biochem. Biophys. Res. Commun.* 240, 213–215.
287. Seeber, R. M., Smith, J. T., and Waddell, B. J. (2002). Plasma leptin-binding activity and hypothalamic leptin receptor expression during pregnancy and lactation in the rat. *Biol. Reprod.* 66, 1762–1767.
288. Mounzih, K., Qiu, J., Ewart-Toland, A., and Chehab, F. F. (1998). Leptin is not necessary for gestation and parturition but regulates maternal nutrition via a leptin resistance state. *Endocrinology* 139, 5259–5262.
289. Rosso, P. (1987). Regulation of food intake during pregnancy and lactation. *Ann. N Y Acad. Sci.* 499, 191–196.
290. Malik, N. M., Carter, N. D., Murray, J. F., Scaramuzzi, R. J., Wilson, C. A., and Stock, M. J. (2001). Leptin requirement for conception, implantation, and gestation in the mouse. *Endocrinology* 142, 5198–5202.
291. Dearth, R. K., Hiney, J. K., and Dees, W. L. (2000). Leptin acts centrally to induce the prepubertal secretion of luteinizing hormone in the female rat. *Peptides* 21, 387–392.
292. Hohmann, J. G., Teal, T. H., Clifton, D. K., Davis, J., Hruby, V. J., Han, G., and Steiner, R. (2000). A. Differential role of melanocortins in mediating leptin's central effects on feeding and reproduction. *Am. J. Physiol. Regul. Integr. Comp. Physiol.* 278, R50–R59.
293. Carro, E., Pinilla, L., Seoane, L. M., Considine, R. V., Aguilar, E., Casanueva, F. F., and Dieguez, C. (1997). Influence of endogenous leptin tone on the estrous cycle and luteinizing hormone pulsatility in female rats. *Neuroendocrinology* 66, 375–377.
294. Hakansson, M. L., Brown, H., Ghilardi, N., Skoda, R. C., and Meister, B. (1998). Leptin receptor immunoreactivity in chemically defined target neurons of the hypothalamus. *J. Neurosci.* 18, 559–572.
295. Magni, P., Vettor, R., Pagano, C., Calcagno, A., Beretta, E., Messi, E., Zanisi, M., Martini, L., and Motta, M. (1999). Expression of a leptin receptor in immortalized gonadotropin-releasing hormone-secreting neurons. *Endocrinology* 140, 1581–1585.
296. Sone, M., Nagata, H., Takekoshi, S., and Osamura, R. Y. (2001). Expression and localization of leptin receptor in the normal rat pituitary gland. *Cell Tissue Res.* 305, 351–356.
297. Caprio, M., Isidori, A. M., Carta, A. R., Moretti, C., Dufau, M. L., and Fabbri, A. (1999). Expression of functional leptin receptors in rodent Leydig cells. *Endocrinology* 140, 4939–4947.
298. Andreelli, F., Hanaire-Broutin, H., Laville, M., Tauber, J. P., Riou, J. P., and Thivolet, C. (2000). Normal reproductive function in leptin-deficient patients with lipoatropic diabetes. *J. Clin. Endocrinol. Metab.* 85, 715–719.
299. Schneider, J. E., Blum, R. M., and Wade, G. N. (2000). Metabolic control of food intake and estrous cycles in Syrian hamsters. I. Plasma insulin and leptin. *Am. J. Physiol. Regul. Integr. Comp. Physiol.* 278, R476–R485.
300. Schneider, J. E., and Zhou, D. (1999). Interactive effects of central leptin and peripheral fuel oxidation on estrous cyclicity. *Am. J. Physiol.* 277, R1020–R1024.
301. Schneider, J. E. (2004). Energy balance and reproduction. *Physiol. Behav.* 81, 289–317.
302. Polonsky, K. S., Given, B. D., and Van Cauter, E. (1988). Twenty-four-hour profiles and pulsatile patterns of insulin secretion in normal and obese subjects. *J. Clin. Invest.* 81, 442–448.
303. Bagdade, J. D., Bierman, E. L., and Porte, D. Jr. (1967). The significance of basal insulin levels in the evaluation of the insulin response to glucose in diabetic and nondiabetic subjects. *J. Clin. Invest.* 46, 1549–1557.
304. Crone, C. (1965). Facilitated transfer of glucose from blood into brain tissue. *J. Physiol.* 181, 103–113.
305. Hom, F. G., Goodner, C. J., and Berrie, M. A. (1984). A (3H)2-deoxyglucose method for comparing rates of glucose metabolism and insulin responses among rat tissues in vivo. Validation of the model and the absence of an insulin effect on brain. *Diabetes* 33, 141–152.
306. Havrankova, J., Schmechel, D., Roth, J., and Brownstein, M. (1978). Identification of insulin in rat brain. *Proc. Natl. Acad. Sci. U S A* 75, 5737–5741.
307. Havrankova, J., Roth, J., and Brownstein, M. (1978). Insulin receptors are widely distributed in the central nervous system of the rat. *Nature* 272, 827–829.
308. Posner, B. I., Kelly, P. A., Shiu, R. P., and Friesen, H. G. (1974). Studies of insulin, growth hormone and prolactin binding: tissue distribution, species variation and characterization. *Endocrinology* 95, 521–531.
309. Werther, G. A., Hogg, A., Oldfield, B. J., McKinley, M. J., Figdor, R., Allen, A. M., and Mendelsohn, F. A. (1987). Localization and characterization of insulin receptors in rat brain and pituitary gland using in vitro autoradiography and computerized densitometry. *Endocrinology* 121, 1562–1570.
310. Baskin, D. G., Sipols, A. J., Schwartz, M. W., and White, M. F. (1993). Immunocytochemical detection of insulin receptor substrate-1 (IRS-1) in rat brain: colocalization with phosphotyrosine. *Regul. Pept.* 48, 257–266.
311. Folli, F., Bonfanti, L., Renard, E., Kahn, C. R., and Merighi, A. (1994). Insulin receptor substrate-1 (IRS-1) distribution in the rat central nervous system. *J. Neurosci.* 14, 6412–6422.
312. Wolf, D. H., Numan, S., Nestler, E. J., and Russell, D. S. (1999). Regulation of phospholipase C gamma in the mesolimbic dopamine system by chronic morphine administration. *J. Neurochem.* 73, 1520–1528.

313. Numan, S., and Russell, D. S. (1999). Discrete expression of insulin receptor substrate-4 mRNA in adult rat brain. *Brain Res. Mol. Brain Res.* 72, 97–102.
314. Unger, J. W., and Betz, M. (1998). Insulin receptors and signal transduction proteins in the hypothalamo-hypophyseal system: a review on morphological findings and functional implications. *Histol. Histopathol.* 13, 1215–1224.
315. Margolis, R. U., and Altszuler, N. (1967). Insulin in the cerebrospinal fluid. *Nature* 215, 1375–1376.
316. Plata-Salaman, C. R. (1991). Insulin in the cerebrospinal fluid. *Neurosci. Biobehav. Rev.* 15, 243–258.
317. Schwartz, M. W., Figlewicz, D. P., Baskin, D. G., Woods, S. C., and Porte, D. Jr. (1992). Insulin in the brain: a hormonal regulator of energy balance. *Endocr. Rev.* 13, 387–414.
318. Baskin, D. G., Woods, S. C., West, D. B., van Houten, M., Posner, B. I., Dorsa, D. M., and Porte, D. Jr. (1983). Immunocytochemical detection of insulin in rat hypothalamus and its possible uptake from cerebrospinal fluid. *Endocrinology* 113, 1818–1825.
319. van Houten, M., Posner, B. I., Kopriwa, B. M., and Brawer, J. R. (1979). Insulin-binding sites in the rat brain: in vivo localization to the circumventricular organs by quantitative radioautography. *Endocrinology* 105, 666–673.
320. van Houten, M., Posner, B. I., Kopriwa, B. M., and Brawer, J. R. (1980). Insulin binding sites localized to nerve terminals in rat median eminence and arcuate nucleus. *Science* 207, 1081–1083.
321. Woods, S. C., and Porte, D. Jr. (1977). Relationship between plasma and cerebrospinal fluid insulin levels of dogs. *Am. J. Physiol.* 233, E331–E334.
322. Strubbe, J. H., Porte, D. Jr., and Woods, S. C. (1988). Insulin responses and glucose levels in plasma and cerebrospinal fluid during fasting and refeeding in the rat. *Physiol. Behav.* 44, 205–208.
323. Steffens, A. B., Scheurink, A. J., Porte, D. Jr., and Woods, S. C. (1988). Penetration of peripheral glucose and insulin into cerebrospinal fluid in rats. *Am. J. Physiol.* 255, R200–R204.
324. Owen, O. E., Reichard, G. A., Jr., Boden, G., and Shuman, C. (1974). Comparative measurements of glucose, beta-hydroxybutyrate, acetoacetate, and insulin in blood and cerebrospinal fluid during starvation. *Metabolism* 23, 7–14.
325. Woods, S. C., Lotter, E. C., McKay, L. D., and Porte, D. Jr. (1979). Chronic intracerebroventricular infusion of insulin reduces food intake and body weight of baboons. *Nature* 282, 503–505.
326. Brief, D. J., and Davis, J. D. (1984). Reduction of food intake and body weight by chronic intraventricular insulin infusion. *Brain Res. Bull.* 12, 571–575.
327. Plata-Salaman, C. R., Oomura, Y., and Shimizu, N. (1986). Dependence of food intake on acute and chronic ventricular administration of insulin. *Physiol. Behav.* 37, 717–734.
328. Ikeda, H., West, D. B., Pustek, J. J., Figlewicz, D. P., Greenwood, M. R., Porte, D. Jr., and Woods, S. C. (1986). Intraventricular insulin reduces food intake and body weight of lean but not obese Zucker rats. *Appetite* 7, 381–386.
329. McGowan, M. K., Andrews, K. M., Kelly, J., and Grossman, S. P. (1990). Effects of chronic intrahypothalamic infusion of insulin on food intake and diurnal meal patterning in the rat. *Behav. Neurosci.* 104, 373–385.
330. Menendez, J. A., and Atrens, D. M. (1991). Insulin and the paraventricular hypothalamus: modulation of energy balance. *Brain Res.* 555, 193–201.
331. Rothwell, N. J., and Stock, M. J. (1988). Insulin and thermogenesis. *Int. J. Obes.* 12, 93–102.
332. Sipols, A. J., Baskin, D. G., and Schwartz, M. W. (1995). Effect of intracerebroventricular insulin infusion on diabetic hyperphagia and hypothalamic neuropeptide gene expression. *Diabetes* 44, 147–151.
333. Stein, L. J., Dorsa, D. M., Baskin, D. G., Figlewicz, D. P., Porte, D., Jr., and Woods, S. C. (1987). Reduced effect of experimental peripheral hyperinsulinemia to elevate cerebrospinal fluid insulin concentrations of obese Zucker rats. *Endocrinology* 121, 1611–1615.
334. Carvalheira, J. B., Ribeiro, E. B., Araujo, E. P., Guimaraes, R. B., Telles, M. M., Torsoni, M., Gontijo, J. A., Velloso, L. A., and Saad, M. J. (2003). Selective impairment of insulin signalling in the hypothalamus of obese Zucker rats. *Diabetologia* 46, 1629–1640.
335. Strubbe, J. H., and Mein, C. G. (1977). Increased feeding in response to bilateral injection of insulin antibodies in the VMH. *Physiol. Behav.* 19, 309–313.
336. Obici, S., Feng, Z., Karkanias, G., Baskin, D. G., and Rossetti, L. (2002). Decreasing hypothalamic insulin receptors causes hyperphagia and insulin resistance in rats. *Nat. Neurosci.* 5, 566–572.
337. Bruning, J. C., Gautam, D., Burks, D. J., Gillette, J., Schubert, M., Orban, P. C., Klein, R., Krone, W., Muller-Wieland, D., and Kahn, C. R. (2000). Role of brain insulin receptor in control of body weight and reproduction. *Science* 289, 2122–2125.
338. Helmreich, D. L., Mattern, L. G., and Cameron, J. L. (1993). Lack of a role of the hypothalamic-pituitary-adrenal axis in the fasting-induced suppression of luteinizing hormone secretion in adult male rhesus monkeys (*Macaca mulatta*). *Endocrinology* 132, 2427–2437.
339. Schreihofer, D. A., Parfitt, D. B., and Cameron, J. L. (1993). Suppression of luteinizing hormone secretion during short-term fasting in male rhesus monkeys: the role of metabolic versus stress signals. *Endocrinology* 132, 1881–1889.
340. Williams, N. I., Lancas, M. J., and Cameron, J. L. (1996). Stimulation of luteinizing hormone secretion by food intake: evidence against a role for insulin. *Endocrinology* 137, 2565–2571.
341. Chen, M. D., O'Byrne, K. T., Chiappini, S. E., Hotchkiss, J., and Knobil, E. (1992). Hypoglycemic "stress" and gonadotropin-releasing hormone pulse generator activity in the rhesus monkey: role of the ovary. *Neuroendocrinology* 56, 666–673.
342. Ohkura, S., Ichimaru, T., Itoh, F., Matsuyama, S., and Okamura, H. (2004). Further evidence for the role of glucose as a metabolic regulator of hypothalamic gonadotropin-releasing hormone pulse generator activity in goats. *Endocrinology* 145, 3239–3246.
343. Clarke, I. J., Horton, R. J., and Doughton, B. W. (1990). Investigation of the mechanism by which insulin-induced hypoglycemia decreases luteinizing hormone secretion in ovariectomized ewes. *Endocrinology* 127, 1470–1476.
344. Medina, C. L., Nagatani, S., Darling, T. A., Bucholtz, D. C., Tsukamura, H., Maeda, K., and Foster, D. L. (1998). Glucose availability modulates the timing of the luteinizing hormone surge in the ewe. *J. Neuroendocrinol.* 10, 785–792.
345. Rodriguez, M., Arias, P., Refojo, D., Feleder, C., and Moguilevsky, J. (1999). Arrest of pulsatile luteinizing hormone (LH) secretion during insulin-induced hypoglycemia (IIH): improvement by intrahypothalamic perfusion with glucose. *Exp. Clin. Endocrinol. Diabetes* 107, 257–261.
346. Goubillon, M. L., and Thalabard, J. C. (1996). Insulin-induced hypoglycemia decreases luteinizing hormone secretion in the castrated male rat: involvement of opiate peptides. *Neuroendocrinology* 64, 49–56.
347. Cagampang, F. R., Cates, P. S., Sandhu, S., Strutton, P. H., McGarvey, C., Coen, C. W., and O'Byrne, K. T. (1997). Hypoglycaemia-induced inhibition of pulsatile luteinizing hormone secretion in female rats: role of oestradiol, endogenous opioids and the adrenal medulla. *J. Neuroendocrinol.* 9, 867–872.

348. Heisler, L. E., Pallotta, C. M., Reid, R. L., and Van Vugt, D. A. (1993). Hypoglycemia-induced inhibition of luteinizing hormone secretion in the rhesus monkey is not mediated by endogenous opioid peptides. *J. Clin. Endocrinol. Metab.* 76, 1280–1285.
349. Bing-You, R. G., and Spratt, D. I. (1992). Serum estradiol but not gonadotropin levels decrease acutely after insulin-induced hypoglycemia in cycling women. *J. Clin. Endocrinol. Metab.* 75, 1054–1059.
350. Wade, G. N., Schneider, J. E., and Friedman, M. I. (1991). Insulin-induced anestrus in Syrian hamsters. *Am. J. Physiol.* 260, R148–R152.
351. McClure, T. J. (1968). Hypoglycaemia, an apparent cause of infertility of lactating cows. *Br. Vet. J.* 124, 126–130.
352. He, D., Funabashi, T., Sano, A., Uemura, T., Minaguchi, H., and Kimura, F. (1999). Effects of glucose and related substrates on the recovery of the electrical activity of gonadotropin-releasing hormone pulse generator which is decreased by insulin-induced hypoglycemia in the estrogen-primed ovariectomized rat. *Brain Res.* 820, 71–76.
353. Burcelin, R., Thorens, B., Glauser, M., Gaillard, R. C., and Pralong, F. P. (2003). Gonadotropin-releasing hormone secretion from hypothalamic neurons: stimulation by insulin and potentiation by leptin. *Endocrinology* 144, 4484–4491.
354. Miller, D. W., Blache, D., and Martin, G. B. (1995). The role of intracerebral insulin in the effect of nutrition on gonadotrophin secretion in mature male sheep. *J. Endocrinol.* 147, 321–329.
355. Hileman, S. M., Schillo, K. K., and Hall, J. B. (1993). Effects of acute, intracerebroventricular administration of insulin on serum concentrations of luteinizing hormone, insulin, and glucose in ovariectomized lambs during restricted and ad libitum feed intake. *Biol. Reprod.* 48, 117–124.
356. Bartak, V., Josifko, M., and Horackova, M. (1975). Juvenile diabetes and human sperm quality. *Int. J. Fertil.* 20, 30–32.
357. Kolodny, R. C., Kahn, C. B., Goldstein, H. H., and Barnett, D. M. (1974). Sexual dysfunction in diabetic men. *Diabetes* 23, 306–309.
358. Fairburn, C. (1981). The sexual problems of diabetic men. *Br. J. Hosp. Med.* 25, 484, 487, 489–491.
359. Schiavi, R. C., Stimmel, B. B., Mandeli, J., and Rayfield, E. J. (1993). Diabetes mellitus and male sexual function: a controlled study. *Diabetologia* 36, 745–751.
360. Betancourt-Albrecht, M., and Cunningham, G. R. (2003). Hypogonadism and diabetes. *Int. J. Impot. Res.* 15 Suppl 4, S14–S20.
361. Griffin, M. L., South, S. A., Yankov, V. I., Booth, R. A., Jr., Asplin, C. M., Veldhuis, J. D., and Evans, W. S. (1994). Insulin-dependent diabetes mellitus and menstrual dysfunction. *Ann. Med.* 26, 331–340.
362. South, S. A., Asplin, C. M., Carlsen, E. C., Booth, R. A., Jr., Weltman, J. Y., Johnson, M. L., Veldhuis, J. D., and Evans, W. S. (1993). Alterations in luteinizing hormone secretory activity in women with insulin-dependent diabetes mellitus and secondary amenorrhea. *J. Clin. Endocrinol. Metab.* 76, 1048–1053.
363. Distiller, L. A., Sagel, J., Morley, J. E., and Seftel, H. C. (1975). Pituitary responsiveness to luteinizing hormone-releasing hormone in insulin-dependent diabetes mellitus. *Diabetes* 24, 378–380.
364. Djursing, H., Hagen, C., Nyholm, H. C., Carstensen, L., and Andersen, A. N. (1983). Gonadotropin responses to gonadotropin-releasing hormone and prolactin responses to thyrotropin-releasing hormone and metoclopramide in women with amenorrhea and insulin-treated diabetes mellitus. *J. Clin. Endocrinol. Metab.* 56, 1016–1021.
365. Dong, Q., Lazarus, R. M., Wong, L. S., Vellios, M., and Handelsman, D. J. (1991). Pulsatile LH secretion in streptozotocin-induced diabetes in the rat. *J. Endocrinol.* 131, 49–55.
366. Sudha, S., Valli, G., Julie, P. M., Arunakaran, J., Govindarajulu, P., and Balasubramanian, K. (2000). Influence of streptozotocin induced diabetes and insulin treatment on the pituitary-testicular axis during sexual maturation in rats. *Exp. Clin. Endocrinol. Diabetes* 108, 14–20.
367. Murray, F. T., Orth, J., Gunsalus, G., Weisz, J., Li, J. B., Jefferson, L. S., Musto, N. A., and Bardin, C. W. (1981). The pituitary-testicular axis in the streptozotocin diabetic male rat: evidence for gonadotroph, Sertoli cell and Leydig cell dysfunction. *Int. J. Androl.* 4, 265–280.
368. Steger, R. W., Amador, A., Lam, E., Rathert, J., Weis, J., and Smith, M. S. (1989). Streptozotocin-induced deficits in sex behavior and neuroendocrine function in male rats. *Endocrinology* 124, 1737–1743.
369. Howland, B. E., and Zebrowski, E. J. (1980). Pituitary response to gonadotropin-releasing hormone in diabetic male rats. *Experientia* 36, 610–611.
370. Bestetti, G. E., Boujon, C. E., Reymond, M. J., and Rossi, G. L. (1989). Functional and morphological changes in mediobasal hypothalamus of streptozocin-induced diabetic rats. In vitro study of LHRH release. *Diabetes* 38, 471–476.
371. King, T. S., and Kang, I. S. (1989). Effects of chronic diabetes on in vitro GnRH release from the hypothalamus of the male rat. *Biomed. Res.* 10, 333–339.
372. Lawrence, A. M., and Contopoulos, A. N. (1960). Reproductive performance in the alloxan diabetic female rat. *Acta. Endocrinol. (Copenh).* 33, 175–184.
373. Kirchick, H. J., Keyes, P. L., and Frye, B. E. (1978). Etiology of anovulation in the immature alloxan-diabetic rat treated with pregnant mare's serum gonadotropin: absence of the preovulatory luteinizing hormone surge. *Endocrinology* 102, 1867–1873.
374. Katayama, S., Brownscheidle, C. M., Wootten, V., Lee, J. B., and Shimaoka, K. (1984). Absent or delayed preovulatory luteinizing hormone surge in experimental diabetes mellitus. *Diabetes* 33, 324–327.
375. Blades, R. A., Bryant, K. R., and Whitehead, S. A. (1985). Feedback effects of steroids and gonadotrophin control in adult rats with streptozotocin-induced diabetes mellitus. *Diabetologia* 28, 348–354.
376. Kovacs, P., Parlow, A. F., and Karkanias, G. B. (2002). Effect of centrally administered insulin on gonadotropin-releasing hormone neuron activity and luteinizing hormone surge in the diabetic female rat. *Neuroendocrinology* 76, 357–365.
377. Gentry, R. T., Wade, G. N., and Blaustein, J. D. (1977). Binding of (3H)estradiol by brain cell nuclei and female rat sexual behavior: inhibition by experimental diabetes. *Brain Res.* 135, 135–146.
378. Karkanias, G. B., Morales, J. C., and Li, C. S. (1997). Deficits in reproductive behavior in diabetic female rats are due to hypoinsulinemia rather than hyperglycemia. *Horm. Behav.* 32, 19–29.
379. Spindler-Vomachka, M., and Johnson, D. C. (1985). Altered hypothalamic-pituitary function in the adult female rat with streptozotocin-induced diabetes. *Diabetologia* 28, 38–44.
380. Bestetti, G., Locatelli, V., Tirone, F., Rossi, G. L., and Muller, E. E. (1985). One month of streptozotocin-diabetes induces different neuroendocrine and morphological alterations in the hypothalamo-pituitary axis of male and female rats. *Endocrinology* 117, 208–216.
381. Kirchick, H. J., Keyes, P. L., and Frye, B. E. (1979). An explanation for anovulation in immature alloxan-diabetic rats treated with pregnant mare's serum gonadotropin: reduced pituitary response to gonadotropin-releasing hormone. *Endocrinology* 105, 1343–1349.

382. Bucholtz, D. C., Chiesa, A., Pappano, W. N., Nagatani, S., Tsukamura, H., Maeda, K. I., and Foster, D. L. (2000). Regulation of pulsatile luteinizing hormone secretion by insulin in the diabetic male lamb. *Biol. Reprod.* 62, 1248–1255.
383. Steger, R. W., and Kienast, S. G. (1990). Effect of continuous versus delayed insulin replacement on sex behavior and neuroendocrine function in diabetic male rats. *Diabetes* 39, 942–948.
384. Kovacs, P., Morales, J. C., and Karkanias, G. B. (2003). Central insulin administration maintains reproductive behavior in diabetic female rats. *Neuroendocrinology* 78, 90–95.
385. Tanaka, T., Nagatani, S., Bucholtz, D. C., Ohkura, S., Tsukamura, H., Maeda, K., and Foster, D. L. (2000). Central action of insulin regulates pulsatile luteinizing hormone secretion in the diabetic sheep model. *Biol. Reprod.* 62, 1256–1261.
386. Burks, D. J., de Mora, J. F., Schubert, M., Withers, D. J., Myers, M. G., Towery, H. H., Altamuro, S. L., Flint, C. L., and White, M. F. (2000). IRS-2 pathways integrate female reproduction and energy homeostasis. *Nature* 407, 377–382.
387. Olson, B. R., Scott, D. C., Wetsel, W. C., Elliot, S. J., Tomic, M., Stojilkovic, S., Nieman, L. K., and Wray, S. (1995). Effects of insulin-like growth factors I and II and insulin on the immortalized hypothalamic GTI-7 cell line. *Neuroendocrinology* 62, 155–165.
388. Arias, P., Rodriguez, M., Szwarcfarb, B., Sinay, I. R., and Moguilevsky, J. A. (1992). Effect of insulin on LHRH release by perifused hypothalamic fragments. *Neuroendocrinology* 56, 415–418.
389. Steiner, R. A., Cameron, J. L., McNeill, T. H., Clifton, D. K., and Bremner, W. J. (1983). Metabolic signals for the onset of puberty. In *Neuroendocrine Aspects of Reproduction* (R. L. Norman, Ed.), pp. 183–227. Academic Press, New York.
390. Kordon, C., Drouva, S. V., Martinez de la Escalera, G., and Weiner, R. I. (1994). Role of classic and peptide neuromediators in the neuroendocrine regulation of luteinizing hormone and prolactin. In *The Physiology of Reproduction* (E. Knobil and J. D. Neill, Eds.), Vol. 2, pp. 1621–1681. Raven Press, New York.
391. Fernstrom, J. D. (1983). Role of precursor availability in control of monoamine biosynthesis in brain. *Physiol. Rev.* 63, 484–546.
392. Adashi, E. Y., Hsueh, A. J., and Yen, S. S. (1981). Insulin enhancement of luteinizing hormone and follicle-stimulating hormone release by cultured pituitary cells. *Endocrinology* 108, 1441–1449.
393. Grizard, G., Fournet, M., Rigaudiere, N., Lombard-Vignon, N., and Grizard, J. (1991). Insulin binding to Leydig cells and insulin levels in testicular interstitial fluid at different stages of development in the rat. *J. Endocrinol.* 128, 375–381.
394. Abele, V., Pelletier, G., and Tremblay, R. R. (1986). Radioautographic localization and regulation of the insulin receptors in rat testis. *J. Recept. Res.* 6, 461–473.
395. Lin, T., Haskell, J., Vinson, N., and Terracio, L. (1986). Characterization of insulin and insulin-like growth factor I receptors of purified Leydig cells and their role in steroidogenesis in primary culture: a comparative study. *Endocrinology* 119, 1641–1647.
396. Rigaudiere, N., Loubassou, S., Grizard, G., and Boucher, D. (1988). Characterization of insulin binding and comparative action of insulin and insulin-like growth factor I on purified Leydig cells from the adult rat. *Int. J. Androl.* 11, 165–178.
397. Poretsky, L., and Kalin, M. F. (1987). The gonadotropic function of insulin. *Endocr. Rev.* 8, 132–141.
398. Cameron, J. L. (1996). Regulation of reproductive hormone secretion in primates by short-term changes in nutrition. *Rev. Reprod.* 1, 117–126.
399. McCann, J. P., and Hansel, W. (1986). Relationships between insulin and glucose metabolism and pituitary-ovarian functions in fasted heifers. *Biol. Reprod.* 34, 630–641.
400. Jones, J. E., and Lubbers, L. S. (2001). Suppression and recovery of estrous behavior in Syrian hamsters after changes in metabolic fuel availability. *Am. J. Physiol. Regul. Integr. Comp. Physiol.* 280, R1393–R1398.
401. Grossman, S. P. (1986). The role of glucose, insulin and glucagon in the regulation of food intake and body weight. *Neurosci. Biobehav. Rev.* 10, 295–315.
402. Miselis, R. R., and Epstein, A. N. (1975). Feeding induced by intracerebroventricular 2-deoxy-D-glucose in the rat. *Am. J. Physiol.* 229, 1438–1447.
403. Balagura, S., and Kanner, M. (1971). Hypothalamic sensitivity to 2-deoxy-D-glucose and glucose: effects on feeding behavior. *Physiol. Behav.* 7, 251–255.
404. Penick, S. B., and Hinkle, L. E. Jr. (1961). Depression of food intake induced in healthy subjects by glucagon. *N. Engl. J. Med.* 264, 893–897.
405. Geary, N., and Smith, G. P. (1982). Pancreatic glucagon and postprandial satiety in the rat. *Physiol. Behav.* 28, 313–322.
406. Penicaud, L., and Le Magnen, J. (1980). Recovery of body weight following starvation or food restriction in rats. *Neurosci. Biobehav. Rev.* 4 Suppl 1, 47–52.
407. Woods, S. C., Schwartz, M. W., Baskin, D. G., and Seeley, R. J. (2000). Food intake and the regulation of body weight. *Annu. Rev. Psychol.* 51, 255–277.
408. Smith, G. P., and Epstein, A. N. (1969). Increased feeding in response to decreased glucose utilization in the rat and monkey. *Am. J. Physiol.* 217, 1083–1087.
409. Thompson, D. A., and Campbell, R. G. (1977). Hunger in humans induced by 2-deoxy-D-glucose: glucoprivic control of taste preference and food intake. *Science* 198, 1065–1068.
410. Ritter, R. C., and Slusser, P. (1980). 5-Thio-D-glucose causes increased feeding and hyperglycemia in the rat. *Am. J. Physiol.* 238, E141–E144.
411. Engeset, R. M., and Ritter, R. C. (1980). Intracerebroventricular 2-DG causes feeding in the absence of other signs of glucoprivation. *Brain Res.* 202, 229–233.
412. Ritter, R. C., Slusser, P. G., and Stone, S. (1981). Glucoreceptors controlling feeding and blood glucose: location in the hindbrain. *Science* 213, 451–452.
413. Slusser, P. G., and Ritter, R. C. (1980). Increased feeding and hyperglycemia elicited by intracerebroventricular 5-thioglucose. *Brain Res.* 202, 474–478.
414. Scharrer, E., and Langhans, W. (1986). Control of food intake by fatty acid oxidation. *Am. J. Physiol.* 250, R1003–R1006.
415. Friedman, M. I., Tordoff, M. G., and Ramirez, I. (1986). Integrated metabolic control of food intake. *Brain Res. Bull.* 17, 855–859.
416. Schneider, J. E., Lazzarini, S. J., Friedman, M. I., and Wade, G. N. (1988). Role of fatty acid oxidation in food intake and hunger motivation in Syrian hamsters. *Physiol. Behav.* 43, 617–623.
417. Friedman, M. I., and Tordoff, M. G. (1986). Fatty acid oxidation and glucose utilization interact to control food intake in rats. *Am. J. Physiol.* 251, R840–R845.
418. Schreihofer, D. A., Renda, F., and Cameron, J. L. (1996). Feeding-induced stimulation of luteinizing hormone secretion in male rhesus monkeys is not dependent on a rise in blood glucose concentration. *Endocrinology* 137, 3770–3776.

419. Steger, R. W., and Rabe, M. B. (1997). The effect of diabetes mellitus on endocrine and reproductive function. *Proc. Soc. Exp. Biol. Med.* 214, 1–11.
420. Lado-Abeal, J., Veldhuis, J. D., and Norman, R. L. (2002). Glucose relays information regarding nutritional status to the neural circuits that control the somatotropic, corticotropic, and gonadotropic axes in adult male rhesus macaques. *Endocrinology* 143, 403–410.
421. Temple, J. L., Schneider, J. E., Scott, D. K., Korutz, A., and Rissman, E. F. (2002). Mating behavior is controlled by acute changes in metabolic fuels. *Am. J. Physiol. Regul. Integr. Comp. Physiol.* 282, R782–R790.
422. Howland, B. E. (1980). Effect of glucoprivation induced by 2-deoxy-D-glucose on serum gonadotropin levels, pituitary response to GnRH and progesterone-induced release of luteinizing hormone in rats. *Horm. Metab. Res.* 12, 520–523.
423. Nagatani, S., Bucholtz, D. C., Murahashi, K., Estacio, M. A., Tsukamura, H., Foster, D. L., and Maeda, K. I. (1996). Reduction of glucose availability suppresses pulsatile luteinizing hormone release in female and male rats. *Endocrinology* 137, 1166–1170.
424. Murahashi, K., Bucholtz, D. C., Nagatani, S., Tsukahara, S., Tsukamura, H., Foster, D. L., and Maeda, K. I. (1996). Suppression of luteinizing hormone pulses by restriction of glucose availability is mediated by sensors in the brain stem. *Endocrinology* 137, 1171–1176.
425. Bucholtz, D. C., Vidwans, N. M., Herbosa, C. G., Schillo, K. K., and Foster, D. L. (1996). Metabolic interfaces between growth and reproduction. V. Pulsatile luteinizing hormone secretion is dependent on glucose availability. *Endocrinology* 137, 601–607.
426. Briski, K. P. (1997). Central opioid receptors mediate glucoprivic inhibition of pituitary LH secretion. *Am. J. Physiol.* 272, E517–E522.
427. Funston, R. N., Roberts, A. J., Hixon, D. L., Hallford, D. M., Sanson, D. W., and Moss, G. E. (1995). Effect of acute glucose antagonism on hypophyseal hormones and concentrations of insulin-like growth factor (IGF)-I and IGF-binding proteins in serum, anterior pituitary, and hypothalamus of ewes. *Biol. Reprod.* 52, 1179–1186.
428. Ohkura, S., Tanaka, T., Nagatani, S., Bucholtz, D. C., Tsukamura, H., Maeda, K., and Foster, D. L. (2000). Central, but not peripheral, glucose-sensing mechanisms mediate glucoprivic suppression of pulsatile luteinizing hormone secretion in the sheep. *Endocrinology* 141, 4472–4480.
429. McClure, T. J., Nancarrow, C. D., and Radford, H. M. (1978). The effect of 2-deoxy-D-glucose on ovarian function of cattle. *Aust. J. Biol. Sci.* 31, 183–186.
430. Schneider, J. E., Friedenson, D. G., Hall, A. J., and Wade, G. N. (1993). Glucoprivation induces anestrus and lipoprivation may induce hibernation in Syrian hamsters. *Am. J. Physiol.* 264, R573–R577.
431. Briski, K. P., and Sylvester, P. W. (1998). Role of endogenous opiates in glucoprivic inhibition of the luteinizing hormone surge and fos expression by preoptic gonadotropin-releasing hormone neurones in ovariectomized steroid-primed female rats. *J. Neuroendocrinol.* 10, 769–776.
432. Schneider, J. E., and Wade, G. N. (1989). Availability of metabolic fuels controls estrous cyclicity of Syrian hamsters. *Science.* 244, 1326–1328.
433. Schneider, J. E., Goldman, M. D., Leo, N. A., and Rosen, M. E. (1997). Central vs. peripheral metabolic control of estrous cycles in Syrian hamsters. II. Glucoprivation. *Am. J. Physiol.* 272, R406–R412.
434. Dickerman, R. W., Li, H. Y., and Wade, G. N. (1993). Decreased availability of metabolic fuels suppresses estrous behavior in Syrian hamsters. *Am. J. Physiol.* 264, R568–R572.
435. Schneider, J. E., and Wade, G. N. (1990). Decreased availability of metabolic fuels induces anestrus in golden hamsters. *Am. J. Physiol.* 258, R750–R755.
436. Oomura, Y. (1981). Chemosensitive neuron in the hypothalamus related to food intake behavior. *Jpn. J. Pharmacol.* 31 Suppl, 1P–12P.
437. Oomura, Y., and Kita, H. (1981). Insulin acting as a modulator of feeding through the hypothalamus. *Diabetologia.* 20 Suppl, 290–298.
438. Singer, L. K., and Ritter, S. (1996). Intraventricular glucose blocks feeding induced by 2-deoxy-D-glucose but not mercaptoacetate. *Physiol. Behav.* 59, 921–923.
439. Levin, B. E., Dunn-Meynell, A. A., and Routh, V. H. (2002). CNS sensing and regulation of peripheral glucose levels. *Int. Rev. Neurobiol.* 51, 219–258.
440. Maekawa, F., Toyoda, Y., Torii, N., Miwa, I., Thompson, R. C., Foster, D. L., Tsukahara, S., Tsukamura, H., and Maeda, K. (2000). Localization of glucokinase-like immunoreactivity in the rat lower brain stem: for possible location of brain glucose-sensing mechanisms. *Endocrinology* 141, 375–384.
441. Obici, S., Feng, Z., Arduini, A., Conti, R., and Rossetti, L. (2003). Inhibition of hypothalamic carnitine palmitoyltransferase-1 decreases food intake and glucose production. *Nat. Med.* 9, 756–761.
442. Obici, S., Feng, Z., Morgan, K., Stein, D., Karkanias, G., and Rossetti, L. (2002). Central administration of oleic acid inhibits glucose production and food intake. *Diabetes* 51, 271–275.
443. Berriman, S. J., Wade, G. N., and Blaustein, J. D. (1992). Expression of Fos-like proteins in gonadotropin-releasing hormone neurons of Syrian hamsters: effects of estrous cycles and metabolic fuels. *Endocrinology* 131, 2222–2228.
444. Shimabukuro, M., Koyama, K., Chen, G., Wang, M. Y., Trieu, F., Lee, Y., Newgard, C. B., and Unger, R. H. (1997). Direct antidiabetic effect of leptin through triglyceride depletion of tissues. *Proc. Natl. Acad. Sci. U S A* 94, 4637–4641.
445. Nagatani, S., Thompson, R. C., and Foster, D. L. (2001). Prevention of glucoprivic stimulation of corticosterone secretion by leptin does not restore high frequency luteinizing hormone pulses in rats. *J. Neuroendocrinol.* 13, 371–377.
446. Livingstone, C., Lyall, H., and Gould, G. W. (1995). Hypothalamic GLUT 4 expression: a glucose- and insulin-sensing mechanism? *Mol. Cell. Endocrinol.* 107, 67–70.
447. Oppenheimer, J. H., Schwartz, H. L., and Surks, M. I. (1974). Tissue differences in the concentration of triiodothyronine nuclear binding sites in the rat: liver, kidney, pituitary, heart, brain, spleen, and testis. *Endocrinology* 95, 897–903.
448. Bradley, D. J., Young, W. S. III, and Weinberger, C. (1989). Differential expression of alpha and beta thyroid hormone receptor genes in rat brain and pituitary. *Proc. Natl. Acad. Sci. U S A* 86, 7250–7254.
449. Harris, A. R., Fang, S. L., Azizi, F., Lipworth, L., Vagenakis, A. G., and Barverman, L. E. (1978). Effect of starvation on hypothalamic-pituitary-thyroid function in the rat. *Metabolism.* 27, 1074–1083.
450. Kinlaw, W. B., Schwartz, H. L., and Oppenheimer, J. H. (1985). Decreased serum triiodothyronine in starving rats is due primarily to diminished thyroidal secretion of thyroxine. *J. Clin. Invest.* 75, 1238–1241.
451. Gardner, D. F., Kaplan, M. M., Stanley, C. A., and Utiger, R. D. (1979). Effect of tri-iodothyronine replacement on the metabolic and pituitary responses to starvation. *N. Engl. J. Med.* 300, 579–584.
452. Beer, S. F., Bircham, P. M., Bloom, S. R., Clark, P. M., Hales, C. N., Hughes, C. M., Jones, C. T., Marsh, D. R., Raggatt, P. R., and Findlay, A. L. (1989). The effect of

a 72-h fast on plasma levels of pituitary, adrenal, thyroid, pancreatic and gastrointestinal hormones in healthy men and women. *J. Endocrinol.* 120, 337–350.
453. Hugues, J. N., Burger, A. G., Pekary, A. E., and Hershman, J. M. (1984). Rapid adaptations of serum thyrotrophin, triiodothyronine and reverse triiodothyronine levels to short-term starvation and refeeding. *Acta. Endocrinol. (Copenh).* 105, 194–199.
454. Miyai, K., Yamamoto, T., Azukizawa, M., Ishibashi, K., and Kumahara, Y. (1975). Serum thyroid hormones and thyrotropin in anorexia nervosa. *J. Clin. Endocrinol. Metab.* 40, 334–338.
455. Moshang, T., Jr., Parks, J. S., Baker, L., Vaidya, V., Utiger, R. D., Bongiovanni, A. M., and Snyder, P. J. (1975). Low serum triiodothyronine in patients with anorexia nervosa. *J. Clin. Endocrinol. Metab.* 40, 470–473.
456. Chopra, I. J., and Smith, S. R. (1975). Circulating thyroid hormones and thyrotropin in adult patients with protein-calorie malnutrition. *J. Clin. Endocrinol. Metab.* 40, 221–227.
457. de Rosa, G., Della Casa, S., Corsello, S. M., Ruffilli, M. P., de Rosa, E., and Pasargiklian, E. (1983). Thyroid function in altered nutritional state. *Exp. Clin. Endocrinol.* 82, 173–177.
458. Tveit, B., and Larsen, F. (1983). Suppression and stimulation of TSH and thyroid hormones in bulls during starvation and refeeding. *Acta. Endocrinol. (Copenh).* 103, 223–226.
459. Unger, J. (1988). Fasting induces a decrease in serum thyroglobulin in normal subjects. *J. Clin. Endocrinol. Metab.* 67, 1309–1311.
460. Legradi, G., Emerson, C. H., Ahima, R. S., Flier, J. S., and Lechan, R. M. (1997). Leptin prevents fasting-induced suppression of prothyrotropin-releasing hormone messenger ribonucleic acid in neurons of the hypothalamic paraventricular nucleus. *Endocrinology* 138, 2569–2576.
461. Hefco, E., Krulich, L., Illner, P., and Larsen, P. R. (1975). Effect of acute exposure to cold on the activity of the hypothalamic-pituitary-thyroid system. *Endocrinology* 97, 1185–1195.
462. Ortega, E., Rodriguez, E., Ruiz, E., and Osorio, C. (1990). Activity of the hypothalamo-pituitary ovarian axis in hypothyroid rats with or without triiodothyronine replacement. *Life Sci.* 46, 391–395.
463. Vriend, J., Bertalanffy, F. D., and Ralcewicz, T. A. (1987). The effects of melatonin and hypothyroidism on estradiol and gonadotropin levels in female Syrian hamsters. *Biol. Reprod.* 36, 719–728.
464. Bruni, J. F., Marshall, S., Dibbet, J. A., and Meites, J. (1975). Effects of hyper- and hypothyroidism on serum LH and FSH levels in intact and gonadectomized male and female rats. *Endocrinology* 97, 558–563.
465. Tamura, K., Hatsuta, M., Watanabe, G., Taya, K., and Kogo, H. (1998). Blockage of gonadotropin-induced first ovulation caused by thyroidectomy and its possible mechanisms in rats. *Am. J. Physiol.* 275, E380–E385.
466. Dellovade, T. L., Zhu, Y. S., Krey, L., and Pfaff, D. W. (1996). Thyroid hormone and estrogen interact to regulate behavior. *Proc. Natl. Acad. Sci. U S A* 93, 12581–12586.
467. Nicholls. T. J., Follett, B. K., Goldsmith, A. R., and Pearson, H. (1988). Possible homologies between photorefractoriness in sheep and birds: the effect of thyroidectomy on the length of the ewe's breeding season. *Reprod Nutr Dev.* 28, 375–385.
468. Webster, J. R., Moenter, S. M., Barrell, G. K., Lehman, M. N., and Karsch, F. J. (1991). Role of the thyroid gland in seasonal reproduction. III. Thyroidectomy blocks seasonal suppression of gonadotropin-releasing hormone secretion in sheep. *Endocrinology* 129, 1635–1643.
469. Moenter, S. M., Woodfill, C. J., and Karsch, F. J. (1991). Role of the thyroid gland in seasonal reproduction: thyroidectomy blocks seasonal suppression of reproductive neuroendocrine activity in ewes. *Endocrinology* 128, 1337–1344.
470. Webster, J. R., Moenter, S. M., Woodfill, C. J., and Karsch, F. J. (1991). Role of the thyroid gland in seasonal reproduction. II. Thyroxine allows a season-specific suppression of gonadotropin secretion in sheep. *Endocrinology* 129, 176–183.
471. Dahl, G. E., Evans, N. P., Thrun, L. A., and Karsch, F. J. (1995). Thyroxine is permissive to seasonal transitions in reproductive neuroendocrine activity in the ewe. *Biol. Reprod.* 52, 690–696.
472. Doufas, A. G., and Mastorakos, G. (2000). The hypothalamic-pituitary-thyroid axis and the female reproductive system. *Ann. N Y Acad. Sci.* 900, 65–76.
473. Meikle, A. W. (2004). The interrelationships between thyroid dysfunction and hypogonadism in men and boys. *Thyroid* 14 Suppl 1, S17–25.
474. Barnes, N. D., Hayles, A. B., and Ryan, R. J. (1973). Sexual maturation in juvenile hypothyroidism. *Mayo Clin. Proc.* 48, 849–856.
475. Jannini, E. A., Ulisse, S., and D'Armiento, M. (1995). Thyroid hormone and male gonadal function. *Endocr. Rev.* 16, 443–459.
476. Morgan, M. A., Dellovade, T. L., and Pfaff, D. W. (2000). Effect of thyroid hormone administration on estrogen-induced sex behavior in female mice. *Horm. Behav.* 37, 15–22.
477. Chandrasekhar, Y., D'Occhio, M. J., Holland, M. K., and Setchell, B. P. (1985). Activity of the hypothalamo-pituitary axis and testicular development in prepubertal ram lambs with induced hypothyroidism or hyperthyroidism. *Endocrinology* 117, 1645–1651.
478. Chandrasekhar, Y., D'Occhio, M. J., and Setchell, B. P. (1986). Delayed puberty caused by hyperthyroidism in ram lambs is not a result of suppression in body growth. *J. Reprod. Fertil.* 76, 763–769.
479. Gellert, R. J., Bakke, J. L., and Lawrence, N. L. (1971). Delayed vaginal opening in the rat following pharmacologic doses of T4 administered during the neonatal period. *J. Lab. Clin. Med.* 77, 410–416.
480. Thomas, R., and Reid, R. L. (1987). Thyroid disease and reproductive dysfunction: a review. *Obstet. Gynecol.* 70, 789–798.
481. Schwartz, H. L., and Oppenheimer, J. H. (1978). Physiologic and biochemical actions of thyroid hormone. *Pharmacol. Ther. (B).* 3, 349–376.
482. Palmero, S., Maggiani, S., and Fugassa, E. (1988). Nuclear triiodothyronine receptors in rat Sertoli cells. *Mol. Cell. Endocrinol.* 58, 253–256.
483. Palmero, S., Bardi, G., Bolla, F., and Fugassa, E. (1996). Influence of thyroid hormone on Sertoli cell protein metabolism in the prepubertal pig. *Boll. Soc. Ital. Biol. Sper.* 72, 163–170.
484. Jansen, H. T., Lubbers, L. S., Macchia, E., DeGroot, L. J., and Lehman, M. N. (1997). Thyroid hormone receptor (alpha) distribution in hamster and sheep brain: colocalization in gonadotropin-releasing hormone and other identified neurons. *Endocrinology* 138, 5039–5047.
485. Maruo, T., Hayashi, M., Matsuo, H., Yamamoto, T., Okada, H., and Mochizuki, M. (1987). The role of thyroid hormone as a biological amplifier of the actions of follicle-stimulating hormone in the functional differentiation of cultured porcine granulosa cells. *Endocrinology* 121, 1233–1241.
486. Channing, C. P., Tsai, V., and Sachs, D. (1976). Role of insulin, thyroxin and cortisol in luteinization of porcine granulosa cells grown in chemically defined media. *Biol. Reprod.* 15, 235–247.

487. Mendis-Handagama, S. M., and Ariyaratne, H. B. (2004). Effects of thyroid hormones on Leydig cells in the postnatal testis. *Histol. Histopathol.* 19, 985–997.
488. Chandrashekar, V., Zaczek, D., and Bartke, A. (2004). The consequences of altered somatotropic system on reproduction. *Biol. Reprod.* 71, 17–27.
489. Schreihofer, D. A., Golden, G. A., and Cameron, J. L. (1993). Cholecystokinin (CCK)-induced stimulation of luteinizing hormone (LH) secretion in adult male rhesus monkeys: examination of the role of CCK in nutritional regulation of LH secretion. *Endocrinology* 132, 1553–1560.
490. Beak, S. A., Heath, M. M., Small, C. J., Morgan, D. G., Ghatei, M. A., Taylor, A. D., Buckingham, J. C., Bloom, S. R., and Smith, D. M. (1998). Glucagon-like peptide-1 stimulates luteinizing hormone-releasing hormone secretion in a rodent hypothalamic neuronal cell line. *J. Clin. Invest.* 101, 1334–1341.
491. Furuta, M., Funabashi, T., and Kimura, F. (2001). Intracerebroventricular administration of ghrelin rapidly suppresses pulsatile luteinizing hormone secretion in ovariectomized rats. *Biochem. Biophys. Res. Commun.* 288, 780–785.
492. Fernandez-Fernandez, R., Tena-Sempere, M., Aguilar, E., and Pinilla, L. (2004). Ghrelin effects on gonadotropin secretion in male and female rats. *Neurosci. Lett.* 362, 103–107.
493. Rezek, M., and Novin, D. (1976). Duodenal nutrient infusion: effects on feeding in intact and vagotomized rabbit. *J. Nutr.* 106, 812–820.
494. Simpson, J. B. (1989). The circumventricular organs and brain barrier systems. In *Textbook of Physiology* (H. D. Patton, A. F. Fuchs, B. Hille, A. M. Scher, and R. Steiner, Eds.), Vol. 2, pp. 1215–1228. WB Saunders, Philadelphia.
495. Baura, G. D., Foster, D. M., Porte, D. Jr., Kahn, S. E., Bergman, R. N., Cobelli, C., and Schwartz, M. W. (1993). Saturable transport of insulin from plasma into the central nervous system of dogs in vivo. A mechanism for regulated insulin delivery to the brain. *J. Clin. Invest.* 92, 1824–1830.
496. Frank, H. J., and Pardridge, W. M. (1983). Insulin binding to brain microvessels. *Adv. Metab. Disord.* 10, 291–302.
497. Pardridge, W. M., Eisenberg, J., and Yang, J. (1985). Human blood-brain barrier insulin receptor. *J. Neurochem.* 44, 1771–1778.
498. Bjorbaek, C., Elmquist, J. K., Michl, P., Ahima, R. S., van Bueren, A., McCall, A. L., and Flier, J. S. (1998). Expression of leptin receptor isoforms in rat brain microvessels. *Endocrinology* 139, 3485–3491.
499. Silverman, A. J., Livne, I., and Witkin, J. W. (1994). The gonadotropin-releasing hormone (GnRH), neuronal systems: immunocytochemistry and in situ hybridization. In *The Physiology of Reproduction* (E. Knobil and J. D. Neill, Eds.), Vol. 1, pp. 1683–1709. Raven Press, New York.
500. Merchenthaler, I., Setalo, G., Petrusz, P., and Negro-Vilar, A., Flerko, B. (1989). Identification of hypophysiotropic luteinizing hormone-releasing hormone (LHRH) neurons by combined retrograde labeling and immunocytochemistry. *Exp. Clin. Endocrinol.* 94, 133–140.
501. Sullivan, S. D., DeFazio, R. A., and Moenter, S. M. (2003). Metabolic regulation of fertility through presynaptic and postsynaptic signaling to gonadotropin-releasing hormone neurons. *J. Neurosci.* 23, 8578–8585.
502. Sullivan, S. D., and Moenter, S. M. (2004). Gamma-aminobutyric acid neurons integrate and rapidly transmit permissive and inhibitory metabolic cues to gonadotropin-releasing hormone neurons. *Endocrinology* 145, 1194–1202.
503. Corp, E. S., Woods, S. C., Porte, D., Jr., Dorsa, D. M., Figlewicz, D. P., and Baskin, D. G. (1986). Localization of 125I-insulin binding sites in the rat hypothalamus by quantitative autoradiography. *Neurosci. Lett.* 70, 17–22.
504. Schneider, J. E., and Zhu, Y. (1994). Caudal brain stem plays a role in metabolic control of estrous cycles in Syrian hamsters. *Brain Res.* 661, 70–74.
505. Panicker, A. K., Mangels, R. A., Powers, J. B., Wade, G. N., and Schneider, J. E. (1998). AP lesions block suppression of estrous behavior, but not estrous cyclicity, in food-deprived Syrian hamsters. *Am. J. Physiol.* 275, R158–R164.
506. I'Anson, H., Sundling, L. A., Roland, S. M., and Ritter, S. (2003). Immunotoxic destruction of distinct catecholaminergic neuron populations disrupts the reproductive response to glucoprivation in female rats. *Endocrinology* 144, 4325–4331.
507. Grill, H. J., and Kaplan, J. M. (2002). The neuroanatomical axis for control of energy balance. *Front Neuroendocrinol.* 23, 2–40.
508. Tatemoto, K., Carlquist, M., and Mutt, V. (1982). Neuropeptide Y—a novel brain peptide with structural similarities to peptide YY and pancreatic polypeptide. *Nature.* 296, 659–660.
509. Chronwall, B. M., DiMaggio, D. A., Massari, V. J., Pickel, V. M., Ruggiero, D. A., and O'Donohue, T. L. (1985). The anatomy of neuropeptide-Y-containing neurons in rat brain. *Neuroscience.* 15, 1159–1181.
510. Morris, B. J. (1989). Neuronal localisation of neuropeptide Y gene expression in rat brain. *J. Comp. Neurol.* 290, 358–368.
511. Gehlert, D. R., Chronwall, B. M., Schafer, M. P., and O'Donohue, T. L. (1987). Localization of neuropeptide Y messenger ribonucleic acid in rat and mouse brain by in situ hybridization. *Synapse.* 1, 25–31.
512. Everitt, B. J., Hokfelt, T., Terenius, L., Tatemoto, K., Mutt, V., and Goldstein, M. (1984). Differential co-existence of neuropeptide Y (NPY)-like immunoreactivity with catecholamines in the central nervous system of the rat. *Neuroscience.* 11, 443–462.
513. Blomqvist, A. G., and Herzog, H. (1997). Y-receptor subtypes—how many more? *Trends Neurosci.* 20, 294–298.
514. Morton, G. J., and Schwartz, M. W. (2001). The NPY/AgRP neuron and energy homeostasis. *Int. J. Obes. Relat. Metab. Disord.* 25 Suppl 5, S56–S62.
515. Clark, J. T., Kalra, P. S., Crowley, W. R., and Kalra, S. P. (1984). Neuropeptide Y and human pancreatic polypeptide stimulate feeding behavior in rats. *Endocrinology* 115, 427–429.
516. Krasnow, S. M., Schuh, S. M., Teklemichael, D. N., Leilabadi, S. N., Baumgartner, J. W., Clifton, D. K., and Steiner, R. A. (2001). Differential roles of the NPY Y5 receptor in the regulation of feeding behavior and gonadotropin secretion in the mouse. *31st Annual Meeting of the Society for Neuroscience*, San Diego.
517. Kulkosky, P. J., Glazner, G. W., Moore, H. D., Low, C. A., and Woods, S. C. (1988). Neuropeptide Y: behavioral effects in the golden hamster. *Peptides* 9, 1389–1393.
518. Miner, J. L., Della-Fera, M. A., Paterson, J. A., and Baile, C. A. (1989). Lateral cerebroventricular injection of neuropeptide Y stimulates feeding in sheep. *Am. J. Physiol.* 257, R383–R387.
519. Larsen, P. J., Tang-Christensen, M., Stidsen, C. E., Madsen, K., Smith, M. S., and Cameron, J. L. (1999). Activation of central neuropeptide Y Y1 receptors potently stimulates food intake in male rhesus monkeys. *J. Clin. Endocrinol. Metab.* 84, 3781–3791.
520. Stanley, B. G., Kyrkouli, S. E., Lampert, S., and Leibowitz, S. F. (1986). Neuropeptide Y chronically injected into the hypothalamus: a powerful neurochemical inducer of hyperphagia and obesity. *Peptides* 7, 1189–1192.

521. Kalra, S. P., Dube, M. G., and Kalra, P. S. (1988). Continuous intraventricular infusion of neuropeptide Y evokes episodic food intake in satiated female rats: effects of adrenalectomy and cholecystokinin. *Peptides* 9, 723–728.
522. Catzeflis, C., Pierroz, D. D., Rohner-Jeanrenaud, F., Rivier, J. E., Sizonenko, P. C., and Aubert, M. L. (1993). Neuropeptide Y administered chronically into the lateral ventricle profoundly inhibits both the gonadotropic and the somatotropic axis in intact adult female rats. *Endocrinology* 132, 224–234.
523. Raposinho, P. D., Pierroz, D. D., Broqua, P., White, R. B., Pedrazzini, T., and Aubert, M. L. (2001). Chronic administration of neuropeptide Y into the lateral ventricle of C57BL/6J male mice produces an obesity syndrome including hyperphagia, hyperleptinemia, insulin resistance, and hypogonadism. *Mol. Cell. Endocrinol.* 185, 195–204.
524. Sawchenko, P. E., Swanson, L. W., Grzanna, R., Howe, P. R., Bloom, S. R., and Polak, J. M. (1985). Colocalization of neuropeptide Y immunoreactivity in brainstem catecholaminergic neurons that project to the paraventricular nucleus of the hypothalamus. *J. Comp. Neurol.* 241, 138–153.
525. Kalra, S. P., Dube, M. G., Sahu, A., Phelps, C. P., and Kalra, P. S. (1991). Neuropeptide Y secretion increases in the paraventricular nucleus in association with increased appetite for food. *Proc. Natl. Acad. Sci. U S A* 88, 10931–10935.
526. Marin-Bivens, C. L., Kalra, S. P., and Olster, D. H. (1998). Intraventricular injection of neuropeptide Y antisera curbs weight gain and feeding, and increases the display of sexual behaviors in obese Zucker female rats. *Regul. Pept.* 75–76, 327–334.
527. Akabayashi, A., Wahlestedt, C., Alexander, J. T., and Leibowitz, S. F. (1994). Specific inhibition of endogenous neuropeptide Y synthesis in arcuate nucleus by antisense oligonucleotides suppresses feeding behavior and insulin secretion. *Brain Res. Mol. Brain Res.* 21, 55–61.
528. Lopez-Valpuesta, F. J., Nyce, J. W., Griffin-Biggs, T. A., Ice, J. C., and Myers, R. D. (1996). Antisense to NPY-Y1 demonstrates that Y1 receptors in the hypothalamus underlie NPY hypothermia and feeding in rats. *Proc. R. Soc. Lond. B. Biol. Sci.* 263, 881–886.
529. Schaffhauser, A. O., Stricker-Krongrad, A., Brunner, L., Cumin, F., Gerald, C., Whitebread, S., Criscione, L., and Hofbauer, K. G. (1997). Inhibition of food intake by neuropeptide Y Y5 receptor antisense oligodeoxynucleotides. *Diabetes* 46, 1792–1798.
530. Tang-Christensen, M., Kristensen, P., Stidsen, C. E., Brand, C. L., and Larsen, P. J. (1998). Central administration of Y5 receptor antisense decreases spontaneous food intake and attenuates feeding in response to exogenous neuropeptide Y. *J. Endocrinol.* 159, 307–312.
531. Sanacora, G., Kershaw, M., Finkelstein, J. A., and White, J. D. (1990). Increased hypothalamic content of preproneuropeptide Y messenger ribonucleic acid in genetically obese Zucker rats and its regulation by food deprivation. *Endocrinology* 127, 730–737.
532. Grove, K. L., Chen, P., Koegler, F. H., Schiffmaker, A., Smith, S. M., and Cameron, J. L. (2003). Fasting activates neuropeptide Y neurons in the arcuate nucleus and the paraventricular nucleus in the rhesus macaque. *Brain Res. Mol. Brain Res.* 113, 133–138.
533. Jones, J. E., Pick, R. R., Dettloff, S. L., and Wade, G. N. (2004). Metabolic fuels, neuropeptide Y, and estrous behavior in Syrian hamsters. *Brain Res.* 1007, 78–85.
534. Brady, L. S., Smith, M. A., Gold, P. W., and Herkenham, M. (1990). Altered expression of hypothalamic neuropeptide mRNAs in food-restricted and food-deprived rats. *Neuroendocrinology* 52, 441–447.
535. Sahu, A., Kalra, P. S., and Kalra, S. P. (1988). Food deprivation and ingestion induce reciprocal changes in neuropeptide Y concentrations in the paraventricular nucleus. *Peptides* 9, 83–86.
536. McShane, T. M., May, T., Miner, J. L., and Keisler, D. H. (1992). Central actions of neuropeptide-Y may provide a neuromodulatory link between nutrition and reproduction. *Biol. Reprod.* 46, 1151–1157.
537. Kaye, W. H., Berrettini, W., Gwirtsman, H., and George, D. T. (1990). Altered cerebrospinal fluid neuropeptide Y and peptide YY immunoreactivity in anorexia and bulimia nervosa. *Arch. Gen. Psychiatry.* 47, 548–556.
538. Lewis, D. E., Shellard, L., Koeslag, D. G., Boer, D. E., McCarthy, H. D., McKibbin, P. E., Russell, J. C., and Williams, G. (1993). Intense exercise and food restriction cause similar hypothalamic neuropeptide Y increases in rats. *Am. J. Physiol.* 264, E279–E284.
539. Malabu, U. H., Kilpatrick, A., Ware, M., Vernon, R. G., and Williams, G. (1994). Increased neuropeptide Y concentrations in specific hypothalamic regions of lactating rats: possible relationship to hyperphagia and adaptive changes in energy balance. *Peptides* 15, 83–87.
540. Smith, M. S. (1993). Lactation alters neuropeptide-Y and proopiomelanocortin gene expression in the arcuate nucleus of the rat. *Endocrinology* 133, 1258–1265.
541. Schwartz, M. W., Seeley, R. J., Campfield, L. A., Burn, P., and Baskin, D. G. (1996). Identification of targets of leptin action in rat hypothalamus. *J. Clin. Invest.* 98, 1101–1106.
542. Wilding, J. P., Gilbey, S. G., Bailey, C. J., Batt, R. A., Williams, G., Ghatei, M. A., and Bloom, S. R. (1993). Increased neuropeptide-Y messenger ribonucleic acid (mRNA) and decreased neurotensin mRNA in the hypothalamus of the obese (ob/ob) mouse. *Endocrinology* 123, 1939–1944.
543. Chua, S. C., Jr., Brown, A. W., Kim, J., Hennessey, K. L., Leibel, R. L., and Hirsch, J. (1991). Food deprivation and hypothalamic neuropeptide gene expression: effects of strain background and the diabetes mutation. *Brain Res. Mol. Brain Res.* 11, 291–299.
544. Erickson, J. C., Hollopeter, G., and Palmiter, R. D. (1996). Attenuation of the obesity syndrome of ob/ob mice by the loss of neuropeptide Y. *Science.* 274, 1704–1707.
545. Stephens, T. W., Basinski, M., Bristow, P. K., Bue-Valleskey, J. M., Burgett, S. G., Craft, L., Hale, J., Hoffmann, J., Hsiung, H. M., and Kriauciunas, A., et al. (1995). The role of neuropeptide Y in the antiobesity action of the obese gene product. *Nature* 377, 530–532.
546. Schwartz, M. W., Baskin, D. G., Bukowski, T. R., Kuijper, J. L., Foster, D., Lasser, G., Prunkard, D. E., Porte, D., Jr., Woods, S. C., Seeley, R. J., and Weigle, D. S. (1996). Specificity of leptin action on elevated blood glucose levels and hypothalamic neuropeptide Y gene expression in ob/ob mice. *Diabetes* 45, 531–535.
547. Mercer, J. G., Hoggard, N., Williams, L. M., Lawrence, C. B., Hannah, L. T., Morgan, P. J., and Trayhurn, P. (1996). Coexpression of leptin receptor and preproneuropeptide Y mRNA in arcuate nucleus of mouse hypothalamus. *J. Neuroendocrinol.* 8, 733–735.
548. Schwartz, M. W., Sipols, A. J., Marks, J. L., Sanacora, G., White, J. D., Scheurink, A., Kahn, S. E., Baskin, D. G., Woods, S. C., and Figlewicz, D. P., et al. (1992). Inhibition of hypothalamic neuropeptide Y gene expression by insulin. *Endocrinology* 130, 3608–3616.
549. Williams, G., Steel, J. H., Cardoso, H., Ghatei, M. A., Lee, Y. C., Gill, J. S., Burrin, J. M., Polak, J. M., and Bloom, S. R. (1988). Increased hypothalamic neuropeptide Y concentrations in diabetic rat. *Diabetes* 37, 763–772.

550. Sergeyev, V., Broberger, C., Gorbatyuk, O., and Hokfelt, T. (2000). Effect of 2-mercaptoacetate and 2-deoxy-D-glucose administration on the expression of NPY, AGRP, POMC, MCH and hypocretin/orexin in the rat hypothalamus. *Neuroreport.* 11, 117–121.
551. Li, A. J., and Ritter, S. (2004). Glucoprivation increases expression of neuropeptide Y mRNA in hindbrain neurons that innervate the hypothalamus. *Eur. J. Neurosci.* 19, 2147–2154.
552. He, B., White, B. D., Edwards, G. L., and Martin, R. J. (1998). Neuropeptide Y antibody attenuates 2-deoxy-D-glucose induced feeding in rats. *Brain Res.* 781, 348–350.
553. Minami, S., Kamegai, J., Sugihara, H., Suzuki, N., Higuchi, H., and Wakabayashi, I. (1995). Central glucoprivation evoked by administration of 2-deoxy-D-glucose induces expression of the c-fos gene in a subpopulation of neuropeptide Y neurons in the rat hypothalamus. *Brain Res. Mol. Brain Res.* 33, 305–310.
554. Dunn-Meynell, A. A., Routh, V. H., Kang, L., Gaspers, L., and Levin, B. E. (2002). Glucokinase is the likely mediator of glucosensing in both glucose-excited and glucose-inhibited central neurons. *Diabetes* 51, 2056–2065.
555. Muroya, S., Yada, T., Shioda, S., and Takigawa, M. (1999). Glucose-sensitive neurons in the rat arcuate nucleus contain neuropeptide Y. *Neurosci. Lett.* 264, 113–116.
556. Fraley, G. S., and Ritter, S. (2003). Immunolesion of norepinephrine and epinephrine afferents to medial hypothalamus alters basal and 2-deoxy-D-glucose-induced neuropeptide Y and agouti gene-related protein messenger ribonucleic acid expression in the arcuate nucleus. *Endocrinology* 144, 75–83.
557. Kalra, S. P., and Kalra, P. S. (1996). Nutritional infertility: the role of the interconnected hypothalamic neuropeptide Y-galanin-opioid network. *Front Neuroendocrinol.* 17, 371–401.
558. Tsuruo, Y., Kawano, H., Kagotani, Y., Hisano, S., Daikoku, S., Chihara, K., Zhang, T., and Yanaihara, N. (1990). Morphological evidence for neuronal regulation of luteinizing hormone-releasing hormone-containing neurons by neuropeptide Y in the rat septo-preoptic area. *Neurosci. Lett.* 110, 261–266.
559. Turi, G. F., Liposits, Z., Moenter, S. M., Fekete, C., and Hrabovszky, E. (2003). Origin of neuropeptide Y-containing afferents to gonadotropin-releasing hormone neurons in male mice. *Endocrinology* 144, 4967–4974.
560. Campbell, R. E., French-Mullen, J. M., Cowley, M. A., Smith, M. S., and Grove, K. L. (2001). Hypothalamic circuitry of neuropeptide Y regulation of neuroendocrine function and food intake via the Y5 receptor subtype. *Neuroendocrinology* 74, 106–119.
561. Li, C., Chen, P., and Smith, M. S. (1999). Morphological evidence for direct interaction between arcuate nucleus neuropeptide Y (NPY) neurons and gonadotropin-releasing hormone neurons and the possible involvement of NPY Y1 receptors. *Endocrinology* 140, 5382–5390.
562. McDonald, J. K., Koenig, J. I., Gibbs, D. M., Collins, P., and Noe, B. D. (1987). High concentrations of neuropeptide Y in pituitary portal blood of rats. *Neuroendocrinology* 46, 538–541.
563. Sutton, S. W., Toyama, T. T., Otto, S., and Plotsky, P. M. (1988). Evidence that neuropeptide Y (NPY) released into the hypophysial-portal circulation participates in priming gonadotropes to the effects of gonadotropin releasing hormone (GnRH). *Endocrinology* 123, 1208–1210.
564. Kalra, S. P., and Crowley, W. R. (1984). Norepinephrine-like effects of neuropeptide Y on LH release in the rat. *Life Sci.* 35, 1173–1176.
565. Allen, L. G., Kalra, P. S., Crowley, W. R., and Kalra, S. P. (1985). Comparison of the effects of neuropeptide Y and adrenergic transmitters on LH release and food intake in male rats. *Life Sci.* 37, 617–623.
566. Khorram, O., Pau, K. Y., and Spies, H. G. (1987). Bimodal effects of neuropeptide Y on hypothalamic release of gonadotropin-releasing hormone in conscious rabbits. *Neuroendocrinology* 45, 290–297.
567. Woller, M. J., and Terasawa, E. (1992). Estradiol enhances the action of neuropeptide Y on in vivo luteinizing hormone-releasing hormone release in the ovariectomized rhesus monkey. *Neuroendocrinology* 56, 921–925.
568. Crowley, W. R., and Kalra, S. P. (1987). Neuropeptide Y stimulates the release of luteinizing hormone-releasing hormone from medial basal hypothalamus in vitro: modulation by ovarian hormones. *Neuroendocrinology* 46, 97–103.
569. Khorram, O., Pau, K. Y., and Spies, H. G. (1988). Release of hypothalamic neuropeptide Y and effects of exogenous NPY on the release of hypothalamic GnRH and pituitary gonadotropins in intact and ovariectomized does in vitro. *Peptides* 9, 411–417.
570. Sabatino, F. D., Collins, P., and McDonald, J. K. (1989). Neuropeptide-Y stimulation of luteinizing hormone-releasing hormone secretion from the median eminence in vitro by estrogen-dependent and extracellular Ca2+-independent mechanisms. *Endocrinology* 124, 2089–2098.
571. Bauer-Dantoin, A. C., Urban, J. H., and Levine, J. E. (1992). Neuropeptide Y gene expression in the arcuate nucleus is increased during preovulatory luteinizing hormone surges. *Endocrinology* 131, 2953–2958.
572. Sahu, A., Crowley, W. R., and Kalra, S. P. (1994). Hypothalamic neuropeptide-Y gene expression increases before the onset of the ovarian steroid-induced luteinizing hormone surge. *Endocrinology* 134, 1018–1022.
573. Watanobe, H., and Takebe, K. (1992). Evidence that neuropeptide Y secretion in the median eminence increases before the luteinizing hormone surge in ovariectomized steroid-primed rats: estimation by push-pull perfusion. *Neurosci. Lett.* 146, 57–59.
574. Wehrenberg, W. B., Corder, R., and Gaillard, R. C. (1989). A physiological role for neuropeptide Y in regulating the estrogen/progesterone induced luteinizing hormone surge in ovariectomized rats. *Neuroendocrinology* 49, 680–682.
575. Mizuno, M., Gearing, M., and Terasawa, E. (2000). The role of neuropeptide Y in the progesterone-induced luteinizing hormone-releasing hormone surge in vivo in ovariectomized female rhesus monkeys. *Endocrinology* 141, 1772–1779.
576. Kerkerian, L., Guy, J., Lefevre, G., and Pelletier, G. (1985). Effects of neuropeptide Y (NPY) on the release of anterior pituitary hormones in the rat. *Peptides* 6, 1201–1204.
577. McDonald, J. K., Lumpkin, M. D., and DePaolo, L. V. (1989). Neuropeptide-Y suppresses pulsatile secretion of luteinizing hormone in ovariectomized rats: possible site of action. *Endocrinology* 125, 186–191.
578. Malven, P. V., Haglof, S. A., and Degroot, H. (1992). Effects of intracerebral administration of neuropeptide-Y on secretion of luteinizing hormone in ovariectomized sheep. *Brain Res. Bull.* 28, 871–875.
579. Thomas, M. G., Gazal, O. S., Williams, G. L., Stanko, R. L., and Keisler, D. H. (1999). Injection of neuropeptide Y into the third cerebroventricle differentially influences pituitary secretion of luteinizing hormone and growth hormone in ovariectomized cows. *Domest. Anim. Endocrinol.* 16, 159–169.
580. Kaynard, A. H., Pau, K. Y., Hess, D. L., and Spies, H. G. (1990). Third-ventricular infusion of neuropeptide Y suppresses luteinizing hormone secretion in ovariectomized rhesus macaques. *Endocrinology* 127, 2437–2444.

581. Pierroz, D. D., Gruaz, N. M., d'Alieves, V., and Aubert, M. L. (1995). Chronic administration of neuropeptide Y into the lateral ventricle starting at 30 days of life delays sexual maturation in the female rat. *Neuroendocrinology* 61, 293–300.
582. Toufexis, D. J., Kyriazis, D., and Woodside, B. (2002). Chronic neuropeptide Y Y5 receptor stimulation suppresses reproduction in virgin female and lactating rats. *J. Neuroendocrinol.* 14, 492–497.
583. Pierroz, D. D., Catzeflis, C., Aebi, A. C., Rivier, J. E., and Aubert, M. L. (1996). Chronic administration of neuropeptide Y into the lateral ventricle inhibits both the pituitary-testicular axis and growth hormone and insulin-like growth factor I secretion in intact adult male rats. *Endocrinology* 137, 3–12.
584. Crowley, W. R., Hassid, A., and Kalra, S. P. (1987). Neuropeptide Y enhances the release of luteinizing hormone (LH) induced by LH-releasing hormone. *Endocrinology* 120, 941–945.
585. Bauer-Dantoin, A. C., McDonald, J. K., and Levine, J. E. (1991). Neuropeptide Y potentiates luteinizing hormone (LH)-releasing hormone-stimulated LH surges in pentobarbital-blocked proestrous rats. *Endocrinology* 129, 402–408.
586. Xu, M., Hill, J. W., and Levine, J. E. (2000). Attenuation of luteinizing hormone surges in neuropeptide Y knockout mice. *Neuroendocrinology* 72, 263–271.
587. Clark, J. T., Kalra, P. S., and Kalra, S. P. (1985). Neuropeptide Y stimulates feeding but inhibits sexual behavior in rats. *Endocrinology* 117, 2435–2442.
588. Corp, E. S., Greco, B., Powers, J. B., Marin Bivens, C. L., and Wade, G. N. (2001). Neuropeptide Y inhibits estrous behavior and stimulates feeding via separate receptors in Syrian hamsters. *Am. J. Physiol. Regul. Integr. Comp. Physiol.* 280, R1061–R1068.
589. Keene, A. C., Jones, J. E., Wade, G. N., and Corp, E. S. (2003). Forebrain sites of NPY action on estrous behavior in Syrian hamsters. *Physiol. Behav.* 78, 711–716.
590. Sainsbury, A., Schwarzer, C., Couzens, M., Jenkins, A., Oakes, S. R., Ormandy, C. J., and Herzog, H. (2002). Y4 receptor knockout rescues fertility in ob/ob mice. *Genes Dev.* 16, 1077–1088.
591. Pralong, F. P., Gonzales, C., Voirol, M. J., Palmiter, R. D., Brunner, H. R., Gaillard, R. C., Seydoux, J., and Pedrazzini, T. (2002). The neuropeptide Y Y1 receptor regulates leptin-mediated control of energy homeostasis and reproductive functions. *FASEB J.* 16, 712–714.
592. Erickson, J. C., Clegg, K. E., and Palmiter, R. D. (1996). Sensitivity to leptin and susceptibility to seizures of mice lacking neuropeptide Y. *Nature* 381, 415–421.
593. Erickson, J. C., Ahima, R. S., Hollopeter, G., Flier, J. S., and Palmiter, R. D. (1997). Endocrine function of neuropeptide Y knockout mice. *Regul. Pept.* 70, 199–202.
594. Hill, J. W., and Levine, J. E. (2003). Abnormal response of the neuropeptide Y-deficient mouse reproductive axis to food deprivation but not lactation. *Endocrinology* 144, 1780–1786.
595. Mains, R. E., Eipper, B. A., and Ling, N. (1977). Common precursor to corticotropins and endorphins. *Proc. Natl. Acad. Sci. U S A* 74, 3014–3018.
596. Smith, A. I., Funder, J. W. (1988). Proopiomelanocortin processing in the pituitary, central nervous system, and peripheral tissues. *Endocr. Rev.* 9, 159–179.
597. Bronstein, D. M., Schafer, M. K., Watson, S. J., and Akil, H. (1992). Evidence that beta-endorphin is synthesized in cells in the nucleus tractus solitarius: detection of POMC mRNA. *Brain Res.* 587, 269–275.
598. Gee, C. E., Chen, C. L., Roberts, J. L., Thompson, R., and Watson, S. J. (1983). Identification of proopiomelanocortin neurones in rat hypothalamus by in situ cDNA-mRNA hybridization. *Nature* 306, 374–376.
599. Palkovits, M., and Eskay, R. L. (1987). Distribution and possible origin of beta-endorphin and ACTH in discrete brainstem nuclei of rats. *Neuropeptides* 9, 123–137.
600. Bugnon, C., Bloch, B., Lenys, D., Gouget, A., and Fellmann, D. (1979). Comparative study of the neuronal populations containing beta-endorphin, corticotropin and dopamine in the arcuate nucleus of the rat hypothalamus. *Neurosci. Lett.* 14, 43–48.
601. Jacobowitz, D. M., and O'Donohue, T. L. (1978). alpha-Melanocyte stimulating hormone: immunohistochemical identification and mapping in neurons of rat brain. *Proc. Natl. Acad. Sci. U S A* 75, 6300–6304.
602. Roselli-Rehfuss, L., Mountjoy, K. G., Robbins, L. S., Mortrud, M. T., Low, M. J., Tatro, J. B., Entwistle, M. L., Simerly, R. B., and Cone, R. D. (1993). Identification of a receptor for gamma melanotropin and other proopiomelanocortin peptides in the hypothalamus and limbic system. *Proc. Natl. Acad. Sci. U S A* 90, 8856–8860.
603. Mountjoy, K. G., Mortrud, M. T., Low, M. J., Simerly, R. B., and Cone, R. D. (1994). Localization of the melanocortin-4 receptor (MC4-R) in neuroendocrine and autonomic control circuits in the brain. *Mol. Endocrinol.* 8, 1298–1308.
604. Mansour, A., Fox, C. A., Burke, S., Meng, F., Thompson, R. C., Akil, H., and Watson, S. J. (1994). Mu, delta, and kappa opioid receptor mRNA expression in the rat CNS: an in situ hybridization study. *J. Comp. Neurol.* 350, 412–438.
605. Cone, R. D., Lu, D., Koppula, S., Vage, D. I., Klungland, H., Boston, B., Chen, W., Orth, D. N., Pouton, C., and Kesterson, R. A. (1996). The melanocortin receptors: agonists, antagonists, and the hormonal control of pigmentation. *Rec. Prog. Horm. Res.* 51, 287–318.
606. Gantz, I., Konda, Y., Tashiro, T., Shimoto, Y., Miwa, H., Munzert, G., Watson, S. J., DelValle, J., and Yamada, T. (1993). Molecular cloning of a novel melanocortin receptor. *J. Biol. Chem.* 268, 8246–8250.
607. Gantz, I., Miwa, H., Konda, Y., Shimoto, Y., Tashiro, T., Watson, S. J., DelValle, J., and Yamada, T. (1993). Molecular cloning, expression, and gene localization of a fourth melanocortin receptor. *J. Biol. Chem.* 268, 15174–15179.
608. Mansour, A., Fox, C. A., Akil, H., and Watson, S. J. (1995). Opioid-receptor mRNA expression in the rat CNS: anatomical and functional implications. *Trends Neurosci.* 18, 22–29.
609. Schwartz, M. W., Seeley, R. J., Woods, S. C., Weigle, D. S., Campfield, L. A., Burn, P., and Baskin, D. G. (1997). Leptin increases hypothalamic pro-opiomelanocortin mRNA expression in the rostral arcuate nucleus. *Diabetes* 46, 2119–2123.
610. Mizuno, T. M., Kleopoulos, S. P., Bergen, H. T., Roberts, J. L., Priest, C. A., and Mobbs, C. V. (1998). Hypothalamic pro-opiomelanocortin mRNA is reduced by fasting and (corrected) in ob/ob and db/db mice, but is stimulated by leptin. *Diabetes* 47, 294–297.
611. Koegler, F. H., Grove, K. L., Schiffmacher, A., Smith, M. S., and Cameron, J. L. (2001). Central melanocortin receptors mediate changes in food intake in the rhesus macaque. *Endocrinology* 142, 2586–2592.
612. McShane, T. M., Petersen, S. L., McCrone, S., and Keisler, D. H. (1993). Influence of food restriction on neuropeptide-Y, proopiomelanocortin, and luteinizing hormone-releasing hormone gene expression in sheep hypothalami. *Biol. Reprod.* 49, 831–839.
613. Hagan, M. M., Rushing, P. A., Schwartz, M. W., Yagaloff, K. A., Burn, P., Woods, S. C., and Seeley, R. J. (1999). Role of the CNS melanocortin system in the response to overfeeding. *J. Neurosci.* 19, 2362–2367.

614. Kaye, W. H., Berrettini, W. H., Gwirtsman, H. E., Chretien, M., Gold, P. W., George, D. T., Jimerson, D. C., and Ebert, M. H. (1987). Reduced cerebrospinal fluid levels of immunoreactive pro-opiomelanocortin related peptides (including beta-endorphin) in anorexia nervosa. *Life Sci.* 41, 2147–2155.
615. Elias, C. F., Aschkenasi, C., Lee, C., Kelly, J., Ahima, R. S., Bjorbaek, C., Flier, J. S., Saper, C. B., and Elmquist, J. K. (1999). Leptin differentially regulates NPY and POMC neurons projecting to the lateral hypothalamic area. *Neuron* 23, 775–786.
616. Cowley, M. A., Smart, J. L., Rubinstein, M., Cerdan, M. G., Diano, S., Horvath, T. L., Cone, R. D., and Low, M. J. (2001). Leptin activates anorexigenic POMC neurons through a neural network in the arcuate nucleus. *Nature* 411, 480–484.
617. Kim, M. S., Small, C. J., Stanley, S. A., Morgan, D. G., Seal, L. J., Kong, W. M., Edwards, C. M., Abusnana, S., Sunter, D., Ghatei, M. A., and Bloom, S. R. (2000). The central melanocortin system affects the hypothalamo-pituitary thyroid axis and may mediate the effect of leptin. *J. Clin. Invest.* 105, 1005–1011.
618. Thornton, J. E., Cheung, C. C., Clifton, D. K., and Steiner, R. A. (1997). Regulation of hypothalamic proopiomelanocortin mRNA by leptin in ob/ob mice. *Endocrinology* 138, 5063–5066.
619. Korner, J., Chua, S. C., J., Williams, J. A., Leibel, R. L., and Wardlaw, S. L. (1999). Regulation of hypothalamic proopiomelanocortin by leptin in lean and obese rats. *Neuroendocrinology* 70, 377–383.
620. Kim, E. M., Grace, M. K., Welch, C. C., Billington, C. J., and Levine, A. S. (1999). STZ-induced diabetes decreases and insulin normalizes POMC mRNA in arcuate nucleus and pituitary in rats. *Am. J. Physiol.* 276, R1320–R1326.
621. Benoit, S. C., Air, E. L., Coolen, L. M., Strauss, R., Jackman, A., Clegg, D. J., Seeley, R. J., and Woods, S. C. (2002). The catabolic action of insulin in the brain is mediated by melanocortins. *J. Neurosci.* 22, 9048–9052.
622. Cheung, C. C., Clifton, D. K., and Steiner, R. A. (1997). Proopiomelanocortin neurons are direct targets for leptin in the hypothalamus. *Endocrinology* 138, 4489–4492.
623. Ibrahim, N., Bosch, M. A., Smart, J. L., Qiu, J., Rubinstein, M., Ronnekleiv, O. K., Low, M. J., and Kelly, M. J. (2003). Hypothalamic proopiomelanocortin neurons are glucose responsive and express K(ATP) channels. *Endocrinology* 144, 1331–1340.
624. Giraudo, S. Q., Kim, E. M., Grace, M. K., Billington, C. J., and Levine, A. S. (1998). Effect of peripheral 2-DG on opioid and neuropeptide Y gene expression. *Brain Res.* 792, 136–140.
625. Vergoni, A. V., and Bertolini, A. (2000). Role of melanocortins in the central control of feeding. *Eur. J. Pharmacol.* 405, 25–32.
626. Khachaturian, H., Lewis, M. E., Haber, S. N., Akil, H., and Watson, S. J. (1984). Proopiomelanocortin peptide immunocytochemistry in rhesus monkey brain. *Brain Res. Bull.* 13, 785–800.
627. Poggioli, R., Vergoni, A. V., and Bertolini, A. (1986). ACTH-(1-24) and alpha-MSH antagonize feeding behavior stimulated by kappa opiate agonists. *Peptides* 7, 843–848.
628. Fan, W., Boston, B. A., Kesterson, R. A., Hruby, V. J., and Cone, R. D. (1997). Role of melanocortinergic neurons in feeding and the agouti obesity syndrome. *Nature* 385, 165–168.
629. Grill, H. J., Ginsberg, A. B., Seeley, R. J., and Kaplan, J. M. (1998). Brainstem application of melanocortin receptor ligands produces long-lasting effects on feeding and body weight. *J. Neurosci.* 18, 10128–10135.
630. Thiele, T. E., van Dijk, G., Yagaloff, K. A., Fisher, S. L., Schwartz, M., Burn, P., and Seeley, R. J. (1998). Central infusion of melanocortin agonist MTII in rats: assessment of c-Fos expression and taste aversion. *Am. J. Physiol.* 274, R248–R254.
631. Kim, M. S., Rossi, M., Abusnana, S., Sunter, D., Morgan, D. G., Small, C. J., Edwards, C. M., Heath, M. M., Stanley, S. A., Seal, L. J., Bhatti, J. R., Smith, D. M., Ghatei, M. A., and Bloom, S. R. (2000). Hypothalamic localization of the feeding effect of agouti-related peptide and alpha-melanocyte-stimulating hormone. *Diabetes* 49, 177–182.
632. Kask, A., Pahkla, R., Irs, A., Rago, L., Wikberg, J. E., and Schioth, H. B. (1999). Long-term administration of MC4 receptor antagonist HS014 causes hyperphagia and obesity in rats. *Neuroreport* 10, 707–711.
633. Raposinho, P. D., Castillo, E., d'Alleves, V., Broqua, P., Pralong, F. P., and Aubert, M. L. (2000). Chronic blockade of the melanocortin 4 receptor subtype leads to obesity independently of neuropeptide Y action, with no adverse effects on the gonadotropic and somatotropic axes. *Endocrinology* 141, 4419–4427.
634. Seeley, R. J., Yagaloff, K. A., Fisher, S. L., Burn, P., Thiele, T. E., van Dijk, G., Baskin, D. G., and Schwartz, M. W. (1997). Melanocortin receptors in leptin effects. *Nature* 390, 349.
635. Yaswen, L., Diehl, N., Brennan, M. B., and Hochgeschwender, U. (1999). Obesity in the mouse model of pro-opiomelanocortin deficiency responds to peripheral melanocortin. *Nat. Med.* 5, 1066–1070.
636. Krude, H., Biebermann, H., Luck, W., Horn, R., Brabant, G., and Gruters, A. (1998). Severe early-onset obesity, adrenal insufficiency and red hair pigmentation caused by POMC mutations in humans. *Nat. Genet.* 19, 155–157.
637. Chen, A. S., Marsh, D. J., Trumbauer, M. E., Frazier, E. G., Guan, X. M., Yu, H., Rosenblum, C. I., Vongs, A., Feng, Y., Cao, L., Metzger, J. M., Strack, A. M., Camacho, R. E., Mellin, T. N., Nunes, C. N., Min, W., Fisher, J., Gopal-Truter, S., MacIntyre, D. E., Chen, H. Y., and Van der Ploeg, L. H. (2000). Inactivation of the mouse melanocortin-3 receptor results in increased fat mass and reduced lean body mass. *Nat. Genet.* 26, 97–102.
638. Butler, A. A., Kesterson, R. A., Khong, K., Cullen, M. J., Pelleymounter, M. A., Dekoning, J., Baetscher, M., and Cone, R. D. (2000). A unique metabolic syndrome causes obesity in the melanocortin-3 receptor-deficient mouse. *Endocrinology* 141, 3518–3521.
639. Huszar, D., Lynch, C. A., Fairchild-Huntress, V., Dunmore, J. H., Fang, Q., Berkemeier, L. R., Gu, W., Kesterson, R. A., Boston, B. A., Cone, R. D., Smith, F. J., Campfield, L. A., Burn, P., and Lee, F. (1997). Targeted disruption of the melanocortin-4 receptor results in obesity in mice. *Cell* 88, 131–141.
640. Vaisse, C., Clement, K., Guy-Grand, B., and Froguel, P. (1998). A frameshift mutation in human MC4R is associated with a dominant form of obesity. *Nat. Genet.* 20, 113–114.
641. Mergen, M., Mergen, H., Ozata, M., Oner, R., and Oner, C. (2001). A novel melanocortin 4 receptor (MC4R) gene mutation associated with morbid obesity. *J. Clin. Endocrinol. Metab.* 86, 3448.
642. Yeo, G. S., Farooqi, I. S., Aminian, S., Halsall, D. J., Stanhope, R. G., and O'Rahilly, S. (1998). A frameshift mutation in MC4R associated with dominantly inherited human obesity. *Nat. Genet.* 20, 111–112.
643. Farooqi, I. S., Yeo, G. S., Keogh, J. M., Aminian, S., Jebb, S. A., Butler, G., Cheetham, T., and O'Rahilly, S. (2000). Dominant and recessive inheritance of morbid obesity associated with melanocortin 4 receptor deficiency. *J. Clin. Invest.* 106, 271–279.
644. Mezey, E., Kiss, J. Z., Mueller, G. P., Eskay, R., O'Donohue, T. L., and Palkovits, M. (1985). Distribution of

the pro-opiomelanocortin derived peptides, adrenocorticotrope hormone, alpha-melanocyte-stimulating hormone and beta-endorphin (ACTH, alpha-MSH, beta-END) in the rat hypothalamus. *Brain Res.* 328, 341–347.

645. Leranth, C., MacLusky, N. J., Shanabrough, M., and Naftolin, F. (1988). Immunohistochemical evidence for synaptic connections between pro-opiomelanocortin-immunoreactive axons and LH-RH neurons in the preoptic area of the rat. *Brain Res.* 449, 167–176.
646. Chen, W. P., Witkin, J. W., and Silverman, A. J. (1989). beta-Endorphin and gonadotropin-releasing hormone synaptic input to gonadotropin-releasing hormone neurosecretory cells in the male rat. *J. Comp. Neurol.* 286, 85–95.
647. Khong, K., Kurtz, S. E., Sykes, R. L., and Cone, R. D. (2001). Expression of functional melanocortin-4 receptor in the hypothalamic GT1-1 cell line. *Neuroendocrinology* 74, 193–201.
648. Celis, M. E. (1985). Release of LH in response to alpha-MSH administration. *Acta. Physiol. Pharmacol. Latinoam.* 35, 281–290.
649. Caballero, C., and Celis, M. E. (1993). The effect of the blockade of alpha-melanocyte-stimulating hormone on LH release in the rat. *J. Endocrinol.* 137, 197–202.
650. Limone, P., Calvelli, P., Altare, F., Ajmone-Catt, P., Lima, T., and Molinatti, G. M. (1997). Evidence for an interaction between alpha-MSH and opioids in the regulation of gonadotropin secretion in man. *J. Endocrinol. Invest.* 20, 207–210.
651. Watanobe, H., Schioth, H. B., Wikberg, J. E., and Suda, T. (1999). The melanocortin 4 receptor mediates leptin stimulation of luteinizing hormone and prolactin surges in steroid-primed ovariectomized rats. *Biochem. Biophys. Res. Commun.* 257, 860–864.
652. Scimonelli, T., and Celis, M. E. (1990). A central action of alpha-melanocyte-stimulating hormone on serum levels of LH and prolactin in rats. *J. Endocrinol.* 124, 127–132.
653. Khorram, O., Bedran de Castro, J. C., and McCann, S. M. (1984). Physiological role of alpha-melanocyte-stimulating hormone in modulating the secretion of prolactin and luteinizing hormone in the female rat. *Proc. Natl. Acad. Sci. U S A* 81, 8004–8008.
654. Khorram, O., DePalatis, L. R., and McCann, S. M. (1984). The effect and possible mode of action of alpha-melanocyte-stimulating hormone on gonadotropin release in the ovariectomized rat: an in vivo and in vitro analysis. *Endocrinology* 114, 227–233.
655. Stanley, S. A., Small, C. J., Kim, M. S., Heath, M. M., Seal, L. J., Russell, S. H., Ghatei, M. A., and Bloom, S. R. (1999). Agouti related peptide (Agrp) stimulates the hypothalamo pituitary gonadal axis in vivo & in vitro in male rats. *Endocrinology* 140, 5459–5462.
656. Gonzalez, M. I., Celis, M. E., Hole, D. R., and Wilson, C. A. (1993). Interaction of oestradiol, alpha-melanotrophin and noradrenaline within the ventromedial nucleus in the control of female sexual behaviour. *Neuroendocrinology* 58, 218–226.
657. Scimonelli, T., Medina, F., Wilson, C., and Celis, M. E. (2000). Interaction of alpha-melanotropin (alpha-MSH) and noradrenaline in the median eminence in the control of female sexual behavior. *Peptides* 21, 219–223.
658. Bodnar, R. J., and Hadjimarkou, M. M. (2003). Endogenous opiates and behavior: 2002. *Peptides* 24, 1241–1302.
659. Hughes, A. M., Everitt, B. J., and Herbert, J. (1988). The effects of simultaneous or separate infusions of some pro-opiomelanocortin-derived peptides (beta-endorphin, melanocyte stimulating hormone, and corticotrophin-like intermediate polypeptide) and their acetylated derivatives upon sexual and ingestive behaviour of male rats. *Neuroscience.* 27, 689–698.
660. Kim, E. M., Grace, M. K., O'Hare, E., Billington, C. J., and Levine, A. S. (2002). Injection of alpha-MSH, but not beta-endorphin, into the PVN decreases POMC gene expression in the ARC. *Neuroreport* 13, 497–500.
661. Levine, A. S., and Billington, C. J. (1989). Opioids. Are they regulators of feeding? *Ann. N Y Acad. Sci.* 575, 209–219 discussion 219–220.
662. Millan, M. J., and Morris, B. J. (1988). Long-term blockade of mu-opioid receptors suggests a role in control of ingestive behaviour, body weight and core temperature in the rat. *Brain Res.* 450, 247–258.
663. Appleyard, S. M., Hayward, M., Young, J. I., Butler, A. A., Cone, R. D., Rubinstein, M., and Low, M. J. (2003). A role for the endogenous opioid beta-endorphin in energy homeostasis. *Endocrinology* 144, 1753–1760.
664. Bruni, J. F., Van Vugt, D., Marshall, S., and Meites, J. (1977). Effects of naloxone, morphine and methionine enkephalin on serum prolactin, luteinizing hormone, follicle stimulating hormone, thyroid stimulating hormone and growth hormone. *Life Sci.* 21, 461–466.
665. Leadem, C. A., and Kalra, S. P. (1985). Effects of endogenous opioid peptides and opiates on luteinizing hormone and prolactin secretion in ovariectomized rats. *Neuroendocrinology* 41, 342–352.
666. Wardlaw, S. L., and Ferin, M. (1990). Interaction between beta-endorphin and alpha-melanocyte-stimulating hormone in the control of prolactin and luteinizing hormone secretion in the primate. *Endocrinology* 126, 2035–2040.
667. Gilbeau, P. M., Almirez, R. G., Holaday, J. W., and Smith, C. G. (1985). Opioid effects on plasma concentrations of luteinizing hormone and prolactin in the adult male rhesus monkey. *J. Clin. Endocrinol. Metab.* 60, 299–305.
668. Leadem, C. A., and Kalra, S. P. (1985). Reversal of beta-endorphin-induced blockade of ovulation and luteinizing hormone surge with prostaglandin E2. *Endocrinology* 117, 684–689.
669. Abs, R., Verhelst, J., Maeyaert, J., Van Buyten, J. P., Opsomer, F., Adriaensen, H., Verlooy, J., Van Havenbergh, T., Smet, M., and Van Acker, K. (2000). Endocrine consequences of long-term intrathecal administration of opioids. *J. Clin. Endocrinol. Metab.* 85, 2215–2222.
670. Cicero, T. J., Schainker, B. A., and Meyer, E. R. (1979). Endogenous opioids participate in the regulation of the hypothalamus-pituitary-luteinizing hormone axis and testosterone's negative feedback control of luteinizing hormone. *Endocrinology* 104, 1286–1291.
671. Gabriel, S. M., Simpkins, J. W., and Kalra, S. P. (1983). Modulation of endogenous opioid influence on luteinizing hormone secretion by progesterone and estrogen. *Endocrinology* 113, 1806–1811.
672. Orstead, K. M., and Spies, H. G. (1987). Inhibition of hypothalamic gonadotropin-releasing hormone release by endogenous opioid peptides in the female rabbit. *Neuroendocrinology* 46, 14–23.
673. Ebling, F. J., Schwartz, M. L., and Foster, D. L. (1989). Endogenous opioid regulation of pulsatile luteinizing hormone secretion during sexual maturation in the female sheep. *Endocrinology* 125, 369–383.
674. Van Vugt, D. A., Lam, N. Y., and Ferin, M. (1984). Reduced frequency of pulsatile luteinizing hormone secretion in the luteal phase of the rhesus monkey. Involvement of endogenous opiates. *Endocrinology* 115, 1095–1101.

675. Moult, P. J., Grossman, A., Evans, J. M., Rees, L. H., and Besser, G. M. (1981). The effect of naloxone on pulsatile gonadotrophin release in normal subjects. *Clin. Endocrinol. (Oxf).* 14, 321–324.
676. Ellingboe, J., Veldhuis, J. D., Mendelson, J. H., Kuehnle, J. C., and Mello, N. K. (1982). Effect of endogenous opioid blockade on the amplitude and frequency of pulsatile luteinizing hormone secretion in normal men. *J. Clin. Endocrinol. Metab.* 54, 854–857.
677. Delitala, G., Devilla, L., and Arata, L. (1981). Opiate receptors and anterior pituitary hormone secretion in man. Effect of naloxone infusion. *Acta. Endocrinol. (Copenh).* 97, 150–156.
678. Blank, M. S., and Roberts, D. L. (1982). Antagonist of gonadotropin-releasing hormone blocks naloxone-induced elevations in serum luteinizing hormone. *Neuroendocrinology* 35, 309–312.
679. Kesner, J. S., Kaufman, J. M., Wilson, R. C., Kuroda, G., and Knobil, E. (1986). The effect of morphine on the electrophysiological activity of the hypothalamic luteinizing hormone-releasing hormone pulse generator in the rhesus monkey. *Neuroendocrinology* 43, 686–688.
680. Rasmussen, D. D., Liu, J. H., Wolf, P. L., and Yen, S. S. (1983). Endogenous opioid regulation of gonadotropin-releasing hormone release from the human fetal hypothalamus in vitro. *J. Clin. Endocrinol. Metab.* 57, 881–884.
681. Kalra, P. S., Crowley, W. R., and Kalra, S. P. (1987). Differential in vitro stimulation by naloxone and K+ of luteinizing hormone-releasing hormone and catecholamine release from the hypothalami of intact and castrated rats. *Endocrinology* 120, 178–185.
682. Goodman, R. L., Parfitt, D. B., Evans, N. P., Dahl, G. E., and Karsch, F. J. (1995). Endogenous opioid peptides control the amplitude and shape of gonadotropin-releasing hormone pulses in the ewe. *Endocrinology* 136, 2412–2420.
683. Karahalios, D. G., and Levine, J. E. (1988). Naloxone stimulation of in vivo LHRH release is not diminished following ovariectomy. *Neuroendocrinology* 47, 504–510.
684. Thind, K. K., and Goldsmith, P. C. (1988). Infundibular gonadotropin-releasing hormone neurons are inhibited by direct opioid and autoregulatory synapses in juvenile monkeys. *Neuroendocrinology* 47, 203–216.
685. Mitchell, V., Prevot, V., Jennes, L., Aubert, J. P., Croix, D., and Beauvillain, J. C. (1997). Presence of mu and kappa opioid receptor mRNAs in galanin but not in GnRH neurons in the female rat. *Neuroreport.* 8, 3167–3172.
686. Sannella, M. I., and Petersen, S. L. (1997). Dual label in situ hybridization studies provide evidence that luteinizing hormone-releasing hormone neurons do not synthesize messenger ribonucleic acid for mu, kappa, or delta opiate receptors. *Endocrinology* 138, 1667–1672.
687. Lagrange, A. H., Ronnekleiv, O. K., and Kelly, M. J. (1995). Estradiol-17 beta and mu-opioid peptides rapidly hyperpolarize GnRH neurons: a cellular mechanism of negative feedback? *Endocrinology* 136, 2341–2344.
688. Briski, K. P., Quigley, K., and Meites, J. (1984). Endogenous opiate involvement in acute and chronic stress-induced changes in plasma LH concentrations in the male rat. *Life Sci.* 34, 2485–2493.
689. Petraglia, F., Vale, W., and Rivier, C. (1986). Opioids act centrally to modulate stress-induced decrease in luteinizing hormone in the rat. *Endocrinology* 119, 2445–2450.
690. Dyer, R. G., Mansfield, S., Corbet, H., and Dean, A. D. (1985). Fasting impairs LH secretion in female rats by activating an inhibitory opioid pathway. *J. Endocrinol.* 105, 91–97.
691. Cagampang, F. R., and Maeda, K. (1991). Effects of intracerebroventricular administration of opiate receptor antagonists on the suppressed pulsatile LH release during acute fasting in ovariectomized estradiol-treated rats. *Life Sci.* 49, 1823–1828.
692. Sirinathsinghji, D. J., and Martini, L. (1984). Effects of bromocriptine and naloxone on plasma levels of prolactin, LH and FSH during suckling in the female rat: responses to gonadotrophin releasing hormone. *J. Endocrinol.* 100, 175–182.
693. Baranowska, B., Rozbicka, G., Jeske, W., and Abdel-Fattah, M. H. (1984). The role of endogenous opiates in the mechanism of inhibited luteinizing hormone (LH) secretion in women with anorexia nervosa: the effect of naloxone on LH, follicle-stimulating hormone, prolactin, and beta-endorphin secretion. *J. Clin. Endocrinol. Metab.* 59, 412–416.
694. Helmreich, D. L., and Cameron, J. L. (1992). Suppression of luteinizing hormone secretion during food restriction in male rhesus monkeys (*Macaca mulatta*): failure of naloxone to restore normal pulsatility. *Neuroendocrinology* 56, 464–473.
695. Cosgrove, J. R., Booth, P. J., and Foxcroft, G. R. (1991). Opioidergic control of gonadotrophin secretion in the prepubertal gilt during restricted feeding and realimentation. *J. Reprod. Fertil.* 91, 277–284.
696. Ohtaki, T., Kumano, S., Ishibashi, Y., Ogi, K., Matsui, H., Harada, M., Kitada, C., Kurokawa, T., Onda, H., and Fujino, M. (1999). Isolation and cDNA cloning of a novel galanin-like peptide (GALP) from porcine hypothalamus. *J. Biol. Chem.* 274, 37041–37045.
697. Jureus, A., Cunningham, M. J., Li, D., Johnson, L. L., Krasnow, S. M., Teklemichael, D. N., Clifton, D. K., and Steiner, R. A. (2001). Distribution and regulation of galanin-like peptide (GALP) in the hypothalamus of the mouse. *Endocrinology* 142, 5140–5144.
698. Cunningham, M. J., Scarlett, J. M., and Steiner, R. A. (2002). Cloning and distribution of galanin-like peptide mRNA in the hypothalamus and pituitary of the macaque. *Endocrinology* 143, 755–763.
699. Jureus, A., Cunningham, M. J., McClain, M. E., Clifton, D. K., and Steiner, R. A. (2000). Galanin-like peptide (GALP) is a target for regulation by leptin in the hypothalamus of the rat. *Endocrinology* 141, 2703–2706.
700. Larm, J. A., and Gundlach, A. L. (2000). Galanin-like peptide (GALP) mRNA expression is restricted to arcuate nucleus of hypothalamus in adult male rat brain. *Neuroendocrinology* 72, 67–71.
701. Kerr, N. C., Holmes, F. E., and Wynick, D. (2000). Galanin-like peptide (GALP) is expressed in rat hypothalamus and pituitary, but not in DRG. *Neuroreport.* 11, 3909–3913.
702. Takatsu, Y., Matsumoto, H., Ohtaki, T., Kumano, S., Kitada, C., Onda, H., Nishimura, O., and Fujino, M. (2001). Distribution of galanin-like peptide in the rat brain. *Endocrinology* 142, 1626–1634.
703. Takenoya, F., Funahashi, H., Matsumoto, H., Ohtaki, T., Katoh, S., Kageyama, H., Suzuki, R., Takeuchi, M., and Shioda, S. (2002). Galanin-like peptide is co-localized with alpha-melanocyte stimulating hormone but not with neuropeptide Y in the rat brain. *Neurosci. Lett.* 331, 119–122.
704. Cunningham, M. J., Shahab, M., Grove, K. L., Scarlett, J. M., Plant, T. M., Cameron, J. L., Smith, M. S., Clifton, D. K., and Steiner, R. A. (2004). Galanin-like peptide as a possible link between metabolism and reproduction in the macaque. *J. Clin. Endocrinol. Metab.* 89, 1760–1766.
705. Fraley, G. S., Scarlett, J. M., Shimada, I., Teklemichael, D. N., Acohido, B. V., Clifton, D. K., and Steiner, R. A. (2004). Effects of diabetes and insulin on the expression of galanin-like peptide in the hypothalamus of the rat. *Diabetes* 53, 1237–1242.

706. Kumano, S., Matsumoto, H., Takatsu, Y., Noguchi, J., Kitada, C., and Ohtaki, T. (2003). Changes in hypothalamic expression levels of galanin-like peptide in rat and mouse models support that it is a leptin-target peptide. *Endocrinology* 144, 2634–2643.
707. Shen, J., and Gundlach, A. L. (2004). Galanin-like peptide mRNA alterations in arcuate nucleus and neural lobe of streptozotocin-diabetic and obese Zucker rats. Further evidence for leptin-dependent and independent regulation. *Neuroendocrinology* 79, 327–337.
708. Seth, A., Stanley, S., Jethwa, P., Gardiner, J., Ghatei, M., and Bloom, S. (2004). Galanin-like peptide stimulates the release of gonadotropin-releasing hormone in vitro and may mediate the effects of leptin on the hypothalamo-pituitary-gonadal axis. *Endocrinology* 145, 743–750.
709. Saito, J., Ozaki, Y., Kawasaki, M., Ohnishi, H., Okimoto, N., Nakamura, T., and Ueta, Y. (2004). Galanin-like peptide gene expression in the hypothalamus and posterior pituitary of the obese fa/fa rat. *Peptides* 25, 967–974.
710. Fraley, G. S., Scarlett, J. M., Teklemichael, D. N., Shimada, I., Baumgartner, J. W., Acohido, B. V., Clifton, D. K., and Steiner, R. A. (2003).Regulation of galanin-like peptide gene expression by insulin and metabolic fuels in the hypothalamus of the rat. 85th Annual Meeting of the Endocrine Society. Philadelphia, PA.
711. Lawrence, C. B., Williams, T., and Luckman, S. M. (2003). Intracerebroventricular galanin-like peptide induces different brain activation compared with galanin. *Endocrinology* 144, 3977–3984.
712. Matsumoto, Y., Watanabe, T., Adachi, Y., Itoh, T., Ohtaki, T., Onda, H., Kurokawa, T., Nishimura, O., and Fujino, M. (2002). Galanin-like peptide stimulates food intake in the rat. *Neurosci. Lett.* 322, 67–69.
713. Lawrence, C. B., Baudoin, F. M., and Luckman, S. M. (2002). Centrally administered galanin-like peptide modifies food intake in the rat: a comparison with galanin. *J. Neuroendocrinol.* 14, 853–860.
714. Seth, A., Stanley, S., Dhillo, W., Murphy, K., Ghatei, M., and Bloom, S. (2003). Effects of galanin-like peptide on food intake and the hypothalamo-pituitary-thyroid axis. *Neuroendocrinology* 77, 125–131.
715. Krasnow, S. M., Fraley, G. S., Schuh, S. M., Baumgartner, J. W., Clifton, D. K., and Steiner, R. A. (2003). A role for galanin-like peptide in the integration of feeding, body weight regulation, and reproduction in the mouse. *Endocrinology* 144, 813–822.
716. Hansen, K. R., Krasnow, S. M., Nolan. M. A., Fraley, G. S., Baumgartner, J. W., Clifton, D. K., and Steiner, R. A. (2003). Activation of the sympathetic nervous system by galanin-like peptide—a possible link between leptin and metabolism. *Endocrinology* 144, 4709–4717.
717. Matsumoto, H., Noguchi, J., Takatsu, Y., Horikoshi, Y., Kumano, S., Ohtaki, T., Kitada, C., Itoh, T., Onda, H., Nishimura, O., and Fujino, M. (2001). Stimulation effect of galanin-like peptide (GALP) on luteinizing hormone-releasing hormone-mediated luteinizing hormone (LH) secretion in male rats. *Endocrinology* 142, 3693–3696.
718. Fraley, G. S., Shimada, I., Baumgartner, J. W., Clifton, D. K., and Steiner, R. A. (2003). Differential patterns of Fos induction in the hypothalamus of the rat following central injections of galanin-like peptide and galanin. *Endocrinology* 144, 1143–1146.
719. Paredes, R. G., and Baum, M. J. (1997). Role of the medial preoptic area/anterior hypothalamus in the control of masculine sexual behavior. *Annu. Rev. Sex. Res.* 8, 68–101.
720. Fraley, G. S., Thomas-Smith, S. E., Acohido, B. V., Steiner, R. A., and Clifton, D. K. (2004). Stimulation of sexual behavior in the male rat by galanin-like peptide. *Horm. Behav.* 46, 551–557.
721. Swanson, L. W., Sawchenko, P. E., Rivier, J., and Vale, W. W. (1983). Organization of ovine corticotropin-releasing factor immunoreactive cells and fibers in the rat brain: an immunohistochemical study. *Neuroendocrinology* 36, 165–186.
722. Hashimoto, K., Ohno, N., Aoki, Y., Kageyama, J., Takahara, J., and Ofuji, T. (1982). Distribution and characterization of corticotropin-releasing factor and arginine vasopressin in rat hypothalamic nuclei. *Neuroendocrinology* 34, 32–37.
723. Williams, G., Bing, C., Cai, X. J., Harrold, J. A., King, P. J., and Liu, X. H. (2001). The hypothalamus and the control of energy homeostasis: different circuits, different purposes. *Physiol. Behav.* 74, 683–701.
724. Arase, K., York, D. A., Shimizu, H., Shargill, N., and Bray, G. A. (1988). Effects of corticotropin-releasing factor on food intake and brown adipose tissue thermogenesis in rats. *Am. J. Physiol.* 255, E255–E259.
725. Egawa, M., Yoshimatsu, H., and Bray, G. A. (1990). Effect of corticotropin releasing hormone and neuropeptide Y on electrophysiological activity of sympathetic nerves to interscapular brown adipose tissue. *Neuroscience.* 34, 771–775.
726. Raber, J., Chen, S., Mucke, L., and Feng, L. (1997). Corticotropin-releasing factor and adrenocorticotrophic hormone as potential central mediators of OB effects. *J. Biol. Chem.* 272, 15057–15060.
727. Gardner, J. D., Rothwell, N. J., and Luheshi, G. N. (1998). Leptin affects food intake via CRF-receptor-mediated pathways. *Nat. Neurosci.* 1, 103.
728. Masaki, T., Yoshimichi, G., Chiba, S., Yasuda, T., Noguchi, H., Kakuma, T., Sakata, T., and Yoshimatsu, H. (2003). Corticotropin-releasing hormone-mediated pathway of leptin to regulate feeding, adiposity, and uncoupling protein expression in mice. *Endocrinology* 144, 3547–3554.
729. Uehara, Y., Shimizu, H., Ohtani, K., Sato, N., and Mori, M. (1998). Hypothalamic corticotropin-releasing hormone is a mediator of the anorexigenic effect of leptin. *Diabetes* 47, 890–893.
730. Sirinathsinghji, D. J. (1987). Inhibitory influence of corticotropin releasing factor on components of sexual behaviour in the male rat. *Brain Res.* 407, 185–190.
731. Sirinathsinghji, D. J., Rees, L. H., Rivier, J., and Vale, W. (1983). Corticotropin-releasing factor is a potent inhibitor of sexual receptivity in the female rat. *Nature* 305, 232–235.
732. Jones, J. E., Pick, R. R., Davenport, M. D., Keene, A. C., Corp, E. S., and Wade, G. N. (2002). Disinhibition of female sexual behavior by a CRH receptor antagonist in Syrian hamsters. *Am. J. Physiol. Regul. Integr. Comp. Physiol.* 283, R591–R597.
733. Heinrichs, S. C., Min, H., Tamraz, S., Carmouche, M., Boehme, S. A., and Vale, W. W. (1997). Anti-sexual and anxiogenic behavioral consequences of corticotropin-releasing factor overexpression are centrally mediated. *Psychoneuroendocrinology* 22, 215–224.
734. Ono, N., Lumpkin, M. D., Samson, W. K., McDonald, J. K., and McCann, S. M. (1984). Intrahypothalamic action of corticotrophin-releasing factor (CRF) to inhibit growth hormone and LH release in the rat. *Life Sci.* 35, 1117–1123.
735. Rivier, C., and Vale, W. (1984). Influence of corticotropin-releasing factor on reproductive functions in the rat. *Endocrinology* 114, 914–921.
736. Olster, D. H., and Ferin, M. (1987). Corticotropin-releasing hormone inhibits gonadotropin secretion in the ovariectomized rhesus monkey. *J. Clin. Endocrinol. Metab.* 65, 262–267.
737. Petraglia, F., Sutton, S., Vale, W., and Plotsky, P. (1987). Corticotropin-releasing factor decreases plasma luteinizing

hormone levels in female rats by inhibiting gonadotropin-releasing hormone release into hypophysial-portal circulation. *Endocrinology* 120, 1083–1088.

738. Rivest, S., Plotsky, P. M., and Rivier, C. (1993). CRF alters the infundibular LHRH secretory system from the medial preoptic area of female rats: possible involvement of opioid receptors. *Neuroendocrinology* 57, 236–246.
739. Nikolarakis, K. E., Almeida, O. F., and Herz, A. (1986). Corticotropin-releasing factor (CRF) inhibits gonadotropin-releasing hormone (GnRH) release from superfused rat hypothalami in vitro. *Brain Res.* 377, 388–390.
740. Gambacciani, M., Yen, S. S., and Rasmussen, D. D. (1986). GnRH release from the mediobasal hypothalamus: in vitro inhibition by corticotropin-releasing factor. *Neuroendocrinology* 43, 533–536.
741. MacLusky, N. J., Naftolin, F., and Leranth, C. (1988). Immunocytochemical evidence for direct synaptic connections between corticotrophin-releasing factor (CRF) and gonadotrophin-releasing hormone (GnRH)-containing neurons in the preoptic area of the rat. *Brain Res.* 439, 391–395.
742. Tellam, D. J., Perone, M. J., Dunn, I. C., Radovick, S., Brennand, J., Rivier, J. E., Castro, M. G., and Lovejoy, D. A. (1998). Direct regulation of GnRH transcription by CRF-like peptides in an immortalized neuronal cell line. *Neuroreport* 9, 3135–3140.
743. Almeida, O. F., Nikolarakis, K. E., and Herz, A. (1988). Evidence for the involvement of endogenous opioids in the inhibition of luteinizing hormone by corticotropin-releasing factor. *Endocrinology* 122, 1034–1041.
744. Gindoff, P. R., and Ferin, M. (1987). Endogenous opioid peptides modulate the effect of corticotropin-releasing factor on gonadotropin release in the primate. *Endocrinology* 121, 837–842.
745. Williams, C. L., Nishihara, M., Thalabard, J. C., Grosser, P. M., Hotchkiss, J., and Knobil, E. (1990). Corticotropin-releasing factor and gonadotropin-releasing hormone pulse generator activity in the rhesus monkey. Electrophysiological studies. *Neuroendocrinology* 52, 133–137.
746. Tsukahara, S., Tsukamura, H., Foster, D. L., and Maeda, K. I. (1999). Effect of corticotropin-releasing hormone antagonist on oestrogen-dependent glucoprivic suppression of luteinizing hormone secretion in female rats. *J. Neuroendocrinol.* 11, 101–105.
747. Maeda, K., Cagampang, F. R., Coen, C. W., and Tsukamura, H. (1994). Involvement of the catecholaminergic input to the paraventricular nucleus and of corticotropin-releasing hormone in the fasting-induced suppression of luteinizing hormone release in female rats. *Endocrinology* 134, 1718–1722.
748. Jeong, K. H., Jacobson, L., Widmaier, E. P., and Majzoub, J. A. (1999). Normal suppression of the reproductive axis following stress in corticotropin-releasing hormone-deficient mice. *Endocrinology* 140, 1702–1708.
749. Kawauchi, H., Kawazoe, I., Tsubokawa, M., Kishida, M., and Baker, B. I. (1983). Characterization of melanin-concentrating hormone in chum salmon pituitaries. *Nature.* 305, 321–323.
750. Naito, N., Kawazoe, I., Nakai, Y., and Kawauchi, H. (1988). Melanin-concentrating hormone-like immunoreactive material in the rat hypothalamus; characterization and subcellular localization. *Cell Tissue Res.* 253, 291–295.
751. Bittencourt, J. C., Presse, F., Arias, C., Peto, C., Vaughan, J., Nahon, J. L., Vale, W., and Sawchenko, P. E. (1992). The melanin-concentrating hormone system of the rat brain: an immuno- and hybridization histochemical characterization. *J. Comp. Neurol.* 319, 218–245.
752. Skofitsch, G., Jacobowitz, D. M., and Zamir, N. (1985). Immunohistochemical localization of a melanin concentrating hormone-like peptide in the rat brain. *Brain Res. Bull.* 15, 635–649.
753. Chambers, J., Ames, R. S., Bergsma, D., Muir, A., Fitzgerald, L. R., Hervieu, G., Dytko, G. M., Foley, J. J., Martin, J., Liu, W. S., Park, J., Ellis, C., Ganguly, S., Konchar, S., Cluderay, J., Leslie, R., Wilson, S., and Sarau, H. M. (1999). Melanin-concentrating hormone is the cognate ligand for the orphan G-protein-coupled receptor SLC-1. *Nature* 400, 261–265.
754. Saito, Y., Nothacker, H. P., Wang, Z., Lin, S. H., Leslie, F., and Civelli, O. (1999). Molecular characterization of the melanin-concentrating-hormone receptor. *Nature* 400, 265–269.
755. Saito, Y., Cheng, M., Leslie, F. M., and Civelli, O. (2001). Expression of the melanin-concentrating hormone (MCH) receptor mRNA in the rat brain. *J. Comp. Neurol.* 435, 26–40.
756. Qu, D., Ludwig, D. S., Gammeltoft, S., Piper, M., Pelleymounter, M. A., Cullen, M. J., Mathes, W. F., Przypek, R., Kanarek, R., and Maratos-Flier, E. (1996). A role for melanin-concentrating hormone in the central regulation of feeding behaviour. *Nature* 380, 243–247.
757. Tritos, N. A., Mastaitis, J. W., Kokkotou, E., and Maratos-Flier, E. (2001). Characterization of melanin concentrating hormone and preproorexin expression in the murine hypothalamus. *Brain Res.* 895, 160–166.
758. Sahu, A. (1998). Evidence suggesting that galanin (GAL), melanin-concentrating hormone (MCH), neurotensin (NT), proopiomelanocortin (POMC) and neuropeptide Y (NPY) are targets of leptin signaling in the hypothalamus. *Endocrinology* 139, 795–798.
759. Segal-Lieberman, G., Bradley, R. L., Kokkotou, E., Carlson, M., Trombly, D. J., Wang, X., Bates, S., Myers, M. G. Jr., Flier, J. S., and Maratos-Flier, E. (2003). Melanin-concentrating hormone is a critical mediator of the leptin-deficient phenotype. *Proc. Natl. Acad. Sci. U S A* 100, 10085–10090.
760. Murray, J. F., Mercer, J. G., Adan, R. A., Datta, J. J., Aldairy, C., Moar, K. M., Baker, B. I., Stock, M. J., and Wilson, C. A. (2000). The effect of leptin on luteinizing hormone release is exerted in the zona incerta and mediated by melanin-concentrating hormone. *J. Neuroendocrinol.* 12, 1133–1139.
761. Bayer, L., Colard, C., Nguyen, N. U., Risold, P. Y., Fellmann, D., and Griffond, B. (2000). Alteration of the expression of the hypocretin (orexin) gene by 2-deoxyglucose in the rat lateral hypothalamic area. *Neuroreport.* 11, 531–533.
762. Bayer, L., Poncet, F., Fellmann, D., and Griffond, B. (1999). Melanin-concentrating hormone expression in slice cultures of rat hypothalamus is not affected by 2-deoxyglucose. *Neurosci. Lett.* 267, 77–80.
763. Anand, B. K., and Brobeck, J. R. (1951). Localization of a "feeding center" in the hypothalamus of the rat. *Proc. Soc. Exp. Biol. Med.* 77, 323–324.
764. Rossi, M., Choi, S. J., O'Shea, D., Miyoshi, T., Ghatei, M. A., and Bloom, S. R. (1997). Melanin-concentrating hormone acutely stimulates feeding, but chronic administration has no effect on body weight. *Endocrinology* 138, 351–355.
765. Della-Zuana, O., Presse, F., Ortola, C., Duhault, J., Nahon, J. L., and Levens, N. (2002). Acute and chronic administration of melanin-concentrating hormone enhances food intake and body weight in Wistar and Sprague-Dawley rats. *Int. J. Obes. Relat. Metab. Disord.* 26, 1289–1295.
766. Gomori, A., Ishihara, A., Ito, M., Mashiko, S., Matsushita, H., Yumoto, M., Tanaka, T., Tokita, S., Moriya, M., Iwaasa, H., and Kanatani, A. (2003). Chronic intracerebroventricular infusion of MCH causes obesity in mice. Melanin-concentrating hormone. *Am. J. Physiol. Endocrinol Metab.* 284, E583–E588.

767. Ludwig, D. S., Tritos, N. A., Mastaitis, J. W., Kulkarni, R., Kokkotou, E., Elmquist, J., Lowell, B., Flier, J. S., and Maratos-Flier, E. (2001). Melanin-concentrating hormone overexpression in transgenic mice leads to obesity and insulin resistance. *J. Clin. Invest.* 107, 379–386.
768. Shimada, M., Tritos, N. A., Lowell, B. B., Flier, J. S., and Maratos-Flier, E. (1998). Mice lacking melanin-concentrating hormone are hypophagic and lean. *Nature* 396, 670–674.
769. Gonzalez, M. I., Vaziri, S., and Wilson, C. A. (1996). Behavioral effects of alpha-MSH and MCH after central administration in the female rat. *Peptides* 17, 171–177.
770. Gonzalez, M. I., Baker, B. I., and Wilson, C. A. (1997). Stimulatory effect of melanin-concentrating hormone on luteinising hormone release. *Neuroendocrinology* 66, 254–262.
771. Tsukamura, H., Thompson, R. C., Tsukahara, S., Ohkura, S., Maekawa, F., Moriyama, R., Niwa, Y., Foster, D. L., and Maeda, K. (2000). Intracerebroventricular administration of melanin-concentrating hormone suppresses pulsatile luteinizing hormone release in the female rat. *J. Neuroendocrinol.* 12, 529–534.
772. Chiocchio, S. R., Gallardo, M. G., Louzan, P., Gutnisky, V., and Tramezzani, J. H. (2001). Melanin-concentrating hormone stimulates the release of luteinizing hormone-releasing hormone and gonadotropins in the female rat acting at both median eminence and pituitary levels. *Biol. Reprod.* 64, 1466–1472.
773. Willie, J. T., Chemelli, R. M., Sinton, C. M., and Yanagisawa, M. (2001). To eat or to sleep? Orexin in the regulation of feeding and wakefulness. *Annu. Rev. Neurosci.* 24, 429–458.
774. Sakurai, T., Amemiya, A., Ishii, M., Matsuzaki, I., Chemelli, R. M., Tanaka, H., Williams, S. C., Richardson, J. A., Kozlowski, G. P., Wilson, S., Arch, J. R., Buckingham, R. E., Haynes, A. C., Carr, S. A., Annan, R. S., McNulty, D. E., Liu, W. S., Terrett, J. A., Elshourbagy, N. A., Bergsma, D. J., and Yanagisawa, M. (1998). Orexins and orexin receptors: a family of hypothalamic neuropeptides and G protein-coupled receptors that regulate feeding behavior. *Cell* 92, 573–585.
775. de Lecea, L., Kilduff, T. S., Peyron, C., Gao, X., Foye, P. E., Danielson, P. E., Fukuhara, C., Battenberg, E. L., Gautvik, V. T., Bartlett, F. S., 2nd, Frankel, W. N., van den Pol, A. N., Bloom, F. E., Gautvik, K. M., and Sutcliffe, J. G. (1998). The hypocretins: hypothalamus-specific peptides with neuroexcitatory activity. *Proc. Natl. Acad. Sci. U S A* 95, 322–327.
776. Peyron, C., Tighe, D. K., van den Pol, A. N., de Lecea, L., Heller, H. C., Sutcliffe, J. G., and Kilduff, T. S. (1998). Neurons containing hypocretin (orexin) project to multiple neuronal systems. *J. Neurosci.* 18, 9996–10015.
777. Cutler, D. J., Morris, R., Sheridhar, V., Wattam, T. A., Holmes, S., Patel, S., Arch, J. R., Wilson, S., Buckingham, R. E., Evans, M. L., Leslie, R. A., and Williams, G. (1999). Differential distribution of orexin-A and orexin-B immunoreactivity in the rat brain and spinal cord. *Peptides* 20, 1455–1470.
778. Trivedi, P., Yu, H., MacNeil, D. J., Van der Ploeg, L. H., and Guan, X. M. (1998). Distribution of orexin receptor mRNA in the rat brain. *FEBS Lett.* 438, 71–75.
779. Marcus, J. N., Aschkenasi, C. J., Lee, C. E., Chemelli, R. M., Saper, C. B., Yanagisawa, M., and Elmquist, J. K. (2001). Differential expression of orexin receptors 1 and 2 in the rat brain. *J. Comp. Neurol.* 435, 6–25.
780. van den Pol, A. N., Gao, X. B., Obrietan, K., Kilduff, T. S., and Belousov, A. B. (1998). Presynaptic and postsynaptic actions and modulation of neuroendocrine neurons by a new hypothalamic peptide, hypocretin/orexin. *J. Neurosci.* 18, 7962–7971.
781. Cai, X. J., Widdowson, P. S., Harrold, J., Wilson, S., Buckingham, R. E., Arch, J. R., Tadayyon, M., Clapham, J. C., Wilding, J., and Williams, G. (1999). Hypothalamic orexin expression: modulation by blood glucose and feeding. *Diabetes* 48, 2132–2137.
782. Hakansson, M., de Lecea, L., Sutcliffe, J. G., Yanagisawa, M., and Meister, B. (1999). Leptin receptor- and STAT3-immunoreactivities in hypocretin/orexin neurones of the lateral hypothalamus. *J. Neuroendocrinol.* 11, 653–663.
783. Iqbal, J., Pompolo, S., Sakurai, T., and Clarke, I. J. (2001). Evidence that orexin-containing neurones provide direct input to gonadotropin-releasing hormone neurones in the ovine hypothalamus. *J. Neuroendocrinol.* 13, 1033–1041.
784. Horvath, T. L., Diano, S., and van den Pol, A. N. (1999). Synaptic interaction between hypocretin (orexin) and neuropeptide Y cells in the rodent and primate hypothalamus: a novel circuit implicated in metabolic and endocrine regulations. *J. Neurosci.* 19, 1072–1087.
785. Lopez, M., Seoane, L., Garcia, M. C., Lago, F., Casanueva, F. F., Senaris, R., and Dieguez, C. (2000). Leptin regulation of prepro-orexin and orexin receptor mRNA levels in the hypothalamus. *Biochem. Biophys. Res. Commun.* 269, 41–45.
786. Muroya, S., Uramura, K., Sakurai, T., Takigawa, M., and Yada, T. (2001). Lowering glucose concentrations increases cytosolic Ca2+ in orexin neurons of the rat lateral hypothalamus. *Neurosci. Lett.* 309, 165–168.
787. Edwards, C. M., Abusnana, S., Sunter, D., Murphy, K. G., Ghatei, M. A., and Bloom, S. R. (1999). The effect of the orexins on food intake: comparison with neuropeptide Y, melanin-concentrating hormone and galanin. *J. Endocrinol.* 160, R7–R12.
788. Dube, M. G., Kalra, S. P., and Kalra, P. S. (1999). Food intake elicited by central administration of orexins/hypocretins: identification of hypothalamic sites of action. *Brain Res.* 842, 473–477.
789. Haynes, A. C., Jackson, B., Overend, P., Buckingham, R. E., Wilson, S., Tadayyon, M., and Arch, J. R. (1999). Effects of single and chronic intracerebroventricular administration of the orexins on feeding in the rat. *Peptides* 20, 1099–1105.
790. Yamada, H., Okumura, T., Motomura, W., Kobayashi, Y., and Kohgo, Y. (2000). Inhibition of food intake by central injection of anti-orexin antibody in fasted rats. *Biochem. Biophys. Res. Commun.* 267, 527–531.
791. Yamanaka, A., Sakurai, T., Katsumoto, T., Yanagisawa, M., and Goto, K. (1999). Chronic intracerebroventricular administration of orexin-A to rats increases food intake in daytime, but has no effect on body weight. *Brain Res.* 849, 248–252.
792. Asakawa, A., Inui, A., Goto, K., Yuzuriha, H., Takimoto, Y., Inui, T., Katsuura, G., Fujino, M. A., Meguid, M. M., and Kasuga, M. (2002). Effects of agouti-related protein, orexin and melanin-concentrating hormone on oxygen consumption in mice. *Int. J. Mol. Med.* 10, 523–525.
793. Wang, J., Osaka, T., and Inoue, S. (2001). Energy expenditure by intracerebroventricular administration of orexin to anesthetized rats. *Neurosci. Lett.* 315, 49–52.
794. Yoshimichi, G., Yoshimatsu, H., Masaki, T., and Sakata, T. (2001). Orexin-A regulates body temperature in coordination with arousal status. *Exp. Biol. Med. (Maywood).* 226, 468–476.
795. Hara, J., Beuckmann, C. T., Nambu, T., Willie, J. T., Chemelli, R. M., Sinton, C. M., Sugiyama, F., Yagami, K., Goto, K., Yanagisawa, M., and Sakurai, T. (2001). Genetic ablation of orexin neurons in mice results in narcolepsy, hypophagia, and obesity. *Neuron* 30, 345–354.
796. Pu, S., Jain, M. R., Kalra, P. S., and Kalra, S. P. (1998). Orexins, a novel family of hypothalamic neuropeptides,

modulate pituitary luteinizing hormone secretion in an ovarian steroid-dependent manner. *Regul. Pept.* 78, 133–136.

797. Kohsaka, A., Watanobe, H., Kakizaki, Y., Suda, T., and Schioth, H. B. (2001). A significant participation of orexin-A, a potent orexigenic peptide, in the preovulatory luteinizing hormone and prolactin surges in the rat. *Brain Res.* 898, 166–170.
798. Kok, S. W., Roelfsema, F., Overeem, S., Lammers, G. J., Frolich, M., Meinders, A. E., and Pijl, H. (2004). Pulsatile LH release is diminished, whereas FSH secretion is normal, in hypocretin-deficient narcoleptic men. *Am. J. Physiol. Endocrinol. Metab.* 287, E630–E636.
799. Furuta, M., Funabashi, T., and Kimura, F. (2002). Suppressive action of orexin A on pulsatile luteinizing hormone secretion is potentiated by a low dose of estrogen in ovariectomized rats. *Neuroendocrinology* 75, 151–157.
800. Tamura, T., Irahara, M., Tezuka, M., Kiyokawa, M., and Aono, T. (1999). Orexins, orexigenic hypothalamic neuropeptides, suppress the pulsatile secretion of luteinizing hormone in ovariectomized female rats. *Biochem. Biophys. Res. Commun.* 264, 759–762.
801. Campbell, R. E., Grove, K. L., and Smith, M. S. (2003). Gonadotropin-releasing hormone neurons coexpress orexin 1 receptor immunoreactivity and receive direct contacts by orexin fibers. *Endocrinology* 144, 1542–1548.
802. Small, C. J., Goubillon, M. L., Murray, J. F., Siddiqui, A., Grimshaw, S. E., Young, H., Sivanesan, V., Kalamatianos, T., Kennedy, A. R., Coen, C. W., Bloom, S. R., and Wilson, C. A. (2003). Central orexin A has site-specific effects on luteinizing hormone release in female rats. *Endocrinology* 144, 3225–3236.
803. Russell, S. H., Small, C. J., Kennedy, A. R., Stanley, S. A., Seth. A., Murphy, K. G., Taheri, S., Ghatei, M. A., and Bloom, S. R. (2001). Orexin A interactions in the hypothalamo-pituitary gonadal axis. *Endocrinology* 142, 5294–5302.
804. World Health Organization. (1997). Obesity: Preventing and Managing the Global Epidemic. Geneva: World Health Organization.
805. Kopelman, P. G. (1988). Neuroendocrine function in obesity. *Clin. Endocrinol. (Oxf).* 28, 675–689.
806. Pasquali, R., Pelusi, C., Genghini, S., Cacciari, M., and Gambineri, A. (2003). Obesity and reproductive disorders in women. *Hum. Reprod. Update* 9, 359–372.
807. Glass, A. R., Swerdloff, R. S., Bray, G. A., Dahms, W. T., and Atkinson, R. L. (1977). Low serum testosterone and sex-hormone-binding-globulin in massively obese men. *J. Clin. Endocrinol. Metab.* 45, 1211–1219.
808. Glass, A. R., Burman, K. D., Dahms, W. T., and Boehm, T. M. (1981). Endocrine function in human obesity. *Metabolism* 30, 89–104.
809. Rogers, J., and Mitchell, G. W. Jr. (1952). The relation of obesity to menstrual disturbances. *N. Engl. J. Med.* 247, 53–55.
810. Hartz, A. J., Barboriak, P. N., Wong, A., Katayama, K. P., and Rimm, A. A. (1979). The association of obesity with infertility and related menstrual abnormalities in women. *Int. J. Obes.* 3, 57–73.
811. Green, B. B., Weiss, N. S., and Daling, J. R. (1988). Risk of ovulatory infertility in relation to body weight. *Fertil. Steril.* 50, 721–726.
812. Grodstein, F., Goldman, M. B., and Cramer, D. W. (1994). Body mass index and ovulatory infertility. *Epidemiology.* 5, 247–250.
813. Rich-Edwards, J. W., Spiegelman, D., Garland, M., Hertzmark, E., Hunter, D. J., Colditz, G. A., Willett, W. C., Wand, H., and Manson, J. E. (2002). Physical activity, body mass index, and ovulatory disorder infertility. *Epidemiology* 13, 184–190.
814. Sebire, N. J., Jolly, M., Harris, J. P., Wadsworth, J., Joffe, M., Beard, R. W., Regan, L., and Robinson, S. (2001). Maternal obesity and pregnancy outcome: a study of 287,213 pregnancies in London. *Int. J. Obes Relat. Metab. Disord.* 25, 1175–1182.
815. Moran, C., Hernandez, E., Ruiz, J. E., Fonseca, M. E., Bermudez, J. A., and Zarate, A. (1999). Upper body obesity and hyperinsulinemia are associated with anovulation. *Gynecol. Obstet. Invest.* 47, 1–5.
816. Douchi, T., Kuwahata, R., Yamamoto, S., Oki, T., Yamasaki, H., and Nagata, Y. (2002). Relationship of upper body obesity to menstrual disorders. *Acta. Obstet. Gynaecol. Scand.* 81, 147–150.
817. Lima, N., Cavaliere, H., Knobel, M., Halpern, A., and Medeiros-Neto, G. (2000). Decreased androgen levels in massively obese men may be associated with impaired function of the gonadostat. *Int. J. Obes. Relat. Metab. Disord.* 24, 1433–1437.
818. Kopelman, P. G., White, N., Pilkington, T. R., and Jeffcoate, S. L. (1981). The effect of weight loss on sex steroid secretion and binding in massively obese women. *Clin. Endocrinol. (Oxf).* 15, 113–116.
819. Clark, A. M., Ledger, W., Galletly, C., Tomlinson, L., Blaney, F., Wang, X., and Norman, R. J. (1995). Weight loss results in significant improvement in pregnancy and ovulation rates in anovulatory obese women. *Hum. Reprod.* 10, 2705–2712.
820. Norman, R. J., Noakes, M., Wu, R., Davies, M. J., Moran, L., and Wang, J. X. (2004). Improving reproductive performance in overweight/obese women with effective weight management. *Hum. Reprod. Update* 10, 267–280.
821. Caro, J. F., Kolaczynski, J. W., Nyce, M. R., Ohannesian, J. P., Opentanova, I., Goldman, W. H., Lynn, R. B., Zhang, P-L., Sinha, M. K., and Considine, R. V. (1996). Decreased cerebrospinal-fluid/serum leptin ratio in obesity: a possible mechanism for leptin resistance. *Lancet* 348, 159–161.
822. Schwartz, M. W., Peskind, E., Raskind, M., Boyko, E. J., and Porte, D. Jr. (1996). Cerebrospinal fluid leptin levels: relationship to plasma levels and to adiposity in humans. *Nat. Med.* 2, 589–593.
823. Tortoriello, D. V., McMinn, J., and Chua, S. C. (2004). Dietary-induced obesity and hypothalamic infertility in female DBA/2J mice. *Endocrinology* 145, 1238–1247.
824. Glass, A. R., Dahms, W. T., Abraham, G., Atkinson, R. L., Bray, G. A., and Swerdloff, R. S. (1978). Secondary amenorrhea in obesity: etiologic role of weight-related androgen excess. *Fertil. Steril.* 30, 243–244.
825. Hollmann, M., Runnebaum, B., and Gerhard, I. (1996). Effects of weight loss on the hormonal profile in obese, infertile women. *Hum. Reprod.* 11, 1884–1891.
826. Baranowska, B., Wasilewska-Dziubinska, E., Radzikowska, M., Plonowski, A., and Roguski, K. (1997). Neuropeptide Y, galanin, and leptin release in obese women and in women with anorexia nervosa. *Metabolism* 46, 1384–1389.
827. Pasquali, R., Cantobelli, S., Casimirri, F., Bortoluzzi, L., Boschi, S., Capelli, M., Melchionda, N., and Barbara, L. (1992). The role of the opioid peptides in the development of hyperinsulinemia in obese women with abdominal body fat distribution. *Metabolism* 41, 763–767.
828. Djursing, H., Andersen, A. N., Hagen, C., and Petersen, K. (1985). Gonadotropin secretion before and during acute and chronic dopamine-receptor blockade in insulin-dependent diabetic patients with amenorrhea. *Fertil. Steril.* 44, 49–55.

829. Schroeder, B., Hertweck, S. P., Sanfilippo, J. S., and Foster, M. B. (2000). Correlation between glycemic control and menstruation in diabetic adolescents. *J. Reprod. Med.* 45, 1–5.
830. Kirchick, H. J., Keyes, P. L., and Frye, B. E. (1982). Restoration of the LH surge and ovulation by insulin in alloxan-diabetic immature rats treated with pregnant mare's serum gonadotrophin. *Acta. Endocrinol. (Copenh).* 100, 266–273.
831. Havel, P. J., Uriu-Hare, J. Y., Liu, T., Stanhope, K. L., Stern, J. S., Keen, C. L., and Ahren, B. (1998). Marked and rapid decreases of circulating leptin in streptozotocin diabetic rats: reversal by insulin. *Am. J. Physiol.* 274, R1482–R1491.
832. Kiess, W., Anil, M., Blum, W. F., Englaro, P., Juul, A., Attanasio, A., Dotsch, J., and Rascher, W. (1998). Serum leptin levels in children and adolescents with insulin-dependent diabetes mellitus in relation to metabolic control and body mass index. *Eur. J. Endocrinol.* 138, 501–509.
833. Kamoda, T., Saitoh, H., Nakahara, S., Izumi, I., Hirano, T., and Matsui, A. (1998). Serum leptin and insulin concentrations in prepubertal lean, obese and insulin-dependent diabetes mellitus children. *Clin. Endocrinol. (Oxf).* 49, 385–389.
834. Tuominen, J. A., Ebeling, P., Stenman, U-H., Heiman, M. L., Stephens, T. W., and Koivisto, V. A. (1997). Leptin synthesis is resistant to acute effects of insulin in insulin-dependent diabetes mellitus patients. *J. Clin. Endocrinol. Metab.* 82, 381–382.
835. Reitman, M. L., Arioglu, E., Gavrilova, O., and Taylor, S. I. (2000). Lipoatrophy revisited. *Trends Endocrinol. Metab.* 11, 410–416.
836. Burant, C. F., Sreenan, S., Hirano, K., Tai, T. A., Lohmiller, J., Lukens, J., Davidson, N. O., Ross, S., and Graves, R. A. (1997). Troglitazone action is independent of adipose tissue. *J. Clin. Invest.* 100, 2900–2908.
837. Moitra, J., Mason, M. M., Olive, M., Krylov, D., Gavrilova, O., Marcus-Samuels, B., Feigenbaum, L., Lee, E., Aoyama, T., Eckhaus, M., Reitman, M. L., and Vinson, C. (1998). Life without white fat: a transgenic mouse. *Genes Dev.* 12, 3168–3181.
838. Shimomura, I., Hammer, R. E., Richardson, J. A., Ikemoto, S., Bashmakov, Y., Goldstein, J. L., and Brown, M. S. (1998). Insulin resistance and diabetes mellitus in transgenic mice expressing nuclear SREBP-1c in adipose tissue: model for congenital generalized lipodystrophy. *Genes Dev.* 12, 3182–3194.
839. El-Haschimi, K., Dufresne, S. D., Hirshman, M. F., Flier, J. S., Goodyear, L. J., and Bjorbaek, C. (2003). Insulin resistance and lipodystrophy in mice lacking ribosomal S6 kinase 2. *Diabetes* 52, 1340–1346.
840. Pardini, V. C., Victoria, I. M., Rocha, S. M., Andrade, D. G., Rocha, A. M., Pieroni, F. B., Milagres, G., Purisch, S., and Velho, G. (1998). Leptin levels, beta-cell function, and insulin sensitivity in families with congenital and acquired generalized lipoatropic diabetes. *J. Clin. Endocrinol. Metab.* 83, 503–508.
841. Jaquet, D., Khallouf, E., Levy-Marchal, C., and Czernichow, P. (1999). Extremely low values of serum leptin in children with congenital generalized lipoatrophy. *Eur. J. Endocrinol.* 140, 107–109.
842. Oral, E. A., Simha, V., Ruiz, E., Andewelt, A., Premkumar, A., Snell, P., Wagner, A. J., DePaoli, A. M., Reitman, M. L., Taylor, S. I., Gorden, P., and Garg, A. (2002). Leptin-replacement therapy for lipodystrophy. *N. Engl. J. Med.* 346, 570–578.
843. Petersen, K. F., Oral, E. A., Dufour, S., Befroy, D., Ariyan, C., Yu, C., Cline, G. W., DePaoli, A. M., Taylor, S. I., Gorden, P., and Shulman, G. I. (2002). Leptin reverses insulin resistance and hepatic steatosis in patients with severe lipodystrophy. *J. Clin. Invest.* 109, 1345–1350.
844. Shimomura, I., Hammer, R. E., Ikemoto, S., Brown, M. S., and Goldstein, J. L. (1999). Leptin reverses insulin resistance and diabetes mellitus in mice with congenital lipodystrophy. *Nature.* 401, 73–76.
845. Ebihara, K., Ogawa, Y., Masuzaki, H., Shintani, M., Miyanaga, F., Aizawa-Abe, M., Hayashi, T., Hosoda, K., Inoue, G., Yoshimasa, Y., Gavrilova, O., Reitman, M. L., and Nakao, K. (2001). Transgenic overexpression of leptin rescues insulin resistance and diabetes in a mouse model of lipoatrophic diabetes. *Diabetes* 50, 1440–1448.
846. Asilmaz, E., Cohen, P., Miyazaki, M., Dobrzyn, P., Ueki, K., Fayzikhodjaeva, G., Soukas, A. A., Kahn, C. R., Ntambi, J. M., Socci, N. D., and Friedman, J. M. (2004). Site and mechanism of leptin action in a rodent form of congenital lipodystrophy. *J. Clin. Invest.* 113, 414–424.
847. Oral, E. A., Ruiz, E., Andewelt, A., Sebring, N., Wagner, A. J., Depaoli, A. M., and Gorden, P. (2002). Effect of leptin replacement on pituitary hormone regulation in patients with severe lipodystrophy. *J. Clin. Endocrinol. Metab.* 87, 3110–3117.
848. Dolberg, B. K., and Lenhard, M. J. (2000). Successful outcome of pregnancy in a patient with generalized lipoatrophic diabetes mellitus. *Endocr. Pract.* 6, 34–36.
849. Morse, A. N., and Whitaker, M. D. (2000). Successful pregnancy in a woman with lipoatrophic diabetes mellitus. A case report. *J. Reprod. Med.* 45, 850–852.
850. Catalano, P. M., Capeless, E. L., Simmons, G. M., Robbins, D. C., and Horton, E. S. (1990). Successful pregnancy outcome in association with lipoatrophic diabetes mellitus. *Obstet. Gynecol.* 76, 978–979.
851. Carmina, E., and Lobo, R. A. (1999). Polycystic ovary syndrome (PCOS): arguably the most common endocrinopathy is associated with significant morbidity in women. *J. Clin. Endocrinol. Metab.* 84, 1897–1899.
852. Asuncion, M., Calvo, R. M., San Millan, J. L., Sancho, J., Avila, S., and Escobar-Morreale, H. F. (2000). A prospective study of the prevalence of the polycystic ovary syndrome in unselected Caucasian women from Spain. *J. Clin. Endocrinol. Metab.* 85, 2434–2438.
853. Azziz, R., Woods, K. S., Reyna, R., Key, T. J., Knochenhauer, E. S., and Yildiz, B. O. (2004). The prevalence and features of the polycystic ovary syndrome in an unselected population. *J. Clin. Endocrinol. Metab.* 89, 2745–2749.
854. Lobo, R. A. (1995). A disorder without identity: "HCA," "PCO," "PCOD," "PCOS," "SLS." What are we to call it?! *Fertil. Steril.* 63, 1158–1160.
855. Yen, S. S. (1980). The polycystic ovary syndrome. *Clin. Endocrinol. (Oxf).* 12, 177–207.
856. Rebar, R., Judd, H. L., Yen, S. S., Rakoff, J., Vandenberg, G., and Naftolin, F. (1976). Characterization of the inappropriate gonadotropin secretion in polycystic ovary syndrome. *J. Clin. Invest.* 57, 1320–1329.
857. Waldstreicher, J., Santoro, N. F., Hall, J. E., Filicori, M., and Crowley, W. F. Jr. (1988). Hyperfunction of the hypothalamic-pituitary axis in women with polycystic ovarian disease: indirect evidence for partial gonadotroph desensitization. *J. Clin. Endocrinol. Metab.* 66, 165–172.
858. Morales, A. J., Laughlin, G. A., Butzow, T., Maheshwari, H., Baumann, G., and Yen, S. S. (1996). Insulin, somatotropic, and luteinizing hormone axes in lean and obese women with polycystic ovary syndrome: common and distinct features. *J. Clin. Endocrinol. Metab.* 81, 2854–2864.
859. Guzick, D. S. (2004). Polycystic ovary syndrome. *Obstet. Gynecol.* 103, 181–193.

860. Franks, S., and White, D. M. (1993). Prevalence of and etiological factors in polycystic ovarian syndrome. *Ann. N Y Acad. Sci.* 687, 112–114.
861. Gambineri, A., Pelusi, C., Vicennati, V., Pagotto, U., and Pasquali, R. (2002). Obesity and the polycystic ovary syndrome. *Int. J. Obes. Relat. Metab. Disord.* 26, 883–896.
862. Burghen, G. A., Givens, J. R., and Kitabchi, A. E. (1980). Correlation of hyperandrogenism with hyperinsulinism in polycystic ovarian disease. *J. Clin. Endocrinol. Metab.* 50, 113–116.
863. Dunaif, A. (1997). Insulin resistance and the polycystic ovary syndrome: mechanism and implications for pathogenesis. *Endocr. Rev.* 18, 774–800.
864. Chang, R. J., Nakamura, R. M., Judd, H. L., and Kaplan, S. A. (1983). Insulin resistance in nonobese patients with polycystic ovarian disease. *J. Clin. Endocrinol. Metab.* 57, 356–359.
865. Dunaif, A., Segal, K. R., Futterweit, W., and Dobrjansky, A. (1989). Profound peripheral insulin resistance, independent of obesity, in polycystic ovary syndrome. *Diabetes* 38, 1165–1174.
866. Azziz, R., Ehrmann, D., Legro, R. S., Whitcomb, R. W., Hanley, R., Fereshetian, A. G., O'Keefe, M., and Ghazzi, M. N. (2001). Troglitazone improves ovulation and hirsutism in the polycystic ovary syndrome: a multicenter, double blind, placebo-controlled trial. *J. Clin. Endocrinol. Metab.* 86, 1626–1632.
867. Ganie, M. A., Khurana, M. L., Eunice, M., Gulati, M., Dwivedi, S. N., and Ammini, A. C. (2004). Comparison of efficacy of spironolactone with metformin in the management of polycystic ovary syndrome: an open-labeled study. *J. Clin. Endocrinol. Metab.* 89, 2756–2762.
868. Lord, J. M., Flight, I. H., and Norman, R. J. (2003). Metformin in polycystic ovary syndrome: systematic review and meta-analysis. *BMJ* 327, 951–953.
869. Kiddy, D. S., Hamilton-Fairley, D., Bush, A., Short, F., Anyaoku, V., Reed, M. J., and Franks, S. (1992). Improvement in endocrine and ovarian function during dietary treatment of obese women with polycystic ovary syndrome. *Clin. Endocrinol. (Oxf).* 36, 105–111.
870. Saladin, R., De Vos, P., Guerre-Millo, M., Leturque, A., Girard, J., Staels, B., and Auwerx, J. (1995). Transient increase in obese gene expression after food intake or insulin administration. *Nature* 377, 527–529.
871. Kolaczynski, J. W., Nyce, M. R., Considine, R. V., Boden, G., Nolan, J. J., Henry, R., Mudaliar, S. R., Olefsky, J., and Caro, J. F. (1996). Acute and chronic effects of insulin on leptin production in humans: studies in vivo and in vitro. *Diabetes* 45, 699–701.
872. Brzechffa, P. R., Jakimiuk, A. J., Agarwal, S. K., Weitsman, S. R., Buyalos, R. P., and Magoffin, D. A. (1996). Serum immunoreactive leptin concentrations in women with polycystic ovary syndrome. *J. Clin. Endocrinol. Metab.* 81, 4166–4169.
873. Chapman, I. M., Wittert, G. A., and Norman, R. J. (1997). Circulating leptin concentrations in polycystic ovary syndrome: relation to anthropometric and metabolic parameters. *Clin. Endocrinol. (Oxf).* 46, 175–181.
874. Rouru, J., Anttila, L., Koskinen, P., Penttila, T. A., Irjala, K., Huupponen, R., and Koulu, M. (1997). Serum leptin concentrations in women with polycystic ovary syndrome. *J. Clin. Endocrinol. Metab.* 82, 1697–1700.
875. Laughlin, G. A., Morales, A. J., and Yen, S. S. (1997). Serum leptin levels in women with polycystic ovary syndrome: the role of insulin resistance/hyperinsulinemia. *J. Clin. Endocrinol. Metab.* 82, 1692–1696.
876. Mantzoros, C. S., Dunaif, A., and Flier, J. S. (1997). Leptin concentrations in the polycystic ovary syndrome. *J. Clin. Endocrinol. Metab.* 82, 1687–1691.

Knobil and Neill's Physiology of Reproduction, Third Edition
edited by Jimmy D. Neill,
Elsevier © 2006

CHAPTER 48

Stress and the Reproductive System

Michel Ferin

What Is Stress?, 2627
The Classical Concept, 2627
The Current Concept of Stress, 2628
The Clinical Consequences of Stress, 2628
The Stress Response, 2629
The Hypothalamic–Pituitary–Adrenal (HPA) Axis, 2629
The Sympathetic Nervous System, 2631
Brain–PVN Interactions, 2632
Interaction between the Immune System and the HPA Axis, 2633
Stress and the Reproductive System, 2633
Effects of Stressors on the HPG Axis, 2634
Mechanisms Whereby the HPA Axis Interferes with the Reproductive System, 2644
Modulation of the HPA and HPG Axes Response to Stressors by Gonadal Steroids, 2656
Stressors and Development, 2660
A Paracrine Role for CRH in the Uterus and Placenta, 2663
Metabolic-Related Suppression of the HPG Axis: Potential Role of HPA, 2665
General Conclusions, 2666
Acknowledgments, 2667
References, 2667

WHAT IS STRESS?

The Classical Concept

A critical advance in the scientific exploration of stress occurred nearly 150 years ago when Claude Bernard introduced the concept of the "milieu intérieur," or the internal physiologic environment, and established the principle of a dynamic internal physiological equilibrium that is essential to sustain the organism in an external environment typified by its variability (1,2). In the 1930s, Cannon defined this internal equilibrium as "homeostasis" and used the physical terms of strain and stress to describe a modification to the initial state that challenges or overwhelms the homeostatic state (3). The mechanism required to maintain homeostasis was defined as one involving the autonomic nervous system through the release of catecholamines and Cannon identified the acute "fight or flight" response to a stress challenge. Selye went further to more specifically identify the response to stress (4–6). He named the strain/stress-causing agents "stressors" and extended the stress concept significantly by including a crucial role for the hypothalamic–pituitary–adrenal (HPA) endocrine axis in the overall response to a stressor. He proposed that, in addition to specific homeostatic and local responses caused by demands to the body, there was a single overall stereotypic stress response common to all stressors which he referred to as the general adaptation syndrome: stress is thus described as the state in which the general adaptation syndrome has been evoked and in which an increased demand requires the body to perform adaptive functions to reestablish normalcy. Selye described this syndrome as unfolding in several stages: first is the alarm reaction, in which the stressor alters the homeostatic process and is rapidly counteracted by the response of the sympathetic–adrenomedullary axis through the release of catecholamines. Second is a resistance

Department of Obstetrics and Gynecology and the Center for Reproductive Sciences, College of Physicians and Surgeons, Columbia University, New York, New York.

stage which involves the activation of the HPA axis and the release of corticosteroids from the adrenal cortex. If the stressor persists, additional mechanisms allow the organism to achieve increased adaptation to the continuing effects of the stressor. In a final stage, exhaustion and eventual death occur if the stress persists.

The Current Concept of Stress

Some aspects of Selye's theory of stress have since been extended. Despite Selye's emphasis on the centrality of the HPA axis in the response to stressors and although an elevation in cortisol has become the benchmark response to a stressor, it is now known that the adaptation response to stressors requires the participation and synchrony of multiple central neuroendocrine and peripheral pathways (1). Furthermore, despite the fact that Selye defines stress as the nonspecific response of the body to any demand through the activation of both the sympathetic-adrenomedullary system and the HPA axis, there are several examples of heterogeneity of the neuroendocrine response to different stressors, particularly in the initial phase of the response and in severe stress, which suggests that responses can to a certain degree be stressor specific (1,7–10).

Chrousos and other investigators (11–15) have further clarified the concept of stress and summarized it as follows: the internal dynamic equilibrium that is *homeostasis* is constantly challenged by intrinsic or extrinsic adverse forces or *stressors*. This state of challenged homeostasis is identified as *stress*. Equilibrium is reestablished by a series of physiological and behavioral *adaptive responses*. These are the results of the integration of a diversity of neurosensory and blood-borne signals which activate distinct central neural pathways and lead to a cluster of time-limited behavioral and physical changes. The brain's ability to mobilize the response is paralleled by activation of the corticotropin-releasing hormone (CRH) and vasopressin (VP) systems in several brain nuclei and by stimulation of the norepinephrine system in the locus coeruleus. These systems perpetuate one another, interact with several other transmitter pathways in the brain and directly activate the HPA axis and the autonomic nervous system. The resultant changes improve the adaptive response and survival by, among others, increasing vigilance and cognition and by redirecting intermediate metabolism and energy to the central nervous system (CNS) or to stressed areas by accelerating cardiovascular tone. Nonurgent tasks are postponed. Successful adaptation is usually specific to a stressor, but may become relatively nonspecific if a stressor of any kind exceeds a threshold in magnitude or chronicity. When the adaptive response is insufficient or inadequate to restore the organism to a healthy state, pathology or disease may ensue.

Some investigators have introduced the model of *allostasis* in an attempt to focus on long term adaptive responses to stressors, distinguishing between systems that maintain homeostasis or stability within the confines of daily routine and those that maintain stability as the environment or life history change leading to adaptation of the organism (16,17). The cumulative physiological behaviors in response to challenges and changing environments are referred to as the allostatic load. The adaptive responses to daily demands and to unpredictable events may lead to allostatic overload that may exceed the capacity of the individual to cope. Serious physiopathology due to the damaging consequences of overproduction or dysregulation of mediators can then occur if overload is not relieved (17). Other investigators prefer to emphasize the notion of a *central neural network* that regulates the stress response and the process of the stressor-activated central neural response, including the need to clearly identify each stressor (15).

The Clinical Consequences of Stress

Over the years, the clinical emphasis on stress has, understandably so, been focused on its dangers and on the damage that it can produce. Overactivity of the HPA axis is a frequent symptom in widespread diseases, such as diabetes mellitus, obesity, hypertension, and psychiatric disorders, perhaps in part reflecting the increased cumulative load of lifetime stress (17–21). Important features of a successful response to stressors are that physiological systems are not only turned on rapidly but that they also be turned off when the stressor ceases, and that these physiological systems become blunted in the response to repeated or persistent stressor(s) (22). A failure to turn off the response or a lack of habituation may result in various stress-related diseases, such as in the posttraumatic stress disorder and other psychosomatic and psychiatric disorders (22–24). In these situations, systems which are not of immediate importance to the organism, such as the reproductive system, may be shut down in order to redirect energy to vital functions.

While biologists and doctors are more concerned with sources of stress that are concrete and observable and can otherwise be considered as resulting in illness, social and behavioral scientists deal more with sources of stress that represent information that arises from outside the person and is mediated by higher centers of the central nervous system (25,26).

The importance of the psychological response to stressors was in fact recognized early on by Cannon (27), and it is clear from the literature that psychological stressors can provoke endocrine alterations that are typical of the stress response and that potentially can also lead to disease, witness for example the deleterious effects of "psychogenic stress" on the reproductive cycle (see below). In these situations, it may become possible in the near future to identify determinants of the marked inter-individual variations that characterize the response to psychogenic stress. For example, studies in a rhesus monkey colony have demonstrated that several specific fearful-anxious behaviors in infant rhesus monkeys are heritable within this colony and that these phenotypes may represent the genetic liability for anxiety and depression which may influence the individual response to stressors (28). Polymorphisms in the glucocorticoid receptor gene have been shown to have an impact on the HPA response to psychosocial stress and may also contribute to individual vulnerability in HPA-related clinical diseases (29).

It is also important to point out that not all effects of stress are nefarious in the long term and, as already pointed out by Selye in his later studies in which he coined the word "eustress," there are also evident benefits of the immediate stress response, such as the typical increased heart rate for speed, increased glucose synthesis for energy, increased breathing for gas exchanges, and increased muscle tension for endurance (6). Even adverse effects on the reproductive system and its eventual shut down under stress may be considered as beneficial, since pregnancy under adverse conditions may evidently not be a desired outcome. It is also worthwhile to consider the stress response in relation to evolution in that it has been shaped by natural selection to increase the ability of organisms to cope with situations that require action or defense, and thereby provides a selective advantage to the individual (30).

THE STRESS RESPONSE

Stressors primarily activate both the HPA endocrine axis and the sympathetic nervous system (11,14,31–34). The response of these two systems induces the body to make the needed physiological and metabolic changes required to cope with the homeostatic challenge. Stressors activate norepinephrine neurons in the cortex, limbic system, and reticular formation which through sympathetic nerves stimulate the adrenal medulla. They also stimulate neuroendocrine pathways which through the release of neuropeptides from the hypothalamus and adrenocorticotropic hormone (ACTH) from the anterior pituitary activate the adrenal cortex to release glucocorticoids. There is also an important interaction between these pathways and the immune system, particularly when stress is produced by foreign substances or organisms (35,36).

The Hypothalamic–Pituitary–Adrenal (HPA) Axis

The Hypothalamic Paraventricular Nucleus (PVN)

PVN plays a pivotal role in the response to stressors. The neuroendocrine activation of the HPA axis by stressors involves the release of two neuropeptides, corticotropin-releasing hormone and vasopressin, from discrete populations of hypophysiotropic neurons in the parvocellular division of PVN (37,38). These two neuropeptides play a primary role in mediating the effects of stressors on the HPA axis.

CRH. CRH is a 41 amino acid residues peptide (39,40), its gene mapped to the long arm of chromosome 8 (41). CRH neurons located in the parvocellular division of PVN give rise to fibers that terminate in the neurohemal zone of the median eminence, from which CRH is secreted into the hypothalamic–pituitary portal circulation. CRH levels in hypophyseal portal blood are within the range shown to stimulate pituitary ACTH and β-endorphin release from the pituitary gland (42). Stressors of several types have been shown to first induce *c-fos* then to stimulate CRH gene expression within neurons of the parvocellular division of PVN (43–51).

In response to stressors, CRH is released rapidly from hypothalamic neurons in vitro in rodents and monkeys (52–54) and its levels in hypophyseal portal blood increase as well demonstrated in the sheep and rodent (35,55–57). In the monkey, the cytokine interleukin-1 (IL-1) also stimulates CRH release into the cerebrospinal fluid (52). Lesions of PVN in the rodent markedly attenuate CRH immunoreactivity in the median eminence and the plasma ACTH response to stress (58). Administration of CRH antagonists or immunoneutralization of endogenous CRH in the monkey or rat block or blunt the ACTH and corticosteroid responses to a variety of stress challenges (33,35,59,60).

Overall, the data demonstrate a critical role of PVN and of CRH in the regulation of ACTH release from the anterior pituitary during stress. Lewis rats, which have a genetic defect in the hypothalamic production of CRH (but also decreased vasopressin) (61), respond to a restraint challenge with a significantly lower ACTH and corticosterone release (62). Adult CRH knockout mice have low glucocorticoid levels and after adrenalectomy their plasma ACTH level fails to rise to that of wild-type animals (63). These animals also have a marked impairment, but not a complete block, of the ACTH and corticosterone response to stressors, such as restraint (64), fasting (65), or insulin-induced hypoglycemia (66), a further proof that CRH is required for the normal pituitary and adrenal responses to stressors (67,68).

VP. There is ample evidence to indicate that VP is co-localized with CRH within perikarya of PVN neurons, as demonstrated both in the human and rodent (69,70). This suggests that a single population of cells can produce both CRH and VP (71–73). The percentage of co-localization varies but can be greater than 70% after adrenalectomy. [VP is the same nonapeptide also known as antidiuretic hormone that is released from the magnocellular system in the hypothalamus and secreted from the posterior pituitary into the peripheral circulation in response to osmotic challenges (74).]

An active role for VP in the control of the adrenal axis and in the stress response process is suggested by increased VP (as well as CRH) mRNA levels and increased number of VP-expressing parvocellular neurons after a stress challenge (75), by the increased co-release of both neuropeptides into hypophyseal portal blood after adrenalectomy and in acute stress (55,76) and by increased cortisol levels following VP administration (77). A marked shift toward VP production is especially apparent after various types of chronic or repeated stress (78–80).

Overall, VP certainly plays a supporting role in the stimulation of the peripheral HPA axis, a role which may differ depending of the type or chronicity of the stress challenge. However, it appears unlikely that VP can entirely take over the function of CRH since, although VP synthesis is increased in the CRH knockout mouse (63), it cannot completely compensate for the loss of CRH (64–66). Administration of a VP antagonist alone in the monkey does not interfere with the normal cortisol response to an immune/inflammatory stress challenge (77). Studies in the VP-deficient Brattleboro rat remain inconclusive: while they appear to suggest that VP may be necessary for a full ACTH response, they do not indicate whether VP is indispensable or not for the normal activation of HPA by stressors (81–83). It is always possible, however, that the long-term absence of VP in these animals is compensated by other mediators.

Potential Effects of Other Proteins on the HPA Axis. The urocortins: The CRH family of peptides also includes urocortin, a 41 amino acid peptide originally cloned from the rat midbrain but also detected in the human and other animal species, as well as urocortin II and urocortin III (both 38 amino acid peptides) (84–90). Urocortin has a 43% homology to CRH, but the urocortins are more potent activators of CRH receptors (CRH-R) than CRH, particularly for CRH-R2 (89,91).Urocortin mRNA is expressed in discrete regions of the rat brain, and is predominant in the Edinger-Westphal nucleus, with lower expression in some motor nuclei and in the lateral hypothalamus and supraoptic nucleus (86). Urocortin has been shown to be involved in modulating different biological functions, such as blood pressure through its sympathomimetic action, cardiac contractility and inflammatory processes (89,92–94).

Urocortin has a more potent action than CRH in releasing ACTH both from rat anterior pituitary cell cultures (86) and from the intact male rat or sheep (95,96). However, although urocortin is a more potent ACTH secretagogue than CRH, it does not appear to be a major endogenous regulator of pituitary ACTH secretion under basal conditions or after stressors: for example, antiserum to urocortin does not prevent the rise in ACTH levels elicited by electrofootshocks (96), although presently it cannot be excluded that the peptide may be involved in the specific response to other stressors. Yet, unlike CRH-R1 or CRH-R2–deficient mice, urocortin-deficient mice exhibit normal anxiety-like behavior as well as autonomic regulation in response to stressors, suggesting that urocortin does not play an essential role in stress-induced behavioral and autonomic responses (97).

Cocaine- and amphetamine-regulated transcript (CART): CART is widely distributed throughout the central nervous system and has been implicated in stress-related behaviors and in feeding (98,99). Although both CART and CRH are found in PVN, few of the CRH-containing cells are immunoreactive to CART (100). However, administration of CART activates central CRH neurons in PVN (101,102), CRH release from hypothalamic explants (102) and the release of ACTH and cortisol (103).The latter effect is exerted centrally and not through direct activation of pituitary corticotropes and can be attenuated by administration of a CRH antagonist (103). Thus, CART appears to regulate HPA activity through CRH-dependent central mechanisms.

The Pituitary Gland

Adrenocorticotropic hormone is the hormone released by the anterior pituitary in response to CRH and VP stimulation. Although CRH and VP have ACTH-releasing properties when administered alone (104–106), CRH and VP clearly act in synergy to promote ACTH release: in the rodent and in the human, combined neuropeptide administration raises plasma

ACTH to levels that are higher than when each compound is given separately (107–109).

ACTH is a 39 amino acid peptide derived from a larger precursor protein, pro-opiomelanocortin (POMC). This precursor molecule is synthesized principally in the corticotropic cells of the anterior pituitary and in the arcuate nucleus of the hypothalamus (110). On processing, POMC yields several products (111), which in the pituitary include β-lipotropin (β-LPH) and ACTH, with β-LPH being further cleaved to yield in varying degrees β-endorphin and γ-LPH (110). In fact, both CRH and VP are secretagogues for the release of multiple products derived from POMC in the pituitary gland (112). It is important to note that POMC processing occurs to different extents in various tissues, which, in turn, determines the diversity of biological activities. As detailed later, hypothalamic posttranslational processing of POMC may represent a step in the processes that control the response of the reproductive axis to stress.

Two G-protein–coupled CRH receptors, CRH-R1 and CRH-R2, were originally cloned and are encoded by distinct genes (113). Both bind CRH and urocortin with high affinity (114) and the CRH-R1 gene is also under the influence of both CRH and urocortin (115). Current consensus is that CRH-R1, which is found on pituitary corticotropes, is the primary receptor that mediates stressor-induced increases in ACTH release and is a key mediator of anxiety in the limbic system (116,117). CRH-R1 knockout mice have a severe impairment of stress-induced ACTH release and display decreased anxiety-like behavior (118–120). The role of CRH-R2 remains more subtle and may be part of a functionally different circuit involved with the urocortins (88). CRH-R2 knockout mice appear to have deficiencies in the homeostatic response to stressors displaying hypersensitivity to stressors and anxiogenic-like behaviors (118,121,122). The endocrine phenotype of mice lacking both CRH-R1 and CRH-R2 is dominated by the functional loss of CRH-R1 (119).

VP is also known to act through two different types of receptors (V1 and V2), but its action on the corticotropes and on ACTH release is thought to occur primarily through V1 receptors (123).

The Adrenal Glands

Circulating ACTH is the key regulator of glucocorticoid synthesis and release by the adrenal gland. In the primate, the synthesis of glucocorticoids occurs mainly in the zona fasciculata (and to a lesser degree the zona reticularis) of the adrenal cortex. Cortisol is the dominant glucocorticosteroid in the primate and corticosterone is in the rodent (34). Glucocorticoids are the final effectors of the HPA axis: hypersecretion of CRH, such as that observed in a transgenic mouse line overexpressing CRH exclusively in neural tissues, results in the hypertrophy of the adrenal gland and considerably elevated basal plasma corticosterone concentrations (124). Glucocorticoids, through activation of their ubiquitous receptors (125), participate at multiple levels in the organization of the body's response to stressors and redirect bodily energy sources through energy mobilization and gluconeogenesis in the liver (126,127).Cardiac output and respiration are accelerated and blood flow is redirected to provide energy to the brain, heart, and muscles (12). These restorative processes are generally catabolic in nature and, if extended, can take a powerful toll on the organism. As mentioned previously, glucocorticoid hypersecretion is known to accompany numerous long-term metabolic, affective, and psychotic states (18–21,128).

Cortisol (or at least the "free" fraction unbound to corticosteroid- binding globulin) or corticosterone act in a negative (or inhibitory) feedback mode to reduce CRH and ACTH release, thereby participating in the body's homeostasis by terminating the stress response (129–131). Administration of glucocorticoids significantly blunts stress-induced CRH mRNA expression in PVN but does not appear to decrease the excitatory neural input to PVN as reflected by c-fos expression, suggesting that the negative feedback loop directly involves cells intrinsic to the HPA axis (132).

The Sympathetic Nervous System

The sympathetic nervous system responds rapidly to stressors and is an integral part of the stress response and of the "fight or flight" reaction that is evoked by its stimulation (12,133). Stimuli evoked by stressors are sensed in various higher levels of the sympathetic nervous system and responses are initiated in the brainstem and hypothalamus (34). One particularly relevant area is the locus coeruleus which gives rise to efferent pathways that provide a major source of norepinephrine to the brain, including the hypothalamus and PVN (134–136). Stressors have been shown to activate norepinephrine neurons in various areas of the brain and, through the sympathetic nerve, to promote epinephrine release from the adrenal medulla (137–140). Catecholamines released from the medulla in response to activation of the sympathetic nervous system act in turn and in concert with the glucocorticoids to increase glucose by activating glycogenolysis and to raise cardiac output thereby facilitating delivery of substrates to tissues that are critical to the immediate defense of the organism (141). Sympathetic neurons richly innervate the vasculature, heart, skeletal muscles and other organs. Adrenergic stimuli to other brain centers induce a general state of arousal and vigilance and behavior useful in the response to stressors (34). [Studies in CRH knockout mice suggest that the HPA axis and CRH may also play an important role in controlling gene expression of catecholamine biosynthesizing enzymes in the adrenal medulla (142).]

Brain–PVN Interactions

The hypothalamic paraventricular nucleus interacts with other parts of the hypothalamus and of the brain through multiple direct and indirect pathways (143).

Direct Pathways to PVN

The parvocellular portion of PVN, where CRH- and VP-containing neurons are located, receives synaptic innervation from a relatively circumspect set of CNS structures, in general within regions known to receive inputs from somatic nocireceptors, visceral afferents and humoral sensory pathways (33,144). A norepinephrine input originates from the brainstem and is a major HPA modulatory pathway (145,146). The brainstem region appears to be a common site of integration of the HPA response as it shows early gene c-fos activation following several types of stressors (44,147). As demonstrated by studies showing norepinephrine release in PVN after different stressors, this input is mostly excitatory (7,76). Reciprocal neural connections also exist between CRH neurons in PVN and norepinephrine neurons in the locus coeruleus (11,13,148,149). These two systems appear to have a mutually reinforcing stimulatory effect: norepinephrine activates CRH release and CRH increases adrenergic discharge. Micro infusion of CRH in the locus coeruleus increases its neuronal discharge rate and norepinephrine release, while that of a CRH antagonist does the opposite (150–152).

Serotonergic input to PVN also stimulates HPA and, although mostly indirect, there is also direct input from a small number of fibers in the raphe nuclei (153). Lesions of these nuclei decrease the HPA response to stressors, suggesting their involvement in the process (154). Additional direct input to PVN derives from areas in the bed nucleus of the stria terminalis which contain mainly γ-aminobutyric acid (GABA)-containing neurons inhibitory to PVN (155). By far the largest stress-activated direct GABA input to PVN, however, derives from inhibitory GABAergic neurons within the hypothalamus (155,156). These neurons are located within the immediate PVN region (156), the medial preoptic nucleus (157,158), and the dorsomedial hypothalamic nucleus (155). The latter also contains glutamate neurons projecting to PVN, which may provide an additional stimulatory input (159).

Indirect Pathways to PVN

There are several indications of indirect interactions between limbic brain structures, such as the prefrontal cortex, the hippocampus, and amygdala, and the HPA axis. This circuitry has been well reviewed (144). For example, neurons from the hippocampus activate GABAergic neurons in the bed nucleus of the stria terminalis which project to hypothalamic neurons that relay information to PVN (160,161). The central amygdaloid nucleus has numerous connections to brainstem structures that innervate PVN (162). The hippocampus and prefrontal cortex have been implicated in the inhibition of the HPA axis, while the amygdala appears to activate HPA (160,163–167). A major role of the hippocampus in regard to HPA may be its action in terminating the HPA response to specific stress challenges (168). Activation of the amygdala is important for the retrieval and emotional analysis of relevant information for any given stressor and for anticipation of negative events, which when chronic may lead to allostatic load (11,144,169). A variety of psychopathologies may be tied to neurohormonal signals activating the amygdala and the linkage between descending limbic pathways and the PVN. However, conflicting data remain which in part reflect varying experimental designs, animal strain differences and/or the subject's past experience, and thus the precise role of these areas in the response to stressors remains to be elucidated.

Overall, as proposed by Herman et al. (144), stressors which represent a genuine homeostatic challenge may primarily activate direct input pathways to PVN, while the influence, for example, of prior experience to a stressor on the HPA response to a new challenge may require polysynaptic input from limbic-associated structures.

Role of CRH in Extrahypothalamic Areas

CRH and its receptors are also present in many extrahypothalamic areas, and data suggest the involvement of extrahypothalamic CRH in the mediation of complementary neuroendocrine, autonomic and behavioral responses to stress (170,171). Although its presence is minimal in PVN, CRH-R1 expression rapidly increases following a stress challenge (172,173). Changes in CRH expression in response to stressors also occur in many extrahypothalamic brain areas, including the cerebral cortex, the limbic system, the cerebellum, and the spinal cord (46,174). CRH neurons in several regions of the bed nucleus of the stria terminalis are known to project to PVN (175–177). CRH itself activates early gene induction and CRH gene transcription in PVN (178,179), while intracerebroventricular injection of a CRH antagonist decreases c-fos mRNA expression in PVN (180). CRH neurons are also found in areas mediating stress-related autonomic and behavioral responses, such as in the basal telencephalon, brain stem, and lateral hypothalamic/perifornical area (181). Thus, extrahypothalamic CRH may also be involved in coordinating the autonomic nervous system response to stressors and evidence suggests a role for CRH as a neurotransmitter in the locus coeruleus that mediates norepinephrine release by stressors (182) (also see above). In a chronic stress-induced model of depression in the rat, CRH and CRH-R mRNA in the locus coeruleus and plasma ACTH are both increased (183).

Brain glucocorticoid receptors also participate in the processes whereby extrahypothalamic regions modulate the response to stressors. Glucocorticoids inhibit expression of their own receptors and of the CRH gene, an action which in the normal negative feedback mode influences the magnitude of HPA suppression after the initial response (161,184). Significantly and in an opposite way, glucocorticoids elevate rather than decrease CRH gene expression in the amygdala and in the bed nucleus of the stria terminalis (as in the placenta) (see below), an action which may signal adversity (169). Thus, overall, glucocorticoids have adaptive effects in the short run but promote allostatic load when there is repeated stress (185). Structural plasticity of several brain areas, including the hippocampus, may play a role in repeated or chronic stress, first as an adaptive mechanism, later as damage occurs when imbalance sets in (186,187): at that point, physiological elevations of glucocorticoids and stress itself may exacerbate hippocampal neuron loss and the attendant degenerative markers (188,189). [It is worth noting also that polymorphisms in the glucoreceptor gene have been found to be associated with individual vulnerability to HPA-related clinical states such as those associated with psychosocial stress (29).]

Interaction between the Immune System and the HPA Axis

Although both the HPA axis and the immune system were long considered to be independent of each other, they are now known to communicate extensively and bidirectionally via common ligands and receptors, multidirectional signals, and feedback regulations of neuroendocrine- and immunocyte-derived mediators (36,190–192). Thus, in addition to the HPA endocrine axis, the immune system also plays a role in restoring homeostasis. A variety of cytokines, released by macrophages and lymphocytes upon activation by an antigen, are essential actors in the early immune response (193,194). Not only peripheral, but also brain and hypothalamic cytokines increase during infection or after endotoxin administration (195–202). During times of infectious challenges or after endotoxin administration, the immune system stimulates the HPA axis and increases the secretion of glucocorticoids (43,60,203–206). The cytokine IL-1 appears at least in part responsible for this effect primarily by stimulating the activity of CRH neurons in the parvocellular PVN and the release of CRH (35,43,46,47,52,57,207–212). Immunoneutralization of CRH or CRH antagonists block this stimulatory effect of IL-1 on the HPA axis (60,213), while IL-1 R1 knockout mice exposed to mild psychological and metabolic stressors display a diminished HPA response (214). A role for VP in this process has also been postulated (215). Another cytokine, IL-6, also activates HPA by stimulating ACTH and especially the glucocorticoid response (216) and may provide a compensatory mechanism in CRH and CRH-1R knockout mice allowing them to respond to an immune challenge (63,96,217,218).

Glucocorticoids play a crucial role in the endogenous mechanisms that prevent exaggerated responses during an immune/infectious challenge by acting in an inhibitory-like feedback fashion to inhibit proinflammatory signaling and gene expression (219–221). This essential control mechanism by the glucocorticoids also exists within the CNS suggesting that glucocorticoids play a major role in protecting the brain. For example, while endotoxin increases the transcription of genes involved in the immune response, this action does not alter neuronal integrity unless animals are also treated with a glucocorticoid receptor antagonist. In this case, the inflammatory reaction is prolonged and toxicity to the brain is increased (222).

Figure 1 schematizes some of the main pathways responsible for the primary response to stressors.

STRESS AND THE REPRODUCTIVE SYSTEM

There are numerous physiological alterations which occur in response to stressors and which are essential in modifying the internal milieu to allow for the body response to the stress challenge.

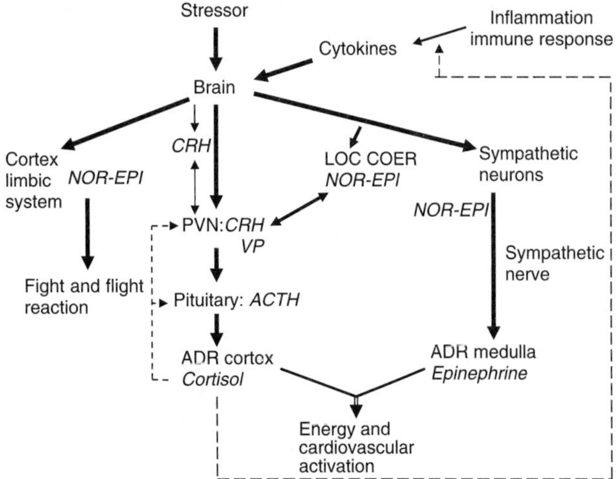

FIG. 1. Some of the primary pathways mediating the response to stress. Various stressors, including immune/inflammatory challenges through the release of cytokines, primarily act on the brain to activate both the HPA endocrine axis and the sympathetic nervous system. The response of these two systems induces the body to make the needed physiological and metabolic changes required to cope with the homeostatic challenge. Acting on the HPA axis, stressors stimulate the paraventricular nucleus of the hypothalamus (PVN) to secrete corticotropin releasing hormone (CRH) and vasopressin (VP) which activate the release of ACTH from the anterior pituitary and glucocorticoids (cortisol in the primate) from the adrenal cortex. In turn, cortisol acts in a negative feedback mode to reduce CRH/VP and ACTH release as well as the inflammatory response, thereby participating in the body's homeostasis by terminating the stress response. Stressors activate the locus coeruleus (LOC COER) and sympathetic neurons to release central norepinephrine (nor-epi) and through sympathetic nerves epinephrine from the adrenal medulla. Catecholamines in concert with cortisol activate energy production and raise cardiac output to facilitate delivery of substrates to tissues that are critical to the immediate defense of the organism. Reciprocal interactions exist between CRH neurons in PVN and nor-epi neurons in the locus coeruleus as well as between extrahypothalamic CRH neurons and PVN.

However, one of the "side effects" of the stress condition is that "nonessential" tasks may be postponed during stress. It is thus that in response to stressors there may be effects on the reproductive system which are usually of an inhibitory nature and which may increase infertility rates (223–225). The pathways and mechanisms, whether central or peripheral, whereby stressors may interfere with normal reproductive function are numerous and intricate and still remain to be completely elucidated.

The reproductive system is under the control of the hypothalamic–pituitary–gonadal (HPG) axis, which is described in considerable details in other chapters of this book (226). In brief, the pituitary–gonadal portion of this endocrine axis is driven by its hypothalamic component responsible for the release of gonadotropin-releasing hormone (GnRH). GnRH is present within several hypothalamic nuclei and released into the hypophyseal portal blood to stimulate the release of the two gonadotropins from the anterior pituitary gland, follicle-stimulating hormone (FSH) and luteinizing hormone (LH). The gonadotropins then act on the ovaries and testes to stimulate gonadal steroid secretion. Significantly, GnRH is released from the hypothalamus in a pulsatile fashion, hence the concept of the GnRH pulse generator and the fact that tonic secretion of gonadotropins is pulsatile in nature. A pulsatile GnRH stimulus is required to increase transcription of the gonadotropin subunit genes and for the proper release of gonadotropins. Conditions that interfere with the function of the GnRH pulse generator will disrupt the pituitary–gonadal axis: complete absence of GnRH pulsatility will result in a dormant reproductive axis while lesser abnormalities, such as a decrease in the GnRH pulse frequency (as occurs for example during stress; see below), may interfere with its proper physiologic function. Constant communication among the several levels of the reproductive axis through feedback loops is essential so that normal function is maintained at all levels of the HPG axis. Important is the negative (or inhibitory) feedback loop, whereby estradiol or testosterone secreted by the target organs, the ovaries or the testes, feed back to the hypothalamic–pituitary unit to adjust GnRH and/or LH and FSH secretion. If this feedback loop is suddenly interrupted, such as after castration, GnRH, FSH, and LH secretion will increase in a (futile) attempt to maintain proper gonadal function. Stressors potentially can influence the HPG axis at any level of control, as will be reviewed below. It is probable, however, that one of the primary and predominant impact is on the brain and specifically on the GnRH pulse generator, although the precise role of extrahypothalamic areas remains to be studied.

Effects of Stressors on the HPG Axis

Although "stress" is a particularly difficult topic to analyze in a systematic fashion, a general overview of the large literature that deals with the relationships between the HPA and the HPG axes allows to define general mechanisms that control this relationship. These mechanisms will be reviewed in the human and in the nonhuman primate (which has a similar reproductive cycle), as well as in other relevant species, such as sheep and rodents. There are obvious difficulties in identifying stressors that are relevant, quantifiable, and especially reproducible (i.e., in which the individuality of the response is minimized), so that comparative experimental protocols can be developed. The most frequently used stressors in laboratory animals have been endotoxin or cytokine administration to induce an inflammatory/immune like challenge, restraint, or insulin-induced hypoglycemia. It should be kept in mind that data obtained with a particular stressor may or may not be applicable to the specific stressor of concern to the reader. Furthermore, in response to a stressor multiple pathways and mechanisms may be activated or inhibited depending upon a large number of variables among which are the specific characteristics of the experimental protocol itself, the age, gender and endocrine status (e.g., the stage of the reproductive cycle) of the subject, and the nature of the challenge, whether it is acute, repetitive, chronic or combined with other stressors. Whenever possible, these variations will be pointed out below.

Inhibitory Effects of Stressors on Tonic Pulsatile Gonadotropin Release

An Inflammatory/Immune Challenge as the Stressor. As briefly reviewed above, there is a firm link between the immune system and the HPA axis: inflammatory/immune signals stimulate the release of a variety of cytokines and other factors which in turn activate the HPA axis. The stressor most frequently used to mimic an inflammatory/immune challenge is endotoxin [also referred to as lipopolysaccharide (LPS)], the toxic moiety of gram negative bacteria which induces physiopathological responses within the body similar to those observed after an acute bacterial challenge, without provoking infection itself (227). Endotoxin administration is widely used to mimic some of the events occurring during sepsis because it causes the endogenous release of both peripheral and central cytokines (198,228) which activate the HPA axis in a manner similar to that of other stressors.

In the rodent, peripheral LPS administration or intracerebroventricular (icv) administration of IL-1α or IL-1β, the major cytokines released by endotoxin, suppresses tonic pulsatile LH release in ovariectomized (ovx) and in orchiectomized rats, resulting in marked reductions in LH pulse frequency and/or LH pulse amplitude (205,229–234). Of interest is that, while both iv and icv injection of IL-1β significantly increases plasma ACTH levels, only icv administration measurably inhibits LH secretion in the rodent (43). IL-1β appears to be relatively more effective than IL-1α in that regard (231). Experiments in orchiectomized rats with specific blockers of IL-1 synthesis indicate that part of the LH suppression caused by icv administration of LPS is mediated via IL-1 synthesized within the brain. However, factor(s) other than IL-1 also participate in LPS-induced effects since administration of any single cytokine does not completely suppress LH secretion and since LH secretion cannot be completely restored to normal levels by administration of a single antagonist (229).

In the ovx and intact sheep, peripheral LPS administration also decreases LH release, LH pulse amplitude and/or LH pulse frequency (235,236). A higher dose of endotoxin is required to affect LH pulse frequency. With this dose of endotoxin, the inflammatory/immune challenge is able to prevent the development of high-frequency LH pulses required for a normal follicular phase (see below) (237). Inhibitory effects of LPS on LH secretion are also observed in rams (238–240) as well as in ovx Holstein heifers (241).

In the ovx monkey, both peripheral endotoxin administration and icv injection of IL-1α stimulate the HPA axis and the central release of CRH (52) and result in an acute inhibition of pulsatile LH and FSH release (60,242,243). However, while concomitant administration of an IL-1 receptor antagonist reverts the inhibitory action of IL-1, the inhibition of LH and stimulation of cortisol that follow LPS administration are not reversed, suggesting that, as in the rodent, the overall endocrine effect of LPS may reflect not only the release of IL-1 but also that of a cascade of other cytokines or factors (242).

Restraint as the Stressor. In the rodent, most studies show significant decreases in plasma LH accompanying increases in ACTH and corticosterone following restraint (whether acute or chronic) (244–249). An effect on FSH is more difficult to detect (perhaps because the longer half-life of the hormone), some studies reporting no change in plasma FSH concentrations, others marginal changes. Restraint also decreases the acute LH, but not FSH, response to the excitatory amino acid N-methyl-D,L aspartic acid (NMDA) (250) and in ovx estrogen-primed rats, suppresses the LH release response to NMDA (251).

In the gonadectomized sheep, pulsatile LH release is inhibited by acute confinement, with both LH pulse frequency and amplitude decreased (252,253). Several days of habituation to the same daily periods of confinement, however, minimize this inhibition and restore normal episodic discharges of LH. Transport (with the coexisting confinement) also decreases LH pulse frequency and/or amplitude in ewes, an effect most marked in the first 4 hours (254,255).

In male rhesus monkeys subjected to restraint in a primate chair, an immediate rise in plasma ACTH and cortisol is accompanied by a rapid fall in LH and testosterone secretion which remain suppressed for several hours after the animals are returned to their home cage (256). However, as shown in ovx estrogen-replaced marmoset monkeys, restraint alone has less effect on plasma LH in habituated subjects (257). In wild baboons, the acute stress of rapid capture and immobilization following anesthesia by phencyclidine darting also suppresses LH and testosterone concentrations (258).

Insulin-Hypoglycemia as the Stressor. In the rodent, insulin-induced hypoglycemia impairs LH secretory patterns and leads to a decrease in LH pulse amplitude and frequency in ovx steroid-primed rats as well as in orchiectomized rats (259–262). This inhibitory effect is readily reversed following restoration of an adequate glucose supply to the hypothalamus (259,263).

In castrated male and female sheep, insulin-induced hypoglycemia reduces LH concentrations. As in the rodent, this inhibitory effect is prevented by iv glucose infusions, suggesting that decreased LH secretion is related to neuroglycopenia (49,264).

In intact and in ovx monkeys, insulin-induced hypoglycemia also produces a decrease in LH (265–267). The stressor also induces a cortisol response, indicating a stimulation of HPA in the primate as also reported in other species.

In man, insulin administration produces a decrement in serum LH and testosterone concentrations, which are restored following glucose infusion, while FSH levels remain unaffected (268,269). The LH decrease appears to be generated by hypoglycemia in glucosensitive areas located inside the blood–brain barrier, while insulin per se is ineffective. In cycling women, estradiol, but strangely not LH, levels are reported to decrease within the first 2 hours after insulin-induced hypoglycemia with increases in cortisol levels evident at 30 minutes and maximal at 120 minutes after insulin injection (270).

Examples of the acute inhibitory effects of the above stressors on pulsatile LH release in the nonhuman primate are illustrated on Fig. 2.

Other Stressors. Other stressors have also been shown to impact on gonadotropin levels. Decreased plasma concentrations of LH and FSH in association with increased plasma concentrations of ACTH and corticosterone are observed in rats after repeated ether stress (271), cold (248,249), and forced muscular exercise (248,249), and in hens exposed to acute thermal stress (272).

Decreased testosterone (and presumably LH) levels are observed in feral baboons during periods of social instability (273). Yet, features of specific clusters of personality traits are difficult to relate to the cortisol response and to sexual behavior (274,275). In the human male, stress related to physical strain during endurance training courses, as identified by increases in plasma cortisol, results in a decrease in levels of LH, FSH, and testosterone (276). Basal tonic levels of FSH, LH, and estradiol are significantly lower in patients with the "functional hypothalamic chronic anovulation" syndrome, a syndrome in which it is assumed that chronic psychogenic stress plays a significant role (further discussed below) (277). Compared to normal cycling women, patients with this syndrome have reduced integrated 24-hour LH and FSH concentrations, reduced LH pulse frequency and increased LH interpulse interval (278,279).

Sites at Which Stressors Inhibit Tonic Pulsatile LH Secretion

Most stressors discussed above inhibit tonic pulsatile LH secretion. This effect could be speculated to be the result of a direct inhibition of the GnRH pulse generator or of a change in the sensitivity of the gonadotrope to GnRH stimulation.

Effects of Stressors on Brain GnRH Release. Direct measurements of GnRH release into hypothalamic-hypophyseal portal blood in ovx ewes provide good evidence that stressors can act at the hypothalamic level to inhibit the activity of the GnRH pulse generator. Indeed, the normal rhythmic release of GnRH and LH of about 1 pulse/hour in controls is either blunted, abolished, or becomes irregular following endotoxin administration at a dose causing significant cortisol release (Fig. 3) (235). (A frank decrease in GnRH pulse frequency, however, occurs only after a high dose of endotoxin.) Multiple unit electrical activity in the mediobasal hypothalamus, which is associated with LH pulses and representative of GnRH pulsatile activity, is also suppressed or shows prolonged intervals between volleys following administration of endotoxin in the ovx goat and rat (280,281). Insulin-induced hypoglycemia as a stressor also reduces pulsatile activity of the GnRH pulse generator both in the ovx monkey and in the estrogen-primed ovx rat, an effect that in the rat is immediately reverted by glucose infusion (263,282,283).

Indirect support to the above data is provided by the observation that foot shock in the anestrous ewe increases GnRH concentration in the median eminence and the medial preoptic area, presumably reflecting an inhibition of GnRH release from the hypothalamus by the stressor (284).

Effects of Stressors at the Pituitary Site. Data from several investigators suggest that stressors do not influence the pituitary LH response to GnRH stimulation. In the rodent, normal LH responses to exogenous GnRH administration are maintained during acute or chronic restraint or after icv injection of IL-1 (233,247,271). In the hen, reduced LH levels in response to heat are not related to a decreased LH response to GnRH (272). In ovx estrogen-primed rats, the suppression of the LH response to NMDA during restraint occurs in the presence of unchanged

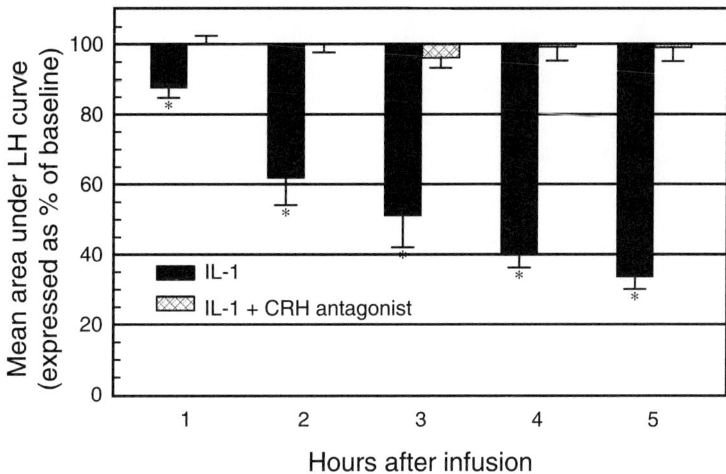

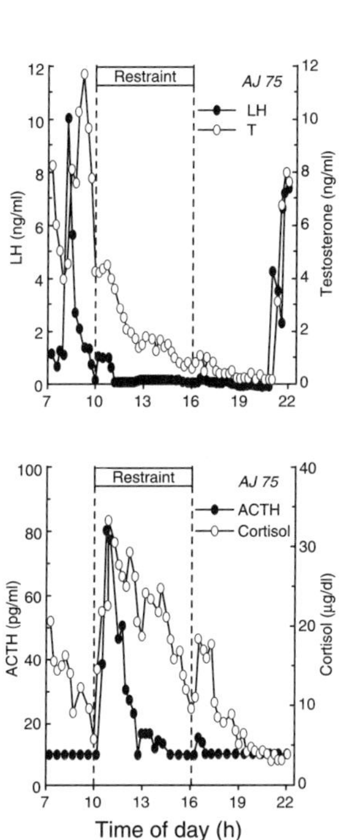

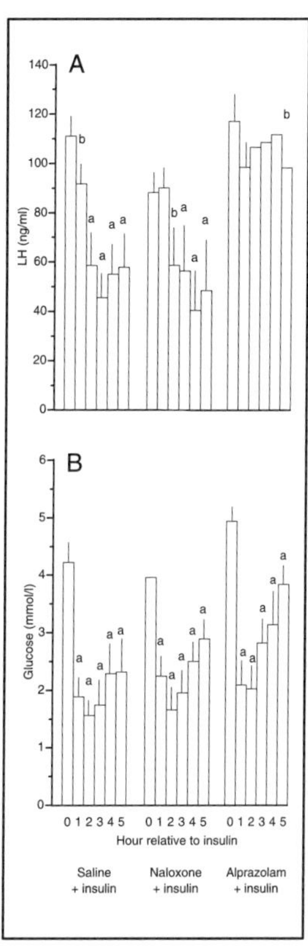

FIG. 2. Examples of acute inhibitory effects of stressors on pulsatile LH release in the nonhuman primate. **Upper panel:** An immune/inflammatory challenge, as reproduced by intracerebroventricular infusion of interleukin-1 (IL-1) (*black bars*). This inhibitory effect on LH is prevented by administration of a CRH antagonist (*gray bars*). [Reprinted with permission from (60), copyright 1991, The Endocrine Society.] **Lower left panel:** Restraint: LH (*closed circles*) and testosterone (*open circles*) release is inhibited *(upper figure)*, with resumption of normal secretion after removal from restraint. Restraint increases ACTH (*closed circles*) and cortisol (*open circles*) release (*lower figure*). [Reprinted with permission from (256), copyright S. Karger AG, Basel.] **Lower right panel:** Insulin-induced hypoglycemia: both LH (*upper figure*) and glucose (*lower figure*) decrease. The inhibitory effect on LH is not prevented by naloxone, an opiate antagonist (*second set of bars*), but is following alprazolam (a compound decreasing CRH activity) (*third set of bars*). [Reprinted with permission from (267), copyright S. Karger AG, Basel.]

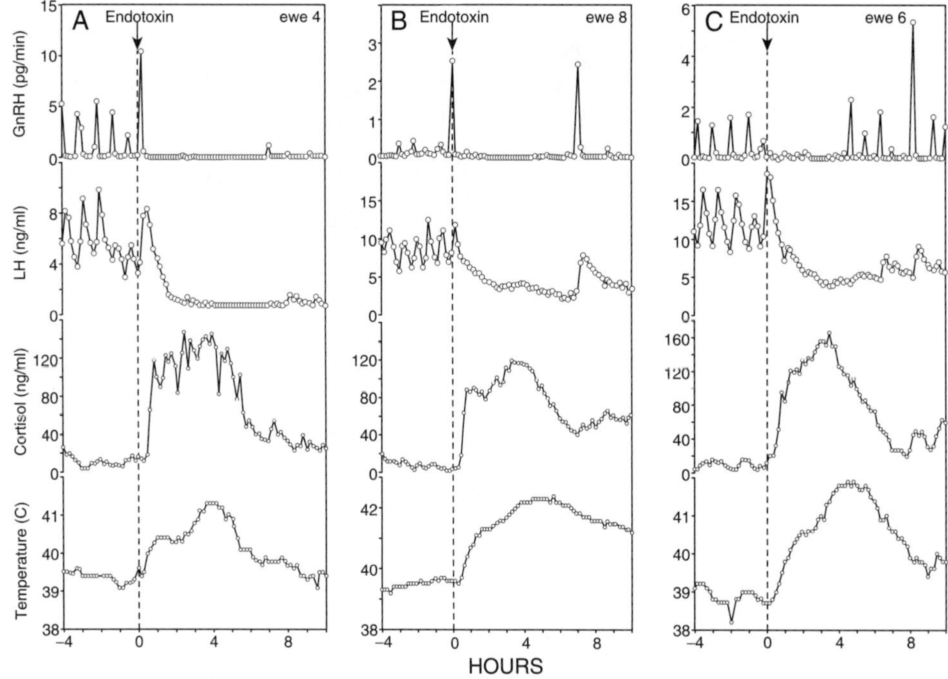

FIG. 3. Immune/inflammatory challenge, as reproduced by peripheral injection of endotoxin in the sheep: the *upper two panels* show the acute inhibitory effect on the release of pulsatile GnRH (in hypophyseal portal blood) and of pulsatile LH (in the jugular vein) in three different ewes. Both cortisol and core body temperature (*lower two panels*) are stimulated by endotoxin. [Reprinted with permission from (235), copyright 1997, The Endocrine Society.]

pituitary LH release in response to GnRH stimulation (251). In the monkey as well, there is no difference in LH responses to varying doses of GnRH between eu- and hypoglycemic animals following insulin administration (285).

In contrast, more recent studies in an ovx ewe model, in which endogenous GnRH secretion is blocked by treatment with ovarian steroids during seasonal anestrus, show that stressors may exert a more subtle influence on the pituitary since endotoxin administration suppresses the amplitude of the LH response to exogenously administered GnRH pulses (286). Furthermore, treatment with a lower dose of endotoxin, a dose that is still capable of inducing cortisol release but not of affecting GnRH release, can disrupt LH pulsatility. Similar reductions in pituitary responsiveness to exogenous GnRH in the sheep are also observed during combined restraint and isolation stressors in the nonbreeding (but not during the breeding) season (287). Observations that in the rodent endotoxin, besides affecting brain GnRH, also reduces GnRH receptor gene expression in the pituitary gland of cycling rats at proestrus also suggest the potential for an action by stressors at a pituitary site in this species (288).

Overall, major evidence suggests that the decrease in tonic pulsatile LH release by stressors is primarily related to an inhibitory action on the hypothalamic GnRH pulse generator. This conclusion is reinforced by observations (reported below) of a direct action of HPA neuropeptides on GnRH pulsatility. It is likely also that stressors may subtly influence pituitary responsiveness to GnRH, an effect that may possibly reflect an action of the corticosteroids (see below).

A Rapid and Transient Stimulatory Effect of Stressors on Tonic LH Release

As reported above, most acute stressors exert an inhibitory action on the HPG axis. However, LH and testosterone secretion during the very early stage of the acute stress response is less coherent. Several investigators have in fact consistently reported a transient *rise* in LH and testosterone in rodents after different stressors, such as handling, restraint, exposure to ether or to a novel environment, or visual, audiogenic, and thermal stressors (249,289–299). An early increase in LH and testosterone is also seen after restraint in cockerels (300). The rise in LH and testosterone is transient and occurs within the first hour of stress. It has been reported to occur in intact but generally not in castrated animals. In the intact male rat, the early increase in plasma LH concentrations parallels a decrease in hypothalamic GnRH content, suggesting that the elevation in LH may reflect an increase in GnRH release, although this has not been demonstrated (247). The increase in LH can be attenuated by prior glucocorticoid treatment, suggesting that this phenomenon may be sensitive to the degree of the HPA response to the stressor (290,301).

Short-term increases in LH and testosterone, not always in parallel, have also been reported in men but not women during physical exercise (302,303) and after surgery (304–306). In wild male baboons, the stress of rapid capture and restraint, in general found to be suppressive of LH and testosterone, can also produce a transient increase in testosterone during the first hour, but only in high ranking males. In contrast, in subordinate animals testosterone declines promptly without prior increase (307). Part of the distinctive pattern between these two types of animals may reflect a lower sensitivity of the testes to the suppressive effects of glucocorticoids in high ranking males. This may be mediated by sympathetic catecholamines, since administration of a sympathetic ganglionic blocker eliminates the transient rise in testosterone (307). In the rat as well, LH does not increase immediately after restraint in animals subjected to prior superior cervical ganglionectomy (308). Inhibition of serotonin synthesis also abolishes LH release caused by restraint (309).

Overall, there is no doubt that under certain circumstances a very transient stimulation of the HPG axis and of LH release may occur within the first hour of stress. The physiological significance of this phenomenon remains however unknown.

Effects of Stressors on the "Surge" Mode of LH Release

Besides acting to inhibit tonic pulsatile LH secretion in its negative feedback loop mode (see above), estradiol can also activate a positive (stimulatory) feedback loop, whereby high concentrations of estradiol stimulate LH release. This phenomenon occurs specifically during the reproductive cycle and is important in that it synchronizes the maturation of the graafian follicles, responsible for the estradiol signal and containing the soon-to-be ovulated oocytes, with the stimulus to ovulation, the preovulatory GnRH/LH surge (226,310,311). One of the potential mechanisms whereby stressors may interfere with the reproductive cycle (see below) would be by inhibiting the proestrus or midcycle estradiol-related gonadotropin surge responsible for inducing ovulation and corpus luteum formation. Investigators have explored the effects of stressors both on the spontaneous GnRH/LH surge at proestrus in the rodent or at the end of the follicular phase in larger animals, and on the estradiol-induced surge in the ovx animal. Indeed, the positive feedback loop can be experimentally mimicked with an infusion of estradiol above a threshold level representative of the species.

The most instructive data were obtained following an immune/inflammatory challenge, both in the rodent and sheep. In the rat, there is substantial evidence to indicate that stressors inhibit the GnRH/LH surge mechanism at proestrus. The percentage of GnRH neurons expressing the c-fos protein in the medial preoptic area (a critical GnRH-containing area of the hypothalamus) during the afternoon of proestrus is significantly decreased after icv infusion of IL-1β or after bilateral micro infusion of the cytokine in this area, as well as after peripheral endotoxin administration (36,312,313). Concurrently, GnRH primary transcript at 15:00 hours on proestrus is significantly attenuated by endotoxin injection (313). GnRH release at proestrus, as measured with a push-pull canula implanted into the median eminence, is markedly suppressed by IL-1 (36,314–316). Endotoxin administration and restraint both inhibit the estradiol-induced LH surge in the ovx rat. This most probably also occurs through a central action since both stressors reduce c-fos expression in GnRH neurons and suppress GnRH mRNA levels in the preoptic area (317–319). Administration of a GnRH agonist concomitantly with the stressor completely restores the LH surge (318). Studies in the ewe also suggest an inhibitory role of endotoxin on the LH surge (Fig. 4) (320), an effect which is exerted centrally since the GnRH surge into the hypophyseal portal

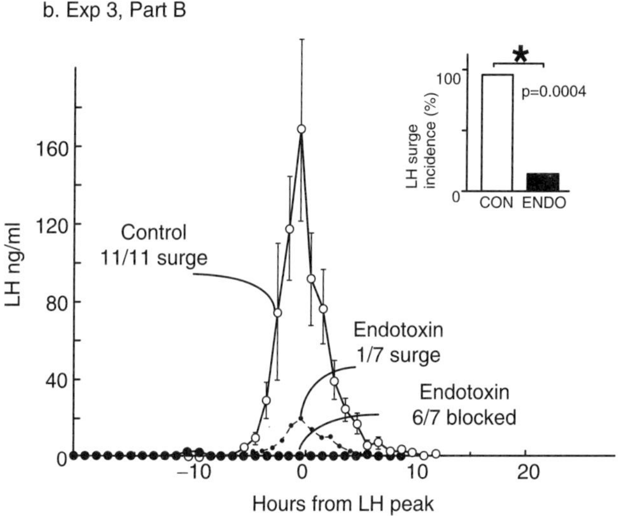

FIG. 4. Blockade of the estradiol-induced LH surge in 6/7 ewes treated with endotoxin, as compared to 11/11 control ewes. The stressor, however, inhibits the LH surge only when given well in advance (10–20 h) of the surge. The inhibitory effect is exerted centrally since the GnRH surge into the hypophyseal portal circulation is also prevented; see (321). [Reprinted with permission from (320), copyright 1999, The Endocrine Society.]

circulation is prevented (321). Significantly however, while endotoxin can rapidly suppress tonic LH release within 1 hour (see above), this stressor inhibits the LH surge in the sheep only when given well in advance (10–20 h) of the surge. Amplitude, duration, and incidence of the LH surge in response to the estradiol stimulus remain unaffected by endotoxin administration at the time of the surge (320). A similar critical period of effectiveness has been reported in the rodent in which IL-1 is most effective when given well ahead of the LH surge period (314). Thus, endotoxin or IL-1 appear to interfere with the early activating effects of the estradiol signal, i.e., during the reading of the signal, but not with the later stages of estradiol signal processing or the actual hormonal surge output (320).

Restraint in the rodent and transport in large animals also result in a suppression of, a reduced or a delayed LH surge (254,255,322–324). In contrast to the above stressors, some investigators failed to demonstrate an inhibition of the estradiol-induced LH surge by insulin-induced hypoglycemia in the rat and monkey (325,326), while others noted a delay in ovulation or a decrease in the magnitude of the LH surge (327,328). In the latter case, the delay is reversed by glucose infusion.

Overall, data in the rodent and ewe suggest that an immune/inflammatory challenge (and possibly other stressors) can directly act most probably at the level of GnRH perikarya to suppress the estradiol-related GnRH/LH surge, provided the challenge occurs ahead of the expected time of the surge, at the initial time of the estradiol signal. These data do not exclude an *indirect* action of stressors on the ovulatory LH surge, since in the intact ewe investigators have also shown that endotoxin can disrupt the spontaneous follicular phase rise in estradiol by preventing the high frequency LH pulses required for normal follicular maturation (and estradiol secretion) and perhaps also by partially impairing follicular responsiveness to gonadotropic stimulation, thereby indirectly leading to a significant *delay* in the timing of the ovulatory LH surge (237,320,329). A similar indirect effect has been observed in the nonhuman primate (330).

Stressors and the Reproductive Cycle

The reproductive cycle in the rodent and larger animals [estrous cycle (331,332)] and in the nonhuman primate and women [menstrual cycle (226,310)] involves a remarkable coordination of hormonal secretion and morphological changes taking place at all levels of the hypothalamic–pituitary–ovarian–uterine axis. A precise synchronized sequence of events is required for a normal reproductive cycle to occur and for fertility. Not surprisingly, there are numerous reports indicating that stress challenges, if sufficiently intense or chronic, disturb the normal estrous or menstrual cycle by interfering with the normal progression of this sequence of events.

Estrous Cycle. In the rat, a chronic inflammatory/immune-like challenge, such as that reproduced by icv IL-1 infusion for 4–6 days, causes a total disruption of the estrous cycle, characterized by persistent diestrus-like vaginal smears (denoting a suppressed estrogen secretion) and decreases in circulating LH and FSH concentrations (333). At the end of the challenge, GnRH gene expression in GnRH neurons distributed between the rostral preoptic area/organum vasculosum of the lamina terminalis and the medial preoptic area is also markedly suppressed. Cyclic disturbances are also reported after a prolonged period of mild unpredictable electric foot shocks, abnormal environmental temperature or overcrowding in the rat or mouse (334–340).

In the ewe, endotoxin administration disrupts the estrous cycle by interrupting normal processes in the preovulatory chain of endocrine events (237,329,341). Indeed, endotoxin is able to prevent the development of high frequency LH pulses which provide the essential stimulus for the preovulatory increase in estradiol secretion from the ovarian follicle during the follicular phase: without this stimulus, the estradiol rise which accompanies follicular maturation is curtailed and estradiol is unable to exert its positive feedback action to release the preovulatory LH surge which is delayed. In the ewe as well, transport and confinement, prolonged mild foot shocks sessions or prolonged exposure to severe climate all have been found to alter the follicular phase, presumably through similar mechanisms, and/or the luteolysis period in the ewe (254,255,336,342,343).

Alterations in follicular or luteal dynamics are also reported after heat or confinement in nonlactating cows and in beef heifers (343–347).

Menstrual Cycle

The Normal Cycle. The menstrual cycle can be divided into two phases, the follicular phase and the luteal phase. At the start of each cycle, there is a hormonal signal (the early follicular phase FSH rise) to recruit a cohort of antral follicles from the large pool of follicles present in the ovary. These follicles will then undergo further development during the follicular phase. Early on, however, through the incompletely understood process of selection, one of these follicles will become a dominant one while the others become atretic. Stimulated by the pulsatile release of the gonadotropins (the optimal pulse frequency is about 1 pulse every 90 min in the human), the dominant follicle grows and creates its own estradiol microenvironment which further contributes to its exponential growth, and releases estradiol in amounts proportional to its growing size. Estradiol restores the endometrium (glandular portion of the uterus) (which through menstruation had been eliminated at the end of the previous cycle). Maturity of the dominant follicle and of its oocyte is marked by a high estradiol level, which when a threshold is reached activates the positive estradiol feedback loop to release a large amount of LH (the preovulatory LH surge) that will induce ovulation (the expulsion of the oocyte from the mature follicle). After ovulation, the cycle enters the luteal phase during which the remaining follicular structure becomes the corpus luteum. Under the stimulation of LH, this structure secretes both estradiol and progesterone. With increasing progesterone, endometrial gland proliferation is arrested and the glands undergo a secretory differentiation, an important event which together with several other factors prepares the uterus for implantation of the fertilized egg. In the absence of fertilization, estradiol and progesterone levels decline, menstruation occurs and a new follicular phase is initiated. [For a detailed review of the menstrual cycle, see (226,310).]

The Functional Hypothalamic Chronic Anovulation Syndrome; Psychogenic Stress. The notion that stressors, by activating the HPA axis, have the potential of interfering with the normal menstrual cycle has been put forward a long time ago. It has been suggested that, if the stressor is severe and/or persistent enough, such condition may result in a complete suppression of the menstrual cycle (amenorrhea). Investigators have focused on the role that "psychogenic" stress, usually the result of overt or latent psychological disturbances, plays in interfering with normal reproductive function. In the 1940s, several investigators called attention to disturbances in the hypothalamus as a common cause of amenorrhea and used the terms "hypothalamic hypoestrinism" (348) or "hypothalamic amenorrhea" (349). Later, the diagnosis in patients showing a silent (but operational, if stimulated) pituitary–ovarian axis with the implication of pathophysiology residing in the central nervous system was referred to as "hypothalamic chronic anovulation" (350). More recently, Berga renamed this syndrome "functional hypothalamic chronic anovulation" (FHCA) (277). When fully established, the syndrome is characterized by ovarian quiescence, amenorrhea and infertility, a tantamount regression to the prepubertal stage. FHCA is estimated to be responsible for up to 35% of cases of secondary amenorrhea (351), although the actual range may well vary from 5% to 70% depending on the population under study.

Several studies in the human underscore the association between FHCA and increased HPA activity. Amenorrheic women diagnosed with FHCA have increased mean basal serum cortisol, reportedly due to higher cortisol levels at night resulting in an amplified circadian excursion, and elevated urinary free cortisol values, as compared to women with other causes of anovulation or eumenorrheic women (278,352–355). These patients may also have a blunted ACTH and cortisol response to CRH administration, perhaps suggesting a reduced sensitivity and expression of the CRH receptor. Recent data also have shown higher cortisol levels in adolescent girls with anorexia nervosa compared with healthy adolescent girls, a consequence of an increased frequency of secretory bursts (356). Women with FHCA who recover from the syndrome have a cortisol level comparable to that in eumenorrheic women and in women with other causes of anovulation (357). Similarly, amenorrheic athletes with normal cortisol levels spontaneously regain menstrual cycles, while those with elevated cortisol remain amenorrheic (358).

Basal tonic levels of FSH, LH, and estradiol are significantly lower in FHCA patients: compared to normal cycling women, these patients have reduced integrated 24-hour LH and FSH concentrations, reduced LH pulse frequency and increased LH interpulse interval (277–279,357).The observation of reduced LH pulse frequency is significant in regard to the progression of the FHCA syndrome, since in the nonhuman primate it has been convincingly demonstrated that even small reductions in the GnRH/LH pulse frequency have profound effects on the quality of follicular development and on cyclic function (359).

A central issue in the diagnosis of FHCA is the appropriate definition of "psychogenic stress." Reifenstein (349) referred to both simple stress and to more severe psychiatric disturbances as potential causes of amenorrhea, and individuals in whom this syndrome is seen were identified as "nurses and students away from home for the first time, debutantes, immigrants, patients with sexual maladjustments, women in fear of pregnancy, women with husbands away at war" and others. Expectedly, investigators describe a variety of settings of onset, suggesting that this syndrome has a greatly varied psychological make-up (360). Because a linear correspondence between the magnitude of a stressor and the extent of psychiatric and physical disability has sometimes been reported (361), a relationship between the type and severity of the challenge and the proportion of women who develop amenorrhea has been assumed (362). Yet, in practice this relationship is usually not recognized and the threshold at which a stressor will interfere with the menstrual cycle is difficult to identify in individual patients. In fact, except for pseudocyesis and anorexia nervosa (363), which are well-established entities representing extreme conditions of FHCA, most patients do not display frank psychiatric disorder or overt neurotic traits. [Several studies reviewing the overall relationship between emotional stress and "infertility" *in general* have not detected significantly greater rates of psychopathology in infertile patients than in the normal population. In particular, there is no general agreement about the role of anxious-depressive symptomatology, in which hyperactivity of the HPA axis is one of the key biological abnormalities in 30% to 50% of patients (364), in the overall pathogenesis of infertility. Yet, it is clear that more sophisticated experimental designs and paradigms are required for investigation (365–369).] Thus, the psychoneuroendocrine evaluation of the most common types of FHCA associated with subtle psychological disturbances is a difficult task. It has been suggested that a "neurotic terrain" must probably preexist for an unremarkable event to precipitate amenorrhea and that thus some subjects may be more "vulnerable" to stress (370). As in the overall response to stressors, a critical factor in evaluating the impact of stress on life events is the individual's perception of them (361). Subtle traits, such as dysfunctional attitudes, perfectionist performance standards and concerns about the judgments of others, tendency to report more depressive symptoms and difficulty coping with daily stress, have been found to discriminate women with FHCA from women with normal menstrual cycles (371). It is also evident that it is difficult if not impossible to demonstrate the independent association between one specific behavior and the initiation of amenorrhea and most women after careful examination display more than one trait or behavior.

Psychosocial stress in males has also been speculated to be associated with the hypogonadal state (372).

On the Road to Amenorrhea and Infertility: The Initial Impact of Stressors on the Cycle. The long-term consequences of chronic stress may include amenorrhea and infertility and result in the full-blown FHCA syndrome described above. This is also at this point that doctors see the patient for the first time. Because of this, there is, unfortunately, little information on the initial impact of a stressor and of an activated HPA axis on the menstrual cycle. It is presumed that intermittent ovarian function reflecting functional abnormalities may occur initially or continue in less severe conditions, but the initial stages of the syndrome are difficult to document in the human. Experiments in the normally cycling nonhuman primate subjected to short-term mild stressors and in the human have begun to address this lack of knowledge. Such studies clearly demonstrate that stressors that are initially of insufficient intensity to induce amenorrhea may however produce different types of temporary cyclic defects which may impact on reproductive fertility.

1. Prolongation of the follicular phase: In the monkey, a 5-day inflammatory/immune challenge, as mimicked by endotoxin administration, during the follicular phase results in a lengthening of the follicular phase and in an interruption of the normal follicular estradiol profile (330). Significantly, in some animals the delay in the resumption of a normal follicular phase greatly exceeds the duration of the stress challenge reflecting a more extended damage to the follicular maturation process. A similar effect of endotoxin has been reported in the ewe, as reported above (237). In both species, the estradiol rise which accompanies follicular maturation is curtailed and estradiol is unable to exert its positive feedback action to release the preovulatory LH surge which is delayed. Similar initial responses to exercise have been reported in women and monkeys (373,374).

2. Inadequate luteal phase: In the monkey, a short-term inflammatory/immune challenge or a psychogenic-like stress (Fig. 5), whether confined to the follicular or luteal phase of the menstrual cycle, invariably results in reduced luteal function, as demonstrated by overall diminished progesterone and inhibin A secretion during the luteal phase (330,375,376). In many animals, these short-term stress challenges continue to interfere with normal luteal function through the first or two subsequent stressor-free cycles. While the decrease in progesterone secretion is most probably related to the observed decrease in LH levels by the stressor (375), studies in the monkey have also shown that chronic cortisol elevations may increase the metabolic clearance rate of progesterone (377). The decrease in luteal progesterone secretion in monkeys subjected to these short-term stress challenges is similar or greater than that reported in clinical studies of patients with the diagnosis of luteal phase deficiency of unspecified origin (378,379). Significantly, these patients may present with a complaint of infertility or recurrent spontaneous abortions, both of which probably reflect decreased progesterone secretion by the corpus luteum resulting in delayed endometrial changes and interference with the process of implantation of the fertilized egg. Thus, impact on fertility may be expected, although this remains to be demonstrated conclusively. Luteal deficiency, which persists even after the return of ovulation, has been described in situations where stress may be suspected, such as in new student nurses after school entry and in practicing nurses (380,381), during a short-term stress endurance exercise (382), and in rural populations (383). The fact that inadequate luteal phase is also seen in monkeys during exercise (374), and in women athletes and runners (373,384–389), suggests that this syndrome may well represent a general initial response to stressors and/or energy-related demands in the normal cyclic female.

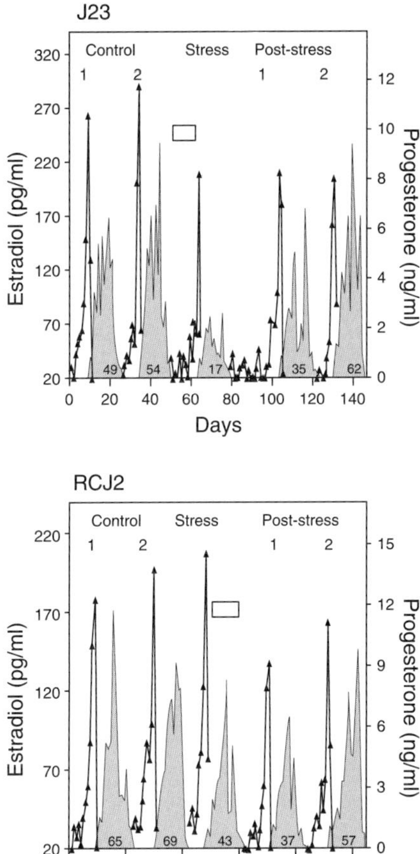

FIG. 5. Inadequate luteal phase in two monkeys following a 10-day psychogenic-like stress during the follicular phase (*upper figure, open rectangle*) or during the luteal phase (*lower figure, open rectangle*). Illustrated in a continuum are daily levels of estradiol during the follicular phase (*lines with triangles*) and of progesterone during the luteal phase (*shaded areas*) throughout two control menstrual cycles (control 1, control 2), the "stress" cycle, and two post-"stress" cycles. Note the significant decrease in luteal progesterone secretion (numbers within the *shaded areas* are integrated luteal progesterone values) in the "stress" cycle and in the first of the two post-stress cycles. [Reprinted with permission from (376), copyright 2002, The Endocrine Society.]

3. Premature LH surge: Studies in the monkey also suggest a novel way by which an acute stressor may interfere with the normal cycle by advancing the timing of the ovulatory LH surge. Activation of the HPA axis by endotoxin or IL-1 in the mid-late follicular phase (but not at other stages of the cycle) results in a large acute LH release, which in some animals resembles the preovulatory LH surge (390). This surge in LH after a short-term immune/inflammatory challenge can be replicated in the ovx monkey pretreated with mid-late follicular estradiol levels (but not in the unreplaced animal) (391), and in post-menopausal women receiving estrogen replacement (Fig. 6) (392). Studies in the monkey suggest that progesterone, which is released in small

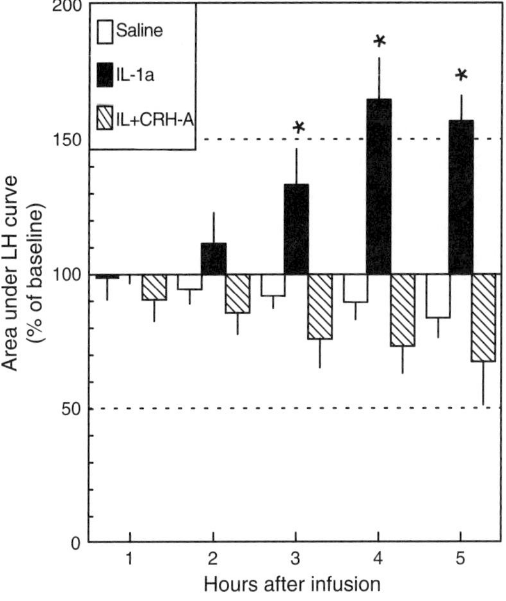

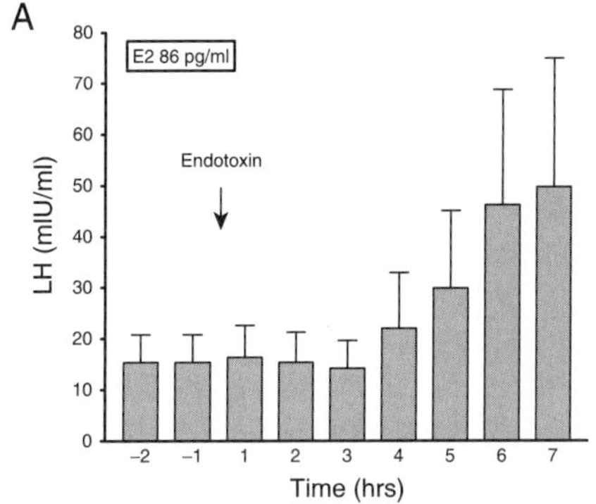

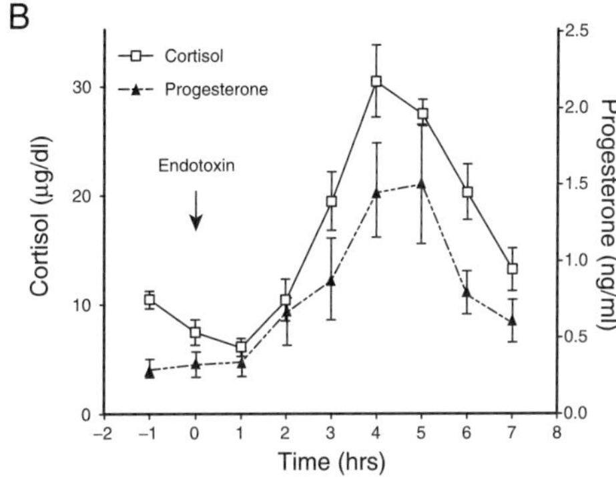

FIG. 6. Examples of LH release by immune/inflammatory stressors in the presence of mid-late follicular phase estradiol levels. **Upper panel:** LH release (*black bars*) following intracerebroventricular IL-1 administration in the estradiol-replaced ovariectomized monkey. This LH release is prevented by co-administration of a CRH antagonist (*hatched bars*). [Reprinted with permission from (391), copyright 1994, The Endocrine Society.] **Lower panel:** LH release (*upper figure, black bars*) following endotoxin administration to estrogen-replaced post menopausal women. Parallel increases in adrenal cortisol (*open squares*) and progesterone (*closed triangles*) in response to activation of the HPA axis are also seen (*lower figure*). [Reprinted with permission from (392), copyright 2000, The Endocrine Society.] It is postulated that such an "ill-timed" surge of LH during the follicular phase may be detrimental to the maturing follicle and to fertility (see text).

amounts from the adrenals in response to the stressor (393), may be responsible for this acute release of LH since this effect is prevented by a progesterone or CRH antagonist (391). This type of LH release can also be simulated by ACTH administration or by blockage of cortisol synthesis with metyrapone in the human or monkey (326,391,392). That progesterone can activate a premature LH surge in the above circumstances may not be surprising since it is well known that this hormone, in amounts that are within the range observed in the immediate pre-LH surge period in the normal cycle, can synergize with circulating estradiol to activate the ovulatory LH surge (394–396). A smaller increase in LH has also been reported in women following exercise during the midfollicular phase (397).

The precise physiopathological significance of a premature acute rise in LH following stressors in the mid- to late follicular phase is not known. However, it may be speculated that, because it occurs at a time when the selected follicle has not attained full growth, the premature surge may have a deleterious effect on the selected follicle: there are indeed several reports indicating that elevated LH concentrations at that stage of the cycle damage the maturing follicle and/or oocyte, desynchronize the ovulatory signal from a timely follicular maturation process and/or result in early pregnancy loss or reduce the fertilization rate in women undergoing in vitro fertilization (398–401). [A similar premature LH surge following stressors has not been reported in the rodent, probably a reflection of the relatively greater biosynthetic capacity of the rat adrenal (402). This would result in progesterone release at levels known to inhibit rather than enhance LH release (403).]

Overall, the existence of a chronic clinical amenorrheic syndrome (FHCA) functionally related to stress, as well as to intensive exercise and diet is well established. Prospective studies of the initial effects of stressors on the cycle have only begun. Data, however, clearly suggest that the HPG axis is closely monitored by the brain and that a stress challenge, even one not sufficient or extended enough to result in amenorrhea, may well provoke endocrine changes that may be too subtle to be appreciated and detected in the clinical environment but that nonetheless may be substantial enough to temporarily interfere with normal cyclic function and fertility (393). Such changes may in fact provide a mechanism whereby pregnancy is temporarily prevented in suboptimal conditions. Since these changes may be induced not only by stressors but also by excessive exercise or deficient energy balance (or a combination of these), it is therefore clear that irregular menstrual bleeding patterns, abnormal luteal phases and unexplained infertility should indeed be considered as part of the symptomatology that *precedes* the fully established FHCA syndrome (277). Clinical attention to these conditions in patients consulting for infertility may well provide for more physiological approaches to the treatment of infertility, since hormonal release can be readily reinstated upon removal of the insult(s). Experimental data should also lead to the study of new and more focused approaches to the treatment of infertility, such as the use of cognitive behavioral therapy (404) or the use of drugs or specific hormone antagonists (see below).

Of interest also is the observation that nonhuman primates in whom stressors produce a more substantial damage to cyclicity are also those who already have a subnormal luteal progesterone profile during the preceding two cycles (330). Thus, an inadequate luteal phase may be viewed as a sign of a stress-sensitive individual.

Mechanisms Whereby the HPA Axis Interferes with the Reproductive System

Notwithstanding the recognition of a link between stressors, and by implication an activated HPA axis, and an inhibition of the HPG axis and of the reproductive cycle, data confirming this association are scant and at best indirect in the human. The most revealing data derive from careful comparisons of cortisol secretion in patients with FHCA to that of women with other causes of anovulation, showing elevated cortisol levels only in FHCA (also see above) (278,352,353). There is, however, emerging evidence of an active role of CRH, the main active neuropeptide controlling HPA activity, and of VP, another neuropeptide which synergizes with CRH to activate HPA, in the inhibition of the reproductive process by stressors. Additional mechanisms may include mediatory effects of central endogenous opioid peptidergic systems as well as potential modulatory roles of secretory products of the peripheral HPA axis.

Role of CRH

CRH is the most potent ACTH secretagogue which is activated by stressors (see above). Direct and indirect data obtained in animals following administration of CRH antagonists or CRH and from CRH measurements provide evidence for a role of this neuropeptide in mediating the inhibitory effects of stressors on gonadotropin release.

Evidence Following Neutralization of Endogenous CRH Activity. A direct line of evidence for a role of CRH in the HPA-HPG link requires the demonstration that neutralization of *endogenous*

CRH activity results in a restoration of normal reproductive activity during a stress challenge, a requirement that has been fulfilled in a few experimental conditions.

In the ovx monkey, concomitant infusion of a CRH receptor antagonist prevents the inhibition of pulsatile LH release, as well as the increase in cortisol, that follow icv injection of IL-1α (Fig. 49-2, upper panel) (60). The CRH-R antagonist also delays the inhibition of multi-unit activity volleys, representative of GnRH pulsatile release, in response to insulin-induced hypoglycemia (282). In ovx monkeys as well, alprazolam, a potent benzodiazepine derivative which binds to the GABA A receptor complex and by increasing GABA activity has been shown to decrease endogenous CRH activity in PVN and the locus coeruleus (94,405–408), also prevents hypoglycemia-induced inhibition of pulsatile LH release while markedly attenuating the ACTH response to hypoglycemia (Fig. 2) (267). This effect is independent of glucose concentrations. In a small study, alprazolam was also shown to acutely restore LH pulses in stress-related anovulatory women (409). Studies in cycling monkeys also show that administration of a CRH-R antagonist prevents the appearance of an inadequate luteal phase after an endotoxin challenge (see above) and accelerates recovery of a normal menstrual cycle (410). Exploratory and sexual behaviors that are normally suppressed in male monkeys exposed to an intense social stressor can be reactivated by a nonpeptide CRH-R1 antagonist (411).

In the rat, the participation of endogenous CRH in the decreased activity of the HPG axis has been shown to vary depending on the type of stressor (36). For example, in the adult castrated male rat exposed to intermittent foot shocks, central administration of a CRH-R antagonist readily reverses the inhibitory action of exposure on pulsatile LH release (412). On the other hand, icv administration of CRH antagonists or of CRH antiserum in gonadectomized male or female rats does not modify the inhibitory effects of IL-1β on LH secretion (36,230,231). This is a marked difference with the monkey, which may reflect varying ability of CRH antagonists to reach different brain structures, species-specific GnRH neuronal organization and/or varying mechanisms mediating stress-induced effects on the HPG axis. Of interest also are studies in the rat that suggest that PVN, which represents the predominant hypothalamic nucleus controlling the response of the HPA axis to a number of stimuli (see above), does not appear to be necessary for the antireproductive effects of an immune/inflammatory challenge: in the rat, bilateral lesion of PVN does not prevent LH decreases after injection of IL-1β, although it attenuates or blocks the increase in HPA axis activity (206). It is possible that decreases in GnRH/LH secretion after IL-1 in the rat may also reflect a direct action of IL on GnRH neurons in the medial preoptic area, a major center of control of the HPG axis in the rodent (36,312).

Evidence from CRH Measurements. In ewes, cortisol release into the peripheral circulation following endotoxin administration parallels that of CRH release into hypophyseal portal blood, which in turn coincides with the suppression of GnRH and LH pulsatile secretion (Fig. 7) (55,235). In the ovx monkey, IL-1α, one of the main cytokines released by endotoxin, not only suppresses pulsatile gonadotropin secretion but also increases in vitro CRH release from monkey hypothalami and CRH release into cerebrospinal fluid (although the precise source of the neuropeptide in CSF remains to be clarified) (52). CRH is also released in sheep in response to a hypoglycemic stressor, which also suppresses LH release (413,414). It should, however, be pointed

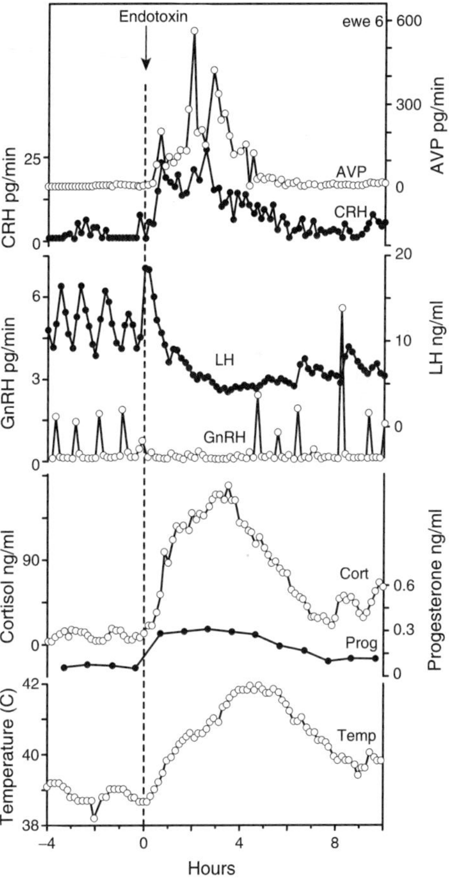

FIG. 7. Simultaneous CRH (*upper panel, closed circles*) and VP (*open circles*) release in hypophyseal portal blood, following endotoxin administration in the ewe. Also illustrated are the simultaneous GnRH (*open circles*) and LH (*closed circles*) inhibition (*second panel*), as well as cortisol release (*third panel, open circles*) and core body temperature increase (*lower panel, open circles*). [Reprinted with permission from (55), copyright 1998, The Endocrine Society.]

out that acute increases in CRH mRNA levels in PVN that follow cortisol synthesis inhibition with metyrapone in the absence of a stressor do not result in an LH inhibition in the ovx monkey (415).

CRH levels in cerebrospinal fluid are elevated in patients with anorexia nervosa, who are hypercortisolemic, hypogonadal, and underweight, but levels become normalized after weight recovery (416). This observation, however, could not be replicated in another group of women with established functional hypothalamic amenorrhea, increased cortisol, and reduced gonadotropin release (417).

Evidence from CRH Injections. In the ovx monkey, administration of CRH clearly results in a rapid inhibition of LH and FSH levels and of large amplitude LH pulses, while concomitantly increasing cortisol release. LH pulsatility resumes immediately at the cessation of the CRH infusion (418). CRH administration also results in a significant decrease in the frequency and duration of GnRH pulses, as documented by hypothalamic multiunit activity measurements, occurring in parallel with a rise in circulating cortisol levels (419). Several reports in the primate suggest that the rapid inhibitory action of CRH on GnRH/LH pulsatile release is not mediated by down-stream activation of the pituitary–adrenal axis: indeed, the inhibitory effects of a CRH infusion on pulsatile LH release cannot be mimicked by ACTH infusions (420), nor does treatment with metyrapone, an inhibitor of adrenal steroidogenesis that prevents the CRH-induced rise in serum cortisol levels, or with dexamethasone, a synthetic steroid known to block the CRH-induced ACTH response, reverse the inhibitory effects of CRH on the frequency or duration of hypothalamic multi-unit activity volleys or on the LH and FSH response (419,421). Similarly, the adrenal glands are not required, since CRH can inhibit pulsatile LH release in the adrenalectomized monkey as well (422). In contrast to the above, an inhibitory effect of CRH on gonadotropin secretion could not be demonstrated in the intact male or female monkey in either the follicular or luteal phase of the cycle, although it increased cortisol in both cases (423,424). In contrast, a significant decrease in plasma LH and FSH levels occurs after peripheral CRH injection in normal women during the midluteal phase of the menstrual cycle (425), although other investigators have been unsuccessful in altering either basal or naloxone-stimulated LH levels in women (426,427). The reasons for these differences remain to be studied in more details, in particular the role of the gonadal steroids in the modulation of the response to stressors (see below). [In the sheep, there are reports of an *increase* in LH secretion and pulsatility after icv CRH administration. This action requires the presence of the gonadal steroids (428,429)].

In the rat, in vitro GnRH release from hypothalamic slices is suppressed in a dose-dependent fashion by CRH (430,431). Intraventricular injection of CRH or direct CRH injection into the median eminence of ovx or ovx/adrenalectomized rats also cause a rapid and prolonged dose-related inhibition of LH secretion together with an increase in ACTH (432–435). However, systemic administration of CRH is without effect on LH release in the rodent (36,432), a difference with the primate (418,436), perhaps reflecting the accessibility of the arcuate nucleus, the presumed site of the GnRH pulse generator, in the latter species. Chronic administration of CRH into the medial preoptic area (the hypothalamic area responsive to the estradiol positive feedback in the rodent) also inhibits GnRH and LH surges at proestrus (230), in accord with an inhibitory role of stressors on the preovulatory gonadotropin surge in the rodent (see above). This action may be exerted via direct synaptic contacts between CRH axon terminals and dendrites of GnRH neurons within the preoptic area (36,437).

Overall, CRH is the most potent ACTH secretagogue which is activated by stressors (see above). Direct and indirect data obtained in animals following administration of CRH antagonists or CRH and from CRH measurements provide evidence for a role of this neuropeptide in mediating the inhibitory effects of stressors on gonadotropin release. Data on the relationship between CRH and gonadotropin release provide direct evidence that *endogenous* CRH can indeed mediate the inhibitory effect of stressors on the HPG axis, although there is a striking difference between the rodent and nonhuman primate in regard to its putative role after an inflammatory/immune stressor. On the other hand, data in CRH-deficient male and female mice indicate that stressors such as restraint or food withdrawal can still inhibit tonic LH secretion or the proestrus LH surge, suggesting that molecules other than CRH (perhaps VP or urocortin II; see below) are also able to suppress LH secretion during the response to stress in the context of a CRH deficiency (65).

Role of VP

Evidence Following Neutralization of Endogenous VP Activity. In the ovx monkey, the inhibitory effects of an icv infusion of IL-1α on pulsatile LH release are prevented by co-administration of an VP antagonist, even though the antagonist is without effect on the cortisol increase (Fig. 8) (77). The same VP antagonist also prevents hypoglycemia-induced LH suppression in ovx and in male monkeys (Fig. 8) (266,438), although other investigators were unable to confirm effects of the same antagonist

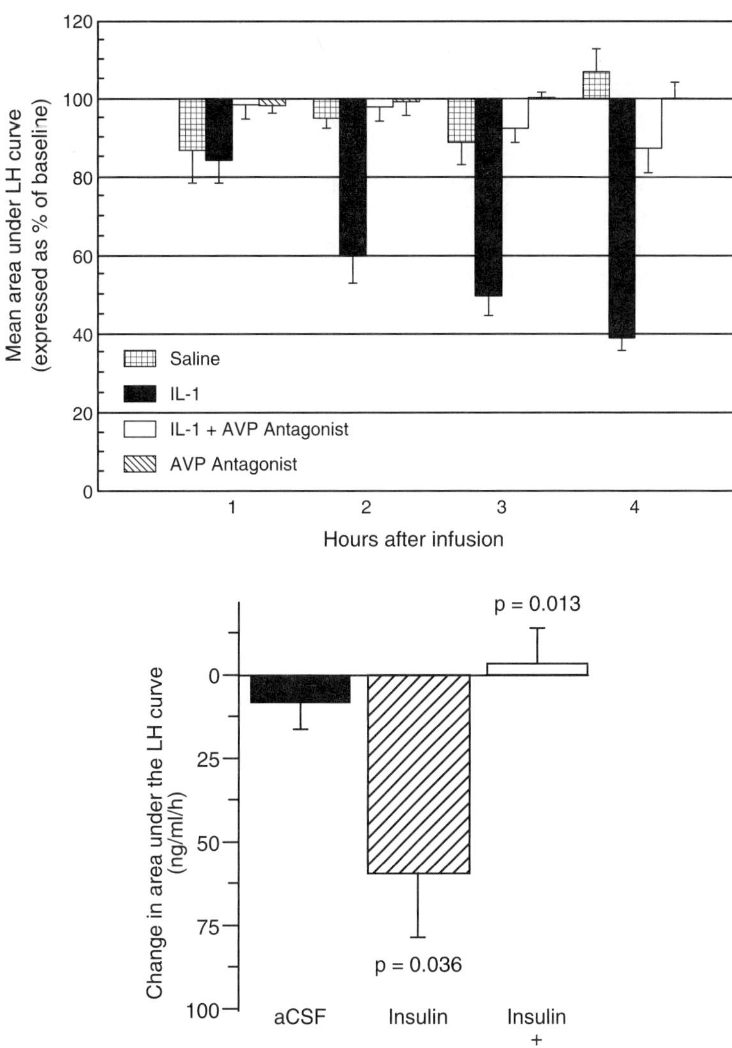

FIG. 8. VP antagonists prevent the inhibitory effects of stressors on LH release in the monkey. **Upper panel:** While icv infusion of IL-1 reduces LH release (*black bars*), the concomitant administration of a VP antagonist (*open bars*) maintains normal LH secretion. [Reprinted with permission from (77), copyright 1992, The Endocrine Society.] **Lower panel:** Insulin-induced hypoglycemia (*hatched bar*) inhibits LH release. Co-administration of a VP antagonist restores normal LH levels. [Reprinted with permission from (266), copyright S. Karger AG, Basel.]

on the GnRH and LH response in ovx monkeys (282). Interestingly, the same VP antagonist is also unable to reverse the inhibitory action of an insulin-induced hypoglycemic challenge on LH release in female primates during the follicular phase (265) and in the ovx estradiol-replaced rat (439), possibly suggesting a modulatory role of estrogen in this process. Antiserum to VP micro-injected into the 3rd ventricle also reduces the immediate increases in ACTH and LH that follow exposure to ether (440).

Evidence from VP Measurements. In ewes, endotoxin potently stimulates VP, as well as CRH, release into hypophyseal portal blood, and the release of both neuropeptides coincides with a significant suppression of pulsatile GnRH and LH secretion in these same animals (Fig. 7) (55). VP is also released in response to insulin-induced hypoglycemia in the sheep, although the release of VP is marked only after the higher insulin doses (413,414), an observation which may explain some of discrepancies reported above on the effects of VP antagonist administration.

Evidence from VP Injections. In the ovx monkey, icv infusion of VP stimulates cortisol release and simultaneously decreases LH levels, an effect that is independent of changes in mean arterial pressure (105,266,282). Co-infusion of a CRH antagonist and VP attenuates the inhibitory effect of VP on LH release but not the VP-induced increase in cortisol, suggesting that CRH is a prerequisite for VP's action on the HPG but not on the HPA axis (441). However, icv VP injections, in contrast to CRH, fail to decrease the frequency of multiple unit activity volleys reflecting the activity of the GnRH pulse generator, even though mean LH concentrations decrease and serum cortisol levels increase (282). In fact, VP unexpectedly appears to accelerate volleys frequency, in some animals to such a degree that the pituitary cannot respond by secreting discrete LH pulses, leading to a decrease in LH pulse frequency. In the ovx rat as well, icv VP injection suppresses LH secretion by decreasing LH pulse amplitude without affecting LH pulse frequency, an effect that is blocked by central administration of a VP antagonist (439). In contrast to the monkey, this inhibitory response is evident only in estradiol-replaced ovx animals, suggesting a sensitizing influence by the gonadal steroid.

In amenorrheic or normal women, a pharmacological dose of desmopressin (an analogue of vasopressin) does not change basal LH secretion (442).

Overall, the above data suggest that VP, a neuropeptide co-localized with CRH within perikarya of PVN neurons and in secretory granules (69,70) and playing an active role in the control of the HPA axis (see above), can also mediate the inhibitory action of stressors on the HPG axis, a role similar to that of CRH but that may involve a different mechanism. More detailed experimentation is required to discriminate between the respective roles of these two neuropeptides in response to various stressors, in different species and under different gonadal status conditions. The observation that VP antagonist administration does not always result in the prevention of the cortisol rise may reflect the primacy of CRH in controlling the HPA axis (77).

A Potential Role for the Urocortins

While CRH has a higher affinity for CRH-R1, urocortin binds with equally high affinity to both CRH-R1 and -R2, while urocortins II and III bind selectively to CRH-R2 and are considered its endogenous ligands (88,89). There is little information on the role of the urocortins in regard to the control of HPG. However, recent evidence suggests a potential role for these peptides in the control of pulsatile LH release. Indeed, administration of a selective CRH-R2 antagonist blocks the suppression of LH pulses induced by restraint in ovx, estrogen-replaced rats, while icv administration of urocortin II induces a dose-dependent decrease in LH pulse frequency (443).

A Modulatory Role by Endogenous Opioid Peptides

Results from retrograde tracings from GnRH neurons show that few CRH neurons are labeled and suggest that the pathway by which CRH mediate inhibitory effects of stressors on LH release may not involve direct CRH projections to GnRH neurons (444). As reviewed now, the endogenous opioid peptides may provide one of such indirect pathways.

The Pro-opiomelanocortin (POMC) System and β-Endorphin Response to Stressors. POMC is a large glycoprotein precursor molecule synthesized primarily in the hypothalamus and pituitary gland. Its posttranslational processing is tissue-specific and results in the production of several peptides with different biological activities (111, 445–447). In the anterior pituitary, POMC cleavage leads primarily to ACTH and β-LPH, while in the hypothalamus it is further processed to yield β-endorphin from β-LPH and α-melanocyte-stimulating hormone (α-MSH) from ACTH. Hormonal regulation of POMC gene expression and of peptide release is also tissue-specific. In the pituitary gland, POMC within the corticotrope is mainly controlled by components of the HPA axis (448,449): ACTH, β-LPH and β-endorphin all are released by CRH and stressors into the peripheral circulation. Their release is controlled by glucocorticoids (450–452). Although POMC has a wide distribution within the brain and brain stem, the highest concentration of β-endorphin–containing neurons is seen within and in the vicinity of the arcuate nucleus of the hypothalamus, an area rich in GnRH neurons (453).

Although effects of CRH and of stressors on pituitary POMC are clear-cut, their role in controlling hypothalamic POMC and β-endorphin levels is less well investigated. There are, however, reports in the rat that CRH stimulates the release of β-endorphin from hypothalamic slices, an effect mediated by CRH receptors since it is antagonized by a CRH-R antagonist (454). Acute foot shocks in the ewe increase β-endorphin release in perfusates from the arcuate nucleus-median eminence region (336,455) and in the rat deplete hypothalamic β-endorphin (456). In contrast, chronic stressors such as prolonged restraint or foot shock in the rat do not appear to affect synthesis or release of POMC-derived peptides from the hypothalamus, perhaps reflecting habituation (457,458).

Role of Endogenous Opioid Peptides in Mediating the Inhibitory Effect of Stressors on Tonic Pulsatile LH Release. Substantial evidence exists to indicate that hypothalamic opioid peptides, principally β-endorphin, are involved in the regulation of GnRH/LH release in the primate (459,460). An acute suppression of pulsatile LH and FSH release is seen following administration of β-endorphin or of morphine in the human (461,462), monkey (463–465), and rodent (466–469). *Endogenous* β-endorphin is a powerful inhibitor of the GnRH pulse generator. For instance, during the luteal phase of the menstrual cycle of the human and nonhuman primate, there is a marked decreased in LH pulse frequency (464,470,471) that parallels elevated endogenous β-endorphin activity (464,472, 473) and that is prevented by the administration of naloxone, a μ-opioid receptor antagonist (462,470, 474,475).

In the primate as well, there is good evidence to suggest a role of hypothalamic endogenous opioid peptides in the inhibitory action of CRH and VP on pulsatile LH release. In ovx monkeys, opioid antagonism by naloxone prevents the CRH- or VP-mediated suppression of gonadotropin release and restores pulsatile LH secretion (Fig. 9) (436,441). Naloxone also blocks the inhibitory effects of CRH on the frequency of multiple unit activity volleys representative of GnRH pulse generator activity (419). In women during the mid-luteal phase of the cycle, co-administration of naloxone also prevents CRH-induced LH and FSH inhibition (425). Furthermore, naloxone also reverses the acute inhibitory effects on LH of several stressors, such as an endotoxin challenge in the ovx monkey (242), restraint in the male monkey or in the estradiol-primed ovx marmoset (256,476), the stress of anesthesia by phencyclidine darting in the wild baboon (258) and the acute stress that follows placement of tethering systems in adolescent male rhesus monkeys (477). Concomitant measurements of GnRH in median eminence perfusate and of LH in plasma of rhesus monkeys during the luteal phase show that infusion of the opioid receptor antagonist nalmefene reinitiates distinctive pulsatile secretion which had been suppressed by the stress of surgery (478). In sharp contrast, naloxone does not appear to prevent the hypoglycemia-induced LH inhibition in neither intact or ovx monkeys (265,267,283,479).

A CRH–opioid interaction in the inhibitory control of LH secretion is also supported in lower species. Naloxone reverses the CRH-induced decrease in LH levels in ovx and estrogen-treated ovx rats (433,480) and in the castrated male rat (481). Other μ-, but not κ-, opioid receptor antagonists are also able to prevent the inhibition of GnRH release that follows the injection of CRH in the medial preoptic

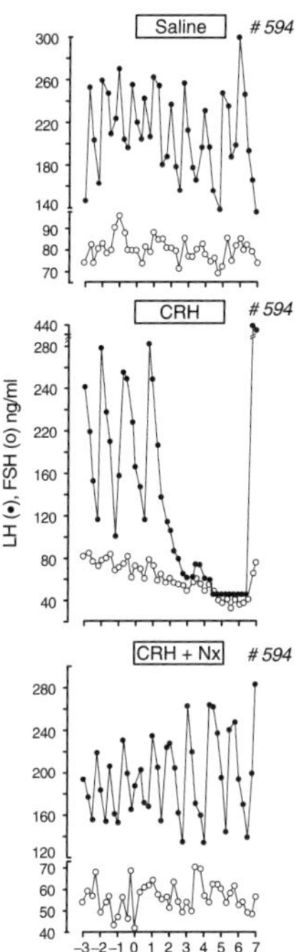

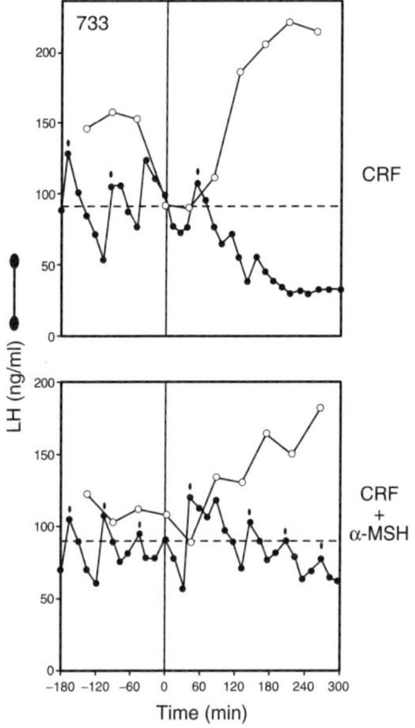

FIG. 9. Opioid peptides derived from the precursor pro-opiomelanocortin (POMC) mediate effects of stressors on LH release in the monkey. **Left panel:** As opposed to a 5-h saline infusion which does not affect pulsatile LH release (*upper graph, closed circles*), infusion of CRH suppresses pulsatile LH release (*middle graph*). (Note the resumption of LH secretion after the infusion has ended.) Co-administration of the opiate antagonist naloxone (Nx) (*lower graph*) prevents the inhibitory effect of CRH on pulsatile LH release. [Reprinted with permission from (436), copyright 1987, The Endocrine Society.] **Right panel:** While CRH (CRF) suppresses pulsatile LH release (*upper graph, closed circles*), co-administration of α-melanocyte stimulating hormone (α-MSH) reverses this effect (*lower graph*), even though stimulation of the peripheral adrenal axis, as shown by similar increases in cortisol (*open circles*) persists. These data suggest that post-translational expression of opioid peptides may represent an important way by which POMC modulates the response of HPG to stressors (see text). [Reprinted with permission from (515), copyright 1992, The Endocrine Society.]

area of the hypothalamus (315). Pretreatment with anti β-endorphin antiserum, β-endorphin antagonist, naloxone, or a more specific μ1-opiate receptor antagonist also reverses the decrease in LH concentrations following several acute stressors in the rat, such as foot shock (482), restraint (293,297,318,483,484), injection of IL-1β or of endotoxin in the rat (231,317) or in ovx heifers (241), ether exposure (293), and, in contrast to the monkey, insulin-induced hypoglycemia in the rat (260) and sheep (262). The role of opioids in chronic stress is less clear, however.

Although it is difficult to specifically separate stress from other etiologies in FHCA (see above), investigators have also reported a role of endogenous opioid peptides in this syndrome in that, to some degree, LH pulsatility and even menstrual cyclicity can be restored by treatment with opiate antagonists. For instance, an amplification of pulsatile LH release is observed after acute opiate antagonist treatment in some (but not all) women with FHCA, weight loss and exercise-related amenorrhea, or anorexia nervosa (475,485–489). Prolonged daily treatment with naltrexone in a large trial of women with FHCA increases both plasma gonadotropin concentrations and the amplitude and frequency of LH pulses, presumably reflecting a disinhibition of the hypothalamic GnRH pulse generator, and in 70% of the patients results in a normalization of menstrual cycles and a cumulative pregnancy rate that appears to closely resemble that of the normal population (490–492). Other investigators, however, were unable to confirm these data (493,494). In monkeys having undergone surgery, menstrual cycles are restored more rapidly following nalmefene treatment than in untreated animals (478).

Role of Endogenous Opioid Peptides in Mediating the Inhibitory Effect of Stressors on "Surge" LH Release. In the primate and other large animals, there is no confirmed observation that stressors act to inhibit the LH "surge" mechanism (see above). In addition, it is known that an increase in opiate tone, such as by morphine administration, does not block the spontaneous or estrogen-induced LH surge in the monkey, pig, or ewe, although a delay in the surge may be observed in some instances (463,495–497). It is also known that naloxone does not enable estradiol to exert its positive feedback action on LH release at a time of high endogenous opiate activity, such as in the luteal phase in the presence of high progesterone levels or after progesterone administration (498). In regard to stressors, naloxone treatment is also unable to prevent the delay in the onset of the spontaneous or exogenous estradiol-induced LH surge caused by transport in cows (322,323,499).

In the rodent, in contrast, endogenous opiates do influence the LH surge since administration of β-endorphin or of morphine blocks the spontaneous preovulatory LH surge at proestrus as well as the estrogen-induced LH and GnRH surges (500–506). These effects are reversed by naloxone administration. Whether endogenous opioid peptides mediate the inhibition of the LH surge by stressors has not been convincingly demonstrated: while some studies report effective antagonism by naloxone of the inhibition of the LH surge by restraint in gonadotropin-treated immature rats or by foot shock in adult rats at proestrus (501,507), others do not confirm these observations (318,508). A role for central opiates in the glucoprivic regulation of surge LH release in the rat has also been proposed (328).

Interaction Between β-Endorphin and α-MSH in the Control of Tonic LH Release. Although there is considerable evidence to suggest that β-endorphin of hypothalamic origin is one of the principal opioid peptide involved in the physiological regulation of LH secretion, another peptide also derived from POMC, α-MSH, produces a variety of physiological effects opposite those of β-endorphin. In the rat, α-MSH antagonizes some of the behavioral effects of β-endorphin, suggesting that it may function as an endogenous opioid antagonist (509). Indeed, α-MSH has been shown to antagonize the inhibitory action of exogenously administered β-endorphin on LH release (as well as on prolactin release) in the primate and rodent (510–513). In support of this notion also are data in women in whom α-MSH is effective in stimulating LH release during the luteal phase, at a time of increased endogenous β-endorphin activity, but not in the early follicular phase when endogenous endorphin activity is lowest (514). α-MSH on its own is similarly unable to increase LH secretion in the ovx monkey, in which endogenous β-endorphin activity is also lowest (515).

Significantly in regard to stress, α-MSH restores pulsatile LH release in ovx monkeys treated with CRH or subjected to a stressor, such as IL-1, in which pulsatile LH release had been suppressed (Fig. 9) (515). This effect presumably reflects its antagonism to endogenously released β-endorphin in response to the stressor. In these α-MSH-treated animals, the stimulation of ACTH and cortisol after CRH persists, indicating that α-MSH does not interfere with the direct action of CRH on the pituitary corticotrope. α-MSH also reverses the inhibitory effects of CRH on pulsatile LH release in women during the luteal phase (516). It can also increase pulsatile LH activity in women with hypothalamic amenorrhea in whom LH pulsatility is reduced (514). BAPa mice with ectopic overexpression of agouti protein (an endogenous antagonist of melanocortin receptors) show exaggerated endocrine and energetic responses to restraint that may be postulated to result from a developmental compensation for chronic inhibition of melanocortin receptors (517).

Overall, there is substantial evidence to indicate that a major inhibitory action of CRH and of most stressors on pulsatile GnRH/LH release occurs at suprapituitary sites that not only involve the neuroendocrine HPA axis but also endogenous opioid peptidergic pathways as mediators. This conclusion is supported by data across species, except in the case of insulin-induced hypoglycemia as a stressor in the monkey. Data also support the notion that posttranslational processing of POMC may represent an important step in these mechanisms and that the "ratio" of β-endorphin to α-MSH may be more significant than the absolute β-endorphin concentration in determining the ultimate effects of stressors on pulsatile LH release.

In regard to the endogenous opioid peptides, two additional features should also be kept in mind. The first is that, although opioid antagonism can reverse the acute inhibitory effects of stressors on LH release, it does not prevent the increase in ACTH and cortisol that follows stressors (242,256, 258,476,477). And indeed, CRH or VP stimulate cortisol release regardless of whether naloxone is administered or not, as a result of their direct stimulatory action on the corticotrope (436,441). The second is that there appears to be an intricate relationship between hypothalamic POMC and the HPG axis in that POMC gene expression in the arcuate nucleus is markedly controlled by gonadal steroids (518–521). In the monkey, basal hypothalamic release of β-endorphin, as measured in hypophyseal portal blood, is magnified in the presence of estradiol and progesterone, such as in the luteal phase of the menstrual cycle, and is significantly decreased after ovariectomy (472,473). Thus, the gonadal steroid background at the time a stressor occurs may also influence the HPG axis response to a stressor (see below).

A Role for ACTH and the Adrenal Glucocorticoids

Activation of the HPA axis by stressors evidently results in elevations of ACTH and glucocorticoids and the question arises whether these hormones play a role in the inhibition of gonadotropin secretion and in the disruption of the reproductive cycle that accompany stress.

A Glucocorticoid-Independent Pathway. There are several lines of experimental evidence obtained in the human, monkey, and other large animals that indicate that activation of the peripheral components of the HPA axis in response to stressors does not play a role in the *acute* inhibition of pulsatile LH release by these stressors. First, the inability of CRH to elevate cortisol release in the adrenalectomized monkey does not prevent the neuropeptide to acutely inhibit pulsatile LH release in a manner comparable to that observed in the nonadrenalectomized animal (422). Second, pharmacological suppression of cortisol synthesis with metyrapone during the acute response to different stressors, such as endotoxin in the ovx sheep, insulin-induced hypoglycemia in the monkey or anesthesia by phencyclidine darting in the wild baboon, does not block or reduce acute stressor-induced suppression of pulsatile GnRH and LH release (258,329,415,522). Third, a short-term ACTH infusion that increases cortisol secretion to levels observed after CRH infusion in the ovx monkey or in deer does not inhibit pulsatile LH release, in contrast to the frank LH inhibition seen after CRH infusion (420,523). Fourth, *short-term* infusions of glucocorticoids in healthy young men and women (524), the intact male monkey (525), the ovx pig (526), or the rodent (527) have no effect on pulsatile LH release. Two experimental examples after infusions of ACTH in the ovx monkey and after CRH infusion in the ovx-adrenalectomized monkey are illustrated in Figure 10.

A Glucocorticoid-Dependent Pathway. Although the preceding data appear to rule out a role of adrenal steroids in mediating the inhibitory effects of stressors on LH secretion, additional data show that glucocorticoids can on their own disrupt gonadotropin release in a manner that appears independent of CRH. For instance, *chronic* administration of natural or synthetic glucocorticoids have been associated with an inhibition of LH and FSH secretion in the castrated male monkey (525,528), in women in the early follicular phase (529,530), in the ovx pig (526,531), in cattle (532–534), and in the rat (535). Chronic ACTH treatment has also been shown to reduce post castration rises in LH in the rodent (536) and LH levels in the bull (537). Cushing's syndrome, with its long-standing elevated cortisol secretion, is also characterized by hypogonadotropism both in men and women (538).

A large number of studies have investigated the mechanism whereby cortisol or corticosterone may inhibit LH secretion and there is a considerable literature that indicates that these steroids act at the pituitary level to affect the gonadotropin response to GnRH stimulation, both in acute and chronic situations. For example, in the ovx sheep cortisol readily decreases the amplitude of the LH response to exogenous hourly GnRH pulses (539).

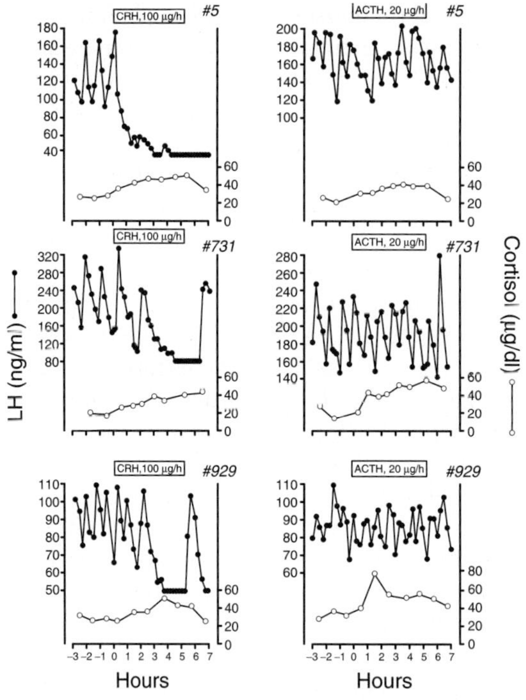

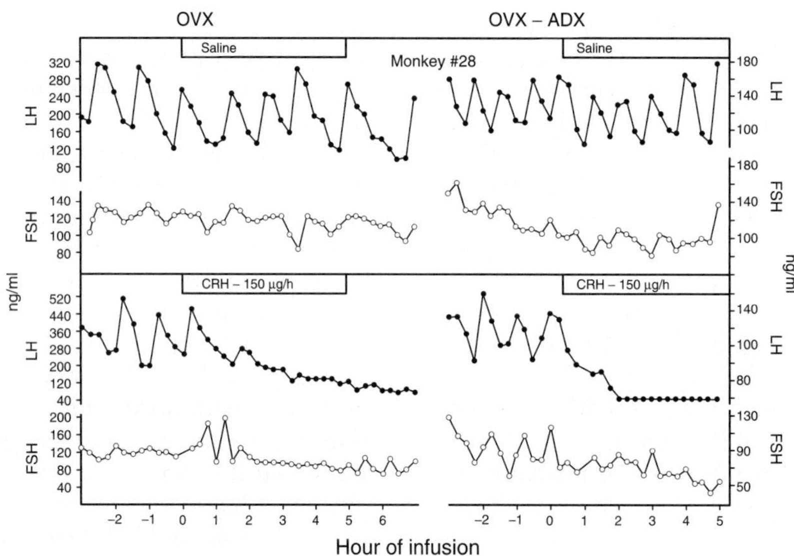

FIG. 10. Although cortisol is able to influence gonadotropin release by decreasing the gonadotropin response to GnRH, activation of the peripheral HPA endocrine axis in the form of increased ACTH and glucocorticoids is not necessary for a stressor to acutely inhibit pulsatile gonadotropin secretion. Data in the ovx monkey show that in the acute situation, ACTH and cortisol do not have a primary role in the decrease in LH that follows activation of the HPA axis by CRH (see also text). **Upper panel:** Comparison in three animals of the effects of CRH infusions (*left graphs*) and of ACTH infusions (*right graphs*) on pulsatile LH release (*closed circles*). While CRH rapidly interrupts pulsatile LH release, ACTH infusions do not, even though they produce similar increases in cortisol as those promoted by CRH. [Reprinted with permission from (420), copyright, The Society for the Study of Reproduction.] **Lower panel:** Comparison of the effects of CRH infusions in the ovariectomized (ovx) (*lower left graph*) and the ovariectomized-adrenalectomized animal (ovx-adx) (*lower right graph*): CRH inhibits pulsatile LH release in the ovx-adx animal as well as in the ovx. [Reprinted with permission from (422), copyright 1989, The Endocrine Society.]

A suppression or decrease of pituitary responsiveness to GnRH by glucocorticoids has also been shown in amenorrheic and fertile women (442), in patients with Cushing's disease (540), in cattle (532,541), and in the rat (542). A similar inhibition of the LH response to GnRH is reported after increases in cortisol following ACTH infusions in the ram (543,544) and in the anoestrous ewe (545). A pituitary effect of glucocorticoids is supported by in vitro observations that corticosterone or cortisol inhibit LH secretion by rat or cow pituitary cells in culture when these are challenged with GnRH (546–548). However, male and female pituitaries appear to respond differently as these results could not be confirmed in dispersed pituitary cells from adult male rats (546,549), although a similar treatment inhibits LH in vivo (542). Glucocorticoid receptors are present in the gonadotrope (550), and, as shown in the ovx ewe, the cortisol-induced suppression of pituitary responsiveness to GnRH may be mediated by type II glucocorticoid receptors (551). An effect of cortisol on the gonadotrope may reflect a reduction of GnRH receptor message and expression, although this mechanism requires the presence of estradiol and may be transient (552–554).

Yet, the literature is complex and results may vary between species, the levels of cortisol attained during an infusion, the chronicity of the cortisol increase, the nature of the stressor, or other circumstances (555,556). The gonadal endocrine status of the animal may also be relevant: gonadal steroids may modulate the site of the response since in estrogen-replaced ovx ewes the stress of isolation preferentially inhibits LH pulse frequency rather than pulse amplitude, suggesting a possible hypothalamic site of action of cortisol (253). The fact that cortisol administered in the follicular phase also inhibits pulse frequency preferentially suggests that estradiol may unveil a specific mechanism allowing cortisol to act at the hypothalamic level (557). Furthermore, in contrast to the preceding data supporting a pituitary effect of cortisol, there are several other studies that have concluded that the ability of the gonadotrope to respond to GnRH is not impaired in the presence of elevated cortisol levels. For instance, in the castrated male monkey and in the pig, the deficit in gonadotropin secretion induced by chronic hypercortisolemia can be avoided by a chronic pulsatile infusion of GnRH (526,528). In addition, in the pig, ram, or ewe replaced with GnRH, inhibitory effects of glucocorticoids on LH release are not seen when the pituitary is surgically disconnected from the hypothalamus (558,559). Such findings suggest that cortisol may also act at a hypothalamic site to decrease GnRH pulse generator activity or that an unknown factor of central origin may mediate cortisol-induced inhibition of pituitary responsiveness to GnRH (539).

It should also be pointed out that, under certain experimental conditions such as after administration of TNFα, a pro-inflammatory cytokine, investigators have shown *protective* effects of glucocorticoids on LH pulsatility by suppressing the excitability of PVN neurons (560).

Overall, it appears safe to conclude that activation of the peripheral HPA endocrine axis in the form of increased ACTH and glucocorticoids is not necessary for a stressor to acutely inhibit pulsatile gonadotropin secretion. At the same time, several reports also suggest that, particularly in the chronic but also in the acute situation, cortisol can influence gonadotropin release by decreasing the gonadotropin response to GnRH. Both conclusions are probably not mutually exclusive: thus, although glucocorticoids may not be required in the primary response to acute stress, their release in response to the activation of the HPA axis by stressors may act to amplify the inhibitory effect of a stressor on the reproductive axis by directly attenuating the response of the gonadotrope to GnRH or indirectly through a central action and/or the release of a putative central factor. The complexity of action of the glucocorticoids and of its interpretation is vividly illustrated by a set of experiments in the ovx ewe using endotoxin as an inflammatory/immune stressor. Investigators have shown that a high dose of endotoxin inhibits LH secretion both at the hypothalamic level, by decreasing pulsatile GnRH release, and at the pituitary level by decreasing responses to GnRH, while a lower dose inhibits LH pulses without alteration in GnRH release (237,286). Suppression of the increase in cortisol that follows endotoxin administration by metyrapone, a drug blocking the crucial enzyme in cortisol synthesis, fails to prevent the inhibitory effects of endotoxin on any aspects of LH pulsatility and suggests that enhanced cortisol release is not essential to suppress the HPG axis (522). Yet, experimental simulation of both low and high endotoxin-induced increases in circulating cortisol by delivering the glucocorticoid in the absence of the endotoxin challenge also results in a rapid suppression of LH secretion at both doses (522). This effect appears to reflect a suppression of pituitary responsiveness rather than a suppression of GnRH (539,551).

Effects of Stressors on Gonadal Steroid Secretion

In the intact animal, stressors are usually accompanied by a decrease in the levels of gonadal steroids. There is evidence that these inhibitory effects are exerted indirectly as well as directly.

An Indirect Action. Studies in the intact animal have shown that stressors diminish gonadal steroid secretion and that parallel decreases in LH and gonadal steroids are observed (244,256,561). Such data clearly suggest that the decrease in ovarian or testicular steroidogenic activity results from a reduced stimulatory input by suppressed circulating gonadotropins in response to stressors.

A Direct Action. There is, however, a significant number of reports suggesting that stressors, potent enough to activate the HPA axis and to elevate endogenous glucocorticoids levels, can reduce the release of testosterone and other testicular steroids as well as steroid testicular content without apparently altering plasma LH concentrations (562–568). Similarly, administration of cortisol in men and dexamethasone in rats is followed by a rapid decrease in serum testosterone levels without apparent changes in LH levels (271,568). These and additional observations below support the possibility that stressors may also disrupt gonadal steroidogenesis via mechanisms that may be independent of changes in circulating levels of gonadotropins. (Importantly however, it should be kept in mind that the reported absence of changes in LH levels in some of the reported experiments may well reflect the paucity of LH data points and especially the absence of pulsatility studies which, if performed, may have revealed significant changes in LH pulse characteristics.) These mechanisms may include a direct effect of glucocorticoids and/or of stressors on the gonads. For example: in the baboon, stressors such as restraint or social instability decrease testosterone secretion to a degree that can be simulated by dexamethasone administration (569). Reduced plasma testosterone levels following restraint in the rat can be partially restored by injections of an antiglucocorticoid (570). In the pubertal carp, administration of a glucocorticoid receptor antagonist reverses the inhibition of the first spermatogenesis waves and the impairment of androgen synthesis that occur following repeated temperature stress or prolonged treatment with cortisol (571,572). It is also worthwhile to mention that during critical illness in the human, at a time of elevated cortisol levels, exogenous GnRH therapy is unable to restore depressed androgen levels (573).

While binding characteristics of LH/hCG receptors appear to be unaffected by a stressor such as restraint (564,574), glucocorticoids have been shown to directly inhibit testicular LH receptor content (575) and a rapid decrease in Leydig cell sensitivity to gonadotropin stimulation occurs during restraint in the rodent (565,576). Decreases in testicular steroid secretion and production after restraint are coupled with enzymatic changes in the testes suggesting that inhibition of enzyme activity may be partially responsible for the disruption of steroidogenesis. Multiple enzymatic steps in steroid synthesis pathways may be affected by stressors. A major impact appears to be on C17,20-lyase, with the result that in vitro production of testosterone from progesterone or 17α-hydroxypregnenolone as substrates is reduced in testicular interstitial cells of stressed rats (563). Restraint has no effect on Km values but reduces the Vmax values of not only C17,20-lyase but also of 17α-hydroxylase and of 3β-hydroxysteroid dehydrogenase (562,563,565). Restraint or glucocorticoids also suppress the translation of transcripts for $P450_{SCC}$ and 3β-hydroxysteroid dehydrogenase (566,577). Another potential contribution to the decrease in circulating androgen levels following stressors follows the report that excessive exposure to corticosterone appears to initiate apoptosis in rat Leydig cells, leading to a decrease in the number of active cells (578).

An inhibitory role of cortisol on the ovary has also been suggested, but data are scant. In the monkey and in the sheep, cortisol or ACTH treatment during the follicular phase interferes with the normal follicular rise of estradiol, delaying the gonadotropin surge and the luteal rise of progesterone (579–582). However, in the sheep, episodic GnRH is shown to override this cortisol-induced delay in follicular maturation (583). Corticosterone and chronic stressors also appear to influence ovarian adrenergic activity (584,585).

A Potential Direct Hypothalamic-Gonadal Pathway. In the rat, testicular responsiveness to hCG is blunted following intraventricular administration of catecholamines or of compounds which release catecholamines, such as CRH or IL-1 in a manner that is independent of changes in LH or corticosterone (586–588). Leydig cell responsiveness to gonadotropin is also reduced after crowding in the adult male rat, a stressor insufficient to alter corticosterone or LH levels (589). These actions have been speculated to reflect the existence of a LH- and glucocorticoid-independent pathway regulating testosterone release (586,587). There is in fact a brain-gonadal neural circuit, the existence of which was postulated by several investigators (590–592). Such a circuit has been recently documented by transsynaptic viral tracing techniques and retrogradely mapped from the testes up the spinal cord to brain stem areas, including the caudal raphe nucleus, the nucleus of the solitary tract and the locus coeruleus, and to the hypothalamus including PVN, the site of CRH, and VP neurons (593,594). PVN is an important functional component of this circuit: for example, microinfusion of CRH or of a β-adrenergic agonist into this hypothalamic nucleus blunts Leydig cell responsiveness to hCG stimulation, while electrolytic PVN lesions abolish the inhibitory effects of intracerebroventricular injections of these compounds on the same endpoint (595). Cells in PVN send projections to the A5 noradrenergic cell group of the medulla which contains neurons transsynaptically linked to the testis (595,596). Spinal cord transection abolishes the ability of intracerebroventricularly injected IL-1β or CRH to blunt the testosterone response to hCG (593). There is also evidence for a brain-ovarian neural circuit and for the regulation of ovarian function by direct neural influences (591, 592,597). The brain structures labeled following the use of transneuronal tracing techniques show considerable overlap with those in the male (594,598). Significantly, retrograde tracing following viral injections in the ovary also highlights the parvocellular portions of PVN, with descending fibers terminating in sympathetic neurons (181,599). Although extrinsic ovarian innervation appears to contribute to ovarian steroidogenesis (600) and corticosterone and chronic stressors may influence ovarian adrenergic activity (584,585), the role of stressors on the ovary through this direct brain-ovarian circuit remains to be fully investigated.

The ovary (and presumably also the testes) has also been shown to contain an intrinsic network of neurons that may play a role in the response of the gonad to stressors. These neurons are phenotypically heterogeneous, with about 20% characterized as catecholaminergic and others as of a peptidergic nature (601,602). Intra-gonadal secretagogues, such as CRH, opiates, catecholamines, cytokines or growth factors, may be activated by stressors and may thereby regulate gonadal steroidogenesis independently of the gonadotropins through a paracrine action (603). For instance, cytokines, such as IL-1 and IL-6, released by an endotoxin challenge and mediating the physiological response to infection can act in vivo and in vitro to inhibit gonadal steroid secretion independently of the presence of the pituitary (233,604–606). A similar paracrine role for testicular β-endorphin in modulating testicular steroidogenesis has been suggested (574,607). A potential local role for CRH in the gonad is reviewed below.

Overall, while the main impact of stressors is most probably exerted on the hypothalamic–pituitary axis and on inhibiting GnRH and LH pulsatile release, thereby reducing gonadal steroid secretion, there is also evidence that several adjunct mechanisms, independent of gonadotropin changes and activated through direct brain-gonadal pathways or through direct effects of stressors or glucocorticoids on the gonads, may enhance or modulate stressor-induced inhibition of gonadal steroids. These latter effects may include a modulation of gonadal steroid enzymatic activity, paracrine effects of local gonadal substances or other mechanisms that remain to be substantiated.

A Paracrine Role for CRH in the Gonads

Besides functioning as the key brain neuropeptide in the HPA cascade, CRH and its receptors have also been localized in the gonads, as well as in other reproductive tissues, such as the uterus and placenta (see below) and has been speculated to play a paracrine role in several reproductive functions (608).

CRH and the Testis. CRH has been shown to be present in rat and ovine testes (609,610) and localized to Leydig cells (609). A major stimulus for Leydig cell-derived CRH appears to be LH which promotes the release of CRH in a dose- dependent fashion. Testicular immunoreactive CRH is drastically reduced after hypophysectomy (609). The action of LH on local testicular CRH appears to be mediated by serotonin pathways (611–613). Rat Leydig cells possess functional high affinity receptors for CRH (614). [In the rat, CRH has also been identified in epididymal spermatozoa (609).]

In cultured adult *rat* Leydig cells, CRH inhibits hCG-induced androgen production in a dose-dependent manner and at sites related to receptor/cyclase coupling or cAMP formation (614). This action possibly results from CRH interaction with local aminergic systems (613). Incubation of adult rat Leydig cells with a CRH antagonist increases basal and hCG-stimulated testosterone secretion and cAMP production, thus possibly revealing a tonic suppression of testosterone production by locally produced CRH (611). In contrast, CRH stimulates cAMP and testosterone production by *mouse* primary cultures of purified Leydig cells as well as by a Leydig tumor cell line, an effect reversed by a CRH antagonist (615). The response to CRH in these mouse cultures is very similar to that seen after LH (611,616,617).

CRH and the Ovary. CRH and its receptors have been detected in the rat and human ovary (618,619). Cytoplasmic CRH immunoreactivity is intense in theca cells surrounding the ovarian follicles, in luteinized cells of the stroma, and in a subpopulation of cells within the corpus luteum. [CRH is not present in oocytes of primordial follicles (618,620).] CRH-R1 have also been detected in theca cells (619,620).

The observation that CRH is significantly higher in premenopausal than in postmenopausal ovaries suggest that ovarian CRH may be related to ovarian function during the reproductive lifespan (621). In fact in the rodent, ovarian CRH fluctuates with the estrous cycle, with higher CRH contents at proestrus (622). The ovary may be locally sensitive to stressors, since an acute stress challenge such as exposure to cold water induces a significant increase in CRH ovarian content, although this occurs only at proestrus (622).

The ovarian CRH system, as the testicular system, may also function as a paracrine regulator of steroid biosynthesis. Several reports have shown that CRH inhibits estrogen production from FSH-stimulated rat and human granulosa cells in culture and androgen production from LH-stimulated theca cells from human ovarian follicles in a dose-dependent fashion (623–627).

Like CRH, urocortin has also been identified in organs which are part of the reproductive system, including the human ovary (628). Intense immunoreactivity is detected in the early degenerating corpus luteum, where it has been speculated to influence steroidogenesis (628).

Overall, findings in cell cultures show that local gonadal CRH may act via functional gonadal receptors to modulate gonadotropin-stimulated gonadal steroidogenesis. This paracrine function may represent an additional mechanism whereby CRH modulates the HPG axis (612).

A Role for Central and Adrenal Medulla Neurotransmitters

Stressors are known to activate central and peripheral catecholaminergic and serotoninergic systems (see above), and in turn it may be speculated that activation of these pathways may also participate in the inhibition of pulsatile LH release and gonadal steroid secretion that is induced by these stressors (629). Indeed, it is well known that catecholaminergic and serotoninergic systems can modulate GnRH and gonadotropin release (630–635). Fibers immunoreactive to enzymes involved in the biosynthesis of catecholamines are seen in proximity to GnRH neurons and have been speculated to affect directly the activity of the GnRH pulse generator (636). In various species, catecholamines or dopamine infusions have also been shown to decrease GnRH/LH release (637–639). In regard to stress, there are a few data suggesting the participation of aminergic pathways in the inhibition of the HPG axis by stressors (640–643).

A role for the adrenal medulla in stress-induced inhibition of LH pulsatility has also been postulated. Circulating catecholamines originating from the adrenal medulla, acting through β-receptors, may partially mediate gonadotropin inhibition by stressors in the rodent, since adrenomedullectomy, while not affecting the increase in corticosterone after restraint, reduces the rise in plasma epinephrine and norepinephrine and diminishes the inhibitory effects of the stressor on LH and testosterone release (641,643). Treatment with an agent toxic to sympathetic neurons decreases epinephrine and norepinephrine responses to restraint and also partially antagonizes its inhibitory effects on testosterone biosynthesis (641). Adrenomedullectomy also prevents the suppression of LH pulses by insulin-induced hypoglycemia, a stressor which also activates the sympathoadrenal axis in the rat (262). Administration of chlorisondamine, a sympathetic ganglionic blocker which attenuates stressor-induced release of epinephrine and norepinephrine, eliminates the inhibitory effect of restraint on testosterone secretion in high ranking wild baboons (307). A role for a descending neural pathway is also suggested by the observation that spinal cord transaction in the rodent or human causes deterioration in spermatogenesis, a testosterone-dependent function (644,645).

Figure 11 outlines some of the main pathways by which stressors affect the HPG axis.

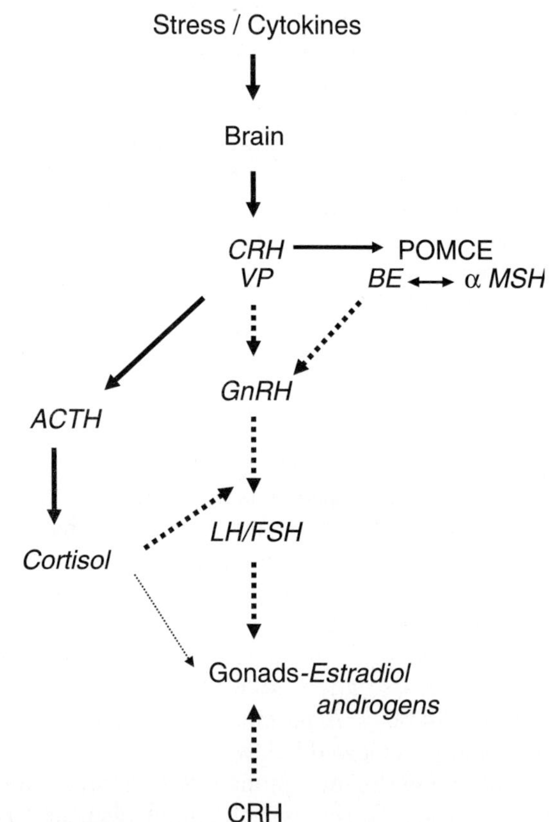

FIG. 11. Schema of the main pathways mediating the HPG endocrine response to stressors. *Continuous lines,* stimulation; *dotted lines,* inhibition. See text for details.

Modulation of the HPA and HPG Axes Response to Stressors by Gonadal Steroids

Gonadal steroids throughout an individual lifetime have been shown to modulate the activity of the HPA endocrine axis and its response to stressors. A large literature, some of which can only briefly be summarized in this chapter, suggests a variety of potential mechanisms.

The mammalian brain is sexually dimorphic (646), a state which presumably in part reflects actions of gonadal steroids during early development (647). Resulting variations in neuroendocrine circuitry may provide an anatomical basis for the numerous reports of differences in basal HPA activity and in the response to stressors between males and females. Some gender differences in the response to stressors may, for instance, reflect the organization of the limbic-HPA neuronal circuits, including PVN, hippocampus, and amygdala, all structures reportedly influenced by gonadal hormone levels during early brain development (648,649). Experimentally, neonatal orchiectomy produces female-like changes within these structures, while neonatal estrogenization results in male-like changes (650,651).

It is also probable that some of the gender differences in the response to stressors result from divergent gonadal hormone secretion in adulthood and, in the female, from gonadal steroid fluctuations throughout the reproductive cycle (556). In fact, one of the major confounding aspects in the analysis of the HPA and HPG axes responses to stressors is the influence that gonadal steroids exert on these responses.

Basal Glucocorticoid Secretion: Gender Dimorphism and Changes with the Reproductive Cycle

In the rodent, there is a well-reported gender dimorphism in basal HPA function. In the unstimulated state, female mice and rats have higher concentrations of plasma corticosterone than males (652–655). Increased corticosterone pulse frequency and pulsatile release are also seen in female compared to male rats (656,657) and are reduced to levels seen in males by ovariectomy (658). The extent of sexual dimorphism in resting plasma corticosterone levels in the rodent, however, is dependent on the estrous cycle stage, the difference between male and female being minimal with females at estrus and highest with females at proestrus (at the peak of estradiol secretion) (655). Indeed, several investigators report higher corticosterone levels at proestrus than at other stages of the estrous cycle (659–661), although others fail to detect such differences (662).

In contrast to the rodent, investigators report slightly lower mean plasma cortisol levels and urinary excretion of total cortisol metabolites in women than men (663–666), although another study finds no differences in urine cortisol production by gender (667). No differences in saliva cortisol levels at different phases of the menstrual cycle were reported (668). It is possible, however, that studies that take into account the overall circadian release of cortisol and the precise stage of the menstrual cycle for comparison may yet reveal subtle gender differences in basal cortisol secretion in the human. That ovarian steroids are able to modulate basal HPA activity in the primate is suggested by the observation that in post-menopausal women the circadian amplitude of serum cortisol can be increased by estrogen replacement therapy (669). In the baboon, as well, a 9-day infusion of estradiol enhances basal ACTH and cortisol release (670), and in the monkey the amplitude of circadian changes and daily mean levels of cortisol are reduced by ovariectomy (671). No apparent effects of gender on serum cortisol levels are found in marsupials (672).

Glucocorticoid Response to Stressors

A Stimulatory Role of Estrogen. A majority of studies, although not all, suggest that estrogen exerts a stimulatory effect on the endocrine HPA response to stressors (673).

Role of gender: There appears to be a gender difference in the intensity of the HPA response to stressors. This has been reasonably well demonstrated in the rodent in which various stressors, such as ether anesthesia (653), psychological stressors (674,675), endotoxin or IL-1 challenges (652,676–678), foot shock (679,680), noise (658,681), and alcohol (682) induce a more robust corticosterone and ACTH responses in the (randomly cycling) female than in the male. The ewe also has a greater cortisol response to an isolation/restraint stressor than the ram (673). Interestingly, gender-related divergences in the response of HPA to an immune-inflammatory challenge agree with otherwise described sexual dimorphism in the immune response itself (676,683–685). The sympatho-adrenal circuit also appears to be more reactive in the female than in the male to stressors like novel environment or foot shock (679). Although there are several reports suggesting that women have a higher incidence of depression and other stress-related disorders than men (686–688), the influence of gender on the response to singular stressors in the human has not been systematically investigated. Some report a greater HPA activation in women than in men in response to exposure to heat, others the opposite after infusion of IL-6 or after a psychological challenge (689,690).

Role of cycle stage: In the rodent, the amplitude of the HPA response to stressors varies with the stage of the estrous cycle, with maximal ACTH and corticosterone release at proestrus compared to the estrus and diestrus phases (338,653,659,662). Other investigators, however, were unable to discern such a difference (680,691). In the ewe, the cortisol response to restraint is lower at stages of the estrus cycle when estradiol secretion is reduced (692). However, in one study in women, the cortisol response to a psychological stress does not appear to vary with menstrual cycle phase (693).

Role of estrogen treatment: Further supporting a stimulatory effect of estrogen on the HPA response to stressors are reports that ACTH and corticosterone responses to restraint, novel environment, foot shock, or exposure to ether are reduced in ovx rats and enhanced following estrogen administration to these animals (659,662,680,694). A similar positive influence of estrogen on pituitary-adrenal function has been inferred to in baboons (670). Results are more variable in the human subjected to psychosocial stress: estrogen treatment in postmenopausal women or in men has been shown to either enhance or attenuate the HPA response (695–700). These variations may reflect gender, age, type of stress as well as duration, dose, and type of estrogen replacement therapy. Differences in salivary free cortisol levels, which are frequently used instead of plasma levels in human studies, may also partly reflect the influence of estradiol on corticosteroid-binding protein levels.

Exceptions to the rule: While estrogen enhances the HPA response to many stressors, estrogen modulation of the HPA response to an immune/inflammatory stressor or to insulin-induced hypoglycemia differs significantly. In the rodent as well as in the non-human primate, the HPA response to endotoxin or IL-1 administration is enhanced by gonadectomy in both sexes and blunted by prolonged estradiol or testosterone replacement (676,677,701–705). Intriguingly in the monkey, the HPA response to IL-1α can be completely blocked by estradiol, but only within a rather narrow range of estradiol concentrations (705). The mechanisms underlying this different response are not completely known. However, endotoxin activates several immune and other cell populations: for instance, it enhances IL-1 release from the pituitary and IL-1 in turn induces IL-6 output from the hypothalamus (706). In the monkey and in the rat, estrogens blunt the IL-6 response to IL-1 (704,707). Since IL-6 is a cytokine which plays a major role in the activation of HPA by endotoxin and IL-1 (708), the data suggest that the blunted HPA response to IL-1 may at least in part be related to estradiol-induced alterations in IL-6 release. A similar mechanism may exist in the human since circulating IL-6 levels are also lower in estrogen-replaced postmenopausal women (709) and there is a parallel attenuation in the response of the cytokines IL-6 and TNF-α to endotoxin and in the response of HPA (710). [Androgens also exert inhibitory effects on IL-6 production and gene expression (711,712).] Ovarian steroids may modulate the HPA response to insulin-induced hypoglycemia in a similar fashion. Rams have a greater cortisol response to insulin-induced hypoglycemia than ewes (692). Ewes have a smaller cortisol response to insulin-induced hypoglycemia in the mid-luteal phase than in the early follicular phase (692), while estradiol treatment reduces the ACTH and cortisol responses to this stressor (713).

Central estrogen pathways: The central nervous system is a major target of estrogen action (714). However, the mechanisms and pathways by which estradiol may modulate the HPA axis response to stressors remain to be completely elucidated. The specific estrogen receptor that mediates this action also remains to be clarified. CRH neurons immunoreactive for ER-α and for ER-β have been identified in the medial preoptic area and in PVN and PVN neurons in the mouse and rat appear to express much greater amounts of ER-β than ER-α (444, 715–717). ER-β in parvocellular PVN neurons is regulated by glucocorticoids: ER-β mRNA expression is up-regulated by stressors and the number of ER-β immunoreactive nuclei is increased following glucocorticoid treatment (an action opposite to that of estrogen which down-regulates ER-β) and ER may thus critically mediate the HPA axis response to stressors (718–720). Data suggest that one of the functions of ER-β may be to modulate the immune response, in that ER-β agonists exert beneficial effects in a transgenic rat model of inflammatory bowel disease and in the Lewis rat adjuvant-induced arthritis model (721). In contrast, ER-α appears to predominate in regions involved with the HPG axis, such as the arcuate nucleus and preoptic area (717). Specific ER patterns may, however, vary between species: in the human for example, ER-α may be the predominant receptor subtype in PVN (722), with apparent greater density in males than females (723).

Findings in the rodent have shown that CRH gene expression is under direct estrogen regulation (724), and that there may be mutual, bidirectional interactions between the HPA and HPG axes (174). Estrogen action is exerted directly on CRH and VP neurons: estrogens stimulate the synthesis of CRH and of VP in PVN, which levels appear to be higher in females than in males (650,724–729). Similar effects of estrogens on CRH are seen in the monkey suggesting that ovarian steroid regulation of CRH synthesis and release may be part of the mechanisms by which ovarian steroids affect the HPA during basal conditions and in its response to stressors (730). Such gender differences have been suggested as an explanation as to why emotional disorders associated with elevated CRH release, such as depression and anxiety (12), have a higher incidence in women than men (731). Ovariectomy decreases hypothalamic CRH content (732) while CRH and VP synthesis in PVN increases following estrogen replacement or an estrogen/progesterone regimen in ovx animals (527,650,730,733). CRH expression in PVN also increases at proestrus, at a time of maximal endogenous estrogen activity during the estrus cycle (48,726). Stressors such as restraint or an endotoxin challenge induce greater CRH and VP transcripts and a greater intensity of c-fos mRNA expression in identical CRH-immunoreactive neurons within parvocellular divisions of PVN on the morning of proestrus, paralleling the high degree of sensitivity of the HPA axis at that time (48,313). Although the above results suggest a stimulatory role for estrogen in the modulation of the CRH and VP neuron, data on peptide synthesis, storage or release do not necessarily reveal simple relationships between these three parameters. It should also be kept in mind that estrogen may also exert an inhibitory effect on CRH activity depending on the duration or strength of the estrogen replacement therapy, as illustrated by results showing that longer chronic administration of high doses of estradiol may reduce hypothalamic CRH content and release (733).

Estrogen effects on the HPA axis may also be mediated by many of the transsynaptic pathways, outlined in the first part of this review (144). PVN is indeed innervated by many regions that include the medial preoptic area, the brain stem, the amygdala, and subdivisions of the bed nucleus of the stria terminalis, all of which express high gonadal steroid receptor activity and are areas which may serve as a potential interface between gonadal steroids and the HPA axis (144,734). It should also be noted here that the glucocorticoid negative (inhibitory) feedback on ACTH and corticosterone responses to stressors appears to be more efficient in females than in males, perhaps the result of higher concentrations of glucocorticoid receptors in hippocampal and hypothalamic regions in the female (735). This would suggest that estrogens can interact with signal transmission by glucocorticoids and possibly impair the glucocorticoid receptor-mediated slow negative feedback (694), indicating that brain corticosteroid receptors are also potential targets for estrogen or other gonadal hormone action (649).

An Inhibitory Role of Androgen. In contrast to the stimulatory effect of estrogens on the HPA, androgens appear to exert opposite effects and to act as restrainers of the HPA and of its response to stressors.

Role of gonadectomy and hormone replacement therapy: In the male rat, gonadectomy enhances both basal HPA activity and the ACTH and corticosterone response to foot shock, noise, or restraint as compared to intact males (649,658,673,736,737). This increased response in male gonadectomized rats appears to be due to a release from androgen-mediated inhibition of the HPA axis, since replacement with testosterone or other androgens restores the HPA response to that seen in the intact animal. Estrogen treatment of male gonadectomized rats also increases post-stress cortisol secretion above that seen prior to treatment (736). In the male hamster, gonadectomy also leads to enhanced levels of glucocorticoids (738). In the baboon, plasma ACTH decreases after several days of androstenedione treatment, although cortisol levels remain unchanged (670).

Central androgen pathways: Orchiectomy increases CRH levels and the number of CRH-immunoreactive cells in PVN compared to those in intact animals, while concomitant dihydrotestosterone treatment prevents these increases, presumably by removing an androgen-dependent repression (739). Orchiectomy also has a stimulatory effect on restraint-induced activation of CRH and VP transcription within medial parvocellular neurons of PVN, probably indicating a requirement for peptide replenishment, and also increases the magnitude of FOS activation within the same area when compared to intact animals (740). Significantly, in the latter study a better correlation is found for VP than CRH, corroborating other data suggesting that changes in HPA activity may conform more with changes in VP than CRH activity in PVN at the transcriptional, synthetic, and secretory levels (741). The targets for androgen, as those for estrogens, may also well involve upstream components, such as for instances the medial preoptic area where testosterone implants decrease the ACTH response to restraint (737,742), and the amygdala and bed nucleus of the stria terminalis (740,743,744).

Overall, there is no doubt that gonadal steroids in part dictate the profile of the HPA response to stressors. Although estrogens appear to enhance this response and androgens to decrease it, this conclusion is valid only for certain stressors, but not all. Effects of gonadal steroids also depend on the absolute levels of the hormones, thus showing variations with the reproductive cycle, the length of a replacement treatment, as well as possibly a myriad of other variations which are difficult to generalize at the present time.

Gonadal Steroid Modulation of the HPG Response to Stressors

There is anecdotal evidence that the response of the HPG axis to stressors or to glucocorticoid exposure is also modulated by gonadal steroids. Estradiol appears to sensitize or amplify the HPG response to several stressors. For examples, while isolation/restraint decreases LH pulse frequency only in ovx ewes, both LH pulse frequency and amplitude are reduced after treatment with estradiol (253). VP decreases LH pulse amplitude in ovx rats, but only after estradiol replacement (439). In the nonhuman primate, the impact of restraint or insulin-induced hypoglycemia on pulsatile GnRH and LH release is greater in the intact than in the ovx monkey and is intensified in the ovx monkey following replacement with estradiol (283,745). In the monkey, however, progesterone does not appear to synergize with estradiol since restraint does not suppress pulsatile LH release in the luteal phase (746). Estradiol replacement in the ovx rat also enhances the suppression of pulsatile LH release by insulin-induced hypoglycemia, while addition of progesterone further potentiates this inhibitory response (262,527).

The precise mechanisms, the pathways that are activated or inhibited by gonadal steroids and the exact relationship to changes in the HPA complex remain to be investigated in-depth. One important point of impact for the gonadal steroids may be the endogenous opioid pathways. Estrogen and progesterone are well-known modulators of endogenous opioid peptide activity (472,473). It may also be that the modulation of the HPG axis response to stressors by gonadal steroids may partly reflect specific effects of the steroids on the HPG axis itself. For instance, estrogens are known to induce significant plastic changes in the GnRH neuron in the primate and rodent (747–751), and it may be speculated that these may lead to configurational changes in its relationship to other neural pathways, including components of the HPA axis.

Stressors and Development

Prenatal life, the early postnatal period, infancy and adolescence are times of active plasticity in the brain and in both central HPA and HPG components. It has been postulated that in these periods individuals may be particularly sensitive to stressors, an effect which in turn may exert more permanent functional alterations, and that therefore early life events may serve to alter HPA responses to stress in adulthood (14).

Prenatal Stress

There are several experimental examples of long-term damage to the reproductive system resulting from prenatal stress. A variety of prenatal stress challenges, such as exposure to phenobarbital, crowding, frequent encounters with another pregnant female or other challenges that result in an activation of HPA, result in a delay in pubertal onset, alter the reproductive cycle or lead to defects in adult sexual and emotional behavior (752–759). In rodents, prenatal maternal stress can demasculinize and feminize the sexual behavior of adult male offspring, an effect simulated by prenatal exposure to ACTH or dexamethasone (760). Such stress results in a decrease in 3β-hydroxysteroid dehydrogenase activity in the fetal testis and lowers basal testosterone concentrations in adult males (761). These results suggest that, notwithstanding the proposed role of ACTH on the fetal testis (see below), suppressed gonadotropin secretion by stressors during critical periods of development can alter fetal testicular function (761).

More generally, prenatal challenges have been shown to induce abnormal or deficient behavior in adulthood, for example defects in attack and threat behavior in male offspring from mice subjected to crowding during late pregnancy, an effect also mimicked by ACTH treatment during the same period of pregnancy (14,762). Such effects have been linked to persistent alterations in the response to stressors that may be associated with lasting changes in the regulation of the HPA axis (763). Significantly, persistent effects of stressors may be found in second generation rats bred from females whose own mothers had been stressed during pregnancy. This second generation is shown to grow more slowly and is smaller than control counterparts, perhaps suggesting a changed genetic programming of first generation ovarian germ cells or a changed uterine environment in the second generation.

Prenatal stressors also affect HPA function later on in life. For example, immature offspring of alcohol-fed pregnant rats show a blunted ACTH response to immune signals, while mature offspring respond to immune or physical stressors with an exaggerated HPA activation (764), although normal responses to acute stressors in adulthood have been observed in other experimental groups (765). There may be a functional relationship between HPA pathways influenced by the prenatal injury and those influenced by female sex steroids (764). Similar effects are seen following exposure of fetal guinea pigs to a synthetic glucocorticoid, a treatment which modifies limbic and hypothalamic glucocorticoid receptor expression in juvenile offspring in a sex-specific fashion (766). This may indicate different windows of vulnerability to prenatal glucocorticoid exposure in fetal life (735). More recent experimental evidence also suggests that prenatal stress appears to alter cognitive and endocrine profiles in a sex-specific fashion, although more systematic studies remain to be performed (767).

In pregnant women, exposure to psychosocial stressors or other stressors have been associated with shortened gestation length and lower newborn body weights. Maternal stress is associated with increased maternal levels of ACTH and cortisol, which may be amplified by placental release of CRH (see below) and increased fetal HPA activity.

A role for ACTH in early gestation?: In the mouse, fetal Leydig cells express the ACTH receptor [melanocortin type 2 receptor (Mc2r)] at 100-fold higher levels than in the adult testis and incubation of fetal or neonatal testes with ACTH stimulates testosterone production. This observation has led to the suggestion that fetal (but not adult) Leydig cells are sensitive to ACTH stimulation (768). In contrast, in the human proliferation and differentiation of Leydig cells after week 10 of gestation appear to be primarily LH/hCG dependent (769). Yet, there are also instances of chronic ACTH stimulation associated with increased androgen secretion or testicular masses, such as in the case of X-linked adrenal hypoplasia and precocious puberty (770) and in patients with congenital adrenal hyperplasia (771). These observations provide an interesting background to speculations of a role for ACTH in human fetal steroidogenesis (772).Yet, their significance to prenatal stress remains to be demonstrated.

Early Neonatal Stress

Maternal separation of neonates from their mothers also alters reproductive behavior during early postnatal development in a gender specific fashion, with males more liable to after effects of neonatal maternal deprivation than females (773,774). Perinatal exposure to daily injections of ACTH or to events or agents that markedly increase ACTH and corticosteroids in rat pups results in a delayed onset of puberty and in deficits in sexual behavior, which appears to be more permanent in the male (775). These results have been related to immediate and long-term changes in central monoamine function, which suggest a hastened maturation and accelerated aging in these systems (776).

Several reports convincingly demonstrate that modulation of early life experience in the neonatal rat permanently influences the function of key pathways that mediate the response to stressors in adulthood (777). Neonatal maternal deprivation (773, 778–782) or exposure to other stress challenges (51,783) lead to a greater sensitivity to further stress in adulthood, with enhanced release of ACTH and corticosterone, especially in females. This has been speculated to be associated with effects on the on-going development of central CRH pathways, specifically those mediated through the activation of brainstem noradrenergic neurons (784). Maternal separation increases PVN norepinephrine responses to restraint in adults compared to nonhandled controls (779). CRH-R1 has been shown to be involved in the release of norepinephrine in PVN induced by conditioned fear, since a CRH-R1 antagonist blocks this fear-induced release of noradrenaline (785). Similarly, the same antagonist inhibits a repertoire of behaviors associated with anxiety and fear but increases exploratory and sexual behaviors that are normally suppressed during stress response in adult male rhesus macaques (411).

In contrast to maternal deprivation, maternal care or daily handling of rat pups by the investigator (which have been related to changes induced by maternal behavior like increased grooming upon the pups return to the cage) result in more modest HPA responses to stress and enhanced cognitive ability which to some degree persists in adulthood (778,786–788). This decrease in the stress response is preceded by a down-regulation of hypothalamic CRH-mRNA levels (780), evident in the rat already by postnatal day 9 and sustained beyond puberty. Thus, maternal care, by regulating systems that normally serve to inhibit CRH expression and release, may influence the development of stress reactivity in the offspring. Systems that have been shown to contribute to this phenomenon may include alterations of glucocorticoid receptors (786,789), of GABA A receptors (a regulatory system that normally serves to inhibit CRH expression and release) (790), and of aminergic systems (791,792). Maternal care in the rat also promotes hippocampal synaptogenesis, and spatial learning and memory through systems known to mediate experience-dependent neural development (779). Maternal care also reduces behavioral fearfulness in response to novelty, thus suggesting that during infancy it serves to "program" behavioral responses to stressors in the offspring by altering the development of the neural systems that mediate fearfulness (778,787,788). Thus, overall maternal care induces changes which can be speculated to make coping more efficient and render the stressful stimulus less noxious. Recent data suggest that differences between two strains of mice in regard to behavioral measures of fear parallel strain differences in the benzodiazepine/GABA (A) receptor system and may reflect, at least in part, the degree of parental care. Such findings may reflect a mammalian example of an indirect genetic effect mediated by maternal care (793). Of interest also is the observation that natural variations in maternal care

are associated with differences in estrogen receptor expression, in that offspring of high caring mothers have higher levels of ERα, which appear to be related to higher estrogen sensitivity, in the medial preoptic area (794).

In the nonhuman primate and human as well, there are several studies which suggest that early postnatal environment and experiences can have profound and prolonged behavioral and physiological consequences and influence the individual response to stressors in later life as well as stress reactivity across generations (795–808).

In the rat, there appears to be a so-called "stress hyporesponsive period" characterized by a reduced capacity to release ACTH and corticosterone in response to stressors during a short period after birth (day 2 to week 2) (809,810). Enhanced sensitivity to the negative (inhibitory) feedback by circulating glucocorticoids, primarily at the pituitary level but perhaps also at the brain, appears to be a critical mediator of this phenomenon (811,812). This hyporesponsive period may be, in part, activated by high corticosterone levels attained in response to the stress of delivery (813). It has been speculated that a hyporesponsive period may ensure the low, stable corticosteroid levels that appear to be optimal for neuronal development in glucocorticoid-sensitive brain regions, thus promoting the adaptive value of the stress hyporesponsive period. Indeed, excessively high or low corticoid levels are associated with abnormal neural and behavioral development (168). A transient increase during the neonatal period in brain 3α-hydroxysteroid dehydrogenase and the resultant increase in neuroactive steroids has also been suggested to contribute to stress hyporesponsiveness (774). Although a hyporesponsive period has also been described in the mouse, it is probably related only to purely neuroendocrine but not to inflammatory stimuli (652). Overall, however, the general relevance of the preceding observations remains questionable since a similar stress hyporesponsiveness period has not been reported in higher species.

Stress at Puberty

Small differences in the HPG response to acute stressors between peripubertal and mature animals have been reported, some of which may reflect differences in the maturation stage of central HPG-related mechanisms. For example, the short-lived acute increases in LH and gonadal steroids in response to acute stressors which are observed in the adult animal (see above) are not seen in animals during the early stages of the pubertal process (814,815). Generally, however, stressors can exert inhibitory effects on the HPG axis of the peripubertal animal as they do in the adult. For instance, repeated restraint or confinement periods applied from prepuberty to early puberty, confirmed by increased plasma ACTH and corticosterone, all cause a decrease in LH and interfere with spermatogenic and androgenic testicular maturation in the rat and pig (245, 816,817). Similar inhibitory effects on gonadotropins, testicular androgen secretion and spermatogenesis are observed during the pubertal period in the carp fed with cortisol-containing food pellets (571). However, closer examinations show that the decrease in LH in response to acute restraint is somewhat smaller in the peripubertal than in the young adult rat (814). Furthermore, in male rats chronic stressors such as cold or restraint do not readily inhibit testosterone in rapidly growing peripubertal rats as they do in young postpubertal rats (815,818). This has led to the hypothesis that the maintenance of fairly constant levels of testosterone in peripubertal male rats could account at least in part for the stability in energy balance, suggesting that rats entering puberty place their resources into ponderal growth while those leaving puberty emphasize maintenance of reproductive capability (815).

In general, basal levels of corticosterone and HPA responses to acute stressors are reported to be weakly influenced by age in rats during the peripubertal period (819,820). There are, however, some observations of gender-dependent differences in the characteristics of HPA responses to specific stressors. For example, during the juvenile age female mice show a hyporesponsive HPA when challenged with endotoxin and the maximal response to this stressor is postponed until after puberty (652). In contrast, in male rats the overall ACTH/corticosterone response to repeated restraint or cold are consistently higher at the beginning than at the end of the pubertal process but a clear habituation occurs more rapidly in the peripubertal rats (815). This particular phenomenon may be due to puberty-related progressive increases in testosterone and its reported effects on the regulation of neural pathways concerned with the HPA response to stressors (see below) (821).

Stressors and the Timing of Pubertal Maturation

In the human, extreme psychological stressors, associated for instance with war, have been implicated in the pathophysiology of puberty and shown to delay the age at menarche (822,823). Intensive physical exercise (824–827) or chronic illness (828) can also delay pubertal maturation. Although there is no direct experimental prospective evidence in the nonhuman primate that psychosocial stress can affect the timing of puberty, a few studies have suggested a negative correlation between low social ranking and the age of puberty in the monkey (829,830). In the above examples, however, it is not always possible to separate effects of the stress from changes in energy intake that may accompany these syndromes. It is indeed believed that nutrition may be one of the crucial factors determining the timing of sexual maturation and that malnutrition is probably a more important mechanism responsible for delayed puberty (828).

In contrast to the above mentioned data in which age at menarche is delayed, other life history models focusing on the role of familial stressors have suggested that psychosocial stressors may accelerate, rather than delay, the timing of puberty (831). Studies have suggested that girls growing up in dysfunctional families in which they were exposed to prolonged socioemotional stressors have earlier menarcheal ages than girls exposed to warm and cohesive family environments (832,833). Some of the data appear to be consistent with the hypothesis that father absence is associated with a developmental trajectory characterized by advanced age at menarche and therefore earlier readiness for reproduction and parenting (834). The subject has been extensively reviewed by Ellis (835).

A Paracrine Role for CRH in the Uterus and Placenta

CRH has been found to be present in the human and rat uterus, where the glandular endometrium is rich in granules positive for CRH, expresses the CRH gene and contains CRH-R1 and CRH-R2 (836–838), and several investigators have postulated paracrine roles for CRH in normal and abnormal uterine function. There are also data of a paracrine role of CRH in the placenta in regard to normal and stress-related function (608,839,840).

Potential Physiological Roles During Early Pregnancy

Data suggest that local CRH may play a role in the physiological events that prepare the uterus during early pregnancy. For example: CRH concentration is higher in secretory than in proliferative endometrium (621) and in endometrial stromal cells that exhibit the decidual reaction (841,842). Progesterone, which has decidualizing properties, increases the production and secretion of immunoreactive CRH from stromal cells and induces the activity of the CRH promoter in human endometrial stroma (843). CRH can decidualize stromal cells in vitro (844). Implantation sites of early pregnant rat uterus have a higher content of immunoreactive CRH and of CRH mRNA than inter-implantation regions (845) and treatment of rats in the first 6 days of gestation with a CRH-R1 specific antagonist results in a decrease in the number of endometrial implantation sites and of live embryos (846). The same treatment also markedly decreases endometrial FasL expression (the ligand for Fas) expression, thereby promoting apoptosis of activated T lymphocytes and suggesting that uterine CRH may participate in local immune early pregnancy phenomena (847).

CRH and the Placenta

The placenta of the human, gorilla and rhesus monkey is a major site of CRH synthesis, displaying CRH mRNA levels >1,000 and >20 times higher than those found in the myometrium and deciduas, respectively, while CRH mRNA is not found in rat, guinea pig and lemur placentae (848–851). The CRH receptor population in placenta appears to consist mainly of CRH-R2 (850), although others report the presence of the CRH-R1α subtype (852). CRH is produced both in syncytiotrophoblast cells as well as in placental deciduas and fetal membranes. Intensity of expression varies according to pregnancy stage, with a dramatic increase during the last 2 months of pregnancy (853–861). Placental CRH appears to be modulated by local steroids. Glucocorticoids enhance CRH gene expression and output in the placenta (859,862) [an action which is in contrast to their inhibitory effects on hypothalamic CRH but in accord with their effect on CRH in two extrahypothalamic sites, the amygdala and the bed nucleus of the stria terminalis (see above)]. Estradiol has an inhibitory effect on CRH gene expression and inhibits CRH gene promoter activity in primary placental cells (863), while progesterone has also been shown to inhibit CRH secretion and production (864). Other endogenous factors that may stimulate local placental CRH production include prostaglandin E2, catecholamines, and the cytokine IL-1 (865,866).

CRH is secreted by the placenta into human maternal and fetal plasma, and amniotic fluid during gestation (849,867). Reflecting changes in CRH expression with the stage of pregnancy, CRH levels in human maternal and fetal plasma rise from undetectable prior to 20 weeks of gestation to reach a peak near term, with an exponential rise during the final 5 weeks of gestation. Within hours of parturition, plasma CRH levels fall and rapidly return to an undetectable baseline (856,867–869). Parallel results are reported in the baboon (870). There is a high maternal-to-fetal concentration difference in plasma CRH levels, reaching 100-fold at term, indicating a greater release into the maternal circulation (871). Most of the circulating CRH, however, is bound to a CRH-binding protein, which is produced by the liver and which neutralizes CRH activity (872). However, concentrations of CRH-binding protein start to decrease drastically in the later stages of pregnancy, thereby increasing free, and presumably bioactive, CRH (873,874). [Urocortin is also present in placenta and secreted in maternal plasma from week 7 of gestation, but unlike CRH its concentrations do not change as pregnancy progresses (875).]

Placental CRH stimulates the pituitary–adrenal endocrine axis and concentrations of ACTH and free cortisol rise in parallel markedly during the third trimester (608,876–881). The addition of CRH to primary trophoblast cell cultures also stimulates local ACTH secretion in a dose-dependent manner, an action mediated by cAMP as second messenger and not inhibited by dexamethasone (880).

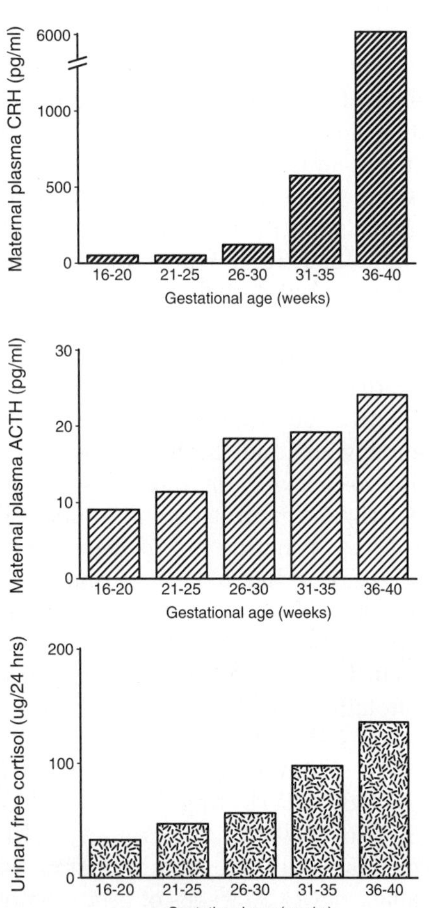

FIG. 12. Secretory profiles of CRH (*upper graph*) and ACTH (*middle graph*) in maternal plasma and of free cortisol (*lower graph*) in urine through gestation. [Reprinted with permission from (876), copyright Elsevier, Inc.]

These effects may be of particular relevance near term, when the maternal and fetal HPA axes are maximally active and the fetus is dependent on elevated glucocorticoid levels for maturation of its organ systems (840,882). In primary cultures of human fetal adrenal cortical cells, CRH increases dehydroepiandrosterone-sulfate (DHEA-S) production and the enzymes required for synthesis of this androgen, which serves as precursor for estrogen synthesis (883). Thus CRH, directly and/or through ACTH, may be necessary to satisfy the high demand on the fetal adrenal to produce DHEA-S and thereby to maintain proper estrogen synthesis in pregnancy (840).CRH exerts a powerful vasodilator effect on the human fetal-placental circulation by activating a nitric oxide-/cGMP-dependent pathway, an effect which is more potent than that of prostacyclin, and thus CRH may also play a role in the control of placental vascular resistance to blood flow in pregnancy (884).

CRH and Stress in Pregnancy

CRH levels are greatly increased in women with pre-eclampsia, a multisystemic disease and the main cause of maternal and neonatal morbidity, and placental release of CRH into the fetal circulation may also be increased in pregnancies complicated by chronic fetal distress resulting in fetal growth retardation and in pre-eclampsia (869,885–887). In these situations, ACTH and cortisol levels parallel the increase in CRH, suggesting that placental CRH, like hypothalamic CRH, is stimulated in conditions of chronic stress and may modulate fetal pituitary-adrenal function in high risk pregnancies (871). In contrast, CRH levels are not significantly altered in patients with hydatidiform mole, polyhydramnios or diabetes. They are, however, elevated in twin pregnancies (888). Higher CRH levels are also reported in products of conception recovered on the 7–9 weeks of gestation from women with a previous history of more than two spontaneous abortions and undergoing elective or spontaneous abortions (889).

Several studies postulate that placental CRH may have a role in the process of parturition and that placental CRH is potentially implicated in the timing of delivery in the human (890–893). CRH production appears to be linked to the length of pregnancy and plasma maternal CRH levels between weeks 28 and 30 of gestation appear to predict gestational length. Highest levels of maternal plasma CRH levels are seen in association with term labor. The positive feedback loop between glucocorticoids and placental CRH (see above) is thought to contribute to this process and thus to parturition (859,862). However, while plasma CRH levels are highest at labor, they do not change significantly during the different stages of labor when evaluated longitudinally (894). Subjects with a pre-term delivery have significantly higher CRH levels in the early third trimester than those who are delivered at term, and pre-term labor is accompanied by inappropriately high levels of CRH for that time of pregnancy (895–897). These levels do not appear to relate to labor itself but may reflect an early activation of the placenta before the onset of pre-term labor (890,898). A significant increase in CRH peptide and mRNA placental content in both spontaneous term and pre-term labor has also been shown in the monkey (899).

CRH may participate in labor through several mechanisms. CRH-R1α and CRH-R2α are present in uterine smooth muscle cells (900) and CRH is a potential local regulator of myometrial contractility (901,902). At term, increases in CRH-R1 mRNA in myometrium in the lower uterine segment (903) and increased CRH-R2 expression in the chorionic mesenchyme (904) appear to be related to the progression of normal labor. CRH however appears unlikely to play a role in the control of myometrial contractility in nonlabor myometrium since the peptide does not affect myometrial tension development in myometrial strips obtained from women undergoing elective nonlabor related caesarean section (905). In vitro observations indicate that CRH and ACTH stimulate the release of prostaglandin-F2α and prostaglandin-E2 from cultured amnion, chorion and placenta (906,907). These compounds are known to play a role in labor in part by stimulating uterine contractility and causing cervix ripening (908). Significantly, addition of a CRH-R antagonist or of a CRH antibody decreases the activity of the enzyme responsible for prostaglandin metabolism into inactive forms. It has been speculated that the rise in CRH in impending pre-term labor contributes to maintain the activity of this enzyme (909).

Overall, while CRH is present in the endometrium, its postulated roles in the decidualization and implantation processes remain to be conclusively demonstrated. Dramatic changes in placental production of CRH and its release in maternal and fetal plasma, especially in the later parts of pregnancy, have been well documented in the primate. These may play a role in preparation to labor and stressful conditions have been postulated to cause premature labor by accelerating the normal CRH profile.

Metabolic-Related Suppression of the HPG Axis: Potential Role of HPA

In many patients, the functional hypothalamic chronic anovulation syndrome (FHCA), which includes amenorrhea and infertility, is thought to be caused primarily by psychogenic stress (see above). FHCA, however, is also associated with lifestyle variables, such as excessive exercise (jogging, athletics) as well as metabolic-related disorders (eating disorders, diets) (277,910–913). Evidently, it may be that exercise or metabolic-related problems are initiated as coping responses to psychosocial stress. However, data in the nonhuman primate also provide evidence that exercise-induced amenorrhea or the acute suppression of LH pulsatility by fasting can be the result of the resultant state of low-energy availability rather than a function of the psychological state (914–916). Similar observations were made in the rodent (917). Since studies have suggested a link between affective disorders and eating disorders (918), it is possible that the combination of mild energy deprivation, such as that induced by nutritional or caloric restriction or exercise, and psychosocial distress, such as performance pressure, unrealistic goals, and negative attributions, may combine synergistically to result in a full-blown FHCA syndrome (277). An indirect support to this theory is the observation that symptoms of disordered eating are shown to discriminate women with FHCA from those with organic amenorrhea or eumenorrheic women (919). Specifically, women with FHCA report more concerns about dieting and weight, fear of weight gain, and tendencies to engage in binge eating, and have higher levels of psychological characteristics associated with disordered eating, such as feelings of general inadequacy, insecurity, and lack of control over life and confusion in identifying and responding to food-related body sensations and emotional states.

Of particular interest to this review is the substantial indirect evidence to suggest the possibility that the HPA axis may be involved in the process whereby low-energy availability affects LH pulsatility. Serum cortisol concentrations are increased in fasted or malnourished human subjects (920). A great number of patients (but not all) in whom FHCA is related to weight loss or excessive exercise also present with hypercortisolemia (277,921–923) which evidently reflects the activation of HPA as in stress-related FHCA (277). Recent data illustrating the dependence of LH pulsatility on energy availability in sedentary, young women in the early follicular phase show a decrease in LH pulse frequency that occurs rather abruptly when energy availability is decreased to a specific level. Significantly, the decrease in LH pulsatility occurs at the same threshold at which cortisol secretion is increased in response to the decrease in energy availability (924). Recent literature has also demonstrated an increase in ACTH and cortisol after ghrelin, neuropeptide Y (NPY) or agouti-related protein (AGRP) administration (925–929). Since ghrelin, or AGRP and NPY (two major neuropeptides mediating the action of ghrelin) (930) are energy-related peptides which have been shown to specifically decrease the frequency of pulsatile LH release (931–933), it may be that one way for these peptides to alter GnRH pulse generator activity in response to deflections from body homeostasis may be through the activation of the HPA axis. It is known that arcuate NPY/AGRP neurons project to PVN (934,935) and thus signals relevant to energy balance may be transmitted by direct innervation of PVN neurons by the arcuate nucleus (144). AGRP increases the release of CRH and VP from hypothalamic explants (936). A role for HPA, and of CRH-R mediated pathways in the action of leptin, another energy-related peptide, on feed intake and energy expenditure has also been postulated (937,938). In the rat, fasting-induced suppression of LH release is observed only in intact or ovx estrogen-replaced rats but not in ovx animals and thus appears to be totally dependent on the presence of estrogen. One of the site at which estradiol appears to act to activate this pathway also involves PVN (939), possibly through an increase in estrogen receptors in that region (940). In support of a role for CRH is the observation that central injection of a CRH antagonist reinstates normal LH release in fasted rats (941).

Besides CRH, urocortin may also be involved in these processes. Injection of urocortin in PVN blocks the stimulatory effects of NPY on feeding, suggesting a possible mediatory role of urocortin (942). One important effect of urocortin administration is to suppress appetite, in an action that is considerably more potent than that of CRH (93,95, 943–945). The anorectic effects of both CRH and urocortin are not blocked by a CRH-R1 antagonist suggesting that effects on food intake by these peptides are mediated primarily by CRH-R2. Feeding is also suppressed following direct injections into PVN or the lateral septum (942,946) and injections in PVN also increase plasma leptin (947). Interestingly in view of observations that decreased caloric intake is generally inhibitory to the HPG axis, central infusion of urocortin in the ovx ewe results in a significant decrease in food intake while concentrations of LH increase (948), although early decreases in LH in the ovx gilt are reported (949).

Overall, the concept that the HPA axis and PVN and CRH or urocortin neurons may mediate some of the inhibitory effects of energy deficits on the HPG axis is intriguing. However, a potential participation of CRH and urocortin in the mechanisms whereby energy balance and nutrition influence the HPG axis remains to be fully investigated in the human and in different animal models. These data should be viewed within the large literature on close interrelationships between feeding, metabolism, energy storage, and glucocorticoid secretion (950).

GENERAL CONCLUSIONS

Although "stress" is difficult to identify in patients with reproductive failure, there is a large literature that documents inhibitory effects of a variety of stressors on normal reproductive function and on the reproductive cycle. It is safe to conclude that one of the major structure affected by stressors is the hypothalamic GnRH pulse generator and that a decline in its ability to properly stimulate the gonadotrope, mainly through a reduction in the frequency of pulsatile GnRH/LH release, is one of the main mechanism involved. This reduction in pulsatility is generally correlated with the clinical spectrum. However, this review also outlines a multiplicity of targets within the central nervous system and within the HPG endocrine axis which stressors can influence and thus a multiplicity of mechanisms are available to modulate the primary organism's response to the stressor.

In a review that strides to provide an overview of the "general" effects of "stress" on the reproductive system, overall statements either represent some consensus in the literature or the fallible judgment of the author. Conflicting data and statements or even the absence of a concluding statement are partly unavoidable because of the difficulty in analyzing, comparing and summarizing data arising from studies using a variety of stressors which, although commonly resulting in the activation of the HPA axis (the common marker for "stress"), not only may influence varying central pathways but may themselves be acute or last varying time periods and to which the subjects may or may not have been habituated. Additional confounding but determining variables such as for example the subject's gender, endocrine status or stage of the reproductive cycle, or the use of gonadal steroid therapy also influence the type of pathways activated and the magnitude of the HPG response. Unfortunately, even within the large existing scientific stress literature, few individual reports have taken such multiple variants into account within identical experimental protocols. The literature in the primate also adds the notion that some subjects may be more "vulnerable" to stressors because of a preexisting "neurotic terrain" or a genetic predisposition, thus making it difficult to compare data between individuals. However, this may explain why some individuals but not others completely cease to have reproductive cycles and become amenorrheic in response to stressors.

Another concept emerging from this review is that, especially in the primate where reproduction requires a large energy investment over a prolonged period of time, stressors at times may have beneficial effects. Indeed, the model may allow females to delay pregnancy and thereby to optimize their lifetime reproductive success by permitting reproduction to occur when conditions for the survival of the conceptus are likely to be sufficiently better than present ones (951). This concept becomes evident in the few prospective studies on the effects of stressors on the menstrual cycle in the nonhuman primate and in women. While amenorrhea, as expressed in the full-blown FHCA syndrome, is clearly the last stage of the inhibitory process, such studies have convincingly demonstrated that a stressor of insufficient strength or duration to produce amenorrhea or affecting a less "vulnerable" individual can interfere with reproductive function in subtle ways without interrupting the menstrual cycle, such as by delaying the maturation of the follicle, inducing an ill-timed ovulatory surge which may damage the maturing follicle or producing an inadequate luteal phase which may interfere with the implantation of the fertilized egg. Because these situations require special efforts to identify in the infertile patient, these are usually not documented and therefore the proportion of infertility due to stress may be underestimated. If this is true, in the future more natural therapeutical approaches to the treatment of infertility could be investigated, since inhibitory effects of stress on HPG are eminently reversible. In addition, the patient may respond to more stress oriented specific therapies, a few examples of which are suggested in the chapter. Finally, in an area untouched in this review, it may also be worthwhile investigating effects besides infertility which stressors may exert, for example by resulting in gonadal hormone deficiencies which may affect nonreproductive systems, such as the cardiovascular system, bone metabolism (356) or the brain. Such studies may reveal that functional reproductive deficits may also become generally detrimental, at least when activated in an environment permissive to the development of chronic disease (225).

A final observation: Although FHCA has been linked to psychogenic stress, the independent association in these patients between the stress and the inhibition of the GnRH pulse generator is difficult to demonstrate. Like in most experimental animal models, such as for example those using restraint, an inflammatory agent, exercise or others as stressors, metabolic insults (such as a decrease in food intake) may accompany the activation of the HPA axis even in the well designed experiment. Thus, in practice it is extremely difficult to separate effects resulting from the activation of energy-related pathways from those resulting from the activation of stress pathways. In fact, there may be an overlap between the two, but it is hoped that, with the recent discovery of new energy related peptides and with future studies of their own effects on the reproductive system, a clearer picture will emerge.

ACKNOWLEDGMENTS

The author would like to thank this section's editor, Dr. Tony Plant, and Dr. Ennian Xiao for reviewing this manuscript and for their helpful comments. The author's research was supported by NIH grants DK 39144 and HD 46715. The author also acknowledges Dr. S. Wardlaw for her long-term and appreciated critical input, as well as W. B. Wehrenberg, D. Van Vugt, N. Y. Lam, D. H. Olster, P. R. Gindoff, E. Xiao, J. Luckhaus, Y. J. Feng, E. Shalts, L. Xia-Zheng, E. Alston, D. Thornell, D. Shanen, A. Barth, J. J. Puder, and N. Vulliemoz for their active contribution to the investigator's studies on the interrelationships between the HPA and HPG axis in the primate.

REFERENCES

1. Mason, J. W. (1975). A historical view of the stress field. *J. Human Stress* 1, 22–36.
2. Kopin, I. J. (1995). Definitions of stress and sympathetic neuronal responses. *Ann. N Y Acad. Sci.* 771, 19–30.
3. Cannon, W. (1935). Stresses and strains of homeostasis. *Am. J. Med.* 89, 1–14.
4. Selye, H. (1936). A syndrome produced by diverse nocuous agents. *Nature* 138, 32–36.
5. Selye, H. (1946). The general adaptation syndrome and the disease of adaptation. *J. Clin. Endocrinol.* 6, 117–230.
6. Selye, H. (1973). The evolution of the stress concept. *Am. Sci.* 61, 692–699.
7. Pacak, K., Palkovits, M., Kvetnansky, R., Yadid, G., Kopin, I. J., and Goldstein, D. S. (1995). Effects of various stressors on in vivo norepinephrine release in the hypothalamic paraventricular nucleus and on the pituitary–adrenocortical axis. *Ann. N Y Acad. Sci.* 771, 115-130.
8. Pacak, K., Palkovits, M., Yadid, G., Kvetnansky, R., Kopin, I. J., and Goldstein, D. S. (1998). Heterogeneous neurochemical responses to different stressors: a test of Selye's doctrine of nonspecificity. *Am. J. Physiol.* 275, R1247–R1255.
9. Pacak, K., and Palkovits, M. (2001). Stressor specificity of central neuroendocrine responses: implications for stress-related disorders. *Endocr. Rev.* 22, 502–548.
10. Mason, J. W. (1971). A re-evaluation of the concept of "non-specificity" in stress theory. *J. Psychiatr. Res.* 8, 323–333.
11. Chrousos, G. P. (1998). Stressors, stress, and neuroendocrine integration of the adaptive response. The 1997 Hans Selye Memorial Lecture. *Ann. N Y Acad. Sci.* 851, 311–335.
12. Chrousos, G. P., and Gold, P. W. (1992). The concepts of stress and stress system disorders. Overview of physical and behavioral homeostasis. *JAMA* 267, 1244–1252.

13. Habib, K. E., Gold, P. W., and Chrousos, G. P. (2001). Neuroendocrinology of stress. *Endocrinol. Metab. Clin. North Am.* 30, 695–728; vii–viii.
14. Charmandari, E., Kino, T., Souvatzoglou, E., and Chrousos, G. P. (2003). Pediatric stress: hormonal mediators and human development. *Horm. Res.* 59, 161–179.
15. Dallman, M. F. (2003). Stress by any other name ... ? *Horm. Behav.* 43, 18–20; discussion 28–30.
16. Sterling, P. and Eyer, J. (1988). Allostasis: a new paradigm to explain arousal pathology. In *Handbook of Life Stress, Cognition, and Health* (S. Fisher and J. Reason, Eds.). Wiley, New York.
17. McEwen, B. S. and Wingfield, J. C. (2003). The concept of allostasis in biology and biomedicine. *Horm. Behav.* 43, 2–15.
18. Perras, B., Schultes, B., Behn, B., Dodt, C., Born, J., and Fehm, H. L. (2004). Intranasal atrial natriuretic peptide acts as central nervous inhibitor of the hypothalamo–pituitary–adrenal stress system in humans. *J. Clin. Endocrinol. Metab.* 89, 4642–4648.
19. Yehuda, R. (2002). Current status of cortisol findings in post-traumatic stress disorder. *Psychiatr. Clin. North. Am.* 25, 341–368, vii.
20. Rosmond, R. (2003). Stress induced disturbances of the HPA axis: a pathway to type 2 diabetes? *Med. Sci. Monit.* 9, RA35–RA39.
21. Gold, P. W., and Chrousos, G. P. (2002). Organization of the stress system and its dysregulation in melancholic and atypical depression: high vs low CRH/NE states. *Mol. Psychiatry* 7, 254–275.
22. McEwen, B. S. (1997). Stress, brain, and behavior: life-long effects upon health and disease. In *Physiology, Stress and Malnutrition. Functional Correlates of Nutritional Intervention* (J. M. Kinney, H. N. Tucker, Eds.), pp. 113–130. Lippincott-Raven, Philadelphia.
23. Yehuda, R. (1999). Risks factors for posttraumatic stress disorder. In *Progress in Psychiatry Series*. American Psychiatry Press, Washington, DC.
24. Chrousos, G. P. (2000). Stress, chronic inflammation, and emotional and physical well-being: concurrent effects and chronic sequelae. *J. Allergy Clin. Immunol.* 106, S275–S291.
25. Hinkle, L. E. Jr. (1987). Stress and disease: the concept after 50 years. *Soc. Sci. Med.* 25, 561–566.
26. Steinberg, A. and Ritzmann, R. F. (1990). A living systems approach to understanding the concept of stress. *Behav. Sci.* 35, 138–146.
27. Cannon, W. (1929). Organization for physiological homeostasis. *Physiol. Rev.* 9, 399–431.
28. Williamson, D. E., Coleman K., Bacanu, S. A., Devlin, B. J., Rogers, J., Ryan, N. D., and Cameron, J. L. (2003). Heritability of fearful-anxious endophenotypes in infant rhesus macaques: a preliminary report. *Biol. Psychiatry* 53, 284–291.
29. Wust, S., Van Rossum, E. F., Federenko, I. S., Koper, J. W., Kumsta, R., and Hellhammer, D. H. (2004). Common polymorphisms in the glucocorticoid receptor gene are associated with adrenocortical responses to psychosocial stress. *J. Clin. Endocrinol. Metab.* 89, 565–573.
30. Nesse, R. M. and Young, E. A. (2000). Evolutionary origins and functions of the stress response. In *Encyclopedia of Stress* (G. Fink, Ed.), Vol. 2, pp. 79–84. Academic Press, San Diego.
31. Miller, D. B. and O'Callaghan, J. P. (2002). Neuroendocrine aspects of the response to stress. *Metabolism* 51, 5–10.
32. Baram, T. Z., Yi, S., Avishai-Eliner, S., and Schultz, L. (1997). Development neurobiology of the stress response: multilevel regulation of corticotropin-releasing hormone function. *Ann. N Y Acad. Sci.* 814, 252–265.
33. Carrasco, G. A., and Van de Kar, L. D. (2003). Neuroendocrine pharmacology of stress. *Eur. J. Pharmacol.* 463, 235–272.
34. Berne, R. M., Levy, M. N., Koeppen, B. M., and Stanton, B. A. (2004). *Physiology*. Mosby, St. Louis.
35. Sapolsky, R., Rivier, C., Yamamoto, G., Plotsky, P., and Vale W. (1987). Interleukin-1 stimulates the secretion of hypothalamic corticotropin-releasing factor. *Science* 238, 522–524.
36. Rivest, S. and Rivier, C. (1995). The role of corticotropin-releasing factor and interleukin-1 in the regulation of neurons controlling reproductive functions. *Endocr. Rev.* 16, 177–199.
37. Swanson, L. W. and Sawchenko, P. E. (1983). Hypothalamic integration: organization of the paraventricular and supraoptic nuclei. *Annu. Rev. Neurosci.* 6, 269–324.
38. Antoni, F. A. (1986). Hypothalamic control of adrenocorticotropin secretion: advances since the discovery of 41-residue corticotropin-releasing factor. *Endocr. Rev.* 7, 351–378.
39. Vale, W., Rivier, C., Brown, M. R., Spiess, J., Koob, G., Swanson, L., Bilezikjian, L., Bloom, F., and Rivier, J. (1983). Chemical and biological characterization of corticotropin releasing factor. *Recent Prog. Horm. Res.* 39, 245–270.
40. Perrin, M. H., and Vale, W. W. (1999). Corticotropin releasing factor receptors and their ligand family. *Ann. N Y Acad. Sci.* 885, 312–328.
41. Arbiser, J. L., Morton, C. C., Bruns, G. A., and Majzoub, J. A. (1988). Human corticotropin releasing hormone gene is located on the long arm of chromosome 8. *Cytogenet. Cell Genet.* 47, 113–116.
42. Gibbs, D. M., and Vale, W. (1982). Presence of corticotropin releasing factor-like immunoreactivity in hypophysial portal blood. *Endocrinology* 111, 1418–1420.
43. Rivest, S., Torres, G., and Rivier, C. (1992). Differential effects of central and peripheral injection of interleukin-1 beta on brain c-fos expression and neuroendocrine functions. *Brain Res.* 587, 13–23.
44. Ericsson, A., Kovacs, K. J., and Sawchenko, P. E. (1994). A functional anatomical analysis of central pathways subserving the effects of interleukin-1 on stress-related neuroendocrine neurons. *J. Neurosci.* 14, 897–913.
45. Imaki, T., Shibasaki, T., Hotta, M., and Demura, H. (1992). Early induction of c-fos precedes increased expression of corticotropin-releasing factor messenger ribonucleic acid in the paraventricular nucleus after immobilization stress. *Endocrinology* 131, 240–246.
46. Rivest, S., and Rivier, C. (1994). Stress and interleukin-1 beta-induced activation of c-fos, NGFI-B and CRF gene expression in the hypothalamic PVN: comparison between Sprague-Dawley, Fisher-344 and Lewis rats. *J. Neuroendocrinol.* 6, 101–117.
47. Suda, T., Tozawa, F., Ushiyama, T., Sumitomo, T., Yamada, M., and Demura, H. (1990). Interleukin-1 stimulates corticotropin-releasing factor gene expression in rat hypothalamus. *Endocrinology* 126, 1223–1228.
48. Nappi, R. E., and Rivest, S. (1995). Ovulatory cycle influences the stimulatory effect of stress on the expression of corticotropin-releasing factor receptor messenger ribonucleic acid in the paraventricular nucleus of the female rat hypothalamus. *Endocrinology* 136, 4073–4083.
49. Adam, C. L., and Findlay, P. A. (1998). Inhibition of luteinizing hormone secretion and expression of c-fos and corticotrophin-releasing factor genes in the paraventricular nucleus during insulin-induced hypoglycaemia in sheep. *J. Neuroendocrinol.* 10, 777–783.
50. Lightman, S. L., Harbuz, M. S., Knight, R. A., and Chowdrey, H. S. (1993). CRF mRNA in normal and stress conditions. *Ann. N Y Acad. Sci.* 697, 28–38.
51. Hatalski, C. G., Guirguis, C., and Baram, T. Z. (1998). Corticotropin releasing factor mRNA expression in the hypothalamic paraventricular nucleus and the central nucleus of

the amygdala is modulated by repeated acute stress in the immature rat. *J. Neuroendocrinol.* 10, 663–669.
52. Xia, L., Matera, C., Ferin, M., and Wardlaw, S. L. (1996). Interleukin-1 stimulates the central release of corticotropin-releasing hormone in the primate. *Neuroendocrinology* 63, 79–84.
53. Navarra, P., Tsagarakis, S., Faria, M. S., Rees, L. H., Besser, G. M., and Grossman, A. B. (1991). Interleukins-1 and -6 stimulate the release of corticotropin-releasing hormone-41 from rat hypothalamus in vitro via the eicosanoid cyclooxygenase pathway. *Endocrinology* 128, 37–44.
54. Zelazowski, P., Patchev, V. K., Zelazowska, E. B., Chrousos, G. P., Gold, P. W., and Sternberg, E. M. (1993). Release of hypothalamic corticotropin-releasing hormone and arginine-vasopressin by interleukin 1 beta and alpha MSH: studies in rats with different susceptibility to inflammatory disease. *Brain Res.* 631, 22–26.
55. Battaglia, D. F., Brown, M. E., Krasa, H. B., Thrun, L. A., Viguie, C., and Karsch, F. J. (1998). Systemic challenge with endotoxin stimulates corticotropin-releasing hormone and arginine vasopressin secretion into hypophyseal portal blood: coincidence with gonadotropin-releasing hormone suppression. *Endocrinology* 139, 4175–4181.
56. Hotta, H., Sato, A., and Sumitomo, T. (1992). Hypothalamic corticotropin-releasing hormone (CRH) secretion into hypophysial portal blood is regulated by cutaneous sensory stimulation in anesthetized rats. *Jpn. J. Physiol.* 42, 515–524.
57. Berkenbosch, F., van Oers, J., del Rey A., Tilders, F., and Besedovsky, H. (1987). Corticotropin-releasing factor-producing neurons in the rat activated by interleukin-1. *Science* 238, 524–526.
58. Bruhn, T. O., Plotsky, P. M., and Vale, W. W. (1984). Effect of paraventricular lesions on corticotropin-releasing factor (CRF)-like immunoreactivity in the stalk-median eminence: studies on the adrenocorticotropin response to ether stress and exogenous CRF. *Endocrinology* 114, 57–62.
59. Li, H. Y., Ericsson A., and Sawchenko, P. E. (1996). Distinct mechanisms underlie activation of hypothalamic neurosecretory neurons and their medullary catecholaminergic afferents in categorically different stress paradigms. *Proc. Natl. Acad. Sci. U S A* 93, 2359–2364.
60. Feng, Y. J., Shalts, E., Xia, L. N., Rivier, J., Rivier, C., Vale, W., and Ferin, M. (1991). An inhibitory effects of interleukin-1a on basal gonadotropin release in the ovariectomized rhesus monkey: reversal by a corticotropin-releasing factor antagonist. *Endocrinology* 128, 2077–2082.
61. Whitnall, M. H., Anderson, K. A., Lane, C. A., Mougey, E. H., Neta, R., and Perlstein, R. S. (1994). Decreased vasopressin content in parvocellular CRH neurosecretory system of Lewis rats. *Neuroreport* 5, 1635–1637.
62. Moncek, F., Kvetnansky, R., and Jezova, D. (2001). Differential responses to stress stimuli of Lewis and Fischer rats at the pituitary and adrenocortical level. *Endocr. Regul.* 35, 35–41.
63. Muglia, L. J., Bethin, K. E., Jacobson, L., Vogt, S. K., and Majzoub, J. A. (2000). Pituitary–adrenal axis regulation in CRH-deficient mice. *Endocr. Res.* 26, 1057–1066.
64. Muglia, L., Jacobson, L., Dikkes, P., and Majzoub, J. A. (1995). Corticotropin-releasing hormone deficiency reveals major fetal but not adult glucocorticoid need. *Nature* 373, 427–432.
65. Jeong, K. H., Jacobson, L., Widmaier, E. P., and Majzoub, J. A. (1999). Normal suppression of the reproductive axis following stress in corticotropin-releasing hormone-deficient mice. *Endocrinology* 140, 1702–1708.
66. Jacobson, L., Muglia, L. J., Weninger, S. C., Markak, K., and Majzoub, J. A. (2000). CRH deficiency impairs but does not block pituitary-adrenal responses to diverse stressors. *Neuroendocrinology* 71, 79–87.
67. Venihaki, M., and Majzoub, J. (2002). Lessons from CRH knockout mice. *Neuropeptides* 36, 96–102.
68. Muglia, L. J., Jacobson, L., Weninger, S. C., Karalis, K. P., Jeong K., and Majzoub, J. A. (2001). The physiology of corticotropin-releasing hormone deficiency in mice. *Peptides* 22, 725–731.
69. Whitnall, M. H., Smyth, D., and Gainer, H. (1987). Vasopressin coexists in half of the corticotropin-releasing factor axons present in the external zone of the median eminence in normal rats. *Neuroendocrinology* 45, 420–424.
70. Mouri, T., Itoi K., Takahashi K., Suda, T., Murakami, O., Yoshinaga K., Andoh, N., Ohtani, H., Masuda, T., and Sasano, N. (1993). Colocalization of corticotropin-releasing factor and vasopressin in the paraventricular nucleus of the human hypothalamus. *Neuroendocrinology* 57, 34–39.
71. Sawchenko, P. E., Swanson, L. W., and Vale, W. W. (1984). Co-expression of corticotropin-releasing factor and vasopressin immunoreactivity in parvocellular neurosecretory neurons of the adrenalectomized rat. *Proc. Natl. Acad. Sci. U S A* 81, 1883–1887.
72. Sawchenko, P. E. (1987). Adrenalectomy-induced enhancement of CRF and vasopressin immunoreactivity in parvocellular neurosecretory neurons: anatomic, peptide, and steroid specificity. *J. Neurosci.* 7, 1093–1106.
73. Sawchenko, P. E., Imaki, T., and Vale, W. (1992). Co-localization of neuroactive substances in the endocrine hypothalamus. *Ciba Found. Symp.* 168, 16–30; discussion 30–42.
74. Zimmerman, E. A. (1981). The organization of oxytocin and vasopressin pathways. *Adv. Biochem. Psychopharmacol.* 28, 63–75.
75. Bartanusz, V., Aubry, J. M., Jezova, D., Baffi, J., and Kiss, J. Z. (1993). Up-regulation of vasopressin mRNA in paraventricular hypophysiotrophic neurons after acute immobilization stress. *Neuroendocrinology* 58, 625–629.
76. Plotsky, P. M. (1987). Facilitation of immunoreactive corticotropin-releasing factor secretion into the hypophysial-portal circulation after activation of catecholaminergic pathways or central norepinephrine injection. *Endocrinology* 121, 924–930.
77. Shalts, E., Feng, Y. J., and Ferin, M. (1992). Vasopressin mediates the interleukin-1 alpha-induced decrease in luteinizing hormone secretion in the ovariectomized rhesus monkey. *Endocrinology* 131, 153–158.
78. de Goeij, D. C., Kvetnansky, R., Whitnall, M. H., Jezova, D., Berkenbosch, F., and Tilders, F. J. (1991). Repeated stress-induced activation of corticotropin-releasing factor neurons enhances vasopressin stores and colocalization with corticotropin-releasing factor in the median eminence of rats. *Neuroendocrinology* 53, 150–159.
79. de Goeij, D. C., Jezova, D., and Tilders, F. J. (1992). Repeated stress enhances vasopressin synthesis in corticotropin releasing factor neurons in the paraventricular nucleus. *Brain Res.* 577, 165–168.
80. Aubry, J. M., Bartanusz, V., Jezova, D., Belin, D., and Kiss, J. Z. (1999). Single stress induces long-lasting elevations in vasopressin mRNA levels in CRF hypophysiotrophic neurones, but repeated stress is required to modify AVP immunoreactivity. *J. Neuroendocrinol.* 11, 377–384.
81. Krieger, D. T., and Liotta, A. (1977). Pituitary ACTH responsiveness in the vasopressin deficient rat. *Life Sci.* 20, 327–335.
82. Zelena, D., Foldes A., Mergl, Z., Barna, I., Kovacs, K. J., and Makara, G. B. (2004). Effects of repeated restraint stress on hypothalamo–pituitary–adrenocortical function in vasopressin deficient Brattleboro rats. *Brain Res. Bull.* 63, 521–530.
83. Makara, G. B., Mergl, Z., and Zelena, D. (2004). The role of vasopressin in hypothalamo–pituitary–adrenal axis activation

during stress: an assessment of the evidence. *Ann. N Y Acad. Sci.* 1018, 151–161.

84. Dautzenberg, F. M. and Hauger, R. L. (2002). The CRF peptide family and their receptors: yet more partners discovered. *Trends Pharmacol. Sci.* 23, 71–77.
85. Hsu, S. Y. and Hsueh, A. J. (2001). Human stresscopin and stresscopin-related peptide are selective ligands for the type 2 corticotropin-releasing hormone receptor. *Nat. Med.* 7, 605–611
86. Donaldson, C. J., Sutton, S. W., Perrin, M. H., Corrigan, A. Z., Lewis, K. A., Rivier, J. E., Vaughan, J. M., and Vale, W. W. (1996). Cloning and characterization of human urocortin. *Endocrinology* 137, 3896.
87. Lewis K., Li, C., Perrin, M. H., Blount A., Kunitake K., Donaldson, C., Vaughan, J., Reyes, T. M., Gulyas, J., Fischer, W., Bilezikjian, L., Rivier, J., Sawchenko, P. E., and Vale, W. W. (2001). Identification of urocortin III, an additional member of the corticotropin-releasing factor (CRF) family with high affinity for the CRF2 receptor. *Proc. Natl. Acad. Sci. U S A* 98, 7570–7575.
88. Reyes, T. M., Lewis K., Perrin, M. H., Kunitake, K. S., Vaughan, J., Arias, C. A., Hogenesch, J. B., Gulyas, J., Rivier, J., Vale, W., and Sawchenko, P. E. (2001). Urocortin II: a member of the corticotropin-releasing factor (CRF) neuropeptide family that is selectively bound by type 2 CRF receptors. *Proc. Natl. Acad. Sci. U S A* 98, 2843–2848.
89. Vaughan, J., Donaldson, C., Bittencourt, J., Perrin, M. H., Lewis K., Sutton, S., Chan, R., Turnbull, A. V., Lovejoy, D., Rivier, C., Sawchenko, P. E., and Vale, W. (1995). Urocortin, a mammalian neuropeptide related to fish urotensin I and to corticotropin-releasing factor. *Nature* 378, 287–292.
90. Iino K., Sasano, H., Oki, Y., Andoh, N., Shin, R. W., Kitamoto, T., Takahashi K., Suzuki, H., Tezuka, F., Yoshimi, T., and Nagura, H. (1999). Urocortin expression in the human central nervous system. *Clin. Endocrinol. (Oxf).* 50, 107–114.
91. Reul, J. M. and Holsboer, F. (2002). Corticotropin-releasing factor receptors 1 and 2 in anxiety and depression. *Curr. Opin. Pharmacol.* 2, 23–33.
92. Turnbull, A. V., Vale, W., and Rivier, C. (1996). Urocortin, a corticotropin-releasing factor-related mammalian peptide, inhibits edema due to thermal injury in rats. *Eur. J. Pharmacol.* 303, 213–216.
93. Spina, M., Merlo-Pich, E., Chan, R. K., Basso, A. M., Rivier, J., Vale, W., and Koob, G. F. (1996). Appetite-suppressing effects of urocortin, a CRF-related neuropeptide. *Science* 273, 1561–1564.
94. Skelton, K. H., Nemeroff, C. B., Knight, D. L., and Owens, M. J. (2000). Chronic administration of the triazolobenzodiazepine alprazolam produces opposite effects on corticotropin-releasing factor and urocortin neuronal systems. *J. Neurosci.* 20, 1240–1248.
95. Parkes, D. G., Vaughan, J., Rivier, J., Vale, W., and May, C. N. (1997). Cardiac inotropic actions of urocortin in conscious sheep. *Am. J. Physiol.* 272, H2115–H2122.
96. Turnbull, A. V., Smith, G. W., Lee, S., Vale, W. W., Lee, K. F., and Rivier, C. (1999). CRF type I receptor-deficient mice exhibit a pronounced pituitary-adrenal response to local inflammation. *Endocrinology* 140, 1013–1017.
97. Wang, X., Su, H., Copenhagen, L. D., Vaishnav, S., Pieri, F., Shope, C. D., Brownell, W. E., De Biasi, M., Paylor, R., and Bradley, A. (2002). Urocortin-deficient mice display normal stress-induced anxiety behavior and autonomic control but an impaired acoustic startle response. *Mol. Cell. Biol.* 22, 6605–6610.
98. Kuhar, M. J., Adams, S., Dominguez, G., Jaworski, J., and Balkan, B. (2002). CART peptides. *Neuropeptides* 36, 1–8.
99. Larsen, P. J., Vrang, N., Petersen, P. C., and Kristensen, P. (2000). Chronic intracerebroventricular administration of recombinant CART(42-89) peptide inhibits and causes weight loss in lean and obese Zucker (fa/fa) rats. *Obes. Res.* 8, 590–596.
100. Larsen, P. J., Seier, V., Fink-Jensen, A., Holst, J. J., Warberg, J., and Vrang, N. (2003). Cocaine- and amphetamine-regulated transcript is present in hypothalamic neuroendocrine neurones and is released to the hypothalamic–pituitary portal circuit. *J. Neuroendocrinol.* 15, 219–226.
101. Vrang, N., Larsen, P. J., Kristensen, P., and Tang-Christensen, M. (2000). Central administration of cocaine-amphetamine-regulated transcript activates hypothalamic neuroendocrine neurons in the rat. *Endocrinology* 141, 794–801.
102. Stanley, S. A., Small, C. J., Murphy, K. G., Rayes, E., Abbott, C. R., Seal, L. J., Morgan, D. G., Sunter, D., Dakin, C. L., Kim, M. S., Hunter, R., Kuhar, M., Ghatei, M. A., and Bloom, S. R. (2001). Actions of cocaine- and amphetamine-regulated transcript (CART) peptide on regulation of appetite and hypothalamo-pituitary axes in vitro and in vivo in male rats. *Brain Res.* 893, 186–194.
103. Smith, S. M., Vaughan, J. M., Donaldson, C. J., Rivier, J., Li, C., Chen A., and Vale, W. W. (2004). Cocaine- and amphetamine-regulated transcript activates the hypothalamic–pituitary– adrenal axis through a corticotropin-releasing factor receptor-dependent mechanism. *Endocrinology* 145, 5202–5209.
104. Salata, R. A., Jarrett, D. B., Verbalis, J. G., and Robinson, A. G. (1988). Vasopressin stimulation of adrenocorticotropin hormone (ACTH) in humans. In vivo bioassay of corticotropin-releasing factor (CRF) which provides evidence for CRF mediation of the diurnal rhythm of ACTH. *J. Clin. Invest.* 81, 766–774.
105. Shalts, E., Xia, L., Xiao, E., and Ferin, M. (1994). Inhibitory effect of arginine-vasopressin on LH secretion in the ovariectomized rhesus monkey. *Neuroendocrinology* 59, 336–342.
106. Rivier, C., Rivier, J., and Vale, W. (1982). Inhibition of adrenocorticotropic hormone secretion in the rat by immunoneutralization of corticotropin-releasing factor. *Science* 218, 377–379.
107. Rivier, C., and Vale, W. (1983). Interaction of corticotropin-releasing factor and arginine vasopressin on adrenocorticotropin secretion in vivo. *Endocrinology* 113, 939–942.
108. Rivier, C., and Vale, W. Modulation of stress-induced ACTH release by corticotropin-releasing factor, catecholamines and vasopressin. *Nature* 305, 325–327.
109. Liu, J. H., Muse K., Contreras, P., Gibbs, D., Vale, W., Rivier, J., and Yen, S. S. (1983). Augmentation of ACTH-releasing activity of synthetic corticotropin releasing factor (CRF) by vasopressin in women. *J. Clin. Endocrinol. Metab.* 57, 1087–1089.
110. Seidah, N. G., Lariviere, N., Boileau, G., Gossard, F., Hsi, K. L., Lu, C. L., and Chretien, M. (1982). Gene expression of peptides: the pro-opiomelanocortin model. *Adv. Biochem. Psychopharmacol.* 33, 19(–26.
111. Krieger, D. T., Liotta, A. S., Brownstein, M. J., and Zimmerman, E. A. (1980). ACTH, beta-lipotropin, and related peptides in brain, pituitary, and blood. *Recent Prog. Horm. Res.* 36, 277–344.
112. Chan, J. S., Lu, C. L., Seidah, N. G., and Chretien, M. (1982). Corticotropin releasing factor (CRF): effects on the release of pro-opiomelanocortin (POMC)-related peptides by human anterior pituitary cells in vitro. *Endocrinology* 111, 1388–1390.
113. Turnbull, A. V., and Rivier, C. (1997). Corticotropin-releasing factor (CRF) and endocrine responses to stress: CRF receptors, binding protein, and related peptides. *Proc. Soc. Exp. Biol. Med.* 215, 1–10.
114. Van Pett K., Viau, V., Bittencourt, J. C., Chan, R. K., Li, H. Y., Arias, C., Prins, G. S., Perrin, M., Vale, W., and

Sawchenko, P. E. (2000). Distribution of mRNAs encoding CRF receptors in brain and pituitary of rat and mouse. *J. Comp. Neurol.* 428, 191–212.

115. Parham, K. L., Zervou, S., Karteris, E., Catalano, R. D., Old, R. W., and Hillhouse, E. W. (2004). Promoter analysis of human corticotropin-releasing factor (CRF) type 1 receptor and regulation by CRF and urocortin. *Endocrinology* 145, 3971–3983.
116. Rivier, C. L., Grigoriadis, D. E., and Rivier, J. E. (2003). Role of corticotropin-releasing factor receptors type 1 and 2 in modulating the rat adrenocorticotropin response to stressors. *Endocrinology* 144, 2396–2403.
117. Gutman, D. A., Owens, M. J., Skelton, K. H., Thrivikraman, K. V., and Nemeroff, C. B. (2003). The corticotropin-releasing factor1 receptor antagonist R121919 attenuates the behavioral and endocrine responses to stress. *J. Pharmacol. Exp. Ther.* 304, 874–880.
118. Bale, T. L., Contarino A., Smith, G. W., Chan, R., Gold, L. H., Sawchenko, P. E., Koob, G. F., Vale, W. W., and Lee, K. F. (2000). Mice deficient for corticotropin-releasing hormone receptor-2 display anxiety-like behaviour and are hypersensitive to stress. *Nat. Genet.* 24, 410–414.
119. Preil, J., Muller, M. B., Gesing A., Reul, J. M., Sillaber, I., van Gaalen, M. M., Landgrebe, J., Holsboer, F., Stenzel-Poore, M., and Wurst, W. (2001). Regulation of the hypothalamic–pituitary–adrenocortical system in mice deficient for CRH receptors 1 and 2. *Endocrinology* 142, 4946–4955.
120. Smith, G. W., Aubry, J. M., Dellu, F., Contarino A., Bilezikjian, L. M., Gold, L. H., Chen, R., Marchuk, Y., Hauser, C., Bentley, C. A., Sawchenko, P. E., Koob, G. F., Vale, W., and Lee, K. F. (1998). Corticotropin releasing factor receptor 1-deficient mice display decreased anxiety, impaired stress response, and aberrant neuroendocrine development. *Neuron* 20, 1093–1102.
121. Coste, S. C., Murray, S. E., and Stenzel-Poore, M. P. (2001). Animal models of CRH excess and CRH receptor deficiency display altered adaptations to stress. *Peptides* 22, 733–741.
122. Bale, T. L., Anderson, K. R., Roberts, A. J., Lee, K. F., Nagy, T. R., and Vale, W. W. (2003). Corticotropin-releasing factor receptor-2-deficient mice display abnormal homeostatic responses to challenges of increased dietary fat and cold. *Endocrinology* 144, 2580–2587.
123. Rivier, C., Rivier, J., Mormede, P., and Vale, W. (1984). Studies of the nature of the interaction between vasopressin and corticotropin-releasing factor on adrenocorticotropin release in the rat. *Endocrinology* 115, 882–886.
124. Groenink, L., Dirks A., Verdouw, P. M., Schipholt, M., Veening, J. G., van der Gugten, J., and Olivier, B. (2002). HPA axis dysregulation in mice overexpressing corticotropin releasing hormone. *Biol. Psychiatry* 51, 875–881.
125. Smith, D. F. and Toft, D. O. (1993). Steroid receptors and their associated proteins. *Mol. Endocrinol.* 7, 4–11.
126. Munck A., Guyre, P. M., and Holbrook, N. J. (1984). Physiological functions of glucocorticoids in stress and their relation to pharmacological actions. *Endocr. Rev.* 5, 25–44.
127. Sapolsky, R. M., Romero, L. M., and Munck, A. U. (2000). How do glucocorticoids influence stress responses? Integrating permissive, suppressive, stimulatory, and preparative actions. *Endocr. Rev.* 21, 55–89.
128. McEwen, B. S. (2003). Mood disorders and allostatic load. *Biol. Psychiatry* 54, 200–207.
129. Keller-Wood, M. E., and Dallman, M. F. (1984). Corticosteroid inhibition of ACTH secretion. *Endocr. Rev.* 5, 1–24.
130. Dallman, M. F., Akana, S. F., Cascio, C. S., Darlington, D. N., Jacobson, L., and Levin, N. (1987). Regulation of ACTH secretion: variations on a theme of B. *Recent Prog. Horm. Res.* 43, 113–173.
131. Swanson, L. W. and Simmons, D. M. (1989). Differential steroid hormone and neural influences on peptide mRNA levels in CRH cells of the paraventricular nucleus: a hybridization histochemical study in the rat. *J. Comp. Neurol.* 285, 413–435.
132. Ginsberg, A. B., Campeau, S., Day, H. E., and Spencer, R. L. (2003). Acute glucocorticoid pretreatment suppresses stress-induced hypothalamic–pituitary–adrenal axis hormone secretion and expression of corticotropin-releasing hormone hnRNA but does not affect c-fos mRNA or fos protein expression in the paraventricular nucleus of the hypothalamus. *J. Neuroendocrinol.* 15, 1075–1083.
133. Stanford, S. C. (1995). Central noradrenergic neurones and stress. *Pharmacol. Ther.* 68, 297–242.
134. Cunningham, E. T. Jr., and Sawchenko, P. E. (1988). Anatomical specificity of noradrenergic inputs to the paraventricular and supraoptic nuclei of the rat hypothalamus. *J. Comp. Neurol.* 274, 60–76.
135. Palkovits, M., Baffi, J. S., and Pacak, K. (1999). The role of ascending neuronal pathways in stress-induced release of noradrenaline in the hypothalamic paraventricular nucleus of rats. *J. Neuroendocrinol.* 11, 529–539.
136. Piekarzewska, A. B., Rosochacki, S. J., and Sender, G. (2000). The effect of acute restraint stress on regional brain neurotransmitter levels in stress-susceptible pietrain pigs. *J. Vet. Med. A. Physiol. Pathol. Clin. Med.* 47, 257–269.
137. Abercrombie, E. D. and Jacobs, B. L. (1987). Single-unit response of noradrenergic neurons in the locus coeruleus of freely moving cats. I. Acutely presented stressful and non-stressful stimuli. *J. Neurosci.* 7, 2837–2843.
138. Cassens, G., Kuruc A., Roffman, M., Orsulak, P. J., and Schildkraut, J. J. (1981). Alterations in brain norepinephrine metabolism and behavior induced by environmental stimuli previously paired with inescapable shock. *Behav. Brain Res.* 2, 387–407.
139. Beck, K. D. and Luine, V. N. (2002). Sex differences in behavioral and neurochemical profiles after chronic stress: role of housing conditions. *Physiol. Behav.* 75, 661–673.
140. Saavedra, J. M. (1982). Changes in dopamine, noradrenaline and adrenaline in specific septal and preoptic nuclei after acute immobilization stress. *Neuroendocrinology* 35, 396–401.
141. Gilbey, M. P. and Spyer, K. M. (1993). Essential organization of the sympathetic nervous system. *Baillieres Clin. Endocrinol. Metab.* 7, 259–278.
142. Kubovcakova, L., Tybitanclova, K., Sabban, E. L., Majzoub, J., Zorad, S., Vietor, I., Wagner, E. F., Krizanova, O., and Kvetnansky, R. (2004). Catecholamine synthesizing enzymes and their modulation by immobilization stress in knockout mice. *Ann. N Y Acad. Sci.* 1018, 458–465.
143. Herman, J. P., Prewitt, C. M., and Cullinan, W. E. (1996). Neuronal circuit regulation of the hypothalamo-pituitary-adrenocortical stress axis. *Crit. Rev. Neurobiol.* 10, 371–394.
144. Herman, J. P., Figueiredo, H., Mueller, N. K., Ulrich-Lai, Y., Ostrander, M. M., Choi, D. C., and Cullinan, W. E. (2003). Central mechanisms of stress integration: hierarchical circuitry controlling hypothalamo–pituitary–adrenocortical responsiveness. *Front. Neuroendocrinol.* 24, 151–180.
145. Cunningham, E. T. Jr., Bohn, M. C., and Sawchenko, P. E. (1990). Organization of adrenergic inputs to the paraventricular and supraoptic nuclei of the hypothalamus in the rat. *J. Comp. Neurol.* 292, 651–667.
146. Plotsky, P. M., Cunningham, E. T. Jr., and Widmaier, E. P. (1989). Catecholaminergic modulation of corticotropin-releasing factor and adrenocorticotropin secretion. *Endocr. Rev.* 10, 437–458.
147. Teppema, L. J., Veening, J. G., Kranenburg A., Dahan, A., Berkenbosch, A., and Olievier, C. (1997). Expression of c-fos

in the rat brainstem after exposure to hypoxia and to normoxic and hyperoxic hypercapnia. *J. Comp. Neurol.* 388, 169–190.
148. Van Bockstaele, E. J., Bajic, D., Proudfit, H., and Valentino, R. J. (2001). Topographic architecture of stress-related pathways targeting the noradrenergic locus coeruleus. *Physiol. Behav.* 73, 273–283.
149. Koob, G. F. (1999). Corticotropin-releasing factor, norepinephrine, and stress. *Biol. Psychiatry* 46, 1167–1180.
150. Curtis, A. L., Lechner, S. M., Pavcovich, L. A., and Valentino, R. J. (1997). Activation of the locus coeruleus noradrenergic system by intracoerulear microinfusion of corticotropin-releasing factor: effects on discharge rate, cortical norepinephrine levels and cortical electroencephalographic activity. *J. Pharmacol. Exp. Ther.* 281, 163–172.
151. Smagin, G. N., Zhou, J., Harris, R. B., and Ryan, D. H. (1997). CRF receptor antagonist attenuates immobilization stress-induced norepinephrine release in the prefrontal cortex in rats. *Brain Res. Bull.* 42, 431–434.
152. Brown, M. R., and Fisher, L. A. (1985). Corticotropin-releasing factor: effects on the autonomic nervous system and visceral systems. *Fed. Proc.* 44, 243–248.
153. Sawchenko, P. E., Swanson, L. W., Steinbusch, H. W., and Verhofstad, A. A. (1983). The distribution and cells of origin of serotonergic inputs to the paraventricular and supraoptic nuclei of the rat. *Brain Res.* 277, 355–360.
154. Jorgensen, H., Knigge, U., Kjaer A., Vadsholt, T., and Warberg, J. (1998). Serotonergic involvement in stress-induced ACTH release. *Brain Res.* 811, 10–20.
155. Cullinan, W. E., Herman, J. P., and Watson, S. J. (1993). Ventral subicular interaction with the hypothalamic paraventricular nucleus: evidence for a relay in the bed nucleus of the stria terminalis. *J. Comp. Neurol.* 332, 1–20.
156. Roland, B. L. and Sawchenko, P. E. (1993). Local origins of some GABAergic projections to the paraventricular and supraoptic nuclei of the hypothalamus in the rat. *J. Comp. Neurol.* 332, 123–143.
157. Silverman, A. J., Hoffman, D. L., and Zimmerman, E. A. (1981). The descending afferent connections of the paraventricular nucleus of the hypothalamus (PVN). *Brain Res. Bull.* 6, 47–61.
158. Cullinan, W. E., Helmreich, D. L., and Watson, S. J. (1996). Fos expression in forebrain afferents to the hypothalamic paraventricular nucleus following swim stress. *J. Comp. Neurol.* 368, 88–99.
159. Ziegler, D. R., Cullinan, W. E., and Herman, J. P. (2002). Distribution of vesicular glutamate transporter mRNA in rat hypothalamus. *J. Comp. Neurol.* 448, 217–229.
160. Herman, J. P., Cullinan, W. E., Morano, M. I., Akil, H., and Watson, S. J. (1995). Contribution of the ventral subiculum to inhibitory regulation of the hypothalamo-pituitary-adrenocortical axis. *J. Neuroendocrinol.* 7, 475–482.
161. Herman, J. P., Schafer, M. K., Young, E. A., Thompson, R., Douglass, J., Akil, H., and Watson, S. J. (1989). Evidence for hippocampal regulation of neuroendocrine neurons of the hypothalamo–pituitary–adrenocortical axis. *J. Neurosci.* 9, 3072–3082.
162. Schwaber, J. S., Kapp, B. S., Higgins, G. A., and Rapp, P. R. (1982). Amygdaloid and basal forebrain direct connections with the nucleus of the solitary tract and the dorsal motor nucleus. *J. Neurosci.* 2, 1424–1438.
163. Sapolsky, R. M., Zola-Morgan, S., and Squire, L. R. (1991). Inhibition of glucocorticoid secretion by the hippocampal formation in the primate. *J. Neurosci.* 11, 3695–3704.
164. Sapolsky, R. M., Armanini, M. P., Sutton, S. W., and Plotsky, P. M. (1989). Elevation of hypophysial portal concentrations of adrenocorticotropin secretagogues after fornix transection. *Endocrinology* 125, 2881–2887.
165. Rubin, R. T., Mandell, A. J., and Crandall, P. H. (1966). Corticosteroid responses to limbic stimulation in man: localization of stimulus sites. *Science* 153, 767–768.
166. Redgate, E. S., and Fahringer, E. E. (1973). A comparison of the pituitary adrenal activity elicited by electrical stimulation of preoptic, amygdaloid and hypothalamic sites in the rat brain. *Neuroendocrinology* 12, 334–343.
167. Knigge, K. M. (1961). Adrenocortical response to stress in rats with lesions in hippocampus and amygdala. *Proc. Soc. Exp. Biol. Med.* 108, 18–21.
168. Sapolsky, R. M., Krey, L. C., and McEwen, B. S. (1986). The neuroendocrinology of stress and aging: the glucocorticoid cascade hypothesis. *Endocr. Rev.* 7, 284–301.
169. Schulkin, J. (1999). Corticotropin-releasing hormone signals adversity in both the placenta and the brain: regulation by glucocorticoids and allostatic overload. *J. Endocrinol.* 161, 349–356.
170. Sawchenko, P. E., Imaki, T., Potter, E., Kovacs, K., Imaki, J., and Vale, W. (1993). The functional neuroanatomy of corticotropin-releasing factor. *Ciba Found. Symp.* 172, 5–21. discussion 21–29.
171. Perrin, M. H., Donaldson, C. J., Chen, R., Lewis, K. A., and Vale, W. W. (1993). Cloning and functional expression of a rat brain corticotropin releasing factor (CRF) receptor. *Endocrinology* 133, 3058–3061.
172. Imaki, T., Naruse, M., Harada, S., Chikada, N., Imaki, J., Onodera, H., Demura, H., and Vale, W. (1996). Corticotropin-releasing factor up-regulates its own receptor mRNA in the paraventricular nucleus of the hypothalamus. *Brain Res. Mol. Brain Res.* 38, 166–170.
173. Mansi, J. A., Rivest, S., and Drolet, G. (1996). Regulation of corticotropin-releasing factor type 1 (CRF1) receptor messenger ribonucleic acid in the paraventricular nucleus of rat hypothalamus by exogenous CRF. *Endocrinology* 137, 4619–4629.
174. Vamvakopoulos, N. C., and Chrousos, G. P. (1994). Hormonal regulation of human corticotropin-releasing hormone gene expression: implications for the stress response and immune/inflammatory reaction. *Endocr. Rev.* 15, 409–420.
175. Silverman, A. J., Hou-Yu, A., and Chen, W. P. (1989). Corticotropin-releasing factor synapses within the paraventricular nucleus of the hypothalamus. *Neuroendocrinology* 49, 291–299.
176. Ju, G., Swanson, L. W., and Simerly, R. B. (1989). Studies on the cellular architecture of the bed nuclei of the stria terminalis in the rat: II. Chemoarchitecture. *J. Comp. Neurol.* 280, 603–621.
177. Dong, H. W., Petrovich, G. D., Watts, A. G., and Swanson, L. W. (2001). Basic organization of projections from the oval and fusiform nuclei of the bed nuclei of the stria terminalis in adult rat brain. *J. Comp. Neurol.* 436, 430–455.
178. Parkes, D., Rivest, S., Lee, S., Rivier, C., and Vale, W. (1993). Corticotropin-releasing factor activates c-fos, NGFI-B, and corticotropin-releasing factor gene expression within the paraventricular nucleus of the rat hypothalamus. *Mol. Endocrinol.* 7, 1357–1367.
179. Imaki, T., Shibasaki, T., Hotta, M., and Demura, H. (1993). Intracerebroventricular administration of corticotropin-releasing factor induces c-fos mRNA expression in brain regions related to stress responses: comparison with pattern of c-fos mRNA induction after stress. *Brain Res.* 616, 114–125.
180. Imaki, T., Shibasaki, T., Wang, X. Q., and Demura, H. (1995). Intracerebroventricular administration of corticotropin-releasing factor antagonist attenuates c-fos mRNA expression in the paraventricular nucleus after stress. *Neuroendocrinology* 61, 445–452.
181. Swanson, L. W., Sawchenko, P. E., Rivier, J., and Vale, W. W. (1983). Organization of ovine corticotropin-releasing factor

immunoreactive cells and fibers in the rat brain: an immunohistochemical study. *Neuroendocrinology* 36, 165–186.
182. Diamant, M., Kashtanov, S. I., Fodor, M., and de Wied, D. (1992). Corticotropin-releasing factor induces differential behavioral and cardiovascular effects after intracerebroventricular and lateral hypothalamic/perifornical injections in rats. *Neuroendocrinology* 56, 750–760.
183. Zeng, J., Kitayama, I., Yoshizato, H., Zhang K., and Okazaki, Y. (2003). Increased expression of corticotropin-releasing factor receptor mRNA in the locus coeruleus of stress-induced rat model of depression. *Life Sci.* 73, 1131–1139.
184. Oitzl, M. S., van Haarst, A. D., and de Kloet, E. R. (1997). Behavioral and neuroendocrine responses controlled by the concerted action of central mineralocorticoid (MRS) and glucocorticoid receptors (GRS). *Psychoneuroendocrinology* 22, Suppl 1, S87–S93.
185. McEwen, B. S. (1998). Protective and damaging effects of stress mediators. *N. Engl. J. Med.* 338, 171–179.
186. McEwen, B. S. (2001). Plasticity of the hippocampus: adaptation to chronic stress and allostatic load. *Ann. N Y Acad. Sci.* 933, 265–277.
187. McEwen, B. S., and Magarinos, A. M. (2001). Stress and hippocampal plasticity: implications for the pathophysiology of affective disorders. *Hum. Psychopharmacol.* 16, S7–S19.
188. Stein-Behrens B., Mattson, M. P., Chang, I., Yeh, M., and Sapolsky, R. (1994). Stress exacerbates neuron loss and cytoskeletal pathology in the hippocampus. *J. Neurosci.* 14, 5373–5380.
189. Stein-Behrens, B. A., Lin, W. J., and Sapolsky, R. M. (1994). Physiological elevations of glucocorticoids potentiate glutamate accumulation in the hippocampus. *J. Neurochem.* 63, 596–602.
190. Chesnokova, V., and Melmed, S. (2002). Minireview: Neuro-immuno-endocrine modulation of the hypothalamic–pituitary–adrenal (HPA) axis by gp130 signaling molecules. *Endocrinology* 143, 1571–1574.
191. Kariagina A., Romanenko, D., Ren, S. G., and Chesnokova, V. (2004). Hypothalamic–pituitary cytokine network. *Endocrinology* 145, 104–112.
192. Tsagarakis, S., and Grossman, A. (1994). Corticotropin-releasing hormone: interactions with the immune system. *Neuroimmunomodulation* 1, 329–334.
193. Dinarello, C. A. (1989). Interleukin-1 and its biologically related cytokines. *Adv. Immunol.* 44, 153–205.
194. Elenkov, I. J., Webster, E. L., Torpy, D. J., and Chrousos, G. P. (1999). Stress, corticotropin-releasing hormone, glucocorticoids, and the immune/inflammatory response: acute and chronic effects. *Ann. N Y Acad. Sci.* 876, 1–11; discussion 11–13.
195. Bauer, J., Berkenbosch, F., Van Dam, A. M., and Dijkstra, C. D. (1993). Demonstration of interleukin-1 beta in Lewis rat brain during experimental allergic encephalomyelitis by immunocytochemistry at the light and ultrastructural level. *J. Neuroimmunol.* 48, 13–21.
196. Higgins, G. C., Foster, J. L., and Postlethwaite, A. E. (1994). Interleukin 1 beta propeptide is detected intracellularly and extracellularly when human monocytes are stimulated with LPS in vitro. *J. Exp. Med.* 180, 607–614.
197. Kotb, M., Ohnishi, H., Majumdar, G., Hackett, S., Bryant A., Higgins, G., and Stevens, D. (1993). Temporal relationship of cytokine release by peripheral blood mononuclear cells stimulated by the streptococcal superantigen pep M5. *Infect. Immun.* 61, 1194–1201.
198. Higgins, G. A., and Olschowka, J. A. (1991). Induction of interleukin-1 beta mRNA in adult rat brain. *Brain Res. Mol. Brain Res.* 9, 143–148.
199. Ma, X. C., Chen, L. T., Oliver, J., Horvath, E., and Phelps, C. P. (2000). Cytokine and adrenal axis responses to endotoxin. *Brain Res.* 861, 135–142.
200. Hopkins, S. J., and Rothwell, N. J. (1995). Cytokines and the nervous system. I: Expression and recognition. *Trends Neurosci.* 18, 83–88.
201. Bandtlow, C. E., Meyer, M., Lindholm, D., Spranger, M., Heumann, R., and Thoenen, H. (1990). Regional and cellular codistribution of interleukin 1 beta and nerve growth factor mRNA in the adult rat brain: possible relationship to the regulation of nerve growth factor synthesis. *J. Cell Biol.* 111, 1701–1711.
202. Laye, S., Gheusi, G., Cremona, S., Combe, C., Kelley K., Dantzer, R., and Parnet, P. (2000). Endogenous brain IL-1 mediates LPS-induced anorexia and hypothalamic cytokine expression. *Am. J. Physiol. Regul. Integr. Comp. Physiol.* 279, R93–R98.
203. Dunn, A. J. (2000). Cytokine activation of the HPA axis. *Ann. N Y Acad. Sci.* 917, 608–617.
204. Bateman A., Singh A., Kral, T., and Solomon, S. (1989). The immune–hypothalamic–pituitary–adrenal axis. *Endocr. Rev.* 10, 92–112.
205. Rivier, C. (1993). Neuroendocrine effects of cytokines in the rat. *Rev. Neurosci.* 4, 223–237.
206. Rivest, S., and Rivier, C. (1991). Influence of the paraventricular nucleus of the hypothalamus in the alteration of neuroendocrine functions induced by intermittent footshock or interleukin. *Endocrinology* 129, 2049–2057.
207. Tsagarakis, S., Gillies, G., Rees, L. H., Besser, M., and Grossman, A. (1989). Interleukin-1 directly stimulates the release of corticotrophin releasing factor from rat hypothalamus. *Neuroendocrinology* 49, 98–101.
208. Barbanel, G., Ixart, G., Szafarczyk, A., Malaval, F., and Assenmacher, I. (1990). Intrahypothalamic infusion of interleukin-1 beta increases the release of corticotropin-releasing hormone (CRH 41) and adrenocorticotropic hormone (ACTH) in free-moving rats bearing a push-pull cannula in the median eminence. *Brain Res.* 516, 31–36.
209. Beishuizen, A. and Thijs, L. G. (2003). Endotoxin and the hypothalamo–pituitary–adrenal (HPA) axis. *J. Endotoxin Res.* 9, 3–24.
210. Peisen, J. N., McDonnell, K. J., Mulroney, S. E., and Lumpkin, M. D. (1995). Endotoxin-induced suppression of the somatotropic axis is mediated by interleukin-1 beta and corticotropin-releasing factor in the juvenile rat. *Endocrinology* 136, 3378–3390.
211. Kakucska, I., Qi, Y., Clark, B. D., and Lechan, R. M. (1993). Endotoxin-induced corticotropin-releasing hormone gene expression in the hypothalamic paraventricular nucleus is mediated centrally by interleukin-1. *Endocrinology* 133, 815–821.
212. van der Meer, M. J., Sweep, C. G., Rijnkels, C. E., Pesman, G. J., Tilders, F. J., Kloppenborg, P. W., and Hermus, A. R. (1996). Acute stimulation of the hypothalamic–pituitary– adrenal axis by IL-1 beta, TNF alpha and IL-6: a dose response study. *J. Endocrinol. Invest.* 19, 175–182.
213. Hagan, P. M., Poole, S., and Bristow, A. F. (1993). Corticotrophin-releasing factor as a mediator of the acute-phase response in rats, mice and rabbits. *J. Endocrinol.* 136, 207–216.
214. Goshen, I., Yirmiya, R., Iverfeldt, K., and Weidenfeld, J. (2003). The role of endogenous interleukin-1 in stress-induced adrenal activation and adrenalectomy-induced adrenocorticotropic hormone hypersecretion. *Endocrinology* 144, 4453–4458.
215. Chikanza, I. C. and Grossman, A. S. (1998). Hypothalamic–pituitary–mediated immunomodulation: arginine vasopressin is a neuroendocrine immune mediator. *Br. J. Rheumatol.* 37, 131–136.
216. Silverman, M. N., Miller, A. H., Biron, C. A., and Pearce, B. D. (2004). Characterization of an interleukin-6- and

adrenocorticotropin-dependent, immune-to-adrenal pathway during viral infection. *Endocrinology* 145, 3580–3589.
217. Bethin, K. E., Vogt, S. K., and Muglia, L. J. (2000). Interleukin-6 is an essential, corticotropin-releasing hormone-independent stimulator of the adrenal axis during immune system activation. *Proc. Natl. Acad. Sci. U S A* 97, 9317–9322.
218. Turnbull, A. V., Prehar, S., Kennedy, A. R., Little, R. A., and Hopkins, S. J. (2003). Interleukin-6 is an afferent signal to the hypothalamo–pituitary–adrenal axis during local inflammation in mice. *Endocrinology* 144, 1894–1906.
219. Nguyen, M. D., Julien, J. P., and Rivest, S. (2001). Induction of proinflammatory molecules in mice with amyotrophic lateral sclerosis: no requirement for proapoptotic interleukin-1beta in neurodegeneration. *Ann. Neurol.* 50, 630–639.
220. McKay, L. I., and Cidlowski, J. A. (1999). Molecular control of immune/inflammatory responses: interactions between nuclear factor-kappa B and steroid receptor-signaling pathways. *Endocr. Rev.* 20, 435–459.
221. Wissink, S., van Heerde, E. C., vand der Burg B., and van der Saag, P. T. (1998). A dual mechanism mediates repression of NF-kappaB activity by glucocorticoids. *Mol. Endocrinol.* 12, 355–363.
222. Nadeau, S., and Rivest, S. (2003). Glucocorticoids play a fundamental role in protecting the brain during innate immune response. *J. Neurosci.* 23, 5536–5544.
223. Dobson, H., Tebble, J. E., Smith, R. F., and Ward, W. R. (2001). Is stress really all that important? *Theriogenology* 55, 65–73.
224. Brown, E., Bain, J., Lerner, P., and Shaul, D. (1983). Psychological, hormonal, and weight disturbances in functional amenorrhea. *Can. J. Psychiatry* 28, 624–628.
225. Kaplan, J. R., and Manuck, S. B. (2004). Ovarian dysfunction, stress, and disease: a primate continuum. *ILAR J.* 45, 89–115.
226. Zeleznik, A. J., and Pohl, C. R. (2006). Control of follicular development, corpus luteum function, and maternal recognition of pregnancy and the neuroendocrine regulation of the menstrual cycle in higher primates. In *Knobil and Neill's Physiology of Reproduction* (J. D. Neill, Ed.). Elsevier, St. Louis.
227. Tilders, F. J., DeRijk, R. H., Van Dam, A. M., Vincent, V. A., Schotanus, K., and Persoons, J. H. (1994). Activation of the hypothalamus–pituitary–adrenal axis by bacterial endotoxins: routes and intermediate signals. *Psychoneuroendocrinology* 19, 209–232.
228. van Dam, A. M., Brouns, M., Louisse, S., and Berkenbosch, F. (1992). Appearance of interleukin-1 in macrophages and in ramified microglia in the brain of endotoxin-treated rats: a pathway for the induction of non-specific symptoms of sickness? *Brain Res.* 588, 291–296.
229. Ebisui, O., Fukata, J., Tominaga, T., Murakami, N., Kobayashi, H., Segawa, H., Muro, S., Naito, Y., Nakai, Y., and Masui, Y., et al. (1992). Roles of interleukin-1 alpha and -1 beta in endotoxin-induced suppression of plasma gonadotropin levels in rats. *Endocrinology* 130, 3307–3313.
230. Rivest, S., and Rivier, C. (1993). Central mechanisms and sites of action involved in the inhibitory effects of CRF and cytokines on LHRH neuronal activity. *Ann. N Y Acad. Sci.* 697, 117–141.
231. Bonavera, J. J., Kalra, S. P., and Kalra, P. S. (1993). Mode of action of interleukin-1 in suppression of pituitary LH release in castrated male rats. *Brain Res.* 612, 1–8.
232. Rettori, V., Gimeno, M. F., Karara A., Gonzalez, M. C., and McCann, S. M. (1991). Interleukin 1 alpha inhibits prostaglandin E2 release to suppress pulsatile release of luteinizing hormone but not follicle-stimulating hormone. *Proc. Natl. Acad. Sci. U S A* 88, 2763–2767.
233. Rivier, C., and Vale, W. (1989). In the rat, interleukin-1 alpha acts at the level of the brain and the gonads to interfere with gonadotropin and sex steroid secretion. *Endocrinology* 124, 2105–2109.
234. Rivier, C., and Vale, W. (1990). Cytokines act within the brain to inhibit luteinizing hormone secretion and ovulation in the rat. *Endocrinology* 127, 849–856.
235. Battaglia, D. F., Bowen, J. M., Krasa, H. B., Thrun, L. A., Viguie, C., and Karsch, F. J. (1997). Endotoxin inhibits the reproductive neuroendocrine axis while stimulating adrenal steroids: a simultaneous view from hypophyseal portal and peripheral blood. *Endocrinology* 138, 4273–4281.
236. Coleman, E. S., Elsasser, T. H., Kemppainen, R. J., Coleman, D. A., and Sartin, J. L. (1993). Effect of endotoxin on pituitary hormone secretion in sheep. *Neuroendocrinology* 58, 111–122.
237. Battaglia, D. F., Krasa, H. B., Padmanabhan, V., Viguie, C., and Karsch, F. J. (2000). Endocrine alterations that underlie endotoxin-induced disruption of the follicular phase in ewes. *Biol. Reprod.* 62, 45–53.
238. Wallgren, M., Kindahl, H., and Larsson, K. (1989). Clinical, endocrinological and spermatological studies after endotoxin in the ram. *Zentralbl. Veterinarmed. A.* 36, 90–103.
239. Daniel, J. A., Whitlock, B. K., Wagner, C. G., and Sartin, J. L. (2002). Regulation of the growth hormone and luteinizing hormone response to endotoxin in sheep. *Domest. Anim. Endocrinol.* 23, 361–370.
240. Daniel, J. A., Abrams, M. S., deSouza, L., Wagner, C. G., Whitlock, B. K., and Sartin, J. L. (2003). Endotoxin inhibition of luteinizing hormone in sheep. *Domest. Anim. Endocrinol.* 25, 13–19.
241. Kujjo, L. L., Bosu, W. T., and Perez, G. I. (1995). Opioid peptides involvement in endotoxin-induced suppression of LH secretion in ovariectomized Holstein heifers. *Reprod. Toxicol.* 9, 169–174.
242. Xiao, E., Xia-Zhang, L., and Ferin, M. (2000). Inhibitory effects of endotoxin on LH secretion in the ovariectomized monkey are prevented by naloxone but not by an interleukin-1 receptor antagonist. *Neuroimmunomodulation* 7, 6–15.
243. Ferin, M. (1995). The antireproductive role of corticotropin releasing hormone and interleukin-1 in the female rhesus monkey. *Ann. Endocrinol. (Paris)* 56, 181–186.
244. Demura, R., Suzuki, T., Nakamura, S., Komatsu, H., Odagiri, E., and Demura, H. (1989). Effect of immobilization stress on testosterone and inhibin in male rats. *J. Androl.* 10, 210–213.
245. Almeida, S. A., Anselmo-Franci, J. A., Rosa e Silva, A. A., and Carvalho, T. L. (1998). Chronic intermittent immobilization of male rats throughout sexual development: a stress protocol. *Exp. Physiol.* 83, 701–704.
246. Lopez-Calderon A., Ariznavarreta, C., Calderon, M. D., and Tresguerres, J. A. (1987). Gonadotropin inhibition during chronic stress: role of the adrenal gland. *J. Steroid Biochem.* 27, 609–614.
247. Lopez-Calderon A., Gonzalez-Quijano, M. I., Tresguerres, J. A., and Ariznavarreta, C. (1990). Role of LHRH in the gonadotrophin response to restraint stress in intact male rats. *J. Endocrinol.* 124, 241–246.
248. Tache, Y., Du Ruisseau, P., Ducharme, J. R., and Collu, R. (1978). Pattern of adenohypophyseal hormone changes in male rats following chronic stress. *Neuroendocrinology* 26, 208–219.
249. DuRuisseau, P. D., Tache, Y., Brazeau, P., and Collu, R. (1978). Pattern of adenohypophyseal hormone changes induced by various stressors in female and male rats. *Neuroendocrinology* 27, 257–271.
250. Srivastava, R. K., Akinbami, M. A., and Mann, D. R. (1995). Acute immobilization stress alters LH and ACTH release in

response to administration of N-methyl-D,L-aspartic acid in peripubertal and adult male rats. *Life Sci.* 56, 1535–1543.

251. Akema, T., Chiba A., Shinozaki, R., Oshida, M., Kimura, F., and Toyoda, J. (1995). Acute stress suppresses the N-methyl-D-aspartate-induced luteinizing hormone release in the ovariectomized estrogen-primed rat. *Neuroendocrinology* 62, 270–276.
252. Rasmussen, D. D., and Malven, P. V. (1983). Effects of confinement stress on episodic secretion of LH in ovariectomized sheep. *Neuroendocrinology* 36, 392–396.
253. Tilbrook, A. J., Canny, B. J., Serapiglia, M. D., Ambrose, T. J., and Clarke, I. J. (1999). Suppression of the secretion of luteinizing hormone due to isolation/restraint stress in gonadectomised rams and ewes is influenced by sex steroids. *J. Endocrinol.* 160, 469–481.
254. Phogat, J. B., Smith, R. F., and Dobson, H. (1999). Effect of transport on pituitary responsiveness to exogenous pulsatile GnRH and oestradiol-induced LH release in intact ewes. *J. Reprod. Fertil.* 116, 9–18.
255. Dobson, H., Tebble, J. E., Phogat, J. B., and Smith, R. F. (1999). Effect of transport on pulsatile and surge secretion of LH in ewes in the breeding season. *J. Reprod. Fertil.* 116, 1–8.
256. Norman, R. L. and Smith, C. J. (1992). Restraint inhibits luteinizing hormone and testosterone secretion in intact male rhesus macaques: effects of concurrent naloxone administration. *Neuroendocrinology* 55, 405–415.
257. O'Byrne, K. T., Lunn, S. F., and Dixson, A. F. (1988). Effects of acute stress on the patterns of LH secretion in the common marmoset (Callithrix jacchus). *J. Endocrinol.* 118, 259–264.
258. Sapolsky, R. M. and Krey, L. C. (1988). Stress-induced suppression of luteinizing hormone concentrations in wild baboons: role of opiates. *J. Clin. Endocrinol. Metab.* 66, 722–726.
259. Rodriguez, M., Arias, P., Refojo, D., Feleder, C., and Moguilevsky, J. (1999). Arrest of pulsatile luteinizing hormone (LH) secretion during insulin-induced hypoglycemia (IIH): improvement by intrahypothalamic perfusion with glucose. *Exp. Clin. Endocrinol. Diabetes* 107, 257–261.
260. Goubillon, M. L. and Thalabard, J. C. (1996). Insulin-induced hypoglycemia decreases luteinizing hormone secretion in the castrated male rat: involvement of opiate peptides. *Neuroendocrinology* 64, 49–56.
261. Cates, P. S. and O'Byrne, K. T. (2000). The area postrema mediates insulin hypoglycaemia-induced suppression of pulsatile LH secretion in the female rat. *Brain Res.* 853, 151–155.
262. Cagampang, F. R., Cates, P. S., Sandhu, S., Strutton, P. H., McGarvey, C., Coen, C. W., and O'Byrne, K. T. (1997). Hypoglycaemia-induced inhibition of pulsatile luteinizing hormone secretion in female rats: role of oestradiol, endogenous opioids and the adrenal medulla. *J. Neuroendocrinol.* 9, 867–872.
263. He, D., Funabashi, T., Sano A., Uemura, T., Minaguchi, H., and Kimura, F. (1999). Effects of glucose and related substrates on the recovery of the electrical activity of gonadotropin-releasing hormone pulse generator which is decreased by insulin-induced hypoglycemia in the estrogen-primed ovariectomized rat. *Brain Res.* 820, 71–76.
264. Clarke, I. J., Horton, R. J., and Doughton, B. W. (1990). Investigation of the mechanism by which insulin-induced hypoglycemia decreases luteinizing hormone secretion in ovariectomized ewes. *Endocrinology* 127, 1470–1476.
265. Lado-Abeal, J., Clapper, J. A., Chen Zhu, B., Hough, C. M., Syapin, P. J., Norman, R. L. (2002). Hypoglycemia-induced suppression of luteinizing hormone (LH) secretion in intact female rhesus macaques: role of vasopressin and endogenous opioids. *Stress* 5, 113–119.
266. Heisler, L. E., Tumber, A. J., Reid, R. L., and van Vugt, D. A. (1994). Vasopressin mediates hypoglycemia-induced inhibition of luteinizing hormone secretion in the ovariectomized rhesus monkey. *Neuroendocrinology* 60, 297–304.
267. Van Vugt, D. A., Washburn, D. L., Farley, A. E., and Reid, R. L. (1997). Hypoglycemia-induced inhibition of LH and stimulation of ACTH secretion in the rhesus monkey is blocked by alprazolam. *Neuroendocrinology* 65, 344–352.
268. Oltmanns, K. M., Fruehwald-Schultes B., Kern, W., Born, J., Fehm, H. L., and Peters, A. (2001). Hypoglycemia, but not insulin, acutely decreases LH and T secretion in men. *J. Clin. Endocrinol. Metab.* 86, 4913–4919.
269. Coiro, V., Volpi, R., Capretti, L., Manfredi, G., Galli, P., Saccani-Jotti, G., and Chiodera, P. (2004). Glucoreceptors located inside the blood–brain barrier mediate hypoglycemia-induced LH inhibition in man. *Horm. Res.* 61, 218–221.
270. Bing-You, R. G., and Spratt, D. I. (1992). Serum estradiol but not gonadotropin levels decrease acutely after insulin-induced hypoglycemia in cycling women. *J. Clin. Endocrinol. Metab.* 75, 1054–1059.
271. Tohei A., Tomabechi, T., Mamada, M., Akai, M., Watanabe, G., and Taya, K. (1997). Effects of repeated ether stress on the hypothalamic–pituitary–testes axis in adult rats with special reference to inhibin secretion. *J. Vet. Med. Sci.* 59, 329–334.
272. Donoghue, D. J., Krueger, B. F., Hargis, B. M., Miller, A. M., and el Halawani, M. (1989). Thermal stress reduces serum luteinizing hormone and bioassayable hypothalamic content of luteinizing hormone-releasing hormone in hens. *Biol. Reprod.* 41, 419–424.
273. Sapolsky, R. (1983). Endocrine aspects of social instability in the olive baboon (Papio anubis). *Am. J. Primatol.* 5, 365–370.
274. Sapolsky, R. M. (1995). Social subordinance as a marker of hypercortisolism. Some unexpected subtleties. *Ann. N Y Acad. Sci.* 771, 626–639.
275. Sapolsky, R. M., Alberts, S. C., and Altmann, J. (1997). Hypercortisolism associated with social subordinance or social isolation among wild baboons. *Arch. Gen. Psychiatry* 54, 1137–1143.
276. Opstad, P. K. (1992). Androgenic hormones during prolonged physical stress, sleep, and energy deficiency. *J. Clin. Endocrinol. Metab.* 74, 1176–1183.
277. Berga, S. L. (1996). Functional hypothalamic chronic anovulation. In *Reproductive Endocrinology, Surgery, and Technology* (E. Y. Adashi, J. A. Rock, and Z. Rosenwaks, Eds.), Vol. 1, pp. 1061–1075. Lippincott-Raven, Philadelphia.
278. Berga, S. L., Mortola, J. F., Girton, L., Suh B., Laughlin, G., Pham, P., and Yen, S. S. (1989). Neuroendocrine aberrations in women with functional hypothalamic amenorrhea. *J. Clin. Endocrinol. Metab.* 68, 301–308.
279. Perkins, R. B., Hall, J. E., and Martin, K. A. (2001). Aetiology, previous menstrual function and patterns of neuro-endocrine disturbance as prognostic indicators in hypothalamic amenorrhoea. *Hum. Reprod.* 16, 2198–2205.
280. Takeuchi, Y., Nagabukuro, H., Kizumi, O., and Mori, Y. (1997). Lipopolysaccharide-induced suppression of the hypothalamic gonadotropin-releasing hormone pulse generator in ovariectomized goats. *J. Vet. Med. Sci.* 59, 93–96.
281. Yoo, M. J., Nishihara, M., and Takahashi, M. (1997). Tumor necrosis factor-alpha mediates endotoxin induced suppression of gonadotropin-releasing hormone pulse generator activity in the rat. *Endocr. J.* 44, 141–148.
282. Chen, M. D., Ordog, T., O'Byrne, K. T., Goldsmith, J. R., Connaughton, M. A., and Knobil, E. (1996). The insulin hypoglycemia-induced inhibition of gonadotropin-releasing hormone pulse generator activity in the rhesus monkey: roles of vasopressin and corticotropin-releasing factor. *Endocrinology* 137, 2012–2021.

283. Chen, M. D., O'Byrne, K. T., Chiappini, S. E., Hotchkiss, J., and Knobil, E. (1992). Hypoglycemic 'stress' and gonadotropin-releasing hormone pulse generator activity in the rhesus monkey: role of the ovary. *Neuroendocrinology* 56, 666–673.
284. Przekop, F., Polkowska, J., and Mateusiak, K. (1988). The effect of prolonged stress on the hypothalamic luteinizing hormone-releasing hormone (LHRH) in the anoestrous ewe. *Exp. Clin. Endocrinol.* 91, 334–340.
285. Lujan, M. E., Krzemien, A. A., and Van Vugt, D. A. (2003). Hypoglycemia does not affect gonadotroph responsiveness to gonadotropin-releasing hormone in rhesus monkeys. *Endocrine* 21, 109–114.
286. Williams, C. Y., Harris, T. G., Battaglia, D. F., Viguie, C., and Karsch, F. J. (2001). Endotoxin inhibits pituitary responsiveness to gonadotropin-releasing hormone. *Endocrinology* 142, 1915–1922.
287. Stackpole, C. A., Turner, A. I., Clarke, I. J., Lambert, G. W., and Tilbrook, A. J. (2003). Seasonal differences in the effect of isolation and restraint stress on the luteinizing hormone response to gonadotropin-releasing hormone in hypothalamo-pituitary disconnected, gonadectomized rams and ewes. *Biol. Reprod.* 69, 1158–1164.
288. Nappi, R. E. and Rivest, S. (1997). Effect of immune and metabolic challenges on the luteinizing hormone-releasing hormone neuronal system in cycling female rats: an evaluation at the transcriptional level. *Endocrinology* 138, 1374–1384.
289. Dunn, J. D., Arimura A., and Scheving, L. E. (1972). Effect of stress on circadian periodicity in serum LH and prolactin concentration. *Endocrinology* 90, 29–33.
290. Euker, J. S., Meites, J., and Riegle, G. D. (1975). Effects of acute stress on serum LH and prolactin in intact, castrate and dexamethasone-treated male rats. *Endocrinology* 96, 85–92.
291. Armario A. and Castellanos, J. M. (1984). Effect of acute and chronic stress on testosterone secretion in male rats. *J. Endocrinol. Invest.* 7, 659–661.
292. Krulich, L., Hefco, E., Illner, P., and Read, C. B. (1974). The effects of acute stress on the secretion of LH, FSH, prolactin and GH in the normal male rat, with comments on their statistical evaluation. *Neuroendocrinology* 16, 293–311.
293. Briski, K. P., Quigley K., and Meites, J. (1984). Endogenous opiate involvement in acute and chronic stress-induced changes in plasma LH concentrations in the male rat. *Life Sci.* 34, 2485–2493.
294. Briski, K. P. and Sylvester, P. W. (1987). Effects of sequential acute stress exposure on stress-induced pituitary luteinizing hormone and prolactin secretion. *Life Sci.* 41, 1249–1255.
295. Briski, K. P. and Sylvester, P. W. (1987). Effects of repetitive daily acute stress on pituitary LH and prolactin release during exposure to the same stressor or a second novel stress. *Psychoneuroendocrinology* 12, 429–437.
296. Siegel, R. A., Weidenfeld, J., Feldman, S., Conforti, N., and Chowers, I. (1981). Neural pathways mediating basal and stress-induced secretion of luteinizing hormone, follicle-stimulating hormone, and testosterone in the rat. *Endocrinology* 108, 2302–2307.
297. Briski, K. P. and Sylvester, P. W. (1992). Differential impact of naltrexone on luteinizing hormone release during single versus repetitive exposure to restraint stress. *Psychoneuroendocrinology* 17, 125–133.
298. Mann, D. R., Evans, D. C., Jacobs, V. L., and Collins, D. C. (1986). Influence of acute intracerebroventricular (i.c.v.) administration of adrenocorticotrophin (ACTH) on LH secretion in male rats: effect of pretreatment (i.c.v.) with ACTH antiserum on the serum LH response to an acute ether stress. *J. Endocrinol.* 108, 275–280.
299. Turpen, C., Johnson, D. C., and Dunn, J. D. (1976). Stress-induced gonadotropin and prolactin secretory patterns. *Neuroendocrinology* 20, 339–351.
300. Heiblum, R., Arnon, E., Gvaryahu, G., Robinzon, B., and Snapir, N. (2000). Short-term stress increases testosterone secretion from testes in male domestic fowl. *Gen. Comp. Endocrinol.* 120, 55–66.
301. Briski, K. P. (1996). Stimulatory vs. inhibitory effects of acute stress on plasma LH: differential effects of pretreatment with dexamethasone or the steroid receptor antagonist, RU 486. *Pharmacol. Biochem. Behav.* 55, 19–26.
302. Dessypris, A., Kuoppasalmi, K., and Adlercreutz, H. (1976). Plasma cortisol, testosterone, androstenedione and luteinizing hormone (LH) in a non-competitive marathon run. *J. Steroid Biochem.* 7, 33–37.
303. Sutton, J. R., Coleman, M. J., Casey, J., and Lazarus, L. (1973). Androgen responses during physical exercise. *Br. Med. J.* 1, 520–522.
304. Aono, T., Kurachi K., Miyata, M., Nakasima A., and Koshiyama, K. (1976). Influence of surgical stress under general anesthesia on serum gonadotropin levels in male and female patients. *J. Clin. Endocrinol. Metab.* 42, 144–148.
305. Carstensen, H., Amer, I., Wide, L., and Amer, B. (1973). Plasma testosterone, LH and FSH during the first 24 hours after surgical operations. *J. Steroid Biochem.* 4, 605–611.
306. Nakashima A., Koshiyama K., Uozumi, T., Monden, Y., and Hamanaka, Y. (1975). Effects of general anaesthesia and severity of surgical stress on serum LH and testosterone in males. *Acta Endocrinol. (Copenh).* 78, 258–269.
307. Sapolsky, R. M. (1986). Stress-induced elevation of testosterone concentration in high ranking baboons: role of catecholamines. *Endocrinology* 118, 1630–1635.
308. Martin, A. I., Lopez-Calderon, A., Tresguerres, J. A., Gonzalez-Quijano, M. I., and Cardinali, D. P. (1995). Restraint-induced changes in serum luteinizing hormone, prolactin, growth hormone and corticosterone levels in rats: effect of superior cervical ganglionectomy. *Neuroendocrinology* 61, 173–179.
309. Armario, A., Marti, O., Gavalda A., and Lopez-Calderon, A. (1993). Evidence for the involvement of serotonin in acute stress-induced release of luteinizing hormone in the male rat. *Brain Res. Bull.* 31, 29–31.
310. Ferin, M. (1996). The menstrual cycle: an integrative view. In *Reproductive Endocrinology, Surgery, and Technology* (E. Y. Adashi, J. A. Rock, and Z. Rosenwaks, Eds.), Vol. 1, pp. 103–122. Lipincott-Raven, Philadelphia.
311. Evans, N. P., Dahl, G. E., Padmanabhan, V., Thrun, L. A., and Karsch, F. J. (1997). Estradiol requirements for induction and maintenance of the gonadotropin-releasing hormone surge: implications for neuroendocrine processing of the estradiol signal. *Endocrinology* 138, 5408–5414.
312. Rivest, S., and Rivier, C. (1993). Interleukin-1 beta inhibits the endogenous expression of the early gene c-fos located within the nucleus of LH-RH neurons and interferes with hypothalamic LH-RH release during proestrus in the rat. *Brain Res.* 613, 132–142.
313. Nappi, R. E., Bonneau, M. J., and Rivest, S. (1997). Influence of the estrous cycle on c-fos and CRH gene transcription in the brain of endotoxin-challenged female rats. *Neuroendocrinology* 65, 29–46.
314. Rivest, S. and Rivier, C. (1993). Centrally injected interleukin-1 beta inhibits the hypothalamic LHRH secretion and circulating LH levels via prostaglandins in rats. *J. Neuroendocrinol.* 5, 445–450.
315. Rivest, S., Plotsky, P. M., and Rivier, C. (1993). CRF alters the infundibular LHRH secretory system from the medial preoptic area of female rats: possible involvement of opioid receptors. *Neuroendocrinology* 57, 236–246.
316. Kalra, P. S., Sahu A., and Kalra, S. P. (1990). Interleukin-1 inhibits the ovarian steroid-induced luteinizing hormone surge and release of hypothalamic luteinizing hormone-releasing hormone in rats. *Endocrinology* 126, 2145–2152.

317. He, D., Sato, I., Kimura, F., and Akema, T. (2003). Lipopolysaccharide inhibits luteinizing hormone release through interaction with opioid and excitatory amino acid inputs to gonadotropin-releasing hormone neurones in female rats: possible evidence for a common mechanism involved in infection and immobilization stress. *J. Neuroendocrinol.* 15, 559–563.
318. Kam, K. Y., Park, Y. B., Cheon, M. S., Kang, S. S., Kim, K., and Ryu, K. (2002). Influence of GnRH agonist and neural antagonists on stress-blockade of LH and prolactin surges induced by 17beta-estradiol in ovariectomized rats. *Yonsei Med. J.* 43, 482–490.
319. Kam, K., Park, Y., Cheon, M., Son, G. H., Kim, K., and Ryu, K. (2000). Effects of immobilization stress on estrogen-induced surges of luteinizing hormone and prolactin in ovariectomized rats. *Endocrine* 12, 279–287.
320. Battaglia, D. F., Beaver, A. B., Harris, T. G., Tanhehco, E., Viguie, C., and Karsch, F. J. (1999). Endotoxin disrupts the estradiol-induced luteinizing hormone surge: interference with estradiol signal reading, not surge release. *Endocrinology* 140, 2471–2479.
321. Breen, K. M., Billings, H. J., Debus, N., and Karsch, F. J. (2004). Endotoxin inhibits the surge secretion of gonadotropin-releasing hormone via a prostaglandin-independent pathway. *Endocrinology* 145, 221–227.
322. Dobson, H. and Smith, R. F. (1995). Stress and reproduction in farm animals. *J. Reprod. Fertil. Suppl.* 49, 451–461.
323. Smart, D., Forhead, A. J., Smith, R. F., and Dobson, H. (1994). Transport stress delays the oestradiol-induced LH surge by a non-opioidergic mechanism in the early postpartum ewe. *J. Endocrinol.* 142, 447–451.
324. Nanda, A. S., Dobson, H., and Ward, W. R. (1990). Relationship between an increase in plasma cortisol during transport-induced stress and failure of oestradiol to induce a luteinising hormone surge in dairy cows. *Res. Vet. Sci.* 49, 25–28.
325. Kawaguchi, M., Funabashi, T., and Kimura, F. (1998). Insulin-induced hypoglycemia does not impair the surge of luteinizing hormone secretion in the proestrous rat. *Neurosci. Lett.* 256, 131–134.
326. Lujan, M. E., MacTavish, P. J., Krzemien, A. A., Bradstock, M. W., and Van Vugt, D. A. (2002). Estrogen-induced gonadotropin surge in rhesus monkeys is not inhibited by cortisol synthesis inhibition or hypoglycemia. *Endocrine* 19, 169–176.
327. Medina, C. L., Nagatani, S., Darling, T. A., Bucholtz, D. C., Tsukamura, H., Maeda, K., and Foster, D. L. (1998). Glucose availability modulates the timing of the luteinizing hormone surge in the ewe. *J. Neuroendocrinol.* 10, 785–792.
328. Briski, K. P. and Sylvester, P. W. (1998). Role of endogenous opiates in glucoprivic inhibition of the luteinizing hormone surge and fos expression by preoptic gonadotropin-releasing hormone neurones in ovariectomized steroid-primed female rats. *J. Neuroendocrinol.* 10, 769–776.
329. Karsch, F. J., Battaglia, D. F., Breen, K. M., Debus, N., and Harris, T. G. (2002). Mechanisms for ovarian cycle disruption by immune/inflammatory stress. *Stress* 5, 101–112.
330. Xiao, E., Xia-Zhang, L., Barth A., Zhu, J., and Ferin, M. (1998). Stress and the menstrual cycle: relevance of cycle quality in the short- and long-term response to a 5-day endotoxin challenge during the follicular phase in the rhesus monkey. *J. Clin. Endocrinol. Metab.* 83, 2454–2460.
331. Freeman, M. (2006). Neuroendocrine control of the ovarian cycle of the rat. In *Knobil and Neill's Physiology of Reproduction* (J. D. Neill, Ed.), pp. 2327–2388. Elsevier, St. Louis.
332. Goodman, R. L., and Inskeep, E. K. (2006). Neuroendocrine control of the ovarian cycle of the sheep. In *Knobil and Neill's Physiology of Reproduction* (J. D. Neill, Ed.), pp. 2389–2447. Elsevier, St. Louis.
333. Rivest, S., Lee, S., Attardi B., and Rivier, C. (1993). The chronic intracerebroventricular infusion of interleukin-1 beta alters the activity of the hypothalamic–pituitary–gonadal axis of cycling rats. I. Effect on LHRH and gonadotropin biosynthesis and secretion. *Endocrinology* 133, 2424–2430.
334. Calhoun, J. B. (1962). Population density and social pathology. *Sci. Am.* 206, 139–148.
335. Christian, J., Lloyd, J. A., and Davis, D. E. (1965). The role of endocrines in the self-regulation of mammalian populations. *Recent Prog. Horm. Res.* 21, 501–578.
336. Domanski, E., Przekop, F., Chomicka, L., and Ostrowska, A. (1989). Effect of stress on the course of oestrous cycle and the release of luteinizing hormone). The role of endorphin in these processes. *Acta Physiol. Pol.* 40, 64–73.
337. Marchlewska-Koj, A., Pochron, E., Galewicz-Sojecka A., and Galas, J. (1994). Suppression of estrus in female mice by the presence of conspecifics or by foot shock. *Physiol. Behav.* 55, 317–321.
338. Pollard, I., White, B. M., Bassett, J. R., and Cairncross, K. D. (1975). Plasma glucocorticoid elevation and desynchronization of the estrous cycle following unpredictable stress in the rat. *Behav. Biol.* 14, 103–108.
339. Barnett, S. A., and Coleman, E. M. (1959). The effect of low environmental temperature on the reproductive cycle of female mice. *J. Endocrinol.* 19, 232–240.
340. Denison, M. E., and Zarrow, M. X. (1955). Changes in estrous cycle of rat during prolonged exposure to cold. *Proc. Soc. Exp. Biol. Med.* 89, 632–634.
341. Smith, R. F., Ghuman, S. P., Evans, N. P., Karsch, F. J., and Dobson, H. (2003). Stress and the control of LH secretion in the ewe. *Reprod. Suppl.* 61, 267–282.
342. Doney, J. M., Gunn, R. G., and Griffiths, J. G. (1973). The effect of premating stress on the onset of oestrus and on ovulation rate in Scottish Blackface ewes. *J. Reprod. Fertil.* 35, 381–384.
343. Liptrap, R. M. (1993). Stress and reproduction in domestic animals. *Ann. N Y Acad. Sci.* 697, 275–284.
344. Guzeloglu A., Ambrose, J. D., Kassa, T., Diaz, T., Thatcher, M. J., and Thatcher, W. W. (2001). Long-term follicular dynamics and biochemical characteristics of dominant follicles in dairy cows subjected to acute heat stress. *Anim. Reprod. Sci.* 66, 15–34.
345. Roth, Z., Meidan, R., Shaham-Albalancy A., Braw-Tal, R., and Wolfenson, D. (2001). Delayed effect of heat stress on steroid production in medium-sized and preovulatory bovine follicles. *Reproduction* 121, 745–751.
346. Burke, J. M., Spiers, D. E., Kojima, F. N., Perry, G. A., Salfen, B. E., Wood, S. L., Patterson, D. J., Smith, M. F., Lucy, M. C., Jackson, W. G., and Piper, E. L. (2001). Interaction of endophyte-infected fescue and heat stress on ovarian function in the beef heifer. *Biol. Reprod.* 65, 260–268.
347. Roth, Z., Meidan, R., Braw-Tal, R., and Wolfenson, D. (2000). Immediate and delayed effects of heat stress on follicular development and its association with plasma FSH and inhibin concentration in cows. *J. Reprod. Fertil.* 120, 83–90.
348. Klinefelter, H. F., Albright, F., and Griswold, G. C. (1943). Experience with a quantitative test for normal or decreased Follicle-stimulating hormone in the urine in endocrinological diagnosis. *J. Clin. Endocrinol.* 3, 529–544.
349. Reifenstein, E. C. (1946). Psychogenic or hypothalamic amenorrhea. *Med. Clin. North Am.* 30, 1103–1114.
350. Lachelin, G. C. and Yen, S. S. (1978). Hypothalamic chronic anovulation. *Am. J. Obstet. Gynecol.* 130, 825–831.
351. Reindollar, R. H., Novak, M., Tho, S. P., and McDonough, P. G. (1986). Adult-onset amenorrhea: a study of 262 patients. *Am. J. Obstet. Gynecol.* 155, 531–543.
352. Biller, B. M., Federoff, H. J., Koenig, J. I., and Klibanski, A. (1990). Abnormal cortisol secretion and responses to

corticotropin-releasing hormone in women with hypothalamic amenorrhea. *J. Clin. Endocrinol. Metab.* 70, 311–317.

353. Meczekalski, B., Tonetti, A., Monteleone, P., Bernardi, F., Luisi, S., Stomati, M., Luisi, M., Petraglia, F., and Genazzani, A. R. (2000). Hypothalamic amenorrhea with normal body weight: ACTH, allopregnanolone and cortisol responses to corticotropin-releasing hormone test. *Eur. J. Endocrinol.* 142, 280–285.
354. Gold, P. W., Gwirtsman, H., Avgerinos, P. C., Nieman, L. K., Gallucci, W. T., Kaye, W., Jimerson, D., Ebert, M., Rittmaster, R., Loriaux, D. L., et al. (1986). Abnormal hypothalamic–pituitary–adrenal function in anorexia nervosa. Pathophysiologic mechanisms in underweight and weight-corrected patients. *N. Engl. J. Med.* 314, 1335–1342.
355. Gwirtsman, H. E., Kaye, W. H., George, D. T., Jimerson, D. C., Ebert, M. H., and Gold, P. W. (1989). Central and peripheral ACTH and cortisol levels in anorexia nervosa and bulimia. *Arch. Gen. Psychiatry* 46, 61–69.
356. Misra, M., Miller, K. K., Almazan, C., Ramaswamy K., Lapcharoensap, W., Worley, M., Neubauer, G., Herzog, D. B., and Klibanski, A. (2004). Alterations in cortisol secretory dynamics in adolescent girls with anorexia nervosa and effects on bone metabolism. *J. Clin. Endocrinol. Metab.* 89, 4972–4980.
357. Berga, S. L., Daniels, T. L., and Giles, D. E. (1997). Women with functional hypothalamic amenorrhea but not other forms of anovulation display amplified cortisol concentrations. *Fertil Steril.* 67, 1024–1030.
358. Ding, J. H., Sheckter, C. B., Drinkwater, B. L., Soules, M. R., and Bremner, W. J. (1988). High serum cortisol levels in exercise-associated amenorrhea. *Ann. Intern. Med.* 108, 530–534.
359. Pohl, C. R., Richardson, D. W., Hutchison, J. S., Germak, J. A., and Knobil, E. (1983). Hypophysiotropic signal frequency and the functioning of the pituitary-ovarian system in the rhesus monkey. *Endocrinology* 112, 2076–2080.
360. Engels, W. D., Pattee, C. J., and Wittkower, E. D. (1964). Emotional settings of functional amenorrhea. *Psychosom. Med.* 26, 682–700.
361. Rabkin, J. G., and Struening, E. L. (1976). Live events, stress, and illness. *Science* 194, 1013–1020.
362. Drew, F. L. (1961). The epidemiology of secondary amenorrhea. *J. Chronic Dis.* 14, 396–407.
363. Newman, M. M., and Halmi, K. A. (1988). The endocrinology of anorexia nervosa and bulimia nervosa. *Neurol. Clin.* 6, 195–212.
364. Scott, L. V., and Dinan, T. G. (2002). Vasopressin as a target for antidepressant development: an assessment of the available evidence. *J. Affect. Disord.* 72, 113–124.
365. Burns, L. H., and Covington, S. H. (1999). Psychology of infertility. In *Infertility Counseling: A Comprehensive Handbook for Clinicians* (L. H. Burns and S. H. Covington, Eds.). Parthenon Publishing, New York.
366. Fassino, S., Piero A., Boggio, S., Piccioni, V., and Garzaro, L. (2002). Anxiety, depression and anger suppression in infertile couples: a controlled study. *Hum. Reprod.* 17, 2986–2994.
367. Greil, A. L. (1997). Infertility and psychological distress: a critical review of the literature. *Soc. Sci. Med.* 45, 1679–1704.
368. Wright, J., Allard, M., Lecours A., and Sabourin, S. (1989). Psychosocial distress and infertility: a review of controlled research. *Int. J. Fertil.* 34, 126–142.
369. Wischmann, T., Stammer, H., Scherg, H., Gerhard, I., and Verres, R. (2001). Psychosocial characteristics of infertile couples: a study by the 'Heidelberg Fertility Consultation Service.' *Hum. Reprod.* 16, 1753–1761.
370. Decourt, J. and Michard, J. P. (1953). Les amenorrhees de cause psychique. *Rev. Prat.* 3, 27–34.
371. Giles, D. E., and Berga, S. L. (1993). Cognitive and psychiatric correlates of functional hypothalamic amenorrhea: a controlled comparison. *Fertil Steril.* 60, 486–492.
372. Nilsson, P. M., Moller, L., and Solstad, K. (1995). Adverse effects of psychosocial stress on gonadal function and insulin levels in middle-aged males. *J. Intern. Med.* 237, 479–486.
373. Loucks, A. B., Mortola, J. F., Girton, L., and Yen, S. S. (1989). Alterations in the hypothalamic–pituitary–ovarian and the hypothalamic–pituitary–adrenal axes in athletic women. *J. Clin. Endocrinol. Metab.* 68, 402–411.
374. Williams, N. I., Caston-Balderrama, A. L., Helmreich, D. L., Parfitt, D. B., Nosbisch, C., and Cameron, J. L. (2001). Longitudinal changes in reproductive hormones and menstrual cyclicity in cynomolgus monkeys during strenuous exercise training: abrupt transition to exercise-induced amenorrhea. *Endocrinology* 142, 2381–2389.
375. Xiao, E., Xia-Zhang, L., and Ferin, M. (1999). Stress and the menstrual cycle: short- and long-term response to a five-day endotoxin challenge during the luteal phase in the rhesus monkey. *J. Clin. Endocrinol. Metab.* 84, 623–626.
376. Xiao, E., Xia-Zhang, L., and Ferin, M. (2002). Inadequate luteal function is the initial clinical cyclic defect in a 12-day stress model that includes a psychogenic component in the Rhesus monkey. *J. Clin. Endocrinol. Metab.* 87, 2232–2237.
377. Kowalski, W., Chatterton, R. T., Jr., Kazer, R. R., and Wentz, A. C. (1993). The impact of subchronic hypercortisolemia on progesterone metabolism and the luteinizing hormone-progesterone axis in the cynomolgus monkey. *J. Clin. Endocrinol. Metab.* 77, 1597–1604.
378. Soules, M. R., Clifton, D. K., Cohen, N. L., Bremner, W. J., and Steiner, R. A. (1989). Luteal phase deficiency: abnormal gonadotropin and progesterone secretion patterns. *J. Clin. Endocrinol. Metab.* 69, 813–820.
379. Soules, M. R., McLachlan, R. I., Ek, M., Dahl, K. D., Cohen, N. L., and Bremner, W. J. (1989). Luteal phase deficiency: characterization of reproductive hormones over the menstrual cycle. *J. Clin. Endocrinol. Metab.* 69, 804–812.
380. Matsumoto, S., Tamada, T., and Konuma, S. (1979). Endocrinological analysis of environmental menstrual disorders. *Int. J. Fertil.* 24, 233–239.
381. Hatch, M. C., Figa-Talamanca, I., and Salerno, S. (1999). Work stress and menstrual patterns among American and Italian nurses. *Scand. J. Work Environ. Health* 25, 144–150.
382. Williams, N. I., Bullen, B. A., McArthur, J. W., Skrinar, G. S., and Turnbull, B. A. (1999). Effects of short-term strenuous endurance exercise upon corpus luteum function. *Med. Sci. Sports Exerc.* 31, 949–958.
383. Nepomnaschy, P. A., Welch K., McConnell, D., Strassmann, B. I., and England, B. G. (2004). Stress and female reproductive function: a study of daily variations in cortisol, gonadotrophins, and gonadal steroids in a rural Mayan population. *Am. J. Hum. Biol.* 16, 523–532.
384. Loucks, A. B. (2003). Introduction to menstrual disturbances in athletes. *Med. Sci. Sports Exerc.* 35, 1551–1552.
385. De Souza, M. J. (2003). Menstrual disturbances in athletes: a focus on luteal phase defects. *Med. Sci. Sports Exerc.* 35, 1553–1563.
386. De Souza, M. J., Miller, B. E., Loucks, A. B., Luciano, A. A., Pescatello, L. S., Campbell, C. G., and Lasley, B. L. (1998). High frequency of luteal phase deficiency and anovulation in recreational women runners: blunted elevation in follicle-stimulating hormone observed during luteal-follicular transition. *J. Clin. Endocrinol. Metab.* 83, 4220–4232.
387. Bonen, A. (1994). Exercise-induced menstrual cycle changes. A functional, temporary adaptation to metabolic stress. *Sports Med.* 17, 373–392.

388. Beitins, I. Z., McArthur, J. W., Turnbull, B. A., Skrinar, G. S., and Bullen, B. A. (1991). Exercise induces two types of human luteal dysfunction: confirmation by urinary free progesterone. *J. Clin. Endocrinol. Metab.* 72, 1350–1358.
389. Insler, V. (1992). Corpus luteum defects. *Curr. Opin. Obstet. Gynecol.* 4, 203–211.
390. Xiao, E., Xia-Zhang, L., Thornell, D., and Ferin, M. (1996). Interleukin-1 stimulates luteinizing hormone release during the midfollicular phase in the rhesus monkey: a novel way in which stress may influence the menstrual cycle. *J. Clin. Endocrinol. Metab.* 81, 2136–2141.
391. Xiao, E., Xia, L., Shanen, D., Khabele, D., and Ferin, M. (1994). Stimulatory effects of interleukin-induced activation of the hypothalamo-pituitary-adrenal axis on gonadotropin secretion in ovariectomized monkeys replaced with estradiol. *Endocrinology* 135, 2093–2098.
392. Puder, J. J., Freda, P. U., Goland, R. S., Ferin, M., and Wardlaw, S. L. (2000). Stimulatory effects of stress on gonadotropin secretion in estrogen-treated women. *J. Clin. Endocrinol. Metab.* 85, 2184–2188.
393. Ferin, M. (1999). Clinical review 105: stress and the reproductive cycle. *J. Clin. Endocrinol. Metab.* 84, 1768–1774.
394. Batista, M. C., Cartledge, T. P., Zellmer, A. W., Nieman, L. K., Merriam, G. R., and Loriaux, D. L. (1992). Evidence for a critical role of progesterone in the regulation of the midcycle gonadotropin surge and ovulation. *J. Clin. Endocrinol. Metab.* 74, 565–570.
395. March, C. M., Goebelsmann, U., Nakamura, R. M., and Mishell, D. R. Jr. (1979). Roles of estradiol and progesterone in eiliciting the midcycle luteinizing hormone and follicle-stimulating hormone surges. *J. Clin. Endocrinol. Metab.* 49, 507–513.
396. Schenken, R. S., Werlin, L. B., Williams, R. F., Prihoda, T. J., and Hodgen, G. D. (1985). Periovulatory hormonal dynamics: relationship of immunoassayable gonadotropins and ovarian steroids to the bioassayable luteinizing hormone surge in rhesus monkeys. *J. Clin. Endocrinol. Metab.* 60, 886–890.
397. Williams, N. I., McArthur, J. W., Turnbull, B. A., Bullen, B. A., Skrinar, G. S., Beitins, I. Z., Besser, G. M., Rees, L. H., Gilbert, I., Cramer, D., et al. (1994). Effects of follicular phase exercise on luteinizing hormone pulse characteristics in sedentary eumenorrhoeic women. *Clin. Endocrinol. (Oxf).* 41, 787–794.
398. Boivin, J. and Takefman, J. E. (1995). Stress level across stages of in vitro fertilization in subsequently pregnant and nonpregnant women. *Fertil Steril.* 64, 802–810.
399. Paulson, R. J., Sauer, M. V., and Lobo, R. A. (1994). Addition of a gonadotropin releasing hormone (GnRH) antagonist and exogenous gonadotropins to unstimulated in vitro fertilization (IVF) cycles: physiologic observations and preliminary experience. *J. Assist. Reprod. Genet.* 11, 28–32.
400. Watson, H., Kiddy, D. S., Hamilton-Fairley, D., Scanlon, M. J., Barnard, C., Collins, W. P., Bonney, R. C., and Franks, S. (1993). Hypersecretion of luteinizing hormone and ovarian steroids in women with recurrent early miscarriage. *Hum. Reprod.* 8, 829–833.
401. Stanger, J. D., and Yovich, J. L. (1985). Reduced in-vitro fertilization of human oocytes from patients with raised basal luteinizing hormone levels during the follicular phase. *Br. J. Obstet. Gynaecol.* 92, 385–393.
402. Resko, J. A. (1969). Endocrine control of adrenal progesterone secretion in the ovariectomized rat. *Science* 164, 70–71.
403. Caligaris, L., Astrada, J. J., and Taleisnik, S. (1971). Biphasic effect of progesterone on the release of gonadotropin in rats. *Endocrinology* 89, 331–337.
404. Berga, S. L., Marcus, M. D., Loucks, T. L., Hlastala, S., Ringham, R., and Krohn, M. A. (2003). Recovery of ovarian activity in women with functional hypothalamic amenorrhea who were treated with cognitive behavior therapy. *Fertil Steril.* 80, 976–981.
405. Kalogeras, K. T., Calogero, A. E., Kuribayiashi, T., Khan, I., Gallucci, W. T., Kling, M. A., Chrousos, G. P., and Gold, P. W. (1990). In vitro and in vivo effects of the triazolobenzodiazepine alprazolam on hypothalamic–pituitary–adrenal function: pharmacological and clinical implications. *J. Clin. Endocrinol. Metab.* 70, 1462–1471.
406. Owens, M. J., Vargas, M. A., Knight, D. L., and Nemeroff, C. B. (1991). The effects of alprazolam on corticotropin-releasing factor neurons in the rat brain: acute time course, chronic treatment and abrupt withdrawal. *J. Pharmacol. Exp. Ther.* 258, 349–356.
407. Owens, M. J., Vargas, M. A., and Nemeroff, C. B. (1993). The effects of alprazolam on corticotropin-releasing factor neurons in the rat brain: implications for a role for CRF in the pathogenesis of anxiety disorders. *J. Psychiatr. Res.* 27, Suppl 1, 209–220.
408. Calogero, A. E., Gallucci, W. T., Chrousos, G. P., and Gold, P. W. (1988). Interaction between GABAergic neurotransmission and rat hypothalamic corticotropin-releasing hormone secretion in vitro. *Brain Res.* 463, 28–36.
409. Judd, S. J., Wong, J., Saloniklis, S., Maiden, M., Yeap B., Filmer, S., and Michailov, L. (1995). The effect of alprazolam on serum cortisol and luteinizing hormone pulsatility in normal women and in women with stress-related anovulation. *J. Clin. Endocrinol. Metab.* 80, 818–823.
410. Xiao, E., Xia-Zhang, L., Vulliemoz, N.R., Rivier, J., and Ferin, M. (2004). Protective effects of a CRH antagonist against the disruption of the menstrual cycle by an inflammatory challenge in the rhesus monkey. The Endocrine Society's 86th annual meeting. Abstract P1-245.
411. Habib, K. E., Weld, K. P., Rice, K. C., Pushkas, J., Champoux, M., Listwak, S., Webster, E. L., Atkinson, A. J., Schulkin, J., Contoreggi, C., Chrousos, G. P., McCann, S. M., Suomi, S. J., Higley, J. D., and Gold, P. W. (2000). Oral administration of a corticotropin-releasing hormone receptor antagonist significantly attenuates behavioral, neuroendocrine, and autonomic responses to stress in primates. *Proc. Natl. Acad. Sci. U S A* 97, 6079–6084.
412. Rivier, C., Rivier, J., and Vale, W. (1986). Stress-induced inhibition of reproductive functions: role of endogenous corticotropin-releasing factor. *Science* 231, 607–609.
413. Caraty, A., Grino, M., Locatelli A., Guillaume, V., Boudouresque, F., Conte-Devolx, B., and Oliver, C. (1990). Insulin-induced hypoglycemia stimulates corticotropin-releasing factor and arginine vasopressin secretion into hypophysial portal blood of conscious, unrestrained rams. *J. Clin. Invest.* 85, 1716–1721.
414. Engler, D., Pham, T., Fullerton, M. J., Ooi, G., Funder, J. W., and Clarke, I. J. (1989). Studies of the secretion of corticotropin-releasing factor and arginine vasopressin into the hypophysial-portal circulation of the conscious sheep. I. Effect of an audiovisual stimulus and insulin-induced hypoglycemia. *Neuroendocrinology* 49, 367–381.
415. Van Vugt, D. A., Piercy, J., Farley, A. E., Reid, R. L., and Rivest, S. (1997). Luteinizing hormone secretion and corticotropin-releasing factor gene expression in the paraventricular nucleus of rhesus monkeys following cortisol synthesis inhibition. *Endocrinology* 138, 2249–2258.
416. Kaye, W. H., Gwirtsman, H. E., George, D. T., Ebert, M. H., Jimerson, D. C., Tomai, T. P., Chrousos, G. P., and Gold, P. W. (1987). Elevated cerebrospinal fluid levels of immunoreactive corticotropin-releasing hormone in anorexia nervosa: relation to state of nutrition, adrenal function, and intensity of depression. *J. Clin. Endocrinol. Metab.* 64, 203–208.

417. Berga, S. L., Loucks-Daniels, T. L., Adler, L. J., Chrousos, G. P., Cameron, J. L., Matthews, K. A., and Marcus, M. D. (2000). Cerebrospinal fluid levels of corticotropin-releasing hormone in women with functional hypothalamic amenorrhea. *Am. J. Obstet. Gynecol.* 182, 776–781; discussion 781–774.
418. Olster, D. H. and Ferin, M. (1987). Corticotropin-releasing hormone inhibits gonadotropin secretion in the ovariectomized rhesus monkey. *J. Clin. Endocrinol. Metab.* 65, 262–267.
419. Williams, C. L., Nishihara, M., Thalabard, J. C., Grosser, P. M., Hotchkiss, J., and Knobil, E. (1990). Corticotropin-releasing factor and gonadotropin-releasing hormone pulse generator activity in the rhesus monkey. Electrophysiological studies. *Neuroendocrinology* 52, 133–137.
420. Xiao, E. N. and Ferin, M. (1988). The inhibitory action of corticotropin-releasing hormone on gonadotropin secretion in the ovariectomized rhesus monkey is not mediated by adrenocorticotropic hormone. *Biol. Reprod.* 38, 763–767.
421. Gindoff, P. R., Xiao EN, Luckhaus, J., and Ferin, M. (1989). Dexamethasone treatment prevents the inhibitory effect of corticotropin-releasing hormone on gonadotropin release in the primate. *Neuroendocrinology* 49, 202–206.
422. Xiao, E., Luckhaus, J., Niemann, W., and Ferin, M. (1989). Acute inhibition of gonadotropin secretion by corticotropin-releasing hormone in the primate: are the adrenal glands involved? *Endocrinology* 124, 1632–1637.
423. Norman, R. L. (1993). Effects of corticotropin-releasing hormone on luteinizing hormone, testosterone, and cortisol secretion in intact male rhesus macaques. *Biol. Reprod.* 49, 148–153.
424. Norman, R. L. (1994). Corticotropin-releasing hormone effects on luteinizing hormone and cortisol secretion in intact female rhesus macaques. *Biol. Reprod.* 50, 949–955.
425. Barbarino A., De Marinis, L., Tofani A., Della Casa, S., D'Amico, C., Mancini A., Corsello, S. M., Sciuto, R., and Barini, A. (1989). Corticotropin-releasing hormone inhibition of gonadotropin release and the effect of opioid blockade. *J. Clin. Endocrinol. Metab.* 68, 523–528.
426. Metzger, D. L., Kerrigan, J. R., and Rogol, A. D. (1997). The short-term infusion of ovine corticotropin-releasing hormone does not alter luteinizing hormone concentrations in young adult men. *J. Clin. Endocrinol. Metab.* 82, 697–700.
427. Fischer, U. G., Wood, S. H., Bruhn, J., Roseff, S. J., Mortola, J., Rivier, J. E., and Yen, S. S. (1992). Effect of human corticotropin-releasing hormone on gonadotropin secretion in cycling and postmenopausal women. *Fertil Steril.* 58, 1108–1112.
428. Caraty, A., Miller, D. W., Delaleu B., and Martin, G. B. (1997). Stimulation of LH secretion in sheep by central administration of corticotrophin-releasing hormone. *J. Reprod. Fertil.* 111, 249–257.
429. Naylor, A. M., Porter, D. W., and Lincoln, D. W. (1990). Central administration of corticotrophin-releasing factor in the sheep: effects on secretion of gonadotrophins, prolactin and cortisol. *J. Endocrinol.* 124, 117–125.
430. Nikolarakis, K. E., Almeida, O. F., and Herz, A. (1986). Corticotropin-releasing factor (CRF) inhibits gonadotropin-releasing hormone (GnRH) release from superfused rat hypothalami in vitro. *Brain Res.* 377, 388–390.
431. Gambacciani, M., Yen, S. S., and Rasmussen, D. D. (1986). GnRH release from the mediobasal hypothalamus: in vitro inhibition by corticotropin-releasing factor. *Neuroendocrinology* 43, 533–536.
432. Rivier, C., and Vale, W. (1984). Influence of corticotropin-releasing factor on reproductive functions in the rat. *Endocrinology* 114, 914–921.
433. Petraglia, F., Sutton, S., Vale, W., and Plotsky, P. (1987). Corticotropin-releasing factor decreases plasma luteinizing hormone levels in female rats by inhibiting gonadotropin-releasing hormone release into hypophysial-portal circulation. *Endocrinology* 120, 1083–1088.
434. Ono, N., Lumpkin, M. D., Samson, W. K., McDonald, J. K., and McCann, S. M. (1984). Intrahypothalamic action of corticotrophin-releasing factor (CRF) to inhibit growth hormone and LH release in the rat. *Life Sci.* 35, 1117–1123.
435. Frias, J., Puertas A., Ruiz, E., and Ortega, E. (1997). Effect of corticotropin releasing factor (CRF) injected into the median eminence on LH secretion in male rats. *Neurochem. Res.* 22, 171–174.
436. Gindoff, P. R., and Ferin, M. (1987). Endogenous opioid peptides modulate the effect of corticotropin-releasing factor on gonadotropin release in the primate. *Endocrinology* 121, 837–842.
437. MacLusky, N. J., Naftolin, F., and Leranth, C. (1988). Immunocytochemical evidence for direct synaptic connections between corticotrophin-releasing factor (CRF) and gonadotrophin-releasing hormone (GnRH)-containing neurons in the preoptic area of the rat. *Brain Res.* 439, 391–395.
438. Lado-Abeal, J., Clapper, J. A., and Norman, R. L. (2001). Antagonism of central vasopressin receptors blocks hypoglycemic stress induced inhibition of luteinizing hormone release in male rhesus macaques. *J. Neuroendocrinol.* 13, 650–655.
439. Cates, P. S., Forsling, M. L., and O'Byrne, K. T. (1999). Stress-induced suppression of pulsatile Luteinising hormone release in the female rat: role of vasopressin. *J. Neuroendocrinol.* 11, 677–683.
440. Ono, N., Bedran de Castro, J., Khorram, O., and McCann, S. M. (1985). Role of arginine vasopressin in control of ACTH and LH release during stress. *Life Sci.* 36, 1779–1786.
441. Xiao, E., Xia-Zhang, L., Thornell, D., Shalts, E., and Ferin, M. (1996). The luteinizing hormone but not the cortisol response to arginine vasopressin is prevented by naloxone and a corticotropin-releasing hormone antagonist in the ovariectomized rhesus monkey. *Neuroendocrinology* 64, 225–232.
442. Monzani, A., Petraglia, F., De Leo, V., Fabbri, G., D'Ambrogio, G., Volpe, A., and Genazzani, A. R. (1989). Glucocorticoids but not vasopressin or oxytocin inhibit luteinizing hormone secretion in patients with psychogenic amenorrhea. *Gynecol. Endocrinol.* 3, 55–62.
443. Li, X. F., Bowe, J. E., Lightman, S. L., and O'Byrne, K. T. (2005). Role of corticotropin-releasing factor receptor-2 in stress-induced suppression of pulsatile luteinizing hormone secretion in the rat. *Endocrinology* 146, 318–322.
444. Hahn, J. D., Kalamatianos, T., and Coen, C. W. (2003). Studies on the neuroanatomical basis for stress-induced oestrogen-potentiated suppression of reproductive function: evidence against direct corticotropin-releasing hormone projections to the vicinity of luteinizing hormone-releasing hormone cell bodies in female rats. *J. Neuroendocrinol.* 15, 732–742.
445. Castro, M. G. and Morrison, E. (1997). Post-translational processing of proopiomelanocortin in the pituitary and in the brain. *Crit. Rev. Neurobiol.* 11, 35–57.
446. Smith, A. I. and Funder, J. W. (1988). Proopiomelanocortin processing in the pituitary, central nervous system, and peripheral tissues. *Endocr. Rev.* 9, 159–179.
447. Gee, C. E., Chen, C. L., Roberts, J. L., Thompson, R., and Watson, S. J. (1983). Identification of proopiomelanocortin neurones in rat hypothalamus by in situ cDNA-mRNA hybridization. *Nature* 306, 374–376.
448. Hollt, V., Przewlocki, R., Haarmann, I., Almeida, O. F., Kley, N., Millan, M. J., and Herz, A. (1986). Stress-induced alterations in the levels of messenger RNA coding for proopiomelanocortin and prolactin in rat pituitary. *Neuroendocrinology* 43, 277–282.

449. Harbuz, M. S. and Lightman, S. L. (1989). Responses of hypothalamic and pituitary mRNA to physical and psychological stress in the rat. *J. Endocrinol.* 122, 705–711.
450. Birnberg, N. C., Lissitzky, J. C., Hinman, M., and Herbert, E. (1983). Glucocorticoids regulate proopiomelanocortin gene expression in vivo at the levels of transcription and secretion. *Proc. Natl. Acad. Sci. U S A* 80, 6982–6986.
451. Vale, W., Spiess, J., Rivier, C., and Rivier, J. (1981). Characterization of a 41-residue ovine hypothalamic peptide that stimulates secretion of corticotropin and beta-endorphin. *Science* 213, 1394–1397.
452. Jackson, R. V., DeCherney, G. S., DeBold, C. R., Sheldon, W. R., Alexander, A. N., Rivier, J., Vale, W., and Orth, D. N. (1984). Synthetic ovine corticotropin-releasing hormone: simultaneous release of proopiolipomelanocortin peptides in man. *J. Clin. Endocrinol. Metab.* 58, 740–743.
453. Zimmerman, E. A. (1981). The distribution of some identified peptide-containing pathways in the central nervous system including the basal ganglia. *Prog. Clin. Biol. Res.* 68, 117–125.
454. Nikolarakis, K. E., Almeida, O. F., and Herz, A. (1986). Stimulation of hypothalamic beta-endorphin and dynorphin release by corticotropin-releasing factor (in vitro). *Brain Res.* 399, 152–155.
455. Tomaszewska, D., Mateusiak K., and Przekop, F. (1999). Changes in extracellular LHRH and beta-endorphin-like immunoreactivity in the nucleus infundibularis-median eminence of anestrous ewes under stress condition. *J. Neural Transm.* 106, 265–274.
456. Millan, M. J., Przewlocki, R., Jerlicz, M., Gramsch, C., Hollt, V., and Herz, A. (1981). Stress-induced release of brain and pituitary beta-endorphin: major role of endorphins in generation of hyperthermia, not analgesia. *Brain Res.* 208, 325–338.
457. Lopez-Calderon, A., Ariznavarreta, C., and Chen, C. L. (1991). Influence of chronic restraint stress on pro-opiomelanocortin mRNA and beta-endorphin in the rat hypothalamus. *J. Mol. Endocrinol.* 7, 197–204.
458. Przewlocki, R., Millan, M. J., Gramsch, C., Millan, M. H., and Herz, A. (1982). The influence of selective adeno- and neurointermedio-hypophysectomy upon plasma and brain levels of beta-endorphin and their response to stress in rats. *Brain Res.* 242, 107–117.
459. Yen, S. S., Quigley, M. E., Reid, R. L., Ropert, J. F., and Cetel, N. S. (1985). Neuroendocrinology of opioid peptides and their role in the control of gonadotropin and prolactin secretion. *Am. J. Obstet. Gynecol.* 152, 485–493.
460. Gindoff, P. and Ferin, M. (1987). Brain opioid peptides and menstrual cyclicity. *Sem. Reprod. Endocrinol.* 5, 123–133.
461. Reid, R. L., Hoff, J. D., Yen, S. S., and Li, C. H. (1981). Effects of exogenous beta h-endorphin on pituitary hormone secretion and its disappearance rate in normal human subjects. *J. Clin. Endocrinol. Metab.* 52, 1179–1184.
462. Ropert, J. F., Quigley, M. E., and Yen, S. S. (1981). Endogenous opiates modulate pulsatile luteinizing hormone release in humans. *J. Clin. Endocrinol. Metab.* 52, 583–585.
463. Ferin, M., Wehrenberg, W. B., Lam, N. Y., Alston, E. J., and Vande Wiele, R. L. (1982). Effects and site of action of morphine on gonadotropin secretion in the female rhesus monkey. *Endocrinology* 111, 1652–1656.
464. Ferin, M., Van Vugt, D., and Wardlaw, S. (1984). The hypothalamic control of the menstrual cycle and the role of endogenous opioid peptides. *Recent Prog. Horm. Res.* 40, 441–485.
465. VanVugt, D. A., Webb, M. Y., and Reid, R. L. (1989). Comparison of the duration of action of nalmefene and naloxone on the hypothalamic–pituitary axis of the rhesus monkey. *Neuroendocrinology* 49, 275–280.
466. Kinoshita, F., Nakai, Y., Katakami, H., Kato, Y., Yajima, H., and Imura, H. (1980). Effect of beta-endorphin on pulsatile luteinizing hormone release in conscious castrated rats. *Life Sci.* 27, 843–846.
467. Sylvester, P. W., Van Vugt, D. A., Aylsworth, C. A., Hanson, E. A., and Meites, J. (1982). Effects of morphine and naloxone on inhibition by ovarian hormones of pulsatile release of LH in ovariectomized rats. *Neuroendocrinology* 34, 269–273.
468. Cicero, T. J., Schainker, B. A., and Meyer, E. R. (1979). Endogenous opioids participate in the regulation of the hypothalamus-pituitary-luteinizing hormone axis and testosterone's negative feedback control of luteinizing hormone. *Endocrinology* 104, 1286–1291.
469. Johnson, J. H., Maughan, G. T., and Anderhub, L. (1982). Inhibition of pulsatile luteinizing hormone release by morphine microinjected into mesencephalic dorsal raphe. *Life Sci.* 30, 1473–1478.
470. Van Vugt, D. A., Lam, N. Y., and Ferin, M. (1984). Reduced frequency of pulsatile luteinizing hormone secretion in the luteal phase of the rhesus monkey. Involvement of endogenous opiates. *Endocrinology* 115, 1095–1101.
471. Filicori, M., Santoro, N., Merriam, G. R., and Crowley, W. F. Jr. (1986). Characterization of the physiological pattern of episodic gonadotropin secretion throughout the human menstrual cycle. *J. Clin. Endocrinol. Metab.* 62, 1136–1144.
472. Wehrenberg, W. B., Wardlaw, S. L., Frantz, A. G., and Ferin, M. (1982). beta-Endorphin in hypophyseal portal blood: variations throughout the menstrual cycle. *Endocrinology* 111, 879–881.
473. Wardlaw, S. L., Wehrenberg, W. B., Ferin, M., Antunes, J. L., and Frantz, A. G. (1982). Effect of sex steroids on beta-endorphin in hypophyseal portal blood. *J. Clin. Endocrinol. Metab.* 55, 877–881.
474. Gindoff, P. R., Jewelewicz, R., Hembree, W., Wardlaw, S., and Ferin, M. (1988). Sustained effects of opioid antagonism during the normal human luteal phase. *J. Clin. Endocrinol. Metab.* 66, 1000–1004.
475. Quigley, M. E. and Yen, S. S. (1980). The role of endogenous opiates in LH secretion during the menstrual cycle. *J. Clin. Endocrinol. Metab.* 51, 179–181.
476. O'Byrne, K. T., Lunn, S. F., and Dixson, A. F. (1989). Naloxone reversal of stress-induced suppression of LH release in the common marmoset. *Physiol. Behav.* 45, 1077–1080.
477. Gilbeau, P. M. and Smith, C. G. (1985). Naloxone reversal of stress-induced reproductive effects in the male rhesus monkey. *Neuropeptides* 5, 335–338.
478. Pau, K. Y., Berria, M., Hess, D. L., and Spies, H. G. (1996). Opiatergic influence on gonadotropin-releasing hormone and luteinizing hormone release during the macaque menstrual cycle. *Biol. Reprod.* 55, 478–484.
479. Heisler, L. E., Pallotta, C. M., Reid, R. L., and Van Vugt, D. A. (1993). Hypoglycemia-induced inhibition of luteinizing hormone secretion in the rhesus monkey is not mediated by endogenous opioid peptides. *J. Clin. Endocrinol. Metab.* 76, 1280–1285.
480. Akema, T., Chiba A., Shinozaki, R., Oshida, M., Kimura, F., and Toyoda, J. (1996). Acute immobilization stress and intraventricular injection of CRF suppress naloxone-induced LH release in ovariectomized estrogen-primed rats. *J. Neuroendocrinol.* 8, 647–652.
481. Almeida, O. F., Nikolarakis, K. E., and Herz, A. (1988). Evidence for the involvement of endogenous opioids in the inhibition of luteinizing hormone by corticotropin-releasing factor. *Endocrinology* 122, 1034–1041.
482. Petraglia, F., Vale, W., and Rivier, C. (1986). Opioids act centrally to modulate stress-induced decrease in luteinizing hormone in the rat. *Endocrinology* 119, 2445–2450.
483. Gonzalez-Quijano, M. I., Ariznavarreta, C., Martin, A. I., Treguerres, J. A., and Lopez-Calderon, A. (1991).

Naltrexone does not reverse the inhibitory effect of chronic restraint on gonadotropin secretion in the intact male rat. *Neuroendocrinology* 54, 447–453.
484. Akinbami, M. A., Taylor, M. F., Collins, D. C., and Mann, D. R. (1994). Effect of a peripheral and a central acting opioid antagonist on the testicular response to stress in rats. *Neuroendocrinology* 59, 343–348.
485. Armeanu, M. C., Lambalk, C. B., Berkhout, G. M., and Schoemaker, J. (1992). Effects of opioid antagonism with naltrexone on pulsatile luteinizing hormone secretion in women with hypothalamic amenorrhea in basal conditions and after discontinuation of treatment with pulsatile LHRH. *Gynecol. Endocrinol.* 6, 3–12.
486. Sauder, S. E., Case, G. D., Hopwood, N. J., Kelch, R. P., and Marshall, J. C. (1984). The effects of opiate antagonism on gonadotropin secretion in children and in women with hypothalamic amenorrhea. *Pediatr. Res.* 18, 322–328.
487. Khoury, S. A., Reame, N. E., Kelch, R. P., and Marshall, J. C. (1987). Diurnal patterns of pulsatile luteinizing hormone secretion in hypothalamic amenorrhea: reproducibility and responses to opiate blockade and an alpha 2-adrenergic agonist. *J. Clin. Endocrinol. Metab.* 64, 755–762.
488. Blankstein, J., Reyes, F. I., Winter, J. S., and Faiman, C. (1981). Endorphins and the regulations of the human menstrual cycle. *Clin. Endocrinol. (Oxf).* 14, 287–294.
489. Baranowska B., Rozbicka, G., Jeske, W., and Abdel-Fattah, M. H. (1984). The role of endogenous opiates in the mechanism of inhibited luteinizing hormone (LH) secretion in women with anorexia nervosa: the effect of naloxone on LH, follicle-stimulating hormone, prolactin, and beta-endorphin secretion. *J. Clin. Endocrinol. Metab.* 59, 412–416.
490. Wildt, L. and Leyendecker, G. (1987). Induction of ovulation by the chronic administration of naltrexone in hypothalamic amenorrhea. *J. Clin. Endocrinol. Metab.* 64, 1334–1335.
491. Wildt, L., Leyendecker, G., Sir-Petermann, T., and Waibel-Treber, S. (1993). Treatment with naltrexone in hypothalamic ovarian failure: induction of ovulation and pregnancy. *Hum. Reprod.* 8, 350–358.
492. Wildt, L., Sir-Petermann, T., Leyendecker, G., Waibel-Treber, S., and Rabenbauer, B. (1993). Opiate antagonist treatment of ovarian failure. *Hum. Reprod.* 8, Suppl. 2, 168–174.
493. Couzinet, B., Young, J., Brailly, S., Chanson, P., and Schaison, G. (1995). Even after priming with ovarian steroids or pulsatile gonadotropin-releasing hormone administration, naltrexone is unable to induce ovulation in women with functional hypothalamic amenorrhea. *J. Clin. Endocrinol. Metab.* 80, 2102–2107.
494. Armeanu, M. C., Moss, R. J., and Schoemaker, J. (1992). Ovulation induction with a single-blind treatment regimen comparing naltrexone, placebo and clomiphene citrate in women with secondary amenorrhea. *Acta Endocrinol. (Copenh).* 126, 410–415.
495. Asanovich, K. M., Johnson B., Chang, W. J., Barb, C. R., Rampacek, G. B., and Kraeling, R. R. (1998). Delay of estradiol-induced surge secretion of LH in gilts by intracerebroventricular injection of morphine. *Domest. Anim. Endocrinol.* 15, 45–53.
496. Currie, W. D., Joseph, I. B., and Rawlings, N. C. (1991). Morphine, naloxone and the gonadotrophin surge in ewes. *J. Reprod. Fertil.* 92, 407–414.
497. Horton, R. J., and Clarke, I. J. (1988). Lack of an effect of morphine or naloxone on the oestrogen-induced LH surge in anoestrous ewes. *J. Endocrinol.* 119, 89–93.
498. Van Vugt, D. A., Heisler, L. E., and Reid, R. L. (1992). Progesterone inhibits the estrogen-induced gonadotropin surge in the rhesus monkey independent of endogenous opiates. *J. Clin. Endocrinol. Metab.* 74, 1312–1319.
499. Nanda, A. S., Ward, W. R., and Dobson, H. (1989). Effects of naloxone on the oestradiol-induced LH surge and cortisol release in transported cows. *J. Reprod. Fertil.* 87, 803–807.
500. Leadem, C. A. and Kalra, S. P. (1985). Reversal of beta-endorphin-induced blockade of ovulation and luteinizing hormone surge with prostaglandin E2. *Endocrinology* 117, 684–689.
501. Hulse, G. K., and Coleman, G. J. (1983). The role of endogenous opioids in the blockade of reproductive function in the rat following exposure to acute stress. *Pharmacol. Biochem. Behav.* 19, 795–799.
502. Ching, M. (1983). Morphine suppresses the proestrous surge of GnrH in pituitary portal plasma of rats. *Endocrinology* 112, 2209–2211.
503. Muraki, T., Tokunaga, Y., and Makino, T. (1977). Effects of morphine and naloxone on serum LH, FSH and prolactin levels and on hypothalamic content of LH-RF in proestrous rats. *Endocrinol. Jpn.* 24, 313–315.
504. Pang, C. N., Zimmermann, E., and Sawyer, C. H. (1977). Morphine inhibition of the preovulatory surges of plasma luteinizing hormone and follicle stimulating hormone in the rat. *Endocrinology* 101, 1726–1732.
505. Ieiri, T., Chen, H. T., Campbell, G. A., and Meites, J. (1980). Effects of naloxone and morphine on the proestrous surge of prolactin and gonadotropins in the rat. *Endocrinology* 106, 1568–1570.
506. Kalra, P. S., Fuentes, M., Sahu, A., and Kalra, S. P. (1990). Endogenous opioid peptides mediate the interleukin-1-induced inhibition of the release of luteinizing hormone (LH)-releasing hormone and LH. *Endocrinology* 127, 2381–2386.
507. Hulse, G., Coleman, G., Nicholas, J., and Greenwood, K. (1982). Reversal of the anti-ovulatory action of stress in rats by prior administration of naloxone hydrochloride. *J. Reprod. Fertil.* 66, 451–456.
508. Roozendaal, M. M., Swarts, J. J., van Maanen, J. C., Wiegant, V. M., and Mattheij, J. A. (1997). Inhibition of the LH surge in cyclic rats by stress is not mediated by opioids. *Life Sci.* 60, 735–742.
509. Contreras, P. C., and Takemori, A. E. (1984). Antagonism of morphine-induced analgesia, tolerance and dependence by alpha-melanocyte-stimulating hormone. *J. Pharmacol. Exp. Ther.* 229, 21–26.
510. Wardlaw, S. L. and Ferin, M. (1990). Interaction between beta-endorphin and alpha-melanocyte-stimulating hormone in the control of prolactin and luteinizing hormone secretion in the primate. *Endocrinology* 126, 2035–2040.
511. Khorram, O. and McCann, S. M. (1986). Interaction of alpha-melanocyte-stimulating hormone with beta-endorphin to influence anterior pituitary hormone secretion in the female rat. *Endocrinology* 119, 1071–1075.
512. Newman, C. B., Wardlaw, S. L., and Frantz, A. G. (1985). Suppression of basal and stress-induced prolactin release and stimulation of luteinizing hormone secretion by alpha-melanocyte-stimulating hormone. *Life Sci.* 36, 1661–1668.
513. Wardlaw, S. L., Smeal, M. M., and Markowitz, C. E. (1986). Antagonism of beta-endorphin-induced prolactin release by alpha-melanocyte-stimulating hormone and corticotropin-like intermediate lobe peptide. *Endocrinology* 119, 112–118.
514. Reid, R. L., Ling, N., and Yen, S. S. (1984). Gonadotropin-releasing activity of alpha-melanocyte-stimulating hormone in normal subjects and in subjects with hypothalamic–pituitary dysfunction. *J. Clin. Endocrinol. Metab.* 58, 773–777.
515. Shalts, E., Feng, Y. J., Ferin, M., and Wardlaw, S. L. (1992). Alpha-melanocyte-stimulating hormone antagonizes the neuroendocrine effects of corticotropin-releasing factor and interleukin-1 alpha in the primate. *Endocrinology* 131, 132–138.

516. Limone, P., Scipioni, T., Calvelli, P., Scaglione, E., Barberis, A. M., Dentelli, P., Isaia, G. C., and Molinatti, G. M. (1994). Alpha-melanocyte-stimulating hormone antagonizes the inhibitory effect of corticotropin-releasing hormone on luteinizing hormone secretion during the luteal phase in normal women. *J. Clin. Endocrinol. Metab.* 79, 571–575.
517. Harris, R. B., Zhou, J., Shi, M., Redmann, S., Mynatt, R. L., and Ryan, D. H. (2001). Overexpression of agouti protein and stress responsiveness in mice. *Physiol. Behav.* 73, 599–608.
518. Wardlaw, S. L. and Blum, M. (1990). Biphasic effect of orchiectomy on pro-opiomelanocortin gene expression in the hypothalamus. *Neuroendocrinology* 52, 521–526.
519. Blum, M., Roberts, J. L., and Wardlaw, S. L. (1989). Androgen regulation of proopiomelanocortin gene expression and peptide content in the basal hypothalamus. *Endocrinology* 124, 2283–2288.
520. Wilcox, J. N. and Roberts, J. L. (1985). Estrogen decreases rat hypothalamic proopiomelanocortin messenger ribonucleic acid levels. *Endocrinology* 117, 2392–2396.
521. Autelitano, D. J., Lundblad, J. R., Blum, M., and Roberts, J. L. (1989). Hormonal regulation of POMC gene expression. *Annu. Rev. Physiol.* 51, 715–726.
522. Debus, N., Breen, K. M., Barrell, G. K., Billings, H. J., Brown, M., Young, E. A., and Karsch, F. J. (2002). Does cortisol mediate endotoxin-induced inhibition of pulsatile luteinizing hormone and gonadotropin-releasing hormone secretion? *Endocrinology* 143, 3748–3758.
523. Bubenik, G. A., Smith, J. H., Pomerantz, D. K., and Schams, D. (1990). Plasma LH, FSH, testosterone, prolactin and androstenedione in male white-tailed deer after ACTH and dexamethasone administration. *Comp. Biochem. Physiol. A.* 95, 163–169.
524. Samuels, M. H., Luther, M., Henry, P., and Ridgway, E. C. (1994). Effects of hydrocortisone on pulsatile pituitary glycoprotein secretion. *J. Clin. Endocrinol. Metab.* 78, 211–215.
525. Hayashi, K. T., and Moberg, G. P. (1987). Influence of acute stress and the adrenal axis on the regulation of LH and testosterone in the male rhesus monkey (macaca mulatta). *Am. J. Primatol.* 12, 263–273.
526. Turner, A. I., Hemsworth, P. H., Canny, B. J., and Tilbrook, A. J. (1999). Inhibition of the secretion of LH in ovariectomised pigs by sustained but not repeated acute elevation of cortisol in the absence but not the presence of oestradiol. *J. Endocrinol.* 163, 477–486.
527. Li, X. F., Mitchell, J. C., Wood, S., Coen, C. W., Lightman, S. L., and O'Byrne, K. T. (2003). The effect of oestradiol and progesterone on hypoglycaemic stress-induced suppression of pulsatile luteinizing hormone release and on corticotropin-releasing hormone mRNA expression in the rat. *J. Neuroendocrinol.* 15, 468–476.
528. Dubey, A. K., and Plant, T. M. (1985). A suppression of gonadotropin secretion by cortisol in castrated male rhesus monkeys (Macaca mulatta) mediated by the interruption of hypothalamic gonadotropin-releasing hormone release. *Biol. Reprod.* 33, 423–431.
529. Saketos, M., Sharma, N., and Santoro, N. F. (1993). Suppression of the hypothalamic–pituitary–ovarian axis in normal women by glucocorticoids. *Biol. Reprod.* 49, 1270–1276.
530. Cunningham, G. R., Goldzieher, J. W., de la Pena, A., and Oliver, M. (1978). The mechanism of ovulation inhibition by triamcinolone acetonide. *J. Clin. Endocrinol. Metab.* 46, 8–14.
531. Frautschy, S. A., Gaon, D., and Liptrap, R. M. (1989). Effect of dexamethasone on gonadotrophin secretion in the gonadectomized sow and boar. *Can. J. Vet. Res.* 53, 490–492.
532. Li, P. S., and Wagner, W. C. (1983). In vivo and in vitro studies on the effect of adrenocorticotropic hormone or cortisol on the pituitary response to gonadotropin releasing hormone. *Biol. Reprod.* 29, 25–37.
533. Stoebel, D. P. and Moberg, G. P. (1982). Effect of adrenocorticotropin and cortisol on luteinizing hormone surge and estrous behavior of cows. *J. Dairy Sci.* 65, 1016–1024.
534. Thibier, M. and Rolland, O. (1976). The effect of dexamethasone (DXM) on circulating testosterone (T) and luteinizing hormone (LH) in young postpubertal bulls. *Theriogenology* 5, 53–60.
535. Baldwin, D. M. (1979). The effect of glucocorticoids on estrogen-dependent luteinizing hormone release in the ovariectomized rat and on gonadotropin secretin in the intact female rat. *Endocrinology* 105, 120–128.
536. Mann, D. R., Evans, D., Edoimioya, F., Kamel, F., and Butterstein, G. M. (1985). A detailed examination of the in vivo and in vitro effects of ACTH on gonadotropin secretion in the adult rat. *Neuroendocrinology* 40, 297–302.
537. Johnson BH, Welsh, T. H. Jr., and Juniewicz, P. E. (1982). Suppression of luteinizing hormone and testosterone secretion in bulls following adrenocorticotropin hormone treatment. *Biol. Reprod.* 26, 305–310.
538. Luton, J. P., Thieblot, P., Valcke, J. C., Mahoudeau, J. A., and Bricaire, H. (1977). Reversible gonadotropin deficiency in male Cushing's disease. *J. Clin. Endocrinol. Metab.* 45, 488–495.
539. Breen, K. M. and Karsch, F. J. (2004). Does cortisol inhibit pulsatile luteinizing hormone secretion at the hypothalamic or pituitary level? *Endocrinology* 145, 692–698.
540. Boccuzzi, G., Angeli A., Bisbocci, D., Fonzo, D., Giadano, G. P., and Ceresa, F. (1975). Effect of synthetic luteinizing hormone releasing hormone (LH-RH) on the release of gonadotropins in Cushing's disease. *J. Clin. Endocrinol. Metab.* 40, 892–895.
541. Chantaraprateep, P. and Thibier, M. (1978). Effects of dexamethasone on the responses of luteinizing hormone and testosterone to two injections of luteinizing hormone releasing hormone in young postpubertal bulls. *J. Endocrinol.* 77, 389–395.
542. Suter, D. E., Schwartz, N. B., and Ringstrom, S. J. (1988). Dual role of glucocorticoids in regulation of pituitary content and secretion of gonadotropins. *Am. J. Physiol.* 254, E595–E600.
543. Matteri, R. L., Watson, J. G., and Moberg, G. P. (1984). Stress or acute adrenocorticotrophin treatment suppresses LHRH-induced LH release in the ram. *J. Reprod. Fertil.* 72, 385–393.
544. Fuquay, J. W. and Moberg, G. P. (1983). Influence of the pituitary-adrenal axis on the induced release of luteinizing hormone in rams. *J. Endocrinol.* 99, 151–155.
545. Dobson, H., Essawy, S. A., and Alam, M. G. (1988). Suppression of LH response to gonadotrophin-releasing hormone or oestradiol by ACTH(1-24) treatment in anoestrous ewes. *J. Endocrinol.* 118, 193–197.
546. Suter, D. E. and Schwartz, N. B. (1985). Effects of glucocorticoids on responsiveness of luteinizing hormone and follicle-stimulating hormone to gonadotropin-releasing hormone by male rat pituitary cells in vitro. *Endocrinology* 117, 855–859.
547. Kamel, F. and Kubajak, C. L. (1987). Modulation of gonadotropin secretion by corticosterone: interaction with gonadal steroids and mechanism of action. *Endocrinology* 121, 561–568.
548. Padmanabhan, V., Keech, C., and Convey, E. M. (1983). Cortisol inhibits and adrenocorticotropin has no effect on luteinizing hormone-releasing hormone-induced release of luteinizing hormone from bovine pituitary cells in vitro. *Endocrinology* 112, 1782–1787.
549. Suter, D. E. and Orosz, G. (1989). Effect of treatment with cortisol in vivo on secretion of gonadotropins in vitro. *Biol. Reprod.* 41, 1091–1096.

550. Antakly, T., and Eisen, H. J. (1984). Immunocytochemical localization of glucocorticoid receptor in target cells. *Endocrinology* 115, 1984–1989.
551. Breen, K. M., Stackpole, C. A., Clarke, I. J., Pytiak, A. V., Tilbrook, A. J., Wagenmaker, E. R., Young, E. A., and Karsch, F. J. (2004). Does the type II glucocorticoid receptor mediate cortisol-induced suppression in pituitary responsiveness to gonadotropin-releasing hormone? *Endocrinology* 145, 2739–2746.
552. Adams, T. E., Sakurai, H., and Adams, B. M. (1999). Effect of stress-like concentrations of cortisol on estradiol-dependent expression of gonadotropin-releasing hormone receptor in orchidectomized sheep. *Biol. Reprod.* 60, 164–168.
553. Daley, C. A., Sakurai, H., Adams, B. M., and Adams, T. E. (1999). Effect of stress-like concentrations of cortisol on gonadotroph function in orchidectomized sheep. *Biol. Reprod.* 60, 158–163.
554. Daley, C. A., Sakurai, H., Adams, B. M., and Adams, T. E. (2000). Effect of stress-like concentrations of cortisol on the feedback potency of oestradiol in orchidectomized sheep. *Anim. Reprod. Sci.* 59, 167–178.
555. Tilbrook, A. J., Turner, A. I., and Clarke, I. J. (2000). Effects of stress on reproduction in non-rodent mammals: the role of glucocorticoids and sex differences. *Rev. Reprod.* 5, 105–113.
556. Tilbrook, A. J., Turner, A. I., and Clarke, I. J. (2002). Stress and reproduction: central mechanisms and sex differences in non-rodent species. *Stress* 5, 83–100.
557. Breen K., Wagenmaker, E. R., and Karsch, F. J. (2003). Insights into a mechanism by which stress interferes with the ovulatory cycle. Annual meeting of the Society for Neurosciences. Abstract 924–929.
558. Estienne, M. J., Barb, C. R., Kesner, J. S., Kraeling, R. R., and Rampacek, G. B. (1991). Luteinizing hormone secretion in hypophysial stalk-transected gilts given hydrocortisone acetate and pulsatile gonadotropin-releasing hormone. *Domest. Anim. Endocrinol.* 8, 407–414.
559. Stackpole, C., Turner, A. I., Clarke, I. J., and Tilbrook, A. J. (2003). Cortisol does not suppress the luteinizing hormone (LH) response to gonadotropin-releasing hormone (GnRH) in hypothalamo-pituitary disconnected rams and ewes. *Biol. Reprod.* 68, Suppl 1, 429, Abstract.
560. Matsuwaki, T., Watanabe, E., Suzuki, M., Yamanouchi K., and Nishihara, M. (2003). Glucocorticoid maintains pulsatile aecretion of luteinizing hormone under infectious stress condition. *Endocrinology* 144, 3477–3482.
561. Rivier, C. (2002). Inhibitory effect of neurogenic and immune stressors on testosterone secretion in rats. *Neuroimmunomodulation* 10, 17–29.
562. Srivastava, R. K., Taylor, M. F., and Mann, D. R. (1993). Effect of immobilization stress on plasma luteinizing hormone, testosterone, and corticosterone concentrations and on 3 beta-hydroxysteroid dehydrogenase activity in the testes of adult rats. *Proc. Soc. Exp. Biol. Med.* 204, 231–235.
563. Orr, T. E. and Mann, D. R. (1990). Effects of restraint stress on plasma LH and testosterone concentrations, Leydig cell LH/hCG receptors, and in vitro testicular steroidogenesis in adult rats. *Horm. Behav.* 24, 324–341.
564. Orr, T. E., Taylor, M. F., Bhattacharyya, A. K., Collins, D. C., and Mann, D. R. (1994). Acute immobilization stress disrupts testicular steroidogenesis in adult male rats by inhibiting the activities of 17 alpha-hydroxylase and 17,20-lyase without affecting the binding of LH/hCG receptors. *J. Androl.* 15, 302–308.
565. Maric, D., Kostic, T., and Kovacevic, R. (1996). Effects of acute and chronic immobilization stress on rat Leydig cell steroidogenesis. *J. Steroid Biochem. Mol. Biol.* 58, 351–355.
566. Akinbami, M. A., Philip, G. H., Sridaran, R., Mahesh, V. B., and Mann, D. R. (1999). Expression of mRNA and proteins for testicular steroidogenic enzymes and brain and pituitary mRNA for glutamate receptors in rats exposed to immobilization stress. *J. Steroid Biochem. Mol. Biol.* 70, 143–149.
567. Charpenet, G., Tache, Y., Forest, M. G., Haour, F., Saez, J. M., Bernier, M., Ducharme, J. R., and Collu, R. (1981). Effects of chronic intermittent immobilization stress on rat testicular androgenic function. *Endocrinology* 109, 1254–1258.
568. Cumming, D. C., Quigley, M. E., and Yen, S. S. (1983). Acute suppression of circulating testosterone levels by cortisol in men. *J. Clin. Endocrinol. Metab.* 57, 671–673.
569. Sapolsky, R. M. (1985). Stress-induced suppression of testicular function in the wild baboon: role of glucocorticoids. *Endocrinology* 116, 2273–2278.
570. Orr, T. E. and Mann, D. R. (1992). Role of glucocorticoids in the stress-induced suppression of testicular steroidogenesis in adult male rats. *Horm. Behav.* 26, 350–363.
571. Consten, D., Lambert, J. G., Komen, H., and Goos, H. J. (2002). Corticosteroids affect the testicular androgen production in male common carp (Cyprinus carpio L.). *Biol. Reprod.* 66, 106–111.
572. Goos, H. J. and Consten, D. (2002). Stress adaptation, cortisol and pubertal development in the male common carp, Cyprinus carpio. *Mol. Cell. Endocrinol.* 197, 105–116.
573. van den Berghe, G., Weekers, F., Baxter, R. C., Wouters, P., Iranmanesh A., Bouillon, R., and Veldhuis, J. D. (2001). Five-day pulsatile gonadotropin-releasing hormone administration unveils combined hypothalamic–pituitary–gonadal defects underlying profound hypoandrogenism in men with prolonged critical illness. *J. Clin. Endocrinol. Metab.* 86, 3217–3226.
574. Mann, D. R. and Orr, T. E. (1990). Effect of restraint stress on gonadal proopiomelanocortin peptides and the pituitary-testicular axis in rats. *Life Sci.* 46, 1601–1609.
575. Bambino, T. H. and Hsueh, A. J. (1981). Direct inhibitory effect of glucocorticoids upon testicular luteinizing hormone receptor and steroidogenesis in vivo and in vitro. *Endocrinology* 108, 2142–2148.
576. Charpenet, G., Tache, Y., Bernier, M., Ducharme, J. R., and Collu, R. (1982). Stress-induced testicular hyposensitivity to gonadotropin in rats. Role of the pituitary gland. *Biol. Reprod.* 27, 616–623.
577. Hales, D. B. and Payne, A. H. (1989). Glucocorticoid-mediated repression of P450scc mRNA and de novo synthesis in cultured Leydig cells. *Endocrinology* 124, 2099–2104
578. Gao, H. B., Tong, M. H., Hu, Y. Q., Guo, Q. S., Ge, R., and Hardy, M. P. (2002). Glucocorticoid induces apoptosis in rat leydig cells. *Endocrinology* 143, 130–138.
579. Hayashi, K. T. and Moberg, G. P. (1990). Influence of the hypothalamic–pituitary–adrenal axis on the menstrual cycle and the pituitary responsiveness to estradiol in the female rhesus monkey (Macaca mulatta). *Biol. Reprod.* 42, 260–265.
580. Macfarlane, M. S., Breen, K. M., Sakurai, H., Adams, B. M., and Adams, T. E. (2000). Effect of duration of infusion of stress-like concentrations of cortisol on follicular development and the preovulatory surge of LH in sheep. *Anim. Reprod. Sci.* 63, 167–175.
581. Kowalski, W. and Chatterton, R. T. Jr. (1992). Peripheral and not central suppression of ovarian function during osmotic pump infusion of adrenocorticotropin-(1-24) for one menstrual cycle in the cynomolgus monkey and its partial compensation by a transitory elevation of sex hormone-binding globulin levels. *Endocrinology* 130, 3582–3592.
582. Kowalski, W. and Chatterton, R. T. Jr. (1993). Effects of subchronic implantation of hydrocortisone phosphate-releasing osmotic pumps on peripheral gonadotropin levels

and estradiol feedback to gonadotropin secretion in the cynomolgus monkey. *Neuroendocrinology* 57, 330–339.
583. Daley, C. A., Macfarlane, M. S., Sakurai, H., and Adams, T. E. (1999). Effect of stress-like concentrations of cortisol on follicular development and the preovulatory surge of LH in sheep. *J. Reprod. Fertil.* 117, 11–16.
584. Aguado, L. I. and Ojeda, S. R. (1986). Prepubertal rat ovary: hormonal modulation of beta-adrenergic receptors and of progesterone response to adrenergic stimulation. *Biol. Reprod.* 34, 45–50.
585. Paredes A., Galvez A., Leyton, V., Aravena, G., Fiedler, J. L., Bustamante, D., and Lara, H. E. (1998). Stress promotes development of ovarian cysts in rats: the possible role of sympathetic nerve activation. *Endocrine* 8, 309–315.
586. Ogilvie, K. M., Held Hales, K., Roberts, M. E., Hales, D. B., and Rivier, C. (1999). The inhibitory effect of intracerebroventricularly injected interleukin 1beta on testosterone secretion in the rat: role of steroidogenic acute regulatory protein. *Biol. Reprod.* 60, 527–533.
587. Turnbull A. and Rivier, C. (1995). Brain-periphery connections: do they play a role in mediating the effect of centrally injected interleukin-1 beta on gonadal function? *Neuroimmunomodulation* 2, 224–235.
588. Turnbull, A. V. and Rivier, C. (1997). Inhibition of gonadotropin-induced testosterone secretion by the intracerebroventricular injection of interleukin-1 beta in the male rat. *Endocrinology* 138, 1008–1013.
589. Armario A. and Lopez-Calderon, A. (1986). Pituitary gonadal function in adult male rats subjected to crowding. *Endocr. Res.* 12, 115–122.
590. Mizunuma, H., DePalatis, L. R., and McCann, S. M. (1983). Effect of unilateral orchidectomy on plasma FSH concentration: evidence for a direct neural connection between testes and CNS. *Neuroendocrinology* 37, 291–296.
591. Nance, D. M., Bhargava, M., and Myatt, G. A. (1984). Further evidence for hypothalamic asymmetry in endocrine control of the ovary. *Brain Res. Bull.* 13, 651–655
592. Gerendai, I. and Halasz, B. (1981). Participation of a pure neuronal mechanism in the control of gonadal functions. *Andrologia* 13, 275–282.
593. Lee, S., Miselis, R., and Rivier, C. (2002). Anatomical and functional evidence for a neural hypothalamic-testicular pathway that is independent of the pituitary. *Endocrinology* 143, 4447–4454.
594. Gerendai, I., Toth, I. E., Boldogkoi, Z., Medveczky, I., and Halasz, B. (2000). Central nervous system structures labelled from the testis using the transsynaptic viral tracing technique. *J. Neuroendocrinol.* 12, 1087–1095.
595. Selvage, D. J., and Rivier, C. (2003). Importance of the paraventricular nucleus of the hypothalamus as a component of a neural pathway between the brain and the testes that modulates testosterone secretion independently of the pituitary. *Endocrinology* 144, 594–598.
596. Banczerowski, P., Csaba, Z., Csernus, V., and Gerendai, I. (2000). The effect of callosotomy on testicular steroidogenesis in hemiorchidectomized rats: a pituitary-independent regulatory mechanism. *Brain Res. Bull.* 53, 227–232.
597. Aguado, L. I. (2002). Role of the central and peripheral nervous system in the ovarian function. *Microsc. Res. Tech.* 59, 462–473.
598. Gerendai, I. and Halasz, B. (2000). Central nervous system structures connected with the endocrine glands: findings obtained with the viral transneuronal tracing technique. *Exp. Clin. Endocrinol. Diabetes.* 108, 389–395.
599. Gerendai, I., Kocsis K., and Halasz, B. (2002). Supraspinal connections of the ovary: structural and functional aspects. *Microsc. Res. Tech.* 59, 474–483.
600. Adashi, E. Y. and Hsueh, A. J. (1981). Stimulation of beta 2-adrenergic responsiveness by follicle-stimulating hormone in rat granulosa cells in vitro and in vivo. *Endocrinology* 108, 2170–2178.
601. D'Albora, H., Anesetti, G., Lombide, P., Dees, W. L., and Ojeda, S. R. (2002). Intrinsic neurons in the mammalian ovary. *Microsc. Res. Tech.* 59, 484–489.
602. Anesetti, G., Lombide, P., D'Albora, H., and Ojeda, S. R. (2001). Intrinsic neurons in the human ovary. *Cell Tissue Res.* 306, 231–237.
603. Saez, J. M. (1994). Leydig cells: endocrine, paracrine, and autocrine regulation. *Endocr. Rev.* 15, 574–626.
604. Calkins, J. H., Guo, H., Sigel, M. M., and Lin, T. (1990). Differential effects of recombinant interleukin-1 alpha and beta on Leydig cell function. *Biochem. Biophys. Res. Commun.* 167, 548–553.
605. Fukuoka, M., Taii, S., Yasuda K., Takakura K., and Mori, T. (1989). Inhibitory effects of interleukin-1 on luteinizing hormone-stimulated adenosine 3′,5′-monophosphate accumulation by cultured porcine granulosa cells. *Endocrinology* 125, 136–143.
606. Tsigos, C., Papanicolaou, D. A., Kyrou, I., Raptis, S. A., and Chrousos, G. P. (1999). Dose-dependent effects of recombinant human interleukin-6 on the pituitary-testicular axis. *J. Interferon Cytokine Res.* 19, 1271–1276.
607. Gerendai, I., Shaha, C., Thau, R., and Bardin, C. W. (1984). Do testicular opiates regulate Leydig cell function? *Endocrinology* 115, 1645–1647.
608. Kalantaridou, S. N., Makrigiannakis A., Mastorakos, G., and Chrousos, G. P. (2003). Roles of reproductive corticotropin-releasing hormone. *Ann. N Y Acad. Sci.* 997, 129–135.
609. Yoon, D. J., Sklar, C., and David, R. (1988). Presence of immunoreactive corticotropin-releasing factor in rat testis. *Endocrinology* 122, 759–761.
610. Audhya, T., Hollander, C. S., Schlesinger, D. H., and Hutchinson, B. (1989). Structural characterization and localization of corticotropin-releasing factor in testis. *Biochim. Biophys. Acta.* 995, 10–16.
611. Fabbri A., Tinajero, J. C., and Dufau, M. L. (1990). Corticotropin-releasing factor is produced by rat Leydig cells and has a major local antireproductive role in the testis. *Endocrinology* 127, 1541–1543.
612. Dufau, M. L., Tinajero, J. C., and Fabbri, A. (1993). Corticotropin-releasing factor: an antireproductive hormone of the testis. *FASEB J.* 7, 299–307.
613. Frungieri, M. B., Zitta K., Pignataro, O. P., Gonzalez-Calvar, S. I., and Calandra, R. S. (2002). Interactions between testicular serotoninergic, catecholaminergic, and corticotropin-releasing hormone systems modulating cAMP and testosterone production in the golden hamster. *Neuroendocrinology* 76, 35–46.
614. Ulisse, S., Fabbri, A., and Dufau, M. L. (1989). Corticotropin-releasing factor receptors and actions in rat Leydig cells. *J. Biol. Chem.* 264, 2156–2163.
615. Huang, B. M., Stocco, D. M., Hutson, J. C., and Norman, R. L. (1995). Corticotropin-releasing hormone stimulates steroidogenesis in mouse Leydig cells. *Biol. Reprod.* 53, 620–626.
616. Huang, B. M., Stocco, D. M., and Norman, R. L. (1997). The cellular mechanisms of corticotropin-releasing hormone (CRH)-stimulated steroidogenesis in mouse Leydig cells are similar to those for LH. *J. Androl.* 18, 528–534.
617. Heinrich, N., Meyer, M. R., Furkert, J., Sasse A., Beyermann, M., Bonigk, W., and Berger, H. (1998). Corticotropin-releasing factor (CRF) agonists stimulate testosterone production in mouse leydig cells through CRF receptor-1. *Endocrinology* 139, 651–658.

618. Mastorakos, G., Webster, E. L., Friedman, T. C., and Chrousos, G. P. (1993). Immunoreactive corticotropin-releasing hormone and its binding sites in the rat ovary. *J. Clin. Invest.* 92, 961–968.
619. Mastorakos, G., Scopa, C. D., Vryonidou A., Friedman TC, Kattis, D., Phenekos, C., Merino, M. J., and Chrousos, G. P. (1994). Presence of immunoreactive corticotropin-releasing hormone in normal and polycystic human ovaries. *J. Clin. Endocrinol. Metab.* 79, 1191–1197.
620. Asakura, H., Zwain, I. H., and Yen, S. S. (1997). Expression of genes encoding corticotropin-releasing factor (CRF), type 1 CRF receptor, and CRF-binding protein and localization of the gene products in the human ovary. *J. Clin. Endocrinol. Metab.* 82, 2720–2725.
621. Zoumakis, E., Chatzaki, E., Charalampopoulos, I., Margioris, A. N., Angelakis, E., Koumantakis, E., and Gravanis, A. (2001). Cycle and age-related changes in corticotropin-releasing hormone levels in human endometrium and ovaries. *Gynecol. Endocrinol.* 15, 98–102.
622. Nappi, R. E., Petraglia, F., Guo, A. L., Criscuolo, M., Trentini, G. P., and Genazzani, A. R. (1996). Estrous cycle- and acute stress-related changes of rat ovarian immunoreactive corticotropin-releasing factor. *Gynecol. Endocrinol.* 10, 75–82.
623. Calogero, A. E., Burrello, N., Negri-Cesi, P., Papale, L., Palumbo, M. A., Cianci A., Sanfilippo, S., and D'Agata, R. (1996). Effects of corticotropin-releasing hormone on ovarian estrogen production in vitro. *Endocrinology* 137, 4161–4166.
624. Calogero, A. E., Barreca A., Burrello, N., Palermo, I., Giordano, G., D'Agata, R., and Vicari, E. (2002). Corticotrophin-releasing hormone inhibits insulin-like growth factor-I release from primary cultures of rat granulosa cells. *J. Endocrinol.* 174, 493–498.
625. Murase, M., Uemura, T., Kondoh, Y., Funabashi, T., and Hirahara, F. (2002). Role of corticotropin-releasing hormone in ovarian steroidogenesis. *Endocrine* 18, 255–260.
626. Ghizzoni, L., Mastorakos, G., Vottero A., Barreca A., Furlini, M., Cesarone A., Ferrari B., Chrousos, G. P., and Bernasconi, S. (1997). Corticotropin-releasing hormone (CRH) inhibits steroid biosynthesis by cultured human granulosa-lutein cells in a CRH and interleukin-1 receptor-mediated fashion. *Endocrinology* 138, 4806–4811.
627. Erden, H. F., Zwain IH, Asakura, H., and Yen, S. S. (1998). Corticotropin-releasing factor inhibits luteinizing hormone-stimulated P450c17 gene expression and androgen production by isolated thecal cells of human ovarian follicles. *J. Clin. Endocrinol. Metab.* 83, 448–452.
628. Muramatsu, Y., Sugino, N., Suzuki, T., Totsune K., Takahashi K., Tashiro A., Hongo, M., Oki, Y., and Sasano, H. (2001). Urocortin and corticotropin-releasing factor receptor expression in normal cycling human ovaries. *J. Clin. Endocrinol. Metab.* 86, 1362–1369.
629. Goldstein, D. S. (2003). Catecholamines and stress. *Endocr. Regul.* 37, 69–80.
630. McCann, S. M., Lumpkin, M. D., Mizunuma, H., Samson, W. K., Ojeda, S. R., and Negro-Vilar, A. (1982). Control of gonadotropin secretion by brain amines and peptides. *Prog. Clin. Biol. Res.* 112 Pt A:15–29.
631. Kamberi, I. A. (1973). The role of brain monoamines and pineal indoles in the secretion of gonadotrophins and gonadotrophin-releasing factors. *Prog. Brain Res.* 39, 261–280.
632. Smith, M. J. and Jennes, L. (2001). Neural signals that regulate GnRH neurones directly during the oestrous cycle. *Reproduction* 122, 1–10.
633. Gore, A. C. and Terasawa, E. (2001). Neural circuits regulating pulsatile luteinizing hormone release in the female guinea-pig: opioid, adrenergic and serotonergic interactions. *J. Neuroendocrinol.* 13, 239–248.
634. Pau, K. Y. and Spies, H. G. (1997). Neuroendocrine signals in the regulation of gonadotropin-releasing hormone secretion. *Chin J. Physiol.* 40, 181–196.
635. Drouva, S. V. and Gallo, R. V. (1977). Further evidence for inhibition of episodic luteinizing hormone release in ovariectomized rats by stimulation of dopamine receptors. *Endocrinology* 100, 792–798.
636. Lehman, M. N., Karsch, F. J., and Silverman, A. J. (1988). Potential sites of interaction between catecholamines and LHRH in the sheep brain. *Brain Res. Bull.* 20, 49–58.
637. Deaver, D. R. and Dailey, R. A. (1982). Effects of dopamine, norepinephrine and serotonin on plasma concentrations of luteinizing hormone and prolactin in ovariectomized and anestrous ewes. *Biol. Reprod.* 27, 624–632.
638. Dailey, R. A., Tsou, R. C., Tindall, G. T., and Neill, J. D. (1978). Direct hypophysial inhibition of luteinizing hormone release by dopamine in the rabbit. *Life Sci.* 22, 1491–1498.
639. Donnelly, P. J. and Dailey, R. A. (1991). Effects of dopamine, norepinephrine and serotonin on secretion of luteinizing hormone, follicle-stimulating hormone and prolactin in ovariectomized, pituitary stalk-transected ewes. *Domest. Anim. Endocrinol.* 8, 87–98.
640. Martin, A. I., Fernandez-Ruiz, J., and Lopez-Calderon, A. (1995). Effects of catecholamine synthesis inhibitors and adrenergic receptor antagonists on restraint-induced LH release. *J. Endocrinol.* 144, 511–515.
641. Collu, R., Gibb, W., and Ducharme, J. R. (1984). Role of catecholamines in the inhibitory effect of immobilization stress on testosterone secretion in rats. *Biol. Reprod.* 30, 416–422.
642. Koh, T., Nakai, Y., Kinoshita, F., Tsukada, T., Tsujii, S., Imura, H., and Maeda, K. (1984). Serotonin involvement in the inhibition of luteinizing hormone (LH) release during immobilization in castrated male rats. *Life Sci.* 34, 1635–1641.
643. Ariznavarreta, C., Calderon, M. D., Tresguerres, J. A., and Lopez-Calderon, A. (1989). Effect of adrenomedullectomy and propranolol treatment on the response of gonadotrophins to chronic stress in male rats. *J. Endocrinol.* 120, 275–279.
644. Chow, S. H., Giglio, W., Anesetti, R., Ottenweller, J. E., Pogach, L. M., and Huang, H. F. (2002). The effects of testicular denervation on spermatogenesis in the Sprague-Dawley rat. *Neuroendocrinology* 72, 37–45.
645. Chen, D., Hartwig, D. M., and Roth, E. J. (1999). Comparison of sperm quantity and quality in antegrade V retrograde ejaculates obtained by vibratory penile stimulation in males with spinal cord injury. *Am. J. Phys. Med. Rehabil.* 78, 46–51.
646. Guillamon, A. and Segovia, S. (1996). Sexual dimorphism in the CNS and the role of steroids. In *CNS Neurotransmitters and Neuromodulators* (T. W. Stone, Ed.), Vol. 1, pp. 127–152. CRC Press, Boca Raton, FL.
647. Madeira, M. D. and Lieberman, A. R. (1995). Sexual dimorphism in the mammalian limbic system. *Prog. Neurobiol.* 45, 275–333.
648. Swaab, D. F., Chun, W. C., Kruijver, F. P., Hofman, M. A., and Ishunina, T. A. (2002). Sexual differentiation of the human hypothalamus. *Adv. Exp. Med. Biol.* 511, 75–100; discussion 100–105.
649. Patchev, V. K. and Almeida, O. F. (1998). Gender specificity in the neural regulation of the response to stress: new leads from classical paradigms. *Mol. Neurobiol.* 16, 63–77.
650. Patchev, V. K., Hayashi, S., Orikasa, C., and Almeida, O. F. (1995). Implications of estrogen-dependent brain organization for gender differences in hypothalamo-pituitary-adrenal regulation. *FASEB J.* 9, 419–423.
651. Patchev, V. K. and Almeida, O. F. (1996). Gonadal steroids exert facilitating and "buffering" effects on glucocorticoid-mediated transcriptional regulation of corticotropin-releasing

hormone and corticosteroid receptor genes in rat brain. *J. Neurosci.* 16, 7077–7084.
652. Spinedi, E., Chisari A., Pralong, F., and Gaillard, R. C. (1997). Sexual dimorphism in the mouse hypothalamic–pituitary–adrenal axis function after endotoxin and insulin stresses during development. *Neuroimmunomodulation* 4, 77–83.
653. Kitay, J. I. (1961). Sex differences in adrenal cortical secretion in the rat. *Endocrinology* 68, 818–824.
654. Critchlow, V., Liebelt, R. A., Bar-Sela, M., Mountcastle, W., and Lipscomb, H. S. (1963). Sex difference in resting pituitary-adrenal function in the rat. *Am. J. Physiol.* 205, 807–815.
655. Atkinson, H. C. and Waddell, B. J. (1997). Circadian variation in basal plasma corticosterone and adrenocorticotropin in the rat: sexual dimorphism and changes across the estrous cycle. *Endocrinology* 138, 3842–3848.
656. Windle, R. J., Wood, S. A., Lightman, S. L., and Ingram, C. D. (1998). The pulsatile characteristics of hypothalamo-pituitary-adrenal activity in female Lewis and Fischer 344 rats and its relationship to differential stress responses. *Endocrinology* 139, 4044–4052.
657. Windle, R. J., Wood, S. A., Kershaw YM, Lightman, S. L., Ingram, C. D., and Harbuz, M. S. (2001). Increased corticosterone pulse frequency during adjuvant-induced arthritis and its relationship to alterations in stress responsiveness. *J. Neuroendocrinol.* 13, 905–911.
658. Seale J. V., Wood, S. A., Atkinson, H. C., Bate, E., Lightman, S. L., Ingram, C. D., Jessop, D. S., and Harbuz, M. S. (2004). Gonadectomy reverses the sexually diergic patterns of circadian and stress-induced hypothalamic–pituitary–adrenal axis activity in male and female rats. *J. Neuroendocrinol.* 16, 516–524.
659. Carey, M. P., Deterd, C. H., de Koning, J., Helmerhorst, F., and de Kloet, E. R. (1995). The influence of ovarian steroids on hypothalamic–pituitary–adrenal regulation in the female rat. *J. Endocrinol.* 144, 311–321.
660. Haim, S., Shakhar, G., Rossene, E., Taylor, A. N., and Ben-Eliyahu, S. (2003). Serum levels of sex hormones and corticosterone throughout 4- and 5-day estrous cycles in Fischer 344 rats and their simulation in ovariectomized females. *J. Endocrinol. Invest.* 26, 1013–1022.
661. Buckingham, J. C., Dohler, K. D., and Wilson, C. A. (1978). Activity of the pituitary-adrenocortical system and thyroid gland during the oestrous cycle of the rat. *J. Endocrinol.* 78, 359–366.
662. Viau, V., and Meaney, M. J. (1991). Variations in the hypothalamic–pituitary–adrenal response to stress during the estrous cycle in the rat. *Endocrinology* 129, 2503–2511.
663. Van Cauter, E., Leproult, R., and Kupfer, D. J. (1996). Effects of gender and age on the levels and circadian rhythmicity of plasma cortisol. *J. Clin. Endocrinol. Metab.* 81, 2468–2473.
664. Tersman, Z., Collins, A., and Eneroth, P. (1991). Cardiovascular responses to psychological and physiological stressors during the menstrual cycle. *Psychosom. Med.* 53, 185–197.
665. Shamim, W., Yousufuddin, M., Bakhai, A., Coats, A. J., and Honour, J. W. (2000). Gender differences in the urinary excretion rates of cortisol and androgen metabolites. *Ann. Clin. Biochem.* 37, Pt. 6, 770–774.
666. Vierhapper, H., Nowotny, P., and Waldhausl, W. (1998). Sex-specific differences in cortisol production rates in humans. *Metabolism* 47, 974–976.
667. Masi, C. M., Rickett, E. M., Hawkley, L. C., and Cacioppo, J. T. (2004). Gender and ethnic differences in urinary stress hormones: the population-based Chicago Health, Aging, and Social Relations Study (CHASRS). *J. Appl. Physiol.* 97, 941–947.
668. McCormick, C. M. and Teillon, S. M. (2001). Menstrual cycle variation in spatial ability: relation to salivary cortisol levels. *Horm. Behav.* 39, 29–38.
669. Gudmundsson, A., Goodman B., Lent, S., Barczi, S., Grace A., Boyle, L., Ershler, W. B., and Carnes, M. (1999). Effects of estrogen replacement therapy on the circadian rhythms of serum cortisol and body temperature in postmenopausal women. *Exp. Gerontol.* 34, 809–818.
670. Giussani, D. A., Farber, D. M., Jenkins, S. L., Yen, A., Winter, J. A., Tame, J. D., and Nathanielsz, P. W. (2000). Opposing effects of androgen and estrogen on pituitary-adrenal function in nonpregnant primates. *Biol. Reprod.* 62, 1445–1451.
671. Smith, C. J. and Norman, R. L. (1987). Influence of the gonads on cortisol secretion in female rhesus macaques. *Endocrinology* 121, 2192–2198.
672. McKenzie, S. and Deane, E. M. (2003). The effects of age, season, and gender on serum cortisol levels in the tammar wallaby, Macropus eugenii. *Gen. Comp. Endocrinol.* 133, 273–278.
673. Young, E. A. (1995). The role of gonadal steroids in hypothalamic–pituitary–adrenal axis regulation. *Crit. Rev. Neurobiol.* 9, 371–381.
674. Lescoat, G., Jego, P., Beraud, G., and Maniey, J. (1970). [Sex influences on the response of the hypothalamo-hypophysio-adrenal axis to emotional and systemic stress in the rat]. *C. R. Seances Soc. Biol. Fil.* 164, 2106–2113.
675. Le Mevel, J. C., Abitbol, S., Beraud, G., and Maniey, J. (1979). Temporal changes in plasma adrenocorticotropin concentration after repeated neurotropic stress in male and female rats. *Endocrinology* 105, 812–817.
676. Spinedi, E., Suescun, M. O., Hadid, R., Daneva, T., and Gaillard, R. C. (1992). Effects of gonadectomy and sex hormone therapy on the endotoxin-stimulated hypothalamo-pituitary-adrenal axis: evidence for a neuroendocrine-immunological sexual dimorphism. *Endocrinology* 131, 2430–2436.
677. Rivier, C. (1994). Stimulatory effect of interleukin-1 beta on the hypothalamic–pituitary–adrenal axis of the rat: influence of age, gender and circulating sex steroids. *J. Endocrinol.* 140, 365–372.
678. Frederic, F., Oliver, C., Wollman, E., Delhaye-Bouchaud, N., and Mariani, J. (1993). IL-1 and LPS induce a sexually dimorphic response of the hypothalamo-pituitary-adrenal axis in several mouse strains. *Eur. Cytokine Netw.* 4, 321–329.
679. Weinstock, M., Razin, M., Schorer-Apelbaum, D., Men, D., and McCarty, R. (1998). Gender differences in sympathoadrenal activity in rats at rest and in response to footshock stress. *Int. J. Dev. Neurosci.* 16, 289–295.
680. Rivier, C. (1999). Gender, sex steroids, corticotropin-releasing factor, nitric oxide, and the HPA response to stress. *Pharmacol. Biochem. Behav.* 64, 739–751.
681. De Boer, S. F., Slangen, J. L., and van der Gugten, J. (1988). Adaptation of plasma catecholamine and corticosterone responses to short-term repeated noise stress in rats. *Physiol. Behav.* 44, 273–280.
682. Rivier, C. (1993). Female rats release more corticosterone than males in response to alcohol: influence of circulating sex steroids and possible consequences for blood alcohol levels. *Alcohol Clin. Exp. Res.* 17, 854–859.
683. Grossman, C. J., Roselle, G. A., and Mendenhall, C. L. (1991). Sex steroid regulation of autoimmunity. *J. Steroid Biochem. Mol. Biol.* 40, 649–659.
684. Spinedi, E., Gaillard, R. C., and Chisari, A. (2002). Sexual dimorphism of neuroendocrine-immune interactions. *Front. Horm. Res.* 29, 91–107.
685. Cutolo, M., Seriolo, B., Villaggio, B., Pizzorni, C., Craviotto, C., and Sulli, A. (2002). Androgens and estrogens modulate the

immune and inflammatory responses in rheumatoid arthritis. *Ann. N Y Acad. Sci.* 966, 131–142.
686. Angold, A., Costello, E. J., and Worthman, C. M. (1998). Puberty and depression: the roles of age, pubertal status and pubertal timing. *Psychol. Med.* 28, 51–61.
687. Piccinelli, M. and Wilkinson, G. (2000). Gender differences in depression. Critical review. *Br. J. Psychiatry* 177, 486–492.
688. Martenyi, F., Dossenbach, M., Mraz, K., and Metcalfe, S. (2001). Gender differences in the efficacy of fluoxetine and maprotiline in depressed patients: a double-blind trial of antidepressants with serotonergic or norepinephrinergic reuptake inhibition profile. *Eur. Neuropsychopharmacol.* 11, 227–232.
689. Silva, C., Ines, L. S., Nour, D., Straub, R. H., and da Silva, J. A. (2002). Differential male and female adrenal cortical steroid hormone and cortisol responses to interleukin-6 in humans. *Ann. N Y Acad. Sci.* 966, 68–72.
690. Traustadottir, T., Bosch, P. R., and Matt, K. S. (2003). Gender differences in cardiovascular and hypothalamic–pituitary–adrenal axis responses to psychological stress in healthy older adult men and women. *Stress* 6, 133–140.
691. Guo, A. L., Petraglia, F., Criscuolo, M., Ficarra, G., Nappi, R. E., Palumbo, M., Valentini A., and Genazzani, A. R. (1994). Acute stress- or lipopolysaccharide-induced corticosterone secretion in female rats is independent of the oestrous cycle. *Eur. J. Endocrinol.* 131, 535–539.
692. Turner, A. I., Canny, B. J., Hobbs, R. J., Bond, J. D., Clarke, I. J., and Tilbrook, A. J. (2002). Influence of sex and gonadal status of sheep on cortisol secretion in response to ACTH and on cortisol and LH secretion in response to stress: importance of different stressors. *J. Endocrinol.* 173, 113–122.
693. Abplanalp, J. M., Livingston, L., Rose, R. M., and Sandwisch, D. (1977). Cortisol and growth hormone responses to psychological stress during the menstrual cycle. *Psychosom. Med.* 39, 158–177.
694. Burgess, L. H. and Handa, R. J. (1992). Chronic estrogen-induced alterations in adrenocorticotropin and corticosterone secretion, and glucocorticoid receptor-mediated functions in female rats. *Endocrinology* 131, 1261–1269.
695. Lindheim, S. R., Legro, R. S., Bernstein, L., Stanczyk, F. Z., Vijod, M. A., Presser, S. C., and Lobo, R. A. (1992). Behavioral stress responses in premenopausal and postmenopausal women and the effects of estrogen. *Am. J. Obstet. Gynecol.* 167, 1831–1836.
696. Komesaroff, P. A., Esler, M. D., and Sudhir, K. (1999). Estrogen supplementation attenuates glucocorticoid and catecholamine responses to mental stress in perimenopausal women. *J. Clin. Endocrinol. Metab.* 84, 606–610.
697. Kirschbaum, C., Schommer, N., Federenko, I., Gaab, J., Neumann, O., Oellers, M., Rohleder, N., Untiedt A., Hanker, J., Pirke, K. M., and Hellhammer, D. H. (1996). Short-term estradiol treatment enhances pituitary-adrenal axis and sympathetic responses to psychosocial stress in healthy young men. *J. Clin. Endocrinol. Metab.* 81, 3639–3643.
698. Kirschbaum, C., Kudielka, B. M., Gaab, J., Schommer, N. C., and Hellhammer, D. H. (1999). Impact of gender, menstrual cycle phase, and oral contraceptives on the activity of the hypothalamus-pituitary-adrenal axis. *Psychosom. Med.* 61, 154–162.
699. Kudielka, B. M., Hellhammer, J., Hellhammer, D. H., Wolf, O. T., Pirke, K. M., Varadi, E., Pilz, J., and Kirschbaum, C. (1998). Sex differences in endocrine and psychological responses to psychosocial stress in healthy elderly subjects and the impact of a 2-week dehydroepiandrosterone treatment. *J. Clin. Endocrinol. Metab.* 83, 1756–1761.
700. Kudielka, B. M., Schmidt-Reinwald A. K, Hellhammer, D. H., and Kirschbaum, C. (1999). Psychological and endocrine responses to psychosocial stress and dexamethasone/corticotropin-releasing hormone in healthy postmenopausal women and young controls: the impact of age and a two-week estradiol treatment. *Neuroendocrinology* 70, 422–430.
701. Lee, S. and Rivier, C. (1993). Effect of exposure to an alcohol diet for 10 days on the ability of interleukin-1 beta to release ACTH and corticosterone in the adult ovariectomized female rat. *Alcohol Clin. Exp. Res.* 17, 1009–1013.
702. Lee, S. and Rivier, C. (1995). Altered ACTH and corticosterone responses to interleukin-1 beta in male rats exposed to an alcohol diet: possible role of vasopressin and testosterone. *Alcohol Clin. Exp. Res.* 19, 200–208.
703. Suescun, M., Chisari A. N., Carino, M., Hadid, R., Gaillard, R. C., and Spinedi, E. (1994). Sex steroid regulation of the hypothalamo-pituitary-adrenal axis activity in middle-aged mice during endotoxic shock. *Neuroimmunomodulation* 1, 315–320.
704. Xiao, E., Xia-Zhang, L., Ferin, M., and Wardlaw, S. L. (2001). Differential effects of estradiol on the adrenocorticotropin responses to interleukin-6 and interleukin-1 in the monkey. *Endocrinology* 142, 2736–2741.
705. Xia-Zhang, L., Xiao, E., and Ferin, M. A. (1995). 5-day estradiol therapy, in amounts reproducing concentrations of the early-mid follicular phase, prevents the activation of the hypothalamo-pituitary-adrenal axis by interleukin-1 alpha in the ovariectomized rhesus monkey. *J. Neuroendocrinol.* 7, 387–392.
706. Spangelo, B. L., and Gorospe, W. C. (1995). Role of the cytokines in the neuroendocrine-immune system axis. *Front. Neuroendocrinol.* 16, 1–22.
707. Mouihate A. and Pittman, Q. J. (2003). Neuroimmune response to endogenous and exogenous pyrogens is differently modulated by sex steroids. *Endocrinology* 144, 2454–2460.
708. Perlstein, R. S., Whitnall, M. H., Abrams, J. S., Mougey, E. H., and Neta, R. (1993). Synergistic roles of interleukin-6, interleukin-1, and tumor necrosis factor in the adrenocorticotropin response to bacterial lipopolysaccharide in vivo. *Endocrinology* 132, 946–952.
709. Straub, R. H., Hense, H. W., Andus, T., Scholmerich, J., Riegger, G. A., and Schunkert, H. (2000). Hormone replacement therapy and interrelation between serum interleukin-6 and body mass index in postmenopausal women: a population-based study. *J. Clin. Endocrinol. Metab.* 85, 1340–1344.
710. Puder, J. J., Freda, P. U., Goland, R. S., and Wardlaw, S. L. (2001). Estrogen modulates the hypothalamic–pituitary–adrenal and inflammatory cytokine responses to endotoxin in women. *J. Clin. Endocrinol. Metab.* 86, 2403–2408.
711. Bellido, T., Jilka, R. L., Boyce, B. F., Girasole, G., Broxmeyer, H., Dalrymple, S. A., Murray, R., and Manolagas, S. C. (1995). Regulation of interleukin-6, osteoclastogenesis, and bone mass by androgens. The role of the androgen receptor. *J. Clin. Invest.* 95, 2886–2895.
712. Kanda, N., Tsuchida, T., and Tamaki, K. (1997). Testosterone suppresses anti-DNA antibody production in peripheral blood mononuclear cells from patients with systemic lupus erythematosus. *Arthritis Rheum.* 40, 1703–1711.
713. Komesaroff, P. A., Esler, M., Clarke, I. J., Fullerton, M. J., and Funder, J. W. (1998). Effects of estrogen and estrous cycle on glucocorticoid and catecholamine responses to stress in sheep. *Am. J. Physiol.* 275, E671–E678.
714. McEwen, B. (2002). Estrogen actions throughout the brain. *Recent Prog. Horm. Res.* 57, 357–384.
715. Alves, S. E., Lopez, V., McEwen, B. S., and Weiland, N. G. (1998). Differential colocalization of estrogen receptor beta (ERbeta) with oxytocin and vasopressin in the paraventricular and supraoptic nuclei of the female rat brain: an immunocytochemical study. *Proc. Natl. Acad. Sci. U S A* 95, 3281–3286.

716. Hrabovszky, E., Kallo, I., Hajszan, T., Shughrue, P. J., Merchenthaler, I., and Liposits, Z. (1998). Expression of estrogen receptor-beta messenger ribonucleic acid in oxytocin and vasopressin neurons of the rat supraoptic and paraventricular nuclei. *Endocrinology* 139, 2600-2604.
717. Mitra, S. W., Hoskin, E., Yudkovitz, J., Pear, L., Wilkinson, H. A., Hayashi, S., Pfaff, D. W., Ogawa, S., Rohrer, S. P., Schaeffer, J. M., McEwen, B. S., and Alves, S. E. (2003). Immunolocalization of estrogen receptor beta in the mouse brain: comparison with estrogen receptor alpha. *Endocrinology* 144, 2055–2067.
718. Somponpun, S. J., Holmes, M. C., Seckl, J. R., and Russell, J. A. (2004). Modulation of oestrogen receptor-beta mRNA expression in rat paraventricular and supraoptic nucleus neurones following adrenal steroid manipulation and hyperosmotic stimulation. *J. Neuroendocrinol.* 16, 472–482.
719. Suzuki, S. and Handa, R. J. (2004). Regulation of estrogen receptor-beta expression in the female rat hypothalamus: differential effects of dexamethasone and estradiol. *Endocrinology* 145, 3658–3670.
720. Isgor, C., Cecchi, M., Kabbaj, M., Akil, H., and Watson, S. J. (2003). Estrogen receptor beta in the paraventricular nucleus of hypothalamus regulates the neuroendocrine response to stress and is regulated by corticosterone. *Neuroscience* 121, 837–845.
721. Harris, H. A., Albert, L. M., Leathurby, Y., Malamas, M. S., Mewshaw, R. E., Miller, C. P., Kharode YP, Marzolf, J., Komm, B. S., Winneker, R. C., Frail, D. E., Henderson, R. A., Zhu, Y., and Keith, J. C. Jr. (2003). Evaluation of an estrogen receptor-beta agonist in animal models of human disease. *Endocrinology* 144, 4241–4249.
722. Osterlund, M. K., Gustafsson, J. A., Keller, E., and Hurd, Y. L. (2000). Estrogen receptor beta (ERbeta) messenger ribonucleic acid (mRNA) expression within the human forebrain: distinct distribution pattern to ERalpha mRNA. *J. Clin. Endocrinol. Metab.* 85, 3840–3846.
723. Kruijver, F. P., Balesar, R., Espila, A. M., Unmehopa, U. A., and Swaab, D. F. (2002). Estrogen receptor-alpha distribution in the human hypothalamus in relation to sex and endocrine status. *J. Comp. Neurol.* 454, 115–139.
724. Vamvakopoulos, N. C. and Chrousos, G. P. (1993). Evidence of direct estrogenic regulation of human corticotropin-releasing hormone gene expression. Potential implications for the sexual dimophism of the stress response and immune/inflammatory reaction. *J. Clin. Invest.* 92, 1896–1902.
725. Greer, E. R., Caldwell, J. D., Johnson, M. F., Prange, A. J. Jr., and Pedersen, C. A. (1986). Variations in concentration of oxytocin and vasopressin in the paraventricular nucleus of the hypothalamus during the estrous cycle in rats. *Life Sci.* 38, 2311–2318.
726. Bohler, H. C., Jr., Zoeller, R. T., King, J. C., Rubin, B. S., Weber, R., and Merriam, G. R. (1990). Corticotropin releasing hormone mRNA is elevated on the afternoon of proestrus in the parvocellular paraventricular nuclei of the female rat. *Brain Res. Mol. Brain Res.* 8, 259–262.
727. Duncko, R., Kiss, A., Skultetyova, I., Rusnak, M., and Jezova, D. (2001). Corticotropin-releasing hormone mRNA levels in response to chronic mild stress rise in male but not in female rats while tyrosine hydroxylase mRNA levels decrease in both sexes. *Psychoneuroendocrinology* 26, 77–89.
728. Frederiksen, S. O., Ekman, R., Gottfries, C. G., Widerlov, E., and Jonsson, S. (1991). Reduced concentrations of galanin, arginine vasopressin, neuropeptide Y and peptide YY in the temporal cortex but not in the hypothalamus of brains from schizophrenics. *Acta Psychiatr. Scand.* 83, 273–277.
729. Forsling, M. L., Kelestimur, H., and Windle, R. (1991). The influence of reproductive status on vasopressin release in the rat. *J. Endocrinol.* 130, 387–393.
730. Roy BN, Reid, R. L., and Van Vugt, D. A. (1999). The effects of estrogen and progesterone on corticotropin-releasing hormone and arginine vasopressin messenger ribonucleic acid levels in the paraventricular nucleus and supraoptic nucleus of the rhesus monkey. *Endocrinology* 140, 2191–2198.
731. Gallucci, W. T., Baum A., Laue, L., Rabin, D. S., Chrousos, G. P., Gold, P. W., and Kling, M. A. (1993). Sex differences in sensitivity of the hypothalamic–pituitary–adrenal axis. *Health Psychol.* 12, 420–425.
732. Haas, D. A. and George, S. R. (1989). Estradiol or ovariectomy decreases CRF synthesis in hypothalamus. *Brain Res. Bull.* 23, 215–218.
733. Haas, D. A. and George, S. R. (1988). Gonadal regulation of corticotropin-releasing factor immunoreactivity in hypothalamus. *Brain Res. Bull.* 20, 361–367.
734. Li, H. Y., Blaustein, J. D., De Vries, G. J., and Wade, G. N. (1993). Estrogen-receptor immunoreactivity in hamster brain: preoptic area, hypothalamus and amygdala. *Brain Res.* 631, 304–312.
735. Owen, D., and Matthews, S. G. (2003). Glucocorticoids and sex-dependent development of brain glucocorticoid and mineralocorticoid receptors. *Endocrinology* 144, 2775–2784.
736. Handa, R. J., Nunley, K. M., Lorens, S. A., Louie, J. P., McGivern, R. F., and Bollnow, M. R. (1994). Androgen regulation of adrenocorticotropin and corticosterone secretion in the male rat following novelty and foot shock stressors. *Physiol. Behav.* 55, 117–124.
737. Viau, V., and Meaney, M. J. (1996). The inhibitory effect of testosterone on hypothalamic–pituitary–adrenal responses to stress is mediated by the medial preoptic area. *J. Neurosci.* 16, 1866–1876.
738. Gaskin, J. H. and Kitay, J. I. (1971). Hypothalamic and pituitary regulation of adrenocortical function in the hamster: effects of gonadectomy and gonadal hormone replacement. *Endocrinology* 89, 1047–1053.
739. Bingaman, E. W., Magnuson, D. J., Gray, T. S., and Handa, R. J. (1994). Androgen inhibits the increases in hypothalamic corticotropin-releasing hormone (CRH) and CRH-immunoreactivity following gonadectomy. *Neuroendocrinology* 59, 228–234.
740. Viau, V., Lee, P., Sampson, J., and Wu, J. (2003). A testicular influence on restraint-induced activation of medial parvocellular neurons in the paraventricular nucleus in the male rat. *Endocrinology* 144, 3067–3075.
741. Kovacs, K. J., Foldes A., and Sawchenko, P. E. (2000). Glucocorticoid negative feedback selectively targets vasopressin transcription in parvocellular neurosecretory neurons. *J. Neurosci.* 20, 3843–3852.
742. McCormick, C. M., Linkroum, W., Sallinen, B. J., and Miller, N. W. (2002). Peripheral and central sex steroids have differential effects on the HPA axis of male and female rats. *Stress* 5, 235–247.
743. Viau, V. (2002). Functional cross-talk between the hypothalamic–pituitary–gonadal and -adrenal axes. *J. Neuroendocrinol.* 14, 506–513.
744. Viau, V., Soriano, L., and Dallman, M. F. (2001). Androgens alter corticotropin releasing hormone and arginine vasopressin mRNA within forebrain sites known to regulate activity in the hypothalamic–pituitary–adrenal axis. *J. Neuroendocrinol.* 13, 442–452.
745. Wilson, R. C., Kesner, J. S., Kaufman, J. M., Uemura, T., Akema, T., and Knobil, E. (1984). Central electrophysiologic correlates of pulsatile luteinizing hormone secretion in the rhesus monkey. *Neuroendocrinology* 39, 256–260.

746. Norman, R. L., McGlone, J., and Smith, C. J. (1994). Restraint inhibits luteinizing hormone secretion in the follicular phase of the menstrual cycle in rhesus macaques. *Biol. Reprod.* 50, 16–26.
747. Witkin, J. W., Ferin, M., Popilskis, S. J., and Silverman, A. J. (1991). Effects of gonadal steroids on the ultrastructure of GnRH neurons in the rhesus monkey: synaptic input and glial apposition. *Endocrinology* 129, 1083–1092.
748. Witkin, J. W., O'Sullivan, H., and Ferin, M. (1995). Glial ensheathment of GnRH neurons in pubertal female rhesus macaques. *J. Neuroendocrinol.* 7, 665–671.
749. Witkin, J. W. (1996). Effects of ovariectomy on GnRH neuronal morphology in rhesus monkey (Macaca mulatta). *J. Neuroendocrinol.* 8, 601–604.
750. Kaur, G., Heera, P. K., and Srivastava, L. K. (2002). Neuroendocrine plasticity in GnRH release during rat estrous cycle: correlation with molecular markers of synaptic remodeling. *Brain Res.* 954, 21–31.
751. Leedom, L., Lewis, C., Garcia-Segura, L. M., and Naftolin, F. (1994). Regulation of arcuate nucleus synaptology by estrogen. *Ann. N Y Acad. Sci.* 743, 61–71.
752. Marchlewska-Koj, A., Kruczek, M., Kapusta, J., and Pochron, E. (2003). Prenatal stress affects the rate of sexual maturation and attractiveness in bank voles. *Physiol. Behav.* 79, 305–310.
753. Harvey, P. W. and Chevins, P. F. (1987). Crowding during pregnancy delays puberty and alters estrous cycles of female offspring in mice. *Experientia* 43, 306–308.
754. Bonner, M. J. (1994). Prenatal and neonatal pharmacologic stress on early behavior and sexual maturation in the hamster. *J. Clin. Pharmacol.* 34, 713–718.
755. Harvey, P. W. and Chevins, P. F. (1984). Crowding or ACTH treatment of pregnant mice affects adult copulatory behavior of male offspring. *Horm. Behav.* 18, 101–110.
756. Herrenkohl, L. R. and Politch, J. A. (1978). Effects of prenatal stress on the estrous cycle of female offspring as adults. *Experientia.* 34, 1240–1241.
757. Rohde, W., Ohkawa, T., Gotz, F., Stahl, F., Tonjes, R., Takeshita, S., Arakawa, S., Kambegawa A., Arai, K., Okinaga, S., et al. (1989). Sex-specific effects on the fetal neuroendocrine system during acute stress in late pregnancy of rat and the influence of a simultaneous treatment by tyrosine. *Exp. Clin. Endocrinol.* 94, 23–42.
758. Gilad, V. H., Rabey, J. M., Eliyayev, Y., and Gilad, G. M. (2000). Different effects of acute neonatal stressors and long-term postnatal handling on stress-induced changes in behavior and in ornithine decarboxylase activity of adult rats. *Brain Res. Dev. Brain Res.* 120, 255–259.
759. Vom Saal, F. S., Even, M. D., and Quadagno, D. M. (1991). Effects of maternal stress on puberty, fertility and aggressive behavior of female mice from different intrauterine positions. *Physiol. Behav.* 49, 1073–1078.
760. Holson, R. R., Gough B., Sullivan, P., Badger, T., and Sheehan, D. M. (1995). Prenatal dexamethasone or stress but not ACTH or corticosterone alter sexual behavior in male rats. *Neurotoxicol. Teratol.* 17, 393–401.
761. Pollard, I., and Dyer, S. L. (1985). Effect of stress administered during pregnancy on the development of fetal testes and their subsequent function in the adult rat. *J. Endocrinol.* 107, 241–245.
762. Harvey, P. W., and Chevins, P. F. (1985). Crowding pregnant mice affects attack and threat behavior of male offspring. *Horm. Behav.* 19, 86–97.
763. Matthews, S. G. (2002). Early programming of the hypothalamo-pituitary-adrenal axis. *Trends Endocrinol. Metab.* 13, 373–380.
764. Lee, S. and Rivier, C. (1996). Gender differences in the effect of prenatal alcohol exposure on the hypothalamic–pituitary–adrenal axis response to immune signals. *Psychoneuroendocrinology* 21, 145–155.
765. Pollard, I. (1986). Prenatal stress effects over two generations in rats. *J. Endocrinol.* 109, 239–244.
766. Dean, F., Yu, C., Lingas, R. I., and Matthews, S. G. (2001). Prenatal glucocorticoid modifies hypothalamo–pituitary–adrenal regulation in prepubertal guinea pigs. *Neuroendocrinology* 73, 194–202.
767. Bowman, R. E., MacLusky, N. J., Sarmiento, Y., Frankfurt, M., Gordon, M., and Luine, V. N. (2004). Sexually dimorphic effects of prenatal stress on cognition, hormonal responses, and central neurotransmitters. *Endocrinology* 145, 3778–3787.
768. O'Shaughnessy, P. J., Fleming, L. M., Jackson, G., Hochgeschwender, U., Reed, P., and Baker, P. J. (2003). Adrenocorticotropic hormone directly stimulates testosterone production by the fetal and neonatal mouse testis. *Endocrinology* 144, 3279–3284.
769. Kremer, H., Kraaij, R., Toledo, S. P., Post, M., Fridman, J. B., Hayashida, C. Y., van Reen, M., Milgrom, E., Ropers, H. H., Mariman, E., et al. (1995). Male pseudohermaphroditism due to a homozygous missense mutation of the luteinizing hormone receptor gene. *Nat. Genet.* 9, 160–164.
770. Domenice, S., Latronico, A. C., Brito, V. N., Arnhold, I. J., Kok, F., and Mendonca, B. B. (2001). Adrenocorticotropin-dependent precocious puberty of testicular origin in a boy with X-linked adrenal hypoplasia congenita due to a novel mutation in the DAX1 gene. *J. Clin. Endocrinol. Metab.* 86, 4068–4071.
771. Stikkelbroeck, N. M, Otten, B. J., Pasic A., Jager, G. J., Sweep, C. G., Noordam K., and Hermus, A. R. (2001). High prevalence of testicular adrenal rest tumors, impaired spermatogenesis, and Leydig cell failure in adolescent and adult males with congenital adrenal hyperplasia. *J. Clin. Endocrinol. Metab.* 86, 5721–5728.
772. Calikoglu, A. S. (2003). Adrenocorticotropic hormone, a new player in the control of testicular steroidogenesis. *Endocrinology* 144, 3277–3278.
773. Rhees, R. W., Lephart, E. D., and Eliason, D. (2001). Effects of maternal separation during early postnatal development on male sexual behavior and female reproductive function. *Behav. Brain Res.* 123, 1–10.
774. Mitev, Y. A., Darwish, M., Wolf, S. S., Holsboer, F., Almeida, O. F., and Patchev, V. K. (2003). Gender differences in the regulation of 3 alpha-hydroxysteroid dehydrogenase in rat brain and sensitivity to neurosteroid-mediated stress protection. *Neuroscience* 120, 541–549.
775. Alves, S. E., Akbari, H. M., Azmitia, E. C., and Strand, F. L. (1993). Neonatal ACTH and corticosterone alter hypothalamic monoamine innervation and reproductive parameters in the female rat. *Peptides* 14, 379–384.
776. Alves, S. E., Akbari H. M., Anderson, G. M., Azmitia, E. C., McEwen, B. C., and Strand, F. L. (1997). Neonatal ACTH administration elicits long-term changes in forebrain monoamine innervation. Subsequent disruptions in hypothalamic–pituitary–adrenal and gonadal function. *Ann. N Y Acad. Sci.* 814, 226–251.
777. Francis, D., Diorio, J., Liu, D., and Meaney, M. J. (1999). Nongenomic transmission across generations of maternal behavior and stress responses in the rat. *Science* 286, 1155–1158.
778. Francis, D. D., Champagne, F. A., Liu, D., and Meaney, M. J. (1999). Maternal care, gene expression, and the development of individual differences in stress reactivity. *Ann. N Y Acad. Sci.* 896, 66–84.

779. Liu, D., Caldji, C., Sharma, S., Plotsky, P. M., and Meaney, M. J. (2000). Influence of neonatal rearing conditions on stress-induced adrenocorticotropin responses and norepinephrine release in the hypothalamic paraventricular nucleus. *J. Neuroendocrinol.* 12, 5–12.
780. Plotsky, P. M., and Meaney, M. J. (1993). Early, postnatal experience alters hypothalamic corticotropin-releasing factor (CRF) mRNA, median eminence CRF content and stress-induced release in adult rats. *Brain Res. Mol. Brain Res.* 18, 195–200.
781. Avishai-Ephner, S., Yi, S. J., Newth, C. J., and Baram, T. Z. (1995). Effects of maternal and sibling deprivation on basal and stress induced hypothalamic–pituitary–adrenal components in the infant rat. *Neurosci. Lett.* 192, 49–52.
782. Smith, M. A., Kim, S. Y., van Oers, H. J., and Levine, S. (1997). Maternal deprivation and stress induce immediate early genes in the infant rat brain. *Endocrinology* 138, 4622–4628.
783. Yi, S. J. and Baram, T. Z. (1994). Corticotropin-releasing hormone mediates the response to cold stress in the neonatal rat without compensatory enhancement of the peptide's gene expression. *Endocrinology* 135, 2364–2368.
784. Francis, D. D., Caldji, C., Champagne, F., Plotsky, P. M., and Meaney, M. J. (199). The role of corticotropin-releasing factor–norepinephrine systems in mediating the effects of early experience on the development of behavioral and endocrine responses to stress. *Biol. Psychiatry* 46, 1153–1166.
785. Otagiri A., Wakabayashi, I., and Shibasaki, T. (2000). Selective corticotropin-releasing factor type 1 receptor antagonist blocks conditioned fear-induced release of noradrenaline in the hypothalamic paraventricular nucleus of rats. *J. Neuroendocrinol.* 12, 1022–1026.
786. Meaney, M. J., Diorio, J., Francis, D., Widdowson, J., LaPlante, P., Caldji, C., Sharma, S., Seckl, J. R., and Plotsky, P. M. (1996). Early environmental regulation of forebrain glucocorticoid receptor gene expression: implications for adrenocortical responses to stress. *Dev. Neurosci.* 18, 49–72.
787. Liu, D., Diorio, J., Tannenbaum, B., Caldji, C., Francis, D., Freedman, A., Sharma, S., Pearson, D., Plotsky, P. M., and Meaney, M. J. (1997). Maternal care, hippocampal glucocorticoid receptors, and hypothalamic–pituitary–adrenal responses to stress. *Science* 277, 1659–1662.
788. Caldji, C., Tannenbaum, B., Sharma, S., Francis, D., Plotsky, P. M., and Meaney, M. J. (1998). Maternal care during infancy regulates the development of neural systems mediating the expression of fearfulness in the rat. *Proc. Natl. Acad. Sci. U S A* 95, 5335–5340.
789. Avishai-Eliner, S., Eghbal-Ahmadi, M., Tabachnik, E., Brunson, K. L., and Baram, T. Z. (2001). Down-regulation of hypothalamic corticotropin-releasing hormone messenger ribonucleic acid (mRNA) precedes early-life experience-induced changes in hippocampal glucocorticoid receptor mRNA. *Endocrinology* 142, 89–97.
790. Caldji, C., Diorio, J., and Meaney, M. J. (2003). Variations in maternal care alter GABA(A) receptor subunit expression in brain regions associated with fear. *Neuropsychopharmacology* 28, 1950–1959.
791. Liu, D., Diorio, J., Day, J. C., Francis, D. D., and Meaney, M. J. (2000). Maternal care, hippocampal synaptogenesis and cognitive development in rats. *Nat. Neurosci.* 3, 799–806.
792. Papaioannou A., Dafni, U., Alikaridis, F., Bolaris, S., and Stylianopoulou, F. (2002). Effects of neonatal handling on basal and stress-induced monoamine levels in the male and female rat brain. *Neuroscience* 114, 195–206.
793. Caldji, C., Diorio, J., Anisman, H., and Meaney, M. J. (2004). Maternal behavior regulates benzodiazepine/GABA(A) receptor subunit expression in brain regions associated with fear in BALB/c and C57BL/6 mice. *Neuropsychopharmacology* 2, 1344–1352.
794. Champagne, F. A., Weaver, I. C., Diorio, J., Sharma, S., and Meaney, M. J. (2003). Natural variations in maternal care are associated with estrogen receptor alpha expression and estrogen sensitivity in the medial preoptic area. *Endocrinology* 144, 4720–4724.
795. Suomi, S. J. (1991). Early stress and adult emotional reactivity in rhesus monkeys. *Ciba Found. Symp.* 156, 171–183; discussion 183–178.
796. Canetti, L., Bachar, E., Galili-Weisstub, E., De-Nour, A. K., and Shalev, A. Y. (1997). Parental bonding and mental health in adolescence. *Adolescence* 32, 381–394.
797. Harlow, H. F., Dodsworth, R. O., and Harlow, M. K. (1965). Total social isolation in monkeys. *Proc. Natl. Acad. Sci. U S A* 54, 90–97.
798. Suomi, S. J., Collins, M. L., Harlow, H. F., and Ruppenthal, G. C. (1976). Effects of maternal and peer separations on young monkeys. *J. Child Psychol. Psychiatry* 17, 101–112.
799. Suomi, S. J., Eisele, C. D., Grady, S. A., and Harlow, H. F. (1975). Depressive behavior in adult monkeys following separation from family environment. *J. Abnorm. Psychol.* 84, 576–578.
800. Suomi, S. J., Harlow, H. F., and Kimball, S. D. (1971). Behavioral effects of prolonged partial social isolation in the rhesus monkey. *Psychol. Rep.* 29, 1171–1177.
801. Mitchell, G. D. and Clark, D. L. (1968). Long-term effects of social isolation in nonsocially adapted Rhesus monkeys. *J. Genet. Psychol.* 113, 117–128.
802. Buchanan A., Ten Brinke, J., and Flouri, E. (2000). Parental background, social disadvantage, public "care," and psychological problems in adolescence and adulthood. *J. Am. Acad. Child Adolesc. Psychiatry* 39, 1415–1423.
803. Dube, S. R., Anda, R. F., Felitti, V. J., Chapman, D. P., Williamson, D. F., and Giles, W. H. (2001). Childhood abuse, household dysfunction, and the risk of attempted suicide throughout the life span: findings from the Adverse Childhood Experiences Study. *JAMA* 286, 3089–3096.
804. Gittleman, M. G., Klein, M. H., Smider, N. A., and Essex, M. J. (1998). Recollections of parental behaviour, adult attachment and mental health: mediating and moderating effects. *Psychol. Med.* 28, 1443–1455.
805. Hall, L. A., Peden, A. R., Rayens, M. K., and Beebe, L. H. (2004). Parental bonding: a key factor for mental health of college women. *Issues Ment. Health Nurs.* 25, 277–291.
806. Heim, C., Newport, D. J., Wagner, D., Wilcox, M. M., Miller, A. H., and Nemeroff, C. B. (2002). The role of early adverse experience and adulthood stress in the prediction of neuroendocrine stress reactivity in women: a multiple regression analysis. *Depress Anxiety* 15, 117–125.
807. Halligan, S. L., Herbert, J., Goodyer, I. M., and Murray, L. (2004). Exposure to postnatal depression predicts elevated cortisol in adolescent offspring. *Biol. Psychiatry* 55, 376–381.
808. Meaney, M. J. (2001). Maternal care, gene expression, and the transmission of individual differences in stress reactivity across generations. *Annu. Rev. Neurosci.* 24, 1161–1192.
809. Guillet, R., Saffran, M., and Michaelson, S. M. (1980). Pituitary-adrenal response in neonatal rats. *Endocrinology* 106, 991–994.
810. Butte, J. C., Kakihana, R., Farnham, M. L., and Noble, E. P. (1973). The relationship between brain and plasma corticosterone stress response in developing rats. *Endocrinology* 92, 1775–1779.
811. Sapolsky, R. M., and Meaney, M. J. (1986). Maturation of the adrenocortical stress response: neuroendocrine control mechanisms and the stress hyporesponsive period. *Brain Res.* 396, 64–76.

812. Walker, C. D., Sapolsky, R. M., Meaney, M. J., Vale, W. W., and Rivier, C. L. (1986). Increased pituitary sensitivity to glucocorticoid feedback during the stress nonresponsive period in the neonatal rat. *Endocrinology* 119, 1816–1821.
813. Hiroshige, T., and Sato, T. (1971). Changes in hypothalamic content of corticotropin-releasing activity following stress during neonatal maturation in the rat. *Neuroendocrinology* 7, 257–270.
814. Armario A., Restrepo, C., Hidalgo, J., and Lopez-Calderon, A. (1987). Differences in prolactin and LH responses to acute stress between peripuberal and adult male rats. *J. Endocrinol.* 112, 9–13.
815. Gomez, F., Houshyar, H., and Dallman, M. F. (2002). Marked regulatory shifts in gonadal, adrenal, and metabolic system responses to repeated restraint stress occur within a 3-week period in pubertal male rats. *Endocrinology* 143, 2852–2862.
816. Almeida, S. A., Petenusci, S. O., Franci, J. A., Rosa e Silva, A. A., and Carvalho, T. L. (2000). Chronic immobilization-induced stress increases plasma testosterone and delays testicular maturation in pubertal rats. *Andrologia* 32, 7–11.
817. Rampacek, G. B., Kraeling, R. R., Fonda, E. S., and Barb, C. R. (1984). Comparison of physiological indicators of chronic stress in confined and nonconfined gilts. *J. Anim. Sci.* 58, 401–408.
818. Gomez, F. and Dallman, M. F. (2001). Manipulation of androgens causes different energetic responses to cold in 60- and 40-day-old male rats. *Am. J. Physiol. Regul. Integr. Comp. Physiol.* 280, R262–273.
819. Lescoat, G., Feliot, J., and Maniey, J. (1978). [Evolution of plasma corticosterone concentration in basal conditions or after psychic aggression during growth in the Rat. Influence of neonatal gonadectomy (author's transl.)]. *J. Physiol. (Paris).* 74, 591–599.
820. Alexandrova, M. and Macho, L. (1976). Plasma corticosterone during postnatal ontogenesis in rats: comparison of protein-binding and fluorometric method. *Endokrinologie* 68, 66–73.
821. Gomez, F., Manalo, S., and Dallman, M. F. (2004). Androgen-sensitive changes in regulation of restraint-induced adrenocorticotropin secretion between early and late puberty in male rats. *Endocrinology* 145, 59–70.
822. Tahirovic, H. F. (1998). Menarchal age and the stress of war: an example from Bosnia. *Eur. J. Pediatr.* 157, 978–980.
823. Prebeg, Z. and Bralic I. (2000). Changes in menarcheal age in girls exposed to war conditions. *Am. J. Human Biol.* 12, 503–508.
824. Brooks-Gunn, J. and Warren, M. P. (1988). Mother-daughter differences in menarcheal age in adolescent girls attending national dance company schools and non-dancers. *Ann. Hum. Biol.* 15, 35–43.
825. Lounana, J., Bantsimba, M., Silou, J., Packa-Tchissambou, B., and Medelli, J. (2002). Effects of intensive and precocious sport activity on the age of menarche in Congolese women population. *Sci. Sports* 17, 17–25.
826. Georgopoulos, N., Markou K., Theodoropoulou A., Paraskevopoulou, P., Varaki, L., Kazantzi, Z., Leglise, M., and Vagenakis, A. G. (1999). Growth and pubertal development in elite female rhythmic gymnasts. *J. Clin. Endocrinol. Metab.* 84, 4525–4530.
827. Klentrou, P. and Plyley, M. (2003). Onset of puberty, menstrual frequency, and body fat in elite rhythmic gymnasts compared with normal controls. *Br. J. Sports Med.* 37, 490–494.
828. Pozo, J. and Argente, J. (2002). Delayed puberty in chronic illness. *Best Pract. Res. Clin. Endocrinol. Metab.* 16, 73–90.
829. Cameron, J. L. (1997). Stress and behaviorally induced reproductive dysfunction in primates. *Semin. Reprod. Endocrinol.* 15, 37–45.
830. Schwartz, S. M., Wilson, M.E., Walker, M. L., and Collins, D. C. (1985). Social and growth correlates of puberty onset in fenale rhesus monkeys. *Nutr. Behav.* 2, 225–232.
831. Belsky, J., Steinberg, L., and Draper, P. (1991). Childhood experience, interpersonal development, and reproductive strategy: and evolutionary theory of socialization. *Child Dev.* 62, 647–670.
832. Hulanicka, B. (1999). Acceleration of menarcheal age of girls from dysfunctional families. *J. Reprod. Infant Psychol.* 17, 119–132.
833. Hulanicka, B., Gronkiewicz, L., and Kotiarek, J. (2001). Effect of familial distress on growth and maturation of girls: a longitudinal study. *Am. J. Hum. Biol.* 13, 771–776.
834. Maestripieri, D., Roney, J. R., DeBias, N., Durante, K. M., and Spaepen, G. M. (2004). Father absence, menarche and interest in infants among adolescent girls. *Dev. Sci.* 7, 560–566.
835. Ellis, B. J. (2004). Timing of pubertal maturation in girls: an integrated life history approach. *Psychol. Bull.* 130, 920–958.
836. Makrigiannakis, A., Zoumakis, E., Margioris, A. N., Theodoropoulos, P., Stournaras, C., and Gravanis, A. (1995). The corticotropin-releasing hormone (CRH) in normal and tumoral epithelial cells of human endometrium. *J. Clin. Endocrinol. Metab.* 80, 185–189.
837. Makrigiannakis, A., Zoumakis, E., Kalantaridou, S., Chrousos, G., and Gravanis, A. (2003). Uterine and embryonic trophoblast CRH promotes implantation and maintenance of early pregnancy. *Ann. N Y Acad. Sci.* 997, 85–92.
838. Karteris, E., Papadopoulou, N., Grammatopoulos, D. K., and Hillhouse, E. W. (2004). Expression and signalling characteristics of the corticotrophin-releasing hormone receptors during the implantation phase in the human endometrium. *J. Mol. Endocrinol.* 32, 21–32.
839. Florio, P., Severi, F. M., Ciarmela, P., Fiore, G., Calonaci, G., Merola A., De Felice, C., Palumbo, M., and Petraglia, F. (2002). Placental stress factors and maternal-fetal adaptive response: the corticotropin-releasing factor family. *Endocrine* 19, 91–102.
840. Majzoub, J. A. and Karalis, K. P. (1999). Placental corticotropin-releasing hormone: function and regulation. *Am. J. Obstet. Gynecol.* 180, S242–246.
841. Di Blasio, A. M., Giraldi, F. P., Vigano, P., Petraglia, F., Vignali, M., and Cavagnini, F. (1997). Expression of corticotropin-releasing hormone and its R1 receptor in human endometrial stromal cells. *J. Clin. Endocrinol. Metab.* 82, 1594–1597.
842. Mastorakos, G., Scopa, C. D., Kao, L. C., Vryonidou, A., Friedman, T. C., Kattis, D., Phenekos, C., Rabin, D., and Chrousos, G. P. (1996). Presence of immunoreactive corticotropin-releasing hormone in human endometrium. *J. Clin. Endocrinol. Metab.* 81, 1046–1050.
843. Makrigiannakis, A., Margioris, A. N., Chatzaki, E., Zoumakis, E., Chrousos, G. P., and Gravanis, A. (1999). The decidualizing effect of progesterone may involve direct transcriptional activation of corticotrophin-releasing hormone from human endometrial stromal cells. *Mol. Hum. Reprod.* 5, 789–796.
844. Ferrari, A., Petraglia, F., and Gurpide, E. (1995). Corticotropin releasing factor decidualizes human endometrial stromal cells in vitro. Interaction with progestin. *J. Steroid Biochem. Mol. Biol.* 54, 251–255.
845. Makrigiannakis, A., Margioris, A. N., Le Goascogne, C., Zoumakis, E., Nikas, G., Stournaras, C., Psychoyos, A., and Gravanis, A. (1995). Corticotropin-releasing hormone (CRH) is expressed at the implantation sites of early pregnant rat uterus. *Life Sci.* 57, 1869–1875.

846. Makrigiannakis, A., Zoumakis, E., Kalantaridou, S., Coutifaris, C., Margioris, A. N., Coukos, G., Rice, K. C., Gravanis, A., and Chrousos, G. P. (2001). Corticotropin-releasing hormone promotes blastocyst implantation and early maternal tolerance. *Nat. Immunol.* 2, 1018–1024.
847. Makrigiannakis, A., Zoumakis, E., Kalantaridou, S., Chrousos, G., and Gravanis, A. (2004). Participation of maternal and fetal CRH in early phases of human implantation: the role of antalarmin. *Curr. Drug Targets Immune Endocr. Metabol. Disord.* 4, 75–78.
848. Robinson, B. G., Arbiser, J. L., Emanuel, R. L., and Majzoub, J. A. (1989). Species-specific placental corticotropin releasing hormone messenger RNA and peptide expression. *Mol. Cell. Endocrinol.* 62, 337–341.
849. Sasaki A., Shinkawa, O., and Yoshinaga, K. (1990). Immunoreactive corticotropin-releasing hormone in amniotic fluid. *Am. J. Obstet. Gynecol.* 162, 194–198.
850. Sehringer, B., Zahradnik, H. P., Simon, M., Ziegler, R., Noethling, C., and Schaefer, W. R. (2004). mRNA expression profiles for corticotrophin-releasing hormone, urocortin, CRH-binding protein and CRH receptors in human term gestational tissues determined by real-time quantitative RT-PCR. *J. Mol. Endocrinol.* 32, 339–348.
851. Hatzoglou A., Margioris, A. N., Bakogeorgou, E., Gravanis, A., and Castanas, E. (1996). Identification, characterization and localization of corticotropin-releasing hormone receptors in human placenta. *Life Sci.* 59, 1871–1879.
852. Karteris, E., Grammatopoulos, D., Dai, Y., Olah, K. B., Ghobara, T. B., Easton, A., and Hillhouse, E. W. (1998). The human placenta and fetal membranes express the corticotropin-releasing hormone receptor 1alpha (CRH-1alpha) and the CRH-C variant receptor. *J. Clin. Endocrinol. Metab.* 83, 1376–1379.
853. Warren, W. B. and Silverman, A. J. (1995). Cellular localization of corticotrophin releasing hormone in the human placenta, fetal membranes and decidua. *Placenta* 16, 147–156.
854. Riley, S. C., Walton, J. C., Herlick, J. M., and Challis, J. R. (1991). The localization and distribution of corticotropin-releasing hormone in the human placenta and fetal membranes throughout gestation. *J. Clin. Endocrinol. Metab.* 72, 1001–1007.
855. Petraglia, F., Tabanelli, S., Galassi, M. C., Garuti, G. C., Mancini, A. C., Genazzani, A. R., and Gurpide, E. (1992). Human decidua and in vitro decidualized endometrial stromal cells at term contain immunoreactive corticotropin-releasing factor (CRF) and CRF messenger ribonucleic acid. *J. Clin. Endocrinol. Metab.* 74, 1427–1431.
856. Perkins, A. V. and Linton, E. A. (1995). Identification and isolation of corticotrophin-releasing hormone-positive cells from the human placenta. *Placenta* 16, 233–243.
857. Frim, D. M., Emanuel, R. L., Robinson, B. G., Smas, C. M., Adler, G. K., and Majzoub, J. A. (1988). Characterization and gestational regulation of corticotropin-releasing hormone messenger RNA in human placenta. *J. Clin. Invest.* 82, 287–292.
858. Okamoto, E., Takagi, T., Azuma, C., Kimura, T., Tokugawa, Y., Mitsuda, N., Saji, F., and Tanizawa, O. (1990). Expression of the corticotropin-releasing hormone (CRH) gene in human placenta and amniotic membrane. *Horm. Metab. Res.* 22, 394–397.
859. Jones, S. A., Brooks, A. N., and Challis, J. R. (1989). Steroids modulate corticotropin-releasing hormone production in human fetal membranes and placenta. *J. Clin. Endocrinol. Metab.* 68, 825–830.
860. Schulte, H. M. and Healy, D. L. (1987). Corticotropin releasing hormone- and adreno-corticotropin-like immunoreactivity in human placenta, peripheral and uterine vein plasma. *Horm. Metab. Res. Suppl.* 16, 44–46.
861. Leung, T. N., Chung, T. K., Madsen, G., Lam, P. K., Sahota, D., and Smith, R. (2001). Rate of rise in maternal plasma corticotrophin-releasing hormone and its relation to gestational length. *BJOG* 108, 527–532.
862. Robinson, B. G., Emanuel, R. L., Frim, D. M., and Majzoub, J. A. (1988). Glucocorticoid stimulates expression of corticotropin-releasing hormone gene in human placenta. *Proc. Natl. Acad. Sci. U S A* 85, 5244–5248.
863. Ni, X., Nicholson, R. C., King, B. R., Chan, E. C., Read, M. A., and Smith, R. (2002). Estrogen represses whereas the estrogen-antagonist ICI 182780 stimulates placental CRH gene expression. *J. Clin. Endocrinol. Metab.* 87, 3774–3778.
864. Karalis, K., Goodwin, G., and Majzoub, J. A. (1996). Cortisol blockade of progesterone: a possible molecular mechanism involved in the initiation of human labor. *Nat. Med.* 2, 556–560.
865. Petraglia, F., Sutton, S., and Vale, W. (1989). Neurotransmitters and peptides modulate the release of immunoreactive corticotropin-releasing factor from cultured human placental cells. *Am. J. Obstet. Gynecol.* 160, 247–251.
866. Petraglia, F., Garuti, G. C., De Ramundo, B., Angioni, S., Genazzani, A. R., and Bilezikjian, L. M. (1990). Mechanism of action of interleukin-1 beta in increasing corticotropin-releasing factor and adrenocorticotropin hormone release from cultured human placental cells. *Am. J. Obstet. Gynecol.* 163, 1307–1312.
867. Goland, R. S., Wardlaw, S. L., Blum, M., Tropper, P. J., and Stark, R. I. (1988). Biologically active corticotropin-releasing hormone in maternal and fetal plasma during pregnancy. *Am. J. Obstet. Gynecol.* 159, 884–890.
868. Campbell, E. A., Linton, E. A., Wolfe, C. D., Scraggs, P. R., Jones, M. T., and Lowry, P. J. (1987). Plasma corticotropin-releasing hormone concentrations during pregnancy and parturition. *J. Clin. Endocrinol. Metab.* 64, 1054–1059.
869. Laatikainen, T., Virtanen, T., Kaaja, R., and Salminen-Lappalainen, K. (1991). Corticotropin-releasing hormone in maternal and cord plasma in pre-eclampsia. *Eur. J. Obstet. Gynecol. Reprod. Biol.* 39, 19–24.
870. Goland, R. S., Stark, R. I., and Wardlaw, S. L. (1990). Response to corticotropin-releasing hormone during pregnancy in the baboon. *J. Clin. Endocrinol. Metab.* 70, 925–929.
871. Ruth, V., Hallman, M., and Laatikainen, T. (1993). Corticotropin-releasing hormone and cortisol in cord plasma in relation to gestational age, labor, and fetal distress. *Am. J. Perinatol.* 10, 115–118.
872. Petraglia, F., Florio, P., Gallo, R., Salvestroni, C., Lombardo, M., Genazzani, A. D., Di Carlo, C., Stomati, M., D'Ambrogio, G., and Artini, P. G. (1996). Corticotropin-releasing factor-binding protein: origins and possible functions. *Horm. Res.* 45, 187–191.
873. Linton, E. A., Perkins, A. V., Woods, R. J., Eben, F., Wolfe, C. D., Behan, D. P., Potter, E., Vale, W. W., and Lowry, P. J. (1993). Corticotropin releasing hormone-binding protein (CRH-BP): plasma levels decrease during the third trimester of normal human pregnancy. *J. Clin. Endocrinol. Metab.* 76, 260–262.
874. Berkowitz, G. S., Lapinski, R. H., Lockwood, C. J., Florio, P., Blackmore-Prince, C., and Petraglia, F. (1996). Corticotropin-releasing factor and its binding protein: maternal serum levels in term and preterm deliveries. *Am. J. Obstet. Gynecol.* 174, 1477–1483.
875. Petraglia, F., Florio, P., Gallo, R., Simoncini, T., Saviozzi, M., Di Blasio, A. M., Vaughan, J., and Vale, W. (1996). Human placenta and fetal membranes express human urocortin mRNA and peptide. *J. Clin. Endocrinol. Metab.* 81, 3807–3810.
876. Goland, R. S., Jozak, S., and Conwell, I. (1994). Placental corticotropin-releasing hormone and the hypercortisolism of pregnancy. *Am. J. Obstet. Gynecol.* 171, 1287–1291.

877. Goland, R. S., Conwell, I. M., Warren, W. B., and Wardlaw, S. L. (1992). Placental corticotropin-releasing hormone and pituitary-adrenal function during pregnancy. *Neuroendocrinology* 56, 742–749.
878. Magiakou, M. A., Mastorakos, G., Rabin, D., Margioris, A. N., Dubbert, B., Calogero, A. E., Tsigos, C., Munson, P. J., and Chrousos, G. P. (1996). The maternal hypothalamic–pituitary–adrenal axis in the third trimester of human pregnancy. *Clin. Endocrinol. (Oxf).* 44, 419–428.
879. Chrousos, G. P., Torpy, D. J., and Gold, P. W. (1998). Interactions between the hypothalamic–pituitary–adrenal axis and the female reproductive system: clinical implications. *Ann. Intern. Med.* 129, 229–240.
880. Petraglia, F., Sawchenko, P. E., Rivier, J., and Vale, W. (1987). Evidence for local stimulation of ACTH secretion by corticotropin-releasing factor in human placenta. *Nature* 328, 717–719.
881. Wadhwa, P. D., Sandman, C. A., Chicz-DeMet, A., and Porto, M. (1997). Placental CRH modulates maternal pituitary adrenal function in human pregnancy. *Ann. N Y Acad. Sci.* 814, 276–281.
882. Waddell, B. J. (1993). The placenta as hypothalamus and pituitary: possible impact on maternal and fetal adrenal function. *Reprod. Fertil. Dev.* 5, 479–497.
883. Smith, R., Mesiano, S., Chan, E. C., Brown, S., and Jaffe, R. B. (1998). Corticotropin-releasing hormone directly and preferentially stimulates dehydroepiandrosterone sulfate secretion by human fetal adrenal cortical cells. *J. Clin. Endocrinol. Metab.* 83, 2916–2920.
884. Clifton, V. L., Read, M. A., Leitch IM, Giles, W. B., Boura, A. L., Robinson, P. J., and Smith, R. (1995). Corticotropin-releasing hormone-induced vasodilatation in the human fetal-placental circulation: involvement of the nitric oxide-cyclic guanosine 3′,5′-monophosphate-mediated pathway. *J. Clin. Endocrinol. Metab.* 80, 2888–2893.
885. Goland, R. S., Jozak, S., Warren, W. B., Conwell, I. M., Stark, R. I., and Tropper, P. J. (1993). Elevated levels of umbilical cord plasma corticotropin-releasing hormone in growth-retarded fetuses. *J. Clin. Endocrinol. Metab.* 77, 1174–1179.
886. Goland, R. S., Conwell, I. M., and Jozak, S. (1995). The effect of pre-eclampsia on human placental corticotrophin-releasing hormone content and processing. *Placenta* 16, 375–382.
887. Florio, P., Imperatore, A., Sanseverino, F., Torricelli, M., Reis, F. M., Lowry, P. J., and Petraglia, F. (2004). The measurement of maternal plasma corticotropin-releasing factor (CRF) and CRF-binding protein improves the early prediction of preeclampsia. *J. Clin. Endocrinol. Metab.* 89, 4673–4677.
888. Wolfe, C. D., Patel, S. P., Linton, E. A., Campbell, E. A., Anderson, J., Dornhorst A., Lowry, P. J., and Jones, M. T. (1988). Plasma corticotrophin-releasing factor (CRF) in abnormal pregnancy. *Br. J. Obstet. Gynaecol.* 95, 1003–1006.
889. Madhappan, B., Kempuraj, D., Christodoulou, S., Tsapikidis, S., Boucher, W., Karagiannis, V., Athanassiou, A., and Theoharides, T. C. (2003). High levels of intrauterine corticotropin-releasing hormone, urocortin, tryptase, and interleukin-8 in spontaneous abortions. *Endocrinology* 144, 2285–2290.
890. Wadhwa, P. D., Porto, M., Garite, T. J., Chicz-DeMet, A., and Sandman, C. A. (1998). Maternal corticotropin-releasing hormone levels in the early third trimester predict length of gestation in human pregnancy. *Am. J. Obstet. Gynecol.* 179, 1079–1085.
891. McLean, M., Bisits A., Davies, J., Woods, R., Lowry, P., and Smith, R. (1995). A placental clock controlling the length of human pregnancy. *Nat. Med.* 1, 460–463.
892. Korebrits, C., Ramirez, M. M., Watson, L., Brinkman, E., Bocking, A. D., and Challis, J. R. (1998). Maternal corticotropin-releasing hormone is increased with impending preterm birth. *J. Clin. Endocrinol. Metab.* 83, 1585–1591.
893. Sandman, C. A., Wadhwa, P. D., Chicz-DeMet, A., Dunkel-Schetter, C., and Porto, M. (1997). Maternal stress, HPA activity, and fetal/infant outcome. *Ann. N Y Acad. Sci.* 814, 266–275.
894. Florio, P., Cobellis, L., Woodman, J., Severi, F. M., Linton, E. A., and Petraglia, F. (2002). Levels of maternal plasma corticotropin-releasing factor and urocortin during labor. *J. Soc. Gynecol. Investig.* 9, 233–237.
895. Erickson, K., Thorsen, P., Chrousos, G., Grigoriadis, D. E., Khongsaly, O., McGregor, J., and Schulkin, J. (2001). Preterm birth: associated neuroendocrine, medical, and behavioral risk factors. *J. Clin. Endocrinol. Metab.* 86, 2544–2552.
896. Smith, R. (1998). Alterations in the hypothalamic pituitary adrenal axis during pregnancy and the placental clock that determines the length of parturition. *J. Reprod. Immunol.* 39, 215–220.
897. McGrath, S., and Smith, R. (2002). Prediction of preterm delivery using plasma corticotrophin-releasing hormone and other biochemical variables. *Ann. Med.* 34, 28–36.
898. Warren, W. B., Patrick, S. L., and Goland, R. S. (1992). Elevated maternal plasma corticotropin-releasing hormone levels in pregnancies complicated by preterm labor. *Am. J. Obstet. Gynecol.* 166, 1198–1204; discussion 1204–1197.
899. Wu, W. X., Unno, S., Giussani, D. A., Mecenas, C. A., McDonald, T. J., and Nathanielsz, P. W. (1995). Corticotropin-releasing hormone and its receptor distribution in fetal membranes and placenta of the rhesus monkey in late gestation and labor. *Endocrinology* 136, 4621–4628.
900. Rodriguez-Linares, B., Linton, E. A., and Phaneuf, S. (1998). Expression of corticotrophin-releasing hormone receptor mRNA and protein in the human myometrium. *J. Endocrinol.* 156, 15–21.
901. Grammatopoulos, D. K. and Hillhouse, E. W. (1999). Role of corticotropin-releasing hormone in onset of labour. *Lancet* 354, 1546–1549.
902. Reis, F. M., Fadalti, M., Florio, P., and Petraglia, F. (1999). Putative role of placental corticotropin-releasing factor in the mechanisms of human parturition. *J. Soc. Gynecol. Investig.* 6, 109–119.
903. Stevens, M. Y., Challis, J. R., and Lye, S. J. (1998). Corticotropin-releasing hormone receptor subtype 1 is significantly up-regulated at the time of labor in the human myometrium. *J. Clin. Endocrinol. Metab.* 83, 4107–4115.
904. Jirecek, S., Tringler, B., Knofler, M., Bauer, S., Topcuoglu, A., and Egarter, C. (2002). Redistribution of corticotropin-releasing hormone receptor R2 using fluorescence immunohistochemistry in fetal membranes of women delivering preterm or at term. *Wien Klin. Wochenschr.* 114, 1008–1012.
905. Simpkin, J. C., Kermani, F., Palmer, A. M., Campa, J. S., Tribe, R. M., Linton, E. A., and Poston, L. (1999). Effects of corticotrophin releasing hormone on contractile activity of myometrium from pregnant women. *Br. J. Obstet. Gynaecol.* 106, 439–445.
906. Benedetto, C., Petraglia, F., Marozio, L., Chiarolini, L., Florio, P., Genazzani, A. R., and Massobrio, M. (1994). Corticotropin-releasing hormone increases prostaglandin F2 alpha activity on human myometrium in vitro. *Am. J. Obstet. Gynecol.* 171, 126–131.
907. Jones, S. A., and Challis, J. R. (1990). Effects of corticotropin-releasing hormone and adrenocorticotropin on prostaglandin output by human placenta and fetal membranes. *Gynecol. Obstet. Invest.* 29, 165–168.
908. Keelan, J. A., Blumenstein, M., Helliwell, R. J., Sato, T. A., Marvin, K. W., and Mitchell, M. D. (2003). Cytokines, prostaglandins and parturition—a review. *Placenta* 24, Suppl A, S33–46.

909. McKeown, K. J. and Challis, J. R. (2003). Regulation of 15-hydroxy prostaglandin dehydrogenase by corticotrophin-releasing hormone through a calcium-dependent pathway in human chorion trophoblast cells. *J. Clin. Endocrinol. Metab.* 88, 1737–1741.
910. Warren, M. P. and Perlroth, N. E. (2001). The effects of intense exercise on the female reproductive system. *J. Endocrinol.* 170, 3–11.
911. Warren, M. P. (1996). Eating disorders and the reproductive axis. In *Reproductive Endocrinology, Surgery, and Technology* (E. Y. Adashi, J. A. Rock, and Z., Rosenwaks, Eds.), Vol. 1, pp. 1039–1060. Lipincott-Raven, Philadelphia.
912. Warren, M. P. and Shantha, S. (2000). The female athlete. *Baillieres Best Pract. Res. Clin. Endocrinol. Metab.* 14, 37–53.
913. Prior, J. C. (1996). Exercise-associated menstrual disturbances. In *Reproductive Endocrinology, Surgery, and Technology* (E. Y. Adashi, J. A. Rock, and Z., Rosenwaks, Eds.). Vol. 1, pp. 1077–1091. Lippincott-Raven, Philadelphia.
914. Williams, N. I., Helmreich, D. L., Parfitt, D. B., Caston-Balderrama, A., and Cameron, J. L. (2001). Evidence for a causal role of low energy availability in the induction of menstrual cycle disturbances during strenuous exercise training. *J. Clin. Endocrinol. Metab.* 86, 5184–5193.
915. Schreihofer, D. A., Parfitt, D. B., and Cameron, J. L. (1993). Suppression of luteinizing hormone secretion during short-term fasting in male rhesus monkeys: the role of metabolic versus stress signals. *Endocrinology* 132, 1881–1889.
916. Cameron, J. L., Helmreich, D. L., and Schreihofer, D. A. (1993). Modulation of reproductive hormone secretion by nutritional intake: stress signals versus metabolic signals. *Hum. Reprod.* 8, Suppl 2, 162–167.
917. Marti, O., Gavalda A., Marti, J., Gil, M., Giralt, M., Lopez-Calderon, A., and Armario, A. (1993). Chronic stress induced changes in LH secretion: the contribution of anorexia associated to stress. *Life Sci.* 52, 1187–1194.
918. Gadpaille, W. J., Sanborn, C. F., and Wagner, W. W. Jr. (1987). Athletic amenorrhea, major affective disorders, and eating disorders. *Am. J. Psychiatry* 144, 939–942.
919. Marcus, M. D., Loucks, T. L., and Berga, S. L. (2001). Psychological correlates of functional hypothalamic amenorrhea. *Fertil Steril.* 76, 310–316.
920. Bergendahl, M., Vance, M. L., Iranmanesh, A., Thorner, M. O., and Veldhuis, J. D. (1996). Fasting as a metabolic stress paradigm selectively amplifies cortisol secretory burst mass and delays the time of maximal nyctohemeral cortisol concentrations in healthy men. *J. Clin. Endocrinol. Metab.* 81, 692–699.
921. McLean, J. A., Barr, S. I., and Prior, J. C. (2001). Cognitive dietary restraint is associated with higher urinary cortisol excretion in healthy premenopausal women. *Am. J. Clin. Nutr.* 73, 7–12.
922. Loucks, A. B. and Horvath, S. M. (1984). Exercise-induced stress responses of amenorrheic and eumenorrheic runners. *J. Clin. Endocrinol. Metab.* 59, 1109–1120.
923. Loucks, A. B. and Horvath, S. M. (1985). Athletic amenorrhea: a review. *Med. Sci. Sports Exerc.* 17, 56–72.
924. Loucks, A. B. and Thuma, J. R. (2003). Luteinizing hormone pulsatility is disrupted at a threshold of energy availability in regularly menstruating women. *J. Clin. Endocrinol. Metab.* 88, 297–311.
925. Nagaya, N., Miyatake K., Uematsu, M., Oya, H., Shimizu, W., Hosoda, H., Kojima, M., Nakanishi, N., Mori, H., and Kangawa, K. (2001). Hemodynamic, renal, and hormonal effects of ghrelin infusion in patients with chronic heart failure. *J. Clin. Endocrinol. Metab.* 86, 5854–5859.
926. Xiao, E., Xia-Zhang, L., Vulliemoz, N. R., Ferin, M., and Wardlaw, S. L. (2003). Agouti-related protein stimulates the hypothalamic–pituitary–adrenal (HPA) axis and enhances the HPA response to interleukin-1 in the primate. *Endocrinology* 144, 1736–1741.
927. Larsen, P. J., Tang-Christensen, M., Stidsen, C. E., Madsen, K., Smith, M. S., and Cameron, J. L. (1999). Activation of central neuropeptide Y Y1 receptors potently stimulates food intake in male rhesus monkeys. *J. Clin. Endocrinol. Metab.* 84, 3781–3791.
928. Inoue, T., Inui A., Okita, M., Sakatani, N., Oya, M., Morioka, H., Mizuno, N., Oimomi, M., and Baba, S. (1989). Effect of neuropeptide Y on the hypothalamic–pituitary–adrenal axis in the dog. *Life Sci.* 44, 1043–1051.
929. Leibowitz, S. F., Sladek, C., Spencer, L., and Tempel, D. (1988). Neuropeptide, Y., epinephrine and norepinephrine in the paraventricular nucleus: stimulation of feeding and the release of corticosterone, vasopressin and glucose. *Brain Res. Bull.* 21, 905–912.
930. Chen, H. Y., Trumbauer, M. E., Chen, A. S., Weingarth, D. T., Adams, J. R., Frazier, E. G., Shen, Z., Marsh, D. J., Feighner, S. D., Guan, X. M., Ye, Z., Nargund, R. P., Smith, R. G., Van der Ploeg, L. H., Howard, A. D., MacNeil, D. J., and Qian, S. (2004). Orexigenic action of peripheral ghrelin is mediated by neuropeptide Y and agouti-related protein. *Endocrinology* 145, 2607–2612.
931. Vulliémoz, N., Xiao, E., Xia-Zhang, L., Wardlaw, S. L., and Ferin, M. (2004). Central infusion of AGRP suppresses pulsatile LH release in the ovariectomized Rhesus monkey. *Endocrinology* 146, 784–789.
932. Kaynard, A. H., Pau, K. Y., Hess, D. L., and Spies, H. G. (1990). Third-ventricular infusion of neuropeptide Y suppresses luteinizing hormone secretion in ovariectomized rhesus macaques. *Endocrinology* 127, 2437–2444.
933. Vulliémoz, N., Xiao, E., Xia-Zhang, L., Rivier, J., and Ferin, M. (2004). Decrease in pulsatile LH frequency during a 5-h peripheral ghrelin infusion in the OVX Rhesus monkey. *J. Clin. Endocrinol. Metab.* 89, 5718–5723.
934. Liposits, Z., Sievers, L., and Paull, W. K. (1988). Neuropeptide-Y and ACTH-immunoreactive innervation of corticotropin releasing factor (CRF)-synthesizing neurons in the hypothalamus of the rat. An immunocytochemical analysis at the light and electron microscopic levels. *Histochemistry* 88, 227–234.
935. Haskell-Luevano, C., Chen, P., Li, C., Chang K., Smith, M. S., Cameron, J. L., and Cone, R. D. (1999). Characterization of the neuroanatomical distribution of agouti-related protein immunoreactivity in the rhesus monkey and the rat. *Endocrinology* 140, 1408–1415.
936. Dhillo, W. S., Small, C. J., Seal, L. J., Kim, M. S., Stanley, S. A., Murphy, K. G., Ghatei, M. A., and Bloom, S. R. (2002). The hypothalamic melanocortin system stimulates the hypothalamo–pituitary–adrenal axis in vitro and in vivo in male rats. *Neuroendocrinology* 75, 209–216.
937. Masaki, T., Yoshimichi, G., Chiba, S., Yasuda, T., Noguchi, H., Kakuma, T., Sakata, T., and Yoshimatsu, H. (2003). Corticotropin-releasing hormone-mediated pathway of leptin to regulate feeding, adiposity, and uncoupling protein expression in mice. *Endocrinology* 144, 3547–3554.
938. Gardner, J. D., Rothwell, N. J., and Luheshi, G. N. (1998). Leptin affects food intake via CRF-receptor-mediated pathways. *Nat. Neurosci.* 1, 103.
939. Maeda, K., Nagatani, S., Estacio, M. A., and Tsukamura, H. (1996). Novel estrogen feedback sites associated with stress-induced suppression of luteinizing hormone secretion in female rats. *Cell. Mol. Neurobiol.* 16, 311–324.
940. Estacio, M. A., Yamada, S., Tsukamura, H., Hirunagi K., and Maeda, K. (1996). Effect of fasting and immobilization stress on estrogen receptor immunoreactivity in the brain in ovariectomized female rats. *Brain Res.* 717, 55–61.

941. Maeda, K., Cagampang, F. R., Coen, C. W., and Tsukamura, H. (1994). Involvement of the catecholaminergic input to the paraventricular nucleus and of corticotropin-releasing hormone in the fasting-induced suppression of luteinizing hormone release in female rats. *Endocrinology* 134, 1718–1722.
942. Currie, P. J., Coscina, D. V., Bishop, C., Coiro, C. D., Koob, G. F., Rivier, J., and Vale, W. (2001). Hypothalamic paraventricular nucleus injections of urocortin alter food intake and respiratory quotient. *Brain Res.* 916, 222–228.
943. Hope, P. J., Turnbull, H., Farr, S., Morley, J. E., Rice, K. C., Chrousos, G. P., Torpy, D. J., and Wittert, G. A. (2000). Peripheral administration of CRF and urocortin: effects on food intake and the HPA axis in the marsupial Sminthopsis crassicaudata. *Peptides* 21, 669–677.
944. Shade, R. E., Blair-West, J. R., Carey, K. D., Madden, L. J., Weisinger, R. S., Rivier, J. E., Vale, W. W., and Denton, D. A. (2002). Ingestive responses to administration of stress hormones in baboons. *Am. J. Physiol. Regul. Integr. Comp. Physiol.* 282, R10–R18.
945. Sinnayah, P., Blair-West, J. R., McBurnie, M. I., McKinley, M. J., Oldfield, B. J., Rivier, J., Vale, W. W., Walker, L. L., Weisinger, R. S., and Denton, D. A. The effect of urocortin on ingestive behaviours and brain Fos immunoreactivity in mice. *Eur. J. Neurosci.* (2003). 18, 373–382.
946. Wang, C. and Kotz, C. M. (2002). Urocortin in the lateral septal area modulates feeding induced by orexin A in the lateral hypothalamus. *Am. J. Physiol. Regul. Integr. Comp. Physiol.* 283, R358–367.
947. Kotz, C. M., Wang, C., Levine, A. S., and Billington, C. J. (2002). Urocortin in the hypothalamic PVN increases leptin and affects uncoupling proteins-1 and -3 in rats. *Am. J. Physiol. Regul. Integr. Comp. Physiol.* 282, R546–551.
948. Holmberg, B. J., Morrison, C. D., and Keisler, D. H. (2001). Endocrine responses of ovariectomized ewes to i.c.v. infusion of urocortin. *J. Endocrinol.* 171, 517–524.
949. Whitley, N. C., Barb, C. R., Kraeling, R. R., Barrett, J. B., Rampacek, G. B., Carroll, J. A., and Keisler, D. H. (2000). Feed intake and serum GH, LH and cortisol in gilts after intracerebroventricular or intravenous injection of urocortin. *Domest. Anim. Endocrinol.* 19, 209–221.
950. Dallman, M. F., la Fleur, S. E., Pecoraro, N. C., Gomez, F., Houshyar, H., and Akana, S. F. (2004). Minireview: glucocorticoids—food intake, abdominal obesity, and wealthy nations in 2004. *Endocrinology* 145, 2633–2638.
951. Wasser, S. K. and Barash, D. P. (1983). Reproductive suppression among female mammals: implications for biomedicine and sexual selection theory. *Q. Rev. Biol.* 58, 513–538.

Knobil and Neill's Physiology of Reproduction, Third Edition
edited by Jimmy D. Neill,
Elsevier © 2006

CHAPTER 49

Aging in the Hypothalamic–Pituitary–Testicular Axis

David J. Handelsman

Introduction, 2697
The Concept of Aging, 2697
Gender Differences in Aging, 2698
Methodological Considerations for Study of Male Reproductive Aging, 2698
Overview of Male Reproductive Function in Young Men, 2699
Aging and the Testicular Exocrine System, 2700
Male Fertility, 2700
Sperm Output, 2703
Testicular Blood Vessels, 2704
Testis Size, 2704
Spermatogenesis, 2705
Sertoli Cells and Hormonal Regulation of Spermatogenesis, 2706
Aging and the Testicular Endocrine System, 2708
Testosterone Production and Leydig Cells, 2708
Hypothalamo–Pituitary Regulation of Gonadotropin Secretion, 2710
Aging and Androgen Action, 2711
Somatic Consequences of Age-Related Decline in Testosterone Production, 2712
Reproductive Consequences of Age-Related Decline in Testosterone Production, 2713
References, 2715

INTRODUCTION

The Concept of Aging

Aging may be defined as the observable physiological effects of the passage of time, on the scale of an organism's life span, ending in death. Between the twin pillars of the irreversibility of time and the inevitability of death, the reproductive system serves the imperative of species propagation. Aging effects are manifest as a process of decay and deterioration with progressive impact, mostly after sexual maturation and reproductive life are completed. Sexual development, a temporal process of growth to sexual maturity that is determined genetically and triggered hormonally, operates effectively to transmit without extinction the spark of life from individuals to their progeny. In that framework, the body can be considered a mortal soma, a superstructural platform on which germ cells can enact their trans-generational immortality (1–3). After genetic and reproductive destiny is fulfilled, the body becomes superfluous and subject to gradual decay and eventual dissolution. An optimistic view of aging is that the underlying mechanism(s) are regulated and susceptible to modulation by physical and biological means. A pessimistic view is that aging is fundamentally a nonregulated, stochastic process of exhaustion whereby wear-and-tear including physical damage and toxic metabolic byproducts of living accumulate as irreparable cellular damage, disability, and ultimately death through terminal diseases, notably cancer and major organ failure.

Director, ANZAC Research Institute and Department of Andrology, Concord Hospital, Professor of Reproductive Endocrinology and Andrology, University of Sydney, Australia.

Yet, mammalian germ line cells are apparently propagated across generations essentially indefinitely, like bacteria or other simple organisms, without the detrimental senescence evident in virtually all somatic tissues. This suggests that mammalian cellular aging may not be an inevitable consequence of living but might involve mechanisms that operate in germ cells to protect them from harmful aging mechanisms that operate without restraint in somatic cells.

Viewed as the prelude to death, aging has been a perpetual human preoccupation and, not surprisingly, theories of aging continue to abound (4) becoming more scientific with the progressive sophistication of our increasing grasp of Nature. Theories differ on whether aging is purposeful and regulated or simply an accidental accumulation of damage that escapes evolutionary selection because its late-onset occurs after reproductive fitness has been tested (1,2,5). The more popular theories highlight oxidative damage from mitochondrial generation of free radicals (reactive oxidative species) causing DNA damage and respiratory chain dysfunction (6), unreliability due to loss of redundancy (7) and older theories of mutation accumulation or programmed cell death (4,7–9). These theories are often not mutually exclusive and serve best to enrich the heuristic search for specific mechanisms of aging susceptible to effective interventions.

Gender Differences in Aging

In understanding aging effects on the human male reproductive axis, a major issue is whether aging effects are simply the consequence of the passage of time and the unfolding of an intrinsic aging program with fixed time course, or whether it results from the cumulative impact of accidents, diseases and other adverse environmental effects that accumulate during living. At one extreme, menopause illustrates an intrinsic age-related process with relatively fixed time course so that ovarian demise is minimally susceptible to the impact of common diseases or adverse environmental effects. Rather the cessation of normal ovarian function is largely predicated on the inexorable attrition of follicles by atresia. Starting in early fetal life this process of apoptosis leads ultimately to follicle exhaustion at a predictable time, denoted as menopause, occurring now in middle of a modern woman's adult life expectancy. In practice, this predictable process of ovarian aging and demise due to atresia appears resistant to all but a very few agents that cause direct ovarian damage. However, in this pattern of predictable senescence well within expected life span, the ovary is distinctive among human organs sharing this pattern of functional obsolescence in early to mid-life only with the thymus. By contrast, the pattern of human testis aging is much closer to that of most other endocrine systems (thyroid, adrenal, pituitary, pancreas, parathyroid) whereby aging is associated with at most modest, slowly progressively and highly variable changes.

Fundamental gender differences in gonadal physiology underlie the wide disparities between men and women in the natural history of human reproductive aging. After the development of spermatogenesis is completed during puberty, the germinal epithelium continues to generate fresh male gametes for the rest of life unless interrupted by disease or its treatment (notably iatrogenic effects of cytotoxic drugs or irradiation) or eventually dampened by aging. This natural history differs radically from that of female gametes whereby the maximal stock is achieved during fetal life, declining inexorably and without significant supplementation until exhaustion at menopause. As a consequence, mature gametes are as old as the body for females whereas in males this applies to the germinal epithelium's stem cells but all progeny are less than 3 months old no matter how old the man, a dichotomy with consequences for age-related accumulation of epigenetic mutations. This sharp disparity in reproductive life span is also illustrated by the fact that, at the average age of vasectomy (~30–35 year), men sacrifice more reproductive life than women ever have. Similarly, while gamete production and steroidogenesis are tightly linked in the ovary, formalized in the didactic concept of the two cell hypothesis for estrogen synthesis, the testis manifests much more flexible linkage between Leydig cell testosterone biosynthesis and sperm production. As a result the natural history of aging effects on spermatogenesis and steroidogenesis need to be considered separately while recognizing that they operate in a joint physical and regulatory environment.

Methodological Considerations for Study of Male Reproductive Aging

Whether the late-life changes shown by all individuals to some degree, can be partitioned into aging as a distinct process distinguishable from the accumulation of specific diseases and concomitants of prolonged existence remains a major and contentious theoretical issue. Consequently, the natural history of the aging of the reproductive system can only be studied meaningfully in animals subject to good general health and environmental conditions. In effect, life expectancy can only be considered to approach its natural limits where there is no significant impact from non-age–related factors such as predation, accidents, disease, or malnutrition. Apart from humans, only domesticated animals achieve this kind of life so studying the effects of aging on the reproductive system in feral animals is at best opportunistic and of limited value. Studies such as of wild-caught animals are unlikely to manifest true aging changes by those criteria. Nor is aging often studies in captive-bred animals such as domestic animals and rodents as the costs of the necessarily prolonged maintenance are generally counter-productive to the original economic purpose for which domestication was forged. This review will therefore focus on aging of the human male reproductive system with animal data contributing where illustrative rather than aiming to provide a comprehensive account of aging in all, or even mammalian or vertebrate, species.

Key methodological issues for studies of human reproductive aging include the choice of cross-sectional vs. longitudinal cohorts, recruitment of representative populations, study size and power, and the handling of co-morbidities. While longitudinal, population-based studies remain the gold standard for public health research, such cohort studies of aging would require generations to answer pressing questions. Hence, telescoping timescales into more tractable dimensions is vital for making progress in analytical aging research. Fortunately, in many instances cross-sectional studies crossing a range of ages can be interpreted as a quasi-longitudinal design, so long as the population characteristics under study and their determinants are relatively stable over time, otherwise significant period-cohort effects become major confounders (10,11). The recruitment of representative study cohorts is vital if findings are to be usefully extrapolated to the general populations such as the source from which study participants are drawn or to other populations. Validity in this setting assumes not only a generally stable population free of major temporal trends in both the characteristic being studied and its major determinants, known and unknown, but also a high participation rate free of (dis)incentives that influence study endpoints. In fact, numerous types of recruitment and participation biases impact on such cohort studies with consequences varying from minor caveats to complete invalidation (10–12). By contrast, in experimental research randomization into groups theoretically ensures balance between both known and unknown influential factors; it is therefore preferable for providing robust answers to well specified questions. But for observational research the search to identify, prevent, eradicate, and adjust for bias remains an open, if not endless, task. Another key validity issue for aging research studies is the optimal handling of co-morbidities, the concomitant diseases that accumulate during aging, particularly if they represent biological confounders because they also influence the study outcomes. Two polar approaches, of exclusion or adjustment, can be considered complementary, although in practice they are largely if not entirely mutually exclusive reflecting a forced choice that must be made. The exclusion of co-morbidities risks weakening study generaliability for an elite population while artificially inflating apparent precision. Alternatively, post-hoc analytical adjustment for co-variables may be difficult due to lower power and uncertainty on the effectiveness of adjustment. These issues remain pivotal for design of prospective studies of reproductive aging and require judicious balance among power, precision, feasibility, and utility.

Another methodological consideration is the difficulty of studying events on a long timescale such as human lifespan or generations. Inevitably, to render such questions tractable, some telescoping of the timescale must be undertaken by introducing surrogate variables. This convenience comes at the cost of creating uncertainty about the validity and especially the temporal extrapolation of such surrogate variables for life expectancy. In addition, the cross-sectional study represents a temporal snap-shot of longitudinal processes that may or may not be in steady state. If the longitudinal processes are in relatively steady-state, then the more tractable and convenient cross-sectional studies may reasonably represent a static snap-shot of the longitudinal process. However, if underlying temporal trends are not at steady state, then the risk of period-cohort confounding arises. In aging studies, this refers to confusion between effects due to the epoch of birth with the effects of aging per se.

Overview of Male Reproductive Function in Young Men

Understanding the impact of aging requires familiarity with the overall functioning of the reproductive axis of healthy young men (for more details, see Chapters 17–25).

The testis serves two physiological functions, spermatogenesis and steroidogenesis, vital for reproductive function and somatic health. Spermatogenesis, the formation of mature male gametes capable of fertilization, is the functional activity of the germinal epithelium; whereas steroidogenesis, the production by Leydig cells of testosterone, is essential both for male sexual function, as the delivery mechanism for the male gamete, as well as for anabolic status of numerous somatic tissues. Accordingly, both testicular functions have well-entrenched biological mechanisms secured by strong evolutionary drive.

The seminiferous tubules are composed of a multilayered germinal epithelium embedded in a scaffold of tightly linked Sertoli cells. Each germinal cell is embedded into, and enshrouded by, the cytoplasm of a Sertoli cell with the whole tubule encased by a layer of peritubular myoid cells situated on a basement membrane. Each seminiferous tubule forms a long, supercoiled loop that is folded into lobules with the ends of each tubule plugged into the rete testis. A convenient functional feature is that, because the seminiferous tubules form the bulk of testicular parenchyma, testis size is a convenient, noninvasive index of spermatogenesis. A terminal specialization of Sertoli cells lining of the tubules forms a one-way valve to ensure forward movement of spermatozoa borne on tubular fluids into the ductular system, propelled by the contractile activity of the peritubular myoid cells. Fully formed but functionally immature sperm undergo maturation and storage in the epididymis and vas deferens where they await ejaculation.

Steroidogenesis occurs in the Leydig cells located in the interstitial space between tubules resulting in the production of testosterone, the major androgen of mature male mammals. The principal regulation of testosterone is by the pulsatile pattern of luteinizing hormone (LH) secreted from pituitary gonadotropes, under tight control by gonadotropin-releasing hormone (GnRH) secreted episodically by hypothalamic neurons into the pituitary portal system. Paracrine regulation of Leydig cell testosterone biosynthesis and secretion represents the fine tuning by local testicular factors from the seminiferous tubules. The close anatomical co-localization reflects that these are functionally separate but inter-dependent compartments of the testis. Although linked, the operation and aging of both exocrine and endocrine compartments of the testis are largely independent of each other. In this review they will be considered separately while recognizing that such analytical convenience should not detract from the overall picture of a well-integrated system. Previous excellent reviews of male reproductive aging (13–15) should be consulted for additional historical details or for a wider zoological survey of vertebrate aging (16).

AGING AND THE TESTICULAR EXOCRINE SYSTEM

Male Fertility

Paternity at an advanced age is well known from older celebrities fathering children, including Pablo Picasso (68 y), Marlon Brando (70 y), Rupert Murdoch (72 y), Charlie Chaplin (73 y), Tony Randall (79 y), Anthony Quinn (81 y), and Saul Bellow (84 y), as well as from published (17) and anecdotal (14) reports of fathers in their 90s. Yet, despite male

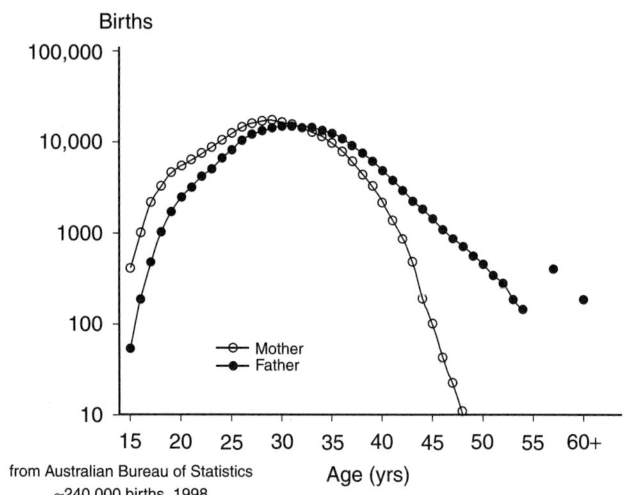

FIG. 1. Plot of births against the age of parents from Australian Bureau of Statistics for ~240,000 births in 1998. Note logarithmic scale on y-axis and that <1% of births are to fathers older than 50 years of age. Similar data has been reported by Rolf and Nieschlag (18) for Germany and Japan (see text).

fertility persisting theoretically until death, <1% of all births are fathered by men over the age of 50 years even in countries where life expectancy exceeds 70 years of age (Fig. 1); see also Fig. 6 in (18). This reflects the fact that couples are mostly close in age and that human female fertility declines steeply with age (19–21) measurably from the age of 30 years (20) and ceasing abruptly by menopause around mid-adult life. This domination of couple fertility by the sharp decline of female fertility can overshadow more subtle contributions from reduced fertility of older men at a population level.

Male fertility is primarily a demographic variable measured by counting conceptions resulting in pregnancy, miscarriage, or birth. The complexity of a biological variable measured for one person by observation of an event occurring in another contributes to the paucity of sound research on this topic, including the effects of age. It is a common fallacy that male fertility can be measured by counting sperm. In fact, sperm output is only a surrogate variable of limited value, especially at a population level (22–25). This is because it is virtually impossible to obtain semen samples from representative samples of men as very few (5%–20%) randomly selected men will participate in studies requiring semen samples to be provided by masturbation. Very low participation rates create potentially serious bias due to participation of subsets of men with concerns or prior knowledge of their fertility.

Demographic and anthropological studies have focused almost exclusively on female fertility (19,26–28), apparently because of simplistic beliefs that only female factors determine population fertility or that paternity is too difficult to determine (29,30). Quantitative methods to estimate male fertility such as the standardized fertility ratio (31,32) based on birth rates or time to pregnancy (33,34) formalizing older demographic methods (27) have been developed but so far have rarely been used to evaluate effects of aging on male fertility (35). The systematic neglect of male fertility, in general, means that few studies are available (36–38), although more focus is developing (39–41).

Ideally, demographic or anthropological studies of male fertility require data from a stable, healthy, noncontracepting ("natural fertility") population where the influence of wife's age and marital duration can be controlled. In practice this largely restricts data to historical populations, such as the classical study of Anderson. She used the 1911 Census of Ireland where sufficient time had elapsed from the 19th century potato famines for restoration of adequate nutrition, while the mass emigration of unemployed younger Irishmen allowed for an unusually wide range of husband's age at marriage. She observed a moderate negative effect (~10%–20% per decade) on annual birth rate as husband's age increased from 40–45 to over 60 years of age (Fig. 2) at fixed marital duration consistently across different strata of wife's age (36). These findings are corroborated by studies showing male age effect on marital fertility in mid-reproductive age (Fig. 3), although toward the end of female reproductive life these effects are overshadowed by female age and marital duration (37,38,42). Using modeling techniques with suitably detailed data from an African tribal village, specific paternal age effects were clearly identified (42). This showed that male fertility effects can be estimated even in a relatively small modern natural fertility population in developing countries. Such observational studies, however, cannot distinguish the cause of the modest but consistent reductions in male fertility with age. These could be due to age-related reductions in sperm numbers and/or function as well as voluntary restraint of fertility, reductions in coital rate or paternally mediated defects in embryo viability, and miscarriage (43). Dissecting these contributions will require further study including more sophisticated modern methods such as use of valid surrogate variables.

Although recent well-designed prospective studies of couples planning a pregnancy in Europe (44), the U.K. (45), and the U.S. (46) are providing information on male fertility, these studies are inevitably restricted to men mostly under 50 years of age. Hence, these studies only provide a modest advances in knowledge of the effects of older age on human male fertility. A large prospective study of nearly 15,000 pregnancies in Avon, U.K., showed a significant effect of older father's age on the time for wives to become pregnant even after correcting for other fertility factors including wife's age, parity and smoking (45). In this study, men younger than 25 years had a

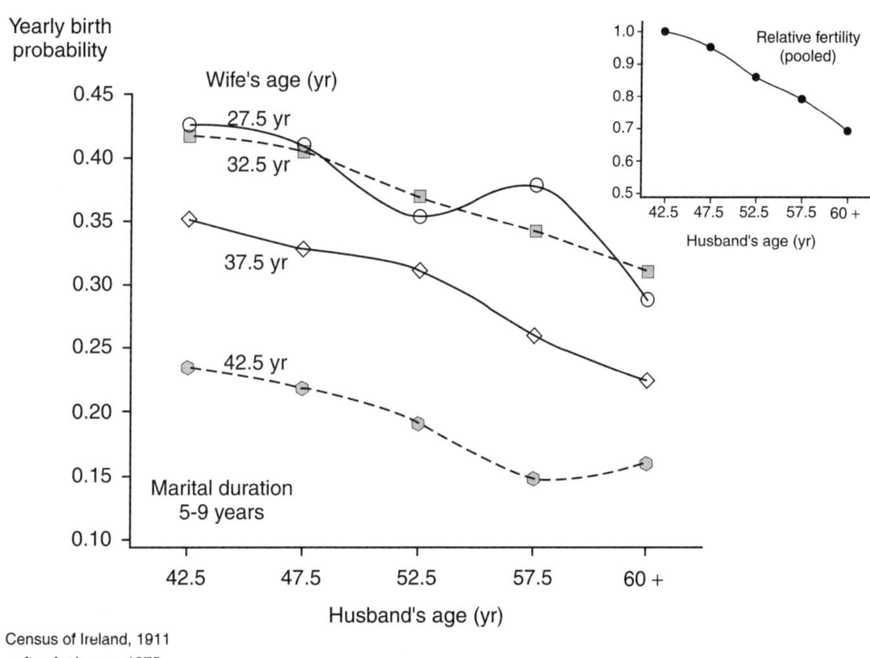

FIG. 2. Plot of annual birth rates according to age of husband from the 1911 Census of Ireland plotted separately for different strata of wife's age between 27.5 and 42.5 years. Inset is the pooled fertility of all husbands according to age across all wives ages. [Data from Anderson, 1975 (36).]

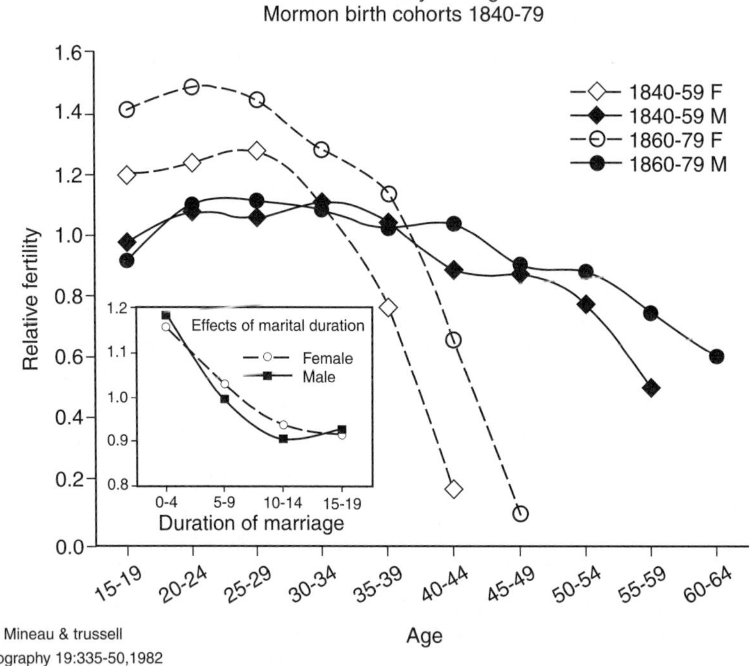

FIG. 3. Plot of male fertility among Mormon birth cohorts in the 19th century, showing effects of male age on birth rates in two different cohorts and distinct from the effects of marital duration (*inset*). [Data from Mineau and Trussell, 1982 (37).]

50% shorter waiting time to conception compared with men older than 35 years (Fig. 4). These findings were corroborated in a study of 2,112 pregnancies in Hull, U.K., which reported a longer time to pregnancy for fathers older than 45 years compared with those under 25 years, an increase consistent across all wife's ages (35). By contrast, two prospective studies of male fertility among 430 first pregnancy planners in Denmark (44) and 210 U.S. couples planning a pregnancy (46) did not identify a male age effect on conception. These discrepancies are most likely due to the smaller sample size and lower power of the latter studies. Alternatively, design differences may lead to systematic differences in outcomes between the studies. While the larger studies can only estimate time to pregnancy from women who became pregnant, their omission of sub-fertile couples leads to a higher estimate of fertility than the smaller studies of pregnancy planners. By contrast, the latter studies omit unplanned pregnancies, thereby eliminating the most fertile couples and producing lower estimates of fertility. Therefore, it remains unclear if the modest decline in male fertility among men under 50 years of age is genuine or reflects subtle limitations in the ability to fully adjust for the highly correlated effects of wife's age and its strong influence on couple fertility.

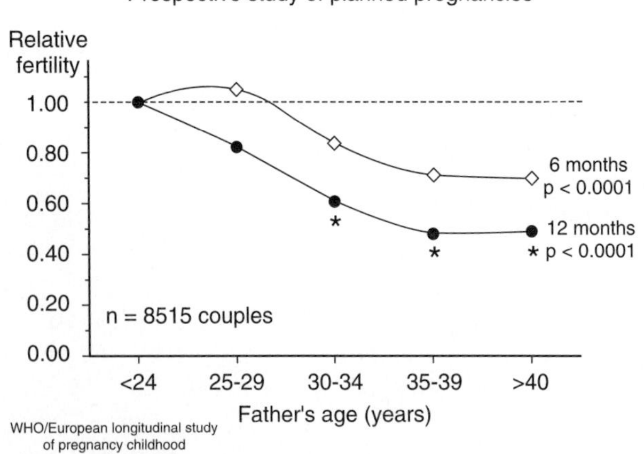

FIG. 4. Plot of time to pregnancy at 6 or 12 months of observation from the ALSPAC prospective study of pregnancy planners showing a male age effect on fertility. [Data from Ford et al., 2000 (45).]

Evaluating age effects on male fertility assumes a relative steady state over recent generations in the relevant male population. If, for example, male fertility was changing at a population level over generations, apparent age-related changes among older men in a cross-sectional study might misattribute to aging those effects actually due to the underlying temporal trends. The available limited evidence suggests male fertility is improving in developed countries. For example, a U.K. national study of time-to-pregnancy shows improved male fertility (40) while a Swedish national fertility study indicates that infertility has declined markedly over recent decades (47), even after discounting truncation bias (48). These observations provide little support for controversial claims that human male fertility might be declining over recent decades (49,50). These claims attribute changes to global pollution by chemicals with estrogenic effects ("endocrine disrupters") that might compromise male fertility by damaging prenatal testis development (51). The hypothesis that prenatal estrogen exposure impairs subsequent male fertility was refuted by a careful follow-up study of a 1950s randomized controlled trial in which women were treated with large doses of diethylstilbestrol (DES) or placebo to prevent pregnancy complications (52). Men prenatally exposed to DES, were more likely to have fathered a child than those of placebo-treated mothers (53). Hence, although the paucity of reliable demographic data precludes firm conclusions, there is little evidence for any decline in male fertility over recent decades.

An additional aspect of male reproductive aging is the increased genetic risk of fatherhood at later age (54,55). Although less well known than maternal age effects on trisomy 21 (56), paternal age effects are observed for a variety of monogenic, most notably autosomal dominant disorders, such as achondroplasia, Apert's, Marfan's, Waardenburg's, Crouzon's, and neurofibromatosis syndromes and X-linked recessive disorders like hemophilia and Duchenne's muscular dystrophy (54,56). In addition, some complex polygenic disorders such as schizophrenia and other psychotic disorders show an excess of older paternal age (57,58), although the responsible mechanisms are not well defined and might include genetic, epigenetic, or environmental factors. Similarly, the risks of reproductive failure due to post-conceptional mechanisms such as miscarriage are also reportedly increased with paternal age in some (59–62) but not all studies (63). The responsible mechanisms may overlap with genetic risks as a paternal age effect in one study (64) was partly due to consanguinity (65); other factors operating remain unclear (66), not least that female and pregnancy factors may not have been fully accounted for.

Sperm Output

Human sperm output is measured by semen analysis which provides measurement of semen volume, sperm density, and total output as well as variables reflecting some aspects of sperm function (sperm vitality, motility, and morphology). Such laboratory techniques remain mainly based on manual methods, although they can be standardized by well-established laboratory methods based on the WHO Semen Manual (67), technician training, and quality control (68,69) with expected difficulties for any manual method in achieving uniform standards (70). While semen analysis is a valuable laboratory tool for investigating infertility of couples, it is a poor epidemiological tool as a surrogate for male fertility in population studies due to its low acceptability, as masturbation is needed to collect samples (23–25). Men's reluctance to provide semen samples unless actively concerned, positively or negatively, about their fertility constitutes an inevitable bias as typically fewer than 20% (71) and as low as 5% (25) of unselected young men will provide semen samples. Such severe selection and participation bias makes valid extrapolation as a representative sample difficult or impossible (23–25). The limited acceptability of semen analysis for older men unconcerned about infertility or limiting their fertility no doubt explains why the very few opportunistic studies of male reproductive aging requiring semen samples are small and population unrepresentative.

Among men younger than 50 years, age has minimal or no consistent detrimental effect on sperm density among healthy noninfertile men (72–74). However, these studies of noninfertile men represent convenience samples of healthy older men recruited from fathers of pregnant women attending an antenatal clinic (72) or from men screened as potential sperm donors (73,74). As such, they may not be representative of the general male community (23,24). Although semen findings from infertility clinics are even less reliable due not only to similar bias from selective recruitment, they also include an unknown proportion of men with defective spermatogenesis. Nevertheless congruent findings have been reported with only minimal effects of age among younger (<50 year) men (75–78).

Few studies have examined sperm output of men older than 50 years (79). Several historical sources indicate consistently the persistence of sperm production to an advanced age in some men. Exner [quoted in Blum (80)] reported that among 165 men over 60 years, sperm were present in the ejaculates of 68.5% of men aged 60–79 years, 59.5% of men aged 70–80 years, and 48% of men aged 80–90 years. Engle reported that more than half of testes and ducts of men over 70 years contained "abundant spermatozoa" (81). Belonoschkin reported sperm present in prostate massage fluids of 55 of 83 "previously healthy" men over 60 years of age (19/26 aged 60–69, 31/45 aged 70–79, 4/11 aged 80–89, and 1/1 aged 92 years) (82). A German study reported that among 463 presumably infertile men older than 50 years, sperm density increased with age up to age 70 years (203 men 50–54 y, 90.8 M/mL; 125 men 55–59 y, 134.4 M/mL; 74 men 60–64 y, 143.3 M/mL; 40 men 65–69 y, 180.2 M/mL) with a decline thereafter (14 men 70–74 y, 62.3 M/mL) (83). An Italian study of 84 generally healthy older men aged 45–81 years,

29 (median age 53 y) provided semen samples with an overall median semen volume (2.5 mL), sperm density (40 M/mL), and total sperm output (108 M per ejaculate), and 25 samples had a sperm density ≥10 M/mL (84).

More formal studies have recruited healthy older men by advertising for vigorous grandfathers (85), from an occupational cohort (86) and from prostate cancer screening (87). None of these convenience samples can be considered population representative and all were relatively small including 23, 42, and 55 older men, respectively. The two larger studies reported that older age was associated with unchanged sperm density but a modest reduction in semen volume and total sperm output (86,87), whereas the smaller study reported increased sperm density and unchanged semen volume (85). If human sperm output was decreasing globally over the last five decades as claimed (49,50), this could influence studies of sperm output of older men. However, evidence for any global decline in sperm output is weak (88–90) and sperm output of domestic animals shows no predicted decline over these decades (91). These observations make it unlikely that observed age effects might be due to temporal trends in human sperm output.

Aging is associated with impaired sperm function manifest as reduced sperm motility and, more frequently, abnormal sperm morphology (85–87), whereas sperm penetration of hamster eggs was unaffected in one study (85). In concert, these limited studies indicate only modest age effects in reducing sperm output with likely minimal consequences for fertility of older men. The small sample size and restriction to elite healthy, potent older men are caveats on interpretation. It remains unclear if these effects are sufficient to contribute significantly to the decline in male fertility among older men described in demographic and anthropological studies in which female subfertility and/or decline in sexual activity with longer marital duration (37,92) appear to be responsible for substantial reduction in couple fertility. Larger studies would be desirable to clarify apparent age effects on human sperm output, function and their determinants.

Testicular Blood Vessels

Vascular changes in the testis microcirculation are particularly important in the pathogenesis of age-related changes in testis structure and function. An elegant study by Regadera et al. correlating angiographic and histological appearances in postmortem testes from 60 men aged 58–88 years compared with 20 young men who died in vehicle accidents showed that regions of arterial nonperfusion corresponded with areas of seminiferous tubular degeneration and fibrosis (93). These findings extended previous observations based on correlating testicular histology with inferred vascular distribution patterns (94) and of micro-angiographic perfusion (95). Such age-related vascular changes in the testis which reduced oxygenation within the tubules where prevailing oxygen availability is marginal, may foster increasing fibrosis of the tunica propria and tunica albuginea as part of a vicious circle resulting in accelerated oxygen deprivation and increasing sclerosis (96). In concert, these findings indicate that vascular defects due to systemic atherosclerosis are an important determinant of age-related degeneration in testicular structure and function further described below.

Testis Size

Biological markers of spermatogenesis would be useful surrogate variables to study age effects on male fertility, given the limitations of demographic estimates of male fertility or of population-based estimates of semen analysis. Testis size is a simple, noninvasive surrogate marker for spermatogenesis because seminiferous tubules comprise the bulk (>80%) of testis volume in mammals (97,98). Testis size can be readily determined from painless, physical examination in vivo as well as from testes excised surgically or at postmortem. Noninvasive measurement of testis size in vivo is most accurately performed by ultrasound measurement of its maximal dimensions allowing calculation of volume as an ellipsoid (99). Orchidometry is a simpler alternative to ultrasound whereby testis size is manually compared against a standard series of solid ellipsoids with known volume. For experienced observers, orchidometry provides a reasonable approximation to sonographic volume, although the between-observer variability is higher (100,101). Measurement of testis size has long been utilized clinically because testicular atrophy, reflected in both reduced volume and palpably increased compressibility, is a reliable indicator of impaired spermatogenesis (102,103) and analogous measurement has long been used in veterinary and agricultural practice.

Ultrasound measurement of testis size has been reported in only one study of healthy younger noninfertile men (99) as well as a single study of older men in which testis volume was decreased by 31% among 115 community-dwelling elderly men aged 76–81 years compared with 42 younger men aged 23–31 years (104). No longitudinal studies of aging effects on human testis size have been reported.

Older studies of testis size across a wide age range have measured testes excised from unselected postmortems or surgery on sex offenders, transgender, or orchidectomy for prostate cancer (105–110). Those with quantitative data showed testis size was unchanged with age (108,109), although one showed reduced testis size in unselected men over 60 years but no change when only testes with normal histology were examined (107). Clinical studies using orchidometry reported unselected older men had reduced testis size (111–113), whereas selected healthy older men had little or no reduction in testis size (85,114,115).

An alternative approach has been to utilize selected postmortem measurements of testis size from men dying suddenly due to accidents, trauma, or otherwise unexpectedly. This selected population of autopsies has the advantage of representing a quasi-random sample of the population. One study of over 1,000 consecutive Australian postmortems, including 857 from men dying suddenly, observed that age and mode of death had a significant and independent influence on testis size (116). Men with chronic, terminal illness or who were malnourished at death had reduced testis weight, whereas diabetes, drug abuse, or pelvic injury had no effect. After adjusting for mode of death, age was associated with reduced testis volume only after 70 years of age. A study of over 20,000 selected forensic postmortems in Tokyo during the 20th century (117) partly confirms findings on age effects. However, as case selection included samples from men with terminal illness and older men only from early decades of the century (118), the Japanese study had confounding between age, period, and terminal illness effects.

Spermatogenesis

Evaluating human spermatogenesis requires access to testicular tissue so the few studies of aging are limited to opportunistic samples from biopsies of living older men or postmortem specimens. The former are available only from highly selected populations, such as men undergoing infertility evaluation, transgender surgery, or orchidectomy for advanced prostate cancer. Such restrictions on the sampling limit the ability to study aging and also overrepresent men with underlying disorders or treatments that impair spermatogenesis. As germinal epithelial damage usually manifests a stereotyped histopathological response, effects of aging may be difficult to distinguish from those due to other common causes of impaired spermatogenesis. Furthermore, testicular biopsies obtained from unfixed testes of living men involve excising small fragments of tubular tissue extruded under pressure from a nick in the tough testicular capsule. This sacrifices natural spatial tubular organization further hampering the identification of specific changes, if any, related to aging.

Consequently, the most reliable quantitative information on aging effects on spermatogenesis arises from relatively small but detailed histological studies of postmortem fixed testis (119). Using an approach developed by Kaler and Neaves (120), the testis is rapidly fixed to avoid artifacts from postmortem autolysis. In addition, these studies focus on men who die suddenly, providing a unique insight into the effects of aging by ensuring correct in situ fixation while avoiding the selection biases of testicular biopsies. These studies calculate daily sperm production rates from numbers of each germinal cell

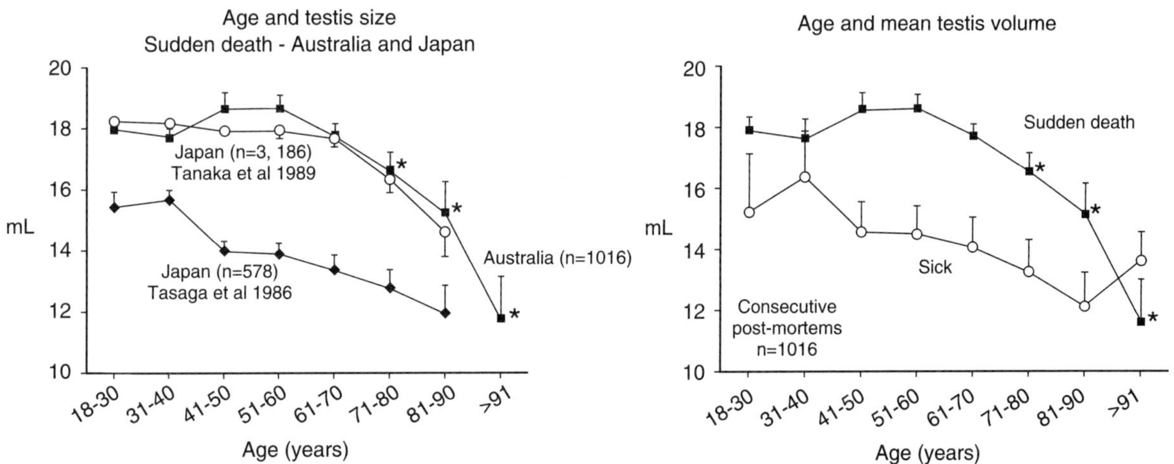

FIG. 5. Plot of testis size against age from two Japanese (118,488) and an Australian (116) postmortem study of men dying suddenly (*left panel*) and of the effects of sudden death compared with dying with a chronic terminal illness or malnutrition (Sick) from the Australian postmortem study.

population by assuming fixed times for each stage of spermatogenesis (121,122). While these time estimates based on limited data are considered axiomatic for healthy men, their validity has not been extensively verified especially for pathological states such as aging, aberrant hormonal states or following germinal cell damage.

Daily sperm production rate was consistently decreased in older men (123,124). The reduced sperm production rates with aging are attributable mainly to increased germ cell loss during premeiotic stages (125) but did not correlate with Leydig cell numbers (124) reflecting a functional excess of these steroidogenic cells at younger ages with the age-related attrition only becoming limiting when a very low threshold is crossed (126). Other quantitative facets of age-related changes in seminiferous tubular structure and function include shortened seminiferous tubular length (126,127), thickening of intertubular (127,128) and capsular (123) tissue, and reduced numbers of Sertoli cells (129), Leydig (120,124,126), myoid (119), and other interstitial (130) cells. Despite the reduced numbers, the functional capacity of remaining nongerminal cells appears intact in aged testes as both the Sertoli cell carrying capacity for germ cells (129) and Leydig cell morphology (120,124,126) appear largely unaffected. Similarly, the average time for sperm to traverse the epididymis is unaffected by age per se, although it is faster with higher daily sperm production rate (131). There is, however, an increased prevalence of pathological features, such as acute and chronic focal inflammatory changes within the parenchyma of testes of older men.

Qualitative information on aging effects on human testicular histology is available from earlier studies (105–110) and has been reviewed in detail (132). Persistence of complete spermatogenesis into the 8th decade of life is reported (106) but more usually there is generally reduced spermatogenesis with tubular lumen dilatation, thickening of the tunica propria (the tubular boundary), a higher prevalence of generalized or focal tubular sclerosis (133), premature exfoliation (sloughing) of germ cells (96), and the appearance of pathological multinucleated germ, Sertoli, and Leydig cells probably due to nonspecific cellular fusions. These nonspecific changes reflect the stereotypical pathological responses to damage or hormone deprivation of the mature testis. Among the characteristic changes associated with aging is the greater variability of histological appearance (106,110). Among the most prominent changes is thickening of the tunica albuginea (123), a feature so consistent it was proposed as a metric for estimating testis age (134). This thickening is due mainly to increases in the middle and inner layers containing more desmin-positive myoid cells (135,136). An alternative categorization of testes from older men, according to whether spermatogenesis was complete or reduced, found that some men had complete spermatogenesis at the most advanced age with persistence of spermatogonial proliferation (137), an increasing proportion of testes showed involuted spermatogenesis by decade of age (127). By the 9th decade even testes with complete spermatogenesis had reduced overall germ cell numbers consistent with the onset of intrinsic age effects on spermatogenesis (116). All germ cell populations decline with age but the greatest decreases occur in premeiotic rather than in post-meiotic germ cells (125,138).

Sertoli Cells and Hormonal Regulation of Spermatogenesis

Sertoli cells have a pivotal role in spermatogenesis (see Chapter 19 for details). These somatic cells regulate, nourish, and embed within their cytoplasm all germinal cells once they become committed to and commence spermatogenic development. Tight junctions between neighboring Sertoli cells form a scaffold for the seminiferous tubules and also create a diffusion-tight, blood–testis barrier forming a highly distinctive fluid environment in the interior adluminal compartment of the tubules where most germinal development takes place. In the healthy mature testis, germ cell-to-Sertoli cell ratios are relatively invariant for any species, consistent with the concept that each mature Sertoli cell has a fixed carrying capacity for germinal cells. Thus, loss of Sertoli cells would dictate a proportionate decrease in germ cell numbers and ultimate output of mature sperm. Consequently, structural or functional effects of aging on Sertoli cells could have numerous and diverse effects on spermatogenesis.

Identifying structural changes in human Sertoli cells due to aging requires access to well-fixed testis histology from representative populations of older men, a constraint that limits the available studies and provides caveats on their interpretation. The most informative studies, from rapidly fixed postmortem testes from men dying suddenly, show a consistent decrease in Sertoli cell numbers with progressive age among older men (127,129). In these studies, the degree of preservation of spermatogenesis (and testis size) is directly proportional to numbers of remaining Sertoli cells, indicating that Sertoli cell loss is an important factor in determining age-related changes in spermatogenesis. This age-related Sertoli cell loss seems likely to be due to testicular ischemia but other contributory causes cannot be excluded. The remaining Sertoli cells demonstrate increased morphological diversity with appearances ranging from normal to immature and features reminiscent of vascular or hormonal deficits. Sertoli cell cytoplasm features increasing lipid and vacuolar inclusions due to phagocytosis of exfoliated degenerating germ cells (139) as well as features of regression to an immature state (140). The latter feature may be responsible for the premature exfoliation of germ cells, a morphological consequence of functional gonadotropin insufficiency (141). These features suggest suboptimal function of the remaining Sertoli cells.

The complex functions of Sertoli cells are difficult to study comprehensively in vivo. Furthermore, the scarce availability of suitable human tissues for in vitro studies means that few functional studies of aging human Sertoli cells are available. A key facet of Sertoli cell function is the fact that they represent the site of hormonal control of spermatogenesis as they express both follicle-stimulating hormone (FSH) (142) and androgen (143–145) receptors, whereas germ cells express neither and spermatogenesis can develop in the absence of germ cell androgen (146,147) or FSH (148,149) receptors. This raises the possibility that aging may affect the Sertoli cell's hormonal regulation of spermatogenesis either through direct effects on Sertoli cells and/or indirectly via effects on gonadotropin secretion.

Androgen stimulation of Sertoli cells is both necessary and sufficient for the initial induction of spermatogenesis (150–154). As a result, an age-related decrease in testicular testosterone production might eventually lead to reduced spermatogenesis, although it is notable that the maintenance of spermatogenesis requires much less testosterone than does initial induction (155). Although human Leydig cell numbers decrease with age (120), sperm production is only impaired when fewer than 10% of Leydig cells remain compared with healthy young men (120,126). This latter observation is quite consistent with evidence from basic (155,156) and clinical (157,158) studies indicating that a comparably low intratesticular testosterone is sufficient to maintain spermatogenesis, although complete androgen withdrawal will fully suppress the germinal epithelium. Why the normal testis has such a surplus of testosterone and in what form it is stored remain intriguing unanswered questions. In practical terms, however such major decreases in testicular testosterone are rarely apparent in older men due to aging alone and crossing such a low threshold requires additional testicular pathology either due to direct damage or indirectly due to gonadotropin deprivation independent of aging per se.

FSH is the second major hormonal stimulus to spermatogenesis (159). Whereas FSH was once believed to be indispensable for induction of mammalian spermatogenesis (160), more recent studies in mice show that qualitatively complete spermatogenesis can be achieved in mice completely lacking FSH due to genetic defects in GnRH 151, FSHβ subunit (161), or the FSH receptor (148,149). In these settings, normal spermatogenesis and fertility occur within a smaller (30%–60%) testis than normal, a gap that can be rectified by FSH replacement through genetic rescue (162–164) or neuronal transplantation (165,166). These findings indicate that the primary role of FSH may be in promoting the proliferation of Sertoli cells during early life prior to sexual maturation, thereby completing the stock of mature cells so as to allow full testis development in maturity. Additional secondary effects of FSH also interact with testosterone to ensure quantitatively complete spermatogenesis. Primate studies indicate some species differences in degree of FSH dependence of spermatogenesis with greater dependence on FSH for induction of spermatogenesis than in rodents (159). On this background, the aging male reproductive system is likely to be relatively impervious to functional loss of FSH in later life with the major impact being manifest on the later meiotic and postmeiotic stages that are androgen-sensitive.

It has long been established that blood FSH concentrations increase progressively with age in older men (85,112,113,115,167,168), although retaining a negative correlation with testis size and, by implication, with spermatogenic efficiency. The magnitude of the rise in FSH in older men varies between studies but is usually to levels less than after orchidectomy, especially in more recent studies using specific two-site FSH immunoassays. The complex carbohydrate side-chains normally present on each FSH subunit creates a series of FSH isoforms in the pituitary and blood with variations in the circulating half-life and biopotency of each FSH isoforms and hence to overall FSH bioactivity. Using an FSH bioassay (169), it is therefore possible that systematic changes in FSH microheterogeneity with aging might systematically change glycosylation side-chain patterns and thereby overall FSH bioactivity. However, from studies using FSH bioassays compared with immunoassays performed on the same blood samples, there is little evidence that bioactive FSH is altered disproportionately with aging (170–173). Hence, the age-related rise in blood FSH is most plausibly interpreted as being a passive reflection, via negative feedback, of a progressive failing of seminiferous tubular function with aging. Verification of this concept is, however, constrained by the limited direct evidence involving sperm production rates of older men.

Sophisticated neuroendocrine studies indicate that the rising FSH secretion in older men is attributable to increases in both basal as well as pulsatile FSH secretion (173–175) with the regularity of FSH secretion being largely preserved with aging (175), consistent with it predominantly reflecting deteriorating Sertoli cells function and spermatogenesis rather than a central hypothalamic mechanism of aging. More recent studies have added evidence that blood concentrations of dimeric inhibin B, a Sertoli cell product known to represent a good marker for Sertoli cell function in young men (176,177), decline with age in older men (104,178–181). Detailed studies suggest that dimeric inhibin B remains a good marker of overall Sertoli cell activity into old age as its correlation with testis volume and inversely with blood FSH remain equally strong (104,179). These findings suggest that blood inhibin B may correspond best to Sertoli cell number rather than function. Earlier studies using an α subunit directed inhibin immunoassay less specific for the bioactive dimeric inhibin B (182,183), provide supportive evidence of the fall in inhibin with age, and its suitability as a marker of overall Sertoli cell activity regardless of age (184–190). Blood levels of another Sertoli cell product from the TGFβ superfamily of proteins, activin A, also increase with aging in older men (180,191,192). Although this homodimer of the inhibin βA subunit demonstrates potent stimulatory effects on FSH secretion in vitro, its major role appears to be in paracrine or autocrine rather than endocrine signaling (193) so the pathophysiological significance of these findings in aging men remains unclear.

Beyond the hormonal regulation of spermatogenesis, the numerous additional functional aspects of the aging Sertoli cell remain harder to study and are consequently much less well understood (194–198).

AGING AND THE TESTICULAR ENDOCRINE SYSTEM

Testosterone Production and Leydig Cells

Steroidogenesis, the other major function of the hypothalamo–pituitary–testicular axis, is geared towards testosterone production by Leydig cells. Located in the interstitial (inter-tubular) space of the testes, Leydig cells actively secrete testosterone during three stages of life—the fetal, perinatal, and post-pubertal periods. During fetal life, Leydig cell testosterone production is gonadotropin independent in rodents (199) while in primates it is stimulated by chorionic gonadotropin (200), an LH homolog secreted from the placenta only in primates (201–204). After fetal life, Leydig cells are tightly regulated by LH secreted episodically from pituitary gonadotropes under burst-like stimulation from hypothalamic GnRH (205–207), including during aging (208). This results in a characteristic, highly pulsatile pattern of blood LH concentrations during the perinatal period and throughout adult life. The tight regulation of testicular testosterone output by pulsatile LH secretion means that changes in hypothalamic–pituitary regulation of pulsatile LH secretion could make a prominent contribution to the age-related decline in testosterone output.

It is now well established that blood testosterone concentrations in men are reduced progressively with age from mid-life onwards (209). The descriptive epidemiology of the aging testicular endocrine system can be considered in four epochs corresponding to the increasing sophistication of their clinical and epidemiological studies. Although blood testosterone was measured with laborious, older chemical methods (15), clinical research studies became feasible following development of steroid immunoassay in 1970 (210,211), using the Nobel prize–winning invention of immunoassay by Berson and Yalow a decade earlier (212,213). The pioneering studies of the endocrinology of male aging examined men across all ages and reported striking decreases in blood testosterone concentrations together with increases in blood sex hormone–binding globulin (SHBG), LH, and FSH (112,113,167,214), findings attributed in a classical interpretation of an endocrine negative feedback system as due to a primary testicular defect. These studies featured convenience samples usually drawing participants indiscriminately from any available source regardless of their health or medical conditions. The second wave of studies was prompted by a reaction to the conclusion that male aging was inevitably associated with a decline in androgen production and featured selective recruitment of healthy aged men (85,168,215). By contrast with the first wave studies, these reports observed minimal, if any, decline in blood testosterone concentrations in the elite healthy older men. As a result, the second wave studies concluded that, as the age-associated decline in blood testosterone was absent or minimal in healthy aging men without major illness or debility, the original reports of age-related decline in blood testosterone was attributable to the diseases that accumulate in unselected elderly men, rather than to aging per se. The third wave of studies recruited more tightly specified samples of older men and found that the age-related decline in blood testosterone was still evident even among distinctive cohorts of community-dwelling healthy older men (216). Finally, the decisive fourth wave of studies developed well-designed cross-sectional and longitudinal epidemiological cohort studies aiming for representative sampling of community-dwelling men across a wide range of ages with defined health status (217–223).

Major methodological reviews have arrived at a consensus that blood testosterone concentrations in men decline gradually with aging from mid-life onwards (209). This decline, of ~1% per year, is progressive with age among groups of men (221–223) with the cross-sectional changes being a reasonably good approximation to the longitudinal changes (223). Nevertheless, for individual men, the changes in blood testosterone concentration on successive measurements in longitudinal studies vary widely without a consistent monotonic trend to decrease with time (224,225). The major known influence on the age-related decline in blood testosterone concentrations is a modest acceleration due to the presence of chronic diseases that accumulate with aging (209).

Testosterone is a nonpolar steroid present in blood at concentrations higher than its solubility in extracellular fluid because it is largely protein bound. Most circulating testosterone (~60%) is bound to SHBG, a hepatic homodimeric glycoprotein with limited, high affinity testosterone binding sites (226–228). The remainder is more loosely bound to serum albumin and a variety of other low-affinity circulating proteins leaving only ~1%–2% of the testosterone in the bloodstream unbound to any protein (229). The latter is the so-called "free" fraction and the combination of the unbound and loosely bound, is termed the "bioavailable" testosterone. It is usually assumed that SHBG-bound testosterone constitutes a biologically inert hormone reservoir, although there is evidence that ligand-bound SHBG can itself bind to cell-surface receptors (230), suggesting SHBG has additional active roles in androgen action. Although immunoassay of blood total testosterone remains the standard measure of testicular testosterone output and of tissue androgen exposure, such derived testosterone measures have become more popular based on the free hormone hypothesis (231–233) which asserts that nonprotein bound, "free" or "bioavailable" testosterone (234,235) may represent a more biologically active fraction of testosterone for its greater accessibility to tissues. Nevertheless, such "free" fractions would also be more accessible to sites of inactivating steroid metabolism so the net effect of whether the "free" fraction identifies a more or less biologically active fraction for any one or all tissues requires empirical evaluation, so far largely lacking. It is notable that rodents have no circulating SHBG (236), further questioning the free hormone hypothesis.

The liver secretes SHBG into the bloodstream so that blood SHBG concentrations are influenced mainly by physiological and pharmacological factors that impact on liver function. These include liver diseases, such as cirrhosis or hepatitis, that increase SHBG or to supraphysiological hepatic exposure to hormones. The latter may arise from oral administration of sex steroids which lead to high, first pass exposure of the liver via the portal route or supraphysiological hormone exposure due to thyroid diseases (thyrotoxicosis, hypothyroidism) or parenteral hormone administration. Blood SHBG concentrations are increased by pharmacological exposure to estrogens or thyroid hormones and reduced by androgens or obesity (237–242).

SHBG concentrations increase consistently with age in virtually all studies of male aging (85,112,113, 167,168,209,214–223,243). Hence its consequences for interpretation of blood testosterone measurement and tissue androgen action warrant careful examination. The consistent increase in blood SHBG with aging excludes the possibility that the age-related reduction in blood testosterone concentrations is a measurement anomaly due to a reduction in blood SHBG such as in obesity (168,239,242,244–249). Conversely, increasing adiposity with age may negate some of the age-related rise in blood SHBG. The rising blood SHBG concentrations may obscure or buffer the fall in testosterone production based on blood total testosterone measurements. Blood SHBG concentration is also a key determinant of whole body metabolic clearance rate for testosterone (250). Hence the rising SHBG with aging would decrease the metabolic clearance rate of testosterone. This counteracts the reduced daily testosterone production rate to better maintain blood testosterone concentrations, although it underestimates the reduced testosterone output in older men. Overall, the consistently rising blood SHBG concentrations in older men means that the progressive decreases in blood testosterone concentrations with male aging must be primarily attributable to reduced LH dependent Leydig cell testosterone production.

Leydig cell attrition with age is estimated as ~1% per annum from early adult life (120). The remaining Leydig cells appear mostly normal but with fewer lipid droplets and more Reinke's crystals (119,251,252). Functionally, the progressive reduction in steroidogenic cells is reflected in proportionately similar age-related reduction in basal (209,223) and in human chorionic gonadotropin (hCG)–stimulated testosterone (85,111,167,168,185) production among healthy aging men. The functional reserve capacity of Leydig cell testosterone synthesis in older men as evaluated by hCG stimulation shows a consistent preservation but quantitative reduction in blood testosterone responses (111,113,168,216,253). These findings have been corroborated by using the more physiological stimulus of recombinant human LH which failed to rectify the dampened testosterone responses of older men during short-term (1 day) studies, despite LH pharmacokinetics being unaffected by age (254,255). This is further supported by observations in older men of impaired Leydig cell testosterone responses to endogenous LH following stimulation by endogenous or exogenous pulsatile GnRH (256–261).

Rodent models have provided insight into mechanistic aspects of Leydig cell aging (262). While subtle effects of aging are evident in interstitial tissues of Sprague Dawley rats at 6 months of age (263), over the last decade the Brown Norway (BN) rat has been more intensely studied for Leydig cell aging. When maintained in disease-free research colonies, it features unusual longevity (>3 years) with relative freedom from obesity and cancers in older age (2–3 years), which confound mechanistic studies of reproductive aging, although the strain is prone to vascular lesions (264) and immune disorders (265,266). No less than in human studies, convenient selection of a rodent model for advantageous features also introduces caveats on interpretation for physiological aspects that may differ from other rodents. For example, while the BN rat shows little reduction in LH with aging, other rodent lines show more consistent lowering of LH. Whether these differences are due to the better housing conditions, superior inherent longevity or strain-specific differences in hypothalamic aging, is unclear. Older studies in various rat strains have shown aging consistently associated with reduced blood testosterone concentrations. However, it remained unexplained whether the primary defect was located in the testis or hypothalamus and pituitary. For example, blood LH concentrations were reported as reduced (15) or unchanged (267) and defective Leydig cell testosterone biosynthesis as being rectified by administration of LH or hCG (268,269) or not (267,270). Hence the nature and functional localization of the major defect in the testis or hypothalamus and pituitary remained uncertain.

Impaired Leydig cell testosterone output in aged BN rats appears attributable to defects in Leydig cell LH receptor signaling. Whereas neither in vitro (271) or in vivo (272,273) stimulation of the LH receptor could rectify these defects, they could be overcome via bypassing the LH receptor cAMP-generating mechanisms with a membrane permeable cAMP analog (274). The functional defect in cAMP generation has been attributed to autogenous mitochondrial generation of reactive oxygen species causing irreversible oxidative damage to signal transduction elements, including membrane fluidity and protein function (275). The functional defects in aging Leydig cell testosterone production are prevented by prolonged suppression of Leydig cell function via sex steroid-induced inhibition of pituitary LH secretion (276), but not caloric restriction (277). Similarly, youthful responsiveness could be restored to Leydig cells following ablation of the resident population of endogenous old Leydig cells with ethane dimethane sulphonate, a specific Leydig cell toxin, that leads to interstitial repopulation with new immature Leydig cells (278). These findings provide striking evidence for age-related defects intrinsic to Leydig cells that cause their own functional demise as a component of the reduction in blood testosterone in aging rodents. A similar mechanism may contribute to the defective function of the surviving Leydig cells in the aging human testis.

Hypothalamo–Pituitary Regulation of Gonadotropin Secretion

Leydig cell testosterone production is highly dependent on physiological LH stimulation. As the age-related decline of blood testosterone is due to reduced Leydig cells testosterone production, the impact of age on pituitary LH secretion is crucial to understanding the functionality of testicular endocrine axis of older men. A monumental series of neuroendocrine studies by Veldhuis have elucidated many aspects of the functional regulation of GnRH, LH and testosterone secretion and its aberration in older men (279). These studies have employed a powerful experimental approach of intensive (10 min) blood sampling over prolonged (usually 24 h) periods to allow detailed characterization of pulsatile LH secretion. They have inferred the patterns of episodic hypothalamic GnRH secretion, its entrainment of pituitary LH and testicular testosterone production with analysis of Leydig cell responsiveness to LH as well as the co-ordination and orderliness of the two hormone pulse trains over time (175,256,257, 261,280). In addition, they have studied responses to perturbations of the closed loop negative feedback system by selective GnRH or stereoidogenic blockers to inhibit endogenous GnRH (254,255), LH, testosterone or its bioactive metabolites (estradiol, dihydrotestosterone) (173,281,282) with or without replacement by exogenous pulsatile GnRH (283,284), LH (254,255), testosterone and its metabolites (174). In concert, these elegant studies have shown that male aging is associated with more sluggish and chaotic operations of the hypothalamo–pituitary–testicular axis with increasing age. Sophisticated mathematical modeling has revealed defects at all levels of the hypothalamo–pituitary–testicular axis in older men including blunted negative testosterone feedback (256), reduced functional GnRH secretion (283), impaired and more chaotic pulsatile LH secretion (282,285), blunted Leydig cells responsiveness to LH (254,255, 257,258,261). In even more intense studies involving sampling every 2.5 minutes for 24 hours, the neuroendocrine defects of the aged male hypothalamus could still be detected when matching basal blood LH and testosterone concentrations (260,285,286). Independent corroboration of an hypothalamic defect in regulation of LH and ultimately testosterone secretion in aging men is provided by the observation that, in healthy young men, the acute stress of prolonged (3.5 days) fasting reduced LH pulse frequency and amplitude, patterns coming to resemble those characteristic of older men among whom fasting had no further effect (287). This versatile array of methodologies has provided concordant findings that the hypothalamo–pituitary regulation of LH secretion in the aging male is distinctively more sluggish, disordered, and disjointed in its secretory responses involving GnRH, LH, and testosterone operating as an ensemble. Although this provides a coherent functional description of the defective hypothalamo–pituitary and testicular dysfunction accounting for the reduced blood testosterone concentrations of older men, the specific neuronal structures and mechanisms responsible for such complex changes remain poorly defined.

Rodent studies provide corroborative evidence for hypothalamic defects in regulation of pituitary LH secretion in aging rats (262,288,289). Strain differences may be important as mean blood LH was reduced in aged male Sprague Dawley rats with reduced LH pulse amplitude and frequency (288,289), whereas in the aged male BN rats mean blood LH did not change with age because the reduced LH pulse frequency was balanced by an increased LH pulse frequency (262). In conjunction with lowered blood testosterone and elevated FSH concentrations, the age-related impairment of BN Leydig cell function does resemble human male reproductive aging in manifesting functional defects at both testicular and hypothalamic levels.

The possibility that the aging human male hypothalamus may have systematic changes in the set points of the negative feedback systems is not yet clearly known. Some evidence suggests that the set point is increased (290), whereas more evidence suggests a decreased (174,256,281,283,291–293) sensitivity to testosterone negative feedback in older men.

Aging and Androgen Action

Androgens have diverse physiological effects on many somatic tissues (294). The androgen receptor is expressed in not only the reproductive organs, but almost ubiquitously in the body's tissues (295,296), apart from germ cells (143–145). Hence, in addition to the classical androgen effects on sexual differentiation, genital development, and induction of spermatogenesis, circulating testosterone is responsible for somatic anabolic effects on many, if not all, tissues and organs. Consequently, the age-related decline in blood testosterone might have impact beyond the reproductive system to involve the anabolic status and function of numerous androgen sensitive somatic tissues.

While the major determinant of androgen action is the blood levels of testosterone, various physiological processes are superimposed to locally modulate androgen effects via pre-receptor, receptor, and post-receptor mechanisms that allows aging to modify the coarse or fine tuning of androgen action. In addition, there is increasing evidence that androgen effects in various tissues are subject to differing dose–response relationships with maximal androgen effects, interpreted as an apparent threshold, occurring at specific endogenous blood testosterone levels or exogenous androgen dosage, which may differ according to age, disease, and even among individuals.

Pre-receptor influence on androgen action theoretically involves three mechanisms: androgen production, activation, or inactivation. Variations in androgen production (LH stimulation of Leydig cell testosterone biosynthesis) or inactivation (via hepatic phase I metabolism by CYP3A4 and related enzymes) can influence blood testosterone levels. As a result, age-related changes in androgen production or inactivation could contribute to the age-related changes in blood testosterone concentrations. By contrast, androgen activation involves two major enzymatic mechanisms governing conversion of testosterone to bioactive metabolites locally within target tissues might be affected by aging independent of changes in blood testosterone concentration. These activation mechanisms involve the metabolism of testosterone to its more potent metabolite DHT via steroid 5α reductase or to estradiol via the cytochrome-P450 enzyme aromatase. By these means testosterone effects are amplified in its effects on the AR or diversified to have effects via estrogen receptors α and/or β. Together, these pre-receptor activating mechanisms broaden and enhance testosterone effects creating distinctive effects in diverse androgen target tissues. There is evidence for reduced steroid 5α reductase in older prostate and other tissues (297–301), but there is no direct evidence of aging effects on aromatase, although the fact that blood estradiol concentrations remain relatively high compared with the steady decline in blood testosterone concentrations, suggesting aromatase may be upregulated in at least some aging tissues (302).

Receptor mechanism can modulate androgen action such as via naturally occurring genetic variation such as the CAG triplet (polyglutamine) repeat polymorphism which is inversely related to intrinsic sensitivity of the androgen receptor (303). This genetic basis for intrinsic androgen sensitivity would be fixed for any individual regardless of age, although differences between individuals may determine susceptibility to age-related changes consequent on a steady decline in blood testosterone concentrations (304–306). Androgen receptor expression may be down regulated in some tissues with age (307,308) and upregulated in some prostate cancers following therapeutic androgen deprivation (309,310). Post-receptor mechanisms remain the least understood potential mediators of tissue androgen sensitivity. Whereas post-receptor transcriptional regulation by the growing list of AR co-regulators (311) may fine tune some tissue androgen effects, the effects of age on the tissue distribution of AR co-regulators have not yet been defined.

The androgen requirement for testosterone effects varies both according to the endpoint and among individuals. For example, male sexual function has a low blood testosterone threshold (312–315), whereas muscle exhibits a linear dose-response from below to well above the physiological range (316). In addition, individuals differ in their own thresholds for androgen deficiency symptoms (317) probably attributable to genetic (303) and environmental (318,319) effects influencing androgen action.

Somatic Consequences of Age-Related Decline in Testosterone Production

A steady decline in blood testosterone concentration with age would be expected to create an androgen deficient state eventually once the decline reaches a sufficient threshold. Together with the resemblance between some nonspecific features of aging with correctable aspects of androgen deficiency in young men, this has long raised the possibility that some aspects of aging may be attributable to the observed decline in blood testosterone concentrations among otherwise healthy older men. This hypothesis has important implications for enhancing healthy aging through the preservation of enjoyable, independent living for as long as possible.

Castration at any age leads to consistent changes in muscle, fat, bone, and psychosexual function, although the best available evidence suggests that life expectancy is not reduced (320–322) [reviewed (323)]. However, the consequences of gradual or less complete reduction in testosterone output remain unclear. For example, an orchidectomy for testis cancer or benign testis disorders (torsion, cryptorchidism) leaving a single healthy testis generally does not usually lead to clinical features of androgen deficiency (324), although reduced testosterone output following unilateral orchidectomy may be partially offset by compensatory reflex increases in LH secretion (325–330). This suggests that for otherwise healthy men, testosterone output may need to be reduced by about 50% to create a genuine androgen deficiency state. From the known rate of decline in blood testosterone concentration, ~1% per annum from mid-life onwards, an androgen deficient state due to aging alone might only be anticipated at advanced age. Nevertheless, the rate of decline in blood testosterone concentration is steeper with concomitant chronic diseases (209,223) and will vary among individuals (317) and for different tissues, such as muscle (316), bone (331), and sexual function (314,315).

If some older men do become functionally androgen deficient, it is possible that some changes attributed to aging may be ameliorated by effective androgen replacement (332). This follows from the fact that testosterone replacement mostly rectifies the consequences of androgen deficiency in young men include decreases in muscle (333), bone (331), mental and psychosexual function, and general sense of energy, endurance, and well being (317). Male aging is associated with progressive decreases in muscle mass (334–344) and strength (345–352), bone mass (353–355) and fractures (355–357), and mental (358) and psychosexual (359–369) functions. Analogous clinical features in older men that impair quality of life include functional loss of muscle (sarcopenia, frailty, unstable gait, falls), fat (obesity, insulin resistance), bone (osteoporosis, fractures), and mental function (mood, cognition, psychosexual). Although these debilities of aging are all individually nonspecific, in concert they may represent a hormone-deficient state analogous to the constellation of clinical features, reflecting the estrogen deficient state of menopause. Whether androgen therapy can significantly ameliorate clinical features of partial, age-related androgen deficiency depends on the magnitude of the deficit and whether older tissues remain equally androgen responsive. Above all, it is crucial to determine whether androgen therapy can be delivered safely to older men (370,371). This refers primarily to the high background rates of cardiovascular and prostate disease, the major predictable androgen-related adverse effects of testosterone in older men, compared with young men with organic androgen deficiency. Even small increases in these major adverse effects are likely to negate morbidity benefits in muscle, bone or mental function.

Age-related changes in androgen production may have important effects on nonclassical androgen target tissue notably the cardiovascular system (323,372). Men have two to three times the rates of cardiovascular disease of women of the same age (373). This clinical observation has long been interpreted as reflecting estrogen protection of pre-menopausal women, an interpretation lacking support from the first well-controlled prospective studies of estrogen therapy for menopause (374,375) and neglecting the alternative possibility that testosterone may have adverse effects on pathogenesis of atherosclerotic cardiovascular disease. Nevertheless, as the best available historical observational evidence suggests that castration does not prolong male lifespan (321,322), androgen effects on fatal cardiovascular disease are presumed limited (323). While numerous retrospective epidemiological studies have linked declining blood testosterone with adverse cardiovascular outcomes (323,372,376), when considered prospectively, blood testosterone concentrations are not predictive of subsequent cardiovascular events (323). This indicates that the lowered blood testosterone concentrations may be a reflection of chronic cardiovascular disease rather than its cause. These considerations raise the possibility that the decline in blood testosterone concentrations in older men may afford a degree of protection against progression of cardiovascular disease (377), a hypothesis warranting careful evaluation for its global implications regarding the most frequent cause of male deaths.

Other idiosyncratic effects of androgen, such as polycythemia (378–385), behavioral disturbance (386–390), sleep disorders (391), and hypertension (392), are additional considerations for older men (371).

The nonspecific features in common between aging and androgen deficiency makes it difficult to diagnose androgen deficiency in individual older men without an underlying hypothalamo–pituitary or testicular disorder. As high-quality clinical trial evidence for such treatment is still lacking, this has led to attempts to define authoritative recommendation for guiding clinical practice for androgen replacement for older men despite the absence of a reliable foundation. Although numerous observational studies link age-related changes in blood testosterone and related hormones with changes in somatic tissues, these cannot establish causality. The pivotal approach to resolve whether age-related reduction in blood testosterone does influence tissue aging is undertaking controlled interventional studies aiming to increase the blood testosterone concentrations in older men within physiological limits to those equivalent to healthy younger or age-matched men. These require well-designed placebo-controlled clinical trials of adequate power to detect effects due to the modest changes in blood testosterone observed in aging men. However, the available evidence in favor of androgen replacement therapy remains weak (393) despite a number of small- to medium-scale placebo-controlled clinical trials of at least 3 months duration involving up to 100 men for as long as 3 years (394–401). These studies have shown small but definite benefits in body composition of older men with reduced fat and increase muscle mass of modest magnitude (mean 2–3 kg) and increased bone mineral density in some (385,398) but not all (402) older, nonosteoporotic men. The 2004 Institute of Medicine report concluded that more convincing evidence of short–medium efficacy was required before even large-scale, long-term clinical trials, similar to those for estrogens in menopause (374,375) were warranted.

Reproductive Consequences of Age-Related Decline in Testosterone Production

Reproductive effects of testosterone that may be affected by aging include androgen effects on the testis as well as behavioral effects on sexual activity. While the testicular effects of androgens support spermatogenesis to create haploid male gametes capable of fertilization, the behavioral effects of testosterone are vital to delivering spermatozoa into the female genital tract where fertilization takes place. The effects of aging on androgen action in supporting spermatogenesis and fertility are considered previously.

Behavioral effects of testosterone are crucial to male sexual activity with the primary role of androgens in human male sexual function being in the maintenance of libido (403,404), the cerebral drive toward interest, initiation, and participation in sexual activity. Androgens are more strongly related to spontaneous nocturnal erections than for psychogenic erections (405). Additional, probably secondary, effects of androgens have been described on the neurovascular mechanisms of erection and ejaculation.

The progressive decline of sexual activity as men age from mid-life onwards has been known throughout human existence. Male sexual activity and its decline with aging have always been of fundamental interest in art and literature, yet the scientific study of sexuality only became feasible in recent decades. Systematic observational and experimental clinical research on human male sexual function was virtually prevented by its social unacceptability until the descriptive work of Kinsey published in 1948 (406). This monumental observational study, based on over 5000 interviews, provided the first comprehensive description of the marked and progressive decline in all aspects of male sexual activity with increasing age. More recently with the greater acceptability of studies of sexuality following Kinsey and the AIDS epidemic, large-scale studies have confirmed the age dependence of male sexual activity in large-scale national surveys of the U.K. (366), France (360), U.S., and Australia (368) as well as in other population-based regional studies in Boston,

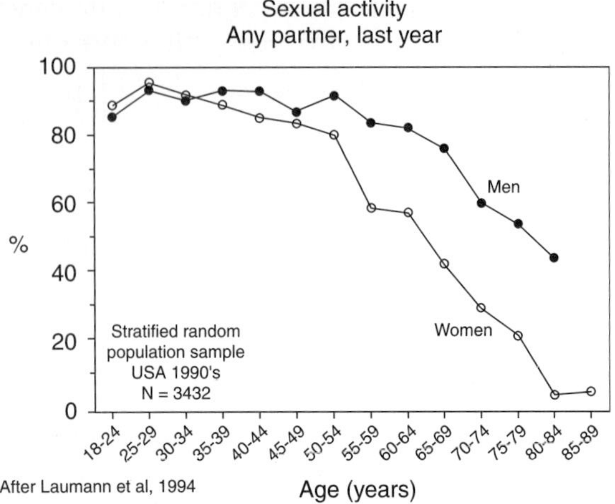

FIG. 6. Plot of male and female sexual activity with any partner in the last year from a stratified random sample of the U.S. population. Note decline in male and female sexual activity from mid-life onwards. [Data from Laumann et al., 1994 (487).]

Massachusetts (361,364,369), France (362), Bristol, U.K. (363), and Cologne, Germany (365), and the U.S. and six European countries (367). While the large-scale national surveys have truncated the ages studied to age 60 years, regional studies have generally studied men up to at least 80 years and include the only longitudinal observations of serial changes in individuals (364,369). The major determinants of sexual activity among older men include his own and his partner's health, notably cardiovascular disease, diabetes, and lower urinary tract symptoms. At the oldest ages, lower sexual activity corresponded to reduced sexual interest (359). As these studies were mostly cross-sectional, interpretation of their findings, as if longitudinal may be subject to bias from confounding period, cohort, or survivorship effects.

The development during puberty of adult male sexual behavior requires exposure to adult male testosterone levels. Once initiated, however, adult male sexual activity no longer depends as strongly on ongoing androgen action. For example, among men who have completed sexual development before castration for sexual offenses (407,408), localized (409), or advanced (410,411) prostate cancer, sexual function continues in some men despite the overall reduction in sexual activity. Similarly, among healthy young men, testosterone effects appear maximal at low blood testosterone concentrations, a threshold well below reference ranges of healthy eugonadal young men (312–315). Consequently, in otherwise healthy men, blood testosterone concentrations must fall to very low levels before sexual function is significantly impaired due to androgen deficiency. Hence, the steep decline of male sexual activity in older men (359–369) is disproportionate to the relatively modest age-related decline in blood testosterone explaining the observation of a poor relationship between sexual activity and blood testosterone concentrations in older men (412). In later life, sexual activity is more predictable from sexual behaviors at earlier ages rather than other biological factors at older age (359). Furthermore, the lowering of blood testosterone in older men may also be a consequence rather than a cause of sexual dysfunction. For example, sexual inactivity itself reduces blood testosterone concentrations (413,414). Similarly, blood testosterone concentrations may also be a consequence of underlying chronic medical disorders that accumulate among older men such as obesity (248,415–417), depression (418–421), physical inactivity (422–424), and other chronic disorders (425). Finally, genuine androgen deficiency is a rare cause of newly diagnosed erectile dysfunction (ED) (426) with fewer than 5% of men presenting with ED due to androgen deficiency from pituitary or testicular pathologies (427). Most men presenting with ED have normal or minimal reduction in blood testosterone concentration. The small reduction in blood testosterone concentration among men with ED who are not frankly androgen deficient may reflect sexual inactivity itself rather being its cause (413). Furthermore, empirical testosterone therapy is ineffective treatment for ED in the absence of overt organic androgen deficiency (428–431). Hence, the steep and progressive decline of sexual activity in older men is unlikely to be primarily, if at all, related to the concomitant age-related decline in blood testosterone concentrations.

The major cause of sexual dysfunction among older men is now understood to be mostly due to organic, primarily vasculogenic, rather than psychogenic factors or androgen deficiency. The neurovascular hydraulic mechanism of penile erection involves nitric oxide (NO) release (432), increased cytosolic cyclic GMP (433) and decreased calcium (434) in smooth muscle cells leading to vasodilatation and cavernosal filling (435). Vasculogenic ED constitutes atherosclerotic vascular disease of the large and small pelvic blood vessels together with endothelial dysfunction (436). Animal models show conclusively that atherosclerosis of the lower aorta, iliac, and penile arteries cause erectile failure in rabbits (437) and monkeys (438). Major risk factors for ED, notably aging, smoking, hyperlipidemia, hypertension, and diabetes, are similar to those for cardiovascular disease (436,439). Consequently, men with ED have higher risk for other atherosclerotic diseases and men with coronary, cerebral, or peripheral atherosclerotic vascular disease are at higher risk for ED (440). As a result, ED may be a sentinel feature of atherosclerotic cardiovascular disease (425,441–443), although direct proof by a well-controlled clinical trial is still lacking.

The medical treatment of male sexual dysfunction has been revolutionized by the recognition that most ED is due to organic rather than psychogenic causes and by the introduction of cavernosal vasodilator therapy which can improve ED of any cause in men of any age (444,445). Cavernosal vasodilator therapy to induce erection was initially achieved by direct injection or topical application of vasodilators such as prostaglandin E1 (446,447), papaverine, and phentolamine (448,449) to relax the vascular smooth muscle of the corpora cavernosae (450). Such treatment had unprecedented effectiveness but the invasiveness of injections was a deterrent with a consequently high attrition rate of users (451–453). Subsequently a more popular and acceptable treatment has become available based on the fortuitous discovery of orally active phosphodiesterase-5 inhibitors. These drugs enhance neural nitric oxide-induced cyclic GMP signaling that triggers the cavernosal vasodilatation to produce an erection (435,454–459) targeting the inhibition of tissue-selective, type 5 phosphodiesterase inhibitor to enhance cyclic GMP action via inhibition of its metabolism (460). The oral phosphodiesterase-5 inhibitors have a good safety record in older men including those with cardiovascular disease (461–463), so long as they are not used together with nitrates, which is an absolute contraindication due to potentially lethal hypotensive drug interaction. Experience indicates that cavernosal vasodilator therapy is as effective and safe in older as for younger men (445,464,465).

Although the primary effect of androgens on human male sexual function is on libido, additional effects on the peripheral neurovascular hydraulic mechanisms of penile erection have been reported to show that testosterone can augment erectile function via peripheral effects on the neurovascular hydraulic mechanisms of erection. In experimental animal models, androgens influence development of pelvic autonomic nerves (466) and penile vasculature and smooth muscle (467). Furthermore, castration causes loss of cavernosal structural and functional features, such as neurons containing nitric oxide synthetase and development of smooth muscle apoptosis, atrophy, and fibrosis with reduced responsiveness to electrical or chemical stimuli (468–484). Whether similar effects apply in castrate men has not been fully evaluated partly due to the limited availability of human tissue. Clinical evidence for peripheral augmentation by testosterone of cavernosal vasodilator effects in men with sexual dysfunction but only mildly reduced blood testosterone concentrations remains inconclusive (485,486).

REFERENCES

1. Charlesworth, B. (2000). Fisher, Medawar, Hamilton and the evolution of aging. *Genetics* 156, 927–931.
2. Bowles, J. (2000). Shattered: Medawar's test tubes and their enduring legacy of chaos. *Med. Hypotheses*. 54, 326–339.
3. Medvedev, Z. A. (1981). On the immortality of the germ line: genetic and biochemical mechanism. A review. *Mech. Ageing Dev.* 17, 331–359.
4. Medvedev, Z. A. (1990). An attempt at a rational classification of theories of ageing. *Biol. Rev. Camb. Philos. Soc.* 65, 375–398.
5. Promislow, D. E., and Pletcher, S. D. (2002). Advice to an aging scientist. *Mech. Ageing Dev.* 123, 841–850.
6. Dufour, E., and Larsson, N. G. (2004). Understanding aging: revealing order out of chaos. *Biochim. Biophys. Acta* 1658, 122–132.
7. Gavrilov, L. A., and Gavrilova, N. S. (2003). The quest for a general theory of aging and longevity. *Sci. Aging Knowledge Environ.* RE5.
8. Kirkwood, T. B., and Holliday, R. (1979). The evolution of ageing and longevity. *Proc. R. Soc. Lond. B. Biol. Sci.* 205, 531–546.
9. Gavrilov, L. A., and Gavrilova, N. S. (2002). Evolutionary theories of aging and longevity. *Sci. World J.* 2, 339–356.
10. Feinstein, A. (1985). *Clinical Epidemiology: The Architecture of Clinical Research*. WB Saunders, Philadelphia.
11. Rothman, K., and Greenland, S. (1998). *Modern Epidemiology*, 2nd ed. Lippincott Willams & Wilkins, Phildelphia.
12. Sackett, D. L. (1979). Bias in analytical research. *J. Chronic Dis.* 32, 51–63.
13. Bishop, M. W. (1970). Ageing and reproduction in the male. *J. Reprod. Fertil. Suppl.* 12, 65–87.
14. Leathem, J. (1977). Aging and the testis. In *The Testis* (A. Johnson and W. Gomes, Eds.), Vol. IV, pp. 547–563. Academic Press, New York.
15. Harman, S., and Talbert, G. (1985). Reproductive aging. In *Handbook of the Biology of Aging* (C. Finch and E. Schneider, Eds.). Van Nostrand Reinhold, New York.
16. vom Saal, F., Finch, C., and Nelson, J. (1994). Natural history and mechanisms of reproductive aging in humans, laboratory rodents and other selected vertebrates. In *The Physiology of Reproduction* (E. Knobil and J. D. Neill, Eds.), Vol. 2, pp. 1213–1314. Raven Press, New York.
17. Seymour, F., Duffy, C., and Koerner, A. (1935). A case of authenticated fertility in a man of 94. *JAMA* 105, 1423–1425.
18. Rolf, C., and Nieschlag, E. (1997). The aging male. In *Andrology: Male Reproductive Health and Dysfunction* (E. Nieschlag and H. M. Behre, Eds.), pp. 419–432. Springer Verlag, Berlin.
19. Leridon, H. (1977). *Human Fertility: The Basic Components*. University of Chicago Press, Chicago.
20. Schwartz, D., and Mayaux, M. J. (1982). Female fecundity as a function of age: results of artificial insemination in 2,193 nulliparous women with azoospermic husbands. Federation CECOS. *N. Engl. J. Med.* 306, 404–406.
21. Menken, J., Trussell, J., and Larsen, U. (1986). Age and infertility. *Science* 233, 1389–1394.
22. Levine, R. J., Blunden, P. B., DalCorso, R. D., Starr, T. B., and Ross, C. E. (1983). Superiority of reproductive histories to sperm counts in detecting infertility at a dibromochloropropane manufacturing plant. *J. Occup. Med.* 25, 591–597.
23. Handelsman, D. J. (1997). Sperm output of healthy men in Australia: magnitude of bias due to self-selected volunteers. *Hum. Reprod.* 12, 2701–2705.
24. Cohn, B. A., Overstreet, J. W., Fogel, R. J., Brazil, C. K., Baird, D. D., and Cirillo, P. M. (2002). Epidemiologic studies of human semen quality: considerations for study design. *Am. J. Epidemiol.* 155, 664–671.
25. Muller A., De La Rochebrochard, E., Labbe-Decleves, C., Jouannet, P., Bujan, L., Mieusset, R., Le Lannou, D., Guerin, J. F., Benchaib, M., Slama, R., and Spira, A. (2004). Selection bias in semen studies due to self-selection of volunteers. *Hum. Reprod.* 19, 2838–2844.
26. Sheps, M., and Menken, J. (1973). *Mathematical Models of Conception and Birth*. University of Chicago Press, Chicago.

27. Shyrock, H. S., Siegel, J. S., and Stockwell, E. G. (1976). *The Methods and Materials of Demography*. Academic Press, New York.
28. Bongaarts, J., and Potter, R. (1983). *Fertility, Biology and Behavior: An Analysis of the Proximate Determinants*. Academic Press, New York.
29. Wood, J. W. (1989). Fecundity and natural fertility in humans. *Oxf. Rev. Reprod. Biol.* 11, 61–109.
30. Campbell, B., and Leslie, P. (1995). Reproductive ecology of human males. *Yearbook of Physical Anthropology* 38, 1–26.
31. Levine, R. J., Symons, M. J., Balogh, S. A., Arndt, D. M., Kaswandik, N. T., and Gentile, J. W. (1980). A method for monitoring the fertility of workers. 1. Method and pilot studies. *J. Occup. Med.* 22, 781–791.
32. Levine, R. J. (1988). Monitoring fertility to detect toxicity to the male reproductive system. *Reprod. Toxicol.* 2, 223–227.
33. Joffe, M. (1997). Time to pregnancy: a measure of reproductive function in either sex. Asclepios Project. *Occup. Environ. Med.* 54, 289–295.
34. Joffe, M. (2003). Invited commentary: the potential for monitoring of fecundity and the remaining challenges. *Am. J. Epidemiol.* 157, 89–93.
35. Hassan, M. A., and Killick, S. R. (2003). Effect of male age on fertility: evidence for the decline in male fertility with increasing age. *Fertil. Steril.* 79, Suppl 3, 1520–1527.
36. Anderson, B. A. (1975). Male age and fertility: results from Ireland prior to 1911. *Population Index* 41, 561–567.
37. Mineau, G. P., and Trussell, J. (1982). A specification of marital fertility by parents' age, age at marriage and marital duration. *Demography* 19, 335–350.
38. Goldman, N., and Montgomery, M. (1989). Fecundability and husband's age. *Soc. Biol.* 36, 146–166.
39. Ratcliffe, A. A., Hill, A. G., and Walraven, G. (2000). Separate lives, different interests: male and female reproduction in the Gambia. *Bull. World Health Organization* 78, 570–579.
40. Joffe, M. (2000). Time trends in biological fertility in Britain. *Lancet* 355, 1961–1965.
41. Ratcliffe, A. A., Hill, A. G., Harrington, D. P., and Walraven, G. (2002). Reporting of fertility events by men and women in rural Gambia. *Demography* 39, 573–586.
42. Strassmann, B. I., and Warner, J. H. (1998). Predictors of fecundability and conception waits among the Dogon of Mali. *Am. J. Phys. Anthropol.* 105, 167–184.
43. Serre, V., and Robaire, B. (1998). Paternal age affects fertility and progeny outcome in the Brown Norway rat. *Fertil. Steril.* 70, 625–631.
44. Bonde, J. P., Ernst, E., Jensen, T. K., Hjollund, N. H., Kolstad, H., Henriksen, T. B., Scheike, T., Giwercman A., Olsen, J., and Skakkebaek, N. E. (1998). Relation between semen quality and fertility: a population-based study of 430 first-pregnancy planners. *Lancet* 352, 1172–1177.
45. Ford, W. C., North, K., Taylor, H., Farrow A., Hull, M. G., and Golding, J. (2000). Increasing paternal age is associated with delayed conception in a large population of fertile couples: evidence for declining fecundity in older men. The ALSPAC Study Team (Avon Longitudinal Study of Pregnancy and Childhood). *Hum. Reprod.* 15, 1703–1708.
46. Zinaman, M. J., Brown, C. C., Selevan, S. G., and Clegg, E. D. (2000). Semen quality and human fertility: a prospective study with healthy couples. *J. Androl.* 21, 145–153.
47. Akre, O., Cnattingius, S., Bergstrom, R., Kvist, U., Trichopoulos, D., and Ekbom, A. (1999). Human fertility does not decline: evidence from Sweden. *Fertil. Steril.* 71, 1066–1069.
48. Jensen, T. K., Keiding, N., Scheike, T., Slama, R., and Spira, A. (2000). Declining human fertility? *Fertil. Steril.* 73, 421–423.
49. Carlsen, E., Giwercman, A., Keiding, N., and Skakkebaek, N. E. (1992). Evidence for decreasing quality of semen during past 50 years. *Br. Med. J.* 305, 609–613.
50. Swan, S. H., Elkin, E. P., and Fenster, L. (2000). The question of declining sperm density revisited: an analysis of 101 studies published 1934–1996. *Environ. Health Perspect.* 108, 961–966.
51. Sharpe, R. M., and Skakkebaek, N. E. (1993). Are oestrogens involved in falling sperm counts and disorders of the male reproductive tract? *Lancet* 341, 1392–1395.
52. Dieckmann, W. J., Davis, M. E., Rynkiewicz, L. M., and Pottinger, R. E. (1953). Does the administration of diethylstilbestrol during pregnancy have therapeutic value? *Am. J. Obstet. Gynecol.* 66, 1062–1081.
53. Wilcox, A. J., Baird, D. D., Weinberg, C. R., Hornsby, P. P., and Herbst, A. L. (1995). Fertility in men exposed prenatally to diethystilbestrol. *N. Engl. J. Med.* 332, 1411–1416.
54. Friedman, J. M. (1981). Genetic disease in the offspring of older fathers. *Obstet. Gynecol.* 57, 745–749.
55. Kuhnert, B., and Nieschlag, E. (2004). Reproductive functions of the ageing male. *Hum. Reprod. Update.* 10, 327–339.
56. Tarin, J. J., Brines, J., and Cano, A. (1998). Long-term effects of delayed parenthood. *Hum. Reprod.* 13, 2371–2376.
57. Malaspina, D. (2001). Paternal factors and schizophrenia risk: de novo mutations and imprinting. *Schizophr. Bull.* 27, 379–393.
58. Sipos A., Rasmussen, F., Harrison, G., Tynelius, P., Lewis, G., Leon, D. A., and Gunnell, D. (2004). Paternal age and schizophrenia: a population based cohort study. *BMJ* 329, 1070.
59. Selvin, S., and Garfinkel, J. (1976). Paternal age, maternal age and birth order and the risk of a fetal loss. *Hum. Biol.* 48, 223–230.
60. Ford, J. H., MacCormac, L., and Hiller, J. (1994). PALS (pregnancy and lifestyle study): association between occupational and environmental exposure to chemicals and reproductive outcome. *Mutat. Res.* 313, 153–164.
61. Wunsch, G., and Gourbin, C. (2002). Parents' age at birth of their offspring and child survival. *Soc. Biol.* 49, 174–184.
62. de la Rochebrochard, E., and Thonneau, P. (2002). Paternal age and maternal age are risk factors for miscarriage; results of a multicentre European study. *Hum. Reprod.* 17, 1649–1656.
63. Resseguie, L. J. (1976). Paternal age, stillbirths and mutation. *Ann. Hum. Genet.* 40, 213–219.
64. al-Ansary, L. A., and Babay, Z. A. (1994). Risk factors for spontaneous abortion: a preliminary study on Saudi women. *J. R. Soc. Health.* 114, 188–193.
65. al-Ansary, L. A., Oni, G., and Babay, Z. A. (1995). Risk factors for spontaneous abortion among Saudi women. *J. Community Health* 20, 491–500.
66. De La Rochebrochard, E., McElreavey, K., and Thonneau, P. (2003). Paternal age over 40 years: the "amber light" in the reproductive life of men? *J. Androl.* 24, 459–465.
67. World Health Organisation. (1999). *WHO Laboratory Manual for the Examination of Human Semen and Sperm–Cervical Mucus Interaction*, 4th ed. Cambridge University Press, Cambridge.
68. Cooper, T. G., Bjorndahl, L., Vreeburg, J., and Nieschlag, E. (2002). Semen analysis and external quality control schemes for semen analysis need global standardization. *Int. J. Androl.* 25, 306–311.
69. McLachlan, R. I., Baker, H. W., Clarke, G. N., Harrison, K. L., Matson, P. L., Holden, C. A., and de Kretser, D. M. (2003). Semen analysis: its place in modern reproductive medical practice. *Pathology (Phila).* 35, 25–33.
70. Keel, B. A., Stembridge, T. W., Pineda, G., and Serafy, N. T. (2002). Lack of standardization in performance of the semen analysis among laboratories in the United States. *Fertil. Steril.* 78, 603–608.
71. Jensen, T. K., Jorgensen, N., Punab, M., Haugen, T. B., Suominen, J., Zilaitiene, B., Horte, A., Andersen, A. G., Carlsen, E., Magnus, O., Matulevicius, V., Nermoen, I.,

Vierula, M., Keiding, N., Toppari, J., and Skakkebaek, N. E. (2004). Association of in utero exposure to maternal smoking with reduced semen quality and testis size in adulthood: a cross-sectional study of 1,770 young men from the general population in five European countries. *Am. J. Epidemiol.* 159, 49–58.
72. MacLeod, J., and Gold, R. Z. (1951). The male factor in fertility and infertility. VII. Semen quality in relation to age and sexual activity. *Fertil. Steril.* 4, 194–209.
73. Schwartz, D., Mayaux, M. J., Spira, A., Moscato, M. L., Jouannet, P., Czyglik, F., and David, G. (1981). Study of a group of 484 fertile men. Part II: relation between age (20–59) and semen characteristics. *Int. J. Androl.* 4, 450–465.
74. Schwartz, D., Mayaux, M. J., Spira A., Moscato, M. L., Jouannet, P., Czyglik, F., and David, G. (1983). Semen characteristics as a function of age in 833 fertile men. *Fertil. Steril.* 39, 530–535.
75. Haidl, G., Jung, A., and Schill, W. B. (1996). Ageing and sperm function. *Hum. Reprod.* 11, 558–560.
76. Rolf, C., Behre, H. M., and Nieschlag, E. (1996). Reproductive parameters of older compared to younger men of infertile couples. *Int. J. Androl.* 19, 135–142.
77. Gallardo, E., Simon, C., Levy, M., Guanes, P. P., Remohi, J., and Pellicer, A. (1996). Effect of age on sperm fertility potential: oocyte donation as a model. *Fertil. Steril.* 66, 260–264.
78. Paulson, R. J., Milligan, R. C., and Sokol, R. Z. (2001). The lack of influence of age on male fertility. *Am. J. Obstet. Gynecol.* 184, 818–822; discussion 822–814.
79. Kidd, S. A., Eskenazi, B., and Wyrobek, A. J. (2001). Effects of male age on semen quality and fertility: a review of the literature. *Fertil. Steril.* 75, 237–248.
80. Blum, V. (1936). Das problem des mannlichen klimakteriums. *Wien Klin. Wochenschr.* 49, 1133–1139.
81. Engle, E. (1952). The male reproductive system. In *Cowdry's Problems of Ageing: Biological and Medical Aspects* (A. Lansing, Ed.), pp. 708–729. Williams & Wilkins, Baltimore.
82. Belonoschkin, B. (1954). Spermiogenesis in elderly men. *Fertil. Steril.* 5, 182–192.
83. Busse, C. (1988). Semen characteristics during senescence. In *Carl Schirren Symposium: Advances in Andrology* (A. Holstein, F. Leidenberger, K. Holzer, and G. Bettendorf, Eds.), pp. 118–119. Diesbach Verlag, Berlin.
84. Natoli A., Riondino, G., and Brancati, A. (1972). Studio della funzione gonadale, ormonica e spermatogenica, nel corso della senescenza maschile. *Giorn. Geront.* 20, 1103–1120.
85. Nieschlag, E., Lammers, U., Freischem, C. W., Langer, K., and Wickings, E. J. (1982). Reproductive function in young fathers and grandfathers. *J. Clin. Endocrinol. Metab.* 55, 676–681.
86. Eskenazi, B., Wyrobek, A. J., Sloter, E., Kidd, S. A., Moore, L., Young, S., and Moore, D. (2003). The association of age and semen quality in healthy men. *Hum. Reprod.* 18, 447–454.
87. Ng, K. K., Donat, R., Chan, L., Lalak, A., Di Pierro, I., and Handelsman, D. J. (2004). Sperm output of older men. *Hum. Reprod.* 19, 1811–1815.
88. Handelsman, D. J. (2001). Estrogens and falling sperm counts. *Reprod. Fertil. Dev.* 13, 317–324.
89. Joffe, M. (2003). Infertility and environmental pollutants. *Br. Med. Bull.* 68, 47–70.
90. Safe, S. (2004). Endocrine disruptors and human health: is there a problem. *Toxicology* 205, 3–10.
91. Setchell, B. P. (1997). Sperm counts in semen of farm animals 1932–1995. *Int. J. Androl.* 20, 209–214.
92. Rao, K. V., and Demaris, A. (1995). Coital frequency among married and cohabiting couples in the United States. *J. Biosoc. Sci.* 27, 135–150.
93. Regadera, J., Nistal, M., and Paniagua, R. (1985). Testis, epididymis, and spermatic cord in elderly men: correlation of angiographic and histologic studies with systemic arteriosclerosis. *Arch. Pathol. Lab. Med.* 109, 663–667.
94. Sasano, N., and Ichijo, S. (1969). Vascular patterns of the human testis with special reference to its senile changes. *Tohoku J. Exp. Med.* 99, 269–280.
95. Suoranta, H. (1971). Changes in the small blood vessels of the adult human testis in relation to age and to some pathological conditions. *Virchows Arch. A. Pathol. Pathol. Anat.* 352, 165–181.
96. Paniagua, R., Nistal, M., Amat, P., Rodriguez, M. C., and Martin, A. (1987). Seminiferous tubule involution in elderly men. *Biol. Reprod.* 36, 939–947.
97. Kothari, L. K., and Gupta, A. S. (1978). The total Leydig cell volume of the testis in some common mammals. *Andrologia* 10, 218–222.
98. van Dop, P. A., Scholtmeijer, R. J., Kurver, P. H., Baak, J. P., Oort, J., and Stolte, L. A. (1980). A quantitative structural model of the testis of fertile males with normal sperm counts. *Int. J. Androl.* 3, 153–169.
99. Lenz, S., Giwercman A., Elsborg A., Cohr, K. H., Jelnes, J. E., Carlsen, E., and Skakkebaek, N. E. (1993). Ultrasonic testicular texture and size in 444 men from the general population: correlation with semen quality. *Eur. Urol.* 24, 231–238.
100. Rivkees, S. A., Hall, D. A., Boepple, P. A., and Crawford, J. D. (1987). Accuracy and reproducibility of clinical measures of testicular volume. *J. Pediatr.* 110, 914–917.
101. Behre, H. M., Nashan, D., and Nieschlag, E. (1989). Objective measurement of testicular volume by ultrasonography: evaluation of the technique and comparison with orchidometer estimates. *Int. J. Androl.* 12, 395–403.
102. Baker, H. W. G. (2001). Male infertility. In *Endocrinology* (L. J. DeGroot, Ed.), Vol. 3, pp. 2308–2328. WB Saunders, Philadelphia.
103. Behre, H., Yeung, C., Holstein, A., Weinbauer, G., Gassner, P., and Nieschlag, E. (1997). Diagnosis of male infertility and hypogonadism. In *Andrology: Male Reproductive Health and Dysfunction* (E. Nieschlag and H. M. Behre, Eds.). Springer Verlag, Berlin.
104. Mahmoud, A. M., Goemaere, S., El-Garem, Y., Van Pottelbergh, I., Comhaire, F. H., and Kaufman, J. M. (2003). Testicular volume in relation to hormonal indices of gonadal function in community-dwelling elderly men. *J. Clin. Endocrinol. Metab.* 88, 179–184.
105. Bothe, A., and Robinson, E. (1933). A histological study of the testicles in one hundred autopsies. *J. Urol.* 29, 425–442.
106. Sniffen, R. (1950). The testis. I. The normal testis. *Arch Pathol.* 50, 259–295.
107. Tillinger, K. G. (1957). Testicular morphology; a histopathological study with special reference to biopsy findings in hypogonadism with mainly endocrine disorders and in gynecomastia. *Acta Endocrinol. (Copenh).* 24, 1–192.
108. Calloway, N. O., Foley, C. F., and Lagerbloom, P. (1965). Uncertainties in geriatric data. II. Organ size. *J. Am. Geriatr. Soc.* 13, 20–28.
109. Harbitz, T. B. (1973). Testis weight and the histology of the prostate in elderly men. An analysis in an autopsy series. *Acta Pathol. Microbiol. Scand. [A].* 81, 148–158.
110. Honore, L. H. (1978). Ageing changes in the human testis: a light-microscopic study. *Gerontology* 24, 58–65.
111. Longcope, C. (1973). The effect of human chorionic gonadotropin on plasma steroid levels in young and old men. *Steroids* 21, 583–592.
112. Stearns, E. L., MacDonnell, J. A., Kaufman, B. J., Padua, R., Lucman, T. S., Winter, J. S. D, and Faiman, C. (1974). Declining testicular function with age: hormonal and clinical correlates. *Am. J. Med.* 57, 761–766.
113. Baker, H. W., Burger, H. G., de Kretser, D. M., Hudson, B., O'Connor, S., Wang, C., Mirovics A., Court, J., Dunlop, M., and Rennie, G. C. (1976). Changes in the pituitary-testicular system with age. *Clin. Endocrinol. (Oxf).* 5, 349–372.

114. Kothari, L. K., and Gupta, A. S. (1974). Effect of ageing on the volume, structure and total Leydig cell content of the human testis. *Int. J. Fertil.* 19, 140–146.
115. Harman, S. M., Tsitouras, P. D., Costa, P. T., and Blackman, M. R. (1982). Reproductive hormones in aging men. II. Basal pituitary gonadotropins and gonadotropin responses to luteinizing hormone-releasing hormone. *J. Clin. Endocrinol. Metab.* 54, 547–551.
116. Handelsman, D. J., and Staraj, S. (1985). Testicular size: the effects of aging, malnutrition and illness. *J. Androl.* 6, 144–151.
117. Mori, C., Hamamatsu, A., Fukata, H., Koh, K. B., Nakamura, N., Takeichi, S., Kusakabe, T., Saito, T., Morita, M., Tanihara, S., Kayama, F., Shiyomi, M., Yoshimura, J., and Sagisaka, K. (2002). Temporal changes in testis weight during the past 50 years in Japan. *Anat. Sci. Int.* 77, 109–116.
118. Tanaka, G., Nakahara, Y., and Nakazima, Y. (1989). [Japanese reference man 1988-IV. Studies on the weight and size of internal organs of normal Japanese]. *Nippon Igaku Hoshasen Gakkai Zasshi.* 49, 344–364.
119. Johnson, L. (1986). Spermatogenesis and aging in the human. *J. Androl.* 7, 331–354.
120. Kaler, L. W., and Neaves, W. B. (1978). Attrition of the human Leydig cell population with advancing age. *Anat. Rec.* 192, 513–518.
121. Heller, C. G., and Clermont, Y. (1963). Spermatogenesis in man: an estimate of its duration. *Science* 140, 184–186.
122. Heller, C. H., and Clermont, Y. (1964). Kinetics of the germinal epithelium in man. *Recent Prog. Horm. Res.* 20, 545–575.
123. Johnson, L., Petty, C. S., and Neaves, W. B. (1984). Influence of age on sperm production and testicular weights in men. *J. Reprod. Fertil.* 70, 211–218.
124. Neaves, W. B., Johnson, L., Porter, J. C., Parker, C. R., and Petty, C. S. (1984). Leydig cell numbers, daily sperm production, and serum gonadotropin levels in aging men. *J. Clin. Endocrinol. Metab.* 55, 756–763.
125. Johnson, L., Petty, C. S., Porter, J. C., and Neaves, W. B. (1984). Germ cell degeneration during postprophase of meiosis and serum concentrations of gonadotropins in young adult and older adult men. *Biol. Reprod.* 31, 779–784.
126. Neaves, W. B., Johnson, L., and Petty, C. S. (1987). Seminiferous tubules and daily sperm production in older adult men with varied numbers of Leydig cells. *Biol. Reprod.* 36, 301–308.
127. Paniagua, R., Martin, A., Nistal, M., and Amat, P. (1987). Testicular involution in elderly men: comparison of histologic quantitative studies with hormone patterns. *Fertil. Steril.* 47, 671–679.
128. Johnson, L., Abdo, J. G., Petty, C. S., and Neaves, W. B. (1988). Effect of age on the composition of seminiferous tubular boundary tissue and on the volume of each component in humans. *Fertil. Steril.* 49, 1045–1051.
129. Johnson, L., Zane, R. S., Petty, C. S., and Neaves, W. B. (1984). Quantification of the human Sertoli cell population: its distribution, relation to germ cell numbers, and age-related decline. *Biol. Reprod.* 31, 785–795.
130. Neaves, W. B., Johnson, L., and Petty, C. S. (1985). Age-related change in numbers of other interstitial cells in testes of adult men: evidence bearing on the fate of Leydig cells lost with increasing age. *Biol. Reprod.* 33, 259–269.
131. Johnson, L., and Varner, D. D. (1988). Effect of daily spermatozoan production but not age on transit time of spermatozoa through the human epididymis. *Biol. Reprod.* 39, 812–817.
132. Paniagua, R., Nistal, M., Saez, F. J., and Fraile, B. (1991). Ultrastructure of the aging human testis. *J. Electron Microsc. Tech.* 19, 241–260.
133. Hatakeyama, S., Takizawa, T., and Kawahara, Y. (1979). Focal atrophy of the seminiferous tubule in the human testis. *Acta Pathol. Jpn.* 29, 901–910.
134. Yoshimura, M., and Fukunishi, M. (1965). Studies on the change with aging of human testicle. 1. Estimation of the age from the thickness of tunica albuginea. *Osaka City Med. J.* 11, 1–5.
135. Arenas, M. I., Bethencourt, F. R., Fraile, B., and Paniagua, R. (1997). Immunocytochemical and quantitative study of the tunica albuginea testis in young and ageing men. *Histochem. Cell. Biol.* 107, 469–477.
136. Arenas, M. I., Bethencourt, F. R., De Miguel, M. P., Fraile, B., Romo, E., and Paniagua, R. (1997). Immunocytochemical and quantitative study of actin, desmin and vimentin in the peritubular cells of the testes from elderly men. *J. Reprod. Fertil.* 110, 183–193.
137. Codesal, J., Santamaria, L., Paniagua, R., and Nistal, M. (1989). Proliferative activity of human spermatogonia from fetal period to senility measured by cytophotometric DNA quantification. *Arch. Androl.* 22, 209–215.
138. Nistal, M., Codesal, J., Paniagua, R., and Santamaria, L. (1987). Decrease in the number of human Ap and Ad spermatogonia and in the Ap/Ad ratio with advancing age. New data on the spermatogonial stem cell. *J. Androl.* 8, 64–68.
139. Paniagua, R., Rodriguez, M. C., Nistal, M., Fraile, B., and Amat, P. (1987). Changes in the lipid inclusion/Sertoli cell cytoplasm area ratio during the cycle of the human seminiferous epithelium. *J. Reprod. Fertil.* 80, 335–341.
140. Paniagua, R., Amat, P., Nistal, M., and Martin, A. (1985). Ultrastructural changes in Sertoli cells in ageing humans. *Int. J. Androl.* 8, 295–312.
141. McLachlan, R. I., O'Donnell, L., Meachem, S. J., Stanton, P. G., de Kretser, D. M., Pratis, K., and Robertson, D. M. (2002). Identification of specific sites of hormonal regulation in spermatogenesis in rats, monkeys, and man. *Recent Prog. Horm. Res.* 57, 149–179.
142. Simoni, M., Gromoll, J., and Nieschlag, E. (1997). The follicle-stimulating hormone receptor: biochemistry, molecular biology, physiology, and pathophysiology. *Endocr. Rev.* 18, 739–773.
143. Bremner, W. J., Millar, M. R., Sharpe, R. M., and Saunders, P. T. (1994). Immunohistochemical localization of androgen receptors in the rat testis: evidence for stage-dependent expression and regulation by androgens. *Endocrinology* 135, 1227–1234.
144. Suarez-Quian, C. A., Martinez-Garcia, F., Nistal, M., and Regadera, J. (1999). Androgen receptor distribution in adult human testis. *J. Clin. Endocrinol. Metab.* 84, 350–358.
145. Zhou, Q., Nie, R., Prins, G. S., Saunders, P. T., Katzenellenbogen, B. S., and Hess, R. A. (2002). Localization of androgen and estrogen receptors in adult male mouse reproductive tract. *J. Androl.* 23, 870–881.
146. Lyon, M. F., Glenister, P. H., and Lamoreaux, M. L. (1975). Normal spermatozoa from androgen-resistant germ cells of chimaeric mice and the role of androgen in spermatogenesis. *Nature* 258, 620–622.
147. Johnston, D. S., Russell, L. D., Friel, P. J., and Griswold, M. D. (2001). Murine germ cells do not require functional androgen receptors to complete spermatogenesis following spermatogonial stem cell transplantation. *Endocrinology* 142, 2405–2408.
148. Dierich A., Sairam, M. R., Monaco, L., Fimia, G. M., Gansmuller A., LeMeur, M., and Sassone-Corsi, P. (1998). Impairing follicle-stimulating hormone (FSH) signalling in vivo: targeted disruption of the FSH receptor leads to aberrant gametogenesis and hormonal imbalance. *Proc. Natl. Acad. Sci. U S A* 95, 13612–13617.
149. Abel, M. H., Wootton, A. N., Wilkins, V., Huhtaniemi, I., Knight, P. G., and Charlton, H. M. (2000). The effect of a null mutation in the follicle-stimulating hormone receptor gene on mouse reproduction. *Endocrinology* 141, 1795–1803.

150. Vanha-Perttula, T., Bardin, C. W., Allison, J. E., Gumbreck, L. C., and Stanley, A. J. (1970). Testicular feminization in the rat: Morphology of the testis. *Endocrinology* 87, 611–619.
151. Singh, J., O'Neill, C., and Handelsman, D. J. (1995). Induction of spermatogenesis by androgens in gonadotropin-deficient (hpg) mice. *Endocrinology* 136, 5311–5321.
152. Holdcraft, R. W., and Braun, R. E. (2004). Androgen receptor function is required in Sertoli cells for the terminal differentiation of haploid spermatids. *Development* 131, 459–467.
153. De Gendt, K., Swinnen, J. V., Saunders, P. T., Schoonjans, L., Dewerchin, M., Devos, A., Tan, K., Atanassova, N., Claessens, F., Lecureuil, C., Heyns, W., Carmeliet, P., Guillou, F., Sharpe, R. M., and Verhoeven, G. (2004). A Sertoli cell-selective knockout of the androgen receptor causes spermatogenic arrest in meiosis. *Proc. Natl. Acad. Sci. U S A* 101, 1327–1332.
154. Chang, C., Chen, Y. T., Yeh, S. D., Xu, Q., Wang, R. S., Guillou, F., Lardy, H., and Yeh, S. (2004). Infertility with defective spermatogenesis and hypotestosteronemia in male mice lacking the androgen receptor in Sertoli cells. *Proc. Natl. Acad. Sci. U S A* 101, 6876–6881.
155. Handelsman, D. J., Spaliviero, J. A., Simpson, J. M., Allan, C. M., and Singh, J. (1999). Spermatogenesis without gonadotropins: maintenance has a lower testosterone threshold than initiation. *Endocrinology* 140, 3938–3946.
156. Zhang, F. P., Pakarainen, T., Poutanen, M., Toppari, J., and Huhtaniemi, I. (2003). The low gonadotropin-independent constitutive production of testicular testosterone is sufficient to maintain spermatogenesis. *Proc. Natl. Acad. Sci. U S A* 100, 13692–13697.
157. Morse, H. C., Horike, N., Rowley, M. J., and Heller, C. G. (1973). Testosterone concentrations in testes of normal men: effects of testosterone propionate administration. *J. Clin. Endocrinol. Metab.* 37, 882–886.
158. McLachlan, R. I., O'Donnell, L., Stanton, P. G., Balourdos, G., Frydenberg, M., de Kretser, D. M., and Robertson, D. M. (2002). Effects of testosterone plus medroxyprogesterone acetate on semen quality, reproductive hormones, and germ cell populations in normal young men. *J. Clin. Endocrinol. Metab.* 87, 546–556.
159. Plant, T. M., and Marshall, G. R. (2001). The functional significance of FSH in spermatogenesis and the control of its secretion in male primates. *Endocr. Rev.* 22, 764–786.
160. Zirkin, B. R., Awoniyi, C., Griswold, M. D., Russell, L. D., and Sharpe, R. (1994). Is FSH required for adult spermatogenesis? *J. Androl.* 15, 273–276.
161. Kumar, T. R., Wang, Y., Lu, N., and Matzuk, M. M. (1997). FSH is required for ovarian follicle maturation but not for male fertility. *Nat. Genet.* 15, 201–204.
162. Mason, A. J., Pitts, S., Nikolics, K., Szonyi, E., Wilcox, J. N., Seeburg, P. H., and Stewart, T. H. (1986). The hypogonadal mouse: reproductive functions restored by gene therapy. *Science* 234, 1372–1378.
163. Kumar, T. R., Low, M. J., and Matzuk, M. M. (1998). Genetic rescue of follicle-stimulating hormone beta-deficient mice. *Endocrinology* 139, 3289–3295.
164. Allan, C. M., Garcia A., Spaliviero, J., Zhang, F. P., Jimenez, M., Huhtaniemi, I., and Handelsman, D. J. (2004). Complete Sertoli cell proliferation induced by follicle-stimulating hormone (FSH) independently of luteinizing hormone activity: evidence from genetic models of isolated FSH action. *Endocrinology* 145, 1587–1593.
165. Krieger, D. T., Perlow, M. J., Gibson, M. J., Davies, T. F., Zimmerman, E. A., Ferin, M., and Charlton, H. M. (1982). Brain grafts reverse hypogonadism of gonadotropin releasing hormone deficiency. *Nature* 298, 468–471.
166. Silverman, A. J., Silverman, R. C., and Gibson, M. J. (1990). Repair of reproductive defects by neural transplantation. *Biol. Reprod.* 42, 29–38.
167. Rubens, R., Dhont, M., and Vermeulen, A. (1974). Further studies on Leydig cell function in old age. *J. Clin. Endocrinol. Metab.* 39, 40–45.
168. Harman, S. M., and Tsitouras, P. D. (1980). Reproductive hormones in aging men I: measurement of sex steroids, basal luteinizing hormone, and Leydig cell response to human chorionic gonadotropin. *J. Clin. Endocrinol. Metab.* 51, 35–40.
169. Wang, C. (1988). Bioassay of follicle-stimulating hormone. *Endocr. Rev.* 9, 374–377.
170. Tenover, J. S., Dahl, K. D., Hsueh, A. J., Lim, P., Matsumoto, A. M., and Bremner, W. J. (1987). Serum bioactive and immunoreactive follicle-stimulating hormone levels and the response to clomiphene in healthy young and elderly men. *J. Clin. Endocrinol. Metab.* 64, 1103–1108.
171. Wide, L. (1989). Follicle-stimulating hormones in anterior pituitary glands from children and adults differ in relation to sex and age. *J. Endocrinol.* 123, 519–529.
172. Matzkin, H., Braf, Z, and Nava, D. (1991). Does age influence the bioactivity of follicle-stimulating hormone in men? *Age Ageing* 20, 199–205.
173. Urban, R. J., Dahl, K. D., Lippert, M. C., and Veldhuis, J. D. (1992). Modulation of immunoradiometric and bioactive follicle stimulating hormone secretion and clearance in young and elderly men during treatment with tamoxifen or flutamide. *J. Androl.* 13, 579–586.
174. Veldhuis, J. D., Iranmanesh A., Samojlik, E., and Urban, R. J. (1997). Differential sex steroid negative feedback regulation of pulsatile follicle-stimulating hormone secretion in healthy older men: deconvolution analysis and steady-state sex-steroid hormone infusions in frequently sampled healthy older individuals. *J. Clin. Endocrinol. Metab.* 82, 1248–1254.
175. Veldhuis, J. D., Iranmanesh A., Demers, L. M., and Mulligan, T. (1999). Joint basal and pulsatile hypersecretory mechanisms drive the monotropic follicle-stimulating hormone (FSH) elevation in healthy older men: concurrent preservation of the orderliness of the FSH release process: a general clinical research center study. *J. Clin. Endocrinol. Metab.* 84, 3506–3514.
176. Anawalt, B. D., Bebb, R. A., Matsumoto, A. M., Groome, N. P., Illingworth, P. J., McNeilly, A. S., and Bremner, W. J. (1996). Serum inhibin B levels reflect Sertoli cell function in normal men and men with testicular dysfunction. *J. Clin. Endocrinol. Metab.* 81, 3341–3345.
177. Meachem, S. J., Nieschlag, E., and Simoni, M. (2001). Inhibin B in male reproduction: pathophysiology and clinical relevance. *Eur. J. Endocrinol.* 145, 561–571.
178. Byrd, W., Bennett, M. J., Carr, B. R., Dong, Y., Wians, F., and Rainey, W. (1998). Regulation of biologically active dimeric inhibin A and B from infancy to adulthood in the male. *J. Clin. Endocrinol. Metab.* 83, 2849–2854.
179. Mahmoud, A. M., Goemaere, S., De Bacquer, D., Comhaire, F. H., and Kaufman, J. M. (2000). Serum inhibin B levels in community-dwelling elderly men. *Clin. Endocrinol. (Oxf).* 53, 141–147.
180. Baccarelli A., Morpurgo P.S., Corsi A., Vaghi, I., Fanelli, M., Cremonesi, G., Vaninetti, S., Beck-Peccoz, P., and Spada, A. (2001). Activin A serum levels and aging of the pituitary–gonadal axis: a cross-sectional study in middle-aged and elderly healthy subjects. *Exp. Gerontol.* 36, 1403–1412.
181. Bohring, C., and Krause, W. (2003). Serum levels of inhibin B in men of different age groups. *Aging Male* 6, 73–78.
182. Robertson, D. M., Tsonis, C. G., McLachlan, R. I., Handelsman, D. J., Leask, R., Baird, D. T., McNeilly, A. S., Hayward, S., Healy, D. L., and Findlay, J. K., et al. (1988). Comparison of

inhibin immunological and in vitro biological activities in human serum. *J. Clin. Endocrinol. Metab.* 67, 438–443.
183. Robertson, D. M., Stephenson, T., Pruysers, E., McCloud, P., Tsigos A., Groome, N., Mamers, P., and Burger, H. G. (2002). Characterization of inhibin forms and their measurement by an inhibin alpha-subunit ELISA in serum from postmenopausal women with ovarian cancer. *J. Clin. Endocrinol. Metab.* 87, 816–824.
184. Tenover, J. S., McLachlan, R. I., Dahl, K. D., Burger, H. G., de Kretser, D. M., and Bremner, W. J. (1988). Decreased serum inhibin levels in normal elderly men: evidence for a decline in Sertoli cell function with aging. *J. Clin. Endocrinol. Metab.* 67, 455–459.
185. Fingscheidt, U., and Nieschlag, E. (1989). The response of inhibin to human chorionic gonadotrophin is decreased in senescent men compared with young men. *J. Endocrinol.* 123, R9–R11.
186. Tenover, J. S., Dahl, K. D., Vale WW, Rivier, J. E., and Bremner, W. J. (1990). Hormonal responses to a potent gonadotropin hormone-releasing hormone antagonist in normal elderly men. *J. Clin. Endocrinol. Metab.* 71, 881–888.
187. Tenover, J. S., and Bremner, W. J. (1991). The effects of normal aging on the response of the pituitary–gonadal axis to chronic clomiphene administration in men. *J. Androl.* 12, 258–263.
188. Tenover, J. S., and Bremner, W. J. (1991). Circadian rhythm of serum immunoreactive inhibin in young and elderly men. *J. Gerontol.* 46, M181–M184.
189. MacNaughton, J. A., Bangah, M. L., McCloud, P. I., and Burger, H. G. (1991). Inhibin and age in men. *Clin. Endocrinol. (Oxf).* 35, 341–346.
190. Haji, M., Tanaka, S., Nishi, Y., Yanase, T., Takayanagi, R., Hasegawa, Y., Sasamoto, S., and Nawata, H. (1994). Sertoli cell function declines earlier than Leydig cell function in aging Japanese men. *Maturitas* 18, 143–153.
191. Harada, K., Shintani, Y., Sakamoto, Y., Wakatsuki, M., Shitsukawa, K., and Saito, S. (1996). Serum immunoreactive activin A levels in normal subjects and patients with various diseases. *J. Clin. Endocrinol. Metab.* 81, 2125–2130.
192. Loria, P., Petraglia, F., Concari, M., Bertolotti, M., Martella, P., Luisi, S., Grisolia, C., Foresta, C., Volpe A., Genazzani, A. R., and Carulli, N. (1998). Influence of age and sex on serum concentrations of total dimeric activin A. *Eur. J. Endocrinol.* 139, 487–492.
193. Welt, C. K., and Crowley, W. F. Jr. (1998). Activin: an endocrine or paracrine agent? *Eur. J. Endocrinol.* 139, 469–471.
194. Lipshultz, L. I., and Norris, J. S. (1977). A successful human Sertoli cell culture technique. *Surg. Forum.* 28, 579–580.
195. Lipshultz, L. I., Murthy, L., and Tindall, D. J. (1982). Characterization of human Sertoli cells in vitro. *J. Clin. Endocrinol. Metab.* 55, 228–237.
196. Holmes, S. D., Lipshultz, L. I., and Smith, R. G. (1983). Effect of cannabinoids on human Sertoli cell function in vitro. *Arch. Androl.* 11, 245–251.
197. Namiki, M., Nakamura, M., Okuyama A., Sonoda, T., Nishimune, Y., Takatsuka, D., and Matsumoto, K. (1987). Influence of temperature on the function of Sertoli and Leydig cells of human testes. *Fertil. Steril.* 47, 475–480.
198. Buch, J. P., Lamb, D. J., Lipshultz, L. I., and Smith, R. G. (1988). Partial characterization of a unique growth factor secreted by human Sertoli cells. *Fertil. Steril.* 49, 658–665.
199. O'Shaughnessy, P. J., Baker, P., Sohnius, U., Haavisto, A. M., Charlton, H. M., and Huhtaniemi, I. (1998). Fetal development of Leydig cell activity in the mouse is independent of pituitary gonadotroph function. *Endocrinology* 139, 1141–1146.
200. Gromoll, J., Eiholzer, U., Nieschlag, E., and Simoni, M. (2000). Male hypogonadism caused by homozygous deletion of exon 10 of the luteinizing hormone (LH) receptor: differential action of human chorionic gonadotropin and LH. *J. Clin. Endocrinol. Metab.* 85, 2281–2286.
201. Licht, P., Papkoff, H., Farmer, S. W., Muller, C. H., Tsui, H. W., and Crews, D. (1976). Evolution of gonadotropin structure and function. *Recent Prog. Horm. Res.* 33, 169–248.
202. Fiddes, J. C., and Goodman, H. M. (1980). The cDNA for the beta-subunit of human chorionic gonadotropin suggests evolution of a gene by readthrough into the 3′-untranslated region. *Nature* 286, 684–687.
203. Fiddes, J. C., and Talmadge, K. (1984). Structure, expression, and evolution of the genes for the human glycoprotein hormones. *Recent Prog. Horm. Res.* 40, 43–78.
204. Albanese, C., Colin, I. M., Crowley, W. F., Ito, M., Pestell, R. G., Weiss, J., and Jameson, J. L. (1996). The gonadotropin genes: evolution of distinct mechanisms for hormonal control. *Recent Prog. Horm. Res.* 51, 23–58; discussion 59–61.
205. Belchetz, P. E., Plant, T. M., Nakai, Y., Keogh, E. J., and Knobil, E. (1978). Hypophysial responses to continuous and intermittent delivery of hypopthalamic gonadotropin-releasing hormone. *Science* 202, 631–633.
206. Plant, T. M. (1986). Gonadal regulation of hypothalamic gonadotropin-releasing hormone release in primates. *Endocr. Rev.* 7, 75–88.
207. Knobil, E. (1992). Remembrance: the discovery of the hypothalamic gonadotropin-releasing hormone pulse generator and of its physiological significance. *Endocrinology* 131, 1005–1006.
208. Veldhuis, J. D., Pincus, S. M., and Keenan, D. M. (2003). Noninvasive in vivo investigative probes of adaptive mechanisms of neuroendocrine regulation: the paradigm of the hypothalamo–pituitary–gonadal axis in the aging male. *J. Endocrinol. Invest.* 26, 2–14.
209. Gray A., Berlin, J. A., McKinlay, J. B., and Longcope, C. (1991). An examination of research design effects on the association of testosterone and male aging: results of a meta-analysis. *J. Clin. Epidemiol.* 44, 671–684.
210. Furuyama, S., Mayes, D. M., and Nugent, C. A. (1970). A radioimmunoassay for plasma testosterone. *Steroids* 16, 415–428.
211. Ismail, A. A., Niswender, G. D., and Midgley, A. R. Jr. (1972). Radioimmunoassay of testosterone without chromatography. *J. Clin. Endocrinol. Metab.* 34, 177–184.
212. Yalow, R. S., and Berson, S. A. (1959). Assay of plasma insulin in human subjects by immunological methods. *Nature* 184, Suppl 21, 1648–1649.
213. Yalow, R. S., and Berson, S. A. (1960). Immunoassay of endogenous plasma insulin in man. *J. Clin. Invest.* 39, 1157–1175.
214. Vermeulen A., Rubens, R., and Verdonck, L. (1972). Testosterone secretion and metabolism in male senscence. *J. Clin. Endocrinol. Metab.* 34, 7305.
215. Sparrow, D., Bosse, R., and Rowe, J. W. (1980). The influence of age, alcohol consumption, and body build on gonadal function in men. *J. Clin. Endocrinol. Metab.* 51, 508–512.
216. Deslypere, J. P., and Vermeulen, A. (1984). Leydig cell function in normal men: effect of age, life-style, residence, diet, and activity. *J. Clin. Endocrinol. Metab.* 59, 955–962.
217. Simon, D., Preziosi, P., Barrett-Connor, E., Roger, M., Saint-Paul, M., Nahoul, K., and Papoz, L. (1992). The influence of aging on plasma sex hormones in men: the Telecom Study. *Am. J. Epidemiol.* 135, 783–791.
218. Barrett-Connor, E., and Khaw, K. T. (1988). Endogenous sex hormones and cardiovascular disease in men. A prospective population-based study. *Circulation* 78, 539–545.
219. Zmuda, J. M., Cauley, J. A., Kriska, A., Glynn, N. W., Gutai, J. P., and Kuller, L. H. (1997). Longitudinal relation between endogenous testosterone and cardiovascular disease risk factors in middle-aged men. *Am. J. Epidemiol.* 146, 609–617.

220. Gray, A., Feldman, H. A., McKinlay, J. B., and Longcope, C. (1991). Age, disease, and changing sex hormone levels in middle-aged men: results of the Massachussetts Male Aging Study. *J. Clin. Endocrinol. Metab.* 73, 1016–1025.
221. Morley, J. E., Kaiser, F. E., Perry, H. M. III, Patrick, P., Morley, P. M., Stauber, P. M., Vellas, B., Baumgartner, R. N., and Garry, P. J. (1997). Longitudinal changes in testosterone, luteinizing hormone, and follicle-stimulating hormone in healthy older men. *Metabolism* 46, 410–413.
222. Harman, S. M., Metter, E. J., Tobin, J. D., Pearson, J., and Blackman, M. R. (2001). Longitudinal effects of aging on serum total and free testosterone levels in healthy men. Baltimore Longitudinal Study of Aging. *J. Clin. Endocrinol. Metab.* 86, 724–731.
223. Feldman, H. A., Longcope, C., Derby, C. A., Johannes, C. B., Araujo, A. B., Coviello, A. D., Bremner, W. J., and McKinlay, J. B. (2002). Age trends in the level of serum testosterone and other hormones in middle-aged men: longitudinal results from the Massachusetts male aging study. *J. Clin. Endocrinol. Metab.* 87, 589–598.
224. Araujo, A. B., O'Donnell, A. B., Brambilla, D. J., Simpson, W. B., Longcope, C., Matsumoto, A. M., and McKinlay, J. B. (2004). Prevalence and incidence of androgen deficiency in middle-aged and older men: estimates from the Massachusetts Male Aging Study. *J. Clin. Endocrinol. Metab.* 89, 5920–5926.
225. Mohr, B. A., Guay, A. T., O'Donnell, A. B., and McKinlay, J. B. (2005). Normal, bound and nonbound testosterone levels in normally ageing men: results from the Massachusetts Male Ageing Study. *Clin. Endocrinol. (Oxf).* 62, 64–73.
226. Siiteri, P. K., Murai, J. T., Hammond, G. L., Nisker, J. A., Raymoure, W. J., and Kuhn, R. W. (1982). The serum transport of steroid hormones. *Recent Prog. Horm. Res.* 38, 457–510.
227. Joseph, D. R. (1994). Structure, function, and regulation of androgen-binding protein/sex hormone-binding globulin. *Vitam. Horm.* 49, 197–280.
228. Hammond, G. L., Avvakumov, G. V., and Muller, Y. A. (2003). Structure/function analyses of human sex hormone-binding globulin: effects of zinc on steroid-binding specificity. *J. Steroid Biochem. Mol. Biol.* 85, 195–200.
229. Dunn, J. F., Nisula, B. C., and Rodbard, D. (1981). Transport of steroid hormones: binding of 21 endogenous steroids to both testosterone-binding globulin and corticosteroid-binding globulin in human plasma. *J. Clin. Endocrinol. Metab.* 53, 58–68.
230. Rosner, W., Hryb, D. J., Khan, M. S., Nakhla, A. M., and Romas, N. A. (1999). Androgen and estrogen signaling at the cell membrane via G-proteins and cyclic adenosine monophosphate. *Steroids* 64, 100–106.
231. Ekins, R. (1990). Measurement of free hormones in blood. *Endocr. Rev.* 11, 5–46.
232. Pardridge, W. M. (1988). Selective delivery of sex steroid hormones to tissues by albumin and by sex hormone-binding globulin. *Oxf. Rev. Reprod. Biol.* 10, 237–292.
233. Mendel, C. M. (1989). The free hormone hypothesis: a physiologically based mathematical model. *Endocr. Rev.* 10, 232–274.
234. Vermeulen, A., Verdonck, L., and Kaufman, J. M. (1999). A critical evaluation of simple methods for the estimation of free testosterone in serum. *J. Clin. Endocrinol. Metab.* 84, 3666–3672.
235. Morley, J. E., Patrick, P., and Perry, H. M. III. (2002). Evaluation of assays available to measure free testosterone. *Metabolism* 51, 554–559.
236. Reventos, J., Sullivan, P. M., Joseph, D. R., and Gordon, J. W. (1993). Tissue-specific expression of the rat androgen-binding protein/sex hormone-binding globulin gene in transgenic mice. *Mol. Cell. Endocrinol.* 96, 69–73.
237. Anderson, D. C. (1974). Sex hormone binding globulin. *Clin. Endocrinol. (Oxf).* 3, 69–96.
238. von Schoultz, B., and Carlstrom, K. (1989). On the regulation of sex-hormone–binding globulin. A challenge of an old dogma and outlines of an alternative mechanism. *J. Steroid Biochem. Mol. Biol.* 32, 327–334.
239. Zumoff, B., Strain, G. W., Miller, L. K., Rosner, W., Senie, R., Seres, D. S., and Rosenfeld, R. S. (1990). Plasma free and non-sex-hormone-binding-globulin-bound testosterone are decreased in obese men in proportion to their degree of obesity. *J. Clin. Endocrinol. Metab.* 71, 929–931.
240. Selby, C. (1990). Sex hormone binding globulin: origin, function and clinical significance. *Ann. Clin. Biochem.* 27, Pt 6, 532–541.
241. Pugeat, M., Crave, J. C., Tourniaire, J., and Forest, M. G. (1996). Clinical utility of sex hormone-binding globulin measurement. *Horm. Res.* 45, 148–155.
242. Tchernof, A., and Despres, J. P. (2000). Sex steroid hormones, sex hormone-binding globulin, and obesity in men and women. *Horm. Metab. Res.* 32, 526–536.
243. Longcope, C., Feldman, H. A., McKinlay, J. B., and Araujo, A. B. (2000). Diet and sex hormone-binding globulin. *J. Clin. Endocrinol. Metab.* 85, 293–296.
244. Glass, A. R., Swerdloff, R. S., Bray, G. A., Dahms, W. T., and Atkinson, R. L. (1977). Low serum testosterone and sex hormone binding globulin in massively obese men. *J. Clin. Endocrinol. Metab.* 45, 1211–1219.
245. Kley, H. K., Deselaers, T., and Peerenboom, H. (1981). Evidence for hypogonadism in massively obese males due to decreased free testosterone. *Horm. Metab. Res.* 13, 639–641.
246. Haffner, S. M. (1996). Sex hormone-binding protein, hyperinsulinemia, insulin resistance and noninsulin-dependent diabetes. *Horm. Res.* 45, 233–237.
247. Pasquali, R., Vicennati, V., Scopinaro, N., Marinari, G., Simonelli A., Flamia, R., Casimirri, F., and Gagliardi, L. (1997). Achievement of near-normal body weight as the prerequisite to normalize sex hormone-binding globulin concentrations in massively obese men. *Int. J. Obes. Relat. Metab. Disord.* 21, 1–5.
248. Niskanen, L., Laaksonen, D. E., Punnonen, K., Mustajoki, P., Kaukua, J., and Rissanen, A. (2004). Changes in sex hormone-binding globulin and testosterone during weight loss and weight maintenance in abdominally obese men with the metabolic syndrome. *Diabetes Obes. Metab.* 6, 208–215.
249. Lewis, J. G., Shand, B. I., Elder, P. A., and Scott, R. S. (2004). Plasma sex hormone-binding globulin rather than corticosteroid-binding globulin is a marker of insulin resistance in obese adult males. *Diabetes Obes. Metab.* 6, 259–263.
250. Petra, P., Stanczyk, F. Z., Namkung, P. C., Fritz, M. A., and Novy, M. L. (1985). Direct effect of sex-steroid binding protein (SBP) of plasma on the metabolic clearance rate of testosterone in the rhesus macaque. *J. Steroid Biochem. Mol. Biol.* 22, 739–746.
251. Mori, H., Fukunishi, R., Fujii, M., Hataji, K., Shiraishi, T., and Matsumoto, K. (1978). Stereological analysis of Reinke's crystals in human Leydig cells. *Virchows Arch. A. Pathol. Anat. Histol.* 380, 1–9.
252. Mori, H., Hiromoto, N., Nakahara, M., and Shiraishi, T. (1982). Stereological analysis of Leydig cell ultrastructure in aged humans. *J. Clin. Endocrinol. Metab.* 55, 634–641.
253. Nieschlag, E., Kley, H. K., Wiegelmann, W., Solbach, H. G., and Kruskemper, H. L. (1973). [Age and endocrine function of the testes in adult man]. *Dtsch. Med. Wochenschr.* 98, 1281–1284.
254. Mulligan, T., Iranmanesh A., and Veldhuis, J. D. (2001). Pulsatile iv infusion of recombinant human LH in leuprolide-suppressed men unmasks impoverished Leydig-cell secretory responsiveness to midphysiological LH drive in the aging male. *J. Clin. Endocrinol. Metab.* 86, 5547–5553.

255. Veldhuis, J. D., Veldhuis, N. J., Keenan, D. M., and Iranmanesh, A. (2005). Age diminishes the testicular steroidogenic response to repeated intravenous pulses of recombinant human LH during acute gnrh-receptor blockade in healthy men. *Am. J. Physiol. Endocrinol. Metab.* 288, E775–E781.
256. Veldhuis, J. D., Iranmanesh A., and Keenan, D. M. (2004). Erosion of endogenous testosterone-driven negative feedback on pulsatile luteinizing hormone secretion in healthy aging men. *J. Clin. Endocrinol. Metab.* 89, 5753–5761.
257. Keenan, D. M., and Veldhuis, J. D. (2004). Divergent gonadotropin-gonadal dose-responsive coupling in healthy young and aging men. *Am. J. Physiol. Regul. Integr. Comp. Physiol.* 286, R381–R389.
258. Keenan, D. M., and Veldhuis, J. D. (2001). Hypothesis testing of the aging male gonadal axis via a biomathematical construct. *Am. J. Physiol. Regul. Integr. Comp. Physiol.* 280, R1755–R1771.
259. Veldhuis, J. D., Iranmanesh A., Godschalk, M., and Mulligan, T. (2000). Older men manifest multifold synchrony disruption of reproductive neurohormone outflow. *J. Clin. Endocrinol. Metab.* 85, 1477–1486.
260. Mulligan, T., Iranmanesh A., Johnson, M. L., Straume, M., and Veldhuis, J. D. (1997). Aging alters feed-forward and feedback linkages between LH and testosterone in healthy men. *Am. J. Physiol.* 273, R1407–R1413.
261. Mulligan, T., Iranmanesh A., Gheorghiu, S., Godschalk, M., and Veldhuis, J. D. (1995). Amplified nocturnal luteinizing hormone (LH) secretory burst frequency with selective attenuation of pulsatile (but not basal) testosterone secretion in healthy aged men: possible Leydig cell desensitization to endogenous LH signaling—a clinical research center study. *J. Clin. Endocrinol. Metab.* 80, 3025–3031.
262. Wang, C., Hikim, A. S., Ferrini, M., Bonavera, J. J., Vernet, D., Leung A., Lue, Y. H., Gonzalez-Cadavid, N. F., and Swerdloff, R. S. (2002). Male reproductive ageing: using the brown Norway rat as a model for man. *Novartis Found. Symp.* 242, 82–95; discussion 95–87.
263. Mendis-Handagama, S. M., and Gelber, S. J. (1995). Signs of aging are apparent in the testis interstitium of Sprague Dawley rats at 6 months of age. *Tissue Cell.* 27, 689–699.
264. Coutard, M., and Osborne-Pellegrin, M. (1997). Genetic susceptibility to experimental cerebral aneurysm formation in the rat. *Stroke* 28, 1035–1041; discussion 1042.
265. Bigazzi, P. E., Kosuda, L. L., and Harnick, L. L. (1977). Sperm autoantibodies in vasectomized rats of different inbred strains. *Science* 197, 1282–1283.
266. Furbeth, C., Hubner, G., and Thoenes, G. H. (1989). Spontaneous immune complex orchitis in brown Norway rats. *Virchows Arch., B., Cell Pathol.* 57, 37–45.
267. Kaler, L. W., and Neaves, W. B. (1981). The androgen status of aging male rats. *Endocrinology* 108, 712–719.
268. Harman, S. M., Danner, R. L., and Roth, G. S. (1978). Testosterone secretion in the rat in response to chorionic gonadotrophin: alterations with age. *Endocrinology* 102, 540–544.
269. Tsitouras, P. D., Kowatch, M. A., Blackman, M. R., and Harman, S. M. (1984). In vivo chorionic gonadotropin administration reverses the testosterone secretory defect of Leydig cells from old rats. *J. Gerontol.* 39, 257–263.
270. Lin, T., Murono, E., Osterman, J., Allen, D. O., and Nankin, H. R. (1980). The aging Leydig cell: 1. Testosterone and adenosine 3′,5′-monophosphate responses to gonadotropin stimulation in rats. *Steroids* 35, 653–663.
271. Chen, H., Hardy, M. P., and Zirkin, B. R. (2002). Age-related decreases in Leydig cell testosterone production are not restored by exposure to LH in vitro. *Endocrinology* 143, 1637–1642.
272. Grzywacz, F. W., Chen, H., Allegretti, J., and Zirkin, B. R. (1998). Does age-associated reduced Leydig cell testosterone production in Brown Norway rats result from understimulation by luteinizing hormone? *J. Androl.* 19, 625–630.
273. Wang, C., Sinha Hikim, A. P., Lue, Y. H., Leung A., Baravarian, S., and Swerdloff, R. S. (1999). Reproductive aging in the Brown Norway rat is characterized by accelerated germ cell apoptosis and is not altered by luteinizing hormone replacement. *J. Androl.* 20, 509–518.
274. Chen, H., Liu, J., Luo, L., and Zirkin, B. R. (2004). Dibutyryl cyclic adenosine monophosphate restores the ability of aged Leydig cells to produce testosterone at the high levels characteristic of young cells. *Endocrinology* 145, 4441–4446.
275. Chen, H., Cangello, D., Benson, S., Folmer, J., Zhu, H., Trush, M. A., and Zirkin, B. R. (2001). Age-related increase in mitochondrial superoxide generation in the testosterone-producing cells of Brown Norway rat testes: relationship to reduced steroidogenic function? *Exp. Gerontol.* 36, 1361–1373.
276. Chen, H., and Zirkin, B. R. (1999). Long-term suppression of Leydig cell steroidogenesis prevents Leydig cell aging. *Proc. Natl. Acad. Sci. U S A* 96, 14877–14881.
277. Chen, H., Irizarry, R. A., Luo, L., and Zirkin, B. R. (2004). Leydig cell gene expression: effects of age and caloric restriction. *Exp. Gerontol.* 39, 31–43.
278. Chen, H., Huhtaniemi, I., and Zirkin, B. R. (1996). Depletion and repopulation of Leydig cells in the testes of aging brown Norway rats. *Endocrinology* 137, 3447–3452.
279. Veldhuis, J. D. (2000). Recent neuroendocrine facets of male reproductive aging. *Exp. Gerontol.* 35, 1281–1308.
280. Veldhuis, J. D., Iranmanesh A., Mulligan, T., and Pincus, S. M. (1999). Disruption of the young-adult synchrony between luteinizing hormone release and oscillations in follicle-stimulating hormone, prolactin, and nocturnal penile tumescence (NPT) in healthy older men. *J. Clin. Endocrinol. Metab.* 84, 3498–3505.
281. Veldhuis, J. D., Urban, R. J., and Dufau, M. L. (1994). Differential responses of biologically active luteinizing hormone secretion in older versus young men to interruption of androgen negative feedback. *J. Clin. Endocrinol. Metab.* 79, 1763–1770.
282. Veldhuis, J. D., and Iranmanesh, A. (2004). Short-term aromatase-enzyme blockade unmasks impaired feedback adaptations in luteinizing hormone and testosterone secretion in older men. *J. Clin. Endocrinol. Metab.* 90, 302–309.
283. Zwart, A. D., Urban, R. J., Odell, W. D., and Veldhuis, J. D. (1996). Contrasts in the gonadotropin-releasing hormone dose–response relationships for luteinizing hormone, follicle-stimulating hormone and alpha-subunit release in young versus older men: appraisal with high-specificity immunoradiometric assay and deconvolution analysis. *Eur. J. Endocrinol.* 135, 399–406.
284. Mulligan, T., Iranmanesh A., Kerzner, R., Demers, L. W., and Veldhuis, J. D. (1999). Two-week pulsatile gonadotropin releasing hormone infusion unmasks dual (hypothalamic and Leydig cell) defects in the healthy aging male gonadotropic axis. *Eur. J. Endocrinol.* 141, 257–266.
285. Pincus, S. M., Mulligan, T., Iranmanesh A., Gheorghiu, S., Godschalk, M., and Veldhuis, J. D. (1996). Older males secrete luteinizing hormone and testosterone more irregularly, and jointly more asynchronously, than younger males. *Proc. Natl. Acad. Sci. U S A* 93, 14100–14105.
286. Liu, P. Y., Pincus, S. M., Keenan, D. M., Roelfsema, F., and Veldhuis, J. D. (2004). Analysis of bidirectional pattern synchrony of concentration-secretion pairs: implementation in the human testicular and adrenal axis. *Am. J. Physiol. Regul. Integr. Comp. Physiol.* 288, R440–R446.
287. Bergendahl, M., Aloi, J. A., Iranmanesh A., Mulligan, T. M., and Veldhuis, J. D. (1998). Fasting suppresses pulsatile luteinizing hormone (LH) secretion and enhances orderliness

of LH release in young but not older men. *J. Clin. Endocrinol. Metab.* 83, 1967–1975.

288. Karpas, A. E., Bremner, W. J., Clifton, D. K., Steiner, R. A., and Dorsa, D. M. (1983). Diminished luteinizing hormone pulse frequency and amplitude with aging in the male rat. *Endocrinology* 112, 788–792.
289. Steiner, R. A., Bremner, W. J., Clifton, D. K., and Dorsa, D. M. (1984). Reduced pulsatile luteinizing hormone and testosterone secretion with aging in the male rat. *Biol. Reprod.* 31, 251–258.
290. Winters, S. J., and Atkinson, L. (1997). Serum LH concentrations in hypogonadal men during transdermal testosterone replacement through scrotal skin: further evidence that ageing enhances testosterone negative feedback. The Testoderm Study Group. *Clin. Endocrinol. (Oxf).* 47, 317–322.
291. Dilman, V. M., and Anisimov, V. N. (1979). Hypothalmic mechanisms of ageing and of specific age pathology-I. Sensitivity threshold of hypothalamo–pituitary complex to homeostatic stimuli in the reproductive system. *Exp. Gerontol.* 14, 161–174.
292. Muta, K., Kato, K., Akamine, Y., and Ibayashi, H. (1981). Age-related changes in the feedback regulation of gonadotrophin secretion by sex steroids in men. *Acta Endocrinol. (Copenh).* 96, 154–162.
293. Veldhuis, J. D., Zwart A., Mulligan, T., and Iranmanesh, A. (2001). Muting of androgen negative feedback unveils impoverished gonadotropin-releasing hormone/luteinizing hormone secretory reactivity in healthy older men. *J. Clin. Endocrinol. Metab.* 86, 529–535.
294. Mooradian, A. D., Morley, J. E., and Korenman, S. G. (1987). Biological actions of androgens. *Endocr. Rev.* 8, 1–28.
295. Kimura, N., Mizokami A., Oonuma, T., Sasano, H., and Nagura, H. (1993). Immunocytochemical localization of androgen receptor with polyclonal antibody in paraffin-embedded human tissues. *J. Histochem. Cytochem.* 41, 671–678.
296. Quigley, C. A., DeBellis, A., Marschke, K. B., El-Awady, M. K., Wilson, E. M., and French, F. F. (1995). Androgen receptor defects: historical, clinical and molecular perspectives. *Endocr. Rev.* 16, 271–321.
297. Tunn, S., Hochstrate, H., Grunwald, I., Fluchter, S. H., and Krieg, M. (1988). Effect of aging on kinetic parameters of 5 alpha-reductase in epithelium and stroma of normal and hyperplastic human prostate. *J. Clin. Endocrinol. Metab.* 67, 979–985.
298. Wu, A. H., Whittemore, A. S., Kolonel, L. N., Stanczyk, F. Z., John, E. M., Gallagher, R. P., and West, D. W. (2001). Lifestyle determinants of 5alpha-reductase metabolites in older African-American, white, and Asian-American men. *Cancer Epidemiol. Biomarkers Prev.* 10, 533–538.
299. Hellwinkel, O. J., Muller, A., Struve, D., and Hiort, O. (2000). Influence of androgens and age on androgen receptor and 5 alpha-reductase II transcription. *Eur. J. Endocrinol.* 143, 217–225.
300. Weisser, H., and Krieg, M. (1997). Kinetic analysis of androstenedione 5 alpha-reductase in epithelium and stroma of human prostate. *Steroids* 62, 589–594.
301. Krieg, M., Weisser, H., and Tunn, S. (1995). Potential activities of androgen metabolizing enzymes in human prostate. *J. Steroid Biochem. Mol. Biol.* 53, 395–400.
302. Vermeulen A., Kaufman, J. M., Goemaere, S., and van Pottelberg, I. (2002). Estradiol in elderly men. *Aging Male* 5, 98–102.
303. Zitzmann, M., and Nieschlag, E. (2003). The CAG repeat polymorphism within the androgen receptor gene and maleness. *Int. J. Androl.* 26, 76–83.
304. Krithivas, K., Yurgalevitch, S. M., Mohr, B. A., Wilcox, C. J., Batter, S. J., Brown, M., Longcope, C., McKinlay, J. B., and Kantoff, P. W. (1999). Evidence that the CAG repeat in the androgen receptor gene is associated with the age-related decline in serum androgen levels in men. *J. Endocrinol.* 162, 137–142.
305. Zitzmann, M., Brune, M., Kornmann, B., Gromoll, J., Junker, R., and Nieschlag, E. (2001). The CAG repeat polymorphism in the androgen receptor gene affects bone density and bone metabolism in healthy males. *Clin. Endocrinol. (Oxf).* 55, 649–657.
306. Harkonen, K., Huhtaniemi, I., Makinen, J., Hubler, D., Irjala, K., Koskenvuo, M., Oettel, M., Raitakari, O., Saad, F., and Pollanen, P. (2003). The polymorphic androgen receptor gene CAG repeat, pituitary–testicular function and andropausal symptoms in ageing men. *Int. J. Androl.* 26, 187–194.
307. Ono, K., Haji, M., Nawata, H., Maki, T., Kato, K., and Ibayashi, H. (1988). Age-related changes in glucocorticoid and androgen receptors of cultured human pubic skin fibroblasts. *Gerontology* 34, 128–133.
308. Ishunina, T. A., Fisser, B., and Swaab, D. F. (2002). Sex differences in androgen receptor immunoreactivity in basal forebrain nuclei of elderly and Alzheimer patients. *Exp. Neurol.* 176, 122–132.
309. Visakorpi, T., Hyytinen, E., Koivisto, P., Tanner, M., Keinanen, R., Palmberg, C., Palotie A., Tammela, T., Isola, J., and Kallioniemi, O. P. (1995). In vivo amplification of the androgen receptor gene and progression of human prostate cancer. *Nat. Genet.* 9, 401–406.
310. Linja, M. J., and Visakorpi, T. (2004). Alterations of androgen receptor in prostate cancer. *J. Steroid Biochem. Mol. Biol.* 92, 255–264.
311. Heinlein, C. A., and Chang, C. (2002). Androgen receptor (AR) coregulators: an overview. *Endocr. Rev.* 23, 175–200.
312. Salmimies, P., Kockott, G., Pirke, K. M., Vogt, H. J., and Schill, W. B. (1982). Effects of testosterone replacement on sexual behaviour in hypogonadal men. *Arch. Sex. Behav.* 11, 345–353.
313. Gooren, L. J. (1987). Androgen levels and sex functions in testosterone-treated hypogonadal men. *Arch. Sex. Behav.* 16, 463–473.
314. Buena, F., Peterson, M. A., Swerdloff, R. S., Pandian, M. R., Steiner, B. S., Galmarini, M., Lutchmansingh, P., and Bhasin, S. (1993). Sexual function does not change when serum testosterone levels are pharmacologically varied within the normal male range. *Fertil. Steril.* 59, 1118–1123.
315. Bagatell, C. J., Heiman, J. R., Rivier, J. E., and Bremner, W. J. (1994). Effects of endogenous testosterone and estradiol on sexual behaviour in normal young men. *J. Clin. Endocrinol. Metab.* 78, 711–716.
316. Woodhouse, L. J., Reisz-Porszasz, S., Javanbakht, M., Storer, T. W., Lee, M., Zerounian, H., and Bhasin, S. (2003). Development of models to predict anabolic response to testosterone administration in healthy young men. *Am. J. Physiol. Endocrinol. Metab.* 284, E1009–E1017.
317. Kelleher, S., Conway, A. J., and Handelsman, D. J. (2004). Blood testosterone threshold for androgen deficiency symptoms. *J. Clin. Endocrinol. Metab.* 89, 3813–3817.
318. Santner, S., Albertson, B., Zhang, G. Y., Zhang, G. H., Santulli, M., Wang, C., Demers, L. M., Shackleton, C., and Santen, R. J. (1998). Comparative rates of androgen production and metabolism in Caucasian and Chinese subjects. *J. Clin. Endocrinol. Metab.* 83, 2104–2109.
319. Jin, B., Turner, L., Zhou, Z, Zhou, E. L., and Handelsman, D. J. (1999). Ethnicity and migration as determinants of human prostate size. *J. Clin. Endocrinol. Metab.* 84, 3613–3619.
320. Hamilton, J. B., and Mestler, G. E. (1969). Mortality and survival: comparison of eunuchs with intact men and women in a mentally retarded population. *J. Gerontol.* 24, 395–411.
321. Nieschlag, E., Nieschlag, S., and Behre, H. M. (1993). Lifespan and testosterone. *Nature* 366, 215.

322. Jenkins, J. S. (1998). The voice of the castrato. *Lancet* 351, 1877–1880.
323. Liu, P. Y., Death, A. K., and Handelsman, D. J. (2003). Androgens and cardiovascular disease. *Endocr. Rev*. 24, 313–340.
324. Stutz, J. A., Barry, B. P., Maslanka, W., Sokal, M., Green, D. J., and Pearson, D. (1998). Bone density: is it affected by orchidectomy and radiotherapy given for stage I seminoma of the testis? *Clin. Oncol. (R. Coll. Radiol.)* 10, 44–49.
325. Laron, Z., and Zilka, E. (1969). Compensatory hypertrophy of testicle in unilateral cryptorchidism. *J. Clin. Endocrinol. Metab*. 29, 1409–1413.
326. Laron, Z, Dickerman, Z, Ritterman, I., and Kaufman, H. (1980). Follow-up of boys with unilateral compensatory testicular hypertrophy. *Fertil. Steril*. 33, 297–301.
327. Hansen, S. W., Berthelsen, J. G., and von der Maase, H. (1990). Long-term fertility and Leydig cell function in patients treated for germ cell cancer with cisplatin, vinblastine, and bleomycin versus surveillance. *J. Clin. Oncol*. 8, 1695–1698.
328. Daehlin, L., Ulstein, M., Thorsen, T., and Hoisaeter, P. A. (1996). Follow-up after torsion of the spermatic cord. *Scand. J. Urol. Nephrol. Suppl*. 179, 139–142.
329. Rybkiewicz, M. (2001). [Long-term and late results of treatment in patients with a history of testicular torsion]. *Ann. Acad. Med. Stetin*. 47, 61–75.
330. Nord, C., Bjoro, T., Ellingsen, D., Mykletun A., Dahl, O., Klepp, O., Bremnes, R. M., Wist, E., and Fossa, S. D. (2003). Gonadal hormones in long-term survivors 10 years after treatment for unilateral testicular cancer. *Eur. Urol*. 44, 322–328.
331. Behre, H. M., Kliesch, S., Leifke, E., Link, T. M., and Nieschlag, E. (1997). Long-term effect of testosterone therapy on bone mineral density in hypogonadal men. *J. Clin. Endocrinol. Metab*. 82, 2386–2390.
332. Liu, P. Y., Swerdloff, R. S., and Veldhuis, J. D. (2004). Clinical review 171: The rationale, efficacy and safety of androgen therapy in older men: future research and current practice recommendations. *J. Clin. Endocrinol. Metab*. 89, 4789–4796.
333. Bhasin, S., Storer, T. W., Berman, N., Yarasheski, K. E., Clevenger, B., Phillips, J., Lee, W. P., Bunnell, T. J., and Casaburi, R. (1997). Testosterone replacement increases fat-free mass and muscle size in hypogonadal men. *J. Clin. Endocrinol. Metab*. 82, 407–413.
334. Novak, L. P. (1972). Aging, total body potassium, fat-free mass, and cell mass in males and females between ages 18 and 85 years. *J. Gerontol*. 27, 438–443.
335. Hughes, V. A., Frontera, W. R., Roubenoff, R., Evans, W. J., and Singh, M. A. (2002). Longitudinal changes in body composition in older men and women: role of body weight change and physical activity. *Am. J. Clin. Nutr*. 76, 473–481.
336. Melton, L. J. III, Khosla, S., Crowson, C. S., O'Connor, M. K., O'Fallon, W. M., and Riggs, B. L. (2000). Epidemiology of sarcopenia. *J. Am. Geriatr. Soc*. 48, 625–630.
337. Frontera, W. R., Hughes, V. A., Fielding, R. A., Fiatarone, M. A., Evans, W. J., and Roubenoff, R. (2000). Aging of skeletal muscle: a 12-yr longitudinal study. *J. Appl. Physiol*. 88, 1321–1326.
338. Baumgartner, R. N., Koehler, K. M., Gallagher, D., Romero, L., Heymsfield, S. B., Ross, R. R., Garry, P. J., and Lindeman, R. D. (1998). Epidemiology of sarcopenia among the elderly in New Mexico. *Am. J. Epidemiol*. 147, 755–763.
339. Dutta, C., and Hadley, E. C. (1995). The significance of sarcopenia in old age. *J. Gerontol. A. Biol. Sci. Med. Sci*. 50, Spec No., 1–4.
340. Kohrt, W. M., Malley, M. T., Dalsky, G. P., and Holloszy, J. O. (1992). Body composition of healthy sedentary and trained, young and older men and women. *Med. Sci. Sports Exerc*. 24, 832–837.
341. Steen, B. (1988). Body composition and aging. *Nutr. Rev*. 46, 45–51.
342. Borkan, G. A., Hults, D. E., Gerzof, S. G., Robbins, A. H., and Silbert, C. K. (1983). Age changes in body composition revealed by computed tomography. *J. Gerontol*. 38, 673–677.
343. Borkan, G. A., and Norris, A. H. (1977). Fat redistribution and the changing body dimensions of the adult male. *Hum. Biol*. 49, 495–513.
344. Cohn, S. H., Vaswani A., Zanzi, I., Aloia, J. F., Roginsky, M. S., and Ellis, K. D. (1976). Changes in body chemical composition with age measured by total-body neutron activation. *Metabolism* 25, 85–95.
345. Lindle, R. S., Metter, E. J., Lynch, N. A., Fleg, J. L., Fozard, J. L., Tobin, J., Roy, T. A., and Hurley, B. F. (1997). Age and gender comparisons of muscle strength in 654 women and men aged 20–93 yr. *J. Appl. Physiol*. 83, 1581–1587.
346. Metter, E. J., Conwit, R., Tobin, J., and Fozard, J. L. (1997). Age-associated loss of power and strength in the upper extremities in women and men. *J. Gerontol. A. Biol. Sci. Med. Sci*. 52, B267–B276.
347. Harris, T. (1997). Muscle mass and strength: relation to function in population studies. *J. Nutr*. 127, 1004S–1006S.
348. Wolfson, L., Judge, J., Whipple, R., and King, M. (1995). Strength is a major factor in balance, gait, and the occurrence of falls. *J. Gerontol. A. Biol. Sci. Med. Sci*. 50, Spec No., 64–67.
349. Skelton, D. A., Greig, C. A., Davies, J. M., and Young, A. (1994). Strength, power and related functional ability of healthy people aged 65–89 years. *Age Ageing*. 23, 371–377.
350. Bassey, E. J., Fiatarone, M. A., O'Neill EF, Kelly, M., Evans, W. J., and Lipsitz, L. A. (1992). Leg extensor power and functional performance in very old men and women. *Clin. Sci. (Lond)*. 82, 321–327.
351. Reed, R. L., Pearlmutter, L., Yochum, K., Meredith, K. E., and Mooradian, A. D. (1991). The relationship between muscle mass and muscle strength in the elderly. *J. Am. Geriatr. Soc*. 39, 555–561.
352. Whipple, R. H., Wolfson, L. I., and Amerman, P. M. (1987). The relationship of knee and ankle weakness to falls in nursing home residents: an isokinetic study. *J. Am. Geriatr. Soc*. 35, 13–20.
353. Riggs, B. L., Wahner, H. W., Seeman, E., Offord, K. P., Dunn, W. L., Mazess, R. B., Johnson, K. A., and Melton, L. J. III. (1982). Changes in bone mineral density of the proximal femur and spine with aging. Differences between the postmenopausal and senile osteoporosis syndromes. *J. Clin. Invest*. 70, 716–723.
354. Looker, A. C., Orwoll, E. S., Johnston, C. C., Jr., Lindsay, R. L., Wahner, H. W., Dunn, W. L., Calvo, M. S., Harris, T. B., and Heyse, S. P. (1997). Prevalence of low femoral bone density in older U.S. adults from NHANES III. *J. Bone Miner. Res*. 12, 1761–1768.
355. Kanis, J. A., Johnell, O., Oden A., De Laet, C., and Mellstrom, D. (2001). Diagnosis of osteoporosis and fracture threshold in men. *Calcif. Tissue Int*. 69, 218–221.
356. Santavirta, S., Konttinen, Y. T., Heliovaara, M., Knekt, P., Luthje, P., and Aromaa, A. (1992). Determinants of osteoporotic thoracic vertebral fracture. Screening of 57,000 Finnish women and men. *Acta Orthopaed. Scand*. 63, 198–202.
357. Anonymous. (2002). Incidence of vertebral fracture in Europe: results from the European Prospective Osteoporosis Study (EPOS). *J. Bone Miner. Res*. 17, 716–724.
358. Salthouse, T. A. (2003). Memory aging from 18 to 80. *Alzheimer Dis. Assoc. Disord*. 17, 162–167.
359. Martin, C. E. (1981). Factor affecting sexual functioning in 60–79-year-old married males. *Arch. Sex. Behav*. 10, 399–420.
360. ACSF. (1992). AIDS and sexual behaviour in France. *Nature* 360, 407–409.
361. Feldman, H. A., Goldstein, I., Hatzichristou, D. G., Krane, R. J., and McKinlay, J. B. (1994). Impotence and its medical and

psychosocial correlates: results of the Massachusetts Male Aging Study. *J. Urol.* (1994). 151, 54–61.

362. Macfarlane, G. J., Botto, H., Sagnier, P. P., Teillac, P., Richard, F., and Boyle, P. (1996). The relationship between sexual life and urinary condition in the French community. *J. Clin. Epidemiol.* 49, 1171–1176.
363. Frankel, S. J., Donovan, J. L., Peters, T. I., Abrams, P., Dabhoiwala, N. F., Osawa, D., and Lin, A. T. (1998). Sexual dysfunction in men with lower urinary tract symptoms. *J. Clin. Epidemiol.* 51, 677–685.
364. Johannes, C. B., Araujo, A. B., Feldman, H. A., Derby, C. A., Kleinman, K. P., and McKinlay, J. B. (2000). Incidence of erectile dysfunction in men 40 to 69 years old: longitudinal results from the Massachusetts male aging study. *J. Urol.* 163, 460–463.
365. Braun, M., Wassmer, G., Klotz, T., Reifenrath, B., Mathers, M., and Engelmann, U. (2000). Epidemiology of erectile dysfunction: results of the 'Cologne Male Survey'. *Int. J. Impot. Res.* 12, 305–311.
366. Johnson, A. M., Mercer, C. H., Erens, B., Copas, A. J., McManus, S., Wellings, K., Fenton, K. A., Korovessis, C., Macdowall, W., Nanchahal, K., Purdon, S., and Field, J. (2001). Sexual behaviour in Britain: partnerships, practices, and HIV risk behaviours. *Lancet* 358, 1835–1842.
367. Rosen, R., Altwein, J., Boyle, P., Kirby, R. S., Lukacs, B., Meuleman, E., O'Leary, M. P., Puppo, P., Robertson, C., and Giuliano, F. (2003). Lower urinary tract symptoms and male sexual dysfunction: the multinational survey of the aging male (MSAM-7). *Eur. Urol.* 44, 637–649.
368. Smith, A. M., Rissel, C. E., Richters, J., Grulich, A. E., and de Visser, R. O. (2003). Sex in Australia. *Aust. N.Z. J. Public Health.* 27, 103–256.
369. Araujo, A. B., Mohr, B. A., and McKinlay, J. B. (2004). Changes in sexual function in middle-aged and older men: longitudinal data from the Massachusetts Male Aging Study. *J. Am. Geriatr. Soc.* 52, 1502–1509.
370. Handelsman, D. J. (1998). The safety of androgens: prostate and cardiovascular disease. In *Male Reproductive Function* (C. Wang, Ed.), pp. 173–190. Kluwer Academic Publishers, Boston.
371. Rhoden, E. L., and Morgentaler, A. (2004). Risks of testosterone-replacement therapy and recommendations for monitoring. *N. Engl. J. Med.* 350, 482–492.
372. Wu, F. C., and von Eckardstein, A. (2003). Androgens and coronary artery disease. *Endocr. Rev.* 24, 183–217.
373. Kalin, M. F., and Zumoff, B. (1990). Sex hormones and coronary disease: a review of the clinical studies. *Steroids* 55, 330–352.
374. Hulley, S., Grady, D., Bush, T., Furberg, C., Herrington, D., Riggs, B., and Vittinghoff, E. (1998). Randomized trial of estrogen plus progestin for secondary prevention of coronary heart disease in postmenopausal women. Heart and Estrogen/progestin Replacement Study (HERS) Research Group. *JAMA* 280, 605–613.
375. Rossouw, J. E., Anderson, G. L., Prentice, R. L., LaCroix AZ, Kooperberg, C., Stefanick, M. L., Jackson, R. D., Beresford, S. A., Howard, B. V., Johnson, K. C., Kotchen, J. M., and Ockene, J. (2002). Risks and benefits of estrogen plus progestin in healthy postmenopausal women: principal results From the Women's Health Initiative randomized controlled trial. *JAMA* 288, 321–333.
376. Alexandersen, P., Haarbo, J., and Christiansen, C. (1996). The relationship of natural androgens to coronary heart disease in male: a review. *Atherosclerosis* 125, 1–13.
377. Snyder, P. J. (2004). Hypogonadism in elderly men—what to do until the evidence comes. *N. Engl. J. Med.* 350, 440–442.
378. Shahidi, N. T. (1973). Androgens and erythropoiesis. *N. Engl. J. Med.* 289, 72–80.
379. Drinka, P. J., Jochen, A. L., Cuisinier, M., Bloom, R., Rudman, I., and Rudman, D. (1995). Polycythemia as a complication of testosterone replacement therapy in nursing home men with low testosterone levels. *J. Am. Geriatr. Soc.* 43, 899–901.
380. Jockenhövel, F., Vogel, E., Reinhardt, W., and Reinwein, D. (1997). Effects of various modes of androgen substitution therapy on erythropoiesis. *Eur. J. Med. Res.* 2, 293–298.
381. Hajjar, R. R., Kaiser, F. E., and Morley, J. E. (1997). Outcomes of long-term testosterone replacement in older hypogonadal males: a retrospective analysis. *J. Clin. Endocrinol. Metab.* 82, 3793–3796.
382. Dobs, A. S., Meikle AW, Arver, S., Sanders, S. W., Caramelli, K. E., and Mazer, N. A. (1999). Pharmacokinetics, efficacy, and safety of a permeation-enhanced testosterone transdermal system in comparison with bi-weekly injections of testosterone enanthate for the treatment of hypogonadal men. *J. Clin. Endocrinol. Metab.* 84, 3469–3478.
383. Viallard, J. F., Marit, G., Mercie, P., Leng, B., Reiffers, J., and Pellegrin, J. L. (2000). Polycythaemia as a complication of transdermal testosterone therapy. *Br. J. Haematol.* 110, 237–238.
384. Siddique, H., Smith, J. C., and Corrall, R. J. (2004). Reversal of polycythaemia induced by intramuscular androgen replacement using transdermal testosterone therapy. *Clin. Endocrinol. (Oxf).* 60, 143–145.
385. Amory, J. K., Watts, N. B., Easley, K. A., Sutton, P. R., Anawalt, B. D., Matsumoto, A. M., Bremner, W. J., and Tenover, J. L. (2004). Exogenous testosterone or testosterone with finasteride increases bone mineral density in older men with low serum testosterone. *J. Clin. Endocrinol. Metab.* 89, 503–510.
386. Freinhar, J. P., and Alvarez, W. (1985). Androgen-induced hypomania. *J. Clin. Psychiatry* 46, 354–355.
387. Su, T. P., Pagliaro, M., Schmidt, P. J., Pickar, D., Wolkowitz, O., and Rubinow, D. R. (1993). Neuropsychiatric effects of anabolic steroids in male normal volunteers. *JAMA* 269, 2760–2764.
388. Weiss, E. L., Bowers, M. B. Jr., and Mazure, C. M. (1999). Testosterone-patch–induced psychotic mania. *Am. J. Psychiatry* 156, 969.
389. Yates, W. R., Perry, P. J., MacIndoe, J., Holman, T., and Ellingrod, V. (1999). Psychosexual effects of three doses of testosterone cycling in normal men. *Biol. Psychiatry* 45, 254–260.
390. Pope, H. G. Jr., Kouri, E. M., and Hudson, J. I. (2000). Effects of supraphysiologic doses of testosterone on mood and aggression in normal men: a randomized controlled trial. *Arch. Gen. Psychiatry* 57, 133–140; discussion 155–136.
391. Liu, P. Y., Yee, B. J., Wishart, S. M., Jimenez, M., Jung, D. G., Grunstein, R. R., and Handelsman, D. J. (2003). The short-term effects of high dose testosterone on sleep, breathing and function in older men. *J. Clin. Endocrinol. Metab.* 88, 3605–3613.
392. Tangredi, J. F., and Buxton, I. L. (2001). Hypertension as a complication of topical testosterone therapy. *Ann. Pharmacother.* 35, 1205–1207.
393. Liverman, C. T., and Blazer, D. G., Eds. (2004). *Testosterone and Aging: Clinical Research Directions. Board on Health Sciences Policy.* Institute of Medicine: The National Academies Press, Washington, DC.
394. Tenover, J. S. (1992). Effects of testosterone supplementation in the aging male. *J. Clin. Endocrinol. Metab.* 75, 1092–1098.
395. Sih, R., Morley, J. E., Kaiser, F. E., Perry, H. M. III, Patrick, P., and Ross, C. (1997). Testosterone replacement in older hypogonadal men: a 12-month randomized controlled trial. *J. Clin. Endocrinol. Metab.* 82, 1661–1667.
396. Clague, J. E., Wu, F. C., and Horan, M. A. (1999). Difficulties in measuring the effect of testosterone replacement therapy on muscle function in older men. *Int. J. Androl.* 22, 261–265.

397. Snyder, P. J., Peachey, H., Hannoush, P., Berlin, J. A., Loh, L., Lenrow, D. A., Holmes, J. H., Dlewati A., Santanna, J., Rosen, C. J., and Strom, B. L. (1999). Effect of testosterone treatment on body composition and muscle strength in men over 65 years of age. *J. Clin. Endocrinol. Metab.* 84, 2647–2653.
398. Kenny, A. M., Prestwood, K. M., Gruman, C. A., Marcello, K. M., and Raisz, L. G. (2001). Effects of transdermal testosterone on bone and muscle in older men with low bioavailable testosterone levels. *J. Gerontol. A. Biol. Sci. Med. Sci.* 56, M266–272.
399. Ly, L. P., Jimenez, M., Zhuang, T. N., Celermajer, D. S., Conway, A. J., and Handelsman, D. J. (2001). A double-blind, placebo-controlled, randomized clinical trial of transdermal dihydrotestosterone gel on muscular strength, mobility, and quality of life in older men with partial androgen deficiency. *J. Clin. Endocrinol. Metab.* 86, 4078–4088.
400. Ferrando, A. A., Sheffield-Moore, M., Yeckel, C. W., Gilkison, C., Jiang, J., Achacosa A., Lieberman, S. A., Tipton, K., Wolfe, R. R., and Urban, R. J. (2002). Testosterone administration to older men improves muscle function: molecular and physiological mechanisms. *Am. J. Physiol. Endocrinol. Metab.* 282, E601–607.
401. Liu, P. Y., Wishart, S. M., and Handelsman, D. J. (2002). A double-blind, placebo-controlled, randomized clinical trial of recombinant human chorionic gonadotropin on muscle strength and physical function and activity in older men with partial age-related androgen deficiency. *J. Clin. Endocrinol. Metab.* 87, 3125–3135.
402. Snyder, P. J., Peachey, H., Hannoush, P., Berlin, J. A., Loh, L., Holmes, J. H., Dlewati A., Staley, J., Santanna, J., Kapoor, S. C., Attie, M. F., Haddad, J. G. Jr., and Strom, B. L. (1999). Effect of testosterone treatment on bone mineral density in men over 65 years of age. *J. Clin. Endocrinol. Metab.* 84, 1966–1972.
403. Bancroft, J. (1995). Are the effects of androgens on male sexuality noradrenergically mediated? Some consideration of the human. *Neurosci. Biobehav. Rev.* 19, 325–330.
404. Wilson, J. D. The role of androgens in male gender role behavior. *Endocr. Rev.* (1999). 20, 726–737.
405. Carani, C., Granata, A. R., Bancroft, J., and Marrama, P. (1995). The effects of testosterone replacement on nocturnal penile tumescence and rigidity and erectile response to visual erotic stimuli in hypogonadal men. *Psychoneuroendocrinology* 20, 743–753.
406. Kinsey, A. C., Pomeroy, W. C., and Martin, C. E. (1948). *Sexual Behaviour in the Human Male.* WB Saunders, Philadelphia.
407. Heim, N. (1981). Sexual behavior of castrated sex offenders. *Arch. Sex. Behav.* 10, 11–19.
408. Zverina, J., Hampl, R., Sulocava, J., and Starka, L. (1990). Hormonal status and sexual behaviour of 16 men after surgical castration. *Arch. Ital. Urol. Nefrol. Androl.* 62, 55–58.
409. Potosky, A. L., Reeve, B. B., Clegg, L. X., Hoffman, R. M., Stephenson, R. A., Albertsen, P. C., Gilliland, F. D., and Stanford, J. L. (2002). Quality of life following localized prostate cancer treated initially with androgen deprivation therapy or no therapy. *J. Natl. Cancer Inst.* 94, 430–437.
410. Herr, H. W., Kornblith, A. B., and Ofman, U. (1993). A comparison of the quality of life of patients with metastatic prostate cancer who received or did not receive hormonal therapy. *Cancer* 71, 1143–1150.
411. Bergman, B., Damber, J. E., Littbrand, B., Sjogren, K., and Tomic, R. (1984). Sexual function in prostatic cancer patients treated with radiotherapy, orchiectomy or oestrogens. *Br. J. Urol.* 56, 64–69.
412. Tsitouras, P. D., Martin, C. E., and Harman, S. M. (1982). Relationship of serum testosterone to sexual activity in healthy elderly men. *J. Gerontol.* 37, 288–293.
413. Jannini, E. A., Screponi, E., Carosa, E., Pepe, M., Lo Giudice, F., Trimarchi, F., and Benvenga, S. (1999). Lack of sexual activity from erectile dysfunction is associated with a reversible reduction in serum testosterone. *Int. J. Androl.* 22, 385–392.
414. Christiansen, K. (2004). Behavioural correlates of testosterone. In *Testosterone: Action Deficiency Substitution* (E. Nieschlag and H. M. Behre, Eds.), pp. 125–172. Cambridge University Press, Cambridge.
415. Strain, G. W., Zumoff, B., Kream, J., Strain, J. J., Deucher, R., Rosenfeld, R. S., Levin, J., and Fukushima, D. K. (1982). Mild hypogonadotropic hypogonadism in obese men. *Metabolism* 31, 871–875.
416. Strain, G. W., Zumoff, B., Miller, L. K., Rosner, W., Levit, C., Kalin, M., Hershcopf, R. J., and Rosenfeld, R. S. (1988). Effect of massive weight loss on hypothalamic–pituitary–gonadal function in obese men. *J. Clin. Endocrinol. Metab.* 66, 1019–1023.
417. Lima, N., Cavaliere, H., Knobel, M., Halpern A., and Medeiros-Neto, G. (2000). Decreased androgen levels in massively obese men may be associated with impaired function of the gonadostat. *Int. J. Obes. Relat. Metab. Disord.* 24, 1433–1437.
418. Unden, F., Ljunggren, J. G., Beck-Friis, J., Kjellman, B. F., and Wetterberg, L. (1988). Hypothalamic–pituitary–gonadal axis in major depressive disorders. *Acta Psychiatr. Scand.* 78, 138–146.
419. Rubin, R. T., Poland, R. E., and Lesser, I. M. (1989). Neuroendocrine aspects of primary endogenous depression VIII. Pituitary–gonadal axis activity in male patients and matched control subjects. *Psychoneuroendocrinology* 14, 217–229.
420. Schweiger, U., Deuschle, M., Weber, B., Korner A., Lammers, C. H., Schmider, J., Gotthardt, U., and Heuser, I. (1999). Testosterone, gonadotropin, and cortisol secretion in male patients with major depression. *Psychosom. Med.* 61, 292–296.
421. Seidman, S. N., Araujo, A. B., Roose, S. P., Devanand, D. P., Xie, S., Cooper, T. B., and McKinlay, J. B. (2002). Low testosterone levels in elderly men with dysthymic disorder. *Am. J. Psychiatry* 159, 456–459.
422. Derby, C. A., Mohr, B. A., Goldstein, I., Feldman, H. A., Johannes, C. B., and McKinlay, J. B. (2000). Modifiable risk factors and erectile dysfunction: can lifestyle changes modify risk? *Urology* 56, 302–306.
423. Blanker, M. H., Bohnen, A. M., Groeneveld, F. P., Bernsen, R. M., Prins A., Thomas, S., and Bosch, J. L. (2001). Correlates for erectile and ejaculatory dysfunction in older Dutch men: a community-based study. *J. Am. Geriatr. Soc.* 49, 436–442.
424. Esposito, K., Giugliano, F., Di Palo, C., Giugliano, G., Marfella, R., D'Andrea, F., D'Armiento, M., and Giugliano, D. (2004). Effect of lifestyle changes on erectile dysfunction in obese men: a randomized controlled trial. *JAMA* 291, 2978–2984.
425. Feldman, H. A., Johannes, C. B., Derby, C. A., Kleinman, K. P., Mohr, B. A., Araujo, A. B., and McKinlay, J. B. (2000). Erectile dysfunction and coronary risk factors: prospective results from the Massachusetts male aging study. *Prev. Med.* 30, 328–338.
426. Buvat, J., and Lemaire, A. (1997). Endocrine screening in 1,022 men with erectile dysfunction: clinical significance and cost-effective strategy. *J. Urol.* 158, 1764–1767.
427. Melman A., and Gingell, J. C. (1999). The epidemiology and pathophysiology of erectile dysfunction. *J. Urol.* 161, 5–11.
428. Carani, C., Zini, D., Baldini, A., Della Casa, L., Ghizzani, A., and Marrama, P. (1990). Effects of androgen treatment in impotent men with normal and low levels of free testosterone. *Arch. Sex. Behav.* 19, 223–234.
429. Guay, A. T., Bansal, S., and Heatley, G. J. (1995). Effect of raising endogenous testosterone levels in impotent men with secondary hypogonadism: double blind placebo-controlled trial with clomiphene citrate. *J. Clin. Endocrinol. Metab.* 80, 3546–3552.

430. Schiavi, R. C., White, D., Mandeli, J., and Levine, A. C. (1997). Effect of testosterone administration on sexual behavior and mood in men with erectile dysfunction. *Arch. Sex. Behav.* 26, 231–241.
431. Rakic, Z, Starcevic, V., Starcevic, V. P., and Marinkovic, J. (1997). Testosterone treatment in men with erectile disorder and low levels of total testosterone in serum. *Arch. Sex. Behav.* 26, 495–504.
432. Rajfer, J., Aronson, W. J., Bush, P. A., Dorey, F. J., and Ignarro, L. J. (1992). Nitric oxide as a mediator of relaxation of the corpus cavernosum in response to nonadrenergic, noncholinergic neurotransmission. *N. Engl. J. Med.* 326, 90–94.
433. Gonzalez-Cadavid, N. F., Ignarro, L. J., and Rajfer, J. (1999). Nitric oxide and the cyclic GMP system in the penis. *Mol. Urol.* 3, 51–59.
434. Archer, S. L. (2002). Potassium channels and erectile dysfunction. *Vascul. Pharmacol.* 38, 61–71.
435. Lue, T. F. (2000). Erectile dysfunction. *N. Engl. J. Med.* 342, 1802–1813.
436. Sullivan, M. E., Keoghane, S. R., and Miller, M. A. (2001). Vascular risk factors and erectile dysfunction. *BJU Int.* 87, 838–845.
437. Azadzoi, K. M., and Goldstein, I. (1992). Erectile dysfunction due to atherosclerotic vascular disease: the development of an animal model. *J. Urol.* 147, 1675–1681.
438. Adams, M. R., Kaplan, J. R., Koritnik, D. R., Weaver, D. S., and Bond, M. G. (1984). Erectile failure in cynomolgus monkeys with atherosclerosis of the arteries supplying the penis. *J. Urol.* 131, 571–573.
439. Virag, R., Bouilly, P., and Frydman, D. (1985). Is impotence an arterial disorder? A study of arterial risk factors in 440 impotent men. *Lancet* 1, 181–184.
440. Michal, V. (1982). Arterial disease as a cause of impotence. *Clin. Endocrinol. Metab.* 11, 725–748.
441. Blumentals, W. A., Gomez-Caminero, A., Joo, S., and Vannappagari, V. (2003). Is erectile dysfunction predictive of peripheral vascular disease? *Aging Male* 6, 217–221.
442. Blumentals, W. A., Gomez-Caminero, A., Joo, S., and Vannappagari, V. (2004). Should erectile dysfunction be considered as a marker for acute myocardial infarction? Results from a retrospective cohort study. *Int. J. Impot. Res.* 16, 350–353.
443. Fung, M. M., Bettencourt, R., and Barrett-Connor, E. (2004). Heart disease risk factors predict erectile dysfunction 25 years later: the Rancho Bernardo Study. *J. Am. Coll. Cardiol.* 43, 1405–1411.
444. Gresser, U., and Gleiter, C. H. (2002). Erectile dysfunction: comparison of efficacy and side effects of the PDE-5 inhibitors sildenafil, vardenafil and tadalafil—review of the literature. *Eur. J. Med. Res.* 7, 435–446.
445. Giuliano, F., Donatucci, C., Montorsi, F., Auerbach, S., Karlin, G., Norenberg, C., Homering, M., Segerson, T., and Eardley, I. (2005). Vardenafil is effective and well-tolerated for treating erectile dysfunction in a broad population of men, irrespective of age. *BJU Int.* 95, 110–116.
446. Linet, O. I., and Ogrinc, F. G. (1996). Efficacy and safety of intracavernosal alprostadil in men with erectile dysfunction. The Alprostadil Study Group. *N. Engl. J. Med.* 334, 873–877.
447. Padma-Nathan, H., Hellstrom, W. J., Kaiser, F. E., Labasky R.F., Lue, T. F., Nolten, W. E., Norwood, P. C., Peterson, C. A., Shabsigh, R., and Tam, P. Y. (1997). Treatment of men with erectile dysfunction with transurethral alprostadil. Medicated Urethral System for Erection (MUSE) Study Group. *N. Engl. J. Med.* 336, 1–7.
448. Leungwattanakij, S., Flynn, V. Jr., and Hellstrom, W. J. (2001). Intracavernosal injection and intraurethral therapy for erectile dysfunction. *Urol. Clin. North Am.* 28, 343–354.
449. Bella, A. J., and Brock, G. B. (2004). Intracavernous pharmacotherapy for erectile dysfunction. *Endocrine* 23, 149–155.
450. Montorsi, F., Salonia A., Deho, F., Cestari A., Guazzoni, G., Rigatti, P., and Stief, C. (2003). Pharmacological management of erectile dysfunction. *BJU Int.* 91, 446–454.
451. Irwin, M. B., and Kata, E. J. (1994). High attrition rate with intracavernous injection of prostaglandin E1 for impotency. *Urology* 43, 84–87.
452. Mulhall, J. P., Jahoda, A. E., Cairney, M., Goldstein, B., Leitzes, R., Woods, J., Payton, T., Krane, R. J., and Goldstein, I. (1999). The causes of patient dropout from penile self-injection therapy for impotence. *J. Urol.* 162, 1291–1294.
453. Purvis, K., Egdetveit, I., and Christiansen, E. (1999). Intracavernosal therapy for erectile failure—impact of treatment and reasons for drop-out and dissatisfaction. *Int. J. Impot. Res.* 11, 287–299.
454. Goldstein, I., Lue, T. F., Padma-Nathan, H., Rosen, R. C., Steers, W. D., and Wicker, P. A. (1998). Oral sildenafil in the treatment of erectile dysfunction. Sildenafil Study Group. *N. Engl. J. Med.* 338, 1397–1404.
455. Porst, H., Rosen, R., Padma-Nathan, H., Goldstein, I., Giuliano, F., Ulbrich, E., and Bandel, T. (2001). The efficacy and tolerability of vardenafil, a new, oral, selective phosphodiesterase type 5 inhibitor, in patients with erectile dysfunction: the first at-home clinical trial. *Int. J. Impot. Res.* 13, 192–199.
456. Brock, G. B., McMahon, C. G., Chen, K. K., Costigan, T., Shen, W., Watkins, V., Anglin, G., and Whitaker, S. (2002). Efficacy and safety of tadalafil for the treatment of erectile dysfunction: results of integrated analyses. *J. Urol.* 168, 1332–1336.
457. Hellstrom, W. J., Gittelman, M., Karlin, G., Segerson, T., Thibonnier, M., Taylor, T., and Padma-Nathan, H. (2002). Vardenafil for treatment of men with erectile dysfunction: efficacy and safety in a randomized, double-blind, placebo-controlled trial. *J. Androl.* 23, 763–771.
458. Hellstrom, W. J., Gittelman, M., Karlin, G., Segerson, T., Thibonnier, M., Taylor, T., and Padma-Nathan, H. (2003). Sustained efficacy and tolerability of vardenafil, a highly potent selective phosphodiesterase type 5 inhibitor, in men with erectile dysfunction: results of a randomized, double-blind, 26-week placebo-controlled pivotal trial. *Urology* 61, 8–14.
459. Stief, C., Porst, H., Saenz De Tejada, I., Ulbrich, E., and Beneke, M. (2004). Sustained efficacy and tolerability with vardenafil over 2 years of treatment in men with erectile dysfunction. *Int. J. Clin. Pract.* 58, 230–239.
460. Corbin, J. D., Francis, S. H., and Webb, D. J. (2002). Phosphodiesterase type 5 as a pharmacologic target in erectile dysfunction. *Urology* 60, 4–11.
461. Padma-Nathan, H., Eardley, I., Kloner, R. A., Laties, A. M., and Montorsi, F. (2002). A 4-year update on the safety of sildenafil citrate (Viagra). *Urology* 60, 67–90.
462. Tran, D., and Howes, L. G. (2003). Cardiovascular safety of sildenafil. *Drug Saf.* 26, 453–460.
463. Brindis, R. G., and Kloner, R. A. (2003). Sildenafil in patients with cardiovascular disease. *Am. J. Cardiol.* 92, 26M–36M.
464. Tsujimura A., Yamanaka, M., Takahashi, T., Miura, H., Nishimura, K., Koga, M., Iwasa A., Takeyama, M., Matsumiya, K., Takahara, S., and Okuyama, A. (2002). The clinical studies of sildenafil for the ageing male. *Int. J. Androl.* 25, 28–33.
465. Montorsi, F., Briganti A., Salonia A., Deho, F., Zanni, G., Cestari A., Guazzoni, G., Rigatti, P., and Stief, C. (2003). The ageing male and erectile dysfunction. *BJU Int.* 92, 516–520.
466. Keast, J. R. Effects of testosterone on pelvic autonomic pathways: progress and pitfalls. *J. Auton. Nerv. Syst.* 79, 67–73.
467. Shabsigh, R. (1997). The effects of testosterone on the cavernous tissue and erectile function. *World J. Urol.* 15, 21–26.
468. Mills, T. M., Wiedmeier, V. T., and Stopper, V. S. (1992). Androgen maintenance of erectile function in the rat penis. *Biol. Reprod.* 46, 342–348.

469. Akingba, A. G., and Burnett, A. L. (2001). Endothelial nitric oxide synthase protein expression, localization, and activity in the penis of the alloxan-induced diabetic rat. *Mol. Urol.* 5, 189–197.
470. Baba, K., Yajima, M., Carrier, S., Akkus, E., Reman, J., Nunes, L., Lue, T. F., and Iwamoto, T. (2000). Effect of testosterone on the number of NADPH diaphorase-stained nerve fibers in the rat corpus cavernosum and dorsal nerve. *Urology* 56, 533–538.
471. Marin, R., Escrig A., Abreu, P., and Mas, M. (1999). Androgen-dependent nitric oxide release in rat penis correlates with levels of constitutive nitric oxide synthase isoenzymes. *Biol. Reprod.* 61, 1012–1016.
472. Mills, T. M., Dai, Y., Stopper, V. S., and Lewis, R. W. (1999). Androgenic maintenance of the erectile response in the rat. *Steroids* 64, 605–609.
473. Dai, Y. T., Stopper, V., Lewis, R., and Mills, T. (1999). Effects of castration and testosterone replacement on veno-occlusion during penile erection in the rat. *Asian J. Androl.* 1, 53–59.
474. Traish, A. M., Park, K., Dhir, V., Kim, N. N., Moreland, R. B., and Goldstein, I. (1999). Effects of castration and androgen replacement on erectile function in a rabbit model. *Endocrinology* 140, 1861–1868.
475. Park, K. H., Kim, S. W., Kim, K. D., and Paick, J. S. (1999). Effects of androgens on the expression of nitric oxide synthase mRNAs in rat corpus cavernosum. *BJU Int.* 83, 327–333.
476. Alcorn, J. F., Toepfer, J. R., and Leipheimer, R. E. (1999). The effects of castration on relaxation of rat corpus cavernosum smooth muscle in vitro. *J. Urol.* 161, 686–689.
477. Bivalacqua, T. J., Rajasekaran, M., Champion, H. C., Wang, R., Sikka, S. C., Kadowitz, P. J., and Hellstrom, W. J. (1998). The influence of castration on pharmacologically induced penile erection in the cat. *J. Androl.* 19, 551–557.
478. Reilly, C. M., Lewis, R. W., Stopper, V. S., and Mills, T. M. (1997). Androgenic maintenance of the rat erectile response via a non-nitric-oxide-dependent pathway. *J. Androl.* 18, 588–594.
479. Reilly, C. M., Zamorano, P., Stopper, V. S., and Mills, T. M. (1997). Androgenic regulation of NO availability in rat penile erection. *J. Androl.* 18, 110–115.
480. Reilly, C. M., Stopper V. S., and Mills, T. M. (1997). Androgens modulate the alpha-adrenergic responsiveness of vascular smooth muscle in the corpus cavernosum. *J. Androl.* 18, 26–31.
481. Garban, H., Marquez, D., Cai, L., Rajfer, J., and Gonzalez-Cadavid, N. F. (1995). Restoration of normal adult penile erectile response in aged rats by long-term treatment with androgens. *Biol. Reprod.* 53, 1365–1372.
482. Vernet, D., Cai, L., Garban, H., Babbitt, M. L., Murray, F. T., Rajfer, J., and Gonzalez-Cadavid, N. F. (1995). Reduction of penile nitric oxide synthase in diabetic BB/WORdp (type I) and BBZ/WORdp (type II) rats with erectile dysfunction. *Endocrinology* 136, 5709–5717.
483. Lugg, J. A., Rajfer, J., and Gonzalez-Cadavid, N. F. (1995). Dihydrotestosterone is the active androgen in the maintenance of nitric oxide-mediated penile erection in the rat. *Endocrinology* 136, 1495–1501.
484. Mills, T. M., Stopper, V. S., and Wiedmeier, V. T. (1994). Effects of castration and androgen replacement on the hemodynamics of penile erection in the rat. *Biol. Reprod.* 51, 234–238.
485. Aversa, A., Isidori, A. M., Spera, G., Lenzi A., and Fabbri, A. (2003). Androgens improve cavernous vasodilation and response to sildenafil in patients with erectile dysfunction. *Clin. Endocrinol. (Oxf).* 58, 632–638.
486. Shabsigh, R., Kaufman, J. M., Steidle, C., and Padma-Nathan, H. (2004). Randomized study of testosterone gel as adjunctive therapy to sildenafil in hypogonadal men with erectile dysfunction who do not respond to sildenafil alone. *J. Urol.* 172, 658–663.
487. Laumann, E. O., Gagnon, J. H., Michael, R. T., and Michaels, S. (1994). *The Social Organization of Sexuality: Sexual Practices in the United States.* University of Chicago Press, Chicago.
488. Tasaka, T., Hiraga, S., Kitamura, M., Iida, T., Kurokawa, J., Hida, M., and Satoh, T. (1986). [First report on studies of normal testicular weight and size in the Japanese]. *Nippon Hinyokika Gakkai Zasshi.* 77, 1506–1510.

Knobil and Neill's Physiology of Reproduction, Third Edition
edited by Jimmy D. Neill,
Elsevier © 2006

CHAPTER 50

Aging in the Hypothalamic–Pituitary–Ovarian Axis

Genevieve Neal-Perry and Nanette F. Santoro

Aging: An Overview, 2729
The Ovary, 2731
Predictors of the Final Menstrual Period, 2731
The Hypothalamic–Pituitary Axis and Reproduction, 2732
The Hypothalamus, 2732
The Pituitary Gland, 2732
Human Menstrual Cycle Overview, 2733
Aging and the Human Hypothalamic–Pituitary Axis, 2735
Age-Related Changes in Gonadotropin Surge Patterns, 2735
Nonhuman Models for Reproductive Aging, 2738
Inhibitory and Excitatory Neurotransmission in Reproductive Aging, 2741
Noradrenergic Neurotransmission and Reproductive Aging, 2743
Changes in Other Neurotransmitters and Neuropeptides as a Function of Reproductive Aging, 2744
Summary, 2745
References, 2745

AGING: AN OVERVIEW

Reproductive aging occurs against a background of generalized, somatic aging. There is increasing evidence that these processes interact with one another, and thus, they must be considered both separately and together.

Generalized aging of tissues is associated with a variety of features. Oxidative stressors affect cellular machinery and lead to DNA damage (1). Persistent oxidative stress causes intracellular damage that is irreversible. Excessive DNA damage can result in mutagenesis—resulting in cancers, or ineffective cellular processes and the loss of "normal" cellular function. In addition to the irreversible cellular damage induced by oxidative stress, chromosomal changes also begin to occur with an increased frequency that seems to be age-dependent. One of these age-related changes includes increased telomere length, a basic and central feature of aging cells (2). Increased telomere length is associated with less efficient transcription, thereby further compromising cellular processes. The ability of cells to buffer the effects of oxidative stressors appears to be a key element in longevity (3,4). Resistance to oxidative stress most likely reflects a combination of inherent attributes of the genomic constitution of an individual organism (5–7) and the ability of exogenous antioxidants to mitigate the detrimental effects of reactive oxygen species (8).

Reduced vascular sufficiency contributes to the process of reproductive and somatic aging by exacerbating the extent of irreversible cell damage and reduced cellular function caused by oxidative stress. Of note, vascular compromise may initiate these events earlier than previously appreciated. For example, in one postmortem study, intramyometrial arteriosclerosis was observed in 37% of women under the age of 30, with an increase to 83% in women over the age of 39 (9). The net effect of these generalized

Department of OB/GYN and Women's Health, Albert Einstein College of Medicine, The Bronx, New York.

processes is reduced energy efficiency at the cellular, organ, and whole organism level—a phenomenon dubbed reduced functional reserve.

In addition, physiological or psychological stress can accelerate tissue aging and alter the onset of morbidity and mortality. Chronically elevated cortisol appears to be a mediator of stress-related tissue aging in human and nonhuman models (10). Increased cortisol is a powerful predictor of loss of functional reserve. The early morning cortisol rise is somewhat blunted in the elderly, and the evening nadir is considerably less pronounced. This lack of a cortisol nadir is linked to sleep disturbances in the elderly (11). The stressors that lead to hypercortisolemia have been collectively named the *allostatic load* (12). Factors that contribute to the "allostatic load" such as insulin resistance, chronic depression, and anxiety states have been posited to predispose an individual to a number of morbidities that include central body fat accumulation, hypertension, and even development of the "syndrome X" phenotype (13).

Reproductive health should therefore be viewed in the context of the influences created by age-dependent nonreproductive endocrine dynamics. For example, insulin resistance and adipose accumulation accompany the aging process in many animals, including humans (14,15). Hyperinsulinemia may directly affect ovarian steroidogenesis and accelerate the process of atherosclerosis. Accumulation of body fat with aging, particularly central adipose tissue, further increases cardiovascular risk and vascular compromise (16). Caloric restriction appears to delay the aging process in animal models (14). However, the effect of caloric restriction on the hypothalamic–pituitary–gonadal axis has not been explored in the context of human reproductive aging.

Adipokines, such as tumor necrosis factor-alpha (TNF-α) and, in some studies, leptin, have been shown to increase with age (17–19). Adipose-derived peptides, such as leptin, have been shown to exert a permissive influence on the reproductive axis in animals and humans (20). In women with hypothalamic amenorrhea, exogenous leptin restores normal cyclicity by reinstating a physiologic pattern of luteinizing hormone (LH) and (by inference) gonadotropin-releasing hormone (GnRH) secretion (21). At this time, a role for leptin in advanced reproductive aging has not been observed (17).

Chronological aging has also been linked to progressive deterioration in the production of growth hormone and insulin-like growth factor-1 (IGF-1) as well as adrenal androgen, the so-called "somatopause" and the "adrenopause," respectively (22,23). The relative loss of growth factors with aging may reduce the potency of endogenous gonadotropins, particularly follicle-stimulating hormone (FSH).

Interest in the physiological machinery that initiates and commits women to the process of reproductive senescence has been driven by the growing trend for women to delay their child bearing into their forties, a time that is most often characterized by reduced fertility. In humans, reproductive aging is associated with a transitional period, the menopausal transition, which is described by changes in gonadal function and altered gonadotropin surge patterns (24,25). Subsequent to the navigation of the menopausal transition, women begin a period of reproductive quiescence known as the menopause. For many years, it was accepted that the fundamental event responsible for reproductive aging was the somatic aging of the ovary, the associated depletion of the fixed quantity of postmeiotic ovarian follicles (26) and the subsequent development of a hypoestrogenic/hypoprogestinemic hormonal milieu (27–33). While it is certain that ovarian failure ultimately defines the menopause, a number of recent studies, albeit primarily of the rodent, have shown that age-related changes are simultaneously occurring within the neuroendocrine axis (34–43) rendering the transition into reproduction senescence to be an irreversible event occurring throughout the hypothalamic–pituitary–ovarian (HPO) axis (44,45).

Epidemiological studies also suggest that there is a common denominator linking the cellular processes that yield reproductive and somatic aging. Women with premature menopause appear to have an increased risk of cardiovascular disease and mortality (46). In one study of women with premature menopause, cardiovascular disease was found to precede the diagnosis of menopause (47). Cardiovascular disease might therefore be viewed as an accelerant of the menopausal process, and not a result. Alternatively, the premature presentation of both gonadal failure and cardiovascular disease may be a manifestation of an early embryonic insult that resulted in the acceleration of both somatic and germ-line mutations. Some of these data have been interpreted to mean that the hypogonadism of premature menopause precipitates the untimely advent of cardiovascular disease. However, in the wake of recent clinical trial information suggesting that estrogen is not an effective cardioprotective agent for postmenopausal women (48,49), alternative explanations for this finding must be considered. Taken together, the available scientific evidence strongly suggests that significant interaction occurs between somatic and reproductive aging and that neither process can be taken out of context or without consideration of the other system.

THE OVARY

All follicles in the ovary, from primordial to preovulatory, are considered part of the total pool, termed *ovarian reserve*. In the human, loss of ovarian reserve is the primary mechanism propelling ovarian senescence. In addition to the loss of follicle numbers, there are age-related decreases in the quality of oocytes contained within them. Aging oocytes are less efficient at undergoing appropriately timed meiosis, consequently increased chromosomal errors, such as nondisjunction, occur. Disruption of the microtubule apparatus critical to chromosomal organization appears to be the source for oocyte abnormalities observed in older reproductive-aged women (50). Oocyte telomere length has a strong inverse relationship with reproductive aging, thus it has been proposed to be a powerful predictor of reduced ovarian reserve, infertility, and the risk for gametogenic errors (51).

This situation differs from the rat and mouse. In these animals, the age-related change in reproductive cyclicity is initially reflected in abnormal gonadotropin release patterns and not coincident with a critical reduction in follicle numbers. Recent data in the mouse indicates that germ cells maintain the potential to proliferate well into advanced reproductive life (52). Relative to observations in humans, age-dependent changes in the reproductive system of rodents appear to have an earlier and larger CNS-based component. However, recent studies in women suggests reproductive aging in women is accompanied by hypothalamic–pituitary dysfunction that, like rodents, also preempt overt gonadal failure (53). The mechanisms involved appear similar to those described in rodents (40,54). Patterns of gonadotropin secretion also appear to become less organized and more chaotic as a result of aging (55,56). Whereas ovarian implants from young mice did not restore estrous cyclicity in aged mice (57), hypothalamic implants of young mice placed into aged, acyclic mice have been reported to restore estrous cyclicity (45), implying that, at least in small mammals, the role of the hypothalamus over-shadows age-related ovarian factors in reproductive aging. Whether these changes reflect specific neuroendocrine alterations or are a by-product of generalized somatic or neurological aging remains to be clarified.

Despite these differences between the pattern of somatic and germ-line aging in humans and rodents, it is still possible to produce changes in the rodent ovary that closely mimic human menopause. One of these models involves the generation of transgenic mice with haploinsufficiency of the FSH receptor. These animals appear to undergo premature oocyte loss, have reduced fertility, and exhibit some features that recapitulate human menopause (58). In another rodent model, exposure to a submaximal dose of a gonadal toxicant, 4-vinylcyclohexene diepoxide (4-VCD), causes an early follicle depletion in which elements of menopause may be able to be viewed over a very short time frame (59). These 4-VCD treated rodents also exhibit features consistent with human menopause, and manifest some of the neuroendocrine features associated with persistent estrus, such as a change in gonadotropin surge patterns and elevated prolactin.

PREDICTORS OF THE FINAL MENSTRUAL PERIOD (FMP)

In the human, ovarian reserve is a major predictor of the age at which natural menopause occurs. Determinants include exposures to agents that are associated with the destruction of normal ovarian follicular cells, such as cigarette smoking (60), radiation, a history of medically treated depression (61), low socio-economic status (62), and possibly galactose consumption (63). Repeated pelvic infections and surgeries for benign diseases such as endometriosis also appear to compromise ovarian reserve (64). Unilateral oophorectomy in women has been estimated to advance the timing of natural menopause by 1–2 years (65), and to be associated with increased circulating FSH (65,66). Lastly, hysterectomy with complete ovarian conservation has been reported to have a similar, though smaller effect (65,67). Thus, both surgical devitalization of pelvic tissues and age-dependent deterioration in ovarian reserve may be responsible for the premature FSH increase and reduced time to menopause observed in patients with a history of pelvic surgery and ovarian disease. On the other hand, long-term administration of oral contraceptives, as well as other treatments or physiological states that prevent the latter stages of follicle maturation and ovulation, has been associated with reduction in FSH levels, most strikingly in women over the age of 45 years (68). This finding implies that the reproductive life span of women taking oral contraceptives may be slightly prolonged. Of note, one of the most powerful predictors of the FMP in epidemiological studies is family history, i.e., the age at which a woman's mother traversed the menopause (68).

Genetic studies are beginning to provide more insight about the factors that determine the timing of the FMP in women. In one large epidemiological cohort, homozygosity for the Pvu II polymorphism for estrogen receptor (ER) alpha was associated with a FMP 0.5 years earlier and a 2.4-fold increase in hysterectomy risk compared to women who lacked a copy of this polymorphism (69). Molecular studies of women with the blepharophimosis/ptosis/epicanthus syndrome—associated with ovarian failure prior to age 40—has led to the discovery of a relationship between the forkhead family of transcription factors and follicle development (70). The FOXL2 gene arrests follicular development and growth. In transgenic null mice lacking FOXL2, primordial follicles do not develop normally (71). In other forkhead family member mutations, granulosa cell growth and development appears to proceed rapidly and without surcease, such that the follicle supply cannot keep pace with follicle development and early menopause results (72). It is possible that similar such mutations, as of yet undiscovered regulators of early follicle growth, are responsible for the clinical phenotype of idiopathic premature ovarian failure. For example, null mice lacking growth and development factor-9 (GDF-9) fail to exhibit normal ovarian reserve (73).

In large, cross-sectional studies in which reproductive phases are grouped into decades, women demonstrate incremental increases in circulating FSH with age (74). Consequently, older ovulatory women have menstrual cycles that become characterized by a distinctive rise in FSH and in some cases an obvious decrement in serum estradiol (25). The follicular phase of older ovulatory women is shortened; a finding that is thought to be the result of accelerated folliculogenesis and earlier ovulation induced by the high FSH levels (25,75–77). Interestingly, the loss of estradiol feedback or receptivity does not account for the age-related increase in FSH levels (77,78).

The available follicle cohort in any given woman is believed to decrease as a function of time, in parallel with the decrease in ovarian reserve (27,79,80). Detailed ultrasound studies of the ovaries in normal women confirm this notion (81–83). Despite the smaller follicle pool, the monotropic rise in FSH should probably be viewed as a compensatory mechanism that permits follicular maturation, ovulatory success, and rescue of reproductive potential in the face of a dwindling supply of follicles and oocytes. FSH levels have been proposed to be a measure of a woman's proximity to menopause. Although elevated FSH levels are predictive of age at menopause in large populations of women, their ability to predict any one woman's trajectory toward her final menstrual period is extremely limited (84,85).

Since FSH is of limited value in predicting the onset of menopause, attention has been focused on ovarian products and their potential to forecast follicular decline. Some have tried to use inhibin as a clinical marker for impending ovarian failure; however, the use of inhibin for this purpose has proved to be inconclusive. Some studies have linked low inhibin levels to poor outcomes of fertility therapies (86–88), but others have not (87,89). At this time, it appears that inhibin quantification does not add useful clinical information about the timing of menopause.

THE HYPOTHALAMIC–PITUITARY AXIS AND REPRODUCTION

The Hypothalamus

As part of the diencephalon, the hypothalamus is one of the most evolutionarily conserved brain structures. It is anatomically located in the base of the brain and contributes to the floor and lateral walls of the third ventricle. The neurophysiological function of the hypothalamus is to integrate and process essential autonomic and endocrine responses central to routine neuroendocrine maintenance, physiological homeostasis, and reproductive competence.

The hypothalamus is divided into three sections: the periventricular, medial, and lateral zones. The periventricular zone houses magnocellular and parvicellular neurons and maintains the preservation of the circadian rhythm and autonomic neuroendocrine responses to stress. The medial zone participates in the control of reproduction, osmoregulation, and thermoregulation. The lateral zone is involved in behavior and arousal. The hypothalamus orchestrates homeostasis and reproductive functions through the synthesis and secretion of a number of regulatory neuropeptides that include gonadotropin-releasing hormone (GnRH), growth hormone-releasing hormone (GHRH), thyrotropin-releasing hormone (TRH), and corticotrophin-releasing hormone (CRH).

The Pituitary Gland

The pituitary gland is derived from the fusion of the neuroectoderm that gives rise to Rathke's pouch (the anterior pituitary) and a portion of the diencephalon (the posterior pituitary). It is located in the sella turcica at the base of the brain, underlying the hypothalamus. The pituitary gland has a specialized blood supply that is derived from tributaries of the internal carotid artery, the superior and inferior hypophyseal arteries. This specialized network of capillaries and arterioles provide blood to each subdivision of the pituitary: the adenohypophysis and neurohypophysis.

The adenohypophysis is composed of the anterior lobe, the pars intermedia and the pars tuberalis, all of which account for approximately 80% of the pituitary gland. Development of the human anterior pituitary begins between the 4th and 5th weeks of pregnancy. By the 9th week of pregnancy, a rudimentary anterior pituitary is recognizable and between the 12th and 17th weeks of pregnancy, a functional hypothalamic–pituitary axis is established. Six major cell types reside in the anterior pituitary: 40%–50% somatotrophs (growth hormone [GH] production), 15%–25% lactotrophs (prolactin [PRL] production), 10%–20% corticotrophs (adrenocorticotropic hormone [ACTH] production), 5% thyrotrophs (thyroid-stimulating hormone [TSH] production), 10% gonadotrophs (follicle-stimulating hormone [FSH] and luteinizing hormone [LH] production) and folliculostellate cells.

The neurohypophysis is organized into the pars nervosa, the infundibular stalk, and the median eminence. The infundibular stalk is surrounded by the pars tuberalis and together they constitute the hypophyseal stalk. The neurohypophysis contains a collection of nerve terminals that arise from magnocellular secretory neurons located within the paraventricular and supraoptic nuclei of the hypothalamus. These axons are in close proximity to the specialized pituitary capillary network. On stimulation, they secrete vasopressin and oxytocin into the hypophyseal veins. Alterations in the concentration of these neuropeptides within the peripheral vasculature result in changes in blood osmolality, blood pressure, and fluid balance.

The pituitary gland is regulated by three interacting elements: hypothalamic inputs, steroid feedback, and paracrine and autocrine actions of the pituitary itself. With regard to the hypothalamus, changing hormonal and physiological cues result in neuropeptide release into the median eminence where the resident arterioles carry neuropeptides to the infundibular stalk and coalesce with the sinusoidal capillaries of the anterior lobe. Venous drainage from the anterior pituitary passes through the posterior pituitary lobe capillary network just prior to draining into the cavernous sinus and the petrosal sinus. An intact and functional hypothalamic–pituitary axis is a critical element in basic reproductive health. Indeed, if hypothalamic–pituitary development goes awry, such as occurs in cases of congenital embryopathy, the disease processes that result include a compromised hypothalamic–pituitary unit in which reproductive competency and gonadal physiology is disrupted.

Several recent studies have proposed that reproductive senescence is a complex process involving but not limited to impaired gonadal steroidogenesis and altered follicular function (24,25,37–40,53, 90–99). Data reported by many of these investigators suggest that age-related changes in the hypothalamic–pituitary axis may critically contribute to the development and maintenance of reproductive senescence. Moreover, these neuroendocrine changes may accelerate the rate of ovarian follicular depletion and the advent of gonadal failure. As a group, these studies have prompted some researchers to hypothesize that reproductive aging involves the disruption of complex interactions amongst gonadal steroids, neuropeptides, and neurotransmitter pathways involved in the regulation and maintenance of the hypothalamic–pituitary–gonadal axis. Furthermore, it has been postulated that even in the absence of gonadal failure, age-related changes in the aging hypothalamic–pituitary axis would continue to progress and yield the physiology associated with reproductive senescence (57). Consistent with these hypotheses, several researchers have shown that the characteristic age-related changes in gonadotropin release precede the advent of irregular reproductive cycles and overt ovarian failure in both rodents (39,93, 97,100,101) and in women (53,102).

Human Menstrual Cycle Overview

Once the menstrual cycle has been well established, an event which on average is achieved within 5 years postmenarche (103,104), a woman enters into her reproductive prime; a period also defined as the "midreproductive" phase. This period is most often coincident with maximal reproductive potential and characterized by the existence of a mature and functional hypothalamic–pituitary axis. The mature hypothalamic–pituitary axis functions to regulate ovarian follicular activity and menses by releasing various neuropeptides and glycoprotein hormones in response to cyclical gonadal peptide and gonadal steroidal feedback.

Hypothalamic regulation of the reproductive cycle is achieved through the pulsatile activity of GnRH neurons. GnRH neurons originate in the epithelium of the medial part of the olfactory placode and migrate during fetal development across the cribiform plate. In primates, the majority of GnRH neuronal cell bodies are located in the mediobasal hypothalamus. With the appropriate hormonal environment and stimulus, GnRH neurons pulse release varying patterns of GnRH into the portal capillary vasculature, which in turn stimulates LH, and to a lesser extent, FSH synthesis and release from the pituitary. A direct linkage between GnRH production by the hypothalamus and peripherally measured LH has been well established by several investigators in numerous animal species, including the human (105–111).

The maintenance of the human menstrual cycle is directly related to changes in the complex and coordinated interactions among patterns of gonadal steroid production, gonadal peptide release, neurotransmitter release, and their composite effects on the rate and pattern of GnRH release and subsequent LH and FSH release. The typical midreproductive-aged female has a biphasic (the follicular and luteal phases) menstrual cycle that lasts between 25–35 days (104) with the definitive physiological event being the release of a mature oocyte for fertilization. In response to changing gonadal steroid levels and other nonsteroidal factors, the intrinsic pulsatile and rhythmic properties of GnRH neurons determine the pattern of gonadotropin release (112–115).

In addition to the classically described negative and positive feedback of estradiol and negative feedback of progesterone on the hypothalamus and pituitary, inhibin production by the ovary and activin production by the pituitary and other sources

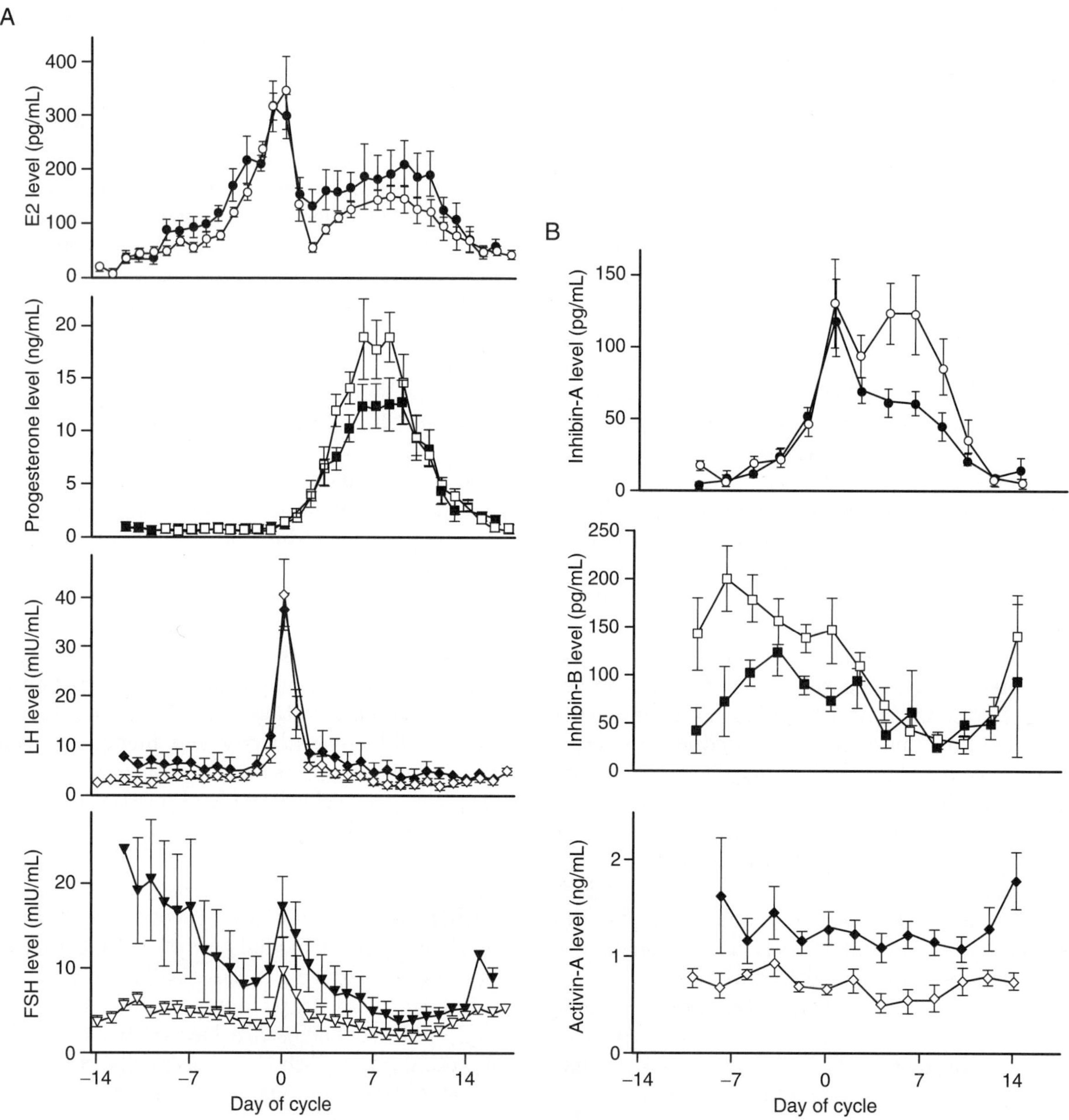

FIG. 1. Relationships between **(A)** estradiol (E2), progesterone, LH, and FSH and **(B)** inhibin A (a product of the dominant follicle and corpus luteum), inhibin B (a product of small, growing follicles), and activin A (a pituitary and adrenal product) over the course of a normal menstrual cycle in midreproductive aged women aged 18–38 (*open symbols*) and older reproductive aged women (43–52) (*closed symbols*). Hormones are standardized around day 0, the day of the LH surge. [Reprinted with permission from (77).]

also appear to play regulatory roles in the normal menstrual cycle. These roles become important when considering the events surrounding reproductive aging, and will be addressed briefly in the following paragraphs. Figure 1 displays the relationships between gonadotropins, sex steroids, and the inhibins in midreproductive and reproductively aged normal volunteers.

Inhibins, dimeric glycoproteins with structural homology consistent with members of the TGF-β family (116), are composed of a common alpha subunit and one of two beta subunits. There are two known types of inhibin, inhibin A (Inh-A) and inhibin B (Inh-B). Inh-B is a product of small, early antral follicles and is secreted primarily during the early to mid-follicular phase of the cycle. Inh-A is a product of the dominant follicle, corpus luteum, and placenta (117). Both types of inhibin exert selective negative feedback on FSH secretion from the pituitary in a manner that differs from the well-established negative feedback of estradiol on LH and FSH secretion.

In the early follicular phase of the menstrual cycle, both Inh-A and -B are at their lowest level, as is estradiol. In other words, FSH is as free from restraint as possible, and therefore rises to a maximal level. This rise in FSH stimulates follicle recruitment, differentiation, and growth. The low steroid and inhibin environment favors an increase in the pulse frequency of GnRH. The initially slow pulsatile pattern of GnRH from the prior luteal phase favors an early follicular increase in FSH secretion (118). By the mid-follicular phase (MFP) of the menstrual cycle, estradiol increases, as does Inh-B, reflecting the selection and growth of small antral follicles. The initiation of the central negative feedback on FSH secretion by both Inh-B and estradiol results in a limitation of available trophic hormone and the selective survival of a single follicle (119). Failure to decrease FSH by as little as 10% at this time in the cycle results in multiple follicle growth (120). Even though mean circulating FSH is decreasing, the small rise in estradiol in the MFP is associated with an increase in the frequency of LH pulsations (121). By the late follicular phase (LFP), Inh-A and estradiol secretion is maximized by the dominant follicle. The rapidly rising levels of serum estradiol in the late follicular phase then exert positive feedback on the hypothalamic–pituitary axis, facilitating the LH surge and ovulation.

The luteal phase (LP) is characterized by high levels of estradiol, progesterone, and Inh-A being secreted by the corpus luteum. Progesterone appears to slow the GnRH pulse generator and exerts negative feedback on gonadotropin secretion in a time-lagged manner (121). The combined negative feedback of estradiol, progesterone, and Inh-A suppress FSH to its lowest level during the entire cycle. With the advent of luteolysis, estradiol, progesterone and Inh-A concentrations all decrease precipitously, releasing the restraint imposed upon FSH secretion. FSH then increases abruptly across the luteal-follicular transition, thereby facilitating the recruitment of the next cohort of pre-antral follicles.

There is a strong and consistent association between the characteristic rise in FSH and absent rise in LH (termed the *monotropic* rise in FSH) and decreased levels of inhibin observed in the menopausal transition. This observation is thought to reflect the diminished number of competent primordial follicles available for follicular recruitment in older reproductive-aged women (77,122–126). The observed decrease in inhibin precedes the rise in FSH. For example, it has clearly been documented that age-dependent changes in ovarian factors such as the production of Inh-B, contributes to the elevation of FSH (127). However, this decrease in Inh-B occurs despite normal circulating estradiol (84,127). This implies that the ovarian follicle is less responsive to FSH. In fact, this early decrease in circulating inhibin levels most likely precipitates the onset of the classical monotropic rise in circulating FSH (84,128,129).

Regulation of pituitary FSH secretion is also achieved with activin, a neuropeptide that is composed of two beta subunits. In response to the changing gonadal steroid environment, this homodimer serves as a paracrine stimulatory factor that directly acts upon the pituitary to *increase* pituitary FSH release (130,131). The existence of cyclical changes in circulating activin and its role in modulating FSH secretion has been controversial. It has been proposed by some that activin levels do not vary across the menstrual cycle (123,132), whereas others report that coincident with FSH surges, total activin levels also increase at mid-cycle and in the late luteal phase, times when FSH rises (121,133–135). The role of activin in reproductive senescence is less clearly defined.

AGING AND THE HUMAN HYPOTHALAMIC–PITUITARY AXIS

Age-Related Changes in Gonadotropin Surge Patterns

The short half life of GnRH (2–4 min) and the dilution factor associated with the entry of GnRH release into the peripheral circulation have rendered the study of GnRH release in humans challenging. Moreover, the existence of an experimental approach to noninvasively record neuronal activity and GnRH release within the median eminence of humans is not currently available.

The difficulties encountered when attempting to measure GnRH release in humans have prompted many researchers to use animal models to investigate the effects of positive and negative gonadal steroid feedback on neuroendocrine peptide and neurotransmitter release. Numerous animal studies have revealed the existence of a tight correlation between GnRH neuronal activity, GnRH release, and peripheral LH levels (108,110,111,136–140). These observations have allowed clinical researchers to study peripheral LH level as an indirect method of monitoring GnRH activity in humans. As a result, most of the scientific literature that characterizes the neuroendocrine aspects of GnRH release and pulsatility in women of mid-reproductive age and beyond report age-related changes in the pulsatility of serum LH levels, namely frequency and amplitude.

The Menopausal Transition

The menopausal transition begins with a skipped menstrual cycle, and in some, the onset of a number of clinical complaints that often include altered sleep patterns and vasomotor symptoms (141–143). Prior to the advent of irregular menstrual cycles, an objective hallmark of the menopausal transition is the monotropic rise in follicular phase FSH. This transitional period has an average duration of 4 years, a median age of onset between 45.5 and 47.5 years of age and lasts until 1 year following the last menstrual period (53,84,103,144,145).

The relative contribution of mounting ovarian exhaustion to age-related changes within the neuroendocrine axis and the initiation of the menopausal transition has been the focal point of many scientific debates. These debates are the result of the intricate interdependence of the various organ systems that define and delineate the functional hypothalamic–pituitary–gonadal axis. In other words, age-related ovarian dysfunction may augment age-related hypothalamic incompetence. Hypothalamic dysfunction may then disrupt the normal physiology of the pituitary.

Alternatively, these observations may represent other somatic changes associated with aging, such as a decreased blood supply to the theca cell layer surrounding the follicle, or a decrease in circulating insulin-like growth factors, which could result in reduced effectiveness of FSH (146). Persistent elevation of FSH without a concomitant change in circulating estradiol also implies that the aging hypothalamic–pituitary axis may be inappropriately sensing "the estrogen signal." This possibility is supported by recent studies in women undergoing the menopausal transition, in which anovulatory cycles were analyzed (53). In this urinary hormone study, intermittent failure of a pituitary LH surge was observed, despite estradiol metabolite excretion that was equivalent to successful ovulatory cycles, implying that selective hypothalamic–pituitary axis failure occurred. Other possibilities, such as fundamental changes in estradiol metabolism accompanying aging, have not been supported by experimental data.

Knowledge of the key role of follicle supply in reproductive aging has been applied to the development of a staging system for reproductive aging. This classification system—Stages of Reproductive Aging (STRAW; Fig. 12)—is a working model that categorizes the menopausal transition in women (147). The number of primordial follicles and their ability to secrete inhibins, protein hormones that are negative feedback regulators of FSH, diminishes gradually over the Early, Peak, and Late Reproductive stages (stages minus 5 through minus 3). Menstrual cycle regularity and quality is conserved during these stages. Eventually, FSH is released from inhibin restraint and rises by the Late Reproductive stage (minus 3). This rise is likely to be intermittent; i.e., varying with the variable size of the recruitable follicle pool each month. The Early Menopausal Transition (stage minus 2) is characterized by skipped or irregular cycles with shortened follicular phases, but overall endocrinologically normal hormonal and follicular dynamics are observed, despite a drastically reduced ability of the resident oocyte to fertilize and grow normally. In the Late Menopausal Transition stage (stage minus 1), menstrual cycles become increasingly irregular (including more than three skipped cycles) and there is an accelerated rise in FSH. It is in this stage that ovarian reserve is critically compromised, and an even greater loss in the number of available follicles have been reported, implying that the process of follicular atresia is accelerated (27). Postmenopausal concentrations of reproductive hormones exhibit different patterns of release. LH and FSH are tonically elevated, and estradiol is tonically low-hypergonadotropic hypogonadism. Although there is some evidence for intermittent estrogen production in the first year after a woman's final menstrual period, progesterone production is almost never observed and ovulation permanently ceases (148). Inhibins are undetectable once STRAW stage zero has been reached (90,149). Thus, many hormonal changes associated with the menopausal transition appear to be explainable by the dynamics of inhibin secretion and ovarian reserve.

STRAW Staging System for Reproductive Aging

Final menstrual period (FMP) 0

Stages	−5	−4	−3	−2	−1	+1		+2
Terminology:	**Reproductive**			**Menopausal transition**		**Postmenopause**		
	Early	Peak	Late	Early	Late*	Early*		Late
				Perimenopause				
Duration of stage:	Variable			Variable		1 yr	4 yrs	Until demise
Menstrual cycles	Variable to regular	Regular	Regular; Length decreases ~ 2 days	Variable cycle length (> 7 days different from normal)	≥2 skipped cycles and an interval of amenorrhea (≥ *60 days*)	†Amenor 12	None	
Endocrine	Normal FSH		↑ FSH	↑ FSH		↑ FSH		

**Stages most likely to be characterized by vasomotor symptoms; †Amenor = Amenorrhea;* ↑ = Elevated

FIG. 2. The STRAW staging system for reproductive aging. See text for details. [Reprinted with permission from (146).]

The Postmenopause and Beyond

Autopsy studies of the hypothalamus in women have suggested that hypothalamic GnRH content is lower in postmenopausal women when compared to reproductive-aged women (150). This reduction in GnRH content is not coincident with reduced numbers of GnRH neurons (151). Interestingly, this period is characterized by hypertrophied hypothalamic neurons. Moreover, reduced GnRH content manifests in a reproductive phase that is most notable for hypergonadotropic hypogonadism. Secretion of GnRH may, however, be increased in the postmenopause. Using a submaximal dose of GnRH antagonist to interrupt endogenous GnRH secretion, it was found that antagonists were less effective with age in two groups of postmenopausal women, 45–55 years and 70–80 years. In contrast, gonadal steroid replacement appears to suppress the hypothalamic– pituitary axis of postmenopausal women more effectively with increasing age, as demonstrated by two different research groups (55,152); however, one study did not find evidence of increased suppression in older women (153). Basal, unstimulated secretion of LH is also lower in older postmenopausal women compared to younger menopausal women (152) or young women with premature menopause (55).

Thus, in advanced reproductive aging, more GnRH is being produced, but less LH secretion is observed. Although this finding could be explained by a slower frequency of GnRH secretion with aging, analysis of pulsatile LH patterns does not support this interpretation. Thus it appears that the ability of the pituitary to secrete LH may become compromised with increased somatic aging. Whether this reflects changes in GnRH receptor physiology or is a manifestation of overall pituitary aging, decreased blood supply or cellular function remains to be elucidated.

Although elevated levels of gonadotropins are reliably associated with the menopausal transition and the early menopausal period (134,154,155), the trend of patterns of gonadotropin release in the late menopause is less well established (55,114,156). When compared to midreproductive aged women, gonadotropin levels throughout the menopause are significantly elevated. However, the effects of extreme aging and duration of the postmenopause on gonadotropin levels have not been sufficiently clarified. It is reported by some that advanced reproductive aging (the late menopause; =70 years of age) is characterized by an attenuation of GnRH release and a reduction in GnRH pulse frequency (92,114, 153,156–159). Others have reported that advanced reproductive aging is associated with increased pulse amplitude with a reduction in GnRH frequency (156). Regardless of how it is achieved, there is agreement that there is a progressive age-related decline in gonadotropin levels (55,114,115,153,156, 158,160). The significance of age-related changes in gonadotropin pulsatility is more controversial (55,114,115,153,156,158,160).

One potential source for the controversy surrounding the concentrations and pattern of serum LH and its role as a marker for GnRH activity in the late menopause may be related to age-related changes in the half-life of LH (156,161,162). Serum LH results may be confounded by other factors that alter LH clearance, such as medications and concurrent diseases. Nonetheless, the current literature strongly suggests that age-related changes in GnRH pulsatility are occurring despite chronic hypoestrogenism and may thus reflect primary neuroendocrine modifications associated with the menopause.

Under physiological conditions, there is a tight correlation between the pulsatile secretion of LH and the glycoprotein free alpha subunit (FAS) throughout the menstrual cycle of normal women (112,121,163). Several recent studies have suggested that when compared to LH, FAS may be a more reliable marker of GnRH activity because it has a shorter half-life (156,164). Consequently, it has been proposed that the study of age-dependent changes in FAS may provide a more accurate and revealing depiction of the impact of chronological and reproductive aging on the performance of the hypothalamic– pituitary axis (156,164), especially with regard to the rapidity of the open-loop pulsatile LH frequency. However, similar to the age-related limitations associated with the use of serum LH levels, there may also be an age-dependent increase in the half-life of FAS (159,165). Moreover, there is a component of FAS secretion that is GnRH independent, therefore potentially limiting the value of this measurement (165–167). Nonetheless, two recent studies with FAS suggest that there is a 40%–50% decline in mean levels of LH and FAS over the postmenopausal life span. The decline in FAS is of interest because each FAS pulse appears to be associated with attenuated pulse amplitudes (92,156). Of note, although a significant reduction in the FAS pulsatility (92,156) was observed as a function of chronological age, there was only a trend for reduced LH pulsatility (156).

As a group, these studies provide compelling evidence for the support of the hypothesis that age-related changes in the neuroendocrine axis continue after the onset of the menopause and well into the late post menopause. Furthermore, postmenopausal age-related changes in the neuroendocrine axis occur in a fashion that is independent of gonadal steroid production such that age-related changes in the neuroendocrine axis most likely reflect hypothalamic dysfunction and not changes in serum estradiol levels beyond the menopausal transition. It is possible that these changes may represent age-dependent changes in the sensitivity of hypothalamic–pituitary axis to positive and/or negative gonadal steroid feedback, thereby altering the regulation of GnRH release and GnRH receptor expression. Alternatively, changes in gonadotropin release patterns may also reflect the manifestation of a programmed, age-related breakdown, in cellular communication within and between the hypothalamus and the pituitary gland. Programmed age-related changes may result in the deterioration in efficiency of hypothalamic afferent input, detrimental changes in GnRH receptor expression or binding kinetics, or changes in the production and/or amplification of second messengers that mediate cellular communication needed for normal hypothalamic–pituitary interactions.

NONHUMAN MODELS FOR REPRODUCTIVE AGING

Nonhuman primates seem to hold the greatest promise with regard to the development of an animal model that closely parallels the process of reproductive senescence in humans (168). The baboon progression into reproductive quiescence most closely mimics the transition into the human menopause. In contrast to that described in other subhuman primates, reproductive senescence in baboons occurs in the latter third of the animal's life span. Similar to humans, baboons also appear to undergo a premenopausal period of compromised ovarian function that manifests as shortened follicular phases with increasing frequencies of anovulatory cycles and menstrual irregularities prior to the cessation of menses (169). Similar to observations in humans, there is also evidence that the baboon perimenopause is characterized by intermittent hypergonadotropic hypogonadism and fluctuating serum gonadotropin and sex steroid concentrations. Consistent with human reproductive aging, wide swings in estrogen production have been observed; again making the baboon a promising model to develop for the generation of studies that may be extrapolated to reproductive aging in humans (24,168,170). Unfortunately, the use of the baboon as a model for reproductive aging has been limited by the availability of facilities equipped to handle baboon colonies and the financial burden that accompanies maintaining baboon colonies (171).

Most nonhuman primate species in captivity typically do not live long enough to undergo ovarian senescence. Similar to the baboon and the human, rhesus monkeys also appear to undergo menopausal transition, but this phenomenon occurs very late in the animal's life span (172). Moreover, when these animals do transit into reproductive senescence, their ovaries are not completely devoid of follicles (172). Furthermore, the monotropic rise in FSH, a sensitive marker for impending ovarian failure in middle-aged regularly cycling women, seems to be absent in older rhesus monkeys with regular cycles (173). Interestingly, skipped cycles, followed by hypergonadotropic hypoestrogenic hypogonadism, appear to occur over a comparatively shorter period of time, implying that the transition into menopause is a relatively rapid process in the rhesus monkey. This transition period is also associated with a transient increase in DHEAS, implying that acute adrenal change is either a cause or a result of the dynamics associated with the changing ovarian milieu (173). Taken together, these data suggest that age-related changes in ovarian reserve initiate the menopausal transition in nonhuman primates as well as humans. It may also be postulated that the reduced pituitary and hypothalamic output associated with advanced reproductive aging are a consequence of age-related changes in the central neuroendocrine axis. In other words, neuroendocrine and ovarian factors appear to differentially contribute to reproductive senescence at differing times in the life cycle of the rhesus monkey.

Although there are certainly many similarities between reproductive aging in primates and rodents, a number of significant differences have led to a relative lack of enthusiasm for the use of rodent models to draw inferences about primate reproduction. Fundamental differences between the reproductive and life cycles of rodents and humans include: (a) rodents have shorter reproductive cycles; (b) reproductive cyclicity in the rodent is tightly linked to the circadian cycle; and (c) rodents do not experience a true "menstrual cycle," therefore by definition they do not have a true "menopause" (94). Lastly, studies in menopausal women demonstrate that menopause is coupled with ovarian follicular depletion (28,174,175). In contrast, studies in rodents suggest that persistent estrous (a rodent's physiological equivalent to menopause) is not characterized by an absolute depletion of the primordial ovarian follicle pool. Instead, the ovaries of persistent estrous rodents contain follicles in various stages of arrested development (101,176,177). Despite these limitations, several key similarities can be capitalized on to produce an acceptable animal model to investigate certain parameters that define reproductive aging. For example, the neuronal pathways and neurotransmitters involved in the regulation of GnRH neurons are conserved across many species (178–181). The conservation of most of the major neurotransmitter pathways involved in the regulation of routine reproductive physiology has permitted the rodent to be a practical model to probe questions that investigate and identify neuronal elements involved in reproductive aging (45,97,182–187). Therefore, for the purpose of this next section, we will discuss the findings of research that has focused on the hypothalamic–pituitary axis in rodents to characterize age-related changes in specific neuronal pathways. When possible, we will draw parallelisms and discuss features of female rodent reproductive aging that are similar to female human reproductive aging.

It is important to understand the reproductive cycle of rodents in order to understand reproductive aging in this animal model. Rodent reproductive physiology is defined by cyclical transitions through three phases, which are easily identified by vaginal smear cytology: proestrous, estrous, and diestrous (for full details see Chapter 43). The estrous cycle affords the rodent the ability to ovulate and potentially to conceive every 4–5 days. Similar to primates, the rodent cycle begins with a follicular phase that is typified by follicular recruitment and increasing serum estradiol levels (diestrous phase lasts for approximately 2–2.5 d). With progression of the cycle, increasing concentrations of estrogen facilitate the selection of dominant follicles, the LH surge, and the initiation of mechanisms that give rise to spontaneous polyovulation (proestrous lasts for approximately 12–14 h). The LH surge facilitates the conversion of post-ovulatory follicles to progesterone-secreting corpora lutea, giving rise to the estrous phase of the cycle (1–1.5 d). In the absence of coital stimulation and fertilization, the luteal-like estrous phase ends quickly with the onset of diestrus. Unlike primates, rodents do not experience menses, and without coital stimulation to produce daily surges of prolactin, they will not develop a fully functional corpus luteum, and will initiate the next reproductive cycle without evidence of bleeding.

Rodent reproductive senescence is primarily characterized by a state of persistent estrous that often evolves into a state of persistent anestrous. The basis for the age-related change in rodent estrous cyclicity seems to reside in the failure to continue to produce progesterone in a cyclic fashion. The ultimate emergence of persistent anestrous (188) generally coincides with significant follicular depletion, a physiological event that takes place long after the onset of central manifestations of reproductive senescence (101,189).

The rodent neuroendocrine reproductive axis thus appears to be the primary site driving the transition into reproductive quiescence (93,97,101, 162,190). The evolution and maintenance of rodent reproductive senescence most likely represents the composite outcome of a number of complex alterations in neuronal cytoarchitecture, neurotransmission, and patterns of neuropeptide release within the hypothalamic–pituitary axis (40,91,97,191–196). Consistent with what is observed in the reproductive cycles of perimenopausal women (24,76), reproductive aging in the female rodent is characterized by a gradual shift from the normal 4–5 day estrous cycles to cycles of variable length (197,198), an increasing proportion of anovulatory cycles, reduced numbers of developing ovarian follicles, compromised fertility (31,98,199), and an increasing proportion of cycles characterized by erratic serum estradiol production and elevated serum FSH. With increasing chronological age, female rodents eventually transit into a period of chronic anovulation, persistent estrus, a phase associated with persistent vaginal cornification (30,101,200,201). This dynamic transition generally begins in the "middle-aged" rodent (approximately 8–12 mo) and persists until complete acyclicity is established between 17–21 months of age (40,202,203).

Prior to the onset of persistent estrous, reproductive aging in rodents is heralded by a decline in preovulatory GnRH release, which manifests as an attenuation, and in most cases, a delay in the peak amplitude of the preovulatory LH surge (41,98,162, 187,190,204–206). Similar to the monotropic rise in FSH that is observed in humans, this age-dependent change in the pattern of LH release and peak amplitude of the rodent preovulatory LH surge emerges in middle-aged rodents that have regular estrous cycles (35,40). This age-related change in LH release does not appear to be related to age-dependent changes in gonadal steroid production. Moreover, the delayed onset and attenuated peak amplitude of the LH surge continues to be observed in gonadectomized middle-aged rats that have been primed with exogenous gonadal steroids for the induction of an LH surge (40,207). In other words, middle-aged rodents continue to exhibit abnormal gonadotropin surge patterns despite the controlled replacement of gonadal steroids with exogenous hormone. These findings clearly implicate the neuroendocrine axis as a primary cause, and not the gonads, in the attenuated and delayed LH surge pattern seen in middle-aged rats. Thus, unlike primates, the aging hypothalamic–pituitary axis in rodents directly contributes to reproductive senescence in a manner that is both distinct from, but concurrent with, gonadal failure (41,190,201,208).

For the most part, when compared to their young counterparts, middle-aged rodents exhibit decreased GnRH expression and reduced GnRH release in the hypothalamus (192,204,207,209). Cycle-specific changes in GnRH expression and release may precipitate the cascade of events that ultimately give rise to the attenuation and delay of the sex steroid induced preovulatory LH surge (40,186, 210,211). However, not all studies regarding age-related changes in GnRH expression and release have been consistent. While some investigators report no change in GnRH expression (99) or release (212), others have observed an age-related increase in GnRH release (213). It has thus been proposed that different populations of GnRH neurons may have specific functions (214,215) in reproduction and are differentially affected by somatic aging (93). It is possible that heterogeneity of GnRH neuronal populations and diffuse organization of GnRH cell bodies located within the preoptic area, the septal nuclei, and other hypothalamic areas are responsible for these somewhat contradictory inter-laboratory observations. For example, if the GnRH population being studied is not the population of neurons being activated at the time of the LH surge or not the population of neurons actively releasing GnRH (216,217), then conclusions drawn about the impact of aging on GnRH neuronal activation and GnRH release have the potential to be inconsistent or erroneous.

Within the medial preoptic area of cycling middle-aged rats killed during the peak or early descending phase of the LH surge, anatomical studies report as much as a 50% reduction in the number of GnRH neurons expressing cFos (as compared to young rats), a nonspecific marker for cellular activation (39,184,218). These studies hypothesized that prior to the LH surge, there is an age-related reduction in the activation of hypothalamic GnRH neurons in middle-aged rodents (219,220). This age-related reduction in GnRH neuronal activation did not reflect an age-related decline in GnRH neuronal density (184) or a diminished capacity for GnRH neurons to synthesize and release GnRH, as evidenced by the finding that potassium chloride-stimulated equivalent GnRH release from hypothalamic fragments of young and old rats (221).

At the time of the LH surge, the anteroventral periventricular preoptic area (AVP), an anatomical site that houses neurons that provide afferent excitatory innervations to GnRH neurons, is proposed to play a critical role in activation of GnRH neurons and GnRH release (222,223). Reduced GnRH neuronal activation has been associated with reduced afferent input from the AVP (39,224). While destruction of the AVP will result in the complete failure of the gonadal steroid-induced LH surge (225–227), lesions lateral to the rostral AVP will not block midcycle GnRH neuronal cFos expression (39). Unilateral lesions of the AVP result in a reduction of GnRH neuronal cFos expression on the side of the lesion and a subsequent reduction in the peak amplitude of the preovulatory LH surge (39). Patterns of

GnRH neuronal activation associated with iatrogenic lesions of the AVP closely resemble the pattern of GnRH neuronal activation observed in reproductive aging (39). An age-related reduction in the ability of the AVP to stimulate GnRH neurons may provide an explanation for both decreased synthesis and secretion of GnRH (39,224,228).

Consistent with the hypothesis that reduced and altered afferent input from the AVP may contribute to reproductive aging, electrical stimulation and the central application of various stimulatory pharmacological agents to the AVP of previously acyclic rodents has been shown to temporarily restore estrous cyclicity (229). It is important to also note that lesions of the rodent AVP do not correlate with or result in reduced GnRH neuronal density. Furthermore, lesions of the rodent AVP also do not compromise GnRH axonal-dendritic integrity (225–227). Other evidence consistent with the hypothesis that reduced GnRH neuronal stimulation results in an age-related reduction in GnRH release is found in studies that show that the nonspecific depolarizing agent potassium chloride stimulates equivalent GnRH release from hypothalamic slices retrieved from young and old rats (221). In other words, when given the appropriate stimulus, the middle-aged hypothalami will respond in a manner that is comparable to the hypothalami of a young reproductive aged rat.

As a group, these studies provide strong evidence that an age-dependent dysfunction in the ability of GnRH neurons to be activated and release GnRH accounts for age-related changes in gonadotropin surge patterns that herald the onset of reproductive senescence. More importantly, these age-related differences in GnRH release precede gonadal failure, thereby strongly implicating that the hypothalamus as the initiating locus of the transition into reproductive senescence.

Many other studies have provided data in strong support of the hypothesis that reproductive senescence in the rodent is a complex physiological event that is driven by changes initiated in the hypothalamic–pituitary axis. For example, studies in which ovaries have been transplanted from old acyclic rodents into young gonadectomized female rats demonstrated a restoration of estrous cyclicity (45,57,182). Additionally estrus cyclicity is not re-instated in young ovariectomized and hypophysectomized mice that receive both ovarian and pituitary explants from acyclic old mice, thereby suggesting that in rodents, the hypothalamus may drive the process of reproductive senescence.

In response to cyclical changes in serum gonadal steroid levels, GnRH pulsatility has direct effects on the sensitivity of the pituitary through the up-regulation of GnRH receptors. With the advent of reproductive senescence, pituitary responsiveness to GnRH has been reported to be unchanged, enhanced (208) or reduced (230,231). Since these studies have used different models with respect to age, gonadal steroid treatment and reproductive status, these data have been difficult to interpret (210,230,231).

Inhibitory and Excitatory Neurotransmission in Reproductive Aging

After migration into the septal and medial preoptic hypothalamic regions, GnRH neuronal network maturation precedes under the influence of both excitatory and inhibitory inputs (192,232–240). Dynamic changes in the patterns of GnRH secretion observed throughout the reproductive cycle reflect the net integration of excitatory and inhibitory input received and processed within the hypothalamus. Although a number of neurotransmitters and their respective receptors participate in the general maintenance of neuroendocrine reproductive axis, the excitatory amino acid neurotransmitters, glutamate (Glu) and aspartate (Asp), and activation of their respective receptors have been reported to play major roles in the regulation of GnRH synthesis and release and the subsequent preovulatory increase in serum LH (192,241–245). Similarly, gamma aminobutyric acid (GABA), an inhibitory amino acid transmitter, norepinephrine and neuropeptide Y have also been identified as key neurotransmitters involved in the regulation of preovulatory GnRH release.

Glutamatergic Neurotransmission and Reproductive Senescence

Within the central nervous system, Glu and Asp are the primary neurotransmitters responsible for excitatory synaptic transmission. Several studies have suggested that excitatory input from glutamatergic neurotransmission contribute to the regulation of GnRH synthesis and release (91,180,241,246–254). The role of glutamatergic neurotransmission in GnRH release and the production of the pre-ovulatory LH surge have been well established (241,255). The putative receptors for Glu and Asp are divided into two groups, the ionotropic and metabotropic receptors. Ionotropic receptors are named for the agonist that has been shown to activate the receptor N-methyl-D-aspartate (NMDA), L-alpha-amino-3-hydroxy-5-methyl-4-isoxazole propionic acid (AMPA), and kainate. Stimulation of the ionotropic Glu receptor is associated with the opening of ion channels and the influx of cations. Metabotropic receptors are coupled to G-proteins and second messenger amplification. Metabotropic receptors are sub-grouped into three categories and each receptor is named according to the second messenger produced and the ligand that best stimulates the receptor (180).

The effect of glutamatergic neurotransmission within the hypothalamic–pituitary axis is primarily related to the activation of hypothalamic GnRH neurons. Neither the infusion of glutamate into the anterior pituitary nor treatment of anterior pituitary cell cultures with glutamate agonist effectively induces the release of LH (256,257). Therefore the reproductive effects of glutamate are believed to reside solely in the hypothalamus. Estrogen and progesterone up-regulate glutamatergic receptor expression in the preoptic area of the hypothalamus. This up-regulation of glutamatergic receptors is proposed to potentiate excitatory neurotransmission and GnRH neuronal activation (228,247,258).

Several neuroanatomical studies have mapped metabotropic (259–261), NMDA (91,99,238,259), AMPA (247,262), and kainate (259,263) receptors to GnRH immunoreactive cells within the medial preoptic area (238,250,259,264–266). Afferent synaptic inputs from glutamatergic neurons to GnRH neurons have also been demonstrated histologically (223,238,267,268) within the hypothalamus.

Activation of the NMDA receptor is primarily associated with GnRH neuronal stimulation within the medial preoptic area (91,250,252,254,269–271). In addition to the induction of the LH surge, medial preoptic area NMDA receptor activation has been associated with preovulatory GnRH synthesis and secretion (254). Likewise, glutamatergic receptor antagonist have been shown to attenuate the estrogenic augmentation of GnRH synthesis and release (91,180,255,272,273). Non-NMDA receptor signaling triggered with AMPA and kainate primarily induces GnRH neuronal activation in the median eminence and the arcuate nucleus 180.

Since rodent reproductive aging is heralded by changes in GnRH neuronal activation, GnRH gene expression and peptide release is associated with the activation of GnRH neurons, it seems logical to question whether age-related changes in excitatory neurotransmission contribute to the mechanisms that yield reproductive senescence. Research focused on the role of glutamatergic receptors in reproductive aging have suggested that the age-dependent attenuation of GnRH/LH secretion may be associated with a loss in the ability of GnRH neurons to respond to NMDA receptor activation (192,228,274). Of note, the etiology of these age-related changes in the response to glutamatergic neurotransmission have not been linked to a reduction in the number of GnRH neurons or a decrease in the density of NMDA receptors located on GnRH immunoreactive neurons (37,99,275,276). Similarly, the age-dependent attenuation in GnRH neuronal activation observed in middle-aged rodents appears to emerge *before* the onset of irregular or persistent estrous cycles. These observations imply that age-related changes in the ability of glutamatergic neurotransmission to activate GnRH neurons may contribute to onset of reproductive senescence and the attenuation of GnRH/LH release and GnRH neuronal activation (37,184,192,212,274–276).

The mechanisms that result in age-related changes in glutamatergic neurotransmission have not been elucidated. It is possible that the age-dependent change in responsiveness to glutamatergic neurotransmission may be related to a reduction in concentration of available neurotransmitter pools of Glu and Asp located within the hypothalamus (274), or a change in glutamate receptor subunit expression and the pharmacokinetics of glutamate receptors (91,99,192,207). For example, it has been proposed that there are age-related changes in the NMDA receptor subunit subtype stoichiometry (99). This change in NMDA receptor stoichiometry could theoretically translate into alterations in receptor function and consequently result in a reduction in the magnitude and the propagation of the pre-ovulatory excitatory impulse needed to precede the preovulatory LH surge (99). A reduction in hypothalamic excitatory neurotransmission may translate into submaximal activation of GnRH neurons. This may result in decreased GnRH release and an inadequate stimulation of the anterior pituitary. This in turn could directly result in a reduced ability of pituitary gonadotrophs to synthesize and secrete LH in a surge-like fashion. Moreover, reduced glutamatergic excitatory neurotransmission could alter the effectiveness of other excitatory neurotransmitters such as norepinephrine (277,278).

In contrast to the NMDA receptor, very little is known about the role of other glutamatergic receptors in the evolution of reproductive senescence. Similar to the age-related changes in the ability of NMDA receptor activation to induce GnRH secretion, it has been proposed that kainate and AMPA also exhibit an age-related compromise in their ability to induce GnRH synthesis and release (192). The role of metabotropic glutamate receptors in reproductive aging has not been explored.

GABAergic Neurotransmission and Reproductive Senescence

The inhibitory amino acid neurotransmitter GABA is the dominant inhibitory neurotransmitter in the hypothalamus. GABA is synthesized in presynaptic neurons from the decarboxylation of Glu by glutamic acid decarboxylase (GAD). GABA synthesis is controlled by the concentration of Glu in GAD-containing elements. Three types of GABA receptors have been identified in the CNS; GABA A, GABA B, and GABA C. The GABA A receptor is a ligand-gated ion channel that is permissive to Cl^- ions upon agonist stimulation. The GABA B receptor is coupled to Ca^{2+} and K^+ channels as well as an intracellular signaling pathway that involve GTP binding proteins. The GABA C receptor is linked to Cl^- conductance.

GABAergic neuronal elements and neurotransmitter pools of GABA, the endogenous ligand, have been mapped within the hypothalamic–pituitary axis (279–281). Morphological studies suggest that there is an abundance of GABA synthesizing neurons in the medial preoptic area in close proximity to GnRH-immunoreactive cells (223,282,283). GABA A receptor–mediated neurotransmission is proposed to have direct (283) and indirect (284) regulatory effects on GnRH release and consequently LH secretion. For example, increased GABA turnover and activation of the GABA A receptor within the preoptic area on the day of diestrus is thought to mediate the inhibition of GnRH neurons (285,286). Conversely, on the day of proestrus and the preovulatory LH surge, preoptic area intracerebral microdialysis studies in ovariectomized, steroid-replaced rodents have suggested that the LH surge is preceded by decreased extracellular GABA levels (38,236,284,287,288). Additional data in support of the hypothesis that the preovulatory LH surge is preceded by reduced GABA mediated neurotransmission can be found in several recent studies that have reported reduced GAD mRNA levels in GABA-containing synaptic elements on the day of proestrus (223,236).

Gamma-vinyl-GABA, an irreversible GABA transaminase inhibitor, and muscimol, a GABA A receptor agonist, as well as the direct infusion of GABA into the preoptic area, have all been observed to block the estrogen and progesterone induced LH surge (160,234,289–291). Conversely, bicuculline, a GABA A antagonist, has been shown to restore the LH surge (288). As a group, these studies imply that in reproductively mature rats GABA imposes an inhibitory effect on GnRH neuronal activation and GnRH release.

Estrogen is proposed to exert dual feedback action on GnRH secretion by virtue of its ability to alter inhibitory and excitatory input into the preoptic area and the mediobasal hypothalamus (238,292–294). Approximately 30%–50% of hypothalamic GABAergic neurons express estrogen receptors (295). In vivo studies in either gonad intact or steroid replaced ovariectomized rodents suggest that estrogen regulates GABA-mediated inhibitory synaptic input by several mechanisms. Estrogen is thought to mediate the induction of ultrastructural changes in GABAergic neuronal elements, thereby changing the synaptic cytoarchitecture and the relationship between GABAergic nerve terminals and post-synaptic contacts on GnRH neurons (296–298). On the day of the LH surge, estrogen has been shown to induce a latent decrease in the expression of GAD, the biosynthetic enzyme for GABA. Reduced GABAergic neurotransmission on the day of proestrus is proposed to facilitate the induction of the LH surge (281,299).

Whereas GABAergic neurotransmission has been clearly implicated in the regulation of GnRH release in the hypothalamus of prepubertal and regularly cycling rats (243,300), few studies have investigated the role of GABA and GABA receptors in the process of reproductive senescence (38,301). When compared to young reproductive aged rodents, increased GABA turnover has been reported in the medial preoptic area and the suprachiasmatic nuclei of older rodents experiencing persistent diestrous or estrous (38). One recent study suggested that reproductive senescence is associated with an increase in GAD mRNA expression on the day of the LH surge and that the diurnal rhythm of anteroventral periventricular nucleus GAD mRNA expression is lost (301). These studies suggest that increased inhibitory tone mediated by GABA contributes to the gonadotropin abnormalities associated with reproductive aging. It is not clear whether these age-related changes in GABA turnover and production are also associated with changes in GABAergic receptor density, GABAergic receptor subtype expression, or changes in the efficiency of GABA receptor conductance of chloride ions. Moreover, it is not clear whether reproductive aging is associated with an altered ability of GABAergic neurons to accumulate estrogen intracellularly, thereby changing the ability of estrogen to regulate GABA-mediated neurotransmission.

Noradrenergic Neurotransmission and Reproductive Aging

Once the pubertal transition has been initiated by a change in the balance between glutamatergic and GABAergic neurotransmission, the adrenergic neurotransmitter pathways starts to contribute to the regulation of GnRH release and the subsequent pre-ovulatory LH surge. The preoptic area is proposed to receive afferent noradrenergic input from three major neuroanatomical regions: the locus ceruleus, the ventrolateral medulla, and nucleus tractus solitarius (302). While noradrenergic terminals have been found in close proximity to GnRH neurons, the existence of synaptic contact by noradrenergic terminals on GnRH soma has been controversial (299).

Adrenergic receptors are categorized into four groups: alpha-1, alpha-2, beta-1, and beta-2. Steroid receptive adrenergic neurons most often express the alpha subtype receptor and less commonly the beta subtype of the estrogen receptor (303,304). Consequently, the hormonal environment significantly impacts the manner by which these adrenergic neurons interact with GnRH neurons (305–307).

Noradrenergic neurons most likely represent interneurons that indirectly contribute to the regulation of GnRH neurons and GnRH release (308–310). The role of noradrenergic neurons in the regulation of GnRH neuronal activation and GnRH release appears to be permissive (311–313). The impact of the adrenergic neurotransmitter on GnRH gene expression and peptide release is dependent on estrogen and progesterone priming (299,314). The gonadal steroid hormonal milieu and the excitability of the GnRH neuron together dictate whether adrenergic neurons cooperate with excitatory or inhibitory input to influence the final pattern of GnRH synthesis and release (224,299,310). Of note, disruptions of noradrenergic input to GnRH neurons only results in the temporary cessation of the estrous cycle in rodents (310).

On the afternoon of proestrous, increased norepinephrine (NE) turnover and release has been observed in the mediobasal hypothalamus and the medial preoptic area (315,316). Diurnal changes in the activation of noradrenergic neurons, increased norepinephrine turnover, and enhanced NE release are thought to be essential to the maintenance of GnRH pulsatility (193,194,305,317). The facilitatory effects of NE on the GnRH release and induction of the LH surge have been linked to alpha-1 adrenergic receptor activation, particularly alpha-1b and activation of the nitric acid synthase/cGMP pathway (314,318–320).

Reproductive aging is associated with changes in the pattern of NE turnover and release (42,43,316,321). Age-related changes in adrenergic neurotransmission have been linked to a variety of changes. For example, it has been postulated that the age-related decrease in NE turnover and release and the associated disruption of downstream second messenger generation contribute to the age-related change in GnRH release and pulsatility (42,43). These age-related changes in NE turnover and release then result in the delayed onset and attenuation of the peak amplitude of the pre-ovulatory LH surge observed in middle-aged rodents (43,94). Interestingly, a recent study by Temel et al. (322) suggested that the age-related decrease in NE turnover was not associated with a significant difference in the number of estrogen positive NE neurons or an age-related impairment in the ability of estradiol to induce cFos production in preoptic area NE immunoreactive elements (322). Instead, Temel et al. hypothesized reproductive aging is associated with an age-related loss in the sensitivity of NE neurons to estrogen positive feedback and this results in a decrease in the expression of tyrosine hydroxylase mRNA, an enzyme critical to NE production (322).

As rodents continue to age and reach an advanced reproductive stage associated with anestrous, a significant age-related reduction in hypothalamic–pituitary noradrenergic tone is maintained (323). Moreover, there appears to be an inverse relationship between tissue concentrations of NE and advanced somatic age (96,324). This reduction in tissue NE levels may reflect reduced production of NE by diminishing numbers of competent hypothalamic noradrenergic neurons (96,316,325) or may be the result of neurotoxicity induced by prolonged exposure to toxic by-products formed during the metabolism of NE (323). Lastly, it has also been recently reported that reproductive aging may be associated with an age-related decrease in the density of synaptic interactions between adrenergic-to-GnRH neuronal elements located within the rodent preoptic area (325).

Changes in Other Neurotransmitters and Neuropeptides as a Function of Reproductive Aging

In rodents, endogenous opioids have been shown to suppress pulsatile GnRH release and LH release whereas opioid antagonists have been shown to reverse the effects of estrogen negative feedback (326). It has been suggested that advanced reproductive aging is associated with increased endogenous opioid tone. This increased opioid tone may exacerbate an already compromised neuroendocrine axis (327). In contrast, studies in women suggest that there may be an age-related decrease in hypothalamic opioid tone (36). Early follicular phase studies in young reproductive-aged women and studies in postmenopausal women indicate sex steroids facilitate the inhibitory effects of opioids on GnRH/LH secretion (328–332). These observations suggest that at least some of the age-related reduction in human opioid tone may be recovered with sex steroid priming.

Vasointestinal peptide (VIP) neurons are located in the ventrolateral aspect of the suprachiasmatic nucleus, a brain region that is associated with the functional maintenance of the circadian rhythm (333). In many small mammals, "middle-age" is associated with central disruption of circadian rhythmicity (334). However, in humans, there is no evidence of altered circadian rhythms until the age of 80 years (335). In rodents, reproductive aging has been associated with an attenuation in the magnitude and change in rhythm of VIP release (336). It is possible that age-related changes in the rhythm of VIP expression may cause the characteristic age-related changes in the LH surge observed in middle-aged rodents (336). Consistent with this hypothesis, Harney et al. described a model in which the infusion of antisense oligonucleotides to VIP into the SCN recreated the characteristic age-dependent changes in the LH surge in young rats (337).

SUMMARY

Reproductive aging is a complex process that involves multiple changes throughout the hypothalamic–pituitary–ovarian axis. Ovarian aging begins as soon as the ovarian follicle supply is maximized, in fetal life, and proceeds continuously until menopause. In humans, clinically apparent changes in cyclicity due to ovarian aging are not manifest until ovarian reserve reaches a critically low level. Once a woman attains her final menstrual period, there is no further evidence of ovarian function or production of gonadal hormones.

Aging of the hypothalamus becomes apparent in middle-aged rodents, and possibly in humans and nonhuman primates as well, when integration of the central nervous system signaling responsible for the induction of the LH surge and ovulation exhibits evidence of disruption. This disruption then contributes to ovulatory failure and acyclicity in rodent and primate models. Eventually, there is a breakdown in the recognition and integration of central circadian signals by the hypothalamus, and the organized, synchronous pattern of GnRH secretion becomes disrupted. In rodents, aberrations in excitatory and inhibitory neurotransmitter pathways appear to mediate age-related disruptions in the hypothalamic–pituitary axis. Finally, due to either pituitary exhaustion or hypothalamic dysfunction, circulating LH and FSH decrease. In humans and higher primates, hypothalamic dysfunction appears to progress at a much slower rate than in rodents, with severe impairment and disruption of the circadian rhythmicity only evident in the terminal stages of old age.

Much has been learned about reproductive aging through the use of animal models. The ability to perform invasive studies in nonhuman models has permitted access to and understanding of neuronal elements and neurochemical changes involved in hypothalamic regulation. These studies have provided a general, unifying working model for reproductive maturation and senescence. While the ovary is primarily responsible for gonadal failure in humans and higher primates, there is accompanying evidence that similar pathway of neuronal dysfunction occur with aging in primate and nonprimate mammals.

It is important to take into account the age of the organism when evaluating HPO axis function. More than likely, changes in central circadian rhythmicity, adrenal androgen production, and the somatotropic axis each impose a negative impact on the basic endocrinological functioning of the HPO axis. Similarly, in experimental models of reproductive aging, producing a menopausal gonadal state without the accompanying changes associated with somatic aging may lead to flawed conclusions. In other words, a young oophorectomized animal is not necessarily the biological equivalent of an aged, menopausal animal. As future models of reproductive aging are developed, this tenet needs to be kept central to the interpretation of results.

REFERENCES

1. Van Remmen, H., Ikeno, Y., and Hamilton, M., et al. (2003). Life-long reduction in MnSOD activity results in increased DNA damage and higher incidence of cancer but does not accelerate aging. *Physiol. Genomics* 16, 29–37.
2. Liu, L., Blasco, M., Trimarchi, J., and Keefe, D. (2002). An essential role for functional telomeres in mouse germ cells during fertilization and early development. *Dev. Biol.* 249, 74–84.
3. Golden, T. R., Hinerfeld, D. A., and Melov, S. (2002). Oxidative stress and aging: beyond correlation. *Aging Cell.* 1, 117–123.
4. Droge, W. (2003). Oxidative stress and aging. *Adv. Exp. Med. Biol.* 543, 191–200.
5. Devi, S. A., and Kiran, T. R. (2004). Regional responses in antioxidant system to exercise training and dietary vitamin E in aging rat brain. *Neurobiol. Aging* 25, 501–508.
6. Heilbronn, L. K., and Ravussin, E. (2003). Calorie restriction and aging: review of the literature and implications for studies in humans. *Am. J. Clin. Nutr.* 78, 361–369.
7. Lane, N. (2003). A unifying view of ageing and disease: the double-agent theory. *J. Theor. Biol.* 225, 531–540.
8. Luczaj, W., Waszkiewicz, E., Skrzydlewska, E., and Roszkowska-Jakimiec, W. (2004). Green tea protection against age-dependent ethanol-induced oxidative stress. *J. Toxicol. Environ. Health A.* 67, 595–606.
9. Naeye, R. L. (1983). Maternal age, obstetric complications, and the outcome of pregnancy. *Obstet. Gynecol.* 61, 210–216.
10. Purnell, J. Q., Brandon, D. D., Isabelle, L. M., Loriaux, D. L., and Samuels, M. H. (2004). Association of 24-hour cortisol production rates, cortisol-binding globulin, and plasma-free cortisol levels with body composition, leptin levels, and aging in adult men and women. *J. Clin. Endocrinol. Metab.* 89, 281–287.
11. Vgontzas, A. N., Zoumakis, M., Bixler, E. O., et al. (2003). Impaired nighttime sleep in healthy old versus young adults is associated with elevated plasma interleukin-6 and cortisol levels: physiologic and therapeutic implications. *J. Clin. Endocrinol. Metab.* 88, 2087–2095.
12. Seeman, T. E., McEwen, B. S., Rowe, J. W., and Singer, B. H. (2001). Allostatic load as a marker of cumulative biological risk: MacArthur studies of successful aging. *Proc. Natl. Acad. Sci. U S A* 98, 4770–4775.
13. Carr, M. C. (2003). The emergence of the metabolic syndrome with menopause. *J. Clin. Endocrinol. Metab.* 88, 2404–2411.

14. Das, M., Gabriely, I., and Barzilai, N. (2004). Caloric restriction, body fat and ageing in experimental models. *Obes. Rev.* 5, 13–19.
15. Moller, N., Gormsen, L., Fuglsang, J., and Gjedsted, J. (2003). Effects of ageing on insulin secretion and action. *Horm. Res.* 60, 102–104.
16. Morley, J. E. (2004). The metabolic syndrome and aging. *J. Gerontol. A. Biol. Sci. Med. Sci.* 59, 139–142.
17. Castracane, V. D., Kraemer, R. R., Franken, M. A., Kraemer, G. R., and Gimpel, T. (1998). Serum leptin concentration in women: effect of age, obesity, and estrogen administration. *Fertil. Steril.* 70, 472–477.
18. Bruunsgaard, H., Pedersen, A. N., Schroll, M., Skinhoj, P., and Pedersen, B. K. (2000). TNF-alpha, leptin, and lymphocyte function in human aging. *Life Sci.* 67, 2721–2731.
19. Ma, X. H., Muzumdar, R., Yang, X. M., Gabriely, I., Berger, R., and Barzilai, N. (2002). Aging is associated with resistance to effects of leptin on fat distribution and insulin action. *J. Gerontol. A. Biol. Sci. Med. Sci.* 57, B225–B231.
20. Meier, U., and Gressner, A. M. (2004). Endocrine regulation of energy metabolism: review of pathobiochemical and clinical chemical aspects of leptin, ghrelin, adiponectin, and resistin. *Clin. Chem.* 50, 1511–1525.
21. Welt, C. K., Chan, J. L., Bullen, J., et al. (2004). Recombinant human leptin in women with hypothalamic amenorrhea. *N. Engl. J. Med.* 351, 987–997.
22. Hoffman, A. R. and Ceda, G. P. (1999). Should we treat the somatopause? *J. Endocrinol. Invest.* 22, 4–6.
23. Labrie, F., Belanger A., Cusan, L., Gomez, J. L., and Candas, B. (1997). Marked decline in serum concentrations of adrenal C19 sex steroid precursors and conjugated androgen metabolites during aging. *J. Clin. Endocrinol. Metab.* 82, 2396–2402.
24. Santoro, N., Brown, J. R., Adel, T., and Skurnick, J. H. (1996). Characterization of reproductive hormonal dynamics in the perimenopause. *J. Clin. Endocrinol. Metab.* 81, 1495–1501.
25. Santoro, N., Isaac, B., Neal-Perry, G., et al. (2003). Impaired folliculogenesis and ovulation in older reproductive aged women. *J. Clin. Endocrinol. Metab.* 88, 5502–5509.
26. Richardson, S. J. (1993). The biological basis of the menopause. *Baillieres Clin. Endocrinol. Metab.* 7, 1–16.
27. Faddy, M. J., Gosden, R. G., Gougeon A., Richardson, S. J., and Nelson, J. F. (1992). Accelerated disappearance of ovarian follicles in mid-life: implications for forecasting menopause. *Hum. Reprod.* 7, 1342–1346.
28. Faddy, M. J. (2000). Follicle dynamics during ovarian ageing. *Mol. Cell. Endocrinol.* 163, 43–48.
29. Finch, C. E., Felicio, L. S., Mobbs, C. V., and Nelson, J. F. (1984). Ovarian and steroidal influences on neuroendocrine aging processes in female rodents. *Endocr. Rev.* 5, 467–497.
30. Mandl, A. M., and Zuckerman, S. (1952). Cyclical changes in the number of medium and large follicles in the adult rat ovary. *J. Endocrinol.* 8, 341–346.
31. Sopelak, V. M., and Butcher, R. L. (1982). Contribution of the ovary versus hypothalamus-pituitary to termination of estrous cycles in aging rats using ovarian transplants. *Biol. Reprod.* 27, 29–37.
32. Nelson, J. F., Bergman, M. D., Karelus, K., and Felicio, L. S. (1987). Aging of the hypothalamo–pituitary–ovarian axis: hormonal influences and cellular mechanisms. *J. Steroid Biochem.* 27, 699–705.
33. Nelson, J. F., and Felicio, L. S. (1986). Radical ovarian resection advances the onset of persistent vaginal cornification but only transiently disrupts hypothalamic–pituitary regulation of cyclicity in C57BL/6J mice. *Biol. Reprod.* 35, 957–964.
34. Cashion, A. B., Smith, M. J., and Wise, P. M. (2003). The morphometry of astrocytes in the rostral preoptic area exhibits a diurnal rhythm on proestrus: Relationship to the luteinizing hormone surge and effects of age. *Endocrinology* 144, 274–280.
35. Cooper, R. L., Conn, P. M., and Walker, R. F. (1980). Characterization of the LH surge in middle-aged female rats. *Biol. Reprod.* 23, 611–615.
36. Coiro, V., Volpi, R., Gramellini, D., et al. (1997). Age-related decrease in the opioid control of LH secretion during reproductive years in normal women. *Gynecol. Obstet. Invest.* 43, 162–165.
37. Gore, A. C., Yeung, G., Morrison, J. H., and Oung, T. (2000). Neuroendocrine aging in the female rat: the changing relationship of hypothalamic gonadotropin-releasing hormone neurons and N-methyl-D-aspartate receptors. *Endocrinology* 141, 4757–4767.
38. Jarry, H., Wise, P. M., Leonhardt, S., and Wuttke, W. (1999). Effects of age on GABA turnover rates in specific hypothalamic areas in female rats. *Exp. Clin. Endocrinol. Diabetes* 107, 59–62.
39. Le W-W, Wise, P. M., Murphy, A. Z., Coolen, L. M., and Hoffman, G. E. (2001). Parallel declines in fos activation of the medial anteroventral periventricular nucleus and LHRH neurons in middle-aged rats. *Endocrinology* 142, 4976–4982.
40. Scarbrough, K., and Wise, P. M. (1990). Age-related changes in pulsatile luteinizing hormone release precede the transition to estrous acyclicity and depend upon estrous cycle history. *Endocrinology* 126, 884–890.
41. Wise, P. M. (1982). Alterations in the proestrous pattern of median eminence LHRH, serum LH, FSH, estradiol and progesterone concentrations in middle-aged rats. *Life Sci.* 31, 165–173.
42. Wise, P. M. (1982). Norepinephrine and dopamine activity in microdissected brain areas of the middle-aged and young rat on proestrus. *Biol. Reprod.* 27, 562–574.
43. MohanKumar, S. M., and MohanKumar, P. S. (2004). Aging alters norepinephrine release in the medial preoptic area in response to steroid priming in ovariectomized rats. *Brain Res.* 1023, 24–30.
44. Granholm, N. H. and Dickens, G. A. (1986). Effects of reciprocal ovary transplantation on reproductive performance of lethal yellow mice (Ay/a; C57BL/6J). *J. Reprod. Fertil.* 78, 749–753.
45. Peng, M. T. and Huang, H. H. (1972). Aging of hypothalamic–pituitary–ovarian function in the rat. *Fertil. Steril.* 23, 535–542.
46. Snowdon, D. A., Kane, R. L., Beeson, W. L., et al. (1989). Is early natural menopause a biologic marker of health and aging? *Am. J. Public Health* 79, 709–714.
47. Luborsky, J. L., Meyer, P., Sowers, M. F., Gold, E. B., and Santoro, N. (2003). Premature menopause in a multi-ethnic population study of the menopause transition. *Hum. Reprod.* 18, 199–206.
48. Anderson, G. L., Limacher, M., Assaf, A. R., et al. (2004). Effects of conjugated equine estrogen in postmenopausal women with hysterectomy: the Women's Health Initiative randomized controlled trial. *JAMA* 291, 1701–1712.
49. Rossouw, J. E., Anderson, G. L., Prentice, R. L., et al. (2002). Risks and benefits of estrogen plus progestin in healthy postmenopausal women: principal results from the Women's Health Initiative randomized controlled trial. *JAMA* 288, 321–333.
50. Battaglia, D. E., Goodwin, P., Klein, N. A., and Soules, M. R. (1996). Influence of maternal age on meiotic spindle assembly in oocytes from naturally cycling women. *Hum. Reprod.* 11, 2217–2222.
51. Liu, L., Franco, S., Spyropoulos, B., Moens, P. B., Blasco, M. A., and Keefe, D. L. (2004). Irregular telomeres impair meiotic synapsis and recombination in mice. *Proc. Natl. Acad. Sci. U S A* 101, 6496–6501.
52. Johnson, J., Canning, J., Kaneko, T., Pru, J. K., and Tilly, J. L. (2004). Germline stem cells and follicular renewal in the postnatal mammalian ovary. *Nature* 428, 145–150.

53. Weiss, G., Skurnick, J. H., Goldsmith, L. T., Santoro, N. F., and Park, S. J. (2004). Menopause and hypothalamic–pituitary sensitivity to estrogen. *JAMA* 292, 2991–2996. [Erratum in *JAMA* (2005) 293, 163.]
54. Wise, P. M., Krajnak, K. M., and Kashon, M. L. (1996). Menopause: the aging of multiple pacemakers. *Science* 273, 67–70.
55. Santoro, N., Banwell, T., Tortoriello, D., Lieman, H., Adel, T., and Skurnick, J. (1998). Effects of aging and gonadal failure on the hypothalamic–pituitary axis in women. *Am. J. Obstet. Gynecol.* 178, 732–741.
56. Pincus, S. M., Veldhuis, J. D., Mulligan, T., Iranmanesh A., and Evans, W. S. (1997). Effects of age on the irregularity of LH and FSH serum concentrations in women and men. *Am. J. Physiol. Endocrinol. Metab.* 273, E989–995.
57. Felicio, L. S., Nelson, J. F., Gosden, R. G., and Finch, C. E. (1983). Restoration of ovulatory cycles by young ovarian grafts in aging mice: potentiation by long-term ovariectomy decreases with age. *Proc. Natl. Acad. Sci. U S A* 80, 6076–6080.
58. Danilovich, N., and Sairam, M. R. (2002). Haploinsufficiency of the follicle-stimulating hormone receptor accelerates oocyte loss inducing early reproductive senescence and biological aging in mice. *Biol. Reprod.* 67, 361–369.
59. Mayer, L. P., Devine, P. J., Dyer, C. A., and Hoyer, P. B. (2004). The follicle-deplete mouse ovary produces androgen. *Biol. Reprod.* 71, 130–138.
60. McKinlay, S. M. (1996). The normal menopause transition: an overview. *Maturitas* 23, 137–145.
61. Harlow, B. L., Wise, L. A., Otto, M. W., Soares, C. N., and Cohen, L. S. (2003). Depression and its influence on reproductive endocrine and menstrual cycle markers associated with perimenopause: the Harvard Study of Moods and Cycles. *Arch. Gen. Psychiatry* 60, 29–36.
62. Luoto, R., Kaprio, J., and Uutela, A. (1994). Age at natural menopause and sociodemographic status in Finland. *Am. J. Epidemiol.* 139, 64–76.
63. Cooper, G. S., Hulka, B. S., Baird, D. D., et al. (1994). Galactose consumption, metabolism, and follicle-stimulating hormone concentrations in women of late reproductive age. *Fertil. Steril.* 62, 1168–1175.
64. Yanushpolsky, E. H., Best, C. L., Jackson, K. V., Clarke, R. N., Barbieri, R. L., and Hornstein, M. D. (1998). Effects of endometriomas on ooccyte quality, embryo quality, and pregnancy rates in in vitro fertilization cycles: a prospective, case-controlled study. *J. Assist. Reprod. Genet.* 15, 193–197.
65. Cooper, G. S., and Thorp, J. M. Jr. (1999). FSH levels in relation to hysterectomy and to unilateral oophorectomy. *Obstet. Gynecol.* 94, 969–972.
66. Beksac, M. S., Kisnisci, H. A., Cakar, A. N., and Beksac, M. (1983). The endocrinological evaluation of bilateral and unilateral oophorectomy in premenopausal women. *Int. J. Fertil.* 28, 219–224.
67. Ahn, E. H., Bai, S. W., Song, C. H., et al. (2002). Effect of hysterectomy on conserved ovarian function. *Yonsei Med. J.* 43, 53–58.
68. Cramer, D. W., Xu, H., and Harlow, B. L. (1995). Does "incessant" ovulation increase risk for early menopause? *Am. J. Obstet. Gynecol.* 172, 568–573.
69. Weel, A. E., Uitterlinden, A. G., Westendorp, I. C., et al. (1999). Estrogen receptor polymorphism predicts the onset of natural and surgical menopause. *J. Clin. Endocrinol. Metab.* 84, 3146–3150.
70. Crisponi, L., Deiana, M., Loi, A., et al. (2001). The putative forkhead transcription factor FOXL2 is mutated in blepharophimosis/ptosis/epicanthus inversus syndrome. *Nat. Genet.* 27, 159–166.
71. Uda, M., Ottolenghi, C., Crisponi, L., et al. (2004). Foxl2 disruption causes mouse ovarian failure by pervasive blockage of follicle development. *Hum. Mol. Genet.* 13, 1171–1181.
72. Richards, J. S., Sharma, S. C., Falender, A. E., and Lo, Y. H. (2002). Expression of FKHR, FKHRL1, and AFX genes in the rodent ovary: evidence for regulation by IGF-I, estrogen, and the gonadotropins. *Mol. Endocrinol.* 16, 580–599.
73. Matzuk, M. M. (2000). Revelations of ovarian follicle biology from gene knockout mice. *Mol. Cell. Endocrinol.* 163, 61–66.
74. Lenton, E. A., Sexton, L., Lee, S., and Cooke, I. D. (1988). Progressive changes in LH and FSH and LH: FSH ratio in women throughout reproductive life. *Maturitas* 10, 35–43.
75. Lenton, E. A., Landgren, B. M., Sexton, L., and Harper, R. (1984). Normal variation in the length of the follicular phase of the menstrual cycle: effect of chronological age. *Br. J. Obstet. Gynaecol.* 91, 681–684.
76. Klein, N. A., Battaglia, D. E., Fujimoto, V. Y., Davis, G. S., Bremner, W. J., and Soules, M. R. (1996). Reproductive aging: accelerated ovarian follicular development associated with a monotropic follicle-stimulating hormone rise in normal older women. *J. Clin. Endocrinol. Metab.* 81, 1038–1045.
77. Santoro, N., Adel, T., and Skurnick, J. H. (1999). Decreased inhibin tone and increased activin A secretion characterize reproductive aging in women. *Fertil. Steril.* 71, 658–662.
78. Klein, N. A., Battaglia, D. E., Miller, P. B., Branigan, E. F., Giudice, L. C., and Soules, M. R. (1996). Ovarian follicular development and the follicular fluid hormones and growth factors in normal women of advanced reproductive age. *J. Clin. Endocrinol. Metab.* 81, 1946–1951.
79. Richardson, S. J., and Nelson, J. F. (1990). Follicular depletion during the menopausal transition. *Ann. N Y Acad. Sci.* 592, 13–20; discussion 44–51.
80. Faddy, M. J., and Gosden, R. G. (1995). A mathematical model of follicle dynamics in the human ovary. *Hum. Reprod.* 10, 770–775.
81. Flaws, J. A., Langenberg, P., Babus, J. K., Hirshfield, A. N., and Sharara, F. I. (2001). Ovarian volume and antral follicle counts as indicators of menopausal status. *Menopause* 8, 175–180.
82. Hansen, K. R., Morris, J. L., Thyer, A. C., and Soules, M. R. (2003). Reproductive aging and variability in the ovarian antral follicle count: application in the clinical setting. *Fertil. Steril.* 80, 577–583.
83. Scheffer, G. J., Broekmans, F. J., Looman, C. W., et al. (2003). The number of antral follicles in normal women with proven fertility is the best reflection of reproductive age. *Hum. Reprod.* 18, 700–706.
84. Burger, H. G. (1999). The endocrinology of the menopause. *J. Steroid Biochem. Mol. Biol.* 69, 31–35.
85. Burger, H. G. (1998). Pituitary dysfunction in recently postmenopausal women. *Clin. Endocrinol. (Oxf).* 48, 124.
86. Loria, P., Petraglia, F., Concari, M., et al. (1998). Influence of age and sex on serum concentrations of total dimeric activin A. *Eur. J. Endocrinol.* 139, 487–492.
87. Ravhon A., Lavery, S., Michael, S., et al. (2000). Dynamic assays of inhibin B and oestradiol following buserelin acetate administration as predictors of ovarian response in IVF. *Hum. Reprod.* 15, 2297–2301.
88. Dzik A., Lambert-Messerlian, G., Izzo, V. M., Soares, J. B., Pinotti, J. A., and Seifer, D. B. (2000). Inhibin B response to EFORT is associated with the outcome of oocyte retrieval in the subsequent in vitro fertilization cycle. *Fertil. Steril.* 74, 1114–1117.
89. Corson, S. L., Gutmann, J., Batzer, F. R., Wallace, H., Klein, N., and Soules, M. R. (1999). Inhibin-B as a test of ovarian reserve for infertile women. *Hum. Reprod.* 14, 2818–2821.
90. Burger, H. G., Dudley, E., Mamers, P., Robertson, D., Groome, N., and Dennerstein, L. (2002). The ageing female

reproductive axis I. *Novartis Found. Symp.* 242, 161–167; discussion 167–171.
91. Gore, A. C. (2001). Gonadotropin-releasing hormone neurons, NMDA receptors, and their regulation by steroid hormones across the reproductive life cycle. *Brain Res. Brain Res. Rev.* 37, 235–248.
92. Gill, S., Lavoie, H. B., Bo-Abbas, Y., and Hall, J. E. (2002). Negative feedback effects of gonadal steroids are preserved with aging in postmenopausal women. *J. Clin. Endocrinol. Metab.* 87, 2297–2302.
93. Rubin, B. S. (2000). Hypothalamic alterations and reproductive aging in female rats: evidence of altered luteinizing hormone-releasing hormone neuronal function. *Biol. Reprod.* 63, 968–976.
94. Wise, P. M., Smith, M. J., Dubal, D. B., et al. (2002). Neuroendocrine modulation and repercussions of female reproductive aging. *Recent Prog. Horm. Res.* 57, 235–256.
95. Weiland, N. G., and Wise, P. M. (1990). Aging progressively decreases the densities and alters the diurnal rhythms of alpha 1-adrenergic receptors in selected hypothalamic regions. *Endocrinology* 126, 2392–2397.
96. Weiland, N. G., Cohen, I. R., and Wise, P. M. (1989). Age-associated alterations in catecholaminergic concentrations, neuronal activity, and alpha 1 receptor densities in female rats. *Neurobiol. Aging* 10, 323–329.
97. Wise, P. M., and Camp, P. (1984). Changes in concentrations of estradiol nuclear receptors in the preoptic area, medial basal hypothalamus, amygdala, and pituitary gland of middle-aged and old cycling rats. *Endocrinology* 114, 92–98.
98. LaPolt, P. S., Matt, D. W., and Lu, J. K. (1998). Progesterone implants delay age-related declines in regular estrous cyclicity and the ovarian follicular reserve in Long-Evans rats. *Biol. Reprod.* 59, 197–201.
99. Gore, A. C., Oung, T., and Woller, M. J. (2002). Age-related changes in hypothalamic gonadotropin-releasing hormone and N-methyl-D-aspartate receptor gene expression, and their regulation by oestrogen, in the female rat. *J. Neuroendocrinol.* 14, 300–309.
100. Wise, P. M., Dueker, E., and Wuttke, W. (1988). Age-related alterations in pulsatile luteinizing hormone release: effects of long-term ovariectomy, repeated pregnancies and naloxone. *Biol. Reprod.* 39, 1060–1066.
101. LaPolt, P. S., and Lu, J. K. (2001). Effects of aging on luteinizing hormone secretion, ovulation, and ovarian tissue-type plasminogen activator expression. *Exp. Biol. Med. (Maywood).* 226, 127–132.
102. Park, S. J., Goldsmith, L. T., and Weiss, G. (2002). Age-related changes in the regulation of luteinizing hormone secretion by estrogen in women. *Exp. Biol. Med.* 227, 455–464.
103. Treloar, A. E. (1974). Menarche, menopause, and intervening fecundability. *Hum. Biol.* 46, 89–107.
104. Treloar, A. E., Boynton, R. E., Behn, B. G., and Brown, B. W. (1967). Variation of the human menstrual cycle through reproductive life. *Int. J. Fertil.* 12, 77–126.
105. Antunes, J. L., Carmel, P. W., Housepian, E. M., and Ferin, M. (1978). Luteinizing hormone-releasing hormone in human pituitary blood. *J. Neurosurg.* 49, 382–386.
106. Clarke, I. J., and Cummins, J. T. (1982). The temporal relationship between gonadotropin releasing hormone (GnRH) and luteinizing hormone (LH) secretion in ovariectomized ewes. *Endocrinology* 111, 1737–1739.
107. Caraty, A., Antoine, C., and Delaleu, B., et al. (1995). Nature and bioactivity of gonadotropin-releasing hormone (GnRH) secreted during the GnRH surge. *Endocrinology* 136, 3452–3460.
108. Karsch, F. J., Dierschke, D. K., Weick, R. F., Yamaji, T., Hotchkiss, J., and Knobil, E. (1973). Positive and negative feedback control by estrogen of luteinizing hormone secretion in the rhesus monkey. *Endocrinology* 92, 799–804.
109. Levine, J. E., and Duffy, M. T. (1988). Simultaneous measurement of luteinizing hormone (LH)-releasing hormone, LH, and follicle-stimulating hormone release in intact and short-term castrate rats. *Endocrinology* 122, 2211–2221.
110. Knobil, E., and Plant, T. M. (1978). The hypothalamic regulation of LH and FSH secretion in the rhesus monkey. *Res. Publ. Assoc. Res. Nerv. Ment. Dis.* 56, 359–372.
111. Handelsman, D. J., Cummins, J. T., and Clarke, I. J. (1988). Pharmacodynamics of gonadotropin-releasing hormone. 1. Effects of gonadotropin-releasing hormone pulse contour on pituitary luteinizing hormone secretion in vivo in sheep. *Neuroendocrinology* 48, 432–438.
112. Reame, N., Sauder, S. E., Kelch, R. P., and Marshall, J. C. (1984). Pulsatile gonadotropin secretion during the human menstrual cycle: evidence for altered frequency of gonadotropin-releasing hormone secretion. *J. Clin. Endocrinol. Metab.* 59, 328–337.
113. Rossmanith, W., Laughlin, G., Mortola, J., Johnson, M., Veldhuis, J., and Yen, S. (1990). Pulsatile cosecretion of estradiol and progesterone by the midluteal phase corpus luteum: temporal link to luteinizing hormone pulses. *J. Clin. Endocrinol. Metab.* 70, 990–995.
114. Rossmanith, W. G., Scherbaum, W. A., and Lauritzen, C. (1991). Gonadotropin secretion during aging in postmenopausal women. *Neuroendocrinology.* 54, 211–218.
115. Rossmanith, W. G. (1995). Gonadotropin secretion during aging in women: Review article. *Exp. Gerontology* 30, 369–381.
116. Baird, D. T., and Smith, K. B. (1993). Inhibin and related peptides in the regulation of reproduction. *Oxf. Rev. Reprod. Biol.* 15, 191–232.
117. Roberts, V. J., Barth, S., el-Roeiy A., and Yen, S. S. (1993). Expression of inhibin/activin subunits and follistatin messenger ribonucleic acids and proteins in ovarian follicles and the corpus luteum during the human menstrual cycle. *J. Clin. Endocrinol. Metab.* 77, 1402–1410.
118. Wildt, L., Hausler A., Marshall, G., et al. (1981). Frequency and amplitude of gonadotropin-releasing hormone stimulation and gonadotropin secretion in the rhesus monkey. *Endocrinology* 109, 376–385.
119. Hodgen, G. D. (1982). The dominant ovarian follicle. *Fertil. Steril.* 38, 281–300.
120. Zeleznik, A. J., Hutchison, J. S., and Schuler, H. M. (1985). Interference with the gonadotropin-suppressing actions of estradiol in macaques overrides the selection of a single preovulatory follicle. *Endocrinology* 117, 991–999.
121. Filicori, M., Santoro, N., Merriam, G. R., and Crowley, W. F. Jr. (1986). Characterization of the physiological pattern of episodic gonadotropin secretion throughout the human menstrual cycle. *J. Clin. Endocrinol. Metab.* 62, 1136–1144.
122. Klein, N. A., Illingworth, P. J., Groome, N. P., McNeilly, A. S., Battaglia, D. E., and Soules, M. R. (1996). Decreased inhibin B secretion is associated with the monotropic FSH rise in older, ovulatory women: a study of serum and follicular fluid levels of dimeric inhibin A and B in spontaneous menstrual cycles. *J. Clin. Endocrinol. Metab.* 81, 2742–2745.
123. Reame, N. E., Wyman, T. L., Phillips, D. J., de Kretser, D. M., and Padmanabhan, V. (1998). Net increase in stimulatory input resulting from a decrease in inhibin B and an increase in activin A may contribute in part to the rise in follicular phase follicle-stimulating hormone of aging cycling women. *J. Clin. Endocrinol. Metab.* 83, 3302–3307.
124. Welt, C. K., McNicholl, D. J., Taylor, A. E., and Hall, J. E. (1999). Female reproductive aging is marked by decreased secretion of dimeric inhibin. *J. Clin. Endocrinol. Metab.* 84, 105–111.

125. Burger, H. G., Dudley, E. C., Hopper, J. L., et al. (1999). Prospectively measured levels of serum follicle-stimulating hormone, estradiol, and the dimeric inhibins during the menopausal transition in a population-based cohort of women. *J. Clin. Endocrinol. Metab.* 84, 4025–4030.
126. Danforth, D. R., Arbogast, L. K., Mroueh, J., et al. (1998). Dimeric inhibin: a direct marker of ovarian aging. *Fertil. Steril.* 70, 119–123.
127. MacNaughton, J., Banah, M., McCloud, P., Hee, J., and Burger, H. (1992). Age related changes in follicle stimulating hormone, luteinizing hormone, oestradiol and immunoreactive inhibin in women of reproductive age. *Clin. Endocrinol. (Oxf).* 36, 339–345.
128. Burger, H. G., Groome, N. P., and Robertson, D. M. (1998). Both inhibin A and B respond to exogenous follicle-stimulating hormone in the follicular phase of the human menstrual cycle. *J. Clin. Endocrinol. Metab.* 83, 4167–4169.
129. Tong, S., Wallace, E. M., and Burger, H. G. (2003). Inhibins and activins: clinical advances in reproductive medicine. *Clin. Endocrinol. (Oxf).* 58, 115–127.
130. Weiss, J., Harris, P. E., Halvorson, L. M., Crowley, W. F. Jr., and Jameson, J. L. (1992). Dynamic regulation of follicle-stimulating hormone-beta messenger ribonucleic acid levels by activin and gonadotropin-releasing hormone in perifused rat pituitary cells. *Endocrinology* 131, 1403–1408.
131. Weiss, J., Crowley, W. F., Jr., Halvorson, L. M., and Jameson, J. L. (1993). Perifusion of rat pituitary cells with gonadotropin-releasing hormone, activin, and inhibin reveals distinct effects on gonadotropin gene expression and secretion. *Endocrinology* 132, 2307–2311.
132. Demura, R., Suzuki, T., Tajima, S., et al. (1993). Human plasma free activin and inhibin levels during the menstrual cycle. *J. Clin. Endocrinol. Metab.* 76, 1080–1082.
133. Muttukrishna, S., Fowler, P. A., George, L., Groome, N. P., and Knight, P. G. (1996). Changes in peripheral serum levels of total activin A during the human menstrual cycle and pregnancy. *J. Clin. Endocrinol. Metab.* 81, 3328–3334.
134. Yen, S. S., Tsai, C. C., Naftolin, F., Vandenberg, G., and Ajabor, L. (1972). Pulsatile patterns of gonadotropin release in subjects with and without ovarian function. *J. Clin. Endocrinol. Metab.* 34, 671–675.
135. Dierschke, D. J., Bhattacharya, A. N., Atkinson, L. E., and Knobil, E. (1970). Circhoral oscillations of plasma LH levels in the ovariectomized rhesus monkey. *Endocrinology* 87, 850–853.
136. Clarke, I. J. (2002). Two decades of measuring GnRH secretion. *Reprod. Suppl.* 59, 1–13.
137. Clarke, I. J., Cummins, J. T., Jenkin, M., and Phillips, D. J. (1989). The oestrogen-induced surge of LH requires a 'signal' pattern of gonadotrophin-releasing hormone input to the pituitary gland in the ewe. *J. Endocrinol.* 122, 127–134.
138. Knobil, E., Plant, T. M., Wildt, L., Belchetz, P. E., and Marshall, G. (1980). Control of the rhesus monkey menstrual cycle: permissive role of hypothalamic gonadotropin-releasing hormone. *Science* 207, 1371–1373.
139. Levine, J. E., Bauer-Dantoin, A. C., and Besecke, L. M., et al. (1991). Neuroendocrine regulation of the luteinizing hormone-releasing hormone pulse generator in the rat. *Recent Prog. Horm. Res.* 47, 97–151; discussion 151–153.
140. Weiss, J., Duca, K. A., and Crowley, W. F. Jr. (1990). Gonadotropin-releasing hormone-induced stimulation and desensitization of free alpha-subunit secretion mirrors luteinizing hormone and follicle-stimulating hormone in perifused rat pituitary cells. *Endocrinology* 127, 2364–2371.
141. Dennerstein, L., Lehert, P., Burger, H., and Dudley, E. (1999). Mood and the menopausal transition. *J. Nerv. Ment. Dis.* 187, 685–691.
142. Dennerstein, L., Dudley, E. C., Hopper, J. L., Guthrie, J. R., and Burger, H. G. (2000). A prospective population-based study of menopausal symptoms. *Obstet. Gynecol.* 96, 351–358.
143. Garamszegi, C., Dennerstein, L., Dudley, E., Guthrie, J. R., Ryan, M., and Burger, H. (1998). Menopausal status: subjectively and objectively defined. *J. Psychosom. Obstet. Gynaecol.* 19, 165–173.
144. Treloar, A. E. (1981). Menstrual cyclicity and the pre-menopause. *Maturitas* 3, 249–264.
145. McKinlay, S. M., Brambilla, D. J., and Posner, J. G. (1992). The normal menopause transition. *Maturitas* 14, 103–115.
146. Adashi, E. Y. (1995). Insulin-like growth factors as determinants of follicular fate. *J. Soc. Gynecol. Investig.* 2, 721–726.
147. Soules, M. R., Sherman, S., and Parrott, E., et al. (2001). Executive summary: Stages of Reproductive Aging Workshop (STRAW). *Fertil. Steril.* 76, 874–878.
148. Metcalf M. G., Donald, R. A., and Livesey, J. H. (1982). Pituitary-ovarian function before, during and after the menopause: a longitudinal study. *Clin. Endocrinol. (Oxf).* 17, 489–494.
149. Hee, J. P., MacNaughton, J., Bangah, M., et al. (1993). Follicle-stimulating hormone induces dose-dependant stimulation of immunoreactive inhibin secretion during the follicular phase of the human menstrual cycle. *J. Clin. Endocrinol. Metab.* 76, 1340–1343.
150. Parker, C. R., Jr., and Porter, J. C. (1984). Luteinizing hormone-releasing hormone and thyrotropin-releasing hormone in the hypothalamus of women: effects of age and reproductive status. *J. Clin. Endocrinol. Metab.* 58, 488–491.
151. Abel, T. W., and Rance, N. E. (2000). Stereologic study of the hypothalamic infundibular nucleus in young and older women. *J. Comp. Neurol.* 424, 679–688.
152. Gill, S., Sharpless, J. L., Rado, K., and Hall, J. E. (2002). Evidence that GnRH decreases with gonadal steroid feedback but increases with age in postmenopausal women. *J. Clin. Endocrinol. Metab.* 87, 2290–2296.
153. Rossmanith, W. G., Reichelt, C., and Scherbaum, W. A. (1994). Neuroendocrinology of aging in humans: Attenuated sensitivity to sex steroid feedback in elderly postmenopausal women. *Neuroendocrinology* 59, 355–362.
154. Hayes, F. J., Hall, J. E., Boepple, P. A., and Crowley, W. F. Jr. (1998). Clinical review 96: Differential control of gonadotropin secretion in the human: endocrine role of inhibin. *J. Clin. Endocrinol. Metab.* 83, 1835–1841.
155. Monroe, S. E., Jaffee, R. B., and Midgley, A. J. (1972). Regulation of human gonadotropins: 13. Changes in serum gonadotropins in menstruating women in response to oophorectomy. *J. Clin. Endocrinol.* 34, 420–422.
156. Hall, J. E., Lavoie, H. B., Marsh, E. E., and Martin, K. A. (2000). Decrease in gonadotropin-releasing hormone (GnRH) pulse frequency with aging in postmenopausal women. *J. Clin. Endocrinol. Metab.* 85, 1794–1800.
157. Hall, J. E., and Gill, S. (2001). Neuroendocrine aspects of aging in women. *Endocrinol. Metab. Clin. North Am.* 30, 631–646.
158. Lambalk, C. B., de Boer, L., Schoute, E., Popp-Snyders, C., and Schoemaker, J. (1997). Post-menopausal and chronological age have divergent effects on pituitary and hypothalamic function in episodic gonadotrophin secretion. *Clin. Endocrinol. (Oxf).* 46, 439–443.
159. Kwekkeboom, D. J., Lamberts, S. W., Blom, J. H., Schroeder, F. H., and De Jong, F. H. (1990). Prolonged treatment with the GnRH analogue buserelin does not affect alpha-subunit production by the pituitary gonadotroph. *Clin. Endocrinol. (Oxf).* 32, 443–451.
160. Lamberts, R., Vijayan, E., Graf, M., Mansky, T., and Wuttke, W. (1983). Involvement of preoptic-anterior hypothalamic

GABA neurons in the regulation of pituitary LH and prolactin release. *Exp. Brain Res.* 52, 356–362.

161. Sharpless, J. L., Supko, J. G., Martin, K. A., and Hall, J. E. (1999). Disappearance of endogenous luteinizing hormone is prolonged in postmenopausal women. *J. Clin. Endocrinol. Metab.* 84, 688–694.
162. Matt, D. W., Gilson M. P., Sales, T. E., et al. (1998). Characterization of attenuated proestrous luteinizing hormone surges in middle-aged rats by deconvolution analysis. *Biol. Reprod.* 59, 1477–1482.
163. Adams, J. M., Taylor, A. E., Schoenfeld, D. A., Crowley, W. F. Jr., and Hall, J. E. (1994). The midcycle gonadotropin surge in normal women occurs in the face of an unchanging gonadotropin-releasing hormone pulse frequency. *J. Clin. Endocrinol. Metab.* 79, 858–864.
164. Hayes, F. J., McNicholl, D. J., Schoenfeld, D., Marsh, E. E., and Hall, J. E. (1999). Free alpha-subunit is superior to luteinizing hormone as a marker of gonadotropin-releasing hormone despite desensitization at fast pulse frequencies. *J. Clin. Endocrinol. Metab.* 84, 1028–1036.
165. Couzinet, B., Lahlou, N., Thomas, G., et al. (1991). Effects of gonadotrophin releasing hormone antagonist and agonist on the pulsatile release of gonadotrophins and alpha-subunit in postmenopausal women. *Clin. Endocrinol. (Oxf).* 34, 477–483.
166. Hall, J. E., Whitcomb, R. W., Rivier, J. E., Vale, W. W., and Crowley, W. F. Jr. (1990). Differential regulation of luteinizing hormone, follicle-stimulating hormone, and free alpha-subunit secretion from the gonadotrope by gonadotropin-releasing hormone (GnRH): evidence from the use of two GnRH antagonists. *J. Clin. Endocrinol. Metab.* 70, 328–335.
167. Roger, M., Lahlou, N., Couzinet, B., Chaussain, J. L., and Scholler, R. (1989). [Free alpha-subunit glycoprotein hormones: physiological and pathological data]. *J. Steroid Biochem.* 33, 763–769.
168. Bellino, F. L., and Wise, P. M. (2003). Nonhuman primate models of menopause workshop. *Biol. Reprod.* 68, 10–18.
169. Martin, L. J., Carey, K. D., and Comuzzie, A. G. (2003). Variation in menstrual cycle length and cessation of menstruation in captive raised baboons. *Mech. Ageing Dev.* 124, 865–871.
170. Shideler, S. E., DeVane, G. W., Kalra, P. S., Benirschke, K., and Lasley, B. L. (1989). Ovarian-pituitary hormone interactions during the perimenopause. *Maturitas* 11, 331–339.
171. VandeBerg, J. L., and Williams-Blangero, S. (1997). Advantages and limitations of nonhuman primates as animal models in genetic research on complex diseases. *J. Med. Primatol.* 26, 113–119.
172. Gilardi, K. V., Shideler, S. E., Valverde, C. R., Roberts, J. A., and Lasley, B. L. (1997). Characterization of the onset of menopause in the rhesus macaque. *Biol. Reprod.* 57, 335–340.
173. Shideler, S. E., Gee, N. A., Chen, J., and Lasley, B. L. (2001). Estrogen and progesterone metabolites and follicle-stimulating hormone in the aged macaque female. *Biol. Reprod.* 65, 1718–1725.
174. Wallace, W. H., and Kelsey, T. W. (2004). Ovarian reserve and reproductive age may be determined from measurement of ovarian volume by transvaginal sonography. *Hum. Reprod.* 19, 1612–1617.
175. Faddy, M. J., and Gosden, R. G. (1996). A model conforming the decline in follicle numbers to the age of menopause in women. *Hum. Reprod.* 11, 1484–1486.
176. Huang, H. H., and Meites, J. (1975). Reproductive capacity of aging female rats. *Neuroendocrinology* 17, 289–295.
177. Anzalone, C. R., Hong, L. S., Lu, J. K., and LaPolt, P. S. (2001). Influences of age and ovarian follicular reserve on estrous cycle patterns, ovulation, and hormone secretion in the Long-Evans rat. *Biol. Reprod.* 64, 1056–1062.
178. Terasawa, E., and Fernandez, D. L. (2001). Neurobiological mechanisms of the onset of puberty in primates. *Endocr. Rev.* 22, 111–151.
179. Downing, J. A., Joss, J., and Scaramuzzi, R. J. (1996). The effects of N-methyl-D,L-aspartic acid and aspartic acid on the plasma concentration of gonadotrophins, GH and prolactin in the ewe. *J. Endocrinol.* 149, 65–72.
180. Brann, D. W., and Mahesh, V. B. (1997). Excitatory amino acids: evidence for a role in the control of reproduction and anterior pituitary hormone secretion. *Endocr. Rev.* 18, 678–700.
181. Fabre-Nys, C., Blache, D., Hinton, M. R., Goode, J. A., and Kendrick, K. M. (1994). Microdialysis measurement of neurochemical changes in the mediobasal hypothalamus of ovariectomized ewes during oestrus. *Brain Res.* 649, 282–296.
182. Aschheim, P. (1968). [Pituitary content of luteinizing hormone (LH) and histophysiologic reaction with circulating LH of ovarian interstitial tissue in different types of senial rats]. *C. R. Acad. Sci. Hebd. Seances Acad. Sci. D.* 267, 1397–1400.
183. Felicio, L. S., Nelson, J. F., and Finch, C. E. (1986). Prolongation and cessation of estrous cycles in aging C57BL/6J mice are differentially regulated events. *Biol. Reprod.* 34, 849–858.
184. Lloyd, J., Hoffman, G., and Wise, P. (1994). Decline in immediate early gene expression in gonadotropin-releasing hormone neurons during proestrus in regularly cycling, middle-aged rats. *Endocrinology* 134, 1800–1805.
185. Gee, C. E., and Roberts, J. L. (1983). In situ hybridization histochemistry: a technique for the study of gene expression in single cells. *DNA* 2, 157–163.
186. Smith, M. S. (1982). Effect of pulsatile gonadotropin-releasing hormone on the release of luteinizing hormone and follicle-stimulating hormone in vitro by anterior pituitaries from lactating and cycling rats. *Endocrinology* 110, 882–891.
187. Wise, P. M. (1987). The role of the hypothalamus in aging of the female reproductive system. *J. Steroid Biochem.* 27, 713–719.
188. Huang, H. H., Steger, R. W., Bruni, J. F., and Meites, J. (1978). Patterns of sex steroid and gonadotropin secretion in aging female rats. *Endocrinology* 103, 1855–1859.
189. Gosden, R. G., Laing, S. C., Felicio, L. S., Nelson, J. F., and Finch, C. E. (1983). Imminent oocyte exhaustion and reduced follicular recruitment mark the transition to acyclicity in aging C57BL/6J mice. *Biol. Reprod.* 28, 255–260.
190. Cooper, R. L., Roberts, B., Rogers, D. C., Seay, S. G., and Conn, P. M. (1984). Endocrine status versus chronologic age as predictors of altered luteinizing hormone secretion in the "aging" rat. *Endocrinology* 114, 391–396.
191. Rubin, B. S., Mitchell, S., Lee, C. E., and King, J. C. (1995). Reconstructions of populations of luteinizing hormone releasing hormone neurons in young and middle-aged rats reveal progressive increases in subgroups expressing Fos protein on proestrus and age-related deficits. *Endocrinology* 136, 3823–3830.
192. Zuo, Z., Mahesh, V., Zamorano, P., and Brann, D. (1996). Decreased gonadotropin-releasing hormone neurosecretory response to glutamate agonists in middle-aged female rats on proestrus afternoon: a possible role in reproductive aging? *Endocrinology* 137, 2334–2338.
193. Pau, F., Lee, C., Cowles, A., Yang, S.-P., Hess, D. L., and Spies, H. (1998). Possible involvement of norepinephrine transporter activity in the pulsatility of hypothalamic gonadotropin-releasing hormone release: influence of the gonad. *J. Neuroendocrinol.* 10, 21–29.
194. Pau, K. Y., Hess, D. L., Kohama, S., Bao, J., Pau, C. Y., and Spies, H. G. (2000). Oestrogen upregulates noradrenaline

release in the mediobasal hypothalamus and tyrosine hydroxylase gene expression in the brainstem of ovariectomized rhesus macaques. *J. Neuroendocrinol.* 12, 899–909.
195. Wise, P. M. (1999). Neuroendocrine modulation of the "menopause": insights into the aging brain. *Am. J. Physiol. Endocrinol. Metab.* 277, E965–E970.
196. Miller, M. M., Bennett, H. P., Billiar, R. B., Franklin, K. B., and Joshi, D. (1998). Estrogen, the ovary, and neutotransmitters: factors associated with aging. *Exp. Gerontol.* 33, 729–757.
197. Lerner, S. P., Meredith, S., Thayne, W. V., and Butcher, R. L. Age-related alterations in follicular development and hormonal profiles in rats with 4-day estrous cycles. *Biol. Reprod.* (1990). 42, 633–638.
198. Butcher, R. L., Inskeep, E. K., and Pope, R. S. (1978). Plasma concentrations of estradiol produced with two delivery systems in ovariectomized rats. *Proc. Soc. Exp. Biol. Med.* 158, 475–477.
199. Mandl, A. M. (1959). The oestrous cycle of the adult rat after x-ray sterilization. *J. Endocrinol.* 18, 426–433.
200. Matt, D. W., Coquelin, A., and Lu, J. K. (1987). Neuroendocrine control of luteinizing hormone secretion and reproductive function in spontaneously persistent-estrous aging rats. *Biol. Reprod.* 37, 1198–1206.
201. Nass, T. E., LaPolt, P. S., Judd, H. L., and Lu, J. K. (1984). Alterations in ovarian steroid and gonadotrophin secretion preceding the cessation of regular oestrous cycles in ageing female rats. *J. Endocrinol.* 100, 43–50.
202. Wise, P. M., Weiland, N. G., Scarbrough, K., Sortino, M. A., Cohen, I. R., and Larson, G. H. (1989). Changing hypothalamo-pituitary function: its role in aging of the female reproductive system. *Horm. Res.* 31, 39–44.
203. Wilkes, M. M., Lu, K. H., Fulton, S. L., and Yen, S. S. (1978). Hypothalamic–pituitary–ovarian interactions during reproductive senescence in the rat. *Adv. Exp. Med. Biol.* 113, 127–147.
204. Rubin, B. S., Lee, C. E., Ohtomo, M., and King, J. C. (1997). Luteinizing hormone-releasing hormone gene expression differs in young and middle-aged females on the day of a steroid-induced LH surge. *Brain Res.* 770, 267–276.
205. Rubin, B. S., and Bridges, R. S. (1989). Alterations in luteinizing hormone-releasing hormone release from the mediobasal hypothalamus of ovariectomized, steroid-primed middle-aged rats as measured by push-pull perfusion. *Neuroendocrinology* 49, 225–232.
206. Nelson, J. F., and Felicio, L. S. (1990). Hormonal influences on reproductive aging in mice. *Ann. N Y Acad. Sci.* 592, 8–12; discussion 44–51.
207. Arias, P., Carbone, S., and Szwarcfarb, B., et al. (1996). Effects of aging on N-methyl-D-aspartate (NMDA)-induced GnRH and LH release in female rats. *Brain Res.* 740, 234–238.
208. Wise, P. M., and Ratner, A. (1980). LHRH-induced LH and FSH responses in the aged female rat. *J. Gerontol.* 35, 506–511.
209. Gruenewald, D. A., and Matsumoto, A. M. (1991). Age-related decreases in serum gonadotropin levels and gonadotropin-releasing hormone gene expression in the medial preoptic area of the male rat are dependent upon testicular feedback. *Endocrinology* 129, 2442–2450.
210. Hogg, B. B., Matt, D. W., and Sayles, T. E. (1992). Attenuated proestrous luteinizing hormone surges in middle-aged rats are associated with decreased pituitary luteinizing hormone-beta messenger ribonucleic acid expression. *Am. J. Obstet. Gynecol.* 167, 303–307.
211. Matt, D. W., Dahl, K. D., Sarkissian A., and Sayles, T. E. (1993). Apparent absence of negative feedback in middle-aged persistent-estrous rats following luteinizing hormone-releasing hormone agonist treatment: relation to plasma inhibin and 17 beta-estradiol. *Biol. Reprod.* 48, 333–339.
212. Sortino, M. A., Weiland, N. G., and Wise, P. M. (1989). Involvement of norepinephrine activity in the regulation of alpha 1 adrenergic receptors in the medial preoptic nucleus of estradiol-treated rats. *Life Sci.* 44, 1223–1229.
213. Rance, N. E., and Uswandi, S. V. (1996). Gonadotropin-releasing hormone gene expression is increased in the medial basal hypothalamus of postmenopausal women. *J. Clin. Endocrinol. Metab.* 81, 3540–3546.
214. Lee, W. S., Smith, M. S., and Hoffman, G. E. (1990). Luteinizing hormone-releasing hormone neurons express Fos protein during the proestrous surge of luteinizing hormone. *Proc. Natl. Acad. Sci. U S A* 87, 5163–5167.
215. Kimura, F., and Funabashi, T. (1998). Two subgroups of gonadotropin releasing hormone neurons control gonadotropin secretion in rats. *News Physiol. Sci.* 13, 225–231.
216. Porkka-Heiskanen, T., Urban, J. H., Turek, F. W., and Levine, J. E. (1994). Gene expression in a subpopulation of luteinizing hormone-releasing hormone (LHRH) neurons prior to the preovulatory gonadotropin surge. *J. Neurosci.* 14, 5548–5558.
217. Petersen, S., McCrone, S., Keller, M., and Shores, S. (1995). Effects of estrogen and progesterone on luteinizing hormone-releasing hormone messenger ribonucleic acid levels: consideration of temporal and neuroanatomical variables. *Endocrinology* 136, 3604–3610.
218. Lee, W. S., Abbud, R., Smith, M. S., and Hoffman, G. E. (1992). LHRH neurons express cJun protein during the proestrous surge of luteinizing hormone. *Endocrinology* 130, 3101–3103.
219. Hoffman, G. E., and Finch, C. E. (1986). LHRH neurons in the female C57BL/6J mouse brain during reproductive aging: no loss up to middle age. *Neurobiol. Aging* 7, 45–48.
220. Funabashi, T., and Kimura, F. (1995). The number of luteinizing hormone-releasing hormone immunoreactive neurons is significantly decreased in the forebrain of old-aged female rats. *Neurosci. Lett.* 189, 85–88.
221. Rubin, B. S. (1992). Isolated hypothalami from aging female rats do not exhibit reduced basal or potassium-stimulated secretion of luteinizing hormone-releasing hormone. *Biol. Reprod.* 47, 254–261.
222. Le, W. W., Berghorn, K. A., Rassnick, S., and Hoffman, G. E. (1999). Periventricular preoptic area neurons coactivated with luteinizing hormone (LH)-releasing hormone (LHRH) neurons at the time of the LH surge are LHRH afferents. *Endocrinology* 140, 510–519.
223. Ottem, E. N., Godwin, J. G., Krishnan, S., and Petersen, S. L. (2004). Dual-phenotype GABA/glutamate neurons in adult preoptic area: sexual dimorphism and function. *J. Neurosci.* 24, 8097–8105.
224. Le, W.-W., Berghorn, K. A., Smith, M. S., and Hoffman, G. E. (1997). Alpha1-adrenergic receptor blockade blocks LH secretion but not LHRH cFos activation. *Brain Res.* 747, 236–245.
225. Wiegand, S. J., Terasawa, E., Bridson, W. E., and Goy, R. W. (1980). Effects of discrete lesions of preoptic and suprachiasmatic structures in the female rat. Alterations in the feedback regulation of gonadotropin secretion. *Neuroendocrinology* 31, 147–157.
226. Wiegand, S. J., and Terasawa, E. (1982). Discrete lesions reveal functional heterogeneity of suprachiasmatic structures in regulation of gonadotropin secretion in the female rat. *Neuroendocrinology* 34, 395–404.
227. Terasawa, E., and Davis, G. A. (1983). The LHRH neuronal system in female rats: relation to the medial preoptic nucleus. *Endocrinol. Jpn.* 30, 405–417.
228. Gu, G., Varoqueaux, F., and Simerly, R. B. (1999). Hormonal regulation of glutamate receptor gene expression in the

anteroventral periventricular nucleus of the hypothalamus. *J. Neurosci.* 19, 3213–3222.

229. Quadri, S. K., Kledzik, G. S., and Meites, J. (1973). Reinitiation of estrous cycles in old constant-estrous rats by central-acting drugs. *Neuroendocrinology* 11, 248–255.
230. Brito, A. N., Sayles, T. E., Krieg, R. J. Jr., and Matt, D. W. (1994). Relation of attenuated proestrous luteinizing hormone surges in middle-aged female rats to in vitro pituitary gonadotropin-releasing hormone responsiveness. *Eur. J. Endocrinol.* 130, 540–544.
231. Krieg, R. J., Jr., Brito, A. N., Sayles, T. E., and Matt, D. W. (1995). Luteinizing hormone secretion by dispersed anterior pituitary gland cells from middle-aged rats. *Neuroendocrinology* 61, 318–325.
232. Carbone, S., Szwarcfarb, B., Otero Losada, M. E., and Moguilevsky, J. A. (1992). Effects of ovarian steroids on the gonadotropin response to N-methyl-D-aspartate and on hypothalamic excitatory amino acid levels during sexual maturation in female rats. *Endocrinology* 130, 1365–1370.
233. Herbison, A. E., Heavens, R. P., and Dyer, R. G. (1990). Endogenous release of gamma-aminobutyric acid from the medial preoptic area measured by microdialysis in the anaesthetised rat. *J. Neurochem.* 55, 1617–1623.
234. Herbison, A. E., Chapman, C., and Dyer, R. G. (1991). Role of medial preoptic GABA neurones in regulating luteinising hormone secretion in the ovariectomised rat. *Exp. Brain Res.* 87, 345–352.
235. Herbison, A. E., and Dyer, R. G. (1991). Effect on luteinizing hormone secretion of GABA receptor modulation in the medial preoptic area at the time of proestrous luteinizing hormone surge. *Neuroendocrinology* 53, 317–320.
236. Herbison, A. E., Augood, S. J., and McGowan, E. M. (1992). Expression of glutamic acid decarboxylase messenger RNA in rat medial preoptic area neurones during the oestrous cycle and after ovariectomy. *Brain Res. Mol. Brain Res.* 14, 310–316.
237. Keen, K. L., Burich, A. J., Mitsushima, D., Kasuya, E., and Terasawa, E. (1999). Effects of pulsatile infusion of the GABA(A) receptor blocker bicuculline on the onset of puberty in female rhesus monkeys. *Endocrinology* 140, 5257–5266.
238. Goldsmith, P. C., Thind, K. K., Perera, A. D., and Plant, T. M. (1994). Glutamate-immunoreactive neurons and their gonadotropin-releasing hormone-neuronal interactions in the monkey hypothalamus. *Endocrinology* 134, 858–868.
239. Thind, K. K., and Goldsmith, P. C. (1995). Glutamate and GABAergic neurointeractions in the monkey hypothalamus: a quantitative immunomorphological study. *Neuroendocrinology* 61, 471–485.
240. Hofmann, G. E., Danforth, D. R., and Seifer, D. B. (1998). Inhibin-B: the physiologic basis of the clomiphene citrate challenge test for ovarian reserve screening. *Fertil. Steril.* 69, 474–477.
241. Ping, L., Mahesh, V. B., Wiedmeier, V. T., and Brann, D. W. (1994). Release of glutamate and aspartate from the preoptic area during the progesterone-induced LH surge: in vivo microdialysis studies. *Neuroendocrinology* 59, 318–324.
242. Bourguignon, J., Gerard A., Mathieu, J., Simons, J., and Franchimont, P. (1989). Pulsatile release of gonadotropin-releasing hormone from hypothalamic explants is restrained by blockade of N-methyl-D,L-aspartate receptors. *Endocrinology* 125, 1090–1096.
243. Donoso, A. O., Lopez, F. J., and Negro-Vilar, A. (1992). Cross-talk between excitatory and inhibitory amino acids in the regulation of luteinizing hormone-releasing hormone secretion. *Endocrinology* 131, 1559–1561.
244. Donoso, A. O., and Banzan, A. M. (1984). Effects of increase of brain GABA levels on the hypothalamic–pituitary–luteinizing hormone axis in rats. *Acta Endocrinol. (Copenh).* 106, 298–304.
245. Donoso, A. O., Lopez, F. J., and Negro-Vilar, A. (1990). Glutamate receptors of the non-N-methyl-D-aspartic acid type mediate the increase in luteinizing hormone-releasing hormone release by excitatory amino acids in vitro. *Endocrinology* 126, 414–420.
246. Bourguignon, J. P., Gerard A., Purnelle, G., et al. (1997). Duality of glutamatergic and GABAergic control of pulsatile GnRH secretion by rat hypothalamic explants: I. Effects of antisense oligodeoxynucleotides using explants including or excluding the preoptic area. *J. Neuroendocrinol.* 9, 183–191.
247. Brann, D. W., and Mahesh, V. B. (1994). Excitatory amino acids: function and significance in reproduction and neuroendocrine regulation. *Front. Neuroendocrinol.* 15, 3–49.
248. Carbone, S., Szwarcfarb, B., Rondina, D., Feleder, C., and Moguilevsky, J. A. (1996). Differential effects of the N-methyl-D-aspartate and non-N-methyl-D-aspartate receptors of the excitatory amino acids system on LH and FSH secretion. Its effects on the hypothalamic luteinizing hormone releasing hormone during maturation in male rats. *Brain Res.* 707, 139–145.
249. Claypool, L. E., Kasuya, E., Saitoh, Y., Marzban, F., and Terasawa, E. (2000). N-methyl D., L-aspartate induces the release of luteinizing hormone-releasing hormone in the prepubertal and pubertal female rhesus monkey as measured by in vivo push-pull perfusion in the stalk-median eminence. *Endocrinology* 141, 219–228.
250. Eyigor, O. and Jennes, L. (2000). Kainate receptor subunit-positive gonadotropin-releasing hormone neurons express c-fos during the steroid-induced luteinizing hormone surge in the female rat. *Endocrinology* 141, 779–786.
251. Ford, H. and Ebling, F. J. P. (2000). Glutamatergic regulation of gonadotropin releasing hormone mRNA levels during development in the mouse. *J. Neuroendocrinol.* 12, 1027–1033.
252. Jarry, H., Hirsch, B., Leonhardt, S., and Wuttke, W. (1992). Amino acid neurotransmitter release in the preoptic area of rats during the positive feedback actions of estradiol on LH release. *Neuroendocrinology* 56, 133–140.
253. Jennes, L., Lin, W., and Lakhlani, S. (2000). Glutamatergic regulation of gonadotropin-releasing hormone neurons. *Prog. Brain Res.* 141, 183–192.
254. Mahachoklertwattana, P., Sanchez, J., Kaplan, S., and Grumbach, M. (1994). N-methyl-D-aspartate (NMDA) receptors mediate the release of gonadotropin-releasing hormone (GnRH) by NMDA in a hypothalamic GnRH neuronal cell line (GT1–1). *Endocrinology* 134, 1023–1030.
255. Brann, D. W. (1995). Glutamate: a major excitatory transmitter in neuroendocrine regulation. *Neuroendocrinology* 61, 213–225.
256. Ondo, J. G. (1981). Effects of the neuroexcitatory amino acids aspartate and glutamate on LH secretion. *Brain Res. Bull.* 7, 333–335.
257. Tal, J., Price, M. T., and Olney, J. W. (1983). Neuroactive amino acids influence gonadotrophin output by a suprapituitary mechanism in either rodents or primates. *Brain Res.* 273, 179–182.
258. Weiland, N. G. (1992). Glutamic acid decarboxylase messenger ribonucleic acid is regulated by estradiol and progesterone in the hippocampus. *Endocrinology* 131, 2697–2702.
259. Eyigor, O., Centers A., and Jennes, L. (2001). Distribution of ionotropic glutamate receptor subunit mRNAs in the rat hypothalamus. *J. Comp. Neurol.* 434, 101–124.
260. Kinoshita A., Shigemoto, R., Ohishi, H., van der Putten, H., and Mizuno, N. (1998). Immunohistochemical localization of metabotropic glutamate receptors, mGluR7a and mGluR7b, in the central nervous system of the adult rat and mouse: a light and electron microscopic study. *J. Comp. Neurol.* 393, 332–352.
261. van den Pol, A. N., Romano, C., and Ghosh, P. (1995). Metabotropic glutamate receptor mGluR5 subcellular

distribution and developmental expression in hypothalamus. *J. Comp. Neurol.* 362, 134–150.

262. Ping, L., Mahesh, V. B., Bhat, G. K., and Brann, D. W. (1997). Regulation of gonadotropin-releasing hormone and luteinizing hormone secretion by AMPA receptors. Evidence for a physiological role of AMPA receptors in the steroid-induced luteinizing hormone surge. *Neuroendocrinology* 66, 246–253.
263. Kiyama, H., Sato, K., and Tohyama, M. (1993). Characteristic localization of non-NMDA type glutamate receptor subunits in the rat pituitary gland. *Brain Res. Mol. Brain Res.* 19, 262–268.
264. Kasuya, E., Nyberg, C. L., Mogi, K., and Terasawa, E. (1999). A role of γ-amino butyric acid (GABA) and glutamate in control of puberty in female rhesus monkeys: effect of an antisense oligodeoxynucleotide for GAD67 messenger ribonucleic acid and MK801 on luteinizing hormone-releasing hormone release. *Endocrinology* 140, 705–712.
265. Herman, J. P., Eyigor, O., Ziegler, D. R., and Jennes, L. (2000). Expression of ionotropic glutamate receptor subunit mRNAs in the hypothalamic paraventricular nucleus of the rat. *J. Comp. Neurol.* 422, 352–362.
266. Ottem, E. N., Godwin, J. G., and Petersen, S. L. (2002). Glutamatergic signaling through the n-methyl-d-aspartate receptor directly activates medial subpopulations of luteinizing hormone-releasing hormone (LHRH) neurons, but does not appear to mediate the effects of estradiol on LHRH gene expression. *Endocrinology* 143, 4837–4845.
267. Pompolo, S., Pereira A., Scott, C. J., Fujiyma, F., and Clarke, I. J. (2003). Evidence for estrogenic regulation of gonadotropin-releasing hormone neurons by glutamatergic neurons in the ewe brain: An immunohistochemical study using an antibody against vesicular glutamate transporter-2. *J. Comp. Neurol.* 465, 136–144.
268. Lin, W., McKinney, K., Liu, L., Lakhlani, S., and Jennes, L. (2003). Distribution of vesicular glutamate transporter-2 messenger ribonucleic acid and protein in the septum-hypothalamus of the rat. *Endocrinology* 144, 662–670.
269. Kuehl-Kovarik, M. C., Pouliot, W. A., Halterman, G. L., Handa, R. J., Dudek, F. E., and Partin, K. M. (2002). Episodic bursting activity and response to excitatory amino acids in acutely dissociated gonadotropin-releasing hormone neurons genetically targeted with green fluorescent protein. *J. Neurosci.* 22, 2313–2322.
270. Herbison, A. E. (1998). Multimodal influence of estrogen upon gonadotropin-releasing hormone neurons. *Endocr. Rev.* 19, 302–330.
271. Miller, B. H., and Gore, A. C. (2002). N-methyl-D-aspartate receptor subunit expression in GnRH neurons changes during reproductive senescence in the female rat. *Endocrinology* 143, 3568–3574.
272. Brann, D., and Mahesh, V. (1991). Endogenous excitatory amino acid involvement in the preovulatory and steroid-induced surge of gonadotropins in the female rat. *Endocrinology* 128, 1541–1547.
273. Nunemaker, C. S., DeFazio, R. A., and Moenter, S. M. (2002). Estradiol-sensitive afferents modulate long-term episodic firing patterns of GnRH neurons. *Endocrinology* 143, 2284–2292.
274. Bonavera, J. J., Swerdloff, R. S., Sinha Hikim, A. P., Lue, Y. H., and Wang, C. (1998). Aging results in attenuated gonadotropin releasing hormone-luteinizing hormone axis responsiveness to glutamate receptor agonist N-methyl-D-aspartate. *J. Neuroendocrinol.* 10, 93–99.
275. Rubin, B. S., and King, J. C. (1994). The number and distribution of detectable luteinizing hormone (LH)-releasing hormone cell bodies changes in association with the preovulatory LH surge in the brains of young but not middle-aged female rats. *Endocrinology* 134, 467–474.
276. Rubin, B. S., Lee, C. E., and King, J. C. (1994). A reduced proportion of luteinizing hormone (LH)-releasing hormone neurons express Fos protein during the preovulatory or steroid-induced LH surge in middle-aged rats. *Biol. Reprod.* 51, 1264–1272.
277. Rettori, V., Belova, N., Dees, W., Nyberg, C., Gimeno, M., and McCann, S. (1993). Role of nitric oxide in the control of luteinizing hormone-releasing hormone release in vivo and in vitro. *Proc. Natl. Acad. Sci. U S A* 90, 10130–10134.
278. Blandina, P., Johnson, D., Walcott, J., and Goldfarb, J. (1992). Release of endogenous norepinephrine from rat hypothalamus by stimulation of N-methyl-D-aspartic acid receptors. *J. Pharmacol. Exp. Ther.* 263, 61–68.
279. Vincent, S. R., Hokfelt, T., and Wu, J. Y. (1982). GABA neuron systems in hypothalamus and the pituitary gland. Immunohistochemical demonstration using antibodies against glutamate decarboxylase. *Neuroendocrinology* 34, 117–125.
280. Herbison, A. E., Augood, S. J., Simonian, S. X., and Chapman, C. (1995). Regulation of GABA transporter activity and mRNA expression by estrogen in rat preoptic area. *J. Neurosci.* 15, 8302–8309.
281. Curran-Rauhut, M. A., and Petersen, S. L. (2002). Regulation of glutamic acid decarboxylase 65 and 67 gene expression by ovarian steroids: Identification of two functionally distinct populations of gaba neurones in the preoptic area. *J. Neuroendocrinol.* 14, 310–317.
282. Jung, H., Shannon, E. M., Fritschy, J. M., and Ojeda, S. R. (1998). Several GABAA receptor subunits are expressed in LHRH neurons of juvenile female rats. *Brain Res.* 780, 218–229.
283. Leranth, C., MacLusky, N. J., Sakamoto, H., Shanabrough, M., and Naftolin, F. (1985). Glutamic acid decarboxylase-containing axons synapse on LHRH neurons in the rat medial preoptic area. *Neuroendocrinology* 40, 536–539.
284. Jarry, H., Leonhardt, S., Schwarze, T., and Wuttke, W. (1995). Preoptic rather than mediobasal hypothalamic amino acid neurotransmitter release regulates GnRH secretion during the estrogen-induced LH surge in the ovariectomized rat. *Neuroendocrinology* 62, 479–486.
285. McCauley, L., and Gee, K. (1995). Influence of the estrus cycle on the discrimination of apparent neuroactive steroid site subtypes on the gamma-aminobutyric acidA receptor complex in the rat. *J. Pharmacol. Exp. Ther.* 275, 1412–1417.
286. Leonhardt, S., Boning, B., Luft, H., Wuttke, W., and Jarry, H. (2000). Activation of gene expression of the gamma-aminobutyric acid rather than the glutamatergic system in the preoptic area during the preovulatory gonadotropin surge of the rat. *Neuroendocrinology* 71, 8–15.
287. Terasawa, E., Luchansky, L. L., Kasuya, E., and Nyberg, C. L. (1999). An increase in glutamate release follows a decrease in gamma aminobutyric acid and the pubertal increase in luteinizing hormone releasing hormone release in the female rhesus monkeys. *J. Neuroendocrinol.* 11, 275–282.
288. Seltzer, A. M., and Donoso, A. O. (1992). Restraining action of GABA on estradiol-induced LH surge in the rat: GABA activity in brain nuclei and effects of GABA mimetics in the medial preoptic nucleus. *Neuroendocrinology* 55, 28–34.
289. Jarry, H., Perschl, A., and Wuttke, W. (1988). Further evidence that preoptic anterior hypothalamic GABAergic neurons are part of the GnRH pulse and surge generator. *Acta Endocrinol. (Copenh).* 118, 573–579.
290. Jarry, H., Leonhardt, S., and Wuttke, W. (1991). Gamma-aminobutyric acid neurons in the preoptic/anterior hypothalamic area synchronize the phasic activity of the gonadotropin-releasing hormone pulse generator in ovariectomized rats. *Neuroendocrinology* 53, 261–267.
291. Scott, C. J., and Clarke, I. J. (1993). Inhibition of luteinizing hormone secretion in ovariectomized ewes during the

breeding season by gamma-aminobutyric acid (GABA) is mediated by GABA-A receptors, but not GABA-B receptors. *Endocrinology* 132, 1789–1796.

292. Silverman, A. J., Krey, L. C., and Zimmerman, E. A. (1979). A comparative study of the luteinizing hormone releasing hormone (LHRH) neuronal networks in mammals. *Biol. Reprod.* 20, 98–110.
293. Witkin, J. W., Ferin, M., Popilskis, S. J., and Silverman, A. J. (1991). Effects of gonadal steroids on the ultrastructure of GnRH neurons in the rhesus monkey: synaptic input and glial apposition. *Endocrinology* 129, 1083–1092.
294. King, J. C., Liu, E., Ronsheim, P. M., Slonimski, M., and Rubin, B. S. (1998). Expression of Fos within luteinizing hormone-releasing hormone neurons, in relation to the steroid-induced luteinizing hormone surge in guinea pigs. *Biol. Reprod.* 58, 316–322.
295. Herbison, A. E. (1997). Noradrenergic regulation of cyclic GnRH secretion. *Rev. Reprod.* 2, 1–6.
296. Leranth, C., Shanabrough, M., and Naftolin, F. (1991). Estrogen induces ultrastructural changes in progesterone receptor-containing GABA neurons of the primate hypothalamus. *Neuroendocrinology* 54, 571–579.
297. Parducz, A., Perez, J., and Garcia-Segura, L. M. (1993). Estradiol induces plasticity of gabaergic synapses in the hypothalamus. *Neuroscience* 53, 395–401.
298. Naftolin, F., Mor, G., Horvath, T. L., et al. (1996). Synaptic remodeling in the arcuate nucleus during the estrous cycle is induced by estrogen and precedes the preovulatory gonadotropin surge. *Endocrinology* 137, 5576–5580.
299. Herbison, A. E. (1997). Estrogen regulation of GABA transmission in rat preoptic area. *Brain Res. Bull.* 44, 321–326.
300. Goroll, D., Arias, P., and Wuttke, W. (1993). Preoptic release of amino acid neurotransmitters evaluated in peripubertal and young adult female rats by push-pull perfusion. *Neuroendocrinology* 58, 11–15.
301. Cashion, A. B., Smith, M. J., and Wise, P. M. (2004). Glutamic acid decarboxylase 67 (GAD67) gene expression in discrete regions of the rostral preoptic area change during the oestrous cycle and with age. *J. Neuroendocrinol.* 16, 711–716.
302. Wright, D. E., and Jennes, L. (1993). Origin of noradrenergic projections to GnRH perikarya-containing areas in the medial septum-diagonal band and preoptic area. *Brain Res.* 621, 272–278.
303. Simonian, S. X., and Herbison, A. E. (1997). Differential expression of estrogen receptor and neuropeptide Y by brainstem A1 and A2 noradrenaline neurons. *Neuroscience* 76, 517–529.
304. Haywood, S. A., Simonian, S. X., van der Beek, E. M., Bicknell, R. J., and Herbison, A. E. (1999). Fluctuating estrogen and progesterone receptor expression in brainstem norepinephrine neurons through the rat estrous cycle. *Endocrinology* 140, 3255–3263.
305. Herbison, A. E., Simonian, S. X., Thanky, N. R., and Bicknell, R. J. (2000). Oestrogen modulation of noradrenaline neurotransmission. *Novartis Found. Symp.* 230, 74–85; discussion 85–93.
306. Simonian, S. X., Spratt, D. P., and Herbison, A. E. (1999). Identification and characterization of estrogen receptor alpha-containing neurons projecting to the vicinity of the gonadotropin-releasing hormone perikarya in the rostral preoptic area of the rat. *J. Comp. Neurol.* 411, 346–358.
307. Conde, G. L., Herbison, A. E., Fernandez-Galaz, C., and Bicknell, R. J. (1996). Estrogen uncouples noradrenergic activation of Fos expression in the female rat preoptic area. *Brain Res.* 735, 197–207.
308. Leranth, C., MacLusky, N. J., Shanabrough, M., and Naftolin, F. (1988). Catecholaminergic innervation of luteinizing hormone-releasing hormone and glutamic acid decarboxylase immunopositive neurons in the rat medial preoptic area. An electron-microscopic double immunostaining and degeneration study. *Neuroendocrinology* 48, 591–602.
309. Gore, A. C., Wu, T. J., Rosenberg, J. J., and Roberts, J. L. (1996). Gonadotropin-releasing hormone and NMDA receptor gene expression and colocalization change during puberty in female rats. *J. Neurosci.* 16, 5281–5289.
310. Clifton, D. K., and Sawyer, C. H. (1979). LH release and ovulation in the rat following depletion of hypothalamic norepinephrine: chronic vs. acute effects. *Neuroendocrinology* 28, 442–449.
311. Jarry, H., Leonhardt, S., and Wuttke, W. (1990). A norepinephrine-dependent mechanism in the preoptic/anterior hypothalamic area but not in the mediobasal hypothalamus is involved in the regulation of the gonadotropin-releasing hormone pulse generator in ovariectomized rats. *Neuroendocrinology* 51, 337–344.
312. Goodman, R. L., Robinson, J. E., Kendrick, K. M., and Dyer, R. G. (1995). Is the inhibitory action of estradiol on luteinizing hormone pulse frequency in anestrous ewes mediated by noradrenergic neurons in the preoptic area? *Neuroendocrinology* 61, 284–292.
313. Legan, S. J., and Callahan, W. H. (1999). Suppression of tonic luteinizing hormone secretion and norepinephrine release near the GnRH neurons by estradiol in ovariectomized rats. *Neuroendocrinology* 70, 237–245.
314. Etgen, A. M. (2003). Ovarian steroid and growth factor regulation of female reproductive function involves modification of hypothalamic alpha 1-adrenoceptor signaling. *Ann. N Y Acad. Sci.* 1007, 153–161.
315. Demling, J., Fuchs, E., Baumert, M., and Wuttke, W. (1985). Preoptic catecholamine, GABA, and glutamate release in ovariectomized and ovariectomized estrogen-primed rats utilizing a push-pull cannula technique. *Neuroendocrinology* 41, 212–218.
316. Mohankumar, P. S., Thyagarajan, S., and Quadri, S. K. (1994). Correlations of catecholamine release in the medial preoptic area with proestrous surges of luteinizing hormone and prolactin: effects of aging. *Endocrinology* 135, 119–126.
317. Wise, M. E., Nieman, D., Stewart, J., and Nett, T. M. (1984). Effect of number of receptors for gonadotropin-releasing hormone on the release of luteinizing hormone. *Biol. Reprod.* 31, 1007–1013.
318. Rosie, R., Sumner, B. E., and Fink, G. (1994). An alpha 1 adrenergic mechanism mediates estradiol stimulation of LHRH mRNA synthesis and estradiol inhibition of POMC mRNA synthesis in the hypothalamus of the prepubertal female rat. *J. Steroid Biochem. Mol. Biol.* 49, 399–406.
319. Chu, H.-P., and Etgen, A. M. (1999). Ovarian hormone dependence of alpha 1-adrenoceptor activation of the nitric oxide-cgmp pathway: relevance for hormonal facilitation of lordosis behavior. *J. Neurosci.* 19, 7191–7197.
320. Karkanias, G. B., Ansonoff, M. A., and Etgen, A. M. (1996). Estradiol regulation of alpha 1b-adrenoceptor mRNA in female rat hypothalamus-preoptic area. *J. Neuroendocrinol.* 8, 449–455.
321. Mohankumar, P. S., Thyagarajan, S., and Quadri, S. K. (1995). Cyclic and age-related changes in norepinephrine concentrations in the medial preoptic area and arcuate nucleus. *Brain Res. Bull.* 38, 561–564.
322. Temel, S., Lin, W., Lakhlani, S., and Jennes, L. (2002). Expression of estrogen receptor-alpha and cFos in norepinephrine and epinephrine neurons of young and middle-aged rats during the steroid-induced luteinizing hormone surge. *Endocrinology* 143, 3974–3983.
323. Meites, J. (1991). Role of hypothalamic catecholamines in aging processes. *Acta Endocrinol. (Copenh).* 125, Suppl 1, 98–103.

324. Robinson, D. S., Nies A., Davis, J. N., et al. (1972). Ageing, monoamines, and monoamine-oxidase levels. *Lancet* 1, 290–291.
325. Miller, M. M., and Zhu, L. (1995). Ovariectomy and age alter gonadotropin hormone releasing hormone-noradrenergic interactions. *Neurobiol. Aging* 16, 613–625.
326. Rubin, B. S. (1993). Naloxone stimulates comparable release of luteinizing hormone-releasing hormone from tissue fragments from ovariectomized, estrogen-treated young and middle-aged female rats. *Brain Res.* 601, 246–254.
327. Kalra, S. P., Sahu A., and Kalra, P. S. (1993). Ageing of the neuropeptidergic signals in rats. *J. Reprod. Fertil. Suppl.* 46, 11–19.
328. Quigley, M. E., Sheehan, K. L., Casper, R. F., and Yen, S. S. (1980). Evidence for an increased opioid inhibition of luteinizing hormone secretion in hyperprolactinemic patients with pituitary microadenoma. *J. Clin. Endocrinol. Metab.* 50, 427–430.
329. Casper, R. F., and Alapin-Rubillovitz, S. (1985). Progestins increase endogenous opioid peptide activity in postmenopausal women. *J. Clin. Endocrinol. Metab.* 60, 34–36.
330. Steele, P. A., and Judd, S. J. (1986). Role of endogenous opioids in reducing the frequency of pulsatile luteinizing hormone secretion induced by progesterone in normal women. *Clin. Endocrinol. (Oxf).* 25, 669–674.
331. Petraglia, F., Porro, C., Facchinetti, F., et al. (1986). Opioid control of LH secretion in humans: menstrual cycle, menopause and aging reduce effect of naloxone but not of morphine. *Life Sci.* 38, 2103–2110.
332. D'Amico, J. F., Greendale, G. A., Lu, J. K., and Judd, H. L. (1991). Induction of hypothalamic opioid activity with transdermal estradiol administration in postmenopausal women. *Fertil. Steril.* 55, 754–758.
333. Albers, H. E., Stopa, E. G., Zoeller, R. T., et al. (1990). Day-night variation in prepro vasoactive intestinal peptide/peptide histidine isoleucine mRNA within the rat suprachiasmatic nucleus. *Brain Res. Mol. Brain Res.* 7, 85–89.
334. Cai A., Scarbrough, K., Hinkle, D. A., and Wise, P. M. (1997). Fetal grafts containing suprachiasmatic nuclei restore the diurnal rhythm of CRH and POMC mRNA in aging rats. *Am. J. Physiol.* 273, R1764–R1770.
335. Van Cauter, E., Leproult, R., and Kupfer, D. J. (1996). Effects of gender and age on the levels and circadian rhythmicity of plasma cortisol. *J. Clin. Endocrinol. Metab.* 81, 2468–2473.
336. Krajnak, K., Kashon, M. L., Rosewell, K. L., and Wise, P. M. (1998). Aging alters the rhythmic expression of vasoactive intestinal polypeptide mRNA but not arginine vasopressin mRNA in the suprachiasmatic nuclei of female rats. *J. Neurosci.* 18, 4767–4774.
337. Harney, J. P., Scarbrough, K., Rosewell, K. L., and Wise, P. M. (1996). In vivo antisense antagonism of vasoactive intestinal peptide in the suprachiasmatic nuclei causes aging-like changes in the estradiol-induced luteinizing hormone and prolactin surges. *Endocrinology* 137, 3696–3701.

Pregnancy

Knobil and Neill's Physiology of Reproduction, Third Edition
edited by Jimmy D. Neill,
Elsevier © 2006

CHAPTER 51

Immunobiology of Human Pregnancy

Joan S. Hunt, Ramsey H. McIntire, and Margaret G. Petroff

Introduction, 2760
The Cycling Human Uterus, 2761
Physiological Features of the Cycling Uterus, 2761
Leukocytes in the Cycling Uterus, 2761
Preimplantation Events, 2762
Immune Protection for the Developing Oocyte, 2762
Immune Protection for the Preimplantation Embryo, 2763
The Maternal–Fetal Interface, 2764
Maternal Cellular Components: Changes and Consequences, 2764
Fetal Cellular Components, 2766
Maternal–Fetal Cell Coexistence and Exchange, 2769
Implantation and Maternal Immune Responses, 2769
Pregnancy as a Th1/Th2-Type Reaction, 2769
Maternal Immune Responses to Implantation, 2770
The Human Implantation Site: Oxygen Regulation of Gene Expression, 2771
Genetic Disparity and Post-Implantation Evasive Strategies, 2771
Foreign Antigens and Immunity: Danger and Self-Nonself Theories, 2771
Mechanisms Preventing Maternal Immune Rejection of the Fetus, 2772
Multifunctionality of Immune System Molecules, 2779
Perspectives, 2779
Acknowledgments, 2780
References, 2780

ABSTRACT

The immune system is designed to protect against both disease and contamination of the host genome by foreign RNA or DNA. Mammalian pregnancy, which features gametes expressing unique antigens and embryos expressing proteins derived from disparate maternal and paternal genes, presents a serious challenge to the maternal immune system. This powerful and highly successful physiological system must be circumvented so as to permit the intimate associations between maternal and fetal cells that characterize follicle development, implantation and, in many mammals, hemochorial placentation. In this chapter, structural features of the human female reproductive organs and specific mechanistic strategies that protect against immune-mediated embryonic loss are described. Critical concepts are that both the mother and the fetus contribute to successful pregnancy, that multiple immune evasion mechanisms are employed, and that adaptations to permit internalization of the fetus are initiated prior to and concurrently with implantation. Major features include (a) seclusion of female gametes within the ovary; (b) changes in the distribution of leukocyte subpopulations in the uterus; (c) production of immune suppressive molecules by both maternal and fetal cells; (d) restrictions on fetal trophoblast cell expression of the HLA molecules that stimulate graft rejection; (e) selection of an inhibitory member of the B7 family of immune cell co-stimulatory proteins for expression on trophoblast cells; and (f) high expression on placenta cells of the regulatory proteins that interfere with complement activation.

Department of Anatomy and Cell Biology, University of Kansas Medical Center, Kansas City, Kansas.

Collectively, these strategies comprise a novel, generally successful circumlocution of the immune system that permits implantation and facilitates semiallogeneic pregnancy.

INTRODUCTION

The unexpected willingness of mothers to permit introduction of genetically different cells during implantation and to harbor the developing mammalian offspring throughout the course of pregnancy has intrigued immunologists for half a century (1). The immune system is designed to repel invaders of all types, from microorganisms to transplanted kidneys, so as to prevent disease and contamination of host genetic material. Yet this powerful system is overcome or bypassed to accommodate the fetus during pregnancy in mammals where the fetus(es) are internalized. This arrangement, which is known as viviparity, provides maximum protection from environmental hazards but introduces a critical requirement for tolerizing or circumventing the mother's immune system. Although imperfect, the mechanisms in place for preventing or bypassing maternal immune cell attack on the fetus are generally successful since species continue to propagate. Furthermore, the strategies appear to be permanently embedded in the genetic makeup of each strain as evidenced by their persistence even when unnecessary, as is the case in inbred strains of laboratory animals.

Multiple overlapping and complementary mechanisms operate for the protection of the fetus. These are most diverse and sophisticated in human pregnancies, where maternal and fetal cells mingle in a type of placentation called hemochorial, and where single pregnancies are the rule. Figure 1 illustrates the critical point that both mother and fetus cooperate to create an appropriate immunological environment for fetal growth and development. This figure also shows some of the major evasive mechanisms that have been identified in human pregnancy. Some are common to maternal and fetal environments. These include both soluble molecules that have immunosuppressive properties (hormones, prostaglandins, cytokines) and those that establish and maintain the maternal–fetal interface as a site of innate immunity (chemokines and cells of the innate immune system). Fetal tissues contribute novel mechanisms that include but are not restricted to restricted expression of molecules that normally mediate graft rejection (human leukocyte antigens [HLA]), high expression

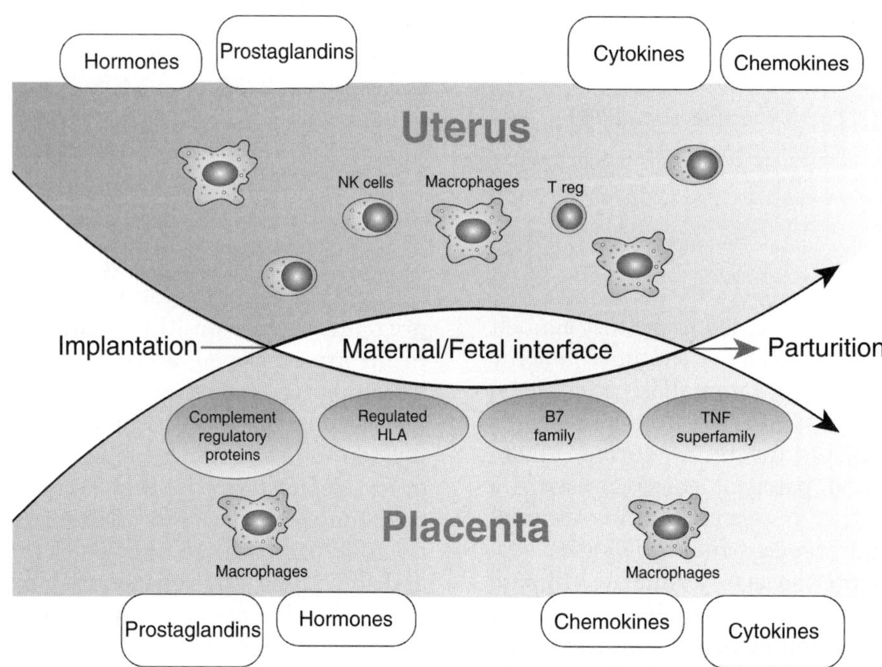

FIG. 1. Multiple mechanisms underlie maternal tolerance of the fetus. Both maternal and fetal mechanisms contribute to maternal immunological acceptance of the genetically different fetal semiallograft. Mothers revise the roster of leukocytes in the decidua to disallow residence of antigen-specific immune cells and promote production of soluble suppressor molecules. Fetal membranes, the placenta and extraplacental membranes, produce many of the same immune suppressive molecules but also exhibit specially selected members of multigene families that bypass or inactivate immune cells. NK cells, natural killer cells; Treg, CD4+ regulatory T cells; HLA, human leukocyte antigen family; TNF superfamily, tumor necrosis factor superfamily.

of proteins that suppress lymphocyte activation (B7) and kill activated lymphocytes (tumor necrosis factor [TNF] superfamily of ligands), and display of proteins that protect against cytotoxic antibodies.

Many of the same evasive strategies prevail in nonhuman primates such as the monkey and the baboon, where placentation is also hemochorial and single pregnancies are usual. In other mammalian species, these two conditions are rarely met; maternal and fetal cells are less intimately associated or multiple fetuses are the rule rather than the exception. Yet many of the mechanisms that ensure human and nonhuman pregnancy are used to a greater or lesser degree in these other species, thus broadening the range of informative experimental models and shedding new light on successful reproduction in a wide range of animals.

This chapter focuses on the spectrum of immune protective mechanisms devised to permit semiallogeneic pregnancy in women, many of which can be explored experimentally in primates. Studies in rodents, which are popular experimental models, are quoted frequently so as to include information derived from transgenic and knockout animals. On occasion, mention is made of other species where particularly novel or interesting findings have been reported.

THE CYCLING HUMAN UTERUS

In terms of immunity, the cycling uterus closely resembles other organs whose surfaces are exposed to environmental pathogens. But unlike other organs, the uterus of women in their child-bearing years undergoes extensive and predictable remodeling due to receptivity to steroid hormones. This remodeling includes changes in the immune cell component.

Physiological Features of the Cycling Uterus

Uterine cells exhibiting specific hormone receptors respond dramatically to cyclic changes in the levels of steroid sex hormones circulating in the blood. As a consequence, major tissue alterations occur in an ordered and chronologically predictable manner. These events are reviewed in detail in this volume (see Chapters 9 and 15). In brief, in the early proliferative phase that follows menstruation, estrogens predominate and stimulate proliferation and partial differentiation of the endometrial epithelial cells. At approximately the time of ovulation, estrogens decline and progesterone from the corpus luteum targets specifically to the mesenchymal cells of the endometrium.

An important aspect of this cycle is that cells providing immune protection may be receptive to the same hormones and/or the cytokines and growth factors produced by hormone-stimulated endometrial cells. Thus, the hormone-driven cyclic changes include alterations in the hematopoietic cell populations in the stroma of the endometrium and changes in their functions.

Leukocytes in the Cycling Uterus

In human uteri, both T and B lymphocytes are established in lymphoid aggregates in the uterus just as they are at other mucosal surfaces. Macrophages and their close relatives, dendritic cells, are randomly distributed through the endometrial stroma and are numerous in the connective tissue in the myometrium. The stroma of cycling uteri also contains natural killer (NK) cells, whose markers vary with the cycle (2). Near the end of the cycle, mast cells are prominent (3). Although most of the leukocytes in the cycling uterus, including the NK cells (4), are believed to function in host defense, the mast cells have been implicated in driving menstruation (5–7). Dependency on specific sex hormones has been extensively tested and verified. It has long been known that estrogens stimulate an influx of leukocytes into the mouse uterus (8) and that human NK cell populations depend on progesterone for maintenance and proliferation (9,10). Estrogens frequently drive production of important cytokines. These include interferon-γ (IFN-γ), a cytokine whose rate of transcription is upregulated by estrogen (11). By contrast, progesterone has a calming effect on immune cells similar to that of corticosteroids and at high levels may act through the glucocorticoid receptor (GR) to downregulate production of inflammatory cytokines (12,13).

Chemokines and chemokine receptors, which are best studied in infection and inflammation (14), operate to facilitate these cyclic changes in uterine leukocytes (15,16). Both steroid hormones and products of the embryo regulate chemokine expression in endometrial epithelial cells and stromal cells (15). For example, progesterone enhances expression of the chemoattractants interleukin-8 (IL-8) and monocyte chemotactic protein-1 (MCP-1) during the window of implantation (16). Two chemokines in particular, IP-10 (CXCL10) and ITAC (CXCL11), have been reported to respond to steroid hormones (17). At times of embryo implantation and menstruation, chemokine patterns are predictable (18). Later, the effects of endometrial ligand–receptor interactions are influenced by products of trophoblast cells in the placenta (15,19–24). The ligands (Table 1) and receptors (Table 2) interact in complex

TABLE 1. *Chemokines in the decidua and villous placenta: identification by cell type [summarized from (19–24)]*

	Decidua			Villous placenta		
Ligands	**Leukocytes**	**Stromal cells**	**xvCTB**	**Macrophages**	**Stromal cells**	**vCTB**
CCL2/MCP-1	+	+/–	–	+	–	–
CCL3/MIP-1α	+	+	+	+	+	–
CCL5/RANTES	–	+	–	–	–	–
CCL7/MCP-3	–	+/–	–	–	+	–
CCL14/HCC-1	–	+	+	–	+	–
CCL21/6Ckine/SLC	–	–	–	–	+/–	–
CXCL1	0	0	0	–	+	–
CXCL6/GCP-2	–	+/–	+	–	+/–	–
CXCL8/IL-8	+	+	–	+	+	+
CXCL10/IP-10	–	+	–	–	+/–	+/–
CXCL11/ITAC	+/–	+	–	–	+	–
CXCL12/SDF-1	+/–	–	+	–	+/–	–
CXCL14/BRAK	–	+/–	–	+/–	–	–
CX3CL1/Fractalkine	–	+	–	–	+	–

+, detected; –, not detected; +/–, inconsistent results; 0, not tested.
xvCTB, extravillous cytotrophoblast cells; vCTB, villous cytotrophoblast cells.

TABLE 2. *Chemokine receptors in decidual leukocytes and placental cytotrophoblast cells [summarized from (19–24)]*

Receptors	**Decidual leukocytes**	**Villous CTB**
CCR1	+	0
CCR2	+	+/–
CCR3	–	+/–
CCR4	+/–	0
CCR5	+	+
CCR7	+	+
CCR8	–	–
US28	–	–
CXCR1	+/–	–
CXCR2	+/–	+/–
CXCR3	+	–
CXCR4	+	+
CXCR5	–	–
CXCR6/STRL33	+	+
CX3CR1/V28	+	+/–
GPR1	+/–	+
GPR15	–	–

+, detected; –, not detected; +/–, inconsistent results; 0, not tested.

TABLE 3. *Chemokines (ligands) and their receptors present in gestational tissues. Receptors fall into two categories, CCR and CXCR, based on their chemokine ligand(s). GPR1 is an orphan receptor whose ligand is yet to be determined.*

Receptors	**Chemokines**
CCR1	HCC-1, MIP-1α, RANTES
CCR2	MCP-1, MCP-3
CCR3	RANTES, MCP-3
CCR4	RANTES
CCR5	MIP-1α, RANTES
CCR7	SLC
CXCR1	IL-8, GCP-2
CXCR2	CXCL1, IL-8, GCP-2
CXCR3	ITAC, IP-10
CXCR4	SDF-1
CXCR6/STRL33	CXCL16
CX3CR1/V28	Fractalkine
GPR1	Unknown

pathways (Table 3) to alter leukocyte subpopulations during pregnancy.

PREIMPLANTATION EVENTS

Immune Protection for the Developing Oocyte

As the ovum develops within the follicle, it is sequestered from immune cells circulating in the blood and resident in the ovarian stroma. Although hematopoietic cells capable of generating antigen-specific immunity, T and B lymphocytes, constantly circulate in the blood stream, the ovary is not a site where these cells naturally migrate and collect as lymphoid aggregates as they do in the secondary lymphoid organs and mucosa. Yet the ovary is not without protection from infection. Macrophages, which are prominent members of the innate or natural immune system, are abundant in the ovarian stroma, where they are thought to play critical roles in folliculogenesis, tissue restructuring during ovulation and both formation and regression of the corpus luteum (25). In cycling women, the macrophages may be in a suppressive mode as a consequence of periodic exposure to progesterone, which is an inhibitor of macrophage activation for priming of the acquired immune system. Macrophages in this mode are common in the pregnant uterus, as described below, and are capable of discouraging lymphocyte migration and proliferation. Alternatively or in addition, ovarian macrophages may be in place to phagocytose debris from atretic follicles and exhausted corpora lutea. Other possibilities remain. Macrophages in the testis, for example, have a central role in directing hormone production by Leydig cells (26,27).

Granulosa cells, a basement membrane, and layers of theca cells comprise the protective casing within which the ovum is sequestered throughout its development. This layering of epithelioid granulosa cells with tight junctions, a thick membrane, and a second layer of overlapping, hormone-producing theca cells effectively prevents trafficking of macrophages and other immune cells into the antrum. The ovum is surrounded by a thick glycoprotein layer called the zona pellucida, which, as with the ovum itself, expresses immunogenic molecules foreign to the mother. Although neither the zona nor the ovum stimulates immunity in normal women, autoimmune disease of the ovary where oocyte antigens are the target may be induced experimentally by treatment with antibodies to zona pellucida (28). In contrast, autoimmune disease of the testis is not uncommon and cytokines stimulated during this process may prevent sperm maturation and interfere with fertility (28,29).

It is likely that under normal circumstances the novel antigens expressed by ova remain unrecognized by the maternal immune system as a consequence of the physical arrangement of follicle components. Regulation of expression of the HLA antigens may also be important. Oocytes lack the highly immunogenic class Ia (HLA-A, B) and class II (HLA-D) major histocompatibility (MHC) antigens (30,31) and instead transcribe a gene in the class Ib family, *HLA-G*, whose protein products are not highly immunogenic (32) and, in many contexts, are uniformly immunosuppressive [reviewed in (33)]. The unique features of the HLA antigens are discussed in greater detail below in Fetal Contributions: Modulation of Cell-Mediated Immunity.

Immune Protection for the Preimplantation Embryo

Following ovulation and until early stages of fertilization, the ovum and its protective covering, the zona pellucida, traverse the Fallopian tube (oviduct), where many immune cells reside in aggregate form (34). In most women, there is no evidence for immune rejection of the ovulated egg. Fertilization occurs, and the new embryo continues in its path toward the hormonally prepared uterus. Nonetheless, clinicians report that 50% or more of potential pregnancies fail. Reproductive immunologists fully recognize the contributions of genetic disorders to these losses but postulate that some of the remaining problems might be due to inadequate immune protective mechanisms in the Fallopian tube or inappropriate maternal immune responses (35,36).

The preimplantation embryo is potentially more susceptible to immune rejection than the oocyte because it expresses a wider array of immunogenic molecules that now include "foreign" (paternal), allogeneic antigens. The assumption is that the zona pellucida and adherent granulosa cells provide protection, but it is also true that the HLA-A, -B, and -D antigens that would stimulate rejection remain unexpressed by human embryos (30), so the maternal immune system might not recognize the foreignness of the embryo.

Another immunoevasive strategy is production of immune suppressive molecules by the preimplantation embryo. For example, both oocytes (37) and embryos (31) transcribe mRNAs specific for a soluble isoform of HLA-G, HLA-G5, and translate the messages into protein. HLA-G5 is readily identified in the human preimplantation embryo using the HLA-G5–specific monoclonal antibody, 1-2C3 (38) (Fig. 2). Soluble HLA-G is present in the media of in vitro cultured embryos, and high levels signal capacity of the embryo

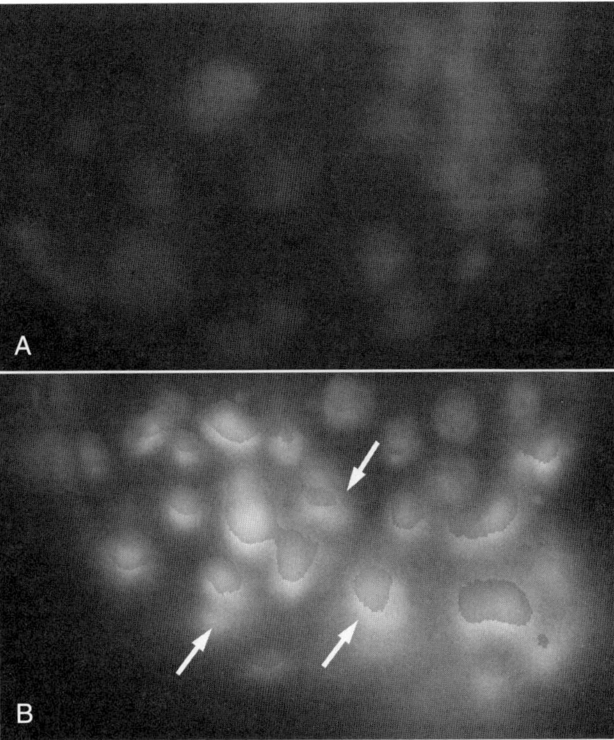

FIG. 2. Expression of HLA-G5, an immunosuppressive member of the HLA class Ib subclass, on a preimplantation embryo. **A:** DAPI stain identifies cell nuclei. **B:** The monoclonal antibody 1-2C3 identifies HLA-G5 in embryonic cells. (Original magnifications ×400; courtesy of C. Simon, University of Valencia, Spain.)

to implant successfully (39–43). Levels and the ultimate success of in vitro fertilization may be related to HLA-G genotype (43). Not surprisingly, soluble HLA-G circulates in mothers during pregnancy (44).

Although HLA-G5 is known to exert profound immune cell suppression [reviewed in (33)], its expression in the embryo could be related to either this immune function or to facilitation of normal embryonic developmental processes, or both, as embryos with HLA-G mRNA exhibit an increased rate of cleavage (31).

Another potential immune cell modulator secreted by the embryo is chorionic gonadotropin (CG). This polypeptide hormone is released prior to implantation, although it is not usually present in sufficient amounts for identification in pregnancy detection kits. Human CG (hCG) is believed to bind to receptors on decidual macrophages but this has been shown only in cells harvested from term placentas and may or may not relate to early human pregnancy (45).

THE MATERNAL–FETAL INTERFACE

Dramatic changes take place in the uterus concurrently with implantation that include restructuring the pregnant uterus as a site of innate immunity and introduction of a new element, fetally derived cells and their products.

Maternal Cellular Components: Changes and Consequences

B and T lymphocytes generate antigen-specific immunity and could therefore, if developed in the mother, comprise a significant danger to the semiallogeneic fetus. Although the cycling uterus is home to several subpopulations of leukocytes with the ability to develop antigen-specific, acquired immunity, major modifications are made in the uterine leukocyte populations concurrently with implantation [reviewed in (46)]. The pregnant uterus is reprogrammed from a mucosal site where acquired immunity is facilitated into a state of innate (natural) immunity. B cells and most T cells disappear from the decidualizing endometrium (47,48).

The most prominent leukocyte subpopulations that remain in the post-implantation uterus are NK cells, antigen presenting cells (APC) that include macrophages and dendritic cells, and a few T lymphocytes. The chemoattractive ligands and receptors that may influence the changes are given in Table 1 (ligands), Table 2 (receptors), and Table 3 (ligand/receptor pairs) (19–24). Monocyte inflammatory protein-1 (MIP-1α) has been identified as a major product of cytotrophoblast cells that attracts monocytes and NK cells into the human uterus (20), whereas IL-8 may influence neutrophil and other leukocyte populations (24). MIP-1α has also been implicated in facilitating cytotrophoblast cell differentiation (19), a finding that illustrates the multifunctional nature of immunological molecules.

The resident leukocytes are environmentally programmed into immune inhibitory profiles consistent with their roles as mediators of the immune privilcged environment required to sustain semiallogeneic pregnancy (49). This is essential; with implantation, genetically dissimilar, highly specialized cytotrophoblast cells from the implanted blastocyst drive directly toward these leukocytes (Fig. 3), and their immediate protection is required.

As a consequence of the cellular changes and their programming for antiinflammatory rather than inflammatory responses, the pregnant uterus is a site where support of embryonic development is preferred to maintenance of host defense. This is in contrast to the cycling uterus where defensins and other antimicrobials are abundant (50). Infections are less vigorously controlled in the pregnant than in the cycling uterus, with infection-associated preterm labor an unfortunate consequence. Caucheteux and colleagues have commented on how repositioning the uterus during pregnancy into a site of innate immunity might relate to fetal loss (51).

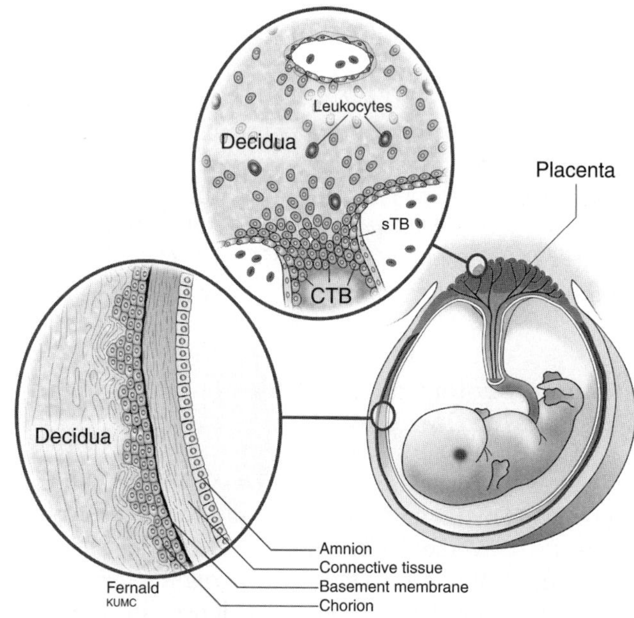

FIG. 3. The maternal–fetal interface. This figure illustrates the point that the fetus is entirely surrounded and encased by trophoblast cells in the placenta and chorion membrane. Mother's cells and fetal cells are intermixed at the border between the placenta and decidua basalis (*upper insert*) and between the chorion membrane and decidua parietalis (*left insert*). CTB, cytotrophoblast cells; sTB, syncytiotrophoblast.

Uterine NK Cells

Decidual NK cells have captured the imagination of many investigators because of their high density in the early human decidua (52) and their unusual structural and functional characteristics. In women these cells have a defined chronological profile where the cells are extremely numerous during the first two trimesters, comprising 20% to 40% of the decidual stromal cells, then decline to undetectable levels as termination of pregnancy approaches [reviewed in (9,10)]. Human uterine NK cells (uNK) are small cells of the lymphoid lineage that contain distinct granules and express a pattern of cell surface markers, $CD56^{bright}CD16^{-}$, which distinguishes them from NK cells in the blood, most of which are $CD56^{dim}CD16^{+}$. In the mouse and rat, the uNK cells are readily identified throughout pregnancy by their exceptionally large size and abundant granules. The life cycle of the mouse uNK cell has been extensively documented, and includes degranulation in the last trimester and replenishment from the spleen [reviewed in (53)].

In other mucosal sites, NK cells serve an immune surveillance function, killing infected and abnormal cells. Unlike other mucosal NK cells, uNK cells do not kill their normal target cells despite having reservoirs of killer molecules such as granzymes and TNF-α in their granules, as demonstrated in the mouse (54). The cytokine profiles of the uNK cells are generally but not invariably consistent with their functional anomalies, being producers of both Th1-type cytokines such as IFN-γ that are associated with inflammation and cell activation, and Th2-type cytokines such as interleukin-10 (IL-10) that drive the immune system into antiinflammatory pathways (55).

NK cells are traditionally known for their surveillance function; these cells recognize and destroy aberrant host cells which fail to exhibit normally expressed HLA. Avoiding this danger may be one reason why cytotrophoblast cells migrating into the uNK-filled decidua choose to exhibit cell surface HLA class I antigens. However, the HLA molecules expressed by these cells, which are HLA-G, -E, and -C [reviewed in (33,56,57) and described in detail below] are likely to have intrinsic protective capacity. Recent studies by King and colleagues suggest that it is binding of trophoblast cell HLA-E to uNK CD94/NKG2 inhibitory receptors which effects loss of killer function in human uNK cells (58) but HLA-G has not been eliminated as a modulator of uNK activity. HLA-C is also present on extravillous trophoblast cells, and Parham has discussed the potential importance of interactions between trophoblast cells expressing this polymorphic HLA class Ia molecule and uNK killer cell immunoglobulin-like (KIR) receptors (59).

Despite considerable experimentation, the role(s) these cells play in pregnancy remain unclear (60). The general principle that uNK activities are unrelated to host defense but are beneficial to pregnancy is well accepted, mainly as a consequence of studies in genetically modified mice. Two popular ideas for which experimental support has been developed in the mouse are that the uNK cells participate in modification of maternal vasculature via production of IFN-γ (61) and promote placental growth and development (62).

Macrophages and Dendritic Cells

APC that include macrophages and dendritic cells are common in the decidua. Macrophages are extremely numerous whereas the dendritic cells are few in number. Although these cells are related, they comprise morphologically distinct subpopulations with defined functions.

Macrophages. In contrast to the fluctuating levels of decidual NK cells, macrophages are a constant feature of the decidua throughout the course of pregnancy, averaging 15% to 20% of the decidual stromal cells (46,47). In human decidua, macrophages are frequently located close to invading cytotrophoblast cells, and are adjacent to the uterine glandular epithelium and proximal to uterine blood vessels. As with macrophages in other locations, these cells are highly versatile and multifunctional. They are proposed to be critical to host defense, uterine homeostasis, and local immune modulation [reviewed in (49,62–64)].

In human decidua, the cells are ubiquitous and comprise valuable contributors to host defense, being fully capable of phagocytosis of microorganisms. In mice, macrophages are scarce in the decidua, and this is believed to cause loss of resistance to infections such as *Listeria monocytogenes* (65). The human decidual macrophages are activated, as evidenced by their expression of HLA class II, CD11c, and CD86 antigens [reviewed in (33)], and could therefore present microbial or other exogenous antigens to T lymphocytes if such cells were nearby. Yet this avenue to acquired immunity appears to be closed; decidual macrophage cytokine profiles indicate that the cells are programmed into an immune suppressive mode, producing powerful immunoinhibitory cytokines such as IL-10 and TGF-β1 [(66,67); McIntire, R. H., and Hunt, J. S., unreported data]. Human endometrial cells also produce TGF-β (68). In the mouse, macrophages are sources of prostaglandin E_2 (PGE_2) (69), a molecule with profound immunosuppressive properties, but human decidual PGE_2 appears to arise from endometrial stromal cells that have differentiated into decidual cells (65). Decidual macrophages have been reported to express B7-H1 (70,71), an inhibitory member of the B7 family of costimulatory molecules, as well as ILT-3, DC-SIGN, MS-1, and factor-13 [reviewed in (33)], all of which are markers associated with immune evasion and activation of macrophages into a suppressive profile.

Under certain conditions, the common condition of preeclampsia being one and gram-negative infections being a second, the environmental factors that drive decidual macrophages into an immune suppressive mode, sometimes referred to as "alternative activation," are overcome and the cells assume a pro inflammatory profile, as proposed in 1989 by Hunt (72). In preeclampsia, decidual macrophages produce TNF-α, which has toxic effects on migrating trophoblast cells and might account for the diminished numbers of viable cytotrophoblast cells in the preeclamptic decidua (73). In gram-negative infections, the decidual macrophages produce extraordinarily high levels of TNF-α, other inflammatory cytokines and PGE_2, which ultimately induce labor [reviewed in (74–76)].

Dendritic cells. The human decidua contains small numbers of another powerful type of APC, the dendritic cell [reviewed in (33,77)]. The dendritic cells fall into two categories: $CD83^+$ mature dendritic cells, which comprise approximately 1% of decidual stromal cells (78), and $CD83^-$ immature macrophage/dendritic cells. As with decidual macrophages, mature $CD83^+$ decidual dendritic cells exhibit an immunosuppressive phenotype (79). These cells secrete less IL-12, a promoter of T cell activation, than monocyte-derived dendritic cells, and induce Th2-producing T cells when co-cultured with naïve $CD4^+$ T cells.

Immature ($CD83^-$) dendritic cells induce T cell anergy. This category appears to be composed of an assortment of related cells, some of which are display classical dendritic cell markers (DC-$SIGN^+CD14^+$) and others of which are DC-$SIGN^-CD14^-$DEC-205^+. The DC-$SIGN^+$ cells may be a precursor population that differentiates into either macrophages or dendritic cells when programmed by certain cytokines. DC-SIGN is utilized for immune evasion by several viral and bacterial pathogens, and binding alters both cytokine production and antigen presentation in ways that benefit the pathogen (80). If this is also a feature of pregnancy, DC-$SIGN^+$ cells could be central players in the development of the pregnant uterus as an immune privileged site benefiting the fetus.

T Cells

Unlike the NK and macrophage subpopulations of hematopoietic cells in the decidua, which are large, T lymphocytes comprise only about 10% of the decidual leukocytes [reviewed in (74a)]. Of the $CD3^+$ subpopulation, most are $CD8^+$, i.e., of the cytotoxic T cell group, but some are mucosal T cells bearing TcRγ/δ and others are $CD4^+$ helper T lymphocytes. Among the $CD3^+CD4^+$ T cells are regulatory T cells (T_{reg}) exhibiting $CD4^+CD25^+$ (82). The cells, whose proliferation is stimulated by estrogen (83), comprise approximately 14% of $CD4^+$ cells in early decidua (82), so altogether constitute only a small fraction of decidual leukocytes.

The decidual T_{reg} cells are similar to those in other locations. Heikkinen and colleagues (82) have reported that the T_{reg} cells express intracellular cytotoxic T lymphocyte antigen-4 (CTLA-4), GITR and OX40, all of which are markers for this subset. The frequency of $CD4^+CD25^+$ cells in the peripheral blood is increased during early pregnancy (84), suggesting that cells circulating in blood migrate into the uterus to increase their numbers during pregnancy. In the mouse, TGF-β, a cytokine that is ubiquitous in the pregnant uterus and placenta (discussed below), induces the $CD4^+CD25^+$ regulatory phenotype from normally $CD4^+CD25^-$ T lymphocytes (85). The pathway of induction is via enhancement of the winged helix/forkhead transcription factor, Foxp3, and down-regulation of Smad 7.

T_{reg} cells are powerful negative regulators of immunity. These cells are responsible in large part for maintaining peripheral tolerance when reactive T cell clones have arisen de novo against self antigens or have escaped from thymic selection (86–88). The cells produce both IL-10 and TGF-β1, now identified as two major immunosuppressive cytokines at the maternal–fetal interface, as discussed below. It is therefore not surprising that $CD4^+CD25^-$ T lymphocytes may ultimately be found to comprise a critical subpopulation of suppressor cells in human decidua, as proposed by Aluvihare and co-workers (89).

Fetal Cellular Components

Figure 3 illustrates the critical point that in normal pregnancies, the placenta and extraplacental membranes form the fetal component of the maternal–fetal interface and therefore carry the major responsibility for preventing the mother from summarily rejecting the "foreign object," the fetus. Special mechanisms allow these cells, which may express "foreign" paternal antigens, to survive and enjoy nourishment from the mother.

Placentas

The placenta, an organ with a limited natural life span, illustrates all of the varied and diverse processes

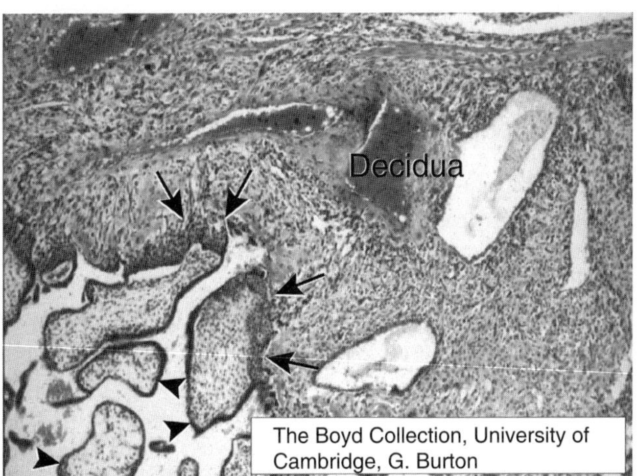

FIG. 4. Hematoxylin and eosin stain of a section of gestation week 5 implantation site. Note the cross-sections of placental villi (*arrowheads*), which are not yet nourished by maternal blood, and the dense accumulations of migrating cytotrophoblast cells at points of attachment to the decidua (*arrows*). (Original magnification ×100; courtesy of G. Burton, the Boyd Collection, University of Cambridge.)

that occur during organogenesis but also illustrates pathways for discarding unnecessary tissue. Thus, studies on the placenta may permit major insights into birth of an organ and death of the same organ from the temporal standpoint. Both phases may be related to immunological processes and involve molecules first identified in the immune system. The maternal–fetal interface of which the placenta is a part is illustrated in Fig. 4, a photomicrograph of a hematoxylin and eosin-stained tissue section taken from a gestation week five implantation site.

Structure of the placenta. From its beginnings in the blastocyst to its natural end at the conclusion of pregnancy the human placenta contains many types of cells and exerts multiple functions. Reproductive immunologists focus on trophoblast cells because they are an entirely unique lineage not found in males or in women other than during pregnancy, and comprise the major fetal cell type interfacing with maternal blood and decidua (Fig. 3). Trophoblast cells arise from the trophectoderm, the outermost layer of the blastocyst. Following implantation, this layer differentiates into several structurally and functionally distinct subpopulations of trophoblast cells (see Chapter 5). These events are easily studied in the mouse, where the sequences of changes are demonstrable hourly and can be examined in histological tissue sections by using light microscopy. Rossant and Cross have made a strong case for parallels between differentiation of trophoblast cells in mice and human placentas that justifies use of the mouse for many if not all types of functional studies (90).

The trophoblast precursor cell in human placentas is the villous cytotrophoblast cell (Fig. 5A), which may differentiate along one of two pathways. In early to middle stages of gestation, the cells may join the syncytium that forms the outermost later of the floating villi or may proliferate, explode from the villus and migrate into the decidua in defined columns. Figures 4 and 5A–C illustrate these three general subpopulations: villous cytotrophoblast cells, syncytiotrophoblast, and migrating extravillous cytotrophoblast cells. Other distinctions include subdivision of the migrating extravillous cytotrophoblast cells

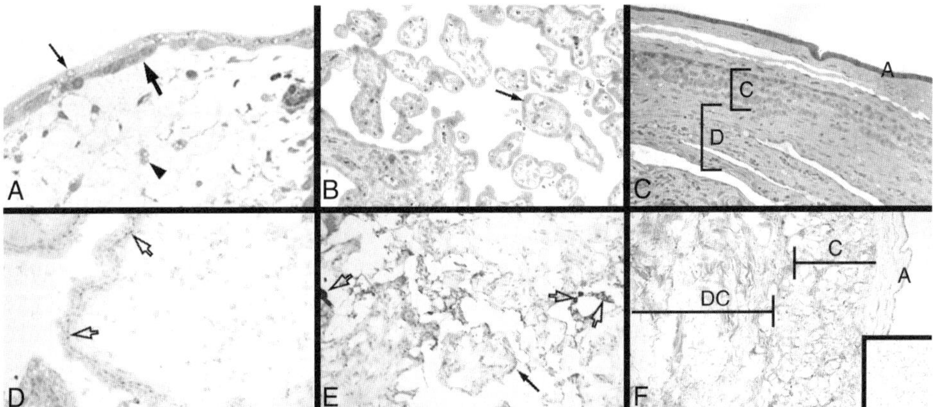

FIG. 5. Histochemical and immunohistochemical stains to show morphologies **(A–C)** and the presence of HLA-G5 **(D–F)**. Toluidine blue stains of plastic-embedded sections of **(A)** a first trimester villous (×200); **(B)** term placental villi (×200); **(C)** extraplacental membranes (×100). Immunohistochemical staining to reveal HLA-G5 in **(D)** a first trimester villus (×200); **(E)** term placenta (×200); and **(F)** term extraplacental membranes (×100). Note that in early villi, staining is intense in villous cytotrophoblast cells underlying the syncytium. In term placentas, HLA-G5 is prominent in maternal blood. In term membranes, HLA-G5 is abundant in both the chorion and the decidua. *Small arrow*, syncytiotrophoblast; *large arrow*, cytotrophoblast cells; *arrowhead*, Hofbauer cell; *C*, chorion membrane; *D*, decidua; *DC*, decidua capsularis.

into those located in the decidual stroma, termed interstitial cytotrophoblast cells, and those lining the maternal spiral arteries, termed the endovascular cytotrophoblast cells.

Later in gestation the villous cytotrophoblast cell population is nearly exhausted and the syncytiotrophoblast layer predominates (Fig. 5B). The extravillous cytotrophoblast cells have regressed and, in the placenta itself, are found only in the basal plate. In the extraplacental membranes that surround the fetus (Fig. 5C), these cells have regressed to form a well-defined chorionic cytotrophoblast cell membrane.

Subpopulation-specific functions of trophoblast cells and relationships with immune evasion strategies. The functions of trophoblast cells are related to their subpopulation status. For example, the villous cytotrophoblast cells must replenish the syncytiotrophoblast cell layer as it expands to accommodate the needs of the growing fetus as well as supply progenitor cells for the development of extravillous trophoblast cells. The syncytiotrophoblast cell layer is responsible for many critical functions: providing defense against immune cells and their potentially destructive products in maternal blood; facilitation of bidirectional transport of nutrients and wastes between the mother and the embryo/fetus; production of pregnancy-specific hormones; serving as a structural barrier to maternal–fetal cell traffic; protecting against environmental toxicants, as discussed by Audus et al. (91). The extravillous cytotrophoblast cells are required to migrate into the decidua, anchor the placenta to the mother and replace the endothelial cells of the spiral arteries. This latter maneuver is essential to opening up the maternal vasculature so that maternal blood can flow over the floating villi, an event that takes place in human placentas around the 10th to 12th week of gestation (92).

All subpopulations of trophoblast cells in humans have developed strategies for avoiding immune rejection by the mother. These include acting as a generally efficient barrier to cell traffic (syncytiotrophoblast), producing immunosuppressive molecules (all subpopulations), and expressing cell surface molecules that avoid or down-regulate cell-mediated immune responses (syncytiotrophoblast, extravillous cytotrophoblast cells). The choices of immune evasion strategies depend in part on anatomical location. Syncytiotrophoblast must defend against blood-born immune cells and inflammatory cytokines; extravillous cytotrophoblast cells must defend themselves from leukocytes resident in the decidua. Some suppressor molecules are used by all subpopulations. HLA-G5 is one of those molecules. Figures 5D and F illustrate immunostaining for this ubiquitous, soluble immunosuppressive molecule in the villous cytotrophoblast cells of early placenta, syncytiotrophoblast and maternal blood in a term placenta, and in the chorion membrane and decidua of a term extraplacental membrane (38).

Certain inner cell mass cells differentiate into the mesenchymal cells that form a supporting structure underlying the trophoblast cell layers. The placental mesenchymal cells continue into the umbilical cord. These cells, which include a substantial macrophage population known in early pregnancy as the Hofbauer cells [reviewed in (93)], become involved in defense of the semiallogeneic embryo when the trophoblast cell layer is breached. This is a common occurrence, called villitis. Maternal cells, mainly macrophages but some lymphocytes, are also found in the denuded villi (94). These invading cells could derive the population of maternal cells that migrate into fetal organs, as discussed below, but there is as yet no experimental evidence for this idea.

Extraplacental Membranes

Figure 3 shows that following implantation and as gestation progresses, the developing fetus resides in a fluid-filled amniotic sac, where it is entirely sequestered from maternal cells. Mammalian amniotic fluid contains numerous cells, among which are macrophages. In the mouse, these cells reportedly migrate into the myometrium and influence pregnancy termination (95). The amnionic sac or membrane is formed of a single epithelial cell layer joined by tight junctions that arise from the inner cell mass of the blastocyst. As with other cells derived from the inner cell mass, amnion cells exhibit HLA class I antigens that can be upregulated by inflammation-associated cytokines such as IFN-γ (96).

Directly beneath the amnion membrane is an area of connective tissue that contains no vasculature but is populated with fetal fibroblasts and macrophages that are interconnected to form a three-dimensional network (97). The purpose of these cells is likely to be communication through the collagenous zone but this idea remains unexplored experimentally.

The trophoblast cell component of the extraplacental membranes, the chorion leave, is composed of regressed elements of the extravillous trophoblast cells that were formed from the trophectoderm (Fig. 5C). By mid to late pregnancy, atrophic villi containing connective tissue and macrophages are evident throughout this four- to six-cell membrane. The chorionic cytotrophoblast cells retain many characteristics of the extravillous cells, although near the conclusion of pregnancy the cells have limited ability to migrate.

The maternal–fetal immunological interface of which the extraplacental membranes comprise an important component is illustrated in Fig. 3 (left insert).

Overall, the immunological interactions between maternal decidual leukocytes and the chorion membrane cytotrophoblast cells are the same as between maternal and fetal cells in the decidua during early to middle stages of pregnancy. The chorionic cytotrophoblast cells exhibit the same immunologically critical cell surface and soluble antigen profiles as their counterparts in early pregnancy. This point is amply illustrated studies on the isoforms of HLA-G. Chorion membrane cytotrophoblast cells strongly express both HLA-G1 and its soluble isoform, HLA-G5, and, in a few cells, HLA-G2/G6, as is the case with their counterparts in the distal column during early pregnancy [(38); reviewed in (33)].

Maternal–Fetal Cell Coexistence and Exchange

The idea that maternal cells and fetal cells might intermingle in organs/tissues other than the maternal–fetal interface and influence one another in unusual ways is fully deserving of exploration. The experimental evidence for peaceful coexistence is presented below, but functional implications remain a matter of speculation.

Coexistence

In species such as humans and nonhuman primates where placentation is hemochorial, maternal and fetal cells are in direct contact and communication. The genetically disparate cells readily co-exist without evidence of inflammation in the decidual stroma and in the maternal spiral arteries undergoing remodeling. In early to middle stages of gestation, stretches of endovascular cytotrophoblast cells and maternal endothelial cells adjoin in the linings of these arteries. Ultimately, most of the endothelial cells proximal to the placenta are entirely replaced by the endovascular trophoblast cells, which have developed strategies for moving through hostile territory as interstitial cytotrophoblast cells then learn to resist cytotoxic cells and substances circulating in maternal blood as endovascular cytotrophoblast cells.

Exchange

Although the maternal–fetal interface is a mixture of mother's cells and fetal cells, little cell transit or exchange normally takes place; maternal cells reside mainly in readily identified maternal tissues and fetal cells remain primarily in fetal tissues. This is not an immutable principle. In both outbred and inbred mice, a few maternal cells are found in fetal tissues such as the thymus (53). In human placentas, maternal macrophages invading placental villi have been identified (94). Furthermore, fetal cells are released into the mother's circulation during pregnancy and at parturition, and some of these reside in maternal tissues for as long as a quarter century [reviewed in (98)].

It is not surprising that maternal cells remain unrecognized and intact in the fetus as the fetal immune system is slow to develop the ability to recognize nonself. The fact that fetal cells remain unrecognized in the mother is likely to be due to failure of the cells that travel into the mother to express a sufficient density of the MHC antigens that stimulate graft rejection. If that were not the case, the "foreign" fetal cells would be summarily rejected. As discussed in detail below, trophoblast cells never express the HLA-A, -B, and -D antigens that serve as targets for cytotoxic T cells, whereas cells derived from the inner cell mass such as those in the mesenchyme of placental villi gradually develop all of these antigens (99). Thus, fetal cells in maternal tissues could be of trophoblast lineage or could comprise a fetal stem cell population. The functions of the traveling cells also remain obscure, but stimulation of maternal "autoimmune" disorders such as scleroderma has been proposed (100). On the positive side, the fetal cells might comprise a reservoir of stem cells for organ regeneration (98,101) and/or induce tolerance in the mother to fetal antigens.

IMPLANTATION AND MATERNAL IMMUNE RESPONSES

Pregnancy as a Th1/Th2-Type Reaction

Pregnancy in mice was first termed a Th2-type phenomenon by Wegmann and co-workers (102). In the immune system, T lymphocytes can be roughly categorized into cytotoxic T cells (CTL) and T helper cells (Th cells). Th cells can be further stratified into Th1 cells, which produce inflammatory cytokines, and Th2 cells, which produce antiinflammatory cytokines and promote B lymphocyte development.

Although appealing in its simplicity, the concept of a Th2 bias in pregnancy is better defined in mice than in women, where many Th1-type cytokines such as IFN-γ and TNF-α have important roles in homeostasis of the pregnant uterus and placenta, as illustrated in a report showing that IFN-γ must be appropriately regulated or preeclampsia ensues (103). Further, the Th1/Th2 paradigm is applicable mainly to the normal, healthy maternal–fetal interface. Infections can overcome the immunosuppressive environment at both the local and systemic levels, with generation of high levels of

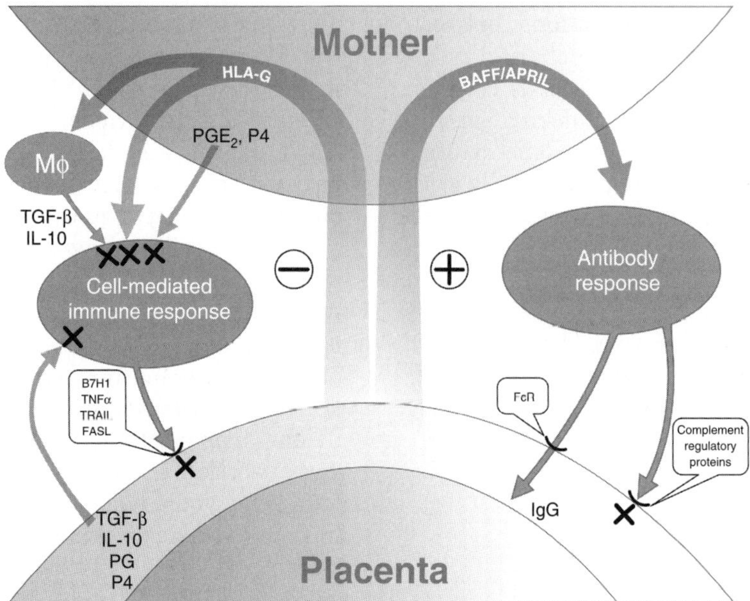

FIG. 6. Proposed pathways by which placental products drive differential expression of the two arms of the immune system, cell-mediated immunity and antibody-mediated immunity. In this scheme, HLA-G along with immunosuppressive cytokines, prostaglandins and progesterone, inhibit maternal cell mediated immunity. Effector cells are impaired by special adaptations on the placental syncytiotrophoblast that include expression of inhibitory cytokines. Antibody-mediated immunity is promoted by nonapoptosis-inducing members of the TNF superfamily of ligands, BAFF and APRIL. Cytotoxic antibodies directed to placental antigens are inhibited by expression of the complement regulatory proteins on syncytiotrophoblast while beneficial IgG is transported into the fetus.

Th1-type responses (72,74,76) and devastating effects on length of gestation.

Yet in normal pregnancies, Th2 cytokine levels are high at the maternal–fetal interface, and antibody responses in the mother are strong. Cells of the innate immune system residing in the pregnant uterus and their Th2-type cytokines undoubtedly contribute to this state, but a substantial body of evidence supports the idea that products of placental cells program these cells, as proposed in Fig. 6. Cell-mediated responses where leukocytes become killers are likely to be dampened by placental antiinflammatory cytokines such as TGF-β1 and IL-10 as well as progesterone and prostaglandins whereas antibody responses may be maintained and even elevated by two nonapoptosis-inducing TNF superfamily ligands produced in placentas, BAFF [also known as BlyS (*B lymphocyte stimulator*) TALL-1, THANK, zTNF4)] and APRIL (*a proliferation inducing ligand*) (104). These elements are discussed in detail below in the section on Fetal Contributions: Modulation of Antibody-Mediated Immunity.

Maternal Immune Responses to Implantation

Studies in mice have generated most of the information that is available on the immunology of early implantation. Concurrent with invasion of the uterine epithelium by the blastocyst, a Th1-type reaction takes place. In mice, this Th1-type reaction is characterized by an influx of leukocytes and synthesis of an array of inflammatory molecules such as IL-1 (105). Numerous observations provide indirect support for the idea that this initial inflammatory response, which could be stimulated in part by maternal immune recognition of foreign antigens, is critical to the success of pregnancy. For example, outbred mice have greater numbers of fetuses than inbred mice, and the fetuses are heavier. In humans, intervals to pregnancy are lengthier when mothers and fathers are genetically similar (106). Thus, the inflammatory response may serve a useful purpose. The practice of immunizing potential mothers with their husband's or third party leukocytes is based on this idea but has been shown in double-blind studies not to be a useful technique and may even be dangerous (107). Because of poor pregnancy outcomes the practice is prohibited in the United States by the Food and Drug Administration (FDA) except under approved experimental protocols.

Following the initial inflammatory response, anti-inflammatory cytokines such as TGF-β1 modulate this inflammatory response and drive nearby maternal cells into an immunosuppressive mode. In mice, TGF-β1 is delivered in the ejaculate, where it is believed to facilitate the initial inflammatory response to implantation (108); in rats, the cytokine is synthesized in diverse uterine cells (109); and in women, TGF-β1 may be produced in uterine macrophages as a consequence of encounter with placental HLA-G5 (67). The TGF-β1 together with IL-10, another powerful lymphocyte inhibiting agent that is produced by uterine macrophages (110), serves to prevent massive infiltration of leukocytes and killer activation of those that are resident in the uterus. In this effort, immunosuppressive cytokines in the uterus are critically aided by progesterone and prostaglandin E_2, both of which are highly immunosuppressive.

The Human Implantation Site: Oxygen Regulation of Gene Expression

Although early implantation is difficult to study in women because of ethical issues, much can be learned from careful histological and immunohistochemical examination of implantation sites in anatomical collections. Figure 4 shows a gestation week 5 implantation site from the Boyd Collection (courtesy of G. Burton, University of Cambridge) where the tissue section was stained with hematoxylin and eosin. Although not apparent in this static tissue section, the decidua is a dynamic, constantly fluctuating environment filled with leukocytes and flooded with cytokines.

Many of the activities of both fetal and maternal cells in this early environment are programmed by oxygen concentration. Prior to approximately 10 weeks of gestation, the implantation site is free of maternal blood and is hypoxic; nutrients are delivered to the fetus via the uterine gland secretions (111). Immune events taking place when fetal cells contact maternal cells may be influenced by hypoxia. For example, levels of HLA-G mRNA in both early and late gestation trophoblast cells are elevated by low oxygen (112), which might assist in creating an immune suppressive environment to the benefit of the fetus. Despite dramatic upregulation of mRNA, limited in vitro experiments on protein expression indicate that neither HLA-G1 nor HLA-G5 protein is increased by low oxygen (112). This observation implies that either transcription and translation are not linked in cytotrophoblast cells or that in vivo, the induction of the antigens is achieved by unknown maternal signals that are absent in in vitro culture systems.

GENETIC DISPARITY AND POST-IMPLANTATION EVASIVE STRATEGIES

Genetic differences that might result in maternal immune cell destruction of the semiallogeneic fetuses are the rule in outbred populations such as humans. It could be that the mother recognizes no danger. Or, it could be that danger is recognized and mechanisms are put into place to shield the semiallogeneic fetus. These arrangements are likely to be programmed into the developmental pathway of fetal cells and, in the case of maternal mechanisms, to be implemented by products of the fetus.

Foreign Antigens and Immunity: Danger and Self-Nonself Theories

The trophoblast cell lineage carries the major responsibility for protecting the embryo/fetus from maternal immune attack. With successful fertilization, the embryo incurs a major risk of rejection because of de novo expression of novel tissue-specific and paternally derived antigens. The embryo itself, which is derived from cells of the inner cell mass, is entirely sequestered from the mother by layers of trophectoderm-derived cells (Fig. 3). Trophoblast cells form the outermost layer of fetal cells in the implanted blastocyst, the placenta and the extraplacental membranes. In general, only trophoblast cells are exposed to maternal blood and maternal tissues, although in some areas of normal and diseased placentas, the syncytiotrophoblast cell layer may be disrupted and placental mesenchymal cells exposed.

The Danger Model. Trophoblast cells express potentially dangerous, organ-specific cell surface and soluble molecules that are unexpressed in other types of cells. These include, for example, the unique human placental isoform of alkaline phosphatase. Surprisingly, these antigens do not pose a risk; maternal cells do not attack the trophoblast cells expressing these antigens and antibodies in the maternal sera with specificity for the antigens are rare to unknown. Collectively, these observations support the "Danger" model of immunity suggested by Matzinger (113), which proposes that immune responses are not initiated by foreignness of an antigen to the host but rather by its capacity to create damage. This is in contrast to the traditional "Self-NonSelf" model where "foreign to the host" is the hallmark of antigens that stimulate immunity. Although immunologists favor the latter model, the former is particularly appealing when dissecting confusing aspects of semiallogeneic pregnancy that include not only the failure of maternal immune cells to attack trophoblast cells expressing tissue-specific antigens but also the role of potential immune modulators in trophoblast cells such as indoleamine 2,3-dioxygenase (IDO), as discussed below.

The Self-NonSelf Model. The Self-NonSelf theory of immunity (1) comes into play when considering the "transplantation" antigens, most of which stimulate graft rejection when foreign to the host. These are mainly the HLA antigens, which are subdivided into HLA class Ia (HLA-A, -B, and -C), HLA class Ib (HLA-E, -F, and -G) and class II (HLA-D) (discussed in detail below). Trophoblast cell selection of specific HLA genes whose antigens are poorly immunogenic from amongst those clustered in the MHC on human chromosome 6p21 is believed by many immunologists to be the major genetic adaptation that permits semiallogeneic pregnancy, as outlined by Loke and King (114). Yet this idea of how the fetal placenta avoids rejection is also challenged by the Danger model of immunity; in and of themselves, in the absence of activated maternal $CD8^+$ cytotoxic T lymphocytes programmed to respond to nonself, paternal HLA antigens expressed on fetal cell surfaces would constitute no risk. If soluble, these antigens would be immunosuppressive rather than immunostimulatory (115). This is in contrast to cell surface and soluble microbial antigens such as lipopolysaccharide (LPS), which bind avidly to receptors on leukocytes and stimulate a major inflammatory response signaling danger.

At present, the most convincing arguments for the Self-Nonself theory include the following points: cytotoxic T lymphocytes generated in the thymus that bear TcR, specific for foreign MHC, are exceptionally numerous but T cells programmed for response to paternal MHC antigens are essentially eliminated from the decidua following implantation; the pregnant uterus and placenta contain a superabundance of additional overlapping and complementary immunoprotective mechanisms that include tightly regulated expression of HLA by trophoblast cells.

Mechanisms Preventing Maternal Immune Rejection of the Fetus

The immune protective mechanisms developed to facilitate semiallogeneic pregnancy in mammals with hemochorial placentation are contributed by both the mother and the developing fetus (Fig. 1). In outbred animals such as humans, these pathways seem never to be an absolute requirement, although that is difficult to ascertain because a natural human knockout would most likely be infertile. In inbred mice, one specific protective strategy has been identified as absolutely essential. This is expression of the complement regulatory proteins that protect trophoblast cells from destruction by maternal cytotoxic antibodies. Yet fertility could be impaired by loss of one or more of the complementary systems of protection and therefore each of these constitutes an important area for research.

Maternal Contributions

Immunological features of the pregnant uterus have been extensively reviewed, most recently by Petroff and Hunt (81) and Kammerer and co-workers (116). The change of leukocyte subpopulations such that cells of the innate immune system predominate (discussed above in section on Maternal Cellular Components: Changes and Consequences) is of critical importance. Although small numbers of T lymphocytes remain, NK cells and mononuclear phagocytes are the predominant subgroups of immune cells in the pregnant uterus.

The maternal portion of this interface is a signaling circuit where the hormone that is absolutely required throughout pregnancy, progesterone, constitutes the driving force as the uterus is reconstructed into an immune privileged site. Progesterone is largely responsible for maintaining that state throughout the course of pregnancy. In women, progesterone is first produced by the corpus luteum. This steroid hormone could have major protective effects during implantation and early pregnancy. It interferes with the NFκB pathway for production of pro-inflammatory molecules and drives cells of the innate immune system into suppressive profiles characterized by lack of production of inflammation-associated cytokines (9). Progesterone also has major effects on T lymphocyte clones, programming the cells for production of both IL-10 and IL-4, as shown by Piccinni et al. (117). Thus, macrophages, NK cells and the few T lymphocytes remaining in the decidua fall into a Th2-type pattern of cytokine secretion under the direction of this powerful hormone.

Although the pregnant uterus is home to many cytokines with immunosuppressive properties, two are identified as of critical importance. These are IL-10 and TGF-β1. The prominence of IL-10, a product of macrophages (62), was first shown in pregnant mice (102) but is now known to be a feature of the human decidua as well. This cytokine not only directs T lymphocytes away from their killer activities but also drives production of HLA-G (118) in a feed-forward pathway that culminates in production of TGF-β1 by macrophages (67). Completing the circuit, HLA-G has profound effects on uNK cells. K-562 cells transfected with HLA-G1 repress synthesis/ release of a Th1-type cytokine (TNF-α) as well as a Th2-type cytokine (IL-10) by first trimester uNK cells (119). By contrast, HLA-E-transfected K-562 cells suppressed only IL-10 production. These data suggest that HLA-G1 stimulates specific immunosuppression that is likely, given the studies quoted above, to include TGF-β1, whereas HLA-E acts as a specific inhibitor of IL-10.

Fetal Contributions: Modulation of Cell-Mediated Immunity

The contributions of fetal cells to maintenance of a safe environment for the fetus fall into three general categories. One is production of soluble molecules (progesterone, prostaglandins, cytokines, growth factors, IDO). A second is production/expression of molecules that may exist in either soluble or cell bound forms (HLA-G, certain cytokines such as FasL). A final group is specifically expressed on fetal cell surfaces (selected members of the HLA and B7 families, complement regulatory proteins). Of these, all but the complement regulatory proteins are designed to interfere with effective cell mediated immunity at the maternal–fetal interface. The complement regulatory proteins interfere with humoral (antibody-mediated) immunity. A complex interplay of the potential effects of placental products on maternal immunity is illustrated in Fig. 6.

Soluble Suppressors: Progesterone, Prostaglandins, and Cytokines. Production of progesterone in the human placenta is well known and extensively documented. This steroid hormone is first produced by the corpus luteum, then production is assumed by trophoblast cells following regression of the corpus luteum. Less well known is the ability of progesterone to calm immune cells, particularly macrophages. In high concentrations, progesterone acts through the GR to diminish Th1-type cytokine production, as shown in mice (12), and may be used in place of dexamethasone to interfere with human preterm labor (120). The placenta and its membranes are also major sources of prostaglandins. Of these, the best known immunosuppressant is prostaglandin E_2, which is synthesized in the amnion membrane (121), and may contribute to both immune privilege and events surrounding parturition.

Identifying production of immunosuppressive cytokines by trophoblast cells is difficult because many preparations of human trophoblast cells are contaminated by macrophages, which are often high producers of pro- and antiinflammatory cytokines. Despite these technical problems, reports of production of IL-10 and TGF-β1 by trophoblast cells have been presented in the scientific literature (122,123). Hanna and co-workers (122) have shown that IL-10 is high in first and second trimester human placentas but diminishes at term. By contrast, TNF-α and IL-1β increase in response to labor-associated conditions. Collectively the observations suggest a cytokine reversal with Th1-type cytokines replacing Th2-type cytokines in relationship to parturition.

IDO. In 1998, Munn et al. presented evidence in a mouse model for the idea that inhibition of tryptophan metabolism by placental cells via production of IDO comprises a major strategy for protecting the embryo from maternal immune cell attack (124). The experiments described a system in which chemical inhibition of IDO, an enzyme which removes tryptophan from the environment and is a known inhibitor of cell proliferation, caused fetal rejection at gestation day 10 that was entirely restricted to allogeneic pregnancies. More recently, the group of investigators who generated the first report on placental IDO demonstrated that in the mouse placenta, IDO is located in trophoblast giant cells (125).

Evaluating the role of IDO in human pregnancy is difficult because women deficient in this ubiquitous enzyme are unknown. Studies on human placentas and decidua have yielded conflicting results, with placental endothelial cells, isolated stretches of syncytiotrophoblast, placental mesenchymal cells, extravillous cytotrophoblast cells and placental macrophages identified as positive by immunohistochemistry (126,127). Whether human decidual macrophages are IDO positive or negative remains in doubt but stage of gestation may be important (110,128). Many other types of leukocytes, including eosinophils (129) and two close relatives of macrophages, i.e., dendritic cells (130) and Langerhans cells (131), also express IDO, but none has been investigated in pregnancy. The assortment of results and their probable contributions to human semiallogeneic pregnancy have been discussed by Entrican (132).

Evaluating IDO functions in mice is also difficult. In contrast to their predictions, Baban and co-workers found that genetically modified $IDO^{-/-}$ mice have entirely normal pregnancies (125). The investigators postulated that other mechanisms of immune suppression compensated for lack of IDO, but the experiments could also be interpreted to mean that IDO is not an immune modulator in pregnancy. Bonney and Matzinger (133) expanded the Danger theory to this concept, raising the question of whether or not, in the original experiments (124), the IDO-reducing chemical simply damaged fetal cells which then emitted the danger signals associated with dead and dying cells. These signals stimulated an influx of paternal antigen-specific and -nonspecific leukocytes that ultimately destroyed the placenta and fetus.

In summary, while intriguing, the importance of this molecule to the immunology of pregnancy remains to be clearly established.

Selective Expression of HLA Family Members: The HLA Class I Genes. The "transplantation antigens" first identified for their roles in graft rejection are now known as the HLA antigens. The telomeric end of human chromosome 6p21 (mouse chromosome 17) contains a cluster of these genes (Fig. 7)

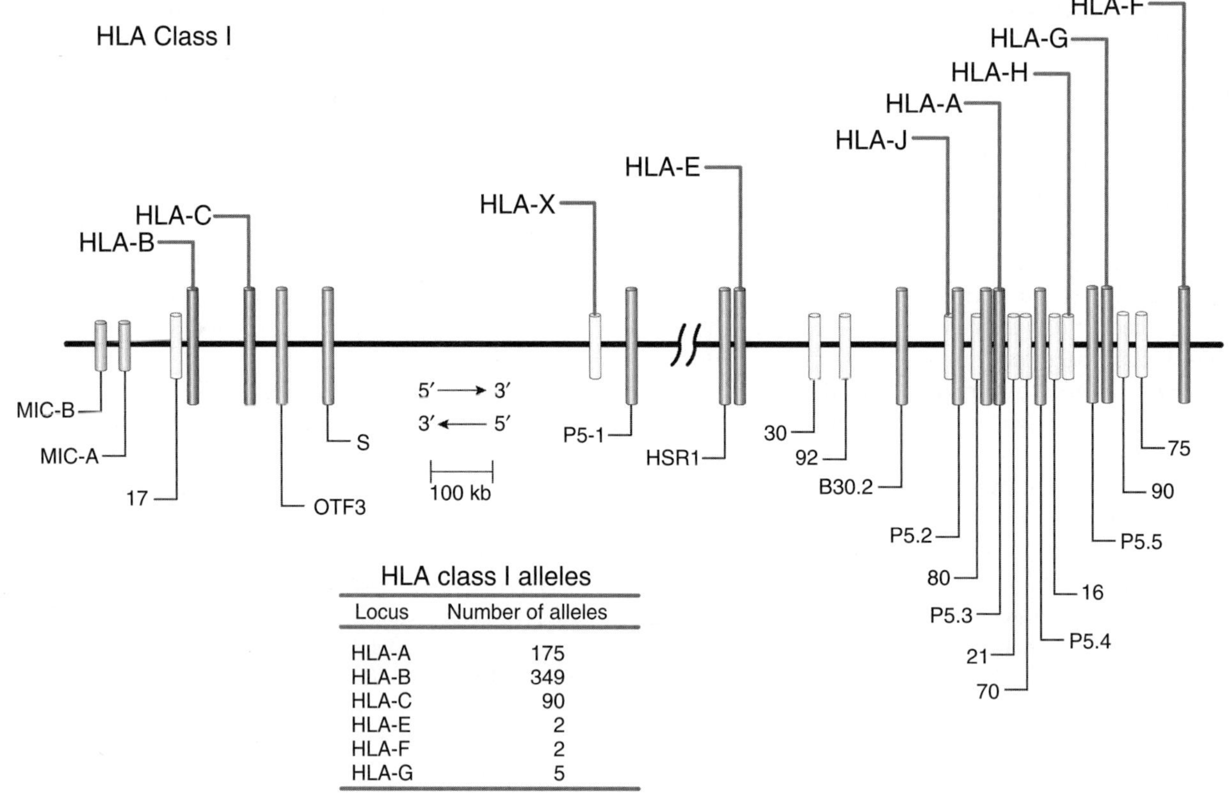

HLA class I alleles

Locus	Number of alleles
HLA-A	175
HLA-B	349
HLA-C	90
HLA-E	2
HLA-F	2
HLA-G	5

FIG. 7. The human major histocompatibility complex, known as HLA, resides on chromosome 6. The HLA class I genes are telomeric to HLA class II. HLA class Ia *(HLA-A, -B, -C)* and Ib *(HLA-E, -F, -G)* genes differ in the numbers of alleles expressed at each locus.

[reviewed in (33,134,135)]. Only a few of the 20 to 25 HLA class I genes in this cluster are transcribed or translated; most are gene fragments or pseudogenes. The class I genes that are expressed by cells are subdivided into the highly polymorphic class Ia, which include *HLA-A, -B,* and *-C* genes, and the poorly polymorphic class Ib genes, which include *HLA-E, -F,* and *-G* genes. HLA class II (*HLA-D*) genes are also highly polymorphic. Expression of the antigens encoded by the three groups differs greatly, being highly cell- and tissue-specific as well as related to cell differentiation and activation. The antigens are functionally versatile, driving either immunostimulatory or immunoinhibitory pathways.

Class I HLA antigens. The protein products of both maternally derived and paternally derived HLA class I genes are expressed on each individual cell. Unless cleaved from the cell surface as cells die and disintegrate, class Ia antigens exist only as membrane bound molecules. Their expression is not cell type-restricted; HLA-A, -B, and probably also HLA-C are expressed by essentially all somatic cells. These antigens are unique markers of individual identity as each gene is highly polymorphic, with nearly 350 alleles known for *HLA-B* (136). By contrast, products of the class Ib genes are held in common amongst most individuals. The numbers of alleles are low for *HLA-E* and *HLA-F,* and of these only two alleles of each gene encode proteins with differing amino acid compositions. *HLA-G* has five such alleles. Proteins derived from the class Ib gene group have restricted distribution, with expression being cell type, tissue or organ-specific and/or related to stage of cell differentiation and activation. The glycoproteins associated with class Ia and class Ib antigens are also significantly different.

Expression of HLA class I antigens by human trophoblast cells. Several decades of investigation have shown that trophoblast cells may either fail entirely to express HLA antigens or may select specific genes with special nonimmunogenic properties for expression [reviewed in (33,134,135)]. The antigens that trophoblast cells never express in vivo are the HLA-A and -B, which, in other contexts, are highly effective stimulators of graft rejection. Trophoblast cells also effectively repress expression of HLA class II antigens (137), which are used for stimulation of the T helper cell subset. Although some types of cells can be induced to express HLA class Ia and class II antigens, trophoblast cells in placental explants cultured in vitro cannot (138). Yet when released from their normal environment in the placenta, antigens can be induced (139). These observations indicate that elements in the placenta have as major role in regulating expression of HLA in trophoblast cells.

Human trophoblast cells express one class Ia molecule, HLA-C, and all three class Ib molecules. The *HLA-C* gene is moderately polymorphic (Fig. 7), and could stimulate maternal anti-fetal acquired immunity if paternal alleles differed from maternal. Yet allelic disparity at the *HLA-C* locus is not usually a causal factor in infertility or termination of pregnancy (59). Hiby and co-workers have reported that special genetic combinations of maternal HLA-C killer inhibitory receptors (KIR) and fetal HLA-C genes may influence the risk of preeclampsia (140).

The other HLA class I antigens expressed by trophoblast cells are HLA-E, -F, and -G. The importance of HLA-E is discussed above in the section on uterine NK cells. Although there was originally much confusion regarding the specific class I HLA that signaled inhibition of uNK cells, it is now clear that HLA-E, which binds to the CD94/NKG2 receptor, is of prime importance in preventing uNK cells from killing their normal targets (58). However, HLA-G might yet prove of value in this setting because the uNK cells express immunoglobulin-like transcript-2 (ILT2) receptors and KIR-2/DL4, which signal inhibition through different pathways. Figure 8 shows the potential immune cell targets of trophoblast cell HLA-G antigens and the receptors known to bind the specific antigens. HLA-F is also present on invading cytotrophoblast cells (141) but its binding patterns and activities remain unknown.

HLA-G structure. The physical properties of HLA-G are unique; two profoundly influence its regulation and functions. First, the gene has a large deletion in its promoter region that precludes enhancement of expression by the usual modulators, i.e., interferons and TNF-α. Second, the HLA-G transcript is alternatively spliced to yield seven different messages of which four are predicted to encode membrane-bound proteins and three are predicted to encode soluble proteins (Fig. 9) [reviewed in (33,134,135)].

Expression of HLA-G by trophoblast cells. HLA-G was the first to be identified in/on trophoblast cells and remains an antigen of great interest and focus of experimental evaluation. The placental expression patterns for HLA-G have been extensively reviewed (33,56,134,135). To summarize, HLA-G was first identified on the remnants of extravillous cytotrophoblast cells that populate the term chorion membrane, then on the same subpopulation forming cytotrophoblast cell columns in early pregnancy. The invasive cytotrophoblast cells categorized as interstitial and endovascular remain HLA-G$^+$. More recent experiments suggest that the isoforms derived from the HLA-G gene by alternative splicing may have similar or different expression patterns. The membrane-bound isoform that is derived from the full-length message, HLA-G1, appears on invasive cells but smaller isoforms that may be either membrane bound or soluble (HLA-G2 and -G6) are also prominent on the extravillous cytotrophoblast cells (31). By contrast, a second soluble isoform, HLA-G5, is ubiquitous in placentas and membranes taken from all three trimesters of pregnancy (38).

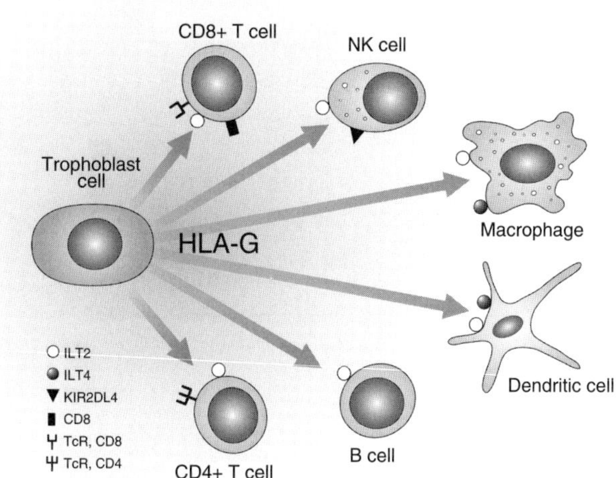

FIG. 8. Potential interactions of trophoblast cell-bound and soluble HLA antigens with cells of the immune system. Each subpopulation of leukocytes expresses its own pattern of receptors for HLA class I, which permits diversity in responses. Note that the immunoglobulin-like transcript (ILT) receptors comprise a large family, and that individual leukocyte subgroups choose specific members for expression.

HLA-G1. The membrane bound form which has been most thoroughly studied is the full length protein, HLA-G1. This isoform is present on invading cytotrophoblast cells in columns distal to the placental villi, on interstitial and endovascular trophoblast cells and on some cells remaining in the term chorion membrane. Its target could be uNK cells through ILT2 or KIR2/DL4 receptors (Fig. 8), although this remains uncertain. There is little information as yet on how this cell-bound molecule might interact with receptors on normal macrophages and/or T_{reg} cells. HLA-G1 as well as HLA-G5 may be expressed by mononuclear phagocytes. In vitro, monocyte HLA-G inhibits autologous CD4 T cell activation and production of both Th1-type (IFN-γ, IL-2) and Th2-type (IL-10) cytokines (142). Interestingly, HLA-G1-transfected APC lines have been shown to display

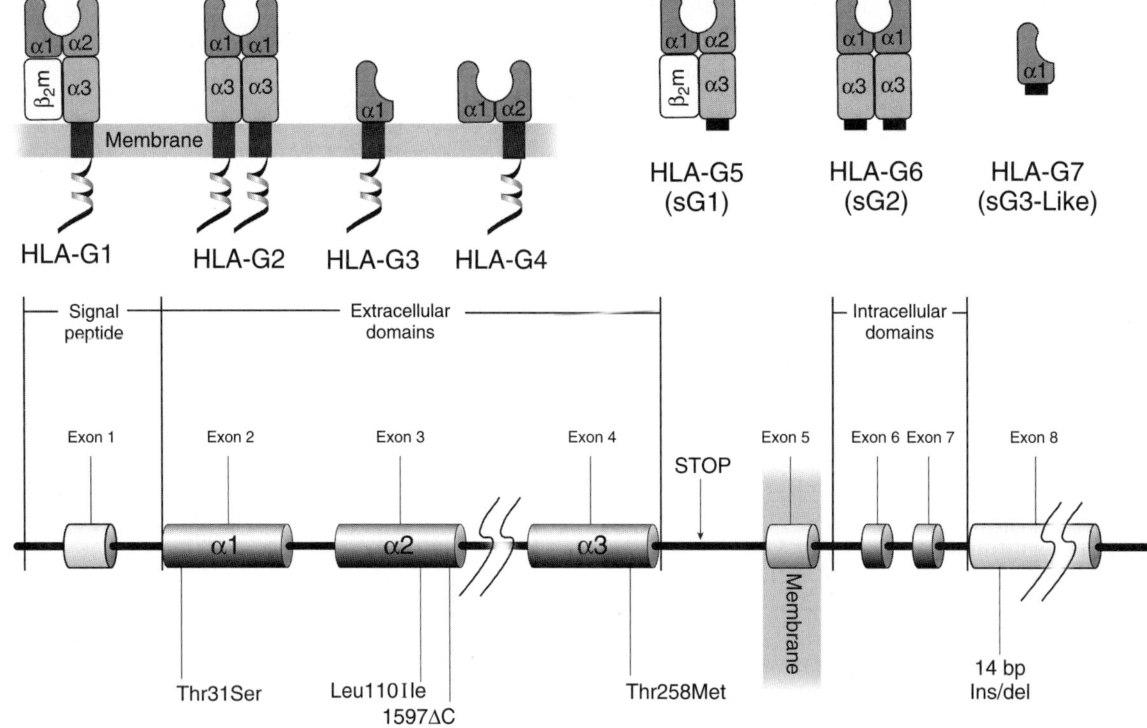

FIG. 9. Illustration of the HLA-G gene and isoforms derived from mRNAs obtained by alternative splicing of a single message. HLA-G1, -G2, -G3, and -G4 molecules are bound to the cell membrane and have short cytoplasmic tails. The soluble isoforms have two sets of identifiers. In this chapter, we use the HLA-G5, -G6, and -G7 designations. These molecules circulate as soluble isoforms because of lacking the transmembrane and cytoplasmic domains present in HLA-G1, -G2, -G3, and -G4. Portions of intron 4 are found tailing the HLA-G5 and -G6 molecules and a portion of intron 2 tails the HLA-G7 molecule.

enhanced levels of receptors such as KIR2DL4, ILT2, ILT3 and ILT4 (143), which could comprise a feed-forward pathway for secretion of cytokines or alteration of other APC activities.

HLA-G5. The functions of HLA-G5 might be similar to those of HLA-G1, as their structures are identical except that HLA-G5 is missing the transmembrane and cytoplasmic domains and is therefore a soluble protein. Binding of HLA-G5 to receptors on leukocytes has profound effects, and these are exerted through multiple pathways. HLA-G5 stimulates death of phytohemagglutinin (PHA)-treated lymphocytes through the Fas/FasL cell death pathway, reduces expression of lymphocyte CD8α, a co-receptor involved in linking CTL with APC, and drives mononuclear phagocytes into an immunosuppressive profile where the cells produce high levels of TGF-β1 [reviewed in (33,67)]. Soluble HLA-G appears to be selective in its enhancement of suppressive cytokines; there is no relation, for example, between levels of HLA-G and IL-10 in serum (144). The first report of binding of HLA-G to macrophages indicated a preference for ILT4, but studies in U937 cells, a histiocytic cell line, indicate that both the ILT2 and the ILT4 receptors on these phagocytes bind HLA-G5 (67).

Other isoforms and isoform-specific functions. Evidence is accumulating for functional distinctions among the isoforms. The argument for similar functions is supported by the finding that in the absence of HLA-G1 and -G5 due to a deletion in the HLA-G gene, women have demonstrable HLA-G proteins in their placentas and produce viable offspring (144,145). These results are consistent with the idea that other isoforms provide functional compensation. Although an early report suggested that the smaller isoforms such as HLA-G6 and -G7 were not synthesized or released in trophoblast cells (146), this study was technically flawed. More recent studies show that these are not only present in placentas but are functional (38,147,148,148a).

The argument for isoform-specific functions is supported by the report that HLA-G2/G6 is expressed exclusively in the extravillous cytotrophoblast cells distal to the villus whereas HLA-G5 is ubiquitously expressed in villous and extravillous cytotrophoblast cells and is prominent in syncytiotrophoblast and maternal blood at term (38,149). At this point in time, the major isoform-specific distinction supported by experimental evidence is concentration-dependency; HLA-G5 is a significantly more effective stimulator of certain responses such as production of TGF-β1 by macrophages (67) than is HLA-G6 [reviewed in (33)]. However, macrophage-like cells demonstrate a significant preference for binding the HLA-G6 isoform via their ILT2 receptors whereas binding of the HLA-G5 isoform is performed equally by ILT2 and ILT4 (67), suggesting that discrete, receptor-driven, isoform-specific functions may yet be uncovered.

The functions of HLA-G2, -G3, and -G4 remain poorly explored with the exception of a report from Riteau and coworkers showing that as with HLA-G1, all of these isoforms protect transfected cells from effector cell lysis (148a). HLA-G7 activities are unknown.

Immunogenicity of HLA-G. The question has been raised as to whether or not HLA-G produced in placentas might be immunogenic to the mother. In approximately 9% of women, anti-HLA-G is produced, but the antibodies are not allele-specific and do not harm the fetus or placenta. Mothers who produce anti-HLA-G have multiple successful pregnancies (32). Interestingly, ELISA titers to HLA-G6 are almost invariably higher than titers to HLA-G5, suggesting that HLA-G6 with its novel homodimeric structure may be more immunogenic than HLA-G5, whose structure closely resembles other HLA class I molecules.

HLA-G and associations with common pregnancy problems. Relationships and potential associations with poor reproductive capacity have been extensively investigated with mixed results that may be dependent upon diagnostic criteria and/or technical differences in the studies.

Unexplained recurrent miscarriage has been tested (150–152), with recent data emerging that relates the condition of miscarriage rates with variation in the HLA-G promoter region (152). Yet Patel and coworkers report that HLA-G on cell membranes is not associated with idiopathic recurrent pregnancy loss (153).

Preeclampsia has been intensively studied; this disease is marked by abnormal invasion of cytotrophoblast cells into the decidua (154) suggesting some role for HLA-G, which appears prominent in invasive cytotrophoblast cells. Levels of HLA-G may be reduced in the invasive cells of preeclamptic women (155,156). Recent studies focusing on HLA-G alleles (157) and isoforms (158) or combinations there of (159) and associations with preeclampsia (160) are now appearing in the scientific literature. Le Bouteiller has reviewed some of the recent observations on HLA-G and preeclampsia (161). Little attention has been paid to intrauterine growth retardation, although it has been definitively shown that this condition is not associated with loss of the ability of a mother to synthesize HLA-G1 and -G5 (158).

In summary, specific DNA sequences, whether in the promoter or coding region, appear to control the amounts of HLA-G (167) that are produced and reduced levels may be associated with disease. Yet when the changes result mainly in reduction of specific isoforms, other isoforms may substitute and compensate, permitting pregnancy to proceed.

Expression of HLA-G by other types of cells. Although HLA-G was originally believed to be expressed only by human trophoblast cells, this is clearly not the case. The first identification of HLA-G in a nontrophoblast cell lineage, human macrophages, was made in 1996 (163). In this study, we reported that expression was linked to activation of the macrophages by IFN-γ. This was subsequently verified in another laboratory where HLA-G5 was identified in activated monocytes using cell ELISA spot assays (164). The scientific literature has since been flooded with additional reports suggesting a role for inflammation in driving expression of HLA-G. It remains to be determined what the functions of the HLA-G molecules may be in nontrophoblastic cells, but evidence in transplant patients suggests that this molecule, which was first identified as a natural immunosuppressant in pregnancy, may exert protective effects in patients undergoing heart transplants (165).

Mesenchymal cells in the placenta gradually develop class Ia and class II antigens during gestation (99). The class Ib molecule, HLA-G, may also be developmentally regulated (163). Stromal cell antigens are fully capable of stimulating antibodies in mothers and have long been known to do so, but as described below, the placenta is competently protected against the cytotoxic activities of these antibodies by proteins that interfere with the complement cascade.

Selective Expression of B7 Family Members on Membranes. The model of lymphocyte activation that is now one of the basic tenets of immunology is one of not only stimulation through the antigen receptor by HLA molecules, but also includes a critical costimulatory signal through the CD28 receptor by B7 molecules, B7-1 or B7-2 (166). Without this costimulatory signal, naïve lymphocytes, instead of becoming activated in response to the antigen/HLA complex, may instead become tolerized to subsequent stimulation (167). With both signals, however, the result of HLA/receptor engagement is lymphocyte activation, proliferation, and cytokine production. Conversely, B7-1 and B7-2 may play a downmodulatory role through the CTLA-4 receptor, which serves to shut off the immune response. The limitation of expression of B7-1 and B7-2 to professional antigen presenting cells ensures that only cells presenting foreign antigen, and not cells expressing only self antigen, can elicit an immune response.

Conditions are different at the maternal–fetal interface, where inhibition rather than the better-described activation of antigen-specific lymphocytes is required of molecules in the B7 family. New studies show that an inhibitory member of the B7 family is expressed in trophoblast cells.

The B7 family. There are currently seven known members of the B7 family, five of which are transcribed and/or translated in the human placenta (168–172). There are also five known CD28 family molecules. Depending in part on the receptor-ligand combination, these interactions have either inhibitory or stimulatory action on immune cells (Fig. 10) (173). The novel feature of these B7 family members is that, whereas B7-1 and B7-2 serve to control outcomes of interaction between antigen presenting cells and lymphocytes, which often occurs within central immune organs such as the thymus and lymph nodes, the newly described B7 family molecules regulate action of immune cells once they have reached the peripheral organs. Thus, in contrast to the narrow cellular distribution of B7-1 and B7-2, the novel B7 family proteins have a much broader tissue distribution, at least at the level of expression of mRNA.

Patterns of expression in placentas and potential regulation. Expression of B7 family member proteins may be restricted and translationally controlled. For example, many tissues seem not to express B7-H1 protein, despite strong tissue signals generated on Northern blots by RNA (174,175). Uniquely, the human placenta expresses abundant quantities of B7-H1 protein, specifically where trophoblast interacts with maternal blood and tissue (168,176). To varying degrees, B7-H1 is expressed by all trophoblast subpopulations, and its regulation may occur by factors that control differentiation, such as epidermal growth factor. In addition, B7-H1 is dramatically upregulated in trophoblast cells at the beginning of the second trimester of pregnancy, which is when maternal blood flow to the placenta begins (92,168). Thus, exposure of fetal trophoblast cells to maternal lymphocytes coincides with the acquisition of high quantities of B7-H1, suggesting that maternal blood signals the presence of potentially dangerous lymphocytes.

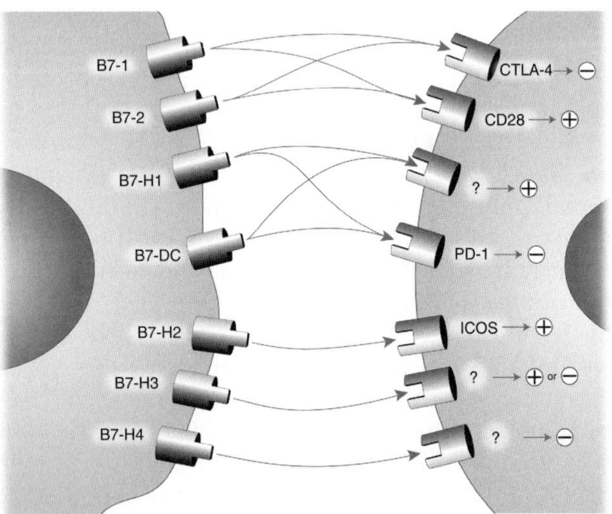

FIG. 10. A schematic representation of the known B7 and CD28 family molecules. B7 family members (*left*) are usually expressed by antigen presenting cells, but many of the novel members may also be expressed by cells within peripheral organs. Expression of CD28 family members (*right*) is restricted to immune cells, especially lymphocytes. *Arrows* between the cells indicate possible ligand-receptor interactions that promote or inhibit pathways leading to immune cell activation. +, stimulatory signal; –, inhibitory signal.

Functions of B7 family members in pregnancy. The function of B7-H1 in the placenta is as yet incompletely understood. However, studies in other models provide important clues as to the overall role of B7-H1 in the immune system. For example, B7-H1 is highly expressed by cancer cells, and facilitates escape of the cells by both inhibiting lymphocyte activation and promoting lymphocyte apoptosis (177,178). Prognosis of cancer patients correlates negatively with the level of B7-H1 expression by cancer cells, suggesting that B7-H1 expression can determine the ultimate outcome of immune surveillance (179). On the other hand, normal tissue expression of B7-H1 and/or its sister ligand seems to be, to some level, required; transgenic mice that cannot respond to these ligands because of a mutation in their receptor, PD-1, undergo severe autoimmune disease. These data, together with the fact that the genetically foreign placenta expresses higher quantities of B7-H1 than any normal other tissue, strongly suggests that the placenta utilizes this protein for survival within the maternal immune environment.

Fetal Contributions: Modulation of Antibody-Mediated Immunity

The question of how antibody levels become elevated in pregnant women has not been thoroughly investigated; researchers have focused mainly on the problem of how cell-mediated responses are diminished. Yet this question is of great importance as newborns need a broad spectrum of antibodies for protection from newly encountered environmental pathogens. Mothers generate antibodies that circulate in the blood and bombard the placenta. Ultimately, noncytotoxic IgG antibodies are delivered to the fetal circulation via special neonatal Fc receptors on syncytiotrophoblast (180,181).

The placenta may help in generating needed antibodies, as proposed in Fig. 6. A recent study (104) showed that human placentas produce two nonapoptosis-inducing TNF superfamily ligands, BAFF and APRIL, with the capacity to sustain B lymphocyte viability. An unfortunate consequence is that mothers with antibody-mediated autoimmune disorders may suffer more during pregnancy. These ligands are elevated in women with autoimmune diseases, and have been suggested to comprise a major underlying defect in their homeostatic mechanisms. The superabundance of maternal antibodies might have detrimental effects, as some could be cytotoxic to the placenta and extraplacental membranes. Yet placental destruction via antibody-mediated immunity does not take place. Mammalian placentas have installed an effective protective mechanism to prevent their destruction by complement-activating cytotoxic antibodies. Human placentas display an array of complement regulatory proteins that interfere with complement deposition and consequent toxicity (182). Mice display only one of the complement regulatory proteins, and a knockout mouse model has shown definitively that in the absence of this protein, the placenta is destroyed (183).

MULTIFUNCTIONALITY OF IMMUNE SYSTEM MOLECULES

Molecules first identified by immunologists serve many other functions in reproduction. Thus, the mere detection of the ligands in the cycling or pregnant uterus does not necessarily demonstrate an immunological role for the ligand. For example, both soluble HLA-G (26), and Qa-2, a mouse MHC molecule with properties similar to HLA-G, are believed to have immunoinhibitory roles at the maternal–fetal interface (184). Instead or in addition these molecules may be, respectively, critical to embryonic cell division and migration of human NK cells across endothelial cells (185). Le Bouteiller and colleagues have commented on how the immune and vascular modification functions of the HLA-G molecules could relate to the common disease of first pregnancy, preeclampsia (161). Primates are expected to be informative in dissecting the functions of these related molecules as preimplantation baboon embryos contain messages encoding the baboon equivalent of HLA-G5, sPaan-AG, and expression of soluble class Ib molecules is high in both baboon and monkey placentas (186,187).

One of the most powerful inflammation-associated cytokines, TNF-α, is synthesized in rat uterus and placenta (188), the human placental syncytiotrophoblast cell layer (189), and mouse uterus and placenta (190). Its immune function is thought to be to protect fetal cells against destruction by binding to activated maternal immune cells expressing high levels of TNF-R1 (191), which signals apoptosis. Yet TNF-α also binds to mouse embryo cells, where it could modulate cell-specific growth patterns. FasL, another member of the TNF superfamily of ligands, has immune functions in pregnancy, protecting against bidirectional maternal–fetal cell traffic across the mouse placenta (192) and, in its soluble form, paving the way for human cytotrophoblast cell migration through potentially toxic decidual leukocytes (193). Yet FasL is widely appreciated for its role in early embryonic development where apoptosis is required for modeling structures such as the notochord (194). The cytokine TNF-related apoptosis inducing ligand (TRAIL) is expressed in the proximal human cytotrophoblast cell column and syncytiotrophoblast, and might provide protection against killer leukocytes (195), but it may also be a factor in placental growth and differentiation because of its impact on cell adhesion molecules such as focal adhesion kinase (196).

Other examples include TGF-β1, which inhibits immune cell function but also has multiple roles in cell to cell interactions and regulation of cell activities (197,198), and colony stimulating factor-1, which promotes macrophage differentiation in the bone marrow and elsewhere but in mouse trophoblast giant cells promotes phagocytosis and is proposed to serve as a member of the innate immune system (199).

These comprise but a few of the known multiple reproduction-paired activities of molecules normally associated with immunity.

PERSPECTIVES

Studies designed to understand the immunological enigma posed by semiallogeneic pregnancy have been extremely revealing. Not only have sophisticated strategies for allowing semi-allogeneic pregnancy been identified but immunologists have found that the natural situation of pregnancy comprises an excellent model for understanding the success of tumors and for devising new strategies that might improve the success rate for transplantation of organs and cells.

Traditional therapies have focused on the host, with administration of cytotoxic drugs to the tumor-bearing patient, and modulation of host immune responses in graft recipients. Yet pregnancy immunology has taught experimentalists and clinicians that both host (mother) and implant (embryo/fetus) cooperate to develop a protected environment. Now, therapies are being devised wherein host immune cells attacking tumors are redesigned to meet the challenge of tumor evasive strategies, and gene-modulated organ and tissue grafts are being considered that would stimulate pregnancy-like host tolerance.

ACKNOWLEDGMENTS

These studies were supported by grants from the National Institutes of Health (HD24212, HD26429, U54 HD33994 [Project III] and PO1 HD39878 [Project III]) to J. S. H., and HD045611 to M. G. P. R. H. M. is supported by a fellowship from the Kansas University Medical Center Biomedical Research Training Program. The authors appreciate the assistance of S. Fernald, Kansas Reproductive Sciences Center, and S. Platt in preparation of the figures. This publication was made possible by NIH grant P20 RR016475 from the INBRE Program of the National Center for Research Resources

REFERENCES

1. Medawar, P. B. (1953). Some immunological and endocrinological problems raised by the evolution of viviparity in vertebrates. In *Symposia of the Society for Experimental Biology* (R. Brown and J. F. Danielli, Eds.), Number VII: Evolution, pp. 320–338. Academic Press, New York.
2. Jones, R. K., Bulmer, J. N., and Searle, R. F. (1998). Phenotypic and functional studies of leukocytes in human endometrium and endometriosis. *Hum. Reprod. Update* 4, 702–709.
3. Jeziorska, M., Salamonsen, L. A., and Woolley, D. E. (1995). Mast cell and eosinophil distribution and activation in human endometrium throughout the menstrual cycle. *Biol. Reprod.* 53, 312–320.
4. Searle, R. F., Jones, R. K., and Bulmer, J. N. (1999). Phenotypic analysis and proliferative responses of human endometrial granulated lymphocytes during the menstrual cycle. *Biol. Reprod.* 60, 871–878.
5. Zhang, J., Nie, G., Jian, W., Woolley, D. E., and Salamonsen, L. A. (1998). Mast cell regulation of human endometrial matrix metalloproteinases: a mechanism underlying menstruation. *Biol. Reprod.* 59, 693–703.
6. Salamonsen, L. A. and Lathbury, L. J. (2000). Endometrial leukocytes and menstruation. *Hum. Reprod. Update* 6, 16–27.
7. Salamonsen, L. A., Zhang, J., and Brasted, M. (2002). Leukocyte networks and human endometrial remodeling. *J. Reprod. Immunol.* 57, 95–108.
8. Hunt, J. S. and Pollard, J. W. (1992). Macrophages in the uterus and placenta. In *Current Topics in Microbiology and Immunology, Macrophages and Macrophage Activation* (S. Gordon and S. W. Russell, Eds.), pp. 39–63. Springer-Verlag, Heidelberg.
9. Moffett-King, A. (2002). Natural killer cells and pregnancy. *Nat. Rev. Immunol.* 2, 656–663.
10. Trundley, A. and Moffett, A. (2004). Human uterine leukocytes and pregnancy. *Tissue Antigens* 63, 1–12.
11. Fox, H. S., Bond, B. L., and Parslow, T. G. (1991). Estrogen regulates the IFN-γ promoter. *J. Immunol.* 146, 4362–4367.
12. Miller, L. and Hunt, J. S. (1998). Regulation of TNF-α production in activated mouse macrophages by progesterone. *J. Immunol.* 160, 5098–5104.
13. Miller, L., Alley, E. W., Murphy, W. J., Russell, S. W., and Hunt, J. S. (1996). Progesterone inhibits inducible nitric oxide synthase gene expression and nitric oxide production in murine macrophages. *J. Leukocyte Biol.* 59, 442–450.
14. Murdoch, C. and Finn, A. (2000). Chemokine receptors and their role in inflammation and infectious diseases. *Blood* 95, 3032–3043.
15. Arcuri, F., Ricci, C., Ietta, F., Cintorino, M., Tripodi, S. A., Cetin, I., Garzia, E., Schatz, F., Klemi, P., Santopietro, R., and Paulesu, L. (2001). Macrophage migration inhibitory factor in the human endometrium: expression and localization during the menstrual cycle and early pregnancy. *Biol. Reprod.* 64, 1200–1205.
16. Caballero-Campo, P., Dominguez, F., Coloma, J., Meseguer, M., Remohi, J., Pellicer, A., and Simon, C. (2002). Hormonal and embryonic regulation of chemokines IL-8, MCP-1 and RANTES in the human endometrium during the window of implantation. *Mol. Hum. Reprod.* 8, 375–384.
17. Sentman, C. L., Meadows, S. K., Wira, C. R., and Eriksson, M. (2004). Recruitment of uterine NK cells: induction of CXC chemokine ligands 10 and 11 in human endometrium by estradiol and progesterone. *J. Immunol.* 173, 6760–6766.
18. Jones, R. L., Hannan, N. J., Kaitu'u, T. J., Zhang, J., and Salamonsen, L. A. (2004). Identification of chemokines important for leukocyte recruitment to the human endometrium at the times of embryo implantation and menstruation. *J. Clin. Endocrinol. Metab.* 89, 6155–6167.
19. Drake, P. M., Red-Horse, K., and Fisher, S. J. (2004). Reciprocal chemokine receptor and ligand expression in the human placenta: implications for cytotrophoblast differentiation. *Dev. Dyn.* 229, 877–885.
20. Drake, P. M., Gunn, M. D, Charo, I. F., Tsou, C.-L., Zhou, Y., Huang, L., and Fisher, S. J. (2001). Human placental cytotrophoblasts attract monocytes and CD56^{bright} natural killer cells via the actions of monocyte inflammatory protein 1α. *J. Exp. Med.* 193, 1199–1212.
21. Garcia-Velasco, J. A. and Arici, A. (1999). Chemokines and human reproduction. *Fertil. Steril.* 71, 983–993.
22. Kayisli, U. A., Mahutte, N. G., and Arici, A. (2002). Uterine chemokines in reproductive physiology and pathology. *Am. J. Reprod. Immunol.* 47, 213–221.
23. Red-Horse, K., Drake, P. M., Gunn, M. D., and Fisher, S. J. (2001). Chemokine ligand and receptor expression in the pregnant uterus: reciprocal patterns in complementary cell subsets suggest functional roles. *Am. J. Pathol.* 159, 2199–2213.
24. Saito, S., Kasahara, T., Sakakura, S., Umekage, H., Harada, N., and Ichijo, M. (1994). Detection and localization of interleukin-8 mRNA and protein in human placenta and decidual tissues. *J. Reprod. Immunol.* 27, 161–172.
25. Wu, R., Van der Hoek, K. H., Ryan, N. K., Norman, R. J., and Robker, R. L. (2004). Macrophage contributions to ovarian function. *Hum. Reprod. Update* 10, 119–133.
26. Hedger, M. P. (2002). Macrophages and the immune responsiveness of the testis. *J. Reprod. Immunol.* 57, 19–34.
27. Hales, D. B. (2002). Testicular macrophage modulation of Leydig cell steroidogenesis. *J. Reprod. Immunol.* 57, 3–18.
28. Hunt, J. S. and Tung, K. K. S. (2001). Immunological aspects of fertility and infertility. In *Samter's Immunologic Diseases* (K. F. Austen, M. M. Frank, J. P. Atkinson, and H. Cantor, Eds.), pp. 771–785. Lippincott Williams & Wilkins, Philadelphia.
29. Lysiak, J. J. (2004). The role of tumor necrosis factor-alpha and interleukin-1 in the mammalian testis and their involvement in testicular torsion and autoimmune orchitis. *Reprod. Biol. Endocrinol.* 2, 9.
30. Hutter, H. and Dohr, G. (1998). HLA expression on immature and mature human germ cells. *J. Reprod. Immunol.* 38, 101–122.
31. Jurisicova, A., Casper, R. F., MacLusky, N. J., Mills, G. B., and Librach, C. L. (1996). HLA-G expression during preimplantation human embryo development. *Proc. Natl. Acad. Sci. U S A* 93, 161–165.
32. Hunt, J. S., Pace, J. L., Morales, P. J., and Ober, C. (2003). Immunogenicity of the soluble isoforms of HLA-G. *Mol. Hum. Reprod.* 9, 729–735.

33. Hunt, J. S., Petroff, M. G., McIntire, R. H., and Ober, C. (2005). HLA-G and immune tolerance in pregnancy. *FASEB J.* 19, 681–693.
34. Givan, A. L., White, H. D., Stern, J. E., Colby, E., Gosselin, E. J., Guyre, P. M., and Wira, C. R. (1997). Flow cytometric analysis of leukocytes in the human female reproductive tract: comparison of fallopian tube, uterus, cervix, and vagina. *Am. J. Reprod. Immunol.* 38, 350–359.
35. Rai, R. and Regan, L. (1997). Antiphospholipid antibodies, infertility and recurrent miscarriage. *Curr. Opin. Obstet. Gynecol.* 9, 279–282.
36. Moffett, A., Regan, L., and Braue, P. (2004). Natural killer cells, miscarriage and infertility. *Brit. Med. J.* 329, 1283–1285.
37. Menicucci, A., Noci, I., Fuzzi, B., Criscuoli, L., Scarselli, G., Baricordi, O., and Mattiuz, P. L. (1999). Non-classic sHLA class I in human oocyte culture medium. *Hum. Immunol.* 60, 1054–1057.
38. Morales, P. J., Pace, J. L., Platt, J. S., Phillips, T. A., Morgan, K., Fazleabas, A. T., and Hunt, J. S. (2003). Placental cell expression of HLA-G2 isoforms is limited to the invasive trophoblast phenotype. *J. Immunol.* 171, 6215–6224.
39. Pfeiffer, K. A., Rebmann, V., Passler, M., van der Ven, K., van der Ven, H., Krebs, D., and Grosse-Wilde, H. (2000). Soluble HLA levels in early pregnancy after in vitro fertilization. *Hum. Immunol.* 61, 559–564.
40. Sher, G., Keskintepe, L., Nouriana, M., Roussev, R., and Batzofin, J. (2004). Expression of sHLA-G in supernatants of individually cultured 46-h embryos: a potentially valuable indicator of 'embryo competency' and IVF outcome. *Reprod. Biomed. Online* 9, 74–78.
41. Noci, I., Fuzzi, B., Rizzo, R., Melchiorri, L., Criscuoli, L., Dabizzi, S., Biagiotti, R., Pellegrini, S., Menicucci, A., and Baricordi, O. R. (2005). Embryonic soluble HLA-G as a marker of developmental potential in embryos. *Hum. Reprod.* 20, 138–146.
42. Yie, S., Balakier, H., Motamedi, B. R. T., and Librach, C. L. (2005). Secretion of human leukocyte antigen-G by human embryos is associated with a higher in vitro fertilization pregnancy rate. *Fertil. Steril.* 83, 30–36.
43. Hviid, T. V. F., Hylenius, S., Lindhard, A., and Christiansen, O. B. (2004). Association between human leukocyte antigen-G genotype and success of in vitro fertilization and pregnancy outcome. *Tissue Antigens* 64, 66–69.
44. Hunt, J. S., Jedhav, L., Chu, W., Geraghty, D. E., and Ober, C. (2000). Soluble HLA-G circulates in mothers during pregnancy. *Am. J. Obstet. Gynecol.* 183, 682–688.
45. Zhang, Y. M., Rao, Ch. V., and Lei, Z. M. (2003). Macrophages in human reproductive tissues contain luteinizing hormone/ chorionic gonadotropin receptors. *Am. J. Reprod. Immunol.* 49, 93–100.
46. Hunt, J. S., Petroff, M. G., and Burnett, T. G. (2000). Uterine leukocytes: key players in pregnancy. *Semin. Cell Dev. Biol.* 11, 127–137.
47. Bulmer, J. N., Pace, D., and Ritson, A. (1988). Immunoregulatory cells in human decidua: morphology, immunohistochemistry and function. *Reprod. Nutr. Dev.* 28, 1599–1613.
48. Vassiliadou, N. and Bulmer, J. N. (1996). Quantitative analysis of T lymphocyte subsets in pregnant and nonpregnant human endometrium. *Biol. Reprod.* 55, 1017–1022.
49. Hunt, J. S. and Robertson, S. A. (1996). Uterine macrophages and environmental programming for pregnancy success. *J. Reprod. Immunol.* 32, 1–25.
50. King, A. E., Critchley, H. O., and Kelly, R. W. (2003). Innate immune defences in the human endometrium. *Reprod. Biol. Endocrinol.* 1, 116.
51. Caucheteux, S. M., Kanellopoiulos-Langevin, C., and Ojcius, D. M. (2003). At the innate frontiers between mother and fetus: linking abortion with complement activation. *Immunity* 18, 169–172.
52. Starkey, P. M., Sargent, I. L., and Redman, C. W. (1988). Cell populations in human early pregnancy decidua: characterization and isolation of large granular lymphocytes by flow cytometry. *Immunology* 65, 129–134.
53. Croy, B. A., He, H., Edsadeg, S., Wei, Q., McCartney, D., Zhang, J., Borzxychowski, A., Ashkar, A. A., Black, G. P., Evans, S. S., Chantakru, S., van den Heuvel, M., Paffaro, Jr., V. A., and Yamada, A. T. (2003). Uterine natural killer cells: insights into their cellular and molecular biology from mouse modeling. *Reprod.* 126, 149–160.
54. Parr, E. L., Chen, H.-L., Parr, M. B., and Hunt, J. S. (1995). Synthesis and granular localization of tumor necrosis factor-alpha in activated NK cells in the pregnant mouse uterus. *J. Reprod. Immunol.* 28, 31–40.
55. Eriksson, M., Meadows, S. K., Wira, C. R., and Sentman, C. L. (2004). Unique phenotype of human uterine NK cells and their regulation by TGF-β. *J. Leukoc. Biol.* 76, 667–675.
56. Le Bouteiller, P. and Mallet, V. (1997). HLA-G and pregnancy. *Rev. Reprod.* 2, 7–13.
57. Hunt, J. S. (2002). Major histocompatibility antigens in reproduction. In *The Endometrium* (S.Glasser, J. Aplin, L. Giudice, and S. Tabibzadeh, Eds.), pp. 405–415. Taylor & Francis, London and New York.
58. King, A., Allan, D. S., Bowen, M., Powis, S. J., Joseph, S., Verma, S., Hiby, S. E., McMichael, A. J., Loke, Y. W., and Braud, V. M. (2000). HLA-E is expressed on trophoblast and interacts with CD94/NKG2 receptors on decidual NK cells. *Eur. J. Immunol.* 30, 1623–1631.
59. Parham, P. (2004). NK cells and trophoblasts: partners in pregnancy. *J. Exp. Med.* 200, 951–955.
60. Bulmer, J. N. and Lash, G. E. (2005). Human uterine natural killer cells: a reappraisal. *Mol. Immunol.* 42, 511–521.
61. Ashkar, A. A., Di Santo, J. P., and Croy, B. A. (2000). Interferon γ contributes to initiation of uterine vascular modification, decidual integrity, and uterine natural killer cell maturation during normal murine pregnancy. *J. Exp. Med.* 192, 259–269.
62. Hunt, J. S. (1990). Current topic: the role of macrophages in the uterine response to pregnancy. *Placenta* 11, 467–475.
63. Vince, G. S., Starkey, P. M., Jackson, M. C., Sargent, I. L., and Redman, C. W. (1990). Flow cytometric characterisation of cell populations in human pregnancy decidua and isolation of decidual macrophages. *J. Immunol. Methods* 132, 181–189.
64. Hunt, J. S. (1994). Immunologically relevant cells in the uterus. *Biol. Reprod.* 50, 461–466.
65. Redline, R. W. and Lu, C. Y. (1988). Specific defects in the anti-listerial immune response in discrete regions of the murine uterus and placenta account for susceptibility to infection. *J. Immunol.* 140, 3947–3955.
66. Heikkinen, J., Mottonen, M., Komi, J., Alanen, A., and Lassila, O. (2003). Phenotypic characterization of human decidual macrophages. *Clin. Exp. Immunol.* 131, 498–505.
67. McIntire, R. H., Morales, P. J., Petroff, M. G., Colonna, M., and Hunt, J. S. (2005). Recombinant HLA-G5 and -G6 drive U937 myelomonocytic cell production of TGF-β1. *J. Leukocyte Biol.* 76, 1220–1228.
68. Kanzaki, H., Hatayama, H., Narukawa, S., Kariya, M., Fujita, J., and Mori, T. (1995). Hormonal regulation in the production of macrophage colony-stimulating factor and transforming growth factor-β by human endometrial cells in culture. *Horm. Res.* 44, Suppl. 2, 30–35.
69. Tawfik, O. T., Hunt, J. S. and Wood, G. W. (1986). Implication of prostaglandin E2 in soluble factor-mediated immune suppression by murine decidual cells. *Am. J. Reprod. Immunol. Microbiol.* 12, 111–117.

70. Scodras, J. M., Parhar. R. S., Kennedy, T. G., and Lala, P. K. (1990). Prostaglandin-mediated inactivation of natural killer cells in the murine decidua. *Cell. Immunol.* 127, 352–367.
71. Petroff, M. G., Chen, L., Phillips, T. A., Azzola, D., Sedlmayr, P., and Hunt, J. S. (2003). B7 family molecules are favorably positioned at the human maternal–fetal interface. *Biol. Reprod.* 68, 1496–1504.
72. Hunt, J. S. (1989). Cytokine networks in the uteroplacental unit: macrophages as pivotal regulatory cells. *J. Reprod. Immunol.* 16, 1–17.
73. Reister, F., Frank, H. G., Kingdom, J. C., Heyl, W., Kaufmann, P., Rath, W., and Huppertz, B. (2001). Macrophage-induced apoptosis limits endovascular trophoblast invasion in the uterine wall of preeclamptic women. *Lab. Invest.* 81, 1143–1152.
74. Gomez, R., Ghezzi, F., Romero, R., Munoz, H., Tolosa, J. E., and Rojas, L. (1995). Premature labor and intra-amniotic infection. Clinical aspects and role of cytokines in diagnosis and pathophysiology. *Clin. Perinatol.* 22, 281–342.
75. Gomez, R., Romero, R., Edwin, S. S., and David, C. (1997). Pathogenesis of preterm labor and preterm premature rupture of membranes associated with intraamniotic infection. *Infect. Dis. Clin. North Am.* 11, 135–176.
76. Ruiz, R. J., Fullerton, J., and Dudley, D. J. (2003). The interrelationship of maternal stress, endocrine factors and inflammation on gestational length. *Obstet. Gynecol. Surv.* 58, 415–428.
77. Gardner, L. and Moffett, A. (2003). Dendritic cells in the human decidua. *Biol. Reprod.* 69, 1438–1446.
78. Kammerer, U., Schoppet, M., McLellan, A. D., Kapp, M., Huppertz, H. I., Kampgen, E., and Dietl, J. (2000) Human decidua contains potent immunostimulatory CD83(+) dendritic cells. *Am. J. Pathol.* 157, 159–169.
79. Miyazaki, S., Tsuda, H., Sakai, M., Hori, S., Sasaki, Y., Futatani, T., Miyawaki, T., and Saito, S. (2003). Predominance of Th2-promoting dendritic cells in early human pregnancy decidua. *J. Leukoc. Biol.* 74, 514–522.
80. van Kooyk, Y. and Geijtenbeek, T. B. (2003). Dendritic cell-SIGN: escape mechanism for pathogens. *Nat. Rev. Immunol.* 3, 697–709.
81. Petroff, M. G. and Hunt, J. S. (2004). Immunity at the maternal–fetal interface. In *Mucosal Immunology,* 3rd ed. (J. Mestecky, Ed.), pp. 1735–1745. Academic Press, New York.
82. Heikkinen, J., Mottonen, M., Alanen, A., and Lassila, O. (2004). Phenotypic characterization of regulatory T cells in the human decidua. *Clin. Exp. Immunol.* 136, 373–378.
83. Polanczyk, M. J., Carson, B. D., Subramanian, S., Afentoulis, M., Vandenbark, A. A., Ziegler, S. F., and Offner, H. (2004). Cutting Edge: estrogen drives expansion of the $CD4^+CD25^+$ regulatory T cell compartment. *J. Immunol.* 173, 2227–2230.
84. Somerset, D. A., Zheng, Y., Kilby, M. D., Sansom, D. M., and Drayson, M. T. (2004). Normal human pregnancy is associated with an elevation in the immune suppressive CD25+CD4+ regulatory T-cell subset. *Immunol.* 112, 38–43.
85. Fantini, M. C., Becker, C., Monteleone, G., Pallone, F., Galle, P. R., and Neurath, M. F. (2004). TGF-β induces a regulatory phenotype in $CD4^+CD25^-$ T cells through Foxp3 induction and down-regulation of Smad7. *J. Immunol.* 172, 5149–5153.
86. Itoh, M., Takahashi, T., Sakaguchi, N., Kuniyasu, Y., Shimizu, J., Otsuka, F., and Sakaguchi, S. (1999). Thymus and autoimmunity: production of $CD25^+CD4^+$ naturally anergic and suppressive T cells as a key function of the thymus in maintaining immunologic self-tolerance. *J. Immunol.* 162, 5317–5326.
87. Sakaguchi, S. (2000). Regulatory T cells: key controllers of immunologic self-tolerance. *Cell* 101, 455–458.
88. Read, S. and Powrie, F. (2001). CD4 (+) regulatory T cells. *Curr. Opin. Immunol.* 13, 644–649.
89. Aluvihare, V. R., Kallikourdis, M., and Betz, A. G. (2004). Regulatory T cells mediate maternal tolerance to the fetus. *Nature Immunol.* 5, 266–271.
90. Rossant, J. and Cross, J. C. (2001). Placental development: lessons from mouse mutants. *Nature Genetics Rev.* 2, 538–548.
91. Audus, K. L., Soares, M. J., and Hunt, J. S. (2002). Characteristics of the fetal/maternal interface with potential usefulness in the development of future immunological and pharmacological strategies. *J. Pharmacol. Exp. Therapeut.* 301, 402–409.
92. Burton, G. J., Hempstock, J., and Jauniaux, E. (2002). Review: Oxygen, early embryonic metabolism and free radical-mediated embryopathies. *Reprod. BioMed. Online* 6, 84–96.
93. Vince, G. S. and Johnson, P. M. (1996). Immunobiology of human uteroplacental macrophages: friend and foe? *Placenta* 17, 191–199.
94. Boyd, T. K., and Redline, R. W. (2000). Chronic histiocytic intervillositis: a placental lesion associated with recurrent reproductive loss. *Hum. Pathol.* 32, 1022–1023.
95. Condon, J. C., Jeyasuria, P., Faust, J. M., and Mendelson, C. R. (2004). Surfactant protein secreted by the maturing mouse fetal lung acts as a hormone that signals the initiation of parturition. *Proc. Natl. Acad. Sci. U S A* 101, 4978–4983.
96. Hunt, J. S. and Wood, G.W. (1986). Gamma interferon induces class I HLA and beta-2-microglobulin expression by human amnion cells. *J. Immunol.* 136, 364–367.
97. Ockleford, C., Malak, T., Hubbard, A., Bracken, K., Burton, S. A., Bright, N., Blakey, G., Goodliff, J., Garrod, D., and d'Lacey, C. (1993). Confocal and conventional immunofluorescence and ultrastructural localization of intracellular strength-giving components of human amniochorion. *J. Anat.* 183, 483–505.
98. Khosrotchram, K., Johnson, K. L., Cha, D. H., Salomon, R. N., and Bianchi, D. W. (2005). Transfer of fetal cells with multilineage potential to maternal tissue. *Obstet. Gynecol. Surv.* 60, 100–101.
99. Lessin, D. L., Hunt, J. S., King, C. R., and Wood, G. W. (1988). Antigen expression by cells near the maternal–fetal interface. *Am. J. Reprod. Immunol. Microbiol.* 16, 1–7.
100. Lambert, N. C., Erickson, T. D., Yan, Z., Pang, J. M., Guthrie, K. A., Furst, D. E., and Nelson, J. L. (2004). Quantification of maternal microchimerism by HLA-specific real-time polymerase chain reaction: studies of healthy women and women with scleroderma. *Arthritis Rheum.* 50, 906–914.
101. Stevens, A. M., McDonnell, W. M., Mullarkey, M. E., Pang, J. M., Leisenring, W., and Nelson, J. L. (2004). Liver biopsies from human females contain male hepatocytes in the absence of transplantation. *Lab. Invest.* 84, 1603–1609.
102. Wegmann, T. G., Lin, H., Guilbert, L., and Mosmann, T. R. (1993). Bidirectional cytokine interactions in the maternal–fetal relationship, is successful pregnancy a Th2 phenomenon? *Immunol. Today* 14, 353–356.
103. Banerjee, S., Smallwood, A., Moorhead, J., Chambers, A. E., Papageorghiou, A., Campbell, S., and Nicolaides, K. (2004). Placental expression of interferon-gamma (IFN-gamma) and its receptor IFN-gamma R2 fail to switch from early hypoxic to late normotensive development in preeclampsia. *J. Clin. Endocrinol. Metab.* 90, 944–952.
104. Phillips, T. A., Ni, J., and Hunt, J. S. (2003). Cell-specific expression of B lymphocyte- (APRIL, BLyS) and Th2-(CD30L/CD153) promoting tumor necrosis factor superfamily ligands in human placentas. *J. Leukocyte Biol.* 74, 81–87.
105. McMaster, M. T., Newton, R. C., Dey, S. K., and Andrews, G. K. (1992). Activation and distribution of inflammatory

cells in the mouse uterus during the preimplantation period. *J. Immunol.* 148, 1699–1705.

106. Ober, C. (1999). Studies of HLA, fertility and mate choice in a human isolate. *Hum. Reprod. Update* 5, 103–107.
107. Ober, C., Karrison, T., Odem, R. R., Barnes, R. B., Branch, D. W., Stephenson, M. D., Baron, B., Walker, M. A., Scott, J. R., and Schreiber, J. R. (1999). Mononuclear-cell immunisation in prevention of recurrent miscarriages: a randomised trial. *Lancet* 354, 365–369.
108. Robertson, S. A., Mau, V. J., Hudson, S. N., and Tremellen, K. P. (1997). Cytokine-leukocyte networks and the establishment of pregnancy. *Am. J. Reprod. Immunol.* 37, 438–442.
109. Chen, H.-L., Yelavarthi, K. K., Hu, X.-L., and Hunt, J. S. (1993). Identification of transforming growth factor–β1 mRNA in virgin and pregnant rats by in situ hybridization. *J. Reprod. Immunol.* 25, 221–233.
110. Heikkinen, J., Mottonen, M., Komi, J., Alanen, A., and Lassila, O. (2003). Phenotypic characterization of human decidual macrophages. *Clin Exp Immunol.* 131, 498.
111. Burton, G. J. and Jaunaiux, E. (2001). Maternal vascularisation of the human placenta: does the embryo develop in a hypoxic environment? *Gynecol. Obstet. Fertil.* 29, 303–508.
112. Nagamatsu, T., Fujii, T., Yamashita, T., Miki, A., Kanai, T., Kusumi, M., Osuga, Y., Kozuma, S., and Taketani, Y. (2004). Hypoxia does not reduce HLA-G expression on extravillous cytotrophoblasts. *J. Reprod. Immunol.* 63, 85–95.
113. Matzinger, P. (1994). Tolerance, danger, and the extended family. *Annu. Rev. Immunol.* 12, 991–1045.
114. Loke, Y. W. and King, A. (2000). Immunology of implantation. *Baillieres Best Pract. Res. Clin. Obstet. Gynaecol.* 14, 827–837.
115. Zavazava, N. (1998). Soluble HLA class I molecules: biological significance and clinical implications. *Mol. Med. Today* 4, 116–121.
116. Kammerer, U., von Wolff, M., and Markert, U. R. (2004). Immunology of human endometrium. *Immunobiol.* 209, 569–574.
117. Piccinni, M. P., Beloni, L., Livi, C., Maggi, E., Scarselli, G., and Romagnani, S. (1998). Defective production of both leukemia inhibitory factor and type 2 T-helper cytokines by decidual T cells in unexplained recurrent abortions. *Nat. Med.* 4, 1020–1024
118. Moreau, P., Adrian-Cabestre, F., Menier, C., Guiard, V., Gourand, L., Dausset, J., Carosella, E. D., and Paul, P. (1999). IL-10 selectively induces HLA-G expression in human trophoblasts and monocytes. *Int. Immunol.* 11, 803–811.
119. Rieger, L., Hfmeister, V., Probe, C., Dietl, J., Weiss, E. H., Steck, T., and Kammerer, U. (2002). Th1- and Th2-like cytokine production by first trimester decidual large granular lymphocytes is influenced by HLA-G and HLA-E. *Mol. Hum. Reprod.* 3, 255–261.
120. Ragusa, A., de Carolis, C., dal Lago, A., Miriello, D., Ruggiero, G., Brucato, A., Pisoni, M. P., Muscara, M., Merati, R., Maccario, L., and Nobili, M. (2004). Progesterone supplement in pregnancy: an immunologic therapy? *Lupus* 13, 639–642.
121. Meadows, J. W., Eis, A. L., Brockman, D. E., and Myatt, L. (2003). Expression and localization of prostaglandin E synthase isoform in human fetal membranes in term and preterm labor. *J. Clin. Endocrinolpi. Metab.* 88, 433–439.
122. Hanna, N., Hanna, I., Hleb, M., Wagner, E., Dougherty, J., Balkundi, D., Padbury, J., and Sharma, S. (2000). Gestational age-dependent expression of IL-10 and its receptor in human placental tissues and isolated cytotrophoblasts. *J. Immunol.* 164, 5721–5728.
123. Lysiak, J. J., Hunt, J., Pringle, G. A., and Lala, P. K. (1995). Localization of transforming growth factor β and its natural inhibitor decorin in the human placenta and decidua throughout gestation. *Placenta* 16, 221–231.
124. Munn, D. H., Zhou, M., Attwood, J. T., Bondarev, I., Conway, S. J., Marshall, B., Brown, C., and Mellor, A. L. (1998). Prevention of allogeneic fetal rejection by tryptophan catabolism. *Science* 281, 1191–1193.
125. Baban, B., Chandler, P., McCool, D., Marshall, B., Munn, D. H., and Mellor, A. L. (2004). Indoleamine 2,3-dioxygenase expression is restricted to fetal trophoblast giant cells during murine gestation and is maternal genome specific. *J. Reprod. Immunol.* 61, 67–77.
126. Santoso, D. I., Rogers, P., Wallace, E. M., Manuelpillai, U., Walker, D., and Subakir, S. B. (2002). Localization of indoleamine 2,3-dioxygenase and 4-hydroxynonenakl in normal and preeclampticx placentae. *Placenta* 23, 373–379.
127. Sedlmayr, P., Blaschitz, A., Wintersteiger, R., Semlitsch, M., Hammer, A., MacKenzie, C. R., Walcher, W., Reich, O., Takikawa, O., and Dohr, G. (2002). Localization of indoleamine 2,3-dioxygenase in human female reproductive organs and the placenta. *Mol. Hum. Reprod.* 8, 385–391.
128. Cupurdija, K., Azzola, D., Hainz, U., Gratchev, A., Heitger, A., Takikawa, O., Goerdt, S., Wintersteiger, R., and Dohr, G. (2004). Macrophages of human first trimester decidua express markers associated to alternative activation. *Am. J. Reprod. Immunol.* 51, 117–122.
129. Odemuyiwa, S. O., Ghahary, A., Li, Y., Puttagunta, L., Lee, J. E., Musat-Marcu, S., Ghahary, A., and Moqbel, R. (2004). Cutting edge: human eosinophils regulate T cell subset selection through idoleamine 2,3-dioxygenase. *J. Immunol.* 173, 5909–5913.
130. Terness, P., Bauer, T. M., Rose, L., Dufter, C., Watzlik, A., Simon, H., and Opelz, G. (2002). Inhibition of allogeneic T cell proliferation by indoleamine 2,3-dioxygenase-expressing dendritix cells: mediation of suppression by tryptophan metabolites. *J. Exp. Med.* 196, 447–457.
131. von Bubnoff, D., Bausinger, H., Matz, H., Koch, S., Hacker, G., Takikawa, O., Bieber, T., Hanau, D., and de la Salle, H. (2004). Human epidermal langerhans express the immunoregulatory enzyme indoleamine 2,3-dioxygenase. *J. Invest. Dermatol.* 123, 298–304.
132. Entrican, G. (2004). IDO: a crossroads of immunology and physiology? *J. Reprod. Immunol.* 61, 63–65.
133. Bonney, E. A. and Matzinger, P. (1998). Much IDO about pregnancy. *Nature Med.* 4, 1128–1129.
134. Hunt, J. S. and Orr, H. T. (1992). HLA and maternal–fetal recognition. *FASEB J.* 6, 2344–2348.
135. Le Bouteiller, P. (1994). HLA class I chromosomal region, genes, and products: facts and questions. *Crit. Rev. Immunol.* 14, 89–129.
136. Meyer, D. and Thomson, G. (2001). How selection shapes variation of the human major histocompatibility complex: a review. *Ann. Hum. Genet.* 65, 1–26.
137. Murphy, S. P., Choi, J. C., and Holtz, R. (2004). Regulation of major histocompatibility complex class II gene expression in trophoblast cells. *Reprod. Biol. Endocrinol.* 2, 52.
138. Hunt, J. S., Andrews, G. K., and Wood, G. W. (1987). Normal trophoblasts resist induction of class I HLA. *J. Immunol.* 138, 2481–2487.
139. Lenfant, F., Fort, M., Rodriguez, A. M., Campan, A., Aguerre-Girr, M., Somm, E., Abbal, M., Ohayon, E., and Le Bouteiller, P. (1998). Absence of imprinting of HLA class Ia genes leads to coexpression of biparental alleles on term human trophoblast cells upon IFN-γ induction. *Immunogenet.* 47, 297–304.
140. Hiby, S. E., Walker, J. J., O'Shaughnessy, K. M., Redman, C. W. G., Carrington, M., Trowsdale, J., and Moffett, A. (2004). Combinations of maternal KIR and fetal HLA-C

genes influence the risk of preeclampsia and reproductive success. *J. Exp. Med.* 200, 957–965.

141. Ishitani, A., Sageshima, N., Lee, N., Dorofeeva, N., Hatake, K., Marquardt, H., and Geraghty, D. E. (2003). Protein expression and peptide binding suggest unique and interacting functional roles for HLA-E, F, and G in maternal–placental immune recognition. *J. Immunol.* 171, 1376–1384.
142. Mitsdoerffer, M., Schreiner, B., Kieseier, B. C., Neuhaus, O., Dichgans, J., Hartung, H.-P., Weller, M., and Wiendl, H. (2005). Monocyte-derived HLA-G acts as a strong inhibitor of autologous CD4 T cell activation and is upregulated by interferon-β in vitro and in vivo: rationale for the therapy of multiple sclerosis. *J. Neuroimmunol.* 159, 155–164.
143. LeMaoult, J., Zafaranloo, K., Le Danff, C., and Carosella, E. D. (2005). HLA-G up-regulates ILT2, ILT3, ILT4, and KIR2DL4 in antigen presenting cells, NK cells and T cells. *FASEB J.* 19, 662–664.
144. Ober, C., Aldrich, C., Rosinsky, B., Robertson, A., Walker, M. A., Willadsen, S., Verp, M. S., Geraghty, D. E., and Hunt, J. S. (1998). HLA-G1 protein expression is not essential for fetal survival. *Placenta* 19, 127–132.

144a. Hviid, T. V., F., Rizzo, R., Christiansen, O. B., Melchiorri, L., Lindhard, A., and Baricordi, O. R. (2004). HLA-G and IL-10 in serum in relation to HLA-G genotype and polymorphisms. *Immunogenet.* 56, 135–141.

145. Castro, M. J., Morales, P., Rojo-Amigo, R., Martinez-Laso, J., Allende, L., Varela, P., Garcia-Berciano, M., Guillen-Perales, J., and Arnaiz-Villena, A. (2000). Homozygous HLA-G*0105N healthy individuals indicate that membrane-anchored HLA-G1 molecule is not necessary for survival. *Tissue Antigens* 56, 232–329.
146. Bainbridge, D. R., Ellis, S. A., and Sargent, I. L. (2000). The short forms of HLA-G are unlikely to play a role in pregnancy because they are not expressed at the cell surface. *J. Reprod. Immunol.* 47, 1–16.
147. Menier, C., Riteau, B., Dausset, J., Carosella, E. D., and Rouas-Freiss, N. (2000). HLA-G truncated isoforms can substitute for HLA-G1 in fetal survival. *Hum. Immunol.* 61, 1118–1125.
148. Riteau, B., Rouas-Freiss, N., Menier, C., Paul, P., Dausset, J., and Carosella, E. D. (2001). HLA-G2, -G3, and -G4 isoforms expressed as nonmature cell surface glycoproteins inhibit NK and antigen-specific CTL cytolysis. *J. Immunol.* 166, 5018–5026.

148a. Moreau, P., Dausset, J., Carosella, E. D., and Rouas-Freiss, N. (2002). Viewpoint on the functionality of the human leukocyte antigen-G null allele at the maternal–fetal interface. *Biol. Reprod.* 67, 1375–1378.

149. Solier, C., Aguerre-Girr, M., Lenfant, F., Campan, A., Berrebi, A., Rebmann, V., Grosse-Wilde, H., and Le Bouteiller, P. (2002). Secretion of pro-apoptotic intron 4-retaining soluble HLA-G1 by human villous trophoblast. *Eur. J. Immunol.* 32, 3576–3586.
150. Aldrich, C. L., Stephenson, M. D., Karrison, T., Odem, R. R., Branch, D. W., Scott, J. R., Schreiber, J. R., and Ober, C. (2001). *HLA-G* genotypes and pregnancy outcome in couples with unexplained recurrent miscarriage. *Mol. Hum. Reprod.* 7, 1162–1172.
151. Pfeiffer, K. A., Fimmers, R., Engels, G., van Der Ven, H., and van Der Ven, K. (2001). The HLA-G genotype is potentially associated with idiopathic recurrent spontaneous abortion. *Mol. Hum. Reprod.* 7, 373–378.
152. Ober, C., Aldrich, C. L., Chervoneva, I., Billstrand, C., Rahimov, F., Gray, H. L., and Hyslop, T. (2003). Variation in the HLA-G promoter region influences miscarriage rates. *Am. J. Hum. Genet.* 72, 1425–1435
153. Patel, R. N., Quack, K. C., Hill, J. A., and Schust, D. J. (2003). Expression of membrane-bound HLA-G at the maternal–fetal interface is not associated with pregnancy maintenance among patients with idiopathic recurrent pregnancy loss. *Mol. Hum. Reprod.* 9, 551–557.
154. Lim, K. H., Zhou, Y., Janatpour, M., McMaster, M., Bass, K., Chun, S. H., and Fisher, S. J. (1997). Human cytotrophoblast differentiation/invasion is abnormal in pre-eclampsia. *Amer. J. Pathol.* 151, 1809–1818.
155. Colbern, G. T., Chiang, M. H., and Main, E. K. (1994). Expression of the nonclassic histocompatibility antigen HLA-G by preeclamptic placenta. *Am. J. Obstet. Gynecol.* 170, 1244–1250.
156. Goldman-Wohl, D. S., Ariel, I., Greenfield, C., Hochner-Celnikier, D., Cross, J., Fisher, S., and Yagel, S. (2000). Lack of human leukocyte antigen-G expression in extravillous trophoblasts is associated with pre-eclampsia. *Mol. Hum. Reprod.* 6, 88–95.
157. Rebmann, V., van der Ven, K., Passler, M., Pfeiffer, K., Krebs, D., and Grosse-Wilde, H. (2001). Association of soluble HLA-G plasma levels with HLA-G alleles. *Tissue Antigens* 57, 15–21.
158. Aldrich, C., Verp, M. S., Walker, M. A., and Ober, C. (2000). A null mutation in HLA-G is not associated with preeclampsia or intrauterine growth retardation. *J. Reprod. Immunol.* 47, 41–48
159. Hviid, T. V., Hylenius, S., Rorbye, C., and Nielsen, L. G. (2003). HLA-G allelic variants are associated with differences in the HLA-G mRNA isoform profile and HLA-G mRNA levels. *Immunogenetics* 55, 63–79.
160. Emmer, P. M., Joosten, I., Schut, M. H., Zusterzeel, P. L. M., Hendriks, J. C. M., and Steegers, E. A. P. (2004). Shift in expression of HLA-G mRNA splice forms in pregnancies complicated by preeclampsia. *J. Soc. Gynecol. Investig.* 11, 220–226.
161. Le Bouteiller, P., Pizzato, N., Barakonyi, A., and Solier, C. (2003). HLA-G, pre-eclampsia, immunity and vascular events. *J. Reprod. Immunol.* 59, 219–234.
162. O'Brien, M., McCarthy, T., Jenkins, D., Paul, P., Dausset, J., Carosella, E. D., and Moreau, P. (2001). Altered HLA-G transcription in pre-eclampsia is associated with allele specific inheritance: possible role of the HLA-G gene in susceptibility to the disease. *Cell. Mol. Life Sci.* 58, 1943–1439.
163. Yang, Y., Chu, W., Geraghty, D. E., and Hunt, J. S. (1996). Expression of HLA-G in human mononuclear phagocytes and selective induction by interferon-γ. *J. Immunol.* 156, 4224–4231.
164. Rebmann, V., Busemann, A., Lindemann, M., and Grosse-Wilde, H. (2003). Detection of HLA-G secreting cells. *Hum. Immunol.* 64, 1017–1024.
165. Lila, N., Amrein, C., Guillemain, R., Chevalier, P., Latremouille, C., Fabiani, J. N., Dausset, J., Carosella, E. D., and Carpentier, A. (2002). Human leukocyte antigen-G expression after heart transplantation is associated with a reduced incidence of rejection. *Circulation* 105, 1949–1954.
166. Greenwald, R. J., Freeman, G. J., and Sharpe, A. H. (2005). The B7 family revisited. *Annu. Rev. Immunol.* 23, 515–548.
167. Finger, E. B. and Bluestone, J. A. (2002). When ligand becomes receptor: tolerance via B7 signaling on DCs. *Nat. Immunol.* 3, 1056–1057.
168. Petroff, M. G., Chen, L., Phillips, T. A., Azzola, D., Sedlmayr, P., and Hunt, J. S. (2003). B7 family molecules are favorably positioned at the human maternal–fetal interface. *Biol. Reprod.* 68, 1496–1504.

169. Ling, V., Wu, P. W., Miyashiro, J. S., Marusic, S., Finnerty, H. F., and Collins, M. (2001). Differential expression of inducible costimulator-ligand splice variants: lymphoid regulation of mouse GL50-B and human GL50 molecules. *J. Immunol.* 166, 7300–7308.
170. Brown, J. A., Dorfman, D. M., Ma, F. R., Sullivan, E. L., Munoz, O., Wood, C. R., Greenfield, E. A., and Freeman, G. J. (2003). Blockade of programmed death-1 ligands on dendritic cells enhances T cell activation and cytokine production. *J. Immunol.* 170, 1257–1266.
171. Chapoval, A. I., Ni, J., Lau, J. S., Wilcox, R. A., Flies, D. B., Liu, D., Dong, H., Sica, G. L., Zhu, G., Tamada, K., and Chen, L. (2001). B7-H3: a costimulatory molecule for T cell activation and IFN-gamma production. *Nat. Immunol.* 2, 269–274.
172. Prasad, D. V., Nguyen, T., Li, Z., Yang, Y., Duong, J., Wang, Y., and Dong, C. (2004). Murine B7-H3 is a negative regulator of T cells. *J. Immunol.* 173, 2500–2506.
173. Wang, S., and Chen, L. (2004). Co-signaling molecules of the B7-CD28 family in positive and negative regulation of T lymphocyte responses. *Microbes Infect.* 6, 759–766.
174. Dong, H., Zhu, G., Tamada, K., and Chen, L. (1999). B7-H1, a third member of the B7 family, co-stimulates T-cell proliferation and interleukin-10 secretion. *Nat. Med.* 5, 1365–1369.
175. Latchman, Y., Wood, C. R., Chernova, T., Chaudhary, D., Borde, M., Chernova, I., Iwai, Y., Long, A. J., Brown, J. A., Nunes, R., Greenfield, E. A., Bourque, K., Boussiotis, V. A., Carter, L. L., Carreno, B. M., Malenkovich, N., Nishimura, H., Okazaki, T., Honjo, T., Sharpe, A. H., and Freeman, G. J. (2001). PD-L2 is a second ligand for PD-1 and inhibits T cell activation. *Nat. Immunol.* 2, 261–268.
176. Petroff, M. G., Chen, L., Phillips, T. A., and Hunt, J. S. (2002). B7 family molecules: novel immunomodulators at the maternal–fetal interface. *Placenta* 23, Suppl. A, S95–S101.
177. Dong, H., Strome, S. E., Salomao, D. R., Tamura, H., Hirano, F., Flies, D. B., Roche, P. C., Lu, J., Zhu, G., Tamada, K., Lennon, V. A., Celis, E., and Chen, L. (2002). Tumor-associated B7-H1 promotes T-cell apoptosis: a potential mechanism of immune evasion. *Nat. Med.* 8, 793–800.
178. Iwai, Y., Ishida, M., Tanaka, Y., Okazaki, T., Honjo, T., and Minato, N. (2002). Involvement of PD-L1 on tumor cells in the escape from host immune system and tumor immunotherapy by PD-L1 blockade. *Proc. Natl. Acad. Sci. U S A* 99, 12293–12297.
179. Thompson, R. H., Gillett, M. D., Cheville, J. C., Lohse, C. M., Dong, H., Webster, W. S., Krejci, K. G., Lobo, J. R., Sengupta, S., Chen, L., Zincke, H., Blute, M. L., Strome, S. E., Leibovich, B. C., and Kwon, E. D. (2004). Costimulatory B7-H1 in renal cell carcinoma patients: Indicator of tumor aggressiveness and potential therapeutic target. *Proc. Natl. Acad. Sci. U S A* 101, 17174–17179.
180. Simister, N. E., Story, C. M., Chen, H.-L., and Hunt, J. S. (1996). An IgG transporting Fc receptor expressed in the syncytiotrophoblast of human placenta. *Eur. J. Immunol.* 26, 1527–1531.
181. Leach, J. L., Sedmak, D. D., Osborne, J. M., Rahill, B., Lairmore, M. D., and Anderson, C. L. (1996). Isolation from human placenta of the IgG transporter, FcRn, and localization to the syncytiotrophoblast: implications for maternal–fetal antibody transport. *J. Immunol.* 157, 3317–3322.
182. Hsi, B.-L., Hunt, J. S., and Atkinson, J. P. (1991). Detection of complement regulatory proteins on subpopulations of human trophoblast. *J. Reprod. Immunol.* 19, 209–223.
183. Xu, C., Mao, D., Holers, V. M., Palanca, B., Cheng, A. M., and Molina, H. (2000). A critical role for murine complement receptor crry in fetomaternal tolerance. *Science* 287, 498–501.
184. Comiskey, M., Goldstein, C. Y., De Fazio, S. R., Mammolenti, M., Newmark, J. A., and Warner, C. M. (2003). Evidence that HLA-G is the functional homologue of mouse Qa-2, the Ped gene product. *Hum. Immunol.* 64, 999–1004.
185. Dorling, A., Monk, N., and Lechler, R. (2000). HLA-G inhibits the transendotheial cell migration of human NK cells: a strategy for inhibiting xenograft rejection. *Transplant. Proc.* 32, 938.
186. Langat, D. K., Morales, P. J., Fazleabas, A. T., Mwenda, J. M., and Hunt, J. S. (2002). Baboon placentas express soluble and membrane-bound Paan-AG proteins encoded by alternatively spliced transcripts of the class Ib major histocompatibility complex gene, *Paan-AG*. *Immunogenet.* 54, 164–173.
187. Slukvin, I. I., Lunn, D. P., Watkins, D. I., and Golos, T. G. (2000). Placental expression of the nonclassical MHC class I molecule Mamu-AG at implantation in the rhesus monkey. *Proc. Natl. Acad. Sci. U S A* 97, 9104–9109.
188. Yelavarthi, K. K., Chen, H.-L., Yang, Y., Fishback, J. L., Cowley, Jr., B. D., and Hunt, J. S. (1991). Tumor necrosis factor-alpha mRNA and protein in rat uterine and placental cells. *J. Immunol.* 146, 3840–3848.
189. Chen, H.-L., Yang, Y., Hu, X.-L., Yelavarthi, K. K., Fishback, J. L., and Hunt, J. S. (1991). Tumor necrosis factor-alpha mRNA and protein are present in human placental and uterine cells at early and late stages of gestation. *Am. J. Pathol.* 139, 327–335.
190. Hunt, J. S., Chen, H.-L., Hu, X.-L., and Pollard, J. W. (1993). Normal distribution of tumor necrosis factor–α messenger ribonucleic acid and protein in virgin and pregnant osteopetrotic (*op*/*op*) mice. *Biol. Reprod.* 49, 441–452.
191. Hunt, J. S., Chen, H.-L., and Miller, L. (1996). Tumor necrosis factors: Pivotal factors in pregnancy? *Biol. Reprod.* 54, 554–562.
192. Hunt, J. S., Vassmer, D., Ferguson, T. A., and Miller, L. (1997). Fas ligand is positioned in mouse uterus and placenta to prevent trafficking of activated leukocytes between the mother and the conceptus. *J. Immunol.* 158, 4122–4128.
193. Abrahams, V. M., Straszewski-Chavez, S. L., Guller, S., and Mor, G. (2004). First trimester trophoblast cells secrete Fas ligand which induces immune cell apoptosis. *Mol. Hum. Reprod.* 10, 55–63.
194. Inui, Y., Nishida, K., Doita, M., Takada, T., Miyamoto, H., Yoshiya, S., and Kurosaka, M. (2004). Fas-ligand expression on nucleus pulposus begins in developing embryo. *Spine* 29, 2365–2369.
195. Phillips, T. A., Ni, J., Pan, G., Ruben, S. M., Wei, Y.-F., and Hunt, J. S. (1999). TRAIL (Apo-2L) and TRAIL receptors in human placentas: implications for tumor protection and immune privilege. *J. Immunol.* 162, 6053–6059.
196. Wen, L. P., Fahrni, J. A., Troie, S., Guan, J. L., Orth, K., and Rosen, G. D. (1997). Cleavage of focal adhesion kinase by caspases during apoptosis. *J. Biol. Chem.* 272, 26056–26061.
197. Letterio, J. J., and A. B. Roberts. (1998). Regulation of immune responses by TGF-beta. *Annu. Rev. Immunol.* 16, 137.
198. Ingman, W. V., and S. A. Robertson. (2002). Defining the actions of transforming growth factor beta in reproduction. *Bioessays* 24, 904–914.
199. Guleria, I., and Pollard, J. W. (2000). The trophoblast is a component of the innate immune system during pregnancy. *Nat. Med.* 6, 589–593.

Knobil and Neill's Physiology of Reproduction, Third Edition
edited by Jimmy D. Neill,
Elsevier © 2006

CHAPTER 52

Placental Transfer

D. E. Atkinson, R. D. H. Boyd, and C. P. Sibley

Conceptual and Methodological Issues in Placental Transfer, 2787
Paraplacental Transport, 2792
Concepts of Transfer, 2793
Common Animal Models of Transfer, 2795
A Comprehensive Model of Placental Transfer, 2795
Driving Forces, 2796
Ontogeny of Placental Transfer Mechanisms, 2796
Blood Flow, 2797
Exchange Barrier Dimensions, 2798
Concentration and Electrical Gradients, 2798
Transporter-Mediated Transfer, 2799
Transport of Selected Substances, 2799
Water, 2799
Protein, 2801
Inorganic Ions, 2802
Organic Ions, 2808
Bulk Nutrients, 2809
Respiratory Gases, 2817
Micronutrients, 2822
Drugs and Xenobiotics, 2824
Acknowledgments, 2827
References, 2827

CONCEPTUAL AND METHODOLOGICAL ISSUES IN PLACENTAL TRANSFER

A circulation separate from that of the mother is present in the embryo from early pregnancy (see section Ontogeny of Placental Transfer Mechanisms). Transfer in both directions across the tissue barrier separating the two circulations must be possible to allow the growth and metabolism of the conceptus. The placenta is the organ in which the two microcirculations, maternal and fetal, are opposed and across which most, but not all (see below) transfer takes place.

The earlier placental transfer literature is reviewed in Villee (1), which provides a useful annotated bibliography and in Hagerman and Villee (2). Schultz (3) is a very comprehensive list of transfer studies to that date. Both are well worth consulting, as in vivo transfer studies of many compounds have not been repeated in recent years. Faber and Thornburg (4) is an outstanding review with an emphasis on theoretical aspects. Morriss et al. (5) summarizes material to 1993 and provides a table of the sometimes striking materno–fetal plasma concentration differences that have been found (Table 1).

Clearly transfer other than that of water, solutes, and gases also takes place between mother and conceptus. Heat is lost from the fetus (6). Cells also cross in both directions (7), presumably via placenta, although the mechanism is not clear, and this may have long-term implications for both mother and child (8,9). Uptake and/or transfer of microorganisms leading to antenatal infection of the fetus may also involve the placenta. These important topics are not considered further here.

The anatomy of the placental barrier at both gross and microscopic levels arguably shows more interspecies difference than is the case for any other mammalian organ (Fig. 1). Enders and Blankenship (10) usefully summarize the anatomy of the most

Academic Unit of Child Health, Human Development and Reproductive Health Division, The Medical School, University of Manchester, St. Mary's Hospital, Manchester, U.K.

TABLE 1. *Human maternal and fetal plasma or serum concentrations of some selected substances (near term unless otherwise stated)*

Solute	Units	Mean maternal concentration (± SE)	Mean fetal concentration (± SE)	Concentration difference *P*<0.05	Comment	Ref.
Ions and Minerals						
Sodium	mmol/kgwater	135.4 ± 2.4	137.0 ± 2.8	*	C-section; venous samples	(160)
Potassium	mmol/kgwater	4.3 ± 0.1	4.8 ± 0.2	NS	Fetal rises during collection	(164)
Chloride	mmol/kgwater	107 ± 2.0	109.0 ± 2.3	*	C-section; venous samples	(160)
Magnesium	mmol/kgwater	0.6 ± 0.1	0.7 ± 0.1	NS	C-section; venous samples	(160)
Calcium	mmol/kgwater	2.1±0.1	2.6 ± 0.1	*	C-section; venous samples	(160)
Phosphate	mmol/kgwater	0.3 ± 0.03	0.5 ± 0.02	*	C-section; venous samples	(160)
Bicarbonate	mmol/kgwater	21.4 ± 0.8	24.4 ± 1.6	*	C-section; venous samples	(160)
Sulfate	mmol/kgwater	0.43 ± 0.02	0.46 ± 0.01	*	Umbilical-cord blood; anion chromatography	(546)
Thiocyanate	mg/dl	0.21 ± 0.01	0.19 ± 0.01	?	Umbilical cord-blood; nonsmokers	(547)
Iron	μg/dl	88.2 ± 9.5	210 ± 38	*	Total serum iron	(548)
Copper	μg/ml	3.16 ± 0.05	0.43 ± 0.02	*	Total copper; umbilical-cord blood; maternal values higher in labor	(549)
Zinc	μg/dl	52 ± 3	74 ± 2	*	Total zinc; umbilical-cord blood; Navajo Indians	(550)
Lead	ng/ml	40.4 ± 4.2	37.1 ± 3.1	NS	Atomic absorption; whole blood	(551)
Cadmium	ng/ml	1.1 ± 0.2	0.4 ± 0.04	*	Atomic absorption; whole blood	(551)
Small Organic Molecules						
Glucose	mmol/l	4.4 ± 0.2	3.6 ± 0.2	*	C-section; venous sample	(341)
Lactate	mmol/l	1.23 ± 0.16	1.25 ± 0.12	NS	Fetal samples at 18–21 weeks	(427)
Lactate	mmol/l	0.66 ± 0.07	1.15 ± 0.08	*	Umbilical-artery; maternal artery; c-section	(552)
Pyruvate	mmol/l	0.05 ± 0.01	0.07 ± 0.01	NS	18–21 weeks; maternal and fetal venous samples	(427)
Free Fatty Acids	mEq/l	0.67 ± 0.04	0.32 ± 0.02		Umbilical-cord veno-arterial difference 0.026 ± 0.01	(385)
Free Fatty Acids	mEq/l	0.9 ± 0.06	0.29 ± 0.02	*	C-section; umbilical-artery lower	(402)
Free Fatty Acids		–	–	–	Most individual FFA lower in fetal blood	(553)
Cholesterol	mg/dl	230 ± 10	60 ± 5	*	Mixed umbilical-cord blood; lipoprotein fractions	(554)
Cholesterol	mg/dl	206 ± 5	69.6 ± 2.0		Umbilical-venous blood; vaginal delivery	(385)
Triglycerides	mg/dl	208 ± 14	28.9 ± 2.8		Umbilical-venous blood; vaginal delivery	(385)
Ketones	mmol/l	0.49 ± 0.06	0.25 ± 0.03	*	18–21 weeks; maternal and fetal venous samples	(427)
Myoinositol	μmol/l	17.9 ± 1.2	60 ± 5	*	C-section umbilical-venous samples lower in umbilical-artery; higher in mid-gestation	(555)
Hypoxanthine	μmol/l	6.1	4.2	NS	Umbilical-cord mixed blood; HPLC	(556)
Urea	mg/dl	17.5 ± 1.3	20.0 ± 1.1	*	Umbilical-arterial samples	(557)
Urea	mmol/kgwater	2.6 ± 0.2	2.7 ± 0.3	NS	C-section; NS at midgestation	(160,300)
Amino acids and amino N	–	–	–	–	Fetal concentrations usually higher than maternal at midgestation or at term	(299–301, 314)
Choline	mmol/l	12.2 ± 0.03	39.8 ± 5.6	*	Adult values fasting; newborn umbilical-cord or capillary blood	(558)
Carnitine	mmol/l	17.4 ± 1.2	25.9 ± 2.7	*	C-section; maternal and fetal levels correlate	(559)

TABLE 1. *Human maternal and fetal plasma or serum concentrations of some selected substances (near term unless otherwise stated)–cont'd*

Solute	Units	Mean maternal concentration (± SE)	Mean fetal concentration (± SE)	Concentration difference $P<0.05$	Comment	Ref.
Bile acids (e.g., conjugated cholic acid)	μmol/l	0.34±0.17(?SD)	0.8±0.4(?SD)	*	Abortus material; immunoassay	(560)
Vitamins						
Vitamin A	μg/dl	37.7±1.5	14.7±0.9	*	Umbilical-cord blood Navajo Indians	(550)
Thiamine	nmol/l	5.9±0.4	13.3±0.9	*	Maternal venous and umbilical-venous plasma	(561)
Pyridoxal phosphate	μg/ml	27.5±4.1	7.1±1.9	*	Fluorescence method; umbilical-cord blood	(562)
Pyridoxal	μg/ml	1.1±0.4	2.2±0.4	?	Fluorescence method; umbilical-cord blood	(562)
B12	pg/ml	340±21	796±34	*	Umbilical-cord blood radioimmunoassay	(563)
B12	pg/ml	370±60	410±60 SE estimated from data	NS	Maternal venous and	(564)
Vitamin D	–	–	–	–	Maternal levels of 3 principal D^3 metabolites higher than fetal	(566–568)
Vitamin E	mg/dl	1.23±0.06	0.58±0.06	*	Brazilian poor; umbilical-cord blood; colorimetric assay	(569)
Vitamin K1	ng/ml	1.52	0.10		Means of 13 Japanese women at term; maternal and umbilical-venous blood	(570)
Riboflavin (free)	μg/dl	0.55	2.14	*	Photofluorimetry; umbilical-cord blood; other fractions given	(571)
Folacin	ng/ml	16±3	25±3	*	Umbilical-cord blood Navajo Indians	(550)
Carotene	μg/dl	151±9	22±3	*	Umbilical-cord blood Navajo Indians	(550)
Nicotinate	μg/ml	4.5±0.1	5.5±0.2	?	No method in paper; SE estimated from data	(572)
Selected Proteins						
Total protein	g/dl	6.7±0.1	5.4±0.1	*	Umbilical-cord blood Navajo Indians electrophoretic values listed	(550)
Albumin	g/dl	3.1±0.1	3.6±0.1	*	C-section; venous sample	(160)
Ferritin	μg/ml	25±6	202±16	*		(548)
Cytoferrin	μg/ml	12±0.7	34±2	*		(548)
Retinol-binding protein	μg/dl	48±3	22±1	*	Radioimmunoassay Navajo Indians	(550)
α-fetoprotein	ng/ml	4.7±3	48,000±35,000	*	Newborn capillary blood; rapid decline in perinatal period; adult values, donor blood	(573)
Complement					Maternal levels approximately 10-fold fetal levels at mid-gestation	(574)
Clotting factors: Factor 8 antigen	%	160±10	39±2	*	Mid-gestation; other clotting factors similar	(575)
IgG	mg/dl	1006±50	1218±35	?	Normal vaginal delivery; immunodiffusion	(576)
IgG1	mg/dl	447±37	748±62	*	IgG_2 and IgG_3 not different; IgG_4 fetal greater than mother	(577)
IgA	mg/dl	198±11	1.2±0.1	*	Normal vaginal delivery; immunodiffusion	(576)

Continued

TABLE 1. *Human maternal and fetal plasma or serum concentrations of some selected substances (near term unless otherwise stated)—cont'd*

Solute	Units	Mean maternal concentration (± SE)	Mean fetal concentration (± SE)	Concentration difference *P*<0.05	Comment	Ref.
IgM	mg/dl	141±14	8.4±1	*	Normal vaginal delivery; immunodiffusion	(576)
IgE	IU/ml	107	0.8	*	Undetectable in 68% of umbilical-cord bloods	(578)
Selected Hormones						
Growth hormone	ng/ml	4.7±0.3	22.2±3.3	*	Umbilical-cord blood; radioimmunoassay; sex difference noted	(579,580)
Somato-mammotrophin	μg/ml	6.34±0.39	0.027±0.003	*	Vaginal delivery; umbilical-cord blood	(581)
Somatostatin	pg/ml	23.1±2.3	27.4±1.8	NS	Umbilical-cord blood radioimmunoassay	(579)
Insulin	μU/ml	10.1±2.1	8.2±1.1	–	C-section; immunoassay not paired	(582)
Glucagon	pg/ml	72.0±8.5	69.4±8.9		C-section; immunoassay not paired	(582)
Cortisol	μg/dl	55.8±3.6	7.0±0.6	*	Radioimmunoassay; C-section; mid-gestation. 11-DOC fetal higher than maternal; other steroids measured	(583,584)
Gastrin	pg/ml	53±2	84±6	*	Umbilical-cord blood radioimmunoassay	(585)
Chorionic gonado-trophin	IU/ml	9.45±0.93	0.027±0.003	*	Vaginal delivery; sex difference noted	(581)
Prolactin	ng/ml	251±35	368±30	*	Vaginal delivery; radioimmunoassay	(586)
Progesterone	nmol/l	245±20	408±35	*	Radioimmunoassay; C-section; umbilical-artery lower	(587)
Estriol	nmol/l	40±4	279±25	*	Umbilical-vein; umbilical-artery lower	(588)
Epinephrine	pg/ml	227±31	770±97	*	C-section; vaginal delivery higher; norepinephrine and dopamine also measured.	(589)

Adapted From Morriss et al. (5).

commonly studied species except sheep; and Benirschke has provided an outstanding Web site (Comparative Placentation http://medicine.ucsd.edu/cpa/). Mossman (11) and Ramsey (12) provide fuller accounts, as does Boyd and Hamilton (13) for the human. Placental anatomy is an issue of importance in understanding the evolution of all vertebrates [e.g., (14)]. In this chapter we restrict ourselves to studies of mammals, and when no species is mentioned, human studies are generally referred to. In addition to anatomy, there are also great differences between species in the synthetic and metabolic capacities of placental tissue [e.g., (1,15,16)]; a systematic review of this topic is long overdue, although there are useful reviews of some aspects of comparative metabolism, e.g., Desoye and Shafrir (17) on the diabetic placenta or Malassiné et al. (18) on placental endocrinology of mouse and human.

Despite anatomical and functional species diversity, there are common conceptual issues that apply to transfer across the placenta of all species. Such issues are considered in this section. In addition, we address the range of methodological approaches that have been applied to the study of maternofetal transfer. The different approaches used have given powerful and useful insights into the understanding of transfer. However, investigations have been limited in number and scope in comparison to the studies of, for example, transfer across the intestinal wall or the renal tubule. Many gaps to understanding remain. Perhaps for this reason, publications reporting results obtained using a specific method are found not infrequently to draw conclusions that ignore issues essential to the overall understanding of transfer in vivo, or at other stages of pregnancy. A difficulty in interpreting findings may

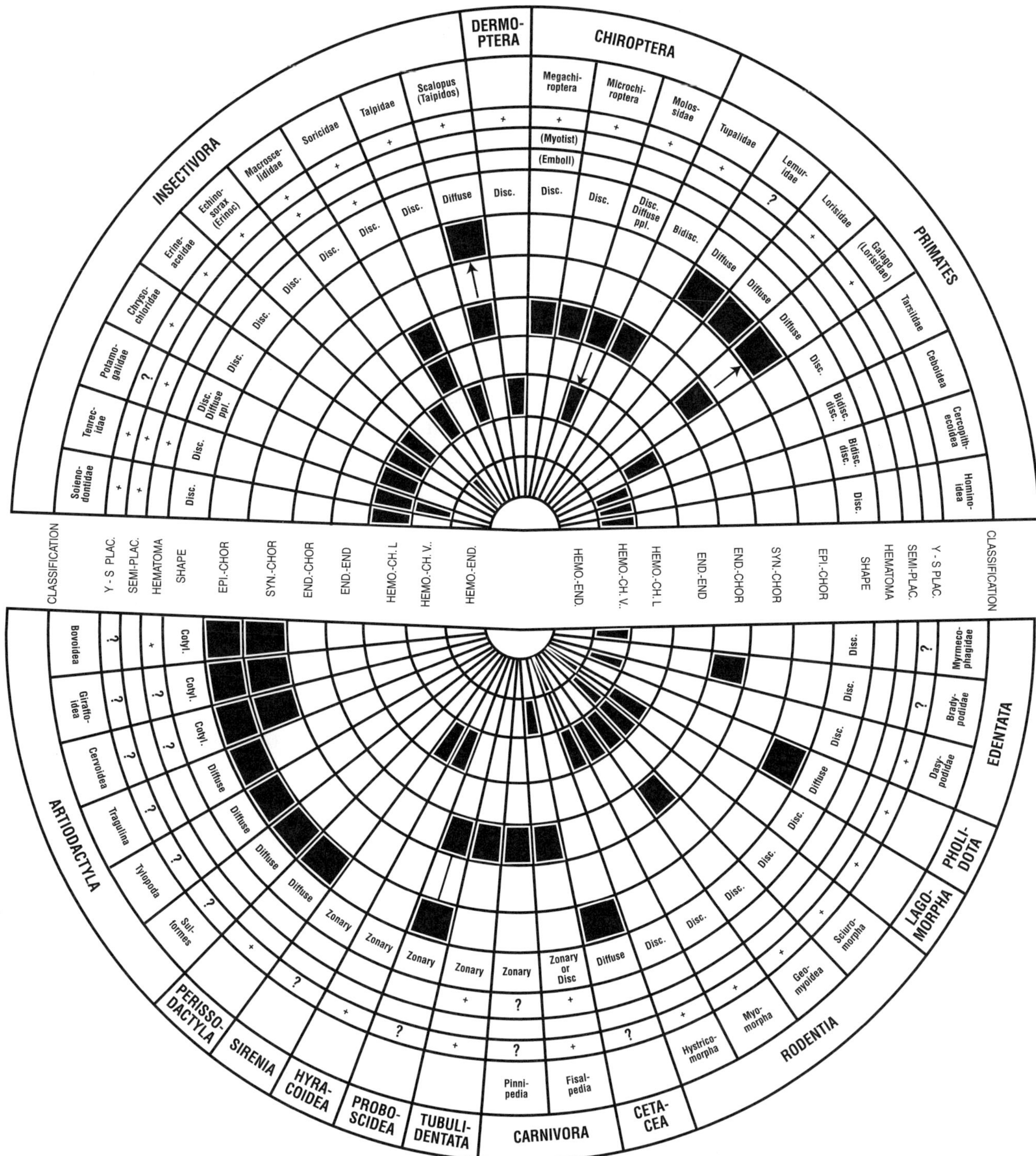

FIG. 1. Interspecies differences in placental structure. A 1962 tabulation of placental characteristics in representative eutherian mammals. All orders are represented and are arranged in systematic sequence in a clockwise direction, beginning with the Insectivora. The *black wedges* designate the definitive organization of the chorioallantoic placental barrier (as per the Grosser classification); two unjoined *shaded wedges* subtending the same arc mean that the indicated placental types have been described in different species within the listed group; *wedges joined by a line* indicate that both conditions prevail within the same placenta; *wedges joined by an arrow* indicate that the condition pointed to is definitive but was preceded at an earlier stage by the other. *Blank areas* generally indicate an absence of the condition indicated; question marks indicate a dubious status or lack of information in cases where the indicated condition might be expected to occur. The diagram does not include all known paraplacental structures. Y-S, yolk-sac; Plac, placenta; Epi, epithelio; Chor, chorial; Syn, Syndesmo; end, endothelio; Ch. L., chorionic labyrinth; Ch. V., chorionic villi. [Redrawn from (670) in (5). A few species have been reassigned; see (11,12,671).]

also be contributed to by the tradition of different sub-disciplines of bioscience using different species in studying placenta.[1]

It follows from the laws of conservation that the molecular and energetic content of the fetus (or conceptus) by term must exactly equal the net amount transferred to it across the placenta (or via other routes). Furthermore, fetal growth rate is not uniform across time. This nonuniformity applies equally but individually differently to the growth of fetal organs such as skeleton, liver, or brain (19), each of which has a different chemical composition. There must thus be a range of control loops linking net transfer of individual solutes to fetal development. The rate-limiting element for transfer of a given "substance" must be modified to match, or perhaps control, fetal development as gestation proceeds. The modification must be either in the transport capacity of placenta (e.g., changes in its surface area for exchange or in the density or behavior of its transporter systems), or in the driving forces across it (e.g., maternofetal plasma water concentration differences at the exchange area, driven in turn by the placental blood flow rates and their composition, or in the hydrostatic or electrical forces across the exchange area). Polarity of transport to or from the fetus leading to net transfers in either direction will either depend on asymmetrical driving forces between the two circulations, and/or on asymmetrical behavior of transporters located in whichever is rate-limiting of the cell membrane lipid bilayers between maternal and fetal blood. Solute movement in response to asymmetrical driving forces may be increased by the presence in this lipid bilayer of unpolarized transport systems which effectively increase permeability, e.g., facilitated transporters (e.g., GLUT I for glucose) or ion channels (e.g., for chloride); or by the presence of shunts/"pores" which bypass the cellular (or syncytial) layers between the two microcirculations. Transporters may behave asymmetrically, that is transport actively, either directly fueled by ATP (e.g., calcium), or by demonstrating co-transport (e.g., certain amino acids whose transporters are driven by the asymmetrical sodium gradient across the cell membrane).

Molecular studies over the last decade have done much to elucidate the presence and behavior of such transporting systems, but understanding of their function is often poorly integrated with studies of transport in vivo or across in vitro systems that maintain placental structural integrity. The presence of a transport system within placenta does not necessarily imply that it contributes importantly, or at all, to *trans*placental transfer. It may translocate internally between different placental cells or intracellular compartments. Alternatively, its function may be quantitatively unimportant compared, for example, to the rate of paracellular transplacental diffusion.

The identification and analysis of *rate-limiting* elements in placental transfer is thus important. Understanding their interplay, their control and their disturbance in abnormal gestation is currently a central concern in placental transfer studies. Control of transfer is limited to nonneural mechanisms. Although a full acetyl choline system (20) and other neurotransmitter receptors [e.g., (21)] are present in placental tissue, a nerve supply to the placenta, although not the subject of detailed recent study, is almost certainly absent (22). In the absence of a nerve supply the organ can only "see" and "sense" the solutes in blood delivered to its maternal and fetal microvascular circulations (or conceivably diffused from the locally adjacent interstitium) and the hydrostatic perfusion pressures to which it is exposed.

The placenta clearly has other functions as well as transfer, i.e., endocrinological, immunological, and metabolic and has a complex developmental cycle. A biological application of Occam's Razor[2] suggests these functions will be ultimately found to be integrated with transfer as well as with each other. Although evidence for such integration is currently scant, the possible involvement of amino acid transporter proteins in trophoblast syncytialization (23), the fact that the cationic amino acid transporter System y^+ is the same protein that acts as a cell-surface receptor for murine ecotropic retrovirus (24), and the possible involvement of tryptophan transport and metabolism in prevention of immunological rejection of the placenta (25), provide hints of shared functions.

Paraplacental Transport

Beyond early gestation, the dominant locus for high volume maternofetal exchange is placenta as

[1]55,645 publications have been listed under "placenta" in PubMed since its inception (673). Of five individual species with over a thousand references, the human accounts for 72% but is under-represented in "transfer" studies (57%). Of the other four, rabbit accounts for 20% of studies on "immunology" but only 10% of those on "infection"; rat for 51% on "hormone" but only 16% on "infection"; sheep 48% of "circulation" but less than 8% on "immunology"; mouse 49% of "infection" but only 14% on "circulation." Of "transfer" studies the mouse is greatly under-, and the sheep over-represented. "Kidney" gives 460,320 publications over the same period! (R. D. H. Boyd, 2004, personal observation.)

[2]The razor can be defined as "Keep it simple stupid"—Chesney's (673) excellent critique of the doctrine does not in our view, when combined with evolutionary considerations, invalidate application of Occam to biological systems.

evidenced by: (a) the high proportion of the microscopic surface area of the interface between mother and fetus which is placental [e.g., (26–28)]: the placental surface area is much less in pig and horse (28) and placental dominance in maternofetal transfer may thus be less in these species; (b) the similar high proportion of total uterine and umbilical blood flow which flows to placenta in sheep (29,30) or, of uterine flow, in monkey (31).

However, at least in some species, some permeants, especially those transferring in quantitatively small amounts, pass predominantly by routes other than through the chorioallantoic placenta. In rodents and rabbits (32) the overwhelmingly dominant route for maternofetal antibody supply is via a separate chorio-vitelline (yolk sac) placenta. Similarly, in several species iron (33) and, in the horse, iron and calcium (34) are supplied to the fetus via a range of paraplacental routes and mechanisms. Furthermore, the conceptus includes, in addition to the fetus itself, one or more extrafetal fluid compartments (amnion, allantois, yolk sac, extraembryonic coelom) surrounded by their individual paraplacental fetal membranes. Exchange between these compartments and the maternal circulation, or maternal interstitial fluid may not be limited by the same barrier as that between fetal and maternal circulations. For example, at least from mid-gestation to term, amniotic fluid protein in humans is, surprisingly, overwhelmingly of maternal rather than fetal in origin (35), implying a route from mother to amnion for protein transfer predominantly across extraplacental fetal membranes rather than across placenta. Transfer routes may be different at different stages of gestation, e.g., extraembryonic coelom may play a role in early human gestation (36).

Concepts of Transfer

At its simplest, placental transfer can be conceptualized as a three-compartment model of two blood pools separated by a barrier (Fig. 2). Such a model has two important properties.

First, net transfer from maternal blood to fetal blood per unit time (J_{net}) is the arithmetical difference between two unidirectional transfer rates (fluxes): that from the maternal blood compartment to the fetal (J_{mf}) and that from fetal to maternal (J_{fm}). The conceptual distinction between J_{net} and J_{mf} has been clear at least since the first use of tracer isotopes to study placental transfer by Flexner [see excellent summary by Faber (37)]. It is a critical distinction, as J_{net} and J_{mf} may differ in quantity by orders of magnitude [e.g., 50-fold for sodium transfer in the rabbit at mid-gestation (38)].

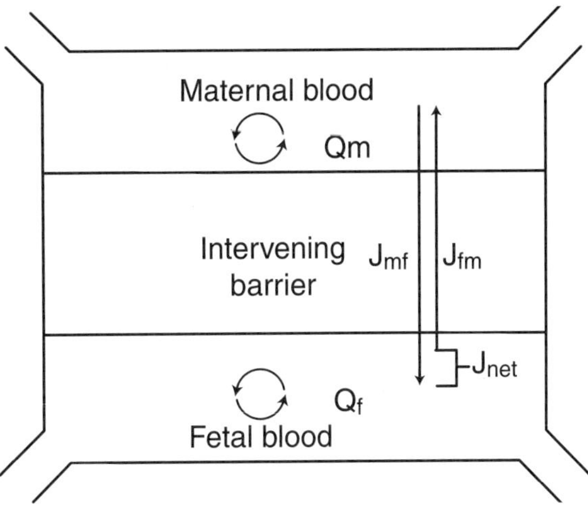

FIG. 2. Three compartment model of placental transfer: maternal blood circulates at a rate Q_m (ml/min-1) through the maternal placental microcirculation (in woman the intervillous space). Circulation through the fetal placental capillary microcirculation is Qf (ml/min-1). At a steady state, the net flux (rate of transfer) to the fetus J_{net} (M/min-1) = $J_{mf} - J_{fm}$ where J_{mf} and J_{fm} are, respectively, the unidirectional fluxes from mother to fetus and fetus to mother. Where the value of J_{fm} exceeds J_{mf}, net flux is from fetus to mother and J_{net} has a negative value (e.g., for fetal excretory products).

Second, the "stirring" of the two blood compartments (a function of the relative flow rates Q_m and Q_f and also of their relative geometries) is important. Transfer rate will depend, for a given permeant, on the balance between the rate of its movement across the barrier and the rate of its delivery or removal from the blood pools by the circulation. If, for a given permeant substance, the flow delivery is relatively low and the barrier permeability relatively high, then the permeant will approach equal concentration in the two microcirculations during a single pass.[3] The transfer of such a permeant will be limited by the rate of blood flows (Q_m and Q_f), the available blood concentrations of the permeant and the geometry of the two flows. Transfer under these conditions is described as being "flow limited." If, by contrast, barrier permeability is relatively low in relation to flow rate, transfer is described as, "diffusion limited"—"barrier limited" might be a better term. There are intermediate cases in which both flow and barrier limitation play a part (39).

Figure 2 is over-simplified but before considering a fuller model, it may be helpful to summarize the broad experimental approaches that have been applied. Even though in vivo human work is limited

[3]How close the concentrations in the maternal and fetal venous outflows approach each other will depend on whether the two flows are concurrent, countercurrent, or follow a more complicated arrangement. Species vary; the subject of an extensive older literature, see section Geometry of Placental Blood Flow.

by ethical concerns, a great strength of human placental studies is, in contrast to work with other human organs, the plentiful supply of material! Experimental approaches that have been used to study the human placenta are listed in Table 2.

Animal study allows more flexibility. The great advantage of large animals is the ease of surgery associated with the ability, through indwelling catheters, to allow sampling from maternal and fetal circulations in the undisturbed state. These techniques have been most powerfully applied to the sheep. Most of the methods listed in Table 2 have also been applied to smaller laboratory animals and some relatively undisturbed studies have also been achieved in them [e.g., (40)], but fetal blood sampling is difficult and changes in solute concentrations during

TABLE 2. *Some methods of placental transfer study applied to the human*

Method	Variable measured	Details	Comment	Example Refs.
Carcass analysis	J_{net}[a]	Analysis of conceptus content at different gestational ages	Only reliable for nometabolized solutes, e.g., Ca. Includes all maternofetal routes. Poor time resolution	(590)
Analysis of experiments of Nature	Evidence of placental impermeability or dysfunction		e.g., hypothyroid or albuminemic fetus; can pose questions on control systems, e.g., hydrops	
Accumulation of exogenous tracer[c]	J_{mf}[b]	Fetal tracer content estimated after delivery	Any metabolism of tracer will complicate analysis	(41,591,592, 564)
Imaging techniques	Solute accumulation in vivo		Potentially interesting; undisturbed fetus (monkey)	(593)
Umbilical AV difference at delivery	J_{net} (J_{mf}[b] if exogenous tracer)	Cord clamped at cesarean section	Non–steady-state artefact likely	(5)
Cordocentesis A-V difference	Maternofetal plasma concentration difference		Mid-gestation; minimal disturbance	(299,314,594)
Sampling embryonic fluid compartments	Lower limit of J_{mf}	Concentration differences or tracer uptake	Possible in early gestation; sum of all routes into embryo	(595)
Cotyledonary perfusion	J_{mf} J_{fm}	Steady-state transfer	Perfusion might alter passive permeability and thus wrongly estimate proportion of specific transport	(596)
Cotyledonary perfusion	Kinetics of uptake and back flux from maternal and fetal circulation	Single pass, indicator dilution	Can demonstrate transport asymmetry. May not reflect rate-limiting element of transfer	(284)
Single villi or villous clumps	Uptake or efflux or electrophysiology		Maintains cell polarity; available through gestation	(62)
Slices; unoriented fragments	Kinetics of individual systems	Uptake or efflux studies	May not be quantitatively important transplacental transport systems	(303,597,598)
Membrane vesicles	Kinetics of individual systems	Microvillous (maternal facing), basal (fetal facing)	May not be transplacental transport systems.Normalization is per mg membrane protein which may not reflect surface area	(146,317,373)
Cultured trophoblast	Uptake or efflux	Primary, differentiating to syncytium, monolayers	Confirmation of cell type important	(599,600)
Receptor studies	Affinity/number/rate constant	Over 25 receptors described	May have immune, metabolic or endocrine as well as transport functions	(601)
Microelectrode and patch clamp studies	Transmembrane PD and ion channel characteristics		Important for ion movements	(62,168,180, 602–604)
Molecular studies	Transport molecules and systems	Expression and reconstitution; sequences and structure	Detailed behavior of transport systems in vitro	See text

[a]See Equation in legend to Fig. 2.
[b]If fetal plasma concentration not close to zero, a hybrid of J_{net} and J_{mf} is measured.
[c]A tracer is any identifiable exogenous solute.
The presence of a transport system within placenta does not, in the absence of other evidence, demonstrate that it plays a role in maternofetal transfer. It may have another function. Conversely, for solutes at high concentration in placenta, the absence of demonstrated transfer in short-term tracer experiments does not exclude the possibility of transfer delayed by the tracer equilibrating with a "middle compartment" containing the solute.
Modified from Sibley and Boyd (149).

sampling may lead to errors of interpretation. However, the easy repeatability of studies is a major advantage of small animals, while the study of knockin and knockout mice is already providing exciting results (41,42).

Common Animal Models of Transfer

Species Differences

Although a wide range of species has contributed to studies of placental transfer, the most coherent bodies of recent evidence have perhaps come from three: human (and closely related primates), sheep, and rat. Their main similarities and differences are summarized in Table 3. The same conceptual issues apply to transfer in each, but it should be clear from the table that quantitative conclusions drawn from one species (e.g., studies of drug kinetics) are likely to be misleading if applied without caution to another.

It can be seen that the transfer mechanisms across placental tissues for solutes that show a high extraction from the uterine circulation, such as glucose or amino acids, are very similar in all three species and indeed in others. The same appears to be true for respiratory gases, which are also highly extracted. By contrast, very different mechanisms are found for the range of solutes that, while important to the fetus or newborn, have a much lower net transfer rate. Iron or immunoglobulin G (IgG) are good examples. Quantitative data suggest that placental transfer mechanisms in the rat and human more closely resemble each other (43) than those of the sheep, which is generally much less permeant. On the other hand, the rate of change in transfer capacity (e.g., for Na^+ or Ca^{2+}) over time in late gestation is much greater in the rodent and other small mammals than in the human. This reflects their much greater daily increase in fetal weight as a proportion of maternal weight as gestation proceeds. Integration between fetal need and placental transfer in small species may thus be predicted to depend on more precise mechanisms than those in human pregnancy.

A Comprehensive Model of Placental Transfer

A full analysis of placental transfer requires a much more complex model than the three compartments of Fig. 2.

For any permeant to cross the placenta it has to traverse, or bypass, several lipid bilayer barriers and several fluid compartments. For some, transfer may also involve release from or uptake by blood cells, plasma complexes, or carrier proteins; or metabolic interconversion within placenta.

The flux (rate of transfer) through each of the barriers is a function of both the ease of movement through it (its permeability) and on the difference in relevant electrical, chemical, hydrostatic, or thermal driving force across it. Movement through the fluid compartments is likely to be a function of diffusion and convection and is usually assumed not to be rate-limiting. Overall, the rate will depend on the perfused surface area of the placenta. Figure 3 illustrates, using human placenta as an example, the broad anatomical

TABLE 3. *Some similarities and differences in nongaseous transfer by human, rat, and sheep placenta in the last third of gestation*

	Amino acid	Glucose	Cl^-J_{mf} (nM/min/g placenta)[a]	Cl^-J_{net} (nM/min/g placenta)	Cl^- Approx. ratio J_{mf}/J_{net}	Aqueous equivalent pore radius nM[b]	IgG	Iron
Human	Numerous transporters: overall active concentrative	Facilitated diffusion GLUT I	3,800	0.8	5,000	~2.5	Placental concentrative transfer	Placental endocytosis from transferrin
Rat	Numerous transporters: overall active concentrative	Facilitated diffusion GLUT I	7,400	11	700	~3.3	(Yolk sac)	From transferrin
Sheep	Numerous transporters: overall active concentrative	Facilitated diffusion GLUT I	320	1.7	200	~0.44	– (Postnatal gut)	– Hemophagus organ-ingestion from red cells

–, nontransferred.
[a]From Boyd et al. (94).
[b]From Faber, quoted in (88) and Boyd et al. (89).

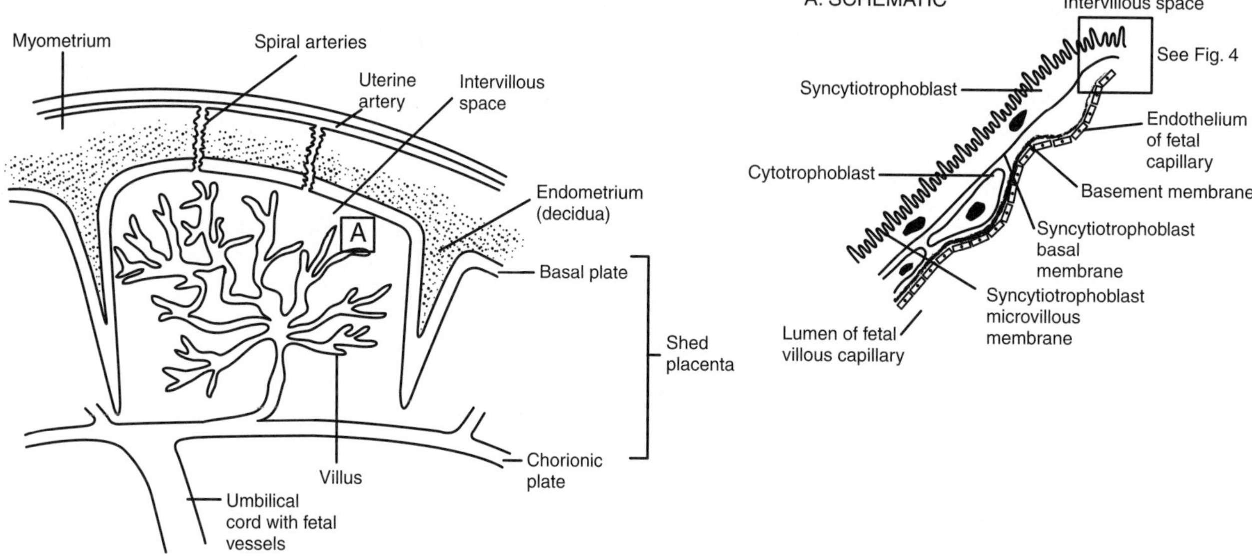

FIG. 3. Human anatomical context of transfer. Schematic cross-section through the human placenta. Maternal blood enters the intervillous space via the spiral arteries; fetal blood enters via the umbilical artery. *Insert A* schematic shows a more detailed section of a single villus. The maternal and fetal circulations are separated by the microvillous and basal membranes of the villous syncytiotrophoblast, the basement membrane, and the fetal capillary endothelium. Together these layers constitute the placental "barrier" to transfer in the human placenta. The section outlined in the schematic is shown in more detail in Fig. 4. [Adapted from (5).]

context of transfer; while Figs. 4 and 5 indicate something of the complexity of the detailed transfer process. Bear in mind that for any given permeant at a given time, only one element is likely to be *rate-limiting* for its transfer. That element will be the likely control point for matching changing fetal need to placental transfer across gestation (see section Ontogeny of Placental Transfer Mechanisms). It will also be the element in which a modest change due to disease is likely to have the biggest impact, while a major change may alter which element is, indeed, rate-limiting. The number of variables in Figs. 4 and 5 and, in many cases, the uncertain measurement of their values, puts the goal of achieving an accurate and comprehensive quantitative model of how the different aspects of placental transport are integrated beyond reach at the present time. It is worth emphasizing that while most work in recent years on membrane limited permeants has concentrated on movement across syncytiotrophoblast, the assumption that it is rate-limiting for all of them may be in error. The endothelium may be a more important barrier for some (44).

Driving Forces

The rates of transfer through the elements listed in the legend to Fig. 4 will depend on the interaction between driving force and the element in question. Cardiac function and the contents of blood ultimately provide the driving forces but, at the local level, rates depend as appropriate, on local energy supply as ATP and on local concentration differences (ΔC); on hydrostatic and osmotic pressure differences (ΔP and $\Delta\pi$) for water and entrained solutes; and on electrical potential difference ($\Delta\psi$) for ions.

These different forces inevitably interact. For example, several amino acids share transporters. An increase in the concentration of one will compete with another and thus increase the driving force required to maintain its flux at the same rate (see section on Amino Acids). Similarly, a change in transmembrane potential difference (PD) will differentially affect the movement of anions and cations into syncytiotrophoblast. Understanding the cross-talk between the transport of different solutes, whether into or out of syncytiotrophoblast or across placenta as a whole, is currently very limited.

ONTOGENY OF PLACENTAL TRANSFER MECHANISMS

The capacity of the placenta to transfer solutes and water has to be sufficient to allow fetal growth. That flux is matched to fetal demand has been illustrated recently in the *Igf2* knockout mouse (41). Evidence that this may also be true in the human

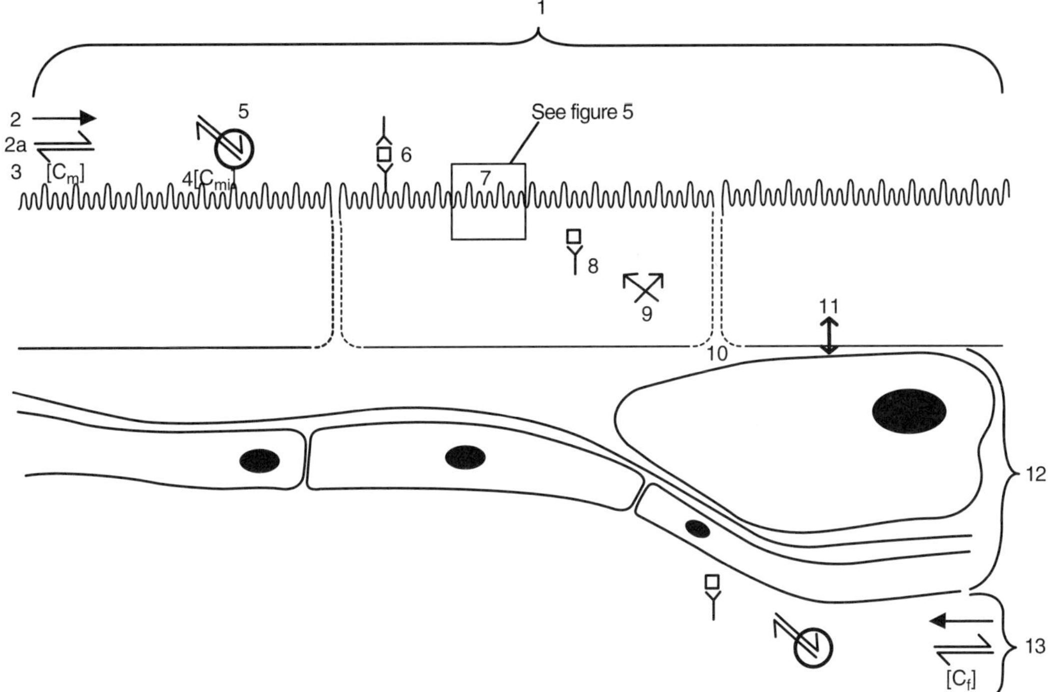

FIG. 4. Elements which may contribute to the overall rate of transfer of different substances in either direction across human placenta. Which elements are quantitatively dominant will be different for different permeants (examples are given in brackets). **1.** Total effective surface area available for exchange (all permeants). The effective surface area is a function of both structure and of whether blood flow on both sides is maintained. **2.** Blood flow rate and (2a) direction and geometry of blood flow (e.g., flow limited permeants such as anesthetic gases). **3.** Concentration (108) in whole blood entering placenta (flow limited permeants). **4.** (Cm) concentration in unstirred plasma water layer adjacent to microvillous membrane (most permeants). Relevant variable is more strictly defined as activity rather than concentration. **5.** Rate of transfer between red cell interior and plasma water (e.g., respiratory gases). **6.** Rate of transfer for bound solutes between plasma carrier and plasma membrane receptor; and release from receptor (e.g., iron). **7.** Mechanisms and rate of transfer across syncytial microvillous membrane from plasma water adjacent to it into placental syncytiosol (see Fig. 5). **8.** Binding molecules putatively involved in movement across syncytiosol (e.g., 9K calcium-binding protein). **9.** Intraplacental metabolic interconversion or catabolism (e.g., glucose, lactate, steroid hormones). **10.** Aqueous parasyncytial pathway across placenta (e.g., a high proportion of salt and water transfer). **11.** A range of transport mechanisms in basal syncytial membrane, probably comparable to those listed in 7. **12.** Transfer across basal lamina, villous core, fetal capillary endothelium including exposure to a range of intravillous cell types including, in earlier gestation, a cytotrophoblast layer. (Importance uncertain but likely to be significant at least for immune function.) **13.** Transfer between placenta and fetal blood and its constituents and involving any or all of elements 1–6 above. These inverse elements on fetal "side" of placenta may, of course, have different properties from their maternal equivalents (e.g., different fetal and maternal red blood cell affinity for O_2).

comes from in vitro data showing that the activity per mg of membrane protein of the System A amino acid transporter on the microvillous plasma membrane is related to birth weight (45–47). Absolute fetal growth rate changes over gestation: it is very slow in first trimester then accelerates in the second trimester to reach the high rate of the third trimester (48) [a rate that is relatively higher in smaller species (49)]. Therefore, it is likely that the transfer capacity of the placenta also changes during the course of pregnancy. Gestational changes in the variables contributing to net flux indicated in Fig. 4 need to be examined in this regard.

Blood Flow

It is now well established that blood flow to the intervillous space of the human placenta does not begin until 8–10 weeks when trophoblast plugging the spiral arteries breaks free (see Chapter 5). It is possible that before this time, nutrients are extracted from the plasma filtrate that does reach the intervillous space and that protein secretions from the uterine glands also contribute in a "histiotrophic" form of nutrition. The fetal circulation of the placenta becomes patent at about 6 weeks (see Chapter 5). It is possible that the yolk sac, which

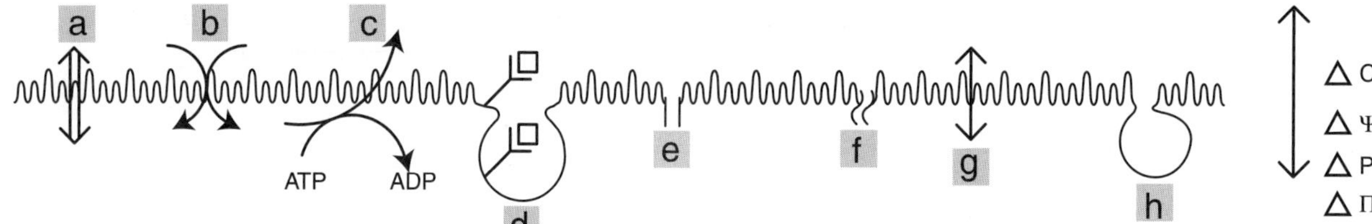

FIG. 5. Transfer mechanisms across syncytial microvillous membrane. (a) Facilitated diffusion (e.g., glucose); (b) co-transport (e.g., alanine and sodium); (c) ATP-driven active transport (e.g., Na^+ extrusion by $3Na^+2K^+$ exchange); (d) receptor mediated endocytosis of receptor bound solute (e.g., iron, IGG); (e) ion channels (e.g., Cl^-); (f) aquaporin (e.g., water); (g) presumed lipophilic diffusion (e.g., respiratory gases or anesthetic gases); (h) nonreceptor mediated endocytosis (e.g., perhaps some nonimmunoglobulin proteins). ΔC, concentration differences; ΔΠ, osmotic pressure differences, ΔP, hydrostatic pressure differences; ΔΨ, electrical potential difference; all as expressed across the microvillous membrane.

is bathed in coelomic fluid on the fetal side of the chorion, is functional in absorbing solute in the human up until this time (see Chapter 5). In other species, such as rodents, the yolk sac certainly has an important role in nutrition of the early embryo/fetus and for at least immunoglobulins is important until term (50). The human yolk sac in contrast degenerates and disappears by late first trimester (13). By the time the fetus has attained its maximal growth rate in second trimester, both uterine and umbilical blood flows are fully developed and are able to deliver and extract, e.g., oxygen in the required amounts.

Exchange Barrier Dimensions

In contrast to the fetus, the human placenta grows rapidly in the first trimester and has reached its maximum size by about 12 weeks (13). Surface area for exchange is also increased by the expansion of the villi, formation of terminal villi, and increased number of microvilli on the maternal facing plasma membrane of the syncytiotrophoblast over the course of gestation (51,52). Furthermore, the harmonic mean thickness of the barrier as estimated from fixed material, decreases over gestation (53). The permeability of the rat placenta to hydrophilic solutes (mannitol, CrEDTA, inulin) increases over gestation, as would be expected for increases in exchange surface area and decreased thickness (54), although Faber takes a contrary view (55). Recent data show that there are changes in the dimensions of the mouse placenta over the last half of gestation (56) and that these are associated with a similar increase in the permeability of the placenta to mannitol, CrEDTA, and inulin. Furthermore, knockout of the placental specific transcript of *Igf2* in the mouse leads to a decrease in placental weight with reduced surface area of the labyrinth exchange area, an increase in the estimated harmonic mean thickness of the barrier and a strongly correlated decrease in permeability to diffusional markers, as well as fetuses which are 30% smaller than wild type (42). Interestingly, similar changes in placental barrier surface area and thickness occur in intrauterine growth restriction in humans (57). These data demonstrate the importance of the normal changes in exchange barrier dimensions and permeability for adequate placental transfer capacity, as well as showing that *Igf2* is a key regulator of these structural and functional developments.

Concentration and Electrical Gradients

Gestational changes do occur in the maternal and fetal plasma concentrations of solutes, so affecting driving forces. For example, glucose and amino acid concentrations in maternal plasma increase over gestation, at least partly due to the effects of insulin resistance in pregnancy and of hormones such as human placental lactogen (see Chapter 54). A maternal-fetal electrical potential difference, if expressed across the placental exchange barrier will have an effect on the exchange of ions. There is data in rat, goat, sheep, human, and probably guinea pig (58) that the PD between mother and fetus (PD_{mf}) declines over the latter part of gestation. In the human PD_{mf} is small in value, being -2.7 ± 0.4 mV fetus negative in mid-gestation (59) and zero or close to it at term (60). A PD between the mother and coelomic cavity was measured in first trimester pregnancies (61). There was no apparent change in the value of this PD (8.7 ± 1 mV fetus negative) between 9 and 13 weeks of gestation, which was somewhat higher than the PD across the syncytiotrophoblast (3 mV villous core negative but with a wide range) measured using microelectrodes, in term placental villi in vitro (62). The in vitro PD across the microvillous membrane of human syncytiotrophoblast decreases between early (median –32 mV) and late first trimester (median –24 mV), with only a small subsequent fall to term (–21 mV) (63). This suggests that the driving force for cation flux into the syncytiotrophoblast decreases and for anions increases as pregnancy progresses.

The relationship between these various potential differences and how they alter over pregnancy is an area that still requires much attention [fully reviewed in (64)]. In sheep (65) and guinea pig (66,67), the endometrium rather than the placenta appears to be the site at which the PD_{mf} is generated, as PD across the placental exchange area (PD_{plac}) is probably small, though in the sheep at least, not zero (68). In rat (69) or pig (70), by contrast, PD_{mf} and PD_{plac} appear to be closely related and PD_{mf} is probably generated by and expressed at the exchange area (64) across which gestational changes in PD are likely to both influence and be influenced by ion fluxes.

Transporter-Mediated Transfer

Gestational changes in the flux of solutes through transporter-mediated processes could result from alterations in the number of transporters in each plasma membrane, or in their behavior, such as their turnover (i.e., rate of binding to and release from the transporter) or their affinity for solute, as well as from changes in the driving forces acting on them such as solute concentrations and electrochemical gradients or ATP availability. There is a variety of evidence that such changes do occur, best documented in relation to the microvillous plasma membrane of the syncytiotrophoblast.

Using the isolated microvillous membrane vesicle technique it has been shown that the V_{max} of the Na^+-dependent system A amino acid transporter (transports small neutral amino acids such as glycine) increases by about fourfold, per mg membrane protein, between first trimester and term (71). The activity of the System y^+ cationic amino acid transporter increases over gestation, whereas the activity of the System y^+L transporter decreases (72,73). This decrease in System y^+L activity is due to a decrease in the affinity of the transporter for substrate and is accompanied by an increased expression of 4F2hc monomer of the dimer protein (72,73). The reason for this decline is not known but could well be associated with a specific fetal need. Glucose transporter, GLUT1, expression in microvillous membrane increases between first trimester and term (74) Na^+/H^+ exchanger activity is lower in first trimester microvillous membrane vesicles as compared to term (71,75), a result borne out by studies on the intrasyncytiotrophoblast pH of isolated placental villi from the two stages in gestation (76). Interestingly, the expression of the NHE1 isoform of this exchanger in the microvillous membrane does not change across gestation, but the expression of both of its NHE2 and NHE3 isoforms increases between weeks 14–18 and term (75). In contrast there is no difference in Cl^-/HCO_3^- exchanger activity or, by Western blotting, expression of its AE1 isoform, between first trimester and term (77).

These data show that there are highly selective changes in transporter expression and activity in the microvillous membrane over gestation: increase, decrease, and no change, dependent on the particular transporter. The regulation of these gestational changes is not yet understood. This may partly be through hormones such as IGF-II from the fetus signaling demand for the nutrients required for growth (78). There may be regulation in response to fetal needs at particular times of gestation. For example, there is a huge increase in the requirement of the fetus for calcium in the late third trimester when bone is being formed (79,80). In the rat, it has been shown that placental mRNA expression of calbindin-D_{9k}, a protein which probably plays a key role in calcium shuttling across the cytoplasmic compartment, increases 18- to 20-fold late in gestation, simultaneously with a rise in maternal-fetal calcium flux; this might be evidence of signaling between fetal bone and placenta, although the nature of the signal is unknown (see section Concentration and Electrical Gradients).

TRANSPORT OF SELECTED SUBSTANCES

Water

As was pointed out long ago by Barcroft (19), the fetus in late pregnancy, even without allowing for the fluid in extra-fetal compartments, requires a net flux to it of more moles of water than of any other substance, exceeding several-fold even the oxygen requirement. Understanding water movement is also central to understanding placental transfer as its filtration under hydrostatic or osmotic forces can lead to entrainment of solutes.

Diffusion of water across placenta is high as evidenced by compartmental modelling of unidirectional fluxes of labeled water (81,82). Direct measurement, e.g., in sheep, shows that water probably comes into equilibrium between the maternal and fetal microcirculations in a single pass (39). Net water movements between the circulations will thus depend on the driving forces between the circulations. The impact of water movement on transfer of plasma solutes will depend on the route(s) through which water moves and particularly on the existence or otherwise of transplacental paracellular channels.

Paracellular Channels

There is strong functional evidence for the existence of an aqueous paracellular route across placenta (83) as inert, unmetabolized molecules (e.g., L-glucose, mannitol, Cr-EDTA) believed to be excluded by the absence of specific transporters from the placental intracellular space nevertheless cross the placenta. Permeability toward such solutes is not saturable or competitively inhibitable, providing evidence that specific transporters for them are indeed absent. In hemochorial species placental permeability for these substances is a function of their free diffusion coefficient in water. Permeability to larger molecules such as albumin is proportionately less, showing "steric hindrance." The morphological site of paracellular pores is uncertain but, in human, may be at the site of so-called trophoblastic denudations (84–86).

The "pore radius" for presumed paracellular channels in hemochorial species ranges from 1.7 nM in rabbit to 3.3 nM in rat, with humans probably around 2.5 nM according to Faber (88). Stulc's estimates are somewhat smaller (83). There is evidence of change in radius through gestation (54) not confirmed by Faber (88), who, however, made the possibly insecure assumption that all water flux was paracellular. Interestingly, Stulc (83) calculates that hemochorial placentas, despite having paracellular pores of radius comparable to that of many capillary beds, have an overall paracellular permeability 10^{-2} to 10^{-3} that of capillary, as the pore area per unit surface area is (assuming an identical pore length) much smaller.

By contrast, the placenta of nonhemochorial species, such as the sheep, is very restrictive to the transplacental diffusion of small solutes (e.g., Cr-EDTA) and its maximum pore radius appears to be at least an order of magnitude less than in hemochorial species (89–91). In passing, this striking difference is the main reason why the use of the sheep for drug disposition studies, though fairly frequent is likely to be useless, at least for hydrophilic compounds, in estimating their quantitative kinetics and availability to the human fetus. On the basis of its permeability to sodium (92), it can be assumed that the pig has channels with an even smaller pore radius than sheep or fewer of them if indeed it has paracellular channels at all. Dogs and cats may be inferred from very early data to be intermediate in paracellular permeability (93). The existence of a paracellular route through channels or "pores" is widely but not universally accepted (94) and there is no doubt that the concept of transplacental paracellular transfer does have some important weaknesses. For example, the pore radius will be overestimated, perhaps grossly, if intracellular exclusion of supposed extracellular permeants is not complete. For example, chloride, often used as an extracellular marker, and which has a high placental permeability in hemochorial species, surprisingly was found to move predominantly transcellularly across rat placenta as judged by inhibition studies (95). There may also be more than one rate-limiting pore radius in series in any paracellular route. The discovery of aquaporins, which may increase permeability to solutes as well as to water, further complicates the issue.

The apparent discrepancy between functional evidence for pores large enough to show no steric hindrance for small molecules crossing the placenta of hemochorial species, contrasted with their placenta's ability to maintain a concentration difference, fetus higher, for such small solutes as amino acids or ionized calcium, is a central challenge to understanding placental transport. Either there is a flaw in the interpretation of the data underpinning the concept of pores, or, the placenta indulges in energetically wasteful maternofetal transport of such solutes, allowing a back flux through the pores of excess supply (83).

Aquaporins

Aquaporins are membrane proteins that facilitate the transmembrane movement of water. Several aquaporins are present in placenta. In the human, AQP8 is present in chorioamnion (96) and placenta and AQP3 and AQP9 are found in microvillous membrane, where they may contribute not only to its water but also perhaps to urea, purine, and pyrimidine permeability (97). Aquaporins are also present in trophoblasts in early embryonic stages (98). AQP1 is found on the chorionic membranes separating the fetal side of the placenta from the amnion (99). The changing water permeability with gestation of human microvillous and basal membranes (100) presumably depends on a variable presence of aquaporins. Fetal development is, however, normal as judged by the birth of a Mendelian ratio of null neonates in at least AQ4 and AQ2 (101,102) null mice. AQP3 and AQP8 mRNA are expressed in the trophoblast and AQP1 in the endothelium of sheep, with changing expression with gestational age (103).

Stulc (83) suggests that there are three paths for water and small hydrophilic solutes across sheep placenta: paracellular, a transcellular porous route which is probably aquaporin dependent, and diffusion through the lipid bilayer of cell membrane as a function of lipid solubility (104).

Control

Regardless of route of permeation, the control of net water transfer is generally assumed to be a function of the difference between the maternal and fetal placental microcirculations in hydrostatic and in osmotic pressures. The osmotic pressure gradients exerted by plasma solute concentration differences across placenta will depend on the reflection coefficient of those solutes; which in turn is a function of pore radius. In osmotic terms, only macromolecular (e.g., protein) concentration differences are thought to be important for hemochorial placentas with their wide pore radii, while small molecules (e.g., ions) will be important in sheep. With regard to hydrostatic pressures, Faber (88) points out that fetal hydration could be controlled by a causal link between fetal cardiovascular function and water transfer through manipulation of placental capillary pressure. However, the situation may be more complex. First, vascular control mechanisms may be different in different species; most of the pressure studies have been done in sheep. Second, very local asymmetries of extracellular solute concentration within placenta may be important in driving water flows, in keeping with the observation that movement of the, apparently untransported, extracellular solute, Cr-EDTA is asymmetrical across the rat placenta (105,106), although Faber (107) interprets the data differently. Similar asymmetry is found in nonplacental circulations (108,109). Kirksey et al.'s old data, (110) demonstrating that in rat fetal plasma and total sodium concentrations are maintained normal in the presence of gross maternal hyponatremia and without fetal dehydration, also suggests that salt and water balance can be controlled at placental level in ways which the simple models do not explain. If they are, this is not always the case, as the lambs of ewes rendered hyponatremic by salivary drainage also become hyponatremic (111); and in the Dancis and Springer study (112), fetal rats were not protected from hyponatremia, although they were substantially protected from hypokalemia when the mother was hypokalemic.

Control of conceptus hydration may not be limited to the control of transplacental water movement. There is a complex interplay of water flows between fetus, mother, and amnion (and in sheep allantois) very well reviewed in a workshop report (87). In sheep, net inflow to amnion is from fetal urine (25% of fetal body weight per day) and lung liquid (5%), while outflow from it occurs through fetal swallowing (20%) and through an important but relatively recently identified direct "intramembranous" flow from amnion to fetal vasculature (10%) (87). In this species, the filtration of water directly between amnion and maternal circulation is sufficiently low that it is immeasurable unless the placenta is experimentally excluded (113). Transplacental maternofetal water movement predominates. In hemochorial species, notably the human, the same assumption is generally made (87), although early compartmental studies (82) suggested that, in early gestation, unidirectional exchanges between amnion and mother are equal to those between mother and fetus via the placenta; the former fall markedly in relation to the latter in later pregnancy.

Protein

With the exception of IgG [and retinol binding protein in the rhesus monkey, (114)], transfer of protein is highly restricted by the placentas of all species. Thus, the human placenta at term is sufficiently restrictive to protein transfer to maintain concentrations of IgA, hCG, or hPL more than 100 times higher in maternal than fetal serum, while alpha-feta protein (AFP) is 500 times lower (115). Earlier in gestation, the fetal serum AFP concentration may be over 10,000 times lower than maternal (116). Nevertheless, a positive correlation is seen between fetal and maternal serum concentrations of these proteins supporting the existence of some quantitatively small permeability to them (115), as demonstrated in earlier kinetic studies using radio-iodinated proteins (117). The presence of normal amniotic fluid AFP concentrations in the majority of cases of raised maternal serum AFP (118), and the presence of more placental infarcts (119) and ultrasound abnormalities in the placenta (120–122) suggests increased placental permeability to AFP in such cases, although this was not borne out by direct study of the perfused shed placenta (86). Permeability to the polysaccharide molecule inulin appears to be only moderately restricted by human (123,124) or rat (54) placenta. Nevertheless, the relatively small macromolecule insulin does show a significant maternofetal concentration difference, fetus lower, in human pregnancy (125) and permeability toward it may depend on transport as an immune complex (126). On the basis of its apparently much smaller pore radius (89), sheep placenta is likely to have low or zero protein, inulin, or insulin permeability.

Placental handling of IgG shows a striking species diversity. In the human, a series of kinetic studies using radio-iodinated proteins (116,127) found IgG transfer from mid-gestation to be much more rapid than that of albumin or other immunoglobulins. By term, fetal plasma total IgG concentrations generally exceed maternal. This is especially true of IgG subclass 1; fetal plasma IgG2 is usually found to be lower than maternal (128). IgG2 is also transferred more slowly across the perfused placental lobule (129). In rabbit (130) and rat (32) IgG transfer occurs via the yolk sac rather than the chorioallantoic placenta. An IgG receptor, FcRn, found in both human placental syncytiotrophoblast (131,132) and (133) and rat and mouse yolk sac (134,135) appears to play a key role in IgG uptake; mutations which impair IgG binding prevent transfer across the perfused placenta (136,137). As emphasized by Junghans (138), binding to the receptor also prevents IgG catabolism. As the receptor only binds at acid pH, it seems probable that IgG enters the syncytiotrophoblast by fluid phase endocytosis, as demonstrated by its co-localization with horseradish peroxidase (139), after which local acidification is hypothesized to allow IgG binding and protection from catabolism (140). After transfer across the syncytiotrophoblast and release from the basal membrane (possibly as a result of exposure of antibody receptor complex to less acidic pH), IgG appears to cross the fetal endothelium transcellularly (139,141,142). Lyden et al. (143) have identified a different Fc receptor (FCgRIIb2) that could play a role in this process, although Simister (140) points out that its antibody affinity is inappropriate. He also suggests that Fc receptors on stromal cells in the villi may play a role in reducing transfer of immune complexes to the fetus, although this does not appear to be entirely prevented (144). Rats and mice obtain IgG in utero and also postnatally through the gut, involving the same receptor as on yolk sac (138). In sheep, antibody transfer is restricted solely to the neonatal gut, which is very briefly extremely permeant to macromolecules including IgG enormously concentrated in sheep colostrum (138).

As mentioned in section Paraplacental Transport, proteins are also found in amniotic fluid where they are predominantly of maternal origin (35,127). The mechanism of materno–amniotic transfer is unknown but appears to be an inverse function of molecular size, suggesting that the process is subject to steric restriction. The largest proteins are excluded (127). Amniotic fluid proteins are swallowed by the fetus and digested and may therefore make a contribution to fetal amino acid requirements (145).

Inorganic Ions

Inorganic ion transfer is reviewed generally by Stulc (83) and the involved transport systems are considered from a functional perspective by Shennan and Boyd (146). Molecular studies in recent years have indicated plausible molecular mechanisms for many of the functions described in the latter review.

Sodium

Plasma Na^+ concentration is higher in fetal than maternal plasma, at least in some studies (Table 1). However, whether this ion is transferred by diffusion, with the asymmetric plasma concentrations the result of the fetal side negative transtrophoblast PD (62) (see section Concentration and Electrical Gradients), or utilizing transporters needs to be analyzed.

The unidirectional maternal–fetal transfer of sodium over the last two thirds of gestation in several species was measured, in a series of now classic studies, by Flexner and colleagues (147,148). The data showed increases in unidirectional flux over gestation. Also there were large differences between species, the hemochorial placentas being the most permeable in agreement with the more recent data on placental permeability to hydrophilic solutes (see section Paracellular Channels). A broad range in the ratio of unidirectional to net flux was found from 1000 in human to only 3.5 in pig.

Evidence that sodium transfer utilizes a transcellular route in addition to the paracellular route is threefold. Firstly the permeability of the human placenta (based on Flexner's data) and rat placenta to sodium is greater than would be expected, relative to its diffusion coefficient in water, by comparison to that of inert hydrophilic solutes (43,149). Secondly, there are a whole range of sodium transporters in the human placenta, including the sodium-amino acid co-transporters (see section on Amino Acids) (Table 5), Na^+/H^+ exchanger (150–152), sodium-phosphate co-transporter (153) and, most importantly, Na^+/K^+ATPase (154). Thirdly, there is direct evidence in the pig and rat for transcellular sodium transport (155,156). The membranous epitheliochorial pig placenta was mounted in an Ussing chamber and a net sodium flux toward the fetal side measured and found to be ouabain (blocker of Na^+/K^+ATPase) inhibitable, as well as catecholamine-stimulable (155). In the rat it was found that 80% of unidirectional maternofetal flux near term is transcellular, being temperature-dependent, saturable, and ouabain-inhibitable (156).

The implication from the pig and rat data described above is that active maternofetal sodium transport takes place by a process which involves entry of the ion down its electrochemical gradient into the trophoblast across the maternal facing plasma membrane, utilizing one of the sodium-dependent co-transporters. Na^+/K^+ATPase would then input energy by pumping the sodium out of the trophoblast across the fetal facing plasma membrane. It does not, however, seem that such a model can be applied to the human placenta. Sodium transfer across the perfused human placenta does not appear to have an ouabain-inhibitable component (157). Furthermore, despite the presence of many sodium dependent co-transporters in the microvillous membrane of the human syncytiotrophoblast, as described above, a detailed study by Johansson and colleagues (154) shows that Na^+/K^+ATPase expression and activity is predominantly localized to the maternal facing microvillous membrane rather than to the fetal facing basal membrane. Further work in this area is clearly required.

Potassium

Fetal plasma potassium concentrations have been reported to be higher than in maternal plasma in studies in guinea pig, dog, human, and rat (112, 158–161). However, other studies report no difference in the two concentrations, or fetal plasma lower, in sheep, rat, and human (160,162,163). The problem is that if plasma sampling is not rapid then potassium may leak from erythrocytes, in which it is in high concentration, artifactually raising plasma concentrations. Such an effect has been demonstrated in human studies (164). It is therefore not possible to make any conclusions regarding mechanisms of placental potassium transfer from these observations.

The steady state permeability of the sheep and guinea pig placenta to potassium is 2–3 times higher than to sodium (165), whereas the difference in diffusion coefficient in water of the two ions is not this great (158). In the dually perfused rat placenta potassium fluxes were asymmetric, being nearly twice as high in the maternofetal than fetomaternal direction; furthermore the unidirectional maternofetal clearance for ^{42}K was between 12–70 times higher than that for ^{51}CrEDTA, used as a paracellular marker (166). It was also found in this study that unidirectional maternofetal clearance was saturable and blockable by ouabain and barium, a general inhibitor of potassium channels. These data therefore provide good evidence that potassium transfer across the rat placenta is transcellular, utilizing potassium channels and the Na^+/K^+ATPase.

Ouabain inhibits ^{86}Rb (used as substitute for ^{42}K) movement into the perfused guinea pig placenta (158) but not the perfused human placenta (157). There is a wide range of potassium channels present in the human syncytiotrophoblast and cytotrophoblast cells (62,167–172). However, these are just as likely to subserve the needs of cell homeostasis, such as the response to alterations in the osmolarity of extracellular fluid (168) or to circulating agonists (171,173) as to transplacental potassium transfer. Therefore, as with sodium, the definitive mechanism of potassium transfer across the human placenta remains to be established.

There is some evidence in the rat that potassium transfer is regulated. When rat mothers were fed a potassium deficient diet and so were rendered hypokalemic, fetal plasma potassium concentrations were maintained (112,161). Also there is a relationship between transplacental potassium transfer and maternofetal PD which breaks down when the mothers are made hypokalaemic (163). The nature of the regulation, at which these data hint, is not known but might be directed towards fetomaternal, rather than maternofetal, fluxes (161).

Chloride

A range of chloride channels and transporters have been identified or inferred to be present in human microvillous and basal syncytiotrophoblast membranes, in cultured cells of trophoblastic lineage, and in intact villi Fig. 6 [reviewed in (94)]. These include a N-K-Cl co-transporter in BeWo cells (174) and an anion exchanger 1 (AE1) DIDS inhibitable chloride: bicarbonate exchanger in microvillous membrane (a) (175–179) but probably not in basal membranes (179). There is a chloride conductance pathway present in both membranes (179). In support of this, a DIDS-inhibitable maxi chloride channel (b) has indeed been found electrophysiologically on the microvillous membrane of intact isolated villi (180) and in cytotrophoblast cells (177). Intermediate DPC-inhibitable (c) (181) and small PKA/ATP/cAMP activated channels (d) (181,182) have also been identified. A DIDS-inhibitable swelling-activated chloride efflux (183) may play a role in cell volume control.

More recently, maxi-chloride channel activity has been studied in a series of papers from Riquelme et al. (184–186). The channel can be reconstituted into liposomes and is found to have an open probability but not a conductance modified by 17-β estradiol and tamoxifen (184), and by arachidonic acid. The ATP effect on chloride channels is in keeping with evidence of purinurgic control being of importance in control of chloride flux in other tissues. Berryman and Bretscher (187) have identified a novel channel CLIC5 (e) in placenta, which is associated with microvillous actin and, like CLIC4, (f) can be localized by immunofluorescence to the microvillous membrane, while CLIC1 (g) is found to be distributed throughout syncytiotrophoblast. Fetuses with chloride channel abnormalities may have polyhydramnios, usually attributed to polyuria (188) but abnormalities in placental chloride transfer might also plausibly play a role. CFTR (h) is expressed in placenta (189,190) though no placental abnormalities have been identified in cystic fibrosis. CFTR may play a role in fluid acquisition by the early embryo (191).

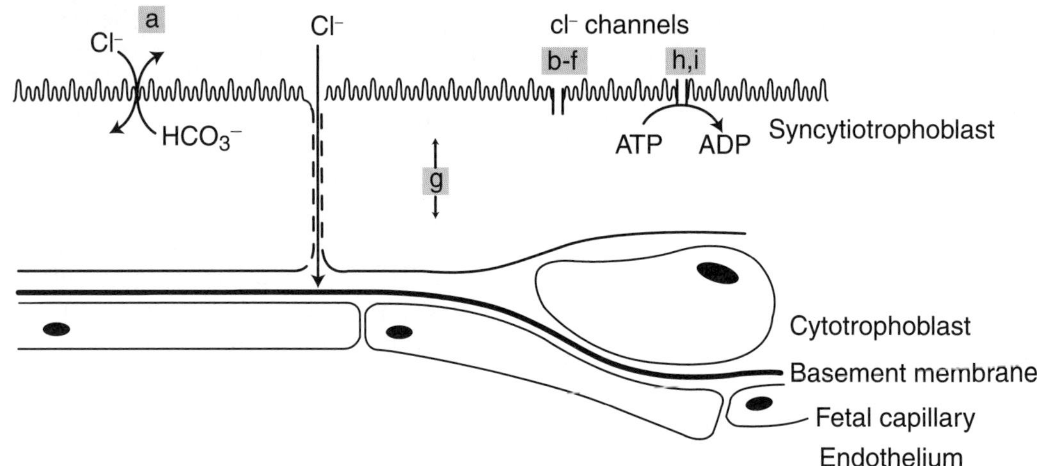

FIG. 6. Mechanisms for placental chloride transport for which there is direct or indirect evidence. (Letters refer to description in text.) (a) AE1 Cl/bicarbonate exchanger; (b) maxichloride channel; (c) intermediate DPC-inhibitable channel; (d) small PKA/ATP/cAMP activated channel; (e) CLIC5; (f) CLIC4; (g) CLIC1; (h) CFTR; (i) MDR1 P-glycoprotein.

MDR1 P-glycoprotein (i) has been shown to act as a Cl^- channel important in volume regulation and is expressed in placenta (189) specifically localized to the microvillous membrane (192). Cultured monolayers from an early pregnancy trophectoderm cell line in the pig demonstrate active transport of chloride as well as of K^+ and Na^+. If these are co-transported, it may explain the effect of chloride withdrawal increasing PD across the in vitro pig placenta in later gestation (193).

There are thus several lines of evidence that chloride movement into and out of the syncytiotrophoblast is probably both subject to control and likely to be important to the integration of syncytial function, e.g., through an effect on transmembrane potential, as Birdsey et al. (168) found that inhibition of chloride channels with DIDS increased the hypopolarization of the syncytiotrophoblast seen in villi in vitro when exposed to hypotonic medium.

The challenge is to clarify whether these various mechanisms, which might move chloride transsyncytially, do so to a quantitatively important extent, thus playing an important role in the overall fetal acquisition of chloride. Plasma chloride concentrations are some 5 mmol higher in fetal than in maternal plasma in the human (183) providing prima facie evidence that transcellular polarized transport towards the fetus is likely to be important. However, in the sheep, the concentration difference is opposite (194). Net transfer to the fetus is approximately 1 nmol/min/g/placenta in both these species while in the rat the figure is nearer 10 nmol. Unidirectional fluxes across the intact placenta in at least one direction have been measured in all three species. Perhaps surprisingly, these are orders of magnitude higher than the net flux.

In the near-term rat unidirectional flux is 700 times greater than the net flux (94). At least 75% of the unidirectional flux in this species is transcellular, showing competition with bromide or nitrate, striking inhibition by DIDS and dependence on the presence of fetal bicarbonate (95). This provides clear evidence that apparently "futile" cycling of chloride across rat placenta utilizes at least some of the molecular mechanisms present in the human placenta. Whether concentration or PD differences require the transport to be active is not known.

In the sheep, unidirectional flux again exceeds net (by about 100-fold) despite a permeability low in relation to most other ions (165). It is unknown what fraction of transfer is transcellular in this species. Assuming a low transcellular component, Faber and Anderson (90) have calculated, based on urinary drainage experiments, a reflection coefficient for chloride movement across this placenta of 0.68.

The most prevalent current assumption in the human is that most of the unidirectional flux (at least 1,000 times the net flux at term) is paracellular (94) on the grounds that, unlike rat, unidirectional flux across the perfused human placenta is only about 16% inhibitable (195). Such a high paracellular chloride permeability is not surprising in view of the much higher paracellular pore radius in this species than in sheep, but the low transcellular component compared to rat is hard to explain as its pore radius is similar to human. If most human placental chloride movement is paracellular, it is difficult to explain human fetal chloride acquisition against a concentration gradient, although Bissonnette et al. (183) suggest that the osmotic gradient generated by amino acid accumulation, which, for example, in liver is balanced by extrusion of K^+ and Cl^-, could account for the fetomaternal concentration difference across placenta. Speculatively, the transcellular route may be more important in earlier gestation, or perhaps filtration may play a greater role than is sometimes supposed, in contributing to asymmetrical paracellular chloride fluxes.

Iodide

Placental iodide transfer is necessary for fetal synthesis of thyroid hormones as these cross the placenta in insufficient quantity to fully meet fetal need—hypothyroid fetuses have delayed growth at birth. Free iodide concentration in fetal plasma does not appear to have been measured but following injection of radioiodide to pregnant guinea pigs, fetal plasma concentration becomes higher than maternal if thyroid uptake of iodine is inhibited (196). Kinetic studies in both sheep (197) and hemochorial species (196) indicate that iodide transfer is predominantly carrier mediated being saturable, inhibitable, e.g., by thiocyanate, and, in sheep, having a substantially higher permeability than that of chloride (165).

These findings are compatible with more recent cellular and molecular work mainly on human material. Thus, the sodium iodide symporter (NIS), one of the sodium-glucose co-transporter family (SLC5) (198), is present in high concentration in JAR cells [a choriocarcinoma cell line (199)] and in placenta (200). NIS transfers $2Na^+$ for each I^- (201) when expressed in oocytes, and iodide movement is thus electrogenic. The placental transmembrane potential and sodium concentration difference may thus be predicted to influence iodide movement. This fits neatly with the kinetic observation in sheep that unidirectional iodide flux is a function of maternofetal electrical potential difference (197).

Pendrin (PDS), an anion exchanger (202), which, in a polarized kidney cell line, plays a role in I^- efflux (203), is also expressed in placenta, especially on the microvillous membrane. Its expression increases as a function of syncytialization and, unlike NIS, also of gestation (204). PDS is also strongly expressed in endometrium where it cycles with menstruation (205).

The placenta may perhaps control fetal iodide acquisition as evidenced by the observation that rats exposed to a reduced iodide intake show both increased fetal thyroid and placental NIS (206). The compensation is, however, not sufficient to achieve normal fetal uptake (207). I^- efflux from cytotrophoblast in culture (208) or from villous explants (209) is stimulated by their exposure to hypotonic media.

Calcium

Both total and ionized calcium concentrations are higher in fetal plasma than in maternal plasma toward term in humans, rats, mice, and sheep (40,160,210–212). Several lines of evidence show that there is asymmetrical movement of calcium across placenta driven by transporter-mediated active transfer of calcium from mother to fetus. Firstly, calcium is concentrated in fetal perfusate against a concentration gradient in perfused human, guinea pig, and sheep placenta (213–215). Secondly, in such perfusion experiments, calcium transfer is reduced by the addition of metabolic inhibitors such as dinitrophenol or cyanide or by lowering perfusate temperature. Thirdly, the placenta expresses transporter proteins known to be involved in active calcium transport in intestine and kidney (see below). Finally, J_{mf} for calcium is greater than J_{fm} across perfused human, guinea pig and rat placenta (216–218). In those species in which fetal circulation is electrically negative to maternal, the PD may also contribute to asymmetrical maternofetal and fetomaternal fluxes favoring fetal acquisition if the PD is expressed across the exchange area (see section Concentration and Electrical Gradients).

Studies in the rat provide the most thorough characterization of fluxes across the intact placenta. In the umbilically perfused rat placenta at 21 days, the unidirectional maternofetal flux of calcium (J_{mf}) measured using ^{45}Ca is about 5 times greater than fetomaternal (J_{fm}) and decreases when the placenta is perfused with blockers of metabolism. Omitting Na^+ from the perfusate has no effect. Doubling maternal plasma concentration of calcium reduces the measured J_{mf} of ^{45}Ca but neither J_{mf} nor J_{net} are affected by varying umbilical calcium concentrations (218). Studies on the dually perfused rat placenta confirm that J_{mf} is saturable and also demonstrate that it is not inhibited by nifedipine, a blocker of voltage gated calcium channels (219). Taken together these data suggest that (a) maternofetal flux of calcium across the rat placenta is by an active, saturable process that does not involve Na^+/Ca^{2+} exchange or voltage gated calcium channels; (b) there is no prompt feedback control of maternofetal flux by the concentration of calcium on the fetal side of the placenta; (c) fetomaternal flux is relatively low and is therefore likely to be accomplished by diffusion alone or by a low capacity transport system.

The conclusions from the rat fit well with data on the perfused human placenta (220).

The mechanism of maternofetal flux of calcium across the placental trophoblast appears to have much in common with transepithelial transport of the ion

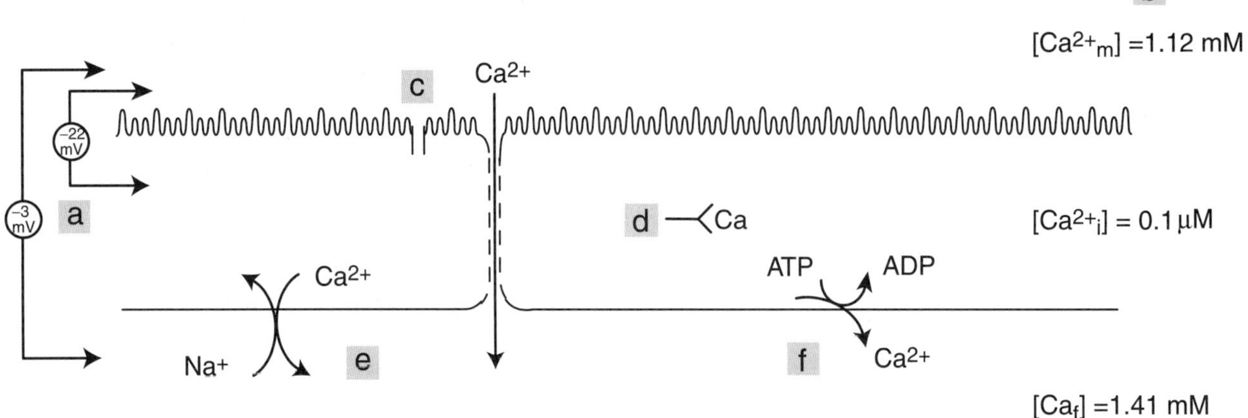

FIG. 7. Syncytiotrophoblast calcium transport. (Letters refer to description in text.) (a) Potential difference; (b) concentration differences; (c) epithelial Ca^{2+} Channels; (d) calcium binding protein; (e) Na^+/Ca^{2+} exchanger; (f) Ca^{2+} ATPase.

across duodenum and intestine (221) in having three components (Fig. 7).

Transport from the Maternal Plasma into the Syncytiotrophoblast Cytosol across the Microvillous Plasma Membrane The electrochemical gradient for calcium diffusion into the syncytiotrophoblast across the microvillous plasma membrane is strongly favorable. The median potential difference (a) across the microvillous membrane measured in vitro, is, at least in intermediate villi, (the only ones studied), –22mV (cytosol negative) (62) and the intrasyncytial ionized calcium concentration (b) is around 10^{-7}M (222,223), four orders of magnitude lower than that in plasma. Surprisingly, perhaps, in view of this strong driving force, studies using isolated microvillous plasma membrane vesicles from human term placenta found that calcium uptake was saturable and involved both high affinity and low affinity systems (224,225), suggesting a facilitated diffusion mechanism increasing microvillous membrane permeability. In agreement with the rat observations, these studies found no effect of verapamil or nifedipine on calcium uptake (225). A family of epithelial Ca^{2+} channels (c) (ECaC1 and ECaC2, recently renamed TRPV5 and TRPV6) have recently been cloned and shown to provide a route of calcium diffusion into duodenal and renal tubular cytoplasm (221). These are expressed in placenta (226) but their contribution to calcium entry across the microvillous plasma membrane has yet to be elucidated.

Translocation across the Syncytiotrophoblast Cytosol Without Markedly Perturbing the Normal Very Low Intracellular Calcium Concentration This is probably achieved predominantly by specific binding proteins buffering calcium. A 9kDa calcium binding protein (calbindin-D_{9k}) (d) appears to be particularly important in this regard. Thus, the expression of calbindin-D_{9k} (227–230) and mRNA for this protein (231–233) increases markedly over the last third of gestation in mice and rats. There is a close temporal association between this increase and the sharp increase in J_{mf} of calcium in late gestation (232), raising the possibility that, at least in this species, calbindin-D_{9k} concentration may be rate-limiting. Further support for this comes from studies on the diabetic rat placenta in which both calbindin-D_{9k} mRNA and J_{mf} for calcium are down regulated (234). In the human placenta, however, evidence is conflicting as to whether mRNA for calbindin-D_{9k} or the protein itself is or is not expressed (235–238). The most recent study found calbindin-D_{9k} mRNA and protein in both the syncytiotrophoblast and in cytotrophoblast cells (239). As extrapolated from studies on the intestine this protein may serve not only as an intracellular calcium buffer (240) but also as a facilitator of calcium movement through the cytosol (241–244).

Efflux of Calcium from the Syncytiotrophoblast across the Basal Plasma Membrane Requiring Active Transport Against the Electrochemical Gradient Transfer of calcium across this membrane in vitro requires ATP (245,246). Curiously, an ATP-independent, saturable mechanism also exists involving facilitated diffusion that may allow back flux into syncytiotrophoblast. A Na^+/Ca^{2+} exchanger (e) has also been identified (246–248). Messenger RNA for the plasma membrane Ca^{2+} ATPase (PMCA) (f) has been found in human and rat placenta (232,244) and PMCA has been localized immunohistochemically to the fetal-facing plasma membrane in the placentas of these species (249) as well as that of cat (250). This calcium "pump" is therefore well positioned to mediate calcium efflux across the basal plasma membrane followed, it is presumed, by diffusion across the capillary endothelium through lateral intercellular spaces into fetal plasma. Ca^{2+} ATPase activity in human basal plasma membrane vesicles increases with gestation during the third trimester, consistent with the increasing demands of the fetus for bone mineralization (251).

Recent data suggest that parathyroid hormone-related peptide (PTHrP), rather than parathyroid hormone (PTH) or the vitamin D metabolite 1,25 $(OH)_2D_3$, is the main regulator of placental calcium transport (252). PTHrP is synthesized by several organs as a 141 amino acid pro-hormone that can be cleaved into different active fragments, each with different actions (253). PTHrP-(1-34) has considerable homology with PTH and activates the common PTH/PTHrP receptor (PPR1) (252,253). Mid-molecule fragments of PTHrP have no homology with PTH and do not act via PPR1: the receptor for these fragments remains unidentified. PTHrP pro-hormone is expressed in both human and mouse placenta, but it is not yet clear which fragments are present (253). PTHrP is also present in the fetal circulation and so could act by endocrine, paracrine, or autocrine actions (253). Studies in sheep first showed that full-length and mid-molecule fragments of PTHrP probably stimulated calcium transport across perfused placentas, whereas PTH had no effect (254). In mouse conceptus from which the PTHrP gene had been knocked out, the fetal plasma ionized calcium concentration was reduced compared to wild-type or heterozygote animals (212). Furthermore, J_{mf} of calcium in the $PTHrP^{-/-}$ animals, estimated by measuring the amount of ^{45}Ca as a ratio to that of $^{51}CrEDTA$ (as a marker of diffusional permeability in the fetuses), following injection of both radioisotopes into anaesthetized mothers, was significantly reduced as compared to wild type or heterozygote animals. Injection of PTHrP-(1-86), PTHrP-(67-86), but not PTHrP-(1-34), or PTH-(1-84), into the $PTHrP^{-/-}$ fetuses returned the $^{45}Ca/^{51}CrEDTA$ ratio, measured 1 hour later, to normal. In similar studies Kovacs and colleagues (255) found that J_{mf} of calcium was not affected in *Hoxa3*-null fetuses that lacked parathyroid glands, although plasma ionized Ca^{2+} in the mutant fetuses was lower than normal. This provides evidence that PTH acts primarily on fetal bone and kidney rather than controlling placental calcium transport. The same group also studied mouse fetuses lacking the PTH/PTHrP receptor gene (PPR1) and found that the $^{45}Ca/^{51}CrEDTA$ ratio was raised in these fetuses as was their plasma concentration of PTHrP (255). These data therefore seem straightforward: in the absence of fetal PTHrP, maternofetal calcium transfer is reduced; when it is normal, even when PTH is absent, transfer is normal and when PTHrP in the fetus is raised then so is J_{mf} of Ca^{2+} (255). However, there are complexities. The calcium content of $PTHrP^{-/-}$ fetuses per g ash weight was not reduced despite their lower $^{45}Ca/^{51}CrEDTA$ transfer ratio, and was significantly lower in $PPR1^{-/-}$ fetuses despite their higher $^{45}Ca/^{51}CrEDTA$ transfer ratio (255). Furthermore, a different group who generated their own $PTHrP^{-/-}$ fetuses report lower fetal plasma ionized Ca^{2+} concentrations, as in the study by Kovacs et al. (212) but significantly higher fetal Ca^{2+} content per g ash weight (256). There are several possible explanations for these apparent inconsistencies. Perhaps gene deletion reduces J_{fm} as well as J_{mf} so that J_{net} does not follow J_{mf}. Alternatively, diffusional permeability to $^{51}CrEDTA$ might be changed leading to an altered $^{45}Ca/^{51}CrEDTA$ transfer ratio: these possibilities require investigation.

Ca^{2+}ATPase activity in human basal plasma membranes is increased by acute exposure to physiological concentrations of PTHrP-(38-94), whereas PTH, PTHrP-(1-34), and PTHrP-67-86) only stimulate at supraphysiological concentrations (257). These effects are therefore broadly consistent with the effects of PTHrP on Ca^{2+} transport reported in the animal studies. Importantly, a further study showed that whereas expression of basal membrane Ca^{2+} ATPase was lower in human intrauterine growth retardation (IUGR), its activity was significantly increased, as were fetal plasma concentrations of mid-molecule PTHrP (257). This is puzzling as raised calcium transporter activity in IUGR is inconsistent with the neonatal hypocalcemia and reduced bone mineral density reported in this condition (258,259). Perhaps the increase in pump activity and of PTHrP mid-molecule fragment fetal plasma concentrations observed are a response to reduced fetal calcium accretion in IUGR.

Magnesium

Total magnesium concentrations are higher in fetal than maternal plasma in rats, rabbits, sheep and humans (260–265). In rats both total and ultrafiltrable concentrations of magnesium in fetal plasma are higher than maternal. In sheep J_{mf} for magnesium is higher than J_{fm} suggesting active transport toward the fetus (263). Studies on the umbilically perfused rat placenta have demonstrated that J_{mf} for magnesium is significantly higher than that of paracellular diffusional markers (266–268) and is reduced by the addition of the metabolic inhibitor cyanide or by lowering the perfusate temperature (266), supporting the view that active transport is involved. These studies also indicate that the carrier mediating magnesium transfer is sodium-dependent as addition of amiloride and ouabain, or the removal of sodium from the perfusate, results in reduced transfer (268). $^{28}Mg^{2+}$ uptake from maternal plasma into rat fetuses is also reduced following maternal injection of amiloride (269). These data are compatible with the involvement of a Na^{+}/Mg^{2+} exchanger in active maternofetal magnesium transfer (270), but no such entity has yet been identified at the molecular level.

Organic Ions

Lactate

Monocarboxylates such as lactate and pyruvate play an important role in the cellular metabolism of all cells, with lactic acid being produced as an end product of glycolysis. The pK of lactic acid is 3.86, which means that at physiological pH it is almost entirely dissociated to the lactate anion. Some tissues such as white skeletal muscle and red blood cells rely on glycolysis for production of the majority of their ATP under normal physiological conditions, while other tissues become dependent on this pathway only during hypoxia. In this situation efflux of lactate and H^{+} from the cell must keep pace with production in order to prevent decrease in intracellular pH. Some tissues such as the brain, heart, and red skeletal muscle can readily oxidize lactate, which can therefore be used as a respiratory fuel under certain circumstances, with lactate being transferred rapidly into the cell. The fetus of many animals is no different in this respect, with fetal heart, brain, and skeletal muscle using lactic acid as a major respiratory fuel and other tissues producing lactic acid as an end point of glycolysis (271). Additionally, studies using perfused isolated cotyledons and radiolabeled glucose, have shown that like other mitochondria-poor tissues the placenta is a lactate producer, 80% of glucose carbon being found in lactate, of which 90% was released, in this system, to the maternal circulation (272).

Studies, both animal and human, suggest that lactate is predominantly transferred across the hemochorial placenta by a stereospecific, inhibitable, carrier mediated, H^{+} dependent co-transport mechanism. Thus, steady state permeability of the guinea pig placenta to L-lactate is 3 times greater than that to D-lactate. The system shows saturability and competition with pyruvate but not with D-glucose (273). The net movement of lactate follows the transplacental concentration gradient (274), which at term is reported to be in the fetomaternal direction (271,275); the fetus being a net lactate producer. Soothill et al. (276,277) report that in early second trimester fetuses the concentration gradient is again in the fetomaternal direction. Although the system is independent of sodium ion gradients there is evidence for co-transport of lactate with hydrogen ions, as a proton gradient induced by infusion of HCl can generate net transfer of lactate against its concentration gradient (273,278). Studies on isolated plasma membrane vesicles from both the rat and human placenta have also demonstrated uptake of radiolabeled L-lactate across both the apical and basal membrane, which is stimulated by an inwardly directed proton gradient. A 1:1 stoichiometry of lactate/H^{+} co-transport is found as is competitive inhibition by other monocarboxylates (278–283). There is a minimal or absent diffusional component to lactate movement across these isolated membranes.

Findings in the perfused human placenta (284,285) and rhesus monkey (286) demonstrate facilitated diffusion of lactate. Observations made on both the maternal and fetal sides of the same placenta in the guinea pig, using the single pass indicator dilution technique, have demonstrated carriers on both sides of the placental trophoblast which can bind both D-and L-lactate. These may both be inhibited by phloretin (287).

In other tissues lactate transfer across plasma membranes has been shown to be predominantly facilitated by a family of monocarboxylate transporters: the MCTs [for reviews see (288,289)]. Initially characterized in the erythrocyte, eight members of the MCT family have now been identified (288,290,291). These may be distinguished by tissue distribution, kinetics, substrate affinities, and inhibitor sensitivities (Table 4). mRNA expression of MCT isoforms in placental homogenates and other tissues has been investigated by Price et al. (291) using Northern blotting. This group studied all the currently known isoforms and showed MCT1 and MCT4 were ubiquitously expressed. High levels of MCT5 and MCT6 mRNA were found in the human placenta in addition to MCT1 and MCT4. At the protein level, however, little information is available on their placental expression. A recent study by Settle et al. (283) using Western blotting of microvillous and basal plasma membranes isolated from the

TABLE 4. *Ki (μm) for inhibitors of MCT 1-4*

Inhibitor	MCT1	MCT2	MCT3	MCT4
CHC	166	24	NI	991
Phloretin	5	14	NI	41
pCMBS	112	NI	NI	21
DIDS	434			

NI, not inhibited.
Data for MCT 1, 2, and 4 from Manning et al. (292) and Bröer et al. (605). Data for MCT 3 from Grollman et al. (606).

human placenta, and immunolocalization in term placental tissue, demonstrated a polarized distribution of MCT1 and MCT4. MCT4 was predominantly localized on the microvillous membrane with MCT1 predominantly on the basal membrane. In keeping with this asymmetrical expression, differences in the kinetics of lactate uptake at the two faces of the placenta were found. K_m of the quantitatively major component of uptake on the microvillous membrane was approximately 6 mM while that on the basal membrane was 25 mM. The fact that these values are in near agreement with the range of K_m values reported for both MCT4 (17–34 mM) and MCT1 (3.5–4.5 mM) (288,292,293) further support the identity of the transporters involved. A direct comparison of the two trophoblast layers in this study show that proton gradient driven lactate transport in the microvillous membrane is 3 times higher than in basal membrane. As a result, in common with other transporters, e.g., GLUT1, the basal membrane appears, on the basis of this data, to be the rate-limiting step in the transtrophoblastic transfer of lactate in the human.

Kirk et al. (294) using co-immunoprecipitation and chemical crosslinking have established a close link between the plasma membrane protein CD147 and MCT1 and MCT4. In heart cells CD147 and MCT1 were shown to co-localize at the t-tubular and intercalated disk regions. In mammalian cell lines, MCT transfection alone caused accumulation of protein in the perinuclear compartment, while co-transfection with CD147 enabled expression of active MCT1 and MCT4 (but not MCT2) protein in the plasma membrane. CD147 is therefore thought to be a chaperone protein for these two transporters, enabling appropriate trafficking to the cell surface. Settle et al. (283) using Western blotting have shown that CD147 is equally expressed on the microvillous and basal plasma membrane of the placenta, which is consistent with such a role in this tissue.

Bulk Nutrients

Amino Acids

Amino acid transfer across the placenta is required both for fetal protein synthesis and to provide metabolic substrates for the fetus. Consequently, although placental transfer of the essential amino acids is absolutely required for fetal development, the transfer of the nonessential amino acids is also very important, both for catabolism, interconversion with other substrates, and oxidation (295,296). Furthermore, placental amino acid metabolism occurs in concert with transport processes. There have been several recent excellent reviews of placental amino acid transfer and metabolism (295–298) and the current overall picture is summarized here.

Fetal plasma concentrations of most amino acids are higher than in maternal plasma, and placental tissue concentrations of amino acids are higher than in either circulation (299–302). Although one needs to be wary of extrapolating tissue concentrations to intracellular concentrations, the indication from these data is that amino acids are transported from the maternal plasma into the syncytiotrophoblast against a concentration gradient, from where they are transferred down a smaller concentration gradient to placental extracellular fluid and thence to fetal plasma. It is unlikely that paracellular diffusion plays an important role in maternofetal amino acid flux, but will contribute to fetomaternal flux: transporter mediated processes are central to maternofetal net flux.

The importance of transporter mediated processes in maternofetal amino acid flux is supported by a wealth of data: uptake of amino acids by placenta is energy dependent (303,304); transfer across the perfused guinea pig placenta and perfused human placental cotyledon is both asymmetric and stereospecific (305,306); a range of transporter proteins has now been identified in both the microvillous and basal plasma membranes of the syncytiotrophoblast (Table 5).

Transporter proteins in the human placenta have mainly been studied using isolated microvillous and basal membrane vesicles and investigating uptake or efflux of tracer radiolabeled amino acids in the presence or absence of other varying concentrations of competing amino acids and in the presence and absence of ions such as Na^+ and Cl^-. This enabled amino acid transporter systems, akin to those described by Christensen in other cell types (307) to be identified on the basis of kinetic and ion-dependence criteria. Recent molecular work has enabled many, but not all, of the genes and proteins associated with these kinetically defined systems to be identified and has often necessitated a change in our view of how they operate. A single kinetically defined transport system may now be associated with more than one gene or protein (e.g., System A) and some systems are the product of covalent, heterodimeric association of two proteins, with families of transporters related by the type of heavy-chain protein/light-chain protein interaction. This is summarized in Table 5.

Data from vesicle studies, indicator dilution work in intact cotyledons and immunocytochemistry also indicate asymmetry between microvillous membrane and basal membrane localization of the transporters (Table 5) (305,308), necessary to effect net maternofetal flux. Jansson (297) has proposed mechanisms of vectorial transfer of different classes of amino acid.

TABLE 5. *Amino acid transporters in placenta*

Transporter	Substrates	Properties	Localization[a]	Protein	Genes[b,c]	Ref.
System A	Neutral Ala, Ser, Gly, MeAIB, AIB	Na^+-dependent pH sensitive	**MVM** BM	Monomeric	ATA1, ATA2 ATA3(1st tri> term)	(607,608,610, 611)
System ASC	Neutral Ala, Ser, Cys	Na^+-dependent	MVM **BM**	Monomeric	ASCT1 ASCT2	(607,608, 612–614)
System B°	Neutral Broad specificity: BCH	Na^+-dependent	?	Monomeric	ATB°	(23,615–618)
System β	Neutral Tau, β-Ala	Na^+-dependent Requires Cl^-	**MVM** BM	Monomeric	TAUT	(308,329, 619–624)
System N	Neutral His, Gln, Asn	Na^+-dependent pH-sensitive	?MVM	Monomeric	mNAT	(625–628)
System Gly	Neutral Gly	Na^+-dependent	MVM	Monomeric	GLYT-1 GLYT-2	(629,630)
System L	Neutral Leu, BCH and other branch chain amino acids	Na^+-independent	MVM BM	Heterodimeric	4F2hc plus LAT-1 or LAT-2	(23,607,631–633)
System $b^{o,+}$	Cationic/ Neutral	Na^+-independent	?BM	Heterodimeric	rBAT plus $b^{o,+}$AT?	(72,73,634)
System y^+L	Cationic/Neutral in the presence of Na^+	Na^+-independent but Na^+ requirement for neutral amino acids	MVM **BM**	Heterodimeric	4F2hc plus y+LAT-1or y+LAT-2	(23,72,635, 636)
System y^+	Cationic Arg, Lys	electrogenic	**MVM** BM	Monomeric	CAT1, CAT2A, CAT2b, CAT4	(72,73, 637–639)
System X_{AG}^-	Anionic Asp, Glu	Na^+/K^+-dependent	MVM BM	Monomeric	EAAT 1-4	(320,640–645)

[a]MVM, microvillous plasma membrane; BM, basal plasma membrane. Bold denotes predominant localization if known; ? denotes localization uncertain.

[b]Genes underlined are expressed by placenta, at least at RNA level.

[c]A more systematic nomenclature for amino acid, and other transporter, genes is now being introduced (see Hediger et al. [645a]).

Glycine transfer is proposed to utilize the System A transporter in the microvillous membrane. This transporter is Na^+-dependent and so will be able to utilize the Na^+ gradient into the cell, maintained by the Na^+/K^+ ATPase, to transport glycine against its concentration gradient; glycine can then either diffuse down its concentration gradient across the basal membrane into extracellular fluid, or could utilize the System L transporter in the basal membrane to increase the rate of such movement. Leucine is transported predominantly by the Na^+-independent System L in both membranes: vectorial transport into the cell against a concentration gradient from the maternal circulation might occur, in the absence of Na^+ dependence by exchange of extracellular leucine for intracellular glycine (pumped in by System A) by System L. Lysine will be transported across the microvillous membrane via System y^+: this could be driven by movement of the cationic amino acid down the electrical gradient [syncytiotrophoblast membrane potential difference being around –22 mV at term: (62)]. On the fetal side lysine transport out of the syncytiotrophoblast may occur via exchange with neutral amino acids on System y^+L.

These models are useful starting places for understanding syncytiotrophoblast amino acid transport but the complexity of the system must not be underrated: there will be multiple interactions among different amino acids and different transporters in vivo because of overlapping substrate specificities; amino acid metabolism by mother, placenta, and fetus also needs to be considered. As regards the latter, it is clear from studies on chronically catheterized sheep fetuses that there are complex fetoplacental amino acid metabolic interactions (296). Two good examples are the glutamate-glutamine cycling and serine-glycine cycling between placenta and fetal liver (295,296). Glutamate is predominantly taken up by the sheep placenta from the umbilical circulation rather than the uterine circulation; within the placenta around 94% is oxidized with the rest being converted to glutamine, which is also taken up from the uterine circulation and transported into the umbilical circulation. In the liver, the reverse fluxes take place with about half of the glutamine taken up being released as glutamate. Serine is taken up from both uterine and umbilical circulations by the sheep placenta, so that fetal serine requirements must be met by fetal synthesis, not transplacental transport. There appears to be no net uptake of glycine from the maternal circulation by the sheep placenta and the glycine delivered into the umbilical circulation appears to all be synthesized within the placenta. Conversely, serine is produced in the fetal liver from glycine. The purpose of this inter-organ exchange of serine and glycine is not known. It could provide a means of fetal/placental signaling in regard to metabolic requirements. More simply, one product of the serine to glycine conversion in the placenta is methyl tetrahydrofolate, which could be used for purine production. It is not yet certain whether such inter-organ cycling occurs in the human; indeed, the activity of serine hydroxymethyltransferase, the enzyme which catalyses serine/glycine conversion, is much lower in the human placenta than in the sheep (R. Lewis et al., personal communication).

Placental amino acid transport does appear to be regulated in relation to fetal growth, but in a complex fashion. System A activity in microvillous membrane (measured as ^{14}C-MeAIB uptake per mg membrane protein) varies inversely with birth weight and abdominal circumference of the baby at term (47) in normal pregnancy. As placental weight increases directly with birth weight, this suggests that overall total placental capacity to transport System A substrates stays fairly constant over the range of normal fetal growth rates. This contrasts with frank IUGR in which System A activity in the microvillous membrane is found to be significantly lower than that in placentas of normally grown babies (45,309–311). In mice where the placental specific insulin-like growth factor II (Igf-II) gene has been knocked out, placental weight is reduced by day 14 of gestation (term is day 20), as compared to wild-type placentas, but fetal growth remains unchanged until day 19 when it is significantly lower (41). Interestingly, maternofetal ^{14}C-MeAIB transfer, per mg placenta, is increased in the knockout at day 16, compared to wild type, but the difference is much less by day 19. This suggests that the ability of the placenta to transport System A substrates is initially upregulated enabling fetal growth to continue normally but that this upregulation is not sustainable, contributing to the decrease in fetal growth rate (41). Further evidence of a strong relationship between placental System A activity and fetal growth was provided by Cramer et al. (312). These authors infused MeAIB into rats over days 7–20 of pregnancy and found that both placental System A activity and fetal weight were reduced: both this and the mouse study described above demonstrate that an alteration in placental transport activity can be the cause of a change in fetal growth.

A series of studies have shown that the activity of other human placental amino acid transporters, as well as that for other solute transporters, are affected in IUGR (Table 6). These data are consistent with the in vivo observations that umbilical plasma amino acid concentrations are significantly lower in IUGR than normally gown fetuses (313,314). Paolini and colleagues (315) have shown directly using stable isotopes that transplacental fluxes of ^{13}C-leucine and ^{13}C-phenylalanine are reduced in IUGR; interestingly, and further emphasizing the intrinsic complexity of the process, the transplacental fluxes of neither ^{13}C-glycine nor ^{13}C-proline were affected in IUGR.

There is also evidence that placental amino acid transport is affected in babies born to diabetic mothers, although two separate studies observed different relationships to birth weight (316,317). Understanding these relationships between amino acid transport and fetal growth will require knowledge of regulatory mechanisms. There are clues from in vitro experiments. Substrate availability may directly regulate the activity of some transporters such as System A, System β, and X^-_{AG} (318–320); insulin, insulin- like growth factor I, epidermal growth factor, and leptin have all been shown to stimulate System A activity (321–325). Tumour necrosis factor-alpha down-regulates System A in the rat (326,327). Taurine transport is affected by protein kinase C and inhibited by calcium (328,329). This is therefore an important area with scope for potentially very exciting new discoveries, particularly with gene-knockout and knockin strategies both in vivo and in vitro.

TABLE 6. *Changes in activity of transporters in the microvillous (MVM) and basal (BM) plasma membrane of placentas from IUGR pregnancies as compared to normal pregnancies*

Transporter	MVM	BM	Ref.
System A	Decreased	No change	(45,311,317)
System L (leucine)	Decreased	Decreased	(646)
System y^+/y^+L (arginine/lysine)	No change	Decreased	(73,646)
System β	Decreased	No change	(623)
Na^+-independent taurine	No change	Decreased	(308,623)
GLUT1	No change	No change	(74)
Na^+/K^+ATPase	Decreased	No change	(647)
$Ca2^+$ATPase	Not present	Increased	(257)
Na^+/H^+ exchanger	Decreased	Activity not present	(311,648)
H^+/lactate	No change	Decreased	(283)

Note that there are increases, decreases, and no change reported for the transporters studied, suggesting there is a discrete pathology associated with IUGR. It could be that this reflects whether a change in transporter activity is a contributor to the cause of IUGR (e.g., System A), is not involved (GLUT1) or changes in response to the disease (Ca^{2+}ATPase).

Glucose

Glucose is the main metabolic fuel used by mammalian cells and is one of the major nutrients transferred from mother to fetus throughout gestation. In the absence of significant gluconeogenesis in the fetus in early gestation (330,331) the importance of maintaining a supply of glucose from mother to fetus across the placenta is apparent. Fetal demand for glucose is high, especially in the third trimester when rapid fetal growth necessitates a rapid, high-capacity system for maternofetal transfer of glucose. Glucose is highly hydrophilic and as a result cannot permeate the cell membranes of the trophoblast to any great extent and its transfer will therefore be membrane limited. The low permeability of the extracellular porous route described previously (see section Conceptual and Methodological Issues in Placental Transfer and section Paracellular Channels) cannot provide sufficient capacity to satisfy both fetal and placental demands (74,332). Observations made in a wide range of species indicate that glucose transfer across the chorioallantoic placenta is by a nonconcentrating, saturable, stereospecific, sodium-independent, carrier mediated process, i.e., facilitated diffusion, a term first coined by Widdas on the basis of placental studies (333,334).

Fetal plasma concentrations have been shown to be consistently lower than maternal plasma concentrations in a wide range of species including sheep (335,336), cattle (337), horse (338,339), monkey (340), and human (341). Thus a maternofetal plasma concentration gradient exists for glucose in all these species.

Crandell et al. (336) and Hay et al. (342) have shown that the net transfer rate of glucose across the sheep placenta in vivo increases with increasing maternal plasma concentrations over the physiological range. However, when maternal concentrations are increased to supraphysiological levels the rate of increase in net glucose flux diminishes. In the human, in vitro perfusion of the term placenta with increasing concentrations of glucose indicate that saturation occurs at maternal concentrations of approximately 20 mM, a value considerably outside the normal range (343). The glucose transport system thus demonstrates saturability, albeit at higher than physiological concentrations.

In the sheep, placental permeability to fructose (whose concentration is high in fetal lamb plasma) (344,345) α-methyl-D-glucopyranoside and L-glucose (346) has been shown to be much lower than that of D-glucose, while other monosaccharides such as 3-0-methyl-D-glucopyranose have a high permeability and compete with D-glucose (346). In the BeWo choriocarcinoma cell line both mannose and 3-0-methyl-glucose compete with D-glucose uptake while fructose and galactose have been shown to be relatively poor inhibitors of glucose uptake (347). Thus D-glucose transfer across the placenta demonstrates chemical- and stereo-specificity. The sugar specificity of the placenta most closely resembles that of the human red blood cell and is different from that of the gut and renal tubule (348).

In human placental microvillous and basal membrane vesicles glucose uptake has been shown to be sodium independent and is inhibited by phloretin and cytochalasin B, which are known inhibitors of the facilitated glucose transporter (349,350). Similarly D-glucose transporter proteins isolated from the microvillous membrane of the human trophoblast and incorporated into phospholipid liposomes, demonstrate selective D-glucose uptake, which is sodium independent (351).

Two families of glucose transporters have been identified in mammalian cells. First the sodium-dependent glucose transporter (SGLT) family, found predominantly in the intestine and the kidney, absorb glucose against a concentration gradient and are dependent on the Na^+ concentration gradient maintained by Na^+/K^+ ATPase. The transfer of glucose across the placenta has been shown to be sodium independent, as discussed above, and therefore involvement of the SGLT transporters is unlikely. Second, a family of facilitated glucose transporters (the GLUT family) has been identified. Initially isolated from human erythrocyte ghosts (352), GLUT1 was the first of a family of 12 (GLUT1-GLUT12) to be identified (353).

The two main isoforms of the GLUT family expressed in the human placental trophoblast are GLUT1 and GLUT3. These two proteins show both spatial and temporal differences in expression. GLUT1 mRNA is found throughout gestation in all placental cells including extravillous trophoblast and cells of the amnion [reviewed in (354)]. Localization of GLUT1 protein in the term placenta is somewhat uncertain. Farrell et al. (355) using immunohistochemistry showed a predominantly basal expression of GLUT1, as did Arnott et al. (356) using confocal and electronmicroscopy. In contrast other studies have shown GLUT1 protein to be expressed on both the microvillous and basal plasma membranes of the term placenta, with a three-fold higher expression on the microvillous as compared to basal membrane (74,357). This has also been demonstrated by immunohistochemistry (358). This asymmetric expression is mirrored in measurements of GLUT1 activity per mg membrane protein in microvillous and basal plasma membranes. Taking into account the difference in submicroscopic surface area of the two trophoblast membranes, Jansson et al. (74) estimate that the total transport capacity of the microvillous membrane at term may be as high as 20 times that of the basal membrane. As a result, concentrations within the syncytiotrophoblast, are, it is suggested, maintained close to maternal plasma concentrations, thereby allowing a maximal concentration gradient toward the fetus. The higher transport capacity of the microvillous membrane may also subserve the high demand for glucose by the trophoblast itself (5,17).

The lower expression of GLUT1 on the basal membrane of the syncytiotrophoblast means that this membrane may be the rate-limiting step in the transplacental transport of glucose. Any change in basal expression therefore has the potential to alter transplacental glucose transport. Changes in GLUT 1 mRNA levels throughout gestation are controversial with some groups reporting increased mRNA levels in first trimester (359), lower levels in first trimester (360), or unchanged levels from mid to late gestation (361). The asymmetric transporter protein distribution first noted in term placenta is present in preterm placental tissue (74). This study reports no change in microvillous expression over the last trimester but a 50% increase in basal membrane expression over the same period. This increase in transporter expression in the rate-limiting barrier parallels the increased transport of glucose to the fetus over the second half of gestation. GLUT1 expression has also been reported in the mononuclear cytotrophoblast cells of the term placenta (362).

GLUT3 mRNA has been identified in term placental trophoblast with a similar distribution to that of GLUT1 (361,363). mRNA for this isoform is also present in the vascular endothelium (363). However, despite unequivocal mRNA expression, the localization of protein expression is again controversial. Hauguel-de Mouzon et al. (363), using immunoblotting, report GLUT3 protein in membranes isolated from whole placentas and in endothelial cells derived from intraplacental microvessels, but not in isolated trophoblast cells. Arnott et al. (356) report a predominantly microvillous location for this protein. Conversely, Barros et al. (364), Jansson et al. (74), and Clarson et al. (362) report no GLUT3 protein expression in either microvillous or basal membranes of the term syncytiotrophoblast. Studies using trophoblast cells in culture have demonstrated GLUT3 protein expression in JAr (362), JEG, and BeWo cells (365) but no expression in cytotrophoblast cells in culture, whether studied 18 or 66 hours after isolation. Protein expression at term is therefore somewhat inconsistent with mRNA expression. In situ hybridization studies have shown that GLUT3 mRNA expression decreases over the course of gestation (360). Ogura et al. (366) have reported GLUT3 expression in trophoblast cells from first trimester tissue. It has been suggested that the expression of GLUT3 may be confined to metabolically active, mitotic trophoblast cells. Thus the terminally differentiated syncytiotrophoblast would not be expected to express this isoform. This hypothesis would be in keeping with the decreasing levels of expression through gestation as the proportion of terminally differentiated syncytiotrophoblasts increases as gestation proceeds (362). GLUT3 has been shown to have a higher affinity than GLUT1. Baumann et al. (367) suggest that early in gestation, before maternal circulation has been established, the presence of higher levels of GLUT3 may be important in maintaining glucose uptake from the maternal blood spaces where the poor circulation at this stage is likely to result in reduced glucose concentration adjacent to the placenta.

GLUT4 is the major insulin responsive isoform of the GLUT transporter family. It is expressed primarily in tissues in which glucose transport is rapidly stimulated by insulin, e.g., adipocytes and muscle cells. In these tissues GLUT4 is localized on cytoplasmic vesicles. Insulin causes the vesicles to fuse with the plasma membrane, thereby rapidly increasing cell surface GLUT4 expression and thence glucose transport activity by these cells. The presence of GLUT4 in the human placenta has been reported at both the RNA level (359) and the protein level (368), although in both cases analysis showed only low levels of expression. Xing et al. (369) report strong staining of intravillous stromal cells for GLUT4 and suggest co-localization of GLUT4 transporters with insulin receptors in these cells. In contrast, Takata et al. (370) and Baros et al. (364) failed to identify GLUT4 expression by immunohistochemistry. Although a large number of insulin receptors have been identified in the human placenta (17,371) glucose transport is generally regarded as insulin insensitive (372,373). The role of the GLUT4 transporter in this tissue is thus still a matter of discussion.

In the sheep placenta GLUT1 expression increases up to 120 days and then decreases to approximately two thirds of its maximal level by term. In contrast GLUT3 expression increases throughout gestation reaching its maximal expression at term (374). Das et al. (375) have shown that GLUT1 is localized on both the apical and basolateral membranes of the trophectoderm and in the epithelial cells lining the placental uterine glands. In contrast GLUT3 is found on only the apical membrane of the trophectoderm.

In the rat Takata et al. (376) and Shin et al. (377) have demonstrated that GLUT1 is abundant in the cell membranes of syncytiotrophoblast layers I and II. These studies have shown that GLUT1 is localized to the maternal side of syncytiotrophoblast layer I and to the fetal side of syncytiotrophoblast layer II. GLUT3 has also been shown to be expressed in the rat placenta (378), however, this isoform is localized to the maternal facing plasma membrane of syncytiotrophoblast layer I. Shin et al. (377) using double fluorescence microscopy have also shown that connexin 26 is present between syncytiotrophoblast layers I and II and, using immunogold electron microscopy, have localized this to the gap junctions between layers I and II. This group hypothesize that GLUT1 and GLUT3 facilitate the entry of glucose into the cytoplasm of syncytiotrophoblast layer I from maternal blood. Connexin 26 then allows transfer from syncytiotrophoblast layer I to syncytiotrophoblast layer II. Glucose then passes into the fetal circulation via GLUT1 on the fetal facing membrane of layer II. In the placentas of diabetic rats expression of GLUT3 is found to be increased, whereas that of GLUT1 is unchanged (379). Similarly in rats with short term hyperglycaemia GLUT3 expression increases with no obvious change in GLUT1. Therefore, GLUT3 in the rat placenta appears to be highly sensitive to changes in maternal blood glucose levels and may play an important role in adaptation to changing maternal glucose status. Zhou et al. (380) have shown that placental GLUT1 mRNA levels in the rat decrease after mid gestation, whereas levels of GLUT3 mRNA are unchanged. This contrasts to data obtained for the mouse in which, GLUT1 mRNA levels in placenta and decidua increase throughout gestation.

Thus glucose transfer across the chorioallantoic placenta demonstrates all the characteristics of a facilitated diffusion mechanism. However, the study of placental glucose transfer is complicated by the metabolism of this substrate by the placenta itself. Simultaneous measurements in vivo of the loss of D-glucose from the uterine circulation and gain of D-glucose by the umbilical circulation indicate that in the sheep the placenta and uterus consume a large fraction of the D-glucose removed from the uterine circulation (335,381). Under basal conditions only 28% of glucose entering placental tissue is transferred across to the fetus in this species. Meschia et al. (382) determined glucose utilization by the ovine uterus and placenta to be 0.205 mmol/min. Carter (383) calculates that in the sheep 87% of the oxygen consumed by the placenta is used for the oxidative phosphorylation of glucose, this producing approximately 3.4 mmol/min ATP, which is in turn used to fuel processes such as placental growth and transport functions.

Lipids

A complex variety of lipids are presented to the placental exchange area. Transport of lipids is inevitably governed by the high hydrophobicity exhibited by these molecules; hence, circulation in aqueous plasma requires mechanisms that circumvent this. Free fatty acids (FFA) in plasma are physically bound to albumin while other lipids, i.e., triglycerides, phospholipids, and cholesterol circulate as lipoproteins. These complexes have the more hydrophobic triglycerides and cholesterol esters at their core with the less hydrophobic phosphatidylcholine towards the periphery, associated with specific apolipoproteins. The lipoproteins are classified on the basis of density, as determined by centrifugation, into: very high-density (VHDL), high-density (HDL), low-density (LDL), very low-density (VLDL) lipoproteins and chylomicrons. Table 7 shows the major classes of lipids in maternal plasma together with the equivalent fetal plasma concentrations.

It has long been recognized that there are marked changes in maternal lipid metabolism throughout pregnancy. During the first and second trimesters, increases in maternal fat depots result from both hyperphagia and increased lipogenesis. The third trimester, in contrast, is a time of accelerated lipolytic activity in maternal adipose tissues, this coinciding with the time of maximal fetal growth. The main consequence of these changes in maternal metabolism is a marked maternal hyperlipidemia, a prominent

TABLE 7. *Major classes of lipids in human plasma*

		Concentration in fetal plasma	
Class of Lipid	**Concentration in maternal plasma**	**Umbilical artery**	**Umbilical vein**
Saturated fatty acids[a] (μg/ml)	1432.1 ± 581.4	435.3 ± 193.7	373.9 ± 120.3
Monounsaturated fatty acids[a] (μg/ml)	1062.8 ± 376.8	279.4 ± 126	245.5 ± 83.0
Polyunsaturated fatty acids[a] (μg/ml)			
n-6	1173.4 ± 354.2	2851.1 ± 95.4	269.8 ± 91.2
n-3	77.6 ± 34.7	32.0 ± 13.2	29.9 ± 91
Cholesterol[b] (mg/dl)	206 ± 5	69.6 ± 2	
Triglycerides[b] (mg/dl)	208 ± 14	28.9 ± 2.8	
Ketones[c] (mMol/l)	0.49 ± 0.06	0.25 ± 0.03	

[a]From Ruyle et al. (649).
[b]From Knopp et al. (385).
[c]From Aynsley-Green et al. (427).

TABLE 8. *Changes in composition of lipoproteins during pregnancy*

Lipoprotein class	Triglyceride		Cholesterol	
	Nonpregnant	Pregnant	Nonpregnant	Pregnant
VLDL	33 ± 14 mg/dl	107 ± 41 mg/dl	11 ± 6 mg/dl	22 ± 9 mg/dl
LDL	14 ± 10 mg/dl	72 ± 21 mg/dl	104 ± 23 mg/dl	161 ± 39 mg/dl
HDL	12 ± 6 mg/dl	29 ± 9 mg/dl	56 ± 12 mg/dl	64 ± 9 mg/dl

From Knopp et al. (385).

feature of which is an increase in plasma triglycerides and cholesterol. The increase in triglycerides results not only from an increase in VLDL but also from an enrichment of triglycerides in all other classes of lipoproteins, see Table 8. [For a review see (384,385).] These maternal metabolic adaptations to pregnancy provide an increased supply of fatty acids, long-chain polyunsaturated fatty acids (LCPFA), triglycerides, cholesterol and, ketones (see below) to the fetus. It has also been suggested that they allow the use of lipid as a maternal oxidative fuel in late gestation, facilitating maximal glucose transfer to the fetus (385).

The hydrophobic nature of lipids means that diffusion through the lipid phase of the placental membranes is likely to be rapid and placental permeability correspondingly high. Their transfer will therefore tend toward being flow limited with both maternal and fetal blood flow rates being likely to have a significant effect on transfer. Albumin bound fatty acid transfer will be affected by albumin concentrations in both the fetal and maternal circulations and by the relative association/dissociation constants for the different fatty acids, which in turn is dependant on chain length (386). The complex nature of lipoproteins and the ability of the placenta to bind and metabolize them complicate understanding of the transfer of their lipid components to the fetus. The fetus also has the capability for de novo synthesis of nonessential fatty acids, this together with requirements for growth, hormone synthesis and ecosanoid production by the placenta itself leads to a very complex picture when considering the mechanisms by which fetal need is met.

Free Fatty Acids and Triglycerides Essential fatty acids (EFA) and their derivatives, long-chain polyunsaturated fatty acids (LCPFA), are required for normal fetal growth and development (387). As essential components of membrane phospholipids, they maintain cellular and organelle integrity, they are precursors for the eicosanoids and, increasingly, they are being recognized as intracellular regulators of gene expression (388). Decosahexanoic acid (DHA), for example, constitutes a large proportion of the phospholipid fatty acids in the brain and retina, and fetal accumulation of this LCPFA in the third trimester of human pregnancy is approximately 21 g per week (389). A reduction in this accumulation of DHA during the last trimester has been associated with impaired retinal function and learning ability (387,390). Nonessential fatty acids are not only important as structural components but are also stored in fetal body fat in those species, such as the human, which have a substantial fetal fat content, and for whom they are an important energy substrate early in postnatal life.

Both the FFA bound to albumin in maternal plasma and the esterified fatty acids, in triglyceride enriched lipoproteins, are potential sources of fatty acids in maternal plasma. There is, however, little evidence for the direct placental transfer of lipoproteins, phospholipids, or triglycerides. EFA derived from maternal lipoprotein triglycerides rather become available to the fetus by virtue of lipoprotein receptors on the maternal facing microvillous membrane of the placental trophoblast. VLDL/apo E receptor, LDL receptor, and LDL receptor-related protein are all expressed in the human placenta (391). Together with lipoprotein lipase (LPL) on the maternal facing plasma membrane of the placenta, they are thought to bind maternal lipoproteins and release their FA intratrophoblastically (392–394).

The net flux of FFA across the placenta varies greatly from one species to another depending on the type of placenta. In the hemochorial placenta of the rabbit (395), guinea pig (396), rat (397), and primate (398) net placental transfer of fatty acids exceeds that needed to fulfil lipid storage requirements of the fetus. By contrast, early studies, which measured venous-arterial differences across the umbilical circulation of the epitheliochorial placenta in the sheep, showed little net transfer of FFA to the fetus (399). Similarly, in the pig (400) and cat (401) net FFA transfer to the fetus is low.

Measurement of individual fatty acid concentrations in human maternal plasma and in umbilical venous and arterial plasma at term reveals a maternal concentration generally higher than fetal with a higher umbilical venous than umbilical arterial concentration for most fatty acids (385,402–404). Two notable exceptions to this are DHA and arachidonic acid (AA). The concentrations of these two LCPFAs in fetal plasma exceed that in maternal plasma. In fetal plasma they also exceed the concentrations of their precursors, α-linoleic acid (ALA) and linoleic acid (LA), a situation only ever seen in fetal life [for review see (405)]. Since placental tissue does not express Δ desaturase activity (406) any LCPFA in fetal plasma is probably of maternal origin, although there is some evidence to suggest that fetal liver and brain may express some Δ desaturase activity (407).

Several studies using the in situ perfused rabbit or guinea pig placenta have shown that net FFA transfer to the fetus correlates with maternal FFA plasma concentrations (408,409) and is a function of the transplacental concentration difference, an observation also made in the human placenta (402). In the guinea pig, an increase in albumin concentration in the fetal perfusate was associated with an increase in net FFA transport, as was an increase in umbilical circulation perfusion rate. These observations together suggest that placental transfer of most fatty acids is by a simple diffusional process that is flow limited. However, in the in situ perfused rabbit placenta, under experimental manipulation to remove the maternofetal concentration gradient, net transfer of fatty acids was still observed (410,409). Also Haggerty et al. (411) have shown preferential transfer of certain fatty acids across the perfused human placenta, i.e., DHA>ALA>LA>oleic>AA. These findings together with the higher fetal plasma concentrations of DHA and AA indicate that, for at least some FFA, active and selective processes must be present in addition to a diffusional pathway.

There is increasing evidence that cellular fatty acid uptake indeed occurs via a process of facilitated diffusion. Campbell et al. (412,413) have identified a plasma membrane fatty acid–binding protein, located on the microvillous membrane of the human placenta, which shows selective binding of fatty acids. This 40kDa protein has also been identified in the placenta of both the sheep (414) and rat (415). It is similar, but not identical, to the ubiquitous plasma membrane fatty acid binding protein ($FABP_{pm}$) first described in rat hepatocytes (416) which showed saturation kinetics consistent with a facilitated transport mechanism for fatty acids (416). Furthermore, Lafond et al. (417) have demonstrated linoleic acid uptake into microvillous membrane vesicles of the human placenta by a "single saturable membrane associated system."

Placental $FABP_{pm}$ ($pFABP_{pm}$) has been shown to have selective affinities for different fatty acids. Campbell et al. (418), using purified $pFABP_{pm}$, showed higher affinities and binding capacities for AA and DHA than for linoleic and oleic acids. The same group (419) have also shown similar binding affinities of fatty acids to human placental membranes, i.e., AA>AL>ALA>oleic acid. $pFABP_{pm}$ is therefore thought to contribute, at least in part, to the observed preferential transfer of LCPFA to the fetus, and therefore to the high level of DHA and AA seen in fetal plasma. In addition to this preferential binding of AA by $pFABP_{pm}$, AA uptake by the syncytiotrophoblast membranes has also been shown to be highly sodium and ATP dependent and an active transport mechanism for this fatty acid has been suggested (420).

In addition to $pFABP_{pm}$, several other fatty acid transport proteins are expressed in the placenta [for review see (421)]. Both membrane and cytoplasmic fatty acid transport/binding proteins have been identified in human (422) and rat (415) placenta. Fatty acid translocase (FAT, 88kDa) and fatty acid transport protein (FATP, 62kDa) are present on both the microvillous and basal plasma membranes of the human placenta (422) and in the junctional and labyrinth zones of the rat placenta, although FAT in the rat placenta shows predominant expression in the labyrinth zone (415). Several cytoplasmic FABPs are also expressed by the placenta, with both heart (h) and liver (l) type FABP present in the trophoblast of the human placenta (422) and hFABP present in the labyrinth zone of the rat placenta (415). Although expression of these proteins has been confirmed, their role in fatty acid uptake and metabolism in the placenta has yet to be elucidated.

Placental metabolism also contributes to the selectivity of fatty acid transfer. A certain proportion of AA is converted to prostaglandins (406) while there is selective incorporation of certain fatty acids into phospholipids (423,424), oxidation of placental fatty acids and placental lipid synthesis (425). Therefore, while the biggest determinant of fatty acid delivery to the fetus is the composition of lipids in and delivery by the maternal circulation, a combination of selective uptake and placental metabolism also influences the rate and selectivity of placental transfer of fatty acids. The importance of metabolism and transport in relation to fetal growth is highlighted by the recent observation that microvillous lipoprotein lipase activity is reduced in preterm IUGR, compared to gestation matched placentas, and increased in placentas from insulin-dependent diabetes mellitus (IDDM) pregnancies; the expression of a FABP (L-FABP) was increased in both IDDM and gestational diabetes (426).

Ketones There are major species differences in the placental transfer of the ketones acetoacetate and β-hydroxybutyrate. In woman the concentration in maternal plasma is about twice that in fetal both at delivery (Table 1) and at mid-gestation (427), although the absolute concentrations are lower in the latter. Fetal and maternal concentrations correlate with each other (428). In anesthetized nonhuman primates the concentrations reported are almost identical with those found in the human during labor (429). The approximate 2:1 maternofetal concentration ratio found in both species is maintained during maternal infusion of β-hydroxybutyrate in the baboon even when this is sufficient to raise maternal concentration from 0.7 ± 0.21 to 5.42 ± 0.93 mM. Such an infusion lead to maternal and fetal lactic acidemia.

In rats permeability appears to be even higher with both β-hyroxybutyrate and acetoacetate concentrations being almost identical in anesthetized mother and fetus. This is true in both fed and fasted animals although concentrations were 50 to 100 times higher in the latter state (430). In this species the high permeability may be due to a co-transport system for β-hydroxybutyrate and protons, which has been found in microvillous vesicles prepared from rat placenta. The activity of this transporter does not change in the last three days of gestation, but is increased by maternal fasting (431).

In the sheep, by contrast, there is little transfer of either β-hydroxybutyrate or acetoacetate with fetal concentrations being about ten times lower than maternal at rest and 40 times lower following maternal infusion of β-hydroxybutyrate (432). However, despite its low placental permeability in this species, fetally infused β-hydroxybutyrate can entirely replace glucose in fetal limb metabolism (433) with a concomitant rise in release of lactate into fetal circulation similar to that seen with its infusion in the baboon.

Respiratory Gases

Oxygen

In addition to its central role as a substrate for oxidative metabolism, oxygen has been suggested to have a plethora of diverse functions, e.g., regulation of placental growth and development (434,435), invasiveness of placental trophoblast (436–438) and gene switching (438,439). These functions, however, are not considered further here. An adequate oxygen supply to the fetus appears to be essential for normal growth and development. In the sheep, restriction of umbilical blood flow and the accompanying fall in fetal oxygen uptake is associated with a fall in fetal growth rate (440,441). Similarly the higher incidence of fetal growth restriction reported in high altitude pregnancies is believed to be the result of chronic hypoxia in these pregnancies (442). Likewise, maternal cigarette smoking causes decreased maternal and fetal O_2 capacities as a result of increased levels of CO and this is believed to be the cause of the fetal growth restriction seen in these cases (443).

The placental permeability towards oxygen as measured by its diffusing capacity is high (4) and is generally believed to be a function of the lipid solubility of the gas, although an old inhibition study suggested the possibility that P450 might play a role (444). The reaction velocities for association/dissociation of oxygen with hemoglobin also appear to be sufficiently high so as to not limit the transfer of oxygen. Transfer across the placental tissue is likely to be fast and maternofetal flux therefore predominantly flow rather than membrane limited. Maternal and fetal blood flow rates, together with the hemoglobin content and oxygen affinities of the two circulations and the geometric arrangement of fetal and maternal blood vessels (discussed below) can therefore be expected to predominate in determining the net flux of oxygen. A weakness is that most functional studies of this flow limitation have been made in a single nonhemochorial species, the sheep.

Evidence that O_2 transfer across the placenta is flow limited has come from several early studies in the sheep [reviewed in (445)]. Partial occlusion of the umbilical vessels for 3–4 minutes resulted in a decrease in O_2 delivery to the fetus (446). Since umbilical venous oxygen content did not change, umbilical venous O_2 delivery to the fetus was linearly related to blood flow. Interestingly in these studies there was a progressive increase in the proportion of delivered O_2 extracted by the fetus as umbilical blood flow was reduced, such that the umbilical flow could be reduced to almost 50% of control values without reducing fetal O_2 consumption. Net flux was maintained by a fall in umbilical arterial oxygen content. These observations were confirmed in a similar series of experiments where umbilical flow was reduced by partially occluding the fetal aorta for 4–8 minutes (447). Again umbilical venous O_2 delivery was linearly related to blood flow while net maternofetal O_2 flux was maintained by an increased fetal fractional O_2 extraction. It has been suggested that this response to short term reductions in umbilical flow provides a mechanism by which the fetus can compensate for short term reductions in flow, e.g., during cord compressions in labor. During chronic reductions in umbilical flow, fetal oxygen uptake was reduced from 286 to 237 μmol (6.4 to 5.3 ml)/min/kg as umbilical blood flow fell from 202 ± 47 to 116 ± 36 ml/min/kg (448). Again PO_2 of umbilical venous blood remained constant. This restriction in flow had previously been shown to be associated with a decrease in fetal growth rate from 4% to 2% per day (440).

Reduction in uterine blood flow by partial occlusion of the uterine artery similarly shows that net flux of oxygen into the umbilical circulation correlates with uterine blood flow (449). Net oxygen flux into the umbilical circulation was independent of the calculated uterine oxygen delivery rate when this was greater than 1.0 mmol/min/kg fetal weight. However, when this

TABLE 9. *PO_2 in sheep, monkey, and human (mm Hg)*

	Sheep	Monkey	Human
Maternal artery	90	108	100
Uterine vein	50	37	55 (intervillous space sample)
Umbilical artery	22	15	–
Umbilical vein	35	22	51

Reproduced from Sibley and Boyd (149).

rate was reduced to below 1.0 mmol (22 ml)/min/kg fetal weight, by reduction of uterine blood flow, the net flux of oxygen into the umbilical circulation became limited by the uterine blood flow rate.

The driving force for net oxygen transfer is the difference in partial pressures for the gas between the maternal intervillous space and the fetal capillary (Table 9). The partial pressures of oxygen in the maternal and fetal bloods are determined not only by blood flow rates and transplacental transfer rates but also by the affinities of hemoglobin for oxygen in both maternal and fetal bloods, i.e., the respective red cell oxyhemoglobin dissociation curves (Fig. 8), a function of adult and fetal hemoglobin structures and of 2:3 DPG concentration in the cell and their interactions which change during gestation (450). In most species the fetal oxyhemaglobin dissociation curve is displaced to the left, the cat being an exception to this.

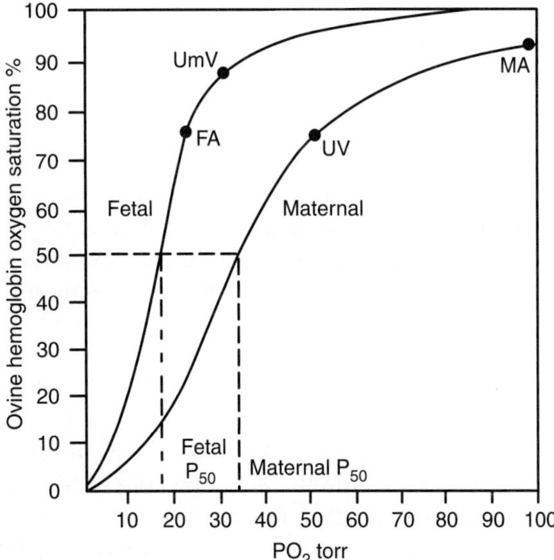

FIG. 8. Maternal and fetal hemoglobin-oxygen dissociation curves for sheep. MA, maternal artery; UV, uterine vein; FA, fetal artery; UmV, umbilical vein. Oxygen transfer from maternal to fetal circulations is associated with progressive decreases in PO_2 from MA to UV to UmV, but the O_2 saturation in UmV is greater and the PO_2 in FA is less than these respective values would be if the fetal hemoglobin affinity for O_2 were not greater than that of maternal hemoglobin. The P_{50} values characterize the relative affinities of hemoglobin for oxygen. [Redrawn from (671) in (5).]

TABLE 10. *Half-saturation pressures (P_{50} torr) of maternal and fetal bloods of several species at pH ~7.4*

Species	Maternal P_{50}	Fetal P_{50}	Difference*
Cat	36	36	0
Llama	21	18	3
Camel	20	17	3
Elephant	24	21	3
Rabbit	31	27	4
Human	26	22	4
Baboon	31	25	6
Dog	28	22	6
Seal	29	21	8
Cow	31	22	9
Rat	38	28	10
Guinea pig	30	19	11
Pig	33	22	11
Goat	30	19	11
Macaque	32	19	13
Sheep	34	17	17

*Maternal pressure–fetal pressure.
From Faber and Thornburg (4).

Table 10 lists the half saturation pressures (P_{50}) for several species at normal body temperature and a pH of around 7.4 (dotted line on Fig. 8). It can be seen that in the cat there is no difference between maternal and fetal P_{50} values, while the sheep is at the opposite end of the spectrum with a 17 torr difference in P_{50} between maternal and fetal bloods. The most important consequence of this left shift in the fetal oxyhemoglobin dissociation curve is that umbilical blood is more highly saturated than it would be if hemoglobin in maternal and fetal red cells had the same affinities for oxygen. Conversely, when the transferred oxygen is extracted by fetal tissues, the end capillary and thus fetal tissue PO_2 will be lower than would be the case were the fetus perfused with maternal blood. This has been suggested to be essential for fetal well being since it contributes to the mechanisms that keep the ductus arteriosus open and the pulmonary vascular bed closed. The functional importance of the left shift in the oxyhemoglobin dissociation curve has been demonstrated in fetal lambs in which fetal blood was replaced with adult blood by exchange transfusion (451,452). Studies in sheep 1–2 days after exchange transfusion showed no change in umbilical blood flow rate. There was, however, a marked decrease in fetal oxygen consumption when fetal oxygen delivery fell to below 0.5 mmol(12 ml)/min/kg fetal weight (452). In individual sheep, the situation is complicated by the fact that their red cells may contain two different hemoglobin types, either a high-affinity (A) or a low-affinity (B) type. The differences in affinity between the two are fully compensated for by a higher rate of perfusion of the uterus, differences in arterial oxygen saturation and a greater O_2 diffusing capacity (453).

The hemoglobin content determines the oxygen carrying capacity of the blood. Since the intra-erythrocytic hemoglobin concentration remains relatively constant this will be determined by hematocrit. The hematocrit is a major determinant of blood viscosity and influences rate of blood flow. Therefore, a hematocrit that gives a compromise between increased carrying capacity and reduced rate of blood flow will result in optimal oxygen supply to the fetus. This has been demonstrated by Edelstone et al. (447) in a series of experiments in sheep in which the fetal hematocrit was manipulated by isovolemic exchange transfusion with packed red cells, or plasma, from a donor fetus. Figure 9 shows that oxygen delivery was maximal at a hematocrit of 33%, which corresponds well to the normal hematocrit in the sheep of 32% ± 3%. As fetal hematocrit was increased above this optimal value O_2 delivery decreased due to a reduction in umbilical flow that was proportionally greater than the rise in umbilical O_2 content. The fetal oxygen uptake remained fairly constant at hematocrits between 16% to 48% due to compensatory changes in fractional oxygen extraction. However, outside this range oxygen uptake by the fetus decreased.

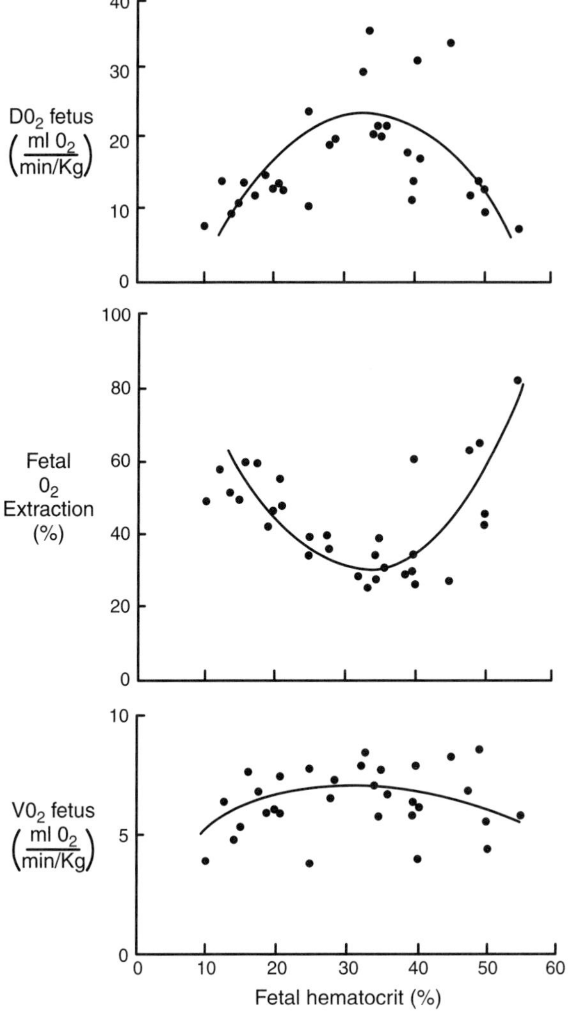

FIG. 9. Fetal oxygen delivery, extraction, and uptake in the sheep fetus as functions of fetal hematocrit. Fetal O_2 delivery (DO_2 defined as umbilical blood flow x umbilical venous $[O_2]$) was greatest at a hematocrit close to the normal value of 32%. However, fetal O_2 uptake (VO_2 defined as umbilical venous $[O_2]$-umbilical arterial $[O_2]$) was maintained, by increased O_2 extraction (defined as umbilical venous $[O_2]$-umbilical arterial $[O_2]$/umbilical venous $[O_2]$), at hematocrits ranging from 16% to 48%, as long as O_2 delivery was >14 ml (0.6 mmol)/min/kg fetus. [Redrawn from (672).]

Geometry of Placental Blood Flow. As well as the rates of maternal and fetal blood flow, the geometric arrangement of flow on either side of the placenta is also likely to be an important determinant of oxygen transfer (Table 11). Several models have been put forward of theoretically possible vascular arrangements in placenta (4,454).

Double Pool Flow. To obtain a functional placenta, the exchange membrane has to be in contact with the maternal and fetal circulations. In the simplest case, the blood is led into chambers that border the exchange area. The blood on each side is well mixed. Such an arrangement is termed a double pool flow exchanger.

Single Pool Flow. In this case, only on one side of the placenta does blood enter a chamber, on the other side, the blood travels in vessels.

Concurrent. Blood on both sides of the exchange area travels in vessels. The maternal and fetal vessels lie parallel to each other and the blood flow in each is in the same direction.

Countercurrent. Again the maternal and fetal vessels are parallel to one another but in this case the blood flows in opposite directions.

Crosscurrent. In this case the vessels are at right angles to one another, therefore flows on maternal and fetal sides of the exchange membrane are perpendicular.

Multivillous. The fetal vessels enter and branch within the maternal intervillous space while maternal blood pools between the villi.

The "efficiency" of each of these models, as defined by Schroder (454) as the arteriovenous concentration difference on one side, divided by the arterial maternofetal concentration difference, is discussed by both Faber and Thornburg (4) and Schroder (454). It is concluded, that under conditions of flow limited transfer the countercurrent and crosscurrent exchangers are more efficient than the other exchangers discussed, apart from the multivillous model, which has been suggested to most closely represent flow geometry in the human placenta, and whose efficiency is thought to lie between the countercurrent and concurrent models. For diffusion to occur from maternal to fetal circulations, it is necessary for maternal plasma partial

TABLE 11. *Estimates of maternal and fetal placental blood flows in several species near term*

	Mean placental blood flows (± SE)				
Species	Maternal (ml/min/kg uterus plus contents)	Ref.	Fetal (ml/min/kg fetus)	Ref.	Method of measurement
Human	124 ± 12	(650)	≤35wk: mean = 120 ≥35wk: mean = 90	(651)	Non–steady-state N_2O equilibration in spinal or epidural anaesthetized woman Pulsed Doppler ultrasound in conscious women
Rhesus monkey	126 ± 12	(652)	208 ± 8	(653)	Radiolabeled microspheres in anaesthetized restrained monkeys
Guinea pig	87%–93% of uterine blood flow 171 ± 15	(654)	–	–	Radiolabeled microspheres in conscious guinea pigs
Sheep	288 ± 55	(655)	256 ± 18	(656)	Radiolabeled microspheres in conscious ewes

pressure, or concentration in the case of nongaseous solutes, to be greater than fetal plasma concentration. The best a concurrent exchanger can do is to allow venous equilibration, in which case, the fetal arteriovenous concentration difference is only half the arterial concentration difference, assuming flows are equal (Fig. 10). In a countercurrent exchanger, the concentration in maternal blood can drop to a level almost as low as the fetal arterial concentration without violating the requirement that for a permeant only crossing by diffusion, its concentration in maternal plasma must exceed fetal everywhere along the exchange vessels or vice versa for fetomaternal flux. In sheep, the umbilical venous PO_2 has been shown to be always less than the maternal uterine venous PO_2. From this observation, a concurrent exchanger has been inferred (455). The horse appears to be similar (456). However, other experimental evidence in sheep suggests that some crosscurrent flow may be also be present (30,457).

Histological examination of the labyrinthine placentas of rabbits and guinea pigs also led to the hypothesis that a countercurrent exchange system operates in these species (458,459) and physiological studies on placental transport of highly diffusable molecules confirmed this histologic finding (460–463). The significance of these different vessel geometries is discussed in detail by (464).

The study of placental flow geometries is complicated by the presence of vascular shunts bypassing the placental exchange area. Studies in the late gestation sheep using radiolabeled microspheres found that 16% of uterine blood flow and 6% of umbilical flow perfuses nonexchange tissues (29,30). It is also possible that due to the close proximity of the fetal arterioles and venules in the placental villi, diffusion of highly permeant molecules between the two may lead to a different category of shunt allowing such permeants to bypass the maternofetal exchange area. This umbilical diffusional shunt has been estimated in the sheep to be up to 25% of umbilical flow (465). For both these reasons, PO_2 of umbilical venous blood may be predicted to be lower than that in capillary blood leaving the exchange area.

Oxygen supply to the human fetus changes dramatically over gestation, as noted in section Blood Flow. Hustin and Schaaps (466), using a variety of in vivo imaging techniques demonstrated that there is little maternal blood flow to the intervillous space prior to 10–12 weeks of gestation. As a consequence the oxygen concentration within the fetoplacental unit is low during early gestation, i.e., 25.6 mm Hg

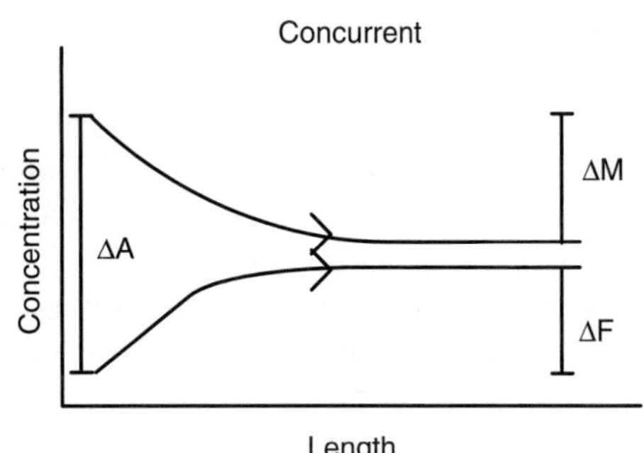

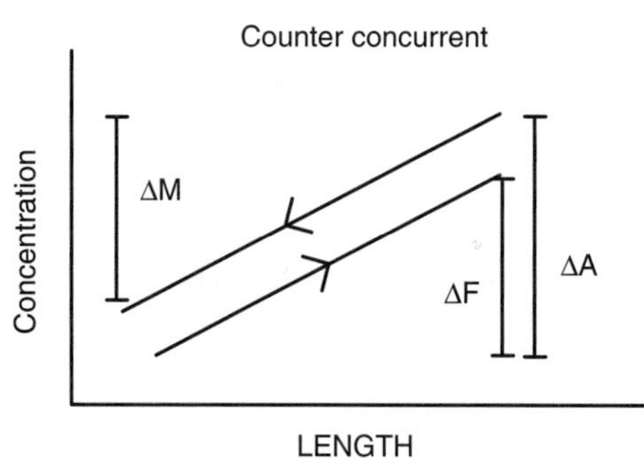

FIG. 10. Efficiency of concurrent and countercurrent models of placental blood flow. △A, maternofetal arterial concentration difference; △M, maternal uterine arteriovenous concentration difference; △F, fetal umbilical arteriovenous concentration difference. [Adapted from (4).]

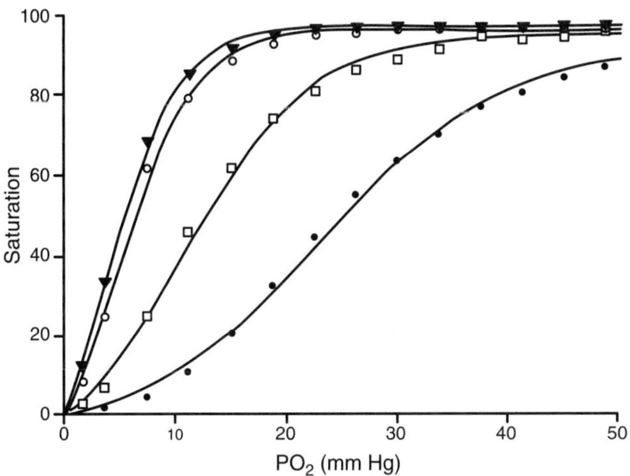

FIG. 11. Oxygen binding to embryonic hemoglobins. The saturation of hemoglobin with oxygen as a function of oxygen partial pressure is shown for adult hemoglobin (*black circles*), embryonic hemoglobin 1 (*black inverted triangles*), 2 (*open squares*), and 3 (*open circles*). [Reproduced from (469).]

at 7–10 weeks, rising to 56.2 mm Hg at the end of the first trimester (467). Furthermore Jauniaux et al. (468) have suggested that the structure of the first trimester gestational sac actually limits rather than facilitates the transfer of oxygen to the fetus. From week 4 of gestation to the establishment of the maternal circulation at the end of the first trimester three embryonic hemoglobins of different oxygen affinities are found. The three types presumably cater for the changing oxygen tensions to which the conceptus is exposed through the first trimester (Fig. 11). The earliest embryonic hemoglobin has a high affinity for oxygen together with a low sensitivity to modulation of oxygen affinity, and has characteristics in common with myoglobin. The second embryonic hemoglobin exhibits a slightly lower oxygen affinity and a greater degree of allosteric modulation. The final form of embryonic hemoglobin is present at the time of the embryonic to fetal transition and has a higher oxygen affinity and an allosteric sensitivity that closely matches fetal hemoglobin [for a review see (469)]. At the start of the second trimester the maternal circulation within the intervillous space becomes well established and oxygen tension rises. This transition period represents a period of oxidative stress for the placenta and fetus. The responses of the fetoplacental tissues to these changing oxygen tensions may play an important role in the success of the pregnancy and are the subject of much current research (470).

The placenta is a highly metabolic organ. In both the sheep and human, placental oxygen consumption accounts for approximately 40% of the oxygen uptake of the uterus. Of this, 30% has been calculated to be utilized in protein synthesis in the human (471) and 40% in the sheep (472). If uterine oxygen supply is limited, then the high oxygen requirements of the placenta will place limitations on the supply to the fetus. During acute alterations in oxygen supply, fetal uptake is reduced, while placental oxygen consumption remains relatively constant at the expense of the fetus (473). However, during long-term reductions in flow/oxygen supply in the sheep, Owens et al. (474) demonstrate that uteroplacental oxygen consumption is reduced to a greater extent than fetal oxygen consumption such that placental consumption accounts for a smaller proportion of total uterine oxygen uptake.

Carbon Dioxide

Carbon dioxide (CO_2) is produced by all cells as an end product of oxidative metabolism. In the fetal sheep approximately 0.94 mol of CO_2 are produced for every mole of oxygen metabolized, this is eliminated via the placenta to the maternal circulation. Quantifying the transfer of CO_2 across the placenta is complicated by the fact that dissolved CO_2 within both fetal and maternal erythrocytes is in equilibrium with bicarbonate ions (HCO_3^-) via the formation of carbonic acid in a chemical reaction catalyzed by carbonic anhydrase. There are also small amounts of undissociated carbonic acid and carbonate present (4). Thus, it is possible that elimination of CO_2 from the fetal circulation involves not only diffusion of the dissolved gas but also transfer of HCO_3^-. The placental membrane is believed to be highly permeable to CO_2 (475) so that transfer of dissolved CO_2 across the placenta is assumed to be flow limited, although experimental evidence for this is scarce. Bicarbonate by contrast is not freely permeable and mammalian cells ubiquitously express bicarbonate transport proteins to facilitate transmembrane bicarbonate flux.

CO_2. Under normal conditions in the sheep, near term, the P_{CO_2} in umbilical venous blood is approximately 43 torr, and P_{CO_2} in uterine venous blood is approximately 38 torr. An induced reduction in umbilical blood flow rate of 50% to 75% resulted in a small increase in fetal arterial P_{CO_2} of 3–5 torr, but no change in umbilical venous P_{CO_2} (446). In contrast, reductions in the uterine blood flow rate were associated with increases in the uterine venous P_{CO_2} and in the uterine venous–maternal arterial P_{CO_2} difference. Furthermore, umbilical venous P_{CO_2} was strongly correlated with uterine venous P_{CO_2} over a range of umbilical venous P_{CO_2} values induced by changing uterine blood flow rate (449).

Mathematical models of placental CO_2 transfer have been used to predict the effect of changing fetal or maternal capillary blood flow rates (476).

Both maternal and fetal flow rates are predicted to correlate with the rate of CO_2 exchange. An increase in either blood flow rate would be predicted to increase CO_2 exchange, while a decrease in flow rate would be expected to decrease it. These predictions are supported by the observed increase in umbilical venous PCO_2 associated with induced reductions in uterine blood flow in the sheep (449) (see above). Similar mathematical models indicate that CO_2 exchange across the placenta is essentially independent of the placental diffusing capacity for CO_2, leading to the belief that placental CO_2 transfer is flow limited. Further experimental validation of these mathematical models is, however, long overdue.

Both oxygen and carbon dioxide bind reversibly to hemoglobin. Because the binding of each gas with hemoglobin is affected by the binding of the other, the exchange of each of these gases across the placenta in opposite directions is enhanced by the exchange of the other. Once again modelling has been used to estimate the contribution that this makes to the exchange of each gas. It is predicted that the contribution of oxygen exchange (by the Haldane effect) to CO_2 transfer is approximately 45% of the total (476). In contrast, the contribution to oxygen transfer by CO_2 exchange (via the Bhor effect) is estimated to be only 8% of the total.

Bicarbonate

Bicarbonate transporters clusters into two separate families the SLC4 family of bicarbonate transporters and the SLC26a family of sulphate transporters, some of which also transport bicarbonate [for reviews see (477–479)]. As described earlier (see section Chloride) the human placenta expresses AE1 (SLC4A1) on the microvillous plasma membrane throughout gestation (77,178). Also recently Lacey et al. (480) have identified mRNA expression for AE2 throughout gestation. The mouse preimplantation embryo has also been shown to express AE1 at all stages.

In erythrocytes AE1 plays a key role in net CO_2 transfer from the systemic tissues to the lungs. The CO_2 generated by metabolism, present in plasma enters the erythrocytes. Here it is converted, by reaction with H_2O, to HCO_3^- and H^+ by the action of carbonic anhydrase (CA). AE1 then exchanges HCO_3^- for chloride, while the deoxygenated hemoglobin (which has a higher affinity for H^+ than the oxygenated form) buffers the H^+. This then allows the erythrocyte to take up more CO_2 (i.e., the Jacobs-Stewart cycle). In the lung the process is reversed. Erythrocytes express a high level of Aquaporin 1, which is also permeable to CO_2, and Sterling et al. (481) suggest that a cluster of four proteins (AQ1, AE1, hemoglobin, and CA) together promote CO_2/HCO_3^- transmembrane movement.

Both CA1 and CA11 isoforms of carbonic anhydrase have been identified by immunocytochemistry in human placental syncytiotrophoblast, fetal erythrocytes, and placental capillaries (482,483). The concentration of both isoforms increases significantly with gestation (484). This, together with the reported AE1/2 and aquaporin expression suggests that the placenta has all the components necessary to use bicarbonate as a means of excreting CO_2, but whether this is a quantitatively important route has yet to be established. Earlier functional studies in the guinea pig (485) suggested two mechanisms for placental bicarbonate transfer in this species. A portion of the bicarbonate is converted to CO_2 and this then diffuses across the placenta, while an anion exchanger transports the rest. Such a system would agree well with the pattern of transporter expression described above.

Micronutrients

Iron

Iron accretion by the fetus accelerates in the last period of gestation but there are major species differences. In humans data is limited but 60% of the total appears to be transferred after about 32 weeks. In guinea pig, transfer rate increases about 50 times between days 25 and 53 but then halves by term (486). In the rabbit, some 90% is transferred in the last third of gestation (487), while iron transport across rat placenta increases several hundred-fold in the last third of gestation (488). Near term total iron concentrations in humans are higher in fetal than maternal plasma being, e.g., 134 ± 8 and 85 ± 8 μg%, respectively (489).

The broad picture for maternofetal transfer of iron, at least in later gestation, has been clear for many years [elegantly reviewed by Morgan (490)] with major differences between those species with hemochorial placentas which transfer iron from maternal transferrin and the remainder which depend on other sources. Understanding of molecular mechanisms has made substantial progress in the last few years in the former group. Because of the high plasma affinity of apotransferrin insufficient iron is present in free solution in plasma for this to be considered a significant source, in either group of species.

Iron Transfer in Hemochorial Species Kinetic studies [e.g., rat, rabbit (491), guinea pig (486), human (117)] following maternal injection of transferrin doubly labeled with ^{59}Fe and ^{125}I indicate both a quantitatively adequate transfer of iron from transferrin and absence of significant transfer or degradation of the transferrin itself. Ligation of vitelline vessels to the yolk sac of the fetal rabbit after mid-gestation does not impair transfer (492), confirming the dominance, in this species, of the chorioalloantoic placenta and the absence of a quantitatively important yolk sac route for iron transfer, at least at this stage of gestation. On the basis of a single experiment, there appears to be little or no fetomaternal flux of iron in the rabbit (487).

Recent studies have increasingly clarified the molecular mechanisms involved in transfer across the syncytiotrophoblast [reviewed by Srai (493)]. Understanding of transfer mechanisms across the fetal endothelium remains, as for most permeants, essentially unexplored. At the microvillous, maternal face, of placental syncytiotrophoblast, maternal plasma Fe^{2+} transferrin binds to the transferrin receptor and the resulting complex is invaginated into coated pits and endosomes. Acidification to around pH 6.0 leads to the release of iron from apotransferrin, which is returned by relatively slow recycling over some 10 minutes to maternal plasma. The transferrin receptor appears to be colocalized with the hereditary hemochromotosis gene product (hFE) and beta 2-microglobulin in the microvillous membrane (494). The divalent metal transporter DMTI (also abbreviated as DMC1 or Nramp2) is found in placenta in both its iron regulatory element (IRE) regulated and in a non-IRE regulated form. It is not co-localized with the transferrin receptor (495) but, by analogy with other cell lines, is hypothesized to play a role in release from the endosome. Intrasyncytial iron must then pass through the basal membrane of syncytiotrophoblast in the fetal direction. Ferroportin, a 62kD protein (also known as IREG1 or MTP1) is found in this membrane (496) which, in the gut, together with the copper-oxidase, hephaestin, is thought to be key to basal iron release, in which copper plays a role. Although hephaestin has been found in whole placenta, it may not be present in trophoblast cell lines in which another copper oxidase may, according to Srai et al. (493), replace it. The expression of copper oxidase in BeWo cells in culture is inversely related to ambient iron and in 5%, but not 20% oxygen, iron release from such cultures is moderately reduced in low copper conditions. (497). Trophoblast monolayer studies (498) demonstrate transferrin uptake at the basal as well as the maternal face although at a slower rate. There is also some evidence of a nontransferrin-dependent iron uptake capacity (499,499).

The feedback control loop between fetal iron acquisition and placental transfer is not clear and may be slow acting (500). Transferrin receptor number increases in several species as growth proceeds (500). Surprisingly, in the rat, radiolabeled iron uptake into placenta continues unchecked, for at least 20 hours after operative fetal removal. The placenta by then contains 10 times its normal iron content—equal to that found in control fetus and placenta combined! In the presence of dietary iron deficiency, however, some compensatory placental changes are seen with upregulated transferrin receptors, increase in placental copper oxidase and in the IRE regulated content of DMTI. BeWo cells cultured in desferrioxamine, to lower ambient iron concentrations, show rather similar changes (501) and syncytiotrophoblast cultured in low iron concentration demonstrates increased release of preloaded iron (502), and of transferrin receptor synthesis (503,504). A comparison of a group of placentas from diabetic pregnancies with low cord serum ferritin concentrations with diabetic controls, found that iron regulatory protein 1 (IRP1) content was an inverse function of cord blood ferritin concentration (505).

Human stable isotope studies confirm that fetal need for iron takes precedence over maternal near term and suggests the possibility of some control at maternal intestinal level in response to fetal demand (506). It could be hypothesized that some examples of neonatal hemochromatosis might follow from a pathological increase in placental iron transport at the expense of the mother (507) who may become anemic (508). However, there appears to have been little direct attention to the placenta in this condition except to note villous edema in some cases (509). No relevant mutations have been found in beta-2-microglobulin or Hfe genes in fetal fibroblasts (508).

Iron Transfer in Nonhemochorial Species Based predominantly on a range of histological evidence, at least three possible pathways have been identified (33,510).

Hemophagus regions. These are a very diverse range of structures, not present in humans, which can be microscopically demonstrated to ingest and degrade maternal red cells in at least some of carnivora, insectivora, kyroptera and ungulata. The ultrastructural as opposed to the anatomical picture is rather uniform. The microvilli of a tall columnar trophoblastic epithelium surround red cells that can be visualized in large underlying vacuoles before apparently becoming degraded into smaller hemoglobin-containing vacuoles, which presumably release their contained iron to the fetal side. Functional confirmation is found in the cat in which, unlike chorioallantoic species, transferrin iron moves only in small quantities to the fetus (511), while transfer of radio-iron in labeled red cells is 100-fold greater and can be calculated to provide enough iron to meet fetal requirement (512). Local ferritin may provide an iron store (513,514).

Endometrial gland secretion. Secretion of the iron-rich protein uteroferrin followed by absorption into placental areolae appears to be the probable route in pig (515) and horse (34).

Yolk sac uptake of transferrin. This may play a part in iron transfer in early gestation in the rat (516) but not later (488) and, plausibly in other species with a major functioning yolk sac during early gestation.

Drugs and Xenobiotics

For many years the placenta was believed to act a barrier to drugs and other foreign compounds (xenobiotics), providing a sheltered environment in which the fetus could grow and develop normally. The thalidomide disaster of the 1960s changed that view dramatically and it is now assumed that any drug or xenobiotic can potentially cross the placenta. As a result the use of drugs during pregnancy is avoided where possible. However, there are certain circumstances where maternal drug therapy is unavoidable, e.g., epilepsy, hypertension, and diabetes. In addition to therapeutic drugs, recreational drugs, drugs of abuse, and environmental xenobiotics, may all be present in maternal plasma and therefore potentially transfer to the fetus. As with other substances, the rate at which a drug or xenobiotic crosses the placenta and the amount delivered to the fetus will depend on issues covered in Fig. 4, e.g., its physicochemical properties, the electrochemical gradient between maternal and fetal circulations, the presence of transporters within the placenta, protein binding in plasma and metabolism within the placenta.

The majority of drugs and xenobiotics are assumed to cross the placenta by diffusion. For lipophilic compounds the rate of transfer is likely to be flow limited (see section Conceptual and Methodological Issues in Placental Transfer and section Respiratory Gases), while for hydrophilic compounds, passage by paracellular diffusion will be dependent on molecular size (see section Conceptual and Methodological Issues in Placental Transfer). Passage of hydrophilic drugs across the placenta may be expected to be lower in species with a small pore radius, e.g., the sheep as compared to rodents or primates. This has been suggested to be the case for digoxin which readily crosses the rodent and human placenta but not that of the sheep (517). It is usually stated that, in general,

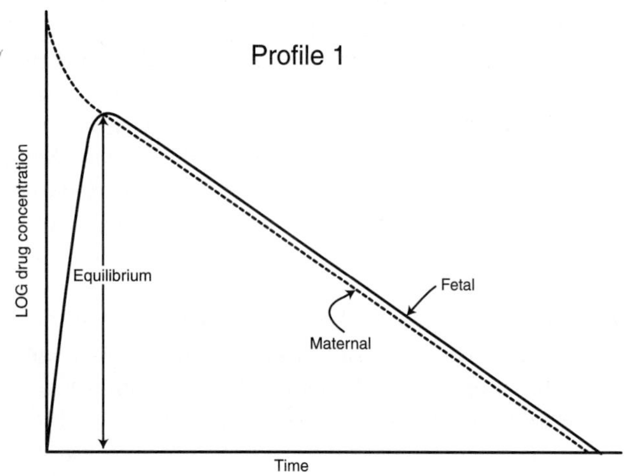

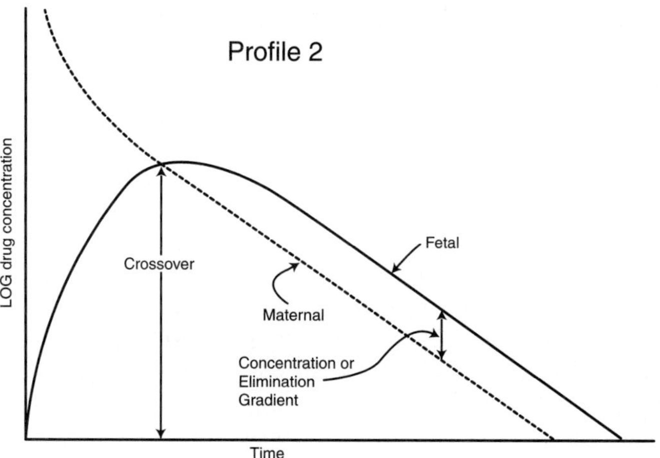

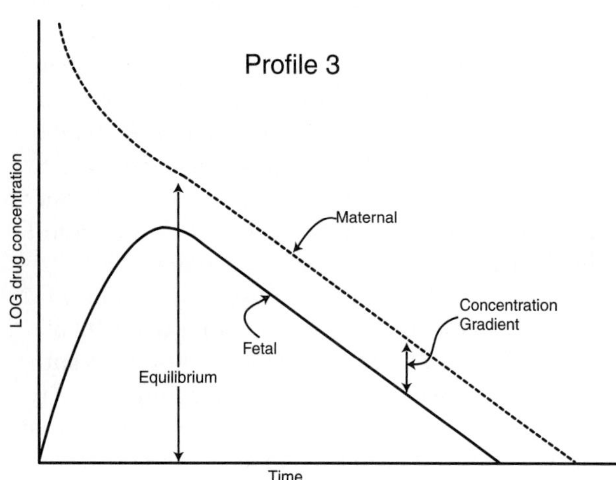

FIG. 12. Suggested pharmokinetic profiles. [Adapted from (519).]

unbound water-soluble drugs with molecular weights under 500 Da can cross the human placenta (see section Paracellular Channels). Protein binding in plasma unless easily reversible, will increase the effective molecular size of the compound and thus reduce its permeability sharply. Binding will also alter lipid solubility which can also have a major influence on placental transfer as may charge through decreasing lipid solubility and thereby permeability. The topic is excellently reviewed by Plonait and Nau (518).

Broadly, three pharmacokinetic profiles can be described (519) (Fig. 12). In the first profile, the drug freely crosses the placenta and rapidly equilibrates between maternal and fetal plasma. There is no binding in either maternal or fetal plasma and the only route of elimination from the fetus is by transfer back across the placenta, e.g., methimazole, carbimazole, in the rat. The second profile is created by the presence of a large fetal compartment that requires longer to equilibrate with maternal plasma, e.g., extensive binding in fetal plasma. The concentration in fetal plasma becomes higher than in maternal circulation due to the faster turnover in the maternal compartment relative to fetal, e.g., valproic acid, diazepam, oxazepam, in the human. The third type of profile is produced when, for example, extensive binding in maternal plasma favors a higher maternal concentration, or when the rate of metabolism, excretion or disposition in the fetus is greater than transfer of drug from the mother, e.g., propylthiouracil in the rat, succinyldicholine in the monkey.

If the foreign compound interacts with transporters or drug metabolizing enzymes within the placenta, this may have a profound effect on its pharmacokinetics. As described throughout this chapter, the role of the placenta in transfer of nutrients to the fetus and elimination of metabolic waste products is facilitated by various transporters expressed differentially on the microvillous and basal plasma membranes of the syncytiotrophoblast. The specificity of these transporters is, however, not restricted to their physiological substrates. Nonendogenous compounds, which bear a structural resemblance to physiological substrates can also interact with several transporters (520). Table 12 shows examples of such transporters together with their physiological substrates and the drugs and/or xenobiotics that have been shown to interact with them. As a consequence, these transporters can mediate the transfer of drugs and xenobiotics across the placenta in addition to, and potentially at the expense of, their physiological substrates (521).

P-glycoprotein (P-gp), encoded by the multi-drug resistance gene MDR1, is of particular interest in this respect. In addition to a few endogenous substrates for P-gp (e.g., cortisol and aldosterone, leukotrienes), it has been shown to extrude a huge variety of structurally diverse drugs and xenobiotics from cells in which it is expressed (522). This transport is ATP dependent, can occur against a concentration gradient, and is independent of electrochemical transmembrane potential or proton gradients (523). In concert with drug metabolizing enzymes such as the cytochrome P450s (discussed later), this active efflux pump has been shown to influence tissue bioavailability and disposition of many drugs and xenobiotics by limiting their uptake from the gastrointestinal tract and facilitating excretion from the kidneys and liver. Moreover, its expression in the blood brain barrier has been shown to protect the brain from the toxic effects of xenobiotics in the circulation (524). Initially identified as the gene that confers multi-drug resistance on cancer cells (525), P-gp has been shown to transport not only the cytotoxic drugs used in cancer chemotherapy but also other therapeutic compounds such as HIV protease inhibitors, anti-convulsants, immunosuppressants, Ca^{2+} channel blockers, cardiac glycosides, and antibiotics (526). This huge range of structurally unrelated compounds transported by P-gp together with its vectorial transport capacity and its expression in excretory tissues and classic pharmacological barriers has led to the belief that this protein has evolved to protect an organism from environmental xenobiotics.

P-gp is expressed in the human placenta throughout gestation (189,192,527–529) and has been shown to be localized predominantly on the microvillous plasma membrane of the syncytiotrophoblast (192,527). Several recent studies have suggested a role for P-gp in placental xenobiotic handling in both the human (192,528,530–532) and the rat (533). Atkinson et al. (192) have demonstrated that reduced expression of MDR1 in human placental trophoblast cells results in a diminished capacity to efflux substrate drugs and that this can be corrected by retroviral transduction of MDR1. Expression of P-gp on the microvillous membrane of the human placenta would therefore be expected to maintain lower levels of those xenobiotics it extrudes within the fetoplacental unit and provide protection to the fetus from those xenobiotics in maternal plasma. This protective role of P-gp has been clearly demonstrated in animal studies, where absence of P-gp in mouse placenta results in 100% susceptibility to ivermectin-induced cleft palate (534) and a seven-fold increase in transfer of substrate drugs such as digoxin, saquinavir, and taxol to the fetus (535). This, incidentally, confirms that the simple assumption of paracellular diffusional transfer of, e.g., digoxin is incorrect, at least in the mouse.

Other xenobiotic/drug efflux proteins expressed by the human placenta include the multi-drug resistance

TABLE 12. *Transport of nonendogenous compounds by placental transporters*

Transporter	Endogenous substrate	Drug/xenobiotic	Comment	Ref.
SERT	Seratonin	Amphetamines, cocaine, tricyclic and nontricyclic anti-depressants	Na^+/Cl^-–coupled active uptake of amphetamine. Binding but no transfer of cocaine and antidepressants	(657)
NET	Norepinephrine Dopamine	Amphetamines, cocaine, tricyclic and nontricyclic anti-depressants	Na^+/Cl^-–coupled active uptake of amphetamine. Affinity much higher than SERT. Binding but no transfer of cocaine and antidepressants	(657)
Monoamine transporter OCT_3	Na/Cl-independent transport of dopamine, norepinephrine, seratonin, and histamine	Amphetamines; 1-methyl-4-phenyl pyridinium (neurotoxin); imipramine, desipramine (anti-depressants) tetraethylammonium (K^+ channel blocker)	OCT3 accepts organic cations; driving force for transport is the membrane potential	(658,659)
Carnitine transporter ($OCTN_2$)	Carnitine	Tetraethylammonium, quinidine, verapamil, pyrilamine, cephaloridine, anticonvulsants	Na^+-dependent transport of zwitterions; Na^+-independent transport of cations	(660–663)
Monocarboxylate transporters MCT1, MCT3, MCT4, MCT5, MCT7	e.g., Lactate, pyruvate	Benzoic acid, acetic acid, acetylsalicytic acid (aspirin),cefdinir	Electroneutral co-transport of H^+ and monocarboxylates	(664,665)
Dicarboxylate transporters ($NaDC_3$)	Succinate α-ketoglutarate	Trans-pyrrolidine-2,4-dicarboxlate (glutamate transporter blocker)		(666)
Sodium/multivitamin transporter (SMVT)	Biotin, pantothenate, lipoate	Anticonvulsant drugs including carbamazepine, primidone	Na^+-dependent electrogenic transport	(520)
Folate receptor (α-isoform)	N-methyl-tetrahydrofolate	Methotrexate	Folate receptor on MVM	(667, 668)
Folate transporter(FOLT1)			Folate transporter on BM	
MDR1 p-glycoprotein P-gp	Binds progesterone, transports steroids, e.g., cortisol, aldosterone	A huge range of structurally unrelated drugs and xenobiotics including anticancer drugs, cytotoxics, HIV protease inhibitors, and drugs of abuse	Unidirectional active transport of drugs and xenobiotics out of cells	(192,527, 530)
Multi-drug resistance protein MRP1	Cystenyl leukotriene LTC_4 Bilirubin	A huge range of structurally unrelated drugs and xenobiotics Heavy metals	Active transport of drugs and xenobiotics out of cells often as glutathione conjugates or co-transported with conjugates	(192,527, 669)
Amino acid transporters	Amino acids	Gabapentin (anticonvulsant) arginine analogs, thyroid hormone mimics		(520)

associated proteins MRP1and MRP2 and the breast cancer resistance protein BCRP. MRP1 is capable of transporting a similar range of drugs and xenobiotics as P-gp. The main difference between the two being that many MRP1 substrates are transported either as glucuronide or glutathione conjugates or co-transported with glutathione (536). This protein has been shown to be expressed on both the basal membrane of the syncytiotrophoblast (192,537) and the fetal endothelium of the human placenta (527). From its known substrate profile it is likely that MRP1 will be involved in the transport of drugs and xenobiotics, but its basolateral position in the placenta means that efflux of xenobiotics mediated by it will presumably be towards the fetus. MRP1, however, may play a role in preventing accumulation of xenobiotics in cells of the trophoblast itself. MRP2 has been shown to be expressed on the microvillous membrane of the human syncytiotrophoblast. The substrate profile of this transporter is less clearly defined than P-gp and MRP1 but its microvillous membrane location would facilitate a role in the ATP-dependent efflux of organic anions from the trophoblast. BCRP has been shown to be involved in the resistance to chemotherapeutic drugs in ovary, breast, colon, and gastric cancers and its substrate profile shows some degree of overlap with P-gp and MRP1. In normal human tissues, BCRP has been shown to be expressed predominantly in placenta, with lower expression in liver small intestine and colon (538,539). Due to its high level of expression in placenta a role in protecting the fetus from drugs and xenobiotics in maternal plasma has been suggested for this protein (540).

In the rat Leazer and Klaassen (541) have identified expression of 16 transporters likely to have a role in xenobiotic transfer in the placenta: Mdr1a, Mdr1b, Mrp1, Mrp5, Oct3, Octn1, Oatp3, Oatp12, four metal transporters, prostaglandin transporter, AbcG8, ENT1, and ENT2. Of the 40 genes screened, these 16 were found to be expressed in placenta at levels similar to or higher than those in maternal liver and kidney.

Drug/xenobiotic metabolism in both maternal and fetal liver and the placenta will affect the exposure of the fetus to such compounds. The cytochrome P450 (CYP) enzymes are known to catalyse oxidative metabolism of a vast number of compounds and about 20 xenobiotic metabolizing forms of CYP are known to exist in humans. These are most abundant in the adult liver, although extrahepatic forms do exist, e.g., in the gut (542), adrenal, lung, (543) and placenta (544,545). The human placenta has been shown to express mRNA for CYP 1A1, 1A2, 2C, 2D6, 2E1, 2F1, 3A4, 3A5, 3A7, and 4B1 during the first trimester and CYP1A1, 2E1, 2F1, 3A3/4, 3A5, and 3A7 at term [for review see (545)]. The fetal liver also has a relatively well-developed metabolism of xenobiotics by comparison to that of the rat or mouse and there is experimental evidence for CYP1A1, 1B1, 2C8, 2D6, 2E1, 3A4, 3A5, and 3A7 [for review see (544)].

ACKNOWLEDGMENTS

Many thanks to Jean French and Stephen Gregory for all their help with reference management.

REFERENCES

1. Villee, C. A. (1960). *The Placenta and Fetal Membranes.* Williams & Wilkins, Baltimore.
2. Hagerman, D. D., and Villee, C. A. (1960). Transport functions of the placenta. *Physiol. Rev.* 40, 313–330.
3. Schultz, R. L. Placental transport: a review. *Ob Gyn Survey.* 25, 979–1020.
4. Faber, J. J., and Thornburg, K. L. (1983). *Placental Physiology.* Raven Press, New York.
5. Morriss, F. H., Boyd, R. D. H., and Mahendran, D. (1994). Placental transport. In *The Physiology of Reproduction* (E. Knobil and J. D. Neill, Eds.), Vol. 2, pp. 813–863. Raven Press, New York.
6. Power, G. G., Blood, A. B., and Hunter, C. J. (2004). Perinatal thermal physiology. In *Fetal and Neonatal Physiology* (R. A. Polin, W. W. Fox, and S. H. Abman, Eds.), Vol. 1, pp. 541–548. WB Saunders, Philadelphia.
7. Lo, Y. M., Lau, T. K., Chan, L. Y., Leung, T. N., and Chang, A. M. (2000). Quantitative analysis of the bidirectional fetomaternal transfer of nucleated cells and plasma DNA. *Clin. Chem.* 46, 1301–1309.
8. Nelson, J. L. (2002). Microchimerism: incidental byproduct of pregnancy or active participant in human health. *Trends Mol. Med.* 8, 109–113.
9. Lo, Y. M. (2000). Fetal DNA in maternal plasma: biology and diagnostic applications. *Clin. Chem.* 46, 1903–1906.
10. Enders, A. C., and Blankenship, T. N. (1999). Comparative placental structure. *Adv. Drug Deliv. Rev.* 38, 3–15.
11. Mossman, H. W. (1986). *Vertebrate and Fetal Membranes.* Macmillan, London.
12. Ramsey, E. M. (1975). The placenta of laboratory animals and man. Holt, Rinehart & Winston, New York.
13. Boyd, J. D., and Hamilton, W. J. (1970). *The Human Placenta.* W. Heffer & Sons, Cambridge, U.K.
14. Mess A., Blackburn, D. G., and Zeller, U. (2003). Evolutionary transformations of fetal membranes and reproductive strategies. *J. Exp. Zoolog. Part A Comp. Exp. Biol.* 299, 3–12.
15. Christie, G. A. (1968). Comparative histochemical studies on carbohydrate, lipid and RNA metabolism in the placenta and foetal membranes. *J. Anat.* 103, 91–112.
16. Ridderstrale, Y., Persson, E., Dantzer, V., and Leiser, R. (1997). Carbonic anhydrase activity in different placenta types: a comparative study of pig, horse, cow, mink, rat, and human. *Microsc. Res. Tech.* 38, 115–124.
17. Desoye, G., and Shafrir, E. (1994). Placental metabolism and its regulation in health and diabetes. *Mol. Aspects Med.* 15, 505–682.
18. Malassine A., Frendo, J. L., and Evain-Brion, D. (2003). A comparison of placental development and endocrine functions between the human and mouse model. *Hum. Reprod. Update.* 9, 531–539.
19. Barcroft, J. (1946). *Researches on Pre-Natal Life.* Blackwell Scientific Publications, Oxford, U.K.
20. Sastry, B. V. (1997). Human placental cholinergic system. *Biochem. Pharmacol.* 53, 1577–1586.
21. Whitsett, J. A., Johnson, C. L., Noguchi A., Darovec-Beckerman, C., and Costello, M. (1980). Beta-adrenergic receptors and catecholamine-sensitive adenylate cyclase of the human placenta. *J. Clin. Endocrinol. Metab.* 50, 27–32.
22. Reilly, R. D., and Russell, P. T. (1977). Neurohistochemical evidence supporting an absence of adrenergic and cholinergic innervation in the human placenta and umbilical cord. *Anat. Rec.* 188, 277–286.
23. Kudo, Y., and Boyd, C. A. (2002). Changes in expression and function of syncytin and its receptor, amino acid transport system B(0) (ASCT2), in human placental choriocarcinoma BeWo cells during syncytialization. *Placenta* 23, 536–541.
24. Wang, H., Paul, R., Burgeson, R. E., Keene, D. R., and Kabat, D. (1991). Plasma membrane receptors for ecotropic murine retroviruses require a limiting accessory factor. *J. Virol.* 65, 6468–6477.
25. Kudo, Y., and Boyd, C. A. (2001). The physiology of immune evasion during pregnancy; the critical role of placental tryptophan metabolism and transport. *Pflugers Arch.* 442, 639–641.
26. Luckhardt, M., Leiser, R., Kingdom, J., Malek A., Sager, R., Kaisig, C., and Schneider, H. (1996). Effect of physiologic perfusion-fixation on the morphometrically evaluated dimensions of the term placental cotyledon. *J. Soc. Gynecol. Investig.* 3, 166–171.
27. Boyd, P. A., Brown, R. A., and Stewart, W. J. (1980). Quantitative structural differences within the normal term human placenta: a pilot study. *Placenta* 1, 337–344.

28. Baur, R. (1977). Morphometry of the placental exchange area. In *Advances in Anatomy, Embryology, and Cell Biology*, Vol. 53/1, pp. 7–64. Springer-Verlag, Berlin/Heidelberg.
29. Makowski, E. L., Meschia, G., Droegemueller, W., and Battaglia, F. C. (1968). Distribution of uterine blood flow in the pregnant sheep. *Am. J. Obstet. Gynecol.* 101, 409–412.
30. Makowski, E. L., Meschia, G., Droegemueller, W., and Battaglia, F. C. (1968). Measurement of umbilical arterial blood flow to the sheep placenta and fetus in utero. Distribution to cotyledons and the intercotyledonary chorion. *Circ. Res.* 23, 623–631.
31. Lees, M. H., Hill, J. D., Ochsner, A. J., 3rd, Thomas, C. L., and Novy, M. J. (1971). Maternal placental and myometrial blood flow of the rhesus monkey during uterine contractions. *Am. J. Obstet. Gynecol.* 110, 68–81.
32. Brambell, F. W. R. (1970). *The Transmission of Passive Immunity from Mother to Young*. North Holland Publishing, Amsterdam.
33. King, B. F. (1992). Compartive-studies of structure and function in mammalian placentas with special reference to maternal–fetal transfer of iron. *Am. Zool.* 32, 331–342.
34. Wooding, F. B., Morgan, G., Fowden, A. L., and Allen, W. R. (2000). Separate sites and mechanisms for placental transport of calcium, iron and glucose in the equine placenta. *Placenta* 21, 635–645.
35. Johnson, A. M., Umansky, I., Alper, C. A., Everett, C., and Greenspan, G. (1974). Amniotic fluid proteins: maternal and fetal contributions. *J. Pediatr.* 84, 588–593.
36. Jauniaux, E., and Gulbis, B. (2000). Fluid compartments of the embryonic environment. *Hum. Reprod. Update* 6, 268–278.
37. Faber, J. J. (1999). Flexner's Na+ clearances and the histological structure and age of the placenta in various species. *Placenta* 20, 141–147.
38. Flexner, L. B., and Pohl, H. A. (1941). *Am. J. Physiol.* 132, 594–606.
39. Meschia, G., Battaglia, F. C., and Bruns, P. D. (1967). Theoretical and experimental study of transplacental diffusion. *J. Appl. Physiol.* 22, 1171–1178.
40. Stulc, J., and Stulcova. B. (1986). Transport of calcium by the placenta of the rat. *J. Physiol.* 371, 1–16.
41. Constancia, M., Hemberger, M., Hughes, J., Dean, W., Ferguson-Smith A., Fundele, R., Stewart, F., Kelsey, G., Fowden A., Sibley, C., and Reik, W. (2002). Placental-specific IGF-II is a major modulator of placental and fetal growth. *Nature* 417, 945–948.
42. Sibley, C. P., Coan, P. M., Ferguson-Smith, A. C., Dean, W., Hughes, J., Smith, P., Reik, W., Burton, G. J., Fowden, A. L., and Constancia, M. (2004). Placental-specific insulin-like growth factor 2 (Igf2) regulates the diffusional exchange characteristics of the mouse placenta. *Proc. Natl. Acad. Sci. U S A* 101, 8204–8208.
43. Sibley, C. P. (1994). Review article: mechanisms of ion transfer by the rat placenta: a model for the human placenta? *Placenta* 15, 675–691.
44. Firth, J. A., and Leach, L. (1996). Not trophoblast alone: a review of the contribution of the fetal microvasculature to transplacental exchange. *Placenta* 17, 89-96.
45. Mahendran, D., Donnai, P., Glazier, J. D., D'Souza, S. W., Boyd, R. D., and Sibley, C. P. (1993). Amino acid (system A) transporter activity in microvillous membrane vesicles from the placentas of appropriate and small for gestational age babies. *Pediatr. Res.* 34, 661–665.
46. Harrington, B., Glazier, J., D'Souza, S., and Sibley, C. (1999). System A amino acid transporter activity in human placental microvillous membrane vesicles in relation to various anthropometric measurements in appropriate and small for gestational age babies. *Pediatr. Res.* 45, 810–814.
47. Godfrey, K. M., Matthews, N., Glazier, J., Jackson A., Wilman, C., and Sibley, C. P. (1998). Neutral amino acid uptake by the microvillous plasma membrane of the human placenta is inversely related to fetal size at birth in normal pregnancy. *J. Clin. Endocrinol. Metab.* 83, 3320–3326.
48. Sibley, C. P., Birdsey, T. J., Brownbill, P., Clarson, L. H., Doughty, I., Glazier, J. D., Greenwood, S. L., Hughes, J., Jansson, T., Mylona, P., Nelson, D. M., and Powell, T. (1998). Mechanisms of maternofetal exchange across the human placenta. *Biochem. Soc. Trans.* 26, 86–91.
49. Flexner, L. B., and Gellhorn, A. (1942). The comparative physiology of placental transfer. *Am. J. Obstet. Gynecol.* 43, 965–974.
50. Thornburg, K. L., and Faber, J. J. (1977). Transfer of hydrophilic molecules by placenta and yolk sac of the guinea pig. *Am. J. Physiol.* 233, C111–C124.
51. Teasdale, F. (1980). Gestational changes in the functional structure of the human placenta in relation to fetal growth: a morphometric study. *Am. J. Obstet. Gynecol.* 137, 560–568.
52. Mayhew, T. M., Jackson, M. R., and Boyd, P. A. (1993). Changes in oxygen diffusive conductances of human placentae during gestation (10–41 weeks) are commensurate with the gain in fetal weight. *Placenta* 14, 51–61.
53. Mayhew, T. M. (1998). Thinning of the intervascular tissue layers of the human placenta is an adaptive response to passive diffusion in vivo and may help to predict the origins of fetal hypoxia. *Eur. J. Obstet. Gynecol. Reprod. Biol.* 81, 101–109.
54. Atkinson, D. E., Robinson, N. R., and Sibley, C. P. (1991). Development of passive permeability characteristics of rat placenta during the last third of gestation. *Am. J. Physiol.* 261, R1461–R1464.
55. Faber, J. J. (1999). Filtration and diffusion across the immature placenta of the anaesthetized rat embryo. *Placenta* 20, 331–337.
56. Coan, P. M., Ferguson-Smith, A. C., and Burton, G. J. (2004). Developmental dynamics of the definitive mouse placenta assessed by stereology. *Biol. Reprod.* 70, 1806–1813.
57. Mayhew, T. M., Ohadike, C., Baker, P. N., Crocker, I. P., Mitchell, C., and Ong, S. S. (2003). Stereological investigation of placental morphology in pregnancies complicated by pre-eclampsia with and without intrauterine growth restriction. *Placenta* 24, 219–226.
58. Boyd, R., Glazier, J., Greenwood, S., and Sibley, C. (1997). Some gestational changes in placental transfer of ions. In *Placental Molecules in Hemodynamics, Transport and Cellular Regulation* (T. Hata, M. Takayama, and I. Taki, Eds.), Vol. 9, pp. 75–85. University of Rochester Press, Rochester, NY.
59. Stulc, J., Svihovec, J., Drabkova, J., Stribrny, J., Kobilkova, J., Vido, I., and Dolezal, A. (1978). Electrical potential difference across the mid-term human placenta. *Acta Obstet. Gynecol. Scand.* 57, 125–126.
60. Mellor, D. J., Cockburn, F., Lees, M. M., and Blagden, A. (1969). Distribution of ions and electrical potential differences between mother and fetus in the human at term. *J. Obstet. Gynaecol. Br. Commonw.* 76, 993–998.
61. Ward, S., Jauniaux, E., Shannon, C., Rodeck, C., Boyd, R., and Sibley, C. (1998). Electrical potential difference between exocelomic fluid and maternal blood in early pregnancy. *Am. J. Physiol.* 274, R1492–R1495.
62. Greenwood, S. L., Boyd, R. D., and Sibley, C. P. (1993). Transtrophoblast and microvillus membrane potential difference in mature intermediate human placental villi. *Am. J. Physiol.* 265, C460–C466.
63. Birdsey, T. J., Boyd, R. D., Sibley, C. P., and Greenwood, S. L. (1997). Microvillous membrane potential (Em) in villi from first trimester human placenta: comparison to Em at term. *Am. J. Physiol.* 273, R1519–R1528.

64. Ward, S., Sibley, C. P. S, and Boyd, R. D. H. (1999). Materno-fetal and transplacenta electrical potential differences. *Trophoblast Res.* 13, 343–361.
65. Thornburg, K. L., Binder, N. D., and Faber, J. J. (1979). Distribution of ionic sulfate, lithium, and bromide across the sheep placenta. *Am. J. Physiol.* 236, C58–C65.
66. Binder, N. D., Faber, J. J., and Thornburg, K. L. (1978). The transplacental potential difference as distinguished from the maternal–fetal potential difference of the guinea-pig. *J. Physiol.* 282, 561–570.
67. Dale, P. S., McNaughton, T. G., Power, L. A., Gilbert, R. D., and Power, G. G. (1990). Transuterine ion movement and electrical potential difference in pregnant guinea pigs. *Am. J. Physiol.* 258, R889–R894.
68. Canning, J. F., Stacey, T. E., Ward, R. H., and Boyd, R. D. (1986). Radioiodide transfer across sheep placenta. *Am. J. Physiol.* 250, R112–R119.
69. Stulc, J., and Stulcova, B. (1996). Maternal-fetal potential difference in the rat is generated by the placenta. *Placenta* 17, 653–659.
70. Ward, S., Jauniaux, E., Shannon, C., Rodeck, C., Boyd, R., and Sibley, C. P. (1998). Electrical potential difference between exocoelomic fluid and maternal blood in early pregnancy. *Am. J. Physiol.* 43, 1492–1495.
71. Mahendran, D., Byrne, S., Donnai, P., D'Souza, S. W., Glazier, J. D., Jones, C. J., and Sibley, C. P. (1994). Na+ transport, H+ concentration gradient dissipation, and system A amino acid transporter activity in purified microvillous plasma membrane isolated from first-trimester human placenta: comparison with the term microvillous membrane. *Am. J. Obstet. Gynecol.* 171, 1534–1540.
72. Ayuk, P. T., Sibley, C. P., Donnai, P., D'Souza, S., and Glazier, J. D. (2000). Development and polarization of cationic amino acid transporters and regulators in the human placenta. *Am. J. Physiol. Cell Physiol.* 278, C1162–C1171.
73. Ayuk, P. T., Theophanous, D., D'Souza, S. W., Sibley, C. P., and Glazier, J. D. (2002). L-arginine transport by the microvillous plasma membrane of the syncytiotrophoblast from human placenta in relation to nitric oxide production: effects of gestation, preeclampsia, and intrauterine growth restriction. *J. Clin. Endocrinol. Metab.* 87, 747–751.
74. Jansson, T., Wennergren, M., and Illsley, N. P. (1993). Glucose transporter protein expression in human placenta throughout gestation and in intrauterine growth retardation. *J. Clin. Endocrinol. Metab.* 77, 1554–1562.
75. Hughes, J. L., Doughty, I. M., Glazier, J. D., Powell, T. L., Jansson, T., D'Souza, S. W., and Sibley, C. P. (2000). Activity and expression of the Na(+)/H(+) exchanger in the microvillous plasma membrane of the syncytiotrophoblast in relation to gestation and small for gestational age birth. *Pediatr. Res.* 48, 652–659.
76. Powell, T. L., and Illsley, N. P. (1996). A novel technique for studying cellular function in human placenta: gestational changes in intracellular pH regulation. *Placenta* 17, 661–668.
77. Doughty, I. M., Glazier, J. D., Powell, T. L., Jansson, T., and Sibley, C. P. (1998). Chloride transport across syncytiotrophoblast microvillous membrane of first trimester human placenta. *Pediatr. Res.* 44, 226–232.
78. Reik, W., Constancia, M., Fowden A., Anderson, N., Dean, W., Ferguson-Smith, A., Tycko, B., and Sibley, C. (2003). Regulation of supply and demand for maternal nutrients in mammals by imprinted genes. *J. Physiol.* 547, 35–44.
79. Widdowson, E. M. (1977). Prenatal nutrition. *Ann. N Y Acad. Sci.* 300, 188–196.
80. Widdowson, E. M. (1982). Nutrition from conception to extreme old age. *Food Nutr. (Roma)* 8, 32–40.
81. Friedman, E. A., Gray, M. J., Hutchinson, D. L., and Plentl, A. A. (1959). The role of the monkey fetus in the exchange of the water and sodium of the amniotic fluid. *J. Clin. Invest.* 38, 961–970.
82. Hutchinson, D. L., Gray, M. J., Plentl, A. A., Alvarez, H., Caldeyro-Barcia, R., Kaplan, B., and Lind, J. (1959). The role of the fetus in the water exchange of the amniotic fluid of normal and hydramniotic patients. *J. Clin. Invest.* 38, 971–980.
83. Stulc, J. (1997). Placental transfer of inorganic ions and water. *Physiol. Rev.* 77, 805–836.
84. Edwards, D., Jones, C. J., Sibley, C. P., and Nelson, D. M. (1993). Paracellular permeability pathways in the human placenta: a quantitative and morphological study of maternal-fetal transfer of horseradish peroxidase. *Placenta* 14, 63–73.
85. Brownbill, P., Edwards, D., Jones, C., Mahendran, D., Owen, D., Sibley, C., Johnson, R., Swanson, P., and Nelson, D. M. (1995). Mechanisms of alphafetoprotein transfer in the perfused human placental cotyledon from uncomplicated pregnancy. *J. Clin. Invest.* 96, 2220–2226.
86. Brownbill, P., Mahendran, D., Owen, D., Swanson, P., Thornburg, K. L., Nelson, D. M., and Sibley, C. P. (2000). Denudations as paracellular routes for alphafetoprotein and creatinine across the human syncytiotrophoblast. *Am. J. Physiol. Regul. Integr. Comp. Physiol.* 278, R677–R683.
87. Ross, M. G., and Brace, R. A. (2001). National institute of child health and development conference summary: Amnioic fluid biology-basic and clinical aspects. *J. Maternal-Fetal Med.* 10, 2–19.
88. Faber, J. J. (1999). Filtration and diffusion across the immature placenta of the anaesthetized rat embryo. *Placenta* 20, 331–337.
89. Boyd, R. D., Haworth, C., Stacey, T. E., and Ward, H. T. (1976). Permeability of the sheep placenta to unmetabolized polar non-electrolytes. *J. Physiol.* 256, 617–634.
90. Faber, J. J., and Anderson, D. F. (1995). Concentrations of Na+ and Cl- in transplacental ultrafiltrate in sheep. *J. Physiol.* 487, Pt 1, 159–167.
91. Stulc, J. (1996). Transfer pathways across the sheep placenta. *Placenta* 17, 377–379.
92. Gellhorn A., Flexner, L. B., and Pohl, H. A. (1941). The transfer of radioactive sodium across the placenta of the sow. *J. Cell. Comp. Physiol.* 18, 393–399.
93. Needham, J. (1931). *Chemical Embryology*, p. 1506. Cambridge University Press, London.
94. Boyd, R. D. H., Doughty, I. M., Glazier, J., Greenwood, S. L., and Sibley, C. P. (1997). Techniques for the study of placental transport-transfer of chloride as an example. In *Nestle Nutrition Workshop Series* (F. C. Battaglia, Ed.), Vol. 39, *Placental Function and Fetal Nutrition*, pp. 1–20. Lippincott-Raven, Philadelphia.
95. Stulc, J., Stulcova, B., Husain, S., and Sibley, C. P. (1996). Transfer of Cl- across placenta of anesthetized rat. *Am. J. Physiol.* 271, R1107–R1114.
96. Wang, S., Kallichanda, N., Song, W., Ramirez, B. A., and Ross, M. G. (2001). Expression of aquaporin-8 in human placenta and chorioamniotic membranes: evidence of molecular mechanism for intramembranous amniotic fluid resorption. *Am. J. Obstet. Gynecol.* 185, 1226–1231.
97. Damiano A., Zotta, E., Goldstein, J., Reisin, I., and Ibarra, C. (2001). Water channel proteins AQP3 and AQP9 are present in syncytiotrophoblast of human term placenta. *Placenta* 22, 776–781.
98. Barcroft, L. C., Offenberg, H., Thomsen, P., and Watson, A. J. (2003). Aquaporin proteins in murine trophectoderm mediate transepithelial water movements during cavitation. *Dev. Biol.* 256, 342–354.

99. Mann, S. E., Ricke, E. A., Yang, B. A., Verkman, A. S., and Taylor, R. N. (2002). Expression and localization of aquaporin1 and 3 in human fetal membranes. *Am. J. Obstet. Gynecol.* 187, 902–907.
100. Jansson, T., Powell, T. L., and Illsley, N. P. (1999). Gestational development of water and non-electrolyte permeability of human syncytiotrophoblast plasma membranes. *Placenta* 20, 155–160.
101. Ma, T., Yang, B., Gillespie A., Carlson, E. J., Epstein, C. J., and Verkman, A. S. (1997). Generation and phenotype of a transgenic knockout mouse lacking the mercurial-insensitive water channel aquaporin-4. *J. Clin. Invest.* 100, 957–962.
102. Yang, B., Gillespie A., Carlson, E. J., Epstein, C. J., and Verkman, A. S. (2001). Neonatal mortality in an aquaporin-2 knock-in mouse model of recessive nephrogenic diabetes insipidus. *J. Biol. Chem.* 276, 2775–2779.
103. Liu, H., Koukoulas, I., Ross, M. C., Wang, S., and Wintour, E. M. (2004). Quantitative comparison of placental expression of three aquaporin genes. *Placenta* 25, 475–478.
104. Stulc, I. (1996). Transfer pathways across the sheep placenta. *Placenta* 17, 377–379.
105. Stulc, J., and Stulcova, B. (1993). Asymmetrical transfer of inert hydrophilic solutes across rat placenta. *Am. J. Physiol.* 265, R670–R675.
106. Stulc, J., and Stulcova, B. (1996). Effect of NaCl load administered to the fetus on the bidirectional movement of 51Cr-EDTA across rat placenta. *Am. J. Physiol.* 270, R984–R989.
107. Faber, J. J. (1997). The non-steady state of pregnancy. *Am. J. Physiol.* 272, 438–440.
108. Adamson, R. H., Lenz, J. F., Zhang, X., Adamson, G. N., Weinbaum, S., and Curry, F. E. (2004). Oncotic pressures opposing filtration across non-fenestrated rat microvessels. *J. Physiol.* 557, 889–907.
109. Michel, C. C. (2004). Fluid exchange in the microcirculation. *J. Physiol.* 557, 701–702.
110. Kirksey, A., Pike, R. L., and Callahan, J. A. (1962). Some effects of high and low sodium intakes during pregnancy in the rat. II. Electrolyte concentrations of maternal plasma, muscle, bone and brain and of placenta, amniotic fluid, fetal plasma and total fetus in normal pregnancy. *J. Nutr.* 77, 43–51.
111. Bell, R. J., McDougall, J. G., Wang, X., and Wintour, E. M. (1985). Effects of maternal sodium depletion on the composition of ovine fetal fluids. *J. Endocrinol.* 107, 177–182.
112. Dancis, J., and Springer, D. (1970). Fetal homeostasis in maternal malnutrition: potassium and sodium deficiency in rats. *Pediatr. Res.* 4, 345–351.
113. Anderson, D. F., Borst, N. J., Boyd, R. D., and Faber, J. J. (1990). Filtration of water from mother to conceptus via paths independent of fetal placental circulation in sheep. *J. Physiol.* 431, 1–10.
114. Vahlquist, A., and Nilsson, S. (1984). Vitamin A transfer to the fetus and to the amniotic fluid in rhesus monkey (Macaca mulatta). *Ann. Nutr. Metab.* 28, 321–333.
115. Malek A., Sager, R., and Schneider, H. (1998). Transport of proteins across the human placenta. *Am. J. Reprod. Immunol.* 40, 347–351.
116. Gitlin, D. (1975). Normal biology of alpha-fetoprotein. *Ann. N Y Acad. Sci.* 259, 7–16.
117. Gitlin, D., Kumate, J., Urrusti, J., and Morales, C. (1964). The selectivity of the human placenta in the transfer of plasma proteins from mother to fetus. *J. Clin. Invest.* 43, 1938–1951.
118. Thomas, R. L., and Blakemore, K. J. (1990). Evaluation of elevations in maternal serum alpha-fetoprotein: a review. *Obstet. Gynecol. Surv.* 45, 269–283.
119. Boyd, P. A., and Keeling, J. W. (1986). Raised maternal serum alpha-fetoprotein in the absence of fetal abnormality-placental findings. A quantitative morphometric study. *Prenat. Diagn.* 6, 369–373.
120. Boyd, P. A. (1992). Why might maternal AFP be high in pregnancies in which the fetus is normally formed. *Br. J. Obstet. Gynaecol.* 99, 93–95.
121. Bernstein, I. M., Barth, R. A., Miller, R., and Capeless, E. L. (1992). Elevated maternal serum alpha-fetoprotein: association with placental sonolucencies, fetomaternal hemorrhage, vaginal bleeding, and pregnancy outcome in the absence of fetal anomalies. *Obstet. Gynecol.* 79, 71–74.
122. Chaddha, V., Alkazaleh, F., Viero, S., Windrim, R., and Kingdom, J. (2004). Placental ultrasound and maternal thrombophilia in patients with combined elevations of AFP and hCG at 15–18 weeks gestation. *Placenta* 25, A, 32.
123. Bain, M. D., Copas, D. K., Taylor, A., Landon, M. J., and Stacey, T. E. (1990). Permeability of the human placenta *in vivo* to four non-metabolized hydrophilic molecules. *J. Physiol.* 431, 505–513.
124. Thornburg, K. L., Burry, K. J., Adams, A. K., Kirk, E. P., and Faber, J. J. (1988). Permeability of placenta to inulin. *Am. J. Obstet. Gynecol.* 158, 1165–1169.
125. Wellik, S. R., de Veciana, M., Morgan, M. A., Berkowitz, K. M., and Arquilla, E. R. (1995). Naturally occurring insulin autoantibodies in neonates of normal pregnancies and their relationship to insulinemia and birth weight. *Am. J. Obstet. Gynecol.* 173, 1878–1884.
126. Menon, R. K., Cohen, R. M., Sperling, M. A., Cutfield, W. S., Mimouni, F., and Khoury, J. C. (1990). Transplacental passage of insulin in pregnant women with insulin-dependent diabetes mellitus. Its role in fetal macrosomia. *N. Engl. J. Med.* 323, 309–315.
127. Gitlin, D., and Gitlin, J. D. (1975). *Fetal and Neonatal Development of Human Plasma Proteins*, pp. 264–317. Academic Press, New York.
128. Malek A., Sager, R., and Schneider, H. (1994). Maternal–fetal transport of immunoglobulin G and its subclasses during the third trimester of human pregnancy. *Am. J. Reprod. Immunol.* 32, 8–14.
129. Malek, A. (2003). Ex vivo human placenta models: transport of immunoglobulin G and its subclasses. *Vaccine* 21, 3362–3364.
130. Brambell, F. W. R., Hemmings, W. A., Henderson, M., Parry, H. J., and Rowlands, W. T. (1949). The route of antibodies passing from the maternal to the fetal circulation of the rabbit. *Proc. Royal. Soc. B.* 136, 131–144.
131. Simister, N. E., Story, C. M., Chen, H. L., and Hunt, J. S. (1996). An IgG-transporting Fc receptor expressed in the syncytiotrophoblast of human placenta. *Eur. J. Immunol.* 26, 1527–1531.
132. Leach, J. L., Sedmak, D. D., Osborne, J. M., Rahill, B., Lairmore, M. D., and Anderson, C. L. (1996). Isolation from human placenta of the IgG transporter, FcRn, and localization to the syncytiotrophoblast: implications for maternal-fetal antibody transport. *J. Immunol.* 157, 3317–3322.
133. Kristoffersen, E. K., and Matre, R. (1996). Co-localization of the neonatal Fc gamma receptor and IgG in human placental term syncytiotrophoblasts. *Eur. J. Immunol.* 26, 1668–1671.
134. Roberts, D. M., Guenthert, M., and Rodewald, R. (1990). Isolation and characterization of the Fc receptor from the fetal yolk sac of the rat. *J. Cell Biol.* 111, 1867–1876.
135. Ahouse, J. J., Hagerman, C. L., Mittal, P., Gilbert, D. J., Copeland, N. G., Jenkins, N. A., and Simister, N. E. (1993). Mouse MHC class I-like Fc receptor encoded outside the MHC. *J. Immunol.* 151, 6076–6088.

136. Firan, M., Bawdon, R., Radu, C., Ober, R. J., Eaken, D., Antohe, F., Ghetie, V., and Ward, E. S. (2001). The MHC class I-related receptor, FcRn, plays an essential role in the maternofetal transfer of gamma-globulin in humans. *Int. Immunol.* 13, 993–1002.
137. Israel, E. J., Patel, V. K., Taylor, S. F., Marshak-Rothstein, A., and Simister, N. E. (1995). Requirement for a beta 2-microglobulin-associated Fc receptor for acquisition of maternal IgG by fetal and neonatal mice. *J. Immunol.* 154, 6246–6251.
138. Junghans, R. P. (1997). Finally! The Brambell receptor (FcRB). Mediator of transmission of immunity and protection from catabolism for IgG. *Immunol. Res.* 16, 29–57.
139. Leach, L., Eaton, B. M., Firth, J. A., and Contractor, S. F. (1991). Immunocytochemical and labelled tracer approaches to uptake and intracellular routing of immunoglobulin-G (IgG) in the human placenta. *Histochem. J.* 23, 444–449.
140. Simister, N. E. (2003). Placental transport of immunoglobulin G. *Vaccine* 21, 3365–3369.
141. King, B. F. (1982). Absorption of peroxidase-conjugated immunoglobulin G by human placenta: an *in vitro* study. *Placenta* 3, 395–406.
142. Bright, N. A., Ockleford, C. D., and Anwar, M. (1994). Ontogeny and distribution of Fc gamma receptors in the human placenta; transport or immune surveillance. *J. Anat.* 184, 297–308.
143. Lyden, T. W., Robinson, J. M., Tridandapani, S., Teillaud, J. L., Garber, S. A., Osborne, J. M., Frey, J., Budde, P., and Anderson, C. L. (2001). The Fc receptor for IgG expressed in the villus endothelium of human placenta is Fc gamma RIIb2. *J. Immunol.* 166, 3882–3889.
144. Casas, R., and Bjorksten, B. (2001). Detection of Fel d 1-immunoglobulin G immune complexes in cord blood and sera from allergic and non-allergic mothers. *Pediatr. Allergy Immunol.* 12, 59–64.
145. Gitlin, D., Kumate, J., Morales, C., Noriega, L., and Arevalo, N. (1972). The turnover of amniotic fluid protein in the human conceptus. *Am. J. Obstet. Gynecol.* 113, 632–645.
146. Shennan, D. B., and Boyd, C. A. R. (1987). Ion transport by the placenta: a review of membrane transport sytems. *Biochim. Biophys. Acta* 906, 437–457.
147. Flexner, L. B., and Gellhorn, A. (1943). Comparitive physiology of the placental transfer. *Am. J. Obstet. Gynecol.* 43, 965–974.
148. Flexner, L. B., Cowie, D. B., Hellman, L. M., Wilde, W. S., and Vosburgh, G. J. (1948). The permeability of the human placenta to sodium in normal and abnormal pregnancies and the supply of sodium to the human fetus as determined with radioactive sodium. *Am. J. Obstet. Gynecol.* 55, 469–480.
149. Sibley, C. P., and Boyd, R. D. H. (2004). Mechanisms of transfer across the human placenta. In *Fetal and Neonatal Physiology* (R. A. Polin, W. W. Fox, and S. H. Abman, Eds.), pp. 111–122. WB Saunders, Philadelphia.
150. Balkovetz, D. F., Leibach, F. H., Mahesh, V. B., Devoe, L. D., Cragoe, E. J. Jr., and Ganapathy, V. (1986). Na^+-H^+ exchanger of human placental brush-border membrane: identification and characterization. *Am. J. Physiol.* 251, C852–C860.
151. Chipperfield, A. R., Langridge-Smith, J. E., and Steele, L. W. (1988). Sodium entry into human placental microvillous (maternal) plasma membrane vesicles. *Q. J. Exp. Physiol.* 73, 399–411.
152. Glazier, J. D., Jones, C. J., and Sibley, C. P. (1988). Purification and Na+ uptake by human placental microvillus membrane vesicles prepared by three different methods. *Biochim. Biophys. Acta* 945, 127–134.
153. Brunette, M. G., and Allard, S. (1985). Phosphate uptake by syncytial brush border membranes of human placenta. *Pediatr. Res.* 19, 1179–1182.
154. Johansson, M., Jansson, T., and Powell, T. L. (2000). Na(+)-K(+)-ATPase is distributed to microvillous and basal membrane of the syncytiotrophoblast in human placenta. *Am. J. Physiol. Regul. Integr. Comp. Physiol.* 279, R287–R294.
155. Sibley, C. P., Ward, B. S., Glazier, J. D., Moore, W. M., and Boyd, R. D. (1986). Electrical activity and sodium transfer across *in vitro* pig placenta. *Am. J. Physiol.* 250, R474–R484.
156. Stulc, J., Stulcova, B., and Sibley, C. P. (1991). Evidence for active transport of Na across the placenta of the rat. *Placenta* 12, 438.
157. Dancis, J., Kammerman, S., Jansen, V., and Levitz, M. (1983). The effect of ouabain on placental transport of 86Rb. *Placenta* 4, 351–359.
158. Bailey, D. J., Bradbury, M. W., France, V. M., Hedley, R., Naik, S., and Parry, H. (1979). Cation transport across the guinea-pig placenta perfused in situ. *J. Physiol.* 287, 45–56.
159. Serrano, C. V., Talbert, L. M., and Welt, L. G. (1964). Potassium deficiency in the pregnant dog. *J. Clin. Invest.* 43, 27–31.
160. Faber, J. J., and Thornburg, K. L. (1981). The forces that drive inert solutes from water across the epitheliochorial placenta of the sheep and the goat and the hemochorial placentae of the rabbit and the guinea pig. In *Placental Transfer: Methods and Interpretations* (M. Young, Ed.), pp. 203–214. WB Saunders, London.
161. Mohammed, T., Stulc, J., Sibley, C. P., and Boyd, R. D. (1992). Effect of maternal hypokalaemia on unidirectional maternofetal and net potassium fluxes across the placenta of the anaesthetized rat. *Placenta* 13, 231–240.
162. Canning, J. F., and Boyd, R. D. H. (1984). Mineral and water exchange between mother and fetus. In *Fetal Physiology and Medicine. The Basis of Perinatology* (R. W. Beard and P. W. Nathanielsz, Eds.), pp. 481–509, Dekker, New York.
163. Fantel, A. G. (1975). Fetomaternal potassium relations in the fetal rat on the twentieth day of gestation. *Pediatr. Res.* 9, 527–530.
164. Mohammed, T., Maresh, M., Sibley, C. P., and Boyd, R. D. (1990). Absent maternofetal potassium gradient at term. *Arch. Dis. Child.* 65, 384–386.
165. Boyd, R. D. H., Canning, J. F., Stacey, T. E., and Ward, R. T. H. (1981). Steady-state feto-maternal ion fluxes across the sheep placenta. In *Placental Transfer: Methods and Interpretations (Placenta Supplement 2)* (M. Young, Ed.), pp. 229–234, WB Saunders, London.
166. Mohammed, T., Stulc, J., Glazier, J. D., Boyd, R. D., and Sibley, C. P. (1993). Mechanisms of potassium transfer across the dually perfused rat placenta. *Am. J. Physiol.* 265, R341–R347.
167. Yamamoto, T., Bkaily, G., and Belisle, S. (1991). Inwardly rectifying currents in human term placental cells. *Pflugers Arch.* 419, 421–423.
168. Birdsey, T. J., Boyd, R. D., Sibley, C. P., and Greenwood, S. L. (1999). Effect of hyposmotic challenge on microvillous membrane potential in isolated human placental villi. *Am. J. Physiol.* 276, R1479–R1488.
169. Mylona, P., Clarson, H., Greenwood, S. L., and Sibley, C. P. (1998). Expression of the Kir2. 1 (inwardly rectifying potassium channel) gene in the human placenta and in cultured cytotrophoblast cells at different stages of differentiation. *Mol. Hum. Reprod.* 4, 195–200.
170. Clarson, L. H., Greenwood, S. L., Mylona, P., and Sibley, C. P. (2001). Inwardly rectifying K(+) current and differentiation of human placental cytotrophoblast cells in culture. *Placenta* 22, 328–236.

171. Clarson, L. H., Roberts, V. H., Greenwood, S. L., and Elliott, A. C. (2002). ATP-stimulated Ca(2+)-activated K(+) efflux pathway and differentiation of human placental cytotrophoblast cells. *Am. J. Physiol. Regul. Integr. Comp. Physiol.* 282, R1077–R1085.
172. Bai, X., Greenwood, S. L., Glazier, J. D., Baker, P. N., Sibley, C. P., Taggart, M. J., and Fyfe, G. K. (2005). Localization of TASK and TREK, two-pore domain K+ channels, in human cytotrophoblast cells. *J. Soc. Gynecol. Investig.* 12, 77–83.
173. Clarson, L. H., Roberts, V. H., Hamark, B., Elliott, A. C., and Powell, T. (2003). Store-operated Ca2+ entry in first trimester and term human placenta. *J. Physiol.* 550, 515–528.
174. Zhao, H., and Hundal, H. S. (2000). Identification and biochemical localization of a Na-K-Cl cotransporter in the human placental cell line BeWo. *Biochem. Biophys. Res. Commun.* 274, 43–48.
175. Shennan, D. B., Davis, B., and Boyd, C. A. (1986). Chloride transport in human placental microvillus membrane vesicles. I. Evidence for anion exchange. *Pflugers Arch.* 406, 60–64.
176. Illsley, N. P., Glaubensklee, C., Davis, B., and Verkman, A. S. (1988). Chloride transport across placental microvillous membranes measured by fluorescence. *Am. J. Physiol.* 255, C789–C797.
177. Byrne, S., Glazier, J. D., Greenwood, S. L., Mahendran, D., and Sibley, C. P. (1993). Chloride transport by human placental microvillous membrane vesicles. *Biochim. Biophys. Acta* 1153, 122–126.
178. Grassl, S. M. (1989). C1/HCO3 exchange in human placental brush border membrane vesicles. *J. Biol. Chem.* 264, 11103–11106.
179. Powell, T. L., Lundquist, C., Doughty, I. M., Glazier, J. D., and Jansson, T. (1998). Mechanisms of chloride transport across the syncytiotrophoblast basal membrane in the human placenta. *Placenta* 19, 315–321.
180. Brown, P. D., Greenwood, S. L., Robinson, J., and Boyd, R. D. (1993). Chloride channels of high conductance in the microvillous membrane of term human placenta. *Placenta* 14, 103–115.
181. Byrne, S., Greenwood, S. L., and Sibley, C. P. (1993). Chloride channels in trophoblast cells cultured from human placentas. In *Clinical Ecology of Cystic Fibrosis* (H. Escobar, F. Baquero, and L. Suarez, Eds.), pp. 3–8. Elsevier Science, New York.
182. Cronier, L., Bois, P., Herve, J. C., and Malassine, A. (1995). Effect of human chorionic gonadotrophin on chloride current in human syncytiotrophoblasts in culture. *Placenta* 16, 599–609.
183. Bissonnette, J. M., Weiner, C. P., and Power, G. G. Jr. (1994). Amino acid uptake and chloride conductances in human placenta. *Placenta* 15, 445–446.
184. Henriquez, M., and Riquelme, G. (2003). 17beta-estradiol and tamoxifen regulate a maxi-chloride channel from human placenta. *J. Membr. Biol.* 191, 59–68.
185. Bernucci, L., Umana, F., Llanos, P., and Riquelme, G. (2003). Large chloride channel from pre-eclamptic human placenta. *Placenta* 24, 895–903.
186. Riquelme, G., Stutzin A., Barros, L. F., and Liberona, J. L. (1995). A chloride channel from human placenta reconstituted into giant liposomes. *Am. J. Obstet. Gynecol.* 173, 733–738.
187. Berryman, M., and Bretscher, A. (2000). Identification of a novel member of the chloride intracellular channel gene family (CLIC5) that associates with the actin cytoskeleton of placental microvilli. *Mol. Biol. Cell.* 11, 1509–1521.
188. Schlingmann, K. P., Konrad, M., Jeck, N., Waldegger, P., Reinalter, S. C., Holder, M., Seyberth, H. W., and Waldegger, S. (2004). Salt wasting and deafness resulting from mutations in two chloride channels. *N. Engl. J. Med.* 350, 1314–1319.
189. Mylona, P., Glazier, J. D., Greenwood, S. L., Sides, M. K., and Sibley, C. P. (1996). Expression of the cystic fibrosis (CF) and multidrug resistance (MDR1) genes during development and differentiation in the human placenta. *Mol. Hum. Reprod.* 2, 693–698.
190. Faller, D. P., Egan, D. A., and Ryan, M. P. (1995). Evidence for location of the CFTR in human placental apical membrane vesicles. *Am. J. Physiol.* 269, C148–C155.
191. Ben-Chetrit, A., Antenos, M., Jurisicova A., Pasyk, E. A., Chitayat, D., Foskett, J. K., and Casper, R. F. (2002). Expression of cystic fibrosis transmembrane conductance regulator during early human embryo development. *Mol. Hum. Reprod.* 8, 758–764.
192. Atkinson, D. E., Greenwood, S. L., Sibley, C. P., Glazier, J. D., and Fairbairn, L. J. (2003). Role of MDR1 and MRP1 in trophoblast cells, elucidated using retroviral gene transfer. *Am. J. Physiol. Cell. Physiol.* 285, C584–C591.
193. Boyd, R. D. H., Glazier, J. D., Moore, W. M. O., Sibley, C. P., and Ward, B. S. (1985). Effect of chloride replacement on the electrical activity of the pig placenta in vitro. *J. Physiol.* 371, 270P.
194. Armentrout, T., Katz, S., Thornburg, K. L., and Faber, J. J. (1977). Osmotic flow through the placental barrier of chronically prepared sheep. *Am. J. Physiol.* 233, H466–H474.
195. Doughty, I. M., Glazier, J. D., Greenwood, S. L., Boyd, R. D., and Sibley, C. P. (1996). Mechanisms of maternofetal chloride transfer across the human placenta perfused *in vitro*. *Am. J. Physiol.* 271, R1701–R1706.
196. Logothetopoulos, J., and Scott, R. F. (1956). Active iodide transport across the placenta of the guinea-pig, rabbit and rat. *J. Physiol.* 132, 365–371.
197. Canning, J. F., Stacey, T. E., Ward, R. H. T., and Boyd, R. D. H. (1986). Radioiodide transfer across sheep placenta. *Am. J. Physiol. Reg. Int. Comp. Physiol.* 250, R112–R119.
198. Wright, E. M., and Turk, E. (2004). The sodium/glucose cotransport family SLC5. *Pflugers Arch. Eur. J. Physiol.* 447, 510–518.
199. Arturi, F., Lacroix, L., Presta, I., Scarpelli, D., Caillou, B., Schlumberger, M., Russo, D., Bidart, J. M., and Filetti, S. (2002). Regulation by human chorionic gonadotropin of sodium/iodide symporter gene expression in the JAr human choriocarcinoma cell line. *Endocrinology* 143, 2216–2220.
200. Mitchell, A. M., Manley, S. W., Morris, J. C., Powell, K. A., Bergert, E. R., and Mortimer, R. H. (2001). Sodium iodide symporter (NIS) gene expression in human placenta. *Placenta* 22, 2–3.
201. Eskandari, S., Loo, D. D. F., Dai, G., Levy, O., Wright, E. M., and Carrasco, N. (1997). Thyroid Na^+/I^- Symporter. *J. Biol. Chem.* 272, 27230–27238.
202. Soleimani, M., Greeley, T., Petrovic, S., Wang, Z. H., Amlal, H., Kopp, P., and Burnham, C. E. (2001). Pendrin: an apical $Cl^-/OH^-/HCO_3^-$ exchanger in the kidney cortex. *Am. J. Physiol. Renal Physiol.* 280, F356–F364.
203. Gillam, M. P., Sidhaye, A. R., Lee, E. J., Rutishauser, J., Stephan, C. W., and Kopp, P. (2004). Fuctional characterization of pendrin in a polarized cell system: evidence for pendrin-mediated apical iodide efflux. *J. Biol. Chem.* 279, 13004–13010.
204. Bidart, J. M., Lacroix, L., Evain-Brion, D., Caillou, B., Lazar, V., Frydman, R., Bellet, D., Filetti, S., and Schlumberger, M. (2000). Expression of Na^+/I^- symporter and pendred syndrome genes in trophoblast cells. *J. Clin. Endocrinol. Metab.* 85, 4367–4372.
205. Suzuki, K., Royaux, I. E., Everett, L. A., Mori-Aoki, A., Suzuki, S., Nakamura, K., Sakai, T., Katoh, R., Toda, S., Green, E. D., and Kohn, L. D. (2002). Expression of PDS/Pds,

the Pendred syndrome gene, in endometrium. *J. Clin. Endocrinol. Metab.* 87, 938–941.

206. Schröder-van der Elst, J. P., van der Heide, D., Kastelijn, J., Rousset, B., and Obregón, M. J. (2001). The expression of the sodium/iodide symporter is up-regulated in the thyroid of fetuses of iodine-deficient rats. *Endocrinology* 142, 3736–3741.
207. Versloot, P. M., Schröder-van der Elst, J. P., van der Heide, D., and Boogerd, L. (1997). Effects of marginal iodine deficiency during pregnancy: iodide uptake by the maternal and fetal thyroid. *Am. J. Physiol. Endocrinol. Metab.* 273, E1121–E1126.
208. Turner, M. A., Sides, M. K., Sibley, C. P., and Greenwood, S. L. (1999). Anion efflux from cytotrophoblast cells derived from normal term human placenta is stimulated by hyposmotic challenge and extracellular A23187 but not by membrane-soluble cAMP. *Exp. Physiol.* 84, 27–40.
209. Shennan, D. B. (1999). Properties of volume-activated taurine and iodide efflux from term human placental tissue. *Placenta* 20, 485–491.
210. Schauberger, C. W., and Pitkin, R. M. (1979). Maternal–perinatal calcium relationships. *Obstet. Gynecol.* 53, 74–76.
211. Delivoria-Papadopoulos, M., Battaglia, F. C., Bruns, P. D., and Meschia G. (1967). Total, protein-bound, and ultrafiltrable calcium in maternal and fetal plasmas. *Am. J. Physiol.* 213, 363–366.
212. Kovacs, C. S., Lanske, B., Hunzelman, J. L., Guo, J., Karaplis, A. C., and Kronenberg, H. M. (1996). Parathyroid hormone-related peptide (PTHrP) regulates fetal-placental calcium transport through a receptor distinct from the PTH/PTHrP receptor. *Proc. Natl. Acad. Sci. U S A* 93, 15233–15238.
213. Twardock, A. R., and Austin, M. K. (1970). Calcium transfer in perfused guinea pig placenta. *Am. J. Physiol.* 219, 540–545.
214. Weatherly, A. J., Ross, R., Rickard, D. W., and Care, A. D. (1983). The transfer of calcium during perfusion of the placenta in intact and thyroparathyroidectomised sheep. *Placenta* 4, 271–278.
215. Abramovich, D. R., Dacke, C. G., Elcock, C., and Page, K. R. (1987). Calcium transport across the isolated dually perfused human placental lobule. *J. Physiol.* 382, 397–410.
216. Sweiry, J. H., and Yudilevich, D. L. (1984). Asymmetric calcium influx and efflux at maternal and fetal sides of the guinea pig placenta: Kinetics and specificity. *J. Physiol. (Lond).* 355, 295–311.
217. Sweiry, J. H., Page, K. R., Dacke, C. G., Abramovich, D. R., and Yudilevich, D. L. (1986). Evidence of saturable uptake mechanisms at maternal and fetal sides of the perfused human placenta by rapid paired-tracer dilution studies with calcium and choline. *J. Dev. Physiol.* 8, 435–445.
218. Štulc, J., and Štulcová, B. (1986). Transport of calcium by the placenta of the rat. *J. Physiol. (Lond).* 371, 1–16.
219. Štulc, J., Štulcová, B., and Svihovec, J. (1990). Transport of calcium across the dually perfused placenta of the rat. *J. Physiol. (Lond).* 420, 295–311.
220. Štulc, J., Štulcová, B., Smid, M., and Sach, I. (1994). Parallell mechanisms of Ca^{++} transfer across the perfused human placental cotyledon. *Am. J. Obstet. Gynecol.* 170, 162–167.
221. Hoenderop, J. G. J, Nilius, B., and Bindels, R. J. M. (2003). Epithelial calcium channels: from identification to function and regulation. *Plugers Arch. Eur. J. Physiol.* 446, 304–308.
222. Bax, C., Bax, B., Bain, M., and Zaidi, M. (1994). Ca^{2+} channnels in human term trophoblast cells in vitro. A study using the Ca^{2+}-sensitive dye Fura 2. *Trophoblast Res.* 8, 573–580.
223. Hellman, P., Ridefelt, P., Juhlin, C., Åkerström, G., Rastad, J., and Gylfe, E. (1992). Parathyroid-like regulation of parathyroid-hormone-related protein release and cytoplasmic calcium in cytotrophoblast cells of human placenta. *Arch. Biochem. Biophys.* 293, 174–180.
224. Brunette, M. G., and Leclerc, M. (1992). Ca^{2+} transport through the brush border membrane of human placenta syncytiotrophoblasts. *Can. J. Physiol. Pharmacol.* 70, 835–842.
225. Kamath, S. G., Kelley, L. K., Friedman, A. F., and Smith, C. H. (1992). Transport and binding in calcium uptake by microvillous membrane of human placenta. *Am. J. Physiol.* 262, C789–C794.
226. Moreau, R., Hamel A., Daoud, G., Simoneau, L., and Lafond, J. (2002). Expression of calcium channels along the differentiation of cultured trophoblast cells from human term placenta. *Biol. Reprod.* 67, 1473–1479.
227. Bruns, M. E. H., Fausto, A., and Avioli, L. (1978). Placental calcium binding protein in rats. Apparent identity with vitamin D-dependent calcium binding protein from rat intestine. *J. Biol. Chem.* 253, 3186–3190.
228. Bruns, M. E. H., Vollmer, S., Wallshein, V., and Bruns, D. E. (1981). Vitamin D-dependent calcium-binding protein. Immunochemical studies and synthesis by placental tissue in vitro. *J. Biol. Chem.* 256, 4649–4653.
229. Delorme, A. C., Marche, P., and Garel, J. M. (1979). Vitamin D-dependent calcium-binding protein. Changes during gestation, prenatal and postnatal development in rats. *J. Dev. Physiol.* 1, 181–194.
230. Mathieu, C. L., Burnett, S. H., Mills, S. E., Overpeck, J. G., Bruns, D. E., and Bruns, M. E. (1989). Gestational changes in calbindin-D9k in rat uterus, yolk sac, and placenta: Implications for maternal–fetal calcium transport and uterine muscle function. *Proc. Natl. Acad. Sci. U S A* 86, 3433–3437.
231. de Maintenant, F., Perret, C., Brun, P., and Thomasset, M., (1985). Expression of cholecalin (9000 MW cholecaliferol-induced CaBP) gene in rat duodenum and placenta during gestation. In *Vitamin D, a Chemical, Biochemical and Clinical Update*, pp. 355–356. Walter de Gruyter, Berlin.
232. Glazier, J. D., Atkinson, D. E., and Thornburg, K. L. (1992). Gestational changes in Ca^{2+} transport across rat placenta and mRNA for calbindin9K and Ca^{2+}-ATPase. *Am. J. Physiol.* 263, R930–R935.
233. Krisinger, J., Dann, J. L., Jeung, E. B., and Leung, P. C. K. (1992). Calbindin-D9k gene expression during pregnancy and lactation in the rat. *Mol. Cell. Endocrinol.* 88, 119–128.
234. Husain, S. M., Birdsey, T. J., Glazier, J. D., Mughal, M. Z., Garland, H. O., Sibley, C. P. (1994). Effect of diabetes mellitus on maternofetal flux of Ca and Mg and calbindin9k mRNA expression in rat placenta. *Pediatr. Res.* 35, 376–381.
235. Brun, P., Dupret, J. M., Perret, C., Thomasset, M., and Mathieu, H. (1987). Vitamin D-dependent calcium-binding proteins (CaBPs) in human fetuses: Comparative distribution of 9K CaBP mRNA and 28K CaBP during development. *Pediatr. Res.* 21, 362–367.
236. Howard A., Legon, S., and Walters, J. R. F. (1992). Plasma membrane calcium pump expression in human placenta and small intestine. *Biochem. Biophys. Res. Commun.* 183, 499–505.
237. Jeung, E. B., Krisinger, J., Dann, J. L., and Leung, P. C. K. (1992). Molecular cloning of the full-length cDNA encoding the human calbindin-D9k. *FEBS Lett.* 307, 224–228.
238. Jeung, E. B., Leung, P. C. K., and Krisinger, J. (1994). The human calbindin-D9k gene. Complete structure and implications on steroid hormone regulation. *J. Mol. Biol.* 235, 1231–1238.
239. Belkacemi, L., Gariepy, G., Mounier, C., Simoneau, L., and Lafond, J. (2004). Calbindin-D9k (CaBP9k) localization and levels of expression in trophoblast cells from human term placenta. *Cell Tissue Res.* 315, 107–117.

240. Christakos, S., Gabrielides, C., and Rhoten, W. B. (1989). Vitamin D-dependent calcium binding proteins: Chemistry, distribution, functional considerations, and molecular biology. *Endocr. Rev.* 10, 3–26.
241. Feher, J. J., Fullmer, C. S., and Fritzsch, G. K. (1989). Comparison of the enhanced steady-state diffusion of calcium by calbindin-D9K and calmodulin: Possible importance in intestinal calcium absorption. *Cell Calcium.* 10, 189–203.
242. Bronner, F. (1988). Vitamin D-dependent active calcium transport: The role of CaBP. *Calcif. Tissue Int.* 43, 133–137.
243. Bronner, F., Pansu, D., and Stein, W. D. (1986). An analysis of intestinal calcium transport across the rat intestine. *Am. J. Physiol.* 250, G561–G569.
244. Feher, J. J., Fullmer, C. S., and Wasserman, R. H. (1992). Role of facilitated diffusion of calcium by calbindin in intestinal calcium absorption. *Am. J. Physiol.* 262, C517–C526.
245. Fisher, G. J., Kelley, L. K., and Smith, C. H. (1987). ATP-dependent calcium transport across basal plasma membranes of human placental trophoblast. *Am. J. Physiol.* 252, C38–C46.
246. Lafond, J., Leclerc, M., and Brunette, M. G. (1991). Characterization of calcium transport by basal plasma membranes from human placental syncytiotrophoblast. *J. Cell. Physiol.* 148, 17–23.
247. Kamath, S. G., and Smith, C. H. (1994). $Na^+/Ca2^+$ exchange, $Ca2^+$ binding, and electrogenic $Ca2^+$ transport in plasma membranes of human placental syncytiotrophoblast. *Pediatr. Res.* 36, 461–467.
248. Kamath, S. G., Haider, N., and Smith, C. H. (1994). ATP independent calcium transport and binding by basal plasma membrane of human placenta. *Placenta* 15, 147–155.
249. Borke, J. L., Caride A., and Verma, A. K. (1989). Calcium pump epitopes in placental trophoblast basal plasma membranes. *Am. J. Physiol.* 257, C341–C346.
250. Champion, E., Glazxier, J., Greenwood, S., Mann, S., Rawlings, J., Sibley, C. P., and Jones, C. (2003). Localisation of alkaline phosphatase and Ca2+-ATPase in the cat placenta. *Placenta* 24, 453–461.
251. Strid, H., and Powell, T. L. (2000). ATP-dependent Ca^{2+} transport is up-regulated during third trimester in human syncytiotrophoblast basal membranes. *Pediatr. Res.* 48, 58–63.
252. Tobias, J. H., and Cooper, C. (2004). PTH/PTHrP activity and the programming of skeletal development in utero. *J. Bone Miner. Res.* 19, 177–185.
253. Philbrick, W. M., Wysolmerski, J. J., Galbraith, S., Holt, E., Orloff, J. J., Yang, K. H., Vasavadar, C., Weir, E. C., Broadus, A. E., and Stewart, A. F. (1996). Defining the roles of parathyroid hormone-related protein in normal physiology. *Physiol. Revs.* 76, 127–173.
254. Care, A. D., Abbas, S. K., Pickard, D. W., Barri, M., Drinkhill, M., Findlay, J. B. C., White, I. R., and Caple, I. W. (1990). Stimulation of ovine placental transport of calcium and magnesium by mid-molecule fragments of human parathyroid hormone-related protein. *Exp. Physiol.* 75, 605–608.
255. Kovacs, C. S., Manley, N. R., Moseley, J. M., Martin, T. J., and Kronenberg, H. M. (2001). Fetal parathyroids are not required to maintain placental calcium transport. *J. Clin. Invest.* 107, 1007–1015.
256. Tucci, J., Hammond, V., Senior PV, Gibson A., and Beck, F. (1996). The role of fetal parathyroid hormone-related protein in transplacental calcium transport. *J. Mol. Endcrinol.* 17, 159–164.
257. Strid, H., Bucht, E., Jansson, T., Wennergren, M., and Powell, T. L. (2003). ATP-dependent Ca2+ transport across basal membrane of human syncytiotrophoblast in pregnancies complicated by intrauterine growth restriction or diabetes. *Placenta* 24, 445–452.
258. Oh, W. (1977). Considerations in neonates with intrauterine growth retardation. In *Clinical Obstetrics and Gynaecology*, 20(4), 991–1003.
259. Namgung, R., Tsang, R. C., Specker, B. L., Sierra, R. I., and Ho, M. L. (1993). Reduced serum osteocalcin and 1, 25-dihydroxyvitamin D concentrations and low bone mineral content in small for gestational age infants: evidence of decreased bone formation rates. *J. Pediatr.* 122, 269–275.
260. Dancis, J., Springer, D., and Cohlan, S. Q. (1971). Fetal homeostasis in maternal malnutrition. II Magneisum deprivation. *Pediatr. Res.* 5, 131–136.
261. Vormann, J., and Gunther, T. (1986). Development of fetal mineral and trace element metabolism in rats with normal as well as magnesium and zinc deficient diets. *Biol. Trace. Elem. Res.* 9, 37–53.
262. Kriesten, K., Palawinskas, R., and Sommer, H. (1984). Magnesium levels in the body fluids of dam and fetal rabbits during the reproductive phase. *Comp. Biochem. Physiol.* 79A, 197–200.
263. Care, A. D., Pickard, D. W., and Weatherley, A. (1979). The measurement of transplacental magnesium fluxes in sheep. *Res. Vet. Sci.* 27, 121–122.
264. Gupta, M. M., Kuppuswamy, G., and Subramanian, A. R. (1982). Transplacental transfer of 25-hydroxy-cholecalciferol. *Postgrad. Med.* 28, 408–410.
265. Paunier, L., Girardin, N. E., P. A., B., and Beguin, F. (1987). Maternal–fetal relationship of extra and intracellular magnesium and potassium concentration. In *Magnesium in Cellular Processes and Medicine.* (B. M. Altura, J. Durlach, and M. S. Seelig, Eds.), pp. 151–155. Karger, Basel.
266. Shaw, A. M., Mughal, M. Z., Mohammed, T., Maresh, M. J., Sibley, C. P. (1990). Evidence for active maternofetal transfer of magnesium across the in situ perfused placenta. *Pediatr. Res.* 27, 622–625.
267. Shaw, A. M., Mughal, M. Z., Maresh, M. J., Sibley, C. P. (1991). Effects of two synthetic parathyroid hormone-related protein fragments on maternofetal transfer of calcium and magnesium and release of cyclic AMP by the in-situ perfused rat placenta. *J. Endocrinol.* 129, 399–404.
268. Shaw, A. M., Mughal, M. Z., Maresh, M. J., Sibley, C. P. (1991). Sodium-dependent magnesium transport across the in situ perfused rat placenta. *Am. J. Physiol.* 261, R369–R372.
269. Günther, T., Vormann, J., and Höllriegl, V. (1988). Effects of amiloride and furosemide on 28Mg transport into fetuses and maternal tissues of rats. *Mag. Bull.* 10, 34–37.
270. Standley, P. R. and Standley, C. A. (2003). Identification of a functional Na+-Mg2+ exchanger in human trophoblast cells. *Am. J. Hypertens.* 16, 95–96.
271. Piquard, F., Schaefer A., Dellenbach, P., and Haberey, P. (1990). Lactate movements in the term human placenta in situ. *Biol. Neonate.* 58, 61–68.
272. Schneider, H. (1981). In *Placental Transfer: Methods and Interpretations (Placenta Supplement 2)* (M. Young, Ed.), pp. 129–138. WB Saunders, Philadelphia.
273. Moll, W., Girard, H., and Gros, G. (1980). Facilitated diffusion of lactic acid in the guinea-pig placenta. *Pflugers Arch.* 385, 229–38.
274. Kastendieck, E., and Moll, W. (1977). The placental transfer of lactate and bicarbonate in the guinea-pig. *Pflugers Arch.* 370, 65–71.
275. Suidan, J. S., Antoine, C., Silverman, F., Lustig, I. D., Wasserman, J. F., and Young, B. K. (1984). Human maternal-fetal lactate relationships. *J. Perinat. Med.* 12, 211–217.
276. Soothill, P. W., Nicolaides, K. H., Rodeck, C. H., and Campbell, S. (1986). Effect of gestational age on fetal and intervillous blood gas and acid-base values in human pregnancy. *Fetal Ther.* 1, 168–175.

277. Soothill, P. W., Nicolaides, K. H., Rodeck, C. H., and Gamsu, H. (1986). Blood gases and acid-base status of the human second-trimester fetus. *Obstet. Gynecol.* 68, 173–176.
278. Balkovetz, D. F., Leibach, F. H., Mahesh, V. B., and Ganapathy, V. (1988). A proton gradient is the driving force for uphill transport of lactate in human placental brush-border membrane vesicles. *J. Biol. Chem.* 263, 13823–13830.
279. Alonso de la Torre, S. R., Serrano, M. A., Caropaton, T., and Medina, J. M. (1991). Proton gradient-dependent active transport of L-lactate in basal plasma membrane vesicles isolated from syncytiotrophoblast human placenta. *Biochem. Soc. Trans.* 19, 409S.
280. Alonso de la Torre, S. R., Serrano, M. A., Alvarado, F., and Medina, J. M. (1991). Carrier-mediated L-lactate transport in brush-border membrane vesicles from rat placenta during late gestation. *Biochem. J.* 278, Pt 2, 535–541.
281. Inuyama, M., Ushigome, F., Emoto A., Koyabu, N., Satoh, S., Tsukimori, K., Nakano, H., Ohtani, H., and Sawada, Y. (2002). Characteristics of L-lactic acid transport in basal membrane vesicles of human placental syncytiotrophoblast. *Am. J. Physiol. Cell. Physiol.* 283, C822–C830.
282. Nakamura, H., Ushigome, F., Koyabu, N., Satoh, S., Tsukimori, K., Nakano, H., Ohtani, H., and Sawada, Y. (2002). Proton gradient-dependent transport of valproic acid in human placental brush-border membrane vesicles. *Pharm. Res.* 19, 154–161.
283. Settle, P., Mynett, K., Speake, P., Champion, E., Doughty, I. M., Sibley, C. P., D'Souza, S. W., and Glazier, J. (2004). Polarized lactate transporter activity and expression in the syncytiotrophoblast of the term human placenta. *Placenta* 25, 496–504.
284. Carstensen, M. H., Leichtweiss, H. P., and Schroder, H. (1983). Lactate carriers in the artificially perfused human term placenta. *Placenta* 4, 165–174.
285. Illsley, N. P., Wootton, R., Penfold, P., Hall, S., and Duffy, S. (1986). Lactate transfer across the perfused human placenta. *Placenta* 7, 209–220.
286. Friedman, E. A., Gray, M. J., Grynfogel, M., Hutchinson, D. L., Kelly, W. T., and Plentl, A. A. (1960). The distribution and metabolism of C14-labeled lactic acid and bicarbonate in pregnant primates. *J. Clin. Invest.* 39, 227–235.
287. Leichtweiss, H. P., and Schroder, H. (1981). L-Lactate and D-Lactate carriers on the fetal and the maternal side of the trophoblast in the isolated guinea pig placenta. *Pflugers Arch.* 390, 80–85.
288. Halestrap, A. P. and Price, N. T. (1999). The proton-linked monocarboxylate transporter (MCT) family: structure, function and regulation. *Biochem. J.* 343, Pt 2, 281–299.
289. Juel, C., and Halestrap, A. P. (1999). Lactate transport in skeletal muscle: role and regulation of the monocarboxylate transporter. *J. Physiol.* 517, Pt 3, 633–642.
290. Garcia, C. K., Goldstein, J. L., Pathak, R. K., Anderson, R. G., and Brown, M. S. (1994). Molecular characterization of a membrane transporter for lactate, pyruvate, and other monocarboxylates: implications for the Cori cycle. *Cell* 76, 865–873.
291. Price, N. T., Jackson, V. N., and Halestrap, A. P. (1998). Cloning and sequencing of four new mammalian monocarboxylate transporter (MCT) homologues confirms the existence of a transporter family with an ancient past. *Biochem. J.* 329, Pt 2, 321–328.
292. Manning Fox, J. E., Meredith, D., and Halestrap, A. P. (2000). Characterisation of human monocarboxylate transporter 4 substantiates its role in lactic acid efflux from skeletal muscle. *J. Physiol.* 529, Pt 2, 285–293.
293. Bonen, A. (2001). The expression of lactate transporters (MCT1 and MCT4) in heart and muscle. *Eur. J. Appl. Physiol.* 86, 6–11.
294. Kirk, P., Wilson, M. C., Heddle, C., Brown, M. H., Barclay, A. N., and Halestrap, A. P. (2000). CD147 is tightly associated with lactate transporters MCT1 and MCT4 and facilitates their cell surface expression. *EMBO J.* 19, 3896–3904.
295. Cetin, I. (2001). Amino acid interconversions in the fetal-placental unit: the animal model and human studies *in vivo*. *Pediatr. Res.* 49, 148–154.
296. Regnault, T. R., de Vrijer, B., and Battaglia, F. C. (2002). Transport and metabolism of amino acids in placenta. *Endocrine.* 19, 23–41.
297. Jansson, T. (2001). Amino acid transporters in the human placenta. *Pediatr. Res.* 49, 141–147.
298. Battaglia, F. C., and Regnault, T. R. (2001). Placental transport and metabolism of amino acids. *Placenta* 22, 145–161.
299. McIntosh, N., Rodeck, C. H., and Heath, R. (1984). Plasma amino acids of the mid-trimester human fetus. *Biol. Neonate.* 45, 218–224.
300. Soltesz, G., Harris, D., Mackenzie, I. Z., and Aynsley-Green, A. (1985). The metabolic and endocrine milieu of the human fetus and mother at 18–21 weeks of gestation. I. Plasma amino acid concentrations. *Pediatr. Res.* 19, 91–93.
301. Young, M., and Prenton, M. A. (1969). Maternal and fetal plasma amino acid concentrations during gestation and in retarded fetal growth. *J. Obstet. Gynaecol. Br. Commonw.* 76, 333–334.
302. Pearse, W. H., and Sornson, H. (1969). Free amino acids of normal and abnormal human placenta. *Am. J. Obstet. Gynecol.* 105, 696–701.
303. Dancis, J., Money, W. L., Springer, D., and Levitz, M. (1968). Transport of amino acids by placenta. *Am. J. Obstet. Gynecol.* 101, 820–829.
304. Litonjua, A. D., Canlas, M., Soliman, J., and Paulino, D. Q. (1967). Uptake of alpha-aminoisobutyric acid in placental slices at term. *Am. J. Obstet. Gynecol.* 99, 242–246.
305. Eaton, B. M., and Yudilevich, D. L. (1981). Uptake and asymmetric efflux of amino acids at maternal and fetal sides of placenta. *Am. J. Physiol.* 241, C106–C112.
306. Schneider, H., Mohlen, K. H., and Dancis, J. (1979). Transfer of amino acids across the *in vitro* perfused human placenta. *Pediatr. Res.* 13, 236–240.
307. Christensen, H. N. (1979). Developments in amino acid transport, illustrated for the blood-brain barrier. *Biochem. Pharmacol.* 28, 1989–1992.
308. Roos, S., Powell, T. L., and Jansson, T. (2004). Human placental taurine transporter in uncomplicated and IUGR pregnancies: cellular localization, protein expression, and regulation. *Am. J. Physiol. Regul. Integr. Comp. Physiol.* 287, R886–R893.
309. Dicke, J. M., and Henderson, G. I. (1988). Placental amino acid uptake in normal and complicated pregnancies. *Am. J. Med. Sci.* 295, 223–227.
310. Dicke, J. M., and Vergas, D. K. (1994). Neutral amino acid uptake by microvillous and basal plasma membrane vesicles from appropriate and small for gestational age human pregnancies. *J. Matern. Fetal Med.* 3, 246.
311. Glazier, J. D., Cetin, I., Perugino, G., Ronzoni, S., Grey, A. M., Mahendran, D., Marconi, A. M., Pardi, G., and Sibley, C. P. (1997). Association between the activity of the system A amino acid transporter in the microvillous plasma membrane of the human placenta and severity of fetal compromise in intrauterine growth restriction. *Pediatr. Res.* 42, 514–519.
312. Cramer, S., Beveridge, M., Kilberg, M., and Novak, D. (2002). Physiological importance of system A-mediated amino acid transport to rat fetal development. *Am. J. Physiol. Cell. Physiol.* 282, C153–C160.

313. Bernardini, I., Evans, M. I., Nicolaides, K. H., Economides, D. L., and Gahl, W. A. (1991). The fetal concentrating index as a gestational age-independent measure of placental dysfunction in intrauterine growth retardation. *Am. J. Obstet. Gynecol.* 164, 1481–1487.
314. Cetin, I., Corbetta, C., Sereni, L. P., Marconi, A. M., Bozzetti, P., Pardi, G., and Battaglia, F. C. (1990). Umbilical amino acid concentrations in normal and growth-retarded fetuses sampled in utero by cordocentesis. *Am. J. Obstet. Gynecol.* 162, 253–261.
315. Paolini, C. L., Marconi, A. M., Ronzoni, S., Di Noio, M., Fennessey P.V., Pardi, G., and Battaglia, F. C. (2001). Placental transport of leucine, phenylalanine, glycine, and proline in intrauterine growth-restricted pregnancies. *J. Clin. Endocrinol. Metab.* 86, 5427–5432.
316. Kuruvilla, A. G., D'Souza, S. W., Glazier, J. D., Mahendran, D., Maresh, M. J., and Sibley, C. P. (1994). Altered activity of the system A amino acid transporter in microvillous membrane vesicles from placentas of macrosomic babies born to diabetic women. *J. Clin. Invest.* 94, 689–695.
317. Jansson, T., Ylven, K., Wennergren, M., and Powell, T. L. (2002). Glucose transport and system A activity in syncytiotrophoblast microvillous and basal plasma membranes in intrauterine growth restriction. *Placenta* 23, 392–399.
318. Steel, R. B., Smith, C. H., and Kelley, L. K. (1982). Placental amino acid uptake. VI. Regulation by intracellular substrate. *Am. J. Physiol. Cell. Physiol.* 243, 46–51.
319. Jayanthi, L. D., Ramamoorthy, S., Mahesh, V. B., Leibach, F. H., and Ganapathy V. (1995). Substrate-specific regulation of the taurine transporter in human placental choriocarcinoma cells (JAR). *Biochim. Biophys. Acta* 1235, 351–360.
320. Novak, D., Quiggle, F., Artime, C., and Beveridge, M. (2001). Regulation of glutamate transport and transport proteins in a placental cell line. *Am. J. Physiol. Cell. Physiol.* 281, C1014–C1022.
321. Karl, P. I., Alpy, K. L., and Fisher, S. E. (1992). Amino acid transport by the cultured human placental trophoblast: effect of insulin on AIB transport. *Am. J. Physiol.* 262, C834–C839.
322. Karl, P. I. (1995). Insulin-like growth factor-1 stimulates amino acid uptake by the cultured human placental trophoblast. *J. Cell. Physiol.* 165, 83–88.
323. Bloxam, D. L., Bax, B. E., and Bax, C. M. (1994). Epidermal growth factor and insulin-like growth factor I differently influence the directional accumulation and transfer of 2-aminoisobutyrate (AIB) by human placental trophoblast in two-sided culture. *Biochem. Biophys. Res. Commun.* 199, 922–929.
324. Bonkobara, M., Thongsong, B., Ohmori, T., Matsuki, N., Inaba, M., and Ono, K. (2002). Effects of epidermal growth factor on placental amino acids uptake in pregnant rats. *J. Vet. Med. Sci.* 64, 689–692.
325. Jansson, N., Greenwood, S. L., Johansson, B. R., Powell, T. L., and Jansson, T. (2003). Leptin stimulates the activity of the system A amino acid transporter in human placental villous fragments. *J. Clin. Endocrinol. Metab.* 88, 1205–1211.
326. Carbo, N., Lopez-Soriano, F. J., and Argiles, J. M. (1995). Administration of tumor necrosis factor-alpha results in a decreased placental transfer of amino acids in the rat. *Endocrinology* 136, 3579–3584.
327. Carbo, N., Lopez-Soriano, F. J., Fiers, W., and Argiles, J. M. (1996). Tumour growth results in changes in placental amino acid transport in the rat: a tumour necrosis factor alpha-mediated effect. *Biochem. J.* 313, Pt 1, 77–82.
328. Kulanthaivel, P., Miyamoto, Y., Mahesh, V. B., Leibach, F. H., and Ganapathy, V. (1991). Inactivation of taurine transporter by calcium in purified human placental brush border membrane vesicles. *Placenta* 12, 327–340.
329. Kulanthaivel, P., Cool, D. R., Ramamoorthy, S., Mahesh, V. B., Leibach, F. H., and Ganapathy, V. (1991). Transport of taurine and its regulation by protein kinase C in the JAR human placental choriocarcinoma cell line. *Biochem. J.* 277, Pt 1, 53–58.
330. Battaglia, F. C., and Meschia, G. (1978). Principal substrates of fetal metabolism. *Physiol. Revs.* 58, 499–527.
331. Jones, C. T. (1991). Control of glucose metabolism in the perinatal period. *J. Dev. Physiol.* 15, 81–89.
332. Illsley, N. P., Hall, S., Penfold, P., and Stacey, T. E. (1985). Diffusional permeability of the human placenta. *Contrib. Gynecol. Obstet.* 13, 92–97.
333. Widdas, W. F. (1952). Inability of diffusion to account for placental glucose transfer in the sheep and consideration of the kinetics of a possible carrier transfer. *J. Physiol.* 118, 23–39.
334. Widdas, W. F. (1951). Inability of diffusion to account for placental glucose transfer in the sheep. *J. Physiol.* 115, 36–37.
335. Simmons, M. A., Battaglia, F. C., and Meschia, G. (1979). Placental transfer of glucose. *J. Dev. Physiol.* 1, 227–243.
336. Crandell, S. S., Palma, P. A., and Morriss, F. H. Jr. (1983). Umbilical glucose and lactate extractions during maternal hyperglycemia in sheep. *Am. J. Physiol.* 244, R882–R887.
337. Comline, R. S., and Silver, M. (1976). Some aspects of foetal and uteroplacental metabolism in cows with indwelling umbilical and uterine vascular catheters. *J. Physiol.* 260, 571–586.
338. Silver, M., and Comline, R. S. (1975). Transfer of gases and metabolites in the equine placenta: a comparison with other species. *J. Reprod. Fertil. Suppl.* 589, 94.
339. Silver, M., and Comline, R. S. (1976). Fetal and placental O2 consumption and the uptake of different metabolites in the ruminant and horse during late gestation. *Adv. Exp. Med. Biol.* 75, 731–736.
340. Chinard, F. P., Danesino, V., Hartmann, W. L., Huggett, A. S., Paul, W., and Reynolds, S. R. (1956). The transmission of hexoses across the placenta in the human and the rhesus monkey (Macaca mulatta). *J. Physiol.* 132, 289–303.
341. Morriss, F. H. Jr., Makowski, E. L., Meschia, G., and Battaglia, F. C. (1974). The glucose/oxygen quotient of the term human fetus. *Biol. Neonate* 25, 44–52.
342. Hay, W. W. Jr., and Meznarich, H. K. (1989). Effect of maternal glucose concentration on uteroplacental glucose consumption and transfer in pregnant sheep. *Proc. Soc. Exp. Biol. Med.* 190, 63–69.
343. Hauguel, S., Desmaizieres, V., and Challier, J. C. (1986). Glucose uptake, utilization, and transfer by the human placenta as functions of maternal glucose concentration. *Pediatr. Res.* 20, 269–273.
344. Huggett, A. S., and Nixon, D. A. (1961). Fructose as a component of the foetal blood in several mammalian species. *Nature* 190, 1209.
345. Huggett, A. S., Warren, F. L., and Warren, N. V. (1951). The origin of the blood fructose of the foetal sheep. *J. Physiol.* 113, 258–275.
346. Stacey, T. E., Weedon, A. P., Haworth, C., Ward, R. H., and Boyd, R. D. (1978). Fetomaternal transfer of glucose analogues by sheep placenta. *Am. J. Physiol.* 234, E32–E37.
347. Shah, S. W., Zhao, H., Low, S. Y., McArdle, H. J., and Hundal, H. S. (1999). Characterization of glucose transport and glucose transporters in the human choriocarcinoma cell line, BeWo. *Placenta* 20, 651–659.
348. Bissonnette, J. M., Hohimer, A. R., Cronan, J. Z., and Black, J. A. (1979). Glucose transfer across the intact guinea-pig placenta. *J. Dev. Physiol.* 1, 415–426.

349. Bissonnette, J. M., Black, J. A., Wickham, W. K., and Acott, K. M. (1981). Glucose uptake into plasma membrane vesicles from the maternal surface of human placenta. *J. Membr. Biol.* 58, 75–80.
350. Johnson, L. W., and Smith, C. H. (1985). Glucose transport across the basal plasma membrane of human placental syncytiotrophoblast. *Biochim. Biophys. Acta* 815, 44–50.
351. Bissonnette, J. M., Black, J. A., Thornburg, K. L., Acott, K. M., and Koch, P. L. (1982). Reconstitution of D-glucose transporter from human placental microvillous plasma membranes. *Am. J. Physiol.* 242, C166–C171.
352. Kasahara, M., and Hinkle, P. C. (1977). Reconstitution and purification of the D-glucose transporter from human erythrocytes. *J. Biol. Chem.* 252, 7384–7390.
353. Joost, H. G., and Thorens, B. (2001). The extended GLUT-family of sugar/polyol transport facilitators: nomenclature, sequence characteristics, and potential function of its novel members (review). *Mol. Membr. Biol.* 18, 247–256.
354. Hahn, T., and Desoye, G. (1996). Ontogeny of glucose transport systems in the placenta and its progenitor tissues. *Early Pregnancy* 2, 168–182.
355. Farrell, C. L., Yang, J., and Pardridge, W. M. (1992). GLUT-1 glucose transporter is present within apical and basolateral membranes of brain epithelial interfaces and in microvascular endothelia with and without tight junctions. *J. Histochem. Cytochem.* 40, 193–199.
356. Arnott, G., Coghill, G., McArdle, H. J., and Hundal, H. S. (1994). Immunolocalization of GLUT1 and GLUT3 glucose transporters in human placenta. *Biochem. Soc. Trans.* 22, 272S.
357. Tadokoro, C., Yoshimoto, Y., Sakata, M., Fujimiya, M., Kurachi, H., Adachi, E., Maeda, T., and Miyake, A. (1996). Localization of human placental glucose transporter 1 during pregnancy. An immunohistochemical study. *Histol. Histopathol.* 11, 673–681.
358. Hahn, T., Hartmann, M., Blaschitz A., Skofitsch, G., Graf, R., Dohr, G., and Desoye, G. (1995). Localisation of the high affinity facilitative glucose transporter protein GLUT 1 in the placenta of human, marmoset monkey (Callithrix jacchus) and rat at different developmental stages. *Cell Tissue Res.* 280, 49–57.
359. Hauguel-de Mouzon, S., Leturque A., Alsat, E., Loizeau, M., Evain-Brion, D., and Girard, J. (1994). Developmental expression of Glut1 glucose transporter and c-fos genes in human placental cells. *Placenta* 15, 35–46.
360. Sakata, M., Kurachi, H., Imai, T., Tadokoro, C., Yamaguchi, M., Yoshimoto, Y., Oka, Y., and Miyake, A. (1995). Increase in human placental glucose transporter-1 during pregnancy. *Eur. J. Endocrinol.* 132, 206–212.
361. Jansson, T., Cowley, E. A., and Illsley, N. P. (1995). Cellular localization of glucose transporter messenger RNA in human placenta. *Reprod. Fertil. Dev.* 7, 1425–1430.
362. Clarson, L. H., Glazier, J. D., Sides, M. K., and Sibley, C. P. (1997). Expression of the facilitated glucose transporters (GLUT1 and GLUT3) by a choriocarcinoma cell line (JAr) and cytotrophoblast cells in culture. *Placenta* 18, 333–339.
363. Hauguel-de Mouzon, S., Challier, J. C., Kacemi A., Cauzac, M., Malek A., and Girard, J. (1997). The GLUT3 glucose transporter isoform is differentially expressed within human placental cell types. *J. Clin. Endocrinol. Metab.* 82, 2689–2694.
364. Barros, L. F., Yudilevich, D. L., Jarvis, S. M., Beaumont, N., and Baldwin, S. A. (1995). Quantitation and immunolocalization of glucose transporters in the human placenta. *Placenta* 16, 623–633.
365. Illsley, N. P., Sellers, M. C., and Wright, R. L. (1998). Glycaemic regulation of glucose transporter expression and activity in the human placenta. *Placenta* 19, 517–524.
366. Ogura, K., Sakata, M., Okamoto, Y., Yasui, Y., Tadokoro, C., Yoshimoto, Y., Yamaguchi, M., Kurachi, H., Maeda, T., and Murata, Y. (2000). 8-bromo-cyclicAMP stimulates glucose transporter-1 expression in a human choriocarcinoma cell line. *J. Endocrinol.* 164, 171–178.
367. Baumann, M. U., Deborde, S., and Illsley, N. P. (2002). Placental glucose transfer and fetal growth. *Endocrine* 19, 13–22.
368. Kainulainen, H., Jarvinen, T., and Heinonen, P. K. (1997). Placental glucose transporters in fetal intrauterine growth retardation and macrosomia. *Gynecol. Obstet. Invest.* 44, 89–92.
369. Xing, A. Y., Challier, J. C., Lepercq, J., Cauzac, M., Charron, M. J., Girard, J., and Hauguel-de Mouzon, S. (1998). Unexpected expression of glucose transporter 4 in villous stromal cells of human placenta. *J. Clin. Endocrinol. Metab.* 83, 4097–4101.
370. Takata, K., Kasahara, T., Kasahara, M., Ezaki, O., and Hirano, H. (1992). Localization of erythrocyte/HepG2-type glucose transporter (GLUT1) in human placental villi. *Cell Tissue Res.* 267, 407–412.
371. Fujita-Yamaguchi, Y., Choi, S., Sakamoto, Y., and Itakura, K. (1983). Purification of insulin receptor with full binding activity. *J. Biol. Chem.* 258, 5045–5049.
372. Challier, J. C., Hauguel, S., and Desmaizieres, V. (1986). Effect of insulin on glucose uptake and metabolism in the human placenta. *J. Clin. Endocrinol. Metab.* 62, 803–807.
373. Johnson, L. W. and Smith, C. H. (1980). Monosaccharide transport across microvillous membrane of human placenta. *Am. J. Physiol.* 238, C160–C168.
374. Currie, M. J., Bassett, N. S., and Gluckman, P. D. (1997). Ovine glucose transporter-1 and -3: cDNA partial sequences and developmental gene expression in the placenta. *Placenta* 18, 393–401.
375. Das, U. G., Sadiq, H. F., Soares, M. J., Hay, W. W. Jr., and Devaskar, S. U. (1998). Time-dependant regulation of rodent and ovine placental glucose transporter (GLUT-1) protein. *Am. J. Physiol.* 274, 339–347.
376. Takata, K., Kasahara, T., Kasahara, M., Ezaki, O., and Hirano, H. (1994). Immunolocalization of glucose transporter GLUT1 in the rat placental barrier: possible role of GLUT1 and the gap junction in the transport of glucose across the placental barrier. *Cell Tissue Res.* 276, 411–418.
377. Shin, B. C., Suzuki, T., Matsuzaki, T., Tanaka, S., Kuraoka A., Shibata, Y., and Takata, K. (1996). Immunolocalization of GLUT1 and connexin 26 in the rat placenta. *Cell Tissue Res.* 285, 83–89.
378. Shin, B. C., Fujikura, K., Suzuki, T., Tanaka, S., and Takata, K. (1997). Glucose transporter GLUT3 in the rat placental barrier: a possible machinery for the transplacental transfer of glucose. *Endocrinology* 138, 3997–4004.
379. Boileau, P., Mrejen, C., Girard, J., and Hauguel-de Mouzon, S. (1995). Overexpression of GLUT3 placental glucose transporter in diabetic rats. *J. Clin. Invest.* 96, 309–317.
380. Zhou, J., and Bondy, C. A. (1993). Placental glucose transporter gene expression and metabolism in the rat. *J. Clin. Invest.* 91, 845–852.
381. Hay, W. W. Jr., Sparks, J. W., Quissell, B. J., Battaglia, F. C., and Meschia, G. (1981). Simultaneous measurements of umbilical uptake, fetal utilization rate, and fetal turnover rate of glucose. *Am. J. Physiol.* 240, E662–E668.
382. Meschia, G., Battaglia, F. C., Hay, W. W., and Sparks, J. W. (1980). Utilization of substrates by the ovine placenta in vivo. *Fed. Proc.* 39, 245–249.
383. Carter, A. M. (2000). Placental oxygen consumption. Part I: in vivo studies—a review. *Placenta* 21, Suppl A, S31–S37.

384. Herrera, E., and Lasuncion, M. A., (2004). Maternal–fetal transfer of lipid metabolites. In *Fetal and Neonatal Physiology* (R. A. Polin, W. W. Fox, and S. H. Abman, Eds.), pp. 376–447. WB Saunders, Philadelphia.
385. Knopp, R. H., Warth, M. R., Charles, D., Childs, M., Li, J. R., Mabuchi, H., and Van Allen, M. I. (1986). Lipoprotein metabolism in pregnancy, fat transport to the fetus, and the effects of diabetes. *Biol. Neonate.* 50, 297–317.
386. Dancis, J. (1981). Placental transport of amino acids, fats and minerals. *Mead Johnson Symp. Perinat. Dev. Med.* 18, 25–31.
387. Innis, S. M. (1991). Essential fatty acids in growth and development. *Prog. Lipid Res.* 30, 39–103.
388. Sellmayer, A., Danesch, U., and Weber, P. C. (1997). Modulation of the expression of early genes by polyunsaturated fatty acids. *Prostaglandins Leukot. Essent. Fatty Acids.* 57, 353–357.
389. Clandinin, M. T., Chappell, J. E., and Heim, T. (1981). Do low weight infants require nutrition with chain elongation-desaturation products of essential fatty acids? *Prog. Lipid Res.* 20, 901–904.
390. Uauy, R., Treen, M., and Hoffman, D. R. (1989). Essential fatty acid metabolism and requirements during development. *Semin. Perinatol.* 13, 118–130.
391. Wittmaack, F. M., Gafvels, M. E., Bronner, M., Matsuo, H., McCrae, K. R., Tomaszewski, J. E., Robinson, S. L., Strickland, D. K., and Strauss, J. F. III. (1995). Localization and regulation of the human very low density lipoprotein/apolipoprotein-E receptor: trophoblast expression predicts a role for the receptor in placental lipid transport. *Endocrinology* 136, 340–348.
392. Rothwell, J. E., and Elphick, M. C. (1982). Lipoprotein lipase activity in human and guinea-pig placenta. *J. Dev. Physiol.* 4, 153–159.
393. Bonet, B., Brunzell, J. D., Gown, A. M., and Knopp, R. H. (1992). Metabolism of very-low-density lipoprotein triglyceride by human placental cells: the role of lipoprotein lipase. *Metabolism* 41, 596–603.
394. Stammers, J. P., Stephenson, T. J., Colley, J. A., and Hull, D. (1996). Fetal lipoproteins and placental release of esterified lipids. *Reprod. Fertil. Dev.* 8, 457–463.
395. Elphick, M. C., Hudson, D. G., and Hull, D. (1975). Transfer of fatty acids across the rabbit placenta. *J. Physiol.* 252, 29–42.
396. Hershfield, M. S., and Nemeth, A. M. (1968). Placental transport of free palmitic and linoleic acids in the guinea pig. *J. Lipid Res.* 9, 460–468.
397. Hummel, L., Schirrmeister, W., and Zimmermann, T. (1974). Origin of the fatty acid moiety of fetal serum phospholipids in rats. *Biol. Neonate.* 24, 292–297.
398. Portman, O. W., Behrman, R. E., and Soltys, P. (1969). Transfer of free fatty acids across the primate placenta. *Am. J. Physiol.* 216, 143–147.
399. James, E., Meschia, G., and Battaglia, F. C. (1971). A-V differences of free fatty acids and glycerol in the ovine umbilical circulation. *Proc. Soc. Exp. Biol. Med.* 138, 823–826.
400. Hull, D., and Stammers, J. P. (1985). Placental transfer of fatty acids. *Biochem. Soc. Trans.* 13, 821–822.
401. Elphick, M. C., and Hull, D. (1984). Transfer of fatty acid across the cat placenta. *J. Dev. Physiol.* 6, 517–525.
402. Elphick, M. C., Hull, D., and Sanders, R. R. (1976). Concentrations of free fatty acids in maternal and umbilical cord blood during elective Caesarean section. *Br. J. Obstet. Gynaecol.* 83, 539–544.
403. Sheath, J., Grimwade, J., Waldron, K., Bickley, M., Taft, P., and Wood, C. (1972). Arteriovenous nonesterified fatty acids and glycerol differences in the umbilical cord at term and their relationship to fetal metabolism. *Am. J. Obstet. Gynecol.* 113, 358–362.
404. Berghaus, T. M., Demmelmair, H., and Koletzko, B. (1998). Fatty acid composition of lipid classes in maternal and cord plasma at birth. *Eur. J. Pediatr.* 157, 763–768.
405. Dutta-Roy, A. K. (1997). Fatty acid transport and metabolism in the feto-placental unit and the role of fatty acid-binding proteins. *J. Nutr. Bochem.* 8, 548–557.
406. Kuhn, D. C., and Crawford, M. (1986). Placental essential fatty acid transport and prostaglandin synthesis. *Prog. Lipid Res.* 25, 345–353.
407. Chambaz, J., Ravel, D., Manier, M. C., Pepin, D., Mulliez, N., and Bereziat, G. (1985). Essential fatty acids interconversion in the human fetal liver. *Biol. Neonate.* 47, 136–140.
408. Thomas, C. R., and Lowy, C. (1983). Placental transfer of free fatty acids: factors affecting transfer across the guinea-pig placenta. *J. Dev. Physiol.* 5, 323–332.
409. Stephenson, T. J., Stammers, J. P., and Hull, D. (1990). Maternal to fetal transfer of free fatty acids in the in situ perfused rabbit placenta. *J. Dev. Physiol.* 13, 117–123.
410. Stephenson, T. J., Stammers, J. P., and Hull, D. (1991). Effects of altering umbilical flow and umbilical free fatty acid concentration on transfer of free fatty acids across the rabbit placenta. *J. Dev. Physiol.* 15, 221–227.
411. Haggarty, P., Page, K., Abramovich, D. R., Ashton, J., and Brown, D. (1997). Long-chain polyunsaturated fatty acid transport across the perfused human placenta. *Placenta* 18, 635–642.
412. Campbell, F. M., and Dutta-Roy, A. K. (1995). Plasma membrane fatty acid-binding protein (FABPpm) is exclusively located in the maternal facing membranes of the human placenta. *FEBS Lett.* 375, 227–230.
413. Campbell, F. M., Taffesse, S., Gordon, M. J., and Dutta-Roy, A. K. (1995). Plasma membrane fatty-acid-binding protein in human placenta: identification and characterization. *Biochem. Biophys. Res. Commun.* 209, 1011–1017.
414. Campbell, F. M., Gordon, M. J., and Dutta-Roy, A. K. (1994). Plasma membrane fatty acid-binding protein (FABPpm) of the sheep placenta. *Biochim. Biophys. Acta* 1214, 187–192.
415. Knipp, G. T., Liu, B., Audus, K. L., Fujii, H., Ono, T., and Soares, M. J. (2000). Fatty acid transport regulatory proteins in the developing rat placenta and in trophoblast cell culture models. *Placenta* 21, 367–375.
416. Stremmel, W., Kleinert, H., Fitscher, B. A., Gunawan, J., Klaassen-Schluter, C., Moller, K., and Wegener, M. (1992). Mechanism of cellular fatty acid uptake. *Biochem. Soc. Trans.* 20, 814–817.
417. Lafond, J., Simoneau, L., Savard, R., and Gagnon, M. C. (1994). Linoleic acid transport by human placental syncytiotrophoblast membranes. *Eur. J. Biochem.* 226, 707–713.
418. Campbell, F. M., Gordon, M. J., and Dutta-Roy, A. K. (1998). Placental membrane fatty acid-binding protein preferentially binds arachidonic and docosahexaenoic acids. *Life Sci.* 63, 235–240.
419. Campbell, F. M., Gordon, M. J., and Dutta-Roy, A. K. (1996). Preferential uptake of long chain polyunsaturated fatty acids by isolated human placental membranes. *Mol. Cell. Biochem.* 155, 77–83.
420. Lafond, J., Moukdar, F., Rioux A., Ech-Chadli, H., Brissette, L., Robidoux, J., Masse A., and Simoneau, L. (2000). Implication of ATP and sodium in arachidonic acid incorporation by placental syncytiotrophoblast brush border and basal plasma membranes in the human. *Placenta* 21, 661–669.
421. Dutta-Roy, A. K. (2000). Cellular uptake of long-chain fatty acids: role of membrane-associated fatty-acid-binding/transport proteins. *Cell. Mol. Life Sci.* 57, 1360–1372.
422. Campbell, F. M., Bush PG, Veerkamp, J. H., and Dutta-Roy, A. K. (1998). Detection and cellular localization

of plasma membrane-associated and cytoplasmic fatty acid-binding proteins in human placenta. *Placenta* 19, 409–415.

423. Shand, J. H., and Noble, R. C. (1985). Incorporation of linoleic and arachidonic acids into ovine placental phospholipids *in vitro*. *Biol. Neonate.* 48, 299–306.
424. Crabtree, J. T., Gordon, M. J., Campbell, F. M., and Dutta-Roy, A. K. (1998). Differential distribution and metabolism of arachidonic acid and docosahexaenoic acid by human placental choriocarcinoma (BeWo) cells. *Mol. Cell. Biochem.* 185, 191–198.
425. Robertson, A., and Sprecher, H. (1967). Human placental lipid metabolism. II. Analysis of fatty acids from placentas after intrauterine death. *Am. J. Obstet. Gynecol.* 99, 247–249.
426. Magnusson, A. L., Waterman, I. J., Wennergren, M., Jansson, T., and Powell, T. L. (2004). Triglyceride hydrolase activities and expression of fatty acid binding proteins in the human placenta in pregnancies complicated by intrauterine growth restriction and diabetes. *J. Clin. Endocrinol. Metab.* 89, 4607–4614.
427. Aynsley-Green, A., Soltesz, G., Jenkins, P. A., and Mackenzie, I. Z. (1985). The metabolic and endocrine milieu of the human fetus at 18–21 weeks of gestation. II. Blood glucose, lactate, pyruvate and ketone body concentrations. *Biol. Neonate* 47, 19–25.
428. Paterson, P., Sheath, J., Taft, P., and Wood, C. (1967). Maternal and foetal ketone concentrations in plasma and urine. *Lancet* 1, 862–865.
429. Paton, J. B., Levitsky, L. L., and Fisher, D. E. (1989). Placental transfer and fetal effects of maternal sodium beta-hydroxybutyrate infusion in the baboon. *Pediatr. Res.* 25, 435–439.
430. Girard, J. R., Ferre, P., Gilbert, M., Kervran A., Assan, R., and Marliss, E. B. (1977). Fetal metabolic response to maternal fasting in the rat. *Am. J. Physiol.* 232, E456–E463.
431. Alonso de la Torre, S. R., Serrano, M. A., and Medina, J. M. (1992). Carrier-mediated beta-D-hydroxybutyrate transport in brush-border membrane vesicles from rat placenta. *Pediatr. Res.* 32, 317–323.
432. Miodovnik, M., Lavin, J. P., Harrington, D. J., Leung LS, Seeds, A. E., and Clark, K. E. (1982). Effect of maternal ketoacidemia on the pregnant ewe and the fetus. *Am. J. Obstet. Gynecol.* 144, 585–593.
433. Harding, J. E., Charlton, V. E., and Evans, P. C. (1992). Effects of beta-hydroxybutyrate infusion on hind limb metabolism in fetal sheep. *Am. J. Obstet. Gynecol.* 166, 671–676.
434. Kingdom, J. C., and Kaufmann, P. (1999). Oxygen and placental vascular development. *Adv. Exp. Med. Biol.* 474, 259–275.
435. Burton, G. J., and Caniggia, I. (2001). Hypoxia: implications for implantation to delivery-a workshop report. *Placenta* 22, Suppl A, S63–S65.
436. Zhou, Y., Genbacev, O., Damsky, C. H., and Fisher, S. J. (1998). Oxygen regulates human cytotrophoblast differentiation and invasion: implications for endovascular invasion in normal pregnancy and in pre-eclampsia. *J. Reprod. Immunol.* 39, 197–213.
437. Lash, G. E., Postovit, L. M., Matthews, N. E., Chung, E. Y., Canning, M. T., Pross, H., Adams, M. A., and Graham, C. H. (2002). Oxygen as a regulator of cellular phenotypes in pregnancy and cancer. *Can. J. Physiol. Pharmacol.* 80, 103–109.
438. Graham, C. H., Postovit, L. M., Park, H., Canning, M. T., and Fitzpatrick, T. E. (2000). Adriana and Luisa Castellucci award lecture 1999: role of oxygen in the regulation of trophoblast gene expression and invasion. *Placenta* 21, 443–450.
439. Kudo, Y., Boyd, C. A., Sargent, I. L., and Redman, C. W. (2003). Hypoxia alters expression and function of syncytin and its receptor during trophoblast cell fusion of human placental BeWo cells: implications for impaired trophoblast syncytialisation in pre-eclampsia. *Biochim. Biophys. Acta* 1638, 63–71.
440. Anderson, D. F., and Faber, J. J. (1984). Regulation of fetal placental blood flow in the lamb. *Am. J. Physiol.* 247, R567–R574.
441. Paulone, M. E., Edelstone, D. I., and Shedd, A. (1987). Effects of maternal anemia on uteroplacental and fetal oxidative metabolism in sheep. *Am. J. Obstet. Gynecol.* 156, 230–236.
442. Zamudio, S. (2003). The placenta at high altitude. *High Alt. Med. Biol.* 4, 171–191.
443. Cole, H. (1986). Studying reproductive risks, smoking. *JAMA* 255, 22–23.
444. Gurtner, G. H., and Burns, B. (1973). The role of cytochrome P-450 of placenta in facilitated oxygen diffusion. *Drug Metab. Dispos.* 1, 368–373.
445. Carter, A. M. (1989). Factors affecting gas transfer across the placenta and the oxygen supply to the fetus. *J. Dev. Physiol.* 12, 305–322.
446. Itskovitz, J., LaGamma, E. F., and Rudolph, A. M. (1983). The effect of reducing umbilical blood flow on fetal oxygenation. *Am. J. Obstet. Gynecol.* 145, 813–818.
447. Edelstone, D. I., Peticca, B. B., and Goldblum, L. J. (1985). Effects of maternal oxygen administration on fetal oxygenation during reductions in umbilical blood flow in fetal lambs. *Am. J. Obstet. Gynecol.* 152, 351–358.
448. Anderson, D. F., Parks, C. M., and Faber, J. J. (1986). Fetal O2 consumption in sheep during controlled long-term reductions in umbilical blood flow. *Am. J. Physiol.* 250, H1037–H1042.
449. Wilkening, R. B., and Meschia, G. (1983). Fetal oxygen uptake, oxygenation, and acid-base balance as a function of uterine blood flow. *Am. J. Physiol.* 244, H749–H755.
450. Delivoria-Papadopoulos, M., and McGowan, J. E., (2004). Oxygen transport and delivery. In *Fetal and Neonatal Physiology* (R. A. Polin, W. W. Fox, and S. H. Abman, Eds.), pp. 880–889. WB Saunders, Philadelphia.
451. Itskovitz, J., Goetzman, B. W., Roman, C., and Rudolph, A. M. (1984). Effects of fetal-maternal exchange transfusion on fetal oxygenation and blood flow distribution. *Am. J. Physiol.* 247, H655–H660.
452. Edelstone, D. I., Darby, M. J., Bass, K., and Miller, K. (1989). Effects of reductions in hemoglobin-oxygen affinity and hematocrit level on oxygen consumption and acid-base state in fetal lambs. *Am. J. Obstet. Gynecol.* 160, 820–826.
453. Wilkening, R. B., Molina, R. D., and Meschia, G. (1988). Placental oxygen transport in sheep with different hemoglobin types. *Am. J. Physiol.* 254, R585–R589.
454. Schroder, H. (1982). Structural and functional organization of the placenta from the physiological point of view. *Bibl. Anat.* 22, 4–12.
455. Rankin, J. H., Meschia, G., Makowski, E. L., and Battaglia, F. C. (1971). Relationship between uterine and umbilical venous PO2 in sheep. *Am. J. Physiol.* 220, 1688–1692.
456. Silver, M., Steven, D. H., and Comline, R. S., (1973). Placental exchange and morphology in ruminants and the mare. In *Foetal and Neonatal Physiology* (R. S. Comline, Ed.), pp. 245–271. Cambridge University Press, Cambridge.
457. Steven, D. H. (1966). Arteriovenous anastomoses in the uterus of the sheep. *J. Physiol.* 187, 18p–19p.
458. Mossman, H. W. (1926). *Am. J. Anat.* 37, 433.
459. Kaufmann, P. (1969). Guinea pig placenta and its development. *Z. Anat. Entwicklungsgesch.* 129, 83–101.
460. Schroder, H., and Leichtweiss, H. P. (1977). Perfusion rates and the transfer of water across isolated guinea pig placenta. *Am. J. Physiol.* 232, H666–H670.

461. Bailey, D. J. (1974). Proceedings: Counter-current flow of maternal and foetal blood-streams in guinea-pig placenta. *J. Physiol.* 242, 104P–105P.
462. Faber, J. J., and Hart, F. M. (1966). The rabbit placenta as an organ of diffusional exchange. Comparison with other species by dimensional analysis. *Circ. Res.* 19, 816–833.
463. Moll, W., and Kastendieck, E. (1977). Transfer of N2O, CO and HTO in the artificially perfused guinea-pig placenta. *Respir. Physiol.* 29, 283–302.
464. Wilkening, R. B., and Meschia, G. (1992). Current topic: comparative physiology of placental oxygen transport. *Placenta* 13, 1–15.
465. Wilkening, R. B., and Meschia, G. (1989). Effect of umbilical blood flow on transplacental diffusion of ethanol and oxygen. *Am. J. Physiol.* 256, H813–H820.
466. Hustin, J., and Schaaps, J. P. (1987). Echographic (corrected) and anatomic studies of the maternotrophoblastic border during the first trimester of pregnancy. *Am. J. Obstet. Gynecol.* 157, 162–168.
467. Jauniaux, E., Watson, A. L., Hempstock, J., Bao, Y. P., Skepper, J. N., and Burton, G. J. (2000). Onset of maternal arterial blood flow and placental oxidative stress. A possible factor in human early pregnancy failure. *Am. J. Pathol.* 157, 2111–2122.
468. Jauniaux, E., Gulbis, B., and Burton, G. J. (2002). The first trimester gestational sac limits rather than facilitates oxygen transfer to the foetus: a review. *Placenta (Supplement A Tropohoblast Research V17)* 24, 86–93.
469. Brittain, T. (2002). Molecular aspects of embryonic hemoglobin function. *Mol. Aspects Med.* 23, 293–342.
470. Graham, C. H., and Burton, G. J. (2004). Oxygen and trophoblast behaviour. A workshop report. *Trophoblast Res.* 18, S90–S92.
471. Bonds, D. R., Crosby, L. O., Cheek, T. G., Hagerdal, M., Gutsche, B. B., and Gabbe, S. G. (1986). Estimation of human fetal–placental unit metabolic rate by application of the Bohr principle. *J. Dev. Physiol.* 8, 49–54.
472. Carroll, M. J., and Young M. (1987). Observations on the energy and redox state and protein synthetic rate in animal and human placentas. *J. Perinat. Med.* 15, 21–30.
473. Gu, W., Jones, C. T., and Parer, J. T. (1985). Metabolic and cardiovascular effects on fetal sheep of sustained reduction of uterine blood flow. *J. Physiol.* 368, 109–129.
474. Owens, J. A., Falconer, J., and Robinson, J. S. (1987). Effect of restriction of placental growth on oxygen delivery to and consumption by the pregnant uterus and fetus. *J. Dev. Physiol.* 9, 137–150.
475. Power, G. G., and Longo, L. D. (1969). Graphical analysis of maternal and fetal exchange of O_2 and CO_2. *J. Appl. Physiol.* 26, 38–47.
476. Hill, E. P., Power, G. G., and Longo, L. D. (1973). A mathematical model of carbon dioxide transfer in the placenta and its interaction with oxygen. *Am. J. Physiol.* 224, 283–299.
477. Sterling, D., and Casey, J. R. (2002). Bicarbonate transport proteins. *Biochem. Cell Biol.* 80, 483–497.
478. Kopito, R. R., Lee, B. S., Simmons, D. M., Lindsey, A. E., Morgans, C. W., and Schneider, K. (1989). Regulation of intracellular pH by a neuronal homolog of the erythrocyte anion exchanger. *Cell* 59, 927–937.
479. Romero, M. F., Fulton, C. M., and Boron, W. F. (2004). The SLC4 family of HCO 3-transporters. *Pflugers Arch.* 447, 495–509.
480. Lacey, H. A., Nolan, T., Greenwood, S. L., Glazier, J. D., and Sibley, C. P. (2004). Gestational profile of Na^+/H^+ exchanger and Cl^-/HCO_3^- anion exchanger mRNA expression in placenta using real-time QPCR. *Placenta* 26, 93–98.
481. Sterling, D., Reithmeier, R. A., and Casey, J. R. (2001). A transport metabolon. Functional interaction of carbonic anhydrase II and chloride/bicarbonate exchangers. *J. Biol. Chem.* 276, 47886–47894.
482. Aliakbar, S., Brown, P. R., and Nicolaides, K. H. (1990). Localization of CAI and CAII isoenzymes in normal term human placenta by immunofluorescence techniques. *Placenta* 11, 35–39.
483. Muhlhauser, J., Crescimanno, C., Rajaniemi, H., Parkkila, S., Milovanov, A. P., Castellucci, M., and Kaufmann, P. (1994). Immunohistochemistry of carbonic anhydrase in human placenta and fetal membranes. *Histochemistry* 101, 91–98.
484. Aliakbar, S., and Brown, P. R. (1996). Measurement of human erythrocyte CAI and CAII in adult, newborn, and fetal blood. *Clin. Biochem.* 29, 157–164.
485. Hatano, H., Leichtweiss, H. P., and Schroder, H. (1989). Uptake of bicarbonate/CO2 in the isolated guinea pig placenta. *Placenta* 10, 213–221.
486. Wong, C. T., and Morgan, E. H. (1973). Placental transfer of iron in the guinea pig. *Q. J. Exp. Physiol. Cogn. Med. Sci.* 58, 47–58.
487. Bothwell, T. H., Pribilla, W. F., Mebust, W., and Finch, C. A. (1958). Iron metabolism in the pregnant rabbit; iron transport across the placenta. *Am. J. Physiol.* 193, 615–622.
488. McArdle, H. J., and Morgan, E. H. (1982). Transferrin and iron movements in the rat conceptus during gestation. *J. Reprod. Fertil.* 66, 529–536.
489. Bogden, J. D., Thind, I. S., Kemp, F. W., and Caterini, H. (1978). Plasma concentrations of calcium, chromium, copper, iron, magnesium, and zinc in maternal and cord blood and their relationship to low birth weight. *J. Lab. Clin. Med.* 92, 455–462.
490. Morgan, E. H. (1982). Placental transfer of iron. *Proc. Aust. Physiol. Pharmacol. Soc.* 13, 11–17.
491. Morgan, E. H. (1964). Passage of transferrin, albumin and gamma globulin from maternal plasma to foetus in the rat and rabbit. *J. Physiol.* 171, 26–41.
492. Davies, J., Brown, E. B. Jr., Stewart, D., Terry, C. W., and Sisson, J. (1959). Transfer of radioactive iron via the placenta and accessory fetal membranes in the rabbit. *Am. J. Physiol.* 197, 87–92.
493. Srai, S. K., Bomford, A., and McArdle, H. J. (2002). Iron transport across cell membranes: molecular understanding of duodenal and placental iron uptake. *Best Pract. Res. Clin. Haematol.* 15, 243–259.
494. Parkkila, S., Waheed A., Britton, R. S., Bacon, B. R., Zhou, X. Y., Tomatsu, S., Fleming, R. E., and Sly, W. S. (1997). Association of the transferrin receptor in human placenta with HFE, the protein defective in hereditary hemochromatosis. *Proc. Natl. Acad. Sci. U S A* 94, 13198–13202.
495. Georgieff, M. K., Wobken, J. K., Welle, J., Burdo, J. R., and Connor, J. R. (2000). Identification and localization of divalent metal transporter-1 (DMT-1) in term human placenta. *Placenta* 21, 799–804.
496. Donovan, A., Brownlie, A., Zhou, Y., Shepard, J., Pratt, S. J., Moynihan, J., Paw, B. H., Drejer, A., Barut, B., Zapata, A., Law, T. C., Brugnara, C., Lux, S. E., Pinkus, G. S., Pinkus, J. L., Kingsley, P. D., Palis, J., Fleming, M. D., Andrews, N. C., and Zon, L. I. (2000). Positional cloning of zebrafish ferroportin1 identifies a conserved vertebrate iron exporter. *Nature* 403, 776–781.
497. Danzeisen, R., Fosset, C., Chariana, Z., Page, K., David, S., and McArdle, H. J. (2002). Placental ceruloplasmin homolog is regulated by iron and copper and is implicated in iron metabolism. *Am. J. Physiol. Cell. Physiol.* 282, C472–C478.

498. Verrijt, C. E., Kroos, M. J., Huijskes-Heins, M. I., Cleton-Soeteman, M. I., van Run, P. R., van Eijk, H. G., and van Dijk, J. P. (1999). Accumulation and release of iron in polarly and non-polarly cultured trophoblast cells isolated from human term placentas. *Eur. J. Obstet. Gynecol. Reprod. Biol.* 86, 3–81.
499. Verrijt, C. E., Kroos, M. J., Huijskes-Heins, M. I., van Eijk, H. G., and van Dijk, J. P. (1998). Non-transferrin iron uptake by trophoblast cells in culture. Significance of a NADH-dependent ferrireductase. *Placenta* 19, 525–530.
500. van Dijk, J. P. (1988). Regulatory aspects of placental iron transfer—a comparative study. *Placenta* 9, 215–226.
501. Gambling, L., Danzeisen, R., Gair, S., Lea, R. G., Charania, Z., Solanky, N., Joory, K. D., Srai, S. K., and McArdle, H. J. (2001). Effect of iron deficiency on placental transfer of iron and expression of iron transport proteins in vivo and in vitro. *Biochem. J.* 356, 883–889.
502. Douglas, G. C., and King, B. F. (1990). Uptake and processing of 125I-labelled transferrin and 59Fe-labelled transferrin by isolated human trophoblast cells. *Placenta* 11, 41–57.
503. Starreveld, J. S., van Dijk, H. P., Kroos, M. J., and van Eijk, H. G. (1993). Regulation of transferrin receptor expression and distribution in *in vitro* cultured human cytotrophoblasts. *Clin. Chim. Acta* 220, 47–60.
504. Kroos, M. J., Starreveld, J. S., Verrijt, C. E, van Eijk, H. G., and Van Dijk, J. P. (1996). Regulation of transferrin receptor synthesis by human cytotrophoblast cells in culture. *Eur. J. Obstet. Gynecol. Reprod. Biol.* 65, 231–234.
505. Georgieff, M. K., Berry, S. A., Wobken, J. D., and Leibold, E. A. (1999). Increased placental iron regulatory protein-1 expression in diabetic pregnancies complicated by fetal iron deficiency. *Placenta* 20, 87–93.
506. O'Brien, K. O., Zavaleta, N., Abrams, S. A., and Caulfield, L. E. (2003). Maternal iron status influences iron transfer to the fetus during the third trimester of pregnancy. *Am. J. Clin. Nutr.* 77, 924–930.
507. Knisely, A. S., Harford, J. B., Klausner, R. D., and Taylor, S. R. (1989). Neonatal hemochromatosis. The regulation of transferrin-receptor and ferritin synthesis by iron in cultured fibroblastic-line cells. *Am. J. Pathol.* 134, 439–445.
508. Kelly, A. L., Lunt, P. W., Rodrigues, F., Berry, P. J., Flynn, D. M., McKiernan, P. J., Kelly, D. A., Mieli-Vergani, G., and Cox, T. M. (2001). Classification and genetic features of neonatal haemochromatosis: a study of 27 affected pedigrees and molecular analysis of genes implicated in iron metabolism. *J. Med. Genet.* 38, 599–610.
509. Murray, K. F. and Kowdley, K. V. (2001). Review article: Neonatal hemochromatosis. *Pediatrics* 108, 960–964.
510. Burton, G. J. (1982). Review article. Placental uptake of maternal erythrocytes: a comparative study. *Placenta* 3, 407–434.
511. Baker, E., and Morgan, E. H. (1973). Placental iron transfer in the cat. *J. Physiol.* 232, 485–501.
512. Wong, C. T., and Morgan, E. H. (1974). Source of fetal iron in the cat. *Aust. J. Exp. Biol. Med. Sci.* 52, 413–416.
513. Dumartin, B., and Canivenc, R. (1992). Placental iron transfer regulation in the haemophagous region of the badger placenta: ultrastructural localization of ferritin in trophoblast and endothelial cells. *Anat. Embryol. (Berl).* 185, 175–179.
514. Dantzer, V., and Nielsen, M. H. (1984). Intracellular pathways of native iron in the maternal part of the porcine placenta. *Eur. J. Cell Biol.* 34, 103–109.
515. Roberts, R. M., Raub, T. J., and Bazer, F. W. (1986). Role of uteroferrin in transplacental iron transport in the pig. *Fed. Proc.* 45, 2513–2518.
516. McArdle, H. J. and Priscott, P. K. (1984). Uptake and metabolism of transferrin and albumin by rat yolk sac placenta. *Am. J. Physiol.* 247, C409–C414.
517. Mirkin, B. L. (1975). Perinatal pharmacology: placental transfer, fetal localization, and neonatal disposition of drugs. *Anesthesiology* 43, 156–170.
518. Plonait, S. L., and Nau, H., (2004). Physicochemical and structural properties regulating placental drug transfer. In *Fetal and Neonatal Physiology* (R. A. Polin, W. W. Fox, and S. H. Abman, Eds.), pp. 197–211. WB Saunders, Philadelphia.
519. Waddell, W. J., and Marlowe, C. (1981). Transfer of drugs across the placenta. *Pharmacol. Ther.* 14, 375–390.
520. Ganapathy, V., Prasad, P. D., Ganapathy, M. E., and Leibach, F. H. (2000). Placental transporters relevant to drug distribution across the maternal-fetal interface. *J. Pharmacol. Exp. Ther.* 294, 413–420.
521. Wu, S. P., Shyu, M. K., Liou, H. H., Gau, C. S., and Lin, C. J. (2004). Interaction between anticonvulsants and human placental carnitine transporter. *Epilepsia* 45, 204–210.
522. Nooter, K., and Stoter, G. (1996). Molecular mechanisms of multidrug resistance in cancer chemotherapy. *Pathol. Res. Pract.* 192, 768–780.
523. Schinkel, A. H. (1997). The physiological function of drugtransporting P-glycoproteins. *Semin. Cancer Biol.* 8, 161–170.
524. Schinkel, A. H., Wagenaar, E., Mol, C. A., and van Deemter, L. (1996). P-glycoprotein in the blood-brain barrier of mice influences the brain penetration and pharmacological activity of many drugs. *J. Clin. Invest.* 97, 2517–2524.
525. Juliano, R. L., and Ling, V. (1976). A surface glycoprotein modulating drug permeability in Chinese hamster ovary cell mutants. *Biochim. Biophys. Acta* 455, 152–162.
526. Brinkmann, U., and Eichelbaum, M. (2001). Polymorphisms in the ABC drug transporter gene MDR1. *Pharmacogenomics J.* 1, 59–64.
527. St-Pierre, M. V., Serrano, M. A., Macias, R. I., Dubs, U., Hoechli, M., Lauper, U., Meier, P. J., and Marin, J. J. (2000). Expression of members of the multidrug resistance protein family in human term placenta. *Am. J. Physiol. Regul. Integr. Comp. Physiol.* 279, R1495–R1503.
528. Nakamura, Y., Ikeda, S., Furukawa, T., Sumizawa, T., Tani A., Akiyama, S., and Nagata, Y. (1997). Function of P-glycoprotein expressed in placenta and mole. *Biochem. Biophys. Res. Commun.* 235, 849–853.
529. MacFarland, A., Abramovich, D. R., Ewen, S. W., and Pearson, C. K. (1994). Stage-specific distribution of P-glycoprotein in first-trimester and full-term human placenta. *Histochem. J.* 26, 417–423.
530. Ushigome, F., Takanaga, H., Matsuo, H., Yanai, S., Tsukimori, K., Nakano, H., Uchiumi, T., Nakamura, T., Kuwano, M., Ohtani, H., and Sawada, Y. (2000). Human placental transport of vinblastine, vincristine, digoxin and progesterone: contribution of P-glycoprotein. *Eur. J. Pharmacol.* 408, 1–10.
531. Utoguchi, N., Chandorkar, G. A., Avery, M., and Audus, K. L. (2000). Functional expression of P-glycoprotein in primary cultures of human cytotrophoblasts and BeWo cells. *Reprod. Toxicol.* 14, 217–224.
532. Ushigome, F., Koyabu, N., Satoh, S., Tsukimori, K., Nakano, H., Nakamura, T., Uchiumi, T., Kuwano, M., Ohtani, H., and Sawada, Y. (2003). Kinetic analysis of P-glycoprotein-mediated transport by using normal human placental brush-border membrane vesicles. *Pharm. Res.* 20, 38–44.
533. Pavek, P., Staud, F., Fendrich, Z., Sklenarova, H., Libra, A., Novotna, M., Kopecky, M., Nobilis, M., and Semecky, V. (2003). Examination of the functional activity of P-glycoprotein in the rat placental barrier using rhodamine 123. *J. Pharmacol. Exp. Ther.* 305, 1239–1250.
534. Lankas, G. R., Wise, L. D., Cartwright, M. E., Pippert, T., and Umbenhauer, D. R. (1998). Placental P-glycoprotein

deficiency enhances susceptibility to chemically induced birth defects in mice. *Reprod. Toxicol.* 12, 457–463.

535. Smit, J. W., Huisman, M. T., van Tellingen, O., Wiltshire HR, and Schinkel, A. H. (1999). Absence or pharmacological blocking of placental P-glycoprotein profoundly increases fetal drug exposure. *J. Clin. Invest.* 104, 1441–1447.
536. Deeley, R. G., and Cole, S. P. (1997). Function, evolution and structure of multidrug resistance protein (MRP). *Semin. Cancer Biol.* 8, 193–204.
537. Nagashige, M., Ushigome, F., Koyabu, N., Hirata, K., Kawabuchi, M., Hirakawa, T., Satoh, S., Tsukimori, K., Nakano, H., Uchiumi, T., Kuwano, M., Ohtani, H., and Sawada, Y. (2003). Basal membrane localization of MRP1 in human placental trophoblast. *Placenta* 24, 951–958.
538. Doyle, L. A., Yang, W., Abruzzo, L. V., Krogmann, T., Gao, Y., Rishi, A. K., and Ross, D. D. (1998). A multidrug resistance transporter from human MCF-7 breast cancer cells. *Proc. Natl. Acad. Sci. U S A* 95, 15665–15670.
539. Doyle, L. A., and Ross, D. D. (2003). Multidrug resistance mediated by the breast cancer resistance protein BCRP (ABCG2). *Oncogene* 22, 7340–7358.
540. Young, A. M., Allen, C. E., and Audus, K. L. (2003). Efflux transporters of the human placenta. *Adv. Drug Deliv. Rev.* 55, 125–132.
541. Leazer, T. M., and Klaassen, C. D. (2003). The presence of xenobiotic transporters in rat placenta. *Drug Metab. Dispos.* 31, 153–167.
542. Lown, K. S., Mayo, R. R., Leichtman, A. B., Hsiao, H. L., Turgeon, D. K., Schmiedlin-Ren, P., Brown, M. B., Guo, W., Rossi, S. J., Benet, L. Z., and Watkins, P. B. (1997). Role of intestinal P-glycoprotein (mdr1) in interpatient variation in the oral bioavailability of cyclosporine. *Clin. Pharmacol. Ther.* 62, 248–260.
543. Nishimura, M., Yaguti, H., Yoshitsugu, H., Naito, S., and Satoh, T. (2003). Tissue distribution of mRNA expression of human cytochrome P450 isoforms assessed by high-sensitivity real-time reverse transcription PCR. *Yakugaku Zasshi.* 123, 369–375.
544. Hakkola, J., Pelkonen, O., Pasanen, M., and Raunio, H. (1998). Xenobiotic-metabolizing cytochrome P450 enzymes in the human feto-placental unit: role in intrauterine toxicity. *Crit. Rev. Toxicol.* 28, 35–72.
545. Pasanen, M. (1999). The expression and regulation of drug metabolism in human placenta. *Adv. Drug Deliv. Rev.* 38, 81–97.
546. Cole, D. E., Baldwin, L. S., and Stirk, L. J. (1984). Increased inorganic sulfate in mother and fetus at parturition: evidence for a fetal-to-maternal gradient. *Am. J. Obstet. Gynecol.* 148, 596–599.
547. Andrews, J. (1973). Thiocyanate and smoking in pregnancy. *J. Obstet. Gynaecol. Br. Commonw.* 80, 810–814.
548. Knisely, A. S., Grady, R. W., Kramer, E. E., and Jones, R. L. (1989). Cytoferrin, maternofetal iron transport, and neonatal hemochromatosis. *Am. J. Clin. Pathol.* 92, 755–759.
549. Kundu, N., Wachs, M., Iverson, G. B., and Petersen, L. P. (1981). Fetal and maternal serum copper levels before and during labor in normal and complicated pregnancies. *Obstet. Gynecol.* 57, 313–319.
550. Butte, N. F., and Calloway, D. H. (1982). Proteins, vitamin, A., carotene, folacin, ferritin and zinc in Navajo maternal and cord blood. *Biol. Neonate.* 41, 273–278.
551. Korpela, H., Loueniva, R., Yrjanheikki, E., and Kauppila, A. (1986). Lead and cadmium concentrations in maternal and umbilical cord blood, amniotic fluid, placenta, and amniotic membranes. *Am. J. Obstet. Gynecol.* 155, 1086–1089.
552. Marconi, A. M., Cetin, I., Ferrazzi, E., Ferrari, M. M., Pardi, G., and Battaglia, F. C. (1990). Lactate metabolism in normal and growth-retarded human fetuses. *Pediatr. Res.* 28, 652–656.
553. McConathy, W. J., and Lane, D. M. (1980). Studies on the apolipoproteins and lipoproteins of cord serum. *Pediatr. Res.* 14, 757–761.
554. Parker, C. R. Jr., Deahl, T., Drewry, P., and Hankins, G. (1983). Analysis of the potential for transfer of lipoprotein-cholesterol across the human placenta. *Early Hum. Dev.* 8, 289–295.
555. Quirk, J. G. Jr., and Bleasdale, J. E. (1983). myo-Inositol homeostasis in the human fetus. *Obstet. Gynecol.* 62, 41–44.
556. O'Connor, M. C., Harkness, R. A., Simmonds, R. J., and Hytten, F. E. (1981). The measurement of hypoxanthine, xanthine, inosine and uridine in umbilical cord blood and fetal scalp blood samples as a measure of fetal hypoxia. *Br. J. Obstet. Gynaecol.* 88, 381–390.
557. Gresham, E. L., Simons, P. S., and Battaglia, F. C. (1971). Maternal–fetal urea concentration difference in man: metabolic significance. *J. Pediatr.* 79, 809–811.
558. Zeisel, S. H., Epstein, M. F., and Wurtman, R. J. (1980). Elevated choline concentration in neonatal plasma. *Life Sci.* 26, 1827–1831.
559. Cederblad, G., Niklasson A., Rydgren, B., Albertsson-Wikland, K., and Olegard, R. (1985). Carnitine in maternal and neonatal plasma. *Acta Paediatr. Scand.* 74, 500–504.
560. Colombo, C., Roda, A., Roda, E., Buscaglia, M., dell' Agnola, C. A., Filippetti, P., Ronchi, M., and Sereni, F. (1985). Correlation between fetal and maternal serum bile acid concentrations. *Pediatr. Res.* 19, 227–231.
561. Dancis, J., Wilson, D., Hoskins, I. A., and Levitz, M. (1988). Placental transfer of thiamine in the human subject: *in vitro* perfusion studies and maternal-cord plasma concentrations. *Am. J. Obstet. Gynecol.* 159, 1435–1439.
562. Contractor, S. F., and Shane, B. (1970). Blood and urine levels of vitamin B6 in the mother and fetus before and after loading of the mother with vitamin B6. *Am. J. Obstet. Gynecol.* 107, 635–640.
563. Giugliani, E. R., Jorge, S. M., and Goncalves, A. L. (1985). Serum vitamin B12 levels in parturients, in the intervillous space of the placenta and in full-term newborns and their interrelationships with folate levels. *Am. J. Clin. Nutr.* 41, 330–335.
564. Willis, D. M., O'Grady, J. P., Faber, J. J., and Thornburg, K. L. (1986). Diffusion permeability of cyanocobalamin in human placenta. *Am. J. Physiol.* 250, R459–R464.
565. Kaminetzky, H. A., Baker, H., Frank, O., and Langer, A. (1974). The effects of intravenously administered water-soluble vitamins during labor in normovitaminemic and hypovitaminemic gravidas on maternal and neonatal blood vitamin levels at delivery. *Am. J. Obstet. Gynecol.* 120, 697–703.
566. Delvin, E. E., Glorieux, F. H., Salle, B. L., David, L., and Varenne, J. P. (1982). Control of vitamin D metabolism in preterm infants: feto-maternal relationships. *Arch. Dis. Child.* 57, 754–757.
567. Steichen, J. J., Tsang, R. C., Gratton, T. L., Hamstra, A., and DeLuca, H. F. (1980). Vitamin D homeostasis in the perinatal period: 1,25-dihydroxyvitamin D in maternal, cord, and neonatal blood. *N. Engl. J. Med.* 302, 315–319.
568. Gray, T. K., Lowe, W., and Lester, G. E. (1981). Vitamin D and pregnancy: the maternal-fetal metabolism of vitamin D. *Endocr. Rev.* 2, 264–274.
569. Martinez, F. E., Goncalves, A. L., Jorge, S. M., and Desai, I. D. (1981). Brief clinical and laboratory observations. Vitamin E in placental blood and its interrelationship to maternal and newborn levels of vitamin E. *J. Pediatr.* 99, 298–300.

570. Hiraike, H., Kimura, M., and Itokawa, Y. (1988). Distribution of K vitamins (phylloquinone and menaquinones) in human placenta and maternal and umbilical cord plasma. *Am. J. Obstet. Gynecol.* 158, 564–569.
571. Lust, J. E., Hagerman, D. D., and Villee, C. A. (1954). The transport of riboflavin by human placenta. *J. Clin. Invest.* 33, 38–40.
572. Baker, H., Frank, O., Thomson, A. D., Langer, A., Munves, E. D., De Angelis, B., and Kaminetzky, H. A. (1975). Vitamin profile of 174 mothers and newborns at parturition. *Am. J. Clin. Nutr.* 28, 59–65.
573. Wu, J. T., Book, L., and Sudar, K. (1981). Serum alpha fetoprotein (AFP) levels in normal infants. *Pediatr. Res.* 15, 50–52.
574. Stabile, I., Nicolaides, K. H., Bach A., Teisner, B., Rodeck, C., Westergaard, J. G., and Grudzinskas, J. G. (1988). Complement factors in fetal and maternal blood and amniotic fluid during the second trimester of normal pregnancy. *Br. J. Obstet. Gynaecol.* 95, 281–285.
575. Forestier, F., Daffos, F., Galacteros, F., Bardakjian, J., Rainaut, M., and Beuzard, Y. (1986). Hematological values of 163 normal fetuses between 18 and 30 weeks of gestation. *Pediatr. Res.* 20, 342–346.
576. Yang, S. L., Chin-Chu, L., River, P., and Moawad, A. H. (1983). Immunoglobulin concentrations in newborn infants associated with intrauterine growth retardation. *Obstet. Gynecol.* 62, 561–564.
577. Einhorn, M. S., Granoff, D. M., Nahm, M. H., Quinn A., and Shackelford, P. G. (1987). Concentrations of antibodies in paired maternal and infant sera: relationship to IgG subclass. *J. Pediatr.* 111, 783–788.
578. Cederqvist, L. L., Caprio, R. E., Rappaport, I., Litwin, S. D., Abdel-Latif, N., and Bhalla, R. B. (1984). Correlation between maternal and fetal immunoglobulin E levels. *Obstet. Gynecol.* 63, 674–676.
579. Furuhashi, N., Takahashi, T., Fukaya, T., Kono, H., Shinkawa, O., Tachibana, Y., Wada, Y., and Suzuki, M. (1983). Plasma somatostatin and growth hormone in the human fetus and its mother at delivery. *Gynecol. Obstet. Invest.* 16, 59–62.
580. Furuhashi, N., Fukaya, T., Kono, H., Shinkawa, O., Tachibana, Y., Takahashi, T., Wada, Y., and Tanaka, M. (1983). Cord serum growth hormone in the human fetus. Sex difference and a negative correlation with birth weight. *Gynecol. Obstet. Invest.* 16, 119–124.
581. Crosignani, P. G., Nencioni, T., and Brambati, B. (1972). Concentration of chorionic gonadotrophin and chorionic somatomammotrophin in maternal serum, amniotic fluid and cord blood serum at term. *J. Obstet. Gynaecol. Br. Commonw.* 79, 122–126.
582. Metzger, B. E., Rodeck, C., Freinkel, N., Price, J., and Young, M. (1985). Transplacental arteriovenous gradients for glucose, insulin, glucagon and placental lactogen during normoglycaemia in human pregnancy at term. *Placenta* 6, 347–354.
583. Kohno, H., Furuhashi, N., Fukaya, T., Shinkawa, O., Tachibana, Y., and Suzuki, M. (1984). Serum cortisol levels in maternal venous, umbilical arterial and umbilical venous blood at delivery in the cases of fetal distress. *Tohoku J. Exp. Med.* 144, 299–304.
584. Partsch, C. J., Sippell WG, MacKenzie, I. Z., and Aynsley-Green, A. (1991). The steroid hormonal milieu of the undisturbed human fetus and mother at 16–20 weeks' gestation. *J. Clin. Endocrinol. Metab.* 73, 969–974.
585. Dokumov, S., Tarkolev, N., Shterev A., and Istatkov, M. (1981). Serum gastrin I concentrations of mother and newborn immediately after birth. *Br. J. Obstet. Gynaecol.* 88, 126–128.
586. Furuhashi, N., Fukaya, T., Kono, H., Shinkawa, O., Takahashi, T., and Suzuki, M. (1983). Simultaneous measurement of human prolactin and growth hormone levels in maternal and fetal serum and amniotic fluid. *Tohoku J. Exp. Med.* 141, 207–210.
587. Farquharson, R. G., and Klopper, A. I. (1984). Progesterone concentrations in maternal and fetal blood. *Br. J. Obstet. Gynaecol.* 91, 133–137.
588. Farquharson, R. G., and Klopper, A. I. (1984). Unconjugated oestriol in the cord blood of infants delivered by caesarean section at term. *Arch. Gynecol.* 236, 77–82.
589. Jones, C. M., 3rd, and Greiss, F. C. Jr. (1982). The effect of labor on maternal and fetal circulating catecholamines. *Am. J. Obstet. Gynecol.* 144, 149–153.
590. Widdowson, E. M., and Spray, C. M. (1951). Chemical development in utero. *Arch. Dis. Child.* 26, 205.
591. Bain, M. D., Copas, D. K., Landon, M. J., and Stacey, T. E. (1988). *In vivo* permeability of the human placenta to inulin and mannitol. *J. Physiol.* 399, 313–319.
592. Flexner, L. B., and Gellhorn, A. (1942). The comparative physiology of placental transfer. *Am. J. Obstet. Gynecol.* 431, 995.
593. Benveniste, H., Fowler, J. S., Rooney, W. D., Moller, D. H., Backus, W. W., Warner, D. A., Carter, P., King, P., Scharf, B., Alexoff, D. A., Ma, Y., Vaska, P., Schlyer, D., and Volkow, N. D. (2003). Maternal-fetal in vivo imaging: a combined PET and MRI study. *J. Nucl. Med.* 44, 1522–1530.
594. Nicolaides, K. H., Soothill, P. W., Rodeck, C. H., and Campbell, S. (1986). Ultrasound-guided sampling of umbilical cord and placental blood to assess fetal wellbeing. *Lancet* 1, 1065–1067.
595. Jauniaux, E., Sherwood, R. A., Jurkovic, D., Boa, F. G., and Campbell, S. (1994). Amino acid concentrations in human embryological fluids. *Hum. Reprod.* 9, 1175–1179.
596. Schneider, H., and Dancis, J. (1985). *In Vitro Perfusion of Human Placetal Tissue.* S Karger, Basel.
597. Longo, L. D., Yuen, P., and Gusseck, D. J. (1973). Anaerobic, glycogen-dependent transport of amino acids by the placenta. *Nature* 243, 531–533.
598. Siman, C. M., Sibley, C. P., Jones, C. J., Turner, M. A., and Greenwood, S. L. (2001). The functional regeneration of syncytiotrophoblast in cultured explants of term placenta. *Am. J. Physiol. Regul. Integr. Comp. Physiol.* 280, R1116–R1122.
599. Furesz, T. C., Smith, C. H., and Moe, A. J. (1993). ASC system activity is altered by development of cell polarity in trophoblast from human placenta. *Am. J. Physiol.* 265, C212–C217.
600. Douglas, G. C., and King, B. F. (1988). Receptor-mediated endocytosis of 125I-labelled transferrin by human choriocarcinoma (JAR) cells. *Placenta* 9, 253–265.
601. Sibley, C. P., and Boyd, R. D. H. (1988). Control of transfer across the mature placenta. In *Oxford Reviews of Reproductive Biology* (J. R. Clarke, Ed.), Vol. 10, pp. 382–435. Oxford University Press, Oxford.
602. Birdsey, T. J. (1995). Microvillous membrane potential (E_m) in isolated placental villi from the first trimester. *Placenta* 16, A6.
603. Bara, M., Challier, J. C., and Guiet-Bara, A. (1988). Membrane potential and input resistance in syncytiotrophoblast of human term placenta in vitro. *Placenta* 9, 139–146.
604. Carstensen, M., Leichtweibb, J. P., and Schroder, H. (1973). [Cellular potentials in the human placenta (author's transl.)]. *Arch. Gynakol.* 215, 299–303.
605. Broer, S., Broer A., Schneider, H. P., Stegen, C., Halestrap, A. P., and Deitmer, J. W. (1999). Characterization of the high-affinity monocarboxylate transporter MCT2 in Xenopus laevis oocytes. *Biochem. J.* 341, Pt 3, 529–535.

606. Grollman, E. F., Philp, N. J., McPhie, P., Ward, R. D., and Sauer, B. (2000). Determination of transport kinetics of chick MCT3 monocarboxylate transporter from retinal pigment epithelium by expression in genetically modified yeast. *Biochemistry* 39, 9351–9357.
607. Johnson, L. W., and Smith, C. H. (1988). Neutral amino acid transport systems of microvillous membrane of human placenta. *Am. J. Physiol.* 254, C773–C780.
608. Hoeltzli, S. D., and Smith, C. H. (1989). Alanine transport systems in isolated basal plasma membrane of human placenta. *Am. J. Physiol.* 256, C630–C637.
609. Novak, D. A., Beveridge, M. J., Malandro, M., and Seo, J. (1996). Ontogeny of amino acid transport system A in rat placenta. *Placenta* 17, 643–651.
610. Wang, H., Huang, W., Sugawara, M., Devoe, L. D., Leibach, F. H., Prasad, P. D., and Ganapathy, V. (2000). Cloning and functional expression of ATA1, a subtype of amino acid transporter A from human placenta. *Biochem. Biophys. Res. Commun.* 273, 1175–1179.
611. Desforges, M., Lacey, H. A., Glazier, J. D., Greenwood, S. L., and Sibley, C. P. (2004). Gestational profile of IGF-II and System A amino acid transporter mRNA expression in human placenta. *Placenta* 25, A. 19.
612. Karl, P. I., Teichberg, S., and Fisher, S. E. (1991). Na(+)-dependent amino acid uptake by human placental microvillous membrane vesicles: importance of storage conditions and preservation of cytoskeletal elements. *Placenta* 12, 239–250.
613. Shafqat, S., Tamarappoo, B. K., Kilberg, M. S., Puranam, R. S., McNamara, J. O., Guadano-Ferraz, A., and Fremeau, R. T. Jr. (1993). Cloning and expression of a novel Na(+)-dependent neutral amino acid transporter structurally related to mammalian Na+/glutamate cotransporters. *J. Biol. Chem.* 268, 15351–15355.
614. Utsunomiya-Tate, N., Endou, H., and Kanai, Y. (1996). Cloning and functional characterization of a system ASC-like Na+-dependent neutral amino acid transporter. *J. Biol. Chem.* 271, 14883–14890.
615. Novak, D. A., and Beveridge, M. J. (1997). Glutamine transport in human and rat placenta. *Placenta* 18, 379–386.
616. Carbo, N., Lopez-Soriano, F. J., and Argiles, J. M. (1997). Neutral amino acid transport in placental plasma membrane vesicles in the late pregnant rat. Evidence for a B0-like transport system. *Eur. J. Obstet. Gynecol. Reprod. Biol.* 71, 85–90.
617. Kekuda, R., Prasad, P. D., Fei YJ, Torres-Zamorano, V., Sinha, S., Yang-Feng, T. L., Leibach, F. H., and Ganapathy, V. (1996). Cloning of the sodium-dependent, broad-scope, neutral amino acid transporter Bo from a human placental choriocarcinoma cell line. *J. Biol. Chem.* 271, 18657–18661.
618. Torres-Zamorano, V., Kekuda, R., Leibach, F. H., and Ganapathy, V. (1997). Tyrosine phosphorylation- and epidermal growth factor-dependent regulation of the sodium-coupled amino acid transporter B0 in the human placental choriocarcinoma cell line JAR. *Biochim. Biophys. Acta* 1356, 258–270.
619. Karl, P. I., and Fisher, S. E. (1990). Taurine transport by microvillous membrane vesicles and the perfused cotyledon of the human placenta. *Am. J. Physiol.* 258, C443–C451.
620. Ramamoorthy, S., Leibach, F. H., Mahesh, V. B., Han, H., Yang-Feng, T., Blakely, R. D., and Ganapathy, V. (1994). Functional characterization and chromosomal localization of a cloned taurine transporter from human placenta. *Biochem. J.* 300, Pt 3, 893–900.
621. Miyamoto, Y., Balkovetz, D. F., Leibach, F. H., Mahesh, V. B., and Ganapathy, V. (1988). Na^{++} Cl^{-}-gradient-driven, high-affinity, uphill transport of taurine in human placental brush-border membrane vesicles. *FEBS Lett.* 231, 263–267.
622. Moyer, M. S., Insler, N., and Dumaswala, R. (1992). The role of chloride in taurine transport across the human placental brush-border membrane. *Biochim. Biophys. Acta* 1109, 74–80.
623. Norberg, S., Powell, T. L., and Jansson, T. (1998). Intrauterine growth restriction is associated with a reduced activity of placental taurine transporters. *Pediatr. Res.* 44, 233–238.
624. Kulanthaivel, P., Leibach, F. H., Mahesh, V. B., and Ganapathy, V. (1989). Tyrosine residues are essential for the activity of the human placental taurine transporter. *Biochim. Biophys. Acta* 985, 139–146.
625. Karl, P. I., Tkaczevski, H., and Fisher, S. E. (1989). Characteristics of histidine uptake by human placental microvillous membrane vesicles. *Pediatr. Res.* 25, 19–26.
626. Kilberg, M. S., Handlogten, M. E., and Christensen, H. N. (1980). Characteristics of an amino acid transport system in rat liver for glutamine, asparagine, histidine, and closely related analogs. *J. Biol. Chem.* 255, 4011–4019.
627. Chaudhry, F. A., Reimer, R. J., Krizaj, D., Barber, D., Storm-Mathisen, J., Copenhagen, D. R., and Edwards, R. H. (1999). Molecular analysis of system N suggests novel physiological roles in nitrogen metabolism and synaptic transmission. *Cell* 99, 769–780.
628. Nakanishi, T., Kekuda, R., Fei, Y. J., Hatanaka, T., Sugawara, M., Martindale, R. G., Leibach, F. H., Prasad, P. D., and Ganapathy, V. (2001). Cloning and functional characterization of a new subtype of the amino acid transport system N. *Am. J. Physiol. Cell. Physiol.* 281, C1757–C1768.
629. Dicke, J. M., Verges, D., Kelley, L. K., and Smith, C. H. (1993). Glycine uptake by microvillous and basal plasma membrane vesicles from term human placentae. *Placenta* 14, 85–92.
630. Borowsky, B., Mezey, E., and Hoffman, B. J. (1993). Two glycine transporter variants with distinct localization in the CNS and peripheral tissues are encoded by a common gene. *Neuron* 10, 851–863.
631. Prasad, P. D., Wang, H., Huang, W., Kekuda, R., Rajan, D. P., Leibach, F. H., and Ganapathy, V. (1999). Human LAT1, a subunit of system L amino acid transporter: molecular cloning and transport function. *Biochem. Biophys. Res. Commun.* 255, 283.
632. Ganapathy, M. E., Leibach, F. H., Mahesh, V. B., Howard, J. C., Devoe, L. D., and Ganapathy, V. (1986). Characterization of tryptophan transport in human placental brush-border membrane vesicles. *Biochem. J.* 238, 201–208.
633. Ramamoorthy, S., Leibach, F. H., Mahesh, V. B., and Ganapathy, V. (1992). Modulation of the activity of amino acid transport system L by phorbol esters and calmodulin antagonists in a human placental choriocarcinoma cell line. *Biochim. Biophys. Acta* 1136, 181–188.
634. Furesz, T. C., and Smith, C. H. (1997). Identification of two leucine-sensitive lysine transport activities in human placental basal membrane. *Placenta* 18, 649–655.
635. Eleno, N., Deves, R., and Boyd, C. A. (1994). Membrane potential dependence of the kinetics of cationic amino acid transport systems in human placenta. *J. Physiol.* 479, Pt 2, 291–300.
636. Furesz, T. C., Moe, A. J., and Smith, C. H. (1995). Lysine uptake by human placental microvillous membrane: comparison of system y+ with basal membrane. *Am. J. Physiol.* 268, C755–C761.
637. Kamath, S. G., Furesz, T. C., Way, B. A., and Smith, C. H. (1999). Identification of three cationic amino acid transporters in placental trophoblast: cloning, expression, and characterization of hCAT-1. *J. Membr. Biol.* 171, 55–62.

638. Furesz, T. C., Heath-Monnig, E., Kamath, S. G., and Smith, C. H. (2002). Lysine uptake by cloned hCAT-2B: comparison with hCAT-1 and with trophoblast surface membranes. *J. Membr. Biol.* 189, 27–33.
639. Speake, P. F., Glazier, J. D., Ayuk, P. T., Reade, M., Sibley, C. P., and D'Souza, S. W. (2003). L-arginine transport across the basal plasma membrane of the syncytiotrphoblast of the human placenta from normal and pre-eclamptic pregnancies. *J. Clin. Endocrinol. Metab.* 88, 4287–4292.
640. Novak, D. A., Kilberg, M. S., and Beveridge, M. J. (1994). Ontogeny and plasma-membrane domain localization of amino acid transport system L in rat liver. *Biochem. J.* 301, Pt 3, 671–674.
641. Malandro, M. S., Beveridge, M. J., Novak, D. A., and Kilberg, M. S. (1996). Rat placental amino acid transport after protein-deprivation-induced intrauterine growth retardation. *Biochem. Soc. Trans.* 24, 839–843.
642. Hoeltzli, S. D., Kelley, L. K., Moe, A. J., and Smith, C. H. (1990). Anionic amino acid transport systems in isolated basal plasma membrane of human placenta. *Am. J. Physiol.* 259, C47–C55.
643. Moe, A. J., and Smith, C. H. (1989). Anionic amino acid uptake by microvillous membrane vesicles from human placenta. *Am. J. Physiol.* 257, C1005–C1011.
644. Matthews, J. C., Beveridge, M. J., Malandro, M. S., Rothstein, J. D., Campbell-Thompson, M., Verlander, J. W., Kilberg, M. S., and Novak, D. A. (1998). Activity and protein localization of multiple glutamate transporters in gestation day 14 vs. day 20 rat placenta. *Am. J. Physiol.* 274, C603–C614.
645. Matthews, J. C., Beveridge, M. J., Dialynas, E., Bartke A., Kilberg, M. S., and Novak, D. A. (1999). Placental anionic and cationic amino acid transporter expression in growth hormone overexpressing and null IGF-II or null IGF-I receptor mice. *Placenta* 20, 639–650.
645a. Hediger, M. A., Romero, M. F., Peng, J. B., Rolfs A., Takanaga, H., and Bruford, E. A. (2004). The ABCs of solute carriers: physiological, pathological and therapeutic implications of human membrane transport proteins Introduction. *Pflugers Arch* 447, 465–468,
646. Jansson, T., Scholtbach, V., and Powell, T. L. (1998). Placental transport of leucine and lysine is reduced in intrauterine growth restriction. *Pediatr. Res.* 44, 532–537.
647. Johansson, M., Karlsson, L., Wennergren, M., Jansson, T., and Powell, T. L. (2003). Activity and protein expression of Na+/K+ ATPase are reduced in microvillous syncytiotrophoblast plasma membranes isolated from pregnancies complicated by intrauterine growth restriction. *J. Clin. Endocrinol. Metab.* 88, 2831–2837.
648. Johansson, M., Glazier, J. D., Sibley, C. P., Jansson, T., and Powell, T. L. (2002). Activity and protein expression of the Na+/H+ exchanger is reduced in syncytiotrophoblast microvillous plasma membranes isolated from preterm intrauterine growth restriction pregnancies. *J. Clin. Endocrinol. Metab.* 87, 5686–5694.
649. Ruyle, M., Connor, W. E., Anderson, G. J., and Lowensohn, R. I. (1990). Placental transfer of essential fatty acids in humans: venous-arterial difference for docosahexaenoic acid in fetal umbilical erythrocytes. *Proc. Natl. Acad. Sci. U S A* 87, 7902–7906.
650. Metcalfe, J., Romney, S. L., Ramsey, L. H., Reid, D. E., and Burwell, C. S. (1955). Estimation of uterine blood flow in normal human pregnancy at term. *J. Clin. Invest.* 34, 1632–1638.
651. Gill, R. W., Trudinger, B. J., Garrett, W. J., Kossoff, G., and Warren, P. S. (1981). Fetal umbilical venous flow measured *in utero* by pulsed-Doppler and B-mode ultrasound. I. Normal pregnancies. *Am. J. Obstet. Gynecol.* 139, 720–725.
652. Novy, M. J., Thomas, C. L., and Lees, M. H. (1975). Uterine contractility and regional blood flow responses to oxytocin and prostaglandin E2 in pregnant rhesus monkeys. *Trans. Pac. Coast Obstet. Gynecol. Soc.* 42, 8–21.
653. Behrman, R. E., Lees, M. H., Peterson, E. N., De Lannoy, C. W., and Seeds, A. E. (1970). Distribution of the circulation in the normal and asphyxiated fetal primate. *Am. J. Obstet. Gynecol.* 108, 956–969.
654. Peeters, L. L., Sparks, J. W., Grutters, G., Girard, J., and Battaglia, F. C. (1982). Uteroplacental blood flow during pregnancy in chronically catheterized guinea pigs. *Pediatr. Res.* 16, 716–720.
655. Morriss, F. H., Rosenfeld, C. R., Crandell, S. S., and Adcock, E. W. III. (1980). Effects of fasting on uterine blood flow and substrate uptake in sheep. *J. Nutr.* 110, 2433–2443.
656. Crandell, S. S., Fisher, D. J., and Morriss, F. H. Jr. (1985). Effects of ovine maternal hyperglycemia on fetal regional blood flows and metabolism. *Am. J. Physiol.* 249, E454–E460.
657. Ramamoorthy, J. D., Ramamoorthy, S., Leibach, F. H., and Ganapathy, V. (1995). Human placental monoamine transporters as targets for amphetamines. *Am. J. Obstet. Gynecol.* 173, 1782–1787.
658. Wu, X., Kekuda, R., Huang, W., Fei YJ, Leibach, F. H., Chen, J., Conway, S. J., and Ganapathy, V. (1998). Identity of the organic cation transporter OCT3 as the extraneuronal monoamine transporter (uptake2) and evidence for the expression of the transporter in the brain. *J. Biol. Chem.* 273, 32776–32786.
659. Kekuda, R., Prasad, P. D., Wu, X., Wang, H., Fei, Y. J., Leibach, F. H., and Ganapathy, V. (1998). Cloning and functional characterization of a potential-sensitive, polyspecific organic cation transporter (OCT3) most abundantly expressed in placenta. *J. Biol. Chem.* 273, 15971–15979.
660. Huang, W., Shaikh, S. N., Ganapathy, M. E., Hopfer, U., Leibach, F. H., Carter, A. L., and Ganapathy, V. (1999). Carnitine transport and its inhibition by sulfonylureas in human kidney proximal tubular epithelial cells. *Biochem. Pharmacol.* 58, 1361–1370.
661. Ohashi, R., Tamai, I., Yabuuchi, H., Nezu, J. I., Oku, A., Sai, Y., Shimane, M., and Tsuji, A. (1999). Na(+)-dependent carnitine transport by organic cation transporter (OCTN2): its pharmacological and toxicological relevance. *J. Pharmacol. Exp. Ther.* 291, 778–784.
662. Wu, X., Huang, W., Prasad, P. D., Seth, P., Rajan, D. P., Leibach, F. H., Chen, J., Conway, S. J., and Ganapathy, V. (1999). Functional characteristics and tissue distribution pattern of organic cation transporter 2 (OCTN2), an organic cation/carnitine transporter. *J. Pharmacol. Exp. Ther.* 290, 1482–1492.
663. Ganapathy, M. E., Huang, W., Rajan, D. P., Carter, A. L., Sugawara, M., Iseki, K., Leibach, F. H., and Ganapathy, V. (2000). Beta-lactam antibiotics as substrates for OCTN2, an organic cation/carnitine transporter. *J. Biol. Chem.* 275, 1699–1707.
664. Tsuji, A., and Tamai, I. (1996). Carrier-mediated intestinal transport of drugs. *Pharm. Res.* 13, 963–977.
665. Utoguchi, N., Magnusson, M., and Audus, K. L. (1999). Carrier-mediated transport of monocarboxylic acids in BeWo cell monolayers as a model of the human trophoblast. *J. Pharm. Sci.* 88, 1288–1292.
666. Ganapathy, V., Ganapathy, M. E., Tiruppathi, C., Miyamoto, Y., Mahesh, V. B., and Leibach, F. H. (1988). Sodium-gradient-driven, high-affinity, uphill transport of succinate in human placental brush-border membrane vesicles. *Biochem. J.* 249, 179–184.

667. Prasad, P. D., Ramamoorthy, S., Moe, A. J., Smith, C. H., Leibach, F. H., and Ganapathy, V. (1994). Selective expression of the high-affinity isoform of the folate receptor (FR-alpha) in the human placental syncytiotrophoblast and choriocarcinoma cells. *Biochim. Biophys. Acta* 1223, 71–75.
668. Prasad, P. D., Ramamoorthy, S., Leibach, F. H., and Ganapathy, V. (1995). Molecular cloning of the human placental folate transporter. *Biochem. Biophys. Res. Commun.* 206, 681–687.
669. Cole, S. P., Sparks, K. E., Fraser, K., Loe, D. W., Grant, C. E., Wilson, G. M., and Deeley, R. G. (1994). Pharmacological characterization of multidrug resistant MRP-transfected human tumor cells. *Cancer Res.* 54, 5902–5910.
670. Wimsatt, W. A. (1962). Some aspects of the comparative anatomy of the mammalian placenta. *Am. J. Obstet. Gynecol.* 84, 1568–1594.
671. Faber, J. J., and Thornburg, K. L. (1983). *Placental Physiology. Structure and Function of Fetomaternal Exchange.* Raven Press, New York.
672. Edelstone, D. I., Caine, M. E., and Fumia, F. D. (1985). Relationship of fetal oxygen consumption and acid-base balance to fetal hematocrit. *Am. J. Obstet. Gynecol.* 151, 844–851.
673. Chesney, R. W. (2004). American Pediatric Society Presidential Address 2003: the failure of Ockham's razor in 21st century pediatrics. *Pediatr. Res.* 55, 903–907.

Knobil and Neill's Physiology of Reproduction, Third Edition
edited by Jimmy D. Neill,
Elsevier © 2006

CHAPTER 53

Placental Endocrine Function

Felice Petraglia, Pasquale Florio, and Michela Torricelli

Introduction, 2848
Hypothalamus-Like Hormones, 2848
Gonadotropin-Releasing Hormone (GnRH), 2848
Corticotropin-Releasing Factor (CRF) and CRF-Related Peptides, 2849
Growth Hormone-Releasing Hormone (GHRH) and Somatostatin (SST), 2851
Pituitary-Like Hormones, 2852
Human Chorionic Gonadotrophin (hCG), 2852
Opioid Peptides, 2853
Placental Growth Hormone (PGH), 2854
Human Placental Lactogen (hPL), 2855
Other Neurohormones, 2855
Neuropeptide Y (NPY), 2855
Chromogranin A (CgA), 2856
Oxytocin and Vasopressin, 2856
Neurokinin B (NKB), 2857
Monoamines, 2858
Growth Factors, 2858
Transforming Growth Factor-β (TGF-β), 2858
Inhibin-Related Peptides, 2859
Insulin-Like Growth Factor and Insulin-Like Growth Factor-Binding Protein (IGF and IGF-BP), 2860
Fibroblast Growth Factor (FGF) and Epidermal Growth Factor (EGF), 2862
Vascular Endothelial Growth Factor (VEGF), 2863
Progesterone and Estrogens, 2864
Vasoactive Peptides, 2865
Adrenomedullin (AM), 2865
Endothelin (ET), 2866
Calcitonin Gene-Related Peptide (CGRP), 2867
Parathyroid Hormone-Related Peptide (PTHrP), 2868
Metabolic Hormones, 2868
Leptin, 2868
Ghrelin, 2869
Paracrine/Autocrine Mechanisms Regulating Placental Hormonogenesis, 2870
GnRH-hCG-Inhibins-Steroid Hormone Axis, 2870
GHRH-PGH-hPL-IGF Axis, 2872
CRF-ACTH-Opioid Axis, 2873
Placental Hormones, Cell Growth, and Differentiation, 2874
Trophoblast Differentiation and Invasion, 2874
Placental Angiogenesis, 2875
Paracrine/Autocrine Actions on Fetal-Placental Blood Flow Regulation, 2876
Effects of CRF and NPY, 2876
Effects of NKB, 2877
Effects of ET-1, 2878
Effects of ADM, 2878
Effect of CGRP, 2878
Effect of PTHrP, 2879
Clinical Usefulness of Placental Hormones Measurement Throughout Pregnancy, 2879
Ectopic Pregnancy, 2879
Hydatidiform Mole, 2879
Down's Syndrome, 2880
Preeclampsia (PE), 2881
Fetal Growth Restriction (FGR), 2882
Preterm Delivery, 2883
References, 2883

Chair of Obstetrics and Gynecology, Department of Pediatrics, Obstetrics and Reproductive Medicine, University of Siena, Siena School of Medicine, Italy.

INTRODUCTION

During the last two decades our concept of the role and functioning of human placenta during human pregnancy has been changing. While initially considered solely for their role in the exchange of gas and transfer of nutrients and waste, now we consider intrauterine tissues [human placenta, fetal membranes (amnion, chorion) and maternal decidua] capable of hormones production and metabolism serving as an endocrine organ. Indeed, a large variety of hormones are produced by intrauterine tissues: brain, pituitary, gonadal, and adrenocortical hormones, chemically identical and biologically active as their counterparts (1,5) (Table 1).

TABLE 1. *Hormones produced by intrauterine tissues (placenta, decidua, amnion, chorion)*

Hypothalamus-like Hormones
Gonadotrophin-releasing hormone (GnRH)
Corticotrophin-releasing factor (CRF)
Urocortin
CRF-binding protein (CRF-BP)
Growth hormone-releasing hormone (GH-RH)
Somatostatin (SST)
Pituitary-like Hormones
Human chorionic gonadotrophin (hCG)
ACTH and opioids
Placental growth hormone (PGH)
Human placental lactogen (hPL)
Other Neurohormones
Neuropeptide Y (NPY)
Chromogranin A (CgA)
Oxytocin (OT)
Vasopressin
Neurokinin B (NKB)
Monoamines
Growth Hormones
Transforming growth factor-β (TGF-β)
Activin A
Follistatin
Follistatin-related gene (FLRG)
Inhibin A and B
Insulin-like growth factor (IGF)
IGF-binding protein (IGF-BP)
Fibroblast growth factor (FGF)
Epidermal growth factor (EGF)
Vascular endothelial growth factor (VEGF) peptides
Progesterone
Estrogens
Vasoactive Peptides
Adrenomedullin (ADM)
Endothelin (ET)
Calcitonin gene-related peptide (CGRP)
Parathyroid hormone-related peptide (PTH-rp)
Metabolic Hormones
Leptin
Ghrelin

A major result derived from the studies of placental physiology is a model of interaction that resembles in many aspects the organization of hypothalamus–pituitary axis, throughout a cell communication (paracrine) and/or within the same cells (autocrine). However, the placenta produces a number of hormones that are not otherwise synthesized in the organism, suggesting that a distinct set of hormones are required to bring about the physiological changes of pregnancy rather than simply producing more of certain hormones. Therefore, placental hormones may also enter the maternal and fetal circulation, where are often present at concentrations far in excess of what is found for similar hormones in the nonpregnant adult, to play a pivotal role in the maternal–fetal hormonal dialog. The placenta, decidua, amnion, and chorion actively participate in human reproductive physiology in allowing implantation, the regulation of fetal-placental blood flow, the growth of the conceptus throughout gestation, and in directing the appropriate signals for the timing of parturition.

HYPOTHALAMUS-LIKE HORMONES

Gonadotropin-Releasing Hormone (GnRH)

GnRH is a decapeptidic hormone produced by the hypothalamus that controls reproduction in vertebrates through the hypothalamic–pituitary–gonadal axis (3).

Expression and Localization

GnRH is produced also by the human placental tissue and by cultured placental cells (4,5) and it is immunologically and chemically identical to hypothalamic GnRH (4–7). Placental GnRH mRNA and protein are expressed in human placenta, since the first trimester till term, both in the cyto- and syncytiotrophoblast (7). The GnRH staining is intense in cytotrophoblast and in the villous stroma from early placentas (8 weeks of pregnancy) (8), and GnRH immunoreactivity is also detected in the syncytiotrophoblast of the normal human placenta from the first half of pregnancy, and in syncytiotrophoblast cells of hydatidiform mole and choriocarcinoma (9). The total placental concentration of immunoreactive GnRH progressively increases during the first 24 weeks of gestation and remains relatively constant in the third trimester (10) while, on the contrary, the mRNA expression remains constant throughout gestation (11).

Receptors

The human placenta contains specific binding sites for GnRH that interact with GnRH agonists and antagonists (12). By in situ hybridization, GnRH receptor (GnRHR) mRNAs (mRNAs) gas been localized to the cytotrophoblast and syncytiotrophoblast cell layers (13). Using oligonucleotide primers specific to the human GnRHR, the predicted polymerase chain reaction (PCR) product has been found in human placenta cells (7,14) and choriocarcinoma cell line (JAR and JEG-3) (15,16). The receptor expressed in human placenta is identical to the counterpart of pituitary (17), and it is coupled to the protein kinase C (PKC) and cAMP/protein kinase A (PKA) pathways (17). Moreover, there is evidence that GnRH induces activation of the mitogen-activated protein kinase (MAPK) signaling pathway in normal and carcinoma cells of the human ovary and placenta (18).

The contemporary presence of GnRH and GnRHR in the same cells strongly suggests an autocrine/paracrine regulation by GnRH in human placenta. On this regard, two classes of placental GnRH binding sites have been described to date: high affinity (Kd = 10–8 mol/L) and low affinity (Kd = 10–5 mol/L) (17,18). From first trimester to term, the human placenta contains low-affinity GnRH binding sites that interact with GnRH agonist or antagonist (17–20). The GnRH receptor levels decrease observed between 10–20 weeks of gestation is probably due to a decreased expression/synthesis (or increased catabolism) of placental GnRHR, or increased occupancy (or down-regulation) of placental GnRHR by an endogenous GnRH-like ligand (19,20). The mRNA of the high affinity GnRH binding site is expressed in human cytotrophoblast and syncytiotrophoblast cell layers (13), and its signal intensity varies with gestational ages and is abundant at 6 weeks, peaks at 9 weeks, declines at 12 and 20 weeks, and is undetectable at term. The GnRH administration to pregnant women increases serum levels of hCG in the first trimester, but not in the third trimester (21), probably due to the decreased number of GnRH receptors in the term placenta (13).

Biological Fluids

GnRH is measurable in the maternal circulation, and levels are significantly higher during pregnancy than in nonpregnant cycling women, in particular in the first half of pregnancy (22). Pulsatile changes of maternal GnRH values have been shown, with highest amplitude in the first trimester and lowest at term (23). Maternal levels at 25–35 weeks of gestation are higher in women who later had post-term pregnancies (24).

Corticotropin-Releasing Factor (CRF) and CRF-Related Peptides

CRF is a 41 amino acid peptide released from the medial eminence of the hypothalamus, acting at the corticotroph cells in the anterior pituitary to stimulate the release of ACTH and related peptides in response to stress events, and to modulate behavioral, vascular, and immune response to stress (25).

Expression and Localization

Human placenta, decidua, chorion and amnion produce CRF (26,27). Immunoreactive CRF was first detected in the extracts of human placenta obtained at full term from spontaneous delivery (28), and results as bioactive as rat hypothalamic CRF or synthetic ovine CRF on the release of immunoreactive ACTH and beta-endorphin from cultures of rat anterior pituitary cells (29). The content of immunoreactive CRF is higher in extracts of placenta obtained at term than in tissue obtained at 10 weeks of gestation (30,31), and a progressive increase of placental CRF content has been described during normal pregnancy, paralleling a similar time course of placental CRF mRNA expression, which starts from early gestation (7–8 weeks) (31). The structure of placental CRF mRNA is similar to that predicted for hypothalamic CRF mRNA (32). As regards the cellular localization, into the placental villi at term CRF is localized in some cytotrophoblast cells (26), as well as in syncytiotrophoblast cells (33). Cytotrophoblast cells are transformed to syncytial cells, which release CRF factor when maintained in culture (34).

CRF is also released from cultured amnion, chorion, and decidual cells at term (34,35) with an output similar to that by the placental cells (34). Immunohistochemical localization of CRF in fetal membranes showed that CRF is distributed in the epithelial cells, in some cells of the subepithelial layer of amnion, and in cells of the reticular layer of chorion (26,33). Immunoreactive CRF is present in decidual cells (27,36), as well as in endometrial cells treated hormonally to achieve in vitro decidualization (37). Increased placental synthesis of CRF is evident at both the mRNA and protein levels in preeclampsia (38,39).

Recently, another component of the CRF family, urocortin, has been described. Its sequence is similar to fish urotensin (63%) and human CRF (45%) (40). Placental and decidual cells collected at 8–11 weeks or 38–40 weeks of gestation express urocortin mRNA and immunohistochemistry localized urocortin staining in syncytiotrophoblast and cytotrophoblast cells of placental villi at term, in amnion, chorion, and decidua, as well as in fetal membranes and maternal decidua (41,42).

CRF-binding protein (CRF-BP) is a 37 kDa protein of 322 amino acids, mainly produced by the human brain and the liver (43) that is able to bind circulating CRF and urocortin. Further sources of CRF-BP during pregnancy are placental trophoblast, decidua and fetal membranes (43,44). In details, the syncytial layer of placental villi at term intensely expresses CRF-BP mRNA and immunoreactivity, whereas rare positively hybridized cells are observed within the cytotrophoblasts and mesenchymal cells. Large decidual cells, amniotic epithelial cells, and chorionic cytotrophoblast stained positively for CRF-BP mRNA and protein (43,44).

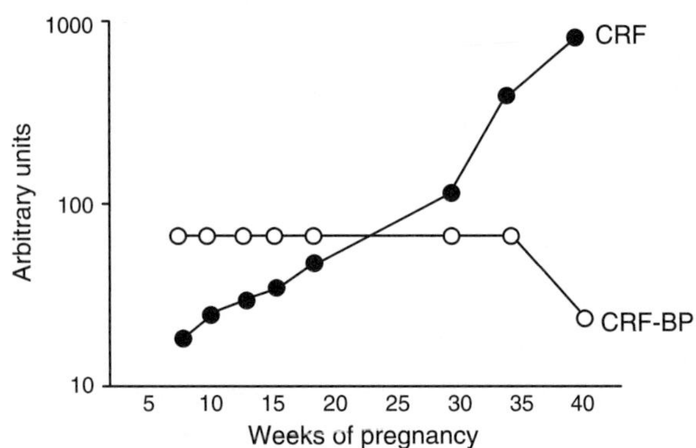

FIG. 1. Changes of corticotropin-releasing factor (CRF) and CRF-binding protein (CRF-BP) in maternal circulation throughout gestation. Levels are inversely correlated at term.

Receptors

CRF and urocortin interact with two distinct receptors: R1 (classified in R1-α, R1-β, R1-c, and R1-d subtypes) and R2 (R2-α, R2-β, and R2-γ subtypes) (45). Fluorescent in situ hybridization and immunofluorescence demonstrated that syncytiotrophoblast cells and amniotic epithelium are the cell types expressing CRF-R1-α, -Rc (45), and -R2-β mRNA (46).

CRF-receptors (mRNA and protein) have been also described in human myometrium. In particular, recent findings show the presence in pregnant myometrium of subtypes 1α, 1β, 2α, and 2β and the variant-Rc, whereas only the 1α, 1β, and 2β receptors are detectable in nonpregnant myometrium (47). Urocortin binds to CRF receptors types 1 and 2, with a particularly high affinity for type 2 receptor (25).

Levels in Biological Fluids

Plasma CRF levels are almost undetectable in nonpregnant women (less than 10 pg/ml) and become higher during the first trimester of pregnancy, then rising steadily till term (48) (Fig. 1). In fetal circulation a linear correlation exists between maternal and fetal plasma CRF levels, despite umbilical cord plasma CRF levels are 20- to 30-fold lower than in maternal circulation (49). In addition, CRF concentrations in umbilical venous plasma are higher than in the umbilical artery, supporting the placenta as a major source of fetal plasma CRF (50). The significant correlation between the amniotic fluid and maternal plasma CRF levels obtained simultaneously (51) suggests a placental source for amniotic CRF.

During spontaneous labor maternal plasma CRF levels progressively rise, reaching the maximum values at the most advanced stages of cervical dilatation (48,52). In addition, subjects who have elective caesarean delivery show plasma and amniotic fluid CRF levels significantly lower than patients after spontaneous vaginal delivery (48,52). Moreover, the amount of CRF in placental extracts obtained at term after spontaneous vaginal delivery is significantly greater than the amount of extracted from placentas obtained after caesarean delivery (124). Women with preterm labor have maternal plasma CRF levels significantly higher than those measured in the course of normal pregnancy (53).

In patients with pregnancy-induced hypertension (PIH) and preeclampsia (PE), maternal CRF levels at all stages of pregnancy are significantly higher than in healthy controls (54). Furthermore, umbilical cord venous plasma CRF concentrations are significantly higher in patients with PE and higher than in cord arterial plasma, indicating the secretion of CRF from the placenta into the fetal circulation (55).

CRF-BP is measurable in maternal plasma, and levels remain stable in nonpregnant women and during gestation until the third trimester of pregnancy (48,54) (Fig. 1). At this time, maternal plasma CRF-BP concentrations significantly and rapidly decrease in the last 4–6 weeks before labor (48,54,56), returning to approximately nonpregnant levels during the first 24 hours postpartum. Thus, opposite changes in concentrations of CRF (higher) and CRF-BP (lower) in maternal plasma occur at term, determining an increased availability of bioactive CRF at labor (Fig. 1). Cord blood CRF-BP levels are higher (57), while amniotic fluid are lower than in maternal plasma and have a similar trend, decreasing until term pregnancy (58). During spontaneous physiological labor a significant decrease in CRF-BP levels in maternal plasma (56), cord blood (57), and amniotic fluid (58) occurs. Furthermore, maternal and fetal plasma CRF-BP levels are low in preterm labor (57,59) resembling the physiologic pattern observed at term. Plasma CRF-BP levels are lower in PIH and PE patients than in normotensive controls, and they are inversely correlated to the increased CRF levels in maternal circulation (54).

Urocortin is undetectable in maternal plasma during pregnancy, with no rise with increasing gestational age (60). This lack of urocortin throughout pregnancy is confirmed by an absence of gestational age-related changes in placental urocortin mRNA expression (42). Plasma levels are high at labor than during pregnancy, but they did not change significantly at the different stages of labor (61).

Growth Hormone-Releasing Hormone (GHRH) and Somatostatin (SST)

GHRH and SST are two hypothalamic releasing factors involved in the regulation of the complex process of growth hormone (GH) synthesis and secretion in the anterior pituitary. While GHRH functions as a specific stimulator of GH secretion, on the contrary, SST acts as the inhibitory counterpart (62).

GHRH is a 44 amino acid peptide, and it is synthesized as a 107 amino acid precursor protein, preprosomatocrinin, with a molecular weight of approximately 13.000 dalton. The human GHRH is encoded by a single gene on chromosome 20 (62). SST is a neuropeptide encoded by a single gene on human chromosome 3. In mammals there are two naturally occurring bioactive somatostatin peptides containing 14 amino acids (SST-14) and the amino terminally extended SST-28 arising from a common precursor molecule (62). SST was the first recognized hypothalamic hormone whose action was to inhibit rather than to stimulate the release of pituitary hormones: it was found to decrease secretion of growth hormone (GH), as it counteracts the growth-promoting effects of pituitary GH by reducing pituitary responsivity to GHRH (62).

Expression and Localization

Two different molecular forms of mouse immunoreactive GHRH were detected in the mouse hypothalamus and one in placenta. GHRH gene is abundantly expressed in human (63), mouse, and rat placenta (64,65), and in situ hybridization revealed that cells expressing the GHRH mRNA had the morphological characteristics of cytotrophoblasts (65).

Although mainly located in the gastrointestinal and central nervous system, SST may be the most widely distributed of all hormones (66), and it is expressed in human placenta and maternal decidua (67). Immunohistochemical studies have revealed the presence of SST in the cytotrophoblastic layer and in the stroma of human placental villi at 6–12 weeks of gestation, while no immunostaining is noted in the syncytiotrophoblast layer (67,68). When gestational age increases, the amount of immunoreactive somatostatin decreases (69).

Receptors

The GHRH receptor is a seven transmembrane G-protein–linked receptor found predominantly in the pituitary gland (62), and significant amounts are found in the placenta (70).

SST effects are mediated through five heptahelical receptors, which are linked to pertussis toxin-sensitive G proteins, activating multiple postreceptor signal transduction pathways (62). The receptors were named chronologically, SSTR1, SSTR2, SSTR3, SSTR4, and SSTR5, according to their publication (62). All receptors identified so far bind SST-14 and SST-28 with high affinity. The similarities in receptor sequence and in the binding profiles of short synthetic SST analogues such as octreotide, MK 678 or RC 160 for SSTR1-5 indicate the existence of two classes of receptors SSTR1/SSTR4 with virtually no or very low affinity, and SSTR2/SSTR3/SSTR5 with intermediate to high affinity for the short SST analogues (62). SSTR1, SSTR4, and SSTR5 mRNA have been found in human placental tissue and villous cytotrophoblast cells, and SSTR4 mRNA only in syncytiotrophoblast cells (70).

Biological Fluids

The presence of GHRH immunoreactivity (GRH-IR) has been described in maternal blood and amniotic fluid. Although its biological function in placental, maternal plasma, and amniotic fluid is not yet clear, the presence of GHRH-IR in these circulating fluids suggests the possibility that it may exert an important role in both fetal and maternal physiology (64).

Plasma SST fluctuates with a significant circadian rhythm, and a nocturnal peak occurs at the first hours of sleep (71). SST levels in maternal circulation do not differ from nonpregnant women suggesting a scarce placental contribution, while levels in fetoplacental circulation originate in the fetus. In fact, umbilical artery SST levels are higher than umbilical vein levels, and even greater differences have been found between umbilical artery levels and levels in the peripheral vein of the mothers (72). In preterm human neonates significantly higher SST plasma levels than those usually found in adults have been discovered (73). SST levels in last trimester of patients with PIH is significantly higher than those in normal pregnancy. Moreover levels of SST in newborns is higher than those in mothers at delivery and the levels of SST in neonates born to mother of PIH is much higher compared with normal pregnancy (74). SST is also measurable in the amniotic fluid, and the highest levels are found in the first trimester of pregnancy (72,74), but to date there are no evidence about a biological action of the SST in the placenta.

PITUITARY-LIKE HORMONES

Human Chorionic Gonadotrophin (hCG)

hCG belongs to the glycoprotein hormones family that also comprises luteinizing hormone (LH), follicle-stimulating hormone (FSH), and thyroid-stimulating hormone (TSH). All members are heterodimers consisting of an α and a β subunit. The α subunit is common to all these four glycoprotein hormones, and contains 92 amino acids, while the β chains determine the biological activity and display extensive homology, that between hCG and LH is being about 80%. The molecular weight of hCG is about 37,500, resulting from 23,500 of β-hCG subunit and 14,000 of α-hCG (75).

Expression and Localization

hCG is produced in large amounts by syncytiotrophoblasts, however, the site of origin of hCG has been the subject of much controversy. Indeed, immunocytochemical localization studies have suggested that it is produced by the syncytiotrophoblasts layer of the placenta rather than the cytotrophoblast, as first believed. It is elaborated by all types of trophoblastic tissue, including that from hydatiform mole, choriocarcinoma. Molecular biology studies have demonstrated that also trisomy 21 trophoblasts show a marked increase in β-hCG mRNA and a smaller increase in α-hCG mRNA, suggesting that one of the causes of high hCG levels in maternal serum is the increased hCG production and secretion by the placenta (76).

Receptors

Human placenta, amniotic epithelium, chorionic cytotrophoblasts, and decidual cells contain hCG/LH receptors. The comparison of immunostaining intensities has suggested that syncytiotrophoblasts contain more receptors than cytotrophoblasts at mid-pregnancy, while mesenchymal cells or blood vessels contained no detectable receptors. Furthermore, there are more receptors in decidua than in fetal membranes at mid- and term pregnancy. While the amniotic epithelial receptors decrease, the receptors in chorionic cytotrophoblasts and decidual cells increase from mid- to term pregnancy (77).

Biological Fluids

In normal pregnancy, the primitive trophoblast produces hCG very early, since hCG is detectable 9 days after the midcycle LH peak, which is 8 days after ovulation and only 1 day after implantation. Therefore, through the evaluation of hCG pregnancy can be detected before the first missed menstrual period. This has clinical utility when it is important to determine the presence of pregnancy at an early stage (78).

Maternal serum hCG peaks at 8–10 week and then declines to reach a plateau at 18–20 week of gestation. Five hCG-related molecules are present in maternal serum: nonnicked hCG, which represents the active hormone; nicked hCG; free β-subunit; free β-subunit; and the nicked free β-subunit (79). The free β-subunit can derive from three sources, i.e., direct trophoblast cell production, dissociation of hCG into free β- and free β-subunits, and by macrophage or neutrophil enzymes nicking the hCG molecule. The free β-hCG circulating in maternal serum corresponds to only about 0.3%–4% of the total hCG (80).

When excreted into urine, much of hCG and β-hCG is degraded to partially cleaved or "nicked" forms and fragments. In nicked hCG and β-hCG, the peptide chain is cleaved between amino acids 44-45 and 47-48. Much of the hCG-immunoreactivity in urine consists of a fragment of β-hCG comprising amino acids 6-40 and 55-92 linked together by disulfide bridges (81). This fragment is called the core fragment of β-hCG. Different degraded forms are recognized to a variable extent by different antibodies causing between-assay-variation especially when determining hCG in urine (79,82). This is of importance mainly for pregnancy tests because serum assays are used for monitoring of hCG during pregnancy and in cancer patients. Indeed, elevated concentrations of hCG are a common finding in the serum of patients with trophoblastic or other germ cell tumors and can be a valuable marker for monitoring these conditions. All trophoblastic tumors produce hCG, and monitoring of therapy is largely based on the determination of hCG in serum (83). In the serum and urine the concentrations of all forms of hCG are lower in cases of ectopic pregnancy and spontaneous miscarriage than in viable pregnancies (84).

Second trimester maternal serum hCG levels are significantly higher in Down's syndrome compared with normal pregnancies, with an average of 2.00 MoM (85,86), and in patients who develop fetal growth restriction (FGR), ranging from 1.21 to 2.5 MoM (87). Patients with overt preeclampsia in the third trimester have increased maternal serum hCG levels (87).

Opioid Peptides

Opioid peptides have a morphine-like activity, and three families are recognized: endorphin (END), enkephalins (ENK), and dynorphins (DYN). They derive from three precursor of similar molecular size and sequence homology. Pro-opiomelanocortin (POMC) is the precursor of ACTH, a-melanocyte-stimulating hormone (a-MSH), β-endorphin (β-END) and lipotropin. Proenkephalin (P-ENK) is the precursor of enkephalin and prodynorphin (P-DYN) of dynorphins, rimorphin, leumorphin and neo-endorphins (88).

Expression and Localization

β-END, methionine-ENK (M-ENK), and dynorphins 1-8 and 1-13 are the main opioid peptides identified in placental extracts. Dynorphin 1-8 seem to be the predominant opioid peptide present in placental villus tissue (89). β-END was the first detected endogenous opioid peptide in the human placenta (90). Immunohistochemical staining of placental tissue for β-END immunoreactivity is positive in the syncytiotrophoblast in both early and term pregnancy (91). Cultures of human placenta cells collected at term release β-END (92), and it is measurable in homogenates of human amnion, chorion, and decidua collected throughout gestation (93).

M-ENK is the major representative of the other family of opioid peptides, the enkephalins. Syncytial and cytotrophoblast cells contain immunoreactive M-ENK, and its de novo synthesis in culture villi at term has been shown (94).

The third family of opioids is represented by DYN, and human placenta is also the source of the multiple forms of DYN despite dynorphins A (1-8) is the major opioid present in the placental extracts (89).

Receptors

Mu (μ), kappa (κ), and delta (δ) are the main opioid receptor types. Each opioid exhibits distinct binding activity towards each type of opioid receptor. The μ receptor has high affinity for M-ENK and β-END (as well as morphine and dynorphin A); the δ receptor for leu-enkephalin; the κ receptor is the main target for the dynorphins (88).

Receptor subtypes for the various endogenous opioid peptides are present on placental cell membranes (95,96), but κ receptors is the more important type present in the placenta, since the order of potency in cells in vitro from term trophoblast tissue is $\kappa >>> \mu > \delta$ (97). Placental content of κ receptors increases with gestational age, and term placental content of κ receptors correlates with route of delivery (90).

Levels in Biological Fluids

Concentration of β-END in maternal plasma during pregnancy have been reported as unchanged (97), decreased (98), or progressively increasing (93,99), so that the question of a possible placental contribution to the circulating pool remains unsolved. However, the finding that (a) β-END concentrations are higher in the placental tissue than in the maternal or cord plasma (1,2); (b) immunoreactive β-END in placental homogenates in the first is significantly higher than in the second trimester; (c) at delivery the β-END content is greater than in the second trimester and; (d) in tissues collected at term, in the absence of labor, β-END levels are higher in tissues collected after vaginal delivery (93) together suggest that gestational tissues are important sources, and that stress of delivery greatly stimulates the placental secretion. On the contrary, amniotic fluid β-END concentrations have a completely different gestational pattern, with no changes at parturition (100,101), suggesting a different source in this compartment.

Moreover, β-END is elevate, approximately twofold higher than circulating plasma levels, in the colostrum and transitional milk of mothers who were vaginally delivered. Therefore, it was hypothesized that β-END may contribute to postnatal fetal adaptation, to overcoming birth stress of natural labor and delivery, and at the same time to the postnatal development of several related biologic functions of breast-fed infants (102).

Maternal plasma levels of M-ENK are not significantly different from those of nonpregnant women and do not change throughout pregnancy (94), supporting a local role of the peptide. DYN is measurable in maternal blood, umbilical vein, and amniotic fluid. No significant change occurs in the plasma levels of DYN in the first and second trimester of pregnancy as compared with plasma obtained from nonpregnant women. However, a 2.2-fold increase in maternal plasma DYN levels occurs during the third trimester, as well as at delivery, suggesting placental contribution (103). High DYN levels are also detectable in amniotic fluid and in umbilical vein plasma (103).

Placental Growth Hormone (PGH)

PGH, the product of the GH-V gene, is a GH-like peptide synthesized by the syncytiotrophoblast and released into the maternal circulation, where it gradually replaces the pituitary GH from the second trimester of gestation. The peptide exhibits both somatogenic and lactogenic activities and is structurally distinguishable from pituitary GH by 13 amino acids (104).

Expression and Localization

In situ hybridization with a GH-V specific probe has demonstrated that the syncytiotrophoblast is the production site of GH-V (105), and that GH-V transcript levels increase with in vitro syncytiotrophoblast formation (106). GH-V mRNA expression is also detected in human trophoblastic neoplasms, choriocarcinoma cells and samples of complete hydatidiform moles (104,107).

Receptors

PGH binds to the GH cell surface receptor (GHR) with similar affinity to pituitary GH (108) and is equipotent to pituitary GH as a ligand for circulating GH-binding protein (GHBP) (109). The GHR gene encodes not only the membrane-associated receptor molecule, but also a truncated peptide that binds GH and is found in the circulation, referred to as GHBP. GHBP is thought to derive from the alternate splicing of GH receptor mRNA or proteolytic processing of the receptor (110).

Both GHR and GHBP are present in human placenta. Strong immunopositive staining is associated with the syncytium of placenta at term and was present as early as 8 weeks gestation, while cytotrophoblast cells are immunonegative. Immunoreactivity is also abundant on adherent decidual cells, but absent from chorionic trophoblasts and the amnion (111).

Biological Fluids

Placental GH is secreted only in the maternal circulation as both free and bound PGH (112), and it is not detected in the fetal blood (113). However, the early stages of pregnancy (up to 15 weeks of amenorrhea) have pituitary GH as the main form present in the maternal circulation, displaying a highly pulsatile 24-hour serum concentration profile (113,114), and only from 10–20 weeks to term PGH gradually replaces pituitary GH, which becomes undetectable (113,114) (Fig. 1). In contrast to pituitary GH, the 24-hour serum concentration profile of PGH is basically nonpulsatile (115). At term after placental removal there is a rapid fall (within 1 h) in the serum PGH concentration (115), confirming the relatively short half-life of the hormone and its placental origin. A drastic fall in PGH is observed with the onset of labor (114), which is likely related to changes in uteroplacental blood flow occurring with labor, and to a release of placental proteases with the initiation of labor (104).

PGH levels are not different in the maternal circulation in cases of fetal anencephalia, which supports the independence of PGH secretion regulation from fetal pituitary axis (116). On the contrary, levels are low in cases of FGR (112,114,116), as well as the mean number of placental cells/area expressing the GH-V mRNA, as evaluated by in situ hybridization (114). McIntyre et al. (112) showed a strong correlation between PGH and glycemia at 28–30 weeks of gestation, and they hypothesized that in long-term regulation PGH levels in diabetic pregnancy are driving increased glycemia.

GHBP levels fall throughout gestation (Fig. 1), and positively correlate with maternal body weight and body mass index. GHBP levels in pregnancies complicated by noninsulin-dependent diabetes mellitus (NIDDM) are substantially elevated at all gestational ages, but reduced at 36–38 weeks in patients with insulin-dependent diabetes mellitus (IDDM) (117). Furthermore, at 36 weeks gestation maternal serum levels of GHBP were significantly elevated in those women whose neonates had low birth weight (<10th percentile) (112).

Human Placental Lactogen (hPL)

The classical placental hormone with GH-like activity is chorionic somatomammotropin, also named hPL (118). The dual nomenclature reflects its mammosomatotropic characteristic, due to the structural homology with GH and prolactin. The hPL molecule is a polypeptide of 191 amino acids with 96% homology with GH, but no more than 3% of the somatotropic activity of GH. hPL is synthesized by the syncytiotrophoblast layer of the placenta. The affinity of hPL for the GH receptor was 10,000-fold lower than GH-V and GH.

The levels of hPL in maternal circulation are very low in early pregnancy and increase progressively (Fig. 2), showing some correlation with placental weight (119). Because the half-life of hPL in maternal plasma is very short, the daily placental production of hPL has to be of great magnitude to maintain circulating levels (119). hPL is demonstrable in placenta within 5 to 10 days after conception; it can be detected in serum as early as 3 weeks after fertilization (Fig. 2), disappears rapidly after delivery, and cannot be detected in serum after the first postpartum day. hPL accounts for 7% to 10% of proteins synthesized by placenta at term, and the production rate of hPL increases as pregnancy progresses, approximately in proportion to placental mass with peak levels being reached during the last 4 weeks, representing one of the major metabolic and biosynthetic activities of the syncytiotrophoblast in humans (about 1g/day production near term). The half-life of hPL in maternal plasma is between 10 and 30 minutes. There may be a tenfold increase in circulating levels of hPL from the first to the third trimester. The concentrations of hPL in maternal peripheral blood is about 300 times that in umbilical vein blood, is less in amniotic fluid than in maternal plasma but greater than in fetal plasma (120,121). Maternal serum hPL levels reflect placental biosynthesis and are positively correlated with the size of the fetus (122), suggesting the possibility of using this hormonal marker for the screening of FGR. Very little of hPL is detected in fetal blood or in the urine of the mother or newborn, but it is found in serum and urine of molar pregnancies. Prolonged maternal starvation in the first half of pregnancy leads to an increase in the placental plasma concentration of hPL (120).

OTHER NEUROHORMONES

Neuropeptide Y (NPY)

Human NPY is a peptide of 36 amino acid residues belonging to a family of regulatory peptides that also includes peptide tyrosine tyrosine (PYY) (123). NPY is abundantly and widely spread expressed in the central and peripheral nervous system, in particular in brain, in sympathetic neurons innervating cardiovascular and respiratory system, gastrointestinal, and genitourinary tract (123). Physiological effects attributed to NPY include the stimulation of food intake and inhibition of anxiety in the central nervous system (CNS); presynaptic inhibition of neurotransmitter release in the CNS and the periphery; vasoconstriction; inhibition of insulin release; regulation of gut motility; and gastrointestinal and renal epithelial secretion. Moreover, there is evidence that NPY is involved in the regulation of anterior pituitary hormone secretion: in particular, NPY plays a critical role in stimulating the basal pattern of LH release (123).

Expression and Localization

NPY is produced by human placenta, maternal decidua and fetal membranes (124). Acidic extracts of human placental tissues collected at term pregnancy contain high immunoreactive NPY (irNPY) concentrations, and the extracted irNPY elutes with the same retention time as synthetic NPY. An intense NPY staining is present in the cytoplasm of the epithelial amnion cells and of the cytotrophoblast cells, and intermediate trophoblast of the chorion. Primary cultures of human placental cells release irNPY into the culture medium and the addition of high K^+ concentrations increase the release of the peptide (124,125).

Receptors

Binding sites for NPY are present in all peripheral cells of placental terminal villi (124,126). All NPY receptors mediate their responses through pertussis toxin sensitive G-proteins of the Gi/0 family, resulting in inhibition of adenylate cyclase activity, but they are also able to increase intracellular Ca^{2+} levels. A variety of receptor subtypes for NPY exists, i.e., Y1, Y2, Y3, Y4, Y5, Y6 receptor, and NPGPR (123). The Y1 and Y3 receptor (126), and NPGPR (128) have been identified within placenta. NPY1R and NPY3R are located on brush-border membranes of syncytiotrophoblastic cells of placental villi (129).

Biological Fluids

During pregnancy, NPY is secreted from human placental tissues in maternal, fetal circulation and in amniotic fluid. NPY levels are higher than in nonpregnant women, without significant changes throughout gestation. During labor maternal plasma NPY levels progressively increased, matching the highest levels at the most advanced stages of cervical dilatation and at the time of vaginal delivery (129). NPY is also measurable in amniotic fluid and umbilical cord serum, and levels are comparable to those found in maternal circulation, being highest at term and mainly during the early or late stages of labor. Plasma NPY levels fall immediately after delivery, supporting its placental origin during pregnancy (129).

A recent study has demonstrated that maternal plasma NPY levels in pregnant women with eclampsia and preeclampsia are significantly elevated with respect to that in normotensive pregnant women. At 6 days after delivery the concentration of plasma NPY was significantly decreased in women with eclampsia and PE, and in women with normotension, compared with the value measured on admission (130).

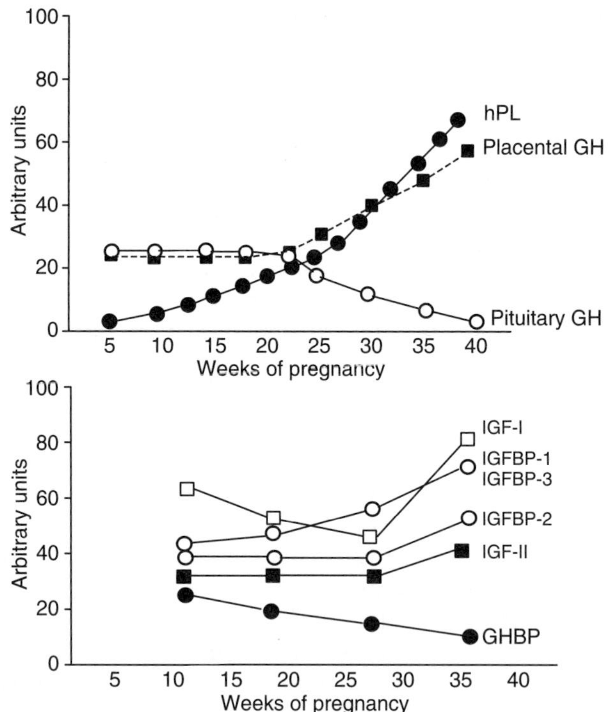

FIG. 2. Relative changes of placental and pituitary GH, GHBP, hPL, IGFs and IGFBPs into maternal circulation throughout pregnancy.

Chromogranin A (CgA)

CgA is a 49kDa glycoprotein of 439 amino acids, belonging to the granin family of regulated secretory proteins identified in secretory granules throughout the neuroendocrine system and in a variety of neurons, both central and peripheral (131). CgA is mainly costored within the granules with catecholamines, enkephalins, vasointestinal peptide (VIP), substance P, and NPY and coreleased by exocytosis with catecholamines and NPY from storage vesicles (131). In plasma, CgA levels are increased in some neuroendocrine tumors, and in response to a large-amplitude stressful events, i.e. hypoxia, physical exercise or other stressful events (131).

Expression and Localization

CgA mRNA and peptide are expressed by placental trophoblast, decidua and membranes (132). Double immunofluorescence of isolated trophoblasts showed colocalization of CgA with hPL and hCG in syncytiotrophoblasts (132).

Biological Fluids

CgA is measurable in maternal and fetal plasma, in umbilical cord blood and in amniotic fluid (133,134). No significant difference in maternal CgA levels during pregnancy compared with levels out of pregnancy, even if median CgA level in maternal sera at term are higher than at 6–11 weeks or in sera from nonpregnant women (134). In umbilical cord sera median CgA level are significantly higher than in term sera, while in amniotic fluid median CgA value was significantly higher at term than in second trimester (134). With respect to labor, umbilical cord plasma and amniotic fluid levels of CgA are higher in women who had spontaneous vaginal delivery than in those delivered by elective cesarean (133,134), suggesting a fetal origin, while no change are observed in maternal circulation (134).

Oxytocin and Vasopressin

Oxytocin (OT) is a neurohypophyseal hormone composed of nine amino acids, synthesized in the hypothalamus and stored in the neurohypophysis, where it acts as a neurotransmitter (135). Vasopressin is a neurohypophyseal hormone similar to oxytocin, since it has a structure close to that of oxytocin and, like oxytocin, is produced in the hypothalamic supraoptic and paraventricular nuclei (135).

Expression and Localization

Northern blot analysis, ribonuclease protection assays and in situ hybridization analysis indicate a local production of OT mRNA in trophoblast, amnion, chorion, and decidua (136,137). In details, the highest abundance is shown in the decidua where the transcript appears to be slightly smaller than that in the hypothalamus and ovary, considerably less in chorion and amnion and very low in trophoblast

(136). With respect to trophoblast localization a large quantity of OT-like substance exists in human placental tissue, mainly in the syncytiotrophoblast (138).

By ribonuclease protection assays, a significantly higher amount of OT mRNA has been detected in tissue obtained after spontaneous labor compared with those obtained at term but before labor onset. OT mRNA levels increase around the time of parturition as a result of either increased transcription of the mRNA or increased stability of the mRNA, thus supporting a role for OT in the mechanism of labor onset. OT peptide concentrations in the decidua are significantly higher than in the amnion or the chorion (137). The content of immunoreactive OT in total placenta extracts increases throughout gestation, in parallel to maternal blood levels (137). Since placental content is approximately fivefold greater than in the posterior pituitary lobe, the main source of OT in pregnancy is the placenta (138).

Receptors

The OT signaling is transduced to physiological actions via the OT receptor (OTR), a 389 amino acid polypeptide with seven transmembrane domains (139). The OTR gene is present in single copy in the human genome and was mapped to the gene locus 3p25-3p26.2, and the human OTR mRNAs shows two different sizes, being of 3.6 Kb in breast and of 4.4 Kb in ovary, endometrium and myometrium (140).

By in situ hybridization and immunohistochemistry OTR mRNA has been detected in decidual cells and in the trophoblast of the chorion leave, but not in the trophoblast into the placenta, suggesting differences in trophoblast expression, depending on the localization. Indeed, the expression of OTR mRNA and protein in the amnion is much lower than in the trophoblasts in chorion leave (141). During pregnancy, significant changes occur in oxytocin receptor gene expression and protein synthesis in target cells. Uterine concentration of oxytocin receptors considerably increases from 20 weeks of pregnancy onwards and undergoes a further dramatic increase around the 30 week (142).

There are at least three distinct myometrial receptors for vasopressin: V1a, V1b, and V2 (143). V1a receptor is likely to be mainly involved in the myometrial contractile response to vasopressin during labor because its tissue mRNA concentrations do not decrease throughout pregnancy and labor (144).

Biological Fluids

A number of technical difficulties have been found in measuring plasma OT in humans, because the pulsatile OT secretion, and the presence in serum of oxytocinase (145). OT is measurable in maternal plasma during pregnancy with a gradual rise of its levels with advancing gestation, but levels do not differ between early labor and late pregnancy (146). OT is secreted in discrete pulses and the frequency of these pulses is significantly higher during spontaneous labor than before the onset of labor (145). After spontaneous vaginal delivery, umbilical arterial plasma levels of OT are consistently higher than those in the umbilical vein, while the fetal arteriovenous difference is less pronounced at elective cesarean section. At spontaneous vaginal delivery, plasma levels from the umbilical cord artery are significantly higher than the maternal levels, and significantly higher than at elective abdominal delivery. Therefore, it is concluded that the human fetus can be an important source of OT (147), and this indirectly supports the hypothesis that locally produced OT may act without being reflected in maternal circulation.

Vasopressin concentrations in maternal peripheral blood do not increase during pregnancy and labor (148), however, fetal production and amniotic fluid are increased at labor (149).

Neurokinin B (NKB)

NKB is a 10 amino acids peptide belonging to the family of tachykinins related peptides, that also comprises substance P (SP) and neurokinin A, that is expressed in the central nervous system and spinal cord, where it plays role as excitatory neurotransmitter (150).

Expression and Localization

Expression of placental NKB mRNA was detected in the first trimester and increased by three- to five-fold at term. In situ hybridization studies on the human placenta at term have shown its localization in the outer syncytiotrophoblasts ideally positioned to secrete its peptide into the maternal blood (150).

Receptors

NKB is able to bind to three G-protein–linked receptors named NK1R, NK2R, and NK3R, of 407, 398, and 465 amino acids, respectively, and it exhibits agonist activity on each receptor subtypes in which its rank order of potency is NK3R>NK2R>NK1R (149). At term, there is predominant expression of NK1R, some expression of NK2R, and absent expression of NK3R mRNA in the whole placental homogenate. NK1R and NK2R, but not NK3R, expressions are detectable in the human umbilical vein endothelial cells (HUVEC) (151). The mRNA of

NK1R is also expressed by the myometrium from pregnant women (152).

Biological Fluids

Significant amounts of immunoreactive human NKB are measurable in early and term placental homogenates, with chromatographic properties that were identical to synthetic NKB. A similar elution pattern is observed on plasma extracts from women at term. Samples taken from nonpregnant women had low (near the sensitivity of the assay) or undetectable levels of NKB (150). However, NKB concentrations increased with advance in gestational week, to decrease during puerperium. The umbilical blood concentration was significantly higher than the maternal blood concentration in late pregnancy, and there was no significant difference between umbilical venous and arterial blood (153). Pregnant women in the third trimester suffering from pregnancy-induced hypertension and preeclampsia have considerably higher levels of NKB (150).

Monoamines

The placenta is a source and target for epinephrine, norepinephrine, dopamine, and 5-hydroxytryptamin (serotonin). The enzymes involved in monoamine synthesis and metabolism as well as monoamine transporters and receptors have been identified in the placenta (154). The placental metabolism and transport of monoamines has a function in determining the availability and bioactivity of biogenic amines to both mother and fetus. In preeclampsia there is an increased activity of tyrosine hydroxylase in placental tissue and this is likely to contribute the higher levels of catecholamines in maternal circulation (155). Placental norepinephrine transporter mRNA expression is reduced in some gestational diseases, resulting in increased norepinephrine levels in fetal circulation (156). The activity of serotonin transporter in placental cells is suppressed by agonistic stimulation of cannabinoid receptors, indicating that placental clearance of serotonin may account for adverse effects of cannabinoid use during pregnancy (157).

GROWTH FACTORS

Transforming Growth Factor-β (TGF-β)

The TGF-β superfamily comprises of a large number of structurally related polypeptides, the majority of which are dimers. In addition to five known TGF-β subtypes, the family also includes inhibins, activins, and bone morphogenic peptides (158).

Expression and Localization

mRNAs for TGF-β1, -β2, and -β3 are expressed by in first trimester decidua, placenta, and trophoblast with TGF-β1 mRNA predominating in the villous placenta and TGF-β2 in the decidua (159,160). In human placenta, mRNA and protein for TGF-β1 and -β2 are expressed throughout pregnancy (161): the protein is particularly evident in the invasive trophoblast in early pregnancy (162), and TGF-β1 placental expression peaks at mid- and late gestation (163).

TGF-β3 mRNA and protein expression increases between weeks 7 and 8 of gestation, then falls dramatically at 9 weeks (161), and the protein is essentially absent by the end of the first trimester (161). In the decidua, TGF-β protein is present in the first and third trimesters, although distribution and abundance of the protein vary with gestational age (162). The expression of TGF-β3 in placental cells is increased by hypoxia (164), and alterations in its expression have been proposed as one causal factor in the failure of trophoblastic invasion and maturation. TGF- β3 is also over-expressed in placentae from pregnancies affected by PE (161).

Receptors

Biological effects of TGF-β are mediated through interactions with cell surface receptors, designated types I, II, and III (158). TGF-β binds the type II but not type I receptor; the type I receptor recognizes the ligand bound type II receptor with which it forms a heteromeric complex. The type I receptor then undergoes phosphorylation, which is probably catalyzed by the type II kinase. The type III receptor, named betaglycan, is a proteoglycan with no cytoplasmic kinase domain and involved in presentation of TGF-β to the type II receptor (158).

Endoglin is a homodimeric glycoprotein that shares regions of sequence identity with betaglycan, and that associates with receptors type II and type I to form a functional receptor complex. While betaglycan recognizes all three TGF-β isoforms, endoglin binds only TGF-β1 and TGF-β3 with high affinity (165).

The presence and localization of TGF-β receptors type I and II has have been evaluated in human term placenta using reverse transcriptase-PCR (RT-PCR) and immunohistochemistry: mRNA for both receptors is present in human term placenta, while an intense staining was reported in the syncytiotrophoblastic layer, chorionic plate, and in cells of the extravillous trophoblasts using immunohistochemistry (166). Betaglycan expression is discussed in the section Inhibin-Related Peptides.

Immunofluorescence staining of sections of first trimester placenta has revealed that endoglin is found at higher levels on the syncytiotrophoblast of samples ranging from 6 to 12 weeks of gestation (167), and in the syncytiotrophoblast of full-term placenta (168).

Biological Fluids

Maternal serum concentration of TGF-β1 in women with PE are not increased and does not correlate with blood pressure (169). On the contrary, maternal serum TGF-β2 levels significantly increase in cases of severe PE (170).

Inhibin-Related Peptides

Inhibins and activins are growth factors belonging to the TGF-β superfamily, and composed by two subunits: inhibins are heterodimers of a α subunit with a β subunit (inhibin A = αβA; inhibin B = αβB) while activins are homodimers of β subunit (activin A = βAβA; activin B = βBβB; and activin AB = βAβB) (159). They were originally identified from gonads as factors acting antagonistically in the endocrine regulation of pituitary FSH production, but successive descriptions of their expression in numerous cells types and tissues outside the hypothalamic–pituitary–gonadal axis indicate different functions, particularly as modulators of cell growth, differentiation, and apoptosis (158).

Expression and Localization

The α subunit mRNA in the human trophoblast is expressed from the first trimester of gestation, with the highest values at term, and it is localized within the structure of placental villi (cyto- and intermediate trophoblast, mesenchymal cells) (171), in maternal decidua (172), amnion and chorionic cells (173). βA subunit is also expressed from the first trimester of gestation, with the highest value at term, while the βB subunit, present in the outer syncytial layer, is detected only at term (171). βA subunit is localized in the external syncytial layer of placental villi, in some structure of the stroma, in maternal decidual cells, and in some amnion and chorionic cells (173) but, in term trophoblast, also in endothelial cells within the placental villi (174).

Placental mRNA expression of both α and βA subunits is increased in preeclampsia, with or without superimposed fetal growth restriction (175).

Receptors and Binding Proteins

Activins signal through a heteromeric complex of receptor serine kinases which include at least two type I (IA and IB) and two type II (IIA and IIB) receptors. These receptors are all transmembrane proteins, composed of a ligand-binding extracellular domain, a transmembrane domain, and a cytoplasmic domain with predicted serine/threonine specificity (159). First and second trimester placentae express the various receptor proteins in the syncytium, whereas at term the distribution is confined to vascular endothelial cells of villous blood vessels. In the fetal membranes they are localized to some epithelial cells, mesenchyme and chorionic trophoblast (174).

The biological function of activin A is tightly regulated by follistatin, a structurally unrelated protein, that binds with high affinity to activin (similar to that for activin receptors) and neutralize its activity, regulating activin bioavailability on target tissues and functions (159). Recently, a new binding protein of 70 amino acids for activin A has been identified, namely follistatin-related gene (FLRG), closely related to follistatin. FLRG interacts physically with activin A and, preventing the binding on ActRs, regulates activin A functions (176). Follistatin and FLRG are both present in trophoblast, deciduas, and fetal membranes. In particular, immunoreactive FLRG is detectable in maternal decidua, in syncytiotrophoblast and in epithelial amniotic and chorionic cells, with the most intense staining in the decidual and placental blood vessels (177). This localization differs from that previously shown for follistatin, which is mainly localized in cyto- and syncytio-trophoblast cells (178) but resembles the activin receptors being present in the lining of blood vessels (174). The finding of FLRG, activin receptors and activin A distribution in vascular endothelial cells of deciduas and villous blood vessels in term placenta supports the hypothesis that FLRG may act by paracrine/autocrine mechanisms as a modulator of activin A vascular actions, as activin A is known to affect the endothelial proliferation and angiogenesis (179).

The transduction of the inhibin A pathway into the cells is mediated through activin receptor type II (180). However, the inhibin A binding to activin receptor type II is facilitated by betaglycan, that allows the formation of a ternary complex including inhibin A, betaglycan, and activin receptor type II. Betaglycan also interacts with activin A, because it prevents the binding of activin A to activin receptor type II, thereby antagonizing the activin signal (180).

Betaglycan is also found in first and third trimester decidual cells and blood vessel walls, in the chorion, in the syncytiotrophoblast (but not cytotrophoblast), and in the placental endothelial cells only at the first trimester (181). Therefore, betaglycan, FLRG, and activin receptors are co-localized in syncytiotrophoblast, and in endothelial cells of trophoblast villi, supporting the hypothesis that these binding sites are involved in modulating the activin A effects on vascular adaptations of pregnancy.

Levels in Biological Fluids

In this trimester of pregnancy, the fetoplacental unit is the main source of circulating activin A and inhibin A, and their maternal serum levels are higher than in nonpregnancy, while inhibin B concentrations do not significantly differ from levels detected during the menstrual cycle [reviewed in (182)]. At this gestational period, coelomic fluid activin A and inhibin B levels are higher than in maternal serum and amniotic fluid. In amniotic fluid, inhibin A is not detectable (182). During the second trimester of pregnancy, inhibin A and activin A further increase in maternal serum and amniotic fluid, while inhibin B increases only in amniotic fluid. At term, maternal serum levels of inhibin A, inhibin B, and activin A and those in amniotic fluid are at their highest (Fig. 3). Inhibins A and B and activin A are also measurable in umbilical cord blood (182). In the second trimester maternal serum inhibin A levels are significantly increased in mothers carrying a Down's syndrome fetuses (183), suggesting a clinical usefulness of inhibin A measurement in the screening of that diseases (see the section Clinical Usefulness of Placental Hormones Measurement Throughout Pregnancy).

In presence of complete miscarriage, a condition associated to the trophoblast dismission, maternal serum levels of activin A and inhibin A are the lowest, indicating the presence of a failed trophoblast (184). Furthermore, patients with threatened abortion (with early pregnancy bleeding and an intrauterine sac with fetal cardiac activity) progressing till term have higher inhibin A and activin A than those with threatened abortion but whose pregnancy failed later (185). Activin A and inhibin A levels are increased in the serum of women with PE compared with normotensive gestational age-matched gravidas (87,183), and very high maternal serum activin A levels are found in presence of FGR (186).

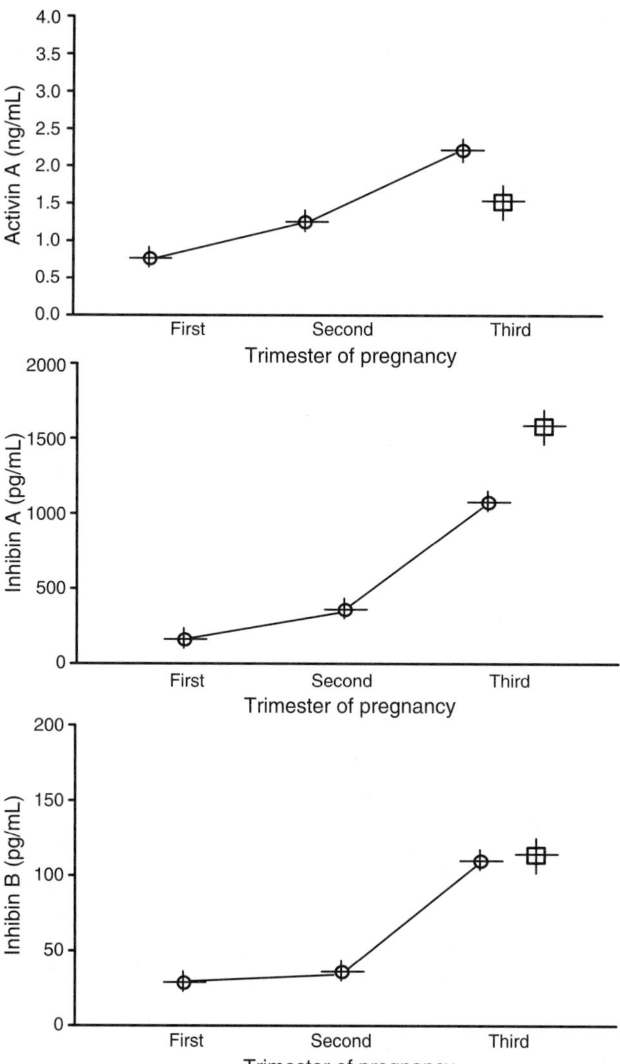

FIG. 3. Changes of activin A, inhibin A, and inhibin B in maternal serum (*open circles*) throughout gestation and in third trimester amniotic fluid.

Insulin-Like Growth Factor and Insulin-Like Growth Factor-Binding Protein (IGF and IGF-BP)

IGFs are single-chain polypeptides of approximately 7 kDa molecular mass that are part of the same family as proinsulin, since they are structurally similar to insulin. The two major forms are IGF-I and IGF-II, which share a 62% sequence homology; both factors have primary structural similarity with insulin (187).

Expression and Localization

IGF-I and IGF-II are produced in human placenta, as early as 9 weeks of gestation (188). In situ hybridization has showed that IGF-I mRNA is expressed primarily by syncytiotrophoblasts, while IGF-II mRNA appears stained in placental fibroblasts and cytotrophoblast localized in the superficial basal plate (189). Immunohistochemistry has showed that IGF-I and -II are localized within syncytiotrophoblast, intermediated trophoblast of the villi and chorion, and amnion and deciduas both in first trimester and term placentas (190). Placental IGF-I mRNA levels are highest at first trimester, slightly decreased by second trimester, and lowest at term. In particular, the second trimester placentas express more IGF-II mRNA than those of first trimester and normal term (190).

Receptors

IGFs act on placental cell membrane receptors for IGF-I and IGF-II (187). The IGF-I receptor is a heterotetrameric protein possessing ligand-activated tyrosine kinase activity, while the IGF-II receptor (IGF-II/mannose-6-phosphate) is a monomer that, in addition to binding IGF-II, binds mannose-6-phosphate and participates in lysosomal enzyme trafficking. Placental IGF-I receptors bind, with decreasing affinity, IGF-I, IGF-II, and insulin, while IGF-II receptors bind IGF-II with higher affinity than IGF-I but do not bind insulin (187). Human placental cells express mRNA for two IGF-I receptors, both detected in placental vessels, stroma, syncytio-, and cytotrophoblast (189), thus suggesting the possibility that tissue-specific isotypes of the IGF-I receptor could mediate the different biological actions of IGF-I. The differential ability of IGFBPs to regulate IGF bioactivity is likely to be related to their association with cell membranes, leading to modifications of IGF-receptor interaction.

In the circulation and tissues, IFGs are found in association with specific IGFBPs, which play a pivotal role in determining the biological stability of IGF-I and -II, inhibiting or enhancing the actions of IGFs at the cellular level. At least, six types of IGFBPs protein have been identified (191): tissue extracts, cell-conditioned culture medium, and blood express IGFBPs during pregnancy (190).

IGFBP-I is a 28 to 30 kDa protein first purified from amniotic fluid with similar affinity IGF-I and IGF-II. It is identical to protein previously termed placental protein 12, or pregnancy-associated endometrial α1-globulin. Syncytiotrophoblast, intermediate trophoblast, and fetal membranes scarcely immunostain for IGFBP-1 both at first trimester and the term (192). Conversely, cultured human decidual cells secrete IGFBP-1 (190) and may explain the high levels of IGFBP-1 circulating in the first trimester. The peak values of maternal IGFBP-1 occur at 12–13 weeks and remain elevated until term (193). Also, amniotic fluid IGFBP-1 levels peak at 9–12 weeks of gestation. In any case, IGFBP-1 levels are much higher in extraembryonic coelom than in amniotic fluid and maternal serum (193) because both chorion and deciduas, or the amnion itself, may contribute to the coelomic pool. Immunoreactive IGFBP-1 in chorion and amnion explain its presence in amniotic fluid (190).

IGFBP-2 is a 34 kDa peptide with greater affinity for IGF-II than IGF-I, present in maternal serum. Immunoreactive IGFBP-2 is associated with the syncytiotrophoblast, intermediate trophoblasts of the fetal villi and chorion, amnion, and deciduas, while weaker staining is present in some but not all cytotrophoblasts (192). In particular, IGFBP-2 mRNA is concentrated in a layer of placental cells within the cytotrophoblast shell opposed to IGF-II mRNA-expressing cells (189). Maternal serum IGFBP-2 levels progressively decrease throughout gestation (190), and they are also measurable in amniotic fluid.

IGFBP-3 mRNA is expressed in both the decidua and certain intermediate trophoblasts of the basal plate and anchoring villi of placenta, and in amnion and chorion laeve of fetal membranes (194). Human IGFBP-3 exists in human serum as a IGF/BP complex (150 kDa), comprising a glycosylated IGF-binding subunit with a greater affinity for IGF-II than IGF-I (IGFBP-3) and an acid-labile subunit that only binds to IGFBP-3 in the presence of IGF-I or IGF-II at neutral pH. No immunoreactive stain has been observed in human placenta for IGFBP-3 (192): this is apparently in contrast with the report of IGFBP-3 immunoreactive stain in the villous mesemchyme (195). This does not imply that IGFBP-3 is not synthesize by placenta, because specific proteases have been identified in maternal circulation of the pregnant rat and human that can selectively destroy the IGF-binding capacity of IGFBP-3 (196). In any case, the 150 kDa complex is found in maternal blood at term, but is absent in second trimester (190), while IGFBP-3 is abundant in fetal serum at this time (197), suggesting that the acid-labile subunit is lacking during early development. IGFBP-3 could act as a reservoir of the IGFs needed for the rapid growth of the placenta. It is possible that placental IGFBP-3 might be confined to the local tissue environment of the placenta, since it does not cross the capillary barrier, while in the circulation it would be rapidly degraded by proteases (190). Circulating levels of IGFBP-3 markedly decrease after 6 weeks of gestation, continuing progressively to term, and returning to nonpregnant levels by 5 days postpartum (198).

Human IGFBP-4 to IGFBP-6 have been cloned from a placental library and their amino acid sequence deduced (199). IGFBP-4 and IGFBP-5 mRNA are expressed in the deciduas, in intermediate trophoblasts of the basal plate and anchoring villi of placenta, in amnion and chorion laeve of fetal membranes, and in low abundance in the chorionic mesoderm (194).

The IGFBP-4 protease, pregnancy-associated plasma protein-A (PAPP-A), is expressed in syncytiotrophoblasts, extravillous cytotrophoblasts and decidualized endometrial stromal cells (200); and its physiological inhibitor, the proform of eosinophil major basic protein (proMBP) is expressed in the extravillous trophoblasts of the anchoring villi (201).

Biological Fluid

Circulating IGF-I levels increase as pregnancy advances both in normal women and in those with GH deficiency (190): the levels in the third trimester are 2.5 times higher than those in the first trimester, while IGF-II levels do not change or only slightly increase (190) (Fig. 2). After delivery, maternal IGF-I levels rapidly decrease, suggesting a major placental contribution. In the fetal compartment, IGF-I levels increase significantly from 18 to 40 weeks of gestation, but IGF-I levels are four to five times lower than those in the maternal circulation. The weak association between maternal and umbilical cord serum IGF-I levels at term and lack of correlation between maternal IGF-I levels and fetal birth weight (202) suggests that the production of IGF-I in the maternal and fetal compartments is independent. Throughout pregnancy, there was a rapid increase in mean IGFBP-1 and IGFBP-3 levels, while IGFBP-2 concentrations fell after conception and started to increase toward term (203) (Fig. 2).

Third trimester IGF-1 and IGF-BP3 levels were lower, and IGFBP-1 higher in PE and eclampsia than controls (204), and reduced levels of maternal IGF-I levels were seen in women with FGR (205). Interestingly, this hormonal pattern is associated with a significant increase in IGFBP-1 but not IGFBP-3 levels (206).

Maternal serum PAPP-A levels are detectable as early as 8 weeks of gestation and then rise throughout pregnancy (207): when measured between 8 and 13 weeks of gestation, levels are significantly reduced in Down's syndrome pregnancies, with an average decrease of 2.5 times (208). Interestingly, maternal serum PAPP-A levels return to normal values between 17 and 19 week of gestation (209).

Fibroblast Growth Factor (FGF) and Epidermal Growth Factor (EGF)

To date, five members of the FGF family have been identified [acid and basic FGF (aFGF and bFGF), int-2, and FGF-5]. The bFGF genes give rise to two polyadenylated mRNA of 3.7 and 7.0 kb, sharing a 55% total amino acid sequence homology with aFGF; both forms are highly conserved among species. bFGF exists as a single-chain nonglycosylated of 146 amino acids and has an ubiquitous distribution. Two cell surface receptors for FGF have been identified on a variety of different cell types, both aFGF and bFGF interacting with the same receptors, also in view of their structural similarity (210).

EGF is a 53 amino acid polypeptide with three disulfide bonds, having a variety of biological activities (211).

Expression and Localization

Human placenta express b-FGF using immunoblot analysis as well as chromatography (212). The expression of specific mRNA for bFGF in trophoblast cells at first trimester and term has been shown (213). In particular, placental villous trophoblast show a perinuclear pattern and intense homogenous staining of the nucleus cytoplasm, observed at different stages of mitosis and cytokinesis, thus suggesting a role for bFGF in cytotrophoblast proliferation in vivo and raising the possibility that bFGF has an intracrine role in trophoblast proliferation (214).

Human placenta contains EGF mRNA and immunoreactive EGF. An intense EGF immunostaining is found in first trimester human placenta. Using rabbit antisera, EGF has been identified in both cytotrophoblast and syncytiotrophoblast, with a more intense staining in the syncytiotrophoblast cells. While light to absent EGF immunostaining has been noted in the villous stroma in first trimester placenta, increased staining is absent in second trimester and term placentas (215). In the perfused placenta model, [^{125}I]EGF does not cross the placenta barrier (216). Human amnion cells also express mRNA for EGF (215).

In human endometrium the immunohistochemical EGF staining is mainly localized to stromal cells, particularly in those surrounding arterioles and which in early human pregnancy are invaded by trophoblast. In decidua, light to moderate EGF immunostaining is noted in the decidualized stromal cells, while intense staining is noted in the surface epithelium (215).

Receptors

Several studies have demonstrated the presence of EGF-R in human placenta, endometrium, and fetal membranes (217). In particular, in very early (4–5 weeks) pregnancy EGF and EGF-R are predominantly on the cytotrophoblast, whereas in early

(6–12 weeks), mid-, and late gestation they are localized to syncytiotrophoblast. In human deciduas both stromal and epithelial cells contain EGF-R. The highest EGF-R concentrations occur in early placenta, which decrease as pregnancy progresses (217).

Biological Fluid

Maternal urinary levels are inversely related to fetal amniotic fluid levels. Amniotic EGF concentration is higher near term than at mid-gestation (218).

Vascular Endothelial Growth Factor (VEGF)

The VEGF family is composed by VEGF (or VEGF-A), VEGF-B, VEGF-C, VEGF-D, and placental growth factor (PlGF). Human VEGF has at least six isoforms of 121, 145, 165, 183, 189, and 206 amino acids that also differ in their bioavailability and biological activities (219). The human genes for VEGF-B span about 4 kb of DNA and are composed of seven exons. Due to alternative splicing of exon 6, two isoforms of VEGF-B (VEGF-B_{167} and VEGF-B_{186}) are generated. The human genes for VEGF-C span over 40 kb of DNA and consist of seven exons. It encodes a protein of 419 amino acid residues with a predicted molecular mass of 46.9 kD (219). VEGF-C is secreted as a disulphide-bonded homodimer that is proteolytically processed from the precursor polypeptide and the secreted form contains the C-terminal silk domain (219). The human gene for VEGF-D is located on chromosome Xp22.31, and the cDNA encodes a protein of 354 amino acid residues (219). PlGF is 149 amino acids long that exists as three isoforms which differ in size and binding properties: PlGF-1 ($PlGF_{131}$), PlGF-2 ($PlGF_{152}$), and PlGF-3 ($PlGF_{203}$) (220).

Expression and Localization

In the human placenta, VEGF expression has been analyzed by *in situ hybridization* and immunohistochemical studies, and has been demonstrated to be localized to the villous trophoblast and macrophages of both fetal and maternal origin, but not in the villous endothelium (221). The expression of VEGF in syncytiotrophoblasts is uneven in the first trimester and most intense at the sprouting sites. In addition, some stromal cells in the villi and decidual cells are also positive in the first trimester (222). In placental and related chorioallantoic tissues VEGF is upregulated by hypoxia and downregulated by hyperoxia (223).

By *Northern blot hybridization* and *immunoblot* low levels of VEGF-C mRNA and protein expression, respectively, are detectable in fibroblasts and cytotrophoblasts isolated from first and second trimester placentas (224). In first trimester chorionic villi very weak VEGF-B immunostaining is present in cytotrophoblast tissues, while there are no changes at any of the gestational ages analyzed, i.e., 6 weeks till term (224). With respect to localization, in the first trimester VEGF-C is in most of the cytotrophoblast populations; in the second trimester the staining tends to be stronger than in the first trimester, with cytotrophoblast stem cells, cells in columns, and cytotrophoblasts within the uterine wall (endovascular cytotrophoblasts) exhibiting similar immunoreactivity levels. By term (38 weeks), little or no staining for VEGF-C is observed (224). Northern blot hybridization shows that fibroblasts and cytotrophoblasts isolated from first and second trimester placentas expressed VEGF-C mRNAs (224). No data are available on the VEGF-D mRNA and protein expression in human placenta.

PlGF is expressed in both villous syncytiotrophoblast (223) and in the media of larger stem vessels (225) in the human placenta, and its expression is diminished in cultured trophoblast by reduced oxygen (226).

Receptors

VEGF exerts its biological effects by binding with high affinity to two tyrosine kinase receptors VEGFR-1/Flt-1 (227) and VEGFR2/KDR (228). PlGF binds Flt-1 but not KDR (229). In the placenta both endothelial and nonendothelial cells express the VEGF receptors. Both Flt-1 and KDR are expressed by human placental villous endothelium (230), and Flt-1 mRNA has also been demonstrated to be highly expressed in the cytotrophoblast shell and columns (231) while the protein localizes to the trophoblast bilayer and extravillous trophoblast (EVT) (232). RT-PCR has demonstrated the presence of KDR both in the human choriocarcinoma cell line, BeWo (231), and in the nontransformed first trimester trophoblast cell line, ED27 (232). KDR immunoreactive protein is localized to the trophoblast in human placenta (232). VEGF, PlGF, and the two receptors are differentially expressed by human placenta throughout gestation: VEGF and KDR are most intense during early gestation and decline as pregnancy advances (222) while PlGF and Flt-1 increase toward term (233).

Neuropilin-1 and VEGFR-3 (Flt-4) are additional VEGF receptors: neuropilin-1 modulates VEGF binding to KDR and subsequent bioactivity (234), but it also functions as the placenta growth factor-2 receptor (235); the protein product of the VEGFR-3 cDNA is structurally similar to the Flt-1 and KDR receptors (236), and it binds VEGF-C but not VEGF (237). To date, the expression of neuropilin-1 is only shown in ovine fetal placental artery endothelial cell line, but not in human placenta (238).

Invasive cytotrophoblasts in early, but not term, gestation are reported to express VEGFR-3 mRNA. Immunoreactive VEGFR-3 is localized in chorionic villi, cell columns, decidua, and myometrium, relatively weaker in the syncytiotrophoblast and strong in endovascular cytotrophoblasts (224). Diminished placental trophoblastic VEGF immunoreactivity, and weaker VEGFR-1, VEGFR-2 expression in decidual endothelium of spontaneous abortions is described. VEGFR-1, but not the other receptors, is over-expressed in placental syncytiotrophoblasts from women with severe PE and FGR when compared to normal placentas (239). In PE elevated levels of soluble VEGFR-1 protein exist in the amniotic fluid and maternal circulation and increased placental mRNA expression (240).

Biological Fluids

Circulating levels of VEGF are elevated and correlated to blood pressure in patients with PE, although decreased concentrations also have been described [reviewed in (241)]. The source of the increased VEGF in PE is unclear; however, VEGF mRNA expression in placental tissue decreases with increasing gestation and is even further decreased in PE. VEGF might also be released by aggregating platelets and vascular smooth muscle cells (241).

Progesterone and Estrogens

Progesterone is one of the main hormones of pregnancy. It is produced by corpus luteum, starting from cholesterol, until about 10 weeks of pregnancy when the placenta becomes the major source and its levels increase throughout pregnancy. Into the placenta, progesterone is produced starting from cholesterol, which is converted by the activity of specific enzymes (mainly *cholesterol side-chain cleavage cytochrome P450*, and 3β-*hydroxysteroid dehydrogenase*) (242). In fact, human placenta by itself is not able to produce directly progesterone, and it converts low-density lipoprotein cholesterol, coming mainly from maternal circulation, first into pregnenolone and then rapidly and efficiently to progesterone (242). The fetal contribute is negligible, as fetal death has no immediate influence on progesterone production (243).

Progesterone levels progressively increase from early to term gestation, and at term concentrations are ten- to 20-fold higher than during luteal phase, determining a *progesterone storm* throughout the entire course of pregnancy (244). Moreover, this rise is independent of factors that normally regulate steroid synthesis and secretion, and whereas exogenous hCG increases progesterone production in pregnancy (245), hypophysectomy has no effect (245). Despite a fall in progesterone concentration is the single most common endocrine event associated with parturition across species (246), in the human there is no evidence of a fall in maternal plasma or uterine tissues progesterone at labor (both term and preterm).

Estrogens also are synthesized by the placenta depending on circulating androgen precursors [dehydroepiandrosterone (DHEA); DHEA-sulphate (DHEA-S)], which are derived from both fetal and maternal androgens (246). Fetal DHEA-S, produced mainly by the fetal adrenals, is converted by placental sulfatase to DHEA and, consequently, to androsterone and testosterone, through the enzymatic pathways common to steroid-producing tissues. In turns, these androgens are aromatized by the placenta to estrone and estradiol, respectively. Finally, fetal DHEA-S is also and mainly metabolized to estriol, the most important estrogen of pregnancy, not detectable out of pregnancy. Thus, maternal serum or urinary estriol concentrations may reflect fetal and placental circulation and metabolism. Henceforth, 90% of the estriol production originates from DHEA-S synthesized by the fetal adrenal glands (247). In fact, unlike total estriol, unconjugated estriol is produced almost entirely by the fetal-placental unit and therefore is a more sensitive indicator of fetal health. Estetrol is another estrogen produced exclusively in pregnancy, whose role has not clear to date (248).

Estrogen levels increase from early to term pregnancy: maternal serum estrone and estradiol concentrations are increased several folds over their pre-pregnancy levels, on the contrary estriol increases from approximately 2 mg/24 h at 26 weeks to 35–45 mg/24 h at term. Estriol is also found in high concentrations in amniotic fluid and in the maternal circulation. However, maternal estriol concentrations are low in pregnancies with placental sulfatase deficiency (249), due to the lack of the enzyme involved in the conversion of fetal DHEA-S, and in the case of the fetal anencephaly (250), because little fetal DHEA-S is produced as fetal adrenal stimulation by ACTH is lacking. Maternal serum unconjugated estriol levels are significantly lower in Down's syndrome pregnancies, ranging from 0.65 (251) to 0.79 MoM (252). Amniotic fluid and placental unconjugated estriol levels are also significantly lower in

Down's syndrome. Whereas placental turnover is not affected by trisomy 21, DHEA-S levels in maternal serum, placental tissue, and fetal liver are decreased, suggesting the possibility that the reduction of estriol is linked to diminished fetal DHEA-S synthesis (249).

Maternal plasma estriol levels at 31–34 weeks are, on average, higher in women destined to initiate spontaneous labor before 37 weeks (253), and on the other hand the measurement of estriol in maternal saliva allows to distinguish women at risk of preterm delivery (254).

VASOACTIVE PEPTIDES

Adrenomedullin (AM)

AM is a 52 amino acid peptide, first isolated from pheochromocytoma tissue, showing some homology with calcitonin gene-related peptide (CGRP) and with amylin. AM is produced from a 185 amino acid prohormone, preproadrenomedullin, from which, after translational processing, derives a 20 amino acid N-terminal fragment (proadrenomedullin-N-terminal 20 peptide, or PAMP) sharing some homology with AM, and two other peptides "45-92 proadrenomedullin" and "153-185 proadrenomedullin," also named adrenotensin, according to its weak vasoconstrictor activity (255).

Expression and Localization

Human placenta produces ADM mRNA in all three trimesters of pregnancy (256). Northern blotting revealed that ADM mRNA is expressed in chorionic villi and basal plate regions, but most abundantly by choriodecidua. By *in situ hybridization*, ADM mRNA has been localized to the syncytiotrophoblasts and the extravillous cytotrophoblasts in the basal plate and choriodecidua regions (256).

Immunohistochemical studies have revealed that AM is expressed in syncytiotrophoblasts (257) and cytotrophoblasts in chorionic villi (258), extravillous trophoblasts (259), and amnion and trophoblasts of the chorion (260). The prevalence of positive trophoblast cells is significantly lower in spontaneous abortions than in controls (261). Conflicting results have been reported also in the expression of AM in fetoplacental tissues in PE: ADM in placentas of women with PE is decreased (259) or unchanged (261), and AM mRNA expression is either decreased (262) or unchanged in the placenta and uterine muscle (263), decreased in fetal membranes and increased in umbilical artery (263). However, the evaluation of the regional distribution and cellular localization of ADM mRNA expression by *Northern blot* and *in situ hybridization* of different regions of the placenta and fetal membranes from pregnancies complicated by severe PE and from normotensive pregnancies revealed that ADM mRNA expression is increased in the choriodecidua, syncytial knots, and cytotrophoblasts in peri-infarct regions in preeclampsia. In chorionic villous explant studies maintained at reduced oxygen tension, ADM mRNA abundance is increased at 12, 24, and 48 hours (256). ADM mRNA are detectable in human placental tissue of 20 term pregnancies complicated by birth asphyxia, and in human choriocarcinoma cells (BeWo) exposed to hypoxia (1% O_2) for 1–24 hours. ADM mRNA levels significantly increase in response to hypoxia within a period of 12 hours in BeWo cells, and high levels of ADM mRNA are present in human placental tissue, suggesting that ADM gene expression in human placenta is up-regulated in presence of hypoxia (264).

Receptors

A complex consisting of receptor activity-modifying protein-2 (RAMP2, a 175-amino acid type I transmembrane protein containing four conserved cysteine residues in the N terminus and a basic-acidic pair of residues) and calcitonin receptor-like receptor (CRLR, a seven transmembrane domain, G-protein–coupled receptor) functions as an ADM receptor, since RAMP2 presents the CRLR at the cell surface, and RAMP2-transported CRLR receptor act as ADM-specific receptors (265).

Makino et al. (266) studying the changes of expression of RAMP2 and CRLR in fetomaternal tissues between normotensive pregnant women and patients with PIH by *Northern blot analysis* found that RAMP2 and CRLR mRNA were significantly decreased in the umbilical artery, increased in the fetal membrane, and unchanged in the placenta of the patients with PIH.

Biological Fluids

The regional and cellular localization of ADM mRNA and peptide in villous and extravillous trophoblasts and in fetal membranes suggests that fetal/placental tissues contribute to the maternal and fetal plasma and amniotic fluid levels of ADM. Maternal plasma ADM levels in pregnant women are three- to fivefold higher than in nonpregnant women and this increase is found in maternal and amniotic fluid samples collected as early as 8 weeks of gestation and remains stable throughout pregnancy (267). Maternal plasma ADM levels rises immediately after implantation, and gradual increase in plasma ADM and amniotic fluid is observed as pregnancy progressed. Concentration of ADM is higher in umbilical cord vein than in arteries (268).

Human labor is not associated with specific changes in ADM concentration in maternal or fetal plasma, nor in ADM expression or concentrations in fetoplacental tissues, both in term or preterm pregnancies (269). In contrast, ADM concentrations are higher in amniotic fluid from women with preterm but not term labor (269). Moreover, plasma ADM concentration peaks at the beginning of the active stage of labor, and decreases with advancing cervical dilatation (269).

ADM levels are increased in fetuses with FGR in response to reduced uteroplacental blood flow and suggests that it may participate in the fetal hemodynamic modifications (270). In pregnant women with PIH maternal circulating ADM has been reported as either increased, decreased, or unchanged, whereas is high in umbilical plasma and amniotic fluid (271). ADM concentrations are also high in amniotic fluid of pregnant women with diabetes, whereas no differences are found in maternal circulation (271).

Endothelin (ET)

ETs are small (21 amino acid) vasoactive peptides produced by endothelial and epithelial cells, macrophages, fibroblasts, and many other cell types (272). There are at least three ET isoforms (ET-1, ET-2, and ET-3) that are encoded by three distinct, independently expressed genes. ET-1, the first described member of the family, is synthesized as a 212 amino acid preproET-1 that is cleaved by a specific endopeptidase into a 38-amino acid big ET-1. Big ET-1 is subsequently cleaved into mature ET-1 by an ET-converting enzyme. ET-3 mRNA encodes a 230 amino acid precursor that includes ET-3 and a 15 amino acid homologous segment called the ET-3–like sequence. ET-2 preproprotein consists of 178 amino acid residues, and the ET-like sequences found in the preproET-1 and ET-3 was conserved in this preproET-2 (272).

Expression and Localization

The trophoblast expresses preproET-1 mRNA that is ultimately translated into the mature peptide. The preproET-1 and ET-3 genes are expressed in the placental villi (273) and in syncytiotrophoblasts in culture (274). With respect to the localization of the peptide, a broad distribution of ET-1 in human placenta, including localization to chorionic villous endothelial cells (275), amnionic epithelium (276), and villous trophoblasts (277) has been shown. Intense immunopositive staining is present in the chorionic trophoblast following vaginal deliveries in term and preterm tissues (275). Placental ET-1 mRNA expression is also high in PE (277).

The processing of human ET-1 requires the action of an endothelin-converting enzyme (ECE-1), a phosphoramidon sensitive metallopeptidase that specifically converts big ET-1 to ET-1 rather than the other two endothelin isopeptides, big ET-2 and big ET-3 (278). As big ET-1 has little or no vasoconstrictor effect itself, ECE-1 is therefore critical in regulating levels of vascular smooth muscle tone and uterine smooth muscle tone, through its production of ET-1 (278). High levels of ECE-1 expression in human placenta by *Northern blot analysis* (279), by *immunohistochemical analysis* and by *immunofluorescence*, is localized in endothelial cells lining the maternal basal plate blood vessels, in the intermediate trophoblasts, in the endothelial cells lining the chorionic villous blood vessels, in the chorionic villous stromal cells and the chorionic villous trophoblasts. The localization of ECE-1 corresponds with the previously reported distribution of ET-1 in the human placenta (280).

With respect to ET-2 and ET-3, the expression of their preproendothelin genes in cultured trophoblast has shown the ET-3 precursor gene expression at low level (detected only after *Southern blotting*) but not ET-2 precursor mRNA (281). In the first trimester (7–12 weeks) and at term the human placenta expresses the genes encoding for preproET-2 and ET-3 (282).

Receptors

There are two distinct ET-R subtypes, ETA (ETA-R) and ETB (ETB-R) receptors, which belong to the superfamily of receptors with seven transmembrane domains that are coupled to heterotrimeric G proteins. ETA-R is selective for ET-1 and ET-2, whereas ETB-R binds to ET-1, ET-2, and ET-3 with equal affinity. An additional subtype (ETc-R), which is present in nonmammalian species and selectively binds ET-3, has been cloned (283).

Specific high-affinity ET-1 binding sites are present on fetoplacental vessels, trophoblasts and plasma membranes purified from the syncytiotrophoblast of term placentas (284). ET-Rs are the most abundant in first trimester placenta, and their concentration gradually decreases toward term (285) but their binding affinities do not significantly change.

Human placenta has regional differences in the relative proportions of ETA-R and ETB-R subtypes, since in early gestation placenta ETB-R represents 72% and ETA-R 28% of total receptors, while in term placentas, ETA-R predominates in the proximal regions of vessels in the chorionic plate and stem villi, whereas ETB-R is concentrated on intermediate and terminal villi and on the decidua (286). ET receptors mRNAs are observed at low level in human amnion epithelial cells and tissue explants tissues probably by the consequence of a negative regulation in response to high concentration of ET-1 in the amniotic fluid (273).

Biological Fluids

High concentration of ET-1 are detectable in amniotic fluid and umbilical cord blood, and considering the potent contractile effect of ET-1 on the myometrium and intramyometrial vessels (284), it is proposed that ET may have a role at and parturition. Although no difference with regard to the mode of delivery is found in ET-1 or big ET-1 concentrations in any of the sampling locations, a role of ET-1 as a regulator of uterine contractions cannot be ruled out. During parturition, an alteration of the ET receptor function might contribute to an enhanced uterotonic effect of ET-1 without any change in the amniotic or intravascular concentration of the peptide (284).

Maternal plasma concentrations of ET-1 have been measured in a number of studies involving pregnant women with PE. Most investigators have found high plasma concentrations of endothelin (two- to threefold) in women with PE, in particular during the latter stage of the disease, suggesting that ET-1 may not be involved in the initiation of PE but rather in the progression of disease into a malignant phase (287,288). In pregnancies complicated by abnormal umbilical flow velocity waveforms, the umbilical ET-1 concentration is increased (289) as well as in umbilical cord blood (289) and in amniotic fluid (290) in cases of FGR.

Calcitonin Gene-Related Peptide (CGRP)

CGRP is a 37 amino acid neuropeptide produced by tissue-specific alternative splicing of the primary transcript of the calcitonin gene. A second gene encoding a similar peptide (β-CGRP) has also been identified in various rat and human tissues, including the central nervous system, heart, and kidney (291). The distribution of CGRP-producing cells and pathways in the brain and other tissues suggests functions for CGRP in nociception, ingestive behavior, and modulation of the autonomic and endocrine systems (291).

Expression and Localization

CGRP mRNA is expressed by human placenta, but mainly by decidual cells (262,292). In general, CGRP gene expression indicated by mRNA amounts is slightly higher in chorionic plate tissue than in placental villi (262). Moreover, in placenta of women with PE reduced CGRP mRNAs in contrast to unchanged mRNA levels of their receptors have been described (262). Decidual cells express both CGRP mRNA and protein, which is secreted by decidual cells in vitro (292).

Receptors

Two classes of CGRP receptors exist: (a) is sensitive to hCGRP (8–37) C-terminal fragment; (b) insensitive to this fragment. CGRP acts at the cellular level by binding to a seven transmembrane domain, G-protein–coupled receptor, and receptors are linked to the activation of adenylate cyclase in several systems and in intracellular calcium level modulation (270). In human placenta there are specific binding sites for CGRP, able to bind α- and β-CGRP in a dose-dependent and saturable manner consistent with a single binding site of high affinity, with a low affinity for calcitonin (269).

CGRP receptors are localized on human syncytiotrophoblast brush-border membrane (facing the mother) and in basal plasma membrane (facing the fetus), and are able to bind CGRP in a specific, rapid, time-dependent, and high-affinity manner (293). The expression of CGRP receptors has been also detected by *Southern blot hybridization* and *RT-PCR* in decidual cells and extravillous trophoblast cells (292). In addition to placental and decidual sites, CGRP receptors are also expressed by human myometrium (294), and the myometrial expression is increased during pregnancy and significantly down-regulated after labor (294). Indeed, CGRP receptors are abundant in myometrial cells of pregnant women who are not in labor and are minimal in uterine specimens from women in labor and in the nonpregnant state (294). Finally, the sensitivity of myometrial tissues to CGRP significantly decreases at term labor (295).

Levels in Biological Fluids

CGRP is secreted in maternal and fetal circulation in increasing amounts from early to term gestation (296).

Pregnant women at term have higher plasma CGRP levels than nonpregnant women and spontaneous labor does not alter maternal CGRP levels, as levels do not differ between vaginal delivery and elective caesarean section and do not correlate with cervical ripening throughout labor (297). Controversial reports exist on maternal plasma CGRP concentrations in PE (298,299).

Parathyroid Hormone-Related Peptide (PTHrP)

PTHrP is a 141 amino acids protein involved, signaling through its receptor PTHR1, on endochondral bone development and epithelial-mesenchymal interactions during the formation of the mammary glands and teeth (300). However, PTHrP is involved in regulating local tissue functions, in contrast to the systemic hormonal function of PTH. PTHrP functions as a polyhormone that gives rise to several biologically active peptides, each of which presumably has it own receptor (300). Eight of the first 13 amino acids in the mature PTHrP peptide are identical to those of PTH, but the sequence diverges completely after amino acid 13, and the subsequent region accounts for the distinctive biological actions of the two peptides (183).

Expression and Localization

The placenta and the mammary glands are the main sources of PTHrP (301). PTHrP (1–34), in fact, its mRNA has been identified in placenta, myometrium, decidua and fetal membranes (302,303) and the peptide is localized in both syncytiotrophoblast and cytotrophoblast cells (304). With respect to mRNA levels, the expression is higher in placental amnion than in reflected amnion (305). By using *immunohistochemistry*, it has been described as a differential localization of ir-PTHrP (1–34) and ir-PTHrP (67–86) in the human placenta and fetal membranes (306). In fact, PTHrP (1–34) is strongly localized to the syncytiotrophoblast of the placenta, while PTHrP (67–86) is present predominantly in the endothelial cells of capillaries in the placental villi. Moreover, the staining for ir-PTHrP (1–34) is less in placenta and membranes obtained from women at the time of labor than at elective caesarean section, whereas ir-PTHrP (67–86) staining is not different (306). In women with PE, the PTHrP expression in placenta and amnion is not increased in association with maternal hypertension, placental insufficiency, and vasoconstriction. PTHrP mRNA expression is low in choriodecidua in association with term but not preterm PE (304,307).

Receptors

PTHrP binds to the same receptor as PTH, a seven transmembrane domain, G-protein–linked receptor which signals via both adenylate cyclase and phospholipase C (300). Using *RT-PCR,* PTH/PTHrP receptor mRNA is expressed in the myometrium and in preterm and term samples of placenta, amnion over placenta, reflected amnion, and choriodecidua (308). In details, PTHrP receptor are detectable in human trophoblast in proximity to sites of PTHrP expression (309), thus suggesting a possible autocrine and paracrine functions of PTHrP in all preterm and term tissues, including amnion, chorodecidua, placenta, and myometrium (308,309).

Levels in Biological Fluids

Plasma levels of PTHrP increase throughout pregnancy with higher levels at term (310) and PTHrP produced in either the fetoplacental unit can reach the circulation of pregnant women in the third trimester and at 1 month postpartum in women with breastfeeding and mixed feeding (310). PTHrP is detectable in fetal blood and concentrations are lower in maternal blood (310). Moreover, PTHrP levels are high in fetal than maternal circulation (311) and, concentrations in the umbilical artery are higher than in the vein, thus suggesting that the fetus is the main source of PTHrP in the cord blood circulation (312). The concentrations in umbilical cord plasma are high in FGR, while unaltered in gestational diabetes (313).

METABOLIC HORMONES

Leptin

Leptin a 16 kDa nonglycosylated protein is produced predominantly by white adipose cells, and the mature protein is encoded by the obese (OB) gene (314). Since circulating leptin levels are proportional to adipose tissue mass, and the hypothalamic nuclei are the target where leptin exerts most of its effects on energy metabolism, leptin can be considered as a signal to the body of its energy reserves (314).

Expression and Localization

By *Northern blot analysis* a single leptin mRNA species of the same size (~4.5 kb) has been identified in placental chorionic tissue. Expression of the leptin gene is abundant in the first trimester chorionic villi, and low in the third trimester chorion laeve and amnion (315). Human leptin mRNA and protein are localized to the syncytiotrophoblast at the maternal interface, and their expressions are comparable to those of the adipose tissues (315). Both syncytiotrophoblasts and cytotrophoblasts stain positively for leptin (315). Leptin gene is also expressed in the villous vascular endothelial cells which are in direct contact with the fetal blood, and human umbilical cord and fetal membranes at term (316).

Leptin mRNA expression is markedly augmented in the placental tissue from women with PE, and in severe PE concentrations are markedly higher than in mild PE. Furthermore, placental leptin mRNA levels are roughly parallel to plasma leptin levels in all the preeclamptic women examined (317). Augmented placental leptin mRNA expression is also reported in pregnancies complicated with diabetes mellitus, especially in pregnant women treated with insulin (318).

Receptors

The leptin receptor (OB-R) is a single membrane spanning receptor with homology to the class I cytokine receptor family. Multiple isoforms of human and mouse OB-R have been identified, including a long form (OB-RL) with a large cytoplasmic domain capable of signal transduction, and several receptor isoforms with short cytoplasmic domains (OB-Rs) lacking signal-transducing capabilities (319). The leptin receptors OB-Rs and OB-RL were found in the murine placenta by *RT-PCR*, *in situ hybridization*, *Western blot analysis*, and *immunohistochemistry* (320). Using quantitative *RT-PCR*, leptin receptors were also detected in the human placenta and a relative decline of the mRNA levels in advanced stages of gestation was noted (321). Both OB-RL and OB-Rs were localized to syncytiotrophoblast cells using *in situ hybridization* and *immunohistochemistry* (322).

Levels in Biological Fluids

Placental leptin content increases with gestational age and positively correlates with estradiol and β-hCG concentrations. The contribution of placental leptin to maternal plasma levels accounts for only 15% (323), and cannot explain the up to two- to fourfold increase in leptin levels of pregnant women at term (324). The residual leptin supply must come from maternal adipose tissue, possibly due to stimulation by placental hormones. The evidence that plasma leptin levels in umbilical arteries and umbilical veins were significantly lower than those in paired maternal plasma and that leptin levels in umbilical artery are significantly lower than those in paired umbilical vein (325), together indicate that leptin is synthesized in the human placenta, and is secreted to both maternal and fetal circulations.

Fetal plasma leptin increases during development in utero (326), and a significant correlation with birth weight exists (327). Gender differences (lower leptin levels in males than in females) are already present at birth, and persist during later life (328,329). Furthermore, concentrations of leptin in human umbilical cord blood at birth correlate positively with placental size (329), and cord blood leptin levels are elevated in infants of diabetic mothers and in large-for-gestational-age newborns (327). It is generally accepted that fetal leptin reflects fetal fat mass, as it does in the adult, however, fetal leptin levels assessed by cordocentesis do not reveal differences between normal and fetuses with FGR, at least before 34 weeks (330).

No significant difference in maternal plasma leptin levels exists between the mild PE group and its gestational age-matched control group. By contrast, plasma leptin levels in the severe PE group are approximatively threefold higher than those in its gestational age matched control group. (331).

Ghrelin

Ghrelin is a 28 amino acid peptide, synthesized predominantly in epithelial cells lining the fundus of the stomach. In vitro and in vivo assays revealed that ghrelin is able to induce GH release in a clear dose-dependent manner, and that it functions to increase hunger though its action on hypothalamic feeding centers and is one of several hormonal signals that communicates the state of energy balance in the body to the brain (332).

Expression and Localization

Ghrelin mRNA was found in first trimester and postpartum normal placentas. Ghrelin peptide is heavily expressed in the first trimester, mainly expressed in cytotrophoblast cells and to a lesser extent in syncytiotrophoblast cells. Ghrelin expression decreases with the evolution of pregnancy, reaching lowest level in full-term placentas (333).

A ghrelin transcript corresponding to the predicted 141 bp size is present in the decidual tissues. Moreover, *immunohistochemical analysis* reveals strong signals in decidual cells, and weak ones in luminal and glandular epithelium. With respect to the first trimester placenta, strong signals are present in the EVT at the tips of the chorionic villi, and weak signals were detected in the cytotrophoblast (334).

Receptors

The growth hormones secretagogue receptor (GHS-R) type 1a, a seven transmembrane domain G-protein–coupled receptor, is shown to transduce the GH-releasing effect of the synthetic GHSs as well as ghrelin. It is expressed in several nuclei of the brain, in particular in the hypothalamus and in the anterior pituitary gland, and at very low levels in the pancreas. A nonspliced, non-functional receptor mRNA variant is also identified (GHS-R1b) that contains all of exon 1 and an additional 74 bases from the intron (332).

The distribution of GHS-R and GHS-R1b mRNA expression are investigated in various human tissues using classical and *real-time RT-PCR*: human placenta expresses GHS-R1b but not GSH-R mRNA (335).

Biological Fluids

Maternal plasma concentration of ghrelin are markedly decreased in normal pregnant women in the third trimester. Plasma ghrelin concentration are significantly higher in patients with PIH than in normal pregnant women, and there is a significant correlation between plasma ghrelin concentration and systemic blood pressure in patients with PIH. Ghrelin concentrations in umbilical artery blood are higher than those in maternal blood normal pregnant women, suggesting that the fetal-placental unit may contribute to its production (336).

No difference between umbilical venous and arterial concentrations exists, suggesting that fetal tissues are a source of ghrelin (337). Small for gestational age neonates have higher umbilical cord ghrelin plasma concentrations than appropriate, and large for gestational age newborns (338).

PARACRINE/AUTOCRINE MECHANISMS REGULATING PLACENTAL HORMONOGENESIS

A major role for the hormones expressed in human placenta is the control of local hormone production through paracrine and/or autocrine mechanisms, since their action may occur in the same tissue where they originate, as well as in the contiguous tissues. The control of local hormonogenesis is one of the major local events that play a role in the physiology of gestational unit, since placental hormones are involved in the endocrine control of the fetoplacental unit. Furthermore, the studies conducted by using primary cultures, cell lines, or tissue explants (incubated or perifused) have suggested that locally produced hormones modulate the release of classic placental peptide and steroid hormones, and that because some of these factors are identical to the hypophysiotropic hormones, a correlation with the regulation of the hypothalamus–pituitary axis may be suggested.

GnRH-hCG-Inhibins-Steroid Hormone Axis

In agreement with the neuroendocrine action on hypothalamic nuclei, depolarizing agents (veratridine or high doses of K) stimulate GnRH release from cultured trophoblast cells, and their effect is reversed by verapamil, cobalt, and EGTA (Ca^{2+} blockers), suggesting that the cell membrane depolarization stimulates the release of placental GnRH by promoting the influx of Ca^{2+} into the cells as occurs in the hypothalamus (339) (Fig. 4). Prostaglandin E_2 (PGE_2) and F_2 (PGF_2), and epinephrine increase the release of GnRH from cultured placental cells, probably acting via the cAMP. The epinephrine-induced GnRH release is reversed by propanol, a β-adrenergic receptor antagonist, suggesting an involvement of β-adrenergic receptors in modulating placental GnRH release (340). Activin A, insulin, and vasoactive intestinal polypeptide (VIP) stimulate the release of GnRH from cultured human placental cells, while the addition of substance P, OT, and NPY have no effects on GnRH release (340,341). Inhibin A, β-END, and dynorphin inhibit GnRH release from placental cells in vitro (340) (Fig. 4). Steroid hormones also modulate GnRH release from human cultured placental cells. In particular, estradiol and estriol potentiate, while progesterone reduces, the action of 8-Br-cAMP on GnRH release from cultured human placental cells (Fig. 4). The role of steroid hormones is also emphasized by the positive action of estriol and the negative effect of progesterone on activin-stimulated GnRH release (341).

Placental GnRH stimulates secretion of hCG in vitro (5), in agreement with the neuroendocrine action of the hypothalamic counterpart. HCG release by first-trimester placenta is pulsatile, and pulses are single and of similar amplitude, with a peak to peak interval of 11–23 min (342). Previous studies have shown that plasma hCG levels in early pregnancy fluctuate periodically (343), reflecting pulsatile secretion of hCG by the placenta. Addition of 1 minute pulses of GnRH analog significantly increases, whereas pulses of progesterone decreases, pulsatile placental hCG secretion (342).

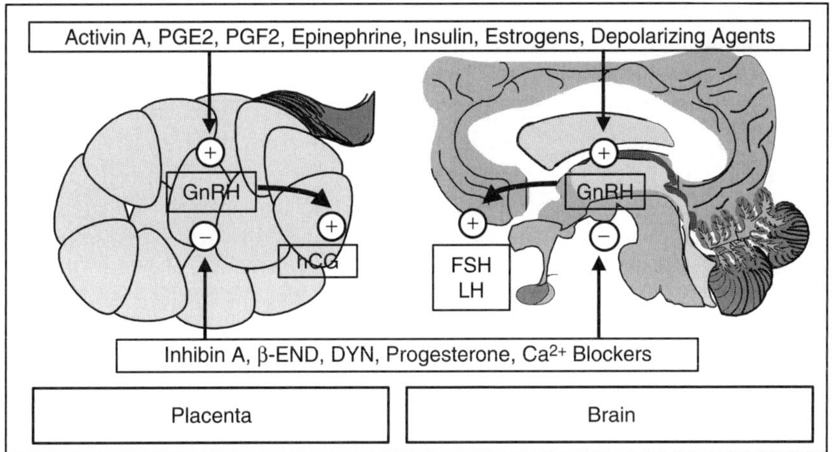

FIG. 4. The mechanisms stimulating GnRH release from medial hypothalamic nuclei are chemically identical to those operating in the human placenta.

Several factors regulating the release of GnRH from hypothalamus also regulates the secretion of placental GnRH, thus having a local role in modulating indirectly the placental hCG release. Indeed, activin A stimulates, while inhibin A inhibits, the placental secretion of hCG, 17β-estradiol and progesterone (340), and their effects on placental hCG release is comparable to pituitary/FSH regulation (Fig. 5). In addition, the cellular co-localization of activin A, inhibin A, and GnRH suggests the occurrence of autocrine events in the regulation of hCG release. The addition of an inhibin antiserum increases the placental hCG release (171,341), while the addition of recombinant inhibin inhibits it secretion (171,341). These effects are mediated at least in part by GnRH, as preincubation with a GnRH antagonist partially reduced the increase of hCG after immunoneutralization of inhibin (171). Activin A increases hCG, and GnRH-induced hCG release from cultured human placental cells collected at first trimester and at term (171,178) (Fig. 5). The effect of activin A on GnRH release is potentiated by estradiol or estriol, reduced by progesterone, and antagonized by tamoxifen or RU486. Progesterone reverses the effect of estriol, thus suggesting that estrogens and progesterone have opposite effects on placental hCG release (Fig. 4). Finally, follistatin inhibits activin A-induced hCG release, as it binds and inactivates activin A (178). Opioid peptides (β-END and DYN) play a role in regulating secretion of hCG release from trophoblast tissue (97): DYN has a significant stimulatory effect on pulsatile hCG secretion in the first trimester placenta cell cultures (344), while β-END in vitro inhibits hCG secretion (345) (Fig. 5).

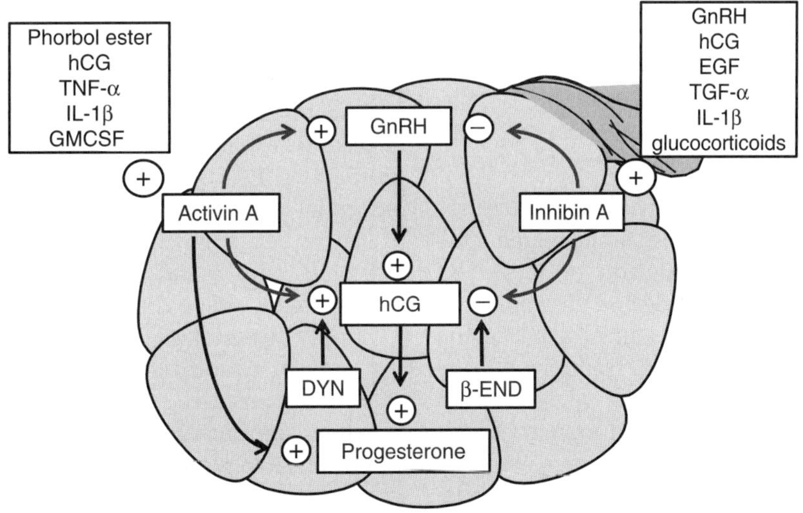

FIG. 5. Factors regulating the release of GnRH, hCG, activin, and inhibin from human placental cells in culture.

With respect to inhibin secretion from cultured placental cells, hCG, PGs, and EGF are potent stimulators, while TGF-β and activin A are suppressors for inhibin production in cultured trophoblast cells (171). Furthermore, the addiction of GnRH and glucocorticoids induces an increase in the release of the inhibin in cultured human trophoblast cells (346). With respect to activin A, the addition of GnRH, inhibin, TGF-β, dexamethasone, cAMP, and interleukin-1α has no effect on its production in cultured trophoblast cells, while phorbol ester (346), tumor necrosis factor (TNF)-α, and interleukin (IL)-1β trigger activin A secretion (Fig. 5).

A limited amount of peptides seems to modulate the release of placental steroid hormones. Increasing doses of GnRH augment placental estrogens and progesterone release (347). The stimulatory effect on progesterone is more pronounced in cultured placental cells of 16/17 weeks of gestation than in placental cultures at term, while the estrone and estradiol release increase only in midgestation cells cultures. The incubation of placental cultures with GnRH antagonist decreases the progesterone and estrogen release (348). A previous observation showed that high doses of GnRH decrease progesterone release from placental explants at various gestational ages, with maximal effect at 12 weeks (349).

A significant dose-related effect of activin A in increasing progesterone release is shown in cultured placental cells at term. The increased release of progesterone in the presence of activin is completely reversed by inhibin (340) or follistatin (178). Nanomolar concentration of inhibin or follistatin block activin-induced progesterone release in cultured of placental cells at term. Both inhibin or follistatin have no significant effect on progesterone release when incubated alone.

The addition of opioids (DYN) significantly decreases progesterone release from placental explants (344).

GHRH-PGH-hPL-IGF Axis

Although the physiological role of placental GHRH is still unknown, it has been suggested that it can play a role in the regulation of fetal GH secretion during the embryonic period as well as in the regulation of placental functions in a paracrine manner (2). A mouse placental cell primary culture system was established to study the regulation of mouse GHRH-IR release. Turnover of mouse GHRH in placental cells was rapid, resulting in a 24-hour media content of 10 times that present in cells. Both 1-oleoyl-2-acetyl-sn-glycerol and 1,2-dioctanoyl-sn-glycerol significantly stimulated the release of mouse GHRH-IR from cultured placental cells into the incubation media but had no effect on total peptide synthesis. These results suggest that the release of mouse GHRH-IR from placental cells is mediated, at least in part, by the activation of protein kinase C (64).

Activin A inhibited and inhibin A activated mGHRH secretion by primary placental cells harvested at day 12 of pregnancy. Follistatin, the binding protein of activin A, completely eliminated the ability of activin to inhibit mGHRH secretion. The steady-state level of mGHRH mRNA, as assessed by *Northern analysis*, was reduced by incubation of placental cells with activin A. TGF-β1 is one of the potent regulators of mouse GHRH secretion with inhibitory effect through autocrine or paracrine mechanisms in mouse placenta in vivo (351).

With respect to PGH, the regulation of its placental secretion is very different from pituitary GH. Indeed, GHRH does not modulate PGH in vivo (352) or in vitro (353), while PGH secretion is inhibited by glucose in vitro (354). In vivo, PGH concentrations decreased during an oral glucose tolerance test in women with gestational diabetes (350), and were increased during a hyperinsulinemic hypoglycemic clamp in pregnant insulin-dependent diabetes mellitus subjects (355), suggesting that the syncytiotrophoblast responds to rapid variations in maternal blood glucose levels by modifying PGH secretion.

The secretion of hPL by the placenta is insensitive to several factors known to affect pituitary GH secretion, since GHRH and somatostatin do not affect its release from placental cells (63,356). Like GH, however, the levels of hPL have been reported to rise following hypoglycemia (357). Indeed, the role of hPL in pregnancy is related to its metabolic properties, rather than somatotropic or lactogenic effects. In the fasting state, the augmented release of hPL coupled with low insulin levels result in increased lipolysis and decreased glucose uptake by the mother, which protects the fetus from hypoglycemia. Thus, the GH-like and anti-insulin effects of hPL would lead to impaired glucose uptake and stimulation of free fatty acids release, with resultant decreased effective insulin. In this manner, maternal metabolism would be directed toward mobilization of maternal sources to furnish substrates for the fetus. There would be a steady source for the fetus, of which glucose would be the major one to ensure an adequate supply of metabolic nutrient.

More substantial effects are observed in the presence of increasing doses of IGF-I (358) or EGF (359), which significantly increase hPL release from human placental explants. Conversely, TGF-β decreases hPL release from cultured placental cells at term. The synthesis of hPL is stimulated by insulin and cAMP. PGE-2α seems to inhibit the hPL secretion (350).

With respect to IGFs, the findings that placental GH and maternal IGF-I levels increase in pregnancies of GH-deficient women (360,361) support the hypothesis of placental GH, and not pituitary GH, being the main regulator of maternal IGF-I during pregnancy. Moreover, cross-sectional studies of a large number of normal and pathological pregnancies have revealed that IGF1 values in the maternal plasma correlate with corresponding hPGH values but not hPL values, regardless of complications and gestational age (362).

CRF-ACTH-Opioid Axis

Some mechanisms stimulating CRF release from medial hypothalamic eminence in the brain are identical to those operating in the human placenta (Fig. 6). In fact, both PGF2 and PGE2 increase the CRF concentration in the culture medium with a dose-dependent effect (34,339). Norepinephrine and acetylcholine are the most active neurotransmitters in increasing CRF release. In particular, the norepinephrine effect is reversed by prazosin, an α1-adrenergic antagonist, or yohimbine, an α2-adrenergic receptor antagonists. The involvement of both adrenergic receptor subtype is further supported by the evidence that methoxamine or clonidine, α1- and α2-adrenergic receptor agonists, respectively, stimulates CRF release from placental cells (363). Acetylcholine acts via a muscarinic receptor: atropine or hexamethonium, specific muscarinic receptor antagonists, reverses the effect of acetylcholine on CRF release. In addition, the human placenta synthesizes acetylcholine and contains acetylcholine concentrations higher than in mammalian brain tissue (1,2). Interestingly, the positive effect of norepinephrine and acetylcholine on placental immunoreactive CRF release agrees with the observation that these neurotransmitters stimulate CRF release from rat hypothalamic tissue in vitro and increase CRF levels in the hypophysial portal circulation (25), suggesting a close correlation between hypothalamic and placental regulation of CRF release (Fig. 6). In agreement with the hypothalamic mechanisms of secretion, some neuropeptides also modulate placental CRF release: angiotensin II and arginine vasopressin increase the release of placental CRF from cultured trophoblasts (363). On the contrary, OT has different effects being inhibitory to CRF/hypothalamus–pituitary–adrenal (HPA) axis (25), and stimulatory on CRF and ACTH secretion from cultured placental cells (34) (Fig. 6).

Placental ACTH, also called chorionic corticotropin (hCC) is a product of the pro-opiomelanocortin (POMC) gene and has the same structure of pituitary ACTH, retaining its immunogenic and biologic activity (25). Placental ACTH is localized to the cytotrophoblast in the first trimester and to the syncytiotrophoblast in the second and third trimesters (364). There is a significant increase of POMC gene expression in the placenta with the advance of gestation, which is manifested by increasing levels of POMC mRNA as well as immunoreactive ACTH (364). The addition of CRF to primary trophoblast cell cultures stimulates ACTH secretion in a dose-dependent manner (34,365), and the concentration

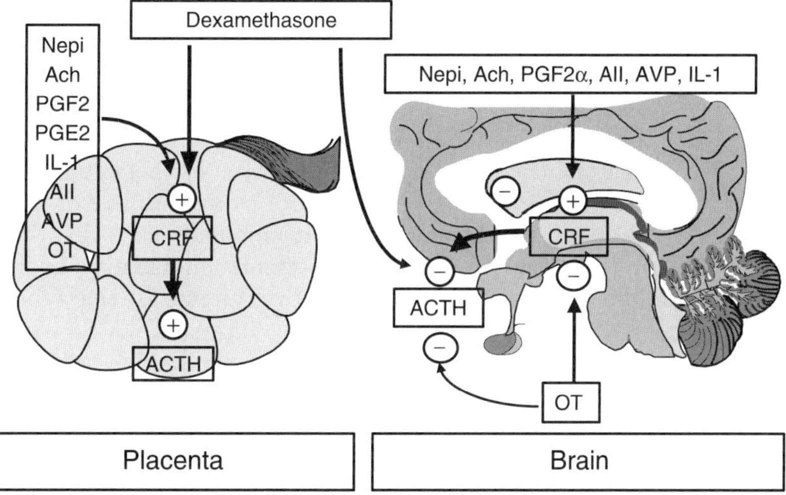

FIG. 6. The mechanisms stimulating CRF release from medial basal hypothalamus are in part chemically identical to those operating in the human placenta. Prostaglandin F2 (PGF2) and E2 (PGE2), norepinephrine (Nepi), acetylcholine (Ach), angiotensin II (AII), arginine vasopressin (AVP), stimulate CRF in hypothalamus, as well as in placental cells. On the contrary, the effect of oxytocin (OT) on CRF and HPA hormones in human placenta, is different being stimulatory. In turns, placental CRF stimulates ACTH secretion from cultured human placental cells.

of CRF required for 50% of maximal stimulation of ACTH secretion is higher than the concentration necessary to release ACTH from cultured anterior pituitary cells (34). Moreover, the addition of a CRF antagonist is able to block the CRF-induced ACTH release from placental cells (34,365). CRF-induced ACTH secretion is mediated by cyclic adenosine monophosphate as second messenger and evidence that this intracellular mechanism operates in placenta comes from the observation that dibutyryl cyclic adenosine monophosphate and forskolin, a diterpene that stimulates adenylate cyclase activity, stimulate ACTH release from cultured trophoblast cells with the same intensity of corticotropin-releasing factor without potentiating the effect of CRF (34).

CRF-BP reverses the CRF-induced ACTH release from placental cells (43), as in the pituitary (25). These findings indicate a similarity between pituitary and placental CRF-induced ACTH release. Furthermore, in contrast to the corticosteroid negative feedback on pituitary ACTH secretion, glucocorticoids stimulate placental CRF secretion and mRNA expression, and dexamethasone does not inhibit the effect of CRF on placental ACTH release (34) (Fig. 6).

Placental CRF is responsible for the high levels of free and bound cortisol circulating in pregnant women (48). However, despite plasma ACTH levels increase throughout pregnancy, they remain within the normal range of nonpregnant women (119), and the injection of exogenous CRF in pregnant women does not induce an increase of circulating ACTH, suggesting that high cortisol levels may desensitize maternal pituitary corticotrophs (366). Moreover, in pregnant women, the diurnal rhythm for plasma ACTH, cortisol, and β-endorphin is maintained, while CRF does not have a circadian rhythm (23,367). These findings, and the fact that the changes of plasma CRF do not correlate with those of ACTH or cortisol throughout normal pregnancy or out of the time of labor (367), support the hypothesis that placental CRF is not the only regulator of maternal ACTH and cortisol levels.

Placental CRF secreted into the fetal circulation may stimulate the production of pituitary ACTH as well as of adrenal hormones. The effect of CRF on fetal pituitary ACTH release is potentiated by arginine vasopressin and possibly mediated by cyclic adenosine monophosphate, and may be antagonized by dexamethasone (25). A direct effect of CRF on DHEA-S release from cultured fetal adrenal cells has been shown (368). Expression of mRNA encoding type-1 CRF receptor is described in midgestation human fetal adrenals (368), suggesting that the fetal adrenal cortex may be directly responsive to CRF. Placenta of humans and higher primates uses DHEA-S supplied by the fetal adrenals as the main substrate for estrogen synthesis, and estrogens produced by the placenta play a pivotal role in the endocrine control of pregnancy and induce many of the key changes involved at parturition (1,2).

Human CRF increased DHEA-S production by cultured human fetal adrenal cortical cells in a dose-dependent fashion, being as effective as ACTH at stimulating DHEA-S production, although it is considerably less potent than ACTH in stimulating cortisol synthesis (249). CRF does not alter cell number, indicating that it is not mitogenic for fetal adrenal cortical cells. Therefore, placental CRF production, which rises exponentially during human pregnancy, may play a key role in promoting DHEA-S production by the fetal adrenals, which could lead to an increase in placental estrogen synthesis.

PLACENTAL HORMONES, CELL GROWTH, AND DIFFERENTIATION

Some of the hormones expressed in human placenta are highly specialized in regulating the cross-talk between the mother and the fetoplacental unit. However, regulation of trophoblast growth and differentiation derives from the harmonized effects of several hormones produced not only by human placenta, but also by maternal decidua, that may stimulate or inhibit placental growth from early to late pregnancy, by influencing the differentiation process from cyto to syncytial cells, and the cells to enter the invasive pathway. This present section will focus only on the hormones produced by human placenta.

Trophoblast Differentiation and Invasion

The human placenta is a highly invasive "tumor-like" structure which invades the uterine endometrium and its vasculature during normal pregnancy in order to establish an exchange of nutrients and other molecules between the fetal and the maternal blood. The availability of in vitro models of human cytotrophoblast cells has allowed to demonstrate that cytotrophoblast cells in anchoring villi either fuse to give rise to the syncytiotrophoblast cells or form columns of EVT, which proliferate, migrate out of chorionic villi, and invade the uterine decidua and the first third of the myometrium reaching the associated spiral arteries (369). This results in the conversion of the narrow caliber arteries into distended uteroplacental arteries which allow increased blood flow to the placenta, thereby supplying oxygen and nutrients to the growing fetus (170).

The control of EVT proliferation and invasion is the result of a complex and coordinated network of several placental hormones, which play a role as stimulating or inhibiting factors. With respect to stimulating factors, IGF-II and IGFBP-1, but not IGF-I, induce EVT cell invasiveness under serum reduced conditions, both in a synergistic manner or independently each other (370) (Fig. 7).

The addition of EGF to first trimester cytotrophoblast cultures has been shown to produce dramatic changes in morphology (extension of pseudopodia and ruffling of the cell membrane, but no syncytium formation) and a several-fold increase in invasive capacity. The effects of EGF on later gestation cytotrophoblasts, whose invasive capacity is diminished, were much less pronounced (371). Human EVT cells also express ETA and ETB receptors, and through the binding to its own receptors, ET-1 has been shown to promote migration, but not proliferation, of EVT (372) (Fig. 7).

The addition of activin A, but not inhibin A, stimulated the differentiation of proliferative cytotrophoblast into extravillous invasive trophoblast cells of the anchoring villi, and also induced the expression of markers characteristic of invasive cytotrophoblast cells, such as metalloproteinase-2. Addition of the activin-binding protein, follistatin, blocked all of the effects of exogenous activin (373). Results of immunocytochemical and RT-PCR analysis have revealed that both protein and mRNA of VEGF, Flt-1, and KDR are expressed by cultured normal EVT cells. Moreover, the addition of VEGF stimulated proliferation, but not migration or invasiveness, of ETV cells

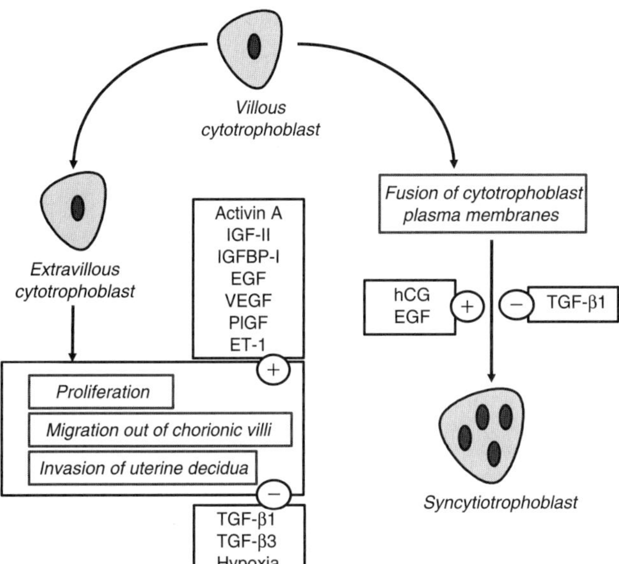

FIG. 7. The control of cytotrophoblast proliferation and differentiation is the result of a complex and coordinated network of several placental hormones, which play a role as stimulating or inhibiting factors. Activin A, IGF-II, IGFBP-I, EGF, VEGF, PIGF, and ET-1 stimulate, and TGF-β1 and -β3 inhibit, extravillous trophoblast proliferation, migration out of chorionic villi and invasion of the uterine decidua. With respect to syncytiotrophoblast formation, hCG and EGF stimulate, while TGF-β1 inhibits, the process of cytotrophoblast aggregation followed by fusion of plasma membranes.

in a concentration-dependent manner, and the effect is abolished with a VEGF-neutralizing antibody (374). The addition of PlGF, a member of the VEGF-family of peptides, stimulates proliferation of first trimester EVT, suggesting that this growth factor and its receptor Flt-1 play a role in the growth and development of the placenta (375) (Fig. 7).

Among the inhibitors, the TGF-β family of peptides is du important modulator of invasion, since TGF-β1 and TGF-β3 inhibit migration and proliferation of EVT (170) (Fig. 7). The human placenta is poorly perfused during the first weeks of pregnancy when trophoblast plugs largely occlude the spiral arteries. These plugs are lost between 9 and 10 weeks of gestation, allowing full maternal blood flow and oxygenation of the placenta (376). The expression of TGF-β3 is under the control of hypoxia inducible factor 1α (HIF-1α), such that hypoxia (as would be found in early pregnancy) induces expression of HIF-1α, which in turn induces expression of TGF-β3 (170). Thus, via HIF-1α and TGF-β3, hypoxia would tend to induce a proliferative rather than an invasive phenotype in trophoblast. In addition, treatment of human villous explants from 5–8 weeks' gestation with antibodies and antisense oligonucleotides to endoglin, the binding protein of TGF-β1 and TGF-β3, stimulated EVT differentiation along the invasive pathway, increasing EVT outgrowth and migration (377).

With respect to syncytiotrophoblast, it is formed from mononuclear cytotrophoblasts through a process of differentiation that involves cellular aggregation followed by fusion of plasma membranes (369). Increasing evidence suggests that the syncytium formation is an autocrine- and paracrine-regulated event involving a number of trophoblast products in placenta (369). Among the trophoblast hormones, perhaps important is hCG, which stimulates the syncytium formation as demonstrated by using biochemical, morphological, and biophysical techniques (378). EGF is another trophoblast product that increases syncytial hormone secretion (hCG and hPL) as well as syncytialization (358), also through the hCG pathway, since preventing the hCG actions blocks the differentiation induced not only by hCG but also by EGF (379). On the contrary, the only inhibitor of this process described has been TGF-β1, which inhibits the EGF-induced syncytial formation and EGF-stimulated secretion of hCG and hPL (380) (Fig. 7).

Placental Angiogenesis

Adequate placental angiogenesis is critical for the establishment of the placental circulation and thus for normal fetal growth and development. Indeed, human gestation is dependent not only on trophoblast differentiation and invasion of maternal decidua, but also on blood vessel growth into the implantation site (through angiogenesis); the development and expansion of villous vasculature; and the remodeling of the maternal uterine circulation near the maternofetal interface.

Little is known on the effects of placental hormones on angiogenesis of human placenta, however, recent observations, including gene knockout studies in mice, have led to the identification of VEGF, FGF, as well as their respective receptors as the major factors regulating the angiogenic process of human placenta (381). VEGF or VEGFR mRNA are associated with the angiogenesis that occurs during development of placental tissues in mice (382). In contrast, TFG-β1 and activin A inhibit both the basal and VEGF-stimulated proliferation of decidual endothelial cells (383). Furthermore, gene knockout studies have shown that in mice, homozygous knockouts of the genes for VEGF receptors (Flt-1 or KDR) led to defects in fetal and placental vasculogenesis and angiogenesis, such as an abnormal vascular formation, organization, and patterning, as well as abnormal endothelial morphology (384,385). Similarly, VEGF-gene knockout homozygous embryos exhibited dramatic cardiovascular defects, such as delayed or abnormal development of the heart, aorta, major vessels, and extra-embryonic vasculature, including the yolk sac and placenta (386,387). Finally, heterozygous VEGF-gene knockout embryos that expressed reduced VEGF levels have defects in fetal and placental angiogenesis (385,387). With respect to FGFs, they stimulate proliferation of both uterine arterial and fetal placental arterial endothelial cells (388).

The remodeling of the maternal uterine circulation has been attributed to the invasive actions of extravillous cytotrophoblast cells, which replace the endothelium of decidual arteries and, therefore, promotes vasodilation. Indeed, adaptation of the uterine vasculature to the rising needs of the fetus occurs through both vasodilation and development of new vessels, and it is a crucial step for normal intrauterine development of the fetus. Recent evidence has suggested that locally produced factors by human placenta may influence angiogenesis of decidual blood vessels. The evidence that: (a) hCG/LH receptors have localized in endothelium and smooth muscle of uterine blood vessels in vivo; (b) their expression level significantly increased in the intramyometrial segment (77,389); and (c) hCG has a direct angiogenic function on hCG/LH receptor-expressing uterine endothelial cells by increasing capillary formation in vitro and on neovascularization (390); together support the hypothesis that hCG may have an important regulatory role in angiogenesis and vascular function in the female reproductive tract, but also in the placental vasculogenesis and angiogenesis in early stages of placental development, as high hCG concentrations were detected in fetal tissue of early pregnancy.

PARACRINE/AUTOCRINE ACTIONS ON FETAL-PLACENTAL BLOOD FLOW REGULATION

The control of fetal-placental blood flow is very important throughout pregnancy, as the nutrients and oxygen to the fetus come from the mother and have to pass the placental barrier. In fact, the fetal vessels of the human placenta are not innerved, so that the control of blood flow in this vascular bed is partly dependent on locally produced and circulating vasoactive factors (391). So, the tight regulation of the blood flow through the uterine arteries (from the mother to the placenta) and the umbilical cord (from the placenta to the fetus) is critical for the growth ad differentiation of the embryo/fetal tissues. In healthy pregnant women an increase in the uterine blood flow and a decrease in uterine vascular resistance are typical features and, the mechanism causing this decreased vascular resistance is probably the result of multiple factors, including a loss of smooth muscle in myometrial resistance vessels (spiral arteries and terminations of radial arteries), an increased angiogenesis, as well as an augmented local uterine artery vasodilation probably related to an increased role of endogenous vasodilators (392).

Effects of CRF and NPY

Several in vitro evidence demonstrated that CRF has vasodilatory effects in a number of species. In fact, both in rat and human i.v. administration of CRF lowers arterial pressure due to peripheral vasodilatation caused by a direct action on vascular smooth muscle (393,394). However, in most animals and in nonpregnant humans, peripheral concentrations of CRF are low (25), which suggest that CRF may play a minimal role in the control of vascular tone. In contrast, in the pregnant human, plasma CRF concentrations rise exponentially, peaking at term (48). CRF, when administered chronically in pregnant rats, causes a decrease in blood pressure (394), and it is also a potent relaxant of the uterine artery of pregnant rats, acting both on the endothelium [mediated by nitric oxide (NO)-cGMP system] and the vascular smooth muscle (395). Animal studies revealed that reduced uterine blood flow and consequently hypoxia induce an increased expression and secretion of CRF, and consequently of ACTH and cortisol (396). As CRF acts as vasodilator in placental circulation, increased CRF could act systemically or be released locally in the placenta as a compensatory mechanisms to reduce uterine resistance to blood flow (397). In humans, an impaired placental CRF secretion has been associated with the lack of uterine artery dilatation and the decrease of the uteroplacental blood flow through uterine arteries, supporting the concept that at mid-gestation placental CRF is involved in the control of the uterine artery tone (398).

Recent investigations have shown that CRF is a potent dilatator of the human fetal-placental circulation (399–401), acting at concentrations comparable to plasma CRF levels in maternal and fetal circulation. Placental CRF may, therefore, have a significant physiological role as a regulator of feto-placental vascular tone. This effect is due to endothelial independent pathways (395), but in some species CRF may also operate via endothelium-dependent mechanism, acting on specific receptors expressed by endothelium, as in the case of HUVEC (402). In the human fetal-placental circulation CRF causes vasodilatation via a NO- and cGMP-mediated pathway, as the addition of a blocker of NO formation and inhibitors of cGMP formation, respectively, cause marked attenuation of CRF-stimulated vasodilatation (Fig. 8). The addition of CRF to preconstricted placental vessels is able to attenuate all constrictor mechanisms without variation in CRF potency as a vasodilator agent. CRF-induced vasodilatation appears to be mediated by a CRF receptor, as the vasodilatory response to CRF is antagonized in the presence of a CRF receptor antagonists (400,401). CRF-induced vasodilatation occurred at concentrations comparable to plasma CRF levels found in the maternal and fetal circulations (46), and CRF is approximately 50 times more potent than prostacyclin (401).

Urocortin has the same effects of CRF: administered i.v in rats is more potent than CRF in causing hypotension (403) and, with respect to placental circulation, it causes vasodilatation (Fig. 8), reducing fetal-placental vascular resistance via CRF type 2 receptors, and being more potent that CRF (404).

As syncytiotrophoblast is the main source of CRF during pregnancy, placental CRF may access the fetal-placental circulation to cause dilatation by paracrine or endocrine mechanisms. It may be released locally to affect the vascular smooth muscle and endothelium, or it may be secreted into the feto-placental circulation and travel to its site of action through the placental vascular system. Finally, CRF may maximize the release of products such as POMC peptides (Fig. 8) or PGs in vivo, by causing vasodilatation of placental vascular tissue (399). NPY is involved in regulation of the local uterine blood flow (405). Since level of NPY in plasma is increased in women with eclampsia, preeclampsia, and hypertension (406), perhaps elevated plasma NPY values may play a key role in the development of these pathologies.

Effects of NKB

In the placental bed, the overall action of NKB is to cause placental vasodilation that is mediated predominantly through the tachykinin NK_1 receptor,

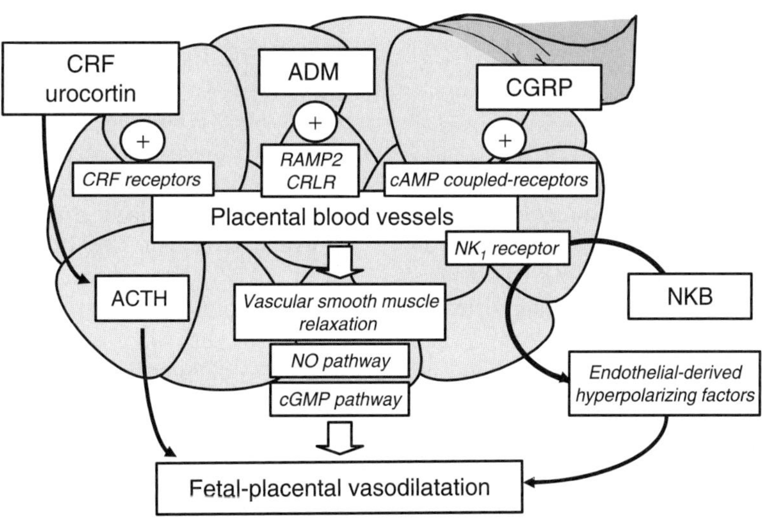

FIG. 8. The control of blood flow from the mother to the fetus through the placenta is mediated by several placental hormones (CRF, urocortin, CGRP, ADM, NKB), that activate several pathways both on placental blood vessels and trophoblast cells.

since the NKB effect was largely diminished response in the presence of the NK_1 receptor antagonist L-732, 138 and no effect with the NK_2 receptor antagonist SR48968. In addition, *RT-PCR* results have shown the predominant expression of NK_1, some expression of NK_2, and absent expression of NK_3 receptor mRNA in the whole placental homogenate. A similar pattern of expression of tachykinin receptor subtypes in the HUVEC cell line suggests a similar role of NKB in the umbilical vein, supporting the notion of a uniform dilatory action of NKB throughout the fetoplacental circulation, downstream of its site of synthesis (151).

In some other vascular beds, vasodilatory effects of NKB are mediated through the release of NO or prostacyclin (407). However, the vasodilatory response of the fetal-placental vasculature was not mediated through NO synthesis, because inhibition of NO synthesis (through the use of L-NAME) failed to inhibit the NKB effect. Similarly, prostacyclin does not appear to mediate the NKB effect, because this was not altered by indomethacin, and similarly, a cocktail of L-NAME and indomethacin failed to inhibit the NKB dilatory response, so mediation of this response is possible through a NO- and prostacyclin-independent mechanism. It remains undetermined whether other endothelial-derived signaling factors might influence smooth muscle relaxation in response to NKB. Therefore, the vasodilatory effect of NKB in the fetal-placental circulation is predominantly due to direct activation of the NK_1 tachykinin receptor, perhaps mediating the release of the putative endothelial-derived hyperpolarizing factors (151) (Fig. 8). Increased syncytiotrophoblast production in preeclampsia could mean that NKB has a significant local role in this condition, vasodilating the fetoplacental circulation in compensation for the reduced blood flow through the uterine circulation (150).

Effects of ET-1

ET-1 is a potent vasoconstrictor of the human fetoplacental vasculature and has the potential to influence placental blood flow either directly or in concert with other vasoactive agents (408). In placental stem villi vessels, which are the major sites of placental vascular resistance, ET-1 exerts a direct vasoconstrictive effect on the smooth muscle layer of these vessels, by binding to its own receptor ETA-R and ETB-R (409). Indeed, significant concentration-dependent increases in fetal-placental perfusion pressure were seen with ET-1 added to single cotyledons of term placentas dually perfused in vitro (410), as well as on uteroplacental arteries (411).

Effects of ADM

ADM is able to decrease vasomotor tone in human fetoplacental vasculature previously constricted with the thromboxane sympathomimetic U46619 in the isolated perfused placental cotyledon (412), and in stem villous arteries dissected and mounted on a wire myograph system (413). This vasodilatation occurs in a dose-dependent manner and suggests that during normal pregnancy, ADM may contribute to the low placental vascular resistance. ADM plays an important role in the regulation of peripheral circulation and blood pressure, directly inducing the relaxation of vascular smooth muscle cells and indirectly acting on endothelial cells through an increase of NO synthesis and release. The localization of ir-ADM at the same sites of NO synthetase in the placenta (275) has suggests that this peptide may also interact with the L-arginine-NO system in regulating placental blood flow. ADM has been shown to exert its effects via stimulation of the NO-cGMP pathway (414) (Fig. 8), activation of adenylate cyclase-camp (415) and through potassium channel activation (413). Furthermore, ADM has been demonstrated to inhibit thrombin- and PDGF-induced ET-1 production (416), through a cAMP-dependent process. These finding suggest that ADM may also modulate vascular tone as a paracrine regulator through the inhibition and/or the stimulation of the vasoactive factors in physiological and pathological state during pregnancy, such as preeclampsia and IUGR.

Effect of CGRP

CGRP is a potent vasodilator in a variety of animal and human systems (291) and, the mode of action of CGRP is still under investigation, as in some tissues it appears to act independently on endothelium, apparently by interacting directly with coupled cAMP receptors on smooth muscle (417), or via mechanisms mediated by the release of vasodilator NO (418). On the basis of this evidence, it has been suggested that CGRP may play a role in the control of vasoactivity of the human fetal-placental circulation (419) (Fig. 8) and, of preconstricted human chorionic plate vessels in vitro through two classes of receptors and also independently of NO pathway (420). CGRP participates in regulation of the vascular adaptations occurring during normal pregnancy, and it appears to be involved in the pathophysiology of preeclampsia (288). In fact, in rats the co-administration of L-NAME (a drug that increases blood pressure) and CGRP prevented the gestational (not the postpartum) L-NAME hypertension and decreased pup mortality to 6.4% but did not reverse the decreased fetal weight (421). The same phenomenon was evident in the presence of adequate levels of progesterone in the postpartum period, as if CGRP regulates vascular adaptations during pregnancy and these effects may be progesterone dependent.

Sex steroid hormones are raised during pregnancy, and the increase of plasma CGRP levels might be related to steroid hormone levels as levels are higher in women than in men, and mainly in women taking contraceptive pills (422). Taken together, we could hypothesize that the relaxant effect of CGRP on the nonpregnant and pregnant uterus, as well as its vasodilatory actions are, at least in part, under progesterone control. In fact, progesterone treatment of ovariectomized mice resulted in a significant increase in the responsiveness of the myometrium to CGRP (423) and, recent studies in female rats indicate that the vasodilator effects of CGRP are sex steroid hormone-dependent, as chronic CGRP administration to pregnant rats increased systolic blood pressure, indicating a role for CGRP in maintaining a normotensive state during pregnancy (424). Furthermore, sex hormones increase both the synthesis of CGRP (425) and the responsiveness to synthetic CGRP in the vasculature (426).

Effect of PTHrP

PTHrP a potent vasodilator (427) through the activation of adenylyl cyclase of myometrial cells, and its expression in vascular smooth muscle cells increases in response to hypertension, vasoconstrictor agents, increased flow, shear stress and stretch. PTHrP has vasodilator effect in the human fetal-placental circulation and is 100 times more potent than PTH (428).

CLINICAL USEFULNESS OF PLACENTAL HORMONES MEASUREMENT THROUGHOUT PREGNANCY

In the last years the understanding of the physiological and pathological functions of human placenta revealed the hypersecretion of hormones in presence of gestational diseases and raised the question whether they are the cause of disorders in pregnancy, or part of an adaptive response of placenta to escape adverse conditions. In any case, there is evidence indicating that placental hormone secretion changes may have a clinical usefulness, since they are measurable in clinically accessible sites (maternal serum, amniotic fluid, saliva, and urine), and may have a value as predictive marker or as prognostic tool.

Ectopic Pregnancy

Ectopic pregnancy causes major maternal morbidity and mortality, with pregnancy loss, and its incidence is increasing worldwide: in Western Countries it increased from 11.2 to 18.8 per 1,000 pregnancies in the last decade (84). Ectopic pregnancy usually present after 7 weeks of amenorrhoea, and the diagnosis can be difficult: any sexually active woman presenting with abdominal pain and vaginal bleeding after an interval of amenorrhoea may have an ectopic pregnancy until proved otherwise. Early diagnosis can be made through serial measurements of hCG, pelvic ultrasonography, measurements of serum progesterone, and uterine curettage; and the accurate diagnosis can be life saving, reduce invasive investigations, and allow conservative treatment. Ectopic pregnancies produce lower concentrations of hCG than normal pregnancies, however, there is still controversy about the concentration of serum hCG that is diagnostic. In the presence of an ectopic mass or fluid in the pouch of Douglas, a cut off point for a serum concentration of hCG of 1,500 IU/L is recommended, but in the absence of any ultrasound signs concentration of 2,000 IU/L should be the cutoff point before an ectopic pregnancy is diagnosed. Absence of an intrauterine gestation with hCG concentrations above this cutoff is diagnostic of an ectopic pregnancy (84).

In a normal pregnancy, serum concentrations of hCG double every 2–3.5 days in the 4th to 8th week of gestation reaching a peak around the 8th to 12th weeks, as calculated from the last menstrual period (20,21). A failure of this increase is suggestive of an ectopic pregnancy although it is also associated with early pregnancy failure and a 2-day sampling interval has been recommended. Progesterone measurements are also useful as a single-time screening device, since a single measurement can exclude ectopic pregnancy with 97.5% sensitivity for concentrations of 79.5 nmol/L (25 ng/mL) or higher (84).

Hydatidiform Mole

Partial hydatidiform mole is a triploid gestational trophoblastic tumor, occurring in about 3 per 1,000 pregnancies (429). Also in the case of this trophoblast disease, the measurement of hCG has reached great relevance not only for establishing the diagnosis, but mainly for determining the response to chemotherapy, and for detecting the rare recurrences. Indeed, hydatidiform mole produces very large amounts of hCG, however, the initial estimation of this placental hormone is not definitive in differentiating between a normal and a molar pregnancy, since the quantity of hCG produced by a normal pregnancy can vary over a wide range (2). As consequence, partial hydatidiform mole is rarely diagnosed in very early pregnancy only according to hCG values, and it is considered when the ultrasound evaluation reveals a mixed pattern of echogenicity (429).

In the last decade it has been shown that some of hormones produced by the human trophoblast are hypersecreted in occurrence of hydatidiform moles (430). However, their measurement did not result to be useful in diagnosing trophoblastic disease, because frequently their serum levels were not elevated when hCG concentrations were already significantly high, or depending on the difficulties encountered in establishing a reliable assay (430). Inhibin A and activin A are expressed by trophoblastic tumors (431) and their detection by immunohistochemistry has been proposed as a useful marker of trophoblastic neoplasia to be included in the diagnostic antibody panel in association with β-hCG (431). In addition, serum levels of activin A and inhibin A in molar pregnancies are high than in normal healthy pregnant controls (seven- to tenfold higher than the 95% confidence interval of values in normal pregnant women), without any considerable overlap with normal pregnancy values, contrary to hCG, thus suggesting to be a reliable tumor markers. Furthermore, serum inhibin A and activin A levels after evacuation decrease to nonpregnant concentrations more rapidly than hCG, showing levels typical of fertile women, since they have a shorter half-life than hCG (432).

Down's Syndrome

Total or partial trisomy of chromosome 21 is responsible for the occurrence of Down's syndrome, which is the most common severe abnormality at birth and has an incidence of 1:700 live births (433). Down's syndrome diagnosis requires invasive testing (chorionic villus sampling in the first trimester, or amniocentesis in the second trimester, or cordocentesis in the second or third trimester), but because they are associated with a risk of pregnancy loss ranging from 0.5% to 1% (433), noninvasive screening tests have acquired much interest. The current approach combines biochemical markers with maternal age (Fig. 9), a variable that is strongly associated with trisomy incidence (433). In fact, in Down's syndrome maternal serum levels of α-fetoprotein (AFP, a glycoprotein produced by the yolk sack and fetal gastrointestinal tract), of unconjugated estriol and of PAPP-A are significantly decreased, while concentrations of hCG and inhibin A are elevated (87).

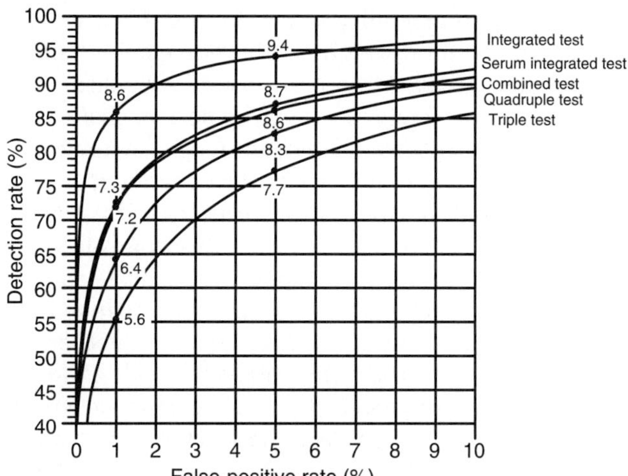

FIG. 9. Down's syndrome detection rates and false-positive rates for specified screening tests. **Triple test:** early second trimester (14–20 completed wk) test based on the measurement of AFP, unconjugated estriol, and hCG together with maternal age. **Quadruple test:** early second trimester (14–20 completed wk) test based on the measurement of AFP, unconjugated estriol, free β-hCG (or total hCG) and inhibin A together with maternal age. **Combined test:** first trimester (10–13 completed wk) test based on combining nuchal translucency measurement with free β-hCG, PAPP-A, and maternal age. **Integrated test:** integration of nuchal translucency measurement and PAPP-A in the first trimester with the Quadruple test in the second. Serum integrated test: a variant of the Integrated test using serum markers only (PAPP-A in the first trimester and the Quadruple test in the second trimester). [Modified from (434).]

The use of maternal age of 35 years or more as the only criterion for performing invasive testing results in identification of only 20% of Down's syndrome pregnancies (433). In the second trimester, the combination of maternal age with AFP elevates the detection rate to approximately 28% (87). The addition of hCG and unconjugated estriol represents a major improvement, as the average detection rate increases to 60% at a fixed false-positive rate of 5% (87). This combination (AFP, hCG, and estriol) is called "triple test." The use of inhibin A as an additional marker can increase substantially the performance of the triple test (434). This combination has been named "quadruple test." Interestingly, the use of β-hCG, AFP, and inhibin A with the exclusion of unconjugated estriol may improve the sensitivity and reduce the false-positive rate of the triple test (433) (Fig. 9). The rate of detection of Down's syndrome increases from 53% to 75% with the same specificity when inhibin A is added to maternal age, AFP, and β-hCG (434,435).

Among the possible markers for first-trimester Down's syndrome screening, the combination of maternal age, PAPP-A, and free β-hCG has produced the best results (436). In this double analyte test, PAPP-A is the most informative marker. In the first trimester ultrasonography scan, embryo nuchal translucency (a physiological space between the back of the fetal neck and the overlying skin) can be evaluated. Several studies have described an association of an increased nuchal translucency thickness with Down's syndrome, with a mean detection rate of 72% (437). Interestingly, the mathematical combination of nuchal translucency ultrasound measurement with maternal age, PAPP-A, and free β-hCG has a hypothetical 80% detection rate for a fixed 5% false-positive rate (87,437) (Fig. 9).

Preeclampsia (PE)

PE is an important cause of fetal and maternal morbidity and mortality, and is clinically defined as hypertension (blood pressure of 140/90 mm Hg or higher) first diagnosed after 20 weeks of gestation and proteinuria (defined as 300 mg or more of urinary protein excretion per 24 h) (438). The etiology of preeclampsia remains uncertain, however recent observations suggest that it originates from an abnormally shallow endovascular cytotrophoblast invasion in the spiral arteries, characterized by an increased apoptotic index of the cytotrophoblast, an exaggerated inflammatory response, an endothelial cell activation, and a relative placental ischemia (439). Another important feature of preeclampsia is that the invasive trophoblasts reach the vicinity of the spiral arteries but fail to penetrate them, precluding the conversion of the spiral arteries into low-resistance channels and inducing the placenta to secrete hypertensive substances (439).

Since trophoblastic abnormalities play a central role in the development of preeclampsia and precede the appearance of clinical signs and symptoms, it is not surprising that some placental hormones change in the maternal circulation, indicating the derangement of placental function. The levels of several placental hormones are elevated in maternal serum long before overt preeclampsia is diagnosed, and these may be considered preclinical manifestations of the earlier stages of the disease. Therefore, such hormones have been proposed as early predictive markers of preeclampsia. The use of hormone markers to predict the risk of preeclampsia may serve more than one objective: first, to be used in a screening program covering all pregnant women indiscriminately; and a second objective may be to identify among high-risk pregnant women, i.e., patients with chronic hypertension or a history of preeclampsia, those with a very low probability of presenting preeclampsia in their present gestation.

HCG, inhibin A, activin A, and CRF are the placental hormones currently investigated as early predictive markers of preeclampsia. There are accumulating data from studies that evaluated whether a single elevated hCG value (usually above 2.0 MoM) between 14 and 24 weeks of gestation is predictive of preeclampsia [reviewed in (87)]. The results of these studies are convergent in suggesting that women with elevated hCG levels in the second trimester are at increased risk for preeclampsia, but there is divergence regarding the accuracy of this test and, by consequence, its predictive value. Indeed, no more than one third of the future cases of preeclampsia can be predicted by a second trimester screening using hCG alone, if one assumes that the false-positive rate must remain within 10%. Lowering the cutoff point from 2.0 to 1.0 MoM would obviously increase the test sensitivity, but the high false-positive rate would render the test useless as an indicator for prophylactic intervention (87,183). Since preeclampsia has a relatively low prevalence in most populations, any new screening test should have a high specificity and positive predictive value to identify only high-risk women and avoid the emotional and social costs imposed by false-positive results.

Furthermore, an other important consideration is related to the fact that values lower than 2.0 MoM (which are detected in 80%–90% of all primigravidas) indicate a very low (<4%) probability of developing preeclampsia, which at first glance may sound useful to predict a good outcome after a negative test. This finding merits cautious interpretation because all studies have been performed on populations whose prevalence of preeclampsia was lower than 5%, and it is probable that in selected populations with a higher prevalence or with specific risk factors the probability of preeclampsia would increase even among subjects with a normal hCG screening test.

The use of activin A and inhibin A is the most recent proposed as useful markers to predict the risk of preeclampsia (87,183) both in low- and high-risk pregnant women, i.e., patients with chronic hypertension or a history of preeclampsia, those with a very low probability of presenting preeclampsia in their present gestation. With respect to the prediction of preeclampsia in low-risk populations, inhibin A and activin A have limited sensitivity and low positive predictive value (87,183). Indeed, inhibin A is elevated several weeks before the onset of clinical signs of preeclampsia and women who had at midtrimester inhibin A levels exceeding 2.0 MoM were more likely to develop preeclampsia: the performance (sensitivity) of that measurement in predicting the onset of the disease ranges from 18% to 48.6% (95% coefficient interval) (87,183). The addition of hCG data to inhibin A does not improve the sensitivity for preeclampsia compared to inhibin A alone (87). Interestingly, inhibin A levels tend to be higher when the onset of preeclampsia occur within a shorter interval after collection of the second-trimester screening sample. These observations suggested that second trimester inhibin A would be more effective in predicting early-onset than later onset disease (87,440,441). Compared to inhibin A, activin A seems to be a more sensitive marker at mid-trimester (441), since in a low-risk population at 15 to 19 weeks its concentrations could discriminate preeclampsia with a sensitivity of 41%, and of 59% at 21 to 25 weeks (441).

With respect to high-risk population, the contribution of inhibin A and activin A measurement to the screening may be modest if it does not accumulate other risk factors, such as impaired uterine arteries blood flow. Indeed, an important feature of preeclampsia is that the invasive trophoblasts reach the vicinity of the spiral arteries but fail to penetrate them, precluding the conversion of the spiral arteries into low-resistance channels (442). The increased vascular resistance is responsible for the typical modifications in the Doppler waveforms and velocimetry of the uterine arteries (442): between 20 and 24 weeks of gestation, the presence of an early diastolic notch and a mean resistance index of 0.65 or greater for the unilateral notch and 0.55 or greater for the bilateral notch has an approximately 65% sensitivity and a false-positive rate of 11% (442). The measurement of maternal serum activin A and inhibin A in women with altered uterine artery Doppler velocimetry at 24 weeks of gestation provides clinically relevant information for the prediction of preeclampsia. Indeed, activin A and inhibin A levels in patients who develop preeclampsia are higher than in those who do not present with preeclampsia at follow-up: activin A at the cutoff 1.7 MoM achieves a sensitivity of 61% and a specificity of 89%, whereas inhibin A at the cutoff 1.8 MoM combines a sensitivity of 39% with a specificity of 92% for prediction of preeclampsia. The combination of both hormones improves the predictive value of the test, since the probability of preeclampsia was 86% and 17% when both activin A and inhibin A levels are elevated or unaltered, respectively (443).

The evidence that increased CRF and decreased CRF-BP levels circulate in maternal plasma of pregnancies complicated by preeclampsia has prompted to investigate whether their evaluation may predict the third trimester onset preeclampsia at-risk pregnant women, consecutively enrolled between 28–29 weeks of pregnancy. CRF and CRF-BP levels are significantly higher and lower, respectively, in the patients who later develop PE than in those who to not present PE at follow-up. CRF at the cutoff 425.95 pmol/L achieves a sensitivity of 94.8% and a specificity of 96.9%, whereas CRF-BP at the cutoff 125.8 nmol/L combines a sensitivity of 92.5% and a specificity of 82.5% as single markers for prediction of PE. The probability of PE was 34.5% in the whole study population, 93.75% when both CRF and CRF-BP levels are changed, and 0% if both hormone markers are unaltered (444). Measuring CRF and CRF-BP levels yields a positive, but overall a negative, predictive value that differs materially from the overall prevalence of PE in the study population. This is an important point of discussion because the use of tests that achieve a high negative predictive value (100%) may prevent unnecessary hospitalization and testing, eliminating unnecessary interventions and providing reassurance for the patient.

Fetal Growth Restriction (FGR)

FGR is a heterogeneous pathology characterized by a disrupted fetal growth, and caused by multiple factors of fetal, placental, and maternal origin, and it has an incidence of 3%–10% in developed countries and 6%–30% in the developing world. Traditionally, working definitions of FGR have been birth weight below 2 SDS of the mean or less than the 10th percentile for the same gestational age (445).

Since FGR is strongly associated with neonatal death, necrotizing enterocolitis, and respiratory distress syndrome, and has a less strict association with intraventricular hemorrhage (445), and it is a significant risk factor for the subsequent development of ischemic heart disease, chronic hypertension, obstructive lung disease, and diabetes in adult life (446), there continues to be a compelling demand for screening tools to predict this heterogeneous pathology.

Placental hormones have been investigated with this purpose, and for a long period in the history of obstetrics, maternal serum and urinary estriol and hPL levels have been tools for the monitoring of fetal welfare and fetal growth (87), and a reduction of their levels was associated with FGR. Indeed, decreased second trimester unconjugated estriol levels (<0.75 MoM) are significantly associated with FGR, and in urine estriol below the 10th percentile is found in about 30% of FGR pregnancies (87).

With respect to hPL, low maternal serum levels have been associated with FGR (87). Thus, since declining levels of hPL would be an early sign of impaired placental function and chronic fetal distress, serum hPL was proposed in the 1970s as a screening test to detect and monitor pregnancies at risk, particularly those with FGR. However, a major limitation of maternal hPL measurement in the screening of FRG is the fact that it has low if any predictive value in the first or second (87) trimester.

In the first trimester women with serum hCG levels below the 10th percentile have an increased risk of subsequent FGR (447). With respect to the association between hCG values and FGR in the second-trimester, it is a matter of controversy (87), probably because FGR is closely associated with preeclampsia, and the latter is a strong confounding factor since the pathological features of preeclampsia include an excessive placental secretion of hCG.

Despite maternal serum and placental GH levels are lower in FGR pregnancies (112), with an average reduction of 50% in the third trimester, its importance in clinical practice is still limited. On the contrary, gene knockout mice have shown the importance of IGF-I and IGF-II related to fetal growth, since disruption of either the IGF-I or IGF-II (190) gene determines severe FGR. Furthermore, both maternal and fetal IGF-I influence placental metabolism and thereby regulate the availability of substrates required for fetal growth, and therefore maternal IGF-I may influence fetal growth only affecting placental function (190). However, a consistent relationship between maternal IGF-I levels and fetal birth weight has not been established. On the contrary, the measurement of IGF-I and IGFBP-1 in cord blood obtained by cordocentesis during the third trimester increases the predictive value of ultrasound in estimating the weight of fetuses affected by FGR, since infant birth weight is positively correlated with cord blood IGF-I concentrations and is inversely correlated with IGFBP-1 (190).

Women with a PAPP-A in the lowest fifth percentile at 8- to 14-week gestation had an increased risk of intrauterine growth restriction [adjusted odds ratio (AOR): 2.9], extremely premature delivery (AOR: 2.9), moderately premature delivery (AOR: 2.4), preeclampsia (AOR: 2.3), and stillbirth (AOR: 3.6), and the strengths of the associations were similar when the test was performed before 13-week gestation or between 13- and 14-week gestation (448).

With respect to leptin, it does not seem to be a suitable antenatal marker of FGR. Indeed, despite umbilical cord serum leptin levels are positively correlated with fetal birth weight (325,326), fetal leptin levels assessed by cordocentesis do not reveal differences between normal and growth restricted fetuses, at least before 34 weeks (330); no significant correlation exists between maternal and fetal leptin levels (330); and maternal leptin concentrations do not predict fetal birth weight.

Preterm Delivery

Preterm delivery is defined as delivery occurring before 37 completed gestational weeks. It complicates nearly 10% of all births and accounts for approximately 70% of all neonatal deaths throughout the world (246). The cascade of parturition, both term and preterm, is activated by endocrine (maternal and fetal) as well as local factors from the uteroplacental unit. Preterm labor is a syndrome that may be initiated by many causes, representing either a breakdown in the mechanisms responsible for maintaining uterine quiescence or an overwhelming of the normal parturition cascade.

Many lines of evidence support the involvement of placental CRF in the mechanisms controlling the onset of labor, either term or preterm (246). Mean maternal plasma CRF levels at 18–20 weeks of gestation are threefold higher in the group of women who subsequently has spontaneous preterm delivery compared with matched controls (449). A large cohort study performed in Hong Kong followed more than 1,000 low-risk women starting at 15–20 weeks of gestation (450). CRF levels resulted higher in the group who has delivered before 34 weeks but not in the whole preterm group, and its positive predictive value is only 3.6% for an outcome whose prevalence in the study population is 1.1%.

The value of mid-trimester CRF measurement to predict the risk of preterm birth (either spontaneous or planned) has been evaluated alone and in combination with AFP and a clinical risk factor score. CRF alone had a positive predictive value of 27%, a fourfold increase compared with the 7.3% prevalence of preterm birth in the study population. The combination of CRF with AFP and a risk factor score enhances the sensitivity of all single markers without losing specificity (451).

The measurement of maternal serum CRF does not satisfy the requisites of a screening test for preterm delivery in a low-risk population, although it should be considered as a potential marker to be used in populations with a higher prevalence of the problem.

REFERENCES

1. Petraglia, F., Florio, P., Nappi, C., and Genazzani, A. R. (1996). Peptide signaling in human placenta and membranes: autocrine, paracrine, and endocrine mechanisms. *Endocr. Rev.* 17, 156–186.
2. Reis, F. M. and Petraglia, F. (2001). The placenta as a neuroendocrine organ. *Front. Horm. Res.* 27, 216–228.
3. Stojilkovic, S. S., Reinhart, J., and Catt, K. J. (1994). Gonadotropin-releasing hormone receptors: structure and signal transduction pathways. *Endocr. Rev.* 15, 462–499.
4. Khodr, G. S. and Siler-Khodr, T. M. (1978). Localization of luteinizing hormone-releasing factor in the human placenta. *Fertil. Steril.* 29, 523–530.
5. Khodr, G. S. and Siler-Khodr, T. M. (1980). Placental luteinizing hormone-releasing and its synthesis. *Science* 207, 315–317.

6. Tan, L. and Rousseau, P. (1982). The chemical identity of the immunoreactive LHRH-like peptide biosynthesized in the human placenta. *Biochem. Biophys. Res. Commun.* 109, 1061–1071.
7. Wolfahrt, S., Kleine, B., and Rossmanith, W. G. (1998). Detection of gonadotrophin releasing hormone and its receptor mRNA in human placental trophoblasts using in-situ reverse transcription-polymerase chain reaction. *Mol. Hum. Reprod.* 4, 999–1006.
8. Miyake, A., Sakumoto, T., Aono, T., Kawamura, Y., Maeda, T., and Kurachi, K. (1982). Changes in luteinizing hormone-releasing hormone in human placenta throughout pregnancy. *Obstet. Gynecol.* 60, 444–449.
9. Seppala, M., Wahlstrom, T., Lehtovirta, P., Lee, J. N., and Leppalouto, J. (1980). Immunohistochemical demonstration of luteinizing hormone-releasing factor-like material in human syncytiotrophoblast and trophoblastic tumours. *Clin. Endocrinol. (Oxf)* 12, 441–451.
10. Siler-Khodr, T. M. and Khodr, G. S. (1978). Content of luteinizing hormone-releasing factor in the human placenta. *Am. J. Obstet. Gynecol.* 130, 216–219.
11. Kelly, A. C., Rodgers, A., Dong, K. W., Barrezueta, N. X., Blum, M., and Roberts, J. L. (1991). Gonadotropin-releasing hormone and chorionic gonadotropin gene expression in human placental development. *DNA Cell Biol.* 10, 411–421.
12. Leung, P. C. and Peng, C. (1996). Gonadotropin-releaing hormone receptor: gene structure, expression and regulation. *Biol. Signals* 5, 63–69.
13. Lin, L. S., Roberts, V. J., and Yen, S. S. (1995). Expression of human gonadotropin-releasing hormone receptor gene in the placenta and its functional relationship to human chorionic gonadotropin secretion. *J. Clin. Endocrinol. Metab.* 80, 580–585.
14. Boyle, T. A., Belt-Davis, D. I., and Duello, T. M. (1998). Nucleotide sequence analyses predict that human pituitary and human placental gonadotropin-releasing hormone receptors have identical primary structures. *Endocrine* 9, 281–287.
15. Yin, H., Cheng, K. W., Hwa, H. L., Peng, C., Auersperg, N., and Leung, P. C. (1998). Expression of the messenger RNA for gonadotropin-releasing hormone and its receptor in human cancer cell lines. *Life Sci.* 62, 2015–2023.
16. Lamharzi, N., Halmos, G., Armatis, P., and Schally, A. V. (1998). Expression of mRNA for luteinizing hormone-releasing hormone receptors and epidermal growth factor receptors in human cancer cell lines. *Int. J. Oncol.* 12, 671–675.
17. Cheng, K. W., Nathwani, P. S., and Leung, P. C. (2000). Regulation of human gonadotropin-releasing hormone receptor gene expression in placental cells. *Endocrinology* 141, 2340–2349.
18. Kang, S. K., Tai, C. J., Cheng, K. W., and Leung, P. C. (2000). Gonadotropin-releasing hormone activates mitogen-activated protein kinase in human ovarian and placental cells. *Mol. Cell. Endocrinol.* 170, 143–151.
19. Bramley, T. A., McPhie, C. A., and Menzies, G. S. (1992). Human placental gonadotrophin-releasing hormone (GnRH) binding sites: I. Characterization, properties and ligand specificity. *Placenta* 13, 555–581.
20. Bramley, T. A., McPhie, C. A., and Menzies, G. S. (1994). Human placental gonadotrophin-releasing hormone (GnRH) binding sites: III. Changes in GnRH binding levels with stage of gestation. *Placenta* 15, 733–745.
21. Iwashita, M., Kudo, Y., Shinozaki, Y., and Takeda, Y. (1993). Gonadotropin-releasing hormone increases serum human chorionic gonadotropin in pregnant women. *Endocr. J.* 40, 539–544.
22. Siler-Khodr, T. M., Khodr, G. S., and Valenzuela, G. (1984). Immunoreactive gonadotropin-releasing hormone level in maternal circulation throughout pregnancy. *Am. J. Obstet. Gynecol.* 150, 376–379.
23. Petraglia, F., Genazzani, A. D., Aguzzoli, L., Gallinelli, A., De Vita, D., Caruso, A., and Genazzani, A. R. (1994). Pulsatile fluctuations of plasma-gonadotropin-releasing hormone and corticotropin-releasing factor levels in healthy pregnant women. *Acta Obstet. Gynecol. Scand.* 73, 284–289.
24. Gohar, J., Mazor, M., and Leiberman, J. R. (1996). GnRH in pregnancy. *Arch. Gynecol. Obstet.* 259, 1-6.
25. Bale, T. L. and Vale, W. W. (2004). CRF and CRF receptors: role in stress responsivity and other behaviors. *Annu. Rev. Pharmacol. Toxicol.* 44, 525–557.
26. Warren, W. B. and Silverman, A. J. (1995). Cellular localization of corticotrophin releasing hormone in the human placenta, fetal membranes and decidua. *Placenta* 16, 147–156.
27. Petraglia, F., Tabanelli, S., Galassi, M. C., Garuti, G. C., Mancini, A. C., Genazzani, A. R., and Gurpide, E. (1992). Human decidua and in vitro decidualized endometrial stromal cells at term contain immunoreactive corticotropin–releasing factor (CRF) and CRF messenger ribonucleic acid. *J. Clin. Endocrinol. Metab.* 74, 1427–1431.
28. Shibasaki, T., Odagiri, E., Shizume, K., and Ling, N. (1982). Corticotropin releasing factor-like activity in human placental extracts. *J. Clin. Endocrinol. Metab.* 55, 384–386.
29. Sasaki, A., Tempst, P., Lotta, A. S., Margioris, A. N., Hood, L. E., Kent, S. B., Sato, S., Shinkawa, O., Yoshinaga, K., and Krieger, D. T. (1988). Isolation and characterization of a corticotropin-releasing hormone-like peptide from human placenta. *J. Clin. Endocrinol. Metab.* 67, 768–773.
30. Schulte, H. M. and Healy, D. L. (1987). Corticotropin releasing hormone- and adreno-corticotropin-like immunoreactivity in human placenta, peripheral and uterine vein plasma. *Horm. Metab. Res.* 16, 44–46.
31. Frim, D. M., Emanuel, R. L., Robinson, B. G., Smas, C. M., Adler, G. K., and Majzoub, J. A. (1988). Characterization and gestational regulation of corticotropin-releasing hormone messenger RNA in human placenta. *J. Clin. Invest.* 82, 287–292.
32. Grino, M., Chrousos, G. P., and Margioris, A. N. (1987). The corticotropin releasing hormone gene is expressed in human placenta. *Biochem. Biophys. Res. Commun.* 148, 1208–1214.
33. Warren, W. B. and Silverman, A. J. (1995). Cellular localization of corticotrophin releasing hormone in the human placenta, fetal membranes and decidua. *Placenta* 16, 147–156.
34. Petraglia, F., Sawchenko, P. E., Rivier, J., and Vale, W. (1987). Evidence for local stimulation of ACTH secretion by corticotropin-releasing factor in human placenta. *Nature* 328, 717–719.
35. Riley, S. C. and Challis, J. R. (1991). Corticotropin-releasing hormone production by the placenta and fetal membranes. *Placenta* 12, 105–119.
36. Florio, P., Rossi, M., Sigurdardottir, M., Ciarmela, P., Luisi, S., Vigano, P., Grasso, D., Fiore, G., Cobellis, L., Di Blasio, A. M., and Petraglia, F. (2003). Paracrine regulation of endometrial function: interaction between progesterone and corticotropin-releasing factor (CRF) and activin A. *Steroids* 68, 801–807.
37. Ferrari, A., Petraglia, F., and Gurpide, E. (1995). Corticotropin releasing factor decidualizes human endometrial stromal cells in vitro. Interaction with progestin. *J. Steroid Biochem. Mol. Biol.* 54, 251–255.
38. Ahmed, I., Glynn, B. P., Perkins, A. V., Castro, M. G., Rowe, J., Morrison, E., and Linton, E. A. (2000). Processing of procorticotropin-releasing hormone (pro-CRH): molecular forms of CRH in normal and preeclamptic pregnancy. *J. Clin. Endocrinol. Metab.* 85, 755–764.
39. Goland, R. S., Conwell, I. M., and Jozak, S. (1995). The effect of pre-eclampsia on human placental corticotrophin-releasing hormone content and processing. *Placenta* 16, 375–382.
40. Vaughan, J., Donaldson, C., Bittencourt, J., Perrin, M. H., Lewis, K., Sutton, S., Chan, R., Turnbull, A. V., Lovejoy, D.,

Rivier, C., and Vale, W. W. (1995). Urocortin, a mammalian neuropeptide related to fish urotensin I and to corticotropin-releasing factor. *Nature* 378, 287–292.

41. Petraglia, F., Florio, P., Gallo, R., Simoncini, T., Giuntini, A., and Genazzani, A. R. (1996). Human placenta and fetal membranes express human urocortin mRNA and peptide. *J. Clin. Endocrinol. Metab.* 81, 3807–3810.
42. Florio, P., Rivest, S., Reis, F. M., Simoncini, T., Martinelli, P., Genazzani, A. R., and Petraglia, F. (1999). Lack of gestational-related changes of urocortin gene expression in human placenta. *Prenat. Neonat. Med.* 4, 296–300.
43. Petraglia, F., Florio, P., Gallo, R., Salvestroni, C., Lombardo, M., Genazzani, A. D., Di Carlo, C., Stomati, M., D'Ambrogio, G., and Artini, P. G. (1996). Corticotropin-releasing factor-binding protein: origins and possible functions. *Horm. Res.* 45, 187–191.
44. Petraglia, F., Potter, E., Cameron, V. A., Sutton, S., Behan, D. P., Woods, R. J., Sawchenko, P. E., Lowry, P. J., and Vale, W. (1993). Corticotropin-releasing factor-binding protein is produced by human placenta and intrauterine tissues. *J. Clin. Endocrinol. Metab.* 77, 919–924.
45. Karteris, E., Grammatopoulos, D., Dai, Y., Olah, K. B., Ghobara, T. B., Easton, A., and Hillhouse, E. W. (1998). The human placenta and fetal membranes express the corticotropin-releasing hormone receptor 1 alpha (CRH-1 alpha) and the CRH-C variant receptor. *J. Clin. Endocrinol. Metab.* 83, 1376–1379.
46. Florio, P., Franchini, A., Reis, F. M., Pezzani, I., Ottaviani, E., and Petraglia, F. (2000). Human placenta, chorion, amnion and decidua express different variants of corticotropin-releasing factor receptor messenger RNA. *Placenta* 21, 32–37.
47. Hillhouse, E. W., and Grammatopoulos, D. K. (2002). Role of stress peptides during human pregnancy and labour. *Reproduction* 124, 323–329.
48. Florio, P. Severi, F. M., Ciarmela, P., Fiore, G., Calonaci, G., Merola, A., De Felice, C., Palumbo, M., and Petraglia, F. (2002). Placental stress factors and maternal-fetal adaptive response: the corticotropin-releasing factor family. *Endocrine* 19, 91–102.
49. Economides, D., Linton, E., Nicolaides, K., Rodeck, C. H., Lowry, P. J., and Chard, T. (1987). Relationship between maternal and fetal corticotrophin-releasing hormone-41 and ACTH levels in human mid-trimester pregnancy. *J. Endocrinol.* 114, 497–501.
50. Goland, R. S., Wardlaw, S. L., Blum, M., Tropper, P. J., and Stark, R. I. (1988). Biologically active corticotropin-releasing hormone in maternal and fetal plasma during pregnancy. *Am. J. Obstet. Gynecol.* 159, 884–890.
51. Laatikainen, T. J., Raisanen, I. J., and Salminen, K. R. (1988). Corticotropin-releasing hormone in amniotic fluid during gestation and labor and in relation to fetal lung maturation. *Am. J. Obstet. Gynecol.* 159, 891–895.
52. Petraglia, F., Giardino, L., Coukos, G., Calza, L., Vale, W., and Genazzani, A. R. (1990). Corticotropin-releasing factor and parturition: plasma and amniotic fluid levels and placental binding sites. *Obstet. Gynecol.* 75, 784–790.
53. Korebrits, C., Ramirez, M. M., Watson, L., Brinkman, E., Bocking, A. D., and Challis, J. R. (1998). Maternal corticotropin-releasing hormone is increased with impending preterm birth. *J. Clin. Endocrinol. Metab.* 83, 1585–1591.
54. Petraglia, F., Florio, P., Benedetto, C., Gallo, C., Woods, R. J., Genazzani, A. R., and Lowry, P. J. (1996). High levels of corticotropin-releasing factor (CRF) are inversely correlated with low levels of maternal CRF-binding protein in pregnant women with pregnancy-induced hypertension. *J. Clin. Endocrinol. Metab.* 81, 852–856.
55. Laatikainen, T., Virtanen, T., Kaaja, R., and Salminen-Lappalainen, K. (1991). Corticotropin-releasing hormone in maternal and cord plasma in pre-eclampsia. *Eur. J. Obstet. Gynecol. Reprod. Biol.* 39, 19–24.
56. Linton, E. A., Perkins, A. V., Woods, R. J., Eben, F., Wolfe, C. D., Behan, D. P., Potter, E., Vale, W. W., and Lowry, P. J. (1993). Corticotropin releasing hormone-binding protein (CRH-BP): plasma levels decrease during the third trimester of normal human pregnancy. *J. Clin. Endocrinol. Metab.* 76, 260–262.
57. Petraglia, F., Florio, P., Simoncini, T., Woods, R. J., Giuntini, A., Gremigni, R., Serra, G. B., Genazzani, A. R., and Lowry, P. J. (1997). Cord plasma corticotropin-releasing factor-binding protein (CRF-BP) in term and preterm labour. *Placenta* 18, 115–119.
58. Florio, P., Woods, R. J., Genazzani, A. R., Lowry, P. J., and Petraglia, F. (1997). Changes in amniotic fluid immunoreactive corticotropin-releasing factor (CRF) and CRF-binding protein levels in pregnant women at term and during labor. *J. Clin. Endocrinol. Metab.* 82, 835–838.
59. Berkowitz, G. S., Lapinski, R. H., Lockwood, C. J., Florio, P., Blackmore Prince, C., and Petraglia, F. (1996). Corticotropin-releasing factor and its binding protein: maternal serum levels in term and preterm deliveries. *Am. J. Obstet. Gynecol.* 174, 1477–1478.
60. Glynn, B. P., Wolton, A., Rodriguez-Linares, B., Phaneuf, S., and Linton, E. A. (1998). Urocortin in pregnancy. *Am. J. Obstet. Gynecol.* 179, 533–539.
61. Florio, P., Cobellis, L., Woodman, J., Severi, F. M., Linton, E. A., and Petraglia, F. (2002). Levels of maternal plasma corticotropin-releasing factor and urocortin during labor. *J. Soc. Gynecol. Investig.* 9, 233–237.
62. Sassolas, G. (2000). Growth hormone-releasing hormone: past and present. *Horm. Res.* 53, 88–92.
63. Berry, S. A., Srivastava, C. H., Rubin, L. R., Phipps, W. R., and Pescovitz, O. H. (1992). Growth hormone-releasing hormone-like messenger ribonucleic acid and immunoreactive peptide are present in human testis and placenta. *J. Clin. Endocrinol. Metab.* 75, 281–284.
64. Mizobuchi, M., Downs, T. R., and Frohman, L. A. (1995). Growth hormone-releasing hormone immunoreactivity in mouse placenta, maternal blood, and amniotic fluid: molecular characterization and secretion from primary cell cultures in vitro. *Endocrinology* 136, 1731–1736.
65. Margioris, A. N., Brockmann, G., Bohler, H. C., Jr., Grino, M., Vamvakopoulos, N., and Chrousos, G. P. (1990). Expression and localization of growth hormone-releasing hormone messenger ribonucleic acid in rat placenta: in vitro secretion and regulation of its peptide product. *Endocrinology* 126, 151–158.
66. Hofland, L. J. and Lamberts, S. W. (2003). The pathophysiological consequences of somatostatin receptor internalization and resistance. *Endocr. Rev.* 24, 28–47.
67. Kumasaka, T., Nishi, N., Yaoi, Y., Kido, Y., Saito, M., Okayasu, I., Shimizu, K., Hatakeyama, S., Sawano, S., and Kokubu, K. (1979). Demonstration of immunoreactive somatostatin-like substance in villi and decidua in early pregnancy. *Am. J. Obstet. Gynecol.* 134, 39–44.
68. Watkins, W. B. and Yen, S. S. (1980). Somatostatin in cytotrophoblast of the immature human placenta: localization by immunoperoxidase cytochemistry. *J. Clin. Endocrinol. Metab.* 50, 969–971.
69. Saito, H., Saito, S., Sano, T., Hosoi, E., and Saito, H. (1983). Fetal and maternal plasma levels of immunoreactive somatostatin at delivery: evidence for its increase in the umbilical artery and its arterio-venous gradient in the feto-placental circulation. *J. Clin. Endocrinol. Metab.* 56, 567–571.
70. Caron, P., Buscail, L., Beckers, A., Esteve, J. P., Igout, A., Hennen, G., and Susini, C. (1997). Expression of somatostatin receptor SST4 in human placenta and absence of octreotide effect on human placental growth hormone concentration during pregnancy. *J. Clin. Endocrinol. Metab.* 82, 3771–3776.

71. Strazzulla, G., Guazzelli, R., Romano, S., Matassi, L., Cotrozzi, G., and Brocchi, A. (1990). Circadian variations of plasma somatostatin levels in healthy subjects. *Chronobiologia* 17, 219–225.
72. Saito, H., Saito, S., Sano, T., Hosoi, E., and Saito, H. (1983). Fetal and maternal plasma levels of immunoreactive somatostatin at delivery: evidence for its increase in the umbilical artery and its arterio-venous gradient in the feto-placental circulation. *J. Clin. Endocrinol. Metab.* 56, 567–571.
73. Sann, L., Chayvialle, J. A., and Descos, F. (1982). Plasma somatostatin concentration in the preterm neonate. *Eur. J. Pediatr.* 139, 148–150.
74. Goldstein, A., Armony-Sivan, R., Rozin, A., and Weller, A. (1995). Somatostatin levels during infancy, pregnancy, and lactation: a review. *Peptides* 16, 1321–1326.
75. Birken, S., Berger, P., Bidart, J. M., Weber, M., Bristow, A., Norman, R., et al. (2003). Preparation and characterization of new WHO reference reagents for human chorionic gonadotropin and metabolites. *Clin. Chem.* 49, 144–154.
76. Eldar-Geva, T., Hochberg, A., deGroot, N., and Weinstein, D. (1995). High maternal serum chorionic gonadotropin level in Downs' sindrome pregnancies is caused by elevation of both subunits messenger ribonucleic acid level in trophoblasts. *J. Clin. Endocrinol. Metab.* 80, 3528–3531.
77. Reshef, E., Lei, Z. M., Rao, C. V., Pridham, D. D., Chegini, N., and Luborsky, J. L. (1990). The presence of gonadotropin receptors in nonpregnant human uterus, human placenta, fetal membranes, and decidua. *J. Clin. Endocrinol. Metab.* 70, 421–430.
78. Davies, S., Byrn, F., and Cole, L. A. (2003). Human chorionic gonadotropin testing for early pregnancy viability and complications. *Clin. Lab. Med.* 23, 257–264.
79. Cole, L. A. (1997). Immunoassay of human chorionic gonadotropin, its free subunits, and metabolites. *Clin. Chem.* 43, 2233–2243.
80. Cole, L. A., Kardana, A., Park, S.-Y., and Braunstein, G. D. (1993). The deactivation of hCG by nicking and dissociation. *J. Clin. Endocrinol. Metab.* 76, 704–713.
81. Birken, S., Armstrong, E. G., Kolks, M. A., Cole, L. A., Agosto, G. M., Krichevsky, A., et al. (1988). Structure of the human chorionic gonadotropin beta-subunit fragment from pregnancy urine. *Endocrinology* 123, 572– 583.
82. O'Connor, J. F., Schlatterer, J. P., Birken, S., Krichevsky, A., Armstrong, E. G., McMahon, D., et al. (1988). Development of highly sensitive immunoassays to measure human chorionic gonadotropin, its beta-subunit, and beta core fragment in the urine: application to malignancies. *Cancer Res.* 48, 1361–1366.
83. Soper, J. T., Mutch, D. G., Schink, J. C., and American College of Obstetricians and Gynecologists. (2004). Diagnosis and treatment of gestational trophoblastic disease: ACOG Practice Bulletin No. 53. *Gynecol. Oncol.* 93, 575–585.
84. Tay, J. I., Moore, J., and Walker, J. J. (2000). Ectopic pregnancy. *BMJ* 320, 916–919.
84. Bogart, M. H., Pandian, M. R., and Jones, O. W. (1987). Abnormal maternal serum chorionic gonadotropin levels in pregnancies with fetal chromosome abnormalities. *Prenat. Diagn.* 7, 623–630.
85. Kellner, L. H., Weiner, Z., Weiss, R. R., Neuer, M., Martin, G. M., Mueenuddin, M., and Bombard, A. (1995). Triple marker (α-fetoprotein, unconjugated estriol, human chorionic gonadotropin) vs. α-fetoprotein plus free-β subunit in second-trimester maternal serum screening for fetal Down syndrome: a prospective comparison study. *Am. J. Obstet. Gynecol.* 173, 1306–1309.
86. Extermann, P., Bischof, P., Marguerat, P., and Mermillod, B. (1998). Second trimester maternal serum screening for Down's syndrome: free β-human chorionic gonadotrophin (HCG) and α-fetoprotein, with or without unconjugated oestriol, compared with total HCG, α-fetoprotein and unconjugated oestriol. *Hum. Reprod.* 13, 220–223.
87. Reis, F. M., D'Antona, D., and Petraglia, F. (2002). Predictive value of hormone measurements in maternal and fetal complications of pregnancy. *Endocr. Rev.* 23, 230–257.
88. Tegeder, I. and Geisslinger, G. (2004). Opioids as modulators of cell death and survival: unraveling mechanisms and revealing new indications. *Pharmacol. Rev.* 56, 351–369.
89. Agbas, A., Ahmed, M. S., Millington, W., Cemerikic, B., Desiderio, D. M., Tseng, J. L., and Dass, C. (1995). Dynorphin A(1-8) in human placenta: amino acid sequence determined by tandem mass spectrometry. *Peptides* 16, 623–627.
90. Ahmed, M. S., Cemerikic, B., and Agbas, A. (1992). Properties and functions of human placental opioid system. *Life Sci.* 50, 83–97.
91. Odagiri, E., Sherrell, B. J., Mount, C. D., Nicholson, W. E., and Orth, D. N. (1979). Human placental immunoreactive corticotropin, lipotropin, and beta-endorphin: evidence for a common precursor. *Proc. Natl. Acad. Sci. U S A* 76, 2027–2031.
92. Liotta, A. S. and Krieger, D. T. (1980). In vitro biosynthesis and comparative posttranslational processing of immunoreactive precursor corticotropin/beta-endorphin by human placental and pituitary cells. *Endocrinology* 106, 1504–1511.
93. Facchinetti, F., Garuti, G., Petraglia, F., Mercantini, F., and Genazzani, A. R. (1990). Changes in beta-endorphin in fetal membranes and placenta in normal and pathological pregnancies. *Acta Obstet. Gynecol. Scand.* 69, 603–607.
94. Sastry, B. V., Barnwell, S. L., Tayeb, O. S., Janson, V. E., and Owens, L. K. (1980). Occurrence of methionine enkephalin in human placental villus. *Biochem. Pharmacol.* 29, 475–478.
95. Porthe, G., Valette, A., and Cros, J. (1981). Kappa opiate binding sites in human placenta. *Biochem. Biophys. Res. Commun.* 101, 1–6.
96. Porthe, G., Valette, A., Moisand, A., Tafani, M., and Cros, J. (1982). Localization of human placental opiate binding sites on the syncytial brush border membrane. *Life Sci.* 31, 2647–2654.
97. Cemerikic, B., Schabbing, R., and Ahmed, M. S. (1992). Selectivity and potency of opioid peptides in regulating human chorionic gonadotropin release from term trophoblast tissue. *Peptides* 13, 897–903.
98. Genazzani, A. R., Facchinetti, F., and Parrini, D. (1981). Beta-lipotrophin and beta-endorphin plasma levels during pregnancy. *Clin. Endocrinol. (Oxf).* 14, 409–418.
99. Newnham, J. P., Tomlin, S., Ratter, S. J., Bourne, G. L., and Rees, L. H. (1983). Endogenous opioid peptides in pregnancy. *Br. J. Obstet. Gynaecol.* 90, 535–538.
100. Genazzani, A. R., Petraglia, F., Parrini, D., Nasi, A., Angioni, G., Facchinetti, F., Facchini, V., and Volpe, A. (1984). Lack of correlation between amniotic fluid and maternal plasma contents of beta-endorphin, beta-lipotropin, and adrenocorticotropic hormone in normal and pathologic pregnancies. *Am. J. Obstet. Gynecol.* 148, 198–203.
101. Mauri, A., Serri, F., Caminiti, F., Mancuso, S., Fratta, W., Gessa, G. L., and Argiolas, A. (1990). Correlation between amniotic levels of alpha-MSH, ACTH and beta-endorphin in late gestation and labour in normal and complicated pregnancies. *Acta Endocrinol. (Copenh).* 123, 637–642.
102. Zanardo, V., Nicolussi, S., Carlo, G., Marzari, F., Faggian, D., Favaro, F., and Plebani, M. (2001). Beta endorphin concentrations in human milk. *J. Pediatr. Gastroenterol. Nutr.* 33, 160–164.
103. Valette, A., Desprat, R., Cros, J., Pontonnier, G., Belisle, S., and Lemaire, S. (1986). Immunoreactive dynorphine in

maternal blood, umbilical vein and amniotic fluid. *Neuropeptides* 7, 145–151.

104. Alsat, E., Guibourdenche, J., Couturier, A., and Evain Brion, D. (1998). Physiological role of human placental growth hormone. *Mol. Cell. Endocrinol.* 140, 121–127.
105. Scippo, M. L., Frankenne, F., Hooghe-Peters, E. L., Igout, A., Velkeniers, B., and Hennen, G. (1993). Syncytiotrophoblastic localization of the human growth hormone variant mRNA in the placenta. *Mol. Cell. Endocrinol.* 92, R7–13.
106. Frendo, J. L., Therond, P., Guibourdenche, J., Bidart, J. M., Vidaud, M., and Evain-Brion, D. (2000). Modulation of copper/zinc superoxide dismutase expression and activity with in vitro differentiation of human villous cytotrophoblasts. *Placenta* 21, 773–781.
107. Lytras, A., Bock, M. E., Dodd, J. G., and Cattini, P. A. (1994). Detection of placental growth hormone variant and chorionic somatomammotropin ribonucleic acid expression in human trophoblastic neoplasms by reverse transcriptase-polymerase chain reaction. *Endocrinology* 134, 2461–2467.
108. Ray, J., Okamura, H., Kelly, P. A., Liebhaber, S. A., and Cooke, N. E. (1990). Alteration in the receptor binding specificity of human growth hormone by genomic exon exchange. *Mol. Endocrinol.* 4, 101–107.
109. Baumann, G., Davila, N., Shaw, M. A., Ray, J., Liebhaber, S. A., and Cooke, N. E. (1991). Binding of human growth hormone (GH)-variant (placental GH) to GH-binding proteins in human plasma. *J. Clin. Endocrinol. Metab.* 73, 1175–1179.
110. Herington, A. C., Ymer, S. I., Stevenson, J. L., and Roupas, P. (1994). Growth hormone receptor/binding protein: physiology and function. *Proc. Soc. Exp. Biol. Med.* 206, 238–242.
111. Hill, D. J., Riley, S. C., Bassett, N. S., and Waters, M. J. (1992). Localization of the growth hormone receptor, identified by immunocytochemistry, in second trimester human fetal tissues and in placenta throughout gestation. *J. Clin. Endocrinol. Metab.* 75, 646–650.
112. McIntyre, H. D., Serek, R., Crane, D. I., Veveris-Lowe, T., Parry, A., Johnson, S., Leung, K. C., Ho, K. K., Bougoussa, M., Hennen, G., Igout, A., Chan, F. Y., Cowley, D., Cotterill, A., and Barnard, R. (2000). Placental growth hormone (GH), GH-binding protein, and insulin-like growth factor axis in normal, growth-retarded, and diabetic pregnancies: correlations with fetal growth. *J. Clin. Endocrinol. Metab.* 85, 1143–1150.
113. Frankenne, F., Closset, J., Gomez, F., Scippo, M. L., Smal, J., and Hennen, G. (1988). The physiology of growth hormones (GHs) in pregnant women and partial characterization of the placental GH variant. *J. Clin. Endocrinol. Metab.* 66, 1171–1180.
114. Mirlesse, V., Frankenne, F., Alsat, E., Poncelet, M., Hennen, G., and Evain-Brion, D. (1993). Placental growth hormone levels in normal pregnancy and in pregnancies with intrauterine growth retardation. *Pediatr. Res.* 34, 439–442.
115. Eriksson, L. (1989). Growth hormone in human pregnancy. Maternal 24-hour serum profiles and experimental effects of continuous GH secretion. *Acta Obstet. Gynecol. Scand.* 147, 1–38.
116. Evain-Brion, D., Alsat, E., Igout, A., Frankenne, F., and Hennen, G. (1994). Placental growth hormone variant: assay and clinical aspects. *Acta Paediatr. Suppl.* 399, 49–51.
117. Barnard, R., Chan, F. Y., and McIntyre, H. D. (1997). Growth hormone-binding protein in normal and pathologic gestation: correlations with maternal diabetes and fetal growth. *J. Clin. Endocrinol. Metab.* 82, 1879–1884.
118. Josimovich, J. B. and MacLaren, J. A. (1962). Presence in human placenta and term serum of highly lactogenic substance immunologically related to pituitary growth hormone. *Endocrinology* 71, 209–215.
119. Jaffe, R. B. (1998). Neuroendocrine-metabolic regulation of pregnancy. In *Reproductive Endocrinology* (S. S. C. Yen, R. B. Jaffe, and R. L. Barbieri, Eds.), pp. 751–784. WB Saunders, Philadelphia.
120. Walker, W. H., Fitzpatrick, S. L., Barrera-Saldana, H. A., Resendez-Perez, D., and Saunders, G. F. (1991). The human placental lactogen genes: structure, function, evolution and transcriptional regulation. *Endocr. Rev.* 12, 316–328.
121. Handwerger, S. and Freemark, M. (2000). The roles of placental growth hormone and placental lactogen in the regulation of human fetal growth and development. *J. Pediatr. Endocrinol. Metab.* 13, 343–356.
122. Jacobsen, G., Hoffman, H. J., and Bakketeig, L. S. (1997). Small-for-gestational-age (SGA) infants born at term: growth and development during the first year of life. *Acta Obstet. Gynecol. Scand.* 165, 93–101.
123. Pedrazzini, T., Pralong, F., and Grouzmann, E. (2003). Neuropeptide Y: the universal soldier. *Cell. Mol. Life Sci.* 60, 350–377.
124. Petraglia, F., Calzà, L., Giardino, L., Sutton, S., Marrama, P., Rivier, J., Genazzani, A. R., and Vale, W. (1989). Identification of immunoreactive neuropeptide-gamma in human placenta: localization, secretion, and binding sites. *Endocrinology* 124, 2016–2022.
125. Petraglia, F., Calzà, L., Giardino, L., Zanni, M., Florio, P., Ferrari, A. R., Nappi, C., and Genazzani, A. R. (1993). Maternal decidua and fetal membranes contain immunoreactive neuropeptide Y. *J. Endocrinol. Invest.* 16, 201–205.
126. Robidoux, J., Simoneau, L., St-Pierre, S., Ech-Chadli, H., and Lafond, J. (1998). Human syncytiotrophoblast NPY receptors are located on BBM and activate PLC-to-PKC axis. *Am. J. Physiol.* 274, E502–E509.
127. Wharton, J., Gordon, L., Byrne, J., Herzog, H., Selbie, L. A., Moore, K., Sullivan, M. H., Elder, M. G., Moscoso, G., Taylor, K. M., Shine, J., and Polak, J. M. (1993). Expression of the human neuropeptide tyrosine Y1 receptor. *Proc. Natl. Acad. Sci. U S A* 90, 687–691.
128. Cikos, S., Gregor, P., and Koppel, J. (1999). Sequence and tissue distribution of a novel G-protein-coupled receptor expressed prominently in human placenta. *Biochem. Biophys. Res. Commun.* 256, 352–356.
129. Petraglia, F., Coukos, G., Battaglia, C., Bartolotti, A., Volpe, A., Nappi, C., Segre, A., and Genazzani, A. R. (1989). Plasma and amniotic fluid immunoreactive neuropeptide-Y level changes during pregnancy, labor, and at parturition. *J. Clin. Endocrinol. Metab.* 69, 324–328.
130. Khatun, S., Kanayama, N., Belayet, H. M., Bhuiyan, A. B., Jahan, S., Begum, A., Kobayashi, T., and Terao, T. (2000). Increased concentrations of plasma neuropeptide Y in patients with eclampsia and preeclampsia. *Am. J. Obstet. Gynecol.* 182, 896–900.
131. Taupenot, L., Harper, K. L., and O'Connor, D. T. (2003). The chromogranin-secretogranin family. *N. Engl. J. Med.* 348, 1134–149.
132. Syversen, U., Waldum, H. L., and O'Connor, D. T. (1992). Rapid, high-yield isolation of human chromogranin A from chromaffin granules of pheochromocytomas. *Neuropeptides* 22, 235–240.
133. Moftaquir-Handaj, A., Barbe, F., Barbarino-Monnier, P., Aunis, D., and Boutroy, M. J. (1995). Circulating chromogranin A and catecholamines in human fetuses at uneventful birth. *Pediatr. Res.* 37, 101–105.
134. Florio, P., Mezzesimi, A., Turchetti, V., Severi, F. M., Ticconi, C., Forconi, S., and Petraglia, F. (2002). High levels of human chromogranin A in umbilical cord plasma

and amniotic fluid at parturition. *J. Soc. Gynecol. Investig.* 9, 32–36.
135. Brownstein, M. J., Russell, J. T., and Gainer, H. (1980). Synthesis, transport, and release of posterior pituitary hormones. *Science* 207, 373–378.
136. Mitchell, B. (1995). Oxytocin synthesis and metabolism in human decidua. *Reprod. Fertil. Dev.* 7, 319-321.
137. Chibbar, R., Miller, F. D., and Mitchell, B. F. (1993). Local synthesis of oxytocin in amnion, chorion and decidua may regulate the timing of parturition. *J. Clin. Invest.* 91, 185–192.
138. Blanks, A. M. and Thornton, S. (2003). The role of oxytocin in parturition. *BJOG* 110, 46–51.
139. Mauri, A., Argiolas, A., Ticconi, C., and Piccione, E. (1995). Oxytocin in human intrauterine tissues at parturition. *Reprod. Fertil. Dev.* 7, 1481–1484.
140. Zingg, H. H. and Laporte, S. A. (2003). The oxytocin receptor. *Trends Endocrinol. Metab.* 14, 222–227.
141. Zeeman, G. G., Khan-Dawood, F. S., and Dawood, M. Y. (1997). Oxytocin and its receptor in pregnancy and parturition: current concepts and clinical implications. *Obstet. Gynecol.* 89, 873–883.
142. Wathes, D. C., Borowick, S. C., Timmons, P. M., Leung, S. T., and Thornton, S. (1999). Oxytocin receptor expression in human term and preterm gestational tissues prior to and following the onset of labour. *J. Endocrinol.* 161, 134–151.
143. Birnbaumer, M. (2000). Vasopressin receptors. *Trends Endocrinol. Metab.* 11, 406–410.
144. Ivanisevic, M., Behrens, O., Helmer, H., Demarest, K., and Fuchs, A.-R. (1989). Vasopressin receptors in human pregnant myometrium and decidua: interactions with oxytocin and vasopressin agonists and antagonists. *Am. J. Obstet. Gynecol.* 161, 1637–1643.
145. Fuchs, A.-R., Romero, R., Keefe, D., Parra, M., Oyarzun, E., and Behnke, E. (1991). Oxytocin secretion and human parturition: pulse frequency and duration increase during spontaneous labor in women. *Am. J. Obstet. Gynecol.* 165, 1515–1523.
146. Otsuki, Y., Yamaji, K., and Fujita, M. (1983). Serial plasma oxytocin concentrations during pregnancy and labor. *Acta Obstet. Gynecol. Scand.* 62, 15–18.
147. Chard, T., Hudson, C. N., Edwards, C. R. W., and Boyd, N. R. H. (1971). Release of oxytocin and vasopressin by the human foetus during labour. *Nature* 234, 352–353.
148. Thornton, S., Baldwin, P. J., Harris, P. A., Harding, F., Davison, J. M., Baylis, P. H., Timmons, P. M., and Wathes, D. C. (2002). The role of arginine vasopressin in human labour: functional studies, fetal production and localisation of V1a receptor mRNA. *Br. J. Obstet. Gynaecol.* 109, 57–62.
149. Nussdorfer, G. G. and Malendowicz, L. K. (1998). Role of tachykinins in the regulation of the hypothalamo-pituitary-adrenal axis. *Peptides* 19, 949–968.
150. Page, N. M., Woods, R. J., Gardiner, S. M., Lomthaisong, K., Gladwell, R. T., Butlin, D. J., Manyonda, I. T., and Lowry, P. J. (2000). Excessive placental secretion of neurokinin B during the third trimester causes pre-eclampsia. *Nature* 405, 797–800.
151. Brownbill, P., Bell, N. J., Woods, R. J., Lowry, P. J., Page, N. M., and Sibley, C. P. (2003). Neurokinin B is a paracrine vasodilator in the human fetal placental circulation. *J. Clin. Endocrinol. Metab.* 88, 2164–2170.
152. Patak, E., Candenas, M. L., Pennefather, J. N., Ziccone, S., Lilley, A., Martin, J. D., Flores, C., Mantecon, A. G., Story, M. E., and Pinto, F. M. (2003). Tachykinins and tachykinin receptors in human uterus. *Br. J. Pharmacol.* 139, 523–532.
153. Sakamoto, R., Osada, H., Iitsuka, Y., Masuda, K., Kaku, K., Seki, K., and Sekiya, S. (2003). Profile of neurokinin B concentrations in maternal and cord blood in normal pregnancy. *Clin. Endocrinol. (Oxf).* 58, 597–600.
154. Nguyen, T. T., Tseng, Y. T., McGonnigal, B., Stabila, J. P., Worrell, L. A., Saha, S., and Padbury, J. F. (1999). Placental biogenic amine transporters: in vivo function, regulation and pathobiological significance. *Placenta* 20, 3–11.
155. Manyonda, I. T., Slater, D. M., Fenske, C., Hole, D., Choy, M. Y., and Wilson, C. (1998). A role for noradrenaline in pre-eclampsia, towards a unifying hypothesis for the pathophysiology. *Br. J. Obstet. Gynaecol.* 105, 641–648.
156. Bzoskie, L., Yen, J., Tseng, Y. T., Blount, L., Kashiwai, K., and Padbury, J. F. (1997). Human placental norepinephrine transporter mRNA: expression and correlation with fetal condition at birth. *Placenta* 18, 205–210.
157. Kenney, S. P., Kekuda, R., Prasad, P. D., Leibach, F. H., Devoe, L. D., and Ganapathy, V. (1999). Cannabinoid receptors and their role in the regulation of the serotonin transporter in human placenta. *Am. J. Obstet. Gynecol.* 181, 491–497.
158. Chang, H., Brown, C. W., and Matzuk, M. M. (2002). Genetic analysis of the mammalian transforming growth factor-beta superfamily. *Endocr. Rev.* 23, 787–823.
159. Ando, N., Hirahara, F., Fukushima, J., Kawamoto, S., Okuda, K., Funabashi, T., Gorai, I., and Minaguchi, H. (1998). Differential gene expression of TGF-beta isoforms and TGF-beta receptors during the first trimester of pregnancy at the human maternal-fetal interface. *Am. J. Reprod. Immunol.* 40, 48–56.
160. Kauma, S., Matt, D., Strom, S., Eierman, D., and Turner, T. (1990). Interleukin-1 beta, human leukocyte antigen HLA-DR alpha, and transforming growth factor-beta expression in endometrium, placenta, and placental membranes. *Am. J. Obstet. Gynecol.* 163, 1430–1437.
161. Caniggia, I., Grisaru-Gravnosky, S., Kuliszewsky, M., Post, M., and Lye, S. J. (1999). Inhibition of TGF-beta 3 restores the invasive capability of extravillous trophoblasts in preeclamptic pregnancies. *J. Clin. Invest.* 103, 1641–1650.
162. Selick, C., Horowitz, G., Gratch, M., Scott, R., Navot, D., and Hofmann G. (1994). Immunohistochemical localization of transforming growth factorbeta in human implantation sites. *J. Clin. Endocrinol. Metab.* 78, 592–596.
163. Dungy, L. J., Siddiqi, T. A., and Khan S. (1991). Transforming growth factor-beta1 expression during placental development. *Am. J. Obstet. Gynecol.* 165, 853–857.
164. Caniggia, I., Mostachfi, H., Winter, J., Gassmann, M., Lye, S. J., Kuliszewski, M., and Post, M. (2000). Hypoxia–inducible factor-1 mediates the biological effects of oxygen on human trophoblast differentiation through TGFbeta(3). *J. Clin. Invest.* 105, 577–587.
165. Cheifetz, S., Bellon, T., Cales, C., Vera, S., Bernabeu, C., Massague, J., and Letarte, M. (1992). Endoglin is a component of the transforming growth factor-β receptor system in human endothelial cells. *J. Biol. Chem.* 267, 19027–19030.
166. Schilling, B. and Yeh, J. (2000). Transforming growth factor-beta(1), -beta(2), -beta(3) and their type I and II receptors in human term placenta. *Gynecol. Obstet. Invest.* 50, 19–23.
167. St-Jacques, S., Forte, M., Lye, S. J., and Letarte, M. (1994). Localization of endoglin, a transforming growth-factor-β binding protein, and of CD44 and integrins in placenta during the first trimester of pregnancy. *Biol. Reprod.* 51, 405–413.
168. Gougos, A., St-Jacques, S., Greaves, A., O'Connell, P. J., d'Apice, A. J., Buhring, H. J., Bernabeu, C., Van Hourik, J. A., and Letarte, M. (1992). Identification of distinct epitopes of endoglin, an RGD-containing glycoprotein of endothelial cells, leukemic cells, and syncytiotrophoblasts. *Int. Immunol.* 4, 83–92.
169. Hennessy, A., Orange, S., Willis, N., Painter, D. M., Child, A., and Horvath, J. S. (2002). Transforming growth factor-beta 1 does not relate to hypertension in pre-eclampsia. *Clin. Exp. Pharmacol. Physiol.* 29, 968–971.

170. Caniggia, I. and Winter, J. L. (2002). Adriana and Luisa Castellucci Award lecture 2001. Hypoxia inducible factor-1: oxygen regulation of trophoblast differentiation in normal and pre-eclamptic pregnancies—a review. *Placenta* 23, S47–S57.
171. Petraglia, F., Sawchenko, P., Lim, A. T., Rivier, J., and Vale, W. (1987). Localization, secretion, and action of inhibin in human placenta. *Science* 237, 187–189.
172. Petraglia, F., Calza, L., Garuti, G. C., Abrate, M., Giardino, L., Genazzani, A. R., Vale, W., and Meunier, H. (1990). Presence and synthesis of inhibin subunits in human decidua. *J. Clin. Endocrinol. Metab.* 71, 487–492.
173. Petraglia, F., Anceschi, M. M., Calza, L., Garuti, G. C., Fusaro, P., Giardino, L., Genazzani, A. R., and Vale, W. (1993). Inhibin and activin in human fetal membranes: evidence for a local effect on prostaglandin release. *J. Clin. Endocrinol. Metab.* 77, 542–548.
174. Schneider-Kolsky, M. E., Manuelpillai, U., Waldron, K., Dole, A., and Wallace, E. M. (2002). The distribution of activin and activin receptors in gestational tissues across human pregnancy and during labour. *Placenta* 23, 294–302.
175. Silver, H. M., Lambert-Messerlian, G. M., Reis, F. M., Diblasio, A. M., Petraglia, F., and Canick, J. A. (2002). Mechanism of increased maternal serum total activin A and inhibin A in preeclampsia. *J. Soc. Gynecol. Investig.* 9, 308–312.
176. Hayette, S., Gadoux, M., Martel, S., Bertrand, S., Tigaud, I., Magaud, J. P., and Rimokh, R. (1998). FLRG (follistatin-related gene), a new target of chromosomal rearrangement in malignant blood disorders. *Oncogene* 16, 2949–2954.
177. Ciarmela, P., Florio, P., Toti, P., Franchini, A., Maguer-Satta, V., Ginanneschi, C., Ottaviani, E., and Petraglia, F. (2003). Human placenta and fetal membranes express follistatin-related gene mRNA and protein. *J. Endocrinol. Invest.* 26, 641–645.
178. Petraglia, F., Gallinelli, A., Grande, A., Florio, P., Ferrari, S., Genazzani, A. R., Ling, N., and DePaolo, L. V. (1994). Local production and action of follistatin in human placenta. *J. Clin. Endocrinol. Metab.* 78, 205–210.
179. Kozian, D. H., Ziche, M., and Augustin, H. G. (1997). The activin-binding protein follistatin regulates autocrine endothelial cell activity and induces angiogenesis. *Lab. Invest.* 76, 267–276.
180. Gray, P. C., Bilezikjian, L. M., and Vale, W. (2002). Antagonism of activin by inhibin and inhibin receptors: a functional role for betaglycan. *Mol. Cell. Endocrinol.* 188, 254–260.
181. Ciarmela, P., Florio, P., Toti, P., Grasso, D., Santopietro, R., Tosi, P., and Petraglia, F. (2003). Expression of betaglycan in pregnant tissues throughout gestation. *Eur. J. Endocrinol.* 149, 433–437.
182. Florio, P., Cobellis, L., Luisi, S., Ciarmela, P., Severi, F. M., Bocchi, C., and Petraglia, F. (2001). Changes in inhibins and activin secretion in healthy and pathological pregnancies. *Mol. Cell. Endocrinol.* 180, 123–130.
183. Florio, P., Luisi, S., Ciarmela, P., Severi, F. M., Bocchi, C., and Petraglia, F. (2004). Inhibins and activins in pregnancy. *Mol. Cell. Endocrinol.* 225, 93–100.
184. Luisi, S., Florio, P., D'Antona, D., Severi, F. M., Sanseverino, F., Danero, S., and Petraglia, F. (2003). Maternal serum inhibin A levels are a marker of a viable trophoblast in incomplete and complete miscarriage. *Eur. J. Endocrinol.* 148, 233–236.
185. Florio, P., Luisi, S., D'Antona, D., Severi, F. M., Rago, G., and Petraglia, F. (2004). Maternal serum inhibin A levels may predict pregnancy outcome in women with threatened abortion. *Fertil. Steril.* 81, 468–470.
186. Wallace, E. M., Schneider-Kolsky, M. E., Edwards, A., Baker, L., and Jenkin, G. (2003). Maternal serum activin A levels in association with intrauterine fetal growth restriction. *BJOG* 110, 306–310.
187. Le Roith, D., Bondy, C., Yakar, S., Liu, J. L., and Butler, A. (2001). The somatomedin hypothesis: 2001. *Endocr. Rev.* 22, 53–74.
188. Graham, C. H., Lysiak, J. J., McCrae, K. R., and Lala, P. K. (1992). Localization of insulin-like growth factor I mRNA in the developing human placenta and in term placenta of diabetics. *Mol. Endocrinol.* 2, 217–221.
189. Zhou, J. and Bondy, C. (1992). Insulin-like growth factor II and its binding proteins in placental development. *Endocrinology* 131, 1230–1240.
190. Nayak, N. R. and Giudice, L. C. (2003). Comparative biology of the IGF system in endometrium, decidua, and placenta, and clinical implications for foetal growth and implantation disorders. *Placenta* 24, 281–296.
191. Firth, S. M. and Baxter, R. C. (2002). Cellular actions of the insulin-like growth factor binding proteins. *Endocr. Rev.* 23, 8248–8254.
192. Hill, D. J., Clemmons, D. R., Riley, S. C., Bassett, N., and Challis, J. R. G. (1993). Immunohistochemical localization of insulin-like growth factors (IGFs) and IGF binding proteins-l, -2, -3 in human placenta and fetal membranes. *Placenta* 14, 1–12.
193. Wathen, N. C., Wang, H. S., Cass, P. L., Campbell, D. J., and Chard, T. (1991). Insulin-like growth factor-l and insulin-like growth factor binding protein-l in early human pregnancy. *Early Hum. Dev.* 28, 105–110.
194. Han, V. K., Bassett, N., Walton, J., and Challis, J. R. (1996). The expression of insulin-like growth factor (IGF) and IGF-binding protein (IGFBP) genes in the human placenta and membranes: evidence for IGF-IGFBP interactions at the feto-maternal interface. *J. Clin. Endocrinol. Metab.* 81, 2680–2693.
195. Fant, M. (1990). Insulin-like growth factor binding proteins (BP) from human placenta are immunologically related to growth hormone dependent binding protein in adult human serum (BP-53). *Placenta* 11, 123–133.
196. Davenport, M. L., Clemmons, D. R., Miles, M. V., Camacho-Hubner, C., D'Ercole, A. J., and Underwood, L. E. (1990). Regulation of serum insulin-like growth factor-I (IGF-I) and IGF binding protein during rat pregnancy. *Endocrinology* 127, 1278–1286.
197. Binoux, M. (1991). Serum insulin-like growth factors and insulin-like growth factor binding proteins in the human fetus. Relationships with growth in normal subjects and in subjects with intrauterine growth retardation. *Pediatr. Res.* 29, 219–225.
198. Giudice, L. C., Farell, E. M., Pham, H., Lamson, G., and Rosenfeld, R. G. (1990). Insulin-like growth factor binding proteins in maternal serum throughout gestation and in the puerperium: effects of a pregnancy-associated serum protease activity. *J. Clin. Endocrinol. Metab.* 71, 806–810.
199. Shimasaki, S., Uchiyama, F., Shimonaka, M., and Ling, N. (1990). Molecular cloning of the cDNAs encoding a novel insulin-like growth factor-binding protein from rat and human. *Mol. Endocrinol.* 4, 1451–1458.
200. Sun I. Y., Overgaard, M. T., Oxvig, C., and Giudice, L. C. (2002). Pregnancy-associated plasma protein A proteolytic activity is associated with the human placental trophoblast cell membrane. *J. Clin. Endocrinol. Metab.* 87, 5235–5240.
201. Bonno, M., Oxvig, C., Kephart, G. M., Wagner, J. M., Kristensen, T., Sottrup-Jensen, L., and Gleich, G. J. (1994). Localization of pregnancy-associated plasma protein-A and colocalization of pregnancy-associated plasma protein-A messenger ribonucleic acid and eosinophil granule major basic protein messenger ribonucleic acid in placenta. *Lab. Invest.* 71, 560–566.
202. Hall, K., Enberg, G., Hellem, E., Lundin, G., Ottosson-Seeberger, A., Sara, V., Trygstad, O., and Ofverholm, U. (1984).

Somatomedin levels in pregnancy: longitudinal study in healthy subjects and patients with growth hormone deficiency. *J. Clin. Endocrinol. Metab.* 59, 587–594.

203. Monaghan, J. M., Godber, I. M., Lawson, N., Kaur, M., Wark, G., Teale, D., and Hosking, D. J. (2004). Longitudinal changes of insulin-like growth factors and their binding proteins throughout normal pregnancy. *Ann. Clin. Biochem.* 41, 220–226.
204. Ingec, M., Gursoy, H. G., Yildiz, L., Kumtepe, Y., and Kadanali, S. (2004). Serum levels of insulin, IGF-1, and IGFBP-1 in pre-eclampsia and eclampsia. *Int. J. Gynaecol. Obstet.* 84, 214–219.
205. Monzavi, R. and Cohen, P. (2002). IGFs and IGFBPs: role in health and disease. *Best Pract. Res. Clin. Endocrinol. Metab.* 16, 433–447.
206. Holmes, R., Montemagno, R., Jones, J., Preece, M., Rodeck, C., and Soothill, P. (1997). Fetal and maternal plasma insulin-like growth factors and binding proteins in pregnancies with appropriate or retarded fetal growth. *Early Hum. Dev.* 49, 7–17.
207. Folkersen, J., Grudzinskas, J. G., Hindersson, P., Teisner, B., and Westergaard, J. G. (1981). Pregnancy-associated plasma protein A: circulating levels during normal pregnancy. *Am. J. Obstet. Gynecol.* 139, 910–914.
208. Canick, J. A. and Kellner, L. H. (1999). First trimester screening for aneuploidy: serum biochemical markers. *Semin. Perinatol.* 5, 359–368.
209. Knight, G. J., Palomaki, G. E., Haddow, J. E., Miller, W., Bersinger, N. A., and Schneider, H. (1993). Pregnancy associated plasma protein A as a marker for Down syndrome in the second trimester of pregnancy. *Prenat. Diagn.* 13, 222–223.
210. Reuss, B. and von Bohlen und Halbach, O. (2003). Fibroblast growth factors and their receptors in the central nervous system. *Cell Tissue Res.* 313, 139–57.
211. Jorissen, R. N., Walker, F., Pouliot, N., Garrett, T. P., Ward, C. W., and Burgess, A. W. (2003). Epidermal growth factor receptor: mechanisms of activation and signalling. *Exp. Cell Res.* 284, 31–53.
212. Gospodarowicz, D., Cheng, J., Lui, G. M., Fujii, D. K., Baird, A., and Bohle, N. P. (1985). Fibroblast growth factor in the human placenta. *Biochem. Biophys. Res. Commun.* 128, 554–562.
213. Di Blasio, A. M., Carniti, C., Vigano, P., Florio, P., Petraglia, F., and Vignali, M. (1997). Basic fibroblast growth factor messenger ribonucleic acid levels in human placentas from normal and pathological pregnancies. *Mol. Hum. Reprod.* 3, 1119–1123.
214. Cattini, P. A., Nickel, B., Bock, M., and Kardami, E. (1993). Immunolocalization of basic fibroblast growth factor (bFGF) in growing and growth-inhibited placental cells: a possible role for bFGF in placental cell development. *Placenta* 12, 341–352.
215. Hofmann, G. E., Scott, R. T., Bergh, P. A., and Deligdisch, L. (1991). Immunohistochemical localization of epidermal growth factor in human endometrium, decidua, and placenta. *J. Clin. Endocrinol. Metab.* 73, 882–886.
216. Eaton, B. M., Richards, R. C., and Contractor, S. F. (1987). Uptake and metabolism of epidermal growth factor in the perifused human placenta. *J. Dev. Physiol.* 9, 69–78
217. Maruo, T., Matsuo, H., Otani, T., and Mochizuki, M. (1995). Role of epidermal growth factor (EGF) and its receptor in the development of the human placenta. *Reprod. Fertil. Dev.* 7, 1465–1470.
218. Hofmann, G. E. and Abramowicz, J. S. (1990). Epidermal growth factor (EGF) concentrations in amniotic fluid and maternal urine during pregnancy. *Acta Obstet. Gynecol. (Scand).* 69, 217–220
219. Cross, M. J., Dixelius, J., Matsumoto, T., and Claesson-Welsh, L. (2003). VEGF-receptor signal transduction. *Trends Biochem. Sci.* 28, 488–494.
220. Autiero, M., Luttun, A., Tjwa, M., and Carmeliet, P. (2003). Placental growth factor and its receptor, vascular endothelial growth factor receptor-1: novel targets for stimulation of ischemic tissue revascularization and inhibition of angiogenic and inflammatory disorders. *J. Thromb. Haemost.* 1, 1356–1370.
221. Vuorela, P., Hatva, E., Lymboussaki, A., Kaipainen, A., Joukov, V., Persico, M. G., Alitalo, K., and Halmesmaki, E. (1997). Expression of vascular endothelial growth factor and placenta growth factor in human placenta. *Biol. Reprod.* 56, 489–494.
222. Shiraishi, S., Nakagawa, K., Kinukawa, N., Nakano, H., and Sueishi, K. (1996). Immunohistochemical localization of vascular endothelial growth factor in the human placenta. *Placenta* 17, 111–121.
223. Shore, V. H., Wang, T. H., Wang, C. L., Torry, R. J., Caudle, M. R., and Torry, D. S. (1997). Vascular endothelial growth factor, placenta growth factor and their receptors in isolated human trophoblast. *Placenta* 18, 657–665.
224. Zhou, Y., McMaster, M., Woo, K., Janatpour, M., Perry, J., Karpanen, T., Alitalo, K., Damsky, C., and Fisher, S. J. (2002). Vascular endothelial growth factor ligands and receptors that regulate human cytotrophoblast survival are dysregulated in severe preeclampsia and hemolysis, elevated liver enzymes, and low platelets syndrome. *Am. J. Pathol.* 160, 1405–1423.
225. Khaliq, A., Li, X. F., Shams, M., Sisi, P., Acevedo, C. A., Whittle, M. J., Weich, H., and Ahmed, A. (1996). Localisation of placenta growth factor (PIGF) in human term placenta. *Growth Factors* 13, 243–250.
226. Nagamatsu, T., Fujii, T., Kusumi, M., Li, Z., Yamashita, T., Osuga, Y., Momoeda, M., Kozuma, S., and Taketani, Y. (2004). Cytotrophoblasts up-regulate soluble Fms-like tyrosine Kinase-1 (sFlt-1) expression under reduced oxygen: an implication for the placental vascular development and the pathophysiology of preeclampsia. *Endocrinology* 145, 4838–4845.
227. de Vries, C., Escobedo, J. A., Ueno, H., Houck, K., Ferrara, N., and Williams, L. T. (1992). The fms-like tyrosine kinase, a receptor for vascular endothelial growth factor. *Science* 255, 989–991.
228. Terman, B. I., Dougher-Vermazen, M., Carrion, M. E., Dimitrov, D., Armellino, D. C., Gospodarowicz, D., and Bohlen, P. (1992). Identification of the KDR tyrosine kinase as a receptor for vascular endothelial cell growth factor. *Biochem. Biophys. Res. Commun.* 187, 1579–1586.
229. Park, J. E., Chen, H. H., Winer, J., Houck, K. A., and Ferrara, N. (1994). Placenta growth factor. Potentiation of vascular endothelial growth factor bioactivity, in vitro and in vivo, and high affinity binding to Flt-1 but not to Flk-1/KDR. *J. Biol. Chem.* 269, 25646–25654.
230. Crescimanno, C., Foidart, J. M., Noel, A., Polette, M., Maquoi, E., Birembaut, P., Baramova, E., Kaufmann, P., and Castellucci, M. (1996). Cloning of choriocarcinoma cells shows that invasion correlates with expression and activation of gelatinase A. *Exp. Cell Res.* 227, 240–251.
231. Charnock-Jones, D. S., Sharkey, A. M., Boocock, C. A., Ahmed, A., Plevin, R., Ferrara, N., and Smith, S. K. (1994). Vascular endothelial growth factor receptor localization and activation in human trophoblast and choriocarcinoma cells. *Biol. Reprod.* 51, 524–530.
232. Ahmed, A., Li, X. F., Dunk, C., Whittle, M. J., Rushton, D. I., and Rollason, T. (1995). Colocalisation of vascular endothelial growth factor and its Flt-1 receptor in human placenta. *Growth Factors.* 12, 235–243.

233. Kumazaki, K., Nakayama, M., Suehara, N., and Wada, Y. (2002). Expression of vascular endothelial growth factor, placental growth factor, and their receptors Flt-1 and KDR in human placenta under pathologic conditions. *Hum. Pathol.* 33, 1069–1077.
234. Soker, S., Gollamudi-Payne, S., Fidder, H., Charmahelli, H., and Klagsbrun, M. (1997). Inhibition of vascular endothelial growth factor (VEGF)-induced endothelial cell proliferation by a peptide corresponding to the exon 7-encoded domain of VEGF165. *J. Biol. Chem.* 272, 31582–31588.
235. Migdal, M., Huppertz, B., Tessler, S., Comforti, A., Shibuya, M., Reich, R., Baumann, H., and Neufeld, G. (1998). Neuropilin-1 is a placenta growth factor-2 receptor. *J. Biol. Chem.* 273, 22272–22278.
236. Pajusola, K., Aprelikova, O., Korhonen, J., Kaipainen, A., Pertovaara, L., Alitalo, R., and Alitalo, K. (1992). FLT4 receptor tyrosine kinase contains seven immunoglobulin-like loops and is expressed in multiple human tissues and cell lines. *Cancer Res.* 52, 5738–5743.
237. Lee, J., Gray, A., Yuan, J., Luoh, S. M., Avraham, H., and Wood, W. I. (1996). Vascular endothelial growth factor-related protein: a ligand and specific activator of the tyrosine kinase receptor Flt4. *Proc. Natl. Acad. Sci. U S A* 93, 1988–1992.
238. Tsoi, S. C., Wen, Y., Chung, J. Y., Chen, D., Magness, R. R., and Zheng, J. (2002). Co-expression of vascular endothelial growth factor and neuropilin-1 in ovine feto-placental artery endothelial cells. *Mol. Cell. Endocrinol.* 196, 95–106.
239. Helske, S., Vuorela, P., Carpen, O., Hornig, C., Weich, H., and Halmesmaki, E. (2001). Expression of vascular endothelial growth factor receptors 1, 2 and 3 in placentas from normal and complicated pregnancies. *Mol. Hum. Reprod.* 7, 205–210.
240. Chung, J. Y., Song, Y., Wang, Y., Magness, R. R., and Zheng, J. (2004). Differential expression of vascular endothelial growth factor (VEGF), endocrine gland derived-VEGF, and VEGF receptors in human placentas from normal and preeclamptic pregnancies. *J. Clin. Endocrinol. Metab.* 89, 2484–2490.
241. VanWijk, M. J., Kublickiene, K., Boer, K., and VanBavel, E. (2000). Vascular function in preeclampsia. *Cardiovasc. Res.* 47, 38–48.
242. Strauss, J. F. 3rd, Martinez, F., and Kiriakidou, M. (1996). Placental steroid hormone synthesis: unique features and unanswered questions. *Biol. Reprod.* 54, 303–311.
243. Aspillaga, M. O., Whittaker, P. G., Grey, C. E., and Lind, T. (1983). Endocrinologic events in early pregnancy failure. *Am. J. Obstet. Gynecol.* 147, 903–908.
244. Yen, S. C., Jaffe, R. B., and Barbieri, R. L. (1998). Endocrinology of pregnancy. In *Reproductive Endocrinology* (S. S. C. Yen, R. B. Jaffe, and R. L. Barbieri, Eds.). WB Saunders, Philadelphia.
245. Pepe, G. and Rothchild, I. (1972). The effect of hypophysectomy on day 12 of pregnancy on the serum progesterone level and time of parturition in the rat. *Endocrinology* 91, 1380–1385.
246. Challis, J. R. G., Matthews, S. G., Gibb, W., and Lye, S. J. (2000). Endocrine and paracrine regulation of birth at term and preterm. *Endocr. Rev.* 21, 514–550.
247. Siiteri, P. K. and MacDonald, P. C. (1966). Placental estrogen biosynthesis during human pregnancy. *J. Clin. Endocrinol. Metab.* 26, 751–761.
248. Smith, R., Mesiano, S., and McGrath, S. (2002). Hormone trajectories leading to human birth. *Regul. Pept.* 108, 159–164.
249. Kashork, C. D., Sutton, V. R., Fonda Allen, J. S., Schmidt, D. E., Likhite, M. L., Potocki, L., O'Brien, W. E., and Shaffer, L. G. (2002). Low or absent unconjugated estriol in pregnancy: an indicator for steroid sulfatase deficiency detectable by fluorescence in situ hybridization and biochemical analysis. *Prenat. Diagn.* 22, 1028–1032.
250. Korte, K., Hemsell, P. G., and Mason, J. I. (1982). Sterol sulfate metabolism in the adrenals of the human fetus, anencephalic newborn, and adult. *J. Clin. Endocrinol. Metab.* 55, 671–675.
251. Newby, D., Aitken, D. A., Howatson, A. G., and Connor, J. M. (2000). Placental synthesis of oestriol in Down's syndrome pregnancies. *Placenta* 21, 263–267.
252. Cuckle, H. S., Wald, N. J., Barkai, G., Fuhrmann, W., Altland, K., Brambati, B., Knight, G., Palomaki, G., Haddow, J. E., and Canick, J. (1988). First-trimester biochemical screening for Down syndrome. *Lancet* 2, 851–852.
253. Ellis, M. J., Livesey, J. H., Inder, W. J., Prickett, T. C., and Reid, R. (2002). Plasma corticotropin-releasing hormone and unconjugated estriol in human pregnancy: gestational patterns and ability to predict preterm delivery. *Am. J. Obstet. Gynecol.* 186, 94–99.
254. Heine, R. P., McGregor, J. A., Goodwin, T. M., Artal, R., Hayashi, R. H., Robertson, P. A., and Varner, M. W. (2000). Serial salivary estriol to detect an increased risk of preterm birth. *Obstet. Gynecol.* 96, 490–497.
255. Kitamura, K., Kangawa, K., Kawamoto, M., Ichiki, Y., Nakamura, S., Matsuo, H., and Eto, T. (1993). Adrenomedullin: a novel hypotensive peptide isolated from human pheochromocytoma. *Biochem. Biophys. Res. Commun.* 192, 553–560.
256. Gratton, R. J., Gluszynski, M., Mazzuca, D. M., Nygard, K., and Han, V. K. (2003). Adrenomedullin messenger ribonucleic acid expression in the placentae of normal and preeclamptic pregnancies. *J. Clin. Endocrinol. Metab.* 88, 6048–6055.
257. Marinoni, E., Di Iorio, R., Letizia, C., Villaccio, B., Scucchi, L., and Cosmi, E. V. (1998). Immunoreactive adrenomedullin in human fetoplacental tissues. *Am. J. Obstet. Gynecol.* 179, 784–787.
258. Moriyama, T., Otani, T., and Maruo, T. (2001). Expression of adrenomedullin by human placental cytotrophoblasts and choriocarcinoma JAr cells. *J. Clin. Endocrinol. Metab.* 86, 3958–3961.
259. Kanenishi, K., Kuwabara, H., Ueno, M., Sakamoto, H., and Hata, T. (2000). Immunohistochemical adrenomedullin expression is decreased in the placenta from pregnancies with pre-eclampsia. *Pathol. Int.* 50, 536–540.
260. Marinoni, E., Di Iorio, R., Lucchini, C., Di Netta, T., Letizia, C., and Cosmi, E. V. (2004). Adrenomedullin and nitric oxide synthase at the maternal-decidual interface in early spontaneous abortion. *J. Reprod. Med.* 49, 153–161.
261. Di Iorio, R., Marinoni, E., Letizia, C., Alò, P., Villaccio, B., and Cosmi, E. V. (1998). Adrenomedullin, a new vasoactive peptide, is increased in preeclampsia. *Hypertension* 32, 758–763.
262. Knerr, I., Dachert, C., Beinder, E., Metzler, M., Dotsch, J., Repp, R., et al. (2002). Adrenomedullin, calcitonin gene-related peptide and their receptors: evidence for a decreased placental mRNA content in preeclapsia and HELLP syndrome. *Eur. J. Obstet. Gynecol. Reprod. Biol.* 101, 47–53.
263. Makino, Y., Shibata, K., Makino, I., Ono, Y., Kangawa, K., and Kawarabayashi, T. (1999). Expression of adrenomedullin in feto-placental circulation of human normotensive pregnant women and pregnancy-induced hypertensive women. *Endocrinology* 140, 5439–5442.
264. Trollmann, R., Schoof, E., Beinder, E., Wenzel, D., Rascher, W., and Dotsch, J. (2002). Adrenomedullin gene expression in human placental tissue and leukocytes: a potential marker of severe tissue hypoxia in neonates with birth asphyxia. *Eur. J. Endocrinol.* 147, 711–716.
265. McLatchie, L. M., Fraser, N. J., Main, M. J., Wise, A., Brown, J., Thompson, N., Solari, R., Lee, M. G., and Foord, S. M. (1998). RAMPs regulate the transport and ligand specificity of the calcitonin-receptor-like receptor. *Nature* 393, 333–339.

266. Makino, Y., Shibata, K., Makino, I., Kangawa, K., and Kawarabayashi, T. (2001). Alteration of the adrenomedullin receptor components gene expression associated with the blood pressure in pregnancy-induced hypertension. *J. Clin. Endocrinol. Metab.* 86, 5079–5082.
267. Di Iorio, R., Marinoni, E., Letizia, C., Villaccio, B., Alberini, A., and Cosmi, E. V. (1999). Adrenomedullin production is increased in normal human pregnancy. *Eur. J. Endocrinol.* 140, 201–206.
268. Di Iorio, R., Marinoni, E., and Cosmi, E. V. (1998). New peptides, hormones and parturition. *Gynecol. Endocrinol.* 12, 429–434.
269. Di Iorio, R., Marinoni, E., Letizia, C., Alo, P., Villaccio, B., Poverini, R., and Cosmi, E. V. (2001). Influence of labor on fetoplacental adrenomedullin concentrations. *Am. J. Obstet. Gynecol.* 185, 697–702.
270. Di Iorio, R., Marinoni, E., Letizia, C., Gazzolo, D., Lucchini, C., and Cosmi, E. V. (2000). Adrenomedullin is increased in the fetoplacental circulation in intrauterine growth restriction with abnormal umbilical artery waveforms. *Am. J. Obstet. Gynecol.* 182, 650–654.
271. Di Iorio, R., Marinoni, E., Letizia, C., and Cosmi, E. V. (2003). Adrenomedullin in perinatal medicine. *Regul. Pept.* 112, 103–113.
272. Pollock, D. M., Keith, T. L., and Highsmith, R. F. (1995). Endothelin receptors and calcium signaling. *FASEB J.* 9, 1196–1204.
273. McMahon, L. P., Redman, C. W. G., and Firth, J. D. (1993). Expression of the three endothelin genes and plasma levels of endothelin in pre-eclamptic and normal gestations. *Clin. Sci.* 85, 417–424.
274. Robert, B., Malassine, A., Bourgeois, C., Mignot, T. M., Cronier, L., Ferre, F., et al. (1996). Expression of endothelin precursor genes in human trophoblast in culture. *Eur. J. Endocrinol.* 134, 490–496.
275. Marinoni, E., Picca, A., Scucchi, L., Cosmi, E. V., and Di Iorio, R. (1995). Immunohistochemical localization of endothelin-1 in placenta and fetal membranes in term and preterm human pregnancy. *Am. J. Reprod. Immunol.* 34, 213–218.
276. Hemsen, A., Gillis, C., Larsson, O., Haegerstrand, A., and Lundberg, J. M. (1991). Characterization, localization and actions of endothelins in umbilical vessels and placenta of man. *Acta Physiol. Scand.* 143, 395–404.
277. Napolitano, M., Miceli, F., Calce, A., Vacca, A., Gulino, A., Apa, R., and Lanzone, A. (2000). Expression and relationship between endothelin-1 messenger ribonucleic acid (mRNA) and inducible/endothelial nitric oxide synthase mRNA isoforms from normal and preeclamptic placentas. *J. Clin. Endocrinol. Metab.* 85, 2318–2323.
278. Matsumura, Y., Ikegawa, R., Tsukahara, Y., Takaoka, M., and Morimoto, S. (1990). Conversion of big endothelin-1 to endothelin-1 by two types of metalloproteinases derived from porcine aortic endothelial cells. *FEBS Lett.* 272, 166–170.
279. Schmidt, M., Kroger, B., Jacob, E., Seulberger, H., Subkowski, T., Otter, R., Meyer, T., Schmalzing, G., and Hillen, H. (1994). Molecular characterization of human and bovine endothelin converting enzyme (ECE-1). *FEBS Lett.* 356, 238–243.
280. Ahmad, Z. and Reznik, S. E. (2000). Immunohistochemical localization of ECE-1 in the human placenta. *Placenta* 21, 226–233.
281. Robert, B., Malassiné, A., Bourgeois, C., et al. (1996). Expression of endothelin precursor genes in human trophoblast in culture. *Eur. J. Endocrinol.* 134, 490–496.
282. Benigni, A., Gaspari, F., Orisio, S., et al. (1991). Human placenta expresses endothelin gene and corresponding protein is excreted in increasing amounts during normal pregnancy. *Am. J. Obstet. Gynecol.* 164, 844–848.
283. Smith, P. L., Lee, C. P., Pullen, M., Ohlstein, E. H., Beck, G., Eddy, E. P., et al. (1995). Non peptide endothelin receptor antagonists. IV. Identification of receptors in rabbit colonic muscosa and smooth muscle and correlation with physiological effects. *J. Pharmacol. Exp. Ther.* 272, 1204–1210.
284. Rutherford, R. A. D., Wharton, J., McCarthy, A., Gordon, L., Sullivan, M. H. F., Elder, M. G., et al. (1993). Differential localization of endothelin ETA and ETB binding sites in human placenta. *Br. J. Pharmacol.* 109, 544–552.
285. Hemsen, A., Gillis, C., Larsson, O., Haegerstrand, A., and Lundberg, J. M. (1991). Characterization, localization and actions of endothelins in umbilical vessels and placenta of man. *Acta Physiol. Scand.* 143, 395–404.
286. Kohnen, G., Mackenzie, F., Collett, G. P., Campbell, S., Davenport, A. P., Cameron, I. T., et al. (1997). Differential distribution of endothelin receptor subtypes in placentae from normal and growth-restricted pregnancies. *Placenta* 18, 173–180.
287. Nova, A., Sibai, B. M., Barton, J. R., Mercer, B. M., and Mitchell, M. D. (1991). Maternal plasma level of endothelin is increased in preeclampsia. *Am. J. Obstet. Gynecol.* 165, 724–727.
288. Schiff, E., Ben Baruch, G., Peleg, E., Rosenthal, T., Alcalay, M., Devir, M., and Mashiach, S. (1992). Immunoreactive circulating endothelin-1 in normal and hypertensive pregnancies. *Am. J. Obstet. Gynecol.* 166, 624–628.
289. Erdem, A., Erdem, M., Himmetoglu, O., Yildirim, G., and Arslan, M. (2003). Maternal and fetal plasma endothelin levels in intrauterine growth restriction: relation to umbilical artery Doppler flow velocimetry. *J. Perinat. Med.* 31, 52–59.
290. Di Iorio, R., Marinoni, E., Anceschi, M. M., Emiliani, S., Letizia, C., and Cosmi, E. V. (1996). Amniotic fluid endothelin-1 levels are increased in pregnancy-induced hypertension and intrauterine growth retardation. *Am. J. Reprod. Immunol.* 36, 260–263.
291. Poyner, D. R., Sexton, P. M., Marshall, I., Smith, D. M., Quirion, R., Born, W., Muff, R., Fischer, J. A., and Foord, S. M. (2002). International Union of Pharmacology. XXXII. The mammalian calcitonin gene-related peptides, adrenomedullin, amylin, and calcitonin receptors. *Pharmacol. Rev.* 54, 233–246.
292. Tsatsaris, V., Tarrade, A., Merviel, P., Garel, J. M., Segond, N., Jullienne, A., and Evain-Brion, D. (2002). Calcitonin gene-related peptide (CGRP) and CGRP receptor expression at the human implantation site. *J. Clin. Endocrinol. Metab.* 87, 4383–4390.
293. Lafond, J., St-Pierre, S., Masse, A., Savard, R., and Simoneau, L. (1997). Calcitonin gene-related peptide receptor in human placental syncytiotrophoblast brush-border and basal plasma membranes. *Placenta* 18, 181–188.
294. Dong, Y. L., Fang, L., Kondapaka, S., Gangula, P. R., Wimalawansa, S. J., and Yallampalli, C. (1999). Involvement of calcitonin gene-related peptide in the modulation of human myometrial contractility during pregnancy. *J. Clin. Invest.* 104, 559–565.
295. Chan, K. K., Robinson, G., and Pipkin, F. B. (1997). Differential sensitivity of human nonpregnant and pregnant myometrium to calcitonin gene-related peptide. *J. Soc. Gynecol. Investig.* 4, 15–21.
296. Yallampalli, C., Chauhan, M., Thota, C. S., Kondapaka, S., and Wimalawansa, S. J. (2002). Calcitonin gene-related peptide in pregnancy and its emerging receptor heterogeneity. *Trends Endocrinol. Metab.* 13, 263–269.
297. Florio, P., Margutti, A., Apa, R., Miceli, F., Pezzani, I., Degli Uberti, E. C., and Petraglia, F. (2001). Maternal plasma calcitonin gene-related peptide levels do not change during

labour and are not influenced by delivery route. *J. Soc. Gynecol. Investig.* 8, 165–168.

298. Schiff, E., Friedman, S. A., Sibai, B. M., Kao, L., and Schifter, S. (1995). Plasma and placental calcitonin gene-related peptide in pregnancies complicated by severe preeclampsia. *Am. J. Obstet. Gynecol.* 173, 1405–1409.
299. Halhali, A., Wimalawansa, S. J., Berentsen, V., Avila, E., Thota, C. S., and Larrea, F. (2001). Calcitonin gene- and parathyroid hormone-related peptides in preeclampsia: effects of magnesium sulfate. *Obstet. Gynecol.* 97, 893–897.
300. Strewler, G. J. (2000). The parathyroid hormone-related protein. *Endocrinol. Metab. Clin. North Am.* 29, 629–645.
301. Ardawi, M. S., Nasrat, H. A., and BA'Aqueel, H. S. (1997). Calcium-regulating hormones and parathyroid hormone-related peptide in normal human pregnancy and postpartum: a longitudinal study. *Eur. J. Endocrinol.* 137, 402–409.
302. Emly, J. F., Gregory, J., Bowden, S. J., Ahmed, A., Whittle, M. J., Rushton, D. I., and Ratcliffe, W. A. (1994). Immunohistochemical localization of parathyroid hormone-related protein (PTHrP) in human term placenta and membranes. *Placenta* 15, 653–660.
303. Curtis, N. E., Ho, P. W., King, R. G., Farrugia, W., Moses, E. K., Gillespie, M. T., Moseley, J. M., Rice, G. E., and Wlodek, M. E. (1997). The expression of parathyroid hormone-related protein mRNA and immunoreactive protein in human amnion and choriodecidua is increased at term compared with preterm gestation. *J. Endocrinol.* 154, 103–112.
304. Clemens, T. L., Cormier, S., Eichinger, A., Endlich, K., Fiaschi-Taesch, N., Fischer, E., Friedman, P. A., Karaplis, A. C., Massfelder, T., Rossert, J., Schluter, K. D., Silve, C., Stewart, A. F., Takane, K., and Helwig, J. J. (2001). Parathyroid hormone-related protein and its receptors: nuclear functions and roles in the renal and cardiovascular systems, the placental trophoblasts and the pancreatic islets. *Br. J. Pharmacol.* 134, 1113–1136.
305. Ferguson, J. E., 2nd, Gorman, J. V., Bruns, D. E., Weir, E. C., Burtis, W. J., Martin, T. J., and Bruns, M. E. (1992). Abundant expression of parathyroid hormone-related protein in human amnion and its association with labor. *Proc. Natl. Acad. Sci. U S A* 89, 8384–8388.
306. Ramirez, M. M., Fraher, L. J., Goltzman, D., Hendy, G. N., Matthews, S. G., Sangha, R., and Challis, J. R. (1995). Immunoreactive parathyroid hormone-related protein: its association with preterm labor. *Eur. J. Obstet. Gynecol. Reprod. Biol.* 63, 21–26.
307. Curtis, N. E., King, R. G., Moseley, J. M., Ho, P. W., Rice, G. E., and Wlodek, M. E. (1998). Intrauterine expression of parathyroid hormone-related protein in normal and pre-eclamptic pregnancies. *Placenta* 19, 595–601.
308. Ferguson, J. E., 2nd, Seaner, R. M., Bruns, D. E., Iezzoni, J. C., and Bruns, M. E. (1998). Expression and specific immunolocalization of the human parathyroid hormone/parathyroid hormone-related protein receptor in the uteroplacental unit. *Am. J. Obstet. Gynecol.* 179, 321–329.
309. Curtis, N. E., King, R. G., Moseley, J. M., Ho, P. W., Rice, G. E., and Wlodek, M. E. (1998). Intrauterine expression of parathyroid hormone-related protein in normal and pre-eclamptic pregnancies. *Placenta* 19, 595–601.
310. Hirota, Y., Anai, T., and Miyakawa, I. (1997). Parathyroid hormone-related protein levels in maternal and cord blood. *Am. J. Obstet. Gynecol.* 177, 702–706.
311. Bucht, E., Rong, H., Bremme, K., Granberg, B., Rian, E., and Torring, O. (1995). Midmolecular parathyroid hormone-related peptide in serum during pregnancy, lactation and in umbilical cord blood. *Eur. J. Endocrinol.* 132, 438–443.
312. Papantoniou, N. E., Papapetrou, P. D., Antsaklis, A. J., Kontoleon, P. E., Mesogitis, S. A., and Aravantinos, D. (1996). Circulating levels of immunoreactive parathyroid hormone-related protein and intact parathyroid hormone in human fetuses and newborns. *Eur. J. Endocrinol.* 134, 437–442.
313. Strid, H., Bucht, E., Jansson, T., Wennergren, M., and Powell, T. L. (2003). ATP dependent Ca(2+) transport across Basal membrane of human syncytiotrophoblast in pregnancies complicated by intrauterine growth restriction or diabetes. *Placenta* 24, 445–452.
314. Ahima, R. S., Saper, C. B., Flier, J. S., and Elmquist, J. K. (2000). Leptin regulation of neuroendocrine systems. *Front. Neuroendocrinol.* 21, 263–307.
315. Masuzaki, H., Ogawa, Y., Sagawa, N., Hosoda, K., Matsumoto, T., Mise, H., Nishimura, H., Yoshimasa, Y., Tanaka, I., Mori, T., and Nakao, K. (1997). Nonadipose tissue production of leptin: leptin as a novel placenta-derived hormone in humans. *Nat. Med.* 3, 1029–1033.
316. Akerman, F., Lei, Z. M., and Rao, C. V. (2002). Human umbilical cord and fetal membranes co-express leptin and its receptor genes. *Gynecol. Endocrinol.* 16, 299–306.
317. Mise, H., Sagawa, N., Matsumoto, T., Yura, S., Nanno, H., Itoh, H., Mori, T., Masuzaki, H., Hosoda, K., Ogawa, Y., and Nakao, K. (1998). Augmented placental production of leptin in preeclampsia: possible involvement of placental hypoxia. *J. Clin. Endocrinol. Metab.* 83, 3225–3229.
318. Lepercq, J., Cauzac, M., Lahlou, N., Timsit, J., Girard, J., Auwerx, J., and Hauguel-de Mouzon, S. (1998). Overexpression of placental leptin in diabetic pregnancy: a critical role for insulin. *Diabetes* 47, 847–850.
319. Tartaglia, L. A. (1997). The leptin receptor. *J. Biol. Chem.* 272, 6093–6096.
320. Hoggard, N., Hunter, L., Duncan, J. S., Williams, L. M., Trayhurn, P., and Mercer, J. G. (1997). Leptin and leptin receptor mRNA and protein expression in the murine fetus and placenta. *Proc. Natl. Acad. Sci. U S A* 94, 11073–11078.
321. Henson, M. C., Swan, K. F., and O'Neil, J. S. (1998). Expression of placental leptin and leptin receptor transcripts in early pregnancy and at term. *Obstet. Gynecol.* 92, 1020–1028.
322. Bodner, J., Ebenbichler, C. F., Wolf, H. J., Muller-Holzner, E., Stanzl, U., Gander, R., Huter, O., and Patsch, J. R. (1999). Leptin receptor in human term placenta: in situ hybridization and immunohistochemical localization. *Placenta* 20, 677–682.
323. Linnemann, K., Malek, A., Sager, R., Blum, W. F., Schneider, H., and Fusch, C. (2000). Leptin production and release in the dually in vitro perfused human placenta. *J. Clin. Endocrinol. Metab.* 85, 4298–4301.
324. Hardie, L., Trayhurn, P., Abramovich, D., and Fowler, P. (1997). Circulating leptin in women: a longitudinal study in the menstrual cycle and during pregnancy. *Clin. Endocrinol. (Oxf).* 47, 101–106.
325. Yura, S., Sagawa, N., Mise, H., Mori, T., Masuzaki, H., Ogawa, Y., and Nakao, K. (1998). A positive umbilical venous-arterial difference of leptin level and its rapid decline after birth. *Am. J. Obstet. Gynecol.* 178, 926–930.
326. Cetin, I., Morpurgo, P. S., Radaelli, T., Taricco, E., Cortelazzi, D., Bellotti, M., Pardi, G., and Beck-Peccoz, P. (2000). Fetal plasma leptin concentrations: relationship with different intrauterine growth patterns from 19 weeks to term. *Pediatr. Res.* 48, 646–651.
327. Shekhawat, P. S., Garland, J. S., Shivpuri, C., Mick, G. J., Sasidharan, P., Pelz, C. J., and McCormick, K. L. (1998). Neonatal cord blood leptin: its relationship to birth weight, body mass index, maternal diabetes, and steroids. *Pediatr. Res.* 43, 338–343.
328. Ertl, T., Funke, S., Sarkany, I., Szabo, I., Rascher, W., Blum, W. F., and Sulyok, E. (1999). Postnatal changes of leptin levels in full-term and preterm neonates: their relation to

intrauterine growth, gender and testosterone. *Biol. Neonate* 75, 167–176.

329. Matsuda, J., Yokota, I., Iida, M., Murakami, T., Naito, E., Ito, M., Shima, K., and Kuroda, Y. (1997). Serum leptin concentration in cord blood: relationship to birth weight and gender. *J. Clin. Endocrinol. Metab.* 82, 1642–1644.
330. Laml, T., Hartmann, B. W., Ruecklinger, E., Preyer, O., Soeregi, G., and Wagenbichler, P. (2001). Maternal serum leptin concentrations do not correlate with cord blood leptin concentrations in normal pregnancy. *J. Soc. Gynecol. Investig.* 8, 43–47.
331. Mise, H., Sagawa, N., Matsumoto, T., Yura, S., Nanno, H., Itoh, H., Mori, T., Masuzaki, H., Hosoda, K., Ogawa, Y., and Nakao, K. (1998). Augmented placental production of leptin in preeclampsia: possible involvement of placental hypoxia. *J. Clin. Endocrinol. Metab.* 83, 3225–3229.
332. Korbonits, M., Goldstone, A. P., Gueorguiev, M., and Grossman, A. B. (2004). Ghrelin—a hormone with multiple functions. *Front. Neuroendocrinol.* 25, 27–68.
333. Gualillo, O., Caminos, J., Blanco, M., Garcia-Caballero, T., Kojima, M., Kangawa, K., Dieguez, C., and Casanueva, F. (2001). Ghrelin, a novel placental-derived hormone. *Endocrinology* 142, 788–794.
334. Tanaka, K., Minoura, H., Isobe, T., Yonaha, H., Kawato, H., Wang D.F., Yoshida, T., Kojima, M., Kangawa, K., and Toyoda, N. (2003). Ghrelin is involved in the decidualization of human endometrial stromal cells. *J. Clin. Endocrinol. Metab.* 88, 2335–2340.
335. Gnanapavan, S., Kola, B., Bustin, S. A., Morris, D. G., McGee, P., Fairclough, P., Bhattacharya, S., Carpenter, R., Grossman, A. B., and Korbonits, M. (2002). The tissue distribution of the mRNA of ghrelin and subtypes of its receptor, GHS-R, in humans. *J. Clin. Endocrinol. Metab.* 87, 2988.
336. Makino, Y., Hosoda, H., Shibata, K., Makino, I., Kojima, M., Kangawa, K., and Kawarabayashi, T. (2002). Alteration of plasma ghrelin levels associated with the blood pressure in pregnancy. *Hypertension* 39, 781–784.
337. Cortelazzi, D., Cappiello, V., Morpurgo, P. S., Ronzoni, S., Nobile De Santis, M. S., Cetin, I., Beck-Peccoz, P., and Spada, A. (2003). Circulating levels of ghrelin in human fetuses. *Eur. J. Endocrinol.* 149, 111–116.
338. Farquhar, J., Heiman, M., Wong, A. C., Wach, R., Chessex, P., and Chanoine, J. P. (2003). Elevated umbilical cord ghrelin concentrations in small for gestational age neonates. *J. Clin. Endocrinol. Metab.* 88, 4324–4327.
339. Petraglia, F., Lim, A. T., and Vale, W. (1987). Adenosine 3′,5′-monophosphate, prostaglandins, and epinephrine stimulate the secretion of immunoreactive gonadotropin-releasing hormone from cultured human placental cells. *J. Clin. Endocrinol. Metab.* 65, 1020–1025.
340. Petraglia, F., Vaughan, J., and Vale, W. (1989). Inhibin and activin modulate the release of gonadotropin-releasing hormone, human chorionic gonadotropin, and progesterone from cultured human placental cells. *Proc. Natl. Acad. Sci. U S A* 86, 5114–5117.
341. Petraglia, F., Vaughan, J., and Vale, W. (1990). Steroid hormones modulate the release of immunoreactive gonadotropin-releasing hormone from cultured human placental cells. *J. Clin. Endocrinol. Metab.* 70, 1173–1178.
342. Barnea, E. R. and Kaplan, M. (1989). Spontaneous, gonadotropin-releasing hormone-induced, and progesterone-inhibited pulsatile secretion of human chorionic gonadotropin in the first trimester placenta in vitro. *J. Clin. Endocrinol. Metab.* 69, 215–217.
343. Owens, O. M., Ryan, K. J., and Tulchinsky, D. (1981). Episodic secretion of human chorionic gonadotropin in early pregnancy. *J. Clin. Endocrinol. Metab.* 53, 1307–1309.
344. Barnea, E. R., Ashkenazy, R., and Sarne, Y. (1991). The effect of dynorphin on placental pulsatile human chorionic gonadotropin secretion in vitro. *J. Clin. Endocrinol. Metab.* 73, 1093–1098.
345. Barnea, E. R., Ashkenazy, R., Tal, Y., Kol, S., and Sarne, Y. (1991). Effect of beta-endorphin on human chorionic gonadotrophin secretion by placental explants. *Hum. Reprod.* 6, 1327–1331.
346. Keelan, J., Song, Y., and France, J. T. (1994). Comparative regulation of inhibin, activin and human chorionic gonadotropin production by placental trophoblast cells in culture. *Placenta* 15, 803–818.
347. Siler-Khodr, T. M., Khodr, G. S., Valenzuela, G., and Rhode, J. (1986). GnRH effects on placental hormones during gestation. II. Progesterone, estrone, estradiol and estriol. *Biol. Reprod.* 34, 245–254.
348. Siler-Khodr, T. M., Khodr, G. S., Rhode, I., Vickerv, B. H., and Nestor, I. R. II. (1987). Gestational age related inhibition of placental hCG, hCG and steroid hormone release in vitro by a GnRH antagonist. *Placenta* 8, 1–14.
349. Wilson, E. A. and Jawad, M. J. (1980). LH-RI-I suppression of human placental progesterone production. *Fertil. Steril.* 33, 91–93.
350. Lacroix, M. C., Guibourdenche, J., Frendo, J. L., Pidoux, G., and Evain-Brion, D. (2002). Placental growth hormones. *Endocrine* 19, 73–79.
351. Yamaguchi, M., Endo, H., Maeda, T., Tahara, M., Ono, M., Miki, N., and Miyake, A. (1994). Transforming growth factor-beta 1 post-transcriptionally inhibits mouse growth hormone releasing factor secretion in placenta. *Biochem. Biophys. Res. Commun.* 204, 1206–1211.
352. de Zegher, F., Vanderschueren-Lodeweyckx, M., Spitz, B., Faijerson, Y., Blomberg, F., Beckers, A., Hennen, G., and Frankenne, F. (1990). Perinatal growth hormone (GH) physiology: effect of GH-releasing factor on maternal and fetal secretion of pituitary and placental GH. *J. Clin. Endocrinol. Metab.* 71, 520–522.
353. Evain-Brion, D., Alsat, E., Mirlesse, V., Dodeur, M., Scippo, M. L., Hennen, G., and Frankenne, F. (1990). Regulation of growth hormone secretion in human trophoblastic cells in culture. *Horm. Res.* 33, 256–259.
354. Patel, N., Alsat, E., Igout, A., Baron, F., Hennen, G., Porquet, D., and Evain-Brion, D. (1995). Glucose inhibits human placental GH secretion, in vitro. *J. Clin. Endocinol. Metab.* 80, 1743–1746.
355. Bjorklund, A. O., Adamson, U. K., Carlstrom, K. A., Hennen, G., Igout, A., Lins, P. E., and Westgren, L. M. (1998). Placental hormones during induced hypoglycemia in pregnant women with insulin-dependent diabetes mellitus: evidence of an active role for placenta in hormonal counter-regulation. *Br. J. Obstet. Gynaecol.* 105, 649–655.
356. Macaron, C., Kynel, M., and Rutsky, L. (1978). Failure of somatostatin to affect human chorionic somaiomammotropin and human chorionic eonadotrouin secretion in vitro. *J. Clin. Endocrinol. Metab.* 47, 141–156.
357. Artenisio, A. C., Volpe, A., Ragonese, F., Maccarrone, G., Forte, F., and Consolo, F. (1980). Behaviour of HPL and GH plasmatic rate in pregnant women at different times of their pregnancy during dynamic tests. *Horm. Metab. Res.* 12, 205–208.
358. Bhaumick, B., Dawson, E. P., and Bala, R. M. (1987). The effects of insulin-like growth factor 1 and insulin on placental lactogen production by human term placental explants. *Biochem. Biophys. Res. Commun.* 144, 674-682.
359. Morrish, D. W., Bhardwaj, D., Dabbagh, L. K., Marusyk, H., and Siy, O. (1987). Epidermal growth factor induces differentiation and secretion of human chorionic gonadotropin

and placental lactogen in normal human placenta. *J. Clin. Endocrinol. Metab.* 65, 1282–1290.

360. Lonberg, U., Damm, P., Andersson, A. M., Main, K. M., Chellakooty, M., Lauenborg, J., Skakkebaek, N. E., and Juul, A. (2003). Increase in maternal placental growth hormone during pregnancy and disappearance during parturition in normal and growth hormone-deficient pregnancies. *Am. J. Obstet. Gynecol.* 188, 247–251.
361. Verhaeghe, J., Bougoussa, M., Van Herck, E., de Zegher, F., Hennen, G., and Igout, A. (2000). Placental growth hormone and IGF-I in a pregnant woman with Pit-1 deficiency. *Clin. Endocrinol. (Oxf).* 53, 645–647.
362. Caufriez, A., Frankenne, F., Hennen, G., and Copinschi, G. (1993). Regulation of maternal IGF-I by placental GH in normal and abnormal human pregnancies. *Am. J. Physiol.* 265, E572–E577.
363. Petraglia, F., Sutton, S., and Vale, W. (1989). Neurotransmitters and peptides modulate the release of immunoreactive corticotropin-releasing factor from cultured human placental cells. *Am. J. Obstet. Gynecol.* 160, 247–251.
364. Cooper, E. S., Greer, I. A., and Brooks, A. N. (1996). Placental proopiomelanocortin gene expression, adrenocorticotropin tissue concentrations, and immunostaining increase throughout gestation and are unaffected by prostaglandins, antiprogestins, or labor. *J. Clin. Endocrinol. Metab.* 81, 4462–4469.
365. Petraglia, F., Florio, P., Benedetto, C., Marozio, L., Di Blasio, A. M., Ticconi, C., Piccione, E., Luisi, S., Genazzani, A. R., and Vale, W. (1999). Urocortin stimulates placental adrenocorticotropin and prostaglandin release and myometrial contractility in vitro. *J. Clin. Endocrinol. Metab.* 84, 1420–1423.
366. Sasaki, A., Shinkawa, O., and Yoshinaga, K. (1989). Placental corticotropin-releasing hormone may be a stimulator of maternal pituitary adrenocorticotropic hormone secretion in humans. *J. Clin. Invest.* 84, 1997–2001.
367. Chan, E. C., Smith, R., Lewin, T., Brinsmead, M. W., Zhang, H. P., Cubis, J., Thornton, K., and Hurt, D. (1993). Plasma corticotropin releasing hormone, β-endorphin and cortisol inter-relationship during human pregnancy. *Acta Endocrinol.* 128, 339–344.
368. Smith, R., Mesiano, S., Chan, E. C., Brown, S., and Jaffe, R. B. (1998). Corticotropin-releasing hormone directly and preferentially stimulates dehydroepiandrosterone sulfate secretion by human fetal adrenal cortical cells. *J. Clin. Endocrinol. Metab.* 83, 2916–2920.
369. Kaufmann, P. and Castellucci, M. (1997). Extravillous trophoblast in the human placenta. A review. *Trophoblast Res.* 10, 21–65.
370. Hamilton, G. S., Lysiak, J. J., Han, V. K. M., and Lala, P. K. (1998). Autocrine-paracrine regulation of human trophoblast invasiveness by IGF-II and IGFBP-1. *Exp. Cell. Res.* 244, 147–156.
371. Bass, K. E., Morrish, D., Roth, I., Bhardwaj, D., Taylor, R., Zhou, Y., and Fisher, S. J. (1994). Human cytotrophoblast invasion is up-regulated by epidermal growth factor: evidence that paracrine factors modify this process. *Dev. Biol.* 164, 550–561.
372. Chakraborty, C., Barbin, Y. P., Chakrabarti, S., Chidiac, P., Dixon, S. J., and Lala, P. K. (2003). Endothelin-1 promotes migration and induces elevation of [Ca2+]i and phosphorylation of MAP kinase of a human extravillous trophoblast cell line. *Mol. Cell. Endocrinol.* 201, 63–73.
373. Caniggia, I., Lye, S. J., and Cross, J. C. (1997). Activin is a local regulator of human cytotrophoblast cell differentiation. *Endocrinology* 138, 3976–3986.
374. Athanassiades, A., Hamilton, G. S., and Lala, P. K. (1998). Vascular endothelial growth factor stimulates proliferation but not migration or invasiveness in human extravillous trophoblast. *Biol. Reprod.* 59, 643–654.
375. Athanassiades, A. and Lala, P. K. (1998). Role of placenta growth factor (PIGF) in human extravillous trophoblast proliferation, migration and invasiveness. *Placenta* 19, 465–473.
376. Foidart, J. M., Hustin, J., Dubois, M., and Schaaps, J. P. (1992). The human placenta becomes haemochorial at the 13th week of pregnancy. *Int. J. Dev. Biol.* 36, 451–453.
377. Caniggia, I., Taylor, C. V., Ritchie, J. W., Lye, S. J., and Letarte, M. (1997). Endoglin regulates trophoblast differentiation along the invasive pathway in human placental villous explants. *Endocrinology* 138, 4977–4988.
378. Shi, Q. J., Lei, Z. M., Rao, C. V., and Lin, J. (1993). Novel role of human chorionic gonadotropin in differentiation of human cytotrophoblasts. *Endocrinology* 132, 1387–1395.
379. Yang, M., Lei, Z. M., and Rao, C. V. (2003). The central role of human chorionic gonadotropin in the formation of human placental syncytium. *Endocrinology* 144, 1108–1120.
380. Morrish, D. W., Bhardwaj, D., and Paras, M. T. (1991). Transforming growth factor beta 1 inhibits placental differentiation and human chorionic gonadotropin and human placental lactogen secretion. *Endocrinology* 129, 22–26.
381. Reynolds, L. P. and Redmer, D. A. (2001). Angiogenesis in the placenta. *Biol. Reprod.* 64, 1033–1040.
382. Dumont, D. J., Fong, G. H., Puri, M. C., Gradwohl, G., Alitalo, K., and Breitman, M. L. (1995). Vascularization of the mouse embryo: a study of flk-1, tek, tie, and vascular endothelial growth factor expression during development. *Dev. Dyn.* 203, 80–92.
383. Grimwood, J., Bicknell, R., and Rees, M. C. (1995). The isolation, characterization and culture of human decidual endothelium. *Hum. Reprod.* 10, 2142–2148.
384. Fong, G. H., Rossant, J., Gertsenstein, M., and Breitman, M. L. (1995). Role of the flt-1 receptor tyrosine kinase in regulating the assembly of vascular endothelium. *Nature* 376, 66–70.
385. Shalabi, F., Rossant, J., Yamaguchi, T. P., Gertenstein, M., Wu X-F, Breitman, M., and Schuh, A. C. (1995). Failure of blood island formation and vasculogenesis in flk-1 deficient mice. *Nature* 376, 62.
386. Carmeliet, P., Ferreira, V., Breier, G., Pollefeyt, S., Kieckens, L., Gertsenstein, M., Fahrig, M., Vandenhoeck, A., Harpal, K., Eberhardt, C., Declercq, C., Pawling, J., Moons, L., Collen, D., Risau, W., and Nagy, A. (1996). Abnormal blood vessel development and lethality in embryos lacking a single VEGF allele. *Nature* 380, 435–439.
387. Ferrara, N., Carver-Moore, K., Chen, H., Dowd, M., Lu, L., O'Shea, K. S., Powell-Braxton, L., Hillan, K. J., and Moore, M. W. (1996). Heterozygous embryonic lethality induced by targeted inactivation of the VEGF gene. *Nature* 380, 439–442.
388. Zheng, J., Vagnoni, K. E., Bird, I. M., and Magness, R. R. (1997). Expression of basic fibroblast growth factor, endothelial mitogenic activity, and angiotensin II type-1 receptors in the ovine placenta during the third trimester of pregnancy. *Biol. Reprod.* 56, 1189–1197.
389. Toth, P., Li, X., and Rao, C. V. (1994). Expression of functional human chorionic gonadotropin (hCG)/luteinizing hormone receptor gene in human uterine arteries. *J. Clin. Endocrinol. Metab.* 79, 307–315.
390. Zygmunt, M., Herr, F., Keller-Schoenwetter, S., Kunzi-Rapp, K., Munstedt, K., Rao, C. V., Lang, U., and Preissner, K. T. (2002). Characterization of human chorionic gonadotropin as a novel angiogenic factor. *J. Clin. Endocrinol. Metab.* 87, 5290–5296.
391. Boura, A. L., Walters, W. A., Read, M. A., and Leitch, I. M. (1994). Autacoids and control of human placental blood flow. *Clin. Exp. Pharmacol. Physiol.* 21, 737–748.

392. Poston, L., McCarthy, A. L., and Ritter, J. M. (1995). Control of vascular resistance in the maternal and feto-placental arterial beds. *Pharmacol. Ther.* 65, 215–239.
393. Hermus, A. R., Pieters, G. F., Willemsen, J. J., Ross, H. A., Smals, A. G., Benraad, T. J., and Kloppenborg, P. W. (1987). Hypotensive effects of ovine and human corticotrophin-releasing factors in man. *Eur. J. Clin. Pharmacol.* 31, 531–534.
394. Jain, V., Shi, S. Q., Vedernikov, Y. P., Saade, G. R., Chwalisz, K., and Garfield, R. E. (1998). In vivo effects of corticotropin-releasing factor in pregnant rats. *Am. J. Obstet. Gynecol.* 178, 186–191.
395. Jain, V., Vedernikov, Y. P., Saade, G. R., Chwalisz, K., and Garfield, R. E. (1999). Endothelium-dependent and -independent mechanisms of vasorelaxation by corticotropin-releasing factor in pregnant rat uterine artery. *J. Pharmacol. Exp. Ther.* 288, 407–413.
396. Sue-Tang, A., Bocking, A. D., Brooks, A. N., Hooper, S., White, S. E., Jacobs, R. A., Fraher, L. J., and Challis, J. R. (1992). Effects of restricting uteroplacental blood flow on concentrations of corticotrophin-releasing hormone, adrenocorticotrophin, cortisol, and prostaglandin E2 in the sheep fetus during late pregnancy. *Can. J. Physiol. Pharmacol.* 70, 1396–1402.
397. Gagnon, R., Murotsuki, J., Challis, J. R., Fraher, L., and Richardson, B. S. (1997). Fetal sheep endocrine responses to sustained hypoxemic stress after chronic fetal placental embolization. *Am. J. Physiol.* 272, E817–E823.
398. Florio, P., Severi, F. M., Fiore, G., Micheli, L., Bocchi, C., Nencini, C., Pezzani, I., Giorgi, G., and Petraglia, F. (2003). Impaired uterine artery blood flow at mid gestation and low levels of maternal plasma corticotropin-releasing factor. *J. Soc. Gynecol. Investig.* 10, 294–297.
399. Clifton, V. L., Read, M. A., Boura, A. L., Robinson, P. J., and Smith, R. (1996). Adrenocorticotropin causes vasodilatation in the human fetal-placental circulation. *J. Clin. Endocrinol. Metab.* 81, 1406–1410.
400. Clifton, V. L., Read, M. A., Leitch, I. M., Boura, A. L., Robinson, P. J., and Smith, R. (1994). Corticotropin-releasing hormone-induced vasodilatation in the human fetal placental circulation. *J. Clin. Endocrinol. Metab.* 79, 666–669.
401. Clifton, V. L., Read, M. A., Leitch, I. M., Giles, W. B., Boura, A. L., Robinson, P. J., and Smith, R. (1995). Corticotropin-releasing hormone-induced vasodilatation in the human fetal-placental circulation, involvement of the nitric oxide-cyclic guanosine 3′,5′-monophosphate-mediated pathway. *J. Clin. Endocrinol. Metab.* 80, 2888–2893.
402. Simoncini, T., Apa, R., Reis, F. M., Miceli, F., Stomati, M., Driul, L., Lanzone, A., Genazzani, A. R., and Petraglia, F. (1999). Human umbilical vein endothelial cells: a new source and potential target for corticotropin-releasing factor. *J. Clin. Endocrinol. Metab.* 84, 2802–2806.
403. Torpy, D. J., Webster, E. L., Zachman, E. K., Aguilera, G., and Chrousos, G. P. (1999). Urocortin and inflammation: confounding effects of hypotension on measures of inflammation. *Neuroimmunomodulation* 6, 182–186.
404. Leitch, I. M., Boura, A. L., Botti, C., Read, M. A., Walters, W. A., and Smith, R. (1998). Vasodilator actions of urocortin and related peptides in the human perfused placenta in vitro. *J. Clin. Endocrinol. Metab.* 83, 4510–4513.
405. Fallgren, B., Edvinsson, L., Ekblad, E., and Ekman, R. (1989). Involvement of perivascular neuropeptide Y nerve fibres in uterine arterial vasoconstriction in conjunction with pregnancy. *Regul. Pept.* 24, 119–130.
406. Fried, G., Hokfelt, T., Lundberg, J. M., Terenius, L., and Hamberger, L. (1986). Neuropeptide Y and noradrenaline in human uterus and myometrium during normal and preeclamptic pregnancy. *Hum. Reprod.* 1, 359–364.
407. Shirahase, H., Kanda, M., Kurahashi, K., and Nakamura, S. (2000). Endothelium-dependent relaxation followed by contraction mediated by NK(1) receptors in precontracted rabbit intrapulmonary arteries. *Br. J. Pharmacol.* 129, 937–942.
408. MacLean, M. R., Templeton, A. G. B., and McGrath, J. C. (1992). The influence of endothelin-1 on human feto-placental blood vessels: a comparison with 5-hydroxytryptamine. *Br. J. Pharmacol.* 106, 937–941.
409. Sabry, S., Mondon, F., Levy, M., Ferré, F., and Dinh-Xuan, A. T. (1995). Endothelial modulation of vasoconstrictor responses to endothelin-1 in human placental stem villi small arteries. *Br. J. Pharmacol.* 115, 1038–1042.
410. Myatt, L., Bewer, A. S., and Brockman, D. E. (1992). The comparative effects of big endothelin-1, endothelin-1, and endothelin-3 in the human placental circulation. *Am. J. Obstet. Gynecol.* 167, 1651–1656.
411. Sand, A. E., Andersson, E., and Fried, G. (2002). Effects of nitric oxide donors and inhibitors of nitric oxide signalling on endothelin- and serotonin-induced contractions in human placental arteries. *Acta Physiol. Scand.* 174, 217–223.
412. Hoeldtke, N. J., Wagner, R. K., Calhoun, B. C., and Hume, R. F. Jr. (2000). Vasodilatory response of fetoplacental vasculature to adrenomedullin after constriction with the thromboxane sympathomimetic U46619. *Am. J. Obstet. Gynecol.* 183, 1573–1578.
413. Jerat, S., Morrish, D. W., Davidge, S. T., and Kaufman, S. (2001). Effect of adrenomedullin on placental arteries in normal and preeclamptic pregnancies. *Hypertension* 37, 227–231.
414. Takahashi, Y., De Vroomen, M., Gournay, V., Roman, C., Rudolph, A. M., and Heymann, M. A. (1999). Mechanisms of adrenomedullin-induced increase of pulmonary blood flow in fetal sheep. *Pediatr. Res.* 45, 276–281.
415. Eguchi, S., Hirata, Y., Kano, H., Sato, K., Watanabe, Y., Watanabe T.X., Nakajima, K., Sakakibara, S., and Marumo, F. (1994). Specific receptors for adrenomedullin in cultured rat vascular smooth muscle cells. *FEBS Lett.* 340, 226–230.
416. Kohno, M., Kano, H., Horio, T., Yokokawa, J., Yasunari, K., and Takeda, T. (1995). Inhibition of endothelin production by adrenomedullin in vascular smooth muscle cells. *Hypertension* 25, 1185–1190.
417. Fiscus, R. R., Wang, X., and Hao, H. (1992). hCGRP8-37 antagonizes vasodilations and cAMP responses to rat CGRP in rat caudal artery. *Ann. N Y Acad. Sci.* 657, 513–515.
418. Fiscus, R. R., Zhou, H. L., Wang, X., Han, C., Ali, S., Joyce, C. D., and Murad, F. (1991). Calcitonin gene-related peptide (CGRP)-induced cyclic AMP, cyclic GMP and vasorelaxant responses in rat thoracic aorta are antagonized by blockers of endothelium-derived relaxant factor (EDRF). *Neuropeptides* 20, 133–143.
419. Mandsager, N. T., Brewer, A. S., and Myatt, L. (1994). Vasodilator effects of parathyroid hormone, parathyroid hormone-related protein, and calcitonin gene-related peptide in the human fetal-placental circulation. *J. Soc. Gynecol. Investig.* 1, 19–24.
420. Firth, K. F. and Pipkin, F. B. (1989). Human alpha- and beta-calcitonin gene-related peptides are vasodilators in human chorionic plate vasculature. *Am. J. Obstet. Gynecol.* 161, 1318–1319.
421. Gangula, P. R., Supowit, S. C., Wimalawansa, S. J., Zhao, H., Hallman, D. M., DiPette, D. J., and Yallampalli, C. (1997). Calcitonin gene-related peptide is a depressor in NG-nitro-L-arginine methyl ester-induced hypertension during pregnancy. *Hypertension* 29, 248–253.
422. Valdermarsson, S., Edvinsson, L., Hedner, P., and Ekman, R. (1990). Hormonal influence on CGRP in man: Effects of sex

differences and contraceptive pills. *Scand. J. Clin Lab. Invest.* 385–388.

423. Naghashpour, M., and Dahl, G. (2000). Sensitivity of myometrium to CGRP varies during mouse estrous cycle and in response to progesterone. *Am. J. Physiol. Cell. Physiol.* 278, C561–C569.
424. Gangula, P. R., Dong, Y. L., Wimalawansa, S. J., and Yallampalli, C. (2002). Infusion of pregnant rats with calcitonin gene-related peptide (CGRP)(8-37), a CGRP receptor antagonist, increases blood pressure and fetal mortality and decreases fetal growth. *Biol. Reprod.* 67, 624–629.
425. Gangula, P. R., Wimalawansa, S. J., and Yallampalli, C. (2000). Pregnancy and sex steroid hormones enhance circulating calcitonin gene-related peptide concentrations in rats. *Hum. Reprod.* 15, 949–953.
426. Gangula, P. R., Zhao, H., Wimalawansa, S. J., Supowit, S. C., DiPette, D. J., and Yallampalli, C. (2001). Pregnancy and steroid hormones enhance the systemic and regional hemodynamic effects of calcitonin gene-related peptide in rats. *Biol. Reprod.* 64, 1776–1783.
427. Nickols, G. A., Nickols, M. A., and Helwig, J. J. (1990). Binding of parathyroid hormone and parathyroid hormone-related protein to vascular smooth muscle of rabbit renal microvessels. *Endocrinology* 126, 721–727.
428. Mandsager, N. T., Brewer, A. S., and Myatt, L. (1994). Vasodilator effects of parathyroid hormone, parathyroid hormone-related protein, and calcitonin gene-related peptide in the human fetal-placental circulation. *J. Soc. Gynecol. Investig.* 1, 19–24.
429. Seckl, M. J., Fisher, R. A., Salerno, G., Rees, H., Paradinas, F. J., Foskett, M., and Newlands, E. S. (2000). Choriocarcinoma and partial hydatidiform moles. *Lancet* 356, 36–39.
430. Lewis, J. L. Jr. (1993). Diagnosis and management of gestational trophoblastic disease. *Cancer* 71, 1639–1647.
431. McCluggage, W. G., Ashe, P., McBride, H., Maxwell, P., and Sloan, J. M. (1998). Localization of the cellular expression of inhibin in trophoblastic tissue. *Histopathology* 32, 252–256.
432. Florio, P., Severi, F. M., Cobellis, L., Danero, S., Bome, A., Luisi, S., and Petraglia, F. (2002). Serum activin A and inhibin A. New clinical markers for hydatidiform mole. *Cancer* 94, 2618–2622.
433. Knight, G. J. and Palomaki, G. E. (2003). Epidemiologic monitoring of prenatal screening for neural tube defects and Down syndrome. *Clin. Lab. Med.* 23, 531–551.
434. Wald, N. J., Rodeck, C., Hackshaw, A. K., and Rudnicka, A. (2004). SURUSS in perspective. *BJOG* 111, 521–531.
435. Aitken, D. A., Wallace, E. M., Crossley, J. A., Swanston, I. A., van Pareren, Y., van Maarle, M., Groome, N. P., Macri, J. N., and Connor, J. M. (1996). Dimeric inhibin A as a marker for Down's syndrome in early pregnancy. *N. Engl. J. Med.* 334, 1231–1236.
436. Haddow, J. E., Palomaki, G. E., Knight, G. J., Williams, J., Miller, W. A., and Johnson, A. (1998). Screening of maternal serum for fetal Down's syndrome in the first trimester. *N. Engl. J. Med.* 338, 955–961.
437. Nicolaides, K. H. (2004). Nuchal translucency and other first-trimester sonographic markers of chromosomal abnormalities. *Am. J. Obstet. Gynecol.* 191, 45–67.
438. National High Blood Pressure Education Program Report of the National High Blood Pressure Education Program Working Group on high blood pressure in pregnancy. (2000). *Am. J. Obstet. Gynecol.* 183, S1–S22.
439. Roberts, J. M., and Cooper, D. W. (2001). Pathogenesis and genetics of pre-eclampsia. *Lancet* 357, 53–56.
440. Lambert-Messerlian, G. M., Silver, H. M., Petraglia, F., Luisi, S., Pezzani, I., Maybruck, W. M., Hogge, W. A., Hanley-Yanez, K., Roberts, J. M., Neveux, L. M., and Canick, J. A. (2000). Second-trimester levels of maternal serum human chorionic gonadotropin and inhibin a as predictors of preeclampsia in the third trimester of pregnancy. *J. Soc. Gynecol. Investig.* 7, 170–174.
441. Muttukrishna, S., North, R. A., Morris, J., Schellenberg, J. C., Taylor, R. S., Asselin, J., Ledger, W., Groome, N., and Redman, C. W. (2000). Serum inhibin A and activin A are elevated prior to the onset of pre-eclampsia. *Hum. Reprod.* 15, 1640–1645.
442. Albaiges, G., Missfelder-Lobos, H., Lees, C., Parra, M., and Nicolaides, K. H. (2000). One-stage screening for pregnancy complications by color Doppler assessment of the uterine arteries at 23 weeks' gestation. *Obstet. Gynecol.* 96, 559–564.
443. Florio, P., Reis, F. M., Pezzani, I., Luisi, S., Severi, F. M., and Petraglia, F. (2003). The addition of activin A and inhibin A measurement to uterine artery Doppler velocimetry to improve the early prediction of pre-eclampsia. *Ultrasound Obstet. Gynecol.* 21, 165–169.
444. Florio, P., Imperatore, A., Sanseverino, F., Torricelli, M., Reis, F. M., Lowry, P. J., and Petraglia, F. (2004). The measurement of maternal plasma corticotropin-releasing factor (CRF) and CRF-binding protein improves the early prediction of preeclampsia. *J. Clin. Endocrinol. Metab.* 89, 4673–4677.
445. Baschat, A. A. (2004). Pathophysiology of fetal growth restriction: implications for diagnosis and surveillance. *Obstet. Gynecol. Surv.* 59, 617–627.
446. Baker, D. J. P. and Robinson, R. J. (1992). Fetal and infant origin of adult disease. *Br. Med. J.* 308, 171–174.
447. Haddad, B., Abirached, F., Louis-Sylvestre, C., Le Blond, J., Paniel, B. J., and Zorn, J. R. (1999). Predictive value of early human chorionic gonadotrophin serum profiles for fetal growth retardation. *Hum. Reprod.* 14, 2872–2875.
448. Smith, G. C. S., Stenhouse, Crossley, J. A., Aitken, D. A., Cameron, A. D., and Connor, J. M. (2002). Early pregnancy levels of pregnancy-associated plasma protein A and the risk of intrauterine growth restriction, premature birth, preeclampsia, and stillbirth. *J. Clin. Endocrinol. Metab.* 87, 1762–1767.
449. Hobel, C. J., Dunkel-Schetter, C., Roesch, S. C., Castro, L. C., and Arora, C. P. (1999). Maternal plasma corticotropin-releasing hormone associated with stress at 20 weeks' gestation in pregnancies ending in preterm delivery. *Am. J. Obstet. Gynecol.* 180, S257–S263.
450. Leung, T. N., Chung, T. K., Madsen, G., McLean, M., Chang, A. M., and Smith, R. (1999). Elevated mid-trimester maternal corticotrophin-releasing hormone levels in pregnancies that delivered before 34 weeks. *Br. J. Obstet. Gynaecol.* 106, 1041–1046.
451. McLean, M., Bisits, A., Davies, J., Walters, W., Hackshaw, A., De Voss, K., and Smith, R. (1999). Predicting risk of preterm delivery by secondtrimester measurement of maternal plasma corticotropin-releasing hormone and α-fetoprotein concentrations. *Am. J. Obstet. Gynecol.* 181, 207–215.

Knobil and Neill's Physiology of Reproduction, Third Edition
edited by Jimmy D. Neill,
Elsevier © 2006

CHAPTER 54

Maternal Adaptation to Pregnancy

Kent L. Thornburg,[1,2] Susan P. Bagby,[2] and George D. Giraud[1–3]

Introduction, 2899
Metabolic Adaptations, 2900
Energy Costs of Pregnancy, 2901
Diet, Pregnancy, and Fetal Programming, 2901
Cardiovascular Adaptations, 2901
Blood Volume, 2901
Circulatory Adaptations, 2904
Respiratory System Adaptations, 2908
Hyperventilation, 2909
Lung Volumes and Function, 2909
Diffusing Capacity, 2910
Renal Adaptations, 2910
Overview, 2910
Recent Progress in Understanding Renal Adaptations to Pregnancy, 2911
The Kidney and Extracellular Fluid Volume/Blood Pressure Regulation in Pregnancy: The Relaxin Paradigm in the Pregnant Rat, 2915
Proposing an Integrated View of the Renal Response to Pregnancy, 2916
References, 2918

INTRODUCTION

Female mammals nurture their developing offspring within the intrauterine environment for at least the minimum period of time required for organ maturation and extrauterine survival. Offspring receive further nurture after birth in the form of milk from the maternal mammary gland that provides all nutritional requirements. Lactation provides the opportunity for the bonding and physical stimulation that is important in postnatal brain maturation (1). This form of reproduction reflects a highly successful reproductive strategy but demands a high energy cost from the mother in the form of remodeling of the maternal body, growing extra mammary tissue, carrying the weight of the developing fetus, consuming excess nutrients during pregnancy and postnatal nursing, as well as providing attention to the offspring for times ranging from weeks to years.

The purpose of this chapter is to cover current knowledge of the major physiological changes that accompany pregnancy in women. Many adaptations of pregnancy are imperceptible even to the woman carrying a child, yet are essential for a successful pregnancy. An incomplete cardiovascular adaptation to pregnancy can lead to improper fetal development or even fetal demise. The physiological processes that remodel the female body in preparation for a 40-week pregnancy have intrigued medical scientists for centuries. Yet, in spite of decades of research on the mechanisms that underlie changes in pregnancy, the most fundamental cellular and molecular mechanisms remain beyond reach. Pregnancy-related changes in the cardiorespiratory and renal systems have been most studied and are best understood. This chapter emphasizes the cardiovascular–respiratory–renal changes as examples of the elaborate modifications of the human body that make childbearing possible, and highlights crucial areas that need further investigation. Because there has been important progress in our understanding of renal adaptations to pregnancy, a large portion of the chapter is devoted

[1]Department of Physiology and Pharmacology;
[2]Department of Medicine, Heart Research Center, Oregon Health & Science University; [3]Portland VA Medical Center, Portland, Oregon.

to these new findings. Much of what we know about pregnancy has come from animal experimentation, and it is likely that many of the cellular details of signals for remodeling will be learned in the upcoming decades through the combined use of whole-animal studies, genetic models, and cell culture experiments.

METABOLIC ADAPTATIONS

Energy Costs of Pregnancy

The total energy cost of a pregnancy over its full-term duration has been estimated to be approximately 80,000 kcal (2,3). The average weight gain during pregnancy is approximately 12.5 kg and includes the increased weights of the fetus, amniotic fluid, uterine tissue, breast tissue, and blood, as well as extravascular and intracellular water. Changes in maternal body composition have been studied (4). Approximately 1 kg of weight gain is protein, 4 kg is fat, and 7 kg is water (5,6). Half of the protein weight accrues in the fetus and placenta. Pregnancy weight gain varies considerably among individuals, with very low or very high weight gains being related to worse fetal outcomes compared with moderate weight gains. Thus, the Institute of Medicine has proposed guidelines for weight gain based on a woman's prepregnancy body mass index (BMI; kg/m^2), with less weight gain recommended with each step increase in their prepregnancy BMI (7).

The maternal basal metabolic rate (BMR) also increases during pregnancy. BMR increases from approximately 1,300 kcals/day to approximately 1,700 kcal/day over gestation in women of average prepregnancy BMI (8). Figure 1 shows the increases in BMR above prepregnant levels over gestation for three groups of women with different BMIs. In reality, the total daily energy expenditure includes not only the BMR but additional energy expended in daily activities. Total energy expenditure ranges from approximately 2,700 to 3,000 kcal/day at 36 weeks, depending on BMI at the time of conception and the degree of activity of the three BMI groups (8). Hytten and Leitch proposed that the increased energy costs of pregnancy could be met by a reduction of activity energy expenditure without an increase in food energy consumption (5). Although this proposal might be correct, it has not been tested. It is clear that most women expend excess energy from a high level of activity during pregnancy and require a significant increase in food intake (8). There is evidence that metabolic rate is regulated differently in pregnant women compared with nonpregnant women. Before pregnancy, metabolic rate is most closely related to fat free body mass, whereas during the third trimester metabolic rate is most closely related to body fat mass (9). This may mean that the maternal metabolic rate is uniquely regulated by adipose reserves during pregnancy.

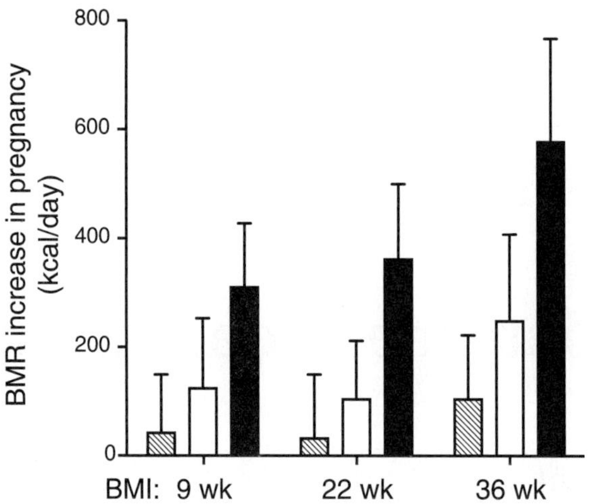

FIG. 1. Changes in basal metabolic rate (BMR) during pregnancy at three time points, 9 weeks, 22 weeks, and 36 weeks, in three groups of women. The first group had a below-average body mass index (BMI) before conception (BMI < 19.8 kg/m^2, *hatched bar*), the second group, a normal BMI (19.8 to 26.0 kg/m^2, *unfilled bar*), and the third group, a high BMI (>26.0 kg/m^2, *filled bar*). (Data from Butte, N. F., Wong, W. W., Treuth, M. S., Ellis, K. J., and O'Brian, S. E. [2004]. Energy requirements during pregnancy based on total energy expenditure and energy deposition. *Am. J. Clin. Nutr.* 79, 1078–1087.)

Oxygen consumption ($\dot{V}O_2$) increases by approximately 50 mL/min in pregnant women by term (10). Figure 2 shows the increases found in individual organs. As expected, the gravid uterus accounts for the largest increase in oxygen consumption. Term women sitting upright show increases in resting ($\dot{V}O_2$)

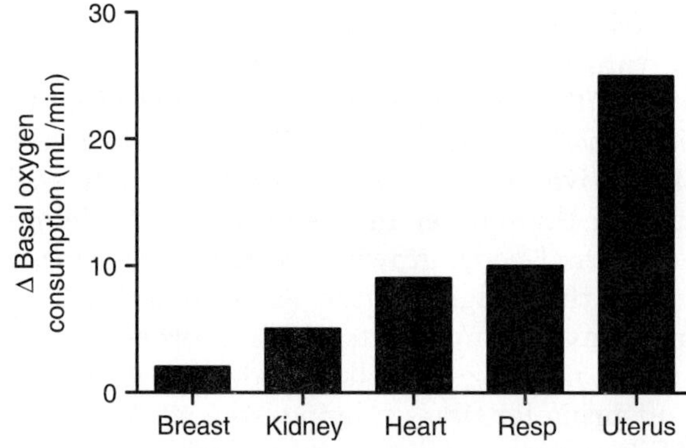

FIG. 2. Increases in basal oxygen consumption (mL/min) for individual organs in near-term pregnancy compared with nonpregnant consumption rates. (Data from Metcalfe, J., McAnulty, J., and Ueland, K. [1986]. *Burwell and Metcalfe's Heart Disease in Pregnancy: Physiology and Management.* Little, Brown, Boston.)

over nonpregnant women by approximately 33%, reflecting increased metabolism. There is evidence of a fundamental change in the regulation of the oxygen consumption threshold during pregnancy in sheep (11). It is not known whether this occurs in humans. However, it is known that physiological conditions affect oxygen consumption in pregnant women. For example, resting ($\dot{V}O_2$) is decreased by 15% in first-trimester women who experience hyperemesis gravidarum, severe "morning sickness," and are unable to consume a normal daily calorie intake (12).

Diet, Pregnancy, and Fetal Programming

There is increasing evidence from animal (13–15) and human (16) studies that maternal diet affects the growth patterns of the placenta, fetus, and offspring. It is also clear that fetuses who experience growth faltering before birth are at risk for chronic diseases as adults (17). Thus, evidence for links between maternal diet, fetal growth, and adult disease is becoming solidified. Shiell and coworkers studied the blood pressures of 626 young men and women of Motherwell, Scotland, whose mothers were recommended to eat a high-meat, low-carbohydrate diet during pregnancy (18). The offspring of mothers who had the highest consumption of meat and fish had the highest mean systolic pressures. These findings point to the importance of maternal diet and body composition in establishing the health of their offspring in later life. Maternal appetite is increased during pregnancy, but little is known about the drive for specific food types.

CARDIOVASCULAR ADAPTATIONS

Blood Volume

Blood volume increases continuously from 6–8 weeks of gestation, reaches a maximum at approximately 32 to 34 weeks, and changes little thereafter (19–23). Bernstein and associates were unable to detect plasma volume expansion until the 6th menstrual week (24). By 12 menstrual weeks, plasma volume had expanded by approximately 14% ± 12% (mean ± SD) over follicular phase estimates. Although the average blood volume increases from approximately 6.5 L before to 8.5 L during pregnancy, increases vary considerably among women, ranging from 30%–50%. Blood volume is known to increase with the number of fetuses (25,26); increases have been estimated in a singleton pregnancy at 1,570 mL, and at 1,960 mL in a twin pregnancy (25). Women show a similar plasma volume expansion pattern in their subsequent pregnancies (21,25). Most of the added volume of blood is found in the uterus, breast, kidney, striated muscle, and skin.

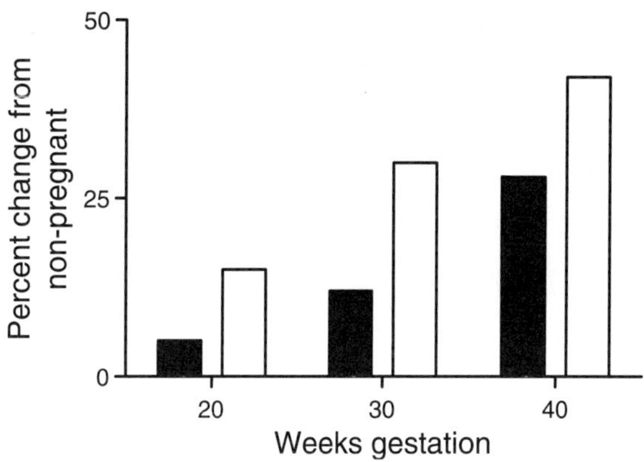

FIG. 3. Average red cell mass (*filled bar*) and blood volume (*open bar*) increase over the course of pregnancy. (Data from Scott, D. E. [1972]. Anemia in pregnancy. *Obstet. Gynecol. Annu.* 1, 219–244.)

The increase in plasma volume (40%–50%) is relatively greater than that of red cell mass (20%–30%), resulting in a decrease in hemoglobin concentration. Figure 3 shows that the decrease in hemoglobin concentration is not the result of simple hemodilution; both red cell mass and blood volume increased. If there is an iron deficiency, supplemental iron and folic acid intakes are necessary to restore hemoglobin levels to normal pregnant levels (~12 g/dL). Leukocyte counts are variable during gestation, but remain at the upper limits of normal; leukocyte production appears to be regulated independently of red cell production. Like red cell concentration, plasma protein concentrations and total plasma osmolality decrease during pregnancy.

Pregnancies complicated by fetal intrauterine growth restriction have lower mean maternal plasma volumes than do pregnancies with normal fetuses (2,976 ± 76 mL versus 3,594 ± 103 mL, respectively) (27). Maternal blood volume expansion supports fetal growth. The increased blood volume serves at least two purposes: (a) it facilitates maternal and fetal exchanges of respiratory gases, nutrients, and metabolites; and (b) it reduces the impact of maternal blood loss at delivery. Blood losses of 300 to 500 mL for vaginal births and 750 to 1,000 mL for cesarean sections are typical and are compensated to some degree by the so-called "autotransfusion" of blood from the contracting uterus. Further blood losses tap the resources of maternal stores.

The mechanism by which blood volume increases is not fully understood, yet there are a number of physiological changes that are thought to be important. The role of the kidney is discussed in detail later. There is evidence that systemic vascular resistance is decreased under the influence of gestational hormones, which leads to activation of the renin–angiotensin–aldosterone system, causing increased sodium retention and blood volume (28–30). There is evidence from experiments in baboons that systemic vasodilation precedes the measured increase in maternal blood volume as pregnancy progresses (31).

The placenta may also contribute to the hormonal milieu that affects maternal volume regulation. Placental estrogen synthesis depends on the estrogen precursor (dehydroepiandrosterone) from the fetal adrenal gland. Estrogen stimulates the renin–angiotensin system and thus promotes aldosterone production. Plasma renin levels, which are high during pregnancy, ensure high levels of angiotensin II (Ang II), which stimulates arginine vasopressin (AVP) release and water retention (28,30,32,33). In the kidney, increased mineralocorticoid activity appears to promote sodium and water retention by the distal renal tubule while, paradoxically, renal glomerular filtration and filtered sodium load are increased (34) and lead to increases in urinary sodium and water excretion (35). Extra-adrenal conversion of progesterone to deoxycorticosterone also increases mineralocorticoid activity (36). There is a net accumulation of sodium during pregnancy of approximately 500 to 900 mEq (37,38). Further studies are needed to integrate these physiological features of pregnancy.

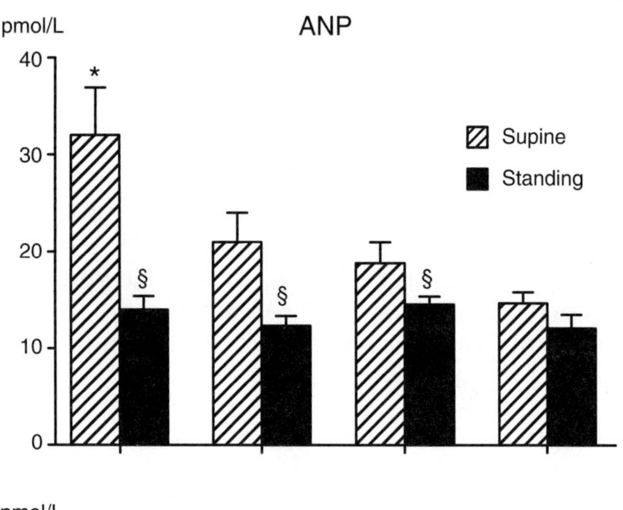

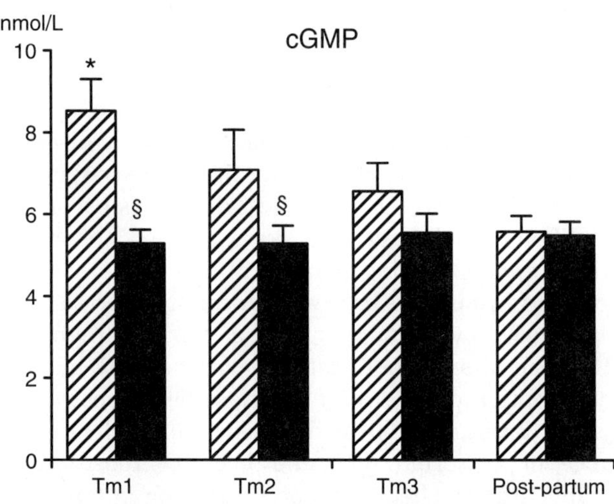

FIG. 4. Plasma atrial natriuretic peptide (ANP) and cyclic guanosine monophosphate (cGMP) at each trimester (Tm) of pregnancy and postpartum in 15 healthy women in the supine and upright positions (mean ± SEM). *p <0.05 vs. postpartum in the same position; §p <0.05 vs. corresponding supine value. (From Sala, C., Campise, M., Ambroso, G., Motta, T., Zanchetti, A., and Morganti, A. [1995]. Atrial natriuretic peptide and hemodynamic changes during normal human pregnancy. *Hypertension* 25, 631–636, with permission.)

Altitude

The hypobaric hypoxia of altitude causes an increase in hematocrit and hyperviscosity in long-term residents. Pregnant women adjust to altitude (~4,400 m) differently from women at sea level in that they show blunted adaptive hematological adaptations in general (39). Hematocrit and viscosity decrease with gestation in women at altitude, but much less than in women at sea level. Thus, women at altitude have a higher blood viscosity than women at sea level even during pregnancy.

Circulating Factors and Exercise

A number of circulating factors are important in blood volume regulation. Many of these are also powerful regulators of vascular tone and of vessel remodeling. Atrial natriuretic peptide (ANP) is maintained at normal levels during pregnancy or is increased depending on posture (40,41). Figure 4 shows plasma ANP levels from 15 healthy women as a function of gestational age and posture (42). In healthy men and nonpregnant women, plasma osmolality, plasma renin activity, and plasma concentrations of AVP, ANP, and aldosterone all increase with increasing exercise intensity. However, in nonpregnant individuals, high-intensity exercise has an antidiuretic and antinatriuretic effect, opposite to the effects expected when plasma ANP levels are elevated (43). Heenan and colleagues studied the responses to exercise in pregnant women, expecting that their hypervolemic state would attenuate the normal increases in Ang II and AVP (44). Thirteen pregnant women of 37 weeks' gestation were studied at rest and at two intensity levels as a percentage of the estimated anaerobic threshold (ventilatory threshold value). Figure 5 shows that plasma Ang II levels were high in these pregnant women but tended to decrease with increasing exercise intensity; in contrast, nonpregnant women had lower Ang II levels that increased with exercise.

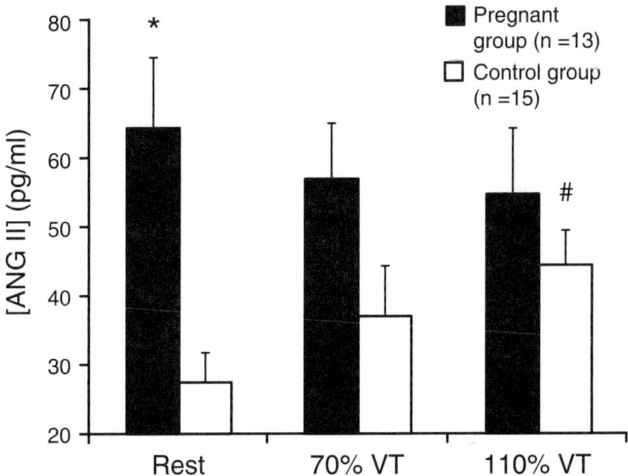

FIG. 5. Plasma angiotensin II concentration ([ANG II]) at rest and at two work rates as a percentage of the estimated anaerobic threshold or ventilatory threshold (VT). *Significant difference ($p < 0.05$) between groups. #Significant change ($p < 0.05$) within group from rest. (From Heenan, A. P., Wolfe, L. A., Davies, G. A., and McGrath, M. J. [2003]. Effects of human pregnancy on fluid regulation responses to short-term exercise. *J. Appl. Physiol.* 95, 2321–2327, with permission.)

The AVP response to exercise was completely blunted in these pregnant women. However, plasma levels of ANP were elevated with exercise.

Verkeste and associates found that the increase in plasma volume during pregnancy in rats results from fluid retention rather than from a shift in the extracellular fluid volume (ECFV) from the interstitium to the intravascular compartment (45). They found a gradual synchronous increase in all fluid compartments, without consistent change in their relative distributions. Chapman and associates found that peripheral vasodilation occurs in the luteal phase of the normal menstrual cycle in association with an increase in renal plasma flow and filtration (46). They found increased activation of the renin–angiotensin–aldosterone system, indicating that vasodilating mediators may be responsible for the observed hemodynamic changes. The authors suggest that hormonal changes during the normal menstrual cycle mimic changes seen in early pregnancy. This would argue for changes in hormonal environment as the primary driving force for changes in blood volume regulation.

Atrial Distention

The role of atrial distention in regulating maternal blood volume has been investigated and is controversial. Some scientists have suggested that the maternal atrium grows to accommodate increased blood volume, which, in turn, decreases the degree of stretch and blunts the stretch–ANP release response. That explanation is not likely to be true for two reasons: (a) an increase in systemic blood volume would not necessarily increase atrial transmural pressure, and (b) an increase in the radius of curvature of the atrial wall would increase wall stress and presumably make the atrium more sensitive to ANP release unless there was a disproportionate increase in wall thickness. Thus far, no study has provided convincing evidence for an increase in atrial stretch during pregnancy, although it may exist. Kaufman used the rat model to study whether pregnancy affects the sensitivity of atrial distention in the release of ANP and whether neural input from the atrial volume receptor is altered (47). Atrial distention was achieved by implanting small balloons positioned at the superior vena caval/right atrial junction of rats; inflation of the balloon did not impede blood flow through the heart. In virgin rats, atrial stretch caused increased urine production and urine sodium and potassium output and decreased free water clearance; plasma ANP levels also increased. No such response was observed in pregnant animals. Distention of isolated atria derived from unmated and 7-day pregnant rats resulted in an increase in secretion of ANP into the perfusate, whereas atria from 14- and 21-day pregnant rats were unresponsive to distention. The authors concluded that there is attenuation of both hormonal and neural responses to atrial distention in the pregnant rat. The sensitivity of ANP release from the heart may accommodate blood volume increases in pregnant women without eliciting homeostatic mechanisms to eliminate the extra fluid.

Deng and Kaufman found that atrial distention caused an increase in c-fos expression in the paraventricular nucleus of the lateral hypothalamus in rats (48). During pregnancy, this response was markedly attenuated. Zhang and Kaufman sought to determine whether nitric oxide (NO) was involved in cardiovascular homeostasis in pregnancy (49). They found that NO synthase (NOS) inhibition selectively reduced plasma volume in pregnant rats. Lo and Kaufman (2001) found that plasma levels of 5α-pregnan-3α-ol-20-one (pregnan), a potential regulator of NOS, is increased during pregnancy in rats (50). One milligram of pregnan given to rats over a 2-day period stimulated NO biosynthesis and an increase in plasma volume in nonpregnant rats. These data are interesting in that whereas NO is known to be a powerful vasodilator in pregnancy, it may also affect cardiac function (51).

During pregnancy in most species, the resting levels of plasma Ang II and adrenocorticotropic hormone (ACTH; corticotropin) are increased (52). Cortisol is undoubtedly important in blood volume regulation; when maternal cortisol is decreased experimentally in sheep, blood volume during pregnancy is reduced (53).

Lindheimer and Davison have shown that the osmotic thresholds for thirst and antidiuretic hormone release simultaneously decrease in pregnancy so that water intake is increased, water is retained, and body fluids are diluted (32). In pregnancy, AVP release is not suppressed at the usual level of body tonicity so that ingested water is retained. With increased body water, plasma osmolality declines until it is beneath the osmotic thirst threshold and a new steady state is established. Pregnancy is characterized by increments in intravascular volume, but volume-sensing mechanisms appear to accommodate as gestation progresses so that each new volume increment is "sensed" as normal. The metabolic clearance of AVP increases severalfold because of rising plasma cystine aminopeptidase (vasopressinase), which is produced by the placenta.

Vasopressin and ACTH responses to hypotension are altered in pregnant ewes (52,54,55) and the relationship between mean arterial pressure and vasopressin or ACTH response is shifted, consistent with a change in setpoint for regulation of mean arterial pressure. The decrease in hormone responses to hypotension is stimulus specific; ACTH responses to hypoglycemia are increased in the pregnant ewe and AVP responses to hyperosmolality are not altered in the pregnant ewe (52,54). The heart rate responses to hypotension are also decreased in pregnant ewes, consistent with the observation that baroreflex responses are decreased in the pregnant rat. The vasopressin and cortisol responses to hypotensive hemorrhage are also altered in the pregnant dog and rabbit (56,57); the relation between mean arterial pressure and hormone response is depressed. These data indicate that pregnant animals have a reduced ability to maintain arterial pressure during hemorrhage compared with nonpregnant animals. In addition, increases in heart rate, renal sympathetic activity, vasopressin, ACTH, and cortisol that are mediated by the baroreceptor reflex are reduced during pregnancy (57).

Circulatory Adaptations

Whole-body remodeling of the maternal circulation begins within a few weeks of conception. Circulatory changes continue through parturition, when most of the changes are reversed. Table 1 summarizes many of the changes that occur. Among the most profound changes are the early decrease in systemic vascular resistance and increase in cardiac output. These changes occur even before there is an increase in metabolic rate (58). The heart is remodeled with increased left ventricular chamber dimensions and increased left ventricular mass, reminiscent of changes observed in athletes (59–62).

TABLE 1. *Hemodynamic changes in pregnancy*

Increased	Decreased
Blood volume	Systemic vascular resistance
Uterine blood flow	Pulmonary vascular resistance
Red cell mass	Hematocrit
Left ventricular diastolic dimension	Arterial blood pressure
Stroke volume	Plasma albumin concentration
Heart rate	Arterial carbon dioxide tension
Arterial oxygen tension	Arterial hydrogen ion concentration
Arterial compliance	Colloid osmotic pressure

Cardiac Output

Cardiac output increases by more than 50% during pregnancy (Fig. 6). It peaks by the middle third trimester, but half of that increase takes place within the first 8 weeks, long before $\dot{V}O_2$ has increased. Thus, the increase in cardiac output ($\dot{Q}_{CO}$) is not the sole result of increased tissue demand:

$$\dot{Q}_{CO} = SV \times HR$$

Because cardiac output is the product of stroke volume (SV) and heart rate (HR), each has the potential to augment cardiac output in pregnancy. Figure 6 shows that both are important. Heart rate and stroke volume increase early in gestation and both reach their peak at mid-gestation. Stroke volume reaches a plateau by approximately 20 weeks and appears to decrease gradually over the remaining course of gestation. The magnitudes of change in stroke volume and heart rate are quite variable between individuals, and some studies have shown somewhat different patterns for these changes.

Stroke volume is regulated through increases in end-diastolic pressure, which changes wall stress (preload, the force/area in the myocardium), through decreases in systolic load (afterload, the force/area

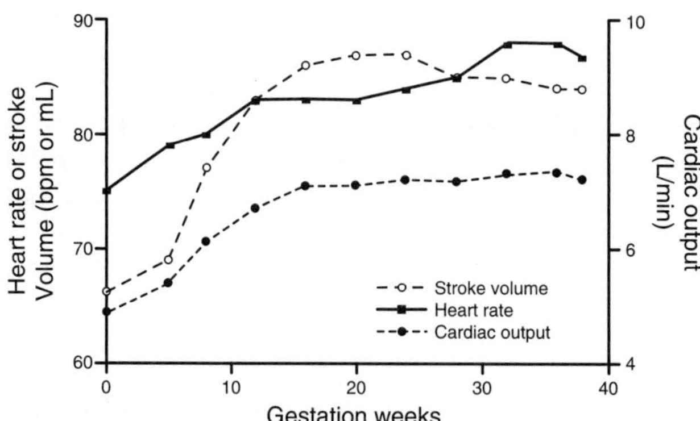

FIG. 6. Cardiac output (*closed circles*) increases over gestation as the product of increased stroke volume (*open circles*) and heart rate (*solid line*). (Data from Robson, S. C., Hunter, S., Boys, R. J., and Dunlop, W. [1989]. Serial study of factors influencing changes in cardiac output during human pregnancy. *Am. J. Physiol.* 256, H1060–H1065. Reproduced from Thornburg, K. L., Jacobson, S. L., Giraud, G. D., and Morton, M. J. [2000]. Hemodynamic changes in pregnancy. *Semin. Perinatol.* 24, 11–14, with permission.)

in the myocardium during ejection) and through changes in contractility, the contractile state in the myocardium. Of these three, only the decrease in afterload is a well-accepted mechanism for increasing stroke volume during pregnancy. Although many investigators have found increases in surrogate measures for preload, an equal number have not seen increases. However, the heart of the pregnant woman may be quite sensitive to changes in preload, so that profound alterations in cardiac anatomy could be driven by preload changes that are hardly measurable. One important factor that must be considered when measuring central venous or end-diastolic pressure is maternal posture. As for all adults, preload increases in the supine position compared with sitting upright. However, in the later stages of pregnancy, the weight of the uterus on the inferior vena cava in the supine posture changes filling pressures of the heart in a dramatic way. The importance of posture in determining plasma ANP levels is shown in Fig. 4.

There is even greater controversy over the role of contractility as an adaptation to pregnancy. Some authors have found increases (63), others have agreed with the Poppas study (64) that there is no change (65), and yet others have found a decrease in contractility (66,67). Differences in contractile state are notoriously difficult to measure in humans using noninvasive methods in which estimates are affected by the loading conditions of the heart. In the Poppas (64) and Mone (66) studies, contractility was estimated from a method believed to be load and heart rate independent, the velocity of circumferential fiber shortening relative to the stress in the wall at end systole. Yet the two groups had different findings. Thus, the degree to which preload and contractility are changed during pregnancy requires further study.

Arterial Blood Pressure

Mean arterial blood pressure decreases over the course of pregnancy and is lowest at mid-gestation. Systolic and diastolic pressures, as well as pulse pressure, drop modestly during pregnancy and reach their lowest values at approximately 20 weeks. This drop in pressure is the result of decreased systemic vascular resistance (Fig. 7). The behavior of the cardiovascular system can be understood in simplified form by applying a form of Ohm's law for direct current in a simple circuit. Under such conditions, current (I) is equal to the ratio of applied voltage (ΔV) divided by total resistance (R):

$$I = \Delta V/R$$

Following the same form for the systemic circuit of the cardiovascular system, the total flow through the system, cardiac output ($\dot{Q}_{CO}$) is equal to the driving

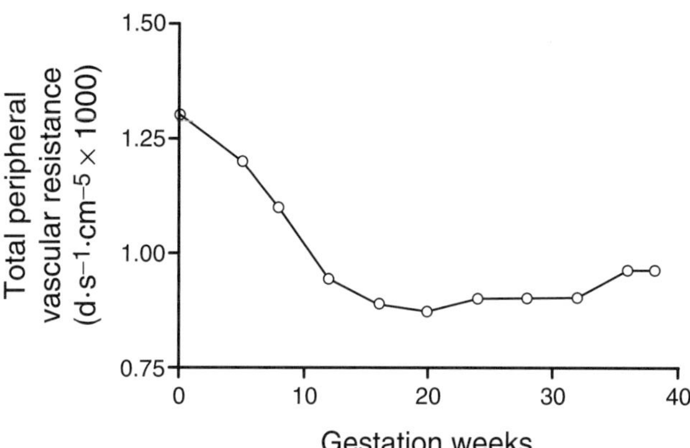

FIG. 7. Total peripheral resistance decreases dramatically over the first 16 weeks of pregnancy. (Data from Robson, S. C., Hunter, S., Boys, R. J., and Dunlop, W. [1989]. Serial study of factors influencing changes in cardiac output during human pregnancy. *Am. J. Physiol.* 256, H1060–H1065. Reproduced from Thornburg, K. L., Jacobson, S. L., Giraud, G. D., and Morton, M. J. [2000]. Hemodynamic changes in pregnancy. *Semin. Perinatol.* 24, 11–14, with permission.)

pressure across the systemic circuit, aortic pressure minus right atrial pressure (Pao − Pra) divided by the total vascular resistance of the circuit (TVR):

$$\dot{Q}_{CO} = (Pao - Pra)/TVR$$

Systemic arterial pressure decreases during the first trimester of pregnancy. Systolic, diastolic, and mean arterial pressure decrease by approximately 10% by the 7th week of pregnancy (23). Because, in an oversimplified view, arterial pressure is determined by the product of cardiac output and total vascular resistance, one or both of these factors must decrease to have a decrease in arterial pressure:

$$BP = \dot{Q}_{CO} \times TVR$$

Because cardiac output is increased early in pregnancy, it cannot account for a decreased arterial pressure in pregnancy. However, total vascular resistance is decreased. Thus, total resistance to flow is decreased to a greater degree than cardiac output is increased, leading to a drop in arterial pressure. It has been argued that the reduction in total vascular resistance is due to the presence of the dilated uterine bed. However, the gravid uterus, at its peak in flow, has been estimated to contribute approximately 20% to the decrease in total vascular resistance. Both systolic and diastolic blood pressure decrease until approximately 28 weeks, and then gradually approach prepregnant values by term (28). Blood pressure values are sensitive to the methods used to make the measurement. Ginsburg and Duncan found that mean systolic and diastolic blood pressures were lower by 6 and 15 mm Hg, respectively, when determined by catheter in the radial artery compared with sphygmomanometry (68). Similarly, Kirshon and associates found a lower systolic blood pressure by automated sphygmomanometry compared with a direct radial arterial measurement, but diastolic pressure was not different (69).

It is well known that hypertensive disorders are a leading cause of maternal and perinatal loss of life (70). It has been estimated that approximately 25% of cases of intrauterine growth retardation are associated with inadequate cardiovascular adaptations to pregnancy and especially insufficient remodeling of the uterine arterial vascular bed (71). There are a number of causes of hypertension in pregnancy, and undoubtedly other causes remain undiscovered. Steer and coworkers have shown that both hypertension and hypotension during pregnancy are associated with infants who are undergrown and who suffer high mortality (72). Among the adaptations to pregnancy are a reduced pressor response to vasoconstrictors such as Ang II, AVP, norepinephrine, and phenylephrine. A number of studies in whole animals and isolated vessels have sought to determine the underlying mechanisms for this adaptation (73–78). D'Angelo and Osol found an increased sensitivity to vasoconstrictors in uterine arcuate arteries in late pregnant rats compared with nonpregnant rats (78). Clearly, more investigation is required for clarification of the sensitivities of resistance vessels to constrictor–dilator substances in the pregnant woman.

Changes in central venous pressures during pregnancy have been difficult to prove, but lower-body venous pressures are known to rise under the influence of mechanical factors. From mid-pregnancy, the enlarged uterus compresses both the inferior vena cava and the lower aorta in the supine position. Obstruction of the inferior vena cava reduces venous return to the heart, leading to a fall in cardiac output by as much as 24% toward term when a woman is lying in the supine position. Most women compensate for the resultant decrease in stroke volume by increasing systemic vascular resistance and heart rate. Obstruction of the lower aorta and its branches can cause diminished blood flow to kidneys, uteroplacental unit, and lower extremities. During the last trimester, maternal kidney function is often lower in the supine than in the lateral position. Rarely, obstruction of the lower aorta and its branches can cause diminished blood flow to the placenta and result in insufficient transplacental gas exchange (10).

Vascular Changes

Changes in vascular behavior with pregnancy have been studied in a number of animal models and in humans. Most studies show that arterial distensibility is increased over the duration of pregnancy. Whether veins behave in the same way is more controversial. The early literature suggests that venous compliance is increased under the influence of sex steroid hormones and that venous pressures in the lower limbs are elevated during pregnancy and prone to stasis (79). Edouard and coworkers found that the distensibility and viscoelastic components of the lower limb veins decreased with duration of pregnancy and that upper limb veins made no such changes (80). Vasodilator substances that are known to be important in pregnancy include estrogen (81), NO (82), prostacyclin (81), relaxin (83), and calcitonin gene–related peptide (84). These vasodilators have effects in a number of vascular beds, including cerebral (85), mesenteric (84), coronary (86), and uterine vessels (81).

In sheep, vasodilatory prostacyclin and associated enzymes for its production are increased in the uterine arteries. Rupnow and colleagues showed that estrogen and progesterone have independent effects on the production of prostacyclin-regulating enzymes in endothelium and vascular smooth muscle of uterine, renal, coronary, and omental vessels (81). However, the combination of increased estrogen and progesterone, as seen with pregnancy, causes an increase in phospholipase 2 and cyclooxygenase in the uterine artery. Thus, changes in vascular remodeling and behavior are vessel specific.

The mechanisms that underlie the remodeling of the uterine vascular bed are the subject of intense investigation because the stakes in terms of fetal health are so profound. It is now well recognized that the uterine arteries of pregnant animals are extensively remodeled, being profoundly dilated in pregnancy compared with the nonpregnant state (76,82,87).

Twin Pregnancies

Maternal cardiovascular responses to pregnancy are exaggerated in twin pregnancies compared with singletons. Twin pregnancies carry a higher risk for gestational hypertension and preeclampsia (88,89). Kametas and coworkers reported that cardiac output is 20% greater, stroke volume is 15% greater, and heart rate is slightly higher in twin pregnancies than in singleton pregnancies (90). Left atrial diameter and left ventricular end-diastolic dimension are also larger. Others have found similar increases in cardiovascular hemodynamics (60,63,88,89).

Pulsatile Characteristics of the Cardiovascular System in Pregnancy

As mentioned previously, the use of Ohm's law to characterize the circulation, as one would on electrical circuit, is the most straightforward way to understand the relationships between steady flow, driving pressure, and resistance to flow. Using these relationships, we are able to understand changes that occur in the cardiovascular system under steady load conditions during pregnancy. However, because an intermittent pump drives blood flow, many of the characteristics of the circulatory system can be understood only by analyzing the relationships of pulsatile pressure and flow.

Poppas and coworkers assessed both steady and oscillatory aspects of the cardiovascular system using noninvasive measurements in 14 pregnant women during each trimester of pregnancy and postpartum (64). They used instantaneous subclavian artery pulse tracings, Doppler flow velocities, and left ventricular M-mode echocardiography (Fig. 8). Their measurements of steady hemodynamics and cardiac anatomy fit with previous reports. They found a pregnancy-related increase in left ventricular muscle mass and end-diastolic chamber dimension. They did not detect changes in myocardial contractility.

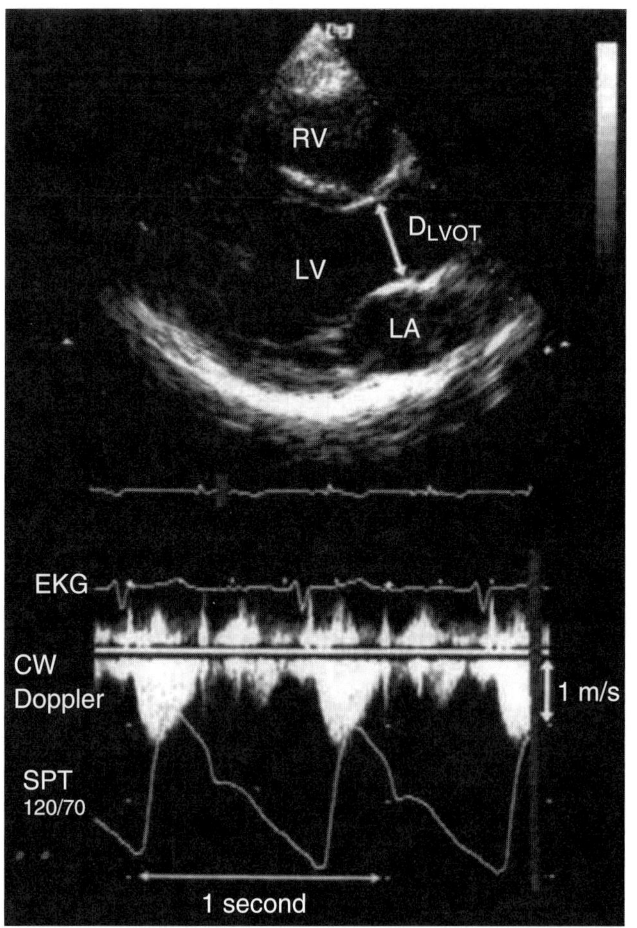

FIG. 8. Two-dimensional parasternal long-axis echocardiographic view showing measurements of left ventricular outflow tract diameter (LVOT). Also shown are simultaneous electrocardiogram (EKG), flow velocity (CW Doppler), and pressure profile from subclavian pulse tracing (SPT) in a pregnant woman. (From Poppas, A., Shroff, S. G., Korcarz, C. E., Hibbard, J. U., Berger, D. S., Lindheimer, M. D., and Lang, R. M. [1997]. Serial assessment of the cardiovascular system in normal pregnancy: role of arterial compliance and pulsatile arterial load. *Circulation* 95, 2407–2415, with permission.)

Poppas and coworkers found that global compliance in the subclavian artery increased with gestational age as total vascular resistance decreased (64). Global arterial compliance is the relationship between volume and pressure change during exponential diastolic pressure decay in an arterial segment ($\Delta V/\Delta P$) (91,92). This finding is undoubtedly related to well-documented changes in distensibility of arteries and veins during pregnancy under the influence of estrogen and relaxin, as discussed previously (80,93–96). There are two mechanisms that could alter the distensibility of the arterial tree at a given transmural pressure: (a) the vascular wall might be remodeled so that matrix elements have an altered behavior, and (b) vascular smooth muscle tone might be reduced. Unfortunately, simply testing the pressure–volume characteristics of specific vessels does not adequately distinguish between these possibilities under in vivo conditions.

One of the most important contributions of the Poppas study was the finding that oscillatory power (time-related work, $\dot{W}_{OSC}$) was progressively increased in pregnancy compared with the nonpregnant state (Fig. 9).

Hematological Factors

Blood volume gradually increases by an average of approximately 50% by term, but with great individual

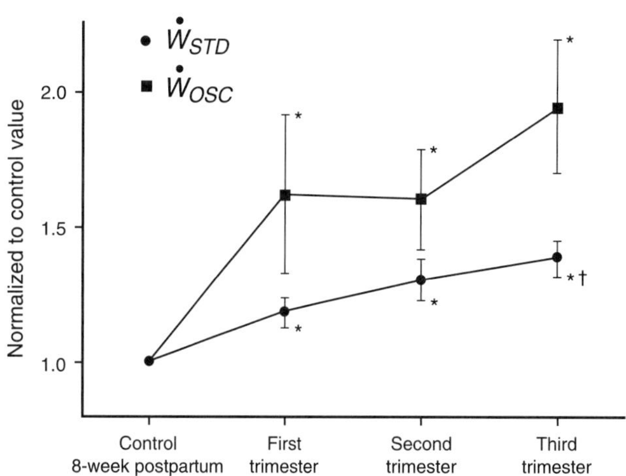

FIG. 9. Increases in steady ($\dot{W}_{STD}$) and oscillatory ($\dot{W}_{OSC}$) power during gestation. Data are normalized to 8-week postpartum control values (mean ± SEM). *$p < 0.05$ vs. control; †$p < 0.05$ vs. first trimester. (From Poppas, A., Shroff, S. G., Korcarz, C. E., Hibbard, J. U., Berger, D. S., Lindheimer, M. D., and Lang, R. M. [1997]. Serial assessment of the cardiovascular system in normal pregnancy: role of arterial compliance and pulsatile arterial load. *Circulation* 95, 2407–2415, with permission.)

variability (21). Pritchard reported an average increase in red blood cell mass of approximately 30% (450 mL) in both singleton and twin pregnancies as determined by the chromium (^{51}Cr)-labeled red blood cell technique (25). Thus, it has been known for decades that plasma erythropoietin levels and consequent erythropoiesis increase during pregnancy (97). However, more recent studies have shown that red cell mass is suppressed in the first trimester, as is erythropoiesis (98,99). Figure 10 shows that hematocrit decreases with gestational age as erythropoietin increases. The "setpoint" for red cell production appears to be altered with pregnancy because the rate of erythropoiesis is never adequate to maintain hematocrit at prepregnant levels. Thus, the increase in red cell mass during pregnancy is less than the increase in plasma volume and leads to the so-called "anemia of pregnancy" (21,100). It has been suggested that the availability of iron may limit erythropoiesis. Iron stores may be suboptimal in up to two thirds of reproductive-age women before pregnancy because of recurrent iron loss from menstruation (101). However, virtually all women have depressed hematocrits during pregnancy, regardless of iron stores (98). Therefore, the iron limitation hypothesis is unlikely to explain the decrease. It is important to emphasize that the decreased hematocrit is not due to simple hemodilution, as is often stated in textbooks. As mentioned, red cell production rate increases during pregnancy, but increases less than required to maintain red cell concentration in the face of increased plasma volume. Like red cell concentration, plasma protein concentrations and total plasma osmolality also decrease during pregnancy, which decreases blood viscosity.

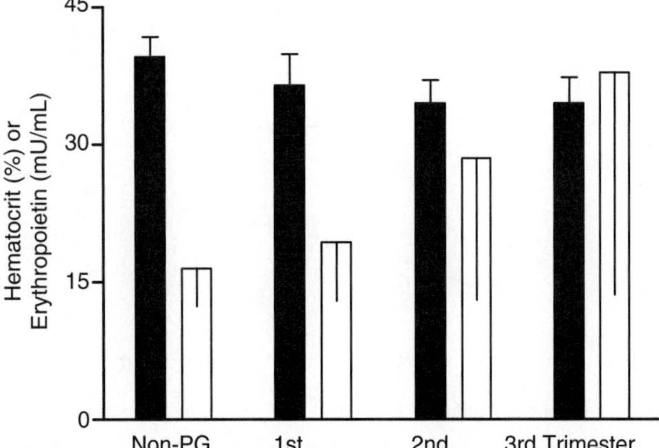

FIG. 10. Hematocrit (*filled bars* ± SD) and plasma erythropoietin concentrations (*unfilled bars*) in nonpregnant women and in pregnant women in each trimester. (Data from Beguin, Y., Lipscei, G., Thoumsin, H., and Fillet, G. [1991]. Blunted erythropoietin production and decreased erythropoiesis in early pregnancy. *Blood* 78, 89–93.)

Hemostasis

There is a commonly held view that maternal blood clots more easily during pregnancy. Some have argued from a teleological vantage that this hypercoagulability is related to the need to prevent excess hemorrhage with delivery of the placenta. However, a more important factor in stemming hemorrhage at birth is contraction of uterine smooth muscle, which occludes blood vessels in the uterus as the placenta is delivered. Regardless, the alteration in hemostasis with pregnancy carries an increased risk of thromboembolism. Thus, pregnant women are thought to be at increased risk for fatal pulmonary embolism and coronary occlusive disease (102,103). The hypercoagulability is related primarily to changes in coagulant and anticoagulant proteins in the blood (104). Table 2 shows coagulation proteins that are increased or decreased with pregnancy. Of note, procoagulant activity in the blood is primarily due to increases in factors VII, VIII, and X, as well as fibrinogen and von Willebrand's factor. Factor VII increases dramatically, up to tenfold, whereas plasma levels of fibrinogen may double. Von Willebrand's factor and factor VII increase late in pregnancy, whereas factor XI usually decreases (105). In spite of the view that pregnancy is a relatively hypercoagulable state, neither clotting nor bleeding times are outside the normal range. Thus, there is some controversy regarding the importance of the known changes in hemostasis. In pregnant women there are a number of conditions that may lead to pathological hypercoagulability and consequent fetal loss (106,107). These thrombophilias include antithrombin III deficiencies, protein C and S deficiencies, dysfibrinogenemias, activated protein C resistance, factor V Leiden mutation, and antiphospholipid syndrome (106).

RESPIRATORY SYSTEM ADAPTATIONS

It is not surprising that the respiratory system must accommodate the energy demands of pregnancy. The growing fetus consumes ever-increasing amounts

TABLE 2. *Coagulation proteins in pregnancy*

	Increased	Decreased
Procoagulants	Factors I, V, VII, VIII, IX, X	Factor XI
Anticoagulants	Soluble thrombomodulin	Protein S
Adhesive proteins	von Willebrand's factor	
Fibrinolytic proteins	Plasminogen activator inhibitors-1 and 2	Tissue plasminogen activator
Placental trophoblast	Tissue factor	Tissue factor pathway inhibitor

Modified from Brenner, B. (2004). Haemostatic changes in pregnancy. *Thromb. Res.* 114, 409–414.

of oxygen, and multiple gestations require even more at any corresponding age. Oxygen flow into the maternal blood must then increase not only because of increasing fetal oxygen consumption but because of increased maternal tissue mass and metabolic demand. Many of the respiratory adaptations have been the subjects of published studies and reviews (108–125). The most profound pregnancy-induced adaptations of the respiratory system arise from two maternal changes:

1. Increases in progesterone/estrogen that drive alveolar ventilation during pregnancy.
2. Reconfiguration of the chest wall and elevation of the diaphragm to accommodate growth of the gravid uterus and decreased mobility of the rib cage (125).

Hyperventilation

Minute ventilation ($\dot{V}_E$) and alveolar ventilation ($\dot{V}_A$) each increase by approximately 50% during pregnancy as a result of the central influence of progesterone, and perhaps depressed osmolality and augmented Ang II and AVP (108,115,117,120, 121,123,126). Hyperventilation begins at least by 8 weeks of gestation and depresses alveolar/arterial carbon dioxide pressure ($P{CO_2}$) from its normal resting prepregnant value of approximately 38 torr (menstrual cycle dependent) to approximately 30 torr at term; even lower values would not be unusual. The increased alveolar ventilation of pregnancy is maintained mostly by an increase in tidal volume without much change in respiratory rate. Depressed alveolar $P{CO_2}$ is accompanied by an increase in the partial pressure of oxygen in arterial blood ($Pa{O_2}$) to at least 108 torr early in gestation. Some investigators report a small decrease in $Pa{O_2}$ to approximately 104 torr at the very end of term. As the fetus grows and produces more carbon dioxide, the mother must excrete the extra carbon dioxide produced by the fetus in addition to her own. The changes in maternal arterial gas pressures improve the transplacental gradients across the placenta for oxygen, facilitating oxygen flux into the fetal compartment and, for carbon dioxide, facilitating the net carbon dioxide flux from the fetus to the mother. The decreased maternal arterial $P{CO_2}$ results in a mild respiratory alkalosis with appropriate renal compensation. Maternal serum pH is maintained at or below 7.45 during pregnancy with a depressed bicarbonate ion level of 18 to 21 mEq/L.

Lung Volumes and Function

Figures 11 and 12 show the primary changes in lung volumes and flows that begin after the second half of pregnancy and become pronounced during the third trimester (108,114,119). These changes include the following:

1. Vital capacity (VC; the amount of gas that can be expelled from the lungs following inspiration to total lung capacity) is unchanged.
2. Tidal volume (V_T) is increased by approximately 40%.
3. Functional residual capacity (FRC; the gas volume remaining in the lungs at rest after exhalation of

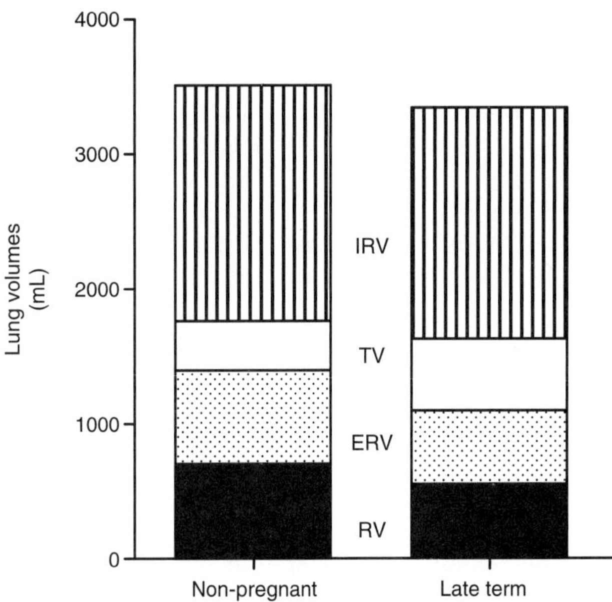

FIG. 11. Lung volumes before and during term pregnancy. IRV, inspiratory reserve volume; TV, tidal volume; ERV, expiratory reserve volume; RV, residual volume. (Data from Elkus, R., and Popovich, J. Jr. [1992]. Respiratory physiology in pregnancy. *Clin. Chest Med.* 13, 555–565.)

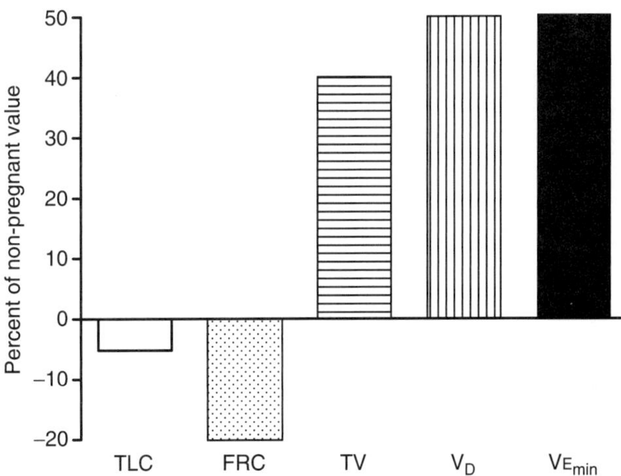

FIG. 12. Maximal percentage changes in respiratory parameters during pregnancy compared with the nonpregnant state. TLC, total lung capacity; FRC, functional residual capacity; TV, tidal volume; VD, dead space volume; VEmin, minute volume, rate of expired gas per minute. (Data from Crapo, R. O. [1996]. Normal cardiopulmonary physiology during pregnancy. *Clin. Obstet. Gynecol.* 39, 3–16.)

a tidal breath) is reduced by 20%. The component volumes of the FRC are the expiratory reserve volume (ERV; the volume of gas remaining after a tidal breath that can be exhaled with effort) and the residual volume (RV; the volume remaining in the lung after maximal expiratory effort). Each is reduced approximately equally by about 20%.

4. The sum of all volumes, the total lung capacity (TLC), is reduced by a modest 5%.
5. Dead space (V_D) increases by approximately 50%, but the resting dead space to tidal volume ratio (V_D/V_T) does not change because of the increase in tidal volume.

Virtually all respiratory adaptations of pregnancy resolve soon after delivery (124). Total airway resistance may be reduced during pregnancy, by unknown mechanisms, but overall lung compliance does not change. There are no major changes in standard respiratory functional indices using spirometry-based measurements. Forced expiratory flow in 1 second (FEV_1), mean peak expiratory flow rate (~450 mL/min), and vital capacity are virtually unchanged in U.S. women at sea level (127). However, ventilatory drive appears to be increased (116,122).

In 2004, McAuliffe and coworkers reported on respiratory physiological measurements in each trimester of women with singleton pregnancies at sea level (Lima, Peru) and at altitude (Cerro de Pasco/4,300 m) (128). They made three interesting findings: (a) nonpregnant and pregnant women alike who lived at altitude had larger lung volumes by approximately 1 L than did sea level women in spite of a 4-cm deficit in height for Cerro de Pasco women; (b) women living at sea level had greater TLC and RV in the third trimester compared with the first trimester; and (c) women living at altitude showed a lower FEV_1 in the third trimester compared with the first. These data suggest that pregnancy at altitude demands additional adaptations in respiratory function that are not needed in populations that have lived for generations at sea level.

In normal women, the oxygen cost of breathing is increased in late gestation by approximately 50%. This is due, in part, to the decreased chest wall compliance that accompanies the reconfiguration of the wall due to the increased intra-abdominal pressure as the growing uterus impinges on abdominal contents and the diaphragm. As pregnancy progresses, the circumference of the abdomen increases, the diaphragm is elevated by approximately 5 cm, and the chest configuration is altered. One aspect of the reconfiguration is the increase in the average costal angle that accompanies pregnancy earlier than can be explained by increased abdominal pressure with uterine growth. The transverse and anteroposterior dimensions of the chest increase, which leads to a circumferential increase in the lower thorax of approximately 5 to 7 cm (112,116,122,123).

Diffusing Capacity

The degree to which diffusing capacity of the lung for oxygen changes during pregnancy is controversial. Diffusing capacity for carbon monoxide (DLCO) measurements have been made in women during pregnancy. Because decreasing hemoglobin concentrations in pregnancy affect the measurement of the DLCO, one would expect decreased DLCO values even without any change in the physical attributes of the pulmonary barrier. Milne and coworkers reported an early increase (11 weeks) and a mid-gestation decrease in DLCO (115). Thus, there may be changes in barrier function, but further investigation is required to clarify this issue.

RENAL ADAPTATIONS

Overview

Early in pregnancy, well before the human embryo is of significant size, the maternal kidney undergoes dramatic changes in structure, hemodynamic regulation, and tubular handling of solute and water. The primary renal responses to pregnancy have been well documented in previous studies and summarized (129–133). These well-established changes in the kidney during pregnancy can be briefly summarized as follows.

Glomerular Filtration Rate and Effective Renal Plasma Flow

Glomerular filtration rate (GFR), as measured by creatinine or inulin clearance, increases by approximately 40%–65% by the end of the first trimester to a peak of approximately 180 mL/min (131,134–136), and is subsequently sustained until term (136). Effective renal plasma flow (RPF) also increases by 50%–85% during early pregnancy, and remains at this level until the final weeks of pregnancy, at which time it declines by 15%–25% (129,130,132–134).

Blood Urea Nitrogen and Creatinine

Blood urea nitrogen and creatinine levels decrease during pregnancy to values averaging 9.0 and 0.5 mg/dL, respectively, from nonpregnant values of 13.0 and 0.8 mg/dL (132). A serum creatinine value of greater than 0.8 mg/dL may therefore indicate abnormal renal function in pregnancy (137).

Protein Excretion

Total urinary nonalbumin protein excretion increases of up to 260 mg/day are normal during pregnancy (138). Urinary albumin excretion variably increases but remains within the normal pregnant range.

Sodium Handling

Renal tubular function is also changed during pregnancy (129,130,133,139). The filtered load of sodium increases significantly because of the increased GFR. Despite this, a concomitant increase in tubular reabsorption of sodium results in a net retention of up to 1 g of sodium per day (140,141).

Renin Release

Renin release and plasma renin activity increase approximately twofold early in pregnancy, are sustained through the second trimester, and plateau (136,142–144). This reflects reduced vascular filling (145), lower blood pressure, sympathetic nervous system (SNS) neural activation (136), and hormonal changes.

Water Metabolism

Pregnancy is also a period of marked water retention (133,140,141). Intravascular volume expands by approximately 1 to 2 L and extravascular volume by approximately 4 to 7 L (139).

Glycosuria, Aminoaciduria

Loss of glucose in the urine (glycosuria) is a normal finding during pregnancy, resulting from increased glomerular filtration and decreased distal tubular glucose reabsorption (146–150). Similarly, aminoaciduria in normal pregnancy is believed to reflect altered tubular reabsorptive activity (151).

Structural Changes

Kidneys increase in length and volume during pregnancy, reflecting increased vascular volume and interstitial fluid as well as nephron hypertrophy (129,133). "Physiological hydronephrosis" describes the typical pelvic and collecting system dilatation of pregnancy commonly attributed to hormonal influences (133).

Recent Progress in Understanding Renal Adaptations to Pregnancy

Recent investigations now address the cellular and molecular mechanisms that underlie renal modifications associated with pregnancy (132,152).

Renal Hemodynamic Changes in Pregnancy

A number of studies in normal rat (153–155), rabbit (156,157), sheep (158), and human (137,159–161) pregnancies have established that RPF and GFR progressively increase over the course of gestation as renal vascular resistance declines. In women, GFR peaks in the first trimester and maintains elevated levels of 40%–60% above nonpregnant levels until delivery (161,162). RPF also increases early but plateaus after 8–12 weeks of gestation. In the rat, both RPF and GFR peak in mid-gestation at approximately 30% above virgin levels (131). These hemodynamic changes are induced by maternal factors because they do not require the presence of the fetoplacental unit. Similar hemodynamic responses are observed in pseudopregnancy (163,164).

Chapman and colleagues reexamined the time course of renal hemodynamic changes in normal human pregnancy (n = 13) through serial measurements of women in the left lateral decubitus position, using gold standard methods (clearances of inulin and para-aminohippuric acid [PAH]) that included a preconception baseline and postconception measurements spanning 6–36 weeks' gestation (136) (Fig. 13). In agreement with prior work, RPF was increased

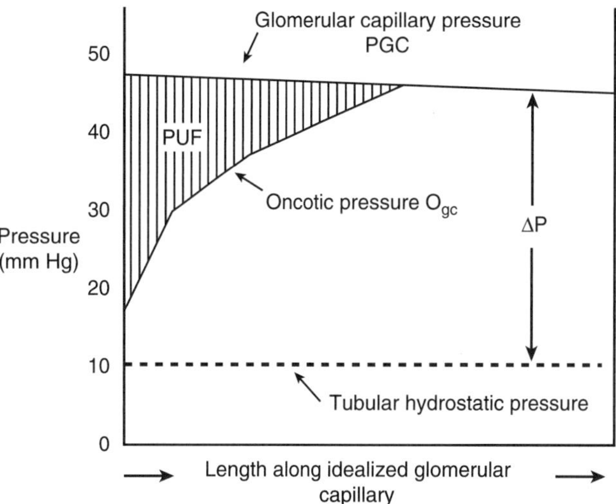

FIG. 13. Changes in renal plasma flow and glomerular filtration rate in normal human pregnancy. Percentage increases in effective renal plasma flow (ERPF; *solid line*) and glomerular filtration rate (GFR; *dashed line*) over the course of gestation are shown relative to the nonpregnant state. (Adapted from Paller, M. S., and Connaire, J. J. [2004]. The kidney and hypertension in pregnancy. In *Brenner and Rector's The Kidney,* 7th ed. B.M. Brenner [Ed.], pp. 1659–1695. Saunders, Philadelphia.)

by 6 weeks, reached a maximum at 8 weeks, and declined slightly by 36 weeks. Renal vascular resistance decreased in a mirror-image pattern, as expected (136). In contrast, GFR rose in parallel to RPF over approximately the first 6 weeks and then continued to increase until 12–24 weeks. Peak RPF increased to approximately 50% above nonpregnant levels at 8 weeks; the GFR increased by an average of approximately 27% by 12 weeks and 55% by 36 weeks. Filtration fraction (fraction of RPF filtered at the glomerulus) remained below prepregnancy values throughout gestation. In a separate study, Roberts and coworkers estimated 72% and 45% increases in RPF at 16 and 36 weeks' gestation, respectively (165). They also mostly agreed with the Chapman study, reporting an approximately 55% increase in GFR at both time points in normal pregnancy (n = 11).

Mechanisms of Increased GFR: Does Hyperfiltration in Pregnancy Convey Risk of Renal Injury?

Understanding the increased GFR in pregnancy is important in considering whether pregnancy can induce injury to the normal kidney (131). GFR is determined by the average net filtration pressure in the glomerular capillary (Fig. 14). An increase in glomerular capillary pressure (PGC) is the typical mechanism underlying glomerular hyperfiltration in pathological states of nephron loss when sustained chronically.

The net filtration pressure at any point along the glomerular capillary is determined by the algebraic sum of hydrostatic and osmotic pressures across the filtration barrier. The net hydrostatic pressure is the pressure difference (ΔP) between PGC and the tubular hydrostatic pressure (P_T); the oncotic pressure component is determined by plasma oncotic pressure (πgc) alone. The net hydrostatic pressure across the capillary wall favors fluid flow out of the capillary and into the renal tubule; at any point along the capillary, ΔP is offset by the oncotic pressure that opposes water flow from the capillary. The formula for GFR can be expressed as:

$$GFR = (\Delta P - \pi gc)\ K_f$$

where K_f is a filtration constant that incorporates the filtration surface area and intrinsic hydraulic conductivity of the glomerular capillary wall. The average ultrafiltration pressure is the summation of pressures along the capillary length, represented in Fig. 14 as the shaded area. An additional factor influencing GFR is the RPF, which determines the rate of rise of πgc along the capillary length (Fig. 14). Thus, an increase in RPF, when changed in isolation, slows the rate of rise of πgc, enhancing the net filtration pressure. When comparing across species, the degree to which changes in RPF affects GFR is believed to differ substantially in rats versus humans. Under basal conditions, the rat kidney is in "filtration pressure equilibrium"—that is, the rise in oncotic pressure leads to cessation of filtration at a point before the end of the capillary length (Fig. 14, solid line for πgc). Plasma flow has a large influence on GFR in the rat because flow increases (e.g., dotted line in Fig. 14) allow filtration to occur along a greater fraction of the capillary length, thereby substantially increasing filtration surface area. In the human kidney, where "filtration pressure disequilibrium" pertains (Fig. 14), filtration occurs along the entire capillary length under basal conditions; thus, increases in RPF cannot recruit additional filtration surface area and have a more modest effect on GFR.

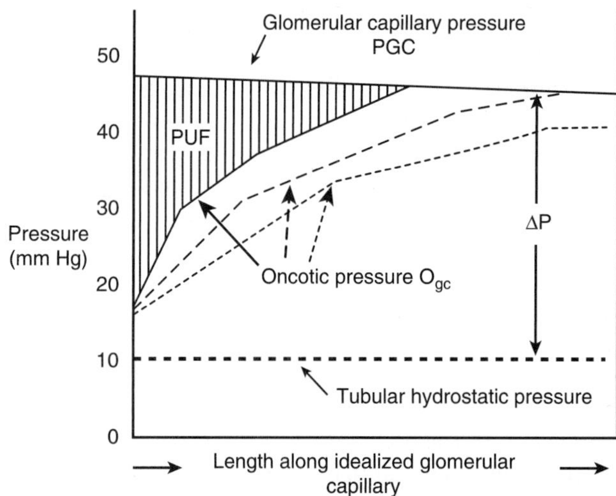

FIG. 14. Regulation of glomerular ultrafiltration. At any point along the length of a glomerular capillary, the force driving filtration is a function of net hydraulic pressure favoring filtration (ΔP = glomerular capillary pressure [PGC] minus intratubular pressure [P_T]) and net colloid osmotic pressures (effectively, glomerular capillary oncotic pressure [ngc] opposing filtration). Increased glomerular filtration rate (GFR) can be induced by an increase in plasma flow, by a decrease in oncotic pressure, or by an increase in glomerular capillary pressure. GFR is proportional to net ultrafiltration pressure, the *shaded area* described between the PGC and πgc profiles. "Filtration equilibrium" refers to an oncotic pressure profile (e.g., *solid line A* or *dashed line B*) where filtration ceases before the end of the capillary. The rat operates in filtration equilibrium under basal conditions. With increased plasma flow rate, the πgc profile can shift to, for example, line B or line C. The parallel increase in PUF area reflects the increase in GFR due to elevation of plasma flow. In "filtration pressure disequilibrium" (πgc profile C), πgc rises sufficiently slowly that it never reaches ΔP; thus filtration continues to occur along the entire capillary length. Humans, in contrast to rat, are believed to operate basally in filtration disequilibrium. An increase in renal plasma flow (RPF) thus cannot increase filtration surface area and thus has much more limited impact on GFR. (Adapted from Maddox, D. A., and Brenner, B. M . [2004]. Glomerular ultrafiltration. In *Brenner and Rector's The Kidney,* 7th ed. B.M. Brenner [Ed.], pp. 353–412. Saunders, Philadelphia.)

Glomerular Hemodynamics in Rat Pregnancy

Micropuncture studies in the rat, where GFR is substantially dependent on RPF, have shown that the pregnancy-associated increase in single-nephron GFR is accounted for by the increase in RPF (166,167). Directly measured pressure in the glomerular capillary, PGC, was unchanged by pregnancy in rats, indicating that (a) hyperfiltration in rat pregnancy does not require glomerular capillary hypertension, and (b) renal vasodilation involves proportional reductions in resistances in the afferent and efferent arterioles (167). Because a value for the filtration coefficient, K_f, could not be determined, a change in K_f could not be excluded in the normal pregnant rat.

To address directly the potential for renal injury with repetitive pregnancies, Baylis and Rennke tested the long-term effect of five consecutive, closely spaced pregnancies in normal Munich-Wistar rats (168). No increase in PGC was apparent compared with age-matched virgins, and no adverse effects of repeated pregnancies on kidney function or structure were observed (168). Although Reckelhoff found a greater tendency for renal functional decline in Sprague-Dawley rats after eight successive pregnancies, no proteinuria or renal structural pathology was noted (169). Studies in the rat indicate that the renal adaptation to pregnancy does not involve intracapillary hypertension, that pregnancy-induced changes fully resolve during the postpartum period, and that renal hyperfiltration during repetitive pregnancies does not appear to impose a risk of renal injury (131).

Intraglomerular Hemodynamics in Human Pregnancy

Renal micropuncture is not possible in humans. Furthermore, because rat intraglomerular hemodynamics differ significantly from the human (see earlier), interspecies inferences are potentially unreliable. Therefore, investigations have attempted to estimate determinants of GFR in human pregnancy using computerized mathematical modeling approaches. RPF, GFR, πgc, and clearances of neutral dextrans across a range of sizes were measured and combined with estimates for unmeasurable parameters like mean pressure of the glomerular capillary and K_f (162,165,170,171). Roberts and coworkers suggested that the mid-gestation increase in GFR of 55% could be fully explained by increased RPF, whereas the late GFR increases are likely due to additional factors, including a reduction from peak RPF as well as a fall in πgc of 5.9% in early gestation and 19.7% in late gestation (165). The authors suggest a lack of evidence supporting glomerular capillary hypertension as the primary driver for increased GFR across a range of feasible K_f and ΔP values. In response to technical criticisms, this group repeated their studies with improved protocols using two different mathematical models (170). They again concluded that the 38% increase in GFR in late pregnancy found in the new study reflects increased RPF, reduced πgc, and increased K_f, with no increase in the net driving pressure across the capillary wall.

Postpartum Glomerular Hemodynamics in Human Pregnancy

Postpartum observations in women add perspective to our understanding of the mechanisms that lead to the resolution of renal adaptive responses during pregnancy and to the timing of these mechanisms. Lafeyette and coworkers studied women immediately after cesarean delivery, using computer models of glomerular parameters (172). They reported a persistent glomerular hyperfiltration without hyperperfusion; the former was attributed entirely to the persistent decrease in plasma oncotic pressure after delivery. The same group reported similar studies at 1 day and 2 weeks postpartum in 22 uncomplicated pregnancies (162). On the first day after delivery, the model again supported reduced capillary oncotic pressure as the predominant factor in maintaining a 41% increase in GFR. By 2 weeks postpartum, GFR was still increased by 20% even though RPF and oncotic pressure had returned to normal nonpregnant levels; thus, only unmeasurable parameters could be postulated to maintain GFR, such as an increase in ΔP, K_f, or both. Milne and coworkers reported that GFR and RPF had returned to prepregnancy values by 4 months after delivery (170).

Despite these careful efforts to enhance our knowledge of renal changes in human pregnancy, caution is in order. Inferences about ΔP necessarily involve modeling based on rat data and require assumptions that cannot be validated at present. Moreover, these theoretical considerations do not address the possibility that the glomerular hypertrophy of pregnancy, through increased glomerular capillary length, might create glomerular filtration pressure equilibrium in the human kidney as in the rat. This would enhance the flow dependence of GFR. Glomerular hypertrophy might also increase filtration surface area. Either the increased capillary length or the increased filtration area could obviate the need for increased PGC. In addition, even if renal adaptations in pregnancy are safe in the setting of normal renal function, even modest impairment of renal function in human pregnancy (GFR < 50 mL/min) imposes a substantial risk of further renal functional loss (173–175). Thus, it is not yet known whether each pregnancy increases the risk for injury to the kidney.

Glomerular Capillary Permselectivity to Neutral Macromolecules in Human Pregnancy

An estimate of glomerular capillary wall porosity to macromolecules in human pregnancy has been made, using fractional clearance of neutral dextrans spanning a wide size range. This measurement is independent of hydraulic permeability. Neutral dextrans are not restricted in their passage across the glomerular filter by charge or configurational factors, nor are they absorbed or secreted by the renal tubule. Hence, they are classically used to assess the glomerular size-selective barrier function. It was a surprise when Lindheimer's group found reduced fractional clearance of dextrans at the lower size ranges during late gestation (165,170). This finding points to an altered structure of the glomerular capillary membrane. Their findings could be explained by (a) an increase in molecular movement through a nondiscriminatory "shunt" pathway, or (b) an increased range of pore sizes in the late-gestation glomerular capillary wall (170).

Urinary Protein and Albumin Excretion in Pregnancy

Protein that is present in normal urine derives from filtered protein that escapes proximal tubular reabsorption or from secreted protein, primarily Tamm-Horsfall protein formed by the distal nephron segments. The urinary albumin excretion rate can increase owing to a decrease in the discrimination of the glomerular capillary wall or to a reduced tubular reabsorption of normally filtered proteins. This leads to an increased excretion of nonalbumin low–molecular-weight proteins—below 10,000 Da—together with small increases in albumin. Several studies report that total urinary protein—but *not* albumin—increases by approximately twofold by late gestation (170,176–178). Higby and coworkers reported an average total protein excretion of 117 mg/day with an upper 95% confidence limit of 260 mg/day in 270 healthy pregnant subjects (138). In normal pregnancy, albumin excretion increased minimally to an average of 12 mg/day with an upper limit of 29 mg/day, thus remaining within the normal nonpregnant range. Collectively, this suggests that proximal tubular reabsorption of normally filterable low–molecular-weight proteins (but not albumin) may be selectively reduced. Alternatively, increases in plasma levels of filterable smaller–molecular-weight proteins or increased tubular synthesis/secretion of Tamm-Horsfall protein could account for the increase.

Changes in albumin excretion in human pregnancy remain controversial. Misiani and coworkers measured albumin by radioimmunoassay in healthy pregnant women and found a reduction in urinary albumin excretion to below nonpregnant levels throughout gestation (178). This was in distinct contrast to gravidas with pregnancy-induced hypertension, where albumin excretion began increasing by 28 weeks' gestation (178). Milne and coworkers found that urinary albumin was higher than nonpregnant levels in 11 healthy women in late gestation and persisted at 4 months postpartum (170). But, once again, these changes were within the normal nonpregnant range. Furthermore, whereas the increase in glomerular permselectivity to small neutral dextrans fully normalizes by 12 weeks postpartum, urinary albumin remains at late-pregnant levels (165,170). These findings suggest that additional factors favoring increased albumin excretion may be operative. Of note, increased urinary amino acid excretion during pregnancy has been documented and could indicate a more generalized impairment of proximal tubular reabsorptive capacities (151).

Structural Changes of Kidney and Genitourinary Tract

During the course of normal pregnancy, the kidneys enlarge by up to 30% (133). In one large autopsy series, kidney weights during or shortly after delivery (range, 300 to 310 g) exceeded the nonpregnant norms; glomerular size, but not cell number, was increased (179). X-ray (129,133) and ultrasound (180) studies have shown that kidneys increase in length by approximately 1 cm during human gestation. In rats, kidney dry weight during pregnancy has been found to increase along with increases in proximal tubular length, changes resembling those in compensatory renal hypertrophy (181,182). In other studies, increased renal weight was attributed to renal water content (183). These changes, seen in more than 90% of gravidas, are associated with dilation of the renal calyces, pelves, and ureters (129,184–186), the latter often more marked on the right because of dextrorotation of the uterus (187). Dilation of the renal calyces is accompanied by ureteral smooth muscle hypertrophy and connective tissue hyperplasia. The mechanisms that lead to these remarkable changes are unknown. Renal structural changes may persist up to 4 months postpartum (186).

It is unknown whether increased compliance of these structures in pregnancy reflects the actions of relaxin on smooth muscle, as described for the vasculature and uterus in the rat (83,188), the actions of relaxin to alter connective tissue structure of the rodent cervix (189), or other hormone-related mechanisms.

The Kidney and Extracellular Fluid Volume/Blood Pressure Regulation in Pregnancy: The Relaxin Paradigm in the Pregnant Rat

Investigators have long struggled to understand how the regulatory system for renal sodium reabsorption and ECFV is co-opted to retain Na^+ and produce a positive Na^+ balance, while avoiding the expected feedback suppression of Na^+ retention. It is not clear, for example, whether observed renal changes are secondary to systemic signals, or whether they are initiators of systemic responses. One view holds that primary systemic vasodilation induces secondary renal Na^+ and water retention, driving the expansion of ECFV and blood volume that is characteristic of pregnancy from its early stages (30,34,132). An alternative view is that the kidney vasodilates early in pregnancy and acts as an arteriovenous shunt, which secondarily activates systemic Na^+/water–retaining pathways (132). Neither view alone fully reconciles renal responses in pregnancy with known principles of hemodynamic regulation. In the nonpregnant state, the introduction of extrarenal vasodilation leads to vasoconstriction of renal circulation as a key feature of its response to circulatory underfilling. In contrast, the renal circulation undergoes a remarkable vasodilation from early in pregnancy, concomitantly with systemic vasodilation (30). We propose another option: that systemic and renal resistances are reduced proportionally by a common stimulus.

Candidate Vasoactive Mediators in Rat Pregnancy

Attempts to identify specific systemic or locally active vasodilators that account for all the hemodynamic changes in pregnancy have failed. Factors such as estrogen, progesterone, prolactin, and vasodilatory prostaglandin species may be important but do not explain all of the vasodilator actions needed for adaptation to pregnancy (129,132). However, in elegant studies in the pregnant rat, Conrad and colleagues have documented a major role for relaxin, a transforming growth factor-β family member with systemic vasodilating capacity (132,190). Relaxin is produced by the corpus luteum and levels rise in temporal association with the early progressive fall in systemic vascular resistance as pregnancy progresses (191). In the nonpregnant rat, Conrad and coworkers found that relaxin reproduces many of the characteristics of the pregnant vasculature: reduced vasomotor tone, increased global arterial compliance, modification of vascular extracellular matrix structure, and reduced vascular stiffness (measured independently of tone) (96,191). These effects of relaxin were found in small renal arteries (96) as well as in numerous arteriolar, capillary, and postcapillary venular sites throughout the body (188). In the pregnant rat, relaxin was shown to mediate maternal renal vasodilation and hyperfiltration (190), to reduce the myogenic activity of small renal arteries, and to reduce their reactivity to Ang II (192). Subsequent work by this group elucidated downstream roles of endothelin (193) and NO (194,195) in mediating the renal vasodilation and hyperfiltration effects of relaxin in rat pregnancy (132). Specifically, the vascular effects of relaxin were mediated by endothelin 1 (ET1), acting through the endothelin B (ETB) receptor to stimulate NO production (96,193,196). Moreover, relaxin was found to increase the ETB-to-endothelin A (ETA) receptor ratio. Jeyabalan and coworkers showed that increases in the vascular gelatinase, matrix metalloproteinase 2 (MMP2), are essential for relaxin-induced renal vascular effects, possibly through conversion of big ET1 to ET1-32 (197). This work establishes a novel mechanism underlying classic pregnancy-induced renal hemodynamic changes in the rat, and concomitantly unveils a unique MMP2-linked vasodilatory pathway with potentially broad relevance for renal vascular regulation.

Renal Tubular Changes during Pregnancy

Before considering the implications of this novel vasodilatory mechanism for pregnancy-associated hemodynamic adaptations, it is important first to consider recently recognized factors acting on tubular sodium handling during pregnancy. Both antinatriuretic (sodium ion retentive) and natriuretic (sodium loss in urine) forces are present during pregnancy. What allows the former to predominate is not clear, but pregnancy is unequivocally associated with renal sodium retention (129,133). The positive sodium balance in women over the course of gestation averages 2 to 6 mEq of sodium daily, with cumulative gains of 500 to 950 mEq (129,133). Studies in humans using lithium clearance as a sodium marker document enhanced proximal tubular sodium reabsorption in pregnancy (198). Micropuncture studies in the rat have also shown that sodium delivery to the end of the accessible proximal tubule is normal in pregnancy, indicating that the excess glomerular filtrate is fully reabsorbed proximally (199). Paradoxically, this increased reabsorptive work in the proximal tubule is not accompanied by increases in renal microsomal Na^+/K^+ adenosine triphosphatase (ATPase) activity, the primary energy source for tubular reabsorption throughout the nephron (200).

A mechanical factor has been shown to influence proximal tubular sodium retention in pregnancy. Khraibi and coworkers documented a reduction in intrarenal interstitial hydrostatic pressure (RIHP) in the pregnant rat (201–203), a feature which should enhance sodium reabsorptive capacity (204). In the nonpregnant rat, increased renal perfusion pressure or volume expansion is associated with an increase in RIHP and subsequently natriuresis (204); if the increase in RIHP is prevented by stripping the renal capsule, no natriuresis ensues. In pregnant rats, basal RIHP was found to be low and failed to increase normally after raised renal perfusion pressure or saline expansion (202,203); however, an expansion of the ECFV with saline did induce a normal natriuretic response, whereas an increase in perfusion pressure did not. It is useful to consider how reduced basal RIHP might influence sodium-retaining forces (elevated Ang II, norepinephrine) versus the natriuretic forces (variably increased ANP) (136). Low RIHP would be expected to reduce the passive intercellular backleak of sodium (with chloride and water) in the leaky proximal tubule, causing more reabsorbate to be retained and thereby enhancing net reabsorptive flux. Sodium reabsorptive activity of Ang II and SNS traffic in the proximal tubule could thereby be substantially amplified. In contrast, the natriuretic response to ANP was blunted in pregnant rats (205). RIHP would seem less likely to influence more distal tubular segments because of their greater water impermeability.

In the distal tubule, additional factors are believed to modulate Na/H_2O retention during pregnancy. The mineralocorticoid hormone, aldosterone, which is secreted from the adrenal cortex, is essential for normal renal sodium and potassium balance and thus indirectly regulates blood volume. The rate at which aldosterone is secreted and consequent plasma levels are increased in normal gestation, frequently approaching levels associated with primary hyperaldosteronism (129,136). In addition, it is known that a specific mutation in the mineralocorticoid receptor leads to enhanced hypertension during pregnancy (206). Experiments in sheep suggest that loss of aldosterone or cortisol leads to a decrease in blood volume during pregnancy (206). The sex steroid progesterone has both agonistic and antagonistic activity at the mineralocorticoid receptor. However, in vivo, efficient renal enzymes metabolize progesterone, effectively protecting the mineralocorticoid receptor (207,208). However, pregnancy also increases intrarenal enzymes which metabolize progesterone (209,210). The net effect of hormone action is not yet clear. Nonetheless, overall, the kidney effectively retains 2 to 6 mEq of sodium daily during gestation, establishing ECFV increases of 5 to 7 L. The increased levels of aldosterone appear necessary to achieve this critically important adaptation.

The relaxin story offers a potential solution to one major pregnancy conundrum: how to entice the kidney to generate and maintain expanded ECFV in anticipation of future fetoplacental needs. The key elements of the solution in the rat are now discernible (132). First, a single systemic hormone (relaxin) simultaneously mediates vasorelaxation at multiple systemic sites and in the kidney. Second, the downstream vasodilatory effect of relaxin is ensured by upregulation of a classic vasoconstrictor system (ET1) while enhancing its vasodilatory activity through an increased ratio of ETB:ETA receptor. Third, the final common pathway of this hormonal system is NO, a potent vasodilator that also has known affects on capillary permeability and structure. Finally, relaxin has been shown to have widespread effects on tissue structure to increase compliance, a recurring theme in tissue responses to pregnancy (132,188,211).

Proposing an Integrated View of the Renal Response to Pregnancy

The actions of relaxin in the pregnant rat create a physiologically unique set of intrarenal conditions. We propose the following canonical model to explain many of the features of mammalian pregnancy and associated changes in the kidney. First, primary vasodilation of both systemic and renal vasculatures generates systemic vascular underfilling through increased vascular capacity, a change likely to activate central venous stretch receptors. The increased SNS traffic and its subsequent stimulation of renal renin release initiates renal salt and water retention and simultaneously stimulates both preglomerular renal vasoconstriction and renal tubular Na/H_2O reabsorptive capacity. Relaxin-ETB receptor–induced NO concomitantly mediates resistance to the pressor effects of Ang II and norepinephrine. This ensures that renal preglomerular vasodilation is sustained. It also ensures that reduced stretch at central venous receptors cannot be relieved through the usual compensatory pressor actions of Ang II and the SNS to reduce volume capacity of the vascular tree. Thus, vascular underfilling is perpetuated even as ECFV increases. In the kidney, this persistent relaxin-mediated preglomerular vasodilation leads to greater transmission of aortic mean arteriolar pressure (MAP) to the afferent arteriole. Thus, the lower systemic blood pressure can now sustain a normal afferent arteriolar pressure, thereby shifting the renal setpoint downward to accept and sustain a lower MAP. In effect, despite systemic vascular underfilling, preglomerular vasodilation that is resistant to Ang II and norepinephrine permits the kidney to accept the lower MAP of early pregnancy as "normal." The concomitant resistance to ANP (see earlier) suggests that its potential to counteract these forces is effectively neutralized.

As outlined previously, the tubular sodium retention limb of the feedback loop remains intact. This may have important effects on the behavior of the macula densa to regulate GFR in pregnancy. Normally, elevated GFR, despite an increased absolute proximal reabsorption (i.e., glomerulotubular balance), will also deliver excess Na/H_2O to the thick ascending limb. The higher thick ascending limb flow rate would limit NaCl capture and create a higher concentration of NaCl sensed by the specialized epithelial cells of the macula densa. This stimulus would normally activate tubular glomerular feedback, causing macula densa cells to release basolateral ATP and activate afferent arteriolar constriction at a discrete sphincter-like site located just within the glomerulus. The outcome would be to reduce PGC and GFR (212). In contrast, the pregnant state is accompanied by normal fluid/sodium delivery measured in the late proximal tubule and is compatible with the view that, despite high GFR, stimulated sodium reabsorption proximally—together with low RIHP/reduced backleak—prevents any increase in [NaCl] at the macula densa (199).The latter is thus rendered blind to the high GFR of pregnancy. This is precisely compatible with the observations of Baylis and Blantz that macula densa function in pregnancy is reset to a higher level of GFR, but operates normally around that setpoint (199).

Another factor of potential importance in promotion of proximal tubular reabsorptive capacity and modification of macula densa activity during pregnancy is proximal tubular hypertrophy. In early chemically induced diabetes in rats, Thompson and coworkers provided compelling data to support the idea that hyperglycemia-mediated proximal tubular hypertrophic growth directly enhances sodium reabsorption, reduces [NaCl] at the macula densa, induces afferent sphincter dilation, and causes a compensatory increase in GFR (213). According to this view, tubular hypertrophy as a primary event directly leads to enhanced proximal reabsorption and glomerular hyperfiltration by activating (rather than silencing) the macula densa. Evidence in the pregnant rat that proximal tubule length is increased lends support to the potential for growth-dependent initiation of glomerular hyperfiltration in pregnancy (181,182).

Mediators of Renal Hemodynamic Adaptations in Human Pregnancy

Available data based on maternal relaxin levels in human pregnancy suggest that relaxin, at least in circulating form, does not mediate the systemic and renal vasodilation (132). However, the similarities in hemodynamic responses of rat and human suggest that there will be homologous systemically active vasodilatory molecules and common downstream pathways in human pregnancy. In support of that, there is limited evidence that NO may be increased in pregnant women (136,214). Goodrum and coworkers found higher arginine and NO production in mid-gestation (214). On the other hand, von Mandach and coworkers found increased NO predominantly in the fetal compartment (215). Similarly, Schiessl and coworkers detected no changes in either NO metabolites or cyclic guanosine monophosphate (cGMP) levels in plasma or urine over the course of gestation in 49 healthy pregnancies (216). In the first study to control carefully for dietary nitrate and nitrite, Conrad and coworkers found no increase in whole-body NO production but marked increases in cGMP production during pregnancy, especially during the period of rapid vasodilatation in the first trimester (217).

Role of Angiotensin 1-7 or Angiotensin AT2 Receptors

If relaxin is not a mediator of pregnancy-associated hemodynamic change in human pregnancy (132), is there a human "relaxin-equivalent?" (One would expect multiple layers of regulation of blood volume and blood pressure during pregnancy.) Brosnihan and coworkers have reported increased levels of the vasodilator Ang 1-7 in plasma and urine of normal pregnant women (218–220).

This vasodilatory metabolite is formed by conversion of Ang II by angiotensin-converting enzyme-2 (ACE2) or by a two-step reaction converting Ang I to Ang 1-9 by ACE2, then to Ang 1-7 by ACE (Fig. 15). Also, in pregnant rats, increased renal and urinary levels of Ang 1-7 were observed, together with

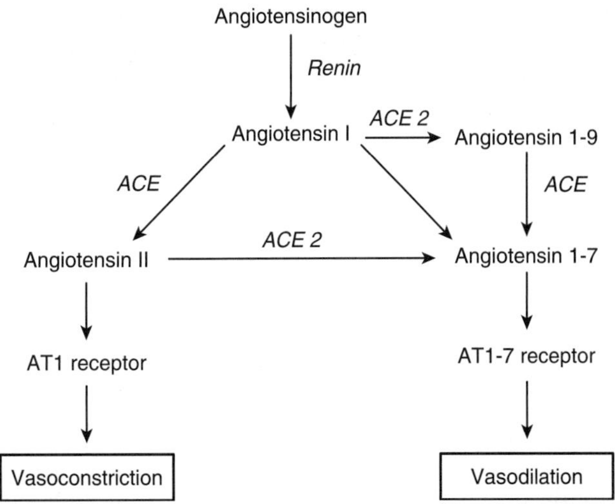

FIG. 15. Enzymatic pathways for formation of the vasodilator angiotensin (Ang) 1-7. Ang 1-7 can be formed from either Ang II by the action of angiotensin-converting enzyme-2 (ACE2) or from angiotensin I through a two-step reaction requiring ACE2 followed by ACE. Ang 1-7 has been reported to increase in the third trimester of normal pregnancy, whereas levels in preeclamptic pregnancies were significantly reduced compared with normal pregnancy. (From Brosnihan, K. B., Neves, L. A., Joyner, J., Averill, D. B., Chappell, M. C., Sarao, R., Penninger, J., and Ferrario, C. M. [2003]. Enhanced renal immunocytochemical expression of ANG-(1-7) and ACE2 during pregnancy. *Hypertension* 42, 749–753.)

enhanced ACE2 expression in kidney (220). These observations suggest a role for vasodilatory components of the renin–Ang II system in human and rodent pregnancy. It is interesting to speculate that, whereas relaxin co-opts the endothelin pressor system and converts it to vasodilatory action in the pregnant rat, the Ang II system might be co-opted to a vasodilatory system in human pregnancy, potentially through increased Ang 1-7, a shift to AT2 receptor dominance compared with AT1 receptor, or enhanced Ang II–induced endothelial vasodilatory factors. Evidence already available indicates functional resistance to AT1 receptor in pregnancy. Like the upregulated endothelin ETB receptor in rat pregnancy, the AT2 receptor is linked to bradykinin, and thus could potentially act through NO or endothelium-derived hyperpolarizing factor (EDHF). The relaxin paradigm in the rat may provide a valuable template for reassessing the human response to pregnancy.

REFERENCES

1. Weaver, I. C., Cervoni, N., Champagne, F. A., D'Alessio, A. C., Sharma, A., Seckl, J. R., Dymov, S., Szyf, M., and Meaney, M. J. (2004). Epigenetic programming by maternal behaviour. *Nat. Neurosci.* 7, 847–854.
2. Hytten, F. E., and Chamberlain, G. (1980). *Clinical Physiology in Obstetrics.* Blackwell Scientific Publications, Oxford.
3. Catalano, P. M., and Hollenbeck, C. (1992). Energy requirements in pregnancy: a review. *Obstet. Gynecol. Surv.* 47, 368–372.
4. McCarthy, E. A., Strauss, B. J., Walker, S. P., and Permezel, M. (2004). Determination of maternal body composition in pregnancy and its relevance to perinatal outcomes. *Obstet. Gynecol. Surv.* 59, 731–742.
5. Hytten, F. E. and Leitch, I. (1971). *The Physiology of Human Pregnancy.* Blackwell Scientific Publications, Oxford.
6. Paxton, A., Lederman, S. A., Heymsfield, S. B., Wang, J., Thornton, J. C., and Pierson, R. N. Jr. (1998). Anthropometric equations for studying body fat in pregnant women. *Am. J. Clin. Nutr.* 67, 104–110.
7. Institute of Medicine—Food and Nutrition Board. (1990). *Nutrition during Pregnancy, Weight Gain and Nutrient Supplements.* National Academy Press, Washington, DC.
8. Butte, N. F., Wong, W. W., Treuth, M. S., Ellis, K. J., and O'Brian, S. E. (2004). Energy requirements during pregnancy based on total energy expenditure and energy deposition. *Am. J. Clin. Nutr.* 79, 1078–1087.
9. Bronstein, M. N., Mak, R. P., and King, J. C. (1996). Unexpected relationship between fat mass and basal metabolic rate in pregnant women. *Br. J. Nutr.* 75, 659–668.
10. Metcalfe, J., McAnulty, J., and Ueland, K. (1986). *Burwell and Metcalfe's Heart Disease in Pregnancy: Physiology and Management* (2nd ed.). Little, Brown, Boston.
11. Evans, W., Capelle, S. C., and Edelstone, D. I. (1996). Lack of a critical cardiac output and critical systemic oxygen delivery during low cardiac output in the third trimester in the pregnant sheep. *Am. J. Obstet. Gynecol.* 175, 222–228.
12. Chihara, H., Otsubo, Y., Yoneyama, Y., Sawa, R., Suzuki, S., Power, G. G., and Araki, T. (2003). Basal metabolic rate in hyperemesis gravidarum: comparison to normal pregnancy and response to treatment. *Am. J. Obstet. Gynecol.* 188, 434–438.
13. Langley, S. C., Seakins, M., Grimble, R. F., and Jackson, A. A. (1994). The acute phase response of adult rats is altered by in utero exposure to maternal low protein diets. *J. Nutr.* 124, 1588–1596.
14. Nishina, H., Green, L. R., McGarrigle, H. H., Noakes, D. E., Poston, L., and Hanson, M. A. (2003). Effect of nutritional restriction in early pregnancy on isolated femoral artery function in mid-gestation fetal sheep. *J. Physiol. (Lond.)* 553, 637–647.
15. Woods, L. L., Weeks, D. A., and Rasch, R. (2004). Programming of adult blood pressure by maternal protein restriction: role of nephrogenesis. *Kidney Int.* 65, 1339–1348.
16. Godfrey, K., Robinson, S., Barker, D. J., Osmond, C., and Cox, V. (1996). Maternal nutrition in early and late pregnancy in relation to placental and fetal growth. *BMJ* 312, 410–414.
17. Barker, D. J. P., Eriksson, J. G., Forsen, T., and Osmond, C. (2002). Fetal origins of adult disease: strength of effects and biological basis. *Int. J. Epidemiol.* 31, 1235–1239.
18. Shiell, A. W., Campbell-Brown, M., Haselden, S., Robinson, S., Godfrey, K. M., and Barker, D. J. (2001). High-meat, low-carbohydrate diet in pregnancy: relation to adult blood pressure in the offspring. *Hypertension* 38, 1282–1288.
19. Caton, W. L., Roby, C. C., Reid, D. E., Caswell, R., Maletskos, C. J., Fluharty, R. G., and Gibson, J. G., II. (1951). The circulating red cell volume and body hematocrit in normal pregnancy and the puerperium by direct measurement using radioactive red cells. *Am. J. Obstet. Gynecol.* 61, 1207–1217.
20. Hytten, F. E. and Paintin, D. B. (1963). Increase in plasma volume during normal pregnancy. *J Obstet. Gynaecol. Br. Emp.* 70, 402–407.
21. Lund, C. J. and Donovan, J. C. (1967). Blood volume during pregnancy: significance of plasma and red cell volumes. *Am. J. Obstet. Gynecol.* 98, 394–403.
22. Scott, D. E. (1972). Anemia in pregnancy. *Obstet. Gynecol. Annu.* 1, 219–244.
23. Clapp, J. F., III, Seaward, B. L., Sleamaker, R. H., and Hiser, J. (1988). Maternal physiologic adaptations to early human pregnancy. *Am. J. Obstet. Gynecol.* 159, 1456–1460.

24. Bernstein, I. M., Ziegler, W., and Badger, G. J. (2001). Plasma volume expansion in early pregnancy. *Obstet. Gynecol.* 97, 669–672.
25. Pritchard, J. A. (1965). Changes in the blood volume during pregnancy and delivery. *Anesthesiology* 26, 393–399.
26. Rovinsky, J. J. and Jaffin, H. (1965). Cardiovascular hemodynamics in pregnancy: I. Blood and plasma volumes in multiple pregnancy. *Am. J. Obstet. Gynecol.* 93, 1–15.
27. Salas, S. P., Rosso, P., Espinoza, R., Robert, J. A., Valdes, G., and Donoso, E. (1993). Maternal plasma volume expansion and hormonal changes in women with idiopathic fetal growth retardation. *Obstet. Gynecol.* 81, 1029–1033.
28. Wilson, M., Morganti, A. A., Zervoudakis, I., Letcher, R. L., Romney, B. M., Von Oeyon, P., Papera, S., Sealey, J. E., and Laragh, J. H. (1980). Blood pressure, the renin-aldosterone system and sex steroids throughout normal pregnancy. *Am. J. Med.* 68, 97–104.
29. Pedersen, E. B., Johannesen, P., Rasmussen, A. B., and Danielsen, H. (1985). The osmoregulatory system and the renin-angiotensin-aldosterone system in pre-eclampsia and normotensive pregnancy. *Scand. J. Clin. Lab. Invest.* 45, 627–633.
30. Duvekot, J. J., Cheriex, E. C., Pieters, F. A., Menheere, P. P., and Peeters, L. H. (1993). Early pregnancy changes in hemodynamics and volume homeostasis are consecutive adjustments triggered by a primary fall in systemic vascular tone. *Am. J. Obstet. Gynecol.* 169, 1382–1392.
31. Phippard, A. F., Horvath, J. S., Glynn, E. M., Garner, M. G., Fletcher, P. J., Duggin, G. G., and Tiller, D. J. (1986). Circulatory adaptation to pregnancy: serial studies of haemodynamics, blood volume, renin and aldosterone in the baboon *(Papio hamadryas). J. Hypertens.* 4, 773–779.
32. Lindheimer, M. D. and Davison, J. M. (1995). Osmoregulation, the secretion of arginine vasopressin and its metabolism during pregnancy. *Eur. J. Endocrinol.* 132, 133–143.
33. Chiodera, P., Volpi, R., Caiazza, A., Giuliani, N., Magotti, M. G., and Coiro, V. (1998). Arginine vasopressin and oxytocin responses to angiotensin II are mediated by AT1 receptor subtype in normal men. *Metabolism* 47, 893–896.
34. Schrier, R. W. and Briner, V. A. (1991). Peripheral arterial vasodilation hypothesis of sodium and water retention in pregnancy: implications for pathogenesis of preeclampsia-eclampsia. *Obstet. Gynecol.* 77, 632–639.
35. Oparil, S., Ehrlich, E. N., and Lindheimer, M. D. (1975). Effect of progesterone on renal sodium handling in man: relation to aldosterone excretion and plasma renin activity. *Clin. Sci. Mol. Med.* 49, 139–147.
36. Winkel, C. A., Milewich, L., Parker, C. R. Jr., Gant, N. F., Simpson, E. R., and MacDonald, P. C. (1980). Conversion of plasma progesterone to deoxycorticosterone in men, nonpregnant and pregnant women, and adrenalectomized subjects. *J. Clin. Invest* 66, 803–812.
37. Seitchik, J. (1967). Total body water and total body density of pregnant women. *Obstet. Gynecol.* 29, 155–166.
38. Lindheimer, M. D., and Katz, A. I. (1973). Sodium and diuretics in pregnancy. *N. Engl. J. Med.* 288, 891–894.
39. Kametas, N. A., Krampl, E., McAuliffe, F., Rampling, M. W., and Nicolaides, K. H. (2005). Pregnancy at high altitude: a hyperviscosity state. *Obstet. Gynecol. Surv.* 60, 78–80.
40. Hatjis, C. G., Kofinas, A. D., Greelish, J. P., Swain, M., and Rose, J. C. (1990). Interrelationship between atrial natriuretic factor concentrations and acute volume expansion in pregnant and nonpregnant women. *Am. J. Obstet. Gynecol.* 163, 45–50.
41. Heenan, A. P. and Wolfe, L. A. (2003). Plasma osmolality and the strong ion difference predict respiratory adaptations in pregnant and nonpregnant women. *Can. J. Physiol. Pharmacol.* 81, 839–847.
42. Sala, C., Campise, M., Ambroso, G., Motta, T., Zanchetti, A., and Morganti, A. (1995). Atrial natriuretic peptide and hemodynamic changes during normal human pregnancy. *Hypertension* 25, 631–636.
43. Freund, B. J., Shizuru, E. M., Hashiro, G. M., and Claybaugh, J. R. (1991). Hormonal, electrolyte, and renal responses to exercise are intensity dependent. *J. Appl. Physiol.* 70, 900–906.
44. Heenan, A. P., Wolfe, L. A., Davies, G. A., and McGrath, M. J. (2003). Effects of human pregnancy on fluid regulation responses to short-term exercise. *J. Appl. Physiol.* 95, 2321–2327.
45. Verkeste, C. M., Slangen, B. F., Dubelaar, M. L., van Kreel, B. K., and Peeters, L. L. (1998). Mechanism of volume adaptation in the awake early pregnant rat. *Am. J. Physiol.* 274, H1662–H1666.
46. Chapman, A. B., Zamudio, S., Woodmansee, W., Merouani, A., Osorio, F., Johnson, A., Moore, L. G., Dahms, T., Coffin, C., Abraham, W. T., and Schrier, R. W. (1997). Systemic and renal hemodynamic changes in the luteal phase of the menstrual cycle mimic early pregnancy. *Am. J. Physiol.* 273, F777–F782.
47. Kaufman, S. (1995). Control of intravascular volume during pregnancy. *Clin. Exp. Pharmacol. Physiol.* 22, 157–163.
48. Deng, Y. and Kaufman, S. (1998). Pregnancy-induced changes in central response to atrial distension mimicked by progesterone metabolite. *Am. J. Physiol.* 275, R1875–R1877.
49. Zhang, Y. and Kaufman, S. (2000). Effect of nitric oxide synthase inhibition on cardiovascular and hormonal regulation during pregnancy in the rat. *Can. J. Physiol. Pharmacol.* 78, 423–427.
50. Lo, F. and Kaufman, S. (2001). Effect of 5 alpha-pregnan-3 alpha-ol-20-one on nitric oxide biosynthesis and plasma volume in rats. *Am. J. Physiol. Regul. Integr. Comp. Physiol.* 280, R1902–R1905.
51. Paulus, W. J., and Shah, A. M. (1999). NO and cardiac diastolic function. *Cardiovasc. Res.* 43, 595–606.
52. Keller-Wood, M. (1995). Reflex regulation of hormonal responses during pregnancy. *Clin. Exp. Pharmacol. Physiol.* 22, 143–151.
53. Jensen, E., Wood, C., and Keller-Wood, M. (2002). The normal increase in adrenal secretion during pregnancy contributes to maternal volume expansion and fetal homeostasis. *J. Soc. Gynecol. Investig.* 9, 362–371.
54. Keller-Wood, M. (1994). Corticotropin responses to hypoglycemia and hypotension during ovine pregnancy. *Am. J. Physiol.* 266, R180–R187.
55. Charles, C. J., Rademaker, M. T., Richards, A. M., and Nicholls, M. G. (2001). Adrenomedullin augments the neurohumoral response to haemorrhage in non-pregnant but not in pregnant sheep. *J. Endocrinol.* 171, 363–371.
56. Brooks, V. L., Quesnell, R. R., Cumbee, S. R., and Bishop, V. S. (1995). Pregnancy attenuates activity of the baroreceptor reflex. *Clin. Exp. Pharmacol. Physiol.* 22, 152–156.
57. Clow, K. A., Giraud, G. D., Ogden, B. E., and Brooks, V. L. (2003). Pregnancy alters hemodynamic responses to hemorrhage in conscious rabbits. *Am. J. Physiol. Heart Circ. Physiol.* 284, H1110–H1118.
58. Spaanderman, M. E., Meertens, M., van Bussel, M., Ekhart, T. H., and Peeters, L. L. (2000). Cardiac output increases independently of basal metabolic rate in early human pregnancy. *Am. J. Physiol. Heart Circ. Physiol.* 278, H1585–H1588.
59. Katz, R., Karliner, J. S., and Resnik, R. (1978). Effects of a natural volume overload state (pregnancy) on left ventricular performance in normal human subjects. *Circulation* 58, 434–441.
60. Veille, J. C., Morton, M. J., and Burry, K. J. (1985). Maternal cardiovascular adaptations to twin pregnancy. *Am. J. Obstet. Gynecol.* 153, 261–263.
61. Robson, S. C., Hunter, S., Boys, R. J., and Dunlop, W. (1989). Hemodynamic changes during twin pregnancy: a Doppler and M-mode echocardiographic study. *Am. J. Obstet. Gynecol.* 161, 1273–1278.

62. Campos, O. (1996). Doppler echocardiography during pregnancy: physiological and abnormal findings. *Echocardiography* 13, 135–146.
63. Robson, S. C., Hunter, S., Boys, R. J., and Dunlop, W. (1989). Serial study of factors influencing changes in cardiac output during human pregnancy. *Am. J. Physiol.* 256, H1060–H1065.
64. Poppas, A., Shroff, S. G., Korcarz, C. E., Hibbard, J. U., Berger, D. S., Lindheimer, M. D., and Lang, R. M. (1997). Serial assessment of the cardiovascular system in normal pregnancy: role of arterial compliance and pulsatile arterial load. *Circulation* 95, 2407–2415.
65. Katz, V. L. (1991). Physiologic changes during normal pregnancy. *Curr. Opin. Obstet. Gynecol.* 3, 750–758.
66. Mone, S. M., Sanders, S. P., and Colan, S. D. (1996). Control mechanisms for physiological hypertrophy of pregnancy. *Circulation* 94, 667–672.
67. Simmons, L. A., Gillin, A. G., and Jeremy, R. W. (2002). Structural and functional changes in left ventricle during normotensive and preeclamptic pregnancy. *Am. J. Physiol. Heart Circ. Physiol.* 283, H1627–H1633.
68. Ginsburg, J. and Duncan, S. (1969). Direct and indirect blood pressure measurement in pregnancy. *J. Obstet. Gynaecol. Br. Commonw.* 76, 705–710.
69. Kirshon, B., Lee, W., Cotton, D. B., and Giebel, R. (1987). Indirect blood pressure monitoring in the postpartum patient. *Obstet. Gynecol.* 70, 799–801.
70. Committee on Technical Bulletins of the American College of Obstetricians and Gynecologists. (1996). ACOG technical bulletin: Hypertension in pregnancy. No. 219, January 1996 (replaces no. 91, February 1986). *Int. J. Gynaecol. Obstet.* 53, 175–183.
71. Creasy, R. K. and Resnik, R. (1989). Intrauterine growth retardation. In *Maternal-Fetal Medicine: Principles and Practice* (R. K. Creasy, R. Resnik, and J. Iams, Eds.), pp. 547–564. WB Saunders, Philadelphia.
72. Steer, P. J., Little, M. P., Kold-Jensen, T., Chapple, J., and Elliott, P. (2004). Maternal blood pressure in pregnancy, birth weight, and perinatal mortality in first births: prospective study. *BMJ* 329, 1312.
73. Chesley, L. C., Talledo, E., Bohler, C. S., and Zuspan, F. P. (1965). Vascular reactivity to angiotensin II and norepinephrine in pregnant women. *Am. J. Obstet. Gynecol.* 91, 837–842.
74. Paller, M. S. (1984). Mechanism of decreased pressor responsiveness to ANG II, NE, and vasopressin in pregnant rats. *Am. J. Physiol.* 247, H100–H108.
75. Magness, R. R. and Rosenfeld, C. R. (1988). Mechanisms for attenuated pressor responses to alpha-agonists in ovine pregnancy. *Am. J. Obstet. Gynecol.* 159, 252–261.
76. Rosenfeld, C. R., and Naden, R. P. (1989). Uterine and nonuterine vascular responses to angiotensin II in ovine pregnancy. *Am. J. Physiol.* 257, H17–H24.
77. Crandall, M. E., Keve, T. M., and McLaughlin, M. K. (1990). Characterization of norepinephrine sensitivity in the maternal splanchnic circulation during pregnancy. *Am. J. Obstet. Gynecol.* 162, 1296–1301.
78. D'Angelo, G. and Osol, G. (1993). Regional variation in resistance artery diameter responses to alpha-adrenergic stimulation during pregnancy. *Am. J. Physiol.* 264, H78–H85.
79. Fawer, R., Dettling, A., Weihs, D., Welti, H., and Schelling, J. L. (1978). Effect of the menstrual cycle, oral contraception and pregnancy on forearm blood flow, venous distensibility and clotting factors. *Eur. J. Clin. Pharmacol.* 13, 251–257.
80. Edouard, D. A., Pannier, B. M., London, G. M., Cuche, J. L., and Safar, M. E. (1998). Venous and arterial behavior during normal pregnancy. *Am. J. Physiol.* 274, H1605–H1612.
81. Rupnow, H. L., Phernetton, T. M., Modrick, M. L., Wiltbank, M. C., Bird, I. M., and Magness, R. R. (2002). Endothelial vasodilator production by uterine and systemic arteries: VIII. Estrogen and progesterone effects on cPLA2, COX-1, and PGIS protein expression. *Biol. Reprod.* 66, 468–474.
82. Rosenfeld, C. R., Cox, B. E., Roy, T., and Magness, R. R. (1996). Nitric oxide contributes to estrogen-induced vasodilation of the ovine uterine circulation. *J. Clin. Invest* 98, 2158–2166.
83. Conrad, K. P. and Novak, J. (2004). Emerging role of relaxin in renal and cardiovascular function. *Am. J. Physiol. Regul. Integr. Comp. Physiol.* 287, R250–R261.
84. Gangula, P. R., Lanlua, P., Bukoski, R. D., Wimalawansa, S. J., and Yallampalli, C. (2004). Mesenteric arterial relaxation to calcitonin gene-related peptide is increased during pregnancy and by sex steroid hormones. *Biol. Reprod.* 71, 1739–1745.
85. Cipolla, M. J., Vitullo, L., and McKinnon, J. (2004). Cerebral artery reactivity changes during pregnancy and the postpartum period: a role in eclampsia? *Am. J. Physiol. Heart Circ. Physiol.* 286, H2127–H2132.
86. Barrett-Connor, E. (1997). Sex differences in coronary heart disease: why are women so superior? The 1995 Ancel Keys Lecture. *Circulation* 95, 252–264.
87. Osol, G. and Cipolla, M. (1993). Pregnancy-induced changes in the three-dimensional mechanical properties of pressurized rat uteroplacental (radial) arteries. *Am. J. Obstet. Gynecol.* 168, 268–274.
88. Sibai, B. M., Hauth, J., Caritis, S., Lindheimer, M. D., MacPherson, C., Klebanoff, M., VanDorsten, J. P., Landon, M., Miodovnik, M., Paul, R., Meis, P., Thurnau, G., Dombrowski, M., Roberts, J., and McNellis, D. (2000). Hypertensive disorders in twin versus singleton gestations. National Institute of Child Health and Human Development Network of Maternal-Fetal Medicine Units. *Am. J. Obstet. Gynecol.* 182, 938–942.
89. Coonrod, D. V., Hickok, D. E., Zhu, K., Easterling, T. R., and Daling, J. R. (1995). Risk factors for preeclampsia in twin pregnancies: a population-based cohort study. *Obstet. Gynecol.* 85, 645–650.
90. Kametas, N. A., McAuliffe, F., Krampl, E., Chambers, J., and Nicolaides, K. H. (2003). Maternal cardiac function in twin pregnancy. *Obstet. Gynecol.* 102, 806–15.
91. McDonald, D. A. (1960). *Blood Flow in Arteries.* Edward Arnold, London.
92. Mackenzie, I. S., Wilkinson, I. B., and Cockcroft, J. R. (2002). Assessment of arterial stiffness in clinical practice. *QJM* 95, 67–74.
93. Williams, J. K., Adams, M. R., Herrington, D. M., and Clarkson, T. B. (1992). Short-term administration of estrogen and vascular responses of atherosclerotic coronary arteries. *J. Am. Coll. Cardiol.* 20, 452–457.
94. Gilligan, D. M., Badar, D. M., Panza, J. A., Quyyumi, A. A., and Cannon, R. O., III. (1995). Effects of estrogen replacement therapy on peripheral vasomotor function in postmenopausal women. *Am. J. Cardiol* 75, 264–268.
95. Page, K. L., Celia, G., Leddy, G., Taatjes, D. J., and Osol, G. (2002). Structural remodeling of rat uterine veins in pregnancy. *Am. J. Obstet. Gynecol.* 187, 1647–1652.
96. Conrad, K. P., Debrah, D. O., Novak, J., Danielson, L. A., and Shroff, S. G. (2004). Relaxin modifies systemic arterial resistance and compliance in conscious, nonpregnant rats. *Endocrinology* 145, 3289–3296.
97. Jepson, J. H. and Lowenstein, L. (1968). Role of erythropoietin and placental lactogen in the control of erythropoiesis during pregnancy. *Can. J. Physiol. Pharmacol.* 46, 573–576.
98. Beguin, Y., Lipscei, G., Thoumsin, H., and Fillet, G. (1991). Blunted erythropoietin production and decreased erythropoiesis in early pregnancy. *Blood* 78, 89–93.
99. Choi, J. W., Im, M. W., and Pai, S. H. (2000). Serum transferrin receptor concentrations during normal pregnancy. *Clin. Chem.* 46, 725–727.

100. Peck, T. M. and Arias, F. (1979). Hematologic changes associated with pregnancy. *Clin. Obstet. Gynecol.* 22, 785–798.
101. Scott, D. E., and Pritchard, J. A. (1967). Iron deficiency in healthy young college women. *JAMA* 199, 897–900.
102. James, K. V., Lohr, J. M., Deshmukh, R. M., and Cranley, J. J. (1996). Venous thrombotic complications of pregnancy. *Cardiovasc. Surg.* 4, 777–782.
103. Roth, A. and Elkayam, U. (1996). Acute myocardial infarction associated with pregnancy. *Ann. Intern. Med.* 125, 751–762.
104. Brenner, B. (2004). Haemostatic changes in pregnancy. *Thromb. Res.* 114, 409–414.
105. Greer, I. (1994). Haemostasis and thrombosis in pregnancy. In *Haemostasis and Thrombosis* (3rd ed. A. L. Bloom, C. D. Forbes, D. P. Thomas, and E. G. C. Tuddenham, Eds.), pp. 987–1015. Churchill Livingstone, Edinburgh.
106. Blumenfeld, Z. and Brenner, B. (1999). Thrombophilia-associated pregnancy wastage. *Fertil. Steril.* 72, 765–774.
107. Robertson, L., Wu, O., and Greer, I. (2004). Thrombophilia and adverse pregnancy outcome. *Curr. Opin. Obstet. Gynecol.* 16, 453–458.
108. Cugell, D. W., Frank, N. R., Gaensler, E. A., and Badger, T. L. (1953). Pulmonary function in pregnancy: I. Serial observations in normal women. *Am. Rev. Tuberc.* 67, 568–597.
109. Goucher, D., Rubin, A., and Russo, N. (1956). The effect of pregnancy upon pulmonary function in normal women. *Am. J. Obstet. Gynecol.* 72, 963–969.
110. Gee, J. B., Packer, B. S., Millen, J. E., and Robin, E. D. (1967). Pulmonary mechanics during pregnancy. *J. Clin. Invest.* 46, 945–952.
111. Gazioglu, K., Kaltreider, N. L., Rosen, M., and Yu, P. N. (1970). Pulmonary function during pregnancy in normal women and in patients with cardiopulmonary disease. *Thorax* 25, 445–450.
112. Bonica, J. J. (1973). Maternal respiratory changes during pregnancy and parturition. In *Parturition and Perinatology* (G. F. Marx, Ed.), pp. 2–19. FA Davis, Philadelphia.
113. Baldwin, G. R., Moorthi, D. S., Whelton, J. A., and MacDonnell, K. F. (1977). New lung functions and pregnancy. *Am. J. Obstet. Gynecol.* 127, 235–239.
114. Alaily, A. B. and Carrol, K. B. (1978). Pulmonary ventilation in pregnancy. *Br. J. Obstet. Gynaecol.* 85, 518–524.
115. Milne, J. A. (1979). The respiratory response to pregnancy. *Postgrad. Med. J.* 55, 318–324.
116. Weinberger, S. E., Weiss, S. T., Cohen, W. R., Weiss, J. W., and Johnson, T. S. (1980). Pregnancy and the lung. *Am. Rev. Respir. Dis.* 121, 559–581.
117. Metcalfe, J. and Bissonnette, J. M. (1987). Gas exchange in pregnancy. In *Handbook of Physiology, Section 3: The Respiratory System* (A. P. Fishman, L. E. Farhi, S. M. Tenney, and S. R. Geiger, Eds.), pp. 341–350. American Physiological Society, Bethesda, MD.
118. Gilroy, R. J., Mangura, B. T., and Lavietes, M. H. (1988). Rib cage and abdominal volume displacements during breathing in pregnancy. *Am. Rev. Respir. Dis.* 137, 668–672.
119. Norregaard, O., Schultz, P., Ostergaard, A., and Dahl, R. (1989). Lung function and postural changes during pregnancy. *Respir. Med.* 83, 467–470.
120. Camann, W. R. and Ostheimer, G. W. (1990). Physiological adaptations during pregnancy. *Int. Anesthesiol. Clin.* 28, 2–10.
121. Rees, G. B., Broughton, P. F., Symonds, E. M., and Patrick, J. M. (1990). A longitudinal study of respiratory changes in normal human pregnancy with cross-sectional data on subjects with pregnancy-induced hypertension. *Am. J. Obstet. Gynecol.* 162, 826–830.
122. Contreras, G., Gutierrez, M., Beroiza, T., Fantin, A., Oddo, H., Villarroel, L., Cruz, E., and Lisboa, C. (1991). Ventilatory drive and respiratory muscle function in pregnancy. *Am. Rev. Respir. Dis.* 144, 837–841.
123. Elkus, R. and Popovich, J. Jr. (1992). Respiratory physiology in pregnancy. *Clin. Chest Med.* 13, 555–565.
124. Crapo, R. O. (1996). Normal cardiopulmonary physiology during pregnancy. *Clin. Obstet. Gynecol.* 39, 3–16.
125. Gordon, M. C. (2002). Maternal physiology in pregnancy. In *Obstetrics: Normal and Problem Pregnancies* (4th ed. S. Gabbe, J. Niebyl, and J. Simpson, Eds.), pp. 63–91. Churchill Livingstone, New York.
126. Dempsey, J. A., Olson, E. B., and Skatrud, J. B. (1986). Hormones and neurochemicals in the regulation of breathing. In *Handbook of Physiology, Section 3: The Respiratory System, Volume II: Control of Breathing* (A. P. Fishman, N. S. Cherniack, J. G. Widdicombe, and S. R. Geiger, Eds.), pp. 181–221. American Physiological Society, Bethesda, MD.
127. Brancazio, L. R., Laifer, S. A., and Schwartz, T. (1997). Peak expiratory flow rate in normal pregnancy. *Obstet. Gynecol.* 89, 383–386.
128. McAuliffe, F., Kametas, N., Espinoza, J., Greenough, A., and Nicolaides, K. (2004). Respiratory function in pregnancy at sea level and at high altitude. *Br. J. Obstet. Gynaecol.* 111, 311–315.
129. Lindheimer, M. D. and Katz, A. I. (1992). Renal physiology and disease in pregnancy. In *The Kidney: Physiology and Pathophysiology* (D. W. Seldin, and G. Giebisch, Eds.), pp. 3371–3432. Raven Press, New York.
130. Monga, J. and Creasy, R. K. (1994). Cardiovascular and renal adaptation to pregnancy. In *Maternal-Fetal Medicine: Principles and Practice* (3rd ed. R. K. Creasy, and R. Resnik, Eds.), pp. 758–767. WB Saunders, Philadelphia.
131. Baylis, C. (1999). Glomerular filtration rate in normal and abnormal pregnancies. *Semin. Nephrol.* 19, 133–139.
132. Conrad, K. P. (2004). Mechanisms of renal vasodilation and hyperfiltration during pregnancy. *J. Soc. Gynecol. Investig.* 11, 438–448.
133. Paller, M. S. and Connaire, J. J. (2004). The kidney and hypertension in pregnancy. In *Brenner and Rector's The Kidney* (7th ed. B. M. Brenner, Ed.), Vol. 2, pp. 1659–1695. Elsevier Saunders, New York.
134. Bucht, H. and Werko, L. (1953). Glomerular filtration rate and renal blood flow in hypertensive toxaemia of pregnancy. *J. Obstet. Gynaecol. Br. Emp.* 60, 157–164.
135. Davison, J. M. and Hytten, F. E. (1974). Glomerular filtration during and after pregnancy. *J. Obstet. Gynaecol. Br. Commonw.* 81, 588–595.
136. Chapman, A. B., Abraham, W. T., Zamudio, S., Coffin, C., Merouani, A., Young, D., Johnson, A., Osorio, F., Goldberg, C., Moore, L. G., Dahms, T., and Schrier, R. W. (1998). Temporal relationships between hormonal and hemodynamic changes in early human pregnancy. *Kidney Int.* 54, 2056–2063.
137. Sims, E. A. and Krantz, K. E. (1958). Serial studies of renal function during pregnancy and the puerperium in normal women. *J. Clin. Invest.* 37, 1764–1774.
138. Higby, K., Suiter, C. R., Phelps, J. Y., Siler-Khodr, T., and Langer, O. (1994). Normal values of urinary albumin and total protein excretion during pregnancy. *Am. J. Obstet. Gynecol.* 171, 984–989.
139. Barron, W. M. and Lindheimer, M. D. (1984). Renal sodium and water handling in pregnancy. *Obstet. Gynecol. Annu.* 13, 35–69.
140. Lindheimer, M. D., and Katz, A. I. (1970). Current concepts: the kidney in pregnancy. *N. Engl. J. Med.* 283, 1095–1097.
141. Cha, S. C., Aberdeen, G. W., Mukaddam-Daher, S., Quillen, E. W. Jr., and Nuwayhid, B. S. (1993). Tubular handling of fluid and electrolytes during ovine pregnancy. *Am. J. Physiol.* 265, F278–F284.
142. August, P., Lenz, T., Ales, K. L., Druzin, M. L., Edersheim, T. G., Hutson, J. M., Muller, F. B., Laragh, J. H., and Sealey, J. E. (1990). Longitudinal study of the renin-angiotensin-aldosterone system in hypertensive

pregnant women: deviations related to the development of superimposed preeclampsia. *Am. J. Obstet. Gynecol.* 163, 1612–1621.

143. Brown, M. A., Zammit, V. C., and Adsett, D. (1990). Stimulation of active renin release in normal and hypertensive pregnancy. *Clin. Sci. (Lond.)* 79, 505–511.
144. Hanssens, M., Keirse, M. J., Spitz, B., and van Assche, F. A. (1991). Angiotensin II levels in hypertensive and normotensive pregnancies. *Br. J. Obstet. Gynaecol.* 98, 155–161.
145. Duvekot, J. J. and Peeters, L. L. (1994). Renal hemodynamics and volume homeostasis in pregnancy. *Obstet. Gynecol. Surv.* 49, 830–839.
146. Davison, J. M. and Hytten, F. E. (1975). The effect of pregnancy on the renal handling of glucose. *Br. J. Obstet. Gynaecol.* 82, 374–381.
147. Bishop, J. H. and Green, R. (1980). Effects of pregnancy on glucose handling by rat kidneys. *J. Physiol. (Lond.)* 307, 491–502.
148. Bishop, J. H. and Green, R. (1983). Glucose handling by distal portions of the nephron during pregnancy in the rat. *J. Physiol. (Lond.)* 336, 131–142.
149. Bishop, J. H., Green, R., and Thomas, S. (1981). Glucose transport by short loops of Henle in the rat. *J. Physiol. (Lond.)* 320, 127–138.
150. Bishop, J. H. and Green, R. (1981). Effects of pregnancy on glucose reabsorption by the proximal convoluted tubule in the rat. *J. Physiol. (Lond.)* 319, 271–285.
151. Hytten, F. E. and Cheyne, G. A. (1972). The aminoaciduria of pregnancy. *J. Obstet. Gynaecol. Br. Commonw.* 79, 424–432.
152. Conrad, K. P. and Lindheimer, M. D. (1999). Renal and cardiovascular alterations. In *Chesley's Hypertensive Disorders in Pregnancy* (2nd ed. M. D. Lindheimer, J. M. Roberts, and F. G. Cunningham, Eds.), pp. 263–326. Appleton & Lange, Stamford, CT.
153. Atherton, J. C. and Pirie, S. C. (1981). The effect of pregnancy on glomerular filtration rate and salt and water reabsorption in the rat. *J. Physiol. (Lond.)* 319, 153–164.
154. Baylis, C. (1994). Glomerular filtration and volume regulation in gravid animal models. *Baillieres Clin. Obstet. Gynaecol.* 8, 235–264.
155. Baylis, C. (1987). Glomerular filtration and volume regulation in gravid animal models. *Baillieres Clin. Obstet. Gynaecol.* 1, 789–813.
156. Woods, L. L., Mizelle, H. L., and Hall, J. E. (1987). Autoregulation of renal blood flow and glomerular filtration rate in the pregnant rabbit. *Am. J. Physiol.* 252, R69–R72.
157. Brown, G. P. and Venuto, R. C. (1991). Renal blood flow response to angiotensin II infusions in conscious pregnant rabbits. *Am. J. Physiol.* 261, F51–F59.
158. Kaufman, C. F. and Bergman, E. N. (1978). Renal function studies in normal and toxemic pregnant sheep. *Cornell Vet.* 68, 124–137.
159. de Alvarez, R. R. (1958). Renal glomerulotubular mechanisms during normal pregnancy: I. Glomerular filtration rate, renal plasma flow, and creatinine clearance. *Am. J. Obstet. Gynecol.* 75, 931–944.
160. Assali, N. S., Dignam, W. J., and Dasgupta, K. (1959). Renal function in human pregnancy: II. Effects of venous pooling on renal hemodynamics and water, electrolyte, and aldosterone excretion during gestation. *J. Lab Clin. Med.* 54, 394–408.
161. Dunlop, W. (1981). Serial changes in renal haemodynamics during normal human pregnancy. *Br. J. Obstet. Gynaecol.* 88, 1–9.
162. Hladunewich, M. A., Lafayette, R. A., Derby, G. C., Blouch, K. L., Bialek, J. W., Druzin, M. L., Deen, W. M., and Myers, B. D. (2004). The dynamics of glomerular filtration in the puerperium. *Am. J. Physiol. Renal Physiol.* 286, F496–F503.
163. Atherton, J. C., Bu'lock, D., and Pirie, S. C. (1982). The effect of pseudopregnancy on glomerular filtration rate and salt and water reabsorption in the rat. *J. Physiol. (Lond.)* 324, 11–20.
164. Baylis, C. (1982). Glomerular ultrafiltration in the pseudopregnant rat. *Am. J. Physiol.* 243, F300–F305.
165. Roberts, M., Lindheimer, M. D., and Davison, J. M. (1996). Altered glomerular permselectivity to neutral dextrans and heteroporous membrane modeling in human pregnancy. *Am. J. Physiol.* 270, F338–F343.
166. Baylis, C. (1987). The determinants of renal hemodynamics in pregnancy. *Am. J. Kidney Dis.* 9, 260–264.
167. Baylis, C. and Reckelhoff, J. F. (1991). Renal hemodynamics in normal and hypertensive pregnancy: lessons from micropuncture. *Am. J. Kidney Dis.* 17, 98–104.
168. Baylis, C. and Rennke, H. G. (1985). Renal hemodynamics and glomerular morphology in repetitively pregnant aging rats. *Kidney Int.* 28, 140–145.
169. Reckelhoff, J. F. (1997). Age-related changes in renal hemodynamics in female rats: role of multiple pregnancy and NO. *Am. J. Physiol.* 272, R1985–R1989.
170. Milne, J. E., Lindheimer, M. D., and Davison, J. M. (2002). Glomerular heteroporous membrane modeling in third trimester and postpartum before and during amino acid infusion. *Am. J. Physiol. Renal Physiol.* 282, F170–F175.
171. Moran, P., Baylis, P. H., Lindheimer, M. D., and Davison, J. M. (2003). Glomerular ultrafiltration in normal and preeclamptic pregnancy. *J. Am. Soc. Nephrol.* 14, 648–652.
172. Lafayette, R. A., Malik, T., Druzin, M., Derby, G., and Myers, B. D. (1999). The dynamics of glomerular filtration after caesarean section. *J. Am. Soc. Nephrol.* 10, 1561–1565.
173. Lindheimer, M. D. and Katz, A. I. (1994). Gestation in women with kidney disease: prognosis and management. *Baillieres Clin. Obstet. Gynaecol.* 8, 387–404.
174. Sanders, C. L. and Lucas, M. J. (2001). Renal disease in pregnancy. *Obstet. Gynecol. Clin. North Am.* 28, 593–600, vii.
175. Baylis, C. (2003). Impact of pregnancy on underlying renal disease. *Adv. Ren. Replace. Ther.* 10, 31–39.
176. Thorpe, C. P. S., Gero, G., and Anthony, F. (1986). Urinary protein excretion in pregnancy-induced hypertension: total protein, albumin and retinal-binding protein as indicators of changing renal function. In *Proceedings of the 5th World Congress of the International Society for the Study of Hypertension in Pregnancy*. Nottingham, p. 122.
177. Beetham, R., Dawnay, A., Menabawy, M., and Silver, A. (1988). Urinary excretion of albumin and retinol-binding protein during normal pregnancy. *J. Clin. Pathol.* 41, 1089–1092.
178. Misiani, R., Marchesi, D., Tiraboschi, G., Gualandris, L., Pagni, R., Goglio, A., Amuso, G., Muratore, D., Bertuletti, P., and Massazza, M. (1991). Urinary albumin excretion in normal pregnancy and pregnancy-induced hypertension. *Nephron* 59, 416–422.
179. Sheehan, H. L. and Lynch, J. P. (1973). *Pathology of Toxaemia of Pregnancy*. Williams & Wilkins, Baltimore.
180. Cietak, K. A. and Newton, J. R. (1985). Serial quantitative maternal nephrosonography in pregnancy. *Br. J. Radiol.* 58, 405–413.
181. Balmer, B. D., Garland, H. O., Green, R., Moriarty, R. J., and Richardson, S. (1977). The effect of pregnancy on proximal tubule lengths and total number of nephrons in the rat kidney [proceedings]. *J. Physiol. (Lond.)* 273, 83P–84P.
182. Garland, H. O., Green, R., and Moriarty, R. J. (1978). Changes in body weight, kidney weight, and proximal tubule length during pregnancy. *Renal Physiol.* 1, 42–47.
183. Davison, J. M. and Lindheimer, M. D. (1980). Changes in renal haemodynamics and kidney weight during pregnancy in the unanaesthetized rat. *J. Physiol. (Lond.)* 301, 129–136.
184. Dure-Smith, P. (1970). Pregnancy dilatation of the urinary tract: the iliac sign and its significance. *Radiology* 96, 545–550.

185. Rasmussen, P. E. and Nielsen, F. R. (1988). Hydronephrosis during pregnancy: a literature survey. *Eur. J. Obstet. Gynecol. Reprod. Biol.* 27, 249–259.
186. Fried, A. M., Woodring, J. H., and Thompson, D. J. (1983). Hydronephrosis of pregnancy: a prospective sequential study of the course of dilatation. *J. Ultrasound Med.* 2, 255–259.
187. Hertzberg, B. S., Carroll, B. A., Bowie, J. D., Paine, S. S., Kliewer, M. A., Paulson, E. K., Weber, T. M., and Gimenez, E. I. (1993). Doppler US assessment of maternal kidneys: analysis of intrarenal resistivity indexes in normal pregnancy and physiologic pelvicaliectasis. *Radiology* 186, 689–692.
188. Sherwood, O. D. (2004). Relaxin's physiological roles and other diverse actions. *Endocr. Rev.* 25, 205–234.
189. Kibblewhite, D., Larrabee, W. F. Jr., and Sutton, D. (1992). The effect of relaxin on tissue expansion. *Arch. Otolaryngol. Head Neck Surg.* 118, 153–156.
190. Novak, J., Danielson, L. A., Kerchner, L. J., Sherwood, O. D., Ramirez, R. J., Moalli, P. A., and Conrad, K. P. (2001). Relaxin is essential for renal vasodilation during pregnancy in conscious rats. *J. Clin. Invest.* 107, 1469–1475.
191. Danielson, L. A. and Conrad, K. P. (2003). Time course and dose response of relaxin-mediated renal vasodilation, hyperfiltration, and changes in plasma osmolality in conscious rats. *J. Appl. Physiol.* 95, 1509–1514.
192. Novak, J., Ramirez, R. J., Gandley, R. E., Sherwood, O. D., and Conrad, K. P. (2002). Myogenic reactivity is reduced in small renal arteries isolated from relaxin-treated rats. *Am. J. Physiol. Regul. Integr. Comp. Physiol.* 283, R349–R355.
193. Conrad, K. P., Gandley, R. E., Ogawa, T., Nakanishi, S., and Danielson, L. A. (1999). Endothelin mediates renal vasodilation and hyperfiltration during pregnancy in chronically instrumented conscious rats. *Am. J. Physiol.* 276, F767–F776.
194. Bani, D., Failli, P., Bello, M. G., Thiemermann, C., Bani, S. T., Bigazzi, M., and Masini, E. (1998). Relaxin activates the L-arginine-nitric oxide pathway in vascular smooth muscle cells in culture. *Hypertension* 31, 1240–1247.
195. Danielson, L. A. and Conrad, K. P. (1995). Acute blockade of nitric oxide synthase inhibits renal vasodilation and hyperfiltration during pregnancy in chronically instrumented conscious rats. *J. Clin. Invest.* 96, 482–490.
196. Failli, P., Nistri, S., Quattrone, S., Mazzetti, L., Bigazzi, M., Sacchi, T. B., and Bani, D. (2002). Relaxin up-regulates inducible nitric oxide synthase expression and nitric oxide generation in rat coronary endothelial cells. *FASEB J.* 16, 252–254.
197. Jeyabalan, A., Novak, J., Danielson, L. A., Kerchner, L. J., Opett, S. L., and Conrad, K. P. (2003). Essential role for vascular gelatinase activity in relaxin-induced renal vasodilation, hyperfiltration, and reduced myogenic reactivity of small arteries. *Circ. Res.* 93, 1249–1257.
198. Atherton, J. C., Bielinska, A., Davison, J. M., Haddon, I., Kay, C., and Samuels, R. (1988). Sodium and water reabsorption in the proximal and distal nephron in conscious pregnant rats and third trimester women. *J. Physiol. (Lond.)* 396, 457–470.
199. Baylis, C. and Blantz, R. C. (1985). Tubuloglomerular feedback activity in virgin and 12-day–pregnant rats. *Am. J. Physiol.* 249, F169–F173.
200. Mahaney, J., Felton, C., Taylor, D., Fleming, W., Kong, J. Q., and Baylis, C. (1998). Renal cortical Na^+-K^+-ATPase activity and abundance is decreased in normal pregnant rats. *Am. J. Physiol.* 275, F812–F817.
201. Khraibi, A. A. (2000). Renal interstitial hydrostatic pressure and pressure natriuresis in pregnant rats. *Am. J. Physiol. Renal Physiol.* 279, F353–F357.
202. Khraibi, A. A., Solhaug, M. J., Dobrian, A. D., and Berndt, T. J. (2002). Renal interstitial hydrostatic pressure and natriuretic responses to volume expansion in pregnant rats. *Am. J. Physiol. Renal Physiol.* 282, F821–F825.
203. Khraibi, A. A. (2002). Renal interstitial hydrostatic pressure and sodium excretion in hypertension and pregnancy. *J. Hypertens.* 20 (Suppl 3), S21–S27.
204. Khraibi, A. A. (1991). Direct renal interstitial volume expansion causes exaggerated natriuresis in SHR. *Am. J. Physiol.* 261, F567–F570.
205. Masilamani, S., Castro, L., and Baylis, C. (1994). Pregnant rats are refractory to the natriuretic actions of atrial natriuretic peptide. *Am. J. Physiol.* 267, R1611–R1616.
206. Geller, D. S., Farhi, A., Pinkerton, N., Fradley, M., Moritz, M., Spitzer, A., Meinke, G., Tsai, F. T., Sigler, P. B., and Lifton, R. P. (2000). Activating mineralocorticoid receptor mutation in hypertension exacerbated by pregnancy. *Science* 289, 119–123.
207. Hayslett, J. P. (2001). A new role for progesterone: an agonist for mineralocorticoid receptor activation and pregnancy-related hypertension. *Am. J. Kidney Dis.* 38, 893–895.
208. Quinkler, M., Meyer, B., Bumke-Vogt, C., Grossmann, C., Gruber, U., Oelkers, W., Diederich, S., and Bahr, V. (2002). Agonistic and antagonistic properties of progesterone metabolites at the human mineralocorticoid receptor. *Eur. J. Endocrinol.* 146, 789–799.
209. Quinkler, M., Johanssen, S., Bumke-Vogt, C., Oelkers, W., Bahr, V., and Diederich, S. (2001). Enzyme-mediated protection of the mineralocorticoid receptor against progesterone in the human kidney. *Mol. Cell. Endocrinol.* 171, 21–24.
210. Quinkler, M., Meyer, B., Oelkers, W., and Diederich, S. (2003). Renal inactivation, mineralocorticoid generation, and 11β-hydroxysteroid dehydrogenase inhibition ameliorate the antimineralocorticoid effect of progesterone in vivo. *J. Clin. Endocrinol. Metab.* 88, 3767–3772.
211. McDonald, G. A., Sarkar, P., Rennke, H., Unemori, E., Kalluri, R., and Sukhatme, V. P. (2003). Relaxin increases ubiquitin-dependent degradation of fibronectin in vitro and ameliorates renal fibrosis in vivo. *Am. J. Physiol. Renal Physiol.* 285, F59–F67.
212. Komlosi, P., Fintha, A., and Bell, P. D. (2004). Current mechanisms of macula densa cell signalling. *Acta Physiol. Scand.* 181, 463–469.
213. Thomson, S. C., Vallon, V., and Blantz, R. C. (2004). Kidney function in early diabetes: the tubular hypothesis of glomerular filtration. *Am. J. Physiol. Renal Physiol.* 286, F8–15.
214. Goodrum, L. A., Saade, G. R., Belfort, M. A., Moise, K. J. Jr., and Jahoor, F. (2003). Arginine flux and nitric oxide production during human pregnancy and postpartum. *J. Soc. Gynecol. Investig.* 10, 400–405.
215. von Mandach, U., Lauth, D., and Huch, R. (2003). Maternal and fetal nitric oxide production in normal and abnormal pregnancy. *J. Matern. Fetal Neonatal Med.* 13, 22–27.
216. Schiessl, B., Strasburger, C. J., Bidlingmaier, M., Spannagl, M., Ugele, B., and Kainer, F. (2003). Decreasing resistance in the maternal uterine and peripheral arterial system is apparently unrelated to plasma and urinary levels of nitrite/nitrate and cyclic-guanosine monophosphate during the course of normal pregnancies. *J. Perinat. Med.* 31, 281–286.
217. Conrad, K. P., Kerchner, L. J., and Mosher, M. D. (1999). Plasma and 24-h NO(x) and cGMP during normal pregnancy and preeclampsia in women on a reduced NO(x) diet. *Am. J. Physiol.* 277, F48–F57.
218. Brosnihan, K. B., Neves, L. A., Joyner, J., Averill, D. B., Chappell, M. C., Sarao, R., Penninger, J., and Ferrario, C. M. (2003). Enhanced renal immunocytochemical expression of ANG-(1-7) and ACE2 during pregnancy. *Hypertension* 42, 749–753.
219. Merrill, D. C., Karoly, M., Chen, K., Ferrario, C. M., and Brosnihan, K. B. (2002). Angiotensin-(1-7) in normal and preeclamptic pregnancy. *Endocrine* 18, 239–245.
220. Brosnihan, K. B., Neves, L. A., Anton, L., Joyner, J., Valdes, G., and Merrill, D. C. (2004). Enhanced expression of Ang-(1-7) during pregnancy. *Braz. J. Med. Biol. Res.* 37, 1255–1262.

Knobil and Neill's Physiology of Reproduction, Third Edition
edited by Jimmy D. Neill,
Elsevier © 2006

CHAPTER 55

Parturition

William Gibb,[1,2] Stephen J. Lye,[2–4] and John R. G. Challis[2–4]

Introduction, 2925
Contraction of the Myometrium, 2926
Phases of Parturition and Their Regulation, 2927
Pregnancy: Phase 0 of Parturition, 2927
Activation: Phase 1 of Parturition, 2930
Stimulation: Phase 2 of Parturition, 2942
The Future, 2955
Acknowledgments, 2955
References, 2955

ABSTRACT

Preterm labor is one of the major problems in obstetrics and its incidence has not decreased since the mid-1980s. There have, however, been considerable advances in our understanding of the mechanisms involved in parturition. The onset of parturition is controlled by endocrine, inflammatory, and mechanical factors. Maturation of the fetal hypothalamic–pituitary–adrenal axis and stretch of the uterus due to growth of the fetus regulate the expression of various genes involved in parturition. During most of pregnancy the myometrium is relatively quiescent, and contractions that do occur are poorly coordinated. Near term, the myometrium becomes responsive to substances that stimulate strong contractions (uterotonins). Many of these uterotonins are synthesized in the uterine tissues. The factors involved in maintaining uterine quiescence during pregnancy are discussed, as is the regulation of the synthesis of contraction-associated proteins that enable the myometrium to contract. The regulation of the synthesis of various uterotonins, including prostaglandins and oxytocin, and the mechanisms involved in cervical ripening and membrane rupture, are reviewed. Regulation of parturition is multifactorial, and when the appropriate expression of contraction-associated proteins and uterotonins occurs, normal labor will ensue. The review discusses studies using various experimental animals, including sheep, rodents, and nonhuman primates, and attempts to relate the findings to studies that have been carried out in human pregnancy.

INTRODUCTION

Parturition is the process by which the fetus is expelled from the uterus into the extrauterine environment. It results from a complex interplay of maternal and fetal factors (1,2). During most of pregnancy the myometrium is a relatively quiescent organ and the cervix is hard and unyielding. In late pregnancy, the uterus undergoes a maturational process in response to mechanical signals resulting from uterine stretch and endocrine or paracrine signals from the mother and fetus. The fetal genome appears to play a major role in this sequence of events through both the endocrine and mechanical pathways. Maturation of the hypothalamic–pituitary–adrenal (HPA) system resulting in increased adrenal cortisol production is

[1]Department of Obstetrics and Gynecology, Department of Cellular and Molecular Medicine, University of Ottawa;
[2]Department of Physiology, University of Toronto;
[3]Department of Obstetrics and Gynecology, University of Toronto;
[4]Division of Child Health and Human Development, Samuel Lunenfeld Research Institute, Mount Sinai Hospital, Toronto, Canada.

particularly relevant in the sheep fetus, and we would suggest that cortisol production from the fetal adrenal or in the fetal membranes also plays an important role in human parturition. When there is the appropriate synchronous coordination of these endocrine/-paracrine and mechanical signals, normal labor ensues.

In the human, preterm labor and delivery is a major problem in obstetrics and occurs in approximately 7% to 10% of all pregnancies. Its incidence has remained relatively unchanged since the mid-1980s (3). It is associated with major neonatal morbidity and mortality and developmental disorders in infants and children (3–5). It is hoped that studies to understand the mechanisms involved in labor both at term and preterm will provide information that can be used to predict better those patients in preterm labor who will deliver, and aid in the development of specific methods to control term and preterm labor. Knowing which patients will in fact deliver will also enable the treatment of mothers with synthetic glucocorticoids to be restricted to the appropriate patients. This is particularly important because studies have shown that glucocorticoid treatment of the fetus can have detrimental effects on neonatal and postnatal development.

In this review, we provide an up-to date summary of studies on parturition in the human and using various experimental models, including primates, sheep, and rodents. Rodents, particularly mice and rats, have provided important models to understand the regulation of parturition, and in the case of the mouse, many knockout animals have been studied. An important limitation with the study of mice and rats as models for human parturition is that throughout pregnancy the ovary is the major source of progesterone, in contrast to human pregnancy, in which the placenta is the major source. Unlike the human, withdrawal of progesterone through luteolysis is a prerequisite for the onset of parturition in these species. Of the species that produce progesterone in the placenta, the sheep is the most widely studied animal model. There are some limited studies with subhuman primates, and for the most part human studies are limited to in vitro approaches. The focus of the review is to relate the findings in these animal models to what is known of parturition in the human studies.

The uterus is relatively quiescent throughout most of pregnancy (6), and the contractions that do occur are poorly synchronized and of low amplitude and cause relatively small increases in uterine pressure. This pattern of contractions has been termed "contractures" in the sheep, baboon, and rhesus monkey, and in women it has been speculated that contractures correspond to Braxton Hicks contractions (6–8). Contractures are associated with transient decreases in fetal Pa_{O_2}, endocrine responses, and mechanical stimulation of the fetus (6). At term, well-coordinated, painful contractions develop in the uterus, and the myometrium becomes highly excitable and generates the high-amplitude, high-frequency contractions of labor (6).

It has been useful to consider that there are different stages in the process of parturition (1,9). Phase 0 of parturition, which corresponds to 95% of gestation, is the stage when the myometrium in relatively quiescent, and the uterine cervix is hard and unyielding. Near-term activation of the myometrium, which corresponds to phase 1 of parturition, occurs with softening of the cervix (10–12). The switch from phase 0 to phase 1 is referred to as *activation*, and results from the effects of uterotropins, removal of inhibitory factors, and generation of mechanical forces due to uterine stretch (10–12). Phase 1 of parturition also depends on endocrine factors that are regulated through activation of the fetal HPA axis in late pregnancy and paracrine factors produced in the intrauterine tissues. Mediators at this stage include changes in estrogen synthesis/action, perhaps in combination with local progesterone withdrawal and the action of glucocorticoids. This activation results in the expression of various contraction-associated proteins (CAPs) in the uterus such as connexin 43, prostaglandin, and oxytocin receptors. Phase 2 of parturition is the contraction phase and corresponds to the forceful contractions of labor when synchronous uterine contractions are stimulated by various uterotonins such as prostaglandins and oxytocin. Finally, phase 3 of parturition corresponds to the postpartum involution of the uterus and is associated with the neuroendocrine release of oxytocin. The designation of these different phases of parturition has helped considerably in developing experimental paradigms to understand the regulation of parturition, but it should be stressed that these phases are in fact a continuum and factors acting in one phase may also be involved in other phases.

Before considering the different phases of parturition, and their regulation, the mechanisms involved in the relaxation and contraction of the myometrium will be reviewed.

CONTRACTION OF THE MYOMETRIUM

The myometrium is the uterine smooth muscle and the mechanisms involved in myometrial contraction have been reviewed in detail. Contraction at term and preterm depends on the interaction of actin and myosin (13–17). The myosin molecule is a hexamer composed of two heavy chains 200 kDa each and two light chains, one of 17 kDa and one of 20 kDa. The heavy chains combine to form α helices with a globular head at the N-terminal. These globular heads contain adenosine triphosphatase (ATPase) activity and an actin binding site. The 20-kDa light chain of myosin contains Ca^{2+} and Mg^{2+} binding sites and a phosphorylation site. The myosin molecules are arranged in one direction, forming filaments 2.2 μm long. The actin molecule is a globular protein that polymerizes into long, thin filaments. For each myosin filament, there are 11 to 15 actin filaments. This structure may enable maximum tension generation during shortening of the muscle cell. Myosin and actin filaments slide over each other, leading to shortening of the myocyte, and this complex is capable of converting the chemical energy from ATP into the mechanical energy of contraction (15).

Myosin light-chain kinase (MLCK) is the central enzyme involved in the process of myometrial contraction (16,18–21). It catalyzes the phosphorylation of the 20-kDa light chain of myosin and is activated by Ca^{2+}. The calcium-binding protein calmodulin binds four Ca^{2+} ions, and Ca^{2+}–calmodulin interacts with the MLCK to induce a conformational change in the enzyme that results in its activation and phosphorylation of the light chain of myosin. MLCK itself can also be phosphorylated, and the phosphorylated enzyme is inactive. The phosphorylation of MLCK is catalyzed by a number of enzymes, including cyclic adenosine monophosphate (cAMP)–dependent protein kinase A (PKA) and protein kinase C (PKC). There is also a myosin phosphatase in the myometrium that removes the phosphate from the phosphorylated myosin, resulting in inactivation. Therefore, the activity of MLCK and subsequently the level of phosphorylated myosin are regulated by calcium and cAMP.

Most of the receptors involved in myometrial activation are G-protein–coupled receptors (22). Receptors coupled to uterotonins such as the oxytocin receptor and prostaglandin $F_{2\alpha}$ receptor (FP) are coupled through $G\alpha_q$ and activate phospholipase C and Ca^{2+} pathways, whereas receptors that act to relax the uterus, such as β-adrenoreceptors, are coupled to $G\alpha_s$ receptors and stimulate cAMP formation and relaxation. Those linked through $G\alpha_i$, such as α-adrenergic receptors, likely act by inhibiting cAMP production and enhancing contraction.

In general, substances that cause the myometrium to contract stimulate Ca^{2+} levels by promoting efflux through membrane channels or stimulating release from intracellular stores, including the sarcoplasmic reticulum (17). Small guanosine triphosphatases (GTPases) of the Rho family have also been shown to be involved in a calcium-independent pathway of myometrial contraction (23,24). The Rho kinases phosphorylate myosin phosphatase and inhibit its action, resulting in increased levels of phosphorylated myosin.

Substances that inhibit myometrial contraction do so by increasing intracellular levels of cAMP or cyclic guanosine monophosphate (cGMP), which in turn inhibit Ca^{2+} release from intracellular stores and reduce MLCK activity (17). Binding to the membrane receptor results in dissociation of the heteromeric G protein G_s into its α, β, and γ subunits. The α subunit activates adenylate cyclase. The cAMP formed activates PKA, which phosphorylates various proteins, including MLCK and other cellular proteins that bind calcium and reduce the intracellular concentrations of free calcium. PKA is composed of two regulatory and two catalytic subunits, and its activity is restricted to specific regions in the cell by anchor proteins, A-kinase anchor proteins (AKAPs) (25). The expression of certain of these anchor proteins is altered in the myometrium during gestation (see later).

Extensive studies have been carried out on the regulation of myometrial calcium levels, and a number of reviews on the role of Ca^{2+} in myometrial contractions are available (17,26,27). During contractions free resting Ca^{2+} concentrations increase from 150 nM to approximately 500 nM through the influx of extracellular calcium or from intracellular calcium stores. Extracellular calcium enters the cell through different Ca^{2+} channels, including the voltage-gated L-type channels (26). This channel is regulated by protein kinases, including PKA, PKC, and PKG (28). Binding of uterotonins to their membrane receptors activates G-protein transducers, resulting in activation of phospholipase C, which hydrolyzes phosphatidylinositol to produce inositol 1,4,5-triphosphate (IP3) and diacylglycerol (29–31). IP3 binds to receptors on the endoplasmic reticulum and releases Ca^{2+} from these intracellular stores. The sustained release of calcium from the intracellular organelles is critical for the maintenance of sustained contractions (27,32,33). IP3 binding in myometrium can be inhibited by calcium, which might be a method of regulating the IP3 response in the myometrium.

PHASES OF PARTURITION AND THEIR REGULATION

Pregnancy: Phase 0 of Parturition

Throughout most of pregnancy the myometrium is unresponsive to various uterotonins, such as prostaglandins and oxytocin, and contractions are poorly coordinated. However, the uterus is not totally inactive. Low-frequency, low-amplitude, long-duration contractions occur that are associated with episodes of increased electromyographic (EMG) activity lasting 2 to 5 minutes. In the sheep this occurs from day 64 of gestation (term, 145 to 150 days) and from day 110 in the rhesus monkey (term, 167 days) (6). In vitro studies indicate that these contractures likely reflect an intrinsic pattern of myometrial activity and do not depend on endocrine or paracrine factors (34,35). The role of these contractures is not well defined. They do not result in cervical changes, nor do they result in delivery of the fetus. They do, however, stimulate behavioral changes in the fetus and endocrine responses (36).

During pregnancy the myometrium undergoes extensive growth and phenotypic changes to allow the growth of the developing fetus and placenta (37,38). To date, few studies have examined myometrial growth in any great detail. Human and rat myometrium is composed principally of muscular tissue and extracellular matrix (ECM) (39). The ECM is composed of structural proteins, such as collagens and elastin, and adhesion molecules such as laminin, fibronectin, and proteoglycans. During late pregnancy important changes occur in the structure of the myometrium because of tissue remodeling, myometrial cell growth, and deposition of ECM (40). The collagen content of the rat uterus increases with advancing gestation, as does the production of elastin. In the human there are increased levels of collagen I, collagen III, and fibronectin messenger RNAs (mRNAs) in pregnant tissues compared with nonpregnant tissues (41). Early studies demonstrated that mechanical stretch promoted uterine hypertrophy and remodeling in nonpregnant, pregnant, and postpartum animals (37,42,43). More recent studies have examined the role of progesterone and mechanical stretch in regulating the ECM components of the rat myometrium and have demonstrated that marked changes occur before labor (see later).

During pregnancy, the myometrium is acted on by one or more different inhibitors, which may include progesterone, prostacyclin (prostaglandin I_2 [PGI_2]), relaxin, parathyroid hormone–related protein (PTHrP), and nitric oxide (NO) (1,44). It is likely that an important contributory mechanism to the onset of labor is the withdrawal of one or more of these inhibitory substances.

Studies have suggested that increased coupling of receptors to $G\alpha_s$ and to adenylate cyclase may be involved in maintaining uterine quiescence during pregnancy (45,46). At least nine isoforms of adenylate cyclase are expressed in the uterus, and all of them are stimulated by $G\alpha_s$ (47). They have been divided into four subfamilies according to their mode of regulation, and there are selective increases in certain of these adenylate cyclases in pregnant myometrium (22). $G\alpha_s$ expression is upregulated during pregnancy and is associated with an increase in $G\alpha_s$-coupled adenylate cyclase activity (46,48). During term and preterm labor the $G\alpha_s$-coupled adenylate cyclase activity in the myometrium is reduced, indicating that expression of $G\alpha_s$ may be involved in maintaining uterine quiescence throughout gestation (48). The regulation of the expression of this protein is complex, involving cAMP and phosphorylated nuclear proteins (49). The phosphodiesterase that hydrolyzes cAMP is inhibited by the high levels of progesterone during pregnancy (50). Once cAMP is produced, PKA can act on various targets to stimulate relaxation, including promoting calcium extrusion and phosphorylation of MLCK. A number of genes involved in parturition, including the β_2-adrenergic receptor, oxytocin receptor, connexin 43, and PGH_2 synthase (PGHS)-2, contain cAMP response elements (CREs) in their promoter regions.

In the rat, AKAP5 is involved in the regulation of uterine contractions (51). In early pregnancy, AKAP5 targets PKA to phospholipase C β3, and the subsequent phosphorylation inhibits the enzyme and the effect of receptors that are coupled to phospholipase C, such as the oxytocin receptor. In late pregnancy, PKA levels decrease and AKAP5 binds to a phosphatase that activates phospholipase C, resulting in increased oxytocin effect at term. Studies have found that expression of CRE modular proteins are altered in the myometrium with advancing gestation and labor (52). In human myometrium there is an increase in the expression of the regulatory RII subunit of PKA in the myometrium with advancing gestation and a decrease in expression at labor. This corresponds to similar changes in particulate PKA activity (53). In addition, two activating transcription factor-2 proteins were described that are differentially expressed in the upper and lower segment of the myometrium (52). How these factors are involved in the maintenance of pregnancy and labor has yet to be determined.

Large-conductance calcium-dependent potassium channels have also been implicated in maintaining uterine quiescence during pregnancy (54–59). A considerable number of these channels are found in human myometrium, and increased intracellular Ca^{2+} results in activation of these channels, causing rapid hyperpolarization. Membrane hyperpolarization reduces Ca^{2+} entry through voltage-activated channels, thereby preventing contractions. It has been shown that the guinea pig chorion releases a factor that causes relaxation of oxytocin-stimulated myometrial contractions by opening these Ca^{2+}-dependent K^+ channels (60). During human labor, there is downregulation of α and β subunits of the calcium-activated potassium channel (61).

PTHrP is synthesized in human myometrium, and its production is increased by transforming growth factor beta (TGF-β) and progesterone (62). PTHrP may act through specific receptors on the myometrium to activate the $G\alpha_s$ subunit of G proteins to increase intracellular levels of cAMP (63,64). In rats, PTHrP has been shown to delay the normal increase in expression of the gap junction–associated protein, connexin 43, and the oxytocin receptor, which normally occurs at term (65).

NO is a potent endogenous relaxant of uterine smooth muscle. It is thought to act through a cGMP-dependent process (66–68), although a study with term human myometrium found that NO-induced relaxation was not dependent on cGMP but was due to activation of calcium-activated K^+ channels (69). Administration of the NO precursor L-arginine or the NO donor nitroprusside, which results in increased endogenous production of NO, inhibits myometrial activity in the human and the rat (68). In vitro, nitroprusside decreased the force of contraction of myometrium and decreased phosphorylation of the myosin light chain. NO is also synthesized in the myometrium and the fetal membranes, and mRNAs for NO synthase (NOS) isoforms have been detected in human fetal membranes and choriodecidua (66,70). Several authors have reported a decrease in levels of inducible NOS (iNOS) mRNA in myometrium collected from patients at term (67). NOS activity in decidua and myometrium decreases before parturition in a variety of species. This would be consistent with a diminution of the inhibitory activity of NO on the uterus at this time. Of particular interest are studies in the rat showing that loss of NOS activity in the myometrium is associated with an increase in NOS activity in the cervix, specifically with increased capacity for NO production by inflammatory cells (68). It has been shown that NO effects on the cervix are mediated by cGMP and prostaglandins (71), which suggests that NO may effect cervical effacement and relaxation at a time when its effects to relax the myometrium are diminished.

There is substantial evidence indicating that relaxin has paracrine/autocrine effects on myometrial activity (72–80). Relaxin is a 6-kDa peptide whose A and B chains are derived from post-translational cleavage of preprorelaxin (78). Relaxin has structural similarities to insulin and nerve growth factor. Administration of relaxin stimulates separation of the pubic symphysis in species in which this is necessary for delivery. Relaxin inhibits spontaneous uterine activity in many species, and, as discussed later, enhances cervical ripening.

The human genome contains two relaxin genes, H1 and H2, that encode two peptides with different amino acid sequences. In the corpus luteum, a major source of relaxin, the H2 relaxin gene is the principal one that is transcribed. Relaxin is also expressed in the fetal membranes, decidua, and placenta (77,80,81). In these tissues both the H1 and H2 genes are transcribed. The peak level of relaxin expression in the decidua and placenta occurs at 12 to 14 weeks' gestation, and at term after labor there was a slight increase in relaxin expression in both tissues.

Relaxin suppresses uterine contractility in a variety of species through elevation of myometrial cAMP levels and inhibition of oxytocin-induced turnover of phosphoinositide (82). Relaxin was shown to suppress spontaneous uterine contractions in the rat and guinea pig but did not alter the sensitivity of the uterus to oxytocin, indicating that relaxin was acting as a frequency modulator (74,75). In the mouse, relaxin increased expression of endothelial NOS but did not alter the expression of iNOS (83). Also, the inhibitory effect of relaxin on myometrial contractility could be reduced by an NOS inhibitor. In the rat uterine artery and myometrium, relaxin was found to inhibit contraction at mid-pregnancy but not at term, and its effects on the uterine artery were mediated by cGMP, cAMP, and NO (84). Relaxin also has important effects on the cervix and fetal membranes (see later).

It is well known that prostacyclin (PGI_2) is involved in maintaining uterine quiescence. PGI_2 infused into the nonpregnant sheep inhibited uterine contractions (85,86), and in the human a similar inhibitory effect was demonstrated (87). In vitro, when human myometrial strips were exposed to PGI_2, initial contraction was followed by relaxation (88), and in vitro PGI_2 has dual actions on the rat myometrium (89). The level of prostacyclin synthase, the enzyme necessary for PGI_2 synthesis, has been shown to decrease with advancing gestation in the human myometrium (90). $PG1_2$ acts through a specific IP receptor to stimulate adenylate cyclase and to increase intracellular levels of cAMP (91). Corticotropin-releasing hormone (CRH) can also stimulate cAMP production in myometrial cells and may act in a synergistic fashion with PGI_2 (92). The role of CRH in parturition is discussed in detail later in this review.

Progesterone maintains uterine quiescence in numerous species, and the presence of progesterone is refractory to stimulation by oxytocin and prostaglandins (93). Administration of exogenous progesterone decreases uterine activity in several species and delays parturition (94,95). Progesterone may act in a number of ways, including uncoupling of excitation– contraction (2,96,97), inhibition of CAP synthesis, and decreasing uterotonin production by stimulating prostaglandin metabolism and inhibiting CRH synthesis. The latter effects are discussed in detail later. Progesterone may also act to inhibit oxytocin release by the pituitary.

Other substances also inhibit uterine activity, including calcitonin gene–related peptide (98,99), vasoactive intestinal peptide, and endogenous β-adrenergic agonists. These compounds act to increase cAMP production or decrease calcium levels or availability. Animal models have provided good evidence for withdrawal of one or more of these inhibitory influences on the myometrium as a prerequisite to the onset of labor (100). In a study that used DNA microarrays, the pattern of gene expression in the rat uterus indicated that more genes were suppressed than activated at labor (101), consistent with the proposal that removal of inhibitory stimuli may play an important role in the onset of labor.

Activation: Phase 1 of Parturition

Near the end of gestation there are major changes in the myometrium resulting in the synthesis of CAP proteins that will enable the tissue to provide strong synchronous contractions when various uterotonins are present. In addition, the cervix, which is rigid throughout most of pregnancy, softens so that it can undergo dilatation and enable delivery of the newborn.

The Uterine Cervix

The cervix is a heterogeneous tissue that is hard and unyielding throughout most of pregnancy. At term it softens (ripens) to allow the infant to be delivered. The cervix has an extensive ECM that is composed of collagen, elastin, and proteoglycans (102–105). Proteoglycans consist of a central protein core substituted with at least one glycosaminoglycan chain such as heparin, hyaluronic acid, or dermatan sulfate. The cervix also contains smooth muscle, fibroblast epithelial cells, and blood vessels. Extensive changes in the composition of the cervix occur during gestation and just before labor, resulting in changes in the interactions between the various components and infiltration of inflammatory cells.

Collagen I and III, rod-shaped molecules that impart tensile strength, are the predominant forms of collagen expressed in the cervix (106). The collagen fibril is composed of tropocollagen molecules orientated in parallel. Each collagen unit has three collagen molecules wrapped around each other in a helical fashion, and cross-linking of the collagen helices increases its tensile strength. The small dermatan sulfate proteoglycan, decorin, coats the collagen fibers and is important in determining cervical structure (107). Elastin provides the tissue with elasticity. Elastin fibers are arranged in parallel between the collagen fibers and enable this tissue to be stretched (108). Patients with incompetent cervices exhibit a decrease in elastin and disruption of the normal cervical structure (109).

In rats, from early gestation there is turnover of fibroblasts and muscle cells in the cervix. With advancing gestation, decorin expression is increased, and the increased decorin-to-collagen ratio results in disruption of the collagen fibers (110–112). With the increase in hyaluronic acid, there is an influx of water into the cervix (113–115). There is also increased apoptosis with advancing gestation, and decorin has been shown to suppress cell proliferation (116). In human pregnancy, there is a decrease in the collagen concentration with advancing gestation (117). The changes in collagen content correlate with changes in proteolytic enzyme activities. Increased collagen degradation occurs during pregnancy and there is a marked increase in collagenase activity during labor (111,118). Also, with the increase in apoptosis there is invasion of the cervix by neutrophils and macrophages, which further stimulate collagen degradation and cervical remodeling. The observation that leukocyte infiltration and degranulation occur in postpartum tissues in a manner similar to an inflammatory reaction supports the idea that cervical ripening is similar to an inflammatory response.

Many factors have been implicated in the regulation of cervical ripening. Relaxin stimulates matrix metalloproteinase (MMP) expression in the cervix. Thus, relaxin would appear to be involved in remodeling cervical connective tissue in addition to its suppressive actions on the myometrium discussed previously (119,120). Similarly, NO, which is involved in maintaining a quiescent myometrium, has also been implicated in the softening of the cervix. Studies in the rat demonstrated that as NOS activity in the myometrium decreased, there was an increase in NOS activity in the cervix, specifically with increased capacity for NO production by inflammatory cells (68). The action of NO on the cervix appears to involve cGMP and prostaglandins (71). PGHS-2 expression is increased and PGE_2 and $PGF_{2\alpha}$ are released when patients are treated with sodium nitroprusside, an NO donor. The morphological changes induced are similar to those seen in normal cervical ripening, namely, separation of collagen fibers, absence of longitudinal collagen bundles, and proinflammatory reactions in fibroblasts, macrophages, and leukocytes. In vitro, NO increases PGE_2 output by cervical tissues.

Cytokines originating from inflammatory cells that invade the cervix at term are also involved in cervical ripening (121–123). For example, IL-6 and IL-1 are found in cervicovaginal fluids and in the human cervix at term labor and there is a marked increase in cytokine concentrations (124). Cytokines act on fibroblasts and smooth muscle cells to release proteases (125). IL-1β and IL-8 increase activity of MMPs, and IL-1β also downregulates tissue inhibitors of MMP (TIMP) expression. IL-8 administration causes cervical ripening in experimental animals and this cytokine, which acts as a neutrophil chemotactic factor, may be involved in the inflammatory cell invasion of the cervix at term (123,126,127). The increased concentration of hyaluronic acid that occurs with advancing gestation and labor may be involved in the process of inflammatory cell invasion and stimulation of proteolysis (103).

Prostaglandins have an important role in the ripening of the cervix (128). Prostaglandin-treated cervices show increased collagen degradation and increased collagenase activity (129). It has also been proposed that PGE_2 may be involved in the production of chemokines that attract neutrophils into the term cervix. Secretion of the chemokine IL-8 is increased in the cervix at term and during labor, and PGE_2 can stimulate IL-8 release from cervical tissues (130–132). Prostaglandins may be produced in the cervix, although there is no increase in prostaglandin content of cervical secretions with advancing gestation or during labor (133,134). Studies have demonstrated decreased expression of 15-hydroxyprostaglandin dehydrogenase (PGDH) mRNA in cervical tissues at term and preterm labor, but there was no change in PGHS-2 mRNA expression (135). Overall, this would be expected to result in decreased metabolism of prostaglandins in the tissues at labor. It is also possible that the prostaglandins acting on the cervix are produced by adjacent tissues such as the decidua or fetal membranes (discussed in detail later). Prostaglandin analogs are used extensively in the clinical situation to ripen the cervix.

The role of progesterone in cervical ripening has not been well defined, but mice deficient in 5α-reductase cannot metabolize progesterone and do not deliver viable fetuses because the cervix does not ripen (136,137). It is therefore possible that progesterone withdrawal occurs in the cervix through changes in its metabolism, or through altered responsiveness of the tissue. This occurs in the myometrium, and may also be important in cervical ripening. In the guinea pig there is a decrease in progesterone levels before parturition that may be indicative of progesterone withdrawal (138). A study in women has provided some evidence of decreased progesterone levels in cervical tissues obtained after labor (139). Also, in the human, estrogen receptor (ER)α and ERβ are expressed in the cervix at term (140), and ERβ has been localized to leukocytes, indicating the potential for estrogen involvement in the regulation of cervical function (141).

Peptidergic sensory neurons have also been implicated in cervical ripening. Substance P release from neurons terminating in the cervix and spinal cord peaks before parturition (142). Studies have shown that mRNA levels of calcitonin gene–related peptide, which colocalizes with substance P, increases at term and its expression is increased by estrogen (143). The human cervix also expresses functional human chorionic gonadotropin receptors (144). Treatment of cervical tissues with human chorionic gonadotropin resulted in increased levels of cAMP and decreased levels of immunoreactive PGHS-2. It is therefore possible that this important placental hormone may regulate cervical function during pregnancy.

Cervical structure has been used to predict patients at a risk for preterm delivery. Using vaginal ultrasonography, patients with small cervices were found to have a higher incidence of preterm delivery than those with cervices of normal length (145,146). The structure of the cervix changes toward term and the onset of labor, and studies have found that these structural changes can be quantified using light-induced fluorescence. This method may be useful to determine which patients are going to deliver (147).

The Myometrium

The evolution of spontaneous contractions in the myometrium is due to changes in its phenotype. As labor approaches, the myometrium becomes more excitable because there is a relative depolarization of the resting membrane potential and it becomes highly responsive to uterotonins. In addition, there is a marked increase in the electrical coupling between myometrial cells. Together, these changes result in the tissue responding to uterotropins by generating high-frequency contractions that rapidly spread over the entire uterus.

The contractile activity of the human myometrium increases with advancing gestation. In the mean, frequency increases from 0.32/h at 25 weeks to 2.33/h at 40 weeks, and in the multipara the increase is from 0.43/h to 2.27/h (148). In the sheep there is a sharp change in the pattern of contractions from "contracture" to "contractions" just before delivery, but in the human and subhuman primate this change occurs over several days before delivery (10,148). During this period there is cycling between contractures and contractions. It is possible that the switch in activities is related to the diurnal rhythm of estrogen levels, and it has been shown that these cycles can be modulated by estrogen, prostaglandin synthesis inhibitors, and even feeding (6,149). During the labor-like activity the myometrium is more responsive to oxytocin, and the switch from contractures to contractions occurs at a time when oxytocin levels are maximal and oxytocin antagonists can inhibit the contractions (150). This pattern of contractions may be responsible for the higher incidence of human deliveries occurring in the early morning compared with during the day (151).

This activation phase is due to the increased expression of various CAPs, such as oxytocin receptors, prostaglandin receptors, calcium channels, and gap junction proteins. The high input resistance of smooth muscle cells during most of pregnancy and the poor coordination of myometrial contractions was shown by Garfield and colleagues to be due to the paucity of gap junctions in the myometrium throughout pregnancy (152,153). With the onset of labor there is a marked increase in the number of gap junctions in the myometrium, enabling the muscle to develop high-amplitude synchronous contractions at labor (154). In all species studied, the concentration of gap junctions in the myometrium increases, indicating that this is a necessary component for labor. The proteins that form gap junctions are connexins, which are classified according to their molecular weights (155). Connexins are arranged in hexameric hemichannels that align across adjacent cells to form an interconnecting pore. These pores allow low-resistance electrical or ionic coupling and metabolic transfer between cells (156). Hundreds of these molecular pores are organized to form a gap junction. Regulation of the synthesis of connexins occurs at the transcriptional and translational level as well as in their transport to the membrane and assembly into connexons to form the functional channels (156–159). This is a complex process that is not well understood, but involves steroid hormones and stretch (see later).

Role of the Fetus

It is evident that for survival of the species the fetus must develop appropriately in utero so that it can survive in the extrauterine environment. It is therefore logical that the fetus be involved in timing the onset of labor, and that this will occur when the appropriate development of the various organ systems has been achieved.

Early studies on the role of the fetus in parturition have been reviewed (160–162). Genetic abnormalities in cattle and ingestion of *Vibrio californicum* by sheep was found to result in prolonged gestation. Marked pituitary and adrenal hypoplasia were found in the fetal sheep. The studies of Liggins and Thorburn, and others provided the initial experimental evidence for the importance of the fetus in timing the onset of labor in sheep and goats, as well as the experimental basis for most of the studies in fetal physiology that were subsequently carried out (160–163). Liggins showed that fetal hypophysectomy resulted in prolonged gestation in sheep. In these animals there was marked hypoplasia of the fetal adrenal. In addition, when the lambs were delivered their lungs were found to be immature. The latter finding led to the use of glucocorticoids in the maturation of the fetal lung in humans delivering prematurely. Subsequent studies by a number of investigators demonstrated that infusion of adrenocorticotropic hormone (ACTH) or glucocorticoids to the sheep fetus could induce preterm labor. These prematurely delivered fetuses were found to be viable, with appropriate organ development. In these experiments and others it was shown that the timing of parturition was unaltered by maternal hypophysectomy or by the administration of ACTH or glucocorticoids to the mother, indicating that the fetal endocrine signals played a different role from maternal signals (161,162). Additional support for the fetus regulating the length of gestation in the sheep was provided by use of embryo transfer techniques. In these studies twin pregnancies were established with one embryo from a shorter-gestation breed (141 days) and the other from a longer-gestation breed (150 days) (164,165). The length of gestation was appropriate for the shorter-gestation twin, with the appropriate increase in fetal cortisol concentrations occurring in this twin. In the other twin there was a delayed rise in cortisol levels, but estrogen and androgen levels increased in both twins at the same time.

In the human and subhuman primate the role of the fetus in timing the length of gestation is less clear. Original studies by Malpas in 1933, in which the gestational length of pregnancies with various fetal anomalies, including anencephaly, were studied, suggested that the fetal pituitary and adrenal provided the trigger for labor (166). A further study from Holland, with a large number of anencephalic fetuses, did not lead to such a definitive conclusion. Swaab et al. (167) found that the median time of delivery for 147 anencephalic pregnancies was shorter than for the control population. When pregnancies complicated with polyhydramnios or twins, or those given injections in which labor was induced, were removed, the median gestational age for the remaining anencephalic pregnancies was the same as that of the control population. However, in the anencephalic pregnancies there was a larger proportion of both preterm and post-term deliveries, indicating that the fetus was involved in timing delivery. Similar results were obtained after experimental anencephaly in rhesus monkeys (168). In monkeys, fetal hypophysectomy resulted in prolonged gestation, but fetal adrenalectomy did not affect the gestational length (169,170). Placental retention after fetectomy in the rhesus monkey was longer than in control animals (171), and fetectomy in the baboon did not affect the delivery of the placenta (172–174). Results with primates are often difficult to interpret because in many instances the numbers of experimental animals are small and the effects of stretch and uterine volume have not been examined as they have in smaller animals. Although studies with humans and subhuman primates would indicate that the fetus is not solely responsible for triggering the onset of labor, as it is in the sheep, it has become clear that the primate fetus still plays an important role in regulating the length of gestation.

More recently, it has become evident that regulation of phase 1 of parturition depends on the fetal genome, which exerts its effects through a number of pathways, including a mechanical growth pathway that influences stretch of the myometrium, as well as an endocrine/paracrine pathway. Activation of these pathways results in upregulation of a cassette of CAPs in the myometrium (1,175,176).

Role of Mechanical Stimuli

In the rat, the synthesis of connexin43 increases markedly during the last 48 hours of pregnancy, and the highest levels are found during delivery itself (177,178). Increases in connexin43 mRNA and protein occur in sheep and human myometrium with the onset of labor (154,178). Garfield demonstrated that cell–cell communication in the myometrium is reduced by elevated $[Ca^{2+}]_i$ and increased by cAMP (152). These factors may regulate the phosphorylation of consensus serine and tyrosine residues in the cytoplasmic domain of connexin43, which regulates the permeability of gap junctions.

Earlier studies had shown the importance of stretch in the regulation of uterine growth (see previously), and more recent studies have examined the role of stretch and mechanical forces in regulating the expression of CAP genes, such as the oxytocin receptor, connexin43, and the FP receptor. Using an approach that involves insertion of an inert tube into the uterine lumen of pregnant rats followed by various endocrine manipulations, investigators have begun to dissect the factors that are involved in the activation of the uterus. They have been able to determine which genes are regulated by mechanical signals, which required hormonal signals, and which required mechanical and hormonal signals.

Garfield demonstrated that progesterone inhibited gap junction formation in the rat uterus and that administration of RU486, a progesterone antagonist, increased connexin43 levels and the number of gap junctions (153), and it was also shown that the mRNA level of connexin43 was reduced by progesterone and increased precociously by RU486 (95). In the rat, the levels of mRNAs for the oxytocin receptor and connexin43 were increased by placing inert tubing in the uterine lumen (179). This response to stretch was diminished in animals treated with progesterone or in late-gestation, unilaterally pregnant animals in which the endogenous systemic progesterone concentration was still high. In contrast to connexin43, connexin26 expression is elevated at mid-gestation in the rat and may be associated with uterine quiescence in this species (11,12). Furthermore, the expression of the mRNA transcripts encoding the $PGF_{2\alpha}$ receptor is temporally associated with progesterone withdrawal but does not depend on mechanical stretch (180). The effect of stretch may be mediated through AP-1 transcription factors (181). In vitro studies demonstrated that mechanical stretch increased expression of the AP-1 family of transcription factors and progesterone attenuates the effect of stretch on the expression of these genes.

It has been proposed that there are three phases of uterine growth during pregnancy (182). The first occurs during the first trimester, results in uterine hyperplasia, and is regulated by endocrine factors. The second phase occurs during the second and third trimesters as the growth of the uterus follows the growth of the fetus. In the final phase the uterus grows at a slower rate than the fetus, resulting in increased uterine stretch and tension. It is evident that progesterone is necessary for stretch-induced hypertrophy, and that near term the drop in progesterone concentrations (or functional withdrawal of progesterone—see later) results in discordance between fetal growth and uterine size, resulting in increased tension and gap junction formation. Studies have examined the role of stretch and progesterone on the synthesis of ECM components of the rat myometrium (183). mRNA levels of laminin B2, fibronectin, and collagen IV were low during gestation but increased markedly at term. Expression of fibrillar collagen types I and III peaked at day 19 and decreased at term. Progesterone was shown markedly to decrease levels of fibronectin and laminin mRNA and prevented the drop in collagen III mRNA levels. The expression of these ECM components could be increased using the progesterone antagonist RU486. Furthermore, most of the changes in ECM components occurred in the gravid horn. Gap junction formation also requires the presence of cell adhesion molecules. This laboratory had demonstrated previously that focal adhesion signaling is abruptly terminated with the onset of labor, likely because of the remodeling required for the tissue to prepare for optimal contractile activity at labor (184).

Teleologically, it would make sense that the size, or growth, of the fetus should have some influence on the timing and process of birth, and may explain the higher incidence of preterm birth in multifetal pregnancies. However, studies could not find any difference in the expression of connexin43 in tissues obtained from twin pregnancies compared with singletons. It is clear from the preceding discussion that there are extensive interactions between the mechanical and endocrine pathways that result in the expression of the various CAPs and enable the uterus to respond to uterotonins in the appropriate manner. The development and regulation of these endocrine pathways that originate in the fetus are discussed next.

Role of the Endocrine/Paracrine Pathways

Fetal Hypothalamic–Pituitary–Adrenal Axis in the Sheep. The sheep has become the experimental animal of choice for the study of fetal development and it allows in vivo studies to be carried out with continuous fetal and maternal monitoring that is not possible in smaller animals or the human. Although such studies would likely be feasible with subhuman primate models, experimentation with these animals is severely restricted. In the discussions that follow it will be seen that there are numerous similarities in the regulation of sheep and human parturition, but also some important differences. We would contend that the similarities outweigh the differences and that underlying molecular mechanisms are the same or similar in ovine and human species.

Hypothalamus/Pituitary. In the ovine fetus, activation of the HPA axis leading to increased production of cortisol by the fetal adrenal occurs at the end of gestation (36,185,186). This increase in cortisol production results in decreased progesterone synthesis and increased estrogen synthesis by the placenta, together with increased production by of prostaglandins by maternal and fetal tissues, resulting in labor and delivery, as discussed later (1).

There is a progressive increase in fetal plasma levels of ACTH 1-39 and cortisol in late gestation of the sheep (187–190). The initial rise in ACTH precedes that of cortisol (188), but in the last 10 days of pregnancy fetal plasma levels of cortisol increase exponentially (187). In the guinea pig there is similar maturation of the HPA axis. Fetal cortisol is responsible for the maturation of many organ systems, including the lung and kidney, and is critical in the normal development and programming of the brain (191–193). This increase in ACTH and cortisol in the fetal sheep is somewhat paradoxical because in the adult, cortisol suppresses pituitary proopiomelanocortin (POMC) gene expression and ACTH output.

The mechanisms of HPA activation in fetal sheep have been studied. CRH is produced by the paraventricular neurons of the hypothalamus. It is secreted into the portal system and acts on the pituitary to stimulate POMC synthesis and ACTH secretion. In the sheep fetus, hypothalamic CRH mRNA and CRH protein are expressed by day 60 of gestation (term, 147 to 150 days) (194). There is a gradual rise in expression until day 120 and then a marked increase in expression during the final 20 days of gestation, with a subsequent decrease in the newborn lamb.

In the fetal pituitary, expression of POMC, the precursor of ACTH, varies in different regions of the gland. POMC is detectable in the inferior region of the pars distalis from 60 days' gestation and POMC mRNA levels increase in both the superior and inferior regions of the pars distalis from days 60 to 120, when there is a further increase to term (195,196). These changes in POMC expression are consistent with the increase in immunoreactive ACTH levels, and increased immunoreactive ACTH immunostaining in the pituitary corticotropes during the same period. POMC mRNA levels in the pars intermedia are relatively high at 60 days' gestation and increase to 100 days' gestation, after which they remain constant to term. The differential changes in expression during gestation may explain the earlier conflicting results using Northern blot analysis (197,198), which cannot of course differentiate expression in different areas of the brain. The lack of negative feedback of cortisol on POMC mRNA levels was clearly demonstrated using fetal pituitary tissues obtained at specific times during late gestation (199). Processing of POMC by the various prohormone convertases (PCs) may also regulate increasing levels of ACTH found with advancing gestation. At term, levels of PC-1 and PC-2 mRNAs were higher in the pars intermedia than in the pars distalis, and PC-1 but not PC-2 mRNA levels increased in the pars distalis with gestational age, but did not change with labor (199). It is possible that the changes in POMC distribution during gestation in the fetal sheep indicate the transition from fetal to adult corticotropes.

The ovine pituitary binds CRH from 100 days' gestation (200), and binding increases up to approximately 135 days and then decreases to term (201). The CRH receptor subtype R1 was shown to be expressed in the fetal pituitary from day 100 of gestation, and there was a decrease in mRNA and protein expression to term (202). Using hypothalamopituitary-disconnected fetuses, the hypothalamus was shown to be necessary for the decrease in CRH R1 levels to occur (203). In vitro studies demonstrated that CRH downregulates its own receptor. Cortisol also decreases CRH binding (201) and, in vivo, cortisol decreases CRH R1 mRNA levels in the pituitary (202). CRH would therefore be expected to have its greatest effect on the pituitary at approximately 120 to 135 days' gestation. Consistent with this proposal, the responsiveness of the fetal pituitary to CRH infusion increases from day 110 to day 125 and then decreases to term (188), and in vitro corticotrope responsiveness to CRH decreases during late pregnancy (204). Therefore, although hypothalamic CRH expression increases until term, the responsiveness of the pituitary to CRH is decreased and therefore is unlikely to be responsible for the sustained function of the HPA axis at term.

Arginine vasopressin (AVP) is produced by the supraoptic and paraventricular nuclei (magnocellular and parvocellular neurons) of the hypothalamus and may be involved in the regulation of the synthesis and secretion of pituitary ACTH (205). AVP is expressed by the parvocellular neurons from day 60 of gestation, and its expression does not change during the rest of gestation (194). In magnocellular neurons, however, AVP mRNA increases with gestational age. In the newborn lamb there are marked increases in AVP mRNA expression in both regions of the paraventricular nucleus. It has been proposed that magnocellular AVP is involved primarily in the regulation of fetal fluid balance, whereas that produced by the parvocellular neurons is involved primarily with the stimulation of pituitary ACTH secretion (206). In vitro studies demonstrated that CRH and AVP stimulated POMC expression and ACTH output by cultures of fetal sheep pituitary cells (207). At 120 to 138 days' gestation, AVP was more potent that CRH, but at term these factors were equipotent. The effect of AVP and CRH on ACTH output was additive. Intrafetal administration of CRH and AVP increased ACTH production throughout gestation. However, the response to CRH decreased and the response to AVP increased (188,208). At approximately 120 days' gestation, AVP and CRH effects were synergistic, whereas at other gestational ages their effects were additive. It is therefore possible that CRH is responsible for the early rise in ACTH and AVP may be more important during the final 20 days of gestation.

A number of mechanisms have been proposed for the lack of feedback inhibition of cortisol on fetal CRH and ACTH production during gestation. Prostaglandins are involved in the maturation of the HPA axis and may be involved in the lack of inhibition of HPA activity seen with advancing gestation. In late gestation the concentrations of PGE_2 and cortisol in the fetal plasma increase with a similar time course (209,210). Infusion of PGE_2 increases plasma levels of cortisol and ACTH (160,211), and inhibition of PGE_2 production decreases fetal plasma cortisol and ACTH levels (212,213). Furthermore, interruption of the hypothalamic–pituitary stalk abrogates the effects of PGE_2 on the HPA axis (214). Prostaglandins may also be synthesized within the fetal pituitary, hypothalamus, and brain stem; there is an exponential increase in the level of PGHS-2 in the hypothalamus and pituitary with advancing gestation (215), and PGE_2 directly stimulates cytochrome P450 C17 expression in the fetal adrenal, resulting in increased cortisol production (216).

It is possible that there is modification of the glucocorticoid threshold in the brain and pituitary. Glucocorticoids increase expression of corticosteroid-binding globulin (CBG) in the fetal sheep liver, which is opposite to the effect found in the adult (217,218). The pattern of glycosylation of CBG is different in the fetus than the adult, and CBG increases in concentration with advancing gestation and maintains a fairly constant level of free cortisol (218,219). Near term, however, there is a marked increase in cortisol production that overcomes the CBG binding capacity, resulting in a marked increase in free cortisol levels just before birth (220). This increase in free cortisol in the fetus is a consistent finding in many species (221).

The levels of free cortisol can also be regulated by metabolism, in particular through the interconversion of cortisol and cortisone catalyzed by the 11β-hydroxysteroid dehydrogenase (11βHSD) enzymes (222). Two isoforms of this enzyme exist: 11βHSD-1, which acts principally as a reductase and converts cortisone to cortisol, and 11βHSD-2, which acts as a dehydrogenase and can act in both directions, but most often oxidizes cortisol to cortisone. 11BHSD is expressed in the fetal pituitary, although in this tissue the enzyme appears to act mainly in the oxidative direction, inactivating cortisol (222). The predominant isoform of 11βHSD in the fetal pituitary, however, remains to be resolved.

The level of glucocorticoid receptor (GR) in the pituitary may also influence glucocorticoid action (223). GR mRNA is present in the fetal pituitary from early gestation and its expression increases with advancing gestation (224). During the course of labor, there is a marked decrease in GR expression in the pars distalis that would be consistent with there being a decrease in glucocorticoid negative feedback during labor (223,225). There is also a decrease in

GR expression in the hypothalamic paraventricular nucleus near term. These changes are specific to CRH- and AVP-positive neurons. In the fetal sheep and guinea pig, GR mRNA levels in the paraventricular nucleus decrease in late gestation, and in fetal sheep GR mRNA levels decrease before labor. All of these finding are consistent with a reduction in the potential for glucocorticoid feedback in brain structures with advancing gestation and labor (226).

Estrogens also influence the HPA axis. Fetal plasma concentrations of estrone sulfate increase with advancing gestation and in some studies estrogen infusion was found to increase ACTH levels in fetal plasma. The fetal hypothalamus, hippocampus, and brain stem express estrogen sulfatase, enabling local estrogen production from sulfated estrogens within these structures (227). It is possible that the effect of estrogens is modulated by CRH because the CRH promoter contains an estrogen response element (228).

Other substances may also be involved in HPA activation and sustaining this activation. Oxytocin is expressed in the parvocellular and magnocellular neurons of the hypothalamus, and its expression increases with gestational age. In vitro it has been shown to stimulate ACTH output by hypothalamic cells, and it can act synergistically with CRH (206). Urocortin is also a potential candidate for maintaining ACTH production in the presence of high cortisol levels (229). Urocortin has been localized to the fetal pituitary in late gestation; urocortin mRNA levels increased with advancing gestation and were temporally associated with increases in the level of POMC mRNA. In vitro studies with pituitary cells demonstrated that inhibition of urocortin expression decreased ACTH output, and when added to the culture media urocortin stimulated ACTH output. Furthermore, in late gestation, fetal infusion of cortisol increased urocortin mRNA levels in the pituitary. These studies taken together would support an important role for urocortin in maintaining stimulation of the HPA axis in the late-gestation fetal sheep even in the presence of increasing cortisol concentrations.

Fetal Adrenal. In the fetal sheep there is a marked increase in cortisol production by the adrenal during the final 20 to 25 days of gestation (189,230,231). This is due both to the increasing levels of ACTH in the fetal plasma and to the maturation of adrenal function (230). In vitro studies with dispersed cells and tissue incubations demonstrated that the responsiveness of the adrenal changed markedly during gestation (232–234). Adrenal responsiveness to ACTH is high in early gestation (approximately 50 days) and at term (235), but is markedly attenuated at mid-gestation (approximately 100 days) (232). This decrease in responsiveness is likely due to the marked decrease in the expression of cytochrome P450 C17 and P450 SCC (side chain cleavage) in the fetal adrenal cortical cells at mid-gestation (236,237). ACTH infusion into the fetus increases expression of the mRNAs for both of these enzymes (238,239). 3β-Hydroxysteroid dehydrogenase (3βHSD) is likely rate limiting for cortisol production in the first half of pregnancy (240), and 3βHSD has been demonstrated in cells of the zona fasciculata of the fetal adrenal from day 50 of gestation to term (241). ACTH receptor mRNA is detected from day 60 of gestation, its level increases slightly until days 126 to 128, when there is a marked increase, followed by another marked increase in expression around days 140 to 141 (242), and ACTH receptor mRNA stability increases with gestation (243). The low level of adrenal responsiveness at day 100 is therefore not due to absence of the receptor, but may be due to the low plasma levels of ACTH (244) at this time or to the antagonism of ACTH action by TGF-β (245). The increase in ACTH receptor levels in the latter part of gestation may be responsible for the increase in adrenal responsiveness seen at this time. A number of substances may regulate ACTH receptor expression in the fetal adrenal, including ACTH itself, cortisol, insulin-like growth factor-II, and TGF-β (246–248).

In vitro and in vivo studies indicate that ACTH can stimulate adrenal maturation (249–252). Infusion of ACTH increases cAMP generation by enhancing the coupling of the ACTH receptor with adenylate cyclase (253–255). ACTH also increases expression of cytochrome P450 C17, C11, and C21 and 3βHSD enzymes (256,257), and the continued presence of ACTH was found to be necessary to maintain increased levels of these enzymes. The pattern of ACTH secretion may determine the steroid output by the adrenal because continuous infusion of ACTH favored corticosterone output, whereas pulsatile infusion favored cortisol production (258,259).

ACTH-induced increases in cytochrome P450 C17 appear to be a key step in the synthesis of cortisol by the fetal adrenal (256,260). However, the obligatory role of increasing ACTH levels to drive the fetal adrenal maturation has been questioned. The infusion of ACTH to maintain constant low levels of plasma ACTH in hypophysectomized fetal sheep resulted in the normal rise in cortisol concentrations (261). Birth occurred around the expected time, with the normal decrease in plasma progesterone concentrations.

There is considerable evidence to support the view that stress, in particular hypoxemia, influences birth by stimulating the HPA axis (262,263). Fetal plasma ACTH and cortisol concentrations are markedly increased in response to even modest reductions in arterial Po_2 (264,265). In fetuses in which the hypothalamus and pituitary are disconnected, release of CRH and AVP into the portal system is abolished and there is no increase in ACTH levels (266). Changes in fetal oxygen levels of only 4 to 5 mm Hg, similar to the hypoxemia that occurs in fetal sheep during contracture activity, also resulted in increased plasma levels of ACTH (267). Therefore, it is possible that myometrial activity in late gestation contributes to the increased levels of ACTH in the fetal plasma. At 135 days' gestation, hypoxia increased CRH mRNA levels in the paraventricular nucleus and POMC mRNA levels in the pars distalis within 6 hours (268). Coinfusion of cortisol inhibited this response, indicating the presence of negative feedback mechanisms. After 48 hours of hypoxia, POMC levels in the pars distalis were elevated but they were decreased in the pars intermedia, consistent with differential regulation of the two zones in the pituitary (269). This would be consistent with inhibition of pituitary POMC synthesis by dopamine, which is exacerbated in hypoxia. Infusion of bromocriptine, a dopamine receptor agonist, during late pregnancy decreased POMC mRNA levels in the pars intermedia but had no effect on levels in the pars distalis (270). Activation of the HPA axis by hypoxemia is clearly involved in preterm birth because a sustained pulsatile hypoxemic stimulus activates HPA gene expression (242). It is therefore likely that sustained hypoxemia together with fetal compromise would result in preterm birth (271).

Studies have also highlighted the importance of peri-implantation insults in the regulation of the fetal HPA axis during the subsequent pregnancy. Undernutrition around the time of conception causes premature activation of the HPA axis and subsequent preterm delivery in approximately 50% of the fetuses from undernourished dams (272). Fetal plasma cortisol concentrations increase earlier in the ewes that delivered prematurely. In this study, although precociously high cortisol levels were found only in 50% of the fetuses, all underfed animals had higher ACTH levels. This may indicate that maturation of the HPA axis in the fetus must reach a threshold level to stimulate the feed-forward loop of continued increase in cortisol and ACTH levels. The role of infection in the process of preterm delivery in these animals was excluded because fetal plasma concentrations of IL-6 and tumor necrosis factor-α (TNF-α) were not altered in the fetuses delivering preterm compared with those delivering at term. If these findings are applicable to human pregnancy, it is clear that insults before and around conception may have a major impact on the incidence of preterm birth.

The HPA axis clearly matures during gestation, resulting in increased levels of fetal cortisol. This cortisol then acts on the placenta to activate the processes required to enable the fetus to be delivered.

The Placenta. Fetal cortisol acts on the ovine placenta to alter steroid production (273–275). Progesterone output decreases and estrogen production increases at term (276–279). These changes are associated with increased expression of the cytochrome P450 C17 enzyme that enables the conversion of C21 steroids to C19 steroids, which can be metabolized by the aromatase enzyme to form estrogens (275,280). The aromatase enzyme is expressed in the ovine placenta and its expression is increased in late gestation. (It should be noted that induction of cytochrome P450 C17 does not occur in the human placenta, and, as discussed elsewhere, the human placenta is an incomplete endocrine organ and requires C19 steroids to be supplied either by the fetal or maternal adrenal for estrogen synthesis.) It had been widely accepted for a number of years that the effect of cortisol on the ovine placenta was to increase P450 C17 expression directly and that the subsequent increase in estrogen production was responsible for the increase in prostaglandin production occurring at term (2,281–283). The increased production of prostaglandins then resulted in contraction of the myometrium and the subsequent delivery of the fetus (162,284–290).

A number of studies, however, led us to question the validity of this scheme of events, and we have recently proposed a new model for the sequence of events that result in parturition in the sheep. These studies led us to believe that a similar mechanism may be occurring in human labor. It had previously been shown that the sheep, like the human, can use fetal adrenal C19 precursors to form estrogens (291). Androstenedione and dehydroepiandrosterone (DHEA) are secreted by the fetal adrenal gland and their secretion is stimulated by ACTH and hypoxia, and androstenedione infused into the fetus is metabolized to estrogens (292,293). Throughout the latter part of gestation there is a progressive increase in conjugated estrogens in maternal plasma (278), and this precedes the sharp rise in free estrogen that occurs just before delivery in response to induction of the cytochrome P450 C17 enzyme (277). In addition, infusion of estradiol into the fetal lamb did not stimulate placental levels of PGHS-2 mRNA (176).

It has also been shown that PGHS-2 expression and activity in sheep placental tissue increase during the final 20 days of gestation before the final surge in estrogens (294–300). At this time, there is a significant increase in the expression of the PGE synthase (PGES) necessary for PGE_2 synthesis (301). The PGHS-2 enzyme is exquisitely localized to the fetal trophoblast of the fetal placenta. In contrast to the PGHS-2 isoform, there is no change in the expression of the PGHS-1 enzyme (296,300). The increase in expression of the PGHS-2 isoform that occurs in the sheep with advancing gestation and labor has also been shown to occur in many other species studied, including the human (see later). The increase in circulating levels of fetal PGE_2 correlates closely with increasing levels of cortisol (209,210) and, as discussed, PGE_2 may play an important role in the development of the HPA axis. Using placental tissues obtained at term before labor, during early labor, and during active labor, we demonstrated that increased expression of PGHS-2 in the fetal placenta during labor preceded the induction the P450 C17 enzyme (302). Furthermore, increased expression of the PGHS-2 enzyme in the maternal endometrium occurred after induction of the placental enzyme (303). This suggested that there was a time-dependent and tissue-specific cascade in the expression of PGHS-2 during labor and that different factors regulated its expression in the fetal and maternal compartments.

Fetal infusion of cortisol into adrenalectomized animals was found to increase expression of PGHS-2 in the ovine placenta, myometrium, and endometrium (304). In a series of studies, the effect of cortisol infusion was examined in the presence and absence of the aromatase inhibitor 4-hydroxyandrostenedione. These demonstrated clearly that PGHS-2 expression in the placenta was cortisol dependent, did not require an increase in estrogen, and was also responsible for the increase in PGE_2 in the fetal circulation (305). In contrast, expression of PGHS-2 in the maternal uterus was estrogen dependent and necessary for the increased plasma levels of $PGF_{2\alpha}$ that occur in the mother (305). Although the PGHS-2 enzyme was markedly increased by cortisol infusion into the fetus, this treatment did not alter the expression of the PGE synthase enzyme (301). Others found that betamethasone infusion increased expression of PGHS-2 but did not increase expression of PGF synthase (306). Because PGE_2 has been shown to stimulate cytochrome P450 C17 expression in other tissues, and in human in vitro cultures cortisol stimulates PGHS-2 expression, we have proposed that in the sheep during gestation increasing fetal cortisol levels are responsible for increasing the expression of PGHS-2 in fetal placental trophoblast cells. The PGE_2 formed is involved in the maturation of the HPA axis (214,216,307–309) and other organs in the fetus and also acts in an autocrine or paracrine fashion to increase expression of the P450 C17 enzyme in the placenta, enabling formation of C19 steroid precursors that are converted to estrogens. This increased estrogen production has its primary effect in the maternal compartment to increase expression of CAP genes in the myometrium and increase expression of PGHS-2, resulting in increased levels of $PGF_{2\alpha}$, which stimulates myometrial contractions. Studies in which estradiol was infused to induce labor are consistent with this model. $PGF_{2\alpha}$ levels were increased with estradiol infusion, but there was no change in the expression of cytochrome P450 C17 or PGHS-2 in the fetal placenta (310). Consistent with this new model it has been shown that GR but not ER is expressed in the ovine placenta, that ER is expressed in the maternal endometrium (311), and at term labor there is increased expression of immunoreactive GR in the placental trophoblast (312). In addition, using ovine placental trophoblast cells in culture, we have shown that the cortisol-induced output of PGE_2 was inhibited by meloxicam, a specific PGHS-2 inhibitor. Furthermore, we have demonstrated that PGE_2 stimulates expression of cytochrome P450 C17 mRNA and protein by ovine trophoblast cells in a concentration- and time-dependent manner, thereby substantiating this key step in the new model of ovine parturition.

Fetal Hypothalamic–Pituitary–Adrenal Axis in the Primate. The development of the primate fetal pituitary and adrenal has been reviewed (313–315). During the second half of gestation in the human and in subhuman primates, the pituitary is essential for appropriate development of the adrenal. In anencephalic fetuses, adrenal development is impaired. In the baboon fetus treated with synthetic glucocorticoids, pituitary POMC mRNA levels are suppressed and the weight of the fetal adrenal is reduced; these findings, together with decreases in ACTH receptor mRNA and 3βHSD mRNA, indicate that pituitary ACTH is necessary for the normal functioning of the primate fetal adrenal (313). CRH- and AVP-like activity is present in the human fetal hypothalamus from about 12 weeks' gestation (316). In the case of CRH, both a high–molecular-weight form and CRH 1-41 are present. AVP and CRH were found to act synergistically in vitro to increase ACTH production by the pituitary. Furthermore, ACTH-like activity has been described in human fetal pituitaries from as early as 5 weeks of gestation [see Challis and Lye (2)]. The regulation of ACTH output by CRH and AVP in the human is similar to that reported with the ovine fetus. In the baboon, POMC mRNA levels in the fetal anterior pituitary increase from mid-gestation to term, and there is a corresponding increase in pituitary cells expressing ACTH (317,318). These increased levels of POMC mRNA may be due to increases in the activity of the CRH receptor. However, estrogen administration at mid-gestation also increased POMC mRNA levels and ACTH-positive corticotropes (318).

It has been argued that the effect of estrogen is secondary to its action on the placental 11βHSD-2 enzyme (319). Estrogen stimulates expression of

11βHSD-2, which under normal circumstances increases with advancing gestation (313,319). At mid-gestation, the lower levels of 11βHSD-2 allow passage of maternal cortisol to the fetus, and this acts to suppress fetal HPA activity. With advancing gestation and higher placental 11βHSD-2 activity, less cortisol has access to the fetus and therefore inhibition of the HPA axis is reduced. Thus, increased ACTH production and fetal adrenal development ensue. These findings are compatible with the observation that the primate adrenal is relatively inactive during much of pregnancy (320,321). Studies have shown that 11βHSD-2, which inactivates cortisol, is higher at term than midterm, and its expression is increased by glucocorticoids (322).

In the primate, the fetal adrenal gland has a large fetal zone that produces substantial amounts of sulfated androgens, which are the precursors for estrogens (323–325). In late gestation there is an increase in 3βHSD activity and ACTH receptor expression in the definitive or adult zone of the adrenal, allowing cortisol synthesis, and a decrease in ACTH receptor and DHEA sulfate (DHEAS) production by the fetal zone (326–328). The expression of cytochrome P450 C17 and SCC does not change significantly. Therefore, in the baboon, 3βHSD expression seems to be rate limiting in cortisol formation, whereas in the sheep cytochrome P450 C17 appears to be rate limiting. In the baboon, ERs are expressed in all zones of the fetal adrenal at mid- and late gestation (329). This may indicate an important role of estrogens in the ACTH-dependent maturation of the fetal adrenal in this species. Estrogen also stimulates cholesterol synthesis by the human fetal liver, thereby increasing fetal levels of low-density lipoprotein (LDL) cholesterol, a major precursor of DHEAS in the human fetus (330).

Estrogen production in the primate during pregnancy depends on the supply of androgen precursors from the adrenal (331,332). In the human, estriol is the principal estrogen formed, and more than 90% of the estriol formed is from fetal precursors. DHEAS is produced by the fetal zone of the fetal adrenal and further hydroxylated in the 16 position in the fetal liver. The 16α-OH-DHEAS is metabolized by the placenta through the sulfatase, 3βHSD, and aromatase enzymes to form estriol. In the case of estradiol/estrone, 50% of the precursor DHEAS comes from the fetal adrenal and 50% from the maternal adrenal. The fetal adrenal is activated with advancing gestation, and this is reflected in the increase in estriol concentrations in maternal plasma. The increase in DHEAS levels in the human fetus are similar to the changes in cortisol seen in the fetal sheep, and both are related to the increase in estrogen production seen in each species, albeit through different mechanisms.

In the primate, the fetal adrenal is divided into three functionally different zones: the outer, adult zone that produces mainly aldosterone; an inner, fetal zone responsible for the production of huge quantities of DHEAS; and a transition zone between these zones that produces mainly cortisol (315,333–335). Consistent with their steroidogenic function, these zones show expression of specific enzymes. For example cytochrome P450 SCC and C21 are expressed in all three zones, whereas C17 is expressed in the transitional and fetal zones but not the definitive (adult) zone (315). At mid-gestation, 3βHSD is not expressed in the adrenal but is expressed in the definitive and transitional zones in late gestation. Cytochrome P450 C11 is found in the transitional zone from mid-gestation and throughout the adrenal cortex in late gestation. ACTH stimulates cortisol secretion from the definitive zone and DHEAS secretion from the fetal zone. CRH, likely of placental origin, has been suggested to play an important role in regulating steroid output by the fetal adrenal. In vitro studies demonstrate that CRH increase DHEAS and cortisol output by cultured fetal zone cells and increases expression of the P450 C17 enzyme (336–338). This has been suggested to result in increased estrogen synthesis through subsequent metabolism of DHEAS by the placenta. The estrogens formed may then act on various intrauterine tissues to stimulate the expression of various CAP genes.

Although there are differences in fetal adrenal structure between the sheep and primate, as well as differences in the pattern of steroid production and the pathway for estrogen synthesis, there are also many similarities. In late gestation, the primate and ovine fetal adrenal produces increasing levels of cortisol (339). It is not possible to make exact correlations between the species in the temporal changes that occur in plasma cortisol concentrations because binding to CGB and metabolism through the 11βHSD enzymes, which can regulate cortisol access to the fetus and its local effects, all need to be taken into account. In both the sheep and primate, C19 steroids from the fetal adrenal precursors are used as precursors of estrogens formed by the placenta. In the human, this is an absolute requirement throughout gestation because the human placenta never expresses the cytochrome P450 C17 enzyme necessary for the formation of C19 steroids. In the sheep at term there is also a major increase in estrogen that results from activation of the sheep placental P450 C17 enzyme; however, we have suggested that the mechanism by which this activation occurs is increased expressed of the PGHS-2 enzyme by cortisol (a stimulatory mechanism previously shown to occur in the human fetal membranes; see later). The major difference in the sheep and human in relation to estrogen synthesis at term is therefore the source of the C19 steroids for estrogen synthesis, although in both species the fetal adrenal is intimately involved in estrogen synthesis.

Steroid Hormones and Parturition. The major steroid hormones produced in human pregnancy are estrogens and progesterone. Progesterone acts to inhibit myometrial contractions and in most species estrogens act to increase contraction of the myometrium. In experimental animals but not in the human or subhuman primate, parturition is preceded by an increase in plasma levels of estrogen and a decrease in plasma levels of progesterone. It is therefore important to bear this difference in mind when extrapolating data from experimental animal models to the human. The site of production of these hormones and the source of steroid precursors for the active steroids also differentiate some experimental animals used in parturition research from the human. For example, in mice and rats progesterone is produced throughout gestation by the corpus luteum of the ovary, and estrogen is produced by the placenta only at the end of gestation, just before delivery. At the time of the increase in plasma estrogen levels there is a marked reduction in plasma progesterone levels because of luteolysis. In the sheep and guinea pig, both progesterone and estrogen are produced by the placenta. Progesterone levels are high until term, when there is a marked decrease in the plasma levels of progesterone and an increase in estrogen levels. In the sheep, as discussed previously, the increased production of estrogen results from induction of the cytochrome P450 C17 enzyme in the fetal placental trophoblasts, which allows C21 steroid to be converted to androgens and subsequently estrogens.

In the human and subhuman primates, the placenta is the major source of progesterone and estrogen during pregnancy (340). In the human, maternal LDL cholesterol is the principal precursor of placental progesterone, and the P450 SCC enzyme is expressed at extremely high levels, resulting in massive amounts of progesterone being produced in an almost unregulated fashion. Increasing amounts of progesterone are produced throughout gestation, and plasma levels tend to plateau but do not decrease at term before delivery. The plasma levels of progesterone in the human reach concentrations that are much higher than those found during the luteal phase of the menstrual cycle. A number of suggestions have been advanced to explain how labor can occur in the presence of the high plasma levels of progesterone (see discussion later).

In the human, there are also large amounts of estrogens, including estrone, estradiol, estriol, estetrol, and their metabolites and conjugates, produced by the fetoplacental unit. Plasma estrogen concentrations increase throughout gestation to term, with the plasma estrogen-to-progesterone ratio increasing during the latter part of gestation (341). Estrogen synthesis depends on C19 precursors produced by the fetal and maternal adrenals. Estriol synthesis depends almost exclusively on androgen precursors provided by the fetal adrenal gland and fetal liver, and estrone and estradiol synthesis depends on androgen precursors (DHEAS) from both the fetal and maternal adrenal.

The human fetal membranes also have important roles in steroid biosynthesis and metabolism directly related to the process of parturition. These tissues are not merely an extension of the placenta, but possess important synthetic and metabolic capabilities with tissue- and cell-specific properties. Steroid sulfatases that can act on various sulfated steroids are present in the membranes (342–345), and the chorion laeve expresses high levels of 3βHSD (346–350). Various steroid dehydrogenases and reductases are also present in the membranes and decidua (351–354). Cholesterol side-chain cleavage (355), aromatase (356), and even cytochrome P450 C17 lyase activities (357) have been described, but their activities are extremely low. Expression of 11βHSD in the fetal membranes has been suggested to play an important role in regulating cortisol production in the tissues (358,359).

Progesterone. Excitability of the myometrium increases before the onset of the strong contractions of labor and, as described, is due to the priming of the uterus that results in the synthesis of various CAP proteins. In the mouse and rat there is progesterone withdrawal resulting from luteolysis, which leads to the onset of labor and delivery. The role of progesterone in preventing gap junction formation in the rat and in inhibiting the expression of other CAPs has already been discussed.

In sheep, as discussed, there is a marked decrease in progesterone production by the placenta, which, together with increasing estrogen production, results in parturition. Inhibition of progesterone production, decreased plasma progesterone concentrations, and increased plasma levels of PGFM (the reduction product of $PGF_{2\alpha}$) resulted in preterm delivery (360–362). Other studies have shown that induction of preterm labor with fetal infusion of glucocorticoids can be blocked by administration of a synthetic progestin. Administration of the progesterone receptor antagonist RU486 induces labor in the sheep and stimulates intense uterine contractions in the rhesus monkey (212,363).

In the human and subhuman primates, in contrast to other animals, plasma progesterone concentrations remain high at term and decrease in maternal plasma only after delivery of the placenta. How labor proceeds in the presence of high levels of progesterone has remained an enigma until very recently. Studies have begun to explain how the functional withdrawal of progesterone in a particular tissue could occur without there being a decrease in circulating levels of the hormone.

It was suggested in the mid-1970s that local effects of steroids occurred in the human fetal membranes (an important source of prostaglandins—see later), and that at term the expression of a specific progesterone-binding protein resulted in the withdrawal of progesterone action in these tissues (364). A role for such a protein was not confirmed experimentally and it was later shown to be identical to CBG (365). To date, no further studies on the role of this protein moiety in the amnion have been carried out. It was also proposed that progesterone withdrawal in the fetal membranes could be achieved by changes in metabolism in the tissues. However, no major changes in the expression or activities of enzymes involved in the synthesis or metabolism of progesterone were found to occur in these tissues around the time of parturition (346,349,353,366).

Evidence for the importance of progesterone metabolism in parturition was produced when it was discovered that 5α-reductase–deficient mice did not deliver (136). This was subsequently shown to be due to the inability of the cervix to ripen (137). This observation would therefore support the proposal that local modulation of progesterone concentrations in the cervix plays an important role in the changes in structure that occur in this tissue at term.

The possibility that an antiprogestin was responsible for progesterone withdrawal in pregnancy was made a number of years ago, and TGF-β was suggested to be the substance involved (367). Using cultured endometrial cells, TGF-β antagonized the effects of progesterone on the expression of preproendothelin-1, connexin43, enkephalinase, and PTHrP. More recently, studies have indicated that glucocorticoids might also act as antiprogestins. Certain effects of progesterone in intrauterine tissues may be exerted through the GR. Increases in endogenous glucocorticoids occur in late gestation, and by displacing progesterone from the GR, glucocorticoid could reverse the activities of progesterone (368–370). This may be involved in the regulation of placental CRH production and the expression of PGDH in the human fetal membranes.

Probably the most exciting idea to explain functional progesterone withdrawal is that changes take place in the response of the myometrium to progesterone through alterations in the response machinery. Progesterone, like other steroids, acts by binding to a nuclear receptor. This receptor binds to response elements in the promoter regions of genes and recruits various coactivators that are involved in histone acetylation, recruitment of chromatin remodeling proteins, and subsequent gene transcription (371,372). Corepressors have also been described that act to inhibit transcription. Studies from a number of laboratories have provided evidence that the relative activities of the different isoforms of the progesterone receptor (PR-A and PR-B) might determine receptor-mediated progesterone action on the uterus (373). In the myometrium during human parturition there are increasing levels of PR-A, a dominant negative repressor of PR-B that may contribute to functional progesterone withdrawal in this tissue (374). PR-A dominance in the myometrium of the rhesus monkey at labor has also been proposed to be involved in functional progesterone withdrawal in this species (375). Declining levels of steroid receptor coactivators (SRCs) may also enhance functional withdrawal of progesterone in the human myometrium and mouse uterus at term (376). Condon et al. (376) reported that in the human myometrium during labor the level of expression of SRC-2, SRC-3, and CBP (cAMP-response element binding [CREB]-binding protein) decreases, and suggested that this may result in decreased progesterone action in this tissue. Similar changes were found to occur in the mouse uterus at parturition, although in this case SRC-1, SRC-3, and CBP levels declined. Furthermore, in the mouse uterus and human myometrium the levels of acetylated histones decreased. Dong et al. (377) identified a progesterone-binding protein in myometrium that interacted with both PR-A and PR-B. This protein was shown to be identical to a previously described RNA splicing factor, PSF, and was found to act as a repressor of progesterone-mediated gene expression. In addition, its expression was shown to increase in the rat uterus with advancing gestation, with the maximum level of expression being seen at the onset of labor. Taken together, these studies indicate that functional withdrawal of progesterone in the myometrium results from a number of alterations in the cellular response mechanisms, including changes in expression of progesterone receptor subtypes and of coactivators and corepressors.

Estrogen. The uterine actions of estrogen are generally antagonistic to progesterone. Administration of estradiol to the pregnant sheep results in premature delivery and an increase in response to oxytocin (378,379). In nonpregnant sheep, estradiol stimulates the myometrium, increases its responsiveness to agonists, and reverses the inhibitory effects of progesterone (380,381). Studies have also shown that in the sheep the increased estrogen production that occurs at term is necessary for the maternal production of $PGF_{2\alpha}$, likely because of increased expression of PGHS-2 in the maternal endometrium (305). When preterm delivery is induced by cortisol or estrogen infusion there is also a marked increase in maternal levels of $PGF_{2\alpha}$ (305,310).

In the human, the principal estrogen produced during pregnancy is estriol, but its role in the process of human parturition is not well defined. Maternal levels of estriol increase exponentially during the latter part of human gestation and have been found to be increased in preterm labor (340,382,383). In patients with a steroid sulfatase deficiency, labor still occurs, although the pregnancy may be prolonged and the cervix does not ripen appropriately (384). Interestingly, in a study by Liggins et al. (385), the efficacy of estrogen given with glucocorticoids to induce labor in post-term pregnancies was greater than that of glucocorticoids alone. This would be consistent with the ability of glucocorticoids to stimulate prostaglandin production (see later) and estrogen to stimulate synthesis of CAPs. Maternal estrogen concentrations reflect the production of DHEAS by the fetal adrenal and subsequent hydroxylation by the fetal liver (340). In the stressed fetus, one would expect fetal adrenal activation and increased production of these estrogen precursors to occur with a subsequent rise in maternal estriol levels. Indeed, maternal plasma and salivary estriol levels are increased in a subset of patients in preterm labor (382). In addition, estriol may increase CAP expression, and the increasing estriol levels during the third trimester may be responsible for the increase in uterine responsiveness at this time. During the final 5 to 6 weeks of gestation in the human there is also a marked increase in the concentration of estrone, estradiol, and estrone sulfate in the amniotic fluid (386). Estrogen sulfatase has been described in the amnion, chorion, and decidua (345,387), and in intact cells or subcellular fractions from the fetal membranes, the hydrolysis of estrone sulfate was markedly increased during labor. This increase in metabolism is probably the most remarkable change in steroid metabolism seen in these tissues around labor. There is also active 17β-hydroxysteroid dehydrogenase in these tissues, enabling the estrone to be converted to estradiol. More recently, estrogens were found to increase expression of oxytocin mRNA in the fetal membranes (388). It is possible that these estrogens are derived from amniotic fluid estrogens and may stimulate oxytocin production through paracrine actions, which can subsequently stimulate myometrial contractions.

In the intact rhesus monkey the increase in maternal estrogen concentrations in late gestation parallels the increase in the fetal adrenal precursors DHEA and DHEAS (313,389). The infusion of androstenedione into the pregnant rhesus monkey at 0.75 gestation resulted in an increase in maternal estrogen concentrations and premature birth (390). This effect could be blocked by the coinfusion of the aromatase inhibitor 4-OHA, which prevented maternal endocrine changes and nocturnal increases in myometrial contractility (391). However, systemic infusion of estrogen to the mother increased myometrial activity but did not result in preterm delivery. The authors therefore suggested that for estrogen to be effective in sustaining uterine contractions it had to be produced locally in the intrauterine tissues, enabling paracrine–autocrine interactions to occur. The infusion of androstenedione into the pregnant monkey and baboon (392,393) reproduced endocrine, biophysical, and biochemical changes similar to those found at normal term labor, including the forward shift in the nocturnal estradiol surge relative to maternal progesterone that results in increased nocturnal myometrial activity.

Studies with human myometrium have described increased expression of ERα at parturition, and this may indicate increased action of estrogen in the myometrium at this time (394). These investigators demonstrated a positive correlation between the levels of ERα and the expression of the CAP genes PGHS-2, and the oxytocin receptor in the myometrium from nonlaboring women. This increase in ERα expression occurred at the same time as the relative increase in PR-A expression over that of PR-B (394).

Interestingly, most of the effects of estrogens that occur in parturition appear to be due to their action on maternal tissues. In the sheep and the human, for example, estrogens have not been found to alter prostaglandin production in the fetal placenta or fetal membranes, respectively, in contrast to their known stimulatory actions in maternal tissues.

Stimulation: Phase 2 of Parturition

This phase of parturition corresponds to the three phases of labor and is characterized by strong, painful uterine contractions. The onset of labor is associated with a marked change in the contractile activity of the myometrium. In sheep this begins 6 to 18 hours before delivery, when the frequency and amplitude of contractions begin to increase and their duration decreases (10). The constant EMG firing and the smooth, regular contraction pressure traces seen during contractures change into shorter bursts and pressure cycles. Single contractions associated with EMG bursts appear and increase in frequency and amplitude as labor progresses (8,10). In phase 2 of parturition, various uterotonins act to stimulate the strong myometrial contractions; these include prostaglandins and oxytocin.

Prostaglandins

Prostaglandins have long been associated with the labor process, at term and preterm. A number of years ago it was proposed that increased prostaglandin production was the consequence of labor rather than a trigger for labor (395,396), but subsequent studies demonstrated that before labor there is an increase in the concentrations of prostaglandins in amniotic fluid, and increased prostaglandin production by fetal membranes (397–400). Numerous studies have shown that inhibitors of prostaglandin synthesis delay delivery (401–403). Prostaglandins contribute to phase 1 of parturition, as previously described, and to phase 2 of parturition.

An important source of prostaglandins in human parturition is the human fetal membranes and decidua, and a large number of in vitro studies have been carried out with these tissues using explant and cell culture systems. The fetal membranes are composed of two tissues, neither of which is vascularized. The amnion has a single layer of epithelial cells in contact with the amniotic fluid and a layer of connective tissue–containing cells of mesenchymal origin that are fibroblast-like. The chorion laeve is in contact with the amnion and is composed of a mesenchymal layer and mononuclear cytotrophoblast cells. The cytotrophoblast cells are in direct contact with maternal decidual cells, and there is no clear separation between these cell types. The various cell types are characterized by the expression of cytokeratin in amnion epithelial cells and chorion leave trophoblast cells, and vimentin in the mesenchymal and decidual cells (404,405). The study of human fetal membranes is limited to in vitro incubation systems. Obviously, these tissues are obtained from human patient samples of diverse genetic backgrounds, the cultured cells are not always pure, and even if the same cell isolation procedures are used, quite different cell populations may be obtained from one patient to another (406). A number of procedures have been described for the isolation of pure populations of the various cell types (358,404,407). In addition, the isolated cells have been removed from the intimate cell–cell contact found in vivo, which may alter their properties. Therefore, although important information has been obtained using these systems, the extrapolation to in vivo circumstances will always be limited and necessitates the use of various experimental animal models, as we have discussed.

Prostaglandins are formed from membrane phospholipids through the obligate precursor arachidonic acid (408). Hydrolysis of arachidonic acid from phosphatidylethanolamine and phosphatidylinositol occurs through the action of one or more isozymes of phospholipase C or phospholipase A_2 (PLA_2). The role of the PLA_2 enzymes in the synthesis of prostaglandins at parturition has received considerable attention. The PLA_2 enzymes hydrolyze the sn-2 acyl ester bond of 1,2-diacyl-sn-3-glycerophospholipids, producing 1-acyl lysophospholipid and free fatty acid (409). The lysophospholipids may serve as a precursor to platelet-activating factor or may have proinflammatory effects. PLA_2 has an intracellular form, cytosolic PLA_2 ($cPLA_2$), and secretory forms ($sPLA_2$). The secretory enzymes have low molecular weights (14 to 19 kDa) and have optimal activity at neutral or alkaline pH in the presence of millimolar concentrations of calcium (410). To date, 10 human $sPLA_2$ isoforms have been described (411). The $cPLA_2$ enzyme family has molecular weighs around 85 kDa and has three isoforms (412,413). These isoforms have high homology but differ in their specificity for arachidonic acid at the sn-2 position (the precursor fatty acid for a number of prostanoids involved in human parturition). The $cPLA_{2\alpha}$ isoform is particularly selective for arachidonic acid–containing phospholipids (414). In contrast to the $sPLA_{2\alpha}$ enzymes, the $cPLA_2$ enzymes are active at low concentrations of calcium (412).

Arachidonic acid may be metabolized through different lipoxygenase pathways to form leukotrienes and hydroxyeicosatetraenoic acids, and through the PGHS pathway to form prostaglandins. Early studies suggested that arachidonic acid metabolism is preferentially directed toward lipoxygenase products during human pregnancy, but that a progressive switch to PGHS products occurs at the time of labor (415). Arachidonic acid is converted to primary prostaglandins such as PGE_2, PGF2α, and prostacyclin (PGI_2) by the action of the PGHS enzyme and specific synthases and isomerases. PGHS catalyses the cyclooxygenation of arachidonic acid followed by isomerization to form the intermediate PGG_2, which then undergoes a two-electron reduction by the peroxidase activity of PGHS to form the committed intermediate in prostaglandin synthesis, PGH_2 (408,416). A number of isoforms of PGHS have been described that are separate gene products: PGHS-1, PGHS-2, and, more recently, PGHS-3 (408,417). The PGHS-1 and PGHS-2 isoforms have been widely studied in relation to parturition, and the PGHS-2 isoform is particularly important for the synthesis of prostaglandins at parturition in many species. The third isoform, PGHS-3, is highly expressed in brain tissues (418,419). To date, there are no reports concerning its expression in intrauterine tissues. The PGHS-1 gene has the properties of a housekeeping gene (408), whereas PGHS-2 gene expression can be stimulated markedly by various growth factors and cytokines.

The major metabolizing enzyme for prostaglandins is a nicotinamide adenine dinucleotide

(NAD^+)–dependent (type 1) PGDH, the activity of which results in the formation of 15-keto and 13, 14-dihydro-15-keto compounds, which have reduced biological activity (420). This step may be important in preventing biologically active prostaglandin derived from amnion or chorion from reaching the underlying decidua and myometrium during most of pregnancy. Failure of this inactivation process in chorion may be one cause of preterm labor.

Prostaglandins act through specific receptors, and these receptors have been cloned. There are four main subtypes of PGE_2 receptors (EP1 to EP4) and one main subtype of FP receptor (421,422). There are also a number of isoforms of each receptor generated by alternate splicing. In the case of the EP3 receptor, for example, there are at least eight isoforms. EP1 receptors stimulate calcium release, as does the FP receptor. The EP3 receptor inhibits cAMP formation. Therefore, in the myometrium activation of these receptors stimulates contraction. In contrast, EP2 and EP4 receptors activate increased intracellular cAMP formation and therefore promote smooth muscle relaxation. Differential expression of the different forms of EP receptor in the fundus and lower segment of the human myometrium may result in contraction of the fundus at the same time as the lower segment relaxes to allow passage of the fetus during labor. During gestation, changes in expression may contribute to altered patterns of myometrial contractility (423). It has been suggested that precocious expression of the FP receptor in the myometrium may be involved in preterm labor (424).

Various gene knockout mice have been studied in relation to the role of prostaglandins in parturition. Evidence from these studies is inconclusive as to whether prostaglandin production is obligatory for labor in this species. Mice constitutively lacking the gene for PGHS-1 have a delayed parturition, and the young are delivered with reduced viability (425). Although oxytocin knockout animals deliver normally, simultaneous deletion of oxytocin and PGHS-1 overcomes the delay in delivery found in the single PGHS-1 knockout animals, but these animals have a protracted labor (426). PGHS-2 knockout animals have impaired fertility and implantation and have not been studied in relation to labor (427,428). Mice lacking the FP receptor or PLA_2 enzyme do not deliver because they do not undergo luteolysis at term (429). Luteolysis allows a fall in progesterone to occur, followed by increased expression of oxytocin receptors and parturition. However, more recent studies have supported the view that FP receptor expression in the uterus is involved in term and preterm labor (424). These various gene deletion studies indicate that many backup systems are likely acting in parturition, and it has been shown that in the mouse birth occurs when at least three of four factors that activate the uterus are induced (424). It is not surprising that such an important process possesses evolutionary redundancy.

Prostaglandins in Human and Subhuman Primate Labor. Prostaglandins have a number of important roles during human parturition. They are involved in cervical ripening (see previously), stimulate MMP expression, which can promote membrane rupture, and can both stimulate and relax myometrium. A number of extensive reviews have been published concerning various aspects of prostaglandin formation and metabolism during gestation and labor (401,430–434). Generation of prostaglandins is discretely compartmentalized within the tissues and cells that make up the human fetal membranes and decidua (1,176,435), and also occurs in the cervix and myometrium.

Prostaglandin Synthesis. The initial step in prostaglandin synthesis is the release of arachidonic acid from phospholipids by the action of phospholipases. During pregnancy, the roles of $sPLA_{2\text{-IIA}}$ and the arachidonic acid–specific $cPLA_{2\alpha}$ have received considerable attention (431,436). However, studies on the expression of phospholipases during gestation and labor are often contradictory.

The $cPLA_2$ enzyme with its specificity for arachidonic acid–containing phospholipids is the most extensively studied phospholipase in relation to parturition. The $cPLA_2$ enzyme is expressed in the placenta, amnion, chorion, and myometrium, with the highest levels in the fetal membranes (437–439). In human amnion, $cPLA_2$ is the principal enzyme involved in arachidonic acid release. $cPLA_2$ activity in amnion increases with gestation but decreases during labor at term and preterm (440).

Secretory PLA_2 is released from the placenta, human fetal membranes, and myometrium during gestation, but its role is undefined (434). In the placenta, type II $sPLA_2$ predominates (437,441). Type II $sPLA_2$ was found to increase in placenta with labor in one study, but not in another (442,443). It has been proposed that the change in $sPLA_2$ activity that occurs in placenta with labor is due to post-translational modification caused by annexin-1, which inhibits the enzyme (443). In preterm labor tissues (amnion and choriodecidua), $sPLA_2$ content is higher than in preterm tissues that were not in labor (444). $sPLA_2$ levels are also elevated in maternal plasma and amniotic fluid in patients delivering preterm (445,446). It has been proposed that this increase is due to the action of bacterial endotoxin in infection-induced preterm labor (434,447,448). In one study, $sPLA_2$ and $cPLA_2$ in the myometrium were unchanged with term or preterm labor (439), but others found that $sPLA_2$ mRNA, although not $cPLA_2$ mRNA, expression increased with gestational age and term labor in both the upper and lower segment of the myometrium (449).

Once arachidonic acid is released from phospholipids, it is acted on by PGHS. The inducible form of PGHS, PGHS-2, appears to play an important role in human parturition, as it does in other species. In amnion, PGHS activity predominates and PGE_2 is the major prostaglandin formed. The levels of PGHS-2 but not PGHS-1 mRNA and protein increase with advancing gestation and during preterm and term labor (450–452). In contrast, membrane-bound PGE isomerase (mPGES) levels do not change with term or preterm labor (453,454). The 3′ untranslated region of PGHS-2 mRNA contains sequences that confer instability in the message. Therefore, the observed increase in mRNA levels could be due to increased transcription or decreased degradation, or a combination of both. It was demonstrated that the level of PGHS-2 mRNA in term amnion is regulated at the transcriptional level (455). Labor did not increase PGHS-2 gene expression or mRNA stability, suggesting that upregulation of transcription occurred before labor, generating stable mRNA that continues to accumulate during labor.

In chorion, PGHS is expressed, but in contrast to the amnion there are very high levels of PGDH (456). The functionally prevalent form of PGHS that increases with advancing gestation and with labor at term and preterm is PGHS-2 (457), although increased levels of PGHS-1 mRNA were also found in the chorion at these times. The high PGDH activity in chorion trophoblasts forms a metabolic barrier preventing the passage of prostaglandins generated in the amnion or chorion from reaching the underlying decidua or myometrium.

Decidua expresses both synthesizing and some metabolizing enzymes, but little overall change occurs in the capacity for prostaglandin synthesis with labor, and the levels of PGHS-2 and PGE isomerase are much lower in decidua compared with levels in amnion and chorion (451,453). Decidual prostaglandin production may in fact result from constitutive expression of PGHS-1 rather than induction of PGHS-2.

At term, if the prostaglandins produced in the amnion and chorion are to access the myometrium, the synthetic activity of amnion/chorion would need to exceed the metabolic potential of chorionic trophoblasts. Certain patients who delivered preterm were found to have a relative deficiency of PGDH in chorion (456). It has been proposed that this deficiency might allow prostaglandins generated in amnion or chorion to pass unmetabolized to the underlying decidua and myometrium and provoke uterine contractility (456). There is also a decrease in PGDH expression during labor in the region of the chorion overlying the cervix, and this may enable prostaglandins from the amnion and chorion to access the cervix (458).

Studies on prostaglandin synthesis in myometrium during gestation are confusing, in part because of the difficulty in obtaining appropriate tissue for examination. It is likely that enzyme expression varies according to the region of the myometrium sampled, or with the proximity of the patient to labor. Increases, decreases, and no changes in PGHS-2 mRNA levels in myometrium at labor have been reported (459–462). Using in situ hybridization, PGHS-2 expression was found to decrease in labor both at term and preterm (459). In contrast, other investigators found that PGHS-2 mRNA increased with labor (460). In myometrial samples obtained from the lower segment, some workers reported that there was no significant change in the levels of PGHS-1 or PGHS-2 during labor at term or preterm (463,464), whereas others found that PGHS-2 mRNA expression increased with advancing gestation but did not change further with labor (461,462). In contrast, PGDH activity in the myometrium decreases during labor (463). It should be noted that PGHS-2 is required for production of PGI_2 (an inhibitory prostaglandin), as well as PGE_2 and $PGF_{2\alpha}$, and PGI synthase expression was found to decrease in the myometrium with advancing gestation, although there was no change in the expression of PGES (90). Taken together, these studies would be consistent with a decrease in production of the inhibitory PGI_2 and a decrease in metabolism of prostaglandins, resulting in increases in local levels of stimulatory prostaglandins, in the myometrium during labor. Furthermore, if regional differences in prostaglandin synthase activity occur at the time of labor, measurements made on tissue obtained from the lower segment may not be representative of changes observed in the fundal region. Also, in vitro studies with isolated human myometrial cells indicate that stretch can upregulate expression of PGI synthase, resulting in increased PGI_2 levels, but had no effect on PGE_2 or $PGF_{2\alpha}$ levels (465). This may therefore be an important way in which the quiescence of the myometrium is maintained during expansion of the uterus with advancing gestation. In placenta, PGHS synthase and PGDH are also expressed, but there are no changes in activity or expression with labor.

The baboon and rhesus monkey are important models for human parturition (466). Because myometrial activity can be monitored continuously in these animals, investigators can define more precisely the stages of labor than is possible in human pregnancy. In the baboon at late gestation and during labor, there is an increase in the PGHS enzyme, specifically the PGHS-2 isoform. The expression of other enzymes involved in prostaglandin synthesis does not change. Thus, the increase in the PGHS-2 isoform would appear to occur in all species studied to date. In the baboon, the earliest change occurs with increased expression of PGHS-2 mRNA in the lower uterine segment and the cervix (467). This increase occurs well before labor. It has been proposed that the increase in forebag prostaglandin concentrations as a result of this activation is responsible for preparing the cervix and lower birth canal at term (466). In the baboon, estrogen concentrations increase approximately 10 days before labor and may be responsible for the induction of the PGHS-2 before labor (468). Further studies have suggested that local paracrine factors may also be involved through stretch-induced mechanisms (466).

Prostaglandin Metabolism. PGDH localizes to chorion trophoblasts in human fetal membranes. Levels of PGDH mRNA, protein, and activity are lower at term spontaneous labor than at elective cesarean section, which suggests a facilitatory role in the labor process. PGDH expression is reduced significantly in a proportion of women in idiopathic preterm labor without infection, and is reduced substantially further in many patients in preterm labor with an underlying infective process (456,469). Loss of PGDH expression is specific for chorion, and levels of PGDH activity in placental tissue from the same groups of patients were similar. Loss of PGDH activity in infection is associated with reduced numbers of chorionic trophoblast cells. In idiopathic preterm labor, chorion trophoblasts are present, but PGDH expression and activity may be reduced. Regional differences in expression of PGDH are also seen in the human uterus (458). At labor, the activity of PGDH in chorion trophoblasts overlying the internal os is reduced compared with that in chorion adjacent to the placenta or midway between these sites. We have speculated that this difference in PGDH expression might allow prostaglandins generated in amnion and chorion in the lower segment to escape metabolism in chorion at the time of labor and to effect effacement and dilatation of the cervix. It is possible that specific downregulation of PGDH in lower segment membranes could occur in response to bacteria and elevated levels of cytokines in the vagina or cervix.

Tissue-specific PGDH distribution, regional variation in expression, and changes in expression with labor have been demonstrated in the baboon (466,470). PGDH was undetectable in the amnion, as in the human, but was expressed in the chorion, decidua, lower uterine segment, and fundal region of the myometrium and in the cervix. PGDH mRNA expression decreases in the decidua and cervix with advancing gestation and decreases in the fundus and decidua during labor. These changes that occur in PGDH expression with advancing gestation and labor may allow prostaglandins from the membranes and decidua to access the myometrium and cervix, as discussed previously for the human. The baboon therefore provides an important model that may enable the significance of the in vitro mechanisms that regulate the human enzyme (see later) to be studied in vivo.

Regulation of Prostaglandin Production In Vitro. Before the enzymes involved in prostaglandin formation and metabolism were cloned, extensive studies were carried out on the regulation of prostaglandin formation in tissue explant cultures and primary cultures of cells derived from human amnion, chorion, and decidua. A large number of substances, including epidermal growth factor, β-adrenergic agonists, Ca^{2+} ionophores, platelet-activating factor, activators of protein kinase A, catecholamines, steroids, and cytokines, have been shown to stimulate prostaglandin output by these cell types (435). The regulation of specific enzymes involved in prostaglandin synthesis and metabolism has been the focus of more recent studies.

Regulation of the $cPLA_{2\alpha}$ enzyme is complex and can occur at transcriptional, translational, and post-translational levels. Rapid changes in activity can be achieved through increases in intracellular levels of calcium that result in translocation of $cPLA_2$ to the nuclear membrane, where it interacts with its phospholipid substrate (471). $cPLA_2$ can be phosphorylated on serine 505 by mitogen-activated protein (MAP) kinase, which results in activation of the enzyme (472–474), and p11, a member of the S110 family of calcium-binding proteins, can inhibit the enzyme (475). More prolonged activation of $cPLA_2$ activity can occur through transcription. The promoter region of the $cPLA_{2\alpha}$ gene contains response elements for a number of transcription factors, including activating protein-1 (AP-1), AP-2, nuclear factor-κB (NF-κB), glucocorticoid response element (GRE), and peroxisome proliferator–activated receptor (PPAR) (476). $sPLA_2$ does not require the membrane-associated Ca^{2+} activation shown by the $cPLA_2$ enzyme, and its transcription is regulated by a number of factors (436). The promoter region of $sPLA_{2\text{-IIA}}$ contains consensus sequences for AP-1, NF-κB, Stat, CRE, GRE, and PPAR.

Using isolated cells and tissue explants, the regulation of $cPLA_2$ and $sPLA_2$ has been studied in intrauterine tissues. Bacterial endotoxins increase the expression and release of type II $sPLA_2$ from placental and choriodecidual explants (434,447,448). Also, in placenta when $sPLA_{2\text{-IIA}}$ activity or release was inhibited, prostaglandin production was reduced by up to 80% (477). It has been suggested that $sPLA_2$

may act through specific receptors to elicit important effects in parturition and that it is also involved in apoptosis and cell migration (434). In amnion-derived WISH cells, IL-1β induces increased $cPLA_2$ expression, and the effect is inhibited by glucocorticoids (478).

Because the promoter regions of $sPLA_{2\text{-IIA}}$ and $cPLA_2$ contain NF-κB binding sites, studies have examined the role of NF-κB in the regulation of both of these enzymes (434,479). In vitro tissue incubation studies with placenta, chorion, and amnion demonstrated that the NF-κB inhibitor sulfasalazine inhibited basal and lipopolysaccharide-stimulated $sPLA_2$ release and $cPLA_2$ protein content. In addition 15d-PGJ_2, which stimulates PPAR-γ, was shown to inhibit the release of $sPLA_{2\text{-IIA}}$ from placental explants, indicating the possibility the PPAR-γ regulates its expression.

The promoter of the human PGHS-2 gene (408) contains many regulatory elements, including a C/EBP (CCAAT/enhancer-binding protein) motif, AP-2 and SP1 (specificity protein-1) sites, NF-κB sites, a CRE motif, and a GRE (480). It is likely that growth factors such as platelet-derived growth factor stimulate PGHS-2 synthesis through functional CREs. However, it should be noted that β-sympathomimetic drugs also stimulate increases in cAMP and increase the output of prostaglandins from uterine tissue in vitro and in vivo (481). It is therefore possible that the reduced efficacy of these drugs to suppress uterine activity may in part be due to their stimulation of prostaglandins.

In the human PGHS-2 mRNA localizes to chorion laeve trophoblast cells, amnion epithelial cells, as well as subepithelial mesenchymal cells, including fibroblasts (451). Primary cultures of these various cells have been used to examine the regulation of PGHS in vitro. Interestingly, glucocorticoids were found to increase expression of PGHS-2 mRNA and protein in amnion cells in culture and increase their output of PGE_2 (405,482–484). The amnion fibroblasts are particularly responsive to the stimulatory effects of glucocorticoids (405), and glucocorticoids also induce expression of $cPLA_2$ but not PGES in these cells (485). In addition, 11βHSD-1 activity has been described in these cells, and its expression can be induced with glucocorticoids and cytokines, raising the possibility of a feed-forward mechanism that stimulates prostaglandin production in these cells (486), as previously proposed in chorion trophoblast cells (487). There is conflicting evidence in the literature as to whether glucocorticoids inhibit or stimulate expression of PGHS-2 in epithelial cells (404,407). It remains a possibility that glucocorticoids have a biphasic effect on PGHS-2 expression in these cells. In addition, both inhibition of prostaglandin production and stimulation of PGHS-2 expression have been reported to occur in chorion laeve trophoblasts (487). The GR is expressed in amnion epithelial cells, amnion/chorion fibroblast cells, and chorion trophoblast cells (488). In these cells, the receptor is localized in the nucleus, whereas in the decidua the receptor is found mainly in the cytoplasm. It is not known whether the stimulatory effect of glucocorticoids is mediated through interaction with a GRE in the PGHS-2 promoter. The effects of glucocorticoids may be exerted through the upregulation of a cell-specific product or by interaction with the NF-κB system.

NF-κB has been localized in the fetal membranes (489) and shown to be to be involved in the IL-1β stimulation of PGE_2 output by amnion cell cultures (490). Transfection studies with human amnion cells have suggested that NF-κB may be involved in functional progesterone withdrawal in this tissue during labor (491).

The amnion epithelial-derived cell line, WISH, has been used extensively as a model system to study the regulation of prostaglandin formation. These cells increase prostaglandin output in response to various cytokines, oxytocin, and various other substances in ways similar to the response of primary cultures of amnion epithelial cells (478,492–494). In addition, NF-κB is clearly required for activation of PGHS-2 expression in WISH cells (495). In WISH cells, glucocorticoids inhibit IL-1β–stimulated prostaglandin production, and PGHS-2 expression occurs through an NF-kB–dependent mechanism (496); estradiol was found to stimulate arachidonic acid and PGE_2 release (494,497), although a similar effect has not been reported in primary cultures of amnion cells. Some studies, however, have raised important questions concerning the validity of WISH cells as an appropriate model for amnion epithelial cells. Kniss et al. (498) reported that polymerase chain reaction analysis of the WISH cell line used in their laboratory indicated that it was identical to HeLa cells. At this time it would appear that extrapolation of the findings with these cell lines to the physiology of human amnion epithelial cells should be made only with considerable caution.

Prostaglandin Metabolism. Regulation of PGDH is multifactorial. In contrast to the stimulatory effects of glucocorticoids on PGHS-2 expression already discussed, they inhibit PGDH activity and decrease levels of PGDH mRNA (1,370,499,500). Synthetic progestins increase PGDH expression and inhibitors of progesterone action such as RU486 or inhibitors of progesterone synthesis such as trilostane all decrease PGDH activity (501). This suggested that locally produced progesterone contributes to maintaining basal levels of PGDH expression. Progesterone was able to reverse the inhibitory effect of trilostane, and could partially reverse the inhibitory effect of cortisol. These studies represent one of the few instances in which an autocrine/paracrine mechanism has been demonstrated to occur in the human fetal membranes. It has therefore been suggested that through normal pregnancy regulation of PGDH expression may reflect a balance between the activities of progesterone and glucocorticoids. Evidence has been presented to indicate that this may occur by competition for binding to the GR, although it is also possible that the progesterone receptor is involved in the regulation of PGDH (370,500,502). Control of PGDH has been studied by transfection of the promoter into myometrium-derived cells and was found to involve progesterone (502). Whether this is the case in chorion leave will require further study. However PGDH is being regulated, in vitro studies with chorion trophoblast demonstrate that glucocorticoids may be acting as an antiprogesterone, in the same fashion as they regulate placental CRH production (see later). The 11β HSD-1 enzyme is expressed in the chorion trophoblast and may regulate cortisol levels in this tissue. Cortisone was found to decrease PGDH activity in chorion because of its conversion to cortisol by 11βHSD-1 (503). The 11βHSD-1 activity in chorion also increases with advancing gestation, and in vitro studies have shown that it is stimulated by prostaglandins and glucocorticoids (359,487). Taken together, these findings provide a feed-forward mechanism for increased prostaglandin production in the fetal membranes at term (1,499). Cortisol can stimulate PGHS-2 expression and decrease PGDH expression, resulting in increased prostag-landin production that could then stimulate the 11βHSD-1 activity, resulting in increased cortisol, further stimulation of 11βHSD-1, and further increases in prostaglandin production. Locally produced CRH in the chorion laeve trophoblast has also been proposed to provide a tonic stimulatory signal to maintain high PGDH activity in this tissue, through a calcium-dependent mechanism (504).

Cytokines such as IL-1 also downregulate PGDH expression (505–507). Thus, proinflammatory cytokines increase PGHS-2 and decrease PGDH expression (493,508), which may then result in a marked increase in prostaglandin production in preterm labor associated with infection (see later).

Prostaglandin Receptors. A number excellent reviews concerning the role of prostaglandin receptors in parturition have been published (422,423). Prostaglandins act through specific receptors that signal various G proteins to stimulate formation of second messengers such as cAMP, IP3, or intracellular calcium release (423). There is little information available concerning the promoter regions of the various prostaglandin receptor genes, although the chromosomal localizations have been mapped in the human (509). The uterine transcription of the EP2 gene has been studied in the mouse, and there appears to be a uterine-specific promoter (510). Various response element sequences such as AP-2, AP-4, and NF-κB have been identified, and a possible progesterone response element has been localized upstream of the transcription start site in the 5′ flanking region.

Mice lacking the FP receptor were unable to deliver fetuses at term, and myometrial PGHS-2 was not induced at parturition (511,512). There was no progesterone withdrawal and oxytocin did not induce delivery. After ovariectomy, the mice do become sensitive to oxytocin and deliver. These results indicated that the FP receptor is necessary for luteolysis in the mouse. However, more recent studies have indicated that expression of the uterine FP receptor may have a key role in parturition, and it has been suggested that precocious expression of the FP receptor in the myometrium may be involved in preterm labor (424). None of the other prostaglandin receptor null mice had altered parturition.

Differential regulation of prostaglandin receptors may contribute to altered patterns of myometrial contractility through the course of gestation (423,513,514). The expression of EP2 receptors has been studied in the pregnant rat. In this species, EP2 mRNA expression was highest at day 16 and decreased toward delivery. Expression of FP receptor mRNA was low at day 16 and increased up to delivery. These findings would be consistent with $PGE_{2\alpha}$ acting to relax the myometrium during gestation; the decrease in EP at term would accentuate the contractile effects of PGF_2 acting through the FP receptors, which are upregulated (513). The increased expression of the FP receptor can be blocked by progesterone, but not by stretch (515). As would be expected, ovariectomy increases FP receptor expression. Interestingly, there appear to be changes in the subcellular localization of FP and EP2 receptors with advancing gestation (44), although the role of these events in parturition has yet to be defined. Nuclear prostaglandin receptors have been described together with expression of $cPLA_2$ and PGHS-2 in the nuclear envelop, potentially emphasizing intracrine actions of prostaglandins (516). An interesting discussion on nuclear prostaglandin receptors in pregnancy and parturition has been published (433).

The expression of prostaglandin receptors has also been examined in human myometrium. The expression of EP3 and FP receptors was found to be decreased in myometrium from pregnant women compared with tissues from nonpregnant patients (517). EP2 mRNA expression was studied in lower segment myometrium obtained during gestation and after labor (514). EP2 mRNA was decreased in term tissues compared with preterm tissues delivered before labor. In these studies, FP receptor mRNA expression also decreased with advancing gestation. There has also been evidence presented for regional differences in prostaglandin receptor expression in myometrium (182,518,519). Both PGE_2 and $PGF_{2\alpha}$ stimulated upper segment myometrium, whereas in lower segment tissues PGE_2 inhibited contractility and $PGF_{2\alpha}$ had no effect (518,519). More recently, Lye et al. (182) found that with the onset of labor there was decreased FP expression and increased EP4 expression in the lower segment. This would be consistent with the proposition that in labor the lower segment may relax to allow passage of the fetal head at the same time that the upper segment contracts to expel the fetus. Expression of connexin43 was found to be higher in the upper segment compared with the lower segment, whereas expression of PGHS isoforms were higher in the lower segment (461). This distribution may facilitate cervical ripening and contraction of the upper segment. However, other studies have not supported this thesis, and no difference was found in the contractile properties of myometrium obtained from the upper and lower segments (520).

Similar findings have been reported for the baboon uterus. PGE_2 contracted the fundus but not the lower segment, whereas $PGF_{2\alpha}$ contracted both areas. Both segments expressed FP and EP4 equally, but EP2 was more highly expressed in the lower segment, whereas the stimulatory EP3 was more highly expressed in the fundus (521). These studies also demonstrated the expression of EP1 through EP4, IP (prostaglandin receptor), FP (prostaglandin $F_{2\alpha}$ receptor), and TP (thromboxane receptor) mRNAs in baboon myometrium, but their expression did not change with advancing gestation (522). Levels of EP1, EP3, and IP mRNAs were highest in the fundus, whereas EP2 mRNA levels were lower. In these animals there was a decrease in expression of EP2 and IP during labor, but no change in the expression of EP1 and EP3. In the sheep, some investigators have reported an increase in expression of the contractile EP3 and FP receptors during labor in the myometrium, whereas other investigators found no change in the level of expression (303,523). Levels of the receptor proteins were not investigated.

Prostaglandin receptors are also expressed in the fetal membranes and decidua, although information is lacking on regulation and changes with pregnancy. Prostaglandins stimulate MMP and 11βHSD expression in the membranes, and this presumably occurs through specific prostaglandin receptors. Studies are ongoing to define further their expression during gestation and labor (454,524). In the chorion, the stimulatory effect of $PGF_{2\alpha}$ on 11βHSD-1 occurs through the FP receptor (358). In the baboon, expression of prostaglandin receptors was defined in fetal membranes and decidua throughout gestation and during labor (525). Chorion and decidua express EP1, EP2, EP3, EP4, and IP mRNAs. In the case of FP, this was localized exclusively to the decidua. Expression of EP1, FP, and EP4 receptors decreased in chorion with advancing gestation. Changes were also seen in the expression of certain receptors during labor. In chorion and decidua, IP expression increased during labor, whereas the expression of EP2 in decidua decreased. With development of specific antibodies to these receptors and more detailed studies on their subcellular distribution, the role of these receptors in the onset and progression of human labor should be an important topic for future studies.

Cytokines

Cytokines play an important role in many cases of preterm labor as well as in normal labor at term (526). They are involved in a number of different aspects of parturition. including cervical ripening, as previously discussed, the regulation of prostaglandin production, and the stimulation of MMP expression in membrane rupture (see later). Cytokines not only are produced by inflammatory cells that invade the intrauterine tissues but are synthesized in the intrauterine tissues themselves. Complementary DNA array studies have identified clusters of inflammation-associated genes upregulated with labor both at term and preterm, and in a suppression subtractive hybridization study with human myometrium at term, a number of genes with immunoregulatory and inflammatory roles were upregulated (527).

Approximately 30% to 40% of preterm labor is believed to be associated with an underlying infective process (435,528). Concentrations of lipoxygenase and PGHS products are elevated in the amniotic fluid of patients with infection, as are levels of cytokines, including IL-1β, IL-6, and TNF-α (528). In animal models, cytokine administration has been shown to induce premature birth, and cytokines and eicosanoids interact in a feed-forward fashion to stimulate each others' production and induce parturition (435). A number of years ago it was shown that the IL-1 receptor antagonist (IL-1RA) prevented IL-stimulated labor in mice, indicating that such an approach may provide a method for preventing preterm delivery in instances when infection is involved (529). However, in many cases of infection one might expect that the fetus needs to be delivered from an inhospitable intrauterine environment to survive.

Extensive studies by Romero and collaborators (528) have suggested that an ascending bacterial infection allows organisms to pass between the membranes and reach the amniotic cavity. Bacteria may release phospholipases that hydrolyze membrane phospholipids to produce arachidonic acid and hence prostaglandins. These organisms also release endotoxins, including lipopolysaccharide, that can act on macrophages to provoke prostaglandin or cytokine release. These cytokines can then act on the fetal membranes and decidua to produce prostaglandins. There is an extensive body of evidence concerning the production and action of cytokines in the fetal membranes and decidua, particularly from Murray Mitchell's laboratory (508,526,530). Cytokines have been shown to influence prostaglandin production in the myometrium itself and may also cause the release of oxytocin and CRH from myometrium, decidua, membranes, or placenta, and these compounds may then affect the myometrium directly or indirectly to stimulate parturition.

Inflammatory cytokine concentrations increase in amniotic fluid at term labor (531–535). TNF-α, IL-1α, IL-6, and IL-8 are present in cervicovaginal fluids before labor at term, and most often in women with bacterial vaginosis or intra-amniotic infection who deliver prematurely (536–538). There is also an increase in the concentrations of inflammatory cytokines in cervicovaginal secretions during term labor (539,540). There is strong evidence to suggest that increased cytokine production in the intrauterine tissues themselves is involved in both preterm and term parturition, and this has been reviewed (508,526). Increased expression of cytokines occurs in the fetal membranes and decidua both in term and preterm labor and is particularly evident in the presence of intrauterine infection.

At term labor there is increased IL-1α production in placenta, amnion, chorion, and decidua, and IL-1β expression is also increased in amnion, chorion, decidua, and possibly the placenta (541–544). In amnion, chorion, and decidua, IL-6 mRNA and protein expression increases with labor (530,534,542). Whether there is an increase in IL-6 expression in the placenta with labor is equivocal (534,545,546). TNF-α mRNA expression is also increased in fetal membranes and decidua during labor (542). Studies have implicated NF-κB in the regulation of IL-6, IL-8, and TNF-α in the fetal membranes (547).

Although antiinflammatory cytokines are produced by intrauterine tissues, there is no evidence for marked changes in their production occurring at labor [see Bowen et al. (526)]. IL-10 secretion from the chorion-decidua was found to decrease at labor in one study (548), but in another study decidual production of IL-10 was not found to change with labor (549). There is also little or no change in the concentrations of IL-10 and IL-4 in the amniotic fluid during labor (550,551). IL-1RA concentrations in amniotic fluid are high at mid- to late gestation, but neither its concentration in amniotic fluid nor its expression in the decidua changes with labor (529,552).

The concentrations of the chemokine IL-8 in amniotic fluid increase with advancing gestation and labor at term, and there are increased levels of IL-8 in the lower uterine segment at labor (528,553). Studies are contradictory as to whether there is increased release of IL-8 by intrauterine tissues at labor (530,554).

IL-1 and TNF-α stimulate prostaglandin output from various intrauterine tissues and cell cultures, including the amnion, chorion, decidua, and myometrium; in most cases this appears to occur through increased expression of PGHS-2 [for reviews, see Bowen et al. (526) and Hansen et al. (555)]. However, studies have shown that TNF-α and IL-1β also regulate expression of PGES. These cytokines increased expression of the microsomal PGES (mPGES) but not the cytosolic PGES enzyme in placental cultures, whereas in chorion trophoblast cultures only TNF-α increased mPGES expression (556). In myometrium, IL-1β was shown to induce phosphodiesterase expression through a PGE_2 and cAMP pathway (557), CRH partially to inhibit IL-1β, and oxytocin to stimulate prostaglandin production (558). IL-6 stimulates prostaglandin production by the amnion and decidua, and MIP-1α acts on the chorion and amnion to stimulate prostaglandin output (559,560). In contrast to their effects on the prostaglandin synthetic enzyme PGHS-2, IL-1β and TNF-α reduce the activity and expression of PGDH in cultured chorion leave cells (see previously). Because PGDH activity has been shown to decrease in the area of the membranes overlying the cervix during labor, it is possible that cytokines in this area of the uterus possibly arising from the forebag are responsible for this downregulation of PGDH.

Some cytokines have also been shown to inhibit prostaglandin production in intrauterine tissues. TGF-β_1 decreases cytokine-stimulated prostaglandin production by the amnion and decidua, as do IL-1RA and TGF-β_1 in the myometrium, and IL-10 has been shown to attenuate the effects of IL-1 in promoting PGHS-2 mRNA accumulation (561). In cultured human myometrial cells, IL-1β induction of calcium transients and PGHS-2 expression can be attenuated by interferon-γ (562). Interestingly, the anti-inflammatory cytokines IL-10 and IL-4 have been shown to stimulate prostaglandin production and cytokine release by the fetal membranes, and IL-1RA increases prostaglandin production by decidua (563–566). It has been proposed that this inflammatory action of usually antiinflammatory cytokines might be an evolutionary adaptation to accelerate parturition and delivery and allow the fetus rapidly to exit a hostile environment (548).

As previously discussed, locally produced cytokines may regulate MMP expression in the cervix and be involved in cervical ripening. These cytokines are likely derived from inflammatory cells that have invaded the tissue (121–123). In addition, the stimulatory effect of cytokines on MMP expression is likely involved in membrane rupture, which is discussed later.

IL-1β also stimulates CRH production by the placenta and endometrium, and this effect appears to be at least in part mediated by prostaglandins (567,568). Therefore, it is possible that cytokines might stimulate prostaglandin production at parturition not only by increasing PGHS-2 expression and decreasing PGDH expression, but through stimulation of CRH production, which may further increase prostaglandin production.

Oxytocin

Oxytocin is a nonapeptide synthesized by the hypothalamic magnocellular neurons located in the supraoptic and paraventricular nuclei (569–571). It is secreted into the circulation by the posterior pituitary. Oxytocin has long been known to be a potent uterotonin (572–574) that stimulates myometrial contraction in late gestation; it is also involved in milk release from the mammary gland during lactation (569,574,575). The synthetic form of oxytocin (Syntocinon) is used clinically to induce contraction of the myometrium. For more detailed discussion of the role of oxytocin in parturition, the reader is referred to a number of reviews (576–579). As is the case with many substances, oxytocin does not appear to be essential for parturition because oxytocin null mice deliver normally, although the pups die postpartum because of the mother's inability to eject milk (580,581). It is possible that in these knockout animals AVP has compensated for the absence of oxytocin (582,583).

In the human, oxytocin is present in maternal and fetal plasma during late gestation, and secretion in these two compartments occurs independently. Oxytocin concentrations do not rise before labor in the human, which raised questions concerning the role of oxytocin in the labor process (574,584). However, the plasma concentrations of oxytocin increase during the course of labor, and a high-amplitude spurt of oxytocin is released into maternal blood at the time of labor (572). In contrast to the modest changes in plasma levels of oxytocin, there is a marked increase in oxytocin receptor number in the myometrium and decidua from patients at term and preterm labor (572). This increase in uterine oxytocin receptor concentration at term occurs in all species studied, including sheep, rabbits, and rodents (579,585–588). In vitro studies have shown that the oxytocin receptor can be induced by estrogens and uterine stretch (576). In vivo, the estrogen that is responsible for increased receptor expression could be derived from the placenta or may be produced locally in choriodecidual tissue (577,589). It has been shown that estrogen sulfatase activity in the fetal membranes increases during labor, and estrogen formed from the hydrolysis of estrogen sulfatase could act to stimulate oxytocin or oxytocin receptor production in the intrauterine tissues (345,388,590).

It is also possible that oxytocin is produced locally in human fetal membranes and decidua (590) and rat uterus (591,592). Mitchell and Chibbar have shown that choriodecidual tissue produces oxytocin and that the rate of oxytocin mRNA synthesis increases at parturition (577). At this time there is also an increase in the concentration of oxytocin receptors. In vitro, estrogens stimulate expression of oxytocin and the oxytocin receptor, and this can be inhibited by administration of the ER antagonist tamoxifen (388,593). The oxytocin gene has an estrogen response element in its promoter, consistent with this effect (594). Oxytocin metabolism does not change in chorion or decidua in relation to labor, which suggests that increases in concentration reflect de novo synthesis. In the sheep, increased oxytocin receptor expression occurs at term as it does in other animals and the human, but there is no evidence for the local production of oxytocin in intrauterine tissues in this species (588).

Oxytocin acts on the myometrium to cause contractions by increasing intracellular Ca^{2+} concentrations. This occurs by increasing Ca^{2+} entry into myocytes as well as release of Ca^{2+} from intracellular stores. These effects are secondary to an increase in production of IP3. The oxytocin receptor couples to phospholipase C through a GTP-binding protein to promote calcium mobilization (82). This pathway can be inhibited by activation of protein kinase A, indicating that cAMP-dependent phosphorylation at a step involving coupling of the GTP-binding protein to phospholipase C can antagonize the stimulatory effect on IP3 formation by oxytocin. Thus, there may be critical interactions between these two pathways. The oxytocin receptor has also been shown to be sensitive to homologous desensitization during human labor (595). It was found that the concentration of oxytocin receptors and oxytocin receptor mRNA decreased in human myometrium during oxytocin-induced labor or prolonged labor in which oxytocin was used.

In the human fetal membranes and decidua, oxytocin also stimulates prostaglandin production and may therefore have a dual action, increasing uterotonins in the uterus and acting directly on the myometrium to stimulate contraction. In addition, oxytocin has been shown to increase PGHS-2 expression in the myometrium itself through G-protein–coupled MAP kinase (596). In general, oxytocin appears to be important during phase 2 of parturition, particularly during the second phase of labor, and during phase 3 of parturition, the involution of the uterus, rather than in phase 1 of parturition.

A number of years ago it was suggested that the oxytocin receptor could be regulated by progesterone in a nongenomic fashion (597). Progesterone was found to bind to the oxytocin receptor and prevent oxytocin binding in the rat uterus. When the human receptor was expressed in CHO cells it did not bind progesterone but did bind the progesterone metabolite 5β-dihydroprogesterone. It was proposed that this binding of progesterone or its metabolite to the oxytocin receptor might be one of the mechanisms by which progesterone maintained myometrial quiescence throughout pregnancy. However, since this initial finding there have been no other studies to examine this possibility in greater detail, and a recent study could not find any effect of 5β-dihydroprogesterone on oxytocin binding to its receptor (598).

Corticotropin-Releasing Hormone

During human pregnancy the placenta produces CRH, and it has been proposed that this has an important role in term and preterm labor (599–602). Increasing levels of CRH mRNA are expressed in the placenta with advancing gestation, resulting in increased concentrations of CRH peptide in the placenta and an exponential increase in CRH concentrations in maternal peripheral plasma (572). In plasma, CRH binds to CRH-binding protein, which is also produced by the placenta, and this reduces its activity. The CRH-binding protein has been shown to inhibit CRH stimulation of ACTH release from pituitary cells and to inhibit CRH effects on prostaglandin production in uterine tissues (572). At term, CRH-binding protein concentrations fall. Combined with the increased production of CRH, this results in a marked increase in the concentration of free CRH (603).

CRH has been localized to placental syncytiotrophoblasts, intermediate trophoblasts, and to the mononuclear trophoblast cells in the chorion leave in the fetal membranes (604). In vitro studies have shown that CRH production is decreased by NO and progesterone and stimulated by prostaglandins, cytokines, vasopressin, and glucocorticoids (368,572). Glucocorticoids therefore appear to act as an antiprogesterone in the regulation of CRH production, overcoming the inhibitory effect of progesterone and resulting in functional withdrawal of progesterone. It has been proposed that these effects are achieved through the GR (368). In vitro studies have demonstrated that cAMP is involved in the stimulation of CRH expression by glucocorticoids, and the CRE regulatory element in the CRH promoter is necessary for this stimulation to occur (605,606). Other studies have shown that overexpression of PR-A or GR decreases CRH promoter activity, whereas the promoter activity is increased when PR-B is overexpressed (607). The CRE regulatory element in the promoter is also involved in progesterone regulation of CRH expression.

Glucocorticoids also simulate placental CRH output in vivo. Plasma CRH concentrations are elevated within 24 hours of the administration of synthetic glucocorticoids given to the mother to promote fetal lung maturity (608,609). As discussed previously, CRH and corticosteroids increase PGHS-2 expression and glucocorticoids decrease PGDH activity. CRH also stimulates the fetal adrenal to produce androgens (336) that can be metabolized by the placenta to form estrogens, which may then act to stimulate the synthesis of CAP proteins. It is therefore not surprising that transient elevations in uterine activity have been reported after the administration of synthetic corticosteroids (610,611).

Maternal perceived stress, anxiety, and depression have been linked to preterm labor (612–614). In stress, women have increased concentrations of plasma CRH (612). High concentrations of CRH in maternal plasma predict preterm labor, and presumably occur in relation to upregulation of the maternal HPA axis (612,615). The fetus also responds to stress such as in utero hypoxemia with HPA activation and increased adrenal cortisol output. This fetal cortisol may increase CRH expression by the placenta (616) and result in increased production of placental ACTH. ACTH and CRH may then act as vasodilators through an NO- and cyclic GTP–dependent pathway. If this series of events does not improve the fetal hypoxemia, continued increases in fetal glucocorticoid could stimulate further CRH output, prostaglandin production, estrogen production (through activation of the fetal adrenal), and premature delivery. It is likely that this scenario provides an explanation for some instances of preterm labor.

It has been suggested that CRH acts as a placental clock regulating the timing of parturition (599,600,603). High levels of plasma CRH in maternal plasma predict patients who will deliver preterm, and lower levels predict patients who will deliver post-term. CRH concentrations in maternal plasma are elevated in women at risk of preterm delivery in the absence of underlying infection (617). From 28 to 36 weeks' gestation, but not 24 to 28 weeks' gestation, maternal plasma CRH concentrations were higher in women with a diagnosis of preterm labor who went on to delivery within 24 to 48 hours than in patients with the same diagnosis of preterm labor but who did not deliver until term (617).

CRH, oxytocin, and prostaglandins have been suggested to act synergistically to stimulate myometrial contractility (572). Activation of PGHS-2 can occur through a CRE in the promoter, and CRH may increase PGHS-2 expression through PKA. Both CRH and urocortin (a CRH-related peptide) are synthesized in the myometrium, and therefore autocrine and paracrine actions of CRH and related peptides may be important in the regulation of myometrial contractility or relaxation. CRH receptors are expressed in the myometrium and CRH may act both to relax and contract the myometrium. In the lower segment, the CRH type 1 receptor is expressed and the levels of CRH type 1 receptor mRNA increase in patients in both term and preterm labor (618). In vitro studies with cultured human myometrial cells found that CRH, but not urocortin II, increases NOS and cGMP production, which may act to relax myometrium (619). The CRH type 2 receptor is also expressed in the myometrium, and treatment of myometrial cells with urocortin II stimulated phosphorylation of the myosin light chain (620). It is therefore possible that CRH might enable relaxation of the lower segment of the myometrium and stimulate contraction in the fundus (602). At present, however, no definitive information concerning these mechanisms is available. However, it is clear that the stimulatory and inhibitory actions of CRH on the myometrium may be similar to those discussed previously for prostaglandins and be regulated by differential spatial expression of receptor subtypes.

There appear to be similarities between the role of CRH in human parturition and the role of PGE_2 in ovine parturition. In the sheep, PGE_2 production is stimulated by cortisol, as is CRH production in the human. In the sheep, PGE_2 produced by the placenta is involved in maturation of the fetal adrenal, resulting in increased production of adrenal androgens and cortisol; in the human, CRH also acts on the adrenal to increase production of cortisol and adrenal androgens.

Glucocorticoids

Considerable evidence discussed throughout this review supports an important role for glucocorticoids in parturition in the human and experimental animals. Glucocorticoids may act as antiprogestins and stimulate or overcome progesterone inhibition of CRH and PGDH expression in the placenta and fetal membranes. Glucocorticoids also stimulate PGHS-2 expression in the fetal membranes. Taken together, these actions may result in increased prostaglandin production at labor. Prostaglandins were also shown to increase activity of 11βHSD-1 in the chorion and amnion, which may result in a further feed-forward loop promoting increased prostaglandin production. Glucocorticoids also stimulate fibronectin production in fetal membrane mesenchymal cells, and increased levels of fetal fibronectin concentrations in cervical and vaginal secretions have been found to be related to the onset of parturition (621–623).

In the sheep, glucocorticoid production by the fetus is clearly essential for normal parturition to occur. In the human an increase in fetal glucocorticoid output is not essential for parturition to occur, but these steroids do regulate of a number of genes involved in parturition. The fact that preterm labor is not stimulated in patients receiving synthetic glucocorticoids to mature the fetal lung has often been used to underline the lack of importance of glucocorticoids in human parturition. However, it should be remembered that even in the sheep maternal infusion of glucocorticoids did not induce labor (161,162). In the human, a few limited studies have reported increased uterine activity in response to synthetic glucocorticoid treatment of the mother. It is possible that these patients do not progress to labor and delivery because the glucocorticoids pass through the placenta into the fetus and inhibit the fetal HPA axis. This would result in decreased production of the androgen precursor of estrogens, DHEAS, by the fetal adrenal, thereby decreasing estrogen levels. This decrease may then interfere with the expression of estrogen-dependent CAPs (611,624) that are necessary for labor.

Interestingly, a number of the effects of glucocorticoids that occur in fetal tissues, such as stimulation of placental CRH, of PGHS-2, and of fetal liver CBG expression, are the opposite of those that occur in adult tissues. The reason for this has not yet been elucidated, but it could be due to the expression of different coregulators, to different GR species, or to regulation by fetal-specific promoters.

Membrane Rupture

Extensive ECM remodeling occurs in many of the processes associated with both term and preterm parturition, including placental detachment, cervical ripening, and membrane rupture. In the case of membrane rupture there do not appear to be widespread structural changes throughout the membranes, but localized changes, particularly in the area overlying the cervix, are seen (625,626). Cytokines acting independently or through the production of prostaglandins may be responsible for stimulating production of MMPs in the membranes that in turn are involved in membrane rupture at both term and preterm. MMPs are inhibited by a family of inhibitory proteins, the TIMPs (627). In intrauterine tissues, alterations in the activity of TIMPs and MMPs are associated with the onset of labor and with membrane rupture (628–631).

MMP-9 concentrations are increased in amniotic fluid with intrauterine infection, and this may be due to increased output of cytokines that have been shown to increase MMP secretion in trophoblasts (631,632). TNF-α stimulates MMPs and decreases TIMP production by chorion, and IL-1α increases MMP-1 synthesis in the chorion (633). IL-8 production by amnion and chorion is stimulated by IL-1β and TNF-α, and in late gestation and labor IL-8 concentrations correlate with the level of MMP-9 in the amnion and chorion during cervical dilatation (553,634,635). Other studies have also suggested that apoptosis may contribute to membrane rupture (636,637).

In the area of the membranes overlying the cervix, pro-MMP-9 was found to be higher before labor compared with the membranes in the midzone, and only after delivery were MMP-2 and MMP-9 detected (638). MMP-2 and MMP-9 expression was studied in human fetal membranes and placenta obtained before and after labor at term and preterm (639). MMP-9 expression was increased in fetal membranes obtained after preterm and term labor compared with nonlabored tissues, and in placental tissue MMP-9 expression was higher in tissues after term labor. No differences were found in the expression of MMP-2. These enzymes were localized to specific cells in the membranes and placenta. Prostaglandins have been implicated in the regulation of MMP-9 production in the fetal membranes, and it was found that the production of endogenous PGE_2 was related to the production of MMP-9 by these tissues (640).

Relaxin receptors have been localized in the decidua and chorionic trophoblast cells. Relaxin acts through these receptors to upregulate expression of various MMPs and is likely involved in premature rupture of the membranes (81,641,642). Both forms of relaxin, H1 and H2, are produced by the decidua, and relaxin increases MMP release and expression in explants of human fetal membranes (81). In tissues obtained after premature rupture of membranes there was overexpression of relaxin in both the placenta and the decidua (643). Similar effects of relaxin on MMPs occur in the cervix (see previously), which suggests that this polypeptide may contribute to remodeling of cervical connective tissue as well as affecting membrane rupture.

An interesting series of studies by Strauss' group investigated the role of MMP polymorphism in premature rupture of the membranes. Polymorphisms in the promoter of MMP-9 were found to be associated with an increased risk of premature rupture of membranes in African Americans (644), and a single-nucleotide polymorphism in the MMP-1 promoter influenced MMP-1 expression in amnion cells (645). In addition, there was a significant association between carriage of the 2Gg allele and premature rupture of membranes, and this allele conferred increased responsiveness of amnion cells to stimuli that induce MMP-1 (645).

Tocolysis

The mechanisms involved in the regulation of term and preterm labor discussed in this review have provided a rational basis for the development of drugs to manage preterm labor. However, at present there is no widely acceptable method to regulate preterm labor that is efficient and devoid of side effects. The regulation of myometrial contraction and relaxation is achieved by altering the intracellular levels of Ca^{2+} and cAMP or by changing the expression of the various agonist and antagonist receptors. Furthermore, for contractions to occur, it is necessary to synthesize the appropriate agonists in the myometrium itself or in adjacent tissues. Tocolysis is therefore based on stimulating the synthesis of antagonists, inhibiting the synthesis of agonists, or blocking the effect of agonists on the myometrium. The major problem with the tocolytics in current use is that they are not specific for the uterus and therefore affect many other tissues in the mother and fetus. A number of reviews have discussed the use of various tocolytics (401,578,646).

Magnesium sulfate is probably the most widely used tocolytic in North America. It is believed to act by competing with Ca^{2+} and thereby interfering with activation of MLCK. Calcium channel blockers have also been used as tocolytics. The calcium channel blocker nifedipine, which blocks voltage-gated Ca^{2+} channels, has been studied in relation to β-adrenergic agonists and found to be comparable with fewer side effects (646,647). β-Adrenergic agonists such as ritrodine were used extensively as tocolytics, but because of their numerous side effects their use is now limited both in North America and the United Kingdom.

Inhibition of prostaglandin synthesis was one of the first methods attempted to reduce the production of uterotonins. Indomethacin, a nonsteroidal anti-inflammatory drug that inhibits PGHS, has been used as a tocolytic. However, it resulted in a number of serious side effects in the fetus, including increased incidence of necrotizing enterocolitis and intraventricular hemorrhage (648,649). Based on studies in animals and humans indicating that the PGHS-2 isoform of PGHS is particularly important in prostaglandin synthesis at labor, the possibility of using PGHS-2–specific inhibitors has been examined. Ongoing studies are being carried out using meloxicam in chronically instrumented animal models to determine the likelihood of potential side effects, particularly in the fetus (212). There are also some limited studies in humans. The fetal side effects of nimesilide were found to be similar to those of indomethacin and sulindac (402), and nimesilide has also been shown to result in oligohydramnios (650,651). However, additional studies have highlighted the fact that PGHS-2 is also expressed in many other fetal and maternal tissues during pregnancy (401,652), which underlines the potential for these compounds to have extensive side effects.

An oxytocin receptor antagonist (atosiban) has received considerable interest and has been studied using various animal models and in human pregnancy. A number of clinical trials have been carried out, and it would appear that atosiban is as efficacious as the β-agonist ritodrine, as well as safer (578,653). Clinical trials using NO donors as inhibitors of myometrial activity are being carried out, although this agent of course can have important cardiovascular side effects. Also, the different activities of NO on the myometrium and cervix, discussed previously, need to be considered.

Studies have also raised the possibility of using a specific PGF receptor antagonist to prevent preterm delivery (654). Expression of the FP receptor is increased in the uterus in a number of species, and ongoing studies with the ovine model indicate that this receptor may be a potential target for tocolysis (424). In animal models, inhibitors of CRH action have been shown to prolong pregnancy, but this effect is more likely to occur at the level of the fetal pituitary than at the level of the myometrium. In addition, CRH may act not only to stimulate but to relax the myometrium.

THE FUTURE

Since the mid-1990s, major advances have been made in understanding the factors involved in the regulation of term and preterm labor. However, the incidence of preterm labor also has increased during this period, and as yet there are no good methods to predict those patients in preterm labor who will deliver, nor are there good tocolytics available to inhibit labor in a safe and specific fashion. We need to be able to identify pregnancies in which the fetus should be delivered from a hostile intrauterine environment, as well as to differentiate threatened preterm labor from actual preterm labor. This will enable antenatal corticosteroid treatment to be limited to patients who are actually in preterm labor and prevent the repeated administration of corticosteroids to patients who are undergoing episodes of uterine contractility during an otherwise normal pregnancy.

With a more detailed understanding of the mechanisms involved in labor, it may be possible to develop uterine-specific tocolytics or methods to target drugs to the uterine tissues. It is clear from this review that preterm labor is multifactorial and likely involves different mechanisms in different patients. This may be an area of medicine where analysis of individual patients with DNA microarrays will enable the treatment of the patient to be individualized. Finally, most studies to date have focused on the mother and neglected the father; perhaps studies investigating the role of paternal genes in the regulation of parturition may provide a new avenue of investigation.

ACKNOWLEDGMENTS

Work in the authors' laboratories was supported by grants from the Canadian Institutes for Health Research (W. G., J. R. G. C., and S. J. L.), the Institute for Child and Youth Health and Human Development (J. R. G. C.), and the National Institutes of Health (S. J. L.). J. R. G. and S. J. L. are members of the Canadian Institutes for Health Research Groups in Fetal and Neonatal Health and Development.

REFERENCES

1. Challis, J., Matthews, S., Gibb, W., and Lye, S. (2000). Endocrine and paracrine regulation of birth at term and preterm. *Endocr. Rev.* 21, 514–550.
2. Challis, J., and Lye, S. (1994). Parturition. In *The Physiology of Reproduction* (E. Knobil and J. D. Neill, Eds.), pp. 985–1031. Raven Press, New York.
3. Creasy, R. (1991). Preventing preterm birth. *N. Engl. J. Med.* 325, 727–729.
4. Joseph, K. S., Kramer, M. S., Marcoux, S., Ohlsson, A., Wen, S. W., Allen, A., and Platt, R. (1998). Determinants of preterm birth rates in Canada from 1981 through 1983 and from 1992 through 1994. *N. Engl. J. Med.* 12, 1434–1439.
5. Meis, P. J., Goldenberg, R. L., Mercer, J. E., Iams, J. D., Moawad, A. H., Miodovnik, M., Menard, M. K., Caritis, S. N., Thurnau, G. R., Bottoms, S. F., Das, A., Roberts, J. M., and McNellis, D. (1998). The preterm prediction study: risk factors for indicated preterm births. Maternal-Fetal Medicine Units Network of the National Institute of Child Health and Human Development. *Am. J. Obstet. Gynecol.* 178, 562–567.
6. Nathanielsz, P., Binlenda, Z., Wimsatt, J., Figueroa, J. P., and Massaman, A. (1987). Patterns of myometrial activity and their regulation in the pregnant monkey. In *The Onset of Labor: Cellular and Integrative Mechanisms* (D. McNellis, J. Challis, P. MacDonald, P. W. Nathanielsz, and J. M. Roberts, Eds.), p. 359. Perinatology Press, Ithaca, NY.

7. Lye, S. (1997). Initiation of parturition. *Anim. Reprod. Sci.* 42, 495.
8. Harding, R., Poore, E. R., Bailey, A., et al. (1982). Electromyographic activity on the nonpregnant and pregnant sheep uterus. *Am. J. Obstet. Gynecol.* 142, 448–457.
9. Challis, J. R. G. (1998). Characteristics of parturition. In *Maternal-Fetal Medicine: Principles and Practice* (R. Creasey and R. Resnik, Eds.), pp. 484–497. WB Saunders, Philadelphia.
10. Lye, S. J., and Freitag, C. L. (1990). Local and systemic control of myometrial contractile activity during labor in the sheep. *J. Reprod. Fertil.* 90, 483–492.
11. Lye, S. J. (1994). The initiation and inhibition of labour: towards a molecular understanding. *Semin. Reprod. Endocrinol.* 12, 284–294.
12. Lye, S., Ou, C.-W., Tech, T., et al. (1998). The molecular basis of labour and tocolysis. *Fetal Matern. Med. Rev.* 10, 121–136.
13. Huszar, G. B., and Walsh, M. P. (1991). Relationship between myometrial and cervical functions in pregnancy and labor. *Semin. Perinatol.* 15, 97–117.
14. MacKenzie, L. W., Word, M. A., Casey, M. L., and Stull, J. T. (1990). Myosin light chain phosphorylation in human myometrial smooth muscle cells. *Am. J. Physiol.* 258, C92–C98.
15. Schoenber, C. F. (1977). The contractile mechanism and ultrastructure of the myometrium. In *Biology of the Uterus* (R. M. Wynn, Ed.), pp. 497–554. Plenum Press, New York.
16. Garfield, R. E. (1988). Structural and functional studies of the control of myometrial contractility and labour. In *The Onset of Labour: Cellular and Integrative Mechanisms* (D. McNellis, J. R. G. Challis, P. C. MacDonald, P. W. Nathanielsz, and J. M. Roberts, Eds.), pp. 55–80. Perinatology Press, Ithaca, NY.
17. Wray, S., Jones, K., Kupittayanant, S., et al. (2003). Calcium signalling and uterine contractility. *J. Soc. Gynecol. Investig.* 10, 252–264.
18. Word, R. A., Tang, D. C., and Kamm, K. E. (1994). Activation properties of myosin light chain kinase during contraction/relaxation cycles of tonic and phasic smooth muscles. *J. Biol. Chem.* 269, 596–602.
19. Hsu, C. J., and Sanborn, B. M. (1986). Relaxin treatment alters the kinetic properties of myosin light chain kinase activity in rat myometrial cells in culture. *Endocrinology* 118, 499–505.
20. MacKenzie, L. W., Word, M. A., Casey, M. L., and Stull, J. T. (1990). Myosin light chain phosphorylation in human myometrial cells. *Am. J. Physiol.* 258, C92–C98.
21. Izumi, H., Ichikara, J., Uchiumi, Y., and Shirakawa, K. (1990). Gestational changes in mechanical properties of skinned muscle tissues of human myometrium. *Am. J. Obstet. Gynecol.* 163, 638C–647C.
22. Lopez Bernal, A. (2003). Mechanisms of labor: biochemical aspects. *Br. J. Obstet. Gynaecol.* 110, 39–45.
23. Moore, F., Da Silva, C., Wilde, J. I., Smarason, A., and Lopez Bernal, A. (2000). Up-regulation of p21-and RhoA-activated protein kinases in human pregnant myometrium. *Biochem. Biophys. Res. Commun.* 269, 322–326.
24. Moran, C. J., Friel, A. M., Smith, T. J., Cairns, M., and Morrison, J. J. (2002). Expression and modulation of Rho kinase in human pregnant myometrium. *Mol. Hum. Reprod.* 8, 196–200.
25. Moss, S. B., and Gerton, G. L. (2001). A-kinase anchor proteins in endocrine systems and reproduction. *Trends Endocrinol. Metab.* 12, 434–440.
26. Sanborn, B. M. (2000). Relationship of ion channel activity to control of myometrial calcium. *J. Soc. Gynecol. Investig.* 7, 4–11.
27. Sanborn, B. M. (2001). Hormones and calcium: mechanisms controlling uterine smooth muscle contractile activity. *Exp. Physiol.* 86, 223–237.
28. Keef, K. D., Hume, J. R., and Zhong, J. (2001). Regulation of cardiac smooth muscles Ca^{2+} channels (Ca(V)1.2a,b) by protein kinases. *Am. J. Physiol.* 281, C1743–C1756.
29. Walsh, M. P. (1991). Calcium-dependent mechanisms of regulation of smooth muscle contraction. *Biochem. Cell Biol.* 69, 771–800.
30. Berridge, M. J. (1995). Capacitative calcium entry. *Biochem. J.* 312, 1–11.
31. Asboth, G., Phaneuf, S., Europe-Finner, G. N., Toth, M., and Lopez, B. (1996). Prostaglandin E2 activates phospholipase c and elevates intracellular calcium in cultured myometrial cells: involvement of EP1 and EP3 receptor subtypes. *Endocrinology* 137, 2572–2579.
32. Sanborn, B. M., Yue, C., Wang, W., and Dodge, K. L. (1998). G protein signalling pathways in myometrium. *Rev. Reprod.* 3, 196–205.
33. Burghardt, R. C., Barhoumi, R., Sanborn, B. M., and Anderson, J. (1999). Oxytocin-induced Ca^{2+} responses in human myometrial cells. *Biol. Reprod.* 60, 777–782.
34. Lye, S., and Freitag, C. (1988). An in vitro model to examine the electromyographic activity of isolated myometrial tissue from pregnant sheep. *J. Reprod. Fertil.* 82, 51–61.
35. Sigger, J. N., Harding, R., and Jenkin, G. (1984). Relationship between electrical activity of the uterus and surgically isolated myometrium in the pregnant and non-pregnant ewe. *J. Reprod. Fertil.* 70, 103–114.
36. Challis, J., and Brooks, A. (1989). Maturation and activation of hypothalamic-pituitary-adrenal function in fetal sheep. *Endocr. Rev.* 10, 182–204.
37. Cullen, B. M., and Harkness, R. D. (1968). Collagen formation and changes in cell population in the rat's uterus after distension with wax. *Q. J. Exp. Physiol. Cogn. Med. Sci.* 53, 33–42.
38. Goldspink, D. F., and Douglas, A. J. (1988). Protein turnover in gravid and nongravid horns of uterus in pregnant rats. *Am. J. Physiol.* 254, E549–E554.
39. Schwalm, H., and Dubrausky, V. (1966). The structure of the musculature of the human uterus-muscles and connective tissue. *Am. J. Obstet. Gynecol.* 94, 391–404.
40. Fata, J. E., Ho, A. T., Leco, K. J., Moorehead, R. A., and Khokha, R. (2000). Cellular turnover and extracellular matrix remodeling in female reproductive tissues: functions of metalloproteinases and their inhibitors. *Cell. Mol. Life Sci.* 57, 77–95.
41. Stewart, E. A., Floor, A. E., Jain, P., and Novak, R. A. (1995). Increased expression of messenger RNA for collagen type I, collagen type III and fibronectin in myometrium of pregnancy. *Obstet. Gynecol.* 86, 417–422.
42. Csapo, A., Erdos, T., De Mattos, C. R., Gramss, E., and Moscowitz, C. (1965). Stretch induced uterine growth, protein synthesis and function. *Nature* 207, 1378–1379.
43. Leung, D. Y., Glagov, S., and Matthews, M. B. (1976). Cyclic stretching stimulates synthesis of matrix components by arterial smooth muscle cells in vitro. *Science* 191, 475–477.
44. Al-Matubsi, H. Y., Eis, A. L., Brodt-Eppley, J., et al. (2001). Expression and localization of the contractile prostaglandin F receptor in pregnant rat myometrium in late gestation, labour and postpartum. *Biol. Reprod.* 65, 1029–1037.
45. Europe-Finner, G. N., Phaneuf, S., Watson, L. P., and Lopez Bernal, A. (1993). Identification and expression of G-proteins in human myometrium: upregulation of $G\alpha_s$ in pregnancy. *Endocrinology* 132, 2484–2490.
46. Europe-Finner, G. N., Phaneuf, S., Tolkovsky, A. M., Watson, S. P., and Lopez Bernal, A. (1994). Down regulation of $G\alpha_s$ in human myometrium in term and preterm labor: a mechanism for parturition. *J. Clin. Endocrinol. Metab.* 79, 1835–1839.
47. Price, S., Pochun, I., Phaneuf, S., and Lopez Bernal, A. (2000). Adenylyl cyclase isoforms in pregnant and non-pregnant human myometrium. *J. Endocrinol.* 164, 21–30.

48. Europe-Finner, G. N., Phaneuf, S., Tolkovsky, A. M., Watson, S. P., and Lopez Bernal, A. (1994). Down regulation of Gα_s in human myometrium in term and preterm labor: a mechanism for parturition. *J. Clin. Endocrinol. Metab.* 79, 1835–1839.
49. Phillips, R., Bailey, J., Robson, S., and Europe-Finner, G. (2002). Differential expression of the adenylyl cyclase-stimulatory guanosine triphosphate-binding protein G(s)α in the human myometrium during pregnancy and labor involves transcriptional regulation by cyclic adenosine 3′5′-monophosphate and binding of phosphorylated nuclear proteins to multiple GC boxes within the promoter. *J. Clin. Endocrinol. Metab.* 87, 5675–5685.
50. Kofinas, A. D., Rose, J. C., Koritinik, D. R., and Meis, P. J. (1990). Progesterone and estradiol concentrations in non-pregnant and pregnant myometrium: effect of progesterone and estradiol on cyclic adenosine monophosphate-phosphodiesterase activity. *J. Reprod. Med.* 35, 1045–1050.
51. Dodge, K. L., Carr, D. W., and Sanborn, B. M. (1999). A role for AKAP (A kinase anchoring protein) scaffolding in the loss of acyclic adenosine 3′5′-monophosphate inhibitory response in late pregnant rat myometrium. *Mol. Endocrinol.* 13, 1977–1987.
52. Bailey, J., Phillips, R. J., Pollard, A. J., et al. (2002). Characterization and functional analysis of cAMP response element modulator protein and activating transcription factor 2 (ATF2) isoforms in the human myometrium during pregnancy and labor: identification of a novel ATF2 species with potent transactivation properties. *J. Clin. Endocrinol. Metab.* 87, 1717–1728.
53. MacDougall, M. W., Europe-Finner, G. N., and Robson, S. C. (2003). Human myometrial quiescence and activation during gestation and parturition involve dramatic changes in the expression and activity of particulate type II (RIIα) protein kinase A haloenzyme. *J. Clin. Endocrinol. Metab.* 88, 2194–2205.
54. Perez, G. J., Toro, L., Erulkar, S. D., and Stefani, E. (1993). Characterization of large conductance calcium-activated potassium channels from human myometrium. *Am. J. Obstet. Gynecol.* 168, 652–660.
55. Anwer, K., Oberti, C., Perez, G. J., et al. (1993). Ca^{2+}-activated K^+ channels as modulators of human myometrial contractility. *Am. J. Physiol.* 265, C976–C985.
56. Khan, R. N., Morrison, J. J., Smith, S. K., and Ashford, M. L. J. (1998). Activation of large conductance potassium channels in pregnant human myometrium by pinacidil. *Am. J. Obstet. Gynecol.* 178, 1027–1034.
57. Khan, R., Mathroo-Ball, B., Arikumaran, S., and Ashford, M. L. J. (2001). Potassium channels in the human myometrium. *Exp. Physiol.* 86, 255–264.
58. Song, M., Zhu, N., Olcese, R., Barila, B., Toro, L., and Stefani, E. (1999). Hormonal control of protein expression and mRNA levels of MaxiK channel alpha subunit in myometrium. *FEBS Lett.* 460, 427–432.
59. Chien, E. K., Zhang, Y., Hiroto, F., and Hara, M. (1999). Expression of adenosine triphosphate-sensitive potassium channel subunits in female rat reproductive tissues: overlapping distribution of messenger ribonucleic acid for weak inwardly rectifying potassium channel subunit 6.1 and sulfonylurea-binding regulatory subunit 2. *Am. J. Obstet. Gynecol.* 180, 1121–1126.
60. Carvajal, J. A., Thompson, L. P., and Weiner, C. P. (2003). Chorion-induced myometrial relaxation is mediated by large conductance Ca^{2+} activated K^+ channel opening in the guinea pig. *Am. J. Obstet. Gynecol.* 188, 84–91.
61. Matharoo-Ball, B., Ashford, M. L., Arulkumaran, S., Khan, R. N. (2003). Down regulation of the alpha- and beta-subunits of the calcium activated potassium channel in human myometrium with parturition. *Biol. Reprod.* 68, 2135–2141.
62. Fergusen, J. E., Gorman, J. V., Bruns, D. E., Weir, E. C., Burtis, W. J., Martin, T. J., and Bruns, M. E. (1992). Abundant expression of parathyroid hormone related protein in human amnion and its association with labor. *Proc. Natl. Acad. Sci. U S A* 89, 8384–8388.
63. Morimoto, T., Devora, G., and Mibe, M. (1997). Parathyroid hormone-related protein and human myometrial cells: Action and regulation. *Mol. Cell. Endocrinol.* 129, 91.
64. Thiede, M. A., Daifotis, A. G., Weir, E. C., Brines, M. L., Burtis, W. J., Ikeda, K., Dreyer, B. E., Garfield, R. E., and Broadus, A. E. (1990). Intrauterine occupancy controls expression of the parathyroid hormone-related peptide gene in preterm rat myometrium. *Proc. Natl. Acad. Sci. U S A* 87, 6969–6973.
65. Mitchell, J., Ting, T., Wong, S., Mitchell, B., and Lye, S. (2003). Parathyroid hormone-related protein treatment of pregnant rat delays the increase in connexin 43 and oxytocin receptor expression in the myometrium. *Biol. Reprod.* 69, 556–562.
66. Dennes, W., Slater, D., and Bennett, P. (1997). Nitric oxide synthase mRNA expression in human fetal membranes: a possible role in parturition. *Biochem. Biophys. Res. Commun.* 233, 276–278.
67. Bansal, R. K., Goldsmith, P. C., and He, Y. (1997). A decline in myometrial nitric oxide synthase expression is associated with labor and delivery. *J. Clin. Invest.* 99, 2502–2508.
68. Chwalisz, K., and Garfield, R. (1997). Regulation of the uterus and cervix during pregnancy and labor: role of progesterone and nitric oxide. *Ann. N Y Acad. Sci.* 828, 238–253.
69. Buxton, I., Kaiser, R., Malmquist, N., and Tichenor, S. (2001). NO-induced relaxation of labouring and non-labouring human myometrium is not mediated by cyclic GMP. *Br. J. Pharmacol.* 134, 206–214.
70. Eis, A. L., Brockman, D. E., and Myatt, L. (1997). Immunolocalization of the inducible nitric oxide synthase isoform in human fetal membranes. *Am. J. Reprod. Immunol.* 38, 289–294.
71. Ekerhovd, E., Weijdegard, B., Brannstrom, M., Mattsby-Baltzer, I., and Norstrom, A. (2002). Nitric oxide induced cervical ripening in the human: involvement of cyclic guanosine monophosphate, prostaglandin $F_{2\alpha}$ and prostaglandin E_2. *Am. J. Obstet. Gynecol.* 186, 745–750.
72. Downing, S. J., and Sherwood, O. D. (1985). The physiological role of relaxin in the pregnant rat II. The influence of relaxin on uterine contractile activity. *Endocrinology* 116, 1206–1214.
73. Porter, D. G., Downing, S. J., and Bradshaw, J. M. C. (1979). Relaxin inhibits spontaneous and prostaglandin driven myometrial activity in anaesthetized rats. *Endocrinology* 83, 183–192.
74. Porter, D. G. (1982). Unsolved problems of relaxin's physiological role. *Ann. N Y Acad. Sci.* 380, 151–162.
75. Porter, D. G., Lye, S. J., Bradshaw, J. M. C., and Kendall, J. Z. (1981). Relaxin inhibits myometrial activity in the ovariectomized non-pregnant ewe. *J. Reprod. Fertil.* 61, 409–414.
76. Porter, D. G., and Watts, A. D. (1986). Relaxin and progesterone are myometrial inhibitors in the ovariectomized non-pregnant mini-pig. *J. Reprod. Fertil.* 76, 205–213.
77. Hansell, D. J., Bryant-Greenwood, G. D., and Greenwood, D. C. (1991). Expression of the human relaxin H1 gene in the decidua trophoblast, and prostate. *J. Clin. Endocrinol. Metab.* 72, 899–904.
78. Bryant-Greenwood, G. D. (1991). The human relaxins: consensus and dissent. *Mol. Cell. Endocrinol.* 79, C125–C132.
79. Castracane, V. D., Lessins, J., Brenner, S., and Weiss, G. (1985). Relaxin in the pregnant baboon: evidence for local production in reproductive tissues. *J. Clin. Endocrinol. Metab.* 60, 133–136.

80. Sakbun, V., Ali, S. M., Greenwood, F. C., and Bryant-Greenwood, G. D. (1990). Human relaxin in the amnion, chorion, decidua parietalis, basal plate and placental trophoblast by immunohistochemistry and northern analysis. *J. Clin. Endocrinol. Metab.* 70, 508–514.
81. Tashima, L., Yamamoto, S., Yasuda, M., Millar, L., and Bryant-Greenwood, G. (2002). Decidual relaxins: gene and protein up-regulation in preterm premature rupture of the membranes by complementary DNA arrays and quantitative immunocytochemistry. *Am. J. Obstet. Gynecol.* 187, 785–797.
82. Sanborn, B., Qian, A., and Ku, C. (1995). Mechanisms regulating oxytocin receptor coupling to phospholipase C in rat and human myometrium. *Adv. Exp. Med. Biol.* 395, 469–479.
83. Bani, D., Baccari, M. C., Nistri, S., et al. (1999). Relaxin up-regulates the nitric oxide biosynthetic pathway in the mouse uterus: involvement in the inhibition of myometrial contractility. *Endocrinology* 140, 4434–4441.
84. Longo, M., Jain, V., Vedernikov, Y., Garfield, R., and Saade, G. (2003). Effects of recombinant human relaxin on pregnant rat uterine artery and myometrium in vitro. *Am. J. Obstet. Gynecol.* 188, 1468–1474.
85. Challis, J. R. G., and Lye, S. J. (1986). Parturition. *Rev. Reprod. Biol.* 8, 61–129.
86. Lye, S. J., and Challis, J. R. G. (1982). Inhibition by PGI_2 of myometrial activity in vivo in non-pregnant ovariectomized sheep. *J. Reprod. Fertil.* 66, 311–315.
87. Omini, C., Folco, G. C., Pasargiklian, R., Fano, M., and Berti, F. (1979). Prostacyclin (PGI_2) in the pregnant human uterus. *Prostaglandins* 17, 113–120.
88. Wikland, M., Lindblom, B., Hammarstrom, S., and Wiqvist, N. (1983). The effect of prostaglandin I_2 on the contractility of the term pregnant human myometrium. *Prostaglandins* 26, 905–1001.
89. Williams, K. I., Tahir, K. E. H., and Marcinkiewicz, E. (1979). Dual actions of prostacyclin (PGI_2) on the rat pregnant uterus. *Prostaglandins* 17, 668–672.
90. Giannoulias, D., Alfaidy, N., Holloway, A. C., Gibb, W., Sun, M., Lye, S. J., and Challis, J. R. (2002). Expression of prostaglandin I2 synthase, but not prostaglandin E synthase, changes in myometrium of women at term pregnancy. *J. Clin. Endocrinol. Metab.* 87, 5274–5282.
91. Negishi, M., Sugimoto, Y., and Ichikawa, A. (1995). Molecular mechanisms of diverse actions of prostanoid receptors. *Biochem. Biophys. Acta* 1259, 109–119.
92. Grammatopoulos, D., and Hillhouse, E. W. (1999). Role of corticotropin releasing hormone in onset of labour. *Lancet* 354, 1546–1549.
93. Csapo, A. I. (1977). The "see saw" theory of parturition. *Ciba Found. Symp.* 47, 159–210.
94. Lye, S. J., and Porter, D. G. (1978). Demonstration that progesterone "blocks" uterine activity in the ewe in vivo by a direct action on the myometrium. *J. Reprod. Fertil.* 52, 87–92.
95. Petrocelli, T., and Lye, S. J. (1993). Regulation of transcripts encoding the myometrial gap junction protein, connexin-43, by estrogens and progesterone. *Endocrinology* 133, 284–290.
96. Csapo, A. I. (1969). The four direct regulatory factors of myometrial function. In *Progesterone: Its Regulatory Effect on the Myometrium* (G. E. W. Wolstenholme, Ed.), pp. 13–19. Churchill, London.
97. Csapo, A. I. (1973). Regulatory interplay of progesterone and $PGF_{2\alpha}$ in the control of the pregnant uterus. In *Uterine Contractions: Side Effects of Steroid Contraceptives* (J. B. Josomovich, Ed.), pp. 223–230. Wiley, London.
98. Dong, Y. L., Wimalawansa, S., and Yallampalli, C. (2003). Effects of steroid hormones on calcitonin gene-related peptide receptors in cultured human myometrium. *Am. J. Obstet. Gynecol.* 188, 466–472.
99. Naghashpour, M., and Dhal, G. (2000). Relaxation of myometrium by calcitonin gene-related peptide is independent of nitric oxide synthase activity in mouse uterus. *Biol. Reprod.* 63, 1421–1427.
100. Liggins, G., and Thorburn, G. (1994). Initiation of parturition. In *Marshall's Physiology of Reproduction*, Vol. 3, Part 2, *Pregnancy and Lactation* (G. Lamming, Ed.), pp. 863–1002. Chapman & Hall, London.
101. Girotti, M., and Zingg, H. (2003). Gene expression profiling of rat uterus at different stages of parturition. *Endocrinology* 144, 2254–2265.
102. Leppert, P. C., Keller, S., Ceretta, J., and Mandl, I. (1982). Conclusive evidence for the presence of elastin in human and monkey cervix. *Am. J. Obstet. Gynecol.* 142, 179–182.
103. Ludmir, J., Sehdev, H., and Harish, M. (2000). Anatomy and physiology of the uterine cervix. *Clin. Obstet. Gynecol.* 43, 433–439.
104. Huzar, G. (1991). Physiology of the uterine cervix in reproduction. *Semin. Perinatol.* 15, 95–105.
105. Leppert, P. C., and Woessner, J. F. (1991). The extracellular matrix of the uterus cervix and fetal membranes: synthesis, degradation and hormonal regulation. In *The Extracellular Matrix of the Uterus Cervix and Fetal Membranes* (C. Phyllis, P. C. Leppert, and J. F. Woessner, Eds.), pp. 68–76. Perinatology Press, Ithaca, NY.
106. Kleisssel, H. P., Van der Rest, M., Naftolin, F., Glorieux, F. K., and de Leon, A. (1978). Collagen changes in the human cervix at parturition. *Am. J. Obstet. Gynecol.* 130, 748–753.
107. Uldbjerg, N., and Danielsen, C. C. (1988). A study of the interaction in vitro between type 1 collagen and a small dermatan sulfate proteoglycan. *Biochem. J.* 251, 643–648.
108. Leppert, P. C., Yu, S. Y., Keller, S., and Cerreta, J. (1986). Orientation of elastic fibers in the human cervix. *Am. J. Obstet. Gynecol.* 155, 219–224.
109. Leppert, P. C., Yu, S. Y., Keller, S., and Cerreta, J. (1987). Decreased elastic fibers and desmosine content in incompetent cervix. *Am. J. Obstet. Gynecol.* 157, 1134–1139.
110. Rechberger, T., Abramson, S. R., and Woessner, J. F., Jr. (1996). Onapristone and prostaglandin E_2 induction of delivery in the rat in late pregnancy: a model for the analysis of cervical softening. *Am. J. Obstet. Gynecol.* 175, 719–723.
111. Rechberger, T., and Woessner, J. F. (1993). Collagenase, its inhibitors, and decorin in the lower uterine segment in pregnant women. *Am. J. Obstet. Gynecol.* 175, 719–723.
112. Leppert, P. C., Kokenyesi, R., Klemenich, C. A., and Fisher, J. (2000). Further evidence of a decorin-collagen interaction in the disruption of cervical collagen fibers during rat gestation. *Am. J. Obstet. Gynecol.* 182, 805–811.
113. Woessner, J. F., Jr., and Rechberger, T. (1993). Collagenase, its inhibitors, and decorin in the lower uterine segment in pregnant women. *Am. J. Obstet. Gynecol.* 175, 719–723.
114. Von Maillot, K., Stuhlsatz, H. W., Mohanaradhakrishnan, V., and Greiling, H. (1979). Changes in glycosaminoglycans distribution pattern in the human uterine cervix during pregnancy and labor. *Am. J. Obstet. Gynecol.* 135, 503–506.
115. Golichowski, A. (1980). Cervical stromal interstitial polysaccharide metabolism in pregnancy. In *Dilatation of the Uterine Cervix* (F. Naftolin and P. G. Stubblefield, Eds.), pp. 99–112. Raven Press, New York.
116. De Lucas, A., Santra, M., Baldi, A., Giordano, A., and Iozzo, R. (1996). Decorin-induced growth suppression is associated with the up-regulation of p21, an inhibitor of cyclin-dependent kinases. *J. Biol. Chem.* 1271, 18961–18965.
117. Uldberg, N., Ekman, G., Malstrom, A., Olsson, K., and Ulmsten, U. (1980). Ripening of the human uterine cervix related changes in collagen, glycosaminoglycans and collagenolytic activity. *Am. J. Obstet. Gynecol.* 147, 662–666.

118. Rajabi, M., Dean, D. D., Beydoun, S. N., and Woessner, J. F. (1988). Elevated tissue levels of collagenase during dilatation of the uterine cervix in human parturition. *Am. J. Obstet. Gynecol.* 159, 971–976.
119. Bryant-Greenwood, G., Yamomoto, S., and Bogic, L. (1997). Relaxin: a role in the premature rupture of the fetal membranes independent of infection. In *Preterm Birth: Etiology, Mechanisms and Prevention* (S. Charleston, Ed.). University of Alabama Press, Birmingham.
120. Goldsmith, L., Palejwala, S., and Weiss, G. (1997). The role of relaxin in preterm labor. In *Preterm Birth: Etiology, Mechanisms and Prevention* (S. Charleston, Ed.). University of Alabama Press, Birmingham.
121. Osman, I., Young, A., Ledingham, M., et al. (2003). Leukocyte density and pro-inflammatory cytokine expression in human fetal membranes, decidua, cervix and myometrium before and during labour at term. *Mol. Hum. Reprod.* 9, 41–45.
122. Yellon, S., Mackler, A., and Kirby, M. (2003). The role of leukocyte traffic and activation in parturition. *J. Soc. Gynecol. Investig.* 10, 323–338.
123. Kelly, R. (2002). Inflammatory mediators and cervical ripening. *J. Reprod. Immunol.* 57, 217–224.
124. Winkler, M., Fischer, D., Hlubek, M., et al. (1998). Interleukin-1β and interleukin-8 concentrations in the lower uterine segment during parturition at term. *Obstet. Gynecol.* 91, 945–949.
125. Watari, M., Watari, H., DiSanto, M., et al. (1999). Pro-inflammatory cytokines induce expression of matrix-metabolizing enzymes in human cervical smooth muscle cells. *Am. J. Pathol.* 154, 1755–1762.
126. El Marandy, E., Kanayma, N., Halim, A., et al. (1995). The effect of interleukin-1 in rabbit cervical ripening. *J. Obstet. Gynecol.* 60, 75–80.
127. Chwalisz, K., Benson, M., Scholz, P., et al. (1994). Cervical ripening with cytokines interleukin 8, interleukin 1β, tumor necrosis factor α in guinea pigs. *Hum Reprod.* 9, 2173–2181.
128. Hertelendy, F., and Zakar, T. (2004). Prostaglandins and the myometrium and cervix. *Prostaglandins Leukot. Essent. Fatty Acids* 70, 207–222.
129. Ekman, E., Uldjerg, N., Malmstrom, A., and Ulmsten, U. (1983). Increased postpartum collagenolytic activity in cervical connective tissue from women treated with prostaglandin E_2. *Gynecol. Obstet. Invest.* 16, 292–298.
130. Sennstrom, B., Ekman, G., Westergren, G. T., et al. (2000). Human cervical ripening, an inflammatory process mediated by cytokines. *Mol. Hum. Reprod.* 6, 375–381.
131. Winkler, M., Fischer, D. C., Hlubek, M., van de Leur, E., Haubeck, H. D., and Rath, W. (1998). Interleukin-1β and interleukin-8 concentrations in the lower uterine segment during parturition at term. *Obstet. Gynecol.* 91, 945–949.
132. Denison, F., Calder, A., Kelly, R. (1999) The action of prostaglandin E2 on the human cervix: stimulation of interleukin 8 and inhibition of secretory leukocyte protease inhibitor. *Am. J. Obstet. Gynecol.* 180, 614–620.
133. Toth, M., Rehnstrom, J., and Fuchs, A. R. (1989). Prostaglandin E and F in cervical mucus of pregnant women. *Am. J. Perinatol.* 6, 142–144.
134. Platz-Christensen, J. J., Pernevi, P., Bokstrom, H., and Wiqvist, N. (1997). Prostaglandin E and $F_{2\alpha}$ concentrations in the cervical mucus and the mechanism of cervical ripening. *Prostaglandins* 53, 253–261.
135. Tornblom, S. A., Patel, F. A., Bystrom, B., et al. (2004). 15-Hydroxyprostaglandin dehydrogenase and cyclooxygenase messenger RNA expression and immunohistochemical localization in human cervical tissue during term and preterm labor. *J. Clin. Endocrinol. Metab.* 89, 2909–2915.
136. Mahendroo, M., Cala, K., and Russell, D. (1996). 5α-Reduced androgens play a key role in murine parturition. *Mol. Endocrinol.* 10, 380–392.
137. Mahendroo, M., Porter, A., Russell, D., and Word, R. (1999). The parturition defect in steroid 5α-reductase type 1 knockout mice is due to impaired cervical ripening. *Mol. Endocrinol.* 13, 981–992.
138. Rodriguez, H. A., Kass, L., Varayoud, J., et al. (2003). Collagen remodeling in the guinea-pig uterine cervix at term is associated with a decrease in progesterone receptor expression. *Mol. Hum. Reprod.* 9, 807–813.
139. Stjerholm-Vladic, Y., Wang, H., Stygar, D., Ekman, G., and Sahlin, L. (2004). Differential regulation of the progesterone receptor A and B in the human uterine cervix at parturition. *Gynecol. Endocrinol.* 18, 41–46.
140. Wang, H., Stjerholm-Vladic, Y., Ekman, G., Eriksson, H., and Sahlin, L. (2001). Differential regulation of oestrogen receptors alpha and beta in the human cervix at term pregnancy. *Mol. Hum. Reprod.* 7, 293–300.
141. Stygar, D., Wang, H., Vladic, Y. S., et al. (2001). Co-localization of oestrogen receptor beta and leukocyte markers in the human cervix. *Mol. Hum. Reprod.* 7, 881–886.
142. Collins, J., Usip, S., McCarson, K., and Papka, R. (2002). Sensory nerves and neuropeptides in uterine cervical ripening. *Peptides* 23, 167–183.
143. Mowa, C., Usip, S., Collins, J., et al. (2003). The effects of pregnancy and estrogen on the expression of calcitonin gene-related peptide (CGRP) in the uterine cervix, dorsal root ganglia and spinal cord. *Peptides* 24, 1163–1174.
144. Lin, P. C., Li, X., Lei, Z. M., and Rao C. V. (2003). Human cervix contains functional luteinizing hormone/human chorionic gonadotropin receptors. *J. Clin. Endocrinol. Metab.* 88, 3409–3414.
145. Iams, J. D. (2003). What we have learned about cervical ultrasound. *Semin. Perinatol.* 27, 194–203.
146. Iams, J. D. (2003). Prediction and detection of preterm labor. *Obstet. Gynecol.* 101, 402–412.
147. Garfield, R., Maul, H., Shi, L., et al. (2001). Methods and devices for the management of term and preterm labor. *Ann. N Y Acad. Sci.* 943, 203–224.
148. Zahn, V. (1984). Uterine contractions during pregnancy. *J. Perinat. Med.* 12, 107–112.
149. Figueroa, J., Honnebier, M., Binlenda, Z., Wimsatt, J., and Nathanielsz, P. (1989). Effect of 48 hours intravenous D-4-androstenedione infusion on the pregnant rhesus monkey during the last third of gestation: changes in maternal plasma estradiol concentrations and myometrial contractility. *Am. J. Obstet. Gynecol.* 161, 481–486.
150. Honnebier, M. B., Figueroa, J. P., Rivier, J., Vale, W., and Nathanielsz, P. W. (1989). Studies on the role of oxytocin in late pregnancy in the pregnant rhesus monkey: plasma concentrations of oxytocin in maternal circulation throughout the 24h-day and the effect of the synthetic oxytocin antagonist [1-β-Mpa (β-(CH2)5)1,(Me (Tyr2, Orn8] oxytocin on spontaneous nocturnal myometrial contractions. *Endocrinology* 125, 1498–1503.
151. Seron-Ferre, M., Prado, J., and Germain, A. M. (1991). Circadian rhythms and delivery in the human. In *Reproduction, Growth and Development* (A. P. G. Negro-Villar, Ed.), pp. 59–67. Raven Press, New York.
152. Garfield, R. E. (1988). Structural and functional studies of the control of myometrial contractility and labour. In *The Onset of Labour: Cellular and Integrative Mechanisms* (D. McNellis, J. R. G. Challis, P. C. MacDonald, P. W. Nathanielsz, and J. M. Roberts, Eds.), pp. 55–80. Perinatology Press, Ithaca, NY.

153. Garfield, R. (1990). Intercellular coupling and modulation of uterine contractility. In *Uterine Contractility* (R. E. Garfield, Ed.), pp. 21–40. Serono Symposia USA, Norwell, MA.
154. Sakai, N., Tabb, T., and Garfield, R. E. (1992). Modulation of cell-to-cell coupling between myometrial cells of the human uterus during pregnancy. *Am. J. Obstet. Gynecol.* 167, 472–480.
155. Beyer, E. C., Kistler, J., Paul, D. L., and Goodenough, D. A. (1989). Antisera directed against connexin-43 peptides react with a 43kD protein localized to gap junctions in myocardium and other tissues. *J. Cell Biol.* 108, 595–605.
156. Risek, B., Guthrie, S., Kumar, N., and Gilula, N. (1990). Modulation of gap junction transcript and protein expression during pregnancy in the rat. *J. Cell Biol.* 108, 595–605.
157. Yu, W., Dahl, G., and Werner, R. (1994). The connexin-43 gene is responsive to oestrogen. *Proc. R. Soc. Lond. B Biol. Sci.* 255, 125–132.
158. Laird, D., Puranam, K., and Revel, J.-P. (1991). Turnover and phosphorylation dynamics of connexin-43 gap junction protein in cultured cardiac myocytes. *Biochem. J.* 273, 67–72.
159. Piersanti, M., and Lye, S. (1995). Increase in messenger ribonucleic acid encoding the myometrial gap junction protein, connexin-43, requires protein synthesis and is associated with increased expression of the activator protein-1, c-fos. *Endocrinology* 136, 3571–3578.
160. Thorburn, G., and Challis, J. (1979). Endocrine control of parturition. *Physiol. Rev.* 59, 863–918.
161. Liggins, G. C. (1969). The foetal role in the initiation of parturition. In *Foetal Autonomy* (G. E. W. Wolstenholme and M. O'Connor, Eds.), Ciba Foundation Symposium, pp. 111–122. Churchill, London.
162. Liggins, G. C., Fairclough, R. J., Grieves, S. A., Kendall, J. Z., and Knox, B. S. (1973). The mechanism of the initiation of parturition in the ewe. *Recent Prog. Horm. Res.* 29, 111–128.
163. Currie, W. B., and Throburn, G. D. (1977). The fetal role in timing the initiation of parturition in the goat. *Ciba Found. Symp.* 47, 49–72.
164. Kitts, D. D., Anderson, G. B., BonDurant, R. H., and Stabenfeldt, G. H. (1984). Temporal patterns of D-4 C-21 steroids in coexisting, genetically dissimilar twin lamb fetuses throughout late gestation. *Endocrinology* 114, 703–710.
165. Kitts, D. D., Anderson, G. B., BonDurant, R. H., and Kindahl, H. (1985). Studies on the endocrinology of parturition: relative steroidogenesis in coexisting genetically dissimilar ovine fetuses concomitant with temporal patterns of maternal C18 and C19 steroids and prostaglandin $F_{2\alpha}$ release. *Biol. Reprod.* 33, 67–75.
166. Malpas, P. (1933). Postmaturity and malformations of the foetus. *J. Obstet. Gynaecol. Br. Emp.* 40, 1046–1050.
167. Swaab, D., Boer, K., and Honnebier, W. (1977). The influence of the fetal hypothalamus and pituitary on the onset and course of parturition in the fetus and birth. *Ciba Found. Symp.* 47, 379–400.
168. Novy, M. J., Walsh, S. W., and Kittinger, G. W. (1977). Experimental adrenal anencephaly in the rhesus monkey: effect on gestational length and fetal and maternal plasma steroids. *J. Clin. Endocrinol. Metab.* 45, 1031–1038.
169. Chez, R. A., Hutchinson, D. L., Salazar, H., and Mintz, D. H. (1970). Some effects of fetal and maternal hypophysectomy in pregnancy. *Am. J. Obstet. Gynecol.* 108, 643–650.
170. Mueller-Heubach, E., Myers, R. E., and Adamson, K. (1972). Effect of adrenalectomy on pregancy length in the rhesus monkey. *Am. J. Obstet. Gynecol.* 112, 221–226.
171. Nathanielsz, P., Figueroa, J. P., and Honnebier, M. B. (1992). In the rhesus monkey placental retention after fetectomy at 121 to 130 days gestation outlasts the normal duration of pregnancy. *Am. J. Obstet. Gynecol.* 166, 1529–1535.
172. Albrecht, E. D., Haskins, A. L., and Pepe, G. J. (1980). The influence of fetectomy at mid gestation upon the peripheral serum concentrations of progesterone, estrone and estradiol in baboons. *Endocrinology* 107, 766–770.
173. Albrecht, E. D., and Pepe, G. J. (1985). The placenta remains functional following fetectomy in baboons. *Endocrinology* 116, 843–845.
174. Lanman, J. T. (1977). Parturition in non-human primates. *Biol. Reprod.* 16, 28–38.
175. Lye, S., Mitchell, J., Nashman, N., et al. (2001). Role of mechanical signals in the onset of term and preterm labor. *Front. Horm. Res.* 27, 165–178.
176. Challis, J., Lye, S., and Gibb, W. (1997). Prostaglandins and parturition. *Ann. N Y Acad. Sci.* 828, 254–267.
177. Winterhager, E., Stutenkemper, R., Traub, O., Beyer, E. C., and Willecke, K. (1991). Expression of different connexin genes in rat uterus during decidualization and at term. *Eur. J. Cell Biol.* 55, 133–142.
178. Chow, L., and Lye, S. J. (1994). Expression of the gap junction protein, connexin-43, is increased in myometrium towards term and with the onset of labor. *Am. J. Obstet. Gynecol.* 170, 788–795.
179. Ou, C.-W., Orsino, A., and Lye, S. (1997). Expression of connexin-43 and connexin-26 in the rat myometrium during pregnancy and labor is differentially regulated by mechanical and hormonal signals. *Endocrinology* 138, 5398–5407.
180. Ou, C.-W., Chen, Z., Qi, S., and Lye, S. (2000). Expression and regulation of the messenger ribonucleic acid encoding the prostaglandin F2α receptor in the rat myometrium during pregnancy and labor. *Am. J. Obstet. Gynecol.* 182, 919–925.
181. Mitchell, J. A., Shynlova, O., Langille, B. L., and Lye, S. J. (2004). Mechanical stretch and progesterone differentially regulate activator protein-1 transcription factors in primary rat myometrial smooth muscle cells. *Am. J. Physiol. Endocrinol. Metab.* 287, 439–445.
182. Lye, S., Ou, C.-W., Teoh, T., et al. (1998). The molecular basis of labour and tocolysis. *Fetal Matern. Med. Rev.* 10, 121–136.
183. Shynlova, O., Mitchell, J. A., Tsampalieros, A., Langille, B. L., and Lye, S. J. (2004). Progesterone and gravidity differentially regulate expression of extracellular matrix components in the pregnant rat myometrium. *Biol. Reprod.* 70, 986–992.
184. MacPhee, D. J., and Lye, S. J. (2000). Focal adhesion signalling in the rat myometrium is abruptly terminated with the onset of labor. *Endocrinology* 141, 274–283.
185. Rose, J., Meis, P., and Morris, M. (1981). Ontogeny of endocrine (ACTH, vasopressin, cortisol) responses to hypotension in lamb fetuses. *Am. J. Physiol.* 240, E656–E661.
186. Challis, J., Lye, S., and Welsh, J. (1986). Ovine fetal adrenal maturation at term and during fetal ACTH administration: evidence that the modulating effect of cortisol may involve cAMP. *Can. J. Physiol. Pharmacol.* 64, 1085–1090.
187. MacIsaac, R., Bell, R., McDougall, J., et al. (1985). Development of the hypothalamic pituitary axis in the ovine fetus: ontogeny of action of ovine corticotropin-releasing factor. *J. Dev. Physiol.* 7, 329–338.
188. Norman, L., Lye, S., Wlodek, M., and Challis, J. R. (1985). Changes in pituitary responses to synthetic ovine corticotrophin-releasing factor in fetal sheep. *Can. J. Physiol. Pharmacol.* 63, 1398–1403.
189. Bassett, J. M., and Thorburn, G. D. (1969). Foetal plasma corticosteroids and the initiation of parturition in sheep. *J. Endocrinol.* 44, 285–286.
190. Magyar, D., Fridshal, D., Elsner, C., et al. (1980). Time trend analysis of plasma cortisol concentrations in the fetal sheep in relation to parturition. *Endocrinology* 197, 155–160.

191. Liggins, G. C. (1994). The role of cortisol in preparing the fetus for birth. *Reprod. Fertil. Dev.* 6, 141–150.
192. Matthews, S. G. (2001). Antenatal glucocorticoids and the developing brain. *Semin. Neonatol.* 6, 309–317.
193. Matthews, S. G. (2002). Early programming of the hypothalamo-pituitary-adrenal axis. *Trends Endocrinol. Metab.* 13, 373–380.
194. Matthews, S., and Challis, J. (1995). Regulation of CRH and AVP mRNA in the developing ovine hypothalamus effects of stress and glucocorticoids. *Am. J. Physiol.* 268, E1096–E1107.
195. Matthews, S., Han, X., Lu, F., and Challis, J. (1994). Developmental changes in the distribution of pro-opiomelanocortin and prolactin mRNA in the pituitary of the ovine fetus and lamb. *J. Mol. Endocrinol.* 13, 175–185.
196. Matthews, S., and Challis, J. (1996). Regulation of the hypothalamo-pituitary-adrenocortical axis in fetal sheep. *Trends Endocrinol. Metab.* 7, 239–246.
197. Merei, J. J., Rao, A., Clarke, I. J., and McMillen, I. C. (1996). Proopiomelanocortin, prolactin and growth hormone messenger ribonucleic acid levels in the fetal sheep pituitary during gestation. *Acta Endocrinol. (Copenh.)* 129, 263–267.
198. McMillen, I. C., Mercer, J. E., and Thorburn, G. D. (1988). Pro-opiomelanocortin mRNA levels fall in the fetal sheep before birth. *J. Mol. Endocrinol.* 1, 141–145.
199. Holloway, A., Gyomorey, S., and Challis, J. (2001). Effects of labor on pituitary expression of proopiomelanocortin, prohormone convertase (PC)-1 PC-2 and glucocorticoid receptor mRNA in fetal sheep. *Endocrine* 13, 17–23.
200. Lu, F., Yang, K., and Challis, J. (1994). Characteristics and developmental changes of corticotrophin-releasing hormone binding sites in the foetal sheep anterior pituitary. *J. Endocrinol.* 130, 223–229.
201. Lu, F., Yang, K., and Challis, J. (1994). Regulation of ovine fetal pituitary function by corticotrophin-releasing hormone, arginine vasopressin and cortisol in vitro. *J. Endocrinol.* 143, 199–208.
202. Green, J. L., Figueroa, J. P., Massman, G. A., Schwartz, J., and Rose, J. C. (2000). Corticotropin-releasing hormone type 1 receptor messenger ribonucleic acid and protein levels in the ovine fetal pituitary: ontogeny and effect of chronic cortisol administration. *Endocrinology* 141, 2870–2876.
203. Young, S. F., Tatter, S. B., Valego, N. K., et al. (2003). The role of hypothalamic input on corticotroph maturation in the fetal sheep. *Am. J. Physiol. Regul. Integr. Comp. Physiol.* 284, R1621–R1630.
204. Young, S. F., and Rose, J. C. (2002). Attenuation of corticotropin-releasing hormone and arginine vasopressin responsiveness during late-gestation pregancy in sheep. *Biol. Reprod.* 66, 1805–1812.
205. Brieu, V., Tonon, M. C., Lutz Bucher, B., and Durand, P. (1989). Corticotropin-releasing factor-like immunoreactivity, arginine vasopressin-like immunoreactivity and ACTH-releasing bioactivity in hypothalamic tissue from fetal and neonatal sheep. *Neuroendocrinology* 49, 164–168.
206. Matthews, S., and Challis, J. (1996). Regulation of the hypothalamo-pituitary-adrenocortical axis in fetal sheep. *Trends Endocrinol. Metab.* 7, 239–246.
207. Durand, P., Cathiard, A., Dacheux, F., Naaman, E., and Saez, J. (1986). In vitro stimulation and inhibition of adrenocorticotropin release by pituitary cells from ovine fetuses and lambs. *Endocrinology* 118, 1387–1394.
208. Carr, G. A., Jacobs, R. A., Young, I. R., et al. (1995). Development of adrenocorticotropin (1-39) and precursor peptide secretory responses in the fetal sheep during the last third of gestation. *Endocrinology* 136, 175–185.
209. Challis, J., Dilley, S., Robinson, J., and Thorburn, G. (1976). Prostaglandins in the circulation of the fetal lamb. *Prostaglandins* 11, 1041–1052.
210. Fowden, A., Harding, R., Ralph, M., and Thorburn, G. (1987). The nutritional regulation of plasma prostaglandin E concentrations in the fetus and pregnant ewe during late gestation. *J. Physiol. (Lond.)* 394, 1–12.
211. Louis, T., Challis, J., Robinson, J., and Thorburn, G. (1976). Rapid increase of fetal corticosteroids after prostaglandin E_2. *Nature* 264, 797–799.
212. McKeown, K., Challis, J., Small, C., et al. (2000). Altered fetal pituitary-adrenal function in the ovine fetus treated with RU486 and meloxicam, an inhibitor of prostaglandin synthase-II. *Biol. Reprod.* 63, 1899–1904.
213. Unno, N., Wu, W., Wong, C., et al. (1998). Prostaglandin regulation of fetal plasma adrenocorticotropin and cortisol concentrations in late-gestation sheep. *Biol. Reprod.* 58, 514–519.
214. Young, I., Deayton, J., Hollingsworth, S., and Thorburn, G. (1996). Continuous intrafetal infusion of prostaglandin E2 prematurely activates the hypothalamus-pituitary-adrenal axis and induces parturition in sheep. *Endocrinology* 137, 2424–2431.
215. Deauseault, D., Giroux, D., and Wood, C. E. (2000). Ontogeny of immunoreactive prostaglandin endoperoxide synthase isoforms in ovine fetal pituitary, hypothalamus and brainstem. *Neuroendocrinology* 71, 287–291.
216. Rainey, W., Naville, D., Cline, N., and Mason, J. (1991). Prostaglandin E2 is a positive regulator of adrenocorticotropin receptors, 3β-hydroxysteroid dehydrogenase, and 17α-hydroxylase expression in bovine adrenocortical cells. *Endocrinology* 129, 1333–1339.
217. Challis, J., Nancekievill, E., and Lye, S. (1985). Possible role of cortisol in the stimulation of cortisol binding capacity in the plasma of fetal sheep. *Endocrinology* 116, 1139–1144.
218. Berdusco, E. T., Hammond, G. L., Jacobs, R., et al. (1993). Glucocorticoid-induced increase in plasma corticosteroid binding globulin levels in fetal sheep is associated with increased biosynthesis and alterations in glycosylation. *Endocrinology* 132, 2001–2008.
219. Ballard, P. L., Kitterman, J. A., Bland, R. D., et al. (1982). Ontogeny and regulation of corticosteroid binding globulin capacity in plasma of fetal and newborn lambs. *Endocrinology* 110, 359–366.
220. Fairclough, R. J., and Liggins, G. C. (1975). Protein binding of plasma cortisol in the foetal lamb near term. *J. Endocrinol.* 67, 333–341.
221. Challis, J. R. G., Berdusco, E. T., Jeffray, T. M., Yang, K., and Hammond, G. I. (1995). Corticosteroid-binding globulin (CBG) in fetal development. *J. Steroid Biochem. Mol. Biol.* 53, 523–527.
222. Yang, K., Matthews, S., and Challis, J. (1995). Developmental and glucocorticoid regulation of pituitary 11-hydroxysteroid dehydrogenase 1 gene expression in the ovine fetus and lamb. *J. Mol. Endocrinol.* 14, 109–116.
223. Yang, K., Hammond, G., and Challis, J. (1992). Characterization of an ovine glucocorticoid receptor cDNA and developmental changes in its mRNA levels in the fetal sheep hypothalamus, pituitary and adrenal gland. *J. Mol. Endocrinol.* 8, 173–180.
224. Matthews, S., Yang, K., and Challis, J. (1995). Changes in glucocorticoid receptor mRNA in the developing ovine pituitary and the effects of exogenous cortisol. *J. Endocrinol.* 144, 483–490.
225. Holloway, A., Gyomorey, S., and Challis, J. (2000). Effects of labor on pituitary expression of proopiomelanocortin, prohormone convertase (PC)-1 PC-2 and glucocorticoid receptor mRNA in fetal sheep. *Endocrine* 13, 17–23.

226. McDonald, T. J., Hoffmann, G. E., Myers, D. A., and Nathanielsz, P. W. (1990). Hypothalamic glucocorticoid implants prevent ovine adrenocorticotropin secretion in response to stress. *Endocrinology* 127, 2862–2868.
227. Purniton, S., Newman, H., Castro, M., and Wood, C. (1999). Ontogeny of estrogen sulfatase activity in ovine fetal hypothalamus, hippocampus and brainstem. *Am. J. Physiol.* 276, R1647–R1652.
228. Wood, C., and Saoud, C. (1997). Influence of estradiol and androstenedione on ACTH and cortisol secretion in the ovine fetus. *J. Soc. Gynecol. Investig.* 4, 279–283.
229. Holloway, A. C., Howe, D. C., Chan, G., et al. (2002). Urocortin: mechanism for the sustained activation of the HPA axis in the late-gestation ovine fetus? *Am. J. Physiol. Endocrinol. Metab.* 283, 165–171.
230. Challis, J. R. G., Manchester, E. L., Mitchell, B. F., and Patrick, J. E. (1982). Activation of adrenal function in fetal sheep by infusion of adrenocorticotropin (ACTH) to the fetus in utero. *Biol. Reprod.* 27, 1026–1032.
231. Hennessy, D. P., Coghlan, J. P., Hardy, K. J., Scoggins, B. A., and Wintour, E. M. (1982). The origin of cortisol in the blood of fetal sheep. *J. Endocrinol.* 95, 71–76.
232. Glickman, J., and Challis, J. (1980). The changing response pattern of sheep fetal adrenal cells throughout the course of gestation. *Endocrinology* 106, 1371–1376.
233. Challis, J., Lye, S. J., Mitchell, B., et al. (1985). Fetal signals for birth. In *Physiological Development of the Fetus and Newborn* (C. Jones and P. W. Nathanielsz, Eds.), pp. 363–370. Academic Press, London.
234. Rose, J., Meis, P., Urban, R., and Greiss, J. F. (1982). In vivo evidence for increased adrenal sensitivity to adrenocorticotrophin-(1-24) in the lamb fetus in late gestation. *Endocrinology* 111, 80–85.
235. Wintour, E., Brown, E., Denton, D., et al. (1975). The ontogeny and regulation of corticosteroid secretion by the ovine foetal adrenal. *Acta Endocrinol. (Copenh.)* 79, 301–316.
236. Manchester, E. L., and Challlis, J. R. G. (1982). The effects of adrenocorticotropin, guanylimidodiphosphate, dibutyryl adenosine 3′,5′-monophosphate and exogenous substrates on corticosteroid output by ovine fetal adrenal cells at different times in pregnancy. *Endocrinology* 111, 889–895.
237. Tangalakis, K., Coglan, J. P., Connell, J., et al. (1989). Tissue distribution and levels of gene expression of three steroid hydroxylases in ovine fetal adrenal glands. *Acta Endocrinol. (Copenh.)* 120, 225–232.
238. Conley, A., and Bird, I. (1997). The role of cytochrome P450 17α-hydroxylase and 3β-hydroxysteroid dehydrogenase in the integration of gonadal and adrenal steroidogenesis via the Δ5 and Δ4 pathways of steroidogenesis in mammals. *Biol. Reprod.* 56, 789–799.
239. Tangalakis, K., Coghlan, J. P., Crawford, R., and Hammond, R. E. (1990). Steroid hydroxylase gene expression in the ovine fetal adrenal gland following ACTH infusion. *Acta Endocrinol. (Copenh.)* 123, 371–377.
240. Riley, S. C., Boshier, D. P., Labrie, F., and Challis, J. R. G. (1992). Immunohistochemical localization of 3β-hydroxysteroid dehydrogenase Δ 5-4 isomerase, tyrosine hydroxylase and phenylethanolamine N-methyl transferase in the adrenal glands of sheep fetuses throughout gestation. *J. Reprod. Fertil.* 96, 127–134.
241. Han, V. K., Lu, F., Bassett, N., et al. (1992). Insulin-like growth factor-II (IGF-II) mRNA is expressed in steroidogenic cells of the developing ovine adrenal: evidence for an autocrine/paracrine role of IGF-II. *Endocrinology* 131, 3100–3109.
242. Fraser, M., Braems, G. A., and Challlis, J. R. (2001). Developmental regulation of corticotropin receptor gene expression in the adrenal gland of the ovine fetus and newborn lamb: effects of hypoxia during late pregnancy. *J. Endocrinol.* 169, 1–10.
243. Wang, J. J., Valego, N. K., Su, Y., and Rose, J. C. (2004). Developmental aspects of ovine adrenal. *J. Soc. Gynecol. Investig.* 11, 27–35.
244. Wintour, E., Coghlan, J., Hardy, K., et al. (1980). Adrenal corticosteroids and immunoreactive ACTH in chronically cannulated ovine fetuses with bilateral adrenalectomy. *Acta Endocrinol. (Copenh.)* 95, 546–552.
245. Holta, M., and Baird, A. (1986). Differential effects of transforming growth factor type β on the growth and function of adrenocortical cells in vitro. *Proc. Natl. Acad. Sci. U S A* 83, 7795–7799.
246. Naaman, E., Chatelain, P., Saez, J., and Durand, P. (1989). In vitro effect of insulin and insulin-like growth factor-I on cell multiplication and adrenocorticotropin responsiveness of fetal adrenal cells. *Biol. Reprod.* 40, 570–577.
247. Penhoat, A., Jaillard, C., and Saez, J. (1989). Synergistic effects of corticotropin and insulin-like growth factor I on corticotropin receptors and corticotropin responsiveness in cultured bovine adrenocortical cells. *Biochem. Biophys. Res. Commun.* 165, 355–359.
248. Durand, P., Cathiard, A., and Saez, J. (1985). Involvement of the regulatory protein N_S in the maturation of ACTH-sensitive adenylate cyclase of ovine fetal adrenal during late gestation. *Mol. Cell. Endocrinol.* 39, 145–150.
249. Boshier, D. P., Holloway, H., and Liggins, G. C. (1981). Effects of cortisol and ACTH on adrenocortical growth and cytodifferentiation in the hypophysectomized sheep. *J. Dev. Physiol.* 3, 355–373.
250. Wintour, E. M. (1984). Developmental aspects of hypothalamic-pituitary adrenal axis. *J. Dev. Physiol.* 6, 291–299.
251. Durand, P., Cathiard, A. M., Morera, A.-M., Dazord, A., and Saez, J. M. (1984). In vitro maturation of steroidogenic capacity of ovine fetal and neonatal adrenal cells. *Endocrinology* 114, 1128–1134.
252. Durand, P., Cathiard, A. M., and Saez, J. M. (1982). In vitro maturation of ovine fetal adrenal cells adenylate cyclase: corticotropin-dependent and independent development of the response to corticotropin. *Biochem. Biophys. Res. Commun.* 106, 8–15.
253. Durand, P., Cathiard, A., Morera, A., Dazord, A., and Saez, J. (1981). Maturation of adrenocorticotropin-sensitive adenylate cyclase of ovine fetal adrenal during late pregnancy. *Endocrinology* 108, 2114–2119.
254. Durand, P., Locatelli, A., Cathiard, A., Dazord, A., and Saez, J. (1981). ACTH induction of the maturation of ACTH-sensitive adenylate cyclase system in the ovine fetal adrenal. *J. Steroid Biochem.* 15, 445–448.
255. Challis, J., Huhtanen, D., Sprague, C., Mitchell, B., and Lye, S. (1985). Modulation by cortisol of adrenocorticotropin-induced activation of adrenal function in fetal sheep. *Endocrinology* 116, 2267–2272.
256. Conley, A., and Bird, I. (1997). The role of cytochrome P450 17α-hydroxylase and 3β–hydroxysteroid dehydrogenase in the integration of gonadal and adrenal steroidogenesis via the delta 5 and delta 4 pathways of steroidogenesis in mammals. *Biol. Reprod.* 56, 789–799.
257. Durand, P., Cathiard, A., Locatelli, A., and Saez, J. (1982). Modifications of the steroidogenic pathway during spontaneous and adrenocorticotropin-induced maturation of ovine fetal adrenal. *Endocrinology* 110, 500–505.
258. Lye, S., Sprague, C., Mitchell, B., and Challis, J. (1983). Activation of ovine fetal adrenal function by pulsatile or continuous administration of ACTH 1-24: I. Effects on fetal plasma corticosteroids. *Endocrinology* 113, 770–782.

259. Manchester, E., Lye, S., and Challis, J. (1983). Activation of ovine fetal adrenal function by pulsatile or continuous administration of adrenocorticotropin-(1-24): II. Effects on adrenal cell responses in vitro. *Endocrinology* 113, 777–782.
260. Rainey, W., Oka, K., Magness, R., and Mason, J. (1991). Ovine fetal adrenal synthesis of cortisol: regulation by adrenocorticotropin, angiotensin II and transforming growth factor-B. *Endocrinology* 129, 1784–1790.
261. Jacobs, R., Young, I., Hollingworth, S., and Thorburn, G. (1994). Chronic administration of low doses of adrenocorticotropin to hypophysectomized fetal sheep leads to normal term labor. *Endocrinology* 134, 1389–1394.
262. Challis, J., Sloboda, D., Matthews, S., et al. (2000). Fetal hypothalamic-pituitary adrenal (HPA) development and activation as a determinant of the timing of birth, and of postnatal disease. *Endocr. Res.* 26, 489–504.
263. Challis, J. R., Sloboda, D., Matthews, S. G., et al. (2001). The fetal placental hypothalamic-pituitary-adrenal (HPA) axis, parturition and post natal health. *Mol. Cell. Endocrinol.* 185, 135–144.
264. Stark, R., Daniel, S., Husain, M., Zubrow, A., and James, L. (1984). Effects of hypoxia on vasopressin concentrations in cerebrospinal fluid and plasma of sheep. *Neuroscience* 38, 453–460.
265. Stark, R., Daniel, S., Husain, M., Tropper, P., and James, L. (1985). Cerebrospinal fluid and plasma vasopressin in the fetal lamb: basal concentration and the effect of hypoxia. *Endocrinology* 116, 65–72.
266. Ozolins, I., Young, I., and McMillen, I. (1992). Surgical disconnection of the hypothalamus from the fetal pituitary abolishes the corticotrophic response to intrauterine hypoglycemia or hypoxemia in the sheep during late gestation. *Endocrinology* 130, 2438–2445.
267. Akagi, K., Berdusco, E. T., and Challis, J. R. G. (1990). Cerebrospinal fluid and plasma vasopressin in the fetal lamb: basal concentration and the effect of hypoxia. *J. Dev. Physiol.* 14, 319–324.
268. Matthews, S. G., and Challis, J. R. G. (1995). Levels of pro-opiomelanocortin and prolactin mRNA in the fetal sheep pituitary following hypoxemia and glucocorticoid treatment in late gestation. *J. Endocrinol.* 147, 139–146.
269. Braems, G. A., Matthews, S. G., and Challis, J. R. G. (1996). Differential regulation of pro-opiomelanocortin mRNA in the fetal pituitary pars distalis and pars intermedia following 48 hours of hypoxia in late gestation. *Endocrinology* 137, 2731–2738.
270. Matthews, S. G., Fraser, M., and Challis, J. R. G. (1996). Dopaminergic regulation of pituitary function in the late gestation fetal sheep. *J. Endocrinol.* 150, 187–194.
271. Murotsuki, J., Challis, J. R. G., and Gagnon, R. (1995). Increased fetal plasma prostaglandin E2 concentrations during fetal embolization in pregnant sheep. *Am. J. Obstet. Gynecol.* 173, 30–35.
272. Bloomfiled, F., Oliver, M., Hawkins, P., et al. (2003). A periconceptional nutritional origin of noninfectious preterm birth. *Science* 300, 606.
273. Flint, A. P. F., Anderson, A. B. M., Steele, P. A., and Turnbull, A. C. (1975). The mechanism by which foetal cortisol controls the onset of parturition in the sheep. *Biochem. Soc. Trans.* 3, 1189–1194.
274. Mason, J. I., France, J. T., Magness, R. R., Murry, B. A., and Rosenfeld, C. R. (1989). Ovine placental steroid 17α-hydroxylase/C-17,20-lyase, aromatase and sulfatase in dexamethasone-induced and natural parturition. *J. Endocrinol.* 122, 351–356.
275. Steele, P. A., Flint, A. P. P., and Turnbull, A. C. (1976). Activity of steroid C-17,20-lyase in the ovine placenta: effect of exposure of foetal glucocorticoid. *Endocrinology* 69, 239–246.
276. Power, S. G. A., Patrick, J. E., Carson, G. D., and Challis, J. R. G. (1982). The fetal membranes as a possible source of progesterone in the amniotic and allantoic fluids of pregnant sheep. *Endocrinology* 110, 481–486.
277. Challis, J. R. G. (1971). Sharp increase in free circulating oestrogen immediately before parturition in sheep. *Nature* 229, 208.
278. Challis, J. R. G., and Patrick, J. E. (1981). Fetal and maternal estrogen concentrations throughout pregnancy in the sheep. *Can. J. Physiol. Pharmacol.* 59, 970–978.
279. Jenkin, G., and Thorburn, G. D. (1985). Inhibition of progesterone secretion by a 3β-hydrosteroid dehydrogenase inhibitor in late pregnant sheep. *Can. J. Physiol. Pharmacol.* 63, 136–142.
280. Steele, P. A., Flint, A. P. P., and Turnbull, A. C. (1976). Increased utero-ovarian androstenedione production before parturition in sheep. *J. Reprod. Fertil.* 46, 443–445.
281. Ma, X. H., Wu, W. X., and Nathanielsz, P. W. (1999). Differential effects of natural and synthetic glucocorticoids on cytochrome 17α-hydroxylase (P45017α) and cytochrome P450 side-chain cleavage (P450scc) messenger ribonucleic acid in the sheep placenta. *Am. J. Obstet. Gynecol.* 180, 1215–1221.
282. Wu, W. X., Owiny, J., Zhang, Q., Ma, X. H., and Nathanielsz, P. W. (1996). Regulation of the estrogen receptor and its messenger ribonucleic acid in the ovariectomized sheep myometrium and endometrium: the role of estradiol and progesterone. *Biol. Reprod.* 55, 762–768.
283. Wu, W. X., Ma, X. H., and Nathanielsz, P. W. (1999). Tissue-specific ontogenic expression of prostaglandin H synthase 2 in the ovine myometrium, endometrium, and placenta during late gestation and at spontaneous term labor. *Obstet. Gynecol.* 181, 1512–1519.
284. Wu, W. X., Ma, X. H., Zhang, Q., Nathanielsz, P. W., and Buchwalder, L. (1997). Regulation of prostaglandin endoperoxide H synthase 1 and 2 by estradiol and progesterone in nonpregnant ovine myometrium and endometrium in vivo. *Endocrinology* 138, 4005–4012.
285. Wu, W. X., Ma, X. H., and Nathanielsz, P. W. (1999). Changes in prostacyclin synthase in pregnant sheep myometrium, endometrium, and placenta at spontaneous term labor and regulation by estradiol and progesterone. *Am. J. Obstet. Gynecol.* 180, 744–749.
286. Wu, W. X., Unno, N., Ma, X. H., and Nathanielsz, P. W. (1998). Inhibition of prostaglandin produced by nimesilide is accompanied by changes in expression of the cassette of uterine labor-related genes in pregnant sheep. *Endocrinology* 139, 3096–3103.
287. Evans, C. A., Kennedy, T. G., Patrick, J. E., and Challis, J. R. G. (1982). The effects of indomethacin on uterine activity and prostaglandin (PG) concentrations during labor induced by administering ACTH to fetal sheep. *Can. J. Physiol. Pharmacol.* 60, 1200–1209.
288. Evans, C. A., Kennedy, T. G., and Challis, J. R. G. (1982). Gestational changes in prostanoid concentration in intrauterine tissues and fetal fluids from pregnant sheep, and the relation to prostanoid output in vitro. *Biol. Reprod.* 127, 1–11.
289. Wimsatt, J., Nathanielsz, P. W., and Sirois, J. (1993). Induction of prostaglandin endoperoxide synthase isoform-2 in ovine cotyledonary tissues during late gestation. *Endocrinology* 133, 1068–1073.
290. Liggins, G. C., and Grieves, S. A. (1971). Possible role for prostaglandin F2 in parturition in sheep. *Nature* 232, 629–631.
291. Davies, I. J., Ryan, K. J., and Petro, Z. (1970). Estrogen synthesis by adrenal-placental tissues of the sheep and the rhesus monkey in vitro. *Endocrinology* 86, 1457–1459.

292. Yu, H. K., Cabalum, T., Jansen, C. A. M., Buster, J. E., and Nathanielsz, P. W. (1983). Androstenedione, testosterone, and estradiol concentrations in fetal and maternal plasma in late pregnant sheep. *Endocrinology* 113, 2216–2220.
293. Mitchell, B. F., Lye, S. J., Lukash, L., and Challis, J. R. G. (1986). Androstenedione metabolism in the late gestation sheep fetus. *Endocrinology* 118, 63–68.
294. Wimsatt, J., and Nathanielsz, P. (1995). Prostaglandin H synthase activity in the sheep placenta during cortisol-induced labor at 128–131 days of gestation and during spontaneous delivery at term. *Biol. Reprod.* 53, 53–58.
295. Boshier, D. P., Jacobs, R. A., Han, V. K., et al. (1991). Immunohistochemical localization of prostaglandin H synthase in the sheep placenta from early pregnancy to term. *Biol. Reprod.* 45, 322–327.
296. Gibb, W., Matthews, S., and Challis, J. (1996). Localization and developmental changes in prostaglandin H synthase (PGHS) and PGHS messenger ribonucleic acid in ovine placenta throughout gestation. *Biol. Reprod.* 54, 654–659.
297. Langlois, D. A., Fraher, I. J., Khalil, M. W., Fraser, M., and Challis, J. R. G. (1993). Preferential increase in cyclooxygenase compared to lipoxygenase activity in sheep placenta and amnion at term pregnancy and after intrafetal glucocorticoid administration. *J. Endocrinol.* 139, 195–204.
298. Rice, G. E., Wong, M. H., and Thorburn, G. D. (1988). Gestational changes in prostaglandin synthase activity in ovine cotyledonary microsomes. *J. Endocrinol.* 139, 195–204.
299. Rice, G. E., Payne, M. J., Wong, M. H., and Thorburn, G. D. (1988). Immunoreactive prostaglandin G/H synthase content increases in ovine cotyledons during late gestation. *Placenta* 13, 429–437.
300. Rice, G. E., Freed, K. A., Aitken, M. A., and Jacobs, R. A. (1995). Gestational- and labour-associated changes in the relative abundance of prostaglandin G/H synthase-1 and -2 mRNA in ovine placenta. *J. Mol. Endocrinol.* 14, 237–245.
301. Martin, R., Whittle, W., Holloway, A., et al. (2002). Ontogeny and regulation of ovine placental prostaglandin E2 synthase. *Biol. Reprod.* 67, 868–873.
302. Gyomorey, S., Gupta, S., Lye, S., J., Gibb, W., and Challis, J. R. G. (2000). Temporal expression of PGHS-2 and P450C17 in ovine placentomes with the natural onset of labor. *Placenta* 21, 478–486.
303. Gyomorey, S., Lye, S., Gibb, W., and Challis, J. (2000). Fetal to maternal progression of prostaglandin H2 synthase-2 expression in ovine intrauterine tissues during the course of labor. *Biol. Reprod.* 62, 797–805.
304. Wu, W. X., Ma, X. H., Unno, N., and Nathanielsz, P. W. (2001). In vivo evidence for stimulation of placental, myometrial, and endometrial prostaglandin G/H synthase 2 by fetal cortisol replacement after fetal adrenalectomy. *Endocrinology* 142, 3857–3864.
305. Whittle, W., Holloway, A., Lye, S., Gibb, W., and Challis, J. (2000). Prostaglandin production at the onset of ovine parturition is regulated by both estrogen-independent and estrogen dependent pathways. *Endocrinology* 141, 3783–3790.
306. Wu, W. X., Ma, X. H., Yoshizato, T., Shimozuko, N., and Nathanielsz, P. W. (2001). Increase in prostaglandin H synthase 2, but not prostaglandin F2α synthase mRNA in intrauterine tissues during betamethasone-induced premature labor and spontaneous term labor in sheep. *J. Soc. Gynecol. Investig.* 8, 69–76.
307. Brooks, A. N., and Gibson, F. (1992). Prostaglandin E2 enhances AVP-stimulated but not CRF-stimulated ACTH secretion from cultured fetal pituitary cells. *Endocrinology* 132, 33–38.
308. Hollingworth, S. A., Deayton, J. M., Young, I. R., and Thorburn, G. D. (1995). Prostaglandin E2 administered to fetal sheep increases the plasma concentration of adrenocorticotropin (ACTH) and the proportion of ACTH in low molecular weight forms. *Endocrinology* 136, 1233–1240.
309. Liggins, G. C., Scroop, G. C., and Haughey, K. G. (1982). Comparison of the effects of prostaglandin E2, prostacyclin and 1-24 adrenocorticotrophin on plasma cortisol levels in fetal sheep. *Endocrinology* 95, 153–162.
310. Wu, W., Ma, X., Coksaygan, T., et al. (2004). Prostaglandin mediates premature delivery in pregnant sheep induced by estradiol at 121 days of gestational age. *Endocrinology* 145, 1444–1452.
311. Leung, S. T., Wathes, D. C., Young, I. R., and Jensen, C. A. M. (1999). Effect of labor induction on the expression of oxytocin receptor, cytochrome P450 aromatase, and estradiol receptor in the reproductive tract of the late-pregnant ewe. *Biol. Reprod.* 60, 814–820.
312. Gupta, S., Gyomorey, S., Lye, S., Gibb, W., and Challis, J. (2003). Effect of labor on glucocorticoid receptor (GR total and GRβ) proteins in ovine intrauterine tissues. *J. Soc. Gynecol. Investig.* 10, 136–144.
313. Pepe, G., and Albrecht, E. (1995). Actions of placental and fetal adrenal steroid hormones in primate pregnancy. *Endocr. Rev.* 16, 608–648.
314. Pepe, G., and Albrecht, E. (1990). Regulation of the primate fetal adrenal cortex. *Endocr. Rev.* 11, 151–176.
315. Mesiano, S., and Jaffe, R. (1997). Developmental and functional biology of the primate fetal adrenal cortex. *Endocr. Rev.* 18, 378–403.
316. Ackland, J. F., Ratter, S. J., Bourne, G. L., and Rees, L. H. (1986). Corticotrophin-releasing factor-like immunoreactivity and bioactivity of human fetal and adult hypothalami. *Endocrinology* 108, 171–180.
317. Berghorn, K. A., Albrecht, E. D., and Pepe, G. J. (1991). Responsivity of the baboon fetal pituitary to corticotropin-releasing hormone in utero at mid-gestation. *Endocrinology* 129, 1424–1428.
318. Pepe, G., Davies, W., and Albrecht, E. (1994). Activation of the baboon fetal pituitary-adrenocortical axis at midgestation by estrogen: enhancement of fetal pituitary proopiomelanocortin messenger ribonucleic acid expression. *Endocrinology* 135, 2581–2587.
319. Pepe, G. J., Babischkin, J. S., Burch, M. G., Leavitt, M. G., and Albrecht, E. D. (1996). Developmental increase in expression of the messenger ribonucleic acid and protein levels of 11β-hydroxysteroid dehydrogenase types 1 and 2 in the baboon placenta. *Endocrinology* 137, 5678–5684.
320. Jaffe, R. B., Seron, F. M., and Mitchell, B. F. (1979). Perinatal regulation of cortisol in the primate. *J. Steroid Biochem.* 11, 549–555.
321. Voutilainen, R., and Miller, W. L. (1987). Coordinate trophic hormone regulation of mRNAs for insulin-like growth factor II and the cholesterol side-chain cleavage enzyme, P450scc in human steroidogenic tissues. *Proc. Natl. Acad. Sci. U S A* 84, 1590–2594.
322. Ma, X. H., Wu, W. X., and Nathanielsz, P. W. (2003). Gestation-related and betamethasone-induced changes in 11 β-hydroxysteroid dehydrogenase types 1 and 2 in the baboon placenta. *Am. J. Obstet. Gynecol.* 188, 13–21.
323. Albrecht, E. D., Henson, M. C., Walker, M. L., and Pepe, G. J. (1990). Modulation of adrenocorticotropin-stimulated baboon fetal adrenal dehydroepiandrosterone formation in vitro by estrogen at mid and late gestation. *Endocrinology* 126, 3083–3088.
324. Pepe, G. J., and Albrecht, E. D. (1985). Regulation of baboon fetal adrenal androgen production by adrenocorticotropic hormone, prolactin and growth hormone. *Biol. Reprod.* 33, 545–550.

325. Pepe, G. J., and Albrecht, E. D. (1985). Prolactin stimulates adrenal androgen secretion in intact baboons. *Endocrinology* 33, 545–550.
326. Aberdeen, G. W., Babischkin, J. S., Davies, W. A., Pepe, G. J., and Albrecht, E. D. (1997). Decline in adrenocorticotropin receptor messenger ribonucleic acid expression in the baboon fetal adrenocortical zone in the second half of pregnancy. *Endocrinology* 138, 1634–1641.
327. Albrecht, E. D., Aberdeen, G. W., Babischkin, J. S., Tilly, J. L., and Pepe, G. J. (1996). Biphasic developmental expression of adrenocorticotropin receptor messenger ribonucleic acid levels in the baboon fetal adrenal gland. *Endocrinology* 137, 1292–1298.
328. Pepe, G. J., Waddel, B. J., and Albrecht, E. D. (1989). Effect of estrogen on pituitary peptide-induced dehydroepiandrosterone secretion in the baboon fetus at mid gestation. *Endocrinology* 125, 1519–1524.
329. Albrecht, E. D., Babischkin, J. S., Davies, W. A., Leavitt, M. G., and Pepe, G. J. (1999). Identification and developmental expression of the estrogen receptor α and β in the baboon fetal adrenal gland. *Endocrinology* 140, 5953–5961.
330. Simpson, E. R., Parker, C. J., and Carr, B. (1985). Role of lipoproteins in the regulation of steroidogenesis by the human fetal adrenal. In *The Endocrine Physiology of Pregnancy and the Peripartal Period* (R. Jaffe and S. Dell'acqua, Eds.), pp. 7–17. Raven Press, New York.
331. Ryan, K. J. (1969). Theoretical basis for endocrine control of gestation: a comparative approach. In *Feto-Placental Unit* (A. Pecile and C. Finzi, Eds.), pp. 120–132. Excerpta Medica, Amsterdam.
332. Novy, M. J. (1977). Endocrine and pharmacological factors which influence the onset of labour in rhesus monkeys. *Ciba Found. Symp.* 47, 259–295.
333. Jaffe, R., Mesiano, S., Smith, R., et al. (1998). The regulation and role of fetal adrenal development in human pregnancy. *Endocr. Res.* 24, 19–26.
334. Ratcliffe, J., Nakanishi, M., and Jaffe, R. (2003). Identification of definitive and fetal zone markers in the human fetal adrenal gland reveals putative developmental genes. *J. Clin. Endocrinol. Metab.* 88, 3272–3277.
335. Coulter, C., and Jaffe, R. (1998). Functional maturation of the primate fetal adrenal in vivo: 3. Specific zonal localization and developmental regulation of CYP21A2 (P450c21) and CYP11B1/CYP11B2 (P450c11/aldosterone synthase) lead to integrated concept of zonal and temporal steroid biosynthesis. *Endocrinology* 139, 144–150.
336. Smith, R., Mesiano, S., Chan, E.-C., Brown, S., and Jaffe, R. (1998). Corticotropin-releasing hormone directly and preferentially stimulates dehydroepiandrosterone sulfate secretion by human fetal adrenal cortical cells. *J. Clin. Endocrinol. Metab.* 83, 2916–2920.
337. Parker, C. J., Stankovic, A., and Goland, R. (1999). Corticotropin-releasing hormone stimulates steroidogenesis in cultured human adrenal cells. *Mol. Cell. Endocrinol.* 155, 19–25.
338. Chakravorty, A., Mesiano, S., and Jaffe, R. (1999). Corticotropin-releasing hormone stimulates P450 17α-hydroxylase/17,20-lyase in human adrenal cells via protein kinase C. *J. Clin. Endocrinol. Metab.* 84, 3732–3738.
339. Yoon, B., Romero, R., Jun, J., et al. (1998). An increase in fetal plasma cortisol but not dehydroepiandrosterone sulfate is followed by the onset of preterm labor in patients with preterm premature rupture of membranes. *Am. J. Obstet. Gynecol.* 179, 1107–1114.
340. Cunningham, F. G., MacDonald, P. C., and Gant, N. F. (1989). The placental hormones. In *Williams Obstetrics* (18th ed.), pp. 67–85. Appleton and Lange, Norwalk, CT.
341. Tulchinsky, D., Hobel, C. J., Yeager, E., and Marshall, J. R. (1972). Plasma estrone, estradiol, progesterone and 17-hydroxy-progesterone in human pregnancy: I. Normal pregnancy. *Am. J. Obstet. Gynecol.* 112, 1195–1199.
342. Chibbar, R., Hobkirk, R., and Mitchell, B. F. (1986). Sulfahydrolase activity for estrone sulfate and dehydroepiandrosterone sulfate in human fetal membranes and decidua around the time of parturition. *J. Clin. Endocrinol. Metab.* 62, 90.
343. Gant, N. F., Milewich, L., Calvert, M. E., and MacDonald, P. C. (1977). Steroid sulfatase activity in human fetal membranes. *J. Clin. Endocrinol. Metab.* 45, 965–972.
344. Milewich, L., and Garcia, R. L. (1986). Steroid sulfatase activity in amnion tissue and amnion calls maintained in monolayer culture. *J. Clin. Endocrinol. Metab.* 61, 812.
345. Mitchell, B. F., Cross, J., Hobkirk, R., and Challis, J. (1984). Formation of unconjugated estrogens from estrone sulfate by dispersed cells from human fetal membranes and decidua. *J. Clin. Endocrinol. Metab.* 58, 845.
346. Gibb, W., Lavoie, J. C., and Roux, J. F. (1978). In vitro conversion of pregnenolone to progesterone by term human fetal membranes. *Am. J. Obstet. Gynecol.* 136, 631–634.
347. Gibb, W., Lavoie, J. C., and Roux, J. F. (1978). 3β-Hydroxysteroid dehydrogenase activity in human fetal membranes. *Steroids* 32, 365–372.
348. Mitchell, B. F., and Powell, W. A. (1984). Progesterone production by human fetal membranes: an in vitro incubation system for studying hormone production and metabolism. *Am. J. Obstet. Gynecol.* 148, 303–308.
349. Mitchell, B. F., Cruickshank, B., McLean, D., and Challis, J. (1982). Local modulation of progesterone production in human fetal membranes. *J. Clin. Endocrinol. Metab.* 55, 1237–1239.
350. Jirasek, J. E., Sulcova, J., Capkova, A., Rohling, S., and Starka, L. (1969). Histochemical and biochemical investigation of 3β-hydroxy-D5-steroid dehydrogenase in the chorion, adrenals and gonads of human foetuses. *Endokrinologie* 54, 173–183.
351. Milewich, L., Gant, N. F., Schwarz, B. E., et al. (1977). Initiation of human parturition: VI. Identification and quantification of progesterone metabolites produced by the components of human fetal membranes. *J. Clin. Endocrinol. Metab.* 45, 400–411.
352. Lopez Bernal, A., Anderson, A. B. M., and Turnbull, A. C. (1982). Cortisol: cortisone interconversion by human decidua in relation to parturition: effect of tissue manipulation on 11β-hydroxysteroid dehydrogenase activity. *J. Endocrinol.* 93, 141–149.
353. Milewich, L., Gant, N. F., Schwarz, B. E., Chen, G. T., and MacDonald, P. C. (1977). Initiation of human parturition: VIII. Metabolism of progesterone by fetal membranes of early and late human gestation. *Obstet. Gynecol.* 50, 45–48.
354. Murphy, B. E. P. (1977). Chorionic membrane as an extra-adrenal source of foetal cortisol in human amniotic fluid. *Nature* 266, 179–181.
355. Gunasegaram, R., Loganath, A., Peh, K. L., et al. (1983). Evidence for cleavage of (26-^{14}C) cholesterol side-chain by human foetal membranes in vitro. *Acta Endocrinol. (Copenh.)* 102, 288–291.
356. Gibb, W., and Lavoie, J. C. (1981). Steroid aromatization by term human fetal membranes. *J. Steroid Biochem.* 14, 401–403.
357. Loganath, A., Peh, K. L., Wong, Y. C., and Ng, S. C. (2001). Enhanced dehydroepiandrosterone synthesis by amnion compared to chorion: a comparative study using the reverse-isotope dilution technique. *Mol. Reprod. Dev.* 58, 276–280.

358. Alfaidy, N., Xiong, Z., Myatt, L., et al. (2001). Prostaglandin F2α potentiates cortisol production by stimulating 11β-hydroxysteroid dehydrogenase 1: a novel feedback loop that may contribute to human labour. *J. Clin. Endocrinol. Metab.* 86, 5585–5592.
359. Alfaidy, N., Li, W., MacIntosh, T., Yang, K., and Challis, J. R. (2003). Late gestation increase in 11-β-hydroxysteroid dehydrogenase 1 expression in human fetal membranes: a novel intrauterine source of cortisol. *J. Clin. Endocrinol. Metab.* 88, 5033–5038.
360. Ledger, W. L., Webster, M. A., Anderson, A. B. M., and Turnbull, A. C. (1985). Effect of inhibition of prostaglandin synthesis on cervical softening and uterine activity during parturition resulting from progesterone withdrawal induced by epostane. *J. Endocrinol.* 105, 227–231.
361. Jenkin, G., and Thorburn, G. D. (1985). Inhibition of progesterone secretion by a 3β-hydrogenase inhibitor in late pregnant sheep. *Can. J. Physiol. Pharmacol.* 63, 136–142.
362. Taylor, M. J., Webb, R., Mitchell, M. D., and Robinson, J. S. (1982). Effect of progesterone withdrawal in sheep during late pregnancy. *J. Endocrinol.* 92, 85–89.
363. Haluska, G. J., and Novy, M. J. (1993). Hormonal modulation of uterine activity during primate parturition. *Reprod. Endocrinol.* 11, 261–271.
364. Schwarz, B. E., Milewich, L., Gant, N. F., et al. (1977). Progesterone binding and metabolism in human fetal membranes. *Ann. N Y Acad. Sci.* 286, 304–312.
365. Kossman, J. C., Bard, H., and Gibb, W. (1982). Characterization of specific steroid binding in human amnion at term. *Biol. Reprod.* 27, 320–326.
366. Mitchell, B. F., Challis, J. R. G., and Lukash, L. (1987). Progesterone synthesis by human amnion, chorion and decidua at term. *Am. J. Obstet. Gynecol.* 157, 349–353.
367. Casey, M., and MacDonald, P. (1996). TGF-β inhibits progesterone-induced enkephalinase expression in human endometrial stromal cells. *J. Clin. Endocrinol. Metab.* 81, 4022–4027.
368. Karalis, K., Goodwin, G., and Majzoub, J. (1996). Cortisol blockade of progesterone: a possible molecular mechanism involved in the initiation of human labor. *Nat. Med.* 2, 556–560.
369. Majzoub, J., and Karalis, K. (1999). Placental corticotropin-releasing hormone: function and regulation. *Am. J. Obstet. Gynecol.* 180, S242–S246.
370. Patel, F., and Challis, J. (2002). Cortisol/progesterone antagonism in regulation of 15-hydroxyprostaglandin dehydrogenase activity and mRNA levels in human chorion and placental trophoblast cells at term. *J. Clin. Endocrinol. Metab.* 87, 700–708.
371. Mckenna, N., and O'Malley, B. (2002). Combinational control of gene expression by nuclear receptors and coregulators. *Cell* 108, 465–474.
372. Lee, K. C., and Kraus, W. L. (2001). Nuclear receptors, coactivators and chromatin: new approaches, new insights. *Trends Endocrinol. Metab.* 12, 191–197.
373. Mesiano, S., Chan, E., Fitter, J., et al. (2002). Progesterone withdrawal and estrogen activation in human parturition are coordinated by progesterone receptor A expression in the myometrium. *J. Clin. Endocrinol. Metab.* 87, 2924–2930.
374. Pieber, D., Allport, V., Hills, F., Johnson, M., and Bennett, P. (2001). Interactions between progesterone receptor isoforms in myometrial cells in human labour. *Mol. Hum. Reprod.* 7, 875–879.
375. Haluska, G., Wells, T., Hirst, J., et al. (2002). Progesterone receptor localization and isoforms in myometrium, decidua, and fetal membranes from rhesus macaques: evidence for functional progesterone withdrawal at parturition. *J. Soc. Gynecol. Investig.* 9, 125–136.
376. Condon, J., Jeyasuria, P., Faust, J., Wilson, J., and Mendelson, C. (2003). A decline in the levels of progesterone receptor coactivators in the pregnant uterus at term may antagonize progesterone receptor function and contribute to the initiation of parturition. *Proc. Natl. Acad. Sci. U S A* 100, 9518–9523.
377. Dong, X., Shylnova, O., Challis, J. R. G., and Lye, S. J. (2004). Identification of a novel repressor of the progesterone receptor which is expressed in the myometrium during labor and may contribute to a functional withdrawal of progesterone [abstract]. *J. Soc. Gynecol. Investig.* 11(Suppl.)
378. Liggins, G. C., Grieves, S. A., Kendall, J. Z., and Knox, B. X. (1972). The physiological roles of progesterone, oestradiol-17β and prostaglandin $F_{2\alpha}$ in the control of parturition. *J. Reprod. Fertil. Suppl.* 16, 85–93.
379. Cahill, L. P., Knee, B. W., and Lawson, R. A. S. (1976). Induction of parturition in ewes with a single dose of oestradiol benzoate. *J. Reprod. Fertil.* 46, 528–533.
380. Lye, S. J., Spargue, C. L., and Challis, J. R. G. (1983). Modulation of ovine myometrial activity by estradiol-17β: the possible involvement of prostaglandins. *Can. J. Physiol. Pharmacol.* 61, 729–734.
381. Lye, S. J., Wathes, D. C., and Porter, D. G. (1983). Oestradiol 17β both inhibits and stimulates myometrial activity in ewes. *J. Reprod. Fertil.* 67, 335–340.
382. Darne, J., McGarrigle, H. H. G., and Lachelin, G. C. L. (1987). Increased saliva oestriol to progesterone ratio before idiopathic preterm delivery: a possible predictor for preterm labour. *BMJ* 294, 270–272.
383. Darne, J., McGarrigle, H. H. G., and Lachelin, G. C. L. (1987). Saliva oestriol, oestradiol, oestrone and progesterone levels in pregnancy: spontaneous labour at term is preceded by a rise in saliva oestriol:progesterone ratio. *Br. J. Obstet. Gynaecol.* 94, 227–235.
384. France, J. T., Seddon, R. J., and Liggins, G. C. (1973). A study of a pregnancy with low estrogen production due to placental sulfatase deficiency. *J. Clin. Endocrinol. Metab.* 36, 1–3.
385. Liggins, G. C., Forster, C. S., Grieves, S. A., and Schwartz, A. L. (1977). Control of parturition in man. *Biol. Reprod.* 16, 39–45.
386. Beling, C. (1977). Estrogens. In *Endocrinology of Pregnancy* (F. Fuchs and A. Klopper, Eds.), pp. 76–136. Harper & Row, Hagerstown, MD.
387. Mitchell, B. F., and Challis, J. R. G. (1988). Estrogen and progesterone metabolism in human fetal membranes. In *The Physiology and Biochemistry of Human Fetal Membranes* (B. F. Mitchell, Ed.), pp. 5–28. Perinatology Press, Ithaca, NY.
388. Chibbar, R., Wong, S., Miller, F., and Mitchell, B. (1995). Estrogen stimulates oxytocin gene expression in human fetal membranes and decidua. *J. Clin. Endocrinol. Metab.* 80, 567–572.
389. Challis, J., Davies, I., Benirschke, K., Henrickx, A., and Ryan, K. (1974). The concentrations of progesterone, estrone and estradiol-17β in the peripheral plasma of the rhesus monkey during the final third of gestation and after the induction of abortion with PGF_2. *Endocrinology* 95, 547–553.
390. Figueroa, J., Honnebier, M., Binlenda, Z., Wimsatt, J., and Nathanielsz, P. (1989). Effect of 48 hours intravenous 4A androstenedione infusion on the pregnant rhesus monkey during the last third of gestation: changes in maternal plasma estradiol concentrations and myometrial contractility. *Am. J. Obstet. Gynecol.* 161, 481–486.
391. Mecenas, C., Giussani, D., Owiny, J., et al. (1996). Production of premature delivery in pregnant rhesus monkeys by androstenedione infusion. *Nat. Med.* 2, 443–448.

392. Bievenue, A. M., Jenkins, S. L., and Nathanielsz, P. W. (2002). The effect of photoperiod on the switching of myometrial contractility patterns of pregnant baboons: relationship to surgery and parturition. *J. Soc. Gynecol. Investig.* 9, 27–31.

393. Giussani, D. A., Jenkins, S. L., Winter, J. A., Tame, J. D., and Nathanielsz, P. W. (2000). Androstenedione treatment of pregnant baboons at 0.7–0.8 of gestation promotes a premature forward shift in the nocturnal maternal plasma estradiol surge relative to progesterone and increases myometrial contraction activity. *Endocrinology* 141, 3296–3303.

394. Mesiano, S., Chan, E., Fitter, J., et al. (2002). Progesterone withdrawal and estrogen activation in human parturition are coordinated by progesterone receptor a expression in the myometrium. *J. Clin. Endocrinol. Metab.* 87, 2924–2930.

395. MacDonald, P., and Casey, M. (1996). Preterm birth. *Sci. Am. Sci. Med.* 3, 42–48.

396. MacDonald, P., and Casey, M. (1993). The accumulation of prostaglandins (PG) in amniotic fluid is an after effect of labor and not indicative of a role for PGE2 or PGF2α in the initiation of human parturition. *J. Clin. Endocrinol. Metab.* 77, 805–810.

397. Romero, R., Munoz, H., Gomez, R., et al. (1995). Increase in prostaglandin bioavailability precedes the onset of human parturition. Prostaglandins Leukot. Essent. Fatty Acids 54, 187–191.

398. Brown, N., Alvi, S., Elder, M., Bennett, P., and Sullivan, M. (1998). A spontaneous induction of fetal membrane prostaglandin production precedes clinical labour. *J. Endocrinol.* 157, R1–R6.

399. Mijovic, J., Zakar, T., Nairn, T., and Olson, D. (1997). Prostaglandin endoperoxidase H synthase-2 expression and activity increases with term labour in the human chorion. *Am. J. Physiol.* 272, 8832–8840.

400. Teixiera, F., Zakar, T., Hirst, J., et al. (1994). Prostaglandin endoperoxides-H synthase (PGHS) activity and immunoreactive PGHS-1 and PGHS-2 levels in human amnion throughout gestation, at term and during labor. *J. Clin. Endocrinol. Metab.* 78, 1396–1402.

401. Mitchell, B. F., and Olson, D. M. (2004). Prostaglandin endoperoxidase H synthase inhibitors and other tocolytics in preterm labor. *Prostaglandins Leukot. Essent. Fatty Acids* 70, 167–187.

402. Sawdy, R., Lye, S., Fisk, N., and Bennet, P. (2003). A double-blind randomised study of fetal side effects during and after the short-term maternal administration of indomethacin, sulindac, and nimesilide for the treatment of preterm labor. *Am. J. Obstet. Gynecol.* 188, 1046–1051.

403. Sawdy, R., Slater, D., Fisk, N., Edmonds, D. K., and Bennet, P. (1997). Use of a cyclooxygenase type-2-selective non-steroidal anti inflammatory agent to prevent delivery. *Lancet* 350, 265–266.

404. Whittle, W., Gibb, W., and Challis, J. (2000). Characterization of human amnion epithelial and mesenchymal cells; the cellular expression, activity and glucocorticoid regulation of prostaglandin output. *Placenta* 21, 394–401.

405. Economopoulos, P., Sun, M., Purgina, B., and Gibb, W. (1996). Glucocorticoids stimulate prostaglandin H synthase type-2 (PGHS-2) in the fibroblast cells in human amnion cultures. *Mol. Cell. Endocrinol.* 117, 141–147.

406. Gibb, W., and Sun, M. (1998). Cellular specificity of interleukin-1b–stimulated expression of type-2 prostaglandin H synthase in human amnion cell cultures. *Biol. Reprod.* 59, 1139–1142.

407. Blumenstein, M., Hansen, W. R., DeVal, D., and Mitchell, M. D. (2000). Differential regulation in human amnion epithelial and fibroblast cells of prostaglandin E2 production and prostaglandin H synthase-2 mRNA expression by dexamethasone but not tumor necrosis factor-α. *Placenta* 21, 210–217.

408. Smith, W., and DeWitt, D. (1996). Prostaglandin endoperoxide H synthases-1 and -2. *Adv. Immunol.* 62, 167–215.

409. Flower, R., and Blackwell, G. (1976). The importance of phospholipase A2 in prostaglandin synthesis. *Biochem. Pharmacol.* 25, 285–291.

410. Bingham, C. O., and Austen, K. F. (1999). Phospholipase A2 enzymes in eicosanoid generation. *Proc. Assoc. Am. Physicians* 111, 524.

411. Murakami, M., and Kudo, I. (2002). Phospholipase A_2. *J. Biochem.* 131, 285–292.

412. Hirabayashi, T., and Shimizu, T. (2004). Localisation and regulation of cytosolic PLA_2. *Biochem. Biophys. Acta* 1488, 124–138.

413. Lappas, M., and Rice, G. E. (2004). Phospholipase A2 isoenzymes in pregnancy and parturition. *Prostaglandins Leukot. Essent. Fatty Acids* 70, 87–100.

414. Clark, J., Lin, L., Kri, R., et al. (1991). A novel arachidonic acid-selective cytosolic PLA2 contains a Ca^{2+} dependent translocation domain with homology to PKC and GAP. *Cell* 65, 1043–10510.

415. Rose, M. P., Myatt, L., and Elder, M. G. (1990). Pathways of arachidonic acid metabolism in human amnion cells at term. *Prostaglandins Leukot. Essent. Fatty Acids.* 39, 303.

416. Smith, W., DeWitt, D., and Garavito, R. (2000). Cyclooxygenases: structural, cellular, and molecular biology. *Annu. Rev. Biochem.* 69, 145–182.

417. Hla, T., and Neilson, K. (1992). Human cyclooxygenase-2 cDNA. *Proc. Natl. Acad. Sci. U S A* 89, 7384–7388.

418. Warner, T. D., and Mitchell, J. A. (2002). Cyclooxygenase-3 (COX-3): filling the gaps towards a COX continuum? *Proc. Natl. Acad. Sci. U S A* 15, 13371–13373.

419. Kis, B., Snipes, A., Bari, F., and Busija, D. (2004). Regional distribution of cyclooxygenase-3 mRNA in the rat central nervous system. *Brain Res. Mol. Brain Res.* 126, 78–80.

420. Matuso, M., Ensor, C., and Tai, H.-H. (1997). Characterization of the generic structure and promoter of the mouse NAD-dependent 15-hydroxyprostaglandin dehydrogenase gene. *Biochem. Biophys. Res. Commun.* 235, 582.

421. Negishi, M., Sugimoto, Y., and Ichikawa, A. (1995). Molecular mechanisms of adverse actions of prostanoid receptors. *Biochem. Biophys. Acta* 1259, 109.

422. Slater, D., Zervou, S., and Thornton, S. (2002). Prostaglandins and prostanoid receptors in human pregnancy and parturition. *J. Soc. Gynecol. Investig.* 9, 118–124.

423. Myatt, L., and Lye, S. (2004). Expression, localization and function of prostaglandin receptors in myometrium. *Prostaglandins Leukot. Essent. Fatty Acids* 70, 137–148.

424. Olson, D., Zaragoza, D., Shallow, M., et al. (2003). Myometrial activation and preterm labor: evidence supporting a role for the prostaglandin F receptor: a review. *Placenta* 24, S47–S54.

425. Langenback, R., Morham, S., and Tiano, H. (1995). Prostaglandin synthase 1 gene disruption in mice reduces arachidonic acid-induced inflammation and indomethacin-induced gastric ulceration. *Cell* 83, 483.

426. Gross, G., Imamura, T., and Luedke, C. (1998). Opposing actions of prostaglandins and oxytocin determine the onset of murine labor. *Proc. Natl. Acad. Sci. U S A* 95, 11871–11875.

427. Morham, S., Langenbach, R., and Loftin, C. (1995). Prostaglandin synthase 2 gene disruption causes severe renal pathology in the mouse. *Cell* 83, 473–482.

428. Lim, H., Paria, B., and Das, S. (1997). Multiple fetal reproductive failures in cyclooxygenase-2-deficient mice. *Cell* 91, 197–208.

429. Sugimoto, Y., Yamasaki, A., and Segi, E. (1997). Failure of parturition in mice lacking the prostaglandin F receptor. *Science* 277, 681–683.
430. Gibb, W. (1998). The role of prostaglandins in human parturition. *Ann. Med.* 30, 235–241.
431. Lindstrom, T., and Bennett, P. (2004). Transcriptional regulation of genes for enzymes of the prostaglandin biosynthetic pathway. *Prostaglandins Leukot. Essent. Fatty Acids* 70, 115–135.
432. Zakar, T., and Hertelendy, F. (2001). Regulation of prostaglandin synthesis in the human uterus. *J. Matern. Fetal Med.* 10, 223–235.
433. Helliwell, R. J. A., Berry, E. B. E., O'Carrol, S. J., and Mitchell, M. D. (2004). Nuclear prostaglandin receptors: role in pregnancy and parturition? *Prostaglandins Leukot. Essent. Fatty Acids* 70, 149–165.
434. Lappas, M., and Rice, G. E. (2004). PhospholipaseA2 isozymes in pregnancy and parturition. *Prostaglandins Leukot. Essent. Fatty Acids* 70, 87–100.
435. Mitchell, M., and Trautman, M. (1993). Molecular mechanisms regulating prostaglandin action. *Mol. Cell. Endocrinol.* 93, C7–C10.
436. Lappas, M., and Rice, G. E. (2004). PhospholipaseA2 isozymes in pregnancy and parturition. *Prostaglandins Leukot. Essent. Fatty Acids* 70, 87–100.
437. Freed, K., Moses, E., Brennecke, S., and Rice, G. E. (1997). Differential expression of type II, IV and cytosolic PLA2 messenger RNA in human intrauterine tissues at term. *Mol. Hum. Reprod.* 3, 493–499.
438. Johansen, B., Rakkestad, K., Balboa, M. A., and Dennis, E. (2000). Expression of cytosolic and secreted forms of PLA2 and COX in human placenta, fetal membranes and chorionic cell lines. *Prostaglandins* 60, 119–125.
439. Skannal, D., Eis, A., Brockman, D., Siddiqi, T., and Myatt, L. (1997). Immunohistochemical localization of phospholipase A2 isoforms in human myometrium during pregnancy and parturition. *Am. J. Obstet. Gynecol.* 176, 878–882.
440. Skannal, D., Brockman, D., Eis, A., et al. (1997). Changes in activity of cytosolic phospholipase A2 in human amnion at parturition. *Am. J. Obstet. Gynecol.* 177, 179–184.
441. Rice, G. E., Wong, M., Farrugia, W., and Scott, K. (1998). Contribution of type II phospholipase A2 to in vitro phospholipase A2 enzymatic activity in human term placenta. *J. Endocrinol.* 157, 25–31.
442. Aitken, M. A., Rice, G. E., and Brennecke, S. P. (1992). Relative abundance of human placental phospholipase A2 messenger RNA in late pregnancy. *Prostaglandins* 43, 361–370.
443. Bennett, P., Slater, D., Berger, I., and Moore, G. (1994). The expression of phospholipase A2 and lipocortins (annexins) I, II and V in human fetal membranes and placenta in association with labour. *Prostaglandins* 48, 81–90.
444. Lappas, M., Permezel, M., Georgiou, H. M., and Rice, G. E. (2001). Immunoreactive type II phospholipase A2 levels in human preterm gestational tissues. *Placenta* 22, 64–69.
445. Rice, G. E., Brennecke, S. P., Scott, K. F., et al. (1992). Elevated maternal plasma immunoreactive phospholipase A2 in human preterm and term labour. *Eicosanoids* 5, 9–12.
446. Koyama, M., Ito, S., Nakajima, A., et al. (2000). Elevations of group II phospholipase A2 concentrations in serum and amniotic fluid in association with preterm labour. *Am. J. Obstet. Gynecol.* 183, 1537–1543.
447. Nguyen, H., Rice, G. E., Farrugia, W., Wong, M., and Brennecke, S. (1994). Bacterial endotoxin releases type II phospholipase A2 immunoreactive content and phospholipase A2 enzymatic activity in human choriodecidua. *Biol. Reprod.* 50, 526–534.
448. Farrugia, W., Nicholls, L., and Rice, G. E. (1999). Effect of bacterial endotoxin on the in vitro release of type II phospholipase-A2 and prostaglandin E2 from human placenta. *J. Endocrinol.* 160, 291–296.
449. Slater, D. M., Myles, L., Sykes, A. L., Poston, L., and Bennett, P. R. (2000). Expression of secretory and cytosolic phospholipase A2 in human myometrium: changes in relation to gestational age and labour onset [abstract]. *J. Soc. Gynecol. Investig.* 303.
450. Hirst, J., Teixiera, F., and Zakar, T. (1995). Prostaglandin endoperoxide-H synthase-1 and -2 messenger ribonucleic acid levels in human amnion with spontaneous labor onset. *J. Clin. Endocrinol. Metab.* 80, 517–523.
451. Gibb, W., and Sun, M. (1996). Localization of prostaglandin H synthase type 2 protein and mRNA in term human fetal membranes and decidua. *J. Endocrinol.* 150, 497–503.
452. Slater, D. M., Berger, L., Newton, R., Moore, G. E., and Bennett, P. R. (1995). Changes in the expression of types 1 and 2 cyclo-oxygenase in human fetal membranes at term. *Am. J. Obstet. Gynecol.* 172, 77–82.
453. Alfaidy, N., Sun, M., Challis, J. R. G., and Gibb, W. (2003). Expression of membrane prostaglandin E synthase in human placenta and fetal membranes and effect of labor. *Endocrine* 20, 219–225.
454. Meadows, J. W., Brockman, D. E., and Myatt, L. (2003). Expression and localization of prostaglandin E synthase isoforms in human fetal membranes in term and preterm labor. *J. Clin. Endocrinol. Metab.* 88, 433–439.
455. Johnson, R., Mitchell, C., Giles, W., Walter, W., and Zakar, T. (2002). The in vivo control of prostaglandin H-synthase messenger ribonucleic acid expression in the human amnion at parturition. *J. Clin. Endocrinol. Metab.* 87, 2816–2823.
456. Sangha, R., Walton, J., and Ensor, C. (1994). Immunohistochemical localization, mRNA abundance and activity of 15-hydroxyprostaglandin dehydrogenase in placenta and fetal membranes during term and preterm labor. *J. Clin. Endocrinol. Metab.* 78, 982–989.
457. Mijovic, J. E., Zakar, T., Nairn, T. K., and Olson, D. M. (1998). Prostaglandin endoperoxide H synthase (PGHS) activity and PGHS-1 and PGHS-2 messenger ribonucleic acid abundance in human chorion throughout gestation and with preterm labor. *J. Clin. Endocrinol. Metab.* 83, 1358–1367.
458. van Meir, C., Matthews, S., and Ramirez, M. (1997). Chorionic prostaglandin catabolism is decreased in the lower uterine segment with term labor. *Placenta* 18, 109–114.
459. Zuo, J., Lei, Z. M., Rao, C. V., Pietrantoni, M., and Cook, V. D. (1994). Differential cyclo-oxygenase-1 and -2 gene expression in human myometria from preterm and term deliveries. *J. Clin. Endocrinol. Metab.* 79, 894–899.
460. Erkinheimo, T.-L., Saukkonen, K., Narko, K., et al. (2000). Expression of cyclooxygenase-2 and prostanoid receptors by human myometrium. *J. Clin. Endocrinol. Metab.* 85, 3468–3475.
461. Sparey, C., Robson, S. C., Bailey, J., Lyall, F., and Europe-Finner, G. N. (1999). The differential expression of myo–metrial connexin-43, cyclooxygenase-1 and -2, and Gs alpha proteins in the upper and lower segments of the human uterus during pregnancy and labor. *J. Clin. Endocrinol. Metab.* 84, 1705–1710.
462. Moore, S. D., Brodt-Eppley, J., Cornelison, L. M., et al. (1999). Expression of prostaglandin H synthase isoforms in human myometrium at parturition. *Am. J. Obstet. Gynecol.* 180, 103–109.
463. Giannoulias, D., Patel, F., Holloway, A., et al. (2002). Differential changes in 15-hydroxyprostaglandin dehydrogenase and prostaglandin H synthase (types I and II) in human pregnant myometrium. *J. Clin. Endocrinol. Metab.* 87, 1345–1352.

464. Slater, D. M., Dennes, W. J., Campa, J. S., Poston, L., and Bennet, P. R. (1999). Expression of cyclo-oxygenase types-1 and -2 in human myometrium throughout pregnancy. *Mol. Hum. Reprod.* 5, 880–884.
465. Korita, D., Sagawa, N., Itoh, H., et al. (2002). Cyclic mechanical stretch augments prostacyclin production in cultured human uterine myometrial cells from pregnant women: possible involvement of up-regulation of prostacyclin synthase expression. *J. Clin. Endocrinol. Metab.* 87, 5209–5219.
466. Nathanielsz, P. W., Smith, G., and Wu, W. (2004). Topographical specialization of prostaglandin function in late pregnancy and at parturition in the baboon. *Prostaglandins Leukot. Essent. Fatty Acids* 70, 199–206.
467. Wu, W. X., Ma, X. H., Smith, G. C. S., Koenen, S. V., and Nathanielsz, P. W. (2000). A new concept on the significance of regional distribution of prostaglandin H synthase 2 throughout the uterus during late pregnancy. *Am. J. Obstet. Gynecol.* 183, 1287–1295.
468. Wilson, L., Parsons, M. T., and Fouret, G. (1991). Forward shift in the initiation of the nocturnal estradiol surge in the pregnant baboon: is this the genesis of labor? *Am. J. Obstet. Gynecol.* 165, 1487–1498.
469. van Meir, C., Matthews, S., and Kierse, M. (1997). 15-Hydroxyprostaglandin dehydrogenase (PGDH): implications in preterm labor with and without ascending infection. *J. Clin. Endocrinol. Metab.* 82, 969–976.
470. Wu, W. X., Ma, X. H., Smith, G. C. S., et al. (2000). Prostaglandin dehydrogenase mRNA in baboon intrauterine tissues in late gestation and spontaneous labor. *Am. J. Physiol.* 279, R1082–R1090.
471. Peters-Golden, M., and McNish, R. (1993). Redistribution of 5-lipoxygenase and cytosolic phospholipase A2 to the nuclear fraction upon macrophage activation. *Biochem. Biophys. Res. Commun.* 196, 147–153.
472. Lin, L., Wartmann, M., Lin, A., et al. (1993). cPLA2 is phosphorylated and activated by MAP kinase. *Cell* 72, 269–278.
473. Durstin, M., Durstin, S., Molski, T., Becker, E., and Sha'afi, R. (1994). Cytoplasmic phospholipase A2 translocates to membrane fraction in human neutrophils activated by stimuli that phosphorylate mitogen-activated protein kinase. *Proc. Natl. Acad. Sci. U S A* 91, 3142–3146.
474. Qiu, Z., De Carvalho, M., and Leslie, C. (1993). Regulation of phospholipase A2 activation by phosphorylation in mouse peritoneal macrophages. *J. Biol. Chem.* 268, 24506–24513.
475. Wu, T., Angus, C. W., Yao, X.-L., Logun, C., and Shellhamer, J. (1997). P11, a unique member of the S100 family of calcium-binding proteins, interacts with and inhibits the activity of the 85kDa cytosolic phospholipase A2. *J. Biol. Chem.* 272, 17145–17153.
476. Tay, A., Maxwell, P., Li, Z., Goldberg, H., and Skorecki, K. (1994). Isolation of promoter for cytosolic phospholipase A2 (cPLA2). *Biochem. Biophys. Acta* 1217, 345–347.
477. Lappas, M., Munns, M., King, R., and Rice, G. E. (2001). Antisense oligonucleotide inhibition of type II phospholipase A2 expression, release and activity in vitro. *Placenta* 22, 418–424.
478. Xue, S., Slater, D., Bennet, P., and Myatt, L. (1996). Induction of both cytosolic phospholipase A2 and prostaglandin H synthase-2 by interleukin-1 beta in WISH cells in inhibited by dexamethasone. *Prostaglandins* 51, 107–124.
479. Lappas, M., Permezel, M., and Georgiou, H. M. (2004). Regulation of phospholipase isozymes by nuclear factor-κB in human gestational tissues in vitro. *J. Clin. Endocrinol. Metab.* 89, 2365–2372.
480. Tazawa, R., Xu, X.-M., Wu, K., and Wang, L.-H. (1994). Characterization of the genomic structure, chromosomal location and promoter of human prostaglandin H synthase-2 gene. *Biochem. Biophys. Res. Commun.* 203, 190–199.
481. Warrick, C., Skinner, K., and Mitchell, B. (1985). Relation between cyclic AMP and prostaglandin output by dispersed cells from human amnion and decidua. *Am. J. Obstet. Gynecol.* 153, 66–71.
482. Mitchell, M., Lytton, F., and Varticovski, L. (1988). Paradoxical stimulation of both lipocortin and prostaglandin production in human amnion cells by dexamethasone. *Biochem. Biophys. Res. Commun.* 151, 137–141.
483. Potestio, F., Zakar, T., and Olson, D. (1988). Glucocorticoids stimulate prostaglandin synthesis in human amnion cells by a receptor-mediated mechanism. *J. Clin. Endocrinol. Metab.* 67, 1205–1210.
484. Zakar, T., and Olson, D. (1989). Dexamethasone stimulates arachidonic acid conversion to prostaglandin E2 in human amnion cells. *J. Dev. Physiol.* 12, 269–272.
485. Sun, K., Ma, R., Cui, X., et al. (2003). Glucocorticoids induce cytosolic phospholipase A2 and prostaglandin H synthase type 2 but not microsomal prostaglandin E synthase (PGES) and cytosolic PGES expression in cultured primary human amnion cells. *J. Clin. Endocrinol. Metab.* 88, 5564–5571.
486. Sun, K., and Myatt, L. (2003). Enhancement of glucocorticoid-induced 11β-hydroxysteroid dehydrogenase type 1 expression by proinflammatory cytokines in cultured human amnion fibroblasts. *Endocrinology* 144, 5568–5577.
487. Sun, K., He, P., and Yang, K. (2002). Intracrine induction of 11β-hydroxysteroid dehydrogenase type 1 expression by glucocorticoid potentiates prostaglandin production in the human chorionic trophoblast. *Biol. Reprod.* 67, 1450–1455.
488. Sun, M., Ramirez, M., and Challis, J. (1996). Immunohistochemical localization of the glucocorticoid receptor in human fetal membranes and decidua at term and preterm delivery. *J. Endocrinol.* 149, 243–248.
489. Yan, X., Sun, M., and Gibb, W. (2002). Localization of nuclear factor-κB (NFκB) and inhibitory factor-κB (IκB) in human fetal membranes and decidua at term and preterm delivery. *Placenta* 23, 288–293.
490. Yan, X., Xiao, C. W., Sun, M., Tsang, B. K., and Gibb, W. (2002). Nuclear factor kappa B (NFκB) activation and regulation of cyclooxygenase type 2 (COX-2) expression in human amnion mesenchymal cells by IL1β. *Biol. Reprod.* 66, 1667–1671.
491. Allport, V. C., Pieber, D., Slater, D. M., et al. (2001). Human labour is associated with nuclear factor-kappa B activity which mediates cyclo-oxygenase-2 expression and is involved with the "functional progesterone withdrawal." *Mol. Hum. Reprod.* 7, 581–586.
492. Xue, S., Brockman, D., Slater, D., and Myatt, L. (1995). Interleukin-1 beta induces the synthesis and activity of cytosolic phospholipase A2 and the release of prostaglandin E2 in human amnion-derived WISH cells. *Prostaglandins* 49, 351–369.
493. Bowen, J. M., Chamley, L., Keelan, J. A., and Mitchell, M. D. (2002). Cytokines of the placenta and extra-placental membranes: roles and regulation during human pregnancy and parturition. *Placenta* 23, 257–273.
494. Pavan, B., Biondi, C., Ferretti, M. E., Lunghi, L., and Paganetto, G. (2001). 17β-Estradiol modulates prostaglandin E2 release from amnion-derived WISH cells. *Biol. Reprod.* 64, 1677–1681.
495. Allport, V. C., Slater, D. M., Newton, R., and Bennet, P. R. (2000). NF-κB and AP-1 are required for cyclo-oxygenase 2 gene expression in amnion epithelial cell line (WISH). *Mol. Hum. Reprod.* 6, 561–565.
496. Wang, Z., and Tai, H.-H. (1998). Interleukin-1β and dexamethasone regulate gene expression of prostaglandin H

synthase-2 via the NF-κB pathway in human amnion derived WISH cells. *Prostaglandins Leukot. Essent. Fatty Acids* 59, 63–69.

497. Fiorini, S., Ferretti, M. E., Biondi, C., et al. (2003). 17β-Estradiol stimulates arachidonate release from human amnion-like WISH cells through a rapid mechanism involving a membrane receptor. *Endocrinology* 144, 3359–3367.
498. Kniss, D. A., Xic, Y., Li, Y., et al. (2002). ED (27) trophoblast-like cells isolated from first-trimester chorionic villi are genetically identical to HeLa cells yet exhibit a distinct phenotype. *Placenta* 23, 32–43.
499. Gibb, W., and Challis, J. (2002). Mechanisms of term and preterm birth. *J. Obstet. Gynaecol. Can.* 24, 874–883.
500. Patel, F., Funder, J., and Challis, J. (2003). Mechanism of cortisol/progesterone antagonism in the regulation of 15-hydroxyprostaglandin dehydrogenase activity and messenger ribonucleic acid levels in human chorion and placental trophoblast cells at term. *J. Clin. Endocrinol. Metab.* 88, 2922–2933.
501. Patel, F., Clifton, V., and Chwalisz, K. (1999). Steroidal regulation of prostaglandin dehydrogenase activity and expression in human term placenta and choriodecidua in relation to labor. *J. Clin. Endocrinol. Metab.* 84, 291.
502. Greenland, K., Jantke, I., Jenatschke, S., et al. (2000). The human NAD^+ dependent 15-hydroxyprostaglandin dehydrogenase gene promoter is controlled by Ets and activating protein 1 transcription factors and progesterone. *Endocrinology* 141, 581–597.
503. Patel, F., Sun, K., and Challis, J. (1999). Local modulation by 11β-hydroxysteroid dehydrogenase of glucocorticoid effects on the activity of 15-hydroxyprostaglandin dehydrogenase in human chorion and placental trophoblast cells. *J. Clin. Endocrinol. Metab.* 84, 395–400.
504. McKeown, K. J., and Challis, J. R. (2003). Regulation of 15-hydroxy prostaglandin dehydrogenase by corticotrophin-releasing hormone through a calcium-dependent pathway in human chorion trophoblast cells. *J. Clin. Endocrinol. Metab.* 88, 1737–1741.
505. Mitchell, M., Goodwin, V., Mesnage, S., and Keelan, J. (2000). Cytokine-induced coordinate expression of enzymes of prostaglandin biosynthesis and metabolism: 15-hydroxyprostaglandin dehydrogenase. *Prostaglandins Leukot. Essent. Fatty Acids* 62, 1–5.
506. Pomini, F., Patel, F., Mancuso, S., and Challis, J. (2000). Activity and expression of 15-hydroxyprostaglandin dehydrogenase in cultured chorionic trophoblast and villous trophoblast cells and in chorionic explants at term with and without spontaneous labor. *Am. J. Obstet. Gynecol.* 182, 221–226.
507. Brown, N., Alvi, S., Elder, M., Bennett, P., and Sullivan, M. (1998). Interleukin-1β and bacterial endotoxin change the metabolism of prostaglandins E2 and F2α in intact term fetal membranes. *Placenta* 19, 625–630.
508. Keelan, J., Blumenstein, M., Helliwell, R., et al. (2003). Cytokines, prostaglandins and parturition: a review. *Placenta* 24, S33–S46.
509. Duncan, A. M., Anderson, L. L., Funk, C. D., Abramovitz, M., and Adam, M. (1995). Chromosomal localization of the human prostanoid receptor gene family. *Genomics* 25, 740–742.
510. Katsuyama, M., Sugimoto, Y., Okano, K., et al. (1998). Characterization of the gene for the mouse prostaglandin E receptor subtype EP2: tissue-specific initiation of transcription in the macrophage and the uterus. *Biochem. J.* 330, 1115–1121.
511. Sugimoto, Y., Yamasaki, A., Segi, E., et al. (1997). Failure of parturition in mice lacking the prostaglandin F receptor. *Science* 277, 681–683.
512. Tsuboi, K., Sugimoto, Y., Iwane, A., et al. (2000). Uterine expression of prostaglandin H2 synthase in late pregnancy and during parturition in prostaglandin F receptor-deficient mice. *Endocrinology* 141, 315–324.
513. Brodt-Eppley, J., and Myatt, L. (1998). Changes in expression of contractile FP and relaxatory EP2 receptors in pregnant rat myometrium during late gestation, at labour, and postpartum. *Biol. Reprod.* 59, 878–883.
514. Brodt-Eppley, J., and Myatt, L. (1999). Prostaglandin receptors in lower segment myometrium during gestation and labor. *Obstet. Gynecol.* 93, 89–93.
515. Ou, C.-W., Chen, Z., Qi, S., and Lye, S. (2000). Expression and regulation of the messenger ribonucleic acid encoding the prostaglandin $F_{2\alpha}$ receptor in the rat myometrium during pregnancy and labour. *Am. J. Obstet. Gynecol.* 182, 919–925.
516. Myatt, L., Sahay, R., Eis, A., and Brockman, D. (1999). Changes in intracellular localization of prostaglandin H synthase-2 and cytosolic phospholipase A2 associated with cytokine and growth factor induced PGE2 synthesis in WISH cells. In the *Proceedings of the 46th Annual Meeting of the Society for Gynecologic Investigation*, Atlanta. 1999, Abstract 326, p. 70A. Elsevier, NY.
517. Matsumoto, T., Sagawa, N., Yoshida, M., et al. (1997). The prostaglandin E2 and $F2_{\alpha}$ receptor genes are expressed in human myometrium and are down-regulated during pregnancy. *Biochem. Biophys. Res. Commun.* 238, 838–841.
518. Wikland, M., Lindblom, B., and Wigvist, N. (1984). Myometrial response to prostaglandins during labour. *Gynecol. Obstet. Invest.* 17, 131–138.
519. Wiqvist, N., Bryman, I., Lindblom, B., Norstrom, A., and Wikland, M. (1985). The role of prostaglandins for the coordination of myometrial forces during labor. *Acta Physiol. Hung.* 65, 313–322.
520. Luckas, M., and Wray, S. (2000). A comparison of the contractile properties of human myometrium obtained from the upper and lower uterine segments. *Br. J. Obstet. Gynaecol.* 107, 1309–1311.
521. Smith, G., Baguma-Nibasheka, M., Wu, W., and Nathanielsz, P. (1998). Regional variations in contractile responses to prostaglandins and prostanoid receptor messenger ribonucleic acid in pregnant baboon uterus. *Am. J. Obstet. Gynecol.* 179, 1545–1552.
522. Smith, G., Wu, W., and Nathanielsz, P. (2001). Effects of gestational age and labor on expression of prostanoid receptor genes in baboon uterus. *Biol. Reprod.* 64, 1131–1137.
523. Ma, X., Wu, W. X., and Nathanielsz, P. W. (1999). Differential regulation of prostaglandin EP and FP receptors in pregnant sheep myometrium and endometrium during spontaneous term labor. *Biol. Reprod.* 61, 1281–1286.
524. Helliwell, R., Mitchell, M. D., and Lan, H. Y. (2004). Presence and regulation of the prostaglandin E2 receptor in human gestational tissues [abstract]. *J. Soc. Gynecol. Investig.* 11.
525. Smith, G. C. S., Wu, W. X., and Nathanielsz, P. W. (2001). Expression of prostanoid receptor genes in baboon chorion and decidua during pregnancy and parturition. *J. Endocrinol.* 168, 263–272.
526. Bowen, J. M., Chamley, L., Keelan, J. A., and Mitchell, M. D. (2002). Cytokines of the placenta and extra-placental membranes: roles and regulation during human pregnancy and parturition. *Placenta* 23, 257–273.
527. Chan, E., Fraser, S., Yin, S., et al. (2002). Human myometrial genes are differentially expressed in labor: a suppression subtractive hybridization study. *J. Clin. Endocrinol. Metab.* 87, 2435–2441.
528. Romero, R., Avila, C., and Brekus, C. (1991). The role of systemic and intrauterine infection in preterm parturition. *Ann. N Y Acad. Sci.* 622, 355–375.

529. Romero, R., Sepulveda, M., Mazor, M., et al. (1992). The natural interleukin-1 receptor antagonist in term and preterm parturition. *Am. J. Obstet. Gynecol.* 167, 863–872.
530. Keelan, J., Marvin, K., Sato, T., et al. (1999). Cytokine abundance in placental tissues: evidence of inflammatory activation in gestational membranes with term and preterm parturition. *Am. J. Obstet. Gynecol.* 181, 1530–1536.
531. Olah, K., Vince, G., Neilson, J., Deniz, G., and Johnson, P. (1996). Interleukin-6, interferon-gamma, interleukin-8, and granulocyte macrophage colony stimulating factor levels in human amniotic fluid at term. *J. Reprod. Immunol.* 32, 89–98.
532. Romero, R., Parvizi, S., Oyarzun, E., et al. (1990). Amniotic fluid interleukin-1 in spontaneous labor at term. *J. Reprod. Med.* 35, 235–238.
533. Romero, R., Mazor, M., Sepulveda, M., et al. (1992). Tumor necrosis factor in preterm and term labor. *Am. J. Obstet. Gynecol.* 166, 1576–1587.
534. Laham, N., Brennecke, S., Bendtzen, K., and Rice, G. E. (1996). Differential release of interleukin-6 from human gestational tissues in association with labour and in vitro endotoxin treatment. *J. Endocrinol.* 149, 431–439.
535. Saito, S., Kasahara, T., Kato, Y., Ishihara, Y., and Ichijo, M. (1993). Elevation of amniotic fluid interleukin 6 (IL-6) IL-8 and granulocyte colony stimulating factor (G-CSF) in term and preterm parturition. *Cytokine* 5, 81–88.
536. Inglis, S. R., Jeremias, J., Kuno, K., et al. (1994). Detection of tumor necrosis factor-α, interleukin-6 and fetal fibronectin in the lower genital tract during pregnancy: relation to outcome. *Am. J. Obstet. Gynecol.* 171, 5–10.
537. Wennerholm, U. B., Holm, B., Mattsby-Baltzer, I., et al. (1998). Interleukin-1α, interleukin-6 and interleukin-8 in cervico/ vaginal secretion for screening preterm birth in twin gestation. *Acta Obstet. Gynecol. Scand.* 77, 508–514.
538. Mattsby-Baltzer, I., Platz-Christensen, J. J., Hosseini, N., and Rosen, P. (1998). IL-1β , IL-6, TNFα, fetal fibronectin, and endotoxin in the lower genital tract of pregnant women with bacterial vaginosis. *Acta Obstet. Gynecol. Scand.* 77, 701–706.
539. Steinborn, A., Kuhnert, M., and Halberstadt, E. (1996). Immunomodulating cytokines induce term and preterm parturition. *J. Perinat. Med.* 24, 381–390.
540. Cox, S., King, M., Casey, M., and MacDonald, P. (1993). Interleukin-1β, -1α, and -6 and prostaglandins in vaginal/-cervical fluids of pregnant women before and during labour. *J. Clin. Endocrinol. Metab.* 77, 805–815.
541. Taniguchi, T., Matsuzaki, N., Kameda, T., et al. (1991). The enhanced production of placental interleukin-1 during labor and intrauterine infection. *Am. J. Obstet. Gynecol.* 165, 131–137.
542. Dudley, D., Collmer, D., Mitchell, M. D., and Trautman, M. (1996). Inflammatory cytokine mRNA in human gestational tissues: Implications for term and preterm labor. *J. Soc. Gynecol. Investig.* 3, 328–335.
543. Laham, N., Brennecke, S., Bendtzen, K., and Rice, G. E. (1996). Labour-associated increase in interleukin-1-α release in vitro by human gestational tissues. *J. Endocrinol.* 150, 515–522.
544. Gunn, L., Hardiman, P., Tharmaratnam, S., Lowe, D., and Chard, T. (1996). Measurement of interleukin-1-α and interleukin-6 in pregnancy-associated tissues. *Reprod. Fertil. Dev.* 8, 1069–1073.
545. Steinborn, A., von Gall, C., Hildenbrand, R., Stutte, H., and Kaufmann, M. (1998). Identification of placental cytokine-producing cells in term and preterm labor. *Obstet. Gynecol.* 91, 329–335.
546. Matsuzaki, N., Taniguchi, T., Shimoya, K., et al. (1993). Placental interleukin-6 production is enhanced in intrauterine infection but not in labor. *Am. J. Obstet. Gynecol.* 168, 94–97.
547. Lappas, M., Permezel, M., Georgiou, H., and Rice, G. E. (2002). Nuclear factor kappa B regulation of pro-inflammatory cytokines in human gestational tissues in vitro. *Biol. Reprod.* 67, 668–673.
548. Simpson, K., Keelan, J., and Mitchell, M. D. (1998). Labor-associated changes in interleukin-10 production and its regulation by immunomodulators in human choriodecidua. *J. Clin. Endocrinol. Metab.* 83, 4332–4337.
549. Jones, C., Finlay-Jones, J., and Hart, P. (1997). Type-1 and type-2 cytokines in human late-gestation decidual tissue. *Biol. Reprod.* 57, 303–311.
550. Dudley, D., Hunter, C., Mitchell, M. D., and Varner, M. (1997). Amniotic fluid interleukin-10 (IL-10) concentrations during pregnancy and with labor. *J. Reprod. Immunol.* 33, 147–156.
551. Dudley, D., Hunter, C., Varner, M., and Mitchell, M. (1996). Elevation of amniotic fluid interleukin-4 concentrations in women with preterm labor and chorioamnionitis. *Am. J. Perinatol.* 13, 443–447.
552. Ammala, M., Nyman, T., Salmi, A., and Rutanen, E. M. (1997). The interleukin-1 system in gestational tissues at term: effect of labour. *Placenta* 18, 717–723.
553. Osmers, R., Blaser, J., Kuhn, W., and Tschesche, H. (1995). Interleukin-8 synthesis and the onset of labor. *Obstet. Gynecol.* 86, 223–229.
554. Laham, N., Brennecke, S. P., and Rice, G. E. (1999). Interleukin-8 release from human gestational tissue explants: effects of gestation, labor, and chorioamnionitis. *Biol. Reprod.* 61, 823–827.
555. Hansen, W. R., Keelan, J. A., Skinner, S. J., and Mitchell, M. D. (1999). Key enzymes of prostaglandin biosynthesis and metabolism: coordinate regulation of expression by cytokines in gestational tissues: a review. *Prostaglandins Other Lipid Mediat.* 57, 243–257.
556. Premyslova, M., Li, W., Alfaidy, N., et al. (2003). Differential expression and regulation of microsomal prostaglandin E_2 synthase in human fetal membranes and placenta with infection and in cultured trophoblast cells. *J. Clin. Endocrinol. Metab.* 88, 6040–6047.
557. Oger, S., Mehats, C., Dallot, E., Ferre, F., and Leroy, M.-J. (2002). Interleukin-1β induces phosphodiesterase 4B2 expression in human myometrial cells through a prostaglandin E2 and cyclic adenosine 3′,5′-monophosphate-dependent pathway. *J. Clin. Endocrinol. Metab.* 87, 5524–5531.
558. Grammatopoulos, D., and Hillhouse, E. (1999). Basal and interleukin-1β-stimulated prostaglandin production from cultured human myometrial cells: differential regulation by corticotropin-releasing hormone. *J. Clin. Endocrinol. Metab.* 84, 2204–2211.
559. Mitchell, M. D., Dudley, D. J., Edwin, S. S., and Schiller, S. L. (1991). Interleukin-6 stimulates prostaglandin production by human amnion and decidual cells. *Eur. J. Pharmacol.* 192, 189–191.
560. Dudley, D. J., Edwin, S. S., and Mitchell, M. D. (1996). Macrophage inflammatory protein-1α regulates prostaglandin and interleukin-6 production by human gestational tissues in vitro. *J. Soc. Gynecol. Investig.* 3, 12–16.
561. Pomini, F., Caruso, A., and Challis, J. (1999). Interleukin-10 modifies the effects of interleukin-1β and tumor necrosis factor-α on the activity and expression of prostaglandin H synthase-2 and the NAD^+-dependent 15-hydroxyprostaglandin dehydrogenase in cultured term human villous trophoblast and chorion trophoblast cells. *J. Clin. Endocrinol. Metab.* 84, 4645–4651.
562. Hertelendy, F., Molnar, M., and Romero, R. (2002). Interferon gamma antagonizes interleukin-1β-induced cyclooxygenase-2 expression and prostaglandin E2 production

in human myometrial cells. *J. Soc. Gynecol. Investig.* 9, 215–219.

563. Mitchell, M. D., Edwin, S. S., Silver, R. M., and Romero, R. (1993). Potential agonist action of the interleukin-1 receptor antagonist protein: implications for treatment of women. *J. Clin. Endocrinol. Metab.* 76, 1386–1388.
564. Adamson, S., Edwin, S. S., and LaMarche, S. M. (1994). Actions of interleukin-4 on prostaglandin biosynthesis by human amnion cells. *Prostaglandins Leukot. Essent. Fatty Acids* 50, 133–135.
565. Simpson, K., Keelan, J., and Mitchell, M. D. (1999). The anti-inflammatory cytokine IL-10 has proinflammatory effects on human amnion [abstract]. *J. Soc. Gynecol. Investig.* 6(Suppl.), 202.
566. Adamson, S., Edwin, S. S., LaMarche, S., and Mitchell, M. D. (1993). Actions of interleukin-4 on prostaglandin biosynthesis at the chorion decidual interface. *Am. J. Obstet. Gynecol.* 169, 1442–1447.
567. Petraglia, F., Garuti, G. C., De Ramundo, B., et al. (1990). Mechanisms of action of interleukin-1 beta in increasing corticotropin-releasing factor and adrenocorticotropin hormone release from cultured placental cells. *Am. J. Obstet. Gynecol.* 163, 1307–1312.
568. Makrigiannakis, A., Margioris, A. N., Zoumakis, E., Stournaras, C., and Gravanis, A. (1999). The transcription of corticotropin-releasing hormone in human endometrial cells is regulated by cytokines. *Neuroendocrinology* 70, 451–459.
569. Chard, T. (1977). Oxytocin. In *Clinical Neuroendocrinology* (I. Martini and G. M. Besser, Eds.), pp. 569–583. Academic Press, New York.
570. Zingg, H. H., and Lefebvre, D. L. (1988). Oxytocin and vasopressin gene expression during gestation and lactation. *Mol. Brain Res.* 4, 1–6.
571. Soloff, M. S. (1988). The role of oxytocin in the initiation of labour and oxytocin-prostaglandin interactions. In *The Onset of Labour: Cellular and Integrative Mechanisms* (D. McNellis, J. R. G. Challis, P. C. MacDonald, P. W. Nathanielsz, and J. M. Roberts, Eds.), p. 203. Perinatology Press, Ithaca, NY.
572. Petraglia, F., Florio, P., and Nappi, C. (1996). Peptide signaling in human placenta and membranes: autocrine, paracrine, and endocrine mechanisms. *Endocr. Rev.* 17, 156–186.
573. Chard, T., Hudson, T., and Edwards, C. (1971). Release of oxytocin and vasopressin by the human fetus during labor. *Nature* 243, 352–354.
574. Chard, T. (1989). Fetal and maternal oxytocin in human parturition. *Am. J. Perinatol.* 2, 145–152.
575. Fuchs, A. R. (1985). Oxytocin secretion and milk ejection in the human. In *Oxytocin in Animal Parturition* (J. A. Amico and A. G. Robinson, Eds.), pp. 200–206. Excerpta Medica, Amsterdam.
576. Mitchell, B., and Schmid, B. (2001). Oxytocin and its receptor in the process of parturition. *J. Soc. Gynecol. Investig.* 8, 122–133.
577. Mitchell, B., and Chibbar, R. (1995). Synthesis and metabolism of oxytocin in late gestation in human decidua. *Adv. Exp. Med. Biol.* 395, 365–380.
578. Lamont, R. (2003). The development and introduction of anti-oxytocic tocolytics. *Br. J. Obstet. Gynaecol.* 110, 108–112.
579. Blanks, A. M., and Thornton, S. (2003). The role of oxytocin in parturition. *Br. J. Obstet. Gynaecol.* 110, 46–51.
580. Young, W., Shepard, E., and Amico, J. (1996). Deficiency in mouse oxytocin prevents milk ejection, but not fertility or parturition. *J. Neuroendocrinol.* 8, 847–853.
581. Nishimori, K., Young, L., Guo, Q., et al. (1996). Oxytocin is required for nursing but is not essential for parturition or reproductive behavior. *Proc. Natl. Acad. Sci. U S A* 93, 11699–11704.
582. Gross, G., Imamura, T., and Luedke, C. (1998). Opposing actions of prostaglandins and oxytocin determine the onset of murine labor. *Proc. Natl. Acad. Sci. U S A* 95, 11871–11875.
583. Muglia, L. (2000). Genetic analysis of fetal development and parturition control in the mouse. *Pediatr. Res.* 47, 443.
584. Fuchs, A., Fuchs, F., Husslein, P., Soloff, M., and Fernstrom, M. (1982). Oxytocin receptors and human parturition: a dual role for oxytocin in the initiation of human labor. *Science* 215, 1396–1398.
585. Soloff, M., Alexandrova, M., and Fernstrom, M. (1979). Oxytocin receptor: triggers for parturition and lactation. *Science* 204, 1313–1315.
586. Riemer, R. K., Goldfien, A. C., Goldfien, A., and Roberts, J. M. (1986). Rabbit uterine oxytocin receptors and in vitro contractile response: abrupt changes at term and the role of eicosanoids. *Endocrinology* 119, 699–709.
587. El, A. A., Onoris, E., Cynober, E., and Germain, G. (1990). Heterogeneity of oxytocin receptors in the pregnant rat myometrium near parturition. *Eur. J. Pharmacol.* 186, 231–238.
588. Wathes, D. C., Smith, H. F., Leung, S. T., et al. (1996). Oxytocin receptor development in ovine uterus and cervix throughout pregnancy and at parturition as determined by in situ hybridization analysis. *J. Reprod. Fertil.* 106, 23–31.
589. Mitchell, B., Fang, X., and Wong, S. (1998). Oxytocin: a paracrine hormone in the regulation of parturition? *Rev. Reprod.* 3, 113–122.
590. Chibbar, R., Miller, F., and Mitchell, B. (1993). Synthesis of oxytocin in amnion, chorion, and decidua may influence the timing of human parturition. *J. Clin. Invest.* 91, 185–192.
591. Lefebvre, D. L., Giaid, A., Bennett, H., Lariviere, R., and Zingg, H. H. (1993). Oxytocin gene expression in rat uterus. *Science* 256, 1553.
592. Lefebvre, D. L., Lariviere, R., and Zingg, H. H. (1993). Rat amnion: a novel site of oxytocin production. *Biol. Reprod.* 48, 632–639.
593. Chibbar, R., Wong, S., Miller, F., and Mitchell, B. (1995). Estrogen stimulates oxytocin gene expression in human chorio-decidua. *J. Clin. Endocrinol. Metab.* 80, 567–572.
594. Richard, S., and Zingg, H. H. (1990). The human oxytocin promoter is regulated by estrogens. *Biol. Reprod.* 265, 6098–6103.
595. Phaneuf, S., Rodriguez Linares, B., TambyRaha, R., and Lopez Bernal, A. (2000). Loss of myometrial oxytocin receptors during oxytocin-induced and oxytocin-augmented labour. *J. Reprod. Fertil.* 120, 91–97.
596. Molnar, M., Rigo, J., Jr., Romero, R., and Hertelendy, F. (1999). Oxytocin activates mitogen-activated protein kinase and up-regulates cyclooxygenase-2 and prostaglandin production in human myometrial cells. *Am. J. Obstet. Gynecol.* 181, 42–49.
597. Grazzini, E., Guillon, G., Mouillac, B., and Zingg, H. H. (1998). Inhibition of oxytocin receptor function by direct binding of progesterone. *Nature* 392, 509–512.
598. Astle, S., Khan, R. N., and Thornton, S. (2003). The effects if a progesterone metabolite, 5-β-dihydroprogesterone, on oxytocin receptor binding in human myometrial membranes. *Br. J. Obstet. Gynaecol.* 110, 589–592.
599. Smith, R. (1999). The timing of birth. *Sci. Am.* 280, 68–75.
600. McLean, M., and Smith, R. (2001). Corticotrophin-releasing hormone and human parturition. *Reproduction* 121, 493–501.
601. Smith, R., Mesiano, S., and McGrath, S. (2002). Hormone trajectories leading to human birth. *Regul. Pept.* 108, 159–164.
602. Hillhouse, E., and Grammatopoulos, D. (2002). Role of stress peptides during human pregnancy and labour. *Reproduction* 124, 323–329.

603. McLean, M., Bisits, A., and Davies, J. (1995). A placental clock controlling the length of human pregnancy. *Nat. Med.* 1, 460–463.
604. Riley, S., Walton, J., and Herlick, J. (1991). The localization and distribution of corticotropin-releasing hormone in the human placental and fetal membranes throughout gestation. *J. Clin. Endocrinol. Metab.* 72, 1001–1007.
605. Cheng, Y.-H., Nicholson, R., King, B., et al. (2000). Glucocorticoid stimulation of corticotropin-releasing hormone gene expression requires a cyclic adenosine 3′,5′-monophosphate regulatory element in human primary placental cytotrophoblast cells. *J. Clin. Endocrinol. Metab.* 85, 1937–1945.
606. Cheng, Y.-H., Nicholson, R., King, B., et al. (2000). Corticotropin-releasing hormone gene expression in primary placental cells is modulated by cyclic adenosine 3′,5′-monophosphate. *J. Clin. Endocrinol. Metab.* 85, 1239–1244.
607. Ni, X., Hou, Y., Yang, R., et al. (2004). Progesterone receptors A and B differentially modulate corticotropin-releasing hormone gene expression through a cAMP regulatory element. *Cell. Mol. Life Sci.* 61, 1114–1122.
608. Mariononi, E., Korebrits, C., Di Iorio, R., Cosmi, E., and Challis, J. (1998). Effect of betamethasone in vivo on placental corticotrophin-releasing hormone (CRH) in human pregnancy. *Am. J. Obstet. Gynecol.* 178, 770–778.
609. Korebrits, C., Yu, D., and Ramirez, M. (1998). Antenatal glucocorticoid administration increases corticotrophin-releasing hormone (CRH) in maternal plasma. *Br. J. Obstet. Gynaecol.* 105, 556–561.
610. Elliott, J., and Radin, T. (1995). The effect of corticosteroid administration on uterine activity and preterm labor in high order multiple gestations. *Obstet. Gynecol.* 85, 250–254.
611. Yeshaya, A., Orvieto, R., Ben-Shem, E., et al. (1996). Uterine activity after betamethasone administration for the enhancement of fetal lung maturation. *Eur. J. Obstet. Reprod. Biol.* 67, 139–141.
612. Hobel, C., Dunkel-Schetter, C., Roesch, S., Castro, L., and Arora, C. (1999). Maternal plasma corticotropin-releasing hormone associated with stress at 20 weeks' gestation in pregnancies ending in preterm delivery. *Am. J. Obstet. Gynecol.* 180, S257–S263.
613. Lockwood, C. (1999). Stress-associated preterm delivery: the role of corticotropin-releasing hormone. *Am. J. Obstet. Gynecol.* 180, S264–S266.
614. Dayan, J., Creveuil, C., Herlicoviez, M., et al. (2002). Role of anxiety and depression in the onset of spontaneous preterm labor. *Am. J. Epidemiol.* 155, 293–301.
615. Hobel, C., Roesch, S., and Dunkel-Schetter, C. (1997). Idiopathic preterm birth: could the "maternal stress reaction" initiate the placental-maternal-fetal cascade? In *Preterm Birth: Etiology, Mechanisms, and Prevention* (S. Charleston, Ed.). University of Alabama Press, Birmingham.
616. Challis, J., and Brooks, A. (1989). Maturation and activation of hypothalamic-pituitary-adrenal function in sheep. *Endocr. Rev.* 10, 182–204.
617. Korebrits, C., Ramirez, M., and Watson, L. (1998). Maternal CRH as a predictor of impending preterm birth. *J. Clin. Endocrinol. Metab.* 83, 1585–1591.
618. Stevens, M., Challis, J., and Lye, S. (1998). Corticotropin-releasing hormone receptor subtype 1 is significantly up-regulated at the time of labor in the human myometrium. *J. Clin. Endocrinol. Metab.* 83, 4107–4115.
619. Aggelidou, E., Hillhouse, E. W., and Grammatopoulos, D. (2002). Up-regulation of nitric oxide synthase and modulation of the guanylate cyclase activity by corticotropin-releasing hormone but not urocortin II or urocortin III in cultured human pregnant myometrial cells. *Proc. Natl. Acad. Sci. U S A* 99, 3300–3305.
620. Karteris, E., Hillhouse, E., and Grammatopoulos, D. (2004). Urocortin II is expressed in human pregnant myometrial cells and regulates myosin light chain phosphorylation: potential role of the type-2 corticotropin-releasing hormone receptor in the control of myometrial contractility. *Endocrinology* 145, 890–900.
621. Lockwood, C. J. (2002). Biochemical predictors of prematurity. *Front. Horm. Res.* 27, 258–268.
622. Lockwood, C. J., Senyei, A. E., Dische, M. R., Casal, D., et al. (1991). Fetal fibronectin in cervical and vaginal secretions as a predictor of preterm delivery. *N. Engl. J. Med.* 325, 669–674.
623. Ma, Y., Lockwood, C. J., Bunim, A. L., et al. (2000). Cell type specific regulation of fetal fibronectin expression in amnion: conservation of glucocorticoid responsiveness in human and nonhuman primates. *Biol. Reprod.* 52, 1812–1817.
624. Elliott, J., and Radin, T. (1995). The effect of corticosteroid administration on uterine activity and preterm labor in high-order multiple gestations. *Obstet. Gynecol.* 85, 250–254.
625. Malak, T., M., and Bell S. C. (1994). Structural characteristics of term human fetal membranes: a novel zone of extreme morphological alteration within the rupture site. *Br. J. Obstet. Gynaecol.* 101, 375–386.
626. McLaren, J., Malak, T., and Bell, S. (1999). Structural characteristics of term human fetal membranes prior to labour: identification of an area of altered morphology overlying the cervix. *Hum. Reprod.* 14, 237–241.
627. Birkedale-Hansen, H., Moore, W. G., Bodden, M. K., et al. (1993). Matrix metalloproteinases: a review. *Crit. Rev. Oral Biol. Med.* 4, 197–250.
628. Bryant-Greenwood, G., and Yamamoto, S. (1995). Control of peripartal collagenolysis in the human chorion-decidua. *Am. J. Obstet. Gynecol.* 172, 63–70.
629. Draper, D., McGregor, J., Hall, J., et al. (1995). Elevated protease activities in human amnion and chorion correlate with preterm premature rupture of membranes. *Am. J. Obstet. Gynecol.* 173, 1506–1512.
630. Riley, S., Leask, R., Denison, F., et al. (1999). Secretion of tissue inhibitors of matrix metalloproteinases by human fetal membranes, decidua and placenta at parturition. *J. Endocrinol.* 162, 351–359.
631. Athayde, N., Edwin, S. S., Romero, R., et al. (1998). A role for matrix metalloproteinase-9 in spontaneous rupture of the fetal membranes. *Am. J. Obstet. Gynecol.* 179, 1248–1253.
632. Meisser, A., Chardonnens, D., Campana, A., and Bischof, P. (1999). Effects of tumour necrosis factor-α, interleukin-1α, macrophage colony stimulating factor and transforming growth factor β on trophoblastic matrix metalloproteinases. *Mol. Hum. Reprod.* 5, 252–260.
633. So, T., Ito, A., Sato, T., Mori, Y., and Hirakawa, S. (1992). Tumor necrosis factor-α stimulates the biosynthesis of matrix metalloproteinases and plasminogen activator in cultured human chorionic cells. *Biol. Reprod.* 46, 772–778.
634. Ito, A., Nakamura, T., Uchiyama, T., et al. (1994). Stimulation of the biosynthesis of interleukin 8 by interleukin 1 and tumor necrosis factor a in cultured human chorionic cells. *Biol. Pharm. Bull.* 17, 1463–1467.
635. Trautman, M., Dudley, D., Edwin, S., Collmer, D., and Mitchell, M. D. (1992). Amnion cell biosynthesis of interleukin-8: regulation by inflammatory cytokines. *J. Cell. Physiol.* 153, 38–43.
636. McLaren, J., Taylor, D., and Bell, S. (1999). Increased incidence of apoptosis in non-labour-affected cytotrophoblast cells in term fetal membranes overlying the cervix. *Hum. Reprod.* 14, 2895–2900.
637. Runic, R., Lockwood, C., LaChapelle, L., et al. (1998). Apoptosis and Fas expression in human fetal membranes. *J. Clin. Endocrinol. Metab.* 83, 660–666.

638. McLaren, J., Taylor D. J., and Bell, S. C. (2000). Increased concentration of pro-matrix metalloproteinase 9 in term fetal membranes overlying the cervix before labor: implications for membrane remodeling and rupture. *Am. J. Obstet. Gynecol.* 182, 409–416.
639. Xu, P., Alfaidy, N., and Challis, J. R. (2002). Expression of matrix metalloproteinase (MMP)-2 and MMP-9 in human placenta and fetal membranes in relation to preterm and term labor. *J. Clin. Endocrinol. Metab.* 87, 1353–1361.
640. McLaren, J., Taylor, D. J., and Bell, S. C. (2000). Prostaglandin E2-dependent production of latent matrix metalloproteinase-9 in cultures of human fetal membranes. *Mol. Hum. Reprod.* 6, 1033–1040.
641. Bryant-Greenwood, G., and Millar, L. (2000). Human fetal membranes: their preterm premature rupture. *Biol. Reprod.* 63, 1575–1579.
642. Qin, X., Chua, P., Ohira, R., and Bryant-Greenwood, G. (1997). An autocrine /paracrine role of human decidual relaxin: II. Stromelysin-1 (MMP-3) and tissue inhibitor of matrix metalloproteinase-1 (TIMP-1). *Biol. Reprod.* 56, 812–820.
643. Bogic, L. V., Yamamoto, S. Y., Millar, L. K., and Bryant-Greenwood, G. D. (1997). Developmental regulation of the human relaxin genes in the decidua and placenta: overexpression in the preterm premature rupture of the fetal membranes. *Biol. Reprod.* 57, 908–920.
644. Ferrand, P., E., Parry, S., Sammel, M., et al. (2002). A polymorphism in the matrix metalloproteinase-9 promoter is associated with increased risk of preterm premature rupture of membranes in African Americans. *Mol. Hum. Reprod.* 8, 494–501.
645. Fujimoto, T., Parry, S., Urbanek, M., et al. (2002). A single nucleotide polymorphism in the matrix metalloproteinase-1 (MMP-1) promoter influences amnion cell MMP-1 expression and risk for preterm premature rupture of the fetal membranes. *J. Biol. Chem.* 277, 6296–6302.
646. Tsatsaris, V., Papatsonis, D., Goffinet, F., Dekker, G., and Carbonne, B. (2001). Tocolysis with nifedipine or -adrenergic agonists: a meta-analysis. *Obstet. Gynecol.* 97, 840–847.
647. King, J. F., Flenady, V. J., Papatsonis, D. N., Dekker, G. A., and Carbonne, B. (2002). Calcium channel blockers for inhibiting preterm labour. *Cochrane Database Syst. Rev.* (2) CD002255.
648. Norton, M. E., Merril, J., Cooper, B. A., Kuller, J. A., and Clyman, R. I. (1993). Neonatal complications after the administration of indomethacin for preterm labor. *N. Engl. J. Med.* 329, 1602–1607.
649. Besinger, R. E., Niebyl, J. R., Keyes, W. G., and Johnson, T. R. (1991). Randomized comparative trial of indomethacin and ritodrine for the long term treatment of preterm labor. *Am. J. Obstet. Gynecol.* 164, 986–988.
650. Holmes, R. P., and Stone, P. R. (2000). Severe oligohydramnios induced by cyclooxygenase-2 inhibitor nimesilide. *Obstet. Gynecol.* 96, 810–811.
651. Locatelli, A., Vergani, P., Bellini, P., Strobelt, N., and Ghidini, A. (2001). Can a cyclooxygenase type-2 selective tocolytic agent avoid the fetal side effects of indomethacin? *Br. J. Obstet. Gynaecol.* 108, 325–362.
652. Olson, D., Mijovic, J., Zaragoza, D., and Cook, J. (2001). Prostaglandin endoperoxide H synthase type 1 and type 2 messenger ribonucleic acid in human fetal tissues throughout gestation and in the newborn infant. *Am. J. Obstet. Gynecol.* 184, 169–174.
653. Moutquin, J., Sherman, D., Cohen, H., et al. (2000). Double-blind randomized, controlled trial of atosiban and ritodrine in the treatment of preterm labor: a multicentre effectiveness and safety study. *Am. J. Obstet. Gynecol.* 182, 1191–1199.
654. Peri, K., Quiniou, C., Hou, X., et al. (2002). THG113: a novel selective FP antagonist that delays preterm labor. *Semin. Perinatol.* 26, 389–397.

Knobil and Neill's Physiology of Reproduction, Third Edition
edited by Jimmy D. Neill,
Elsevier © 2006

CHAPTER 56

Developmental Origins of Health and Disease: Theoretical Considerations, Mechanisms, and Implications

Peter D. Gluckman[1] and Mark A. Hanson[2]

Introduction, 2975
Background, 2975
The Scope of the Problem, 2976
Experimental Observations, 2977
Altered Tissue Differentiation, 2977
Epigenetic Processes, 2979
Mitochondrial Function, 2979
Evolutionary and Theoretical Perspectives, 2980
Maternal Constraint, 2983
Transitions, 2984
Maternal Health, 2985
Transgenerational Influences, 2986
Postnatal Factors, 2986
Interventions, 2986
References, 2987

INTRODUCTION

There is now considerable interest in the concept of developmental origins of health and disease (DOHaD). This has arisen partly as a result of retrospective epidemiological studies showing the links between the early environment and risk of later disease, partly as a result of research in experimental animals that shows that perturbation of developmental processes can indeed produce pathophysiological and epigenetic changes in the adult offspring, and partly as a result of theoretical work that brings together insights from developmental, evolutionary, and life history biology. In addition, prospective epidemiological studies have now been launched and detailed clinical investigations are being undertaken. The implications of the DOHaD concept are being realized and possible interventions to prevent this component of the etiology of chronic disease are being actively considered. In this chapter, we discuss the empirical and theoretical basis for the DOHaD concept, examine current ideas about the underlying mechanisms, and explore the implications of the concept for the global burden of chronic disease.

Background

In the late 1980s, retrospective studies of cohorts in the United Kingdom revealed a striking statistical correlation between size at birth and subsequent mortality from heart disease (1). The smaller the infant at birth, the greater the risk of coronary heart disease through life. The observations were soon extended to demonstrate inverse relationships between birth size (most often birth weight) and the risks for development of hypertension (2), type 2 diabetes mellitus, or glucose intolerance (3), and to

[1]Liggins Institute, University of Auckland, Auckland, New Zealand;
[2]Centre for Developmental Origins of Health and Disease, University of Southampton, Southampton, U.K.

show that the effects were demonstrable in other populations (4–6). These observations led to the proposition that these adult life-style diseases had a "fetal origin," and this was termed the "fetal origins of adult disease" concept or the "Barker hypothesis," David Barker being the epidemiologist who made these seminal observations. Many subsequent epidemiological studies essentially confirmed these results, but they were not without controversy. Arguments about confounding variables such as smoking or social class and considerations of the plausibility of a causal relationship between fetal growth and adult disease arose (7). The debate became heated on the issue of the effect size. The absence of a clear understanding of the possible underlying mechanisms fueled these debates.

The original study had used mortality as the outcome measure but virtually all subsequent studies had to use surrogates of disease such as measures of blood pressure. For a while, the epidemiology became dominated by the relationship between blood pressure and birth weight—which was based on a misunderstanding of the original findings. In fact, blood pressure per se had been used only as a surrogate for the development of heart disease—further thinking rapidly revealed that birth weight was not causal to the development of heart disease, but itself a surrogate for some environmental influences on fetal development [discussed in Barker (8) and Jackson (9)]. Studies on the progeny of women who were pregnant at the time of the Dutch winter famine of 1944 and 1945, in which a previously well-nourished population became acutely undernourished, showed long-term consequences to those adults who were exposed as fetuses without alterations in their birth weight (10). From the outset, epidemiological studies showed an interaction between birth size, adult measures of obesity (e.g., body mass index [BMI] or ponderal index), and disease risk. Those most at risk were born small and were heavy as adults. Further observations on Finnish cohorts born in the 1930s suggested that the pattern of infant growth could also influence lifetime disease risk for diabetes, with both growth faltering in infancy (11) and early adiposity rebound with rapid weight gain before adolescence (12) contributing to disease risk. As a result of these complexities, and with the emergence of a body of relevant experimental data, the general model became known as "the developmental origins of adult disease." Essentially, this states that environmental factors acting in early life, often in embryonic or fetal life, affect the structural and functional development of the individual so that its risks for development of disease in later life are altered.

The Scope of the Problem

The first issue to arise is which diseases have their origins in early life. There is unequivocal evidence for several forms of cardiovascular disease (i.e., heart disease, stroke, and hypertension) (13), for type 2 diabetes (14,15), and for osteoporosis (16). In addition, there is very strong evidence for early life effects on cognitive decline in the elderly (17) and some forms of affective disorders (18). Evidence for certain forms of cancer, particularly breast (19) and testicular cancer [see Akre et al. (20) and Montgomery et al. (21) and references therein], being associated with enhanced growth in early life is accumulating but at this stage is not unequivocal.

The cost implications of the aforementioned major classes of disease are enormous. It is estimated that 15 million people die of cardiovascular disease across the globe annually; nearly 1 million people died in the United States of coronary heart disease in 2002 (22). Over 100 million people in the world have type 2 diabetes, and this is estimated to rise to 250 million people by 2025 (23). Each year it is estimated that 1.5 million people suffer an osteoporosis-related fracture. In senior citizens with hip fracture, the subsequent decline in physical and mental health leads to 20% dying within 1 year (24). One of every two women older than age 50 years will have an osteoporosis-related fracture in their lifetime, with the number of hip fractures in the United States doubling or even tripling by 2020. Worldwide, there were 1.7 million reported hip fractures in 1990, and this number is expected to increase to 6 million by 2050 (25). Differences in adult life-style go only a small way to explaining why one person experiences such diseases when another does not. In addition, life-style does not account for the higher rates of these diseases among poorer people and ethnic minorities.

As noted, a range of epidemiological studies have now established unequivocally that small infants have a higher risk as adults of hypertension, raised serum cholesterol (26), raised plasma fibrinogen and clotting factors (27), raised C-reactive protein (28), impaired glucose tolerance and type 2 diabetes (3), and obesity (29). Additional studies have shown endothelial dysfunction is related to low birth weight (30). This cluster of symptoms is often grouped under the heading of the *metabolic syndrome*, and Fig. 1 shows that the incidence of this syndrome is greater in lower–birth-weight infants. It is important to note that the effect is graded across the normal birth weight range, and is not just a feature of infants who are small for gestational age or exhibit intrauterine growth retardation. In fact, given that

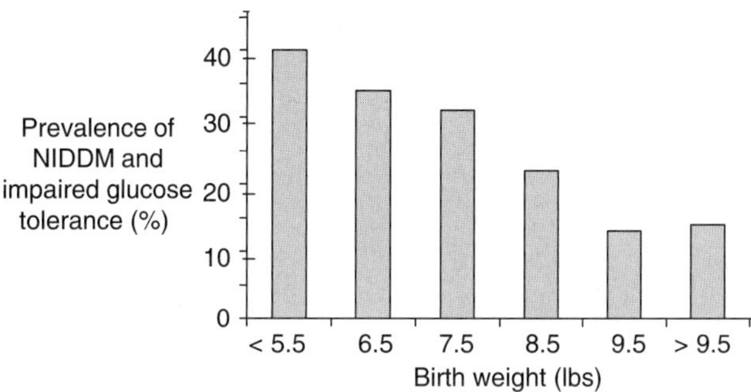

FIG. 1. Prevalence of type 2 diabetes (non–insulin-dependent diabetes [NIDDM]) and impaired glucose tolerance in 370 men aged 59 to 70 years. (Data from Godfrey, K. M. [1998]. Maternal regulation of fetal development and health in adult life. *Eur. J. Obstet. Gynecol. Reprod. Biol.* 78, 141–150.)

these data relate to a retrospective study of adults born in the county of Hertfordshire (United Kingdom) in the middle of the last century, the survival rate of growth-retarded infants would have been low and they probably form only a very small proportion of this data set.

The second issue that arises concerns the magnitude of this early life effect—how important is it to the etiology of these life-style diseases? One study (31) demonstrates that there is a nearly fourfold greater risk of coronary heart disease in adult men who were of low ponderal index at birth and whose BMI exceeded 18 at 12 years of age. Calculations from the same study suggest that if lower–birth-weight infants could be brought to the mean for the population and if growth until 1 year of life postnatally could similarly be optimized, then the risk of later hypertension would be reduced by 25%, and that of type 2 diabetes by 57% (32). The magnitude of the effects apparent from this study alone makes the phenomenon one that clearly has to be taken very seriously. Questions arise about how representative this Finnish population is of current westernized societies, but even if the effects are an order of magnitude greater than currently operating, they still indicate that early development provides a greater component to risk than does, for example, smoking. A further fundamentally important point to emerge from the Helsinki studies is that the effect of smaller size at birth is amplified by poor growth in the early years of childhood followed by a tendency to obesity in the preadolescent years. Clearly, we cannot single out a single period of development in determining this risk: a life course approach needs to be adopted.

Until recently, a considerable degree of research into the risks of cardiovascular disease concerned genetic predisposition. The Finnish cohort provides an alternative view by considering early environmental effects. For example, the Pro12Pro polymorphism in the peroxisome proliferator–activated receptor (PPAR)-γ2 gene, which confers an elevated risk of type 2 diabetes in adulthood, has been examined in the context of birth weight. It was found (33) that adult plasma fasting insulin or homeostasis model assessment (HOMA) index were elevated in those individuals with a combination of the Pro12Pro polymorphism and low birth weight. There were no effects of birth weight on fasting insulin or HOMA index in members of the population who were Pro12Ala or Ala12Ala. Hence, individuals with the polymorphism in the PPAR-γ2 gene have a predisposition to type 2 diabetes that is brought out by a reduction in fetal growth. The concept of gene–environment interactions now underpins DOHaD and makes it clear that neither genetic factors nor environmental factors alone can explain the phenomenon.

EXPERIMENTAL OBSERVATIONS

Data obtained from experimental approaches have confirmed that transient environmental exposures, such as unbalanced maternal nutrition or glucocorticoid administration, during early development can have consequences for adult physiology that are compatible with the epidemiological observations. Many studies have focused on the control of glucose homeostasis (34), insulin secretion and action (35), blood pressure regulation (36), and renal (37) and endothelial function (38), but there is also an increasing focus on changes in neural and endocrine functions, such as the regulation of appetite (39) and control of the hypothalamic–pituitary–adrenal axis (40). These studies have not only demonstrated changes in systems suspected to be involved in disease, but have pointed to the fundamental underlying mechanisms by which developmental plasticity leads to irreversible changes in developmental trajectory. Much attention has been focused on epigenetic processes, such as DNA methylation and histone acetylation, as the processes underlying the developmental origins phenomenon (41,42). These may become manifest as altered tissue differentiation or altered regulatory function.

Altered Tissue Differentiation

During the rat perinatal period, the cells of the developing pancreatic islet undergo apoptosis, likely reflecting the transition from a fetal form of islet cell to a postnatal form with a different regulatory profile (43). In rat pups exposed to low-protein diets in utero, the rate of beta cell apoptosis is increased (44), perhaps because of an increased sensitivity to cytokines (45). Inhibition of this apoptosis in the neonatal period prevents the development of subsequent diabetes (46). In rats, maternal protein undernutrition during pregnancy also leads to offspring that initially demonstrate an increased insulin response to a glucose challenge, followed later by reduced insulin secretion and degranulation of the islets. This suggests that the homeostatic capability of the pancreas has been exceeded, and may parallel the development of tissue insulin resistance. Similar effects have been reported for the offspring of dams fed a high-fat diet, and these have also been linked to disturbances in mitochondrial function (47).

Adult nephron number is reduced both in humans who are born small and in animals experimentally manipulated in utero, leading to the development of hypertension (48,49). According to the Brenner hypothesis (50), it is the increased filtration performed by the remaining nephrons that compromises their overall function and leads to hypertension of renal origin.

In the rat, maternal undernutrition leads to reduced vascularity in various tissues, including the pancreas and skeletal muscle (51,52). Such reduced peripheral vascular density may contribute indirectly to the peripheral insulin resistance seen in the adult offspring.

In addition, programming in utero alters the ratio of periportal to perivenous hepatocytes, leading to an altered ratio of phosphoenolpyruvate carboxykinase (PEPCK) to glucokinase-enriched cells and relatively greater hepatic glucose production (53). Expression of lipid pathway enzymes, including carnitine palmitoyl transferase, is also suppressed (54), and expression of insulin-like growth factor (IGF)–binding protein-1 is increased, possibly reflecting concurrent insulin resistance (55). Skeletal muscle from programmed rat pups is resistant to both IGF-1 and insulin (56), and demonstrates mitochondrial dysfunction and reduced expression of protein kinase C-ζ, which may lead to altered glucose transporter-4 (GLUT4) mobilization (57,58). Cardiac muscle from such rats also demonstrates reduced expression of GLUT1, GLUT4, and hexokinase II (59).

Insulin resistance or deficiency appears central to the clinical observations. Glucose intolerance and type 2 diabetes were one of the first correlates found in the epidemiological observations. In general, there appears to be a change in body composition with a reduction in skeletal muscle relative to fat—both in adults with metabolic syndrome and in infants born small. Recent data have shown that growth-retarded infants preserve visceral fat at the expense of subcutaneous fat and muscle (60).

The endothelium is another target of particular focus. Vascular endothelial cells play an important role in normal life, regulating (61)

- Vessel caliber
- Vascular remodeling
- Tissue and organ growth
- Tissue metabolism
- Immune responses
- Blood fluidity
- Platelet and white cell stickiness
- Vascular permeability

Endothelial function has been shown to be disturbed in hypertension (61), atherogenesis (61), type 2 diabetes (62,63), coronary heart disease (61), in the metabolic syndrome and obesity (64,65), and in diseases of pregnancy, including preeclampsia (66).

Hyperglycemia (through poor control of blood glucose) induces hepatic overproduction of superoxide through the mitochondrial respiratory change. This in turn inhibits glyceraldehyde-3-phosphate dehydrogenase (GAPDH) and leads to hepatic mitochondrial damage. This is followed by vascular and other pathophysiological effects (67). Endothelial function was disturbed in both male and female offspring of dams fed the high-fat diet during pregnancy (68); however, if the offspring of these high-fat-fed dams were also fed the high-fat diet, their endothelial function outcome was no different from control animals (68). Similar observations were made by Norman and LeVeen (69) in piglets whose mothers were fed an atherogenic diet and who themselves were fed an atherogenic or controlled diet. Only animals fed an atherogenic diet prenatally, followed by a normal diet postnatally, showed evidence of atherosclerotic plaques in their coronary arteries. These results illustrate nicely a hypothesis we term the *predictive adaptive response*. We discuss predictive adaptive responses in more detail later on.

Endothelial function, as measured by flow-mediated vasodilation, is inversely related to birth weight (70) and is also associated with preterm birth (71). In animal models it can be assessed using wire myography (72). Feeding a low-protein diet to rats during pregnancy reduces endothelial function in the uterine arteries of those rats (73) and also produces a reduction in both endothelial function and smooth muscle vasodilator responses to nitric oxide in the offspring (74,75). How early these effects develop is not known in the rat, although they have been shown to be evident by mid-gestation in the sheep fetus (76). A range of epidemiological and animal data now suggests that the effect of dietary imbalance during pregnancy depends on the timing of the insult. Interest in the early gestational period comes from studies of the Dutch "hunger winter" of 1944 and 1945. Fetuses whose mothers were exposed to famine during the first trimester had a greater risk of cardiovascular disease as adults, whereas those exposed to famine during the third trimester had an increased risk of type 2 diabetes (10). Interestingly, the former group were not smaller at birth. We have made parallel observations in the guinea pig in which global nutrition was reduced by 50% in the first or the second half of pregnancy (77). Offspring of dams nutrient restricted during the first half of pregnancy were not smaller at birth but had elevated blood pressure and left ventricular hypertrophy as adults. Conversely, offspring of dams nutrient restricted in late pregnancy were smaller at birth but had normal blood pressure and left ventricular mass as adults. More detailed consideration of environmental effects on the early embryo comes from the work of Kwong and colleagues (78), who exposed pregnant rats to a low-protein diet for the first 4.25 days of gestation (i.e., before implantation). Offspring had elevated blood pressure and changes in embryonic cell allocation.

Epigenetic Processes

Such effects in early gestation focus attention on mechanisms that involve epigenetic processes. The best-described concern imprinted genes that have an underlying role in fetal overgrowth or large offspring syndrome (79). The work of Constancia et al. has explored the mechanisms for this process in the mouse through differential effects on the IGF-2/H19 cluster (80) and through maternal imprinting of the IGF-2 receptor versus paternal imprinting of IGF-2 peptide (81). In addition, they have shown that a specific transcript, IGF-2 P0, which is expressed in the placenta, is also paternally imprinted with growth-promoting effects (81). The IGF-2 receptor is not imprinted in humans, so whether analogous processes operate is unclear.

The DNA methylation processes underlying such imprinting, and indeed epigenetic control of the genome depends on the supply of one-carbon groups from dietary amino acid. Methyl groups are predominantly supplied from glycine, a nonessential amino acid but one for which the fetal requirements are very large in late gestation (82). In humans, fetal glycine provision was thought to arise primarily from system A transport and serine-to-glycine interconversion mediated by the enzyme serine hydroxymethyl transferase (SHMT). However, we have shown (83) that although this may operate in the sheep, the activity of SHMT is considerably less in the human placenta, suggesting that other sources of one-carbon groups must therefore be more relevant in the human fetus. In the rat, supplementation of the low-protein diet with glycine rectifies the endothelial dysfunction of the pregnant dam (84) and also prevents the onset of hypertension in the offspring (85).

Because the provision of methyl groups depends on cofactors, including folate, it may be that maternal dietary folate provision can also have a key effect on epigenetic processes in the offspring. It is therefore of interest that folate supplementation of the low-protein diet in the rat prevented the elevated blood pressure, vascular dysfunction, and reduced endothelial nitric oxide synthase expression in the offspring (C. Torrens et al., unpublished data). Other examples of epigenetic processes that produce long-term effects on the offspring through changes in DNA methylation/histone acetylation are the effects of uterine blood flow restriction on renal P53 gene expression in the kidneys (86); the effects of maternal licking and grooming behavior on the methylation of the nerve growth factor-1A binding region of the glucocorticoid receptor promoter in the hippocampus (42); and the well-established effects of micronutrient supplementation in preventing the induction of the agouti mutation in the mouse (87).

Specific target genes that may be affected in early life to produce later metabolic syndrome are now being researched. Prominent among these are the PPARs (88). In the fasting state, PPAR-α promotes fatty acid uptake by the liver and muscle, where it undergoes peroxisomal and mitochondrial β-oxidation. Gluconeogenesis is also promoted by PPAR-α, and this cluster of effects is glucose sparing. In the fed state, PPAR-γ promotes lipoprotein lipase activity, and lipoproteins are taken up by adipocytes and stored as triglycerides. In addition, PPAR-γ sensitizes insulin-mediated glucose uptake (88). In the liver of offspring of protein-restricted rats, we have shown increased expression of PPAR-α coupled with reduced methylation of the promoter region of the gene (89). These processes were both reversed by dietary folate supplementation of the dam. No effects were seen on PPAR-γ expression in the liver. In addition, the changes in PPAR-α were mirrored by effects on acyl co-oxidase, the rate-limiting step of fatty acid β-oxidation, which is stimulated by PPAR-α. Finally, there were also effects on hepatic glucocorticoid receptor expression, which is one of the modulators of PPAR-α expression.

Mitochondrial Function

The effects on mitochondrial function have formed the focus of recent interest. Mitochondria are central to metabolic control, and mitochondrial DNA is susceptible to environmental effects that could produce changes in mitochondrial copy number.

Such epigenetic effects may operate by changes in mitochondrial DNA methylation, by the effects of proinflammatory cytokines, and through the effects of reactive oxygen species (90). These effects are proposed to form part of the basis of the metabolic syndrome itself. Changes in mitochondrial DNA are passed through the female line to future generations, so they offer the possibility of a transgenerational process for induction of phenotype. Support for this has been gained through studies in which animals were bred over 11 generations to select for reduced exercise tolerance (91). The animals then showed all the components of the human metabolic syndrome and underlying defects in mitochondrial function. Mitochondria are also thought to be a key target in the etiology of the metabolic syndrome because mitochondrial damage builds up over time. This therefore offers a possible mechanism involved in diseases developing over the life course. We have shown impaired mitochondrial function in the offspring of rats fed a high-fat diet during pregnancy, coupled with insulin and leptin resistance and relative insulin depletion of the pancreatic islets (47).

EVOLUTIONARY AND THEORETICAL PERSPECTIVES

From the first epidemiological observations of the developmental origins phenomenon, teleological and theoretical arguments were put forward to explain these relatively surprising observations. Barker and Hales proposed that the fetus had selected a "thrifty phenotype"—that is, to adapt to the deprived fetal environment, the fetus had selected a developmental trajectory involving insulin resistance and small size that, persisting after birth, would be more suited to a deprived ("thrifty") environment than the one to which it was actually exposed (92). The choice of the term "thrifty" followed an earlier use by Neel (93), who proposed that modern humans had evolved with "thrifty genes" to be able to conserve energy as hunter-gatherers—these genes were now more likely to induce obesity and insulin resistance and be disadvantageous in a modern environment. Indeed, some had argued that such thrifty genes could explain the epidemiological observations linking low birth weight and later disease risk. It was suggested, for example, that genetic variation in insulin sensitivity could explain a relationship between birth size and diabetes risk (94,95). However, such models could not explain findings such as those made on the offspring off the Dutch winter famine or the extensive data derived from experimental animals. The thrifty phenotype model persisted as a favored explanation, although it was difficult to see how it could adequately explain the epidemiological data, particularly those in larger infants. It was subsequently modified to suggest that individual fetal organs could respond to periods of maternal environmental change during critical periods and adapt their physiology to a deprived environment (96). But some induction of the phenomenon had to occur at higher birth sizes, too, given the continuous relationship between birth size and disease risk, so a "thrifty model" seemed inappropriate.

Experimental data had clearly shown that there were interactions between the setting of the prenatal environment and the postnatal nutritional environment (97) (Fig. 2), suggesting that the magnitude of the environmental transition between early and later life might be the key factor and could occur against any baseline nutritional state in development. This was also consistent with the later clinical observations on the Finnish cohort referred to earlier. Bateson (98) and we (99) have suggested that the prenatal settings establish the developmental trajectory, not so much for immediate advantage, but rather for later advantage in life. In other words, the developing organism predicts its future according to its assessment of the maternal environment and sets its physiological trajectories for that predicted environment. Disease is likely when there is a mismatch

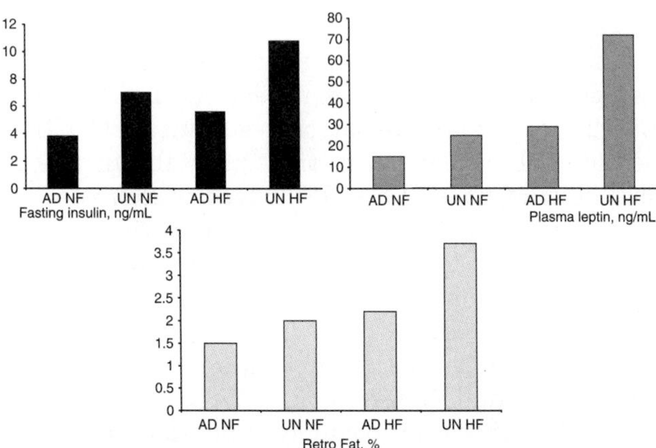

FIG. 2. Evidence that the prenatal gene–environmental interaction determines the impact of postnatal nutrition. Pregnant rats were fed ad libitum (AD) or undernourished throughout pregnancy (UN; 30% of an ad libitum diet). At birth, their pups were cross-fostered to ad libitum ("normal") rats until weaning, at which time the pups were placed on normal-fat (NF) or high-fat (HF) diets. Fasting insulin, leptin levels, and retroperitoneal fat pads were measured on day 125. Both prenatal undernutrition and postnatal high-fat diets have independent effects on all three measures (p <0.001), and for each there is a marked interaction between the prenatal and postnatal diet, demonstrating a synergistic effect (p <0.001). (Data from Vickers, M. H., Breier, B. H., Cutfield, W. S., Hofman, P. L., and Gluckman, P. D. [2000]. Fetal origins of hyperphagia, obesity and hypertension and its postnatal amplification by hypercaloric nutrition. *Am. J. Physiol.* 279, E83–E87.)

between the predicted and actual postmature environments. Modeling has demonstrated that this is an evolutionarily advantageous strategy (100). This is better understood when we consider our evolutionary history.

Early hominids developed an estimated 2,000,000 years ago, although *Homo sapiens* did not arise until approximately 200,000 years ago. For much of our evolutionary history since then, we were predominantly hunter-gatherers, with expansion beyond Africa only in the last 65,000 years, a cultural explosion some 30,000 years ago, and primitive agriculture and settlement appearing in the last 10,000 years. These developments resulted in a change in food supply and exposure to pathogens that constituted a major transition in our life-styles; in terms of evolutionary time, however, these developments have been recent. The changes that occurred in the industrial revolution (approximately 250 years ago) are even more recent, as of course are the enormous changes in diet and life-style that have occurred over the last 20 years or so. Thus, the bulk of human evolution occurred in a distinct and no-longer-extant environment [also termed the "environment of evolutionary adaptedness" (101)], and our capacity as a species to tolerate our current environment is constrained by this evolutionary past.

Table 1 shows the composition of diet believed to have been consumed by hunter-gatherers, and for which we are adapted, contrasted to the current North American dietary composition. Three things are apparent: first, the protein content of the contemporary diet has more than halved, and the meat sources of protein are now much higher in fat than they were during our evolutionary past; second, fruit and vegetable intake has also halved; and third, more than half of the contemporary diet is made up of carbohydrates and other high–glycemic-index sources of energy. These considerations highlight one of the major ways in which environmental transitions underpin the etiology of much current chronic disease.

There has been much research on the processes by which the developing organism is able to respond to its environment, and indeed on how it gets information about this environment. Much of this research focuses on "maternal effects." There are multiple examples of this process across plant and animal phyla (102). To draw one example from insects, the phenotypic shift from the solitary to the gregarious form of the desert locust is apparent at the larval stage after hatching. It is produced by chemical signals in the fluid surrounding the eggs laid by the mother, and these relate to the levels of nutrition and population density around her (103,104). When the population density is high the maternally transduced signal is transmitted to the eggs and a higher proportion of offspring are born in the gregarious, swarming phenotypes. These animals are capable of longer-duration flights, are able to consume a broader range of diets, and fly in swarms even during daylight hours. When population density is lower, there is no need for swarming to move the population to new sites, and a higher proportion of eggs hatch into the solitary form, which is nocturnal, green in color, and more selective about diet. There are many examples in mammals of equivalent maternal signaling effects. For example, in the meadow vole, maternal photoperiod before conception determines coat thickness in the offspring (105,106). When day length decreases in the autumn, offspring are born with a thicker coat than in the spring, when the day is lengthening. This provides a clear survival

TABLE 1. *Comparison of a late Paleolithic diet, a contemporary American diet, and U.S. dietary recommendations*

	Late Paleolithic diet[a]	Current American diet[b]	U.S. Senate Select Committee recommendations[b]
Total dietary energy (%)			
Protein	34	12	12
Carbohydrate	45	46	58
Fat	21	42	30
P:S ratio	1.41	0.44	1.00
Cholesterol (mg)	591	600	300
Fiber (g)	45.7	19.7[c]	30-60
Sodium (mg)	690	2,300-6,900	1,100-3,300
Calcium (mg)	1,580	740[d]	800-1,200[e]
Ascorbic acid (mg)	392.3	87.7[d]	45[e]

P:S, polyunsaturated:saturated fats.

[a] Assuming the diet contained 35% meat and 65% vegetables.

[b] Select Committee on Nutrition and Human Needs, United States Senate (1977). Dietary Goals for the United States. U.S. Government Printing Office, Washington, DC.

[c] British National Food Survey, 1976.

[d] U.S. Department of Agriculture Food Consumption Survey, 1977–1978.

[e] Recommended Daily Dietary Allowance, Food and Nutrition Board, National Academy of Sciences – National Research Council.

advantage for them during the winter or summer climatic conditions they will experience. However, the signal for adopting this development strategy is clearly detected prenatally and must come from the mother—intrauterine temperature is similar throughout the year. In these two examples, it is clear that phenotypic changes are made in response to a maternal signal received during development. Moreover, the change is made not in response to the developmental environment per se, but to provide an adaptive advantage in a later environment that the offspring expects to inhabit during its postlarval phase of independent life.

The concept of developmental "choices" is of course not a new one. Waddington (107) used the model of a ball rolling down a gradient in a valley and subsequently being channeled down separate pathways depending on which subsidiary valley it entered. At each juncture when it could go one way or another, a "decision" was made that was irrevocable. Thus, after a series of decisions its developmental path had been determined. This concept was adopted more recently under the heading of "fetal programming." The term is now less favored because, like Waddington's landscape, it implies that the landscape is fixed and the only variable is the choices the ball takes in deciding which path to take. This suits a purely genetically driven concept of development, and indeed the term *genetic program* was first coined by Jacob and Monod (108). Programming also implies that the developmental trajectory is ballistic (i.e., once initiated, it does not change), and in this sense the use is similar to a computer program that either runs to completion or crashes. These deterministic overtones make the term *programming* somewhat inappropriate for the gene–environment interactions that are now known to underpin much of DOHaD. Finally, from a developmental biology perspective the term *programming* is unhelpful because it suggests that the cue that initiates the program contains within itself the instructions for the program. Clearly this is not the case, as for example where a change in circadian melatonin levels in the maternal blood induces a change in offspring coat thickness in the meadow vole (105).

In refining these models we have stressed the importance of graded phenotypic changes that involve interactions with the environment during development and constitute a strategy for surviving rapid or short-duration environmental changes (109). These strategies have been termed *predictive adaptive responses* (PARs), and their characteristics may be summarized as follows:

- PARs are induced by an environmental factor acting in early life.
- PARs manifest in irreversible change in physiology/structure.
- There may be multiple pathways to PARs induction.
- PARs confer a survival advantage in the predictive reproductive environment.
- PARs may be pathophysiological when predictive reproductive or postreproductive environmental boundaries are exceeded.

Aspects of these paths are discussed in detail in the following. At this stage, however, two additional points need to be noted, namely:

- PARs are likely to be transgenerationally exhibited.
- PARs are likely to be neo-darwinian adaptations that permit the species to survive short-term environmental challenges while preserving maximum genotypic variation.

These last points are important because phenotypic change induced by mutation in the genotype is an extremely risky event and the likelihood that it could occur within the time frame of a rapid environmental change is extremely small indeed. In the face of such a rapid change in environment (as may occur with climatic change in animals, or migration in humans), rapid phenotypic changes need to be made if the species is to survive. Even if a helpful mutation had occurred over one generation, the survival advantage could be realized only if the individual carrying that mutation was extremely fertile. We therefore see PARs as constituting adaptive processes that allow a species to survive a rapidly occurring, but short-term environmental transition, and working in conjunction with the darwinian processes of natural selection for genotypic changes and indeed speciation over a longer period of time.

The effects of the prenatal environment on subsequent adult survival may be divided into several categories (110), summarized in Fig. 3. Very poor prenatal environments produce a directly disruptive effect on development, rather akin to teratogenesis. Similar considerations may apply to the effects of teratogens or pharmacological interventions during pregnancy. For example, the effects of single high doses of exogenous glucocorticoids on the development of the fetal kidney (49) may fall into this category. The fetal kidney receives only 3% of the cardiac output (compared with over 20% in the adult) and thus changes in renal growth provide relatively little survival advantage to the fetus in utero. Nonetheless, dramatic effects of environmental changes have been shown for the kidney if they occur at sensitive periods during nephrogenesis.

In the face of a slightly better, but nonetheless challenging prenatal environment, the fetus may adopt a "trade-off" strategy, reducing growth to promote immediate chances of survival but then having to cope with the results in later life. This effect underlies the "thrifty phenotype" concept (92), in which reduced birth weight was associated with later risk

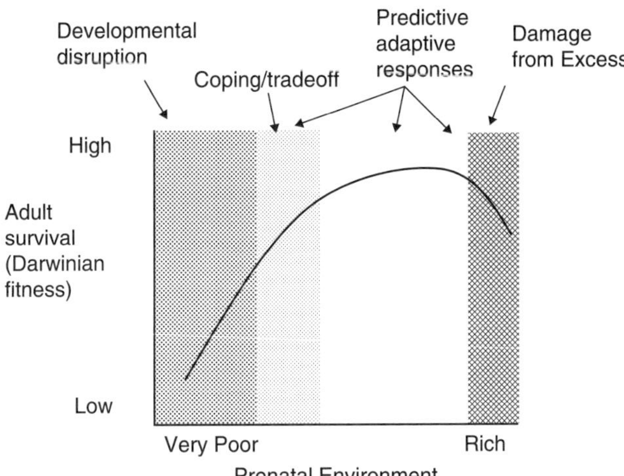

FIG. 3. Proposed relationship between the environmental (nutritional) plane to which the developing organism is exposed during the phase of developmental plasticity and the consequences for its darwinian fitness in adult life. The relationship is determined by different mechanisms according to the degree of shift in the environmental state from the optimal, as suggested by the *shaded columns*. (Data from Gluckman, P. D., Hanson, M. A., Spencer, H. G., and Bateson, P. [2005]. Environmental influences during development and their later consequences for health and disease: implications for the interpretation of empirical studies. *Proc. Biol. Sci.* 272, 671.)

of disease. It is therefore important to understand that PARs may overlap with this range of prenatal environment, but in addition extend beyond it. By definition, PARs involve a response made by the developing organism to cues from its mother and subsequent changes in phenotype made in prediction of a later environment, rather than in response to an immediate environmental challenge. Thus, a prenatal environment of relative abundance may also exert detrimental effects on development, such as shown for the macrosomic fetuses of diabetic pregnancies.

Thus, the PARs model emphasizes the importance of the transition between prenatal and postnatal environment, or more specifically the mismatch between the postnatal environment predicted on the basis of prenatal environmental cues, and the actual postnatal environment experienced (Fig. 4). The precise shape of the relationship shown in Fig. 3 is not known, although it is assumed to show a plateau at the higher end of the prenatal environment. Across the relationship there must be a range of appropriate PARs (i.e., when the actual postnatal environment accurately reflects the prenatal prediction), and this is associated with health. Outside this range PARs are inappropriate and the risk of disease is enhanced. The most obvious example is that of a fetus from a relatively deprived prenatal environment that predicts it will be born into a deprived postnatal environment: if instead this fetus is born into a rich postnatal environment, where food is plentiful and energy expenditure is low, then risk of chronic disease is accentuated. PARs are therefore a normal component of mammalian development.

Maternal Constraint

Maternal constraint is a term that refers to the physiological, rather than pathological, influences from the mother on fetal growth. Mammalian fetal growth must be constrained to allow vaginal birth—otherwise the potential for a large male mating a much smaller female would put both mother and fetus at risk and potentially threaten the continuance of the germ line. Hence, fetal growth cannot be solely controlled by the fetal genotype. The issue is not as prominent in polytocous species, but there is still a negative relationship between fetal number and mean birth weight, suggesting that there is a limit on the amount of nutritional support a mother can provide. However, in monotocous species maternal constraint is much more critical—in many monotocous species, the fetus grows to a limit such that it is just able to pass through the pelvic canal. In humans, the evolution of the upright posture made maternal constraint a critical process for species survival. On the one hand, the size of the cranium increased, and on the other the dimensions of the pelvic canal became reduced with the adoption of upright posture. We can see the operation of maternal constraint today in the correlation between a mother's birth weight and those of her progeny (with no such relation being apparent for paternal birth weight), and indeed in the relation between size at birth and pelvic dimensions (111,112). The precise mechanisms of maternal constraint are not known, but are thought to involve at least two processes. One concerns the supply of nutrients to the growing fetus, in turn dependent not only on maternal diet and metabolism but on uterine blood flow. Because this flow increases during a first pregnancy to a smaller degree than in subsequent pregnancies, birth size is smaller in primiparous than multiparous pregnancies. Greater maternal constraint also seems to operate at the extremes of maternal reproductive age. In adolescent pregnancy it has been proposed that the growth of the mother takes precedence for nutrient supply over her conceptus (113), and in older women uterine blood flow and placentation may be less efficient (114). These considerations are highly relevant to contemporary society because in many developed societies women choose to start families later and to have smaller families. A higher percentage of infants born in developed societies are from primiparous pregnancies and at extremes of maternal age. In countries such as China, where family size

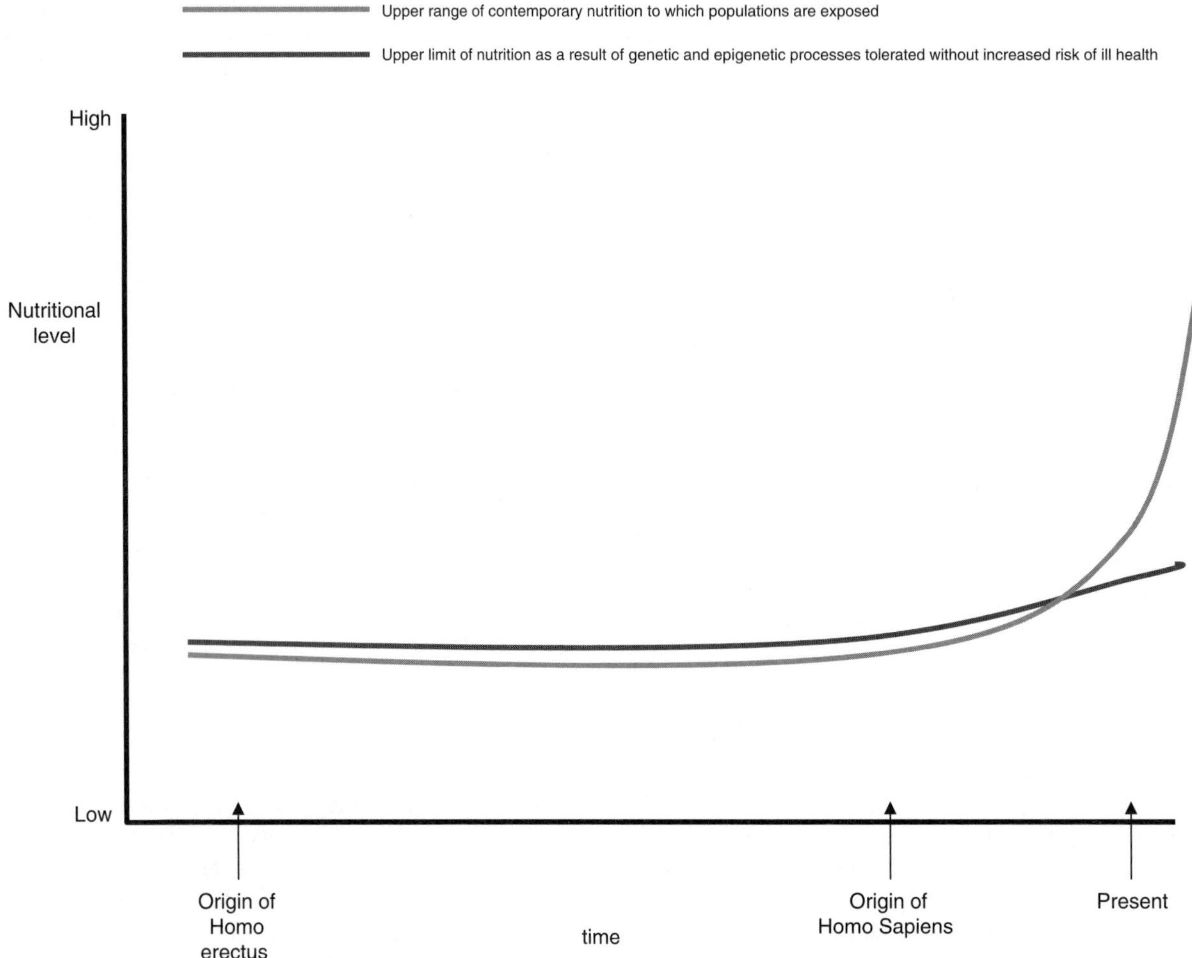

FIG. 4. Schema showing how the level of nutrition tolerated without ill health, set by evolutionary, genetic, and epigenetic processes, is now exceeded. (Data from Gluckman, P., and Hanson, M. [2004]. Echoes of the past: evolution, development, health and disease. *Discov. Med.* 4, 401–407.)

has been limited by legislation, the effects are greater still. All these processes of maternal constraint are accentuated further by imbalance of maternal diet or body composition, by maternal placental disease, and by maternal smoking.

A seemingly obvious situation where prenatal growth is constrained is that of multiple pregnancy. Clearly, several fetuses are likely to have less individual nutrient supply than would a singleton fetus. Concern has been raised about the long-term consequences of multiple pregnancies with the increasing use of assisted reproductive technology. However, the situation may not be quite that simple because it has been shown that the death of one twin in early pregnancy is not followed by a greatly increased growth of the surviving twin: it appears that fetal growth trajectory is set earlier in gestation than was formerly thought.

This consideration relates to the other major potential mechanism for maternal constraint, namely, epigenetic control of the fetal genome. This idea has been discussed in the context of imprinted genes, especially the IGF-2/H19 cluster. Maternal imprinting of the IGF-2 receptor permits control of growth velocity even when it is driven by paternally expressed IGF-2 production (115,116). The epigenetic control of offspring gene expression by environmental factors is now thought to extend substantially beyond simply effects on growth and imprinted genes, however, and this is discussed later.

Transitions

Development should be seen as a continuum, or perhaps as a progression of transitional steps. Environmental changes can induce alterations in this development when they occur during sensitive or plastic periods. Once the period of plasticity has passed, the opportunity for a phenotypic change is lost and the phenotypic attributes are fixed. It is thought to be too expensive in terms of the utilization of resources by a species for plasticity to be maintained beyond a certain period. Thus, the environment poses a threat to those members of a species not appropriately adapted. Recent concepts of the biology of aging reflect this idea, and again use the "trade-off" concept. Maturation may be traded off against growth to promote reproduction, and this will be traded off against biological repair processes (117–119). The increase in degenerative disease associated with the increasing longevity in many societies underscores this effect.

Human migration and economic transition can provide substantial transitions in environment between prenatal and postnatal life, almost invariably in terms of greater access to westernized nutrition and a more sedentary lifestyle. The impact of such transitions has been shown in several animal species (120), and the effect of a transition in the reverse direction in promoting longevity has also been demonstrated (121). Experimental models in which a pregnant animal is fed an unbalanced or restricted diet and the offspring a normal or enriched diet after weaning, have shown that it is possible to produce, in those offspring, phenotypic changes that correlate with each aspect of the human metabolic syndrome—namely, hypertension (36,97), glucose intolerance (97) and insulin resistance (97,122), leptin resistance (97), dyslipidemia, endothelial dysfunction (38), obesity (123), reduced aerobic exercise capacity and mitochondrial dysfunction (47), increased appetite (97), preference for high-fat and high-sugar foods (124), and lethargy (125). Each of these aspects may be seen to be part of a "survival" phenotype, better suited to life in an environment where food supply is uncertain; for example, the reduced blood supply and capillarity in skeletal muscle and the reduced skeletal muscle mass and insulin resistance all provide mechanisms for economizing on metabolic fuel usage. Hominids evolved as hunter-gatherers, consuming a high-protein, low-carbohydrate diet in an episodic way (126). Depositing truncal fat gave a survival advantage, and it is believed that maternal constraint processes induced this pattern of fat deposition (109,127). Modern diets are, of course, higher in carbohydrates and lower in protein and need little exercise to acquire (Fig. 4). To preserve protein balance, we ingest too much carbohydrate and deposit it as truncal fat. Research has shown that portion sizes for many meals have increased two- to fivefold since approximately the mid-20th century, and that the energy density of the typically Western diet has increased (often exceeding 1,000 kJ per 100 g versus the recommended 500 kJ per 100 g). Although it is widely thought that fast food consumption is detrimental to health, there are in fact very few long-term data on the consequences of such consumption. One of the CARDIA (Coronary Artery Risk Development in Young Adults) studies conducted a 15-year prospective analysis of the consequences of fast food consumption (128). They concluded that visiting fast food outlets more than twice a week was associated with increased weight gain and insulin resistance, after correcting for life-style factors.

Maternal Health

The focus on epigenetic processes operating in early gestation raises the question of how aspects of the maternal environment in early pregnancy or preconception may influence later outcome. Research in this area has necessitated adopting a prospective approach, and the most comprehensive study in this respect is the Southampton Women's Survey. Early data to emerge from this study examined effects on the distribution of venous return from the umbilical cord to the fetal liver, or through the ductus venosus to the fetal heart (129). Godfrey et al. showed no relation between mother's prepregnancy subscapular skinfold thickness and umbilical venous flow, but a striking relation with ductus venosus shunting and an inverse relation with liver blood flow (129). Thus, perfusion of the fetal liver was greater in thin women and shunting through the ductus venosus was correspondingly less. These effects were independent of maternal age and parity, social class, education, and smoking. In addition, it was shown that a similar pattern was produced by the mother consuming an "imprudent" diet (low in fruit and vegetables, high in red meat, white bread, sugar, and potato chips) (129). These observations are interesting for two reasons. First, they illustrate that fetal cardiovascular adaptations occur even within the normal range of fetal growth and that they are not necessarily associated with impaired growth or "brain-sparing" phenomena associated with hypoxia or placental insufficiency. Indeed, a novel category of "liver-sparing" responses appears to operate in the normal fetal adaptive range. Second, the results demonstrate that fetal adaptations in late gestation are influenced by aspects of maternal body composition and life-style even within what is considered to be a normal range in a developed society.

The consequences of these observations are wide-reaching. It is known that thinner women have offspring with a higher risk of type 2 diabetes. In the Southampton study, roughly 25% of women reported dieting to lose weight at around the time of conception (130). In Japan, the BMI distribution of women of reproductive age has shifted downward since the early 1990s, and there is an increasing tendency to limit weight gain during pregnancy (131).

It is now becoming increasingly clear that the epigenetic changes induced in offspring prenatally can be passed over several subsequent generations, not just one. Retrospective studies have shown in humans that the risk of diabetes in adult men is related to the diets of their grandfathers during their prepubescent years (132). And there is evidence that risk of diabetes in women is transmitted over several generations when associated with gestational diabetes mellitus (133,134). Animal studies reinforce these concepts.

Transgenerational Influences

Administration of a low-protein diet in pregnancy produces endothelial dysfunction and hypertension in the F_2 generation in the rat (135), even without any additional nutritional manipulation of the offspring in the F_1 generation. Dexamethasone administration in pregnancy similarly produces effects on glucose tolerance and insulin resistance in the F_2 offspring in the rat (136).

The mechanisms by which such multigenerational transmission occurs are not known, although there are several possibilities. First is that the epigenetic change (e.g., in DNA methylation) is induced in the gametes that will form the F_2 generation while the F_1 parents are in utero. Such effects on the developing oocyte have been reported in other contexts. Alternatively, it may be that the DNA "marked" in the F_1 generation by nutritional or environmental manipulation of the F_0 mothers can be passed to the F_2 offspring because such changes in DNA methylation need not be entirely erased at meiosis. Finally, it may be that the adaptations to pregnancy of the F_1 generation have been impaired by their experience in utero, and that this affects the delivery of nutrients to their offspring, the F_2 generation, when they are in utero. Whatever the mechanisms, such nongenomic inheritance mechanisms impose important questions about the longer-term transmission of health and disease through the population. In addition, such processes limit the interpretation that can be placed on traditional genetic methods, such as family linkage analysis, for disease traits.

Postnatal Factors

The Helsinki study and other retrospective cohorts have shown that growth trajectory during infancy and childhood also plays an important part in the risk of later disease. Indeed, it is clear that a life course approach needs to be adopted in assessing the factors that contribute to such disease (137). Examination of the Helsinki data reveals that those men who had coronary heart disease as adults were not only smaller at birth but showed reduced growth in relation to the cohort mean at 1 year of age. Subsequently, their growth caught up and both BMI and weight (although not height) approached the cohort mean or exceeded it by approximately 12 years of age (31,138). Indeed, viewed from this perspective the hazard ratio for coronary heart disease in both the Hertfordshire and Helsinki cohorts is highly significant in its relationship to weight at 1 year of age. A similar picture is seen for the risk of impaired glucose tolerance for type 2 diabetes in young subjects from New Delhi (139). After a fall below the cohort mean at approximately 3 years of age, those subjects with impaired glucose tolerance or diabetes at 28 to 30 years of age showed a faster adiposity rebound. In this group of subjects where the adiposity rebound occurred before 6 years of age, the incidence of obesity was 14% to 21%, of diabetes, 13% to 20% and of blood pressure on the high side of normal, 24% to 25%. The scale of such effects in a developing country in rapid economic transition is striking and suggests that urgent action is needed to address the problem.

INTERVENTIONS

The foregoing discussion raises the issue of possible interventions. A particular focus here has been on micronutrient balance before and during pregnancy. The evidence that folate supplementation may be beneficial has been discussed previously, although this is not universally recognized as being desirable. Vitamins C and E improve endothelial function in animals by decreasing vascular superoxide anion production (140), and trials are ongoing of this regimen for the prevention of preeclampsia (141). However, the questions of safety and feasibility of such interventions in pregnancy or before pregnancy need to be urgently discussed. Follow-up studies in countries such as Chile, where folate supplementation of flour was introduced in 2000, might give valuable insight into the pros and cons of such a strategy. Similar considerations apply to interventions in infancy. There is little doubt that breast feeding is superior to feeding infant formulas in terms of longer-term health of the offspring. Singhal and Lucus and their colleagues, based on a prospective study of preterm infants, have gone further to suggest that rapid catch-up growth of such infants should be avoided (142,143). However, the feeding strategy that would need to be adopted to achieve this, and its possible consequences, have not been examined. Our use of the PARs model may suggest that different strategies will be needed in populations at different stages in the nutritional transition. M.A.H. is supported by the British Heart Foundation.

REFERENCES

1. Barker, D. J., and Osmond, C. (1986). Infant mortality, childhood nutrition, and ischaemic heart disease in England and Wales. *Lancet* 1, 1077–1081.
2. Barker, D. J., and Osmond, C. (1988). Low birth weight and hypertension. *BMJ* 297, 134–135.
3. Ravelli, A. C., van der Meulen, J. H., Michels, R. P., Osmond, C., Barker, D. J., Hales, C. N., et al. (1998). Glucose tolerance in adults after prenatal exposure to famine. *Lancet* 351, 173–177.
4. Forrester, T. E., Wilks, R. J., Bennett, F. I., Simeon, D., Osmond, C., Allen, M., et al. (1996). Fetal growth and cardiovascular risk factors in Jamaican schoolchildren. *BMJ* 312, 156–160.
5. Rich-Edwards, J. W., Stampfer, M. J., Manson, J. E., Rosner, B., Hankinson, S. E., Colditz, G. A., et al. (1997). Birth weight and risk of cardiovascular disease in a cohort of women followed up since 1976. *BMJ* 315, 396–400.
6. Stein, C. E., Fall, C. H., Kumaran, K., Osmond, C., Cox, V., and Barker, D. J. (1996). Fetal growth and coronary heart disease in South India. *Lancet* 348, 1269–1273.
7. Huxley, R., Neil, A., and Collins, R. (2002). Unravelling the fetal origins hypothesis: is there really an inverse association between birthweight and subsequent blood pressure? *Lancet* 360, 659–665.
8. Barker, D. J. (2004). The developmental origins of well-being. *Philos. Trans. R. Soc. Lond. B Biol. Sci.* 359, 1359–1366.
9. Jackson, A. A. (2000). Nutrients, growth, and the development of programmed metabolic function. *Adv. Exp. Med. Biol.* 478, 41–55.
10. Roseboom, T. J., van der Meulen, J. H. P., Ravelli, A. C. J., Osmond, C., Barker, D. J. P., Bleker, O. P. (2001). Effects of prenatal exposure to the Dutch Famine on adult disease in later life: an overview. *Twin Res.* 4, 293–298.
11. Eriksson, J. G., Forsen, T. J., Osmond, C., and Barker, D. J. P. (2003). Pathways of infant and childhood growth that lead to type 2 diabetes. *Diabetes Care* 26, 3006–3010.
12. Eriksson, J. G., Forsen, T., Tuomilehto, J., Osmond, C., and Barker, D. J. (2003). Early adiposity rebound in childhood and risk of type 2 diabetes in adult life. *Diabetologia* 46, 190–194.
13. Wintour, E. M., Johnson, K., Koukoulas, I., Mortiz, K., Tersteeg, M., and Dodic, M. (2003). Programming the cardiovascular system, kidney and the brain: a review. *Placenta* 24(Suppl. A), S65–S71.
14. Fowden, A. L., and Hill, D. J. (2001). Intra-uterine programming of the endocrine pancreas. *Br. Med. Bull.* 60, 123–142.
15. Ong, K. K., and Dunger, D. B. (2004). Birth weight, infant growth and insulin resistance. *Eur. J. Endocrinol.* 151, U131–U139.
16. Cooper, C., Javaid, M. K., Taylor, P., Walker-Bone, K., Dennison, E., and Arden, N. (2002). The fetal origins of osteoporotic fracture. *Calcif. Tissue Int.* 70, 391–394.
17. Gale, C. R., Walton, S., and Martyn, C. N. (2003). Foetal and postnatal head growth and risk of cognitive decline in old age. *Brain* 126, 2273–2278.
18. Wahlbeck, K., Forsen, T., Osmond, C., Barker, D. J. P., and Eriksson, J. G. (2001). Association of schizophrenia with low maternal body mass index, small size at birth, and thinness during childhood. *Arch. Gen. Psychiatry* 58, 48–52.
19. Ahlgren, M., Melbye, M., Wohlfahrt, J., and Sorensen, T. I. A. (2004). Growth patterns and the risk of breast cancer in women. *N. Engl. J. Med.* 351, 1619–1626.
20. Akre, O., Ekbom, A., Hsieh, C. C., Trichopoulos, D., and Adami, H. O. (1996). Testicular nonseminoma and seminoma in relation to perinatal characteristics. *J. Natl. Cancer Inst.* 88, 883–889.
21. Montgomery, S. M., Granath, F., Ehlin, A., Sparen, P., and Ekbom, A. (2005). Germ-cell testicular cancer in offspring of Finnish immigrants to Sweden. *Cancer Epidemiol. Biomarkers Prev.* 14, 280–282.
22. American Heart Association. (2005). *Heart Disease and Stroke Statistics: 2005 Update.* American Heart Association, Dallas, TX.
23. King, H., Aubert, R. E., and Herman, W. H. (1998). Global burden of diabetes, 1995–2025. *Diabetes Care* 21, 1414–1431.
24. Johnell, O., Kanis, J. A., Odén, A., Sernbo, I., Redlund-Johnell, I., Petterson, C., et al. (2004). Mortality after osteoporotic fractures. *Osteoporos. Int.* 15, 38–42.
25. Genant, H. K., Cooper, C., Poor, G., Reid, I., Ehrlich, G., Kanis, J., et al. (1999). Interim report and recommendations of the World Health Organization task-force for osteoporosis. *Osteoporos. Int.* 10, 259–264.
26. Barker, D. J., Martyn, C. N., Osmond, C., Hales, C. N., and Fall, C. H. (1993). Growth in utero and serum cholesterol concentrations in adult life. *BMJ* 307, 1524–1527.
27. Barker, D. J. P., Meade, T. W., Fall, C. H. D., Lee, A., Osmond, C., Phipps, K., et al. (1992). Relation of fetal and infant growth to plasma fibrinogen and factor VII concentrations in adult life. *BMJ* 304, 148–152.
28. Sattar, N., McConnachie, A., O'Reilly, D., Upton, M. N., Greer, I. A., Davey Smith, G., et al. (2004). Inverse association between birth weight and C-reactive protein concentrations in the MIDSPAN Family Study. *Arterioscler. Thromb. Vasc. Biol.* 24, 583–587.
29. Ravelli, A. C., van der Meulen, J. H., Osmond, C., Barker, D. J., and Bleker, O. P. (1999). Obesity at the age of 50 y in men and women exposed to famine prenatally. *Am. J. Clin. Nutr.* 70, 811–816.
30. Leeson, C. P. M., Kattenhorn, M., Morley, R., Lucas, A., and Deanfield, J. E. (2001). Impact of low birth weight and cardiovascular risk factors on endothelial function in early adult life. *Circulation* 103, 1264–1268.
31. Eriksson, J. G., Forsen, T., Tuomilehto, J., Osmond, C., and Barker, D. J. (2001). Early growth and coronary heart disease in later life: longitudinal study. *BMJ* 322, 949–953.
32. Barker, D. J. P., Eriksson, J. G., Forsén, T., and Osmond, C. (2002). Fetal origins of adult disease: strength of effects and biological basis. *Int. J. Epidemiol.* 31, 1235–1239.
33. Eriksson, J. G., Lindi, V., Uusitupa, M., Forsen, T. J., Laakso, M., Osmond, C., et al. (2002). The effects of the Pro12Ala polymorphism of the peroxisome proliferator-activated receptor-γ2 gene on insulin sensitivity and insulin metabolism interact with size at birth. *Diabetes* 51, 2321–2324.
34. Ozanne, S. E., Smith, G. D., Tikerpae, J., and Hales, C. N. (1996). Altered regulation of hepatic glucose output in the male offspring of protein-malnourished rat dams. *Am. J. Physiol.* 270, E559–E564.
35. Ozanne, S. E. (1999). Programming of hepatic and peripheral tissue insulin sensitivity by maternal protein restriction. *Biochem. Soc. Trans.* 27, 94–97.
36. Langley-Evans, S., and Jackson, A. A. (1994). Increased systolic blood pressure in adult rats induced by fetal exposure to maternal low protein diet. *Clin. Sci.* 86, 217–222.
37. Langley-Evans, S. C., Welham, S. J., and Jackson, A. A. (1999). Fetal exposure to a maternal low protein diet impairs

nephrogenesis and promotes hypertension in the rat. *Life Sci.* 64, 965–974.

38. Brawley, L., Poston, L., and Hanson, M. (2003). Mechanisms underlying the programming of small artery dysfunction: review of the model using low protein diet in pregnancy in the rat. *Arch. Physiol. Biochem.* 111, 25–35.
39. El-Haddad, M. A., Desai, M., Gayle, D., and Ross, M. G. (2004). In utero development of fetal thirst and appetite: potential for programming. *J. Soc. Gynecol. Investig.* 11, 123–130.
40. Bloomfield, F. H., Oliver, M. H., Hawkins, P., Holloway, A. C., Campbell, M., Gluckman, P. D., et al. (2004). Periconceptional undernutrition in sheep accelerates maturation of the fetal hypothalamic-pituitary-adrenal axis in late gestation. *Endocrinology* 145, 4278–4285.
41. Drake, A. J., and Walker, B. R. (2004). The intergenerational effects of fetal programming: non-genomic mechanisms for the inheritance of low birth weight and cardiovascular risk. *J. Endocrinol.* 180, 1–16.
42. Weaver, I. C. G., Cervoni, N., Champagne, F. A., D'Alessio, A. C., Sharma, S., Seckl, J. R., et al. (2004). Epigenetic programming by maternal behavior. *Nat. Neurosci.* 7, 847–854.
43. Hill, D. J., and Duvillie, B. (2000). Pancreatic development and adult diabetes. *Pediatr. Res.* 48, 269–274.
44. Petrik, J., Reusens, B., Arany, E., Remacle, C., Coelho, C., Hoet, J. J., et al. (1999). A low protein diet alters the balance of islet cell replication and apoptosis in the fetal and neonatal rat and is associated with a reduced pancreatic expression of insulin-like growth factor-II. *Endocrinology* 140, 4861–4873.
45. Merezak, S., Hardikar, A. A., Yajnik, C. S., Remacle, C., and Reusens, B. (2001). Intrauterine low protein diet increases fetal beta-cell sensitivity to NO and IL-1 beta: the protective role of taurine. *J. Endocrinol.* 171, 299–308.
46. Stoffers, D. A., Desai, B. M., DeLeon, D. D., and Simmons, R. A. (2003). Neonatal exendin-4 prevents the development of diabetes in the intrauterine growth retarded rat. *Diabetes* 52, 734–740.
47. Taylor, P. D., McConnell, J., Khan, I. Y., Holemans, K., Lawrence, K. M., Asare-Anane, H., et al. (2005). Impaired glucose homeostasis and mitochondrial abnormalities in offspring of rats fed a fat rich diet in pregnancy. *Am. J. Physiol.* 288, R134–R139.
48. Mackenzie, H. S., and Brenner, B. M. (1995). Fewer nephrons at birth: a missing link in the etiology of essential hypertension? *Am. J. Kidney Dis.* 26, 91–98.
49. Wintour, E. M., Moritz, K. M., Johnson, K., Ricardo, S., Samuel, C. S., and Dodic, M. (2003). Reduced nephron number in adult sheep, hypertensive as a result of prenatal glucocorticoid treatment. *J. Physiol. (Lond.)* 549, 929–935.
50. Brenner, B. M., Garcia, D. L., and Anderson, S. (1988). Glomeruli and blood pressure: less of one, more of the other? *Am. J. Hypertens.* 1, 335–347.
51. Boujendar, S., Arany, E., Hill, D., Remacle, C., and Reusens, B. (2003). Taurine supplementation of a low protein diet fed to rat dams normalizes the vascularization of the fetal endocrine pancreas. *J. Nutr.* 133, 2820–2825.
52. Duvillie, B., Currie, C., Chrones, T., Bucchini, D., Jami, J., Joshi, R. L., et al. (2002). Increased islet cell proliferation, decreased apoptosis, and greater vascularization leading to beta-cell hyperplasia in mutant mice lacking insulin. *Endocrinology* 143, 1530–1537.
53. Desai, M., Byrne, C. D., Zhang, J., Petry, C. J., Lucas, A., Hales, C. N., et al. (1997). Programming of hepatic insulin-sensitive enzymes in offspring of rat dams fed a protein-restricted diet. *Am. J. Physiol.* 272, G1083–G1090.
54. Lane, R. H., Kelley, D. E., Gruetzmacher, E. M., and Devaskar, S. U. (2001). Uteroplacental insufficiency alters hepatic fatty acid-metabolizing enzymes in juvenile and adult rats. *Am. J. Physiol.* 280, R183–R190.
55. El-Khattabi, I., Gregoire, F., Remacle, C., and Reusens, B. (2003). Isocaloric maternal low-protein diet alters IGF-1, IGFBPs, and hepatocyte proliferation in the fetal rat. *Am. J. Physiol.* 285, E991–E1000.
56. Simmons, R. A., Flozak, A. S., and Ogata, E. S. (1993). The effect of insulin and insulin-like growth factor-I on glucose transport in normal and small for gestational age fetal rats. *Endocrinology* 133, 1361–1368.
57. Ozanne, S. E., Olsen, G. S., Hansen, L. L., Tingey, K. J., Nave, B. T., Wang, C. L., et al. (2003). Early growth restriction leads to down regulation of protein kinase C zeta and insulin resistance in skeletal muscle. *J. Endocrinol.* 177, 235–241.
58. Selak, M. A., Storey, B. T., Peterside, I., and Simmons, R. A. (2003). Impaired oxidative phosphorylation in skeletal muscle of intrauterine growth-retarded rats. *Am. J. Physiol.* 285, E130–E137.
59. Tsirka, A. E., Gruetzmacher, E. M., Kelley, D. E., Ritov, V. H., and Devaskar, S. U. (2001). Myocardial gene expression of glucose transporter 1 and glucose transporter 4 in response to uteroplacental insufficiency in the rat. *J. Endocrinol.* 169, 373–380.
60. Yajnik, C. S., Fall, C. H. D., Coyaji, K. J., Hirve, S. S., Rao, S., Barker, D. J. P., et al. (2003). Neonatal anthropometry: the thin-fat Indian baby. The Pune Maternal Nutrition Study. *Int. J. Obes.* 27, 173–180.
61. Endemann, D. H., and Schiffrin, E. L. (2004). Endothelial dysfunction. *J. Am. Soc. Nephrol.* 15, 1983–1992.
62. Hsueh, W. A., Lyon, C. J., and Quinones, M. J. (2004). Insulin resistance and the endothelium. *Am. J. Med.* 117, 109–117.
63. Sjoholm, A., and Nystrom, T. (2005). Endothelial inflammation in insulin resistance. *Lancet* 365, 610–612.
64. Caballero, A. E. (2003). Endothelial dysfunction in obesity and insulin resistance: a road to diabetes and heart disease. *Obes. Res.* 11, 1278–1289.
65. Fornoni, A., and Raij, L. (2005). Metabolic syndrome and endothelial dysfunction. *Curr. Hypertens. Rep.* 7, 88–95.
66. Rodie, V. A., Freeman, D. J., Sattar, N., and Greer, I. A. (2004). Pre-eclampsia and cardiovascular disease: metabolic syndrome of pregnancy? *Atherosclerosis* 175, 189–202.
67. Vanhorebeek, I., De Vos, R., Mesotten, D., Wouters, P. J., De Wolf-Peeters, C., and Van den Berghe, G. (2005). Protection of hepatocyte mitochondrial ultrastructure and function by strict blood glucose control with insulin in critically ill patients. *Lancet* 365, 53–59.
68. Khan, I., Dekou, V., Hanson, M., Poston, L., and Taylor, P. (2004). Predictive adaptive responses to maternal high-fat diet prevent endothelial dysfunction but not hypertension in adult rat offspring. *Circulation* 110, 1097–1102.
69. Norman, J. F., and LeVeen, R. F. (2001). Maternal atherogenic diet in swine is protective against early atherosclerosis development in offspring consuming an atherogenic diet postnatally. *Atherosclerosis* 157, 41–47.
70. Leeson, C. P., Whincup, P. H., Cook, D. G., Donald, A. E., Papacosta, O., Lucas, A., et al. (1997). Flow-mediated dilation in 9- to 11-year-old children: the influence of intrauterine and childhood factors. *Circulation* 96, 2233–2238.
71. Singhal, A., Kattenhorn, M., Cole, T. J., Deanfield, J., and Lucas, A. (2001). Preterm birth, vascular function, and risk factors for atherosclerosis. *Lancet* 358, 1159–1160.
72. Schiotz, L., Buus, C. L., Hessellund, A., and Mulvany, M. J. (2000). Effect of mitogens on growth and contractile responses of rat small arteries: In vitro studies. *Acta Physiol. Scand.* 169, 103–113.
73. Itoh, S., Brawley, L., Wheeler, T., Anthony, F. W., Poston, L., and Hanson, M. A. (2002). Vasodilation to vascular endothelial

growth factor in the uterine artery of the pregnant rat is blunted by low dietary protein intake. *Pediatr. Res.* 51, 485–491.
74. Brawley, L., Itoh, S., Torrens, C., Barker, A., Bertram, C., Poston, L., et al. (2003). Dietary protein restriction in pregnancy induces hypertension and vascular defects in rat male offspring. *Pediatr. Res.* 54, 83–90.
75. Brawley, L., Torrens, C., Itoh, S., Barker, A. C., Poston, L., Clough, G. F., et al. (2002). Dietary protein restriction attenuates acetylcholine-induced nitric oxide levels in small mesenteric arteries from pregnant rat dams. *FASEB J.* 16, A104.
76. Nishina, H., Green, L. R., Mcgarrigle, H. H. G., Noakes, D. E., Poston, L., and Hanson, M. A. (2003). Effect of nutritional restriction in early pregnancy on isolated femoral artery function in mid gestation fetal sheep. *J. Physiol. (Lond.)* 553, 637–647.
77. Khan, O. A., Loades, C., Boullin, J., Bertram, C., Ohri, S. K., and Hanson, M. A. (2004). Early prenatal undernutrition causes hypertrophy and hypertension. *Circulation* 110, 486 suppl. S.
78. Kwong, W. Y., Wild, A. E., Roberts, P., Willis, A. C., and Fleming, T. P. (2000). Maternal undernutrition during the preimplantation period of rat development causes blastocyst abnormalities and programming of postnatal hypertension. *Development* 127, 4195–4202.
79. Young, L. E., Fernandes, K., McEvoy, T. G., Butterwith, S. C., Gutierrez, C. G., Carolan, C., et al. (2001). Epigenetic change in IGF2R is associated with fetal overgrowth after sheep embryo culture. *Nat. Genet.* 27, 153–154.
80. Lopes, S., Lewis, A., Hajkova, P., Dean, W., Oswald, J., Forne, T., et al. (2003). Epigenetic modifications in an imprinting cluster are controlled by a hierarchy of DMRs suggesting long-range chromatin interactions. *Hum. Mol. Genet.* 12, 295–305.
81. Constancia, M., Hemberger, M., Hughes, J., Dean, W., Ferguson-Smith, A., Fundele, R., et al. (2002). Placental-specific IGF-II is a major modulator of placental and fetal growth. *Nature* 417, 945–948.
82. Widdowson, E. M. (1979). Nutritional requirement and its assessment, with special reference to energy, protein and calcium. *Bibl. Nutr. Dieta* 148–154.
83. Lewis, R. M., Godfrey, K. M., Jackson, A. A., Cameron, I. T., and Hanson, M. A. (2005). Low serine hydromethyltransferase activity in the human placenta has important implications for fetal glycine supply. *J. Clin. Endocrinol. Metab.* 90, 1594–1598.
84. Brawley, L., Torrens, C., Anthony, F. W., Itoh, S., Wheeler, T., Jackson, A. A., et al. (2004). Glycine rectifies vascular dysfunction induced by dietary protein imbalance during pregnancy. *J. Physiol. (Lond.)* 554, 497–504.
85. Jackson, A. A., Dunn, R. L., Marchand, M. C., and Langley-Evans, S. C. (2002). Increased systolic blood pressure in rats induced by maternal low-protein diet is reversed by dietary supplementation with glycine. *Clin. Sci.* 103, 633–639.
86. Pham, T. D., MacLennan, N. K., Chiu, C. T., Laksana, G. S., Hsu, J. L., and Lane, R. H. (2003). Uteroplacental insufficiency increases apoptosis and alters p53 gene methylation in the full-term IUGR rat kidney. *Am. J. Physiol.* 285, R962–R970.
87. Waterland, R. A., and Jirtle, R. L. (2003). Transposable elements: targets for early nutritional effects on epigenetic gene regulation. *Mol. Cell. Biol.* 23, 5293–5300.
88. Ferre, P. (2004). The biology of peroxisome proliferator-activated receptors: relationship with lipid metabolism and insulin sensitivity. *Diabetes* 53, S43–S50.
89. Lillycrop, K. A., Phillips, E. S., Jackson, A. A., Hanson, M. A., and Burdge, G. C. (2005). Dietary protein restriction of pregnant rats and folic acid supplementation prevents epigenetic modification of hepatic gene expression in the offspring. *J. Nutr.* 135, 1382–1386.
90. McConnell, J. L., and Petrie, L. (2004). Mitochondrial DNA turnover occurs during preimplantation development and can be modulated by environmental factors. *Reprod. Biomed. Online* 9, 418–424.
91. Wisloff, U., Najjar, S. M., Ellingsen, O., Haram, P. M., Swoap, S., Al-Share, Q., et al. (2005). Cardiovascular risk factors emerge after artificial selection for low aerobic capacity. *Science* 307, 418–420.
92. Hales, C. N., and Barker, D. J. (1992). Type 2 (non-insulin-dependent) diabetes mellitus: the thrifty phenotype hypothesis. *Diabetologia* 35, 595–601.
93. Neel, J. V. (1962). Diabetes mellitus: a "thrifty" genotype rendered detrimental by "progress?" *Am. J. Hum. Genet.* 14, 353–362.
94. Hattersley, A. T., Beards, F., Ballantyne, E., Appleton, M., Harvey, R., and Ellard, S. (1998). Mutations in the glucokinase gene of the fetus result in reduced birth weight. *Nat. Genet.* 19, 209–210.
95. Hattersley, A. T., and Tooke, J. E. (1999). The fetal insulin hypothesis: an alternative explanation of the association of low birthweight with diabetes and vascular disease. *Lancet* 353, 1789–1792.
96. Hales, C. N., and Barker, D. J. (2001).The thrifty phenotype hypothesis. *Br. Med. Bull.* 60, 5–20.
97. Vickers, M. H., Breier, B. H., Cutfield, W. S., Hofman, P. L., and Gluckman, P. D. (2000). Fetal origins of hyperphagia, obesity and hypertension and its postnatal amplification by hypercaloric nutrition. *Am. J. Physiol.* 279, E83–E87.
98. Bateson, P. (2001). Fetal experience and good adult design. *Int. J. Epidemiol.* 30, 928–934.
99. Gluckman, P. D., and Hanson, M. A. (2004). Living with the past: evolution, development and patterns of disease. *Science* 305, 1773–1776.
100. Jablonka, E., Oborny, B., Molnar, I., Kisdi, E., Hofbauer, J., and Czaran, T. (1995). The adaptive advantage of phenotypic memory in changing environments. *Philos. Trans. R. Soc. Lond. B Biol. Sci.* 350, 133–141.
101. Irons, W. (1998). Adaptively relevant environments versus the environment of evolutionary adaptedness. *Evol. Anthropol.* 6, 194–204.
102. Agrawal, A. A., Laforsch, C., and Tollrian, R. (1999). Transgenerational induction of defences in animals and plants. *Nature* 401, 60–63.
103. Simpson, S. J., McCaffery, A. R., and Hagelle, B. F. (1999). A behavioural analysis of phase change in the desert locust. *Biol. Rev.* 74, 461–480.
104. Simpson, S. J., Raubenheimer, D., Behmer, S. T., Whitworth, A., and Wright, G. A. (2002). A comparison of nutritional regulation in solitarious- and gregarious-phase nymphs of the desert locust *Schistocerca gregaria*. *J. Exp. Biol.* 205, 121–129.
105. Lee, T. M., Spears, N., Tuthill, C. R., and Zucker, I. (1989). Maternal melatonin treatment influences rates of neonatal development of meadow vole pups. *Biol. Reprod.* 40, 495–502.
106. Lee, T. M., and Zucker, I. (1988). Vole infant development is influenced perinatally by maternal photoperiodic history. *Am. J. Physiol.* 255, R831–R838.
107. Waddington, C. H. (1957). *The Strategy of Genes: A Discussion of Some Aspects of Theoretical Biology*. George Allen & Unwin, London.
108. Jacob, F., and Monod, J. (1961). Genetic regulatory mechanisms in the synthesis of proteins. *J. Mol. Biol.* 3, 318–356.
109. Gluckman, P. D., and Hanson, M. A. (2004). *The Fetal Matrix: Evolution, Development, and Disease*. Cambridge University Press, Cambridge.
110. Bateson, P., Barker, D., Clutton-Brock, T., Deb, D., D'Udine, B., Foley, R. A., et al. (2004). Developmental plasticity and human health. *Nature* 430, 419–421.

111. Morton, N. E. (1955). The inheritance of human birth weight. *Ann. Hum. Genet.* 20, 125–134.
112. Robson, E. B. (1953). The genetics of human birth weight. *Heredity* 7, 149.
113. Wallace, J., Bourke, D., Da Silva, P., and Aitken, R. (2001). Nutrient partitioning during adolescent pregnancy. *Reproduction* 122, 347–357.
114. Naeye, R. L. (1983). Maternal age, obstetric complications, and the outcome of pregnancy. *Obstet. Gynecol.* 61, 210–216.
115. Constancia, M., Kelsey, G., and Reik, W. (2004). Resourceful imprinting. *Nature* 432, 53–57.
116. Haig, D., and Graham, C. (1991). Genomic imprinting and the strange case of the insulin-like growth factor II receptor. *Cell* 64, 1045–1046.
117. Metcalfe, N. B., and Monaghan, P. (2003). Growth versus lifespan: perspectives from evolutionary ecology. *Exp. Gerontol.* 38, 935–940.
118. Miller, R. A., Harper, J. M., Dysko, R. C., Durkee, S. J., and Austad, S. N. (2002). Longer life spans and delayed maturation in wild-derived mice. *Exp. Biol. Med.* 227, 500–508.
119. Promislow, D. E. L. (1998). Longevity and the barren aristocrat. *Nature* 396, 719–720.
120. Ozanne, S. E., and Hales, C. N. (2004). Lifespan: catch-up growth and obesity in male mice. *Nature* 427, 411–412.
121. Sayer, A. A., Cooper, C., and Barker, D. J. P. (1997). Is lifespan determined in utero? *Fetal Neonat. Med.* 77, F162–F164.
122. Simmons, R. A., Templeton, L. J., and Gertz, S. J. (2001). Intrauterine growth retardation leads to the development of type 2 diabetes in the rat. *Diabetes* 50, 2279–2286.
123. Breier, B. H., Vickers, M. H., Ikenasio, B. A., Chan, K. Y., and Wong, W. P. (2001). Fetal programming of appetite and obesity. *Mol. Cell. Endocrinol.* 185, 73–79.
124. Bellinger, L., Lilley, C., and Langley-Evans, S. C. (2004). Prenatal exposure to a maternal low-protein diet programmes a preference for high-fat foods in the young adult rat. *Br. J. Nutr.* 92, 513–520.
125. Vickers, M., Breier, B. H., McCarthy, D., and Gluckman, P. (2003). Sedentary behaviour during postnatal life is determined by the prenatal environment and exacerbated by postnatal hypercaloric nutrition. *Am. J. Physiol.* 285, R271–R273.
126. Eaton, S. B., and Konner, M. (1985). Paleolithic nutrition: a consideration of its nature and current implications. *N. Engl. J. Med.* 312, 283–289.
127. Gluckman, P. D., and Hanson, M. A. (2004). Maternal constraint of fetal growth and its consequences. *Semin. Fetal. Neonatal. Med.* 9, 419–425.
128. Pereira, M. A., Kartashov, A. I., Ebbeling, C. B., Van Horn, L., Slattery, M. L., Jacobs, D. R., et al. (2005). Fast-food habits, weight gain, and insulin resistance (the CARDIA study): 15-year prospective analysis. *Lancet* 365, 36–42.
129. Haugen, G., Hanson, M., Kiserud, T., Crozier, S., Inskip, H., and Godfrey, K. M. (2005). Fetal liver-sparing cardiovascular adaptations linked to mother's slimness and diet. *Circ. Res.* 96, 12–14.
130. Robinson, S. M., Crozier, S. R., Borland, S. E., Hammond, J., Barker, D. J., and Inskip, H. M. (2004). Impact of educational attainment on the quality of young women's diets. *Eur. J. Clin. Nutr.* 58, 1174–1180.
131. Ohmi, H., Hirooka, K., Hata, A., and Mochizuki, Y. (2001). Recent trend of increase in proportion of low birthweight infants in Japan. *Int. J. Epidemiol.* 30, 1269–1271.
132. Kaati, G., Bygren, L. O., and Edvinsson, S. (2002). Cardiovascular and diabetes mortality determined by nutrition during parents' and grandparents' slow growth period. *Eur. J. Hum. Genet.* 10, 682–688.
133. Boney, C. M., Verma, A., Tucker, R., and Vohr, B. R. (2005). Metabolic syndrome in childhood: association with birth weight, maternal obesity, and gestational diabetes mellitus. *Pediatrics* 115, 290–296.
134. McCarron, P., Davey Smith, G., Hattersley, A. T., and the ALSPAC Study Team. (2004). Type 2 diabetes in grandparents and birth weight in offspring and grandchildren in the ALSPAC study. *J. Epidemiol. Community Health* 58, 517–522.
135. Torrens, C., Brawley, L., Dance, C. S., Itoh, S., Poston, L., and Hanson, M. A. (2002). First evidence for transgenerational vascular programming in the rat protein restriction model. *J. Physiol. (Lond.)* 543, 41P–42P.
136. Drake, A. J., Walker, B. R., and Seckl, J. R. (2005). Intergenerational consequences of fetal programming by in utero exposure to glucocorticoids in rats. *Am. J. Physiol.* 288, R34–R38.
137. World Health Organization/U. N. Food and Agriculture Organization. (2003). *Diet, Nutrition and the Prevention of Chronic Diseases*. Technical Report Series 916. World Health Organization, Geneva.
138. Eriksson, J. G., Forsen, T., Tuomilehto, J., Winter, P. D., Osmond, C., and Barker, D. J. (1999). Catch-up growth in childhood and death from coronary heart disease: longitudinal study. *BMJ* 318, 427–431.
139. Bhargava, S. K., Sachdev, H. S., Fall, C. H. D., Osmond, C., Lakshmy, R., Barker, D. J. P., et al. (2004). Relation of serial changes in childhood body-mass index to impaired glucose tolerance in young adulthood. *N. Engl. J. Med.* 350, 865–875.
140. Franco, M. C., Akamine, E. H., Aparecida de Oliveira, M., Fortes, Z. B., Tostes, R. C., Carvalho, M. H., et al. (2003). Vitamins C and E improve endothelial dysfunction in intrauterine-undernourished rats by decreasing vascular superoxide anion concentration. *J. Cardiovasc. Pharmacol.* 42, 211–217.
141. Poston, L., Raijmakers, M., and Kelly, F. (2004). Vitamin E in preeclampsia. *Ann. N. Y. Acad. Sci.* 1031, 242–248.
142. Fewtrell, M. S., Morley, R., Abbott, R. A., Singhal, A., Stephenson, T., MacFadyen, U. M., et al. (2001). Catch-up growth in small-for-gestational-age term infants: a randomized trial. *Am. J. Clin. Nutr.* 74, 516–523.
143. Singhal, A., Cole, T. J., Fewtrell, M., and Lucas, A. (2004). Breastmilk feeding and lipoprotein profile in adolescents born preterm: follow-up of a prospective randomised study. *Lancet* 363, 1571–1578.

Lactation

Knobil and Neill's Physiology of Reproduction, Third Edition
edited by Jimmy D. Neill,
Elsevier © 2006

CHAPTER 57

Lactation and Its Hormonal Control

Margaret C. Neville

Introduction, 2994
Developmental Aspects of Milk Secretion, 2994
Anatomical Development, 2994
Functional Differentiation, 2995
Lactation, 2996
Involution, 2998
Endocrine Regulation of Functional Development, 2998
Regulation of Alveolar Proliferation Requires Paracrine Signaling, 2998
Regulation of Secretory Differentiation, 2998
Secretory Activation, 3003
Problems with Secretory Differentiation Lead to Defective Secretory Activation, 3004
Mammary Gland Involution, 3004
Milk Synthesis and Secretion, 3005
Glucose and Amino Acid Metabolism in the Lactating Mammary Alveolar Cell, 3007
The Exocytic Pathway for Secretion of the Aqueous Components of Milk, 3010
Secretion of Lipid into Milk, 3019
Secretion of the Monovalent Ions Na^+, K^+, and Cl^- into Milk, 3023
Transcytosis, 3026
Regulation of Junctional Complexes in the Mammary Gland, 3027
Regulation of Milk Secretion, 3032
Prolactin, 3032
Milk Removal, 3034
Glucocorticoids, 3034
Milk Ejection, 3035
The Let-Down Reflex, 3035
Characteristics of Oxytocin, 3035
The Neuroendocrinology of Oxytocin Synthesis and Release, 3036
Oxytocin Receptors, 3036
Interaction of Oxytocin with Myoepithelial Cells, 3037
Oxytocin and Maternal Behavior, 3037
Lactation in Monotremes and Marsupials, 3037
Acknowledgments, 3038
References, 3039

ABSTRACT

The mammary gland undergoes its final development in the adult animal, filling the mammary fat pad with a ductal tree and rudimentary alveoli during puberty, and expanding and fully differentiating during pregnancy. The hormones of pregnancy, prolactin, placental lactogen, growth hormone, and progesterone, cooperate to produce a gland that is fully developed but nonfunctional. Secretory activity commences around parturition, with the withdrawal of progesterone and maintained prolactin and glucocorticoid. The lactating gland produces milk of a composition defined for the species using several specialized pathways, including (a) exocytosis for the secretion of milk proteins, lactose, and divalent ions; (b) a unique lipid secretion pathway that produces membrane-bound milk fat globules; (c) transport systems for monovalent ions, glucose, and amino acids; and (d) transcytosis for the secretion of immunoglobulins and other milk components. Tight junctions form a gasket around the apical surface of the epithelial cells that is open to traffic of large and small molecules in pregnancy, but tightly closed in lactation. During lactation, prolactin provides a comprehensive signal that fosters synthesis and

University of Colorado Health Sciences Center at Fitzsimmons, Aurora, Colorado.

secretion of milk components and the survival of the alveolar cell. The volume of milk produced is determined by milk removal from the gland, a function dependent on oxytocin secretion by the posterior pituitary and contraction of myoepithelial cells to force milk out of the alveoli. These considerations provide a framework for the application of a myriad of new molecular technologies that are beginning to bring our understanding of these complex phenomena to a new level.

Abbreviations: ADPH, adipophilin; Btn, Butyrophilin; CLA, *trans*-10, *cis*-12 conjugated linoleic acid; Cx, connexin; GH, growth hormone; GHR, growth hormone receptor; GR, glucocorticoid receptor; LPL, lipoprotein lipase; PL, placental lactogen; PR, progesterone receptor; PRL, prolactin; SREBP, sterol response element binding protein; XOR, xanthine oxidoreductase.

INTRODUCTION

Sometime in the 1950s, Jim Linzell, working at the Agricultural Research Council of the Institute of Animal Physiology at Babraham, Cambridge, began to examine the blood flow in the mammary gland of the goat. Along with a young researcher, Malcolm Peaker, he soon began to combine blood flow measurements with new techniques that allowed metabolic reactions to be traced with radioactive isotopes in the animal. These experiments, summarized in *Physiological Reviews* in 1971 (1), provided insights into the mechanisms of milk secretion that had been undreamed of even a decade earlier. Around the same time, a world away in Berkeley, California, another researcher, Dorothy Pitelka, was applying her skills with the new miracle of microscopy, the electron microscope, to an examination of the morphology of the mammary gland at the minutest level (2). Pitelka used her skills with transmission and freeze-fracture electron microscopy to publish a series of papers that essentially defined cell–cell contacts in the mammary gland. She then went on to establish tissue culture models that would allow study of the regulation of these contacts and other aspects of the molecular control of the synthesis of milk proteins even to the present (3). One of the exciting moments in the field was when Linzell and Pitelka both presented their work at a Gordon Conference in 1973, where each discovered that the other was working on complementary aspects of tight junction regulation in the mammary gland, providing a framework for the work on tight junction regulation in the mammary gland that continues today. Many other researchers, too numerous to mention in this introduction, have contributed in the interim to our current understanding of the physiology of lactation, and some of their work is described in this chapter, which focuses on the mechanisms of mammary development and milk secretion in the animal. Many of the molecular aspects are dealt with more completely in many literature articles recently summarized by Hennighausen and Robinson (4).

We are again at a strategic junction in our understanding of the biology of milk secretion. The biochemistry was largely defined in the 1970s before the molecular revolution that consumed the 1980s and 1990s. Now with the human, mouse, and other genomes firmly in hand along with a variety of techniques that allow dissection of minute molecular interactions, high-throughput technologies that allow us to put those interactions into pathways, and genetically altered mice that allow us to test the hypotheses we generate, we are at the brink of another great leap in our understanding of the biology of milk secretion and its regulation. The goal of this chapter is to provide a scaffold of past and current knowledge on which to build the future.

DEVELOPMENTAL ASPECTS OF MILK SECRETION

Anatomical Development

Unlike most secretory organs, the capacity for milk secretion develops in the adult in concert with other reproductive events and is directed by reproductive hormones (5,6). Although the mammary anlagen and the adipose fat pad into which the gland will develop are specified during embryonic development, further growth and differentiation must await the hormones of puberty and pregnancy (7–10). The developmental process has been best described in the rodent, where the ductal development that occurs during early puberty has received extensive study. Ductal development is thought to require estrogen, insulin-like growth factor (IGF)-1, and growth hormone (GH) (11). It terminates with a branching ductal tree expanded to the margins of the fat pad (Fig. 1A). Depending in part on the strain (compare Fig. 1A, B, and C) and age (compare Fig. 1C and 1D) of the animal and its reproductive history, the ducts may be decorated with regularly spaced side branches arranged perpendicularly to the duct and somewhat narrower in diameter (Fig. 1B and D) (12). Cross-transplantation experiments by Naylor and Ormandy showed that strain-specific patterns of side branching are a function of the mammary stroma rather than the epithelium (12). In the rat, the side branches are generally more complex but tend to be variable (13). In the human, the lobular units that are derived from the side branches are complex in character. In this species, the so-called

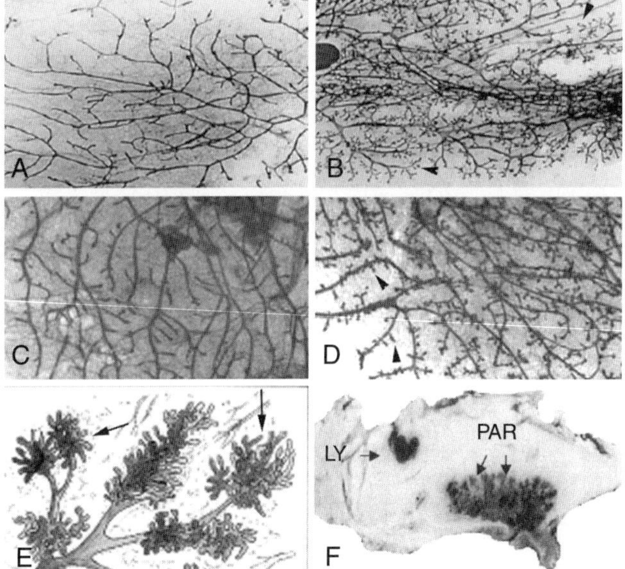

FIG. 1. Whole mounts of mammary glands from virgin animals of different species. **A–D:** Mouse. **A:** C57Bl/6J; **B:** 129 Ola/Sv/Pas; **C:** FVB; **D:** 77-day-old FVB. *Arrowheads* mark side branches. **E:** Human; drawing of a subgross preparation of a mammary gland from a 22-year-old nulliparous woman. *Arrows* point to terminal duct lobular units (TDLUs). **F:** Ewe; pubertal gland. PAR, parenchyma; LY, lymph node. The *white area* surrounding the parenchyma is the adipose fat pad, which will not fill with mammary tissue until pregnancy. [**A** and **B** from Naylor and Ormandy (12), reprinted with permission of Wiley-Liss; **C** and **D** from Cardiff et al. (15); **E** from Cardiff and Wellings (14); **C–E** reprinted with kind permission of Springer Science and Business Media; **F** from Akers (16), with permission.]

terminal duct lobular unit (TDLU) consists of a straight extralobular duct that branches into several intralobular ducts terminating in alveolar or acinar complexes (14,15) (Fig. 1E). In most agricultural species, the ductal structure does not reach the margin of the fat pad until the hormones of pregnancy stimulate both ductal elongation and branching (16) (Fig. 1F). Whatever the pattern in the virgin animal, these side branches proliferate during pregnancy into the complex alveoli that eventually become the milk secretory unit. For a more complete description of these morphological changes, the reader is referred to a number of excellent reviews (13–15,17) and an excellent online tutorial at http://mammary.nih.gov.

Functional Differentiation

Functional differentiation of the gland can be divided into three phases that have been most thoroughly studied in the mouse: (a) the *proliferative* phase of early pregnancy; (b) *secretory differentiation* starting in mid-pregnancy, during which the gland becomes competent to secrete milk; and (c) *secretory activation* around parturition, when milk secretion commences.

Proliferative Phase

The *proliferative phase* starts immediately after conception, reaching a peak at approximately day 5, when an astonishing 25% of the cells are labeled with [^{3}H]-thymidine in the mouse (18) (Fig. 2A). It tapers off gradually through the remainder of pregnancy to reach quiescence just before parturition. During this period, the content of messenger RNA (mRNA) for epithelial markers cytokeratin 19 and claudin 7 increases by nearly three orders of magnitude (19) (Fig. 2B), indicating a remarkable expansion of the alveolar compartment. Whether these markers reflect cell number, cell size, or a denser cytoplasm is not currently clear. Three paracrine factors, Rank ligand, Wnt-4, and amphiregulin, show an expression pattern that parallels proliferative activity (Fig. 2C). In the mouse a single, coordinated round of cell division takes place immediately after parturition (20) (Fig. 2D) that may result in the high proportion of binucleate cells present in the lactating gland (21). Note that none of the three paracrine factors elevated during early pregnancy is elevated during the round of cell division in early lactation, indicating that proliferation is regulated by a different mechanism at this stage. It is not clear whether the proliferative activity in early lactation is specific to the mouse or occurs in other species as well.

Secretory Differentiation

Secretory differentiation, sometimes referred to as lactogenesis I (22), begins at approximately mid-pregnancy and is signaled by changes that depend on the species and the experimental paradigm used for their study. For example, Akers (16) showed that α-lactalbumin increased in mid-pregnancy in the secretion product of the bovine mammary gland (Fig. 3A); Mellenberger and Bauman (23) showed a biphasic increase in the enzymes of lipid synthesis in the rabbit mammary gland (Fig. 3C); Hartmann and colleagues (24) showed that lactose appeared in the plasma and urine in the women at mid-pregnancy; and McManaman and colleagues (25,26) have documented the accumulation of lipid droplets in the mammary alveolar cells of the mouse starting around days 8 to 10 of pregnancy (Fig. 3B). Milk protein gene expression increases approximately fivefold during this process (Fig. 3D), and the proportion of adipose tissue and connective tissue stroma decreases markedly, as shown by a decrease in expression of adipocyte-specific genes and most collagen genes (27) (Fig. 3D). Whether these changes represent active stromal remodeling or simply dilution of stromal components by the expanding epithelium is not clear. During late pregnancy the gland begins to produce

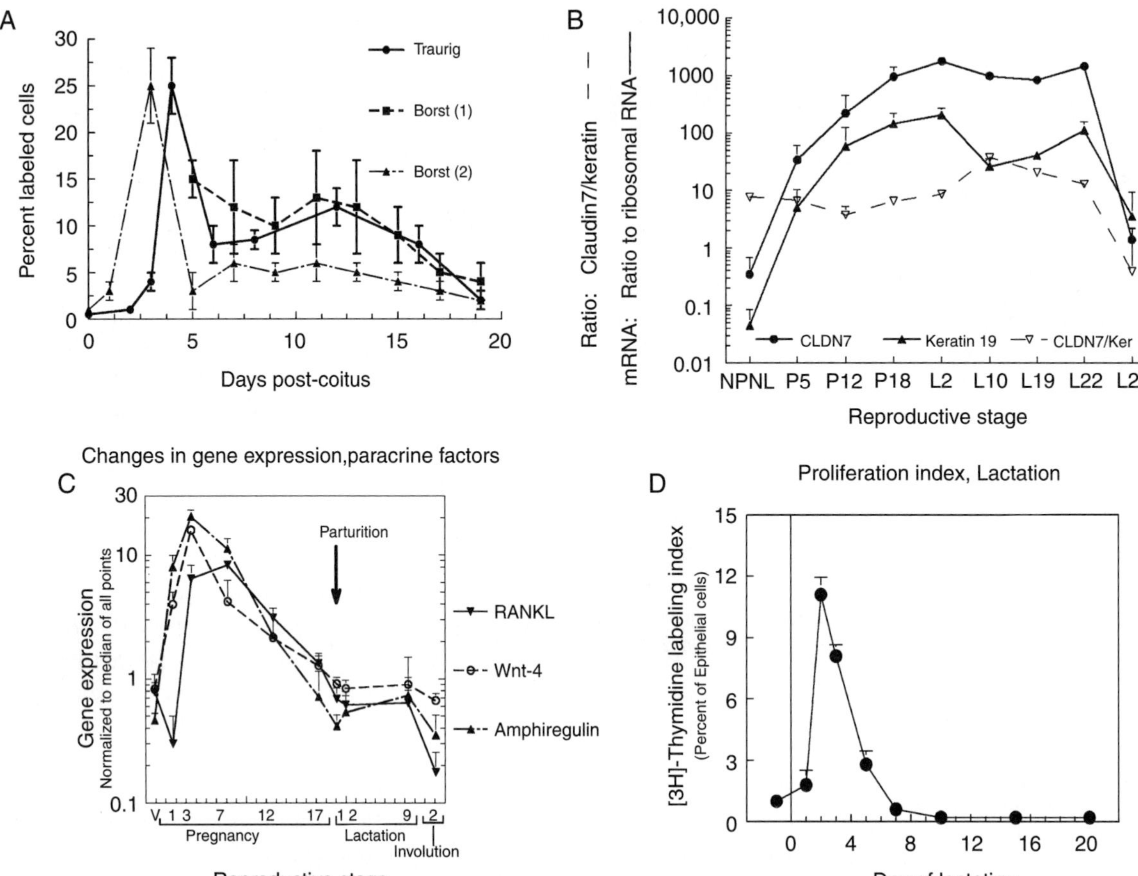

FIG. 2. Proliferative activity during pregnancy and lactation in the mouse. **A:** Proliferation through pregnancy. **B:** Expression of mRNA for keratin 19 and claudin 7, epithelial cell markers determined by real-time reverse transcriptase polymerase chain reaction. *Dashed line* is the ratio of claudin7 to keratin 19. **C:** Changes in expression of mRNA for Rank ligand (RankL), Wnt-4, and amphiregulin. **D:** [^{3}H]-Thymidine labeling index during lactation. The synchronized round of DNA synthesis probably results in the production of the binucleate cells that are numerous in the mouse mammary gland. [**A** and **D:** Data replotted from Traurig (18,20) and Borst, D. W., and Mahoney, W. B. (1982). Mouse mammary gland DNA synthesis during pregnancy. *J. Exp. Zool.* 221, 245–250. **B:** Figure from Blackman et al. (19), used by permission. **C:** Unpublished data from Rudolph, M., and Neville, M. C.]

small amounts of secretion product. This product can escape the gland through the junctional complexes between the alveolar cells, which are highly permeable during pregnancy (28,29). Copious milk secretion is inhibited by the high concentrations of circulating progesterone present in the plasma at this time. Potential paracrine mediators of this effect are discussed in the section on Potential Paracrine Factors in the Pregnant Mammary Gland.

Secretory Activation

Secretory activation is the onset of copious milk secretion; it is sometimes referred to as lactogenesis II or the initiation of lactation (22). This process is set in motion by the fall of progesterone around parturition in all species that have been studied (30–36). It is accompanied by profound changes in the synthetic and metabolic activities of many classes of molecules, including milk proteins and the enzymes of lipid and cholesterol synthesis (Fig. 3D). The molecular regulation of these processes is currently poorly understood. We deal with secretory differentiation and activation in some detail in this chapter because they are they are crucial to the development of the functional gland.

Lactation

The continuous production of milk, known as *lactation* or, formerly, *galactopoiesis*, consists of two phases in placental mammals, a colostral phase during which the secretion product contains large concentrations of immunoglobulins and other protective

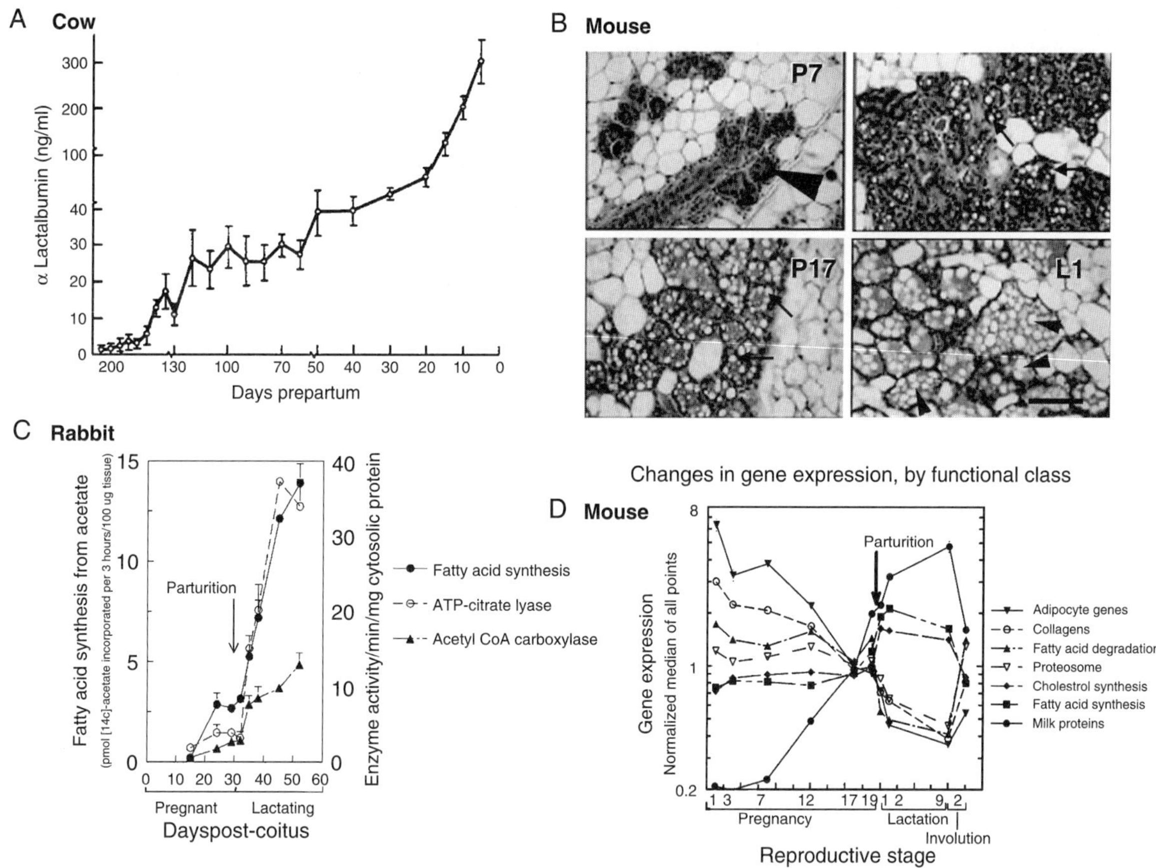

FIG. 3. Changes in the differentiated activity of the mammary gland in pregnancy and lactation. **A:** Cow. The change in α-lactalbumin in the secretion product that can be extracted from a cow mammary gland during pregnancy. (From Akers, R. M. [2000]. Selection for milk production from a lactation biology viewpoint. *J. Dairy Sci.* 83, 1151–1158, used by permission.) **B:** Mouse. Morphology of secretory development in the mouse. Note the condensed alveolar complexes on pregnancy day 7 (P7, *large arrowhead*), small fat droplets in the alveoli at P12 that are clearly enlarged at pregnancy day 17 (P17, *arrows*). By lactation day 1 (L1), the fat droplets have moved into the lumen (*arrowheads*). Bar = 200 μm. (Modified from Rudolph, M., Phang, T., McManaman, J. L., Hunter, L., and Neville, M. C. [2003]. Microarray analysis of secretory activation in the mammary gland. *J. Mammary Gland Biol. Neoplasia* 8, 287–307, with kind permission of Springer Science and Business Media.) **C:** Rabbit. Changes in fatty acid synthesis and the activity of some fatty acid synthesis enzymes during pregnancy and lactation in the rabbit. Note the increase in synthesis and enzyme activity at approximately day 20 of pregnancy. (Redrawn from Mellenberger, R. W., and Bauman, D. E. [1974]. Metabolic adaptations during lactogenesis: fatty acid synthesis in rabbit mammary tissue during pregnancy and lactation. *Biochem. J.* 138, 373–379.) **D:** Mouse. Changes in gene expression from microarray studies. Expression of adipocyte specific genes and collagens decreases six- to eightfold during pregnancy and another twofold at parturition, whereas the genes for fatty acid degradation and many components of the proteosome are level during pregnancy and decrease approximately twofold at parturition. Milk protein genes, on average increase approximately fivefold during pregnancy and another threefold around parturition, whereas the genes for fatty acid and cholesterol synthetic enzymes increase approximately twofold just after parturition. Normalized data for each class were averaged to produce the lines on this graph. (From Rudolph, M., and Neville, M. C., unpublished data.)

substances (37), and a mature secretion phase during which large quantities of milk are produced. Mature milk provides all the nutrients required by the offspring of the particular species for growth during a period that varies from 7 days in the guinea pig to a year or longer in the human. Some adjustments in milk composition occur during this process (38). For example, in humans the concentrations of protein, sodium, potassium, chloride, and calcium decline significantly between 3 weeks and 6 months postpartum, whereas lipid, lactose, and glucose increase (38). In general, these changes are small, usually not more than 20%. For a relatively complete description of milk composition in various species, the reader is referred to works by Jenness (39,40) and Jensen (41). In marsupials, in which the young are born at an extremely immature stage and develop in an abdominal pouch, the changes in milk composition during lactation are much more extreme, as described in the section on Lactation in Monotremes and Marsupials.

Involution

After weaning, most of the secretory elements undergo apoptosis, and the gland remodels and returns nearly to the prepartum state (42,43). The process is highly programmed, proceeding through at least two distinct phases; the first of these is reversible and protease independent. The second depends on expression of proteases and involves considerable connective tissue remodeling (44). Patterns of gene expression during involution in the mouse have been exhaustively described by two groups using microarray technology (45,46). Although there is little firm evidence for invasion by immune cells during involution (Monks, J., personal communication), there is nevertheless a strong upregulation of immune components, death receptors, and acute-phase response proteins at this time. How these components interact to restore the gland to its prepregnant state remains to be determined by future research.

ENDOCRINE REGULATION OF FUNCTIONAL DEVELOPMENT

Having set the stage with a description of the developmental stages of the mammary gland, in this section the focus is on the hormonal and paracrine regulation of proliferation and differentiation in pregnancy and during secretory activation.

Regulation of Alveolar Proliferation Requires Paracrine Signaling

Although alveolar proliferation is not a primary focus of this chapter, we summarize the rather extensive data that now exist on this topic because it sets the stage for our discussion of the endocrine regulation of functional differentiation. It has become clear from studies using genetic manipulation that progesterone (47,48) and prolactin (PRL) (49) are the major hormones that promote side branching and the formation of alveoli (7). Importantly, *the receptors for both progesterone and PRL are expressed in only a subset of cells in the mammary epithelium* (50–54). (Fig. 4). Further, there are extensive data showing that those cells that possess progesterone receptors (PRs) do not proliferate (50,51) (Fig. 4A). Nonetheless, in early pregnancy the tips of the side branches show very heavy PR, at least in the LacZ knockin mouse (Fig. 4C). This means that progesterone- and likely PRL-dependent proliferation must be directed by paracrine signaling (55). There is very good evidence that three paracrine factors mediate the effects of progesterone and PRL on proliferation in early alveolar development, Rank ligand, Wnt-4, and amphiregulin (56–63). Other factors such as leukemia inhibitory factor and various fibroblast growth factors may also be involved, as reviewed by Anderson et al. (50). The reader is also referred to excellent reviews by Brisken (7) and Conneely and colleagues (64) for more detailed information about the roles of these factors.

Regulation of Secretory Differentiation

The hormones that have been implicated in secretory differentiation are PRL, progesterone, and the lactogenic product of the placenta, placental lactogen (PL) (65). There are likely species differences in the roles of these hormones, particularly PRL and PL, but even in the rodent, which has received the most attention, the relative roles of these factors, as well as the receptors involved, are still unclear (5). In addition to these hormones, GH is lactogenic in many species (66). However, a role in differentiation was generally discounted because the GH receptor (GHR) was thought to be present only in the stroma. A recent study (67) showed, however, that GHR is indeed present in the epithelium as well, at least in the mouse, and therefore GH may be a direct participant in differentiation or function of the luminal epithelial cells. In addition, lactation occurred but was not normal in the GHR$^{-/-}$ mouse (68). Finally, it is increasingly recognized that GH is secreted by the mammary epithelium, its secretion being stimulated by progesterone in both the dog (69) and the human (70). The downstream mediators of lactogenic stimulation are also little known with the exception of Stat5 (signal transducer and activator of transcription-5), which has received extensive attention from the Hennighausen laboratory (49,71,72) and others (73), and CAAT/enhancer-binding protein-β (C/EBPβ) (74,75).

Prolactin

PRL remains high throughout human pregnancy and women with low PRL levels during pregnancy have difficulty lactating (76), suggesting that PRL is involved in secretory differentiation in humans (77). In addition, there does seem to be a need for PRL in late pregnancy in gilts (78). However, the story appears to be different in rodents, which have received more extensive study. It is likely that a PRL-like signaling is crucial for secretory differentiation, but PRL

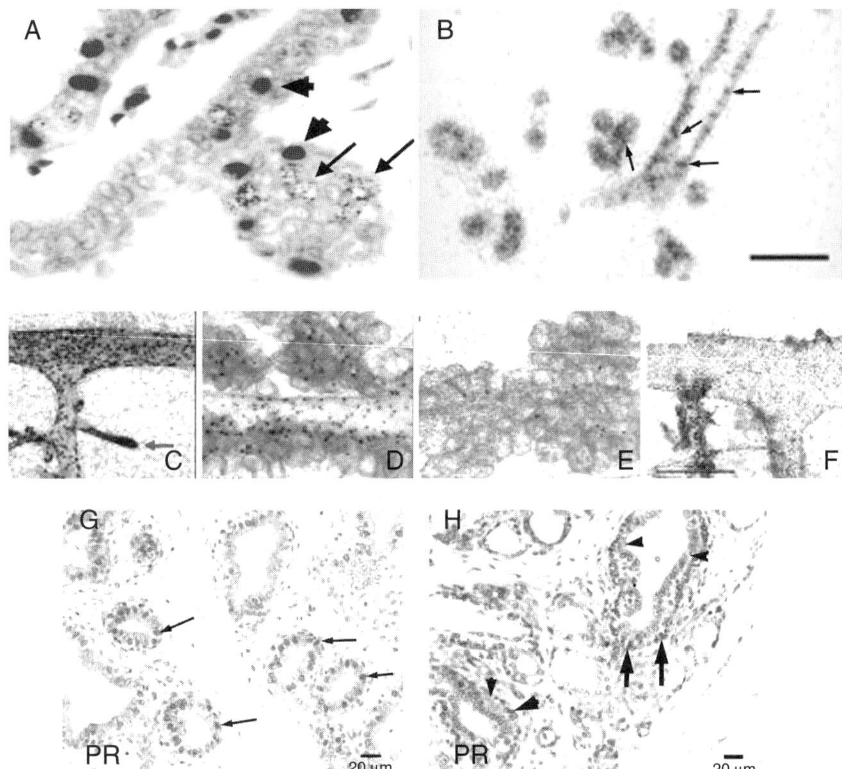

FIG. 4. Localization of the progesterone receptor (PR) and the prolactin receptor (PRLR) in the mammary epithelium. **A:** Human mammary gland. *Arrowheads* indicate immunohistochemistry for PR showing nuclear localization; *arrows* indicate nuclei labeled with tritiated thymidine. Note that PR is present in only a subset of the epithelial cells and that PRs stain and proliferating cells do not overlap. (From Anderson, E., Clarke, R. B., and Howell, A. [1998]. Estrogen responsiveness and control of normal human breast proliferation. *J. Mammary Gland Biol. Neoplasia* 3, 23–35, with kind permission of Springer Science and Business Media.) **B:** Mouse mammary gland, in situ hybridization for PRLR (*arrows*). PRLR is present in only a subset of the epithelial cells. Bar = 200 µm. (From Grimm, S. L., Seagroves, T. N., Kabotyanski, E. B., et al. [2002]. Disruption of steroid and prolactin receptor patterning in the mammary gland correlates with a block in lobuloalveolar development. *Mol. Endocrinol.* 16, 2675–2691, copyright 2002, The Endocrine Society.) **C–F:** Whole mounts of the mouse mammary gland showing LacZ knockin for PR expression. **C:** Day 3 of pregnancy. Note concentration of LacZ at the tips of the forming side branches (*arrow*). (From Ismail, P. M., Amato, P., Soyal, S. M., Demayo, F. J., Conneely, O. M., O'Malley, B. W., and Lydon, J. P. [2003]. Progesterone involvement in breast development and tumorigenesis: as revealed by progesterone receptor "knockout" and "knockin" mouse models. *Steroids* 68, 779–787, with permission of Elsevier.) **D:** Pregnancy day 18. PR (*black dots*) expression is much sparser than in early pregnancy in both the ducts and alveoli. (From Schams, D., Kohlenberg, S., Amselgruber, W., Berisha, B., Pfaffl, M. W., and Sinowatz, F. [2003]. Expression and localisation of oestrogen and progesterone receptors in the bovine mammary gland during development, function, and involution. *J. Endocrinol.* 177, 305–317, with permission.) **E:** PR expression in the lactating gland (day 7) is extremely sparse. **F:** PR begins to return during involution. Bar = 200 µm. **G, H:** Immunostaining for PR in the bovine mammary gland. (Schams, D., Kohlenberg, S., Amselgruber, W., Berisha, B., Pfaffl, M. W., and Sinowatz, F. [2003]. Expression and localisation of oestrogen and progesterone receptors in the bovine mammary gland during development, function and involution. *J. Endocrinol.* 177, 305–317. Copyright 2003, Society for Endocrinology. Reproduced by permission.) **G:** Nonpregnant heifer. **H:** Pregnant heifer, 194 days. Note nuclear stain in **G** (*arrows*) and diffuse cytoplasmic stain (*arrowheads*) of alveolar cells in **H.** Curiously, the nuclei of the basal or myoepithelial cells surrounding the alveoli show heavy stain as well (*large arrows*).

itself is not necessary in mid-pregnancy in rodents and may even be deleterious by inducing premature secretory activation. For example, PRL is found at very low levels after day 8 of pregnancy in the rat (Fig. 5A); it does not return to high levels until just before parturition (79) (Fig. 5B). Further, the PRL receptor (PRLR) is downregulated in pregnancy, increasing markedly just at parturition (80) (Fig. 5D). Finally, if a constitutively active PRLR is expressed under the control of the β-lactoglobulin promoter, alveolar development is increased in late pregnancy, but the animals fail to lactate (81). Significant evidence suggests that the PRLR must be involved in some way in mammary development during pregnancy because PRLR$^{-/-}$ homozygous or heterozygous mice express little or no β-casein (72,82). In addition,

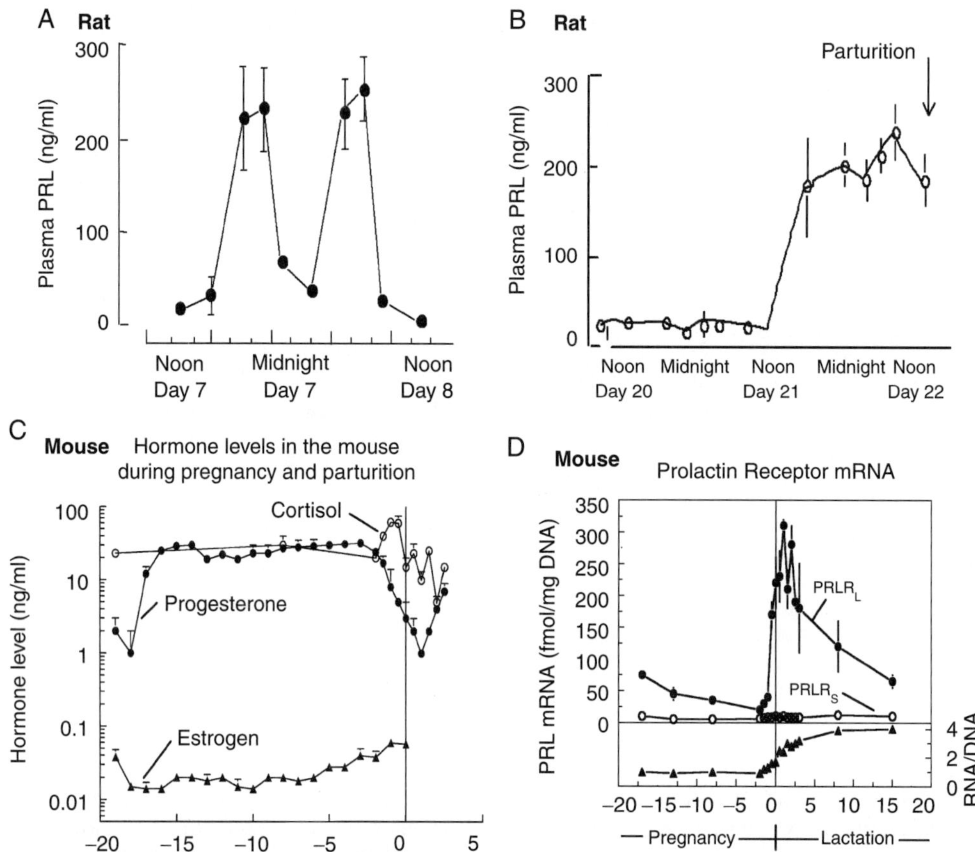

FIG. 5. Hormone profiles during pregnancy in rats and mice. **A:** Pulsatile secretion of prolactin (PRL) during early pregnancy in the rat. At 8 days postcoitus, this pulsatile activity ceases. **B:** Return of PRL secretion 24 hours before parturition in the rat. (A and B, replot of data from Andrews, Z. B., Kokay, I. C., and Grattan, D. R. [2001]. Dissociation of prolactin secretion from tuberoinfundibular dopamine activity in late pregnant rats. *Endocrinology* 142, 2719–2724.) **C:** Steroid hormone profiles in the mouse during pregnancy. Cortisol remains approximately constant with a slight elevation at parturition, possibly due to stress. Progesterone rises early in pregnancy and falls 1 day before parturition. Estradiol rises approximately threefold over the course of pregnancy. (Data replotted from Mizoguchi, Y., Yamaguchi, H., Aoki, F., Enami, J., and Sakai, S. [1997]. Corticosterone is required for the prolactin receptor gene expression in the late pregnant mouse mammary gland. *Mol. Cell. Endocrinol.* 132, 177–183.) **D:** Expression of mRNA for the long (PRL_L) and short (PRL_S) forms of the prolactin receptor during the transition from pregnancy to lactation. (From Mizoguchi, Y., Kim, J. Y., Sasaki, T., Hama, T., Sasaki, M., Enami, J., and Sakai, S. [1996]. Acute expression of the PRL receptor gene after ovariectomy in midpregnant mouse mammary gland. *Endocr. J.* 43, 537–544, with permission.)

terminal differentiation of the mammary gland and lactation failed in mice deficient in Stat5a (71). Thus, based on the low levels of PRL and its receptor in late pregnancy and the data on overexpression of the activated PRLR (81), it seems likely that PRL and its receptor are necessary for the proliferative but not the differentiation phase of alveolar development. Interestingly, as shown in Fig. 4B, like PR the PRLR is expressed in only a subset of epithelial cells in early pregnancy (52,83); whether these cells are the same cells as those expressing PR is not yet clear. Once the milk-secreting cells have been established, secretory differentiation may be under the control of other hormones and receptors such as PL and the GHR, as described next.

Placental Lactogen and the Growth Hormone Receptor

PL is present at high levels only during secretory differentiation (84–87). The hormone has evolved at least twice, in rodents and ruminants from the PRL gene and in primates from the GH gene (65,88,89). Women with deletions in the GH/PL gene complex and undetectable PL concentrations in plasma have been reported to have quite normal lactations (90,91), suggesting that PRL is the hormone responsible for secretory differentiation in humans. Several rodent PLs have been fully characterized (mouse, rat, hamster) and their patterns of secretion in pregnancy measured. The twice-daily surges of pituitary PRL in rats are replaced in mid-pregnancy by, first, a surge in PL I and then by PL II, which continues to increase until term. There is currently no evidence for a specific PL receptor, and PL is thought to act through the GHR or PRLR, or both (92). In earlier experiments using membranes from sheep mammary glands, PRL and oPL binding were not mutually inhibitory, and GH displaced oPL from its receptor. More recent work shows that ruminant PL does not homodimerize the homologous GHR of ruminants, but suggests that a heterodimer of the PRLR and GHR may be the most effective in PL signaling (92). In a more recent and particularly elegant experiment, a heterodimer of the ovine GHR and the PRLR was shown to phosphorylate and activate both Stat1 and Stat3 (93). Whether such signaling is important in secretory differentiation remains for future investigation.

Progesterone

Although it is not clear whether progesterone plays a role in differentiation of the mammary epithelium in late pregnancy, the evidence is incontrovertible that progesterone inhibits the activation of secretion during this time (30,94). Thus, secretion and tight junction closure are triggered in the mammary epithelium of late pregnancy by withdrawal of progesterone. Glucocorticoids and PRL are necessary for this effect (94), which can be simulated an inhibitor of progesterone action, RU486. Because RU486 acts through the PR, it is likely that progesterone, acting through the small number of PRs present in late pregnancy (Fig. 4D), is responsible for the inhibition of secretory activity in all the alveolar epithelial cells. This action implies a paracrine factor, just as a paracrine factor is responsible for alveolar proliferation in early pregnancy. On the other hand, uniform cytoplasmic staining has been observed under some circumstances (95), particularly in the bovine mammary gland during late pregnancy and lactation (96) (Fig. 4H). Evidence suggests that at least some of the actions of progesterone are mediate by a subpopulation of PR that interacts with signal transduction pathways in the cytoplasm (97). The question of what role direct versus indirect mechanisms play in inhibition of secretory activation by progesterone acting through the PR requires future investigation. Whatever the case in late pregnancy, PR is generally considered to be absent or expressed at extremely low levels during lactation (53,95,96,98,99) (Fig. 4E).

Potential Paracrine Factors in the Pregnant Mammary Gland

Paracrine signaling has been shown to play a significant role in embryonic development in the mammary gland (100,101), ductal elongation in the pubertal animal (102), and proliferative activity in the early pregnant animal, as summarized previously. As shown in, Fig. 6 a number of paracrine mediators, including Wnt-5b, IGF-binding protein-5 (IGFBP5), transforming growth factor (TGF)-β2, and TGF-β3

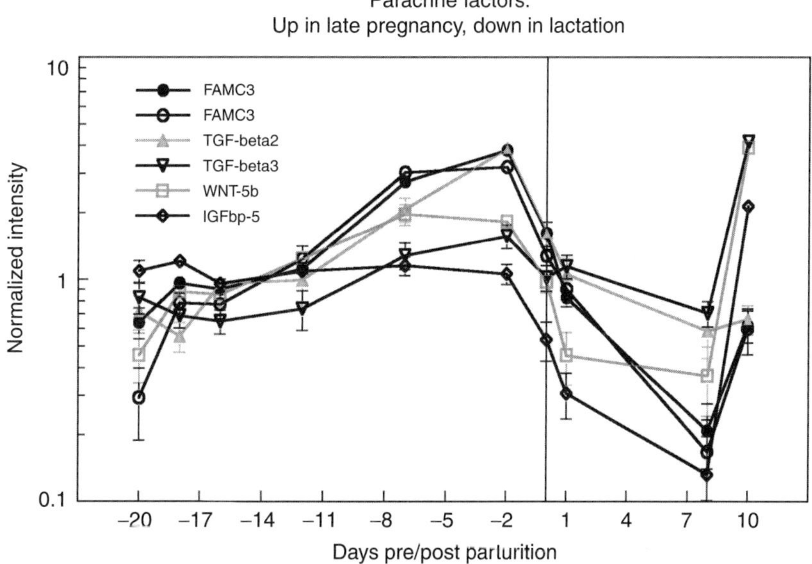

FIG. 6. Profiles of regulatory factors potentially important in secretory differentiation. Microarray results showing paracrine and transcription factors that increase through pregnancy and peak just before parturition, falling as progesterone falls around parturition. (Rudolph, M., and Neville, M. C., unpublished data.)

are present at reasonably high concentrations in late pregnancy, although only the role of TGF-β has received significant attention. Microarray data also implicate a cytokine, Fam3c, about which sequence information and some expression data are available (103,104). Our current understanding of the role of these factors is summarized here.

Transforming Growth Factor-β and Mammary Development. Charles Daniel and his colleagues first showed that TGF-β is responsible for ductal spacing in the pubertal gland by suppressing lateral buds (105,106), and that the various isoforms of TGF-β have distinct developmental regulatory roles. TGF-β1 is expressed in epithelial cells in actively growing end buds in the virgin animal, whereas TGF-β3 is found in the myoepithelial cells. Using antibodies that can distinguish active TGF-β1 from the latent form of the growth factor, Barcellos-Hoff and colleagues showed that only latent TGF-β1 was present in the stroma, whereas both active and latent TGF-β were associated with the epithelium (107). The situation changed during pregnancy, when both TGF-β2 and TGF-β3 were expressed at increasing levels, localized to the ductal and alveolar epithelium. Both fell rapidly at lactation (105,108,109), a pattern also observed in microarrays (Rudolph, M., and Neville, M. C., unpublished data). TGF-β3 expression increases to very high levels at involution (108,110), when it likely participates in epithelial apoptosis and remodeling of the gland (108). To assess the functional relevance of TGF-β in pregnancy, Daniel and colleagues used explant cultures (106) to show that active TGF-β at a dose of 5 ng/mL suppressed β-casein protein synthesis and secretion, but not mRNA expression. On the basis of this finding, they hypothesized that *TGF-b2 and TGF-b3, upregulated by progesterone, inhibit milk secretion in late pregnancy.*

A great deal of data support this hypothesis (110). Sudlow et al. (110) showed β-casein inhibition in explants from pregnant, but not lactating mammary glands, and Mieth et al. (111) obtained the same results in HC11 cells. Yamamoto et al. (112) showed inhibition of β-lactalbumin by TGF-β in a similar model. Muc4, a mucin that is highly expressed during lactation in the rat, has been shown to be inhibited by TGF-β by a mechanism involving inhibition of processing of the protein (113,114). In addition to these experiments in cell culture, data from three types of transgenic mice also support the idea that TGF-β is a progesterone-regulated paracrine factor. When TGF-β was expressed from the WAP promoter, which turns on in late pregnancy, alveolar development was inhibited (115) and the stem cell pool available for alveolar morphogenesis was reduced (116). These experiments highlight the well-known inhibitory effects of TGF-β on normal cell proliferation (117). Moses and colleagues found that targeting a dominant negative TGF-β type II receptor (mouse mammary tumor virus [MMTV]-DNTGRII) to the mammary gland inhibited TGF-β signaling and resulted in alveolar hyperplasia and premature secretory activation in pregnancy, followed by lactation failure (118–120), a result expected if TGF-β is inhibiting both alveolar proliferation and secretory activation. There was, however, no lactation phenotype associated with a DNRII in another laboratory (121). Expression of a constitutively active subunit of the type I receptor did lead to apoptosis in pregnancy and decreased proliferation in a constitutively active receptor I–overexpressing mouse (121). Thus, most of the available data are in accord with the hypothesis that TGF-β suppresses ductal and alveolar proliferation as well as milk secretion in late pregnancy.

If this hypothesis is correct, TGF-β should be upregulated directly or indirectly by progesterone. Although there are no direct data from the mammary gland showing that progesterone stimulates TGF-β production, there is some evidence bearing on this point from other organs. For example, progesterone upregulates TGF-βs in osteoblasts (122). Progesterone, hydrocortisone, and oxytocin all increase TGF-β1 output from human fetal membranes (123). The mechanisms are not clear at this point, in part because there appear to be few data on the transcriptional regulation of TGF-β, and none directly relevant to the mammary gland.

If TGF-β inhibits secretory activation, it should inhibit tight junction closure as a part of these actions. There are precedents for effects of TGF-β on tight junctions. The tight junction barrier in Sertoli cells must disassemble and reassemble periodically to allow developing spermatocytes to pass into the luminal compartment of the seminiferous tubules. Tight junction assembly in Sertoli cells in culture was accompanied by a decline in TGF-β2 and TGF-β expression (124) and was prevented by application of TGF-β3 in the medium at a concentration of 3 ng/mL for a maximal effect. Lui et al. (125) found that the p38 mitogen-activated protein (MAP) kinase pathway rather than the SMAD (Sma/mothers against decapentaplegic homolog) pathway was involved in these responses. Experiments in cell culture also suggest that TGF-β might suppress the closure of the tight junctions that occurs at parturition. Firestone and colleagues (126), using 31EG4 cells, a derivative of the IM-2 cell line of Reichmann et al. (127), showed that TGF-β added with dexamethasone to cultures grown on permeable filter supports prevented dexamethasone-stimulated tight junction closure, suggesting that TGF-β could also play a role in maintaining the permeable tight junction of pregnancy.

Insulin-like Growth Factor–binding Protein-5. IGFBP5 is elevated during pregnancy and involution, where it is expressed by alveolar and ductal epithelium at high levels and at low levels in the stroma (128). The protein persists through early lactation, falling after day 2 to moderate amounts associated mostly with the ductal epithelium (128). Its role is poorly understood, but seems unlikely that it sequesters IGF-1 or IGF-2 at this time because these molecules are thought to be involved in alveolar proliferation or milk secretion. It has been postulated to have alternate roles, including independent proliferative effects and a role in binding extracellular matrix (129), but the exact nature of its actions in pregnancy and early lactation is a matter of speculation (130). The timing of the postparturition fall in IGFBP5, even if it does sequester IGFs, is not consistent with a role as the mediator of progesterone inhibition of secretory activation. IGFBP5 has been firmly associated with involution, where it has been postulated to withdraw IGF-1, inhibiting alveolar cell survival (131,132).

Wnt-5b. The first indication that the Wnt family might play a role in mammary gland development was published in 1992 (133), when Gavin and McMahon showed in Northern blots that Wnt-5b is expressed at rising levels during late pregnancy, falling sharply with secretory activation. This result has been confirmed by us and by the Dale laboratory (27,134). Along with the fall in Wnt-5b gene expression at parturition, we found a profound downregulation of many members of its downstream signaling pathway, including β-catenin (27), suggesting that the fall in Wnt-5b has functional consequences. In line with this reasoning, Renou and colleagues expressed a stabilized form of β-catenin in the mammary gland, finding that it inhibited differentiation and brought about keratinization (135). Other studies dealing with the significance of Wnt and β-catenin signaling have generally focused on the Wnt-4–dependent alveolar proliferation of early pregnancy (63,136). Such models shed little light on the role of the Wnt pathway in differentiation.

Fam3c. This gene is part of a cytokine-like gene family Fam3 first reported in 2002 (104). Fam3C is thought to be ubiquitously expressed and appears to play a role in inner ear development (103). A related family member, Fam3b, is found in the pancreas, where overexpression brings about apoptosis (137). Interestingly, Fam3c (also known as D15Wsu77e) has been identified in a proteomic screen of HC11 cells as a protein downregulated by PRL (138).

Secretory Activation

Secretory activation (also called the initiation of lactation, or lactogenesis II) is accompanied by a plethora of changes in the mammary epithelium, all of which are integrated and coordinated to produce milk. In many species, such as cows, goats, mice, and rats, secretory activation occurs before birth of the young, brought about by the sharp decline in plasma progesterone that also triggers parturition (5,139). In humans, progesterone does not decline sharply prepartum, but falls by approximately two orders of magnitude during the first 4 days after birth, accompanied by a programmed transformation of the mammary epithelium that leads to secretion of approximately 500 mL/day of milk by day 5 postpartum (140). Both the concentration of milk constituents and milk volume can be monitored in women, allowing a dramatic illustration of the changes in the activity of the gland over the period of secretory activation (141–144) (Fig. 7). The increase in milk volume, which takes place primarily between days 2 and 4 postpartum in women and earlier in many other species, is brought about by the coordinated increase in the activity of pathways for ion (citrate, phosphate, potassium, magnesium, calcium, sodium, and chloride) secretion and for lipid, lactose, and protein synthesis and secretion. A number of other changes take place before this coordinated increase in milk secretion. As detailed in the section on Mammary Tight Junctions and Their Regulation, tight junctions close at the onset of lactation, leading to closure of the direct pathway between the interstitium and the milk space. These changes are accompanied by a decrease in the sodium and chloride flux from the interstitial space into the lumen as a well as a decrease in lactose flux out of the gland; the result is the rapid increase in lactose concentration and fall in sodium and chloride shown in Fig. 8A. At the same time there is a substantial secretion of protective molecules into the milk including secretory immunoglobulin A (sIgA), lactoferrin (145,146) (Fig. 8B), and oligosaccharides (147) (not shown). Secretion of an early milk rich in immunoglobulins is characteristic of milk secretion in all species where it has been examined (148,149); however, the mechanisms are poorly understood.

In all species yet studied, a fall in progesterone, in the presence of PRL and corticosteroid, triggers secretory activation after mid-pregnancy when the gland is prepared for lactation (5). The patterns of hormone secretion around parturition are shown in Fig. 5B and C. The fall in progesterone has been discussed above, and the evidence is unequivocal that this fall must occur if secretory activation is to take

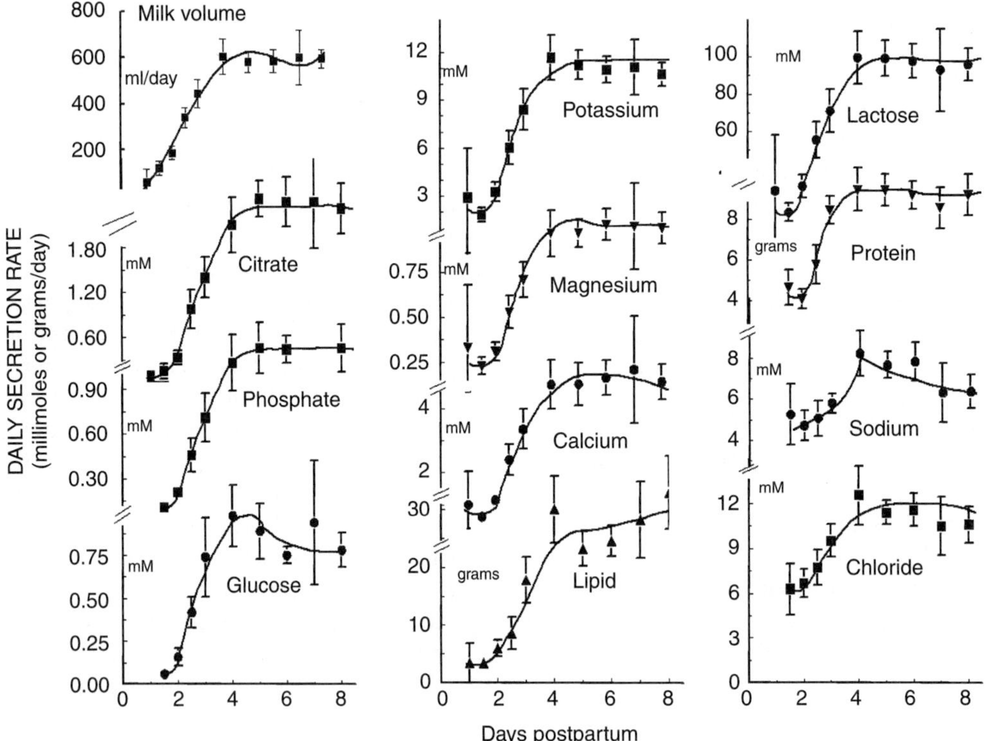

FIG. 7. Milk volume secretion and the rate of secretion of several milk components during the first week postpartum in women. [Data from Neville et al. (141); used by permission from Neville, M. C., and Morton, J. (2001). Physiology and endocrine changes underlying human lactogenesis II. *J. Nutr.* 131, 3005S–3008S. Used by permission of the American Society for Nutritional Sciences.]

place (30–36). Endocrine ablation experiments (150) and studies in organ and cell culture systems (7,151–153) point to a role for glucocorticoids in mammary differentiation and lactation. Although glucocorticoids increase around parturition in most or all species (Fig. 5C) (154), probably as a reaction to the stress of labor, there is no firm evidence that this increase is the trigger for lactation. As described in the section on Glucocorticoids, our understanding of the molecular role of glucocorticoids in the maintenance of lactation is rudimentary. On the other hand, it is well known that if pituitary PRL secretion is inhibited with the dopamine-like agent bromocriptine, secretory activation is delayed until bromocriptine is withdrawn, providing compelling evidence for a necessary role for PRL in secretory activation as well as continued milk synthesis and secretion (94,155,156). PRL action is discussed extensively in the section on Prolactin.

Problems with Secretory Differentiation Lead to Defective Secretory Activation

Disturbances in the carefully programmed series of developmental changes that occur in pregnancy appear to lead to inadequate lactation. As mentioned previously, Djiane and colleagues showed that overexpression of a constitutively active form of the PRLR under the sheep β-lactoglobulin promoter led to florid alveolar development during pregnancy followed by lactation failure, as indicated by the failure of the pups to gain weight in the postpartum period (157). Interestingly, overexpression of a dominant negative form of the TGF-β type II receptor under the MMTV promoter led to a very similar phenotype (119,158). Work from our group shows that precocious formation of lipid droplets in the alveoli during pregnancy (25) in mice overexpressing an activated form of Akt-1 in the mammary also have inadequate lactation. Lactation failure was also seen in hypoxia-inducible factor (HIF)-1α–deficient mammary glands, but in this case the phenotype appeared to be due to inadequate alveolar development during late pregnancy rather than to precocious differentiation (159). All these findings demonstrate that a carefully balanced program of differentiation during pregnancy is critical to successful lactation.

Mammary Gland Involution

After weaning of the pups, involution is characterized by the extensive death of mammary epithelial

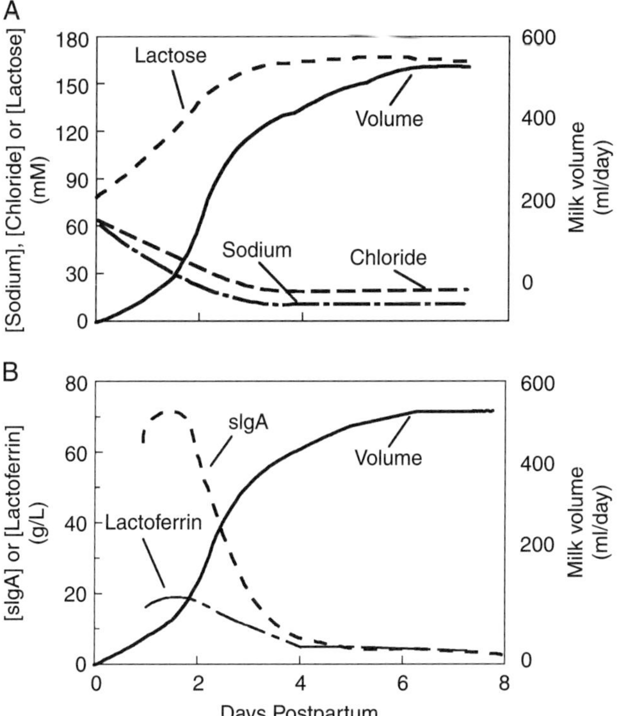

FIG. 8. Changes in human milk composition during the colostrum-forming stage. **A:** Sodium and chloride concentrations fall rapidly and lactose increases as the tight junctions between the epithelial cells close (141). Not that these changes are well underway before the increase in milk volume secretion beginning on day 2. **B:** Secretory IgA (sIgA) and lactoferrin are found at very high concentrations in the mammary secretion during the first 3 days postpartum, the period when colostrum formation is at its peak (145). (Replotted from Neville, M. C. [1995]. Lactogenesis in women: evidence for a cascade of cellular events. In *Handbook of Milk Composition* [R. G. Jensen, Ed.], pp. 87–98. Academic Press, San Diego, with permission of Elsevier.)

cells and remodeling of the mammary gland (160,161). Although it does not completely mimic the changes seen in the mammary alveolar cell (11), weaning is often experimentally induced by pup withdrawal. In this model an early involution phase lasting 2 to 4 days is characterized by cessation of secretory function and disorganization of the Golgi (162,163), followed by apoptosis of mammary epithelial cells. These changes are reversible if pups are returned within 2 days (163). A late phase starting at 3 to 4 days after pup removal is characterized by the activation of metalloproteinases and tissue remodeling (160,161,163). The peak of apoptosis occurs between 3 and 4 days after pup withdrawal in most strains of mice. This first stage of involution is also characterized by increased levels of IGFBP5, Bax, TGF-β3, phosphorylated Stat3, tissue inhibitor of metalloproteinase-1, and decreases in the amount of phosphorylated Stat5 and Bcl-xL (163–165). It has been suggested that changes in the relative levels of proapoptotic and antiapoptotic Bcl2 family members may be critical in regulating involution (164,166–168). Consistent with this hypothesis, overexpression of Bcl2 in the mammary glands of transgenic mice delayed involution (167), as did the overexpression of activated Akt1 (169). Conditional deletion of Stat3 from the mammary gland also delayed involution (170,171), whereas increased TGF-β3 increased apoptosis (108). There is no clear sense of which factor precedes the other and how the entire process is initiated except for some interesting new data implicating Stat5. Deletion of Stat5 in pregnancy brought about significant apoptosis at a time when the gland is expressing significant TGF-β and IGFBP5, as described previously, suggesting that Stat5 may be upstream of other elements of the apoptotic pathways. The fact that so many factors appear to be involved in various aspects of mammary involution indicates both the complexity of the process and the susceptibility of the epithelium to apoptotic stimuli. Thus, in tissue culture it is well known that to study apoptosis, growth factors and cell–cell or cell–matrix attachments often have to be minimized before the cell responds. The mammary cell in lactation seems "poised for death" because PRL is absolutely necessary for its survival in the animal, where serum factors and cell matrix should be optimal; further, sustained stretch of the alveolar cells results in significant apoptosis within 2 days.

MILK SYNTHESIS AND SECRETION

Milk secretion is a complex process involving a myriad of synthetic and secretory mechanisms, all of which must be coordinated to produce milk of a composition characteristic of the species. The complexity of the process is reflected in the structure of the mammary alveolar cell, illustrated in the beautiful electron micrograph by Clermont and colleagues (172) shown in Fig. 9. The major routes for milk secretion are illustrated in Fig. 10. Much of the aqueous fraction of milk is secreted by the *exocytotic pathway* (I, Fig. 10) within which proteins, lactose, calcium, phosphate, and many other compounds are synthesized or processed for secretion. The *lipid secretory pathway* (II, Fig. 10) synthesizes and secretes lipids by a mechanism, unique to the mammary cell, in which lipid droplets, synthesized in the endoplasmic reticulum, move toward the plasma membrane and are budded into the milk space, surrounded by a modified plasma membrane (173). A few milk components, mostly water, urea, ions, and small organic compounds such as monosaccharides and amino acids, are transported intact from the interstitial space into the milk, as illustrated by pathway III, the *transport pathway* (174). The *transcytotic pathway* engulfs IgA and other compounds in the interstitial space,

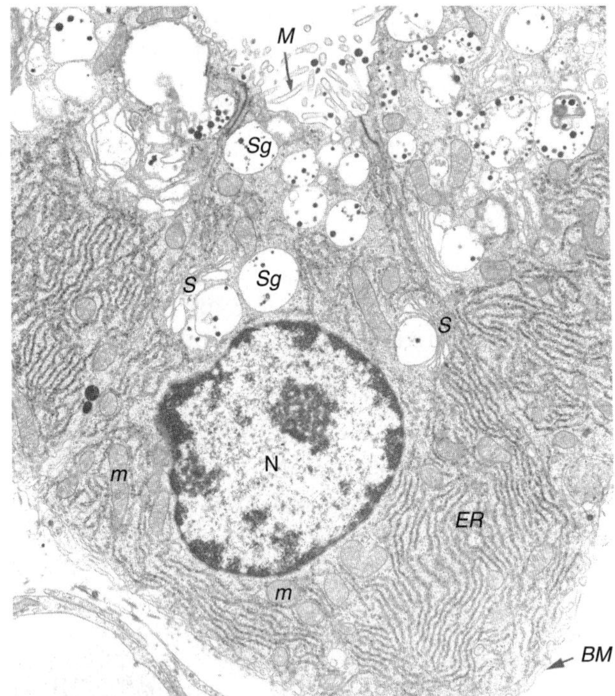

FIG. 9. The anatomy of the alveolar cell in lactation. The apical borders of the central secretory cell are marked by two *black arrows* perpendicular to the lumen that represent the tight junction plaques that seal the lumen (M) from the interstitial space at the bottom of the figure. Abundant endoplasmic reticulum (ER) and mitochondria (m) can be seen at the base of the cell, which also possesses a large, basally located nucleus (N). Secretory vesicles (Sg) containing abundant casein micelles are visible throughout the apical pole. The cell on the left displays a beautiful expanded Golgi stack within which the casein micelles can be seen forming. The empty space at the pole of this cell represents a lipid droplet, fallen out, which is in the process of being secreted. (From Clermont, Y., Xia, L., Rambourg, A., Turner, J. D., and Hermo, L. [1993]. Transport of casein submicelles and formation of the secretion granules in the Golgi apparatus of epithelial cells of the lactating mammary gland of the rat. *Anat. Rec.* 235, 363–373, reprinted with permission of Wiley-Liss.)

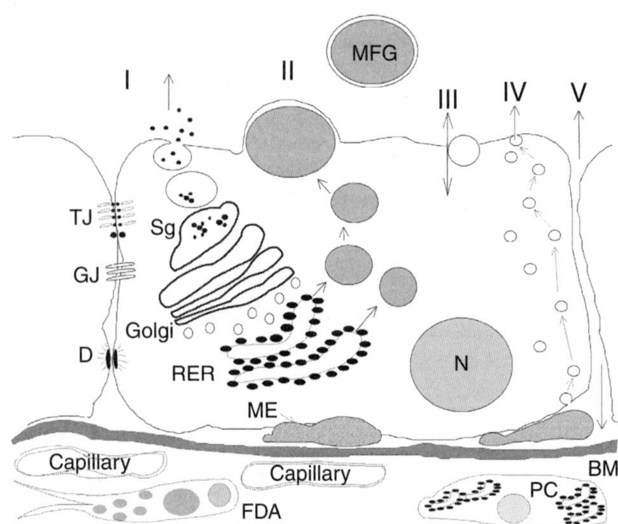

FIG. 10. Diagram of the mammary alveolar cell showing the major routes for secretion of milk components. The exocytic route (I) begins in the rough endoplasmic reticulum (RER), where proteins destined either for the milk (milk proteins such as lactoferrin, casein, and α-lactalbumin) or for the Golgi compartment (transporters that bring small molecules into the Golgi compartment or enzymes involved in the modification of milk components in this compartment) are modified. It continues through the Golgi and terminates in the fusion of specialized secretory vesicles with the apical membrane, where their contents are secreted by exocytosis. The lipid secretion route (II) begins in the smooth endoplasmic reticulum, where lipid droplets are synthesized; these are transported to the apical pole of the cell, where they contact the apical membrane, fuse with additional lipid droplets, and are budded into the milk space. The membrane transport route (III) involves passage of small molecules such as water, Na^+, K^+, Cl^-, glucose, and urea through channels or transporters directly from the cytoplasm to the milk space. These milk components also pass unchanged across the basal membrane through similar or different channels or transporters. The transcytotic route (IV) transfers proteins and probably other molecules from the interstitial space through the cell to the apical membrane, where they are secreted by exocytosis. The paracellular pathway (V) is open in pregnancy, mastitis, and involution and allows exchange of small and large molecules directly between the luminal and interstitial spaces. It is tightly closed in the lactating gland. MFG, milk fat globule; TJ, tight junction; GJ, gap junction; D, desmosome; SG, secretory granule; ME, process of myoepithelial cell; BM, basement membrane; FDA, fat-depleted adipocyte; PC, plasma cell, for immunoglobulin secretion. (Modified from Neville, M. C. [1983]. In *Lactation: Physiology, Nutrition and Breast-feeding* [M. C. Neville and M. A. Neifert, Eds.], pp. 49–104. Plenum Press, New York.)

transfers them across the epithelium in transcytotic vesicles, and secretes them at the apical surface by exocytosis (pathway IV) (175). Finally, in pregnancy, mastitis, and involution, components of the interstitial space pass freely through the *paracellular pathway* (pathway V), which admits molecules as large as immunoglobulins (29). Conversely, this pathway is tightly closed during lactation, as shown on the left side of the diagram.

Some milk components are secreted by more than one pathway. For example, proteins (casein, α-lactalbumin, lactoferrin, whey acidic protein [WAP], and β-lactoglobulin) can be secreted by (a) the exocytotic pathway (b) the transcytotic pathway [immunoglobulins in all species that have been examined (173,175) and albumin in rodents; (176)], and (c) as components of the milk fat globule (MFG) membrane (MFGM) [butyrophilin, xanthine oxidoreductase, adipophilin, MFG-E8, and others (173)]. In some species a small fragment of cytoplasm, the cytoplasmic crescent (177), is engulfed along with the lipid droplet, providing a conduit for any cytoplasmic protein to reach the milk. Monovalent ions, water, and glucose can be secreted by both apical transport pathways and as a part of the contents of the secretory vesicle. Although we cannot deal with the subject in any detail, drugs can reach milk through any of these pathways, depending largely on their lipid solubility or their ability to use carriers already present in the plasma membranes for transfer across the cell. The interested reader is referred to reviews for more information (178–180).

Before describing each of these pathways in detail, glucose and amino acid transport and metabolism in the lactating mammary gland are summarized because these substrates contribute to the synthesis of all the major nutrients in milk—lactose, lipid, and proteins—in a variety of ways. An understanding of the mechanisms by which glucose and amino acids are metabolized is of great importance as we head into an era where genetic modifications in regulatory pathways provide powerful tools for elucidation of metabolic control mechanisms.

Glucose and Amino Acid Metabolism in the Lactating Mammary Alveolar Cell

Balanced Aerobic Glycolysis in Milk Synthesis

Glycolysis is the anaerobic conversion of glucose to pyruvate, generating adenosine triphosphate (ATP) and, in the absence of oxidative phosphorylation, lactate (Fig. 11). The term *aerobic glycolysis* was coined to describe the increased glycolysis displayed by tumor cells, even in the presence of adequate oxygen (181). The mammary alveolar cell also upregulates glycolysis in the absence of anoxia but does so in a way that leads not to the production of lactate, but to the balanced production of lactose and lipid, as shown in Fig. 11, a phenomenon that we like to call *balanced aerobic glycolysis*.

Glucose is a key molecule in milk synthesis because it is the foremost substrate for the synthesis of two of the major components of milk, lactose and lipid. Although the proportion of lipid to lactose varies among species, together these two substances provide at least 80% of the caloric content of the milk of all species. In humans, where the proportion of milk lactose is the highest of any species except the horse (39), lactose is present at 200 mM or approximately 70 g/L, providing 37% of the calories. In addition, lactose provides most of the osmotic pressure that draws water into milk (1). In rodents, lactose is present at approximately 25 g/L or 70 mM, but still provides the osmotic driving force for milk secretion, as shown by the observation that when α-lactalbumin, a key coenzyme for lactose synthesis, is missing, milk secretion fails (182). Milk lipid, present at approximately 40 g/L, provides approximately 53% of the calories of human and cow's milk. Among commonly studied species, the mouse is a champion in terms of milk lipid synthesis. In the common FVB and CD-1 strains, approximately 2 g of mammary tissue are estimated to secrete approximately 5 mL of milk containing approximately 30% lipid, or 1.5 g of lipid per day. Over the course of a 20-day lactation the dam therefore secretes approximately 30 g of milk lipid,

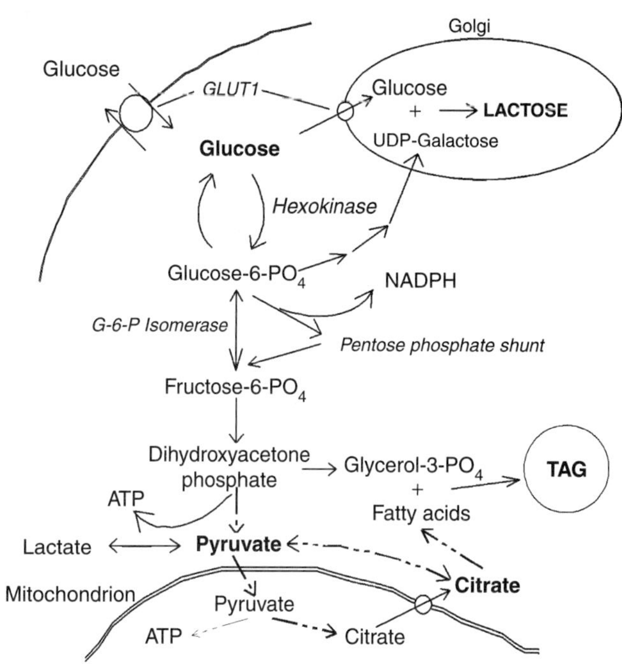

FIG. 11. Pathways for glucose entry and utilization in the lactating mammary alveolar cell. Glucose enters the cell via GLUT-1, a non-insulin-sensitive transporter. Free glucose enters the Golgi by GLUT-1 or other glucose transporters in the Golgi membrane. Here it is combined with uridine diphosphate (UDP)-galactose, also derived from glucose, to make the disaccharide, lactose. Because the Golgi membrane is not permeable to disaccharides, lactose draws water osmotically into the Golgi compartment. Glucose is also converted to glucose-6-phosphate (glucose-6-PO_4) by hexokinase. The glucose-6-PO_4 isomerizes to fructose-6-PO_4, which is then converted by dihydroxyacetone phosphate to glycerol-3-phosphate or to pyruvate through several reactions (not shown) into pyruvate, yielding two molecules of adenosine triphosphate (ATP). Glucose-6-PO_4 may also enter the pentose phosphate shunt, a major source of nicotinamide adenine dinucleotide phosphate (NADPH) for lipid synthesis. Pyruvate enters the mitochondrion, where two major products are ATP and citrate. Citrate has three fates: it serves as the substrate for fatty acid synthesis by conversion to malonyl-CoA; it can be converted to pyruvate through the malate shunt, which provides additional NADPH; and it can be transported into the Golgi for secretion into milk (not shown). NADPH, glycerol-3-phosphate, and pyruvate all contribute to triacylglycerol (TAG) synthesis.

equivalent to her entire body weight! In the mouse eating a normal chow diet, at least half the lipid is derived from glucose. All these considerations indicate that the gland has a major requirement for glucose.

One of the characteristic features of the lactating mammary alveolar cell is its high cytoplasmic glucose concentration. Most cells maintain glucose concentrations in the range of 0.1 to 0.4 mM. As shown first by Kuhn and White in rats (183) and confirmed by Neville et al. using stable isotope studies in women (184), the concentration of glucose in milk is equal to that in the mammary alveolar cell, providing an easily accessible window on the concentration of this substrate in the alveolar cell. The glucose concentration in human milk increases markedly during the initiation of lactation, from approximately 0.34 mM to 1.5 mM (38,185). During a hyperglycemic clamp, milk glucose did respond to plasma glucose, increasing from a steady-state value of 1.7 mM to 4 mM as plasma glucose went from 4 to 8 mM. Importantly, this increase was not associated with any changes in the rate of milk lactose secretion or lipid synthesis, implying that during lactation the concentration of glucose is far above the K_m for lactose and glucose-6-phosphate formation. There was no effect of insulin on any parameters, including milk glucose, when glucose was maintained constant as insulin was increased systematically (i.e., a euglycemic clamp) (186). However, milk glucose did fall as milk secretion fell during weaning (141), consistent with the concept first elucidated by Faulkner and Peaker (187) that milk glucose is proportional to milk volume secretion. Together, these considerations show that cytoplasmic glucose is regulated by the secretory activity of the cell in a manner different from glucose regulation in insulin-dependent tissues. We have only a few clues as to how this regulation is accomplished.

Glucose enters the mammary alveolar cell through glucose transporter-1 (GLUT1), a non–insulin-dependent glucose transporter whose basal plasma membrane localization is significantly increased during the onset of lactation (155,188,189). Once in the cell, glucose is either converted to lactose in the Golgi or phosphorylated in the cytoplasm to glucose-6-phosphate, which enters either the glycolytic chain to be converted to pyruvate, or the pentose phosphate shunt, where it contributes to nicotinamide adenine dinucleotide phosphate (NADPH) synthesis. Pyruvate in turn enters the mitochondrion, where it is ultimately converted to ATP or citrate (Fig. 11). Citrate is an important intermediate in lipid synthesis, providing both malonyl CoA for fatty acid synthesis and oxaloacetic acid that is metabolized by the malate shuttle to produce additional NADPH. The large amount of NADPH required for fatty acid synthesis (190) is thus derived either from the pentose phosphate shunt or as a side product of the malate shuttle. We have found that the enzymes involved in creation of NADPH are coordinately regulated with the enzymes of fatty acid synthesis in the mammary gland (27), as they are in all tissues that synthesize fatty acids. This coordination is achieved by the transcription factor sterol response element–binding protein-1c (SREBP-1c) in most tissues (191). However, lactation is compromised only slightly in SREBP-1c null mice (Rudolph, M., and Neville, M. C., unpublished data), suggesting alternative modes of regulation in the mammary gland.

An important question is: *How is balanced aerobic metabolism maintained so glucose is properly partitioned between lactose and lipid synthesis?* Insight into this process has come from observations of the effects of overexpressing a constitutively active form of the serine/threonine kinase Akt (also known as protein kinase B or PKB) in mouse mammary epithelial cells (25). This kinase is normally located in the cytoplasm. Stimulation of cells with growth factors and hormones can activate phosphatidylinositol 3-kinase (IP_3), thereby increasing plasma membrane inositol phosphates, phosphatidylinositol 4,5-bisphosphate (PIP_2) and phosphatidylinositol 3,4,5-trisphosphate (PIP_3). Akt binds to these compounds through its pleckstrin homology domain, translocating to the plasma membrane, where it is phosphorylated and activated (192,193). Activated Akt stimulates glucose transport and lipid synthesis in a wide variety of cell types (194) and also promotes cell survival (195,196). Most of the studies have been carried out on insulin-sensitive tissues, where Akt2 appears to play a critical role in the translocation of the transporter GLUT4 to the plasma membrane (197,198). However, in the mammary gland, Akt2 is present at very low levels, whereas Akt1 is upregulated more than tenfold at both the mRNA and protein level during the initiation of lactation (25,27,199). Expression of constitutively active myristoylated Akt1 (Myr-Akt) under the control of the MMTV promoter, which directs expression to the mammary gland (25), increased epithelial triglyceride accumulation, elevated milk fat content from its normal level of 30% to 70%, and led to decreased pup growth rates. Although the phenotype is complex, the increased fat content of the milk in the face of decreased pup growth suggests imbalances in the synthetic and secretion processes for lactose and lipid. Importantly, corresponding cell culture studies linked increased lipid accumulation in mammary epithelial cells transfected with Myr-Akt to increased glucose uptake (25). Thompson and colleagues (200) showed, using a hematopoietic cell line, that Akt1 increased glucose transport by stimulating movement of GLUT1 to the plasma membrane, and that elevated Akt1 increased the activity of hexokinase and increased NADPH production through the pentose phosphate shunt. In the mammary gland we found (Rudolph, M., Neville, M. C., and Anderson, S. M., unpublished data) that both hexokinase and transketolase gene expression were increased by expression of MMTV-Myr-Akt1.

Putting all these observations together, we hypothesize that Akt1 acts normally during the initiation of lactation to stimulate balanced aerobic glycolysis by increasing glucose transport, hexokinase activity, and NADPH production by the alveolar cells of the lactating mammary gland. Overexpression would increase the activity of hexokinase, decreasing the availability of glucose for lactose synthesis and shifting substrate to lipid synthesis. Although considerable work needs to be done to test this hypothesis, the use of genetically altered mice in this context represents an important approach to understanding metabolic control in the lactating mammary gland. That this pathway may be linked to mammary cell survival was also shown in these mice, in which involution was significantly delayed by expression of the MMTV-Myr-Akt1 transgene (169).

Amino Acid Transport and Metabolism in the Lactating Mammary Gland

The differences between arterial and venous levels of amino acids, particularly the essential ones, are very large, necessitated by the very high rate of protein synthesis during lactation. For example, the normal lactating dairy cow is estimated to require 1 kg of amino acids per day to support milk protein synthesis (174); because protein breakdown is not generally thought to occur in the mammary gland, this large quantity of amino acids must be transported into the mammary alveolar cell from the bloodstream. Lactating mammary alveolar cells have been shown to possess at least six different amino acid transport systems that have been defined by uptake or efflux experiments with explants (174). These include sodium-dependent transporters for glutamate and aspartate (system X^-_{AG}) (201,202), a sodium and chloride–dependent system for taurine uptake (possibly system β) (203,204), sodium-dependent and -independent systems for the uptake of α-aminoisobutyric acid (systems A and L, respectively) (205), sodium-dependent methionine uptake (system ASC) (206,207), and sodium-independent 2-aminobicyclo (2,2,1) heptane-2-carboxylic acid (BCH)–insensitive uptake of tyrosine (208). Gene expression studies in porcine mammary gland added cationic amino acid transporters CAT-1, CAT-2B (207), and system B0,+ (207).

The molecular identity of some of these transport systems is available, but no molecular or genetic studies have been done in mammary tissue. Such work should be greatly facilitated by the recent availability of a set of excellent reviews based on the solute transporters (the SLCs) identified in the course of the human genome project (209). In one of these reviews we find, for example, that the X^-_{AG} family is coded by five genes in the SLCA1 family (210); it is not currently clear which of these is highly expressed in mammary tissue, although the pattern of the effects of inhibitors suggests that SLCA1, SLCA2, and SLCA3 are not involved (174). System ASC, which mediates sodium-dependent transport of neutral amino acids, corresponds to two genes, SLC1A4 and SLC1A5 (210), both of which are expressed in mammary tissue. From their expression profiles, SLC1A4 is likely expressed in adipose tissue and SLC1A5 in the mammary alveolar cell (27), where it corresponds to the sodium-independent L system of neutral amino acid exchangers. In this role, SLC1A5 likely functions to exchange intracellular amino acids taken up by a sodium-dependent transporter, allowing a wider variety of amino acids to be transported into the cell. CAT-1 (SLC7A1, a so-called system y+ transporter) and CAT-2B (SLC7A2) are involved in the arginine-requiring nitrous oxide–generating system in many tissues (211), and the amino acid transporter B0,+ corresponds to SLC7A9, which is thought to transport cystine and cationic amino acids with high affinity.

Taurine, a β-amino acid that is not incorporated into proteins, is found at high concentrations in the milks of many, but not all, species (212). Taurine transport has been studied in the mammary gland of a number of species and it appears to be sodium-, and sometimes chloride-mediated (204,213). The molecular characteristics of the transporter in the mammary gland are not available at this writing. Interestingly, a peptide transporter, PEPT2, is expressed at all stages of development in the mammary epithelium (214). Although its role is unknown, failure to demonstrate uptake of peptides from the bloodstream in rats (174) led the authors of the study to speculate that it could be involved in reuptake of peptides from hydrolyzed milk proteins in the lumen of the gland or prevent drug secretion into milk. What is needed at this juncture is a thorough analysis of the expression of amino acid transporters in the mammary gland, the use of highly specific antibodies to localize their proteins and transport activity, and the use of mutant mouse models to determine the functional significance of their expression. The availability of the structure of the genome from many species, the existence of high-throughput technologies for the analysis of mRNA, proteins, and even metabolites, and the ability to produce genetically altered mice, where the gene defect can be turned off and on at will, mean that the stage is set for rapid progress in understanding the molecular basis for the contribution of amino acid transport to lactational function.

Within the cell, amino acids are used for protein synthesis, are secreted free into the milk, or are deaminated and the products used in the glycolytic pathway or secreted as urea and ammonium. Milk contains rather large quantities of free amino acids, up to 1.6 mM of glutamine and glutamate and 0.6 mM of taurine (215), as well as significant amounts of urea. The urea reflects the blood concentration of the mother, suggesting that it diffuses from the blood; however, the presence of free amino acids in milk implies the presence of amino acid transporters on both the apical and basal membranes of the alveolar cell. It has long been known that amino acids are deaminated and their carbon skeletons funneled into pyruvate, acetyl CoA, and several citric acid cycle intermediates, including α-ketoglutarate, succinyl CoA, fumarate, and oxaloacetate (1). Direct evidence that milk production depends in part on the amino acid supply for constituents other than milk protein comes from an experiment in which Finnish cows were fed excess glucose, starch, and casein along with an adequate supply of essential amino acids (216). Only the supplemental casein increased milk and lactose outputs, suggesting that catabolism of amino acids not only contributes to synthesis of several milk components but that it may, in some way, regulate milk output. The new science of metabolomics, which allows accurate and simultaneous measurement of the concentrations of many metabolites as well as their fluxes in tissues and even in the animal, may allow new insight into these processes and their regulation (217).

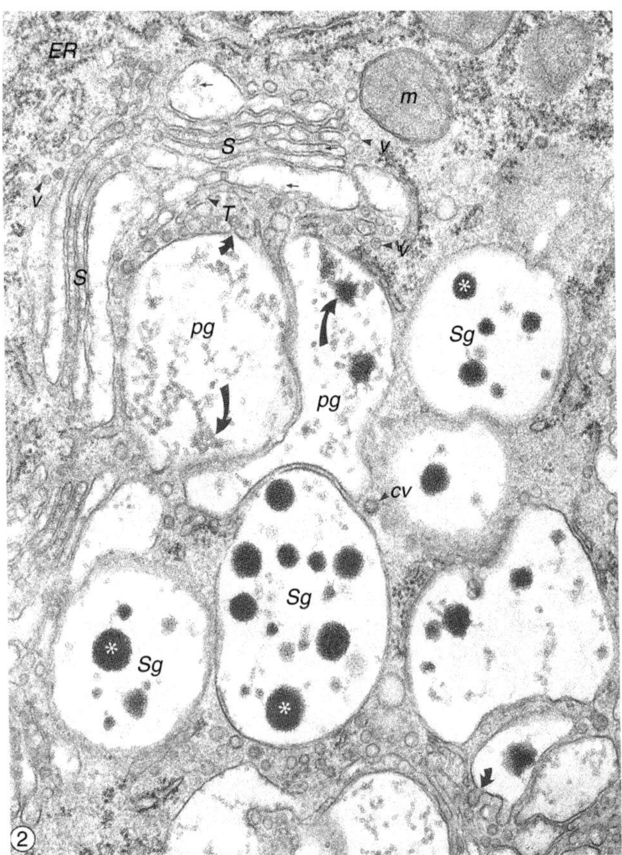

FIG. 12. The specialized *trans*-Golgi compartment of the mammary secretory cell. Portion of a rat mammary alveolar cell in lactation. The Golgi stacks are shown at the upper left (S) giving rise to small vesicles at their lateral borders (V). Wispy strands of casein can be seen in the slightly swollen vesicle of the *trans*-Golgi. As the vesicles continue to swell because of the osmotic effect of the lactose being secreted, they become prosecretory granules (Pg), where additional casein submicelles (*large curved arrows*) are beginning to coalesce into fully mature micelles (*asterisks*). Coated vesicles (cv) can be observed budding off some of the cisternae (*small curved arrows*) and in the interstitial space. (From Clermont, Y., Xia, L., Rambourg, A., Turner, J. D., and Hermo, L. [1993]. Structure of the Golgi apparatus in stimulated and nonstimulated acinar cells of mammary glands of the rat. *Anat. Rec.* 237, 308–317, reprinted with permission of Wiley-Liss.)

The Exocytic Pathway for Secretion of the Aqueous Components of Milk

A large number of milk components are secreted by exocytosis, including most milk proteins, lactose and oligosaccharides, calcium, citrate, and phosphate. As in all exocrine cells, protein synthesis begins in the endoplasmic reticulum, from which proteins are transferred to the Golgi compartment for further processing and packaged into secretory vesicles that move constitutively or in a regulated fashion to fuse with the apical membrane of the cell, where they are secreted into the lumen. However, some special features of protein secretion are present in the mammary alveolar cell. First, the process goes from a basal level in late pregnancy to high activity with the onset of lactation. During this process, the rough endoplasmic reticulum expands twofold, the Golgi expands five- to tenfold, and there is a twofold increase in the number of cisternae (218–220). In addition, secretory vesicles increase in size and the content of casein micelles increases over the 2 to 4 days required for onset of secretory function. Second, the *trans*-Golgi is the site of massive synthesis of a disaccharide, lactose, in most species resulting in a highly expanded compartment in which the secretory granules, in this case made of casein, are surrounded by a large clear space (Fig. 12). Third, massive casein phosphorylation and binding of calcium take place in this compartment. Two questions arise from these observations: first, how do secreted proteins progress from one Golgi stack to the next and on to the secretory vesicles/granules? Second, what are the detailed changes that take place in the mammary alveolar cell that allow the dramatic upregulation of the exocytotic pathway at the onset of lactation? A partial answer to the first question is provided by the process of cisternal maturation, elegantly described in a 1997 hypothesis article by Glick and colleagues (221), for which considerable evidence has been obtained in the interim, as described later. We have only hints about the answer to the second question, provided by a proteomic analysis of Golgi membranes isolated from late pregnant and lactating rats (220).

Cisternal Maturation as a Mechanism for Formation of Secretory Vesicles

The progression of changes in the Golgi cisternae and secretory vesicles as one traverses the stack from the condensed *cis* side to the expanded *trans* side is elegantly illustrated in the electron micrograph shown in Fig. 12, published by Louis Hermo and colleagues in 1993 (172). This picture was difficult to reconcile with the hypothesis popular at the time that transport of secreted proteins from one stack of the Golgi to the next proceeded by a distinct population of membrane vesicles and that each stack formed a stable compartment with a particular complement of processing enzymes (221,222). The question was: Why, if the forming casein micelles observed in the *trans*-Golgi were carried forward to secretory vesicles by anterograde vesicles, was casein never observed in the vesicles in the region of the Golgi? Similar questions were being asked about glycoprotein scales in certain algae as well as lipoprotein particles in the liver [reviewed in Glick et al. (221)]. The findings that Golgi resident enzymes turned over rapidly, that they could be packed in COPI (coat protein I) vesicles, and that they could be observed traveling from one compartment to another led some investigators to reconsider the cisternal maturation hypothesis. As explained by Glick, according to this hypothesis a Golgi cisterna is formed at the *cis* face of the stack, where it acquires the processing enzymes characteristic of that stack both from COPII vesicles coming forward from the endoplasmic reticulum and from COPI vesicles traveling in a retrograde manner from the medial Golgi (Fig. 13). As the cisterna proceeds up the stack, it loses the enzymes characteristic of the previous position and the enzymes of the next position are concentrated by retrograde transport from more forward compartments. At the leading edge of the stack, constitutive secretory vesicles would arise from the *trans*-Golgi network by selective removal of material targeted to forming cisternae and other destinations. Considerable evidence has accumulated in favor of this hypothesis (223,224). However, it also seems likely that this mechanism is too simple (225). For example, three-dimensional tomographic studies of the mammalian Golgi complex predict that the mechanisms of transport through, and exit from, the Golgi vary with physiological state, and in different cells and tissues (226). Such analyses of the lactating mammary epithelium are not yet available, but are likely greatly to inform our understanding once they are.

At this writing we do not have any understanding of the molecular details that govern the transition from the nonsecreting Golgi to the secreting Golgi. A proteomic analysis of 30 proteins from the isolated Golgi (Table 1) undoubtedly represents only a partial listing of the species of proteins that are upregulated (227). Of interest are the number of membrane fusion proteins, regulatory proteins (Rabs), actin remodeling proteins, and microtubule motor proteins identified.

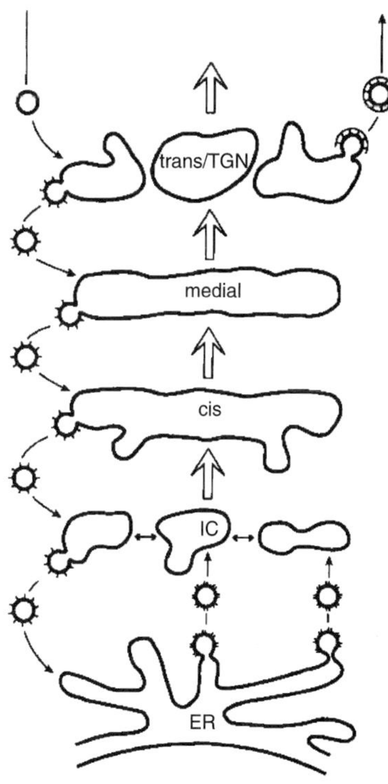

FIG. 13. A model for cisternal maturation. The intermediate compartment (IC) represents a nascent Golgi cisterna forming from vesicles (COPII coated) budding off the endoplasmic reticulum (ER). Several ICs fuse to form the *cis* cisterna. Other COPI-coated vesicles recycle enzymes from the *cis* cisterna as it migrates through the stack, gradually acquiring the characteristics of the medial and *trans*-Golgi as it proceeds. At the *trans*-Golgi network (TGN), the Golgi membranes reorganize into secretory vesicles, which in the case of the mammary gland would proceed to the apical membrane largely in a constitutive manner while *trans*-Golgi–specific enzymes like β-galactosyltransferase are moved back into the *trans* compartment by COPI-coated vesicles. (From Glick, B. S., Elston, T., and Oster, G. [1997]. A cisternal maturation mechanism can explain the asymmetry of the Golgi stack. *FEBS Lett.* 414, 177–181. Used by permission of Elsevier.)

The interaction of these molecules to bring about massive milk secretion is a fascinating problem for the future.

The Cell Biology of Lactose Synthesis

Lactose has long been known to be synthesized in a membrane compartment that can be isolated in good quantity from the lactating mammary alveolar cell (228,229). The enzyme responsible, β1, 4-galactosyltransferase (230), is found in the lumen of the last two or three *trans*-Golgi cisternae, where it catalyzes the transfer of galactose to glycoprotein-bound acetylglucosamines. In the mammary gland, α-lactalbumin serves as a cofactor that allows transfer of galactose to glucose, leading to the formation of lactose (229). The enzyme is highly upregulated in the lactating gland by a mechanism that was elegantly shown by Shaper and colleagues (231,232)

TABLE 1. *Proteins identified in purified Golgi fractions from 18-day pregnant (P18) and 10-day lactating (L10) rats using two-dimensional electrophoresis and mass spectrometry*

Map ID	Identification[a] (abbrevation)	P18	L10	Mr (kDa)	pI
2	Albumin[b]	+	++	68.7	6.20
3	Bip[b]	++	++	74.2	4.81
6	Lactoferrin	+	++	76.3	7.12
7	O–acetyl GD3 ganglioside synthase	+	++	47.4 (65)	6.78
8	Arrestin 2[c]	–	++	45.0	6.30
11	Tubulin[b]	++	++	47.9	5.17
12	Actin[b]	++	++	41.5	5.22
13	Somatin like protein	–	++	38.7	6.40 (6.1)
14	GBB 1[b]	++	++	38.7	6.40 (6.1)
16	14-3-3 β[c]	+	++	29.1	4.55 (6.0)
17	14-3-3 ε[c]	+	++	27.8	4.80 (6.1)
22	gp25LP	–	++	25.1	6.40 (7.0)
26	Ca^{2+} binding protein/p22	+	++	19.0	5.06
28	Rap 1B	–	++	19.0 (29)	4.63 (6.3)
31	Rap 1A/B	++	+	19.0 (29)	4.63 (6.4)
Membrane fusion proteins					
9	Synaptotagmin	–	++	42.6	6.71
10	Annexin I	+	++	35.0	7.77
15	Annexin V[c]	+	++	35.6	4.70
18	α-SNAP	–	++	33.2	5.23
19	Annexin IV	–	++	35.9	5.80 (8.0)
20	Annexin III	–	++	32.5	5.85 (8.0)
Regulatory proteins					
24	Rab6	–	++	23.5	5.42
25	Rab18	–	++	23.0	5.24
29	Rab11b	–	+	24.5	5.98
30	Rab11e	–	+	23.4	6.44
33	Rab14	–	++	23.8	5.85 (0.5)
34	Rab2	–	++	23.5	6.08 (6.5)
Actin remodeling proteins					
21	G25K/cdc42	+	++	21.3	6.16 (6.7)
27	RhoGAP	–	++	245.0 (28)	8.53 (6.2)
32	Hep27[c]	+	++	22.8	6.12 (6.5)
35	RhoA	–	++	20.5	4.99 (6.5)
Microtubule motor proteins					
1	Myosin heavy chain[c]	+	++	226.0 (100)	5.49
4	Kinesin[c]	–	++	125.2	7.36
5	Dynein intermediate chain[c]	–	++	68.3 (80)	6.66
23	Dynein light chain[c]	–	++	28.6	6.96 (6.1)

Numbers in the Mr and pI columns are theoretical. Numbers in parentheses are the actual and were given if the experimental number deviated substantially from the theoretical. Deviations in Mr are most likely due to proteolysis (ID#1 and 27), glycosylation (ID #7), and interactions with the cytoskeleton (ID #5, 28, and 31). Deviations in pI are most likely due to post-translational modifications.
[a]All protein identifications listed were determined by tandem mass spectometry.
[b]Identified proteins also used as fiducial markers or landmarks.
[c]Proteins also identified from the TX114 detergent insoluble phase.
Data from Wu, C. C., Howell, D. K., Neville, M. C., Rates, J. R., III, and McManaman, J. L. (2000). Proteomics reveal a link between the endoplasmic reticulum and lipid secretory mechanisms in mammary epithelial cells. *Electrophoresis* 21, 3470–3482.

to involve utilization of an alternate promoter start site that is regulated by a strong mammary-specific factor. The mRNA transcript formed lacks approximately 180 nucleotides of the 5′ untranslated region (UTR) of the transcript in other cells, leading, in addition, to much more efficient transcription. The specific reaction is

$$\text{UDP-galactose} + \text{glucose} \rightarrow \text{lactose} + \text{UDP}$$

Both uridine diphosphate (UDP)-galactose and glucose must be translocated into the Golgi compartment by transporters. It has long been known that translocation of UDP-galactose from the cytosol into the Golgi is mediated by an antiporter that transports uridine monophosphate in the opposite direction (233,234). More recently, Haney and coworkers (235,236) have published evidence suggesting that glucose is transported to the Golgi by the glucose transporter, GLUT1. Expression studies showing significant upregulation of GLUT1at the onset of lactation (188) are consistent with its role in this processes, but others have found that both GLUT8 (237) and GLUT12 (188), are also upregulated in the mammary gland at the onset of lactation. Macheda et al. observed GLUT12 both on the apical membrane and in the cytoplasm (188). The same group found that an antibody to GLUT1 strongly stained the basal membrane as well as an undefined intracellular compartment. To add to the current confusion, there is considerable evidence for a sodium-dependent glucose transporter, SLGT1, leading Obermeier et al. (238) to speculate that glucose transport in the lactating mammary epithelium is sodium dependent. The conclusion must be that until we have highly specific antibodies for the relevant glucose transporters, and coimmunolocalization with compartmental markers by confocal microscopy is available, we will not be able to understand which transporters in the mammary gland are contributing to the very large flux of glucose into the gland for lactose and lipid synthesis (184). Importantly, the insulin-regulated glucose transporter, GLUT4, is also found in the mammary gland but is likely confined to the adipose compartment.

Proteins Synthesized and Processed in the Exocrine Pathway

In humans, the major milk proteins, casein, lactoferrin, and α-lactalbumin, are secreted by the exocrine pathway, whereas IgA is secreted by the transcytotic pathway, as described later. In rodents, camels, and marsupials (239), whey acidic protein (WAP) is a major milk protein (240,241). Ruminants secrete little lactoferrin in mature milk, but do secrete β-lactoglobulin. WAP and lactoferrin are secreted by exocytosis. Casein is known to transfer large quantities of calcium and phosphate to the neonate, lactoferrin has widely studied anti-infectious and anti-immune properties (242), and the role of α-lactalbumin as a cofactor in lactose synthesis has been discussed previously. The function of WAP is unknown, although it was earlier suggested to function as a protease (239). More recently, overexpression of WAP in the mammary gland has been found to inhibit proliferation of alveolar cells; it also decreased growth of a normal mouse clonal cell line (HC11) by decreasing the rate of progression through the cell cycle (240,241). In the mouse, an increase in WAP gene expression occurs while proliferation is decreasing (compare milk proteins in Fig. 3D with proliferation rates in Fig. 2A), consistent with a role for the protein in downregulation of proliferation. β-Lactoglobulin is a lipocalin, one of a large class of molecules known to bind hydrophobic molecules, and has been postulated to have immunoregulatory functions, among other things. However, Kontopidis and colleagues, noting its similarity to an endometrial protein, glycodelin, postulate that β-lactoglobulin arose by fortuitous gene duplication from glycodelin and that it serves merely as a source of amino acids for the neonate (243). It is considered to be the major allergen causing cow's milk allergy in human infants (244). Thus, even though milk proteins have been the subject of intensive study for a long time, we are still in doubt about the function of some of them.

Secretion of the Major Divalent Ions: Calcium, Magnesium, Citrate, and Phosphate

The divalent ions, calcium, magnesium, citrate, and phosphate, are the second most abundant ionic components of milk, next to the monovalent ions we discussed previously. Because of their importance to the formation of bones and teeth, they vary considerably from species to species, the concentration being roughly proportional to the growth rate of the young. The concentrations of these entities in human and bovine milk are given in Table 2. For complete species information on these milk elements, the reader is referred to work by Jenness (39). In contrast to the monovalent ions, the divalent ions interact with each other and with the casein component of milk in a complex equilibrium that depends on the milk pH. For that reason, we begin this section by discussing the major hydrogen ion buffers in milk. We then consider the major structural compartments in milk, and end with a discussion of the mechanism of secretion of each of the ions listed, to the extent known. Because calcium has received the most attention, its secretion is described in the greatest detail.

The pH of Milk. Because CO_2 passes freely from the interstitial space into the milk space of the mammary gland, the following reaction is at equilibrium in the milk (245):

$$CO_2 + H_2O \Leftrightarrow H_2CO_3 \Leftrightarrow H^+ + HCO_3^- \Leftrightarrow 2H^+ + CO_3^{2-}$$

TABLE 2. *Major calcium and magnesium binding constituents of human and bovine milk*[a]

Component	Units	Human milk	Bovine milk
Magnesium, total	mM	1.8	5.1
Calcium, total	mM	7.5	29.4
Calcium, ionized	mM	3.0	2.0[b]
Citrate	mM	2.6	9.2
Phosphate, free	mM	1.8	11.2
Bicarbonate	mM	6.0	4.8
Casein	g/L	18	261
pH		6.8	6.7

[a]Values for human milk at 90 days postpartum from Allen et al. (38), except for casein, which comes from Casey and Hambidge (256). Values for bulk herd bovine milk are from White and Davies (257).

[b]Value obtained by calculation using a series of assumed equilibrium constants (258). If the murexide method is used, a value closer to 3.0 is obtained (258).

In all mammals, the body fluids, including milk, are in equilibrium with 5% CO_2 (246). When milk is removed from the mammary gland by milking, although not by the neonate, and exposed to the air, CO_2 is lost and the equation shifts to the left, leading to an increase in the pH of milk and, paradoxically, an increase in CO_3^{2-}, which binds calcium, decreasing the ionized calcium in the milk. This means that to obtain the equilibrium as it exists in the gland, ionized calcium must be measured on fresh milk, equilibrated with 5% CO_2. The problem is particularly acute for human milk, where bicarbonate is the major buffer. As can be seen from Table 2, cow's milk has a much higher phosphate concentration, and phosphate is a more important buffer in this milk and stabilizes the bicarbonate equilibrium. In human milk the native pH is 6.6 to 6.8, but it often rises to 7.2 if the milk is allowed to stand for any period (245). The pH given for cow's milk, 6.8, does not represent CO_2-equilibrated milk.

Distribution of the Divalent Ions among the Structural Compartments of Milk. Milk is a complex fluid that can be separated into several compartments by centrifugation at increasing speeds (245). Centrifugation of fresh milk at 6,000 × g for 30 minutes produces a layer of milk lipid droplets at the upper surface and a small cellular pellet. Subsequent high-speed centrifugation (100,000 × g for 2 hours) of the aqueous middle fraction produces a complex pellet composed of a layer of large casein micelles topped by a layer of smaller casein and then a loose "fluff" layer containing a variety of membranous and cellular debris (247). In human milk we found that approximately 72% of the calcium is associated with the aqueous fraction and the rest with the casein pellet (245) and α-lactalbumin, whereas in cow's milk 65% of the calcium is associated with casein (248). Little or no calcium was associated with the MFG or cellular fractions in fresh milk (245,249), although after freezing and thawing, which distorts the MFGM and may lead to production of free fatty acids, calcium can be associated with membrane or lipid fractions (250,251). Magnesium is largely present in the aqueous fraction (83%), but significant amounts are associated with the lipid (7%) and membranous fractions (10%). Citrate is found only in the aqueous fraction in human milk but is partly associated with the casein fraction of bovine milk (252). The phosphate value given in Table 1 represents free phosphate; the total phosphate in milk is much higher because of association of significant quantities of phosphate with casein, both due to actual phosphorylation of the molecule and association of casein with calcium phosphate complexes (253). The milk protein α-lactalbumin (254), which is present at a concentration of approximately 1 mM in human milk, binds approximately 1 molecule of calcium per molecule of protein (255) and likely accounts for approximately 1 mM of calcium in human milk and 0.5 mM in bovine milk (245).

Ionic Equilibria in Milk. To determine the electrochemical interactions of any ionic species in milk, it is necessary to know the ionic activity, the concentrations of all potential ligands, and the equilibrium constants for the relevant interactions (245,259). We have already given the concentrations of the relevant ionic species in Table 2. The relevant equilibrium constants can be found in Neville et al. (245) for the ionic equilibria illustrated in Fig. 14 (251). The concentration of ionized calcium can be determined using either ion-selective electrodes or equilibrium dialysis. A highly specific electrode is available for measurement of ionized calcium. However, as described earlier, great care must be taken with its use because the equilibria are extremely sensitive to changes in pH and CO_2 (260). When the ionic equilibria were calculated for both human and bovine milk, taking the pH and bicarbonate concentrations into account, the ionic species that contribute significantly to the calcium equilibria in human and bovine milk were found to be restricted to calcium citrate and calcium phosphate (Table 3). As Fig. 15. As shows, for most species the calcium concentration is roughly proportional to the casein concentration.

The Casein Micelle. The casein micelle is an amazingly elegant structure composed of amphipathic casein subunits held in submicelles by colloidal calcium phosphate and hydrophobic bonding (262,263). The formation of the casein micelle from submicelles in the trans-Golgi and secretory granules/vesicles can be clearly observed in Fig. 12 (162). The process can be studied in the test tube (259) using saturated calcium phosphate solutions and

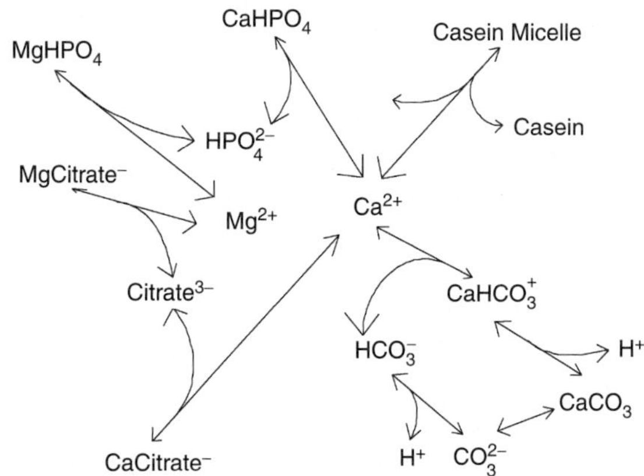

FIG. 14. Equilibria between divalent ions and bicarbonate in milk. (Used by permission from Neville, M. C., Allen, J. C., and Zhang, P. [1995]. Ionic interactions in milk. In *Handbook of Milk Composition* [R. G. Jensen, Ed.], pp. 577–592. Academic Press, San Diego. Used by permission of Elsevier.)

TABLE 3. *Calculated values for major forms of calcium and magnesium in human and bovine milk*

Binding species	Human milk (mmol/L)[a]	Bovine milk (mmol/L)[b]
Calcium		
$[Ca^{2+}]$	3.0	2.0
$[CaCit^-]$	2.0	6.9
$[CaPO_4]$	0.4	0.6
Casein	1.1	19.4
α-lactalbumin	1.0	0.5
Magnesium		
$[Mg^{2+}]$	0.94	0.8
$[MgCit^-]$	0.82	2.0
$[MgPO_4]$	0.03	0.3

[a]Values for human milk from Neville et al. (245) and unpublished data from Neville and Allen, and calculated using data from human milk at 3 months postpartum (38). Ionized calcium was measured on anaerobic samples using an ionized calcium electrode, as described. All other values were calculated from the ionic equilibria as described in Neville et al. (245)

[b]Values for bovine milk were taken from Holt et al. (258). These values may not be entirely accurate because ionized calcium was derived by calculation, and loss of CO_2 from the milk was not considered. Calcium bound to α-lactalbumin in bovine milk was calculated from the value of 10 mg/L α-lactalbumin given in Swaisgood (261).

casein phosphopeptides, where nanoparticles comprising a very stable spherical core of approximately 355 $CaHPO_4 \cdot 2\ H_2O$ units, surrounded by approximately 50 peptide chains, make a structure with a diameter in excess of 4 nm. Although the calcium remains exchangeable, the peptides prevent further

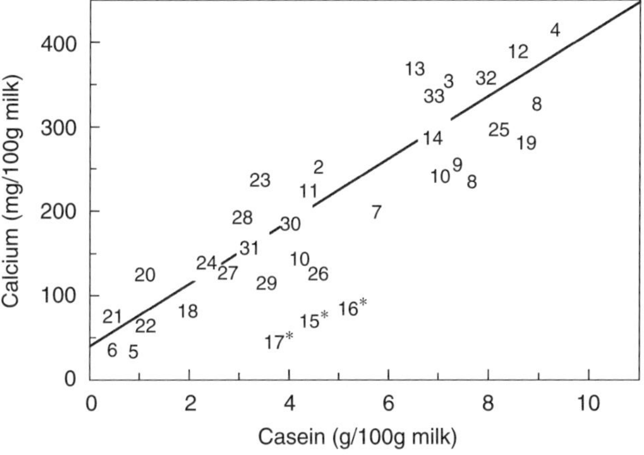

FIG. 15. Relation between calcium and casein concentrations in the milks of many species. 1, Long-tailed bat; 2, little brown bat; 3, tree-tailed bat; 4, rabbit; 5, baboon, 6, human; 7, hamster; 8, rat; 9, mouse; 10, guinea pig; 11, dog; 12; black bear; 13, grizzly bear; 14, polar bear; 15, fur seal; 16, elephant seal; 17, harp seal; 18, Indian elephant; 19, aardvark; 20, horse; 21, burro; 22, rhinoceros; 23, pig; 24, camel; 25, reindeer; 26, giraffe; 27, cow; 28, buffalo; 29, goat, 30, sheep; 31, pygmy sperm whale; 32, fin whale; 33, blue whale. *Starred points* were omitted from the regression. The best-fitting line had a slope of 8.7 mM/g casein and an intercept of 9.1 mM/kg milk. (Redrawn from Jenness, R. [1979]. Comparative aspects of milk proteins. *J. Dairy Res.* 46, 197–210.)

growth of the calcium phosphate crystals, allowing milk to maintain large quantities of calcium and phosphate in the absence of precipitation of the salt. The submicelles aggregate to make a micellar structure 40 to 200 nm in diameter stabilized by a special glycosylated form of casein, κ-casein (264,265). Thus, the casein micelle is a highly efficient package for the delivery of protein, calcium, and phosphate to the neonate. The difference in proportion of calcium bound to casein between human and bovine milks is due largely to the almost tenfold difference in milk casein content between the two species. However, in addition, the citrate associated with the casein micelle in bovine milk enhances its calcium-binding capacity (266). Further, human casein is not fully phosphorylated (267), and the calcium-binding capacity of human milk is only approximately 14 mol per mol of casein (245), as opposed to 20 mol of calcium per mol of casein (40) in the milks of most species.

Changes in Divalent Ions with Duration of Lactation. The total concentrations of calcium, phosphate, and citrate in human milk vary significantly with the stage of lactation (Fig. 3), whereas pH, magnesium, and ionized calcium remain relatively constant (141). To determine the basis for the changes in milk calcium, we calculated the equilibria among calcium-binding milk components using these data (38). The results (Fig. 16C) show that the increase of calcium during initiation of lactation is entirely due to an increase of citrate- and protein-bound calcium during this period. Because casein is not present in human milk until approximately day 3 of lactation (37,268) we can assume that much of the increase is due to the onset of casein secretion into milk. Similarly, the decline in late lactation (Fig. 16D) appears to be due to the loss of both protein- and citrate-bound calcium with no change in $[Ca^{2+}]$.

Mechanisms for Secretion of Divalent Ions into Milk. Because of the complex equilibria among the milk divalent ions, it might be inferred that the mechanisms by which they are secreted into milk are also complex, but, in fact, on a superficial level, they are likely quite simple, although the carriers and transporters involved are only now beginning to be identified. There is evidence stretching back three decades that calcium is pumped from the cytoplasm into the Golgi compartment (269–271), where it interacts with whatever calcium-binding molecules are present, equilibrating with these molecules as predicted by the equilibria in Fig. 14 and fostering the formation of the casein micelle. Both ionized and bound calcium are then secreted by exocytosis. The difficulty comes in understanding how the large amount of calcium necessary can be transferred into the Golgi and secretory compartments while cytosolic calcium is maintained in the micromolar range.

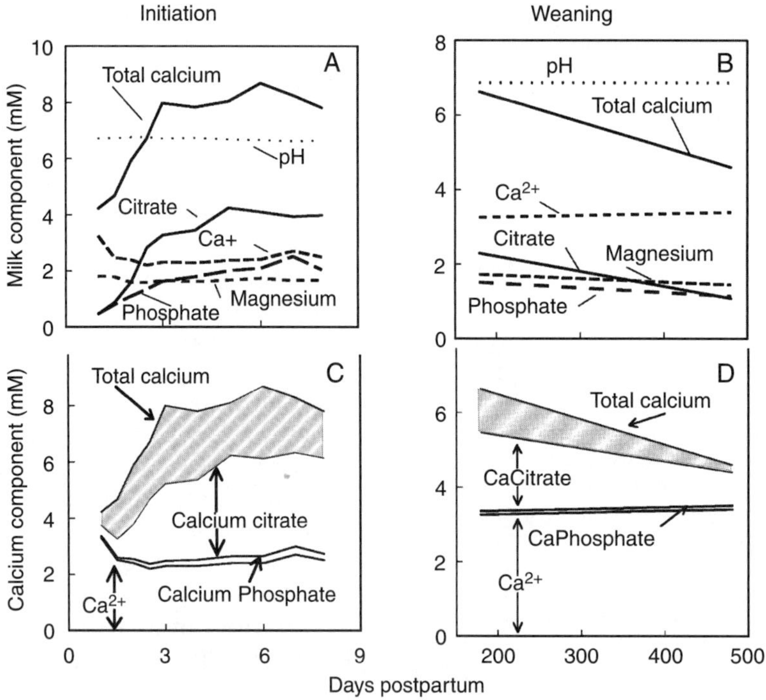

FIG. 16. Changes in divalent ions and calcium equilibria at initiation of lactation and weaning in women. **A, B:** Concentrations of milk components taken from studies of American women during initiation of lactation (12 subjects) and weaning (6 subjects) (141). **C, D:** Calcium binding to phosphate and citrate calculated from these data as described in Neville et al. (245). The *shaded areas* equal calcium bound to proteins, presumably casein and α-lactalbumin. (Figures taken with permission from Neville, M. C., Keller, R. P., Casey, C. E., and Allen, J. C. [1994]. Calcium partitioning in human and bovine milk. *J. Dairy Sci.* 77, 1964–1975.)

Nothing is known about magnesium transport. The mammary gland has the capacity to extract very large quantities of phosphate from the plasma (174), but the mechanism by which it enters the Golgi compartment is not yet understood. Citrate is an important metabolic intermediate (see the section on Balanced Aerobic Glycolysis in Milk Synthesis); however, the mechanism by which it is transferred into milk is also poorly understood. Our current understanding of calcium, phosphate, and citrate transport in the mammary gland is summarized next.

Divalent Ion Secretion into Milk. In the 1970s, Neville and Peaker made the important observation that isotopic calcium infused through the mammary artery into goats requires 1.5 hours to equilibrate with the calcium in the milk, a duration consistent with secretion by the Golgi–secretory vesicle system (272). These authors showed that isotopic calcium infused through the teat canal was entirely recovered in subsequent milkings, indicating that calcium does not enter the alveolar cell directly through the apical membrane. Further, milk calcium failed to increase when the milk in the gland was diluted for several hours with an infusion of isotonic sucrose, suggesting that there is no functional calcium pump in the apical membrane. Similar experiments indicated that phosphate and citrate are also secreted by the Golgi–secretory vesicle system (272,273). All these experiments, plus the observation that the casein micelle, a structure dependent on the presence of millimolar concentrations of calcium (274), forms in the Golgi–secretory vesicle system (Fig. 12) (172), are consistent with the hypothesis that the apical membrane of the mammary alveolar cell is impermeable to calcium, phosphate, and citrate and that these milk components are secreted solely by the Golgi–secretory vesicle system.

Figure 17 shows a model for calcium transport into milk. Crucial steps are the pumping of calcium into the Golgi against a concentration gradient of approximately three orders of magnitude, the transport of phosphate, and the synthesis and transport of citrate into the Golgi compartment. Casein micelles form in the swollen *trans*-Golgi [called *prosecretory granules* by Hermo and colleagues (172)] and mature as the secretory vesicles are formed (Fig. 12). The secretory vesicles fuse with the apical membrane and release their contents into the milk space. Not shown in Fig. 17 is the basal membrane, which must participate by transferring large quantities of calcium and phosphate from the interstitial space to the cytoplasm. The presence of a calcium pump on the apical

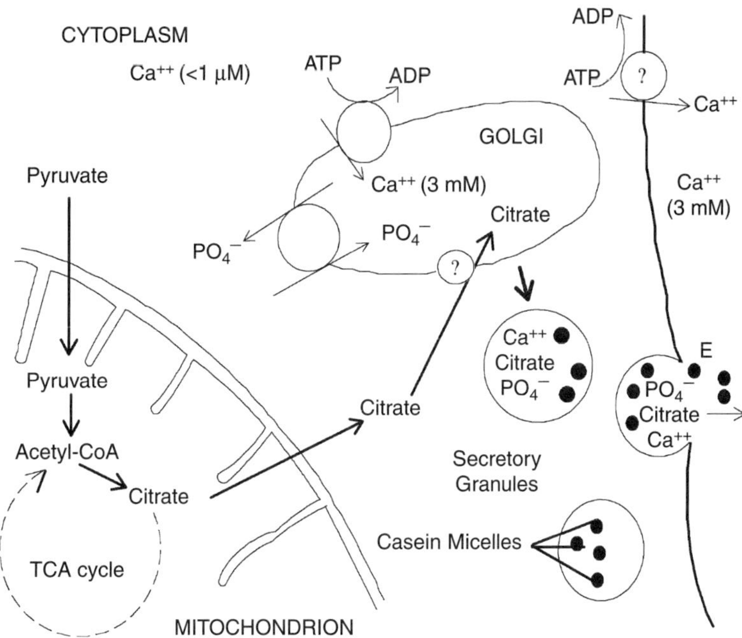

FIG. 17. Model for the secretion of divalent ions into milk. Calcium is pumped from the cytoplasm, where its concentration is less than 1 μM, into the Golgi by a calcium adenosine triphosphatase (ATPase) to achieve a concentration of approximately 3 mM. Within the Golgi, it interacts with citrate, phosphate, and casein. As the secretory granules move toward the apical membrane, they contain 3 mM [Ca^{2+}], mature casein micelles, and calcium citrate, the latter two at a concentration determined by the synthetic processes of the cell. Phosphate and citrate are also transported into the Golgi and are secreted both as the free anion and as part of the calcium complexes. E, secretion by exocytosis. A plasma membrane calcium ATPase has also been proposed by Reinhardt et al. (275). (Modified from Atkinson, S., Alston-Mills, B., Lonnerdal, B., and Neville, M. C. [1995]. Major minerals and ionic constituents of human and bovine milks. In *Handbook of Milk Composition* [R. G. Jensen, Ed.], pp. 593–622. Academic Press, San Diego.)

membrane of the alveolar cell that participates in the transport of calcium directly into the milk compartment has been proposed by Reinhardt and colleagues (275,276). The physiological relevance of such a pump is, however, questionable because calcium secretion into goat milk was not increased when isotopic sucrose was placed in the milk space, as described earlier (272). A good bit of work over the years has been devoted to identification of the calcium ATPases in the mammary alveolar cell and their functional significance. However, the mechanisms by which calcium traverses the basolateral membrane are totally unknown.

Calcium ATPases in the Mammary Alveolar Cell. Evidence for a calcium pump in membrane fractions of the bovine mammary gland was first obtained by Baumrucker and Keenan (271). Subsequent work in other laboratories confirmed the present of ATP- and magnesium-dependent calcium accumulation in Golgi-derived vesicles from the mammary gland of mice and rats (269,270,277). The calcium-sensitive ATPase activity in these vesicles had a K_m for [Ca^{2+}] in the submicromolar range and therefore could be active in the low-calcium environment of the cytoplasm. As pointed out by Taylor and associates (278), the problem is complicated by the potential presence in the Golgi of both resident calcium ATPases and ATPases in transit to the plasma membrane. Three categories of calcium ATPases have been identified in isolated "Golgi fractions": sarcoplasmic reticulum/endoplasmic reticulum calcium ATPases (SERCAs), which are inhibited by thapsigargin; plasma membrane calcium transporters (PMCAs); and secretory pathway calcium ATPase or SPCA, an evolutionarily ancient homolog of the yeast NMR1 calcium (279) that is resident in the Golgi of many organs. The work of Bingham and colleagues (280) showing that the biochemical properties of the calcium sensitive ATPases in bovine mammary Golgi membrane fractions differ significantly from those of the microsomes suggests that SERCAs may not play an important role in Golgi calcium transport in the mammary gland, a conclusion consistent with more recent experimental results.

Reinhardt and Horst examined mRNA from various stages of mammary development and found expression of six calcium ATPases by Northern blot: two SERCAs, three PMCAs, and the mammalian homolog (an SPCA) to a yeast calcium ATPase, NMR1 (281). The endoplasmic reticulum transporters SERCA2 and SERCA3 were upregulated during lactation, as might be expected from the massive increase in endoplasmic reticulum in the lactating gland. Studies of protein levels confirmed that the three PMCAs found in Northern blots were also present as proteins (275). Of the two housekeeping isoforms, PMCA1 was moderately upregulated during lactation and PMCA4b protein was downregulated. However, the specialized isoform PMCA2b was present at a level 2,000 times that found in pregnancy on day 1 and was upregulated more than fivefold between days 1 and 18 of lactation. It was present as a specialized splice variant, PMCA2bw, containing a 45–amino-acid insertion, and was found on the MFGM, leading Reinhardt and colleagues to suggest that it may participate in calcium transport across the apical membrane (275). Studies in knockout mice show that milk calcium is significantly reduced when this transporter is absent (276), confirming that it plays an important role in mammary calcium transport during lactation.

Interestingly, SPCA was upregulated approximately sixfold in mid-pregnancy in the rat and was stated to be upregulated 7 days before lactation in the cow, suggesting a possible role in mammary calcium transport during lactation (275). This calcium ATPase is generally thought to be present in Golgi membranes (279), where it transports not only Ca^{2+} but Mn^{2+}, a cation necessary for the activation of the major galactosyltransferase in the mammary Golgi (282,283). Its subcellular localization in the mammary gland has not yet been defined. However, it is found at very high levels on mammary microsomes, and unlike PMCA2bw, at much lower levels on MFGMs (276).

In summary, two calcium ATPases have been found at very high levels in the lactating rat mammary gland, PMCA2bw and SPCA. Both are associated with membrane fractions from the mammary gland, but only PMCA2bw is found in significant quantities on the MFGM. It seems likely that SPCA is involved in the transfer of both calcium and manganese into the Golgi compartment. The role of PMCA2bw in Golgi and apical plasma membrane calcium transport is ripe for further investigation.

Phosphate Transport in the Mammary Alveolar Cell. Phosphate is an important ion in milk, being present as the free ion, complexed with calcium, and complexed with casein. Flux into milk is rapid; for example, it has been estimated that the entire plasma pool of phosphate turns over every 10 minutes in the lactating rat (284). Using mammary explants and in vivo perfusion techniques, Shillingford et al. (285) obtained strong evidence that phosphate transport into mammary tissue is sodium dependent. A sodium phosphate transporter, Na-Pi2b (classified as SLC34A2), was found to be upregulated in late pregnancy and highly expressed in the lactating mammary gland (286) (Rudolph, M., and Neville, M. C., unpublished data). This transporter is thought to have a $Na^{+}:PO_4^{2-}$ ratio of 3:1, making it an electrogenic transporter. It has been detected in a number of tissues, including small intestine, lung, mammary glands, testis, and liver (287), and has been clearly shown to be present on the apical border of enterocytes, alveolar type II cells, and mammary alveolar cells (286,288,289). With this localization, it seems unlikely that it is responsible for the transport of phosphate across the basolateral or Golgi membranes, and its exact function at the apical membrane is not clear.

Shillingford et al. were also able to detect PO_4^{2-} transport by a DIDS (4,4′-diisothiocyanostilbene-2,2′-disulfonic acid)–sensitive anion exchanger, distinct from the anion exchanger that transports Cl^- and SO_4^{2-} in mammary explants (285). No other information about this transporter is available at this writing, and its physiological role is still unclear as is the mechanism by which phosphate is transferred into the Golgi. Microarray studies from this laboratory show that SLC20A1, a sodium phosphate transporter that is also a viral receptor called Pit-1, is upregulated fourfold at the onset of lactation (Neville, M. C., and Rudolph, M., unpublished data). Because this transporter is known to form assemblies at the basal surfaces of cells (290), it may represent the basal phosphate transporter. Because no phosphate transporter has been identified in the Golgi membranes, the possibility exists that the phosphate in milk could be derived from the transport and hydrolysis of UDP-galactose during lactose synthesis. However, it is not clear that all species transport sufficient UDP-galactose to satisfy the phosphate requirements of the casein micelle. Thus, many unanswered questions surround phosphate transport into milk.

Citrate Transport in the Mammary Alveolar Cell. In contrast to the other divalent anions described in this section, citrate is synthesized from pyruvate in the mitochondria of the mammary alveolar cell. It reaches the cytoplasm through the mitochondria citrate transporter (tricarboxylic acid carrier, SLC25A1). This carrier is highly upregulated during lactation in the mouse (27), presumably to provide citrate for synthesis of malonyl CoA and NADPH, the substrate and energy source, respectively, for fatty acid synthesis. In this species, unlike humans and ruminants, citrate does not enter the milk. The mechanism by which citrate enters the Golgi–secretory granule compartment is not known.

Calcium as a Regulator of Milk Secretory Activity. Considerable circumstantial evidence exists that calcium also plays a regulatory role in milk secretion (291,292). The problem has been studied at the cellular level by the laboratory of Burgoyne, who examined the determinants of casein secretion in isolated mammary acini (293,294). They found that much of the casein secretion is constitutive. However, ionomycin, a calcium ionophore, increased casein secretion, providing evidence for the existence of a calcium-regulated pathway (294). Hypotonic solutions, an in vitro surrogate for stretch of mammary alveolar cells, also showed increased $[Ca^{2+}]_i$, albeit transiently (295), suggesting that changes in cytoplasmic calcium may be part of the signal that downregulates milk secretion during milk stasis (296). In addition, calcium appears to be necessary at various points along the secretory pathway for protein synthesis and phosphorylation. Thus, calcium depletion led to a loss of ribosomes from the endoplasmic reticulum in isolated mammary organelles, leading in turn to inhibition of protein synthesis (296). In the same series of experiments, calcium was found to be necessary for the second phase, but not the first, of casein phosphorylation. It is now feasible to use confocal or multiphoton microscopy to visualize cells or compartments labeled with fluorescent dyes in the mammary gland of the living animal (297,298). Although the experiment has not yet been carried out with calcium-sensitive dyes, such dyes could likely be loaded into the alveolar cells by intraductal injection (299) and visualized in situ in the live mouse to monitor cytoplasmic calcium activity as the cells are stretched by, for example, overfilling the mammary lumen by intraductal injection of isotonic sucrose. Low-affinity calcium-sensitive dyes such as mag-fura-2 have been shown to accumulate in a thapsigargin-sensitive compartment of fibroblasts and used to monitor the effects of various agents on endoplasmic reticulum calcium (300,301). Adaptation of such techniques to the in vivo mammary gland should allow definitive studies of the role of endoplasmic reticulum and Golgi calcium in the regulation of milk secretion.

Secretion of Lipid into Milk

The composition of milk lipid varies among species (302,303) primarily in the proportion of medium-chain fatty acids, which are synthesized from glucose in the mammary gland (Fig. 11), and long-chain fatty acids, which are derived from both dietary lipid and adipose tissue stores (Fig. 18). Triacylglycerol synthesis takes place in the endoplasmic reticulum using fatty acids derived either from the plasma or by endogenous synthesis and glycerol-3-phosphate derived from glucose-6 phosphate or by phosphorylation of plasma-derived glycerol. Lipid droplets form in the endoplasmic reticulum and gradually move toward the plasma membrane, where they interact with a complex complement of proteins, are engulfed by the membrane, and budded into the milk space as a membrane-bound particle called the *MFG* (173).

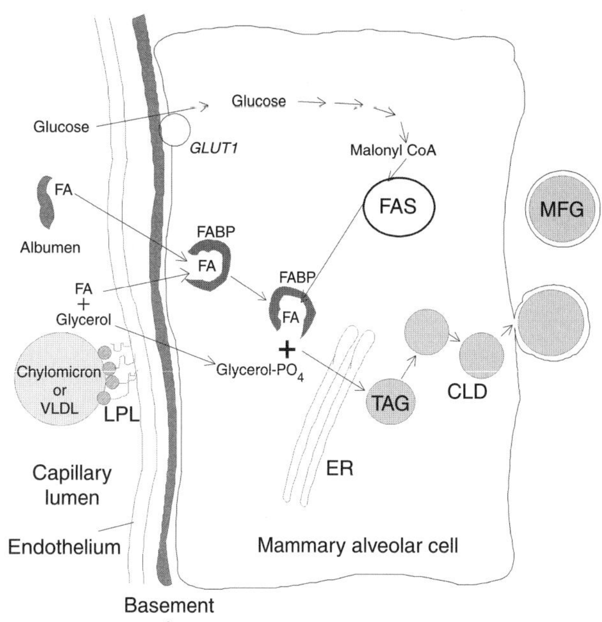

FIG. 18. Diagram depicting the sources of fatty acids for lipid synthesis and the secretion of the milk fat globule. Glucose transporter-1 (GLUT-1) brings glucose into the cell for conversion to malonyl CoA, where fatty acid synthase (FAS) acts on malonyl CoA to convert it to medium-chain fatty acids (FA) that are immediately bound by a fatty acid–binding protein (FABP); FA enter the cell either from albumen, to which they are bound during fasting, or from hydrolysis of triacylglycerol (TAG) in chylomicrons or very–low-density lipoprotein (VLDL) by the enzyme lipoprotein lipase (LDL). Glycerol also enters the cell after release from plasma TAG and is converted to glycerol-3-phosphate. Fatty acids and glycerol-3-phosphate are combined into TAG by the action of acyltransferases and formed immediately into cytoplasmic lipid droplets (CLD), which move toward the apical membrane, where they coalesce into larger droplets, are engulfed by plasma membrane, and secreted as the membrane-bound milk fat globule (MFG). (Modified with permission from Neville, M. C., and Picciano, M. F. [1997]. Regulation of milk lipid synthesis and composition. *Annu. Rev. Nutr.* 17, 159–184, with permission of Annual Reviews, http://annualreviews.org.)

The source of plasma-derived fatty acids used for triglyceride synthesis in the mouse and human mammary gland depends on the dietary state. In the fed state, triglycerides in chylomicrons or very–low-density lipoprotein are broken down by lipoprotein lipase to glycerol and fatty acids, which are transported into the cell possibly by glycerol transporters such as the aquaporins (304) or fatty acid transporters such as FAT/CD36 (305,306). In the fasting state, the plasma concentration of nonesterified fatty acids rises considerably, allowing transfer of lipid from body stores to the alveolar cell. These substrate transfers have been best studied in the lactating goat (307), although new technology (308) should soon allow studies in smaller animals such as rodents. Once in the mammary alveolar cells, fatty acids are either bound to a fatty acid–binding protein (FABP; Fig. 18) (309) or activated by esterification to coenzyme A and then bound to an acyl CoA–binding protein (310). Activated fatty acids are transesterified to glycerol-3-phosphate in a series of reactions catalyzed by specific transacylases located in the membrane of the endoplasmic reticulum, eventually forming triglycerides (311). Triglycerides and cholesterol esters are then packaged into phospholipid- and protein-bound structures called *cytoplasmic lipid droplets* (CLDs). The mechanism for fat secretion by the mammary gland (Fig. 18) is not duplicated in any other organ (173). Although the details of this process are not well understood, CLDs travel to the apex of the cell, where they are enveloped by the plasma membrane and eventually bud off as bilayer-coated structures known as MFGs (173,311).

The Milk Fat Globule

The MFG consists of a lipid core that is approximately 99% triacylglycerol (TAG) and 1% cholesterol esters. The MFGM contains phospholipids and cholesterol, which are present at concentrations approximately 0.8% and 0.3% of the total lipid, respectively (312). The MFGM membrane is a complex structure whose formation is not entirely understood. The major proteins in MFGM have been the subject of a careful review by Mather (313), so their functions are only briefly summarized here.

1. Butyrophilin (Btn) is the most abundant protein in the MFGM and is expressed only in the mammary gland (313). It is a type I transmembrane protein related to the immunoglobulin superfamily and is expressed only by mammary epithelial cells (313). Studies in Btn null mice show that it is essential for the secretion of CLDs (314).
2. Xanthine oxidoreductase (XOR) is a metabolic enzyme that catalyzes the reduction of xanthine and hypoxanthine to urea, but is probably a structural component of the MFGM (26,315). It is also essential for milk lipid droplet secretion (316).
3. Adipophilin (ADPH) is a member of the PAT (perilipin, ADPH, TIP47) family of lipid storage droplet–associated proteins. It is associated with the CLDs in many cell types, including the mammary alveolar cell, and is found at high levels on the MFGM (317–319). It does not seem to be essential for milk lipid droplet formation because ADPH null mice apparently lactate normally; the possibility that the congener TIP47 takes over the function of ADPH in lipid droplet formation is currently under investigation (McManaman, J. L., personal communication).
4. FAT/CD36 is a widely expressed protein with diverse functions, including fatty acid translocation (320) and binding to thrombospondin and collagen (313). It is not clear whether its presence in the MFGM is accidental or functional.
5. MFGM-E8 has been found to be important in the clearance of apoptotic cells in many systems (321,322). Whether it plays a role in clearance of apoptotic cells during mammary gland involution is still a matter of investigation (323).
6. MUC1 is a heavily glycosylated glycoprotein that is present at very high levels in human milk (almost 1 g/L) and lower concentrations (0.04 g/L) in bovine milk (313). Despite intensive investigation its function is not understood, although it may be a protective factor, binding pathogens and preventing their entry into the intestinal mucosa (147).
7. FABPs are expressed at very high concentrations in the mammary cytoplasm, where they presumably prevent high concentrations of free fatty acids from having deleterious detergent effects on mammary membranes (324). Like FAT/CD36, their presence in MFGs may be related more to association with CLDs than to a specific function in these structures.

Of the seven proteins listed, only three have been clearly implicated in the secretion of the MFG: Btn, XOR, and ADPH (Fig. 19). Mather and Keenan proposed that butyrophilin acts as a transmembrane scaffold for interaction with other proteins of the MFGM (173,313). The cytoplasmic tail has a remarkably conserved "30.6" domain of approximately 170 amino acids containing approximately 15 β strands folded into two immunoglobulin-like regions. It interacts specifically with XOR, an interaction that depends on sulfhydryl bonds as shown by the gel filtration experiment in Fig. 19 (26,325), in which a high–molecular-weight complex containing Btn, XOR, and ADPH is shown to be dissociated by dithiothreitol (DTT) into a complex containing Btn and ADPH, but no XOR. Interestingly, in the cytoplasm of the alveolar cell, XOR is in the reduced xanthine dehydrogenase form, whereas most of the enzyme in milk is present as the oxidized xanthine oxidase form (326). A sulfhydryl oxidase associated with the MFGM has been postulated to be responsible for the conversion, which likely also mediates an interaction between XOR, Btn, and ADPH (326,327). Although both XOR and Btn are found in the apical plasma membrane of the lactating mammary gland (26), ADPH is associated with cytoplasmic lipid droplets as they arise from the endoplasmic reticulum and are transported through the cell to the apical membrane. The molecule does not possess a distinctive hydrophobic domain as is characteristic of lipid storage droplet proteins in plants; instead, two lipid droplet targeting domains with complex structural properties are found in its N- and C-terminal regions (318).

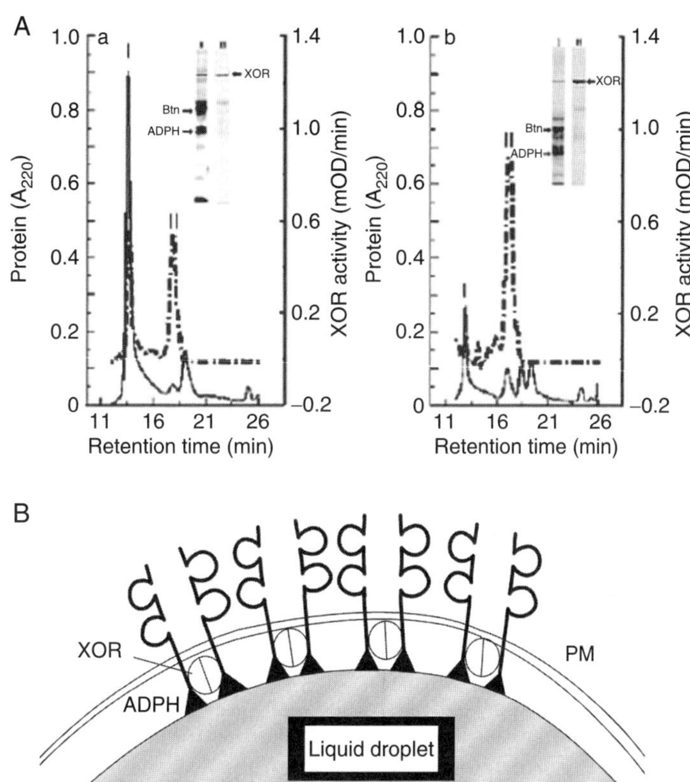

FIG. 19. Adipophilin (ADPH), butyrophilin (Btn), and xanthine oxidoreductase (XOR) form a complex in milk fat globule membranes that depends on sulfhydryl interactions. **A:** Proteins were extracted from washed milk fat globule membranes using Triton X-100 and subjected to size-exclusion chromatography. Panel *a* shows the elution profiles of XOR enzymatic activity (*dashed line*) and 220-nm absorbing material (*continuous line*). The *insets* show silver-stained gels of proteins in peaks I and II separated by sodium dodecyl sulfate–polyacrylamide gel electrophoresis. The identities of XOR, Btn, and ADPH bands were confirmed by N-terminal microsequence analysis. Panel *b* shows the elution properties of XOR, Btn, and ADPH after incubating peak I with 5 mM dithiothreitol (DTT). XOR no longer interacts with ADPH and Btn when sulfhydryl bonds are broken. (From McManaman, J. L., Palmer, C. A., Wright, R. M., and Neville, M. C. [2002]. Functional regulation of xanthine oxidoreductase expression and localization in the mouse mammary gland: evidence of a role in lipid secretion. *J. Physiol. [Lond].* 545, 567–579.) **B:** Model for interactions of Btn, XOR, and ADPH with the lipid droplet of the milk fat globule. (Model after Mather, I. H., and Keenan, T. S. [1998]. Origin and secretion of milk lipids. *J. Mammary Gland Biol. Neoplasia* 3, 259–274, with permission of Elsevier.)

Thus, ADPH is implicated both in formation of lipid droplets and their targeting to the apical membrane complex. How the latter is accomplished is not known.

Regulation of Milk Lipid Synthesis and Secretion

Several well-defined conditions alter either the rate of secretion of milk lipid or the fatty acid composition of milk, including the initiation of lactation, fasting and starvation, and hormones. In addition, a dietary fatty acid, conjugated linoleic acid, has been shown to produce a low–milk-fat syndrome in dairy cows (302) and to reduce the lipid content of human milk (328). Our current understanding of each of these areas is summarized in turn.

Initiation of Lactation. Around parturition, there is a coordinated increase in many of the enzymes of lipid synthesis in the mammary gland, most completely studied in the rabbit (23) until our current microarray studies defined the time course in the mouse (27). Although the increased lipid synthesis around parturition was initially interpreted as arising largely from changes in phosphorylation of key enzymes (329), it is now clear from studies of liver and other lipid-synthesizing tissues as well as the mammary gland (330,331) that changes in the activity of lipid synthetic enzymes largely occur at the level of gene expression (191,332). The 1c form of the SREBP, a member of a family of basic helix–loop helix (bHLH) transcription factors important in the regulation of fatty acid and cholesterol synthesis (191,333), has been suggested to be responsible for transcriptional regulation of lipid synthesis, although it is likely not the only molecule involved because peroxisome proliferator–activated receptors (PPARs) (334), liver X receptors (335), and carbohydrate response element–binding proteins (336) have also been implicated in the regulation of the enzymes of fatty acid synthesis in other tissues. So far, however, only SREBP-1c has been implicated in the mammary gland (302,337).

A transcriptional regulator, SREBP is stored in the membranes of the endoplasmic reticulum bound to a sterol-sensitive escort protein, SREBP cleavage–activating protein (SCAP). When sterol levels fall, SCAP undergoes a conformation change (338–340), probably decreasing its binding to an endoplasmic reticulum retention protein, INSIG-1 (341), and moves to the Golgi, taking SREBP with it. Two proteinases in the Golgi cleave SREBP, releasing the bHLH transcription factor that makes up its N-terminus. The bHLH protein moves to the nucleus and binds to sterol response elements (SREs) in the promoters of many of the genes involved in lipid synthesis (342,343). A distinct form of SREBP, SREBP-2, appears to be responsible for regulation of cholesterol synthesis (344). Another member of the SREBP family, SREBP-1, has been shown to regulate the enzymes specific to fatty acid synthesis in liver (345,346), adipose tissue (347), muscle (348), and pancreatic beta cells (349,350). It comes in two variants, SREBP-1a and SREBP-1c, determined by alternate start sites. SREBP-1a is the most active and predominates in highly proliferative tissue culture cells (338,351); however, SREBP-1c is the major form found most organs in vivo, where it appears to be the isoform that specifically regulates the genes for fatty acid synthesis (333,338,352).

The mechanisms by which SREBP-1 is activated and regulated are less well understood than those of SREBP-2. Although it seems clear that the molecule is activated by cleavage of the bHLH region (191,353), the nature of the molecular factors that promote movement from the endoplasmic reticulum to the Golgi are currently unclear, although there is evidence that SCAP is involved (354). What is clear is that SREBP-1c mRNA, protein, and activity all decrease in response to fasting (351) and long-chain fatty acids (355,356) and increase in response to carbohydrate feeding, refeeding after fasting, and insulin (351). However, the best-understood regulation of SREBP-1c expression is through a feedback loop whereby the bHLH region of SREBP-1c activates its own transcription (357), interacting with a combined SRE Sp1 site approximately 50 nucleotides upstream from the start site. The SRE elements in the promoters of the genes of fatty acid synthesis are more variable than those in the promoters of genes for cholesterol synthesis and, in some cases, appear to be replaced or supplemented by E-boxes (358), a more general bHLH response element (353).

Work is beginning to appear on the role of SREBP-1c in the mammary gland in the initiation of lactation. Barber and coworkers showed that the mature form of SREBP-1 is upregulated ninefold in nuclear extracts from the mammary gland of lactating sheep (359). To determine the basis for this increase, they studied the promoter of acetyl CoA carboxylase (ACC), an enzyme that catalyzes the first committed step in fatty acid synthesis by converting acetyl CoA to malonyl CoA. They had previously shown that ACC-α has three promoter sequences, two of which are significantly upregulated in the lactating sheep enzyme at the onset of lactation (360). Promoter III leads to transcription of a unique exon upstream of exon 5 that they have termed exon E5A. Studies of promoter III–reporter constructs in transfected cells showed that transcription factors NF-Y and USF-1 and -2 interact functionally with SREBP-1 at a DNAase-hypersensitive site corresponding to an E-box in promoter III, leading to increased transcription of the E5A variant of the enzyme. They also form an SREBP-1–interacting site in promoter II of the same gene, providing an explanation for the coordinate upregulation of both forms of the enzyme at the initiation of lactation in sheep. Promoter III and exons E5A are found in humans (361), but not in rats and mice, suggesting alternate regulation of ACC in these species. That SREBP-1c may be the relevant form of the enzyme in the mammary gland is suggested by the finding from real-time reverse transcriptase polymerase chain reaction that SREBP1-c is the major form in the mammary gland of the lactating mouse (Rudolph, M., and Neville, M. C., unpublished data). SREBP-1c was also found to activate fatty acid synthase in a MCF7 cells, a human breast cancer cell line (362).

Dietary Regulation. In animals with little metabolic reserve compared with the rate of milk synthesis (e.g., rats and mice and dairy cows), 24 hours of food deprivation leads to a state of starvation with profound effects on the rate of synthesis of all milk components, including lipids (363). Fasting and refeeding has often been used as an experimental paradigm for the study of lipid synthesis in the rodent mammary gland (363–368), and we have used a fasting paradigm in this laboratory to study the regulation of mammary lipoprotein lipase (369). However, it is not a very satisfactory protocol in the lactating mouse, who reacts to more than 10 hours of food deprivation by eating her pups. It is also not a very clean protocol for studying the details of regulation of fatty acid synthesis because many plasma parameters, including insulin, fatty acids, and glucagons, are significantly altered on refeeding, and it is difficult to ascertain which is responsible for observed responses. This point is well illustrated by a consideration of studies of the effects of insulin.

Insulin, in part because of its in vitro effects on milk protein synthesis, has long been considered to be a regulator of lipid synthesis in the mammary gland. In addition, an apparent role of this hormone, derived from studies using a fasting/refeeding paradigm in rats (368,370–372), may have been confused with responses to circulating free fatty acids (303). In more recent experiments using the glucose clamp paradigm, which allows independent regulation of insulin and glucose, no effects of insulin on the synthesis of milk components were demonstrated (186,373). This finding is in marked contrast to the liver, where insulin appears to be a major regulator of the SREBP-1 pathway (374). In fact, it would be counterproductive for the mammary gland, which produces an energy-containing fluid for export at a relatively constant rate (375), to depend on insulin, whose levels vary with food intake, for activation of lipid synthesis. Finally, except in very–high-producing species where food intake directly provides the calories for milk production (363), feeding and fasting have little effect on the rate of milk production (303,376,377).

It seems likely that dietary paradigms designed to alter the proportion of fat and carbohydrate in the diet may be useful experimental paradigms to study the regulation of lipogenesis in the mammary gland. Medium-chain fatty acids are synthesized only in the mammary gland, using glucose (or acetate in ruminants) as substrate (363). Synthesis of this class of fatty acids has been shown to vary with the lipid content of the diet in cows (378–380), rats (366,381), and humans (382). However, these paradigms have not yet been put to use in studies of the molecular mechanisms by which lipogenesis is regulated in the mammary gland except for some recent studies of the effect of a very interesting nutrient, conjugated linoleic acid, described next.

Effect of Conjugated Linoleic Acid on Milk Lipid Synthesis. Trans fatty acids, particularly trans-10, cis-12 conjugated linoleic acid (t10,c12 CLA), which can be found in hydrogenated fats and is increasingly used as a weight-loss supplement, have been shown to inhibit milk lipid synthesis in cows, both acutely and when fed on a chronic basis (337,383). Ingestion of CLA by pigs has been shown to alter the composition of body fat of the suckling (384), and diets high in CLAs decrease milk lipid content in women (328,385), suggesting that similar effects are present in other species. The effect in dairy cows, called *low milk fat syndrome*, was initially attributed to changes in plasma insulin with certain types of feed supplement (385). However, the glucose clamp studies described previously (186,373) and other data (303) indicate that this is not the mechanism. Bauman and coworkers (330,337) showed that the mRNA for many of the genes of lipid synthesis, including ACC, fatty acid synthase, FABP, delta-9 sterol desaturase, lipoprotein lipase, glycerol phosphate acyltransferase, and acylglycerol phosphate acyltransferase, were downregulated in dairy cows ingesting t10,c12 CLA, resulting in corresponding or larger decreases in lipid synthetic reactions. They have also shown that effects of t10,c12 CLA on fatty acid synthesis in a bovine mammary cell line are associated with a decrease in the mature nuclear form of SREBP-1, clearly implicating this transcription factor in low milk fat syndrome (386). Whether SREBP-1 will turn out to be the sole regulatory mechanism is a matter for further research because there is considerable evidence from tissues other than the mammary gland that mechanisms such as changes in cyclooxygenase activity and eicosanoid production or PPAR activity may be involved (387).

Summary. Mammary lipid synthesis has been the subject of detailed study for many decades, and indeed studies in the mammary gland were instrumental in elucidating many of the pathways of lipid metabolism (388). However, the molecular mechanisms by which these pathways are regulated are only now beginning to be subjected to the tool of modern molecular biology. Recent studies have led to a clear role for the transcription factor SREBP-1 in regulation of lipogenesis; however, it is likely that many other factors will be involved. Some interesting times lie ahead in this field.

Secretion of the Monovalent Ions Na^+, K^+, and Cl^- into Milk

Sodium, potassium, and chloride ions are among the most prevalent minerals in milk, comprising approximately 10% of the osmolarity of human milk, 25% of the osmolarity of bovine milk, and 65% of the osmolarity of rabbit milk (389). Because the total concentration of sodium and potassium in milk is equal whether determined by ion-selective electrodes or flame photometry (258,390), they are considered to be free and uncomplexed in milk. The same assumption is usually made for chloride (266). Thus, we have only to consider how the free ions traverse the mammary epithelia cell from the interstitial fluid and how the electrochemical gradients of these ions between the cell and its surroundings are maintained. In most species, the paracellular pathway between epithelial cells is tightly closed during lactation and the

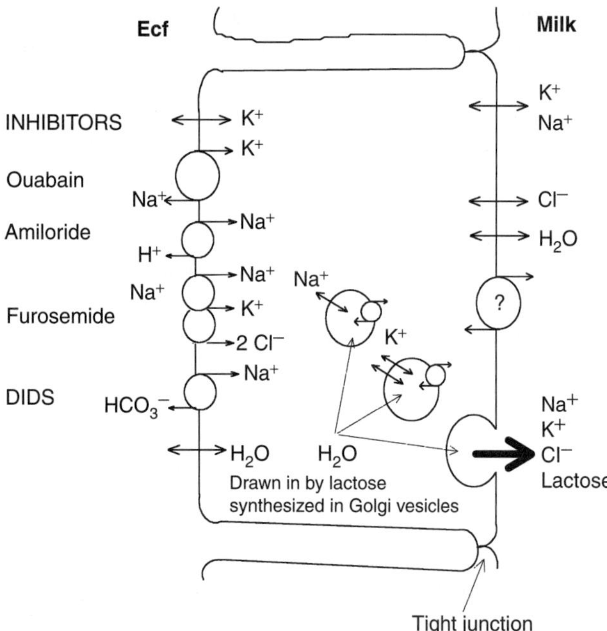

FIG. 20. Model for monovalent ion transport in the lactating mammary epithelium. Transporters on the basal surface of the cell move sodium, potassium, hydrogen ion, chloride, and bicarbonate to control the composition of the cytoplasm. In the cell, lactose synthesis in the Golgi draws water and ions osmotically into these compartments, which form into secretory vesicle and are secreted by exocytosis. Channels in the apical membrane equilibrate sodium and potassium according to their electrochemical gradient. Chloride channels are also present, but chloride is out of equilibrium, its low concentration in the milk presumably maintained by a chloride pump or exchanger whose identity has not yet been established. During lactation the tight junctions are impermeable to ions, so that all concentration gradients must be established by transcellular mechanisms. (Model redrawn from Neville, M. C., Allen, J. C., and Zhang, P. [1995]. Ionic interactions in milk. In *Handbook of Milk Composition* [R. G. Jensen, Ed.], pp. 577–592. Academic Press, San Diego.)

concentrations of monovalent ions must be strictly controlled by the transport activities of the alveolar cell, which are, unfortunately, only partially understood. Linzell and Peaker proposed a scheme for ion transport across the mammary epithelium more than 30 years ago that remains largely unchallenged today (1,174,391). The major features of this scheme, diagrammed in Fig. 20, were:

1. All mammary membranes are freely permeable to water.
2. The ducts have the same ionic permeability properties as the alveolar cells, and milk composition is unchanged as it passes through the mammary ducts.
3. Ion concentrations are established in the Golgi and secretory vesicles, which possess ion channels and pumps similar to those on the apical membrane of the cell.
4. The ionic concentration in the cytoplasm is maintained by pumps and exchangers present on the basolateral membranes of the cells.
5. The apical membrane is permeable to sodium, potassium, and chloride.
6. Sodium and potassium are at electrochemical equilibrium across the apical membrane, whereas the concentration of chloride in the milk is lower than predicted by passive distribution.

The major evidence for the free permeability of the mammary membranes to water is that milk is iso-osmotic with plasma under any circumstances in which it has been measured (174,251). An interesting example of this principle is that during Ramadan, when women fast, both their blood and their milk become slightly hypertonic (392). Water channels called *aquaporins* are thought to be responsible for the water permeability of cell membranes. Aquaporin 5 was found to be present on the apical membranes of the mammary ducts of virgin animals (289) and possibly intracellularly during lactation, although the authors claim that the intracellular stain observed during this time is nonspecific. Aquaporin 1 was observed in capillaries of breast tissue and possibly myoepithelial cells or basal membranes on tissue microarrays (393); this aquaporin was not observed in normal murine mammary epithelial cells. Preliminary data from microarray studies suggest that both aquaporins 2 and 8 are upregulated during lactation (27).

Ample evidence for point 2, that milk does not change ionic composition as it passes through the milk ducts, is present in the literature. However, firm evidence for point 3 has not been obtained because Golgi membranes have not been studied for their complement of monovalent ion transporters, exchangers, and channels.

Basolateral Membrane Transporters

There is evidence for a number of pumps and exchangers in the basolateral membrane. The functional and immunohistochemical evidence for a ouabain-sensitive Na/K-ATPase in the basolateral membranes of the lactating mammary alveolar cells has been obtained in many laboratories [reviewed in Shennan and Peaker (174)] and it is widely accepted that the sodium and potassium levels in the mammary cell are maintained by this enzyme. However, sodium levels in the cell [e.g., 43 mM in the guinea pig, 26 mM in the mouse (394)] are higher than in many other cells. The mechanism is not clear, but many sodium-dependent exchangers are thought to be present on the basolateral membrane (174), which could account for excess entry of this ion into the cell (Fig. 20). Evidence for an amiloride-sensitive Na^+/H^+ exchanger as well as a DIDS-sensitive Na/HCO_3^- cotransporter was obtained in the 31EG4 cell line (395), a nontransformed mouse mammary cell line derived from the IM-2 cell line (127). Shennan (396) obtained physiologic evidence for a presence of electroneutral $Na^+/K^+/2Cl^-$ in mammary explants that was inhibitable by furosemide. The electroneutral sodium-potassium-chloride cotransporter NKCC1 was detected by immunohistochemistry in the luminal epithelial cells distributed along the basolateral membranes (286,289,397). Levels were highest in the ducts of the virgin gland, but the protein was detected throughout pregnancy and lactation, generally localized to basolateral membranes. Mice with a null mutation of this transporter showed no obvious defects in morphology during lactation; however, the pups failed to thrive (397). What could not be determined from the experiments was whether the lactation failure was due to a defect endogenous to the mammary gland or to some other deficiency, because the mice have defects in a number of tissues.

Apical Membrane Transport

Apical membrane transport has also been characterized in a number of laboratories. The earliest experimental evidence was obtained in goats in 1974. Linzell and Peaker found that NaCl and KCl solutions infused up the teat were absorbed into the bloodstream and that isotonic sucrose solutions infused in the same way drew ions into the milk space (398). In addition, measurement of the effect of changes in the Na^+, K^+, and Cl^- concentration in the milk space of the goat mammary gland on the transepithelial potential difference provided evidence for the presence of a nonselective cation channel and a chloride channel in the apical membrane (399). Additional evidence for K^+ and Cl^- channels comes from studies of ion transport across membrane vesicles isolated from MFGMs (400,401). The question is what role these channels play in the apical membrane. In the guinea pig (391), the rat, and the rabbit, the ratio of the concentrations of Na^+ and K^+ in milk and mammary gland was approximately equal at 3, and both were at approximately electrochemical equilibrium across the apical membrane, suggesting that they are passively distributed (402). This was also the case in the mouse, where careful measurements of the transepithelial and basolateral potential in vivo gave a transepithelial potential of –35 mV and a basolateral potential of –49 mV. By subtraction, the apical potential should be in the range of –14 mV, not too different from the equilibrium potentials of Na^+ and K^+ across the apical membrane of –26 and –15 mV, respectively (394,403).

The chloride concentration is much higher in the cell than in the milk of all species where it has been measured (402), and the ion is obviously out of electrochemical equilibrium. The mechanism by which milk chloride is maintained at a low level is not clear, but experiments in tissue culture models are beginning to show how these transporters may function and the mechanisms by which the electrical potential across the apical membrane is maintained. There is no molecular information on the nature of apical monovalent transporters or channels from in vivo systems at this writing.

Cell Culture Models for Monovalent Ion Transport

Experiments in HC11 cells showed that furosemide-sensitive Cl^- transport is increased by PRL and that this increase was accompanied by phosphorylation of NKCC1, suggesting that PRL activates NKCC1 in this manner (404). However, although a good model for PRL action, HC11 cells are an unsatisfactory model for mammary transport function because they do not form a tight monolayer that allows for study of transepithelial transport. The 31EG4 cell line, mentioned previously, is a much more satisfactory model because it forms a tight monolayer when grown on a Transwell support. The resistance of this monolayer is substantially increased by glucocorticoid in a manner reminiscent of the in vivo mammary gland (405). 31EG4 cells were found to possess both the amiloride-sensitive epithelial sodium channel (ENaC) and the cystic fibrosis transmembrane conductance regulator (CFTR), a chloride channel that is stimulated by cyclic adenosine monophosphate (cAMP) and inhibited by diphenylamine-2-carboxylate (DPC) (405). Both were localized to the apical border of 31EG4 cells by immunostaining, a location that was confirmed by electrophysiological experiments. The CFTR was shown to be present in ducts in the human mammary gland. Elegant electrophysiological experiments in this cell line showed that fluid absorption is mainly controlled by the "transfer of sodium down its electrochemical gradient through apical ENaC, while fluid secretion is regulated by cell-to-lumen movement of Cl^- and/or HCO_3^- through cAMP-stimulated CFTR." In this epithelium, counterions move though the apparently "leaky" tight junctions, and the authors make the point that the ability of an epithelium to exhibit both fluid secretion and absorption depends on the presence of such junctions. Whether this principle is applicable to the lactating mammary gland with its very tight junctions is a question for future research. It seems more likely that counterions for transport by these two channels are transferred by transporters that may not be expressed in this cell line. In addition, it seems unlikely that the CFTR plays an essential role in milk secretion because women with cystic fibrosis have no difficulty with lactation (406).

Mammary tumor cells were found to secrete the nucleotides uridine triphosphate and adenosine diphosphate after mechanical stimulation; these compounds in turn increased intracellular calcium in the tumor cells (407). The receptor responsible for these nucleotide effects was found to be the metabotropic P2 purinergic receptor, P2Y. Blaug and colleagues (408) have used 31EG4 cells for a second set of experiments in which they show that ATP and UTP stimulate apical chloride movement by acting on P2Y to increase Ca^{2+} release from the endoplasmic reticulum (Fig. 21). The possibility that these ATP-mediated effects are important during the letdown reflex requires investigation.

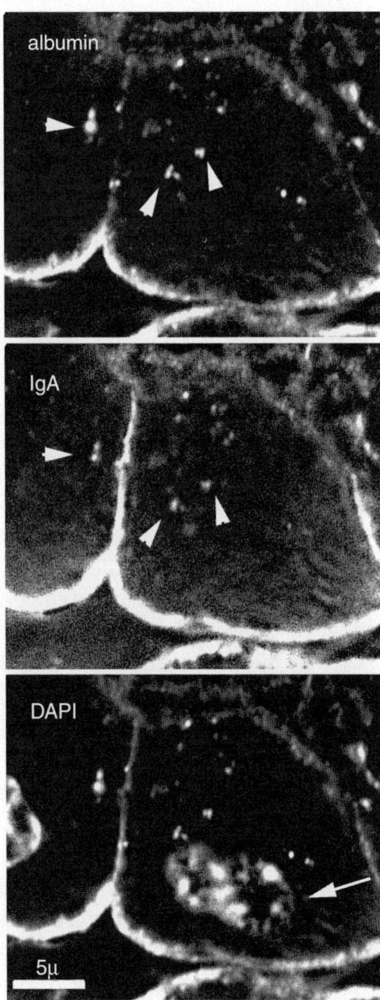

FIG. 21. Fluorescence micrograph illustrating transcytosis of albumin and IgA across the mammary epithelium. Frozen sections of the mammary gland of a lactating mouse were stained with antibodies to murine albumin and IgA as well as DAPI to visualize nuclei. Digital confocal microscopy using filters for FITC (albumin), IgA (Cy3), and DAPI was used to obtain images specific to each color probe. The bottom image is merged from all three colors to show the morphology of the cell. (From Monks, J., and Neville, M. C. [2004]. Albumin transcytosis across the epithelium of the lactating mouse mammary gland. *J. Physiol. [Lond].* 560, 267–280.)

Transcytosis

In 1976, the Gitlins measured the transfer of iodinated proteins across the mammary epithelium of lactating mice (409), finding that transmammary protein transfer was highly selective, with IgGs and IgA readily traversing the epithelium. Many other proteins have been found in the intervening years to cross the rodent mammary epithelium, including albumin (176), ApoB (410), hormones such as insulin and PRL (411), cytokines (412), and lipoprotein lipase (413). In addition, exogenous substances such as proteins from the mother's diet (414) and drugs or toxins (6) ingested by the mother are excreted into the milk. Indeed, in the Gitlin study all the labeled proteins used were transferred in some degree (409). Soon after the Gitlin study, it became apparent that the source of the IgA in milk was antibody-secreting cells, which, having been primed in the gut and respiratory tract, homed to the interstitial space of the mammary gland, where they sat down and began to secrete IgA (415,416). The IgA in the milk was bound to a peptide, secretory component, that made it resistant to digestion in the gastrointestinal tract (417). This compound, secretory IgA (sIgA), is thought to be a major factor in the protection of the breast-fed infant against diarrheal disease (418).

Secretory IgA Transport

The molecular mechanisms by which sIgA secretion is regulated have been the subject of research in many laboratories. Wilson and Butcher have defined an epithelial chemokine, CCL28, upregulated in the mammary gland during lactation, as the mediator of the homing response of antibody secreting cells (419). However, experiments using a transgenic mouse line in which the polymeric immunoglobulin receptor (pIgR) was expressed under the control of a milk protein promoter show that the limiting factor for IgA transport is pIgR expression (420), not the concentration of IgA available from the antibody-secreting cells. More recently, polymeric IgA expression has been shown, like milk protein genes, to be regulated by PRL and glucocorticoids in vitro and in vivo (421). In vitro it was upregulated by interferon-γ (422). However, the in vivo significance of this finding is not yet clear because a role for interferon in milk secretion has not been defined.

From a cell biological standpoint, the early questions were: How did sIgA reach the milk from the interstitial space, and where did the secretory component originate? These questions were answered first by studies with mRNA from a rabbit mammary gland showing that secretory component was part of the pIgR complex, released by proteolytic processing (423). MDCK cells, a dog kidney cell line, form a monolayer when cultured on membrane filters. Mostov and Deitcher (424) transfected these cells with the pIgR and were able to verify that secretory component is derived from proteolysis of a fragment of the pIgR. This system has served as a very useful model for dissecting the molecular mechanisms of IgA endocytosis, sorting, and transcytosis (425). In the meantime, Solari and Kraehenbuhl (426) performed pulse-chase experiments with isolated mammary acini to study the synthesis and processing of the pIgR in lactating rabbit mammary gland minilobules. The results of these experiments verified that pIgR is synthesized as a transmembrane protein that is glycosylated and matures in the Golgi, is subsequently routed to the basolateral membrane, binds ligand, and is transcytosed across the cell, with final cleavage of secretory component and liberation at the apical membrane. Molecular understanding of immunoglobulin transport across the mammary epithelium is, therefore, quite advanced, as detailed in an outstanding review by Kraehenbuhl and Hunziker (175). One question that remains open is how the very large quantities of IgG in the colostrum of species like ruminants that lack transplacental immunity are transferred across the mammary epithelium.

Transferrin Transport

Mammary epithelial cell endocytosis, recycling, and transport of transferrin has been studied in sheep (427), rabbits (428,429), and mice (430). Transferrin is internalized in coated pits with accumulation in endosomes, and the majority of the transferrin is recycled back to the cell surface. However, some follows a transcytotic pathway to the milk (429,430). In isolated acini, the transferrin was observed in multivesicular bodies, the Golgi and secretory vesicles containing casein (429). In vivo, however, we found that iodinated transferrin injected into the bloodstream was transferred into milk with a steady-state whey-to-serum ratio of 0.12 (430), similar to the fluid-phase transport ratio expected on the basis of the transport of IgA into the milk of J-chain–deficient mice. Because the J-chain is essential for IgA binding to the pIgR (431), IgA transport in these mice sets an upper limit for fluid-phase transfer from the interstitial space (176). Using an in vivo incubation technique in which fluorescent transferrin could be visualized at high resolution in the mammary alveolar cell, we were unable to find any transferrin associated with the casein secretory compartment in mice (176), leading us to conclude that the transcytotic and exocytic compartments are completely separate.

Serum Albumin Transport

Murine milk contains large amounts of serum albumin; in fact, its concentration equal to that in the serum of the lactating mouse (176,432). All this albumin originates in the plasma. Using an in vivo incubation technique, Monks and Neville observed albumin in cytoplasmic vesicles in the alveolar cell that were devoid of caveolins, distinguishing transcytosis across the mammary alveolar cell from that observed in capillary walls (176). The albumin was localized to vesicles containing IgA (Fig. 21) and transport was disrupted by agents that interfere with clathrin-mediated endocytosis. The one-to-one ratio of whey to plasma albumin is five times the ratio for fluid-phase transport inferred from J-chain–deficient mice, suggesting that uptake of albumin for transcytosis into milk is carrier mediated. However, the mechanism must be different from those described for the capillary (433) and lung (434) epithelium.

It is not clear to what extent the findings in mouse with respect to albumin transport apply to other species. In human and cow milk the albumin blood-to-milk ratio is approximately 0.01, suggesting that the concentrative transport mechanism may not be present. This ratio also sets an upper limit for fluid-phase transcytosis in these species. Nonetheless, the finding of dietary proteins in human milk suggests that the transcytotic pathway is important in women as well. Whatever the mechanism, mediated or nonspecific, the notion that transcytotic pathways carry valuable nutrients and protective and potentially detrimental substances to the neonate that sensitize or tolerize the neonatal immune system to antigens present in the maternal environment requires additional investigation.

Regulation of Junctional Complexes in the Mammary Gland

Cell–cell junctions in the mammary gland were first described in detail by Dorothy Pitelka in a series of classic electron micrographs, both transmission and freeze fracture (2,435). She described four types of junctions, starting at the apical border of the cell:

- *Tight junctions (zonula occludens).* The tight junctions form a continuous boundary around the apical borders of the epithelial cells in the mammary epithelial monolayer. These junctions regulate the flux of molecules between the interstitial and milk spaces and are open to rather large molecules in pregnancy and tightly closed during lactation (28,29). They have received much more attention than the other types of junctions, in part because their permeability can be measured by a number of in vivo techniques and in part because their permeability changes in a remarkable manner during the initiation of lactation.

- *Adherens junctions.* Adherens junctions are found directly below the tight junctions and in fact share some of the cytoplasmic scaffold of the tight junctions. Their function seems to be to provide mechanical stability to the cell monolayer, but they may also function to organize the tight junction complex. They have received considerable attention in the mammary gland because the major transmembrane component, E-cadherin, is often downregulated in breast cancer (436). Adherens junctions appear to be a constitutive component of the mammary epithelium and the protein does not appear, by immunohistochemistry, to change remarkably between pregnancy and lactation (Nguyen, D.-A. D., and Neville, M. C., unpublished observation).
- *Gap junctions.* Gap junctions provide cell–cell communication between epithelial cells and between epithelial and myoepithelial cells (403). Their composition changes markedly between pregnancy and lactation, an observation that potentially has profound consequences for the coordination of milk secretory processes during lactation. (437).
- *Desmosomes.* Desmosomes provide stable mechanical interactions between the epithelial cells and well as between the cell and the extracellular matrix. They are abundant in the major ducts of the lactating mammary gland, where they can be found on the lateral borders of epithelial cells, connecting them one to the other, as well as between myoepithelial and epithelial cells (2,435). Hemidesmosomes can be found at the basal surface of the myoepithelial cells, linking them to the basal lamina. Alveoli and small ducts in the lactating mouse lack desmosomes and most hemidesmosomes, possibly providing the flexibility for changes in cell shape as the milk lumen fills and empties.

More recently, extensive attention has been devoted to the interactions of mammary epithelial cells with components of the extracellular matrix, studies that have been facilitated by the ability of mammary epithelial cells to form hollow, acinar-like structures when cultured on a reconstituted basement membranes (Matrigel®) (438,439). These culture systems have added greatly to our understanding of the role of cell–matrix interactions in the branching morphogenesis that leads to development of the mammary gland as well as breast cancer progression. Many excellent reviews of this subject are available (440–443). Unfortunately, although changes in epithelium–matrix interactions certainly take place in the late pregnant and lactating glands, as indicated by profound changes in expression of genes for components of the stroma (27), these changes are not currently well defined and are not further discussed here. Instead, we focus on the regulation of tight and gap junctions.

Mammary Tight Junctions and Their Regulation

The earliest experiments on the permeability of the mammary tight junctions were carried out by Linzell and Peaker in the 1970s (398,444), measuring the transfer into the bloodstream of [^{14}C]-sucrose injected through the teat canal into the lumen of the mammary gland of the goat. The tracer appeared immediately in the blood in the pregnant goat, but no tracer was measured in the blood during lactation. They also observed that [^{14}C]-sucrose moved from the bloodstream into the milk during pregnancy but not lactation. Higher concentrations of the isotope were observed in milk fractions derived from the alveoli than from the ducts, suggesting that most of the "leaky" tight junctions are in the alveolar epithelium and not the ducts. Additional evidence of a profound change in epithelial permeability between pregnancy and lactation is that no electrical potential can be measured between electrodes in the bloodstream and milk space during pregnancy, but blood–milk potentials of approximately –20 mV and –35 mV were found during lactation in goats (402) and mice (403), respectively. This very high potential difference across the mammary epithelium indicates that the epithelial tight junctions are highly impermeable during lactation.

More recently, advances in fluorescence microscopy have allowed examination of transfer of macromolecules labeled with fluorescent tracers across the epithelium, either from the lumen to the interstitial space or in the opposite direction (176,298). These experiments, illustrated in Fig. 22, show that during pregnancy molecules as large as IgA permeate the junctional complexes (29), but that the junctions are completely impermeable in the lactating gland. Together, the findings from isotope tracer, blood–milk potential, and fluorescent tracer experiments indicate that the mammary tight junctions, particularly in the alveoli, are leaky during pregnancy and close around parturition to form a tight barrier that prevents paracellular movement of molecules across the mammary epithelium.

Freeze-fracture studies carried out by Pitelka (2,435) and confirmed in this laboratory (Nguyen, D.-A. D.,

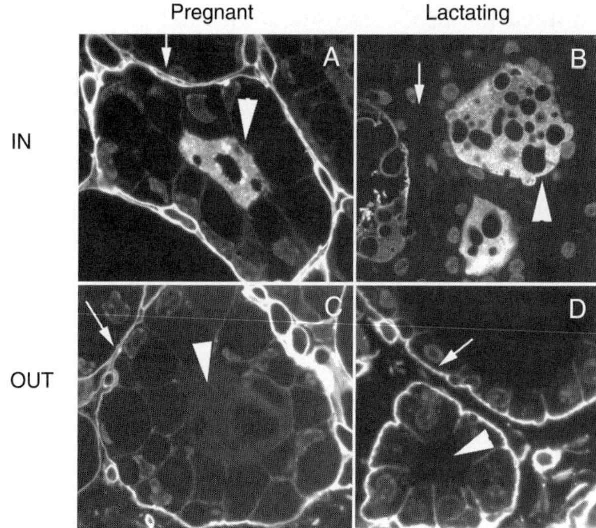

FIG. 22. FITC-labeled albumin passage does not enter the paracellular pathway. **A, B:** A glass pipette containing the tracer (3 mg/mL) was inserted intraductally into the mammary gland and 10 μL of solution injected into the lumen of the mammary gland, followed immediately by injection of paraformaldehyde to fix the gland. **C, D:** The fourth mammary gland of an anesthetized mouse was exposed and overlaid with a solution containing the tracer. After 2 hours, the gland was perfusion fixed. Sections of glands from both procedures were examined by digital confocal microscopy. *Large arrowheads* indicate glandular lumina; *arrows* indicate interstitial space. (From Nguyen, D. D., Parlow, A. F., and Neville, M. C. [2001]. Hormonal regulation of tight junction closure in the mouse mammary epithelium during the transition from pregnancy to lactation. *J. Endocrinol.* 170, 347–356. Copyright 2003, Society for Endocrinology. Reproduced by permission.)

and Neville, M. C., unpublished data) showed that the tight junction strands in the gland during pregnancy are highly disordered, suggesting that molecules permeate between the strands. In lactation the strands are highly ordered and show no breaks that might allow passage of large or small molecules. Although claudins in low-resistance epithelia permit the passage of monovalent ions (445,446), the high transepithelial potential maintained in these glands despite substantial transcellular permeability to these ions suggest that the complement of claudins in the mammary epithelium is such that ionic permeation does not occur. However, no published information is available at present about the complement of claudins in the mammary epithelium.

Hormone Dependence of the Change in Tight Junction Permeability between Pregnancy and Lactation. To study the transition between pregnancy and lactation, we ovariectomized mice on day 17 of pregnancy, a procedure that removes progesterone and sets the transition between pregnancy and lactation in motion. Figure 23 shows that the sucrose permeability of the epithelium progressively decreases over the subsequent 20 hours (94). We found that the injection of a progesterone antagonist, RU486, produced the same effect and that the effect could be delayed by injection of progesterone at intervals, firmly establishing that progesterone withdrawal was the trigger for tight junction closure. A series of endocrine ablation experiments showed

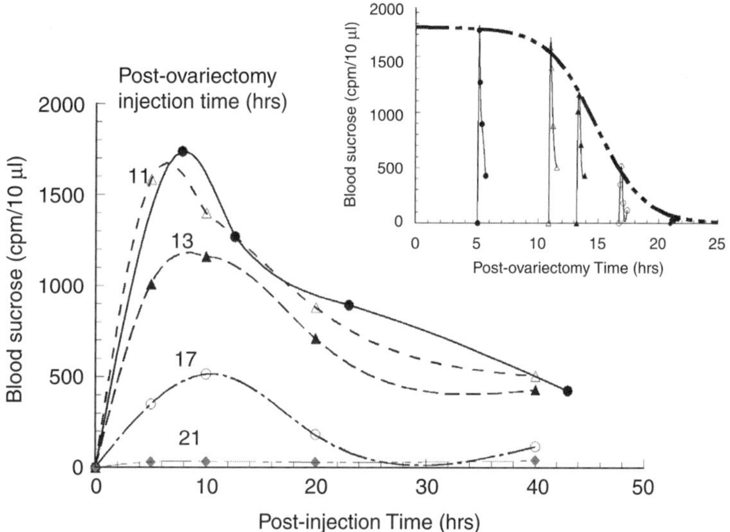

FIG. 23. Injection of [^{14}C]-sucrose intraductally at intervals after ovariectomy of a 17-day pregnant mouse. The counts of sucrose in the blood are shown at intervals after the intraductal injection at the times, in hours, indicated by the numbers above the graph. *Inset:* The same curves are shown plotted as a function of time after ovariectomy to show the decline in permeability between 10 and 20 hours, indicating closure of the tight junctions. (From Nguyen, D.-A. D., Beeman, N., and Neville, M. C. [2001]. Regulation of tight junction permeability in the mammary gland. In *Tight Junctions* [M. Cerejido and J. M. Anderson, Eds.], pp. 395–414. CRC Press, New York, with permission of CRC Press.)

that both PRL and glucocorticoid are necessary for mammary tight junction closure (447). These are the endocrine conditions that promote activation of milk secretion as well.

Changes in Milk Composition Due to Tight Junction Closure at the Initiation of Lactation. The direct passage of interstitial molecules such as sodium and chloride across the mammary epithelium during pregnancy leads to an increase in the sodium and chloride concentration of colostrum and the appearance of milk components such as lactose and α-lactalbumin in the bloodstream, as illustrated by the diagram in Fig. 24. Such changes in milk composition have been fully documented during the onset of lactation in both goats (398) and humans (141,448), and can serve as an indicator of tight junction status under conditions where direct tracer measurement of the permeability of the mammary epithelium is not possible.

Epithelial Permeability during Mastitis and Involution. Considerable evidence indicates that tight junction permeability increases during involution (449–451) and with mastitis (452,453). Much less research has been devoted to these stages and it is not clear whether the changes result from a return of the tight junction strands to the disordered condition of pregnancy or simply from a loss of integrity of the epithelium as epithelial cells are lost by apoptosis. Stelwagen and colleagues showed that increased milk components such as lactose and α-lactalbumin passed into the bloodstream if goats and cows were left unmilked for 18 hours (451,454). The authors interpreted these results as showing an effect of milk stasis on tight junction permeability. However, they did not inject tracers into the gland or bloodstream, as was done by Peaker and Linzell (398), to distinguish between opening of tight junctions and aberrant, basally directed exocytosis in the extended mammary epithelium. We have found, using intraductal injection of [^{14}C]-sucrose, that the permeability of the mammary epithelium in the mouse does not increase until 48 hours after forced weaning, at a time when apoptosis reaches a peak (Beeman, N., and Neville, M. C., unpublished data). This finding suggests that closure of junctions at the initiation of lactation may be an irreversible process and that loss of epithelial cells is responsible for changes during involution and mastitis. However, the matter requires additional investigation.

Connexins in the Mammary Epithelium

Direct cell–cell signaling is mediated by the passage of molecules with molecular weights less than 1,000 kDa through gap junctions. These junctions are formed from six connexin molecules that aggregate to form a hemichannel or connexon in each cell; when these hemichannels in the opposing membranes of two cells are aligned, they form a channel with a pore that allows intercellular passage of signaling molecules, metabolites, vitamins, and other

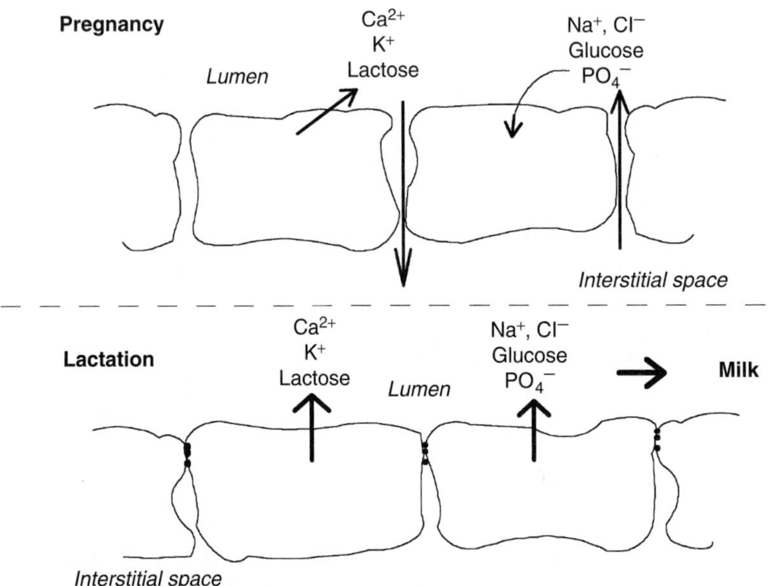

FIG. 24. The paracellular pathway is open in pregnancy and closed in lactation. The diagram illustrates the net flux of several small molecules during pregnancy when the junctional complexes are very leaky, and in lactation when they are tightly closed. (From Neville, M. C. [1995]. Lactogenesis in women: evidence for a cascade of cellular events. In *Handbook of Composition of Milks* [R. G. Jensen, Ed.], pp. 87–98. Academic Press, San Diego, with permission of Elsevier.)

small molecules, but not macromolecules. Freeze-fracture studies showed that gap junctions between epithelial cells in the mammary gland are composed of an aggregate of many connexons (435). Later studies using lucifer yellow dye showed extensive coupling between cells of the mammary alveolus as well as suggesting dye transfer between alveolar and myoepithelial cells (403).

Four connexins have been shown to be expressed in rodent mammary glands (455). Connexin26 (Cx26) and connexin32 (Cx32) are expressed at all stages of mammary development (455,456) but are markedly upregulated in pregnancy (Cx26) (437,457,458) and lactation (Cx26 and Cx32) (437,458). Connexin30 (Cx30) was expressed after day 15 of pregnancy and peaked at the onset of lactation (455). By microarray analysis, the mRNA for Cx30 was low, but upregulated during lactation (437). However, no other information is available about mammary Cx30 at this time. Connexin43 (Cx43) is the only isoform that has been positively identified in the myoepithelium (459). This protein was dramatically upregulated on the first day of lactation in rats, when two bands of equal intensity were observed at 43 and 45 kDa on Western blots (459). As lactation progressed, the mRNA decreased (437), but the protein persisted (459). However, the 43-kDa protein band disappeared, leaving only the higher–molecular-weight form (459). Alkaline phosphatase treatment decreased the molecular weights of both forms, suggesting that the phosphorylation state of Cx43 may be related to its function during lactation, as expected from earlier studies showing that Cx43 phosphorylation is essential for its function (460). Although expression of Cx43 in the lactating mammary gland is low compared with Cx26 and Cx32, it may be important for the myoepithelial cell response to oxytocin because pups suckling dams heterozygous for a Cx43/30 knockin at the Cx43 locus showed poor survival (461). Conditional knockout of Cx26 caused disorganization of the ductal epithelium in early pregnancy but had no effect in late pregnancy or lactation, suggesting that gap junctions are essential for coordinated growth of the epithelium in early pregnancy (462). Cx32 may take over the function of Cx26 in lactation, rendering Cx26 superfluous.

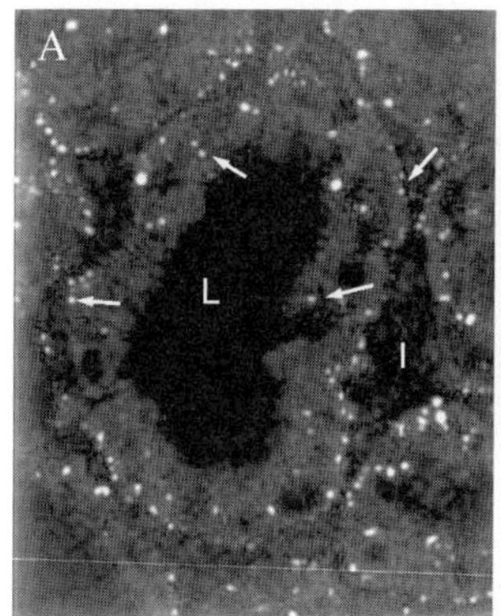

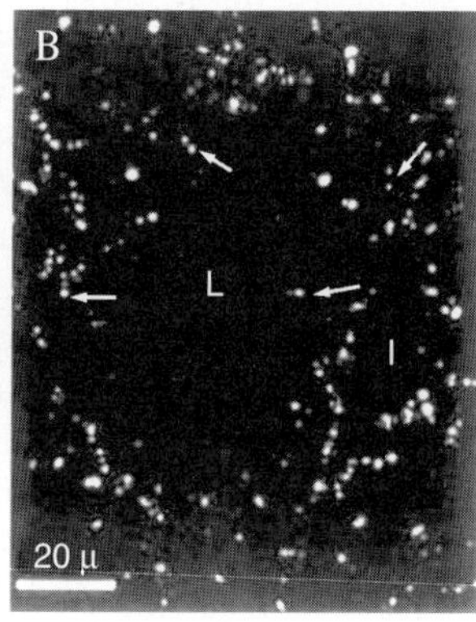

FIG. 25. Connexin26 (**A;** visualized with TRITC) and connexin32 (**B;** visualized with FITC) on day 3 of lactation in the rat mammary gland. At the level of the light microscope, almost complete overlap between the two connexins (*arrows*) is observed. L, lumen; I, interstitial space. The epithelial layer can be visualized from the autofluorescence of the image. Note that much of the connexin expression is found in plaques at the base of the cell. (From Yamanaka, I., Kuraoka, A., Inai, T., Ishibashi, T., and Shibata, Y. [2001]. Differential expression of major gap junction proteins, connexins 26 and 32, in rat mammary glands during pregnancy and lactation. *Histochem. Cell Biol.* 115, 277–284, with kind permission of Springer Science and Business Media.)

Figure 25 shows the localization of Cx26 and Cx32 in the epithelium of the day 3 lactating rat. There was almost complete overlap between the two forms of connexin, which appeared from electron micrographs to be localized to the luminal epithelium, albeit mainly between epithelial cell projections at the basal surface. These data raise two questions: do the two forms of connexin come together in the same channel, or are the channels between the epithelial cells homomeric? And, how does the presence of one or another form of connexin relate to the function of the epithelium? These questions are admirably addressed in a paper by Locke et al. (437) in which analytical size-exclusion chromatography was used to resolve homomeric from heteromeric Cx26 and Cx32 hemichannels. When Cx32 first appeared in isolated plasma membranes from mouse mammary gland, 4 to 8 hours postpartum, it was exclusively found in heteromeric channels with Cx26. As its membrane concentration increased, both homomeric and heteromeric channels began to appear and were maintained throughout lactation. Using transport-specific fractionation with liposomes to examine the permeability properties of the various channel configurations to sugars of increasing sizes, Locke et al. (437) showed that the size of the channel increased in the following order: homomeric Cx26 ≤ heteromeric Cx26Cx32 < homomeric Cx32 < homomeric Cx43. Surprisingly, when a series of signaling intermediates was used in the same system, only the heteromeric Cx26Cx32 hemichannel was impermeant to cAMP and IP_3, and, to a lesser extent, to cyclic guanosine monophosphate, suggesting that the state of the channels may regulate cell–cell flux of signaling molecules. At the very least, the permeability studies suggest that the wider Cx30 homomeric hemichannels may allow for very efficient coupling between mammary epithelial cells in the lactating mouse. That such coupling is necessary for efficient milk secretion is suggested by the finding in a cultured mammary epithelial cell line that gap junction communication was necessary for β-casein expression (463).

REGULATION OF MILK SECRETION

The factors acting directly on the mammary gland that are important in the regulation of milk secretion are PRL and milk removal. PRL is critically necessary for alveolar proliferation (9,464,465), synthesis of milk proteins, and survival of mammary epithelial cells in the lactating animal (466–468). Milk removal prevents the cell death that begins 2 days or more after milk stasis; studies in dairy animals and women show that it is a critical factor in determining the volume of milk secreted. Many metabolically active hormones have been implicated in the regulation of milk secretion, including insulin and glucocorticoids. We have argued against a direct role for insulin in regulation of lipid secretion in the section on Dietary Regulation; although a large amount of data implicates insulin in the maintenance of mammary differentiation in tissue culture (8,152), the possibility that IGF-1 or IGF-2 plays a maintenance role in vivo requires serious consideration (128,469). However, because the functions of the IGFs in lactation have not been clearly defined, we do not consider these potentially important factors here. The role of glucocorticoids has received much less attention in recent years, although classic endocrine ablation and replacement studies clearly demonstrated that glucocorticoids were necessary for milk secretion (150,470,471). More recent evidence from knockout mice, detailed later, suggests that the actions of glucocorticoids in lactation may not require the epithelial glucocorticoid receptor (GR). Oxytocin is critical to the maintenance of milk secretion (472) because of the role of myoepithelial contraction in milk ejection, a phenomenon that is discussed in the section on Milk Ejection. We elaborate here on the mechanisms by which PRL and milk removal exert their effects on milk secretion because of their unequivocal importance in the maintenance of lactation. The most recent evidence on glucocorticoids also is summarized.

Prolactin

Prolactin Secretion during Lactation

PRL is secreted in a pulsatile fashion during lactation stimulated by the suckling of the young (Fig. 26). Plasma levels are influenced by day length in seasonal mammals like ruminants (473) (Fig. 26A), and in women by the interval between feeds and their duration as well as by the time postpartum (474) (Fig. 26B). Bromocriptine, a dopamine analog that inhibits pituitary PRL secretion, completely suppresses milk yields in rats (475) and women (476), but has variable effects in ruminants. Although PRL is clearly necessary for the maintenance of lactation (466,467), it has long been known that plasma PRL levels do not correlate with the rate of milk secretion (150). To give only one of a myriad of examples, Hagasawa and Yanai (477) found no correlation between litter growth rates and either PRLR levels or plasma PRL in two strains of mice. Observations like these lead to the suggestion that PRL may be a permissive or survival factor rather than a regulatory factor for milk secretion. Two mechanisms have been proposed for PRL's survival effects: (a) stimulation of IGF-1 secretion from the parenchyma with concomitant inhibition of the mammary production of IGFBP5 (466); and (b) the action of the PRL signal transduction intermediate Stat5 (49). Withdrawal of PRL in the lactating mouse led to a rapid decrease in the expression of genes for milk synthesis, total inhibition of milk secretion, and induction of changes in the mammary gland associated with involution (Russell, T., Anderson, S. M., and Neville, M. C., unpublished data), suggesting that PRL is required for maintenance of all alveolar cell processes associated with milk secretion as well as cell survival.

Prolactin Receptor

The PRLR belongs to the cytokine receptor superfamily. PRL binding results in receptor dimerization and activation of the Jak/Stat pathway, leading to transcription of milk-specific proteins (478) and inhibition of involution (479). The existence of several forms of the PRLR has been recognized in the mammary glands of rodents and ruminants for some time, but in general only the long form has been implicated in the effects of PRL on differentiation, with a possibility that at least one of the short forms is inhibitory (480–483). Interestingly, like PRL, the PRLR is downregulated in the mammary glands of mice and rats during pregnancy, reappearing near the end of pregnancy (Fig. 5C) (481,484). A four- to tenfold change in PRLR mRNA expression between late pregnancy and the day of birth was seen for the long form of the receptor in the mouse (484), with the short-form mRNA remaining virtually constant (Fig. 5D) (480).

Prolactin Receptor Signal Transduction

The main elements of PRLR signal transduction were worked out in the mid- to late 1990s (485–488), and are reviewed only briefly here. Briefly, PRL binding results in receptor dimerization, activation of the associated protein tyrosine kinase Jak2 and phosphorylation and translocation to the nucleus of Stat5.

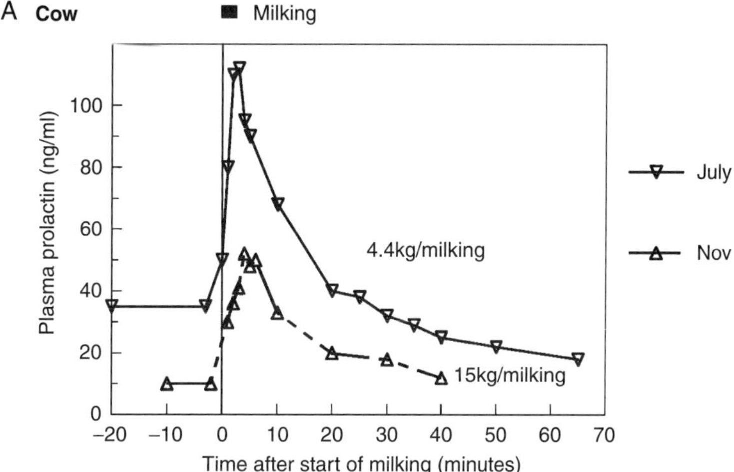

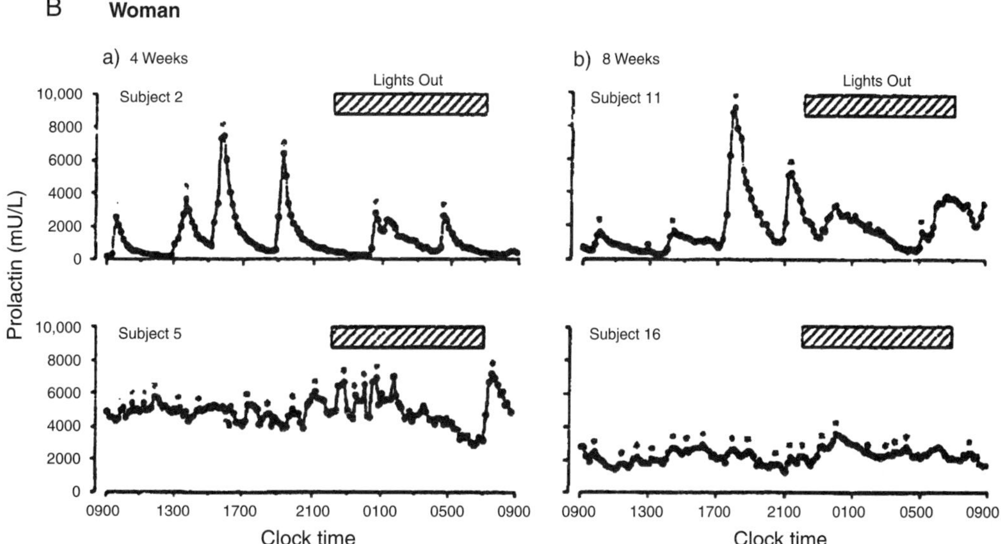

FIG. 26. The release of prolactin as a function of milking or suckling in cows and women. **A:** Two lactating Swiss brown cows were stimulated by a 3- to 6-minute machine milking interval in the summer (e.g., July) and in the winter (e.g., November). Note that the prolactin baseline and peak are much higher in the summer. (Data replotted from Schams, D. [1972]. Prolactin levels in bovine blood, influenced by milking manipulation, genital stimulation and oxytocin administration with specific consideration of the seasonal variations. *Acta Endocrinol.* 71, 684–696.) **B:** Prolactin secretion over daily suckling episodes (*asterisks*) in four women. (From Tay, C. C. K., Glasier, A. F., and McNeilly, A. S. [1996]. Twenty-four hour patterns of prolactin secretion during lactation and the relationship to suckling and the resumption of fertility in breast-feeding women. *Hum. Reprod.* 11, 950–955. Copyright European Society of Human Reproduction and Embryology. Reproduced with permission of Oxford University Press/Human Reproduction.)

Downstream effects of Stat5 have been most extensively studied because they affect the β-casein promoter. Studies with a conditional knockout of Jak2 confirm a role of the Jak/Stat pathway in both alveolar morphogenesis and survival of mammary epithelial cells (489). However, PRL has additional modes of action. For example, it specifically activates the binding of the cytoplasmic poly(A) elongation element–binding protein (CPEB) to an element in the 3′UTR of the β-casein message, leading to increased polyadenylation of this message and a substantial increase in casein translation (490). The linkage between PRLR activation and CPEB binding is discussed in the next section. Both interactions with the extracellular matrix (443,491) and the cytoskeleton (492) appear to be important in PRL signaling and have received extensive attention. A number of other signaling molecules are activated by PRL, including extracellular signal–related kinase (Erk) (493), c-Jun N-terminal kinase (Jnk) (494), cbl (495), and, potentially, Akt/PKB (25,169). As described in the section on Balanced Aerobic Glycolysis in

Milk Synthesis, Akt1 is significantly upregulated at the beginning of lactation, and overexpression alters both lipid synthesis and survival of mammary epithelial cells.

Milk Removal

The rate of milk secretion is determined largely by milk removal from the gland (496,497) by mechanisms that are very poorly understood. Using an ovariectomy model like that described for studies of tight junction regulation (see the section on Hormone Dependence of the Change in Tight Junction Permeability between Pregnancy and Lactation), Sakai and colleagues (480) showed that milk removal effected by supplying foster pups 24 hours after ovariectomy had no effect on the pattern of expression of $PRLR_L$, but was required to maintain β-casein gene expression beyond 36 hours. Although it is clear that milk removal from the human breast is not necessary for the early changes in milk composition in women, it is equally clear that milk removal by day 3 or 4 is necessary for the successful initiation of breastfeeding (143). Further, suckling has been shown to be necessary for full development of the enzymes of lipid synthesis in rats (332) and rabbits (23). Once lactation is established, the rate of milk secretion is adjusted to the demands of the offspring, primarily by feedback mechanisms thought to be related to the amount of milk remaining in the alveoli at the termination of suckling (496). The mechanisms involved are not well understood, but likely involve both chemical factors in the milk itself (496) and stretching of the epithelial cells (294). In some species where maximal milk production is achieved, primarily dairy species and rodents with large litters, the amount of mammary tissue (498) and the ability of the body to supply nutrients to this tissue (466,498) are limiting to milk production. In women and other animals with small numbers of offspring, the demands of the offspring set the rate of milk production by a given individual (497).

Two local mechanisms have been implicated in the regulation of milk volume production. Peaker and Wilde (496) have summarized the extensive evidence (499) for the existence of a feedback inhibitor of lactation in milk. This factor is thought to build up as milk accumulates in the lumen of the mammary gland and inhibit milk secretion. A factor with the appropriate physiological properties has also been identified in human (500) and wallaby milk (501). Isolation of the factor has been extremely difficult, and no reliable compounds have emerged from considerable effort. One group has provided evidence that casein hydrolysates could be involved (502).

The second candidate for local regulation is the alveolar cell stretch that accompanies filling of the alveolar lumina. Thus, hyperosmotic media, which bring about significant shrinkage of mammary alveoli, were shown significantly to decrease the level of casein mRNA in rat mammary explants (503). More recently, the poly(A) tail of β-casein was shown to be much longer in the mammary gland from lactating compared with weaning rats (504); a decrease in the length of this tail destabilized the mRNA (505). In a very elegant analysis, Choi et al. (490), using CID9 cells (506), a mouse mammary cell line derived from the well-differentiated comma 1D cell line (507), carefully dissected this response, showing that PRL and insulin rapidly synergize to increase initiation of β-casein synthesis by more than 50-fold in the absence of changes in the mRNA levels. They further showed that in the presence of insulin, PRL specifically actuates the binding of CPEB to an element in the 3′UTR of the β-casein message, leading to a tenfold elongation of the poly(A) tail of the mRNA and a substantial increase of casein mRNA–associated polysomes. If this elegant mechanism were to be inhibited by a shape change associated with overfilling of the alveoli, it could provide at least a partial explanation for the effects of both milk removal and PRL withdrawal on the rate of milk synthesis.

Glucocorticoids

Glucocorticoid levels are often increased during lactation (154), but this finding is generally confined to dairy species and rodents, species whose nutrient stores are diminished to a state of partial starvation during lactation. If anything, glucocorticoid levels are lower in the lactating woman (508) in whom body fat stores are usually not depleted. Thus, changes in systemic glucocorticoid levels probably do not play a direct role in the day-to-day regulation of milk secretion. Nonetheless, extensive endocrine ablation experiments carried out almost four decades ago (150), as well as studies in organ and cell culture systems (7,151–153), clearly point to a role for adrenal steroids in mammary differentiation and lactation. Corticosterone was shown to be necessary for the increase in PRLR expression in the late pregnant mouse mammary gland (481) as well as tight junction closure around parturition (94). Further, the effects of Stat5 and C/EBPβ on β-casein and WAP production have been shown to depend on glucocorticoid acting through the GR (74,509–511).

The question is whether glucocorticoids have a direct effect on the mammary gland during lactation. In an attempt to answer this question, mammary glands from the $GR^{-/-}$ mouse were transplanted into cleared mammary fat pads (512). Although there were some defects in ductal development, surprisingly, development in pregnancy appeared to be normal, as did secretory activation on the first day of lactation. In confirmation of these findings, the GR was specifically knocked out in the late pregnant and lactating mammary gland using Cre-Lox technology (513). Although mammary growth during late pregnancy was decreased, the mice lactated, with the pups achieving a growth rate only 20% lower than in the wild type, and alveolar morphology appeared to be normal in lactation. Both experiments suggest that glucocorticoid effects on lactation are mediated through another receptor or are indirect. Rosen and colleagues suggested that the mineralocorticoid receptor could substitute for GR in the $GR^{-/-}$ mouse (512). Other possibilities are that glucocorticoids act either through the stroma, where the receptor was intact in the Cre-Lox experiment, or through systemic effects. Because glucocorticoids are involved in nutrient flux—maintaining blood glucose during starvation, for example—the potential for indirect effects in lactation is great and requires additional investigation.

MILK EJECTION

Milk ejection is effected by the contraction of myoepithelial cells, whose processes form a basket-like network around the alveoli where milk is stored. Stimulation of nerve terminals in the nipple or teat produces afferent impulses that reach hypothalamic neurons whose axons terminate in the posterior pituitary (514). Released oxytocin is carried through the bloodstream to the lactating mammary gland, where it interacts with specific receptors on myoepithelial cells, initiating a coordinated contraction that expels milk from the alveoli into the ducts and subareolar sinuses or udder. Milk ejection is essential for lactation, as shown by the observation that mice genetically deficient in oxytocin have a severe lactation deficit (472,515).

The Let-Down Reflex

The rat nurses her litter on the average approximately half of each hour. Let-down is delayed for at least 15 minutes after the attachment of the pups (514). Thereafter, increases in mammary pressure corresponding to oxytocin-induced milk ejection can be measured every 5 to 12 minutes (Fig. 27A). In the woman, ejection was measured as a rise in pressure sensed with a small catheter placed in a mammary duct or noted subjectively by the mother

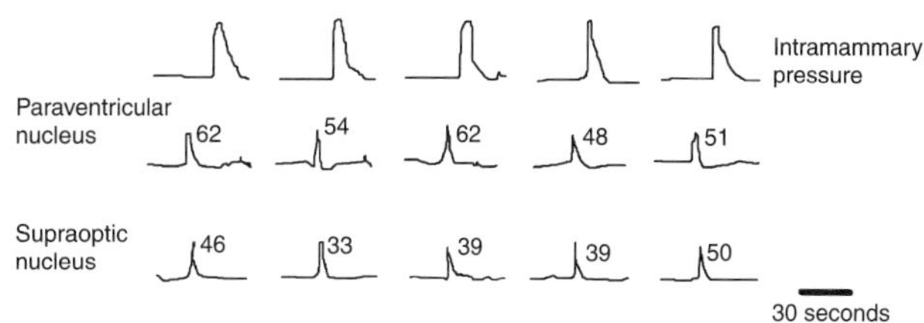

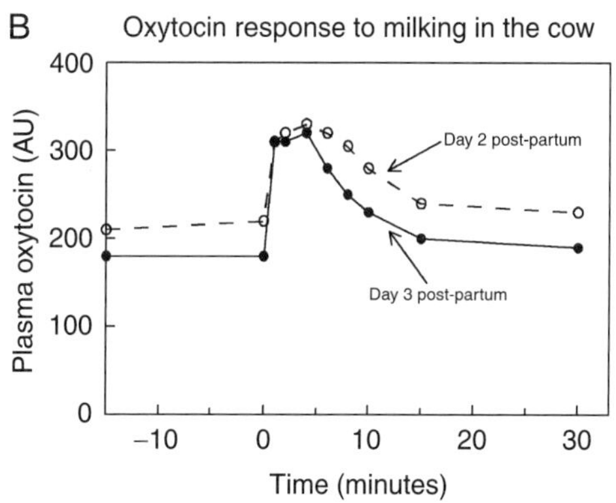

FIG. 27. Oxytocin secretion. **A:** Recordings from a pair of oxytocin-releasing neurons in the anesthetized rat, one in the supraoptic nucleus of the hypothalamus and the other in the paraventricular nucleus, made simultaneously with recordings of intramammary pressure. Bursts of neuronal activity with firing rates indicated by the numbers above each peak are spaced at 5- to 12-minute intervals in the rat. Both neurons fire simultaneously. In fact, the bursts of firing of all oxytocin neurons appears to be closely synchronized, even from opposite sides of the brain (520). Coordinated firing of the oxytocin neurons leads to a pulse of oxytocin release from nerve terminals in the posterior pituitary and a rise in the plasma oxytocin level followed shortly by a rise in intramammary pressure, as shown. (Modified from Crowley, W. R., and Armstrong, W. E. [1992]. Neurochemical regulation of oxytocin secretion in lactation. *Endocrine Rev.* 13, 33–65.) **B:** Oxytocin release in the cow showing the prolonged pulse of oxytocin in the plasma during milking on days 2 and 3 postpartum. (Akers, R. M., and Lefcourt, A. M. [1982]. Milking- and suckling-induced secretion of oxytocin and prolactin in parturient dairy cows. *Horm. Behav.* 16, 87–93, with permission of Elsevier.)

as a "tingling sensation" in the breast before or shortly after the start of suckling (516). The contractions lasted approximately 1 minute and occurred with a frequency of 4 to 10 contractions per 10 minutes. When properly measured (e.g., from blood samples drawn at 2-minute intervals), the pulses of oxytocin in the bloodstream correspond to these contractile episodes (517,518). In cows, a more sustained release of oxytocin has been measured (Fig. 27B) and is absolutely essential for proper milk removal from the gland (519).

Characteristics of Oxytocin

Oxytocin is a member of a family of nonapeptides with disulfide bridges between Cys residues 1 and 8 that includes vasopressin and a number of widely distributed similar peptides. The hormone holds the distinction of being the first naturally occurring peptide hormone to be synthesized (520), a feat for which du Vigneaud received the Nobel Prize in 1955. Oxytocin is found in all placental mammals, whereas marsupials have a variant called mesotocin in which Ile is substituted for Leu at position 8 (521,522). This hormone has all the activities of oxytocin in placental mammals and its presence is thought to be the result of a neutral mutation rather than functional evolution. Regulation of oxytocin gene expression has been extensively studied and multiple potential regulatory sites, including an estrogen response element, have been identified in the promoter of the oxytocin gene [reviewed in Gimpl and Fahrenholz (523)]. The exact physiological role of this and other promoter elements has yet to be defined.

The Neuroendocrinology of Oxytocin Synthesis and Release

Immunostaining techniques were used to show that oxytocin is synthesized mainly in specialized magnocellular neurons in the supraoptic and paraventricular nucleus of the hypothalamus, separate from the vasopressin-synthesizing neurons (523). Fibers from the paraventricular nucleus course to the level of the supraoptic nucleus, where they are joined by fibers from this nucleus as well. Both sets of fibers then travel down the median eminence to the posterior pituitary.

Brownstein and coworkers (524) used pulse-chase experiments to show that oxytocin is synthesized as part of a 30-kDa prohormone in the hypothalamus. After cleavage to smaller subunits, the prohormone is packaged into secretory granules and transported down the axonal processes of the magnocellular neurons to the posterior pituitary. There it is further cleaved to oxytocin and its binding protein, a 10-kDa neurophysin. The complex between oxytocin and neurophysin is stable at pH 5.5, but dissociates at pH 7.4, facilitating its dissociation as the complex is released to plasma (514,525). Like the release of neurotransmitters and peptide hormones, oxytocin release involves exocytosis of secretory granules after a burst of impulses carried to the neurohypophysis from the oxytocin neurons in the hypothalamus (514).

Measurements of suckling-released oxytocin have been made in a variety of species. The response is intermittent, with a return to baseline between milk ejections, and correlates with the bursting activity of oxytocin neurons (Fig. 27A), which continues even during continued suckling. The neural controls underlying oxytocin release during lactation in the rat have received extensive investigation using an experimental paradigm in which lactating rats are separated from their pups for several hours, anesthetized with urethane, and then reunited with the pups. The reader is referred to excellent reviews by Crowley and Armstrong (514), Leng (526), and Gimpl and Fahrenholz (522) for more detail.

Oxytocin release is modulated by many agents, including progesterone and estrogen (514,522,526), and the mechanisms involved remain the object of intense study by neurophysiologists. Endogenous opioid peptide systems interact with oxytocin secretion at several levels to inhibit both basal hormone secretion and the response to various physiological stimuli. Thus, opioids inhibit release of oxytocin at the cell bodies of the magnocellular neurons in the hypothalamus as well as their neurosecretory terminals in the posterior pituitary. They likely also have an inhibitory effect on the cell bodies of afferent input cells that stimulate magnocellular neuron activity (527). It is likely that this system mediates the inhibitory effect of stress on milk ejection in women (528).

Oxytocin Receptors

The oxytocin receptor belongs to a family of nonapeptide hormone receptors that includes the vasopressin receptors and that belong to the seven membrane span rhodopsin-like class I G-protein–coupled receptors (529). The human oxytocin receptor has 389 amino acids and a core molecular mass of approximately 40 to 45 kDa that can be increased by glycosylation. The oxytocin receptor gene has been cloned from a number of species (522,530), and a number of regulatory elements have been identified in the promoter, including a palindromic estrogen response element and a half-steroid response element (530) that may be responsible for the stimulation of oxytocin receptor mRNA by estrogen. In an elegant study, Olins and Bremel (531) found that oxytocin stimulated the influx of extracellular calcium ions, leading to phosphorylation of the myosin light chain in rat mammary myoepithelial cells. Influx of extracellular calcium ions regulated the duration of the response. There is also evidence that intracellular calcium stores are involved.

Cholesterol is an essential allosteric modulator of oxytocin receptor function, as is Mg^{2+} (522), although whether either plays a role in modulation of receptor activity in the mammary gland is unknown. Progesterone has little effect on mRNA expression but exerts a powerful inhibitory action on receptor activity at a nongenomic level. Whether this effect is mediated by interaction of progesterone directly with the oxytocin receptor, as reported by Grazzini et al. (532), or in some way alters the essential interaction of cholesterol with the receptor (533) remains to be elucidated. An interesting question is whether this interaction with the oxytocin receptor plays a role in the elusive mechanism by which secretory activity is downregulated by progesterone in late pregnancy.

Interaction of Oxytocin with Myoepithelial Cells

The processes of the myoepithelial cells lie in the basement membrane of the mammary alveolus and along the interlobular ducts. Autoradiographic studies showed that oxytocin binding sites are similarly localized (534). A LacZ knockin mouse (535) confirmed this finding, but also showed oxytocin receptors localized to mammary adipocytes. Soloff and colleagues (534) found a tenfold increase in the concentration of oxytocin receptors in the rat mammary gland during pregnancy. The gradual increase in mammary receptor concentration contrasted sharply to the sudden increase in oxytocin receptors in the uterus on the day of parturition. Oxytocin appears to stimulate the growth of myoepithelial cells both in vitro and in vivo. When administered with estrogen and progesterone through implantable pellets into a virgin mouse mammary gland, it also enhanced myoepithelial differentiation of the cap cells surrounding the terminal end bud (535). However, in the mammary gland its more characteristic and physiological action is to promote contraction of the myoepithelial cells.

Oxytocin and Maternal Behavior

In the late 1970s, Pederson and Prange and others noted that intraventricular administration of oxytocin induced maternal behavior in virgin rats (536). This observation led to extensive studies of the role of oxytocin in maternal behavior in rats [summarized in Gimpl and Fahrenholz (522)]. Similar observations have been made in sheep, suggesting that these effects may be relevant in many species (537). However, this notion is called into question by the finding that mice lacking oxytocin show normal maternal behavior (515).

LACTATION IN MONOTREMES AND MARSUPIALS

Space does not allow a full discussion of the evolution of mammary glands, a topic that has been beautifully and thoroughly discussed in the work of Olaf Oftedal (538,539). He concludes that milk-secreting glands originated from apocrine/sebaceous glands, secretory structures that are associated with hair follicles and secrete quite varied products often under control of steroid hormones. He further proposes that "mammary secretion first evolved as a means of supplying water to eggs and was essential to the evolution of endothermy among the egg-laying cynodonts that were ancestral to mammals" (539). The mammary glands of echidnas, monotremes that are widely distributed in Australia and New Zealand, are flat secretory structures with many alveoli opening directly onto an enlarged (2-cm) areola in a specialized pouch that is used both for incubating the eggs and for milk secretion (540,541). According to Oftedal, both monotremes and marsupials secrete a milk whose composition changes as lactation progresses (538), although limited information is available on the composition of the milk of monotremes.

In contrast, the milk of marsupials, particularly that of the wallaby (542,543) and, to a lesser extent, the opossum, has been extensively characterized (544,545). The lactation cycle in the tammar wallaby (Fig. 28) was well summarized by Trott et al. (543). It has been divided into four phases, the first being the 26.5-day developmental cycle during pregnancy when the gland becomes competent to secrete casein. At birth, the altricial young climb from the birth canal into the pouch and attach themselves permanently to a teat, where they remain for 100 to 125 days (phase 2A). During phase 2B, which lasts 75 to 100 days, the young remains in the pouch where it is intermittently attached to the teat. Phase 3 begins when the young leaves the pouch and begins to eat grass. At approximately day 250, the young leaves the pouch permanently, but continues to suckle until weaning, 10 to 12 months after birth.

As Fig. 28 shows, milk composition changes significantly during this cycle (546), with the largest change occurring when the young leaves the pouch at approximately 200 days (Fig. 28A). At this time the concentration of carbohydrate in the milk declines drastically, whereas the protein and lipid increase three- to fourfold. In fact, significant changes in gene expression occur at the boundary between phases 2A and 2B when WAP takes over from early lactation protein (542). When the young leave the pouch, special late lactation proteins (547) are substituted for many of the more common milk proteins such as the caseins, WAP, β-lactoglobulin, and α-lactalbumin (Fig. 28B). Interestingly, these changes, which meet the nutritional needs of the young while not compromising maternal nutrition, are not mandated by milk removal by the young. This was shown in a series of well-designed studies in which 60-day-old

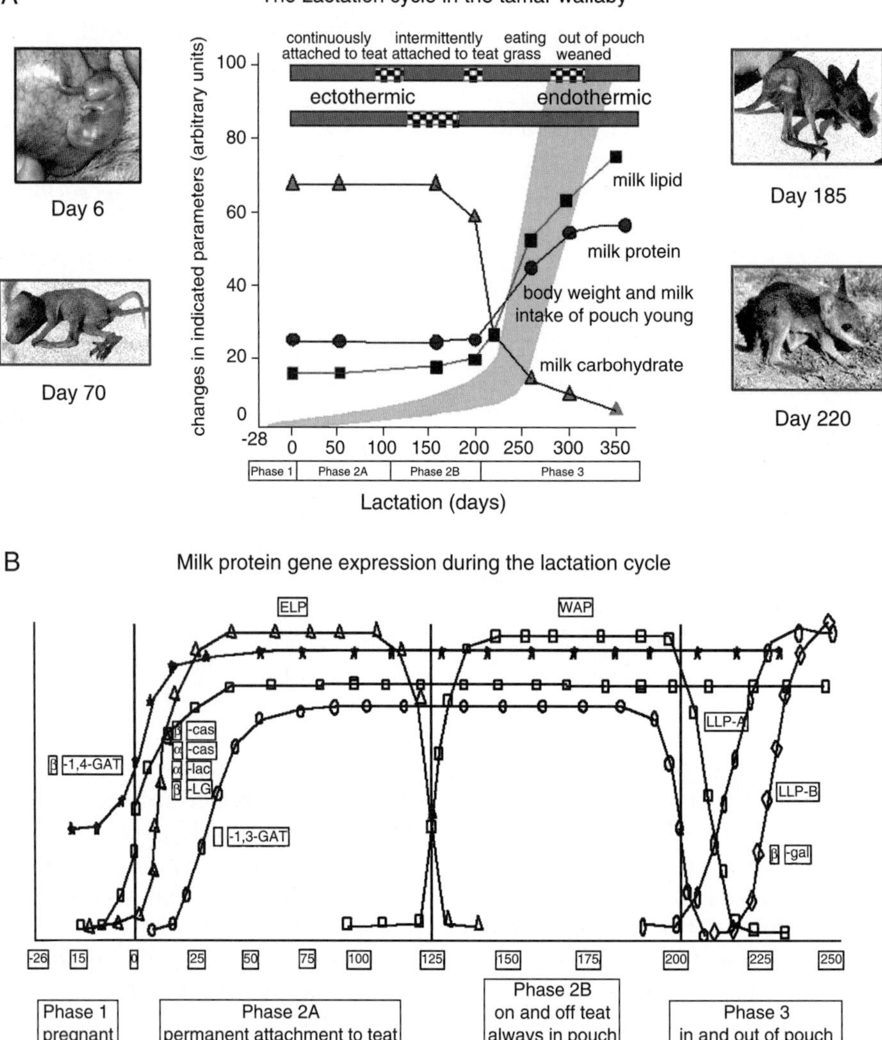

FIG. 28. The lactation cycle and milk protein genes in the tammar wallaby. **A:** The phases of lactation cycle correlate with the development of the young. **B:** Changes in the expression of milk proteins and synthetic enzymes in the mammary gland. β1,4 GAT, β1,4-galactosyltransferase; β1,3 GAT, β1,3-galactosyltransferase; β-cas, β-casein; α-cas, α-casein; α-lac, α-lactalbumin; β-LG, β-lactoglobulin; ELP, early lactation protein; WAP, whey acidic protein; LL-A, late lactation protein A; LL-B, late lactation protein B. (Graphs kindly provided by K. Nicholas, University of Melbourne, Australia.)

pouch young were fostered onto a group of host mothers in such a way that the lactational stage progressed 42 days ahead of the age of the young (543). The unexpected result was that once the young have attached and commenced nursing, the stages of lactation are developmentally fixed, rather than being regulated either by hormones or by milk removal. This developmental mechanism has two consequences: it allows for maternal regulation of milk secretion in such a way that the mother may be protected from the significant extremes of the Australian climate, and it means that the mother can be suckling two young of very different developmental stages simultaneously (548).

ACKNOWLEDGMENTS

The author thanks Michael Rudolph and James McManaman for critical reading of parts of the manuscript. She also expresses her gratitude to Kevin Nicholas for providing Fig. 28. The work was supported in part by National Institutes of Health grant PO1-HD38129. This chapter is dedicated to the late Dorothy Pitelka, to Malcolm Peaker, and to Jeffrey Rosen, whose guidance was invaluable as I gradually came to understand the marvels of the developing mammary gland. Many others taught me much along the way; their outstanding contributions are recognized in this manuscript.

REFERENCES

1. Linzell, J. L., and Peaker, M. (1971). Mechanism of milk secretion. *Physiol. Rev.* 51, 564–597.
2. Pitelka, D. R., Hamamoto, S. T., Duafala, J. G. and Nemanic, M. K. (1973). Cell contacts in the mouse mammary gland: I. Normal gland in postnatal development and the secretory cycle. *J. Cell Biol.* 56, 797–818.
3. Emerman, J. T., and Pitelka, D. R. (1977). Maintenance and induction of morphological differentiation in dissociated mammary epithelium on floating collagen membranes. *In Vitro* 13, 316–328.
4. Hennighausen, L., and Robinson, G. W. (2005). Information networks in the mammary gland. Nature Reviews: *Molecular Cell Biology* 6, 715–725.
5. Neville, M. C., McFadden, T. B., and Forsyth, I. A. (2002). Hormonal regulation of mammary differentiation and lactation. *J. Mammary Gland Biol. Neoplasia* 7, 49–66.
6. McManaman, J. L., and Neville, M. C. (2003). Mammary physiology and milk secretion. *Adv. Drug Deliv. Rev.* 55, 625–641.
7. Brisken, C. (2002). Hormonal control of alveolar development and its implications for breast carcinogenesis. *J. Mammary Gland Biol. Neoplasia* 7, 39–48.
8. Hovey, R. C., Trott, J. F., and Vonderhaar, B. K. (2002). Establishing a framework for the functional mammary gland: from Endocrinology to morphology. *J. Mammary Gland Biol. Neoplasia* 7, 17–38.
9. Kelly, P. A., Bachelot, A., Kedzia, C., Hennighausen, L., Ormandy, C. J., Kopchick, J. J., and Binart, N. (2002). The role of prolactin and growth hormone in mammary gland development. *Mol. Cell. Endocrinol.* 197, 127–131.
10. Hennighausen, L., and Robinson, G. W. (2001). Signaling pathways in mammary gland development. *Dev. Cell* 1, 467–475.
11. Hadsell, D. L., Bonnette, S. G., and Lee, A. V. (2002). Genetic manipulation of the IGF-1 axis to regulate mammary gland development and function. *J. Dairy Sci.* 85, 365–377.
12. Naylor, M. J., and Ormandy, C. J. (2002). Mouse strain-specific patterns of mammary epithelial ductal side branching are elicited by stromal factors. *Dev. Dyn.* 225, 100–105.
13. Masso-Welch, P. A., Darcy, K. M., Stangle-Castor, N. C., and Ip, M. M. (2000). A developmental atlas of rat mammary gland histology. *J. Mammary Gland Biol. Neoplasia* 5, 165–185.
14. Cardiff, R. D., and Wellings, S. R. (1999). The comparative pathology of human and mouse mammary glands. *J. Mammary Gland Biol. Neoplasia* 4, 105–122.
15. Cardiff, R. D., Wagner, U., and Hennighausen, L. G. (2000). Mammary cancer in humans and mice: a tutorial for comparative pathology [CD-ROM]. *J. Mammary Gland Biol. Neoplasia* 5, 243–244.
16. Akers, R. M. (2000). Selection for milk production from a lactation biology viewpoint. *J. Dairy Sci.* 83, 1151–1158.
17. Richert, M. M., Schwertfeger, K. L., Ryder, J., and Anderson, S. M. (2000). An atlas of mouse mammary gland development. *J. Mammary Gland Biol. Neoplasia* 5, 227–241.
18. Traurig, H. H. (2001). A radioautographic study of cell proliferation in the mammary gland of the pregnant mouse. *Anat. Rec.* 159, 239–248.
19. Blackman, B., Russell, T., Nordeen, S., Medina, D., and Neville, M. C. (2005). Claudin-7 expression and localization in the normal murine mammary gland and murine mammary tumors. *Breast Cancer Res.* 7, R248–R255.
20. Traurig, H. H. (1967). Cell proliferation in the mammary gland during late pregnancy and lactation. *Anat. Rec.* 157, 489–504.
21. Steiner, M. R., Altenburg, B., Richards, C. S., Dudley, J. P., Medina, D., and Butel, J. S. (1978). Differential response of cultured mouse mammary cells of varying tumorigenicity to cytochalasin B. *Cancer Res.* 38, 2719–2721.
22. Hartmann, P. E. (1973). Changes in the composition and yield of the mammary secretion of cows during the initiation of lactation. *J. Endocrinol.* 59, 231–247.
23. Mellenberger, R. W., and Bauman, D. E. (1974). Metabolic adaptations during lactogenesis: fatty acid synthesis in rabbit mammary tissue during pregnancy and lactation. *Biochem. J.* 138, 373–379.
24. Arthur, P. G., Smith, M., and Hartmann, P. (1989). Milk lactose, citrate and glucose as markers of lactogenesis in normal and diabetic women. *J. Pediatr. Gastroenterol. Nutr.* 9, 488–496.
25. Schwertfeger, K. L., McManaman, J. L., Palmer, C. A., Neville, M. C., and Anderson, S. M. (2003). Expression of constitutively activated Akt in the mammary gland leads to excess lipid synthesis during pregnancy and lactation. *J. Lipid Res.* 44, 1100–1112.
26. McManaman, J. L., Palmer, C. A., Wright, R. M., and Neville, M. C. (2002). Functional regulation of xanthine oxidoreductase expression and localization in the mouse mammary gland: evidence of a role in lipid secretion. *J. Physiol. (Lond).* 545, 567–579.
27. Rudolph, M., Phang, T., McManaman, J. L., Hunter, L., and Neville, M. C. (2003). Microarray analysis of secretory activation in the mammary gland. *J. Mammary Gland Biol. Neoplasia* 8, 287–307.
28. Nguyen, D.-A. D., and Neville, M. C. (1998). Tight junction regulation in the mammary gland. *J. Mammary Gland Biol. Neoplasia* 3, 233–246.
29. Nguyen, D.-A. D., Beeman, N., and Neville, M. C. (2001). Regulation of tight junction permeability in the mammary gland. In *Tight Junctions* (M. Cerejido and J. M. Anderson, Eds.), pp. 395–414. CRC Press, New York.
30. Kuhn, N. J. (1969). Progesterone withdrawal as the lactogenic trigger in the rat. *J. Endocrinol.* 44, 39–54.
31. Loizzi, R. F. (1985). Progesterone withdrawal stimulates mammary gland tubulin polymerization in pregnant rats. *Endocrinology* 116, 2543–2547.
32. Deis, R. P., and Delouis, C. (1983). Lactogenesis induced by ovariectomy in pregnant rats and its regulation by oestrogen and progesterone. *J. Steroid Biochem.* 18, 687–690.
33. Martin, C. E., Hartmann, P., and Gooneratne, A. (1978). Progesterone and corticosteroids in the initiation of lactation in the sow. *Aust. J. Biol. Sci.* 31, 517–525.
34. Hartmann, P., Trevethan, P., and Shelton, J. N. (1973). Progesterone and oestrogen and the initiation of lactation in ewes. *J. Endocrinol.* 59, 249–259.
35. Rosen, J. M., O'Neal, D. L., McHugh, J. E., and Comstock, J. P. (1978). Progesterone-mediated inhibition of casein mRNA and polysomal casein synthesis in the rat mammary gland during pregnancy. *Biochemistry* 17, 290–297.
36. Halban, J. (1905). Die innere Secretion von Ovarium und Placenta und ihre Bedeutung für die Function der Milchdruese. *Arch. Gynaekol.* 75, 353–441.
37. Patton, S., Huston, G. E., Montgomery, P. A., and Josephson, R. V. (1986). Approaches to the study of colostrum: the onset of lactation. In *Human Lactation 2: Maternal and Environmental Factors* (M. Hamosh and A. S. Goldman, Eds.), pp. 231–240. Plenum Press, New York.
38. Allen, J. C., Keller, R. P., Archer, P. C., and Neville, M. C. (1991). Studies in human lactation: 6. Milk composition and daily secretion rates of macronutrients in the first year of lactation. *Am. J. Clin. Nutr.* 54, 69–80.
39. Jenness, R. (1974). The composition of milk. In *Lactation: A Comprehensive Treatise,* vol. 3 (B. L. Larson and V. R. Smith, Eds.), pp. 3–107. Academic Press, New York.
40. Jenness, R. (1979). Comparative aspects of milk proteins. *J. Dairy Res.* 46, 197–210.
41. Jensen, R. G. (1995). *Handbook of Milk Composition.* Academic Press, San Diego.

42. Furth, P. A. (1999). Mammary involution and apoptosis. *J. Mammary Gland Biol. Neoplasia* 4, 123–240.
43. Lascelles, A. K., and Lee, C. S. (1978). Involution of the mammary gland. In *Lactation*, vol. IV (B. L. Larson, Ed.), pp. 115–177. Academic Press, New York.
44. Lund, L. R., Romer, J., Thomasset, N., Solberg, H., Pyke, C., Bissell, M. J., Dano, K., and Werb, Z. (1996). Two distinct phases of apoptosis in mammary gland involution: proteinase-independent and -dependent pathways. *Development* 122, 181–193.
45. Clarkson, R. W. E., Wayland, M. T., Lee, J., Freeman, T., and Watson, C. J. (2004). Gene expression profiling of mammary gland development reveals putative roles for death receptors and immune mediators in post-lactational regression. *Breast Cancer Res.* 6, 9–109.
46. Stein, T., Morris, J. S., Davies, C. R., Weber-Hall, S. J., Duffy, M., Heath, V. J., Bell, A. K., Ferrier, R. K., Sandilands, G. P., and Gusterson, B. (2004). Involution of the mouse mammary gland is associated with an immune cascade and an acute-phase response, involving LBP, CD14 and STAT3. *Breast Cancer Res.* 6, 75–91.
47. Atwood, C. S., Hovey, R. C., Glover, J. P., Chepko, G., Ginsburg, E., Robison, J., and Vonderhaar, B. K. (2000). Progesterone induces side-branching of the ductal epithelium in the mammary glands of peripubertal mice. *J. Endocrinol.* 167, 39–52.
48. Plaut, K., Maple, R., Ginsburg, E., and Vonderhaar, B. K. (1999). Progesterone stimulates DNA synthesis and lobulo-alveolar development in mammary glands in ovariectomized mice. *J. Cell. Physiol.* 180, 298–304.
49. Cui, Y., Riedlinger, G., Miyoshi, K., Tang, W., Li, C., Deng, C. X., Robinson, G. W., and Hennighausen, L. (2004). Inactivation of Stat5 in mouse mammary epithelium during pregnancy reveals distinct functions in cell proliferation, survival, and differentiation. *Mol. Cell. Biol.* 24, 8037–8047.
50. Anderson, E., Clarke, R. B., and Howell, A. (1998). Estrogen responsiveness and control of normal human breast proliferation. *J. Mammary Gland Biol. Neoplasia* 3, 23–35.
51. Clarke, R. B., Howell, A. B., Potten, C. S., and Anderson, E. (1997). Dissociation between steroid receptor expression and cell proliferation in the human breast. *Cancer Res.* 57, 4987–4991.
52. Grimm, S. L., Seagroves, T. N., Kabotyanski, E. B., Hovey, R. C., Vonderhaar, B. K., Lydon, J. P., Miyoshi, K., Hennighausen, L., Ormandy, C. J., Lee, A. V., Stull, M. A., Wood, T. L., and Rosen, J. M. (2002). Disruption of steroid and prolactin receptor patterning in the mammary gland correlates with a block in lobuloalveolar development. *Mol. Endocrinol.* 16, 2675–2691.
53. Ismail, P. M., Amato, P., Soyal, S. M., Demayo, F. J., Conneely, O. M., O'Malley, B. W., and Lydon, J. P. (2003). Progesterone involvement in breast development and tumorigenesis: as revealed by progesterone receptor "knockout" and "knockin" mouse models. *Steroids* 68, 779–787.
54. Lee, A. V., Zhang, P., Ivanova, M., Bonnette, S., Oesterreich, S., Rosen, J. M., Grimm, S., Hovey, R. C., Vonderhaar, B. K., Kahn, C. R., Torres, D., George, J., Mohsin, S., Allred, D. C., and Hadsell, D. L. (2003). Developmental and hormonal signals dramatically alter the localization and abundance of insulin receptor substrate proteins in the mammary gland. *Endocrinology* 144, 2683–2694.
55. Brisken, C., Park, S., Vass, T., Lydon, J. P., O'Malley, B. W., and Weinberg, R. A. (1998). A paracrine role for the epithelial progesterone receptor in mammary gland development. *Proc. Natl. Acad. Sci. U S A* 95, 5076–5081.
56. Conneely, O. M., Mulac-Jericevic, B., Demayo, F., Lydon, J. P., and O'malley, B. W. (2002). Reproductive functions of progesterone receptors. *Recent Prog. Horm. Res.* 57, 339–355.
57. Cao, Y., Bonizzi, G., Seagroves, T. N., Greten, F. R., Johnson, R., Schmidt, E. V., and Karin, M. (2001). IKKα provides an essential link between RANK signaling and cyclin D1 expression during mammary gland development. *Cell* 107, 763–775.
58. Mulac-Jericevic, B., Lydon, J. P., Demayo, F. J., and Conneely, O. M. (2003). Defective mammary gland morphogenesis in mice lacking the progesterone receptor B isoform. *Proc. Natl. Acad. Sci. U S A* 100, 9744–9749.
59. Suchyta, S. P., Sipkovsky, S., Halgren, R. G., Kruska, R., Elftman, M., Weber-Nielson, M., Vandehaar, M. J., Xiao, L., Temperman, R. J., and Coussens, P. M. (2004). Bovine mammary gene expression profiling using a cDNA microarray enhanced for mammary-specific transcripts. *Physiol. Genomics* 16, 8–18.
60. Forsyth, I. A., Taylor, J. A., Keable, S., Turvey, A., and Lennard, S. (1997). Expression of amphiregulin in the sheep mammary gland. *Mol. Cell. Endocrinol.* 126, 41–48.
61. Kenney, N. J., Smith, G. H., Rosenberg, K., Cutler, M. L., and Dickson, R. B. (1996). Induction of ductal morphogenesis and lobular hyperplasia by amphiregulin in the mouse mammary gland. *Cell Growth Differ.* 7, 1769–1781.
62. Bradbury, J. M., Edwards, P. A. W., Niemeyer, C. C., and Dale, T. C. (1995). Wnt-4 expression induces a pregnancy-like growth pattern in reconstituted mammary glands in virgin mice. *Dev. Biol.* 170, 553–563.
63. Brisken, C., Heineman, A., Chavarria, T., Elenbaas, B., Tan, J., Dey, S. K., McMahon, J. A., McMahon, A. P., and Weinberg, R. A. (2000). Essential function of Wnt-4 in mammary gland development downstream of progesterone signaling. *Genes Dev.* 14, 650–654.
64. Conneely, O. M., Jericevic, B. M., and Lydon, J. P. (2003). Progesterone receptors in mammary gland development and tumorigenesis. *J. Mammary Gland Biol. Neoplasia* 8, 205–214.
65. Talamantes, F., and Ogren, L. (1988). The placenta as an endocrine organ: polypeptides. In *The Physiology of Reproduction* (E. Knobil and J. D. Neill, Eds.), pp. 2093–2144. Raven Press, New York.
66. Caron, R. W., Jahn, G. A., and Deis, R. P. (1994). Lactogenic actions of different growth hormone preparations in pregnant and lactating rats. *J. Endocrinol.* 142, 535–545.
67. Ilkbahar, Y. N., Thordarson, G., Camarillo, I. G., and Talamantes, F. (1999). Differential expression of the growth hormone receptor and growth hormone-binding protein in epithelia and stroma of the mouse mammary gland at various physiological stages. *J. Endocrinol.* 161, 77–87.
68. Zhou, Y., Xu, B. C., Maheshwari, H. G., He, L., Reed, M., Lozykowski, M., Okada, S., Cataldo, L., Coschigamo, K., Wagner, T. E., Baumann, G., and Kopchick, J. J. (1997). A mammalian model for Laron syndrome produced by targeted disruption of the mouse growth hormone receptor/binding protein gene (the Laron mouse). *Proc. Natl. Acad. Sci. U S A* 94, 13215–13220.
69. Mol, J. A., Lantinga-Van Leeuwen, I., Van Garderen, E., and Rijnberk, A. (2000). Progestin-induced mammary growth hormone (GH) production. *Adv. Exp. Med. Biol.* 480, 71–76.
70. Gregoraszczuk, E. L., Milewicz, T., Kolodziejczyk, J., Krzysiek, J., Basta, A., Sztefko, K., Kurek, S., and Stachura, J. (2001). Progesterone-induced secretion of growth hormone, insulin-like growth factor I and prolactin by human breast cancer explants. *Gynecol. Endocrinol.* 15, 251–258.
71. Liu, X., Robinson, G. W., Wagner, K. U., Garrett, L., Wynshaw-Boris, A., and Hennighausen, L. (1997). Stat5a is mandatory for adult mammary gland development and lactogenesis. *Genes Dev.* 11, 179–186.
72. Gallego, M. I., Binart, N., Robinson, G. W., Okagaki, R., Coschigano, K. T., Perry, J., Kopchick, J. J., Oka, T., Kelly, P. A., and Hennighausen, L. (2001). Prolactin, growth hormone, and epidermal growth factor activate Stat5 in different

compartments of mammary tissue and exert different and overlapping developmental effects. *Dev. Biol.* 229, 163–175.
73. Doppler, W., Windegger, M., Soratroi, C., Tomasi, J., Lechner, J., Rusconi, S., Cato, A. C., Almlof, T., Liden, J., Okret, S., Gustafsson, J., Richard-Foy, H., Starr, D. B., Klocker, H., Edwards, D., and Geymayer, S. (2001). Expression level-dependent contribution of glucocorticoid receptor domains for functional interaction with STAT5. *Mol. Cell. Biol.* 21, 3266–3279.
74. Wyszomierski, S. L., and Rosen, J. M. (2001). Cooperative effects of stat5 (signal transducer and activator of transcription 5) and C/EBP β (CAAT/enhancer-binding protein-β) on β-casein gene transcription are mediated by the glucocorticoid receptor. *Mol. Endocrinol.* 15, 228–240.
75. Zahnow, C. A. (2002). CCAAT/enhancer binding proteins in normal mammary development and breast cancer. *Breast Cancer Res.* 4, 113–121.
76. Martin, R. H., and Oakey, R. E. (1982). The role of antenatal oestrogen in post-partum human lactogenesis: evidence from oestrogen-deficient pregnancies. *Clin. Endocrinol.* 17, 403–408.
77. Rigg, L. A., Lein, A., and Yen, S. S. (1977). Pattern of increase in circulating prolactin levels during human gestation. *Am. J. Obstet. Gynecol.* 129, 454–456.
78. Farmer, C., and Petitclerc, D. (2003). Specific window of prolactin inhibition in late gestation decreases mammary parenchymal tissue development in gilts. *J. Anim. Sci.* 81, 1823–1829.
79. Andrews, Z. B., Kokay, I. C., and Grattan, D. R. (2001). Dissociation of prolactin secretion from tuberoinfundibular dopamine activity in late pregnant rats. *Endocrinology* 142, 2719–2724.
80. Mizoguchi, Y., Kim, J. Y., Enami, J., and Sakai, S. (1997). The regulation of the prolactin receptor gene expression in the mammary gland of early pregnant mouse. *Endocr. J.* 44, 53–58.
81. Gourdou, I., Paly, J., Hue-Beauvais, C., Pessemesse, L., Clark, J., and Djiane, J. (2004). Expression by transgenesis of a constitutively active mutant form of the prolactin receptor induces premature abnormal development of the mouse mammary gland and lactation failure. *Biol. Reprod.* 70, 718–728.
82. Brisken, C., Kaur, S., Chavarria, T. E., Binart, N., Sutherland, R. L., Weinberg, R. A., Kelly, P. A., and Ormandy, C. J. (1999). Prolactin controls mammary gland development via direct and indirect mechanisms. *Dev. Biol.* 210, 96–106.
83. Hovey, R. C., Trott, J. F., Ginsburg, E., Goldhar, A. S., Sasaki, M. M., Fountain, S. J., Sundararajan, K., and Vonderhaar, B. K. (2001). Transcriptional and spatiotemporal regulation of prolactin receptor mRNA and cooperativity with progesterone receptor function during ductal branch growth in the mammary gland. *Dev. Dyn.* 222, 192–205.
84. Thordarson, G., Villalobos, R., Colosi, P., Southard, L., Ogren, L., and Talamantes, F. (1986). Lactogenic response of cultured mouse mammary epithelial cells to mouse placental lactogen. *J. Endocrinol.* 109, 263–274.
85. Buttle, H. L., Cowie, A. T., Jones, E. A., and Turvey, A. (1979). Mammary growth during pregnancy in hypophysectomized or bromocriptine-treated goats. *J. Endocrinol.* 80, 343–351.
86. Forsyth, I. A. (1986). Variation among species in the endocrine control of mammary growth and function: the roles of prolactin, growth hormone, and placental lactogen. *J. Dairy Sci.* 69, 886–903.
87. Whitelaw, C. B. (1995). Regulation of ovine beta-lactoglobulin gene expression during the first stage of lactogenesis. *Biochem. Biophys. Res. Commun.* 209, 1089–1093.
88. Forsyth, I. A. (1994). Comparative aspects of placental lactogens: structure and function. *Exp. Clin. Endocrinol.* 102, 244–251.
89. Gootwine, E. (2004). Placental hormones and fetal-placental development. *Anim. Reprod. Sci.* 82–83, 551–566.
90. Gaede, P., Trolle, D., and Pedersen, H. (1978). Extremely low placental lactogen hormone (hPL) values in an otherwise uneventful pregnancy preceding delivery of a normal baby. *Acta Obstet. Gynecol. Scand.* 57, 203–209.
91. Nielson, P. V., Pederson, H., and Kampmann, E. (1979). Absence of placental lactogen in an otherwise uneventful pregnancy. *Am. J. Obstet. Gynecol.* 135, 322–330.
92. Herman, A., Bignon, C., Daniel, N., Grosclaude, J., Gertler, A., and Djiane, J. (2000). Functional heterodimerization of prolactin and growth hormone receptors by ovine placental lactogen. *J. Biol. Chem.* 275, 6295–6301.
93. Biener, E., Martin, C., Daniel, N., Frank, S. J., Centonze, V. E., Herman, B., Djiane, J., and Gertler, A. (2003). Ovine placental lactogen-induced heterodimerization of ovine growth hormone and prolactin receptors in living cells is demonstrated by fluorescence resonance energy transfer microscopy and leads to prolonged phosphorylation of signal transducer and activator of transcription (STAT)1 and STAT3. *Endocrinology* 144, 3532–3540.
94. Nguyen, D. D., Parlow, A. F., and Neville, M. C. (2001). Hormonal regulation of tight junction closure in the mouse mammary epithelium during the transition from pregnancy to lactation. *J. Endocrinol.* 170, 347–356.
95. Silberstein, G. B., Van Horn, K., Shyamala, G., and Daniel, C. W. (1996). Progesterone receptors in the mouse mammary duct: distribution and developmental regulation. *Cell Growth Differ.* 7, 945–952.
96. Schams, D., Kohlenberg, S., Amselgruber, W., Berisha, B., Pfaffl, M. W., and Sinowatz, F. (2003). Expression and localisation of oestrogen and progesterone receptors in the bovine mammary gland during development, function and involution. *J. Endocrinol.* 177, 305–317.
97. Boonyaratanakornkit, V., and Edwards, D. P. (2004). Receptor mechanisms of rapid extranuclear signalling initiated by steroid hormones. *Essays Biochem.* 40, 105–120.
98. Haslam, S. Z., and Shyamala, G. (1980). Progesterone receptors in normal mammary gland: Receptor modulations in relation to differentiation. *J. Cell Biol.* 86, 730–737.
99. Shyamala, G., Schneider, W., and Schott, D. (1990). Developmental regulation of murine mammary progesterone receptor gene expression. *Endocrinology* 126, 2882–2889.
100. Foley, J., Dann, P., Hong, J., Cosgrove, J., Dreyer, B., Rimm, D., Dunbar, M., Philbrick, W. M., and Wysolmerski, J. J. (2001). Parathyroid hormone-related protein maintains mammary epithelial fate and triggers nipple skin differentiation during embryonic breast development. *Development* 128, 513–525.
101. Dunbar, M., and Wysolmerski, J. J. (1999). Parathyroid hormone-related protein: a developmental regulatory molecule necessary for mammary gland development. *J. Mammary Gland Biol. Neoplasia* 4, 21–34.
102. Kleinberg, D. L. (1997). Early mammary development: growth hormone and IGF-1. *J. Mammary Gland Biol. Neoplasia* 2, 49–57.
103. Pilipenko, V. V., Reece, A., Choo, D. I., and Greinwald, J. H., Jr. (2004). Genomic organization and expression analysis of the murine Fam3c gene. *Gene* 335, 159–168.
104. Zhu, Y., Xu, G., Patel, A., Mclaughlin, M. M., Silverman, C., Knecht, K., Sweitzer, S., Li, X., McDonnell, P., Mirabile, R., Zimmerman, D., Boyce, R., Tierney, L. A., Hu, E., Livi, G. P., Wolf, B., Abdel-Meguid, S. S., Rose, G. D., Aurora, R., Hensley, P., Briggs, M., and Young, P. R. (2002). Cloning, expression, and initial characterization of a novel cytokine-like gene family. *Genomics* 80, 144–150.
105. Daniel, C. W., Silberstein, G. B., Van Horn, K., Strickland, P., and Robinson, S. (1989). TGF-β1–induced inhibition of mouse mammary ductal growth: developmental specificity and characterization. *Dev. Biol.* 135, 20–30.

106. Robinson, S. D., Roberts, A. B., and Daniel, C. W. (1993). TGFβ suppresses casein synthesis in mouse mammary explants and may play a role in controlling milk levels during pregnancy. *J. Cell Biol.* 120, 245–251.
107. Barcellos-Hoff, M. H., and Ewan, K. B. (2000). Transforming growth factor-beta and breast cancer: mammary gland development. *Breast Cancer Res.* 2, 92–99.
108. Nguyen, A. V., and Pollard, J. W. (2000). Transforming growth factor β3 induces cell death during the first stage of mammary gland involution. *Development* 127, 3107–3118.
109. Pollard, J. W. (2001). Tumour-stromal interactions: Transforming growth factor-β isoforms and hepatocyte growth factor/scatter factor in mammary gland ductal morphogenesis. *Breast Cancer Res.* 3, 230–237.
110. Sudlow, A. W., Wilde, C. J., and Burgoyne, R. D. (1994). Transforming growth factor-β1 inhibits casein secretion from differentiating mammary-gland explants but not from lactating mammary cells. *Biochem. J.* 304, 333–336.
111. Mieth, M., Boehmer, F. D., Ball, R., Groner, B., and Grosse, R. (1990). Transforming growth factor-beta inhibits lactogenic hormone induction of beta-casein expression in HC11 mouse mammary epithelial cells. *Growth Factors* 4, 9–15.
112. Yamamoto, T., Komura, H., Morishige, K., Tadokoro, C., Sakata, M., Kurachi, H., and Miyake, A. (1994). Involvement of autocrine mechanism of transforming growth factor-beta in the functional differentiation of pregnant mouse mammary gland. *Eur. J. Endocrinol.* 130, 302–307.
113. Price-Schiavi, S. A., Carraway, C. A., Fregien, N., and Carraway, K. L. (1998). Post-transcriptional regulation of a milk membrane protein, the sialomucin complex (ascites sialoglycoprotein [ASGP]-1/ASGP-2, rat muc4), by transforming growth factor beta. *J. Biol. Chem.* 273, 35228–35237.
114. Price-Schiavi, S. A., Zhu, X., Aquinin, R., and Carraway, K. L. (2000). Sialomucin complex (rat Muc4) is regulated by transforming growth factor beta in mammary gland by a novel post-translational mechanism. *J. Biol. Chem.* 275, 17800–17807.
115. Jhappan, C., Geiser, A. G., Kordon, E. C., Bagheri, D., Hennighausen, L., Roberts, A. B., Smith, G. H., and Merlino, G. (1993). Targeting expression of a transforming growth factor beta 1 transgene to the pregnant mammary gland inhibits alveolar development and lactation. *EMBO J.* 12, 1835–1845.
116. Kordon, E. C., McKnight, R. A., Jhappan, C., Hennighausen, L., Merlino, G., and Smith, G. H. (1995). Ectopic TGFβ1 expression in the secretory mammary epithelium induces early senescence of the epithelial stem cell population. *Dev. Biol.* 168, 47–61.
117. Ewan, K. B., Shyamala, G., Ravani, S. A., Tang, Y., Akhurst, R., Wakefield, L., and Barcellos-Hoff, M. H. (2002). Latent transforming growth factor-beta activation in mammary gland: regulation by ovarian hormones affects ductal and alveolar proliferation. *Am. J. Pathol.* 160, 2081–2093.
118. Gorska, A. E., Joseph, H., Derynck, R., Moses, H. L., and Serra, R. (1998). Dominant-negative interference of the transforming growth factor beta type II receptor in mammary gland epithelium results in alveolar hyperplasia and differentiation in virgin mice. *Cell Growth Differ.* 9, 229–238.
119. Gorska, A. E., Jensen, R. A., Shyr, Y., Aakre, M. E., Bhowmick, N. A., and Moses, H. L. (2004). Transgenic mice expressing a dominant-negative mutant type II transforming growth factor-beta receptor exhibit impaired mammary development and enhanced mammary tumor formation. *Am. J. Pathol.* 163, 1539–1549.
120. Lenferink, A. E. G., Magoon, J., Pepin, M. C., Guimond, A., and O'Connor-McCourt, M. D. (2003). Expression of TGF-beta type II receptor antisense RNA impairs TGF-beta signaling in vitro and promotes mammary gland differentiation in vivo. *Int. J. Cancer* 107, 919–928.
121. Siegel, P. M., Shu, W., Cardiff, R. D., Muller, W. J., and Massague, J. (2003). Transforming growth factor beta signaling impairs Neu-induced mammary tumorigenesis while promoting pulmonary metastasis. *Proc. Natl. Acad. Sci. U S A* 100, 8430–8435.
122. Luo, X. H., Liao, E. Y., and Su, X. (2002). Progesterone upregulates TGF-β isoforms (β1, β2, and β3) expression in normal human osteoblast-like cells. *Calcif. Tissue Int.* 71, 329–334.
123. Zicari, A., Ticconi, C., Realacci, M., Cela, O., Santangelo, C., Pietropoli, A., Russo, M. A., and Piccione, E. (2002). Hormonal regulation of cytokine release by human fetal membranes at term gestation: effects of oxytocin, hydrocortisone and progesterone on tumour necrosis factor-alpha and transforming growth factor-beta 1 output. *J. Reprod. Immunol.* 56, 123–136.
124. Lui, W.-L., Lee, W. M., and Cheng, C. Y. (2001). Transforming growth factor β3 perturbs the intersertoli tight junction permeability barrier in vitro possibly mediated via its effects on occludin, zonula occludens-1 and claudin-11. *Endocrinology* 142, 1865–1877.
125. Lui, W. Y., Lee, W. M., and Cheng, C. Y. (2003). Transforming growth factor beta3 regulates the dynamics of Sertoli cell tight junctions via the p38 mitogen-activated protein kinase pathway. *Biol. Reprod.* 68, 1597–1612.
126. Woo, P. L., Cha, H. H., Singer, K. L., and Firestone, G. L. (1996). Antagonistic regulation of tight junction dynamics by glucocorticoids and transforming growth factor-beta in mouse mammary epithelial cells. *J. Biol. Chem.* 271, 404–412.
127. Reichmann, E., Ball, R., Groner, B., and Friis, R. R. (1989). New mammary epithelial and fibroblastic cell clones in coculture differentiate functionally. *J. Cell Biol.* 108, 1127–1138.
128. Allar, M. J., and Wood, T. L. (2004). Expression of insulin-like growth factor binding proteins during postnatal development of the murine mammary gland. *Endocrinology* 145, 2467–2477.
129. Schneider, M. R., Wolf, E., Hoeflich, A., and Lahm, H. (2002). IGF-binding protein-5: flexible player in the IGF system and effector on its own. *J. Endocrinol.* 172, 423–440.
130. Flint, D. J., Beattie, J., and Allan, G. J. (2003). Modulation of the actions of IGFs by IGFBP-5 in the mammary gland. *Horm. Metab. Res.* 35, 809–815.
131. Tonner, E., Barber, M. C., Allan, G. J., Beattie, J., Webster, J., Whitelaw, C. B. A., and Flint, D. J. (2002). Insulin-like growth factor binding protein-5 (IGFBP-5) induces premature cell death in the mammary glands of transgenic mice. *Development* 129, 4547–4557.
132. Marshman, E., Green, K. A., Flint, D. J., White, A., Streuli, C. H., and Westwood, M. (2003). Insulin-like growth factor binding protein 5 and apoptosis in mammary epithelial cells. *J. Cell Sci.* 116, 675–682.
133. Gavin, B. J., and McMahon, A. P. (1992). Differential regulation of the Wnt gene family during pregnancy and lactation suggests a role in postnatal development of the mammary gland. *Mol. Cell. Biol.* 12, 2418–2423.
134. Weber-Hall, S. J., Phippard, D. J., Neimeyer, C. C., and Dale, T. C. (1994). Developmental and hormonal regulation of Wnt gene expression in the mouse mammary gland. *Differentiation* 57, 205–214.
135. Renou, J. P., Bierie, B., Miyoshi, K., Cui, Y., Djiane, J., Reichenstein, M., Shani, M., and Hennighausen, L. (2003). Identification of genes differentially expressed in mouse mammary epithelium transformed by an activated beta-catenin. *Oncogene* 22, 4594–4610.
136. Hsu, W., Shakya, R., and Costantini, F. (2001). Impaired mammary gland and lymphoid development caused by inducible expression of axin in transgenic mice. *J. Cell Biol.* 155, 1055–1064.

137. Cao, X., Gao, Z., Robert, C. E., Greene, S., Xu, G., Xu, W., Bell, E., Campbell, D., Zhu, Y., Young, R., Trucco, M., Markmann, J. F., Naji, A., and Wolf, B. A. (2003). Pancreatic-derived factor (FAM3B), a novel islet cytokine, induces apoptosis of insulin-secreting beta-cells. *Diabetes* 52, 2296–2303.
138. Desrivieres, S., Prinz, T., Castro-Palomino Laria, N., Meyer, M., Boehm, G., Bauer, U., Schafer, J., Neumann, T., Shemanko, C., and Groner, B. (2003). Comparative proteomic analysis of proliferating and functionally differentiated mammary epithelial cells. *Mol. Cell. Proteomics* 2, 1039–1054.
139. Neville, M. C., Casey, C. E., Keller, R. P., and Archer, P. (1986). Human lactation 2: effects of maternal and environmental factors. In *Changes in Milk Composition after Six Months of Lactation: The Effects of Duration of Lactation and Gradual Weaning* (M. Hamosh and A. Goldman, Eds.), pp. 141–154. Plenum Press, New York.
140. Neville, M. C., Keller, R. P., Seacat, J., Lutes, V., Neifert, M., Casey, C. E., Allen, J. C., and Archer, P. (1988). Studies in human lactation: milk volumes in lactating women during the onset of lactation and full lactation. *Am. J. Clin. Nutr.* 48, 1375–1386.
141. Neville, M. C., Allen, J. C., Archer, P. C., Casey, C. E., Seacat, J., Keller, R. P., Lutes, V., Rasbach, J., and Neifert, M. (1991). Studies in human lactation: milk volume and nutrient composition during weaning and lactogenesis. *Am. J. Clin. Nutr.* 54, 81–92.
142. Fleet, I. R., Goode, J. A., Hamon, M. H., Laurie, M. S., Linzell, J. L., and Peaker, M. (1975). Secretory activity of goat mammary glands during pregnancy and the onset of lactation. *J. Physiol. (Lond).* 251, 763–773.
143. Neville, M. C., and Morton, J. (2001). Physiology and endocrine changes underlying human lactogenesis II. *J. Nutr.* 131, 3005S–3008S.
144. Neville, M. C., Morton, J., and Umemora, S. (2001). Lactogenesis: the transition from pregnancy to lactation. *Pediatr. Clin. North Am.* 48, 35–52.
145. Lewis-Jones, D. I., Lewis-Jones, M. S., Connolly, R. C., Lloyd, D. C., and West, C. R. (1985). Sequential changes in the antimicrobial protein concentrations in human milk during lactation and its relevance to banked human milk. *Pediatr. Res.* 19, 561–565.
146. McClelland, D. B. L., McGrath, J., and Samson, R. R. (1978). Antimicrobial factors in human milk: studies of concentration and transfer to the infant during the early stages of lactation. *Acta Paediatr. Scand. Suppl.* 271, 1–20.
147. Newburg, D. S. (1996). Oligosaccharides and glycoconjugates of human milk: their role in host defense. *J. Mammary Gland Biol. Neoplasia* 1, 271–283.
148. Lascelles, A. K. (1977). Role of the mammary gland and milk in immunology. In *Comparative Aspects of Lactation* (M. Peaker, Ed.), pp. 241–260. Academic Press, London.
149. Guy, M. A., McFadden, T. B., Cockrell, D. C., and Besser, T. E. (1994). Regulation of colostrum formation in beef and dairy cows. *J. Dairy Sci.* 77, 3002–3007.
150. Cowie, A. T., Forsyth, I. A., and Hart, I. C. (1980). Hormonal control of lactation. *Monogr. Endocrinol.* 15(I–XIV), 1–275.
151. Hovey, R. C., Trott, J., and Vonderhaar, B. K. (2001). Establishing a framework for the functional mammary gland: from endocrinology to morphology. *J. Mammary Gland Biol. Neoplasia* 7, 17–38.
152. Topper, Y. J., and Freeman, C. S. (1980). Multiple hormone interactions in the developmental biology of the mammary gland. *Physiol. Rev.* 60, 1049–1106.
153. Vonderhaar, B. K., and Ziska, S. E. (1989). Hormonal regulation of milk protein gene expression. *Annu. Rev. Physiol.* 51, 641–652.
154. Feng, Z., Marti, A., Jehn, B., Altermatt, H. J., Chicaiza, G., and Jaggi, R. (1995). Glucocorticoid and progesterone inhibit involution and programmed cell death in the mouse mammary gland. *J. Cell Biol.* 131, 1095–1103.
155. Camps, M., Vilaro, S., Testar, X., Palacin, M., and Zorzano, A. (1994). High and polarized expression of GLUT1 glucose transporters in epithelial cells from mammary gland: acute down-regulation of GLUT1 carriers by weaning. *Endocrinology* 134, 924–934.
156. Kulski, J. K., Hartmann, P. E., Martin, J. D., and Smith, M. (1978). Effects of bromocriptine mesylate on the composition of the mammary secretion in non-breast-feeding women. *Obstet. Gynecol.* 52, 38–42.
157. Gourdou, I., Paly, J., Hue-Beauvais, C., Pessemesse, L., Clark, J., and Djiane, J. (2003). Expression by transgenesis of a constitutively active mutant form of the prolactin receptor induces premature abnormal development of the mouse mammary gland and lactation failure. *Biol. Reprod.* 70, 718–728.
158. Bailey, J. P., Nieport, K., Herbst, M. P., Srivastava, S., Serra, R., and Horseman, N. D. (2004). Prolactin and transforming growth factor-β signaling exert opposing effects on mammary gland morphogenesis, involution, and the Akt-forkhead pathway. *Mol. Endocrinol.* 18, 1171–1184.
159. Seagroves, T. N., Hadsell, D., McManaman, J., Palmer, C., Liao, D., McNulty, W., Welm, B., Wagner, K. U., Neville, M., and Johnson, R. S. (2003). HIF1α is a critical regulator of secretory differentiation and activation, but not vascular expansion, in the mouse mammary gland. *Development* 130, 1713–1724.
160. Strange, R., Li, F., Friis, R. R., Reichmann, E., Haenni, B., and Burri, P. H. (1991). Mammary epithelial differentiation in vitro: minimum requirements for a functional response to hormonal stimulation. *Cell Growth Differ.* 2, 549–559.
161. Strange, R., Li, F., Saurer, S., Burkhardt, A., and Friis, R. R. (1992). Apoptotic cell death and tissue remodeling during mouse mammary gland involution. *Development* 115, 49–58.
162. Clermont, Y., Xia, L., Rambourg, A., Turner, J. D., and Hermo, L. (1993). Structure of the Golgi apparatus in stimulated and nonstimulated acinar cells of mammary glands of the rat. *Anat. Rec.* 237, 308–317.
163. Li, M. L., Liu, X. W., Robinson, G., Bar-Peled, U., Wagner, K. U., Young, W. S., Hennighausen, L., and Furth, P. A. (1997). Mammary-derived signals activate programmed cell death during the first stage of mammary gland involution. *Proc. Natl. Acad. Sci. U S A* 94, 3425–3430.
164. Walton, K. D., Wagner, K. U., Rucker, E. B., 3rd, Shillingford, J. M., Miyoshi, K., and Hennighausen, L. (2001). Conditional deletion of the bcl-x gene from mouse mammary epithelium results in accelerated apoptosis during involution but does not compromise cell function during lactation. *Mech. Dev.* 109, 281–293.
165. Tonner, E., Barber, M. C., Travers, M. T., Logan, A., and Flint, D. J. (1997). Hormonal control of insulin-like growth factor-binding protein-5 production in the involuting mammary gland of the rat. *Endocrinology* 138, 5101–5107.
166. Schorr, K., Li, M., Krajewski, S., Reed, J. C., and Furth, P. A. (1999). Bcl-2 gene family and related proteins in mammary gland involution and breast cancer. *J. Mammary Gland Biol. Neoplasia* 4, 153–164.
167. Schorr, K., Li, M., Bar-Peled, U., Lewis, A., Heredia, A., Lewis, B., Knudson, C. M., Korsmeyer, S. J., Jager, R., Weiher, H., and Furth, P. A. (1999). Gain of Bcl-2 is more potent than bax loss in regulating mammary epithelial cell survival in vivo. *Cancer Res.* 59, 2541–2545.
168. Pullan, S., Wilson, J., Metcalfe, A., Edwards, G. M., Goberdhan, N., Tilly, J., Hickman, J. A., Dive, C., and Streuli, C. H. (1996). Requirement of basement membrane

for the suppression of programmed cell death in mammary epithelium. *J. Cell Sci.* 109, 631–642.

169. Schwertfeger, K. L., Richer, M. M., and Anderson, S. M. (2001). Mammary gland involution is delayed by activated Akt in transgenic mice. *Mol. Endocrinol.* 15, 867–881.
170. Humphreys, R. C., Bierie, B., Zhao, L., Raz, R., Levy, D., and Hennighausen, L. (2002). Deletion of Stat3 blocks mammary gland involution and extends functional competence of the secretory epithelium in the absence of lactogenic stimuli. *Endocrinology* 143, 3641–3650.
171. Chapman, R. S., Lourenco, P. C., Tonner, E., Flint, D. J., Selbert, S., Takeda, K., Akira, S., Clarke, A. F., and Watson, C. J. (1999). Suppression of epithelial apoptosis and delayed mammary gland involution in mice with a conditional knockout of Stat3. *Genes Dev.* 13, 2604–2616.
172. Clermont, Y., Xia, L., Rambourg, A., Turner, J. D., and Hermo, L. (1993). Transport of casein submicelles and formation of the secretion granules in the Golgi apparatus of epithelial cells of the lactating mammary gland of the rat. *Anat. Rec.* 235, 363–373.
173. Mather, I. H., and Keenan, T. S. (1998). Origin and secretion of milk lipids. *J. Mammary Gland Biol. Neoplasia* 3, 259–274.
174. Shennan, D. B., and Peaker, M. (2000). Transport of milk constituents by the mammary gland. *Physiol. Rev.* 80, 925–951.
175. Kraehenbuhl, J.-P., and Hunziker, W. (1998). Epithelial transcytosis of immunoglobulins. *J. Mammary Gland Biol. Neoplasia* 3, 289–304.
176. Monks, J., and Neville, M. C. (2004). Albumin transcytosis across the epithelium of the lactating mouse mammary gland. *J. Physiol. (Lond).* 560, 267–280.
177. Huston, G. E., and Patton, S. (1990). Factors related to the formation of cytoplasmic crescents on milk fat globules. *J. Dairy Sci.* 73, 2061–2066.
178. Ito, S., and Lee, A. (2003). Drug excretion into breast milk: overview. *Adv. Drug Deliv. Rev.* 55, 617–627.
179. Fleishaker, J. C. (2003). Models and methods for predicting drug transfer into human milk. *Adv. Drug Deliv. Rev.* 55, 643–652.
180. Ito, S., and Alcorn, J. (2003). Xenobiotic transporter expression and function in the human mammary gland. *Adv. Drug Deliv. Rev.* 55, 653–665.
181. Elstrom, R. L., Bauer, D. E., Buzzai, M., Karnauskas, R., Harris, M. H., Plas, D. R., Zhuang, H., Cinalli, R. M., Alavi, A., Rudin, C. M., and Thompson, C. B. (2004). Akt stimulates aerobic glycolysis in cancer cells. *Cancer Res.* 64, 3892–3899.
182. Soulier, S., Lepourry, L., Stinnakre, M. G., Mercier, J. C., and Vilotte, J. L. (1997). Expression of a bovine α-lactalbumin transgene in α-lactalbumin-deficient mice can rescue lactation: in vivo relationship between bovine α-lactalbumin expression content and milk composition. *J. Dairy Res.* 64, 145–148.
183. Kuhn, N. J., and White, A. (1975). Milk glucose as an index of the intracellular glucose concentration of rat mammary gland. *Biochem. J.* 152, 153–155.
184. Neville, M. C., Hay, W. W., Jr., and Fennessey, P. (1990). Physiological significance of the concentration of human milk glucose. *Protoplasma* 159, 118–128.
185. Neville, M. C. (1995). Lactogenesis in women: evidence for a cascade of cellular events. In *Handbook of Composition of Milks* (R. G. Jensen, Ed.), pp. 87–98. Academic Press, San Diego.
186. Neville, M. C., Sawicki, V., and Hay, W. W., Jr. (1993). Effect of fasting, elevated plasma glucose and plasma insulin concentrations on milk secretion in women. *J. Endocrinol.* 139, 165–173.
187. Faulkner, A., and Peaker, M. (1987). Regulation of mammary glucose metabolism in lactation. In *The Mammary Gland: Development, Regulation and Function* (M. C. Neville and C. W. Daniel, Eds.), pp. 536–562. Plenum Press, New York.
188. Macheda, M. L., Williams, E. D., Best, J. D., Wlodek, M. E., and Rogers, S. (2003). Expression and localization of GLUT1 and GLUT12 glucose transporter in the pregnant and lactating rat mammary gland. *Cell Tissue Res.* 311, 91–97.
189. Prosser, C. G., and Topper, Y. J. (1986). Changes in the rate of carrier-mediated glucose transport by mouse mammary epithelial cells during ontogeny: hormone dependence delineated in vitro. *Endocrinology* 119, 91–96.
190. Smith, S. (1994). The animal fatty acid synthase: one gene, one polypeptide, seven enzymes. *FASEB J.* 8, 1248–1259.
191. Horton, J. D., Goldstein, J. L., and Brown, M. S. (2002). SREBPs: activators of the complete program of cholesterol and fatty acid synthesis in the liver. *J. Clin. Invest.* 109, 1125–1135.
192. Filippa, N., Sable, C. L., Hemmings, B. A., and Van Obberghen, E. (2000). Effect of phosphoinositide-dependent kinase 1 on protein kinase B translocation and its subsequent activation. *Mol. Cell. Biol.* 20, 5712–5721.
193. Scheid, M. P., and Woodgett, J. R. (2001). PKB/AKT: functional insights from genetic models. *Nat. Rev. Mol. Cell. Biol.* 2, 760–768.
194. Whiteman, E. L., Cho, H., and Birnbaum, M. J. (2002). Role of Akt/protein kinase B in metabolism. *Trends Endocrinol. Metab.* 13, 444–451.
195. Downward, J. (2004). PI 3–kinase, Akt and cell survival. *Semin. Cell Dev. Biol.* 15, 177–182.
196. Marte, B. M., and Downward, J. (1997). PKB/Akt: connecting phosphoinositide 3-kinase to cell survival and beyond. *Trends Biochem. Sci.* 22, 355–358.
197. Jiang, Z. Y., Zhou, Q. L., Coleman, K. A., Chouinard, M., Boese, Q., and Czech, M. P. (2003). Insulin signaling through Akt/protein kinase B analyzed by small interfering RNA-mediated gene silencing. *Proc. Natl. Acad. Sci. U S A* 100, 7569–7574.
198. Katome, T., Obata, T., Matsushima, R., Masuyama, N., Cantley, L. C., Gotoh, Y., Kishi, K., Shiota, H., and Ebina, Y. (2003). Use of RNA interference-mediated gene silencing and adenoviral overexpression to elucidate the roles of AKT/protein kinase B isoforms in insulin actions. *J. Biol. Chem.* 278, 28312–28323.
199. Chodosh, L. A., Gardner, H. P., Rajan, J. V., Stairs, D. B., Marquis, S. T., and Leder, P. A. (2000). Protein kinase expression during murine mammary development. *Dev. Biol.* 219, 259–276.
200. Rathmell, J. C., Fox, C. J., Plas, D. R., Hammerman, P. S., Cinalli, R. M., and Thompson, C. B. (2003). Akt-directed glucose metabolism can prevent Bax conformation change and promote growth factor-independent survival. *Mol. Cell. Biol.* 23, 7513–7328.
201. Millar, I. D., Calvert, D. T., Lomax, M. A., and Shennan, D. B. (1996). The mechanism of L-glutamate transport by lactating rat mammary tissue. *Biochim. Biophys. Acta* 1282, 200–206.
202. Millar, I. D., Calvert, D. T., Lomax, M. A., and Shennan, D. B. (1997). Substrate specificity of the mammary tissue anionic amino acid carrier operating in the cotransport and exchange modes. *Biochim. Biophys. Acta* 1326, 92–102.
203. Shennan, D. B., and McNeillie, S. A. (1994). High affinity (Na^+ + Cl^-)–dependent taurine transport by lactating mammary tissue. *J. Dairy Res.* 61, 335–343.
204. Shennan, D. B. (1995). Identification of a high affinity taurine transporter which is not dependent on chloride. *Biosci. Rep.* 15, 231–239.
205. Neville, M. C., Lobitz, C. J., Ripoll, E. A., and Tinney, C. (1980). The sites for α-aminoisobutyric acid uptake in normal mammary gland and ascites tumor cells. *J. Biol. Chem.* 255, 7311–7316.

206. Verma, N., and Kansal, V. K. (1993). Characterisation of the routes of methionine transport in mouse mammary glands. *Indian J. Med. Res.* 98, 297–304.
207. Laspiur, J. P., Burton, J. L., Weber, P. S., Kirkwood, R. N., and Trottier, N. L. (2004). Short communication: amino acid transporters in porcine mammary gland during lactation. *J. Dairy Sci.* 87, 3235.
208. Kansal, R., and Kansal, V. K. (1996). Lack of functional correlation between gamma-glutamyl transpeptidase and amino acid transport in the lactating mouse mammary gland. *Indian J. Exp. Biol.* 34, 267–269.
209. Hediger, M. A., Romero, M. F., Peng, J. B., Rolfs, A., Takanaga, H., and Bruford, E. A. (2004). The ABCs of solute carriers: physiological, pathological and therapeutic implications of human membrane transport proteins. Introduction. *Pflugers Arch.* 447, 465–468.
210. Kanai, Y., and Hediger, M. A. (2004). The glutamate/neutral amino acid transporter family SLC1: molecular, physiological and pharmacological aspects. *Pflugers Arch.* 447, 469–479.
211. Verrey, F., Closs, E. I., Wagner, C. A., Palacin, M., Endou, H., and Kanai, Y. (2004). CATs and HATs: the SLC7 family of amino acid transporters. *Pflugers Arch.* 447, 532–542.
212. Rassin, D. K., Sturman, J. A., and Guall, G. E. (1978). Taurine and other free amino acids in milk of man and other mammals. *Early Hum. Dev.* 2, 1–13.
213. Shennan, D. B., McNeillie, S. A., and Cliff, M. J. (1995). The relationship between volume-sensitive taurine efflux and the anion-exchanger in lactating mammary tissue. *Biochem. Soc. Trans.* 23, 392S.
214. Groneberg, D. A., Doring, F., Theis, S., Nickolaus, M., Fischer, A., and Daniel, H. (2002). Peptide transport in the mammary gland: expression and distribution of PEPT2 mRNA and protein. *Am. J. Physiol. Endocrinol. Metab.* 282, E1172–1179.
215. Atkinson, S. A., and Lonnerdal, B. (1995). Nonprotein nitrogen fractions of human milk. In *Handbook of Milk Composition* (R. G. Jensen, Ed.), pp. 369–387. Academic Press, San Diego.
216. Vanhatalo, A., Varvikko, T., and Huhtanen, P. (2003). Effects of various glucogenic sources on production and metabolic responses of dairy cows fed grass silage-based diets. *J. Dairy Sci.* 86, 3249–3259.
217. Serkova, N., Bendrick-Peart, J., Alexander, B., and Tissot Van Patot, M. C. (2003). Metabolite concentrations in human term placentae and their changes due to delayed collection after delivery. *Placenta* 24, 227–235.
218. Wooding, R. B. P. (1977). Comparative mammary fine structure. In *Comparative Aspects of Lactation* (M. Peaker, Ed.) pp. 1–41. Academic Press, London.
219. Hollman, K. H. (1974). Cytology and fine structure of the mammary gland. In *Lactation* (B. H. Larson and V. R. Smith, Eds.), Vol. 1, pp. 3–95. Academic Press, New York.
220. Wu, C. C., Yates, J. R., III, Neville, M. C., and Howell, K. E. (2000). Proteomic analysis of two functional states of the Golgi complex in mammary epithelial cells. *Traffic* 1, 769–782.
221. Glick, B. S., Elston, T., and Oster, G. (1997). A cisternal maturation mechanism can explain the asymmetry of the Golgi stack. *FEBS Lett.* 414, 177–181.
222. Rothman, J. E., Miller, R. L., and Urbani, L. J. (1984). Intercompartmental transport in the Golgi complex is a dissociative process: facile transfer of membrane protein between two Golgi populations. *J. Cell Biol.* 99, 260–271.
223. Pelham, H. R. (2001). Traffic through the Golgi apparatus. *J. Cell Biol.* 155, 1099–1101.
224. Mironov, A. A., Beznoussenko, G. V., Nicoziani, P., Martella, O., Trucco, A., Kweon, H. S., Di Giandomenico, D., Polishchuk, R. S., Fusella, A., Lupetti, P., Berger, E. G., Geerts, W. J., Koster, A. J., Burger, K. N., and Luini, A. (2001). Small cargo proteins and large aggregates can traverse the Golgi by a common mechanism without leaving the lumen of cisternae. *J. Cell Biol.* 155, 1225–1238.
225. Kweon, H. S., Beznoussenko, G. V., Micaroni, M., Polishchuk, R. S., Trucco, A., Martella, O., Di Giandomenico, D., Marra, P., Fusella, A., Di Pentima, A., Berger, E. G., Geerts, W. J., Koster, A. J., Burger, K. N., Luini, A., and Mironov, A. A. (2004). Golgi enzymes are enriched in perforated zones of Golgi cisternae but are depleted in COPI vesicles. *Mol. Biol. Cell* 15, 4710–4724.
226. Mogelsvang, S., Marsh, B. J., Ladinsky, M. S., and Howell, K. E. (2004). Predicting function from structure: 3D structure studies of the mammalian Golgi complex. *Traffic* 5, 338–345.
227. Wu, C. C., Howell, D. K., Neville, M. C., Yates, J. R., III, and McManaman, J. L. (2000). Proteomics reveal a link between the endoplasmic reticulum and lipid secretory mechanisms in mammary epithelial cells. *Electrophoresis* 21, 3470–3482.
228. Brew, K., and Hill, R. L. (1975). Lactose biosynthesis. *Rev. Physiol. Biochem. Pharmacol.* 72, 103–158.
229. Kuhn, N. J., and White, A. (1975). The topography of lactose synthesis. *Biochem. J.* 148, 77–84.
230. Strous, G. J. (1986). Golgi and secreted galactosyltransferase. *CRC Crit. Rev. Biochem.* 21, 119–151.
231. Russo, R. N., Shaper, N. L., and Shaper, J. H. (1990). Bovine β1,4-galactosyltransferase: two sets of mRNA transcripts encode two forms of the protein with different aminoterminal domains: in vitro translation experiments demonstrate that both the short and the long forms of the enzyme are type II membrane-bound glycoproteins. *J. Biol. Chem.* 265, 3324–3331.
232. Shaper, J. H., Harduin-Lepers, A., Rajput, B., and Shaper, N. L. (1995). Murine beta 1,4-galactosyltransferase: analysis of a gene that serves both a housekeeping and a cell specific function. *Adv. Exp. Med. Biol.* 376, 95–104.
233. Kuhn, N. J., and White, A. (1976). Evidence for specific transport of uridine diphosphate galactose across the Golgi membrane of rat mammary gland. *Biochem. J.* 154, 243–244.
234. Abeijon, C., Mandon, E. C., and Hirschberg, C. B. (1997). Transporters of nucleotide sugars, nucleotide sulfate and ATP in the Golgi apparatus. *Trends Biochem. Sci.* 22, 203–207.
235. Nemeth, B. A., Tsang, S. W. Y., Geske, R. S., and Haney, P. M. (2000). Golgi targeting of the GLUT1 glucose transporter in lactating mouse mammary gland. *Pediatr. Res.* 47, 444–450.
236. Haney, P. M. (2001). Localization of the GLUT1 glucose transporter to brefeldin A sensitive vesicles of differentiated CIT3 mouse mammary epithelial cells. *Cell Biol. Int.* 25, 277–288.
237. Zhao, F.-Q., Miller, P. J., Wall, E. H., Zheng, U.-C., Dong, M. S., Neville, M. C., and McFadden, T. B. (2004). Bovine glucose transporter GLUT8: cloning, expression, and developmental regulation in mammary gland. *Biochim. Biophys. Acta* 1680, 103–113.
238. Obermeier, S., Huselweh, B., Tinel, H., Kinne, R. H., and Kunz, C. (2000). Expression of glucose transporters in lactating human mammary gland epithelial cells. *Eur J. Nutr.* 39, 194–200.
239. Simpson, K. J., Ranganathan, S., Fisher, J. A., Janssens, P. A., Shaw, D. C., and Nicholas, K. R. (2000). The gene for a novel member of the whey acidic protein family encodes three four-disulfide core domains and is asynchronously expressed during lactation. *J. Biol. Chem.* 275, 23074–23081.
240. Nukumi, N., Ikeda, K., Osawa, M., Iwamori, T., Naito, K., and Tojo, H. (2004). Regulatory function of whey acidic protein in the proliferation of mouse mammary epithelial cells in vivo and in vitro. *Dev. Biol.* 274, 31–44.
241. Ikeda, K., Nukumi, N., Iwamori, T., Osawa, M., Naito, K., and Tojo, H. (2004). Inhibitory function of whey acidic

protein in the cell-cycle progression of mouse mammary epithelial cells (EpH4/K6 cells). *J. Reprod. Dev.* 50, 87–96.

242. Legrand, D., Elass, E., Pierce, A., and Mazurier, J. (2004). Lactoferrin and host defence: an overview of its immuno-modulating and anti-inflammatory properties. *Biometals* 17, 225–229.
243. Kontopidis, G., Holt, C., and Sawyer, L. (2004). Invited review: beta-lactoglobulin: binding properties, structure, and function. *J. Dairy Sci.* 87, 785–796.
244. Rytkonen, J., Karttunen, T. J., Karttunen, R., Valkonen, K. H., Jenmalm, M. C., Alatossava, T., Bjorksten, B., and Kokkonen, J. (2002). Effect of heat denaturation on beta-lactoglobulin-induced gastrointestinal sensitization in rats: denatured βLG induces a more intensive local immunologic response than native βLG. *Pediatr. Allergy Immunol.* 13, 269–277.
245. Neville, M. C., Keller, R. P., Casey, C. E., and Allen, J. C. (1994). Calcium partitioning in human and bovine milk. *J. Dairy Sci.* 77, 1964–1975.
246. Linzell, J. L., and Peaker, M. (1975). The distribution and movements of carbon dioxide, carbonic acid and bicarbonate between blood and milk in the goat. *J. Physiol. (Lond).* 244, 771–782.
247. Huston, G. E., and Patton, S. (1986). Membrane distribution in human milks throughout lactation as revealed by phospholipid and cholesterol analyses. *J. Pediatr. Gastroenterol. Nutr.* 5, 602–607.
248. Griffin, M. C. A., Lyster, R. L., and Price, J. C. (1988). The disaggregation of calcium-depleted casein micelles. *Eur. J. Biochem.* 174, 339–343.
249. Blake, H. H., and Henning, S. J. (1988). Absorption and transport of milk calcium by infant rats. *Am. J. Physiol.* 254, G12–G19.
250. Ruegg, M., and Blanc, B. (1982). Structure and properties of the particulate constituents of human milk: a review. *Food Microstruc.* 1, 25–47.
251. Neville, M. C., Allen, J. C., and Zhang, P. (1995). Ionic interactions in milk. In *Handbook of Milk Composition* (R. G. Jensen, Ed.), pp. 577–592. Academic Press, San Diego.
252. Farrell, H. M. (1988). Physical equilibria: proteins. In *Fundamentals of Dairy Chemistry* (N. P. Wong, Ed.), pp. 461–509. Van Nostrand Reinhold, New York.
253. Holt, C., Timmins, P. A., Errington, N., and Leaver, J. (1998). A core-shell model of calcium phosphate nanoclusters stabilized by beta-casein phosphopeptides, derived from sedimentation equilibrium and small-angle X-ray and neutron- scattering measurements. *Eur. J. Biochem.* 252, 73–78.
254. Lonnerdal, B., and Lien, E. L. (2003). Nutritional and physiologic significance of α-lactalbumin in infants. *Nutr. Rev.* 61, 295–305.
255. Segawa, T., and Sugai, S. (1983). Interactions of divalent metal ions with bovine, human, and goat α-lactalbumins. *J. Biochem. (Tokyo)* 93, 1321–1328.
256. Casey, C. E., and Hambidge, K. M. (1983). Nutritional aspects of human lactation. In *Lactation: Physiology, Nutrition and Breast-feeding* (M. C. Neville and M. A. Neifert, Eds.), pp. 199–248. Plenum Press, New York.
257. White, J. C. D., and Davies, D. T. (1958) The relation between the chemical composition of milk and the stability of the caseinate complex: I. General introduction, description of samples, methods, and chemical composition of samples. *J. Dairy Res.* 30, 171–189.
258. Holt, C., Dalgleish, D. G., and Jenness, R. (1981). Calculation of the ion equilibria in milk diffusate and comparison with experiment. *Anal. Biochem.* 113, 154–163.
259. Little, E. M., and Holt, C. (2004). An equilibrium thermodynamic model of the sequestration of calcium phosphate by casein phosphopeptides. *Eur. Biophys. J.* 33, 435–447.
260. Allen, J. C., and Neville, M. C. (1983). Ionized calcium in human milk determined with a calcium-selective electrode. *Clin. Chem.* 29, 858–861.
261. Swaisgood, H. E. (1995). Protein and amino acid composition of bovine milk. In *Handbook of Milk Composition* (R. G. Jensen, Ed.), pp. 464–467. Academic Press, San Diego.
262. Slattery, C. W. (1976). Review: Casein micelle structure: an examination of models. *J. Dairy Sci.* 59, 1547–1556.
263. Schmidt, D. G. (1982). Association of caseins and casein micelle structure. In *Developments in Dairy Chemistry: 1. Proteins* (P. F. Fox, Ed.), pp. 61–86. Applied Science Publishers, London.
264. Kumosinski, T. F., Brown, E. M., and Farrell, H. M., Jr. (1993). Three-dimensional molecular modeling of bovine caseins: a refined, energy-minimized kappa-casein structure. *J. Dairy Sci.* 76, 2507–2520.
265. Kakalis, L. T., Kumosinski, T. F., and Farrell, H. M., Jr. (1990). A multinuclear, high-resolution NMR study of bovine casein micelles and submicelles. *Biophys. Chem.* 38, 87–98.
266. Holt, C., and Jenness, R. (1984). Interrelationships of constituents and partition of salts in milk samples from eight species. *Comp. Biochem. Physiol. A* 77, 275–282.
267. Groves, M. L., and Gordon, W. G. (1970) The major component of human casein: a protein phosphorylated at different levels. *Arch. Biochem. Biophys.* 140, 47–51.
268. Chen, D. C., Nommsen-Rivers, L., Dewey, K. G., and Lonnerdal, B. (1998). Stress during labor and delivery and early lactation performance. *Am. J. Clin. Nutr.* 68, 335–344.
269. Neville, M. C., Selker, F., Semple, K., and Watters, C. (1981). ATP-dependent calcium transport by a Golgi-enriched membrane fraction from mouse mammary gland. *J. Membr. Biol.* 61, 97–105.
270. West, D. W. (1981). Energy-dependent calcium sequestration activity in a Golgi apparatus fraction derived from lactating rat mammary gland. *Biochim. Biophys. Acta* 673, 374–386.
271. Baumrucker, C. R., and Keenan, T. W. (1975). Membranes of the mammary gland: X. ATP-dependent calcium accumulation by a Golgi apparatus-rich fraction from bovine mammary gland. *Exp. Cell Res.* 90, 253–260.
272. Neville, M. C., and Peaker, M. (1979). The secretion of calcium and phosphorus into milk. *J. Physiol. (Lond).* 313, 561–570.
273. Linzell, J. L., Mepham, T. B., and Peaker, M. (1976). The secretion of citrate into milk. *J. Physiol. (Lond).* 260, 739–750.
274. Waugh, D. F., and Talbot, B. (1971) Equilibrium casein micelle systems. *Biochemistry* 10, 4153–4162.
275. Reinhardt, T. A., Filoteo, A. G., Penniston, J. T., and Horst, R. L. (2000). Ca(2+)-ATPase protein expression in mammary tissue. *Am. J. Physiol. Cell Physiol.* 279, C1595–C1602.
276. Reinhardt, T. A., Lippolis, J. D., Shull, G. E., and Horst, R. L. (2004). Null mutation in the gene encoding plasma membrane Ca^{2+}-ATPase isoform 2 impairs calcium transport into milk. *J. Biol. Chem.* 279, 42369–42373.
277. Watters, C. (1984). A Ca^{2+}-stimulated adenosine triphosphatase in Golgi-enriched membranes of lactating murine mammary tissue. *Biochem. J.* 224, 39–45.
278. Taylor, R. S., Jones, S. M., Dahl, R. H., Nordeen, M. H., and Howell, K. E. (1997). Characterization of the Golgi complex cleared of proteins in transit and examination of calcium uptake activities. *Mol. Biol. Cell* 8, 1911–1931.
279. Wuytack, F., Raeymaekers, L., and Missiaen, L. (2003). PMR1/SPCA Ca^{2+} pumps and the role of the Golgi apparatus as a Ca^{2+} store. *Pflugers Arch.* 446, 148–153.
280. Bingham, E. W., McGranaghan, M. B., Wickham, E. D., Leung, C. T., and Farrell, H. M., Jr. (1993). Properties of $[Ca^{2+} + Mg^{2+}]$-adenosine triphosphatases in the Golgi apparatus and microsomes of the lactating mammary glands of cows. *J. Dairy Sci.* 76, 393–400.

281. Reinhardt, T. A., and Horst, R. L. (1999). Ca^{2+}-ATPases and their expression in the mammary gland of pregnant and lactating rats. *Am. J. Physiol.* 276, C796–802.
282. Witsell, D. L., Casey, C. E., and Neville, M. C. (1990). Divalent cation galactosyltransferase in native mammary golgi vesicles. *J. Biol. Chem.* 265, 15731–15737.
283. Kuhn, N. J., Ward, S., and Leong, W. S. (1991). Submicromolar manganese dependence of Golgi vesicular galactosyltransferase (lactose synthetase). *Eur. J. Biochem.* 195, 243–250.
284. Brommage, R. (1989). Measurement of calcium and phosphorus fluxes during lactation in the rat. *J. Nutr.* 119, 428–438.
285. Shillingford, J. M., Calvert, D. T., Shennan, D. B., and Beechey, R. B. (1996). Phosphate transport via Na^+-Pi cotransport and anion exchange in lactating rat mammary tissue. *Exp Physiol* 81, 273–284.
286. Miyoshi, K., Shillingford, J. M., Smith, G. H., Grimm, S. L., Wagner, K. U., Oka, T., Rosen, J. M., Robinson, G. W., and Hennighausen, L. (2001). Signal transducer and activator of transcription (Stat) 5 controls the proliferation and differentiation of mammary alveolar epithelium. *J. Cell Biol.* 155, 531–542.
287. Shillingford, J. M., Miyoshi, K., Robinson, G. W., Grimm, S. L., Rosen, J. M., Neubauer, H., Pfeffer, K., and Hennighausen, L. G. (2002). Jak2 is an essential tryosine kinase involved in pregnancy-mediated development of mammary secretory epithelium. *Molecular Endo.* 16(3), 563–570.
288. Murer, H., Forster, I., and Biber, J. (2004). The sodium phosphate cotransporter family SLC34. *Pflugers Arch.* 447, 763–767.
289. Shillingford, J. M., Miyoshi, K., Robinson, G. W., Bierie, B., Cao, Y., Karin, M., and Hennighausen, L. (2003). Proteotyping of mammary tissue from transgenic and gene knockout mice with immunohistochemical markers: a tool to define developmental lesions. *J. Histochem. Cytochem.* 51, 555–565.
290. Salaun, C., Gyan, E., Rodrigues, P., and Heard, J. M. (2002). Pit2 assemblies at the cell surface are modulated by extracellular inorganic phosphate concentration. *J. Virol.* 76, 4304–4311.
291. Vanhouten, J., Dann, P., McGeoch, G., Brown, E. M., Krapcho, K., Neville, M. C., and Wysolmerski, J. J. (2004). The calcium-sensing receptor regulates mammary gland parathyroid hormone related protein production and calcium transport. *J. Clin. Invest.* 113, 598–608.
292. Vanhouten, J. (2005). Calcium sensing by the mammary gland. *J. Mammary Gland Biol. Neoplasia* 10, 129–139.
293. Turner, M. D., Rennison, M. E., Handel, S. E., Wilde, C. J., and Burgoyne, R. D. (1992). Proteins are secreted by both constitutive and regulated secretory pathways in lactating mouse mammary epithelial cells. *J. Cell Biol.* 117, 269–278.
294. Burgoyne, R. D., Duncan, J. S., and Sudlow, A. W. (1998). Role of calcium in the pathway for milk protein secretion and possible relevance for mammary gland physiology. *Biochem. Soc. Symp.* 63, 91–100.
295. Shennan, D. B., Grant, A. C., and Gow, I. F. (2002). The effect of hyposmotic and isosmotic cell swelling on the intracellular [Ca^{2+}] in lactating rat mammary acinar cells. *Mol. Cell. Biochem.* 233, 91–97.
296. Sudlow, A. W., and Burgoyne, R. D. (1997). A hypo-osmotically induced increase in intracellular Ca2+ in lactating mouse mammary epithelial cells involving Ca^{2+} influx. *Pflugers Arch.* 433, 609–616.
297. Wang, W., Wyckoff, J. B., Frohlich, V. C., Oleynikov, Y., Huttelmaier, S., Zavadil, J., Cermak, L., Bottinger, E. P., Singer, R. H., White, J. G., Segall, J. E., and Condeelis, J. S. (2002). Single cell behavior in metastatic primary mammary tumors correlated with gene expression patterns revealed by molecular profiling. *Cancer Res.* 62, 6278–6288.
298. Nguyen, D.-A. D., Beeman, N. G., Lewis, M. T., Schaack, J., and Neville, M. C. (2000). Intraductal injection into the mouse mammary gland. In *Methods in Mammary Gland Biology and Breast Cancer Research* (M. M. Ip and B. B. Asch, Eds.), pp. 259–270. Kluwer Academic/Plenum, New York.
299. Russell, T., Fischer, A., Beeman, N., Freed, E., Neville, M. C., and Schaack, J. (2003). Transduction of the mouse mammary epithelium with adenoviral vectors in vivo. *J. Virol.* 77, 5801–5809.
300. Hofer, A. M., Landolfi, B., Debellis, L., Pozzan, T., and Curci, S. (1998). Free [Ca^{2+}] dynamics measured in agonist-sensitive stores of single living intact cells: a new look at the refilling process. *EMBO J.* 17, 1986–1995.
301. Caroppo, R., Colella, M., Colasuonno, A., Deluisi, A., Debellis, L., Curci, S., and Hofer, A. M. (2003). A reassessment of the effects of luminal [Ca^{2+}] on inositol 1,4,5–trisphosphate-induced Ca^{2+} release from internal stores. *J. Biol. Chem.* 278, 39503–39508.
302. Bauman, D. E., and Griinari, J. M. (2003). Nutritional regulation of milk fat synthesis. *Annu. Rev. Nutr.* 23, 203–227.
303. Neville, M. C., and Picciano, M. F. (1997). Regulation of milk lipid synthesis and composition. *Annu. Rev. Nutr.* 17, 159–184.
304. Borgnia, M., Nielson, S., Engel, A., and Agre, P. (1999). Cellular and molecular biology of the aquaporin water channels. *Annu. Rev. Biochem.* 68, 425–458.
305. Aoki, N., Ishii, T., Ohira, S., Yamaguchi, Y., Negi, M., Adachi, T., Nakamura, R., and Matsuda, T. (1997). Stage specific expression of milk fat globule membrane glycoproteins in mouse mammary gland: Comparison of MFG-E8, butyrophilin, and CD36 with a major milk protein, β-casein. *Biochim. Biophys. Acta* 1334, 182–190.
306. Abumrad, N. A., El-Maghrabi, M. R., Amri, E.-Z., Lopez, E., and Grimaldi, P. (1993). Cloning of a rat adipocyte membrane protein implicated in binding or transport of long-chain fatty acids that is induced during preadipocyte differentiation. *J. Biol. Chem.* 268, 17665–17669.
307. West, C. E., Bickerstaffe, R., Annison, E. F., and Linzell, J. L. (1972) Studies on the mode of uptake of blood triglycerides by the mammary gland of the lactating goat: the uptake and incorporation into milk fat and mammary lymph of labelled glycerol, fatty acids and triglycerides. *Biochem. J.* 126, 477–490.
308. Oakes, N. E., and Furler, S. M. (2002). Evaluation of free fatty acid metabolism in vivo. *Ann. NY Acad. Sci.* 967, 158–175.
309. Bansal, M. P., and Medina, D. (1993). Expression of fatty acid-binding proteins in the developing mouse mammary gland. *Biochem. Biophys. Res. Commun.* 191, 61–69.
310. Knudsen, J. (1990). Acyl-CoA-binding protein (ACBP) and its relation to fatty acid-binding protein (FABP): an overview. *Mol. Cell. Biochem.* 98, 217–223.
311. Deeney, J. T., Valivullay, H. M., Dapper, C. H., Dylewski, D. P., and Keenan, T. W. (1985). Microlipid droplets in milk secreting mammary epithelial cells: Evidence that they originate form endoplasmic reticulum and are precursors of milk lipid globules. *Eur. J. Cell Biol.* 38, 16–26.
312. Jensen, R. G., Bitman, J., Carlson, S. E., Couch, S. C., Hamosh, M., and Newburg, D. S. (1995). Milk lipids: A. Human milk lipids. In *Handbook of Milk Composition* (R. G. Jensen, Ed.), pp. 495–542. Academic Press, San Diego.
313. Mather, I. H. (2000). A review and proposed nomenclature for major proteins of the milk-fat globule membrane. *J. Dairy Sci.* 83, 203–247.
314. Ogg, S. L., Weldon, A. K., Dobbie, L., Smith, A. J., and Mather, I. H. (2004). Expression of butyrophilin (Btn1a1) in lactating mammary gland is essential for the regulated secretion of milk-lipid droplets. *Proc. Natl. Acad. Sci. U S A* 101, 10084–10089.
315. McManaman, J. L., and Bain, D. L. (2002). Structural and conformational analysis of the oxidase to dehydrogenase conversion of xanthine oxidoreductase. *J. Biol. Chem.* 277, 21261–21268.

316. Vorbach, C., Scriven, A., and Capecchi, M. R. (2002). The housekeeping gene xanthine oxidoreductase is necessary for milk fat droplet enveloping and secretion: gene sharing in the lactating mammary gland. *Genes Dev.* 16, 3223–3235.
317. Heid, H. W., Schnolzer, M., and Keenan, T. W. (1996). Adipocyte differentiation-related protein is secreted into milk as a constituent of milk lipid globule membrane. *Biochem. J.* 320, 1025–1030.
318. McManaman, J. L., Zabaronick, W., Schaack, J., and Orlicky, D. J. (2003). Lipid droplet targeting domains of adipophilin. *J. Lipid Res.* 44, 668–673.
319. Heid, H. W., Moll, R., Schwetlick, I., Rackwitz, H. R., and Keenan, T. W. (1998). Adipophilin is a specific marker of lipid accumulation in diverse cell types and diseases. *Cell Tissue Res.* 294, 309–321.
320. Bonen, A., Campbell, S. E., Benton, C. R., Chabowski, A., Coort, S. L., Han, X. X., Koonen, D. P., Glatz, J. F., and Luiken, J. J. (2004). Regulation of fatty acid transport by fatty acid translocase/CD36. *Proc. Nutr. Soc.* 63, 245–249.
321. Hanayama, R., Tanaka, M., Miyasaka, K., Aozasa, K., Koike, M., Uchiyama, Y., and Nagata, S. (2004). Autoimmune disease and impaired uptake of apoptotic cells in MFG-E8-deficient mice. *Science* 304, 1147–1150.
322. Hanayama, R., Tanaka, M., Miwa, K., and Nagata, S. (2004). Expression of developmental endothelial locus-1 in a subset of macrophages for engulfment of apoptotic cells. *J. Immunol.* 172, 3876–3882.
323. Monks, J., Rosner, D., Geske, J., Lehman, L., Hanson, L., Neville, M. C., and Fadok, V. A. (2005). Epithelial cells as phagocytes: apoptotic epithelial cells are engulfed by mammary alveolar epithelial cells and repress inflammatory mediator release. *Cell Death Differ.* 12, 107–114
324. Zschiesche, W., Kleine, A. H., Spitzer, E., Veerkamp, J. H., and Glatz, J. F. C. (1996). Histochemical localization of heart-type fatty-acid binding protein in human and murine tissues. *Histochem. Cell Biol.* 103, 147–156.
325. Ishii, T., Aoki, N., Noda, A., Adachi, T., Nakamura, R., and Matsuda, T. (1995). Carboxy-terminal cytoplasmic domain of mouse butyrophilin specifically associates with a 150-kDa protein of mammary epithelial cells and milk fat globule membrane. *Biochim. Biophys. Acta* 1245, 285–292.
326. McManaman, J. L., Neville, M. C., and Wright, R. M. (1999). Mouse mammary gland xanthine oxidoreductase: purification, characterization and regulation. *Arch. Biochem. Biophys.* 371, 308–316.
327. Clare, D. A., Horton, H. R., Stabel, T. J., Swaisgood, H. E., and Lecce, J. G. (1984). Tissue distribution of mammalian sulfhydryl oxidase. *Arch. Biochem. Biophys.* 230, 138–145.
328. Masters, N., McGuire, M. A., Beerman, K. A., Dasgupta, N., and McGuire, M. K. (2002). Maternal supplementation with CLA decreases milk fat in humans. *Lipids* 37, 133–138.
329. Barber, M. C., Clegg, R. A., Travers, M. T., and Vernon, R. G. (1997). Lipid metabolism in the lactating mammary gland. *Biochim. Biophys. Acta* 1347, 101–126.
330. Baumgard, L. H., Matitashvili, E., Corl, B. A., Dwyer, D. A., and Bauman, D. E. (2002). Trans 10, cis 12 conjugated linoleic acid decreases lipogenic rates and expression of genes involved in fatty acid synthesis in dairy cows. *J. Dairy Sci.* 85, 2155–2163.
331. Vernon, R. G., and Flint, D. J. (1983). Control of fatty acid synthesis in lactation. *Proc. Nutr. Soc.* 42, 315–331.
332. Martyn, P., and Hansen, I. A. (1981). Initiation of lipogenic enzyme activities in rat mammary glands. *Biochem. J.* 198, 187–192.
333. Shimomura, I., Shimano, H., Horton, J. D., Goldstein, J. L., and Brown, M. S. (1997). Differential expression of exons 1a and 1c in mRNAs for sterol regulatory element binding protein-1 in human and mouse organs and cultured cells. *J. Clin. Invest.* 99, 838–845.
334. Matsuzaka, T. Q., Shimano, H., Yahagi, N., Ameniya-Kudo, M., Yoshikawa, T., Hasty, A. H., Tamura, Y., Osuga, J., Okazaki, H., Iizuka, Y., Takahashi, A., Sone, H., Gotoda, T., Ishibashi, S., and Yamada, N. (2002). Dual regulation of mouse Δ5 and Δ6-desaturase gene expression by SREBP-1 and PPARα. *J. Lipid Res.* 43, 107–114.
335. Tobin, K. A., Ulven, S. M., Schuster, G. U., Steineger, H. H., Andreson, S. M., Gustafsson, J. A., and Nebb, H. (2002). Liver X receptors as insulin mediating factors in fatty acid and cholesterol biosynthesis. *J. Biol. Chem.* 277, 10691–10697.
336. Uyeda, K., Yamashita, H., and Kawaguchi, A. (2002). Carbohydrate responsive element-finding protein (ChREBP), a key regulator of glucose metabolism and fat storage. *Biochem. Pharmacol.* 63, 2075–2080.
337. Peterson, D. R., Matitashvili, E. A., and Bauman, D. E. (2003). Diet-induced milk fat depression in dairy cows results in increased trans-10, cis-12 CLA in milk fat and coordinate suppression of mRNA abundance for mammary enzymes involved in milk fat synthesis. *J. Nutr.* 133, 3098–3102.
338. Shimano, H., Horton, J. D., Shimomura, I., Hammer, R. E., Brown, M. S., and Goldstein, J. L. (1997). Isoform 1c of sterol regulatory element binding protein is less active than isoform 1a in livers of transgenic mice and in cultured cells. *J. Clin. Invest.* 96, 846–854.
339. Nohrturfft, A., Brown, M. S., and Goldstein, J. L. (2002). Sterols regulate processing of carbohydrate chains of wild-type SREBP cleavage activating protein (SCAP) but not sterol-resistant mutants Y298C or D443N. *Proc. Natl. Acad. Sci. U S A* 95, 12848–12853.
340. Nohturfft, A., Yabe, D., Goldstein, J. L., and Brown, M. S. (2000). Regulated step in cholesterol feedback localized to budding of SCAP from ER membranes. *Cell* 102, 315–323.
341. Yang, T., Espenshade, P. J., Wright, M. E., Yabe, D., Gong, Y., Aebersold, R., Goldstein, J. L., and Brown, M. S. (2002). Crucial step in cholesterol homeostasis: sterols promote binding of SCAP to INSIG-1, a membrane protein that facilitates retention of SREBP's in ER. *Cell* 110, 489–500.
342. Lopez, J. M., Bennett, M. K., Sanchez, H. B., Rosenfeld, J. M., and Osborne, T. F. (1996). Sterol regulation of acetyl coenzyme A carboxylase: a mechanism for coordinate control of cellular lipid. *Proc. Natl. Acad. Sci. U S A* 93, 1049–1053.
343. Streicher, R., Kotaka, J., Muller-Wieland, D., Siemeister, G., Munck, M., Avci, H., and Krone, W. (1996). SREBP-1 mediates activation of the low density lipoprotein receptor promoter by insulin and insulin-like growth factor-1. *J. Biol. Chem.* 271, 7128–7133.
344. Horton, J. D., Shimomura, I., Brown, M. S., Hammer, R. E., Goldstein, J. L., and Shimano, H. (1998). Activation of cholesterol synthesis in preference to fatty acid synthesis in liver and adipose tissue of transgenic mice overproducing sterol regulatory element-binding protein-2. *J. Clin. Invest.* 101, 2331–2339.
345. Shimomura, I., Shimano, H., Korn, B. S., Bashmakov, Y., and Horton, J. D. (1998). Nuclear sterol regulatory element-binding proteins activate genes responsible for the entire program of unsaturated fatty acid biosynthesis in transgenic mouse liver. *J. Biol. Chem.* 273, 35299–35306.
346. Yahagi, N., Shimano, H., Hasty, A. H., Ameniya-Kudo, M., Okazaki, H., Tamura, Y., Iizuka, Y., Shionoiri, F., Ohashi, K., Osuga, J., Harada, K., Gotoda, T., Nagai, J., Ishibashi, S., and Yamada, N. (1999). A crucial role of sterol regulatory element-binding protein-1 in the regulation of lipogenic gene expression by polyunsaturated fatty acids. *J. Biol. Chem.* 274, 35840–35844.

347. Morrison, R. F., and Farmer, S. R. (1999). Insights into the transcriptional control of adipocyte differentiation. *J. Cell. Biochem. Suppl.* 32/33, 59–67.
348. Guillet-Deniau, I., Mieulet, V., Lelay, S., Achouri, Y., Carre', D., Girard, J., Foufelle, F., and Ferré, P. (2002). Sterol regulatory element binding protein-1c expression and action in rat muscles: insulin-like effects on the control of glycolytic and lipogenic enzymes and UCP3 gene expression. *Diabetes* 51, 1722–1728.
349. Andreolas, C., Xavier, G. D., Diraison, F., Zhao, C., Varadi, A., Lopez-Casillas, F., Ferré, P., Foufelle, F., and Rutter, G. A. (2002). Stimulation of acetyl-CoA carboxylase gene expression by glucose requires insulin release and sterol regulatory element binding protein 1c in pancreatic MIN6β-cells. *Diabetes* 51, 2536–2545.
350. Medvedev, A. V., Ribidoux, J., Bai, X., Cao, W., Floering, L. M., Daniel, K. W., and Collins, S. (2002). Regulation of the uncoupling protein-2 gene in INS-1 β-cells by oleic acid. *J. Biol. Chem.* 277, 42639–42644.
351. Foufelle, R., and Ferré, P. (2002). New perspectives in the regulation of hepatic glycolytic and lipogenic genes by insulin and glucose: a role for the transcription factor SREBP-1c. *Biochem. J.* 366, 377–391.
352. Shimano, H., Yahagi, N., Ameniya-Kudo, M., Hasty, A. H., Osuga, J., Tamura, Y., Shionoiri, F., Iizuka, Y., Ohashi, K., Harada, K., Gotoda, T., Ishibashi, S., and Yamada, N. (1999). Sterol regulatory element-binding protein-1 as a key transcription factor for nutritional induction of lipogenic enzyme genes. *J. Biol. Chem.* 274, 35832–35839.
353. Brown, M. S., and Goldstein, J. L. (1997). The SREBP pathway: regulation of cholesterol metabolism by proteolysis of a membrane-bound transcription factor. *Cell* 89, 331–340.
354. Azzout-Marniche, D., Becard, D., Guichard, C., Feretz, M., Ferre, P., and Foufelle, F. (2000). Insulin effects on sterol regulatory-element-binding protein 1-c (SREBP-1c) transcriptional activity in rat hepatocytes. *Biochem. J.* 350, 289–293.
355. Xu, J., Nakamura, M. T., Cho, H. K., and Clarke, S. D. (1999). Sterol regulatory element binding protein-1 expression is suppressed by dietary polyunsaturated fatty acids. *J. Biol. Chem.* 274, 23577–23583.
356. Xu, J., Teran-Garcia, M., Park, J. H. Y., Nakamura, M. T., and Clarke, S. D. (2001). Polyunsaturated fatty acids suppress hepatic sterol regulatory element-binding protein-1 expression by accelerating transcript decay. *J. Biol. Chem.* 276, 9800–9807.
357. Ameniya-Kudo, M., Shimano, H., Yoshikawa, T., Yahagi, N., Hasty, A. H., Okazaki, H., Tamura, Y., Shionoiri, F., Iizuka, Y., Ohashi, K., Osuga, J., Harada, K., Takanari, G., Sato, R., Kimura, S., Ishibashi, S., and Yamada, N. (2000). Promoter analysis of the mouse sterol regulatory element-binding protein-1c gene. *J. Biol. Chem.* 275, 31078–31085.
358. Ameniya-Kudo, M., Shimano, H., Hasty, A. H., Yahagi, N., Yoshikawa, T., Matsuzaka, T., Okazaki, H., Tamura, Y., Iizuka, Y., Ohashi, K., Osuga, J., Harada, K., Gotoda, T., Sato, R., Kimura, S., Ishibashi, S., and Yamada, N. (2002). Transcriptional activities of nuclear SREBP-1a, -1c and 2 to different target promoters of lipogenic and cholesterolgenic genes. *J. Lipid Res.* 43, 1220–1235.
359. Barber, M. C., Vallance, A. J., Kennedy, H. T., and Travers, M. T. (2003). Induction of transcripts derived from promoter III of the acetyl-CoA carboxylase-alpha gene in mammary gland is associated with recruitment of SREBP-1 to a region of the proximal promoter defined by a DNase I hypersensitive site. *Biochem. J.* 375, 489–501.
360. Barber, M. C., and Travers, M. T. (1996). Induction of a variant acetyl-CoA carboxylase mRNA in ovine mammary gland during lactation. *Biochem. Soc. Trans.* 24, 360S.
361. Travers, M. T., Vallance, A. J., Clegg, R. A., Thomson, R., Price, N. T., and Barber, M. C. (2003). Characterisation of an N-terminal variant of acetyl-CoA carboxylase-α: expression in human tissues and evolutionary aspects. *Biochim. Biophys. Acta* 1634, 97–106.
362. Yang, Y. A., Morin, P. J., Han, W. F., Chen, T., Bornman, D. M., Gabrielson, E. W., and Pizer, E. S. (2003). Regulation of fatty acid synthase expression in breast cancer by sterol regulatory element binding protein-1c. *Exp. Cell Res.* 282, 132–137.
363. Williamson, D. H. (1990). The lactating mammary gland of the rat and the starved refed transition: a model system for the study of the temporal regulation of substrate utilization. *Biochem. Soc. Trans.* 18, 853–856.
364. Mercer, S. W., and Williamson, D. H. (1986). Time course of changes in plasma glucose and insulin concentrations and mammary-gland lipogenesis during re-feeding of starved conscious lactating rats. *Biochem. J.* 239, 489–492.
365. Mercer, S. W., and Williamson, D. H. (1987). The regulation of lipogenesis in vivo in the lactating mammary gland of the rat during the starved-refed transition. *Biochem. J.* 242, 235–243.
366. Munday, M. R., and Williamson, D. H. (1987). Insulin activation of lipogenesis in isolated mammary acini from lactating rats fed a high-fat diet: evidence that acetyl-CoA carboxylase is a site of action. *Biochem. J.* 242, 905–911.
367. Munday, M. R., and Hardie, D. G. (1987). Role of protein phosphorylation in the regulation of fatty acid synthesis in the mammary gland of the lactating rat. In *The Mammary Gland: Development, Regulation and Function* (M. C. Neville and C. W. Daniel, Eds.), pp. 563–613. Plenum Press, New York.
368. Hagopian, K., and Munday, M. R. (1997). The role of pyruvate dehydrogenase, phosphofructo-1-kinase and acetyl-CoA carboxylase in the regulation of fatty acid synthesis in the lactating rat mammary gland during the starved to re-fed transition. *Biochim. Biophys. Acta* 1336, 474–484.
369. Jensen, D. R., Gavigan, S., Sawicki, V., Witsell, D., Eckel, R. H., and Neville, M. C. (1994). Regulation of lipoprotein lipase activity in the mammary gland of the lactating mouse. *Biochem. J.* 298, 321–327.
370. Jones, R. G., Ilic, V., and Williamson, D. H. (1984). Physiological significance of altered insulin metabolism in the conscious rat during lactation. *Biochem. J.* 220, 455–460.
371. Threadgold, L. C., and Kuhn, N. J. (1984). Monosaccharide transport in the mammary gland of the intact lactating rat. *Biochem. J.* 218, 213–219.
372. Williamson, D. H. (1990). The lactating mammary gland of the rat and the starved-refed transition: a model system for the study of the temporal regulation of substrate utilization. *Biochem. Soc. Trans.* 18, 853–856.
373. Griinari, J. M., McGuire, M. A., Dwyer, D. A., Bauman, D. E., Barbano, D. M., and House, W. E. (1997). The role of insulin in the regulation of milk protein synthesis in dairy cows. *J. Dairy Sci.* 80, 2361–2371.
374. Fleischmann, M., and Iynedjian, P. B. (2000). Regulation of sterol regulatory-element binding protein 1 gene expression in liver: role of insulin and protein kinase B/cAkt. *Biochem. J.* 349, 13–17.
375. Bauman, D. E., and Currie, W. B. (1980). Partitioning of nutrients during pregnancy and lactation: a review of mechanisms involving homeostasis and homeorrhesis. *J. Dairy Sci.* 63, 1514–1529.
376. Hartmann, P. E., Kulski, J. K., Rattigan, S., and Saint, L. (1980). Lactation in Australian women. *Proc. Nutr. Soc. Aust.* 5, 104–110.
377. Oftedal, O. T. (1993). The adaptation of milk secretion to the constraints of fasting in bears, seals, and baleen whales. *J. Dairy Sci.* 76, 3234–3246.

378. Rigout, S., Lemosquet, S., Bach, A., Blum, J. W., and Ruiquin, H. (2002). Duodenal infusion of glucose decreases milk fat production in grass silage-fed dairy cows. *J. Dairy Sci.* 85, 2541–2550.
379. Moore, J. H., and Steele, W. (1968). Dietary fat and milk fat secretion in the cow. *Proc. Nutr. Soc.* 27, 66–70.
380. Gooden, J. M., and Lascelles, A. K. (1973). Effect of feeding protected lipid on the uptake of precursors of milk fat by the bovine mammary gland. *Aust. J. Biol. Sci.* 26, 1201–1210.
381. Souza, P. F., and Williamson, D. H. (1993). Effects of feeding medium-chain triacylglycerols on maternal lipid metabolism and pup growth in lactating rats. *Br. J. Nutr.* 69, 779–787.
382. Insull, W., Jr., Hirsch, J., James, T., and Ahrens, E. H. (1958) The fatty acids of human milk. II. Alterations produced by manipulation of caloric balance and exchange of dietary fats. *J. Clin. Invest.* 38, 443–450.
383. Bauman, D. E., and Grinari, J. M. (2000). Regulation and nutritional manipulation of milk fat: low-fat milk syndrome. *Livestock Prod. Sci.* 70, 15–29.
384. Bee, G. (2000). Dietary conjugated linoleic acid consumption during pregnancy and lactation influences growth and tissue composition in weaned pigs. *J. Nutr.* 130, 2981–2989.
385. Palmquist, D. L., Beaulieu, A. D., and Barbano, D. M. (1993). Feed and animal factors influencing milkfat composition. *J. Dairy Sci.* 76, 1753–1771.
386. Peterson, D. G., Matitashvili, E. A., and Bauman, D. E. (2004). The inhibitory effect of trans-10, cis-12 CLA on lipid synthesis in bovine mammary epithelial cells involves reduced proteolytic activation of the transcription factor SREBP-1. *J. Nutr.* 134, 2523–2527.
387. Belury, M. A. (2002). Dietary conjugated linoleic acid in health: Physiological effects and mechanisms of action. *Annu. Rev. Nutr.* 22, 505–531.
388. Dils, R. R. (1983). Milk fat synthesis. In *Biochemistry of Lactation* (T. B. Mepham, Ed.), pp. 141–157. Elsevier, Amsterdam.
389. Peaker, M. (1977). The aqueous phase of milk: ion and water transport. *Symp. Zool. Soc. Lond.* 41, 113–134.
390. Neville, M. C., Keller, R. P., Seacat, J., Casey, C. E., Allen, J. C., and Archer, P. (1984). Studies on human lactation: I. Within-feed and between-breast variation in selected components of human milk. *Am. J. Clin. Nutr.* 40, 635–646.
391. Linzell, J. L., and Peaker, M. (1971) Intracellular concentrations of sodium, potassium and chloride in the lactating mammary gland and their relation to the secretory mechanism. *J. Physiol. (Lond).* 216, 683–700.
392. Prentice, A. M., Prentice, A., Lamb, W. H., Lunn, P. G., and Austin, S. (1983). Metabolic consequences of fasting during Ramadan in pregnant and lactating women. *Hum. Nutr. Clin. Nutr.* 37C, 283–294.
393. Mobasheri, A., and Marples, D. (2004). Expression of the AQP-1 water channel in normal human tissues: a semiquantitative study using tissue microarray technology. *Am. J. Physiol. Cell Physiol.* 286, C529–537.
394. Berga, S. E., and Neville, M. C. (1985). Sodium and potassium distribution in the lactating mouse mammary gland in vivo. *J. Physiol. (Lond).* 361, 219–230.
395. Sjaastad, M. D., Zettl, K. S., Parry, G., Firestone, G. L., and Machen, T. E. (1993). Hormonal regulation of the polarized function and distribution of Na/H exchange and Na/HCO_3 cotransport in cultured mammary epithelial cells. *J. Cell Biol.* 122, 589–600.
396. Shennan, D. B. (1990). Efflux of chloride from lactating rat mammary tissue slices. *Comp. Biochem. Physiol. A* 95, 367–371.
397. Shillingford, J. M., Miyoshi, K., Flagella, M., Shull, G. E., and Hennighausen, L. (2002). Mouse mammary epithelial cells express the Na-K-Cl cotransporter, NKCC1: characterization, localization, and involvement in ductal development and morphogenesis. *Mol. Endocrinol.* 16, 1309–1321.
398. Linzell, J. L., and Peaker, M. (1974). Changes in colostrum composition and in the permeability of the mammary epithelium at about the time of parturition in the goat. *J. Physiol. (Lond).* 243, 129–151.
399. Blatchford, D. R., and Peaker, M. (1988). Effect of ionic composition of milk on transepithelial potential in the goat mammary gland. *J. Physiol. (Lond).* 402, 533–541.
400. Shennan, D. B. (1992). K^+ and Cl^- transport by mammary secretory cell apical membrane vesicles isolated from milk. *J. Dairy Res.* 59, 339–348.
401. Smith, K. M., McNeillie, S. A., and Shennan, D. B. (1990). K^+ (Rb^+) transport by a mammary secretory cell apical membrane fraction isolated from goats' milk. *Exp. Physiol.* 75, 349–358.
402. Peaker, M. (1977). Mechanism of milk secretion: milk composition in relation to potential difference across the mammary epithelium. *J. Physiol. (Lond).* 270, 489–505.
403. Berga, S. E. (1984). Electrical potentials and cell-to-cell dye movement in mouse mammary gland during lactation. *Am. J. Physiol.* 247, C20–C25.
404. Selvaraj, N. G., Omi, E., Gibori, G., and Rao, M. C. (2000). Janus kinase 2 (JAK2) regulates prolactin-mediated chloride transport in mouse mammary epithelial cells through tyrosine phosphorylation of Na^+- K^+-2 Cl^- cotransporter. *Mol. Endocrinol.* 14, 2054–2065.
405. Blaug, S., Hybiske, K., Cohn, J., Firestone, G. L., Machen, T. E., and Miller, S. S. (2001). ENaC- and CFTR-dependent ion and fluid transport in mammary epithelia. *Am. J. Physiol. Cell Physiol.* 281, C633–648.
406. Shiffman, M. L., Seale, T. S., Flux, M., Rennert, O. R., and Swender, P. T. (1989). Breast-milk composition in women with cystic fibrosis: report of two cases and a review of the literature. *Am. J. Clin. Nutr.* 49, 612–617.
407. Enomoto, K., Furuya, K., Yamagishi, S., Oka, T., and Maeno, T. (1994). The increase in the intracellular Ca^{2+} concentration induced by mechanical stimulation is propagated via release of pyrophosphorylated nucleotides in mammary epithelial cells. *Pflugers Arch.* 427, 533–542.
408. Blaug, S., Rymer, J., Jalickee, S., and Miller, S. S. (2003). P2 purinoceptors regulate calcium-activated chloride and fluid transport in 31EG4 mammary epithelia. *Am. J. Physiol. Cell Physiol.* 284, C897–909.
409. Gitlin, J. D., Gitlin, J. I., and Gitlin, D. (1976). Selective transfer of plasma proteins across mammary gland in lactating mouse. *Am. J. Physiol.* 230, 1594–1602.
410. Monks, J., Heuy, P., Hanson, L., Eckel, R. H., Neville, M. C., and Gavigan, S. (2001). A lipoprotein particle in mouse milk. *J. Lipid Res.* 42, 686–696.
411. Koldovsky, O. (1996). The potential physiological significance of milk-borne hormonally active substances for the neonate. *J. Mammary Gland Biol. Neoplasia* 1, 317–323.
412. Goldman, A. S., Chheda, S., Garofalo, R., and Schmalstieg, F. C. (1996). Cytokines in human milk: properties and potential effects upon the mammary gland and the neonate. *J. Mammary Gland Biol. Neoplasia* 1, 251–258.
413. Jensen, D. R., Bessesen, D. H., Etienne, J., Eckel, R. H., and Neville, M. C. (1991). Distribution and source of lipoprotein lipase in mouse mammary gland. *J. Lipid Res.* 32, 733–742.
414. Kilshaw, P. J., and Cant, A. J. (1984). The passage of maternal dietary proteins into human breast milk. *Int. Arch. Allergy Appl. Immunol.* 75, 8–15.
415. Goldblum, R. M., Ahlstedt, S., Carlsson, B., Hanson, L. A., Jodal, U., Lidin-Janson, G., and Sohl-Akerlund, A. (1975).

Antibody-forming cells in human colostrum after oral immunisation. *Nature* 257, 797–798.

416. Tanneau, G. M., Hibrand-Saint Oyant, L., Chevaleyre, C. C., and Salmon, H. P. (1999). Differential recruitment of T- and IgA B-lymphocytes in the developing mammary gland in relation to homing receptors and vascular addressins. *J. Histochem. Cytochem.* 47, 1581–1592.
417. Lindh, E. (1975). Increased resistance of immunoglobulin A dimers to proteolytic degradation after binding of secretory component. *J. Immunol.* 114, 284–286.
418. Feachem, R. G., and Koblinsky, M. A. (1984). Interventions for the control of diarrhoeal diseases among young children: promotion of breast-feeding. *Bull. World Health Org.* 62, 271–291.
419. Wilson, E., and Butcher, E. C. (2004). CCL28 controls immunoglobulin (Ig)A plasma cell accumulation in the lactating mammary gland and IgA antibody transfer to the neonate. *J. Exp. Med.* 200, 805–809.
420. De Groot, N., Van Kuik-Romeijn, P., Lee, S. H., and De Boer, H. A. (2000). Increased immunoglobulin A levels in milk by over-expressing the murine polymeric immunoglobulin receptor gene in the mammary gland epithelial cells of transgenic mice. *Immunology* 101, 218–224.
421. Rincheval-Arnold, A., Belair, L., and Djiane, J. (2002). Developmental expression of pIgR gene in sheep mammary gland and hormonal regulation. *J. Dairy Res.* 69, 13–26.
422. Rincheval-Arnold, A., Belair, L., Cencic, A., and Djiane, J. (2002). Up-regulation of polymeric immunoglobulin receptor mRNA in mammary epithelial cells by IFN-gamma. *Mol. Cell. Endocrinol.* 194, 95–105.
423. Mostov, K. E., Kraehenbuhl, J.-P., and Blobel, G. (1980). Receptor-mediated transcellular transport of immunoglobulin: Synthesis of secretory component as multiple and larger transmembrane forms. *Proc. Natl. Acad. Sci. U S A* 77, 7257–7261.
424. Mostov, K. E., and Deitcher, D. L. (1986). Polymeric immunoglobulin receptor expressed in MDCK cells transcytoses IgA. *Cell* 46, 613–621.
425. Mostov, K. (1993). Protein traffic in polarized epithelial cells: the polymeric immunoglobulin receptor as a model system. *J. Cell Sci. Suppl.* 17, 21–26.
426. Solari, R., and Kraehenbuhl, J.-P. (1984). Biosynthesis of the IgA antibody receptor: a model for the transepithelial sorting of a membrane glycoprotein. *Cell* 36, 61–71.
427. Sanchez, L., Lujan, L., Oria, R., Castillo, H., Perez, D., Ena, J. M., and Calvo, M. (1992). Synthesis of lactoferrin and transport of transferrin in the lactating mammary gland of sheep. *J. Dairy Sci.* 75, 1257–1262.
428. Seddiki, T., Delpal, S., and Ollivier-Bousquet, M. (1992). Endocytosis and intracellular transport of transferrin across the lactating rabbit mammary epithelial cell. *J. Histochem. Cytochem.* 40, 1501–1510.
429. Ollivier-Bousquet, M. (1998). Transferrin and prolactin transcytosis in the lactating mammary epithelial cell. *J. Mammary Gland Biol. Neoplasia* 3, 303–313.
430. Zhang, P., Sawicki, V., Lewis, A., Hanson, L., Monks, J., and Neville, M. C. (2000). The effect of serum iron concentration on iron secretion into mouse milk. *J. Physiol. (Lond).* 522 Pt 3, 479–491.
431. Johansen, F. E., Braathen, R., and Brandtzaeg, P. (2001). The J chain is essential for polymeric Ig receptor-mediated epithelial transport of IgA. *J. Immunol.* 167, 5185–5192.
432. Halsey, J. F., Mitchell, C., Meyer, R., and Cebra, J. J. (1982). Metabolism of immunoglobulin A in lactating mice: origins of immunoglobulin A in milk. *Eur. J. Immunol.* 12, 107–112.
433. Vogel, S. M., Minshall, R. D., Pilipovic, M., Tiruppathi, C., and Malik, A. B. (2001). Albumin uptake and transcytosis in endothelial cells in vivo induced by albumin-binding protein. *Am. J. Physiol. Lung Cell. Mol. Physiol.* 281, L1512–L1522.
434. Kim, K.-J., and Malik, A. B. (2003). Protein transport across the lung epithelial barrier. *Am. J. Physiol. Lung Cell. Mol. Physiol.* 264, L247–L259.
435. Pitelka, D. R. (1978). Cell contacts in the mammary gland. In *Lactation* (B. L. Larson, Ed.), Vol. IV, pp. 41–66. Academic Press, New York.
436. Lenferink, A. E. G., Magoon, J., Cantin, C., and O'Connor-McCourt, M. D. (2004). Investigation of three new mouse mammary tumor cell lines as models for transforming growth factor (TGF)-β and Neu pathway signaling studies: identification of a novel model for TGF-β-induced epithelial-to-mesenchymal transition. *Breast Cancer Res.* 6, R514–R530.
437. Locke, D., Stein, T., Davies, C., Morris, J., Harris, A. L., Evans, W. H., Monaghan, P., and Gusterson, B. (2004). Altered permeability and modulatory character of connexin channels during mammary gland development. *Exp. Cell Res.* 298, 643–660.
438. Neville, M. C., Stahl, L., Brozo, L. A., and Lowe-Lieber, J. (1991). Morphogenesis and secretory activity of mouse mammary cultures on EHS biomatrix. *Protoplasma* 160, 110–123.
439. Barcellos-Hoff, M. H., Aggeler, J., Ram, T. G., and Bissell, M. J. (1989). Functional differentiation and alveolar morphogenesis of primary mammary epithelial cells cultured on a reconstituted basement membrane. *Development* 105, 223–235.
440. Fata, J. E., Werb, Z., and Bissell, M. J. (2004). Regulation of mammary gland branching morphogenesis by the extracellular matrix and its remodeling enzymes. *Breast Cancer Res.* 6, 1–11.
441. Streuli, C. H., and Edwards, G. M. (1998). Control of normal mammary epithelial phenotype by integrins. *J. Mammary Gland Biol. Neoplasia* 3, 151–163.
442. Haslam, S. Z., and Woodward, T. L. (2004). Host microenvironment in breast cancer development: epithelial-cell–stromal-cell interactions and steroid hormone action in normal and cancerous mammary gland. *Breast Cancer Res.* 5, 208–215.
443. Bissell, M. J., Radisky, D. C., Rizki, A., Weaver, V. M., and Petersen, O. W. (2002). The organizing principle: microenvironmental influences in the normal and malignant breast. *Differentiation* 70, 537–546.
444. Linzell, J. L., and Peaker, M. (1971) The permeability of mammary ducts. *J. Physiol. (Lond).* 216, 701–716.
445. Schneeberger, E. E. (2003). Claudins form ion-selective channels in the paracellular pathway. Focus on claudin extracellular domains determine paracellular charge selectively and resistance but not tight junction fibril architecture. *Am. J. Physiol.* 284, C1331–C1333.
446. Colegio, O. R., Van Itallie, C. M., Rahner, C., and Anderson, J. M. (2003). Claudin extracellular domains determine paracellular charge selectivity and resistance but not tight junction fibril architecture. *Am. J. Physiol. Cell Physiol.* 284, C1331–C1354.
447. Stelwagen, K., McFadden, H. A., and Demmer, J. (1999). Prolactin, alone or in combination with glucocorticoids, enhances tight junction formation and expression of the tight junction protein occludin in mammary cells. *Mol. Cell. Endocrinol.* 156, 55–61.
448. Neville, M. C. (1995). Lactogenesis in women: evidence for a cascade of cellular events. In *Handbook of Composition of Milks* (R. G. Jensen, Ed.), pp. 87–98. Academic Press, San Diego.
449. Hartmann, P. E., and Kulski, J. K. (1978). Changes in the composition of the mammary secretion of women after abrupt termination of breast feeding. *J. Physiol. (Lond).* 275, 1–11.
450. Atwood, C. S., and Hartmann, P. E. (1995). Assessment of mammary gland metabolism in the sow: III. Cellular

metabolites in the mammary secretion and plasma following weaning. *J. Dairy Res.* 62, 221–236.
451. Stelwagen, K., Davis, S. R., Farr, V. C., Prosser, C. G., and Sherlock, R. A. (1994). Mammary epithelial cell tight junction integrity and mammary blood flow during an extended milking interval in goats. *J. Dairy Sci.* 77, 426–432.
452. Linzell, J. L., and Peaker, M. (1972) Day-to-day variations in milk composition in the goat and cow as a guide to the detection of subclinical mastitis. *Br. Vet. J.* 128, 284–295.
453. Linzell, J. L., and Peaker, M. (1971) Early detection of mastitis. *Vet. Rec.* 89, 393–394.
454. Stelwagen, K., Farr, V. C., McFadden, H. A., Prosser, C. G., and Davis, S. R. (1997). Time course of milk accumulation-induced opening of mammary tight junctions, and blood clearance of milk components. *Am. J. Physiol.* 273, R379–R386.
455. El-Sabban, M. E., Abi-Mosleh, L. F., and Talhouk, R. S. (2003). Developmental regulation of gap junctions and their role in mammary epithelial cell differentiation. *J. Mammary Gland Biol. Neoplasia* 8, 463–473.
456. Locke, D., Perusinghe, N., Newman, T., Jayatilake, H., Evans, W. H., and Monaghan, P. (2000). Developmental expression and assembly of connexins into homomeric and heteromeric gap junction hemichannels in the mouse mammary gland. *J. Cell. Physiol.* 183, 228–237.
457. Tu, Z. J., Kollander, R., and Kiang, D. T. (1998). Differential up-regulation of gap junction connexin 26 gene in mammary and uterine tissues: the role of Sp transcription factors. *Mol. Endocrinol.* 12, 1931–1938.
458. Yamanaka, I., Kuraoka, A., Inai, T., Ishibashi, T., and Shibata, Y. (2001). Differential expression of major gap junction proteins, connexins 26 and 32, in rat mammary glands during pregnancy and lactation. *Histochem. Cell Biol.* 115, 277–284.
459. Yamanaka, I., Kuraoka, A., Inai, T., Ishibashi, T., and Shibata, Y. (1997). Changes in the phosphorylation states of connexin43 in myoepithelial cells of lactating rat mammary glands. *Eur J. Cell Biol.* 72, 166–173.
460. Musil, L. S., Cunningham, B. A., Edelman, G. M., and Goodenough, D. A. (1990). Differential phosphorylation of the gap junction protein connexin43 in junctional communication-competent and -deficient cell lines. *J. Cell Biol.* 111, 2077–2088.
461. Plum, A., Hallas, G., Magin, T., Dombrowski, F., Hagendorff, A., Schumacher, B., Wolpert, C., Kim, J., Lamers, W. H., Evert, M., Meda, P., Traub, O., and Willecke, K. (2000). Unique and shared functions of different connexins in mice. *Curr. Biol.* 10, 1083–1091.
462. Bry, C., Maass, K., Miyoshi, K., Willecke, K., Ott, T., Robinson, G. W., and Hennighausen, L. (2004). Loss of connexin 26 in mammary epithelium during early but not during late pregnancy results in unscheduled apoptosis and impaired development. *Development* 267, 418–429.
463. El-Sabban, M. E., Sfeir, A. J., Daher, M. H., Kalaany, N. Y., Bassam, R. A., and Talhouk, R. S. (2003). ECM-induced gap junctional communication enhances mammary epithelial cell differentiation. *J. Cell Sci.* 116, 3531–3541.
464. Ormandy, C. J., Naylor, M. J., Harris, J., Robertson, F., Horseman, N. D., Lindeman, G. J., Visvader, J., and Kelly, P. A. (2003). Investigation of the transcriptional changes underlying functional defects in the mammary glands of prolactin receptor knockout mice. *Recent Prog. Horm. Res.* 58, 297–323.
465. Grimm, S. L., Seagroves, T. N., Kabotyansky, E., Hovey, R. C., Vonderhaar, B. K., Lydon, J. P., Miyoshi, K., Hennighausen, L. G., Ormandy, C. J., Lee, A. V., Stull, J. A., Wood, T. L., and Rosen, J. M. (2002). Disruption of steroid and prolactin receptor patterning in the mammary gland correlates with a block in lobuloalveolar development. *Mol. Endocrinol.* 16, 2675–2691.
466. Flint, D. J., and Knight, C. H. (1997). Interactions of prolactin and growth hormone in the regulation of mammary gland function and epithelial cell survival. *J. Mammary Gland Biol. Neoplasia* 2, 41–48.
467. Travers, M. T., Barber, E., Tonner, E., Quarrie, L., Wilde, C. J., and Flint, D. J. (1996). The role of prolactin and growth hormone in the regulation of casein gene expression and mammary cell survival: relationships to milk synthesis and secretion. *Endocrinology* 137, 1530–1539.
468. Rosen, J. M., Wysolmerski, S. L., and Hadsell, D. (1999). Regulation of milk protein gene expression. *Annu. Rev. Nutr.* 19, 407–436.
469. Wood, T. L., Richert, M. M., Stull, M. A., and Allar, M. A. (2000). The insulin-like growth factors (IGFs) and IGF binding proteins in postnatal development of murine mammary glands. *J. Mammary Gland Biol. Neoplasia* 5, 31–42.
470. Cowie, A. T., and Lyons, W. R. (1959) Mammogenesis and lactogenesis in hypophysectomized, ovariectomized, adrenalectomized rats. *J. Endocrinol.* 19, 29–32.
471. Cowie, A. T., Hartmann, P. E., and Turvey, A. (1969). The maintenance of lactation in the rabbit after hypophysectomy. *J. Endocrinol.* 43, 651–662.
472. Young, W. S., Iii, Shepard, E., Amico, J., Hennighausen, L., Wagner, K. U., Lamarca, M. E., McKinney, C., and Ginns, E. I. (1996). Deficiency in mouse oxytocin prevents milk ejection, but not fertility or parturition. *J. Neuroendocrinol.* 8, 847–853.
473. Schams, D. (1972). Prolactin levels in bovine blood, influenced by milking manipulation, genital stimulation and oxytocin administration with specific consideration of the seasonal variations. *Acta Endocrinol.* 71, 684–696.
474. Tay, C. C. K., Glasier, A. F., and McNeilly, A. S. (1996). Twenty-four hour patterns of prolactin secretion during lactation and the relationship to suckling and the resumption of fertility in breast-feeding women. *Hum. Reprod.* 11, 950–955.
475. Flint, D. J., and Vernon, R. G. (1998). Effects of food restriction on the responses of the mammary gland and adipose tissue to prolactin and growth hormone in the lactating rat. *J. Endocrinol.* 156, 299–305.
476. Benedek-Jaszmann, L. J., and Sternthal, V. (1976). Late suppression of lactation with bromocriptine. *Practitioner* 216, 450.
477. Nagasawa, H., and Yanai, R. (1978). Mammary gland prolactin receptor and pituitary prolactin secretion in lactating mice with different lactational performance. *Acta Endocrinol.* 88, 94–98.
478. Bole-Feysot, C., Goffin, V., Edery, M., Binart, N., and Kelly, P. A. (1998). Prolactin (PRL) and its receptor: Actions, signal transduction pathways and phenotypes observed in PRL receptor knockout mice. *Endocr. Rev.* 19, 225–268.
479. Cassy, S., Charlier, M., Belair, L., Guillomot, M., Charron, G., Bloch, B., and Djiane, J. (1998). Developmental expression and localization of the prolactin receptor (PRL-R) gene in ewe mammary gland during pregnancy and lactation: estimation of the ratio of the two forms of PRL-R messenger ribonucleic acid. *Biol. Reprod.* 58, 1290–1296.
480. Mizoguchi, Y., Kim, J. Y., Sasaki, T., Hama, T., Sasaki, M., Enami, J., and Sakai, S. (1996). Acute expression of the PRL receptor gene after ovariectomy in midpregnant mouse mammary gland. *Endocr. J.* 43, 537–544.
481. Mizoguchi, Y., Yamaguchi, H., Aoki, F., Enami, J., and Sakai, S. (1997). Corticosterone is required for the prolactin receptor gene expression in the late pregnant mouse mammary gland. *Mol. Cell. Endocrinol.* 132, 177–183.
482. Hovey, R. C., Trott, J. F., and Vonderhaar, B. K. (2002). Establishing a framework for the functional mammary gland: from endocrinology to morphology. *J. Mammary Gland Biol. Neoplasia* 7, 17–38.

483. Trott, J. F., Hovey, R. C., Koduri, S., and Vonderhaar, B. K. (2004). Multiple new isoforms of the human prolactin receptor gene. *Adv. Exp. Med. Biol.* 554, 495–499.
484. Nishikawa, S., Moore, R. C., Nonomura, N., and Oka, T. (1994). Progesterone and EGF inhibit mouse mammary gland prolactin receptor and β-casein gene expression. *Am. J. Physiol.* 267, C1467–C1472.
485. Groner, B., and Gouilleux, F. (1995). Prolactin-mediated gene activation in mammary epithelial cells. *Curr. Opin. Genet. Dev.* 5, 587–594.
486. Liu, X., Robinson, G. W., Gouilleux, F., Groner, B., and Hennighausen, L. (1995). Cloning and expression of Stat5 and an additional homologue (Stat5b) involved in prolactin signal transduction in mouse mammary tissue. *Proc. Natl. Acad. Sci. U S A* 92, 8831.
487. Hennighausen, L. G., and Robinson, G. W. (1998). Think globally, act locally: the making of a mouse mammary gland. *Genes Dev.* 12, 449–455.
488. Kazansky, A. V., Raught, B., Lindsey, S. M., Wang, Y. F., and Rosen, J. M. (1995). Regulation of mammary gland factor Stat5a during mammary gland development. *Mol. Endocrinol.* 9, 1598–1609.
489. Wagner, K. U., Krempler, A., Triplett, A. A., Qi, Y., George, N. M., Zhu, J., and Rui, H. (2004). Impaired alveologenesis and maintenance of secretory mammary epithelial cells in Jak2 conditional knockout mice. *Mol. Cell. Biol.* 24, 5510–5520.
490. Choi, K. M., Barash, I., and Rhoads, R. E. (2004). Insulin and prolactin synergistically stimulate beta-casein messenger ribonucleic acid translation by cytoplasmic polyadenylation. *Mol. Endocrinol.* 18, 1670–1686.
491. Fata, J. E., Werb, Z., and Bissell, M. J. (2004). Regulation of mammary gland branching morphogenesis by the extracellular matrix and its remodeling enzymes. *Breast Cancer Res.* 6, 1–11.
492. Zoubiane, G. S., Valentijn, A., Lowe, E. T., Akhtar, N., Bagley, S., Gilmore, A. P., and Streuli, C. H. (2004). A role for the cytoskeleton in prolactin-dependent mammary epithelial cell differentiation. *J. Cell Sci.* 117, 271–280.
493. Lee, R. C., Walters, J. A., Reyland, M. E., and Anderson, S. M. (1999). Constitutive activation of the prolactin receptor results in the induction of growth factor-independent proliferation and constitutive activation of signaling molecules. *J. Biol. Chem.* 274, 10024–10034.
494. Schwertfeger, K. L., Hunter, S., Heasley, L. E., Levresse, V., Leon, R. P., Degregori, J., and Anderson, S. M. (2000). Prolactin stimulates activation of c-jun N-terminal kinase (JNK). *Mol. Endocrinol.* 14, 1592–1602.
495. Hunter, S., Koch, B. L., and Anderson, S. M. (1997). Phosphorylation of cbl after stimulation of Nb2 cells with prolactin and its association with phosphatidylinositol 3-kinase. *Mol. Endocrinol.* 11, 1213–1222.
496. Peaker, M., and Wilde, C. J. (1996). Feedback control of milk secretion from milk. *J. Mammary Gland Biol. Neoplasia* 1, 307–315.
497. Neville, M. C. (1995). Volume and caloric density of human milk. In *Handbook of Milk Composition* (R. G. Jensen, Ed.), pp. 101–113. Academic Press, San Diego.
498. Hammond, K. A., Lloyd, K. C., and Diamond, J. (1996). Is mammary output capacity limiting to lactational performance in mice. *J. Exp. Biol.* 199, 337–349.
499. Wilde, C. J., Addey, C. V. P., Boddy, L. M., and Peaker, M. (1995). Autocrine regulation of milk secretion by a protein in milk. *Biochem. J.* 305, 51–58.
500. Prentice, A., Addey, C. P., and Wilde, C. J. (1989). Evidence for local feedback control of human milk secretion. *Biochem. Soc. Trans.* 16, 122.
501. Hendry, K. A., Simpson, K. J., Nicholas, K. R., and Wilde, C. J. (1998). Autocrine inhibition of milk secretion in the lactating tammar wallaby (*Macropus eugenii*). *J Mol. Endocrinol.* 21, 169–177.
502. Shamay, A., Shapiro, F., Leitner, G., and Silanikove, N. (2003). Infusions of casein hydrolyzates into the mammary gland disrupt tight junction integrity and induce involution in cows. *J. Dairy Sci.* 86, 1250–1258.
503. Millar, I. D., Barber, M. C., Lomax, M. A., Travers, M. T., and Shennan, D. B. (1997). Mammary protein synthesis is acutely regulated by the cellular hydration state. *Biochem. Biophys. Res. Commun.* 230, 351–355.
504. Kuraishi, T., Sun, Y., Aoki, F., Imakawa, K., and Sakai, S. (2000). The poly(A) tail length of casein mRNA in the lactating mammary gland changes depending upon the accumulation and removal of milk. *Biochem. J.* 347, 579–583.
505. Kuraishi, T., Mizoguchi, Y., Sun, Y., Aoki, F., Imakawa, K., and Sakai, S. (2002). The casein mRNA decay changes in parallel with the poly(A) tail length in the mouse mammary gland. *Mol. Cell. Endocrinol.* 190, 101–107.
506. Schmidhauser, C., Bissell, M. J., Myers, C. J., and Casperson, G. F. (1991). Extracellular matrix and hormones transcriptionally regulate bovine β-casein 5′ sequences in stably transfected mouse mammary cells. *Proc. Natl. Acad. Sci. U S A* 87, 9118–9122.
507. Danielson, K. G., Oborn, C. J., Durban, E. M., Butel, J. S., and Medina, D. (1984). Epithelial mouse mammary cell line exhibiting normal morphogenesis in vivo and functional differentiation in vitro. *Proc. Natl. Acad. Sci. U S A* 81, 3756–3760.
508. Butte, N. F., Hopkinson, J. M., Moon, J. K., Mehta, N., and Smith, E. O. (1999). Adjustments in energy expenditure and substrate utilization during late pregnancy and lactation. *Am. J. Clin. Nutr.* 69, 299–307.
509. Mukhopadhyay, S. S., Wyszolmierski, S. L., Gronostajski, R. M., and Rosen, J. M. (2001). Differential interactions of specific nuclear factor I isoforms with the glucocorticoid receptor and STAT5 in the cooperative regulation of WAP gene transcription. *Mol. Cell. Biol.* 21, 6859–6869.
510. Doppler, W., Groner, B., and Ball, R. K. (1989). Prolactin and glucocorticoid hormones synergistically induce expression of transfected rat á-casein gene promoter constructs in a mammary epithelial cell line. *Proc. Natl. Acad. Sci. U S A* 86, 104–108.
511. Stoecklin, E., Wissler, M., Gouilleux, F., and Groner, B. (1996). Functional interactions between Stat5 and the glucocorticoid receptor. *Nature* 383, 726–728.
512. Kingley-Kallesen, M., Mukhopadhyay, S. S., Wyszomierski, S. L., Schanler, S., Schutz, G., and Rosen, J. M. (2002). The mineralocorticoid receptor may compensate for the loss of the glucocorticoid receptor at specific stages of mammary gland development. *Mol. Endocrinol.* 16, 2008–2018.
513. Wintermantel, T. M., Bock, D., Fleig, V., Greiner, E. F., and Schutz, G. (2005). The epithelial glucocorticoid receptor is required for the normal timing of cell proliferation during mammary lobuloalveolar development, but dispensable for milk production. *Mol. Endocrinol.* 19, 340–349.
514. Crowley, W. R., and Armstrong, W. E. (1992). Neurochemical regulation of oxytocin secretion in lactation. *Endocr. Rev.* 13, 33–65.
515. Nishimori, K., Young, L. J., Guo, Q., Wang, Z., Insel, T. R., and Matzuk, M. M. (1996). Oxytocin is required for nursing but is not essential for parturition or reproductive behavior. *Proc. Natl. Acad. Sci. U S A* 93, 11699–11704.
516. Cobo, E., Debernal, M., Gaitan, E., and Quintero, C. A. (1967). Neurohypophysial hormone release in the human: II. Experimental study during lactation. *Am. J. Obstet. Gynecol.* 97, 519–529.

517. McNeilly, A. S., Robinson, I. C., Houston, M. J., and Howie, P. W. (1983). Release of oxytocin and prolactin in response to suckling. *BMJ Clin. Res.* 286, 257–259.
518. Ueda, T., Yokoyama, Y., Irahara, M., and Aono, T. (1994). Influence of psychological stress on suckling-induced pulsatile oxytocin release. *Obstet. Gynecol.* 84, 259–262.
519. Bruckmaier, R. M., and Schams, D. (1998). Oxytocin release and milk removal in ruminants. *J. Dairy Sci.* 81, 939–949.
520. Duvigneaud, B., Ressler, C., Swan, J. M., Roberts, C. W., and Katsoyannis, P. G. (1954) The synthesis of oxytocin. *J. Am. Chem. Soc.* 76, 3115–3121.
521. Bathgate, R. A., Parry, L. J., Fletcher, T. P., Shaw, G., Renfree, M. B., Gemmell, R. T., and Sernia, C. (1995). Comparative aspects of oxytocin-like hormones in marsupials. *Adv. Exp. Med. Biol.* 395, 639–655.
522. Gimpl, G., and Fahrenholz, F. (2001). The oxytocin receptor system: Structure, function and regulation. *Physiol. Rev.* 81, 629–683.
523. Zimmerman, E. A., and Defendini, R. (1977). Hypothalamic pathways containing oxytocin, vasopressin and associated neurophysins. In *International Conference on the Neurohypophysis* (A. M. Moses and L. Share, Eds.), pp. 22–29. Karger, New York.
524. Brownstein, M. J., Russell, J. T., and Gainer, H. T. (1980). Synthesis, transport and release of posterior pituitary hormones. *Science* 207, 373–378.
525. Rose, J. P., Wu, C. K., Hsiao, D. C., Breslow, E., and Wang, B. C. (1996). Crystal structure of the neurophysin-oxytocin complex. *Nat. Struct. Biol.* 3, 163–169.
526. Leng, G. (2000). Steroidal influences on oxytocin neurones. *J. Physiol. (Lond).* 524.2, 315.
527. Brown, C. H., Russell, J. A., and Leng, G. (2000). Opioid modulation of magnocellular neurosecretory cell activity. *Neurosci. Res.* 36, 97–120.
528. Newton, M., and Newton, N. R. (1948). The let-down reflex in human lactation. *J. Pediatr.* 33, 698–704.
529. Gould, B. R., and Zingg, H. H. (2003). Mapping oxytocin receptor gene expression in the mouse brain and mammary gland using an oxytocin receptor-LacZ reporter mouse. *Neuroscience* 122, 155–167.
530. Rozen, F., Russo, C., Banville, D., and Zingg, H. H. (1995). Structure, characterization and expression of the rat oxytocin receptor gene. *Proc. Natl. Acad. Sci. U S A* 92, 200–204.
531. Olins, G. M., and Bremel, R. D. (1984). Oxytocin-stimulated myosin phosphorylation in mammary myoepithelial cells: roles of calcium ions and cyclic nucleotides. *Endocrinology* 114, 1617–1626.
532. Grazzini, E., Guillon, G., Mouillac, B., and Zingg, H. H. (1998). Inhibition of oxytocin receptor function by direct binding of progesterone. *Nature* 392, 509–512.
533. Burger, K., Fahrenholz, F., and Gimpl, G. (1999). Non-genomic effects of progesterone on the signaling function of G protein-coupled receptors. *FEBS Lett.* 464, 25–29.
534. Soloff, M. S., Alexandrova, M., and Fernstrom, M. J. (1979). Oxytocin receptors: triggers for parturition and lactation? *Science* 204, 1313–1315.
535. Sapino, A., Macri, L., Tonda, L., and Bussolati, G. (1993). Oxytocin enhances myoepithelial cell differentiation and proliferation in the mouse mammary gland. *Endocrinology* 133, 838–842.
536. Pedersen, C. A., Caldwell, J. D., Walker, C., Ayers, G., and Mason, G. A. (1994). Oxytocin activates the postpartum onset of rat maternal behavior in the ventral tegmental and medial preoptic areas. *Behav. Neurosci.* 108, 1163–1171.
537. Keverne, E. B., and Kendrick, K. M. (1994). Maternal behaviour in sheep and its neuroendocrine regulation. *Acta Paediatr. Suppl.* 397, 47–56.
538. Oftedal, O. T. (2002). The mammary gland and its origin during synapsid evolution. *J. Mammary Gland Biol. Neoplasia* 7, 225–252.
539. Oftedal, O. T. (2002). The origin of lactation as a water source for parchment-shelled eggs. *J. Mammary Gland Biol. Neoplasia* 7, 253–266.
540. Schmid, J., Andersen, N. A., Speakman, J. R., and Nicol, S. C. (2003). Field energetics of free-living, lactating and non-lactating echidnas (*Tachyglossus aculeatus*). *Comp. Biochem. Physiol. A Mol. Integr. Physiol.* 2003, 903–909.
541. Griffiths, M. (1978). *Biology of the Monotremes.* Academic Press, New York.
542. Bird, P. H., Hendry, K. A., Shaw, D. C., Wilde, C. J., and Nicholas, K. R. (1994). Progressive changes in milk protein gene expression and prolactin binding during lactation in the tammar wallaby (*Macropus eugenii*). *J Mol. Endocrinol.* 13, 117–125.
543. Trott, J. F., Simpson, K. J., Moyle, R. L. C., Mearn, C. M., Shaw, G., Nicholas, K. R., and Renfree, M. B. (2003). Maternal regulation of milk composition, milk production and pouch young development during lactation in the tammar wallaby. *Biol. Rep.* 68, 929–936.
544. Nicholas, K., Simpson, K., Wilson, M., Trott, J., and Shaw, D. (1997). The tammar wallaby: a model to study putative autocrine-induced changes in milk composition. *J. Mammary Gland Biol. Neoplasia* 2, 299–310.
545. Grigor, M. R., Bennett, B. L., Carne, A., and Cowan, P. E. (1991). Whey proteins of the common brushtail possum (*Trichosurus vulpecula*): isolation, characterization and changes in concentration in milk during lactation of transferrin, α-lactalbumin and serum albumin. *Comp. Biochem. Physiol. B* 98, 451–459.
546. Dove, H., and Cork, S. J. (1989). I. Milk consumption and the algebraic description of the lactation curve. *J. Zool. (Lond).* 219, 385–397.
547. Collet, C., and Joseph, R. (1993). A novel member of the lipocalin superfamily: tammar wallaby late-lactation protein. *Biochim. Biophys. Acta* 1167, 219–222.
548. Nicholas, K. R. (1988). Asynchronous dual lactation in a marsupial, the tammar wallaby (*Macropus eugenii*). *Biochem. Biophys. Res. Commun.* 154, 529–536.

Knobil and Neill's Physiology of Reproduction, Third Edition
edited by Jimmy D. Neill,
Elsevier © 2006

CHAPTER 58

Oxytocin: Synthesis, Secretion, and Reproductive Functions

J. Peter H. Burbach,[1] Larry J. Young,[2] and John A. Russell[3]

Introduction, 3055
Oxytocin Systems, 3056
Endocrine Oxytocin System, 3056
Central Oxytocin System, 3065
Paracrine Oxytocin Systems in Peripheral Organs, 3067
Gene Structure and Biosynthesis, 3068
Gene Structure and Gene Products, 3068
Oxytocin Promoter, 3070
Posttranscriptional Regulation of Oxytocin mRNA, 3071
Biosynthetic Processing and Posttranslational Modification, 3072
Regulation of Oxytocin Gene Expression in Reproduction, 3073
Physiological Functions of Oxytocin in Reproduction, 3075
Role for Oxytocin in Parturition, 3075
Other Reproductive Roles of Oxytocin, 3077
Secretory Activity of Oxytocin Neurons in Pregnancy and Parturition, 3090
Pathways to Parturition, 3090
Pulsatile Oxytocin Secretion, 3096
Somatodendritic Oxytocin Release, 3097
Preventing Premature Oxytocin Secretion in Pregnancy, 3098
Concluding Remarks, 3104
Acknowledgments, 3105
References, 3105

INTRODUCTION

Oxytocin was the first peptide hormone whose chemical identity could be coupled to a reproductive function. As early as 1906, it was recognized by Sir Henry Dale that extracts of the posterior lobe of the pituitary gland contained activity with powerful capacity to stimulate contractile activity of the uterus (1). He named this activity "oxytocin" after the Greek for "rapid birth." Ott and Scott (2) and Oliver and Schäfer (3) discovered that similar pituitary extracts had "galactagogic" actions on the mammary gland. In the early 1950s, the technology to separate substances and chemically identify peptides was advanced enough to analyze the primary structure of this activity. Du Vigneaud et al. (4) elucidated the nonapeptide structure of oxytocin, Cys-Tyr-Ile-Gln-Asn-Cys-Pro-Leu-GlyNH_2, and that of vasopressin and chemically synthesized both peptides. Oxytocin and vasopressin were generally referred to as the neurohypophysial peptides, based on their source of isolation and prime storage site of the body. Since then, we have learned more about the physiology and pharmacology of these neurohypophysial

[1]Department of Pharmacology and Anatomy, University Medical Center Utrecht, Rudolf Magnus Institute of Neuroscience, Utrecht, The Netherlands;
[2]Department of Psychiatry, Center for Behavioral Neuroscience, Yerkes Research Center, Emory University, Atlanta, Georgia;
[3]Laboratory of Neuroendocrinology, Centre for Integrative Physiology, School of Biomedical Sciences, University of Edinburgh, Edinburgh, U.K.
Parts of this review were adapted from Burbach, J. P. H., Luckman, S. M., Murphy, D., and Gainer, H. (2001). Gene regulation in the magnocellular hypothalamo-neurohypophysial system. *Physiol. Rev.* 81, 1197–1267.

peptides, and this knowledge has contributed importantly to establishing general physiological principles in endocrinology, cell biology, and neuroscience.

Now, 100 years after its initial discovery, the major endocrine functions of oxytocin are well recognized in reproductive biology. As known for about a century, oxytocin acts on the uterus and has powerful uterotonic activity to assist during delivery. After parturition, oxytocin stimulates milk ejection by action on the mammary gland. We also know that the role of oxytocin in reproduction is much more diverse and subtle. Oxytocin plays a role in much of the orchestration of reproduction as an integrated process of behavior and endocrinology. Oxytocin participates in behaviors that lead to partnership and mating, controls maternal endocrinology to prepare delivery and feeding of the newborn, and also sets maternal behaviors to ensure care of the offspring.

This chapter concerns the peptide hormone oxytocin and its reproductive functions. We distinguish the different sources of oxytocin biosynthesis and secretion and link these to the potential sites of action, characterized by the expression of the oxytocin receptor gene. Many effects, roles, or functions of oxytocin in relation to reproduction have been reported. Here we attempt to provide an integrative view on oxytocin and its participation in reproduction.

OXYTOCIN SYSTEMS

Communication between chemical signals and receptors is the basis of a physiological system. When considering oxytocin systems, three issues are pertinent. First, it is important to know the sites where oxytocin, the chemical signal, is synthesized and released. Second, we need to know on which receptors oxytocin acts and where these receptors are located. Third, the origin of oxytocin and the localization of its receptors are linked together to make up a "system."

For oxytocin there is one major endocrine system. The massive storage of oxytocin in the posterior lobe of the pituitary gland (over 1 nmol in the rat) is the origin of oxytocin levels in the peripheral blood circulation, which typically are intermittently and transiently massively increased during copulation, parturition, and suckling. The oxytocin receptors on the uterus during the latest phases of pregnancy, on the mammary gland and other cells, are readily accessible to oxytocin in blood. Furthermore, there are sites of oxytocin production and release within the nervous system. These concern neurons that do not necessarily deliver to the posterior pituitary gland but innervate multiple brain regions and reach the many oxytocin receptors in the brain and brainstem. These constitute central oxytocin systems, which serve different functions sometimes coherent with those of peripheral oxytocin systems.

Finally, several types of peripheral cells and tissues produce and release oxytocin in relatively minute quantities for local action. These are the paracrine oxytocin systems.

Endocrine Oxytocin System

Magnocellular Hypothalamo-Neurohypophysial System

Anatomical Organization. Classically, the hypothalamo-neurohypophysial neurosecretory system (HNS) is defined as the neuronal complex that consists of magnocellular neurons with a 20- to 40-μm cell body diameter located in the supraoptic (SON) and paraventricular (PVN) nuclei of the hypothalamus that have axonal fiber tracts terminating on the blood capillaries of the posterior lobe of the pituitary gland, also termed the posterior pituitary or neural lobe (5) (Fig. 1). The magnocellular neurons have been identified as the production sites of oxytocin and vasopressin. The use of immunocytochemistry (6,7) and injection of retrograde tracers into the neural lobe (8,9) have since defined accessory magnocellular cell groups scattered across the hypothalamus. Oxytocin is also present in accessory magnocellular neurons of the preoptic area, the lateral hypothalamus, the anterior commissural nucleus (that does not produce vasopressin), the nucleus circularis, the perifornical nucleus, and near the third ventricle. The SON is wholly magnocellular, whereas the PVN is divided into a lateral magnocellular subdivision and a more medial parvocellular subdivision.

Oxytocin and vasopressin magnocellular neurons are found intermingled in the magnocellular nuclei, although there is some topographical segregation (6,7). Within the SON, oxytocin magnocellular neurons are mainly rostral and dorsal. The unmyelinated varicose axons of the magnocellular neurons leave the SON dorsomedially and then turn caudally to pass through the dorsal portions of the internal layer of the median eminence and pituitary stalk before reaching the neural lobe (10). As they leave the SON, the axons give rise to collaterals, many of which terminate close to the nucleus where they may contact lateral cholinergic neurons (11–13). There are also local interneurons, at least some of which are GABAergic (14).

The magnocellular PVN is more complex. In the magnocellular subdivision of the PVN, oxytocin

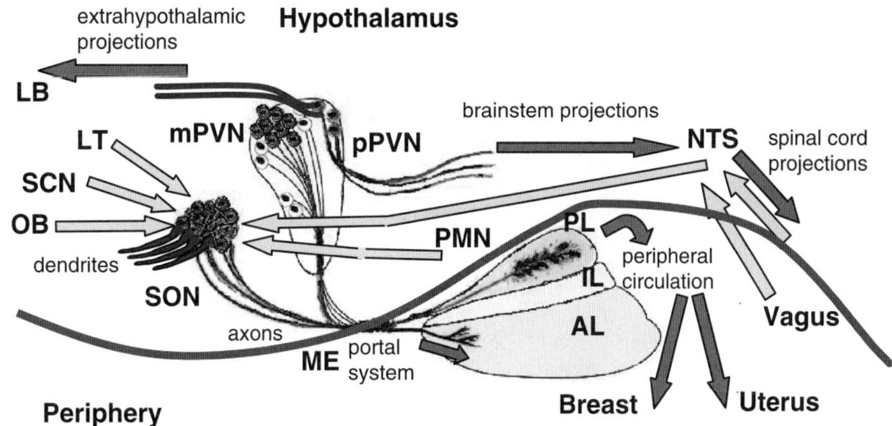

FIG. 1. Anatomical organization of the oxytocin-producing neurons of the hypothalamo-neurohypophysial system. Magnocellular neurons located in the supraoptic nucleus (SON) and the magnocellular part of the paraventricular nucleus (mPVN) send axons to the posterior or neural lobe (PL) of the pituitary gland and secrete oxytocin into blood capillaries of the systemic blood circulation to act as a classical hormone on peripheral target organs, especially the lactating breast and parturient uterus. Oxytocin is also released from these axons at axonal swellings at the level of the median eminence (ME) into the portal vascular system that supplies the anterior lobe of the pituitary gland (AL) (804). These swellings are illustrated in Fig. 2. In this way oxytocin acts as a releasing hormone on the synthesis and release of specific pituitary hormones. Magnocellular neurons may influence extrahypothalamic brain regions through secretion from their dendrites. Parvocellular oxytocin neurons in a subdivision of the PVN (pPVN) project caudally to brainstem regions (especially the nucleus tractus solitarius [NTS] and spinal cord and also rostrally to limbic brain [LB] regions) and release oxytocin as a neuropeptide in the regulation of neural systems. Inputs to the magnocellular oxytocin neurons are (1) caudal projections, including spinal and vagal afferents that relay in the NTS; (2) hypothalamic projections, including those from the premammillary nuclei (PMN); and (3) rostral projections, including those from the olfactory bulbs (OB), suprachiasmatic nuclei (SCN) and lamina terminalis (LT). IL, intermediate lobe of the pituitary gland.

magnocellular neurons are predominantly rostral, whereas vasopressin magnocellular neurons are more caudal and lateral and are surrounded by a shell of oxytocin neurons. Their axons leave the PVN laterally, then course ventrally and medially over the SON, and pass through the ventral portion of the median eminence and pituitary stalk toward the posterior pituitary (10). Collaterals of PVN magnocellular neurons terminate within the nucleus or in the nearby perifornical region (11,15). There is electrophysiological evidence to suggest that some magnocellular axons that terminate in the neural lobe may also give off collaterals to the median eminence, medial amygdaloid nucleus, and lateral septum (16). Oxytocin terminals (17) have been demonstrated to contact magnocellular neurons containing the same peptide, although the origin of these terminals, and their incidence, remains unclear. The major source of local release of oxytocin within the magnocellular nuclei is the dendrites of the neurons, which project ventrally from the SON, to the ventral glial lamina, and are gathered together, and medially from the PVN (Figs. 1 and 2) (see section on Somatodendritic Oxytocin Release).

Immunohistochemical examination at the light microscope level indicates that oxytocin fibers enter the periphery of the posterior pituitary gland (18), whereas vasopressin fibers enter the central region. However, the fibers become dispersed and terminals of the two cell types can be found intermingled throughout the neural lobe, with emphasis on the outer region for the highest density of oxytocin fibers and terminals (19). The neuronal elements of the pituitary are, as in the hypothalamic nuclei, interposed by the processes of resident astroglia, the pituicytes (20). The interrelationships of the magnocellular nerve endings, pituicytes, and capillaries also change dramatically during times of increased hormone secretion (21,22).

The soma and dendrites of the magnocellular neurons are sometimes directly adjacent to those of other magnocellular neurons, although normally interposed between them are thin astrocytic glial processes. This organization undergoes marked changes during times of high hormone demand. These alterations occur in the architecture of the magnocellular nuclei, bringing cell bodies and dendrites into direct contact and increasing synaptic inputs to

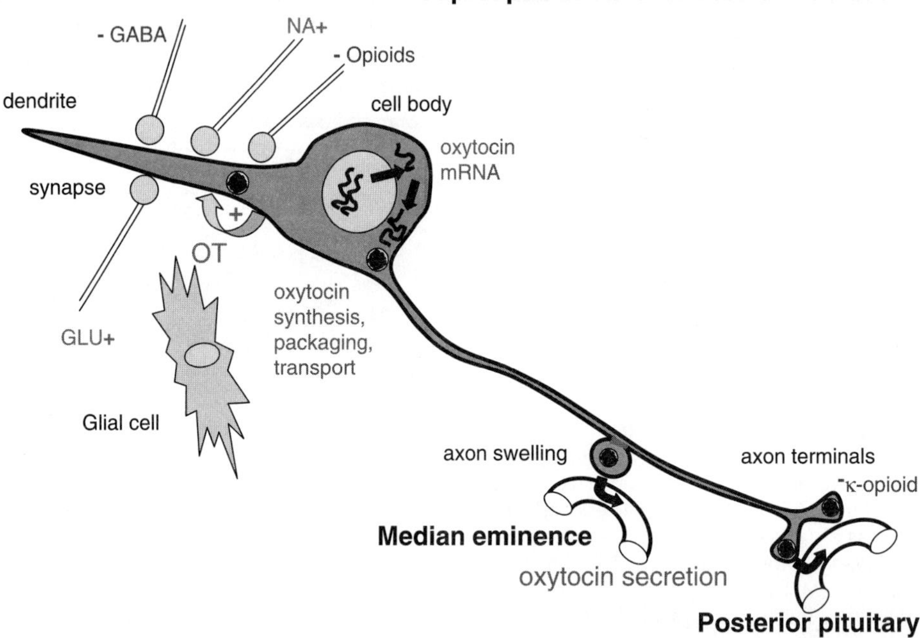

FIG. 2. Schematic of main features of a magnocellular oxytocin neuron. *Oxytocin synthesis and transport.* The cell body (soma, perikaryon) is the site of transcription of the oxytocin gene, translation into the oxytocin-neurophysin precursor and packaging in granules for transport within the axon, at 8 mm/h, for storage in the secretory terminals (5,000 per axon) and then release in the posterior pituitary gland and entry into the systemic circulation. Some oxytocin/neurophysin-containing granules are transported into the dendrites. Secretory granules accumulate in axon swellings (Herring bodies) in the internal zone of the median eminence, providing oxytocin for carriage in the hypothalamohypophysial portal system to the anterior pituitary gland. *Oxytocin secretion*. Oxytocin is released by Ca^{2+}-dependent exocytosis. Oxytocin release from axon swellings or terminals is triggered by action potentials (spikes) produced by excitation of the cell body. This is a result of greater activity of glutamate (GLU) or noradrenergic (NA) excitatory synapses over γ-aminobutyric acid (GABA) or opioid inhibitory synapses. Oxytocin released by dendrites is also excitatory. Glial cells provide the extracellular environment to which the activity of oxytocin neurons can respond, for example through the amino acid transmitter taurine.

the neurons (21,22). Magnocellular neurons make somatic and dendritic contacts via gap junctions, thus producing cytoplasmic coupling. The extent of this coupling varies with the structural plasticity within the magnocellular nuclei (21).

The varicose appearance of magnocellular axons is due to the presence of large swellings (Fig. 2), which, in the neural lobe of the pituitary, bud to form palisades of terminals along the pericapillary basal lamina (23,24). This, together with the fact that axons and swellings are themselves capable of releasing hormone (25), has left the definition of terminals in the neural lobe dependent only on their relative size and the presence of translucent microvesicles in addition to the hormone-containing electron-dense granules.

Abundance and Overlap of Oxytocin Gene Expression. The oxytocin magnocellular neuron (Fig. 2) can be distinguished from related vasopressin magnocellular neurons on the basis of the expression of the oxytocin gene as the major output gene, the oxytocin gene products, as well as the morphological and electrical properties. The level of expression of the oxytocin or vasopressin gene is very high in magnocellular neurons. A rat hypothalamus contains about 7,000 oxytocin neurons. Most estimates on the oxytocin mRNA level in these neurons agree on roughly 10^6 molecules of oxytocin mRNA per neuron as a steady-state level (26–31).

The earliest studies of the magnocellular neurons, using immunohistochemistry or in situ hybridization, identified only two distinct cellular phenotypes in the HNS, the oxytocin- and the vasopressin-expressing magnocellular neurons (6,26,27,32,32–40). These observations and others led to the generalization that expression of the oxytocin and vasopressin genes was mutually exclusive (38). However, reports of an oxytocin- and vasopressin-coexpressing magnocellular neuron phenotype (41,42), which increased in amount under specific physiological conditions (42,43), argued strongly against this simple proposition. This oxytocin- and vasopressin-coexpressing phenotype is found at a frequency of ~1%–3% in the HNS neuronal population under basal physiological conditions but increases to 17% in the SON after 2 days of lactation (42), as with a chronic osmotic stimulus or adrenalectomy (41,43).

Further studies using a sensitive reverse transcriptase polymerase chain reaction (RT-PCR) analysis of single magnocellular neurons have confirmed the existence of an oxytocin- and vasopressin-coexpressing magnocellular neuron phenotype in the HNS. Coexpressing magnocellular neurons comprise up to 17%–18% of the cell population during lactation. Furthermore, virtually all the magnocellular neurons of the oxytocin phenotype contain some vasopressin mRNA (44–46). Those of the vasopressin phenotype also contain oxytocin mRNA at low levels. Separate quantitative RT-PCR analyses of magnocellular neurons in normal female rats showed that in the oxytocin and vasopressin magnocellular neuron phenotypes, the major nonapeptide mRNA species was ~100-fold greater than the minor peptide mRNA. In contrast, in the oxytocin- and vasopressin-coexpressing magnocellular neuron phenotypes, the ratio of the two peptide mRNAs is around 2 (47). These observations show that expression of the oxytocin gene and vasopressin gene is not entirely mutually exclusive in the magnocellular neurons as originally thought. Magnocellular neurons are highly plastic neurons that have a preferred, but not exclusive, expression of either the oxytocin gene or vasopressin gene. Both genes in the same neuron are subject to regulation in response to changes in physiological status. The physiological significance of this change in peptide phenotype and its underlying mechanisms have not been clarified, however.

Coexpression of Other Biologically Active Peptides. Several other peptide genes are also expressed in oxytocin magnocellular neurons at far lower levels. However, these levels are strongly regulated by specific experimental conditions. Immunohistochemistry and in situ hybridization have shown that corticotropin-releasing hormone, thyrotropin-releasing hormone, cholecystokinin (CCK), and the human intestinal trefoil factor (48) are exclusively coexpressed in oxytocin magnocellular neurons (49–56). Other peptide genes are expressed in both oxytocin and vasopressin magnocellular neurons: pro-enkephalin, chromogranin, RESP18, and CART (50,51,57–62). For pro-enkephalin products, it has been shown by electron microscopy that opioid peptides are present in the same secretory granules as oxytocin or vasopressin and are thus cosecreted (51,63) (see section on Negative Feedback in Parturition and section on Endogenous Opioids).

Secretion of the coexpressed peptides can have modulatory effects on the secretion of oxytocin (64). Galanin released from vasopressin cells has a marked stimulatory effect on CCK release from oxytocin cells, and CCK can stimulate both vasopressin and oxytocin secretion (65). Dynorphin, on the other hand, selectively inhibits oxytocin secretion (66). The copeptides can have additional paracrine functions on pituicytes of the neural lobe. Pituicytes are known to express functional receptors for several of the copeptides and display a markedly plastic morphology in association with secretory activity of the peptidergic terminals (67,68).

Afferent Inputs to Magnocellular Oxytocin Neurons. The magnocellular neurons provide hormonal output of the nervous system that responds to changes in physiological status. The magnocellular neuron is a center of integration, receives neuronal afferent inputs from other brain systems, and is responsive to humoral signals from the periphery (Fig. 1).

Physiological conditions that require systemic release of oxytocin from the posterior lobe, such as during parturition and suckling in lactation in the female, relay afferent neural signals, autonomic from the uterus and somatic from the mammary glands, via the brainstem to oxytocin magnocellular neurons. The nucleus tractus solitarii (NTS) appears to act as an important relay for these pathways ascending to oxytocin magnocellular neurons. Whereas electrical stimulation of the NTS leads to indirect activation of vasopressin neurons via the A1 cell group of the ventrolateral medulla, there is a direct excitatory input to oxytocin neurons that is mediated, at least in part, by the A2 noradrenergic cell group lying within the NTS (69,70) (Fig. 1). One stimulus that has been used to investigate this pathway is systemic administration of the peptide CCK (71). Physiologically, this octapeptide activates CCK-A receptors on endings of the vagus nerve after distension of the gut after a meal, leading to the release, at least in the rat, of oxytocin but not vasopressin into the bloodstream from the HNS (72,73). With the use of *c-fos* and tyrosine hydroxylase immunocytochemistry combined with tract tracing, it has been possible to show that this information is relayed in A2 neurons of the NTS that project directly to the magnocellular nuclei of the hypothalamus (74–77).

Activation of catecholaminergic neurons in the NTS has been associated with a number of other stimuli that cause the release of oxytocin, including administration of nausea-producing agents (78), stress, interleukin-1 (79,80), and nicotine (81). Furthermore, A2 neurons of the NTS are activated during parturition (82), but not by suckling (83) and are thus possibly part of the pathway from the uterus to the HNS (84) (see section on Pathways to Parturition). These observations suggest that catecholaminergic neurons of the NTS may integrate information from a number of sources. However, this likely involves different subsets of NTS neurons. Neuropeptide Y- and somatostatin-containing neurons are activated after CCK (85), whereas inhibin neurons appear to mediate at least part of the suckling stimulus (86).

There are a number of other inputs to the HNS, although any direct relevance of these to reproductive functions is largely unclear (Fig. 1). They include scattered inputs from other hypothalamic nuclei, the preoptic area, septum, and limbic structures (87–91). The SON receives direct and indirect afferent information from both the suprachiasmatic nucleus and the retina (92,93). These inputs, together with the influence of the pineal gland on hormone release (94), provide substrates for the diurnal regulation of HNS function. There are well-documented direct projections from histamine neurons of the tuberomammillary nuclei to the HNS (95,96). These involve inputs to vasopressin magnocellular neurons via histamine H_1 receptors that are electrically excitatory and inputs to oxytocin magnocellular neurons via H_2 receptors that are electrically inhibitory (97). Interestingly, histamine administered intracerebroventricularly induces *c-fos* in both oxytocin and vasopressin magnocellular neurons and increases the expression of both neurohormone genes (98,99), but this may not be a direct effect. Roles for the histaminergic input, which arises from the tuberomammillary nuclei (Fig. 1), have been implied during pregnancy and parturition (100), lactation (101), dehydration (102,103), and novelty stress (104) (see section on Inhibitory Restraint: Endogenous Opioid, GABA, Histamine, Nitric Oxide). Similarly, direct afferents from the main and accessory olfactory bulbs to the SON have been described (105,106) (Fig. 1), the electrical stimulation of which can, like stimulation of the histaminergic input, lead to increased dye coupling between magnocellular neurons (21). Rostral input from the lamina terminalis (Fig. 1) conveys information about plasma osmolarity and signals from circulating relaxin and angiotensin II, which in the rat regulate oxytocin secretion in the context of its actions in salt balance (see Other Reproductive Roles of Oxytocin: Oxytocin and Hydromineral Balance).

Development. The present ideas about the mechanisms underlying the developmental origin and differentiation of the HNS follow the general notion that neurons are committed early on by an interaction between endogenous gene expression programs and extracellular signals provided by growth factors and contact with neighboring cells. These forces also mediate migratory direction and axonal pathfinding. Although anatomical descriptions of the developing HNS are limited, details on the developmental events rolling out in magnocellular neurons have been contributed recently by studies on mice in which specific transcription factors were inactivated. These studies provide insights into the molecular mechanisms of magnocellular neuron development.

Magnocellular neurons have been traced back to embryonic day (E) 10.5 using [^{3}H]thymidine incorporation and histochemical methods. Magnocellular progenitors arise in the neuroepithelium of the ventricular region around E10.5 in the mouse and E12.5 in the rat (107–113) and migrate laterally. Using an antiserum against the calcium-binding protein calbindin D-28K, Yamakuni et al. (114) found that these neurons can be detected at E11.5 in the anterior hypothalamus of the mouse as a cluster of cells separated from the ventricular surface. Two streams of cells migrate from this cluster toward the hypothalamus, one following a lateral–ventral direction that will form the SON and the other following the midline that will form the PVN. Migration takes place between E10.5 and E14.5, after which axonal extensions are formed. The first expression of neurophysin (NP), the coproduct of the oxytocin and vasopressin precursors, has been reported on E13 as the earliest time point in the rat (115,116), and an anatomical distribution on E17 has been provided (117). Innervation of the rat neurohypophysis has been reported to start around E16 (118,119). In the mouse, an antibody against the glycosylated C-terminal peptide of the vasopressin precursor (see section on Gene Structure and Gene Products) revealed immunoreactive cell bodies and fibers at E14.5. These fibers start to innervate the neurohypophysis from E14.5 onward in the mouse (120,121). As early as E15.5, vasopressin and oxytocin transcripts are present in different groups of neurons. Oxytocin expression is seen first in the PVN at E15.5 and in the SON at E18.5 (122).

There are several intriguing questions about the development of the HNS, three of which are relevant for the physiological functions that the HNS has to exert during adulthood. First, what is the molecular program required for progenitor neurons to develop into magnocellular neurons? All data on this aspect come from studies on mouse mutants in which transcription factors were inactivated by homologous recombination. Four factors and their phenotypic relationships have been studied in detail. These studies can now be fitted in a hierarchical scheme, suggesting a sequence of transcriptional events intrinsic to the development of magnocellular neurons. The knockout of the POU homeobox gene *Brn2* results in a defect in migration of early magnocellular neurons around E12.5 (113,123). Consequently, no magnocellular neurons are present at birth, and no

HNS has been established. A very similar phenotype has been described for a mouse mutant in which the basic-helix-loop-helix–PAS gene *Sim1* was inactivated (113), again caused by early apoptosis of magnocellular neurons that fail to migrate normally. On the basis of analysis of *Brn2* expression in *Sim1* mutants, it has been proposed that *Sim1* may be required for induction or maintenance of expression of *Brn2*. This would explain the common phenotype and indicates that *Sim1* is upstream of *Brn2*. In relation to these findings, another basic-helix-loop-helix–PAS gene, *Arnt2*, has been studied in a natural mutant mouse strain (124). This strain, C^{112K}, fails to express the *Arnt2* gene due to a deletion in the albino locus (125). *Arnt2* maps within the C^{112K} deletion. Several basic-helix-loop-helix–PAS proteins need to heterodimerize for transcriptional action. The phenotype of the C^{112K} strain appears identical to the *Sim1* null mutant regarding the hypothalamic defects. Moreover, *Sim1* and *Arnt2* are coexpressed in magnocellular neurons and physically interact in vitro (124). These data show that *Sim1–Arnt2* heterodimerization forms the transcriptionally active complex that is required for development of the magnocellular neurons.

A fourth gene for a transcription factor, *Otp*, has been inactivated resulting in a phenotype very similar to those obtained by null mutation of *Brn2* or *Sim1* (124,126). This gene is expressed in magnocellular neurons. The homeobox *Orthopedia (Otp)* null mice lack the morphological features of the PVN and SON and do not express vasopressin and magnocellular oxytocin. The detailed analysis of the mutant mice lacking one of the four transcription factors has allowed the proposition of a genetic pathway that outlines the hierarchy of these factors (127,128). Upstream is *Otp* that is already active during the proliferation phase of progenitor neuroblasts of magnocellular neurons. *Otp* is required for the maintenance of *Brn2* expression. Parallel to this interaction, *Sim1* and *Arnt2* act upstream of *Brn2* and also affect its maintenance. One other mouse mutant has been described from which defects in the development of the HNS may be deduced. This concerns a mouse mutant for the winged helix/forkhead transcription factor *Mf3*, also termed *Fkh5* and *HFH-e5.1*, which is expressed in the hypothalamus (129,130). This mutant displays severe developmental defects in the diencephalon, mammillary body region, and midbrain. Interestingly, $Mf3^{-/-}$ females cannot eject milk but can be stimulated to give milk after intraperitoneal administration of oxytocin (129). This defect may be due to lack or reduction in oxytocin synthesis or release from magnocellular neurons. It has not been established whether there is a developmental defect in the HNS or whether this phenotype is due to altered circuitry elsewhere in the brain. Another mutant with defective milk ejection and reduced number of oxytocin-producing magnocellular neurons is the paternally expressed zinc finger gene $Peg3^{-/-}$ mouse (131). Mutant females display a dramatic reduction in maternal behavior as well.

The second question is whether vasopressin- and oxytocin-producing neurons have a common origin and, if so, how is their terminal specification rolled out? The many similarities of vasopressin- and oxytocin-producing magnocellular neurons argue in favor of a common progenitor cell for both cell types and a fate-determining event in the course of development. Direct evidence for a common progenitor cell is lacking, however. This is partly because the discriminating property, namely expression of the vasopressin gene or oxytocin gene, is only visible relatively late in development, that is, after E14.5. With the use of double-label in situ hybridization, it was demonstrated that at E16.5 vasopressin and oxytocin transcripts exist in separate populations of neurons (122). The observation that a single cluster of calbindin D-28K positive cells is present in the anterior hypothalamus (113) suggests that a common progenitor may exist. The relatedness of the two magnocellular types is also illustrated in adulthood by the plastic overlap in expression of the vasopressin and oxytocin genes in certain physiological conditions (see section on Gene Structure and Gene Products).

Third, how are axons of the magnocellular neurons guided to the formation of the neurohypophysis? It is assumed that the area that forms the neurohypophysis secretes growth factors to attract and guide outgrowing axons from magnocellular neurons (123). Four mouse mutants have been described in which magnocellular neurons are absent at birth. In one of them, the $Brn2^{-/-}$ mouse, it was reported that the neurohypophysis was extremely hypoplastic (113,123). Nakai et al. (113) reported that pituicytes were present in $Brn2^{-/-}$ newborn animals, but Schonemann et al. (123) described the total absence of pituicytes leaving a vacated neurohypophysis and an infolding of the intermediate lobe. This discrepancy has not been solved as yet. In the $Brn2^{-/-}$ mice described by Schonemann et al. (123), a normal population of pituicytes is present at E16 but starts to disappear concurrently (132). This finding suggests that magnocellular neuron axons elaborate a trophic effect on the pituicytes of the neurohypophysis.

In the $Sim1^{-/-}$ and $Arnt2^{-/-}$ mice, there is a similar absence of magnocellular neurons at birth (124,128). For these mice, the neurohypophysis was reported to be ~50% smaller and to contain 50% less pituicytes.

It was noted that the pituicytes in Sim1$^{-/-}$ mice were more densely packed (128).

A mouse mutant was created in which regulated secretion in neurons is impaired through null mutation of Munc18-1, a key protein in the core complex of synaptic vesicle docking and secretion (133). In these mice there is hardly any axonal innervation by magnocellular neurons of the neurohypophysis, which appears expanded and with disorganized pituicytes (121). Taken together, available studies indicate that the pituicytes of the neurohypophysis develop independently but are organized by the axonal innervation from the hypothalamus. Whether this organizing influence includes maintenance and/or proliferation is uncertain. It will be highly relevant to identify the trophic factors involved in the organizing events, which are possibly growth factors or cell surface molecules. In this respect, the expression of "7365," a protein with epidermal growth factor and follistatin modules, in the HNS is notable (134) and deserves further attention.

Although the expression of vasopressin and oxytocin genes is relatively low during late embryonic development, the expression is strongly enhanced in the first 2 weeks postnatally (135,136). This enhancement follows the recruitment of physiological control mechanisms in the newborn, such as the regulation of normal diuresis by the end of the second week of age (137). Both the vasopressin gene and oxytocin gene display a similar enhancement, and both SON and PVN magnocellular neurons are involved. After the first month there is still a mild increase in levels of vasopressin mRNA and oxytocin mRNA, up to ~3 months. This increase in mRNA levels that ranges between 1.2- to twofold in Wistar rats is correlated with a twofold increase in pituitary peptide levels, whereas plasma peptide levels remain unaltered (136,138). There may be strain differences, with respect to postnatal development, because the spontaneously hypertensive rat displays a more rapid increase in postnatal vasopressin and oxytocin gene expression (138,139).

The mature magnocellular neurons respond to perturbations in water balance by releasing large amounts of stored vasopressin and oxytocin into the general circulation. This is accompanied by functional remodeling of magnocellular neuron activity, the molecular basis of which is still far from understood. The snapshot of epigenetic plasticity discussed here is mediated by receptors and coupled transduction systems as well as by transcription factors of which identified components are described in Gene Structure and Biosynthesis.

Oxytocin Receptor and Expression in Peripheral Target Tissues

Oxytocin Receptor. An in-depth review of the oxytocin receptor can be found elsewhere (140). There is a single receptor gene for oxytocin in the known mammalian genomes (141,142). The structure of the receptor, molecular modeling of ligand binding and subsequent activation (143,144), and the post–oxytocin receptor signaling mechanisms in the myometrium (145) are well described. The human oxytocin receptor is 389 amino acids long and is located on chromosomal position 3p25 (contig NT 022517.17, locus OXTR). The oxytocin receptor gene consists of four exons, two of which contain the coding region. Different transcripts from the single oxytocin receptor gene have been found in the mammary gland (146). These result from alternative polyadenylation. Editing of nucleotides in the 3′-untranslated region and the coding region has been suggested (147).

Site-directed mutagenesis has revealed details of the key residues involved in extracellular ligand binding, the conformation of the receptor, and intracellular $G_{\alpha q/11}$ coupling to phospholipase C (143,148). Modeling studies based on the mutagenesis data show a three-dimensional image of the docking of oxytocin and how the altered conformation of the activated receptor forms a docking pocket that allows the binding of G protein to the intracellular surface of the receptor (143,149). Within the target cell membrane, the oxytocin receptors are more concentrated in cholesterol-rich caveolin-containing regions. This milieu protects high affinity receptors against thermal denaturation, evidently by interactions of cholesterol with specific residues in transmembrane domains of the receptor (140).

The cellular responses to oxytocin are mediated by the oxytocin receptor protein, as a typical seven-transmembrane domain G-protein–coupled receptor. The oxytocin receptor is structurally related to the vasopressin receptor subtypes, V1a, V1b, and V2, and together these receptors form a subfamily of G-protein–coupled receptors with amino acid homologies ranging from 40% to 50%. Within this receptor subfamily for the neurohypophysial hormones, the oxytocin receptor displays the least specificity. It can bind vasopressin with a tenfold lower affinity than oxytocin, and for this reason the oxytocin receptor has been indicated as a "nonselective" receptor (150). The V1a, V1b, and V2 receptors are selective for vasopressin; the V1b receptor is activated only at very high concentrations of oxytocin, which may be relevant for observed effects of oxytocin on the release of proopiomelanocortin peptides from the anterior

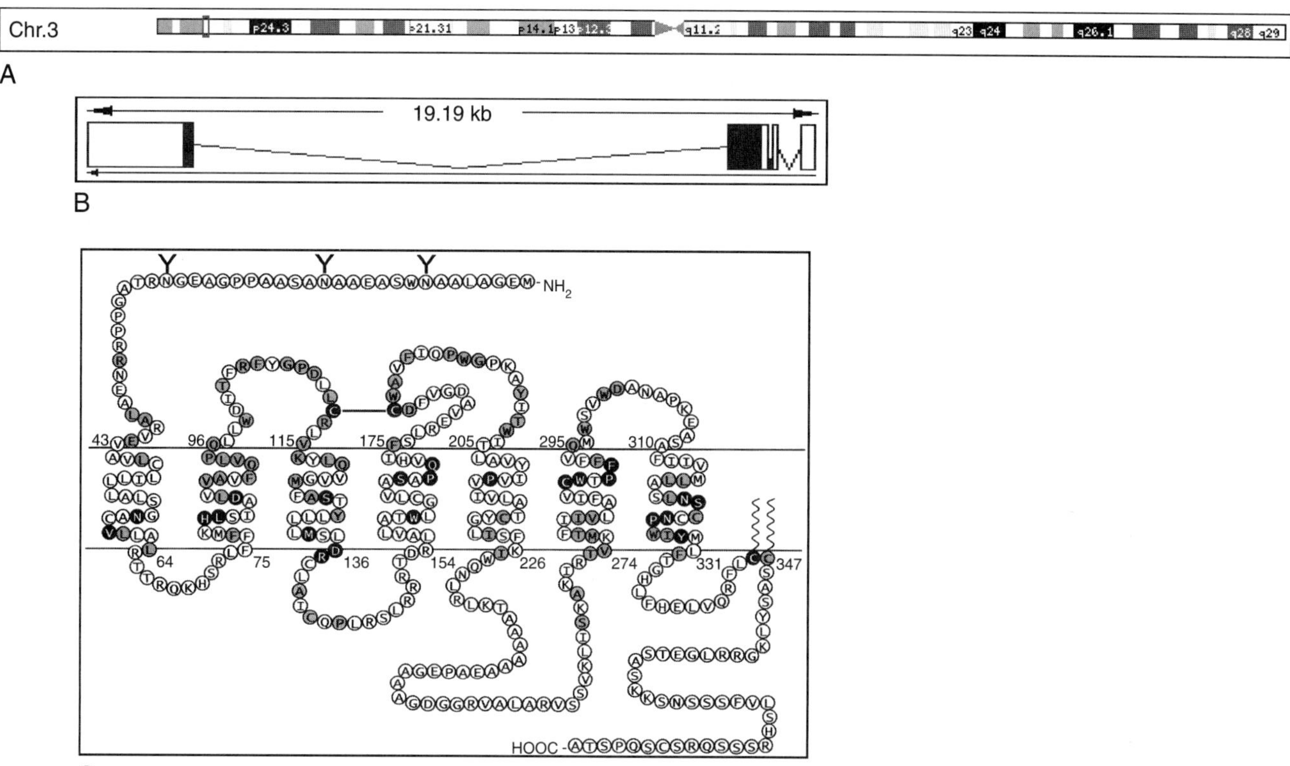

FIG. 3. Gene and protein structure of the oxytocin receptor. The oxytocin receptor in humans is a single gene located on chromosome 3p25 (A). It spans almost 20 kB having four exons (B). The receptor protein is 389 amino acids long and has a predicted structure of seven transmembrane domains that reside in the cell membrane, typical of the G-protein–coupled receptor family (C). The three-dimensional structure provides extracellular access to a binding pocket formed by critical transmembrane residues and the N-terminal tail [reviewed in (140)]. The intracellular part couples to G protein G_{q11}, which triggers mobilization of intracellular Ca^{2+}. The biochemical pathways of postreceptor signaling are delineated in Fig. 7. Elements of the figure are from http://www.ensembl.org and (140).

pituitary gland (see section on Other Reproductive Roles of Oxytocin: Stress).

The postreceptor signal transduction events triggered by oxytocin receptor activation have mostly been derived from studies of the myometrium (145, 151,152) (see section on Control of Cyclic Uterine Oxytocin Receptor Expression in Uterus and Mammary Gland). The receptor is coupled to G_{q11}, which stimulates the activation of phospholipase, which results in an accumulation of inositol trisphosphate (IP_3) and diacylglycerol. IP_3 triggers the release of Ca^{2+} from intracellular stores, whereas diacylglycerol results in the activation of protein kinase C (PKC), leading to protein phosphorylation. In smooth muscle tissue, such as myometrium or myoepithelial cells, calcium–calmodulin complex formation stimulates myosin light chain kinase activity, leading to smooth muscle contraction. In neurons, intracellular Ca^{2+} accumulation may result in increased neurotransmitter release. The oxytocin receptor is expressed in many tissues throughout the body, including brain, pituitary, kidney, ovary, testis, thymus, heart vascular endothelium, osteoclasts, myoblasts, pancreas, and adipose tissue.

Female Reproductive Tissues

Uterus. Oxytocin has two distinct functions in the uterus. Activation of oxytocin receptors in the myometrium results in the classic uterine contractions associated with parturition. In contrast, in the endometrium/decidua, oxytocin stimulates the production of prostaglandin (PG) $F_{2\alpha}$ (153,154), which stimulates luteolysis (see section on Luteolysis). There may also be physiological effects of oxytocin on the uterus mediated by paracrine effects of oxytocin synthesized locally. Effects of oxytocin are dynamically regulated by changes in oxytocin and/or oxytocin receptor expression during pregnancy and at parturition. In the rat, oxytocin is synthesized in the uterine epithelium and oxytocin mRNA increases approximately 100-fold at term and reaches levels 70-fold higher than in the hypothalamus (155). In addition, the sensitivity of the uterus to oxytocin increases markedly at the onset of parturition (156).

In humans, circulating concentrations of oxytocin in maternal plasma are not increased early in labor, and neither is oxytocin gene expression in the uterus. Oxytocin receptors are found in both the myometrium and glandular endometrial in the human uterus. However, the oxytocin receptor is much more abundant in the myometrium at parturition compared with endometrium (157,158). In both rat and human uterus, oxytocin receptor protein and mRNA increase dramatically at parturition (158–161). In rat uterus, oxytocin receptor mRNA is increased 25-fold during gestation: 4.5-fold during the first 21 days and then sixfold within the 24 hours between day 21 and the onset of parturition. Uterine oxytocin receptor mRNA levels fall rapidly by 85% within 24 hours of parturition. In that study, oxytocin receptor mRNA was concentrated in the myometrium and not detected by in situ hybridization in the endometrium. After parturition, the density of oxytocin receptor and oxytocin receptor mRNA decline, presumably to avoid continued uterine contractions during lactation. In the human myometrium, both oxytocin receptor protein and mRNA increase dramatically at the time of parturition.

During the rat estrous cycle, oxytocin receptor mRNA levels in the uterus increase about twofold between metestrus and proestrus, whereas oxytocin binding increases tenfold during that time (161). The hypothalamic oxytocin mRNA levels are increased 1.5- to twofold at estrus, whereas circulating oxytocin levels rise during proestrus and peak at estrus (161–163). The changes in uterine oxytocin receptor expression are likely due to changes in estradiol concentration because estrogen treatment also significantly increases oxytocin receptor binding and expression in the uterus (161). In the human uterus, oxytocin receptor gene expression in epithelial cells is elevated at the time of ovulation relative to other phases of the cycle (157).

Ovary. Oxytocin and oxytocin receptors have been found in the corpora lutea of several species (164,165). Oxytocin appears to have both luteotropic and luteolytic effects in the porcine corpus luteum. In vitro studies suggest that oxytocin has direct inhibitory effects on progesterone secretion. Simultaneously, however, in vivo studies demonstrate that oxytocin stimulates estrogen secretion, which in turn facilitates progesterone induction. It has been suggested that in young luteal cells, which are rich in oxytocin receptors, oxytocin is luteotropic due to its indirect effects on progesterone secretion. Later, however, oxytocin receptor densities decrease and the inhibitory effects on progesterone synthesis predominate, and luteolysis ensues (164).

Mammary Tissues. The most well-established physiological role for oxytocin is in the regulation of the milk ejection reflex during lactation. Oxytocin secreted in a pulsatile burst-like manner into the circulation from the neurohypophysis in response to suckling stimulates contraction of the myoepithelial cells, resulting in the ejection of milk from the secretory alveoli, through the lactiferous ducts, sinuses, and nipple of the lactating breast. The oxytocin antagonist atosiban inhibits milk ejection, and oxytocin gene knockout mice fail to eject milk accumulated in the lactating mammary gland (166,167), demonstrating this essential function of oxytocin. In rats, there is a 100-fold increase in oxytocin receptor expression in the mammary gland between the end of pregnancy and late lactation (168), and a similar increase in oxytocin receptor mRNA has been reported (169). In contrast, another study failed to find these differences when comparing lactating and nonlactating human breast samples (170). Immunocytochemical analysis reveals unexpectedly that oxytocin receptors are located at highest levels in the ductal/glandular epithelium compared with myoepithelial cells (170).

Male Reproductive Tissues

Penis and Penile Erection. Oxytocin participates, together with other transmitters like dopamine, glutamate, and nitric oxide, in the central regulation of penile erection (171). The parvocellular paraventricular hypothalamic oxytocin neurons that project to extrahypothalamic brain areas and the spinal cord are involved in this regulation (172,173). This should be considered a central function of oxytocin that acts in coordination with hormonal oxytocin in peripheral male organs to promote reproduction (see section on Paracrine Oxytocin Systems in Peripheral Organs). A multiple involvement of oxytocin in reproductive functions in the male has been proposed. Oxytocin has been implicated in the control of erectile function and copulation (171,174,175). In several species, a peripheral oxytocin pulse is secreted with ejaculation and may be involved in stimulating smooth muscle cells associated with ejaculation, although evidence for this is not clear (176). Several studies have found a significant rise in plasma oxytocin concentration during sexual arousal and orgasm in men and women (177–179). Oxytocin contracts the corpus cavernosum of the penis, which expresses oxytocin receptors (180,181). So the oxytocin secreted by the posterior pituitary at ejaculation may, after promoting sperm transport, lead to postcoital detumescence. The oxytocin sensitivity of the corpus cavernosum is estrogen dependent (180).

Testis. In several species, including humans, oxytocin receptors are present in the testis, epididymis, and prostate (182). Two functions have been associated with oxytocin in the testis: regulation of seminiferous tubule contractility and the modulation of steroidogenesis. Seminiferous tubules are spontaneously contractile in vitro, and this contractile activity increases with the addition of oxytocin (183). This function of oxytocin could serve in enhancing sperm shedding and transport, because oxytocin has been reported to increase the numbers of sperm leaving the testis in the pubertal rat, mouse, and ram (184–186). Oxytocin receptors are not found on the smooth muscle myoid contractile cells of the seminiferous tubules but are found instead on Sertoli cells (176,187,188), which perhaps could modulate contractile activity of the seminiferous tubules through an unknown mechanism. Oxytocin receptors are also located on the testosterone producing Leydig cells, and oxytocin treatment results in increased testosterone secretion (189).

The testes of the mature rat also produce oxytocin, principally in the Leydig cells, but oxytocin production also requires a functioning seminiferous tubule epithelium (190). Luteinizing hormone (LH) stimulates not only testosterone secretion but also oxytocin production (191). Despite these local intratesticular modulatory actions of oxytocin in the rat and decreased circulating testosterone concentration in mice overexpressing an oxytocin gene in the testes (192), male transgenic mice with inactivation of the oxytocin gene are fertile (165,193), but male mice with inactivation of the oxytocin receptor gene may have reduced fertility (194,195).

Prostate. Oxytocin is also synthesized in the prostate (184), and oxytocin increases the resting tone of prostate tissue in a variety of species and evoked contractile activity in prostate tissue in vitro (184,196,197). Oxytocin released from the posterior pituitary gland in males with ejaculation may have stimulatory actions on duct contractility and testosterone metabolism in the prostate. This suggests that oxytocin is involved in the expulsion of prostatic secretions at the time of ejaculation. Oxytocin also stimulates growth in the rat prostate (198). When treated with oxytocin, 5α-reductase activity in the prostate is increased, resulting in an increase in the conversion of testosterone into dihydrotestosterone (184). The action of oxytocin in increasing 5α-dihydrotestosterone production might be involved in prostatic hypertrophy (184). Oxytocin content is increased in the human prostate with benign hypertrophy, and in primary culture androgens stimulate oxytocin release, in contrast with inhibitory actions of androgens in the rat prostate (199,200). In view of the stimulatory actions of oxytocin on prostatic growth and 5α-reductase activity, oxytocin may have a role in prostatic hypertrophy (200).

Breast Cancer. The oxytocin receptor is expressed in human breast cancer cells in most primary breast tumors (201,202). Receptor expression is regulated by estradiol and progesterone in breast cancer cell lines (203). Primary endometrial adenocarcinoma cells also express the oxytocin receptor (202). Oxytocin evidently inhibits proliferation in vitro in these different types of tumor cells in vitro and inhibits growth of a mammary tumor cell line in mice (202,204). However, the growth of some other types of tumor cell lines is stimulated by oxytocin (202). The direction of effect on proliferation is related to the disposition of the oxytocin receptor in the cell membrane and the consequent differences in postreceptor signaling pathway activation (202). Possible applications of these findings to target radiolabels or chemotherapy to oxytocin receptor–expressing tumors have been proposed (202).

Central Oxytocin System

Parvocellular and Magnocellular Hypothalamic Neurons

The hypothalamus also harbors parvocellular oxytocin neurons; these typically have smaller cell bodies (10–15 μm cell body diameter) than the magnocellular neurons (20–40 μm). Parvocellular neurons have different functions from magnocellular neurons, and it is not known whether their major genes are regulated by the same mechanisms. The phenotype of null mutants for several transcription factors indicated that parvocellular neurons expressing corticotropin- releasing hormone, thyrotropin releasing hormone, and somatostatin may have a developmental origin common to magnocellular neurons. For oxytocin-producing neurons, parvocellular and magnocellular neurons contribute to innervation of other brain nuclei. In this way oxytocin is implicated in the physiological control of brain function, some of which actions concern reproduction. There are axonal projections to brainstem and spinal cord (34) as well as to limbic and olfactory areas (205–207).

An important parvocellular oxytocin-expressing cell group is present in the caudal portion of the PVN. These neurons project to the hindbrain, brainstem, and spinal cord. Projection areas include the NTS and the rostroventrolateral medulla (208). This pathway has been implicated in the transmission of leptin signals from the hypothalamus to the NTS, which receives satiety signals from the gastrointestinal tract (209), and in the regulation of the autonomic nervous system.

A second oxytocin system innervates the hypothalamic ventromedial nucleus. This region receives oxytocin from parvocellular and magnocellular neurons (210). Oxytocin can trigger reproductive behavior, particularly lordosis behavior, in this nucleus (see section on Female Sexual Receptivity) (211,212).

Distribution of the Oxytocin Receptor in the Brain

The distribution and regulation of oxytocin receptors in the central nervous system have been most extensively studied in the rat. Here we focus primarily on the forebrain receptor systems related to reproduction. Oxytocin receptors were first described in the rat brain using receptor autoradiography, and subsequently in situ hybridization and PCR techniques largely confirmed these observations. In the rat forebrain, specific oxytocin receptor binding sites are concentrated in the anterior olfactory nucleus; the shell of the nucleus accumbens; islands of Calleja; lateral septum; medial, central, and basomedial nuclei of the amygdala; and ventromedial nucleus of the hypothalamus (213–216). Oxytocin receptor binding sites have also been found in the PVN and SON using receptor autoradiography in lactating females, although these sites are typically not detected in other studies (215). In addition, in vitro binding assays have detected oxytocin receptors in the medial preoptic area and ventral tegmental area, and these sites are up-regulated at the time of pregnancy and are thought to be involved in the regulation of maternal behavior (217). The distribution of oxytocin receptor mRNA in the brain generally matches that of binding sites, including high levels of expression in the PVN and SON (146,218–220).

The distribution of oxytocin receptor binding sites has been described in the brain of several species, including mouse, vole, and human (221–223), and distribution of oxytocin receptors is highly species specific (Fig. 2). For example, the mouse has much higher concentrations of oxytocin receptor in the septum and hippocampus and lower concentrations in the bed nucleus of the stria terminalis than the rat. Oxytocin receptor distribution even differs dramatically between closely related species, and these differences are associated with different mating patterns. For example, the monogamous prairie vole has high levels of oxytocin receptor in the nucleus accumbens, whereas this region is devoid of receptors in the nonmonogamous montane vole (223). Interestingly, the distribution of oxytocin receptor binding sites changes dramatically during development in the rat, with receptors being more broadly distributed before postnatal day 22 than after (224).

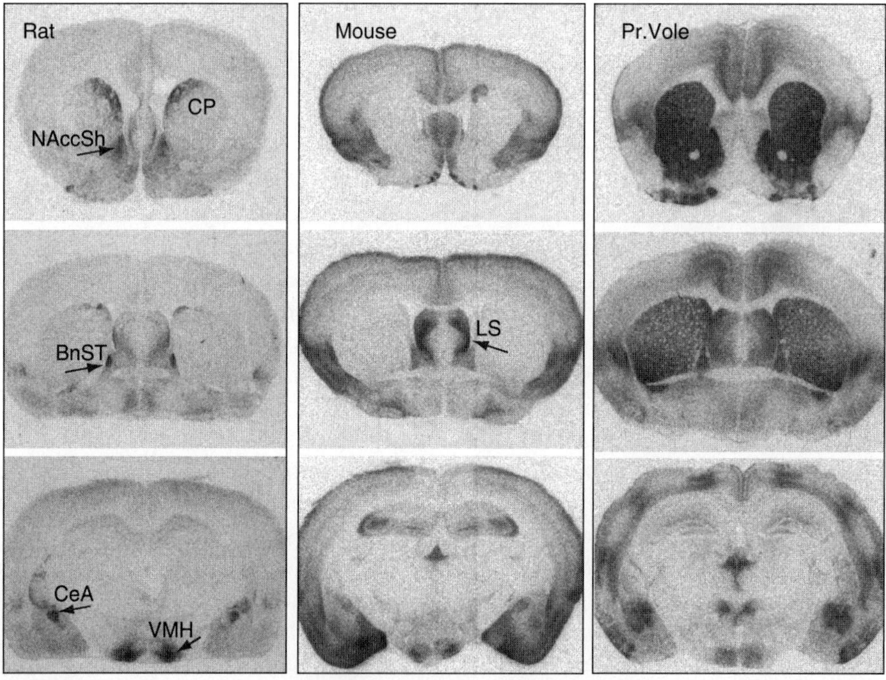

FIG. 4. Receptor autoradiograms illustrating the distribution of oxytocin receptor binding sites in the forebrain of rat, mouse, and prairie vole. Note that although the distribution of receptors is conserved in some brain regions, such as the lateral septum (LS), bed nucleus of the stria terminalis (BnST), central nucleus of the amygdala (CeA), and the ventromedial nucleus of hypothalamus (VMH), several striking species differences are apparent. For example, in the rat striatum, oxytocin receptors are restricted to the dorsal caudate putamen (CP) and the shell of the nucleus accumbens (NAccSh), whereas oxytocin receptors are abundant throughout the prairie vole striatum and absent in the mouse striatum.

For example, before weaning, high levels of oxytocin receptor are detected in the nucleus accumbens but nearly disappear thereafter.

Regulation by Gonadal Steroids. Certain populations of brain oxytocin receptors are exquisitely regulated by gonadal steroids. In male rats, castration dramatically reduces and testosterone or estrogen replacement increases oxytocin receptor binding in the islands of Calleja and the ventromedial nucleus of the hypothalamus but not in other oxytocin receptor populations (225–227). Similarly, estradiol increases oxytocin receptor binding density in the ventromedial nucleus of the hypothalamus of ovariectomized females (228–230). In situ hybridization studies suggest that this is due to transcriptional regulation by the estrogen receptor (ER). In castrated male rats, both testosterone propionate and estradiol benzoate increase oxytocin receptor mRNA in the ventromedial nucleus of the hypothalamus, whereas dihydrotestosterone was ineffective (231). Furthermore, oxytocin receptor mRNA levels are elevated during the morning of proestrus in the cycling female rat (232). These changes in oxytocin receptor expression are thought to play a critical role in the regulation of female sexual behavior (see section on Female Sexual Receptivity).

Interestingly, in the male mouse, testosterone increases oxytocin receptor binding in the lateral septum but decreases binding in the ventromedial nucleus of the hypothalamus (222). However, estrogen significantly increases oxytocin receptor density in the amygdala of mice, and this effect is mediated by ER-α, because the effect disappears in ER-α knockout mice (233).

Central Oxytocin Systems and Release

Parvocellular hypothalamic oxytocin neurons give rise to numerous central projections, including the septal region, nucleus accumbens, hippocampus, amygdala, mediobasal hypothalamus, and the autonomic zones of the brainstem and spinal cord (234). The release of oxytocin from central projections has been poorly characterized directly, although it is inferred, based on the effectiveness of oxytocin antagonists infused into various extrahypothalamic nuclei at modulating behaviors. Oxytocin release within the central nervous system has been most extensively studied in the sheep in relation to the regulation of maternal and sexual behavior.

In vivo microdialysis studies have demonstrated an increase in extracellular oxytocin in the olfactory bulb at parturition as well as by artificial vaginocervical stimulation (235). Retrodialysis of oxytocin into the olfactory bulb resulted in an increase in acetylcholine, noradrenaline, and γ-aminobutyric acid (GABA), suggesting that oxytocin release during parturition may play a role in the formation of the olfactory memory in the ewe for her lamb. Other studies have demonstrated that oxytocin is released within the mediobasal hypothalamus in receptive female sheep after an intromission. Further studies suggest that this release may serve to inhibit receptive behavior in the sheep after multiple intromissions (236). Vaginocervical stimulation has also been shown to stimulate oxytocin release within the spinal cord of the rat (237). This study found that oxytocin release in the spinal cord after vaginocervical stimulation was higher in estrogen-treated females than in vehicle-treated females. In addition, oxytocin levels in the cerebrospinal fluid were found to increase several-fold in male rats after ejaculation (238). Oxytocin release within the septum and hippocampus has also been documented in the female rat during lactation and suckling (239).

In vivo microdialysis studies have also documented the release of oxytocin within the PVN nucleus as a result of various physiological stimuli [reviewed in (240)]. Stimuli increasing oxytocin release in the PVN include parturition, suckling, and maternal defense.

Paracrine Oxytocin Systems in Peripheral Organs

Although by far the most abundant expression of the oxytocin gene takes place in magnocellular neurons of the hypothalamus, additional sites of oxytocin gene expression have been reported. These concern several peripheral organs and tissues that are often related to, or part of, the endocrine system. Minute production of oxytocin in peripheral organs is considered to serve paracrine functions, if the appropriate receptor is also present. Nonhypothalamic sites of oxytocin gene expression have been assigned to the following organs and tissues that also express the oxytocin receptor:

- *Uterus:* Although the uterus is one of the main target tissues of oxytocin during delivery, the uterus expresses the oxytocin gene locally in some species. The oxytocin gene was found to be expressed in the rat uterine epithelium at term. The estrogen-induced elevation of oxytocin mRNA levels was restricted to 3 days (241). Immunohistochemistry localized oxytocin to chorionic trophoblast, decidual stroma, and glandular epithelium (242).
- *Placenta:* Oxytocin gene expression was shown to be present in rat placenta and amnion (243) and in humans in amnion, chorion, and decidua (140,244).

- *Ovary:* Soon after elucidation of the cDNA and gene for oxytocin, it was recognized that the bovine corpus luteum abundantly expressed the oxytocin gene (165,245). This occurs not only in ruminants, but also to a lesser extent in primates and rodents (165,246). In the bovine, expression concerns granulosa cells and preovulatory follicles (247). In women, oxytocin is produced in the corpus luteum (248). In the marmoset monkey, in vivo and in vitro studies suggest that oxytocin can act as a follicular luteinization factor. It was observed that after human chorionic gonadotropin treatment, almost all granulosa cell layers in antral follicles showed oxytocin immunoreactivity. Oxytocin was produced only by granulosa cells derived from preovulatory follicles (249). In women oxytocin is also expressed in cumulus cells that surround the oocytes (250).
- *Testis and prostate:* Oxytocin transcripts have been identified in the testis and prostate. Both the Sertoli cells and the Leydig cells have been indicated as site of local oxytocin production (251,252).
- *Heart and vasculature:* The heart evidently produces locally acting oxytocin (253). Relatively high oxytocin concentrations were measured in the atrium, but there was no direct evidence for oxytocin mRNA. Recent studies by RT-PCR indicated expression of the oxytocin gene in the aorta and vena cava (254) and showed that oxytocin and oxytocin receptor expression increased strongly in the first postnatal weeks. These authors suggested from these observations and from in vitro studies that oxytocin can be involved in developmental formation of the coronary vessels (255).
- *Pituitary gland:* One report indicated expression of oxytocin in the anterior lobe of the pituitary gland (256). The anterior pituitary contains prolactin cells that are targets for oxytocin. Although the expression of "copeptides" is common in the peptidergic cells of this gland, local oxytocin production could contribute to the control of hormone production and release in a paracrine fashion. Oxytocin mRNA has also been detected in the mouse neurohypophysis (257) and in the bovine neurointermediate lobe (258).
- *Pancreas:* Oxytocin has been detected in the pancreas and interpreted as derived from local production (259).
- *Thymus:* The presence of oxytocin gene products in the thymus has been indicated (260,261), but the precise nature of the detected materials and local synthesis could not be substantiated.

GENE STRUCTURE AND BIOSYNTHESIS

Gene Structure and Gene Products

The organization of the oxytocin gene is peculiar. The oxytocin gene is small and located within the same chromosomal locus as the vasopressin gene on human chromosome 20p13 (contig NT 011387.8, locus OXT). The distance between the two genes is short (3–11 kb), and they are in a head-to-head orientation (Fig. 5) (262–265). Although the exon organization is simple, the genomic control regions are complex and likely dependent on large parts of the locus (266). Notably, in the vasopressin–oxytocin locus of the rat, long interspersed repeated DNA elements are present and transcribed. Expression of long interspersed repeated DNA elements is predominantly in the brain (267). A role of these long interspersed repeated DNA elements in the expression of the vasopressin and oxytocin genes is not known.

The oxytocin gene consists of three exons and comprises less than 1 kb. The structural organization of the vasopressin gene is very similar to the oxytocin gene. The first exon of the oxytocin gene, exon A, codes for a signal peptide, the hormone (oxytocin), a three-amino acid spacer (-Gly-Lys-Arg-), and the first nine NH_2-terminal amino acids of NP. Exon B codes for the highly conserved mid-portion of NP (amino acids 10–76), and exon C of the oxytocin gene codes for the remaining COOH-terminal amino acids of NP. The vasopressin gene has an additional COOH-terminal glycopeptide of 39 amino acids. In total, the mammalian oxytocin prohormones contain 106 amino acids; the vasopressin precursor is 145 amino acids long.

The close linkage and homologies between the oxytocin and vasopressin genes suggest that they are derived from duplication of an ancestral gene. The similarities extend to the structures of the hormonally active nonapeptides. Still, these differences are the structural basis for selectivity for their respective receptors. About 17 variations in nonapeptide structures are found in the animal kingdom (140). Oxytocin and vasopressin are closely related to each other in primary structure and differ in only two amino acids at position 3 and 8.

The strong conservation of the NP domain of the different oxytocin/vasopressin-related prohormones further strengthens this assumption. The NP domain is remarkably well conserved and exists in all vertebrate and invertebrate species. It is generally believed that the mammalian oxytocin/vasopressin family developed from a common ancestral molecule by gene duplication about 450 million years ago (268).

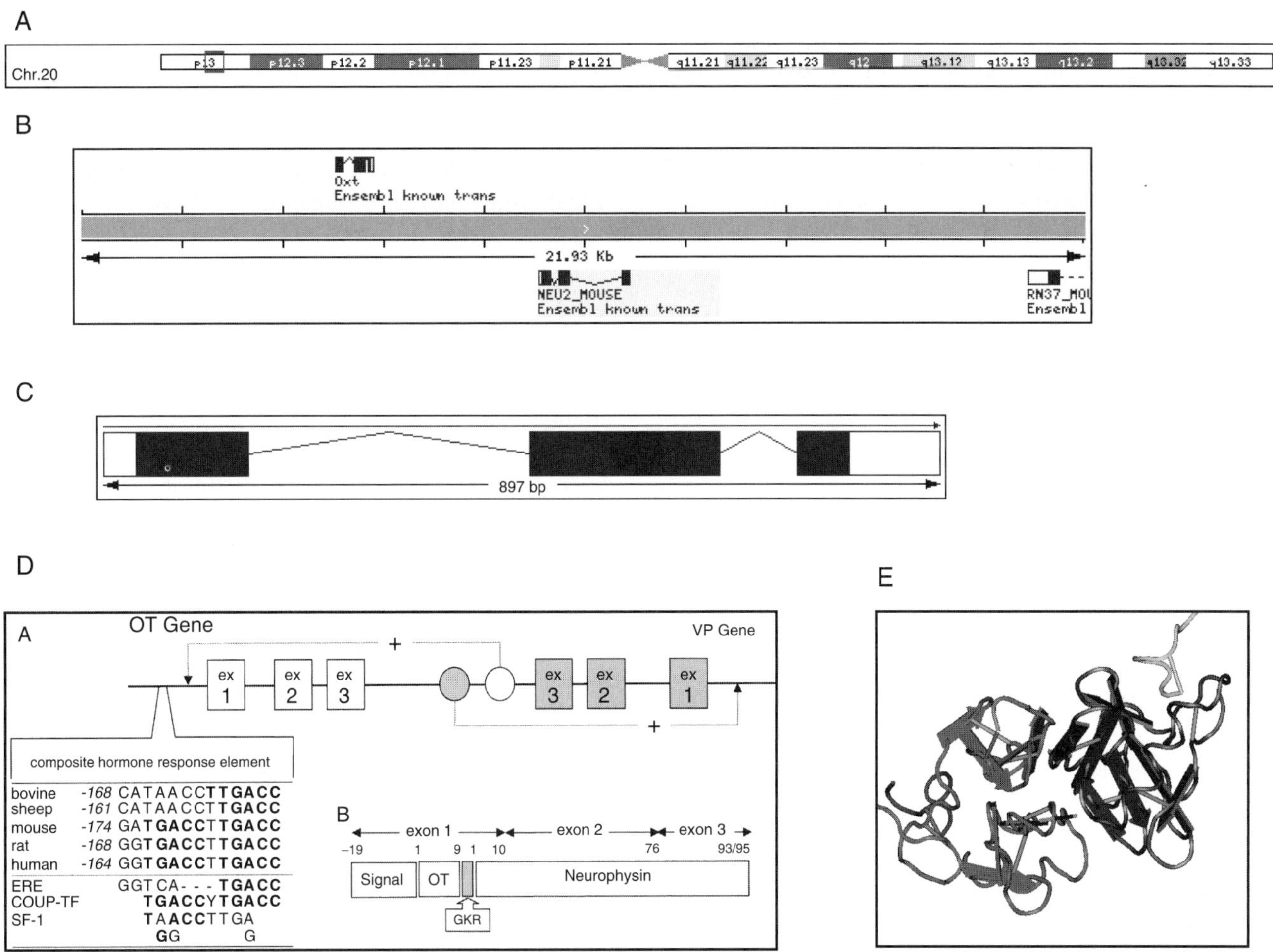

FIG. 5. Organization and expression of the oxytocin gene. *Gene.* The oxytocin gene is part of a small gene cluster together with the vasopressin gene, located on chromosome 20p13 in humans (contig NT 011387.8, locus OXT; **A**). The two genes are positioned in opposite orientation to each other with a distance that is variable between species; in humans about 11 kb, in mouse about 3 kb (**B**). Both genes contain three exons (**C**). There is significant influence of the intergenic region on the expression of both genes. The proximal 5′ flanking region of the oxytocin gene displays significant homology between species in a regulatory motif that serves as a composite hormone response element (**D**). *Protein.* The oxytocin gene encodes a 106 amino acid precursor composed of the oxytocin hormone (9 amino acids) and the oxytocin-associated neurophysin (93–95 amino acids) separated by the sequence -Gly-Lys-Arg- (**D**). This amino acid sequence provides the cleavage site for the prohormone convertase and the amidation substrate for the amidating enzyme PAM. In secretory granules the oxytocin nonapeptide binds to the neurophysin protein under the acidic intragranular conditions, but dissociation is favored by the neutral environment after release. *Protein complex.* The three-dimensional structure of a dimer of the oxytocin–neurophysin complex is shown (**E**). Two oxytocin molecules are shown. Two molecules of neurophysin (249) contact each other and each carry one oxytocin molecule. The view on the neurophysin molecule reveals the two-domain structure of this protein. The complex exists at the acidic pH provided by the intragranular environment of the secretory granules that transport and store the oxytocin precursor products. Elements of the figure were obtained from http://www.ensembl.org and (140) and modeling with the program Cn3D of NCBI.

In particular, the exon B-encoded NP portion ($NP_{10\text{-}76}$) shows a considerable homology between different species. Two segments in the NP sequence, which fall within the region encoded by exon B, share 58% homology with each other. This indicates that the primordial NP gene itself arose from a partial gene duplication that extended an initially smaller structure (269). The secondary structure of NP also indicates a two-domain organization of the molecule in terms of the disulfide pairing that clearly divides the molecule into an NH_2-terminal and a COOH-terminal domain (249,270) (Fig. 5E). NP contains 14 cysteine residues, which are conserved with almost equal interdistances. The NH_2- and COOH-terminal domains, each with three Cys residues, are held together by the extremely well-conserved amino acids Cys^{44}-Gln-Glu-Glu-Asn-Tyr-Leu-Pro-Ser-Pro-Cys^{54} and a disulfide bridge between Cys^{10} and Cys^{54}. It has been shown that NP serves an intracellular function in the sorting of the hormone precursor and in the transport of biosynthetic precursor products. Although it has been proposed that NPs also have biological activity (271,272), such a role has not been firmly established.

Oxytocin Promoter

Comparison of the proximal 5′-flanking region of the oxytocin genes of different species reveals only a few homologous motifs, which is striking, because many promoters share large regions of homology as an indication that their mechanisms of regulation are conserved among species (273). For the oxytocin gene, this indicates common principles in the regulation of oxytocin genes through proximal regulatory elements but in particular suggests that the transcriptional control must involve much more complex mechanisms involving distal elements. Homologous proximal promoter sequences particularly involve potential regulatory elements of nuclear receptors. These are reviewed below. Evidence for the importance of distal elements has been obtained in studies on transgenic mice.

Studies in transgenic mice and organotypic slice cultures point to the oxytocin–vasopressin intergenic region as the site of a critical enhancer element (195,274–276). The DNA sequences in the intergenic region responsible for both oxytocin and vasopressin gene expression are located in a 178-base pair domain immediately downstream of exon 3 of the vasopressin gene. In addition, another domain in the intergenic region, 430 base pairs immediately downstream of exon 3 of the oxytocin gene, contains a positive regulatory element for oxytocin gene expression in the hypothalamus. Alignment of the DNA sequences in the 178- and 430-base pair domains reveals four common sequences (motifs) that may be candidates for the putative enhancers in the intergenic region that regulate oxytocin and vasopressin gene hypothalamic-specific expression.

Nuclear Receptors

Early physiological observations on the stimulation by estrogens of the HNS (277,278) prompted initial studies on the oxytocin promoter. The studies aimed to test the potential of the oxytocin promoter to respond to estrogen and have led to the elucidation of a broad regulatory potential of the oxytocin gene promoter to respond to a wide variety of nuclear receptors and related orphan receptors (Fig. 5). Nuclear orphan receptors are transcription factors of the nuclear hormone receptor family for which no ligands have been identified or which have no ligands and act constitutively. Less attention has been given to other regulatory mechanisms based on proximal elements, basically because clues provided by species conserved homologies are lacking in this region.

Multiple estrogen response element motifs and variations thereof are present in the proximal 5′ flanking region of cloned oxytocin genes. This suggests that many members of the nuclear receptor family may have the potential to interact with the oxytocin gene and regulate its expression. This family consists of true hormone receptors as well as orphan receptors (279) and potentially contains many members able to act and to interact on the oxytocin promoter, some of which have been characterized in the HNS.

The human and rat oxytocin promoters can be stimulated by the ligand-activated ER-α, ER-β, thyroid hormone receptor α, retinoic acid receptors α and β in transient transfection experiments using a variety of cell types (280–284). In the appropriate cellular context, the strongest activators of oxytocin transcription in transfected cell lines are ER-α and ER-β. These studies pinpointed regulatory sequences for nuclear hormone receptors in the proximal region from nucleotides –195 to –70. Overall, this region appears to be a complex regulatory domain of the oxytocin gene providing separate as well as synergistic actions of nuclear receptors. Each receptor appears to prefer a different combination of AGGTCA-like motifs in this promoter region and displays a different activating potential. The core element for all the receptors investigated so far is the region between nucleotides –172 and –148 of the rat oxytocin gene. This region contains an imperfect variant of the palindromic estrogen response element, but also integrates a direct repeat with no spacing (DR0) (281). Deletion results in complete loss of triiodothyronine responsiveness and most of the responsiveness to estradiol and RA. The –172/–148 hormone response element synergizes with the proximal elements for the estradiol responsiveness (285) and is essential for the positive regulation by RA (282,286). ER-α has a higher affinity for the oxytocin estrogen response element than ER-β in transient transfection studies (287) but activates the oxytocin estrogen response element less effectively than a perfect estrogen response element, like the vitellogenin A_2 estrogen response element. This has been explained by DNA-induced modulation of ER-β conformation on the imperfect oxytocin estrogen response element that influences coactivator recruitment (288). It has been proposed, but not tested on the oxytocin promoter, that cross-talk between ER isoforms and thyroid hormone receptor isoforms provides a mechanism for neuroendocrine regulation (289).

Investigations of the oxytocin promoter in particular have pointed toward a role of nuclear orphan receptors in the regulation of this gene (279,289). Specifically, COUP-TFI, ARP-1 (COUP-TFII), SF1/ELP, Ear2, ROR/RZRα, and TR4 can act on the composite response element of the oxytocin gene. They can have two functions: first, to exert transcriptional control by their own and, second, to modulate hormone-stimulated activation of the oxytocin gene. The orphan receptor SF1 is one particular isoform of the SF1/ELP gene that codes at least for three isoforms, addressed here is ELP1, ELP2, and ELP3 (290). In the hypothalamus ELP3/SF1 is the predominant form (291,292). All forms can activate the rat oxytocin promoter, but the splice variant ELP1 is the most potent activator and acts through multiple elements in the –195 to –70 region of the rat oxytocin gene. TR4 can also enhance oxytocin promoter activity significantly, via sequences in the –112/–77 region for this action. This orphan and the orphan receptor TR4 are factors with constitutive activating properties on the oxytocin gene. ROR/RZRα activates the oxytocin promoter through the –172/–148 composite hormone response element (293). Other orphan receptors display repressing activity. The related orphan receptors COUP-TFI, COUP-TFII (ARP-1), and Ear2 lack intrinsic activity on the rat oxytocin promoter, but they repress the induction by estradiol, triiodothyronine, and RA (294). The mechanism of repression of hormone responses of the rat oxytocin gene by COUP-TF I is the competitive binding of COUP-TFI and hormone receptors to the –172/–148 element.

Despite the great regulatory potential of nuclear receptors on the oxytocin promoter, knowledge of the endogenous presence of nuclear receptors in hypothalamic oxytocin neurons is still limited. ER-β is expressed in SON and PVN, but ER-α is not (295,296). More detailed analyses (208,297,298) showed that oxytocin magnocellular neurons of the SON, PVN, and accessory nuclei contain low levels of ER-β in ovariectomized rats. High levels of ER-β were found in parvocellular oxytocin neurons of the ventral and dorsal subdivisions of the medial parvocellular subnucleus and lateral parvocellular subnucleus (297). ER-β was also found in subsets of hypothalamic oxytocin neurons in humans (297). By the use of an ER-β–null mutant it has been shown that ER-β mediates physiological effects on oxytocin gene expression in the hypothalamus (299). Thyroid hormone receptors have been localized in SON and PVN (300,301). Indeed, a small but significant influence of thyroid hormones on oxytocin gene expression in vivo has been found in euhydrated rats (302). However, a significant interaction between effects of thyroid hormones and estrogens has been described (303).

Other Transcription Factors

The 5′ flanking region of the rat oxytocin gene contains a sequence at –1535 to –1270 that binds POU class III transcription factors identified by band shift assays. All POU class III proteins (304), Brn-1, Brn-2, Brn-4, or Oct-6, bind to this distal region but display a different and marked cell-type dependent effect on oxytocin promoter activity. These data are relevant in view of the expression of Brn2 and Brn4 in hypothalamic oxytocin neurons and the requirement of Brn2 for development of oxytocin and other hypothalamic neurons. Moreover, heterozygous Brn2 null mutants display a reduced expression of oxytocin mRNA in the hypothalamus, indicating that Brn2 can control basal expression of the oxytocin gene. Although a direct action of POU class III transcription factors has not been demonstrated, these proteins may contribute to the basal expression of the oxytocin gene and may participate in regulatory responses.

Posttranscriptional Regulation of Oxytocin mRNA

Gene expression in oxytocin magnocellular neurons is regulated at the posttranscriptional level in two unusual ways. First, the poly(A) tail lengths of the oxytocin mRNAs is modulated by physiological stimuli. Second, selected gene transcripts may be subject to axonal transport.

Poly(A) Tail Length Regulation

First discovered for the vasopressin gene transcripts in magnocellular neurons during osmotic stimulation, the poly(A) tail length of the rat vasopressin mRNA appeared to increase in length from ~200 to 400 residues (305). This process is not limited to the vasopressin mRNA and also occurs for the oxytocin mRNA, but it does not affect coexpressed messages (306). The osmotic effect on poly(A) tail lengths is species specific, having being observed in rat (305) and chicken (307) but not in mouse (308).

Still, the increases in vasopressin mRNA abundance and poly(A) tail length can occur independently of each other (306,309–311) and are probably separately regulated. The increase in vasopressin and oxytocin mRNA poly(A) tail length is an extremely rapid response to an osmotic stimulus, being apparent within a couple of hours of its onset (306) and coinciding with a burst of transcriptional activity (308). These data suggest that old vasopressin and oxytocin mRNAs, bearing short poly(A) tails, are rapidly degraded after the onset of an osmotic stimulus and are replaced, through de novo transcription, by a homogeneous class of mRNAs with elongated poly(A) tails.

The mechanics of poly(A) tail regulation in the SON are not understood. The poly(A) tail length of the vasopressin mRNA appears to be regulated in the cell nucleus (312), and thus only newly synthesized transcripts are subject to this mechanism. There is evidence that the protein kinase pathway can mediate an increase in the vasopressin mRNA poly(A) tail length (313). Stimulation of the PKC pathway also increased the length of the vasopressin mRNA poly(A) tail but had no effect on vasopressin mRNA abundance (313).

The physiological role of the poly(A) tail length shift of oxytocin mRNA is not clear. It has been suggested that translational efficiency and/or message stability may be affected, thus contributing to an optimal capacity to synthesize hormone (305,310).

Biosynthetic Processing and Posttranslational Modification

The goal of gene transcription and transcript modifications is to produce gene products. The primary product of the oxytocin gene is the oxytocin prehormone. The presequence is immediately lost on translation. The oxytocin and vasopressin prohormones are extremely rich in disulfide-bonded loops (314), that is, eight disulfide bridges exist. Both the biological functions of the peptides and their structural interactions with the NPs are completely dependent on the integrity of these disulfide bonds (314). The three-dimensional structures of NP–peptide complexes have been elucidated by crystallography (315). This structure emphasizes the function of the disulfide bridges for tight folding of the NP protein and for allowing a "peptide-binding" pocket to accommodate vasopressin or oxytocin by multiple intermolecular contacts.

A characteristic feature of the oxytocin and vasopressin prohormones is reflected in the binding of the nonapeptides to the NPs. For many years, the binding by the NPs of oxytocin and vasopressin has been something of a mystery, and its presumed role as a "carrier protein" (316) has been questioned, because no other prohormone is known to contain such a molecule. Nevertheless, both theoretical considerations (314) and experimental evidence (317) give increasing evidence to the view that the NPs serve as "chaperone-like" molecules supporting intracellular transport. Experimental evidence for a role of NP–hormone interaction in sorting to the regulated secretory pathway includes the missorting of a mutant vasopressin prohormone encoding a vasopressin peptide that is unable to bind to NP (318,319). When expressed in neuronal cells, this prohormone is partly retained in the endoplasmic reticulum (ER) and mostly secreted unprocessed via the constitutive secretory pathway. These data favor a mechanism in which the nonapeptide domain of vasopressin and oxytocin precursors is crucial for correct trafficking of the prohormone through the secretory pathway. It suggests that the mechanism involves an association between the nonapeptide domain and the NP pocket, leading to correct prohormone folding in the endoplasmic reticulum. The misrouting of natural vasopressin prohormone mutants that cause familial neurohypophysial diabetes insipidus supports such a model (320–323). The prohormones are selected from the Golgi and packaged in large dense-cored vesicles. After sorting, subsequent posttranslational processing steps generally involve a variety of proteolytic as well as nonproteolytic modifications. These steps occur within membrane-bound compartments in the cell, starting within the *trans*-Golgi network and then primarily in the large dense-cored vesicle (Fig. 2). The first processing step is to proteolytically cleave the intact prohormone into its correct peptide fragments. For the oxytocin prohormone, this involves one single cleavage site that splits the prohormone into an NH_2-terminal product containing oxytocin and a COOH-terminal product containing NP. Oxytocin is generated by carboxypeptidase H/E trimming of the paired-basic amino acids and conversion of the COOH-terminal glycine to an amide (Fig. 5D). There are at least eight species of prohormone (precursor) convertases found in mammalian cells, which are in the subtilisin-like (Kex-2) endoprotease family, which can process precursors of secreted proteins (316,324–329).

The prohormone (precursor) convertases that have been reported to be present in the HNS are PC1 (330), PC2, PC5, and PACE4 (330–333). Carboxypeptidase H/E (334–338) and PAM (339–343) have similarly been demonstrated to exist in the HNS. In only a few studies has there been an effort to specifically assign expression of the prohormone (precursor) convertase subtypes to either the oxytocin or the vasopressin magnocellular neurons. In situ hybridization studies have indicated that PC5 is the predominant prohormone (precursor) convertase expressed in the vasopressin and oxytocin magnocellular neurons (330,333).

An intriguing differential processing mechanism is apparent during development of the oxytocin magnocellular neuron. In the oxytocin magnocellular neurons there appears to be a cleavage of the oxytocin prohormone to oxytocin COOH-terminally extended peptide intermediates, but processing of the extended oxytocin forms by carboxypeptidase E/H does not occur until postnatal periods (116,344). Although the processing of the vasopressin precursor is complete in all investigated stages of development (from E19 onward), the processing of the oxytocin precursor lags behind significantly (116,117). At E19, no fully processed oxytocin is found but rather intermediates of processing such as oxytocin-Gly-Lys-Arg, oxytocin-Gly-Lys, and oxytocin-Gly, indicating that initial cleavage of the precursor occurs but that trimming by carboxypeptidase H/E and amidation are not fully operative yet. By postnatal day 14, native oxytocin is the end product of processing (116). It is not known if the extended forms of oxytocin serve a physiological function. The COOH-terminally extended forms of oxytocin do not exhibit any known biological activities. There are also reports of the secretion of the extended forms of oxytocin in primates when they are stimulated by estrogen (345,346). An N-terminally acetylated form of oxytocin has been isolated from bovine pineal extracts and identified by mass spectrometry (347). Although evidence for an N-acetylated form of vasopressin was also obtained (348), N-acetylation of oxytocin and vasopressin is not an abundant posttranslational modification, and these forms are at best minor products of biosynthesis.

In the HNS, oxytocin is stored in large dense-cored vesicles located in the nerve terminals in the neurohypophysis where they are mobilized for secretion by electrical activity (349,350). However, some large dense-cored vesicles can also be stored in cell bodies and dendrites, and calcium-dependent secretion can also occur in the dendrites (351,352). An example of this dual mode of oxytocin secretion occurs in the hypothalamus of lactating animals. As part of the suckling reflex, these neurons secrete large boluses of oxytocin from nerve terminals in the pituitary into the bloodstream to act on their distant targets, the mammary glands. They also secrete oxytocin from their dendrites into the hypothalamus itself for a paracrine function, that is, to modulate the synchrony of the oxytocin neuron population's electrical response to suckling to generate bursts of oxytocin release (351,353) (see section on Somatodendritic Oxytocin Release and Chapter 59).

Regulation of Oxytocin Gene Expression in Reproduction

Because the magnocellular oxytocin system is a major neuroendocrine output system of the brain, it is strictly regulated in response to physiological stimuli. The main levels of control are the regulation of gene expression and the regulation of release. Although release of oxytocin results in rapid almost immediate availability of hormone in the bloodstream or at central synapses, changes in gene expression become evident over days or weeks. It is generally believed that release and exhaustion of stored pools of peptide are coupled to gene expression to replenish the peptide. Many studies have shown parallels between oxytocin release and gene expression under various physiological conditions, particularly those that involve long-term changes. The secretory activity of oxytocin neurons and its control are addressed in the section on Secretory Activity of Oxytocin Neurons in Pregnancy and Parturition. Here, the responsiveness of oxytocin gene expression to sex steroids and during reproductive states is dealt with. Concerning the characteristics of oxytocin gene expression changes, hyper- and hypo-osmolality have been used to demonstrate the regulatory potential of oxytocin neurons.

It was recognized early that oxytocin is released from the posterior pituitary gland upon hyperosmolar stimulation, in parallel to vasopressin (354–356). Studies on the osmotic regulation of gene expression in magnocellular neurons have benefited from the exploitation of well-established paradigms for the modulation of the activity of vasopressin magnocellular neurons. Experimentally, plasma osmolality can be increased acutely, generally by the intraperitoneal injection of 1.5 M NaCl or chronically by complete fluid deprivation for up to 3 days, usually referred to as "dehydration," or replacement of the normal drinking water with 2% (wt/vol) NaCl for up to 12 days, usually referred to as "salt loading." Experimental animals can be rendered hyponatremic by a combination of the chronic administration of the vasopressin agonist des-amino[D-Arg^8] vasopressin (357) via osmotic minipumps and the provision of a dilute liquid diet as the only source of nutrients (357).

Vasopressin and oxytocin release as a consequence of an increase in plasma osmolality leads to a functional demand for increased biosynthesis to replace the depleted stores. A hierarchy of transcriptional and posttranscriptional processes achieves this. Hyperosmotic stimuli such as dehydration or salt loading result in an increase in transcription of the oxytocin gene (358–360), similar to the vasopressin gene (308,361), and a concomitant increase in vasopressin mRNA abundance (359,362–366). The oxytocin and vasopressin genes are normally expressed in distinct magnocellular neurons within the SON and PVN (38), but chronic osmotic stimulation increases the proportion of neurons expressing both oxytocin and vasopressin (41).

Sex Steroid Sensitivity and the Role of the HNS in Reproduction

Concerning the role of sex steroids in the reproductive functions of oxytocin, of critical debate at the moment is whether the effects of sex steroid hormones during reproductive cycles are directly on the magnocellular neurons themselves or whether they are indirect via steroid-sensitive interneurons. Although it is generally accepted that magnocellular neurons do not possess ER-α, evidence has been provided for the presence of ER-β in both oxytocin and vasopressin magnocellular neurons (see section on Preventing Premature Oxytocin Secretion in Pregnancy: Estrogens). The oxytocin gene promoter, but not the vasopressin gene promoter, possesses estrogen responsive elements (see section on The Oxytocin Gene Promoter). However, in most cases, both genes in the HNS are regulated similarly during reproductive cycles (see below), as is the sensitivity of both oxytocin and vasopressin neurons to other stimuli (see section on Oxytocin and Hydromineral Balance), evidence that may favor an indirect effect of sex steroids on the HNS.

Although areas of the brain that are known to project to the HNS contain estrogen-receptive neurons, only one study has combined retrograde tracing from the SON with immunostaining for ER-α (367). In summary, this study shows that most forebrain estrogen-receptive neurons that project to the SON are located in the medial preoptic area and lamina terminalis but with scattered neurons throughout the hypothalamus. Interestingly, few of the neurons in the medulla oblongata, which project to the SON, contain ER-α (367). It will be important to investigate estrogen-receptive neurons in other parts of the brain that project to the HNS and that may play a role in reproductive function, such as the olfactory lobes or the tuberomammillary nuclei (100,368), as well as to map possible innervation by neurons that contain ER-β.

ER-β has been found to be expressed in magnocellular neurons of the SON and PVN (295), indicating that estrogens have the potential to act directly on magnocellular neurons by regulating transcription. Alternatively, sex steroids may be having direct but nongenomic effects on magnocellular neurons. Sex steroids can cause rapid increases in magnocellular cell electrical activity (369,370), the release of oxytocin and vasopressin into the circulation (61,371–373), and intrahypothalamic dendritic release (374).

Sex Differences and the Ovarian Cycle

Gonadectomy largely, but not totally, prevents the increase in oxytocin gene expression that is associated with the onset of puberty in rats (375,376). In addition, treatment of prepubertal rats with either estrogen or testosterone has no effect on oxytocin mRNA. Therefore, factors other than gonadal steroids appear to be involved in the maturation of the HNS at the time of puberty. There is a sexual dimorphism in the size of, primarily, vasopressin magnocellular neurons (larger in males) that is related to body weight and does not seem to be dependent on gonadal steroids (377). As the volume of magnocellular neurons increases with age and body weight, it is likely that this is a reflection of increased hormone demand associated with water balance.

Plasma concentrations of oxytocin and vasopressin are reported to vary during the ovarian cycle in both rats (61,372,378,379) and humans (371,380), with highest levels corresponding to peak estrogen levels. There is less compelling evidence for changes in magnocellular cell synthesis over the 4-day rat ovarian cycle. Van Tol et al. (163) demonstrated an increase in oxytocin mRNA during estrus as measured in extracts of SON micropunches. Changes in HNS peptide content have also been reported (61,381). However, short-term (2–7 days) estrogen treatment of ovariectomized rats had no effect on oxytocin mRNA in the SON as determined by in situ hybridization or Northern analysis (382,383), although a significant increase was noted after 2 months of treatment (383). It has been speculated initially that gonadal steroids may affect oxytocin magnocellular neurons indirectly (285), because no ER expression could be detected. However, once ER-β had been discovered (284,384), it was shown to be present in magnocellular neurons and to confer direct actions of estrogens on gene expression in magnocellular neurons (295).

Control of Oxytocin during Pregnancy and Parturition

There is a large accumulation of oxytocin peptide in the HNS during pregnancy, and this is secreted from the pituitary at the time of parturition (385–390). The general consensus is that this buildup of oxytocin peptide is due to a combination of increased gene expression (100,163,391,392), transcript polyadenylation (393), and an inhibition of hormone release (386,394). There is little doubt that oxytocin magnocellular neurons are activated at the time of parturition (395–398), although it has been postulated that oxytocin produced in the uterus itself, in some species, may contribute to the hormone's peripheral effects at term (155,399). Although the rat uterus contains 70-fold more oxytocin mRNA than the hypothalamus at the end of pregnancy, and this disappears with parturition, the actual oxytocin peptide content of the uterus is small compared with that in the neurohypophysis. Furthermore, the increase in oxytocin mRNA reported in the rat and human uterus does not occur in all animals, such as the pig, cow, mouse, and sheep (400–403).

In a series of elegant studies, Crowley and coworkers (404,405) manipulated sex steroids in ovariectomized virgin rats to mimic the changes that occur during pregnancy: a rise in both estrogen and progesterone, followed by an abrupt decline in progesterone (normally 48 hours preterm), while estrogen levels were maintained (406). In this way, they demonstrated that this pattern of estrogen priming followed by progesterone withdrawal can enhance oxytocin mRNA abundance in the SON and magnocellular PVN (405). Maintenance of progesterone artificially at the end of pregnancy not only delays birth (407), it also prevents the normal rise in oxytocin mRNA (405).

The activity of oxytocin magnocellular neurons and plasma oxytocin secretion is pulsatile during birth, although there is an increase in circulating levels that occurs before fetal passage through the birth canal (398,408–410). Pulsatile delivery of oxytocin to rats the day before expected delivery leads to uterine contractions and a pattern of neuronal activity similar to that seen during normal parturition (82,396,411). The significance of oxytocin pulsatility is probably species dependent (409,412,413), although it is likely that the HNS plays a critical role in allowing birth to progress normally (396,414).

Lactation

Oxytocin gene expression in magnocellular neurons falls immediately postpartum (391) and then increases during early lactation to cope with the demand for milk of the growing young (100,163, 359,392). The level of oxytocin mRNA during early lactation is maintained by the suckling stimulus, suggesting an important role for afferent stimulation (415). Removal of pups during early lactation leads to a decline in oxytocin mRNA expression that can be attenuated by intracerebroventricular oxytocin infusion, implying an action of centrally released hormone. Oxytocin may be released from dendrites and the cell bodies in response to suckling to have this effect (353,416). In contrast, the expression of oxytocin mRNA in the rat during late lactation shows increasing dependence on the steroidal milieu (391). Estrogen levels in the rat are suppressed by suckling for the first week postpartum and then rise gradually as lactation continues, whereas progesterone levels increase during early lactation to peak at about day 10 before declining again (417). Thus, around days 10–12 of lactation or ~48 hours after interruption of nursing by pup removal (418), the steroid status of the rat is similar to that of the late-pregnant animal, that is, rising estrogen and falling progesterone levels. Both of these situations are accompanied by further increases in oxytocin gene expression, which can be abolished by maintaining progesterone artificially (419).

The highly pulsatile activity of oxytocin magnocellular neurons induced by suckling is critical for successful lactation (420). A discussion of the mechanisms that might allow this distinctive patterning to be displayed by the whole population of oxytocin neurons in synchrony is beyond the realms of this review and can be found elsewhere (22,421,422) (see Chapter 59). However, it is clear that a number of changes in neuron structure and membrane properties will require complex alterations in gene expression by magnocellular neurons that have yet to be elucidated.

PHYSIOLOGICAL FUNCTIONS OF OXYTOCIN IN REPRODUCTION

Role for Oxytocin in Parturition

Oxytocin is a distinctly mammalian neuropeptide, though its forebears are phylogenetically ancient and involved in egg laying [see (423)], which is analogous to its role in promoting the birth of young in mammals. There have been doubts about the importance of oxytocin in parturition, based on difficulty in measuring increases in circulating concentrations in women in labor and on the finding that mice with targeted deletion of the oxytocin gene, or the oxytocin receptor gene, give birth (167,193,194). However, the fact that oxytocin infusion can initiate labor and that an oxytocin antagonist is effective in delaying threatened preterm labor in women (424) or delaying the start of parturition or impeding its progress in several species (425,426) indicates an important role for oxytocin in driving parturition. This role seems to be normally to ensure that the uterus is stimulated to contract sufficiently powerfully in established labor to expel the fetus(es) and, in women, to ensure also that when the placenta separates the severed spiral arterioles are "ligated" by the contracted myometrium to avoid torrential hemorrhage. Hence, in human and veterinary obstetric practice, labor is often supported by oxytocin infusion, and in women oxytocin is routinely given (with an ergot) as the infant is delivered to ensure powerful uterine contraction and so prevent postpartum hemorrhage (427). Increased circulating oxytocin concentrations during parturition have been measured with radioimmunoassay in several species (409,410,428–430). In women, this is most evident in the second stage of labor, although the rapid breakdown of oxytocin by circulating placental aminopeptidase and the secretion of oxytocin in discrete pulses make such measurements difficult (408).

In the tammar wallaby, a typical Australian macropodid marsupial, mesotocin (differing from oxytocin by Ile replacing Leu at position 8) functions like oxytocin and has the same actions on the oxytocin receptor as oxytocin; conversely, mesotocin and oxytocin act similarly on the wallaby mesotocin/oxytocin receptor (431). There are low circulating levels of mesotocin before birth, but mesotocin stored in the posterior pituitary increases approximately sixfold across pregnancy, and sensitivity of the gravid uterus to contractile effects of mesotocin develop only in the last 3 days of pregnancy, with a corresponding fourfold increase in receptor expression. With the birth of the fetus there is a sharp large increase in mesotocin secretion, lasting about 15 minutes. This surge begins 10–30 seconds before the fetus is ejected from the birth canal. Infusion of the oxytocin receptor antagonist Atosiban in the last few days of pregnancy delays birth by about 2 days.

The oxytocin gene is expressed in the uterus, placenta, and fetal membranes in human and rat pregnancy (155,432,433), which has been taken to indicate that oxytocin from this source may be important in promoting parturition. However, little oxytocin is produced, and local release does not seem to be onto the myometrial cells (434); furthermore, the evidence is against oxytocin gene expression in the pregnant uterus in many other species, including the mouse (402). In the bovine, oxytocin produced by the corpus luteum (see section on Luteolysis) at the end of pregnancy may contribute to circulating oxytocin at term (430).

There is abundant evidence from studies, in particular in the rat, that leads inexorably to the conclusion that oxytocin secretion by the posterior pituitary gland is increased during parturition, that oxytocin is important in driving the normal progress of parturition, and that this increased oxytocin secretion is stimulated by neural signals from the contracting uterus and the birth canal as the fetus(es) pass through it (435). Because oxytocin stimulates further uterine contractions, which, until each fetus is born, generates further neural stimuli, this process involves *positive* feedback regulation of oxytocin secretion. This body of evidence is summarized as follows.

Increased oxytocin secretion from the posterior pituitary gland during parturition is shown by loss of a substantial portion of the increased store that accumulates in pregnancy (388) and by increased circulating oxytocin concentration. The stimulation of the magnocellular oxytocin neurons during parturition is shown by finding, with immunocytochemistry in rats and mice, increased expression of Fos, the transcription factor protein translated from the immediate–early gene *c-fos* (82,429,436). There is also a rapid increase during parturition in expression in supraoptic neurons of the oxytocin gene immediate transcript (heteronuclear RNA) (437). The release of oxytocin from the dendrites and cell bodies of supraoptic magnocellular neurons is increased during parturition (353,438), which provides further evidence for activation, but, also as discussed below, plays a role in autocrine stimulation of oxytocin neurons. In a still unique study, in rats with a chronically implanted recording electrode in a PVN, presumptive oxytocin neurons increased their firing rate during parturition, with peak ("bursting") activity correlated with the birth of a pup (398).

The fact that mice genetically engineered to inactivate the oxytocin, or oxytocin receptor, gene give birth in the light of all the evidence is interpreted to mean that there is redundancy in the mechanisms that bring about parturition, but not that oxytocin is unimportant (423,435). The importance of oxytocin secreted by the posterior pituitary during parturition in driving the process is shown by the consequences of inhibiting oxytocin secretion or action.

Opiates

In rats, opiate drugs inhibit oxytocin secretion, including during parturition; this includes actions of μ-opioid receptor agonists (morphine and pethidine [meperidine]) (439,440) and the κ-agonist U50,488 (441). These opiates slow or halt established parturition in rats; the actions of morphine are reversed by administration of naloxone, an opioid antagonist, or, crucially, by oxytocin infusions (396,439). Pethidine actions are not reversed in this way because it also directly inhibits uterine contractions (440).

Stress in Parturition

Exposing pigs, rats, or mice during parturition to mild stress, such as moving to a novel environment or exposing to an air puff, also slows, in pigs and rats, or stops, in mice, the births of further young (428,429,442,443). Such interruption of parturition by stress reduces oxytocin secretion and is reversed by oxytocin administration; hence, in the circumstance of parturition stress reduces oxytocin secretion, and this is responsible for the interruption of parturition. In rats and pigs the inhibition of oxytocin secretion is mediated by endogenous opioids, as naloxone prevents the stress-induced delay in births and the reduced oxytocin secretion (428,442). In mice, endogenous opioids seem not to be involved as naloxone has no effect; instead, in this species a β-adrenergic receptor antagonist prevents the stress-induced suspension of births, indicating that activation of adrenergic mechanisms inhibits uterine contractions (429). In both cases it seems that stress interrupts the positive feedback neural loop from the parturient uterus to the oxytocin neurons, by a central opioid mechanism (pigs and rats) or by a peripheral adrenergic mechanism (mice). The interruption of parturition and reduced oxytocin secretion seem paradoxical in view of the stimulation of oxytocin secretion by stressors at other times (see below). However, interruption of pulsatile secretion by stress can explain this apparent paradox; such an effect is exemplified by the dual actions of relaxin in lactating rats, which stimulates steady oxytocin secretion but stops milk ejection pulses (444,445).

Oxytocin Receptor Antagonists

Further evidence for the important role of oxytocin in driving parturition comes from studies of the effects of an oxytocin receptor antagonist. In rats and mice, treatment with an oxytocin antagonist slows parturition (426,429). The antagonist atosiban (Tractocile® in Europe), recently introdued into obstetric practice, reduces the number of premature deliveries for more than a week in women with threatened preterm labor, without immediate adverse maternal or fetal effects, which may be seen with other tocolytic treatments, such as β-agonists (424). However, in rats prenatal atosiban treatment impairs subsequent lactation performance, likely a consequence of impaired milk transfer (446). It should be noted that atosiban is a V1a receptor antagonist as well as an oxytocin receptor antagonist (447). It has been shown that this antagonist reduces uterine hypertonus during labor, indicating excess oxytocin or uterine hypersensitivity to oxytocin underlies this state (448). A more recently developed peptide (FE 200 440) has similar in vitro potency to atosiban in inhibiting oxytocin-induced contractions of parturient human myometrial strips in vitro, without the antagonist actions on V1 receptors shown by atosiban (449).

Nonpeptide oxytocin antagonists with varying selectivity for oxytocin receptors over the V1a receptor have recently been described; these antagonists are effective in suppressing uterine contractions in late pregnancy (450) and in delaying parturition in pregnant rats and mice (429,451). Caution is needed in any clinical application of such compounds if they cross the blood–brain barrier because of likely antagonist actions on the expression of maternal behavior (see section on Maternal Behavior) or on central actions of oxytocin that promote the milk ejection reflex (see Chapter 59).

Mice are not little rats. No naturally occurring mutation of the oxytocin peptide or receptor genes has yet been described, so the only gene inactivation models to date are mice with targeted deletion of these genes (195). As mentioned already, these mice give birth, though a full quantitative description of the birth process in these engineered mutants has not yet been reported. Whether spontaneous inactivating mutation of the oxytocin peptide or receptor gene in other species would result in abnormal parturition would be difficult to discover, as the essential role of oxytocin in milk transfer means that offspring of a mother homozygous for such a mutation will not survive without cross-fostering, and the phenotype will be lost. Mice heterozygous for targeted inactivation of the oxytocin gene do not show a phenotype and so could not be identified in a wild population. Although magnocellular oxytocin neurons are clearly activated and secrete oxytocin, during parturition, the mouse and rat show major differences in the regulation of oxytocin neurons in pregnancy and parturition, as noted above: For instance, in mice stress interrupts parturition by an adrenergic mechanism, and endogenous opioids are not involved (429).

Other Reproductive Roles of Oxytocin

Oxytocin and the Anterior Pituitary Gland

The oxytocin receptor is expressed in the anterior pituitary gland, but only in lactotrophs, and expression is greatly increased at the end of pregnancy and is stimulated by estrogen treatment (452,453). Oxytocin actions on corticotrophs are via the V1b receptor (454) (see section on Stress). Because oxytocin has an essential role in lactation in effecting the milk ejection reflex, a complementary role in regulating milk production through control of prolactin secretion has long been considered (455). Likewise, as ovulation is suppressed in lactation, it might be expected that oxytocin is involved in suppressing gonadotropin secretion. There is evidence for a prolactin releasing action of oxytocin, but also for a stimulatory action on LH secretion. Oxytocin is secreted into the hypothalamohypophysial portal system, from swellings on magnocellular neuronal axons en passant in the internal zone of the median eminence and perhaps from parvocellular axon terminals in the external zone (456–459); the possibility of oxytocin passing from the posterior pituitary via short portal vessels to the anterior pituitary is not supported by measurement of oxytocin content after suckling (460).

LH Secretion. Oxytocin can stimulate gonadotropin-releasing hormone release in the marmoset (461) and advance the ovulatory LH surge in women (462), whereas an oxytocin antagonist (atosiban) reduces the ovulatory LH surge in women (463). In vitro, oxytocin stimulates LH secretion by rat anterior pituitary and augments the action of subsequent gonadotropin-releasing hormone exposure (464). Oxytocin also stimulates in vitro LH release by anterior pituitary cells from pigs treated with estrogen (465). A problem with interpretation is that oxytocin receptor expression has not been detected in rat gonadotrophs (452). Moreover, female mice with targeted inactivation of the oxytocin gene are fertile, indicating adequate gonadotropin secretion (167,193). A physiological role for oxytocin in regulating LH secretion is not clear; this contrasts with the actions of oxytocin on or in the ovary and on the endometrium in the regulation of the corpus luteum in ruminants (see section on Luteolysis).

Prolactin. Concentrations of oxytocin in hypothalamohypophysial portal blood and prolactin in peripheral blood are correlated (379), and in vitro and in vivo oxytocin stimulates prolactin secretion from dispersed anterior pituitary cells but not in the presence of dopamine, the endogenous prolactin release inhibiting factor (466). The oxytocin receptor is expressed only by a subpopulation of lactotrophs, which also express receptor for vasoactive intestinal peptide (another hypothalamic prolactin-releasing factor) (455). The action of oxytocin on lactotrophs involves increasing intracellular [Ca^{2+}] (467) but does not involve IP_3, so the typical post–oxytocin receptor pathway is not used (468). Immunoneutralization of oxytocin indicates a modulatory role for oxytocin in the regulation of prolactin secretion stimulated by suckling (469), although other studies do not support this finding (470). These direct actions of oxytocin contrast with the inhibitory effects on prolactin secretion of oxytocin given centrally, which evidently acts by reducing vasoactive intestinal peptide secretion from the hypothalamus (471). A role for oxytocin in driving daily mating-induced surges of prolactin secretion in the female rat is indicated by suppression of the surges by a systemic oxytocin antagonist (472). In this model, oxytocin neurons are regulated by vasoactive intestinal peptide input from the suprachiasmatic nuclei, as shown by the effects of local vasoactive intestinal peptide antisense oligonucleotide microinjection (467). Other studies on the relationship between oxytocin secretion and oxytocin neuron activity after mating, in females, have deduced that oxytocin secretion does not cause post-mating prolactin surges (473).

In women, intravenous oxytocin infusion does not affect prolactin secretion (474). In mice with targeted inactivation of the oxytocin gene, milk is secreted and is ejected upon exogenous oxytocin treatment (193), but without oxytocin treatment the mammary gland secretory epithelium involutes (475). This is interpreted as a consequence of lack of milk removal rather than lack of prolactin.

Luteolysis

Nonfertile Ovarian Cycle (Ruminants). In the sheep and other ruminants, oxytocin is important in inducing luteolysis in the absence of conception (476,477). Oxytocin acts on endometrial oxytocin receptors, induced by the actions of estradiol (476), to stimulate $PGF_{2\alpha}$ synthesis and release (478,479). $PGF_{2\alpha}$ reaches the ipsilateral ovary in high concentration as a result of countercurrent exchange between the uterine vein and ovarian artery and triggers luteolysis (476). The nonpregnant uterus thus provides the signal for a new ovarian cycle to start (Fig. 6).

The source of the oxytocin is the corpus luteum itself (480,481) and the posterior pituitary (477). A marked up-regulation of expression of the oxytocin gene in bovine granulosa cells follows the ovulatory pulse of LH; the consequently increased cAMP formation and insulin-like growth factor I (after FSH stimulation) promote oxytocin gene expression as the follicle becomes the corpus luteum (482).

In intact cycling sheep, large long-duration (~1 hour) pulses of oxytocin are secreted during the luteolytic stage. Surgical removal of the corpus luteum shows that perhaps 90% of the oxytocin released is from the corpus luteum; oxytocin secretion from the posterior pituitary at this time, driven by estradiol and progesterone, may have a supplemental role in stimulating $PGF_{2\alpha}$ production by the endometrium, and hence luteolysis (477). Thus, in estrogen-treated ovariectomized sheep, intermittent

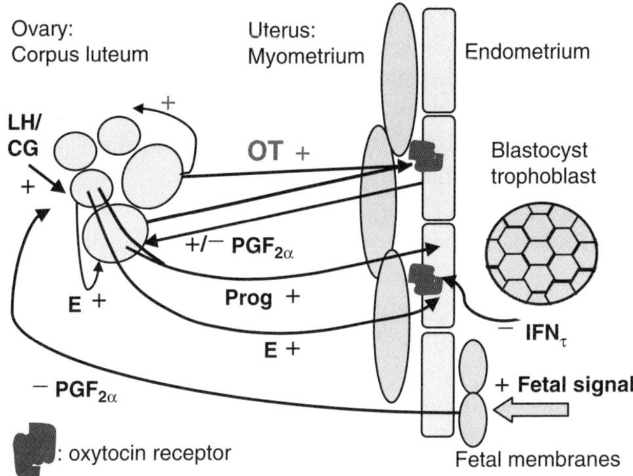

FIG. 6. Changing roles of oxytocin in interactions between the corpus luteum and endometrium in the reproductive cycle. Composite diagram based on studies in several species. The ovulatory LH signal initiates the conversion of the follicle cells to secrete more progesterone and form the corpus luteum. Progesterone (Prog) has essential actions on the endometrium to permit implantation and sustenance of pregnancy; in the event of pregnancy luteal function is sustained by luteotrophic hormones (e.g., chorionic gonadotropin, CG). The oxytocin gene is expressed in large luteal cells and locally released oxytocin has trophic actions via oxytocin receptors in the small luteal cells, promoting gap junction formation between luteal cells [in primates (487), mice (495)]. Oxytocin expression is strongly up-regulated in the *ruminant* corpus luteum (related to estrogen production and cAMP signaling by LH action) (482). This oxytocin acts on oxytocin receptors in the *endometrium*, where oxytocin receptor expression is stimulated by estrogen (E) (476). Oxytocin stimulates $PGF_{2\alpha}$ production (see Fig. 7) and, reaching the corpus luteum via a local vascular system, stimulates further oxytocin secretion and inhibits progesterone; this positive feedback $PGF_{2\alpha}$–oxytocin interaction leads to luteolysis. In *ruminants*, pregnancy depends on suppression of oxytocin receptor expression in the endometrium by interferon-τ (IFN-τ) produced by trophoblast of the blastocyst (476). In species dependent on progesterone from the corpus luteum throughout pregnancy (nonprimates), luteolysis is essential to allow parturition: $PGF_{2\alpha}$ from the fetal membranes, stimulated by a fetal signal is the luteolytic signal, removing intraluteal trophic actions of oxytocin and halting progesterone secretion (494). +, stimulatory action; –, inhibitory action.

uterine contractions that are blocked by oxytocin antagonist correspond to pulses of oxytocin from the posterior pituitary, measured in jugular venous blood (477).

The luteolytic action of $PGF_{2\alpha}$ from the endometrium involves a positive feedback loop, because its action via receptors in the corpus luteum include stimulation of further oxytocin secretion (483), which in turn acts on the endometrium to further stimulate $PGF_{2\alpha}$ production in a series of large bursts that cause luteolysis and to terminate oxytocin secretion from this source (484). The rapid stimulation of oxytocin secretion by $PGF_{2\alpha}$ likely involves activation of PKC and Ca^{2+} entry, but the much slower suppression of progesterone production involves endothelin-1 and tumor necrosis factor-α (485).

Oxytocin has a similar luteolytic role in swine, though the endometrium may be the source of oxytocin (486). In primates, a uterine signal is not necessary for luteolysis (477). In contrast to ruminants, oxytocin has a trophic action on the corpus luteum in primates. In the baboon, and evidently in the human and other primates, oxytocin produced and secreted by the large cells of the corpus luteum, maximally in the midluteal phase, is not directly involved in regulation of steroidogenesis. Instead, oxytocin may act locally, via oxytocin receptors on the small steroidogenic luteal cells, to regulate direct communication between luteal cells by stimulating expression of connexin-43 (487).

Pregnancy. Pregnancy requires that progesterone production is sustained and thus that luteolysis is prevented. In ruminants, this requires suppression of the luteolytic mechanism involving stimulation by oxytocin of $PGF_{2\alpha}$ production by the endometrium. This is achieved by suppression of endometrial oxytocin receptor by the trophoblast of the blastocyst, which produces interferon-τ, and this strongly suppresses oxytocin receptor expression in the endometrium (476). This breaks the positive feedback loop between the corpus luteum and the endometrium and allows the corpus luteum (and the embryo) to survive.

Pregnancy at Term. At term in nonprimate species, as exemplified by ruminants in which the corpus luteum remains the source of progesterone throughout pregnancy, luteolysis is an essential step in the initiation of parturition; this has been considered to be induced by $PGF_{2\alpha}$ from the uterus, driven by fetal signals (488). Doubt has been cast on the role of uterine $PGF_{2\alpha}$ because of lack of correlation between circulating progesterone and $PGF_{2\alpha}$ levels in the last few days of pregnancy (489). Nonetheless, interactions between $PGF_{2\alpha}$, progesterone and oxytocin have a critical role in other species. The importance of $PGF_{2\alpha}$-induced luteolysis in permitting parturition is shown in rats by $PGF_{2\alpha}$ injection in late pregnancy, which induces parturition; the progesterone withdrawal precipitated by $PGF_{2\alpha}$ injection advances the induction of oxytocin receptor mRNA expression in the myometrium, provided estrogen is available (490). In mice with targeted deletion of the prostaglandin ($PGF_{2\alpha}$) receptor gene, at term oxytocin receptor mRNA expression in the myometrium is not up-regulated, progesterone secretion does not decrease, and parturition does not occur (491). The maternal phenotype is rescued by performing ovariectomy in late pregnancy, removing the inhibitory actions of progesterone on the myometrium, which allows delivery to proceed (492).

Similarly, in mice lacking the cyclooxygenase-1 gene, essential for $PGF_{2\alpha}$ synthesis by the endometrium, the corpora lutea do not regress and progesterone secretion does not collapse at term, so parturition is delayed and the newborn die (493,494); however, in mice with inactivation of both the cyclooxygenase-1 and oxytocin genes, parturition occurs normally. Hence, it seems that in the absence of a $PGF_{2\alpha}$ luteolytic signal from the uterus, oxytocin has a luteotrophic action, and without this luteolysis can occur (493). Accordingly, infusion of low doses of oxytocin in late pregnant mice with oxytocin gene inactivation delays parturition and luteolysis, whereas larger doses of oxytocin, acting on the myometrium, induced parturition (495).

Control of Cyclical Uterine Oxytocin Receptor Expression

There are massive changes in oxytocin receptor expression in the uterus and mammary gland as a result of pregnancy. These changes are out of phase, as myometrial oxytocin receptor expression peaks at the end of pregnancy and declines rapidly postpartum, whereas increased myoepithelial oxytocin receptor in the mammary gland remains increased through lactation. Less dramatic changes in oxytocin receptor expression in the brain occur during the reproductive cycle (see section on Central Control of Behaviors Related to Reproduction). Expression of oxytocin receptor mRNA increases in the SON near the end of pregnancy (496), indicating possible increased sensitivity to effects of locally released oxytocin.

Oxytocin Receptor Gene Regulation. Although the structure of the oxytocin receptor gene in several mammalian species is known, the mechanisms of regulation of the oxytocin receptor remain unclear (140,149,497). According to species, the major upregulation of uterine oxytocin receptor expression may be in the myometrium (e.g., rat, human) or in the endometrium (e.g., bovine), reflecting different interactions between endometrium and myometrium in the regulation of myometrial contractile activity (144). Methylation of the oxytocin receptor gene may be important in regulation of tissue specific expression (498).

Myometrium. Many studies have pointed to the importance of estrogen in inducing up-regulation of myometrial oxytocin receptor expression (149,498, 499) and of progesterone in inhibiting expression (490). Despite the evident importance of estrogen in stimulating oxytocin receptor expression, the mechanisms are not clear. Direct actions on the oxytocin receptor gene are unlikely (149,500). Relaxin seems important, because myometrial oxytocin receptor mRNA expression does not increase in late pregnant mice with targeted relaxin gene deletion; ER-α mRNA expression also does not increase, which may explain the lack of increase in oxytocin receptor mRNA (501). However, stretch of the uterus and fetoplacental factors are implicated in regulating oxytocin receptor expression, independently of gonadal sex steroids (502).

The oxytocin receptor gene contains potential interleukin response elements in the 5′ flanking regions (141,142), suggesting the possibility that infection might up-regulate expression and contribute to the association between vaginouterine infection and preterm labor. However, interleukin-1β and interleukin-6 negatively regulate the oxytocin receptor gene in myometrial cells through these elements (503–505). The pattern of protein binding to the 5′ flanking region of the oxytocin receptor gene in the myometrium is altered at term, suggestive of term-specific regulation (498).

Endometrium. Also, in bovine endometrium, proinflammatory cytokines inhibit oxytocin receptor gene expression (506). Endometrial oxytocin receptor expression in early ruminant pregnancy is inhibited by interferon-τ, which is essential to allow pregnancy by preventing luteolysis (476). This acts via an interferon-responsive element in the receptor gene and indeed is the only physiological mechanism shown to regulate expression of the gene at this level so far (500).

In the ovine endometrium, oxytocin receptor mRNA shows heterogeneity, partly due to alternative polyadenylation sites but also due to editing of the coding region. Hence, oxytocin receptors with different amino acid sequences are expected to be products of the same single gene but with modified properties (147).

Mammary Gland. In the mammary gland, the myoepithelial cells express the oxytocin receptor gene (146). The oxytocin receptor mRNA is found as two species, one like that in the uterus and one that is shorter. The coding regions are identical, and the difference results from a polyadenylation site specific to the mammary gland. In striking contrast with oxytocin receptor expression in the uterus, estrogen has no effect on oxytocin receptor mRNA expression in the mammary gland (169). In women, immunocytochemical studies show that the receptor protein is more strongly expressed on the secretory epithelial cells than on the myoepithelial cells; the level of expression is not increased in lactation (170).

Oxytocin Postreceptor Signaling. The oxytocin receptor is a seven-transmembrane G protein–coupled receptor that increases the binding to G proteins when oxytocin binds into the receptor pocket (see section on The Oxytocin Receptor and Fig. 3). Depending on which cell type it is expressed in, a number of postreceptor signaling events have been delineated (Fig. 7).

Contractions. In the myometrium, the oxytocin receptor increases binding to G protein when

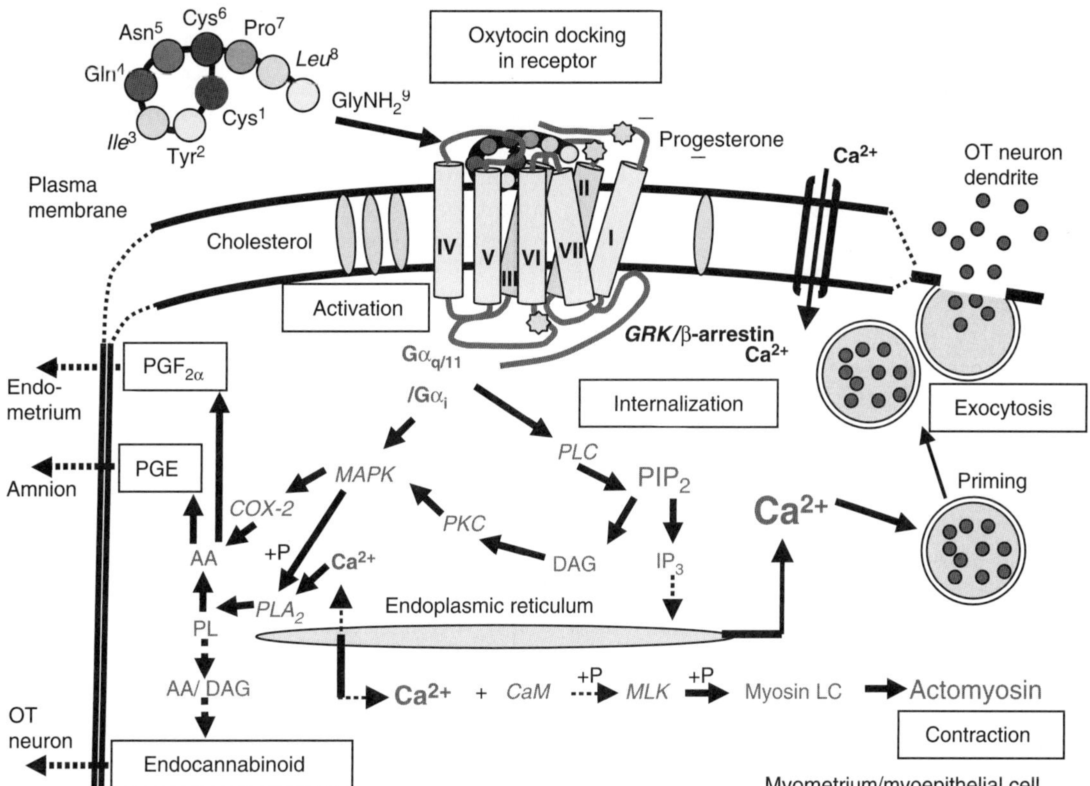

FIG. 7. Receptor function: binding and postreceptor signaling mechanisms. Summary of oxytocin receptor mechanisms in target cells (*enzymes* are in italics). Oxytocin can cause cell contraction (mammary myoepithelial cells; myometrium), secretion by exocytosis (oxytocin neurons; lactotrophs; corticotrophs; atrial cells), production of prostanoids (endometrium, fetal membranes), and cannabinoids (oxytocin neurons). The oxytocin receptor is a seven transmembrane domain G protein–coupled receptor. The high affinity state of the receptor requires local high density of membrane cholesterol. Progesterone reduces cholesterol availability (805) and antagonizes oxytocin binding (757). Oxytocin binding to the human oxytocin receptor is inhibited by 5β-dihydroprogesterone. Oxytocin, depicted as peptide, docks in an extracellular pocket: oxytocin residues 3 and 6 distinguish the oxytocin molecule from vasopressin; key receptor residues in extracellular domains 1 and 2 are indicated by stars [e.g., (149)]. Receptor occupancy leads to coupling to $G\alpha_{q11}$ and $G\alpha_i$; a key residue in G protein binding is indicated by a star on the third intracellular loop (143,149). Stimulation of phospholipase C (PLC) follows, which triggers several pathways leading to target cell responses. Oxytocin occupancy of the receptor also triggers interaction with G-protein–coupled receptor kinase (GRK), allowing binding to β-arrestin, leading to receptor internalization (806). *In contractile cells* (myometrium, myoepithelium), PLC catalyses phosphatidylinositide 4,5-bisphosphate (PIP_2) breakdown to diacylglycerol (DAG) and inositol 1,4,5-trisphosphate (IP_3). IP_3 triggers Ca^{2+} release from smooth endoplasmic reticulum; Ca^{2+} also enters via plasma membrane channels, L-type or capacitative, and is also PLC-dependent (145). Ca^{2+} activates calmodulin (CaM), leading to phosphorylation (P) of myosin light chain kinase (MLK) that phosphorylates myosin light chains (myosin LC), facilitating actin–myosin interaction, and contraction (145). *In secretory cells* (e.g., oxytocin neurons), increased $[Ca^{2+}]_i$, released from IP_3 sensitive stores and entering from outside, primes oxytocin secretory granules, and triggers exocytosis (742–744). *In prostaglandin (PG) secreting cells* (endometrium, decidua, amnion), Ca^{2+} causes phospholipase A_2 (PLA_2) binding to the endoplasmic reticulum; mitogen-activated protein kinase (MAPK) phosphorylates PLA_2, and membrane glycerophospholipid (PL) is broken down to yield arachidonic acid (AA), and prostanoids. Cyclooxygenase 2 (COX-2, induced by MAPK) action forms PGH_2; *synthases* form PGE_2 [amnion; see (510)] or $PGF_{2\alpha}$ [nonpregnant/pregnant ruminant endometrium (478,807)]. In magnocellular oxytocin neurons, oxytocin autostimulates production of endocannibinoids (748). These are not yet identified but are glycerophospholipid (e.g., AA, DAG) derivatives: e.g., anandamide (*N*-arachidonoylethanolamine) and/ or arachidonoylglycerol (2-AG) (808). Synthesis is Ca^{2+} dependent and probably involves N-acyltransferase/aPLD and PLC/diacylglycerol lipases, respectively (808).

activated by oxytocin. This results in activation of phospholipase C and increase in cytoplasmic [Ca^{2+}] by release from intracellular stores and opening of plasma membrane channels (145).

The receptor binds $G_{\alpha q/11}$ and also $G_{\alpha i3}$, with increased binding at term (507). Activation of phospholipase C stimulates phosphoinositide hydrolysis to produce IP_3, which releases Ca^{2+} from the endoplasmic reticulum (145,508). Oxytocin also increases Ca^{2+} entry, either via L-type channels or as a result of capacitative entry, perhaps via TrP proteins, and consequent on phospholipase C–mediated depletion of the intracellular store (145). The increase in cytoplasmic [Ca^{2+}] then leads to contraction via calmodulin and stimulation of actin–myosin interactions (145). Activation of phospholipase C by oxytocin is inhibited by cAMP, which can contribute to inhibition of uterine contractions by β-adrenergic agonists (145).

Prostaglandins. In the ovine endometrium, oxytocin action leads to stimulation of $PGF_{2\alpha}$ production. Here, the oxytocin receptor is not coupled to a G_i protein, and stimulation of $PGF_{2\alpha}$ production is mediated by phosphorylation of extracellular signal regulated protein kinase (ERK1/2), a mitogen-activated protein kinase (479), and involves rapid stimulation of PG synthase-2 mRNA expression (509).

In the rabbit amnion oxytocin stimulates PGE_2 synthesis. This involves increased expression of oxytocin receptors near the end of pregnancy and, when oxytocin binds, to coupling via $G_{\alpha q/11}$ and phospholipase C to mechanisms increasing cytoplasmic [Ca^{2+}], which causes phospholipase A_2 to associate with intracellular membranes. $G_{\alpha q/11}$ also activates PKC and hence ERK, which activates phospholipase A_2; arachidonic acid subsequently released from membrane phospholipid provides the substrate for PG endoperoxide-H synthase 1 to produce PGE_2 (510).

Mammary Gland. Myoepithelial cells around the secretory alveoli contract on exposure to oxytocin, effecting the essential function of milk ejection (see Chapter 59). Oxytocin action includes stimulating increased intracellular [Ca^{2+}] but also increasing cAMP formation, which might be expected to be inhibitory (511). ATP is released by secretory epithelial cells in the lactating mammary gland when they are compressed by the surrounding myoepithelial cells, contracting in response to oxytocin (512). This has the interesting consequence of potentiating the contractile response of myoepithelial cells to oxytocin, so that the alveoli most distended with milk are preferentially emptied (512). Oxytocin also has a trophic action on myoepithelial cells (513).

Receptor Desensitization. Continuous exposure of myometrium to oxytocin leads to reduced oxytocin binding and effectiveness, like that seen for other G-protein–coupled receptors, except that the process is slow for myometrial oxytocin receptors (514). The mechanism is proposed to involve G-protein–coupled receptor kinases, which phosphorylate only agonist-activated receptors, allowing binding to arrestins, uncoupling of G protein, and endocytosis (515). It is not yet clear whether this mechanism explains oxytocin receptor desensitization (514).

Oxytocin and Hydromineral Balance

Natriuresis. In rats, oxytocin secretion from the posterior pituitary is stimulated by hyperosmolarity of extracellular fluid, as occurs with water deprivation or salt ingestion (Fig. 8). Oxytocin secreted in these conditions has a natriuretic action in rats (516), but it acts directly (517) and indirectly via the heart (518). In the heart, acting via oxytocin receptors, oxytocin stimulates the secretion of atrial natriuretic peptide (ANP) by the atria, which then acts on the kidneys (519). Oxytocin is also produced and released by the atria but acts in a paracrine manner (253); blood volume expansion evidently also stimulates oxytocin neurons and secretion (520,521). The natriuretic action of oxytocin can be viewed as complementary to the water-retaining actions of vasopressin in regulating extracellular fluid osmolarity. In late pregnancy plasma ANP level is low, as is oxytocin receptor mRNA content in the left atrium and kidney (522,523). Together with reduced activity of oxytocin neurons in late pregnancy, these changes indicate reduced activity of the oxytocin–ANP axis and hence Na^+ retention, which aids plasma volume expansion.

Stimulation of ANP secretion by oxytocin released during lactation, increased oxytocin receptor mRNA and ANP mRNA expression in the atria, and increased ANP secretion postpartum (523) and direct oxytocin action on the kidney enhance natriuresis. This aids osmoregulation as milk secretion represents a net water loss because milk has a lower [Na^+] than plasma.

Relaxin. In pregnancy blood volume is increased, with plasma volume increasing by ~40%, but osmolarity and plasma [Na^+] concentration are decreased (524). These changes are due to increased drinking and increased vasopressin secretion. Drinking increases as pregnancy progresses, and this is partly due to relaxin from the corpora lutea of pregnancy (525); relaxin is dipsogenic via actions on the subfornical organ (SFO), where neurons expressing relaxin receptors are excited by relaxin (526,527). Central angiotensin mechanisms are involved, and in pregnancy AT1A receptor expression in the SFO is increased (528,529). Stimulation of vasopressin and oxytocin neurons by relaxin is mediated by the

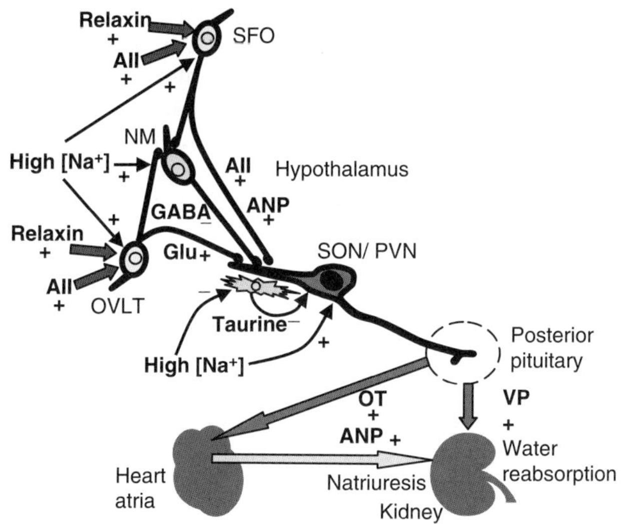

FIG. 8. Regulation of magnocellular oxytocin neurons by afferent inputs from the lamina terminalis involving the influence of osmolarity, angiotensin II, and relaxin. In the rat, oxytocin and vasopressin secretion from the posterior pituitary are regulated by afferent inputs from the lamina terminalis. In the rat, oxytocin is natriuretic, and vasopressin conserves water. In pregnancy, plasma volume is increased and osmolarity is decreased; relaxin resets the control mechanisms. Hyperosmotic stimulation (high [Na^+], *black arrows*) acts (1) directly on magnocellular oxytocin (OT) and vasopressin (VP) neurons as they are osmosensitive, increasing activity with increased extracellular [Na^+] (534); (2) by inhibiting release of taurine, an inhibitory amino acid of astrocytic glia in the supraoptic nuclei (SON) (535); and (3) on osmosensitive neurons in the subfornical organ (SFO) and organum vasculosum of the lamina terminalis (OVLT) that project to oxytocin and vasopressin neurons, directly and via nucleus medianus (NM; also directly osmosensitive). These projections use glutamate (Glu), angiotensin II (AII), and atrial natriuretic peptide (ANP) as excitatory transmitters and GABA as an inhibitory transmitter (536,540). Increased circulating angiotensin II reflects decreased renal blood flow and stimulates oxytocin and vasopressin neurons via the SFO and OVLT (538). Oxytocin acts on oxytocin receptors in the atria of the heart, stimulating ANP secretion (519); ANP acts on the kidneys, causing natriuresis. Vasopressin acts on the kidneys stimulating water reabsorption and excretion of more concentrated urine. These actions restore normal extracellular fluid osmolarity, which removes the initial feedback signal. The control of oxytocin secretion in this context is via negative feedback. In pregnancy the corpora lutea secrete relaxin. This hormone acts via the SFO to stimulate drinking and via the OVLT and relays in the nucleus medianus (NM) to stimulate oxytocin and vasopressin secretion, involving central angiotensin (AII) synapses (527). In pregnancy these actions of relaxin lead to water retention, expansion of blood volume, and reduced extracellular fluid osmolarity.

organum vasculosum of the lamina terminalis, which contains neurons expressing relaxin receptors (526) and from which neurons project indirectly via the nucleus medianus to the magnocellular neurons (527) (Fig. 8). Consequently, relaxin also stimulates the electrical activity of oxytocin neurons and hence secretion of oxytocin, partly via central angiotensin II actions (444,530), which may stimulate Na^+ excretion, but the action on vasopressin neurons, and subsequent water retention, evidently predominates in pregnancy (531). In late pregnancy exogenous relaxin no longer stimulates oxytocin secretion (531). It has been proposed that the action of relaxin on oxytocin neurons in pregnancy leads to induction of inhibitory endogenous opioid tone on these neurons and hence in time reduces oxytocin secretion (532). This would aid the expansion of blood volume in pregnancy.

Osmoregulation of Oxytocin Neurons. The increase in oxytocin secretion with extracellular fluid hyperosmolarity in rats is a consequence of a sustained increase in firing rate of the neurons, linearly related to the osmotic stimulus, without any bursting activity (533). This contrasts with the burst firing seen during parturition (see section on Pulsatile Oxytocin Secretion) and during suckling in lactation (see Chapter 59). There are three ways in which oxytocin neurons are activated by osmotic stimulation, and these are the same as the mechanisms that osmotically stimulate vasopressin neurons (Fig. 8). First, these neurons are directly osmosensitive. This property is conferred by stretch-inactivated cationic conductances, that is, when the neurons shrink with increased extracellular osmolarity, these conductances open and the neurons depolarize (534). Second, the release of taurine by glial cells in the magnocellular nuclei is reduced when extracellular osmolarity increases, and this reduces an inhibitory restraint via glycine receptors and associated Cl^- channels on SON neurons (535). However, the depolarization that results from increased extracellular osmolarity is not sufficient alone to initiate spikes, and for this, excitatory synaptic activity is required (533). The third mechanism is through activation of inputs from the lamina terminalis, which are from two circumventricular organs, the SFO and the organum vasculosum of the lamina terminalis, whose central projections are mainly via relays in the nucleus medianus and from which neurons project to the magnocellular oxytocin (and vasopressin neurons) (536).

Angiotensin II. The SFO and organum vasculosum of the lamina terminalis contain neurons that are osmosensitive (537) and neurons that are excited by angiotensin II (538). Circulating angiotensin II concentrations are increased after renin release by the kidneys when renal perfusion is reduced, and the SFO mediates the stimulation of oxytocin secretion by circulating angiotensin II; also, SFO neurons projecting to magnocellular neurons use angiotensin II as a transmitter (539). Angiotensin II acts on magnocellular neurons via angiotensin I receptors linked to the stretch-inactivated channels that confer osmosensitivity; this provides a mechanism for integration of blood osmotic and volume information at the level of the effector cell (534).

Atrial Natriuretic Peptide. The lamina terminalis and SFO contain neurons that are immunoreactive for ANP, and these project to the magnocellular nuclei (540). ANP in these neurons is evidently involved in the hyperosmotic stimulation of oxytocin secretion, because intracerebroventricular injection of ANP antibody inhibits osmotically stimulated oxytocin secretion (540). In pregnancy, the expression of ANP mRNA decreases in rostral regions but is increased in the SON, and ANP content in the hypothalamus increases also; these changes suggest a changing role for central ANP in oxytocin neuron regulation in pregnancy (541).

In pregnancy, oxytocin secretion is weakly stimulated by a hyperosmotic stimulus that strongly stimulates oxytocin secretion in virgin rats (542). This is explained by the reduced extracellular fluid osmolarity in pregnancy due to water retention resulting from stimulation of vasopressin secretion by relaxin, as discussed above. Evidently, although extracellular fluid osmolarity and [Na^+] are reduced in pregnancy, the threshold for osmotic stimulation of oxytocin neurons is not fully adjusted, as a stronger osmotic stimulus does stimulate oxytocin neuron firing rate and oxytocin secretion (542–544). Increased taurine release by glia in the magnocellular nuclei is expected as a result of the decreased extracellular osmolarity in pregnancy and hence increased inhibition of oxytocin neurons (535). This local mechanism, together with closure of osmosensitive stretch-inactivated channels in oxytocin neurons, could explain why electrical stimulation of the lamina terminalis, which has a key role in osmoregulation of oxytocin neurons, is less effective in stimulating oxytocin neurons in pregnancy (545).

Stress

In the rat oxytocin is a stress hormone. Activation of oxytocin neurons and stimulation of secretion is increased by physical stressors such as immune stimulation [interleukin-1β (546)], by systemic CCK [a stressor as it activates the hypothalamic–pituitary–adrenal (HPA) axis (547)], and by emotional stressors, such as conditioned fear (548), restraint (549), or shaker stress (550), or mixed stressors, such as forced swimming (551), but not social defeat stress in male rats (552). Oxytocin is not a stress hormone in primates (553).

The stimulation of oxytocin neurons by these stressors involves the brainstem noradrenergic A2 (NTS) projection in particular, so that neurotoxic lesion of the noradrenergic input to the SON prevents activation of supraoptic neurons by unconditioned fear (548,554). In addition, prolactin-releasing peptide (PrRP) expressing NTS neurons projecting to the SON are also activated by CCK and by conditioned fear, and intracerebroventricular injection of anti-PrRP antibody reduces oxytocin release in response to the fear stressor (548,555). Interleukin-1β may act directly on the posterior pituitary via prostaglandins (556,557) but evidently not in pregnancy (558).

Activation of oxytocin secretion by stress during pregnancy might be expected to predispose to preterm labor, in particular by interleukin-1β, produced by immune cells as result of infection. However, as discussed above, studies in the rat have shown that powerful mechanisms normally inhibit stress-induced oxytocin secretion in pregnancy. If these mechanisms are overwhelmed, then parturition could be triggered, given myometrial sensitivity to oxytocin. Otherwise, actions of oxytocin released in response to stress may be as follows.

Peripheral Roles of Oxytocin in Stress

Pro-opiomelanocortin Peptide Secretion. Oxytocin can act via V1b receptors on anterior pituitary corticotrophs to augment the action of corticotropin-releasing hormone in rats but is not very effective alone (454). Here, as discussed above, oxytocin enters the primary capillaries of the hypothalamohypophysial portal system after release in the median eminence. In humans oxytocin inhibits rather than stimulates adrenocorticotropic hormone (ACTH) secretion, blocking the response to corticotropin-releasing factor in men (559,560). A study with porcine pituitary cells in vitro indicated that high concentrations of oxytocin stimulated release of β-endorphin and elevated cellular levels of cAMP and cGMP (465), suggesting that oxytocin can modulate release of pro-opiomelanocortin peptides through a cAMP- and cGMP-coupled mechanism.

The fact that ACTH secretion is not increased during parturition in the rat (443,561) when oxytocin secretion is increased suggests that at least the increased level of circulating oxytocin in parturition is not sufficient to stimulate ACTH secretion. Oxytocin secretion does not seem to be increased by stress in mice (novel environment or swimming) (562). In mice with targeted deletion of the oxytocin gene, corticosterone secretion in response to stress is greater than in wild-type mice (563), suggesting that ACTH secretion is not compromised by lack of oxytocin.

Cardiovascular and Natriuretic Actions. Stimulation of renal Na^+ excretion is discussed in section on Oxytocin and Hydromineral Balance. It is not clear how this action, and consequent hemodynamic changes, might be adaptive in stress.

Central Roles of Oxytocin in Stress

Modulation of HPA Axis Responses. Somatodendritic release of oxytocin is increased within the

SON and PVN during stress (564–566) and after central administration of interleukin-1β (567). As well as effects of central oxytocin discussed above, central oxytocin administration reduces HPA axis restraint stress responses in female rats, evidently reducing activation in forebrain circuits, including the PVN (568). The possibility that centrally released oxytocin attenuates HPA axis activity is supported by finding increased ACTH and corticosterone secretion after intracerebroventricular oxytocin antagonist infusion in male and virgin rats under basal conditions (569). In female mice with targeted oxytocin gene inactivation, the hyperthermic and corticosterone secretory responses to acute stress and the activation of the amygdala are greater than in wild-type mice (563). In male rats, local oxytocin antagonist administration by microdialysis indicates that oxytocin acts within the PVN (but not the amygdala), possibly reducing corticotropin-releasing factor neuron activity (569). This mechanism is evidently not active in pregnant rats as intracerebroventricular oxytocin antagonist has no effect on HPA axis activity (570).

In contrast to inhibitory actions of oxytocin on basal HPA axis activity, central administration of an oxytocin antagonist during stress decreases the HPA axis response, indicating a facilitatory, not inhibitory, role for central oxytocin in the PVN (571). In late pregnancy oxytocin release in the PVN during stress is inhibited by endogenous opioids, as naloxone further increases oxytocin release; this is a reversal of the actions of naloxone in virgin rats (566). These findings further show that central oxytocin release by magnocellular neurons in pregnancy is not responsible for attenuated HPA axis stress responses in pregnancy, because naloxone reverses attenuation of HPA axis stress responses (572) while increasing central oxytocin release. Similarly in parturition, naloxone increases HPA axis activity, which is otherwise not activated during parturition (561), yet naloxone does not increase central oxytocin release in parturition (353). Furthermore, an oxytocin antagonist given centrally does not increase HPA axis activity in the presence of a stressor (443).

Anxiolytic. Central administration of oxytocin is anxiolytic in rats and mice (573,574), but oxytocin antagonist given centrally does not alter anxious behaviors on the elevated plus maze in rats (570), except in pregnancy and lactation when anxious behavior is increased by the antagonist, partly acting in the amygdala (571). However, in mice intracerebroventricular oxytocin antagonist does increase anxious behavior on the plus maze, and this behavior is greater in transgenic mice with targeted oxytocin gene inactivation and is reduced by intracerebroventricular oxytocin administration (563).

Maternal Aggression. The aggressive behavior of lactating rats toward male or female intruders provides a means to study social defeat in females as well as female aggressive behavior. Oxytocin release is increased in the PVN (but not in the amygdala) in the intruder and in the aggressor lactating rat (575). Any role for central oxytocin in the maternal aggression that emerges in lactation is contraindicated by finding that central administration of oxytocin antagonist does not modify this behavior in normal rats (576).

Central Role of Oxytocin in Salt Appetite

Within the brain, oxytocin is proposed to suppress salt appetite. In studies on rats, central oxytocin has been shown to be involved in reducing stimulated salt intake, for example, during hypovolemia. Such an action is consonant with the natriuretic actions of oxytocin secreted into the systemic circulation (577). As a corollary, stimuli that increase oxytocin secretion from the posterior pituitary, such as extracellular fluid hyperosmolarity, also reduce salt intake. Several approaches to block oxytocin action, or release, or to ablate central neurons bearing oxytocin receptors support a role for oxytocin in suppressing salt appetite (578). Increased salt appetite is revealed after overnight water deprivation in transgenic mice with targeted deletion of the oxytocin gene, indicating that oxytocin suppresses salt intake in mice (579).

Central Role of Oxytocin in Food Intake

Oxytocin has central anorexic effects (580). It reduces stimulated food intake, for example, in fasted rats (577). Such an action has been viewed as consonant with the central and systemic actions of oxytocin regulating salt balance. There is a close linkage between intake of food, which delivers a solute load, and water intake.

Central oxytocin, in neurons projecting from the parvocellular PVN to the dorsal vagal complex, also inhibits gastrointestinal function, reducing gastric motility (581,582). Inhibition of gastric motility is evidently an appropriate accompaniment to suppression of food intake, and this provides the context for the stimulatory action of systemic CCK, which is secreted after a meal, on oxytocin secretion, and its inhibitory action on food intake.

Notably, oxytocin given by intracerebroventricular injection activates Fos expression in NTS neurons, as does food intake (78). Furthermore, parvocellular PVN oxytocin neurons that project to the dorsal vagal complex are activated by intravenous CCK (583).

The NTS is thus evidently a site of modulation of neurons projecting caudally to regulate gastric motility and rostrally to influence appetite (584) and the activity of oxytocin neurons. Notably, an oxytocin antagonist given into the fourth ventricle attenuates the acute reduction in food intake caused by CCK (585).

CCK is a signal that functions to regulate meal size. Leptin has longer range actions on appetite and regulation of metabolism. Central oxytocin evidently plays a key role in integrating leptin and CCK actions on appetite. Centrally administered leptin reduces food intake and activates parvocellular PVN oxytocin neurons that project to the NTS, and leptin thereby activates NTS neurons and potentiates CCK actions (209). Furthermore, oxytocin mediates the central effects of leptin on food intake by these actions in the NTS, presumably via ascending NTS projections (209,584). Whether NTS neurons that are primarily concerned with appetite or gastric motility control are among those showing altered oxytocin receptor expression in pregnancy (496), and thus may be more sensitive to oxytocin, remains to be studied.

It may be, as pointed out by Verbalis et al. (577), that the central inhibitory actions of oxytocin on eating and salt intake give priority to the procreative activities through the other central actions of oxytocin. Put simply, central actions of oxytocin direct behavior toward reproduction and away from satisfying hunger or salt appetite. The mechanisms that focus the magnocellular neurons to respond to inputs from the birth canal rather than inputs related to stress, digestive activity, or osmoregulation are discussed in the section on Preventing Premature Oxytocin Secretion in Pregnancy.

Central Role of Oxytocin in Memory and Learning Processes

The central administration of oxytocin in rats tested for behavioral paradigms related to memory and learning gave the first demonstration that the hormone acted directly on the brain and, furthermore, emphasized that central processes related to memory and learning functions were modulated by oxytocin (586,587). The long-lasting effects of the peptide and the effects on retrieval processes indicated that oxytocin had memory attenuating effects, in contrast to vasopressin (588). Thus, oxytocin was proposed to be a neuropeptide with amnesic activity (589). Amnesic activity of oxytocin was found by application of the peptide in the nucleus basalis of Meynert in rats exposed to a spatial learning paradigm (590). However, memory-enhancing activity of oxytocin has also been reported. Particularly, in mice oxytocin improved long-lasting spatial memory during motherhood (591). It was shown that this effect was mediated by the hippocampus and involved the mitogen-activated protein kinase signal transduction pathway (591). More recently, it was shown that null mutation of the oxytocin gene in mice results in social amnesia due to the lack of development of social memory (592). The roles of oxytocin and vasopressin in memory and learning processes were recently reviewed (593).

In the human, amnesic activity of oxytocin has been reported. Early studies concerned women that received oxytocin as part of the treatment to induce second trimester therapeutic abortion (594,595). Oxytocin given as nasal spray to healthy volunteers also reduced memory functions (596,597). These studies pointed out that the type of test stimuli and the psychobiological context play a role in the responsiveness to oxytocin. It has been speculated that the memory effects of oxytocin may play a biological role in woman to cope with stress during parturition and to adapt to a nursing role after delivery (598).

Central Control of Behaviors Related to Reproduction

Based on behavioral studies, oxytocin has been dubbed an "affiliative" neuropeptide because of its prominent role in regulating a number of sociosexual behaviors in a variety of species (599). Pharmacological studies have demonstrated that oxytocin is involved in regulating male and female sexual behaviors, affiliation and the neural processing of social cues, maternal behavior, and pair bonding in monogamous species. Some of these topics are covered in more detail in other chapters in this volume. Here we summarize some of the findings relating oxytocin to behaviors associated with reproduction.

Social Recognition and Affiliation. Social recognition is necessary for the development of all social relationships. Social recognition is the process by which an individual recognizes another individual as familiar based on social cues. This process involves the perception and storage of the relevant cues as well as the recall of those cues in the appropriate context and is essential for the recognition of kin, mates, or offspring. Social recognition in rodents is olfactory based and can be evaluated in rodents by measuring the degree of social investigation on subsequent exposures to the same individual stimulus animal: Less investigation indicates recognition (600). Low doses of oxytocin infused into the brain facilitate social recognition in rats (601), and both male and female oxytocin knockout mice display a profound disruption of social recognition (296,592). This disruption is not related to general impairment in olfaction or cognition because oxytocin knockout mice easily recognize nonsocial odors. Pharmacological studies in oxytocin knockout mice demonstrate that oxytocin acts in the medial amygdala in the initial processing of olfactory signals. A single injection of oxytocin into the medial amygdala before but not after the initial social exposure to a stimulus female completely rescues social recognition in oxytocin knockout mice (602).

Oxytocin also appears to be important for mate selection in mice. Female mice display an aversion to males with a parasite load and mate preferentially with nonparasitized male mice. In one study, wild-type female mice displayed a preference for males not infected with lice and an aversion to males infected with lice (603). This preference is presumably based on olfactory cues disseminating from the infected males. In contrast, female oxytocin knockout mice failed to display this preference; however, they did display aversions to nonsocial aversive odors (603). These studies suggest that oxytocin is involved in the neural processing of socially salient olfactory cues, which is critical for recognizing familiar individuals.

In addition to social recognition, oxytocin has been shown to increase social contact as well. When chronic intracerebroventricular infusions of oxytocin were given to male rats paired with either ovariectomized or estrous females, the time that the males spent in contact with the females doubled, without affecting sexual behavior (604). In addition, the highly gregarious bonnet macaque has higher levels of oxytocin in the cerebrospinal fluid than the less affiliative pigtail macaque (605). Thus, oxytocin may act to promote affiliative social interaction in a range of mammalian species.

Female Sexual Receptivity. Estrogen and progesterone act sequentially to stimulate female sexual behavior (see Chapter 34). Central administration of oxytocin facilitates female sexual behavior in estrogen- and estrogen- plus progesterone-primed female rats (212,606–608). Furthermore, selective oxytocin receptor antagonists decrease both proceptive and receptive components of female sexual behavior, although female sexual behavior was never eliminated (609–611). There is evidence that oxytocin acts both in the ventromedial nucleus of the hypothalamus and medial preoptic area to stimulate female sexual behavior. The case for the ventromedial nucleus of the hypothalamus is quite strong. First, estradiol stimulates up to a fourfold increase in oxytocin receptor expression in the ventromedial nucleus of the hypothalamus (225,227). Estrogen also increases the density of oxytocin fibers in a region just lateral to the ventromedial nucleus of the hypothalamus (212). Progesterone treatment results in an apparent spread of the oxytocin receptor to the more lateral aspects of the the posterior ventromedial nucleus of the hypothalamus, presumably by translocation on neural processes (212,612).

Finally, oxytocin stimulates lordosis behavior in estrogen plus progesterone–treated but not estrogen alone–treated rats when infused laterally to the posterior but not anterior ventromedial nucleus of the hypothalamus (612). This illustrates the permissive role of gonadal steroids in oxytocin-mediated induction of female sexual behaviors. Furthermore, infusions of oxytocin receptor antisense oligonucleotides into the ventromedial nucleus of the hypothalamus reduce the frequency and amplitude of lordosis in female rats (613). In contrast to these findings, Caldwell et al. (609,614) found that oxytocin increased and oxytocin antagonist suppressed lordosis quotients when directly infused into the medial preoptic area. All the evidence points to the conclusion that oxytocin modulates the expression of female proceptive and receptive behavior in rats. However, the effects of oxytocin on female sexual behavior vary across species; oxytocin knockout mice display apparently normal female sexual behavior (167), and oxytocin released into the mediobasal hypothalamus of sheep inhibits sexual behavior (236). Interestingly, plasma levels of oxytocin increases in women after orgasm compared with baseline measurements (177,179). Also, the permissive role of progesterone in oxytocin-stimulated sexual behavior has been documented in women in a case report (615). This report describes anecdotally a woman who resumed breast-feeding after an intermission of 6–8 months. While she was on a progestogen contraceptive pill (levonorgestrel), she was given a synthetic oxytocin spray (Syntocinon) to stimulate milk letdown. Two hours after the use of the oxytocin nasal spray, she experienced intense sexual responses, including secretion of vaginal transudate, sexual desire, cervical contraction, and orgasm (615).

Male Sexual Behavior. In males, peripherally released oxytocin affects genitalia and seminal emission (see section on Male Reproductive Tissues), whereas centrally released oxytocin modulate the expression of sexual behavior as well as sympathetic and parasympathetic outflow to internal and external reproductive organs. The role of oxytocin in male sexual behavior has not been studied as extensively as in female sexual behavior. An early study found that intravenous injections of purified pituitary extract of oxytocin stimulated male sexual behavior in rabbits as measured by an increase in the number of ejaculations in a 30-minute test (616). Both peripheral and central infusions of oxytocin were found to shorten the latency to ejaculation and postejaculation interval in male rats but had no effects on other parameters of sexual behavior (617). A selective oxytocin receptor antagonist given intracerebroventricularly significantly decreased the number of mounts and intromissions and nearly abolished ejaculation at all doses studied (618). This effect could possibly be due to its effect on penile erection because this dose of antagonist also prevented oxytocin-induced penile erection (619). Repeated episodes of penile erection and yawning can be induced in rats by central administration of oxytocin (620,621), and this effect is blocked by a selective oxytocin antagonist (619). The PVN is the most sensitive brain area for the induction of penile erection by oxytocin (622). Parvocellular projections from this area to extrahypothalamic regions and to the spinal cord modulate penile erection together with other transmitters, particularly dopamine, glutamate, and nitric oxide (623). In humans, circulating plasma concentrations of oxytocin increase during sexual arousal and ejaculation (177). In contrast, male oxytocin knockout mice are fertile, and no obvious disruptions in sexual function have been reported (167).

Maternal Behavior. The initial observations suggesting a role for oxytocin in regulating maternal behavior were not without controversy. In the early studies, Pedersen and colleagues (624,625) found that intracerebroventricular infusions of oxytocin stimulated a rapid onset of full maternal behavior in estrogen primed females. Although some groups replicated these observations (626), others failed to find a significant effect on maternal behavior in virgin rats (627,628). Later studies reported that oxytocin infusions only facilitated rapid onset of maternal behavior in virgin females when tested in a novel cage (629) or were rendered anosmic (630). This suggested that the effects of oxytocin on maternal behavior depended highly on the testing conditions and were most evident when maternal behavior performance was compromised. However, the large number of studies substantiating a role for oxytocin in maternal behavior provides strong evidence that oxytocin neurotransmission is indeed involved in regulating the onset of maternal behavior. The rapid onset of maternal behavior is delayed by reducing oxytocin neurotransmission by injecting a specific receptor antagonist (631), infusing an anti-oxytocin antibody (632), or lesioning the PVN of the hypothalamus (633). Remarkably, once maternal behavior has been induced, these same interventions have no effect. Thus, oxytocin has a role in the transition from pup avoidance to maternal behavior, but, once established, oxytocin has generally been found to be unnecessary for the maintenance or performance of maternal behavior. However, recent studies with centrally administered oxytocin antagonist indicate a role for oxytocin in the brain in the maintenance of established maternal behavior (611,634).

Central oxytocin appears to modulate maternal behavior through actions in at least three brain regions: the medial preoptic area, the ventral tegmental area, and the olfactory bulb. Infusions of oxytocin antagonist into the medial preoptic area and the ventral tegmental area prevented parturient dams from retrieving pups or assuming a nursing posture over pups (217). Oxytocin receptor expression in both of these areas is increased at the time of parturition, and the levels of oxytocin receptor mRNA are elevated in the medial preoptic area during pregnancy (217,220). Some evidence suggests that oxytocin in the olfactory bulb may reduce olfactory processing, thus facilitating approach and reducing avoidance of pups. In one study, a low concentration of a specific oxytocin receptor antagonist infused directly into the olfactory bulb delayed the onset of maternal behavior in parturient rats, whereas oxytocin infused directly into the bulb induced maternal behavior in half of the animals within 2 hours of pup exposure (635). Furthermore, electrophysiological stimulation of the PVN inhibited the firing of olfactory bulb mitral cells (636). This effect was blocked by intrabulbar infusion of oxytocin receptor antagonist and mimicked by intrabulbar infusion with oxytocin. Thus, oxytocin in the olfactory bulb may decrease olfactory processing, potentially modulating the olfactory aversion to pup odors.

Oxytocin gene expression in the hypothalamus increases during pregnancy and lactation (163). Estrogen or progesterone alone do not increase oxytocin synthesis, but prolonged estrogen and progesterone followed by progesterone withdrawal results in a marked increase in oxytocin mRNA expression (637). Perhaps even more important, oxytocin receptors in select brain regions increase as well. Oxytocin receptor mRNA in the preoptic area is elevated during mid-pregnancy, and oxytocin receptor binding is elevated in the medial preoptic area, the bed nucleus of the stria terminalis, ventromedial nucleus of the hypothalamus, and the ventral tegmental area on postpartum day 1 (217,220,638).

The data discussed above are based on studies in rats. However, there is evidence that oxytocin also plays a role in regulating maternal behavior in mice. Many laboratory strains of mice display spontaneous maternal behavior. However, mice recently derived from wild-caught mice exhibit high levels of infanticide. Intracerebroventricular infusion of oxytocin was found to significantly decrease infanticide in these wild-type strains (639). In contrast, oxytocin knockout mice display apparently normal maternal behavior, and this was initially explained by the fact that although in rats oxytocin is required

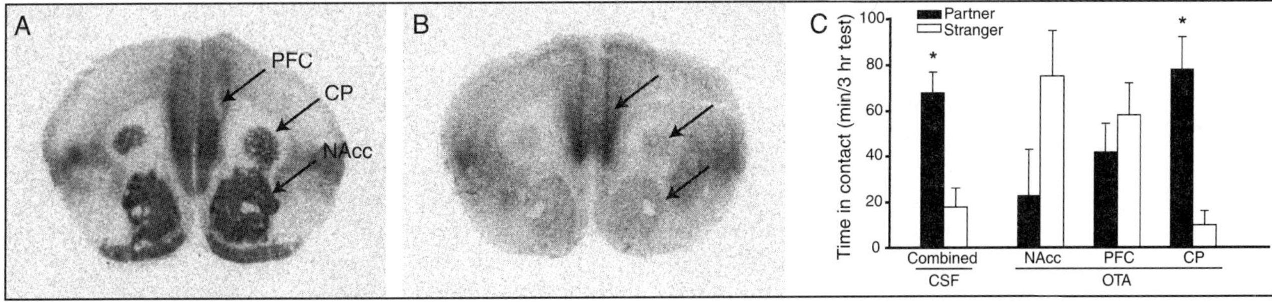

FIG. 9. Oxytocin receptor regulation of pair bonding in female prairie voles. Prairie voles (A) have high densities of oxytocin receptors in the nucleus accumbens (NAcc) and caudate putamen (CP) compared with nonmonogamous montane voles (B). Both species have oxytocin receptors in the prefrontal cortex (PFC). (C) A selective oxytocin receptor antagonist (OTA; 1 ng per side in 200 nl) infused bilaterally into the NAcc or PFC, but not the CP, blocks partner preference formation in female prairie voles (653).

for the initiation of maternal behavior, laboratory strains of mice are spontaneously maternal (167,640). However, recent studies on oxytocin receptor knockout mice suggest that functional oxytocin receptors are essential for the expression of maternal behavior even in spontaneously maternal strains of mice (641).

A remarkable body of research has implicated central oxytocin pathways in the regulation of maternal behavior and bonding with offspring in sheep (642). Infusions of oxytocin into the brain of estrogen-primed ovariectomized ewes induced short latency maternal responses in less than 30 seconds (643). Maternal behavior develops naturally at the onset of parturition in sheep or can be stimulated artificially with vaginocervical stimulation. In either case, the onset of maternal behavior is associated with an increase in oxytocin release in several brain structures, including the olfactory bulb, medial preoptic area, and the PVN (644–647). An epidural anesthetic prevents both the increase in oxytocin release and maternal behavior. Administration of oxytocin directly into the PVN by retrodialysis appears to be as potent for inducing maternal behavior as vaginocervical stimulation (647). These results suggest that oxytocin released within the PVN during parturition may, via positive feedback, coordinate the release of oxytocin in several other central sites, thus facilitating the onset of maternal behavior. Oxytocin released in the olfactory bulb is thought to play a role in the olfactory recognition of the lamb via modulation of the release of other neurotransmitters that play a role in olfactory learning.

Pair Bonding in Monogamous Species. Given the role of oxytocin in maternal bonding with offspring, it is not surprising that oxytocin also plays a critical role in the development of pair bonds in monogamous species. Prairie voles are monogamous rodents that form lifelong pair bonds with their mates (648,649). Mating normally facilitates pair bond formation in prairie voles; however, intracranial infusions of oxytocin in female prairie voles facilitate the formation of partner preferences even in the absence of mating (650). Furthermore, selective oxytocin antagonists block the development of partner preferences when given before mating (650). The effects of oxytocin are somewhat gender specific, because oxytocin antagonists are more effective at blocking pair bond formation in the female than in the male [(651), but see (652)].

Approximately 5% of mammals display social monogamy (as opposed to sexual exclusivity), and several vole species are promiscuous breeders and consequently do not form pair bonds. Comparisons of the distribution of oxytocin receptors in monogamous and nonmonogamous species suggest an association between oxytocin receptor distribution in the brain and social organization (223). For example, monogamous prairie voles have high levels of oxytocin receptor in the nucleus accumbens and caudate putamen, whereas nonmonogamous montane voles do not (Fig. 9). Infusion of oxytocin antagonist into the nucleus accumbens and prelimbic cortex, but not into the caudate putamen, inhibits partner preference formation in female prairie voles (653). Interestingly, the nucleus accumbens and the prelimbic cortex are part of the mesolimbic dopamine reward circuit. Infusions of dopamine D2 antagonist into the nucleus accumbens of female prairie voles also prevent partner-preference formation (654). In fact, activation of both dopamine D2 receptors and oxytocin receptors are necessary for pair bond formation in the female prairie vole (655). Given the role of oxytocin in social recognition and the neural processing of social signals, this has led to the hypothesis that pair bond formation results from the convergence of the neural pathways involved in processing the olfactory signatures of the partner and reward resulting from sexual activity (649). This convergence may result in an association of the olfactory signature of the partner and the rewarding nature of sex, culminating in the development of a conditioned partner preference. Oxytocin is less effective at stimulating partner preferences in male prairie voles. Instead, vasopressin facilitates and vasopressin V1a receptor antagonist blocks partner preferences in male prairie voles (656). Interestingly, V1a receptors are concentrated in the ventral pallidum of prairie monogamous voles but not of nonmonogamous montane voles, and blockade of vasopressin receptors in this region blocks pair bond formation in male prairie voles (657). The ventral pallidum is a major output of the nucleus accumbens, suggesting that parallel neural mechanisms underlie pair bonding in male and female prairie voles even though the process is regulated by different peptides.

SECRETORY ACTIVITY OF OXYTOCIN NEURONS IN PREGNANCY AND PARTURITION

Pathways to Parturition

Two questions arise from the earlier conclusion that oxytocin secreted by the posterior pituitary has an important role in parturition (see Role for Oxytocin in Parturition). First, what are the pathways from the contracting uterus and the distended birth canal that lead to excitation of the magnocellular neurons in the SON and PVN and hence to oxytocin secretion from the posterior pituitary during parturition? Second, how is premature activation of this positive feedback mechanism prevented?

Positive Feedback in Parturition

The stimulation of oxytocin secretion by neural signals from the birth canal was proposed by Ferguson (658). Measurement of circulating oxytocin concentrations show that oxytocin secretion is not increased in the rat until births begin, so it is evident that oxytocin secretion is stimulated by the birth process, which is initiated by local mechanisms (410). In anesthetized rats in late pregnancy, intravenous infusions of oxytocin stimulate uterine contractions and increase the electrical activity of putative oxytocin neurons in the SON (659). In conscious late pregnant rats, intravenous infusions of oxytocin advance the onset of parturition and activate magnocellular oxytocin neurons, as indicated by Fos expression (411).

Afferent pathways from the uterus and cervix include the vagal and pelvic nerves, relaying in the spinal cord (660,661). Electrical activity in afferents from the uterus, including the pelvic nerves, increases during parturition (661), and if these nerves are lesioned, the fetus expulsion reflex is prevented and oxytocin secretion is reduced (660). These afferents relay in the brainstem on neurons projecting to the hypothalamus.

Magnocellular oxytocin neurons are densely innervated: there are some 37×10^6 synapses per mm^3 in the SON (662). Oxytocin neurons express receptors for a wide variety of excitatory and inhibitory neurotransmitters (663,664), and how many of these are critically involved in driving, or regulating, oxytocin neurons in parturition is not yet clear. Approaches to answer this include the following: tracing and phenotyping projections to the oxytocin neurons that are activated in parturition, seeking increased expression in neurons projecting to oxytocin neurons of genes for neurotransmitter-synthesising enzymes or neuropeptide genes in pregnancy, measuring the release of neurotransmitters in the vicinity of the oxytocin neurons, and testing actions on oxytocin neuron activity and parturition of antagonists for receptors for putative key transmitters. The activity of oxytocin neurons is regulated by input from distant projections, by more local neurons, and by autocrine or paracrine mechanisms. The latter have a key role in generating the patterning of oxytocin neuron firing and hence in the shaping of the profile of oxytocin secretion.

Organization and Signaling Properties of Oxytocin Neurons. Like other neurons, oxytocin cells have four compartments (Fig. 10):

1. *Dendrites,* which receive most synapses from axon terminals of other neurons and conduct electrical events after action of neurotransmitters on their receptors to the soma (perikaryon, cell body), but also are actively involved in retrograde chemical signaling to modulate the presynaptic inputs. There are also membrane receptors (including for oxytocin) that regulate local biochemical signaling pathways.
2. The *soma,* which integrates electrical signals from synapses on the soma and the dendrites, triggering an action potential at the axon hillock if depolarization reaches threshold; the soma is also the location of most of the cell's mRNA translation, protein synthesis, and processing machinery and all of the cell's nonmitochondrial genome. The soma contains intracellular ligand-operated transcription factors (including some steroid receptors).
3. The nonmyelinated *axon,* which conducts action potentials rapidly through its plasma membrane (~5,000 mm/sec) to the neurosecretory terminals in the posterior pituitary gland and slowly transports (~8 mm/h) neurosecretory granules in

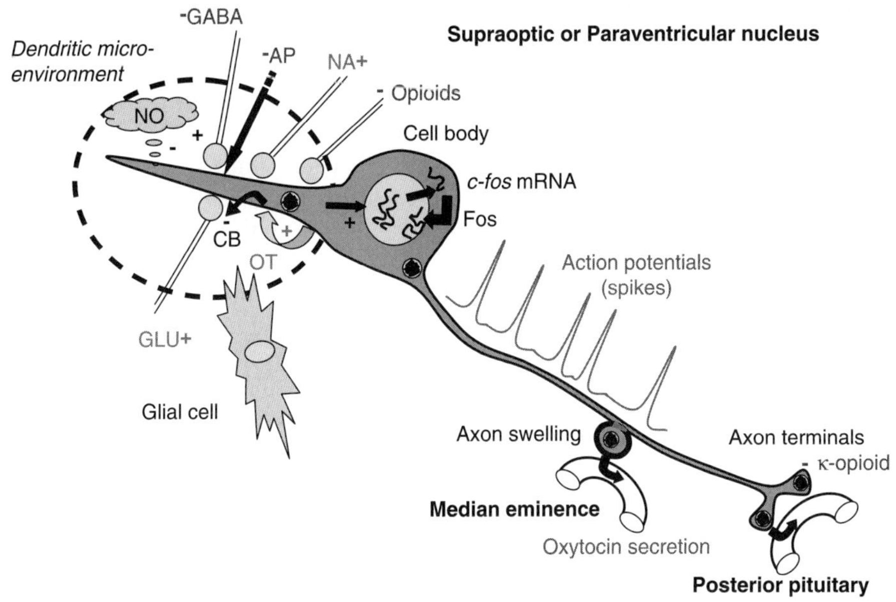

FIG. 10. Schematic of regulation of electrical activity of a magnocellular oxytocin neuron. Action potentials (spikes) are initiated at the axon hillock (origin of the axon from the cell body), propagated along the axon at about 5 $m \cdot s^{-1}$. Synchronized burst firing of oxytocin neurons in parturition (Fig. 11) and lactation (see Chapter 59) causes secretion of oxytocin pulses. *Dendritic signaling.* The dendrites (about three per neuron) and cell body have excitatory (+; e.g., glutamate, GLU; noradrenaline, NA) and inhibitory (–; e.g., GABA, opioid) synapses. Summation of inhibitory and excitatory postsynaptic potentials determines action potential initiation. Somadendrite depolarization, and/or increased $[Ca^{2+}]_i$, activates auto- and paracrine regulatory mechanisms and alters gene expression (e.g., *c-fos* and hence Fos protein). Auto- and paracrine mechanisms include generation of nitric oxide (NO), which is autoinhibitory directly and indirectly by presynaptically increasing GABA release (711); oxytocin stimulates further oxytocin release (731) and cannabinoid (CB) production (see Fig. 7), which inhibits glutamate synapses (748). Oxytocin also inhibits GABA release (749). *Axon terminals.* Release of oxytocin can be inhibited by endogenous opioid acting on κ-receptors (794). *Glia.* Allopregnanolone (AP), a neuroactive progesterone metabolite produced by glia cells, enhances GABA inhibition by allosteric modulation of $GABA_A$ receptors (708). The glial environment also regulates the ionic and excitatory and inhibitory amino acid content of the extracellular fluid environment of the cell body and dendrites (see Fig. 8).

its cytoplasm to the axon terminals in the posterior pituitary gland.

4. The *axon terminals*, of which the axon of each neuron has about 5,000, abut the fenestrated capillaries in the posterior pituitary gland, contain much of the oxytocin stored in the posterior pituitary (there are also preterminal swellings with stored oxytocin as well, and axonal swellings—Herring bodies—in the median eminence), and release oxytocin by exocytosis when action potentials arrive and trigger Ca^{2+} entry.

Glial Interactions with Oxytocin Neurons. Among the perikarya and particularly the dendrites in the magnocellular nuclei are *glial* cells, and similarly there are modified glial cells, *pituicytes*, around the axon terminals in the posterior pituitary. These *astrocytic* glial cells have important roles in regulating the immediate extracellular environment of the neurons, with respect to amino acids that act as excitatory or inhibitory transmitters and $[K^+]$ (which, if increased, depolarizes neurons) (422,535,665) (Figs. 8 and 10). Glial cells can act more directly on SON neurons by releasing taurine, an inhibitory amino acid that acts via glycine receptors on the neurons; taurine release by glia is decreased by increased extracellular osmolarity and increased by hypoosmolarity, reflecting a role of taurine in cell volume regulation. Consequently, glial cells contribute to the osmoregulation of oxytocin neurons (535). The morphology of the glial cells is *plastic*, and when the magnocellular oxytocin system is strongly stimulated, as in lactation (see Chapter 59), withdrawal of their processes evidently permits closer contact between the dendrites and somata of adjacent neurons and the sharing of synaptic inputs, which may have a role in synchronized firing (422,666,667). At the end of pregnancy, before parturition, there is already closer contact between dendrites in the ventral zone of the SON, and more oxytocin neurons share glutamate and GABA synaptic terminals (435,668,669).

The withdrawal of glial processes also changes the dynamic regulation of the immediate extracellular environment of the neurons (535).

In the posterior pituitary, retraction of pituicyte processes in response to local chemical signals released with strong stimulation of the oxytocin system increases the extent of contact of axon terminals with the blood capillaries and is envisaged as facilitating secretion. Pituicytes express receptors for many neuropeptides and neurotransmitters, including products, except oxytocin, of oxytocin neurons (422,535). Glial cells express enzymes that generate *neuroactive steroids* (670), and may contribute to the intensification of GABA inhibition of oxytocin neurons in pregnancy, discussed below.

Action Potential Generation and Regulation. As in other neurons, an oxytocin neuron discharges an action potential from its cell body, at the axon hillock, when it is depolarized above threshold for opening Na^+ channels, to produce the large depolarization that can be recorded extracellularly as a *spike*. This excitation is propagated as a regenerating wave along the axon to the terminals. The depolarization of the cell body is primarily the result of summation of *excitatory* and *inhibitory postsynaptic potentials*, brief events recorded only with an intracellular electrode that follow the release of packets of excitatory or inhibitory neurotransmitter from synaptic terminals on the dendrites and soma. These events can alternatively be recorded with patch-clamp electrodes as *excitatory* or *inhibitory postsynaptic currents* and as currents through individual channels. Whenever the sum of ongoing excitatory and inhibitory postsynaptic potentials exceeds threshold, an action potential is triggered. If the neuron is still recovering from a previous action potential it is briefly *refractory* to further stimulation and cannot respond. Thus, postspike excitability of an oxytocin neuron, like other neurons, is regulated by intrinsic mechanisms, which involve activity of specific K^+ channels (conductances) that are activated by influxing Ca^{2+} during an action potential, and allow K^+ to leave the neuron, which is consequently hyperpolarized and inexcitable (671–673). These "*after*-potential" mechanisms govern the maximum firing rate of an oxytocin neuron. Other longer lasting events follow excitation of an oxytocin neuron that modulate the effectiveness of further excitatory stimulation and are discussed below.

Inputs: Brainstem Neurons. During parturition, neurons in the NTS and ventrolateral medulla are activated (i.e., increase Fos expression). Noradrenergic neurons from these areas project directly to the magnocellular oxytocin neurons and provide about 10% of the synapses on oxytocin neurons (674) (Fig. 11). Noradrenaline depolarizes oxytocin neurons via α_1-adrenoreceptors (675,676). Retrograde labeling of these neurons by injection of fluorescent microspheres into the SON before pregnancy allows their identification after parturition; combined immunocytochemistry for Fos and tyrosine hydroxylase (the enzyme that catalyses conversion of tyrosine to dihydroxyphenylalanine, the precursor of noradrenaline) has enabled identification of putative noradrenergic brainstem neurons that project to the oxytocin neurons and are activated during parturition (677). Further evidence for the involvement of a noradrenergic relay from the birth canal is provided by using microdialysis to measure increased noradrenaline release in the SON during parturition (678). Furthermore, intravenous infusion of oxytocin in anesthetized late pregnant rats activates Fos expression in NTS neurons and leads to increased noradrenaline release in the SON (659). Because circulating oxytocin does not enter the brain and there are no oxytocin receptors in the circumventricular organs, the central effects of oxytocin infusions, which are seen only in late pregnancy, are deduced to be due to stimulation of uterine contractions. As further evidence of the importance of the noradrenergic input in stimulating oxytocin neurons, local application of benoxathian, an α_1-adrenergic antagonist, into a SON in late pregnant rats blocks Fos activation that, in controls, follows pulsatile intravenous oxytocin infusion (659).

The release of noradrenaline into the SON in response to intravenous pulsatile oxytocin secretion itself shows intermittent increases. However, the temporal resolution of the noradrenaline measurements is not adequate to show precise relationship to uterine contractions. There is a correlation between oxytocin, and also vasopressin, neuron firing rate and uterine contractile activity induced by oxytocin infusions, but without synchrony between uterine contractions and firing rate change. Thus, there is not, in general, a simple relationship between an oxytocin-stimulated uterine contraction and increased oxytocin neuron firing rate (659). However, these studies were performed on anesthetized late pregnant rats, and births rarely follow oxytocin infusions in these conditions, so the above mechanisms concern interactions before and between births. Stimuli arising from the passage of a fetus through the birth canal are probably important in triggering the bursts of oxytocin neuron firing (Fig. 11) as recorded during natural parturition in rats (398) and as triggered by mechanical vaginocervical stimulation in late pregnant rats (679) (Fig. 11). Synthesizing this information indicates that uterine contractile activity before the first delivery stimulates oxytocin secretion and hence further uterine contractions that move the first fetus along the birth canal. As the fetus

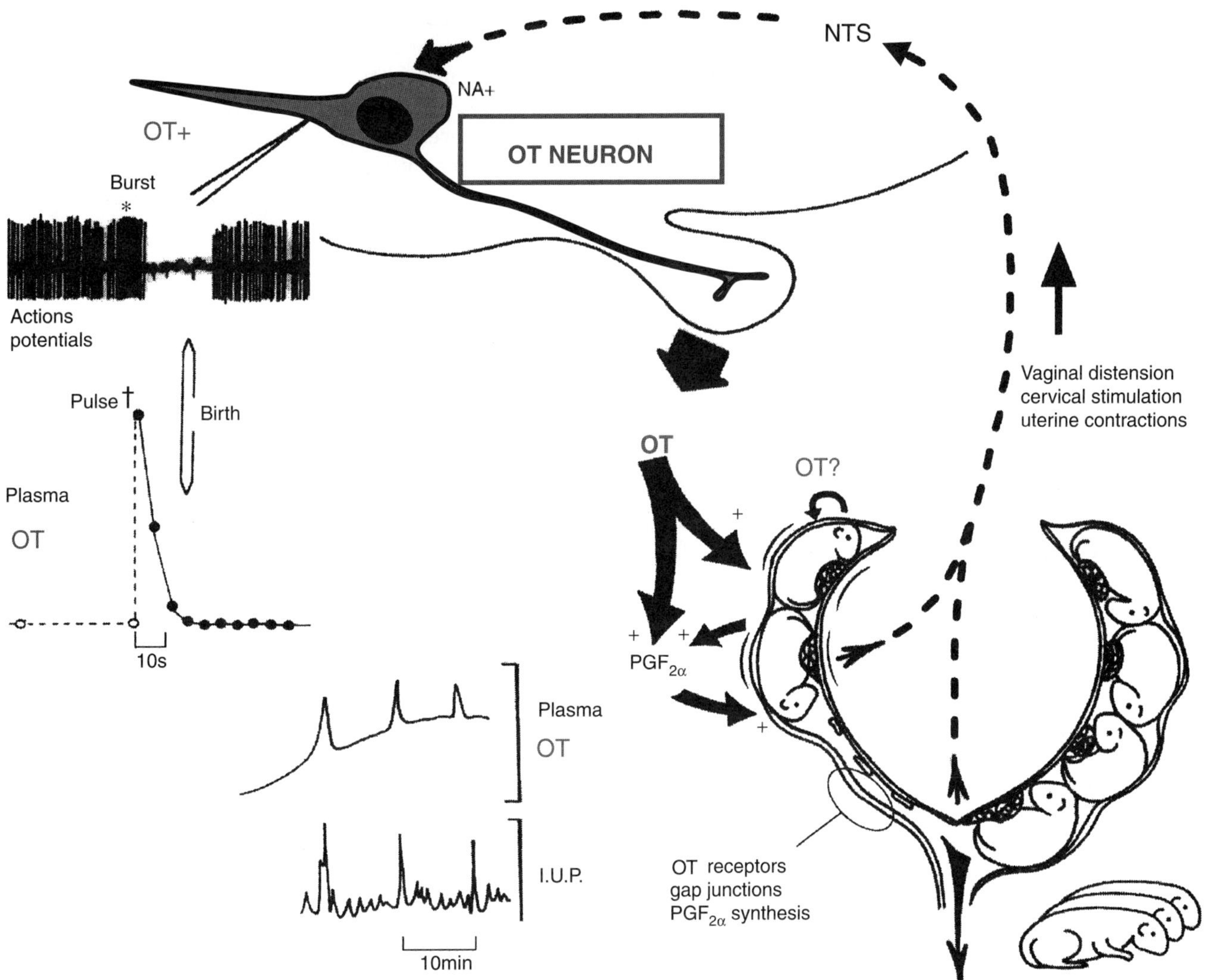

FIG. 11. Oxytocin neuron activation in parturition: the Ferguson reflex pathway. Oxytocin secretion increases with uterine contractions (659): Electrophysiological recording shows burst firing (*) of presumptive oxytocin neurons, and blood sampling shows a consequent pulse (†) of oxytocin secretion just before the birth of a rat pup, indicating triggering of a pulse by stretching of the birth canal with passage of a fetus (Ferguson reflex) (398,410). The pathway from the uterus/birth canal relays in the nucleus tractus solitarius (NTS) to noradrenergic neurons that project to oxytocin cells (677). This stimulation of oxytocin neurons by uterine (intrauterine pressure: I.U.P.)/birth canal signals, shown by Fos expression (396,429,436) and oxytocin secretion, reveals a positive feedback loop. This depends on myometrial sensitivity to oxytocin, i.e., up-regulation of oxytocin receptor expression late in pregnancy (490), and is facilitated by local oxytocin-stimulated $PGF_{2\alpha}$ production (see Fig. 7), possibly by locally produced oxytocin (155) and estrogen-stimulated myometrial gap junction formation. Oxytocin released in the supraoptic nucleus can have autocrine actions (438) (see Figs. 2,10). Endogenous opioid peptide (EOP), probably in the supraoptic nucleus, restrains oxytocin neuron excitation in parturition (781,783). Local oxytocin release probably facilitates excitation (438), as in lactation (725) and by desensitizing $GABA_A$ receptors to allopregnanolone (809). [Adapted from (423) with permission.]

completes this journey, vaginocervical distension triggers a burst of oxytocin neuron firing and a pulse of oxytocin is secreted. This pulse may support the expulsion of the fetus, placental separation, and passage of the next fetus.

An α_1-adrenergic agonist more effectively stimulates oxytocin secretion in late gestation than at other times (680). This may reflect the greater store of oxytocin in the posterior pituitary or strengthened noradrenergic synaptic connections onto oxytocin neurons in pregnancy.

Vaginocervical mechanical stimulation has been used to mimic the stimulation of the birth canal during parturition. In proestrous rats, direct vaginocervical stimulation increases release of several neurotransmitters in the NTS (noradrenaline, glutamate, aspartate, and GABA) but not after section of the abdominal vagus or pelvic and hypogastric nerves (681).

Systemic administration of CCK has been used as another stimulus to explore the NTS–oxytocin neuron pathway (see section on Afferent Inputs to Magnocellular Oxytocin Neurons). CCK activates noradrenergic NTS neurons, as shown by Fos expression, and noradrenaline release in the SON as well as activating oxytocin neurons and oxytocin secretion from the posterior pituitary (76,682,683). Neurotoxic lesion of the noradrenergic neurons projecting to the SON prevents oxytocin neuron activation by systemic CCK (76).

In vivo electrophysiological studies show that vaginocervical stimulation and intravenous CCK administration activate different NTS neurons in late pregnant rats (684). Nonetheless, systemic CCK administration has been used to explore changes in the regulation of oxytocin neurons by brainstem noradrenergic input in pregnancy (see section on Endogenous Opioids) (435).

Descending Oxytocin Projections. Some parvocellular PVN neurons project to the NTS. The expression of oxytocin receptor mRNA by NTS neurons is increased at the end of pregnancy, and NTS neurons expressing oxytocin receptor and projecting to the SON are activated in parturition (496). These findings indicate increased oxytocin sensitivity of NTS neurons relaying afferent information from the birth canal. The electrophysiological responsiveness of dorsal vagal complex neurons to oxytocin is decreased by estrogen and increased by progesterone, but the projections of these neurons are not known (685).

There is an oxytocin neural projection from parvocellular PVN neurons to the sympathetic outflow neurons in the spinal cord (686). Vaginocervical stimulation also activates descending oxytocin projections, from the PVN, causing estrogen-dependent release of oxytocin in the lumbosacral spinal cord (237). Actions of oxytocin on spinal cord sympathetic preganglionic neurons seem to be mediated by V1 rather than oxytocin receptors (687). A sympathetic reflex response to vaginocervical stimulation is pupillary dilation, and this depends on oxytocin-mediated transmission in the spinal cord (686). This descending oxytocin projection is evidently involved in regulating uterine contractions, as shown by studies in nonpregnant rats in which oxytocin was applied intrathecally, at the lumbosacral or thoracolumbar levels, and found to increase contractile activity, but only at proestrus or estrus (688). Thus, vaginocervical stimulation not only stimulates, via the NTS, activation of magnocellular oxytocin neurons, but also stimulates, likely via the same pathway, activation of parvocellular PVN neurons that project to preganglionic sympathetic neurons in the spinal cord. Whether this is important in parturition remains to be explored, but another positive feedback loop, involving activation of parvocellular nucleus oxytocin neurons by births, and further stimulation of uterine contractions via actions of sympathetic outflow regulated by oxytocin released in the spinal cord seems possible.

Local Modulation. Without pregnancy, indications for roles of other neurotransmitters in the regulation of the noradrenergic input to oxytocin neurons come from several studies of the effects of central administration of candidate modulators. In female rats, a neurokinin 3 receptor agonist given centrally, including by microdialysis into the PVN, increased oxytocin secretion into the systemic circulation, and this was prevented by an α_1-adrenergic antagonist, suggesting that tachykinins modulate noradrenaline release onto oxytocin neurons (689).

In the SON local microdialysis of oxytocin enhances the release of noradrenaline stimulated by locally increased extracellular [K^+], evidently by actions via oxytocin receptors on noradrenergic terminals; furthermore, centrally administered oxytocin antagonist reduces stress-induced oxytocin secretion, which is stimulated by noradrenergic input to oxytocin neurons (690).

In lactating rats, oxytocin secretion in response to central angiotensin II administration is reduced by an α_1-adrenergic antagonist, suggesting a role for noradrenaline in mediating angiotensin II action, but angiotensin II antagonists do not reduce suckling-induced oxytocin secretion (690a). Whether this applies in parturition is not known, but it seems likely in view of other findings indicating that lamina terminalis inputs are not involved in stimulating oxytocin secretion in parturition (702).

Other Transmitters in Brainstem Neurons. Noradrenaline is not the only transmitter involved in the brainstem projection to oxytocin neurons, because intracerebroventricular infusion of an α_1-antagonist does not impact on parturition, though in anesthetized late pregnant rats the antagonist prevents Fos activation in the SON induced by intravenous oxytocin infusion (659). The brainstem noradrenergic neurons produce other transmitters, including neuropeptide Y and somatostatin (691). Neuropeptide Y or somatostatin given intracerebroventricularly excite oxytocin neurons, stimulating oxytocin secretion, although the local action of somatostatin on oxytocin neurons is inhibitory (692,693). Neuropeptide Y acts via Y-1 receptors and potentiates the action of an α-adrenergic agonist on oxytocin secretion in vivo and in vitro from a hypothalamic explant (692,694). Hence, corelease of neuropeptide Y with noradrenaline is expected to enhance the effectiveness of noradrenaline. It is not clear whether these neuropeptides are involved in regulating oxytocin neurons during parturition; however, the expression of somatostatin mRNA in NTS neurons is decreased in late pregnancy but increases during parturition (693).

PrRP is coproduced in NTS A2 neurons, and PrRP terminals contact oxytocin neurons (695). PrRP given centrally stimulates oxytocin secretion (696). However, electrophysiological studies indicate that stimulatory actions on oxytocin neurons of PrRP given alone are indirect (697). PrRP mRNA expression in the NTS is increased by estrogen or progesterone treatment and in pregnancy (698), which indicates a possibly enhanced role in regulating oxytocin neurons in pregnancy.

The release of other transmitters in the SON has been measured with microdialysis, including aspartate, dopamine, and serotonin, but only glutamate increases and only before the first birth (678). Analysis of retrogradely labeled inputs to magnocellular oxytocin neurons has shown activation (Fos expression) during parturition in rostral inputs only in the olfactory bulb mitral cells (677), and neurotransmitter release in the olfactory bulbs is increased after vaginocervical stimulation in proestrous rats (699). The olfactory bulbs provide a glutamate innervation of the SON and might be the source of the increased glutamate release before parturition (678,700). There is no activation during parturition in other rostral regions that are known sources of glutamate and GABA input and involved in osmoregulation of oxytocin neurons, including glutamate neurons in the median preoptic nucleus (677,701). Electrolytic lesion of this rostral lamina terminalis input during parturition has no effect on the process (702).

Negative Feedback in Parturition

Inhibitory Restraint

Endogenous Opioid. The opioid antagonist naloxone given during parturition hastens delivery and increases oxytocin secretion in rats and pigs (428,703). This disrupts the normal pattern of immediate maternal care in these polytocous species, evidently because the interval between births is too brief (704). Clearly, endogenous opioid restraint of oxytocin secretion is sustained during parturition and is involved in the optimal patterning of births. Endogenous opioids do not seem to be involved in regulating oxytocin secretion during parturition in women or mice (429,705).

GABA. Inhibitory GABA input might be similarly important in regulating the activity of oxytocin neurons in parturition, as about one-third of their synapses are GABAergic (662). Part of this GABA input comes from GABA neurons located around the magnocellular nuclei (706). When oxytocin neurons are driven by hyperosmotic stimulation, GABA input has an essential role in regulating the firing of oxytocin neurons (533). However, GABA release in the SON is not increased in parturition, and there is no Fos activation in other sources of GABA input to the oxytocin cells, such as the lamina terminalis or arcuate nucleus (677,707).

The effectiveness of GABA input to oxytocin neurons in late pregnancy is enhanced because the density of GABA synapses is increased (707a) and because of positive allosteric modulation of $GABA_A$ receptors by allopregnanolone (3α-hydroxy dihydroprogesterone), formed from progesterone (708). Allopregnanolone enhances the effectiveness of $GABA_A$ receptor activation by prolonging the opening time of the Cl^- channel in the receptor complex (709). Because brain allopregnanolone level decreases after progesterone withdrawal at the end of pregnancy (710), it is likely that the GABA input to oxytocin neurons is less effective during parturition than during late pregnancy. In addition, oxytocin acts on oxytocin neurons to reduce the amplitude of spontaneous inhibitory postsynaptic currents, an effect due to $GABA_A$ receptor phosphorylation via PKC, stimulated via the oxytocin receptor. Oxytocin and allopregnanolone compete with each other via the $GABA_A$ receptor, and allopregnanolone evidently overrides PKC-mediated metabotropic actions of oxytocin, aided by a phosphatase, and prevents premature activation of oxytocin neurons (709).

Histamine. Without pregnancy, the histamine input to oxytocin neurons from the tuberomammillary nucleus is inhibitory, via H_2 receptors (97). Histidine decarboxylase mRNA expression in the tuberomammillary nucleus is increased at the end of pregnancy, although the direction of histamine actions on oxytocin neurons in parturition is not clear as an H_1 antagonist (but not an H_2 antagonist) slows births, indicating a stimulatory action (100).

Nitric Oxide. Oxytocin neurons express neuronal nitric oxide synthase and, when stimulated, generate nitric oxide, which inhibits their activity (711). Accordingly, the nitric oxide donor sodium nitroprusside given centrally slows established parturition in rats (712). A histochemical marker of nitric oxide synthase (NADPH-diaphorase) indicates increased activity toward the end of pregnancy (713). However, neuronal nitric oxide synthase mRNA expression in magnocellular neurons is reduced at the end of pregnancy, and an inhibitor of neuronal nitric oxide synthase ($N\omega$-L-arginine) no longer increases oxytocin secretion after osmotic stimulation (544). Evidently, endogenous nitric oxide generation by oxytocin neurons is suppressed at the end of pregnancy. At other times, nitric oxide acts both on oxytocin neurons, as an autoinhibitor, and presynaptically to enhance release of GABA, which inhibits oxytocin neurons (711,714). The inhibition of oxytocin secretion by nitric oxide actions in the SON is evidently not via a cGMP pathway as has been described for nitric oxide actions elsewhere (715), and indeed cGMP application excites oxytocin neurons (716). Nitric oxide also acts in the posterior pituitary to inhibit stimulated oxytocin release (717).

Pulsatile Oxytocin Secretion

The now classic studies of Wakerley and Lincoln (718) (see Chapter 59) showed that during continuous suckling in lactating rats, a pulse of oxytocin is secreted every few minutes to cause ejection of milk and that this pulse of oxytocin follows a brief high frequency spike discharge of the cell bodies of oxytocin neurons, synchronized among the four magnocellular nuclei (421). Summerlee (398) then showed that a similar high frequency bursting discharge of presumptive oxytocin neurons occurs during parturition, closely after the birth of each pup, each burst leading to the secretion of a pulse of oxytocin. Each pulse then stimulates uterine contractions to expel the next fetus. The secretion of a pulse of oxytocin after the birth of each fetus has been observed also in the pig (409), and in women oxytocin secretion is pulsatile during labor (408). This pulsatile pattern of secretion is sufficient and efficient, that is, requiring less oxytocin to achieve births, as shown in rats and women (396,719). Pulsatile oxytocin secretion, and consequent intermittent stimulation of the uterus, not only conserves the store of oxytocin in the posterior pituitary (of which about one-third is used during a normal parturition) for possible further secretion in the event of dystocia, but also prevents receptor desensitization in the myometrium by continual oxytocin exposure (see section on Receptor Desensitization) (720).

The processes that convert continual stimulation by suckling into the intermittent synchronized pattern of oxytocin neuron firing have been extensively investigated (see Chapter 59). These studies have led to the conclusion that burst firing of oxytocin neurons, which is only naturally seen during parturition and lactation, reflects properties of the oxytocin neurons and their local paracrine interactions with their excitatory and inhibitory synaptic inputs. The mechanisms of burst firing of oxytocin neurons in parturition have been much less intensively investigated (the transient nature and relatively unpredictable timing of parturition and the interruption of births by general anesthesia are important reasons). Nonetheless, comparison of available mechanisms at the end of pregnancy, when the burst firing of oxytocin neurons is expressed for the first time, with the mechanisms deduced to be important in lactation, and when adaptations to chronic stimulation are superimposed, reveals essential mechanisms for burst firing (435). Oxytocin neurons are already capable of displaying burst firing in response to suckling and vaginocervical stimulation in the last few days of pregnancy (679,721), so it seems that either of these two types of afferent input have the characteristics to determine burst firing. No other natural stimuli induce burst firing of oxytocin neurons: Strong hyperosmotic stimulation or morphine withdrawal excitation never induce burst firing but instead cause a sustained increase in firing rate, to a maximum of ~15 spikes/sec (533,722), compared with the peak rate of ~100 spikes/sec over a few seconds during suckling, though the bursts are less intense during parturition (723,724). Nonetheless, the generation of synchronized burst firing in parturition results from the properties of the oxytocin neurons in interaction with their excitatory and inhibitory synaptic inputs.

Key features from studies in rats are as follows. Just before the start of births, the firing rate of oxytocin neurons increases, from relative quiescence at the end of pregnancy, and then a burst is seen as each pup is born (724). Presumably, these bursts are synchronized among the magnocellular oxytocin neurons, as during suckling in lactation (725), causing the pulse of oxytocin secretion. In lactation, some oxytocin neurons lead coordination of firing among oyxtocin neurons just before bursts (726). Also in lactation, neurons in the ventromedial medulla oblongata that project to both SON are important in coordinating burst firing of oxytocin neurons (726), but whether these neurons are among those activated during parturition and their neurochemical identity are not yet known. However, in vitro in hypothalamic slices, albeit from lactating rats, burst firing of putative supraoptic oxytocin neurons is triggered by superfusion with an α_1-adrenergic receptor agonist (if the medium Ca^{2+} concentration is kept low), indicating the potential for noradrenergic input to drive burst firing of oxytocin neurons in vivo (727). In vitro studies of organotypic cultures of slices of neonatal hypothalamus show bursting activity in identified oxytocin neurons that can be triggered by oxytocin, and importantly, bursting follows volleys of excitatory postsynaptic potentials from glutamate synapses (728). Hence, there may be a hierarchy of mechanisms that ensure synchronized, or coordinated, bursting of oxytocin neurons during parturition and later in lactation. These are potentially a noradrenergic input from the brainstem projecting bilaterally to oxytocin neurons, with activation of a local glutamate network, and a crucial role for auto- or paracrine oxytocin action. This oxytocin is released from the dendrites or cell bodies of oxytocin neurons and has presynaptic as well as postsynaptic actions.

The retraction of glial processes from between oxytocin neuron somata and dendrites at the end of pregnancy facilitates interactions among the transmitters and modulators produced by the oxytocin neurons and the synapses on them. These changes are expected to facilitate at parturition the stimulation by noradrenaline of glutamate release and the inhibition of GABA release, both by removing a physical barrier to diffusion, but also by distancing the active clearance of neurotransmitters by glial processes (729,730). Blocking uptake of glutamate by glia has the effect of decreasing glutamate and GABA release by synapses, as the increased glutamate level acts on presynaptic terminals of both types through inhibitory metabotropic glutamate receptors (730). Glial retraction at the end of pregnancy may similarly enhance local inhibitory feedback presynaptic actions of glutamate, with the consequence of increasing the signal-to-noise ratio, allowing only the most salient and strong stimuli to drive oxytocin neurons, that is, the release of noradrenaline signaling uterine contractions and vaginocervical distension.

Somatodendritic Oxytocin Release

The release of oxytocin from the dendrites and cell bodies of oxytocin neurons was first identified in the context of the milk ejection reflex, from the finding that burst firing of oxytocin neurons during suckling in lactation depends on a central stimulatory action of oxytocin (416). The most important source of this oxytocin was deduced to be the magnocellular nuclei, from measurements of release of oxytocin in vivo and in vitro (731).

There are very few oxytocin-containing axon terminals in the nuclei (17), but there is a dense ventral layer of oxytocin neuron dendrites containing oxytocin in secretory vesicles, from which regulated exocytotic oxytocin secretion can be stimulated (732). The dendrites, and the cell bodies, are now accepted as an important secretory compartment of the neurons, releasing oxytocin to have local, and maybe more distant, autocrine and paracrine actions in the brain (733,734). Perhaps 90% of the oxytocin in the SON is in the dendrites, in secretory granules, and the content of oxytocin in the nucleus increases by ~75% in pregnancy (385,735). This oxytocin is produced and packaged in the cell body, as the dendrites lack Golgi membrane stacks, and the oxytocin-containing secretory granules are transported into the dendrites. The dendrites do contain mRNA species for membrane receptors and for proteins involved in exocytosis (732).

Extracellular concentrations of oxytocin in the magnocellular nuclei, measured by microdialysis and radioimmunoassay, can be 100- to 1,000-fold greater than circulating concentrations (351). During parturition, the extracellular concentration of oxytocin in the SON and PVN increases 2.5-fold (353). The importance of dendritic oxytocin release in regulating parturition is indicated by the disruption of the progress of parturition that follows application by microdialysis of an oxytocin antagonist into a magnocellular nucleus (438). Moreover, intracerebroventricular injection of oxytocin can facilitate burst firing of oxytocin neurons during suckling in late pregnancy (736).

Explanations are needed for the stimuli that cause somatodendritic oxytocin release and its actions. The time course of somatodendritic release of oxytocin is protracted, permitting sustained local actions.

Regulation of Somatodendritic Oxytocin Release and Postsynaptic Autocrine Actions

Whereas the release of oxytocin from axon terminals in the posterior pituitary by exocytosis is consequent on the arrival of action potentials that activate voltage-gated Ca^{2+} channels (737), release of oxytocin from dendrites is also Ca^{2+} dependent but is partly independent of invasion by action potentials (738). An α_1-adrenergic agonist stimulates oxytocin release within the PVN, though this is more effective in mid-gestation than at other times (680). Recording membrane capacitance changes in isolated SON neuron somata as an indicator of exocytotic events has shown that depolarization increases the rate of vesicular release, that this depends on Ca^{2+} entry, and that a train of simulated action potentials of more than 13 spikes/sec increases exocytosis (739).

Once it is released, oxytocin, acting via oxytocin receptors that oxytocin neurons express (140,220, 496), causes release of further oxytocin from the dendrites (731). Close interaction between adjacent neurons is facilitated by closer apposition of adjacent dendrites, as glia reorganize under the influence of local oxytocin, at the end of pregnancy (669,740,741). Oxytocin liberates Ca^{2+} from intracellular IP_3-sensitive stores in isolated oxytocin neurons (742). This may be more related to the regulation of further oxytocin release, through "priming" of the exocytotic mechanism, than to immediate effects on electrical activity. The positive feedback effect of oxytocin on the dendrites primes release mechanisms, and similar priming by exposure to thapsigargin, which mobilizes Ca^{2+} from intracellular stores, indicates that priming by oxytocin involves mobilization of Ca^{2+} (743). Electron microscopy shows that priming involves secretory granules in dendrites moving closer to the plasma membrane (744). After priming, invasion of the dendrites by action potentials can now cause further oxytocin release, and this has local excitatory actions that lead to burst firing (743).

There are but modest changes in the electrophysiological properties of the neurons in pregnancy (435): More oxytocin neurons show a depolarizing after spike potential (DAP) and briefer after hyperpolarizing potentials (AHP), which together potentially increase excitability (745). However, burst firing is not a direct consequence of increased excitability of oxytocin neurons in pregnancy (see section Reduced Responses of Oxytocin Neurons in Pregnancy). Oxytocin also acts on oxytocin neurons to oppose the effectiveness of inhibitory synaptic input via $GABA_A$ receptors by phosphorylation of the receptors through PKC action (see section on Inhibitory Restraint: Endogenous Opioid, GABA, Histamine, Nitric Oxide) (709).

Presynaptic Actions of Somatodendritically Released Oxytocin

Oxytocin inhibits, via oxytocin receptors, the $[K^+]$-stimulated release of excitatory amino acids (glutamate, aspartate) in the SON (746). Electrophysiological studies, using nystatin perforated patch whole cell recording in hypothalamic slices, have shown that oxytocin, exogenous or endogenous, acts presynaptically to inhibit excitatory synapses, reducing the amplitude of excitatory stimulus-evoked postsynaptic currents in oxytocin neurons by reducing glutamate release (747). This action is via N and P/Q type Ca^{2+} channels, but oxytocin does not act directly on the terminals; rather, oxytocin stimulates release of cannabinoid by oxytocin neurons that then acts on presynaptic receptors to inhibit glutamatergic input (748). Oxytocin also acts presynaptically to suppress inhibitory GABA input to oxytocin neurons, and this presynaptic action involves adenosine (749). The presynaptic actions of oxytocin, released when the oxytocin neuron dendrites are stimulated, ensure that the neurons "regulate afferent input unto themselves using their own firing rate as a gauge" (747).

From these studies it is proposed that the actions of oxytocin in suppressing presynaptic excitatory and inhibitory input allow synchronized bursts to emerge. Altogether, the local pre- and postsynaptic actions of oxytocin weakly couple the oxytocin cells, enabling a major net increase in excitatory synaptic input to drive synchronized bursts (750).

In contrast, there are local excitatory actions of oxytocin in the magnocellular nuclei; in sheep oxytocin release in the PVN is increased during parturition (647), which may increase local noradrenaline release, as in lactating rats (751). In rats, microdialysis of oxytocin into the SON augments noradrenaline release stimulated by high $[K^+]$, via a direct action through oxytocin receptors on noradrenergic terminals (690). Increased release of noradrenaline in the SON with noxious stimulation and oxytocin secretion into the circulation are decreased by central oxytocin antagonist administration (548,690). The expression of oxytocin receptor mRNA in NTS neurons is increased in late pregnancy (496), which is expected to lead to increased availability of receptors after transport to the nerve terminals on the oxytocin neurons. Noradrenaline can reduce, via α_2 receptors, GABA release (752) and increase glutamate release (753).

These findings suggest that the noradrenergic input to oxytocin neurons that is activated by afferents from the uterus and birth canal during parturition may have an important role in the initial stimulation of somatodendritic oxytocin release. This then triggers the local actions of oxytocin that prepare for burst firing and secretion of a pulse of oxytocin, ensuring fetal expulsion.

Preventing Premature Oxytocin Secretion in Pregnancy

Timing of Uterine Oxytocin Receptor Up-Regulation

The positive feedback stimulation of oxytocin secretion requires that the myometrium is responsive to oxytocin. The expression of oxytocin receptors in the myometrium increases late in pregnancy, strikingly in the rat only in the last day, after progesterone withdrawal (156,490). Accordingly, the myometrium is most sensitive to oxytocin late in pregnancy (156). Indeed, before this, in the rat oxytocin infusions do not induce parturition or activate brainstem or magnocellular oxytocin neurons (426). In women, oxytocin receptor–binding concentration increases 12-fold from 13–17 weeks to 37–41 weeks of pregnancy, with a further increase with the onset of labor (154,754). An increase in oxytocin receptor binding in human myometrium across the last 7 weeks of pregnancy, including labor, has not been supported in recent studies, although oxytocin receptor mRNA expression does increase over this period (755).

Progesterone and Metabolites

The power of progesterone is shown by preventing progesterone withdrawal in the rat by injection of progesterone; parturition is then delayed and oxytocin infusions at the normal time that parturition starts are ineffective, both in inducing births and in activating the brainstem and oxytocin neurons (756). Progesterone has multiple inhibitory actions. Study of the time course of emergence of response to exogenous oxytocin in the Ferguson reflex pathway in rats treated with a single injection of progesterone in late pregnancy indicated that progesterone withdrawal is progressive, first at the myometrium, then at the brainstem (NTS) relay, and finally at the oxytocin neurons (756).

Uterus. Oxytocin receptor expression in the myometrium increases only after progesterone withdrawal in the rat (156,490), so progesterone prevents premature contractile responses to oxytocin. Progesterone antagonizes oxytocin binding to oxytocin receptors in membranes from rat parturient myometrium or expressed in CHO cells and consequently attenuates postreceptor signaling. Progesterone does not have this effect on oxytocin binding to the human oxytocin receptor, but binding is similarly inhibited by 5β-dihydroprogesterone, a progesterone metabolite (757). This has not been replicated for the bovine oxytocin receptor (144), although progesterone suppresses oxytocin binding to sheep endometrial membranes (758).

Brain. There are three ways in which progesterone can act on neurons. It can act via cytoplasmic/nuclear progesterone receptors to regulate gene expression, via progesterone receptors in the plasma membrane, or after metabolism to the neuroactive steroid allopregnanolone. Unlike effects on the uterus, direct actions of progesterone on oxytocin binding to oxytocin receptors in the brain have not been found. Expression of progesterone receptor in the hypothalamus, brainstem, and elsewhere in the brain is dependent on stimulation by estrogen; progesterone receptor is not detectable after ovariectomy, reappears after estrogen replacement, and varies with estrous cycle stage (759–761).

Magnocellular Oxytocin Neurons. In virgin rats, progesterone receptor protein or mRNA expression is not detectable in magnocellular neurons (759,762,763), though progesterone receptor mRNA has been detected in the PVN (764). Despite the increased secretion of estrogen near the end of pregnancy, expression of progesterone receptor is not induced in magnocellular oxytocin neurons, nor in any groups of paraventricular parvocellular neurons; this contrasts with substantially increased numbers of progesterone receptor–expressing neurons in the medial preoptic nucleus at the end of pregnancy in the same study (759). Lack of progesterone receptors in oxytocin neurons clearly indicates that progesterone cannot act via the classical genomic mechanism. Activation of Fos expression in SON neurons in late pregnant rats treated with progesterone and then given the progesterone/glucocorticoid antagonist RU486 (756) indicates that progesterone inhibits activation of magnocellular oxytocin neurons through progesterone receptors either in afferent neurons or the uterus. Whether oxytocin neurons express membrane progesterone receptors has not yet been reported.

Allopregnanolone (3α,5α-Tetrahydroprogesterone). In contrast with the lack of evidence for inhibitory actions of progesterone on oxytocin neurons through classical progesterone receptor mechanisms, there is abundant evidence for direct inhibitory actions of allopregnanolone, the progesterone neuroactive metabolite. Allopregnanolone acts as detailed above (see section Inhibitory Restraint: Endogenous Opioid, GABA, Histamine, Nitric Oxide). It is a positive allosteric modifier of $GABA_A$ receptors on oxytocin neurons, prolonging opening time of the Cl^- channels and interfering with autostimulatory actions of oxytocin (709). Allopregnanolone is produced by sequential actions of 5α-reductase (converting progesterone to 5α-dihydroprogesterone) and 3β-hydroxysteroid dehydrogenase (converting 5α-dihydroprogesterone to allopregnanolone), which are expressed in the brain, predominantly in glia, including in the hypothalamus; allopregnanolone can also be produced from progesterone made in the brain (670). The content of progesterone in the brain (measured in cerebral cortex) in pregnancy is increased in parallel with circulating concentrations and decreases sharply after luteolysis; brain allopregnanolone concentration peaks nearer the end of pregnancy but also declines sharply with luteolysis (710). Consequently, the effectiveness of GABA inhibition of oxytocin neurons decreases near the end of pregnancy. Decreased effectiveness of allopregnanolone in potentiating GABA actions on oxytocin neurons at the end of pregnancy is a consequence of postreceptor changes affecting $GABA_A$ receptor phosphorylation rather than, as initially proposed, due to changes in $GABA_A$ receptor α-subunit expression (709,765). However, the decreased production of allopregnanolone after progesterone withdrawal would seem to be normally the primary means by which GABA inhibition of oxytocin neurons is reduced as term approaches.

Allopregnanolone is important in pregnancy in preventing inadvertent premature stimulation of the release of the posterior pituitary store of oxytocin (Fig. 12). This is indicated by finding that the failure of systemic interleukin-1β to stimulate oxytocin in pregnant rats (see section on Reduced Responses of Oxytocin Neurons in Pregnancy) is overcome by prior treatment with finasteride, a 5α-reductase inhibitor (765a).

Pregnancy, Progesterone Withdrawal, Parturition, and Oxytocin Gene Expression. There has been confusion about whether oxytocin gene expression, measured as oxytocin mRNA expression, in magnocellular neurons is increased during pregnancy. Different techniques have been used to address this question. Review of the several reports (435) indicates that studies confined to measurements on supraoptic magnocellular neurons have shown either that oxytocin mRNA or hnRNA expression is not altered or is increased by just 27%, compared with virgin rats or very early pregnancy. Measurements in the other reports generally included the PVN, which contains parvocellular as well as magnocellular oxytocin neurons. If the increases in oxytocin gene expression are as great as some of these studies indicate (392), then efficiency of translation would be greatly reduced, as the posterior pituitary content of oxytocin increases by less than 50%, and secretion is not increased during pregnancy. A mere 10% greater oxytocin synthesis rate, or a similar reduction in secretion, would be sufficient to accumulate the measured amount of oxytocin. The store of oxytocin in the dendrites increases by a similar percentage in pregnancy, but this represents a much smaller total store than in the posterior pituitary.

In contrast, when oxytocin secretion is strongly stimulated during lactation and the posterior pituitary is depleted (766), there is agreement in all studies that

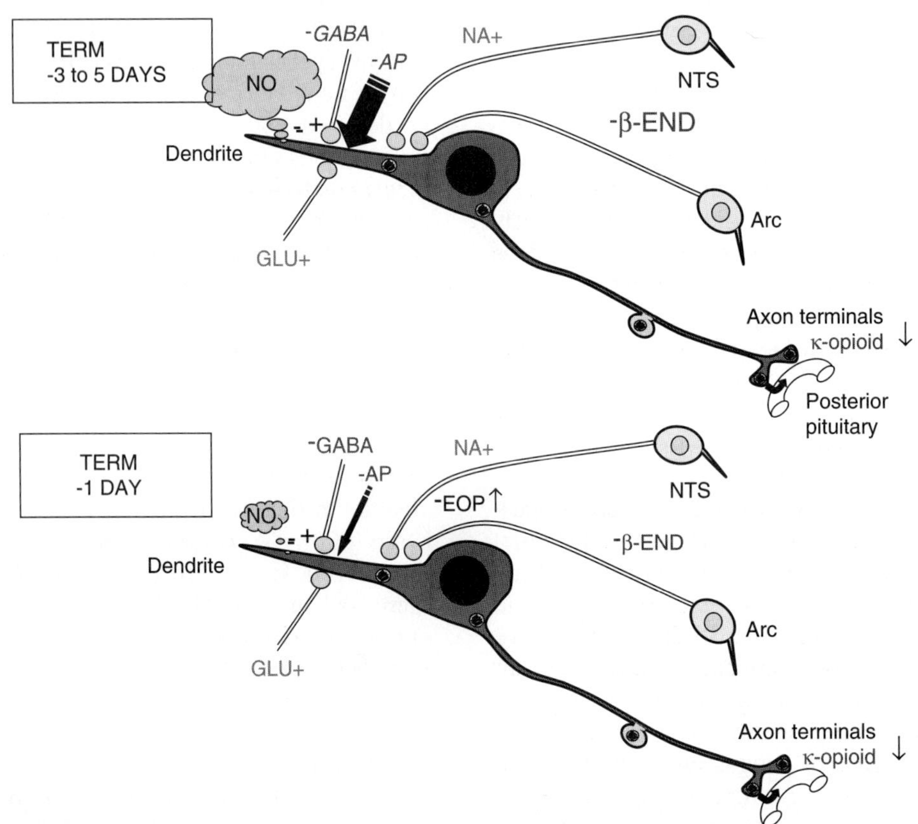

FIG. 12. Mechanisms restraining oxytocin neurons in pregnancy. A few days before expected parturition, inhibition of oxytocin neurons by GABA is enhanced by pre- and postsynaptic actions of nitric oxide (NO) (544) and by the high levels of allopregnanolone (AP) (710), which potentiates GABA actions via $GABA_A$ receptors (708). κ-Opioid mechanisms on oxytocin terminals in the posterior pituitary are now down-regulated (795) (*upper panel*). Nearer term, NO mechanisms in the supraoptic nucleus are down-regulated (544), progesterone withdrawal is followed by decreased brain allopregnanolone content (710), and so GABA inhibition is less effective. Central endogenous μ-opioid peptide (EOP) inhibition is now predominant (783). The endogenous opioid may be β-endorphin, because its fiber density in the supraoptic nucleus is now increased (782), or met-enkephalin coproduced in nucleus tractus solitarius (NTS) neurons in which gene expression is now up-regulated (791). +, excitatory; –, inhibitory; Arc, arcuate nucleus.

after several days oxytocin mRNA content in the magnocellular neurons is substantially increased, as it is after sustained hyperosmotic stimulation (437). Measurement of increased expression of oxytocin hnRNA in supraoptic neurons after parturition indicates that the oxytocin gene is rapidly stimulated as a result of the strong activation of the neurons in parturition (437). The link between stimulation of the neurons by their excitatory inputs, via induced expression of transcription factor genes, and increased oxytocin gene expression is not understood (663). Because parturition follows the withdrawal of progesterone, the question of possible linkage between the consequent withdrawal of allopregnanolone and increased oxytocin gene expression has been investigated (764,767). These studies indicate that progesterone may act via allopregnanolone and $GABA_A$ receptors to reduce oxytocin mRNA abundance in PVN neurons. Hence, withdrawal of progesterone/allopregnanolone at the end of pregnancy might initiate increased oxytocin gene expression.

Estrogens

Oxytocin Gene Expression. Given the increased production of estrogens in pregnancy, in particular in late pregnancy in the rat, estrogens are natural candidate regulators of oxytocin gene expression and oxytocin neuron activity. Unfortunately, it is not possible to make a consistent simple statement about estrogen actions on oxytocin neurons. Despite the extensive analysis of regulatory elements in the oxytocin gene (see section on The Oxytocin Gene Promoter), mixed effects of estrogen have been reported on oxytocin gene expression in magnocellular neurons in vivo. 17β-Estradiol treatment of ovariectomized rats rapidly (6 hours) *decreases* oxytocin mRNA expression in the PVN, followed by return to basal over 48 hours (768), and treatment over 7 days had no effect (382). Negative findings might be due to thyroid status, because triiodothyronine has negative effects on oxytocin gene expression, competing with estrogen (303). However, in the stimulated state due to hyperosmolarity, oxytocin mRNA expression in magnocellular neurons requires estrogen in female rats, indicating a stimulatory role for estrogen on oxytocin gene expression (769). In mice, estrogen treatment does not alter oxytocin mRNA expression in supraoptic neurons, but it increases expression in the PVN, but not in mice with targeted deletion of the ER-β gene (770).

Oxytocin Secretion. In vitro estrogen inhibits via a genomic action the secretion of oxytocin by hypothalamic explants stimulated by NMDA but acts on the cell membrane to inhibit hyperosmotically stimulated oxytocin release (771–773). However, estrogen has an opposite stimulatory action via membrane receptors on oxytocin release by magnocellular neuron dendrites (374), and stress-induced oxytocin secretion in female rats is potentiated by estrogen treatment (774).

ER Expression in Oxytocin Neurons. Oxytocin neurons in rats do not express ER-α but do express ER-β (298). ER-β is expressed in human oxytocin neurons (297), but only ER-α is expressed in the guinea pig SON and PVN (775), and in mice few supraoptic neurons express ER-β, but many PVN neurons do (770). Because ER-β has an affinity for phytoestrogens (776), variations in dietary intake may confound in vivo studies of actions of endogenous estrogens on oxytocin neurons.

The expression of ER-β mRNA and protein in magnocellular neurons shows remarkable variation, inversely related to strong stimulation of oxytocin secretion (777), but little affected by pregnancy or sex steroid manipulation (773,778). ER-β mRNA expression in the PVN is most prominent in the centrally projecting parvocellular neurons, which include oxytocin neurons (298,777).

The variable results from studies of estrogen actions on oxytocin gene expression could be interpreted to indicate that such actions may be indirect and to involve cofactors or intermediate transcription factors that interact with ER-β (779). It is also possible that ER-β functions as a transcription factor activated by phosphorylation rather than by estrogen (780).

Progesterone and Neurons Projecting to Oxytocin Neurons. In the brainstem, the number of NTS neurons immunoreactive for progesterone receptor is not increased at term and is decreased at the end of parturition. As very few tyrosine hydroxylase-expressing neurons also express progesterone receptor at the end of pregnancy, it is unlikely that progesterone can inhibit these neurons through classical genomic mechanisms (759).

Progesterone receptor expression is prominent in several hypothalamic groups of neurons that project to the SON, including the arcuate, ventromedial, periventricular, and medial preoptic nuclei (759). However, these neurons are not activated in parturition (677). Whether progesterone regulates β-endorphin–producing POMC arcuate neurons that may provide endogenous opioid restraint of oxytocin neurons in pregnancy is not clear (781,782).

Estrogen and Neurons Projecting to Oxytocin Neurons. In vivo, like progesterone, actions of estrogens on oxytocin neurons may be indirect via their inputs. Many of the inputs that project to oxytocin neurons express ER-α, though only some neurons in these regions do so. Especially important in the context of parturition, none of the NTS noradrenergic neurons projecting to the SON expresses ER-α; the input neurons that express ER-α are predominantly in rostral regions concerned with osmoregulation (367). Of interest in relation to progesterone/allopregnanolone actions on oxytocin neurons via $GABA_A$ receptors is that GABA neurons adjacent to the SON, in the perinuclear zone, express ER-α, which suggests genomic estrogen actions regulate these neurons (706).

Reduced Responses of Oxytocin Neurons in Pregnancy

Oxytocin neurons are less sensitive to a range of stimuli in pregnancy. This can be proposed to ensure that the storage of oxytocin in the posterior pituitary is increased. Thus, the direct stimulation by CCK or oxytocin of supraoptic oxytocin neurons in vitro is substantially reduced in pregnancy (684). Likewise, the excitation of oxytocin neurons by intracerebroventricular neuropeptide Y is essentially abolished in pregnancy (691a). There is also in late pregnancy reduced stimulation of oxytocin secretion by electrical stimulation of the lamina terminalis input, which is essential for hyperosmotic stimulation of oxytocin neurons (545). Moreover, modest hyperosmotic stimulation is also less effective in stimulating oxytocin neurons in pregnancy (542), and this likely reflects reduced extracellular fluid basal osmolarity in pregnancy as a consequence of stimulation of drinking and vasopressin secretion and consequent water retention induced by relaxin, produced by the corpora lutea of pregnancy (see section Relaxin) (527,531). Excitatory responses of oxytocin neurons to systemic CCK, which activates brainstem noradrenergic neurons, are also reduced in pregnancy (783).

Cytokines. Interleukin-1β, a proinflammatory cytokine released by activated macrophages, given systemically activates magnocellular oxytocin neurons, increasing oxytocin secretion in female rats (79,558). The major site of transduction of systemic interleukin-1β to regulate neuroendocrine responses is in the medulla, involving action on receptors in vascular endothelium, local prostaglandin synthesis, and excitation of noradrenergic neurons (784).

The strong excitation of oxytocin neurons by systemic interleukin-1β administration in virgin rats is essentially abolished in late pregnant rats (558). Because interleukin-1β acts by stimulation of brainstem (NTS) noradrenergic neurons, in virgin rats it increases noradrenaline release in the PVN and excites the firing rate of magnocellular oxytocin neurons but has no such effects in late pregnant rats (558). Hence, reduced activation of oxytocin neurons by systemic interleukin-1β in late pregnancy seems to be the result of failure to stimulate the release of noradrenaline in the magnocellular nuclei. There are two mechanisms involved in the suppression of interleukin-1β signaling to oxytocin neurons in pregnancy, which may be interrelated. First is the action of allopregnanolone, which is revealed by showing that pretreatment of rats in late pregnancy with finasteride inhibits 5α-reductase, and hence conversion of progesterone to allopregnanolone restores an oxytocin secretory response to systemic interleukin-1β (765a). Conversely, treatment of virgin rats with allopregnanolone suppresses stimulation of oxytocin secretion by interleukin-1β. It is likely that allopregnanolone acts by enhancing GABA actions via $GABA_A$ receptors on oxytocin neurons (see section on Inhibitory Restraint: Endogenous Opioid, GABA, Histamine, Nitric Oxide and section on Progesterone and Metabolites). The second mechanism involves the inhibitory action of endogenous opioids, revealed by restoration of excitation of oxytocin neurons to interleukin-1β by administration of the opioid antagonist naloxone, as discussed below (558).

Endogenous Opioids

In late pregnancy in rats and pigs (but not mice), a tonic inhibitory tone on oxytocin neurons emerges, which is revealed by the increase in oxytocin neuron activity that follows administration of the opioid receptor antagonist naloxone (386,429,435,572,783). As described above, this opioid restraint continues through parturition and regulates the pace of births. In pregnancy this opioid tone importantly restricts activation of oxytocin secretion by numerous stimuli and can be considered as a mechanism that prevents inadvertent premature discharge of the accumulating store of oxytocin in the posterior pituitary gland. Such premature secretion of oxytocin would not only deplete the store but may desensitise the myometrium (426) or induce preterm labor.

The findings that naloxone evokes Fos expression in magnocellular neurons, increases somatodendritic oxytocin release, and increases the firing rate responses of oxytocin neurons to systemic CCK or interleukin-1β in pregnant, but not virgin, rats reveal that the opioid mechanisms activated in pregnancy act centrally to regulate oxytocin neurons (Fig. 12) (435,783). This central opioid inhibitory restraint is only evident in late pregnant rats and is mediated by μ- and not κ-opioid receptors (783).

Magnocellular oxytocin neurons coexpress enkephalins and dynorphins, respectively, products of the pro-enkephalin A and pro-dynorphin genes, and are secreted with oxytocin, though in much smaller amounts (785). Secreted by oxytocin axon terminals in the posterior pituitary gland, coreleased opioids act to restrain oxytocin secretion (see below). Magnocellular oxytocin neuron cell bodies express μ- and κ-opioid receptors, and agonists selective for these receptors (e.g., μ, morphine and endomorphin; κ, U50488) inhibit both electrical and secretory activity of oxytocin neurons (786,787). Electrically stimulated somatodendritic release of oxytocin without pregnancy is enhanced by naloxone, which might be due to opioid peptides coreleased with oxytocin (788).

Endogenous opioids are also coproduced by several of the inputs to oxytocin neurons, including brainstem neurons that produce β-endorphin and enkephalins (789), which might after release in the magnocellular nuclei act on the oxytocin neurons or on autoreceptors on noradrenergic terminals. Intracerebroventricular administration of a μ-opioid receptor antagonist in late pregnant rats increases plasma oxytocin concentration and the noradrenaline content of the SON and PVN, indicating opioid restraint of the noradrenaline input to oxytocin neurons (790). Moreover, naloxone increases the electrical and secretory excitatory responses of oxytocin neurons to intravenous CCK in late pregnancy (783) and permits excitatory responses to intravenous interleukin-1β (558). Hence, for these stimuli that act via the brainstem noradrenergic input, it is evident that responses of oxytocin neurons are reduced by the emergence of an inhibitory mechanism on this input in pregnancy. This may be a consequence of the up-regulated pro-enkephalin A and μ-opioid receptor gene expression in NTS neurons in late pregnancy (791), which would provide, after axonal transport, μ-opioid receptor and enkephalin for preterminal inhibition of noradrenaline release in the magnocellular nuclei.

Another source of endogenous opioid that may contribute to the regulation of oxytocin neurons in pregnancy is the β-endorphin (POMC) neurons in the arcuate nucleus. There are β-endorphin–containing axons in the vicinity of oxytocin neurons in the SON, and this innervation is more prominent at the end of pregnancy (782). Actions of opioids on many of the sources of inputs to oxytocin neurons are also possible as many of these neurons express opioid receptors on their cell bodies; these include brainstem, lamina terminalis, and olfactory bulb neurons (792).

Opioids and the Posterior Pituitary Gland. The first descriptions of endogenous opioid actions on oxytocin secretion identified inhibitory actions at the level of the axon terminals in the posterior pituitary. These studies showed that the release of oxytocin from electrically stimulated isolated posterior pituitary glands in vitro was enhanced by treatment with the opioid antagonist naloxone (793). Further studies with agonists and antagonists selective for the different types of opioid receptor (δ, κ, μ) led to the conclusion that the opioid restraint of oxytocin secretion in the posterior pituitary gland is mediated through κ receptors, on the neurosecretory oxytocin axon terminals (794,795). The source of the opioid might be adjacent vasopressin terminals, which contain dynorphin (394), but dynorphin and extended enkephalins (Met-enkephalin-Arg^{6}Phe^7, and Met-enkephalin-Arg^{6}Gly^{7}Leu^8), which are also κ selective agonists, are produced by oxytocin neurons (796). The close correlation between depletion of the posterior pituitary gland content of these peptides and oxytocin leads to the conclusion that these opioids produced by oxytocin neurons autoinhibit oxytocin secretion from the posterior pituitary (794). Essentially, this mechanism automatically limits oxytocin secretion, despite increased rate of arrival of action potentials.

Some studies indicate that μ-receptor mediated actions on oxytocin terminals are also involved. Oxytocin release stimulated from isolated posterior pituitary nerve terminals by increasing extracellular [K^+] is inhibited by a μ-selective agonist (DAMGO); this action involves inhibition of Ca^{2+} entry. The effects of DAMGO are reversed by a μ-selective antagonist (CTOP) (797). However, the source of endogenous μ-opioid is still the terminals of oxytocin neurons, in particular, the enkephalins that they produce.

The posterior pituitary opioid mechanism is weakened as pregnancy progresses (386,572,795), and accordingly the central inhibitory opioid mechanism becomes predominant. Down-regulation of the posterior pituitary mechanism involves reduced dynorphin content, reduced κ-opioid receptor binding, and reduced effectiveness of the κ-agonist, U50,488, and antagonist (nor-binaltorphimine) (572,795). Because opioid inhibition at the posterior pituitary does not discriminate between different stimuli acting on oxytocin neurons, this shift in site of opioid action allows for opioid restraint to be applied selectively to specific inputs to the cell bodies.

Neurosecretory Terminal Modulation by Firing Pattern. The firing pattern of oxytocin neurons has an important influence on the effectiveness of each action potential arriving at the axon terminals in the posterior pituitary (750). With high frequency firing, as during a burst in parturition, each action potential releases more oxytocin

(798,799); this is frequency facilitation. It may be due to prolongation of action potentials with sustained firing (800), to prolonged terminal depolarization as a result of extracellular K^+ accumulation (703,798), or to production or release of a stimulatory modulator that has a local positive feedback action. With the down-regulation of the autoinhibitory opioid mechanism in the posterior pituitary as pregnancy progresses, action potentials arriving in bursts, as in parturition, can act through frequency facilitation with greater fidelity to cause high amplitude pulsatile oxytocin release.

Induction of Changes in Opioid Regulation of Oxytocin Neurons in Pregnancy

Central Up-Regulation. Ovariectomized pregnant rats given estrogen and progesterone replacement show a smaller increase in oxytocin secretion after naloxone than intact pregnant rats, indicating that the central opioid tone depends on relaxin (532). However, sex steroids may also be important, as in virgin rats given estrogen and progesterone implants to model exposure to pregnancy levels of these steroids, naloxone increases the oxytocin secretory response to forced swimming, pointing to induction of inhibitory opioid tone on oxytocin neurons (774). The source of this opioid is not established: It might be coproduced opioid in afferent neurons driving the excitation of oxytocin neurons or the β-endorphin input from the arcuate nucleus. These β-endorphin neurons are more active in pregnancy (782).

Posterior Pituitary Down-Regulation. Reduced effectiveness of naloxone, κ-agonist, and antagonist and reduced content of dynorphin and κ-receptor may be attributable to initial stimulation of oxytocin neurons by relaxin in pregnancy and consequent endogenous opioid-induced receptor down-regulation (724,801).

CONCLUDING REMARKS

We have celebrated the 100th birthday of oxytocin in reproductive physiology. From its first described actions as a uterotonic and galactogogic principle early in the 20th century, oxytocin has been established as a peptide hormone with diverse and versatile functions. Notably, most of these functions are coherent in reproductive physiology and allow the hormone to control reproductive functions at multiple levels. This leads to a highly integrated role of oxytocin and allows many factors, for example steroids, neural activity and growth factors, to interact with the actions of oxytocin at specific levels.

The integrative reproductive functions are based on multiple modes of action. First, oxytocin acts classically as a hormone on peripheral target tissues, like the uterus and mammary gland. The oxytocinergic magnocellular neurons of the hypothalamus express the oxytocin gene at a high rate and store the mature peptide hormone in the posterior pituitary gland. One level of control is the synthesis and secretion of oxytocin. Whereas the transcription of the oxytocin gene is affected by external factors and physiological conditions, the regulatory range of gene expression and biosynthesis is limited to a few-fold, due to the high basal oxytocin mRNA content. Only after chronic repression of oxytocin gene transcription, as during long-lasting hyponatremia, can gene expression be up-regulated dramatically. In contrast, the release of oxytocin from the posterior pituitary gland is highly dynamic with large amplitude. The synchronized bursting of oxytocin neurons results in pulses of massively released oxytocin. These are required for milk ejection on demand and the promotion of parturition. The activities of oxytocin as a releasing hormone for prolactin and LH in the anterior pituitary gland are likely linked to the same hypothalamic neurons and augment milk production indirectly through stimulated synthesis and secretion of prolactin.

Second, other hypothalamic neurons innervate brain systems. The central oxytocinergic systems are multifold, as are the central effects of oxytocin administered to the brain. Central oxytocin exerts an important permissive or modulatory role in the control of behaviors that are required for appropriate reproduction. These include social interactions between individual animals, pair bonding between male and female, lordosis behavior and sexual activity, and the induction and maintenance of nursing and other maternal behaviors. Both female and male behaviors are influenced by oxytocin. These functions seem to be controlled mainly at two levels: the secretion of oxytocin locally by neuronal systems involved in behavioral control and the expression of the oxytocin receptor on neurons receiving oxytocinergic afferents. The studies of Kendrick, Neumann, and their respective colleagues in sheep and rodents have demonstrated that the local secretion of oxytocin in the brain is responsive to various physiological conditions of the animal (240,443,566,571,575,647,802,803). Expression of the oxytocin receptor in the brain is highly regulated and displays marked differences between species displaying different patterns of reproductive behaviors. Thus, oxytocin is an important neuropeptide in regulatory neural networks underlying these behaviors. Now, 100 years after its discovery, oxytocin may serve as an example of how a peptide factor, acting as a hormone and a neuropeptide, encoded by a single gene and acting on a single-gene receptor can establish a highly integrated regulatory power on one of the main physiological functions in life to reproduce and maintain offspring.

ACKNOWLEDGMENTS

This chapter was made possible by the expert secretarial assistance of Mrs. Ria van Vlaardingen, who is gratefully acknowledged.

REFERENCES

1. Dale, H. H. (1906). On some physiological actions of ergot. *J. Physiol. (Lond).* 34, 165–206.
2. Ott, I., and Scott, J. C. (1911). The action of infundibulin upon mammary secretion. *Proc. Soc. Exp. Biol. Med.* 8, 48–49.
3. Oliver, G., and Schäfer, E. A. (1895). On the physiological actions of extracts of pituitary body and other glandular organs. *J. Physiol. (Lond).* 18, 277–299.
4. Du Vigneaud, V., Ressler, C., and Swan, J. M., et al. (1953). The synthesis of an octapeptide amide with the hormonal activity of oxytocin. *J. Am. Chem. Soc.* 75, 4879–4880.
5. Scharrer, E., and Scharrer, B. (1940). Secretory cells within the hypothalamus. The Hypothalamus and Central Levels of Autonomic Function, pp. 170–194. Williams & Wilkins, Baltimore.
6. Rhodes, C. H., Morrell, J. I., and Pfaff, D. W. (1981). Immunohistochemical analysis of magnocellular elements in rat hypothalamus: distribution and numbers of cells containing neurophysin, oxytocin, and vasopressin. *J. Comp. Neurol.* 198, 45–64.
7. Sofroniew, M. V., and Glasmann, W. (1981). Golgi-like immunoperoxidase staining of hypothalamic magnocellular neurons that contain vasopressin, oxytocin or neurophysin in the rat. *Neuroscience* 6, 619–643.
8. Kelly, J., and Swanson, L. W. (1980). Additional forebrain regions projecting to the posterior pituitary: preoptic region, bed nucleus of the stria terminalis, and zona incerta. *Brain Res.* 197, 1–9.
9. Sherlock, D. A., Field, P. M., and Raisman, G. (1975). Retrograde transport of horseradish peroxidase in the magnocellular neurosecretory system of the rat. *Brain Res.* 88, 403–414.
10. Alonso, G., and Assenmacher, I. (1981). Radioautographic studies on the neurohypophysial projections of the supraoptic and paraventricular nuclei in the rat. *Cell Tissue Res.* 219, 525–534.
11. Hatton, G. I., Cobbett, P., and Salm, A. K. (1985). Extranuclear axon collaterals of paraventricular neurons in the rat hypothalamus: intracellular staining, immunocytochemistry and electrophysiology. *Brain Res. Bull.* 14, 123–132.
12. Leng, G. (1982). Lateral hypothalamic neurones: osmosensitivity and the influence of activating magnocellular neurosecretory neurones. *J. Physiol. (Lond).* 326, 35–48.
13. Mason, W. T., Ho, Y. W., and Hatton, G. I. (1984). Axon collaterals of supraoptic neurones: anatomical and electrophysiological evidence for their existence in the lateral hypothalamus. *Neuroscience* 11, 169–182.
14. Bains, J. S., and Ferguson, A. V. (1997). Nitric oxide depolarizes type II paraventricular nucleus neurons in vitro. *Neuroscience* 79, 149–159.
15. Mason, W. T., Hatton, G. I., Ho, Y. W., Chapman, C., and Robinson, I. C. (1986). Central release of oxytocin, vasopressin and neurophysin by magnocellular neurone depolarization: evidence in slices of guinea pig and rat hypothalamus. *Neuroendocrinology* 42, 311–322.
16. Pittman, Q. J., Blume, H. W., and Renaud, L. P. (1981). Connections of the hypothalamic paraventricular nucleus with the neurohypophysis, median eminence, amygdala, lateral septum and midbrain periaqueductal gray: an electrophysiological study in the rat. *Brain Res.* 215, 15–28.
17. Theodosis, D. T. (1985). Oxytocin-immunoreactive terminals synapse on oxytocin neurones in the supraoptic nucleus. *Nature* 313, 682–684.
18. van Leeuwen, F. W., de Raay, C., Swaab, D. F., and Fisser, B. (1979). The localization of oxytocin, vasopressin, somatostatin and luteinizing hormone releasing hormone in the rat neurohypophysis. *Cell Tissue Res.* 202, 189–201.
19. Luckman, S. M., and Bicknell, R. J. (1990). Morphological plasticity that occurs in the neurohypophysis following activation of the magnocellular neurosecretory system can be mimicked in vitro by beta-adrenergic stimulation. *Neuroscience* 39, 701–709.
20. Wittkowski, W. (1986). Pituicytes. In *Astrocytes* (S. Fedoroff and A. Vernadakis, Eds.), pp. 173–208. Academic Press, Orlando, FL.
21. Hatton, G. I., and Yang, Q. Z. (1996). Synaptically released histamine increases dye coupling among vasopressinergic neurons of the supraoptic nucleus: mediation by H1 receptors and cyclic nucleotides. *J. Neurosci.* 16, 123–129.
22. Theodosis, D. T., and MacVicar, B. (1996). Neurone-glia interactions in the hypothalamus and pituitary. *Trends Neurosci.* 19, 363–367.
23. Morris, J. F., Chapman, D. B., and Sokol, H. W. (1987). Anatomy and function of the classic vasopressin-secreting hypothalamus-neurohypophysial system. In *Vasopressin: Principles and Properties* (D. M. Gash and G. J. Boer, Eds.), pp. 1–89. Plenum Press, London.
24. Tweedle, C. D., Smithson, K. G., and Hatton, G. I. (1989). Neurosecretory endings in the rat neurohypophysis are en passant. *Exp. Neurol.* 106, 20–26.
25. Morris, J. F., and Pow, D. V. (1988). Capturing and quantifying the exocytotic event. *J. Exp. Biol.* 139, 81–103.
26. Burbach, J. P., van Tol, H. H., Bakkus, M. H., Schmale, H., and Ivell, R. (1986). Quantitation of vasopressin mRNA and oxytocin mRNA in hypothalamic nuclei by solution hybridization assays. *J. Neurochem.* 47, 1814–1821.
27. Young, W. S., Mezey, E., and Siegel, R. E. (1986). Vasopressin and oxytocin mRNAs in adrenalectomized and Brattleboro rats: analysis by quantitative in situ hybridization histochemistry. *Brain Res.* 387, 231–241.
28. Kim, J. K., Soubrier, F., and Michel, J. B., et al. (1990). Arginine vasopressin gene regulation in the homozygous Brattleboro rat. *J. Clin. Invest.* 86, 14–16.
29. LeMoullec, J. M., Jouquey, S., Corvol, P., and Pinet, F. (1997). A sensitive reverse transcriptase polymerase chain reaction assay for measuring the effects of dehydration and gestation on rat amounts of vasopressin and ocytocin mRNAs. *Mol. Cell Endocrinol.* 128, 151–159.
30. Yoshimura, R., Kiyama, H., and Kimura, T., et al. (1993). Localization of oxytocin receptor messenger ribonucleic acid in the rat brain. *Endocrinology* 133, 1239–1246.
31. Xi, D., Chin, H., and Gainer, H. (1999). Analysis of synaptotagmin I-IV messenger RNA expression and developmental regulation in the rat hypothalamus and pituitary. *Neuroscience* 88, 425–435.
32. Swaab, D. F., Nijveldt, F., and Pool, C. W. (1975). Distribution of oxytocin and vasopressin in the rat supraoptic and paraventricular nucleus. *J. Endocrinol.* 67, 461–462.

33. Aspeslagh, M. R., Vandesande, F., and Dierickx, K. (1976). Electron microscopic immunocytochemical demonstration of separate neurophysin-vasopressinergic and neurophysin-oxytocinergic nerve fibres in the neural lobe of the rat hypophysis. *Cell Tissue Res.* 171, 31–37.
34. Buijs, R. M., De Vries, G. J., van Leeuwen, F. W., and Swaab, D. F. (1983). Vasopressin and oxytocin: distribution and putative functions in the brain. *Prog. Brain Res.* 60, 115–122.
35. Dierickx, K., and Vandesande, F. (1977). Immunocytochemical localization of the vasopressinergic and the oxytocinergic neurons in the human hypothalamus. *Cell Tissue Res.* 184, 15–27.
36. Vandesande, F., and Dierickx, K. (1979). The activated hypothalamic magnocellular neurosecretory system and the one neuron—one neurohypophysial hormone concept. *Cell Tissue Res.* 200, 29–33.
37. Hou-Yu, A., Lamme, A. T., Zimmerman, E. A., and Silverman, A. J. (1986). Comparative distribution of vasopressin and oxytocin neurons in the rat brain using a double-label procedure. *Neuroendocrinology* 44, 235–246.
38. Mohr, E., Bahnsen, U., Kiessling, C., and Richter, D. (1988). Expression of the vasopressin and oxytocin genes in rats occurs in mutually exclusive sets of hypothalamic neurons. *FEBS Lett.* 242, 144–148.
39. Mohr, E., Hillers, M., Ivell, R., Haulica, I. D., and Richter, D. (1985). Expression of the vasopressin and oxytocin genes in human hypothalami. *FEBS Lett.* 193, 12–16.
40. van Leeuwen, F. W., and Swaab, D. F. (1977). Specific immunoelectronmicroscopic localization of vasopressin and oxytocin in the neurohypophysis of the rat. *Cell Tissue Res.* 177, 493–501.
41. Kiyama, H., and Emson, P. C. (1990). Evidence for the coexpression of oxytocin and vasopressin messenger ribonucleic acids in magnocellular neurosecretory cells: simultaneous demonstration of two neurophysin messenger ribonucleic acids by hybridization histochemistry. *J. Neuroendocrinol.* 2, 257–259.
42. Mezey, E., and Kiss, J. Z. (1991). Coexpression of vasopressin and oxytocin in hypothalamic supraoptic neurons of lactating rats. *Endocrinology* 129, 1814–1820.
43. Baldino, F. J., O'Kane, T. M., Fitzpatrick-McElligott, S., and Wolfson, B. (1988). Coordinate hormonal and synaptic regulation of vasopressin messenger RNA. *Science* 241, 978–981.
44. Gainer, H. (1998). Cell-specific gene expression in oxytocin and vasopressin magnocellular neurons. *Adv. Exp. Med. Biol.* 449, 15–27.
45. Glasgow, E., Kusano, K., Chin, H., and Gainer, H. (1998). Analysis of differential gene expression in magnocellular oxytocin and vasopressin neurons. Society for Neuroscience Annual Meeting.
46. Glasgow, E., Kusano, K., and Chin, H., et al. (1999). Single cell reverse transcription-polymerase chain reaction analysis of rat supraoptic magnocellular neurons: neuropeptide phenotypes and high voltage-gated calcium channel subtypes. *Endocrinology* 140, 5391–5401.
47. Xi, D., Kusano, K., and Gainer, H. (1999). Quantitative analysis of oxytocin and vasopressin messenger ribonucleic acids in single magnocellular neurons isolated from supraoptic nucleus of rat hypothalamus. *Endocrinology* 140, 4677–4682.
48. Griepentrog, T., Bauer, M., Hornstein, C., Sauer, H., and Jirikowski, G. F. (2000). Coexistence of intestinal trefoil factor (hITF) and oxytocin in magnocellular neurons in the human hypothalamus. *Horm. Metab. Res.* 32, 121–124.
49. Burlet, A., Tonon, M. C., Tankosic, P., Coy, D., and Vaudry, H. (1983). Comparative immunocytochemical localization of corticotropin releasing factor (CRF-41) and neurohypophysial peptides in the brain of Brattleboro and Long-Evans rats. *Neuroendocrinology* 37, 64–72.
50. Martin, R., Geis, R., Holl, R., Schafer, M., and Voigt, K. H. (1983). Co-existence of unrelated peptides in oxytocin and vasopressin terminals of rat neurohypophyses: immunoreactive methionine-enkephalin-, leucine-enkephalin- and cholecystokinin-like substances. *Neuroscience* 8, 213–227.
51. Martin, R., and Voigt, K. H. (1981). Enkephalins co-exist with oxytocin and vasopressin in nerve terminals of rat neurohypophysis. *Nature* 289, 502–504.
52. Mezey, E., Reisine, T. D., Skirboll, L., Beinfeld, M., and Kiss, J. Z. (1986). Role of cholecystokinin in corticotropin release: coexistence with vasopressin and corticotropin-releasing factor in cells of the rat hypothalamic paraventricular nucleus. *Proc. Natl. Acad. Sci. U S A* 83, 3510–3512.
53. Piekut, D. T., and Joseph, S. A. (1985). Relationship of CRF-immunostained cells and magnocellular neurons in the paraventricular nucleus of rat hypothalamus. *Peptides* 6, 873–882.
54. Vanderhaeghen, J. J., Goldman, S., and Lotstra, F., et al. (1985). Co-existence of cholecystokinin- or gastrin-like peptides with other peptides in the hypophysis and the hypothalamus. *Ann. N Y Acad. Sci.* 448, 334–344.
55. Whitnall, M. H., Mezey, E., and Gainer, H. (1985). Co-localization of corticotropin-releasing factor and vasopressin in median eminence neurosecretory vesicles. *Nature* 317, 248–250.
56. Tsuruo, Y., Ceccatelli, S., and Villar, M. J., et al. (1988). Coexistence of TRH with other neuroactive substances in the rat central nervous system [published erratum appears in *J. Chem. Neuroanat.* 1989;2(3):177]. *J. Chem. Neuroanat.* 1, 235–253.
57. Majdoubi, M. E., Metz-Boutigue, M. H., Garcia-Sablone, P., Theodosis, D. T., and Aunis, D. (1996). Immunocytochemical localization of chromogranin A in the normal and stimulated hypothalamo-neurohypophysial system of the rat. *J. Neurocytol.* 25, 405–416.
58. El Majdoubi, M., Poulain, D. A., and Theodosis, D. T. (1996). The glutamatergic innervation of oxytocin- and vasopressin secreting neurons in the rat supraoptic nucleus and its contribution to lactation-induced synaptic plasticity. *Eur. J. Neurosci.* 8, 1377–1389.
59. Shen, P. J., and Gundlach, A. L. (1996). Chromogranin mRNA levels in the brain as a marker for acute and chronic changes in neuronal activity: effect of treatments including seizures, osmotic stimulation and axotomy in the rat. *Eur. J. Neurosci.* 8, 988–1000.
60. Vanderhaeghen, J. J., Lotstra, F., Liston, D. R., and Rossier, J. (1983). Proenkephalin, [Met]enkephalin, and oxytocin immunoreactivities are colocalized in bovine hypothalamic magnocellular neurons. *Proc. Natl. Acad. Sci. U S A* 80, 5139–5143.
61. Crowley, W. R., O'Donohue, T. L., George, J. M., and Jacobowitz, D. M. (1978). Changes in pituitary oxytocin and vasopressin during the estrous cycle and after ovarian hormones: evidence for mediation by norepinephrine. *Life Sci.* 23, 2579–2585.
62. Vrang, N., Larsen, P. J., Clausen, J. T., and Kristensen, P. (1999). Neurochemical characterization of hypothalamic cocaine—amphetamine-regulated transcript neurons [in process citation]. *J. Neurosci. (Online).* 19, RC5.
63. Whitnall, M. H., Gainer, H., Cox, B. M., and Molineaux, C. J. (1983). Dynorphin-A-(1-8) is contained within vasopressin neurosecretory vesicles in rat pituitary. *Science* 222, 1137–1139.
64. Meister, B. (1993). Gene expression and chemical diversity in hypothalamic neurosecretory neurons. *Mol. Neurobiol.* 7, 87–110.
65. Bondy, C. A., Jensen, R. T., Brady, L. S., and Gainer, H. (1989). Cholecystokinin evokes secretion of oxytocin and vasopressin from rat neural lobe independent of external calcium. *Proc. Natl. Acad. Sci. U S A* 86, 5198–5201.

66. Bondy, C. A., Gainer, H., and Russell, J. T. (1988). Dynorphin A inhibits and naloxone increases the electrically stimulated release of oxytocin but not vasopressin from the terminals of the neural lobe. *Endocrinology* 122, 1321–1327.
67. Boersma, C. J., Sonnemans, M. A., and van Leeuwen, .F W. (1993). Immunocytochemical localization of neuropeptide FF (FMRF amide-like peptide) in the hypothalamo-neurohypophyseal system of Wistar and Brattleboro rats by light and electron microscopy. *J. Comp. Neurol.* 336, 555–570.
68. Theodosis, D. T., El Majdoubi, M., Pierre, K., and Poulain, D. A. (1998). Factors governing activity-dependent structural plasticity of the hypothalamoneurohypophysial system. *Cell Mol. Neurobiol.* 18, 285–298.
69. Day, T. A., and Sibbald, J. R. (1989). A1 cell group mediates solitary nucleus excitation of supraoptic vasopressin cells. *Am. J. Physiol.* 257, R1020–R1026.
70. Raby, W. N., and Renaud, L. P. (1989). Dorsomedial medulla stimulation activates rat supraoptic oxytocin and vasopressin neurones through different pathways. *J. Physiol. (Lond).* 417, 279–294.
71. Day, H. E., McKnight, A. T., Poat, J. A., and Hughes, J. (1994). Evidence that cholecystokinin induces immediate early gene expression in the brainstem, hypothalamus and amygdala of the rat by a CCKA receptor mechanism. *Neuropharmacology* 33, 719–727.
72. Luckman, S. M., Hamamura, M., Antonijevic, I., Dye, S., and Leng, G. (1993). Involvement of cholecystokinin receptor types in pathways controlling oxytocin secretion. *Br. J. Pharmacol.* 110, 378–384.
73. Renaud, L. P., Tang, M., McCann, M. J., Stricker, E. M., and Verbalis, J. G. (1987). Cholecystokinin and gastric distension activate oxytocinergic cells in rat hypothalamus. *Am. J. Physiol.* 253, R661–R665.
74. Buller, K. M., and Day, T. A. (1996). Involvement of medullary catecholamine cells in neuroendocrine responses to systemic cholecystokinin. *J. Neuroendocrinol.* 8, 819–824.
75. Leng, G., Dyball, R. E., and Luckman, S. M. (1992). Mechanisms of vasopressin secretion. *Horm. Res.* 37, 33–38.
76. Onaka, T., Luckman, S. M., Antonijevic, I., Palmer, J. R., and Leng, G. (1995). Involvement of the noradrenergic afferents from the nucleus tractus solitarii to the supraoptic nucleus in oxytocin release after peripheral cholecystokinin octapeptide in the rat. *Neuroscience* 66, 403–412.
77. Rinaman, L., Hoffman, G. E., and Dohanics, J., et al. (1995). Cholecystokinin activates catecholaminergic neurons in the caudal medulla that innervate the paraventricular nucleus of the hypothalamus in rats. *J. Comp. Neurol.* 360, 246–256.
78. Olson, B. R., Freilino, M., and Hoffman, G. E., et al. (1993). C-Fos expression in rat brain and brainstem nuclei in response to treatments that alter food intake and gastric motility. *Mol. Cell Neurosci.* 4, 93–106.
79. Ericsson, A., Kovacs, K. J., and Sawchenko, P. E. (1994). A functional anatomical analysis of central pathways subserving the effects of interleukin-1 on stress-related neuroendocrine neurons. *J. Neurosci.* 14, 897–913.
80. Larsen, P. J., and Mikkelsen, J. D. (1995). Functional identification of central afferent projections conveying information of acute "stress" to the hypothalamic paraventricular nucleus. *J. Neurosci.* 15, 2609–2627.
81. Matta, S. G., Foster, C. A., and Sharp, B. M. (1993). Nicotine stimulates the expression of cFos protein in the parvocellular paraventricular nucleus and brainstem catecholaminergic regions. *Endocrinology* 132, 2149–2156.
82. Luckman, S. M. (1995). Fos expression within regions of the preoptic area, hypothalamus and brainstem during pregnancy and parturition. *Brain Res.* 669, 115–124.
83. Li, C., Chen, P., and Smith, M. S. (1999). Neural populations in the rat forebrain and brainstem activated by the suckling stimulus as demonstrated by cFos expression. *Neuroscience* 94, 117–129.
84. Ortega-Villalobos, M., Garcia-Bazan, M., and Solano-Flores, L. P., et al. (1990). Vagus nerve afferent and efferent innervation of the rat uterus: an electrophysiological and HRP study. *Brain Res. Bull.* 25, 365–371.
85. Verbalis, J. G., Hoffman, G. E., and Sherman, T. G. (1995). Use of immediate early genes as markers of oxytocin and vasopressin neuronal activation. *Curr. Opin. Endocrinol Diab.* 2, 157–168.
86. Sawchenko, P. E., Plotsky, P. M., and Pfeiffer, S. W., et al. (1988). Inhibin beta in central neural pathways involved in the control of oxytocin secretion. *Nature* 334, 615–617.
87. Leng, G., Yamashita, H., Dyball, R. E., and Bunting, R. (1988). Electrophysiological evidence for a projection from the arcuate nucleus to the supraoptic nucleus. *Neurosci. Lett.* 89, 146–151.
88. Oldfield, B. J., Hou-Yu, A., and Silverman, A. J. (1985). A combined electron microscopic HRP and immunocytochemical study of the limbic projections to rat hypothalamic nuclei containing vasopressin and oxytocin neurons. *J. Comp. Neurol.* 231, 221–231.
89. Saphier, D., and Feldman, S. (1987). Electrophysiology of supraoptico-paraventricular nucleus connections in the rat. *Exp. Brain Res.* 69, 60–66.
90. Sawchenko, P. E., Swanson, L. W., Steinbusch, H. W., and Verhofstad, A. A. (1983). The distribution and cells of origin of serotonergic inputs to the paraventricular and supraoptic nuclei of the rat. *Brain Res.* 277, 355–360.
91. Thellier, D., Moos, F., Richard, P., and Stoeckel, M. E. (1994). Evidence for connections between a discrete hypothalamic dorsochiasmatic area and the supraoptic and paraventricular nuclei. *Brain Res. Bull.* 34, 261–274.
92. Cui, L. N., Saeb-Parsy, K., and Dyball, R. E. (1997). Neurones in the supraoptic nucleus of the rat are regulated by a projection from the suprachiasmatic nucleus. *J. Physiol. (Lond).* 502 (Pt 1), 149–159.
93. Cui, L. N., Jolley, C. J., and Dyball, R. E. (1997). Electrophysiological evidence for retinal projections to the hypothalamic supraoptic nucleus and its perinuclear zone. *J. Neuroendocrinol.* 9, 347–353.
94. Windle, R. J., Luckman, S. M., Stoughton, R. P., and Forsling, M. L. (1996). The effect of pinealectomy on osmotically stimulated vasopressin and oxytocin release and Fos protein production within the hypothalamus of the rat. *J. Neuroendocrinol.* 8, 747–753.
95. Ericson, H., Watanabe, T., and Kohler, C. (1987). Morphological analysis of the tuberomammillary nucleus in the rat brain: delineation of subgroups with antibody against L-histidine decarboxylase as a marker. *J. Comp. Neurol.* 263, 1–24.
96. Weiss, M. L., Yang, Q. Z., and Hatton, G. I. (1989). Magnocellular tuberomammillary nucleus input to the supraoptic nucleus in the rat: anatomical and in vitro electrophysiological investigations. *Neuroscience* 31, 299–311.
97. Yang, Q. Z., and Hatton, G. I. (1994). Histamine mediates fast synaptic inhibition of rat supraoptic oxytocin neurons via chloride conductance activation. *Neuroscience* 61, 955–964.
98. Kjaer, A., Larsen, P. J., Knigge, U., Moller, M., and Warberg, J. (1994). Histamine stimulates c-fos expression in hypothalamic vasopressin-, oxytocin-, and corticotropin-releasing hormone-containing neurons. *Endocrinology* 134, 482–491.
99. Kjaer, A., Larsen, P. J., Knigge, U., and Warberg, J. (1994). Histaminergic activation of the hypothalamic–pituitary–adrenal axis. *Endocrinology* 135, 1171–1177.

100. Luckman, S. M., and Larsen, P. J. (1997). Evidence for the involvement of histaminergic neurones in the regulation of the rat oxytocinergic system during pregnancy and parturition *J. Physiol. (Lond).* 501 (Pt 3), 649–655.
101. Schagen, F. H., Knigge, U., Kjaer, A., Larsen, P. J., and Warberg, J. (1996). Involvement of histamine in suckling-induced release of oxytocin, prolactin and adrenocorticotropin in lactating rats. *Neuroendocrinology* 63, 550–558.
102. Kjaer, A., Knigge, U., Rouleau, A., Garbarg, M., and Warberg, J. (1994). Dehydration-induced release of vasopressin involves activation of hypothalamic histaminergic neurons. *Endocrinology* 135, 675–681.
103. Kjaer, A., Larsen, P. J., Knigge, U., and Warberg, J. (1995). Dehydration stimulates hypothalamic gene expression of histamine synthesis enzyme: importance for neuroendocrine regulation of vasopressin and oxytocin secretion. *Endocrinology* 136, 2189–2197.
104. Yagi, K. (1994). Effects of a histamine H2-receptor antagonist, ranitidine on the vasopressin and oxytocin responses to novelty stress in the rat. *Neurosci. Res.* 19, 357–364.
105. Smithson, K. G., Weiss, M. L., and Hatton, G. I. (1989). Supraoptic nucleus afferents from the main olfactory bulb. I. Anatomical evidence from anterograde and retrograde tracers in rat. *Neuroscience* 31, 277–287.
106. Smithson, K. G., Weiss, M. L., and Hatton, G. I. (1989). Accessory olfactory bulb inputs to the rat supraoptic nucleus: an anterograde and retrograde study. *Anat. Rec.* 220.
107. Altman, J., and Bayer, S. A. (1978). Development of the diencephalon in the rat. III. Ontogeny of the specialized ventricular linings of the hypothalamic third ventricle. *J. Comp. Neurol.* 182, 995–1015.
108. Altman, J., and Bayer, S. A. (1978). Development of the diencephalon in the rat. II. Correlation of the embryonic development of the hypothalamus with the time of origin of its neurons. *J. Comp. Neurol.* 182, 973–993.
109. Altman, J., and Bayer, S. A. (1978). Development of the diencephalon in the rat. I. Autoradiographic study of the time of origin and settling patterns of neurons of the hypothalamus. *J. Comp. Neurol.* 182, 945–971.
110. Daikoku, S., Tsuruo, Y., and Hashimoto, T., et al. (1989). Hypothalamic neurons from a developmental aspect. *Arch. Histol. Cytol.* 52(Suppl), 217–223.
111. Daikoku, S., Okamura, Y., and Kawano, H., et al. (1984). Immunohistochemical study on the development of CRF-containing neurons in the hypothalamus of the rat. *Cell Tissue Res.* 238, 539–544.
112. Karim, M. A., and Sloper, J. C. (1980). Histogenesis of the supraoptic and paraventricular neurosecretory cells of the mouse hypothalamus. *J. Anat.* 130, 341–347.
113. Nakai, S., Kawano, H., and Yudate, T., et al. (1995). The POU domain transcription factor Brn-2 is required for the determination of specific neuronal lineages in the hypothalamus of the mouse. *Genes Dev.* 9, 3109–3121.
114. Yamakuni, T., Kuwano, R., and Odani, S., et al. (1987). Molecular cloning of cDNA to mRNA for a cerebellar spot 35 protein. *J. Neurochem.* 48, 1590–1596.
115. Sinding, C., Robinson, A. G., Seif, S. M., and Schmid, P. G. (1980). Neurohypophyseal peptides in the developing rat fetus. *Brain Res.* 195, 177–186.
116. Altstein, M., and Gainer, H. (1988). Differential biosynthesis and posttranslational processing of vasopressin and oxytocin in rat brain during embryonic and postnatal development. *J. Neurosci.* 8, 3967–3977.
117. Whitnall, M. H., Key, S., Ben-Barak, Y., Ozato, K., and Gainer, H. (1985). Neurophysin in the hypothalamo-neurohypophysial system. II. Immunocytochemical studies of the ontogeny of oxytocinergic and vasopressinergic neurons. *J. Neurosci.* 5, 98–109.
118. Galabov, P., and Schiebler, T. H. (1978). The ultrastructure of the developing neural lobe. *Cell Tissue Res.* 189, 313–329.
119. Galabov, P. G., and Schiebler, T. H. (1983). Development of the capillary system in the neurohypophysis of the rat. *Cell Tissue Res.* 228, 685–696.
120. Korteweg, N. (2000). Molecular Dissection of Peptide Secretion: Analysis of the Secretion Machinery in Neurons and Endocrine Cells. Utrecht University.
121. Korteweg, N., Maia, A. S., Verhage, M., and Burbach, J. P. (2004). Development of the mouse hypothalamo-neurohypophysial system in the munc18-1 null mutant that lacks regulated secretion. *Eur. J. Neurosci.* 19, 2944–2952.
122. Jing, X., Ratty, A. K., and Murphy, D. (1998). Ontogeny of the vasopressin and oxytocin RNAs in the mouse hypothalamus. *Neurosci. Res.* 30, 343–349.
123. Schonemann, M. D., Ryan, A. K., and McEvilly, R. J., et al. (1995). Development and survival of the endocrine hypothalamus and posterior pituitary gland requires the neuronal POU domain factor Brn-2. *Genes Dev.* 9, 3122–3135.
124. Michaud, J. L., DeRossi, C., May, N. R., Holdener, B. C., and Fan, C. (2000). ARNT2 acts as the dimerization partner of SIM1 for the development of the hypothalamus. *Mech. Dev.* 90, 253–261.
125. Wines, M. E., Tiffany, A. M., and Holdener, B. C. (1998). Physical localization of the mouse aryl hydrocarbon receptor nuclear translocator-2 (Arnt2) gene within the c112K deletion. *Genomics* 51, 223–232.
126. Simeone, A., D'Apice, M. R., and Nigro, V., et al. (1994). Orthopedia, a novel homeobox-containing gene expressed in the developing CNS of both mouse and Drosophila. *Neuron.* 13, 83–101.
127. Acampora, D., Postiglione, M. P., and Avantaggiato, V., et al. (1999). Progressive impairment of developing neuroendocrine cell lineages in the hypothalamus of mice lacking the Orthopedia gene. *Genes Dev.* 13, 2787–2800.
128. Michaud, J. L., Rosenquist, T., May, N. R., and Fan, C. M. (1998). Development of neuroendocrine lineages requires the bHLH-PAS transcription factor SIM1. *Genes Dev.* 12, 3264–3275.
129. Labosky, P. A., Winnier, G. E., and Jetton, T. L., et al. (1997). The winged helix gene, Mf3, is required for normal development of the diencephalon and midbrain, postnatal growth and the milk-ejection reflex. *Development* 124, 1263–1274.
130. Wehr, R., Mansouri, A., de Maeyer, T., and Gruss, P. (1997). Fkh5-deficient mice show dysgenesis in the caudal midbrain and hypothalamic mammillary body. *Development* 124, 4447–4456.
131. Li, L., Keverne, E. B., and Aparicio, S. A., et al. (1999). Regulation of maternal behavior and offspring growth by paternally expressed Peg3. *Science* 284, 330–333.
132. Treier, M., and Rosenfeld, M. G. (1996). The hypothalamic–pituitary axis: co-development of two organs. *Curr. Opin. Cell Biol.* 8, 833–843.
133. Verhage, M., Maia, A. S., and Plomp, J. J., et al. (2000). Synaptic assembly of the brain in the absence of neurotransmitter secretion. *Science* 287, 864–869.
134. Eib, D. W., and Martens, G. J. (1996). A novel transmembrane protein with epidermal growth factor and follistatin domains expressed in the hypothalamo–hypophysial axis of *Xenopus laevis*. *J. Neurochem.* 67, 1047–1055.
135. Almazan, G., Lefebvre, D. L., and Zingg, H. H. (1989). Ontogeny of hypothalamic vasopressin, oxytocin and somatostatin gene expression. *Brain Res. Dev. Brain Res.* 45, 69–75.
136. van Tol, H. H., Snijdewint, F. G., Boer, G. J., and Burbach, J. P. (1986). Postnatal development of vasopressin mRNA

content in supraoptic and paraventricular nucleus of the Wistar rat. *Neurosci. Lett.* 65, 1–6.

137. Zicha, J., Krecek, J., and Dlouha, H. (1982). Age-dependent salt hypertension in Brattleboro rats: a hemodynamic analysis. *Ann. N Y Acad. Sci.* 394, 330–342.
138. van Tol, H. H., van den, B. M., de Jong, W., and Burbach, J. P. (1988). Vasopressin and oxytocin gene expression in the supraoptic and paraventricular nucleus of the spontaneously hypertensive rat (SHR) during development of hypertension. *Brain Res.* 464, 303–311.
139. Burbach, J. P., Liu, B., Voorhuis, T. A., and van Tol, H. H. (1988). Diurnal variation in vasopressin and oxytocin messenger RNAs in hypothalamic nuclei of the rat. *Brain Res.* 464, 157–160.
140. Gimpl, G., and Fahrenholz, F. (2001). The oxytocin receptor system: structure, function, and regulation. *Physiol. Rev.* 81, 629–683.
141. Inoue, T., Kimura, T., and Azuma, C., et al. (1994). Structural organization of the human oxytocin receptor gene. *J. Biol. Chem.* 269, 32451–32456.
142. Rozen, F., Russo, C., Banville, D., and Zingg, H. H. (1995). Structure, characterization, and expression of the rat oxytocin receptor gene [published erratum appears in *Proc. Natl. Acad. Sci. U S A* 1996;93(21):12051]. *Proc. Natl. Acad. Sci. U S A* 92, 200–204.
143. Chini, B., and Fanelli, F. (2000). Molecular basis of ligand binding and receptor activation in the oxytocin and vasopressin receptor family. *Exp. Physiol.* 85 (Spec No), 59S–66S.
144. Ivell, R., Kimura, T., and Muller, D., et al. (2001). The structure and regulation of the oxytocin receptor. *Exp. Physiol.* 86, 289–296.
145. Sanborn, B. M. (2001). Hormones and calcium: mechanisms controlling uterine smooth muscle contractile activity. The Litchfield Lecture. *Exp. Physiol.* 86, 223–237.
146. Gould, B. R., and Zingg, H. H. (2003). Mapping oxytocin receptor gene expression in the mouse brain and mammary gland using an oxytocin receptor-LacZ reporter mouse. *Neuroscience* 122, 155–167.
147. Feng, H. C., Bhave, M., and Fairclough, R. J. (2000). Regulation of oxytocin receptor gene expression in sheep: tissue specificity, multiple transcripts and mRNA editing. *J. Reprod. Fertil.* 120, 187–200.
148. Zhong, M., Navratil, A. M., Clay, C., and Sanborn, B. M. (2004). Residues in the hydrophilic face of putative helix 8 of oxytocin receptor are important for receptor function. *Biochemistry* 43, 3490–3498.
149. Zingg, H. H., and Laporte, S. A. (2003). The oxytocin receptor. *Trends Endocrinol. Metab.* 14, 222–227.
150. Schoots, O., Hernando, F., Knoers, N. V., and Burbach, J. P. (1999). Vasopressin receptors: structural functional relationships and role in neural and endocrine regulation. *Results Probl. Cell Differ.* 26, 107–133.
151. Sanborn, B. M., Dodge, K., and Monga, M., et al. (1998). Molecular mechanisms regulating the effects of oxytocin on myometrial intracellular calcium. *Adv. Exp. Med. Biol.* 449, 277–286.
152. Sanborn, B. M., Qian, A., and Ku, C. Y., et al. (1995). Mechanisms regulating oxytocin receptor coupling to phospholipase C in rat and human myometrium. *Adv. Exp. Med. Biol.* 395, 469–479.
153. Chan, W. Y., Chen, D. L., and Manning, M. (1993). Oxytocin receptor subtypes in the pregnant rat myometrium and decidua: pharmacological differentiations. *Endocrinology* 132, 1381–1386.
154. Fuchs, A. R., Fuchs, F., Husslein, P., Soloff, M. S., and Fernstrom, M. J. (1982). Oxytocin receptors and human parturition: a dual role for oxytocin in the initiation of labor. *Science* 215, 1396–1398.
155. Lefebvre, D. L., Giaid, A., Bennett, H., Lariviere, R., and Zingg, H. H. (1992). Oxytocin gene expression in rat uterus. *Science* 256, 1553–1555.
156. Fuchs, A. R., Periyasamy, S., Alexandrova, M., and Soloff, M. S. (1983). Correlation between oxytocin receptor concentration and responsiveness to oxytocin in pregnant rat myometrium: effects of ovarian steroids. *Endocrinology* 113, 742–749.
157. Takemura, M., Nomura, S., and Kimura, T., et al. (1993). Expression and localization of oxytocin receptor gene in human uterine endometrium in relation to the menstrual cycle. *Endocrinology* 132, 1830–1835.
158. Kimura, T., Tanizawa, O., Mori, K., Brownstein, M. J., and Okayama, H. (1992). Structure and expression of a human oxytocin receptor. *Nature* 356, 526–529.
159. Tence, M., Guillon, G., Bottari, S., and Jard, S. (1990). Labelling of vasopressin and oxytocin receptors from the human uterus. *Eur. J. Pharmacol.* 191, 427–436.
160. Takemura, M., Kimura, T., and Nomura, S., et al. (1994). Expression and localization of human oxytocin receptor mRNA and its protein in chorion and decidua during parturition. *J. Clin. Invest.* 93, 2319–2323.
161. Larcher, A., Neculcea, J., and Breton, C., et al. (1995). Oxytocin receptor gene expression in the rat uterus during pregnancy and the estrous cycle and in response to gonadal steroid treatment. *Endocrinology* 136, 5350–5356.
162. Windle, R. J., and Forsling, M. L. (1997). Renal responses to oxytocin during the 4 days of the oestrous cycle in the rat. *J. Endocrinol.* 154, 347–353.
163. van Tol, H. H., Bolwerk, E. L., Liu, B., and Burbach, J. P. (1988). Oxytocin and vasopressin gene expression in the hypothalamo-neurohypophyseal system of the rat during the estrous cycle, pregnancy, and lactation. *Endocrinology* 122, 945–951.
164. Wuttke, W., Jarry, H., Knoke, I., Pitzel, L., and Spiess, S. (1995). Luteotropic and luteolytic effects of oxytocin in the porcine corpus luteum. *Adv. Exp. Med. Biol.* 395, 495–506.
165. Ivell, R., and Richter, D. (1984). The gene for the hypothalamic peptide hormone oxytocin is highly expressed in the bovine corpus luteum: biosynthesis, structure and sequence analysis. *EMBO J.* 3, 2351–2354.
166. Bruckmaier, R. M., Wellnitz, O., and Blum, J. W. (1997). Inhibition of milk ejection in cows by oxytocin receptor blockade, alpha-adrenergic receptor stimulation and in unfamiliar surroundings. *J. Dairy Res.* 64, 315–325.
167. Nishimori, K., Young, L. J., and Guo, Q., et al. (1996). Oxytocin is required for nursing but is not essential for parturition or reproductive behavior. *Proc. Natl. Acad. Sci. U S A* 93, 11699–11704.
168. Soloff, M. S., and Wieder, M. H. (1983). Oxytocin receptors in rat involuting mammary gland. *Can. J. Biochem. Cell Biol.* 61, 631–635.
169. Breton, C., Scala-Guenot, D., and Zingg, H. H. (2001). Oxytocin receptor gene expression in rat mammary gland: structural characterization and regulation. *J. Mol. Endocrinol.* 27, 175–189.
170. Kimura, T., Ito, Y., and Einspanier, A., et al. (1998). Expression and immunolocalization of the oxytocin receptor in human lactating and non-lactating mammary glands. *Hum. Reprod.* 13, 2645–2653.
171. Argiolas, A., and Melis, M. R. (2004). The role of oxytocin and the paraventricular nucleus in the sexual behaviour of male mammals. *Physiol. Behav.* 83, 309–317.
172. Chen, K., and Chang, L. S. (2001). Oxytocinergic neurotransmission at the hippocampus in the central neural regulation of penile erection in the rat. *Urology* 58, 107–112.

173. Veronneau-Longueville, F., Rampin, O., and Freund-Mercier, M. J., et al. (1999). Oxytocinergic innervation of autonomic nuclei controlling penile erection in the rat. *Neuroscience* 93, 1437–1447.
174. Argiolas, A., Melis, M. R., and Gessa, G. L. (1986). Oxytocin: an extremely potent inducer of penile erection and yawning in male rats. *Eur. J. Pharmacol.* 130, 265–272.
175. Filippi, S., Vignozzi, L., and Vannelli, G. B., et al. (2003). Role of oxytocin in the ejaculatory process. *J. Endocrinol. Invest.* 26, 82–86.
176. Ivell, R., Balvers, M., Rust, W., Bathgate, R., and Einspanier, A. (1997). Oxytocin and male reproductive function. *Adv. Exp. Med. Biol.* 424, 253–264.
177. Carmichael, M. S., Humbert, R., and Dixen, J., et al. (1987). Plasma oxytocin increases in the human sexual response. *J. Clin. Endocrinol. Metab.* 64, 27–31.
178. Ogawa, S., Kudo, S., Kitsunai, Y., and Fukuchi, S. (1980). Increase in oxytocin secretion at ejaculation in male. *Clin. Endocrinol. (Oxf).* 13, 95–97.
179. Blaicher, W., Gruber, D., and Bieglmayer, C., et al. (1999). The role of oxytocin in relation to female sexual arousal. *Gynecol. Obstet. Invest.* 47, 125–126.
180. Vignozzi, L., Filippi, S., and Luconi, M., et al. (2004). Oxytocin receptor is expressed in the penis and mediates an estrogen-dependent smooth muscle contractility. *Endocrinology* 145, 1823–1834.
181. Vignozzi, L., Vannelli, G. B., and Morelli, A., et al. (2004). Identification, characterization and biological activity of oxytocin receptor in the developing human penis. *Mol. Hum. Reprod.*
182. Frayne, J., and Nicholson, H. D. (1998). Localization of oxytocin receptors in the human and macaque monkey male reproductive tracts: evidence for a physiological role of oxytocin in the male. *Mol. Hum. Reprod.* 4, 527–532.
183. Niemi, M., and Kormano, M. (1965). Contractility of tye seminiferous tubele of the postnatal rat testis and its response to oxytocin. *Ann. Med. Exp. Biol. Fenn.* 43, 40–42.
184. Nicholson, H. D., and Jenkin, L. (1995). Oxytocin and prostatic function. *Adv. Exp. Med. Biol.* 395, 529–538.
185. Assinder, S. J., Rezvani, A., and Nicholson, H. D. (2002). Oxytocin promotes spermiation and sperm transfer in the mouse. *Int. J. Androl.* 25, 19–27.
186. Whittington, K., Assinder, S. J., Parkinson, T., Lapwood, K. R., and Nicholson, H. D. (2001). Function and localization of oxytocin receptors in the reproductive tissue of rams. *Reproduction* 122, 317–325.
187. Einspanier, A., and Ivell, R. (1997). Oxytocin and oxytocin receptor expression in reproductive tissues of the male marmoset monkey. *Biol. Reprod.* 56, 416–422.
188. Bathgate, R. A., and Sernia, C. (1994). Characterization and localization of oxytocin receptors in the rat testis. *J. Endocrinol.* 141, 343–352.
189. Frayne, J., and Nicholson, H. D. (1995). Effect of oxytocin on testosterone production by isolated rat Leydig cells is mediated via a specific oxytocin receptor. *Biol. Reprod.* 52, 1268–1273.
190. Nicholson, H. D., Greenfield, H. M., and Frayne, J. (1994). The effect of germ cell complement on the presence of oxytocin in the interstitial and seminiferous tubule fluid of the rat testis. *J. Endocrinol.* 143, 471–478.
191. Frayne, J., and Nicholson, H. D. (1994). Regulation of oxytocin production by purified adult rat Leydig cells in vitro: effects of LH, testosterone and lipoproteins. *J. Endocrinol.* 143, 325–332.
192. Ang, H. L., Ivell, R., and Walther, N., et al. (1994). Over-expression of oxytocin in the testes of a transgenic mouse model. *J. Endocrinol.* 140, 53–62.
193. Young, W. S., Shepard, E., and Amico, J., et al. (1996). Deficiency in mouse oxytocin prevents milk ejection, but not fertility or parturition. *J. Neuroendocrinol.* 8, 847–853.
194. Nishimori, K., Kawamata, M., and Takayanagi, Y. (2002). Functional analysis of oxytocin and oxytocin receptor with the genes deficient mice. *Nogeikagaku Kaishi.* 76, 1086–1089.
195. Young, W. S. III, and Gainer, H. (2003). Transgenesis and the study of expression, cellular targeting and function of oxytocin, vasopressin and their receptors. *Neuroendocrinology* 78, 185–203.
196. Bodanszky, M., Sharaf, H., Roy, J. B., and Said, S. I. (1992). Contractile activity of vasotocin, oxytocin, and vasopressin on mammalian prostate. *Eur. J. Pharmacol.* 216, 311–313.
197. Jenkin, L., and Nicholson, H. D. (1994). The effect of testosterone on prostatic oxytocin in the rat. *J. Endocrinol.* 143 (Suppl), 43.
198. Popovic, A., Plecas, B., Milicevic, Z., Hristic, M., and Jovovic, D. (1990). Stereologic analysis of ventral prostate of oxytocin-treated rats. *Arch. Androl.* 24, 247–253.
199. Jenkin, L., and Nicholson, H. D. (1999). Evidence for the regulation of prostatic oxytocin by gonadal steroids in the rat. *J. Androl.* 20, 80–87.
200. Assinder, S. J., and Nicholson, H. D. (2004). Effects of steroids on oxytocin secretion by the human prostate in vitro. *Int. J. Androl.* 27, 12–18.
201. Ito, Y., Kobayashi, T., and Kimura, T., et al. (1996). Investigation of the oxytocin receptor expression in human breast cancer tissue using newly established monoclonal antibodies. *Endocrinology* 137, 773–779.
202. Cassoni, P., Sapino, A., Marrocco, T., Chini, B., and Bussolati, G. (2004). Oxytocin and oxytocin receptors in cancer cells and proliferation. *J. Neuroendocrinol.* 16, 362–364.
203. Amico, J. A., Rauk, P. N., and Cai, H. M. (2002). Estradiol and progesterone regulate oxytocin receptor binding and expression in human breast cancer cell lines. *Endocrine* 18, 79–84.
204. Bussolati, G., Cassoni, P., Ghisolfi, G., Negro, F., and Sapino, A. (1996). Immunolocalization and gene expression of oxytocin receptors in carcinomas and non-neoplastic tissues of the breast. *Am. J. Pathol.* 148, 1895–1903.
205. Buijs, R. M., and Swaab, D. F. (1979). Immuno-electron microscopical demonstration of vasopressin and oxytocin synapses in the limbic system of the rat. *Cell Tissue Res.* 204, 355–365.
206. Hermes, M. L., Buijs, R. M., Masson-Pevet, M., and Pevet, P. (1988). Oxytocinergic innervation of the brain of the garden dormouse (*Eliomys quercinus* L.). *J. Comp. Neurol.* 273, 252–262.
207. Sofroniew, M. V., and Schrell, U. (1981). Evidence for a direct projection from oxytocin and vasopressin neurons in the hypothalamic paraventricular nucleus to the medulla oblongata: immunohistochemical visualization of both the horseradish peroxidase transported and the peptide produced by the same neurons. *Neurosci. Lett.* 22, 211–217.
208. Stern, J. E., and Zhang, W. (2003). Preautonomic neurons in the paraventricular nucleus of the hypothalamus contain estrogen receptor beta. *Brain Res.* 975, 99–109.
209. Blevins, J. E., Schwartz, M. W., and Baskin, D. G. (2004). Evidence that paraventricular nucleus oxytocin neurons link hypothalamic leptin action to caudal brain stem nuclei controlling meal size. *Am. J. Physiol. Regul. Integr. Comp. Physiol.* 287, R87–R96.
210. Flanagan, L. M., Pfaus, J. G., Pfaff, D. W., and McEwen, B. S. (1993). Induction of FOS immunoreactivity in oxytocin neurons after sexual activity in female rats. *Neuroendocrinology* 58, 352–358.

211. Schulze, H. G., and Gorzalka, B. B. (1991). Oxytocin effects on lordosis frequency and lordosis duration following infusion into the medial pre-optic area and ventromedial hypothalamus of female rats. *Neuropeptides* 18, 99–106.
212. Schumacher, M., Coirini, H., Frankfurt, M., and McEwen, B. S. (1989). Localized actions of progesterone in hypothalamus involve oxytocin. *Proc. Natl. Acad. Sci. U S A* 86, 6798–6801.
213. Kremarik, P., Freund-Mercier, M. J., and Stoeckel, M. E. (1993). Histoautoradiographic detection of oxytocin- and vasopressin-binding sites in the telencephalon of the rat. *J. Comp. Neurol.* 333, 343–359.
214. Elands, J., Beetsma, A., Barberis, C., and De Kloet, E. R. (1988). Topography of the oxytocin receptor system in rat brain: an autoradiographical study with a selective radioiodinated oxytocin antagonist. *J. Chem. Neuroanat.* 1, 293–302.
215. Freund-Mercier, M. J., Stoeckel, M. E., and Palacios, J. M., et al. (1987). Pharmacological characteristics and anatomical distribution of [3H]oxytocin-binding sites in the Wistar rat brain studied by autoradiography. *Neuroscience* 20, 599–614.
216. Tribollet, E., Barberis, C., Jard, S., Dubois-Dauphin, M., and Dreifuss, J. J. (1988). Localization and pharmacological characterization of high affinity binding sites for vasopressin and oxytocin in the rat brain by light microscopic autoradiography. *Brain Res.* 442, 105–118.
217. Pedersen, C. A., Caldwell, J. D., Walker, C., Ayers, G., and Mason, G. A. (1994). Oxytocin activates the postpartum onset of rat maternal behavior in the ventral tegmental and medial preoptic areas. *Behav. Neurosci.* 108, 1163–1171.
218. Adan, R. A., van Leeuwen, F. W., and Sonnemans, M. A., et al. (1995). Rat oxytocin receptor in brain, pituitary, mammary gland, and uterus: partial sequence and immunocytochemical localization. *Endocrinology* 136, 4022–4028.
219. Vaccari, C., Lolait, S. J., and Ostrowski, N. L. (1998). Comparative distribution of vasopressin V1b and oxytocin receptor messenger ribonucleic acids in brain. *Endocrinology* 139, 5015–5033.
220. Young, L. J., Muns, S., Wang, Z., and Insel, T. R. (1997). Changes in oxytocin receptor mRNA in rat brain during pregnancy and the effects of estrogen and interleukin-6. *J. Neuroendocrinol.* 9, 859–865.
221. Loup, F., Tribollet, E., Dubois-Dauphin, M., and Dreifuss, J. J. (1991). Localization of high-affinity binding sites for oxytocin and vasopressin in the human brain. An autoradiographic study. *Brain Res.* 555, 220–232.
222. Insel, T. R., Young, L., Witt, D. M., and Crews, D. (1993). Gonadal steroids have paradoxical effects on brain oxytocin receptors. *J. Neuroendocrinol.* 5, 619–628.
223. Insel, T. R., and Shapiro, L. E. (1992). Oxytocin receptor distribution reflects social organization in monogamous and polygamous voles. *Proc. Natl. Acad. Sci. U S A* 89, 5981–5985.
224. Tribollet, E., Charpak, S., Schmidt, A., Dubois-Dauphin, M., and Dreifuss, J. J. (1989). Appearance and transient expression of oxytocin receptors in fetal, infant, and peripubertal rat brain studied by autoradiography and electrophysiology. *J. Neurosci.* 9, 1764–1773.
225. Tribollet, E., Audigier, S., Dubois-Dauphin, M., and Dreifuss, J. J. (1990). Gonadal steroids regulate oxytocin receptors but not vasopressin receptors in the brain of male and female rats. An autoradiographical study. *Brain Res.* 511, 129–140.
226. Johnson, A. E., Coirini, H., Insel, T. R., and McEwen, B. S. (1990). The regulation of oxytocin receptor binding in the ventromedial hypothalamic nucleus by testosterone and its metabolites. *Endocrinology* 128, 891–896.
227. Johnson, A. E., Coirini, H., Ball, G. F., and McEwen, B. S. (1989). Anatomical localization of the effects of 17 beta-estradiol on oxytocin receptor binding in the ventromedial hypothalamic nucleus. *Endocrinology* 124, 207–211.
228. Coirini, H., Johnson, A. E., and McEwen, B. S. (1989). Estradiol modulation of oxytocin binding in the ventromedial hypothalamic nucleus of male and female rats. *Neuroendocrinology* 50, 193–198.
229. De Kloet, E. R., Voorhuis, T. A., and Elands, J. (1985). Estradiol induces oxytocin binding sites in rat hypothalamic ventromedial nucleus. *Eur. J. Pharmacol.* 118, 185–186.
230. De Kloet, E. R., Voorhuis, D. A., Boschma, Y., and Elands, J. (1986). Estradiol modulates density of putative "oxytocin receptors" in discrete rat brain regions. *Neuroendocrinology* 44, 415–421.
231. Bale, T. L., and Dorsa, D. M. (1995). Regulation of oxytocin receptor messenger ribonucleic acid in the ventromedial hypothalamus by testosterone and its metabolites. *Endocrinology* 136, 5135–5138.
232. Bale, T. L., Dorsa, D. M., and Johnston, C. A. (1995). Oxytocin receptor mRNA expression in the ventromedial hypothalamus during the estrous cycle. *J. Neurosci.* 15, 5058–5064.
233. Young, L. J., Wang, Z., Donaldson, R., and Rissman, E. F. (1998). Estrogen receptor alpha is essential for induction of oxytocin receptor by estrogen. *Neuroreport* 9, 933–936.
234. Lim, M. M., Murphy, A. Z., and Young, L. J. (2004). Ventral striatopallidal oxytocin and vasopressin V1a receptors in the monogamous prairie vole (*Microtus ochrogaster*). *J. Comp. Neurol.* 468, 555–570.
235. Levy, F., Kendrick, K. M., Goode, J. A., Guevara-Guzman, R., and Keverne, E. B. (1995). Oxytocin and vasopressin release in the olfactory bulb of parturient ewes: changes with maternal experience and effects on acetylcholine, gamma-aminobutyric acid, glutamate and noradrenaline release. *Brain Res.* 669, 197–206.
236. Kendrick, K. M., Fabre-Nys, C., Blache, D., Goode, J. A., and Broad, K. D. (1993). The role of oxytocin release in the mediobasal hypothalamus of the sheep in relation to female sexual receptivity. *J. Neuroendocrinol.* 5, 13–21.
237. Sansone, G. R., Gerdes, C. A., and Steinman, J. L., et al. (2002). Vaginocervical stimulation releases oxytocin within the spinal cord in rats. *Neuroendocrinology* 75, 306–315.
238. Hughes, A. M., Everitt, B. J., Lightman, S. L., and Todd, K. (1987). Oxytocin in the central nervous system and sexual behaviour in male rats. *Brain Res.* 414, 133–137.
239. Neumann, I., and Landgraf, R. (1989). Septal and hippocampal release of oxytocin, but not vasopressin, in the conscious lactating rat during suckling. *J. Neuroendocrinol.* 1, 305–308.
240. Landgraf, R., and Neumann, I. D. (2004). Vasopressin and oxytocin release within the brain: a dynamic concept of multiple and variable modes of neuropeptide communication. *Front Neuroendocrinol.* 25, 150–176.
241. Lefebvre, D. L., Farookhi, R., Giaid, A., Neculcea, J., and Zingg, H. H. (1994). Uterine oxytocin gene expression. II. Induction by exogenous steroid administration. *Endocrinology* 134, 2562–2566.
242. Blanks, A. M., Vatish, M., and Allen, M. J., et al. (2003). Paracrine oxytocin and estradiol demonstrate a spatial increase in human intrauterine tissues with labor. *J. Clin. Endocrinol. Metab.* 88, 3392–3400.
243. Lefebvre, D. L., Giaid, A., and Zingg, H. H. (1992). Expression of the oxytocin gene in rat placenta. *Endocrinology* 130, 1185–1192.
244. Mitchell, B. F., and Chibbar, R. (1996) Synthesis and metabolism of oxytocin in late gestation in human decidua. *Adv Exp. Med. Biol.* 395, 365–380.

245. Fehr, S., Ivell, R., and Koll, R., et al. (1987). Expression of the oxytocin gene in the large cells of the bovine corpus luteum. *FEBS Lett.* 210, 45–50.
246. Jones, D. S., and Flint, A. P. (1988). Concentrations of oxytocin-neurophysin prohormone mRNA in corpora lutea of sheep during the oestrous cycle and in early pregnancy. *J. Endocrinol.* 117, 409–414.
247. Fortune, J. E., and Voss, A. K. (1993). Oxytocin gene expression and action in bovine preovulatory follicles. *Regul. Pept.* 45, 257–261.
248. Ivell, R., Furuya, K., Brackmann, B., Dawood, Y., and Khan-Dawood, F. (1990). Expression of the oxytocin and vasopressin genes in human and baboon gonadal tissues. *Endocrinology* 127, 2990–2996.
249. Einspanier, A., Ivell, R., and Hodges, J. K. (1995). Oxytocin: a follicular luteinisation factor in the marmoset monkey. *Adv Exp. Med. Biol.* 395, 517–522.
250. Furuya, K., Mizumoto, Y., Makimura, N., Mitsui, C., Murakami, M., Tokuoka, S., Ishikawa, N., Imaizumi, E., Katayama, E., and Seki, K. (1995). Gene expressions of oxytocin and oxytocin receptor in cumulus cells of human ovary. *Horm. Res.* 44, 47–49.
251. Ungefroren, H., Davidoff, M., and Ivell, R. (1994). Post-transcriptional block in oxytocin gene expression within the seminiferous tubules of the bovine testis. *J. Endocrinol.* 140, 63–72.
252. Foo, N. C., Carter, D., Murphy, D., and Ivell, R. (1991). Vasopressin and oxytocin gene expression in rat testis. *Endocrinology* 128, 2118–2128.
253. Jankowski, M., Hajjar, F., and Kawas, S. A., et al. (1998). Rat heart: a site of oxytocin production and action. *Proc. Natl. Acad. Sci. U S A* 95, 14558–14563.
254. Jankowski, M., Wang, D., and Hajjar, F., et al. (2000). Oxytocin and its receptors are synthesized in the rat vasculature. *Proc. Natl. Acad. Sci. U S A* 97, 6207–6211.
255. Jankowski, M., Danalache, B., and Wang, D., et al. (2004). Oxytocin in cardiac ontogeny. *Proc. Natl. Acad. Sci. U S A* 101, 13074–13079.
256. Dave, J. R., Culp, S. G., Liu, L., Tabakoff, B., and Hoffman, P. L. (1988). Regulation of vasopressin and oxytocin synthesis in anterior pituitary and peripheral tissues. *Adv. Alcohol Subst. Abuse* 7, 231–234.
257. Murphy, D., Levy, A., Lightman, S., and Carter, D. (1989). Vasopressin RNA in the neural lobe of the pituitary: dramatic accumulation in response to salt loading. *Proc. Natl. Acad. Sci. U S A* 86, 9002–9005.
258. Murphy, D., Si-Hoe, S. L., Brenner, S., and Venkatesh, B. (1998). Something fishy in the rat brain: molecular genetics of the hypothalamo-neurohypophysial system. *Bioessays* 20, 741–749.
259. Amico, J. A., Finn, F. M., and Haldar, J. (1988). Oxytocin and vasopressin are present in human and rat pancreas. *Am. J. Med Sci.* 296, 303–307.
260. Argiolas, A., Gessa, G. L., Melis, M. R., Stancampiano, R., and Vaccari, A. (1990). Effects of neonatal and adult thyroid dysfunction on thymic oxytocin. *Neuroendocrinology* 52, 556–559.
261. Geenen, V., Legros, J. J., and Franchimont, P., et al. (1986). The neuroendocrine thymus: coexistence of oxytocin and neurophysin in the human thymus. *Science* 232, 508–511.
262. Ivell, R., and Richter, D. (1984). Structure and comparison of the oxytocin and vasopressin genes from rat. *Proc. Natl. Acad. Sci. U S A* 81, 2006–2010.
263. Mohr, E., Schmitz, E., and Richter, D. (1988). A single rat genomic DNA fragment encodes both the oxytocin and vasopressin genes separated by 11 kilobases and oriented in opposite transcriptional directions. *Biochimie* 70, 649–654.
264. Riddell, D. C., Mallonee, R., and Phillips, J. A., et al. (1985). Chromosomal assignment of human sequences encoding arginine vasopressin-neurophysin II and growth hormone releasing factor. *Somat. Cell Mol. Genet.* 11, 189–195.
265. Summar, M. L., Phillips, J. A., and Battey, J., et al. (1990). Linkage relationships of human arginine vasopressin-neurophysin-II and oxytocin-neurophysin-I to prodynorphin and other loci on chromosome 20. *Mol. Endocrinol.* 4, 947–950.
266. Murphy, D., and Ho, M. Y. (1995). Oxytocin transgenic mice. *Adv. Exp. Med. Biol.* 395, 67–78.
267. Schmitz, E., Mohr, E., and Richter, D. (1991). Rat vasopressin and oxytocin genes are linked by a long interspersed repeated DNA element (LINE): sequence and transcriptional analysis of LINE. *DNA Cell Biol.* 10, 81–91.
268. Urano, A., Hyodo, S., and Suzuki, M. (1992). Molecular evolution of neurohypophysial hormone precursors. *Prog. Brain Res.* 92, 39–46.
269. Capra, J. D., Kehoe, J. M., Kotelchuck, D., Walter, R., and Breslow, E. (1972). Evolution of neurophysin proteins: the partial sequence of bovine neurophysin-I (vasopressin-oxytocin-carrier proteins-automated amino-acid-sequence analysis-homology-protein evolution). *Proc. Natl. Acad. Sci. U S A* 69, 431–434.
270. Burman, S., Wellner, D., Chait, B., Chaudhary, T., and Breslow, E. (1989). Complete assignment of neurophysin disulfides indicates pairing in two separate domains. *Proc. Natl. Acad. Sci. U S A* 86, 429–433.
271. Schally, A. V., Guoth, J. G., and Redding, T. W., et al. (1991). Isolation and characterization of two peptides with prolactin release-inhibiting activity from porcine hypothalami. *Proc. Natl. Acad. Sci. U S A* 88, 3540–3544.
272. Foss, I., Sletten, K., and Trygstad, O. (1973). Studies on the primary structure and biological activity of a human neurophysin. *FEBS Lett.* 30, 151–156.
273. Burbach, J. P. (2002). Regulation of gene promoters of hypothalamic peptides. *Front. Neuroendocrinol.* 23, 342–369.
274. Fields, R. L., House, S. B., and Gainer, H. (2003). Regulatory domains in the intergenic region of the oxytocin and vasopressin genes that control their hypothalamus-specific expression in vitro. *J. Neurosci.* 23, 7801–7809.
275. Ho, M. Y., Carter, D. A., Ang, H. L., and Murphy, D. (1995). Bovine oxytocin transgenes in mice. Hypothalamic expression, physiological regulation, and interactions with the vasopressin gene. *J. Biol. Chem.* 270, 27199–27205.
276. Murphy, D., and Wells, S. (2003). In vivo gene transfer studies on the regulation and function of the vasopressin and oxytocin genes. *J. Neuroendocrinol.* 15, 109–125.
277. Robinson, A. G., Haluszczak, C., Wilkins, J. A., Huellmantel, A. B., and Watson, C. G. (1977). Physiologic control of two neurophysins in humans. *J. Clin. Endocrinol. Metab.* 44, 330–339.
278. Robinson, A. G. (1974). Elevation of plasma neurophysin in women on oral contraceptives. *J. Clin. Invest.* 54, 209–212.
279. Lopes, D. S., and Burbach, J. P. (1995). The nuclear hormone-receptor family in the brain: classics and orphans. *Trends Neurosci.* 18, 542–548.
280. Adan, R. A., Cox, J. J., Beischlag, T. V., and Burbach, J. P. (1993). A composite hormone response element mediates the transactivation of the rat oxytocin gene by different

classes of nuclear hormone receptors. *Mol. Endocrinol.* 7, 17–57.

281. Burbach, J. P., van Schaick, H., Lopes, D. S., Asbreuk, C. H., and Smidt, M. P. (1998). Hypothalamic transcription factors and the regulation of the hypothalamo-neurohypophysial system. *Adv. Exp. Med. Biol.* 449, 29–37.
282. Richard, S., and Zingg, H. H. (1991). Identification of a retinoic acid response element in the human oxytocin promoter. *J. Biol. Chem.* 266, 21428–21433.
283. Richard, S., and Zingg, H. H. (1990). The human oxytocin gene promoter is regulated by estrogens. *J. Biol. Chem.* 265, 6098–6103.
284. Mosselman, S., Polman, J., and Dijkema, R. (1996). ER beta: identification and characterization of a novel human estrogen receptor. *FEBS Lett.* 392, 49–53.
285. Burbach, J. P., and Adan, R. A. (1993). The rat oxytocin gene. Physiological changes in expression in the hypothalamo-neurohypophysial system and responsiveness of promoter activity. *Ann. N Y Acad. Sci.* 689, 34–49.
286. Lipkin, S. M., Nelson, C. A., Glass, C. K., and Rosenfeld, M. G. (1992). A negative retinoic acid response element in the rat oxytocin promoter restricts transcriptional stimulation by heterologous transactivation domains. *Proc. Natl. Acad. Sci. U S A* 89, 1209–1213.
287. Loven, M. A., Wood, J. R., and Nardulli, A. M. (2001). Interaction of estrogen receptors alpha and beta with estrogen response elements. *Mol. Cell Endocrinol.* 181, 151–163.
288. Loven, M. A., Likhite, V. S., Choi, I., and Nardulli, A. M. (2001). Estrogen response elements alter coactivator recruitment through allosteric modulation of estrogen receptor beta conformation. *J. Biol. Chem.* 276, 45282–45288.
289. Vasudevan, N., Koibuchi, N., Chin, W. W., and Pfaff, D. W. (2001). Differential crosstalk between estrogen receptor (ER)alpha and ERbeta and the thyroid hormone receptor isoforms results in flexible regulation of the consensus ERE. *Brain Res. Mol. Brain Res.* 95, 9–17.
290. Ninomiya, Y., Okada, M., and Kotomura, N., et al. (1995). Genomic organization and isoforms of the mouse ELP gene. *J. Biochem. (Tokyo)* 118, 380–389.
291. Ikeda, Y., Luo, X., Abbud, R., Nilson, J. H., and Parker, K. L. (1995). The nuclear receptor steroidogenic factor 1 is essential for the formation of the ventromedial hypothalamic nucleus. *Mol. Endocrinol.* 9, 478–486.
292. Luo, X., Ikeda, Y., Schlosser, D. A., and Parker, K. L. (1995). Steroidogenic factor 1 is the essential transcript of the mouse Ftz-F1 gene. *Mol. Endocrinol.* 9, 1233–1239.
293. Chu, K., and Zingg, H. H. (1999). Activation of the mouse oxytocin promoter by the orphan receptor RORalpha. *J. Mol. Endocrinol.* 23, 337–346.
294. Burbach, J. P., Lopes, D. S., and Cox, J. J., et al. (1994). Repression of estrogen-dependent stimulation of the oxytocin gene by chicken ovalbumin upstream promoter transcription factor I. *J. Biol. Chem.* 269, 15046–15053.
295. Shughrue, P. J., Komm, B., and Merchenthaler, I. (1996). The distribution of estrogen receptor-beta mRNA in the rat hypothalamus. *Steroids* 61, 678–681.
296. Choleris, E., Gustafsson, J. A., and Korach, K. S., et al. (2003). An estrogen-dependent four-gene micronet regulating social recognition: a study with oxytocin and estrogen receptor-alpha and -beta knockout mice. *Proc. Natl. Acad. Sci. U S A* 100, 6192–6197.
297. Hrabovszky, E., Kallo, I., and Steinhauser, A., et al. (2004). Estrogen receptor-beta in oxytocin and vasopressin neurons of the rat and human hypothalamus: immunocytochemical and in situ hybridization studies. *J. Comp. Neurol.* 473, 315–333.
298. Isgor, C., Shieh, K. R., Akil, H., and Watson, S. J. (2003). Colocalization of estrogen beta-receptor messenger RNA with orphanin FQ, vasopressin and oxytocin in the rat hypothalamic paraventricular and supraoptic nuclei. *Anat. Embryol. (Berl).* 206, 461–469.
299. Patisaul, H. B., Scordalakes, E. M., Young, L. J., and Rissman, E. F. (2003). Oxytocin, but not oxytocin receptor, is regulated by oestrogen receptor beta in the female mouse hypothalamus. *J. Neuroendocrinol.* 15, 787–793.
300. Bradley, D. J., Young, W. S., and Weinberger, C. (1989). Differential expression of alpha and beta thyroid hormone receptor genes in rat brain and pituitary. *Proc. Natl. Acad. Sci. U S A* 86, 7250–7254.
301. Swanson, L. W. (1992). Spatiotemporal patterns of transcription factor gene expression accompanying the development and plasticity of cell phenotypes in the neuroendocrine system. *Prog. Brain Res.* 92, 97–113.
302. Adan, R. A., and Burbach, J. P. (1992). Regulation of vasopressin and oxytocin gene expression by estrogen and thyroid hormone. *Prog. Brain Res.* 92, 127–136.
303. Dellovade, T. L., Zhu, Y. S., and Pfaff, D. W. (1999). Thyroid hormones and estrogen affect oxytocin gene expression in hypothalamic neurons. *J. Neuroendocrinol.* 11, 1–10.
304. He, X., Treacy, M. N., and Simmons, D. M., et al. (1989). Expression of a large family of POU-domain regulatory genes in mammalian brain development [published erratum appears in *Nature* 1989;340(6235):662]. *Nature* 340, 35–41.
305. Carrazana, E. J., Pasieka, K. B., and Majzoub, J. A. (1988). The vasopressin mRNA poly(A) tract is unusually long and increases during stimulation of vasopressin gene expression in vivo. *Mol. Cell Biol.* 8, 2267–2274.
306. Carter, D. A., and Murphy, D. (1991). Rapid changes in poly (A) tail length of vasopressin and oxytocin mRNAs form a common early component of neurohypophyseal peptide gene activation following physiological stimulation. *Neuroendocrinology* 53, 1–6.
307. Chaturvedi, C. M., Zheng, Z., Shimada, K., Cornett, L. E., and Koike, T. I. (1996). Changes in poly(A) tail length of arginine vasotocin messenger ribonucleic acid in the hypothalamus of water-deprived chickens. *Gen. Comp. Endocrinol.* 103, 316–322.
308. Murphy, D., and Carter, D. (1990). vasopressin gene expression in the rodent hypothalamus: transcriptional and post-transcriptional responses to physiological stimulation. *Mol. Endocrinol.* 4, 1051–1059.
309. Ang, H. L., Carter, D. A., and Murphy, D. (1993). Neuron-specific expression and physiological regulation of bovine vasopressin transgenes in mice. *EMBO J.* 12, 2397–2409.
310. Carter, D. A., and Murphy, D. (1989). Diurnal rhythm of vasopressin mRNA species in the rat suprachiasmatic nucleus: independence of neuroendocrine modulation and maintenance in explant culture. *Brain Res. Mol. Brain Res.* 6, 233–239.
311. Carter, D. A., and Murphy, D. (1993). Regulation of vasopressin (VP) gene expression in the bed nucleus of the stria terminalis: Gonadal Steroid-dependent changes in VP mRNA accumulation are associated with alterations in mRNA poly (A) tail length but are independent of the rate of VP gene transcription. *J. Neuroendocrinol.* 5, 509–515.
312. Waller, S. J., Carter, D. A., and Ang, H. L., et al. (1993). Regulation of the extent of polyadenylation of vasopressin and growth hormone mRNAs in response to physiological stimuli. *Regul. Pept.* 45, 37–41.
313. Emanuel, R. L., Girard, D. M., Thull, D. L., and Majzoub, J. A. (1992). Regulated expression of vasopressin gene by cAMP and phorbol ester in primary rat fetal hypothalamic cultures. *Mol. Cell Endocrinol.* 86, 29–36.

314. Breslow, E., and Burman, S. (1990). Molecular, thermodynamic, and biological aspects of recognition and function in neurophysin-hormone systems: a model system for the analysis of protein-peptide interactions. *Adv. Enzymol. Relat. Areas Mol. Biol.* 63, 1–67.
315. Rose, J. P., Wu, C. K., Hsiao, C. D., Breslow, E., and Wang, B. C. (1996). Crystal structure of the neurophysin-oxytocin complex. *Nat. Struct. Biol.* 3, 163–169.
316. Gainer, H., and Wray, S. (1994). The cellular and molecular biology of oxytocin and vasopressin. In *The Physiology of Reproduction* (E. Knobil and J. D. Neill, Eds.), pp. 1099–1129. Raven Press, New York.
317. de Bree, F. M., and Burbach, J. P. (1998). Structure-function relationships of the vasopressin prohormone domains. *Cell Mol. Neurobiol.* 18, 173–191.
318. Burbach, J. P., van Schaick, H. S., de Bree, F. M., Lopes, D. S., and Adan, R. A. (1995). Functional domains in the oxytocin gene for regulation of expression and biosynthesis of gene products. *Adv. Exp. Med. Biol.* 395, 9–21.
319. de Bree, F. M., van der Kleij, A. A., and Nijenhuis, M., et al. (2003). The hormone domain of the vasopressin prohormone is required for the correct prohormone trafficking through the secretory pathway. *J. Neuroendocrinol.* 15, 1156–1163.
320. Christensen, J. H., Siggaard, C., and Corydon, T. J., et al. (2004). Impaired trafficking of mutated AVP prohormone in cells expressing rare disease genes causing autosomal dominant familial neurohypophyseal diabetes insipidus. *Clin. Endocrinol. (Oxf).* 60, 125–136.
321. Christensen, J. H., Siggaard, C., and Rittig, S. (2003). Autosomal dominant familial neurohypophyseal diabetes insipidus. *APMIS Suppl.* 92–95.
322. Nijenhuis, M., van den Akker, E. L., and Zalm, R., et al. (2001). Familial neurohypophysial diabetes insipidus in a large Dutch kindred: effect of the onset of diabetes on growth in children and cell biological defects of the mutant vasopressin prohormone. *J. Clin. Endocrinol. Metab.* 86, 3410–3420.
323. Friberg, M. A., Spiess, M., and Rutishauser, J. (2004). Degradation of wild-type vasopressin precursor and pathogenic mutants by the proteasome. *J. Biol. Chem.* 279, 19441–19447.
324. Gainer, H., and Chin, H. (1998). Molecular diversity in neurosecretion: reflections on the hypothalamo-neurohypophysial system. *Cell Mol. Neurobiol.* 18, 211–230.
325. Rouille, Y., Duguay, S. J., and Lund, K., et al. (1995). Proteolytic processing mechanisms in the biosynthesis of neuroendocrine peptides: the subtilisin-like proprotein convertases. *Front Neuroendocrinol.* 16, 322–361.
326. Seidah, N. G., and Chretien, M. (1992). Pro-protein and pro-hormone convertases of the subtilsin family. Recent developments and future perspectives. *Trends Endocrinol. Metab.* 3, 133–140.
327. Seidah, N. G., Chretien, M., and Day, R. (1994). The family of subtilisin/kexin like pro-protein and pro-hormone convertases: divergent or shared functions. *Biochimie* 76, 197–209.
328. Seidah, N. G., Day, R., Marcinkiewicz, M., and Chretien, M. (1993). Mammalian paired basic amino acid convertases of prohormones and proproteins. *Ann. N Y Acad. Sci.* 680, 135–146.
329. Steiner, D. F., Smeekens, S. P., Ohagi, S., and Chan, S. J. (1992). The new enzymology of precursor processing endoproteases. *J. Biol. Chem.* 267, 23435–23438.
330. Birch, N. P., Hakes, D. J., Dixon, J. E., and Mezey, E. (1994). Distribution and regulation of the candidate prohormone processing enzymes SPC2 and SPC3 in adult rat brain. *Neuropeptides* 27, 307–322.
331. Day, R., Schafer, M. K., and Cullinan, W. E., et al. (1993). Region specific expression of furin mRNA in the rat brain. *Neurosci. Lett.* 149, 27–30.
332. Day, R., Schafer, M. K., Watson, S. J., Chretien, M., and Seidah, N. G. (1992). Distribution and regulation of the prohormone convertases PC1 and PC2 in the rat pituitary. *Mol. Endocrinol.* 6, 485–497.
333. Dong, W., Marcinkiewicz, M., and Vieau, D., et al. (1995). Distinct mRNA expression of the highly homologous convertases PC5 and PACE4 in the rat brain and pituitary. *J. Neurosci.* 15, 1778–1796.
334. Birch, N. P., Rodriguez, C., Dixon, J. E., and Mezey, E. (1990). Distribution of carboxypeptidase H messenger RNA in rat brain using in situ hybridization histochemistry: implications for neuropeptide biosynthesis. *Brain Res. Mol. Brain Res.* 7, 53–59.
335. Fricker, L. D. (1988). Carboxypeptidase E. *Annu. Rev. Physiol.* 50, 309–321.
336. Fricker, L. D. (1991). *Peptide Biosynthesis and Processing.* CRC Press, Boca Raton, FL.
337. Lynch, D. R., Braas, K. M., Hutton, J. C., and Snyder, S. H. (1990). Carboxypeptidase E (CPE): immunocytochemical localization in the rat central nervous system and pituitary gland. *J. Neurosci.* 10, 1592–1599.
338. MacCumber, M. W., Snyder, S. H., and Ross, C. A. (1990). Carboxypeptidase E (enkephalin convertase): mRNA distribution in rat brain by in situ hybridization. *J. Neurosci.* 10, 2850–2860.
339. Braas, K. M., Harakall, S. A., Ouafik, L., Eipper, B. A., and May, V. (1992). Expression of peptidylglycine alpha-amidating monooxygenase: an in situ hybridization and immunocytochemical study. *Endocrinology* 130, 2778–2788.
340. Grino, M., Guillaume, V., and Boudouresque, F., et al. (1990). Glucocorticoids regulate peptidyl-glycine alpha-amidating monooxygenase gene expression in the rat hypothalamic paraventricular nucleus. *Mol. Endocrinol.* 4, 1613–1619.
341. Lew, R. A., Clarke, I. J., and Smith, A. I. (1992). Distribution and characterization of peptidylglycine alpha-amidating monooxygenase activity in the ovine brain and hypothalamo–pituitary axis. *Endocrinology* 130, 994–1000.
342. May, V., Ouafik, L., Eipper, B. A., and Braas, K. M. (1990). Immunocytochemical and in situ hybridization studies of peptidylglycine alpha-amidating monooxygenase in pituitary gland. *Endocrinology* 127, 358–364.
343. Schafer, M. K., Stoffers, D. A., Eipper, B. A., and Watson, S. J. (1992). Expression of peptidylglycine alpha-amidating monooxygenase (EC 1.14.17.3) in the rat central nervous system. *J. Neurosci.* 12, 222–234.
344. Morris, M., Castro, M., and Rose, J. C. (1992). Alterations in oxytocin prohormone processing during early development in the fetal sheep. *Am. J. Physiol.* 263, R738–R740.
345. Amico, J. A., Ervin, M. G., and Leake, R. D., et al. (1985). A novel oxytocin-like and vasotocin-like peptide in human plasma after administration of estrogen. *J. Clin. Endocrinol. Metab.* 60, 5–12.
346. Amico, J. A., and Hempel, J. (1990). An oxytocin precursor intermediate circulates in the plasma of humans and rhesus monkeys administered estrogen. *Neuroendocrinology* 51, 437–443.
347. Liu, B., Poulter, L., Neacsu, C., and Burbach, J. P. (1988). Isolation and identification of vasopressin- and oxytocin-immunoreactive substances from bovine pineal gland. Presence of N alpha-acetyloxytocin. *J. Biol. Chem.* 263, 72–75.
348. Lopes, D. S., de Bree, F. M., Evans, D. A., van Leeuwen, F. W., and Burbach, J. P. (1993). Structure and expression of the vasopressin gene: analysis of mutations, novel genes, and gene products. *Ann. N Y Acad. Sci.* 689, 492–503.
349. Castel, M., Gainer, H., and Dellmann, H. D. (1984). Neuronal secretory systems. *Int. Rev. Cytol.* 88, 303–459.

350. Morris, J. F., Nordmann, J. J., and Dyball, R. E. (1978). Structure-function correlation in mammalian neurosecretion. *Int. Rev. Exp. Pathol.* 18, 1–95.
351. Ludwig, M. (1998). Dendritic release of vasopressin and oxytocin. *J. Neuroendocrinol.* 10, 881–895.
352. Pow, D. V., and Morris, J. F. (1989). Dendrites of hypothalamic magnocellular neurons release neurohypophysial peptides by exocytosis. *Neuroscience* 32, 435–439.
353. Neumann, I., Russell, J. A., and Landgraf, R. (1993). Oxytocin and vasopressin release within the supraoptic and paraventricular nuclei of pregnant, parturient and lactating rats: a microdialysis study. *Neuroscience* 53, 65–75.
354. Cobo, E., and Quintero, C. A. (1969). Milk-ejecting and antidiuretic activities under neurohypophyseal inhibition with alcohol and water overload. *Am. J. Obstet. Gynecol.* 105, 877–887.
355. Jones, C. W., and Pickering, B. T. (1969). Changes in the hormone content of the rat neurohypophysis induced by substituting 2 per cent saline for drinking water. *J. Endocrinol.* 43, vi.
356. Young, T. K., and Van Dyke, H. B. (1968). Repletion of vasopressin and oxytocin in the posterior lobe of the pituitary gland of the rat. *J. Endocrinol.* 40, 337–342.
357. Robinson, A., G., Roberts, M. M., Evron, W. A., Verbalis, J. G., and Sherman, T. G. (1990). Hyponatremia in rats induces downregulation of vasopressin synthesis. *J. Clin. Invest.* 86, 1023–1029.
358. Carter, D. A., and Murphy, D. (1989). Independent regulation of neuropeptide mRNA level and poly(A) tail length. *J. Biol. Chem.* 264, 6601–6603.
359. Lightman, S. L., and Young, W. S. III. (1987). Vasopressin, oxytocin, dynorphin, enkephalin and corticotrophin-releasing factor mRNA stimulation in the rat. *J. Physiol. (Lond).* 394, 23–39.
360. van Tol, H. H., Voorhuis, D. T., and Burbach, J. P. (1987). Oxytocin gene expression in discrete hypothalamic magnocellular cell groups is stimulated by prolonged salt loading. *Endocrinology* 120, 71–76.
361. Herman, J. P., Schafer, M. K., Watson, S. J., and Sherman, T. G. (1991). In situ hybridization analysis of arginine vasopressin gene transcription using intron-specific probes. *Mol. Endocrinol.* 5, 1447–1456.
362. Burbach, J. P., De Hoop, M. J., and Schmale, H., et al. (1984). Differential responses to osmotic stress of vasopressin-neurophysin mRNA in hypothalamic nuclei. *Neuroendocrinology* 39, 582–584.
363. Sherman, T. G., Civelli, O., Douglass, J., Herbert, E., and Watson, S. J. (1986). Coordinate expression of hypothalamic pro-dynorphin and pro-vasopressin mRNAs with osmotic stimulation. *Neuroendocrinology* 44, 222–228.
364. Sherman, T. G., and Watson, S. J. (1988). Differential expression of vasopressin alleles in the Brattleboro heterozygote. *J. Neurosci.* 8, 3797–3811.
365. Sherman, T. G., Day, R., and Civelli, O., et al. (1988). Regulation of hypothalamic magnocellular neuropeptides and their mRNAs in the Brattleboro rat: coordinate responses to further osmotic challenge. *J. Neurosci.* 8, 3785–3796.
366. Zingg, H. H., Lefebvre, D., Almazan, G. (1986). Regulation of vasopressin gene expression in rat hypothalamic neurons. Response to osmotic stimulation. *J. Biol. Chem.* 261, 12956–12959.
367. Voisin, D. L., Simonian, S. X., and Herbison, A. E. (1997). Identification of estrogen receptor-containing neurons projecting to the rat supraoptic nucleus. *Neuroscience* 78, 215–228.
368. Swaab, D. F., Hofman, M. A., Lucassen, P. J., Purba, J. S., Raadsheer, F. C., and Van de Nes, J. A. (1993). Functional neuroanatomy and neuropathology of the human hypothalamus. *Anat. Embryol. (Berl).* 187, 317–330.
369. Akaishi, T., and Sakuma, Y. (1985). Estrogen excites oxytocinergic, but not vasopressinergic cells in the paraventricular nucleus of female rat hypothalamus. *Brain Res.* 335, 302–305.
370. Negoro, H., Vissessuwan, S., and Holland, R. C. (1973). Reflex activation of paraventricular nucleus units during the reproductive cycle and in ovariectomized rats treated with oestrogen and progesterone. *J. Endocrinol.* 59, 545–567.
371. Amico, J. A., Seif, S. M., and Robinson, A. G. (1981). Oxytocin in human plasma: correlation with neurophysin and stimulation with estrogen. *J. Clin. Endocrinol. Metab.* 52, 988–993.
372. Skowsky, W. R., Swan, L., and Smith, P. (1979). Effects of sex steroid hormones on arginine vasopressin in intact and castrated male and female rats. *Endocrinology* 104, 105–108.
373. Yamaguchi, K., Akaishi, T., and Negoro, H. (1979). Effect of estrogen treatment on plasma oxytocin and vasopressin in ovariectomized rats. *Endocrinol. Jpn.* 26, 197–205.
374. Wang, H., Ward, A. R., and Morris, J. F. (1995). Oestradiol acutely stimulates exocytosis of oxytocin and vasopressin from dendrites and somata of hypothalamic magnocellular neurons. *Neuroscience* 68, 1179–1188.
375. Chibbar, R., Toma, J. G., Mitchell, B. F., and Miller, F. D. (1990). Regulation of neural oxytocin gene expression by gonadal steroids in pubertal rats. *Mol. Endocrinol.* 4, 2030–2038.
376. Miller, F. D., Ozimek, G., Milner, R. J., and Bloom, F. E. (1989). Regulation of neuronal oxytocin mRNA by ovarian steroids in the mature and developing hypothalamus. *Proc. Natl. Acad. Sci. U S A* 86, 2468–2472.
377. Madeira, M. D., Sousa, N., Cadete-Leite, A., Lieberman, A. R., and Paula-Barbosa, M. M. (1993). The supraoptic nucleus of the adult rat hypothalamus displays marked sexual dimorphism which is dependent on body weight. *Neuroscience* 52, 497–513.
378. Forsling, M. L., and Peysner, K. (1988). Pituitary and plasma vasopressin concentrations and fluid balance throughout the oestrous cycle of the rat. *J. Endocrinol.* 117, 397–402.
379. Sarkar, D. K., and Gibbs, D. M. (1984). Cyclic variation of oxytocin in the blood of pituitary portal vessels of rats. *Neuroendocrinology* 39, 481–483.
380. Forsling, M. L., Akerlund, M., and Stromberg, P. (1981). Variations in plasma concentrations of vasopressin during the menstrual cycle. *J. Endocrinol.* 89, 263–266.
381. Greer, E. R., Caldwell, J. D., Johnson, M. F., Prange, A. J. J., and Pedersen, C. A. (1986). Variations in concentration of oxytocin and vasopressin in the paraventricular nucleus of the hypothalamus during the estrous cycle in rats. *Life Sci.* 38, 2311–2318.
382. Burbach, J. P., Adan, R. A., and van Tol, H. H., et al. (1990). Regulation of the rat oxytocin gene by estradiol. *J. Neuroendocrinol.* 2, 633–639.
383. Chung, S. K., McCabe, J. T., and Pfaff, D. W. (1991). Estrogen influences on oxytocin mRNA expression in preoptic and anterior hypothalamic regions studied by in situ hybridization. *J. Comp. Neurol.* 307, 281–295.
384. Kuiper, G. G., Enmark, E., Pelto-Huikko, M., Nilsson, S., and Gustafsson, J. A. (1996). Cloning of a novel receptor expressed in rat prostate and ovary. *Proc. Natl. Acad. Sci. U S A* 93, 5925–5930.
385. Caldwell, J. D., Greer, E. R., Johnson, M. F., Prange, A. J. J., and Pedersen, C. A. (1987). Oxytocin and vasopressin immunoreactivity in hypothalamic and extrahypothalamic sites in late pregnant and postpartum rats. *Neuroendocrinology* 46, 39–47.

386. Douglas, A. J., Dye, S., Leng, G., Russell, J. A., and Bicknell, R. J. (1993). Endogenous opioid regulation of oxytocin secretion through pregnancy in the rat. *J. Neuroendocrinol.* 5, 307–314.
387. Fuchs, A. R. (1985). Oxytocin in animal parturition. In *Oxytocin: Clinical and Laboratory Studies* (J. Amico and A. G. Robinson, Eds.), pp. 207–235. Excerpta Medica, Amsterdam.
388. Fuchs, A. R., and Saito, S. (1971). Pituitary oxytocin and vasopressin content of pregnant rats before, during and after parturition. *Endocrinology* 88, 574–578.
389. Kumaresan, P., Subramanian, M., Anandarangam, P. B., and Kumaresan, M. (1979). Radioimmunoassay of plasma and pituitary oxytocin in pregnant rats during various stages of pregnancy and parturition. *J. Endocrinol. Invest.* 2, 65–70.
390. Schriefer, J. A. (1991). Diethylstilbesterol- and pregnancy-induced changes in rat neurointermediate lobe oxytocin, arginine vasopressin, methionine enkephalin and dynorphin. *Neuroendocrinology* 54, 185–191.
391. Crowley, R. S., Insel, T. R., O'Keefe, J. A., and Amico, J. A. (1993). Cytoplasmic oxytocin and vasopressin gene transcripts decline postpartum in the hypothalamus of the lactating rat. *Endocrinology* 133, 2704–2710.
392. Zingg, H. H., and Lefebvre, D. L. (1988). Oxytocin and vasopressin gene expression during gestation and lactation. *Brain Res.* 464, 1–6.
393. Zingg, H. H., and Lefebvre, D. L. (1989). Oxytocin mRNA: increase of polyadenylate tail size during pregnancy and lactation. *Mol. Cell Endocrinol.* 65, 59–62.
394. Hartman, R. D., Rosella-Dampman, L. M., Emmert, S. E., and Summy-Long, J. Y. (1986). Inhibition of release of neurohypophysial hormones by endogenous opioid peptides in pregnant and parturient rats. *Brain Res.* 382, 352–359.
395. Fenelon, V. S., Poulain, D. A., and Theodosis, D. T. (1993). Oxytocin neuron activation and Fos expression: a quantitative immunocytochemical analysis of the effect of lactation, parturition, osmotic and cardiovascular stimulation. *Neuroscience* 53, 77–89.
396. Luckman, S. M., Antonijevic, I., and Leng, G., et al. (1993). The maintenance of normal parturition in the rat requires neurohypophysial oxytocin. *J. Neuroendocrinol.* 5, 7–12.
397. Luckman, S. M., Dye, S., and Cox, H. J. (1996). Induction of members of the Fos/Jun family of immediate-early genes in identified hypothalamic neurons: in vivo evidence for differential regulation. *Neuroscience* 73, 473–485.
398. Summerlee, A. J. (1981). Extracellular recordings from oxytocin neurones during the expulsive phase of birth in unanaesthetized rats. *J. Physiol. (Lond).* 321, 1–9.
399. Lefebvre, D. L., Lariviere, R., and Zingg, H. H. (1993). Rat amnion: a novel site of oxytocin production. *Biol. Reprod.* 48, 632–639.
400. Boulton, M. I., McGrath, T. J., Goode, J. A., Broad, K. D., and Gilbert, C. L. (1996). Changes in content of mRNA encoding oxytocin in the pig uterus during the oestrous cycle, pregnancy, at parturition and in lactational anoestrus. *J. Reprod. Fertil.* 108, 219–227.
401. Ivell, R., Rust, W., and Einspanier, A., et al. (1995). Oxytocin and oxytocin receptor gene expression in the reproductive tract of the pregnant cow: rescue of luteal oxytocin production at term. *Biol. Reprod.* 53, 553–560.
402. Murphy, D., and Ho, M. Y. (1995). Oxytocin transgenic mice. *Adv. Exp. Med. Biol.* 395, 67–78.
403. Wathes, D. C., Smith, H. F., and Leung, S. T., et al. (1996). Oxytocin receptor development in ovine uterus and cervix throughout pregnancy and at parturition as determined by in situ hybridization analysis. *J. Reprod. Fertil.* 106, 23–31.
404. Crowley, R. S., and Amico, J. A. (1993). Gonadal steroid modulation of oxytocin and vasopressin gene expression in the hypothalamus of the osmotically stimulated rat. *Endocrinology* 133, 2711–2718.
405. Crowley, R. S., Insel, T. R., O'Keefe, J. A., Kim, N. B., and Amico, J. A. (1995). Increased accumulation of oxytocin messenger ribonucleic acid in the hypothalamus of the female rat: induction by long term estradiol and progesterone administration and subsequent progesterone withdrawal. *Endocrinology* 136, 224–231.
406. Bridges, R. S. (1984). A quantitative analysis of the roles of dosage, sequence, and duration of estradiol and progesterone exposure in the regulation of maternal behavior in the rat. *Endocrinology* 114, 930–940.
407. Antonijevic, I., Bicknell, R.J., Leng, G., and Russell, J. A. (1995). Progesterone suppression of activation of supraoptic nucleus (SON) neurones in late pregnancy in the rat. *Adv. Exp. Med. Biol.* 395, 195–196.
408. Fuchs, A. R., Romero, R., and Keefe, D., et al. (1991). Oxytocin secretion and human parturition: pulse frequency and duration increase during spontaneous labor in women. *Am. J. Obstet. Gynecol.* 165, 1515–1523.
409. Gilbert, C. L., Goode, J. A., and McGrath, T. J. (1994). Pulsatile secretion of oxytocin during parturition in the pig: temporal relationship with fetal expulsion. *J. Physiol. (Lond).* 475, 129–137.
410. Higuchi, T., Tadokoro, Y., Honda, K., and Negoro, H. (1986). Detailed analysis of blood oxytocin levels during suckling and parturition in the rat. *J. Endocrinol.* 110, 251–256.
411. Antonijevic, I. A., Leng, G., and Luckman, S. M., et al. (1995). Induction of uterine activity with oxytocin in late pregnant rats replicates the expression of c-fos in neuroendocrine and brain stem neurons as seen during parturition. *Endocrinology* 136, 154–163.
412. Fuchs, A. R., and Dawood, M. Y. (1980). Oxytocin release and uterine activation during parturition in rabbits. *Endocrinology* 107, 1117–1126.
413. Higuchi, T., Uchide, K., Honda, K., and Negoro, H. (1987). Pelvic neurectomy abolishes the fetus-expulsion reflex and induces dystocia in the rat. *Exp. Neurol.* 96, 443–455.
414. Randolph, G. W., and Fuchs, A. R. (1989). Pulsatile administration enhances the effect and reduces the dose of oxytocin required for induction of labor. *Am. J. Perinatol.* 6, 159–166.
415. Spinolo, L. H., Raghow, R., and Crowley, W. R. (1992). Oxytocin messenger RNA levels in hypothalamic, paraventricular, and supraoptic nuclei during pregnancy and lactation in rats. Evidence for regulation by afferent stimuli from the offspring. *Ann. N Y Acad. Sci.* 652, 425–428.
416. Moos, F., Poulain, D. A., and Rodriguez, F., et al. (1989). Release of oxytocin within the supraoptic nucleus during the milk ejection reflex in rats. *Exp. Brain Res.* 76, 593–602.
417. Smith, M. S., and Neill, J. D. (1977). Inhibition of gonadotropin secretion during lactation in the rat: relative contribution of suckling and ovarian steroids. *Biol. Reprod.* 17, 255–261.
418. Lee, L. R., Haisenleder, D. J., Marshall, J. C., and Smith, M. S. (1989). Expression of alpha-subunit and luteinizing hormone (LH) beta messenger ribonucleic acid in the rat during lactation and after pup removal: relationship to pituitary gonadotropin-releasing hormone receptors and pulsatile LH secretion. *Endocrinology* 124, 776–782.
419. Thomas, A., Crowley, R. S., and Amico, J. A. (1995). Effect of progesterone on hypothalamic oxytocin messenger ribonucleic acid levels in the lactating rat. *Endocrinology* 136, 4188–4194.

420. Poulain, D. A., and Wakerley, J. B. (1982). Electrophysiology of hypothalamic magnocellular neurones secreting oxytocin and vasopressin. *Neuroscience* 7, 773–808.
421. Belin, V., and Moos, F. (1986). Paired recordings from supraoptic and paraventricular oxytocin cells in suckled rats: recruitment and synchronization. *J. Physiol. (Lond).* 377, 369–390.
422. Hatton, G. I. (1997). Function-related plasticity in hypothalamus. *Annu. Rev. Neurosci.* 20, 375–397.
423. Russell, J. A., and Leng, G. (1998). Sex, parturition and motherhood without oxytocin? *J. Endocrinol.* 157, 343–359.
424. Tsatsaris, V., Carbonne, B., and Cabrol, D. (2004). Atosiban for preterm labor. *Drugs* 64, 375–382.
425. Schellenberg, J. C. (2002). Oxytocin in parturition of guinea pigs, humans, and other species. *Prog. Brain Res.* 139, 299–308.
426. Antonijevic, I. A., Douglas, A. J., and Dye, S., et al. (1995). Oxytocin antagonists delay the initiation of parturition and prolong its active phase in rats. *J. Endocrinol.* 145, 97–103.
427. Villar, J., Gulmezoglu, A. M., Hofmeyr, G. J., and Forna, F. (2002). Systematic review of randomized controlled trials of misoprostol to prevent postpartum hemorrhage. *Obstet. Gynecol.* 100, 1301–1312.
428. Lawrence, A. B., Petherick, J. C., and McLean, K., et al. (1992). Naloxone prevents interruption of parturition and increases plasma oxytocin following environmental disturbance in parturient sows. *Physiol. Behav.* 52, 917–923.
429. Douglas, A. J., Leng, G., and Russell, J. A. (2002). The importance of oxytocin mechanisms in the control of mouse parturition. *Reproduction* 123, 543–552.
430. Fuchs, A. R., Ivell, R., Ganz, N., Fields, M. J., and Gimenez, T. (2001). Secretion of oxytocin in pregnant and parturient cows: corpus luteum may contribute to plasma oxytocin at term. *Biol. Reprod.* 65, 1135–1141.
431. Parry, L. J., and Bathgate, R. A. (2000). The role of oxytocin and regulation of uterine oxytocin receptors in pregnant marsupials. *Exp. Physiol.* 85(Spec No), 91S–99S.
432. Chibbar, R., Miller, F. D., and Mitchell, B. F. (1993). Synthesis of oxytocin in amnion, chorion, and decidua may influence the timing of human parturition. *J. Clin. Invest.* 91, 185–192.
433. Lefebvre, D. L., Giaid, A., and Zingg, H. H. (1992). Expression of the oxytocin gene in rat placenta. *Endocrinology* 130,1185–1192.
434. Morel, G., Pechoux, C., Raccurt, M., and Zingg, H. H. (2001). Intrauterine oxytocin system. Compartmental distribution of oxytocin and oxytocin receptors in rat endometrial epithelium. *Cell Tissue Res.* 304, 377–382.
435. Russell, J. A., Leng, G., and Douglas, A. J. (2003). The magnocellular oxytocin system, the fount of maternity: adaptations in pregnancy. *Front Neuroendocrinol.* 24, 27–61.
436. Lin, S. H., Miyata, S., and Weng, W., et al. (1998). Comparison of the expression of two immediate early gene proteins, FosB and Fos in the rat preoptic area, hypothalamus and brainstem during pregnancy, parturition and lactation. *Neurosci. Res.* 32, 333–341.
437. Douglas, A. J., Meeren, H. K., and Johnstone, L. E., et al. (1998). Stimulation of expression of the oxytocin gene in rat supraoptic neurons at parturition. *Brain Res.* 782, 167–174.
438. Neumann, I., Douglas, A. J., Pittman, Q. J., Russell, J. A., and Landgraf, R. (1996). Oxytocin released within the supraoptic nucleus of the rat brain by positive feedback action is involved in parturition-related events. *J. Neuroendocrinol.* 8, 227–233.
439. Russell, J. A., Gosden, R. G., and Humphreys, E. M., et al. (1989). Interruption of parturition in rats by morphine: a result of inhibition of oxytocin secretion. *J. Endocrinol.* 121, 521–536.
440. Russell, J. A., Leng, G., and Coombes, J. E., et al. (1991). Pethidine (meperidine) inhibition of oxytocin secretion and action in parturient rats. *Am. J. Physiol.* 261, R358–R368.
441. Douglas, A. J., Clarke, G., and MacMillan, S. J., et al. (1993). Effects of the kappa-opioid agonist U50,488 on parturition in rats. *Br. J. Pharmacol.* 109, 251–258.
442. Leng, G., Mansfield, S., and Bicknell, R. J., et al. (1988). Endogenous opioid actions and effects of environmental disturbance on parturition and oxytocin secretion in rats. *J. Reprod. Fertil.* 84, 345–356.
443. Neumann, I. D., Bosch, O. J., Toschi, N., Torner, L., and Douglas, A. J. (2003). No stress response of the hypothalamo–pituitary–adrenal axis in parturient rats: lack of involvement of brain oxytocin. *Endocrinology* 144, 2473–2479.
444. Way, S. A., and Leng, G. (1992). Relaxin increases the firing rate of supraoptic neurones and increases oxytocin secretion in the rat. *J. Endocrinol.* 132, 149–158.
445. Summerlee, A. J., O'Byrne, K. T., and Poterski, R. S. (1998). Relaxin inhibits the pulsatile release of oxytocin but increases basal concentrations of hormone in lactating rats. *Biol. Reprod.* 58, 977–981.
446. McAnulty, P. A., and Burns, L. M. (2004). Evaluation of the maternal and neonatal effects of the oxytocin antagonist, atosiban, in a cross-fostering study in rats. *Reprod. Toxicol.* 19, 65–70.
447. Akerlund, M., Bossmar, T., and Brouard, R., et al. (1999). Receptor binding of oxytocin and vasopressin antagonists and inhibitory effects on isolated myometrium from preterm and term pregnant women. *Br. J. Obstet Gynaecol.* 106, 1047–1053.
448. Lurie, S., Sadan, O., Ben Aroya, Z., and Glezerman, M. (2004). Atosiban treatment for uterine hyperactivity during active labor: a pilot study. *J. Perinat Med.* 32, 137–139.
449. Nilsson, L., Reinheimer, T., Steinwall, M., and Akerlund, M. (2003). FE 200 440: a selective oxytocin antagonist on the term-pregnant human uterus. *Br. J. Obstet. Gynaecol.* 110, 1025–1028.
450. Cirillo, R., Gillio, T. E., and Schwarz, M. K., et al. (2003). Pharmacology of (2S,4Z)-N-[(2S)-2-hydroxy-2-phenylethyl]-4-(methoxyimino)-1-[(2'-methyl[1,1'-biphenyl]-4-yl) carbonyl]-2-pyrrolidinecarboxamide, a new potent and selective nonpeptide antagonist of the oxytocin receptor. *J. Pharmacol. Exp. Ther.* 306, 253–261.
451. Serradeil-Le Gal, C., Valette, G., and Foulon, L., et al. (2004). SSR126768A (4-chloro-3-[(3R)-(+)-5-chloro-1-(2,4-dimethoxybenzyl)-3-methyl-2-oxo-2,3- dihydro-1H-indol-3-yl]-N-ethyl-N-(3-pyridylmethyl)-benzamide, hydrochloride): a new selective and orally active oxytocin receptor antagonist for the prevention of preterm labor. *J. Pharmacol. Exp. Ther.* 309, 414–424.
452. Breton C, Pechoux C, Morel G, and Zingg HH. Oxytocin receptor messenger ribonucleic acid: characterization, regulation, and cellular localization in the rat pituitary gland. *Endocrinology* 1995;136, 2928–2936.
453. Quinones-Jenab, V., Jenab, S., and Ogawa, S., et al. (1997). Effects of estrogen on oxytocin receptor messenger ribonucleic acid expression in the uterus, pituitary, and forebrain of the female rat. *Neuroendocrinology* 65, 9–17.
454. Schlosser, S. F., Almeida, O. F., Patchev, V. K., Yassouridis, A., and Elands, J. (1994). Oxytocin-stimulated release of adrenocorticotropin from the rat pituitary is mediated by arginine vasopressin receptors of the V1b type. *Endocrinology* 135, 2058–2063.
455. Samson, W. K., and Schell, D. A. (1995). Oxytocin and the anterior pituitary gland. *Adv. Exp. Med. Biol.* 395, 355–364.

456. Buma, P., and Nieuwenhuys, R. (1988). Ultrastructural characterization of exocytotic release sites in different layers of the median eminence of the rat. *Cell Tissue Res.* 252, 107–114.
457. Sheward, W. J., Coombes, J. E., Bicknell, R. J., Fink, G., and Russell, J. A. (1990). Release of oxytocin but not corticotrophin-releasing factor-41 into rat hypophysial portal vessel blood can be made opiate dependent. *J. Endocrinol.* 124, 141–150.
458. Swanson, L. W., and Sawchenko, P. E. (1983). Hypothalamic integration: organization of the paraventricular and supraoptic nuclei. *Annu. Rev. Neurosci.* 6, 269–324.
459. Zimmerman, E. A., Nilaver, G., Hou-Yu, A., and Silverman, A. J. (1984). Vasopressinergic and oxytocinergic pathways in the central nervous system. *Fed. Proc.* 43, 91–96.
460. Vecsernyes, M., Nagy, G., and Meszaros, L., et al. (2000). Suckling-induced change in oxytocin but not in alpha-MSH concentrations of the median eminence, the neural-, intermediate- and anterior lobes of the pituitary gland. *Endocr. Res.* 26, 333–345.
461. O'Byrne, K. T., Lunn, S. F., and Coen, C. W. (1990). Central oxytocin stimulates luteinizing hormone release in the marmoset, a primate which fails to show lactationally-induced fertility. *J. Neuroendocrinol.* 2, 419–421.
462. Hull, M. L., Reid, R. A., Evans, J. J., Benny, P. S., and Aickin, D. R. (1995). Pre-ovulatory oxytocin administration promotes the onset of the luteinizing hormone surge in human females. *Hum. Reprod.* 10, 2266–2269.
463. Evans, J. J., Reid, R. A., Wakeman, S. A., Croft, L. B., and Benny, P. S. (2003). Evidence that oxytocin is a physiological component of LH regulation in non-pregnant women. *Hum. Reprod.* 18, 1428–1431.
464. Evans, J. J., Pragg, F. L., and Mason, D. R. (2001). Release of luteinizing hormone from the anterior pituitary gland in vitro can be concurrently regulated by at least three peptides: gonadotropin-releasing hormone, oxytocin and neuropeptide Y. *Neuroendocrinology* 73, 408–416.
465. Bogacka, I., Siawrys, G., Okrasa, S., Kaminski, T., and Przala, J. (2002). The influences of GnRH, oxytocin and vasoactive intestinal peptide on LH and PRL secretion by porcine pituitary cells in vitro. *J. Physiol. Pharmacol.* 53, 439–451.
466. Samson, W. K., Alexander, B. D., Skala, K. D., Huang, F. L., and Fulton, R. J. (1992). Ricin-cytotoxin conjugate administration reveals a physiologically relevant role for oxytocin in the control of gonadotropin secretion. *Ann. N Y Acad. Sci.* 652, 411–422.
467. Egli, M., Bertram, R., Sellix, M. T., and Freeman, M. E. (2004). Rhythmic secretion of prolactin in rats: action of oxytocin coordinated by vasoactive intestinal polypeptide of suprachiasmatic nucleus origin. *Endocrinology* 145, 3386–3394.
468. Chadio, S. E., and Antoni, F. A. (1993). Specific oxytocin agonist stimulates prolactin release but has no effect on inositol phosphate accumulation in isolated rat anterior pituitary cells. *J. Mol. Endocrinol.* 10, 107–114.
469. Samson, W. K., Lumpkin, M. D., and McCann, S. M. (1986). Evidence for a physiological role for oxytocin in the control of prolactin secretion. *Endocrinology* 119, 554–560.
470. Freeman, M. E., Kanyicska, B., Lerant, A., and Nagy, G. (2000). Prolactin: structure, function, and regulation of secretion. *Physiol. Rev.* 80, 1523–1631.
471. Mogg, R. J., and Samson, W. K. (1990). Interactions of dopaminergic and peptidergic factors in the control of prolactin release. *Endocrinology* 126, 728–735.
472. Arey, B. J., and Freeman, M. E. (1990). Oxytocin, vasoactive-intestinal peptide, and serotonin regulate the mating-induced surges of prolactin secretion in the rat. *Endocrinology* 126, 279–284.
473. Polston, E. K., Centorino, K.M., and Erskine, M. S. (1998). Diurnal fluctuations in mating-induced oxytocinergic activity within the paraventricular and supraoptic nuclei do not influence prolactin secretion. *Endocrinology* 139, 4849–4859.
474. Chiodera, P., Volpi, R., and Gramellini, D., et al. (1991). Oxytocin does not modify the prolactin response to metoclopramide in normal women. *J. Endocrinol. Invest.* 14, 463–468.
475. Li, M., Liu, X., and Robinson, G., et al. (1997). Mammary-derived signals activate programmed cell death during the first stage of mammary gland involution. *Proc. Natl. Acad. Sci. U S A* 94, 3425–3430.
476. Flint, A. P., Riley, P. R., Kaluz, S., Stewart, H. J., and Abayasekara, D. R. (1995). The sheep endometrial oxytocin receptor. *Adv. Exp. Med. Biol.* 395, 281–294.
477. McCracken, J. A., Custer, E. E., and Lamsa, J. C. (1999). Luteolysis: a neuroendocrine-mediated event. *Physiol. Rev.* 79, 263–323.
478. Asselin, E., Drolet, P., and Fortier, M. A. (1997). Cellular mechanisms involved during oxytocin-induced prostaglandin F2alpha production in endometrial epithelial cells in vitro: role of cyclooxygenase-2. *Endocrinology* 138, 4798–4805.
479. Burns, P. D., Mendes, J. O. Jr., and Yemm, R. S., et al. (2001). Cellular mechanisms by which oxytocin mediates ovine endometrial prostaglandin F2alpha synthesis: role of G(i) proteins and mitogen-activated protein kinases. *Biol. Reprod.* 65, 1150–1155.
480. Wathes, D. C., and Swann, R. W. (1982). Is oxytocin an ovarian hormone? *Nature* 297, 225–227.
481. Ivell, R. (1999). The physiology of ovarian oxytocin. *Reprod. Med. Rev.* 7, 11–25.
482. Meidan, R., Altstein, M., and Girsh, E. (1992). Biosynthesis and release of oxytocin by granulosa cells derived from preovulatory bovine follicles: effects of forskolin and insulin-like growth factor-I. *Biol. Reprod.* 46, 715–720.
483. Flint, A. P., and Sheldrick, E. L. (1982). Ovarian secretion of oxytocin is stimulated by prostaglandin. *Nature* 297, 587–588.
484. Lamsa, J. C., Kot, S. J., Eldering, J. A., Nay, M. G., and McCracken, J. A. (1989). Prostaglandin F2 alpha-stimulated release of ovarian oxytocin in the sheep in vivo: threshold and dose dependency. *Biol. Reprod.* 40, 1215–1223.
485. Ohtani, M., Takase, S., Wijayagunawardane, M. P., Tetsuka, M., and Miyamoto, A. (2004). Local interaction of prostaglandin F2alpha with endothelin-1 and tumor necrosis factor-alpha on the release of progesterone and oxytocin in ovine corpora lutea in vivo: a possible implication for a luteolytic cascade. *Reproduction* 127, 117–124.
486. Hu, J., Ludwig, T. E., Salli, U., Stormshak, F., and Mirando, M. A. (2001). Autocrine/paracrine action of oxytocin in pig endometrium. *Biol. Reprod.* 64, 1682–1688.
487. Khan-Dawood, F. S., and Dawood, M. Y. (1998). Comparative aspects of oxytocin in baboon (Papio hamadryus anubis) and human corpora lutea. *Hum. Reprod. Update* 4, 371–382.
488. Currie, W. B., and Thorburn, G. D. (1977). Parturition in goats: studies on the interactions between the foetus, placenta, prostaglandin F and progesterone before parturition, at term or at parturition induced prematurely by corticotrophin infusion of the foetus. *J. Endocrinol.* 73, 263–278.
489. Ford, M. M., Thorburn, G. D., Caddy, D. J., and Young, I. R. (1999). Pulsatile output of prostaglandin F(2alpha) does not increase around the time of luteolysis in the pregnant goat. *Biol. Reprod.* 61, 411–415.
490. Murata, T., Murata, E., and Liu, C. X., et al. (2000). Oxytocin receptor gene expression in rat uterus: regulation by ovarian steroids. *J. Endocrinol.* 166, 45–52.

491. Sugimoto, Y., Yamasaki, A., and Segi, E., et al. (1997). Failure of parturition in mice lacking the prostaglandin F receptor. *Science* 277, 681–683.
492. Sugimoto, Y., Segi, E., Tsuboi, K., Ichikawa, A., and Narumiya, S. (1998). Female reproduction in mice lacking the prostaglandin F receptor. Roles of prostaglandin and oxytocin receptors in parturition. *Adv. Exp. Med. Biol.* 449, 317–321.
493. Gross, G. A., Imamura, T., and Luedke, C., et al. (1998). Opposing actions of prostaglandins and oxytocin determine the onset of murine labor. *Proc. Natl. Acad. Sci. U S A* 95, 11875–11879.
494. Gross, G., Imamura, T., and Muglia, L. J. (2000). Gene knockout mice in the study of parturition. *J. Soc. Gynecol. Invest.* 7, 88–95.
495. Imamura, T., Luedke, C. E., Vogt, S. K., and Muglia, L. J. (2000). Oxytocin modulates the onset of murine parturition by competing ovarian and uterine effects. *Am. J. Physiol. Regul. Integr. Comp. Physiol.* 279, R1061–R1067.
496. Meddle, S. L., Bishop, V. R., Gkoumassi, E., Van Leeuwen, F. W., and Douglas, A. (2003). Parturition activates oxytocin receptor expressing cells within the brainstem of the rat. *Auton. Neurosci.* 106, 53.
497. Siebel, A. L., Gehring, H. M., and Parry, L. J. (2004) Steroid-independent regulation of uterine oxytocin receptors. *J. Neuroendocrinol.* 16, 398–402.
498. Kimura, T., Saji, F., Nishimori, K., Ogita, K., Nakamura, H., Koyama, M., and Murata, Y. (2003). Molecular regulation of the oxytocin receptor in peripheral organs. *J. Mol. Endocrinol.* 30, 109–115.
499. Ivell, R., and Walther, N. (1999). The role of sex steroids in the oxytocin hormone system. *Mol. Cell Endocrinol.* 151, 95–101.
500. Telgmann, R., Bathgate, R. A., Jaeger, S., Tillmann, G., and Ivell, R. (2003). Transcriptional regulation of the bovine oxytocin receptor gene. *Biol. Reprod.* 68, 1015–1026.
501. Siebel, A. L., Gehring, H. M., Reytomas, I. G., and Parry, L. J. (2003). Inhibition of oxytocin receptor and estrogen receptor-alpha expression, but not relaxin receptors (LGR7), in the myometrium of late pregnant relaxin gene knockout mice. *Endocrinology* 144, 4272–4275.
502. Ou, C. W., Chen, Z. Q., Qi, S., and Lye, S. J. (1998). Increased expression of the rat myometrial oxytocin receptor messenger ribonucleic acid during labor requires both mechanical and hormonal signals. *Biol. Reprod.* 59, 1055–1061.
503. Rauk, P. N., and Friebe-Hoffmann, U. (2000). Interleukin-1 beta down-regulates the oxytocin receptor in cultured uterine smooth muscle cells. *Am. J. Reprod. Immunol.* 43, 85–91.
504. Schmid, B., Wong, S., and Mitchell, B. F. (19). Transcriptional regulation of oxytocin receptor by interleukin-1beta and interleukin-6. *Endocrinology* 142, 1380–1385.
505. Helmer, H., Tretzmuller, U., and Brunbauer, M., et al. (2002). Production of oxytocin receptor and cytokines in primary uterine smooth muscle cells cultivated under inflammatory conditions. *J. Soc. Gynecol. Invest.* 9, 15–21.
506. Leung, S. T., Cheng, Z., and Sheldrick, E.L., et al. (2001). The effects of lipopolysaccharide and interleukins-1alpha, -2 and -6 on oxytocin receptor expression and prostaglandin production in bovine endometrium. *J. Endocrinol.* 168, 497–508.
507. Strakova, Z., and Soloff, M. S. (1997). Coupling of oxytocin receptor to G proteins in rat myometrium during labor: Gi receptor interaction. *Am. J. Physiol.* 272, E870–E876.
508. Phaneuf, S., Europe-Finner, G. N., and Varney, M., et al. (1993). Oxytocin-stimulated phosphoinositide hydrolysis in human myometrial cells: involvement of pertussis toxin-sensitive and -insensitive G-proteins. *J. Endocrinol.* 136, 497–509.
509. Burns, P. D., Tsai, S. J., and Wiltbank, M. C., et al. (1997). Effect of oxytocin on concentrations of prostaglandin H synthase-2 mRNA in ovine endometrial tissue in vivo. *Endocrinology* 138, 5637–5640.
510. Soloff, M. S., Jeng, Y. J., Copland, J. A., Strakova, Z., and Hoare, S. (2000). Signal pathways mediating oxytocin stimulation of prostaglandin synthesis in select target cells. *Exp. Physiol.* 85(Spec No), 51S–58S.
511. Olins, G. M., and Bremel, R. D. (1984). Oxytocin-stimulated myosin phosphorylation in mammary myoepithelial cells: roles of calcium ions and cyclic nucleotides. *Endocrinology* 114, 1617–1626.
512. Furuya, K., Akita, K., and Sokabe, M. (2004). [Extracellular ATP mediated mechano-signaling in mammary glands]. *Nippon Yakurigaku Zasshi.* 123, 397–402.
513. Sapino, A., Macri, L., Tonda, L., and Bussolati, G. (1993). Oxytocin enhances myoepithelial cell differentiation and proliferation in the mouse mammary gland. *Endocrinology* 133, 838–842.
514. Plested, C. P., and Bernal, A. L. (2001). Desensitisation of the oxytocin receptor and other G-protein coupled receptors in the human myometrium. *Exp. Physiol.* 86, 303–312.
515. Oakley, R. H., Laporte, S. A., Holt, J. A., Barak, L. S., and Caron, M. G. (2001). Molecular determinants underlying the formation of stable intracellular G protein-coupled receptor-beta-arrestin complexes after receptor endocytosis. *J. Biol. Chem.* 276, 19452–19460.
516. Verbalis, J. G., Mangione, M. P., and Stricker, E. M. (1991). Oxytocin produces natriuresis in rats at physiological plasma concentrations. *Endocrinology* 128, 1317–1322.
517. Soares, T. J., Coimbra, T. M., and Martins, A. R., et al. (1999). Atrial natriuretic peptide and oxytocin induce natriuresis by release of cGMP. *Proc. Natl. Acad. Sci. U S A* 96, 278–283.
518. McCann, S. M., Antunes-Rodrigues, J., Jankowski, M., and Gutkowska, J. (2002). Oxytocin, vasopressin and atrial natriuretic peptide control body fluid homeostasis by action on their receptors in brain, cardiovascular system and kidney. *Prog. Brain Res.* 139, 309–328.
519. Gutkowska, J., Jankowski, M., and Lambert, C., et al. (1997). Oxytocin releases atrial natriuretic peptide by combining with oxytocin receptors in the heart. *Proc. Natl. Acad. Sci. U S A* 94, 11704–11709.
520. Randolph, R. R., Li, Q., Curtis, K. S., Sullivan, M. J., and Cunningham, J. T. (1998). Fos expression following isotonic volume expansion of the unanesthetized male rat. *Am. J. Physiol.* 274, R1345–R1352.
521. Haanwinckel, M. A., Elias, L. K., and Favaretto, A. L., et al. (1995). Oxytocin mediates atrial natriuretic peptide release and natriuresis after volume expansion in the rat. *Proc. Natl. Acad. Sci. U S A* 92, 7902–7906.
522. Breton, C., Neculcea, J., and Zingg, H. H. (1996). Renal oxytocin receptor messenger ribonucleic acid: characterization and regulation during pregnancy and in response to ovarian steroid treatment. *Endocrinology* 137, 2711–2717.
523. Mukaddam-Daher, S., Jankowski, M., Wang, D., Menaouar, A., and Gutkowska, J. (2002). Regulation of cardiac oxytocin system and natriuretic peptide during rat gestation and postpartum. *J. Endocrinol.* 175, 211–216.
524. Atherton, J. C., Dark, J. M., and Garland, H. O., et al. (1982). Changes in water and electrolyte balance, plasma volume and composition during pregnancy in the rat. *J. Physiol.* 330, 81–93.
525. Omi, E. C., Zhao, S., Shanks, R. D., and Sherwood, O. D. (1997). Evidence that systemic relaxin promotes moderate water consumption during late pregnancy in rats. *J. Endocrinol.* 153, 33–40.

526. Osheroff, P. L., and Phillips, H. S. (1991). Autoradiographic localization of relaxin binding sites in rat brain. *Proc. Natl. Acad. Sci. U S A* 88, 6413–6417.
527. Sunn, N., Egli, M., and Burazin, T. C., et al. (2002). Circulating relaxin acts on subfornical organ neurons to stimulate water drinking in the rat. *Proc. Natl. Acad. Sci. U S A* 99, 1701–1706.
528. Summerlee, A. J., and Robertson, G. F. (1995). Central administration of porcine relaxin stimulates drinking behaviour in rats: an effect mediated by central angiotensin II. *Endocrine* 3, 377–381.
529. Hornsby, D. J., Wilson, B. C., and Summerlee, A. J. (2001). Relaxin and drinking in pregnant rats. *Prog. Brain Res.* 133, 229–240.
530. Geddes, B. J., Parry, L. J., and Summerlee, A. J. (1994). Brain angiotensin-II partially mediates the effects of relaxin on vasopressin and oxytocin release in anesthetized rats. *Endocrinology* 134, 1188–1192.
531. Parry, L. J., Poterski, R. S., and Summerlee, A. J. (1994). Effects of relaxin on blood pressure and the release of vasopressin and oxytocin in anesthetized rats during pregnancy and lactation. *Biol. Reprod.* 50, 622–628.
532. Way, S. A., Douglas, A. J., and Dye, S., et al. (1993). Endogenous opioid regulation of oxytocin release during parturition is reduced in ovariectomized rats. *J. Endocrinol.* 138, 13–22.
533. Leng, G., Brown, C. H., and Bull, P. M., et al. (2001). Responses of magnocellular neurons to osmotic stimulation involves coactivation of excitatory and inhibitory input: an experimental and theoretical analysis. *J. Neurosci.* 21, 6967–6977.
534. Bourque, C. W., Voisin, D. L., and Chakfe, Y. (2002). Stretch-inactivated cation channels: cellular targets for modulation of osmosensitivity in supraoptic neurons. *Prog. Brain Res.* 139, 85–94.
535. Hussy, N. (2002). Glial cells in the hypothalamo-neurohypophysial system: key elements of the regulation of neuronal electrical and secretory activity. *Prog. Brain Res.* 139, 95–112.
536. McKinley, M. J., Mathai, M. L., and McAllen, R. M., et al. (2004). Vasopressin secretion: osmotic and hormonal regulation by the lamina terminalis. *J. Neuroendocrinol.* 16, 340–347.
537. Anderson, J. W., Washburn, D. L., and Ferguson, A. V. (2000). Intrinsic osmosensitivity of subfornical organ neurons. *Neuroscience* 100, 539–547.
538. Sunn, N., McKinley, M. J., and Oldfield, B. J. (2003). Circulating angiotensin II activates neurones in circumventricular organs of the lamina terminalis that project to the bed nucleus of the stria terminalis. *J. Neuroendocrinol.* 15, 725–731.
539. Ferguson, A. V., and Washburn, D. L. (1998). Angiotensin II: a peptidergic neurotransmitter in central autonomic pathways. *Prog. Neurobiol.* 54, 169–192.
540. Chriguer, R. S., Antunes-Rodrigues, J., and Franci, C. R. (2003). Atrial natriuretic peptide mediates oxytocin secretion induced by osmotic stimulus. *Brain Res. Bull.* 59, 505–511.
541. Jankowski, M., Marcinkiewicz, M., Bouanga, J. C., and Gutkowska, J. (2004). Changes of atrial natriuretic peptide in rat supraoptic neurones during pregnancy. *J. Neuroendocrinol.* 16, 441–449.
542. Russell, J. A., Douglas, A. J., and Bull, P. M., et al. (1992). Pregnancy and opioid interactions with the anterior perithird ventricular input to magnocellular oxytocin neurones. *Prog. Brain Res.* 91, 41–53.
543. Koehler, E. M., McLemore, G. L., Tang, W., and Summy-Long, J. Y. (1993). Osmoregulation of the magnocellular system during pregnancy and lactation. *Am. J. Physiol.* 264, R555–R560.
544. Srisawat, R., Ludwig, M., and Bull, P. M., et al. (2000). Nitric oxide and the oxytocin system in pregnancy. *J. Neurosci.* 20, 6721–6727.
545. Bull, P. M., Douglas, A. J., and Russell, J. A. (1994). Opioids and coupling of the anterior peri-third ventricular input to oxytocin neurones in anaesthetized pregnant rats. *J. Neuroendocrinol.* 6, 267–274.
546. Naito, Y., Fukata, J., and Shindo, K., et al. (1991). Effects of interleukins on plasma arginine vasopressin and oxytocin levels in conscious, freely moving rats. *Biochem. Biophys. Res. Commun.* 174, 1189–1195.
547. Onaka, T., Luckman, S. M., and Guevara-Guzman, R., et al. (1995). Presynaptic actions of morphine: blockade of cholecystokinin-induced noradrenaline release in the rat supraoptic nucleus. *J. Physiol. (Lond).* 482(Pt 1), 69–79.
548. Onaka, T. (2004). Neural pathways controlling central and peripheral oxytocin release during stress. *J. Neuroendocrinol.* 16, 308–312.
549. Miyata, S., Itoh, T., and Lin, S. H., et al. (1995). Temporal changes of c-fos expression in oxytocinergic magnocellular neuroendocrine cells of the rat hypothalamus with restraint stress. *Brain Res. Bull.* 37, 391–395.
550. Hashiguchi, H., Ye, S. H., Morris, M., and Alexander, N. (1997). Single and repeated environmental stress: effect on plasma oxytocin, corticosterone, catecholamines, and behavior. *Physiol. Behav.* 61, 731–736.
551. Douglas, A. J., Johnstone, H. A., and Wigger, A., et al. (1998). The role of endogenous opioids in neurohypophysial and hypothalamo–pituitary–adrenal axis hormone secretory responses to stress in pregnant rats. *J. Endocrinol.* 158, 285–293.
552. Engelmann, M., Ebner, K., Landgraf, R., Holsboer, F., and Wotjak1, C. T. (1999). Emotional stress triggers intrahypothalamic but not peripheral release of oxytocin in male rats. *J. Neuroendocrinol.* 11, 867–872.
553. Kalin, N. H., Gibbs, D. M., Barksdale, C. M., Shelton, S. E., and Carnes, M. (1985). Behavioral stress decreases plasma oxytocin concentrations in primates. *Life Sci.* 36, 1275–1280.
554. Zhu, L., and Onaka, T. (2002). Involvement of medullary A2 noradrenergic neurons in the activation of oxytocin neurons after conditioned fear stimuli. *Eur. J. Neurosci.* 16, 2186–2198.
555. Zhu, L. L., and Onaka, T. (2003). Facilitative role of prolactin-releasing peptide neurons in oxytocin cell activation after conditioned-fear stimuli. *Neuroscience* 118, 1045–1053.
556. Christensen, J. D., Hansen, E. W., and Fjalland, B. (1990). Influence of interleukin-1 beta on the secretion of oxytocin and vasopressin from the isolated rat neurohypophysis. *Pharmacol. Toxicol.* 67, 81–83.
557. Yasin, S. A., Costa, A., Forsling, M. L., and Grossman, A. (1994). Interleukin-1 beta and interleukin-6 stimulate neurohypophysial hormone release in vitro. *J. Neuroendocrinol.* 6, 179–184.
558. Brunton, P. J., Sabatier, N., Leng, G., and Russell, J. A. (2004). Stimulation of oxytocin neurones by systemic interleukin-1 beta is suppressed in pregnant rats through actions of endogenous opioids. Soc. Neurosci. Abstracts 30: 662·3
559. Legros, J. J., Chiodera, P., Geenen, V., and von Frenckell, R. (1987). Confirmation of the inhibitory influence of exogenous oxytocin on cortisol and ACTH in man: evidence of reproducibility. *Acta Endocrinol. (Copenh).* 114, 345–349.
560. Page, S. R., Ang, V. T., Jackson, R., and Nussey, S. S. (1990). The effect of oxytocin on the plasma glucagon response to insulin-induced hypoglycaemia in man. *Diabete Metab.* 16, 248–251.

561. Wigger, A., Lorscher, P., and Oehler, I., et al. (1999). Nonresponsiveness of the rat hypothalamo–pituitary–adrenocortical axis to parturition-related events: inhibitory action of endogenous opioids. *Endocrinology* 140, 2843–2849.
562. Douglas, A. J., Brunton, P. J., Bosch, O. J., Russell, J. A., and Neumann, I. D. (2003). Neuroendocrine responses to stress in mice: hyporesponsiveness in pregnancy and parturition. *Endocrinology* 144, 5268–5276.
563. Amico, J. A., Mantella, R. C., Vollmer, R. R., and Li, X. (2004). Anxiety and stress responses in female oxytocin deficient mice. *J. Neuroendocrinol.* 16, 319–324.
564. Nishioka, T., Anselmo-Franci, J. A., Li, P., Callahan, M. F., and Morris, M. (1998). Stress increases oxytocin release within the hypothalamic paraventricular nucleus. *Brain Res.* 781, 56–60.
565. Wotjak, C. T., Ganster, J., and Kohl, G., et al. (1998). Dissociated central and peripheral release of vasopressin, but not oxytocin, in response to repeated swim stress: new insights into the secretory capacities of peptidergic neurons. *Neuroscience* 85, 1209–1222.
566. Wigger, A., and Neumann, I. D. (2002). Endogenous opioid regulation of stress-induced oxytocin release within the hypothalamic paraventricular nucleus is reversed in late pregnancy: a microdialysis study. *Neuroscience* 112, 121–129.
567. Landgraf, R., Neumann, I., Holsboer, F., and Pittman, Q. J. (1995). Interleukin-1 beta stimulates both central and peripheral release of vasopressin and oxytocin in the rat. *Eur. J. Neurosci.* 7, 592–598.
568. Windle, R. J., Kershaw, Y. M., and Shanks, N., et al. (2004). Oxytocin attenuates stress-induced c-fos mRNA expression in specific forebrain regions associated with modulation of hypothalamo-pituitary-adrenal activity. *J. Neurosci.* 24, 2974–2982.
569. Neumann, I. D., Wigger, A., Torner, L., Holsboer, F., and Landgraf, R. (2000). Brain oxytocin inhibits basal and stress-induced activity of the hypothalamo–pituitary–adrenal axis in male and female rats: partial action within the paraventricular nucleus. *J. Neuroendocrinol.* 12, 235–243.
570. Neumann, I. D., Torner, L., and Wigger, A. (2000). Brain oxytocin: differential inhibition of neuroendocrine stress responses and anxiety-related behaviour in virgin, pregnant and lactating rats. *Neuroscience* 95, 567–575.
571. Neumann, I. D. (2002). Involvement of the brain oxytocin system in stress coping: interactions with the hypothalamo–pituitary–adrenal axis. *Prog. Brain Res.* 139, 147–162.
572. Douglas, A. J., and Russell, J. A. (2001). Endogenous opioid regulation of oxytocin and ACTH secretion during pregnancy and parturition. *Prog. Brain Res.* 133, 67–82.
573. Windle, R. J., Shanks, N., Lightman, S. L., and Ingram, C. D. (1997). Central oxytocin administration reduces stress-induced corticosterone release and anxiety behavior in rats. *Endocrinology* 138, 2829–2834.
574. McCarthy, M. M., McDonald, C. H., Brooks, P. J., and Goldman, D. (1996). An anxiolytic action of oxytocin is enhanced by estrogen in the mouse. *Physiol. Behav.* 60, 1209–1215.
575. Bosch, O. J., Kromer, S. A., Brunton, P. J., and Neumann, I. D. (2004). Release of oxytocin in the hypothalamic paraventricular nucleus, but not central amygdala or lateral septum in lactating residents and virgin intruders during maternal defence. *Neuroscience* 124, 439–448.
576. Neumann, I. D., Toschi, N., Ohl, F., Torner, L., and Kromer, S. A. (2001). Maternal defence as an emotional stressor in female rats: correlation of neuroendocrine and behavioural parameters and involvement of brain oxytocin. *Eur. J. Neurosci.* 13, 1016–1024.
577. Verbalis, J. G., Blackburn, R. E., Hoffman, G. E., and Stricker, E. M. (1995). Establishing behavioral and physiological functions of central oxytocin: insights from studies of oxytocin and ingestive behaviors. *Adv. Exp. Med. Biol.* 395, 209–225.
578. Blackburn, R. E., Samson, W. K., Fulton, R. J., Stricker, E. M., and Verbalis, J. G. (1993). Central oxytocin inhibition of salt appetite in rats: evidence for differential sensing of plasma sodium and osmolality. *Proc. Natl. Acad. Sci. U S A* 90, 10380–10384.
579. Amico, J. A., Morris, M., and Vollmer, R. R. (2001). Mice deficient in oxytocin manifest increased saline consumption following overnight fluid deprivation. *Am. J. Physiol. Regul. Integr. Comp. Physiol.* 281, R1368–R1373.
580. Hillebrand, J. J., de Wied, D., and Adan, R. A. (2002). Neuropeptides, food intake and body weight regulation: a hypothalamic focus. *Peptides* 23, 2283–2306.
581. Rogers, R. C., and Hermann, G. E. (1987). Oxytocin, oxytocin antagonist, TRH, and hypothalamic paraventricular nucleus stimulation effects on gastric motility. *Peptides* 8, 505–513.
582. Flanagan, L. M., Dohanics, J., Verbalis, J. G., and Stricker, E. M. (1992). Gastric motility and food intake in rats after lesions of hypothalamic paraventricular nucleus. *Am. J. Physiol.* 263, R39–R44.
583. Olson, B. R., Hoffman, G. E., Sved, A. F., Stricker, E. M., and Verbalis, J. G. (1992). Cholecystokinin induces c-fos expression in hypothalamic oxytocinergic neurons projecting to the dorsal vagal complex. *Brain Res.* 569, 238–248.
584. Luckman, S. M., and Lawrence, C. B. (2003). Anorectic brainstem peptides: more pieces to the puzzle. *Trends Endocrinol. Metab.* 14, 60–65.
585. Blevins, J. E., Eakin, T. J., Murphy, J. A., Schwartz, M. W., and Baskin, D. G. (2003). Oxytocin innervation of caudal brainstem nuclei activated by cholecystokinin. *Brain Res.* 993, 30–41.
586. Walter, R., Van Ree, J. M., and de Wied, D. (1978). Modification of conditioned behavior of rats by neurohypophyseal hormones and analogues. *Proc. Natl. Acad. Sci. U S A* 75, 2493–2496.
587. Schultz, H., Kovacs, G. L., and Telegdy, G. (1974). Effect of physiological doses of vasopressin and oxytocin on avoidance and exploratory behaviour in rats. *Acta Physiol. Acad. Sci. Hung.* 45, 211–215.
588. Bohus, B., Urban, I., Wimersma Greidanus, T. B., and de Wied, D. (1978). Opposite effects of oxytocin and vasopressin on avoidance behaviour and hippocampal theta rhythm in the rat. *Neuropharmacology* 17, 239–247.
589. Kovacs, G. L., and de Wied, D. (1994). Peptidergic modulation of learning and memory processes. *Pharmacol. Rev.* 46, 269–291.
590. Wu, W., and Yu, L. C. (2004). Roles of oxytocin in spatial learning and memory in the nucleus basalis of Meynert in rats. *Regul. Pept.* 120, 119–125.
591. Tomizawa, K., Iga, N., and Lu, Y. F., et al. (2003). Oxytocin improves long-lasting spatial memory during motherhood through MAP kinase cascade. *Nat. Neurosci.* 6, 384–390.
592. Ferguson, J. N., Young, L. J., and Hearn, E. F., et al. (2000). Social amnesia in mice lacking the oxytocin gene. *Nat. Genet.* 25, 284–288.
593. McEwen, B. B. (2004). Advances in Pharmacology. Elsevier/Academic Press, San Diego.
594. Ferrier, B. M., Kennett, D.J., and Devlin, M. C. (1980). Influence of oxytocin on human memory processes. *Life Sci.* 27, 2311–2317.

595. Kennett, D. J., Devlin, M. C., and Ferrier, B. M. (1982). Influence of oxytocin on human memory processes: validation by a control study. *Life Sci.* 31, 273–275.
596. Heinrichs, M., Meinlschmidt, G., Wippich, W., Ehlert, U., and Hellhammer, D. H. (2004). Selective amnesic effects of oxytocin on human memory. *Physiol. Behav.* 83, 31–38.
597. Fehm-Wolfsdorf, G., Born, J., Voigt, K. H., and Fehm, H. L. (1984). Human memory and neurohypophyseal hormones: opposite effects of vasopressin and oxytocin. *Psychoneuroendocrinology* 9, 285–292.
598. de Wied, D. (1980). Behavioural actions of neurohypophysial peptides. *Proc. R. Soc. Lond. B. Biol. Sci.* 210, 183–195.
599. Insel, T. R. (1992). Oxytocin—a neuropeptide for affiliation: evidence from behavioral, receptor autoradiographic, and comparative studies. *Psychoneuroendocrinology* 17, 3–35.
600. Ferguson, J. N., Young, L. J., and Insel, T. R. (2002). The neuroendocrine basis of social recognition. *Front Neuroendocrinol.* 23, 200–224.
601. Popik, P., Vos, P. E., and Van Ree, J. M. (1992). Neurohypophyseal hormone receptors in the septum are implicated in social recognition in the rat. *Behav. Pharmacol.* 3, 351–358.
602. Ferguson, J. N., Aldag, J.M., Insel, T. R., and Young, L. J. (2001). Oxytocin in the medial amygdala is essential for social recognition in the mouse. *J. Neurosci.* 21, 8278–8285.
603. Kavaliers, M., Colwell, D. D., and Choleris, E., et al. (2003). Impaired discrimination of and aversion to parasitized male odors by female oxytocin knockout mice. *Genes Brain Behav.* 2, 220–230.
604. Witt, D. M., Winslow, J. T., and Insel, T. R. (1992). Enhanced social interactions in rats following chronic, centrally infused oxytocin. *Pharmacol. Biochem. Behav.* 43, 855–861.
605. Rosenblum, L. A., Smith, E. L., and Altemus, M., et al. (2002). Differing concentrations of corticotropin-releasing factor and oxytocin in the cerebrospinal fluid of bonnet and pigtail macaques. *Psychoneuroendocrinology* 27, 651–660.
606. Arletti, R., and Bertolini, A. (1985). Oxytocin stimulates lordosis behavior in female rats. *Neuropeptides* 6, 247–253.
607. Caldwell, J. D., Prange, A. J. Jr., and Pedersen, C. A. (1986). Oxytocin facilitates the sexual receptivity of estrogen-treated female rats. *Neuropeptides* 7, 175–189.
608. Gorzalka, B. B., and Lester, G. L. (1987). Oxytocin-induced facilitation of lordosis behaviour in rats is progesterone-dependent. *Neuropeptides* 10, 55–65.
609. Caldwell, J. D., Johns, J. M., Faggin, B. M., Senger, M. A., and Pedersen, C. A. (1994). Infusion of an oxytocin antagonist into the medial preoptic area prior to progesterone inhibits sexual receptivity and increases rejection in female rats. *Horm. Behav.* 28, 288–302.
610. Witt, D. M., and Insel, T. R. (1991). A selective oxytocin antagonist attenuates progesterone facilitation of female sexual behavior. *Endocrinology* 128, 3269–3276.
611. Pedersen, C. A., and Boccia, M. L. (2002). Oxytocin maintains as well as initiates female sexual behavior: effects of a highly selective oxytocin antagonist. *Horm. Behav.* 41, 170–177.
612. Schumacher, M., Coirini, H., Pfaff, D. W., and McEwen, B. S. (1990). Behavioral effects of progesterone associated with rapid modulation of oxytocin receptors. *Science* 250, 691–694.
613. McCarthy, M. M., Kleopoulos, S. P., Mobbs, C. V., and Pfaff, D. W. (1994). Infusion of antisense oligodeoxynucleotides to the oxytocin receptor in the ventromedial hypothalamus reduces estrogen-induced sexual receptivity and oxytocin receptor binding in the female rat. *Neuroendocrinology* 59, 432–440.
614. Caldwell, J. D., Jirikowski, G. F., Greer, E. R., and Pedersen, C. A. (1989). Medial preoptic area oxytocin and female sexual receptivity. *Behav. Neurosci.* 103, 655–662.
615. Anderson-Hunt, M., and Dennerstein, L. (1994). Increased female sexual response after oxytocin. *BMJ* 309, 929.
616. Melin, P., and Kihlstroem, J. E. (1963). Influence of oxytocin on sexual behavior in male rabbits. *Endocrinology* 73, 433–435.
617. Arletti, R., Bazzani, C., Castelli, M., and Bertolini, A. (1985). Oxytocin improves male copulatory performance in rats. *Horm. Behav.* 19, 14–20.
618. Argiolas, A., Collu, M., Gessa, G. L., Melis, M. R., and Serra, G. (1988). The oxytocin antagonist d(CH2)5Tyr(Me)-Orn8-vasotocin inhibits male copulatory behaviour in rats. *Eur. J. Pharmacol.* 149, 389–392.
619. Argiolas, A., Melis, M. R., Vargiu, L., and Gessa, G. L. (1987). d(CH2)5Tyr(Me)-[Orn8]vasotocin, a potent oxytocin antagonist, antagonizes penile erection and yawning induced by oxytocin and apomorphine, but not by ACTH-(1-24). *Eur. J. Pharmacol.* 134, 221–224.
620. Argiolas, A., Melis, M. R., Stancampiano, R., and Gessa, G. L. (1989). Penile erection and yawning induced by oxytocin and related peptides: structure-activity relationship. *Peptides* 10, 559–563.
621. Argiolas, A., Melis, M. R., and Gessa, G. L. (1985). Intraventricular oxytocin induces yawning and penile erection in rats. *Eur. J. Pharmacol.* 117, 395–396.
622. Melis, M. R., Argiolas, A., and Gessa, G. L. (1986). Oxytocin-induced penile erection and yawning: site of action in the brain. *Brain Res.* 398, 259–265.
623. Argiolas, A., and Melis, M. R. (2005). Central control of penile erection: Role of the paraventricular nucleus of the hypothalamus. *Prog. Neurobiol.* 76, 1–21.
624. Pedersen, C. A., and Prange, A. J. Jr. (1979). Induction of maternal behavior in virgin rats after intracerebroventricular administration of oxytocin. *Proc. Natl. Acad. Sci. U S A* 76, 6661–6665.
625. Pedersen, C. A., Ascher, J. A., Monroe, Y. L., and Prange, A. J. Jr. (1982). Oxytocin induces maternal behavior in virgin female rats. *Science* 216, 648–650.
626. Fahrbach, S. E., Morrell, J. I., and Pfaff, D. W. (1985). Possible role for endogenous oxytocin in estrogen-facilitated maternal behavior in rats. *Neuroendocrinology* 40, 526–532.
627. Rubin, B. S., Menniti, F. S., and Bridges, R. S. (1983). Intracerebroventricular administration of oxytocin and maternal behavior in rats after prolonged and acute steroid pretreatment. *Horm. Behav.* 17, 45–53.
628. Bolwerk, E. L., and Swanson, H. H. (1984). Does oxytocin play a role in the onset of maternal behaviour in the rat? *J. Endocrinol.* 101, 353–357.
629. Fahrbach, S. E., Morrell, J. I., and Pfaff, D. W. (1986). Effect of varying the duration of pre-test cage habituation on oxytocin induction of short-latency maternal behavior. *Physiol. Behav.* 37, 135–139.
630. Wamboldt, M. Z., and Insel, T. R. (1987). The ability of oxytocin to induce short latency maternal behavior is dependent on peripheral anosmia. *Behav. Neurosci.* 101, 439–441.
631. van Leengoed, E., Kerker, E., and Swanson, H. H. (1987). Inhibition of post-partum maternal behaviour in the rat by injecting an oxytocin antagonist into the cerebral ventricles. *J. Endocrinol.* 112, 275–282.
632. Pedersen, C. A., Caldwell, J. D., Johnson, M. F., Fort, S. A., and Prange, A. J. Jr. (1985). Oxytocin antiserum delays onset of ovarian steroid-induced maternal behavior. *Neuropeptides* 6, 175–182.
633. Insel, T. R., and Harbaugh, C. R. (1989). Lesions of the hypothalamic paraventricular nucleus disrupt the initiation of maternal behavior. *Physiol. Behav.* 45, 1033–1041.

634. Pedersen, C. A., and Boccia, M. L. (2003). Oxytocin antagonism alters rat dams' oral grooming and upright posturing over pups. *Physiol. Behav.* 80, 233–241.
635. Yu, G. Z., Kaba, H., Okutani, F., Takahashi, S., and Higuchi, T. (1996). The olfactory bulb: a critical site of action for oxytocin in the induction of maternal behaviour in the rat. *Neuroscience* 72, 1083–1088.
636. Yu, G. Z., Kaba, H., and Okutani, F., et al. (1996). The action of oxytocin originating in the hypothalamic paraventricular nucleus on mitral and granule cells in the rat main olfactory bulb. *Neuroscience* 72, 1073–1082.
637. Amico, J. A., Crowley, R. S., Insel, T. R., Thomas, A., and O'Keefe, J. A. (1995). Effect of gonadal steroids upon hypothalamic oxytocin expression. *Adv. Exp. Med. Biol.* 395, 23–35.
638. Insel, T. R. (1990). Regional changes in brain oxytocin receptor post-partum: time-course and relationship to maternal behavior. *J. Neuroendocrinol.* 2, 539–545.
639. McCarthy, M. M. (1990). Oxytocin inhibits infanticide in female house mice (*Mus domesticus*). *Horm. Behav.* 24, 365–375.
640. Young, L. J., Winslow, J. T., and Wang, Z., et al. (1997). Gene targeting approaches to neuroendocrinology: oxytocin, maternal behavior, and affiliation. *Horm. Behav.* 31, 221–231.
641. Olazabal, D. E., Ross, H. E., Nishimori, K., and Young, L. J. (2005). Female oxytocin receptor knockout mice show a deficit in "spontaneous" maternal behavior [abstract]. *Soc. Neurosci.*
642. Kendrick, K. M., Da Costa, A. P., and Broad, K. D., et al. (1997). Neural control of maternal behaviour and olfactory recognition of offspring. *Brain Res. Bull.* 44, 383–395.
643. Kendrick, K. M., Keverne, E. B., and Baldwin, B. A. (1987). Intracerebroventricular oxytocin stimulates maternal behaviour in the sheep. *Neuroendocrinology* 46, 56–61.
644. Kendrick, K. M., Keverne, E. B., Chapman, C., and Baldwin, B. A. (1988). Microdialysis measurement of oxytocin, aspartate, gamma-aminobutyric acid and glutamate release from the olfactory bulb of the sheep during vaginocervical stimulation. *Brain Res.* 442, 171–174.
645. Kendrick, K. M., Keverne, E. B., Chapman, C., and Baldwin, B. A. (1988). Intracranial dialysis measurement of oxytocin, monoamine and uric acid release from the olfactory bulb and substantia nigra of sheep during parturition, suckling, separation from lambs and eating. *Brain Res.* 439, 1–10.
646. Kendrick, K. M., Keverne, E. B., Hinton, M. R., and Goode, J. A. (1992). Oxytocin, amino acid and monoamine release in the region of the medial preoptic area and bed nucleus of the stria terminalis of the sheep during parturition and suckling. *Brain Res.* 569, 199–209.
647. Da Costa, A. P., Guevara-Guzman, R. G., Ohkura, S., Goode, J. A., and Kendrick, K. M. (1996). The role of oxytocin release in the paraventricular nucleus in the control of maternal behaviour in the sheep. *J. Neuroendocrinol.* 8, 163–177.
648. Insel, T. R., and Young, L. J. (2001). The neurobiology of attachment. *Nat. Rev. Neurosci.* 2, 129–136.
649. Young, L. J., and Wang, Z. (2004). The neurobiology of pair bonding. *Nat. Neurosci.* 7, 1048–1054.
650. Williams, J. R., Insel, T. R., Harbaugh, C. R., and Carter, C. S. (1994). Oxytocin administered centrally facilitates formation of a partner preference in female prairie voles (*Microtus ochrogaster*). *J. Neuroendocrinol.* 6, 247–250.
651. Insel, T. R., and Hulihan, T. J. (1995). A gender-specific mechanism for pair bonding: oxytocin and partner preference formation in monogamous voles. *Behav. Neurosci.* 109, 782–789.
652. Cho, M. M., DeVries, A. C., Williams, J. R., and Carter, C. S. (1999). The effects of oxytocin and vasopressin on partner preferences in male and female prairie voles (*Microtus ochrogaster*). *Behav. Neurosci.* 113, 1071–1079.
653. Young, L. J., Lim, M. M., Gingrich, B., and Insel, T. R. (2001). Cellular mechanisms of social attachment. *Horm. Behav.* 40, 133–138.
654. Gingrich, B., Liu, Y., Cascio, C., Wang, Z., and Insel, T. R. (2000). Dopamine D2 receptors in the nucleus accumbens are important for social attachment in female prairie voles (*Microtus ochrogaster*). *Behav. Neurosci.* 114, 173–183.
655. Liu, Y., and Wang, Z. X. (2003). Nucleus accumbens oxytocin and dopamine interact to regulate pair bond formation in female prairie voles. *Neuroscience* 121, 537–544.
656. Winslow, J. T., Hastings, N., Carter, C. S., Harbaugh, C. R., and Insel, T. R. (1993). A role for central vasopressin in pair bonding in monogamous prairie voles. *Nature* 365, 545–548.
657. Lim, M. M., and Young, L. J. (2004). Vasopressin-dependent neural circuits underlying pair bond formation in the monogamous prairie vole. *Neuroscience* 125, 35–45.
658. Ferguson, J. K. (1941). A study of the motility of the intact uterus at term. *Surgery Gynecol Obstet.* 73, 359–366.
659. Douglas, A., Scullion, S., and Antonijevic, I., et al. (2001). Uterine contractile activity stimulates supraoptic neurons in term pregnant rats via a noradrenergic pathway. *Endocrinology* 142, 633–644.
660. Higuchi, T., Uchide, K., Honda, K., and Negoro, H. (1986). Oxytocin release during parturition in the pelvic-neurectomized rat. *J. Endocrinol.* 109, 149–154.
661. Peters, L. C., Kristal, M. B., and Komisaruk, B. R. (1987). Sensory innervation of the external and internal genitalia of the female rat. *Brain Res.* 408, 199–204.
662. El Majdoubi, M., Poulain, D. A., and Theodosis, D. T. (1997). Lactation-induced plasticity in the supraoptic nucleus augments axodendritic and axosomatic GABAergic and glutamatergic synapses: an ultrastructural analysis using the disector method. *Neuroscience* 80, 1137–1147.
663. Burbach, J. P., Luckman, S. M., Murphy, D., and Gainer, H. (2001). Gene regulation in the magnocellular hypothalamo-neurohypophysial system. *Physiol. Rev.* 81, 1197–1267.
664. Mutsuga, N., Shahar, T., and Verbalis, J. G., et al. (2004). Selective gene expression in magnocellular neurons in rat supraoptic nucleus. *J. Neurosci.* 24, 7174–7185.
665. Oliet, S. H., Piet, R., and Poulain, D. A. (2001). Control of glutamate clearance and synaptic efficacy by glial coverage of neurons. *Science* 292, 923–926.
666. Langle, S. L., Poulain, D. A., and Theodosis, D. T. (2002). Neuronal-glial remodeling: a structural basis for neuronal-glial interactions in the adult hypothalamus. *J. Physiol. Paris* 96, 169–175.
667. Theodosis, D. T. (2002). Oxytocin-secreting neurons: a physiological model of morphological neuronal and glial plasticity in the adult hypothalamus. *Front Neuroendocrinol.* 23, 101–135.
668. Perlmutter, L. S., Tweedle, C. D., and Hatton, G. I. (1984). Neuronal/glial plasticity in the supraoptic dendritic zone: dendritic bundling and double synapse formation at parturition. *Neuroscience* 13, 769–779.
669. Montagnese, C. M., Poulain, D. A., Vincent, J. D., and Theodosis, D. T. (1987). Structural plasticity in the rat supraoptic nucleus during gestation, post-partum lactation and suckling-induced pseudogestation and lactation. *J. Endocrinol.* 115, 97–105.
670. Compagnone, N. A., and Mellon, S. H. (2002). Neurosteroids: biosynthesis and function of these novel neuromodulators. *Front Neuroendocrinol.* 21, 1–56.

671. Roper, P., Callaway, J., Shevchenko, T., Teruyama, R., and Armstrong, W. (2003). AHP's, HAP's and DAP's: how potassium currents regulate the excitability of rat supraoptic neurones. *J. Comput. Neurosci.* 15, 367–389.
672. Greffrath, W., Magerl, W., and Disque-Kaiser, U., et al. (2004). Contribution of Ca^{2+}-activated K^{+} channels to hyperpolarizing after-potentials and discharge pattern in rat supraoptic neurones. *J. Neuroendocrinol.* 16, 577–588.
673. Leng, G., Reiff-Marganiec, A., Ludwig, M., and Sabatier, N. (2004). Generating quantitatively accurate, but computational concise, models of single neurons. In Computational Neuroscience: A Comprehensive Approach (J. Feng, Ed.), pp. 185–213. CRC Press, Boca Raton, FL.
674. Michaloudi, H. C., El Majdoubi, M., Poulain, D. A., Papadopoulos, G. C., and Theodosis, D. T. (1997). The noradrenergic innervation of identified hypothalamic magnocellular somata and its contribution to lactation-induced synaptic plasticity. *J. Neuroendocrinol.* 9, 17–23.
675. Randle, J. C., Bourque, C. W., and Renaud, L. P. (1986). Alpha 1-adrenergic receptor activation depolarizes rat supraoptic neurosecretory neurons in vitro. *Am. J. Physiol.* 251, R569–R574.
676. Yamashita, H., Inenaga, K., and Kannan, H. (1987). Depolarizing effect of noradrenaline on neurons of the rat supraoptic nucleus in vitro. *Brain Res.* 405, 348–352.
677. Meddle, S. L., Leng, G., Selvarajah, J. R., Bicknell, R. J., and Russell, J. A. (2000). Direct pathways to the supraoptic nucleus from the brainstem and the main olfactory bulb are activated at parturition in the rat. *Neuroscience* 101, 1013–1021.
678. Herbison, A. E., Voisin, D. L., Douglas, A. J., and Chapman, C. (1997). Profile of monoamine and excitatory amino acid release in rat supraoptic nucleus over parturition. *Endocrinology* 138, 33–40.
679. Jiang, Q. B., and Wakerley, J. B. (1995). Analysis of bursting responses of oxytocin neurones in the rat in late pregnancy, lactation and after weaning. *J. Physiol.* 486, 237–248.
680. Lipschitz, D. L., Crowley, W. R., and Bealer, S. L. (2004). Differential sensitivity of intranuclear and systemic oxytocin release to central noradrenergic receptor stimulation during mid- and late gestation in rats. *Am. J. Physiol. Endocrinol. Metab.* 287, E523–E528.
681. Guevara-Guzman, R., Buzo, E., and Larrazolo, A., et al. (2001). Vaginocervical stimulation-induced release of classical neurotransmitters and nitric oxide in the nucleus of the solitary tract varies as a function of the oestrus cycle. *Brain Res.* 898, 303–313.
682. Luckman, S. M. (1992). Fos-like immunoreactivity in the brainstem of the rat following peripheral administration of cholecystokinin. *J. Neuroendocrinol.* 4, 149–152.
683. Kendrick, K. M., Leng, G., and Higuchi, T. (1991). Noradrenaline, dopamine and serotonin release in the paraventricular and supraoptic nuclei of the rat in response to intravenous cholecystokinin injections. *J. Neuroendocrinol.* 3, 139–144.
684. Carter, S. J., Richardson, C. M., and Wakerley, J. B. (2003). Excitatory effects of oxytocin and cholecystokinin on oxytocin neurones: differences between virgin, pregnant and lactating rats. *Neurosci. Lett.* 351, 13–16.
685. Tolchard, S., and Ingram, C. D. (1993). Electrophysiological actions of oxytocin in the dorsal vagal complex of the female rat in vitro: changing responsiveness during the oestrous cycle and after steroid treatment. *Brain Res.* 609, 21–28.
686. Sansone, G. R., and Komisaruk, B. R. (2001). Evidence that oxytocin is an endogenous stimulator of autonomic sympathetic preganglionics: the pupillary dilatation response to vaginocervical stimulation in the rat. *Brain Res.* 898, 265–271.
687. Sermasi, E., and Coote, J. H. (1994). Oxytocin acts at V1 receptors to excite sympathetic preganglionic neurones in neonate rat spinal cord in vitro. *Brain Res.* 647, 323–332.
688. Benoussaidh, A., Maurin, Y., and Rampin, O. (2004). Spinal effects of oxytocin on uterine motility in anesthetized rats. *Am. J. Physiol. Regul. Integr. Comp. Physiol.* 287, R446–R453.
689. Bealer, S. L., and Flynn, F. W. (2003). Central neurokinin 3 receptors increase systemic oxytocin release: interaction with norepinephrine. *Exp. Neurol.* 184, 1027–1033.
690. Onaka, T., Ikeda, K., Yamashita, T., and Honda, K. (2003). Facilitative role of endogenous oxytocin in noradrenaline release in the rat supraoptic nucleus. *Eur. J. Neurosci.* 18, 3018–3026.
690a. Bealer, S. L., and Crowley, W. R. (2003) Angiotensin II-induced release of oxytocin: Interaction with noepinephrine and role in lactation. *Reg. Peptides.* 111, 41–46.
691. Sawchenko, P. E., and Pfeiffer, S. W. (1995). Ultrastructural localization of inhibin beta- and somatostatin-28-immunoreactivities in the paraventricular and supraoptic nuclei. *Brain Res.* 694, 233–245.
691a. Brunton, P. J., Wylie, L., McGown, C. C., and Russell, J. A. (2004). Central neuropeptide Y (NPY) stimulation of neuroendocrine stress responses, but not feeding behaviour, is ablated in pregnant rats [Abstract]. 12th Int. Congr. Endocrinol. p522.
692. Parker, S. L., and Crowley, W. R. (1993). Central stimulation of oxytocin release in the lactating rat: interaction of neuropeptide Y with alpha-1-adrenergic mechanisms. *Endocrinology* 132, 658–666.
693. Meddle, S. L., Russell, J. A., Leng, G., Ludwig, M., and Bull, P. M. (2004). Somatostatin influences supraoptic nucleus activity in the rat [abstract]. *Soc. Neurosci. Abstr.* 30, 660.19.
694. Kapoor, J. R., and Sladek, C. D. (2001). Substance P and NPY differentially potentiate ATP and adrenergic stimulated vasopressin and oxytocin release. *Am. J. Physiol. Regul. Integr. Comp. Physiol.* 280, R69–R78.
695. Ibata, Y., Iijima, N., and Kataoka, Y., et al. (2000). Morphological survey of prolactin-releasing peptide and its receptor with special reference to their functional roles in the brain. *Neurosci. Res.* 38, 223–230.
696. Maruyama, M., Matsumoto, H., and Fujiwara, K., et al. (1999). Central administration of prolactin-releasing peptide stimulates oxytocin release in rats. *Neurosci. Lett.* 276, 193–196.
697. Honda, K., Narita, K., Murata, T., and Higuchi, T. (2004). Prolactin releasing peptides modulate background firing rate and milk-ejection related burst of oxytocin cells in the supraoptic nucleus. *Brain Res. Bull.* 63, 315–319.
698. Kataoka, Y., Iijima, N., and Yano, T., et al. (2001). Gonadal regulation of PrRP mRNA expression in the nucleus tractus solitarius and ventral and lateral reticular nuclei of the rat. *Brain Res Mol. Brain Res.* 87, 42–47.
699. Guevara-Guzman, R., Barrera-Mera, B., De La, R. C., and Kendrick, K.M. (2000). Release of classical transmitters and nitric oxide in the rat olfactory bulb, evoked by vaginocervical stimulation and potassium, varies with the oestrus cycle. *Eur. J. Neurosci.* 12, 80–88.
700. Yang, Q. Z., Smithson, K. G., and Hatton, G. I. (1995). NMDA and non-NMDA receptors on rat supraoptic nucleus neurons activated monosynaptically by olfactory afferents. *Brain Res.* 680, 207–216.
701. Csaki, A., Kocsis, K., Kiss, J., and Halasz, B. (2002). Localization of putative glutamatergic/aspartatergic neurons

projecting to the supraoptic nucleus area of the rat hypothalamus. *Eur. J. Neurosci.* 16, 55–68.
702. Russell, J. A., Blackburn, R. E., and Leng, G. (1989). Ablation of the region anterior and ventral to the third ventricle (AV3V region) does not impede parturition in rats. *J. Endocrinol.* 121, 109–115.
703. Bicknell, R. J., Leng, G., and Russell, J. A., et al. (1988). Hypothalamic opioid mechanisms controlling oxytocin neurones during parturition. *Brain Res. Bull.* 20, 743–749.
704. Leng, G., Mansfield, S., and Bicknell, R. J., et al. (1985). Central opioids: a possible role in parturition? *J. Endocrinol.* 106, 219–224.
705. Lindow, S. W., van der Spuy, Z. M., and Hendricks, M. S., et al. (1992). The effect of morphine and naloxone administration on plasma oxytocin concentrations in the first stage of labor. *Clin. Endocrinol.* (Oxf). 37, 349–353.
706. Herbison, A. E. (1994). Immunocytochemical evidence for oestrogen receptors within GABA neurones located in the perinuclear zone of the supraoptic nucleus and GABAA receptor beta 2/beta 3 subunits on supraoptic oxytocin neurones. *J. Neuroendocrinol.* 6, 5–11.
707. Fenelon, V. S., and Herbison, A. E. (2000). Progesterone regulation of GABAA receptor plasticity in adult rat supraoptic nucleus. *Eur. J. Neurosci.* 12, 1617–1623.
707a. Brussaard, A. B., Devay, P., Leyting-Vermeulen, J. L., and Kits, K. S. (1999). Changes in properties and neurosteroid regulation of GABAergic synapses in the supraoptic nucleus during the mammalian female reproductive cycle. *J Physiol.* 516, 513–524.
708. Brussaard, A. B., and Herbison, A. E. (2000). Long-term plasticity of postsynaptic GABAA-receptor function in the adult brain: insights from the oxytocin neurone. *Trends Neurosci.* 23, 190–195.
709. Brussaard, A. B., and Koksma, J. J. (2002). Short-term modulation of GABAA receptor function in the adult female rat. *Prog. Brain Res.* 139, 31–42.
710. Concas, A., Mostallino, M. C., and Porcu, P., et al. (1998). Role of brain allopregnanolone in the plasticity of gamma-aminobutyric acid type A receptor in rat brain during pregnancy and after delivery. *Proc. Natl. Acad. Sci. U S A* 95, 13284–13289.
711. Stern, J. E., and Ludwig, M. (2001). NO inhibits supraoptic oxytocin and vasopressin neurons via activation of GABAergic synaptic inputs. *Am. J. Physiol. Regul. Integr. Comp. Physiol.* 280, R1815–R1822.
712. Okere, C. O., Wang, Y. F., and Higuchi, T., et al. (1996). The effect of systemic and central nitric oxide administration on milk availability in lactating rats. *Neuroreport* 8, 243–247.
713. Popeski, N., Amir, S., and Woodside, B. (1999). Changes in NADPH-d staining in the paraventricular and supraoptic nuclei during pregnancy and lactation in rats: role of ovarian steroids and oxytocin. *J. Neuroendocrinol.* 11, 53–61.
714. Ozaki, M., Shibuya, I., and Kabashima, N., et al. (2000). Preferential potentiation by nitric oxide of spontaneous inhibitory postsynaptic currents in rat supraoptic neurones. *J. Neuroendocrinol.* 12, 273–281.
715. Terrell, M. L., Salas, N., Bui, V., Summy-Long, J. Y., and Kadekaro, M. (2003). NO inhibition of the magnocellular neuroendocrine system in rats is independent of cGMP signaling pathway. *Exp. Neurol.* 184, 846–856.
716. Yang, Q. Z., and Hatton, G. I. (1999). Nitric oxide via cGMP-dependent mechanisms increases dye coupling and excitability of rat supraoptic nucleus neurons. *J. Neurosci.* 19, 4270–4279.
717. Kadowaki, K., Kishimoto, J., Leng, G., and Emson, P. C. (1994). Up-regulation of nitric oxide synthase (NOS) gene expression together with NOS activity in the rat hypothalamo-hypophysial system after chronic salt loading: evidence of a neuromodulatory role of nitric oxide in arginine vasopressin and oxytocin secretion. *Endocrinology* 134, 1011–1017.
718. Wakerley, J. B., and Lincoln, D. W. (1973). The milk-ejection reflex of the rat: a 20- to 40-fold acceleration in the firing of paraventricular neurones during oxytocin release. *J. Endocrinol.* 57, 477–493.
719. Dawood, M. Y. (1995). Novel approach to oxytocin induction-augmentation of labor. Application of oxytocin physiology during pregnancy. *Adv. Exp. Med. Biol.* 395, 585–594.
720. Robinson, C., Schumann, R., Zhang, P., and Young, R. C. (2003). Oxytocin-induced desensitization of the oxytocin receptor. *Am. J. Obstet. Gynecol.* 188, 497–502.
721. Wakerley, J. B., Jiang, Q. B., Housham, S. J., Terenzi, M. G., and Ingram, C. D. (1995). Influence of reproductive state and ovarian steroids on facilitation of the milk-ejection reflex by central oxytocin. *Adv. Exp. Med. Biol.* 395, 117–132.
722. Brown, C. H., Stern, J. E., and Jackson, K. L. M., et al. (2005) Morphine withdrawal increases intrinsic excitability of oxytocin neurones in morphine-dependent rats. *Eur. J. Neurosci.* 21, 501–512.
723. Dyball, R. E., and Leng, G. (1986). Regulation of the milk ejection reflex in the rat. *J. Physiol.* 380, 239–256.
724. Summerlee, A. J. (1989). Relaxin, opioids and the timing of birth in rats. In Brain Opioid Systems in Reproduction (R. G. Dyer and R. J. Bicknell, Eds.), pp. 257–270. Oxford University Press, Oxford.
725. Moos, F., and Richard, P. (1989). Paraventricular and supraoptic bursting oxytocin cells in rat are locally regulated by oxytocin and functionally related. *J. Physiol.* 408, 1–18.
726. Moos, F., Fontanaud, P., Mekaouche, M., and Brown, D. (2004). Oxytocin neurones are recruited into co-ordinated fluctuations of firing before bursting in the rat. *Neuroscience* 125, 391–410.
727. Wang, Y. F., and Hatton, G. I. (2004). Milk ejection burst-like electrical activity evoked in supraoptic oxytocin neurons in slices from lactating rats. *J. Neurophysiol.* 91, 2312–2321.
728. Israel, J. M., Le Masson, G., Theodosis, D. T., and Poulain, D. A. (2003). Glutamatergic input governs periodicity and synchronization of bursting activity in oxytocin neurons in hypothalamic organotypic cultures. *Eur. J. Neurosci.* 17, 2619–2629.
729. Tasker, J. G., Di, S., and Boudaba, C. (2002). Functional synaptic plasticity in hypothalamic magnocellular neurons. *Prog. Brain Res.* 139, 113–119.
730. Oliet, S. H., and Piet, R. (2004). Anatomical remodelling of the supraoptic nucleus: changes in synaptic and extrasynaptic transmission. *J. Neuroendocrinol.* 16, 303–307.
731. Moos, F., Freund-Mercier, M. J., and Guerne, Y., et al. (1984). Release of oxytocin and vasopressin by magnocellular nuclei in vitro: specific facilitatory effect of oxytocin on its own release. *J. Endocrinol.* 102, 63–72.
732. Morris, J. F., and Ludwig, M. (2004). Magnocellular dendrites: prototypic receiver/transmitters. *J. Neuroendocrinol.* 16, 403–408.
733. Landgraf, R. (1995). Intracerebrally released vasopressin and oxytocin: measurement, mechanisms and behavioural consequences. *J. Neuroendocrinol.* 7, 243–253.
734. Ludwig, M., and Pittman, Q. J. (2003). Talking back: dendritic neurotransmitter release. *Trends Neurosci.* 26, 255–261.
735. Buijs, R. M., De Vries, G. J., and Van Leeuwen, F. W. (1985). The distribution and synaptic release of oxytocin in the central nervous system. In Oxytocin: Clinical and Laboratory Studies (J. A. Amico and A. G. Robinson AG, Eds.), pp. 77–86. Excerpta Medica, Amsterdam.
736. Jiang, Q., and Wakerley, J. B. (1997). The milk-ejection reflex in the peri-partum rat: effects of oestradiol and progesterone on basal milk-ejection frequency and the

facilitatory response to central oxytocin. *J. Neuroendocrinol.* 9, 9–16.
737. Wang, G., Dayanithi, G., Newcomb, R., and Lemos, J. R. (1999). An R-type Ca(2+) current in neurohypophysial terminals preferentially regulates oxytocin secretion. *J. Neurosci.* 19, 9235–9241.
738. Scala-Guenot, D., Strosser, M. T., and Richard, P. (1987). Electrical stimulations of perifused magnocellular nuclei in vitro elicit Ca^{2+}-dependent, tetrodotoxin-insensitive release of oxytocin and vasopressin. *Neurosci. Lett.* 76, 209–214.
739. Soldo, B. L., Giovannucci, D. R., Stuenkel, E. L., and Moises, H. C. (2004). $Ca(^{2+})$ and frequency dependence of exocytosis in isolated somata of magnocellular supraoptic neurones of the rat hypothalamus. *J. Physiol.* 555, 699–711.
740. Hatton, G. I., and Tweedle, C. D. (1982). Magnocellular neuropeptidergic neurons in hypothalamus: increases in membrane apposition and number of specialized synapses from pregnancy to lactation. *Brain Res. Bull.* 8, 197–204.
741. Theodosis, D. T., Montagnese, C., Rodriguez, F., Vincent, J. D., and Poulain, D. A. (1986). Oxytocin induces morphological plasticity in the adult hypothalamo-neurohypophysial system. *Nature* 322, 738–740.
742. Lambert, R. C., Dayanithi, G., Moos, F. C., and Richard, P. (1994). A rise in the intracellular Ca2+ concentration of isolated rat supraoptic cells in response to oxytocin. *J. Physiol.* 478, 275–287.
743. Ludwig, M., Sabatier, N., and Bull, P. M., et al. (2002). Intracellular calcium stores regulate activity-dependent neuropeptide release from dendrites. *Nature* 418, 85–89.
744. Tobin, V. A., Hurst, G., and Norrie, L., et al. (2004). Thapsigargin-induced mobilization of dendritic dense-cored vesicles in rat supraoptic neurons. *Eur. J. Neurosci.* 19, 2909–2912.
745. Teruyama, R., and Armstrong, W. E. (2002). Changes in the active membrane properties of rat supraoptic neurones during pregnancy and lactation. *J. Neuroendocrinol.* 14, 933–944.
746. Curras-Collazo, M. C., Gillard, E. R., Jin, J., and Pandika, J. (2003). Vasopressin and oxytocin decrease excitatory amino acid release in adult rat supraoptic nucleus. *J. Neuroendocrinol.* 15, 182–190.
747. Kombian, S. B., Hirasawa, M., Mouginot, D., and Pittman, Q. J. (2002). Modulation of synaptic transmission by oxytocin and vasopressin in the supraoptic nucleus. *Prog. Brain Res.* 139, 235–246.
748. Hirasawa, M., Schwab, Y., and Natah, S., et al. (2004). Dendritically released transmitters cooperate via autocrine and retrograde actions to inhibit afferent excitation in rat brain. *J. Physiol.* 559, 611–624.
749. de Kock, C. P., Wierda, K. D., and Bosman, L. W., et al. (2003). Somatodendritic secretion in oxytocin neurons is upregulated during the female reproductive cycle. *J. Neurosci.* 23, 2726–2734.
750. Leng, G., and Brown, D. (1997). The origins and significance of pulsatility in hormone secretion from the pituitary. *J. Neuroendocrinol.* 9, 493–513.
751. Bealer, S. L., and Crowley, W. R. (1998). Noradrenergic control of central oxytocin release during lactation in rats. Am. *J. Physiol.* 274, E453–E458.
752. Wang, Y. F., Shibuya, I., and Kabashima, N., et al. (1998). Inhibition of spontaneous inhibitory postsynaptic currents (IPSC) by noradrenaline in rat supraoptic neurons through presynaptic alpha2-adrenoceptors. *Brain Res.* 807, 61–69.
753. Daftary, S. S., Boudaba, C., Szabo, K., and Tasker, J. G. (1998). Noradrenergic excitation of magnocellular neurons in the rat hypothalamic paraventricular nucleus via intranuclear glutamatergic circuits. *J. Neurosci.* 18, 10619–10628.
754. Fuchs, A. R., Fuchs, F., Husslein, P., and Soloff, M. S. (1984). Oxytocin receptors in the human uterus during pregnancy and parturition. *Am. J. Obstet. Gynecol.* 150, 734–741.
755. Wathes, D. C., Borwick, S. C., Timmons, P. M., Leung, S. T., and Thornton, S. (1999). Oxytocin receptor expression in human term and preterm gestational tissues prior to and following the onset of labor. *J. Endocrinol.* 161, 143–151.
756. Antonijevic, I. A., Russell, J. A., Bicknell, R. J., Leng, G., and Douglas, A. J. (2000). Effect of progesterone on the activation of neurones of the supraoptic nucleus during parturition. *J. Reprod. Fertil.* 120, 367–376.
757. Grazzini, E., Guillon, G., Mouillac, B., and Zingg, H. H. (1998). Inhibition of oxytocin receptor function by direct binding of progesterone. *Nature* 392, 509–512.
758. Dunlap, K. A., and Stormshak, F. (2004). Nongenomic inhibition of oxytocin binding by progesterone in the ovine uterus. *Biol. Reprod.* 70, 65–69.
759. Francis, K., Meddle, S. L., Bishop, V. R., and Russell, J. A. (2002). Progesterone receptor expression in the pregnant and parturient rat hypothalamus and brainstem. *Brain Res.* 927, 18–26.
760. Haywood, S. A., Simonian, S. X., van der Beek, E. M., Bicknell, R. J., and Herbison, A. E. (1999). Fluctuating estrogen and progesterone receptor expression in brainstem norepinephrine neurons through the rat estrous cycle. *Endocrinology* 140, 3255–3263.
761. Hagihara, K., Hirata, S., Osada, T., Hirai, M., and Kato, J. (1992). Distribution of cells containing progesterone receptor mRNA in the female rat di- and telencephalon: an in situ hybridization study. *Brain Res. Mol. Brain Res.* 14, 239–249.
762. Rainbow, T. C., McGinnis, M. Y., Krey, L. C., and McEwen, B. S. (1982). Nuclear progestin receptors in rat brain and pituitary. *Neuroendocrinology* 34, 426–432.
763. Shughrue, P. J., Lane, M. V., and Merchenthaler, I. (1997). Regulation of progesterone receptor messenger ribonucleic acid in the rat medial preoptic nucleus by estrogenic and antiestrogenic compounds: an in situ hybridization study. *Endocrinology* 138, 5476–5484.
764. Thomas, A., Shughrue, P. J., Merchenthaler, I., and Amico, J. A. (1999). The effects of progesterone on oxytocin mRNA levels in the paraventricular nucleus of the female rat can be altered by the administration of diazepam or RU486. *J. Neuroendocrinol.* 11, 137–144.
765. Herbison, A. E. (2001). Physiological roles for the neurosteroid allopregnanolone in the modulation of brain function during pregnancy and parturition. *Prog. Brain Res.* 133, 39–47.
765a. Russell, J. A., and Brunton, P. J. (In press) Neuroactive steroids attenuate oxytocin stress responses in late pregnancy. *Neuroscience.*
766. Higuchi, T., Bicknell, R. J., and Leng, G. (1991). Reduced oxytocin release from the neural lobe of lactating rats is associated with reduced pituitary content and does not reflect reduced excitability of oxytocin neurons. *J. Neuroendocrinol.* 3, 297–302.
767. Blyth, B. J., Hauger, R. L., Purdy, R. H., and Amico, J. A. (2000). The neurosteroid allopregnanolone modulates oxytocin expression in the hypothalamic paraventricular nucleus. *Am. J. Physiol. Regul. Integr. Comp. Physiol.* 278, R684–R691.
768. Shughrue, P. J., Dellovade, T. L., and Merchenthaler, I. (2002). Estrogen modulates oxytocin gene expression in regions of the rat supraoptic and paraventricular nuclei that contain estrogen receptor-beta. *Prog. Brain Res.* 139, 15–29.
769. O'Keefe, J. A., Crowley, R. S., Hrivnak, P., Kim, N. B., and Amico, J. A. (1995). Effect of testosterone and its metabolites

upon the level of vasopressin messenger ribonucleic acid in the hypothalamus of the hyperosmotically stimulated male rat. *Neuroendocrinology* 61, 405–411.

770. Nomura, M., McKenna, E., Korach, K. S., Pfaff, D. W., and Ogawa, S. (2002). Estrogen receptor-beta regulates transcript levels for oxytocin and arginine vasopressin in the hypothalamic paraventricular nucleus of male mice. *Brain Res. Mol. Brain Res.* 109, 84–94.
771. Swenson, K. L., and Sladek, C.D. (1997). Gonadal steroid modulation of vasopressin secretion in response to osmotic stimulation. *Endocrinology* 138, 2089–2097.
772. Somponpun, S., and Sladek, C. D. (2002). Role of estrogen receptor-beta in regulation of vasopressin and oxytocin release in vitro. *Endocrinology* 143, 2899–2904.
773. Somponpun, S. J., and Sladek, C. D. (2003). Osmotic regulation of estrogen receptor-beta in rat vasopressin and oxytocin neurons. *J. Neurosci.* 23, 4261–4269.
774. Douglas, A. J., Johnstone, H., Brunton, P., and Russell, J. A. (2000). Sex-steroid induction of endogenous opioid inhibition on oxytocin secretory responses to stress. *J. Neuroendocrinol.* 12, 343–350.
775. Warembourg, M., and Leroy, D. (2004). Comparative distribution of estrogen receptor alpha and beta immunoreactivities in the forebrain and the midbrain of the female guinea pig. *Brain Res.* 1002, 55–66.
776. Kuiper, G. G., Carlsson, B., and Grandien, K., et al. (1997). Comparison of the ligand binding specificity and transcript tissue distribution of estrogen receptors alpha and beta. *Endocrinology* 138, 863–870.
777. Somponpun, S. J., Johnson, A. K., Beltz, T., and Sladek, C. D. (2004). Osmotic regulation of estrogen receptor-beta expression in magnocellular vasopressin neurons requires lamina terminalis. *Am. J. Physiol. Regul. Integr. Comp. Physiol.* 286, R465–R473.
778. Greco, B., Lubbers, L. S., and Blaustein, J. D. (2003). Estrogen receptor beta messenger ribonucleic acid expression in the forebrain of proestrous, pregnant, and lactating female rats. *Endocrinology* 144, 1869–1875.
779. Stedronsky, K., Telgmann, R., Tillmann, G., Walther, N., and Ivell, R. (2002). The affinity and activity of the multiple hormone response element in the proximal promoter of the human oxytocin gene. *J. Neuroendocrinol.* 14, 472–485.
780. Coleman, K. M., Dutertre, M., and El Gharbawy, A., et al. (2003). Mechanistic differences in the activation of estrogen receptor-alpha (ER alpha)- and ER beta-dependent gene expression by cAMP signaling pathway(s). *J. Biol. Chem.* 278, 12834–12845.
781. Bridges, R. S., and Ronsheim, P. M. (1987). Immunoreactive beta-endorphin concentrations in brain and plasma during pregnancy in rats: possible modulation by progesterone and estradiol. *Neuroendocrinology* 45, 381–388.
782. Douglas, A. J., Bicknell, R. J., Leng, G., Russell, J. A., and Meddle, S. L. (2002). Beta-endorphin cells in the arcuate nucleus: projections to the supraoptic nucleus and changes in expression during pregnancy and parturition. *J. Neuroendocrinol.* 14, 768–777.
783. Douglas, A. J., Neumann, I., and Meeren, H. K., et al. (1995). Central endogenous opioid inhibition of supraoptic oxytocin neurons in pregnant rats. *J. Neurosci.* 15, 5049–5057.
784. Ericsson, A., Arias, C., and Sawchenko, P. E. (1997). Evidence for an intramedullary prostaglandin-dependent mechanism in the activation of stress-related neuroendocrine circuitry by intravenous interleukin-1. *J. Neurosci.* 17, 7166–7179.
785. Meister, B., Cortes, R., Villar, M. J., Schalling, M., and Hokfelt, T. (1990). Peptides and transmitter enzymes in hypothalamic magnocellular neurons after administration of hyperosmotic stimuli: comparison between messenger RNA and peptide/protein levels. *Cell Tissue Res.* 260, 279–297.
786. Doi, N., Brown, C. H., Cohen, H. D., Leng, G., and Russell, J. A. (2001). Effects of the endogenous opioid peptide, endomorphin 1, on supraoptic nucleus oxytocin and vasopressin neurones in vivo and in vitro. *Br. J. Pharmacol.* 132, 1136–1144.
787. Brown, C. H., Russell, J. A., and Leng, G. (2000). Opioid modulation of magnocellular neurosecretory cell activity. *Neurosci. Res.* 36, 97–120.
788. Ingram, C. D., Kavadas, V., Thomas, M. R., and Threapleton, J. D. (1996). Endogenous opioid control of somatodendritic oxytocin release from the hypothalamic supraoptic and paraventricular nuclei in vitro. *Neurosci. Res.* 25, 17–24.
789. Sawchenko, P. E., Arias, C., and Bittencourt, J. C. (1990). Inhibin beta, somatostatin, and enkephalin immunoreactivities coexist in caudal medullary neurons that project to the paraventricular nucleus of the hypothalamus. *J. Comp. Neurol.* 291, 269–280.
790. Kutlu, S., Yilmaz, B., and Canpolat, S., et al. (2004). Mu opioid modulation of oxytocin secretion in late pregnant and parturient rats. Involvement of noradrenergic neurotransmission. *Neuroendocrinology* 79, 197–203.
791. Brunton, P. J., Meddle, S. L., and Ma, S., et al. (2005). Endogenous opioids and attenuated hypothalamo–pituitary–adrenal axis responses to immune challenge in pregnant rats. *J. Neurosci.* 25, 5117–5126.
792. Mansour, A., Fox, C. A., and Burke, S. et al. (1994). Mu, delta, and kappa opioid receptor mRNA expression in the rat CNS: an in situ hybridization study. *J. Comp. Neurol.* 350, 412–438.
793. Bicknell, R. J., and Leng, G. (1982). Endogenous opiates regulate oxytocin but not vasopressin secretion from the neurohypophysis. *Nature* 298, 161–162.
794. Leng, G., Bicknell, R. J., and Brown, D., et al. (1994). Stimulus-induced depletion of pro-enkephalins, oxytocin and vasopressin and pro-enkephalin interaction with posterior pituitary hormone release in vitro. *Neuroendocrinology* 60, 559–566.
795. Leng, G., Dye, S., and Bicknell, R. J. (1997). Kappa-opioid restraint of oxytocin secretion: plasticity through pregnancy. *Neuroendocrinology* 66, 378–383.
796. Panula, P., and Lindberg, I. (1987). Enkephalins in the rat pituitary gland: immunohistochemical and biochemical observations. *Endocrinology* 121, 48–58.
797. Ortiz-Miranda, S. I., Dayanithi, G., and Coccia, V., et al. (2003). mu-Opioid receptor modulates peptide release from rat neurohypophysial terminals by inhibiting Ca$^{(2+)}$ influx. *J. Neuroendocrinol.* 15, 888–894.
798. Bondy, C. A., Gainer, H., and Russell, J. T. (1987). Effects of stimulus frequency and potassium channel blockade on the secretion of vasopressin and oxytocin from the neurohypophysis. *Neuroendocrinology* 46, 258–267.
799. Leng, G., Shibuki, K., and Way, S. A. (1988). Effects of raised extracellular potassium on the excitability of, and hormone release from, the isolated rat neurohypophysis. *J. Physiol.* 399, 591–605.
800. Jackson, M. B., Konnerth, A., and Augustine, G. J. (1991). Action potential broadening and frequency-dependent facilitation of calcium signals in pituitary nerve terminals. *Proc. Natl. Acad. Sci.* U S A 88, 380–384.
801. Jones, S. A., and Summerlee, A. J. (1986). Relaxin acts centrally to inhibit oxytocin release during parturition: an effect that is reversed by naloxone. *J. Endocrinol.* 111, 99–102.
802. Kendrick, K. M. (2000). Oxytocin, motherhood and bonding. *Exp. Physiol.* 85(Spec No), 111S–124S.

803. Levy, F., Kendrick, K. M., Keverne, E. B., Piketty, V., and Poindron, P. (1992). Intracerebral oxytocin is important for the onset of maternal behavior in inexperienced ewes delivered under peridural anesthesia. *Behav. Neurosci.* 106, 427–432.
804. Holmes, M. C., Antoni, F. A., Aguilera, G., and Catt, K. J. (1986). Magnocellular axons in passage through the median eminence release vasopressin. *Nature* 319, 326–329.
805. Gimpl, G., and Fahrenholz, F. (2002). Cholesterol as stabilizer of the oxytocin receptor. *Biochim. Biophys. Acta* 1564, 384–392.
806. Hasbi, A., Devost, D., Laporte, S. A., and Zingg, H. H. (2004). Real-time detection of interactions between the human oxytocin receptor and G protein-coupled receptor kinase-2. *Mol. Endocrinol.* 18, 1277–1286.
807. Fuchs, A. R., Rust, W., and Fields, M. J. (1999). Accumulation of cyclooxygenase-2 gene transcripts in uterine tissues of pregnant and parturient cows: stimulation by oxytocin. *Biol. Reprod.* 60, 341–348.
808. Di, M. V., Bifulco, M., and De Petrocellis, L. (2004). The endocannabinoid system and its therapeutic exploitation. *Nat. Rev. Drug Discov.* 3, 771–784.
809. Koksma, J. J., van Kesteren, R. E., and Rosahl, T. W., et al. (2003). Oxytocin regulates neurosteroid modulation of GABA(A) receptors in supraoptic nucleus around parturition. *J. Neurosci.* 23, 788–797.

Knobil and Neill's Physiology of Reproduction, Third Edition
edited by Jimmy D. Neill,
Elsevier © 2006

CHAPTER 59

Milk Ejection and Its Control

J. B. Wakerley

Anatomical Adaptations Related to Transfer of Milk to the Offspring, 3130
General Structure of the Lactating Mammary Gland, 3130
Storage of Milk within the Mammary Gland, 3131
Contractile Tissue of the Mammary Gland, 3132
Structural Adaptations of the Nipples for Milk Retention and Removal, 3133
Pattern of Milk Transfer, 3134
Maternal Nursing Behavior, 3134
Sucking Behavior and the Dynamics of Milk Removal, 3134
Functional Significance of Active Milk Ejection in Different Species, 3136
Role of Oxytocin in the Milk-Ejection Reflex, 3137
Stimuli for Activation of Oxytocin Release, 3137
Plasma Oxytocin Concentrations during Suckling, 3139
Mode of Action of Oxytocin on the Myoepithelium, 3140
Factors Governing the Intramammary Pressure Response to Oxytocin, 3141
Mechanically Induced Contractions of the Myoepithelium, 3142
Hypothalamic Regulation of Oxytocin Release during Milk Ejection, 3142
Organization of Hypothalamic Oxytocin Neurons and Some Methodological Considerations, 3142
Firing Patterns of Oxytocin Neurons during Suckling, 3145
Molecular Correlates of the Activation of Oxytocin Neurons during Suckling: Fos Expression, 3147
Neural Mechanisms Underlying Synchronous Bursting Activity in Oxytocin Neurons, 3148
Autoregulatory Role of Oxytocin in the Bursting Activity of Oxytocin Neurons, 3156
Control of Release from Oxytocinergic Neurosecretory Terminals at the Level of the Neurohypophysis, 3161
Adaptations of Oxytocin Neurons for Lactation, 3163
Ascending and Descending Neural Pathways Involved in the Regulation of Oxytocin Release during Milk Ejection, 3165
Sensory Receptors and Primary Afferent Fibers, 3165
Transmission of the Sucking Stimulus through the Spinal Cord, 3166
Transmission of the Sucking Stimulus at the Supraspinal Level, 3167
Descending Pathways Mediating Conditioning and Inhibition of the Milk-Ejection Reflex, 3168
Neurotransmitters and Other Neuromodulators in the Milk-Ejection Reflex, 3170
Acetylcholine, 3170
Monoamines and Related Transmitters, 3170
Excitatory and Inhibitory Amino Acids, 3172
Neuropeptides, 3173
Nitric Oxide, 3174
Conclusion, 3175
Acknowledgments, 3176
References, 3176

The basic components of the milk-ejection reflex were first described in detail by Ely and Petersen over 60 years ago (1), and since then there has been a steady expansion in our knowledge of this fascinating neuroendocrine mechanism. This expansion has, if anything, accelerated over the last few years, and substantial new information has been added concerning all aspects of the milk-ejection reflex since the previous edition of this volume. However, perhaps it is in relation to the functional organization of

Department of Anatomy, University of Bristol, School of Medical Sciences, Bristol, U.K.

oxytocin neurons within the hypothalamus where the most spectacular advances have been achieved. As will become evident in this chapter, the advent of new methodological approaches, particularly the development of in vitro preparations for studying oxytocin neurons in complete or relative isolation, has lead to an explosion of information in this area. Doubtless, some readers might argue that the whole of this chapter should be focused on the hypothalamus to reflect this concentration of effort. However, in this age of increasing scientific specialization and the tendency toward the "reductionist" approach, we should be careful not to lose sight of the bigger picture, namely the physiological context in which the microdissected elements operate within the whole animal. Hence, although the larger part of this chapter is concerned with the central neural mechanisms involved in the regulation of oxytocin released during suckling, some space is also devoted to more general information concerning the physiological importance and comparative aspects of the milk-ejection reflex as well as the organization of its peripheral components.

ANATOMICAL ADAPTATIONS RELATED TO TRANSFER OF MILK TO THE OFFSPRING

General Structure of the Lactating Mammary Gland

The mammary glands probably have their evolutionary beginnings in the endothermic therapsids, which had specialized area in the skin of the abdomen for incubating the young (2). The sweat glands of this incubation area became specialized and enlarged, their secretions originally serving to prevent desiccation of the eggs and of the newly hatched offspring. The mammary glands in present-day mammalian forms show considerable diversity in structure but nevertheless retain their obvious anatomical association with the skin and share the cutaneous supply of blood vessels and nerves (3). Moreover, the mammary glands of all species retain the relatively simple tubuloalveolar arrangement of the secretory tissue reminiscent of the tubular secretory units of the sweat glands from which they originated (2).

In one of the most primitive of living mammals, the echidna, there are two mammary glands, each comprising 100 to 150 lobules, drained by separate ducts that open into invaginations of skin from which a mammary hair projects (4). The mammary ducts open onto a specialized areola region to which the young appear to firmly attach; they do not simply lick the mammary hairs as once supposed. In higher mammals this relatively primitive arrangement has become modified by (a) the development of a prehensile nipple, varying in size and prominence, and (b) the development of a complex and extensively branching duct system with a corresponding reduction in the number of ducts opening onto the surface. Thus, in all Metatherian and Eutherian mammals, the lobules of the mammary glands are arranged in lobes, and each lobe is drained by its own duct system. The arrangement of the ducts varies according to whether the ducts from each lobe join together before opening onto the surface of the nipple or remain separate (Fig. 1).

Detailed accounts of the variation in form, relative size, number, and position of the mammary glands found among present-day mammals have been provided by other authors (5,6). The present-day variation in the structure and position of the mammary glands has arisen from evolutionary pressures related to the number and relative maturity of the offspring. Other important factors of the mammary glands are to cause least impairment to maternal locomotion and to be optimally placed for attachment of the young yet minimally exposed to damage from the physical environment. The number of mammary glands ranges from a minimum of 2, as in humans, up to the 26, found in the opossum. In species with numerous glands, these are distributed throughout the entire axillary-inguinal embryonic milk line (e.g., rat, pig), but otherwise they are restricted to the thoracic (elephant, gorilla), abdominal (whales, seals), or

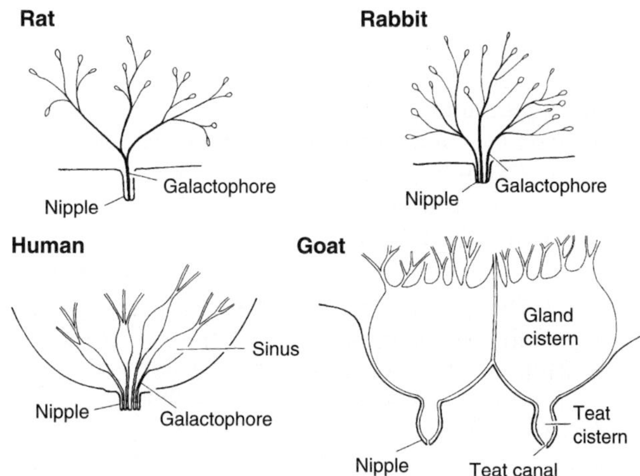

FIG. 1. Examples of the species variation in the arrangement of the milk ducts. In the rat, the milk ducts join to form a single galactophore within the nipple, whereas in the rabbit the ducts remain separate, such that there are some six to eight galactophores opening onto the surface. In humans some 12 to 20 galactophores are present, and these have a further modification in the form of an enlarged lumen (sinus) close to the nipple. In the goat (as in other ruminants) there is extreme specialization of the duct system to increase storage capacity, with the milk ducts opening into a large gland cistern. The gland cistern connects with a smaller teat cistern, which opens onto the surface via the teat canal. [Based on (6).]

inguinal regions. In the horse there are four inguinal mammary glands but only a single pair of teats. The cow, like other ruminants, has four inguinal mammary glands, which are grouped together into a pendulous udder. The coypu, with a mainly aquatic life-style, has its mammary glands on the dorsal surface. In the order Cetacea (whales, dolphins, and porpoises), the mammary glands are elongated narrow flat organs on either side of the ventral midline between the anus and the umbilicus, the nipples being situated on either side of the genitalia. In the order Pinnipediae (sea lions, walruses, and seals) there are usually two pairs of mammary glands, one pair lying caudal and the other cranial to the umbilicus. The glands lie beneath the blubber, and the nipples are hidden beneath a special flap of skin called the pilage.

The secretory tissue of the mammary gland is made up of numerous alveoli, each consisting of a hollow ball of secretory cells surrounded by a network of contractile myoepithelium and capillaries. The alveoli are embedded in a stroma of loose intralobular connective tissue and dense interlobular connective tissue. Thick septa lie between the lobes, and in those forms with pendulous udders, the septa connect with suspensory ligaments anchored to the abdominal wall and skeleton, which help to support the mammary glands (7). The small ducts are lined by secretory cells, whereas the larger ducts and sinuses are lined by a nonsecretory two-layered cuboidal epithelium. The development and growth of the mammary gland are considered in detail elsewhere in Chapter 57.

Storage of Milk within the Mammary Gland

It is important to appreciate that milk secretion is a continuous process that is not restricted to the episodes of nursing. Apart from certain exceptions, such as in marsupials where the youngest offspring are continuously attached to the nipples, nursing occurs intermittently and there has to be an adequate means of storing milk within the mammary glands. If the alveoli become grossly overdistended with milk, this initiates biochemical changes triggering involution of the secretory tissue (8). Hence, it is critically important that the storage capacity of the mammary gland is adequate to avoid any buildup of excess pressure between the nursing episodes. The mammals can be broadly divided into two groups: (a) those species without obvious modification (other than the alveoli) to increase storage and (b) those species in which the duct system has become modified to provide storage space. As we will see, the site of milk storage is an important factor in determining how easily milk can be removed from the mammary glands.

The first group represents the most common situation and includes species such as the rat, rabbit, pig, and cat. In these species the ducts are relatively narrow in diameter, and the alveoli provide by far the largest storage compartment, probably accounting for over 70% to 80% of the milk stored within the gland (9). The alveoli are capable of considerable distension, increasing their diameter perhaps more than tenfold in the engorged gland. In the absence of any special adaptations for milk storage, the capacity is more or less proportional to the total number of alveoli within the mammary gland. In species without substantial alternative storage sites, the alveoli can provide considerable storage capacity. The rat, for example, normally nurses her offspring frequently (every 1 to 2 hours) but, if forcibly separated from her litter, can store some 12 g of milk in her mammary glands, equivalent to about 30% of the daily milk yield (10). The rabbit routinely accumulates a full day's milk production (150 to 200 g) in her mammary glands before nursing her young (11).

The most striking examples of the other group (those with modification for storage space) are provided by the ruminants. In the cow there is a special gland enlargement immediately above the teat that constitutes the so-called cistern of the gland. This cistern holds 100 to 400 ml of milk, and the 5 to 20 large sinus-like ducts that feed into it are also distensible to provide additional storage space (7). As the milk is secreted, it initially fills the alveoli before the cistern and ducts gradually become distended. Movement of milk to the cisterns has been studied using ultrasonography and is accelerated by injections of oxytocin (12). In the cow only about 20% of the stored milk can be held in the cistern and ducts, but in the goat, which has a similar anatomical arrangement, over 80% may be thus held (7,9). Another storage modification in ruminants is that the nipple and its duct system have evolved to form a teat containing a teat cistern. However, the teat cistern when distended can hold only about 25 ml of milk (7) and is probably more important in relation to the mechanism of suckling.

Aside from ruminants, other groups also show evidence of anatomical specialization for milk storage. In the primates (including women), the ducts leading into the nipple are dilated to form sinuses. The African elephant has a teat with several milk ducts, and in the proximal portions of these ducts the walls are greatly folded, suggesting they are capable of considerable dilation (13). In the Cetacea (whales, dolphins, and porpoises) there is a single milk duct that forms an expanded sinus as it approaches the nipple (5). However, these specializations are probably more significant in relation to the mechanics of suckling than for the long-term storage of the milk between nursing episodes. Thus, in most groups apart from ruminants, the alveoli form the most important storage site.

Contractile Tissue of the Mammary Gland

The contractile tissue within the substance of the mammary gland comprises (a) specialized myoepithelial cells that invest the alveoli and small ducts and (b) smooth muscle surrounding the larger milk ducts and blood vessels. Although the smooth muscle around the ducts may play some part in regulating the flow of milk from the gland, there is no evidence that it contributes to the expulsive force during milk ejection, which is provided entirely by the myoepithelial cells. Richardson (14), studying thick (100 μm) sections of goat mammary tissue stained by silver impregnation, provided the first detailed description of the abundant network of myoepithelial cells around the alveoli, ducts, and cisterns, and these have since been studied by scanning electron microscopy (15) (Fig. 2). Linzell (16) extended these observations to the mammary gland of the cat, dog, rat, and humans. Myoepithelial cells have also been described in the African elephant (13). At the time of their discovery, the arrangement of myoepithelial cells around the alveoli indicated that these cells had a role in the expulsion of milk, and this was confirmed when Linzell (17) observed directly the contraction of mouse mammary alveoli in response to topical application of oxytocin. Myoepithelial cells also surround the small ducts of the mammary gland, tending to be spindle- rather than star-shaped and arranged such that their contraction increases the lumen of the ducts (17). This has an obvious benefit in facilitating the passage of milk, especially because mammary flow resistance can be a significant factor in milk transfer (18). As well as their contractile function, myoepithelial cells also serve a structural supportive role, particularly in the involuting gland (19).

Myoepithelial cells are present in the mammary glands of virgin females, but their numbers increase during pregnancy, their proliferation being regulated partly by sex steroids, especially estrogens (20,21). The cells do not complete morphological differentiation until around the onset of lactation, when they become packed with myofilaments, and this correlates with the appearance of contractile activity. Fully developed myoepithelial cells have an elongated lobular nucleus and numerous mitochondria, which may become displaced to the perinuclear area as the abundant myofilaments develop. Pinocytotic vesicles are present on all surfaces, but whether these are involved in the contraction mechanism or transport of materials is uncertain. Desmosomes form attachment

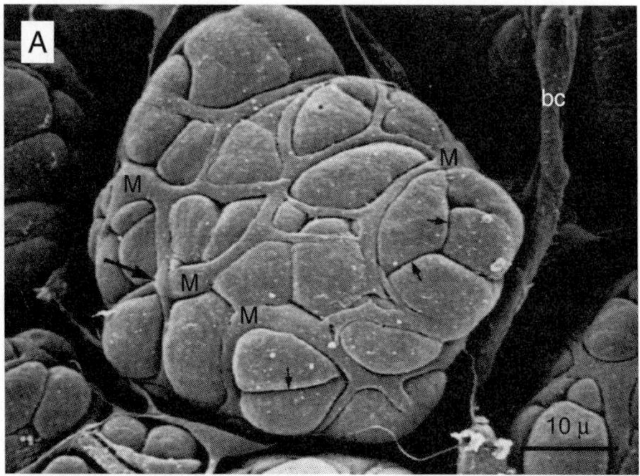

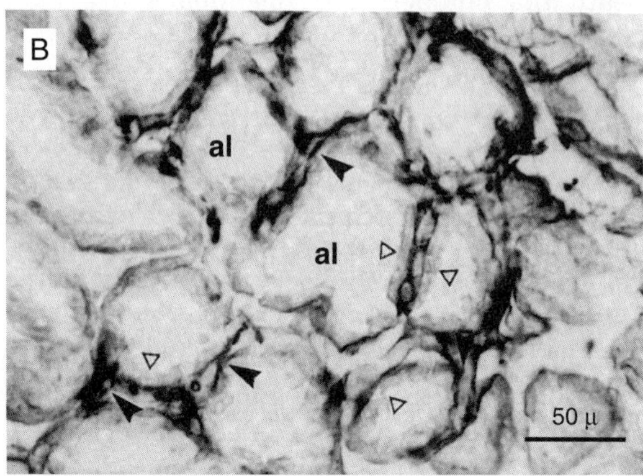

FIG. 2. A: Scanning electron micrograph of rat mammary gland to show the network of myoepithelial cells (each separate cell marked by M) surrounding an alveolus. The *large arrow* indicates overlap of two adjacent myoepithelial cells, and the *small arrows* mark the boundaries of adjacent secretory cells (bc) blood capillary. **B:** Light microscope immunocytochemistry of oxytocin receptors within the alveoli (al). These are located on the myoepithelial cells (*large filled arrows*) but not the secretory cells (*small open arrows*). [A is reproduced from (15) and B from (22), with permission.]

between adjacent myoepithelial cells, which occasionally have overlapping processes, but other specialized close contacts are seen infrequently. The myoepithelium does not form a true syncytium, but membranes of adjacent cells are closely apposed, perhaps forming a basis for electrotonic coupling. Myofilaments in myoepithelial processes around ducts are long and straight and arranged in clear bundles, whereas those in processes around alveoli are more closely packed and not arranged as discrete bundles (20,21).

Both oxytocin receptors (22) (Fig. 2) and oxytocin receptor mRNA have been located in the mammary gland (23). Curiously, in the human mammary gland the oxytocin receptor is expressed to a greater degree in alveolar epithelial cells than in the myoepithelial cells (24). Oxytocin receptors begin to be expressed on the myoepithelial cells shortly before the onset of parturition and disappear during weaning (25,26). The timing of the appearance of oxytocin receptors correlates with the development of sensitivity to exogenous oxytocin in late pregnancy (27). However, unlike oxytocin receptors in the uterus, the mammary oxytocin receptors do not appear to be directly regulated by estrogen (23) so that the factors governing the peripartum increase in mammary oxytocin receptors remain to be clarified.

The mammary gland has also been reported to contain vasopressin receptors of the V1 type, which may mediate stimulation of the intracellular messenger inositol triphosphate, possibly within the vasculature (28). It has been reported in the goat that vasopressin at physiological concentrations may act on the myoepithelium to induce milk ejection responses (29). However, there is no evidence in the literature that suckling evokes vasopressin release in any species.

Structural Adaptations of the Nipples for Milk Retention and Removal

Well-defined nipples are present in all species of Metatherian and Eutherian mammals but are not present in the Monotremes. Where the nipple contains a storage cistern, as in the ruminants, it is usually referred to as a teat. The nipple subserves three interrelated functions: (a) to provide a prehensile structure for attachment of the offspring, with a conducting channel for the milk to be passed into their buccal cavity; (b) to restrict leakage from the gland between nursing episodes; and (c) to provide sensory input for activation of the milk-ejection reflex by the sucking stimulus. The nipple consists of a specialized covering of glabrous skin and a core of connective tissue in which are embedded blood vessels, nerves, and one or more milk ducts. In some species (e.g., rat) the skin around the base of the nipple is adapted to form a small pouch, into which the nipple retracts between nursing episodes.

The human nipple, which has been extensively described (30,31) is covered by a stratified squamous epithelium, with numerous papillae extending from the dermis up to epidermis, enabling the dermal nerve endings to be close to the surface. Deep within the nipple are the 10 to 12 lactiferous ducts, also termed galactophores. These are lined by a stratified squamous epithelium, which changes to the double-layered columnar epithelium lining the sinuses and large ducts of the gland. There may be a plug of keratin in the resting gland, secreted by the cells lining the ducts. At the tip of the nipple (but not the sides) are groups of sebaceous glands associated with the openings of the galactophores. The substance of the nipple consists of connective tissue (with a framework of elastic fibers) and smooth muscle, of significance in the erection of the nipples for helping attachment by the offspring. The blood vessels of the nipple may become engorged with blood during sucking, so contributing to the erection process. As in most other species, there is no obvious sphincter for retention of the milk in the human, and it is well known that milk will leak from an unsuckled nipple during milk ejection. Surrounding the base of the nipple is the pigmented areola region containing raised tubercles, which are called Montgomery's glands. These tubercles are now known (31) to contain the openings of galactophores, which, as at the tip of the nipple, are associated with sebaceous glands. The secretory alveoli opening into these areolar galactophores are identical to those in the main part of the mammary gland. The areola also contains sweat glands but few hair follicles. Smooth muscle fibers are arranged radially and circularly within the areola around the base of the nipple, and, when contracted, these serve to elevate the nipple.

The teat of the cow has also been studied in detail (32,33). The skin on the teat is thin stratified squamous epithelium and devoid of hair, thus ensuring that a good "milk-tight" seal can be formed within the mouth of the calf and that the nerve endings in the skin are maximally sensitive to the mechanical and thermal stimuli associated with sucking. In fact, this probably explains the absence of hair on the nipple in all species, but this adaptation has one disadvantage: the tendency for the skin of the nipple to become ulcerated or chapped, which may seriously impair nursing. In the teat wall there is an outer network of smooth muscle arranged longitudinally, whereas in the deeper portions the fibers are arranged circularly and intermingle with bundles of collagen and elastic fibers. The teat cistern is lined by a double-layered epithelium and opens onto the surface via a single muscular-walled canal, known as the papillary duct or streak canal, lined by stratified squamous epithelium. Retention of the milk is aided by folds of mucosa at the inner end of the streak canal, which press down and occlude its opening, helping to prevent the escape of milk under the influence of gravity. The muscle wall of the streak canal is under autonomic control (34,35), and adrenergic agonists alter milk flow through the teat canal (36). The teat sphincter muscle can undergo rhythmic contractions that may help move milk from the teat cistern (37).

Nipples usually develop in both sexes, but in a few species (e.g., rat, mouse) sex differences occur in the pattern of fetal mammary growth such that no nipples are formed in the male (38). Subsequent enlargement of the nipple in the female is controlled by ovarian and placental hormones, particularly estrogens.

During pregnancy there is an increase in the protractility of the nipple, and this is also under hormonal control (39). Protractility of the nipple can be increased by mechanical stretching (40), and mechanical factors could be significant in the increased self-licking of the nipple lines observed in the rat during pregnancy (41). Mechanical factors may also underlie the differential development of adjacent nipples, as occurs in polytocous species with abnormally small litters, where the inguinal nipples receive less attention from the young and are poorly developed.

PATTERN OF MILK TRANSFER

Maternal Nursing Behavior

Mammals nurse their offspring in a variety of environmental conditions and circumstances and display enormous diversity in the pattern of their nursing behavior [see review in (42)]. For example, some marine mammals, such as whales, manatees, and dugongs, nurse their young while swimming, whereas others, such as seals, heave themselves out of water and onto dry land to nurse (43). Terrestrial mammals that nurse in the bottom of burrows (e.g., rabbits) tend to lie down or crouch over their young, those that nurse on the open Savannah (wildebeest, zebra) are usually standing, and those that nurse in trees (monkeys, opossums, and lemurs) tend to sit and clutch their young to the nipple with an upright posture. Some species, such as the rat, nurse their young in a somnolent state (44,45), whereas others, such as the rabbit (46) and the pig (47), appear to be alert and aroused during nursing. Ruminants seem almost indifferent to nursing and often feed their young without interrupting their grazing activity.

There is also much variation in the frequency and duration of nursing among the different species. In some marsupials (e.g., agile wallaby, red kangaroo) the young joey becomes continuously attached to the nipple for many weeks before becoming mature enough to leave the pouch to return intermittently to suck (48). Marsupials are particularly interesting in that they show concurrent but asynchronous lactation in their mammary glands, whereby one offspring may be continuously attached to the nipple while the other is suckling intermittently. Laboratory rats nurse their young for a total of 12 to 18 hours per day with frequent periods of nursing lasting 20 to 30 minutes (10). At the other extreme are rabbits, which feed their offspring only once per day (49) for just 4 to 5 minutes, and tree shrews, which nurse every 2 days (50). Between these two extremes are species such as the domestic pig, which nurses approximately once per hour for 10-minute periods (51). In the horse suckling occurs for 1 to 2 minutes every 20 minutes (52). The bottle-nosed whale and the dolphin feed their calves at half-hourly intervals (43). Women nurse approximately every 4 hours, although this varies in different cultures: in the !Kung tribe of Africa, the infant may suckle every few minutes night and day (53). Studies in the rabbit (54) suggest that motivation to nurse the young depends on several interacting factors, including the degree of mammary distension and how much time has elapsed since the previous nursing episode.

Sucking Behavior and the Dynamics of Milk Removal

Nipple seeking, although helped by encouragement from the mother, is an innate behavioral pattern of the neonate (55). Newborn human offspring display an inborn sequence of behavior, including massaging and licking the nipple, before they first attach to the nipple (56). In animals it has been shown that the continuation of nipple seeking after the first few days of life is facilitated by positive reinforcement. After about 20 days of artificial rearing by force feeding, young kittens lose the ability to locate and attach to the nipples when returned to a lactating dam (57). Successful location of the nipple depends on tactile sensation, but olfactory cues may also be important. Young rats attach more rapidly to a nipple if it has recently been sucked, and thus is protracted and coated with milky saliva, than one that has not been sucked for some hours (58). Location of the nipple or teat is assisted by erection of the nipple, which involves contraction of smooth muscle under sympathetic control in a process similar to piloerection. Erection of the nipple for attachment of the young would seem especially important in some marine mammals, where the teats are recessed to aid streamline performance (6). Once the offspring begins to suck, then mechanical stretching and vascular engorgement cause further enlargement of the nipple so that the young can become completely attached. In marsupials, such as the red kangaroo, the offspring are born in a very immature state, and attachment to the nipple represents a most remarkable feat involving considerable help from the mother, who licks a track within her fur toward the pouch (58,59). Once the young have the nipple within their mouths, continuous attachment is further secured within a few days by enlargement of the nipple and anatomical changes in the buccal cavity of the offspring.

The mechanism of sucking has been studied by measuring pressure changes within the buccal cavity (58,60), by ultrasonographic imaging (61,62), and by cineradiographic methods using radiopaque markers injected into the mammary gland (63) or smeared onto the nipple (64). In ruminants it has been shown that suction is of relatively minor importance during the actual transfer of milk to the offspring. Instead, transfer of milk is aided by compression of the teat cistern from its base toward its tip (stripping), using the tongue and hard palate. When the pressure at the base of the teat is relaxed, this enables the teat cistern to refill (possibly aided by suction), and the process is then repeated. Considerable quantities of milk can be removed without milk ejection in ruminants, so that the transfer of milk is probably a more prolonged and continuous process than in species totally reliant on intermittent activation of the milk-ejection reflex. The human infant forms a "teat" by drawing both the nipple and areola region (including the underlying milk sinuses) into the buccal cavity. The infant shows bursts of sucking activity lasting 1 to 2 minutes with pauses of 20 to 30 seconds between bursts. This involves a sequential pattern of rhythmical sucking, swallowing, and breathing. The related contributions of intraoral negative pressure and of nipple compression in milk transfer are still controversial, but both tongue and jaw movements are important (62). Milk flow from the breast correlates with sucking activity, and the fact that the bursts become shorter and the pauses longer is one reason why milk intake declines as the feed progresses (65,66). Sucking and gulping activity noticeably increase during periods of milk ejection. Most of the milk is obtained within the first few minutes of the nursing episode (64). It should be noted that both humans and animals rapidly adapt their sucking behavior when fed from a bottle, where suction is more important (62).

The improved survival rates for premature infants have led to investigations of how prematurity affects sucking behavior during breast-feeding (67). Feeding behavior of premature infants matures considerably between 33 and 36 weeks postconception, with an increase in sucking frequency and pressure. Infants under 33 weeks have difficulty in coordinating feeding and respiration, and swallowing is associated with respiratory pauses. After 35 weeks, swallowing infrequently interrupts respiration and occurs at the end of the inspiratory phase. However, premature infants at 40 weeks postconception still display differences in their sucking behavior compared with normal term infants of the same conceptual age (68).

Apart from ruminants and humans, the mechanism of sucking has been little studied in other forms, but it is generally assumed that in most species suction is more important for maintaining a grasp of the nipple and activating the milk-ejection reflex than for removing the milk. This is supported by observations (69) on rat pups that have had their salivary glands extirpated such that they can only suck hard on the nipple if an alternative sealant (paraffin jelly) is applied to the nipple. Such pups can obtain adequate quantities of milk provided they are reared in small litters. However, when reared in large litters they do not survive because they are too easily dislodged from the nipple by their littermates.

In many species, the transfer of milk is restricted to periods of active milk ejection, and the young usually display a marked change in their behavior when they receive the milk. The characteristic stretch reaction of the rat pup (Fig. 3) provides a dramatic example, consisting of cessation of movement with extension of the forelimbs and elevation of the tail (58). Between each milk ejection, the rat pup is more or less somnolent on the nipple, showing bursts of sucks every 15 to 20 seconds, but during the stretch reaction sucking increases dramatically. Such changes in behavior can be used to evaluate the timing of milk ejection (10). In addition to sucking, the offspring may

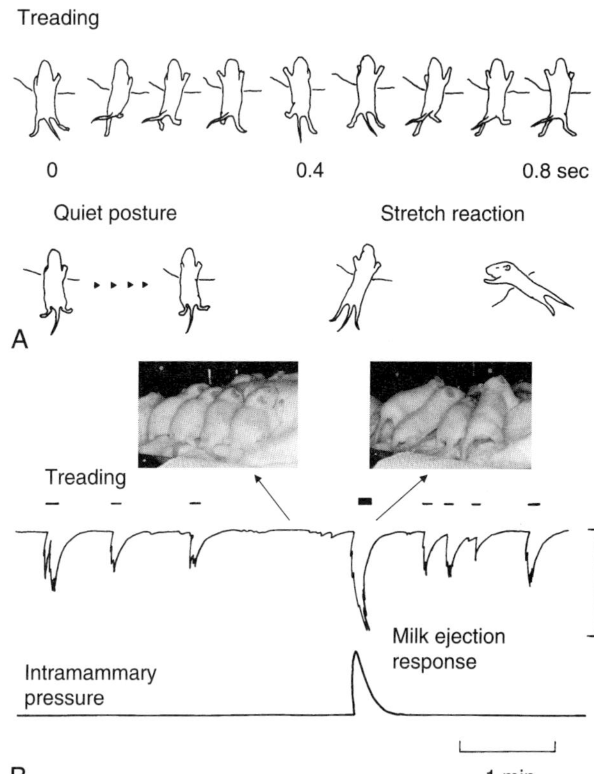

FIG. 3. Sucking-related behavior in the rat pup. **A:** Three types of behavior seen during suckling (drawings taken from cinegraphic film). **B:** Record of sucking activity (recorded through a small cannula opening at the tip of the nipple with the mother anesthetized) and the intramammary pressure changes during suckling. The pup was attached to the nipple throughout the recording. For most of the time the pup adopts a quiet posture, with no detectable sucking activity. Intermittently there are periods of treading combined with sucking activity. At the time of reflex milk ejection the pup adopts the highly characteristic stretch reaction with intense sucking. [Redrawn from (58) with permission.]

show treading activity (e.g., kittens) or may butt the mammary gland (e.g., calves). Such activity may help to increase positive pressure within the gland or stimulate mechanically induced contractions.

Functional Significance of Active Milk Ejection in Different Species

Oxytocin-induced contractions of the mammary myoepithelium, providing circumstantial evidence for a functional milk-ejection reflex, have been observed in a range of terrestrial eutherian mammals, including the cow, goat, guinea pig, mouse, pig, rat, sheep, and human [see review in (9)]. Far less is known about the milk-ejection reflex in marine forms, except that in the whale the mammary gland contains myoepithelial cells, and milk has been observed to be forcibly expelled from the nipples (70). The brevity of nursing in aquatic forms would seem to necessitate an active milk-ejection process.

Among the metatherian mammals, the wallaby has been studied in considerable detail (71), and in this species the milk-ejection reflex shows a fascinating adaptation for allowing two young at very different stages of maturity to be nursed simultaneously. In early lactation the mammary gland is extremely sensitive and shows rhythmic contractions during basal oxytocin release, whereas the mammary gland at later stages of lactation is only responsive to larger pulses of oxytocin. Hence, the release of a trickle of oxytocin enables continuous milk ejection from the gland supporting the youngest joey without producing inappropriate milk ejection in the other gland supporting the older joey. Evidence for the existence of a milk-ejection reflex within the monotremes has been provided by observations in the echidna (4) and the platypus (72), where injection of oxytocin caused visible contraction of the mammary lobules and milk to ooze onto the surface of the areola region. The existence of oxytocin receptors indicative of an active milk-ejection process has been demonstrated in the mammary gland of the brushtail possum (73).

The existence of a milk-ejection reflex does not, of course, mean that it is essential for normal transfer of milk to the young. However, definitive evidence that the reflex is essential has been provided by showing that damaging the hypothalamoneurohypophyseal system prevents milk removal in the rat, rabbit, cat, and dog (6,9). In the guinea pig, passive immunization against oxytocin impairs milk removal (74), and in the mouse, oxytocin knockouts are unable to feed their young unless they are given oxytocin replacement (75). Absence of oxytocin release in the domestic pig during suckling is associated with failure of the young to obtain milk (76).

The role of milk ejection in ruminants has been discussed in detail by Bruckmaier and Blum (77). In the cow, only the cisternal milk (about 20% of the stored milk) is available for removal in the absence of milk ejection (6) and oxytocin release associated with premilking teat stimulation is important for moving milk from the alveoli to the udder cistern (77). Further oxytocin release is then needed during the milking process to maintain the flow of milk into the cistern and ensure complete emptying of the gland. Milking in the cow causes a sustained low-level oxytocin release, and it has been proposed (78) that only low levels of oxytocin are required for milk ejection in this species. However, pulsatile release of oxytocin has also been reported during milking (79). The importance of oxytocin in the cow has been confirmed by showing that in cows injected with the oxytocin antagonist, atosiban, milk ejection is prevented and peak flow rates during milking are reduced (80,81). The situation in the goat and sheep is somewhat different because a far greater amount of milk is stored in the cistern, and premilking stimulation of milk ejection is therefore less important for enabling milk removal than in the cow. Nevertheless, oxytocin release has been reported during milking in the goat (82,83).

In women, powerful suction using a gland pump will allow some milk removal, apparently without active milk ejection, but the amounts are greatly increased by giving buccal oxytocin (84). Newton and Newton (85) reported that young infants can obtain only a small fraction of the milk without the involvement of the milk-ejection process. Women who successfully breast-feed show more symptoms of active milk ejection (e.g., uterine cramps, milk dripping from opposite breast) than do unsuccessful breast-feeders (86). Several workers have detected pulsatile oxytocin release during nursing in the human, but the levels were highly variable, and some women appear to show no oxytocin release despite a successful nursing episode (87–89). Lucas et al. (88) simultaneously measured milk flow and oxytocin release and found that 90% of the milk was removed within 4 minutes of the onset of suckling, even though oxytocin release was often absent during this period. Uvnas-Moberg et al. reported (90) no correlation between milk yield during breast-feeding and oxytocin levels. A possible explanation for these inconsistent results has been provided by McNeilly et al. (89), who found that many women showed a significant oxytocin release in response to cues such as the sight or sound of the infant, whereas the response to suckling was small or absent altogether. McNeilly et al. concluded that conditioned oxytocin release is particularly common in women, and failure to detect oxytocin release during the actual suckling period should not be taken as evidence against a role for the milk-ejection reflex in human lactation (90).

Further evidence for the importance of milk ejection in women has been provided by Ramsay and coworkers (91) using ultrasound imaging of a milk duct as a means of monitoring milk ejection (Fig. 4). Imaging was performed on the unsuckled side during nursing, and the occurrence of milk ejection was signaled by a transitory increase in the diameter of the milk duct as it dilated. Responses were observed in virtually 100% of successful breast-feeds and occurred either as a single large dilation (26% of cases) or as multiple dilations (74% of cases). There was an average of 2.5 ductal responses per breast-feed (average duration approximately 7 minutes), although some women showed more frequent responses. Ductal responses were associated with the sensation of milk ejection in the mother and gulping in the infant, and the number of responses was correlated with milk intake. Relating the ductal response pattern to plasma oxytocin levels may be complicated by possible recoil effects that can result in secondary contractions to a single oxytocin pulse. Nevertheless, the study showed good agreement between the frequency of ductal responses and previous reports of the pattern of oxytocin release (88,89). These findings confirm the importance of milk ejection in the nursing woman and demonstrate the usefulness of ultrasonography as a means of noninvasively monitoring milk ejection during breast-feeding.

ROLE OF OXYTOCIN IN THE MILK-EJECTION REFLEX

Stimuli for Activation of Oxytocin Release

The crucial importance of somatosensory stimuli in triggering the release of oxytocin has been demonstrated by experimental procedures such as local anesthesia of the nipples (92) or section of the spinal cord (93), which block the milk-ejection reflex by disturbing the flow of nerve impulses from the mammary glands. Further evidence is provided by the readiness with which milk ejection can be evoked by electrical stimulation of the nipples (94), the mammary nerves (95,96), or the spinal cord (97). In domestic animals, hand milking and machine milking are also potent stimuli for oxytocin release (78,79,82), and in women, tactile stimulation of the nipples (98) or with a breast pump (99,100) is effective. Postpartum women also release oxytocin in response to massaging and licking of the nipple by the neonate (56). The readiness with which mechanical stimuli can replace the sucking stimulus suggests that it is the mechanical

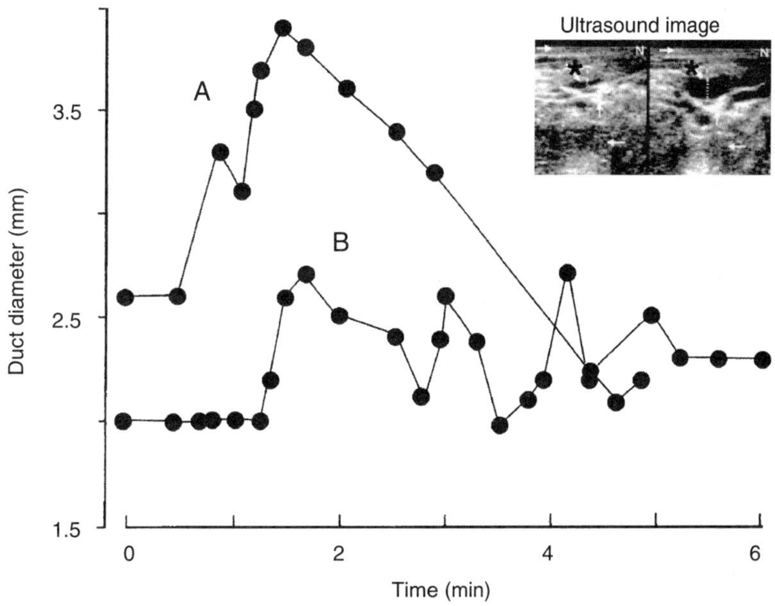

FIG. 4. Monitoring of human milk ejection using ultrasound scanning of a milk duct. This technique depends on changes in the diameter of the milk duct (*inset, asterisked*) caused by milk ejection, with continuous scanning being performed on the nonsuckled mammary gland over the nursing period. Case A shows a large milk ejection at the start of nursing, and case B shows repetitive milk ejections. [Reproduced and modified with permission from (91).]

stimulation associated with sucking, rather than the thermal stimulation, that is more important for activating the milk-ejection reflex.

Considerable discussion in the past has revolved around the possible existence of receptors within the parenchyma of the mammary glands. Such receptors would act to signal the distension of the gland, either to trigger the reflex or to facilitate the response to suckling. However, the evidence for such receptors is flimsy, and most morphological studies have failed to identify receptors within the parenchyma (101,102). The electrophysiological evidence is also against parenchymal receptors being important, because most sensory fibers responding to mammary distension appeared to relate to the stretching of the surface skin (103). A similar pattern of pulsatile oxytocin release was observed in suckled rats regardless of whether the mammary glands were distended or empty (104). Hence, the effects of the butting and treading behavior of the offspring in contributing to milk ejection seem to relate more to mechanical stimulation of the myoepithelium to triggering oxytocin release by stimulation of parenchymal receptors.

Oxytocin release is stimulated by coitus in women (105), and vaginal distension is effective in the ewe (106). In the anesthetized rat, vaginal distension during suckling triggers milk ejection, although this stimulus is ineffective when given alone (107). The ability of stimulation of the reproductive tract to cause milk ejection has been known for considerable time, and this phenomenon has been exploited in dairying practice in some societies (108).

It should be noted that the effectiveness of somatosensory stimuli in triggering oxytocin release is influenced by the reproductive state and stage of lactation. Leake et al. (99) found that the oxytocin response during nipple stimulation in women occurred during the luteal but not the follicular phase of the menstrual cycle. In the rat (27), rabbit (109), and cow (110), the oxytocin response to suckling or milking increases as lactation progresses.

Stimuli Related to Conditioned Oxytocin Release

The ability of the sight and sound of the calf to trigger the milk-ejection reflex of the cow has been known for many years, forming the basis of certain husbandry practices whereby the calf is allowed to be present during milking (108). Later came scientific reports that various visual or auditory stimuli, once they have become recognized to be associated with suckling or milking, can trigger milk ejection as a conditioned response. For example, Deis (111) reported that milk ejection occurred in a rat put within sound of another mother suckling her young. Newton and Newton (85) reported that in the human the milk-ejection reflex could be conditioned to respond to the sound of the infant crying or to activities normally associated with preparing for breast-feeding. In a detailed study in women by McNeilly and coworkers (89), it was found that oxytocin release most frequently occurred before the infant was attached to the nipple. Lactating women also display spontaneous milk ejections between feeding episodes (112), conceivably triggered by imagery of nursing, although this has not been analyzed. Cobo (113) reported that around 80% of women displayed such spontaneous milk ejections, and these became more pronounced as lactation progressed.

In the cow, Peeters and coworkers (114) found that simply showing the calf to the mother was effective in eliciting milk ejection responses in the cow, although this was more effective when combined with other stimuli such as washing the udder with warm water. Cleverley and Folley (79) found that in cows accustomed to a strict milking routine, entry of the milker was often followed by transient appearance of oxytocin in the bloodstream even before milking began. In the sheep, nontactile stimuli emanating from the lamb caused oxytocin levels to double before suckling began (115), although this response required several days of lactation to be initiated. However, others have reported that conditioned oxytocin release does not depend on the lactation state, because it is also seen in nonlactating animals (116).

Stimuli for conditioned oxytocin release interact with somatosensory stimuli in an additive way and ensure the maximal activation of the milk-ejection reflex during suckling (117). Many animals display a complex behavioral repertoire involving the cleaning and grooming of their offspring before nursing. Doubtless this also provides additional conditioning stimuli (tactile, olfactory, and visual) and also evokes the right emotional background or context for triggering activation of the milk-ejection reflex. The evolutionary advantage of developing a mechanism for conditioning of the milk-ejection reflex is not entirely clear, but it perhaps serves to shorten the nursing episode, thereby reducing danger from predators. In ruminants, conditioned milk-ejection responses may help move milk from the alveoli to the gland cisterns, where it can be readily obtained by the young.

Other Stimuli for Oxytocin Release

The list of functions for oxytocin is growing [see reviews in (118,119)], and, not surprisingly, the hormone is released by a variety of stimuli unrelated to reproduction. Electrophysiological recordings or hormone measurements have shown that oxytocin neurons are excited by osmotic stimulation, dehydration, and hypovolemia (120) as well as stimulation of the vagus (121), gut distension (122), and injections of cholecystokinin (123). There is evidence that oxytocin is released by stress (124), although in the lactating animal the oxytocin response to stress is reduced compared with the response in the nonlactating state (125). Fluctuating levels of ovarian steroids may also be significant in determining basal release of oxytocin (126). Although such factors are not directly concerned with the milk-ejection reflex, they influence the background tonic levels of oxytocin on which suckling-induced oxytocin release is superimposed. Background levels of oxytocin are important for determining the occurrence of "spontaneous" rhythmic mammary responses as well as the response of the mammary gland to oxytocin pulses (127). Prolactin may have a stimulatory effect on oxytocin release, acting both centrally (128) and at the level of the posterior pituitary (129).

Plasma Oxytocin Concentrations during Suckling

There are certain difficulties in obtaining an accurate profile of plasma oxytocin during suckling. The plasma oxytocin concentration necessary for milk ejection is relatively low, and degrading enzymes in the plasma may cause rapid breakdown of oxytocin, giving the hormone a very short (less than 2 to 3 minutes) half-life. Another problem is that oxytocin release during suckling is pulsatile, so to obtain an accurate picture it is essential to take frequent blood samples.

In early studies in the rabbit, Fuchs et al. (109) found that oxytocin levels rose from >30 to 200–300 pg/ml[1] within 3 to 4 minutes of suckling and remained elevated for 10 to 20 minutes. Summerlee et al. (130), using frequent blood sampling, found basal oxytocin titers below the limit of sensitivity of the assay (5 pmol/l), but 10 to 20 seconds after the onset of suckling levels started to rise at a rate of 2 to 3 pmol/l per second. Peak levels of 300 to 400 pmol/l were attained toward the end of the 4- to 5-minute suckling period, and the levels then slowly declined at a rate of about 2 pmol/l per second.

Analysis of the variation in hormone levels during the oxytocin peak suggested that it was produced by release of several individual pulses of oxytocin. In some rabbits the decline in oxytocin concentrations after suckling was noticeably protracted, and this relates to secondary activation of oxytocin neurons by grooming of the nipples. In the rat, oxytocin release can be monitored in the anesthetized animal using on-line intramammary pressure recordings, which have provided unequivocal evidence of pulsatile release (10) (Fig. 5). Higuchi and colleagues (131,132) determined oxytocin in randomly collected samples in the anesthetized rat and found significant elevations from 20 pmol/l up to 50 pmol/l in samples that happened to be collected within 30 seconds of the rise in intramammary pressure at milk ejection (Fig. 5). In the guinea pig, Robinson (133) found that levels increased during suckling from less than 5 to over 20 to 40 pg/ml, usually as a single peak lasting 2 to 4 minutes. Oxytocin release was always followed by milk ejection, indicated by behavioral changes in the offspring.

Most studies of oxytocin release in larger species relate to the farm animals. In the goat (134), oxytocin pulses (20–60 μU/ml jugular venous plasma) were detected at the start of both suckling and hand milking. In the pig (76), plasma oxytocin levels rose from <0.2 to a peak of 4 to 20 μU/ml within a few minutes after the onset of suckling. The transient appearance of oxytocin in the blood was foreshadowed by a characteristic grunting pattern in the mother. Comparison of intramammary pressure waves during natural milk ejection and in response to rapid oxytocin injections have provided further evidence that

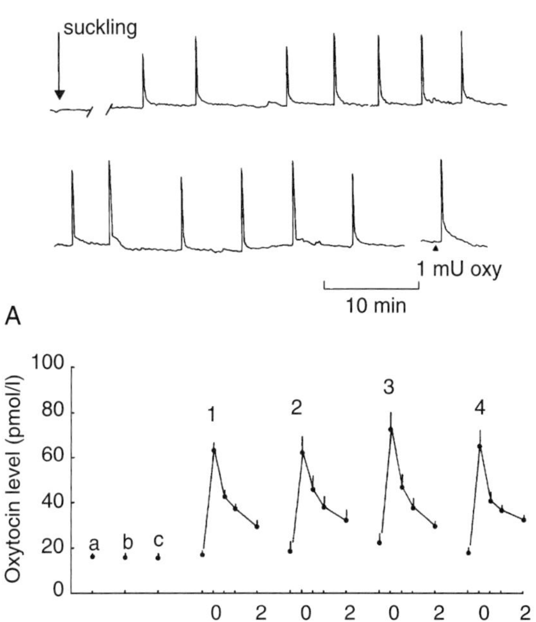

FIG. 5. A: The pattern of intramammary pressure during suckling in the anesthetized rat. Note the intermittent brief milk-ejection responses mimicked by the rapid intravenous injection of 1 mU oxytocin. **B:** Levels of oxytocin in the suckled rat at different times in relation to the occurrence of milk ejection. Samples were collected before suckling (a), when 4–5 pups were attached (b), and when 9–10 pups were attached (c). Thereafter, samples were collected surrounding the first to the fourth milk ejection (1–4). Note the transient pulse of oxytocin that occurs at each milk ejection. [Redrawn from (131), with permission.]

[1]For conversion: 1mU=2.2 ng/ml=2.2 pmole oxytocin.

the release of oxytocin in the suckled pig is pulsatile (135). In the horse, peak oxytocin levels reached around 16 pmol/l of plasma, and the intramammary pressure wave could be mimicked by giving a single injection of oxytocin (52). Oxytocin release usually occurred during a quiet phase after initial stimulation by the foal. Folley and Knaggs (82) found that in cows during machine milking, application of the teat cups was the stimulus most commonly evoking a transient oxytocin release. However, oxytocin levels were highly variable (15–2,000 μU/ml in jugular venous plasma), and sometimes there were additional oxytocin pulses within the subsequent few minutes, unrelated to any well-defined triggering event. Later studies by Schams and coworkers (78) produced different results in that oxytocin levels consistently rose at the onset of manual stimulation (washing udder and premilking) and remained elevated throughout the period of machine milking. Mean peak levels were 26 pmol/l compared with basal levels of 1.5 pmol/l, and the hormone profiles varied, being smooth in some cows and fluctuating in others. Whether these profiles represented closely spaced pulses of oxytocin or a more sustained release was uncertain.

Studies of oxytocin levels during nursing in the human began with circumstantial evidence of a pulsatile release pattern obtained by comparing intramammary pressure responses to endogenous and injected oxytocin (136). Subsequent studies with radioimmunoassay (88,89,137) confirmed this pattern, with seemingly random pulses (up to 50 pg/ml) occurring during suckling, such that hormone profiles varied markedly between individuals (Fig. 6). On average, around two pulses are observed per 20 minutes of nursing (138). Oxytocin pulses often occur in women before the onset of suckling, usually in response to the sight or sound of the infant (89). Oxytocin release during suckling in women occurs without detectable vasopressin release (139).

Mode of Action of Oxytocin on the Myoepithelium

The rich capillary supply to the mammary alveoli ensures that oxytocin gains ready access to the myoepithelium, at least under normal unstressed conditions, when peripheral inhibition of the milk-ejection reflex is not operating. Once in the vicinity of the myoepithelial cells, oxytocin exerts its action by attachment to specific G-protein–coupled receptors on the cell membrane. The oxytocin receptor gene has been sequenced and is identical in all peripheral oxytocin-responsive tissues (140). The three-dimensional structure of the oxytocin receptor has been characterized in detail, including the conformational change that follows binding of the agonist and leads to docking with the G protein (141). The action of oxytocin is mediated via phospholipase C and production of inositol triphosphate (29), which in turn liberates intracellular calcium stores (142). The rise in calcium activates a calcium-dependent myosin light chain kinase (143) that catalyzes phosphorylation of a 20-kDa myosin light chain and so increases myosin ATPase activity. The activity of the myosin light chain kinase is probably regulated by calmodulin (144). Calcium prevents rapid dephosphorylation of the myosin light chain, and if the increase in calcium is inhibited by incubation with cobalt or magnesium, then the myoepithelial cells show only a transient phosphorylation of the myosin light chain in response to oxytocin (143). Oxytocin also stimulates a rise in

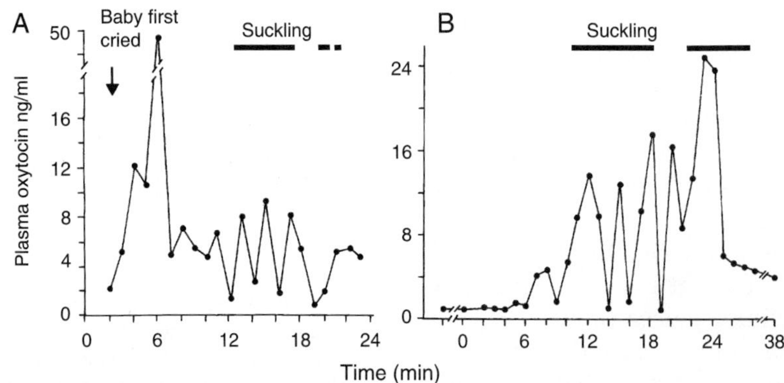

FIG. 6. Plasma oxytocin profiles in the human during suckling. In case A (representing the most typical pattern) there is a release of oxytocin before the infant was applied to the breast, presumably related to a conditioned release triggered by the sound of the infant crying. Case B shows an example in which oxytocin release occurred only after the onset of suckling. The marked fluctuations in oxytocin levels seen in both examples probably reflect the pulsatile pattern of hormone release. [Redrawn from (89), with permission.]

cAMP within myoepithelial cells, but this does not seem to be directly involved in the regulation of myosin phosphorylation (143). However, it has recently been shown that ATP, acting through purine receptors, augments the calcium response to oxytocin in myoepithelial cells. Interestingly, Nakano et al. (145) proposed that ATP is released from alveolar epithelial cells by mechanical stimulation, as would occur during a rise of intramammary pressure. This could provide a mechanism for potentiating the response to oxytocin under conditions of mammary distension. The brevity of the milk-ejection response from a physiological bolus of oxytocin suggests that elimination of the hormone from its receptors and cessation of the contraction mechanism occur rapidly (within seconds) once the extracellular concentrations of oxytocin fall.

There are still aspects of the action of oxytocin that are poorly understood, particularly in relation to the response to slow infusions of oxytocin, which evoke rhythmic contractions. It remains unclear how the activity of the whole population of myoepithelial cells is coordinated such that they contract in synchrony. One possibility is that there is some form of electrotonic coupling between the myoepithelial cells, perhaps mediated by close apposition of their cell membranes (20,21). Presumably, there must also be a pacemaker mechanism for determining the frequency of contractions, as in other smooth muscle tissue, but this has yet to be characterized.

It has been found that oxytocin not only initiates contractions of myoepithelial cells, but may also induce myoepithelial cell proliferation and differentiation (146). This might provide a basis for the ability of oxytocin to slow down involution of the mammary gland after weaning (147).

Myoepithelial cells also respond to other agents, such as acetylcholine, bradykinin, and prostaglandins (148–150). Such agents are unlikely to interact with oxytocin receptors, and it is clear they do not produce the same intracellular effects as oxytocin (143). Their potency in molar terms is considerably less than that of oxytocin.

Factors Governing the Intramammary Pressure Response to Oxytocin

Most observations on the intramammary pressure changes during milk ejection have been made with pressure transducers attached to an intramammary cannula, forming a "closed" recording system (10,76,95,132). This arrangement allows little movement of fluid out of the gland, so that unless the ducts can expand to accommodate the alveolar milk, the myoepithelial contraction tends to be isometric (i.e., occurs without muscle shortening). The closed recording system may result in the generation of artificially high intramammary pressures and may produce recoil when the myoepithelium subsequently relaxes and the pressure wave dissipates. Elastic recoil within the recording system may stimulate the secondary contractions that often follow the initial pressure wave. Hence, such recordings provide a somewhat distorted view of the dynamic pressure changes within the mammary gland during milk ejection.

The amplitude and shape of the intramammary pressure response to a given amount of oxytocin depend to a considerable extent on the rate at which the hormone enters the bloodstream. Slow injection or infusions of oxytocin in the rat evoke rhythmic asynchronous contractions with a period of 1 to 4 minutes (10,127). Each contraction causes a sinusoidal pressure wave with an amplitude of 10 to 20 mm Hg. This type of activity may also be evoked by low-frequency stimulation of the neurohypophysis, which causes low basal output of oxytocin instead of the usual pulsatile release. Sometimes such rhythmic activity is seen spontaneously but can be abolished by lesioning the neural stalk, again suggesting that it depends on basal oxytocin release (127). The physiological importance of this type of activity is uncertain, but in the wallaby, the sensitivity of the mammary gland to a basal tonic output of oxytocin appears to be particularly high in the first 100 days of lactation, providing a means for the differential control of milk ejection in adjacent mammary glands at different lactational stages (71).

If oxytocin enters the bloodstream as a short bolus, the mammary response consists of a single pressure wave (latency 8–10 seconds), perhaps with one or more secondary contractions (10,127). The close relationship between the amplitude of the first pressure wave and the amount of oxytocin injected has long formed the basis for the bioassay of oxytocin (148). In the rat there is a sigmoid dose–response curve, and the relationship between the response and the log of the injected dose is linear over the range 0.5 to 1.5 mU oxytocin. The response to a bolus injection of oxytocin represents the most efficient means of milk removal for the smallest dose of hormone, and pulsatile oxytocin release is thus a highly efficient means of ensuring adequate milk removal during suckling. Another benefit of pulsatile release is that it avoids receptor down-regulation and loss of sensitivity to oxytocin that can arise when the mammary gland is exposed continuously to high levels of oxytocin (104,151).

A further important factor influencing the response of the mammary gland to oxytocin is the amount of milk it contains. In the rat, De Nuccio and Grosvenor (152) found that progressive distension of previously emptied mammary glands caused the slope of their oxytocin dose–response curve to be steepened and shifted to the left. This resulted in a reduction in the dose of oxytocin required to elicit a given response and was probably related to the fact that the force of contraction was increased by the higher resting tension of the myoepithelium. Distended mammary glands also tend to show more protracted secondary contractions after a bolus oxytocin injection. The mammary oscillations evoked by basal oxytocin release also depend on mammary distension and disappear if the mammary gland is emptied (127).

The response of the mammary gland to oxytocin is also regulated by the autonomic nervous system. Sympatheticoadrenal activation causes vasoconstriction of the vessels supplying the mammary glands and prevents oxytocin from reaching the myoepithelium, forming the basis for the peripheral blockade of the milk-ejection reflex during severe stress (11,153). There is also evidence for sympathetic control of the duct system of the mammary gland. Grosvenor and coworkers (154,155) found that the rate at which fluid would flow into the rat mammary gland under gravity was increased by denervation or administration of sympatheticolytic agents, suggesting a tonic constriction of the duct system. Changes in the diameter of the ducts would obviously influence the dynamics of the intramammary pressure response to oxytocin and the movement of milk out of the gland.

Finally, the response of the mammary gland to oxytocin is influenced by the stage of lactation (27,110), reflecting the progressive development and regression of the myoepithelium, coupled with changes in the number of oxytocin receptors. In marsupials with asynchronously lactating mammary glands, the smallest gland (i.e., the one in the earliest stage of lactation) has a much higher sensitivity to oxytocin than the larger gland (71). This difference relates to the concentration of oxytocin receptors: There is a progressive decline in oxytocin receptor number as the weight of the mammary gland increases (73).

Mechanically Induced Contractions of the Myoepithelium

The mammary myoepithelium contracts when mechanical stimulation (sharp tap or pinch of the gland) is applied to the skin overlying the mammary tissue (156,157). Such contractions, sometimes termed tap responses, occur after a latency of 1 to 2 seconds and consist of a single sinusoidal pressure wave lasting some 10 to 15 seconds. The amplitude of the contraction is usually much less than the response to a physiological dose of oxytocin. There is no summation with repeated taps of the skin and little evidence of fatigue provided the stimuli are given less frequently than one per 10 seconds. Tap responses are not prevented by general or local anesthesia, cord section, or intense vasoconstriction of the mammary glands. These mechanically induced contractions result from stretching of the myoepithelium and are easiest to elicit when the mammary gland is distended with milk.

Tap responses could occur as a result of certain forms of suckling-related behavior such as butting (calves) or treading (kittens) of the mammary glands. However, it is clear that in most species oxytocin release is essential for milk removal so that although these mechanical stimuli may be able to supplement the milk-ejection reflex, they certainly cannot replace it.

HYPOTHALAMIC REGULATION OF OXYTOCIN RELEASE DURING MILK EJECTION

Organization of Hypothalamic Oxytocin Neurons and Some Methodological Considerations

The organization and function of the magnocellular nuclei have been the subject of several excellent reviews (158–162). The magnocellular oxytocin neurons responsible for controlling milk ejection lie within the supraoptic and paraventricular nuclei of the hypothalamus. The paraventricular nuclei are positioned within the medial hypothalamus, being located within the lateral walls of the third ventricle. The supraoptic nuclei are in the lateral hypothalamus, lying along the lateral borders of the optic chiasma. The supraoptic nuclei have a relatively simple organization, consisting only of oxytocin and vasopressin neurons projecting to the posterior pituitary, as well as their associated glial astrocytes, and a few interneurons. Both neuronal types are found throughout the nuclei, although there is a tendency for the oxytocin neurons to be distributed mainly in the rostrodorsal regions. The paraventricular nuclei have a more complex organization in which a number of anatomical divisions are recognized. In the rat, there are three magnocellular divisions of the paraventricular nuclei (rostral magnocellular, medial magnocellular and caudal magnocellular), that are surrounded by groups of parvocellular neurons (162). Oxytocin neurons tend to be more concentrated in the rostral than the caudal magnocellular divisions. The parvocellular divisions of the paraventricular nuclei contain neurons that project to the median eminence as well as to the brainstem and spinal cord.

This includes some parvocellular oxytocin and vasopressin neurons, but many other peptidergic cell types are also to be found within these parvocellular divisions (162). Within both the paraventricular and supraoptic nuclei, magnocellular oxytocin neurons tend to lie in clusters, and this may be important in enabling interaction between the neurons (158).

Although oxytocin and vasopressin neurons are found in both the paraventricular and supraoptic nuclei, there are species-dependent variations in the distribution of each cell type. In the rat, oxytocin neurons make up approximately 50% of the magnocellular cells in both nuclei (163). However, the supraoptic nuclei contain substantially more magnocellular neurons, so that in the rat around 80% of the oxytocin neurons contributing to milk ejection are located in this nucleus (120), and this would explain why the supraoptic nuclei alone can support milk ejection (164,165). In other species such as the monkey, there may be a higher percentage of oxytocin cells in the paraventricular than in the supraoptic nucleus (163), and a similar situation exists in the human hypothalamus (166). Thus, it would seem that in these species the paraventricular nucleus has the principle role in the regulation of oxytocin secretion. However, again it should be remembered that because of its larger population of magnocellular neurons, the contribution of the supraoptic nucleus to oxytocin release may still be significant.

Some oxytocin neurons are also found within small accessory nuclei lying between the paraventricular and supraoptic nuclei, and these may also be involved in the milk-ejection reflex (167). Additional populations of oxytocin-containing cells are to be found in the rostral basal forebrain, such as those in the anterior commissural nuclei (168). These cells, which may innervate the bed nucleus of the stria terminalis, change under different reproductive conditions, being particularly prominent in the peripartum period, probably under the influence of estradiol (169). They may be involved in maternal behavior and facilitation of the milk-ejection reflex.

Studies combining in situ mRNA hybridization with immunocytochemistry to investigate coexpression of oxytocin and vasopressin have revealed that under control conditions some 2% to 3% of magnocellular cells produce both hormones, but early in lactation this increases to 17%, mainly because of the appearance of vasopressin in cells previously containing only oxytocin (170). Some magnocellular neurons also display the electrophysiological characteristics of both oxytocin and vasopressin in neurons (171). It has been proposed (170) that this serves to enhance antidiuretic hormone secretion, a possible adaptation for maintaining water balance during lactation. Both the paraventricular and supraoptic nuclei contain other neuropeptides besides oxytocin and vasopressin (159,172,173). Most notably, oxytocin cells coproduce corticotropin-releasing factor, cholecystokinin, enkephalin (172), and dynorphin (174). Whether these are secreted in significant amounts during suckling is not clear, but the fact that these tend to be expressed at far lower levels than oxytocin might suggest they are mainly concerned with local modulatory effects.

The magnocellular oxytocin neurons have a two or three short simple dendrites with a single unmyelinated axon arising from the soma or root of a dendrite (158). The dendrites both receive synaptic inputs and release oxytocin and other modulators (175,176). The axons from the paraventricular nuclei make their way toward the infundibulum (pituitary stalk) by two routes, either following a ventrally directed course close to the third ventricle or passing dorsolaterally in a diffuse band arching over the fornix (159). Fibers that follow the dorsolateral route pass near the supraoptic nucleus before coursing mediocaudally across the optic chiasm intermingled with the supraopticohypophyseal fibers. After passing through the external zone of the median eminence, the combined pathways form the infundibulum (pituitary stalk) before terminating within the pars nervosa (neural lobe) of the pituitary gland. Within the neural lobe the neurosecretory axons are characterized by numerous swellings or dilatations (Herring bodies), which are sites for storage and degradation of neurosecretory granules (177). The neurosecretory terminals contact fenestrated capillaries (thus being outside the blood–brain barrier) and are invested by specialized glial-like cells termed pituicytes. Electrical activity generated at the level of the cell bodies of the oxytocin cells is conducted down their axons to the posterior lobe of the pituitary gland. Depolarization of these terminals causes discharge of neurosecretory granules through calcium-mediated exocytosis (159,178). From what is know of exocytosis in peptidergic neurons (179), it can be expected that significant release of oxytocin requires short interspike intervals and is not restricted to any specialized zone on the terminal membrane. Within the posterior pituitary, the oxytocin terminals are surrounded by the processes of specialized glial cells, the pituicytes. Although the precise role of pituicytes is unclear, it is thought they can influence release from the oxytocin terminals through their control of the extracellular ionic environment (especially potassium) (180,181). Pituicyte processes form a barrier around the oxytocin terminals, limiting the ease with which newly released oxytocin can gain access to the basement membrane of the neurohypophysial capillaries (159). The posterior pituitary is innervated by GABAergic, dopaminergic, and noradrenergic fibers (181,182). The oxytocinergic and vasopressinergic neurosecretory terminals contain opioid peptides that are coreleased with the neurohypophysial hormones (183). The role of these various agents in the control of oxytocin release at the level of the posterior pituitary is considered elsewhere (see section on Control of Release from Oxytocinergic Neurosecretory Terminals at the Level of the Neurohypophysis).

A commonly used way of detecting functional activity in oxytocin neurons in vivo is to measure expression of immediate early genes such as cFos, using in situ hybridization to detect mRNA (184), or immunocytochemistry for the protein products (185). Such approaches have the advantage of enabling an instant sampling of a large population of neurons, but the disadvantage is that they only provide a "snapshot" of activity at the moment the tissue is collected. To monitor second by second changes in activity in real time, it is necessary to undertake extracellular recordings, either in conscious freely moving rats (186) or, more usually, in anesthetized preparations (187–189). Identification of the recorded neuron is usually based on two criteria. The first criterion is that the cell has an axon projecting to the posterior pituitary gland, using the technique of antidromic identification (Fig. 7). Electric shocks are repetitively applied to the neural stalk or posterior pituitary, causing antidromic action potentials with a latency of 8–12 msec to occur in all the magnocellular neurons of the paraventricular and supraoptic nuclei. The second criterion is that the neuron has specific electrophysiological characteristics of oxytocin neurons, based either on their firing pattern and response to secretory stimuli such as suckling (120) or on cholecystokinin injection (160). Oxytocin neurons typically display "spontaneous" background firing in the range of one to five spikes per second, driven by inputs related to osmoregulation and hemodynamic changes (120,160,161,190), gastric distension (191), and stress (192). Thus, these neurons perform other functions besides the regulation of parturition and milk ejection. However, it is noteworthy that in the mouse at least, evidence from oxytocin knockouts (75) suggests that loss of the milk-ejection reflex, leading to death of the offspring, is the only really crucial consequence of the absence of oxytocin.

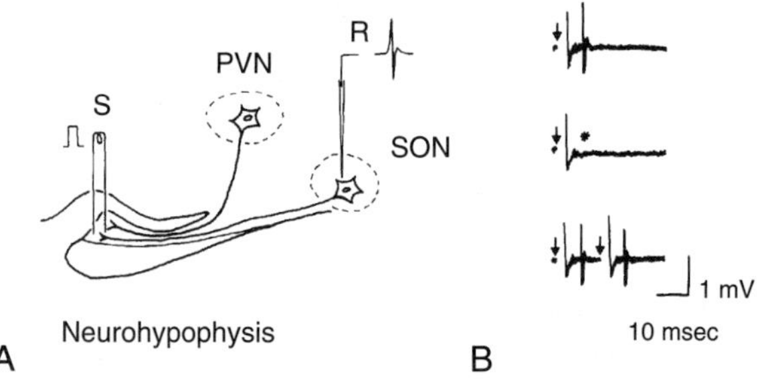

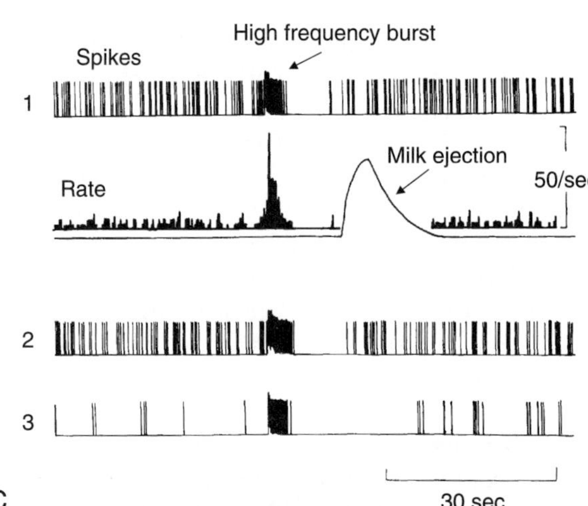

FIG. 7. Analysis of the electrophysiological behavior of oxytocin neurons in relation to milk ejection in the anesthetized rat. **A:** Technique for antidromic identification of neurosecretory cells within the paraventricular (PVN) and supraoptic (SON) nuclei before their analysis in relation to milk ejection. A stimulating electrode (S) is located in the neurohypophysis, and when electrical pulses are applied, all the neurons projecting to this structure are antidromically activated. Individual antidromically activated neurons are monitored by a microelectrode (R) stereotactically positioned within one or another of the magnocellular nuclei. **B:** Examples of the oscilloscope records obtained during antidromic identification of an SO cell. The *upper trace* shows five superimposed sweeps demonstrating constant latency. The *middle trace* shows the collision test, whereby the cell is antidromically activated immediately after the occurrence of a spontaneous (orthodromic) action potential so that the antidromic and orthodromic potentials collide and cancel. The *lower trace* shows frequency following to two shocks delivered at 50 Hz. **C:** Examples of unit firing activity from three different oxytocin neurons during reflex milk ejection evoked by suckling. For clarity the records of firing rate and intramammary pressure have been omitted from the lower two traces. See text for further explanation. [Redrawn from (120) with permission.]

In vitro preparations, most commonly hypothalamic slices, have been extensively used in electrophysiological studies of oxytocin neurons (193,194). In vitro preparations offer far more physical stability, thereby enabling intracellular recording and patch clamping techniques, and make it possible to maintain the extracellular environment. Cell identification by intracellular labeling and subsequent immunocytochemistry (195) is another advantage. However, it should be remembered that neurons in slices may have suffered mechanical damage and retain only their local inputs. Intracellular recordings indicate that oxytocin neurons display electrophysiological characteristics similar to other neurons of the central nervous system, with membrane potentials in the range of –60 to –80 mV (158,161,196). One unusual feature is that the dendritic action potential is mainly calcium rather than sodium mediated, and this could relate to a mechanism for dendritic release of oxytocin (197). Calcium is also a key player in controlling the activation of particular intrinsic inhibitory and excitatory potentials displayed by oxytocin neurons (196); these are discussed elsewhere (see section on Role of Intrinsic Properties of Oxytocin Neurons in Generating Bursts).

An important fact to emerge from in vitro studies is the importance of local interneuronal circuitry in controlling oxytocin neurons. Anatomical evidence suggests that up to 80% of the inputs to the magnocellular neurons arise from local interneurons (158,159), and their existence has been confirmed electrophysiologically, using topical application of glutamate to evoke synaptic responses in magnocellular neurons (198–200). The local projections comprise either excitatory glutamatergic or inhibitory GABAergic neurons (198–200), reflecting the predominance of these transmitters in synaptic terminals within the magnocellular nuclei (159). Inputs to the supraoptic nuclei arise mainly from the perinuclear zone, whereas inputs to the paraventricular nuclei arise from perifornical region and dorsomedial nucleus (200). It is evident that these local interneurons may be important in relaying the effects of more distant connections of the magnocellular nuclei. For example, excitation of supraoptic neurons evoked by stimulation of noradrenergic inputs from the brainstem may involve presynaptic modulation of local glutamatergic afferents (201). Similarly, extrahypothalamic cholinergic inputs may excite supraoptic neurons by presynaptically inhibiting of GABA release from local GABAergic neurons (202). Moreover, these local glutamatergic and GABAergic neurons may interact with each other, through presynaptic GABA (203,204) and glutamate (200) receptors.

Firing Patterns of Oxytocin Neurons during Suckling

Most of the available information concerning the electrophysiological effects of suckling on oxytocin neurons has been derived from studies of the milk-ejection reflex in the anesthetized lactating rat (120,205). This experimental model has several features that make it ideal for electrophysiological studies. First, the milk-ejection reflex of the rat continues to operate under surgical levels of anesthesia, making it possible to undertake physiological studies using anesthetized animals. Second, the rat displays unusually long nursing episodes such that after overnight separation of the mother and pups, milk-ejection response can be monitored over several hours. As already described (see above), suckling in the rat evokes intermittent milk ejection responses that begin about 10 minutes after the onset of suckling and then recur at regular intervals of 5–6 minutes (10). However, to achieve the uninterrupted operation of the reflex, it is important to avoid unfavorable conditions (e.g., deep anesthesia, stress, or arousal of electroencephalogram) that cause the milk ejections to become infrequent or absent altogether. Conversely, certain pharmacological treatments, such as administration of the α_2 agonist, clonidine, result in bursts that are less than 1 minute apart (206). Each milk-ejection response is signaled by a stretch response from the pups and an abrupt rise of intramammary pressure that can be mimicked by the rapid injection of 0.5–0.8 mU (approximately 1–2 ng) oxytocin. Studies of diuresis rates during milk ejection suggest that very little (less than 10 μU) vasopressin is released with each oxytocin pulse (207).

Recordings of antidromically identified magnocellular neurons (189,208–210) indicated that approximately 50% were strongly correlated with milk ejection and were presumed to be oxytocin neurons, whereas the remainder (putative vasopressin neurons) was unresponsive. The oxytocin neurons displayed no immediate change in activity when the pups first started suckling but instead continued their ongoing slow irregular or continuous firing, with rates of two to four spikes per second. About 10–15 seconds before each milk-ejection response, the neurons displayed a brief high frequency burst of firing lasting approximately 2–4 seconds (i.e., involving about 70 to 100 action potentials) that was followed by a period of inhibition lasting around 15 seconds (Fig. 7). Stimulation of the neurohypophysis evoked a mammary response after a similar latency, indicating that this delay represented the time required for oxytocin to circulate and initiate contraction of the myoepithelium.

Similar neuronal responses can be recorded within both the paraventricular and supraoptic nuclei (120,211). Because the supraoptic nuclei contain approximately threefold more oxytocin neurons than the paraventricular nuclei, they are the larger contributor to each oxytocin pulse, at least in the rat (177). The 9,000 oxytocin neurons that project to the rat neurohypophysis fire approximately 500,000 action potentials to release the 0.5 mU (1 ng) oxytocin required for milk ejection. This equates to 2 to 3 fg (the contents of about 40 granules) per action potential, which is some 50–100 times more than occurs during the background activity of the oxytocin neurons (177). This is because the clustering of action potentials during the burst favors the process of "facilitation" of peptide release. Facilitation occurs when the increments in calcium within the neurosecretory terminals during successive spikes summate, so maximizing the release process.

Whereas the amplitude of the high frequency bursts displayed by oxytocin neurons may be quite variable, the overall pattern of their response is highly stereotyped. Moreover, the consistent temporal relationship between their response and the rise of intramammary pressure at milk ejection indicates that the oxytocin neurons are activated in close synchrony. This has been confirmed by double recordings taken simultaneously from the paraventricular and supraoptic nuclei, when the difference in time of activation of oxytocin cells in the two nuclei was always less than 400 msec (212). The order in which each neuron of the pair was activated changed randomly from one milk ejection to the next, with no indication that some neurons always triggered their burst ahead of others. The proportion of neurons activated at milk ejection corresponds closely with what would be predicted from immunocytochemical studies of the numbers of oxytocin-containing neurons within the magnocellular nuclei (120). Hence, the whole pool of oxytocin neurons is simultaneously involved in generating each oxytocin pulse. This high degree of coordination is maintained even during a reduction in the suckling input, which causes a decline in the magnitude and frequency of the bursts, but little change in the number of neurons activated (188,213). However, double recordings by Belin and Moos (214) (Fig. 8) indicated that this tight coordination of the oxytocin neurons may take time to be established when suckling commences. They found that some oxytocin cells did not respond at milk ejection early in suckling but were later recruited into the pool of activated cells; thereafter, they were consistently involved in every response. Once the milk-ejection reflex has become fully established, it seems that only hemisection of the brain (215,216) will disrupt this remarkable synchronization of bursting in the oxytocin neurons.

Although early electrophysiological studies reported that application of the pups to the nipples had no obvious effect on the background firing of oxytocin neurons, subsequent studies have shown that suckling does have subtle effects on basal activity (217), and the background firing rate tends to be lower when fewer pups are suckling (213). These changes are unlikely, themselves, to alter the output of oxytocin from the posterior pituitary gland, but they may be important for enabling the bursting responses.

Electrical recordings taken from conscious unrestrained rats (186) have shown that the observations made in the anesthetized preparation provide a true picture of the normal behavior of oxytocin neurons. Continuous recordings have been obtained from the same oxytocin neuron for up to 24 hours, spanning the period of parturition and subsequent suckling episodes. Suckling evoked high frequency bursts identical to those observed under anesthesia, and each was followed by a milk-ejection response, indicated by a stretch reaction from the pups. Parturition itself caused an increase in overall neuronal activity, with episodes of high frequency firing superimposed on an elevated background activity. The episodes of high frequency firing were related to cervicovaginal passage of the fetus or placenta but were far less organized than the highly stereotyped bursts caused by suckling. This could be because the oxytocin neurons are fundamentally incapable of synchronous high frequency bursts until after they have undergone essential neuroplastic changes associated with birth. However, synchronous high frequency bursts have been recorded during late pregnancy (218), so the more likely explanation is that the differences in neuronal responses between birth and suckling are related to the different nature of the stimuli.

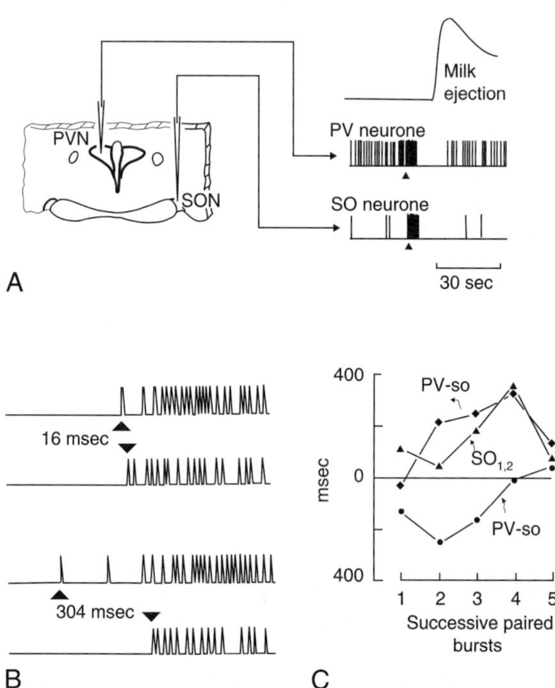

FIG. 8. Paired recordings from oxytocin neurons demonstrating the close synchrony of their activation during suckling in the anesthetized rat. **A:** Technique whereby recording electrodes were simultaneously inserted into the paraventricular (PVN) and contralateral supraoptic (SON) nuclei. The traces show the simultaneous recordings obtained from a PV and contralateral SO unit during suckling. **B:** Successive high frequency discharges from these same two units are shown on an expanded time scale to indicate the short but somewhat variable time difference in the onset of their responses, typical of such double recordings. **C:** Time difference (in milliseconds) in the onset of the high-frequency discharges for five successive responses is plotted for three different pairs of cells. Two cell pairs involved one PV and one contralateral SO cell (marked Pv-so), and the other pair involved adjacent SO cells (marked $SO_{1,2}$). In each case, one of the cells served as the reference cell (the PV cell in the case of the PV-so recordings). If the high frequency burst for the reference cell began first, then the values were above zero; otherwise, values fell below zero. Note the lack of consistency in the time difference and order in which these three different pairs of cells were activated (see text). [Redrawn from (212), with permission.]

The only other species in which the electrophysiological behavior of oxytocin neurons has been studied during suckling is the rabbit (219). This species exhibits a much shorter period of nursing than the rat, and because the milk-ejection reflex is abolished by anesthesia, only recordings in conscious freely moving rabbits have been undertaken. During the 3- to 5-minute nursing period, the oxytocin cells show several closely spaced bursts of firing, superimposed on their background activity (130) (Fig. 9). The bursts of firing are shorter (1–2 seconds) than those recorded in the rat, as are the intervals between the bursts (range 30–60 seconds). Cross-correlation analysis of simultaneous double recordings suggests that, as in the rat, bursting responses are synchronized between adjacent cells so that apart from time scale, there are close similarities between the two species. The closely spaced bursts observed in the rabbit result in a complex profile of oxytocin (130), with the rise of hormone starting shortly after the first burst and lasting several minutes, depending on the pattern of subsequent responses. One notable difference from the rat is that additional bursts are sometimes seen after the mother has left the nest, possibly triggered by grooming of the mammary glands. There was no evidence that oxytocin cells were activated before suckling (e.g., as the mother approached her nest), and electrophysiological correlates of conditioned oxytocin release have yet to be obtained in either the rabbit or the rat.

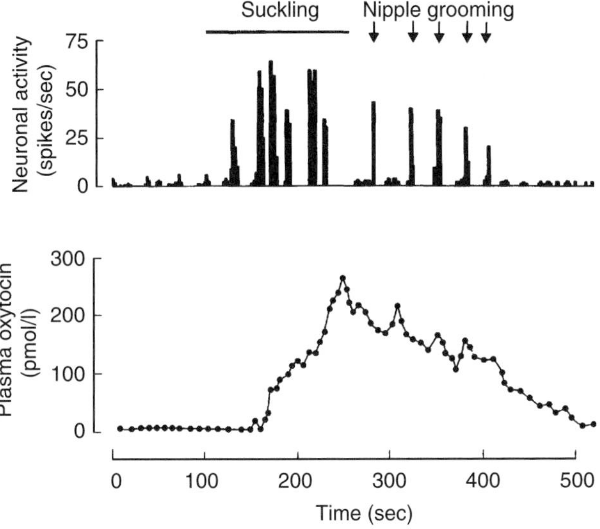

FIG. 9. The relationship between oxytocin neuron activity and plasma oxytocin levels in a conscious rabbit in association with suckling. Before suckling oxytocin neuronal activity was slow and irregular, and plasma oxytocin was undetectable. After the start of suckling the neuron gave six bursts of high frequency firing, with further bursts when the mother moved away from her litter and groomed her nipples. These bursts of firing precipitated a protracted rise in oxytocin in the plasma. Note the bursts are more closely spaced than those seen in the rat. [Redrawn from (130), with permission.]

Although direct electrophysiological evidence is lacking in species other than the rat and rabbit, the wide occurrence of pulsatile oxytocin release during suckling suggests that oxytocin neurons are similarly organized in many other species. Given this apparent generality, it is interesting to speculate on the factors that have resulted in evolution of this type of neuronal organization. In fact, there are several benefits that derive from evolving this type of response. First, as we have already seen, bursts are highly efficient at discharging oxytocin from the neurohypophysial oxytocin terminals (220). Second, release of a pulse of oxytocin causes the hormone to be released as a discreet bolus, and this maximizes the concentration of the peptide as it reaches the myoepithelial receptors. Third, the contraction of the myoepithelium is synchronized both within and between mammary glands, allowing an organized rather than random pattern of milk ejection within the nursing episode.

A fascinating example of the functional implications of low level continuous firing versus high frequency bursting is provided by the milk-ejection reflex of the Agile Wallaby. This species displays concurrent asynchronous lactation, whereby one mammary gland is supporting the youngest offspring permanently attached to the nipple, whereas the other gland is supporting an older offspring that attaches intermittently for an episode of nursing (48). Low level stimulation of the neural stalk, mimicking continuous firing, causes sufficient oxytocin release to cause rhythmical contractions of the early-lactating mammary gland but leaves the late-lactating gland unaffected, because this gland is relatively insensitive oxytocin (71). Only when the stalk is stimulated at higher frequencies, mimicking the burst firing of oxytocin neurons, does the more "mature" gland display a response.

Molecular Correlates of the Activation of Oxytocin Neurons during Suckling: Fos Expression

Activation of oxytocin cells by various stimuli (e.g., osmotic stimulation, injection of cholecystokinin) has been studied using immunocytochemistry or in situ hybridization to detect expression of the immediate early gene c-Fos as a marker of neuronal excitation, and the neurons readily show Fos induction to such stimuli (221,222). Oxytocin neurons also show increased production of Fos during parturition (185,223). However, studies examining the effect of suckling on Fos activation have produced variable results. Fenelon et al. (185) were unable to detect any evidence that suckling induced Fos in oxytocin neurons, whereas Li et al. (224), using a longer separation period of 48 hours, did observe that suckling induced Fos activation in the supraoptic nucleus. Positive results were also obtained by Lin et al. (439), who found that Fos and Fos B were activated in the both the paraventricular and supraoptic nuclei of 24-hour postpartum rats, compared with virgin animals. Removal of the pups immediately after birth eliminated this increased expression, suggesting that it was suckling dependent. However, an additional complication is that Lonstein et al. (225), using similar prolonged separation before the suckling test, reported that the incidence of Fos labeling in the supraoptic nucleus was no greater than when the mothers were in contact with their pups but not allowed to suckle. Based on these variable results, it seems that suckling is far less effective at inducing Fos in oxytocin neurons than other stimuli. One explanation might be that the brief bursts evoked by suckling cause insufficient change in the minute-by minute firing of the cells compared with the protracted increase in activity seen after, for example, intravenous injection of cholecystokinin or during parturition.

Neural Mechanisms Underlying Synchronous Bursting Activity in Oxytocin Neurons

Rhythmical patterns of neuronal activity are widely encountered within the central nervous system, for example, in the neural circuits concerned with respiration (226) and locomotion (227). Pattern generation is also a common feature of hypothalamic neuroendocrine systems, another prime example, other than the oxytocinergic system currently under discussion, being the pulsatile release of gonadotropin-releasing hormone (228). At its simplest, rhythmical firing can be generated within individual neurons, as in the case of the phasic activity displayed by vasopressin neurons (158,159). However, it is self-evident that such a simple mechanism is unlikely to account for the complex behavior of oxytocin neurons during suckling. As we see in this section, it is more likely that the bursting activity of oxytocin neurons is controlled by the interplay of several different mechanisms, including their intrinsic electrophysiological properties, coordination of their synaptic input, and their mutual excitation.

General Overview: Localization of Circuitry Involved in Generating Bursting Activity in Oxytocin Neurons and the Concept of a Hypothalamic Burst Generator

Transmission of the suckling input from the nipples up to the hypothalamus involves afferent pathways that pass through many synaptic relays within the spinal cord and brainstem before reaching the level of the hypothalamus (229). The delayed and intermittent nature of the bursting responses in oxytocin neurons that are evoked by suckling lead to the suggestion that these pathways include "gating mechanisms" that are capable of regulating the flow of ascending impulses transmitted to the oxytocin neurons, whereby excitation is only allowed through at certain times when the bursts are activated (229). Results from stimulation studies at the level of the mammary nerve (95), spinal cord (97), or dorsal vagal complex of the brainstem (230) indicated that short (30–60 second) pulse trains could evoke immediate milk-ejection responses without the long delay characteristic of the response to suckling, as if the gate had been bypassed. Such results encouraged the view that the pulsatility might be organized downstream of the hypothalamus, perhaps in the spinal cord or medulla. The idea of gating below the level of the hypothalamus was reenforced by studies showing that lesions of the midbrain ventral tegmentum altered the pattern of milk ejection (231), a result that was interpreted as indicating that some components of the gating mechanism might also be located at this level (229). Also, the gating function was not ascribed only to the ascending pathways of the milk-ejection reflex. Recordings from the bed nucleus showed that some neurons displayed oscillations in firing that correlated with milk-ejection bursts (232), as if they might be controlling the flow of excitation to the oxytocin neurons. However, lesioning the bed nucleus does not prevent pulsatile oxytocin release during suckling (233), and the fact that the activity of bed nucleus neurons correlates with milk-ejection bursts does not automatically mean that they are directly involved in generating bursting activity. A more likely interpretation for this correlation is that the bed nucleus shares common excitatory inputs with the oxytocin neurons.

Although it remains possible that all aspects of the bursts are fashioned downstream of the hypothalamus, evidence has focused attention on whether the bursting pattern might be organized in local circuitry within the hypothalamus itself. The main line of evidence concerning the ability of local hypothalamic circuitry to fashion bursting responses has emerged from studies of neonatal hypothalamic explants maintained in vitro (234,235). These explants comprise a 400-μm hypothalamic slice prepared from 4- to 6-day-old rat pups and maintained in culture for 3 weeks. Such conditions favor the survival of magnocellular oxytocin neurons, which persist within the explant as clumps of cells that are innervated by local GABA and glutamate neurons, which might seem to resemble hypothalamic circuitry. Electrophysiological recordings from oxytocin neurons in these explants (234,235) have shown that, quite unexpectedly, they display patterns of firing that are remarkably similar to the behavior of oxytocin neurons in the suckled rat. Intracellular recordings from such explants have revealed that most oxytocin neurons are spontaneously active and display repetitive bursts of firing that have many similarities with the high frequency bursts associated with suckling in vivo, including synchronization (236) and autofaciliation with oxytocin (235). From intracellular recordings it is evident that this bursting activity is driven by repetitive volleys of excitatory postsynaptic potentials (EPSPs) that are blocked by glutamate antagonists, indicating that they derive from local glutamatergic neurons. From their observations, Israel et al. (236) concluded that oxytocin neurons in these explants are regulated by a local hypothalamic network of glutamergic neurons capable of generating a rhythmic excitatory input and driving intermittent bursting activity.

Although a degree of caution is required in extrapolating from cell cultures to the in vivo situation, the above observations provide strong support for the idea that high frequency bursting could be organized locally within the hypothalamus. The most direct way of testing this would be to investigate the firing pattern of oxytocin neurons in the deafferented hypothalamus, but because busting activity depends on the suckling input, such an approach is not possible. Nevertheless, there are several lines of indirect evidence that the bursting pattern may be generated at the hypothalamic level. First, it is known that bursting can be modified by microinjections of agents such as GABA (237) or oxytocin itself (238) applied in the immediate proximity of the hypothalamic magnocellular nuclei, presumably by interacting with local circuitry. Second, it is evident that the basic pulsatile nature of oxytocin release persists after experimental interventions, for example, midbrain lesions (231) or administration of pharmacological agents (239,240), acting at different levels of the milk-ejection reflex. This strongly suggests that bursting activity is organized near the output pathway, that is, the oxytocin neurons themselves. Third, as we have already seen when discussing the organization of oxytocin neurons, there is now much evidence that the magnocellular nuclei are surrounded by complex local glutamatergic and GABAergic circuits that serve to process and regulate the flow of excitation to the oxytocin neurons (200). This would seem to provide the necessary local substrate for generating bursting activity close to the oxytocin neurons.

The conclusion that bursting is fundamentally organized within the hypothalamus should not be taken to indicate that other structures do not contribute to timing the bursts. Indeed, we know from stimulation (241) and lesioning studies (231) that the timing of the bursts can be modulated at extrahypothalamic sites, probably through changing the basic excitatory drive impinging on hypothalamic burst generator. However, although such manipulations can lead to a wide range in the frequency of milk-ejection response, it is notable that it is rare to observe milk-ejection intervals of less than 2 minutes under any circumstances. This suggests that that hypothalamic burst generator not only fashions the bursting responses, but also sets a minimum interval between them.

Role of Intrinsic Properties of Oxytocin Neurons in Generating Bursts

Most neurons possess distinct regularity mechanisms that arise from the inherent properties of their cell membranes, and oxytocin neurons are no exception in this regard. These intrinsic regulatory mechanisms arise from the presence of ion channels within the cell membrane that can generate membrane currents distinct from those underlying the action potential. However, because the ion channels mediating intrinsic control are voltage and time dependent and may also be calcium dependent, their opening is often linked to spike activity. Intrinsic properties can be both excitatory and inhibitory and can give rise to complex firing patterns, including rhythmical activity (242). The intrinsic properties of neurons can also profoundly alter the way they respond to their synaptic inputs (243,244).

Although intrinsic properties are doubtless important, perhaps it should be reiterated that they are unlikely to be the sole mechanism for controlling bursting activity in oxytocin neurons. Thus, this pattern is highly stimulus specific (i.e., related only to birth and suckling), contrary to what would be expected if intrinsic currents were driving the bursts. Intraperitoneal injection of hypertonic sodium chloride, for instance, causes supraoptic oxytocin neurons to show a sustained high level of firing, but bursting activity is never observed (190). Furthermore, suckling-evoked bursts cannot be triggered, at least in any immediate way, by antidromic stimulation (187,245), and this characteristic is incompatible with a purely intrinsic regulatory mechanism. Finally, studies in hypothalamic explants have shown that brief depolarizing or hyperpolarizing current pulses have no effect on the periodicity of bursts (234), again indicating they are not under direct intrinsic control. However, the foregoing observations do not preclude the idea that the intrinsic properties of oxytocin neurons indirectly influence bursting activity, and as we will see, there is increasing evidence to support this concept. This section describes the major intrinsic currents that have been identified, but the reader is referred to other reviews (196,246) for a more detailed account of the membrane properties of oxytocin neurons.

One intrinsic excitatory mechanism that has been identified in oxytocin neurons is the depolarizing afterpotential (Fig. 10). Depolarizing afterpotentials in oxytocin neurons occur after single or clusters of spikes and are characterized by a slight depolarization that may last several seconds. The depolarizing afterpotential is activated by calcium influx and may involve both potassium (247) and sodium currents (248). Although the depolarizing afterpotential in oxytocin neurons is rarely powerful enough, on its own, to evoke spike activity (249), it increases the probability of spiking in response to synaptic inputs. Hence, the presence of the depolarizing afterpotential mechanism favors the generation of spike clusters.

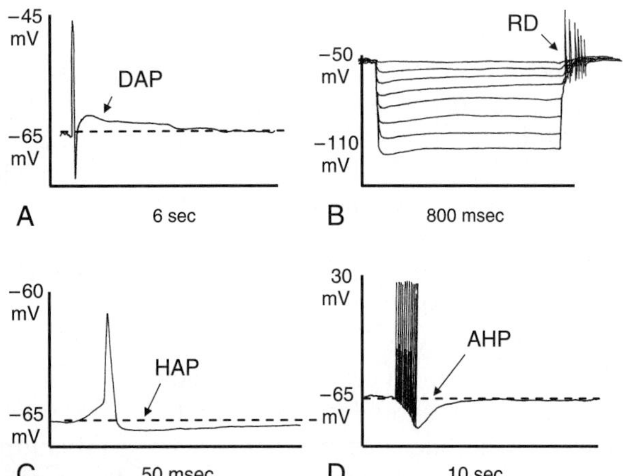

FIG. 10. Electrophysiological recordings showing intrinsic properties that have been detected in magnocellular supraoptic neurons. **A:** Depolarizing afterpotential (DAP) following a single action potential. **B:** Superimposed traces showing rebound depolarization (RD) after prolonged hyperpolarization pulses of different magnitudes. (The cell was held in a slightly depolarized state prior to pulse application.) **C:** Hyperpolarizing afterpotential (HAP) following a single action potential. **D:** Afterhyperpolarization (AHP) that occurs following a train of action potentials. These properties are found in both oxytocin and vasopressin neurons, apart from the RD, which is specific for oxytocin neurons. [A, C, and D are withdrawn from (253), B from (246), with permission.]

As we see in the next section, irregularity of background activity, with spike clustering, seems to be causally linked to bursting in oxytocin neurons (217). Thus, although the depolarizing afterpotential cannot directly drive bursts, it can contribute indirectly through its influence on basal firing. It is notable that the number of oxytocin neurons showing depolarizing afterpotentials is increased to around 50% in lactating rats, as compared with 20% in virgin animals (249).

Oxytocin neurons also display a more subtle form of intrinsically generated excitation that has been referred to as "rebound depolarization" (158,246). The conditions required to observe the rebound depolarization are complex, in that the neuron must be held in a slightly depolarized state and then exposed to a transient hyperpolarizing pulse lasting around 1–2 seconds. Termination of the end of the hyperpolarizing pulse is followed by one or several action potentials over the next 100–200 msec (Fig. 10). This rebound excitation arises from deactivation of a potassium current (referred to as a sustained outward rectification current) during the inhibition, and the brief excitation arises from the time required to reactivate this current. Interestingly, the sustained outward rectifying current and accompanying rebound excitatory phenomenon are particular hallmarks of oxytocin neurons and can be used to identify this cell type (158). As with the depolarizing afterpotential, the rebound excitation is more pronounced in oxytocin neurons from lactating rats (250), and its significance may lie in the requirement for an oscillating background activity for burst generation (217). As pointed out by Armstrong and Stern (251), it could, during periods when there are alternating inhibitory and excitatory inputs, lead to a redistribution of spike activity and an increase in oscillatory firing.

The potential importance of intrinsic excitation in the generation of bursting in oxytocin neurons was further highlighted in a recent article by Wang and Hatton (252). They made recordings from hypothalamic slices and found that around 30% of supraoptic neurons (including some immunocytochemically identified oxytocin neurons) displayed bursting activity quite similar to bursts observed in vivo, when the neurons were excited by phenylephrine in the presence of low calcium medium. These bursts were not driven by excitatory synaptic inputs, nor was there any evidence that they were related to the depolarizing afterpotential or rebound depolarization phenomena described above. Wang and Hatton (252) proposed that the powerful excitation driving the bursts might have arisen through a combination of a sudden increase in sodium influx and a reduction in potassium efflux caused by the special incubation conditions (low calcium) and the presence of phenylephrine. These observations perhaps indicate the presence of an as yet undefined but very powerful intrinsic excitatory mechanism in oxytocin neurons. However, it remains to be established how they relate to in vivo bursting and whether they involved a specific property of oxytocin neurons or a more general property arising from the unusual recording conditions.

As well as possessing intrinsic excitatory mechanisms, oxytocin neurons also possess intrinsic inhibitory mechanisms. Thus, action potentials in oxytocin neurons are followed by a hyperpolarization that arises from activation of calcium-dependent potassium channels (159,161). There are two interrelated components of this postspike hyperpolarization: a fast transient hyperpolarization, called the hyperpolarizing after potential, and a more pronounced long-lasting hyperpolarization, called the afterhyperpolarization, that is generated by trains of spikes (196,246,253) (Fig. 10). The hyperpolarizing afterpotential arises as a result of single spikes and tends to set a lower limit on the interspike interval. This is around 30 msec or so, so that the maximum spiking frequency that might be expected from an oxytocin neuron is around 30–40 spikes/sec. The hyperpolarizing afterpotential tends to interrupt clusters of spikes and so "smooth" the firing rate, making it more regular (254). The significance of the hyperpolarizing after potential for bursting is that (a) it tends to oppose the development of irregular background firing that is linked to bursting, and (b) it tends to impose an upper limit on maximal firing, well below that observed during burst activation. It is evident then that the hyperpolarizing afterpotential must be removed, or over-ridden, before bursting can occur in oxytocin neurons.

The other component of the postspike hyperpolarization, the afterhyperpolarization, is activated after prolonged spike trains and is probably responsible for terminating the burst and imposing the postburst inhibitory period (usually around 10–15 seconds). There is little evidence for recurrent inhibitory circuitry within the magnocellular system (255), so this provides the most likely means of terminating the bursts. This idea is supported by the fact that the postburst inhibitory period can be mimicked by antidromic activation (177). Given their likely role in suppressing bursting activity, it might seem curious that these postspike hyperpolarization currents are enhanced in oxytocin neurons from lactating rats (249). It has been suggested (249) that this may be to limit the excitatory effects of other nonsuckling inputs to oxytocin neurons. At the same time, the impact of removal of this intrinsic inhibition during suckling would be more pronounced, perhaps even providing a means of activating a burst.

To summarize, our current knowledge suggests that the most likely importance of intrinsic control may be to organize the background firing of oxytocin neurons, especially the irregular firing that is linked to bursting activity. Furthermore, it seems clear that intrinsic inhibitory mechanisms (in particular the hyperpolarizing after potential) must be modified in some way during suckling if they are not to dampen burst activation. The results of Wang and Hatton (252) indicate that under specialized in vitro conditions, bursts can be intrinsically generated. However, the relationship of this finding to the in vivo situation is currently unclear.

Insights into the Mechanism of Burst Generation from Analysis of Background Activity

Initial reports concerning the behavior of oxytocin neurons during suckling paid rather little attention to their background activity, other than to note that it was not altered by suckling (208,209). The fact that oxytocin neurons showing bursting responses displayed a range of background rates, and indeed some showed very little background firing at all, tended to reinforce the view that basal activity was unimportant for the generation of bursting responses. However, it was then noticed (256) that giving intracerebroventricular oxytocin to facilitate bursting activity was accompanied by an increase in background firing rate of oxytocin neurons (see example in Fig. 11). This change only occurred if the pups were attached to the nipples (257). This was the first indication of a relationship between background activity and burst generation. Subsequently, Brown and Moos (258) reported that the probability of bursting correlated with the proportion of oxytocin neurons with a firing rate of between one and three spikes per second. The suckling stimulus was reported to alter basal firing, with cells either speeding up or slowing down, such that more of them were brought into this critical range.

Subsequently, Brown et al. (217) investigated whether there was any correlation between bursting activity and the patterning (as opposed to simple frequency) of background firing. Giving hypertonic saline during suckling reduced the incidence of bursting, and this correlated with a reduction in the variability of background firing within individual neurons. In other words, the neurons adopted a more regular pattern, with a narrowing of their interspike interval distribution. On the other hand, giving intracerebroventricular oxytocin to facilitate bursting activity had the opposite effect, causing the oxytocin neurons to adopt a more irregular firing pattern with increased clustering of activity and a broadening of the interspike interval distribution (Fig. 11).

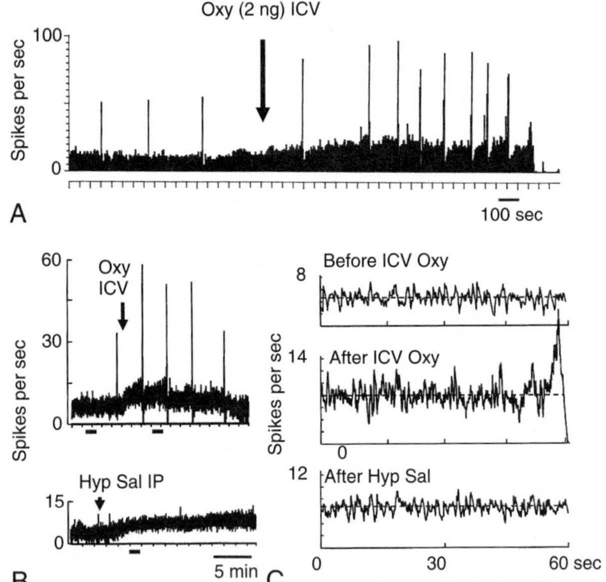

FIG. 11. A: Effect of intracerebroventricular (ICV) injection of 2 ng oxytocin (Oxy) on the background firing and bursting activity of an oxytocin neuron during suckling. Note this increased the background firing of the neuron and facilitated bursting activity, with an increase in both the frequency and amplitude of the bursts. **B:** Recordings comparing the effect of intracerebroventricular oxytocin (*upper*) with an intraperitoneal injection of hypertonic saline (Hyp Sal IP) (*lower*). Note that oxytocin was followed by the onset of bursting activity, but this was not observed following hypertonic saline. **C:** Sections of the recordings in B (indicated by the *bars*) displayed on a wider time base. Note the increased irregularity of the background firing after intracerebroventricular oxytocin. Hypertonic saline induced a similar increase in the rate of background firing of the oxytocin neuron, but the firing pattern did not become irregular. (See text for further explanation.) (Wakerley, previously unpublished data.)

Because both hypertonic saline and intracerebroventricular oxytocin were equally effective in increasing mean firing rate, it seemed that bursting activity was far more strongly related to the irregularity of firing than to the frequency of firing. Brown et al. (217) went on to further suggest that giving intracerebroventricular oxytocin caused individual oxytocin neurons to express the full effect of suckling on their background firing rate (i.e., suckling acts to increase variability in firing), a change that was a prerequisite for enabling bursting responses. They argued that during irregular firing, the oxytocin cells were undergoing cyclic fluctuations of excitation and inhibition. During its excitatory peak, each cell would be both providing greater excitation to other neurons with which it was functionally connected and would, itself, be more receptive to excitation. In other words, irregular firing increased the efficiency of excitatory interaction between oxytocin neurons, driving them into coordinated and resonant cycles of high and low activity. Assuming that the oxytocin neurons behaved as a single network, then such excitatory interaction would make the network increasingly unstable, a situation that could eventually lead to burst activation. Hypertonic saline reduced bursting activity because it tended to regularize background activity, and regular firing dampens down such excitatory interactions.

Fundamental to the above concept was that the oscillations in activity associated with irregular firing were synchronized between adjacent oxytocin neurons. This crucial point was subsequently confirmed by Moos et al. (259), who undertook a detailed analysis of the firing pattern and degree of coordination (or "cross-correlation") in background firing between simultaneously recorded oxytocin neurons. Under conditions where bursts were absent or infrequent, Moos et al. found that oxytocin neurons invariably displayed a high degree of "negative autocorrelation" (i.e., a tendency for short intervals to be followed by long intervals over a time scale of 0.5–2 seconds) and cross-correlation between simultaneously recorded neurons was low. The negative autocorrelation provided the basis for the regular firing pattern typically encountered in nonbursting oxytocin neurons. However, after pharmacological manipulations to increase bursting activity (i.e., administration of intracerebroventricular oxytocin) negative autocorrelation weakened, leading to more irregular firing and simultaneously recorded neurons began to show substantial cross-correlation. Moreover, analysis of oxytocin neurons leading up to burst activation showed that these changes became increasingly evident in the final few seconds before the onset of a burst. This increasing degree of cross-correlation before a burst was evident even between neurons recorded on the opposite sides of the brain, providing striking confirmation that the disparate populations of oxytocin neurons within the hypothalamus do indeed behave as a single coordinated network. Moreover, Moos et al. (259) found that not all oxytocin neurons displayed the above changes to the same extent; some appeared to be profoundly altered well ahead of the burst, whereas others only displayed the changes rather weakly or only in the very last few seconds of the preburst period. This suggested a degree of hierarchical organization within the population of oxytocin neurons, with some behaving as leader neurons and having a dominant role in organizing the background activity. Others (follower cells) were more subordinate, perhaps only being recruited once the network had become sufficiently organized. The functional distinction between the leader and follower cells might arise because of differences in the arrangement of their synaptic input or their anatomical relationship with other oxytocin neurons.

The above analysis of the background activity of oxytocin neurons has thus provided important insights into the way that their bursting activity may be generated. The special feature of the suckling stimulus, and the reason that it evokes bursting activity in oxytocin neurons, is that it induces irregular background firing. This favors excitatory interactions between the oxytocin neurons, and they begin to show coordinated fluctuations in firing. As more oxytocin neurons become recruited into coordinated activity, so the mutual excitatory drive between them increases. This provides positive feedback within the system, further amplifying the fluctuations in firing and recruiting more and more neurons. Eventually, a point is reached when the network rapidly becomes more and more oscillatory, until there is a sudden loss of stability in the system, leading to a full-blow burst. The neuronal basis of the sudden loss of stability is unclear, but such behavior is recognized in other neuronal networks (260). As Moos et al. (259) pointed out, this scheme also reveals the potential importance of intrinsic properties in governing bursting activity. We have already discussed how intrinsic inhibition serves to regularize firing (i.e., increases negative autocorrelation) and it therefore tends to dampen down oscillatory activity. One of the effects of the suckling input may be to diminish this dampening effect. The finding of Moos et al. that negative autocorrelation declines rapidly in the last few seconds before the burst further supports the idea that suppression of intrinsic inhibition may be a crucial factor in destabilizing oxytocin neurons and allowing burst activation.

If oxytocin neurons alter their excitability and show increasing excitatory interaction as a burst approaches, then this might be expected to be accompanied by augmentation of their response to excitatory inputs, particularly those involved in the milk-ejection reflex. Results obtained by Cosgrave et al. (261) (Fig. 12) indicated that this may indeed be the case. They studied the response of oxytocin neurons to stimulation of the ventral tegmentum, a region implicated in the milk-ejection reflex (231). They found that this response was greatly potentiated in the presence of intracerebroventricular oxytocin, and more importantly in the present context, the response increased dramatically in the period leading up to the burst. These observations are consistent

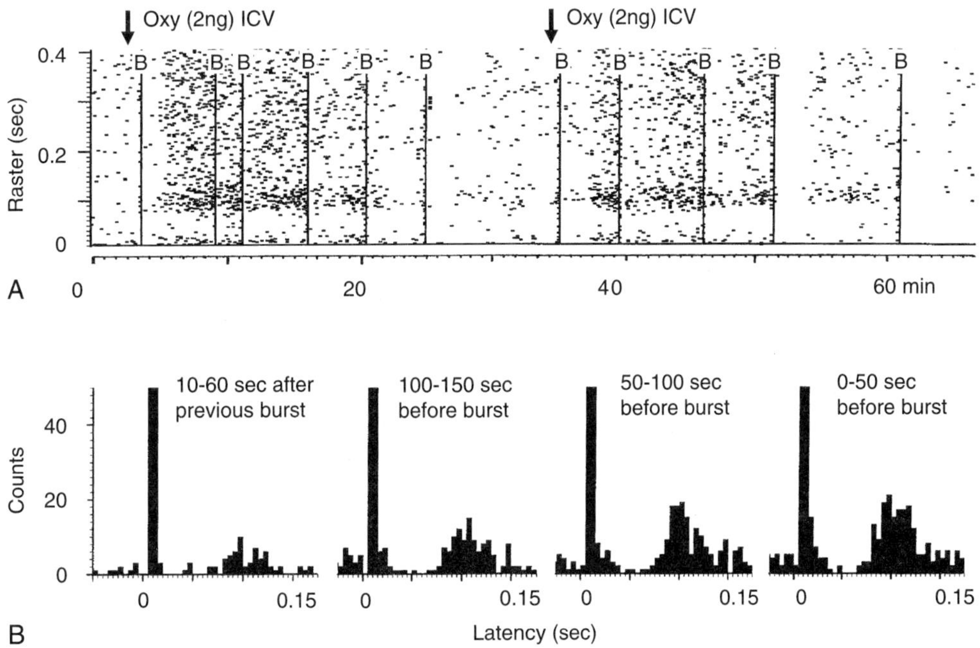

FIG. 12. Changes in the response of an oxytocin neuron to stimulation of the ventral tegmentum of the midbrain, illustrating how the excitatory response is augmented in the run up to burst activation during suckling. **A:** A raster display of the response of the neuron to ventral tegmental stimulation during two intracerebroventricular (ICV) injections of 2 ng oxytocin (Oxy) given to facilitate the milk ejection reflex and increase bursting activity of the neuron. Single shock stimulation was applied at 0.3 Hz, and the dots indicate occurrence of action potentials in each 0.4-second epochs after the stimulus. The occurrence of each milk ejection related burst (B) is indicated by a vertical line. **B:** Peristimulus time interval histograms showing the response to ventral tegmental stimulation in four 50-second time epochs surrounding the bursts, as indicated. Note how the response to ventral tegmental stimulation is small immediately after a burst but increases as the next burst approaches. (See text for further explanation.) [Redrawn from (261), with permission.]

with the idea that the preburst period is characterized by increasing mutual excitation between oxytocin neurons whereby a modest excitatory input can become greatly amplified. They are also compatible with a declining effectiveness of intrinsic inhibition, thereby allowing an increase in the number of spikes evoked by each stimulus shock.

In summary, analysis of background firing has suggested that before burst activation, oxytocin neurons behave in an increasingly coordinated fashion. This may depend on changes in their intrinsic properties that favor their mutual interaction, eventually leading to a synchronous burst. The oscillations in firing that lead to a burst may depend on a patterned synaptic drive that is itself amplified as the burst approaches. An illustration of these points is provided in Fig. 13.

How Are Oxytocin Neurons Coordinated?

In the previous section, we saw how oxytocin neurons can behave as a coordinated network during suckling, not withstanding their anatomical distribution into four different hypothalamic nuclei. This applies as much to their fluctuating background activity leading up to the bursts as it does to the bursts themselves. Coordination between the oxytocin neurons becomes particularly strong in the last few seconds before the burst, when increasing numbers of cells become drawn into the oscillating network. Clearly, understanding how synchronization of activity within this network is accomplished is fundamental to explaining the behavior of oxytocin neurons during suckling.

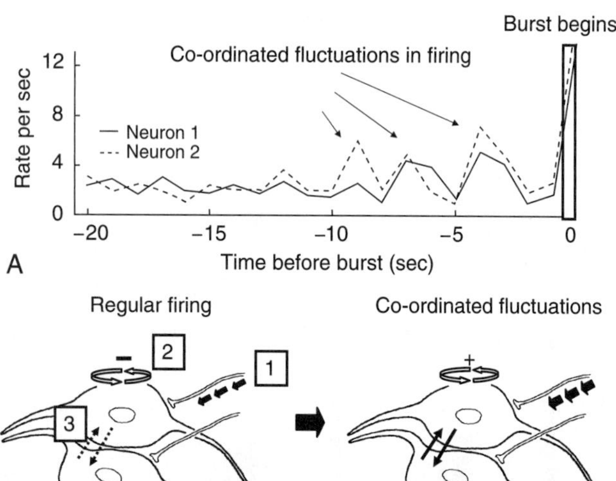

FIG. 13. Coordination of background firing leading up to burst activation of oxytocin neurons during suckling. **A:** An idealized representation of how oxytocin neurons alter their background activity in relation to burst activation. This graph represents two oxytocin neurons (*solid* and *dotted lines*) to illustrate how oxytocin neurons tend to show increasing coordinated fluctuations in background firing in the lead up to burst activation. **B:** Schematic outline of possible mechanisms involved in the change from random regular firing to coordinated fluctuating activity. This may involve a strengthening of an episodic input (1), a change from a predominance of inhibitory to excitatory intrinsic control mechanisms (2), and increased mutual excitation of the oxytocin neurons (3). This scheme is based on the findings of Moos et al. (259). (See text for further explanation.)

One obvious mechanism for coordinating the activity of oxytocin neurons is synchronization of their synaptic input, an idea that is supported by studies of oxytocin neurons in hypothalamic explants that were shown to receive synchronous excitatory inputs (235). Most of the inputs to oxytocin neurons probably arise locally from glutamatergic and GABAergic neurons in the perinuclear zone (200), and burst synchronization within the individual magnocellular nuclei could occur at this level. Furthermore, there is evidence that these local glutamatergic and GABAergic neurons can regulate each other's activity (200,204,262), and such interaction might provide a basis for generating the alternating excitatory and inhibitory synaptic drive necessary for the oscillating activity leading up to burst activation. This idea that oxytocin neurons are coordinated by local interneurons would be strengthened if it were possible to demonstrate that neurons adjacent to the magnocellular nuclei displayed activity that was correlated with bursting activity in oxytocin cells. In fact, a study by Dyball and Leng (187) of perinuclear neurons above the supraoptic nucleus reported that rather few cells displayed activity correlated with milk ejection. However, this study was undertaken before we became aware of importance of coordination of the basal firing for bursting activation, and there is clearly a need for further studies to look for more subtle correlations between the activity of perinuclear and oxytocin neurons.

Within the magnocellular nuclei themselves, it is known that around 5%–10% of oxytocin neurons are the recipients of shared synapses, that is, synapses that simultaneously abut onto two, or sometimes three, oxytocin neurons (263). Many of these shared synapses involve glutamatergic and GABAergic terminals (264). Although the percentage of oxytocin neurons contacted by shared synapses is quite small, their synchronizing effect could be considerably amplified if, as has been argued by Moos et al. (259), coordination of the population of cells with shared synapses leads to further neurons being drawn into the coordinated activity. Such recruitment could involve additional coupling mechanisms that link oxytocin neurons (see below). Furthermore, oxytocin neurons in lactating rats possess intrinsic properties that favor irregular spike activity (251), and these properties may serve to further reinforce and amplify coordinated fluctuations in basal firing.

Evidence for the existence of excitatory coupling between oxytocin neurons that would help coordinate of their activity was provided by McKenzie et al. (265). They found that stimulation of the neural stalk produced an excitation (latency of 25–50 msec) in identified oxytocin neurons that occurred even when the antidromic spike from the stimulation was eliminated by collision with a spontaneous spike. This ruled out simple recurrent excitation and suggested that the excitation was arising from adjacent (presumably oxytocin) neurons. The effect was only seen in oxytocin (rather than vasopressin) neurons and occurred more significantly in lactators as compared with virgin females and males. There are several possible explanations for this mutual excitation. In lactating rats, about 30% of oxytocin neurons have portions of their membrane in direct contact with other oxytocin cells, enabling direct electrotonic coupling (266,267). Oxytocin neurons may also be connected by gap junctions (268,269), although their role in synchronizing bursting is uncertain (236). Mutual excitation could also involve release of a local excitatory neuromodulator, oxytocin being the most likely candidate (see below). Another mechanism, proposed by Coles and Poulain (270), is that increased firing in oxytocin neurons causes a rise in extracellular potassium that tends to depolarize adjacent neurons, although this only operates during intense excitation. Finally, axons of oxytocin neurons give rise to collateral branches (271), and it has been suggested that these might terminate on excitatory interneurons projecting back to other oxytocin neurons (265).

Perhaps the aspect of synchronization that we least understand is how coordination is achieved *between* the magnocellular nuclei. Wang et al. (215) found that after deep parasagittal sections separated the two halves of the hypothalamus, bursting still occurred in response to contralateral suckling but rarely in response to ipsilateral suckling. They proposed that each side of the hypothalamus had its own synchronizing mechanism that could function independently of the opposite side of the hypothalamus. Concerning potential pathways for ipsilateral coordination, there is evidence that neurons of the supraoptic nucleus project to the paraventricular nuclei (272), and electrophysiological studies (273) suggest the existence of ipsilateral connections between these two nuclei. Lesion studies (274) indicate that synchronous bursting may not require projections from the paraventricular nucleus to the supraoptic nucleus, but the importance of projections in the opposite direction has not been studied. Thellier et al. (275) identified neurons in the dorsochiasmatic area of the hypothalamus that projected ipsilaterally both to the supraoptic and paraventricular nuclei, and such common inputs could provide the basis for synchronization of the two nuclei within one-half of the hypothalamus. Another region of potential importance for ipsilateral synchronization of oxytocin neurons is the hypothalamic dorsomedial nucleus. A small proportion of neurons in this region project to the supraoptic nucleus and display bursting responses identical to those found in oxytocin neurons. Moreover, lesions involving the dorsomedial nucleus were found to block milk-ejection responses (276).

Superimposed upon mechanisms coordinating the ipsilateral paraventricular and supraoptic nuclei must be a means of synchronizing the two sides of the hypothalamus so that all four nuclei fire bursts simultaneously. Wang et al. (277) found that after complete disruption of all ascending inputs on one side of the midbrain, synchronous bursts still occurred in both the ipsilateral and contralateral supraoptic nuclei, indicating that decussating pathways for bilateral coordination were located within or close to the hypothalamus. One possibility is that bilateral coordination might involve the supraventricular gray commissure (the region connecting the two paraventricular nuclei through the roof of the third ventricle), especially because the dendrites from the two paraventricular nuclei intermingle in this region. However, Moos and Richard (238) studied the effects of deep parasagittal cuts sectioning this commissure and found that such cuts did not prevent bilateral synchronization of bursting activity. More recently, studies by Moos et al. (216) showed that only very deep interhemispheric cuts that involve commissural fibers passing over the supraoptic chiasm (the supraoptic decussation) disrupt bilateral synchronization of bursting activity. The supraoptic decussations contain several different fiber pathways, including axons of the somatosensory spinohypothalamic tracts (278) and ascending projections from the locus ceruleus that decussate at this level (279). Interestingly, Moos et al. (216) identified a population of neurons in the ventrolateral medulla that projected bilaterally to both supraoptic nuclei, via fibers passing through the supraoptic decussations. Neurons in the ventrolateral medulla also displayed increased Fos expression during suckling (216), supporting the idea that this region might by important for bilateral synchronization of oxytocin neurons. However, whether the ventrolateral medulla has a primary role in the timing of the bursts is not entirely clear, especially in the light of increasing evidence (see above) that the periodicity of bursting is determined within the hypothalamus.

Significance of the Suckling Pattern in Timing Burst Activation

Analysis of the sucking behavior of rat pups indicated that, individually, they display episodes of sucking every 20–30 seconds (58,280). Adjacent pups may display simultaneous sucking episodes, but using behavioral observations, Drewett et al. (58) could not detect any change in the collective activity of the litter leading up to milk ejection. However, using a strain gauge technique, Karyakin and Alekseev (280) reported that the incidence of simultaneous sucking was increased 10–30 seconds before milk ejection. Although this later work might suggest that activation of bursts is triggered by cues from the suckling pups, other evidence suggests that the periodicity is an inherent property of the oxytocin pulse generator mechanism. Thus, continuous unvarying stimulation of the nipples at a frequency 10 Hz evoked a sequence of intermittent milk-ejection responses similar in time course to those observed during natural suckling (281). The pattern of burst activation in oxytocin neurons can be profoundly altered by centrally administered drugs (189,237) or midbrain lesions (231), and this also argues against the idea that burst activation is purely regulated by fluctuations in the intensity of the suckling stimulus. Aside from the rat, some insight into the relationship between the suckling stimulus and activation of oxytocin neurons can also be gained from studies in other species, where plasma oxytocin or milk-ejection responses have been monitored during suckling (52,76,79,82,88,91). Although the onset of milk ejection may be correlated with particular behavioral events in the mother and her offspring during the nursing episode, there is little evidence in these studies that the timing of the individual pulses of oxytocin is linked to a sudden change in suckling intensity.

The finding that suckling causes a particular type of background activity in oxytocin neurons, not seen with other stimuli (217), might indicate that even if the sucking episodes of individual pups do not trigger bursts, they might be essential for causing the irregular oscillatory activity that culminates in burst activation. However, it should be noted that similar irregular activity can be seen in oxytocin neurons recorded in hypothalamic explants (236). This suggests that the characteristic pattern of background firing observed during suckling is not fashioned by patterning of the peripheral stimulus. Rather, it arises because of the unique ability of suckling to cause an organized pattern of excitation within hypothalamic circuitry controlling the oxytocin neurons.

The conclusion that the milk-ejection reflex does not rely on patterning of the suckling input to generate periodic activation of bursts in the oxytocin neurons does not mean that burst activation is never influenced by sudden changes in the somatosensory input. On the contrary, bursts can be prematurely precipitated during suckling by several procedures, namely applying or removing a pup from one of the nipples (188), pumping a mammary gland with isotonic saline (188), and vaginocervical stimulation (107,218). Such procedures are only effective within a particular time window in relation to milk ejection and do not work if given to close to a preceding burst (107,218). It is notable that burst activation after mammary or vaginocervical stimulation is more or less instantaneous, and even when vaginocervical probing fails to precipitate a burst, it usually evokes a small cluster of spikes within 100 msec of applying the stimulus (218). Hence, in certain circumstances there seems to be a direct route by which changes in somatosensory input can be signaled up to the oxytocin neurons.

The fact that cervicovaginal stimulation only precipitates burst activation in the presence of the suckling stimulus (107,218) reflects the requirement for the network controlling burst activation to be in an appropriate state of "readiness" before the sudden increased excitatory drive can precipitate a burst. In other words, the oxytocin neurons probably need display coordinated fluctuations in background firing, as they do during suckling (259), so that an abrupt excitatory input can advance the moment when the oscillating network loses stability and a burst is generated. The idea that ongoing suckling is essential before it is possible to trigger bursts by additional stimulation might seem to be dispelled by reports of milk-ejection responses evoked by stimulation of the mammary nerve in the absence of suckling (95). However, it is uncertain whether mammary nerve stimulation evokes bursting responses comparable with those during suckling, rather than repetitive stimulus-linked firing similar to that seen when milk ejection is evoked by septal stimulation (282).

Autoregulatory Role of Oxytocin in the Bursting Activity of Oxytocin Neurons

It is well established that oxytocin is released within the brain, where it serves as a neuromodulator regulating a range of reproductive and nonreproductive functions (283). Oxytocin does not cross the blood–brain barrier, so the central and peripheral effects of oxytocin can operate independently. Early studies concerning the possible role of oxytocin as a neuromodulator in the milk-ejection reflex first addressed the question of whether the intermittent nature of oxytocin release might arise as a result of negative feedback. However, attempts to demonstrate negative feedback, either by intravenous administration of oxytocin (284) or by electrical stimulation of the neural stalk to evoke additional neurosecretory bursts (285), failed to support this idea, because the periodicity of milk ejection was not depressed. Indeed, stalk stimulation has since been shown to increase milk-ejection frequency (245).

Investigation of the effects of centrally administered oxytocin in the milk-ejection reflex was first undertaken by Freund-Mercier and Richard (286). They found that intracerebroventricular injection of oxytocin had an extraordinarily powerful effect in facilitating bursting activity, and this has been confirmed in many subsequent studies (238,256,283, 287–289) (see example in Fig. 11). The dose of oxytocin required was 1–2 ng, and within a few minutes of making the intracerebroventricular injection, the frequency of the intermittent bursts in the oxytocin cells increased from once every 5–6 minutes to once every 2–3 minutes. The number of action potentials in each burst also increased (from 50–60 spikes to 80–90 spikes). Administration of intracerebroventricular oxytocin to suckled rats that previously failed to display bursting responses precipitated a sequence of bursts. Intracerebroventricular oxytocin also induced recruitment of previously unresponsive oxytocin cells into the pool of activated neurons (214). The facilitatory effect of oxytocin lasted around 15–20 minutes and was followed by a period of reduced bursting activity that may relate to delayed electroencephalogram arousal (290). The physiological significance of these observations was supported by the finding that intracerebroventricular administration of an oxytocin antagonist blocked ongoing bursting activity (256). Furthermore, intrahypothalamic injection of an oligonucleotide encoding the antisense sequence of oxytocin mRNA caused a rapid reduction in milk-ejection frequency, consistent with a requirement for ongoing dendritic oxytocin synthesis for maintaining the normal response of oxytocin neurons during suckling (291). The one caveat is that virtually all the investigations undertaken so far have been in the rat, apart from one study in the goat showing that intracerebroventricular oxytocin increases reflex oxytocin release (292).

Neural Mechanisms Underlying the Facilitatory Action of Oxytocin on the Milk-Ejection Reflex: Intrahypothalamic Effects

Early observations showed that the facilitatory effect of oxytocin on the milk-ejection reflex was most effective when intracerebroventricular injections were made into the third ventricle, and this suggested that a likely site of action was the hypothalamus (256). The most obvious target for oxytocin within the hypothalamus was the magnocellular nuclei themselves, and this was investigated by Moos and Richard (293), who made injections of oxytocin into the paraventricular and supraoptic nuclei while recording bursting responses in a contralateral nucleus (Fig. 14). Localized oxytocin injections made into either nuclei were found to facilitate bursting recorded contralaterally, and this effect was lost after deep hemisection of the brain. Subsequently, the effects of injections of oxytocin antagonist into the magnocellular nuclei were investigated, and, as predicted, these reduced both burst amplitude and frequency. From their collective results, Lambert et al. (289) concluded that oxytocin exerted a local action within the magnocellular nuclei paraventricular and supraoptic nuclei that, in turn, could influence the whole neuronal population. It has been argued that a local action for oxytocin is unlikely because the effects of intracerebroventricular oxytocin are restricted to facilitation of suckling evoked bursting, whereas responses to other stimuli are unaffected (294). We now know that oxytocin can exert subtle

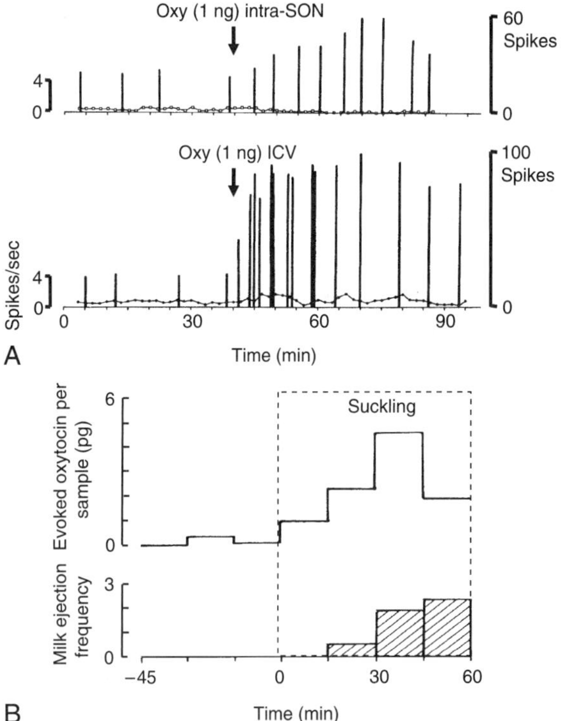

FIG. 14. Evidence for a local autoregulatory effect of oxytocin on bursting activity of oxytocin neurons during suckling. **A:** *Upper record* shows the effect of a microinjection of 1 ng oxytocin (Oxy) into the supraoptic nucleus (SON) on the background firing and bursting activity of an oxytocin neuron recorded in the contralateral nucleus. The *lower record* shows the effect of a similar dose of oxytocin given by intracerebroventricular injection. Although the response was less than seen after intracerebroventricular administration, the intranuclear injection was effective in increasing both the frequency and amplitude of the bursting responses. **B:** Levels of oxytocin in the supraoptic nucleus during suckling, measured by a push–pull cannula. Note a small rise of oxytocin was observed before the onset of reflex milk ejection responses. [A is redrawn from (293), B from (324), with permission.]

effects on excitatory responses in oxytocin neurons that may be selective for particular inputs (261), so such specificity should not be surprising. The conclusions from the in vivo microinjection studies of Moos et al., that oxytocin acts as an autoregulator within the magnocellular nuclei, have received support from electrophysiological studies using hypothalamic explants in vitro. Oxytocin neurons in these explants display spontaneous bursting, and this is facilitated if oxytocin is added to the incubation medium (234,235), just as in vivo oxytocin could also induce bursting in neurons in which such responses were otherwise absent (235).

Another line of evidence that has been used to support a local autoregulatory action for oxytocin within the magnocellular nuclei concerns the effects of electrical stimulation of the neural stalk. Neural stalk stimulation causes antidromic activation of oxytocin neurons and, if the oxytocin dendrites have been appropriately primed (295), should cause a rise in oxytocin within the magnocellular nuclei. As predicted, neural stalk stimulation applied during suckling leads, after a delay of several minutes, to significant facilitation of bursting activity (187,245). Facilitation can also be induced by stimulating within the magnocellular nuclei, although this is far less effective (296). Some, including the present author (294), have raised doubts about whether facilitation after antidromic stimulation does indeed operate via local oxytocin release on the grounds that other procedures that might be expected to increase intranuclear oxytocin, such as osmotic stimulation, do not facilitate bursting. However, it has been shown (217) that bursting requires a particular pattern of background firing that is disrupted by osmotic stimulation, thereby masking any potential facilitation caused by elevated intranuclear oxytocin.

Assuming that oxytocin operates in or near the magnocellular nuclei, what could be the mechanism of this action? Oxytocin neurons have been show to possess oxytocin receptors (297), and one possibility is that oxytocin exerts a direct excitatory effect, thereby mediating the proposed excitatory interaction between the oxytocin neurons. Oxytocin does increase the firing rate of oxytocin neurons on slices in vitro (298,299), an effect that is more pronounced in slices from lactating compared with virgin rats (300). By analogy with the known actions of oxytocin elsewhere in the nervous system (301), this may involve opening of nonspecific cation channels or closing of potassium channels. However, it is important to emphasize that the excitation caused by oxytocin is nowhere near as powerful as that associated with the bursting responses. Moreover, neither patch clamp studies in slices (302,303) nor intracellular recordings studies in explants (235) reported any clear effects of oxytocin on membrane current-voltage relations that might be conducive to a direct depolarizing effect sufficient to evoke a burst. Thus, to date, there is little evidence to support the idea that oxytocin functions simply as an excitatory transmitter causing mutual burst activation in oxytocin neurons.

As discussed in a previous section, facilitation of bursting activity by oxytocin involves a change in the pattern of basal firing in the oxytocin neurons (217), and this might arise through modulation of their intrinsic properties. After binding to its receptor, oxytocin acts through the inositol triphosphate intracellular signaling cascade to cause a rise of intracellular calcium in oxytocin neurons (304). It is known that the depolarizing afterpotential is triggered by spike-related calcium influx, and the afterhyperpolarization similarly operates through a calcium-activated potassium current (196,246). Thus, it is quite likely that by altering calcium dynamics inside the cell, oxytocin might modify these intrinsic regulatory mechanisms. Evidence in support of this idea has been provided by Ludwig et al. (295). They used constant current stimulation to evoke intranuclear oxytocin release, having first dialyzed thapsagargin (a liberator of intracellular calcium stores) into the nucleus to prime the dendrites for peptide release. They found that constant current stimulation caused a change in the patterning of basal firing, whereby the activity became less regular with occasional small bursts of firing. The same effects were obtained when constant current stimulation was applied after giving intracerebroventricular oxytocin to similarly prime local oxytocin release. These effects were observed in virgin female rats in the absence of any suckling stimulus. They provide the first good evidence that oxytocin does indeed act locally to cause reorganization of spike patterning, characterized by more irregular activity of the type previously associated with burst activation (217). It remains to be proven that the irregular activity observed by Ludwig et al. during constant current stimulation arose from a change in the intrinsic properties of oxytocin neurons, but this seems the most likely mechanism.

Another way in which oxytocin probably influences bursting activity is by altering the efficiency of synaptic inputs to oxytocin neurons, especially from local GABAergic and glutamatergic interneurons. This could involve both retrograde presynaptic effects (i.e., modulation of transmitter release) and postsynaptic effects on the responsivity or excitability of the neurons. It known that oxytocin reduces GABAergic transmission, both by presynaptic reduction in GABA release (305) and by postsynaptic reduction in the inhibitory postsynaptic currents (302). The latter effect may involve a change in the phosphorylation state of the GABA receptors, after calcium-mediated activation of a protein kinase (302). The effect of oxytocin on glutamate transmission is less clear-cut. On the one hand there is evidence in other systems, such as the olfactory bulb (306) and spinal cord (307), that oxytocin increases glutamate-mediated excitatory postsynaptic currents. Similarly, in hypothalamic explants oxytocin increases bursting activity in oxytocin neurons by increasing the frequency of glutamate-mediated synaptic potentials (236). However, Kombian et al. (303) found that oxytocin suppressed glutamatergic excitatory postsynaptic currents in oxytocin neurons, probably through endocannabinoid release (308). If the effects of endocannabinoids are blocked, then oxytocin causes a small increase in excitatory postsynaptic current frequency (308). The story is further complicated by that fact that oxytocin induces release of noradrenalin within the supraoptic nucleus (309), and noradrenalin exerts a presynaptic facilitation of glutamatergic inputs to magnocellular neurons (including oxytocin neurons) (310). On balance, the effects of oxytocin may be to increase glutamate excitation of oxytocin neurons, even if this is partly opposed by retrograde inhibition of glutamatergic terminals. This would be consistent with in vivo data showing that intracerebroventricular oxytocin potentiates long-latency excitatory responses of oxytocin neurons to stimulation of the ventral tegmentum (261) (Fig. 15) and dorsomedial hypothalamus (311), inputs that are probably mediated through local glutamatergic interneurons. A scheme illustrating the possible effects of locally released oxytocin in the autoregulation of oxytocin neurons is provided in Fig. 16.

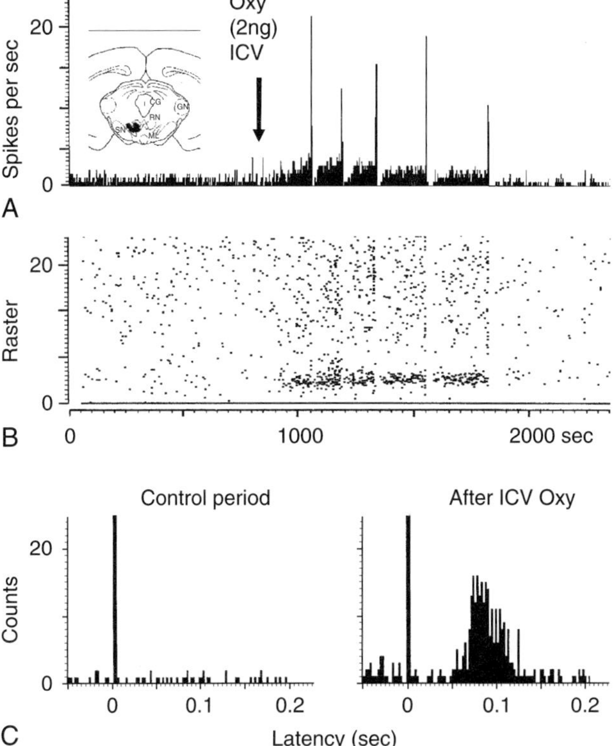

FIG. 15. Recording from an oxytocin neuron illustrating the effect of an intracerebroventricular (ICV) injection of 2 ng oxytocin (Oxy) on the responses to stimulation of the ventral tegmentum of the midbrain during suckling. **A:** Ratemeter recording of the activity of the cell during the experiment. Single shock stimulation of the ventral tegmental was applied at 0.3 Hz throughout the recording to detect changes in the neuronal response. The time of oxytocin administration is indicated by the arrow. Note the increase in background firing and occurrence of repetitive high frequency bursts after oxytocin administration. (*Inset* shows the localization of the stimulation sites for all the experiments in the study.) **B:** A raster display (see Fig. 12) of the response of the oxytocin neuron to the ventral tegmentum stimulation displayed on the same time scale as the rate meter record above. Note the appearance of an excitatory response (indicated by the *cluster of dots*) coincident with the facilitation of bursting after intracerebroventricular administration of oxytocin. **C:** Peristimulus time-interval histograms for a 10-minute control period (*left*) and a 10-minute period after giving intracerebroventricular oxytocin (*right*) showing how the response increased after giving oxytocin. [Redrawn from (261), with permission.]

Effects of Oxytocin at Other Sites in the Milk-Ejection Reflex

The autoradiographic localization of oxytocin binding within the brain (312,313) has revealed a number of extrahypothalamic targets where oxytocin facilitation of the milk-ejection reflex might be mediated. The fact that intraventricular injections of oxytocin are still effective in facilitating the milk-ejection reflex even after lesioning the paraventricular nucleus (274) might suggest that oxytocin can

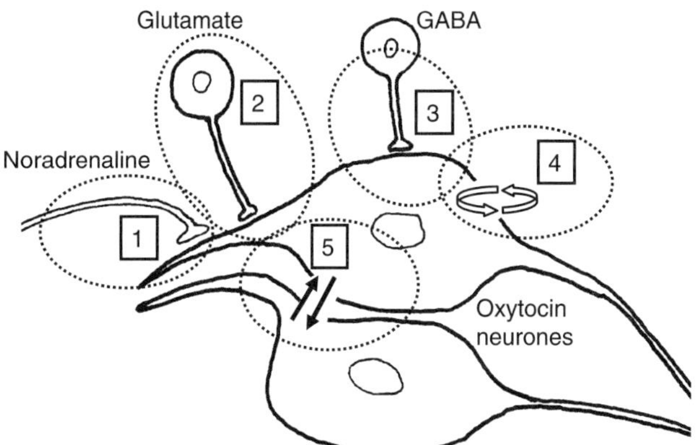

FIG. 16. Scheme indicting the potential mechanisms by which locally released oxytocin can alter the bursting activity of oxytocin neurons during suckling. Oxytocin may increase suckling-evoked release of noradrenalin (1), increase excitation by local glutamatergic neurons (2), reduce GABA inhibition (3), alter the intrinsic regulation of neuronal firing (4), or promote mutual excitation of the oxytocin neurons (5). See text for further explanation.

have effects at other periventricular sites. Attention has especially been focused on the bed nucleus of the stria terminalis, an area also implicated in the effects of oxytocin on maternal behavior (314). Oxytocin binding sites in the bed nucleus increase over the peripartum period (315), probably in response to changing steroids, particularly estradiol (316). Up to 50% of neurons in the bed nucleus respond to oxytocin (317), and their sensitivity to oxytocin is enhanced during lactation, compared with pregnancy and weaning (318). Microinjections of oxytocin into the bed nucleus reliably cause a facilitatory increase in oxytocin cell bursting (319) similar to that observed with intracerebroventricular injections of the peptide. The importance of the bed nucleus in regulating the milk-ejection reflex is further substantiated by the fact that some of its neurons show rhythmic oscillations in firing during suckling that correlate with the activity of oxytocin cells (232).

We know from other systems, for example, in relation to the role of oxytocin in pair bonding behavior (320), that oxytocin interacts with dopaminergic pathways, and dopamine strongly facilitates the milk-ejection reflex (189). Interaction between oxytocin and the dopamine system has been thought to involve effects on mesolimbic dopamine neurons in the ventral tegmental area (321), but there have so far been no reports that oxytocin can excite neurons in this region. A potentially more significant area is the adjacent supramamillary nucleus, a region lying in the caudal hypothalamus that is involved in dopaminergic regulation of reward behavior (322) and contains oxytocin-responsive neurons (323).

Regulation of Central Release of Oxytocin during Suckling

As evidence concerning the local autoregulatory action of oxytocin in the milk-ejection reflex emerged, the question arose as to whether suckling evoked a rise of oxytocin within the magnocellular nuclei. It was first confirmed that oxytocin release occurred within the magnocellular nuclei in vitro (197). An in vivo study was then undertaken by Moos and coworkers (324) using a push–pull cannula inserted into the supraoptic nucleus to sample oxytocin release over 15-minute epochs during suckling (Fig. 14). During a 40-minute sequence of milk-ejection responses evoked by suckling, there was a progressive rise in intranuclear oxytocin. The rise in oxytocin occurred slightly before milk ejections began, suggesting that the initial rise was a consequence of suckling rather than the onset of milk-ejection responses. Release of oxytocin in both the paraventricular and supraoptic nuclei during suckling was later confirmed by Neumann et al. (325), using microdialysis.

Attention then turned to which neural elements might be involved in oxytocin release within the magnocellular nuclei. Studies using tannic acid fixation to capture exocytotic profiles in magnocellular neurons (326) showed that exocytosis of oxytocin-containing granules can occur over multiple compartments of the cell, including both the soma and the dendrites. This is consistent with the view (327) that exocytosis in peptidergic neurons is not restricted to a specialized presynaptic zone, as in "conventional" neurons. Although there may be a small number of oxytocin-containing synaptic terminals within the magnocellular nuclei (328), the most likely origin of intranuclear oxytocin release are the oxytocin-containing dendrites (175,176,295). Dendritic release of oxytocin is a calcium-dependent process that is induced by membrane depolarization (e.g., exposure to high potassium medium or direct electrical stimulation). This would suggest that, physiologically, the main trigger for dendritic release is action potentials initiated in the soma and retrogradely propagated along the dendrites. On this basis, central release of oxytocin in vivo might be expected to parallel peripheral release from neurohypophysial terminals. However, when Ludwig (327) studied dendritic release, using a microdialysis probe placed over the supraoptic nucleus after ventral exposure of the brain, he found that a number of stimuli that were highly effective in causing rapid release of oxytocin from neurohypophysial terminals (intraperitoneal sodium chloride injection or stimulation of the neural stalk) caused only a sluggish delayed release within the supraoptic nuclei. Hence, it was concluded that dendritic release of oxytocin during suckling involved additional factors, for example, the action of a particular neuromodulator rather than just an increase in neuronal firing.

So what special factor enables dendritic release of oxytocin during suckling? It has been established for a number of years that oxytocin itself can induce release of oxytocin from hypothalamic slices in vivo (329). Ludwig et al. (330) proposed that perhaps the important effect of oxytocin was to prime the release process, so enabling subsequent spike activity to trigger dendritic release. Oxytocin was known (304) to increase intracellular calcium in oxytocin neurons by releasing so-called thapsigargin-sensitive calcium stores held within the smooth endoplasmic reticulum. Using microdialysis of the exposed supraoptic nucleus in anesthetized rats, Ludwig et al. (330) showed that thapsigargin pretreatment dramatically potentiated intranuclear (i.e., dendritic) oxytocin release in response to osmotic stimuli or antidromic activation.

Thapsigargin also potentiated oxytocin release from in vitro hypothalamic slices, and, crucially, an oxytocin agonist caused the same priming action. Ludwig et al. (330) suggested that the priming effect of oxytocin might involve movement of oxytocin granules toward the cell membrane, where they were held within a readily releasable pool.

To translate these observations into a physiological context, it would seem that oxytocin is involved in a positive feedback loop whereby oxytocin release from just a few magnocellular neurons at the start of suckling can prime other cells to discharge their oxytocin in response to ongoing spike activity. In turn, further cells undergo priming such that dendritic release is progressively activated within the whole population of neurons. Unfortunately, current methodology only allows measurement of dendritic release over a relatively long time course, so the extent to which intranuclear concentrations of oxytocin fluctuate with oscillations in background firing or in association with high frequency bursts is currently unknown. Such information would have an enormous impact on our understanding of the role of dendritic oxytocin release in bursting activity.

As well as oxytocin itself, conventional neurotransmitters also regulate dendritic oxytocin release. Suckling increases the turnover of noradrenalin within the magnocellular nuclei (331), and this transmitter regulates intranuclear oxytocin release during suckling (332). Dendritic release of oxytocin from in vitro slices can be stimulated by *N*-methyl-D-aspartate (NMDA) agonists, and this occurs even in the absence of action potentials (305). Thus, dendritic release may be regulated by glutamatergic synapses on the dendrites of oxytocin neurons, acting quite independently of neuronal firing. The effects of noradrenalin and glutamate are particularly important at the onset of suckling, before intranuclear oxytocin rises sufficiently to stimulate its own release.

It is also necessary to consider how the levels of oxytocin may be regulated at other sites outside of the magnocellular nuclei, for example, in the bed nucleus of the stria terminalis, where oxytocin acts to facilitate the milk-ejection reflex (319). Microdialysis studies (333) have revealed that suckling causes a rise of oxytocin within the bed nucleus. The bed nucleus lies adjacent to the lateral ventricle, and it is possible that the cerebrospinal fluid is the normal route by which oxytocin reaches this target structure. Indeed, intracerebroventricular injection of oxytocin is effective in activating bed nucleus neurons (232). Oxytocin could reach the cerebrospinal fluid from the paraventricular nuclei, although this has been disputed (334). An oxytocinergic projection to the bed nucleus from the paraventricular nucleus has been demonstrated (335), and this could be part of an important circuitry for providing the appropriate input during suckling. Additionally, oxytocin-containing perikarya have been found in the anterior commissural nucleus during late pregnancy (168,169), and these cells are well positioned to release oxytocin adjacent to the bed nucleus. In fact, Ingram and Moos (335) proposed that the bed nucleus receives its oxytocinergic innervation from a continuum of oxytocin-containing perikarya running from the anterior parvocellular hypothalamic paraventricular nucleus to the anterior commissural nucleus and perifornical region. Oxytocinergic fibers project caudally from the parvocellular regions of the paraventricular nucleus, providing a potential source of oxytocin release in the brainstem and spinal cord, although it has yet to be shown that parvocellular oxytocin neurons are activated by suckling.

Control of Release from Oxytocinergic Neurosecretory Terminals at the Level of the Neurohypophysis

Although the main determinant of oxytocin release during milk ejection is the firing of the oxytocin neurons within the hypothalamus, there may be further "last chance" regulation of oxytocin release at the level of the posterior pituitary (180–182). A range of neurotransmitters and neuropeptides have been implicated in the regulation of oxytocin release from the neurosecretory terminals of the neurohypophysis, but on current evidence, by far the most important are the opioid peptides (183,336,337). This opioidergic regulation was revealed when it was found that morphine, or enkephalin, inhibited oxytocin release from the neurohypophysis both in vivo (338,339) and in vitro (340). In both cases this effect was reversed by the opiate antagonist, naloxone. Furthermore, even in the absence of exogenously applied drugs, release of oxytocin from the rat neurohypophysis was potentiated by naloxone, indicating the existence of endogenous opioid activity (340). Since these early observations, there has been considerable work to determine the identity of the active opioids in the neurohypophysis as well as the type and location of their receptors. The consensus view is that oxytocin terminals possess κ opiate receptors (341) that bind dynorphin. Dynorphin coexists within both oxytocin (342) and vasopressin (174) neurons and is probably released from their axon terminals. κ Agonists are highly effective at inhibiting oxytocin release in vivo (343,344) and dynorphin suppresses calcium influx, as well as exocytosis, in isolated neurosecretory terminals (345). Because pituicytes also possess κ type receptors (346), their involvement is not essential for the inhibitory actions of dynorphin (347).

Although dynorphin is the main opioid regulating oxytocinergic terminals, enkephalin has also been identified in fibers projecting to the neurohypophysis (348), being colocalized within oxytocin-containing nerve terminals (349), and enkephalin can inhibit oxytocin release (340). Another opioid that could influence oxytocin release is β-endorphin, which is present within melanotropes of the pars intermedia, and might reach the neurohypophysis by diffusion (350). Oxytocin release from isolated terminals is depressed by μ receptor agonists (of which β-endorphin could be the endogenous ligand) and, as in the case of κ type receptors, this may operate by reducing calcium influx (351).

Most of the work concerning the physiological significance of opioids in controlling oxytocin release at the level of the neurohypophysis has been undertaken in relation to the regulation of birth, where steroidal modulation of endogenous opioid tone has an important role (337,352). In relation to milk ejection, Clarke et al. (338) reported that morphine administration caused uncoupling of the bursting activity of oxytocin neurons from the associated release of an oxytocin pulse. This suggested that opioids could inhibit the milk-ejection reflex at the level of the neurohypophysis, and a role in stress inhibition of the milk-ejection reflex was proposed. However, it has since been found that opioid action within the neurohypophysis is predominantly mediated by κ receptors, rather than morphine-binding μ receptors, casting doubt on whether such effects are truly at level of the neural lobe rather than the hypothalamus (183). Moreover, attempts to demonstrate an effect of the opiate antagonist naloxone on oxytocin release or intramammary pressure responses during suckling or milking in the cow (353), pig (354), and human (355) have proved unsuccessful. Morphine can inhibit oxytocin release during milking in the cow (353), but naloxone cannot reverse stress-induced blockade of the milk-ejection reflex in this species (356,357). Hence, a role for neurohypophysial opioid action in stress-induced inhibition of the milk-ejection reflex in species other than the rat has still to be established.

An alternative role for opioidergic inhibition of oxytocinergic terminals within the neurohypophysis is to prevent release of oxytocin in response to osmotic stimulation and hemorrhage. In the lactating rat, both oxytocin and vasopressin neurons show a sustained increase in firing during osmotic stimulation and dehydration (120). Opioid antagonists increase oxytocin, but not vasopressin, release in responses to such stimuli (358), suggesting the presence of selective opioid inhibition of the oxytocin terminals. It is possible that coreleased dynorphin, either from oxytocin or vasopressin terminals, may help to limit the release of oxytocin in response to osmotic stimuli and thereby conserve oxytocin stores for milk ejection. The fact that dynorphin in oxytocin neurons is up-regulated during lactation (359) is consistent with this suggestion.

Another potential modulator of oxytocin release at the level of the neurohypophysis is dopamine, which is present in the posterior lobe in nerve terminals originating from the arcuate nucleus (360). Pituicytes possess dopamine receptors (180,181), and the content of dopamine within the lobe fluctuates in synchrony with physiological demands for oxytocin release (361). However, inconsistent effects, either stimulatory (362) or inhibitory (363), of dopamine agonists on oxytocin release have been observed in vitro. Studies on the effects of dopamine agonists on the milk-ejection reflex have reported both stimulatory (189) and inhibitory (331) effects. Bromocriptine did not prevent milk ejection when given in doses that reduced milk secretion (364), so that if neurohypophysial dopamine does have a physiological role it is most likely to be stimulatory.

Norepinephrine-containing fibers are also present in the neural lobe, some contacting pituicytes (181) and others terminating in the perivascular spaces (365). Intriguingly, some of these noradrenergic fibers arise from the A2 noradrenergic cell group in the brainstem (366), and this cell group has also been implicated in the regulation of oxytocin release at the hypothalamic level (160). Noradrenalin stimulates movement of pituicyte processes in vitro (367), so it is conceivable that this pathway might control the morphological adaptations that occur in the neurohypophysis during birth and lactation (see next section). Zhao et al. (368), using an in vitro approach, concluded that noradrenalin, acting via both β and α_1 receptors, facilitated oxytocin release. Direct evidence concerning the role of neurohypophysial noradrenaline in the milk-ejection reflex in vivo is sparse. It has been reported that the β antagonist, propranolol, potentiates milk ejection evoked by neural stalk stimulation (97). This might point to an inhibitory effect of noradrenalin (i.e., the opposite effect to that reported by Zhao et al. in vitro). However, it is possible that the effects of propranolol related to a subtle change in responsivity of the mammary gland to oxytocin. Noradrenalin may also regulate neurohypophysial blood flow (369), an effect that would be absent in vitro.

GABA is present in the neurohypophysis (370), and under conditions of strong stimulation, GABA inhibits oxytocin release from the neurohypophysis (371). However, under milder stimulation GABA has a small potentiating effect (372). GABA acts directly on neurosecretory terminals, and the complex effects of GABA probably relate to the fact that, unexpectedly, GABA depolarizes neurosecretory terminals (178). This can lead either to conduction block, giving reduced release, or to influx of calcium, giving enhanced release (372). GABA receptors in the neurohypophysis also bind taurine, an amino acid released from pituicytes (372). GABAergic regulation of oxytocin release within the neurohypophysis could provide a means whereby steroid metabolites modulate oxytocin release (373). Concerning the milk-ejection reflex, Voisin et al. (374) reported no effect of intracerebroventricular injection of the GABA agonist, bicuculline, on the intramammary response to neural stalk stimulation, although it should be noted that the route of administration may have limited access of the drug to the neurohypophysis.

Other potential modulators of oxytocin release within the neurohypophysis include nitric oxide and ATP. Nitric oxide inhibits neuropeptide release from the isolated rat neural lobe (375), and nitric oxide synthetase is present in neurosecretory nerve terminals (376) and pituicytes (377). ATP is coreleased from neurosecretory terminals and exerts complex effects that include an initial stimulation followed by inhibition, mediated by its metabolite, adenosine (378).

Adaptations of Oxytocin Neurons for Lactation

The increased secretory demands associated with lactation lead to hypertrophy of the oxytocin neurons, with development of the intracellular organelles concerned with hormone synthesis (379) and an increase in oxytocin mRNA (380). Oxytocin neurons also undergo a change in their morphology, with a reduction in total dendritic length and fewer dendritic branches (251). It is unknown whether this leads to an increase in the density of synaptic contacts on the dendrites or whether there is a compensatory loss of synapses. In any case, the shorter dendrites provide a stronger electrotonic contact with the soma, with less dissipation of current arising in the dendrites during integration of excitatory or inhibitory inputs. Oxytocin neurons also change their electrophysiological characteristics, displaying broader action potentials as well as a lower initial firing rate and more rapid adaptation during a sustained depolarization (249). The afterhyperpolarization following a train of spikes is increased, and a greater number of oxytocin neurons display a depolarizing afterpotential. These adaptations in the intrinsic properties of oxytocin neurons during lactation may increase their ability to switch from regular to irregular background firing during suckling, a change that has been implicated in the onset of bursting activity (217). The increased spike width results in greater calcium influx for the activation of intracellular processes involved in the synthesis and release of oxytocin (249).

Another important modification that occurs in oxytocin neurons during lactation concerns their morphological relationship with astroglial cells (381–384) (Fig. 17). This adaptation has provided a useful model for studying neuroplasticity within the adult brain and has been the subject of recent reviews (266,267). In the supraoptic nuclei of nonlactating rats, neurosecretory neurons are usually separated by astroglial processes, and fewer than 10% of cells are involved in direct soma–soma contacts or soma–dendritic contacts. In the lactating animal, cell contacts (soma–soma, soma–dendrite, and dendrite–dendrite) become more numerous and involve some 30% of the cells. The mean length of the membrane appositions increases in lactation from 1.9 to 3.1 μm, attachment plaques are formed between the oxytocin cells, and their dendrites become arranged in bundles. There is also an increased incidence of dye coupling between oxytocin neurons, indicative of the presence of gap junctions (268,269). The change in the glia/neuron relations of the oxytocin neurons is accompanied by reorganization of their synaptic contacts. In the nonlactating animal, fewer than

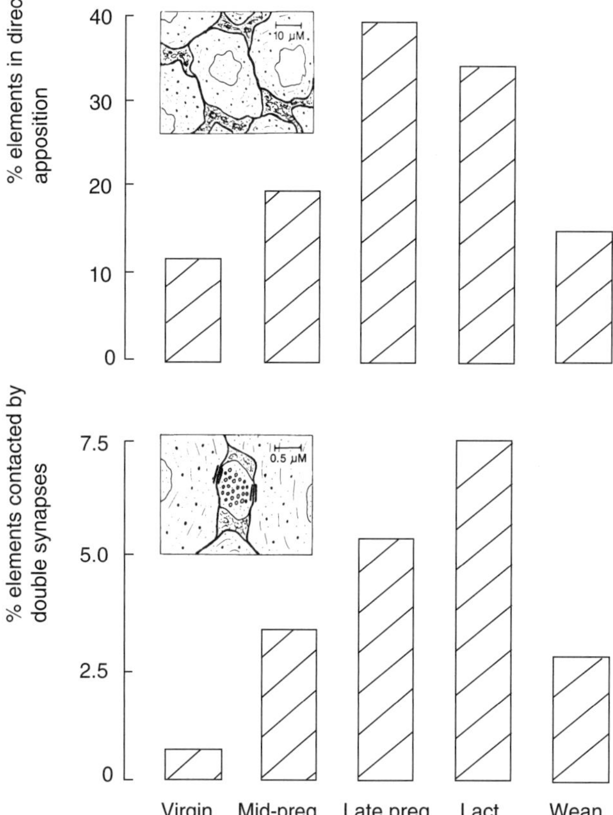

FIG. 17. Structural reorganization of the supraoptic nucleus during lactation in the rat. The histograms show the changes in membrane apposition of the neurosecretory cells (*upper*) and the occurrence of "double" synapses (*lower*) in different reproductive states. The histograms were obtained by analysis of electron micrographs, and the insets give a diagrammatic representation of the morphological features that were noted. See text for further explanation. [Redrawn from (383), with permission.]

1% of oxytocin cells are contacted by double synapses (i.e., terminals simultaneously contacting two adjacent cells), but in the lactating animal, over 7% of cells receive such synapses. Both glutamate and GABAergic synapses are involved in this reorganization (264). These morphological changes occur between the 15th and 21st days of gestation and regress within a month of weaning (383).

It is tempting to relate the above structural adaptations to changes in the ability of oxytocin neurons to display bursting responses, especially because the incidence of bursting activity increases between late pregnancy and lactation (218), that is, over the period of structural reorganization of the hypothalamoneurohypophysial system. However, it should not be forgotten that these changes are not restricted to lactation but are also evoked by osmotic stimuli (385); hence, it is possible that they are of more general functional significance. Nevertheless, the fact that the adaptations occur only in oxytocin neurons and are never observed in vasopressin neurons under identical conditions (386) might suggest that they have some significance in determining the specific firing characteristics of oxytocin neurons during lactation. Thus, the increased numbers of double synapses might serve a generalized function in maintaining synaptic density under conditions of cell hypertrophy but could also have a more specific role in enhancing the ability of oxytocin neurons to fire in synchrony under the particular conditions of parturition and suckling (266).

One important consequence of the restructuring of the magnocellular nuclei may to alter ambient glutamate concentrations around the oxytocin neurons (387). Astrocytes possess a specific glutamate uptake mechanism and have a role in glutamate clearance (388). Hence, it might be predicted that a consequence of the retraction of astroglial processes from between oxytocin neurons would be a buildup of excess glutamate within the extracellular space. Oliet et al. (389) showed that there is indeed a reduced clearance of glutamate within the supraoptic nucleus during lactation. Furthermore, it was found that the buildup of excess glutamate resulted in activation of inhibitory metabotropic receptors on glutamate terminals, which suppressed further glutamate release. This would seem to contradict other reports indicating an increase in synaptic glutamate release in the supraoptic nucleus during lactation (390) and might seem paradoxical in that it would reduce the efficiency of excitatory transmission. However, Oliet and Piet (387) proposed that this homosynaptic inhibition might only operate on low frequency inputs, whereas high frequency inputs, involving more powerful activation of the glutamate terminals, would not be inhibited. In other words, the remodeling could create a high pass band filter, allowing high frequency inputs arising from suckling but filtering out low frequency inputs from other stimuli. Oliet and Piet (387) further proposed that during intense excitation, the excess glutamate may inhibit distant GABAergic terminals, also known to possess metabotropic glutamate receptors. This heterosynaptic inhibition of GABA inputs would further potentiate excitation of oxytocin neurons, particularly during the intense high frequency inputs activated by suckling. Glutamate is regarded as a key transmitter in the excitation of oxytocin neurons during suckling (391,392) highlighting the potential importance of these effects.

It has also been proposed that the restructuring of the magnocellular nuclei during lactation would promote enhanced diffusion of local modulators around the oxytocin neurons. This would arise partly because of the removal of the physical diffusion barrier created by the astrocytic processes, but the associated reduction in extracellular fluid volume within the intercellular clefts might also be important. Any locally released molecule, particularly oxytocin, a major determinant of bursting activity in oxytocin neurons (238,288), would therefore be more effective during lactation. Along the same lines, removal of astrocytes from around the oxytocin neurons may also allow an increased buildup of potassium during neuronal activity, and this could increase the excitability of the oxytocin neurons (266).

The morphological restructuring of the magnocellular nuclei is a dynamic process that can occur within a matter of hours in vitro (393). The increased membrane apposition of the oxytocin neurons is thought to involve active retraction of the astroglia rather than arise as a passive consequence of the expanding size of the oxytocin neurons Similarly, the changing arrangement of the synapses requires disensheathement of presynaptic terminals by retraction of astrocytic processes (159,394). The ability of the astroglia to undergo rapid remodeling has been linked to the expression of an embryonic pattern of surface adhesion molecules, including neural cell adhesion molecule (NCAM) and Tenascin-C (266,267), although the role of the former may not be essential because remodeling occurs in NCAM-deficient mice (395). These adhesion molecules do not themselves initiate the remodeling, but their presence means that the hypothalamoneurohypophysial system retains the ability to undergo reorganization when an appropriate signal is received. One such signal is oxytocin that, if infused into the third ventricle in pseudopregnant rats, has been shown to induce increased membrane apposition (396). Oxytocin also induces remodeling in acute hypothalamic slices in vitro, provided the activity of the oxytocin neurons is simultaneously enhanced by osmotic stimulation or blockade of GABA transmission (393). This suggests that in the presence of oxytocin, astrocytes can sense an increase in the level of activity of the oxytocin neurons and respond by undergoing morphologic reorganization. Consistent with the idea of oxytocin being essential for the lactational adaptation of oxytocin neurons, central blockade of oxytocin receptors during late pregnancy disrupts the systemic release of oxytocin during the subsequent lactation (397).

The action of oxytocin in inducing remodeling of the hypothalamoneurohypophysial system in vitro is potentiated by estrogen (393), and oxytocin is only effective in vivo in the presence of a background of appropriate steroidal hormones, namely, exposure to progesterone treatment followed by estrogen (398). The importance of steroids in the adaptations needed for lactation is further indicated by the ability of estrogen pretreatment to increase and progesterone treatment to reduce the incidence of bursting activity in oxytocin neurons during suckling (399). Progesterone withdrawal over the prepartum period removes GABA inhibition and increases the excitability of oxytocin neurons (400), and this could potentiate intranuclear release of oxytocin and morphological reorganization. Dye coupling of supraoptic magnocellular cells can be induced within hours by electrical stimulation of the lateral olfactory tract (401), indicating that olfactory inputs related to the birth of the pups might also be important in preparing the magnocellular nuclei for lactation.

Lactation is also associated with structural changes at the level of the neurohypophysis, particularly in the relationship between the pituicytes and neurosecretory processes (267,402). In nonlactating animals, pituicytes commonly ensheathe several neurosecretory axons, but at the onset of lactation there is an unwrapping of the neurosecretory axons such that there is a threefold fall in the number of enclosed axons per 100 μm^2 of pituicyte cytoplasm. Unwrapping of axons might allow greater buildup of extracellular potassium during activation of release, causing further depolarization of axonal membranes, although this has yet to be demonstrated. Pituicyte processes also retract from the perivascular space at the onset of lactation, leading to an increase in the number of neurosecretory processes contacting the capillary basement membrane and thus greater access to the neurohypophyseal capillaries. As in the hypothalamus, the changes in the neurohypophysis appear to involve active movement of pituicyte processes. Pituicytes in cell culture show a similar rapid retraction of their processes when β-adrenoceptor agonists are applied (403).

ASCENDING AND DESCENDING NEURAL PATHWAYS INVOLVED IN THE REGULATION OF OXYTOCIN RELEASE DURING MILK EJECTION

Sensory Receptors and Primary Afferent Fibers

The density of nerve endings in the nipples exceeds that of most other areas of the skin surface (404), reflecting the important function of these structures in providing the sensory input for activating the milk-ejection reflex. Detailed descriptions of the sensory innervation of the nipples are available for the rabbit (101,102), cow (405), and human (406,407). Innervation of the nipple in the rabbit (102,103) is derived from nerve bundles that branch from the respective cutaneous nerves and approach the nipple from deep within the mammary tissue (102). At the base of the nipple, these nerve bundles divide into plexuses that lie either within the dermis of the nipple (dermal plexus) or within the connective tissue between the ducts (interlactiferous plexus). At the summit of the nipple these plexuses reunite to form a dense apical network of fibers. As the dermal plexus ascends toward the tip of the nipple, fibers are given off to form a subepidermal network of endings. However, the most profuse network of subepidermal endings arises from the apical plexus, it being generally agreed that the tip of the nipple rather than the sides is the most richly innervated region. Nerve endings also arise within the interlactiferous plexus, but some of these endings may form the motor supply to smooth muscle important in erection of the nipple. The skin surrounding the base of the nipple (specialized to form the areola region in the human) is mainly innervated from more superficial branches of the cutaneous nerves passing beneath the adjacent skin overlying the mammary glands. Innervation of the parenchyma of the gland is sparse and consists mainly of sympathetic fibers to blood vessels plus a few sensory fibers to the larger ducts, a fact that correlates with the limited sensibility of the deep mammary tissue.

Most sensory receptors in the nipple comprise unmyelinated free nerve endings usually grouped together in bundles and embedded within a Schwann cell. In the human nipple these appear under light microscopy as finger-like blobs within the dermis or dermal papillae of the skin and are particularly abundant at the tip of the nipple around the openings of the lactiferous ducts (406). Similar endings are also present within the interlactiferous connective tissue. The areola region has a pattern of nerve endings more akin to that of the adjacent hairy skin than to the nipple. The sensitivity of the human nipple changes during the menstrual cycle and shows a striking increase after childbirth (408), presumably reflecting the influence of sex steroids on the development of the nerve endings.

A number of peptides are localized in nerve fibers of the nipples. Fibers containing calcitonin gene-related peptide have been identified in the nipple dermis of the rat and human and may serve a mechanoreceptive function (409,410). The nipple is also innervated by substance P–containing fibers (409), and lactating rats in which substance P has been depleted by neonatal capsaicin treatment show reduced milk yield, although this could relate to inadequate release of prolactin as well as oxytocin (411). Other neuropeptides present in the nipple include vasoactive intestinal peptide, neuropeptide Y and galanin (409).

Electrophysiological recordings from afferent fibers in the rabbit showed that nipple receptors were mainly sensitive to mechanical stimuli (pressure or stretch during suction of the nipple), but many were also temperature sensitive (103). Most fiber responses were abolished by removal of the skin overlying the mammary glands or infiltration of lidocaine around the nipples, again arguing against the existence of sensory receptors within the parenchyma. Recordings from the dorsal horn of the spinal cord in the rat (412,413) have confirmed the effectiveness of mechanical stimuli, particularly stretching, in activating the nipple receptors. The responses were fast adapting, such that a maintained stretch produced only a transient (1–2 seconds) barrage of impulses, whereas rhythmic stretching designed to mimic the sucking stimulus evoked more sustained and intense activity. Fibers derived from the nipples of the rat have a range of diameters (1–25 μm) and conduction velocities (<6 to >36 m/sec), but most slowly conducting unmyelinated fibers of the C type (414). Recordings from afferent nerves during milk ejection (415) showed that activity was low before milk ejection, but bursts of spikes occurred that correlated with cheek movements of the pups. There was a large increase in firing when the pups increased their sucking at the time of milk ejection.

The sensory input from the mammary glands is conducted to the spinal cord through the segmental spinal nerves, the level of input obviously relating to whether the mammary glands are located in the thoracic, abdominal, or inguinal region. Tracer studies (416) have shown that the afferents from each nipple enter two or three adjacent spinal segments, such that each spinal segment receives an overlapping input from several adjacent nipples. Within the spinal cord, most afferents from the nipples terminate in the substantia gelatinosa of the dorsal horn, and many contain glutamate (417). A few fibers end in the lateral spinal nucleus (nucleus of the dorsolateral fasciculus) and Clarke's column and some also ascend directly to the gracile and cuneate nuclei of the medulla oblongata (416).

Transmission of the Sucking Stimulus through the Spinal Cord

Analysis of evoked responses after electrical stimulation of the nipples in the rat has confirmed that the sensory input is widely dispersed through multisynaptic relays within the substantia gelatinosa and deeper laminae in the dorsal horn of the spinal cord (412). Neurons responding to stretching the nipples could be detected over a wide area of the dorsal horn, and most showed considerable sensory convergence in that they responded to other stimuli, such as stroking of the skin overlying the mammary glands. Dorsal horn neurons received convergent input from adjacent ipsilateral (but not contralateral) nipples and responded to sucking with rhythmic trains of impulses at intervals of 15 to 20 seconds, reflecting the periodic sucking of the pup. Recordings taken during reflex milk ejection (413) confirmed that spinal neurons received an ungated input from the nipples such that their activity correlated with sucking movements rather than activation of oxytocin neurons. However, the fact that the milk-ejection reflex is interrupted by intrathecal opiates (418) suggests the existence of control mechanisms for interrupting transmission of the suckling stimulus at the spinal level. Interestingly, the dorsal horn is innervated by oxytocinergic fibers that control pain transmission (419), but their role in the milk-ejection reflex has not been studied.

Transmission of the suckling stimulus up the spinal cord was studied by Fukuoka et al. (420), who found that lesions of the lateral funiculi blocked the milk-ejection reflex, whereas lesions of the dorsal and ventral funiculi were ineffective. Electrical stimulation of the lateral funiculus was found to evoke milk-ejection responses similar to those seen during suckling (97). The importance of the lateral funiculus in milk ejection indicates involvement of the spinothalamic tract, which runs in this part of cord, and has been implicated in the milk-ejection reflex in other studies (421,422). However, the fibers of the spinothalamic tract decussate and ascend the spinal cord contralaterally, whereas Fukuoka et al. (420) found that the sucking stimulus ascended the lateral funiculus mainly ipsilaterally. Thus, the more probable route for the sucking stimulus is via the spinocervical tract, which ascends ipsilaterally within the lateral funiculus, relaying within the lateral cervical nucleus in the cervical segments of the cord. Lesions of the lateral cervical nucleus block milk ejection in the rat (423). A role for the spinocervicothalamic tract was also suggested from spinal ablation studies by Richard et al. in the sheep (424), where milk ejection was prevented by sectioning the dorsal fasciculus in the thoracic but not cervical region. In the sheep the spinocervical tract runs in the dorsal fasciculus up to the cervical region, where it swings ventrally before contacting the lateral cervical nucleus.

The dorsal column system does not appear to participate in the ascending transmission of suckling stimulus, because studies in the rat (93,420), rabbit (425), and sheep (424) showed that lesioning the dorsal column system did not block the milk-ejection reflex. However, there has been one report suggesting involvement of the dorsal columns in suckling-evoked responses (426).

Transmission of the Sucking Stimulus at the Supraspinal Level

There are comparatively few direct somatosensory inputs to the hypothalamus (427), although a spinohypothalamic pathway has been identified that projects to both the lateral and medial hypothalamus (428). This pathway is involved in the transmission of noxious stimuli arising from C fiber primary afferents, and some of its neurons contain substance P (429). It is unknown whether spinohypothalamic neurons respond to suckling, but it may be significant that this input is also is mediated via unmyelinated C fibers (103) similar to those involved in nociception. Moreover, the spinohypothalamic fibers traverse the supraoptic decussation (428), a region implicated in bilateral coordination of oxytocin neurons (216).

There are other indirect routes through which the suckling stimulus could reach the hypothalamus. Detailed accounts of the ascending neuroanatomical projections to the forebrain are available elsewhere, and this section focuses only on pathways that have been specifically implicated in the milk-ejection reflex by techniques such as lesioning (231), electrical stimulation (421), neuroanatomical tracing (430), or Fos studies (224). An additional complexity in this area is the role played by well-defined neurochemical cell groups that project to the hypothalamus, for example, the noradrenergic pathways originating from the dorsal vagal complex and ventrolateral medulla that have well-defined roles in modulating the activity of the magnocellular nuclei (158,159,161,431). Rather than discuss these neurochemically defined inputs in detail here, they are covered later (see section on Neurotransmitters and Other Neuromodulators in the Milk-Ejection Reflex).

Early work examining the effects of brainstem lesions on the transmission of impulses from the mammary glands to the pituitary stalk (432) or on reflex milk ejection (433) led to the conclusion that the suckling stimulus traverses the lower brainstem through diffuse pathways in the reticular formation and cannot be ascribed to any compact somatosensory pathway. Dubois-Dauphin et al. (430) used a different approach in which they injected neuroanatomical tracers into the mesencephalic lateral tegmentum at a site that was known (231) to be important in the milk-ejection reflex. They found retrogradely labeled cells within the contralateral dorsal horn of the spinal cord (nucleus proprius), the contralateral dorsal column nuclei, and the ipsilateral lateral cervical nucleus. Because other evidence indicated that the dorsal column and spinothalamic pathways were not important for milk-ejection reflex (93,420), Dubois-Dauphin et al. concluded that the suckling stimulus passed through the spinocervical tract and then ascended the caudal brainstem along axons from the lateral cervical nucleus. Anterograde labeling from injection of tracer into the lateral cervical nucleus showed that these axons leave the lateral cervical nucleus to cross the midline beneath the central canal and ascend in the ventrolateral medulla, joining the lateral aspect of the medial lemniscus. At the level of the mesencephalon, the fibers turn dorsally to pass into the mesencephalic lateral tegmentum via the spinal lemniscus.

Stimulation (421,422) and lesioning (231,430) studies have confirmed that the suckling stimulus traverses the mesencephalic lateral tegmentum, in a region adjacent to the brachium of the inferior colliculus and medial geniculate body. To completely block milk ejection, lateral tegmental lesions need to involve the intercollicular area and parts of the external nucleus of the inferior colliculus (430). Unilateral lesions of the lateral tegmentum prevent the response to contralateral, but not ipsilateral, suckling, confirming decussation of the pathway below this level (231). From the work of Dubois-Dauphin et al. (423,430), it is probable that the effects of lateral tegmental lesions relate to damage to ascending projections from the lateral cervical nucleus, particularly those relaying in the external nucleus of the inferior colliculus. However, the lateral tegmental region contains a number of other cell groups and fiber tracts involved in somatosensory functions (430) that might also be important for the milk-ejection reflex.

The route of the suckling stimulus above the midbrain was first studied by Tindal and colleagues using electrical stimulation and knife cuts (6,434).

They reported that projections from the lateral tegmentum ascended toward the hypothalamus in a position medioventral to the medial geniculate body. Near the hypothalamus, the pathway was found to divide into dorsal and ventral components, with the ventral pathway being the more important. This ventral pathway entered the hypothalamus via the zona incerta and fields of Forel, before joining the medial forebrain bundle from where it reached the magnocellular nuclei. Anterograde tracing from the mesencephalic lateral tegmentum and external nucleus of the inferior colliculus revealed terminal labeling within the zona incerta and fields of Forel (430), agreeing with Tindal and Knaggs (434) that these regions were important in the milk-ejection reflex. Moreover, injection of neurotoxin into the peripeduncular region (encompassing the zona incerta and field of Forel) has been found (435) to block the milk-ejection reflex, indicating that this region contained cell bodies of neurons relaying the sucking stimulus up to the hypothalamus and not just axons of passage. Because neurons in the peripeduncular region are not labeled by injection of retrograde tracers into the magnocellular nuclei (436), there must be further relays before the sucking stimulus reaches the oxytocin neurons. This is entirely consistent with the other evidence concerning the importance of local interneurons for relaying inputs to the oxytocin neurons (159,198), as discussed in an earlier section.

Information concerning the path of the suckling stimulus through the brainstem has also been obtained using Fos labeling to detect neurons activated by suckling. A detailed study by Li et al. (224) revealed Fos-labeled neurons in the ventrolateral medulla, locus ceruleus, lateral parabrachial nucleus periaqueductal gray, and paralemniscal nucleus. Some activated neurons in the ventrolateral medulla and locus ceruleus were tyrosine hydroxylase positive (i.e., catecholaminergic neurons), and this is consistent with physiological evidence for involvement of noradrenergic pathways in the regulation of oxytocin release during suckling (437). Moos et al. (216) detected Fos labeling evoked by suckling in both the dorsomedial and ventrolateral medulla, and again many of the cells activated in the ventrolateral medulla were tyrosine hydroxylase positive. Fos activation during suckling has also been detected in the nucleus tractus solitarius of the dorsal vagal complex (438), although the neurochemical identity of the neurons has not been established. Similarly, Lin et al. (439) reported Fos labeling in the nucleus tractus solitarius in lactating rats, although this was significantly lower than that seen in response to parturition. On the other hand, Li et al. (224) were unable to detect any suckling-evoked Fos labeling in this nucleus, despite previous evidence for projections to oxytocin neurons from

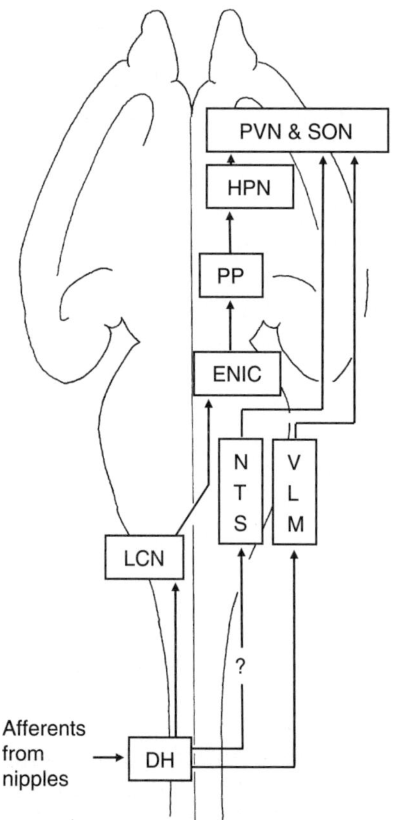

FIG. 18. Summary of the ascending pathways implicated in the regulation of oxytocin release during suckling. Only the main pathways are indicated, and inputs concerned with synchronizing the activity of the different magnocellular nuclei are not included. DH, dorsal horn of spinal cord; ENIC, external nucleus of inferior colliculus; HPN, hypothalamic perinuclear regions surrounding magnocellular nuclei; LCN, lateral cervical nucleus; NTS, nucleus tractus solitarius of the dorsal medulla; PP, peripeduncular region; PVN, paraventricular nucleus, SON supraoptic nucleus, VLM, ventrolateral medulla. See text for further explanation.

this region (431,440), so there is still some doubt concerning its importance in the milk-ejection reflex. A scheme illustrating the ascending pathways involved in the regulation of oxytocin release is presented in Fig. 18.

Descending Pathways Mediating Conditioning and Inhibition of the Milk-Ejection Reflex

The available evidence suggests that activation of oxytocin release by suckling does not require integrity of structures above the diencephalon. Thus, Voloschin and Tramezzani (164) found that the milk-ejection reflex persisted after surgical disruption of rostral, dorsal, and lateral hypothalamic afferents, and only caudal afferents were essential for oxytocin release. Descending pathways from the forebrain can, nevertheless, exert powerful facilitatory and inhibitory influences on the milk-ejection reflex. Such pathways provide the neural basis for conditioned triggering of the milk-ejection reflex in the absence of suckling (89) and are also responsible for inhibition of the reflex during stress (77,138). Such phenomena are likely to be under the control of the limbic system, a phylogenetically ancient group of structures (including the amygdala, septum, and bed nucleus of the stria terminalis) that are concerned with coordinating emotional and visceral responses (441). The reader is referred to other reviews (158–161,431,442) for detailed accounts of the forebrain connections of the magnocellular nuclei. Afferents to the region of the paraventricular and supraoptic nuclei arise from the diagonal band of Broca, bed nucleus of the stria terminalis, hippocampus, ventral subiculum, amygdala, and lateral septum as well as from many areas of the medial and lateral hypothalamus (159, 431,442). As in the case of ascending inputs, it is likely that most of these structures relay close to the magnocellular nuclei, but direct projections to oxytocin neurons have been shown to derive from the dorsomedial nucleus of the hypothalamus, the median preoptic nucleus, and the bed nucleus of the stria terminalis (442). Interestingly, the bed nucleus has been shown to receive indirect projections from the mammary gland (443). However, in a Fos study, the only regions of the forebrain that appeared to show activation specifically in response to suckling were the septum and medial preoptic area, although several other forebrain structures, including the bed nucleus, were activated when the pups were present but not allowed to suckle (224).

A number of forebrain structures have been implicated in conditioning of the milk-ejection reflex on the basis that electrical stimulation of these areas evoked oxytocin release. These mostly lie within or adjacent to the forebrain periventricular system rostral to the paraventricular nuclei and include the cingulate and pyriform cortex, nucleus accumbens, diagonal band of Broca, bed nucleus of the stria terminalis, septum, hippocampus, and fimbria of the fornix (282,444,445). Electrophysiological studies showed that magnocellular cells of the paraventricular or supraoptic nucleus were excited by stimulation of the amygdala, hippocampus, medial septum, dorsomedial thalamus, and olfactory bulb (446–448). However, it cannot be certain whether all the reported effects involved oxytocin-releasing neurons. More direct evidence for the involvement of the bed nucleus of stria terminalis and adjacent septum in controlling oxytocin neurons was obtained by Lambert et al. (232). They found that some bed nucleus neurons displayed cyclic fluctuations in their firing rate during suckling in association with milk ejection. Such neurons may provide a facilitatory influence on oxytocin release, although the fact that lesions of the bed nucleus do not block milk ejection (233) indicates that their input is not essential.

Deis and Prilusky (449) examined the role of the hippocampus in the facilitatory effect on the milk-ejection reflex of an auditory stimulus emanating from the pups and mother. Sectioning of the tractus corticohypothalamicus medialis (carrying fibers from the hippocampus to the hypothalamus) prevented the facilitation, which indicated an excitatory role for this pathway, whereas lesions of stria terminalis had no effect. However, the role of the hippocampal formation is complex because there is also evidence that another major pathway from the hippocampus, the subiculum, mediates central inhibition of oxytocin release (see below).

Early attempts to delineate pathways concerned with central inhibition of oxytocin release involved the technique of dual stimulation, whereby stimulation was simultaneously applied to a second area of the brain to prevent oxytocin release evoked by stimulating a known facilitatory pathway (450). Inhibitory areas detected by this means included the superior colliculus, tectoreticular tract, periaqueductal gray, pontine and mesencephalic raphe, posterior commissural nuclei, pretectum, centromedian parafascicular complex, dorsomedial nucleus of the thalamus, and the pyriform cortex. Electrophysiological recording studies have identified inhibitory pathways to the magnocellular neurosecretory cells from several structures, including the nucleus accumbens, amygdala, and lateral septum (451–453). The pathway from the lateral septum has been shown to impinge on identified oxytocin-releasing neurons (454). Moreover, prolonged electrical stimulation of the septum was found to inhibit the intermittent milk-ejection responses evoked by suckling (241,455). The role of the septum in the milk-ejection reflex is complex because brief low-frequency stimulation of the septum has an opposite stimulatory effect on milk ejection (282).

Central inhibition of oxytocin release may also involve the hippocampal formation. Lesions of the subiculum, a pathway for information from the hippocampal formation to the hypothalamus, have been found to result in "spontaneous" oxytocin release, causing prolonged contractions of the mammary gland (456). Hence, the subiculum may transmit a tonic inhibitory input to the oxytocin cells, which become excited when this is removed. The inhibitory influence of the hippocampus may itself be regulated by pathways from the midbrain, because transverse midbrain transections involving the ventral part of the midbrain central gray, the medial longitudinal fasciculus, and neighboring reticular formation also produced spontaneous oxytocin release (457). From the known neuroanatomical projections from this midbrain area and the fact that transection of the mediodorsal septum (an area that relays brainstem inputs to the hippocampus) also caused spontaneous oxytocin release (458), it seems that an inhibitory control emanates from the ascending reticular formation and ascends via a septohippocamposubicular route to the hypothalamus. The involvement of the reticular formation provides a means whereby the occurrence of milk ejection is linked to the sleep–wake cycle (10,45).

An important consideration in relation to the influence of descending pathways on the milk-ejection reflex is that several of the structures implicated in the regulation of oxytocin release contain receptors for steroid and peptide hormones. A prime example of this is provided by the bed nucleus, which contains estradiol and progesterone receptors (459) as well as receptors for oxytocin (314). Structures such as the bed nucleus provide one potential route for the observed effects of steroids on the milk-ejection reflex (399) and, in relation to its complement of oxytocin receptors, an additional route for the autoregulatory effects of oxytocin (232).

NEUROTRANSMITTERS AND OTHER NEUROMODULATORS IN THE MILK-EJECTION REFLEX

There has been considerable attention focused on the role of various transmitters and other neuromodulators in the magnocellular nuclei [see reviews in (158,161,460,461)]. It should be noted that much of the more recent information has been derived from using in vitro preparations, where it is possible to have at least some degree of certainty concerning the sites at which applied drugs are having their effects. Inevitably, this is not always the case with in vivo experiments, so that unless drugs are injected directly into the magnocellular nuclei, an element of caution should be applied when matching up observations made using in vitro hypothalamic preparations with in vivo data. It should also be noted that some of the in vivo results are derived from nonlactating female animals and male animals. This could be important, given the extent of reorganization within the hypothalamoneurohypophysial system associated with lactation (249,251,266).

Acetylcholine

High levels of choline acetyltransferase are present within the supraoptic nucleus (462), and this nucleus receives an innervation from local cholinergic interneurons (463). Acetylcholine directly excites identified oxytocinergic neurons of the paraventricular and supraoptic nuclei via nicotinic receptors (464). Interestingly, acetylcholine also acts presynaptically to potentiate glutamatergic inputs (465) and attenuate GABAergic inputs (202) to magnocellular neurons, providing an additional means by which this transmitter can excite oxytocin neurons. Consistent with these observations, acetylcholine injected into the supraoptic nucleus increases the firing rate of identified oxytocin neurons in vivo (238). Intracerebroventricularly administered cholinomimetics have been shown to increase the concentration of oxytocin in plasma and to cause a prolonged rise in intramammary pressure (466,467). This effect was elicited with acetylcholine and carbachol, but nicotine was only weakly effective. The evoked release was blocked by atropine and other muscarinic antagonists but not by nicotinic antagonists such as mecamylamine. The lack of involvement of nicotinic receptors would argue against an effect at the level of the oxytocin neurons in these studies. In fact, it has been proposed that centrally injected cholinomimetics do not act on the milk-ejection pathway but on other afferent pathways to the oxytocin cells, perhaps concerned with hemodynamic or osmotic regulation of their activity (467). This conclusion has been borne out in a recent study by Moos et al. (259), who reported that intracerebroventricularly injected acetylcholine increased the firing of oxytocin neurons but inhibited bursting activity, probably by disrupting the irregular background pattern needed for burst activation. Overall, then, it would appear that the excitatory effects of acetylcholine of oxytocin neurons is concerned with inputs other than this involved in the milk-ejection reflex.

Monoamines and Related Transmitters

Noradrenalin

Intraventricular injection of norepinephrine in vivo powerfully excites oxytocin neurons, causing a large intramammary pressure response. This effect is mimicked by α_1-adrenoceptor agonists such as phenylephrine and is blocked by α_1-adrenoceptor antagonists (468). Because the milk-ejection reflex is similarly blocked by these α_1-antagonists, as is osmotically evoked release of oxytocin (167), it seems reasonable that an excitatory input mediated by α_1-adrenoceptors is located close to the oxytocin neurons. Oxytocin neurons of the supraoptic nuclei are directly contacted by noradrenergic terminals (469), and noradrenaline has been shown to excite supraoptic neurons via α_1-adrenoceptors (470). Noradrenalin also potentiates excitatory glutamatergic inputs to supraoptic neurons (310). Moreover, in slices incubated in low calcium medium, phenylephrine induces bursting activity in oxytocin neurons, similar to that evoked by suckling (252). Suckling causes increased turnover of noradrenalin in the paraventricular and supraoptic nuclei and inhibition of noradrenalin synthesis prevents the release of oxytocin during suckling (471). Bealer and Crowley (332) showed that there is a rise of noradrenalin in the paraventricular nucleus during suckling, and this operates via α_1-adrenoceptors to stimulate intranuclear oxytocin release, itself a crucial link in activating of the milk-ejection reflex.

Although the important role of noradrenalin in the milk-ejection reflex is thus very well established, there is still some uncertainty as to which noradrenergic cell group(s) are responsible for the rise of noradrenalin within the magnocellular nuclei during suckling. The main noradrenergic inputs to the magnocellular nuclei arise from the Al (in the caudal ventrolateral medulla) and A2 (in the caudal region of the nucleus of the tractus solitarius) cell groups (472,473). The balance of evidence from both neuroanatomical and electrophysiological studies [see reviews in (159,161,337)] suggests that it is the A2 group of the nucleus of the tractus solitarius that is the most significant for controlling oxytocin release. The fact that the nucleus of the tractus solitarius receives a somatosensory input via the spinosolitary tract (474) and that electrical stimulation of this region evokes milk-ejection responses (440) supports a role for the A2 group in the milk-ejection reflex. The A2 group has also been strongly implicated in the control of oxytocin release during parturition, and in this case there is further supporting evidence, from Fos studies, showing that noradrenergic neurons of the A2 group are activated during labor (337). However, studies involving double labeling for Fos and tyrosine hydroxylase (216,224) suggest that brainstem noradrenergic neurons activated by suckling are localized within the A1 group of the ventrolateral medulla and in the locus ceruleus rather than within the A2 group. One explanation is that neurons of the A2 noradrenergic group are also excited by suckling but with insufficient intensity to induce Fos activation. Alternatively, it may be that the suckling activates a different set of noradrenergic neurons (i.e., the A1 group) as compared with the visceral inputs associated with parturition.

Work has also focused on the possible role of α_2-adrenoceptors in the milk-ejection reflex. α_2 Mechanisms do not seem to be important in controlling oxytocin release at the level of the magnocellular nuclei (475) but probably have effects on afferent pathways at lower levels of the milk-ejection reflex. The α_2 antagonist, iadozoxan, administered to rats suppresses the bursting activity of oxytocin cells, whereas clonidine, an agonist, causes facilitation of the reflex (240). The effect of clonidine is complex and dose dependent in that high doses are inhibitory, whereas lower doses cause a change in the patterning of milk ejections, which become clustered with intervals of less than 1 minute. Clonidine binds to noradrenergic neurons in the brainstem (476), and its effect may be at this level.

In addition to these excitatory actions, there is evidence that noradrenergic transmission may also mediate inhibitory actions on the milk-ejection reflex via β-adrenoceptors. Thus, Moos and Richard (477,478) found that administration of the β agonist isoproterenol inhibited suckling-induced bursting responses, although their background firing was unaffected. β-Adrenoceptor agonists also reduced plasma oxytocin (479). Conversely, the β-adrenoceptor antagonist propranolol can be used to initiate milk ejection in animals apparently blocked by stress (480), and propranolol augments suckling-induced oxytocin release in the conscious rat (479). As with the effects of β agonists, β antagonists such as propranolol do not change the background firing of the oxytocin cells, again suggesting an indirect action (480). In addition to alleviating the central suppression of oxytocin release, propranolol also improves the mammary response to oxytocin (481). Thus, β-adrenergic mechanisms may underlie both the peripheral and central inhibition of the milk-ejection reflex originally described by Cross (11,153). Crowley et al. (471) have shown that β-adrenoceptor inhibition of the reflex may also involve peripheral catecholamines from the adrenal gland.

Dopamine

Another monoamine with a potential role in controlling the milk-ejection reflex is dopamine, which is present in synaptic terminals of the supraoptic nuclei and derives from the A14 and A15 dopaminergic cell groups (482). On the isolated hypothalamoneurohypophyseal system, dopamine has an inhibitory action (483), but in the intact animal dopamine antagonists inhibit rather than facilitate the milk-ejection reflex (189,468). This indicates the existence of a dopamine-mediated excitatory component in the milk-ejection pathway, and this is consistent with increased turnover of dopamine in the forebrain during suckling (484). Moreover, dopamine injected centrally at low concentrations facilitates the bursting activity of oxytocin neurons during suckling, an effect that occurs without changing their background firing rate (189). Crowley et al. (485) found that the release of oxytocin during suckling was blocked by dopamine D1 receptor antagonists, again supporting an excitatory role for this transmitter in the milk-ejection reflex, and this role is exerted within the magnocellular nuclei (486). Dopamine directly excites supraoptic neurons in vitro, but this effect involves D2 rather than D1 receptors (487). In vitro studies indicate that dopamine also modulates local inputs to the supraoptic nucleus, acting presynaptically through dopamine D4 receptors to inhibit glutamate (488) and GABA (489) transmission to supraoptic neurons, but the effect of dopamine D4 agonists or antagonists on the milk-ejection reflex have yet to be investigated.

Serotonin (5-Hydroxytryptamine, 5-HT)

The magnocellular nuclei receive a modest innervation by serotonin-containing fibers that terminate predominantly in regions rich in oxytocinergic neurons and originate from the B7–B9 serotoninergic groups of the midbrain (490). There is surprisingly little electrophysiological data concerning the effects of 5-HT on magnocellular neurons, although early work pointed to a predominantly inhibitory effect (161). On the other hand, 5-HT agonists have recently been shown to up-regulate oxytocin mRNA in the hypothalamus (491), suggesting a stimulatory effect, and 5-HT given intracerebroventricularly causes neurohypophyseal hormone (i.e., both oxytocin and vasopressin) release (492). Regarding effects on the milk-ejection reflex, intracerebroventricular 5-HT administration in conscious rats inhibits the milk-ejection reflex (239), but this may relate to the ability of 5-HT to cause sympatheticoadrenal activation (493). Opposite effects on the milk-ejection reflex were observed in anesthetized rats, whereby administration of 5-HT inhibited the bursting activity of oxytocin neurons (493). Interestingly, administration of 5-HT antagonists exerted a mild facilitatory effect on the milk-ejection reflex, associated with disturbance of the normal sequence of bursts in oxytocin neurons, with the occurrence of "dual" bursts. This correlated with the effects of bilateral lesions of the mesencephalic ventral tegmentum, involving destruction of the B9 serotoninergic cell group that caused clustering of milk-ejection responses (231), paralleling the electrophysiological data. Further evidence for involvement of 5-HT was provided by studies showing that destruction of serotonin neurons in the medial raphe of the caudal mesencephalon caused prolonged desynchronization of the electroencephalogram, with slowing of the periodicity of milk ejection (494). Additionally, some milk ejections occurred in the absence of electroencephalogram synchronization, which was never observed in the intact animal. Finally, injection of 5-HT into the subarachnoid space of the spinal cord caused inhibition of the milk-ejection reflex, indicating a role in regulating transmission of the suckling stimulus at the level of the dorsal horn (495).

Histamine

The magnocellular neurons of the paraventricular and supraoptic nuclei are contacted by histamine-containing fibers, and these nuclei contain histamine receptors (460). Using hypothalamic slices, Yang and Hatton (496) investigated the histaminergic input to the supraoptic nucleus from the tuberomammillary nucleus in the posterior hypothalamus and found that an electrical stimulation of this input evoked inhibitory responses in oxytocinergic neurons. In contrast, most of the available in vivo data points to histamine being stimulatory to oxytocin release (497–500), and blockade of histamine receptors prevented the release of oxytocin during sucking (497). Interestingly, histamine has been shown to cause oxytocin release within the paraventricular nucleus, and blockade of either H_1 or H_2 receptors prevents intranuclear oxytocin release evoked by suckling (499). Bealer and Crowley (498) found that the stimulatory effect of histamine on central and peripheral oxytocin release requires functional α_1-noradrenergic receptors and concluded that histamine stimulated oxytocin via the release of noradrenalin, perhaps through presynaptic receptors on noradrenergic terminals. In further work, Bealer and Crowley (499) showed that suckling caused release of histamine within the paraventricular nucleus so that, overall, there is increasing evidence indicating an excitatory role for this transmitter in the milk-ejection reflex. One explanation for the contradiction between the in vivo hormone release data versus the in vitro electrophysiological evidence suggesting that histamine has an inhibitory effect (496) might be that histamine exerts both an indirect excitatory action, involving a presynaptic action on noradrenergic inputs, and a direct inhibitory action, involving a postsynaptic effect on the oxytocin neurons themselves. These dual effects might involve different subsets of histaminergic neurons, with only those mediating excitation being activated during suckling.

Excitatory and Inhibitory Amino Acids

It is recognized that glutamate and GABA are major transmitters within the paraventricular and supraoptic magnocellular nuclei (158,159,161), and only work directly related to their role in the milk-ejection reflex is considered here. Oxytocin neurons are contacted by glutamatergic synapses (482), and of the two ionotropic receptor subtypes (i.e., α-amino-3-hydroxy-5-methyl-4-isoxazolepropionic acid [AMPA] receptors vs. n-methyl-d-aspartate [NMDA] receptors), it seems that AMPA receptors may be more important in controlling oxytocin neurons. Thus, AMPA receptors are enriched in oxytocinergic neurons (501), and glutamate excitation of oxytocin neurons in vitro has been reported to involve predominantly AMPA receptors (502). However, it has been argued that NMDA receptor excitation of oxytocin neurons is voltage dependent and so can be masked unless the neurons are partly depolarized (158). It should also be noted that NMDA receptors have been implicated in dendritic release of oxytocin (305), which itself can excite oxytocin neurons (299).

In vivo evidence for the importance of AMPA receptors in systemic oxytocin release was obtained by Parker and Crowley (392), who showed that AMPA receptor agonists injected into the supraoptic nucleus evoked a rise in plasma oxytocin and AMPA receptor antagonists prevented oxytocin release in response to suckling. However, subsequent work (391) established that functional NMDA receptors are required for AMPA receptor agonists to induce a full oxytocin response. Electrophysiological studies point to the involvement of AMPA rather than NMDA receptors in the activation of oxytocin neurons during suckling. Thus, Moos et al. (503) found that local application of NMDA antagonists reduced the basal firing of oxytocin neurons but did not prevent their bursting activity. Recordings from cultured hypothalamic explants showed that the bursting activity that oxytocin neurons display in this preparation is blocked by AMPA antagonists (236).

As might be predicted, there is good evidence that GABA has a direct role in controlling the activity of oxytocin neurons during suckling. Voisin et al. (374) found that both muscimol, a $GABA_A$ agonist, and bicuculline, a $GABA_A$ antagonist, injected intracerebroventricularly inhibited the milk-ejection reflex. Similar results were obtained when these drugs were bilaterally infused into the supraoptic nuclei. Electrophysiological recordings after intracerebroventricular injections of the drugs indicated that, in both cases, the inhibition of bursting activity in oxytocin neurons was accompanied by a reduction in their background firing. Similar results were obtained with bilateral infusions of these drugs into the supraoptic nuclei. An explanation for the unexpected blockade of bursting activity with both $GABA_A$ agonists and antagonists was provided by Moos (237). She found that local microinjections of GABA into the supraoptic nucleus reduced the background firing of the oxytocin neurons but facilitated their bursting activity and concluded that the oxytocin neurons required a critical level of GABA "tone" to display bursts. This might be to ensure that they have the appropriate level of background firing (258) or to suppress other inputs that would otherwise interfere with bursting. $GABA_B$ receptors have also been ascribed a role in the milk-ejection reflex. Intracerebroventricular injections of the $GABA_B$ agonist baclofen inhibited the oxytocin release evoked by suckling (504), but because there was no effect when the drug was bilaterally infused into the supraoptic nuclei, this effect appeared to operate outside of the magnocellular nuclei. Hence, the physiological importance of the local presynaptic $GABA_B$ inhibition of glutamate and GABA inputs to oxytocin neurons (203,204) still needs to be established.

Neuropeptides

Opioids

There is considerable evidence to implicate opioid peptides in the regulation of oxytocin secretion [see reviews in (183,336,337)], and their role in controlling release at the level of the pituitary has already been discussed in an earlier section. At the level of the magnocellular nuclei, several sources of opioids are recognized, including corelease of dynorphin and enkephalin from oxytocin cell dendrites and corelease of enkephalin from noradrenergic terminals arising from the A2 cell group of the nucleus of the tractus solitarius (183,337). The magnocellular nuclei also receive a β-endorphinergic projection from the arcuate nucleus (505). Oxytocin fibers innervate β-endorphin neurons of the arcuate nucleus, and this may form the basis for reciprocal interactions between oxytocinergic and β-endorphinergic pathways within the hypothalamus (506). Oxytocin neurons are directly inhibited by both μ and κ opioid agonists (507,508), and μ agonists also inhibit the excitatory effects of noradrenalin, an important excitatory transmitter in the milk-ejection reflex (see above), by presynaptically suppressing noradrenalin release within the magnocellular nuclei (183).

Both μ and κ agonists given centrally or systemically inhibit the firing of oxytocin neurons, and μ agonists cause inhibition of the milk-ejection reflex that is reversed by the opioid antagonist, naloxone (338,509). The doses required are not dissimilar from those needed to produce analgesia (510). Whereas Clarke et al. (338) concluded that the inhibitory effects of opioids occurred mainly at the level of the posterior pituitary, Russell et al. (183) argued that opioids operating through μ receptors also act at the level of the hypothalamus to disrupt bursting activity. However, not all oxytocin neurons are affected, perhaps because the excitation associated with burst activation is sufficiently powerful to overcome opioid inhibition (183).

Physiologically, opioids are known to be important in the control of oxytocin release during parturition, and there is increasing tonic opioid inhibition of oxytocin neurons toward the end of pregnancy that wanes at the time of birth and stays low through lactation (511). Naloxone alone has little effect on milk ejection in healthy rats that are already responding well to suckling but increases oxytocin release in animals that, because of poor health or unfavorable environmental circumstances, are showing a poor response to suckling (338). Thus, the main role of opioids during lactation may relate to stress-induced inhibition of the milk-ejection reflex. However, it should be noted that most of this work relates to the rat, and studies in several other species have failed to observe reversal of stress-induced inhibition of oxytocin release by naloxone (354,356,357).

Prolactin-Releasing Peptide

Prolactin-releasing peptide was originally identified by its ability to release prolactin but is now recognized as a central neuromodulator present within brainstem neurons of the A1 and A2 noradrenergic cell groups as well as in neurons of the dorsomedial nucleus of the hypothalamus (438). Prolactin peptide fibers innervate the magnocellular nuclei, and intracerebroventricular injections of prolactin-releasing peptide cause the release of oxytocin in female rats (512). Moreover, central injections of prolactin peptide administered during suckling increases the background firing and bursting activity of oxytocin neurons (513). However, prolactin-releasing peptide has no effect on oxytocin neurons in vitro (513), so it is probable that the effects in vivo involve actions other than within the magnocellular nuclei. Morales and Sawchenko (438) found no evidence that the prolactin-releasing peptide neurons of the brainstem are activated by suckling, but whether suckling activates prolactin-releasing peptide neurons of the dorsomedial nucleus has not yet been studied.

Substance P and Neuropeptide Y

Substance P is a member of the tachykinin family of peptides that exert their effects via neurokinin receptors, and in the mouse, neurokinin receptors have been localized to oxytocin neurons (514). The substance P innervation of the magnocellular nuclei may derive from the A1 and A2 noradrenergic neurons of the brainstem, some of which may contain colocalized substance P (515). Some fibers ascending directly to the hypothalamus from the spinal cord have been found to contain substance P (429), although it is not known whether these project to the magnocellular paraventricular and supraoptic nuclei. Centrally administered substance P increased the firing of magnocellular neurons (516), and Bealer and Flynn (517) found that substance P also evoked systemic release of oxytocin. Because this effect was blocked by phentolamine, they assumed it involved interaction with ascending noradrenergic pathways to the magnocellular nuclei (517). However, substance P also evokes oxytocin release from the hypothalamoneurohypophysial complex in vitro (518), indicating a possible effect at the hypothalamic level.

Neuropeptide Y is another peptide that, like substance P, is colocalized within noradrenergic neurons, particularly the A1 group of the ventrolateral medulla (519). Parker and Crowley (475) found that neuropeptide Y injected into the paraventricular or supraoptic nuclei increased plasma oxytocin, and this effect was prevented in the presence of an α_1-adrenoceptor antagonist. This suggested that neuropeptide Y was interacting with the noradrenergic input to the oxytocin neurons, and this was confirmed by showing that neuropeptide Y markedly potentiated the release of oxytocin in response to microinjections of the α_1 agonist phenylephrine into the magnocellular nuclei. Along similar lines, Kapoor and Sladek (520) found that although neuropeptide Y did not itself cause release of oxytocin from hypothalamoneurohypophysial explants, it significantly potentiated the oxytocin response induced by phenylephrine. Hence, there is good evidence that coreleased neuropeptide Y augments the well-established excitatory role of noradrenalin in the activation of oxytocin neurons during suckling.

Nitric Oxide

Nitric oxide is a gaseous neuromodulator synthesized from L-arginine by the enzyme nitric oxide synthetase, an enzyme that is found in several hypothalamic cell groups, including the magnocellular supraoptic and paraventricular neurons (521). The activity of nitric oxide synthetase tends to be upregulated under certain conditions of increased oxytocin secretion, suggesting it has a role in controlling the activity of oxytocin neurons (229,522). Srisawat et al. (523) found that application of a nitric oxide donor (sodium nitroprusside) to the supraoptic nucleus by in vivo microdialysis inhibited the firing of oxytocin neurons, whereas a nitric oxide synthetase inhibitor increased their electrical activity. The inhibitory action of nitric oxide involves both direct effects and a retrograde action to increase in GABA transmission (524). A role for nitric oxide in inducing gap junction formation between supraoptic oxytocin neurons has also been proposed (525). Consistent with the inhibitory effect on neuronal activity, administration of a nitric oxide synthetase inhibitor to lactating rats was found to potentiate the bursting activity of oxytocin neurons during suckling and to increase milk ejection. Conversely, central administration of nitroprusside inhibited bursting activity (526).

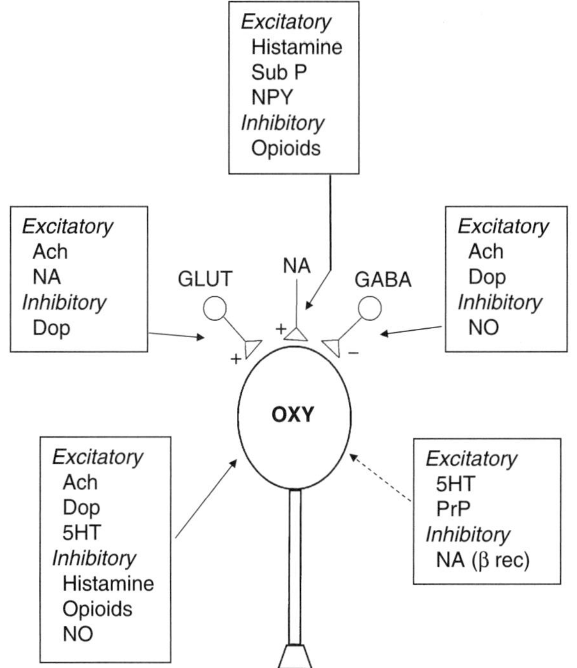

FIG. 19. Summary of the role of neurotransmitters and neuromodulators in the regulation of oxytocin neurons during suckling. These agents may be excitatory or inhibitory to the oxytocin neurons, and their effect may operate directly on the oxytocin neurons or may involve interaction with local glutamate or GABA neurons. Some agents may also operate by modulation of intranuclear noradrenalin release. For some agents the site of action is uncertain but probably involves effects on circuitry quite remote from the oxytocin neurons (indicated by *dotted arrow*). Ach, acetylcholine; β rec, β-adrenoceptors; Dop, dopamine; 5HT, 5-hydroxytryptamine; NA, noradrenalin; NO, nitric oxide; NPY, neuropeptide Y; PrP, prolactin-releasing peptide; Sub P, substance P. See text for further explanation.

However, somewhat unexpectedly, it has been reported that lactation is associated with up-regulation of nitric oxide synthetase within the magnocellular nuclei (527), a change that occurs rapidly in response to suckling (528). One explanation for this up-regulation is that nitric oxide serves as a compensatory mechanism to control hyperexcitation of oxytocin neurons during suckling or to limit their response to other stimuli not involved in the milk-ejection reflex. Another suggestion (529) is that nitric oxide mediates neuron–glia interactions within the magnocellular nuclei, and this would be in line with the observations of Yang and Hatton (525) concerning the role of nitric oxide in gap junction formation. A summary of the various neurotransmitters and neuromodulators implicated in the regulation of oxytocin neurons during milk ejection is presented in Fig. 19.

CONCLUSION

As is evident from this chapter, our understanding of the neuroendocrine control of milk ejection is well advanced, particularly in relation to the peripheral mechanisms that come in to play once the hypothalamic oxytocin neurons have been activated by suckling. However, most of our information relating to the milk-ejection reflex is based on laboratory and domesticated animals, and there is still a need for further work to explore how the milk-ejection reflex operates in other species. On the basis of our present knowledge of the anatomy and physiology of the mammary gland, it appears safe to assume that the milk-ejection reflex is of crucial importance for successful reproduction in the vast majority of species. More information about how stressful environmental factors, such as noise or territorial competition, can influence operation of the milk-ejection reflex in the wild may well reveal hitherto unknown pressures on reproduction that would have obvious implications, for example, in the field of conservation. When it comes to the neural components of the milk-ejection reflex and the central mechanisms that regulate oxytocin neurons, it is immediately obvious that our information is more or less completely derived from one species, the rat. We like to infer that fundamental mechanisms, such as the autoregulatory role of central oxytocin in the milk-ejection reflex, can be extrapolated to other species, but this needs to be confirmed.

Much of this chapter focused on the central neural mechanisms that regulate pulsatile oxytocin release, and no one can fail to be impressed by the recent advances in this area. One highlight is the development of in vitro preparations, especially the cultured hypothalamic explant, that enable bursting activity to be studied within isolated hypothalamic circuitry. Another would be the expansion of our knowledge about the intrinsic regulation of oxytocin neurons and how this controls patterns of spiking at the cellular level. A third important area of achievement has been the insights derived from analysis of the background activity of the oxytocin neurons, showing that these cells begin to behave as a coordinated network many seconds before burst activation. Finally, elucidation of the regulation of dendritic oxytocin release, and the many potential effects that oxytocin can have on the oxytocin neurons, also represents a significant breakthrough. Collectively, these advances have focused our attention very much on the role of local networks in fashioning burst firing and, by analogy with the regulation of gonadotropin-releasing hormone, we can now reasonably use the term "hypothalamic oxytocin pulse generator" to encompass both the oxytocin neurons and the circuitry that controls their periodic activation. From the current viewpoint, it seems that the suckling stimulus has little role in fashioning the periodicity of this pulse generator but is essential for its expression.

There are still many things we do not know about the control of oxytocin neurons during suckling. In particular, we do not yet understand how bursting activity is synchronized within the oxytocin neurons, although some information about the common inputs to the four magnocellular nuclei is beginning to emerge. Within the nuclei themselves, we know a great deal about the remarkable neuroplasticity displayed by the oxytocin neurons and the end of pregnancy. However, the precise way in which these glial changes contribute to synchronous bursting activity is still under investigation. Moreover, it is unknown whether the glial cells themselves are solely involved in passively controlling the level of interneuronal communication or whether they actively contribute to intercellular signaling, as is beginning to emerge from studies in other regions of the brain. A further unknown area concerns how the intrinsic mechanisms that regulate oxytocin neurons are themselves regulated. Models of bursting activity imply that these mechanism (particular those exerting inhibition) must be altered to enable burst activation, but how the various neuromodulators and neurotransmitters implicated in the milk-ejection reflex can modify these intrinsic currents has still to studied. Doubtless, answers to these various questions will emerge over the next few years and will be the subject of future chapters on the control of milk ejection.

ACKNOWLEDGMENTS

I thank Miss Katie Ashford for help with the preparation of this manuscript.

REFERENCES

1. Ely, F., and Petersen, W. E. (1941). Factors involved in the ejection of milk. *J. Dairy Sci.* 24, 211–223.
2. Long, C. A. (1969). The origin and evolution of mammary glands. *Bioscience* 19, 519–523.
3. Linzell, J. L. (1953). The blood and nerve supply to the mammary glands of the cat, and other laboratory animals. *Br. Vet. J.* 109, 427–433.
4. Griffiths, M., McIntosh, D. L., and Coles, R. E. A. (1969). The mammary gland of the echidna, *Tachyglossus aculeatus*, with observations on the incubation of the egg and on the newly hatched young. *J. Zool. (Lond).* 158, 371–386.
5. Turner, C. W. (1952). The mammary gland 1. The anatomy of the udder of cattle and domestic animals. Lucas Brothers, Columbia, MO.
6. Cowie, A. T., and Tindal, J. S. (1971). The physiology of lactation. Edward Arnold, London.
7. Smith, V. R. (1959). Physiology of lactation. Constable and Co., London.
8. Jones, E. A. (1969). Recent developments in the biochemistry of the mammary glands. *J. Dairy Sci.* 36, 145–167.
9. Cross, B. A. (1977). Comparative physiology of milk removal. *Symp. Zool. Soc. Lond.* 41, 193–210.
10. Lincoln, D. W., Hill, A., and Wakerley, J. B. (1973). The milk-ejection reflex of the rat: an intermittent function not abolished by surgical levels of anaesthesia. *J. Endocrinol.* 57, 459–476.
11. Cross, B. A. (1955). Neurohormonal mechanisms in emotional inhibition of milk ejection. *J. Endocrinol.* 12, 29–37.
12. Bruckmaier, R. M., and Jurg, W. B. (1992). B-mode ultrasonography of mammary glands of cows, goats and sheep during alpha- and beta-adrenergic agonists and oxytocin administration. *J. Dairy Res.* 59, 151–159.
13. Welsch, U., Feuerhake, F., van Aarde, R., Buchheim, W., and Patton, S. (1998). Histo- and cytophysiology of the lactating mammary gland of the African elephant (*Loxodonta africana*). *Cell Tissue Res.* 294, 485–501.
14. Richardson, K. C. (1949). Contractile tissues in the mammary gland with special reference to myoepithelium in the goat. *Proc. R. Soc. Lond. (Biol).* 136, 30–45.
15. Nagato, T., Yoshida, H., Yoshida, A., and Uehara, Y. (1980). A scanning electron microscope study of myoepithelial cells in exocrine glands. *Cell Tissue Res.* 209, 1–10.
16. Linzell, J. L. (1952). The silver staining of myoepithelial cells, particularly in the mammary gland, and their relation to the ejection of milk. *J. Anat.* 86, 49–57.
17. Linzell, J. L. (1955). Some observations on the contractile tissue of the mammary glands. *J. Physiol. (Lond).* 130, 257–267.
18. Lau, C., and Henning, S. J. (1987). Mammary resistance: a possible controlling factor in milk ejection. *J. Endocrinol.* 112, 379–385.
19. Verley, J. M., and Hollmann, K. H. (1967). La regression de la glande mammaire a l'arret de la lactation. I. Etude au microscope optique. *Z. Zellforsch. Mikrosk. Anat.* 82, 212–221.
20. Sala, N. L., and Freire, F. (1974). Relationship between ultrastructure and response to oxytocin of the mammary myoepithelium throughout pregnancy and lactation: effect of estrogen and progesterone. *Biol. Reprod.* 11, 7–17.
21. Radnor, C. J. P. (1972). Myoepithelium in the pre-lactating and lactating mammary glands of the rat. *J. Anat.* 112:337–353.
22. Adan, R. A., Van Leeuwen, F. W., Sonnemans, M. A., Brouns, M., Hoffman, G., Verbalis, J. G., and Burbach, J. P. (1995). Rat oxytocin receptor in brain, pituitary, mammary gland, and uterus: partial sequence and immunocytochemical localization. *Endocrinology* 136, 4022–4028.
23. Breton, C., Di Scala-Guenot, D., and Zingg, H. H. (2001). Oxytocin receptor gene expression in rat mammary gland: structural characterization and regulation. *J. Mol. Endocrinol.* 27, 175–189.
24. Kimura, T., Ito, Y., Einspanier, A., Tohya, K., Nobunaga, T., Tokugawa, Y., Takemura, M., Kubota, Y., Ivell, R., Matsuura, N., Saji, F., and Murata, Y. (1998). Expression and immunolocalization of the oxytocin receptor in human lactating and non-lactating mammary glands. *Hum. Reprod.* 13, 2645–2653.
25. Soloff, M. S., Alexandrova, M., and Fernstrom, M. J. (1979). Oxytocin receptors: triggers for parturition and lactation. *Science* 204, 1313–1315.
26. Soloff, M. S., and Wieder, M. H. (1983). Oxytocin receptors in rat involuting mammary gland. *Can. J. Biochem. Cell Biol.* 61, 631–635.
27. Sutherland, R. C., Aizlewood, E. S., and Wakerley, J. B. (1986). Changing characteristics of the milk-ejection reflex during pregnancy, lactation and after weaning in the rat. *J. Reprod. Fertil.* 76, 123–130.
28. Soloff, M. S., Fernstrom, M. A., and Fernstrom, M. J. (1989). Vasopressin and oxytocin receptors on plasma membranes from rat mammary gland. Demonstration of vasopressin receptors by stimulation of inositol phosphate solution and

oxytocin regulators by study of a specific oxytocin agonist. *Biochem. Cell Biol.* 67, 152–162.
29. Pettibone, D. J., Woyden, C. J., and Totaro, J. A. (1990). Identification of functional oxytocin receptors in lactating rat mammary gland in vitro. *Eur. J. Pharmacol.* 188, 235–241.
30. Giacometti, L., and Montagna, W. (1962). The nipple and areola of the human female breast. *Anat. Rec.* 144, 191–198.
31. Montagna, W., and Macpherson, E. E. (1974). Some neglected aspects of the anatomy of the human breasts. *J. Invest. Dermatol.* 63, 10–16.
32. Foust, H. L. (1941). The surgical anatomy of the teat of the cow. *J. Am. Vet. Assoc.* 98, 143.
33. Ponden, W. D., and Grossman, J. D. (1950). Wall structure and closing mechanism of the bovine teat. *Am. J. Vet. Res.* 11, 349–354.
34. Bernabe, J., and Peeters, G. (1980). Study of the function of the teat sphincter in lactating cows. *Arch. Int. Pharmacodyn. Ther.* 246, 162–164.
35. Vandeputte-Van Messom, G., Burvenich, C., and Peeters, G. (1984). Action of epinephrine on the function of the teat sphincter in the lactating cow. *Am. J. Vet. Res.* 45, 2145–2149.
36. Butler, M. C., Hillerton, J. E., and Grindal, R. J. (1991). The control of milk flow through the teats of dairy cows. *J. Dairy Sci.* 75, 1019–1024.
37. Lefcourt, A. M. (1982). Rhythmic contractions of the teat sphincter in bovines: an expulsion mechanism. *Am. J. Physiol.* 242, R181–R184.
38. Raynaud, A. (1971). Foetal development of the mammary gland and hormonal effects on its morphogenesis. In Lactation. Proceedings of an International Symposium (Seventeenth Easter School in Agricultural Science, University of Nottingham) (I. R. Falconer, Ed.), pp. 3–29. Butterworths, London.
39. Hytten, F. E., and Baird, D. (1958). The development of the nipple in pregnancy. *Lancet* 1, 1201–1204.
40. Waller, H. (1950). The early yield of milk, and its relation to the security of lactation. *Lancet* 1, 53–56.
41. Roth, L. L., and Rosenblatt, J. S. (1968). Self-licking and mammary development during pregnancy in the rat. *J. Endocrinol.* 42, 363–378.
42. Shillito-Walser, E. (1977). Maternal behaviour in mammals. *Symp. Zool. Soc. Lond.* 41, 313–331.
43. Harrison, R. J. (1969). Reproduction and reproductive organs. In *The Biology of Marine Mammals* (H. T. Andersen, Ed.). Academic Press, New York.
44. Lincoln, D. W., Hentzen, K., Hin, T., Van der Schoot, P., Clarke, G., and Summerlee, A. J. S. (1980). Sleep: a prerequisite for milk-ejection in the rat. *Exp. Brain Res.* 38, 151–162.
45. Voloschin, L. M., and Tramezzani, N. (1979). Milk ejection reflex linked to slow-wave sleep in nursing rats. *Endocrinology* 105, 1202–1207.
46. Neve, H. A., Paisley, A. C., and Summerlee, A. J. S. (1982). Arousal: a prerequisite for suckling in the rabbit? *Physiol. Behav.* 28, 213–217.
47. Poulain, D. A., Rodrigues, F., and Ellendorff, F. (1981). Sleep is not a prerequisite for the milk-ejection reflex in the pig. *Exp. Brain Res.* 43, 107–110.
48. Findlay, L., and Renfree, M. (1984). Growth, development and secretion of the mammary gland of macropodid marsupials. *Symp. Zool. Soc. Lond.* 51, 403–432.
49. Zarrow, M. X., Denenberg, V. H., and Anderson, C. O. (1965). Rabbit: frequency of suckling in the pup. *Science* 150, 1835–1836.
50. D'Souza, F., and Martin, R. D. (1974). Maternal behaviour and the effects of stress in the tree shrew. *Nature* 251, 309–311.
51. Gill, J. C., and Thompson, W. (1956). Observations on the behaviour of suckling pigs. *Br. J. Anim. Behav.* 4, 46–51.
52. Ellendorff, E., and Schams, D. (1988). Characteristics of milk ejection, associated intramammary pressure changes and oxytocin release in the mare. *J. Endocrinol.* 119, 219–227.
53. Konner, M., and Worthman, C. (1980). Nursing frequency, gonadal function and birth spacing among the !Kung hunter gatherers. *Science* 207, 788–791.
54. Findlay, A. L. R. (1971). Neural and behavioural interactions with lactation. In Lactation. Proceedings of an International Symposium (Seventeenth Easter School in Agricultural Science, University of Nottingham) (I. R. Falconer, Ed.), pp. 75–91. Butterworths, London.
55. Hall, W. G., Cramer, C. P., and Blass, E. M. (1975). Developmental changes in suckling of rat pups. *Nature* 258, 318–320.
56. Matthiesen, A. S., Ransjo-Arvidson, A. B., Nissen, E., and Uvnas-Moberg, K. (2001). Postpartum maternal oxytocin release by newborns: effects of infant hand massage and sucking. *Birth* 28, 13–19.
57. Kovach, J. K., and Kling, A. (1967). Mechanisms of neonate sucking behaviour in the kitten. *Anim. Behav.* 15, 91–101.
58. Drewett, R. F., Statham, C., and Wakerley, J. B. (1974). A quantitative analysis of the feeding behaviour of suckling rats. *Anim. Behav.* 22, 907–913.
59. Sharman, G. B., and Pilton, P. E. (1964). The life history and reproduction of the red kangaroo (*Megaleia rufa*). *Proc. Zool. Soc. Lond.* 142, 20–48.
60. Ardran, G. M., and Kemp, F. H. (1959). A correlation between suckling pressures and the movements of the tongue. *Acta Paedriatr.* 48, 261–272.
61. Weber, F., Woolridge, M. W., and Baum, J. D. (1986). An ultrasonographic study of the organisation of sucking and swallowing by newborn infants. *Dev. Med. Child Neurol.* 28, 19–24.
62. Bu'lock, F., Woolridge, M. W., and Baum, J. D. (1990). Development of coordination of sucking, swallowing and breathing: ultrasound study of term and preterm infants. *Dev. Med. Child Neurol.* 32, 669–678.
63. Ardran, G. M., Cowie, A. T., and Kemp, F. H. (1957). A cineradiographic study of the teat sinus during suckling in the goat. *Vet. Rec.* 69, 1100–1101.
64. Ardran, G. M., Kemp, F. H., and Lind, J. (1958). A cineradiographic study of breast feeding. *Br. J. Radiol.* 31, 156–162.
65. Bowen-Jones, A., Thompson, C., and Drewett, R. E. (1982). Milk-flow and suckling rates during breast-feeding. *Dev. Med. Child Neurol.* 24, 626–633.
66. Woolridge, M. W., Drewett, R. F., Rolfe, P., and Baum, J. D. (1982). The continuous measurement of milk intake at a feed in breast-fed babies. *Early. Hum. Dev.* 6, 365–373.
67. Mizuno, K., and Ueda, A. (2003). The maturation and coordination of sucking, swallowing, and respiration in preterm infants. *J. Pediatr.* 142, 36–40.
68. Medoff-Cooper, B., McGrath, J. M., and Shults, J. (2002). Feeding patterns of full-term and preterm infants at forty weeks postconceptional age. *J. Dev. Behav. Pediatr.* 23, 231–236.
69. Epstein, A. N., Blass, E. M., Batshaw, M. L., and Parks, A. D. (1970). The vital role of saliva as a mechanical sealant for suckling in the rat. *Physiol. Behav.* 5, 1395–1398.
70. Slijper, E. J. (1962). Whales. Hutchinson, London.
71. Lincoln, D. W., and Renfree, M. B. (1981). Mammary gland growth and milk ejection in the agile wallaby, *Macropus agilis*, displaying concurrent asynchronous lactation. *J. Reprod. Fertil.* 63, 193–203.
72. Griffiths, M., Elliott, M. A., Leckie, R. M. C., and Schoefl, G. I. (1973). Observations of the comparative anatomy and ultrastructure of mammary glands and on the fatty acids of the triglycerides in platypus and echidna milk fats. *J. Zool. Lond.* 169, 255–279.

73. Sernia, C., Thomas, W. G., and Gemnell, R. T. (1991). Oxytocin receptors in the mammary gland and reproductive tract of a marsupial, the bushtail possum (*Trichosorus vulpecula*). *Biol. Reprod.* 45, 673–679.
74. Robinson, I. C. A. F., and Parsons, J. A. (1981). Suckling in the guinea-pig: the effects of passive immunization with an antiserum to oxytocin. *J. Endocrinol.* 90, 237–244.
75. Young, W. S. III, Shepard, E., DeVries, A. C., Zimmer, A., LaMarca, M. E., Ginns, E. I., Amico, J., Nelson, R. J., Hennighausen, L., and Wagner, K. U. (1998). Targeted reduction of oxytocin expression provides insights into its physiological roles. *Adv. Exp. Med. Biol.* 449, 231–240.
76. Ellendorf, F., Forsling, M. L., and Poulain, D. A. (1982). The milk-ejection reflex in the pig. *J. Physiol. (Lond).* 333, 577–594.
77. Bruckmaier, R. M., and Blum, J. W. (1998). Oxytocin release and milk removal in ruminants. *J. Dairy Sci.* 81, 939–949.
78. Schams, D., Mayer, H., Prokopp, A., and Worstorff, H. (1984). Oxytocin secretion during milking in dairy cows with regard to the variation and importance of a threshold level for milk removal. *J. Endocrinol.* 102, 337–343.
79. Cleverley, J. D., and Folley, S. J. (1970). The blood levels of oxytocin during machine milking in cows with some observations on its half-life in the circulation. *J. Endocrinol.* 46, 347–361.
80. Bruckmaier, R. M., Wellnitz, O., and Blum, J. W. (1997). Inhibition of milk ejection in cows by oxytocin receptor blockade, alpha-adrenergic receptor stimulation and in unfamiliar surroundings. *J. Dairy Res.* 64, 315–325.
81. Wellnitz, O., Bruckmaier, R. M., Albrecht, C., and Blum, J. W. (1999). Atosiban, an oxytocin receptor blocking agent: pharmacokinetics and inhibition of milk ejection in dairy cows. *J. Dairy Res.* 66, 1–8.
82. Folley, S. J., and Knaggs, G. S. (1966). Milk ejection activity (oxytocin) in the external jugular vein blood of the cow, goat and sow, in relation to the stimulus of milking or suckling. *J. Endocrinol.* 34, 197–214.
83. Bruckmaier, R., and Ritter, D. (1994). Machine Milking of dairy goats during lactation: udder anatomy, Milking characteristics, and blood concentrations of oxytocin and prolactin. *J. Dairy Res.* 61, 457–466.
84. Ruis, H., Rolland, R., Doesbury, W., Broeders, G., and Corbey, R. (1981). Oxytocin enhances onset of lactation among mothers delivering prematurely. *Br. Med. J.* 283, 340–342.
85. Newton, N., and Newton, M. (1967). Psychological aspects of lactation. *N. Engl. J. Med.* 277, 1179–1188.
86. Newton, N. (1992). The relation of the milk-ejection reflex to the ability to breast feed. *Ann. N Y Acad. Sci.* 652, 484–486.
87. Dawood, M. Y., Khan-Dawood, F. S., Waki, R. S., and Fuchs, F. (1981). Oxytocin release and plasma anterior pituitary and gonadal hormones in women during lactation. *J. Clin. Endocrinol. Metab.* 52, 678–683.
88. Lucas, A., Drewett, R. B., and Mitchell, M. D. (1980). Breast-feeding and plasma oxytocin concentrations. *Br. Med. J.* 281, 834–835.
89. McNeilly, A. S., Robinson, I. C. A. F., Houston, M. J., and Howie, P. W. (1983). Release of oxytocin and prolactin in response to suckling. *Br. Med. J.* 286, 257–259.
90. Uvnas-Moberg, K., Widstrom, A. M., Werner, S., Matthieson, A. S., and Winberg, J. (1990). Oxytocin and prolactin levels in breast feeding women. Correlation with milk yield and duration of breast feeding. *Acta Obstet. Gynaecol. Scand.* 69, 301–306.
91. Ramsay, D. T., Kent, J. C., Owens, R. A., and Hartmann, P. E. (2004). Ultrasound imaging of milk ejection in the breast of lactating women. *Pediatrics* 113, 361–367.
92. Findlay, A. L. R. (1968). The effect of teat anaesthesia on the milk ejection reflex in the rabbit. *J. Endocrinol.* 40, 127–128.
93. Eayrs, J. T., and Baddeley, R. M. (1956). Neural pathways in lactation. *J. Anat.* 90, 161–171.
94. Juss, T. S., Sutherland, R. C., and Wakerley, J. B. (1984). Intermittent milk ejection during electrical stimulation of the nipples: evidence for further gating of the milk ejection pathway. *J. Anat.* 139, 734.
95. Mena, F., Pacheco, P., Aguayo, D., Clapp, C., and Grosvenor, C. E. (1978). A rise in intramammary pressure follows electrical stimulation of a mammary nerve in anaesthetized rats. *Endocrinology* 103, 1929–1936.
96. Eriksson, M., and Uvnas-Moberg, K. (1990). Plasma levels of vasoactive intestinal polypeptide and oxytocin in response to suckling, electrical stimulation of mammary nerve and oxytocin infusion in rats. *Neuroendocrinology* 51, 237–240.
97. Poulain, D. A., and Dyer, R. G. (1984). Reproducible increases in intramammary pressure after spinal cord stimulation in lactating rats. *Exp. Brain Res.* 55, 313–316.
98. Christensson, K., Nilsonn, B. A., Stock, S., Matthiesen, A. S., and UnvasMoberg, K. (1989). Effect of nipple stimulation on uterine activity and on plasma levels of oxytocin in full term, healthy, pregnant women. *Acta Obstet. Gynaecol. Scand.* 68, 205–210.
99. Leake, R. D., Buster, J. E., and Fischer, D. A. (1984). The oxytocin secretory response to breast stimulation in women during the menstrual cycle. *Am. J. Obstet. Gynecol.* 148, 457–460.
100. Zinaman, M. J., Hughes, V., Queenan, J. T., Labbok, M. H., and Albertson, B. (1992). Acute prolactin and oxytocin responses and milk yield to infant suckling and artificial methods of expression in lactating women. *Pediatrics* 89, 437–440.
101. Hebb, C. O., and Linzell, J. L. (1966). A histological study of the innervation of the mammary gland. *J. Physiol. (Lond).* 186, 82P–83P.
102. Ballantyne, B., and Bunch, G. A. (1966). The neurohistology of quiescent mammary tissue in *Lepus albus*. *J. Comp. Neurol.* 127, 471–487.
103. Cross, B. A., and Findlay, A. L. R. (1969). Comparative and sensory aspects of milk ejection. In *Lactogenesis: The Initiation of Milk Secretion at Parturition* (M. Reynolds and S. J. Folley, Eds.), pp. 245—252. University of Pennsylvania Press, Philadelphia.
104. Riggs, C. M., Sutherland, R. C., and Wakerley, J. B. (1985). Reappraisal of the influence of mammary distension on the frequency of milk ejection in the rat. *J. Endocrinol.* 105, 127–132.
105. Fox, C. A., and Knaggs, G. S. (1969). Milk ejection activity (oxytocin) in peripheral venous blood in man during lactation and in association with coitus. *J. Endocrinol.* 45, 145–146.
106. Roberts, J. S., and Share, L. (1968). Oxytocin in plasma of pregnant, lactating and cycling ewes during vaginal stimulation. *Endocrinology* 83, 272–278.
107. Negoro, H., Uchide, K., Tadokoro, Y., Honda, K., and Higuchi, T. (1987). Vaginal distension induces milk ejection-related burst of oxytocin neurones interacting with suckling stimuli in lactating rats. *Brain Res.* 404, 331–374.
108. Amoroso, E. C., and Jewell, P. A. (1963). The exploitation of the milk-ejection reflex by primitive peoples. Occasional paper no. 18, pp. 126–137. Royal Anthropological Institute, London.
109. Fuchs, A. R., Cubile, L., Dawood, M. Y., and Jorgensen, F. S. (1984). Release of oxytocin and prolactin by suckling in rabbits throughout lactation. *Endocrinology* 114, 462–469.
110. Mayer, H., Bruckmaier, R., and Schams, D. (1991). Lactational changes in oxytocin release, intramammary pressure and milking characteristics in dairy cows. *J. Dairy Res.* 58, 159–169.
111. Deis, R. P. (1968). The effect of an exteroceptive stimulus on milk ejection in lactating rats. *J. Physiol. (Lond).* 197, 37–46.

112. McNeilly, A. S., and McNeilly, J. P. (1978). Spontaneous milk ejection during lactation and its possible relevance to success of breast feeding. *Br. Med. J.* 2, 466–468.
113. Cobo, E. (1993). Characteristics of the spontaneous milk ejecting activity occurring during human lactation. *J. Perinat. Med.* 21, 77–85.
114. Peeters, G., Stormorken, H., and Vanschoubroek, F. (1960). The effect of different stimuli on milk ejection and diuresis in the lactating cow. *J. Endocrinol.* 20, 163–172.
115. Fuchs, A. R., Ayromlooi, J., and Rasmussen, A. B. (1987). Oxytocin response to conditioned and nonconditioned stimuli in lactating ewes. *Biol. Reprod.* 37, 301–305.
116. Gorewit, R. C., Wachs, E. A., Sagi, R., and Merrill, W. C. (1983). Current concepts on the role of oxytocin in milk ejection. *J. Dairy Sci.* 66, 2236–2250.
117. Ackers, R. M., and Lefcourt, L. M. (1984). Effect of presence of calf on milking induced release of prolactin and oxytocin during early lactation of dairy cows. *J. Dairy Sci.* 67, 115–122.
118. Gimpl, G., and Fahrenholz, F. (2001). The oxytocin receptor system: structure, function, and regulation. *Physiol. Rev.* 81, 629–683.
119. Lippert, T. H., Mueck, A. O., Seeger, H., and Pfaff, A. (2003). Effects of oxytocin outside pregnancy. *Horm. Res.* 60, 262–271.
120. Poulain, D. A., and Wakerley, J. B. (1982). Electrophysiology of hypothalamic magnocellular neurones secreting oxytocin and vasopressin. *Neuroscience* 7, 773–808.
121. Stock, S., and Uvnas-Moberg, K. (1988). Increased plasma levels of oxytocin in response to afferent electrical stimulation of the sciatic and vagal nerves and in response to touch and pinching in anaesthetised rats. *Acta Physiol. Scand.* 132, 29–34.
122. Verbalis, J. G., McCann, M. J., McHale, C. M., and Stricker, E. M. (1986). Oxytocin secretion in response to cholecystokinin and food: differentiation of nausea from satiety. *Science* 232, 1417–1419.
123. Hamamura, M., Leng, G., Emson, P. C., and Kiyama, H. (1991). Electrical activation and c fos mRNA expression in rat neurosecretory neurones after systemic administration of cholecystokin. *J. Physiol. (Lond).* 444, 51–63.
124. Lang, R. E., Heil, J. W. E., Ganten, D., Hermann, K., Unger, T., and Rascher, W. (1983). Oxytocin unlike vasopressin is a stress hormone in the rat. *Neuroendocrinology* 37, 314–316.
125. Higuchi, T., Honda, K., Takano, S., and Negoro, H. (1988). Reduced oxytocin response to osmotic stimuli and immobilisation stress in lactating rats. *J. Endocrinol.* 116, 225–230.
126. Windle, R. J., and Forsling, M. L. (1993). Variations in oxytocin secretion during the 4-day oestrous cycle of the rat. *J. Endocrinol.* 136, 305–311.
127. Poulain, D. A., and Tasker, J. G. (1985). Recurrent mammary gland contractions induced by a low tonic release of oxytocin in rats. *J. Endocrinol.* 107, 89–96.
128. Sarkar, D. K. (1989). Evidence for prolactin feedback actions on hypothalamic oxytocin, vasoactive intestinal peptide and dopamine secretion. *Neuroendocrinology* 49, 520–524.
129. Parker, S. L., Armstrong, W. E., Sladek, C. D., Grosvenor, C. E., and Crowley, W. R. (1991). Prolactin stimulates the release of oxytocin in lactating rats: evidence for a physiological role via an action at the neural lobe. *Neuroendocrinology* 53, 503–510.
130. Summerlee, A. J. S., Paisley, A. C., O'Byrne, K. T., Fairhall, K. M., Robinson, I. C. A. F., and Fletcher, J. (1986). Aspects of the neuronal and endocrine components of reflex milk ejection in conscious rabbits. *J. Endocrinol.* 108, 143–149.
131. Higuchi, T., Tadokoro, Y., Honda, K., and Negoro, H. (1986). Detailed analysis of blood oxytocin levels during suckling and parturition in the rat. *J. Endocrinol.* 110, 251–256.
132. Higuchi, T., Honda, K., Fukuoka, T., Negoro, H., and Wakabayasyi, K. (1985). Release of oxytocin during suckling and parturition in the rat. *J. Endocrinol.* 105, 339–346.
133. Robinson, I. C. A. F. (1986). Oxytocin and the milk-ejection reflex. *Curr. Top. Neuroendocrinol.* 6, 153–172.
134. McNeilly, A. S. (1972). The blood levels of oxytocin during suckling and hand-milking in the goat with some observations on the pattern of release. *J. Endocrinol.* 52, 177–188.
135. Kent, J. C., Kennaugh, L. M., and Hartmann, P. E. (2003). Intramammary pressure in the lactating sow in response to oxytocin and during natural milk ejections throughout lactation. *J. Dairy Res.* 70, 131–138.
136. Cobo, E., Bernal, M. H., de Gaitan, E., and Quintero, C. A. (1967). Neurohypophyseal hormone release in the human. *Am. J. Obstet. Gynecol.* 97, 519–529.
137. Leake, R. D., Waters, C. B., Rubin, R. T., Buster, J. E., and Fisher, D. A. (1983). Oxytocin and prolactin responses in long-term breast-feeding. *Obstet. Gynecol.* 62, 565–568.
138. Ueda, T., Yokoyama, Y., Irahara, M., and Aono, T. (1994). Influence of psychological stress on suckling-induced pulsatile oxytocin release. *Obstet. Gynecol.* 84, 259–262.
139. Johnston, J. M., and Amico, J. A. (1986). A prospective longitudinal study of the release of oxytocin and prolactin in response to infant suckling in long term lactation. *J. Clin. Endocrinol. Metab.* 62, 653–657.
140. Zingg, H. H., and Laporte, S. A. (2003). The oxytocin receptor. *Trends Endocrinol. Metab.* 14, 222–227.
141. Chini, B., and Fanelli, F. (2000). Molecular basis of ligand binding and receptor activation in the oxytocin and vasopressin receptor family. *Exp. Physiol.* 85 Spec No., 59S–66S.
142. Nakano, H., Furuya, K., Furuya, S., and Yamagishi, S. (1997). Involvement of P2-purinergic receptors in intracellular Ca2+ responses and the contraction of mammary myoepithelial cells. *Pflugers Arch.* 435, 1–8.
143. Olins, G. M., and Bremel, R. D. (1984). Oxytocin-stimulated myosin phosphorylation in mammary myoepithelial cells: roles of calcium ions and cyclic nucleotides. *Endocrinology* 114, 1617–1626.
144. Conti, M. A., and Adelstein, R. S. (1981). The relationship between calmodulin binding and phosphorylation of smooth muscle myosin kinase by the catalytic subunit of 3':S'cAMP-dependent protein kinase. *J. Biol. Chem.* 256, 3178–3181.
145. Nakano, H., Furuya, K., and Yamagishi, S. (2001). Synergistic effects of ATP on oxytocin-induced intracellular Ca2+ response in mouse mammary myoepithelial cells. *Pflugers Arch.* 442, 57–63.
146. Sapino, A., Macri, L., Tonda, L., and Bussolati, G. (1993). Oxytocin enhances myoepithelial cell differentiation and proliferation in the mouse mammary gland. *Endocrinology* 133, 838–842.
147. Benson, G., and Folley, S. J. (1957). The effect of oxytocin on mammary gland involution in the rat. *J. Endocrinol.* 16, 189–201.
148. Bisset, G. W., Clark, B. J., Haldar, J., Harris, M. C., Lewis, G. P., and Rocha e Silva, M. (1967). The assay of milk-ejecting activity in the lactating rat. *Br. J. Pharmacol. Chemother.* 31, 537–549.
149. Houvenaghel, A., and Peeters, G. (1968). Influence de la bradykinine sur 1'ejection du lait chez la brebis et la chevre. *Arch. Physiol. Biochim.* 76, 647–657.
150. McNeilly, A. S., and Fox, C. A. (1971). The effect of prostaglandins on the guinea-pig mammary gland. *J. Endocrinol.* 51, 603–604.
151. Macuhova, J., Tancin, V., and Bruckmaier, R. M. (2004). Effects of oxytocin administration on oxytocin release and milk ejection. *J. Dairy Sci.* 87, 1236–1244.

152. De Nuccio, D. J., and Grosvenor, C. E. (1971). Effects of volume and distribution of milk on the oxytocin-induced contraction of the lactating rat mammary gland in vivo. *J. Endocrinol.* 51, 437–446.
153. Cross, B. A. (1955). The hypothalamus and the mechanism of sympathetico-adrenal inhibition of milk ejection. *J. Endocrinol.* 12, 15–28.
154. Grosvenor, C. E., and Findlay, A. L. R. (1968). Effect of denervation on fluid flow into rat mammary gland. *Am. J. Physiol.* 214, 820–824.
155. Mena, F., Pacheco, P., Aguayo, D., Martinex, G., and Grosvenor, C. E. (1979). Reflex regulation of autonomic influences upon the oxytocin induced contractile response of the mammary gland in the anaesthetised rat. *Endocrinology* 104, 751–756.
156. Grosvenor, C. E. (1965). Contraction of lactating rat mammary gland in response to direct mechanical stimulation. *Am. J. Physiol.* 208, 214–218.
157. Cross, B. A. (1954). Milk ejection resulting from mechanical stimulation of mammary myoepithelium in the rabbit. *Nature* 173, 450.
158. Armstrong, W. E. (1995). Morphological and electrophy siological classification of hypothalamic supraoptic neurons. *Prog. Neurobiol.* 47, 291–339.
159. Hatton, G. I. (1990). Emerging concepts of structure-function dynamics in adult brain: the hypothalamo-neurohypophysial system. *Prog. Neurobiol.* 34, 437–504.
160. Leng, G., Brown, C. H., and Russell, J. A. (1999). Physiological pathways regulating the activity of magnocellular neurosecretory cells. *Prog. Neurobiol.* 57, 625–655.
161. Renaud, L. P., and Bourque, C. W. (1991). Neurophysiology and *Neuropharmacology.* of hypothalamic magnocellular neurons secreting vasopressin and oxytocin. *Prog. Neurobiol.* 36, 131–169.
162. Swanson, L. W., and Sawchenko, P. E. (1983). Hypothalamic integration: organization of the paraventricular and supraoptic nuclei. *Annu. Rev. Neurosci.* 6, 269–324.
163. Dierickx, K. (1980). Immunocytochemical localization of the vertebrate cyclic nonapeptide neurohypophyseal hormones and neurophysins. *Int. Rev. Cytol.* 62, 119–185.
164. Voloschin, L. M., and Tramezzani, J. H. (1973). The neural input of the milk ejection reflex in the hypothalamus. *Endocrinology* 92, 973–983.
165. Wakerley, J. B., Juss, T. S., Farrington, R., and Ingram, C. D. (1990). Role of the paraventricular nucleus in controlling the frequency of milk ejection and the facilitatory effect of centrally-administered oxytocin in the suckled rat. *J. Endocrinol.* 125, 467–475.
166. Dierickx, K., and Vandesande, F. (1977). Immuno-cytochemical localization of the vasopressinergic and the oxytocinergic neurons in the human hypothalamus. *Cell Tissue Res.* 184, 15–27.
167. Clarke, G., and Merrick, L. P. (1985). Electrophysiological studies of the magnocellular neurones. In *Current Topics of Neuroendocrinology*, vol. 4 (D. Ganten and D. Pfaff, Eds.), pp. 17–59. Springer Verlag, New York, Heidelberg.
168. Caldwell, J. D., Greer, E. R., Johnson, M. F., Prange, A. J., and Pedersen, C. A. (1987). Oxytocin and vasopressin immunoreactivity in hypothalamic and extrahypothalamic sites in late pregnant and postpartum rat. *Neuroendocrinology* 46, 39–47.
169. Jirikowski, G. F., Caldwell, J. D., Pilgrim, C., Stumpf, W. F., and Pedersen, C. A. (1989). Changes in immunostaining for oxytocin in the forebrain of the female rat during late pregnancy, parturition and early lactation. *Cell Tissue Res.* 256, 411–417.
170. Mezey, E., and Kiss, J. Z. (1991). Co-expression of vasopressin and oxytocin in hypothalamic supraoptic neurons of lactating rats. *Endocrinology* 129, 1814–1820.
171. Moos, F. C., and Ingram, C. D. (1995). Electrical recordings of magnocellular supraoptic and paraventricular neurons displaying both oxytocin- and vasopressin-related activity. *Brain Res.* 669, 309–314.
172. Sawchenko, P. E., Imaki, T., and Vale, W. (1992). Co-localization of neuroactive substances in the endocrine hypothalamus. In *CIBA Foundation Symposium 168* (D. J. Chadwick and J. Marsh, Eds.), pp. 16–42. Wiley, Chichester.
173. Glasgow, E., Kusano, K., Chin, H., Mezey, E., Young, W. S. 3rd, and Gainer, H. (1999). Single cell reverse transcription-polymerase chain reaction analysis of rat supraoptic magnocellular neurons: neuropeptide phenotypes and high voltage-gated calcium channel subtypes. *Endocrinology* 140, 5391–5401.
174. Gaymann, W., and Martin, R. (1987). A re-examination of the localization of immunoreactive dynorphin(1-8), (Leu)enkephalin and (Met)enkephalin in the rat neurohypophysis. *Neuroscience* 20, 1069–1080.
175. Ludwig, M., and Pittman, Q. J. (2003). Talking back: dendritic neurotransmitter release. *Trends Neurosci.* 26, 255–261.
176. Morris, J. F., and Ludwig, M. (2004). Magnocellular dendrites: prototypic receiver/transmitters. *J. Neuroendocrinol.* 16, 403–408.
177. Cross, B. A., Dyball, R. E., Dyer, R. G., Jones, C. W., Lincoln, D. W., Morris, J. F., and Pickering, B. T. (1975). Endocrine neurons. *Rec. Prog. Horm. Res.* 31, 243–294.
178. Branchaw, J. L., Hsu, S. F., and Jackson, M. B. (1998). Membrane excitability and secretion from peptidergic nerve terminals. *Cell Mol. Neurobiol.* 18, 45–63.
179. Zupanc, G. K. (1996). Peptidergic transmission: from morphological correlates to functional implications. *Micron* 27, 35–91.
180. Hussy, N. (2002). Glial cells in the hypothalamo-neurohypophysial system: key elements of the regulation of neuronal electrical and secretory activity. *Prog. Brain Res.* 139, 95–112.
181. Boersma, C. J., and Van Leeuwen, F. W. (1994). Neuron-glia interactions in the release of oxytocin and vasopressin from the rat neural lobe: the role of opioids, other neuropeptides. and their receptors. *Neuroscience* 62, 1003–1020.
182. Falke, N. (1991). Modulation of oxytocin and vasopressin release at the level of the neurohypophysis. *Prog. Neurobiol.* 36, 465–484.
183. Russell, J. A., Leng, G., and Bicknell, R. J. (1995). Opioid tolerance and dependence in the magnocellular oxytocin system: a physiological mechanism? *Exp. Physiol.* 80, 307–340.
184. Luckman, S. M., Dye, S., and Cox, H. J. (1996). Induction of members of the Fos/Jun family of immediate-early genes in identified hypothalamic neurons: in vivo evidence for differential regulation. *Neuroscience* 73, 473–485.
185. Fenelon, V. S., Poulain, D. A., and Theodosis, D. T. (1993). Oxytocin neuron activation and fos expression: a quantitative immunocytochemical analysis of the effect of lactation, parturition, osmotic and cardiovascular stimulation. *Neuroscience* 53, 77–89.
186. Summerlee, A. J. S., and Lincoln, D. W. (1981). Electrophysiological recordings from oxytocinergic neurones during suckling in the unanaesthetised lactating rat. *J. Endocrinol.* 90, 255–265.
187. Dyball, R. E., and Leng, G. (1986). Regulation of the milk ejection reflex in the rat. *J. Physiol.* 380, 239–256.
188. Lincoln, D. W., and Wakerley, J. B. (1975). Factors governing the periodic activation of supraoptic and paraventricular neurosecretory cells during suckling in the rat. *J. Physiol. (Lond).* 250, 443–461.

189. Moos, F., and Richard, P. (1982). Excitatory effect of dopamine on oxytocin and vasopressin reflex releases in the rat. *Brain Res.* 241, 249–260.
190. Brimble, M. J., and Dyball, R. E. (1977). Characterization of the responses of oxytocin- and vasopressin-secreting neurones in the supraoptic nucleus to osmotic stimulation. *J. Physiol.* 271, 253–271.
191. Renaud, L. P., Tang, M., McCann, M. J., Stricker, E. M., and Verbalis, J. G. (1987). Cholecystokinin and gastric distension activate oxytocinergic cells in rat hypothalamus. *Am. J. Physiol.* 253, R661–R665.
192. Onaka, T. (2004). Neural pathways controlling central and peripheral oxytocin release during stress. *J. Neuroendocrinol.* 16, 308–312.
193. Hatton, G. I. (1983). The hypothalamic slice approach to neuroendocrinology. *Q. J. Exp. Physiol.* 68, 483–489.
194. Dudek, F. E., Tasker, J. G., and Wuarin, J. P. (1989). Intrinsic and synaptic mechanisms of hypothalamic neurons studied with slice and explant preparations. *J. Neurosci. Methods* 28, 59–69.
195. Armstrong, W. E., Smith, B. N., and Tian, M. (1994). Electrophysiological characteristics of immunochemically identified rat oxytocin and vasopressin neurones in vitro. *J. Physiol.* 475, 115–128.
196. Hatton, G. I., and Li, Z. H. (1998). Neurophysiology of magnocellular neuroendocrine cells: recent advances. *Prog. Brain Res.* 119, 77–99.
197. Mason, W. T., Hatton, G. I., Ho, Y. W., Chapman, C., and Robinson, I. C. A. F. (1986). Central release of oxytocin, vasopressin and neurophysin by magnocellular neurone depolarization: evidence in slices of guinea pig and rat hypothalamus. *Neuroendocrinology* 42, 311–322.
198. Boudaba, C., Schrader, L. A., and Tasker, J. G. (1997). Physiological evidence for local excitatory synaptic circuits in the rat hypothalamus. *J. Neurophysiol.* 77, 3396–3400.
199. Tasker, J. G., and Dudek, F. E. (1993). Local inhibitory synaptic inputs to neurones of the paraventricular nucleus in slices of rat hypothalamus. *J. Physiol.* 469, 179–192.
200. Tasker, J. G., Boudaba, C., and Schrader, L. A. (1998). Local glutamatergic and GABAergic synaptic circuits and metabotropic glutamate receptors in the hypothalamic paraventricular and supraoptic nuclei. *Adv. Exp. Med. Biol.* 449, 117–121.
201. Daftary, S. S., Boudaba, C., Szabo, K., and Tasker, J. G. (1998). Noradrenergic excitation of magnocellular neurons in the rat hypothalamic paraventricular nucleus via intranuclear glutamatergic circuits. *J. Neurosci.* 18, 10619–10628.
202. Li, D. P., Pan, Y. Z., and Pan, H. L. (2001). Acetylcholine attenuates synaptic GABA release to supraoptic neurons through presynaptic nicotinic receptors. *Brain Res.* 920, 151–158.
203. Shibuya, I., Kabashima, N., Ibrahim, N., Setiadji, S. V., Ueta, Y., and Yamashita, H. (2000). Pre- and postsynaptic modulation of the electrical activity of rat supraoptic neurones. *Exp. Physiol.* 85 Spec No., 145S–151S.
204. Pittman, Q. J., Mouginot, D., and Kombian, S. B. (1998). GABAB receptors and supraoptic neuronal activity. *Adv. Exp. Med. Biol.* 449, 107–115.
205. Lincoln, D. W., and Paisley, A. C. (1982). Neuroendocrine control of milk ejection. *J. Reprod. Fertil.* 65, 571–586.
206. Bailey, A. R. T., Clarke, G., and Wakerley, J. B. (1993). Effect of clonidine on the activity of oxytocin neurones in the suckled lactating rat. *J. Physiol. (Lond).* 459–485.
207. Wakerley, J. B., Dyball, R. E., and Lincoln, D. W. (1973). Milk ejection in the rat: the result of a selective release of oxytocin. *J. Endocrinol.* 57, 557–558.
208. Wakerley, J. B., and Lincoln, D. W. (1973). The milk-ejection reflex of the rat: a 20- to 40-fold acceleration in the firing of paraventricular neurones during oxytocin release. *J. Endocrinol.* 57, 477–493.
209. Lincoln, D. W., and Wakerley, J. B. (1974). Electrophysiological evidence for the activation of supraoptic neurones during the release of oxytocin. *J. Physiol.* 242, 533–554.
210. Moos, F., and Richard, P. (1988). Characteristics of early- and late-recruited oxytocin bursting cells at the beginning of suckling in rats. *J. Physiol.* 399, 1–12.
211. Belin, V., Moos, F., and Richard, P. (1984). Synchronization of oxytocin cells in the hypothalamic paraventricular and supraoptic nuclei in suckled rats: direct proof with paired extracellular recordings. *Exp. Brain Res.* 57, 201–203.
212. Belin, V., and Moos, F. (1986). Paired recordings from supraoptic and paraventricular oxytocin cells in suckled rats, recruitment and synchronization. *J. Physiol.* 377, 369–390.
213. Wang, Y. F., Negoro, H., and Honda, K. (1996). Milk ejection bursts of supraoptic oxytocin neurones during bilateral and unilateral suckling in the rat. *J. Neuroendocrinol.* 8, 427–431.
214. Belin, V., and Moos, F. (1984). Paired recordings from supraoptic and paraventricular oxytocin cells in suckled rats: recruitment and synchronisation. *J. Physiol. (Lond).* 377, 369–390.
215. Wang, Y. F., Negoro, H., and Honda, K. (1997). Effect of interhemispheric sections of the hypothalamus on milk-ejection bursts of supraoptic oxytocin neurones during bilateral and unilateral suckling in the rat. *Neurosci. Lett.* 227, 17–20.
216. Moos, F., Marganiec, A., Fontanaud, P., Guillou-Duvoid, A., and Alonso, G. (2004). Synchronization of oxytocin neurons in suckled rats: possible role of bilateral innervation of hypothalamic supraoptic nuclei by single medullary neurons. *Eur. J. Neurosci.* 20, 66–78.
217. Brown, D., Fontanaud, P., and Moos, F. C. (2000). The variability of basal action potential firing is positively correlated with bursting in hypothalamic oxytocin neurones. *J. Neuroendocrinol.* 12, 506–520.
218. Jiang, Q. B., and Wakerley, J. B. (1995). Analysis of bursting responses of oxytocin neurones in the rat in late pregnancy, lactation and after weaning. *J. Physiol.* 486 (Pt 1), 237–248.
219. Paisley, A. C., and Summerlee, A. J. S. (1984). Activity of putative oxytocin neurones during reflex milk ejection in conscious rabbits. *J. Physiol. (Lond).* 347, 465–478.
220. Lincoln, D. W. (1974). Dynamics of oxytocin secretion. In *Neurosection, the Final Neuroendocrine Pathway* (F. G. W. Knowles and L. Wollrath, Eds.), pp. 129–133. Springer Verlag, Berlin.
221. Humamura, M., Leng, G., Emson, P. C., and Kiyama, H. (1991). Electrical activation and c fos mRNA expression in rat neurosecretory neurones after systemic administration of cholecystokinin. *J. Physiol. (Lond).* 444, 51–63.
222. Giovannelli, L, Shiromani, P. J., Jirikowski, G. F., and Bloom, F. E. (1990). Oxytocin neurons in the rat hypothalamus exhibit c fos-immunoreactivity upon osmotic stress. *Brain Res.* 531, 299–303.
223. Luckman, S. M., Antonijevic, I., and Leng, G. (1993). The maintenance of normal parturition in the rat requires neurohypophysial oxytocin. *J. Neuroendocrinol.* 5, 7–12.
224. Li, C., Chen, P., and Smith, M. S. (1999). Neural populations in the rat forebrain and brainstem activated by the suckling stimulus as demonstrated by cFos expression. *Neuroscience* 94, 117–129.
225. Lonstein, J. S., Simmons, D. A., Swann, J. M., and Stern, J. M. (1998). Forebrain expression of c-fos due to active maternal behaviour in lactating rats. *Neuroscience* 82, 267–281.

226. Ezure, K. (2004). Reflections on respiratory rhythm generation. *Prog. Brain Res.* 143, 67–74.
227. Streit, J., Tscherter, A., Heuschkel, M. O., and Renaud, P. (2001). The generation of rhythmic activity in dissociated cultures of rat spinal cord. *Eur. J. Neurosci.* 14, 191–202.
228. Moenter, S. M., DeFazio, A. R., Pitts, G. R., and Nunemaker, C. S. (2003). Mechanisms underlying episodic gonadotropin-releasing hormone secretion. *Front. Neuroendocrinol.* 24, 79–93.
229. Higuchi, T., and Okere, C. O. (2002). Role of the supraoptic nucleus in regulation of parturition and milk ejection revisited. *Microsc. Res. Tech.* 56, 113–121.
230. Plotsky, P. M., Sawchenko, P. E., and Vale, W. W. (1988). Evidence for inhibin-chain-like peptide mediation of suckling-induced oxytocin secretion. *Soc. Neurosci. Abstr.* 14, 627.
231. Juss, T. S., and Wakerley, J. B. (1981). Mesencephalic areas controlling pulsatile oxytocin release in the suckled rat. *J. Endocrinol.* 91, 233–244.
232. Lambert, R. C., Moos, F. C., Ingram, C. D., Wakerley, J. B., Kremarik, P., Guerne, Y., and Richard, P. (1993). Electrical activity of neurons in the ventrolateral septum and bed nuclei of the stria terminalis in suckled rats: statistical analysis gives evidence for sensitivity to oxytocin and for relation to the milk-ejection reflex. *Neuroscience* 54, 361–376.
233. Ingram, C. D., Adams, T. S., Jiang, Q. B., Terenzi, M. G., Lambert, R. C., Wakerley, J. B., and Moos, F. (1995). Mortyn Jones Memorial Lecture. Limbic regions mediating central actions of oxytocin on the milk-ejection reflex in the rat. *J. Neuroendocrinol.* 7, 1–13.
234. Jourdain, P., Dupouy, B., Bonhomme, R., Theodosis, D. T., Poulain, D. A., and Israel, J. M. (1998). Electrophysiological studies of oxytocin neurons in organotypic slice cultures. *Adv. Exp. Med. Biol.* 449, 135–145.
235. Jourdain, P., Israel, J. M., Dupouy, B., Oliet, S. H., Allard, M., Vitiello, S., Theodosis, D. T., and Poulain, D. A. (1998). Evidence for a hypothalamic oxytocin-sensitive pattern-generating network governing oxytocin neurons in vitro. *J. Neurosci.* 18, 6641–6649.
236. Israel, J. M., Le Masson, G., Theodosis, D. T., and Poulain, D. A. (2003). Glutamatergic input governs periodicity and synchronization of bursting activity in oxytocin neurons in hypothalamic organotypic cultures. *Eur. J. Neurosci.* 17, 2619–2629.
237. Moos, F. C. (1995). GABA-induced facilitation of the periodic bursting activity of oxytocin neurones in suckled rats. *J. Physiol.* 488 (Pt 1), 103–114.
238. Moos, F., and Richard, P. (1989). Paraventricular and supraoptic bursting oxytocin cells in rat are locally regulated by oxytocin and functionally related. *J. Physiol.* 408, 1–18.
239. Moos, F., and Richard, P. (1983). Serotonergic control of oxytocin release during suckling in the rat: opposite effects in conscious and anesthetized rats. *Neuroendocrinology* 36, 300–306.
240. Bailey, A., Clarke, G., and Wakerley, J. (1997). The role of alpha-2 adrenoceptors in the regulation of oxytocin neurones in the suckled rat. *Brain Res. Bull.* 44, 193–197.
241. Lebrun, C. J., Poulain, D. A., and Theodosis, D. T. (1983). The role of the septum in the control of the milk ejection reflex in the rat: effects of lesions and electrical stimulation. *J. Physiol.* 339, 17–31.
242. Arshavsky, Y. I. (2003). Cellular and network properties in the functioning of the nervous system: from central pattern generators to cognition. *Brain Res. Brain Res. Rev.* 41, 229–267.
243. Binder, M. D. (2002). Integration of synaptic and intrinsic dendritic currents in cat spinal motoneurons. *Brain Res. Brain Res. Rev.* 40, 1–8.
244. Grillner, P., and Mercuri, N. B. (2002). Intrinsic membrane properties and synaptic inputs regulating the firing activity of the dopamine neurons. *Behav. Brain Res.* 130, 149–169.
245. Negoro, H., Uchide, K., Honda, K., and Higuchi, T. (1985). Facilitatory effect of antidromic stimulation on milk ejection-related activation ofoxytocin neurons during suckling in the rat. *Neurosci. Lett.* 59, 21–25.
246. Armstrong, W. E., and Stern, J. E. (1998). Electrophysiological distinctions between oxytocin and vasopressin neurons in the supraoptic nucleus. *Adv. Exp. Med. Biol.* 449, 67–77.
247. Li, Z., and Hatton, G. I. (1997). Reduced outward K+ conductances generate depolarizing after-potentials in rat supraoptic nucleus neurones. *J. Physiol.* 505 (Pt 1), 95–106.
248. Inenaga, K., Nagatomo, T., Kannan, H., and Yamashita, H. (1993). Inward sodium current involvement in regenerative bursting activity of rat magnocellular supraoptic neurones in vitro. *J. Physiol.* 465, 289–301.
249. Teruyama, R., and Armstrong, W. E. (2002). Changes in the active membrane properties of rat supraoptic neurones during pregnancy and lactation. *J. Neuroendocrinol.* 14, 933–944.
250. Stern, J. E., and Armstrong, W. E. (1996). Changes in the electrical properties of supraoptic nucleus oxytocin and vasopressin neurons during lactation. *J. Neurosci.* 16, 4861–4871.
251. Armstrong, W. E., and Stern, J. E. (1998). Phenotypic and state-dependent expression of the electrical and morphological properties of oxytocin and vasopressin neurones. *Prog. Brain Res.* 119, 101–113.
252. Wang, Y. F., and Hatton, G. I. (2004). Milk ejection burst-like electrical activity evoked in supraoptic oxytocin neurons in slices from lactating rats. *J. Neurophysiol.* 91, 2312–2321.
253. Roper, P., Callaway, J., Shevchenko, T., Teruyama, R., and Armstrong, W. (2003). AHP's, HAP's and DAP's: how potassium currents regulate the excitability of rat supraoptic neurones. *J. Comput. Neurosci.* 15, 367–389.
254. Russell, J. A., and Leng, G. (2002). Editorial commentary: ready to burst? *J. Neuroendocrinol.* 14, 927–928.
255. Leng, G., and Dyball, R. E. (1984). Recurrent inhibition: a recurring misinterpretation. *Q. J. Exp. Physiol.* 69, 393–395.
256. Freund-Mercier, M. J., and Richard, P. (1984). Electrophysiological evidence for facilitatory control of oxytocin neurones by oxytocin during suckling in the rat. *J. Physiol.* 352, 447–466.
257. Meyer, C., Freund-Mercier, M. J., and Richard, P. (1987). Facilitatory effect of oxytocin on oxytocin cell background activity in the rat is suckling-dependent. *Neurosci. Lett.* 75, 80–84.
258. Brown, D., and Moos, F. (1997). Onset of bursting in oxytocin cells in suckled rats. *J. Physiol.* 503 (Pt 3), 625–634.
259. Moos, F., Fontanaud, P., Mekaouche, M., and Brown, D. (2004). Oxytocin neurones are recruited into co-ordinated fluctuations of firing before bursting in the rat. *Neuroscience* 125, 391–410.
260. Suffczynski, P., Kalitzin, S., and Lopes Da Silva, F. H. (2004). Dynamics of non-convulsive epileptic phenomena modeled by a bistable neuronal network. *Neuroscience* 126, 467–484.
261. Cosgrave, A. S., Richardson, C. M., and Wakerley, J. B. (2000). Permissive effect of centrally administered oxytocin on the excitatory response of oxytocin neurones to ventral tegmental stimulation in the suckled lactating rat. *J. Neuroendocrinol.* 12, 843–852.
262. Schrader, L. A., and Tasker, J. G. (1997). Presynaptic modulation by metabotropic glutamate receptors of excitatory and inhibitory synaptic inputs to hypothalamic magnocellular neurons. *J. Neurophysiol.* 77, 527–536.

263. Theodosis, D. T. (2002). Oxytocin-secreting neurons: a physiological model of morphological neuronal and glial plasticity in the adult hypothalamus. *Front. Neuroendocrinol.* 23, 101–135.
264. El Majdoubi, M., Poulain, D. A., and Theodosis, D. T. (1997). Lactation-induced plasticity in the supraoptic nucleus augments axodendritic and axosomatic GABAergic and glutamatergic synapses: an ultrastructural analysis using the disector method. *Neuroscience* 80, 1137–1147.
265. McKenzie, D. N., Leng, G., and Dyball, R. E. (1995). Electrophysiological evidence for mutual excitation of oxytocin cells in the supraoptic nucleus of the rat hypothalamus. *J. Physiol.* 485 (Pt 2), 485–492.
266. Theodosis, D. T., Piet, R., Poulain, D. A., and Oliet, S. H. (2004). Neuronal, glial and synaptic remodeling in the adult hypothalamus: functional consequences and role of cell surface and extracellular matrix adhesion molecules. *Neurochem. Int.* 45, 491–501.
267. Miyata, S., and Hatton, G. I. (2002). Activity-related, dynamic neuron-glial interactions in the hypothalamo-neurohypophysial system. *Microsc. Res. Tech.* 56, 143–157.
268. Hatton, G. I., and Zhao Yang, Q. (2002). Peripartum interneuronal coupling in the supraoptic nucleus. *Brain Res.* 932, 120–123.
269. Hatton, G. I., Yang, Q. Z., and Cobbett, P. (1987). Dye coupling among immunocytochemically identified neurons in the supraoptic nucleus: increased incidence in lactating rat. *Neuroscience* 21, 923–930.
270. Coles, J. A., and Poulain, D. A. (1991). Extracellular K+ in the supraoptic nucleus of the rat during reflex bursting activity by oxytocin neurones. *J. Physiol.* 439, 383–409.
271. Hatton, G. I., Cobbett, P., and Salm, A. K. (1985). Extranuclear axon collaterals of paraventricular neurons in the rat hypothalamus: intracellular staining, immunocytochemistry and electrophysiology. *Brain Res. Bull.* 14, 123–132.
272. Silverman, A. J., Hoffman, D. L., and Zimmerman, E. A. (1981). The descending afferent connections of the paraventricular nucleus of the hypothalamus (PVN). *Brain Res. Bull.* 6, 47–61.
273. Saphier, D., and Feldman, S. (1987). Electrophysiology of supraoptico-paraventricular nucleus connections in the rat. *Exp. Brain Res.* 69, 60–66.
274. Wakerley, J. B., Juss, T. S., and Farrington, R., and lngram C. D. (1990). Role of the paraventricular nucleus in controlling the frequency of milk ejection and the facilitatory effect of centrally administered oxytocin in the suckled rat. *J. Endocrinol.* 125, 467–475.
275. Thellier, D., Moos, F., Richard, P., and Stoeckel, M. E. (1994). Evidence for connections between a discrete hypothalamic dorsochiasmatic area and the supraoptic and paraventricular nuclei. *Brain Res. Bull.* 34, 261–274.
276. Takano, S., Negoro, H., Honda, K., and Higuchi, T. (1992). Lesion and electrophysiological studies on the hypothalamic afferent pathway of the milk ejection reflex in the rat. *Neuroscience* 50, 877–883.
277. Wang, Y. F., Negoro, H., and Honda, K. (1995). Effects of hemitransection of the midbrain on milk-ejection burst of oxytocin neurones in lactating rat. *J. Endocrinol.* 144, 463–470.
278. Zhang, X., Kostarczyk, E., and Giesler, G. J. Jr. (1995). Spinohypothalamic tract neurons in the cervical enlargement of rats: descending axons in the ipsilateral brain. *J. Neurosci.* 15, 8393–8407.
279. Gitler, M. S., and Barraclough, C. A. (1988). Identification of the hypothalamic site through which locus coeruleus axons decussate to reach and stimulate contralateral LH-RH neurons. *Brain Res.* 447, 205–214.
280. Karyakin, M. G., and Alekseev, N. P. (1992). The influence of tactile stimulation of nipples on the milk ejection reflex in the rat. *J. Comp. Physiol. (A)* 170, 645–650.
281. Sutherland, R. C., Juss, T. S., and Wakerley, J. B. (1987). Prolonged electrical stimulation of the nipples evokes intermittent milk ejection in the anaesthetised lactating rat. *Exp. Brain Res.* 66, 29–34.
282. Ingram, C. D., Sutherland, R. C., and Wakerley, J. B. (1988). Oxytocin release evoked by electrical stimulation of the medial forebrain in the rat: analysis of stimulus parameters and supraoptic neuronal activity. *Neuroscience* 27, 597–605.
283. Richard, P., Moos, F., and Freund-Mercier, M. J. (1991). Central effects of oxytocin. *Physiol. Rev.* 71, 331–370.
284. Lincoln, D. W. (1974). Does a mechanism of negative feedback determine the intermittent release of oxytocin during suckling. *J. Endocrinol.* 60, 193–194.
285. Wakerley, J. B., and Deverson, B. M. (1975). Stimulation of the supraopticohypophysial tract in the rat during suckling: failure to alter the inherent periodicity of oxytocin release. *J. Endocrinol.* 66, 439–440.
286. Freund-Mercier, M. J., and Richard, P. (1981). Excitatory effects of intraventricular injections of oxytocin on the milk-ejection reflex in the rat. *Neurosci. Lett.* 23, 193–198.
287. Freund-Mercier, M. J., Moos, F., Guerne, Y., and Richard, P. (1983). Possible control by oxytocin of periodical and synchronous neurosecretory bursts of oxytocin cells. *Prog. Brain Res.* 60, 197–201.
288. Freund-Mercier, M. J., Moos, F., Poulain, D. A., Richard, P., Rodriguez, F., Theodosis, D. T., and Vincent, J. D. (1988). Role of central oxytocin in the control of the milk ejection reflex. *Brain Res. Bull.* 20, 737–741.
289. Lambert, R. C., Moos, F. C., and Richard, P. (1993). Action of endogenous oxytocin within the paraventricular or supraoptic nuclei: a powerful link in the regulation of the bursting pattern of oxytocin neurons during the milk-ejection reflex in rats. *Neuroscience* 57, 1027–1038.
290. Wakerley, J. B., Foreman, C. T., and Ingram, C. D. (1989). Effect of centrally administered oxytocin on the association between cortical electro encephalogram and milk ejection in the rat. *J. Neuroendocrinol.* 1, 173–178.
291. Neumann, I., Porter, D. W., Landgraf, R., and Pittman, Q. J. (1994). Rapid effect on suckling of an oxytocin antisense oligonucleotide administered into rat supraoptic nucleus. *Am. J. Physiol.* 267, R852–R858.
292. Seckl, J. R., and Lightman, S. L. (1988). Potentiation of lactation-induced oxytocin secretion by intracerebroventricular oxytocin in the conscious goat. *J. Endocrinol.* 116, 273–277.
293. Moos, F., and Richard, P. (1989). Paraventricular and supraoptic bursting oxytocin cells in rat are locally regulated by oxytocin and functionally related. *J. Physiol. (Lond).* 408, 1–18.
294. Wakerley, J. B., and Ingram, C. D. (1988). Regulation of oxytocin pulse frequency in the suckled rat. *Elsevier Int. Cong. Ser.* 797, 49–56.
295. Ludwig, M., Sabatier, N., Dayanithi, G., Russell, J. A., and Leng, G. (2002). The active role of dendrites in the regulation of magnocellular neurosecretory cell behavior. *Prog. Brain Res.* 139, 247–256.
296. Ingram, C. D., Davison, I., and Wakerley, J. B. (1990). Facilitatory effect of hypothalamo-neurohypophysial tract stimulation on milk ejection frequency in the lactating rat. *Neuroscience Lett.* 118, 67–70.
297. Freund-Mercier, M. J., and Stoeckel, M. E. (1995). Somatodendritic autoreceptors on oxytocin neurones. *Adv. Exp. Med. Biol.* 395, 185–194.

298. Inenaga, K., and Yamashita, H. (1986). Excitation of neurones in the rat paraventricular nucleus in vitro by vasopressin and oxytocin. *J. Physiol. (Lond).* 370, 165–180.
299. Yamashita, H., Okuya, S., and Inenaga, K. (1987). Oxytocin predominantly excites putative oxytocin neurons in the rat supraoptic nucleus in vitro. *Brain Res.* 417, 364–368.
300. Carter, S. J., Richardson, C. M., and Wakerley, J. B. (2003). Excitatory effects of oxytocin and cholecystokinin on oxytocin neurones: differences between virgin, pregnant and lactating rats. *Neurosci. Lett.* 351, 13–16.
301. Raggenbass, M. (2001). Vasopressin- and oxytocin-induced activity in the central nervous system: electrophysiological studies using in-vitro systems. *Prog. Neurobiol.* 64, 307–326.
302. Brussaard, A. B. (1995). Oxytocin suppresses the GABAergic synaptic input in supraoptic neurones from the rat. *Adv. Exp. Med. Biol.* 395, 105–115.
303. Kombian, S. B., Hirasawa, M., Mouginot, D., and Pittman, Q. J. (2002). Modulation of synaptic transmission by oxytocin and vasopressin in the supraoptic nucleus. *Prog. Brain Res.* 139, 235–246.
304. Lambert, R. C., Dayanithi, G., Moos, F. C., and Richard, P. (1994). A rise in the intracellular Ca2+ concentration of isolated rat supraoptic cells in response to oxytocin. *J. Physiol.* 478 (Pt 2), 275–287.
305. de Kock, C. P., Burnashev, N., Lodder, J. C., Mansvelder, H. D., and Brussaard, A. B. (2004). NMDA receptors induce somatodendritic secretion in hypothalamic neurones of lactating female rats. *J. Physiol.* 561, 53–64.
306. Osako, Y., Otsuka, T., Taniguchi, M., Oka, T., and Kaba, H. (2001). Oxytocin enhances presynaptic and postsynaptic glutamatergic transmission between rat olfactory bulb neurones in culture. *Neurosci. Lett.* 299, 65–68.
307. Jo, Y. H., Stoeckel, M. E., Freund-Mercier, M. J., and Schlichter, R. (1998). Oxytocin modulates glutamatergic synaptic transmission between cultured neonatal spinal cord dorsal horn neurons. *J. Neurosci.* 18, 2377–2386.
308. Hirasawa, M., Schwab, Y., Natah, S., Hillard, C. J., Mackie, K., Sharkey, K. A., and Pittman, Q. J. (2004). Dendritically released transmitters cooperate via autocrine and retrograde actions to inhibit afferent excitation in rat brain. *J. Physiol.* 559, 611–624.
309. Onaka, T., Ikeda, K., Yamashita, T., and Honda, K. (2003). Facilitative role of endogenous oxytocin in noradrenaline release in the rat supraoptic nucleus. *Eur. J. Neurosci.* 18, 3018–3026.
310. Boudaba, C., Di, S., and Tasker, J. G. (2003). Presynaptic noradrenergic regulation of glutamate inputs to hypothalamic magnocellular neurones. *J. Neuroendocrinol.* 15, 803–810.
311. Cosgrave, A. S., and Wakerley, J. B. (2002). Central oxytocin potentiates excitatory responses of oxytocin neurones to stimulation of the dorsomedial hypothalamic nucleus in the suckled rat. *Neurosci. Lett.* 317, 47–49.
312. Tribollet, E., Barberis, C., Jard, S., Dubois-Dauphin, M., and Dreifuss, I. I. (1988). Localization and pharmacological characterization of high affinity binding sites for vasopressin and oxytocin in the rat brain by light microscopic autoradiography. *Brain Res.* 442, 105–118.
313. Freund-Mercier, M. J., Stoeckel, M. E., Dietl, M. M., Palacios, J. M., and Richard, P. (1988). Quantitative autoradiographic mapping of neurohypophysial-hormone-binding sites in the rat forebrain and pituitary gland. I. Characterization of different types of binding sites and their distribution in the Long Evans strain. *Neuroscience* 26, 261–272.
314. Insel, T. R. (1992). Oxytocin—a neuropeptide for affiliation: evidence from behavioral, receptor autoradiographic, and comparative studies. *Psychoneuroendocrinology* 17, 3–35.
315. Insel, T. R. (1990). Regional changes in brain oxytocin receptors postpartum: time course and relationship to maternal behaviour. *J. Neuroendocrinol.* 2, 539–545.
316. Det, E. R., Voorhis, D. A. M., Boschma, Y., and Elands, J. (1986). Estradiol modulates density of putative oxytocin receptors in discrete rat brain regions. *J. Neuroendocrinology* 44, 415–421.
317. Ingram, C. D., Cutler, K. L., and Wakerley, J. B. (1990). Oxytocin excites neurones in the bed nucleus of the stria terminalis of the lactating rat in vitro. *Brain Res.* 527, 167–170.
318. Ingram, C. D., and Wakerley, J. B. (1993). Post-partum increase in oxytocin induced excitation of neurones in the bed nuclei of the stria terminalis in vitro. *Brain Res.* 602, 325–330.
319. Moos, F., Ingram, C. D., Wakerley, J. B., Guerne, Y., Freund-Mercier, M. J., and Richard, P. (1991). Oxytocin in the bed nucleus of the stria terminalis and lateral septum facilitated bursting of hypothalamic oxytocin neurons in suckled rats. *J. Neuroendocrinol.* 3, 163–171.
320. Young, L. J., and Wang, Z. (2004). The neurobiology of pair bonding. *Nat. Neurosci.* 7, 1048–1054.
321. Stivers, J. A., Kaltwasser, M. T., Hill, P. S., Hruby, V. J., and Crawley, J. N. (1988). Ventral tegmental oxytocin induces grooming. *Peptides* 9 Suppl 1, 223–231.
322. Ikemoto, S., Witkin, B. M., Zangen, A., and Wise, R. A. (2004). Rewarding effects of AMPA administration into the supramammillary or posterior hypothalamic nuclei but not the ventral tegmental area. *J. Neurosci.* 24, 5758–5765.
323. Wakerley, J. B., Crabtree, J. W., and Cumbers, M. R. (2003). Facilitation of the milk ejection reflex by oxytocin injections into the ventral tegmentum: possible involvement of the supramammillary nucleus. 33rd Annual Meeting of the Society of Neuroscience. Abstract No. 613.1.
324. Moos, F., Poulain, D. A., Rodriguez, F., Guerne, Y., Vincent, J. D., and Richard, P. (1989). Release of oxytocin within the supraoptic nucleus during the milk ejection reflex in rats. *Exp. Brain Res.* 76, 593–602.
325. Neumann, I. D., Russell, J. A., and Landgraf, R. 1993). Oxytocin and vasopressin release within the supraoptic and paraventricular nuclei of pregnant, parturient and lactating rats: a microdialysis study. *Neuroscience* 53, 65–75.
326. Pow, D. V., and Morris, J. F. (1989). Dendrites of hypothalamic magnocellular neurons release neurohypophysial peptides by exocytosis. *Neuroscience* 32, 435–439.
327. Ludwig, M. (1998). Dendritic release of vasopressin and oxytocin. *J. Neuroendocrinol.* 10, 881–895.
328. Theodosis, D. T. (1985). Oxytocin-immunoreactive terminals synapse on oxytocin neurones in the supraoptic nucleus. *Nature* 313, 682–684.
329. Moos, F., Freund-Mercier, M. J., Guerne, J. M., Stoeckel, M. E., and Richard, P. (1984). Release of oxytocin and vasopressin by magnocellular nuclei in vitro: specific facilitatory effect of oxytocin on its own release. *J. Endocrinol.* 102, 63–72.
330. Ludwig, M., Sabatier, N., Bull, P. M., Landgraf, R., Dayanithi, G., and Leng, G. (2002). Intracellular calcium stores regulate activity-dependent neuropeptide release from dendrites. *Nature* 418, 85–89.
331. Crowley, W. R., Shyr, S. W., Kacsoh, B., and Grosvenor, C. E. (1987). Evidence for stimulatory noradrenergic and inhibitory dopaminergic regulation of oxytocin release in the lactating rat. *Endocrinology* 121, 14–20.
332. Bealer, S. L., and Crowley, W. R. (1998). Noradrenergic control of central oxytocin release during lactation in rats. *Am. J. Physiol.* 274, E453–E458.
333. Landgraf, R., Neumann, I., Russell, J. A., and Pittman, Q. J. (1992). Push-pull perfusion and microdialysis studies of central oxytocin and vasopressin release in freely moving rats

during pregnancy, parturition, and lactation. *Ann. N Y Acad. Sci.* 652, 326–339.

334. Robinson, I. C. A. F. (1983). Neurohypophysial peptides in cerebrospinal fluid. In *The Neurohypophysis: Structure, Function and Control* (B. A. Cross and G. Leng, Eds.), pp. 129–145. Elsevier, Amsterdam.
335. Ingram, C. D., and Moos, F. (1992). Oxytocin-containing pathway to the bed nuclei of the stria terminalis of the lactating rat brain, immunocytochemical and in vitro electrophysiological evidence. *Neuroscience* 47, 439–452.
336. Brown, C. H., Russell, J. A., and Leng, G. (2000). Opioid modulation of magnocellular neurosecretory cell activity. *Neurosci. Res.* 36, 97–120.
337. Russell, J. A., Leng, G., and Douglas, A. J. (2003). The magnocellular oxytocin system, the fount of maternity: adaptations in pregnancy. *Front. Neuroendocrinol.* 24, 27–61.
338. Clarke, G., Wood, P., Merrick, L., and Lincoln, D. W. (1979). Opiate inhibition of peptide release from the neurohumoral terminals of the hypothalamic neurones. *Nature* 282, 746–748.
339. Clarke, G., and Patrick, G. (1983). Differential inhibitory action by morphine on the release of oxytocin and vasopressin from the isolated neural lobe. *Neurosci. Lett.* 39, 175–180.
340. Bicknell, R. J., and Leng, G. (1982). Endogenous opiates regulate oxytocin but not vasopressin secretion from the neurohypophysis. *Nature* 198, 161–162.
341. Zhao, B-G., Chapman, C., and Bicknell, R. J. (1988). Functional kappa opioid receptors on oxytocin and vasopressin nerve terminals isolated from rat neurohypophysis. *Brain Res.* 462, 62–66.
342. Levin, M. C., and Sawchenko, P. E. (1993). Neuropeptide co-expression in the magnocellular neurosecretory system of the female rat: evidence for differential modulation by estrogen. *Neuroscience* 54, 1001–1018.
343. Summy-Long, J. Y., Rosella-Dampman, L. M., M'Lemore, G. L., and Kachler, E. (1990). Kappa opiate receptors inhibit release of oxytocin from the magnocellular system during dehydration. *Neuroendocrinology* 51, 376–384.
344. Leng, G., Dye, S., and Bicknell, R. J. (1997). Kappa-opioid restraint of oxytocin secretion: plasticity through pregnancy. *Neuroendocrinology* 66, 378–383.
345. Rusin, K. I., Giovannucci, D. R., Stuenkel, E. L., and Moises, H. C. (1997). Kappa-opioid receptor activation modulates Ca2+ currents and secretion in isolated neuroendocrine nerve terminals. *J. Neurosci.* 17, 6565–6574.
346. Bunn, S. J., Hanley, M. R., and Wilkin, G. P. (1985). Evidence for a kappa-opioid receptor on pituitary astrocytes: an autoradiographic study. *Neurosci. Lett.* 55, 317–323.
347. Kato, M., Chapman, C., and Bicknell, R. J. (1992). Activation of kappa-opioid receptors inhibits depolarisation-evoked exocytosis but not the rise in intracellular Ca2+ in secretory nerve terminals of the neurohypophysis. *Brain Res.* 574, 138–146.
348. Rossier, J., Battenberg, E., and Pittman, Q., et al. (1979). Hypothalamic enkephalin neurones may regulate the neurohypophysis. *Nature* 277, 653–654.
349. Martin, R., and Voigt, K. H. (1981). Enkephalin co-exists with oxytocin and vasopressin in nerve terminals of rat neurohypophysis. *Nature* 289, 502–504.
350. O'Donohue, T. L., and Dorsa, D. M. (1982). The opiomelanotropinergic neuronal and endocrine systems. *Peptides* 3, 353–395.
351. Ortiz-Miranda, S. I., Dayanithi, G., Coccia, V., Custer, E. E., Alphandery, S., Mazuc, E., Treistman, S., and Lemos, J. R. (2003). mu-Opioid receptor modulates peptide release from rat neurohypophysial terminals by inhibiting Ca(2+) influx. *J. Neuroendocrinol.* 15, 888–894.
352. Leng, G., and Mansfield, S., Bicknell, R. J., Dean, A. D., Ingram, C. D., Marsh, M. I., Yates, J. O., and Dyer, R. G. (1985). Central opioids: a possible role in parturition? *J. Endocrinol.* 106, 219–224.
353. Tancin, V., Kraetzl, W. D., and Schams, D. (2000). Effects of morphine and naloxone on the release oxytocin and on milk ejection in dairy cows. *J. Dairy Res.* 67, 13–20.
354. Rushen, J., Nay, T. S., Wright, L. R., Payne, D. C., and Foxcroft, G. R. (1995). Stress and nursing in the pig: role of HPA axis and endogenous opioid Peptides. *Physiol. Behav.* 58, 43–48.
355. Johnson, M. R., Andrews, M. A., Seckl, J. R., and Lightman, S. L. (1990). Effect of naloxone on neurohypophyseal peptide responses to breast feeding and breast stimulation in man. *Clin. Endocrinol. (Oxf).* 33, 81–86.
356. Wellnitz, O., Bruckmaier, R. M., and Blum, J. W. (1997). Naloxone and adrenergic blocking agents fail to abolish central inhibition of milk ejection in cows. *J. Dairy Res.* 64, 627–631.
357. Kraetzl, W. D., Tancin, V., and Schams, D. (2001). Inhibition of oxytocin release and milk let-down in postpartum primiparous cows is not abolished by naloxone. *J. Dairy Res.* 68, 559–568.
358. Summy-Long, J. Y., Miller, D. S., Rosella-Dampman, L. M., Hartman, R. D., and Emmert, S. E. (1984). A functional role for opioid peptides in the differential secretion of vasopressin and oxytocin. *Brain Res.* 309, 362–366.
359. Eriksson, M., Ceccatelli, S., Uvnas-Moberg, K., Iadarola, M., and Hokfelt, T. (1996). Expression of Fos-related antigens, oxytocin, dynorphin and galanin in the paraventricular and supraoptic nuclei of lactating rats. *Neuroendocrinology* 63, 356–367.
360. Baumgarten, H. G., Bjorklund, A., Holstein, A. F., and Nobin, A. (1972). Organization and ultrastructural identification of the catecholamine nerve terminals in the neural lobe and pars intermedia of the rat pituitary. *Z. Zellforsch. Mikrosk. Anat.* 126, 483–517.
361. Holzbauer, M., Muscholl, E., Racke, K., and Sharman, D. F. (1983). Evidence that dopamine is a neurotransmitter in the neurointermediate lobe of the hypophysis. In *The Neurohypophysis: Structure, Function and Control* (B. A. Cross and G. Leng, Eds.), pp. 357–364. Elsevier, Amsterdam.
362. Crowley, W. R., Parker, S. L., Armstrong, W. E., Wang, W., and Grosvenor, C. E. (1991). Excitatory and inhibitory dopaminergic regulation of oxytocin secretion in the lactating rat: evidence for respective mediation by D-1 and D-2 dopamine receptor subtypes. *Neuroendocrinology* 53, 493–502.
363. Barnes, P. R. J., and Dyball, R. E. J. (1982). Inhibition of neurohypophysial hormone release by dopamine in the rat. *J. Physiol. (Lond).* 327, 85P–86P.
364. Russell, J. A., Harrison, D. J., and McNeilly, A. S. (1981). Bromocriptine and alpha-ergocryptine do not inhibit oxytocin secretion in the lactating rat. *J. Endocrinol.* 89, 91–98.
365. Alonso, G., Runquist, M., Hussy, N., Duvoid, A., and Moos, F. (2003). Age-related modifications of the morphological organization of pituicytes are associated with alteration of the GABAergic and dopaminergic innervation afferent to the neurohypophysial lobe. *Eur. J. Neurosci.* 18, 1889–1903.
366. Garten, L. L., Sofroniew, M. W., and Dyball, R. E. J. (1989). A direct catecholaminergic projection from the brainstem to the neurohypophysis of the rat. *Neuroscience* 33, 149–155.
367. Hatton, G. I., Luckman, S. M., and Bicknell, R. J. (1991). Adrenalin activation of beta 2-adrenoceptors stimulates morphological changes in astrocytes (pituicytes) cultured from adult rat neurohypophyses. *Brain Res. Bull.* 26, 765–769.
368. Zhao, B-B., Chapman, C., Brown, D., and Bicknell, R. J. (1988). Opioid-noradrenergic interactions in the neurohypophysis.

II. Does noradrenaline mediate the actions of endogenous opioids on oxytocin and vasopressin release? *Neuroendocrinology* 48, 25–31.

369. Livingston, A., and Morris, B. (1986). Localisation of (3H) clonidine binding in rat neurohypophysis by means of electron-microscopic autoradiography. *Cell Tissue Res.* 244, 467–469.
370. Buijs, R. M., van Vulpen, E. H., and Geffard, M. (1987). Ultrastructural localization of GABA in the supraoptic nucleus and neural lobe. *Neuroscience* 20, 347–355.
371. Dyball, R. E., and Shaw, F. D. (1979). Inhibition by GABA of hormone release from the neurohypophysis in the rat (proceedings). *J. Physiol.* 289, 78P–79P.
372. Song, Z., and Hatton, G. I. (2003). Taurine and the control of basal hormone release from rat neurohypophysis. *Exp. Neurol.* 183, 330–337.
373. Zhang, S. J., and Jackson, M. B. (1994). Neuroactive steroids modulate GABAA receptors in peptidergic nerve terminals. *J. Neuroendocrinol.* 6, 533–538.
374. Voisin, D. L., Herbison, A. E., and Poulain, D. A. (1995). Central inhibitory effects of muscimol and bicuculline on the milk ejection reflex in the anaesthetized rat. *J. Physiol.* 483 (Pt 1), 211–224.
375. Lutz-Bucher, B., and Koch, B. (1994). Evidence for an inhibitory effect of nitric oxides on neuropeptide secretion from isolated neural lobe of the rat pituitary gland. *Neurosci. Lett.* 165, 48–50.
376. Bredt, D. S., Hwang, P. M., and Snyder, S. H. (1990). Localization of nitric oxide synthase indicating a neural role for nitric oxide. *Nature* 347, 768–770.
377. Kjeldsen, T. H., Rivier, C., Lee, S., Hansen, E. W., Christensen, J. D., and Moesby, L. (2003). Inducible nitric oxide synthase is responsible for nitric oxide release from murine pituicytes. *J. Neuroendocrinol.* 15, 250–255.
378. Lemos, J. R., and Wang, G. (2000). Excitatory versus inhibitory modulation by ATP of neurohypophysial terminal activity in the rat. *Exp. Physiol.* 85 Spec No., 67S–74S.
379. Kalimo, H. (1975). Ultrastructural studies on the hypothalamic neurosecretory neurones of the rat. III. Paraventricular and supraoptic neurones during lactation and dehydration. *Cell Tissue Res.* 163, 151–168.
380. Spin, L. H., Raghow, R., and Crowley, W. R. (1992). Afferent stimulation regulates oxytocin messenger ribonucleic acid during early lactation in rats. *J. Neuroendocrinol.* 4, 181–187.
381. Hatton, G. I., and Tweedle, C. D. (1982). Magnocellular neuropeptidergic neurons in hypothalamus: increases in membrane apposition and number of specialised synapses from pregnancy to lactation. *Brain Res. Bull.* 8, 197–204.
382. Theodosis, D. T., Poulain, D. A., and Vincent, J. D. (1981). Possible morphological bases for synchronization of neuronal firing in the rat supraoptic nucleus during lactation. *Neuroscience* 6, 919–929.
383. Theodosis, D. T., and Poulain, D. A. (1984). Evidence for structural plasticity in the supraoptic nucleus of the rat hypothalamus in relation to gestation and lactation. *Neuroscience* 11, 83–193.
384. Theodosis, D. T., and Poulain, D. A. (2001). Maternity leads to morphological synaptic plasticity in the oxytocin system. *Prog. Brain Res.* 133, 49–58.
385. Chapman, D. B., Theodosis, D. T., Montagnese, C., Poulain, D. A., and Morris, J. F. (1986). Osmotic stimulation causes structural plasticity of neurone-glia relationships of the oxytocin but not vasopressin secreting neurones in the hypothalamic supraoptic nucleus. *Neuroscience* 17, 679–686.
386. Theodosis, D. T., Chapman, D. B., Montagnese, C., Poulain, D. A., and Morris, J. F. (1986). Structural plasticity in the hypothalamic supraoptic nucleus at lactation affects oxytocin-, but not vasopressin-secreting neurones. *Neuroscience* 17, 661–678.
387. Oliet, S. H., and Piet, R. (2004). Anatomical remodelling of the supraoptic nucleus: changes in synaptic and extrasynaptic transmission. *J. Neuroendocrinol.* 16, 303–307.
388. Danbolt, N. C. (2001). Glutamate uptake. *Prog. Neurobiol.* 65, 1–105.
389. Oliet, S. H., Piet, R., and Poulain, D. A. (2001). Control of glutamate clearance and synaptic efficacy by glial coverage of neurons. *Science* 292, 923–926.
390. Stern, J. E., Hestrin, S., and Armstrong, W. E. (2000). Enhanced neurotransmitter release at glutamatergic synapses on oxytocin neurones during lactation in the rat. *J. Physiol.* 526 (Pt 1), 109–114.
391. Parker, S. L., and Crowley, W. R. (1995). Central stimulation of oxytocin release in the lactating rat by N-methyl-D-aspartate: requirement for coactivation through non-NMDA glutamate receptors or the glycine coagonist site. *Neuroendocrinology* 62, 467–478.
392. Parker, S. L., and Crowley, W. R. (1993). Stimulation of oxytocin release in the lactating rat by central excitatory amino acid mechanisms: evidence for specific involvement of R,S-alpha-amino-3-hydroxy-5-methylisoxazole-4-propionic acid-sensitive glutamate receptors. *Endocrinology* 133, 2847–2854.
393. Langle, S. L., Poulain, D. A., and Theodosis, D. T. (2003). Induction of rapid, activity-dependent neuronal-glial remodelling in the adult rat hypothalamus in vitro. *Eur. J. Neurosci.* 18, 206–214.
394. Theodosis, D. T., and Poulain, D. A. (1992). Neuronal-glial and synaptic remodelling in the adult hypothalamus in response to physiological stimuli. In *Functional Anatomy of the Neuroendocrine Hypothalamus* (D. J. Chadwick and J. Marsh, Eds.), pp. 209–232. Wiley, Chichester.
395. Theodosis, D. T., Schachner, M., and Neumann, I. D. (2004). Oxytocin neuron activation in NCAM-deficient mice: anatomical and functional consequences. *Eur. J. Neurosci.* 20, 3270–3280.
396. Theodosis, D. T., Montagnese, C., Rodriguez, F., Vincent, J. D., and Poulain, D. A. (1986). Oxytocin induces morphological plasticity in the adult hypothalamo-neurohypophysial system. *Nature* 322, 738–740.
397. Lipschitz, D. L., Crowley, W. R., and Bealer, S. L. (2003). Central blockade of oxytocin receptors during late gestation disrupts systemic release of oxytocin during suckling in rats. *J. Neuroendocrinol.* 15, 743–748.
398. Montagnese, C., Poulain, D., and Theodosis, D. T. (1990). Influence of ovarian steroids on the ultrastructural plasticity of the adult supraoptic nucleus induced by central administration of oxytocin. *J. Neuroendocrinol.* 2, 225–231.
399. Jiang, Q., and Wakerley, J. B. (1997). The milk-ejection reflex in the peri-partum rat: effects of oestradiol and progesterone on basal milk-ejection frequency and the facilitatory response to central oxytocin. *J. Neuroendocrinol.* 9, 9–16.
400. Brussaard, A. B., and Herbison, A. E. (2000). Long-term plasticity of postsynaptic GABAA-receptor function in the adult brain: insights from the oxytocin neurone. *Trends Neurosci.* 23, 190–195.
401. Hatton, G. I., and Yang, Q. Z. (1990). Activation of excitatory amono acid inputs to supraoptic neurones. I. Induced increases in dye coupling in lactating but not virgin or male rats. *Brain Res.* 513, 264–269.
402. Tweedle, C. D. (1983) Ultrastructural manifestations of increased hormone release in the neurohypophysis. In *The Neurohypophysis: Structure, Function and Control* (B. A. Cross and G. Leng, Eds.), pp. 259–272. Elsevier, Amsterdam.

403. Bicknell, R. J., Luckman, S. M., and Mason, W. T. (1989). Beta-adrenergic and opioid receptors in pituicytes cultured from adult rat neurohypophysis: regulation of cell morphology. *Brain Res.* 22, 379–388.
404. Cathcart, E. P., Gairns, F. W., and Garven, H. S. D. (1948). The innervation of the human quiescent nipple, with notes on pigmentation, erection and hyperneury. *Trans. R. Soc. (Edinb).* 61, 699–717.
405. Niggli-Stokar von, U. (1961). Faseranalyse der eutemerven and die nervenformationen in der zitzenhaut des rindes. *Acta Anat.* 46, 104–126.
406. Montagna, W. (1970). Histology and cytochemistry of human skin. XXXV. The nipple and areola. *Br. J. Dermatol.* 83, 2–13.
407. Miller, M. R., and Kasahara, M. (1959). The cutaneous innervation of the human female breast. *Anat. Rec.* 135, 153–167.
408. Robinson, J. E., and Short, R. V. (1977). Changes in breast sensitivity at puberty, during the menstrual cycle and at parturition. *Br. Med. J.* 1, 1188–1191.
409. Eriksson, M., Lindh, B., Uvnas-Moberg, K., and Hokfelt, T. (1996). Distribution and origin of peptide-containing nerve fibres in the rat and human mammary gland. *Neuroscience* 70, 227–245.
410. Thulesen, J., Rasmussen, T. N., Schmidt, P., Holst, J. J., and Poulsen, S. S. (1994). Calcitonin gene-related peptide (CGRP) in the nipple of the rat mammary gland. *Histochemistry* 102, 437–444.
411. Traurig, H., Papka, R. E., Saria, A., and Lembeck, F. (1984). Substance P immunoreactivity in the rat mammary nipple and the effects of capsaicin treatment on lactation. *Naunyn Schmiedebergs Arch. Pharmacol.* 328, 1–8.
412. Poulain, D. A., and Wakerley, J. B. (1986). Afferent projections from the mammary glands to the spinal cord in the lactating rat: electrophysiological responses of spinal neurons during stimulation of the nipples, including suckling. *Neuroscience* 7, 773–808.
413. Fenelon, V. S., and Poulain, D. A. (1992). Electrical activity of dorsal horn neurons during the suckling-induced milk ejection reflex in the lactating rat. *J. Neuroendocrinol.* 4, 575–584.
414. Haller, E. W. (1985). Neural and anatomical characteristics of peripheral afferent fibres in the milk-ejection reflex. *Brain Res. Bull.* 15, 563–567.
415. Voloschin, L. M., Decima, E., and Tramezzani, J. H. (1988). Electronmicroscopic and electrophysiological studies of the teat branch of the XIII thoracic nerve: relationship with lactation in the rat. *J. Endocrinol.* 118, 471–483.
416. Tasker, J. G., Theodosis, D. T., and Poulain, D. A. (1986). Afferent projections from the mammary glands to the spinal cord in the lactating rat I. A neuroanatomical study using the transganglionic transport of horseradish peroxidase-wheatgerm agglutinin. *Neuroscience* 19, 495–509.
417. Rousselot, P., Poulain, D. A., and Theodosis, D. T. (1994). Ultrastructural visualization and neurochemical characterization of spinal projections of primary sensory afferents from the nipple: combined use of transganglionic transport of HRP-WGA and glutamate immunocytochemistry. *J. Histochem. Cytochem.* 42, 115–123.
418. Wright, D. M. (1985). Evidence for a spinal site at which opioids may act to inhibit the milk-ejection reflex. *J. Endocrinol.* 106, 401–407.
419. Robinson, D. A., Wei, F., Wang, G. D., Li, P., Kim, S. J., Vogt, S. K., Muglia, L. J., and Zhuo, M. (2002). Oxytocin mediates stress-induced analgesia in adult mice. *J. Physiol.* 540, 593–606.
420. Fukuoka, T., Negoro, H., Honda, K., Higuchi, T., and Nishida, E. (1984). Spinal pathway of the milk-ejection reflex in the rat. *Biol. Reprod.* 30, 74–81.
421. Tindal, J. S., Knaggs, G. S., and Turvey, A. (1969). The afferent path of the milk ejection reflex in the brain of the rabbit. *J. Endocrinol.* 43, 663–671.
422. Tindal, J. S., Knaggs, G. S., and Turvey, A. (1967). The afferent path of the milk ejection reflex in the brain of the guinea pig. *J. Endocrinol.* 38, 337–349.
423. Dubois-Dauphin, M., Armstrong, W. E., Tribollet, E., and Dreifuss, J. J. (1985). Somatosensory systems and the milk-ejection reflex in the rat. II. The effects of lesions in the ventroposterior thalamic complex, dorsal columns and lateral cervical nucleus-dorsolateral fasciculus. *Neuroscience* 15, 1131–1140.
424. Richard, P., Urban, I., and Denamur, R. (1970). The role of the dorsal tracts of the spinal cord and of the mesencephalic and thalamic lemniscal system in the milk-ejection reflex during milking in the ewe. *J. Endocrinol.* 47, 45–53.
425. Mena, F., and Beyer, C. (1968). Effect of spinal cord lesions on milk ejection in the rabbit. *Endocrinology* 83, 615–617.
426. Stern, J. M., Yu, Y. L., and Crockett, D. P. (2002). Dorsolateral columns of the spinal cord are necessary for both suckling-induced neuroendocrine reflexes and the kyphotic nursing posture in lactating rats. *Brain Res.* 947, 110–121.
427. Palkovits, M., and Zaborszky, L. (1979). Neural connections of the hypothalamus. In *Handbook of the Hypothalamus* (P. J. Morgane and J. Panksepp, Eds.), pp. 379–510. Marcel Dekker, New York.
428. Cliffer, K. D., Burstein, R., and Giesler, G. J. Jr. (1991). Distributions of spinothalamic, spinohypothalamic, and spinotelencephalic fibers revealed by anterograde transport of PHA-L in rats. *J. Neurosci.* 11, 852–868.
429. Li, J. L., Kaneko, T., Shigemoto, R., and Mizuno, N. (1997). Distribution of trigeminohypothalamic and spinohypothalamic tract neurons displaying substance P receptor-like immunoreactivity in the rat. *J. Comp. Neurol.* 378, 508–521.
430. Dubois-Dauphin, M., Armstrong, W. E., Tribollet, E., and Dreifuss, J. J. (1985). Somatosensory systems and the milk-ejection reflex in the rat. I. Lesions of the mesencephalic lateral tegmentum disrupt the reflex and damage mesencephalic somatosensory connections. *Neuroscience* 15, 1111–1129.
431. Cunningham, E. T., and Sawchenko, P. (1991). Reflex control of magnocellular vasopressin and oxytocin secretion. *Trends Neurosci.* 14, 406–411.
432. Richard, P. (1970). An electrophysiological study in the ewe of the tracts which transmit impulses from the mammary glands to the pituitary stalk. *J. Endocrinol.* 47, 37–44.
433. Voloschin, L. M., and Dottaviano, E. J. (1976). The channelling of natural stimuli that evoke the ejection of milk in the rat. Effect of transections in the midbrain and hypothalamus. *Endocrinology* 99, 49–58.
434. Tindal, J. S., and Knaggs, G. S. (1971). Determination of the detailed hypothalamic route of the milk-ejection reflex in the guinea-pig. *J. Endocrinol.* 50, 135–152.
435. Hansen, S., and Kohler, C. (1984). The importance of the peripeduncular nucleus in the neuroendocrine control of sexual behavior and milk ejection in the rat. *Neuroendocrinology* 39, 563–572.
436. Tribollet, E., Armstrong, W. E., Dubois-Dauphin, M., and Dreifuss, J. J. (1985). Extra-hypothalamic afferent inputs to the supraoptic nucleus area of the rat as determined by retrograde and anterograde tracing techniques. *Neuroscience* 15, 135–148.
437. Crowley, W. R., and Armstrong, W. E. (1992). Neurochemical regulation of oxytocin secretion in lactation. *Endocr. Rev.* 13, 33–65.

438. Morales, T., and Sawchenko, P. E. (2003). Brainstem prolactin-releasing peptide neurons are sensitive to stress and lactation. *Neuroscience* 121, 771–778.
439. Lin, S. H., Miyata, S., Matsunaga, W., Kawarabayashi, T., Nakashima, T., and Kiyohara, T. (1998). Metabolic mapping of the brain in pregnant, parturient and lactating rats using fos immunohistochemistry. *Brain Res.* 787, 226–236.
440. Sawchenko, P. E., Plotsky, P. M., Pfeiffer, S. W., Cunningham, E. T. Jr., Vaughan, J., Rivier, J., and Vale, W. (1988). Inhibin beta in central neural pathways involved in the control of oxytocin secretion. *Nature* 334, 615–617.
441. LeDoux, J. (2003). The emotional brain, fear, and the amygdala. *Cell. Mol. Neurobiol.* 23, 727–738.
442. Sawchenko, P. E., and Swanson, L. W. (1983). The organization of forebrain afferents to the paraventricular and supraoptic nuclei of the rat. *J. Comp. Neurol.* 218, 121–144.
443. Gerendai, I., Toth, I. E., Kocsis, K., Boldogkoi, Z., Medveczky, I., and Halasz, B. (2001). Transneuronal labelling of nerve cells in the CNS of female rat from the mammary gland by viral tracing technique. *Neuroscience* 108, 103–118.
444. Woods, W. H., Holland, R. C., and Powell, E. W. (1969). Connections of cerebral structures functioning in neurohypophysial hormone release. *Brain Res.* 12, 26–46.
445. Aulsebrook, L. H., and Holland, R. C. (1969). Central regulation of oxytocin release with and without vasopressin release. *Am. J. Physiol.* 216, 818–829.
446. Saphier, D., and Feldman, S. (1987). Effects of septal and hippocampal stimuli on paraventricular nucleus neurons. *Neuroscience* 20, 749–755.
447. Ferguson, A. V., Day, T. A., and Renaud, L. P. (1984). Connections of hypothalamic paraventricular neurons with the dorsal medial thalamus and neurohypophysis: an electrophysiological study in the rat. *Brain Res.* 299, 376–379.
448. Pittman, Q. J., Blume, H. W., and Renaud, L. P. (1981). Connections of the hypothalamic paraventricular nucleus with the neurohypophysis, median eminence, amygdala, lateral septum and midbrain periaqueductal gray: an electrophysiological study in the rat. *Brain Res.* 215, 15–28.
449. Deis, R. P., and Prilusky, J. (1984). Participation of the hippocampus in the facilitatory effect of an exteroceptive stimulus on milk ejection. *Exp. Brain Res.* 55, 177–179.
450. Aulsebrook, H. (1969). Central inhibition of oxytocin release. *Am. J. Physiol.* 216, 830–842.
451. Cirino, M., and Renaud, L. P. (1985). Influence of lateral septum and amygdala stimulation on the excitability of hypothalamic supraoptic neurons. An electrophysiological study in the rat. *Brain Res.* 326, 357–361.
452. Poulain, D. A., Lebrun, C. J., and Vincent, J. D. (1981). Electrophysiological evidence for connections between septal neurones and the supraoptic nucleus of the hypothalamus of the rat. *Exp. Brain Res.* 42, 260–268.
453. Shibuki, K. (1984). Supraoptic neurosecretory cells: synaptic inputs from the nucleus accumbens in the rat. *Exp. Brain Res.* 53, 341–348.
454. Poulain, D. A., Ellendorff, F., and Vincent, J. D. (1980). Septal connections with identified oxytocin and vasopressin neurones in the supraoptic nucleus of the rat. An electrophysiological investigation. *Neuroscience* 5, 379–387.
455. Boudaba, C., and Poulain, D. A. (1991). Further evidence that the septum is not part of the main pathway of the milk ejection reflex in the rat. *J. Neuroendocrinol.* 3, 199–204.
456. Tindal, J. S., and Blake, L. A. (1984). Central inhibition of milk ejection in the rabbit: involvement of hippocampus and subiculum. *J. Endocrinol.* 100, 125–129.
457. Tindal, J. S., and Blake, L. A. (1986). Central inhibition of oxytocin release in the rabbit: role of the midbrain. *J. Endocrinol.* 109, 405–409.
458. Tindal, J. S., and Blake, L. A. (1980). A neural basis for central inhibition of milk ejection in the rabbit. *J. Endocrinol.* 86, 525–531.
459. Greco, B., Allegretto, E. A., Tetel, M. J., and Blaustein, J. D. (2001). Coexpression of ER beta with ER alpha and progestin receptor proteins in the female rat forebrain: effects of estradiol treatment. *Endocrinology* 142, 5172–5181.
460. Bealer, S. L., and Crowley, W. R. (1999). Neurotransmitter interaction in release of intranuclear oxytocin in magnocellular nuclei of the hypothalamus. *Ann. N Y Acad. Sci.* 897, 182–191.
461. Sladek, C. D., and Kapoor, J. R. (2001). Neurotransmitter/neuropeptide interactions in the regulation of neurohypophyseal hormone release. *Exp. Neurol.* 171, 200–209.
462. Kimura, H., McGeer, P. L., Peng, J. H., and McGeer, E. G. (1981). The central cholinergic system studied by choline acetyltransferase immunohistochemistry in the cat. *J. Comp. Neurol.* 200, 151–201.
463. Hatton, G. I., and Yang, Q. Z. (2002). Synaptic potentials mediated by alpha 7 nicotinic acetylcholine receptors in supraoptic nucleus. *J. Neurosci.* 22, 29–37.
464. Zaninetti, M., Tribollet, E., Bertrand, D., and Raggenbass, M. (2002). Nicotinic cholinergic activation of magnocellular neurons of the hypothalamic paraventricular nucleus. *Neuroscience* 110, 287–299.
465. Li, D. P., and Pan, H. L. (2001). Potentiation of glutamatergic synaptic input to supraoptic neurons by presynaptic nicotinic receptors. *Am. J. Physiol. Regul. Integr. Comp. Physiol.* 281, R1105–R1113.
466. Kuhn, E. R., and McCann, S. M. (1971). Release of oxytocin and vasopressin in lactating rats after injection of carbachol into the third ventricle. *Neuroendocrinology* 8, 48–58.
467. Clarke, G., Fall, C. H., Lincoln, D. W., and Merrick, L. P. (1978). Effects of cholinoceptor antagonists on the suckling-induced and experimentally evoked release of oxytocin. *Br. J. Pharmacol.* 63, 519–527.
468. Clarke, G., Lincoln, D. W., and Merrick, L. P. (1979). Dopaminergic control of oxytocin release in lactating rats. *J. Endocrinol.* 83, 409–420.
469. Michaloudi, H. C., el Majdoubi, M., Poulain, D. A., Papadopoulos, G. C., and Theodosis, D. T. (1997). The noradrenergic innervation of identified hypothalamic magnocellular somata and its contribution to lactation-induced synaptic plasticity. *J. Neuroendocrinol.* 9, 17–23.
470. Yamashita, H., Inenaga, K., and Kannan, H. (1987). Depolarizing effect of noradrenaline on neurons of the rat supraoptic nucleus in vitro. *Brain Res.* 405, 348–352.
471. Crowley, W. R., Parker, S. L., Armstrong, W. E., Spinolo, L. H., and Grosvenor, C. E. (1992). Neurotransmitter and neurohormonal regulation of oxytocin secretion in lactation. *Ann. N Y Acad. Sci.* 652, 286–302.
472. Cunningham, E. T. Jr., and Sawchenko, P. E. (1988). Anatomical specificity of noradrenergic inputs to the paraventricular and supraoptic nuclei of the rat hypothalamus. *J. Comp. Neurol.* 274, 60–76.
473. Sawchenko, P. E., and Swanson, L. W. (1982). The organization of noradrenergic pathways from the brainstem to the paraventricular and supraoptic nuclei in the rat. *Brain Res.* 257, 275–325.
474. Menetrey, D., and Basbaum, A. I. (1987). Spinal and trigeminal projections to the nucleus of the solitary tract: a possible substrate for somatovisceral and viscerovisceral reflex activation. *J. Comp. Neurol.* 255, 439–450.
475. Parker, S. L., and Crowley, W. R. (1993). Central stimulation of oxytocin release in the lactating rat: interaction of neuropeptide Y with alpha-1-adrenergic mechanisms. *Endocrinology* 132, 658–666.

476. Unnerstall, J. R., Kopajtic, T. A., and Kuhar, M. J. (1984). Distribution of alpha 2 agonist binding sites in the rat and human central nervous system: analysis of some functional, anatomic correlates of the pharmacologic effects of clonidine and related adrenergic agents. *Brain Res.* 319, 69–101.
477. Moos, F., and Richard, P. (1980). (Dual noradrenergic control of oxytocin release during the milk ejection reflex in the rat). *C. R. Seances Acad. Sci. D.* 290, 1261–1264.
478. Moos, F., and Richard, P. (1979). The inhibitory role of beta-noradrenergic receptors in oxytocin release during suckling. *Brain Res.* 169, 595–599.
479. Song, S. L., Crowley, W. R., and Grosvenor, C. E. (1988). Evidence for involvement of an adrenal catecholamine in the beta-adrenergic inhibition of oxytocin release in lactating rats. *Brain Res.* 457, 303–309.
480. Tribollet, E., Clarke, G., Dreifuss, J. J., and Lincoln, D. W. (1978). The role of central adrenergic receptors in the reflex release of oxytocin. *Brain Res.* 142, 69–84.
481. Vorherr, H. (1971). Catecholamine antagonism to oxytocin-induced milk-ejection. *Acta Endocrinol. Suppl. (Copenh).* 154, 3–38.
482. Jourdain, P., Dupouy, B., Bonhomme, R., Poulain, D. A., Israel, J. M., and Theodosis, D. T. (1999). Visualization of local afferent inputs to magnocellular oxytocin neurons in vitro. *Eur. J. Neurosci.* 11, 1960–1972.
483. Bridges, T. E., Hillhouse, E. W., and Jones, M. T. (1976). The effect of dopamine on neurohypophysial hormone release in vivo and from the rat neural lobe and hypothalamus in vitro. *J. Physiol.* 260, 647–666.
484. Voogt, J. L., and Carr, L. (1974). Plasma prolactin levels and hypothalamic catecholamine synthesis during suckling. *Neuroendocrinology* 16, 108–118.
485. Crowley, W. R., Parker, S. L., Armstrong, W. E., Wang, W., and Grosvenor, C. E. (1991). Excitatory and inhibitory dopaminergic regulation of oxytocin secretion in the lactating rat: evidence for respective mediation by D-1 and D-2 dopamine receptor subtypes. *Neuroendocrinology* 53, 493–502.
486. Parker, S. L., and Crowley, W. R. (1992). Activation of central D-1 dopamine receptors stimulates oxytocin release in the lactating rat: evidence for involvement of the hypothalamic paraventricular and supraoptic nuclei. *Neuroendocrinology* 56, 385–392.
487. Yang, C. R., Bourque, C. W., and Renaud, L. P. (1991). Dopamine D2 receptor activation depolarizes rat supraoptic neurones in hypothalamic explants. *J. Physiol.* 443, 405–419.
488. Price, C. J., and Pittman, Q. (2001). Dopamine D4 receptor activation inhibits presynaptically glutamatergic neurotransmission in the rat supraoptic nucleus. *J. Neurophysiol.* 86, 1149–1155.
489. Azdad, K., Piet, R., Poulain, D. A., and Oliet, S. H. (2003). Dopamine D4 receptor-mediated presynaptic inhibition of GABAergic transmission in the rat supraoptic nucleus. *J. Neurophysiol.* 90, 559–565.
490. Sawchenko, P. E., Swanson, L. W., Steinbusch, H. W., and Verhofstad, A. A. (1983). The distribution and cells of origin of serotonergic inputs to the paraventricular and supraoptic nuclei of the rat. *Brain Res.* 277, 355–360.
491. Jorgensen, H., Kjaer, A., Knigge, U., Moller, M., and Warberg, J. (2003). Serotonin stimulates hypothalamic mRNA expression and local release of neurohypophysial *Peptides J. Neuroendocrinol.* 15, 564–571.
492. Saydorff, J. A., Rittenhous, R. A., Van de Kar, L. D., and Brownfield, M. S. (1991). Enhanced serotonergic transmission stimulates oxytocin secretion in conscious male rats. *J. Pharmacol. Exp. Ther.* 257, 95–99.
493. Krstic, M. K., and Djurkovic, D. (1980). Analysis of cardiovascular responses to central administration of 5-hydroxytryptamine in rats. *Neuropharmacology* 19, 455–463.
494. Clarke, G., Juss, T. S., and Wakerley, J. B. (1985). Effect of lesions of the medial raphe on EEG patterns and reflex milk ejection in the urethane anaesthetized rat. *J. Physiol. (Lond).* 364, 57P.
495. Clarke, G., and Wright, D. M. (1985). Effect of 5-hydroxytryptamine on oxytocin release in anaesthetised lactating rats. *J. Physiol. (Lond).* 369, 128P.
496. Yang, Q. Z., and Hatton, G. I. (1994). Histamine mediates fast synaptic inhibition of rat supraoptic oxytocin neurons via chloride conductance activation. *Neuroscience* 61, 955–964.
497. Schagen, F. H., Knigge, U., Kjaer, A., Larsen, P. J., and Warberg, J. (1996). Involvement of histamine in suckling-induced release of oxytocin, prolactin and adrenocorticotropin in lactating rats. *Neuroendocrinology* 63, 550–558.
498. Bealer, S. L., and Crowley, W. R. (1999). Stimulation of central and systemic oxytocin release by histamine in the paraventricular hypothalamic nucleus: evidence for an interaction with norepinephrine. *Endocrinology* 140, 1158–1164.
499. Bealer, S. L., and Crowley, W. R. (2001). Histaminergic control of oxytocin release in the paraventricular nucleus during lactation in rats. *Exp. Neurol.* 171, 317–322.
500. Knigge, U., Willems, E., Kjaer, A., Jorgensen, H., and Warberg, J. (1999). Histaminergic and catecholaminergic interactions in the central regulation of vasopressin and oxytocin secretion. *Endocrinology* 140, 3713–3719.
501. Ginsberg, S. D., Price, D. L., Blackstone, C. D., Huganir, R. L., and Martin, L. J. (1995). The AMPA glutamate receptor GluR3 is enriched in oxytocinergic magnocellular neurons and is localized at synapses. *Neuroscience* 65, 563–575.
502. Richardson, C. M., and Wakerley, J. B. (1997). Glutamate excitation of oxytocin neurones in vitro involves predominantly non-NMDA receptors. *Brain Res.* 767, 158–161.
503. Moos, F. C., Rossi, K., and Richard, P. (1997). Activation of N-methyl-D-aspartate receptors regulates basal electrical activity of oxytocin and vasopressin neurons in lactating rats. *Neuroscience* 77, 993–1002.
504. Voisin, D. L., Herbison, A. E., Chapman, C., and Poulain, D. A. (1996). Effects of central GABAB receptor modulation upon the milk ejection reflex in the rat. *Neuroendocrinology* 63, 368–376.
505. Sawchenko, P. E., Swanson, L. W., and Joseph, S. A. (1982). The distribution and cells of origin of ACTH(1-39)-stained varicosities in the paraventricular and supraoptic nuclei. *Brain Res.* 232, 365–374.
506. Csiffary, A., Ruttner, Z., Toth, Z., and Palkovits, M. (1992). Oxytocin nerve fibers innervate beta-endorphin neurons in the arcuate nucleus of the rat hypothalamus. *Neuroendocrinology* 56, 429–435.
507. Wakerley, J. B., Noble, R., and Clarke, G. (1983). Effects of morphine and D-Ala, D-Leu enkephalin on the electrical activity of supraoptic neurosecretory cells in vitro. *Neuroscience* 10, 73–81.
508. Pittman, Q. J., Hatton, J. D., and Bloom, F. E. (1980). Morphine and opioid Peptides. reduce paraventricular neuronal activity: studies on the rat hypothalamic slice preparation. *Proc. Natl. Acad. Sci. U S A* 77, 5527–5531.
509. Haldar, J., and Sawyer, W. H. (1978). Inhibition of oxytocin release by morphine and its analogs. *Proc. Soc. Exp. Biol. Med.* 157, 476–480.
510. Clarkek, G., and Wright, D. M. (1984). A comparison of analgesia and suppression of oxytocin release by opiates. *Br. J. Pharmacol.* 83, 799–806.

511. Bicknell, R. J., Leng, G., Russell, J. A., Dyer, R. G., Mansfield, S., and Zhao, B. G. (1988). Hypothalamic opioid mechanisms controlling oxytocin neurones during parturition. *Brain Res. Bull.* 20, 743–749.
512. Maruyama, M., Matsumoto, H., Fujiwara, K., Noguchi, J., Kitada, C., Hinuma, S., Onda, H., Nishimura, O., Fujino, M., Higuchi, T., and Inoue, K. (1999). Central administration of prolactin-releasing peptide stimulates oxytocin release in rats. *Neurosci. Lett.* 276, 193–196.
513. Honda, K., Narita, K., Murata, T., and Higuchi, T. (2004). Prolactin releasing peptides modulate background firing rate and milk-ejection related burst of oxytocin cells in the supraoptic nucleus. *Brain Res. Bull.* 63, 315–319.
514. Liu, H. L., Cao, R., Jin, L., and Chen, L. W. (2002). Immunocytochemical localization of substance P receptor in hypothalamic oxytocin-containing neurons of C57 mice. *Brain Res.* 948, 175–179.
515. Bittencourt, J. C., Benoit, R., and Sawchenko, P. E. (1991). Distribution and origins of substance P-immunoreactive projections to the paraventricular and supraoptic nuclei: partial overlap with ascending catecholaminergic projections. *J. Chem. Neuroanat.* 4, 63–78.
516. Clarke, G., Kirby, P. C., and Thomson, A. M. (1980). Effects on vasopressinergic and oxytocinergic neurones of intraventricular substance P. *J. Physiol. (Lond).* 307, 59P–60P.
517. Bealer, S. L., and Flynn, F. W. (2003). Central neurokinin 3 receptors increase systemic oxytocin release: interaction with norepinephrine. *Exp. Neurol.* 184, 1027–1033.
518. Juszczak, M., Furykiewicz-Nykis, K., and Stempniak, B. (2004). Role of tachykinin receptors and melatonin in oxitocin secretion from isolated rat hypothalmo-neurohypophysial system. *J. Physiol. Pharmacol.* 55, 739–749.
519. Sawchenko, P. E., Swanson, L. W., Grzanna, R., Howe, P. R., Bloom, S. R., and Polak, J. M. (1985). Colocalization of neuropeptide Y immunoreactivity in brainstem catecholaminergic neurons that project to the paraventricular nucleus of the hypothalamus. *J. Comp. Neurol.* 241, 138–153.
520. Kapoor, J. R., and Sladek, C. D. (2001). Substance P and NPY differentially potentiate ATP and adrenergic stimulated vasopressin and oxytocin release. *Am. J. Physiol. Regul. Integr. Comp. Physiol.* 280, R69–R78.
521. Vanhatalo, S., and Soinila, S. (1995). Nitric oxide synthase in the hypothalamo-pituitary pathways. *J. Chem. Neuroanat.* 8, 165–173.
522. Kadekaro, M. (2004). Nitric oxide modulation of the hypothalamo-neurohypophyseal system. *Braz. J. Med. Biol. Res.* 37, 441–450.
523. Srisawat, R., Ludwig, M., Bull, P. M., Douglas, A. J., Russell, J. A., and Leng, G. (2000). Nitric oxide and the oxytocin system in pregnancy. *J. Neurosci.* 20, 6721–6727.
524. Stern, J. E., and Ludwig, M. (2001). NO inhibits supraoptic oxytocin and vasopressin neurons via activation of GABAergic synaptic inputs. *Am. J. Physiol. Regul. Integr. Comp. Physiol.* 280, R1815–R1822.
525. Yang, Q. Z., and Hatton, G. I. (1999). Nitric oxide via cGMP-dependent mechanisms increases dye coupling and excitability of rat supraoptic nucleus neurons. *J. Neurosci.* 19, 4270–4279.
526. Okere, C. O., Wang, Y. F., Higuchi, T., Negoro, H., Okutani, F., Takahashi, S., and Murata, T. (1996). The effect of systemic and central nitric oxide administration on milk availability in lactating rats. *Neuroreport* 8, 243–247.
527. Ceccatelli, S., and Eriksson, M. (1993). The effect of lactation on nitric oxide synthase gene expression. *Brain Res.* 625, 177–179.
528. Otukonyong, E. E., Okere, C. O., Johnstone, L. E., Murata, T., Kaba, H., and Higuchi, T. (2000). Effect of suckling on NADPH-diaphorase (nitric oxide synthase, NOS) reactivity and NOS gene expression in the paraventricular and supraoptic nuclei of lactating rats. *J. Neuroendocrinol.* 12, 1001–1008.
529. Stern, J. E., and Zhang, W. (2004). Cellular sources, targets and actions of constitutive nitric oxide in the magnocellular neurosecretory system of the rat. *J. Physiol.* 562, 725–744.

Index

Page numbers in **bold** refer to Volume 1; page numbers followed by *f* indicate a figure; page numbers followed by *t* indicate tabular material.

A

Acanthopagrus schlegeli, 2022
ACE. *See* Angiotensin-converting enzyme
Acetylcholine
  hormonal effects, 1840–1841
  male sexual behavior and, 1754–1755
  milk ejection and, 3170
  as neurotransmitter action on LH secretion, **1430*t***
  reflex ovulation and, 2308–2309
ACh, effect on male sexual behavior, 1775
Achondroplasia, 2703
Acomys cahirinus, 2028
Acrbp, **17*t***
Acrocephalus sechellensis, 2025
Acrogranin, **17*t***
Acromegaly, **1290**
Acrosin, **17*t***
Acrosome, **4, 14–18**
  components, **17*t***
  contents, **16–18**
  reaction, **62–65**
    description, **62–64, 63*f***
    significance, **64**
    site, **64–65**
    zona pellucida and, **72–78, 73*f*, 74*t*, 76*t***
  structure, **14–16**
ACRV1, **17*t***
ACTH. *See* Adrenocorticotropic hormone
Actin, **8–9**
Actitis macularia, 2015
Activins
  expression, **1529–1530**
  menstrual cycle and, 2455
  mouse models, **408–409**
  as physiological regulator of gonadotropin biosynthesis and secretion, **1658–1659**
  regulation of GnRH receptors, **1528**
Actl7B, **8*t***
Actrt1, **8*t***
ADAMTS-1, **431*f*, 433–435, 434*f*, 481–482, 647–648**
  enzymes, **439–440, 443**
Adenohypophysis, **1316–1329**
  development, **1319–1322, 1320*f***
  epithelial cells, **1316–1319, 1317–1318*f*, 1319*f***
  organization, **1322–1329**
    pars distalis, **1322–1327, 1322*f***
    pars intermedia, **1327–1328**
    pars tuberalis, **1328–1329**
Adenylate cyclase, **1710–1711**
  sperm capacitation and, **59–60**
Adipokines, 2730
Adrenal glands
  HPA axis and, 2629–2631
  puberty and, 2089–2090, 2179–2182
Adrenergic transmitters, female reproductive behavior and, 1842–1844
Adrenocorticotropic hormone (ACTH), **1292**
  adrenal glucocorticoids and, 2651–2653, 2652*f*
  male sexual behavior and, 1783
Adrenodoxin, **443**
Adrenomedullin (AM)
  biological fluids, 2865–2866
  expression and localization, 2865
  placenta and, 2865–2866
  receptors, 2865
African pygmy hedgehog, **130**
Age
  epididymis, **1071–1148**
  gonadotropes and, **1541–1542**
  persistent estrus in rats and, 2313
Aggression
  female reproductive behavior and, 1868–1869
  maternal, 1936–1937, 3085
    hormonal bases, 1931–1932
    motivation, 1936–1937
  parental, 2029
Aging
  chronological, 2730
  definition, 2697–2698
  gender differences, 2698
  human hypothalamic–pituitary axis, 2735–2738
    age-related changes in gonadotropin surge patterns, 2735–2738, 2737*f*
  in the hypothalamic–pituitary–ovarian axis, 2729–2755
    hypothalamic–pituitary axis and reproduction, 2732–2735
      human menstrual cycle overview, 2733–2735, 2734*f*
      hypothalamus, 2732
      pituitary gland, 2732–2733
    ovary, 2731
    overview, 2729–2730
    predictors of the final menstrual period (FMP), 2731–2732
  in the hypothalamic–pituitary–testicular axis, 2697–2728
Aging *(Continued)*
  male reproduction and, 2698–2699
    in young men, 2699–2700
  nonhuman models for reproductive aging, 2738–2745
    changes in other neurotransmitters and neuropeptides as a function of reproductive aging, 2744–2745
    inhibitory and excitatory neurotransmission in reproductive aging, 2741–2743
    noradrenergic neurotransmission and reproductive aging, 2743–2744
  sexual behavior, 1734
  somatic, 2730
  testicular endocrine system and, 2707–2715
    aging and androgen action, 2711
    hypothalamo–pituitary regulation of gonadotropin secretion, 2710–2711
    reproductive consequences of age-related decline in testosterone production, 2713–2715, 2714*f*
    somatic consequences of age-related decline in testosterone production, 2711–2713
    testosterone production and Leydig cells, 2707–2710
  testicular exocrine system and, 2700–2707
    male fertility, 2700–2703, 2700*f*, 2701*f*, 2702*f*
    Sertoli cells and hormonal regulation of spermatogenesis, 2706–2707
    spermatogenesis, 2705–2706
    sperm output, 2703–2704
    testicular blood vessels, 2704
    testis size, 2704–2705
AHR. *See* Aryl hydrocarbon receptor
AKAP3, **27*t***
AKAP4, **27*t***
AKAPs, gonadotropin signaling in the ovary and, **551–552**
Akt substrates
  follicle-stimulating hormone and, **558–566, 558*f*, 562*f*, 563*f*, 565*f***
  luteinizing hormone and insulin/IGF-1, **574–575**
Alectoris rufa, 2009
Allergic responses, relaxin and, **735**
Allomothering, 1957–1958
Allopregnanolone, 3099–3101, 3100*f*

Allostatic load, 2730
Alprostadil, **1184**
AM. *See* Adrenomedullin
AM50, **17*t***
Amenorrhea. *See also* Menstruation cycle
stress and, 2642
AMH. *See* Anti-Müllerian hormone
Amino acids
female reproductive behavior and, 1846
male sexual behavior and, 1783
milk ejection and, 3172–3173
placental transfer, 2809–2811, 2810*t*, 2811*t*
transport and metabolism in lactating mammary gland, 3009–3010
Aminoaciduria, 2911
5-Aminolevulinate synthase, **444**
Amphiregulin, **444**
Ampulla, sperm migration, **131–132**
Amygdala, male sexual behavior and, 1762–1766, 1762*f*, 1763*f*, 1765*f*
Amygdaloid efferents, effect on male sexual behavior, 1766
Analgesia, female reproductive behavior and, 1868
Androgen receptor (AR), **353, 595, 596*f*, 1558–1559**
aging and, 2711
alternate signaling pathways and, **999**
distribution in the granulosa cells, **649*f***
effect on male reproduction, **1221–1229, 1222*f*, 1225*t*, 1226*f***
effect on spermatogenesis, **1047**
endogenous and exogenous ligands, **601*t***
epididymis and, **1098–1099, 1117**
gene expression regulation and, **997–999, 998*f***
intraovarian role in ovarian function, **650–653**
nuclear receptor family and, **995–997, 996–997*f***
ovarian phenotypes in murine models of disrupted signaling, **649–650, 649*t***
ovary and, **536*f***
responses to androgen that require more than one cell type, **999**
selective ablation in specific cell types of the testes, **1000–1001, 1001*t***
signaling
in ovarian function, **648–653**
in uterine function, **619**
spermatogenesis and, **1023–1024**
regulation, **1026–1032, 1027*t*, 1029*f*, 1030*t***
stressors and, 2659
studies in the testis, **1000**
testicular steroidogenesis and, **995–1001, 996–997*f*, 998*f*, 1001*t***
Androstenedione, secretion in sheep, 2396
Anejaculation, **1187**
Anemia, **221–222**
Anesthetics, mating-induced ovulation and, 2312
Angiogenesis, **356–358**
luteal, **482–484, 484*f***
matrix remodeling and, **170–172**
menstruation cycle and, 2462, 2462*f*
placental, 2875–2876
Angiogenesis *(Continued)*
vascular growth during the menstrual cycle, **356–357**
Angiogenic growth factors, receptors and, **346–347**
Angiopoietins, corpus luteum and, **483–484, 484*f***
Angiotensin, as neuropeptide action on LH secretion, **1431*t***
Angiotensin-converting enzyme (ACE), **123–124**
Angiotensin II
effect of microinjections on male sexual behavior, 1766
male sexual behavior and, 1759
oxytocin and, 3083
Anorgasmia, **1187**
Antechinus agilis, **777, 785**
Antechinus stuarti, 2018–2019
Antibodies, **7**
Antiestrogens, effect on lordosis behavior, 1839
Antimicrobial peptides, **1110–1111**
Anti-Müllerian hormone (AMH), **389**
Antiprogestins, effect on lordosis behavior, 1839–1840
Antral follicles, **385–386**
Antrum formation, **636–637**
Anxiety, female reproductive behavior and, 1868
Apert's syndrome, 2703
Apexin, **17*t***
Aphelocoma coerulescens, 2025
Aphelocoma ultramarina, 2025
Apical cells, **1078**
Apodemus sylvaticus, **114**
Apolipoprotein-E, **444**
Apoptosis, **392**
cascade, **206**
cellular, **280–282, 281*f***
in the rat ovary, 2065
regulation, **1543–1546**
Apposition, **148**
Apteronotus albifrons, 2019
Apteronotus leptorhynchus, 2019
AP-1 transcription factor family, **994**
Aquaporins, placental transfer, 2800
AR. *See* Androgen receptor
ARC, **8*t***
Armadillos, **114**
Aromatase
androgen receptor signaling and, **648*f***
estradiol synthesis and, **637–638**
spermatogenesis and, **1040, 1041*t***
Arousal, female, 1882–1891
background and operational definition, 1883–1886
BBURP theory, 1882–1883, 1883*f*
biophysics of arousal transmitters affecting hypothalamic neurons that control sexual behavior, 1887–1891
development of the concept, 1884–1885
hormonal influences, 1885–1886
hormone-dependent genes influencing, 1886–1887, 1886*f*, 1887*f*, 1888–1889*f*, 1890*f*, 1891*f*
response potentiation, 1883, 1883*f*
Arpm2, **8*t***
Arp-T1, **8*t***
Arp-T2, **8*t***
Artificial insemination, **116–117**
in cows, **116–117**
in humans, **117**
ARX, **323**
Arx, **254**
Arylhydrocarbon nuclear translocator (ARNT), **993–994**
Aryl hydrocarbon receptor (AHR), **444**
Arylhydrocarbon receptor (AhR), **993–994**
ASP, **27*t***
Asthenozoospermia, **31**
Atelerix albiventris, **130**
Atosiban (Tractocile), 3077
Atrial natriuretic peptide, oxytocin and, 3084
Attachment, **148–149, 149*f*, 150*f***
Auditory communication, 2002
Autocrine
epidermal growth factor and, **1540**
menstrual cycle regulation and, 2465–2466
Autosomal dominant disorders, 2703
Axis axis, **775**
Axoneme, **4, 20**
Aythya valisinaria, 2009

B

Baboons
estrogen and implantation, **152**
pregnancy, **711**
timing of implantation, **149*t***
Balaenoptera acutorostrata, isolation of relaxin, **684**
Balaenoptera edeni, isolation of relaxin, **684**
Barker hypothesis, 2976. *See also* Health and disease, developmental origins
Basal cells, **1079–1080**
Basal plate, **224**
Bat, **775**
Bathygobius soporator, 2012
BBURP. *See* Bilaterally symmetric, Bipolar (bidirectional), Universal Response-Potentiating system
Bears, **114**
Bed nucleus of the stria terminalis (BST), male sexual behavior and, 1766–1767
Behavior
estrous, 2390–2391
macrophage–Leydig cell interactions and effects of testosterone on, **1220*f***
male sexual, 1734
maternal, 3088–3089
nursing, 3134
recognition of pregnancy, 2470–2475, 2472*f*, 2474*f*
ovarian cyclicity in the ewe and, 2426, 2427*f*, 2428*f*
oxytocin and control of behaviors related to reproduction, 3086–3090, 3089*f*
reproductive material, 1921–1993

Behavior *(Continued)*
- sociosexual, 1870–1871, 1872*f*
- territorial, 2014–2015

β-endorphin
- effect of estradiol on, 2405*t*
- female reproductive behavior and, 1855–1856
- as neuropeptide action on LH secretion, **1431*t***
- as regulator of the estrous cycle, 2347
- reproduction and, 2590–2591
- as steroid-responsive neuron, 2399*t*

Betacellulin (BTC), **444**
Betta splendens, 2019–2020
β-galactosidase, **17*t***
Bicarbonate, placental transfer, 2822
Bilaterally symmetric, Bipolar (bidirectional), Universal Response-Potentiating (BBURP) system, 1882–1883, 1883*f*
Binding proteins, insulin-like growth factors and, **537–539**
Biochemical reactions, during ovulation, **431–440, 431*f***
Bipotential gonad
- formation, **314**
- genes essential for development, **317–320**

Births
- age of parents and, 2700, 2700*f*
- annual rates, 2701, 2701*f*

Blarina brevicauda, 2286–2287*t*, 2292, 2295*t*
Blepharophimosis ptosis, **390**
Blood–epididymis barrier, **1080–1081**
Blood–testis barrier, immunological response, **1235–1236**
Blood vessels, testicular, 2704
BMP, **391–395**
BMP15, **391–395**
Boar, semen characteristics, **117*t***
Bombesin, as neuropeptide action on LH secretion, **1431*t***
Bone morphogenetic protein 15 (Bmp15), **393–394**
Bradykinin, as neuropeptide action on LH secretion, **1431*t***
Brain
- distribution of oxytocin receptor in the brain, 3066–3067
- mammalian, 1826–1827
- oxytocin and, 3099
- prolactin and, **1717**
- PVN interactions, 2632–2633

Brain grafts, fertility and induced ovulation and, 2313–2314
Brainstem, reflex ovulation and, 2305–2307, 2305*f*, 2306*f*, 2307*f*
Brca 1 gene, **278*t***
Breast cancer, **278*t***
- oxytocin and, 3065

Breeding strategies, 2016–2027
- temporal pattern, 2016–2019
- variation, 2023–2027

Brienomyrus brachyistius, 2019
Brood parasitism, 2026–2027
BST. *See* Bed nucleus of the stria terminalis
BTC. *See* Betacellulin
Bufo japonicus, 2012
Bulbourethral (Cowper's) glands, **1165**
Bull
- scrotum, **807*f***
- semen characteristics, **117*t***
- sperm binding, **126**
- swimming patterns of sperm, **128*f***

C

Ca^{2+}
- follicle-stimulating hormone and, **556–557**
- luteinizing hormone increase and, **574**

Cabassous unicinctus, **114**
Caenorhabditis elegans, **527–528**
Calbindins, role in implantation, **168–169**
Calcarius ornatus, 2005
Calcitonin
- prolactin and, **1713**
- role in implantation, **168**

Calcitonin gene-related peptide (CGRP)
- expression and localization, 2867
- levels in biological fluids, 2867–2868
- as neuropeptide action on LH secretion, **1431*t***
- placenta and, 2867–2868
- receptors, 2867

Calcium
- -associated signaling molecules, **168–169**
- intracellular, **1646–1647, 1646*f***
- lactation and, 3013, 3013*t*, 3015*t*
- placental transfer, 2805–2807, 2806*f*
- as regulator of milk secretory activity, 3018–3019
- sperm motility and, **33–34**

Calicin, **8*t***
Calories, as metabolic stressor that impairs reproduction, 2559
Camel bactrianus, 2286–2287*t*, 2295*t*
Camelus dromedaries, 2286–2287*t*, 2292, 2295*t*
cAMP
- gonadotropin signaling in the ovary and, **550–551, 551*f***
- model of possible signal transduction, **701*f***
- signaling in preovulatory cells, **572–573, 573*f***
- sperm motility and, **34–35**

cAMP-response element binding protein (CREB), **444**
- follicle-stimulating hormone and, **558**

cAMP-response element modulator (CREM), **444–445**
Cancer. *See also* Carcinoma
- breast, **278*t***
  - oxytocin and, 3065
- ovarian, insulin-like growth factors and, **537**

Canis familiaris, isolation of relaxin, **684**
Cannabinoids, as neurotransmitter action on LH secretion, **1430*t***
Capacitation, sperm, **56–62, 58*t*, 61*f*, 62*f*, 63*f***
- regulation of sperm–egg interaction, **62**

Capsulin. *See* POD-1
Capza3, **8*t***
Carassius auratus, 2012
Carbohydrates
- deglycosylated forms, **1611–1613**
- gonadotropin glycosylation sites, **1605–1607, 1605*f*, 1606*f***
- sperm binding, **125–127**
- structure and function, **1603–1613, 1608*f*, 1612*f***

Carbon dioxide, placental transfer, 2821–2822
Carbon monoxide (CO), as regulator of the estrous cycle, 2349
Carbonyl reductase (CBR), **445**
Carcinoma. *See also* Cancer
- of the prostate, **1161–1163**

Cardiovascular system, maternal adaptation during pregnancy, 2901–2908
- blood volume, 2901–2904, 2901*f*, 2902*f*, 2903*f*
- circulatory adaptations, 2904–2908, 2904*f*, 2905*f*, 2907*f*, 2908*f*, 2908*t*

CART. *See* Cocaine- and amphetamine-regulated transcript
CASK, **8*t***
Castration, **1500*f*, 1501*t***
- effects of hormones on male sexual behavior, 1779–1780
- effects on ex copula penile responses, 1746–1747, 1747*f*
- time course of changes in copulation following, 1744

Catecholamines, seasonal rhythms of reproduction and, 2257–2258, 2258*f*
Catechol estrogens
- role in estrous cycle, 2352–2353, 2352*f*
- role in implantation, **159**

Cathelicidins, **1111**
Cathepsin L (CatL), **445**
CatL. *See* Cathepsin L
Cats
- hypothalamus and median eminence, **1349, 1350*f***
- pregnancy, **713**
- semen characteristics, **117*t***
- sperm transport, **118**
- testes, **803*f***

Cavia porcellus, **114**
- pregnancy, **713–714**

CBR. *See* Carbonyl reductase
CCAAT/Enhancer-binding protein-β, **445**
CCAAT/Enhancer binding proteins, **993**
- epididymis and, **1106**

Ccin, **8*t***
CCK. *See* Cholecystokinin
CD63 (cell surface antigen), **445**
Cdk5.p35, **24*t***
CE11, **1110**
Cebus apella, 2029
Cell columns, **224–225**
Cell cycle regulation, **169–170**
- in cleavage stage embryos, **285*f***
- in early mitoses, **282–286**
- progression in one-cell and two-cell embryos, **282–284, 282*f***

Cell cycle regulation *(Continued)*
progression through first cell cycles, **282f**
regulators of later cell divisions, **284–286, 285f, 287t**
Cell-enclosed oocyte complex (COC), **425–426**
expansion prior to and following ovulation, **427f**
Cell islands, **224–225**
Cells
apical, **1078**
basal, **1079–1080**
clear, **1078–1079**
fractionation, 1835–1836
halo, **1080**
narrow, **1078**
preimplantation polarity and fate determination, **293–296**
Central nervous system (CNS)
insulin and, 2575
routes of entry, 2584
Central signaling molecules, reproduction and, 2585–2596
Cerebrospinal fluid, male sexual behavior and chemical changes in, 1788–1789
Ceropithecus aethiops, **836**
Cervix
anatomy, **338f**
bovine, **119f**
immune responses, **118**
parturition and, 2930–2931
patency, **5–6**
preferential sperm pathways, **119**
relaxin and, **714–717, 716f**
sperm storage, **119–120**
sperm transport, **117–120**
C-FABP. *See* Cutaneous fatty-acid binding protein
CgA. *See* Chromogranin A
CGRP. *See* Calcitonin gene-related peptide
Chemical stimulation, 2003
effects on male sexual behavior, 1770, 1770*f*
Chemokines, pregnancy and, 2761–2762, 2762*t*
Chemotaxis, **132**
sperm capacitation and, **61–62, 62f**
Chlamydia trachomatis, orchitis and, **1234**
Chloride, placental transfer, 2803–2805, 2804*f*
6-Chloro-6-deoxy-glucose, **1118**
α-Chlorohydrin, **1118**
Cholecystokinin (CCK)
male sexual behavior and, 1759
as neuropeptide action on LH secretion, **1431t**
Cholesterol, sources for steroidogenesis in Leydig cells, **978–979**
Choriocarcinoma, **1047t**
Chorionic gonadotropin, **1585f**
Chorionic plate, **222–224**
Chromatin
genomic methylation, **273–276, 274f, 274t, 275t**
remodeling and epigenetics in early embryos, **272–278**
Chromatin *(Continued)*
remodeling factors, **251**
X-chromosome dosage compensation, **276–278, 276f, 278t**
ZGA control, **272–273, 273t**
Chromis dispilus, 2019
Chromogranin A (CgA)
biological fluids, 2856
expression and localization, 2856
placental function and, 2856
Chromosomal sex, **245–248, 246f**
sex chromosome defects, **246–248, 247f**
Chromosomes, relaxin locations, **692–693, 692f**
Cigarette smoke, interference with embryo transport, **136**
CINC. *See* Cytokine-induced neutrophil chemoattractant
Circadian clocks
in control of the ovarian cycle, 2339–2340, 2339*f*
induced ovulation in rats and, 2312–2313
Citellis tridecemlinatus, 2292
Citellus tridecemlineatus, 2286–2287*t*, 2290, 2295*t*
Cithriomomys glareolus, 2295*t*
Citrate, transport in the mammary alveolar cell, 3018
CKGF. *See* Cystine knot growth factor
Clythriomomys glareolus, 2286–2287*t*, 2292
Cnemidophorus gularis, 2022
Cnemidophorus inornatus, 2022
Cnemidophorus uniparens, 2022
CNS. *See* Central nervous system
CO. *See* Carbon monoxide
COC. *See* Cell-enclosed oocyte complex
Cocaine- and amphetamine-regulated transcript (CART), 2630
as neuropeptide action on LH secretion, 1431*t*
Coccyzus americanus, 2026
Cochrane Menstrual Disorders and Subfertility Study Group, **529**
Coitus-induced ovulation, 2294–2302
Colinus virginianus texanus, 2005
Collagen, modulation, **713f**
Columba livia, 2009
Communication
auditory, 2002
chemical, 2003
electrical, 2003
modalities, 2002–2003
reproduction and, 2027–2030
tactile, 2002
vibrations (seismic stimuli), 2002
visual, 2002
Congestive heart failure, relaxin and, **736f**
Connexins
gap junction membrane channel proteins and, **401–402**
in mammary epithelium, 3030–3031, 3031*f*
role in implantation, **169**
Contractions
myometrial, **121**
uterine, **121**
Cooperative breeding, 2025–2026
Copulation. *See also* Ejaculation; Male sexual function
extra-pair, 2024–2025
intromissions, 1731
male behavior, 1730–1731, 1731*f*
mounts, 1730–1731, 1731*f*
postejaculatory behavior and satiety, 1731–1732
precopulatory behaviors, 1730
pseudocopulation, 2022
role of T metabolites in maintaining and restoring, 1745–1746
time course of changes following castration, 1744
Copulatory gels, **114–116**
Copulatory plugs, **114–116, 116f**
Corpus luteum
classification of types of corpora lutea, **478–480, 478f**
long-lived, **478**
short-lived, **478–479**
ultrashort-lived, **479–480**
domestic animals, **502–503**
formation and function, 2461–2475
autocrine and paracrine regulation, 2465–2466
cellular regulation, 2466–2468, 2466*f*, 2467*f*
development, 2461–2462, 2462*f*
endocrine secretions, 2463
luteotropic hormones, 2463–2465, 2464*f*
maternal recognition and pregnancy, 2470–2475, 2472*f*, 2474*f*
regression of the primate corpus luteum, 2468–2470, 2468*f*
historical perspective, **476–477, 476f**
luteal cell types and their regulation, **505–508**
function and regulation of luteal cells, **506–508, 508f**
small versus large luteal cells, **505–506**
luteal–placental shift, 2473–2475
luteinization, **480–490, 480f**
differentiation of steroid-secreting cells, **484–489, 487f**
luteal angiogenesis, regulation and role, **482–484, 484f**
morphological changes, **481–482**
luteolysis, **489–491**
multiple cell types, **508f**
overview, **475–476**
primates, **503–505**
in the rat, 2332–2333, 2333*f*
regulation of luteal structure–function, **491–505**
luteal function in early pregnancy, **502–505**
luteolytic factors, **496–502, 498f, 500f**
luteotropic factors, **491–496, 491f, 495f**
"rescue" during the active luteal phase of the rat, 2363–2364
structure, function, and regulation, **475–526**

Corticosterones, **989**
onset of maternal behavior and, 1927–1928
Corticotropin-releasing factor (CRF)
expression and localization, 2849–2850
female reproductive behavior and, 1855
levels in biological fluids, 2850–2851, 2850*f*
as neuropeptide action on LH secretion, **1431*t***
placenta and, 2849–2851, 2850*f*
receptors, 2850
reproductive axis and, 2593–2594
Corticotropin-releasing hormone receptor (CRHR), **445–446**
aging and, 2732
male sexual behavior and, 1759
paracrine role
in the gonads, 2655–2656
in the uterus and placenta, 2663–2665
parturition and, 2952–2953
stress and, 2644–2646, 2645*f*
pregnancy and, 2664–2665
Cortisol
aging and, 2730
gonadotropin release and, 2651, 2652*f*
onset of maternal behavior and, 1927–1928
Courtship behavior, 1871–1882, 1875*f*
of female rats, 1877–1880
preoptic neurons and, 1862–1877, 1876*f*, 1880–1881
sexual partner preferences, 1877
social recognition and affiliation, 1881–1882, 1881*f*
Cowper's (bulbourethral) glands, **1165**
Cows
artificial insemination, **116–117**
cervix, **119*f***
ejaculated sperm, **127*f***
epididymal sperm, **127*f***
gonadotropin secretion, 2522–2525
estrus and ovulation, 2522
gonadotropins, 2523–2524
induction of ovulation, 2525
luteinizing hormone, 2524
ovarian activity, 2522–2523, 2523*f*
response to estradiol and ovariectomy, 2524–2525, 2524*f*
placenta, **194*f***
reproductive tract, **120*f***
sperm, **126*f***
timing of implantation, **149*t***
uterotubal junction, **122*f*, 123*f***
uterotubal junction and caudal isthmus, **122*f***
COX-2. *See* Cyclooxygenase-2
CPα3, ***8t***
CPβ3, ***8t***
CREB. *See* cAMP-response element binding protein
CREM. *See* cAMP-response element modulator
CRF. *See* Corticotropin-releasing factor
CRH. *See* Corticotropin-releasing hormone
CRHR. *See* Corticotropin-releasing hormone receptor
Cricetulus griceus, **114**
CRISP-2 (TPX1), **17*t***
Crocidura russula, oviduct features, **130*f***
Crocidura russula monacha, 2286–2287*t*, 2295*t*
Crocidure russula, **130**
Crocuta crocuta, 2023
relaxin and, **714**
c-Ros, epididymis and, **1104–1105**
Crouzon's disorder, 2703
Cryptotis parva, 2286–2287*t*, 2292
flagellating sperm, **130*f***
Cuculus canorus, 2026
Cumulus granulosa cells, **430**
Cumulus oophorus, **65–67**
Cushing's syndrome, **1047*t*, 1050**
Cutaneous fatty-acid binding protein (C-FABP), **446**
Cyclooxygenase-2 (COX-2), **436–438, 436*f*, 437*f*, 446**
implantation and, **155–156**
transport, **1108, 1109*f***
Cyclophosphamide, epididymis and, **1119**
Cylc1, ***8t***
Cylc2, ***8t***
Cylicin I, ***8t***
Cylicin II, ***8t***
CYPT1, ***8t***
Cypt1, ***8t***
Cystine knot growth factor (CKGF), **1581–1582**
Cystine knot proteins, **1582–1583, 1582*f*, 1584–1585*f***
Cytochrome P450
aromatase, **984**
chain cleavage, **980–982**
estrogen receptors and, **1101–1102**
side chain cleavage (P450scc), **446**
Cytokine-induced neutrophil chemoattractant (CINC), **446**
Cytokines
balance, **1206*f***
effects of progesterone receptor antagonist, **346**
in the endometrium, **367–368**
inflammatory and immunoregulatory, **1216–1221, 1217*f*, 1220*f***
oxytocin and, 3102
parturition and, 2949–2951
role in implantation, **165–167**
signaling in the immune system, **1205*f***
Cytoskeletal proteins, sperm head domains and, ***8t***
Cytotrophoblast (Langhans' cells), **203–205, 204*f***

D

DA, effects on male sexual behavior, 1788
Dasyuroides byrnei, **781*f***
Dasyurus hallucatus, 2019
DAX1, **322–324**
DAX-1 (Nr0b1), **250**
Defensins, **1111**
Delayed puberty, 2212–2213. *See also* Puberty
Delta sleep-inducing peptides (DSIP), **1426–1427, 1427–1428**
Dendritic cells, pregnancy and, 2765–2766
Denys-Drash syndrome, **318**
Desert Hedgehog (DHH), **254, 323**
DFS. *See* Dysplasia of the fibrous sheath
DHH. *See* Desert Hedgehog
DHT. *See* Dihydrotestosterone
Diabetes mellitus
developmental origins, 2976–2977, 2977*f*
reproduction and, 2596–2597
Diacylglycerol, signal transduction and, **1647–1648**
Diapause, **153–156, 154*f***
embryo and, **156**
environmental, neural, and pituitary regulation of embryonic diapause, **155**
ovarian regulation, **155**
uterine factors and embryonic diapause, **155–156**
Dicerorhinus sumatrensis, **114**
Dicrostonyx groenlandicus, 2286–2287*t*, 2295*t*
Didelphis albviventris, **806**
Didelphis virginiana, **114**
Diestrus
definition, 2328
in the rat, 2329*f*
Diet. *See* Nutrition
Dihydrotestosterone (DHT)
epididymis and, **1117–1118**
male sexual function and, **1180**
in various species, **1027*t***
Dimethyl disulfide (HVD), 2045
DMDS. *See* Dimethyl disulfide
DMRT1, **321**
DNA
nuclear proteins, **12**
of the sperm nucleus, **11**
Dogs
isolation of relaxin, **684**
pituitary gland, **1310, 1310*f***
placenta, **194*f***
pregnancy, **713**
semen characteristics, **117*t***
sperm transport, **118**
uterotubal junction, **122*f***
DOHaD. *See* Health and disease, developmental origins
Dolphin, testes, **793*f***
Domestic animals
corpus luteum, **502–503**
estrogen receptor-α, **621**
estrogen receptor-β, **625**
estrogen receptor signaling in uterine function, **609**
reproduction and, 2556–2557
Dopamine, **1335**
effect of estradiol on, 2405*t*
effect on male sexual behavior, 1748–1752, 1749*f*, 1772–1774, 1772*f*, 1773*f*
inhibition, **1554–1555**
milk ejection and, 3171–3172

Dopamine *(Continued)*
as neurotransmitter action on LH secretion, **1430t**
prolactin and, **1710–1713, 1711f**
receptors, **1710**
role in estrous cycle, 2362–2363
seasonal rhythms of reproduction and, 2257
as steroid-responsive neuron, 2399t
stimulation, **1555**
Doppler ultrasound, placenta and, **230–232**
Down's syndrome, placental hormones and, 2880–2881, 2880f
Drosophila empty spiracles gene, **320, 325–326**
Drosophila melanogaster, **527–528**
Drugs. *See also* individual drug names
effects of microinjections on male sexual behavior, 1766
effects of systemically or intracerebroventricularly administered drugs, 1748–1759
effects on male sexual behavior, 1772, 1787–1788
placental transfer, 2824–2827, 2824f, 2826t
DSIP. *See* Delta sleep-inducing peptides
Duchenne's muscular dystrophy, 2703
Ductus deferens, innervation, **805, 806f**
Dyneins, **20–21**
Dynorphin, **1335–1336**
as neuropeptide action on LH secretion, **1431t**
as steroid-responsive neuron, 2399t
Dysplasia of the fibrous sheath (DFS), **31**

E

Early growth response protein-1 (Egr-1), **446–447**
E-cadherin, role in implantation, **161–162**
ECM. *See* Extracellular matrix
ECM metalloproteinase inducer (EEMPRIN), role in implantation, **162**
ECM substrates, role in implantation, **161**
Ectopic pregnancy. *See also* Pregnancies
placental hormones and, 2879
ED. *See* Erectile dysfunction
EEMPRIN (ECM metalloproteinase inducer), role in implantation, **162**
Efferent ductules, effects of estrogen exposure and insufficiency, **1042, 1042t**
EGF. *See* Epidermal growth factor
Egg–cumulus complex, **135f**
Egr-1. *See* Early growth response protein-1
Ejaculation, **1186–1187,** 1731. *See also* Copulation
anatomy, **1186**
autonomic and somatic control of, 1740–1741, 1741f
clinical application of physiological principles, **1187**
delayed, **1187**
disorders, **1187**

Ejaculation *(Continued)*
male sexual behavior and, 1791–1793, 1792f
neural control, **1186–1187, 1187f**
postejaculatory behavior and satiety, 1731–1732
retrograde, **1187**
Electrical stimulation, 2003
effects on male sexual behavior, 1770, 1786
Electron microscopy
in evaluation of acrosome, **15**
immunolabeling, **1546, 1546f, 1547f**
Electrophysiological recordings
female reproductive behavior and, 1841–1842
male sexual behavior and, 1775
Elephants, relaxin and, **714**
Elephantus myurus, **783–784**
Elephas maximas, relaxin and, **714**
Embryo
diapause and, **156**
ethics and, **296–297**
immune protection for pre-implantation, 2763–2764, 2763f
knockout models postulated to compromise genomic integrity, **287t**
model of events occurring during early stages of embryonic implantation, **364f**
molecular aspects of uterine interactions, **157–173, 157f**
morphological and metabolic changes in early cleavage, **288–293**
physiological and morphological aspects of embryo–uterine interactions, **148–153**
polarity in mammalian, **293–294, 294f**
remodeling and epigenetics in early, **272–278**
Embryogenesis
cellular apoptosis, **280–282, 281f**
destruction, **278–282**
preimplantation, **261–310**
RNA degradation, **278–279**
ubiquitination of proteins and organelles, **279–280, 279f, 280t**
Embryology, of the oviduct and endometrium, **338–340**
EMX2, **320, 329**
Endocannabinoids, **172**
Endocrine system
immune system and, **1203–1209, 1203t, 1204t, 1205f, 1207f**
menstrual cycle and, 2463
oxytocin and, 3056–3065, 3057f, 3058f, 3063f
parturition and, 2934–2942
regulation and immunophysiology, **1208–1209**
regulation of spermatogenesis, **1017–1069**
spermatogenesis
cell–cell adhesion systems involved in, **1022**

Endocrine system *(Continued)*
spermatogenesis
germ cell development, **1019–1021, 1020f**
spermatogenic cycle organization, **1021–1022**
Endocytosis, male sexual behavior and, 1779
Endometrial glands, secretions, **153**
Endometrial stromal cells, phases of decidualization, **365f**
Endometrial vasculature, **356**
Endometrium, **13–33**
anatomy, **338–340, 338f**
cell constituents, **338–340**
cyclic changes, **337–381**
in primates, **341–343, 342f**
cytokines in, **367–368**
embryology, **338–340**
estrogen and progesterone receptors, **343–345**
growth factor and cytokine expression and action, **345–347**
historical perspective and functions, **340–341**
immune cells, **366–367**
immune environment, **366–368**
molecular mechanisms underlying endometrial function, **354–369, 359f, 360t–361t, 362t–363t, 364f, 365f**
phases of menstrual cycle and endometrial histology, **349–352, 350f, 351f**
regulation of endometrial vessels, **357–358**
secretions and oviductin, **347–349**
steroid hormone receptors, **352–354, 352f**
tissue desquamation and menstruation, **368**
tissue regeneration, **368–369**
ultrastructural characteristics of secretory, **351t**
Endothelin (ET)
biological fluids, 2867
expression and localization, 2866
placenta and, 2866–2867
receptors, 2866–2867
Endothelin-1 (ET-1), **499–500, 500f**
expression, **1530**
as neuropeptide action on LH secretion, **1431t**
prolactin and, **1713**
Endozepines, **988**
Energy, reproduction and, 2553–2558
Enkephalin, as neuropeptide action on LH secretion, **1431t**
Entactin, **160**
Environment
classification of factors regulating reproduction, 2001–2002, 2001f
control of ovarian cyclicity in the ewe and, 2421–2426
gene–environmental interaction, 2980f
influence on reproduction, 2554
puberty and, 2092–2093
seasonal rhythms of reproduction, 2233–2234

Eosinophils, immunophysiology, **1214**
EPI. *See* Epiregulin
Epicanthus inversus syndrome, **390**
Epicardin. *See* POD-1
Epidermal growth factor (EGF), **355**
biological fluid, 2863
expression and localization, 2862
implantation and, **155**
as a paracrine or autocrine regulator, **1540**
placenta and, 2862–2863
receptors and, **345–346,** 2862–2863
as regulator of the estrous cycle, 2349–2350
Epididymal retinoic acid–binding protein (E-RABP), **1111–1112**
Epididymis, **1071–1148**
aging, **1114–1117**
effects on appearance of the epididymal epithelium, **1114–1115**
mechanisms underlying, **1115–1116**
molecular and functional effects, **1115**
spermatozoa changes during, **1116–1117**
anatomy, **782–785, 1076, 1076*f***
development, **1073–1076**
formation of the mesonephric/Wolffian/nephric duct and tubules, **1073–1074**
postnatal, **1074–1076, 1075*f***
historical perspective, **1072–1073**
immune cells, **1214–1215**
immunological response, **1244–1247**
innervation, **805**
luminal compartment functions, **1081–1092**
spermatozoa
maturation, **1083–1087, 1084*f***
microenvironment for maturation, protection, and storage, **1089–1092, 1090*f***
protection of, **1088–1089**
storage, **1087–1088**
transport, **1081–1083, 1082*f***
overview, **1071–1072**
perspective and future directions, **1120**
protein families and their regulation, **1095–1114**
functional families, **1098–1114**
antimicrobial peptides, **1110–1111**
lipocalins, **1111–1114**
receptors and their ligands, **1098–1105**
transcription factors, **1106–1107**
transporters and channels, **1107–1110, 1109*f***
general principles, **1095–1097, 1096*f*, 1097*f***
regulation of epididymal functions, **1092–1095**
hormones, **1092–1093**
testicular factors, **1093–1095, 1094*f***
structural organization, **1076–1081**
anatomy, **782–785, 1076, 1076*f***
blood–epididymis barrier, **1080–1081**
Epididymis *(Continued)*
structural organization
cell types and specific markers, **1076–1080, 1077*f*, 1078*f*, 1079*f***
as target for xenobiotics, **1117–1119**
cyclophosphamide, **1119**
fungicides, **1119**
hormonal regulation, **1117–1118**
metabolism, **1118–1119**
methyl chloride, **1119**
neuronal regulation, **1118**
sulfonates, **1119**
Epinephrine, as neurotransmitter action on LH secretion, **1430*t***
Epiregulin (EPI), **447**
Epithelial–stromal interactions, **354**
Epithelium, sperm trapped by binding, **125**
Eptesicus fuscus, 1744
Equus caballus, **114**
isolation of relaxin, **683–684**
pregnancy, **711–712**
ER-α
maintenance of progesterone actions in absence of, **616**
maintenance of selective estrogen actions in absence of, **615–616**
E-RABP. *See* Epididymal retinoic acid–binding protein
Erectile dysfunction (ED)
PDE5 inhibitors for treatment of, **1182–1184**
targets for development of therapeutic agents for men with, **1185–1186**
Erinaceus europeaus, 2286–2287*t*
ERKs, gonadotropin signaling in the ovary and, **554–556, 554*f*, 555*f*, 556*f***
E. coli, **697, 1111**
orchitis and, **1234**
Essential fatty acids, puberty and, 2090–2091
Estradiol, **135–136, 494–495**
aging and, 2734*f*
aromatase and synthesis, **637–638**
biphasic uterine response to, **612–614, 612*f*, 612*t*, 613*f*, 614*f***
control of secretion, 2410–2411
development of response to, 2130*f*
effect of concentration on secretion in ewes, 2401*f*
effects of different concentrations, 2397*f*
estrous cycle
of the rat and, 2334*f*
of sheep and, 2410–2411
intraovarian roles in ovarian function, **633–634**
mechanisms of action, 2404–2406, 2405*f*, 2405*t*
-mediated negative feedback on thecal cell function, **640–641**
mice lacking synthesis, **632–633**
negative feedback action, 2402–2404, 2403*f*
ovariectomy in cows and, 2524–2525, 2524*f*
ovulation and, **641–642**
puberty and, 2134–2138, 2135*f*, 2136*f*, 2137*f*, 2138*f*
Estradiol *(Continued)*
seasonal rhythms of reproduction and, 2257
secretion in sheep, 2396
sites of action, 2483–2484
synergistic action of follicle-stimulating hormone and, **639*f***
Estradiol benzoate, 1866*f*
Estrogen
in the adult, **1043–1044**
amplification by progesterone, 1868
antiprogestin effects and cellular function, **345**
binding studies, **343**
biphasic response of rodent uterus to, **612*t***
correlation of receptors with lordosis behavior, 1836–1837
development of positive-feedback, 2073*f*
disrupted signaling in humans, **643–644**
effect on GnRH receptor expression, **1523*f***
effects of exposure and insufficiency, **1042, 1042*t***
epididymis and, **1118**
female reproductive behavior and, 1829–1830
folliculogenesis and, **399–401**
growth factor cross-talk in the uterus, **614–615, 615*f***
in human implantation, **152–153**
implant effect on lordosis behavior, 1838
importance in efferent ductule development and function, **1044–1045**
importance in implantation, **151–153**
influence on arousal, 1885
interactions within the GnRH network, **1446–1452, 1446*f*, 1450*f*, 1452*f***
estrogen-negative feedback, **1450–1452**
estrogen-positive feedback, **1448–1450**
modes of action, **1447–1448**
lactotrope response, **1551–1552**
as a luteolysin, 2469
maintenance of selective actions in the absence of ER-α, **615–616**
onset of maternal behavior and, 1925–1927
ovarian phenotypes in murine models of disrupted signaling, **627–633, 627*f*, 628*t*, 631*f***
oxytocin gene expression, 3101
oxytocin neurons and, 3101–3102
oxytocin secretion, 3101
parturition and, 2941–2942
as physiological regulator of gonadotropin biosynthesis and secretion, **1660–1661**
placenta and, 2864–2865
production in the testis, **1040**
profile, **1715*f***
progesterone receptors and, **343–345**
prolactin and, **1713–1714**

Estrogen *(Continued)*
receptor isoforms, **344–345**
receptors, **1557–1558**
receptor signaling in uterine function, **608–616**
regulation and the estrous cycle, **1536–1538, 1537*f***
regulation during the neonatal period, **1041–1043**
regulation of prolactin synthesis, **656–657**
role in estrous cycle, 2352–2353, 2352*f*
stressors and, 2657–2659
Estrogen receptor (ER), **352–353, 352*f*, 595–597, 597*f***
cytochrome P450 aromatase and, **1101–1102**
endogenous and exogenous ligands, **601*t***
receptor-α, **620–622**
receptor-β, **622–626, 624*f***
signaling in ovarian function, **619–644**
spermatogenesis and, **1024–1025**
Estrogen signaling pathways, **399–401**
Estrous cycle
aging in rats and, 2313
estrogen regulation and, **1536–1538, 1537*f***
gonadotropes and lactotropes, **1503–1508, 1503*f*, 1504*f*, 1506*f*, 1508*f***
gonadotrophin and, 2338–2339, 2338*f*
mating-induced ovulation and, 2312
neural events during, in rats, 2339–2350
biogenic amine neurotransmitters controlling the estrous cycle, 2345–2347
circadian control of the ovarian cycle, 2339–2340, 2339*f*
gases as possible regulators of the estrous cycle, 2348–2349
growth factors as possible regulators of the estrous cycle, 2349–2350
peptidergic neurohormones controlling the estrous cycle, 2341–2345, 2342*f*, 2343*f*
peptide transmitters as possible regulators of the estrous cycle, 2347–2348
role of the hypothalamus, 2340–2341, 2341*f*
ovine, 2390–2397
control of ovarian function, 2408–2418
estradiol secretion control, 2410–2411
follicular development control, 2408–2410, 2409*f*
inhibin secretion control, 2411–2412
luteal progesterone secretion control, 2412–2416, 2413*f*
ovulation rate control, 2410
environmental control of ovarian cyclicity in the ewe, 2421–2426
effects of nutrition, 2425–2426

Estrous cycle *(Continued)*
environmental control of ovarian cyclicity in the ewe
effects of pheromones and social interactions, 2426, 2427*f*, 2428*f*
effects of stressors, 2423–2425, 2424*f*
photoperiodic control of seasonal breeding, 2421–2423, 2422*f*, 2423*f*
estrous behavior, 2390–2391
hormonal control, 2397–2398, 2397*f*
hypothalamo–hypophyseal function control, 2398–2408, 2398*f*, 2399*t*, 2401*f*, 2403*f*, 2405*f*, 2405*t*, 2408*f*
hypothalamo–hypophyseal unit, 2391, 2392*f*
neural control, 2398
ovary, 2391–2396, 2394*f*
resynthesis model, 2418–2421, 2419*f*
uterine prostaglandins, 2396–2397
pattern of follicle-stimulating hormone secretion throughout, 2337–2338
pattern of luteinizing hormone secretion throughout, 2336–2337, 2337*f*, 2338*f*
pattern of prolactin secretion throughout, 2338
patterns of pituitary hormone secretion during the estrous cycle, 2335–2339, 2337*f*, 2338*f*
as physiological regulator of gonadotropin biosynthesis and secretion, **1655–1656, 1656*f***
prolactin and, **1715, 1715*f***
in rats, 2327–2388
secretion of ovarian steroids during, 2333–2335, 2334*f*
stressors and, 2640
Estrus
definition, 2328
in the rat, 2329*f*
ET. *See* Endothelin
ET-1. *See* Endothelin-1
Ethics, embryo and, **296–297**
Ethnicity, male hormonal contraceptives and, **1052–1053**
Eusocial systems, 2026
Eutherians
accessory sex glands, **786–788, 789*f***
testes, **771–772**
anatomy, **783–785**
size variation, **774–777**
Ewes
distribution according to genotype, 2265, 2265*f*
plasma GnRH concentrations, **1440*f***
seasonal rhythms of reproduction, 2235–2236, 2235*f*
Excitatory amino acids
female reproductive behavior and, 1848–1850
seasonal rhythms of reproduction and, 2259–2260
Excitatory neurons, 2081–2082
Excurrent ducts
anatomy, **781–785**
blood flow, **801**

Excurrent ducts *(Continued)*
blood vessels, **800–801**
efferent ducts, **781–782, 785*f***
immunophysiology, **1201**
lymphatic vessels, **801**
vasculature, **800–801**
Exercise
as cause of hypogonadotropic hypogonadism, **1047*t***
circulating factors and, 2902–2903, 2902*f*, 2903*f*
as metabolic stressor that impairs reproduction, 2562–2564, 2564*f*
Extracellular matrix (ECM), **481–482**
Extra-pair copulation, 2024–2025
Extravillous trophoblast, **225**

F

Fabp9, **8*t***
Factor in the germline (Figla), **387**
Fallopian tube, anatomy, **338*f***
Fam3c, 3003
Farm animals
growth hormone and, **528–529**
insulin-like growth factors and, **533–534, 533*f***
Fas ligand, **1238**
Felis catus, **118**
Felis catus, 2295*t*
Felis catus, 2286–2287*t*
Fertility. *See also* Male reproductive system
brain grafts and induced ovulation in mouse mutants, 2313–2314
growth hormone and, **529**
inflammation and seminal leukocytes, **1247**
male aging and, 2700–2703, 2700*f*, 2701*f*, 2702*f*
maturation of spermatozoa and, **1083–1086, 1084*f***
Fertilization, in mammals, **55–112**
acquisition of fertilization competence by gametes, **56–65**
pre-ovulatory acquisition of egg's ability to undergo activation, **56**
sperm capacitation, **56–62, 58*t*, 61*f*, 62*f*, 63*f***
applications of fertilization biology, **93**
overview, **55–56**
sperm–egg interaction, **65–93**
early events preceding gamete membrane fusion, **65–79, 65*f*, 68*f*, 69*f*, 71*f*, 73*f*, 74*t*, 76*t***
events following sperm–egg membrane fusion, **83–93, 84*f*, 84*t*, 85*f*, 86*f*, 87*f*, 88*f*, 88*t*, 91*f***
incorporation of fertilizing sperm into the egg and fate of sperm components, **82–83, 83*f***
plasma membrane interactions, **79–82, 81*f*, 82*f***
Fertilization biology, applications, **93**
Fetal development
female, 2063–2064
male, 2084–2089, 2086*f*

Fetal growth restriction (FGR), placental hormones and, 2882–2883
Fetal vessels, **209–212, 210*f*, 211*f***
Fetus
maternal tolerance, 2760, 2760*f*
mechanisms preventing maternal immune rejection of, 2772–2779, 2774*f*, 2775*f*, 2776*f*, 2778*f*
paracrine/autocrine actions on fetal–placental blood flow regulation, 2876–2879
parturition and, 2932–2933
Fetus papyraceous, **199**
FGFs. *See* Fibroblast growth factors
FGR. *See* Fetal growth restriction
Fibrinoid, **226–227**
Fibroblast growth factors (FGFs), **322, 355**
biological fluid, 2863
epididymis and, **1074**
expression and localization, 2862
placenta and, 2862–2863
receptors, 2862–2863
Fibronectin, **160**
Fibrosis, relaxin and, **731–734**
Fibrous sheath, **4, 26–30**
composition, **26–29**
formation, **26**
function, **29–30**
structure, **26**
Figla. *See* Factor in the germline
Final menstrual period (FMP), predictors, 2731–2732
Finasteride, **1180–1181**
Fish, comparison of second alpha-subunit proteins, **1604*f***
Fixed connective tissue, **209**
Flagellar dyskinesia, **31**
Flagellar motion, **32–37**
activation, **33**
calcium and motility, **33–34**
cAMP and motility, **34–35**
flagellar wave, **32–33**
phosphorylation and motility, **35–37**
PKA and motility, **35**
regulation of sperm motility, **33**
Flagellum, **4, 18–37**
abnormal, **30–32**
axoneme, **20**
connecting piece, **19–20, 19*f***
cytoskeletal components, **20*f***
fibrous sheath, **26–32**
flagellar motion, **32–37**
mitochondrial sheath, **21–22**
outer dense fibers, **22–26**
Flehmen, 2042–2043
FLIP. *See* Fluorescence loss in photobleaching
Fluorescence loss in photobleaching (FLIP), **10**
Fluorescence recovery after photobleaching (FRAP), **10**
Fluorogold, effect on male sexual behavior, 1767, 1767*f*
FMP. *See* Final menstrual period
Follicle
estrous cycle of sheep and, 2408–2410, 2409*f*
maturation, **547–548**
Follicle atresia, **653**
Follicle-stimulating hormone (FSH), **397, 1325**
in the adult, **1035–1037**
Akt and, **558–566, 558*f*, 562*f*, 563*f*, 565*f***
beta-subunit gene expression, **1669–1672**
Ca^{2+} and, **556–557**
control of secretion, 2407–2408, 2408*f*
during the estrous cycle of the rat, 2357–2360, 2358*f*
CREB and, **558**
effect on spermatogenesis, **1047–1048**
electron microscopy, **1493*f***
FOXO1 and, **560–561**
gonadotropins and, **1603, 1604*f***
gonadotropin secretion and, 2514
gonadotropin signaling, **548–550, 549*f***
GSK3β and, **558*f*, 561–562**
menstruation cycle and, 2465
mitochondrial-dependent apoptotic pathway, **563*f***
pattern of plasma and estrogen concentrations, 2457–2458, 2457*f*
pattern of secretion throughout the estrous cycle, 2337–2338
PI-3K and, **557–558, 558*f***
preovulatory follicular development, 2455–2456
receptors, **1619**
regulation of gene targets, **566*f***
Sertoli cells and, **956–957**
SGK and, **562–563**
spermatogenesis and, **1025, 1032–1039, 1033*t*, 1037*f***
synergistic action of estradiol, **639*f***
synergistic action of sex steroids, **638*f***
threshold model for selection of preovulatory follicle, 2460, 2460*f*
tuberin and, **558–560**
Follicular apex
at the beginning of the ovulatory process, **427–430, 428*f*, 429*f*, 430*f***
at less than five minutes before rupture, **430–431, 431*f***
at less than one hour before rupture, **430, 430*f*, 431*f***
Follicular development, 2451–2461
characteristics of development, 2451–2453, 2451*f*, 2452*f*
follicular estrogen biosynthesis, 2453–2455
gonadotropic control of preovulatory, 2455–2456
preovulatory follicle selection, 2456–2461, 2457*f*, 2459*f*, 2460*f*
Folliculogenesis, **383–423**
antral follicle development, **395–402**
estrogen signaling pathways and, **399–401**
gap junction membrane channel proteins and connexins, **401–402**
Folliculogenesis *(Continued)*
antral follicle development
hypothalamus and pituitary roles, **395–399**
classification of mouse follicles, **385*f***
factor in the germline (Figla), **387**
female reproductive phenotypes in genetic mouse models, **406*t***
GDF9, BMP15, and BMP receptors in preantral follicle and granulosa cell physiology, **391–395**
BMP receptor type IB receptor (Bmpr 1b, Alk6), **394–395**
bone morphogenetic protein 15 (Bmp15), **393–394**
growth differentiation factor 9, **392–393**
human mutations causing reproductive axis defects in females, **408*t***
metabolism and, **402–404**
growth hormone, **386–387**
insulin-like growth factor signaling system, **404**
leptin, **402–403**
morphology, **384–387**
human, **386–387**
murine, **384–386**
in the mouse, **386*f***
mouse models, **404–410**
activins, **408–409**
follistatin, **409–410**
inhibins, **404–408**
oocyte–granulosa cell interactions, **391**
ovarian histology for mouse knockout models, **388*f***
overview, **383–384**
primordial follicle formation, **387**
anti-Müllerian hormone, **389**
forkhead box L2 (Foxl2), **390–391**
kit ligand (Kitl) and kit receptor (Kit), **389**
nerve growth factor and receptors, **389–390**
newborn ovary homeobox gene (Nobox), **388**
primordial to primary follicle transition, **387–391**
in the rat, 2330–2332, 2331*f*
sheep mutations causing ovarian defects, **394*f***
transcription factors in preantral stage, **395**
Foxo3a, **395**
$TAF_{II}105$, **395**
Follistatin, **409–410**
expression, **1529–1530**
as physiological regulator of gonadotropin biosynthesis and secretion, **1659**
Follitropin, **1598–1599**
Food
availability and influence on reproduction, 2554–2555
puberty and, 2211
Forkhead box L2 (Foxl2), Kit ligand (Kitl) and kit receptor (Kit), **389**
FOS-like antigen-2 (Fra-2), **447**

FOXO1, follicle-stimulating hormone and, **560–561**
Foxo3a, **395**
Fra-2. *See* FOS-like antigen-2
FRAP. *See* Fluorescence recovery after photobleaching
Frasier syndrome, **318**
Free fatty acids, placental transfer, 2815–2816
Freeze-etch. *See* Freeze-fracture
Freeze-fracture, **6**
Frizzled G-protein–coupled receptors (Fzs), **447**
Frizzled-related protein-4 (sFRPs), **447, 448*f***
FS39, **27*t***
Fscn3, **8*t***
FSH. *See* Follicle-stimulating hormone
FSH receptor (FSHR), **398**
FSIP1, **27*t***
FSIP2, **27*t***
Funambulus pennanti, 2286–2287*t*, 2295*t*
"Functional hypothalamic chronic anovulation" syndrome, 2636
  stressors and, 2641
Fungicides, epididymis and, **1119**
Fzs. *See* Frizzled G-protein–coupled receptors

G

GABA. *See* Gamma amino butyric acid
GABAergic neurons, 2082
Galactopoiesis. *See* Lactation
Galago crasicaudatus, cloning of relaxin, **686**
Galanin, **1427**
  male sexual behavior and, 1758
  as neuropeptide action on LH secretion, **1431*t***
Galanin-like peptide (GALP)
  as neuropeptide action on LH secretion, **1431*t***
  reproductive axis and, 2591–2593, 2591*f*, 2592*f*, 2593*f*
GALP. *See* Galanin-like peptide
Gamete transport, **113–145**
  acquisition of fertilization competence, **56–65**
  early events preceding membrane fusion, **65–79, 65*f*, 68*f*, 69*f*, 71*f*, 73*f*, 74*t*, 76*t***
  semen disposition site, **114–117**
  sperm competition and cooperation, **113–114**
  sperm transport through the cervix, **117–120**
    cervical patency, **117–118**
    immune responses in the cervix, **118**
    immunological responses in the uterus, **121–122**
    mucous barrier, **118**
    preferential pathways through the cervix, **119**
    rate of transport, **120**
    sperm storage in the cervix, **119–120**
    uterine peristalsis, **120–121**
Gamete transport *(Continued)*
  sperm transport through the oviduct, **125–132**
    chemotaxis, **132**
    nonfertilizing sperm in the oviduct, **132**
    oviductal sperm reservoir, **125–131**
    sperm migration into the ampulla, **131–132**
  sperm transport through the uterotubal junction, **122–125**
    junction as barrier to sperm migration, **122–123**
    junction as filter, **123–124**
    rapid transit through the junction, **124–125**
  sperm transport through the uterus, **120–122**
  timing of sperm deposition with ovulation, **114**
  vaginal defenses against sperm, **117**
    leukocytic response in the vagina, **117**
    vaginal pH, **117**
  vaginal semen deposition
    artificial insemination, **116–117**
    copulatory plugs and gels, **114–116**
    uterine semen deposition, **116**
    vaginal semen deposition, **114**
  in vitro differentiation, **294*f***
Gamma amino butyric acid (GABA), **1335**
  effect on male sexual behavior, 1755, 1774, 1788
  estrogen interactions, **1449**
  female reproductive behavior and, 1846–1848
  inhibitory influence on GnRH neuronal activity, **1433*f***
  as neurotransmitter action on LH secretion, **1430*t***
  puberty and, 2069
  as steroid-responsive neuron, 2399*t*
GAP. *See* GnRH-associated peptide
GAPD-S, **27*t***
Gases, as possible regulators of the estrous cycle, 2348–2349
GATA4, **321**
GATA4/FOG2, **251**
GDF9. *See* Growth differentiation factor 9
Gender
  differences in aging, 2698
  dimorphism, 2657
  effects on immune function, **1209**
Gene expression
  effect on female reproductive behaviors, 1829–1832
  organization of oxytocin gene, 3069*f*
  regulation of oxytocin in reproduction, 3073–3074
  Sertoli cells, **951–956, 954*t***
Gene knockouts, affecting mouse sperm structure and function, **32**
Gene products, **264–268, 265*f***
Genes, **360–361*t*, 362–363*t***
  catalog of current, **443–455**
  categorization
    by nature/function of protein product, **457–458**
Genes *(Continued)*
  categorization
    by relevance to functional processes/pathways, **458**
    by spatial distribution of expression, **457**
    by temporal pattern of expression, **456–457**
  cloning of relaxin, **684–687, 685*f*, 686*f***
  disruption technologies, **577**
  essential for development, **317–320**
  essential for development of bipotential gonad, **317–320**
  essential for testes development, **317*t*, 320–325**
  expression in mid-secretory phase, **358–364, 359*f***
  implicated in sex determination and associated defects, **249–251, 249*f***
  maternal effect, **265–266, 265*f*, 267**
  ovarian expression during ovulation, **440–458**
  ovulation-specific, **441–442*t***
  relaxin structure, **687**
  RLN2 schematic representation, **688*f***
Genetics
  control of placental development, **228–230**
  control of seasonality and, 2264–2266
    evidence for genetic control, 2264, 2264*f*
    human reproductive seasonality, 2266
    individual genetic variation and photoneuroendocrine pathways, 2264–2266, 2265*f*
  puberty genes and, 2208–2209
Genitals, somatosensory input from, 1791
Genomic methylation, **273–276, 274*f*, 274*t*, 275*t***
Germ cells, **254**
  chromosomal events, **326**
  degeneration, **899–900**
  development, **1028**
    and spermatogenesis, **1034–1035**
  effects of estrogen exposure and insufficiency, **1042–1043, 1042*t***
  effects of gonadotropin on cell types in humans, **1048, 1049*f***
  migration, **326, 878*f***
  origin, **325–326**
  primordial, **325–326**
  Sertoli cells and, **831*f*, 958–959**
  spermatogenesis and, **1019–1021, 1020*f***
  testis transplantation and, **1243–1244**
  transplantation, **960–961**
GermOnline database, **384**
Gestational sac, **203*f***
γ-GCS. *See* γ-Glutamylcysteine synthetase
γ-Glutamylcysteine synthetase (γ-GCS), **447**
GH. *See* Growth hormone
Ghrelin
  biological fluids, 2870
  expression and localization, 2869–2870
  as neuropeptide action on LH secretion, **1431*t***
  placenta and, 2869–2870
  receptors, 2870

GHRH. *See* Growth hormone–releasing hormone
Glaucomys volans, **114**
Glial cells
glia-to-neuron communication, 2069, 2070*f*
GnRH network and, **1439–1440**
puberty and, 2082–2083
Glucocorticoid receptor (GR)
lactation and, 3034–3035
signaling in uterine function, **619**
Glucocorticoids, 2953
lactation and, 3034–3035
regulation of immunity, **1208–1209**
response to stressors, 2657–2659
Glucocorticoid secretion, basal, 2657
Glucose
placental transfer, 2812–2814
relevance of maternal hyperglycemia and hyperinsulinemia, **291–292**
Glucose related protein 78 (GRP 78), **988**
Glutamate, **1430, 1432**
effect on male sexual behavior, 1775
estrogen interactions, **1449**
–GABA interactions and puberty, 2095–2100, 2096*f*, 2099*f*
as neurotransmitter action on LH secretion, **1430*t***
as steroid-responsive neuron, 2399*t*
Glutamatergic neurons, 2081
puberty and, 2068
Glutamate transporters, **1110**
Glutathione S-transferase (GST), **447–448**
Glycine, female reproductive behavior and, 1848
Glycoprotein hormone
alpha-subunit gene expression, **1661–1666, 1662*f***
biosynthetic pathway, **1623*f***
sequence accession numbers, **1615–1617*t***
subunit sequences accession numbers, **1587–1590*t***
Glycosuria, 2911
GnRH. *See* Gonadotropin-releasing hormone
GnRH-associated peptide (GAP), **1428**
Goats, milk ducts, 3130*f*
Golgi complex, **1316**
GnRH receptor expression and, **1533–1534**
production of monohormonal granules and, **1506*f***
Gonadal sex, **248–254**
genes implicated in sex determination and associated defects, **249–251, 249*f***
sex determination, **248–249, 248*f***
testis development, **251–254, 252*f*, 253*f***
Gonadal steroids
modulation of HPA and HPG axes, 2656–2660
oxytocin regulation and, 3067
puberty and, 2100
Gonadectomy, 3074
as physiological regulator of gonadotropin biosynthesis and secretion, **1659–1660**
Gonadotropes, **1483–1579**
autocrine and paracrine influences, **1539–1540**
activin, inhibin, and follistatin and nonparallel release, **1539–1540**
epidermal growth factor as regulator, **1540**
cell type development, **1488–1490**
discovery and historical perspective, **1484**
diversity, **1513–1530**
division of labor, **1514–1519, 1515*f*, 1517*f*, 1518*f***
as a multihormonal cell, **1519–1527, 1520*f*, 1521*f*, 1522*f*, 1523*f*, 1524*f*, 1526*f***
as a regulatory cell, **1528–1530, 1529*f***
roles for monohormonal gonadotropes, **1527–1528**
tumorigenesis, **1513–1514**
early identification, **1484–1485**
early studies, **1492–1497, 1493*f*, 1494*f*, 1495*f*, 1496*f***
effect of age and reproductive hormones, **1541–1542**
effects of growth factors, **1542–1543, 1543*f*, 1544*f***
embryological origin, **1486–1492**
GnRH receptor expression, **1530–1539**
detection of ligand–receptor binding, **1530–1531, 1531*f***
early internalization events, **1532–1533**
Golgi complex, granule, and lysosomal domains, **1533–1534**
identification of hormone stores in GnRH target cells, **1534–1535, 1535*f*, 1536*f***
membrane binding and microaggregation, **1531–1532, 1532*f***
pituitary GnRH versus the GnRH–receptor complex, **1534, 1534*f***
regulation of GnRH receptors, **1535–1539, 1537*f***
immunolabeling, **1485–1486**
integrative studies of lactotropes and, **1556–1562**
communication between, **1559–1560**
detecting receptor populations, **1556–1559**
how past lessons affect future experiments, **1560–1562**
regulating transition from birth to adult, **1556**
monohormonal and multihormonal cells, **1492**
origin and formation of anterior pituitary, **1486–1488, 1487*f***
proliferation of differentiated lactotropes and, **1540–1541**
regulation of apoptosis, **1543–1546**
Gonadotropes *(Continued)*
responses to different physiological states, **1497–1513**
after castration, **1501*t***
changing levels of gonadal steroids, **1499–1503, 1500*f*, 1501*t*, 1502*f*, 1503*f***
during the estrous cycle, **1503–1508, 1503*f*, 1504*f*, 1506*f*, 1508*f***
maturation of the gonadotrope, **1497–1499, 1498*f*, 1499*f***
role of matrix and gap junction proteins, **1512–1513**
in vitro studies, **1508–1511, 1509*f*, 1510*f*, 1511*f*, 1512*f***
spatial organization of transcription factors, **1490–1492**
Gonadotropic hormone (GTH), **1595**
estrous cycle and, 2338–2339, 2338*f*
follicular estrogen biosynthesis and, 2453
Gonadotropin-releasing hormone (GnRH), **396–397**
afferent inputs to GnRH neurons, 2585
biological fluids, 2849
brain grafts and induced ovulation in mouse mutants, 2313–2314
coitus-induced ovulation and, 2296–2297
competency of the hypothalamus before puberty, 2133–2134
effect on germ cell types in humans, **1048, 1049*f***
effect on spermatogenesis, **1047**
expression and localization, 2848
female reproductive behavior and, 1868
GnRH neuron, **1417–1429**
coexpressed neurotransmitters, **1426–1428**
electrical properties, **1421–1422, 1422*f***
GnRH biosynthesis, **1422–1426, 1423*f***
neuroanatomy, **1417–1421, 1418*f*, 1420*f***
GnRH-1 neuronal network, **1429–1456**
components, **1429–1440, 1430*t*, 1431*t*, 1433*f*, 1435*f*, 1438*f***
estrous cycle plasticity, **1446–1455, 1446*f*, 1447*f*, 1452*f*, 1454*f***
heterogeneity with the neuronal phenotype, **1455–1456**
synchronization and pulsatility, **1440–1446, 1440*f*, 1442*f***
GnRH receptor gene expression, **1672–1677**
GnRH regulation of gonadotropin gene expression, **1661–1677**
follicle-stimulating hormone beta-subunit gene expression, **1669–1672**
glycoprotein alpha-subunit gene expression, **1661–1666, 1662*f***
luteinizing hormone beta-subunit gene expression, **1666–1669**
inhibitory influence of GABA, **1433*f***
male sexual behavior and, 1757
melatonin regulation in breeding seasonality, 2256–2260, 2256*f*, 2258*f*

Gonadotropin-releasing hormone (GnRH) *(Continued)*
multiple GnRH molecules for neuroendocrine control of reproduction, **1416–1417**
GnRH-II, **1416–1417**
GnRH-III, **1417**
negative feedback control of secretion, **654–655**
neurobiological bases of juvenile brake on pulsatile release, 2194–2201, 2194*f*, 2195*f*, 2196*f*, 2197*f*, 2198*f*, 2199*f*, 2200*f*
neuronal network, **1415–1482**
neuron synchronization models, **1442*f***
as neuropeptide action on LH secretion, **1431*t***
as neuropeptide transmitter, **1428–1429**
peptide degradation, **1426**
physiological regulators of gonadotropin biosynthesis and secretion, **1653–1661**
activin, **1658–1659**
estrogen and progesterone, **1660–1661**
estrous cycle, **1655–1656, 1656*f***
follistatin, **1659**
GnRH pulsatility, **1653–1655, 1653*f*, 1654*f*, 1655*f***
gonadectomy, **1659–1660**
inhibin, **1657**
testosterone, **1661**
placenta and, 2848–2849
positive feedback control of secretion, **655–656**
receptors, 2849
reflex ovulation and, 2303–2305
regulation, **1426**
of gonadotropin biosynthesis and secretion, **1635–1701**
structure and function of GnRH receptor, **1636–1643**
signal transduction and gonadotropin secretion, **1643–1652**
diacylglycerol, **1647–1648**
gonadotropin secretion, **1651–1652, 1652*f***
guanosine triphosphate–binding proteins, **1643, 1644*f***
immediate–early gene expression, **1650–1651, 1650*f***
intracellular calcium, **1646–1647, 1646*f***
mitogen-activated protein kinase cascade, **1649–1650, 1649*f***
phospholipase C and phosphoinositides, **1643–1646, 1645*f***
protein kinase A (PKA) pathways, **1651**
protein kinase C (PKC) pathways, **1648**
sites of metabolic cue detection, 2584–2585

Gonadotropin-releasing hormone (GnRH) *(Continued)*
structure and function, **1636–1643**
desensitization, **1638–1640, 1639*f*, 1640*f***
effects of mutations, **1640–1641**
internalization, **1638**
ligand affinity and expression in the cell membrane, **1637–1638**
physical characteristics, **1636–1637, 1637*f***
type II GnRH receptor, **1641–1643, 1642*f***
Gonadotropins, **1581–1634**
acute actions of, 2467
alpha/common subunit, **1583, 1586–1590*t*, 1591–1595, 1591*f***
biosynthesis, **1619–1625, 1623*f***
carbohydrate structure and function, **1603–1613, 1608*f*, 1610–1611*f*, 1612*f***
deglycosylated forms, **1611–1613**
gonadotropin glycosylation sites, **1605–1607, 1605*f*, 1606*f***
control of secretion, 2476–2485, 2479*f*, 2482*f*
curvilinear pattern of ovulatory gonadotropin surges, **1298**
genetic relationships of the subunits, **1602–1603**
alpha-subunit gene, **1601–1603, 1602*f***
chorionic gonadotropin and luteinizing hormone beta-subunit gene families, 1603
follicle-stimulating hormone beta-subunit gene, **1603, 1604*f***
orphan subunit genes, **1603, 1604*f***
thyroid-stimulating hormone beta-subunit gene, **1603**
hormone receptors, **1613–1619**
follicle-stimulating hormone receptors, **1619**
luteinizing hormone/chorionic gonadotropin receptors, **1613–1619, 1614*f*, 1615–1617*t***
thyroid-stimulating hormone and orphan leucine-rich repeat/G-protein–coupled receptors, **1619**
long-term actions, 2467–2468, 2467*f*
preovulatory follicular development and, 2455–2456
preovulatory surge, 2081
progesterone influence on secretion, **1299**
secretion during the menstrual cycle, 2475–2476
steroids and, **1300–1301**
structural features, **1582–1601**
aberrant gonadotropin molecules, **1601**
beta/hormone-specific subunit, **1595–1599, 1596–1597*f***
cystine knot proteins, **1582–1583, 1582*f*, 1584–1585*f***
function allocation of glycoprotein hormones, **1599–1601**
three-dimensional structure, **1583, 1585*f***

Gonadotropin secretion
age-related changes in surge patterns, 2735–2738, 2737*f*
cortisol and, 2651, 2652*f*
in the cow, 2522–2525
estrus and ovulation, 2522
gonadotropins, 2523–2524
induction of ovulation, 2525
luteinizing hormone, 2524
ovarian activity, 2522–2523, 2523*f*
response to estradiol and ovariectomy, 2524–2525, 2524*f*
role of prolactin, opioids, and negative energy balance, 2525
in humans, 2528–2533
gonadotropins, 2531, 2531*f*
influence of nutritional status, 2533
luteinizing hormone, 2531–2532
ovarian activity, 2530, 2530*f*
prolactin and suckling, 2532–2533
hypothalamo–pituitary regulation, 2710–2711
mechanisms of suckling-induced ovarian inactivity, 2533–2536, 2534*f*
in nonhuman primates, 2526–2528
gonadotropins, 2526–2528, 2527*f*
ovarian activity, 2526
prolactin, 2528, 2529*f*
in the pig, 2516–2519
estrus and ovulation, 2516–2517
gonadotropins, 2517–2518, 2517*f*
induction of ovulation during lactation, 2518–2519
ovarian activity, 2517
prolactin, opioids, and nutritional status, 2519
in the rat, 2512–2516
estrus and ovulation, 2512–2513, 2512*f*, 2513*f*
ovarian activity, 2513–2514
pattern, 2514–2515, 2514*f*
prolactin and energy balance, 2516
role of ovarian steroids and opioids, 2515
in the sheep, 2519–2521
estrus, ovulation, and gonadotropins, 2519–2521, 2520*f*
prolactin, 2521
suckling and control of, 2511–2551
Gonadotropin signaling, **547–592**
follicle maturation, **547–548**
follicle-stimulating hormone and receptor, **548–550, 549*f***
granulosa cell differentiation, **550–566**
LH signaling pathways, **572–575**
luteinizing hormone and receptor, **570–572**
molecular basis of target gene expression, **575–576**
overview, **547**
ovulation and luteinization, **568–570**
puberty and, 2063–2064
receptors, **549*f***
regulation of follicle-stimulating hormone-responsive gene targets, **566*f***

Gonadotropin signaling *(Continued)*
target gene expression regulation, **566–568**
Gonads
and ducts, **313–336**
anatomy and cell biology of sexual differentiation, **314–315, 314*f*, 315*f*, 316*f***
embryology of ducts of the reproductive tract, **326–330**
formation of bipotential gonad, **314**
genes essential for development of bipotential gonad, **317–320**
genes essential for testes development, **317*t*, 320–325**
ovary formation, **316–317**
primordial germ cells, **325–326**
testes formation, **315–316, 316*f***
embryonic development in the rat, 2065*f*
puberty and, 2179–2182
relaxin and, **741, 742*f***
Gossypol, **1118–1119**
GPR54. *See* G-protein–coupled receptor-54
G-protein–coupled receptor-54 (GPR54), **448**
GR. *See* Glucocorticoid receptor
Graafian follicle, **385–386**
Grafting, intracerebral, 1779
Granulin, **17*t***
Granulosa cell
differentiation, **636–640, 651–652**
estrogens in proliferation, **634–635**
gonadotropin signaling and, **550–566**
proliferation, **650–651**
Grn, **17*t***
Growth differentiation factor 9 (GDF9), **392–393**
Growth factor receptors, epididymis and, **1104**
Growth factors
acting via receptors with tyrosine kinase activity, 2070–2072, 2071*f*
acting via serine-threonine kinase receptors, 2070
cytokine expression and action, **345–347**
glial signaling and, **1439–1440**
immunoregulation, **1219–1221, 1220*f***
implantation and, **164–165**
as possible regulators of the estrous cycle, 2349–2350
signaling molecules and, **329–330**
Growth hormone (GH), **386–387, 403–404**
effect of estrogen on, **1523*f***
growth hormone receptor and, **528**
ovary and, **527–546**
in rodents and farm animals, **528–529**
puberty and, 2203–2205, 2204*f*
Growth hormone receptors, growth hormone and, **528**
Growth hormone–releasing hormone (GHRH)
aging and, 2732
biological fluids, 2851–2852
Growth hormone–releasing hormone (GHRH) *(Continued)*
expression, **1530,** 2851
localization, 2851
placenta and, 2851–2852
receptors, 2851
GRP 78. *See* Glucose related protein 78
GSK3β, follicle-stimulating hormone and, **558*f*, 561–562**
GST. *See* Glutathione S-transferase
GSTM5, **27*t***
GTH. *See* Gonadotropic hormone
Guanosine triphosphate–binding proteins, **1643, 1644*f***
Guinea pig, **114**
Leydig cells, **907*f*, 908*f***
pregnancy, **713–714**
spermatogonia, **829*f***
spermatozoa, **15*f***
steroids and phasic release of gonadotropins, **1300**
timing of implantation, **149*t***
ultrastructure, **903*f***

H

Halo cells, **1080**
Hamsters, **114**
oocyte–cumulus complex, **133*f***
para-acrosomal cytoskeleton, **13**
pheromones and, 2044–2046
relaxin and, **713**
seasonal rhythms of reproduction and, 2232, 2233*f*, 2256–2257, 2256*f*
semen deposition, **114**
spermatozoa, **15*f***
steroids and phasic release of gonadotropins, **1300**
timing of implantation, **149*t***
Hamster vaginal discharge (HVD), 2045
Hand-Foot-Genital syndrome, **328**
hCG. *See* Human chorionic gonadotropin
HE6, epididymis and, **1105**
Health and disease, developmental origins (DOHaD), 2975–2990
cost implications, 2976
evolutionary and theoretical perspectives, 2980–2986
diet comparison, 2981*t*
gene–environmental interaction, 2980*f*
maternal constraint, 2983–2984
maternal health, 2985–2986
postnatal factors, 2986
transgenerational influences, 2986
transitions, 2984–2985
experimental observations, 2977–2980
altered tissue differentiation, 2977–2979
epigenetic processes, 2979
mitochondrial function, 2979–2980
interventions, 2986–2987
overview, 2975–2976
scope of the problem, 2976–2977, 2977*f*
Heart
paracrine oxytocin systems and, 3068
relaxin in cardiovascular function, **722–723, 725–726**
Heart rate, relaxin and, **701**
Hedgehog genes, **323**
Helogale parvula, 2025
Hemophilia, 2703
Hepatocyte growth factor (HGF), **355–356**
Hermaphroditism, 2021–2022
Heterocephalus glaber, 2026
Heteronetta atricapilla, 2026–2027
Hexarelin analog peptides, male sexual behavior and, 1783
HGF. *See* Hepatocyte growth factor
HH. *See* Hypogonadotropic hypogonadism
HIFs. *See* Hypoxia-inducible factors
Hirundo rustica, 2027
Histamine
milk ejection and, 3172
as neurotransmitter action on LH secretion, **1430*t***
role in implantation, **162–163**
Histocompatibility (HLA) complex, 2773–2778, 2774*f*, 2775*f*, 2776*f*, 2778*f*
Histone H3 methylation in one-cell zygotes, **270*f***
HK1-S, **27*t***
Hofbauer cells, **209, 210*f*, 211**
HOMA. *See* Homeostasis model assessment
Homeobox genes, **328–329**
Homeodomain-containing transcription factors, **1106–1107**
Homeostasis model assessment (HOMA), 2977
Hormone–behavior interactions, 1996, 1997*t*
during development phase of reproduction life history stage, 2007–2010
mate choice and pair bonding, 2009–2010
photoperiod, 2008–2009
in the ewe, 2397–2398, 2397*f*
finite-state machine theory and testing predictions, 2027
influencing mature capability of reproduction life history stage, 2010–2014
cycling onset, 2010–2012
parental care, 2012–2013
recycling (renesting), 2013–2014
transitions between sexual and parental phases of reproduction, 2013
support functions for reproductive life history stage, 2014–2015
social status, 2015
territorial behavior, 2014–2015
termination phase of reproduction life history cycle, 2015–2016
effects of offspring and mate, 2016
interactions with molt, 2016
testosterone "costs," 2027
Hormone regulation
epididymal functions and, **1092–1093**
epididymis and, **1117–1118**
hypothalamus-like, 2848–2852
lactation and, 2993–3054

Hormones. *See also* individual names
action on embryonic ducts, **327–330**
as determinants of onset of maternal behavior, 1925–1931, 1926*t*, 1929*f*, 1939*f*
effect on female reproductive behaviors, 1828–1863, 1828*f*
comparisons across species, 1828–1829
electrophysiological actions, 1858–1863
gene memory, 1829–1832
neuropeptides influenced, 1850–1858
neurotransmitters influenced, 1840–1850, 1844*t*
effects of hormone implants on male sexual behavior, 1764–1765, 1765*f*
effects of hormone replacement on ex copula penile responses, 1746–1747, 1747*f*
effects on male sexual behavior, 1767
effects on robustness of maternal experience, 1960–1961
female reproductive behavior and, 1828, 1828*f*
gonadotropes and, **1541–1542**
hypothalamic, **1295**
influence on sexual arousal, 1885–1886
interference actions, 2005
luteotropic, 2463–2465, 2464*f*
male sexual behavior and, 1785
maternal behavior and, 1924–1932
maternal effects, 2006
metabolic, 2868–2870
ovarian, 2481–2485
persistence actions, 2005–2006
pituitary-like, 2852–2855
produced by intrauterine tissues, 2848*t*
requirements for implantation, **151–153**
steroid, **157–159**
suppressive actions, 2004–2005
Horses, **114**
embryo transport in the oviduct, **135**
isolation of relaxin, **683–684**
pregnancy, **711–712**
Hoxa-10, role in implantation, **167**
Hoxa-11, role in implantation, **167**
Hoxd13, **328**
Hox genes, **1155**
Hox (homeobox), **1106–1107**
HPG. *See* Hypothalamic–pituitary–gonadal axis
hPL. *See* Human placental lactogen
HPT axis, estrogen and spermatogenesis in the adult, **1043**
3β-HSD. *See* 3β-Hydroxysteroid dehydrogenase
11β-HSD. *See* 11β-Hydroxysteroid dehydrogenase
17β-HSD. *See* 17β-Hydroxysteroid dehydrogenase, type 4
HT5-HT, effects on male sexual behavior, 1787–1788
Human chorionic gonadotropin (hCG), **538, 538*f***
biological fluids, 2852–2853
expression and localization, 2852
Human chorionic gonadotropin (hCG) *(Continued)*
placenta and, 2852–2853
receptors, 2852
Human placental lactogen (hPL), 2855
Humans
anatomy of prostate, **1153*f***
approach-avoidance in postpartum mothers, 1935–1936
biosynthesis of GnRH and GAP, **1423*f***
cloning of relaxin, **685–686**
developmental origins of health and disease, 2975–2990
disrupted estrogen signaling in, **643–644**
egg–cumulus complex, **135*f***
embryology of prostate, **1150*f***
endocrine regulation, **1045–1053, 1046*f*, 1047*t*, 1049*f*, 1050*f***
hormonal dependency, **1045–1048, 1046*f*, 1047*t*, 1049*f*, 1050*f***
hypogonadotropic hypogonadism, **1047*t*, 1049–1051**
male hormonal contraceptive development, **1051–1053**
estrogen
in the adult, **1043–1044**
and implantation, **152–153**
female reproductive tract, **115*f***
flagellar beating patterns of sperm, **119*f***
gonadotropin oligosaccharide heterogeneity, **1612*f***
gonadotropin secretion, 2528–2533
gonadotropins, 2531, 2531*f*
influence of nutritional status, 2533
luteinizing hormone, 2531–2532
ovarian activity, 2530, 2530*f*
prolactin and suckling, 2532–2533
growth hormone and, **529**
heritable effects on sperm structure, **30–31**
implantation, timing, **149*t***
individual recognition, 1962
isolation of relaxin, **684**
Leydig cells, **909*f*, 914*f***
maternal deprivation and isolation studies, 1969–1971, 1970*f*, 1973, 1973*f*
maternal experience and parity effects in adult females, 1957
mechanisms of maternal experience, 1962–1963
menstrual cycle, **708, 709*f***
milk ducts, 3130*f*
milk ejection monitoring, 3137, 3137*f*
mutations causing reproductive axis defects in females, **408*t***
neuroanatomical bases of maternal experience and olfactory learning, 1965–1966
onset of maternal behavior
with cortisol and corticosterone, 1928, 1929*f*
with steroids, 1927
oviduct, **341**
oviductal lumen, **131*f***
pituitary gland, **1312, 1312*f***
placenta in situ, **198*f***
Humans *(Continued)*
placental transfer. *See* Placental transfer
postpartum maternal experience, 1960
pregnancy. *See* Pregnancies
relaxin sources, **710**
semen coagulation, **116**
sensory mechanisms mediating maternal behavior, 1940
sexual response cycle, **1174–1175, 1174*f***
classification of sexual disorders, **1175*t***
clinical implications, **1174–1175, 1175*t***
sibling experience and female allomothering, 1958
spermatozoa, **15*f***
testes, **783*f*, 786*f***
uterotubal junction, **122*f***
HVD. *See* Hamster vaginal discharge
Hyal, **17*t***
Hyaluronic acid, **133–134**
Hyaluronidase, **17*t***
Hydatidiform mole, 2879–2880
17α-hydroxylase/17-20 lyase, **982–983**
11β-Hydroxysteroid dehydrogenase (11β-HSD), **449**
3α-HSD. *See* 3α-Hydroxysteroid dehydrogenase
3β-hydroxysteroid dehydrogenase, **982**
17β-Hydroxysteroid dehydrogenase, **983**
17β-Hydroxysteroid dehydrogenase, type 4 (17β-HSD), **449**
3β-Hydroxysteroid dehydrogenase (3β-HSD), **449**
3α-Hydroxysteroid dehydrogenase (3α-HSD), **448–449**
Hyena, relaxin and, **714**
Hyperandrogenism
insulin and, **535–537**
insulin-like growth factors and, **535–537**
Hypercortisolemia, 2730
Hyperglycemia, relevance of maternal, **291–292**
Hyperinsulinemia, relevance of maternal, **291–292**
Hypocretins (orexin)
estrous cycle and, 2348
male sexual behavior and, 1784
as neuropeptide action on LH secretion, **1431*t***
reproductive axis and, 2595–2596
Hypoglycemia, as stressor, 2636
Hypogonadotropic hypogonadism (HH), **1045, 1047*t*, 1049–1051**
causes, **1047*t***
congenital, **1046*f***
etiology and effects on spermatogenesis, **1047*t*, 1049–1050**
restoration of spermatogenesis and, **1050–1051**
Hypophysectomy, **1291**
Hypothalamic paraventricular nucleus (PVN), 2629–2630
Hypothalamic–pituitary–adrenal axis, parturition and, 2938–2940

Hypothalamic–pituitary–gonadal (HPG) axis, **386, 393*f***
effects of estrogen exposure and insufficiency, **1042, 1042*t***
effects of stressors on, 2634–2644
energy deficits and, 2559
hormonal control, 2086–2088
maturation, 2072–2073, 2073*f*
mechanisms of interference with the reproductive system, 2644–2660
metabolic-related suppression, 2665–2666
neural control, 2088
ovarian cyclicity in the ewe and, 2398–2408, 2398*f*, 2399*t*, 2401*f*, 2403*f*, 2405*f*, 2405*t*, 2408*f*
postnatal development, 2085–2089, 2086*f*
pubertal activation, 2089
puberty in the rat and, 2066–2068, 2067*f*
Hypothalamic–pituitary–ovarian axis, aging and, 2729–2755
Hypothalamic–pituitary–testicular axis, aging and, 2697–2728
Hypothalamo–hypophysial complex
adenohypophysis, **1316–1329**
development, **1319–1322, 1320*f***
epithelial cells, **1316–1319, 1317–1318*f*, 1319*f***
organization, **1322–1329**
pars distalis, **1322–1327, 1322*f***
pars intermedia, **1327–1328**
pars tuberalis, **1328–1329**
afferents to the MPOA-HTA, **1375*t***
anatomy, **1309–1413**
efferents from MPOA-HTA, **1364*t***
hypothalamus, **1351–1373**
cell bodies, **1351–1355, 1352*f***
neurosecretory systems
development, **1355–1356**
organization, **1356–1373, 1357*t*, 1358*f*, 1363*f*, 1364*t*, 1366*f***
input into the periventricular brain, **1373–1380, 1375*t*, 1378*f***
neurohypophysis, **1329–1343**
development, **1331–1333, 1334*f***
nerve terminals, **1329–1331, 1330*f***
organization, **1333–1343**
infundibular process, **1333–1337**
infundibular stem, **1337**
infundibulum (median eminence), **1337–1343, 1338*f***
overview, **1309–1316, 1310*f*, 1312*f*, 1313*f***
portal system, **1343–1351**
capillaries of the hypophysis, **1343, 1343*f*, 1344*f***
development, **1343–1344**
organization, **1344–1351, 1346*f*, 1347*f*, 1348*f*, 1349*f*, 1350*f***
synapses on neurons, **1357*t***
Hypothalamus, **1351–1373.** *See also* Pituitary gland
aging and, 2732
cell bodies, **1351–1355, 1352*f***
circadian periodicity of the luteinizing hormone surge mechanism, **1296**
competency to produce GnRH before puberty, 2133–2134
Hypothalamus *(Continued)*
folliculogenesis and, **395–399**
GnRH pulse generator and, 2477–2481, 2479*f*
hormone implants effect on lordosis behavior, 1838
hormones, **1295**
male sexual behavior and, 1784, 1784*f*
neurosecretory systems
development, **1355–1356**
organization, **1356–1373, 1357*t*, 1358*f*, 1363*f*, 1364*t*, 1366*f***
oxytocin and, 3065–3066
pharmacologic blockade, **1295–1296**
pseudopregnancy, delayed, **1296–1297**
reflex ovulators, **1296**
role in rats during the estrous cycle, 2340–2341, 2341*f*
role in reproductive behavior control, 1862
"sexual center," **1293**
spontaneous ovulators, **1296**
Hypothalamus-like hormones, placenta and, 2848–2852
Hypothalamic–pituitary–adrenal axis, inflammatory reaction, **1202–1203, 1202*f***
Hypoxia-inducible factors (HIFs), implantation and, **172**

I

ICS. *See* Immotile-cilia syndrome
Idiopathic hypogonadotropic hypogonadism (IHH), **1049**
IGF. *See* Insulin-like growth factor
IGF-binding proteins, **449**
IGF-BP. *See* Insulin-like growth factor-binding protein
IGF-I. *See* Insulin-like growth factor-I
IGF-II. *See* Insulin-like growth factor-II
IHH. *See* Idiopathic hypogonadotropic hypogonadism
Immature intermediate villi, **212*f*, 214**
Immotile-cilia syndrome (ICS), **30–31**
Immune cells, male reproductive tract, **1209–1216**
quantification, **1210*t***
Immune response
in the cervix, **118**
effect on male reproduction, **1216–1232**
in the uterus, **121–122**
Immune system, interaction between HPA axis and, 2633, 2634*f*
Immunobiology, human pregnancy and, 2759–2785
Immunocontraception, **1249**
Immunocytochemistry
effects on male sexual behavior, 1786
male sexual behavior and, 1789
Immunophysiology, male reproductive tract, **1195–1286**
activation of adaptive immune response, **1207*f***
characteristics of innate immunity versus adaptive immunity, **1203*t***
cluster designation markers, **1204*t***
Immunophysiology, male reproductive tract *(Continued)*
effects of inflammation, infection, and immunity on male reproduction, **1216–1232**
androgen production, **1221–1229, 1222*f*, 1225*t*, 1226*f***
inflammatory and immunoregulatory cytokines and testis function, **1216–1221, 1217*f*, 1220*f***
role of macrophages in Leydig cell development and steroidogenesis, **1231–1232**
spermatogenesis, **1229–1231, 1230*f***
endocrine regulation, **1208–1209**
future, **1249–1250, 1250*f***
historical aspects, **1197–1199**
immune cells in male reproductive tract, **1209–1216**
immune cells of the epididymis, **1214–1215**
immune cells of the vas deferens, accessory glands, and urethra, **1215–1216, 1216*f***
lymphocytes and polymorphonuclear cells in the testis, **1213–1214**
quantification, **1210*t***
testicular macrophages, **1210–1213, 1212*f***
immune system and endocrine control, **1203–1209**
characteristics, **1203–1208, 1203*t*, 1204*t*, 1205*f*, 1206*f*, 1207*f***
immunocontraception, **1249**
immunological responses, **1232–1247**
blood–testis barrier, **1235–1236**
epididymis and genital tract, **1244–1247**
germ cell and testis transplantation, **1243–1244**
immune privilege as "testis-specific" phenomenon, **1244**
orchitis, **1233–1235**
sperm antibodies, **1232–1233, 1232*f***
transplantation and immunoregulation, **1236–1243, 1239*t*, 1241*f*, 1242*f***
infection and tumor response, **1247–1248**
overview, **1195–1197, 1196*f*, 1198*f***
structural organization and functions, **1199–1203**
endocrine regulation, **1201–1203, 1202*f***
excurrent ducts and accessory glands, **1201**
testis, **1199–1201, 1200*f***
transplantation medicine implications, **1249**
Immunoregulation, male transplantation and, **1236–1243, 1239*t*, 1241*f*, 1242*f***
Implantation, **147–188**
delayed (diapause), **153–156, 154*f***
embryo and, **156**
environmental, neural, and pituitary regulation of embryonic diapause, **155**

Implantation *(Continued)*
delayed (diapause)
ovarian regulation, **155**
uterine factors and embryonic diapause, **155–156**
effects of hormone implants on male sexual behavior, 1764–1765, 1765*f*
endometrial glands and secretions, **153**
future/new techniques, **173–174**
gene expression in mid-secretory phase, **358–364, 359*f***
hormonal requirements, **151–153**
microarray technology and, **173–174**
mid-secretory endometrium and embryonic, **350*f*, 358**
model of events occurring during early stages of embryonic, **364*f***
molecular aspects of embryo–uterine interactions, **157–173, 158*f***
adhesion molecules, **159–162.** *See also* molecule names
calcium-associated signaling molecules, **168–169**
cell cycle regulation, **169–170**
connexins, **169**
cytokines, **165–167**
endocannabinoids, **172**
growth factors, **164–165**
matrix remodeling and angiogenesis, **170–172**
morphogens, **172–173**
steroid hormones, **157–159**
transcription factors, **167–168**
vasoactive factors, **162–164.** *See also* Vasoactive agents
overview, **147–148**
physiological and morphological aspects of embryo–uterine interactions, **148–153**
relaxin and, **728–730**
stages, **148–150**
apposition, **148**
attachment, **148–149, 149*f*, 150*f***
penetration and decidualization, **149–150**
timing in various species, **149*t***
types, **150–151**
window of, **156–157**
Imprinting, **229–230**
loci with parent-of-origin, **274*t***
mouse models of methylation and defects, **275*t***
Infections, effect on male reproduction, **1216–1232**
Infertility. *See also* Sterility
stress and, 2642–2643, 2643*f*
Inflammation, effect on male reproduction, **1216–1232**
Inhibin-related peptides
expression and localization, 2859
levels in biological fluids, 2860, 2860*f*
placenta and, 2859–2860, 2860*f*
receptors and binding proteins, 2859–2860
Inhibins, **404–408**
aging and, 2734*f*, 2805
control of secretion, 2411–2412
Inhibins *(Continued)*
estrous cycle of sheep and, 2395–2396, 2411–2412
expression, **1529–1530**
menstrual cycle and, **7,** 2463
negative feedback actions, 2485
as physiological regulator of gonadotropin biosynthesis and secretion, **1657**
regulation of GnRH receptors, **1538**
Inhibitory neurons, 2082–2083
Inorganic ions, placental transfer, 2802–2805
Insectivores, oviductal sperm storage, **129–130**
INSL3, **689*f*, 698, 737*f***
primary structure, **699*f***
INSL4, **690**
physiological role, **742–744**
in primates, **693**
INSL5, **690–691**
physiological roles, **743–744**
INSL6, **690–691**
physiological roles, **744**
Insulin
as an adiposity signal to the brain, 2575
central nervous system and, 2575
covalent structure of porcine, **683*f***
energy homeostasis and, 2575–2576
impaired action and reproduction, 2577–2578, 2577*f*
ovarian hyperandrogenism and, **535–537**
reproductive axis and, 2576–2577
sites and mechanisms of action, 2578
as stressor, 2636
Insulin-like growth factor-binding protein (IGF-BP), placenta and, 2860–2862
Insulin-like growth factor (IGF)
binding proteins and the ovary, **537–539**
biological fluids, 2862
expression and localization, 2860–2861
insulin and ovarian hyperandrogenism, **535–537**
mammary development and, 3003
menstrual cycle and, 2454–2455
ovarian cancer and, **537**
ovary and, **527–546, 531–535**
placenta and, 2860–2862
proteolysis, follicular selection, and luteolysis, **539**
puberty in the rat and, 2065
receptors, **322–323, 530–531, 530*f*,** 2861–2862
as regulator of the estrous cycle, 2349–2350
structure–function, **530**
Insulin-like growth factor-I (IGF-1), **354–355**
puberty and, 2091
Insulin-like growth factor-II (IGF-II), **355**
Insulin-like growth factor receptors
folliculogenesis and, **404**
Insulin receptor family, **250–251**
Integrins, role in implantation, **160–161**
Interferons
biological properties, **1225*t***
immunoregulation, **1225, 1225*t***
Interleukin-1
immunoregulation, **1216–1219, 1217*f***
regulation, **1222*f*, 1223**
Interleukin-6, immunoregulation, **1216–1219, 1217*f***
Interleukin-4 receptor-α, **449**
Interleukins, **449**
Intracerebral grafts, effect on male sexual behavior, 1779
Intratesticular ducts, cytology, **828–829**
Intrauterine growth restriction (IUGR), **222**
Intromissions, 1731. *See also* Copulation
In vitro fertilization (IVF)
in humans, **152–153**
maturation of spermatozoa and, **1083–1086, 1084*f***
Involution, 2998
Iodide, placental transfer, 2805
Ion channels, **1711–1712**
Ions, secretion into milk, 3016–3017, 3017*f*, 3023–3026, 3024*f*, 3026*f*
Iron, placental transport, 2822–2824
IUGR. *See* Intrauterine growth restriction
IVF. *See* In vitro fertilization

J

Jackson Laboratories, **384**
Jacobson's organ (vomeronasal organ), 2042
JAK2. *See* Jaus kinase-2
Janus kinase-2 (JAK2), **1708–1709**
JunDE proto-oncogene, **449–450**

K

KA36K-A, **27*t***
Kallmann's syndrome, **1047*t***
female reproductive behavior and, 1870
Kartagener's syndrome, **30–31, 135**
Ketones, placental transfer, 2816–2817
Kidney
extracellular fluid volume/blood pressure regulation in pregnancy, 2915–2916
maternal adaptation during pregnancy, 2914–2915
relaxin and, **723–724, 724*f***
KiSS-1/GPR-54 system, 2096*f*, 2097, 2198–2199, 2199*f*
estrous cycle and, 2348
Kisspeptin, as neuropeptide action on LH secretion, **1431*t***
KLC3, **24*t***
Klinefelter syndrome, **247–248**
Knockout models, in early embryos, **287*t***
Koalas, **777**
Sertoli cells, **873*f***
ultrastructure of seminiferous epithelium, **873*f***

L

Lactate, placental transfer, 2808–2809, 2808*t*

Lactation. *See also* Mammary gland; Milk
dietary regulation, 3022–3023
hormonal control, 2993–3054
developmental aspects of milk secretion, 2994–2998, 2995*f*, 2996*f*, 2997*f*
endocrine regulation of functional development, 2998–3005, 3000*f*, 3001*f*, 3004*f*, 3005*f*
lactation in monotremes and marsupials, 3037–3038, 3038*f*
milk ejection, 3035–3037, 3035*f*
milk synthesis and secretion, 3005–3031, 3006*f*, 3007*f*, 3010*f*, 3011*f*, 3012*t*, 3013*t*, 3014*f*, 3015*f*, 3015*t*, 3016*f*, 3017*f*, 3019*f*, 3021*f*, 3024*f*, 3026*f*, 3029*f*, 3030*f*, 3031*f*
overview, 2993–2994
regulation of milk secretion, 3032–3035, 3033*f*
initiation, 3021–3022
lactotropes and, **1553**
leptin and, 2573
as metabolic stressor that impairs reproduction, 2564–2565
oxytocin and, 3075
prolactin and, **1715**
relaxin and, **726–728**
mammary gland, **727–728**
nipple development, **726–727, 727*f***
Lactotropes, **1483–1579**
autocrine and paracrine influences, **1539–1540**
activin, inhibin, and follistatin and nonparallel release, **1539–1540**
epidermal growth factor as regulator, **1540**
cell type development, **1488–1490**
discovery and historical perspective, **1484**
diversity, **1513–1530**
division of labor, **1514–1519, 1515*f*, 1517*f*, 1518*f***
as a multihormonal cell, **1519–1527, 1520*f*, 1521*f*, 1522*f*, 1523*f*, 1524*f*, 1526*f***
as a regulatory cell, **1528–1530, 1529*f***
roles for monohormonal gonadotropes and, **1527–1528**
tumorigenesis, **1513–1514**
early identification, **1484–1485**
early studies, **1492–1497, 1493*f*, 1494*f*, 1495*f*, 1496*f***
to define population, **1546–1548**
differentiation of small and large prolactin cells, **1546–1548**
discovery of bihormonal somatomammotropes, **1548**
immunolabeling and electron microscopy, **1546, 1546*f*, 1547*f***
effect of age and reproductive hormones, **1541–1542**
effects of growth factors, **1542–1543, 1543*f*, 1544*f***
embryological origin, **1486–1492**
functional heterogeneity, **1548–1551**
dependence on somatotropes and gonadotropes, **1548–1550, 1549*f***

Lactotropes *(Continued)*
functional heterogeneity
significance of prolactin cells, **1550–1551**
single-cell secretion assays, **1551**
GnRH receptor expression, **1530–1539**
detection of ligand–receptor binding, **1530–1531, 1531*f***
early internalization events, **1532–1533**
Golgi complex, granule, and lysosomal domains, **1533–1534**
identification of hormone stores in GnRH target cells, **1534–1535, 1535*f*, 1536*f***
membrane binding and microaggregation, **1531–1532, 1532*f***
pituitary GnRH versus the GnRH–receptor complex, **1534, 1534*f***
regulation of GnRH receptors, **1535–1539, 1537*f***
immunolabeling, **1485–1486**
integrative studies of gonadotropes and, **1556–1562**
communication between, **1559–1560**
detecting receptor populations, **1556–1559**
how past lessons affect future experiments, **1560–1562**
regulating transition from birth to adult, **1556**
monohormonal and multihormonal cells, **1492**
neurointermediate lobe prolactin regulatory factors, **1555–1556**
origin and formation of anterior pituitary, **1486–1488, 1487*f***
prolactin cells as model for studies of cell secretory pathway, **1551–1553**
responses during estrous cycle and estrogen treatment, **1552–1553**
responses to lactation and weaning, **1553**
response to pregnancy, **1551–1552**
proliferation of differentiated gonadotropes and, **1540–1541**
regulation by dopamine, **1554–1555**
inhibition, **1554–1555**
stimulation, **1555**
regulation of apoptosis, **1543–1546**
responses to different physiological states, **1497–1513**
after castration, **1501*t***
changing levels of gonadal steroids, **1499–1503, 1500*f*, 1501*t*, 1502*f*, 1503*f***
during the estrous cycle, **1503–1508, 1503*f*, 1504*f*, 1506*f*, 1508*f***
maturation of the gonadotrope and, **1497–1499, 1498*f*, 1499*f***
role of matrix and gap junction proteins, **1512–1513**
in vitro studies, **1508–1511, 1509*f*, 1510*f*, 1511*f*, 1512*f***
spatial organization of transcription factors, **1490–1492**

Lactotropes *(Continued)*
thyroid-releasing hormone regulation, **1553–1554**
Lama glama, 2286–2287*t*, 2292, 2295*t*
Lama pacos, 2286–2287*t*, 2292, 2295*t*
Laminin, **160**
Langhans' cells (cytotrophoblast), **203–205, 204*f***
Laron syndrome, **403**
Larus occidentalis, 2015
Lasiorhinus latifrons, **777**
L-Carnitine transporters, **1109–1110**
Lcn2. *See* Lipocalin 2
Leaf-eating monkey, placenta, **194*f***
Learning processes, oxytocin and, 3086
Lebistes reticulatus, 2012
Lectins, **6**
Lepomis macrochirus, 2024
Leptin, **402–403, 450**
energy homeostasis and, 2569
expression and localization, 2868–2869
levels in biological fluids, 2869
as metabolic signal to the brain, 2568–2569
Ob-R mutations in humans, 2571–2572
placenta and, 2868–2869
pregnancy/lactation and, 2573
puberty and, 2091, 2202–2203, 2572–2573, 2573*f*
receptors, **450,** 2869
reproduction and, 2569, 2570*f*
rodent models of impaired leptin action, 2569–2571, 2570*f*, 2571*f*
sites of actions, 2573–2574
Lepus americanus, 2028
Lepus europaeus, 2286–2287*t*, 2295*t*
Lesions
cell-body, 1766, 1768–1770, 1779, 1781, 1785, 1786
electrolytic, 1786
male sexual behavior and, 1787
Let-down reflex, 3035, 3035*f*
Leucine, **1333**
Leukemia inhibitory factor (LIF), **346**
implantation and, **155**
Leukocytes, pregnancy and, 2761–2762, 2762*t*
Leydig cells
cholesterol sources for steroidogenesis, **978–979**
cytology, **900–920**
cytoplasm, **907–913, 907*f*, 908*f*, 909*f*, 910*f*, 912*f*, 914*f***
differentiation in testis development, **253–254**
effects of estrogen exposure and insufficiency, **1042, 1042*t***
enzyme involvement, **979–984, 980–981*f***
ER-α–mediated repression, **627*f*, 642–643**
estrogen and spermatogenesis in the adult, **1043**
historical background, **900–901**
from human fetus, **832*f***
life history, **914–918**

Leydig cells *(Continued)*
macrophage cell interactions and effects of testosterone on behavior, **1198f**
morphology, **906t**
nucleus, 905, **907f**
organization of intertubular tissue, **901–905, 901f, 902f, 903f, 904f**
response to testicular damage, **918–920**
role of macrophages in development and steroidogenesis, **1231–1232**
role of StAR protein, **989–992**
spermatogenesis and, **1035**
steroid biosynthesis, **980–981f**
steroidogenesis, **1221**
regulation, **984–995, 987–988f**
ultrastructure, **905, 905t, 906t**
LH. *See* Luteinizing hormone
LHβ, **398–399**
LH receptor, **399**
LHRH. *See* Luteinizing hormone–releasing hormone
LHX1, **319, 329**
LHX9, **319–320**
Libido, male, **1179–1180**
LIF. *See* Leukemia inhibitory factor
Life history, stages, 1998*f*
Ligands, receptors and, **1098–1105**
Lipids
placental transfer, 2814–2817, 2814*t*, 2815*t*
secretion into milk, 3019–3023, 3019*f*, 3021*f*
sperm plasma membrane, **9–10**
changes with maturation and capacitation, **10–11**
composition, **9–10**
diffusion, **10**
rafts, **10**
Lipocalin 2 (Lcn2, NGAL), **1112–1113**
Lipocalins, **1111–1114**
function, **1113–1114**
Lipocalin-type prostaglandin D synthase, **1112**
Lipodystrophies, reproduction and, 2597–2598
Lipopolysaccharides, inflammation and spermatogenesis, **1229, 1230f**
Litter, effects of size and suckling activity, 2513, 2513*f*
Littre glands, **785**
Liver receptor homolog-1 (LRH-1), **450**
Lordosis, 1863–1871. *See also* Reproductive behavior, female
correlation of estrogen receptors and, 1836–1837
feeding behaviors and, 1845–1846
neural circuit for producing, 1864*f*
second messenger systems and, 1858
LRH-1. *See* Liver receptor homolog-1
Luteal cells
development, 2412
function and regulation, **506–508, 508f**
puberty and, 2139
regression, 2413–2416, 2413*f*
small versus large, **505–506**
support, 2412–2413
types and regulation, **505–508**
Luteinization, **480–490, 480f**
differentiation of steroid-secreting cells, **484–489, 487f**
luteal angiogenesis, regulation and role, **482–484, 484f**
morphological changes, **481–482**
ovulation and, **568–570**
Luteinizing hormone (LH)
beta-subunit gene expression, **1666–1669**
control of secretion during the estrous cycle of the rat, 2350–2357, 2352*f*
gonadotropins and, **1603**
secretion in cows and, 2524
secretion in humans and, 2531–2532
signaling and, **570–572**
inhibition of secretion, 2350–2351
oxytocin and, 3078
pattern of secretion throughout the estrous cycle, 2336–2337, 2337*f*, 2338*f*
preovulatory follicular development, 2456
profile, **1715f**
receptors, **570–572, 1613–1619, 1614f, 1615–1617t**
signaling pathways, **572–575**
spermatogenesis and, **1025–1026**
stimulation of secretion, 2351–2352
Luteinizing hormone–releasing hormone (LHRH)
coitus-induced ovulation and, 2297–2300, 2298*f*, 2299*f*
gonadotropin secretion and, 2084
intracellular mechanisms underlying preovulatory discharge, 2083–2084
mechanisms of hormonal control, 1850–1852
neurons, 2068
pheromones and, 2044
reproductive behavior, 1853
transcription, 1852–1853
Luteinzing hormone-like luteotropins, **493–494**
Luteolysin, estrogen as, 2469
Luteolysis, **489–491,** 3078–3080, 3079*f*
insulin-like growth factor and, **539**
Luteotropic hormones, menstrual cycle and, 2463–2465, 2464*f*
Lutra canadensis, 2286–2287*t*
Lycaon pictus, 2025
Lynx canadensis, 2028

M

M33, **321**
Macaca mulatta, 2009
pregnancy, **711**
Macacus rhesus, **836**
α2-Macroglobulin (α2-M), **450**
Macrophage migration inhibitory factor (MIF), **450**
Macrophages
Leydig cell interactions and effects of testosterone on behavior, **1198f**
maturation of the rat testis, **1212f**
pregnancy and, 2765–2766
role in Leydig cell development and steroidogenesis, **1231–1232**
Macropus agilis, **777**
Macropus eugenii, **777**
Magnesium
lactation and, 3013, 3013*t*, 3015*t*
placental transfer, 2807–2808
Magnetic resonance imaging (MRI), placenta and, **232**
Major histocompatibility complex (MHC), 2046–2047
Male accessory organs, **1149–1172**
immune cells, **1215–1216, 1216f**
Male hormonal contraceptive (MHC)
determinants of sperm suppression, **1052–1053**
development, **1045, 1051–1053**
ethnic and genetic factors, **1052–1053**
GnRH antagonists and, **1053**
gonadotropic suppression and coadministration of progestins, **1052**
mechanism of action on spermatogenesis, **1051–1052**
role of 5α-reductase, **1053**
Male reproductive system
accessory sex glands, **785–789, 789f**
aging and, 2698–2699
cytology of the testis and intrinsic control mechanisms, **827–947**
endocrine regulation of spermatogenesis, **1017–1069**
excurrent ducts, **781–785**
immunophysiology, **1195–1286**
male accessory sex structure, physiology, **1149–1172**
male sexual function, **1173–1194**
neurobiology of male sexual behavior, 1729–1823
penis, **789–790, 790f**
reproductive tract, **771–825**
anatomy, **771–791**
innervation, **803–807**
vasculature, **791–803**
scrotum, **790–791**
testes, **771–781, 779f, 780f, 781f, 782f, 783f**
testicular steroidogenesis, **977–1016**
in young men, 2699–2700
Male sexual function, **1173–1194**
ejaculation and emission, **1186–1187**
anatomy, **1186**
clinical application of physiological principles, **1187**
neural control, **1186–1187, 1187f**
human sexual response cycle, **1174–1175, 1174f**
classification of sexual disorders, **1175t**
clinical implications, **1174–1175, 1175t**
mechanisms of penile erection, **1175–1179**
biochemical mechanisms, **1177–1179, 1178f, 1179f**
blood supply to the penis, **1176, 1176f**
hemodynamic changes during, **1177**
neural input into the penis, **1176–1177, 1176t**
penile anatomy, **1175–1176, 1175f**

Male sexual function *(Continued)*
overview, **1173–1174**
physiological principles, **1181–1186**
intracavernosal injection of vasoactive agents, **1184–1185**
intraurethral therapies, **1184**
PDE5 inhibitors for treatment of ED, **1182–1184**
prostatectomy, nerve-sparing, **1185**
targets for development of therapeutic agents for men with ED, **1185–1186**
vacuum devices for inducing erection, **1185**
regulation by sex steroids, **1179–1181**
testosterone effects on mood, **1181**
Malnutrition, as cause of hypogonadotropic hypogonadism, **1047*t***
α2-M. *See* α2-Macroglobulin
Mammalian Reproductive Genetics database, **384**
Mammals
anatomy of interstitial tissue of testes, **779*f***
androgen regulation, **1027*t***
breeding seasons, 2234*f*
classical and non-neuropeptidergic neurotransmitter actions, **1430*t***
comparison of second alpha-subunit proteins, **1604*f***
efferent ducts, **785*f***
evolution of maternal behavior, 1956*f*
fertilization, **55–112**
classification based on mechanisms of polyspermy block, **83–85, 84*t***
early stages of sperm–egg interaction, **83*f***
model of early events, **65*f***
food availability, 2211
hormonal profiles and latencies to ovulation after coitus in induced ovulators, 2295*t*
male sexual patterns, 1730–1738
maternal adaptation to pregnancy, 2899–2923
melatonin rhythm generating system, 2237, 2237*f*
ovarian synchrony from pheromonal stimulation, 2050*t*
pattern of follicular development, 2452, 2452*f*
pheromones, 2041–2057
phylogenetic representation of mating-induced ovulation, 2288
puberty. *See also* Puberty
onset, 2047*t*
reproduction, circannual versus non-circannual species, 2234–2236, 2235*f*
reproductive behavior, female, 1826–1827
seasonality, 2231–2281
social, seasonal, and behavioral parameters affecting induction of ovulation in reflex ovulators, 2286–2287*t*
sources of relaxin in females, **705*t***
suckling and control of gonadotropin secretion, 2511–2551
Mammals *(Continued)*
testes, **772*f***
timing of ovulation relative to onset of mating, 2290*f*
Mammary gland. *See also* Lactation
changes in differentiated activity in pregnancy and lactation, 2997*f*
contractile tissue, 3132–3133, 3132*f*
general structure of lactating, 3130–3131, 3130*f*
glucose and amino acid metabolism in the lactating mammary alveolar cell, 3007–3010, 3007*f*
involution, 3004–3005
milk storage within, 3131–3132
oxytocin and, 3064
paracrine factors in pregnancy, 3001–3003, 3001*f*
regulation of junctional complexes, 3027–3032, 3029*f*, 3030*f*, 3031*f*
relaxin and, **727–728**
from virgin animals of different species, 2995*f*
Manganese superoxide dismutase (Mn-SOD), **454**
MAPK. *See* Mitogen activated protein kinase
Marfan's syndrome, 2703
Marmoset, pregnancy, **711**
Marmota marmota, relaxin and, **714**
Marmota monax, 2286–2287*t*, 2295*t*
Marmots, relaxin and, **714**
Marsupials
accessory sex glands, **788**
egg–cumulus complex, **135*f***
epididymis, **785**
lactation, 3037–3038, 3038*f*
oviductal sperm storage, **129–130**
sperm, **129*f***
in isthmic crypt, **130*f***
testes, **772, 782*f*, 784*f***
size variation, **777**
Martes pennanti, **154**
Mast cells, immunophysiology, **1214**
Mate choice, 2009–2010
pheromone signaling, 2046–2047
sex attractants, 2044–2046
Maternal effect genes, **265–266, 265*f*, 267**
Maternal–fetal barrier, **195–196, 195*f***
Maternal–fetal interdigitation, **194, 195*f***
Maternal–fetal interface, 2764–2769
Mating systems, 2016–2027
variation, 2019–2023
Matrix metalloproteinases (MMPs), **438, 439, 450, 481–482**
Mature intermediate villi, **211*f*, 212*f*, 213**
MC41, **17*t***
MCH. *See* Melanin-concentrating hormone
Mcl-1. *See* Myeloid cell leukemia
MC4R, **1186**
Mechanical stimuli, parturition and, 2933–2934
Medial preoptic area (MPOA), male sexual behavior and, 1768–1779, 1770*f*, 1772*f*, 1773*f*, 1776*f*, 1777*f*
Meiosis
androgen regulation, **1030–1031**
spermatogenesis and, **1021**
Melanin-concentrating hormone (MCH), reproductive axis and, 2594–2595
Melanocortin receptor, **1186**
α-Melanocyte-stimulating hormone, male sexual behavior and, 1757–1758
Melanocyte-stimulating hormone (MSH), **1292**
as neuropeptide action on LH secretion, **1431*t***
Melatonin
biosynthesis and its regulation, 2240–2243, 2241*f*, 2242*f*
puberty and, 2205, 2206*f*
regulation of the hypothalamic–pituitary–gonadal axis in breeding seasonality, 2245–2263
interneurons connecting to GnRH, 2256–2260, 2256*f*, 2258*f*
melatonin control of reproduction, 2253–2256, 2255*f*
melatonin receptors and seasonal breeding, 2250–2253, 2252*f*
model for analyzing melatonin, 2253
role of pineal gland and melatonin in transducing photoperiodic information, 2245–2248, 2247*f*
sites of action of melatonin, 2248–2250, 2249*f*
thyroid hormones as potential mediators of seasonal plasticity, 2260–2263, 2261*f*, 2262*f*
transduction of light–dark control of synthesis, 2243–2245
transport and catabolism, 2245
Membrane-intercalating agents, **6–7**
Membrane rupture, parturition and, 2953–2954
Membrane type-1 matrix metalloproteinase (MT-1-MMP), **450–451**
Memory, oxytocin and, 3086
Menopause
postmenopause, 2737–2738
transition, 2736, 2737*f*
Menstruation cycle, **708, 709*f***
aging and, 2733–2735, 2734*f*
angiogenesis and vascular growth during, **356–357**
average age of girls, 2557, 2558*f*
final menstrual period, 2731–2732
phases and endometrial histology, **349–352, 350*f*, 351*f***
in primates, 2449–2510
secretory patterns, 2475–2476
stressors and, 2640–2644, 2643*f*
tissue desquamation and, **368**
ultrastructural characteristics of secretory endometrium, **351*t***
MENT. *See* 7α-Methyl-19-nortestosterone
Mephitius mephitis, 2286–2287*t*
Mesenchymal villi, **212*f*, 214**
Mesocorticolimbic DA tract, male sexual behavior and, 1779–1781, 1781*f*

Mesocricetus auratus, **114,** 2015, 2044, 2091
  pregnancy, **713**
Met-1. *See* Metallothionein
Metabolic disorders and reproductive function, 2596–2599
Metabolic fuels
  availability
    and feeding, 2579–2580
    and reproductive function, 2580–2581, 2581*f*
  reflection of energy availability, 2579
  reproduction and, 2579–2582
  sites of action, 2581–2582
Metabolic hormones, placenta and, 2868–2870
Metabolic syndrome, 2976–2977
Metabolism
  epididymis and, **1118–1119**
  folliculogenesis and, **402–404**
  maternal adaptation during pregnancy, 2900–2901
    diet, pregnancy, and fetal programming, 2901
    energy costs of pregnancy, 2900–2901, 2900*f*
  physiological mechanisms integrating reproduction and, 2553–2625
  reproduction and, 2555–2557
  stressors that impair reproduction, 2558–2566
Metallothionein-1 (Met-1), **451**
Metastin-producing neurons, 2083
  puberty and, 2068
Methionine, **1333**
7α-Methyl-19-nortestosterone, **1180**
Methyl chloride, epididymis and, **1119**
MFG. *See* Milk fat globule
MHC. *See* Major histocompatibility complex; Male hormonal contraceptive
Mice, **114**
  biosynthesis of GnRH and GAP, **1423*f***
  brain grafts and induced ovulation in mutants, 2313–2314
  cellular composition of seminiferous tubules of WT and SCARKO male mice, **1001*t***
  classification of follicles, **385*f***
  compaction of eight-cell embryo, **291*f***
  distribution of GnRH neurons in the brain, **1418*f***
  embryo, **314*f*, 315*f***
  endocrine and ovarian phenotypes in mice null for progesterone signaling, **646*t***
  epididymal duct, **1074, 1075*f***
  estradiol-filled cannulae implant studies, **1450**
  external os of cervix, **116*f***
  female reproductive phenotypes in genetic models, **406*t***
  fertility, **134**
  folliculogenesis and, **386*f*, 404–410**
  gene knockouts affecting sperm structure and function, **32**
  generation of mutations, **264*f***
Mice *(Continued)*
  genetic background, **250**
  GnRH
    firing patterns, **1422*f***
    neuron morphology, **1420*f***
  heritable effects on sperm structure, **31–32**
  hormone profile during pregnancy, 3000*f*
  lacking ER-α, **607, 627–629, 627*f*, 628*t***
    and ER-β, **608, 630–632, 631*f***
  lacking ER-β, **607, 629–630**
  lacking estradiol synthesis, **608, 632–633**
  Leydig cells, **910*f***
  models of methylation and imprinting defects, **275*t***
  models of X-inactivation, **278*t***
  molecular neurobiology and onset of maternal behavior, 1952–1955
  ovarian histology for knockout models, **388*f***
  oviduct, **123*f***
  oviductal colonization comparison, **124*f***
  ovotestis, **249*f***
  possessing mutant ER-α, **633**
  pregnancy, **711**
  proliferative activity during pregnancy and lactation, 2996*f*
  relaxin bioactivity, **711**
  select CPE-containing mRNAs, **268*t***
  sperm, **128*f***
  spermatogonia, **832*f***
  spermatozoa, **15*f***
  spermatozoon, general features, **5*f***
  sperm frozen section, **123*f***
  SRY and, **248*f***
  steroid hormones
    actions on endometrial cells, **353–354**
    in implantation, **157–159, 158*f***
  steroid signaling in null mice, **607–608**
  testes, **792*f***
  timing of implantation, **149*t***
  transgenic by pronuclear microinjection, **263*f***
  transgenic technology, **263–264, 263*f*, 264*f***
  uterotubal junction, **122*f***
  Wobbler, **32**
Microarray technology, implantation and, **173–174**
Microdialysis
  effect of chemical changes on male sexual behavior, 1780–1781, 1781*f*
  male sexual behavior and, 1775–1778, 1776*f*, 1777*f*, 1783, 1784, 1784*f*
Micronutrients, placental transfer, 2822–2824
Microtubules, **20**
Microtus agrestis, 2286–2287*t*, 2295*t*
Microtus californicus, 2009, 2286–2287*t*, 2292, 2295*t*
Microtus montanus, 2286–2287*t*, 2292, 2295*t*
Microtus montebelli, 2286–2287*t*, 2295*t*
Microtus ochrogaster, 2010, 2051, 2286–2287*t*, 2292, 2295*t*
Microtus pennsylvanicus, 2286–2287*t*, 2289, 2292, 2295*t*
Microtus pinetorum, 2051, 2286–2287*t*
Microtus rossiaemerionalis, 2286–2287*t*
Microtus townsendii, 2286–2287*t*, 2289
Midbrain periaqueductal gray (PAG), male sexual behavior and, 1785–1787
MIF. *See* Macrophage migration inhibitory factor
Milk. *See also* Lactation
  changes in composition due to tight junction closure, 3030, 3030*f*
  changes in human milk composition during the colostrum-forming stage, 3005*f*
  developmental aspects of milk secretion, 2994–2998, 2995*f*, 2996*f*, 2997*f*
  ejection, 3035–3037, 3129–3190
    anatomical adaptations related to transfer of milk to offspring, 3130–3134
      contractile tissue of mammary gland, 3132–3133, 3132*f*
      general structure of lactating mammary gland, 3130–3131, 3130*f*
      storage of milk within mammary gland, 3131–3132
      structural adaptations of nipples for milk retention and removal, 3133–3134
    ascending and descending neural pathways in regulation of oxytocin release during, 3165–3170
      descending pathways mediating conditioning and inhibition of the milk-ejection reflex, 3168–3170
      sensory receptors and primary afferent fibers, 3165–3166
      transmission of sucking stimulus at the supraspinal level, 3167–3168, 3168*f*
      transmission of sucking stimulus through the spinal cord, 3166–3167
    characteristics of oxytocin, 3035–3036
    excitatory and inhibitory amino acids, 3172–3173
    hypothalamic regulation of oxytocin release during, 3142–3165
      adaptations of oxytocin neurons for lactation, 3163–3165, 3163*f*
      autoregulatory role of oxytocin in bursting activity of oxytocin neurons, 3156–3161, 3157*f*, 3159*f*
      firing patterns of oxytocin neurons during suckling, 3145–3147, 3146*f*, 3147*f*
      hypothalamic oxytocin neurons, 3142–3145, 3144*f*
      molecular correlates of activation of oxytocin neurons during suckling, 3147–3148

Milk *(Continued)*
ejection
hypothalamic regulation of oxytocin release during
neural mechanisms underlying synchronous bursting activity in oxytocin neurons, 3148–3156, 3150*f*, 3152*f*, 3153*f*, 3154*f*
interaction of oxytocin with myoepithelial cells, 3037
let-down reflex, 3035, 3035*f*
maternal behavior and oxytocin, 3037
neuroendocrinology of oxytocin synthesis and release, 3036
acetylcholine, 3170
dopamine, 3171–3172
histamine, 3172
monoamines and related transmitters, 3170–3172
neuropeptides, 3173–3174
nitric oxide, 3174–3175, 3175*f*
noradrenalin, 3170–3171
opioids, 3173–3174
prolactin-releasing peptide, 3174
serotonin (5-hydroxytryptamine, 5-HT), 3172
substance P, 3174
neurotransmitters and other neuromodulators in the milk-ejection reflex, 3170–3175, 3175*f*
overview, 3129–3130
oxytocin in milk-ejection reflex, 3137–3142
intramammary pressure response to oxytocin, 3141–3142
mechanically induced contractions of the myoepithelium, 3142
mode of action of oxytocin on the myoepithelium, 3140–3141
plasma oxytocin concentrations during suckling, 3139–3140, 3139*f*, 3140*f*
stimuli for activation of oxytocin release, 3137–3139
oxytocin receptors, 3036–3037
pattern of milk transfer, 3134–3137
functional significance of active milk ejection in different species, 3136–3137, 3137*f*
maternal nursing behavior, 3134
sucking behavior and dynamics of milk removal, 3134–3136, 3135*f*
regulation of secretion, 3032–3035
glucocorticoids, 3034–3035
milk removal, 3034
prolactin, 3032–3034, 3033*f*
synthesis and secretion, 3005–3035, 3006*f*
exocytic pathway for secretion of aqueous components of milk, 3010–3019, 3010*f*, 3011*f*, 3012*t*, 3013*t*, 3014*f*, 3015*f*, 3015*t*, 3016*f*, 3017*f*
glucose and amino acid metabolism in the lactating mammary alveolar cell, 3007–3010, 3007*f*
regulation of junctional complexes in the mammary gland, 3027–3032, 3029*f*, 3030*f*, 3031*f*

Milk *(Continued)*
synthesis and secretion
secretion
of lipid into milk, 3019–3023, 3019*f*, 3021*f*
of monovalent ions Na^+, K^+, and Cl^- into milk, 3023–3026, 3024*f*, 3026*f*
transcytosis, 3026–3027
volume secretion, 3005*f*
Milk-ejection reflex
oxytocin and, 3137–3142
stimuli related to conditioned oxytocin release, 3138–3139
Milk fat globule (MFG), 3020–3021, 3021*f*
Mimicry, 2023–2024
Miniopterus, **775**
Miopithecus talapoin, 2011
Mitochondrial sheath, **21–22**
Mitogen activated protein kinase (MAPK), **451**
MMPs. *See* Matrix metalloproteinases
Mn-SOD. *See* Manganese superoxide dismutase
Molecular markers, for blastocyst dormancy and activation, ***154f***
Molecules. *See also* molecule names
adhesion, **159–162**
calcium-associated signaling, **168–169**
Molothrus ater, 2026
Monkeys
age at menarche, 2181*t*
cloning of relaxin, **686**
hypothalamic–pituitary vascular cast, **1345, 1346*f***
peripubertal developmental changes, 2193, 2193*f*
placenta, ***194f***
pregnancy, **711**
timing of implantation, ***149t***
uterotubal junction, ***122f***
Monoamines
milk ejection and, 3170–3172
placenta and, 2858
Monocyte chemotactic protein-1, **501–502**
Monodelphis domestica, **114**
Monogamy, 2020–2021
Monotremes
accessory sex glands, **788–789**
epididymis, **785**
lactation, 3037–3038, 3038*f*
testes, **772–773**
size variation, **777–778**
Morphine
male sexual behavior and, 1783
role in estrous cycle, 2357
Morphogens, **172–173**
Morula, compaction, **290–291**
MPOA. *See* Medial preoptic area
MRI. *See* Magnetic resonance imaging
mRNAs
select CPE-containing, ***268t***
translational control of maternal, **267–268**
MSH. *See* Melanocyte-stimulating hormone
α-MSH, reproduction and, 2589–2590
MT-1-MMP. *See* Membrane type-1 matrix metalloproteinase

Muc-1, role in implantation, **160**
Mucous barrier, **118**
Müllerian ducts
AMH and, **327–328**
dimorphic development, ***316f***
Multifunctional transcription factors, **994–995**
Muntiacus reevesi, **775**
Mus musculus, 2008
pregnancy, **711**
Mustela furo, 2286–2287*t*, 2290, 2295*t*
Mustela putoris, **155**
Mustela vision, **154**
Mustela vison, 2286–2287*t*, 2295*t*
Myeloid cell leukemia-1 (Mcl-1), **451**
Myometrial contractile activity, relaxin and, **717–719, 718*f***
Myometrium
parturition and, 2931–2932
Myometrium, parturition and, 2926–2927
Myosorex varius, **130**

N

N-acetylglucosaminidase, ***17t***
National Center for Biotechnology Information, **384**
Natriuresis, 3082
Natriuretic peptides, as neuropeptide action on LH secretion, ***1431t***
NE, effect on male sexual behavior, 1774, 1788
Neisseria gonorrhoeae, orchitis and, **1234**
Neonates
development of primordial follicles in the neonatal rat ovary, 2064*f*
estrogen regulation, **1041–1043**
stressors and development, 2661–2662
Nerve growth factor (NGF), **451, 1542**
receptors and, **389–390**
as regulator of the estrous cycle, 2349–2350
tyrosine kinase receptor, **451**
Nervous system, sexual differentiation, **255–256**
Neural–immune–endocrine axis, control of testicular steroidogenesis, **1202*f*, 1227–1229**
Neuroanatomy
as bases of maternal experience and olfactory learning, 1963–1966
maternal behavior and, 1937–1956, 1941*f*, 1943*f*, 1944*f*, 1945*f*, 1947*f*, 1948*f*, 1956*f*
Neurobiology, of male sexual behavior, 1729–1823
Neurochemistry
as bases of maternal experience, 1963
maternal behavior and, 1937–1956, 1941*f*, 1943*f*, 1944*f*, 1945*f*, 1947*f*, 1948*f*, 1956*f*
Neuroendocrine system
control of gonadotropin secretion, 2476–2485, 2479*f*, 2482*f*
control of mating-induced ovulation, 2283–2326

Neuroendocrine system *(Continued)*
control of the ovarian cycle of sheep, 2389–2447
control of the ovarian cycle of the rat, 2327–2388
control system and timing of the menstrual cycle, 2486–2487
frequency changes of the GnRH pulse generator during the menstrual cycle, 2485–2486
gonadotropin secretion during the menstrual cycle, 2475–2476
Neurohormones, control of the estrous cycle and, 2327*f*, 2341–2345, 2342*f*
Neurohypophysis, **1329–1343**
development, **1331–1333, 1334*f***
nerve terminals, **1329–1331, 1330*f***
organization, **1333–1343**
infundibular process, **1333–1337**
infundibular stem, **1337**
infundibulum (median eminence), **1337–1343, 1338*f***
Neurokinin B (NKB)
biological fluids, 2858
expression and localization, 2857
placenta and, 2857–2858
receptors, 2857–2858
as steroid-responsive neuron, 2399*t*
Neuronal regulation, epididymis and, **1118**
Neurons, female reproductive behavior and, 1835
Neuropeptides, **1342–1343**
effect on female reproductive behaviors, 1850–1858
milk ejection and, 3173–3174
Neuropeptide Y (NPY), **1436–1437**
biological fluids, 2855–2856
expression and localization, 2855
male sexual behavior and, 1758
neurons, 2083
oxytocin and, 3094–3095
placenta and, 2855–2856
puberty and, 2069
receptors, 2855
as regulator of the estrous cycle, 2347–2348
regulatory actions and reflex ovulation, 2307–2308, 2307*f*, 2308*f*
reproduction and, 2585–2589, 2588*f*
seasonal rhythms of reproduction and, 2259
as steroid-responsive neuron, 2399*t*
Neurotensin
estrogen interactions, **1449**
expression, **1530**
as neuropeptide action on LH secretion, **1431*t***
Neurotransmitters
biogenic amine controlling the estrous cycle, 2345–2347
effect on female reproductive behaviors, 1840–1850, 1844*t*
male sexual behavior and, 1779, 1787–1788
Neutrophils, vascular inflammation models and, **1229–1231**
Newborn ovary homeobox gene (Nobox), **388**
NF-κβ, follicle-stimulating hormone and, **564–566, 565*f***
NGAL. *See* Lipocalin 2
NGF. *See* Nerve growth factor
Nigrostriatal DA tract, male sexual behavior and, 1781–1782
Nipple
relaxin and, **727–728, 727*f***
structural adaptations for milk retention and removal, 3133–3134
Nitric oxide
effect on male sexual behavior, 1756–1757, 1774–1775
male sexual dysfunction and, **1178**
milk ejection and, 3174–3175, 3175*f*
oxytocin and, 3095–3096
Nitric oxide synthases (NOSs), **452**
Nitrogen, immunoregulation, **1225–1227, 1226*f***
NKB. *See* Neurokinin B
NMDA. *See N*-methyl-D-aspartate
N-methyl-D-aspartate (NMDA), **1425**
male sexual behavior and, 1783
seasonal rhythms of reproduction and, 2260
stressors and, 2635
NO, effects on male sexual behavior, 1788
Nobox. *See* Newborn ovary homeobox gene
Noradrenalin, milk ejection and, 3170–3171
Noradrenergic neurons, 2083
puberty and, 2068–2069
Norepinephrine, **1335, 1434–1435**
effect of estradiol on, 2405*t*
male sexual behavior and, 1752
as neurotransmitter action on LH secretion, **1430*t***
reflex ovulation and, 2305–2307, 2305*f*, 2306*f*, 2307*f*
as steroid-responsive neuron, 2399*t*
NOS, **566**
NOSs. *See* Nitric oxide synthases
Notopthalmus viridescens, 2024
NPY. *See* Neuropeptide Y
NRs. *See* Nuclear receptors
Nuclear factor-κB, **455–456**
Nuclear lamina, **12–13**
Nuclear receptors (NRs), **594.** *See also* individual names
co-activators and co-repressors, **998*f***
coregulators, **598*t***
modular structure, **996–997*f***
response elements, **599*f***
Nucleus
envelope, **12**
proteins, **12**
Nursing, maternal behavior, 3134
Nutrients, placental transfer, 2809–2817
Nutrition
dietary regulation during lactation, 3022–3023
evolutionary diet comparison, 2981*t*
gonadotropin secretion in pigs and, 2519
influence on gonadotropin secretion in humans, 2533
Nutrition *(Continued)*
maternal adaptation during pregnancy, 2901
as metabolic stressor that impairs reproduction, 2559–2561, 2560*f*
ovarian cyclicity in the ewe and, 2425–2426
oxytocin in food intake, 3085–3086
reproduction and, 2566–2584
transgenerational influences on diet, 2986

O

Obesity
as cause of hypogonadotropic hypogonadism, **1047*t***
reproductive function and, 2596
Ob-R, leptin and, 2571–2572
ODF1, **24*t***
ODF2, **24*t***
ODF3, **24*t***
ODF4, **24*t***
Odontaspis taurus, isolation of relaxin, **684**
Offspring, conflict, 2029–2030
OIP1, **24*t***
Olfactory learning, neuroanatomy and, 1963–1966
Olfactory system
accessory, 2042–2043, 2043*f*, 2043*t*
main, 2042
Oligosaccharide structures, **1607–1608, 1608*f***
Oncorhynchus keta, **1595**
Oncorhynchus nerka, 2017
Oncorhynchus tshawytscha, 2018
Oocyte
cigarette smoke interference with embryo transport, **136**
embryo transport in the oviduct, **132–136**
immune protection for developing, 2762–2763
microenvironment, **430**
Oocyte–granulosa cell interactions, **391**
Oogenesis, in the rat, 2329–2330, 2330*f*
Opiatergic neurons, 2082
Opiates
as cause of hypogonadotropic hypogonadism, **1047*t***
effects on male sexual behavior, 1788
male sexual behavior and, 1783
oxytocin and, 3076
Opioid peptides, **1333–1335, 1437–1439, 1438*f***
expression and localization, 2853
female reproductive behavior and, 1855
levels in biological fluids, 2853–2854
modulatory role and stress, 2648–2651, 2649*f*
placenta and, 2853–2854
puberty and, 2069
receptors, 2853
Opioids
effect on male sexual behavior, 1755, 1774

Opioids *(Continued)*
- gonadotropin secretion and
  - in cows, 2525
  - in pigs, 2519
  - in rats, 2515
- milk ejection and, 3173–3174
- onset of maternal behavior and, 1951–1952
- oxytocin and, 3102–3104
- reflex ovulation and, 2308–2309
- role in estrous cycle, 2357
- seasonal rhythms of reproduction and, 2259

Opossum, 806
Orchidectomy, regulation of epididymal functions and, **1092–1093**
Orchitis, immunological response, **1233–1235**
Oreochromis mossambicus, 2019
Orexin (hypocretins)
- estrous cycle and, 2348
- male sexual behavior and, 1784
- as neuropeptide action on LH secretion, **1431*t***
- reproductive axis and, 2595–2596

Orexin receptors, **1105**
Organic ions, placental transfer, 2808–2809, 2808*t*
Organic solute transporters, **1109–1110**
Orphanin FQ (Prepronociceptin), **453**
- as neuropeptide action on LH secretion, **1431*t***

Oryctolagus cuniculus, **114,** 2286–2287*t*, 2295*t*
Oryzias latipes, 2011
Outer dense fibers, **22–26**
- associated proteins, **24*t***
- composition, **23–25**
- formation, **22–23**
- function, **25–26**
- general features, **22**
- satellite fibers, **25**
- structure, **23**

Ovarian cancer, insulin-like growth factors and, **537**
Ovarian cycle
- active luteal phase of the rat, 2363–2367
  - control of mating-induced prolactin secretion, 2364–2367
  - prolactin secretion, 2364, 2364*f*
  - "rescue" of the corpus luteum, 2363–2364
  - short luteal phase significance, 2367
- neuroendocrine control
  - in rats, 2327–2388
  - in sheep, 2389–2447

Ovarian hormones, 2481–2485. *See also* Estradiol; Inhibins; Progesterone
- negative feedback actions, 2481–2482, 2482*f*
  - of ovarian inhibin, 2485
- positive feedback actions, 2482–2483
- sites of action
  - of estradiol, 2483–2484
  - of progesterone, 2484–2485
- surge-blocking action of progesterone, 2482

Ovarian Kaleidoscope database, **384**
Ovariectomy, estradiol in cows and, 2524–2525, 2524*f*
Ovariectomy cell, **1500*f***
Ovaries
- activation, 2079–2081, 2080*f*
- aging and, 2729–2730
- androgen receptor signaling function, **648–653**
- competency before puberty, 2131, 2131*f*
- CRH and, 2655–2656
- diapause and, **155**
- embryonic development and puberty, 2064
- endocrine and ovarian phenotypes in mice null for progesterone signaling, **646*t***
- estrogen receptor signaling function, **619–644**
- extraovarian roles of sex steroids in ovarian function, **653–657**
- formation, **316–317**
- gonadotropin signaling, **547–592**
- gross anatomy of ovarian tumor and cyst formation in mouse models, **393*f***
- growth hormone and, **527–529**
- hormonal control, 2073–2076, 2074*f*
- insulin-like growth factor and, **529–539**
- intraovarian role of androgen in ovarian function, **650–653**
- intraovarian role of progesterone in ovarian function, **646–648**
- intraovarian roles of estradiol in function of, **633–634**
- mechanisms of development, **323**
- nerves and follicular steroidogenesis, 2076–2077
- neural control, 2076
- oxytocin and, 3064
- paracrine oxytocin systems and, 3068
- phenotypes
  - in murine models of disrupted androgen signaling, **649–650, 649*t***
  - in murine models of disrupted estrogen signaling, **627–633, 627*f*, 628*t*, 631*f***
  - in murine models of disrupted progesterone signaling, **646, 646*t***
- pheromones and, 2050–2051, 2050*t*
- premature ovarian failure, **407–408**
- progesterone receptor signaling function, **644–648**
- puberty and, 2180–2181, 2181*t*
- sex steroid receptor
  - actions, **656*f***
  - function, **619**
- steroid hormone receptors in, **593–678**

Oviduct
- anatomy, **338–340, 338*f***
- cell constituents, **338–340**
- cyclic changes, **337–381**
  - in primate, **341–343, 342*f***
- embryology, **338–340**
- estrogen and progesterone receptors, **343–345**

Oviduct *(Continued)*
- extracellular signaling pathways in early development, **288, 289*t*–290*t*, 291*f***
- extracellular signals influencing early development, **286–288**
- growth factor and cytokine expression and action, **345–347**
- historical perspective and functions, **340–341**
- human, **341**
- immunohistochemical localization of steroid receptors in, **343–344**
- microenvironment, **286–288**
- nonfertilizing sperm, **132**
- nonhuman primate, **341–343, 342*f***
- secretions and oviductin, **347–349**
- sperm reservoir, **125–131**

Oviductin, secretions and, **349–352, 350*f*, 351*f***
Ovis ammon, 2127
Ovis musimon, 2127
Ovis orientalis, 2127
Ovulation, **425–474, 652–653,** 2283–2326
- biochemical reactions in, **431–440, 431*f***
  - progesterone, **432–435, 432*f*, 433*f*, 434*f*, 435*f***
  - prostaglandins, **435–438, 436*f*, 437*f***
  - proteases, **438–440**
- control in sheep, 2410
- control of rats, 2410
- estradiol and, **641–642**
- follicular rupture in sheep, 2394–2395
- luteinization and, **568–570**
- morphological features, **426–431, 426*f*, 427*f***
  - follicular apex
    - at the beginning of the ovulatory process, **427–430, 428*f*, 429*f*, 430*f***
    - at less than five minutes before rupture, **430–431, 431*f***
    - at less than one hour before rupture, **430, 430*f*, 431*f***
- neuroendocrine control of mating-induced, 2283–2326
  - endocrine correlates of coitus-induced ovulation, 2294–2302
    - gonadotropin-releasing hormone, 2296–2297
    - hormonal profiles and latencies, 2295*t*
    - luteinizing hormone and follicle-stimulating hormone, 2297–2300, 2298*f*, 2299*f*
    - ovarian progestins, 2301–2302
    - prolactin, 2300–2301
  - evolution, adaptive function, and phylogenetic distribution, 2285–2289, 2286–2287*t*, 2288*f*
  - how spontaneous begets induced ovulation, 2312–2314
    - aging and persistent estrous in rats, 2313
    - anesthetics and estrous cycles, 2312

Ovulation *(Continued)*
neuroendocrine control of mating-induced
how spontaneous begets induced ovulation
brain grafts restore fertility and induced ovulation in mouse mutants lacking GnRH, 2313–2314
circadian clocks and induced ovulation in rats, 2312–2313
male-facilitated ovulation and puberty, 2309–2312
ovulation advance in seasonally breeding ungulates, 2309–2310
in rodents that normally exhibit spontaneous cycles, 2310–2311, 2311*f*
neural pathways and neurotransmitters mediating control of reflex ovulation, 2302–2309
acetylcholine, opioids, and other putative neuroregulatory factors, 2308–2309
brainstem component and role of norepinephrine, 2305–2307, 2305*f*, 2306*f*, 2307*f*
hypothalamic component and GnRH neurons, 2303–2305
regulatory actions of neuropeptide Y, 2307–2308, 2307*f*, 2308*f*
overview, 2284–2285
social, seasonal, and behavioral parameters affecting induction of ovulation in reflex ovulators, 2286–2287*t*
stimuli requirements, 2289–2294, 2290*f*
behavioral estrus of reflex ovulators, 2289–2291
exogenous stimuli for inducing the LH surge and ovulation, 2291–2293
nontactile cues influencing ovulation in reflex ovulators, 2293–2294
ovarian gene expression during, **440–458**
catalog of current genes, **443–455**
comments about ovarian gene expression, **455–458**
complexity of integrating gene expression data, **440–443, 441–442*t***
overview, **425–426, 426*f***
progesterone receptor signaling and, **647–648, 647*f***
in the rat, 2332
timing of sperm deposition and, **114**
Oxygen
immunoregulation, **1225–1227, 1226*f***
placental transfer, 2817–2821, 2818*f*, 2818*t*, 2819*f*, 2820*f*, 2820*t*, 2821*f*
Oxytocin, **1333**
adaptations of neurons for lactation, 3163–3165, 3163*f*
autoregulatory role in bursting activity of neurons, 3156–3161, 3157*f*, 3159*f*
biological fluids, 2857
breast cancer and, 3065

Oxytocin *(Continued)*
characteristics, 3035–3036, 3171–3172
effects on male sexual behavior, 1788
electrophysiological action, 1857
expression and localization, 2856–2857
female reproductive behavior and, 1856–1857
hypothalamic neurons, 3142–3145, 3144*f*
interaction with myoepithelial cells, 3037
lactation and, 3035–3036, 3075
male sexual behavior and, 1755–1756, 1782, 1782*f*
maternal behavior and, 3037
menstruation and, 2463
milk-ejection reflex and, 3137–3142, 3168–3170
molecular synthesis, 1856–1857
neuroendocrinology of synthesis and release, 3036
neuron activation in parturition, 3093*f*
as neuropeptide action on LH secretion, **1431*t***
onset of maternal behavior and, 1949–1951
parturition and, 2951–2952
placenta and, 2856–2857
receptor antagonists, 3077, 3080
receptors, **1103–1104,** 1857, 2857, 3036–3037, 3062–3063, 3081*f*
synthesis, secretion, and reproductive functions, 3055–3128
biosynthetic processing and posttranslational modification, 3072–3073
central oxytocin system, 3065–3067, 3066*f*
control of cyclical uterine oxytocin receptor expression, 3080–3082, 3081*f*
endocrine oxytocin system, 3056–3065, 3057*f*, 3058*f*, 3063*f*
gene structure and gene products, 3068–3070, 3069*f*
luteolysis, 3078–3080, 3079*f*
overview, 3055–3056
oxytocin and anterior pituitary gland, 3077–3078
oxytocin and control of behaviors related to reproduction, 3086–3090, 3089*f*
oxytocin and hydromineral balance, 3082–3084, 3083*f*
oxytocin in food intake, 3085–3086
oxytocin in memory and learning processes, 3086
oxytocin in salt appetite, 3085
oxytocin promoter, 3070–3071
paracrine oxytocin systems in peripheral organs, 3067–3068
posttranscriptional regulation of oxytocin mRNA, 3071–3072
regulation of oxytocin gene expression in reproduction, 3073–3075
role for oxytocin in parturition, 3075–3077

Oxytocin *(Continued)*
synthesis, secretion, and reproductive functions
secretory activity of oxytocin neurons in pregnancy and parturition, 3090–3104
pathways to parturition, 3090–3096, 3091*f*, 3093*f*
premature oxytocin secretion in pregnancy prevention, 3098–3104, 3100*f*
pulsatile oxytocin secretion, 3096–3097
somatodendritic oxytocin release, 3097–3098
stress, 3084–3085

P

p38, **573–574**
p50, **17*t***
PA. *See* Plasminogen activators
PACAP. *See* Pituitary adenylate cyclase activating polypeptide
PAF. *See* Platelet-activating factor
PAG. *See* Midbrain periaqueductal gray
PAI-1. *See* Plasminogen activator inhibitor-1
Pair bonding, 2009–2010
oxytocin and, 3089–3090, 3089*f*
Pancreas, paracrine oxytocin systems and, 3068
Pancreatitis associated protein-III (PAP-III), **452**
Papaverine, male sexual function and, **1185**
PAP-III. *See* Pancreatitis associated protein-III
Papio anubis, 2015
Papio ursinus, pregnancy, **711**
Parablennius sanguinolentus parvicornis, 2025
Paracrine
epidermal growth factor and, **1540**
menstrual cycle regulation and, 2465–2466
parturition and, 2934–2942
Paracrine/autocrine function
actions on fetal–placental blood flow regulation, 2876–2879
mechanisms regulating placental hormonogenesis, 2870–2874
Parathyroid hormone-related peptide (PTHrP)
expression and localization, 2868
levels in biological fluids, 2868
placenta and, 2868
receptors, 2868
Paraventricular nucleus of the hypothalamus, male sexual behavior and, 1782–1784, 1782*f*
Parents
aggression, 2029
births against age of, 2700, 2700*f*
offspring conflict, 2029–2030
Pars distalis, **1322–1327, 1322*f***
epithelial cell, **1316, 1317–1318*f***
Pars intermedia, **1327–1328**

Pars tuberalis, **1328–1329**
Parthenogenesis, 2022–2023
Parturition, 2925–2974. *See also* Pregnancies
  contraction of the myometrium, 2926–2927
  endocrine changes associated with maternal behavior and, 1924–1925, 1924*f*
  future advances, 2955
  overview, 2925–2926
  oxytocin and, 3075–3077
    neuron activation in, 3093*f*
  phases and their regulation, 2927–2955
    phase 0, 2927–2930
    phase 1, 2930–2942
    phase 2, 2942–2955
  relaxin and, **714–726**
  secretory activity of oxytocin neurons in, 3090–3104
  stress and, 3077
  timing and duration, **720–721, 721*t***
Passer domesticus, 2015
Pathogens, signaling in the immune system, **1205*f***
Pax-2, **328–329**
Pax-8, **328–329**
PBR. *See* Peripheral benzodiazepine receptor
PCD. *See* Primary ciliary dyskinesia
PDE5, for treatment of ED, **1182–1184**
PDEF, cardiovascular and hemodynamic effects, **1183–1184**
PDGF. *See* Platelet derived growth factor
PE. *See* Preeclampsia
PEA3 family, **1106**
Penis
  anatomy, **789–790, 790*f***
    and mechanics of erection, 1736–1738, 1737*f*
  cellular mediators of erection, 1741–1743, 1742*f*, 1743*f*
  erections, **1175–1179**
    anatomy, **1175–1176, 1175*f***
    biochemical mechanisms, **1177–1179, 1178*f*, 1179*f***
    blood supply to the penis, **1176, 1176*f***
    drug-induced, 1735
    hemodynamic changes during, **1177**
    innervation of the penis, **1176*t***
    neural input into the penis, **1176–1177, 1176*t***
    nocturnal, **1180**
    noncontact, 1735
    spontaneous, 1735
    touch-based, 1735, 1736*f*
    vacuum devices for inducing, **1185**
  innervation, **807**
  measures of penile function, 1734–1736
    ex copula, 1735–1736
    observations during copulation, 1734–1735
  neural innervation, 1738–1739, 1739*f*, 1740*f*
  oxytocin and, 3064
  relaxation, 1742*f*, 1743*f*
  role in male sexual behavior, 1736–1743
Penis *(Continued)*
  tactile receptors, 1741
  vasculature, **802–803**
PEPCK. *See* Phosphoenolpyruvate carboxykinase
Peptidergic neurohormones, estrous cycle and, 2341–2345, 2342*f*, 2343*f*
Peptides
  antimicrobial, **1110–1111**
  atrial natriuretic, 3084
  effects on testicular blood flow, **798*t***
  female reproductive behavior and, 1834–1835
  inhibin-related, 2859–2860, 2860*f*
  male sexual behavior and, 1757–1758
Peptide-secreting cells, **487–489**
PERF 15, **8*t***
Perforatorium, **13**
Perinuclear theca, **13**
Peripheral benzodiazepine receptor (PBR), **988**
Peristalsis, uterine, **120–121**
Peritubular tissue, cytology, **870*f*, 889**
Persistent vaginal estrus (PVE), 2310
PGC. *See* Primordial germ cells
PGE_1. *See* Prostaglandin E_1
PGE_2. *See* Prostaglandin E_2
PGH. *See* Placental growth hormone
Phagocytosis, of surplus sperm, **132*f***
Phalaropus tricolor, 2015
Phascolarctos cinereus, **777**
Phasogale calura, 2018
Phenacomys longicaudus, 2286–2287*t*
Phenotypic sex, **254–256**
  sexual differentiation
    of the nervous system, **255–256**
    of reproductive organs, **254–255, 255*f***
Phentolamine, male sexual function and, **1185**
Pheromones, 2041–2057
  flehmen, 2042–2043
  ovarian cyclicity in the ewe and, 2426, 2427*f*, 2428*f*
  overview, 2041–2042
  priming, 2047–2052
    ovarian cyclicity, 2050–2051, 2050*t*
    pregnancy block, 2051–2052
    puberty, 2092
      acceleration, 2047–2048, 2047*t*
      inhibition, 2048–2050
  reception of chemosignals, 2042–2044
    olfactory system
      accessory, 2042–2043, 2043*f*, 2043*t*
      main, 2042
    secondary chemical receptors, 2043–2044
  signaling, 2044–2047
    individual identification and mate selection, 2046–2047
    sex attractants, 2044–2046
Phocaena phocaena, isolation of relaxin, **684**
Phocoena phocoena, **776**
Phosphate, transport in the mammary alveolar cell, 3018
Phosphodiesterase, **452**
Phosphoenolpyruvate carboxykinase (PEPCK), **403–404,** 2978
Phospholipase C, **1643–1646, 1645*f***
Phospholipase/phosphoinositols, **1711**
Phosphorylation, sperm motility and, **35–37**
Photoperiods, seasonal regulation of reproduction and, 2234
Photoreceptors, seasonal rhythms of reproduction and, 2237, 2237*f*
Physalaemus pustulosus, 2021
PIF. *See* Prolactin-inhibiting factor
Pigs
  covalent structure of relaxin, **683*f***
  ductus deferens, **806*f***
  estrogen and implantation, **152**
  estrous cycle, **706**
  gonadotropin secretion, 2516–2519
    estrus and ovulation, 2516–2517
    gonadotropins, 2517–2518, 2517*f*
    induction of ovulation during lactation, 2518–2519
    ovarian activity, 2517
    prolactin, opioids, and nutritional status, 2519
  isolation of relaxin, **682–683**
  placenta, **194*f***
  pregnancy, **706–707, 706*f*, 707*f***
  relaxin sources, **707**
  timing of implantation, **149*t***
  vesicular gland, **806*f***
Pig tail monkey, uterotubal junction, **122*f***
PI-3K pathway, follicle-stimulating hormone and, **557–558, 558*f***
Pineal gland
  melatonin regulation in breeding seasonality, 2239–2240, 2240*f*, 2245–2248, 2247*f*
  puberty and, 2091–2092
Pituicytes, **1336–1337**
Pituitary adenylate cyclase activating polypeptide (PACAP), **435, 452–453**
Pituitary gland, **1289–1307.** *See also* Hypothalamus
  aging and, 2732–2733
  anterior cytology, **1292**
  blood supply, 2341*f*
  canine, **1310, 1310*f***
  competency to secrete gonadotropins before puberty, 2132–2133, 2132*f*, 2133*f*, 2134*f*
  folliculogenesis and, **395–399**
  function during puberty, 2183–2184
  gonadal secretions on pituitary secretion, **1298–1301**
    biphasic action of progesterone, **1299**
    episodic gonadotropin release, **1299–1300**
    nonsteroid gonadal feedback, **1301**
    steroids and phasic release of gonadotropins, **1300–1301**
  human, **1312, 1312*f***
  hypophysis terminology, **1290*t***
  neural control of pars distalis secretion, **1292–1298**
    brain stimulation of gonadotropin in rats, **1297–1298**

Pituitary gland *(Continued)*
neural control of pars distalis secretion
curvilinear pattern of ovulatory gonadotropin surges, **1298**
hypothalamic-pituitary control of ovulation, **1295–1297**
neurohumoral regulation, **1294–1295**
pituitary portal vessels, **1293–1294**
onset of puberty and, 2077–2079, 2078*f*
overview, **1289–1292**
oxytocin and, 3077–3078
paracrine oxytocin systems and, 3068
Pituitary hormone receptors, spermatogenesis and, **1025–1026**
Pituitary hormones
secretion during the estrous cycle, 2335–2339, 2337*f*, 2338*f*
of the rat, 2350–2363
secretion during the estrous cycle of the rat
control of follicle-stimulating hormone secretion, 2357–2360, 2358*f*
control of LH secretion, 2350–2357, 2352*f*
control of prolactin secretion, 2360–2363
Pituitary-like hormones, placenta and, 2852–2855
Pitymys subterraneus, 2286–2287*t*, 2295*t*
PKA. *See* Protein kinase A
signaling in preovulatory cells, **572–573, 573*f***
sperm motility and, **35**
PKC. *See* Protein kinase C
Placenta
anatomy and genesis, **189–243**
basic villous structure, **203–213**
cytotrophoblast (Langhans' cells), **203–205, 204*f***
fetal vessels, **209–212, 210*f*, 211*f***
fixed connective tissue, **209**
Hofbauer cells, **209**
integrity of the placental barrier, **212–213**
syncytial fusion, **205–207**
syncytiotrophoblast, **207–208**
trophoblast turnover, **208–209**
villous trophoblast, **203**
comparative placentation, **193–197**
maternal–fetal barrier, **195–196, 195*f***
maternal–fetal interdigitation, **194, 195*f***
placental evolution, **197**
placental shape, **194**
vascular arrangement, **196**
CRH and, 2663–2664
early development, **199–203**
early villous stage, **202–203, 203*f***
lacunar stage, **201–202**
prelacunar stage, **199–201, 200*f***
genetic control of placental development, **228–230**
genetic control, **229**
imprinting and placental development, **229–230**
mouse placenta, **228–229**
Placenta *(Continued)*
historical survey, **189–193, 190*f***
immunobiology and, 2766–2768, 2767*f*
luteal–placental shift, 2473–2475
macroscopy of delivered, **197–199**
morphological correlates of noninvasive diagnostic methods, **230–232**
Doppler ultrasound, **231–232**
magnetic resonance imaging, **232**
ultrasound, **230–231**
multiple pregnancies, **199, 199*f***
nonvillous parts, **222–228**
basal plate and uteroplacental vessels, **224**
cell columns and cell islands, **224–225**
chorionic plate, secondary yolk sac, umbilical cord, and membranes, **222–224**
composition and origin, **227**
extravillous trophoblast, **225**
fibrinoid, **226–227**
functional aspects, **227–228**
placental septa, **224**
uteroplacental vessels, **225–226**
paracrine oxytocin systems and, 3067
in situ, **198*f***
stereological data for normal term, **220*t***
villous trees, **213–222**
development, **214–215**
fetal angioarchitecture, **215–216, 216*f***
intervillous space, **216–218**
oxygen as regulator of villous development, **218–220**
stereological data, **220–222**
structure of villous types, **211*f*, 212*f*, 213–214**
Placental barrier, **204*f***
integrity, **212–213**
Placental endocrine function, 2847–2897
chromogranin A (CgA), 2856
clinical usefulness of placental hormones measurement throughout pregnancy, 2879–2883
Down's syndrome, 2880–2881, 2880*f*
ectopic pregnancy, 2879
fetal growth restriction (FGR), 2882–2883
hydatidiform mole, 2879–2880
preeclampsia (PE), 2881–2882
preterm delivery, 2883
growth factors, 2858–2865
estrogens, 2864–2865
fibroblast growth factor (FGF) and epidermal growth factor (EGF), 2862–2863
inhibin-related peptides, 2859–2860, 2860*f*
insulin-like growth factor and insulin-like growth factor-binding protein (IGF and IGF-BP), 2860–2862
progesterone, 2864–2865
transforming growth factor-β (TGF-β), 2858–2859
vascular endothelial growth factor (VEGF), 2863–2864
hormones produced by intrauterine tissues, 2848*t*
Placental endocrine function *(Continued)*
hypothalamus-like hormones, 2848–2852
corticotropin-releasing factor and CRF-related peptides, 2849–2851, 2850*f*
gonadotropin-releasing hormone, 2848–2849
growth hormone-releasing hormone and somatostatin, 2851–2852
metabolic hormones, 2868–2870
ghrelin, 2869–2870
leptin, 2868–2869
monoamines, 2858
neurokinin B (NKB), 2857–2858
neuropeptide Y (NPY), 2855–2856
overview, 2848, 2848*t*
oxytocin, 2856–2857
paracrine/autocrine actions on fetal–placental blood flow regulation, 2876–2879
ADM effects, 2878
CGRP effects, 2878–2879
CRF and NPY effects, 2876–2877, 2877*f*
ET-1 effects, 2878
NKB effects, 2877–2878
PTH-rP effect, 2879
paracrine/autocrine mechanisms regulating placental hormonogenesis, 2870–2874
CRF-ACTH-opioid axis, 2873–2874, 2873*f*
GHRH-PGH-hPL-IGF axis, 2872–2873
GnRH-hCG-inhibins-steroid hormone axis, 2870–2872, 2871*f*
pituitary-like hormones, 2852–2855
human chorionic gonadotrophin (hCG), 2852–2853
human placental lactogen (hPL), 2855
opioid peptides, 2853–2854
placental growth hormone (PGH), 2854
placental hormones, cell growth, and differentiation, 2874–2876
placental angiogenesis, 2875–2876
trophoblast differentiation and invasion, 2874–2875, 2875*f*
vasoactive peptides, 2865–2868
adrenomedullin (AM), 2865–2866
calcitonin gene-related peptide (CGRP), 2867–2868
endothelin (ET), 2866–2867
parathyroid hormone-related peptide (PTHrP), 2868
vasopressin, 2856–2857
Placental growth hormone (PGH), 2854
biological fluids, 2854
expression and localization, 2854
receptors, 2854
Placental septa, 224
Placental transfer, 2787–2846
amino acids transport, 2809–2811, 2810*t*, 2811*t*
animal models, 2795, 2795*t*
bicarbonate transport, 2822
bulk nutrients transport, 2809–2817

Placental transfer *(Continued)*
calcium transport, 2805–2807, 2806*f*
carbon dioxide transport, 2821–2822
chloride transport, 2803–2805, 2804*f*
comprehensive model, 2795–2796, 2797*f*, 2798*f*
concentration and electrical gradients, 2798–2799
concepts of transfer, 2793–2795, 2793*f*, 2794*t*
driving forces, 2796
drugs transport, 2824–2827, 2824*f*, 2826*t*
free fatty acids transport, 2815–2816
glucose transport, 2812–2814
human maternal and fetal plasma or serum concentrations of selected substances, 2788–2790*t*
inorganic ions transport, 2802–2805
interspecies differences in placental structure, 2787, 2791*f*, 2792
iodide transport, 2805
iron transport, 2822–2824
ketones transport, 2816–2817
lactate transport, 2808–2809, 2808*t*
lipids transport, 2814–2817, 2814*t*, 2815*t*
magnesium transport, 2807–2808
micronutrients transport, 2822–2824
ontogeny of transfer mechanisms, 2796–2799
blood flow, 2797–2798
exchange barrier dimensions, 2798
transporter-mediated transfer, 2799
organic ions transport, 2808–2809, 2808*t*
oxygen transport, 2817–2821, 2818*f*, 2818*t*, 2819*f*, 2820*f*, 2820*t*, 2821*f*
paraplacental transport, 2792–2793
potassium transport, 2803
protein transport, 2801–2802
respiratory gases transport, 2817–2822
sodium transport, 2802–2803
studies, 2794*t*
triglycerides transport, 2815–2816
water transport, 2799–2801
aquaporins, 2800
control, 2801
paracellular channels, 2800
xenobiotics transport, 2824–2827, 2824*f*, 2826*t*
Plasma membrane, sperm capacitation and, **60**
Plasma osmolality, regulation, **721–722**
Plasminogen activator inhibitor-1 (PAI-1), **453**
Plasminogen activators (PA), **438–439**
Platelet-activating factor (PAF), **135**
Platelet-derived growth factor (PDGF), **252**
Pleurodeles waltlii, **1293–1294**
Plocepasser mahali, 2025
PMSG. *See* Pregnant mare serum gonadotropin
POD-1, **323**
Poecilia formosa, 2022–2023
Poecilia reticulata, 2012
Poeciliopsis lucida, 2023
Poeciliopsis monacha, 2023
Poephila guttata, 2004
POF. *See* Premature ovarian failure
Pogonomys macrourus, **777**
Polyandry, 2020–2021
Polycystic ovary syndrome (PCOS), **535–537, 536*f*, 537*f***
reproduction and, 2598–2599
Polygyny, 2020–2021
Polyspermy, **125**
POMC. *See* Pro-opiomelanocortin peptides; Pro-opiomelanocortin system
Porpoise, isolation of relaxin, **684**
Portal system, **1343–1351**
capillaries of the hypophysis, **1343, 1343*f*, 1344*f***
development, **1343–1344**
organization, **1344–1351, 1346*f*, 1347*f*, 1348*f*, 1349*f*, 1350*f***
Potassium, placental transfer, 2803
PPARδ, role in implantation, **168**
ppN. *See* Prepronociceptin
Precocious puberty, 2211–2212. *See also* Puberty
Predators, 2028–2029
Preeclampsia (PE), placental hormones and, 2881–2882
Pregnancies. *See also* Parturition
allomothering, 1957–1958
block, pheromones and, 2051–2052
cats, **713**
clinical usefulness of placental hormones measurement throughout, 2879–2883
coagulation proteins in, 2908, 2908*f*
CRH and stress, 2664–2665
diagrammatic representation of gestational sac, **203*f***
dogs, **713**
ectopic, 2879
extension of luteal function in early, **502**
guinea pig, **713–714**
hamsters, **713**
hemodynamic changes, 2904*t*
human, **708–709, 709*f***
immunobiology of, 2759–2785
cycling human uterus, 2761–2762
leukocytes in, 2761–2762, 2762*t*
physiological features, 2761
genetic disparity and post-implantation evasive strategies, 2771–2779
foreign antigens and immunity, 2771–2772
mechanisms preventing maternal immune rejection of the fetus, 2772–2779, 2774*f*, 2775*f*, 2776*f*, 2778*f*
implantation and maternal immune responses, 2769–2771
maternal immune responses to implantation, 2770–2771
oxygen regulation of gene expression, 2771
pregnancy as a Th1/Th2-type reaction, 2769–2770, 2770*f*
Pregnancies *(Continued)*
immunobiology of
maternal–fetal interface, 2764–2769
cell coexistence and exchange, 2769
maternal cellular components, 2764–2769, 2764*f*, 2767*f*
multifunctionality of immune system molecules, 2779
overview, 2759–2761, 2760*f*
preimplantation events, 2762–2764
immune protection for developing oocyte, 2762–2763
immune protection for the pre-implantation embryo, 2763–2764, 2763*f*
leptin and, 2573
luteal function in early, **502–505**
maternal adaptation to, 2899–2923
cardiovascular adaptations, 2901–2908
blood volume, 2901–2904, 2901*f*, 2902*f*, 2903*f*
circulatory adaptations, 2904–2908, 2904*f*, 2905*f*, 2907*f*, 2908*f*, 2908*t*
metabolic adaptations, 2900–2901
diet, pregnancy, and fetal programming, 2901
energy costs of pregnancy, 2900–2901, 2900*f*
overview, 2899–2900
renal adaptations, 2910–2918
aminoaciduria, 2911
angiotensin 1-7/AT2 receptors, 2917–2918, 2918*f*
blood urea nitrogen and creatinine, 2910
glomerular capillary permselectivity to neutral macromolecules in human pregnancy, 2914
glomerular filtration rate and effective renal plasma flow, 2910
glomerular hemodynamics in rat pregnancy, 2913
glycosuria, 2911
intraglomerular hemodynamics in human pregnancy, 2913
kidney and extracellular fluid volume/blood pressure regulation in pregnancy, 2915–2916
mechanisms of increased GFR, 2912, 2912*f*
mediators of renal hemodynamic adaptations in human pregnancy, 2917
postpartum glomerular hemodynamics in human pregnancy, 2913–2914
protein excretion, 2911
renal hemodynamic changes in pregnancy, 2911–2912, 2911*f*
renin release, 2911
response to pregnancy, 2916–2918, 2918*f*
sodium handling, 2911

Pregnancies *(Continued)*
  maternal adaptation to
    renal adaptations
      structural changes, 2911
      structural changes of kidney and genitourinary tract, 2914–2915
      urinary protein and albumin excretion in pregnancy, 2914
      water metabolism, 2911
    respiratory system adaptations, 2908–2910
      diffusing capacity, 2910
      hyperventilation, 2909
      lung volumes and function, 2909–2910, 2909*f*
  maternal constraints, 2983–2984
  maternal health, 2985–2986
  maternal recognition, 2470–2475, 2472*f*, 2474*f*
  multiple, **199, 199*f*.** *See also* Twins
  oxytocin and, 3079
  paracrine role during early pregnancy, 2663
  parturition, **714–726.** *See also* Parturition
  pigs, **706–707, 706*f*, 707*f***
  plot of time to, 2702*f*
  postpartum maternal experience, 1958–1960, 1958*f*, 1959*f*, 1960*f*
  preterm delivery, 2883
  prevention, 3098–3104, 3100*f*
  primates, **710–711**
  prolactin cells and, **1551–1552**
  rabbits, **713**
  rats, **707–708, 708*f***
  secretory activity of oxytocin neurons in, 3090–3104
  sources of relaxin in female mammals, **705*t***
  stereological data for normal term, **220*t***
  at term, 3079–3080
Pregnant mare serum gonadotropin (PMSG), 2310
Pregnenolone, conversion, **982**
Preimplantation embryogenesis, **261–310**
  cell cycle control in early mitoses, **282–286**
    cell cycle progression in one-cell and two-cell embryos, **282–284, 282*f***
    regulators of later cell divisions, **284–286, 285*f*, 287*t***
  chromatin remodeling and epigenetics in early embryos, **272–278**
    genomic methylation, **273–276, 274*f*, 274*t*, 275*t***
    X-chromosome dosage compensation, **276–278, 276*f*, 278*t***
    ZGA control, **272–273, 273*t***
  creation destruction, **278–282**
    cellular apoptosis, **280–282, 281*f***
    RNA degradation, **278–279**
    ubiquitination of proteins and organelles, **279–280, 279*f*, 280*t***
  early postfertilization development and maternal effect gene products, **264–268, 265*f***
    maternal effect genes, **265–266, 265*f***
Preimplantation embryogenesis *(Continued)*
  early postfertilization development and maternal effect gene products
    quantity of early acting maternal effect genes, **267**
    select CPE-containing mRNAs, **268*t***
    translational control of maternal mRNAs, **267–268**
  ethics and the embryo, **296–297**
  morphological and metabolic changes in early cleavage embryos, **288–293**
    glucose import and relevance of maternal hyperglycemia and hyperinsulinemia, **291–292**
    morula compaction, **290–291**
    TE sidedness and blastocoel development, **292–293, 292*f***
  overview, **261–263, 262*f***
  oviductal factors and extracellular signals influencing early development, **286–288**
    extracellular signaling pathways in early development, **288, 289*t*–290*t*, 291*f***
    oviductal microenvironment, **286–288**
  preimplantation polarity and cell fate determination, **293–296**
    polarity in mammalian embryos, **293–294, 294*f***
    totipotency within the embryo and cell fate determination, **294–296, 294*f*, 295*f***
  transgenic mouse technology, **263–264, 263*f*, 264*f***
  zygote genome activation, **268–272, 269*f*, 270*f***
    products of early transcription, **270–271, 271*t***
    transcription factors, **271–272**
Premature ovarian failure (POF), **407–408**
Prenatal development, stressors and, 2660–2661
Preproacrosin, **17*t***
Prepronociceptin (ppN), **453**
  as neuropeptide action on LH secretion, **1431*t***
Presbytis entellus, 2010
Preterm delivery, placental hormones and, 2883
Priapism, **1185**
Primary ciliary dyskinesia (PCD), **30–31**
Primates
  approach-avoidance in postpartum mothers, 1935
  cloning of relaxin, **686**
  corpus luteum, **503–505**
  estrogen receptor-α, **621–622**
  estrogen receptor-β, **625–626**
  estrogen receptor signaling in uterine function, **609–610**
  gonadotropin secretion, 2526–2528
    gonadotropins, 2526–2528, 2527*f*
    ovarian activity, 2526
    prolactin, 2528, 2529*f*
  insulin-like growth factors and, **534–535, 534*f***
Primates *(Continued)*
  maternal deprivation and isolation studies, 1968–1969, 1969*f*, 1972–1973, 1974
  maternal experience and parity effects in adult females, 1957
  mechanisms of maternal experience, 1962
  menstrual cycle
    corpus luteum formation and function, 2461–2475
      autocrine and paracrine regulation, 2465–2466
      cellular regulation, 2466–2468, 2466*f*, 2467*f*
      development, 2461–2462, 2462*f*
      endocrine secretions, 2463
      luteotropic hormones, 2463–2465, 2464*f*
      maternal recognition of pregnancy, 2470–2475, 2472*f*, 2474*f*
      regression of the primate corpus luteum, 2468–2470, 2468*f*
    follicular development and its control, 2451–2461
      characteristics of development, 2451–2453, 2451*f*, 2452*f*
      follicular estrogen biosynthesis, 2453–2455
      gonadotropic control of preovulatory, 2455–2456
      preovulatory follicle selection, 2456–2461, 2457*f*, 2459*f*, 2460*f*
    history, 2449–2451
    neuroendocrine control, 2475–2487
      control of gonadotropin secretion, 2476–2485, 2479*f*, 2482*f*
      control system and timing of the menstrual cycle, 2486–2487
      frequency changes of the GnRH pulse generator during the menstrual cycle, 2485–2486
      gonadotropin secretion during the menstrual cycle, 2475–2476
  menstrual/estrous cycle, **710–711**
  menstruation cycle, 2449–2510
  neuroanatomical bases of maternal experience and olfactory learning, 1965–1966
  onset of maternal behavior
    with steroids, 1927
    with vaginocervical stimulation, 1930
  opioids and onset of maternal behavior, 1952
  postpartum maternal experience, 1959–1960
  pregnancy, **711**
  prostate, **1151*f***
  pup experience and female allomothering, 1958
  reproductive aging, 2738–2739
Primordial germ cells (PGC), **325–326**
  chromosomal events, **326**
  migration, **326**
  origin, **325–326**
PRL. *See* Prolactin
Proacrosin, **17*t***

Procreation, **147.** *See also* Implantation
Procyon lotor, 2286–2287*t*, 2292
Proestrus
  definition, 2328
  in the rat, 2329*f*
Progesterone
  aging and, 2734*f*
  antiprogestin effects and cellular function, **345**
  biochemical reaction in ovulation, **432–435, 432*f*, 433*f*, 434*f*, 435*f***
  endocrine and ovarian phenotypes in mice null for progesterone signaling, **646*t***
  enhancement of production, **485–487, 487*f***
  estrogen and, 1868
  estrous cycle
    of the rat and, 2334*f*
    of sheep and, 2412–2416, 2413*f*
  female reproductive behavior and, 1831–1832
  history, **477**
  implant effect on lordosis behavior, 1838–1839
  influence on gonadotropin secretion, **1299**
  intraovarian role in ovarian function, **646–648**
  lactation and, 3001
  maintenance of actions in the absence of ER-α, **616**
  male sexual behavior and, 1748
  mechanism of action, 2406
  menstruation cycle and, 2463
  negative feedback action, 2402
  onset of maternal behavior and, 1925–1927
  oxytocin neurons and, 3101
  parturition and, 2940–2941
  as physiological regulator of gonadotropin biosynthesis and secretion, **1660–1661**
  placenta and, 2864–2865
  puberty and, 2163, 2163*f*
  at puberty in the sheep, 2129*f*
  in the rat, 2333, 2333*f*
  receptor isoforms, **344–345**
  receptors, **1558**
  receptor signaling in the uterus, **616–618, 617*f*, 618*t***
  regulated genes in the mammalian uterus, **618*t***
  regulation of GnRH neurons, **1452–1453, 1453*f***
    negative feedback, **1453**
    positive feedback, **1453–1454**
  secretion in sheep, 2396
  signaling in ovarian function, **644–648**
  sites of action, 2484–2485
  suckling and control of gonadotropin secretion, 2515–2516
  surge-blocking action, 2482
Progesterone receptors, **353, 453, 595**
  endogenous and exogenous ligands, **601*t***
  female reproductive behavior and, 1837
Progestin, coitus-induced ovulation and, 2301–2302
Prolactin-inhibiting factor (PIF), **1295**
Prolactin-like luteotropic hormones, **491–493, 491*f***
Prolactin (PRL), **1703–1726**
  biochemistry, **1705–1707**
    models, **1706*f***
    variants, **1706–1707**
  cleaved, **1707**
  coitus-induced ovulation and, 2300–2301
  control of mating-induced secretion, 2364–2367
  control of secretion during the estrous cycle of the rat, 2360–2363
  deficiency, **1718**
  energy balance and, 2516
  episodic release, **1716*f***
    associated with female reproduction, **1714–1715**
      estrous cycle, **1715, 1715*f***
      lactation, **1715**
  estrous cycle of the rat and, 2334*f*
  evolution, **1703–1705**
    extrapituitary PRL, **1705, 1705*f***
    nonlactogenic PRL-like peptides, **1704–1705**
    pituitary, **1703–1704**
    placental lactogens, **1704**
  glycosylated, **1706**
  gonadotropin secretion and
    in cows, 2525
    in humans, 2532–2533
    in nonhuman primates, 2528, 2529*f*
    in pigs, 2519
    in rats, 2516
    in sheet, 2521
  hypersecretion, **1717–1718, 1718*f***
  lactation and, 2998–3000, 3000*f*, 3032–3034, 3033*f*
  male sexual behavior and, 1759
  multimeric forms, **1707**
  nocturnal and diurnal surges, 2364*f*
  onset of maternal behavior and, 1930–1931
  oxytocin and, 3078
  pattern of secretion throughout the estrous cycle, 2338
  phosphorylated, **1706–1707**
  profile, **1715*f***
  receptors and signaling, **1707–1709, 1708*f*,** 3032–3034
    signal transduction, **1708–1709**
  regulators of secretion, **1709–1714, 1710*f***
    calcitonin, **1713**
    dopamine, **1710–1713, 1711*f***
    endothelin-1, **1713**
    estrogen, **1713–1714**
    PRL-releasing factors, **1714**
    somatostatin, **1713**
    transforming growth factor-β 1, **1713**
  reproductive behavior, 1854
  reproductive physiology and pathophysiology in mammals, **1716–1718**
    brain and behavior, **1717**
    female reproductive tissues, **1716–1717**
Prolactin (PRL) *(Continued)*
  reproductive physiology and pathophysiology in mammals
    hypersecretion of PRL, **1717–1718, 1718*f***
    isolated PRL deficiency, **1718**
    male reproductive tissues, **1717**
    physiological levels of PRL, **1716, 1716*f***
  seasonal rhythms of reproduction and, 2250
  secretion
    during the active luteal phase of the rat, 2364, 2364*f*
    during lactation, 3032, 3033*f*
  synthesis, 1854
  transcription, 1854
Prolactin receptor, **453**
  research, **1291–1292**
  synthesis regulation, **656–657**
Prolactin-releasing peptide (PrRP)
  milk ejection and, 3174
  role in estrous cycle, 2363
Pro-opiomelanocortin peptides (POMC), **1528–1529, 1529*f*,** 2648
  reproduction and, 2589–2591
Prostaglandin E_1 (PGE_1), male sexual function and, **1177–1178, 1178*f*, 1185**
Prostaglandin E_2 (PGE_2), effect on male sexual behavior, 1758, 1775
Prostaglandin F_{2a}, **497–499, 498*f***
Prostaglandins
  biochemical reaction in ovulation, **435–438, 436*f*, 437*f***
  menstruation and, 2469
  metabolism, 2946, 2947–2948
  ovarian cyclicity in the ewe and, 2396–2397
  parturition and, 2943–2949
  receptors, 2948–2949
  regulation of production in vitro, 2946–2947
  role in implantation, **163–164**
  synthesis, 2944–2946
Prostatectomy, nerve-sparing, **1185**
Prostate gland, **1150–1161**
  adult, **1152–1153, 1153*f***
  anatomy of human, **1153*f***
  branching morphogenesis expression and effects, **1156*t***
  carcinoma, **1161–1163**
  diseases of, **1161–1163**
  embryology and fetal development of human and primate, **1150–1152, 1150*f*, 1151*f***
  function, **1161**
  histology of mature, **1158–1161, 1160*f***
  issues and research, **1165–1166**
  oxytocin and, 3065
  paracrine oxytocin systems and, 3068
  pubertal maturation, **1152**
  rodent, **1153–1158, 1154*f*, 1156*t***
Protandry, 2022
Proteases
  biochemical reaction in ovulation, **438–440**
  role in implantation, **170–171**

Protein
  phosphorylation, **60**
  placental transfer, 2801–2802
  pregnancy and, 3012*t*
Protein hormones, menstruation and, 2466
Protein kinase A (PKA) pathways, **1641**
Protein kinase C (PKC) pathways, **1648**
Proteins
  cystine knot, **1582–1583, 1582*f*, 1584–1585*f***
  cytoskeletal, **8*t***
  epididymal luminal fluid and, **1092**
  fibrous sheath-associated, **27*t***
  nuclear, **12**
  outer dense fiber-associated, **24*t***
  outer dense fibers and, **24*t***
  regulation of testicular, **1032**
  ubiquitination, **279–280, 279*f*, 280*t***
Proteoglycan M (versican), **455**
Protogyny, 2021–2022
PrRP. *See* Prolactin-releasing peptide
Prss2, **17*t***
P450scc. *See* Cytochrome P450 side chain cleavage
Pseudocopulation, 2022
Pseudomonas aeruginosa, **1111**
Pseudomys apodemoides, **777, 778**
Pseudomys delicatulus, **777, 778**
Pseudomys shortridgei, **777**
Pseudopregnancy, delayed, **1296–1297**
Psychogenic stress, 2641–2642
Pteropus alecto, **775**
Pteropus poliocephalus, **775**
Pteropus scapulatus, **775**
PTHrP. *See* Parathyroid hormone-related peptide
p53 tumor suppressor, **452**
Puberty
  activation of the ovary, 2079–2081, 2080*f*
  effect of undernutrition on timing of, 2560, 2560*f*
  growth and timing, 2141*f*
  leptin and, 2572–2573, 2573*f*
  male-facilitated ovulation and, 2309–2312
    ovulation advance in seasonally breeding ungulates, 2309–2310
    in rodents that normally exhibit spontaneous cycles, 2310–2311, 2311*f*
  in nonhuman primates and humans, 2177–2230
    definition, 2177–2178
    disorders of pubertal development, 2211–2213
      delayed puberty, 2212–2213
      precocious puberty, 2211–2212
    exteroceptive cues and timing of puberty, 2209–2211
    fetal programming and onset of puberty, 2209
    genetics and puberty genes, 2208–2209
    gonadal and adrenal function, 2179–2182
      the adrenal cortex, 2181–2182

Puberty *(Continued)*
  in nonhuman primates and humans
    gonadal and adrenal function
      the ovary, 2180–2181, 2181*t*
      the testis, 2179–2180
    historical perspectives, 2178–2179
    neurobiological bases of the juvenile brake on pulsatile GnRH release, 2194–2201, 2194*f*, 2195*f*, 2196*f*, 2197*f*, 2198*f*, 2199*f*, 2200*f*
    neuroendocrine control system governing gonadotropin secretion, 2184–2194, 2187*f*, 2189*f*, 2191*f*, 2192*f*, 2193*f*, 2194*f*
    pituitary gonadotrope function during puberty, 2183–2184
    pubertal staging, 2182–2183
    recognition, 2178
    secular trends, 2183
    theoretical physiological controls governing duration of prepubertal hiatus in GnRH secretion, 2201–2208, 2204*f*, 2206*f*
  phenotypic changes, 2079
  pheromones and
    acceleration, 2047–2048, 2047*t*
    inhibition, 2048–2050
  in the rat, 2061–2126
    female, 2063–2084
      fetal development, 2063–2064
      postnatal development, 2064–2084, 2064*f*, 2065*f*, 2067*f*, 2070*f*, 2071*f*, 2073*f*, 2074*f*, 2078*f*, 2080*f*
    historical aspects, 2061–2062
    male, 2084–2089
      fetal development, 2084–2089, 2086*f*
    mechanism of onset, 2093–2100
      activation of excitatory inputs, 2095
      glutamate–GABA interactions and, 2095–2100, 2096*f*, 2099*f*
      gonad-independent increase in central drive, 2093–2094, 2093*f*
      "gonadostat" hypothesis, 2093
      loss of central restraint, 2094–2095
    as model, 2062–2063
    modulatory influences regulating timing, 2089–2093
      adrenal gland, 2089–2090
      environmental factors and toxins, 2092–2093
      pheromones, 2092
      pineal gland, 2091–2092
      somatic growth, 2090–2091
  seasonality, 2209–2210
  sexual behavior, 1734
  in sheep, 2127–2176
    definitions, 2128–2129, 2129*f*
    growth based on body weight, 2128
    model, 2167–2168, 2167*f*
    neuroendocrine sequence during transition into adulthood, 2129–2140, 2130*f*, 2131*f*, 2132*f*, 2133*f*, 2134*f*, 2135*f*, 2136*f*, 2137*f*, 2138*f*, 2140*f*
    overview, 2127–2128, 2128*f*

Puberty *(Continued)*
  in sheep
    sex differences in timing the transition into adulthood, 2160–2167, 2160*f*, 2161*f*, 2162*f*, 2163*f*, 2165*f*, 2166*f*
    sexual maturity versus puberty versus onset of breeding season, 2159–2160, 2160*f*
    transition into adulthood, 2140–2160, 2141*f*, 2142*f*, 2144*f*, 2145*f*, 2146*f*, 2147*f*, 2148*f*, 2149*f*, 2150*f*, 2151*f*, 2152*f*, 2154*f*, 2155*f*, 2156*f*, 2157*f*, 2158*f*, 2159*f*
  social factors, 2210–2211
  sterility and, 2140, 2140*f*
  stressors and development, 2662
  testes formation and development, **951**
  timing of pubertal maturation, 2662–2663
Pubic symphysis, relaxin and, **714, 715*f***
PVE. *See* Persistent vaginal estrus
PVN. *See* Paraventricular nucleus of the hypothalamus
Pyknosis, 2331–2332
Pytila melba, 2027

R

Rabbits
  embryo transport in the oviduct, **136**
  fibroblasts in tunica albuginea, **429*f***
  fluorescent dopaminergic neurons, **1366, 1366*f***
  median eminence, **1343, 1343*f*, 1344*f*, 1347, 1348, 1348*f*, 1349*f***
  milk ducts, 3130*f*
  moment of rupture of follicle, **426*f***
  onset of maternal behavior
    with prolactin, 1931
    with steroids, 1926
  ovulation, **1293**
  pars distalis, **1322, 1322*f***
    epithelial cell, **1316, 1317–1318*f***
  pars distalis epithelial cell, **1322, 1322*f***
  relaxin and, **713**
  semen deposition, **114**
  spermatozoon, **15*f***
    general features, **5*f***
  testes, **792*f***
  third ventricle, **1338*f***
  timing of implantation, **149*t***
  ultrastructure of apex of follicle, **428*f*, 431**
  uterotubal junction, **122*f***
Radiofrequency
  effects on male sexual behavior, 1766
Radiofrequency, effects on male sexual behavior, 1763
Raja erinacea, isolation of relaxin, **684**
Ram (buck)
  seasonal rhythms of reproduction, 2232, 2232*f*, 2233*f*
  semen characteristics, **117*t***
  testes, **783*f*, 794*f***
Rana pipiens, 2011

Rats. *See also* Rodents
- accessory sex gland, **802*f***
- aging and persistent estrous, 2313
- apoptosis, 2065*f*
- approach-avoidance in postpartum mothers, 1934–1935
- biosynthesis of GnRH and GAP, **1423*f***
- biphasic response of uterus to estrogens, **612*t***
- circadian clocks and induced ovulation, 2312–2313
- corpus luteum, 2332–2333, 2333*f*
- early preweaning influences on adult maternal behavior, 1966–1967, 1971–1972, 1972*f*
- egg–cumulus complex, **135*f***
- embryonic gonad development, 2065*f*
- epididymal epithelium, **1075, 1075*f***
- estrogen receptor signaling in uterine function, **608–609**
- estrous cycle, 2327–2388
  - and pseudopregnancy, **707**
- folliculogenesis, 2330–2332, 2331*f*
- gonadotropin secretion, 2512–2516
  - estrus and ovulation, 2512–2513, 2512*f*, 2513*f*
  - ovarian activity, 2513–2514
  - pattern, 2514–2515, 2514*f*
  - prolactin and energy balance, 2516
  - role of ovarian steroids and opioids, 2515
- growth hormone and, **528–529**
- hormonal effects on robustness of maternal experience, 1960–1961
- hormone profile during pregnancy, 3000*f*
- implantation, **148**
- insulin-like growth factors and ovary, **531–533, 531*f*, 532*f***
- isolation of relaxin, **683**
- lymphatic vessels, **800*f***
- mammary gland, 3132*f*
- maternal aggression, 1931–1932
- maternal deprivation and isolation studies, 1967–1968, 1967*f*, 1973–1974
- maternal experience and parity effects in adult females, 1956–1957
- maturation of resident macrophage population of testis, **1212*f***
- mechanisms of maternal experience, 1962
- milk ducts, 3130*f*
- neural events during the estrous cycle, 2339–2350
  - biogenic amine neurotransmitters controlling, 2345–2347
  - circadian control of the ovarian cycle, 2339–2340, 2339*f*
  - gases as possible regulators, 2348–2349
  - growth factors as possible regulators, 2349–2350
  - peptidergic neurohormones controlling, 2341–2345, 2342*f*, 2343*f*
  - peptide transmitters as possible regulators, 2347–2348

Rats *(Continued)*
- neural events during the estrous cycle
  - role of the hypothalamus, 2340–2341, 2341*f*
- neural terminalis, **1330, 1330*f***
- neuroanatomical bases of maternal experience and olfactory learning, 1963–1964, 1964*f*
- neurochemical bases of maternal experience, 1963
- neuroendocrine control of the ovarian cycle, 2327–2388
- onset of maternal behavior
  - with opioids, 1951–1952
  - with oxytocin, 1949–1950
  - with prolactin, 1930–1931, 1930*f*
  - with steroids, 1926
  - with vaginocervical stimulation, 1928
- oogenesis, 2329–2330, 2330*f*
- ovulation, 2332
- patterns of pituitary hormone secretion during the estrous cycle, 2335–2339, 2337*f*, 2338*f*
- pineal gland, 2240*f*
- placental transfer, 2795*t*
- platelet-activating factor, **135**
- postpartum maternal experience, 1958–1959, 1958*f*, 1959*f*
- pregnancy, **707–708, 708*f***
- progesterone actions in the uterus, **617–618, 617*f*, 618*t***
- puberty, 2061–2126
- pup experience and female allomothering, 1957–1958
- relaxin sources, **708**
- secretion of ovarian steroids during the estrous cycle, 2333–2335, 2334*f*
- semen deposition, **114**
- seminiferous cycle, **868*f*, 869*f*, 891*f***
- sensitization of maternal behavior in nulliparous females, 1933–1934, 1934*f*
- sensory and psychological features of maternal experience, 1961
- Sertoli cells, **830*f***
- sexual maturity, 2567, 2567*f*
- spermatogenesis, **1020*f***
- spermatogenic cycle, **866*f*, 872*f*, 874*f*, 881*f***
- spermatozoa, **15*f***
- steroids and phasic release of gonadotropins, **1300**
- sucking-related behavior, 3135, 3135*f*
- testes, **792*f*, 885*f***
- timing of implantation, **149*t***
- uterotubal junction, **122*f***

Rattus norvegicus, **114**

Receptor blockers, effect on lordosis behavior, 1839

Receptor interacting protein-140 (RIP-140), **453**

5α-Reductase, **983–984**
- androgen regulation, **1031–1032**
- spermatogenesis and, **1028–1029, 1029*f***

Regulator of G-protein signaling protein-2, **453–454**

Relaxin, **679–768**
- binding and signaling, **698–702**
  - binding sites for other relaxin family peptides, **700**
  - model, **701*f***
  - signaling by other relaxin family peptides, **702**
  - sites, **698–700, 699*f***
- chemistry of relaxin peptides, **694–698**
  - biosynthesis and processing, **694–695**
  - chemical and recombinant synthesis, **696–698**
  - phylogenetic tree, **694*f***
  - structure–activity relationships, **698**
  - structure of relaxin peptides, **696, 697*f***
- chromosomal locations, **692–693, 692*f***
- clinical implications of serum relaxin levels, **709–710**
- covalent structure of porcine, **683*f***
- discovery, **681–691**
  - cloning of the genes, **684–687**
  - isolation of peptide, **681–684**
  - of novel relaxin family genes, **687–691**
- evolution of relaxin peptide family, **691–694**
  - chromosomal locations, **692–693, 692*f***
  - in primates, **693**
- family peptide receptors, **702–704, 702*f*, 703*f*, 704*f***
- gene families, **688*t***
- gene structure, **687**
- heart rate and, **701**
- history, **680–681**
- menstruation and, 2463
- overview, **680**
- oxytocin and, 3082–3083, 3083*f*
- physiological role of INSL4, **742–744**
- physiological role of INSLs 4 to 6, **742**
- physiological role of relaxin-3, **741–742**
- physiological roles in female and male reproductive processes, **714–731**
  - implantation, **728–730**
  - intraovarian effects, **730**
  - lactation, **726–728**
  - male, **730–731**
  - pregnancy and parturition, **714–726**
- physiological roles in nonreproductive processes, **731–735**
  - cardiac protection, **734–735**
  - fibrosis, **731–734**
  - wound healing, **734**
- physiological roles of INSL3, **735–741**
  - expression of the INSL3 gene, **735–737**
  - gonadal functions, **741, 742*f***
  - in testis descent, **738–740, 739*f***
  - testis expression and secretion, **738**
- sources and secretion in the female, **705–714**
  - in horses, **711–712**
  - in humans, **708–710**
  - in mammals, **705*t***
  - in mice, **711**
  - in pigs, **706–707, 706*f***
  - in primates, **710–711**

Relaxin *(Continued)*
sources and secretion in the female
in rats, **707–708**
in ruminants, **712–713**
in various species, **713–714**
Relaxin-3, discovery, **691, 692*f***
Renal system
maternal adaptation during pregnancy, 2910–2918
aminoaciduria, 2911
angiotensin 1-7/AT2 receptors, 2917–2918, 2918*f*
blood urea nitrogen and creatinine, 2910
glomerular capillary permselectivity to neutral macromolecules in human pregnancy, 2914
glomerular filtration rate and effective renal plasma flow, 2910
glomerular hemodynamics in human pregnancy, 2913
glycosuria, 2911
intraglomerular hemodynamics in human pregnancy, 2913
kidney and extracellular fluid volume/blood pressure regulation in pregnancy, 2915–2916
mechanisms of increased GFR, 2912, 2912*f*
mediators of renal hemodynamic adap- tations in human pregnancy, 2917
postpartum glomerular hemodynamics in human pregnancy, 2913–2914
protein excretion, 2911
renal hemodynamic changes in pregnancy, 2911–2912, 2911*f*
renin release, 2911
response to pregnancy, 2916–2918, 2918*f*
sodium handling, 2911
structural changes, 2911
of kidney and genitourinary tract, 2914–2915
urinary protein and albumin excretion in pregnancy, 2914
water metabolism, 2911
Renin–angiotensin system, expression, 1529
Reproduction
adult function, 2561–2562, 2561*f*
aging, 2738–2745
diabetes mellitus and, 2596–2597
energetics and, 2553–2558
environmental influences, 2554
impaired insulin action and, 2577–2578, 2577*f*
lipodystrophies and, 2597–2598
metabolic disorders and, 2596–2599
metabolic stressors that impair, 2558–2566
obesity and, 2596
oxytocin and control of behaviors related to, 3086–3090, 3089*f*
physiological mechanisms integrating metabolism and, 2553–2625
polycystic ovary syndrome and, 2598–2599
Reproduction *(Continued)*
role of food availability, 2554–2555
seasonal influences, 2555
sensitivity to metabolic challenges, 2555–2557
stress and, 2627–2696
Reproduction, in nonhuman primates and humans. *See also* Menstruation cycle; Primates
seasonality, 2209–2210
Reproduction, in sheep
photoperiod, 2150–2158, 2150*f*, 2151*f*, 2152*f*, 2154*f*, 2155*f*, 2156*f*, 2157*f*, 2158*f*
seasonality, 2148–2150, 2149*f*
Reproduction, in vertebrates, 1995–2040
classification of environmental factors regulating reproduction, 2001–2002, 2001*f*
communication in a capricious environment, 2027–2030
finite-stage machine theory and reproductive life history stage, 1998–2001, 1998*f*, 1999*f*, 2000*f*
hormone–behavior interactions, 1996, 1997*t*
during development phase of reproduction life history stage, 2007–2010
mate choice and pair bonding, 2009–2010
photoperiod, 2008–2009
finite-state machine theory and testing predictions, 2027
influencing mature capability of reproduction life history stage, 2010–2014
cycling onset, 2010–2012
parental care, 2012–2013
recycling (renesting), 2013–2014
transitions between sexual and parental phases of reproduction, 2013
support functions for reproductive life history stage, 2014–2015
social status, 2015
territorial behavior, 2014–2015
termination phase of reproduction life history cycle, 2015–2016
effects of offspring and mate, 2016
interactions with molt, 2016
testosterone "costs," 2027
hormones, behavior, and social interactions, 2003–2006
mating systems, breeding strategies, and alternative reproductive tactics, 2016–2027
variation in breeding strategy, 2023–2027
variation in mating systems, 2019–2023
variation in temporal pattern of breeding, 2016–2019
modalities of communication, 2002–2003
ontogeny versus development of adult life history stages, 1996–1998, 1997*f*, 1998*f*
Reproduction, in vertebrates *(Continued)*
synchronizing and integrating factors in the reproductive process, 2006–2007, 2006*f*
Reproductive behavior, female, 1825–1920
arousal, 1882–1891
background and operational definition, 1883–1886
BBURP theory, 1882–1883, 1883*f*
biophysics of arousal transmitters affecting hypothalamic neurons that control sexual behavior, 1887–1891
development of the concept, 1884–1885
hormonal influences, 1885–1886
hormone-dependent genes influencing, 1886–1887, 1886*f*, 1887*f*, 1888–1889*f*, 1890*f*, 1891*f*
response potentiation, 1883, 1883*f*
axiomatic requirements for reproduction, 1827, 1827*f*
background, 1825–1862
hormone actions, 1828–1863
behavior and, 1828, 1828*f*
comparisons across species, 1828–1829
electrophysiological actions, 1858–1863
gene memory, 1829–1832
hormone receptors in brain, 1832–1840, 1833*t*
neuropeptides influenced, 1850–1858
neurotransmitters influenced, 1840–1850, 1844*t*
lordosis, 1863–1871
mammalian brain, 1826–1827
molecular mechanisms, 1865–1871
underlying primary mating behavior, 1865–1871, 1867*f*, 1872*f*, 1873–1874*f*
neural circuit for producing lordosis behavior, 1864*f*
sexual motivation and courtship behaviors, 1871–1882, 1875*f*
courtship and sexual behaviors of female rats, 1877–1880
preoptic area neurons and, 1872–1877, 1876*f*, 1880–1881
social recognition and affiliation, 1881–1882, 1881*f*
sexual partner preferences, 1877
Reproductive physiology, maternal behavior, 1921–1993
early preweaning influences on adult maternal behavior, 1966–1974, 1967*f*, 1969*f*, 1970*f*, 1972*f*, 1973*f*
endocrine changes associated with parturition, 1924–1925, 1924*f*
experiential influences, 1956–1966, 1958*f*, 1959*f*, 1960*f*, 1964*f*
hormonal bases, 1924–1932
hormonal determinants of onset of maternal behavior, 1925–1931, 1926*t*, 1929*f*, 1939*f*
maternal aggression
hormonal bases, 1931–1932
motivation, 1936–1937
motivational analysis, 1932–1937, 1933*f*

Reproductive physiology, maternal behavior *(Continued)*
neural circuits regulating, 1940–1949, 1941*f*
neuroanatomy and neurochemistry, 1937–1956, 1941*f*, 1943*f*, 1944*f*, 1945*f*, 1947*f*, 1948*f*, 1956*f*
overview, 1922–1923
Reproductive tract
of a cow, **120*f***
embryology of ducts, **326–330**
human female, **115*f***
Respiratory gases, placental transfer, 2817–2822
Respiratory system
maternal adaptations during pregnancy, 2908–2910
diffusing capacity, 2910
hyperventilation, 2909
lung volumes and function, 2909–2910, 2909*f*
Rete testis, **828–829, 829*f***
Retinohypothalamic tract, seasonal rhythms of reproduction and, 2237
Retinohypothalamopineal pathways, 2237–2239, 2237*f*, 2238*f*
Retinoic acid receptors, epididymis and, **1102–1103**
Rhesus monkey
distribution of GnRH neuronal network, 2194
glutamate release, 2196, 2196*f*
orchidectomized, 2190–2191, 2191*f*
ovulatory ovarian cycles, 2191, 2192*f*
puberty, 2187–2190, 2187*f*
responses in agonadal GnRH-primed juvenile males, 2194–2195, 2195*f*
steroids and phasic release of gonadotropins, **1300**
timing of implantation, **149*t***
Rhinoceros, **114**
Rho kinase inhibitors, **1185–1186**
Rhophilin, **27*t***
Rhox, **1107**
RIP-140. *See* Receptor interacting protein-140
Rivulus marmoratus, 2021
RNA
degradation, **278–279**
protein structure, **1604*f***
Rodents. *See also* Rats
accessory sex gland, **789*f***
androgen regulation, **1030*t***
approach-avoidance in postpartum mothers, 1935
circadian inputs, **1446**
estrogen receptor-α, **620–621**
estrogen receptor-β, **622–625, 624*f***
maternal aggression, 1931–1932, 1936–1937
onset of maternal behavior
with prolactin, 1930–1931, 1930*f*
with steroids, 1926–1927
penis, **790*f***
prostate, **1154*f***
reproductive aging, 2739–2741
reproductive events, 2043*t*
Rodents *(Continued)*
seasonal rhythms of reproduction and, 2242*f*, 2243
sensory mechanisms mediating maternal behavior, 1938–1939
spermatogenesis, **1033*t***
structure of testes, **778**
testes, **780*f***
Ropporin, **27*t***

S

Saguinus oedipus oedipus, 2049
Saimiri scireus, 2023
Sak57, **24*t***
Salmon, **1595**
Salt, oxytocin and, 3085
SAMP14, **17*t***
SAMP32, **17*t***
SAP. *See* Steroidogenesis activator polypeptide
Saxicola torquata, 2028–2029
SCARKO. *See* Sertoli cell-selective androgen receptor knockout
Sceloporus jarrovi, 2014–2015
Sceloporus undulatus hyacinthinus, 2020
Schizophrenia, 2703
Scrotum
anatomy, **790–791**
innervation, **807, 807*f***
vasculature, **803, 807*f***
Seasonality, in breeding
breeding season, 2159–2160, 2160*f*
in mammals, 2231–2281
genetic control of seasonality, 2264–2266
evidence for genetic control, 2264, 2264*f*
human reproductive seasonality, 2266
individual genetic variation and photoneuroendocrine pathways, 2264–2266, 2265*f*
melatonin regulation of the hypothalamic–pituitary–gonadal axis, 2245–2263
interneurons connecting to GnRH, 2256–2260, 2256*f*, 2258*f*
melatonin control of reproduction, 2253–2256, 2255*f*
melatonin receptors and seasonal breeding, 2250–2253, 2252*f*
model for analyzing melatonin, 2253
role of pineal gland and melatonin in transducing photoperiodic information, 2245–2248, 2247*f*
sites of action of melatonin, 2248–2250, 2249*f*
thyroid hormones as potential mediators of seasonal plasticity, 2260–2263, 2261*f*, 2262*f*
overview, 2231–2232
seasonal rhythms of reproduction, 2232–2237
circannual versus noncircannual species, 2234–2236, 2235*f*
Seasonality, in breeding *(Continued)*
in mammals
seasonal rhythms of reproduction
description, 2232–2233, 2232*f*, 2233*f*, 2234*f*
environmental factors, 2233–2234
importance of photoperiodic history, 2236–2237
photoperiod, 2234
transduction of light–dark information into a rhythm of melatonin secretion, 2237–2245
light–dark control of melatonin synthesis, 2243–2245
melatonin biosynthesis and its regulation, 2240–2243, 2241*f*, 2242*f*
melatonin transport and catabolism, 2245
pineal gland, 2239–2240, 2240*f*
retinohypothalamopineal pathways, 2237–2239, 2237*f*, 2238*f*
in nonhuman primates and humans, 2209–2210
ovarian cyclicity in the ewe and, 2421–2423, 2422*f*, 2423*f*
role in reproduction, 2555
Sertoli cells and, **886–887, 963–965**
in sheep, 2148–2150, 2149*f*
Second messenger systems, 1858
correlation with effects on lordosis, 1858
hormone effects, 1858
Selectins, role in implantation, **162**
Semelparity, 2017–2019
Semen
characteristics in domestic animals, **117*t***
disposition site, **114–117**
expulsion, 1740–1741
Semicossphus pulcher, 2021
Seminal plasma, immunosuppression, **1245–1247**
Seminal vesicles, **1163–1165**
development, **1163–1164, 1163*f***
embryology, **1163*f***
function, **1164–1165**
structure of mature, **1164**
Seminiferous epithelium, estrogen and spermatogenesis in the adult, **1043–1044**
Serinus canarius, 2004
Serotonin (5-hydroxytryptamine, 5-HT), **1335**
effect on male sexual behavior, 1752–1754, 1776
estrous cycle and, 2346
female reproductive behavior and, 1844–1845
milk ejection and, 3172
as neurotransmitter action on LH secretion, **1430*t***
seasonal rhythms of reproduction and, 2258–2259
Serranus subligarius, 2021
Sertoli cells, **949–975**
adult, **1019*f***
changes associated with seasonal breeding and age, **886–887**
comparative biology, **961–965**
vertebrate and invertebrate, **961–963**
cytology, **831*f*, 864–889**

Sertoli cells *(Continued)*
effects of estrogen exposure and insufficiency, **1042, 1042*t***
established cell lines, **959–960**
follicle-stimulating hormones and spermatogenesis, **956–957, 1033–1034**
function, **1035–1036**
gene expression, **951–956, 954*t***
in functionally mature cells, **952–953**
gene array analysis, **953, 955–956**
during testes formation, **952**
transgenic mice and cell-specific expression, **953**
germ cells and, **831*f***
culture and, **958–959**
transplantation and, **960–961**
hormonal regulation of spermatogenesis and, 2706–2707
from human fetus, **832*f***
inter-Sertoli cell junction, **877*f*, 883*f***
morphology, **864*t***
overview, **949–950**
phagocytosis, **875*f***
proliferation and maturation, **1028**
response to injury, **887–889, 888*f*, 889*f*, 890*f***
rodent spermatogenesis, **1030*t***
seasonal breeders, testicular regression, and recrudescence, **963–965**
shape, **864–866, 865*f*, 866*f*, 867*f*, 868*f*, 869*f***
spermatids and, **861–862, 863*f***
stages, **955*t***
surface specializations, **876–886, 877*f*, 878*f*, 881*f*, 883*f*, 884*f*, 885*f***
testes formation and development, **950–951**
cell fates and embryonic cells, **950–951**
functional maturation at puberty, **951**
late embryonic and early postnatal maturation, **951**
testosterone and spermatogenesis, **957–958**
ultrastructural features, **866–876, 867*f*, 868*f*, 869*f*, 870*f*, 871*f*, 872*f*, 873*f*, 874*f*, 875*f***
Sertoli cell-selective androgen receptor knockout (SCARKO), **1000–1001, 1001*t***
Serum albumin, transport, 3027
Sex
attractants, 2044–2046
determination and differentiation, **245–260**
chromosomal sex, **245–248, 246*f***
gonadal sex, **248–254**
phenotypic sex, **254–256**
maturation, 2141–2146, 2141*f*, 2142*f*, 2144*f*, 2145*f*
receptivity, 2139–2140
sensitivity to metabolic changes and reproduction, 2555–2556
Sex glands, male
anatomy, **785–789, 789*f***
innervation, **805–807**
vasculature, **801–802, 802*f***
Sex hormone–binding globulin (SHBG), 2708–2710
Sex steroid receptors
extraovarian roles in ovarian function, **653–657**
follicle-stimulating hormones and, **638*f***
in ovarian function, **619**
ovarian model, **656*f***
Sexual behavior, female. *See also* Behavior
receptivity, 3087
Sexual behavior, male. *See also* Behavior
brain areas and circuitry in the control of sexual behavior, 1759–1789
major integrative sites, 1768–1785
lateral hypothalamus (LH), 1784, 1784*f*
medial preoptic area (MPOA), 1768–1779, 1770*f*, 1772*f*, 1773*f*, 1776*f*, 1777*f*
mesocorticolimbic DA tract, 1779–1781, 1781*f*
nigrostriatal DA tract, 1781–1782
paraventricular nucleus (PVN) of the hypothalamus, 1782–1784, 1782*f*
ventromedial hypothalamus (VMH), 1784–1785
major motor outputs, 1785–1786
midbrain periaqueductal gray (PAG), 1785–1787
ventral premammillary nucleus, 1785
spinal cord, 1787–1789
chemical changes detected by cerebrospinal fluid analysis, 1788–1789
effects of direct application of drugs selective for a neurotransmitter, 1787–1788
effects of lesions or transections, 1787
immunocytochemistry, 1789
presence of steroid receptors, 1789
circuitry and anatomical interconnections, 1789–1794
chemosensory circuits, 1790–1791, 1790*f*
efferents from the MPOA, 1793–1794, 1794*f*
ejaculation-related circuit, 1791–1793, 1792*f*
sexual behavior in the context of mammalian social behavior, 1794
somatosensory input from the genitals, 1791
effects of systemically or intracerebroventricularly administered drugs, 1748–1759
acetylcholine, 1754–1755
alpha-melanocyte–stimulating hormone and related peptides, 1757–1758
angiotensin II, 1759
cholecystokinin (CCK), 1759
corticotropin-releasing hormone (CRH), 1759
Sexual behavior, male *(Continued)*
effects of systemically or intracerebroventricularly administered drugs
dopamine, 1748–1752, 1749*f*
galanin and galanin-like peptide, 1758
gamma amino butyric acid (GABA), 1755
gonadotropin-releasing hormone, 1757
neuropeptide Y (NPY), 1758
nitric oxide, 1756–1757
norepinephrine, 1752
opioids, 1755
oxytocin, 1755–1756
prolactin, 1759
prostaglandin E_2 (PGE_2), 1758
serotonin, 1752–1754
vasoactive intestinal peptide (VIP), 1758
future directions, 1794–1795
neurobiology of male, 1729–1823
other brain areas, 1786–1787
overview, 1730
oxytocin and, 3087–3088
patterns of male mammals, 1730–1738
copulatory behavior, 1730–1731, 1731*f*
measures of penile function, 1734–1736
ex copula, 1735–1736
observations during copulation, 1734–1735
postejaculatory behavior and satiety, 1731–1732
precopulatory behaviors, 1730
role of gonadal steroids in control of male sexual behavior, 1744–1748
effects of castration and hormone replacement on ex copula penile responses, 1746–1747, 1747*f*
role of progesterone, 1748
role of T metabolites in maintaining and restoring copulation, 1745–1746
steroid action on steroid hormone receptors, 1747–1748
time course of changes in copulation following castration, 1744
time course of changes in copulation following T restoration, 1744–1745
role of the penis in male sexual behavior, 1736–1743
anatomy of the penis and mechanics of erection, 1736–1738, 1737*f*
autonomic and somatic control of ejaculation, 1740–1741, 1741*f*
cellular mediators of erection, 1741–1743, 1742*f*, 1743*f*
neural innervation of the penis, 1738–1739, 1739*f*, 1740*f*
tactile receptors on the penis, 1741
sensory inputs, 1759–1768
amygdala, 1762–1766, 1762*f*, 1763*f*, 1765*f*
bed nucleus of the stria terminalis (BST), 1766–1767
central tegmental field/subparafascicular nucleus of the thalamus, 1767–1770, 1767*f*
olfactory bulbs, 1759–1762, 1760*f*, 1761*f*

Sexual behavior, male *(Continued)*
- sexual behavior throughout lifespan, 1734
- sexual experience, 1733
- sexual motivation versus performance, 1732–1733
- spinal cord, 1787–1789
- throughout male lifespan, 1734

Sexual differentiation, anatomy and cell biology, **314–315, 314*f*, 315*f*, 316*f***
Sexual disorders, classification, **1175*t***
Sexual function, male, **1173–1194**
Sexual maturity
- as metabolic stressor that impairs reproduction, 2559–2561, 2560*f*
- versus puberty versus onset of breeding season, 2159–2160, 2160*f*
- in rats, 2567, 2567*f*

Sexual motivation, 1871–1882, 1875*f*
- courtship and sexual behaviors of female rats, 1877–1880
- preoptic area neurons and, 1880–1881
- preoptic neurons, 1862–1877, 1876*f*
- sexual partner preferences, 1877
- social recognition and affiliation, 1881–1882, 1881*f*

sFRPs. *See* Frizzled-related protein-4
SGK, follicle-stimulating hormone and, **562–563**
Shark, isolation of relaxin, **684**
SHBG. *See* Sex hormone–binding globulin
Sheep
- approach-avoidance in postpartum mothers, 1935
- estradiol-filled cannulae implant studies, **1449–1450**
- estrous cycle, 2390–2397
  - control of ovarian function, 2408–2418
    - estradiol secretion control, 2410–2411
    - follicular development control, 2408–2410, 2409*f*
    - inhibin secretion control, 2411–2412
    - luteal progesterone secretion control, 2412–2416, 2413*f*
    - ovulation rate control, 2410
  - environmental control of ovarian cyclicity in the ewe, 2421–2426
    - effects of nutrition, 2425–2426
    - effects of pheromones and social interactions, 2426, 2427*f*, 2428*f*
    - effects of stressors, 2423–2425, 2424*f*
    - photoperiodic control of seasonal breeding, 2421–2423, 2422*f*, 2423*f*
  - estrous behavior, 2390–2391
  - hormonal control, 2397–2398, 2397*f*
  - hypothalamo–hypophyseal function control, 2398–2408, 2398*f*, 2399*t*, 2401*f*, 2403*f*, 2405*f*, 2405*t*, 2408*f*
  - hypothalamo–hypophyseal unit, 2391, 2392*f*
  - neural control, 2398
  - ovary, 2391–2396, 2394*f*
  - resynthesis model, 2418–2421, 2419*f*
  - uterine prostaglandins, 2396–2397

Sheep *(Continued)*
- GnRH neuron morphology, **1420*f***
- gonadotropin secretion and, 2519–2521
  - estrus, ovulation, and gonadotropins, 2519–2521, 2520*f*
  - prolactin, 2521
- individual recognition, 1961–1962
- maternal experience and parity effects in adult females, 1957
- mechanisms of maternal experience, 1962
- mutations causing ovarian defects, **394*f***
- neural circuits regulating maternal behavior, 1941, 1941*f*
- neuroanatomical bases of maternal experience and olfactory learning, 1964–1965
- neuroendocrine control of the ovarian cycle, 2389–2447
- onset of maternal behavior
  - with cortisol and corticosterone, 1928
  - with opioids, 1952
  - with prolactin, 1931
  - with steroids, 1927
  - with vaginocervical stimulation, 1928, 1930
- ovaries, **476*f***
- oxytocin and onset of maternal behavior, 1950–1951
- placental transfer, 2795*t*
- postpartum maternal experience, 1959, 1960*f*
- puberty, 2127–2176
- pup experience and female allomothering, 1958
- sensory mechanisms mediating maternal behavior, 1939–1940
- timing of implantation, **149*t***
- uterotubal junction, **122*f***

Shrews, **114**
- egg–cumulus complex, **135*f***

Sigmodon hispidis, **777**
Signal peptide, **1593**
Signal transduction, gonadotropin secretion and, **1643–1652**
Sildenafil (Viagra), **1182–1183**
Size, reproduction and, 2556
Skene's gland, **1165**
Skunks, regulation of embryonic diapause, **155**
SLLP1, **17*t***
Sluice-flow principle, **217**
Sminthopsis crassicuadata, **130**
Smithsopsis crassicaudata, 2019
Social behavior
- female reproductive behavior and, 1868–1869
- male sexual behavior and, 1794
- recognition and affiliation, 3086–3087

Sociosexual behavior, 1870–1871, 1872*f*
Sodium, placental transfer, 2802–2803
Soluble guanylate cyclase activators, 1185–1186
Somatic growth, puberty and, 2090–2091
Somatoluteotropes, **1521, 1524*f***
Somatometer, 2201–2202
Somatostatin
- effect of estradiol on, 2405*t*
- as neuropeptide action on LH secretion, **1431*t***
- placenta and, 2851–2852
- prolactin and, **1713**
- as steroid-responsive neuron, 2399*t*

Somatotropin, **528**
SOX9, **320–321**
Sp-10, **17*t***
Sp17, **27*t*, 993**
sp32, **17*t***
SPACA1, **17*t***
SPACA3, **17*t***
SPACA4, **17*t***
Spag4, **24*t***
Spag5, **24*t***
sp56 (AM67), **17*t***
Sparisoma viride, 2019
Sperm
- antibodies, immunological response, **1232–1233, 1232*f***
- capacitation, **56–62, 58*t*, 61*f*, 62*f*, 63*f***
- competition and cooperation, **113–114**
- egg factors implicated in sperm chromatin remodeling, **88*t***
- functional changes associated with capacitation, **60–62, 61*f*, 62*f***
- leukocytic response in the vagina, **117**
- migration into the ampulla, **131–132**
- nonfertilizing, in the oviduct, **132**
- output and aging, 2703–2704
- preferential pathways through the cervix, **119**
- storage in the cervix, **119–120**
- timing of deposition with ovulation, **114**
- transport
  - rate, **120**
  - through the cervix, **117–120**
  - through the oviduct, **125–132**
  - through the uterotubal junction, **122–125**
  - through the uterus, **120–122**
- vaginal defenses against, **117**

Spermatic cord
- anatomy, **791–794, 792*f*, 793*f***
- blood vessels, **791–792, 791*f*, 792*f*, 793*f***
- lymphatic vessels, **794**
- variocele, **792–794**

Spermatids, rodent spermatogenesis, **1030*t***
Spermatocytes
- cytology, **839–843, 840*f*, 841*f***
- primary, **840–842, 840*f*, 841*f***
- rodent spermatogenesis, **1030*t***
- secondary, **840*f*, 842–843**

Spermatogenesis, **829–864, 829*f*, 830*f*, 831*f*, 832*f***
- aging and, 2705–2706
- androgen regulation, **1026–1032, 1027*t*, 1029*f*, 1030*t***
  - in the adult, **1029–1032, 1030*t***
  - during initiation of spermatogenesis, **1027–1029, 1029*f***
- developmental considerations, **831–836, 832*f***

Spermatogenesis *(Continued)*
developmental time-line, **1196*f***
duration, **892–894**
effects of inflammation, infection, and immunity and male reproduction, **1229–1231, 1230*f***
endocrine regulation, **1017–1069**
cell–cell adhesion systems involved in, **1022**
germ cell development, **1019–1021, 1020*f***
organization of spermatogenesis, **1018–1022, 1019*f***
overview, **1017–1018**
spermatogenic cycle organization, **1021–1022**
endocrine regulation in humans, **1045–1053, 1046*f*, 1047*t*, 1049*f*, 1050*f***
hormonal dependency, **1045–1048, 1046*f*, 1047*t*, 1049*f*, 1050*f***
hypogonadotropic hypogonadism, **1047*t*, 1049–1051**
male hormonal contraceptive development, **1051–1053**
estrogen regulation of, **1039–1045, 1041*t*, 1042*t***
in the adult, **1043–1044**
estrogen production in the testis, **1040**
importance of estrogen in efferent ductule development and function, **1044–1045**
during the neonatal period, **1041–1043**
FSH regulation of, **1032–1039, 1033*t*, 1037*f***
FSH and T cooperativity, **1037–1039, 1037*f***
role during initiation of spermatogenesis, **1033–1035**
role in the adult, **1035–1037**
future directions, **1053–1054**
hormonal regulation and Sertoli cells, 2706–2707
hormone receptor expression, **1022–1026, 1023*t***
pituitary hormone receptors, **1025–1026**
steroid receptors, **1023–1025**
lipopolysaccharide-induced inflammation and, **1229, 1230*f***
quantitative studies, **862–864, 863*t***
Sertoli cells and, **957–958**
spermatocytes, **839–843, 840*f*, 841*f***
spermatogonia, **836–839, 837*f*, 838*f*, 839*f***
spermiogenesis, **843–864, 843*f*, 844*f*, 845*f*, 846*f*, 847*f*, 848*f*, 850*f*, 851*f*, 852*f*, 853*f*, 854*f*, 856*f*, 858*f*, 859*f*, 860*f*, 861*f*, 862*f*, 863*f*, 863*t***
thyroid hormone and, **1039**
Spermatogonia, **828*f*, 836–839, 837*f*, 838*f*, 839*f*, 1030**
development, **1036**
effect of gonadotropin on cell types in humans, **1048**
rodent spermatogenesis, **1030*t***
Spermatogonial stem cell, renewal, **895–898, 896*f***
Spermatogonial wave, **898–899, 898*f***
Spermatozoa
changes during aging of epididymis, **1116–1117**
maturation, **1083–1087, 1084*f***
microenvironment for maturation, protection, and storage, **1089–1092, 1090*f***
protection of, **1088–1089**
storage, **1087–1088**
transport, **1081–1083, 1082*f***
Spermatozoon, **3–54**
acrosome, **4**
axoneme, **4**
components, **4**
fibrous sheath, **4**
flagellum, **4, 18–37**
mammalian, general features, **4*f***
motility, **33–37**
mouse, **5*f***
nucleus, **4**
overview, **3**
rabbit, **5*f***
spermatogenesis and, **3–4**
sperm head, **4, 11–18**
sperm plasma membrane, **4–11**
Sperm–egg interaction, **65–93.** *See also* Fertility; Fertilization, mammals
Sperm fertility
preservation during storage, **127**
release of sperm, **127–129**
Sperm head, **4, 11–18**
acrosome, **14–18**
cytoskeleton, **13–14**
distribution of regions and membranes, **11*f***
domains and associated cytoskeletal proteins, **8*t***
flagellar beating patterns of human sperm, **119*f***
infundibular cilia with cumulus oophorus, **134*f***
nucleus, **11–13**
Spermiation, **1021**
effect of gonadotropin in humans, **1048, 1050*f***
Spermiogenesis, **843–864, 843*f*, 844*f*, 845*f*, 846*f*, 847*f*, 848*f*, 850*f*, 851*f*, 852*f*, 853*f*, 854*f*, 856*f*, 858*f*, 859*f*, 860*f*, 861*f*, 862*f*, 863*f*, 863*t***
acrosome formation, **844–847, 845*f*, 846*f*, 847*f***
androgen regulation, **1030–1031**
nuclear changes, **847*f*, 848–850, 848*f*, 850*f***
relationships between Sertoli cell and spermatids, **861–862, 863*f***
reorganization of cytoplasm and organelles and spermiation, **859–861, 861*f*, 862*f***
spermatogenesis and, **1021**
tail formation, **845*f*, 850–859, 851*f*, 852*f*, 853*f*, 854*f*, 856*f*, 858*f*, 859*f*, 860*f***
testosterone and, **1036–1037**
Spermophilus columbianus, 2028
Sperm plasma membrane, **4–11**
domain
formation and maintenance, **8–9**
identification, **6–7**
lipids, **9–11**
localized sperm membrane components, **7–8**
mosaic sperm surface, **4–6**
Spilogale putorius, **155**
Spinal cord
male sexual behavior and, 1787–1789
transmission of sucking stimulus through, 3166–3167
Sptrx-1, **27*t***
Sptrx-2, **27*t***
Squalus acanthias, isolation of relaxin, **684**
Squirrels, **114**
SREBP, **993**
SRY, **246–247, 247*f*, 320**
Stages of Reproductive Aging (STRAW), 2736, 2737*f*
Stallion, semen characteristics, **117*t***
StAR. *See* Steroidogenic acute regulatory protein; Steroidogenic acute regulatory protein-1
Star, **391**
STAT4, **8*t***
Stat4, **8*t***
STAT5 transcription factors, **1709**
Stem cells
adult, in the prostate, **1159–1160, 1160*f***
proliferation and disordered growth, **1163**
Stem villi, **212*f*, 213**
Sterility. *See also* Infertility
postpubertal, 2140, 2140*f*
Steroid 5α-reductases, epididymis and, **1099–1101**
Steroid hormone receptors, **352–354**
binding arrangements, **600*f***
biphasic uterine response to estradiol, **612–614, 612*f*, 612*t*, 613*f*, 614*f***
D domain, **600**
E domain, **600–601**
endogenous and exogenous ligands, **601*t***
estrogen receptor signaling function, **608–616**
extraovarian roles of sex steroids in ovarian function, **653–657**
F domain, **602**
hinge region, **600**
LBD, **600–601**
male sexual behavior and, 1789
mechanisms of action, **602–607, 602*f*, 603*f*, 604*f***
classic model, **602–604, 602*f*, 603*f***
cross-talk (ligand independent), **603*f*, 604–605**
plasma membrane–associated steroid signaling, **605–606**
receptor antagonists, **606–607**
tethered (HRE independent), **605**
in the ovary and uterus, **593–678**

Steroid hormone receptors *(Continued)*
progesterone receptor signaling in the uterus, **616–618, 617f, 618t**
review of the literature, **595t**
spermatogenesis and, **1023–1025**
steroid signaling null mice, **607–608**
structure, **597–602, 598t**
in uterine function, **608**
uterine phenotypes in murine models of disrupted estrogen signaling, **610–611, 610t, 611f**
Steroid hormones
follicular estrogen biosynthesis and, 2453–2454
implantation and, **157–159**
menstruation and, 2463, 2465–2466
parturition and, 2940–2942
receptors, **352–354, 352f**
Steroidogenesis
Leydig cells, **1221**
neural–immune–endocrine axis and, **1202f, 1227–1229**
role of macrophages in Leydig cell development and, **1231–1232**
Steroidogenesis activator polypeptide (SAP), **988**
Steroidogenic acute regulatory protein (StAR), **986–988**
role of protein in Leydig cells, **989–992**
transcription factors involved in gene expression, **992–995**
Steroidogenic acute regulatory protein-1 (StAR), **454**
Steroidogenic cell proliferation, **484–485**
Steroidogenic factor-1, **318–319, 992–993**
Steroids
effects on male sexual behavior, 1770–1772
gonadal, 2100
male sexual behavior and, 1783
male sexual function and, **1179–1181**
onset of maternal behavior and, 1925–1931
phasic release of gonadotropins and, **1300–1301**
Steroid-secreting cells, **487–489**
differentiation, **484–489, 487f**
Steroid sex hormones, 1871, 1873–1874*f*
Stratum granulosum, **428f, 429–430, 430f**
STRAW. *See* Stages of Reproductive Aging
Streptopelia risoria, 2008
Stress
classical concept, 2627–2628
clinical consequences, 2628–2629
current concept, 2628
effects of stressors on HPG axis, 2634–2644
effects on the "surge" mode of LH release, 2639–2640, 2639*f*
inhibitory effects of stressors on tonic pulsatile gonadotropin release, 2635–2636, 2637*f*
sites at which stressors inhibit tonic pulsatile LH secretion, 2636, 2638, 2638*f*
stimulatory effect of stressors on tonic LH release, 2638–2639
Stress *(Continued)*
effects of stressors on HPG axis
stressors and the reproductive cycle, 2640–2644, 2643*f*
mechanisms whereby the HPA axis interferes with the reproductive system, 2644–2660
mechanisms whereby the HPA axis interferes with the reproductive system
ACTH and adrenal glucocorticoids, 2651–2653, 2652*f*
central and adrenal medulla neurotransmitters, 2656, 2656*f*
effects of stressors on gonadal steroid secretion, 2653–2655
modulatory role by endogenous opioid peptides, 2648–2651, 2649*f*
paracrine role for CRH in the gonads, 2655–2656
role of CRH, 2644–2646, 2645*f*
role of VP, 2646–2648, 2647*f*
urocortins, 2648
metabolic-related suppression of the HPG axis, 2665–2666
modulation of HPA and HPG axes response to stressors by gonadal steroids, 2656–2660
basal glucocorticoid secretion, 2657
glucocorticoid response to stressors, 2657–2659
gonadal steroid modulation of the HPG response to stressors, 2659–2660
oxytocin and, 3084–3085
paracrine role for CRH in the uterus and placenta, 2663–2665
CRH and stress in pregnancy, 2664–2665
CRH and the placenta, 2663–2664
physiological roles during early pregnancy, 2663
psychogenic, 2641–2642
reproduction system and, 2627–2696
response, 2629–2633
brain–PVN interactions, 2632–2633
direct pathways to PVN, 2632
indirect pathways to PVN, 2632
role of CRH in extrahypothalamic areas, 2632–2633
hypothalamic–pituitary–adrenal (HPA) axis, 2629–2631
adrenal glands, 2631
hypothalamic paraventricular nucleus (PVN), 2629–2630
pituitary gland, 2630–2631
interaction between immune system and HPA axis, 2633, 2634*f*
sympathetic nervous system, 2631–2632
stressors and development, 2660–2663
early neonatal, 2661–2662
prenatal, 2660–2661
puberty, 2662
timing of pubertal maturation, 2662–2663
Stressors
aging and, 2730
development and, 2660–2663
early neonatal, 2661–2662
prenatal, 2660–2661
puberty, 2662
timing of pubertal maturation, 2662–2663
ovarian cyclicity in the ewe and, 2423–2425, 2424*f*
restraint as, 2635–2636
SubH2Bv, **8t**
Substance P
female reproductive behavior and, 1854–1855
milk ejection and, 3174
Sucking
behavior and dynamics of milk removal, 3134–3136, 3135*f*
firing patterns of oxytocin neurons during, 3145–3147, 3146*f*, 3147*f*
transmission of stimulus at the supraspinal level, 3167–3168, 3168*f*
transmission of stimulus through the spinal cord, 3166–3167
Suckling
control of gonadotropin secretion and, 2511–2551
mechanisms of suckling-induced ovarian inactivity, 2533–2536, 2534*f*
Sulfonates, epididymis and, **1119**
Suncus murinus, 2009, 2286–2287*t*, 2290, 2292, 2295*t*
Suncus vulgaris, 2009
Suprachiasmatic nuclei, seasonal rhythms of reproduction and, 2237–2239, 2238*f*
Surface charge, **6**
Surface replica studies. *See* Freeze-fracture
Sus scrofa, **116**
isolation of relaxin, **682–683**
Sympathetic nervous system, 2631–2632
Synaptotagmin IV, **17t**
Syncytial fusion, **205–207**
Syncytiotrophoblast, **207–208**
Syndrome X, 2730
SYT6, **17t**

T

Tachykinins, as neuropeptide action on LH secretion, **1431t**
Tactile communication, 2002
T-actin 2, **8t**
TAF_{II}105, **395**
TAKAP-80, **27t**
Tanycytes, 1337, **1339**
Target gene expression
molecular basis, **575–576**
regulation, **566–568**
Taricha granulosa, 2011
Taurine transporters, **1110**
T cells, pregnancy and, 2766
Tcf21. *See* POD-1
TDLU. *See* Terminal duct lobular unit
TE. *See* Trophectoderm cells

Tektins, **21**
Temperature, as metabolic stressor that impairs reproduction, 2565–2566
Terminal deoxynucleotidyl transferase-mediated dUTP nick end labeling (TUNEL), **392**
Terminal duct lobular unit (TDLU), 2995
Terminal villi, **210*f*, 211*f*, 212*f*, 213–214**
TESP1, **17*t***
Tesp1, **17*t***
TESP2, **17*t***
Tesp2, **17*t***
TESP4, **17*t***
Testes
  anatomy, **771–781, 779*f*, 780*f*, 781*f*, 782*f*, 783*f***
  blood flow, **797–798, 798*t***
  blood vessels, **794–797, 795*f***
  CRH and, 2655
  cycle of seminiferous epithelium, **889–900, 891*f*, 891*t*, 892*f*, 893*f***
    coordination within, **894–895**
    duration of spermatogenesis, **892–894**
    germ cell degeneration, **899–900**
    spermatogonial stem cell renewal, **895–898, 896*f***
    spermatogonial wave, **898–899, 898*f***
  cytology, **827–947**
  development, **251–254**
    cell proliferation, **252**
    cellular changes, **252*f***
    cellular pathway association, **252**
    germ cells, **254**
    Leydig cell differentiation, **253–254**
    male-specific vascular formation, **252*f*, 253**
    mesonephric cell migration, **252–253, 252*f***
    Sertoli cell development, **251–252**
    testicular cord formation, **253, 253*f***
  diagrammatic representation, **1076*f***
  estrogen production, **1040**
  formation, **315–316, 316*f***
    and development, **950–951**
  historical aspects, **827–828**
  human, **783*f*, 786*f***
  immune privilege as "testis-specific" phenomenon, **1244**
  immunological compartmentalization, **1200*f***
  immunoregulation comparison, **1215*f***
  immunoregulatory factors, **1239*t***
  inflammatory and immunoregulatory cytokine function, **1216–1221, 1217*f*, 1220*f***
  innervation, **803–805, 803*f***
  intratesticular ducts, **828–829**
    rete testis, **828–829, 829*f***
    transitional distal segment of seminiferous tubule, **828, 829*f***
  Leydig cells
    cytology, **900–920**
    cytoplasm, **907–913, 907*f*, 908*f*, 909*f*, 910*f*, 912*f*, 914*f***
    historical background, **900–901**
    life history, **914–918**
Testes *(Continued)*
  Leydig cells
    nucleus, **905, 907*f***
    organization of intertubular tissue, **901–905, 901*f*, 902*f*, 903*f*, 904*f***
    response to testicular damage, **918–920**
    ultrastructure, **905, 905*t*, 906*t***
  lymphatic vessels, **799–800, 800*f***
  lymph flow, **800**
  lymphocytes and polymorphonuclear cells, **1213–1214**
  macrophages, **1210–1213, 1212*f***
  mechanism of descent, **773–774**
  oxytocin and, 3064–3065
  paracrine oxytocin systems and, 3068
  peritubular tissue, **870*f*, 889**
  puberty and, 2084, 2179–2180
  relaxin and, **738–740, 739*f***
  Sertoli cells, **831*f*, 864–889**
    changes associated with seasonal breeding and age, **886–887**
    morphology, **864*t***
    response to injury, **887–889, 888*f*, 889*f*, 890*f***
    shape, **864–866, 865*f*, 866*f*, 867*f*, 868*f*, 869*f***
    surface specializations, **876–886, 877*f*, 878*f*, 881*f*, 883*f*, 884*f*, 885*f***
    ultrastructural features, **866–876, 867*f*, 868*f*, 869*f*, 870*f*, 871*f*, 872*f*, 873*f*, 874*f*, 875*f***
  Sertoli cells and, **950–951, 951–956, 954*t***
  size variation, **774–778, 2704–2705**
  spermatogenesis, **829–864, 829*f*, 830*f*, 831*f*, 832*f***
    developmental considerations, **831–836, 832*f***
    spermatocytes, **839–843, 840*f*, 841*f***
    spermatogonia, **828*f*, 836–839, 837*f*, 838*f*, 839*f***
    spermiogenesis, **843–864, 843*f*, 844*f*, 845*f*, 846*f*, 847*f*, 848*f*, 850*f*, 851*f*, 852*f*, 853*f*, 854*f*, 856*f*, 858*f*, 859*f*, 860*f*, 861*f*, 862*f*, 863*f*, 863*t***
  structure, **778–781, 779*f*, 780*f*, 781*f*, 828, 828*f***
  testis cords and vascular development, **315–316**
  vascular permeability and interstitial extracellular fluid, **798–799**
  vasculature, **794–800, 795*f*, 798*t***
Testicles, regulation of testicular proteins, **1032**
Testicular endocrine system, aging and, 2707–2715
  aging and androgen action, 2711
  hypothalamo–pituitary regulation of gonadotropin secretion, 2710–2711
  reproductive consequences of age-related decline in testosterone production, 2713–2715, 2714*f*
  somatic consequences of age-related decline in testosterone production, 2711–2713
Testicles, regulation of testicular proteins *(Continued)*
  testosterone production and Leydig cells, 2707–2710
Testicular exocrine system, aging and, 2700–2707
  male fertility, 2700–2703, 2700*f*, 2701*f*, 2702*f*
  Sertoli cells and hormonal regulation of spermatogenesis, 2706–2707
  spermatogenesis, 2705–2706
  sperm output, 2703–2704
  testicular blood vessels, 2704
  testis size, 2704–2705
Testicular factors, regulation of epididymal functions and, **1093–1095, 1094*f***
Testicular interstitium, immune–endocrine interactions, **1221–1223**
Testicular steroidogenesis, **977–1016**
  androgen action, **995–1001, 996–997*f*, 998*f*, 1001*t***
    androgen receptor and alternate signaling pathways, **999**
    androgen receptor and gene expression regulation, **997–999, 998*f***
    androgen receptor and the nuclear receptor family, **995–997, 996–997*f***
    responses to androgen that require more than one cell type, **999**
    selective ablation of the androgen receptor in specific cell types of the testes, **1000–1001, 1001*t***
    studies in the testis, **1000**
  cholesterol sources for steroidogenesis in Leydig cells, **978–979**
  enzyme involvement in androgen synthesis in Leydig cells, **979–984, 980–981*f***
    17α-hydroxylase/17-20 lyase, **982–983**
    5α-reductase, **983–984**
    3β-hydroxysteroid dehydrogenase, **982**
    17β-hydroxysteroid dehydrogenase, **983**
    cytochrome P450 aromatase, **984**
    cytochrome P450 chain cleavage, **980–982**
  overview, **977–978**
  regulation of Leydig cell steroidogenesis, **984–995, 987–988*f***
    history, **984–986, 987*f***
    other proteins involved in acute regulation of steroidogenesis, **988–989**
    role of StAR protein in Leydig cells, **989–992**
    steroidogenic acute regulatory protein (StAR), **986–988**
    transcription factors involved in StAR regulation gene expression, **992–995**
Testis-determining factor, 2084

Testis fascin, **8t**
Testosterone
effects on mood, **1181**
Leydig cells and production, 2707–2710
macrophage–Leydig cell interactions and effects on behavior, **1198f**
as physiological regulator of gonadotropin biosynthesis and secretion, **1661**
regulation of biosynthesis, **987f**
reproductive consequences of age-related decline in production, 2713–2715, 2714*f*
Sertoli cells and, **957–958**
somatic consequences of age-related decline in production, 2711–2713
as stressor, 2636
testicular steroidogenesis and, **978**
in various species, **1027t**
Wolffian duct differentiation, **327**
TGF-ß
immunoregulation, **1219–1221, 1220f**
pathways in somatic cells, **291f**
Thalassoma bifasciatum, 2022
Theca externa, **428f, 429**
Theca interna, **429, 431f**
Thecal cell steroidogenesis, **652**
Thryonomys swinderianus, **114**
Th1/Th2-type pregnancy, 2769–2770, 2770*f*
Thyroid hormones
influence on sexual arousal, 1885–1886
as potential mediators of seasonal plasticity, 2260–2263, 2261*f*, 2262*f*
reproduction and, 2582–2583
seasonal rhythms of reproduction and, 2260–2263, 2261*f*, 2262*f*
spermatogenesis and, **1026, 1039**
Thyroid-releasing hormone (TRH), **1295**
gonadotropins and, **1603**
lactotropes and, **1553–1554**
Thyroid-stimulating hormone (TSH), **1292, 1325**
orphan leucine-rich repeat/G-protein–coupled receptors, **1619**
Thyrotropin-releasing hormone (TRH), **1315**
aging and, 2732
Tilapia melanocephala, 2011
TIMPs. *See* Tissue inhibitors of metalloproteinases
Tissue, regeneration in the endometrium, **368–369**
Tissue inhibitors of metalloproteinases (TIMPs), **454, 481–482**
Tissue plasminogen activator (tPA), **453**
T metabolites, role in maintaining and restoring copulation, 1745–1746
Tocolysis, parturition and, 2954–2955
Toxins, puberty and, 2092–2093
tPA. *See* Tissue plasminogen activator
Tpx-1, **24t**
Trachemys scripta, 2004
Tractocile (atosiban), 3077
Transcription factors, **254, 295f**
epididymis and, **1106–1107**
homeobox genes and, **328–329**
Transcription factors *(Continued)*
implantation and, **167–168**
in preantral stage, **395**
Transcytosis, 3026, 3026*f*
Transections, male sexual behavior and, 1787
Transferrin, transport, 3027
Transforming growth factor-β 1, **1713**
Transforming growth factor-β (TGF-β), **346**
biological fluids, 2859
expression and localization, 2858
mammary development and, 3002–3003
placenta and, 2858–2859
receptors, 2858–2859
Transplantation
immunoregulation and, **1236–1243, 1239t, 1241f, 1242f**
medicine implications, **1249**
testis, germ cell, **1243–1244**
TRH. *See* Thyroid-releasing hormone; Thyrotropin-releasing hormone
expression, **1530**
Triglycerides, placental transfer, 2815–2816
Trophectoderm (TE) cells, schematic of ion channels, **292f**
Trophinin, role in implantation, **161**
Trophoblast turnover, **208–209**
differentiation and invasion, 2874–2875, 2875*f*
TSC. *See* Tuberous sclerosis
TSG-6. *See* Tumor necrosis factor-stimulated gene-6
Tsga10, **27t**
TSH. *See* Thyroid-stimulating hormone
Tuberin, follicle-stimulating hormone and, **558–560**
Tuberous sclerosis (TSC), **558–559**
Tumor necrosis factor-α (TNF-α), **346, 454, 500–501**
aging and, 2730
regulation, **1223–1224**
Tumor necrosis factor-stimulated gene-6 (TSG-6), **454–455**
Tumors, as cause of hypogonadotropic hypogonadism, **1047t**
TUNEL. *See* Terminal deoxynucleotidyl transferase-mediated dUTP nick end labeling
Tunica albuginea, **428–429, 428f, 429f**
Turner syndrome, **247**
Twins, **199, 199f**
maternal adaptation to, 2906
monozygotic, **199f**

U

UGKO. *See* Uterine gland knockout
Ultrasound, placenta and, **230–231**
Umbilical cord, **222–224**
Unkei-to, **446**
Urethra, immune cells, **1215–1216, 1216f**
Urethrogenital reflex, 1735–1736
Urocortins, 2630, 2648
Ursus americanus, **114**
Uta stansburiana, 2004
Uterine gland knockout (UGKO), **153**
Uteroplacental vessels, **224, 225–226**
Uterotubal junction
as barrier to sperm migration, **122–123**
of a cow, **122f**
as filter, **123–124**
sperm transport, **122–125**
rapid, **124–125**
Uterus
anatomy, **338f**
androgen receptor signaling function, **619**
biphasic response to estradiol, **612–614, 612f, 612t, 613f, 614f**
contractions, **121**
embryonic diapause and, **155–156**
estrogen growth factor cross-talk in, **614–615, 615f**
estrogen receptor signaling function, **608–616**
glucocorticoid receptor signaling function, **619**
immunological responses, **121–122**
molecular aspects of embryonic interactions, **157–173, 158f**
oxytocin and, 3063–3064, 3099
paracrine oxytocin systems and, 3067
peristalsis, **120–121**
physiological and morphological aspects of embryo–uterine interactions, **148–153**
progesterone receptor signaling in, **616–618, 617f, 618t**
progesterone regulated genes in mammals, **618t**
relaxin and, **717–720**
myometrial contractile activity, **717–719, 718f**
uterine growth and development, **719–720, 720f**
semen deposition, **116**
sex steroid receptor function and, **608**
sperm transport through, **120–122**
steroid hormone receptors in, **593–678**
uterine phenotypes in murine models of disrupted estrogen signaling, **610–611, 610t, 611f**

V

Vacuum devices, for inducing erection, **1185**
Vagina
defenses against sperm, **117**
pH, **117**
relaxin and, **720, 720f**
semen deposition, **114, 116**
Vaginocervical stimulation, onset of maternal behavior and, 1928–1930
Varecia variegata, cloning of relaxin, **686**
Vascular endothelial growth factor (VEGF), **346–347, 455**
biological fluids, 2864
corpus luteum and, **482–483**
expression and localization, 2863
placenta and, 2863–2864
receptors, 2863–2864
role in implantation, **164**
signaling, **171–172**

Vas deferens, immune cells, **1215–1216, 1216*f***
Vasoactive agents. *See also* individual agents
  implantation and, **162–164**
  male sexual function and, **1184–1185**
  placenta and, 2865–2868
Vasoactive intestinal peptide (VIP), **435**
  male sexual behavior and, 1758
  as neuropeptide action on LH secretion, **1431*t***
Vasopressin, **1333, 1341–1342**
  biological fluids, 2857
  expression and localization, 2856–2857
  as neuropeptide action on LH secretion, **1431*t***
  placenta and, 2856–2857
  receptors, 2857
VDAC2, **24*t***
VDAC3, **24*t***
VEGF. *See* Vascular endothelial growth factor
Ventral premammillary nucleus, male sexual behavior and, 1785
Ventromedial hypothalamus (VMH), male sexual behavior and, 1784–1785
Versican (proteoglycan M), **455**
Vertebrates. *See also* individual names
  reproduction, 1995–2040
Viagra (sildenafil), **1182–1183**
Vibrations (seismic stimuli), 2002
Vidua paradisea, 2027
Villi
  immature intermediate, **212*f*, 214**
  mature intermediate, **211*f*, 212*f*, 213**
  mesenchymal, **212*f*, 214**
  stem, **212*f*, 213**
  terminal, **210*f*, 211*f*, 212*f*, 213–214**
Villous trees, **213–222**
  development, **214–215**
  fetal angioarchitecture, **215–216, 216*f***
  intervillous space, **216–218**
  oxygen as regulator of villous development, **218–220**
  stereological data, **220–222, 220*t***
  structure of villous types, **211*f*, 212*f*, 213–214**
Villous trophoblast, **203**
VIP. *See* Vasoactive intestinal peptide
Visual communication, 2002
VMH. *See* Ventromedial hypothalamus
Voles, **114**
Vomeronasal organ (Jacobson's organ), 2042
Von Hippel-Lindau tumor, **559**

W

Waardenburg's syndrome, 2703
WAGR syndrome, **318**
Wallabia bicolor, **777**
Wallaby, testes, **784*f*, 791*f*, 795*f***
Water
  influence on formation of the luminal microenvironment, **1089–1090**
  ion transport and, **1107–1109, 1109*f***
  placental transfer, 2799–2801
Weaning
  early preweaning influences on adult maternal behavior, 1956–1974, 1967*f*, 1969*f*, 1970*f*, 1972*f*, 1973*f*
  lactotropes and, **1553**
Weibel-Palade bodies, **445**
Whales, isolation of relaxin, **684**
WHO. *See* World Health Organization
Wilms' tumor, **249–250**
Wilms' tumor related 1 (WT1), **249–250**
  genes essential for development of the bipotential gonad and, **317–318**
Wingless-type MMTV integration site family member-4, **455**
WNT4, **324–325, 330**
WNT7a, **329**
Wnt-5b, 3003
Wolffian ducts
  dimorphic development, **316*f***
  testosterone and, **327**
Women, sperm storage, **130–131, 131*f***
World Health Organization (WHO), **1247**
Wound healing, relaxin and, **734**

X

Xanthocephalus xanthocephalus, 2027
X-chromosome, dosage compensation in preimplantation embryogenesis, **276–278, 276*f*, 278*t***
X-dosage, **247–248**
Xenobiotics
  epididymis as target for, **1117–1119**
  placental transfer, 2824–2827, 2824*f*, 2826*t*
Xenopus laevis, **265, 1328**
Xist, **277–278, 278*t***
X-linked recessive disorders, 2703
XO chromosomes, **247**
XXY chromosomes, **247–248**

Y

Yolk sac, secondary, **222–224**
Y-specific genes, **247**
YY1, **994–995**

Z

ZAN, **17*t***
ZGA, control, **272–273, 273*t***
Zonadhesin, **17*t***
Zona pellucida
  induction of acrosome reaction and, **72–78, 73*f*, 74*t*, 76*t***
  sperm adhesion, **67–72, 68*f*, 69*f*, 71*f***
Zonotrichia leucophrys gambelii, 2005, 2016
Zp3r, **17*t***
Zygote genome activation, **268–272, 269*f*, 270*f***
  early products, **271*t***
  methylation in one-cell zygotes, **270*f***
  products of early transcription, **270–271, 0811*t***
  timeline, **269*f***
  transcription factors, **271–272**
  transcript persistence in early development, **273*t***
Zygote markers, **294*f***
Zygote transport, **261–293**